Discover Biology, Develop Skills, and Make Connections

Since it's trailblazing First Edition, *Biological Sciences* has delivered numerous biology teaching innovations that emphasize higher-order thinking skills and conceptual understanding rather than an encyclopedic grasp of what is known about biology. Central to this shift is a student-centered approach that provides support for mastering core content and developing skills that help students learn and practice biology.

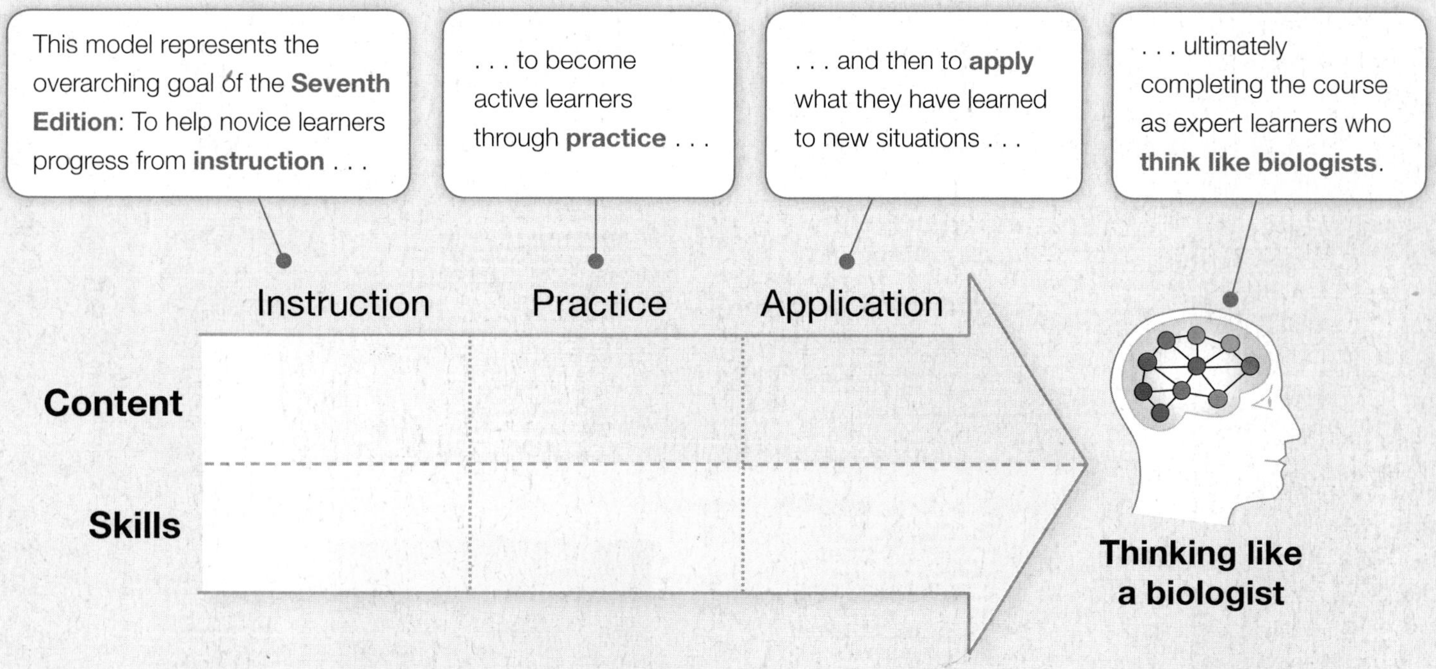

Making Connections Through

NEW Integrative End of Unit Case Study is introduced following Chapter 1. Each unit concludes with a 2-page spread that continues the story, guiding students through an exploration of key biological elements and scientific data. A unifying story about the evolutionary arms race between newts and garter snakes unfolds to illustrate how biology concepts and the various sub-disciplines of biology are connected across multiple levels from molecules, cells, and genetics to evolution and diversity, physiology, and ecology. Materials in Mastering Biology support in-class and out-of-class activities.

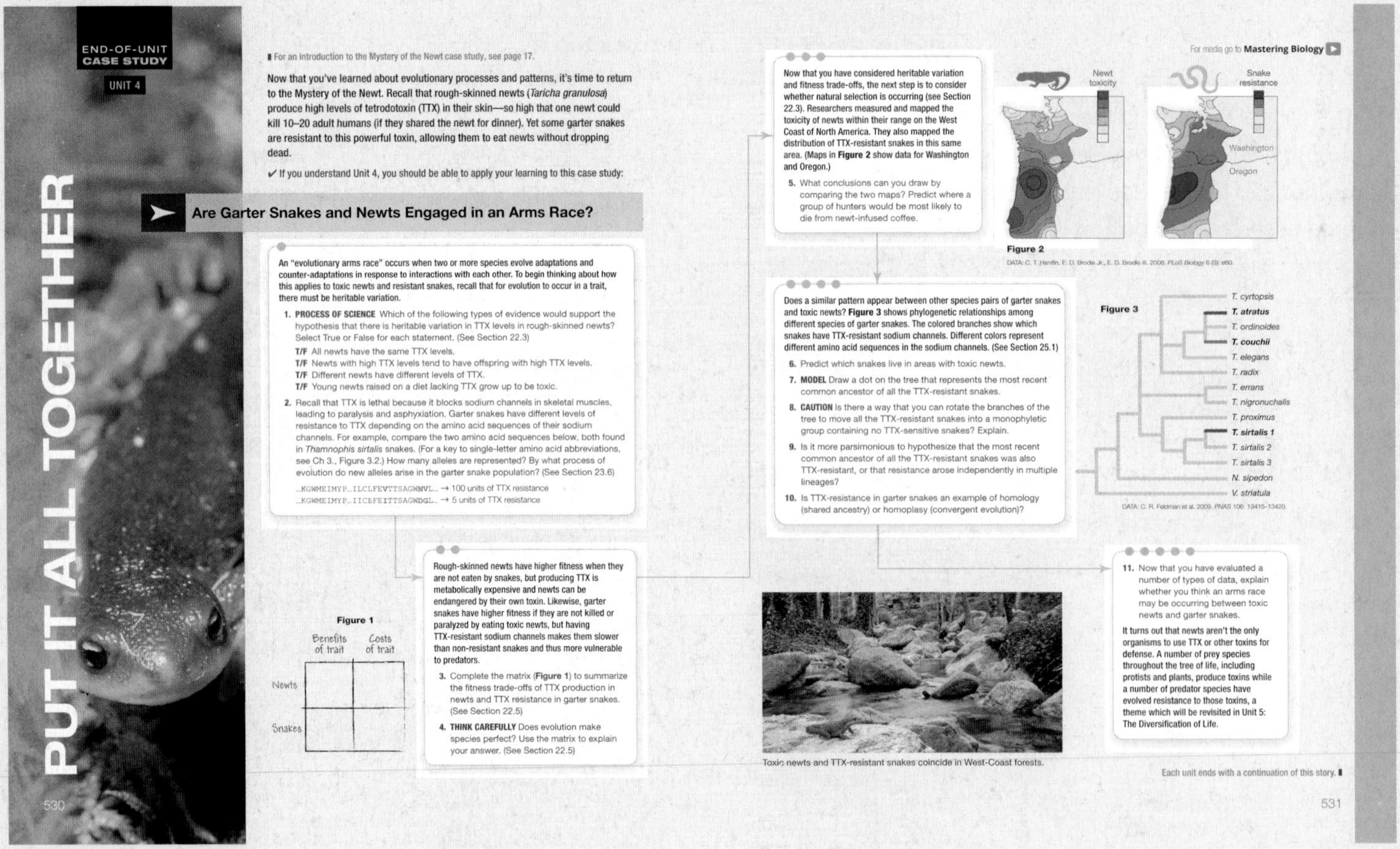

END-OF-UNIT CASE STUDY

UNIT 4

PUT IT ALL TOGETHER

For an introduction to the Mystery of the Newt case study, see page 17.

Now that you've learned about evolutionary processes and patterns, it's time to return to the Mystery of the Newt. Recall that rough-skinned newts (*Taricha granulosa*) produce high levels of tetrodotoxin (TTX) in their skin—so high that one newt could kill 10–20 adult humans (if they shared the newt for dinner). Yet some garter snakes are resistant to this powerful toxin, allowing them to eat newts without dropping dead.

✔ If you understand Unit 4, you should be able to apply your learning to this case study:

Are Garter Snakes and Newts Engaged in an Arms Race?

An "evolutionary arms race" occurs when two or more species evolve adaptations and counter-adaptations in response to interactions with each other. To begin thinking about how this applies to toxic newts and resistant snakes, recall that for evolution to occur in a trait, there must be heritable variation.

1. **PROCESS OF SCIENCE** Which of the following types of evidence would support the hypothesis that there is heritable variation in TTX levels in rough-skinned newts? Select True or False for each statement. (See Section 22.3)
   T/F All newts have the same TTX levels.
   T/F Newts with high TTX levels tend to have offspring with high TTX levels.
   T/F Different newts have different levels of TTX.
   T/F Young newts raised on a diet lacking TTX grow up to be toxic.
2. Recall that TTX is lethal because it blocks sodium channels in skeletal muscles, leading to paralysis and asphyxiation. Garter snakes have different levels of resistance to TTX depending on the amino acid sequences of their sodium channels. For example, compare the two amino acid sequences below, both found in *Thamnophis sirtalis* snakes. (For a key to single-letter amino acid abbreviations, see Ch 3., Figure 3.2.) How many alleles are represented? By what process of evolution do new alleles arise in the garter snake population? (See Section 23.6)
   ...KGWMEIMYP...ILCLFEVTTSAGWNVL... → 100 units of TTX resistance
   ...KGWMEIMYP...IICLFEITTSAGWDGL... → 5 units of TTX resistance

Rough-skinned newts have higher fitness when they are not eaten by snakes, but producing TTX is metabolically expensive and newts can be endangered by their own toxin. Likewise, garter snakes have higher fitness if they are not killed or paralyzed by eating toxic newts, but having TTX-resistant sodium channels makes them slower than non-resistant snakes and thus more vulnerable to predators.

3. Complete the matrix (**Figure 1**) to summarize the fitness trade-offs of TTX production in newts and TTX resistance in garter snakes. (See Section 22.5)
4. **THINK CAREFULLY** Does evolution make species perfect? Use the matrix to explain your answer. (See Section 22.5)

Figure 1

	Benefits of trait	Costs of trait
Newts		
Snakes		

530

For media go to **Mastering Biology**

Now that you have considered heritable variation and fitness trade-offs, the next step is to consider whether natural selection is occurring (see Section 22.3). Researchers measured and mapped the toxicity of newts within their range on the West Coast of North America. They also mapped the distribution of TTX-resistant snakes in this same area. (Maps in **Figure 2** show data for Washington and Oregon.)

5. What conclusions can you draw by comparing the two maps? Predict where a group of hunters would be most likely to die from newt-infused coffee.

Newt toxicity

Snake resistance

Washington

Oregon

Figure 2

DATA: C. T. Hanifin, E. D. Brodie Jr., E. D. Brodie III. 2008. *PLoS Biology* 6 (3): e60.

Does a similar pattern appear between other species pairs of garter snakes and toxic newts? **Figure 3** shows phylogenetic relationships among different species of garter snakes. The colored branches show which snakes have TTX-resistant sodium channels. Different colors represent different amino acid sequences in the sodium channels. (See Section 25.1)

6. Predict which snakes live in areas with toxic newts.
7. **MODEL** Draw a dot on the tree that represents the most recent common ancestor of all the TTX-resistant snakes.
8. **CAUTION** Is there a way that you can rotate the branches of the tree to move all the TTX-resistant snakes into a monophyletic group containing no TTX-sensitive snakes? Explain.
9. Is it more parsimonious to hypothesize that the most recent common ancestor of all the TTX-resistant snakes was also TTX-resistant, or that resistance arose independently in multiple lineages?
10. Is TTX-resistance in garter snakes an example of homology (shared ancestry) or homoplasy (convergent evolution)?

Figure 3

T. cyrtopsis
T. atratus
T. ordinoides
T. couchii
T. elegans
T. radix
T. errans
T. nigronuchalis
T. proximus
T. sirtalis 1
T. sirtalis 2
T. sirtalis 3
N. sipedon
V. striatula

DATA: C. R. Feldman et al. 2009. *PNAS* 106: 13415–13420.

Toxic newts and TTX-resistant snakes coincide in West-Coast forests.

11. Now that you have evaluated a number of types of data, explain whether you think an arms race may be occurring between toxic newts and garter snakes.

It turns out that newts aren't the only organisms to use TTX or other toxins for defense. A number of prey species throughout the tree of life, including protists and plants, produce toxins while a number of predator species have evolved resistance to those toxins, a theme which will be revisited in Unit 5: The Diversification of Life.

Each unit ends with a continuation of this story.

531

Every Chapter and Unit

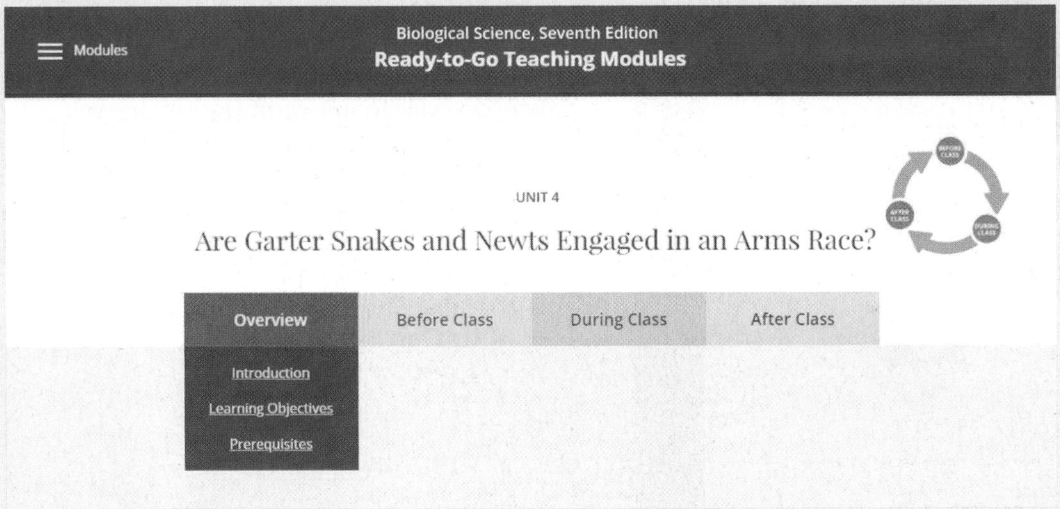

NEW The **End-of-Unit Case Studies** are supported by **Ready-to-Go Teaching Modules** in Mastering Biology that provide pre- and post-class assignments as well as a wealth of ideas for in-class activities. These resources will help enliven your class time and provide students with opportunities to apply what they are learning.

Updated **"Put It All Together" Case Studies** appear at the end of every chapter and provide a sample of contemporary biology research in action. Each case study poses questions that help students connect what they learn in class with current, real-world biology research. At least one question requires students to analyze real data or apply quantitative skills.

✔ PUT IT ALL TOGETHER: Case Study

Are toucans important to tropical forests?

Human activities are causing the fragmentation of the Brazilian Atlantic rain forest. One result is that toucans have become extinct or nearly extinct in some of the forest fragments. Does the absence of toucans affect the forest?

11. Toucans disperse seeds of key forest species such as juçara palms by eating the fruit and defecating the seeds in new locations, sometimes more than a kilometer away. If there are no toucans, is the genetic diversity of palms likely to increase or decrease within forest fragments? Why?
    a. increase (due to increased genetic drift)
    b. decrease (due to decreased gene flow)
    c. decrease (due to decreased mutation rate)
    d. decrease (due to decreased natural selection)
12. **QUANTITATIVE** Toucans can eat fruits with large seeds because their large bills can open wide. Most other birds in the same forest can only eat small seeds. Ecologist Mauro Galetti and his colleagues measured the seed sizes of palms in forest fragments with and without toucans. The graphs show two of the forest populations they studied. What is the take-home message of the data?

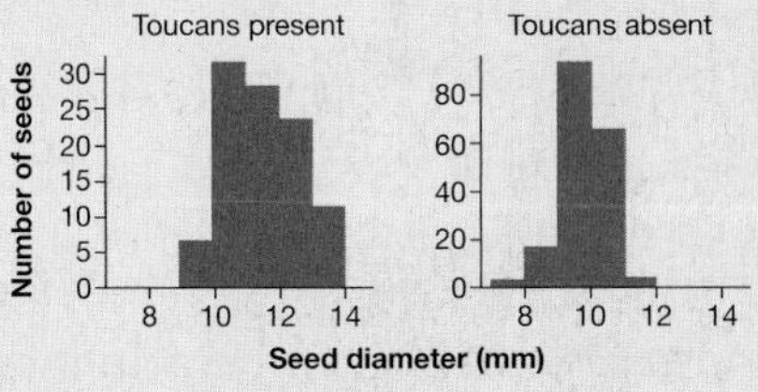

Source: M. Galetti, R. Guevara, and M. C. Côrtes, et al. 2013. *Science* 340: 1086–1090.

13. Do these data illustrate directional, stabilizing, disruptive, or balancing selection? Justify your answer in terms of fitness.
14. Large seeds carry more resources than small seeds and tend to have a higher rate of survival, especially after being dispersed by a bird. Predict how the local extinction of toucans will affect the palm population over time.
15. **PROCESS OF SCIENCE** The data in the graphs are from two of the 22 forest fragments studied by the researchers: 7 with toucans present, 15 with toucans absent. Why do you think the researchers bothered to study so many forest fragments?
16. **SOCIETY** If you were a journalist covering this story, how could you use data from this study to respond to the following social media post? "Evolution is a slow process. Humans do not cause evolution in other organisms."

Developing Skills with

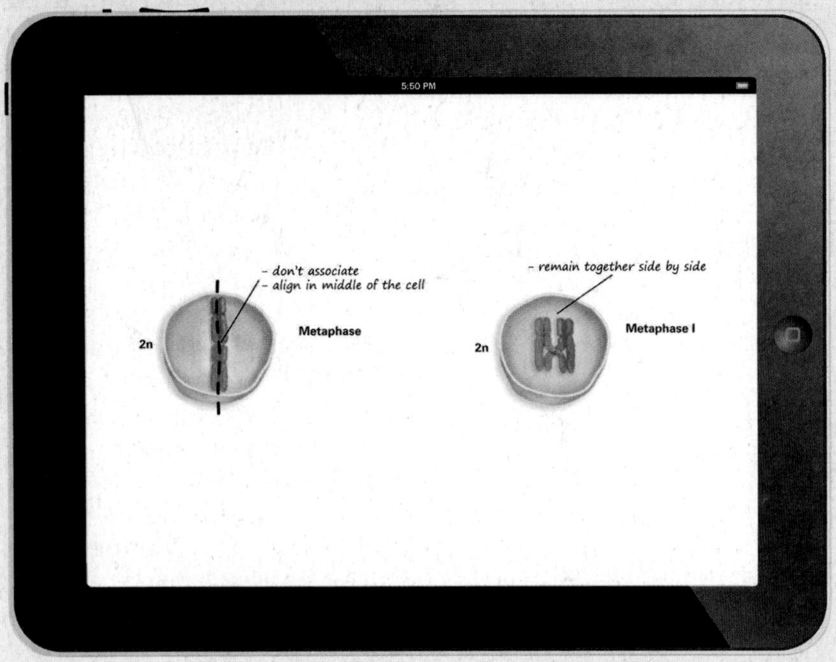

NEW 24 Interactive Figures with Walkthrough Videos help students develop skills to interpret figures, as well as develop a better understanding of key concepts. Figure Walkthrough Videos are embedded in Pearson eText for viewing at the initial point of learning and also assignable in Mastering with questions that help students practice working with visuals.

The **BioSkills** reference section appears between Chapters 1 and 2, drawing attention to key skills students need to succeed in biology. This compendium of easy-to-find reference material supports skill development throughout the course. Each BioSkill includes practice exercises in the book, questions in the Study Area of Mastering, and assignable, skill-reinforcing activities in Mastering.

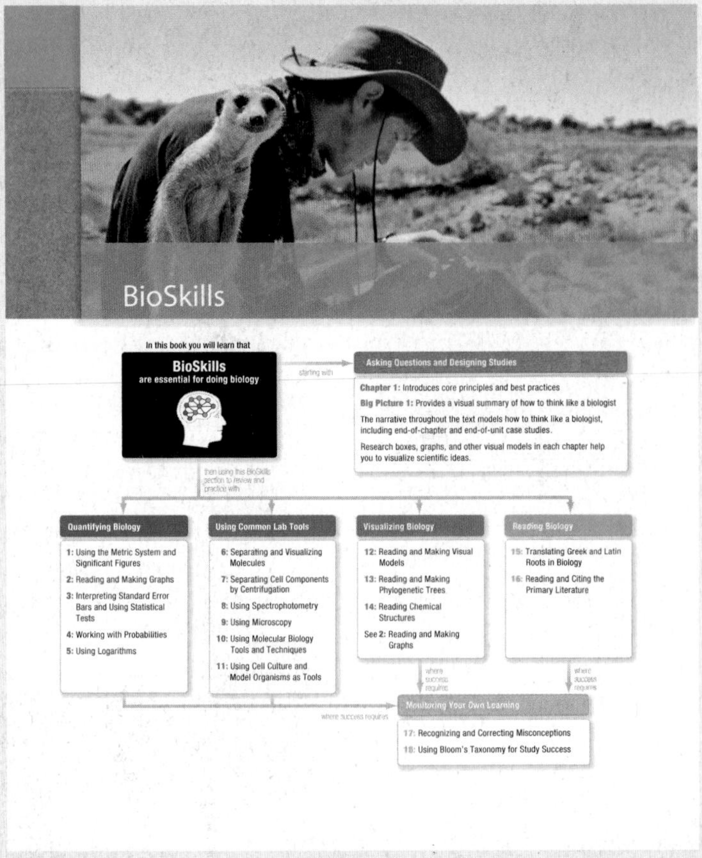

Interactive and Engaging Content

Making Models 5.1 Tips on Drawing Carbohydrates

Drawing simple models is the best way to understand the structures of monosaccharides and glycosidic linkages. In these models, focus on the overall shape of each monomer and how the monomers' carbons are numbered. You can keep the drawings simple by showing only the hydroxyl groups on the carbons being linked together, as in these examples based on α-glucose:

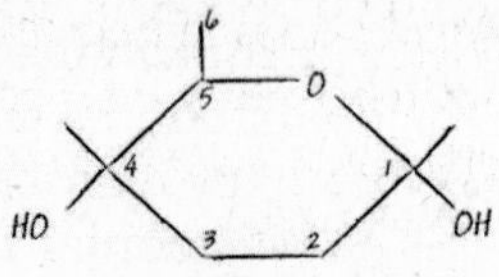

α-Glucose

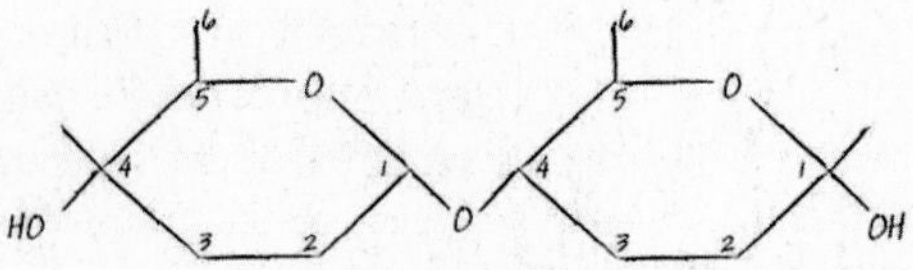

α-1,4-Glycosidic linkage

MODEL Use the examples above and Figure 5.4b to draw simplified models of a β-glucose monosaccharide and a β-glucose disaccharide with a β-1,4-glycosidic linkage.

To see this model in action, go to the Study Area of **Mastering Biology**

Making Models boxes explicitly teach students how to use visual models to learn and do biology. 45 boxes throughout the book guide students in deepening their understanding of modeling and of biology concepts. Making Models are also available for self-study in the Study Area and assignable with questions in Mastering.

3 NEW Making Models boxes are:

- Ch. 5: Tips on Drawing Carbohydrates
- Ch. 40: Tips on Drawing Arrows
- Ch. 48: Tips on Drawing Immune System Processes

Dynamic whiteboard videos support each **Making Models** box, bringing the modeling activity to life and helping students better understand how to interpret and build models. The videos are embedded in the eText, available in the Study Area, and assignable as homework in Mastering Biology.

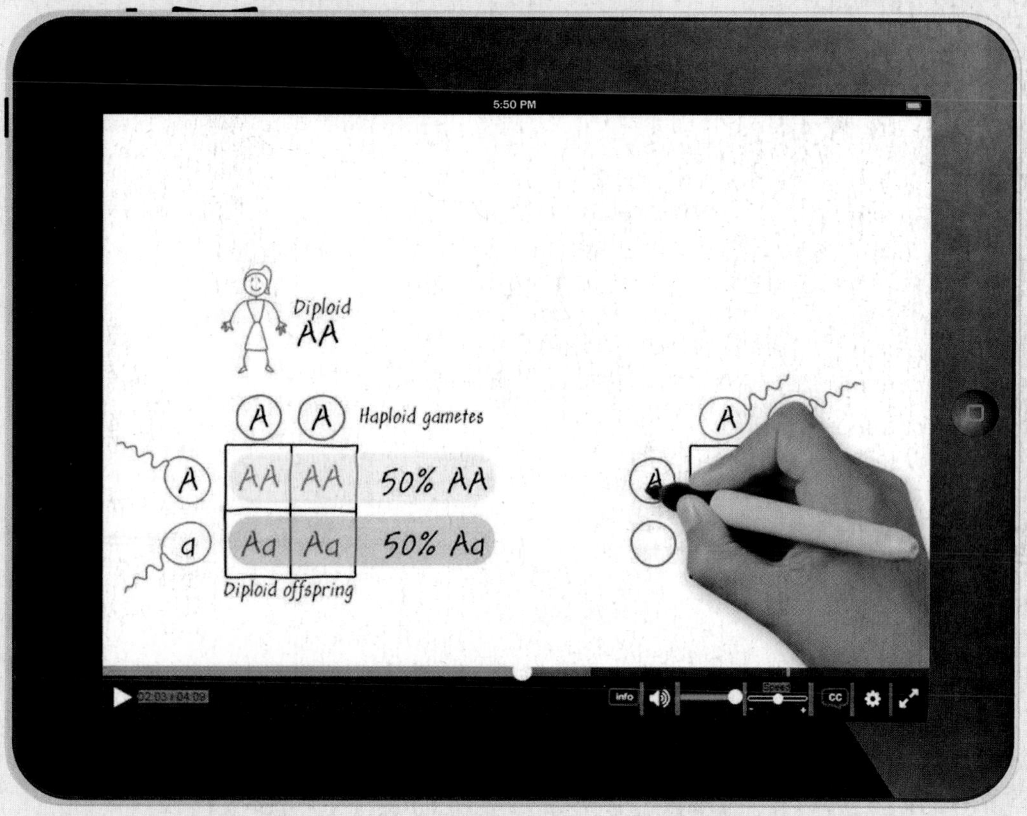

Guiding Students to Learning

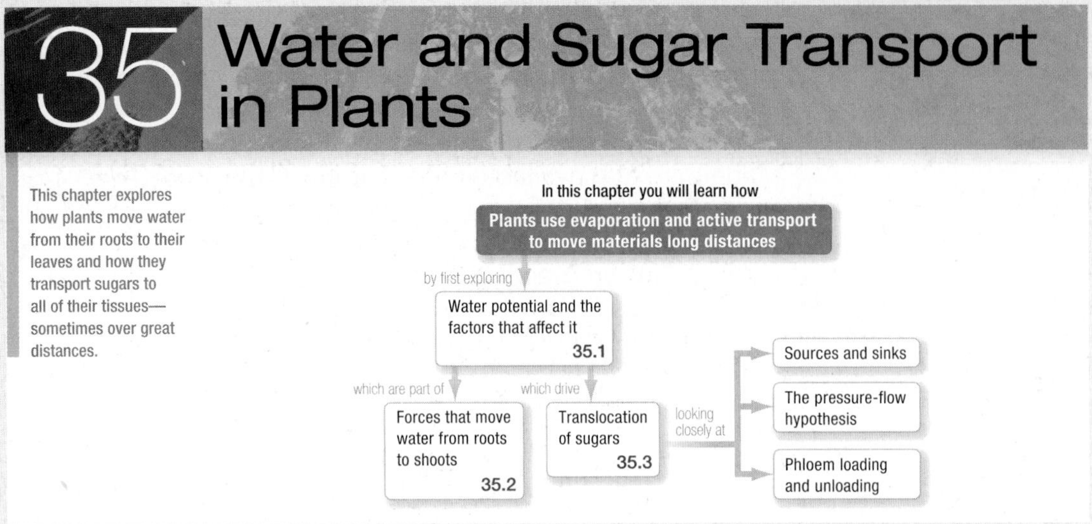

Unique **Chapter-Opening Roadmaps** set the table for learning by visually grouping and organizing information to help students anticipate key ideas as well as recognize meaningful relationships and connections that are explored in the chapter that follows.

Big Picture Concept Maps help students review key ideas. Words and visuals are integrated in these 2-page spreads to help students synthesize information about challenging topics that span multiple chapters or units. Accompanying question sets encourage students to analyze important patterns within each Big Picture. Mastering Biology provides related mapping activities and questions to help students work on higher order problems.

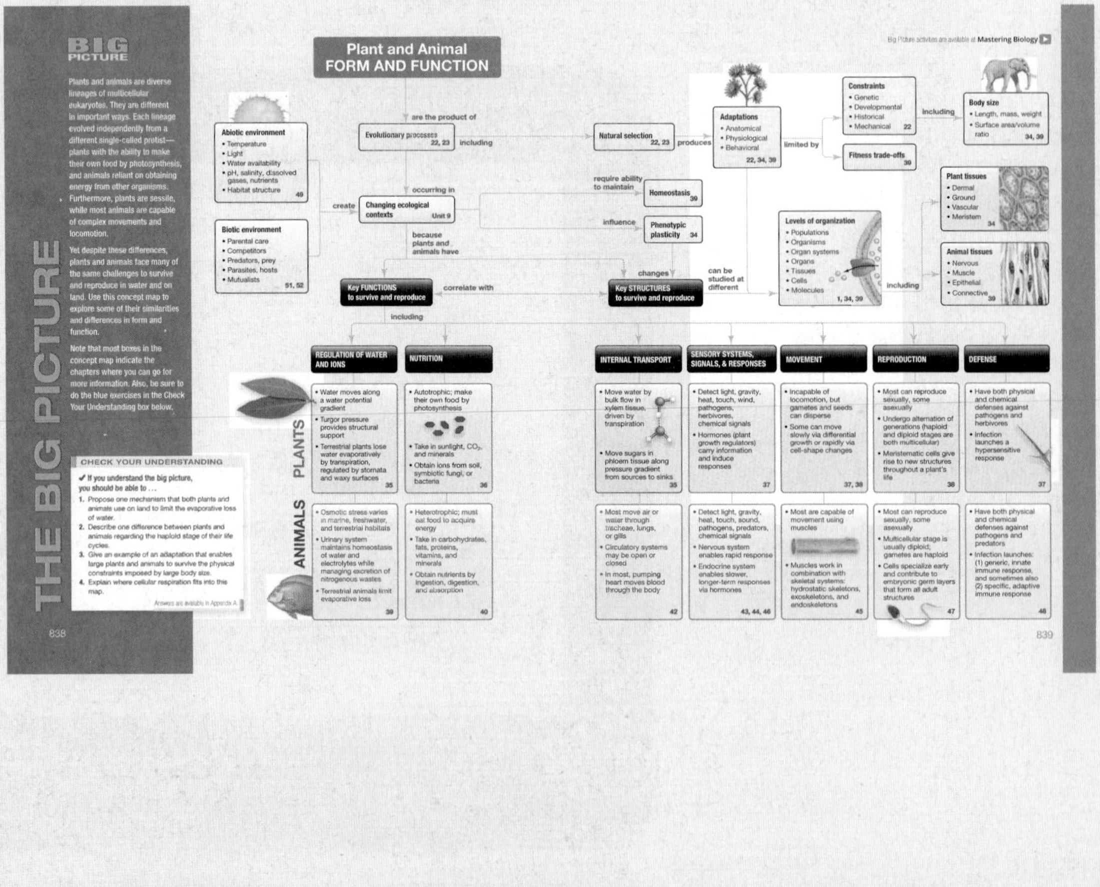

and Increasing Engagement

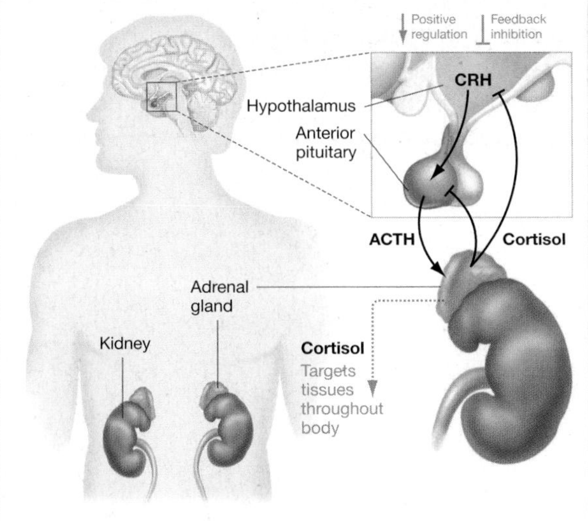

Figure 46.14 The Interaction between Cortisol, ACTH, and CRH Is an Example of Negative Feedback.

✔ **PROCESS OF SCIENCE** Use the figure to devise a test for adrenal failure in humans.

Hallmark **Blue-Thread** questions throughout the text encourage students to engage with content, think like biologists, and monitor their learning. There are a variety of question types throughout the text to help students retrieve and apply information and practice skills at all cognitive levels of Bloom's taxonomy.

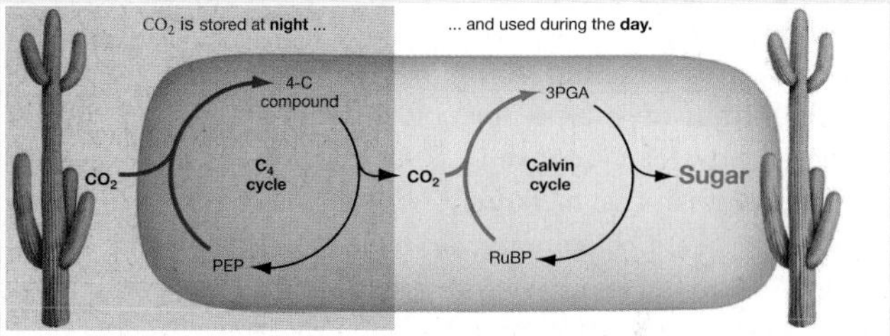

Figure 10.25 **In CAM Plants, Carbon Fixation Occurs at Night and the Calvin Cycle Occurs during the Day.**

✔ At what part of the day would there be the highest concentration of four-carbon organic acids in the vacuoles of CAM plants?

Hear from every student with **Learning Catalytics**. Utilizing a variety of question types, students recall ideas, apply concepts, and develop critical-thinking skills. Students respond using smartphones, tablets, or laptops. Responses are monitored in real-time and allow you to see what your students do—and don't—understand. Instructors can create their own questions, draw from community content, or access Pearson's library of question clusters. Focused on key topics, the clusters consist of 2-5 questions about a single data set or scenario.

Multiple Levels of Assessment

24.1 How Are Species Defined and Identified?

If your friend tells you she's planning to study polar bears and grizzly bears for her summer research project, you'd likely know that these animals are distinct species. But what if your friend is going to compare forest elephants and savanna elephants of Africa? Are they the same species or two different species?

Evolutionary biologists have been wrangling with the definition of species for decades—how can you reliably distinguish two or more species of bears, elephants, or bacteria in the field or fossil record? Although there is no single, universal answer, scientists do agree there is a distinction between the *general definition* of a species and the criteria used in the *practical identification* of species in particular cases.

After you complete this section, you should be able to . . .

- Compare mechanisms of reproductive isolation.
- Compare the advantages and disadvantages of different species concepts.

NEW Learning Objectives at the beginning of every section make it clear what fundamental content students should expect to learn and how they should be able to apply that knowledge.

Check Your Understanding Questions at the end of every section are tightly aligned to the learning objectives for the section.

CHECK YOUR UNDERSTANDING

✔ **If you understood this section, you should be able to . . .**

1. Predict which mechanism of reproductive isolation played a role in trumpeter speciation in the Amazon basin. Note: Trumpeters cannot fly across large rivers.
2. Determine which species concept(s) could be used to identify the number of trumpeter species in the Amazon.

Answers are available in Appendix A.

Steps to Building Understanding

Each chapter ends with three groups of questions that build in difficulty

✔ **TEST YOUR KNOWLEDGE**

Begin by testing your basic knowledge of new information.

✔ **TEST YOUR UNDERSTANDING**

Once you're confident with the basics, demonstrate your deeper understanding of the material.

✔ **TEST YOUR PROBLEM-SOLVING SKILLS**

Work towards mastery of the content by answering questions that challenge you at the highest level of competency.

End-of-Chapter Questions are organized in three levels–Test Your Knowledge, Test Your Understanding, and Test Your Problem-Solving Skills–so students can build from lower- to higher-order cognitive levels of assessment.

Help Students Learn and Practice

Blue Thread questions, throughout the text and figures, help students gauge their learning.

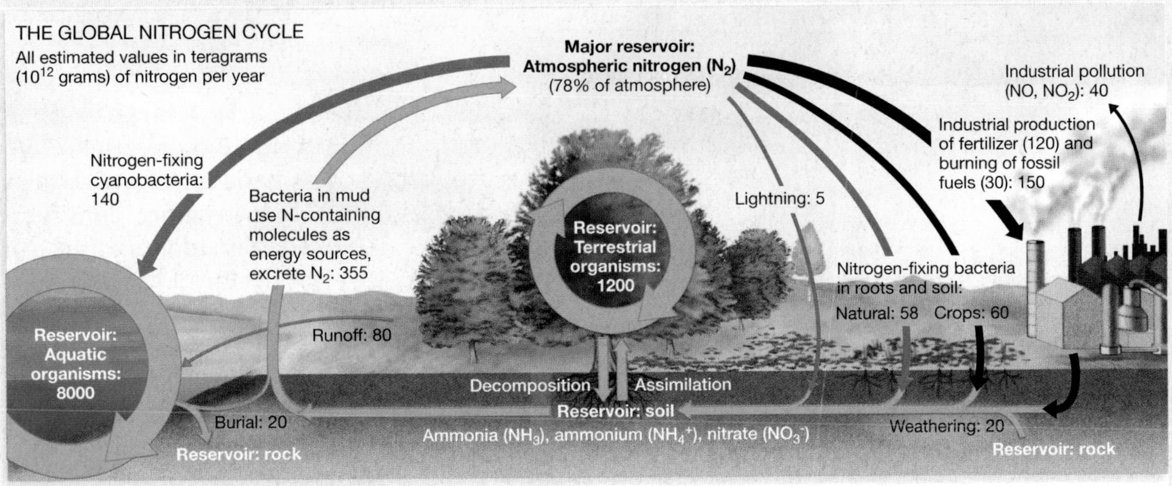

Figure 53.15 The Global Nitrogen Cycle. Nitrogen enters ecosystems as ammonia or nitrate via fixation from atmospheric nitrogen. It is exported in runoff and as nitrogen gas given off by bacteria.

DATA: D. Fowler et al. 2013. *Philosophical Transactions of the Royal Society B* 368 (1621): 20130165.

✔ **QUANTITATIVE** Calculate the percentage of total nitrogen fixation (all downward-pointing arrows) that is caused by human activities (black arrows).

Chapter Assessment Grids help instructors quickly identify suitable assessment questions in the text according to learning outcomes, Bloom's taxonomy level, core concepts and core competencies discussed in the *Vision and Change in Undergraduate Biology Education* report, and when, applicable, common misconceptions.

BLOOMS TAXONOMY RANKING

"Blue Thread" questions, including end-of-chapter problems, are ranked according to **Bloom's taxonomy** and are assignable in Mastering Biology.

LEARNING OUTCOMES

Each question is tagged to a publisher-provided **Learning Outcome**. Instructors may also track their own Learning Outcomes using Mastering Biology.

MISCONCEPTIONS

When applicable, **common student misconceptions** are addressed and identified with targeted questions.

VISION & CHANGE CORE CONCEPTS

Each question that covers a **Core Concept** from the *Vision and Change in Undergraduate Biology Education* report is noted in the chapter assessment grid and in Mastering Biology.

VISION & CHANGE CORE COMPETENCIES

Core Competencies from the *Vision and Change in Undergraduate Biology Education* report are indicated in the chapter assessment grid and in Mastering Biology.

Succeeding with Mastering Biology

Mastering Biology is the teaching and learning platform that empowers you to reach every student. By combining trusted author content with digital tools developed to engage students and emulate the office-hour experience, Mastering personalizes learning and improves results for each student.

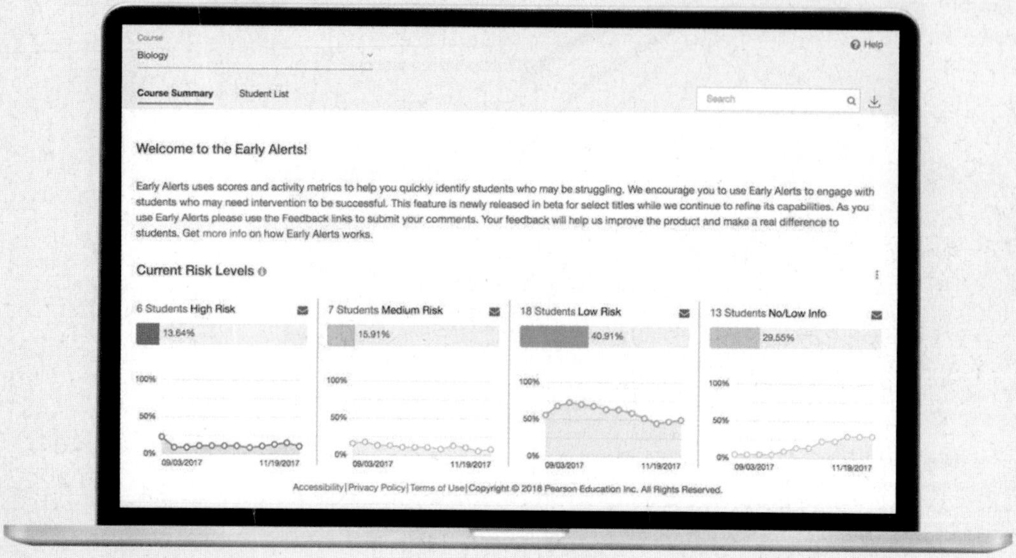

NEW Early Alerts in Mastering Biology uses scores and behavioral data to help instructors identify individual students at risk of not performing well in the course. This insight enables instructors to provide informed feedback and support at the moment struggling students need it so they can stay—and succeed—in the course.

Ready-To-Go Teaching Modules offer prepared teaching tools for use before, during, and after class, including ideas for in-class active learning. The modules incorporate the best that the text, Mastering Biology, and Learning Catalytics have to offer and can be accessed through the Instructor Resources area of Mastering Biology.

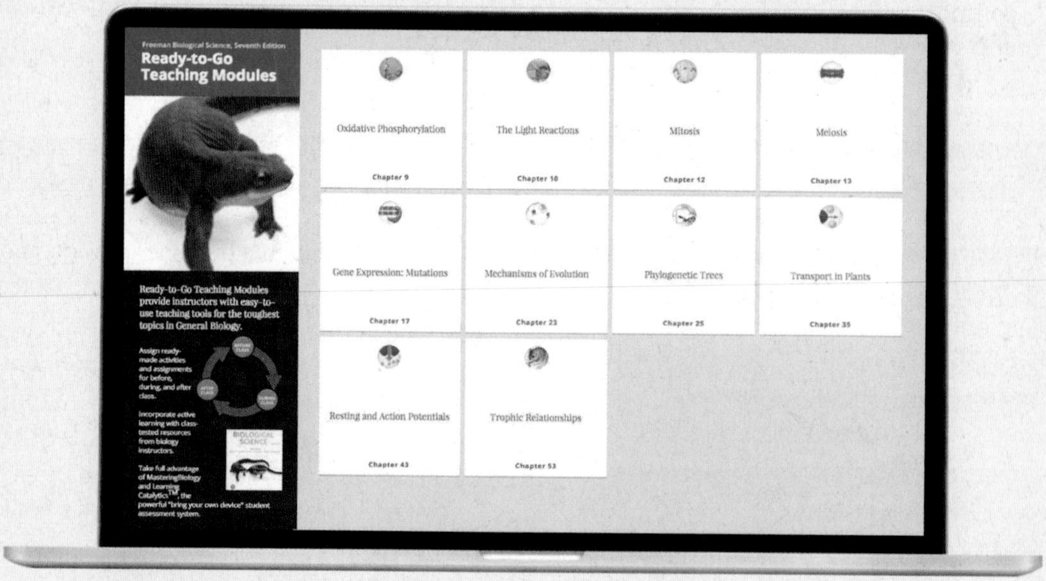

Personalizing Learning and the Classroom

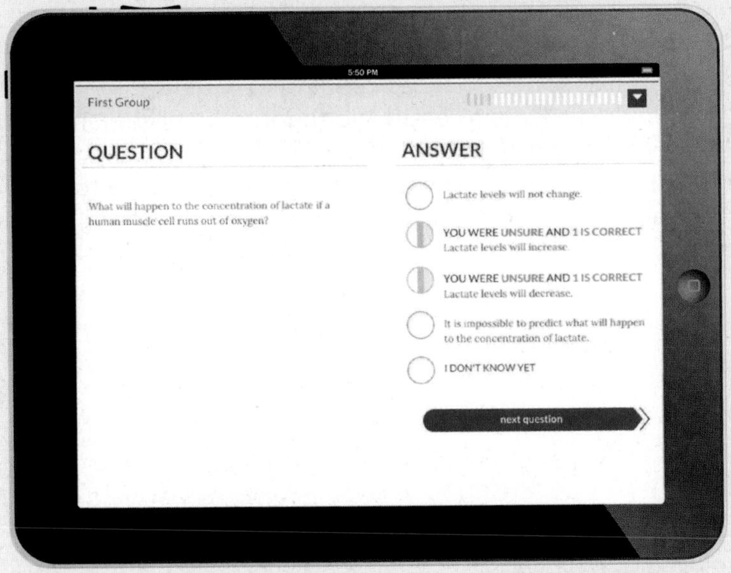

Dynamic Study Modules, based on the latest developments in cognitive science, adapt to student performance in real time to help students study course topics. As a result, students build the confidence they need to deepen their understanding, participate meaningfully, and perform better—in and out of class. Available on smartphones, tablets, and computers.

Adaptive Follow-Up Assignments provide each student with targeted question sets that address the specific concepts and skills he or she struggled with in the original homework assignment.

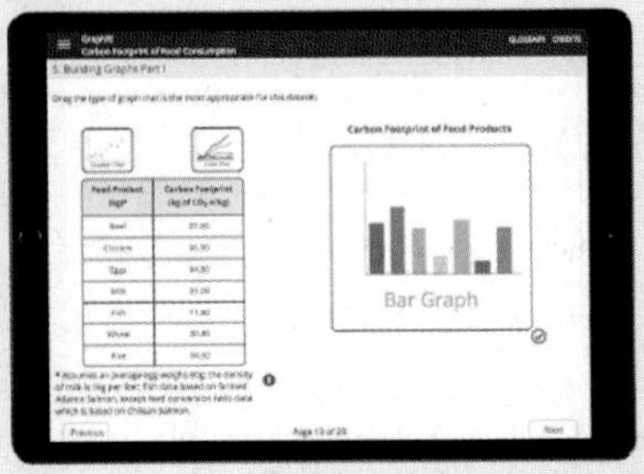

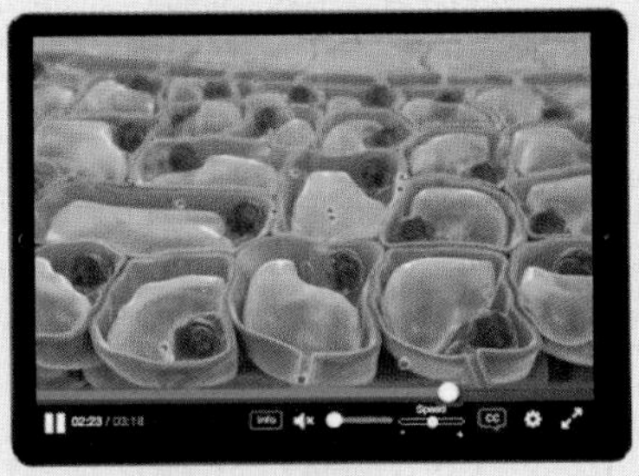

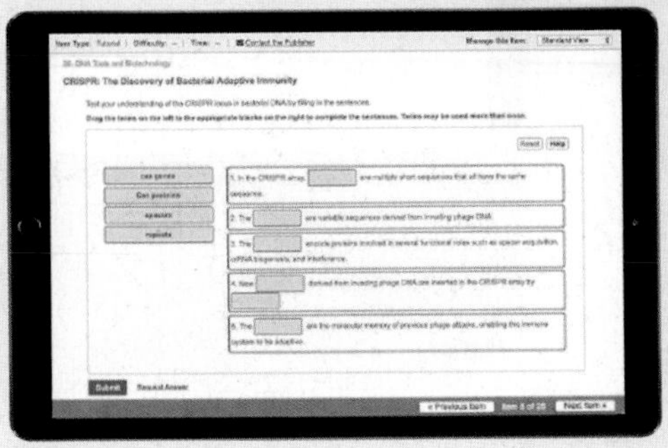

Additional Mastering Resources include: BioFlix, GraphIt! activities, HHMI videos, animations, concept maps, new tutorials, and many other tools to engage students and bring concepts to life. Available for self-study and assignment.

A Whole New Learning Experience with Pearson eText

NEW Pearson eText is a simple-to-use, mobile-optimized, personalized reading experience. It allows students to easily highlight, take notes, and review key vocabulary all in one place. Instructors can share notes from their eText with students to help focus student attention on important ideas.

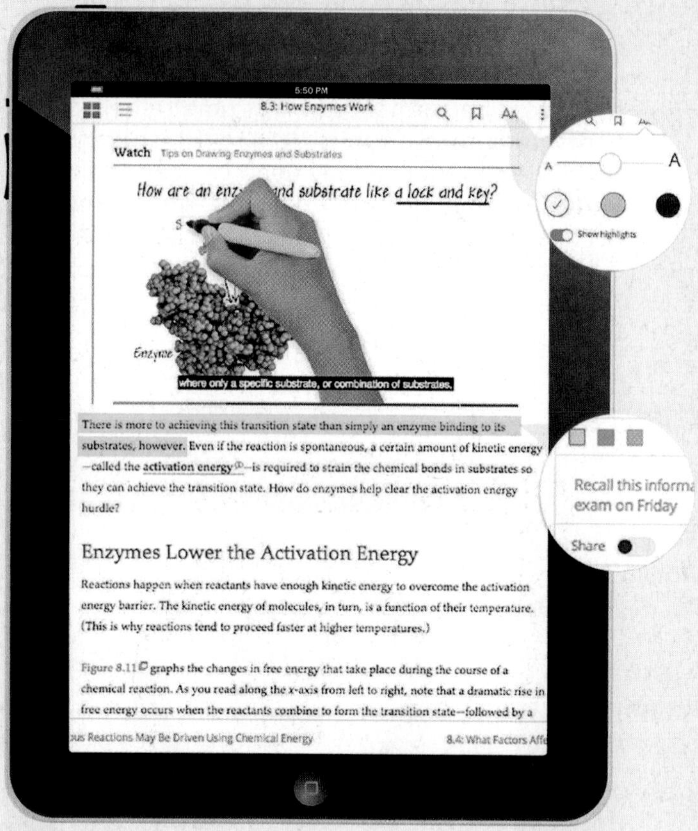

The **7th edition eText** is accessible on computers, tablets, and smart phones. To engage students, it includes embedded multimedia carefully selected or created to support key ideas in the text, including 45 Making Models videos, 25 Figure Walkthrough videos, 12 interactive graphs, and over 150 additional animations and videos.

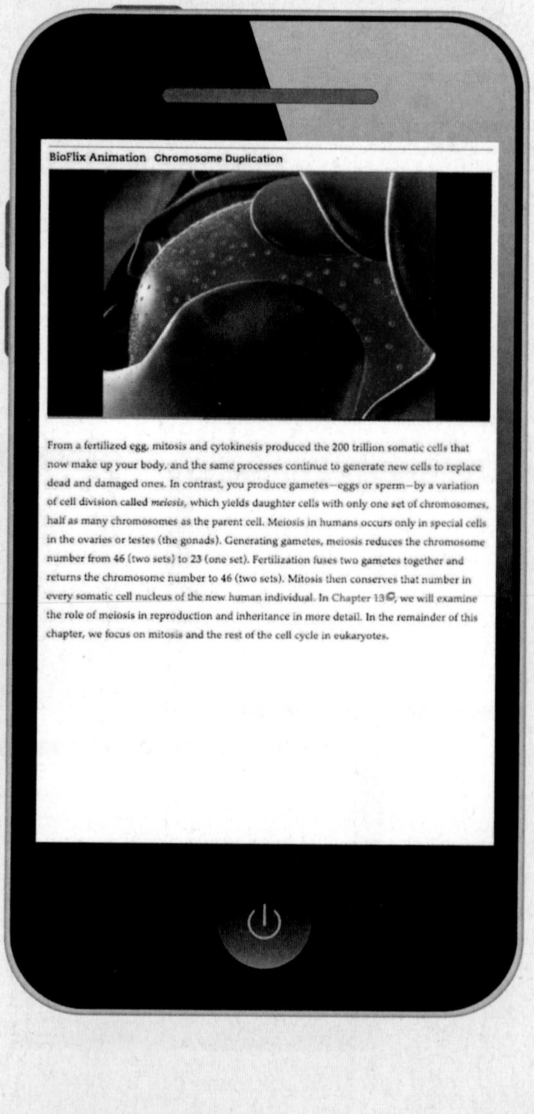

BIOLOGICAL SCIENCE

This garter snake, *Thamnophis sirtalis*, is a fearsome predator. It devours whatever it can easily overpower, including snails, slugs, earthworms, frogs, bird nestlings, mice—and newts. Can it counter the deadly defense of the rough-skinned newt (see back cover)? You'll find out the answer in this edition of *Biological Science*, while exploring an ongoing "evolutionary arms race" between snakes and newts (see the case study at the end of Unit 4 on pages 530–531).

BIOLOGICAL SCIENCE

SEVENTH EDITION

SCOTT FREEMAN
University of Washington

KIM QUILLIN
Salisbury University

LIZABETH ALLISON
College of William & Mary

MICHAEL BLACK
California Polytechnic State University in San Luis Obispo

GREG PODGORSKI
Utah State University

EMILY TAYLOR
California Polytechnic State University in San Luis Obispo

JEFF CARMICHAEL
University of North Dakota

Pearson

Director, Courseware Portfolio Management: Beth Wilbur
Courseware Portfolio Management Specialist: Michael Gillespie
Content Producer: Laura Perry
Managing Producer: Michael Early
Courseware Director, Content Development: Ginnie Simione Jutson
Courseware Senior Analysts: Sonia DiVittorio, Suzanne Olivier, Hilair Chism
Senior Development Editor: Mary Hill
Courseware Editorial Assistant: Summer Giles, Ashley Fallon
Rich Media Content Producer: Nicole Constantine
Full-Service Vendor: Integra Software Services
Copyeditor: Christianne Thillen
Compositor: Integra Software Services
Art Coordinator: Imagineering Media Services
Design Manager: Mark Ong/Side By Side
Interior and Cover Designer: tani hasegawa
Illustrators: Imagineering Media Services
Rights & Permissions Project Manager: Eric Schrader
Rights & Permissions Management: Ben Ferrini
Photo Researcher: Kristin Piljay
Manufacturing Buyer: Stacey J. Weinberger
Director of Field Marketing: Tim Galligan
Director of Product Marketing: Allison Rona
Executive Field Marketing Manager: Kelly Galli
Product Marketing Manager: Alysun Estes
Front and Back Cover Photo Credits: Joel Sartore/National Geographic/Getty Images
Title Page Photo Credit: Dave King ©Dorling Kindersley

Library of Congress Cataloging-in-Publication Data

Names: Freeman, Scott, author. | Quillin, Kim, author. | Allison, Lizabeth A., author. | Black, Michael, author. | Podgorski, Greg, author. | Taylor, Emily, author. | Carmichael, Jeff, author.
Title: Biological science / Scott Freeman, Kim Quillin, Lizabeth Allison, Michael Black, Greg Podgorski, Emily Taylor, Jeff Carmichael.
Description: Seventh edition. | New York, NY: Pearson, [2020] | Includes index.
Identifiers: LCCN 2018031078 | ISBN 9780134678320 (hardcover)
Subjects: LCSH: Biology.
Classification: LCC QH308.2 .F73 2020 | DDC 570—dc23
LC record available at https://lccn.loc.gov/2018031078

Pearson

www.pearson.com

ISBN 10: 0-134-67832-X; ISBN 13: 978-0-134-67832-0 (Student edition)
ISBN 10: 0-135-27280-7; ISBN 13: 978-0-135-27280-0 (Looseleaf Edition)

Detailed Contents

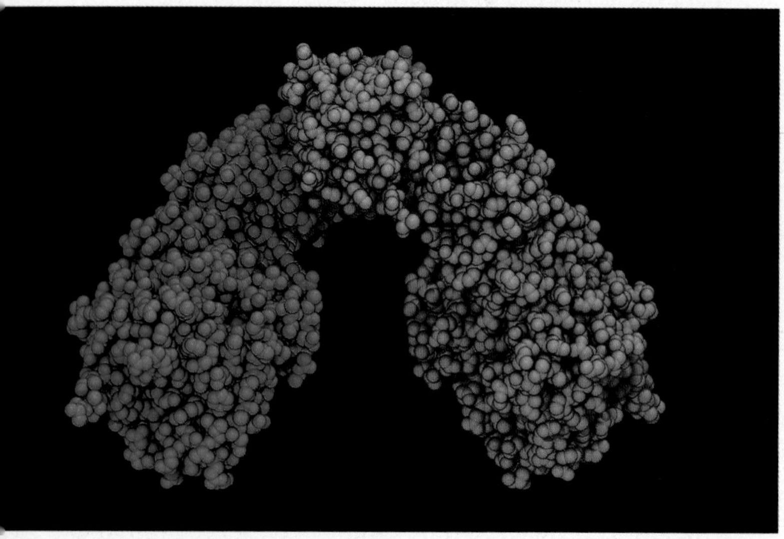

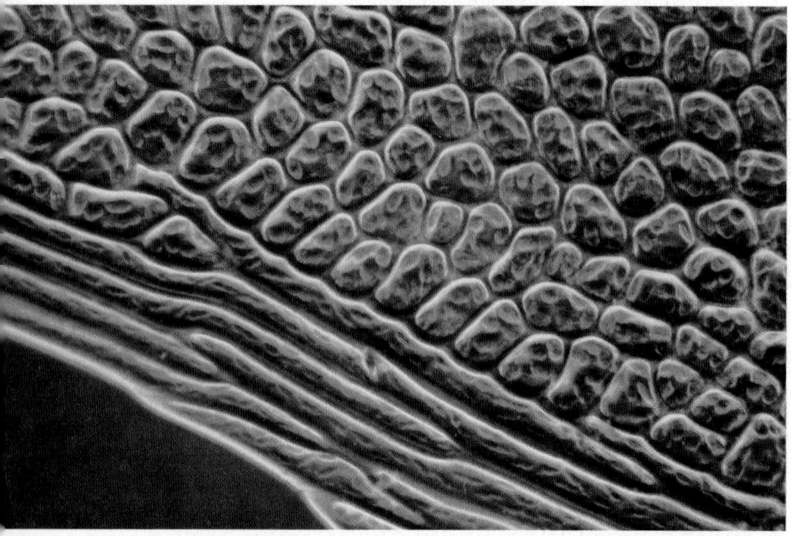

UNIT 4 EVOLUTIONARY PATTERNS AND PROCESSES 448

About the Authors

A Letter from Scott

I started working on *Biological Science* in 1997 with a simple goal: To help change the way biology is taught. After just shy of 20,000 hours of work on four editions of this text, that goal still gets me out of bed in the morning. But instead of focusing my energies on textbook writing, I've decided to devote myself full-time to research on student learning and developing new courses for undergraduate and graduate students at the University of Washington.

I have passed the torch to an all-star cast of leading scientists and educators who have enthusiastically taught from, and contributed to, previous editions of *Biological Science*. The new team brings their passion, talent, and creativity to the book, with expertise that spans the breadth of the life sciences. Just as important, they work beautifully together because they think alike. They are driven by a commitment to the craft of writing and a background in evidence-based teaching.

These pages offer a brief introduction to Liz Allison, Michael Black, Greg Podgorski, Kim Quillin, Jeff Carmichael, and Emily Taylor. As a group, they've built on the book's existing strengths and infused this edition with fresh energy, perspective, and ideas. I'm full of admiration for what they have accomplished, and I'm excited about the impact this edition will have on biology students from all over the world.

—Scott Freeman

Scott Freeman received a PhD in Zoology from the University of Washington and was subsequently awarded an Alfred P. Sloan Postdoctoral Fellowship in Molecular Evolution at Princeton University. He has done research in evolutionary biology on topics ranging from nest parasitism to the molecular systematics of the blackbird family and is coauthor, with Jon Herron, of the standard-setting undergraduate text *Evolutionary Analysis*. Scott is the recipient of a Distinguished Teaching Award from the University of Washington and is currently a Principal Lecturer in the UW Department of Biology, where he teaches introductory biology for majors, a writing-intensive course for majors called The Tree of Life, and a graduate seminar in college science teaching. Scott's current research focuses on how active learning affects student learning and academic performance.

Lizabeth A. Allison is Chancellor Professor of Biology at the College of William & Mary. She received her PhD in Zoology from the University of Washington, specializing in molecular and cellular biology. Before coming to William & Mary, she spent eight years as a faculty member at the University of Canterbury in New Zealand. Liz teaches introductory biology for majors and upper division molecular biology courses. She has mentored graduate students and more than 120 undergraduate research students, many of them coauthoring papers with her on intracellular trafficking of the thyroid hormone receptor in normal and cancer cells. The recipient of numerous awards, including a State Council for Higher Education in Virginia (SCHEV) Outstanding Faculty Award in 2009, Liz received one of the three inaugural Arts and Sciences Faculty Awards for Teaching Excellence in 2011, and a Plumeri Award for Faculty Excellence in 2012. In addition to her work on this text, she is author of *Fundamental Molecular Biology*, now in its second edition, with a third edition under way.

Lead Author; Chapters 1, 33, 48 and BioSkills

laalli@wm.edu

Michael Black received his PhD in Microbiology and Immunology from Stanford University School of Medicine as a Howard Hughes Predoctoral Fellow. After graduation, he studied cell biology as a Burroughs Wellcome Postdoctoral Fellow at the MRC Laboratory of Molecular Biology in Cambridge, England. His current research focuses on the use of molecules to identify and track the transmission of microbes in the environment. Michael is a professor of Cell and Molecular Biology at California Polytechnic State University in San Luis Obispo, where he teaches introductory and advanced classes for majors in cell biology and microbiology. In addition to his teaching and research activities, Michael serves as the director of the Undergraduate Biotechnology Lab, where he works alongside undergraduate technicians to integrate research projects and inquiry-based activities into undergraduate classes.

Chapters 2–12

mblack@calpoly.edu

Greg Podgorski received his PhD in Molecular and Cellular Biology from Penn State University and has been a postdoctoral fellow at the Max Plank Institute for Biochemistry and Columbia University. His research interests are in biology education, developmental genetics, and computational biology. Greg's most recent work has been in mathematical modeling of how patterns of different cell types emerge during development and how tumors recruit new blood vessels in cancer. Greg has been teaching at Utah State University for more than 20 years in courses that include introductory biology for majors and for nonmajors, genetics, cell biology, developmental biology, and microbiology, and he has offered courses in nonmajors biology in Beijing and Hong Kong. He has won teaching awards at Utah State University and has been recognized by the National Academies as a Teaching Fellow and a Teaching Mentor.

Chapters 13–21

greg.podgorski@usu.edu

Jeff Carmichael received his BS in Biology from Slippery Rock University in Pennsylvania and his PhD in Plant Biology from the University of Georgia. As an undergraduate student, he spent some time studying enzyme kinetics through a fellowship at Oak Ridge National Laboratory in Tennessee. His graduate work focused on sexual reproduction in an intriguing group of seed plants. He has been teaching and coordinating Introductory Biology at the University of North Dakota (UND) for more than 20 years. He also works with the campus-wide Teaching Transformation and Development Academy, where he helps other faculty members incorporate evidence-based best teaching practices in their courses. He has received excellence in teaching awards at UND and as a graduate student in Georgia.

Chapters 26–29 and 34–38

Jeffrey.Carmichael@und.edu

Kim Quillin received her BA in Biology at Oberlin College *summa cum laude* and her PhD in Integrative Biology from the University of California, Berkeley, as a National Science Foundation Graduate Fellow. Kim has worked in the trenches with Scott Freeman on every edition of *Biological Science*, starting with the ground-up development of the illustrations in the first edition in 1999 and expanding her role in each edition. Kim currently serves as the Curriculum Coordinator for Introductory Biology for Salisbury University (SU), a member of the University System of Maryland, where she is actively involved in the ongoing student-centered reform of the course for biology majors. She also serves as the Curriculum Facilitator for the biology department, focusing on programmatic review and alignment of the SU biology curriculum to the *Vision and Change* core concepts and competencies.

Chapters 22–25, 30–32, 49–54

kxquillin@salisbury.edu

Emily Taylor earned a BA in English at the University of California, Berkeley. She then earned a PhD in Biological Sciences from Arizona State University, where she conducted research in the field of environmental physiology as a National Science Foundation Graduate Research Fellow. She is currently a professor of Biological Sciences at the California Polytechnic State University in San Luis Obispo. Her student-centered research program focuses on the endocrine and reproductive physiology of free-ranging reptiles, especially rattlesnakes. She teaches numerous undergraduate and graduate courses, including introductory biology, anatomy and physiology, endocrinology, and herpetology, and received the California Faculty Association's Distinguished Educator Award in 2010 and Cal Poly's Distinguished Teaching Award in 2012.

Chapters 39–47

etaylor@calpoly.edu

Preface to Instructors

Since its inception, *Biological Science*'s unique emphasis on the process of scientific discovery and guiding students to think like biologists has placed this book at the forefront of change in the way we teach biology. The Seventh Edition embraces this legacy and continues to exemplify the principles outlined in the *Vision and Change in Undergraduate Biology Education* report. As in previous editions, the cutting-edge biology in the Seventh Edition is pitched at exactly the right level for introductory students and is as accurate and exciting as ever for instructors and students alike. New findings from education research about ways to increase student engagement in the learning process continue to inform and inspire the coauthor team. The Seventh Edition introduces novel approaches that will allow instructors to attain an appropriate blend of both content and skill development in the classroom. Innovative features new to this edition offer students even more opportunities to actively apply concepts in new situations; evaluate experimental design, hypotheses, and data; synthesize results; and make and interpret models. For instructors, additional resources are provided to help align course activities and learning goals with their assessment strategies.

Core Values

In the Seventh Edition, the coauthor team has strived to extend the vision and maintain the core values of *Biological Science*—to provide a book and online resources for dedicated instructors who embrace the challenge of helping students achieve higher levels of learning, and to provide a book that helps students each step of the way in learning to think like scientists, regardless of their starting point in the process. The Seventh Edition provides tools to help students build their cognitive mastery in both transferrable skills and biology content—to learn at the level called for by the National Academy of Sciences, the Howard Hughes Medical Institute, the American Association of Medical Academies, and the National Science Foundation. Reports such as *Biology 2010, Scientific Foundations for Future Physicians*, and *Vision and Change* all place a premium on fundamental skills and concepts as well as connecting core ideas across all levels of biology.

What's New in This Edition

Many of the new or expanded features in the Seventh Edition are designed to seamlessly intertwine delivery of cutting-edge biology content with development of students' skills in data analysis and quantitative model-based reasoning, in an engaging and relevant way. Students will receive initial instruction in biology content and transferrable skills, followed by opportunities for lots of practice in applying knowledge and skills to new contexts. The ultimate goal is for students to learn to construct their own knowledge and think like biologists.

- **Integrative End-of-Unit Case Study** Introduced in Chapter 1 and then revisited throughout the book, the end-of-unit case study "Mystery of the Newt" features a cliff-hanger tale of poisonous newts and resistant garter snakes. The unfolding story illustrates how biology concepts and the various subdisciplines of biology (genetics, evolution, physiology, ecology, etc.) are connected across multiple levels from molecules and cells to ecology and evolution. Each unit concludes with a two-page spread that focuses on contemporary biological elements and poses questions relevant to the unit and the case study. By highlighting the practice of science, we hope to encourage students to envision themselves as scientists. The end-of-unit case study is supported by Ready-to-Go teaching modules with instructional resources that include in-class clicker questions and activities and related Mastering™ Biology questions for pre- and post-class assignments to promote critical thinking and student engagement in the classroom.
- **Interactive Figure Walk-through Videos** Twenty-five figure walk-through videos have been added throughout the text to help students with unpacking complex figures and to cultivate better engagement with the text and other scientific sources. The videos will help students to better understand quantitative and modeling skills, as well as key concepts, by breaking down the information in figures step-by-step. Interactive figures are embedded in Pearson's new and engaging eText for practice at the initial point of learning and included in assignable Mastering Biology activities to help students practice working with visuals.
- **Three New Making Models Boxes** Reports like *Vision and Change* cite the importance of developing model-based reasoning skills. To help attain this goal, Making Models boxes were introduced in the Sixth Edition to explicitly teach students how to use visual models in learning and doing biology. In the Seventh Edition, three new Making Models boxes have been added: Tips on Drawing Carbohydrates (Ch. 5); Tips on Drawing Arrows (Ch. 40); and Tips on Drawing Immune System Processes (Ch. 48). Each Making Models box has three components: instruction in interpreting or creating a specific type of model, an example of that type of model, and an application question so that students can immediately practice their skills. Additionally, whiteboard-style videos allow students to watch and interact with a dynamic presentation of modeling. These videos are embedded in the eText, accessible in the Study Area of Mastering Biology, and integrated in assignable Mastering Biology activities that test students' understanding of models. Lastly, there are test bank questions aligned with the Making Models activities.
- **Enhanced Interactive eText** New and expanded features in the Seventh Edition have been designed not only to enhance

the print text but also to make full use of the new, embedded Pearson eText platform. The eText presents over 250 videos and animations that have been carefully selected to support content in the text. Resources include whiteboard Making Models videos, interactive Figure Walk-through videos, HHMI Biointeractive videos, and BioFlix® Animations and Tutorials that engage students, help them learn, guide them in completing assignments, and bring biology to life.

- **In-Text Learning Objectives** To help students navigate content, a Learning Objective now appears at the beginning of each section of a chapter. Each Learning Objective makes it clear up front what fundamental content students should expect to learn in a particular section and what they should be able to do with what they've learned.
- **Key New Content** Introduced in the Sixth Edition, end-of-chapter Put It All Together case studies engage students by asking them to connect what they've learned in the chapter to an example of contemporary, relevant research. New case studies have been added that are updated for more recent developments, relevancy, and alignment with chapter content: Ch 30: "Why aren't comb jellies the most diverse animals on Earth?; Ch 31: "How could you count all the bees, beetles, and butterflies in one forest?; and Ch. 33: "How does Zika virus cause birth defects?" Also new to the Seventh Edition is a heavily revised Chapter 20, which includes expanded coverage of CRISPR-Cas gene editing and synthetic biology. In addition, coverage of climate change and other human impacts, as an important context for biological systems, is woven throughout relevant chapters, especially in the ecology unit. There is also an increased emphasis on systems biology and big data; for example, in the ecology unit students explore social networks and community networks.
- **Expanded Assessment Matrix** Introduced in the Sixth Edition, an assessment matrix for each chapter identifies for instructors how each question is related to learning objectives, Bloom's level, common misconceptions, and *Vision and Change* core concepts and competencies. New to the Seventh Edition, all in-text and test bank questions are integrated and tightly aligned to two levels of learning objectives: broader "big picture" objectives and more granular supporting objectives. We hope this tool will assist instructors in selecting the most appropriate assessment items to align with the goals of their course.

Hallmark Features of the Text

While we are excited to introduce the new features of the Seventh Edition, we are committed to strengthening the hallmark features that make this book unique.

- **Road Maps** Each chapter opens with a concept map that visually groups and organizes information to help students anticipate key ideas as well as recognize meaningful relationships and connections among ideas.
- **Big Picture Spreads** While the Road Maps help students look forward as they engage with a chapter, Big Picture concept maps, typically found at the end of a unit, help students review content. Words and visuals are integrated to help students synthesize information about challenging topics that span multiple chapters or units. Related activities are available in Mastering Biology to help students work on higher-order problems while synthesizing key ideas.
- **BioSkills** Instructors recognize that biology students need to develop foundational science skills in addition to content knowledge. Since the Third Edition, *Biological Science* has provided a unique, robust set of materials and activities to guide students who need extra help with the skills emphasized in the book. The BioSkills materials reside between Chapters 1 and 2 to emphasize their importance as a resource for success in doing biology, and to make it easier for students to access them throughout the course. The BioSkills are presented in five broad categories: Quantifying Biology, Using Common Lab Tools, Visualizing Biology, Reading Biology, and Monitoring Your Own Learning. BioSkills include practice questions, are cross-referenced throughout the text, and can be assigned online in Mastering Biology.
- **Opportunities for Practice** "Blue Thread" questions, integrated throughout the text, are designed to help students identify what they do and do not understand. Students are encouraged to think like biologists and practice engaging with content at a higher level. The idea is that if students really understand a piece of information or a concept, they should be able to do something with it. As in the Sixth Edition, all questions in the text are assigned a Bloom's taxonomy level to help both students and instructors understand whether a question requires higher-order or lower-order cognitive skills. A substantial proportion of questions are higher order. In addition, skill-based question tags help students and instructors identify opportunities to practice key skills. Questions are tagged to indicate the following: *Process of Science* questions explore the application of the scientific process; *Model* questions ask students to interpret or construct visual models; *Society* questions explore the relationship between science and society; *Quantitative* questions help students perform quantitative analysis and use mathematical reasoning; and *Think Carefully* questions address topics that students often have common misconceptions about, and the answers to these questions include information that addresses the misconception.
    - **In-text "You Should Be Able To" questions** These questions focus on topics and concepts that professors and students have identified as most key or difficult in each chapter.
    - **Caption questions and exercises** Students are challenged to examine the information in a figure or table critically—not just absorb it.
    - **Check Your Understanding boxes** One to three tasks are presented that students should be able to complete in order to demonstrate a mastery of learning objectives.
    - **End-of-chapter questions** Questions are organized in three levels: Test Your Knowledge, Test Your Understanding, and Test Your Problem-Solving Skills—so students can build from lower- to higher-order cognitive levels of assessment.

- **Put It All Together Case Studies** Case studies added in the Sixth Edition briefly introduce contemporary biology research in action, followed by questions that ask students to apply the chapter's content and skills to the research topic. Instructor resources include clicker questions to give instructors the opportunity to use the case studies as discussion prompts in the classroom. An enduring feature of this text is its emphasis on experimental evidence—on teaching how we know what we know. The case studies expand this emphasis, requiring students to evaluate real data and to see how ongoing scientific research is related to core biological ideas.
- **Focus on Real Data** Students now have expanded opportunities to develop skills in working with real data from the primary literature. Sources of the data presented in research boxes, graphs, end-of-chapter case studies, and the end-of-unit case studies are cited to model good practice for students and to provide a resource for students and instructors who wish to evaluate the original data more deeply.
- **Illustrated Summary Tables** The art program is enhanced by illustrated summary tables that deliver content in a streamlined way and facilitate comparison and analysis by students. Photographic summary tables in Unit 5, for instance, illustrate the diversity of life. These tables make subject areas more accessible to visual learners and reinforce a chapter's key concepts.

Integration of Media

The textbook continues to be supported by Mastering Biology, the most powerful online homework, tutorial, and assessment system available. Tutorials follow the Socratic method, coaching students to the correct answer by offering feedback specific to a student's errors or misconceptions as well as supplying hints that students can access if they get stuck. Instructors can associate content with publisher-provided learning outcomes or create their own. Highlights include the following:

- **eText** The new eText provides an engaging learning experience for students. We have integrated over 250 videos and animations, including Making Models videos, Interactive Figure Walk-through videos, BioFlix animations, and data visualizations. The eText is accessible on computers, tablets, and smart phones.
- **Interactive Figure Walk-through Videos** Twenty-five figures that students find challenging to interpret are now enhanced with brief videos that break down the art and provide explanatory narration to help students better understand key concepts and to develop critical thinking skills. The videos are in the eText and included in assignable activities in Mastering Biology that allow students to apply what they've learned.
- **Making Models Activities** Forty-six whiteboard videos bring the Making Models feature from the book to life to help students develop their visual modeling skills. The videos are in the Study Area and are also included in assignable activities that allow students to practice modeling and apply their understanding to new situations.
- **Case Study Questions** Put It All Together case study questions from the end of each chapter are assignable in Mastering Biology.
- **Solve It Tutorials** These activities allow students to act like scientists in simulated investigations. Each tutorial presents an interesting, real-world question that students will answer by analyzing and interpreting data.
- **Experimental Inquiry Tutorials** The call to teach students about the process of science has never been louder. To support such teaching, there are 10 interactive tutorials on classic scientific experiments—ranging from Meselson–Stahl on DNA replication to the Grants' work on Galápagos finches and Connell's work on competition. Students who use these tutorials should be better prepared to think critically about experimental design and evaluate the wider implications of the data—preparing them to do the work of real scientists in the future.
- **BioFlix Animations and Tutorials** BioFlix are movie-quality, 3-D animations that focus on the most difficult core topics and are accompanied by in-depth, online tutorials that provide hints and feedback to guide student learning. Eighteen BioFlix animations and tutorials tackle topics such as meiosis, mitosis, DNA replication, photosynthesis, homeostasis, and the carbon cycle.
- **HHMI BioInteractive Short Films Activities** Documentary-quality movies from HHMI are available in Mastering Biology with assignable questions to make sure students understand key ideas.
- **Galápagos Evolution Video Activities** These incredible videos, filmed on the Galápagos Islands by Peter and Rosemary Grant, bring to life the dynamic evolutionary processes that have an impact on Darwin's finches on Daphne Major Island. Six videos explore important concepts and data from the Grants' field research, and assignable activities keep students focused on the important take-home points.
- **GraphIt! Activities** These activities use real data to help students explore important topics while developing skills to read, interpret, and create graphs.
- **End-of-Chapter Questions** A broad range of end-of-chapter questions are available to assign in Mastering Biology.
- **Blue Thread Questions** Over 500 questions based on the Blue Thread questions in the textbook are assignable in Mastering Biology.
- **Big Picture Concept Map Tutorials** An engaging concept mapping tool is the basis for highly interactive, challenging concept map activities based on the Big Picture figures in the textbook. Students build their own concept maps, which are automatically graded, and then answer questions to make sure they understand key ideas and make important connections.
- **BioSkills Activities** Activities based on the BioSkills content in the textbook are assignable in Mastering Biology, including activities to support the new BioSkills.
- **Reading Quiz Questions** Every chapter includes reading quiz questions that can be assigned to ensure students read

the textbook and understand the basics. These quizzes are perfect as a pre-lecture assignment to get students into the content before class, allowing instructors to use class time more effectively.

In addition, a few tools are of particular help to busy instructors.

- **Ready-to-Go Teaching Modules** Based on key topics that students struggle with, these modules provide instructors with assignments to use before class, in class, and after class. New modules will support the integrative end-of-unit case study about newts to help instructors engage students with a range of activities for in class and outside of class.
- **Early Alerts** A powerful algorithm looks at student performance and behavior to help identify struggling students as early as possible in the course. Instructors are notified and can easily e-mail students from the Early Alerts dashboard to offer additional help to get students back on track.
- **Vision and Change Filters** All content in Mastering Biology is tagged to the *Vision and Change* core concepts and core competencies, so instructors can identify assessment items in every chapter that are related to *Vision and Change*. Instructors can also sort items by the learning outcomes for this book.

Serving a Community of Teachers

All members of the coauthor team are motivated by a deep commitment to students and to supporting the efforts of dedicated teachers. Our passion in life is doing and teaching biology. At various points along our diverse paths, we have been inspired by our own teachers when we were students, and now we are inspired by our colleagues as we strive to become even better teacher-scholars. In the tradition of all previous editions of *Biological Science*, we have tried to infuse this textbook with the spirit and practice of evidence-based teaching. We welcome your comments, suggestions, and questions.

Content Highlights of the Seventh Edition

Chapter 1 Biology—The Study of Life The new title and reorganized sections reflect a greater emphasis on the theme of five characteristics of life, within a framework of three unifying theories: the cell theory, the theory of evolution, and the chromosome theory of inheritance. Coverage of "life requires energy" is expanded to include more examples of energy acquisition and use and emphasize its importance in the diversification of life.

Chapter 2 Water and Carbon—The Chemical Basis of Life Macromolecules are now introduced in this chapter to provide flexibility in how instructors organize their presentation of content in Unit 1. The description of thermodynamic systems is expanded, including a new figure that illustrates the types of systems based on the exchange of energy and matter with the surrounding environment. To sharpen the focus on foundational concepts that support other chapters in the unit, coverage of the models for chemical evolution is condensed.

Chapter 3 Protein Structure and Function The presentation of how electron sharing gives peptide bonds characteristics similar to double bonds is improved. Discussion of molecular chaperones is expanded, and a new figure illustrates their role in the protein folding process. New assessment questions reinforce student understanding of peptide bonds, how proteins fold into different tertiary structures, and the relationship between protein structure and function.

Chapter 4 Nucleic Acids and the RNA World The description of DNA secondary structure and the accompanying figures are updated to include the concept of base stacking and more accurately represent the geometry of nitrogenous bases relative to the sugar–phosphate backbone. The varied functional roles of RNA are now discussed with reference to the Central Dogma, which is introduced in Chapter 1. A new figure illustrates how tertiary structures are formed from secondary structures in RNA.

Chapter 5 An Introduction to Carbohydrates The introduction of monosaccharides is revised to emphasize how variations in the ring structure of the same monosaccharide will have profound functional consequences. A new Making Models figure is included to help students identify the effect of the ring structure on glycosidic linkages between monosaccharides. Several new questions in the text address how carbohydrate structure is related to function of the polymers.

Chapter 6 Lipids, Membranes, and the First Cells A new figure illustrates how the chemical bonds within fatty acids are responsible for the large amount of potential energy stored in fats. The discussion of selective permeability in membranes is streamlined to emphasize the impact of solute polarity, size, and charge. The section on facilitated diffusion is reorganized, and differences between how channel proteins and carrier proteins participate in this process are clarified.

Chapter 7 Inside the Cell Figures of generalized animal and plant cells are replaced with new and vibrant illustrations to represent the diversity of organelle structures. References to the central dogma, introduced in Chapter 1, are included when describing the functions of the nucleoid, ribosome, and nuclear envelope. A new figure shows how cytosolic proteins are imported into the nucleus after being experimentally modified to include a nuclear localization signal. The research box is replaced with a new one that demonstrates how the secretory pathway was identified using a pulse–chase assay. The description of mitochondrial structure is expanded to include the existence of dynamic networks that undergo fusion and fission between these organelles. Discussion of lysosomes and recycling is clarified and includes a new section on autophagy.

Chapter 8 Energy and Enzymes—An Introduction to Metabolism A new figure illustrates the concept of entropy. Discussion of how temperature and concentration affect chemical reactions is expanded, and the research box includes new data and a description of how the "iodine clock" experiment was performed. The discussion of how enzymes affect chemical reactions is reorganized to emphasize the relationship between catalyzed and uncatalyzed reactions.

Chapter 9 Cellular Respiration and Fermentation The discussion of pyruvate oxidation and the citric acid cycle is revised to improve clarity. Figures illustrating electron transport are updated to clearly show how electrons are transferred to ubiquinone via both complexes I and II, and illustrations in the research box are improved to clearly show the coupling of electron transport and oxidative phosphorylation. A revised figure comparing cellular respiration and fermentation emphasizes how these pathways are related in terms of regenerating electron carriers. Discussion of uncoupling proteins, and how they affect electron transport and oxidative phosphorylation efficiency, is expanded.

Chapter 10 Photosynthesis The revised introduction to the light-capturing reactions and Calvin cycle places greater emphasis on the interdependence of these two pathways. The role of proteins in organizing and tuning pigments in light-harvesting complexes is expanded. Material on carbon fixation and the reduction of sugars is reorganized to emphasize their distinct roles and clarify the relationship between C_3 and C_4 pathways.

Chapter 11 Cell–Cell Interactions The chapter opening image is replaced with a vibrant fluorescent micrograph that represents the chapter's key concepts regarding intercellular connections. The introduction of plant and animal fiber composites is reorganized to streamline discussion of animal extracellular matrix (ECM) and intercellular connections. The figure on animal cell ECM is supplemented with a new illustration on the assembly of collagen proteins into fibrils. Discussion of second messengers is expanded to emphasize their roles in diversification and amplification of intracellular signals.

Chapter 12 The Cell Cycle Centrosome replication and participation in forming the mitotic spindle is now included in the discussion of mitosis. The summary table is revised to include microtubules and microtubule motor proteins as structures involved in mitosis. A new section on the discovery of proteins responsible for M-phase-promoting factor (MPF) activity is added, including a new research box on the expression of cyclins in the cell cycle that illustrates how scientists distinguish between correlation and causation. The introduction to the G1 checkpoint is revised to clarify how Rb and E2F are regulated.

Chapter 13 Meiosis Figures are updated to more accurately show how chiasma are maintained throughout metaphase I of meiosis. Discussion of human aneuploid conditions is expanded, and sections of the chapter are streamlined for a sharper focus on essential content: for example, by reducing the discussion of details of synapsis and crossing over during meiosis I.

Chapter 14 Mendel and the Gene Coverage of epistasis is included, and material on gene linkage is expanded. Revisions to improve clarity and understanding were made to the research box; to discussions of quantitative inheritance and multiple alleles; and to the figures showing the genotypes and phenotypes of the ABO blood group and quantitative inheritance.

Chapter 15 DNA and the Gene—Synthesis and Repair The description of the polarity of DNA strands and how the antiparallel nature of DNA leads to a lagging strand and challenges in DNA replication is now more student friendly. The content and figures are updated throughout: for example, by clarifying the sliding clamp's relationship to DNA polymerase, discussing the role of DNA helicase in nucleotide excision repair, and showing an exonuclease site in DNA polymerases that is distinct from the active site for polymerization.

Chapter 16 How Genes Work New assessment questions are provided to enhance student understanding of the central dogma of molecular biology. The description of how the genetic code was cracked is expanded. Updated content includes a new figure on chromosome structural alterations and their functional consequences.

Chapter 17 Transcription, RNA Processing, and Translation Content is updated, including new findings on the prevalence of coupled transcription and translation in bacteria; modifications to the figure on splicing to better illustrate how the spliceosome assembles; enhanced discussion on the need for accuracy of aminoacyl tRNA synthetases; and new information on translation elongation factors, the role of GTP in elongation, and termination of transcription in bacteria.

Chapter 18 Control of Gene Expression in Bacteria A broader view of gene expression in bacteria is provided, including use of the CAP protein in *lac* operon regulation as an example of positive control, new coverage of the *trp* operon, and increased coverage of global gene regulation. Figures are updated for clarity, and a new figure on the SOS regulon is added.

Chapter 19 Control of Gene Expression in Eukaryotes Sections were reordered to improve understanding and logical flow. Coverage of epigenetic inheritance and RNA interference is expanded. Extensive updates to the figures and accompanying text better illustrate key concepts, such as how promoter-proximal elements and enhancers are composed of multiple regulatory sequences; how base-pair projections in major and minor grooves of DNA helix can be recognized by transcription factors; the number of alternative splicing possibilities for a single gene; and how RISC binds and separates double-stranded RNA. A new summary table highlights the components of transcriptional regulation.

Chapter 20 The Molecular Revolution—Biotechnology, Genomics, and New Frontiers This chapter is extensively revised to reflect the rapid advances in biotechnology. Highlights include extensive coverage of CRISPR-Cas genome editing, including applications in agriculture and gene drives; many new examples of the applications of biotechnology across all areas of the chapter; expanded coverage of Next-Gen sequencing, GMOs, RNA-seq, and gene therapy; a new section on synthetic biology; and a reorganization of the chapter to present techniques first, followed by insights gained from these new methodologies and finally by coverage of emerging areas in biology.

Chapter 21 Genes, Development, and Evolution Chapter sections are reorganized and updated to improve comprehension and flow, and they now move up an organizational ladder from single cell properties to organizing multiple cells to creating the body plan and finally to the link between development and evolution. New figures illustrate the stem cell concept, the creation of induced pluripotent stem cells, and the genetic regulatory cascades in *Drosophila* development.

Chapter 22 Evolution by Natural Selection More structure and practice are provided for students in the section on evidence of evolution. The mockingbird phylogeny is replaced with a finch phylogeny to align with the finch case study. Content is updated, for example, by including reference to CRISPR technology, new data on pesticide and herbicide resistance, and organisms experiencing climate change. The summary table on common misconceptions is enlarged to help students identify and modify their conceptions of evolution.

Chapter 23 Evolutionary Processes The introduction to the Hardy–Weinberg principle is more student friendly, including a sea turtle example comparing alleles, genotypes, and phenotypes; an aligned new figure of the gene pool concept; and an aligned new figure helping students bridge the use of Punnett squares in Mendelian genetics and their use in understanding the calculations of the Hardy–Weinberg principle. A few examples have been either updated or replaced to align with examples used in other chapters (e.g., finches and whitefish). Discussion of sexual selection is revised to be more culturally sensitive to stereotypes in human gender identity and sexual orientation.

Chapter 24 Speciation A new figure offers a friendly introduction to the topic of speciation. Mechanisms of reproductive isolation are reformatted in an illustrated summary table, followed by application questions to give students more practice. The seaside sparrow case study is replaced by an expanded case study on the number of elephant species and includes a research box. The apple maggot fly case study is replaced with a case study on sympatric speciation in killer whales. New whitefish data on fusion is added to align with the new disruptive selection example used in Chapter 23.

Chapter 25 Phylogenies and the History of Life Chapter structure is clarified, explaining how phylogenetics and fossil evidence can be used in combination or separately to reveal insights about the history of life. The phylogenetics section is streamlined and updated. Dates and events in the history of life are updated.

Discussion of the Cretaceous extinction is updated to include evidence of volcanism in the Deccan Traps, including a new figure that supplies geological context for the continental configuration at that time. Discussion of the sixth extinction is expanded, including new data showing current extinction rates for vertebrates.

Chapter 26 Bacteria and Archaea The list of key terms is expanded to help students navigate content. Discussion of microbial communities is enhanced, including a new figure on human microbiomes. New figures illustrating DNA transfer by transformation and transduction are included, along with accompanying text to clarify these processes.

Chapter 27 Diversification of Eukaryotes Although protists are still the primary focus, the title is changed to reflect the evolutionary emphasis of the chapter. To support this evolutionary emphasis, figures are modified to better illustrate morphological diversity and key life-cycle events of protists. A new figure is added to illustrate bioluminescence and its important role in protist defense mechanisms.

Chapter 28 Green Algae and Land Plants Several figures and the accompanying text are modified to emphasize morphological diversity and key life-cycle events of green algae and land plants. A new figure is added that illustrates antheridia and archegonia in seedless vascular plants. The fern life-cycle figure is modified to better show how fern gametophytes facilitate cross-fertilization.

Chapter 29 Fungi The text and some figures are updated and improved to help clarify key concepts related to morphological diversity and key life-cycle events of fungi. The sections on reproductive structures in fungi are streamlined to minimize repetition.

Chapter 30 An Introduction to Animals Revisions include clarifications and updates, such as the removal of *Xenoturbellida* from the deuterostomes; a clearer distinction between germ layers and tissues derived from germ layers; and an addition of the terms "deep homology" and "biradial symmetry." A new case study gives students practice in distinguishing traits that were important to the origin of animal phyla versus traits that were important in the diversification within phyla.

Chapter 31 Protostome Animals More references to ecology are added for context, such as the ecological implications of the water-to-land transition and the increasing threat of extinction to protostomes during the Anthropocene. In addition, a new case study emphasizes sampling methods and measures of diversity in tropical forests.

Chapter 32 Deuterostome Animals *Xenoturbellida* is removed from the deuterostome phylogeny (they are no longer considered deuterostomes). The walk-through of vertebrate evolution includes a new figure showing the homology of placoderm head shields to human skull and jaw, as well as an expanded explanation of the origin of lungs. The human evolution section is updated, including *Homo naledi* and increased evidence of interbreeding among species of *Homo*.

Chapter 33 Viruses New content is included on the decimation of native populations in the Americas during the sixteenth to eighteenth centuries by viral epidemics. Content on different modes of viral replication is expanded. Coverage of the reemergence of Zika virus as a current significant international health problem is included, along with a new case study on how Zika virus causes birth defects.

Chapter 34 Plant Form and Function Several additional key terms are added to expand coverage of fundamental concepts and help students navigate this updated content. Figures are enhanced to help clarify common misconceptions about plant form and function. Discussion of experimental studies on phenotypic plasticity in plants is expanded to give students greater exposure to the process of science.

Chapter 35 Water and Sugar Transport in Plants Figures and accompanying text are modified and improved to help clarify key concepts in water and sugar transport. Previous sections on forces that move water from roots to shoots and features that reduce water loss are streamlined and merged.

Chapter 36 Plant Nutrition The text is updated and streamlined for clarity throughout the chapter to emphasize key concepts in plant nutrition. Students will get more practice at interpreting data with the help of a new research box that addresses the question of whether legumes regulate root nodule development based on nitrogen availability.

Chapter 37 Plant Sensory Systems Figures and accompanying text are enhanced to help clarify key concepts in plant sensory systems. To help engage students in the process of science, discussion of experiments on the relationship between photoperiodism and flowering is expanded. Key terms are added to help students navigate updated content.

Chapter 38 Plant Reproduction and Development Key figures and corresponding text are enhanced to help clarify important concepts in plant reproduction and development. Students will benefit from greater exposure to the process of science and practice with interpreting data thanks to a newly added discussion of experiments on the relationship between flower color and pollinator preference.

Chapter 39 Animal Form and Function The study on reproduction and immune function in crickets has been simplified to better illustrate trade-offs. Content is updated, including a description of the genetic basis of adaptation to high elevation in Tibet and enhanced descriptions of the composition of connective tissues, bone, and cartilage. The treatment of epithelia has been expanded with an updated figure that shows simple and stratified epithelia side by side. The mathematical explanation of surface-area-to-volume ratio is simplified to clarify this concept. A new figure shows homeostatic thermoregulation in dogs.

Chapter 40 Animal Water and Electrolyte Balance A new Making Models box shows tips for modeling osmoregulatory challenges and solutions faced by animals. A new figure compares osmoregulation in marine bony fishes and osmoconformation in marine cartilaginous fishes. The sections on marine and freshwater fishes are combined to better compare osmoregulation in the two groups. In addition to filtration and reabsorption in the kidney, secretion of substances into the nephron is now briefly discussed.

Chapter 41 Animal Nutrition Further detail is added on the actions of digestive enzymes in humans, including clarification of the function of lingual lipase; introduction of gastric lipase; clarification of carbohydrate digestion by salivary amylase and pancreatic amylase; introduction of brush border enzymes for digestion of disaccharides and dipeptides; and standardization of the name "enteropeptidase." Information on how fat-soluble and water-soluble vitamins are absorbed is included. A new section covers the human gut microbiome.

Chapter 42 Animal Gas Exchange and Circulation The section on insect tracheae is revised for clarity. A brief discussion of oxygen-carrying pigments other than hemoglobin is added. New material and a new research box are added discussing the special anatomy of the alligator heart. Discussion of the mechanism by which blood pressure drops as blood moves away from the heart is revised to clarify key concepts.

Chapter 43 Animal Nervous Systems Discussion of how the resting membrane potential is established is clarified. Coverage of the peripheral nervous system is expanded, specifically on the autonomic (sympathetic and parasympathetic) nervous system. Content on the vertebrate brain is augmented by couching the main components (cerebrum, cerebellum, diencephalon, and brainstem) in terms of the three region (forebrain, midbrain, hindbrain) schema from Chapter 32. Discussion on learning and memory is revised to highlight how little scientists actually know about these processes.

Chapter 44 Animal Sensory Systems A brief discussion of touch is now included in the mechanosensation section. A detailed discussion and new figure on the inner ear's sensation of equilibrium are added. The section about the "dark current" associated with stimulation of rods in the eye is revised to describe the process at an appropriate level.

Chapter 45 Animal Movement Language is added throughout to make the chapter more interactive (e.g., put your hand on your biceps and flex it). The figure on muscle cross-bridge cycling is updated to show it as a circular process. The description of cardiac and smooth muscle is expanded, including discussions of autorhythmicity and the reason smooth muscle appears smooth.

Chapter 46 Animal Chemical Communication Coverage of positive feedback is added to complement the discussion of negative feedback. The section on Berthold's discovery of testosterone is updated to affirm that he discovered that chemical signal(s) affect rooster development, although the terms "hormone" and "testosterone" were not coined until many years later. Discussion of hormone receptors is updated and streamlined by removing detail on how the estradiol receptor was discovered and focusing on key content.

Chapter 47 Animal Reproduction and Development The discussion of asexual reproduction in animals is updated to show that it does not always result in clones. For clarity, the process of cell migration during gastrulation is now described as "movement" rather than "migration." The concept of the evolution of viviparity in reptiles and mammals is expanded.

Chapter 48 The Immune System in Animals Content is expanded to help students visualize how innate and adaptive immunity work together, including a new Making Models box, "Tips on Drawing Immune Cells." Updates to coverage of antigen presentation and the activation of B and T cells take a more comparative approach.

Chapter 49 An Introduction to Ecology The introductory section on levels of ecology increases the focus on variables that ecologists measure. Section 49.2 is restructured as a case study on açaí palms, including a new niche model comparing açaí palms and coconut palms; a new summary table comparing variables in the present abiotic, present biotic, past abiotic, and past biotic environments; and a new research box focusing on the effect of climate change on the distribution of açaí palms. The chapter now includes altitude as an abiotic factor, the effect of organisms on climate, and a definition of landscape ecology. Other changes improve clarity and timeliness of information.

Chapter 50 Behavioral Ecology Revisions increase the focus on human behavior, including a new case study on consolation behavior in humans, elephants, and prairie voles. Behaviors are categorized in a summary table representing a continuum between innate and learned behaviors. A new figure shows how behaviors are learned via social networks in bees. The section on optimal foraging includes a new research box that looks at cuttlefish counting behavior and an introduction to game theory. The section on mate choice includes a new summary table on mating strategies, including language that helps students to interpret the complexity of human mate choice and sexual identity in this simplistic framework. There is increased emphasis on the importance of variability in behavior in the context of climate change.

Chapter 51 Population Ecology The chapter includes a new summary table of dispersion patterns with an emphasis on proximate and ultimate causation. Increased emphasis on life-history patterns includes a new figure comparing the life-history patterns of two lizard populations. The section on population dynamics includes a new case study on the crash of the reindeer population on St. Paul island to clarify the common misconception that the human population will gently peak at its carrying capacity, and the hare/lynx research box provides a more accessible data set. The section on human population is updated. Application to conservation biology is now more integrated throughout the chapter.

Chapter 52 Community Ecology Content now includes a new summary table comparing consumption interactions, a new section referencing the hare/lynx data in Chapter 51, a new figure illustrating how mutualists can be generalists or specialists, a new figure of interaction networks to show examples other than food webs, a Making Models box on species richness (brought forward from Chapter 54), a resurrection of Paine's keystone species data (moved from Chapter 32), new data on latitudinal diversity, and increased emphasis throughout on human impacts, including climate change.

Chapter 53 Ecosystems and Global Ecology Updated content includes a new opening image of global CO_2 data, a new figure showing solar-induced fluorescence measured via satellite, a new figure helping students to analyze the productivity pyramid while also understanding the impacts of their food choices, increased emphasis that the greenhouse effect and ozone layer are different phenomena, new reference to the Paris climate agreement, and new reference to ocean deoxygenation.

Chapter 54 Biodiversity and Conservation Content is updated throughout, including a new summary table comparing species, phylogenetic and functional diversity; a revised map showing where global vertebrate biodiversity is highest; an updated map of biodiversity hotspots; an updated data set of the distribution of threats to terrestrial, freshwater, and saltwater ecosystems; the inclusion of disease in the topic of invasive species; and a new map illustrating the principle of "Nature Needs Half" as a conservation goal for a sustainable future.

Acknowledgments

Reviewers

The peer review system is the key to quality and clarity in science publishing. Besides providing a filter, the investment that respected individuals make in vetting the material—catching errors or inconsistencies and making suggestions to improve the presentation—gives authors, editors, and readers confidence that the text meets rigorous professional standards.

Peer review plays the same role in textbook publishing. The time and care invested by this book's reviewers are tributes to their professional integrity, their scholarship, and their concern for the quality of teaching. Virtually every page in this edition has been revised and improved based on insights from the following individuals.

S. Tariq Ahmad, *Colby College*
Dan Ardia, *Franklin & Marshall College*
Kim Atwood, *Cumberland University*
Claudia Barreto, *University of New Mexico*
Geoff Benn, *University of California, Davis*
Sarah Bissonnette, *California State University, Stanislaus*
Mary Boggs, *University of Delaware*
Martha Caron, *Arizona State University—Polytechnic*
Paula Checchi, *Marist College*
Ann Cleveland, *Maine Maritime Academy*
Alistair Cullum, *Creighton University*
Karen Curto, *University of Pittsburgh*
Candice Damiani, *University of Pittsburgh*
Christine Davis, *University of Florida*
Abby Drake, *Cornell University*
Miles Engell, *North Carolina State University*
Ben England, *The University of Tennessee, Knoxville*
Lisa Elfring, *The University of Arizona*
Wayne Fagerberg, *University of New Hampshire*
Larry Feinstein, *University of Maine at Presque Isle*
Kenneth Filchak, *University of Notre Dame*
Mark Flood, *Fairmont State University*
Jason Flores, *North Carolina State University*
Vanessa Fox, *DePauw University*
Juliet Fuhrman, *Tufts University*
Jason Fuller, *Bellevue College*
Michael Fultz, *Morehead State University*
Caitlin Gabor, *Texas State University*
Grant Gardner, *Middle Tennessee State University*
Kathryn Gardner, *University of Pittsburgh*
Elise Gervais, *University at Albany, State University of New York*
Mita Ghosh, *Florida State College at Jacksonville*
Brian Gibbens, *University of Minnesota*
Melissa Goldsmith, *Chapman University*
Steven Gorsich, *Central Michigan University*
Linda Green, *Georgia Institute of Technology*
Suzanna Gribble, *University of Pittsburgh*
Scott Grubbs, *Western Kentucky University*
Nancy Guild, *University of Colorado*
David Hanson, *University of New Mexico*
Deborah Harris, *Case Western Reserve University*
Chris Haynes, *Shelton State Community College*
Bethany Henderson-Dean, *University of Findlay*
Mar-Elise Hill, *Northern Arizona University*
Arlene Hoogewerf, *Calvin College*
Justin Hoshaw, *Waubonsee Community College*
Dan Husband, *Florida State College at Jacksonville—Kent*
Ilko Iliev, *Southern University at Shreveport*
Diana Ivankovic, *Anderson University*
Susan Jorstad, *The University of Arizona*
Martin Kapper, *Central Connecticut State University*
Amy Keagy, *University of North Florida*
Tess Killpack, *Wellesley College*
Katarzyna Konior, *Concordia University Chicago*
Carey Krajewski, *Southern Illinois University*
Pat Krug, *California State University, Los Angeles*
Katy Lack, *Wake Forest University*
Nathan Lanning, *California State University, Los Angeles*
Kari Lavalli, *Boston University*
Mark Lazzaro, *College of Charleston*
Abby Levitt, *North Central State College*
Debra Linton, *Central Michigan University*
Janet Loxterman, *Idaho State University*
Jose Maldonado, *El Paso Community College*
Linda Mayerhofer, *University at Albany, State University of New York*
Thaddeus McRae, *Broward College*
Mitch McVey, *Tufts University*
Brad Mehrtens, *University of Illinois*
Jamie Moon, *University of North Florida*
Anna Mosser, *University of Minnesota*
Deborah Muldavin, *Central New Mexico Community College*
Jennifer Nauen, *University of Delaware*
Sarah O'Malley, *Maine Marine Academy*
Joanne Odden, *Pacific University of Oregon*
Jennifer Osterhage, *University of Kentucky*
Lisa Parks, *North Carolina State University*
Shelley Penrod, *Lone Star College*
David Pindel, *Corning Community College*
Marianne Poxleitner, *Gonzaga University*
David Puthoff, *Frostburg State University*
Narayanen Rajendran, *Kentucky State University*
Sami Raut, *University of Alabama at Birmingham*
Mamta Rawat, *California State University, Fresno*
Melissa Reedy, *University of Illinois*
Karen Resendes, *Westminster College*
Ana Ribeiro, *College of Mount Saint Vincent*
Nancy Rice, *Western Kentucky University*
Linda Richardson, *Blinn College*
Laurel Roberts, *University of Pittsburgh*
Melissa Rowland-Goldsmith, *Chapman University*
Thomas Sasek, *University of Louisiana at Monroe*

Judy Schoonmaker, *Colorado School of Mines*
Ryan Sessions, *Florida State College at Jacksonville*
Christine Simmons, *Southern Illinois University Edwardsville*
Jennifer Smith, *Triton College*
Michael Smith, *Western Kentucky University*
Thomas Sobat, *Ivy Tech Community College–Columbus*
Phil Sokolove, *University of Maryland, Baltimore County*
Nancy Solomon, *Miami University*
Karen Stancil, *College of Central Florida*
Cassy Summerlin, *Holy Cross Catholic Academy*
Zuzana Swigonova, *University of Pittsburgh*
Munir Syed, *Hartwick College*
Steve Takata, *University of California, Berkeley*
Noel Takeuchi, *University of South Florida*
Heidi Tarus, *Colby Community College*
Mackenzie Taylor, *Creighton University*
Monica Tischler, *Benedictine University*
Jonathan Titus, *State University of New York Fredonia*
Sara Tolsma, *Northwestern College*
Christopher Vitek, *The University of Texas Rio Grande Valley*
Rebecca Waggett, *University of Tampa*
Alexander Wait, *Missouri State University*
Kira Wennstrom, *Shoreline Community College*
Peter White, *Michigan State University*
Antonia Wijte, *Irvine Valley College*
John Willford, *University of Wyoming*
Denise Williams, *Caldwell Community College and Technical Institute*
Janet Wolkenstein, *Hudson Valley Community College*
Phil Yeager, *Fairmont State University*
Marilyn Yoder, *University of Missouri Kansas City*
Jenn Yost, *California Polytechnic State University—San Luis Obispo*
Brenda Zink, *Northeastern Junior College*

Student Reviewers:
James Hilliard, *Southern Illinois University*
Paige McCaleb, *Southern Illinois University*

Correspondents

We enjoy hearing from instructors, researchers, and students who generously take the time to share feedback with us or to pose questions about material in the book. We appreciate conversations we have had about content coverage with Dylan Schwilk of *Texas Tech University*, Amanda Simcox of *The Ohio State University*, and Jenny McFarland and Gwen Shlichta of *Edmonds Community College*. Maria Bontrager, a student at *Texas State University*, shared helpful insights as well. We thank James Hilliard and Paige McCaleb, two students working with Carey Krajewski at *Southern Illinois University*, who provided thoughtful feedback on chapter content and assessment items.

We also want to offer special thanks to a few individuals who helped our work on the integrative end-of-unit case studies about the rough-skinned newt. To Edmund "Butch" Brodie, Jr., of *Utah State University*, for sharing stories about the mystery of the newt and for a lifetime of research that helped unveil many of the newt's secrets. To Chris Feldman of *University of Nevada, Reno*, and his graduate student Robert del Carlo for assisting us in brainstorming and editing newt–garter snake content. Dr. Feldman's expertise and eager engagement were especially valuable for creating this new multi-unit feature.

Contributors

We are grateful for the hard work and creativity of the contributors who worked on an impressive array of print and online support materials.

Ana Araya-Anchetta, *Northern Arizona University*
Andrea Aspbury, *Texas State University, San Marcos*
Brian Bagatto, *University of Akron*
Jay Brewster, *Pepperdine University*
Warren Burggren, *University of North Texas*
Patrick Cafferty, *Emory University*
Tim Christensen, *East Carolina University*
David Coughlin, *Widener University*
Karen Curto, *University of Pittsburgh*
Candice Damiani, *University of Pittsburgh*
Clarissa Dirks, *The Evergreen State College*
Caitlin Gabor, *Texas State University, San Marcos*
Kathy Gillen, *Kenyon College*
Nancy Ann Guild, *University of Colorado, Boulder*
Jutta Heller, *University of Washington–Tacoma*
Laurel Hester, *Keuka College*
Mar-Elise Hill, *Northern Arizona University*
Mark Holbrook, *University of Iowa*
Jacob L. Kerby, *University of South Dakota*
Craig Lending, *SUNY Brockport*
Cindy Malone, *California State University, Northridge*
Jim Manser, *Harvey Mudd College*
Brad Mehrtens, *University of Illinois at Urbana–Champaign*
Oyenike Olabisi, *University of Delaware*
Jennifer Osterhage, *University of Kentucky*
Melissa Murray Reedy, *University of Illinois at Urbana–Champaign*
Anthony Rossi, *University of North Florida*
Christina T. Russin, *Northwestern University*
Joan Sharp, *Simon Fraser University*
Cara Shillington, *Eastern Michigan University*
Nancy Solomon, *Miami University*
Anna Soper, *University of Massachusetts Amherst*
Chrissy Simmons, *Southern Illiniois University—Edwardsville*
David Sovic, *The Ohio State University*
Catherine Ueckert, *Northern Arizona University*

We also offer a special thanks to Stephen Thomas of *Michigan State University* for bringing the Making Models videos to life.

Book Team

Anyone who has been involved in a major production knows that many people work behind the scenes to make it all happen. The coauthor team is indebted to the many talented individuals who have made this book possible.

The final version of the text was copyedited with mastery by Chris Thillen and expertly proofread by Pete Shanks. Art was rendered by Imagineering Media Services, while Kristen Piljay researched hundreds of new photographs for the Seventh Edition. The book's clean, innovative design was developed by Tani Hasegawa. Text and art were skillfully set in the design by Integra.

Creating Mastering™ Biology tutorials and activities requires a large team. Media content development was overseen by Sarah Jensen. Nicole Constantine, Mireille Pfeffer, and Summer Giles served as media producers, developing and coordinating new and revised media content. We also benefited from the media guidance of Laura Tommasi, Sarah Young-Dualan, Tod Regan, and Kaitlin Smith. Sara Finnigan and Tacha Gennarino were instrumental in working with us on a revised hierarchy of learning outcomes.

Pearson's talented sales reps, who listen to professors, advise the editorial staff, and get the book into students' hands, are supported by the boundless energy of the marketing team: Kelly Galli and Alysun Burns. Marketing materials were produced by Jane Campbell.

The vision and resources required to run this entire enterprise are the responsibility of Courseware Portfolio Management Director Beth Wilbur, who provided wise, inspirational, and focused leadership, and Adam Jaworski, vice president of Pearson Science Courseware Porfolio Management, who displays unwavering commitment to high-quality science publishing.

The editorial team was skillfully directed by Courseware Portfolio Content Development Director Ginnie Simione Jutson. Content Producer Laura Perry and Senior Project Manager Kim Fletcher at Integra efficiently kept the mammoth project steadily rolling forward during development and production. Finally, we are deeply grateful for the key drivers of the Seventh Edition. Courseware Senior Analyst Sonia DiVittorio's remarkable vision and creativity; keen attention to detail, level, and clarity; and absolute insistence on excellence set the standard for us all. Courseware Senior Analyst Suzanne Olivier brought her deep insights and vast experience to the task of improving many chapters, as did Senior Development Editor Mary Hill, who also worked magic to come up with visually aesthetic page layouts. And Courseware Portfolio Management Specialist Michael Gillespie's unstoppable enthusiasm, invaluable skills at team building, upbeat attitude, and sharp intellect have energized and united the team while guiding the book through the hurdles to existence. The coauthor team gives many thanks to all these exceptional people for making the art and science of book writing a productive and exhilarating process.

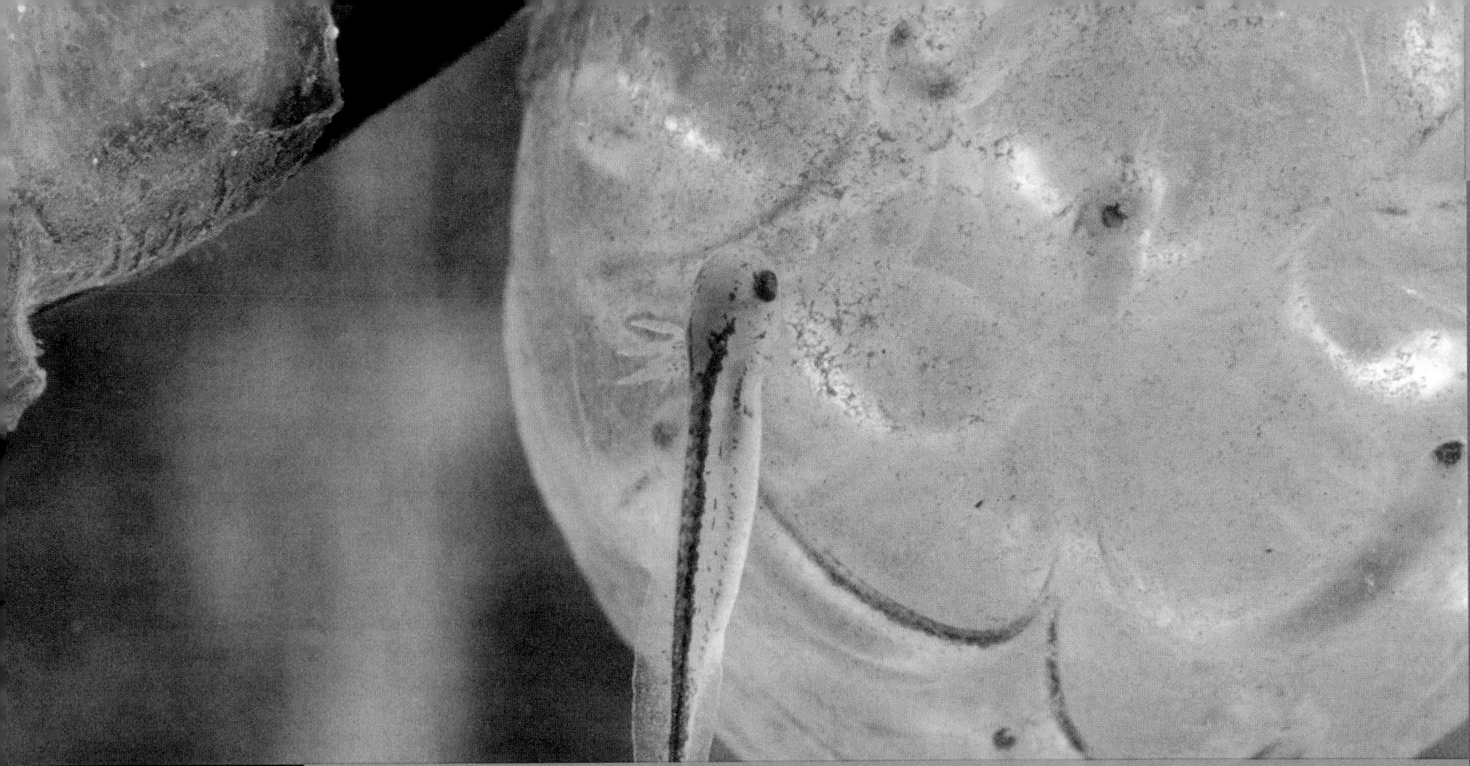

1 Biology: The Study of Life

These newt babies are hatching, exploring their new world, and learning how to find food and stay alive. They represent one of the key characteristics of life introduced in this chapter—replication.

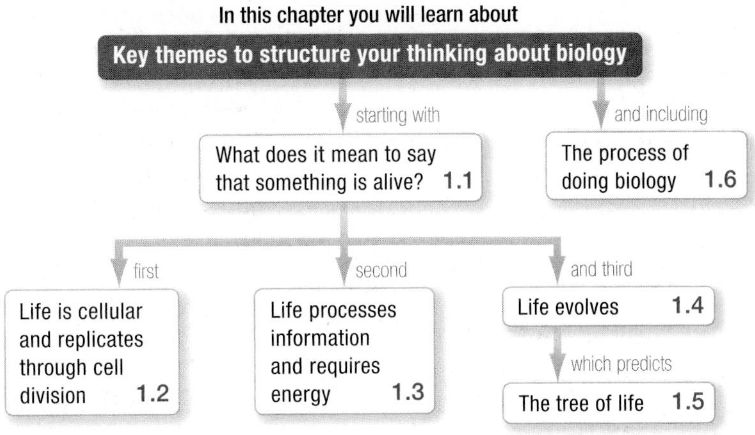

In essence, biological science is the study of life. It searches for ideas and observations that unify our understanding of the diversity of life—from bacteria living in hot springs to humans and majestic sequoia trees.

The goals of this chapter are to introduce the nature of life and explore how biologists go about studying it. In this chapter, you will also encounter themes that will resonate throughout your study of biology:

- Analyzing how organisms work at the molecular level.
- Understanding organisms in terms of their evolutionary history.
- Helping you learn to think like a biologist.

Let's begin with what may be the most fundamental question of all: What is life?

BIG PICTURE

This chapter is part of the Big Picture. See how on pages 18–19.

1.1 What Does It Mean to Say that Something Is Alive?

An **organism** is a life-form—a living entity made up of one or more cells. Although there is no simple definition of life that is endorsed by all biologists, most agree that organisms share a suite of five fundamental characteristics. You can think of this text as one long exploration of these five traits.

> **After you complete this section, you should be able to . . .**
> - Explain what it means to say that an organism is "alive."

1. ***Cells*** Organisms consist of membrane-bound units called **cells**. The membrane of a cell regulates the passage of materials between exterior and interior spaces.
2. ***Replication*** One of the great biologists of the twentieth century, François Jacob, said that the "dream of a bacterium is to become two bacteria." Almost everything an organism does contributes to one goal: replicating itself.
3. ***Information*** Organisms process hereditary, or genetic, information encoded in units called **genes**. Organisms also respond to information from the environment and adjust to maintain stable internal conditions. Right now, cells throughout your body are using information stored in your genes to make the substances, or **molecules**, that keep you alive; your eyes and brain are decoding information on this page that will help you learn some biology, and if your room is too hot you might be sweating to cool off.
4. ***Energy*** To stay alive and reproduce, organisms have to acquire and use energy. To give just two examples: plants absorb sunlight; animals ingest food.
5. ***Evolution*** Organisms are the products of evolution, and their populations continue to evolve today.

Three of the greatest unifying ideas in all of science, which depend on the five characteristics just listed, laid the groundwork for modern biology: **(1)** the cell theory, **(2)** the chromosome theory of inheritance, and **(3)** the theory of evolution. Formally, scientists define a **theory** as an explanation for a very broad class of observed phenomena that is supported by a wide body of evidence. Note that this definition contrasts sharply with the everyday usage of the word "theory," which often carries meanings such as "speculation" or "guess."

The cell theory, the chromosome theory of inheritance, and the theory of evolution address fundamental questions: What are organisms made of, and where do they come from? How is hereditary information transmitted from one generation to the next? How are organisms related to one another?

When these theories emerged in the mid-1800s, they revolutionized the way biologists think about the world. None of these insights came easily, however. The cell theory, for example, emerged after some 200 years of work. Let's examine some of the pivotal discoveries made along the way, starting with the first observation of cells by microscopy.

> **CHECK YOUR UNDERSTANDING**
>
> ✔ **If you understood this section, you should be able to . . .**
> 1. Determine whether the virus that causes the flu is alive, based on the following information: The flu virus is not made of cells and depends on your cells for replication and information processing.
> 2. **MODEL** Diagram the relationship between the three unifying theories of biology and the five characteristics of life.
>
> Answers are available in Appendix A.

1.2 Life Is Cellular and Replicates through Cell Division

In 1665 the Englishman Robert Hooke devised a crude microscope to examine the structure of cork (a bark tissue) from an oak tree. The instrument magnified objects to just 30× (30 times) their normal size, but it allowed Hooke to see something extraordinary. In the cork he observed small, empty-looking compartments that were invisible to the naked eye (**Figure 1.1a**). Hooke coined the term "cells" for these structures because he thought they resembled the cells inhabited by monks in a monastery.

> **After you complete this section, you should be able to . . .**
> - Explain what organisms are made of and how cells come to be.

Soon after Hooke published his results, the Dutch scientist Anton van Leeuwenhoek developed much more powerful microscopes, some capable of magnifications up to 300× (**Figure 1.1b**). With these instruments, van Leeuwenhoek inspected samples of pond water and made the first observations of a dazzling collection of single-celled organisms that he called "animalcules."

In the 1670s an Italian researcher who was studying the leaves and stems of plants with a microscope concluded that plant tissues were composed of many individual cells. By the early 1800s, enough data had accumulated for a German biologist to claim that *all* organisms consist of cells. Did this claim hold up?

All Organisms Are Made of Cells

Advances in microscopy have made it possible to examine the amazing diversity and complexity of cells at higher and higher magnifications. Microscopes tens of thousands of times more powerful than van Leeuwenhoek's have revealed that cells are highly organized compartments separated from their environment by a membrane barrier. With these instruments, biologists have described over a million new species. The basic conclusion made in the 1800s remains intact: All organisms are made of cells.

The smallest organisms known today are bacteria that are barely 200 nanometers wide, or 200 *billionths* of a meter. It would take 5000 of these organisms lined up side by side to span a millimeter. This is the distance between the smallest hash marks on

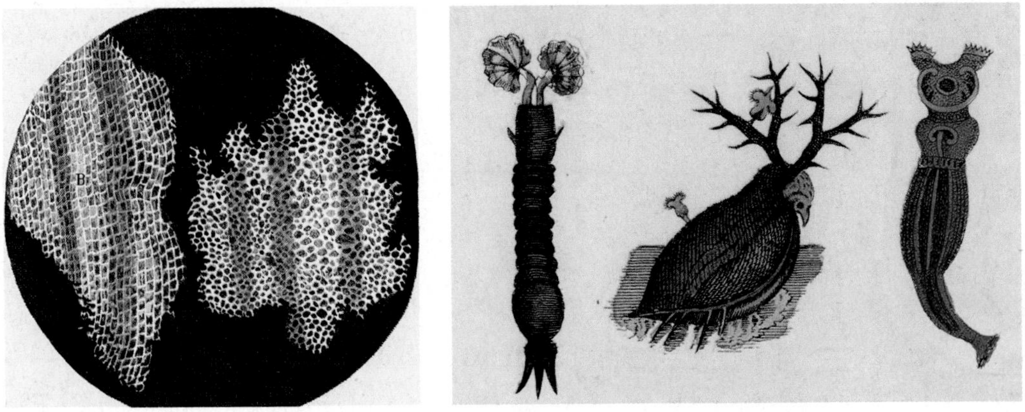

Figure 1.1 What Hooke and van Leeuwenhoek Saw through Their Microscopes.

a metric ruler. In contrast, sequoia trees can be over 100 meters tall, the equivalent of a 20-story office building. Bacteria and sequoias are composed of the same fundamental building block, however—the cell. Bacteria consist of a single cell; sequoias are made up of trillions of cells. (See **BioSkills 1** to review the metric system.[1])

The realization that all organisms consist of cells was fundamentally important, but it formed only the first part of the cell theory. Besides understanding what organisms are made of, scientists wanted to understand how cells come to be.

Where Do Cells Come From?

In 1858, a German scientist named Rudolph Virchow proposed that all cells arise from cells already in existence. The complete **cell theory** builds on this concept: All organisms are made of cells, and all cells come from preexisting cells.

Two Hypotheses The cell theory directly challenged the idea of spontaneous generation—the prevailing explanation of where cells come from. In the mid-1800s, most biologists believed that organisms could arise spontaneously under certain conditions. The bacteria and fungi that spoil foods such as milk and wine were thought to appear in these nutrient-rich media of their own accord—springing to life from nonliving materials. In contrast, the cell theory maintained that cells do not arise spontaneously but are produced only when preexisting cells grow and divide. The all-cells-from-cells explanation was a **hypothesis**: a testable statement to explain a set of observations. Spontaneous generation was an alternative hypothesis to explain the same phenomenon.

Biologists typically use the word "theory" to refer to proposed explanations for broad patterns in nature and "hypothesis" when referring to explanations for more tightly focused questions. A theory serves as a framework for developing new hypotheses.

An Experiment to Settle the Question In 1862, soon after Virchow's all-cells-from-cells hypothesis appeared in print, a French scientist named Louis Pasteur set out to test its predictions in an **experiment**. Experiments are a powerful scientific tool because they allow researchers to test the effect of a single, well-defined factor on a particular phenomenon. An experimental **prediction** describes a measurable or observable result that must be correct if a hypothesis is valid.

Pasteur wanted to determine whether organisms could arise spontaneously in a nutrient broth or whether they appear only when a broth is exposed to a source of preexisting cells. To address the question, he created two treatment groups that were identical in every respect but one: the factor being tested—in this case, a broth's exposure to preexisting cells.

Both treatments used glass flasks filled with the same amount of the same nutrient broth (**Figure 1.2** on page 4). Both flasks were boiled for the same amount of time to kill any existing organisms. After sterilization by boiling, however, any bacteria and fungi clinging to dust particles in the air could drop into the broth in the flask shown in Figure 1.2a because the neck of this flask was straight.

In contrast, in the flask with a long swan neck (Figure 1.2b), water would condense in the crook of the swan neck after boiling and this pool of water would trap any bacteria or fungi that entered on dust particles. Thus, the contents of the swan-necked flask were isolated from any source of preexisting cells even though they were still open to the air. The spontaneous generation hypothesis predicted that cells would appear in both treatment groups. The all-cells-from-cells hypothesis predicted that cells would appear only in the treatment exposed to a source of preexisting cells.

And Pasteur's results? The broth in the straight-necked flask exposed to preexisting cells quickly filled with bacteria and

[1]BioSkills are located after Chapter 1. They focus on general skills that you'll use throughout this course. More than a few students have found them to be a lifesaver. Please use them!

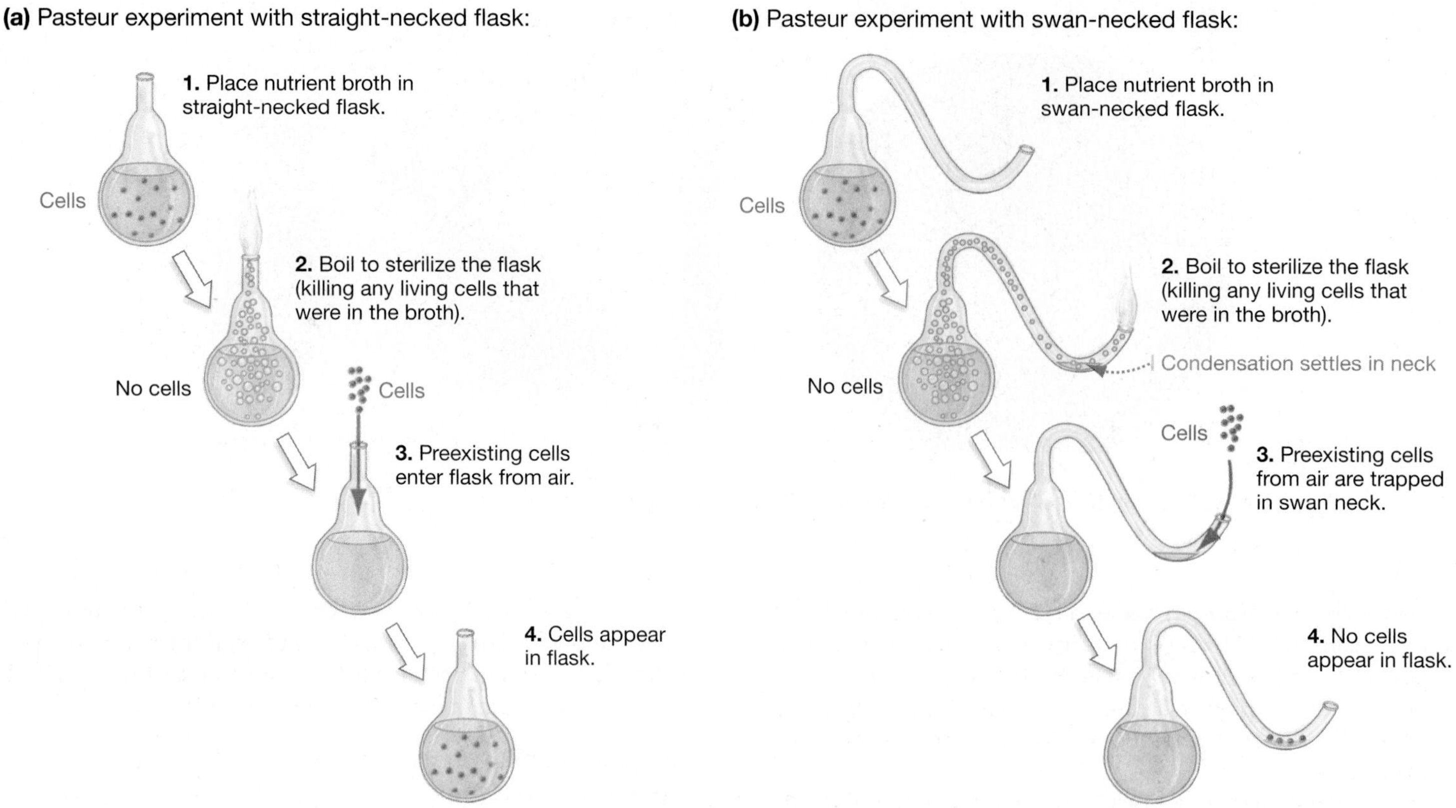

Figure 1.2 The Spontaneous Generation and All-Cells-from-Cells Hypotheses Were Tested Experimentally.

✔ **PROCESS OF SCIENCE** What problem would arise in interpreting these results if Pasteur had (1) put different types of broth in the two treatments, or (2) used a ceramic flask for one treatment and a glass flask for the other?

fungi. This observation was important because it showed that the sterilization step had not altered the nutrient broth's capacity to support growth. The broth in the swan-necked flask remained sterile, however. Even when the flask was left standing for months, no organisms appeared in it. This result was inconsistent with the hypothesis of spontaneous generation.

Because Pasteur's data were so conclusive—meaning that there was no other reasonable explanation for them—the results persuaded most biologists that the all-cells-from-cells hypothesis was correct. But if all cells come from existing cells, where did the first cells come from? Biologists now have evidence that life arose from nonlife early in Earth's history, through a process called **chemical evolution** (Ch. 2, Section 2.4).

Life Replicates through Cell Division

For life on Earth to continue to exist, cells must replicate. Most cells are capable of reproducing by dividing—in effect, by making a copy of themselves. As predicted by the cell theory, all the cells present in your body and in most other multicellular individuals are descended from preexisting cells, tracing back to a fertilized egg. A fertilized egg is a cell created by the fusion of sperm and egg—cells that formed in individuals of the previous generation.

New cells arise when preexisting cells split. In multicellular organisms they become specialized for particular functions by intricate processes. In this way the cells in a multicellular organism may differ, but they are all connected by a common lineage.

But how is the information for making diverse organisms, such as bacteria or sequoias, stored and transmitted from one generation to the next? The second unifying idea—the chromosome theory of inheritance proposed by Walter Sutton and Theodor Boveri in 1902—provided the foundation for biologists to answer this question.

CHECK YOUR UNDERSTANDING

✔ **If you understood this section, you should be able to …**

1. Discuss whether the cell theory states that all cells in a multicellular organism are identical.
2. **MODEL** Diagram how life replicates (use a circle to represent a cell).

Answers are available in Appendix A.

1.3 Life Processes Information and Requires Energy

After Sutton and Boveri proposed the **chromosome theory of inheritance**, the pieces of the puzzle of life began to fall into place. The key point? Inside cells, hereditary or genetic information is encoded in units called genes that are located on chromosomes.

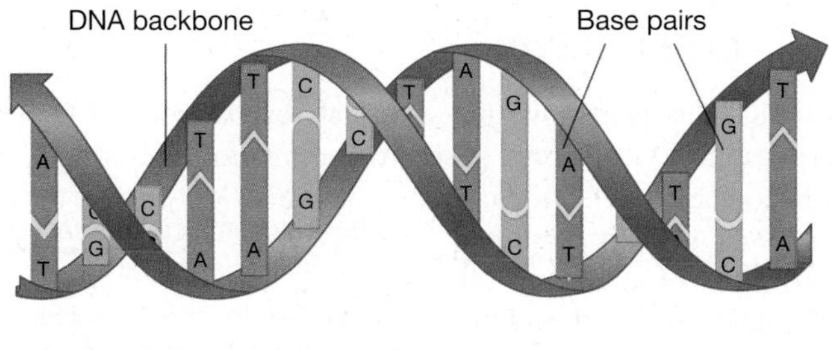

Figure 1.3 **DNA Is a Double Helix.**

But it wasn't until experiments were carried out in the 1950s that biologists figured out the molecular nature of the genetic material: A **chromosome** consists of a molecule of **deoxyribonucleic acid**, or **DNA**. To sum up, DNA is the hereditary material. Genes consist of specific segments of DNA that code for products in the cell.

After you complete this section, you should be able to ...

- Explain how organisms process hereditary information and acquire and use energy.

The Central Dogma

In what is considered one of the greatest scientific breakthroughs of biology, in 1953 James Watson and Francis Crick proposed that DNA is a double-stranded helix (**Figure 1.3**). Crucial insights that led to this model came from structural analyses performed by Rosalind Franklin in Maurice Wilkins' laboratory.

Each strand of the **double helix** is made up of varying sequences of four molecular building blocks, each containing a different kind of base (Ch. 4, Section 4.2). In terms of structure, on each strand of the double helix the building blocks of DNA are connected one to another linearly. In terms of function, they are like letters of the alphabet—the four different kinds of bases in the building blocks are symbolized by the letters A, T, C, and G. A sequence of this letter code is like the sequence of letters in a word—it has meaning. In this way DNA carries, or encodes, the information required for an organism's growth and reproduction.

The two strands of the double helix are joined by interactions between pairs of bases. Base pairing occurs only between certain letters: A always pairs with T, and C always pairs with G (see Figure 1.3). The pairs are arranged much like the rungs on a ladder, with the backbones of strands acting as the sides of the ladder. Base pairing is key: it permits DNA to be copied and faithfully preserves the information encoded within the DNA.

How is the encoded information decoded and put to work? The **central dogma**—first articulated by Crick—describes the flow of information in cells. In this context, the term "dogma" means a framework for understanding. Put simply, DNA codes for RNA, which codes for proteins (**Figure 1.4**).

Molecular machinery in cells makes a copy of a particular gene's information in the form of a closely related molecule called **ribonucleic acid**, or **RNA**. RNA molecules carry out a number of specialized functions in cells. For example, molecular machinery reads a messenger RNA molecule to determine what molecular building blocks to use to make a **protein**. Proteins are crucial to most tasks required for a cell to exist, from forming structural components to promoting the chemical reactions that sustain life.

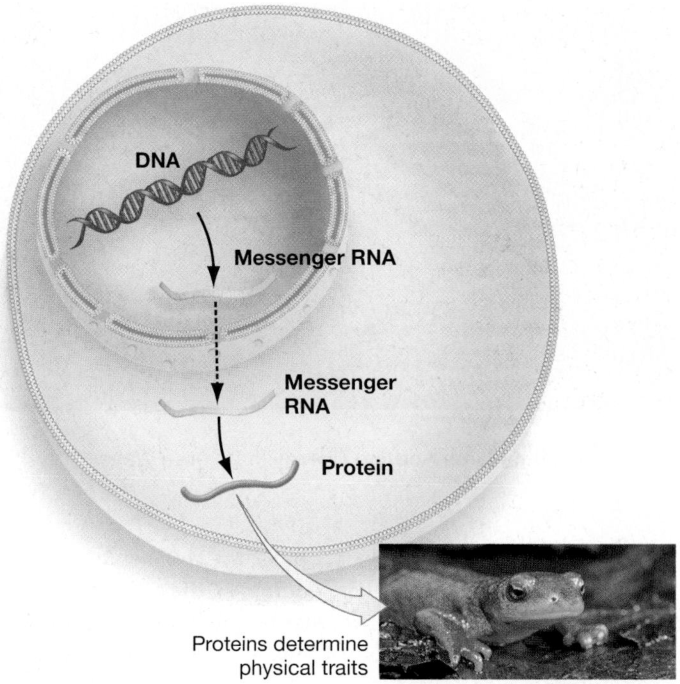

INTERACTIVE Figure 1.4 **The Central Dogma Describes the Flow of Genetic Information.** Genetic information flows from DNA to RNA to proteins. Differences in proteins encoded by different DNA sequences may lead to different physical traits.

Understanding the structure of DNA also provided insight into how genetic information is passed on when cells replicate, or passed from one organism to its offspring. Making a copy of DNA in a cell is a highly accurate process, but mistakes can occur. What happens when a mistake is made? Differences in DNA sequences may lead to differences in the sequence of the building blocks in proteins.

The implications are profound: The outward appearance of an organism is a product of the proteins produced by its molecular machinery, so differences in DNA sequences might lead to a difference, for example, in finch beak size and shape, or in the length of a giraffe's neck. Changes in sequence lead to the heritable variations that underlie the diversity of life.

Life Requires Energy

The chemical reactions that sustain life in all its diversity take place inside cells. Transmitting genetic information, and the other work carried out by cells, requires energy. Organisms—whether single-celled or multicellular—are capable of living in a wide array of environments because they vary in cell structure and in how they acquire and use energy.

Organisms have two fundamental nutritional needs—chemical energy in the form of a molecule called **ATP** (or **adenosine triphosphate**) and molecules that can be used as building blocks

Figure 1.5 **Organisms Acquire Energy in Diverse Ways.**

for the synthesis of DNA, RNA, proteins, the cell membrane, and other large, complex compounds required by the cell. How organisms obtain these materials—whether by using energy in sunlight or from ingested chemical compounds—is central to the tremendous diversification of life.

Plants and some bacteria can produce sugar using energy from sunlight, and either use the sugar to make ATP or store it in other energy-rich molecules. When plants are eaten or decompose, their stored energy-rich molecules are obtained by animals, fungi, archaea, and other bacteria. The overall masters of meeting nutritional needs, however, are bacteria and archaea. Taken together, they can subsist on almost anything from ammonia to crude oil.

While bacteria and archaea can use a wide array of molecules absorbed from their environment as food, one of the most important stories in the diversification of life is in the novel methods different organisms have for securing food. For example, you use your teeth to chew food into pieces small enough to be swallowed, and butterflies drink nectar from flowers through straw-like mouthparts (Figure 1.5).

Recall that all the cells in a multicellular organism are connected by a common lineage. Is the tremendous diversity among organisms also related to common ancestry? The third great founding idea in biology, published the same year as the all-cells-from-cells hypothesis, provided an answer. This was the realization, arrived at independently by the English scientists Charles Darwin and Alfred Russel Wallace, that all diverse **species**—all distinct, identifiable types of organisms—are connected by common ancestry.

CHECK YOUR UNDERSTANDING

✔ If you understood this section, you should be able to ...

1. **MODEL** Use words and arrows to depict the central dogma and show where a gene would be located.
2. Explain the importance for a flamingo of being able to capture aquatic organisms floating in the water.

Answers are available in Appendix A.

1.4 Life Evolves

In 1858, short papers written separately by Darwin and Wallace were read to a small group of scientists attending a meeting of the Linnean Society of London. A year later, Darwin published a book that expanded on the idea summarized in those brief papers. The book was called *On the Origin of Species*. The first edition sold out in a day.

After you complete this section, you should be able to ...

- Explain the relationships between species and how evolution occurs.

What Is Evolution?

Darwin and Wallace's theory made two important claims concerning patterns that exist in the natural world.

1. Species are related by common ancestry (Figure 1.6). This idea contrasted with the prevailing view in science at the time, which was that species represent independent entities created separately by a divine being.
2. The characteristics of species can be modified from generation to generation. Darwin called this process descent with modification. This claim argued against the popular view at the time that species do not change.

Evolution is a change in the characteristics of a population over time. A **population** is defined as a group of individuals of the same species living in the same area at the same time. To put it another way, species are related to one another and can change through time.

What Is Natural Selection?

Several other scientists had already come to the same conclusions as Darwin and Wallace about the relationships between species. The great insight by Darwin and Wallace was in proposing a process, called **natural selection**, that explained how evolution occurs.

Two Conditions of Natural Selection Natural selection occurs whenever two conditions are met.

1. Individuals within a population vary in characteristics that are **heritable**—meaning, traits that can be passed on to offspring.
2. In a particular environment, certain versions of these heritable traits help individuals survive better and reproduce more than do other versions.

If certain heritable traits lead to increased success in producing offspring, then those traits become more common in the population over time. In this way, the population's characteristics change as a result of natural selection acting on individuals. This is a key insight: Natural selection acts on individuals, but evolutionary change occurs in populations.

Groups of individuals within a species can change through time in response to natural selection. But biologists have also observed and documented dozens of cases in which natural

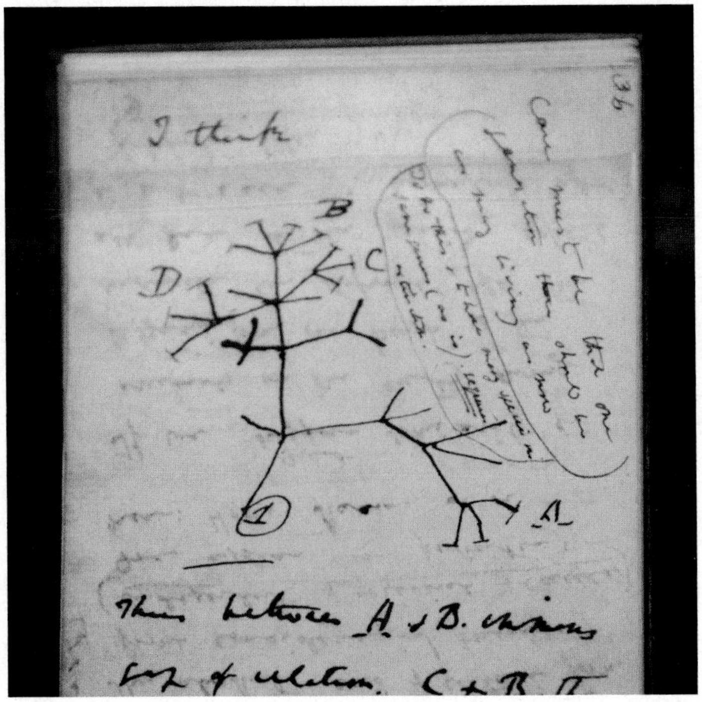

Figure 1.6 Sketch of Lineage Tree from Darwin's Notebook Dated 1837. Below the sketch, Darwin wrote: "... Thus genera would be formed—bearing relation to ancient types."

selection has caused populations of one species to diverge and form new species. This divergence process is called **speciation**.

Research on speciation has two important implications: All species come from preexisting species, and all species, past and present, trace their ancestry back to a single common ancestor.

Fitness and Adaptation Darwin also introduced some new terminology to identify what happens during natural selection.

- In everyday English, "fitness" means "health and well-being." But in biology, **fitness** means "an individual's ability to produce viable offspring relative to that ability in other individuals in the population." Individuals with high fitness produce many surviving offspring.
- In everyday English, "adaptation" means that an individual is adjusting and changing to function in new circumstances. But in biology, an **adaptation** is a heritable trait that increases the fitness of an individual in a particular environment relative to individuals lacking that trait.

Darwin and Wallace's ideas arose from their observations of nature. For example, in finches from the Galápagos Islands Darwin noted the remarkable variation in beak size and shape in species that otherwise appeared similar. He proposed that the birds on different islands in the chain were similar because they descended from a common ancestor—the finch populations that colonized different islands had changed through time and formed new species with distinct beaks.

Long-term observational studies by biologists beginning in the 1970s have recorded dramatic changes in a population of finches on one of the Galápagos Islands. (You will learn more about this study in Ch. 22, Section 22.4.) When small, soft seeds were abundant due to increased rainfall on the island, finches with small, pointed beaks produced more offspring and had higher fitness than individuals with large, deep beaks. In this population and with this food source, a small, pointed beak was an adaptation that allowed certain individuals to thrive, and the incidence of finches with such beaks increased in the population.

Note that during this process, the beak shape of any individual finch did not change within its lifetime—the change occurred in the characteristics of the population over time. Darwin's finches continue to evolve today in response to changes in the environment.

CHECK YOUR UNDERSTANDING

✔ **If you understood this section, you should be able to ...**

Discuss this statement: "Various species of Galápagos finches are adapted to their particular habitats."

1.5 The "Tree of Life" Depicts Evolutionary History

The theory of evolution by natural selection predicts that biologists should be able to construct a **tree of life**—a family tree of organisms. If life on Earth arose just once, then such a diagram would describe the genealogical relationships between species with a single, ancestral species at its base. Has this task been accomplished? If the tree of life exists, what does it look like?

After you complete this section, you should be able to ...

- Describe the significance of the tree of life.

Using Genetic Sequences to Understand the Tree of Life

One of the great breakthroughs in research on the tree of life occurred when American biologist Carl Woese (pronounced *woze*) and colleagues began analyzing the molecular components of organisms as a way to understand their evolutionary relationships. Their goal was to understand the **phylogeny** of all organisms—their actual genealogical relationships. Translated literally, "phylogeny" means "tribe-source."

To understand which organisms are closely versus distantly related, Woese and co-workers needed to study a molecule found in all organisms. They selected a particular RNA molecule, one that codes for an essential part of the machinery that all cells use to grow and reproduce. The researchers based their initial work on the sequence of building blocks observed in this RNA molecule. At that time it was not possible to easily analyze DNA sequences. With advances in technology, biologists now use DNA sequences to investigate phylogenetic relationships.

Making Models 1.1 Tips on Drawing DNA Sequences

In models focused on the information content in DNA, structural details can be left out and the double-stranded DNA double helix simplified to show the letter code on one strand only. Sequences can then be compared for similarities and differences.

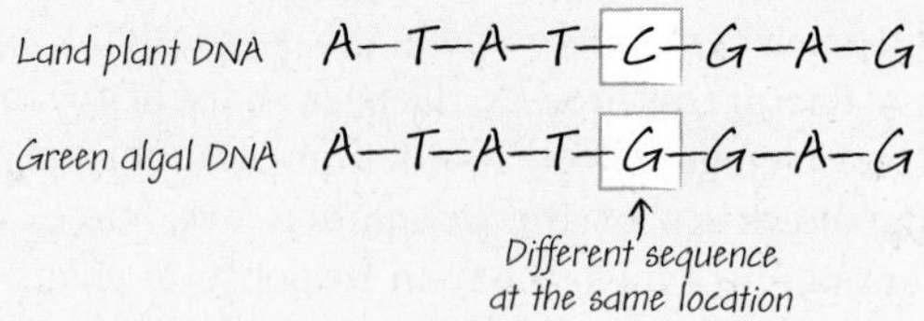

MODEL Suppose that in the same section of DNA, molds and other fungi have the sequence A-T-A-T-**A**-G-A-**C**. Draw a model that compares the sequences. Are fungi closely related to either green algae or land plants? Explain your logic.

To see this model in action, go to the Study Area of **Mastering Biology**

Analyzing Genetic Variation Why might DNA (or RNA) be useful for understanding the relationships between organisms? Because the sequence of building blocks in DNA is a trait that can change during the course of evolution. Although a gene may code for an RNA or protein molecule that performs the same function in all organisms, the corresponding DNA sequence is not identical among species.

How is such genetic variation analyzed? Recall that the building blocks in DNA are symbolized by the letters A, T, C, and G (see Section 1.3). Biologists use this letter code to describe DNA sequences (**Making Models 1.1**). In land plants, for example, a section of DNA might start with the sequence A-T-A-T-**C**-G-A-G. In green algae, which are closely related to land plants, the same section of the molecule might contain A-T-A-T-**G**-G-A-G. But in brown algae, which are not closely related to green algae or to land plants, the same part of the molecule might consist of A-**A**-A-T-**G**-G-A-**C**.

The next step in analyzing genetic variation is to consider what the similarities and differences in the sequences imply about relationships between species. The goal is to produce a diagram that describes the phylogeny of the organisms being compared.

A diagram that depicts evolutionary history in this way is called a **phylogenetic tree**. (For help in learning how to read a phylogenetic tree, see **BioSkills 13**.) Just as a family tree shows relationships between individuals, a phylogenetic tree shows relationships between species. On a phylogenetic tree, branches that share a recent common ancestor—that is, an ancestral population—represent species that are closely related; branches that don't share recent common ancestors represent species that are more distantly related.

The Tree of Life Estimated from Genetic Data To construct a phylogenetic tree like the one shown in **Figure 1.7**, researchers use sophisticated computer programs to find the arrangement of branches that is most consistent with the similarities and differences observed in the genetic data.

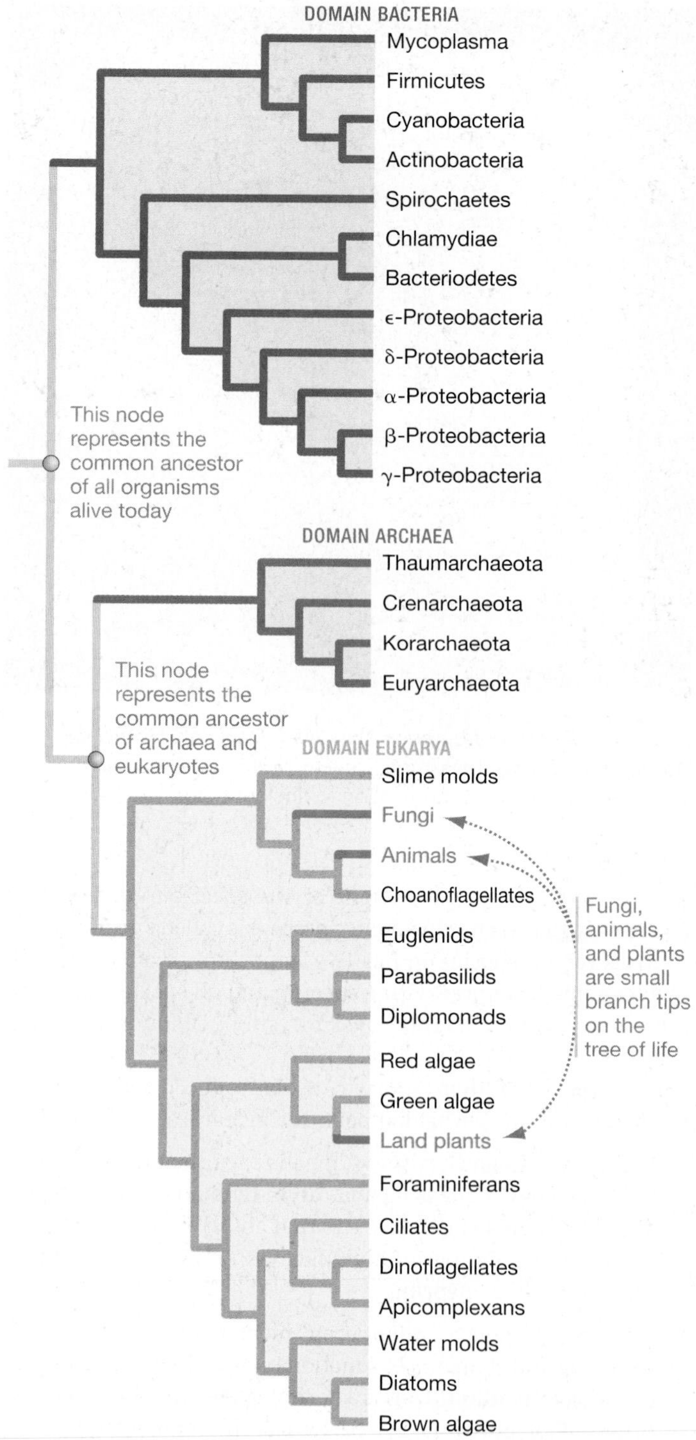

Figure 1.7 The Tree of Life Was Produced by Comparing Genetic Sequence Data. The three domains of life revealed by the analysis are labeled. Common names are given for lineages in the domains Bacteria and Eukarya. Phyla names are given for lineages in the domain Archaea, because most of them have no common names.

Because this tree includes such a diverse array of species, it is often called "the universal tree," or "the tree of life." Notice that the tree's main node is the common ancestor (ancestral population) of all living organisms. Researchers who study the origin of life propose that the tree's root extends all the way back to a "*l*ast *u*niversal *c*ommon *a*ncestor" of cells, or **LUCA**.

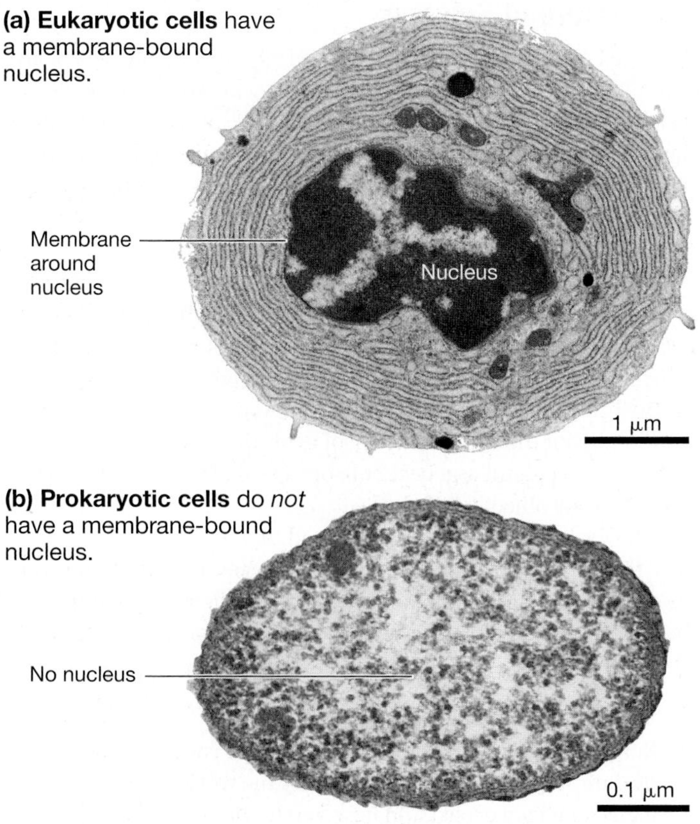

Figure 1.8 Eukaryotic and Prokaryotic Cells Differ in Structure.

✔ **QUANTITATIVE** How many times larger than the prokaryotic cell is the eukaryotic cell in this figure? (Hint: Study the scale bars.)

The tree of life implied by genetic sequence data established that there are three fundamental groups or lineages of organisms: **(1)** the Bacteria, **(2)** the Archaea, and **(3)** the Eukarya. In all **eukaryotes** (literally, "true-kernel"), cells have a prominent component called the nucleus (Figure 1.8a). Because the vast majority of bacterial and archaeal cells lack a nucleus, they are referred to as **prokaryotes** ("before-kernel"; see Figure 1.8b). Bacteria and archaea are almost exclusively unicellular ("one-celled"); many eukaryotes are multicellular ("many-celled").

When results based on genetic data were first published, biologists were astonished. For example:

- Before Woese's work and follow-up studies, biologists thought that the most fundamental division among organisms was between prokaryotes and eukaryotes. The Archaea were virtually unknown, much less recognized as a major and highly distinctive branch on the tree of life.
- Fungi were thought to be closely related to plants. Instead, they are much more closely related to animals.
- Traditional approaches for classifying organisms—including the system of five kingdoms divided into various classes, orders, and families that you may have learned in high school—are inaccurate in many cases, because they do not reflect the actual evolutionary history of the organisms involved.

The Tree of Life Is a Work in Progress Just as researching your family tree can help you understand who you are and where you came from, so the tree of life helps biologists understand the relationships between organisms and the history of species. Discovery of the Archaea and the accurate placement of lineages such as fungi qualify as exciting breakthroughs in our understanding of evolutionary history and life's diversity.

Work on the tree of life continues at a furious pace, however, and the location of certain branches on the tree is hotly debated. As databases expand and as techniques for analyzing data improve, the shape of the tree of life will undoubtedly change. Our understanding of the tree of life, like our understanding of every other topic in biological science, is dynamic.

How Should We Name Branches on the Tree of Life?

In science, the effort to name and classify organisms is called **taxonomy**. Any named group at any level of a classification system is called a **taxon** (plural: **taxa**). Currently, biologists are working to create a taxonomy, or naming system, that accurately reflects the phylogeny of all organisms. Based on the tree of life, Woese proposed a new taxonomic category called the **domain**. He designated the Bacteria, Archaea, and Eukarya as the three domains of life.

Biologists today often use the term **phylum** (plural: **phyla**) to refer to major lineages within each domain. Although the designation is somewhat arbitrary, each phylum is considered a major branch on the tree of life. Within the eukaryotic lineage called animals, biologists currently name 30–35 phyla—each of them distinguished by distinctive aspects of body structure as well as by distinctive gene sequences. For example, the mollusks (clams, squid, octopuses) constitute an animal phylum, as do the chordates (vertebrates and their close relatives).

Because the tree of life is so new, though, naming systems are still being worked out. For example, recent genetic data have fueled a debate over the domains of life. Some argue that rather than three domains, there are only two: Bacteria and the rest of life. (For background on the debate, see the end-of-chapter Case Study in Chapter 27.) One thing that hasn't changed for centuries, however, is the naming system for individual species.

Scientific (Latin) Names In 1735, a Swedish botanist named Carolus Linnaeus established a system for naming species that is still in use today. Each type of organism is assigned a two-part name.

- ***Genus*** The first part of the two-part name indicates the organism's **genus** (plural: **genera**). A genus is made up of a closely related group of species. For example, Linnaeus put humans in the genus *Homo*. Although humans are the only living species in this genus, at least six extinct organisms, all of which walked upright and made extensive use of tools, were later also assigned to *Homo*.
- ***Species name*** The second term in the name identifies the organism's species. Linnaeus gave humans the species name *sapiens*. A species name is always preceded by its genus.

An organism's genus and species designation is called its **scientific name**, or Latin name. Scientific names are always italicized. Genus names are always capitalized, but species names are not—as in *Homo sapiens*. Linnaeus maintained that different types of organisms should not be given the same genus and species names. Other species may be assigned to the genus *Homo* (from the Latin for "man"), and members of other genera may be named *sapiens* (from the Latin for "wise" or "knowing"), but only humans are named *Homo sapiens*. Each scientific name is unique.

Scientific Names Are Often Descriptive Scientific names and terms are often based on Latin or Greek word roots that are descriptive. For example, consider the yeast *Saccharomyces cerevisiae*. *Saccharomyces* is aptly named—the domesticated strains of yeast used in brewing and baking are often fed sugar (Greek root *saccharo*), and yeast is a fungus (Greek root *myces*). The species name of this organism, *cerevisiae*, is Latin for "beer." Loosely translated, then, the scientific name of brewer's yeast means "sugar-fungus for beer."

Scientific names and terms often seem daunting at first glance. So, most biologists find it extremely helpful to memorize some of the common Latin and Greek roots. To aid you in this process, new terms in this text are often accompanied by a translation of their Latin or Greek word roots in parentheses. (A glossary of common root words with translations and examples is also provided in **BioSkills 15**.)

CHECK YOUR UNDERSTANDING

✔ **If you understood this section, you should be able to...**

Examine the following DNA sequences and determine which two species would be closest on a phylogenetic tree.

Species A: A A C T A G C G C G A T

Species B: A A C T A G C G C C A T

Species C: T T C T A G C G G T A T

Answers are available in Appendix A.

1.6 Doing Biology

After you complete this section, you should be able to . . .

- Explain the scientific process.

In this chapter you have read about some of the great ideas in biology. Development of the cell theory, the chromosome theory of inheritance, and the theory of evolution provided cornerstones when the science was young. The central dogma explained the flow of information from DNA to physical traits of an organism, and the more recent insights of the tree of life have revolutionized our understanding of life's diversity.

These are considered great ideas because they explain fundamental aspects of nature, and because they have consistently been shown to be correct. They are considered correct because they have withstood extensive testing. How do biologists go about testing their ideas? Before answering this question, let's step back a bit and consider the types of questions that researchers can and cannot ask.

The Nature of Science

Biologists ask questions about organisms, just as physicists and chemists ask questions about the physical world or geologists ask questions about Earth's history and the processes that shape landforms. No matter what their field, all scientists ask questions that can be answered by observing or measuring things—by collecting data. Conversely, scientists cannot address questions that can't be answered by observing or measuring things.

This distinction is important. It is at the root of continuing controversies about teaching evolution in publicly funded schools. In some parts of the United States and in Turkey, fundamentalist Christian and Islamic leaders have been particularly successful in pushing their claim that evolution and religious faith are in conflict. Even though the theory of evolution is considered one of the most successful and best-substantiated ideas in the history of science, they object to teaching it.

The vast majority of biologists and many religious leaders reject this claim; they see no conflict between evolution and religious faith. Their view is that science and religion are compatible because they address different types of questions.

- Science is about formulating hypotheses and finding evidence that supports or conflicts with those hypotheses.
- Religious faith addresses questions that cannot be answered by data. The questions addressed by the world's great religions focus on why we exist and how we should live.

So how do biologists go about answering questions? After formulating hypotheses, biologists perform studies that yield observational or experimental data, such as observing a behavior, characterizing a structure within a cell by microscopy, or sequencing DNA. Let's consider two examples of this process.

An Introduction to Hypothesis Testing: Why Do Giraffes Have Long Necks?

If you were asked why giraffes have long necks, you might say based on your observations that long necks enable giraffes to reach food that is unavailable to other mammals. This hypothesis is expressed in African folktales and has traditionally been accepted by many biologists. The food competition hypothesis is so plausible, in fact, that for decades no one thought to test it.

In the mid-1990s, however, Robert Simmons and Lue Scheepers assembled data suggesting that the food competition hypothesis is only part of the story. Their analysis of observational data supports an alternative hypothesis: Long necks allow giraffes to use their heads as effective weapons for battering their opponents, and longer-necked giraffes have a competitive advantage in fights.

Before exploring these alternative explanations, it's important to recognize that hypothesis testing is a two-step process:

Step 1 State the hypothesis as precisely as possible and list the predictions it makes.

Step 2 Design an observational or experimental study that is capable of testing those predictions.

If the predictions are accurate, the hypothesis is supported. If the predictions are not met, then researchers do further tests, modify the original hypothesis, or search for alternative explanations. But the process does not end here. Biologists communicate their results to the scientific community and beyond; for example, via informal conversations, scientific meetings, or publications. (See the Big Picture of Doing Biology on pages 18–19.)

Now that you understand more about hypothesis testing, let's return to the giraffes. How did biologists test the food competition hypothesis? What data support their alternative explanation?

The Food Competition Hypothesis: Predictions and Tests The food competition hypothesis claims that giraffes compete for food with other mammals. When food is scarce, as it is during the dry season, giraffes with longer necks can reach food that is unavailable to other species and to giraffes with shorter necks. As a result, the longest-necked individuals in a giraffe population survive better and produce more young than do shorter-necked individuals, and average neck length of the population increases with each generation.

To use the terms introduced earlier, long necks are adaptations that increase the fitness of individual giraffes during competition for food. This type of natural selection has gone on so long that the population has become extremely long necked.

The food competition hypothesis makes several explicit predictions. For example, it predicts that

- neck length is variable among giraffes;
- neck length in giraffes is heritable; and
- giraffes feed high in trees, especially during the dry season, when food is scarce and the threat of starvation is high.

The first prediction is correct. Studies in zoos and natural populations confirm that neck length is variable among individuals. The researchers were unable to test the second prediction, however, because they studied giraffes in a natural population and could not do breeding experiments. As a result, they simply had to accept this prediction as an assumption. In general, though, biologists prefer to test every assumption behind a hypothesis.

What about the prediction regarding feeding high in trees? According to Simmons and Scheepers, this is where the food competition hypothesis breaks down.

Consider, for example, data collected by a different research team on the amount of time that giraffes spend feeding in vegetation of different heights (**Figure 1.9a**). Note that this graph plots the height of vegetation on the *x*-axis, starting from ground level (0 meters on the graph) and continuing up to 7 meters. The percentage of bites taken by a giraffe is plotted on the *y*-axis, for males and for females from the same population in Kenya. A dashed line on each graph indicates the average height of a male or female in this population. (For more help on reading graphs, see **BioSkills 2**.)

Note that the average height of a giraffe in this population is much greater than the height where most feeding takes place. In this population, both male and female giraffes spend most of their feeding time eating vegetation that averages just 60 percent of their full height. Studies on other populations of giraffes, during both the wet and dry seasons, are consistent with these data. Giraffes usually feed with their necks bent (**Figure 1.9b**).

These data cast doubt on the food competition hypothesis, because one of its predictions does not appear to hold. Biologists have not abandoned this hypothesis completely, though, because feeding high in trees may be particularly valuable during extreme droughts, when a giraffe's ability to reach leaves far above the ground could mean the difference between life and death. Still, Simmons and Scheepers have offered an alternative explanation for why giraffes have long necks. The new hypothesis is based on the mating system of giraffes.

(a) Most feeding is done at about shoulder height.

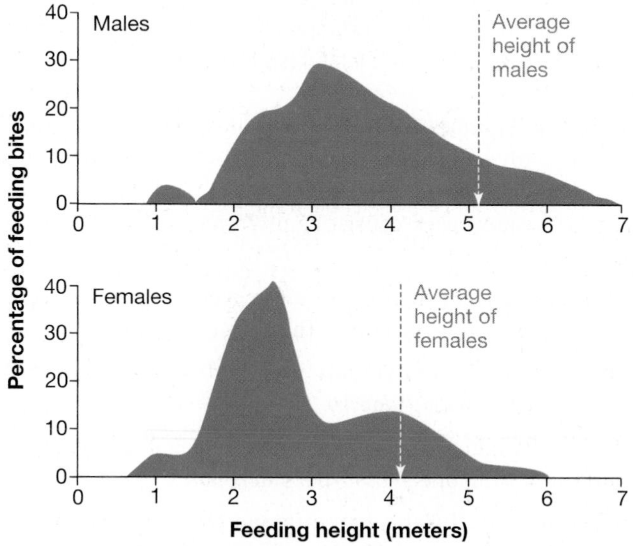

(b) Typical feeding posture in giraffes

Figure 1.9 Giraffes Do Not Usually Extend Their Necks Upward to Feed.
DATA: T. P. Young and L. A. Isbell. 1991. *Ethology* 87: 79–89.

✔ **QUANTITATIVE** At what height of vegetation do male and female giraffes spend most of their time feeding?

The Sexual Competition Hypothesis: Predictions and Tests Giraffes have an unusual mating system. Breeding occurs year-round rather than seasonally. Once a female giraffe enters estrus, males may fight among themselves for the opportunity to mate, though confrontation often is resolved by the males standing very tall and staring hard at each other until one male turns and runs away. When combat does occur, it is spectacular. The bulls stand next to one another, swing their necks, and strike thunderous blows with their heads. Researchers have seen males knocked unconscious for up to 20 minutes after being hit and have cataloged numerous instances in which the loser died.

These observations inspired a new explanation for why giraffes have long necks. The sexual competition hypothesis is based on the idea that longer-necked giraffes can strike harder blows during combat than shorter-necked giraffes. In engineering terms, longer necks provide a longer "moment arm." A long moment arm increases the force of an impact. (Think about the type of sledgehammer you'd choose to bash down a concrete wall—one with a short handle or one with a long handle?)

The idea here is that longer-necked males should win more fights and, as a result, father more offspring than shorter-necked males do. If neck length in giraffes is inherited, then the average neck length in the population should increase over time. Under the sexual competition hypothesis, long necks are adaptations that increase the fitness of males during competition for females.

Although several studies have shown that long-necked males are more successful in fighting and that the winners of fights gain access to estrous females, the question of why giraffes have long necks is not closed. With the data collected to date, many biologists would probably conclude that both the food competition hypothesis and the sexual competition hypothesis need further testing and refinement. It could be that both hypotheses are correct. For our purposes, the important take-home message is that all hypotheses must be tested rigorously.

In many cases in biological science, testing hypotheses rigorously involves experimental manipulation rather than measuring just what can be observed in nature. Experimenting on giraffes is difficult. But in the case study considered next, biologists could test an interesting hypothesis experimentally.

An Introduction to Experimental Design: How Do Ants Navigate?

Let's consider a question that is easier to test than the one about factors that determine giraffe neck length: When ants leave their nest to search for food, how do they find their way back?

The Saharan desert ant lives in colonies and makes a living by scavenging the carcasses of dead insects. Individuals leave the burrow and wander about searching for food at midday, when temperatures at the surface can reach 60°C (140°F) and predators are hiding from the heat. Foraging trips can take the ants hundreds of meters—an impressive distance when you consider that the ants are only about a centimeter long. But when an ant returns, it doesn't follow the same wandering route it took away from the nest. Instead, its return path is a straight line (Figure 1.10). Once individuals are close to the nest, they engage in a characteristic set of back-and-forth U-turns until they find their nest hole. How do they do know how far they are from the nest?

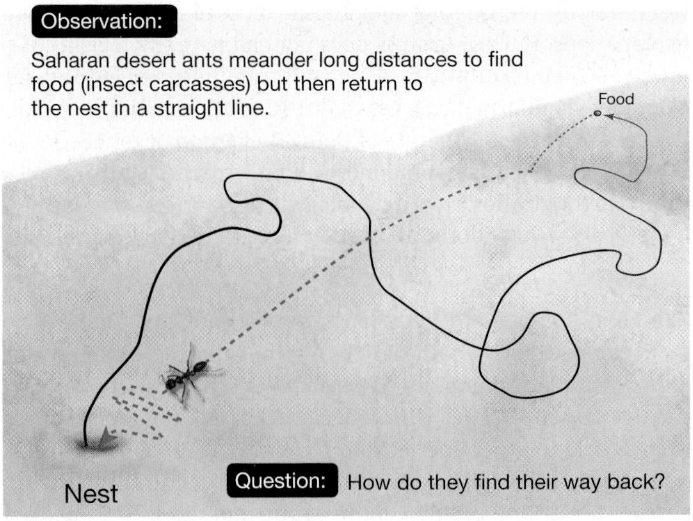

Figure 1.10 Foraging Desert Ants Can Navigate.

The Pedometer Hypothesis Early work on navigation in desert ants showed that they use the Sun's position as a compass—meaning that they always know the approximate direction of the nest relative to the Sun. But how do they know how far to go? Experiments had shown that the ants do not use landmarks to navigate, so Matthias Wittlinger and co-workers set out to test a novel idea. The biologists proposed that these ants know how far they are from the nest by using information from leg movements.

According to this pedometer hypothesis, the ants always know how far they are from the nest because they track the number of steps they have taken and their stride length. The idea is that they can make a beeline back toward the burrow because they integrate information on the angles they have traveled *and* the distance they have gone—based on step number and stride length.

If the pedometer hypothesis is wrong, however, then stride length and step number should have no effect on the ant's ability to get back to its nest. This latter possibility is called a **null hypothesis**. A null hypothesis specifies what should be observed when the hypothesis being tested isn't correct.

Testing the Hypothesis To test their idea, Wittlinger's group allowed ants to walk from a nest to a feeder through a channel—a distance of 10 m. Then they caught ants at the feeder and created three test groups, each consisting of 25 individuals (Figure 1.11 and Figure 1.12):

- ***Stumps*** By cutting off the lower legs of some individuals, the biologists created ants with shorter-than-normal legs.
- ***Normal*** Some individuals were left alone, meaning that they had normal leg length.
- ***Stilts*** By gluing pig bristles onto each leg, the biologists created ants with longer-than-normal legs.

Next they put the ants in a different channel for their return home and recorded how far they traveled in a direct line before starting their nest-searching behavior. To see the data they collected, look at the graph on the left side of the "Results" section in Figure 1.11.

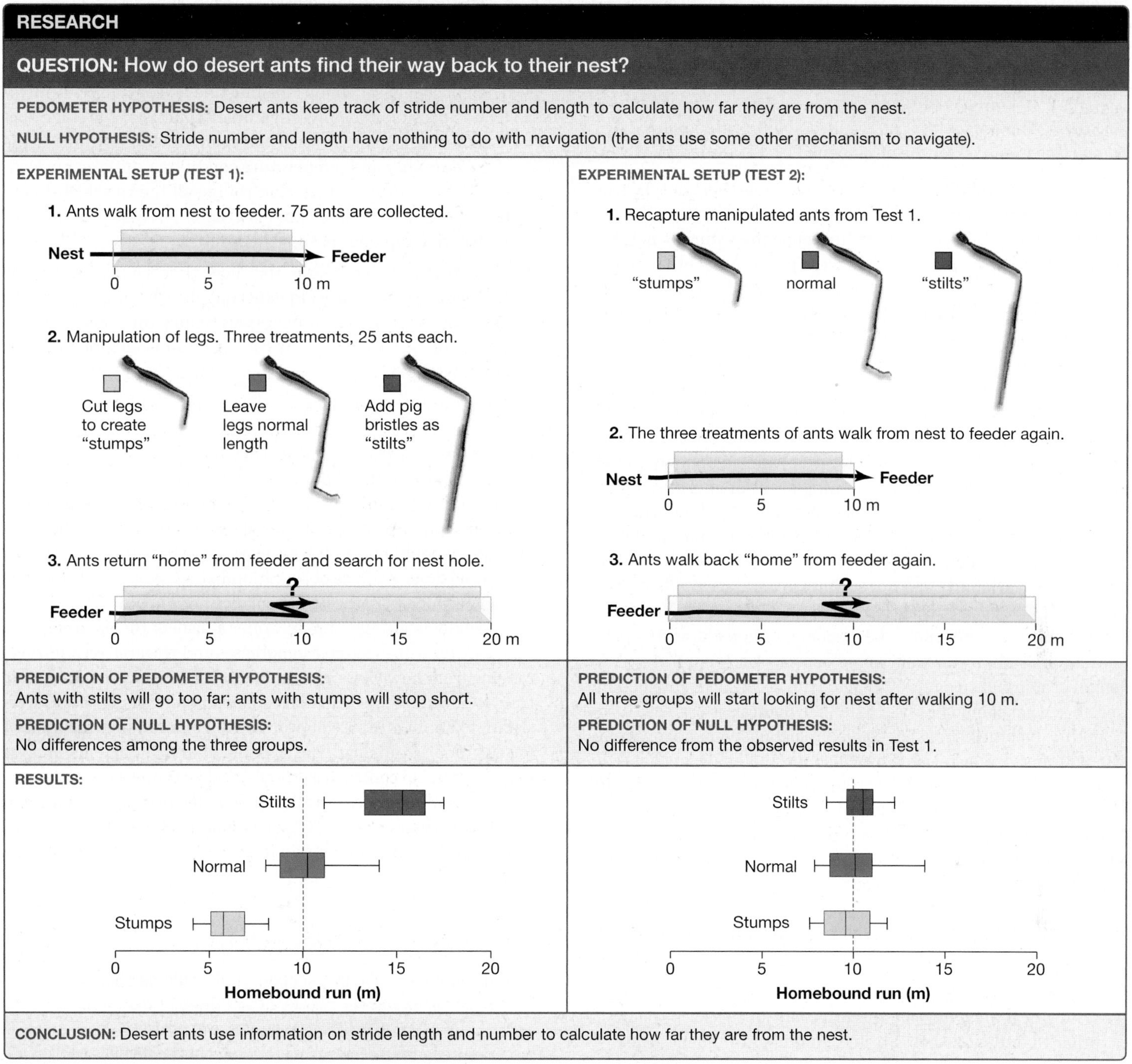

INTERACTIVE Figure 1.11 An Experimental Test: Do Desert Ants Use a "Pedometer"?

SOURCE: M. Wittlinger, R. Wehner, and H. Wolf. 2006. The ant odometer: Stepping on stilts and stumps. *Science* 312: 1965–1967.

✔ **PROCESS OF SCIENCE** What is the advantage of using 25 ants in each group instead of just one?

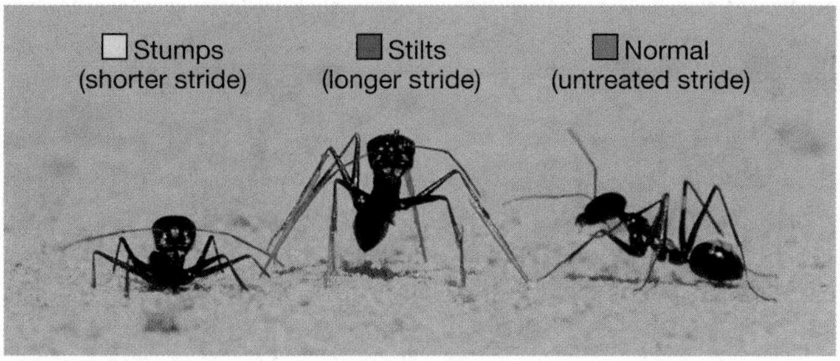

Figure 1.12 **Manipulating Desert Ant Legs Changes Stride Length.**

- ***Stumps*** The ants with stumps stopped short, by about 5 m, before starting to search for the nest opening.
- ***Normal*** The normal ants walked the correct distance—about 10 m.
- ***Stilts*** The ants with stilts walked about 5 m too far before starting to search for the nest opening.

To check the validity of this result, the researchers put the test ants back in the nest and recaptured them one to several days later, when they had walked to the feeder on their stumps, normal legs, or stilts. Now when the ants were put into the other channel to "walk back," they all traveled the correct distance—10 m—before starting to search for the nest (see the graph on the right side of the "Results" section in Figure 1.11).

The graphs in the "Results" display "box-and-whisker" plots that allow you to easily see where most of the data fall. Each box indicates the range of distances where 50 percent of the ants stopped to search for the nest. The whiskers indicate a lower extreme (stopping short of the nest location) and an upper extreme (going too far) of where the other 50 percent of ants stopped to search. The vertical line inside each box indicates the median—meaning that half the ants stopped above this distance and half below. (For more details on how biologists report medians and indicate the variability and uncertainty in data, see **BioSkills 3**.)

Interpreting the Results The pedometer hypothesis predicts that an ant's ability to walk home depends on the number and length of steps taken on its outbound trip. Recall that a prediction specifies what should be observed if a hypothesis is correct. Good scientific hypotheses make testable predictions—predictions that can be supported or rejected by collecting and analyzing data. In this case, the researchers tested the prediction by altering stride length and recording the distance traveled on the return trip. Under the null hypothesis in this experiment, all the ants—altered and unaltered—should have walked 10 m in the first test before they started looking for their nest.

Important Characteristics of Good Experimental Design This study illustrates several important points related to designing effective experiments:

- It is critical to include a **control**. A control checks for factors, other than the one being tested, that might influence the experiment's outcome. In this case, there were two controls. Including a normal, unmanipulated individual controlled for the possibility that switching the individuals to a new channel altered their behavior. The researchers also had to control for the possibility that the manipulation itself—and not the change in leg length—affected the behavior of the stilts and stumps ants. This is why they did the second test, where the ants did outbound and return runs with the same legs.
- The experimental conditions must be as constant or equivalent as possible. The investigators used ants of the same species, from the same nest, at the same time of day, under the same humidity and temperature conditions, at the same feeders, in the same channels. Controlling all the variables except one—leg length in this case—is crucial because it eliminates alternative explanations for the results.
- Repeating the test is essential. It is almost universally true that larger sample sizes in experiments are better. By testing many individuals, researchers can reduce the amount of distortion or "noise" in the data caused by unusual individuals or circumstances.

From the outcomes of these experiments, the researchers concluded that desert ants use stride length and number to measure how far they are from the nest. They interpreted their results as strong support for the pedometer hypothesis.

The giraffe and ant studies demonstrate a vital point: Biologists practice evidence-based decision making. They ask questions about how organisms work, pose hypotheses to answer those questions, and use observational or experimental evidence to decide which hypotheses are correct.

The data on giraffes and ants offer a taste of things to come. In this text you will encounter hypotheses and research on questions ranging from how water gets to the top of 100-meter-tall sequoia trees to how the bacterium that causes tuberculosis has become resistant to antibiotics. As you work through the chapters, you'll get lots of practice thinking about hypotheses and predictions, analyzing the nature of control treatments, and interpreting graphs.

A commitment to tough-minded hypothesis testing and sound experimental design is a hallmark of biological science. Understanding their value is an important first step in becoming a biologist.

CHECK YOUR UNDERSTANDING

✔ If you understood this section, you should be able to . . .

1. **PROCESS OF SCIENCE** Design an experiment to test the hypothesis that desert ants feed during the hottest part of the day because it allows them to avoid being eaten by lizards.
2. **PROCESS OF SCIENCE** Explain how experimental conditions in your ant-predation experiment are controlled or standardized in a way that precludes alternative explanations of the data.

Answers are available in Appendix A.

1.1 What Does It Mean to Say that Something Is Alive?

- There is no single, well-accepted definition of life. Instead, biologists point to five characteristics that organisms share.
- Three of the greatest unifying ideas in biology are (1) the cell theory, (2) the chromosome theory of inheritance, and (3) the theory of evolution.

1.2 Life Is Cellular and Replicates through Cell Division

- The cell theory identified the fundamental structural unit common to all life.
- Most cells are capable of reproducing by dividing.

1.3 Life Processes Information and Requires Energy

- The chromosome theory of inheritance states that genes are located on chromosomes.
- A chromosome consists of a molecule of DNA—the hereditary material. Genes, located on chromosomes, consist of specific segments of DNA that code for products in the cell.
- The flow of information from DNA to RNA to protein is called the central dogma.
- Organisms are highly diverse in how they acquire and use energy.

1.4 Life Evolves

- The theory of evolution states that all organisms are related by common ancestry.
- Natural selection is a well-tested explanation for why species change through time and why they are so well adapted to their habitats.

1.5 The "Tree of Life" Depicts Evolutionary History

- The theory of evolution predicts that all organisms are part of a genealogy of species, and that all species trace their ancestry back to a single common ancestor.
- To construct this phylogeny, biologists have analyzed the sequences in an array of genetic material found in all cells.
- A tree of life, based on similarities and differences in these molecules, has three fundamental lineages, or domains: the Bacteria, the Archaea, and the Eukarya.

1.6 Doing Biology

- Biology is a hypothesis-driven, experimental science.

Answers are available in Appendix A.

✔ TEST YOUR KNOWLEDGE

1. Anton van Leeuwenhoek made an important contribution to the development of the cell theory. How?
   a. He articulated that all organisms are made of cells.
   b. He articulated that all cells come from preexisting cells.
   c. He invented the first microscope and saw the first cell.
   d. He invented more powerful microscopes and was the first to describe the diversity of cells.
2. **PROCESS OF SCIENCE** What does it mean to say that experimental conditions are controlled? Select True or False for each statement.
   T/F The test groups consist of the same individuals.
   T/F The null hypothesis is correct.
   T/F There is no difference in outcome between the control and experimental treatment.
   T/F All physical conditions except for one are identical for all groups tested.
3. What does it mean to say that a characteristic is heritable?
   a. The characteristic evolves.
   b. The characteristic can be passed on to offspring.
   c. The characteristic is advantageous to the organism.
   d. The characteristic does not vary in the population.
4. Could both the food competition hypothesis and the sexual competition hypothesis explain why giraffes have long necks? Why or why not?

✔ TEST YOUR UNDERSTANDING

5. What would researchers have to demonstrate to convince you that they had discovered life on another planet?
6. What did Linnaeus' system of naming organisms ensure?
   a. Two different organisms never end up with the same genus and species name.
   b. Two different organisms have the same genus and species name if they are closely related.
   c. The genus name is different for closely related species.
   d. The species name is the same for each organism in a genus.
7. What is "selected" during natural selection? Explain your answer.
8. **PROCESS OF SCIENCE** Explain why researchers formulate a null hypothesis in addition to a hypothesis when designing an experimental study.

✔ TEST YOUR PROBLEM-SOLVING SKILLS

9. **THINK CAREFULLY** A friend tells you that the theory of evolution is just an educated guess by biologists about how things work. Evaluate this statement.
10. Some humans have genes that make them resistant to infection by HIV. Would human populations likely evolve differently in areas of the world where HIV infection rates are high? Explain your logic.

✔ PUT IT ALL TOGETHER: Case Study

Can a plant act like a chameleon?

You may be familiar with chameleons turning different colors to blend in with their environment. Now biologists have observed that *Boquila trifoliolata,* a climbing vine found in the rain forest of southern Chile, can mimic the leaves of a dozen host tree species. When the vine climbs up a leafy tree, it adjusts the size, shape, and color of its own leaves to match that tree's leaves. But when a vine climbs up a bare tree trunk, it looks the same as one that creeps along the rain forest floor.

11. Outline the flow of information from the genetic material to the physical appearance of the vine pictured earlier.

12. What does the species name of *Boquila trifoliolata* mean? Why is this name appropriate? (Hint: Look carefully at the Case Study photos and see **BioSkills 15**.)

13. **QUANTITATIVE** Researchers hypothesized that leaf mimicry by *B. trifoliolata* protects it from plant-eating animals (herbivores). Results of a study of 45 individual vines are shown in the following graph. Light conditions were very similar in all cases. Researchers compared the level of leaf damage by plant eaters (herbivory index) in vines climbing leafy host trees, vines creeping on the ground with no support, and vines climbing on bare tree trunks. Use the *P* values provided to determine if the differences are significant or not (*** means $P < 0.001$, see **BioSkills 3**). What conclusion, if any, can be drawn about leaf mimicry from this study? What might the researchers do next to further explore the role of leaf mimicry?

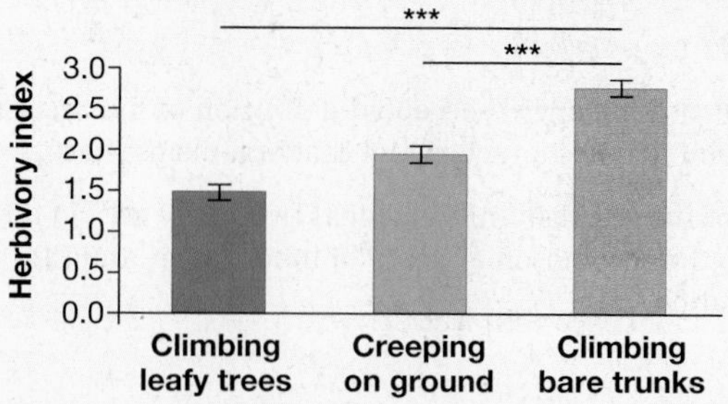

Source: S. N. Vignieri, J. G. Larson, and H. E. Hoekstra. 2010. *Evolution* 64: 2153–2158.

14. **PROCESS OF SCIENCE** If the researchers had compared vines growing under variable light conditions, how might this have changed their interpretation of the data?

15. **PROCESS OF SCIENCE** What was the purpose of including bare tree trunks in the study?

16. By avoiding being eaten, *B. trifoliolata* individuals would have increased fitness. In biology, what does the term "fitness" mean?

Mastering Biology

Students Go to Mastering™ Biology for assignments, the eText, and the Study Area with animations, practice tests, and activities.

Professors Go to Mastering™ Biology for automatically graded tutorials and questions that you can assign to your students, plus Instructor Resources.

Back in the 1950s, so the story goes, three hunters were reported missing in the Oregon Coast Range. Police discovered them dead, around their camp fire, coffee cups in hand, next to a pot of coffee on the now cold embers.

There was no sign of foul play and nothing had been stolen from the campsite. But, then the police found another dead body. Curled up at the bottom of the pot was a coffee-colored newt, later identified as *Taricha granulosa*, the rough-skinned newt.

Mystery of the Newt

Enter undergraduate Edmund D. "Butch" Brodie, Jr. Told this local legend in 1962 by "Doc" Walker (Dr. Kenneth Walker), a biology professor at the Oregon College of Education, Brodie became intrigued. So, he designed a research project to find out whether there was a connection between the rough-skinned newt and the dead hunters.

Little did people know at that time that solving this mystery would encompass all levels of biology, from molecules and cells to ecology and evolution. At the end of each unit, look for more clues as the "Mystery of the Newt" unfolds.

As for Butch Brodie, Jr., he continued to study the newts through his undergraduate degree, a Master's, and Ph.D. from nearby Oregon State University, and in a life-long career doing biology.

Each unit ends with a continuation of this story.

1. **PROCESS OF SCIENCE** Propose a hypothesis for how the hunters died.
2. **PROCESS OF SCIENCE** Design an experiment to test your hypothesis.

This rough-skinned newt is crossing a roadway in western Oregon, as it migrates to its breeding area in the spring.

BIG PICTURE

THE BIG PICTURE

Biologists study the characteristics of life. The cell theory, the theory of evolution by natural selection, the chromosome theory of inheritance, and the tree of life are some of the great ideas in biology that came about by biologists asking questions that can be answered by observing or measuring things—that is, by collecting data.

Notice that the study of life is not a series of linear steps with a beginning and an end. Instead, the process of doing biology is dynamic and ongoing. The answer to one question may lay the foundation for 20 more questions. Working together, biologists from different disciplines integrate data across many levels, from atoms to the biosphere.

Note that the gray numbers in boxes tell you where to go in the book for more information. Also, be sure to do the blue exercises in the Check Your Understanding box below.

Characteristics of living things

- Cells
- Replication
- Information
- Energy
- Evolution

1.1

focuses on

Text section where you can find more information

Scientists regularly integrate across many of these levels

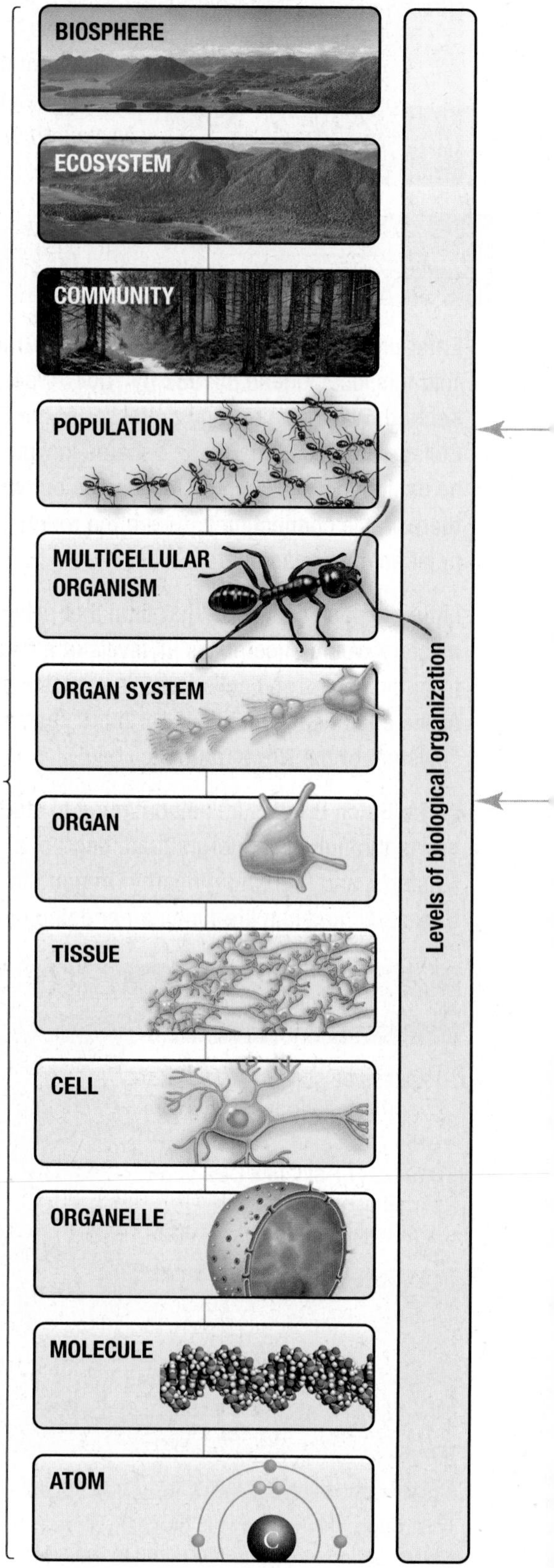

CHECK YOUR UNDERSTANDING

✔ If you understand the big picture, you should be able to ...

1. **PROCESS OF SCIENCE** Describe how biologists go about testing their ideas.
2. Provide an example of how an experimental study could span more than one level of biological organization.
3. Compare and contrast a hypothesis with a theory.
4. **PROCESS OF SCIENCE** Propose the next step to take if data support the hypothesis you are testing.

Answers are available in Appendix A.

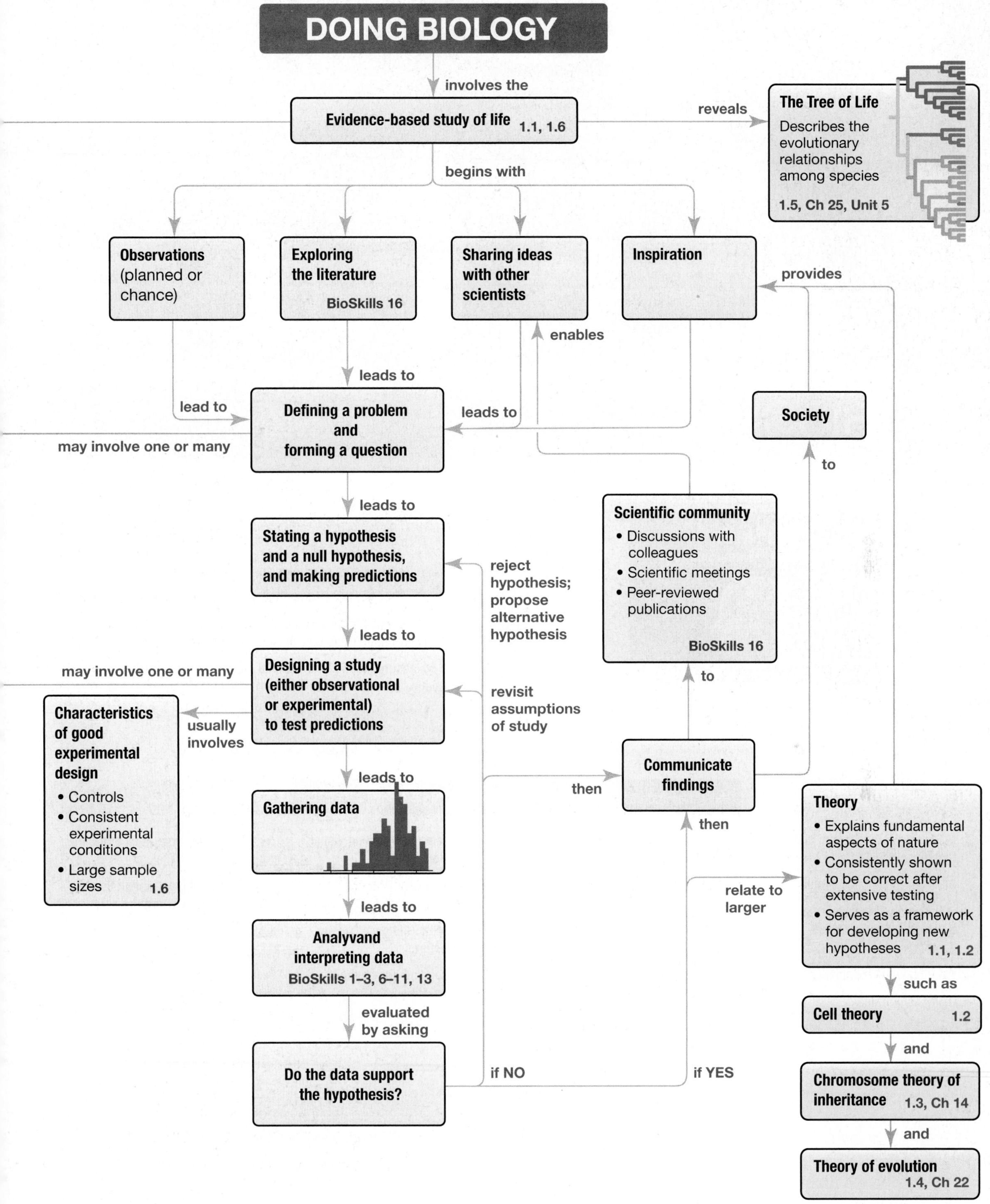
DOING BIOLOGY
involves the
Evidence-based study of life 1.1, 1.6
reveals
The Tree of Life
Describes the evolutionary relationships among species
1.5, Ch 25, Unit 5
begins with
Observations (planned or chance)
Exploring the literature
BioSkills 16
Sharing ideas with other scientists
Inspiration
provides
enables
leads to
lead to
Defining a problem and forming a question
leads to
may involve one or many
Society
to
leads to
Stating a hypothesis and a null hypothesis, and making predictions
reject hypothesis; propose alternative hypothesis
Scientific community
• Discussions with colleagues
• Scientific meetings
• Peer-reviewed publications
BioSkills 16
leads to
may involve one or many
Designing a study (either observational or experimental) to test predictions
revisit assumptions of study
to
Characteristics of good experimental design
• Controls
• Consistent experimental conditions
• Large sample sizes 1.6
usually involves
leads to
Gathering data
then
Communicate findings
then
Theory
• Explains fundamental aspects of nature
• Consistently shown to be correct after extensive testing
• Serves as a framework for developing new hypotheses 1.1, 1.2
leads to
relate to larger
Analyvand interpreting data
BioSkills 1–3, 6–11, 13
evaluated by asking
such as
Cell theory 1.2
and
Do the data support the hypothesis?
if NO
if YES
Chromosome theory of inheritance 1.3, Ch 14
and
Theory of evolution 1.4, Ch 22

BioSkills

In this book you will learn that

BioSkills
are essential for doing biology

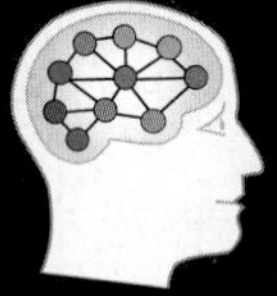

starting with

Asking Questions and Designing Studies

Chapter 1: Introduces core principles and best practices

Big Picture 1: Provides a visual summary of how to think like a biologist

The narrative throughout the text models how to think like a biologist, including end-of-chapter and end-of-unit case studies.

Research boxes, graphs, and other visual models in each chapter help you to visualize scientific ideas.

then using this BioSkills section to review and practice with

Quantifying Biology

- **1:** Using the Metric System and Significant Figures
- **2:** Reading and Making Graphs
- **3:** Interpreting Standard Error Bars and Using Statistical Tests
- **4:** Working with Probabilities
- **5:** Using Logarithms

Using Common Lab Tools

- **6:** Separating and Visualizing Molecules
- **7:** Separating Cell Components by Centrifugation
- **8:** Using Spectrophotometry
- **9:** Using Microscopy
- **10:** Using Molecular Biology Tools and Techniques
- **11:** Using Cell Culture and Model Organisms as Tools

Visualizing Biology

- **12:** Reading and Making Visual Models
- **13:** Reading and Making Phylogenetic Trees
- **14:** Reading Chemical Structures
- See **2:** Reading and Making Graphs

Reading Biology

- **15:** Translating Greek and Latin Roots in Biology
- **16:** Reading and Citing the Primary Literature

where success requires

Monitoring Your Own Learning

- **17:** Recognizing and Correcting Misconceptions
- **18:** Using Bloom's Taxonomy for Study Success

BioSkills 1 **Using the Metric System and Significant Figures**

After you complete BioSkills 1–5, you should be able to . . .

- Demonstrate the quantitative skills needed to succeed in biology.

Scientists ask questions that can be answered by observing and measuring things—by collecting data. What units are used to make measurements? When measurements are reported, how can you tell how reliable the data are?

Metric System Units and Conversions

The metric system is the system of measure used in every country of the world but three (Liberia, Myanmar, and the United States). It is also the basis of the SI system—the International System of Units (abbreviated from the French, *Système international d'unités*)—used in scientific publications.

The popularity of the metric system is based on its consistency and ease of use. These attributes arise from the system's use of base 10 for its units of measure. For example, each unit of length in the system is related to all other measures of length in the system by a multiple of 10. There are 10 millimeters in a centimeter, 100 centimeters in a meter, and 1000 meters in a kilometer.

Measures in the English system, in contrast, do not relate to each other in a regular way. For example, inches are routinely divided into 16ths; there are 12 inches in a foot, 3 feet in a yard, and 5280 feet (or 1760 yards) in a mile.

If you have grown up in the United States and are accustomed to using the English system, it is extremely important that you begin developing a working familiarity with metric units and values. **Table B1.1** and **Table B1.2** on page 22 should help you get started with this process.

As an example, consider the following question: An American football field is 120 yards long, while rugby fields are 144 meters long. In meters, how much longer is a rugby field than an American football field? To solve this problem, first convert yards to meters:

$$120 \text{ yards} \times \frac{\text{meters}}{1.09 \text{ yards}} = 110 \text{ meters}$$

Note that the unit "yards" cancels out. The difference in meters is thus $144 - 110 = 34$ m. If you did the unit conversion calculation on a calculator, you might have come up with 110.09 m.

Table B1.1 **Metric System Units and Conversions**

Measurement	Unit of Measurement and Abbreviation	Metric System Equivalent	Converting Metric Units to English Units
Length	kilometer (km)	$1 \text{ km} = 1000 \text{ m} = 10^3 \text{ m}$	1 km = 0.62 mile
	meter (m)	1 m = 100 cm	1 m = 1.09 yards = 3.28 feet = 39.37 inches
	centimeter (cm)	$1 \text{ cm} = 0.01 \text{ m} = 10^{-2} \text{ m}$	1 cm = 0.3937 inch
	millimeter (mm)	$1 \text{ mm} = 0.001 \text{ m} = 10^{-3} \text{ m}$	1 mm = 0.039 inch
	micrometer (μm)	$1 \ \mu\text{m} = 10^{-6} \text{ m} = 10^{-3} \text{ mm}$	
	nanometer (nm)	$1 \text{ nm} = 10^{-9} \text{ m} = 10^{-3} \ \mu\text{m}$	
Area	hectare (ha)	$1 \text{ ha} = 10{,}000 \text{ m}^2$	1 ha = 2.47 acres
	square meter (m^2)	$1 \text{ m}^2 = 10{,}000 \text{ cm}^2$	$1 \text{ m}^2 = 1.196$ square yards
	square centimeter (cm^2)	$1 \text{ cm}^2 = 100 \text{ mm}^2 = 10^{-4} \text{ m}^2$	$1 \text{ cm}^2 = 0.155$ square inch
Volume	liter (L)	1 L = 1000 mL	1 L = 1.06 quarts
	milliliter (mL)	$1 \text{ mL} = 1000 \ \mu\text{L} = 10^{-3} \text{ L}$	1 mL = 0.034 fluid ounce
	microliter (μL)	$1 \ \mu\text{L} = 10^{-6} \text{ L}$	
Mass	kilogram (kg)	1 kg = 1000 g	1 kg = 2.20 pounds
	gram (g)	1 g = 1000 mg	1 g = 0.035 ounce
	milligram (mg)	$1 \text{ mg} = 1000 \ \mu\text{g} = 10^{-3} \text{ g}$	
	microgram (μg)	$1 \ \mu\text{g} = 10^{-6} \text{ g}$	
Temperature	Kelvin (K)*		K = °C + 273.15
	degrees Celsius (°C)		$°\text{C} = \frac{5}{9}(°\text{F} - 32)$
	degrees Fahrenheit (°F)		$°\text{F} = (\frac{9}{5} \times °\text{C}) + 32$

* Absolute zero is −273.15°C = 0 K.

Table B1.2 Prefixes Used in the Metric System

Prefix	Abbreviation	10^n	Decimal	English expression
pico–	p	10^{-12}	0.000000000001	one trillionth
nano–	n	10^{-9}	0.000000001	one billionth
micro–	μ	10^{-6}	0.000001	one millionth
milli–	m	10^{-3}	0.001	one thousandth
centi–	c	10^{-2}	0.01	one hundredth
deci–	d	10^{-1}	0.1	one tenth
–	–	10^{0}	1	one
deca–	da	10^{1}	10	ten
hecto–	h	10^{2}	100	one hundred
kilo–	k	10^{3}	1000	one thousand
mega–	M	10^{6}	1,000,000	one million
giga–	G	10^{9}	1,000,000,000	one billion
tera–	T	10^{12}	1,000,000,000,000	one trillion

Why was the number of meters rounded off in the equation above? The answer lies in significant figures. Let's take a closer look.

Significant Figures

Significant figures, or "sig figs," are critical when reporting scientific data. The number of significant figures in a measurement, such as 3.524, is the number of digits that are known with some degree of confidence (3, 5, and 2) plus the last digit (4), which is an estimate or approximation. How do scientists know how many digits to include when reporting a measurement?

Rules for Working with Significant Figures Here are the rules for counting significant figures in a reported measurement:

- Nonzero numbers are always significant.
- Embedded zeros (any zero between nonzero digits) are always significant.
- Leading zeros (any zero preceding the first nonzero digit) are not significant; these zeros do nothing but set the decimal point.
- Trailing zeros (any zero following the last nonzero digit) are significant *only* if the decimal point is specified. (Hint: Changing the number to scientific notation makes it easier to see "trailing" zeros that are significant.)

Table B1.3 shows examples of how to apply these rules. The bottom line? Significant figures indicate the accuracy of measurements.

Using Scientific Notation When dealing with numbers that are very large or very small, biologists use scientific notation. For example, instead of writing 0.0000000027, you could write this value as the product of two numbers: 2.7 (the digit term) and 10^{-9} (the exponential term), or 2.7×10^{-9}. The digit term shows the number of significant figures. The exponential term places the decimal point. A negative exponent of value *x* shows that to write the number in long form, you will shift the decimal point *x* number of places to the left. A positive exponent shows that you will shift the decimal point *x* places to the right.

Precision versus Accuracy If biologists count the number of bird eggs in a nest, they report the data as an exact number—say,

Table B1.3 Rules for Working with Significant Figures

Rule	Example	Scientific Notation	Number of Significant Figures
Nonzero numbers are always significant.	35,214	3.5214×10^4	5
Embedded zeros are always significant.	1.035	$1.035 \times (10^0)$	4
Leading zeros are never significant.	0.00352	3.52×10^{-3}	3
Trailing zeros are significant only if the decimal point is specified.	200 vs. 200.0	2×10^2 vs. 2.000×10^2	1 vs. 4

3 eggs. But if the same biologists are measuring the diameter of the eggs, the numbers will be inexact. Just how inexact depends on the equipment used to make the measurements.

If you measure the width of your textbook with a ruler several times, you'll get more or less the same measurement again and again. See **Figure B1.1** for a graphical representation of this idea. Precision refers to how closely individual measurements agree with each other.

You may have determined the book's width with precision, but how do you know if your ruler is accurate? Accuracy refers to how closely a measured value agrees with the correct value. You don't know the accuracy of a measuring device unless you calibrate it. For instance, you could calibrate your ruler by comparing it against a ruler that you know is accurate. As the sensitivity of equipment used to make a measurement increases, the number of significant figures increases. For example, if you used a kitchen scale to weigh some sodium chloride, you might obtain a weight of 3 ± 1 g (an accuracy of 1 significant figure); but an analytical balance in the lab might give a value of 3.524 ± 0.001 g (an accuracy of 4 significant figures).

It is important to follow the "sig fig rules" when reporting a measurement, so that data do not appear to be more accurate than the equipment allows.

Combining Measurements How do you deal with combining measurements with different degrees of accuracy and precision? A simple rule to follow when combining measurements is that the accuracy of the final answer can be no greater than the least accurate measurement (the measurement with the least significant figures).

When you add or subtract measurements, the number of significant figures doesn't matter. The only digit that matters is the last one—the digit furthest to the right. The answer can have *no more decimal places* than the least accurate measurement. For example, consider adding these three measurements: 5.9522, 2.065, and 3.6. If you add these numbers with your calculator, it gives you the answer 11.6172. But this is incorrect—now you must round your answer off to 11.6, which has one decimal place, the least number of decimal places in your data. Note that your rounded answer, in this case, has three significant figures, while your least accurate starting measurement had only two significant figures.

When you are multiplying or dividing measurements, on the other hand, the number of significant figures does matter. Multiplying significant figures will always result in a solution that has the same significant figures as the smallest significant figures you started with. When you multiply or divide measurements, the answer can have *no more significant figures* than the least accurate measurement. For instance, if you multiply 4.268 × 5.3, your calculator will give you 22.6204. But since 4.268 has four significant figures and 5.3 has two significant figures, your solution must only have two significant figures. So, the solution will be 22 instead of 22.6204.

It is important to practice working with metric units and scientific notation, and with nailing down the concept of significant figures. The Check Your Understanding questions in this BioSkills should help you get started with this process.

CHECK YOUR UNDERSTANDING

✔ **If you understood BioSkills 1, you should be able to...**

1. **QUANTITATIVE** Calculate how many miles a runner completes in a 5.0-kilometer run.
2. **QUANTITATIVE** Calculate your normal body temperature in degrees Celsius. (Normal body temperature is 98.6 °F.)
3. **QUANTITATIVE** Calculate your current weight in kilograms.
4. **QUANTITATIVE** Rewrite the following two numbers, using scientific notation: 0.00000092 and 23,000,000.
5. **QUANTITATIVE** Multiply the measurements 2.8723 and 1.6. How many significant figures does your answer have? Why?

Answers are available in Appendix A.

BioSkills 2 Reading and Making Graphs

There's a reason graphs are the most common way to report data. Compared to reading raw numerical values in a table or list, graphs make it much easier to understand what the data mean.

Learning how to read and interpret graphs is one of the most basic skills you'll need to acquire as a biology student. As when learning piano or soccer or anything else, you need to understand a few key ideas to get started and then have a chance to

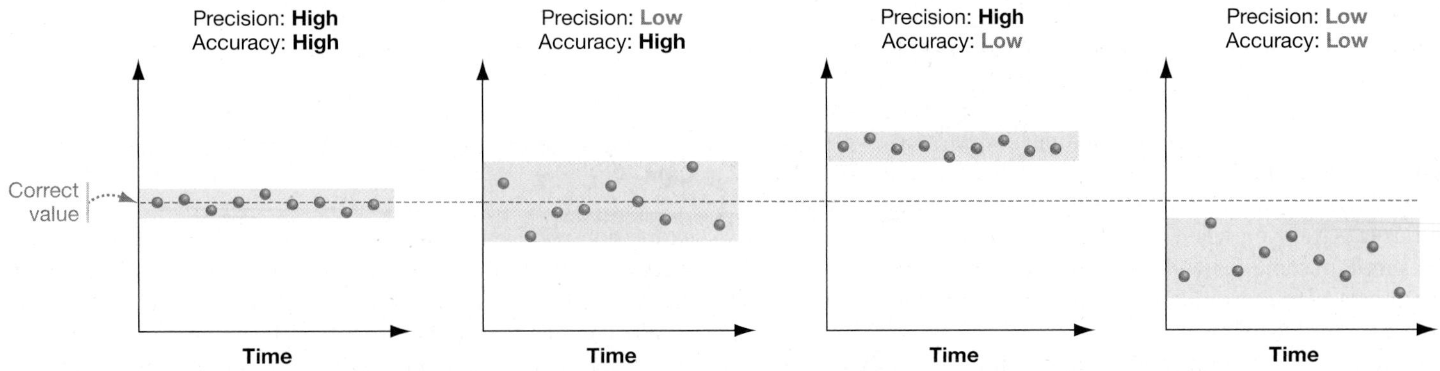

Figure B1.1 Precision versus Accuracy in Measurement. The dashed line shows the correct value (in the text example, the actual width of your textbook). The red dots indicate measurements made over time.

practice—a lot—with some guidance and feedback. At the same time, you'll be developing the skills you need to make your own graphs.

Getting Started

To start reading a graph, you need to do three things: read the axes, figure out what the data points represent—that is, where they came from—and think about the overall message of the data. Let's consider each step in turn.

What Do the Axes Represent? Most graphs have two axes: one horizontal and one vertical. The horizontal axis of a graph is called the *x*-axis. The vertical axis of a graph is called the *y*-axis. Each axis represents a variable that takes on a range of values. These values are indicated by the tick marks and labels on the axis. Note that each axis should *always* be clearly labeled with the unit or treatment it represents.

Figure B2.1 shows the steps in reading a graph where continuous data are plotted against each axis. Continuous data report an array of values over a range. In contrast, discrete data report only a restricted set of values. To grasp the difference, consider a graph of the average height of men and women in your class; in such a graph, height is a continuous variable, but sex is a discrete variable.

To create a graph, researchers plot the independent variable on the *x*-axis and the dependent variable on the *y*-axis (see Figure B2.1a). The terms "independent" and "dependent" are used because the values plotted on the *y*-axis *depend* on the *x*-axis values. In this figure, the researchers wanted to show how the protein content of maize (corn) kernels in a study population changed over time. Thus, the protein concentration plotted on the *y*-axis depended on the generation plotted on the *x*-axis. The value on the *y*-axis always depends on the value on the *x*-axis, but not vice versa.

In many graphs in biology, the independent variable is either time or the various treatments used in an experiment. In these cases, the *y*-axis records how some quantity changes as a function of time or as the outcome of the treatments applied to the experimental cells or organisms.

What Do the Data Points Represent? Once you've read the axes, you need to figure out what each data point is. In our maize kernel example, the data point in Figure B2.1b represents the average percentage of protein (15 percent) found in a sample of kernels from a study population in a particular generation (the 20th).

If it's difficult to figure out what the data points are, ask yourself where they came from—meaning how the researchers got them. You can do this by understanding how the study was done and by understanding what is being plotted on each axis. The *y*-axis will tell you what the researchers measured; the *x*-axis will usually tell you when they measured it or what group they measured. In some cases—for example, in a plot of average body size versus average brain size in primates—the *x*-axis will report a second variable that was measured.

In other cases, graphed data may represent a relative or arbitrary unit of measurement, such as the amount of gene

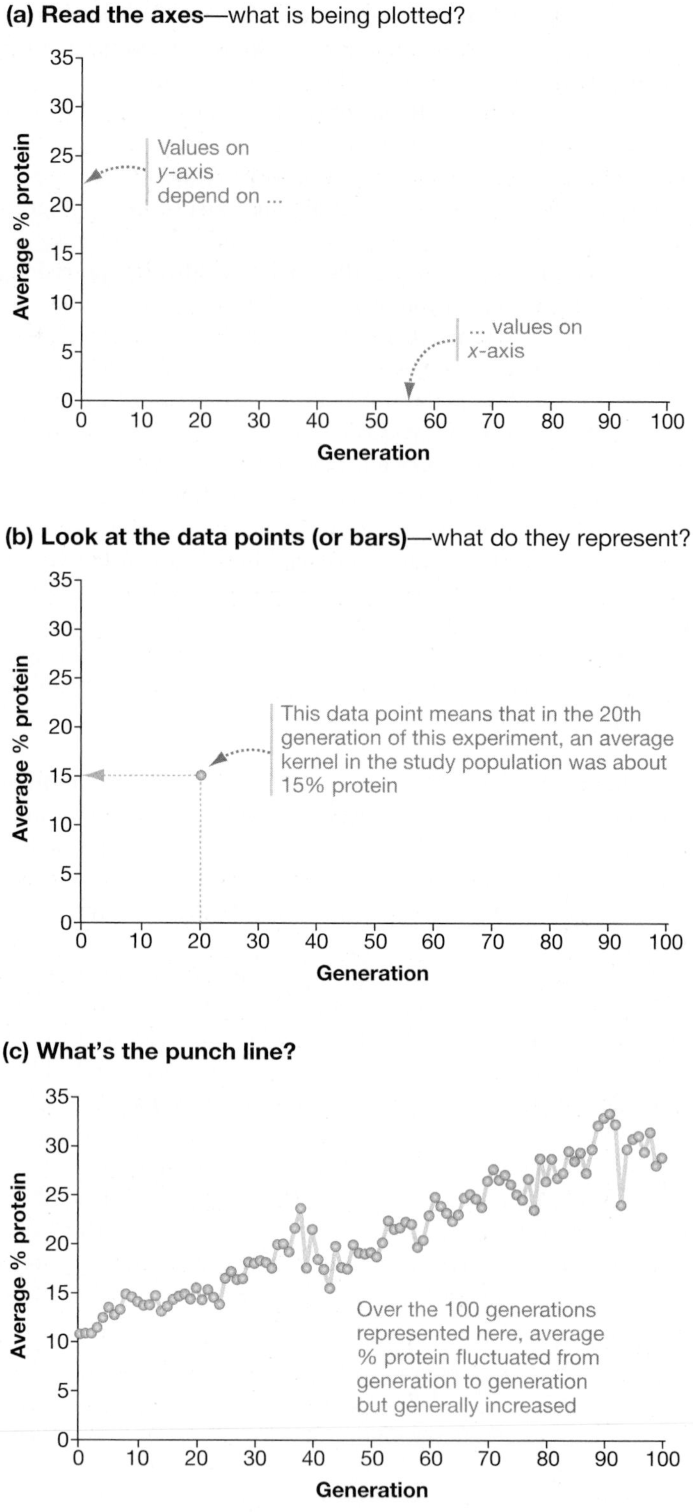

Figure B2.1 **Scatterplots Are One Way to Graph Continuous Data.**

expression *relative* to a control, with the control set at an arbitrary value of 1.0 (for an example, see Ch. 19, Figure 19.6). A data point may show the ratio of the amount of a substance, intensity, or other quantity, relative to a predetermined reference measurement. For example, the *y*-axis might show the percentage of relative activity of an enzyme—the rate of the enzyme-catalyzed reaction, scaled to the highest rate of activity observed (100 percent)—in experiments conducted under conditions that are identical except for one variable, such as pH or temperature (see Ch. 8, Figure 8.15).

What Is the Overall Trend or Message? Look at the data as a whole, and figure out what they mean. Figure B2.1c suggests an interpretation of the maize kernel example. If the graph shows how some quantity changes over time, ask yourself if that quantity is increasing, decreasing, fluctuating up and down, or staying the same. Then ask whether the pattern is the same over time or whether it changes over time.

When you're interpreting a graph, it's extremely important to limit your conclusions to the data presented. Don't extrapolate beyond the data, unless you are explicitly making a prediction based on the assumption that present trends will continue. For example, you can't say that the average percentage of protein content was increasing in the population before the experiment started, or that it will continue to increase in the future. You can say only what the data tell you.

Types of Graphs

Many of the graphs in this text are *scatterplots* like the one shown in Figure B2.1c. But you will also see other types of graphs in this text. When creating your own graphs, you'll want to think carefully about which type of graph is the most appropriate to use for a particular data set.

Scatterplots, Lines, and Curves Some scatterplots, like the one in Figure B2.1c, have data points that are connected by dot-to-dot lines to help make the overall trend clearer. In other scatterplots, the data points are unconnected or have a smooth line drawn through them.

A *smooth line* through data points—sometimes straight, sometimes curved—is a "line of best fit." It represents a mathematical function that summarizes the relationship between the x and y variables. It is "best" in the sense of fitting the data points most accurately. The line may intersect with some of the points, none of the points, or all of the points.

Curved lines often take on characteristic shapes depending on the relationships between the x and y variable. For example, bell-shaped curves typically fit data from studies on enzyme kinetics (Ch. 8, Section 8.3), while sigmoid, or S-shaped, curves fit data from many studies on oxygen–hemoglobin dissociation (Ch. 42, Section 42.4) and population growth (Ch. 51, Section 51.3).

Bar Charts, Histograms, and Box-and-Whisker Plots Scatterplots, or line-of-best-fit graphs, are the most appropriate type of graph when the data have a continuous range of values and you want to show individual data points. But other types of graphs are used to represent different types of data distributions:

- *Bar charts* plot data that have discrete or categorical values instead of a continuous range of values. In many cases the bars might represent different treatment groups in an experiment, as in Figure B2.2a. To interpret the graph, ask yourself how different the values are. If the bar chart reports means (averages) over discrete ranges of values, ask what trend is implied—as you would for a scatterplot.

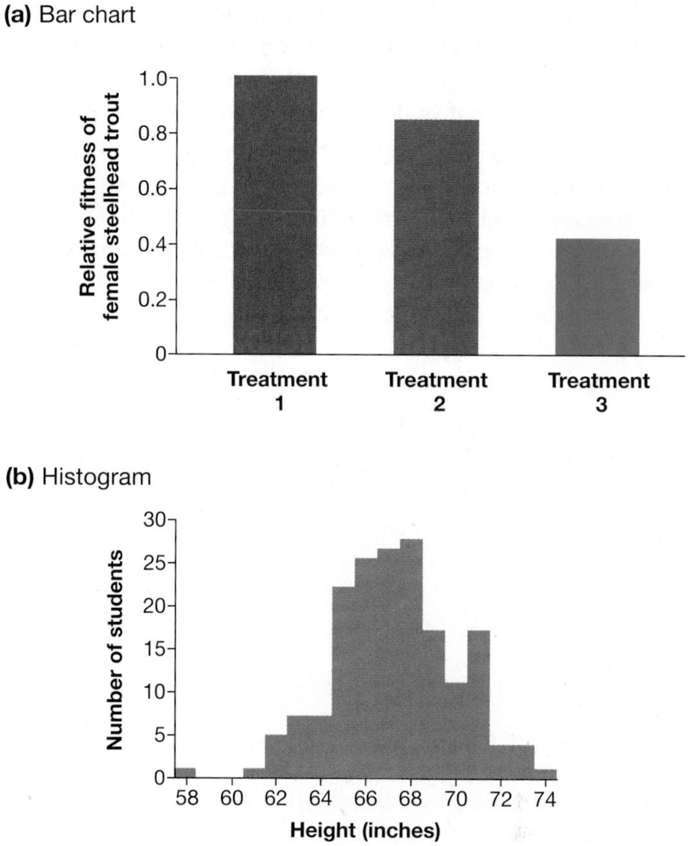

Figure B2.2 **Bar Charts and Histograms. (a)** Bar charts are used to graph data that are discontinuous or categorical. **(b)** Histograms show the distribution of frequencies or values in a population.

- *Histograms* illustrate frequency data and can be plotted as numbers or percentages. Figure B2.2b shows an example where height (in inches) is plotted on the x-axis, and the number of students in a population in the United States is plotted on the y-axis. Each bar indicates the number of individuals in each interval of height, which reflects the relative frequency, in this population, of people whose heights are in that interval. The measurements could also be recalculated so that the y-axis reported the percentage of the population in each interval. Then the total percentage for all the bars would equal 100 percent. Note that if you were to draw a smooth curve connecting the tops of the bars in this histogram, the curve would be roughly bell shaped. To interpret a histogram, ask whether there is a "hump" in the data—indicating a group of values on the x-axis that are more frequent than others. If so, what does it mean? Is the hump in the center of the distribution of values, toward the left, or toward the right?
- *Box-and-whisker plots* allow you to easily see where most of the data in a test set fall (see Ch. 1, Figure 1.11 for an example). Each box indicates where half of the data numbers are. The whiskers indicate the lower extreme and the upper extreme of the data. The vertical line inside each box indicates the median—meaning that half of the data are greater than

this value and half are less. To interpret a box-and-whisker plot, ask yourself what information the graph gives you. What is the range of values for the data? Where are half the data points? Below what value is three quarters of the data?

In all types of graphs, statistical tests can be used to determine whether a difference between treatment groups, or a difference in the relationship between two continuous ranges of values, is "significant." If differences are *statistically significant* (see **BioSkills 3**), it means they are not likely to have occurred by chance, but rather are likely to be attributable to a specific variable.

Getting Practice

Working with this textbook will give you lots of practice reading and interpreting graphs—they appear in almost every chapter. In many graphs, gray arrows and labels have been added to suggest an interpretation or draw your attention to an important point on the graph. In other graphs, you should be able to figure out what the data mean on your own or with the help of other students or your instructor.

CHECK YOUR UNDERSTANDING

✔ If you understood BioSkills 2, you should be able to . . .

1. **QUANTITATIVE** Refer to Figure B2.1 and determine the total change in average percentage of protein in maize kernels, from the start of the experiment until the end.
2. **QUANTITATIVE** Determine the trend in average percentage of protein in maize kernels between generation 37 and generation 42 in Figure B2.1.
3. Explain whether the conclusions from the bar chart in Figure B2.2a would be different if the data and label for Treatment 3 were put on the far left and the data and label for Treatment 1 on the far right.
4. **QUANTITATIVE** Determine the most common height in the class graphed in Figure B2.2b. Convert your answer to centimeters (cm).
5. **MODEL** Make a bar graph from the data in this table that shows how 16 children with central nervous system leukemia responded to treatment with the anti-tumor drug topotecan. Being free of leukemia was considered a complete response. Which is the dependent variable? Which is the independent variable?

Response	Percentage of Children
Complete response	37.5
Stable disease	50
Progressive disease	12.5

DATA: S. L. Potter et al. 2012. *Pediatric Blood Cancer* 58: 362–365.

Answers are available in Appendix A.

BioSkills 3 Interpreting Standard Error Bars and Using Statistical Tests

When biologists do an experiment, they collect data on individuals or samples in a treatment group and a control group, or several such comparison groups. Then they typically test whether the mean (average) values of the dependent variable are different in two (or more) of the groups.

Standard Error Bars

For example, in one experiment, student researchers measured how fast a product formed when they set up a reaction with three concentrations of reactants (see Ch. 8, Figure 8.4). Each treatment—meaning each combination of reactant concentrations—was replicated many times.

Figure B3.1 graphs the mean reaction rate for each of the three treatments in the students' experiment. Note that Treatments 1, 2, and 3 represent increasing concentrations of reactants. The thin "I-beams" at the top of each bar indicate the standard error of each mean. The standard error of the mean is a quantity that indicates the uncertainty in a calculated mean. In effect, it quantifies how confident you are that the mean you've calculated is the mean you'd observe if you did the experiment under the same conditions an extremely large number of times. It is a measure of precision (see **BioSkills 1**).

Note that sometimes error bars represent the confidence interval of the mean. Typically, researchers state in a graph's caption what the error bars represent. A *confidence interval* gives an estimated range of values that is likely to include the population parameter being studied, such as the survival rate of animals after exposure to a pathogen. The estimated range is calculated from a given set of sample data. A 95 percent confidence level means that 95 percent of the confidence intervals would include the population parameter.

You might also have heard the term "standard deviation." How does standard deviation differ from standard error? When biologists calculate the standard deviation of a sample, they are using it as an estimate of the variability of the population that the sample was taken from. For data with a normal distribution, about 95 percent of individuals will have values within two standard deviations of the mean. The standard deviation will not tend to change as sample size increases.

In contrast, the standard error of the mean (SEM) depends on both the standard deviation (SD) and the sample size:

$$\text{SEM} = \frac{\text{SD}}{\sqrt{\text{sample size}}}$$

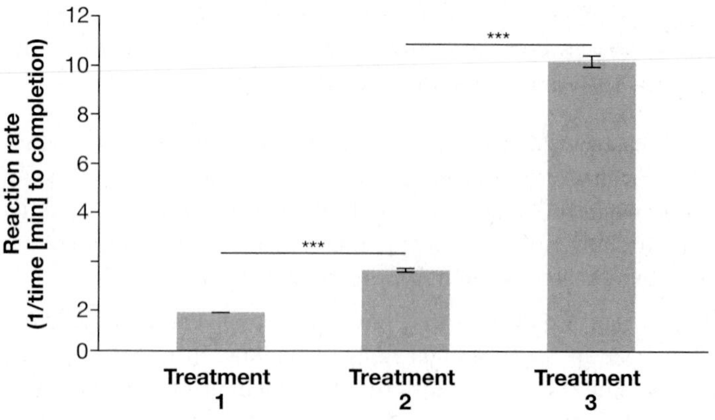

Figure B3.1 Standard Error Bars Indicate the Uncertainty in a Mean.

The standard error of the mean decreases as the sample size increases, because the extent of chance variation is reduced.

Let's consider again the experiment carried out by the student researchers (see Figure B3.1). Suppose two trials with the same concentration of reactants had a reaction rate of 0.075, and two other trials had a reaction rate of 0.025. The mean reaction rate of all four trials would be 0.050. In this case, the standard error would be large. But what if two trials had a reaction rate of 0.051 and two had a reaction rate of 0.049? The mean would still be 0.050, but the standard error would be small.

After calculating these means and standard errors, the student researchers wanted to answer a question: Does reaction rate increase when reactant concentration increases? In looking at the data, you might conclude that the answer is yes. But how could you come to a conclusion like this objectively, instead of subjectively? The answer is to use a statistical test to determine, for example, whether the difference between the rate at the highest reactant concentration and the rate at the lowest reactant concentration is significant. If the difference is found to be statistically significant, then it is not likely to have occurred by chance—it's likely to be attributable to the change in reactant concentration. Let's take a closer look at using and interpreting statistical tests.

Using Statistical Tests

If you take a statistics course, you'll learn which statistical tests are most appropriate for analyzing different types of data. Three commonly used statistical tests are the chi-square test, t-test, and analysis of variance. Other tests examine correlation and regression:

- *Chi-square tests* are used to compare observed data with data you would expect to obtain according to a specific hypothesis. For example, if, according to Mendel's principles (Ch. 14, Sections 14.2 and 14.3), you expected equal numbers of male and female offspring from a cross but you observed 9 males and 23 females, you might want to know whether the difference between the observed and expected numbers was due to chance or to other factors. How much of a difference can occur before you must conclude that something other than chance is at work? The chi-square test always tests the null hypothesis (Ch. 1, Section 1.6), which states that there is no significant difference between the observed and expected results.
- *T-tests* are used to determine if there is a significant difference between the mean values of two groups, such as the mean body sizes of mainland and island tortoises (Ch. 39, end-of-chapter Case Study). In this case, the null hypothesis would be that there is no significant difference between the means of the two data sets.
- *Analysis of variance (ANOVA)* compares the means of two or more sets of data by calculating how widely individual values in each data set vary. If they vary greatly from the mean, the variance is large, and vice versa. When applied to only two data sets, ANOVA will give the same result as a t-test. ANOVA is a powerful statistical test because it allows you to test for each factor while controlling for others and to detect whether one variable affects another. For example, if you were comparing the activity of a particular enzyme in mainland and island tortoises, you might want to determine whether sex affects enzyme activity, so you could also separate the data sets by sex.
- *Correlation analysis* is done when a researcher wants to know whether there is a relationship or correlation between two variables and, if so, is it positive (positive slope) or negative (negative slope). For example, when patients are given increasing amounts of a drug, does their blood pressure increase or decrease proportionally? Correlation is a way to express the relationship between two variables.
- *Regression analysis* is done when a researcher wants to evaluate the scatter of data points around a "line of best fit"(see **BioSkills 2**) in a graph, to determine how well the data fit that line. Does the best-fit line account well for changes in the dependent variable? When a best-fit line accounts for more of the variance, the data points are closer to the line. There are several "goodness-of-fit" statistics for regression analysis. One commonly used statistic is called *R*-squared (R^2). R^2 evaluates the scatter of the data points around a fitted line. R^2 is always between zero and one hundred percent. In general, the higher the R^2, the better the line fits your data. A line that accounts for none of the variation in the dependent variable around its mean would have an R^2 value of zero percent. A line that accounts for all of the variation in the dependent variable around its mean would have an R^2 value of 100 percent. In **Figure B3.2**, the R^2 for the line of best fit on the left, where the data points are highly scattered, is 15 percent. For the line of best fit on the right, where the data points are tightly clustered, R^2 is 85 percent. Even if you have a low R^2 value, when the independent variables are statistically significant, you may still be able to draw important conclusions about the relationships between the variables.

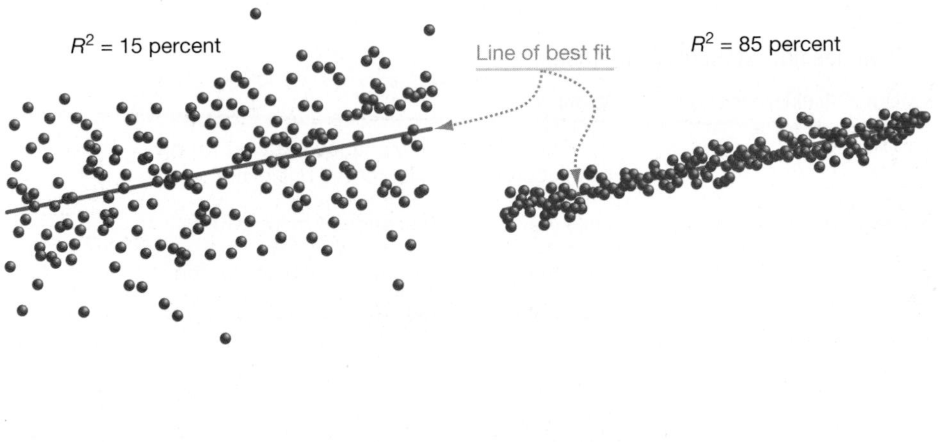

Figure B3.2 R^2 Evaluates the Scatter of the Data Points Around a Line of Best Fit.

You'll likely do statistical tests early in your undergraduate career. To use this textbook, though, you need to know only what statistical testing does and how to interpret a test statistic—a number that characterizes the size of the difference among the data sets.

Interpreting Differences: *P* Values and Statistical Significance

How do you use a statistical test to determine if differences are significant? Let's return to the experiment shown in Figure B3.1 and work through a three-step process:

1. Specify the null hypothesis, which is that reactant concentration has no effect on reaction rate.
2. Calculate a test statistic. In this experiment, the test statistic compares the actual differences in reaction rates at the three reactant concentrations to the difference predicted by the null hypothesis. The null hypothesis predicts that there should be no difference.
3. Determine the probability (see **BioSkills 4**) of getting by chance a test statistic at least as large as the one calculated. This probability, called the ***P* value**, comes from a reference distribution—a mathematical function that specifies the probability of getting various values of the test statistic if the null hypothesis is correct. The *P* value is the estimated probability of rejecting the null hypothesis when that hypothesis is correct. For example, a *P* value of 0.01 means that there is a 1 percent chance that the null hypothesis has been rejected when it is actually correct. One percent is considered a very small chance of making such an error; thus, very small *P* values indicate that researchers have high confidence in the significance of differences in their data.

By convention, most researchers consider a difference among treatment groups to be *statistically significant* if there is less than a 5 percent probability (P) of observing it by chance, or $P < 0.05$. When presenting *P* values in the scientific literature, researchers often use an asterisk rating system as well as quoting the *P* values (**Table B3.1**).

You are very likely to see small differences among treatment groups just by chance. If you flipped a coin ten times, for example, you are unlikely to get exactly five heads and five tails, even if the coin is fair. A reference distribution tells you how likely you are to get, by chance, each of the possible outcomes, such as six heads and four tails.

In the case of the student researchers' experiment (see Figure B3.1), the reference distribution indicated that if the null hypothesis of no difference in reaction rates is correct, you would see differences at least as large as those observed only 0.01 percent of the time by chance ($P < 0.0001$). Because 0.0001 is less than 0.05, the students were able to conclude that the null hypothesis—that reactant concentration has no effect on reaction rate—is not correct. According to their data, the reaction they studied really does happen more rapidly when reactant concentration increases.

What does a result that is not statistically significant mean ($P > 0.05$)? You can conclude that no effect of the treatment was detected in the experiment. However, this doesn't necessarily mean there was no underlying effect. If the sample size in a study is small—particularly in a population with lots of natural variability—researchers may not detect an effect of a particular treatment, even when an effect is actually there.

When reading graphs in this book, you should take care to inspect the standard error bars. As a *very* rough rule of thumb, means often turn out to be significantly different, according to an appropriate statistical test, if there is no overlap between two times the standard errors. But when you are asked to make conclusions about the significance of data shown in a graph, you will be provided with *P* values to interpret.

Evaluating Causation versus Correlation

If the results of an experiment show that data are significantly different between a control and treatment groups, you might be tempted to conclude that one variable caused the observed effect; for example, that consuming lots of *trans* fat causes heart disease (Ch. 6, end-of-chapter Case Study). But when interpreting the implications of the data, biologists must also evaluate whether the results identify the cause of an observed effect or only show a correlation. A correlation might be coincidental, or it might be due to a third factor, or multiple factors. In the case of heart disease, you would need to take into account an individual's age, weight, whether they exercise regularly, genetic factors, and so on.

Causation means that one event is responsible for the occurrence of the other (e.g., studying and doing well on your biology exam), while *correlation* means that two events appear to occur together (wearing your lucky hat and doing well on a biology exam). The take-home message? Correlation does not imply causation.

Table B3.1 Asterisk Rating System for *P* Values and Statistical Significance

P Value	Asterisk Rating	Statistical Significance Level	Meaning
$P > 0.05$	None	Not significant	Greater than a 1 in 20 chance of being wrong (i.e., incorrect rejection of the null hypothesis)
$P < 0.05$	*	Statistically significant	Less than a 1 in 20 chance of being wrong
$P < 0.01$	**	Statistically significant	Less than a 1 in 100 chance of being wrong
$P < 0.001$	***	Statistically significant	Less than a 1 in 1000 chance of being wrong

CHECK YOUR UNDERSTANDING

✔ If you understood BioSkills 3, you should be able to . . .

1. **QUANTITATIVE** Determine which of the following tests used to estimate the average height of individuals in a class is likely to have the smaller standard error, and why.
   - Test 1: Measuring the height of two individuals chosen at random
   - Test 2: Measuring the height of every student who showed up for class on a particular day
2. Interpret data from a recent study in which researchers investigated the evolution of sweet taste perception in hummingbirds. Captive hummingbirds were presented with a control solution (sucrose) and a test solution (either water, the artificial sweetener aspartame, or erythritol, a substance that stimulates the sweet taste receptor). The length of time the birds spent drinking each solution was recorded. What can you conclude from the data shown in the graph below? (Hint: Consult Table B3.1 on the asterisk rating system for *P* values).

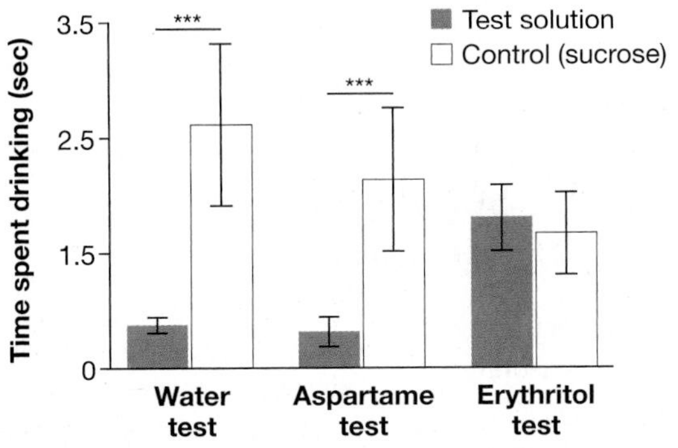

DATA: M. W. Baldwin et al. 2014. *Science* 345: 929–933.

Answers are available in Appendix A.

BioSkills 4 **Working with Probabilities**

What is probability? Probability is the chance or likelihood that an event will occur or that a hypothesis or scientific prediction is correct. In biology, just two exampes of using probability are to evaluate the significance of experimental results and to predict the outcome of genetic crosses.

To answer certain questions, biologists sometimes need to combine the probabilities of different events. You'll encounter examples of this when you solve genetics problems (Ch. 14, Section 14.2) and analyze changes in allele frequencies using the Hardy–Weinberg principle (Ch. 23, Section 23.1).

Two fundamental rules apply when combining probabilities. Each rule pertains to a distinct situation.

The "Both-And" Rule

The both-and rule—also known as the product rule or multiplication rule—applies when you want to know the probability of two or more independent events occurring together. Let's use the rolling of two dice as an example. What is the probability of rolling two sixes? These two events are independent, because the probability of rolling a six on one die has no effect on the probability of rolling a six on the other die.

The probability of rolling a six on the first die is $1/6$. The probability of rolling a six on the second die is also $1/6$. The probability of rolling a six on *both* dice, then, is $1/6 \times 1/6 = 1/36$. In other words, if you rolled two dice 36 times, on average you would expect to roll two sixes once.

In the case of a cross between two parents heterozygous for the *R* gene (genotype *Rr*), the probability of getting a gamete (egg or sperm) with allele *R* from one parent has no effect on the probability of getting a gamete with allele *R* from the other parent. Gametes fuse randomly. The probability of a child getting allele *R* from the father is $1/2$, and the probability of the child getting allele *R* from the mother is $1/2$. Thus, the probability of getting both *R* alleles and having the genotype *RR* is $1/2 \times 1/2 = 1/4$.

The "Either-Or" Rule

The either-or rule—also known as the sum rule or addition rule—applies when you want to know the probability of an event happening when there are two or more alternative ways for that event to occur. In this case, the probability that the event will occur is the sum of the probabilities of each way that it can occur.

For example, suppose you wanted to know the probability of rolling either a one or a six when you toss a die. The probability of rolling each number is $1/6$ so the probability of rolling one or the other is $1/6 + 1/6 = 1/3$. If you rolled a die three times, on average you'd expect to roll a one or a six once.

In the case of a cross between two parents heterozygous for the *R* gene, the probability of getting an *R* allele from the father and an *r* allele from the mother is $1/2 \times 1/2 = 1/4$. Similarly, the probability of getting an *r* allele from the father and an *R* allele from the mother is $1/2 \times 1/2 = 1/4$. Thus, the combined probability of getting the *Rr* genotype in either of the two ways is $1/4 + 1/4 = 1/2$.

CHECK YOUR UNDERSTANDING

✔ If you understood BioSkills 4, you should be able to . . .

1. **QUANTITATIVE** Calculate the probability of getting four "tails" if four students each toss a coin.
2. **QUANTITATIVE** Calculate the probability of getting a two, a three, or a six after a single roll of a die.

Answers are available in Appendix A.

BioSkills 5 **Using Logarithms**

You will encounter logarithms at several points in this text. Logarithms are a way of working with powers—numbers that are multiplied by themselves one or more times.

Logarithms are useful when you are studying something that can have a large range of values, like the concentration of hydrogen ions in a solution or the intensity of sound that the human ear can detect. In cases like these, it's convenient to express the numbers involved using exponents. Using exponents makes a large range of numbers more manageable. For example, instead of saying that the hydrogen ion concentration in a solution can range from 0 to 10^{-14}, the logarithmic pH scale allows you to simply say that it ranges from 0 to 14. Instead of giving the actual value, the pH scale expresses a concentration in terms of an exponent.

Scientists use exponential notation to represent powers. For example,

$$a^x = y$$

means that if you multiply a by itself x times, you get y. In exponential notation, a is called the base and x is called the exponent. The entire expression is called an exponential function.

What if you know y and a, and you want to know x? This is where logarithms come in. You can solve for exponents by using logarithms:

$$x = \log_a y$$

This equation reads that x is equal to the logarithm of y to the base a. Logarithms are a way of working with exponential functions. They are important because so many processes in biology (and in chemistry and physics, for that matter) are exponential. To understand what's going on, you have to describe the process with an exponential function and then use logarithms to work with that function.

Although a base can be any number, most scientists use just two bases when they employ logarithmic notation: 10 and e (sometimes called Euler's number after Swiss mathematician Leonhard Euler). What is e? It is the limit of $(1 + 1/n)^n$ as n tends to infinity. Mathematicians have shown that the base e is an irrational number (like π) that is approximately equal to 2.718. Like 10, e is just a number; $10^0 = 1$ and, likewise, $e^0 = 1$. Both 10 and e have qualities that make them convenient bases to use in science.

Logarithms to the base 10 are so common that they are usually symbolized in the form $\log y$ instead of $\log_{10} y$. A logarithm to the base e is called a natural logarithm and is symbolized as ln (pronounced *EL-EN*) instead of log. You write "the natural logarithm of y" as $\ln y$.

Most scientific calculators have keys that allow you to solve problems involving base 10 and base e. For example, if you know y, they'll tell you what $\log y$ or $\ln y$ are—meaning that they'll solve for x in our first example equation. They'll also allow you to find a number when you know its logarithm to base 10 or base e. Stated another way, they'll tell you what y is if you know x, and y is equal to e^x or 10^x. This process is called finding an antilog. In most cases, you'll use the inverse or second function button on your calculator to find an antilog (above the log or ln key).

To get some practice using a scientific calculator, consider this equation:

$$10^2 = 100$$

If you press the calculator's "log" function key, enter 100, and then press the "equals" key, the readout should say 2. The logarithm tells you what the exponent is. If you enter 10, followed by the antilog function key (caret symbol, ^), followed by 2, and press the "equals" key, the readout should say 100. The antilog solves the exponential function, given the base and the exponent.

If your background in algebra isn't strong, you'll want to get more practice working with logarithms because you'll see them frequently during your undergraduate career. Remember that once you understand the basic notation, there's nothing mysterious about logarithms. They are simply a way of working with exponential functions, which describe what happens when something is multiplied by itself a number of times—like cells that replicate and then replicate again and then again.

CHECK YOUR UNDERSTANDING

✔ **If you understood BioSkills 5, you should be able to . . .**

For questions 1 and 2, use the equation $N_t = N_0 e^{rt}$ (see Ch. 51, Quantitative Methods 51.3).

1. Identify the type of function this equation describes.
2. **QUANTITATIVE** Rewrite this equation after taking the natural logarithm of both sides.

For questions 3 and 4, use the equation $\text{pH} = -\log[\text{H}^+]$ (see Ch. 2, Section 2.2).

3. **QUANTITATIVE** Calculate the pH of a solution whose $[\text{H}^+]$ is 2.75×10^{-4}.
4. **QUANTITATIVE** Determine the $[\text{H}^+]$ of a solution whose pH is 5.43.

Answers are available in Appendix A.

BioSkills 6 Separating and Visualizing Molecules

After you complete BioSkills 6–11, you should be able to . . .

- Describe the key principles underlying common techniques used for scientific inquiry.

To study a molecule, you have to be able to isolate it. Isolating a molecule is a two-step process: The molecule has to be separated from other molecules in a mixture and then physically picked out or located in a purified form. Let's explore some of the techniques that biologists use to separate proteins and nucleic acids and then find the particular molecule they are interested in.

Using Electrophoresis to Separate Molecules

In molecular biology, a standard technique for separating proteins and nucleic acids is called gel electrophoresis or, simply, electrophoresis (literally, "electricity-moving"). You may be using electrophoresis in a lab for this course, and you will be analyzing data derived from electrophoresis in this text.

The principle behind electrophoresis is simple. Nucleic acids carry a negative charge, as do proteins when they are denatured and coated with a charged (ionic) detergent. As a result, these molecules move when placed in an electric field. Negatively charged molecules move toward the positive electrode; positively charged molecules move toward the negative electrode.

An Example "Run" **Figure B6.1** shows an electrophoresis setup. To separate a mixture (sample) of macromolecules so that each one can be isolated and analyzed, researchers load the sample onto a gelatinous substrate ("gel") consisting of long molecules that form a matrix of fibers. The matrix has pores that act like a sieve through which the molecules in the sample can pass.

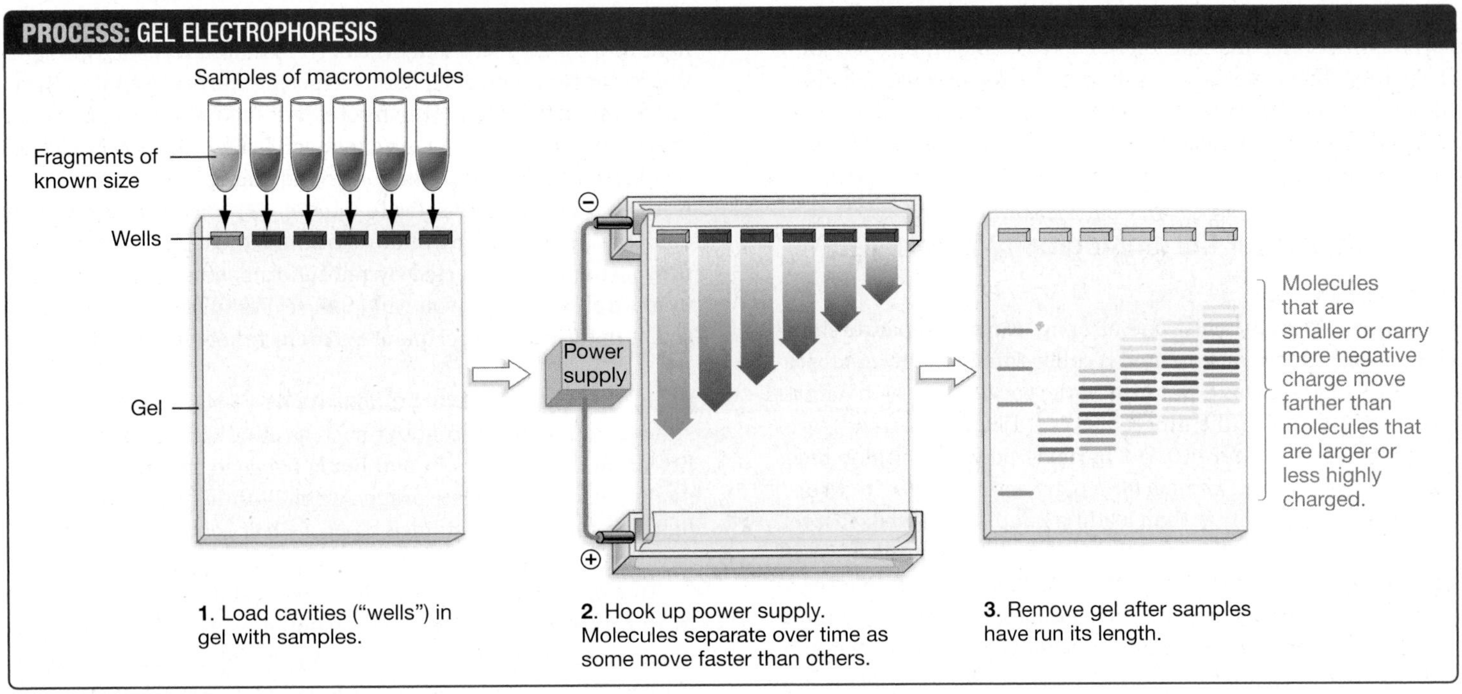

Figure B6.1 Macromolecules Can Be Separated via Gel Electrophoresis.

✔ DNA and RNA move toward the positive electrode. What makes these molecules negatively charged?

As shown in step 1, each sample is placed in a slot ("well") in a sheet or slab of the gel. In many cases, researchers also fill a well with a sample containing proteins or DNA molecules of known size. When the gel is run, these molecules will form a size standard or "ladder."

In step 2, the gel is immersed in a solution that conducts electricity. When an electric field is applied across the gel, the molecules in each well move through the gel toward the positive electrode, forming a lane. Molecules that are smaller or more highly charged for their size move faster through the sieve than do larger or less highly charged molecules. As they move, then, the molecules separate by size or by charge. Small or highly charged molecules end up near the bottom of the gel; large or less-charged molecules remain near the top.

Once molecules of different size or charge have separated from one another, the electric field is turned off and the lanes are analyzed (step 3).

A combination of charge and size is important in separating molecules by electrophoresis. Nucleic acids have a fixed amount of charge for a given length of molecule, so the size of the molecule determines how fast it runs on the gel. The same is true for proteins that are treated with a charged detergent before they are run on a gel. For proteins that are run without being treated with a charged detergent, size and charge work together to determine how fast they separate on a gel.

Why Do Separated Molecules Form Bands? When researchers visualize a particular molecule on a gel, using techniques described later in BioSkills 6, the image that results consists of bands: lines of varying thickness that are as wide as a lane in the gel. Why?

To understand the answer, study Figure B6.2. The left panel shows the original mixture of molecules. In this diagram, the size of each dot represents the size of each molecule. The key is

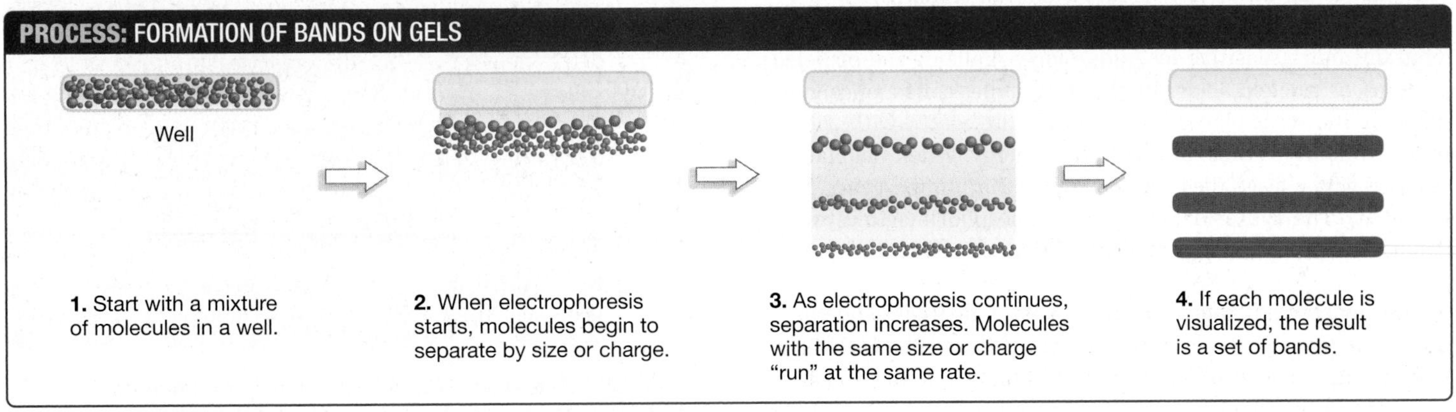

Figure B6.2 On a Gel, Molecules That Are Alike Form Bands.

to realize that the original sample contains many copies of each specific molecule, and that these copies run down the length of the gel together—meaning, at the same rate—because they have the same size or charge.

It's that simple: Molecules that are alike form a band because they stay together.

Using Thin Layer Chromatography to Separate Molecules

Gel electrophoresis is just one of many ways to separate molecules. Another common method is called thin layer chromatography. This method was developed in the early 1900s by botanists who were analyzing the different-colored pigments from leaves of a plant (Ch. 10, Figure 10.5). The name chromatography comes from the Greek words *khroma* for "color" and *graphein*, "to write."

In this method, rather than loading samples into wells in a gel, the samples are deposited or "spotted" near the bottom of a stiff support, either glass or plastic, that is coated with a thin layer of silica gel, cellulose, or a similar porous material. The coated support is then placed in a solvent. As the solvent moves up through the coating by capillary action, it carries the molecules in the samples with it. Molecules are carried different distances, based on their size and solubility in the solvent.

Visualizing Molecules

Once molecules have been separated using electrophoresis or thin layer chromatography, they have to be detected. Although plant pigments are colored, nucleic acids and most proteins are invisible unless they are labeled in some way.

Using Radioactive Isotopes In the early days of molecular biology, the first types of labels in common use were radioactive isotopes—forms of atoms that are unstable and release energy in the form of radiation. Radioactive atoms can be incorporated into proteins or nucleic acids during synthesis, and the radiation then can be used to detect the labeled macromolecules.

Once electrophoresis is complete, the labeled proteins or nucleic acids can be visualized by laying X-ray film over the gel. Because radiation exposes film, a black dot appears wherever a radioactive atom is located in the gel. So many black dots occur so close together that they form a dark band. This technique for visualizing macromolecules is called autoradiography.

Advances in technology have led to the development of a related but more sensitive technique for visualizing radiolabeled proteins or nucleic acids. In this technique, called phosphorimaging, the gel is placed on a special plate coated with "phosphor" compounds that store radiation energy. When the plate is scanned with a laser, the energy is released as luminescence (visible light). The light is detected by the phosphorimaging system and used to produce a digital image of the gel.

Using Fluorescent Tags Starting in the late 1990s and early 2000s, it became common to label macromolecules with fluorescent tags instead of radioactive isotopes. A "tag" is a molecule that is attached chemically to proteins or nucleic acids to allow them to be visualized. In this case, the tag is a fluorescent molecule that absorbs light at one wavelength and reemits the light at a longer wavelength. Once electrophoresis is complete, fluorescence can be detected by exposing the gel to an appropriate wavelength of light; the fluorescent tag fluoresces, or glows, in response. (Fluorescence is explained in Ch. 10, Section 10.2.)

Fluorescent tags have important advantages over radioactive isotopes: **(1)** They are safer to handle. **(2)** They are faster—you don't have to wait hours or days for the radioactive isotope to expose a film or be detected by phosphorimaging. **(3)** They come in multiple colors, so you can label several different macromolecules in the same experiment and detect them independently.

Using Nucleic Acid and Protein Stains DNA and RNA can be stained with a fluorescent dye such as ethidium bromide (EtBr). Ethidium bromide fits in and binds between the bases, causing nucleic acids to fluoresce orange when illuminated by ultraviolet light. Proteins can be stained using a silver solution or dyes such as Coomassie blue that bind to proteins in the gel.

An example of an EtBr-stained gel is shown in **Figure B6.3**. In this experiment, researchers wanted to determine the optimal temperature for annealing (complementary base pairing) of primers to a particular target DNA in a polymerase chain reaction (PCR; see **BioSkills 10**). The far-left lane contains DNA fragments of known size; this lane (the size standard) is used to estimate the size of the molecules in the other lanes, which are numbered: Lane 1 is a control sample containing no target DNA; lanes 2 through 8 are samples in which the primer annealing temperature was varied incrementally from 71 °C to 51 °C. Primer was present in all test samples and the control.

Reading a Gel One of the keys to interpreting, or "reading," a gel or an image of a gel is to realize that brighter or more intense bands contain more of the stain or label, indicating a greater amount of the stained or labeled molecule. Fainter bands contain less of the molecule.

To read a gel, then, you look for **(1)** the presence or absence of bands in some lanes—meaning some experimental samples—and

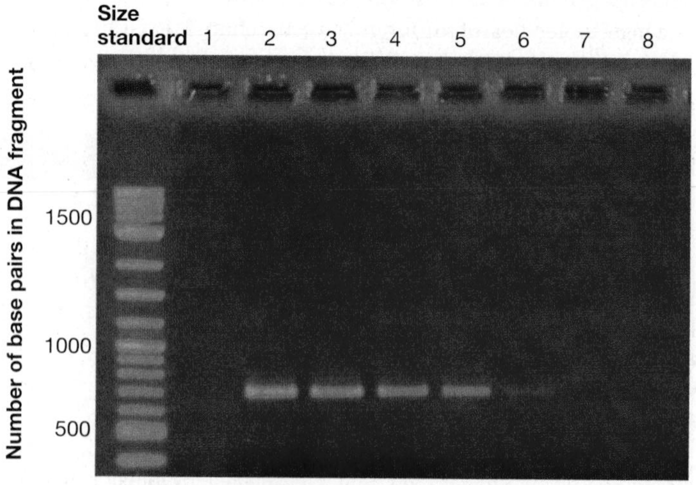

Figure B6.3 Ethidium Bromide Staining Is a Technique for Visualizing Nucleic Acids. The DNA molecules in this gel were stained with ethidium bromide and illuminated by ultraviolet light.

(2) differences in the intensity of the bands—reflecting differences in the amount of DNA or RNA or protein present.

For example, several conclusions can be drawn from the data in Figure B6.3. Note that a DNA fragment (band on gel) containing about 700 base pairs was amplified over a range of primer annealing temperatures. Based on the intensity of the bands, you can conclude that lane 6 contained less of this fragment than lanes 2–5, and lanes 7 and 8 contained none at all. In the absence of the target DNA (lane 1), no 700-base-pair fragment was amplified. The absence of a band in lane 1 indicates that the primers were specific for the target DNA used in all the other lanes.

Using Nucleic Acid Probes In many cases, researchers want to find one specific molecule—a certain DNA sequence, for example—in a collection of molecules. How is this possible? The answer hinges on using a particular molecule as a probe. A probe is a labeled molecule that binds specifically to the molecule of interest. The label is often a radioactive atom, a fluorescent tag, or a tag that can be recognized by an antibody with an attached enzyme that catalyzes a color-forming or light-emitting reaction.

For example, a nucleic acid probe is a single-stranded fragment of DNA or RNA that will bind to a particular single-stranded complementary sequence in a mixture of DNA or RNA molecules. By binding to the target sequence, the probe marks the fragment containing that sequence, distinguishing it from all the other nucleic acid fragments in the mixture. As **Figure B6.4** shows, a nucleic acid probe—in this case a labeled DNA probe—can be found after it has bound to the complementary sequence in the large mixture of fragments.

Before applying your probe to locate a particular DNA or RNA sequence on a gel, you will first need to transfer the nucleic acids to a special membrane by a technique called blotting.

- *Southern blotting*, invented by Edwin Southern, is a technique for identifying DNA segments of interest in a mixed sample. This involves making DNA fragments that have been run on a gel single stranded, transferring them from the gel to a blotting membrane, and then exposing the membrane to a single-stranded probe that binds to the target sequence by complementary base pairing. Once the probe has bound, you can detect the band that contains it through autoradiography, a light-emitting reaction, fluorescence, or a color change.
- *Northern blotting* is a technique for detecting target RNA segments. It involves transferring RNA fragments from a gel to a blotting membrane and then probing them to detect the segment of interest. The name is a play on Southern blotting, the protocol that it was derived from.

Using Antibodies as Probes How can researchers pick a particular protein out of a large collection of different proteins? The answer is to use an antibody. An antibody is an immune system protein that binds specifically to a section of a protein of interest (see Ch. 48, Section 48.2 for more details on antibodies).

To use an antibody as a probe, investigators attach a tag molecule—often an enzyme that catalyzes a color-forming or light-emitting reaction—to the antibody and then add the tagged antibody to the collection of proteins. The antibody will bind to its target protein and can be visualized thanks to the tag it carries.

PROCESS: USING A DNA PROBE

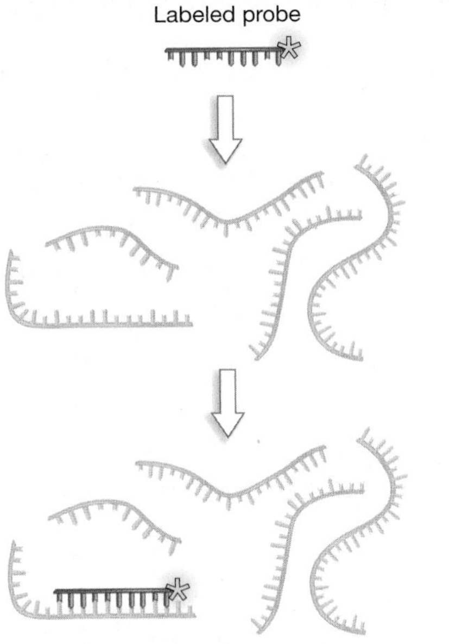

Figure B6.4 DNA Probes Bind to and Identify Specific Target Sequences.

✔ If you understand the concept of a DNA probe, you should be able to explain why the probe must be single stranded and labeled in order to work, and why it binds to just one specific fragment. You should also be able to indicate where a probe with the sequence AATCG will bind to a target DNA strand with the sequence TTTTACCCATTTACGATTGGCCT. (Recall that sequences are always written 5′ → 3′.)

If the proteins in question have been separated by gel electrophoresis and transferred to a membrane, the result is called a western blot. The name is an extension of the naming pattern for Southern and northern blots.

Using Radioimmunoassay and ELISA to Measure Amounts of Molecules Another important method that makes use of antibodies is called a radioimmunoassay. This method is used when investigators want to measure tiny amounts of a molecule, such as a hormone in the blood. In this case, a known quantity of a hormone is labeled with a radioactive isotope. This labeled hormone is then mixed with a known amount of antibody specific for the hormone, and the two bind to one another. Next, a sample of blood, containing an unknown quantity of the same hormone, is added. The hormone from the blood and the radiolabeled hormone compete for antibody binding sites.

As the concentration of unlabeled hormone increases, more of it binds to the antibody, displacing more of the radiolabeled hormone. The amount of unbound radiolabeled hormone is then measured. Using known standards as a reference, the amount of hormone in the blood can be determined.

A commonly used technique based on similar principles is called enzyme-linked immunosorbent assay (ELISA). In ELISA, the amount of a particular molecule is measured using colorimetric signals instead of a radioactive signal.

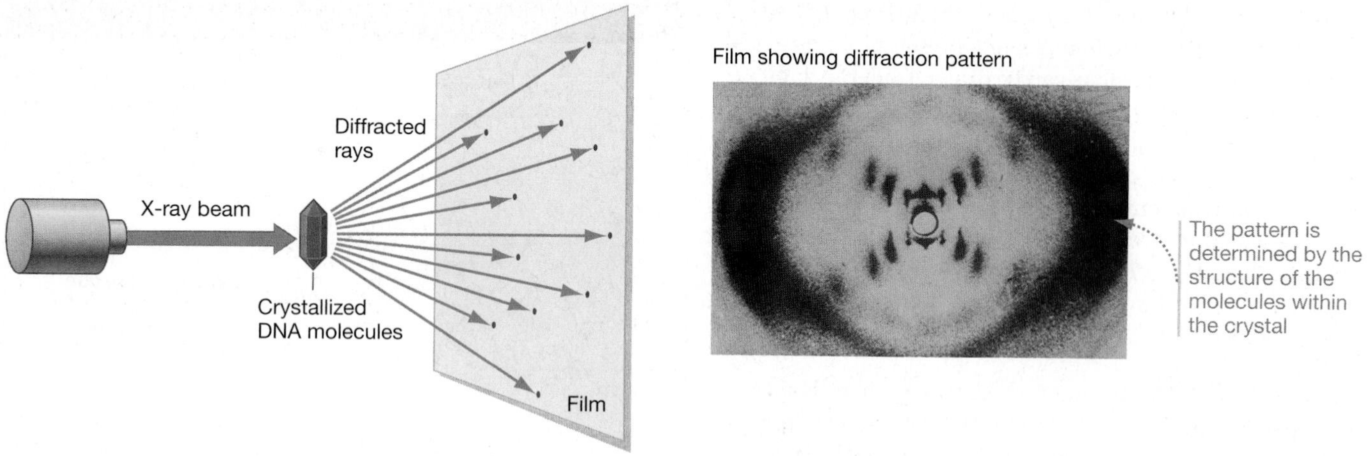

Figure B6.5 X-Ray Crystallography. When crystallized molecules are bombarded with X-rays, the radiation is scattered in distinctive patterns. The photograph at the right, obtained by Rosalind Franklin in 1953, shows an X-ray film that recorded the pattern of scattered radiation from DNA molecules.

Using X-ray Crystallography to Visualize Macromolecules To understand what the 3-D structure of individual macromolecules looks like, researchers use a technique called X-ray crystallography, or X-ray diffraction analysis. The procedure is based on bombarding crystals of a molecule with X-rays. X-rays are scattered in precise ways when they interact with the atoms in a crystal, producing a diffraction pattern that can be recorded on X-ray film or other types of detectors (Figure B6.5).

By varying the orientation of the X-ray beam as it strikes a crystal and documenting the diffraction patterns that result, researchers can construct a map representing the density of electrons in the crystal. Relating these electron-density maps to information about the primary structure of the nucleic acid or protein allows researchers to build a 3-D model of the molecule. Virtually all of the molecular models used in this book were built from X-ray crystallographic data.

CHECK YOUR UNDERSTANDING

✔ **If you understood BioSkills 6, you should be able to . . .**

1. Consider a gel that has been stained for DNA products from a polymerase chain reaction using ethidium bromide. One lane contains no bands. Two lanes have a band in the same location, but one of the bands is very faint and the other is extremely bright. Interpret these results.
2. Explain why the effort to understand the structure of biological molecules is worthwhile even though X-ray crystallography is time-consuming and technically difficult. What's the payoff?

Answers are available in Appendix A.

BioSkills 7 **Separating Cell Components by Centrifugation**

Biologists use a technique called differential centrifugation to isolate specific cell components. A centrifuge accomplishes this task by spinning a cell sample in a solution that allows cell components to separate according to their density or size and shape. The individual parts of the cell can then be purified and studied in detail, in isolation from other parts of the cell.

The first step in preparing a cell sample for centrifugation is to release the cell components by breaking the cells apart. This can be done by putting them in a hypotonic solution (Ch. 6, Figure 6.14), by exposing them to high-frequency vibration, by treating them with a detergent, or by grinding them up. Each of these methods breaks apart plasma membranes and releases the contents of the cells.

The resulting pieces of plasma membrane quickly reseal to form small vesicles, often trapping cell components inside. The suspension that results from the homogenization step is a mixture of these vesicles, free-floating macromolecules released from the cells, and organelles. A suspension like this is called a cell extract or cell homogenate.

When a cell homogenate is placed in a centrifuge tube and spun at high speed, the suspended components move toward the bottom of the tube, along the red arrows in Figure B7.1a. The effect is similar to a merry-go-round, which seems to push you away from the spinning platform. At the same time, the solution in the tube exerts a centripetal (literally, "center-seeking") force that pushes the homogenate away from the bottom of the tube. Larger, denser components resist this inward force more readily than do smaller, less dense ones and so reach the bottom of the tube faster.

To separate the different components of a cell extract, researchers often perform a series of centrifuge runs. Steps 1 and 2 of Figure B7.1b illustrate how an initial run at low speed causes larger, heavier parts of the homogenate to move below smaller, lighter parts. The material that collects at the bottom of the tube is called the pellet, and the solution and components left behind form the supernatant ("above-swimming"). The supernatant is placed in a fresh tube and centrifuged at increasingly higher speeds and longer durations. Each centrifuge run continues to separate cell components, or "fractions" of a cell homogenate, based on their size and density.

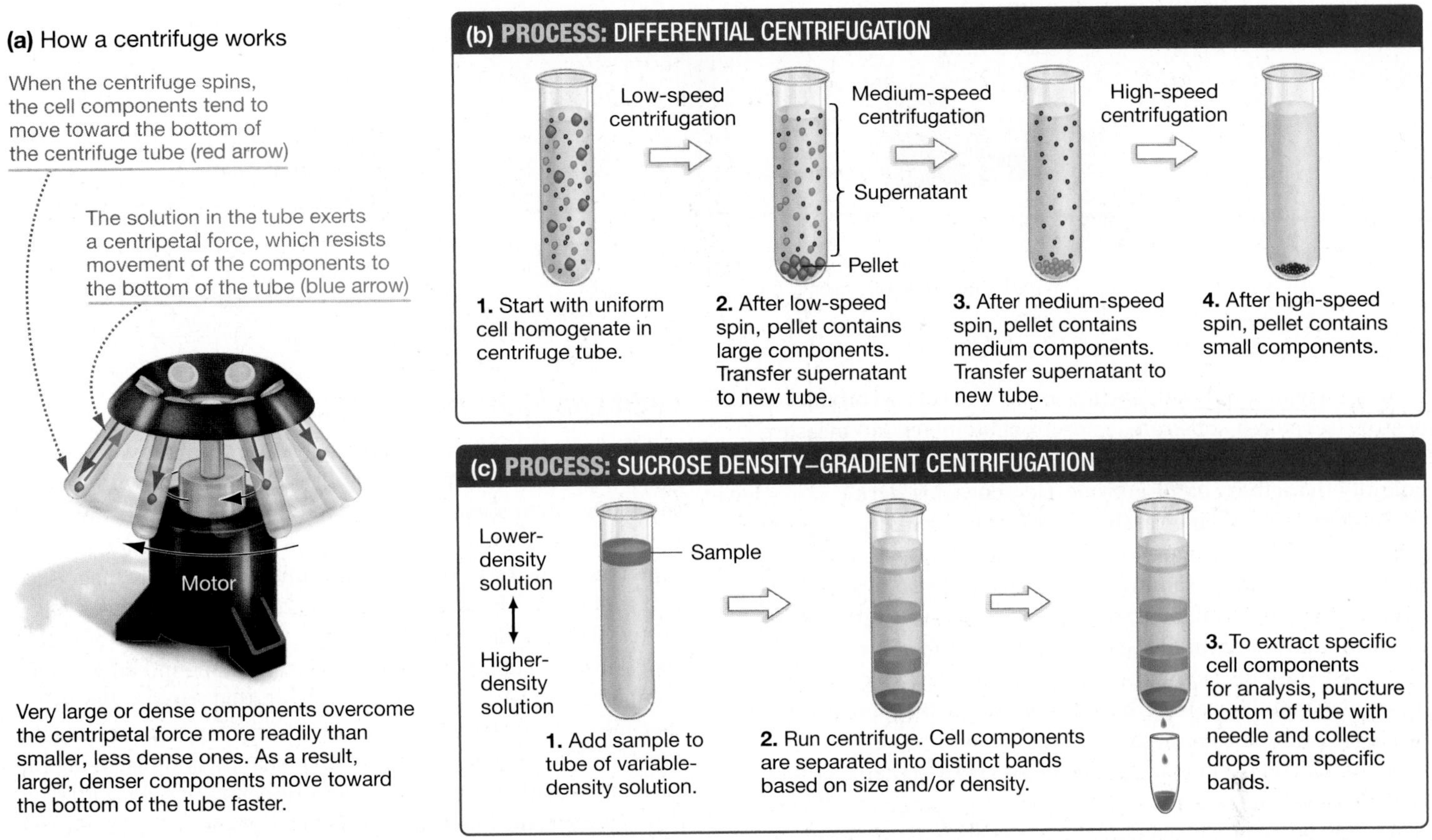

Figure B7.1 Cell Components Can Be Separated by Centrifugation. (a) The forces inside a centrifuge tube allow cell components to be separated. **(b)** Through a series of centrifuge runs made at increasingly higher speeds, an investigator can separate "fractions" of a cell homogenate by size via differential centrifugation. **(c)** A high-speed centrifuge run can achieve extremely fine separation among cell components by sucrose density–gradient centrifugation.

To separate macromolecules or organelles (for a list of eukaryotic cell components, see Ch. 7, Summary Table 7.2), researchers may also fill the centrifuge tube with a series of sucrose solutions of decreasing density, starting with the highest density at the bottom of the tube (Figure B7.1c). The resulting density gradient allows cell components to separate on the basis of small differences in size, shape, and density. When the centrifuge run is complete, each cell component occupies a distinct band of material in the tube, based on where that component settled in the density gradient. A researcher can collect the material in each band for further study.

CHECK YOUR UNDERSTANDING

✔ **If you understood BioSkills 7, you should be able to . . .**

1. List the physical properties of molecules or cell components that allow their separation via centrifugation.
2. State which cell component—ribosomes or mitochondria—you would expect to form a pellet more quickly when you centrifuge a cell homogenate at medium speed using the method shown in Figure B7.1b. Explain why.

Answers are available in Appendix A.

BioSkills 8 Using Spectrophotometry

Spectrophotometry is a versatile technique in which an instrument called a spectrophotometer measures light absorbance by a substance (Ch. 10, Section 10.2, and Ch. 39, Section 39.1). This measurement can be used to determine the concentration of the substance. In the spectrophotometer, light is passed from a lamp through a prism or diffraction grating, which splits the light into individual wavelengths (Figure B8.1 on page 36). A moveable slit is then positioned to allow only light of a single wavelength to reach the sample, which is placed in the light path in a transparent cuvette or test tube. On the other side of the sample holder is a detector that measures the amount of transmitted light (T) that got through the sample. This value is then converted into the amount of absorbed light (A).

In the lab, you can use spectrophotometry in the following tasks: **(1)** calculating the concentration of DNA, RNA, or proteins in a solution; **(2)** following the growth of bacterial cells; **(3)** quantifying the amount of photosynthesis occurring in chloroplasts, or **(4)** determining the rate of an enzyme-catalyzed reaction. Let's examine the last example in more detail.

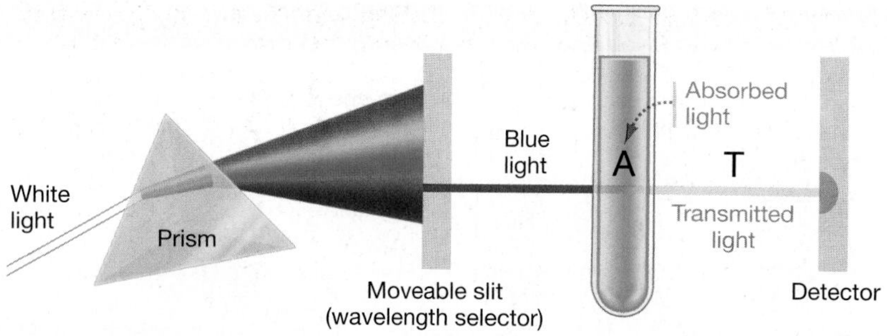

Figure B8.1 How a Spectrophotometer Works.

If an enzyme-catalyzed reaction produces a colored product or destroys a colored substrate, a spectrophotometer can measure how much of that product or substrate is present and thereby quantify the activity of the enzyme. How does this work?

Suppose an enzyme-catalyzed reaction produces a green product that absorbs light best at a wavelength of 475 nm (blue light). If a solution containing this product is placed in a spectrophotometer and illuminated with light of this wavelength, the solution will absorb most of the blue light and transmit only some of it (see Figure B8.1). A more concentrated solution will absorb more blue light than a less concentrated one, so the more colored product a solution contains, the darker it looks.

Even solutions that appear colorless may actually absorb specific wavelengths of light. For instance, a solution of DNA absorbs light best at a wavelength of 260 nm, which is in the ultraviolet range. There are no units for absorbance, but you should always state the wavelength used, for example, "absorbance at 260 nm."

CHECK YOUR UNDERSTANDING

✔ **If you understood BioSkills 8, you should be able to . . .**

Explain the relationship between absorbance and transmittance of light through the sample in the spectrophotometer shown in Figure B8.1.

Answers are available in Appendix A.

BioSkills 9 **Using Microscopy**

A lot of biology happens at levels that can't be detected with the naked eye. Biologists use an array of microscopes to study whole embryos, tissues, small multicellular organisms or individual cells, and the contents of cells.

You'll probably use dissecting microscopes and compound light microscopes to view specimens during your labs for this course, and throughout this text you'll see images generated from other types of microscopy. The key is to recognize that each approach for visualizing microscopic structures has strengths and weaknesses. As a result, each technique is appropriate for studying certain types or aspects of cells or molecules.

Light and Fluorescence Microscopy

If you use a dissecting microscope during labs, you'll recognize that it works by magnifying light that bounces off a whole specimen—often a live organism. You'll be able to view the specimen in three dimensions, which is why these instruments are sometimes called stereomicroscopes, but the maximum magnification possible is only about 20× to 40× (20–40 times normal size).

To view smaller objects, such as wet mounts or prepared slides of specimens, you'll probably use a compound microscope. Compound microscopes magnify light that passes *through* a specimen. The instruments used in introductory labs are usually capable of 400× magnification. This magnification is enough to view individual bacterial or eukaryotic cells and see large structures inside cells, like condensed chromosomes (Ch. 12, Section 12.2). The most sophisticated compound microscopes available can achieve magnifications of up to 2000×.

To prepare a specimen for viewing under a compound light microscope, researchers may need to slice the tissues or cells to create a section thin enough for light to pass through. The section is often stained to increase contrast and make structures visible. In many cases, different types of dyes are used to highlight different types of structures.

To visualize the location of specific proteins, such as structural or regulatory proteins, or to visualize organelles, such as mitochondria, researchers use a technique called immunostaining. After tissues or cells are prepared for viewing, the specimen is stained with fluorescently tagged antibodies. In this case, the cells are viewed under a fluorescence microscope. The fluorescing tag emits visible light when ultraviolet or other wavelengths of light are passed through the specimen. The result? Beautiful cells that glow green, red, or blue.

Electron Microscopy

Until the 1950s, the compound microscope was the biologist's only tool for viewing cells directly. But the invention of the electron microscope provided a new way to view specimens. Two basic types of electron microscopy are now available: one type allows researchers to examine very thin cross sections of cells at extremely high magnification (transmission electron microscopy), the other offers a view of surfaces at somewhat lower magnification (scanning electron microscopy). Let's look at each type in turn.

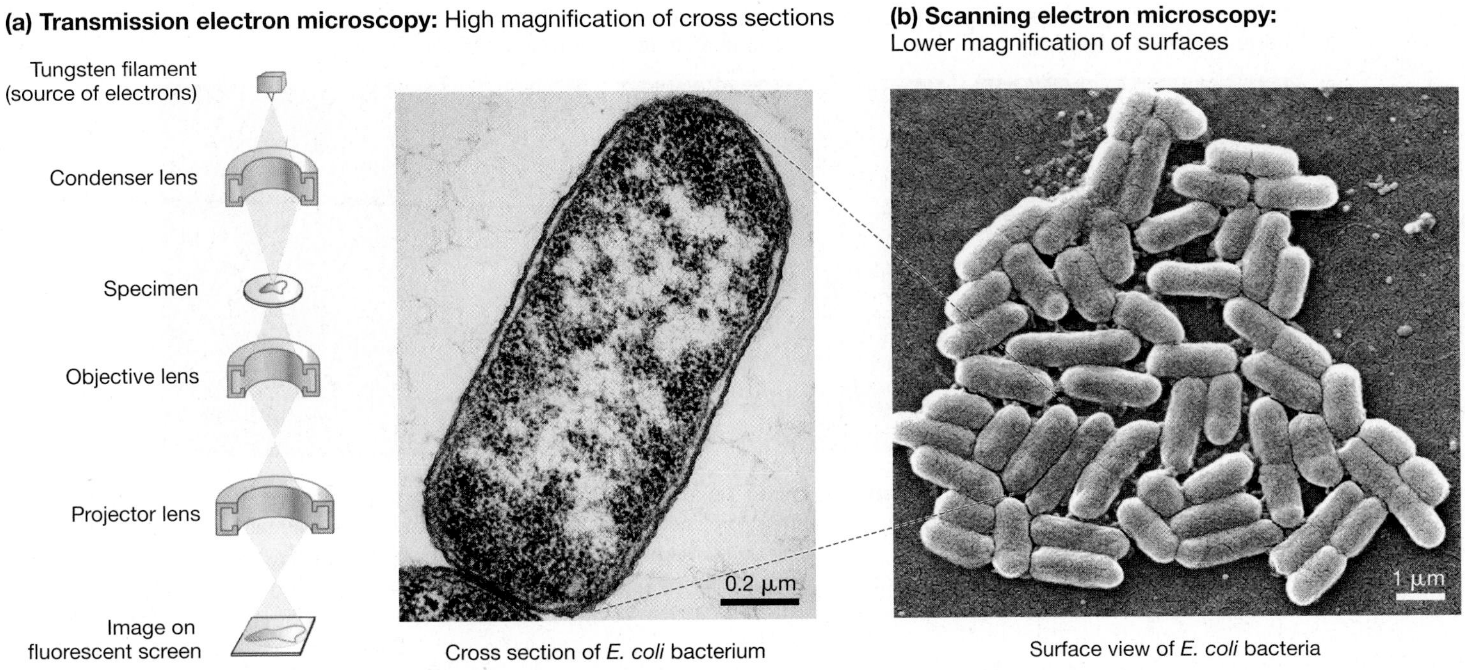

Figure B9.1 There Are Two Basic Types of Electron Microscopy.

Transmission Electron Microscopy The transmission electron microscope (TEM) is an extraordinarily effective tool for viewing cell structure at high magnification. TEM forms an image from electrons that pass through a specimen, just as a compound light microscope forms an image from light rays that pass through a specimen.

Biologists who want to view a cell under a transmission electron microscope begin by "fixing" the cell, which means they treat it with a chemical agent that stabilizes the cell's structure and contents while disturbing them as little as possible. Then the researcher permeates the cell with an epoxy plastic to stiffen it. Once this epoxy hardens, the cell can be cut into extremely thin sections with a glass or diamond knife. Finally, the sectioned specimens are saturated with a metal—often lead. (The reason for this last step is explained shortly.)

Figure B9.1a outlines how the transmission electron microscope works. A beam of electrons is produced by a tungsten filament at the top of a column and directed downward. (All of the air is pumped out of the column, so that the electron beam isn't scattered by collisions with air molecules.) The electron beam passes through a series of lenses and through the specimen. The lenses are electromagnets, which alter the path of the beam much like a glass lens in a dissecting or compound microscope bends light. The electromagnetic lenses magnify and focus the image on a screen at the bottom of the column. There the electrons strike a coating of fluorescent crystals, which emit visible light in response. The light can be detected by a digital camera; the result is a micrograph—a photograph of an image produced by microscopy.

The image itself is created by electrons that pass through the specimen. If no specimen were in place, all the electrons would pass through and the screen (and micrograph) would be uniformly bright. However, cell materials by themselves would also appear fairly uniform and bright. This is because an atom's ability to deflect electrons depends on its mass, and the hydrogen, carbon, oxygen, and nitrogen atoms that dominate biological molecules have low masses. It is why cell biologists must saturate cell sections with solutions containing heavy metals such as lead. These metals have high atomic masses and scatter electrons effectively. Different macromolecules take up the metal atoms in different amounts, so the metals function as "stains" that produce contrast for different structures. With TEM, areas that take up the most metal atoms scatter the electron beam most, producing dark areas in micrographs.

The advantage of TEM is that it can magnify objects up to 250,000×, making intracellular structures clearly visible. The downsides? Researchers are restricted to observing dead, sectioned material, and they must take care not to distort the specimen during the preparation process.

Scanning Electron Microscopy The scanning electron microscope (SEM) is the most useful tool available for looking at the surfaces of biological structures. Materials are prepared for scanning electron microscopy by coating their surfaces with a layer of metal atoms. To create an image of this surface, the microscope scans the surface with a narrow beam of electrons. Electrons that are reflected back from the surface or that are emitted by the metal atoms in response to the beam then strike a detector. The detector counts these electrons and sends the signals to an amplifier. The final image is built up from the number of electrons emitted from each spot on the sample and is displayed on a screen, magnified up to 50,000×. The image is captured directly in a computer.

Because SEM records shadows and highlights, it provides images with a three-dimensional appearance (**Figure B9.1b**). It cannot magnify objects nearly as much as TEM can, however.

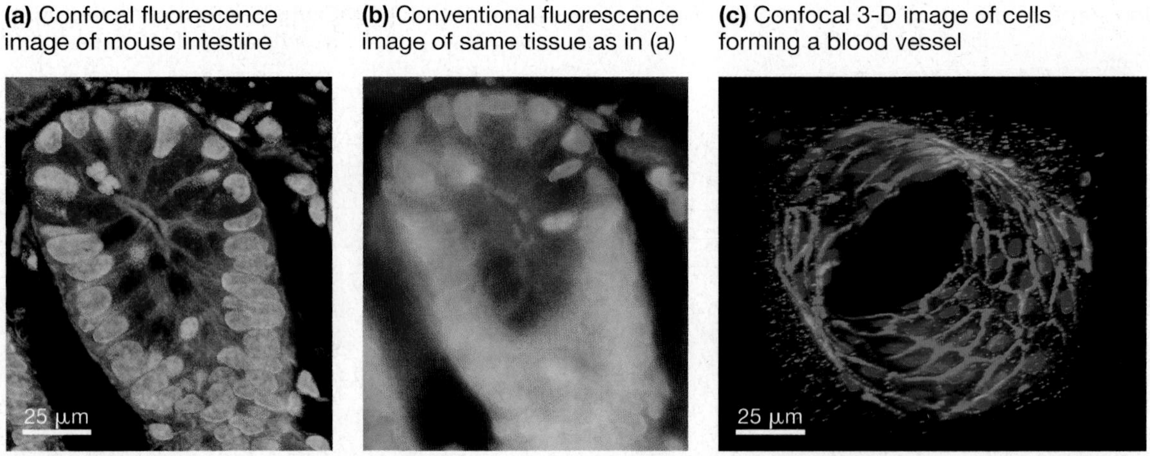

Figure B9.2 **Confocal Microscopy Provides Sharp Images of Living Tissues.** **(a)** This confocal image of a mouse intestine is sharp because it results from light emitted at a single plane, or "section," within the tissue. **(b)** A conventional image of this same tissue is blurred because it results from light emitted by the entire tissue rather than a single plane. **(c)** This 3-D confocal image was reconstructed from optical sections of cells forming a blood vessel.

Studying Live Cells and Real-Time Processes

Until the 1960s, biologists were unable to get clear, high-magnification images of living cells. But a series of innovations in the years since has made it possible to observe organelles and subcellular structures in action.

The development of digital imaging proved revolutionary. It allowed specimens to be viewed at higher magnification, because digital cameras are more sensitive to small differences in contrast than are the human eye. It also made it easier to keep live specimens functioning normally, because the increased light sensitivity of digital cameras allows them to be used with low illumination, so specimens don't overheat. Digital imaging also made possible the use of computers to remove out-of-focus background material and increase image clarity.

A more recent innovation was the use of a fluorescent molecule called green fluorescent protein, or GFP, that allows researchers to tag specific molecules or structures and follow their movement in live cells over time. This was a major advance over immunostaining, in which cells have to be fixed. GFP is naturally synthesized in certain species of jellyfish. By affixing GFP to another protein (using genetic engineering techniques described in Ch. 20, Section 20.1) and expressing that protein in a live cell, investigators can follow the protein's fate over time and record its movement. For example, researchers have made video recordings of GFP-tagged proteins being transported from the rough ER through the Golgi apparatus and out to the plasma membrane. This is *Cell Biology: The Movie*.

GFP's influence has been so profound that the researchers who developed its use in microscopy were awarded the 2008 Nobel Prize in Chemistry. Many other fluorescent proteins have since been developed with colors ranging from cyan (greenish blue) to yellow to red.

Visualizing Cellular Structures in 3-D The world is three-dimensional. To understand how microscopic structures work, it is essential to understand their shapes and spatial relationships. Consider two techniques currently being used to analyze the 3-D structure of cells and organelles.

- *Confocal scanning microscopy* is carried out by mounting a specimen that has been treated with one or more fluorescent tags on a microscope slide and then passing a beam of light at a certain wavelength through a pinhole to focus it at a specific depth within the specimen. The fluorescent tag in the specimen emits light at a different wavelength in response, and a detector records the emitted light. The focused spot of light is moved horizontally across the sample to achieve a sharp image of a precise 2-D plane, or "section," in the tissue being studied (**Figure B9.2a**). Note that if you viewed the same specimen under a conventional fluorescence microscope, the image would be blurry because it results from light emitted by the entire specimen (**Figure B9.2b**). By altering the focal plane, a researcher can record images from an array of depths in the specimen; a computer can then be used to piece the 2-D data together to generate a 3-D image of a cell or tissue (**Figure B9.2c**).
- *Electron tomography* uses a transmission electron microscope to generate a 3-D image of an organelle or other subcellular structures. The specimen is rotated around a single axis while the researcher collects many individual "snapshots." The images are then pieced together with a computer. This technique has provided a much more accurate view of mitochondrial structure than was possible using traditional TEM (see Ch. 9, Figure 9.8).

CHECK YOUR UNDERSTANDING

✔ **If you understood BioSkills 9, you should be able to . . .**

Interpret whether the absence of mitochondria in a transmission electron micrograph of a cancerous human liver cell means that the cell lacks mitochondria.

Answers are available in Appendix A.

BioSkills 10 **Using Molecular Biology Tools and Techniques**

The basic tools and techniques of molecular biology are revolutionary. Molecular biology is the study of the structure and function of cellular molecules—such as nucleic acids and proteins—and how they carry out the biological processes essential for life, including the molecular basis of gene expression and its regulation. The field of molecular biology involves many other areas of biology such as biochemistry, genetics, and cell biology. (Some of the major breakthroughs achieved by molecular biologists are highlighted in Chapter 20). But unless you are doing research in the lab, it can be difficult to fully appreciate all the details underlying the methods. The key is to understand some of the basic principles and steps in each technique, and then to recognize how biologists can use each technique to answer distinct questions. Additional molecular biology tools can be found online (**Table B10.1**).

Often in a molecular biology study, some of the first tools and techniques used are associated with cloning. Biologists clone a DNA region of interest to obtain millions of copies of that region for further analysis. One traditional approach is to make a collection of DNA sequences called a DNA library. Let's take a closer look.

Making and Using cDNA Libraries

A collection of DNA sequences, each of which is inserted into a vector, is called a DNA library. DNA libraries are made up of cloned genes or portions of genes. Each gene sequence can be produced in large quantity and isolated in pure form. If the sequences are fragments of DNA from the genome of an individual, the library is called a genomic library.

Often the sequences are complementary DNA (cDNA)—DNA copies of mRNAs made by a particular cell type or tissue—and the library is called a cDNA library. How is a cDNA library made?

Creating a cDNA Library The viral enzyme reverse transcriptase catalyzes the synthesis of cDNA from an RNA template (see Ch. 33, Figure 33.12). This cDNA can then be used to make a cDNA library, as shown in **Figure B10.1**. The end result, shown in step 5, is a collection of transformed bacterial cells. Each of the cells contains a plasmid with one cDNA from the initial mRNAs isolated from a particular cell type or tissue.

cDNA libraries are important because they give researchers a way to store cDNA fragments from a particular cell type or tissue

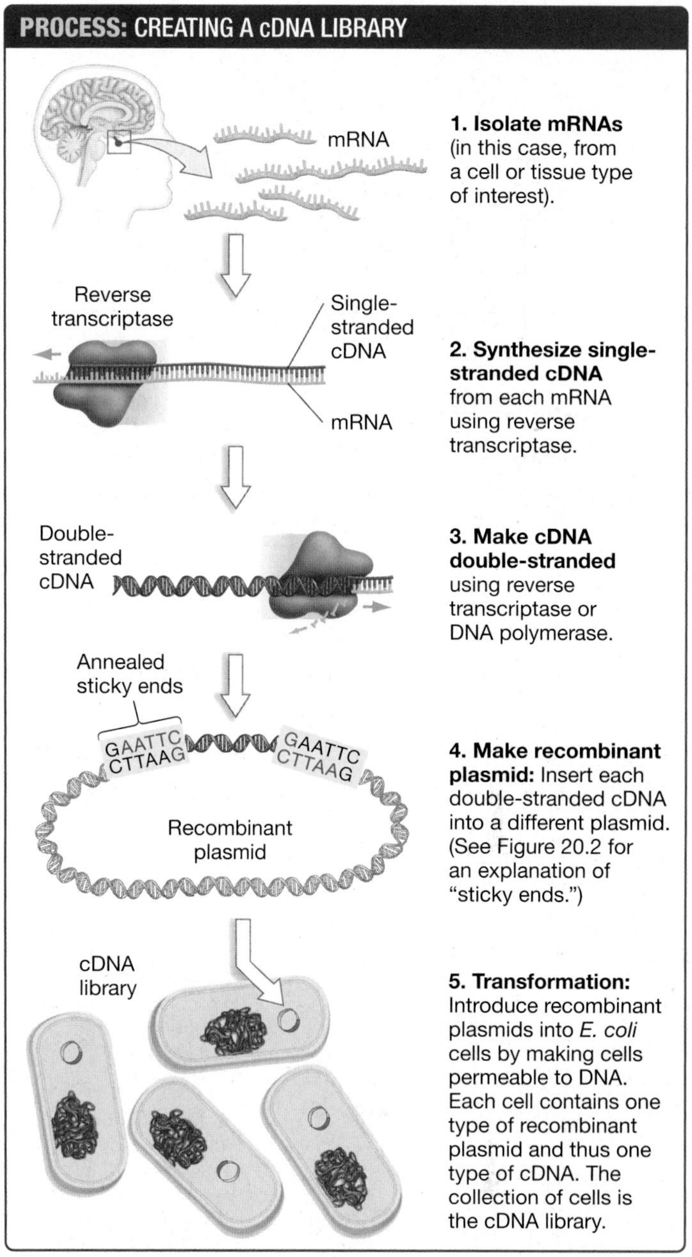

Figure B10.1 Complementary DNA (cDNA) Libraries Represent a Collection of the mRNAs in a Cell.

✔ Would each type of cDNA in the library be represented just once? Why or why not?

Table B10.1 Online Molecular Biology Tools

Name	Description	URL
Basic Local Alignment Search Tool (BLAST)	Many uses, including comparison of nucleotide or protein sequences to sequence databases, matching DNA microarray profiles, designing primers for PCR, searching markers for phylogenetic analyses	https://blast.ncbi.nlm.nih.gov/Blast.cgi
Protein Data Bank (PDB)	Information about the three-dimensional shapes of proteins, nucleic acids, and complex assemblies; tools and resources for education and research	www.rcsb.org/
GenBank	The National Institutes of Health genetic sequence database, containing an annotated collection of all publicly available DNA sequences	https://www.ncbi.nlm.nih.gov/genbank/

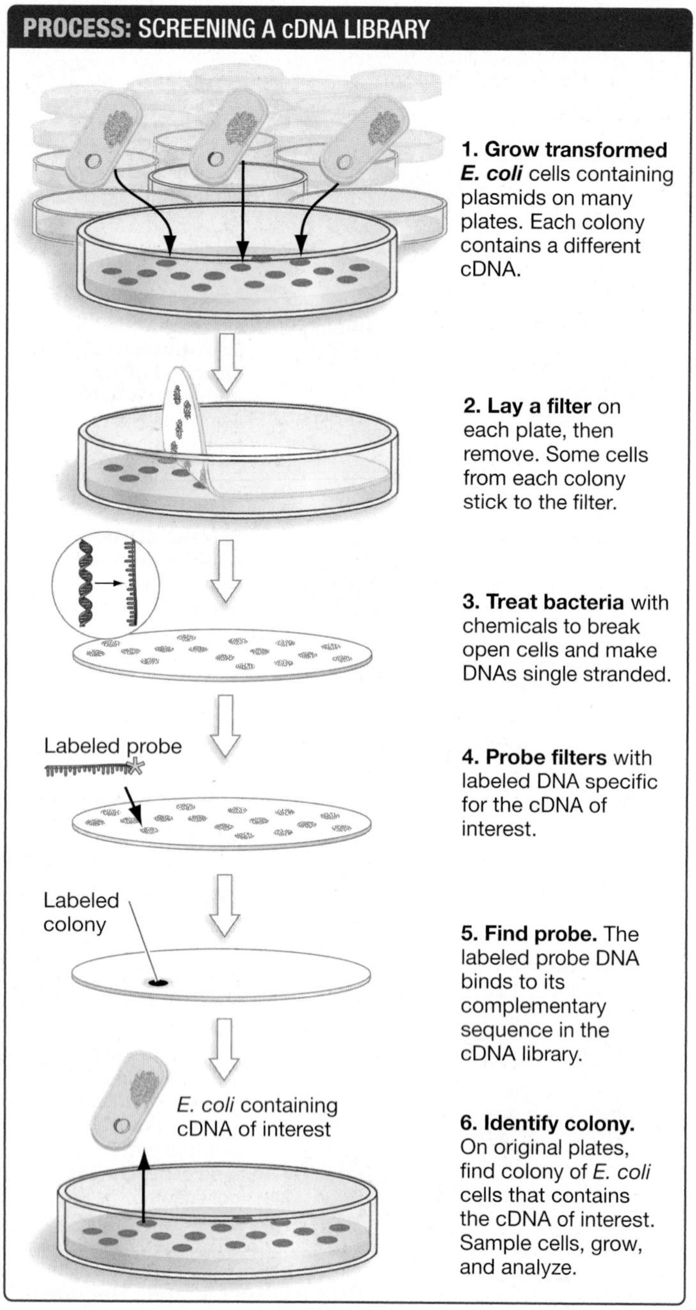

Figure B10.2 Finding a Specific cDNA by Probing a cDNA Library.

in a form that is accessible for gene cloning. But like a college library, a cDNA library isn't very useful unless there is a way to retrieve specific pieces of information. At your school's library, you use call numbers or computer searches to retrieve a particular book or article. How do you go about retrieving a particular cDNA from a library?

Finding a Particular cDNA in a Library Molecular biologists often face the task of finding one specific cDNA in a large collection of DNA fragments. To do this requires a probe—a labeled molecule that binds to the molecule the biologist is looking for (see BioSkills 6).

Figure B10.2 shows how researchers use a probe to find a particular plasmid in a cDNA library. The labeled probe will bind to

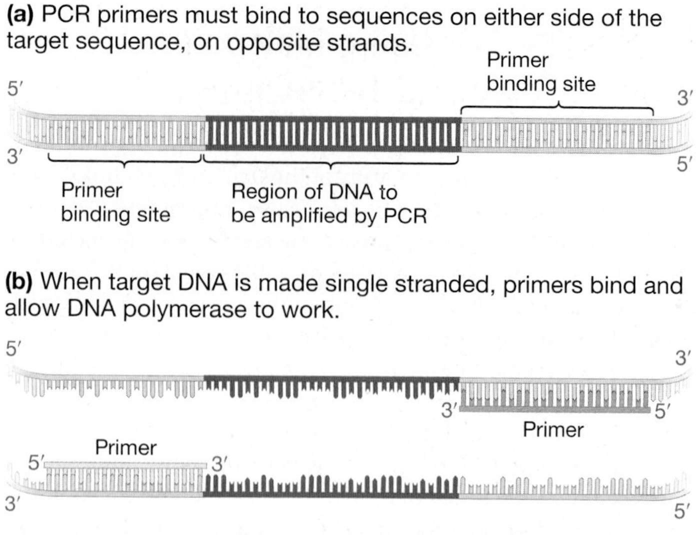

Figure B10.3 The Polymerase Chain Reaction Requires Appropriate Primers. (a) To design an appropriate primer, the base sequences at the primer binding sites must be known. **(b)** The primers bind by complementary base pairing to single-stranded target DNA.

✔ **MODEL** Indicate where DNA polymerase would begin to work on each strand; add an arrow indicating the direction of DNA synthesis.

its complementary sequence in the library. In this way, the recombinant cell that contains the specific cDNA of interest can be identified by the researchers.

Amplifying DNA Using the Polymerase Chain Reaction (PCR)

Another powerful technique for making lots of identical copies of (amplifying) a particular region of DNA is the polymerase chain reaction. The polymerase chain reaction (PCR) is an in vitro DNA synthesis reaction that uses DNA polymerase to replicate a specific section of DNA over and over. The amplified DNA can be used for cloning into a plasmid vector or for many other types of analyses (Ch. 20, Section 20.2).

DNA polymerase cannot work without a primer (Ch. 15, Section 15.3). As Figure B10.3a shows, the primer sequences used must be complementary to bases on either side of the target region—the DNA you want to copy. One primer is complementary to a sequence directly downstream (toward the 3′ end) of the target DNA on one of the double strands; the other primer is complementary to a sequence on the opposite strand of DNA, also directly downstream of the target region. If the target DNA molecule is made single stranded, then the primers can bind to their complementary sequence, as shown in Figure B10.3b. Once the primers are bound, DNA polymerase will extend each new strand of DNA in the 5′ → 3′ direction.

Figure B10.4 shows the steps involved in the polymerase chain reaction.

Step 1 The researcher creates a reaction mixture containing an abundant supply of the four deoxyribonucleoside triphosphates (dNTPs; Ch. 15, Section 15.3), a DNA sample that includes the target DNA of interest, many copies of the two primers, and a heat-resistant DNA polymerase.

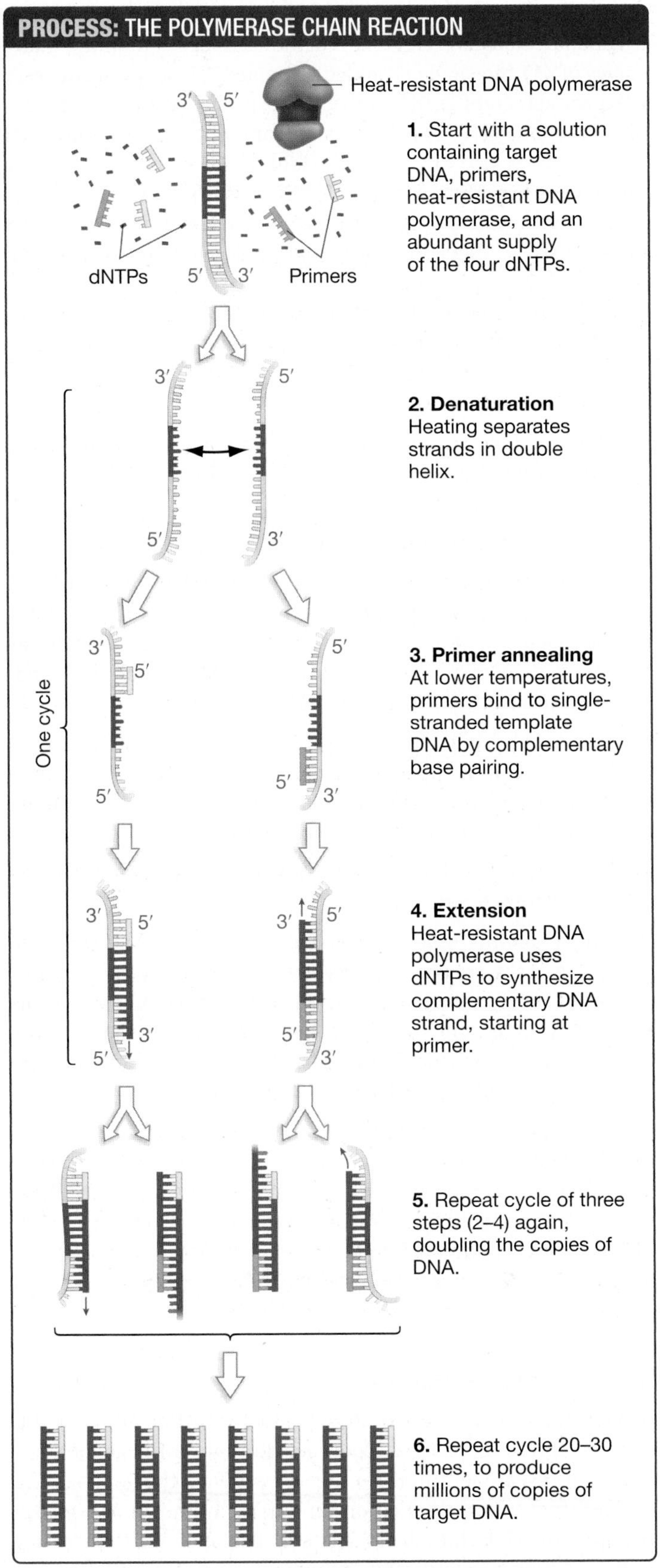

Figure B10.4 The Polymerase Chain Reaction Produces Many Copies of a Specific Sequence. Each PCR cycle (denaturation, primer annealing, and extension) results in a doubling of the number of target sequences.

Step 2 **Denaturation.** The reaction mixture is heated to 95°C–98°C. At this temperature, double-stranded DNA denatures. This means that the two DNA strands separate, forming single-stranded templates.

Step 3 **Primer annealing.** The mixture is allowed to cool to 50°C–72°C, depending on the polymerase and the primer sequences used. In this temperature range, the primers bind, or anneal, to complementary portions of the single-stranded template DNA.

Step 4 **Extension.** The reaction mixture is heated to 72°C. At this temperature, the heat-resistant DNA polymerase efficiently synthesizes the complementary DNA strand from the dNTPs, starting at the primer.

Step 5 Repeat steps 2 through 4.

Step 6 Continue repeating steps 2 through 4 until the necessary number of copies is obtained.

The temperature changes required in each step are controlled by automated PCR machines, and there is no need to add more components to the reaction mixture once the reaction starts.

The denaturation, primer annealing, and extension steps constitute a single PCR cycle. If one copy of a target sequence existed in the original sample, then two copies are present at the end of the first cycle (see step 4 in Figure B10.4). These two copies then act as templates for the second cycle—another round of denaturation, primer annealing, and extension—after which four copies of the target DNA are present (see step 5).

Each time the cycle repeats, the number of copies of target DNA in the reaction mixture doubles (step 6). Doubling occurs because the strands of each newly synthesized segment of DNA serve as templates in the subsequent cycle, along with the previously synthesized segments. Starting with a single target copy, successive cycles result in the production of 2, 4, 8, 16, 32, 64, 128, 256 copies, and so on. A total of n cycles can generate 2^n copies. In just 20 cycles, one double-stranded molecule can be amplified to over a million copies.

Dideoxy Sequencing

After cloning a gene or amplifying a region of DNA by PCR, molecular biologists often want to determine the DNA's base sequence. One way to do this, called dideoxy sequencing, is a clever variation on the basic in vitro DNA synthesis reaction.

The key is to use monomers for DNA synthesis called dideoxyribonucleoside triphosphates (ddNTPs) along with the normal deoxyribonucleoside triphosphates (dNTPs; Ch. 15, Section 15.3) in the reaction mixture. The ddNTPs are identical to dNTPs, except they lack a hydroxyl group at their 3′ carbon. Four types of ddNTPs are used in dideoxy sequencing, each named according to whether it contains adenine (ddATP), thymine (ddTTP), cytosine (ddCTP), or guanine (ddGTP). The use of ddNTPs inspired the name dideoxy sequencing.

If a ddNTP is added to a growing DNA strand, it terminates synthesis. Why? After a ddNTP is added, no hydroxyl group is available on a 3′ carbon to link to the 5′ carbon on an incoming dNTP monomer. As a result, DNA polymerization stops once a ddNTP is added.

Every time a ddNTP is added to a growing strand, the result is a fragment with a length corresponding to the position in the

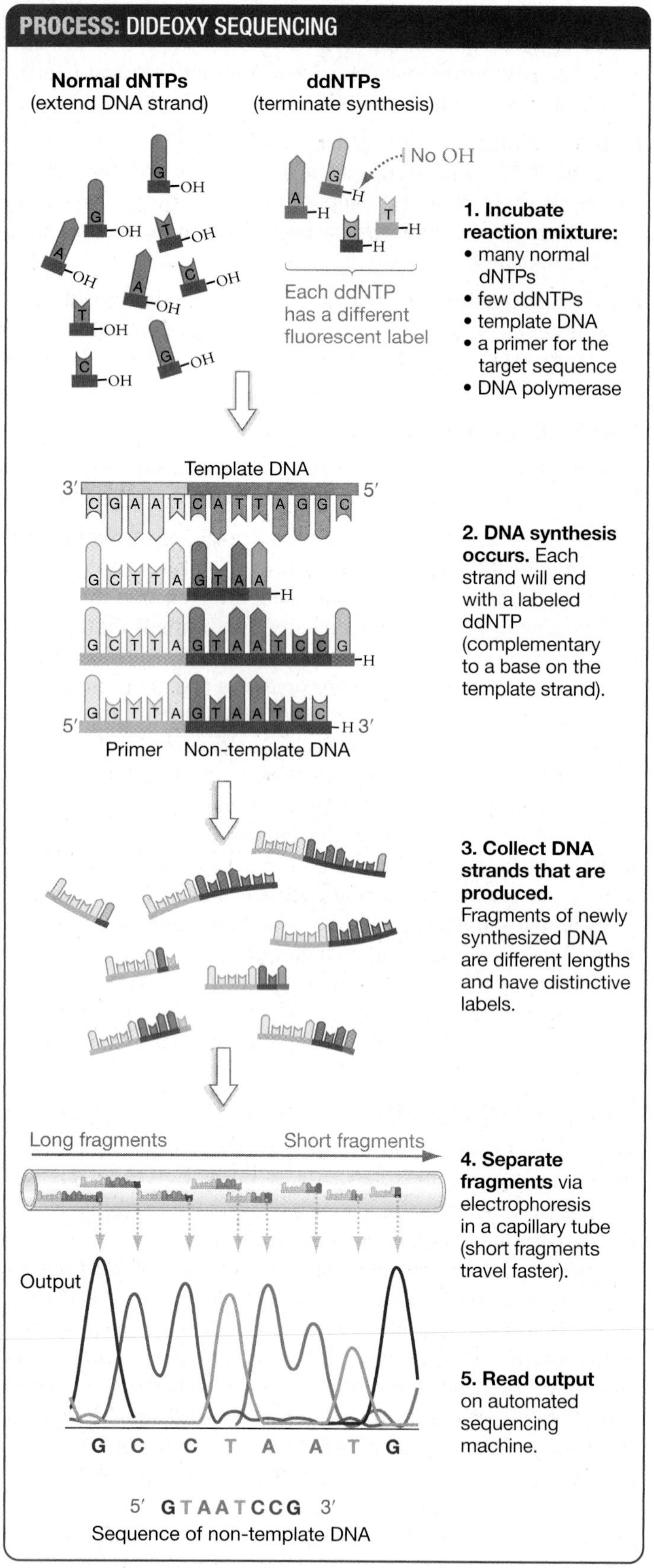

Figure B10.5 Dideoxy Sequencing Can Determine the Base Sequence of DNA.

template of a base complementary to the ddNTP. To produce these fragments, biologists create a reaction mixture containing many copies of **(1)** the single-stranded template DNA to be sequenced, **(2)** a primer, and **(3)** DNA polymerase, as well as **(4)** a large supply of the four dNTPs and **(5)** a small amount of the four ddNTPs (**Figure B10.5**, step 1). Each of the four ddNTPs carries a different fluorescent tag. Fluorescent molecules absorb light at one wavelength and reemit the light at a longer wavelength. As described in **BioSkills 6**, they provide a very sensitive way of detecting molecules.

Under these conditions, many daughter strands of different lengths are synthesized. All fragments that are the same length end in the same kind of ddNTP.

Step 2 in Figure B10.5 shows why:

- DNA polymerase synthesizes a complementary strand from each template strand in the reaction mixture.
- The synthesis of each one of these complementary strands starts at the same point—the primer.
- Because there are many dNTPs and relatively few ddNTPs in the reaction mixture, dNTPs are usually incorporated opposite each complementary base on the template strand as DNA polymerase works its way along the template strand. Incorporating a dNTP allows DNA synthesis to continue.
- Occasionally, one of the few ddNTPs is incorporated into the growing strand, opposite the corresponding base in the template. The complementary base in the template strand pairs randomly with either a ddNTP or a dNTP.
- The addition of the ddNTP stops further elongation of the newly synthesized strand.
- "Stops" of this kind happen for each base in the template strand. As a result, the overall reaction produces a collection of newly synthesized strands (fragments) whose various lengths correspond to the location of each base in the template strand (see step 3 in Figure B10.5). Each fragment will fluoresce in the color of its terminal ddNTP.

Using gel electrophoresis in a capillary tube (step 4 in Figure B10.5), biologists can line up the fragments in order of size. When the fragments are lined up by size, the incorporated dideoxy nucleotides on the successive fragments reveal the sequence of bases in the template DNA. As step 5 shows, a machine can read out the pattern of fluorescence, indicating the sequence of bases in the newly synthesized strand.

Shotgun Sequencing

Sequencing a cDNA clone or PCR product is technically straightforward, but how can researchers sequence an entire genome?

When researchers first set out to sequence the genome of a species, they usually relied on an approach known as shotgun sequencing. Molecular biologists still use shotgun sequencing, although newer sequencing technologies are streamlining the process. Let's take a look at key steps in the traditional method:

Step 1 Application of high-frequency sound waves, or sonication, is used to break a genome randomly into pieces about 160 kilobases (kb) long (1 kb = 1000 bases).

Step 2 Each 160-kb piece is inserted into an engineered version of a bacterial chromosome, a vector called a bacterial artificial chromosome (BAC). BACs permit the cloning of large segments of DNA. Each BAC is then inserted into a different bacterial cell. By allowing each cell to grow into a colony, researchers can isolate large numbers of each 160-kb fragment.

Step 3 After many copies of each 160-kb fragment have been produced, each clone is in turn broken into fragments—but this time, the fragments are about 1 kb long.

Step 4 These small fragments are then inserted into plasmids and placed inside bacterial cells. The plasmids are copied many times as each cell grows into a large population.

Step 5 Next, the cloned 1-kb fragments from each 160-kb BAC clone are sequenced, and computer programs analyze regions where the ends of different 1-kb fragments overlap. Overlaps occur because many copies of each 160-kb segment were made and then fragmented randomly by sonication.

Step 6 Based on the overlaps between 1-kb fragments from a single BAC clone, the computer stitches the sequences together until a continuous sequence across the BAC has been reconstructed.

Step 7 The ends of the reconstructed BACs are analyzed in a similar way. The goal is to link sequences from each 160-kb segment based on regions of overlap until the sequence of an entire genome is assembled.

In essence, the shotgun strategy consists of breaking a genome into many small fragments, sequencing each fragment, and then putting the sequence data back in the correct order. Whether the approach is traditional or modern, this principle holds.

Once genes are cloned and sequenced, researchers can then begin to address important questions about how genes function. (Chapter 20 covers some exciting recent discoveries.) Let's next examine one method that allows researchers to find out how and when all the genes in an organism are expressed.

DNA Microarrays

A DNA microarray lets researchers study the expression of thousands of genes at a time. A microarray consists of as many as 1 million different single-stranded DNA segments that are permanently attached at one end to a glass slide or silicon chip. The DNA sequence of each segment is known, as is its location on the slide or chip. Each segment serves as a probe for a specific RNA transcript.

A typical experiment done with a DNA microarray follows the steps outlined in Figure B10.6. For example, suppose researchers wanted to learn how gene expression in a certain kind of cell is altered to meet the challenges of heat stress. They would begin by isolating mRNAs produced in control cells functioning at normal temperature and in cells of the same kind exposed to high temperatures (step 1).

Once they purified mRNAs from the two populations of cells, the researchers would use the enzyme reverse transcriptase (Ch. 33, Section 33.2) to make a single-stranded cDNA version of each RNA in the two samples. One of the nucleotides in the cDNA would carry a fluorescent tag (step 2). The tag used for the control cells would fluoresce one color (let's say green), while the tag for the heat-stressed cells would fluoresce another color (let's say red). The labeled cDNAs of both colors would then be added to the microarray, where they would bind to complementary DNA probes (step 3). This step is called hybridization because hybrids form between probe DNAs and cDNAs.

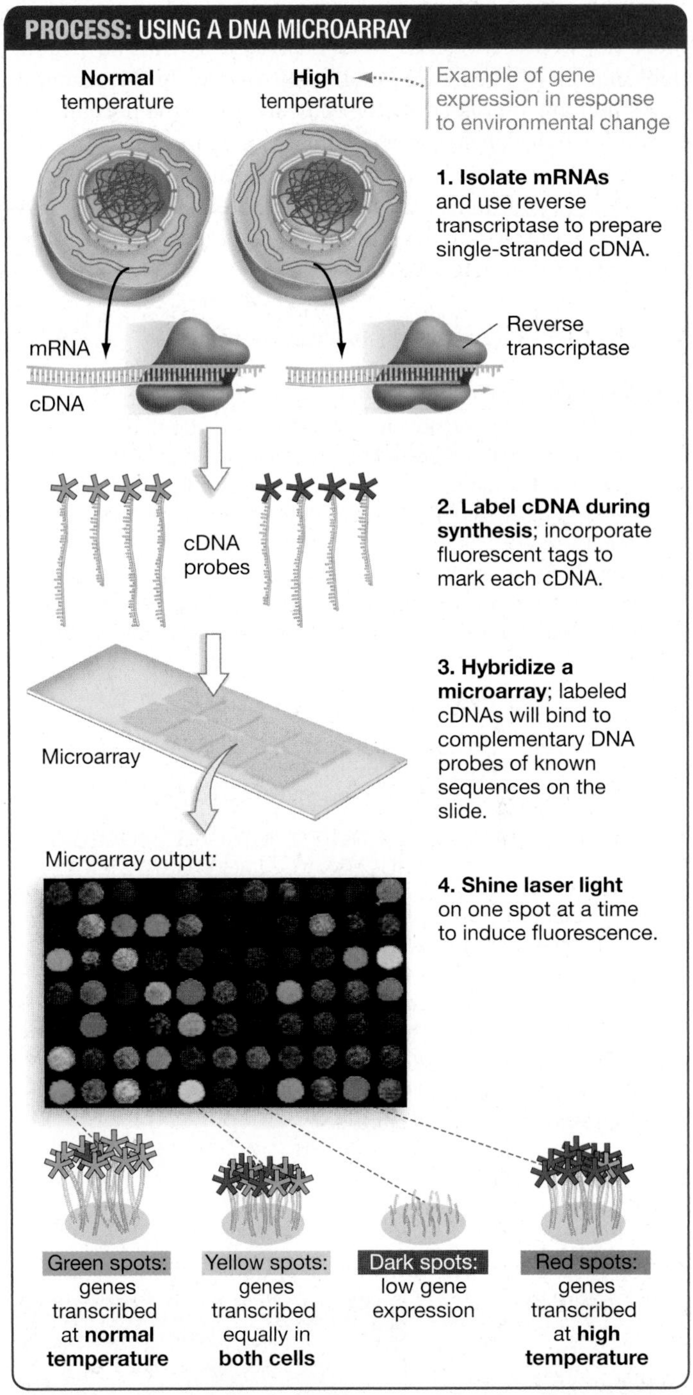

Figure B10.6 DNA Microarrays Can Be Used to Study Changes in Gene Expression. By hybridizing a microarray with labeled cDNAs synthesized from mRNAs, researchers can identify which sequences are being transcribed. Here cDNAs made from cells growing at normal temperature have a green tag, while cDNAs made from cells growing at high temperature have a red tag.

Out of all the probes present on the microarray, then, only those that represent genes being expressed by the two populations of cells will be labeled on the microarray. In this example, genes that are expressed by the control cells at the normal temperature will be labeled green, while those expressed by the cells during heat stress will be labeled red. If a gene is expressed under both sets of conditions, then both green- and red-labeled cDNAs will bind to the DNA in that spot on the microarray, and the spot will appear yellow (step 4).

CHECK YOUR UNDERSTANDING

✔ If you understood BioSkills 10, you should be able to . . .

1. Explain why no further nucleotides can be added after a ddNTP molecule is added to a growing chain of DNA.
2. Consider the following scenario: Suppose a friend of yours is doing a series of PCRs and comes to you for advice. She purchased two sets of primers, hoping that one set would amplify the template sequence shown here. (The dashed lines in the template sequence stand for a long sequence of unspecified bases in the target gene.) Neither of the primer pairs produced any product DNA, however.

	Primer a		Primer b
Primer Pair 1:	5′-CAAGTCC-3′	and	5′-GCTGGAC-3′
Primer Pair 2:	5′-GGACTTG-3′	and	5′-GTCCAGC-3′
Template:	5′-ATTCGGACTTG—GTCCAGCTAGAGG-3′ 3′-TAAGCCTGAAC—CAGGTCGATCTCC-5′		

   a. Explain why each primer pair didn't work. Indicate whether both primers are at fault, or just one of them.
   b. Your friend doesn't want to buy new primers. She asks you whether she can salvage this experiment. What should you tell her to do?
3. Explain how you would use a DNA microarray to compare the genes expressed in human brain cells with those expressed in human liver cells.

Answers are available in Appendix A.

BioSkills 11 Using Cell Culture and Model Organisms as Tools

Research in biological science starts with a question. In most cases, the question is inspired by an observation about a cell or an organism. To answer it, biologists often study cells or tissues in culture. At other times, they perform experiments on model organisms—aptly named because these organisms are intended to serve as models for what is going on in a wide array of species. Let's look at each approach in turn.

Cell and Tissue Culture Methods

For researchers, there are important advantages to culturing plant and animal cells and tissues. Culturing involves growing a cell or tissue outside the organism itself. Cell and tissue cultures produce large populations of a single type of cell or tissue and enable biologists to control experimental conditions precisely.

Animal Cell Culture In 1907, a researcher successfully cultivated amphibian nerve cells in a drop of fluid from the spinal cord, but biologists were not able to routinely culture animal cells in the laboratory until the 1950s and 1960s. It took years to figure out how to re-create the conditions that exist in the intact organism precisely enough for cells to grow normally.

To grow in culture, animal cells must be provided with a liquid mixture containing the nutrients, vitamins, and hormones that stimulate growth. Typically, this mixture is serum, the liquid portion of blood. Serum-free media that are much more precisely defined chemically are available for certain cell types.

Moreover, many types of animal cells will not grow in culture unless they are on a solid surface that mimics the types of surfaces they would adhere to in the intact animal. As a result, animal cells are typically cultured in flasks with special coatings (**Figure B11.1a**, left).

(a) Animal cell culture: immortal HeLa cancer cells

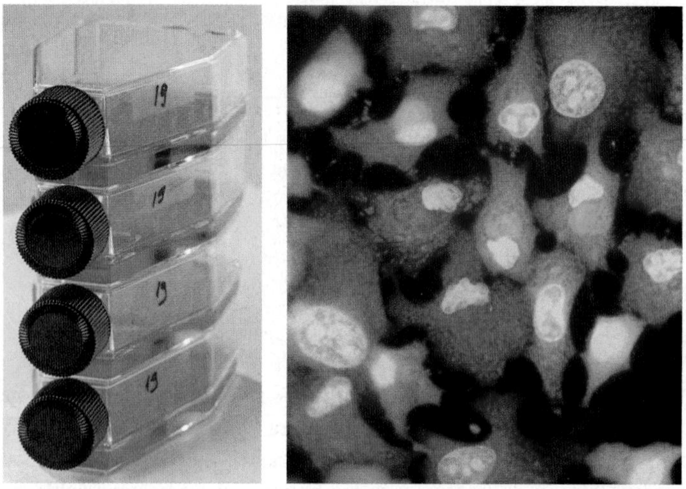

(b) Plant tissue culture: tobacco callus

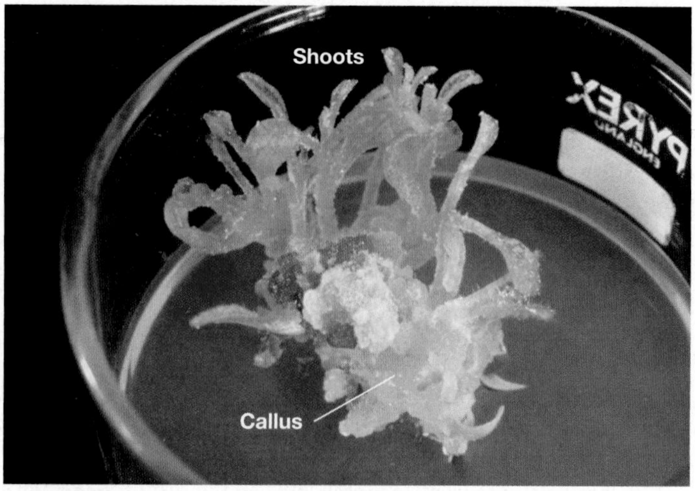

Figure B11.1 Animal and Plant Cells Can Be Grown in the Lab.

Even under optimal conditions, though, normal cells have a finite life span in culture. In contrast, many cultured cancerous cells grow indefinitely. This characteristic correlates with a key feature of cancerous cells in organisms: Their growth is continuous and uncontrolled.

The first human cell type to be grown in culture was isolated in 1951 from a woman with a malignant tumor of the uterine cervix. These cells are called HeLa cells in recognition of the woman, Henrietta Lacks, who died soon after from her cancer. HeLa cells continue to grow in laboratories around the world (see the micrograph on the right in Figure B11.1a).

Because of their immortality and relative ease of growth, cultured cancer cells are commonly used in research on basic aspects of cell structure and function.

Plant Tissue Culture Certain cells found in plants are totipotent—meaning that they retain the ability to divide and differentiate into new cell and tissue types. These cells, called parenchyma cells, are important in wound healing and asexual reproduction. They also allow researchers to grow complete adult plants in the laboratory, starting with a small number of parenchyma cells.

Biologists who grow plants in tissue culture begin by placing parenchyma cells in a liquid or solid medium containing all the nutrients required for cell maintenance and growth. In the early days of plant tissue culture, as for animal cells, investigators found that successful growth and differentiation depended not only on the presence of specific hormones but also on their relative abundance.

The earliest experiments on hormone interactions in plant tissue cultures were done with tobacco cells in the 1950s. Researchers found that when they added roughly equal amounts of the hormones auxin and cytokinin to the cells, the cells began to divide and eventually formed an undifferentiated mass of parenchyma cells called a callus. By varying the proportion of auxin to cytokinin in different parts of the callus and over time, researchers could stimulate the growth and differentiation of root and shoot systems and produce whole new plants (**Figure B11.1b**).

The ability to grow a whole plant in tissue culture from a single cell has been instrumental in the development of genetic engineering (Ch. 20, Section 20.1). Researchers insert recombinant genes into target cells, test the cells to identify those that successfully express the recombinant genes, and then use tissue culture techniques to grow those cells into adult individuals with novel genotypes and phenotypes.

Model Organisms

Certain organisms are chosen as model organisms because they are convenient to study, and because they each have attributes that make them appropriate for the particular type of research proposed.

They tend to have some common characteristics:

- ***Short generation time and rapid reproduction*** This trait is important because it makes it possible to produce offspring quickly and perform many experiments in a short amount of time—you don't have to wait long for individuals to grow.
- ***Large numbers of offspring*** This trait is particularly important in genetics, where many offspring phenotypes and genotypes need to be assessed to get a large sample size.
- ***Small size and simple feeding and habitat requirements*** These attributes make it relatively cheap and easy to maintain individuals in the lab.

The following examples highlight just a few model organisms supporting current work in biological science.

***Escherichia coli* and Its Viruses** Of all model organisms in biology, perhaps none has been more important than the bacterium *Escherichia coli*—a common inhabitant of the human gut. The strain that is most commonly worked on today, called K-12 (**Figure B11.2a** on page 46), was originally isolated from a hospital patient in 1922.

During the last half of the twentieth century, key results in molecular biology originated in studies of *E. coli* and its viruses, or bacteriophage (Ch. 15, Section 15.2, and Ch. 33, Section 33.2). These results include the discovery of enzymes such as DNA polymerase, RNA polymerase, DNA repair enzymes, and restriction endonucleases; the elucidation of ribosome structure and function; and the initial characterization of promoters, regulatory transcription factors, regulatory sites in DNA, and operons. In many cases, initial discoveries made in *E. coli* and bacteriophage allowed researchers to confirm that homologous enzymes and processes existed in an array of organisms, ranging from other bacteria to yeast, mice, and humans.

The success of *E. coli* as a model for other species inspired Jacques Monod's claim that "Once we understand the biology of *Escherichia coli*, we will understand the biology of an elephant." The genome of *E. coli* K-12 was sequenced in 1997, and the strain continues to be a workhorse in studies of gene function, biochemistry, and particularly biotechnology.

In the lab, *E. coli* is usually grown in suspension culture, where cells are introduced to a liquid nutrient medium, or on plates containing agar—a gelatinous mixture of polysaccharides. Under optimal growing conditions—meaning before cells begin to get crowded and compete for space and nutrients—a cell takes just 30 minutes on average to grow and divide. At this rate, a single cell can produce a population of over a million descendants in just 10 hours. Unless they develop new mutations, all of the descendant cells are genetically identical.

Dictyostelium discoideum The cellular slime mold *Dictyostelium discoideum* is not always slimy, and it's not a mold—which is a type of fungus. Instead, it is an amoeba. Amoeba is a general term that biologists use to characterize a unicellular eukaryote that lacks a cell wall and is extremely flexible in shape. *Dictyostelium* has long fascinated biologists because it is a social organism. Independent cells sometimes aggregate to form a multicellular structure (see Ch. 11, Figure 11.18).

Under most conditions, *Dictyostelium* cells are haploid (n) and move about in decaying vegetation on forest floors or other habitats. They feed on bacteria by engulfing them whole. When *Dictyostelium* cells reproduce, they can do so sexually by fusing with another cell and then undergoing meiotic cell division, or asexually by mitotic cell division, which is more common. If food begins to run out, the cells begin to aggregate. In many cases, tens

(a) Bacterium *Escherichia coli* (strain K-12)

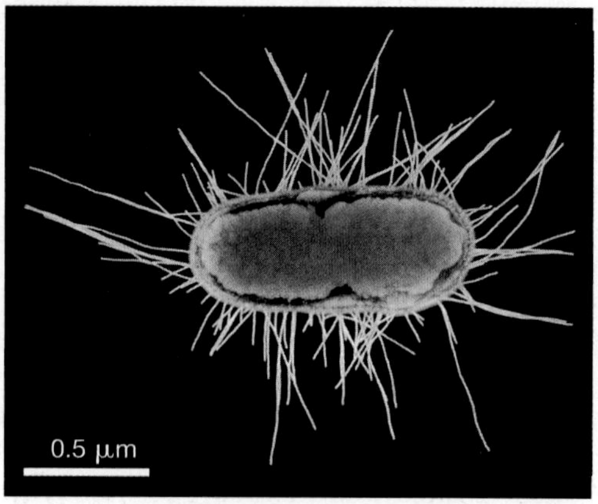

(b) Slime mold *Dictyostelium discoideum*

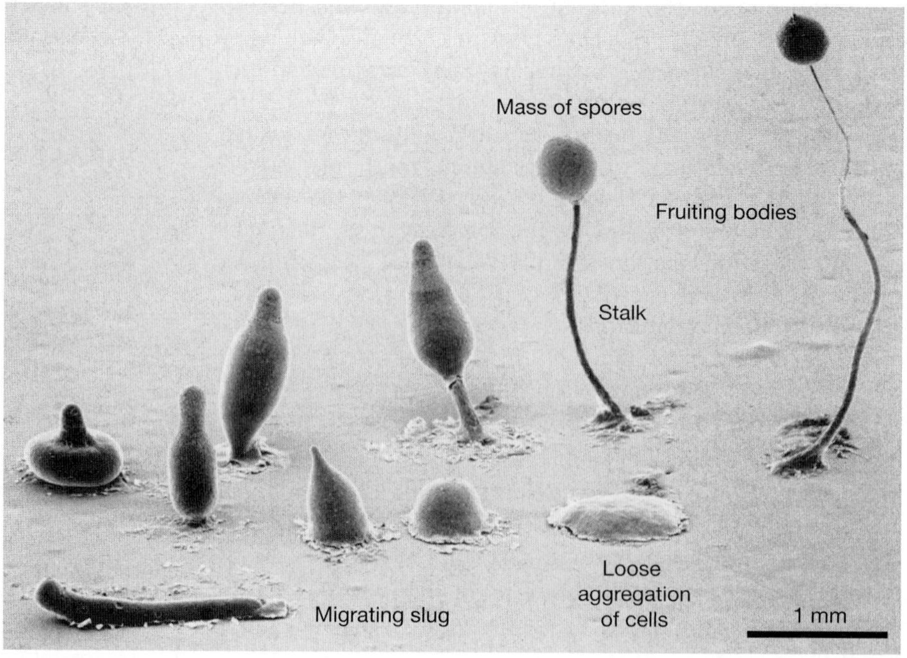

(c) Thale cress *Arabidopsis thaliana*

(d) Yeast *Saccharomyces cerevisiae*

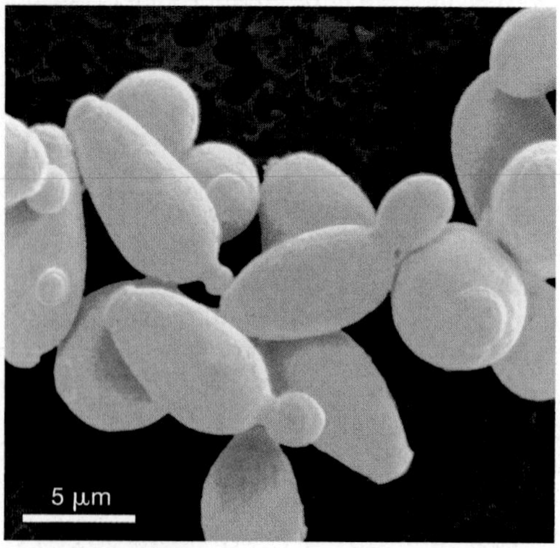

(e) Fruit fly *Drosophila melanogaster*

(f) Roundworm *Caenorhabditis elegans*

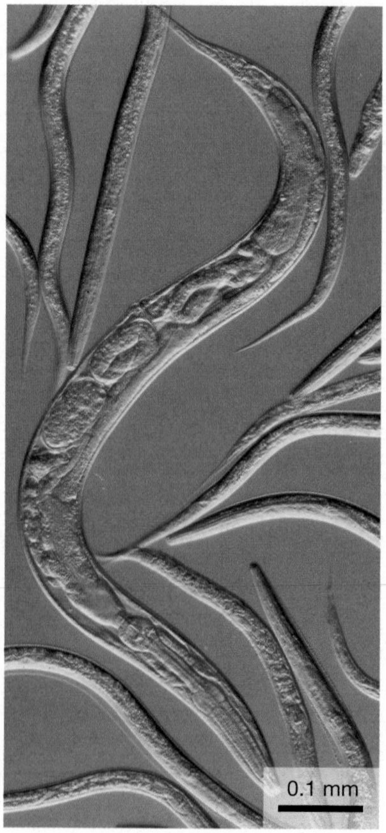

(g) Mouse *Mus musculus*

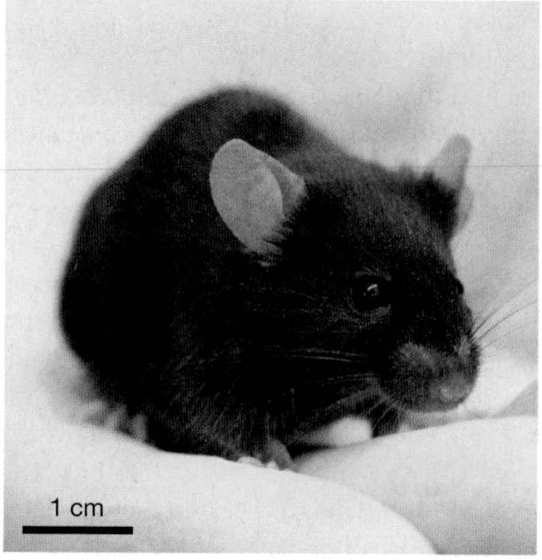

Figure B11.2 Model Organisms.

of thousands of cells cohere to form a 2-mm-long mass called a slug (**Figure B11.2b**). (Note that this is not a slug that is related to snails.)

After migrating to a sunlit location, the slug stops, and individual cells differentiate according to their position in the slug. Some form a stalk; others form a mass of spores at the tip of the stalk. (A spore is a single cell that develops into an adult organism, but it is not formed from gamete fusion like a zygote is.) The entire structure, stalk plus mass of spores, is called a fruiting body.

Cells of *Dictyostelium* that form spores secrete a tough coat and represent a durable resting stage. The fruiting body eventually dries out, and the wind disperses the spores to new locations, where more food might be available.

Dictyostelium has been an important model organism for investigating questions about eukaryotes:

- Cells in a slug are initially identical in morphology but then differentiate into distinctive stalk cells and spores. Studying this process helped biologists better understand how cells in plant and animal embryos differentiate into distinct cell types.
- The process of slug formation has helped biologists study how animal cells move and how they aggregate as they form specific types of tissues.
- When *Dictyostelium* cells aggregate to form a slug, they stick to each other. The discovery of membrane proteins responsible for this cell–cell adhesion helped biologists understand some of the general principles of multicellular life (highlighted in Ch. 11, Section 11.4).

Arabidopsis thaliana In the early days of biology, the best-studied plants were agricultural varieties such as maize (corn), rice, and garden peas. When biologists began to unravel the mechanisms responsible for oxygenic photosynthesis in the early to mid-1900s, they relied on green algae that were relatively easy to grow and manipulate in the lab—often the unicellular species *Chlamydomonas reinhardii*.

Although crop plants and green algae continue to be the subject of considerable research, a new model organism emerged in the 1980s and now serves as the preeminent experimental subject in plant biology. That organism is *Arabidopsis thaliana*, commonly known as thale cress or wall cress (**Figure B11.2c**).

Arabidopsis is a member of the mustard family, or Brassicaceae, so it is closely related to radishes and broccoli. In nature it is a weed—meaning a species that is adapted to thrive in habitats where soils have been disturbed.

One of the most attractive aspects of working with *Arabidopsis* is that individuals can grow from a seed into a mature, seed-producing plant in just four to six weeks. Several other attributes add to its effectiveness as a subject for study:

- It has just five chromosomes and a relatively small genome, which has been completely sequenced and has limited numbers of repetitive sequences.
- It can self-fertilize as well as undergo cross-fertilization.
- It can be grown in a relatively small amount of space and with a minimum of care in the greenhouse.
- It produces up to 10,000 seeds per individual per generation.

Arabidopsis has been instrumental in a variety of studies in plant molecular genetics and development, and it is becoming increasingly popular in ecological and evolutionary studies.

Saccharomyces cerevisiae When biologists want to answer basic questions about how eukaryotic cells work, they often turn to a fungus, the yeast *Saccharomyces cerevisiae*.

S. cerevisiae is unicellular and relatively easy to culture and manipulate in the lab (**Figure B11.2d**). In good conditions, yeast cells grow and divide almost as rapidly as bacteria. As a result, the species has become the organism of choice for experiments on control of the cell cycle and regulation of gene expression in eukaryotes. For example, research has confirmed that several of the genes controlling cell division and DNA repair in yeast have homologs in humans; when mutated, these homologs contribute to cancer in humans. Strains of yeast that carry the homologous mutations are now being used to test drugs that might be effective against cancer.

S. cerevisiae was the first eukaryote with a sequenced genome, and it has become even more important in efforts to interpret the genomes of organisms like rice, mice, zebrafish, and humans. It is much easier to investigate the function of particular genes in *S. cerevisiae* by creating mutants or transferring specific alleles among individuals than it is to do the same experiments in mice or zebrafish. Once the function of a gene has been established in yeast, biologists can look for the homologous gene in other eukaryotes. If such a gene exists, they can usually infer that it has a function similar to its role in *S. cerevisiae*.

Drosophila melanogaster If you walk into a biology building on any university campus around the world, you are almost certain to find at least one lab where the fruit fly *Drosophila melanogaster* is being studied (**Figure B11.2e**).

Drosophila has been a key experimental subject in genetics since the early 1900s. T. H. Morgan was the first to choose this organism as a focus for study, because it can be reared in the laboratory easily and inexpensively, matings can be arranged, the life cycle is completed in less than two weeks, and females lay a large number of eggs. These traits made fruit flies valuable subjects for breeding experiments designed to test hypotheses about how traits are transmitted from parents to offspring (Ch. 14, Section 14.4).

More recently, *Drosophila* has also become a key model organism in the field of developmental biology. The use of flies in developmental studies was inspired largely by the work of Christiane Nüsslein-Volhard and Eric Wieschaus, who in the 1980s isolated flies with genetic defects in early embryonic development. By investigating the nature of these defects, researchers have gained valuable insights into how various gene products influence the development of eukaryotes (Ch. 21, Section 21.4). The complete genome sequence of *Drosophila* has been available to investigators since the year 2000.

Caenorhabditis elegans The roundworm *Caenorhabditis elegans* emerged as a model organism in developmental biology in the 1970s, due largely to work by Sydney Brenner and colleagues. (*Caenorhabditis* is pronounced *see-no-rab-DIE-tiss*.)

C. elegans was chosen for three reasons:

1. Its cuticle (soft outer layer) is transparent, making individual cells relatively easy to observe (**Figure B11.2f**).
2. Adults have exactly 959 nonreproductive cells.

3. Most important, the fate of each cell in an embryo can be predicted because cell fates are invariant among individuals. For example, when researchers examine a 33-cell *C. elegans* embryo, they know exactly which of the 959 cells in the adult will be derived from each of those 33 embryonic cells.

In addition, *C. elegans* is small (less than 1 mm long), can self-fertilize or cross-fertilize, and undergoes early development in just 16 hours. The genome of *C. elegans* was fully sequenced by 2002.

Mus musculus The house mouse *Mus musculus* is the most important model organism among mammals. That's why it is especially prominent in biomedical research, where researchers need to work on animals with strong genetic and developmental similarities to humans.

The house mouse was an intelligent choice for a model organism in mammals: It is small and thus relatively inexpensive to maintain in captivity, and it breeds rapidly. A litter can contain 10 offspring, and generation time is only 12 weeks—meaning that several generations can be produced in a year. Descendants of wild house mice have been selected for docility and other traits that make them easy to handle and rear; these populations are referred to as laboratory mice (**Figure B11.2g** on page 46).

Some of the most valuable laboratory mice are strains with distinctive, well-characterized genotypes. Inbred strains are virtually homogenous genetically and are useful in experiments where gene interactions or environmental effects (Ch. 14, Section 14.5) have to be controlled. Other populations carry mutations that knock out genes and cause diseases similar to those observed in humans. These populations are useful for identifying the causes of genetic diseases and for testing drugs or other types of therapies.

CHECK YOUR UNDERSTANDING

✔ **If you understood BioSkills 11, you should be able to . . .**

1. Identify a limitation in interpreting experiments on HeLa cells.
2. Determine which model organisms described in **BioSkills 11** would be the best choice for the following studies. In each case, explain your reasoning: (a) a study of how specific cells in an embryo become specialized for a particular fate at certain points in normal development; (b) a study of proteins that are required for cells to adhere to each other in multicellular organisms.

Answers are available in Appendix A.

BioSkills 12 Reading and Making Visual Models

After you complete BioSkills 12–14, you should be able to . . .

- Use models to visualize scientific ideas.

Biology is a visual discipline—it's hard to imagine learning biology without illustrations, such as the figures in this textbook. While some illustrations are designed to show structures as realistically as possible (such as a bluebird in a field guide), most illustrations in biology are *visual models*—drawings that are purposely simplified or abstracted to show a concept, process, or relationship clearly. Visual models help you see what is too small, too large, too complex, too hidden, or too conceptual to be seen otherwise.

Since the topics in biology are diverse, the types of visual models are diverse too. They include graphs (see **BioSkills 2**), phylogenetic trees (see **BioSkills 13**), and chemical structures (see **BioSkills 14**). Scientists make models to help them form hypotheses, design experiments, visualize data, and communicate with others. Making your own visual models is a skill that will help you learn biology and become a biologist. But like interpreting models, making models takes practice.

Tips for Interpreting Models

Each figure in this text was designed to focus on a specific take-home message about biology, and many are, by necessity, abstractions of reality. **Table B12.1** lists some basic tips to help you practice interpreting these models. As you examine the table, take your time and ask yourself these five questions:

1. ***Have I read the figure legend and the section of text where the figure is called out? Do I understand what the words mean?*** The text provides additional information that will help walk you through a complex model, step by step. Look for definitions of key terms and consult the glossary.
2. ***Are all the parts of the figure at the same scale, or different scales? At one time, or at different times?*** Different elements of a model may be shown at different scales to make them visible. A double strand of uncondensed DNA is so thin that it would not be visible in a model of a cell if not adjusted in size. Sometimes, one part of a model is an enlarged section of another part. Look for arrows and lines that indicate these connections.
3. ***Do I understand what the symbols in the figure represent?*** Sometimes different symbols are used for the same concept in different contexts due to the amount (or type) of detail needed—DNA, lipid bilayers, and cells are good examples. Or sometimes symbols such as arrows can have different meanings in different contexts. Take your time in mentally "translating" these symbols to make sure that you are not carrying a previous meaning into a new context.
4. ***Do I understand the use of color?*** Artificial coloration is an important aspect of models. For example, DNA and RNA are colorless in real life, but in this book, the default color for DNA is red and the default color for RNA is yellow. Color can also be used for emphasis to draw your attention to certain features. Bright red is more likely to grab your attention than pale gray, so red is often used for important information and gray for supporting information.
5. ***What seems to be the most important part of the figure? What is the take-home message of the figure?*** Many of the figures in this book have blue questions in the captions. You can use these questions (and the answers in Appendix A) to check that you are interpreting the figures appropriately.

Table B12.1 Tips for Interpreting Visual Models

Tips for Interpreting Visual Models	Example
Simplicity Visual models simplify reality to focus your attention on a specific concept. **1.** What is the focus of this model? What is *not* shown for simplicity?	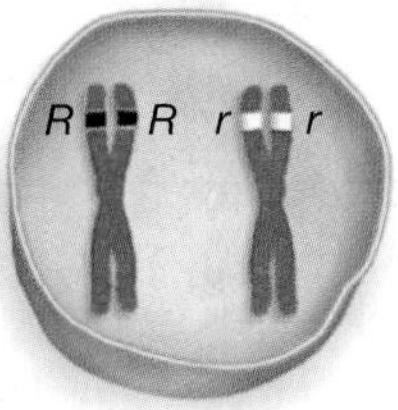
Color use Color coding helps to identify structures and emphasize certain parts of models. **2.** What do the red balls represent? Are they red in real life?	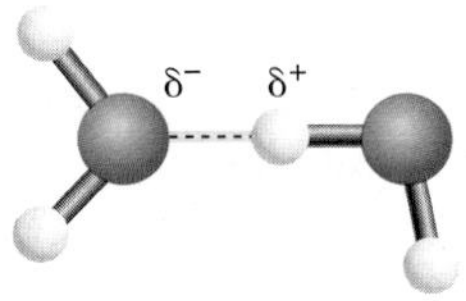
Same concept, different symbols Different symbols may be used to represent the same concept in different contexts, such as at different scales. **3.** What do these three symbols represent?	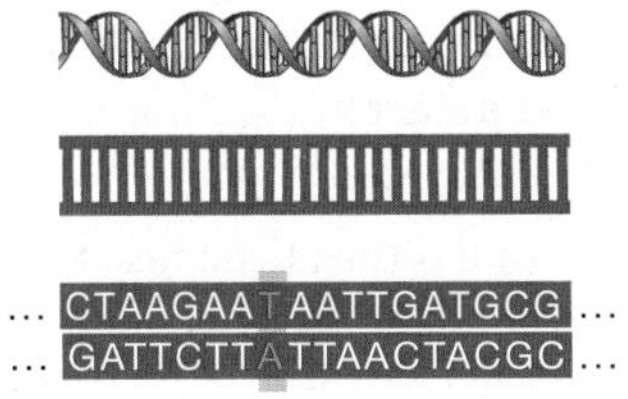
Same symbol, different meanings Some symbols, such as arrows, may have very different meanings in different contexts. **4.** Does this arrow represent a transformation of one thing to another, a movement, a passage of time, or something else?	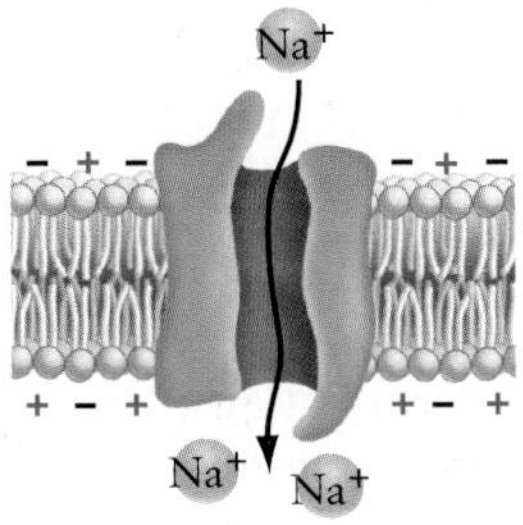
Scale flexibility Different components of a model may be shown at different scales to make them more visible. **5.** Estimate the size of the lizard relative to the human in real life.	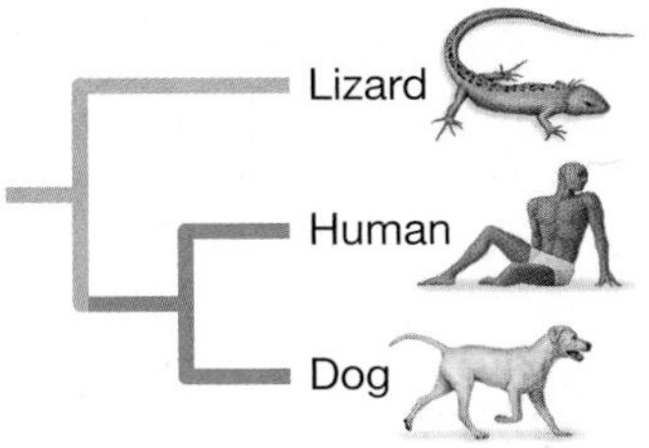
Scale connections Lines and arrows may be used to show that one part of a figure is an enlargement of another. **6.** Why are the stomata cells not shown directly on the leaf?	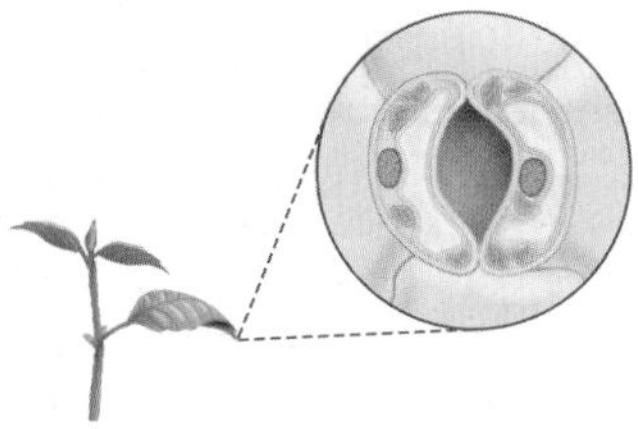

Tips for Making Your Own Models

You may be excited by the opportunity to draw models, or you may feel intimidated by the task. Either way, you have lots of company. The **Making Models** boxes throughout the book give you specific tips on how to make and use different types of visual models in biology—such as pedigrees, membrane models, phylogenetic trees, graphs, and food webs. (See the inside back cover for a complete list of Making Models boxes.) Here, let's consider four general tips to help you get started:

1. ***You don't have to be an artist to make good models.*** Models are often most effective when they are simple. Remember, the goal is to make tools to think with, not realistic art. The figure below shows an example of how your hand-drawn models may differ from typical textbook figures.

Textbook figure

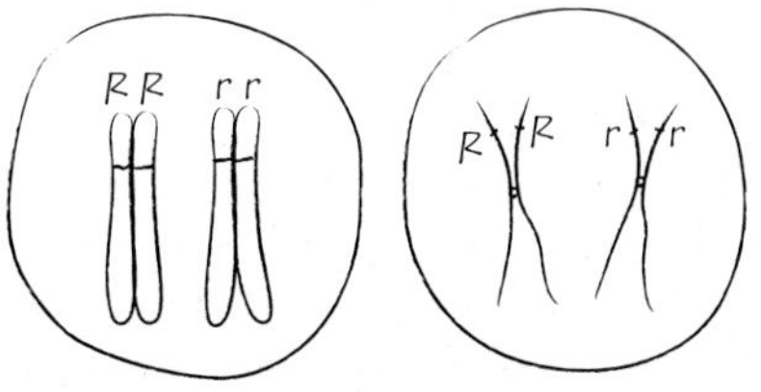

Hand-drawn models

2. ***Focus on the important ideas.*** If you are drawing a model to show the relationship between cellular respiration and photosynthesis in an ecosphere that includes shrimp, algae, and bacteria, don't spend a lot of time drawing shrimp legs. You could simply use an oval containing the word "shrimp."

3. ***Practice using standard symbols.*** Translating from verbal to visual and visual to verbal is like translating between two languages. Learn the meaning of commonly used symbols in biology, such as the circles and squares in pedigrees, the axes in graphs, and the lines in phylogenetic trees.

4. ***Process is as important as product.*** Don't be afraid to change your model or start over. Since models are thinking tools, go ahead and make as many drafts of a model as you need to until you understand an idea or solve a problem.

Let's apply these ideas to one kind of model that you will be prompted to draw in this book: a concept map.

Concept Maps

Concept maps are devices for organizing and expressing what you know about a topic. They have been proven to be an effective studying and learning tool.

Concept maps have two main elements: **(1)** concepts that are identified by words or short phrases and placed in a box or circle (you can think of these as the "nouns"), and **(2)** labeled arrows that physically link two or more concepts and explain the relationship between them (you can think of these as the "verbs"). The concepts can be arranged in different patterns on the page depending on the content. Concept maps may be organized as a hierarchy starting with the big idea at the top and moving down to details, or they may be organized in a time sequence or cycle.

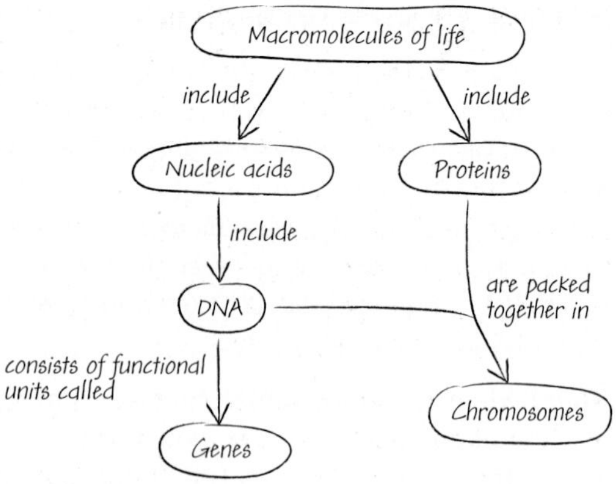

Figure B12.1 A Concept Map of the Relationships between Nucleic Acids and Proteins.

One example of a simple concept map is shown in **Figure B12.1**. Other types of concept maps that you will encounter in this textbook are the road maps that introduce each chapter and the Big Picture concept maps that show how material in multiple chapters is related. When you draw your own concept maps, note how the general principles of model making apply:

- Since concept maps are built using geometric shapes, arrows, and words, you don't need to be an artist to make them.
- The key to good concept maps is focusing on the important ideas and how they are related. Make sure that your connections are accurate.
- It may take time for you to get used to "concept map language," but you will become fluent with practice.
- Your first draft of a map might be a mess. Once you get your ideas down on paper, you may need to create a cleaner draft so that you can read the map easily.

Remember, interpreting and making visual models are skills that can be learned. These skills are valuable not only in your biology class but also in your other science classes and beyond.

CHECK YOUR UNDERSTANDING

✔ If you understood BioSkills 12, you should be able to . . .

1–6. Answer the questions in Table B12.1.

7. **MODEL** Find a way to add "alleles" to the concept map in Figure B12.1.

Answers are available in Appendix A.

BioSkills 13 **Reading and Making Phylogenetic Trees**

Phylogenetic trees show the hypothesized evolutionary relationships among species or other taxa. A taxon is any named group of organisms, such as a population, a species, or a larger group (Ch. 1, Section 1.5). Phylogenetic trees are a type of visual model (see **BioSkills 12**) that can take some practice to interpret and draw correctly.

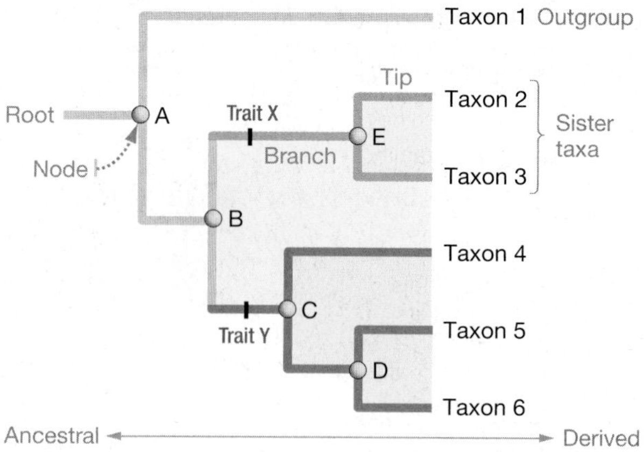

Figure B13.1 A Phylogenetic Tree Has a Root, Branches, Nodes, and Tips

Anatomy of a Phylogenetic Tree

At first glance, a phylogenetic tree may look like just a bunch of lines. But the lines have very specific meanings that include some types of information and not others.

Let's start by examining the tree in **Figure B13.1** and considering each component.

- A *root* is the most ancestral population in a phylogenetic tree—where the tree originates. In this text, most of the trees are rooted on the left with the branches facing right because this orientation is easy to label and read. You may see other orientations in other sources.

- *Branches* represent populations through time. For ease of interpretation in this text, most branches are drawn as horizontal lines with vertical connectors, and the branches have arbitrary lengths. You may also see trees whose branch lengths are proportional to time or to the extent of genetic difference among populations (indicated by a scale bar at the bottom of the tree; Ch. 25, Section 25.1). In other sources, branches may be drawn as diagonal or curved lines.

- *Nodes* (also called forks) occur where a hypothetical ancestral group splits into two or more descendant groups. In such cases, each node represents the most recent common ancestor of the descendant populations that emerge from it.
- *Tips* (also called terminal nodes) are the tree's endpoints. Each name on a tip represents a taxon of organisms living today or in the past. Taxa on tips connected by a single node are called sister taxa.
- A *monophyletic group* (also called a lineage or clade) consists of an ancestral species and all of its descendants. Monophyletic groups can be identified using the "one-snip test": If you cut

any branch on a phylogenetic tree, all of the branches and tips that fall off represent a monophyletic group. Using the one-snip test, you should be able to convince yourself that each of the tips is itself a monophyletic group.

- A *trait* is indicated by a black bar. An ancestral trait is a characteristic that existed in an ancestor; a derived trait is a characteristic that is a modified form of the ancestral trait, found in a descendant. A shared, derived trait (called a synapomorphy) occurs in all the branches to the right of the black bar. A trait can be a characteristic that is *gained* (such as the origin of hair in mammals), or it can be a characteristic that is *lost* (such as legs in snakes).
- An *outgroup* is a taxon that is known to have diverged before the rest of the taxa shown in the tree. Outgroups are used to establish whether a trait is ancestral or derived.

How to Read a Phylogenetic Tree

The key to reading relationships in a phylogenetic tree is to examine which groups share most recent common ancestors. In Figure B13.1, for example, taxa 5 and 6 are more closely related to each other than either is to taxon 4, because taxa 5 and 6 share common ancestor D. Consider two important characteristics of phylogenetic trees:

1. ***Branches can rotate at each node.*** Imagine that a tree is like a mobile hanging by its root from the ceiling. If you were to look up at it, you would see that the taxa at the tips can spin around but the connections—which represent the relationships—don't change. For example, the following two trees are equivalent to the tree in Figure B13.1—although the pattern looks different, the relationships are the same:

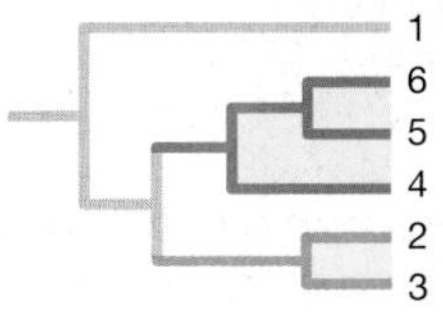

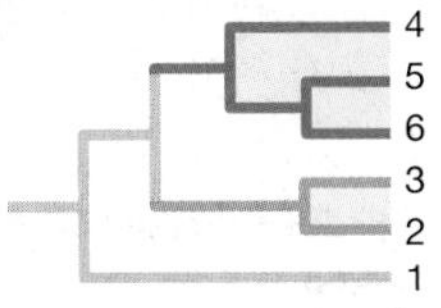

2. ***Branches can be added or removed without changing the relationships of the other branches.*** The taxa included in a tree often represent just a sample of the diversity of life that exists. Individual branches can be "expanded" to include more taxa or "collapsed" to include fewer taxa. For example, the following two trees include different numbers of taxa: 1 beetle + 5 butterflies , or 5 beetles + 1 butterfly :

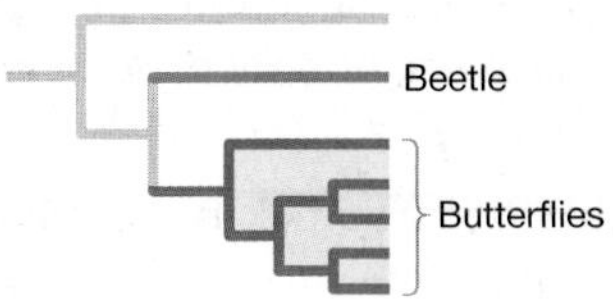

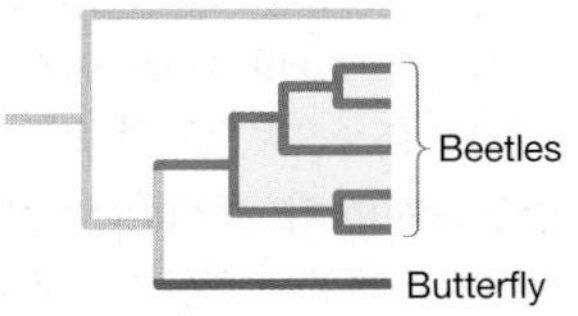

This example shows that counting the number of nodes is not a valid way to decide how closely or distantly related two taxa are—the number of nodes will depend on the number of taxa included in a particular part of a tree.

Let's apply these insights to a question regarding Figure B13.1: Is taxon 3 more closely related to taxon 4, 5, or 6?

- Even though taxa 3 and 4 are closest together on the page, this does not mean that they are more closely related. Remember, the branches can be rotated.
- Even though there are only three nodes between taxa 3 and 4 and four nodes between taxon 3 and taxon 5 or 6, this does not mean that taxa 3 and 4 are more closely related. Remember, the removal of taxon 5 or 6 from the analysis would remove node D.
- The correct answer: Taxon 3 is equally related to taxa 4, 5, and 6 because they all share a common ancestor, B. The branching events that occurred after this ancestor do not affect the underlying relationship.

It takes practice to use these visual models as tools and to draw them correctly. Drawing your own phylogenetic trees can help you improve your skill at interpreting phylogenetic trees.

How to Draw a Phylogenetic Tree

Some of the **Making Models** boxes throughout this book give you specific tips on how to make and use phylogenetic trees. (See the list of Making Models boxes on the inside back cover; also, Ch. 25, Section 25.1 demonstrates how to build a phylogenetic tree based on a data matrix.) Note that many trees in the book are color-coded or include icons of the organisms to make them easier to read. Your hand-drawn trees can be much simpler and still be effective.

CHECK YOUR UNDERSTANDING

✔ If you understood BioSkills 13, you should be able to . . .

Use the tree below to answer the questions.

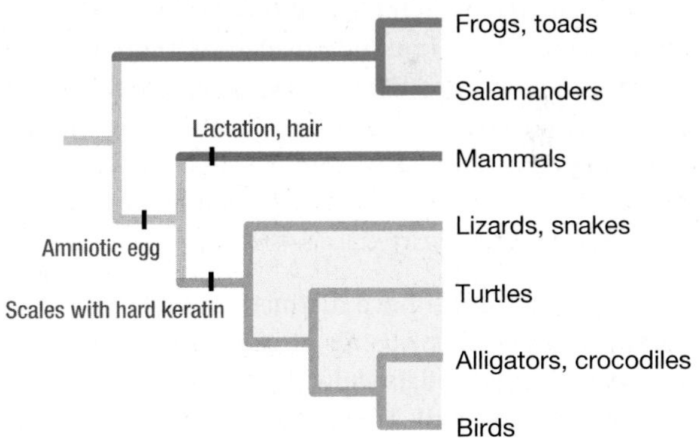

1. **MODEL** Draw a node on the tree that represents the most recent common ancestor of frogs and birds.
2. Determine which taxa have amniotic eggs.
3. **MODEL** Add a trait black bar showing the origin of limbs. How could you modify the tree to show the loss of limbs in snakes?
4. **MODEL** Draw what the tree would look like if you collapsed all the reptile branches into one branch and included two branches of mammals: placentals and marsupials.
5. Determine whether mammals are more closely related to lizards or turtles.

Answers are available in Appendix A.

BioSkills 14 **Reading Chemical Structures**

If you haven't had much chemistry yet, learning basic biological chemistry can be a challenge. One stumbling block is simply being able to read chemical structures and understand what they mean. This task will become much easier once you have a little notation under your belt and you understand some basic symbols. Atoms are the basic building blocks of everything in the universe, just as cells are the basic building blocks of your body. Every type of atom that has been identified on Earth has a one- or two-letter element symbol (see Appendix B, Periodic Table of Elements, for a full list).

Table B14.1 shows the symbols for most of the elements you'll encounter in this book. You should memorize these symbols. The table also offers details on the number of bonds each atom can form, as well as how the atoms are represented in some visual models.

When atoms attach to each other by covalent bonding, a molecule forms. Table B14.1 also includes atoms such as chlorine and potassium that can be joined by ionic bonds to form ionic compounds (Ch. 2, Section 2.1), but the focus here is on atoms that form molecules. Biologists have a couple of different ways of representing molecules—you'll see each of these in the book and in class.

- *Molecular formulas* like the one for glycine (an amino acid; Ch. 3, Section 3.1) in **Figure B14.1a** simply list the atoms present in a molecule. Subscripts indicate how many of each atom are present. If an element symbol has no subscript, only one atom of that element is present. A methane (natural gas) molecule, for example, is written as CH_4. It consists of one carbon atom and four hydrogen atoms.
- *Structural formulas* like the one for glycine in **Figure B14.1b** show which atoms in a molecule are bonded to each other. Each bond in glycine is indicated by a dash. Single covalent bonds are symbolized by a single dash, as in the bonds between the hydrogen atoms and the nitrogen atom. Double bonds are indicated by two dashes, as in the covalent bond between a carbon atom and an oxygen atom. Triple bonds are indicated by three dashes, as in the structural formula for molecular nitrogen (N_2), which is written as N≡N.

Even simple molecules have distinctive shapes, because different atoms form covalent bonds at different angles. Ball-and-stick and space-filling models show the geometry of the bonds in a molecule accurately, while ribbon models are used to depict the way large molecules fold.

- *Ball-and-stick models* are not as realistic as space-filling models, but they make the bonding arrangement of atoms (colored balls) easier to see because the bonds are represented as sticks (**Figure B14.1c** and **Figure B14.2a**). Ball-and-stick models provide information on the three-dimensional relationships among bonded atoms. In some cases, the relative sizes of the atoms may be indicated.
- *Space-filling models* are the most realistic way to represent a chemical structure (**Figure B14.1d** and **Figure B14.2b**). Spheres representing atoms show their true relative sizes. These models depict spatial relationships between atoms more realistically than ball-and-stick models—for example, how closely two atoms can approach each other when they are not linked by a covalent bond.

Table B14.1 Some Attributes of Elements Found in Organisms

Element	Symbol	Number of Bonds an Atom Can Form	Standard Color Code*
Hydrogen	H	1	white
Carbon	C	4	black
Nitrogen	N	3	blue
Oxygen	O	2	red
Sodium	Na	1	—
Magnesium	Mg	2	—
Phosphorus	P	5	orange or purple
Sulfur	S	2	yellow
Chlorine	Cl	1	—
Potassium	K	1	—
Calcium	Ca	2	—

*In ball-and-stick or space-filling models.

(a) Molecular formula: NH_2CH_2COOH **(glycine)**

(b) Structural formula:

(c) Ball-and-stick model:

(d) Space-filling model:

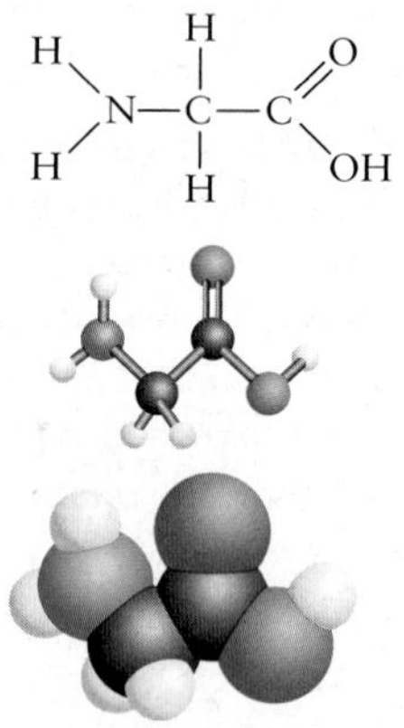

Figure B14.1 A Molecule Can Be Represented in Several Different Ways. The amino acid glycine consists of one nitrogen atom, two carbon atoms, five hydrogen atoms, and two oxygen atoms.

- *Ribbon models* are very schematic (**Figure B14.2c**). They are commonly used to highlight a few major features of protein structure. For example, α-helices are depicted like corkscrews, β-pleated sheets as flat arrows, and loops as simple lines (Ch. 3, Section 3.2).

To learn more about a molecule when you look at a chemical structure, ask yourself three questions:

1. ***Is the molecule polar—meaning are some parts more negatively or positively charged than others?*** Molecules that contain nitrogen or oxygen atoms are often polar because these atoms are very electronegative (Ch. 2, Section 2.1). This trait is important because polar molecules dissolve in water.

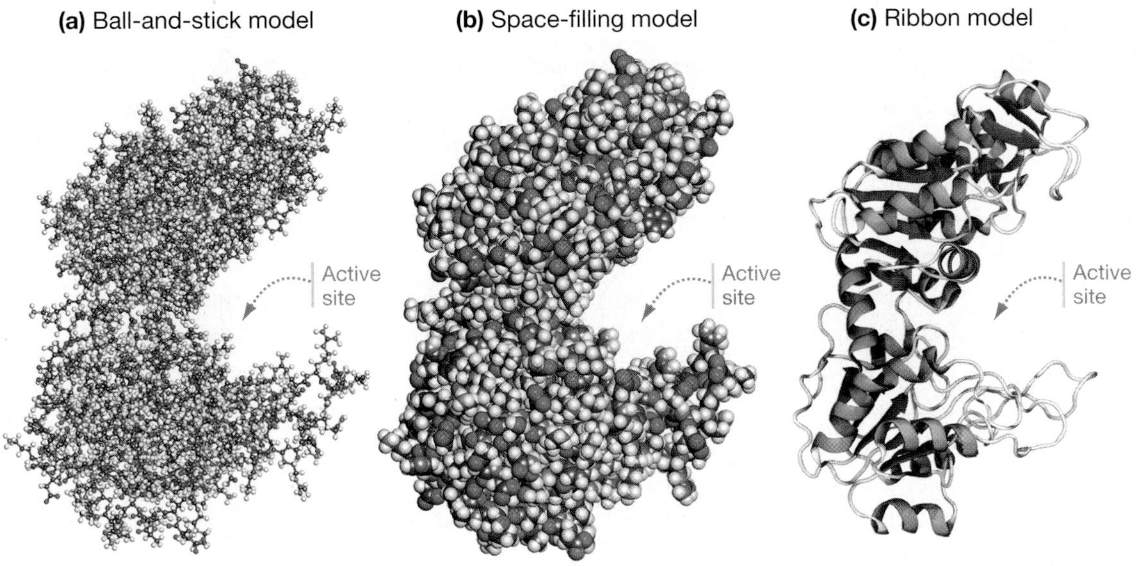

Figure B14.2 Three Different Models of an Enzyme. The enzyme phosphoglycerate kinase catalyzes a step in glycolysis (Ch. 9, Section 9.2). The enzyme's active site appears as a deep cleft.

2. ***Does a structural formula show atoms that might participate in chemical reactions?*** For example, are there charged atoms or amino (—NH_2) or carboxyl (—COOH) functional groups that might make the molecule act as a base or an acid?
3. ***In ball-and-stick and especially space-filling models of large molecules, are there interesting aspects of overall shape?*** For example, is there a groove where a protein might bind to DNA, or a cleft (as shown in Figure B14.2) where a substrate might undergo a reaction in an enzyme?

CHECK YOUR UNDERSTANDING

✔ **If you understood BioSkills 14, you should be able to . . .**

MODEL Write carbon dioxide's molecular formula and then draw its structural formula, a ball-and-stick model, and a space-filling model. (Hint: Carbon dioxide consists of a carbon atom that forms a double bond with each of two oxygen atoms, for a total of four bonds. It is a linear molecule.)

Answers are available in Appendix A.

BioSkills 15 Translating Greek and Latin Roots in Biology

After you complete BioSkills 15–16, you should be able to . . .

- Demonstrate the ability to read the primary literature.

Use the common Greek and Latin roots that are shown in **Table B15.1** on page 54 to help interpret unfamiliar biological terms and come up with a literal translation that will help you remember its meaning.

CHECK YOUR UNDERSTANDING

✔ **If you understood BioSkills 15, you should be able to . . .**

Provide literal translations of the following terms:
1. heterozygote **2.** glycolysis **3.** morphology **4.** trisomy

Answers are available in Appendix A.

BioSkills 16 Reading and Citing the Primary Literature

As part of the process of doing science, biologists communicate their results to the scientific community through publications in scientific journals that report on their original research discoveries (see The Big Picture: Doing Biology on pages 18–19). These published reports are referred to, interchangeably, as the primary literature, research papers, or primary research articles.

What Is the Primary Literature?

Scientists publish "peer-reviewed" papers. This means that several experts in the field have carefully read the paper and considered its strengths and weaknesses. Reviewers write a critique of the paper and make a recommendation to the journal editor as to whether the paper should be published. Often reviewers will suggest additional experiments that need to be completed before a paper is considered acceptable for publication. The peer review process means that research discoveries are carefully vetted before they go to press.

You can distinguish a primary research paper from secondary sources—such as review articles, textbooks, and magazine articles—by looking for key characteristics. A primary research paper includes a detailed description of methods and results, written by the researchers who did the work. As described in **Table B16.1** on page 54, it is typically divided into several sections, including the Title, Abstract, Introduction, Materials and Methods (or Experimental Design), Results and Discussion, and References (or Literature Cited), although the order and name of the sections varies among journals.

Getting Started

At first, trying to read the primary literature may seem like a daunting task. A paper may be peppered with unfamiliar terms and abbreviations. If you tried to read a research paper from start to finish, like you might read a chapter in this textbook, it would be a frustrating experience. But, with practice, the scientific literature becomes approachable, and reading it is well worth

Table B15.1 Some Common Greek and Latin Roots

Greek or Latin Root	English Translation	Example Term	Greek or Latin Root	English Translation	Example Term
a, an	not	anaerobic	*hyper*	over, more than	hypertonic
aero	air	aerobic	*hypo*	under, less than	hypotonic
allo	other	allopatric	*inter*	between	interspecific
amphi	on both sides	amphipathic	*intra*	within	intraspecific
anti	against	antibody	*iso*	same	isotonic
auto	self	autotroph	*logo, logy*	study of	morphology
bi	two	bilateral symmetry	*lyse, lysis*	loosen, burst	glycolysis
bio	life, living	bioinformatics	*macro*	large	macromolecule
blast	bud, sprout	blastula	*meta*	change, beyond	metamorphosis
co	with	cofactor	*micro*	small	microfilament
cyto	cell	cytoplasm	*morph*	form	morphology
di	two	diploid	*oligo*	few	oligopeptide
ecto	outer	ectoparasite	*para*	beside	parathyroid gland
endo	inner, within	endoparasite	*photo*	light	photosynthesis
epi	outer, upon	epidermis	*poly*	many	polymer
exo	outside	exothermic	*soma*	body	somatic cells
foli	leaf	foliage	*sym, syn*	together	symbiotic, synapsis
glyc	sugary	glycolysis	*trans*	across	translation
hetero	different	heterozygous	*tri*	three	trisomy
homo	alike	homozygous	*zygo*	yoked together	zygote
hydro	water	hydrolysis	*zym*	ferment	enzyme

the effort. The primary literature is the cutting edge, the place to read firsthand about the process of doing science. Becoming skilled at reading and evaluating scientific reports is a powerful way to learn how to think critically—to think like a biologist.

To get started, try breaking down the process of reading a primary research article into a series of steps:

1. ***Read the authors' names.*** Where are they from? Are they working as a team or alone? After you delve into the literature, certain familiar names will crop up again and again. You'll begin to recognize the experts in a particular field.

2. ***Read the title.*** It should summarize the key finding of the research and tell you what you can expect to learn from the paper.

3. ***Read the abstract.*** The abstract summarizes the entire paper in a short paragraph. At this point, it might be tempting to stop reading. But sometimes the abstract understates or

Table B16.1 Sections of a Primary Research Paper

Section	Characteristics
Title	Short, succinct, descriptive
Abstract	Summary of Methods, Results, Discussion. Explains why the research was done and why the results are significant.
Introduction	Background information (what past work was done, why the work was important). States the objectives and hypotheses of the study and explains why the study is important.
Materials and Methods	Explains how and where the work was done.
Results	Presents the data from the study and explains what they show.
Discussion	Explains why the data show what they show, how the analysis relates to the objectives stated in the Introduction, the significance of the findings, and how they advance the field.
References	A list of all the literature cited in the research paper. All sources that served as a foundation for the work, including all the ideas and findings that are not those of the authors, should be cited.

overstates the significance and conclusions of the research. You should never cite an article as a reference after having read only the abstract.

4. ***Read the Introduction.*** The first couple of paragraphs should make it clear what the objectives or hypotheses of the research were; the remaining paragraphs will give you the background information you need to understand the point of the paper.
5. ***Flip through the article.*** Look at the figures and tables, including reading the legends.
6. ***Read the Results section carefully.*** Ask yourself these questions: Does this section accurately describe the data presented in the paper? Were all the appropriate controls carried out in an experiment? Are there additional experiments that you think should have been performed? Are the figures and tables clearly labeled?
7. ***Consult the Materials and Methods section.*** It can help you understand the research design and the techniques used.
8. ***Read the Discussion.*** The first and last paragraphs usually summarize the key findings and state their significance. The Discussion is the part of the paper where the results are explained in the context of the scientific literature. The authors should explain what their results mean.
9. ***You don't need to read the reference list.*** Use it as a directory when you want to learn more about certain topics.

Citing Sources

Whether you're writing lab reports, poster presentations, or papers, it's important to get in the habit of citing your sources in the work you do for your biology classes. To avoid plagiarism, you need to credit the information that came from someone else. For those who read your work, citing references also assures them that the information you've used comes from reliable sources.

Your instructor may specify the exact format to use in citing references, but you can also see examples of how to format the reference list as you read the scientific literature. Note that the accepted style varies among journals. Take a moment to look at examples of citations in this textbook—wherever data are shown, a source for the data is provided (see Ch. 1, Figure 1.11 for one example).

Getting Practice

The best way to get practice is to read the scientific literature as often as possible. You could begin by reading some of the references cited in this textbook. You can get an electronic copy of most articles through online databases such as PubMed, ScienceDirect, or Google Scholar, or through your institution's library.

After reading a primary research paper, you should be able to paraphrase the significance of the paper in a few sentences, free of technical jargon. As you become more familiar with reading the scientific literature, you're likely to start thinking about what questions remain to be answered. You may even come up with "the next experiment."

CHECK YOUR UNDERSTANDING

✔ **If you understood BioSkills 16, you should be able to . . .**

Choose a primary research paper on a topic in biology that you would like to know more about, select one figure in the Results section that reports on a specific experiment, and construct a Research box (like the ones in this textbook) depicting that experiment.

Answers are available in Appendix A.

BioSkills 17 **Recognizing and Correcting Misconceptions**

After you complete BioSkills 17–18, you should be able to . . .

- Monitor your own learning.

Have you ever felt confident that you understood a concept in a science class, and then been surprised to get a question wrong on a quiz? To avoid such unpleasant surprises, you can begin by mastering a certain learning skill—recognizing when everyday ways of thinking and everyday uses of words are not a good fit in a scientific context.

Table B17.1 on page 56 summarizes some common ways that your experiences outside of your biology class might hinder your understanding of biology, leading to misconceptions—incorrect ideas, which can be very common among students. With some practice, though, you can learn to identify the common types of misconceptions and steer your thinking accordingly. Use the following three steps as a starting place:

1. ***Study Table B17.1 to familiarize yourself with common categories of misconceptions.*** Are any of them familiar to you already?
2. ***Answer in-chapter and end-of-chapter questions that have a* THINK CAREFULLY *tag.*** You will see this tag on some questions throughout the book. The tag is a prompt to alert you that the question addresses a concept that is often the source of a misconception.
3. ***Check your answers in Appendix A. If you missed a question with a* THINK CAREFULLY** tag, look at Table B17.1 and identify the category of misconception that made it challenging to answer the question.

Becoming aware of how experiences in everyday life can hinder learning in biology will help you learn to think like a biologist. By learning to think like a biologist, you can maximize your study success.

CHECK YOUR UNDERSTANDING

✔ **If you understood BioSkills 17, you should be able to . . .**

1. **THINK CAREFULLY** Identify the type of misconception in this statement: Fish evolved legs so that they could walk on land.
2. Look up the meaning of "theory" in science and compare it to the meaning in everyday use. Describe how this difference in meaning could cause confusion.

Answers are available in Appendix A.

Table B17.1 Common Categories of Misconceptions

Category	Description	Examples
Different meaning of terms	Some terms have a different meaning in everyday use than they do in biology.	theory, fitness, adaptation, selection, "using up" energy, "producing" energy, climate, germ
Goal-oriented thinking	Humans have a natural tendency to assign purpose or reason to events, but most biological processes are not goal driven.	"Plants produce oxygen so that animals can breathe." "Roses evolved thorns so that herbivores wouldn't eat them."
Human-centered thinking	Humans have a natural tendency to view the world from a human perspective, but our personal experiences often do not apply to other biological contexts.	"Plants suck up food through their roots." "Mutations are bad." "Disturbance in communities is bad."
Simplistic thinking	Humans have a natural tendency to mentally simplify complex systems to make sense of them, but this can lead to oversimplification.	"Eye genes occur only in eye cells." "All members of a species are alike."

Bio Skills 18 Using Bloom's Taxonomy for Study Success

At one time or another, you may have wondered why a particular question on an exam seemed so hard while others seemed easy. The explanation has much to do with the type of cognitive skill required to answer the question.

Let's take a closer look.

Categories of Human Cognition

Bloom's Taxonomy is a classification system that instructors use to identify the cognitive skill levels at which they are asking students to work, particularly on practice problems and exams. Bloom's Taxonomy is also a very useful tool for you to know—it can help you figure out the appropriate level at which you should be studying to succeed in a course.

Bloom's Taxonomy distinguishes six categories of human thinking: *Remember, Understand, Apply, Analyze, Evaluate*, and *Create*. One of the most useful distinctions lies not in the differences among these six categories, but rather in the difference between what are classified as high-order cognitive (HOC) skills and low-order cognitive (LOC) skills. **Figure B18.1** shows how the six categories of the taxonomy can be broken into HOC and LOC skills.

Types of thinking that hallmark LOC skills include recall, explanation (*Remember, Understand*), and use of knowledge in the exact way that you have used it before (*Apply*). Note, however, that when knowledge is used in a new way, *Apply* is considered an HOC skill.

Thinking levels that typify HOC skills include the breakdown, critique, and creation of information (*Analyze, Evaluate, Create*). Most college instructors will assume you are proficient at solving LOC questions and will expect you to frequently work at the HOC levels. The good news is that HOC problems usually require use of basic vocabulary and applying knowledge in a new way. Thus, working at the HOC levels has an added benefit—it also helps you to master the LOC levels.

Six Study Steps to Success

You can use Bloom's Taxonomy along with questions in this textbook to prepare for an exam. Follow these six study steps to success:

1. ***Answer in-chapter questions while reading the chapter.*** All questions in this book have been assigned Bloom's levels in answers provided in Appendix A, so you can review the answers and the Bloom's level while you study.
2. ***Identify the Bloom's level(s) of the questions that you are having greatest difficulty answering.*** While working through the text, take note of the content and Bloom's level(s) that you find the most challenging.
3. ***Use the Bloom's Taxonomy Study Guide* (Table B18.1) *to focus your study efforts at the appropriate Bloom's level.*** Table B18.1 lists specific study methods that can help you practice your understanding of material at both the LOC and HOC levels, whether you are studying alone or with a study group.

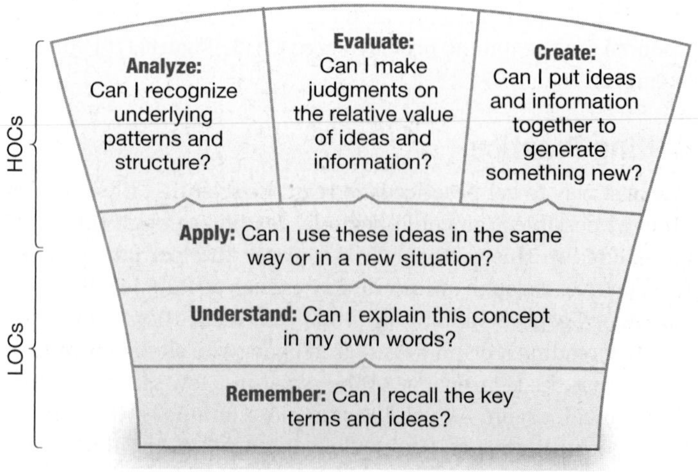

Figure B18.1 Bloom's Taxonomy

4. ***Complete the end-of-chapter questions as if you're taking an exam, without looking for the answers.*** If you look at the chapter text or jump to the answers, then you really aren't testing your ability to work with the content and have reduced the questions to the lowest Bloom's level of *Remember*.
5. ***Grade your answers to the end-of-chapter questions and note the Bloom's level of the questions you got wrong.*** At what level of Bloom's Taxonomy were the questions you missed?
6. ***Use the Bloom's Taxonomy Study Guide to focus your study efforts at the appropriate Bloom's level.*** If you missed a lot of questions, then revisit Table B18.1 to review suggestions for study activities, spend more time studying the material, and find other resources for quizzing yourself, such as online problem sets.

By following these six steps and studying at both the LOC and HOC levels, you should succeed in answering questions on in-class exams.

CHECK YOUR UNDERSTANDING

✔ If you understood BioSkills 18, you should be able to . . .

1. Explain the six tips to study success.
2. Determine the Bloom's level of question 1.

Answers are available in Appendix A.

Table B18.1 Bloom's Taxonomy Study Guide

Skills Level(s)	Individual Study Activities	Group Study Activities
Create (HOC) *Generate something new*	• Generate a hypothesis or design an experiment based on information you are studying. • Create a model based on a given data set. • Create summary sheets that show how facts and concepts relate to each other. • Create questions at each level of Bloom's Taxonomy as a practice test and then take the test.	• Have each student put forward a hypothesis about a biological process and design an experiment to test it, then have your peers critique the hypotheses and experiments. • Create a new model/summary sheet/concept map that integrates each group member's ideas.
Evaluate (HOC) *Defend or judge a concept or idea*	• Provide a written assessment of the strengths and weaknesses of your peers' work or understanding of a given concept based on previously determined criteria.	• Provide a verbal assessment of the strengths and weaknesses of your peers' work or understanding of a given concept based on previously described criteria, and have your peers critique your assessment.
Analyze (HOC) *Distinguish parts and make inferences*	• Analyze and interpret data in primary literature or a textbook without reading the author's interpretation and then compare the authors' interpretation with your own. • Analyze a situation and then identify the assumptions and principles of the argument. • Compare and contrast two ideas or concepts. • Construct a map of the main concepts by defining the relationships of the concepts using one- or two-way arrows.	• Work together to analyze and interpret data in primary literature or a textbook without reading the author's interpretation, and defend your analysis to your peers. • Work together to identify all of the concepts in a paper or textbook chapter, construct individual maps linking the concepts together with arrows and words that relate the concepts, and then grade each other's concept maps.
Apply (HOC or LOC) *Use information or concepts in new ways (HOC) or in the same ways (LOC)*	• Review each process you have learned and then ask yourself: What would happen if you increase or decrease a component in the system, or what would happen if you alter the activity of a component in the system? • If possible, graph a biological process and create scenarios that change the shape or slope of the graph.	• Practice writing out answers to old exam questions on the board, and have your peers check to make sure you don't have too much or too little information in your answer. • Take turns teaching your peers a biological process while the group critiques the content.
Understand (LOC) *Explain information or concepts*	• Describe a biological process in your own words without copying it from a book or another source. • Provide examples of a process. • Write a sentence using the word. • Give examples of a process.	• Discuss content with peers. • Take turns quizzing each other about definitions, and have your peers check your answers.
Remember (LOC) *Recall information*	• Practice labeling diagrams. • List characteristics. • Identify biological objects or components from flash cards. • Quiz yourself with flash cards. • Take a self-made quiz on vocabulary. • Draw, classify, select, or match items. • Write out the textbook definitions.	• Check a drawing that another student labeled. • Create lists of concepts and processes that your peers can match. • Place flash cards in a bag and take turns selecting one for which you must define a term. • Do the preceding activities, and have peers check your answers.

2 Water and Carbon: The Chemical Basis of Life

These deep-sea hydrothermal vents produce hydrogen-rich, highly basic fluids at temperatures that range from 40 to 90 degrees C. It has been proposed that life emerged from similar seafloor chimneys early in Earth's history via chemical evolution.

In this chapter you will learn that

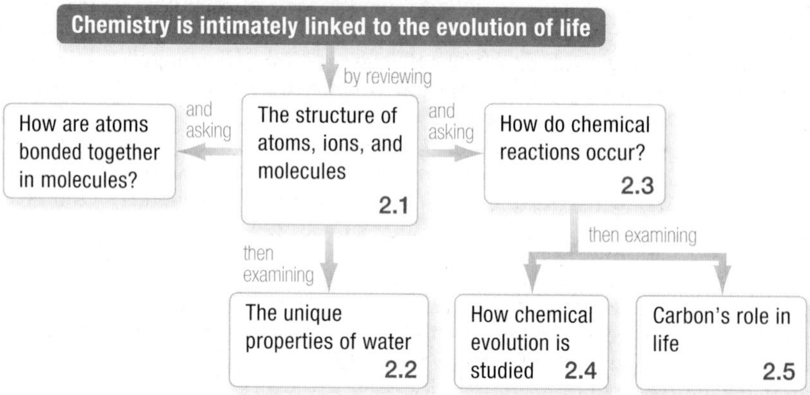

A classic experiment on spontaneous generation by Louis Pasteur tested the idea that organisms arise from nonliving materials (Ch. 1, Section 1.2). The results of this work helped build a consensus that spontaneous generation does not occur. But for life to exist, it must have evolved from nonliving materials at least once early in Earth's history.

How did life begin? This simple query has been called "the mother of all questions." In this chapter you will examine the fundamental principles of chemistry that are necessary to understand the leading scientific explanation for life's origin—the theory of **chemical evolution**. The theory maintains that the formation of increasingly complex carbon-containing substances culminated in a molecule that could replicate itself. At this point, there was a switch from chemical evolution to biological evolution.

As the original molecule multiplied, the process of evolution by natural selection took over. Eventually a descendant of the original molecule became metabolically active and acquired a membrane. When this occurred, the five characteristics of life (discussed in Ch. 1, Section 1.1) were fulfilled. Life had begun.

BIG PICTURE

This chapter is part of the Big Picture. See how on pages 144–145.

At first glance, the theory of chemical evolution may seem implausible. But is it? Let's start by reviewing the structure of key atoms and molecules that would have combined to get chemical evolution started.

2.1 Atoms, Ions, and Molecules: The Building Blocks of Chemical Evolution

An **atom** is the smallest identifiable unit of matter. Just four types of atoms—hydrogen (H), carbon (C), nitrogen (N), and oxygen (O)—make up 96 percent of all matter found in organisms today. Many of the molecules found in your cells contain thousands, or even millions, of these atoms bonded together. But early in Earth's history, H, C, N, and O existed only in simple substances such as water and carbon dioxide, which contain just three atoms apiece.

> **After you complete this section, you should be able to ...**
>
> - Analyze the relationship between atomic structure and how atoms interact in simple molecules.

Two questions are fundamental to understanding how these simple substances could have evolved into the more complex molecules found in living cells:

1. What are the physical structures of hydrogen, carbon, nitrogen, and oxygen atoms?
2. What are the structures of the simple molecules—water, carbon dioxide, and others—that served as the building blocks of chemical evolution?

The focus on structure follows from one of the most central themes in biology: *Structure affects function*. To understand how a molecule affects your body or the role it played in chemical evolution, you have to understand how it is put together.

Basic Atomic Structure

Figure 2.1a shows a simple way of depicting the structure of an atom, using hydrogen and carbon as examples. Extremely small particles called electrons orbit an atomic nucleus made up of larger particles called protons and neutrons. Every element except hydrogen has one or more neutrons in its nucleus. Figure 2.1b provides a sense of scale at the atomic level.

Protons have a positive electric charge (+1), neutrons are electrically neutral, and electrons have a negative electric charge (−1). When the number of protons and the number of electrons in an atom are the same, the charges balance and make the entire atom electrically neutral.

Figure 2.2 shows a segment of the periodic table of the elements. **Elements** are defined as substances that consist entirely of a single type of atom. Notice that each atom of a given element contains a characteristic number of protons, called its **atomic number**. The atomic number is written as a subscript to the left of an element's symbol in Figure 2.2. The sum of the protons and neutrons in an atom is called its **mass number** and is written as a superscript to the left of its symbol.

Although the masses of protons, neutrons, and electrons can be measured in grams, the numbers involved are so small that biologists prefer to use a special unit called the **dalton (Da)**. This unit of measure was named after John Dalton, who was responsible for formulating modern atomic theory in the seventeenth century. The masses of protons and neutrons are virtually identical and are routinely rounded to 1 Da each. The mass of an electron is so small that it is normally ignored. So, the mass of an atom is equal to its mass number.

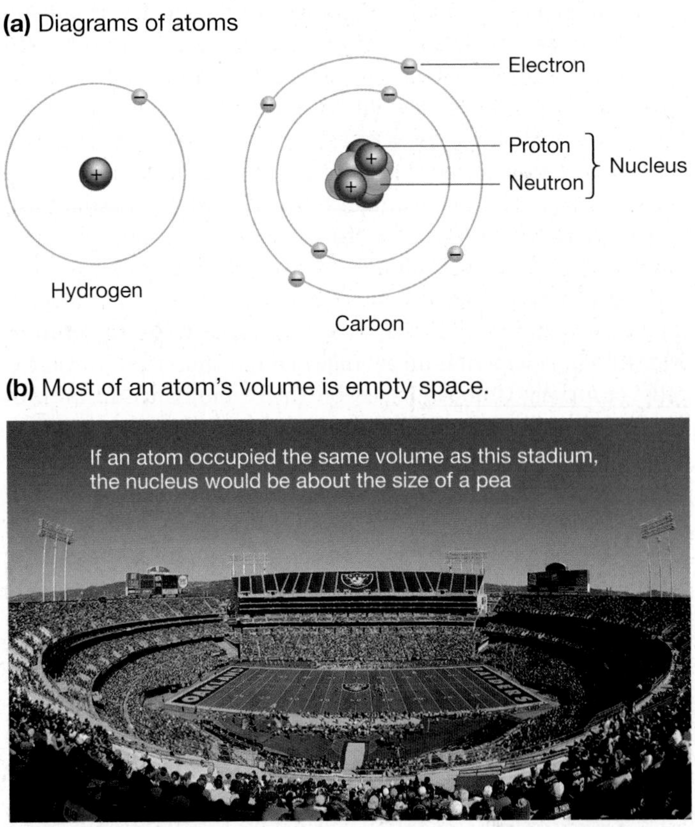

Figure 2.1 Parts of an Atom. A simplified model of an atom with its nucleus, made up of protons and neutrons—or a single proton in the case of hydrogen—surrounded by orbiting electrons. In reality, electrons are not evenly spaced, nor do they orbit the nucleus in concentric circles; their actual orbits are complex.

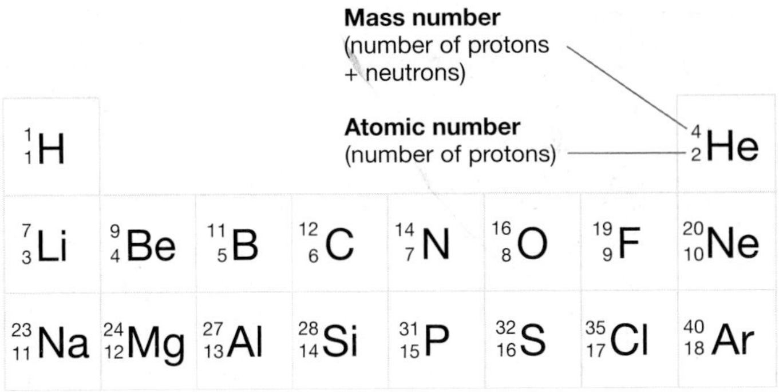

Figure 2.2 A Portion of the Periodic Table. Each element has a unique atomic number and is represented by a unique one- or two-letter symbol. The mass numbers given here are the most common for each element. (Appendix B provides a complete periodic table.)

The number of protons in an element does not vary—if the atomic number of an atom changes, then it is no longer the same element. The number of neutrons present in an element can vary, however. Forms of an element with different numbers of neutrons are known as **isotopes** (literally "equal-places," because all the forms occupy the same position in the periodic table).

Isotopes of the same element have different masses because they have different numbers of neutrons. All atoms of the element carbon have 6 protons, for example, but naturally occurring isotopes of carbon can have 6, 7, or even 8 neutrons, giving them masses of 12, 13, or 14 Da, respectively. The **atomic weight** of an element is an average of all the masses of the naturally occurring isotopes based on their abundance in nature. This is why the atomic weight of an element is often slightly different from its mass number. For example, carbon's atomic weight is 12.01 rather than 12, which reflects that while the most abundant isotope of carbon has 6 neutrons and a mass of 12 daltons (^{12}C), there are also less abundant isotopes with greater atomic weights.

Most common isotopes are stable, but not all. For example, ^{14}C, with 8 neutrons, represents an unstable **radioactive isotope**. Its nucleus will eventually decay and release energy (in the form of radiation). When ^{14}C decays, one of its neutrons changes into a proton, converting ^{14}C to the stable ^{14}N isotope of nitrogen, with 7 protons and 7 neutrons. Timing of decay is specific to each radioactive isotope, a fact that has been very useful in estimating the dates of key events in Earth's history (Ch. 22, Section 22.2).

To understand how the structures of atoms differ, take a moment to study **Figure 2.3**. This chart highlights in blue the elements that are most abundant in living cells. The elements C, H, N, O, P, and S make up over 99 percent of the atoms in your body.

The arrangement of electrons around the nucleus is key to understanding how different elements behave.

- Electrons move around atomic nuclei in specific regions called **orbitals**. Each orbital can hold up to two electrons (i.e., a pair).
- Orbitals are grouped into levels called **electron shells**.
- Electron shells are numbered 1, 2, 3, and so on, to indicate their relative distance from the nucleus. Smaller numbers are closer to the nucleus.
- Each electron shell contains a specific number of orbitals. Each orbital in a shell is loaded with one electron before any orbital is filled with a second, paired electron.
- The electrons of an atom fill the innermost shells first, before filling outer shells.

Now focus on the outermost shell of each element. This is the atom's **valence shell**. The electrons found in this shell are referred to as **valence electrons**. Note that in each of the highlighted elements, the outermost electron shell is not full—it contains at least one orbital with an unpaired valence electron. The number of unpaired valence electrons varies among elements. Carbon, for example, has four valence electrons, all unpaired. Oxygen has six valence electrons; four are paired, two are not. The number of unpaired electrons found in an atom's valence shell is referred to as its **valence**. Carbon's valence is four, oxygen's is two.

The observations just made are significant because an atom is most stable when its valence shell is filled. One way that valence shells can be filled is through the formation of **chemical bonds**—attractions that bind atoms together. When two atoms share electrons, the chemical bond is called a **covalent bond**, and the connected atoms are termed a **molecule**.

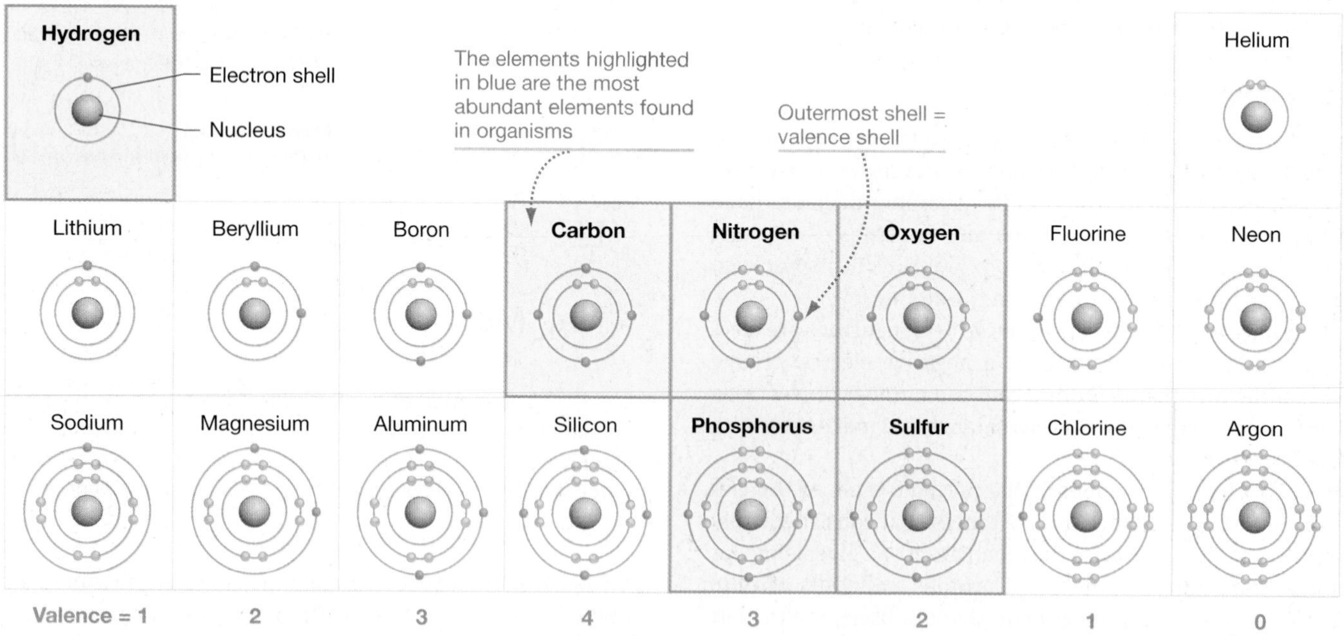

Figure 2.3 The Atomic Structure of the First 18 Elements. The most abundant elements in organisms are highlighted in blue.

✔ **MODEL** Use the atomic structure of phosphorus to determine the number of electrons, electron shells, and orbitals present in the outer shell of this element. Which, if any, of these values would differ between different isotopes?

How Does Covalent Bonding Hold Molecules Together?

To understand how atoms can become more stable by making covalent bonds, consider hydrogen. The hydrogen atom has just one electron, which resides in a valence shell that can hold two electrons.

Because it has an unpaired valence electron, the hydrogen atom does not have a full valence shell and is not very stable. But when two atoms of hydrogen come into contact, their two electrons become shared by their two nuclei as shown in Figure 2.4. This in effect gives each atom a filled outer shell. Together, the hydrogen atoms are more stable than the two individual hydrogen atoms.

Shared electrons "glue" atoms together into molecules. In the case of two hydrogen atoms, the bonded atoms form a single molecule of hydrogen, written as H—H or H_2.

Nonpolar and Polar Covalent Bonds In Figure 2.5a, the covalent bond between hydrogen atoms is represented by a dash and the electrons are drawn as dots halfway between the two nuclei. This depiction shows that the electrons are shared equally between the two hydrogen atoms, resulting in a covalent bond that is symmetrical.

It's important to note, though, that the electrons participating in a covalent bond are not always shared equally between the atoms involved. This may occur in substances called **compounds**, in which atoms of different elements are bonded together. When atoms of different elements form a bond, they may pull shared electrons toward their nuclei with varying strengths. Chemists call this property of an atom its **electronegativity**.

What is responsible for an atom's electronegativity? It's a combination of two things—the number of protons in the nucleus and the distance between the nucleus and the valence shell. If you return to the periodic table in Figure 2.3 and move your finger along a full row from left to right, you will be moving toward elements that increase in number of protons and in electronegativity (ignoring those with full outer shells in the far right column). Each row in the table represents a shell of electrons. As your finger moves down the table, it passes over elements with more shells and less electronegativity. In Figure 2.3, fluorine would have the highest electronegativity and sodium would have the lowest.

Oxygen, which has eight protons and only two electron shells, is among the most electronegative of all elements. It attracts covalently bonded electrons more strongly than does any other atom commonly found in organisms. Nitrogen, which has one fewer proton, has a somewhat lower electronegativity than oxygen. Sulfur, carbon, hydrogen, and phosphorus, in turn, have relatively low and approximately equal electronegativities. Thus, the electronegativities of the six most abundant elements in organisms are related as follows: $O > N > S \cong C \cong H \cong P$.

Because carbon and hydrogen have approximately equal electronegativity, the electrons in a C—H bond are shared equally or symmetrically. A bond that involves equally shared electrons is called a **nonpolar covalent bond**. In contrast, asymmetric sharing of electrons results in a **polar covalent bond**. The electrons in a polar covalent bond spend most of their time close to the nucleus of the more electronegative atom. Why is this important?

Polar Bonds Produce Partial Charges on Atoms To understand the consequences of differences in electronegativity and the formation of polar covalent bonds, consider the water molecule. Water consists of an oxygen atom bonded to two hydrogen atoms, and is written H_2O. As Figure 2.5b illustrates, electrons involved in the covalent bonds in water are not shared equally but are held much more tightly by the oxygen nucleus than by the hydrogen nuclei. Hence, both bonds in a water molecule, between each of the hydrogen atoms and the oxygen atom, are polar covalent bonds.

Here's the key observation: Because electrons are shared unequally in each O—H bond, they spend more time near the oxygen atom, giving it a partial negative charge, and less time near the hydrogen atoms, giving them a partial positive charge. These partial charges are symbolized by the lowercase Greek letter delta (δ), together with a plus (+) or minus (−) sign.

As you will see (Section 2.2), the partial charges on water molecules—due simply to the difference in electronegativity between oxygen and hydrogen—are one of the primary reasons that life exists.

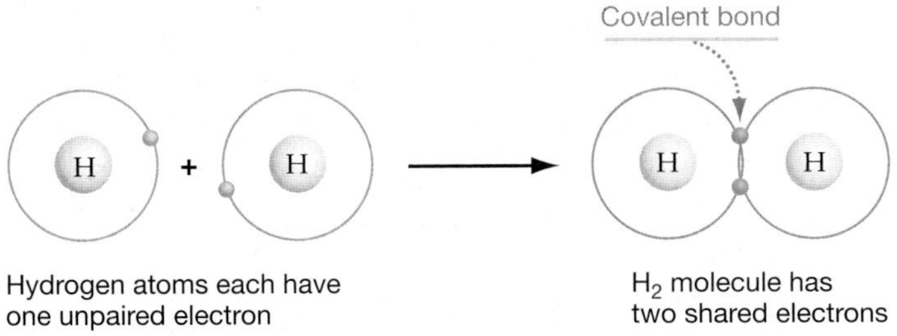

Figure 2.4 Covalent Bonds Result from Electron Sharing. When two hydrogen atoms form a covalent bond, their unpaired valence electrons are shared by each nucleus.

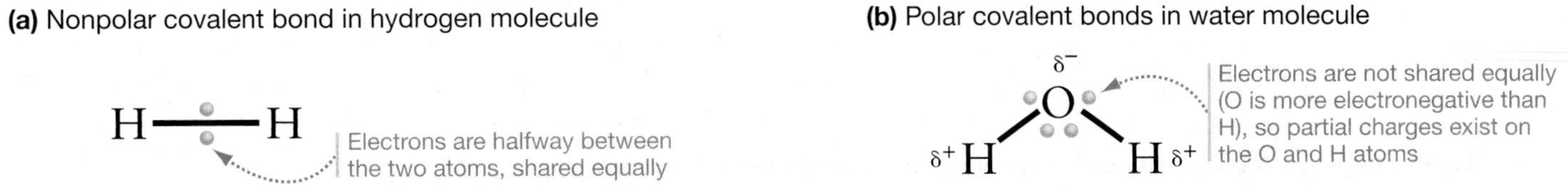

Figure 2.5 Electron Sharing and Bond Polarity. Electrons in a covalent bond can be **(a)** shared equally, resulting in nonpolar bonds, or **(b)** shared unequally, resulting in polar bonds. Delta symbols δ^+ and δ^- associated with polar covalent bonds refer to partial charges that arise owing to unequal electron sharing.

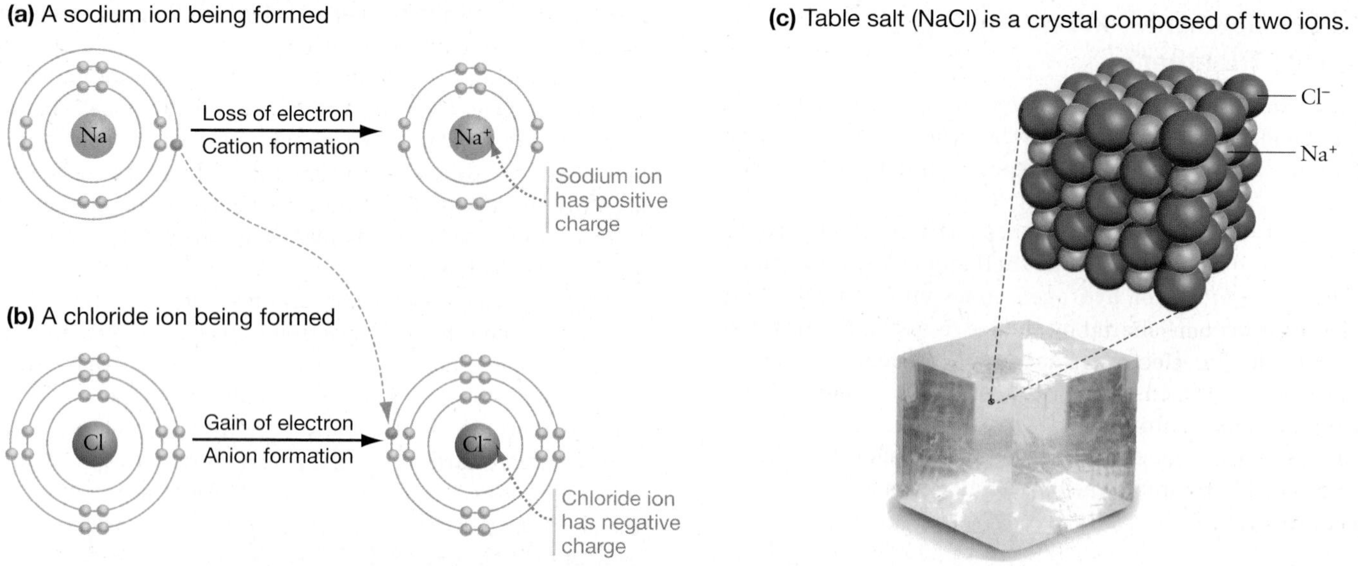

Figure 2.6 Ion Formation and Ionic Bonding. The sodium ion (Na^+) and the chloride ion (Cl^-) are stable because they have full valence shells. In table salt (NaCl), sodium and chloride ions pack into a crystal structure held together by electrical attraction between their positive and negative charges.

Ionic Bonding, Ions, and the Electron-Sharing Continuum

Ionic bonds are similar in principle to covalent bonds, but instead of being shared between two atoms, the electrons in ionic bonds are completely transferred from one atom to the other. The electron transfer occurs because it gives each of the two resulting atoms a full valence shell.

A sodium atom (Na), for example, has three electron shells with a lone electron in its valence shell. Sodium atoms tend to lose an electron, leaving them with a full second shell—a much more energetically stable arrangement (**Figure 2.6a**). The atom that results has a net electric charge of +1, because it has one more proton than it has electrons.

An atom or molecule that carries a full charge, rather than the partial charges that arise from polar covalent bonds, is called an **ion**. The sodium ion is written Na^+ and, like other positively charged ions, is called a **cation** (pronounced *KAT-eye-un*).

Chlorine atoms (Cl), in contrast, tend to gain an electron, filling their outermost shell (**Figure 2.6b**). The resulting ion has a net charge of −1, because it has one more electron than protons. This negatively charged ion, or **anion** (pronounced *AN-eye-un*), is written Cl^- and is called chloride.

When sodium cations and chloride anions combine to form sodium chloride (NaCl, common table salt), they pack into a crystal structure held together by electrical attraction between opposite charges on the ions (**Figure 2.6c**). The electrical attraction is so strong that salt crystals are difficult to break apart.

Based on the previous discussions of covalent and ionic bonds, **Figure 2.7** illustrates an important general observation concerning the role of electrons. The degree to which electrons are shared in chemical bonds forms a continuum from equal sharing

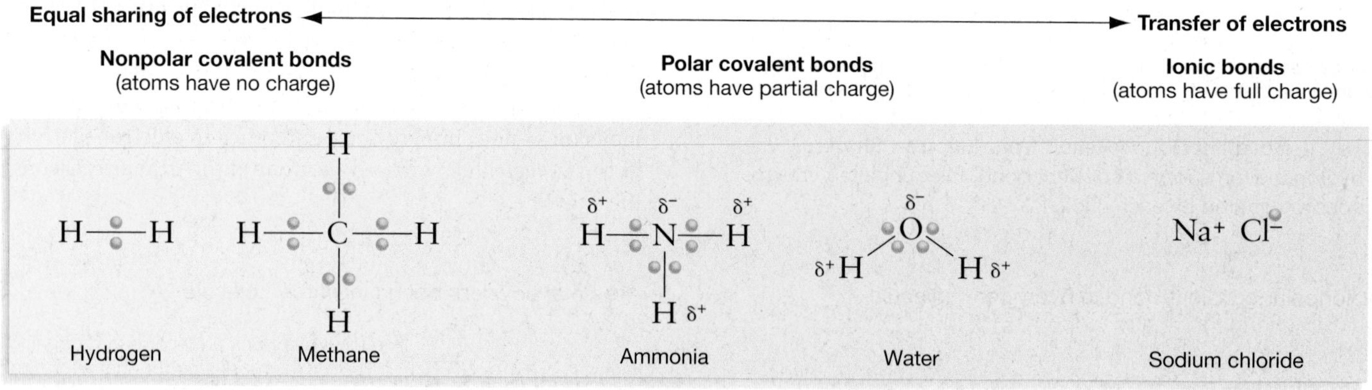

Figure 2.7 The Electron-Sharing Continuum. The degree of electron sharing in chemical bonds can be thought of as a continuum, from equal sharing in nonpolar covalent bonds to complete electron transfer in ionic bonds.

✔ Why do most polar covalent bonds involve nitrogen or oxygen?

in nonpolar covalent bonds to unequal sharing in polar covalent bonds to the transfer of electrons in ionic bonds. Most of the compounds that are present in living organisms are formed from either nonpolar or polar covalent bonds.

Let's look next at how covalent bonds hold atoms together in molecules.

Some Simple Molecules Formed from C, H, N, and O

Look back at the periodic table in Figure 2.3 and count the number of unpaired electrons in the valence shells of carbon, nitrogen, oxygen, and hydrogen. Each unpaired electron in a valence shell can make up half of a covalent bond. It should make sense to you that a carbon atom can form a total of four covalent bonds; nitrogen can form three; oxygen can form two; and hydrogen, one.

When each of the four unpaired electrons of a carbon atom covalently bonds with a hydrogen atom, the molecule that results is written CH_4 and is called methane (**Figure 2.8a**). Methane is the most common molecule found in natural gas. When a nitrogen atom's three unpaired electrons bond with three hydrogen atoms, the result is NH_3, or ammonia. Similarly, an atom of oxygen can form covalent bonds with two atoms of hydrogen, resulting in a water molecule (H_2O). As Figure 2.4 showed, a hydrogen atom can bond with another hydrogen atom to form hydrogen gas (H_2).

Atoms with more than one unpaired electron in the valence shell can also form double bonds or triple bonds. **Figure 2.8b** shows how carbon forms double bonds with oxygen atoms to produce carbon dioxide (CO_2). Triple bonds result when three pairs of electrons are shared. **Figure 2.8c** shows the structure of molecular nitrogen (N_2), which forms when two nitrogen atoms establish a triple bond.

Note that in each of the molecules in Figure 2.8, the single, double, or triple covalent bonds have the effect of giving each atom a full outer shell. Each nitrogen atom in N_2, for example, has one unshared pair of electrons and three shared electron pairs to fill its valence shell with a total of eight electrons.

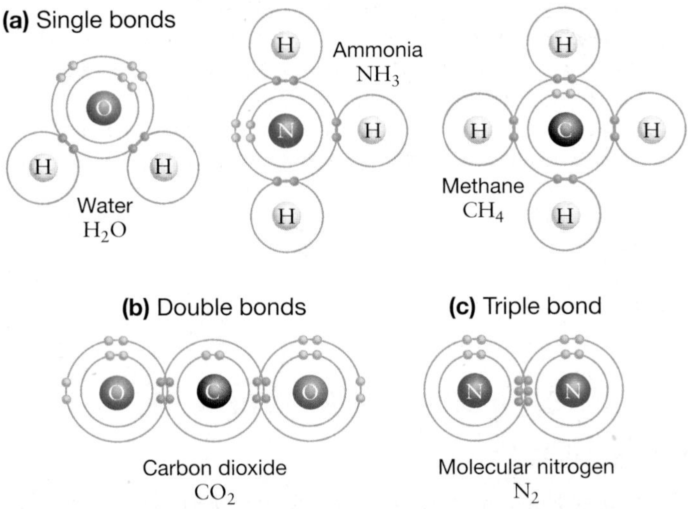

Figure 2.8 Unpaired Electrons in the Valence Shell Participate in Covalent Bonds. Covalent bonding is based on sharing of electrons in the outermost shell. Covalent bonds can be **(a)** single, **(b)** double, or **(c)** triple.

✔ Draw arrows between the atoms in each molecule to indicate the relative position of the shared electrons. If they are equally shared, then draw a double-headed arrow.

The Geometry of Simple Molecules

In many cases, the overall shape of a molecule dictates how it *behaves*. In chemistry and in biology, function is based on structure.

The shapes of the simple molecules you've just learned about are governed by the geometry of their bonds. The position of these bonds results from the repulsive forces between the negative charges in shared and unshared electron pairs in the valence shell.

- Nitrogen (N_2) and carbon dioxide (CO_2) have linear structures (see Figure 2.8). There are only two atoms in N_2, so the molecule can only be linear. The three atoms in CO_2 are linear because the electrons in the two C=O bonds repel one another and are thus 180° apart, which maximizes the distance between them.
- Methane (CH_4) has a tetrahedral structure (**Figure 2.9a**). The tetrahedron forms because the repulsive forces between electrons push the four C—H bonds as far apart as they can get, such that each bond to the central carbon atom is 109.5° away from its neighboring bonds.
- Water (H_2O) has a planar (flat) structure, with a bond geometry that is bent rather than linear (**Figure 2.9b**). Why? The electrons in the four orbitals of oxygen's valence shell repulse each other, just like the electrons in the C—H bonds in methane do. But in water, two of the orbitals in the central oxygen atom are filled with unshared electron pairs, which push the O—H bonds closer together than what is observed in methane's C—H bonds. The result is a flat, V-shaped molecule with O—H bonds that are 104.5° apart.

Later in this chapter you will explore how water's shape, in combination with the partial charges on the oxygen and hydrogen atoms, makes it the most important molecule on Earth (Section 2.2).

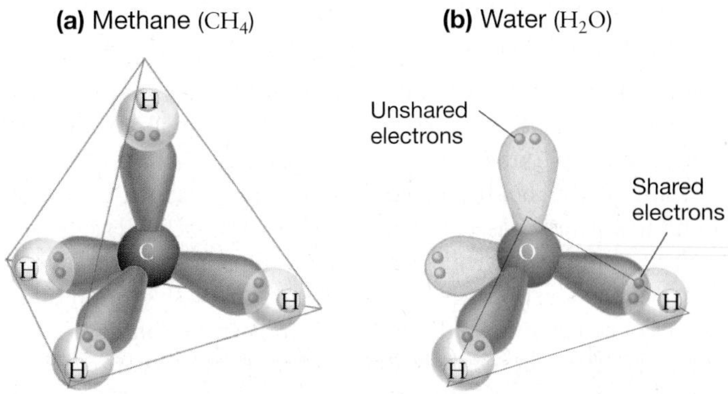

Figure 2.9 The Geometry of Methane and Water.

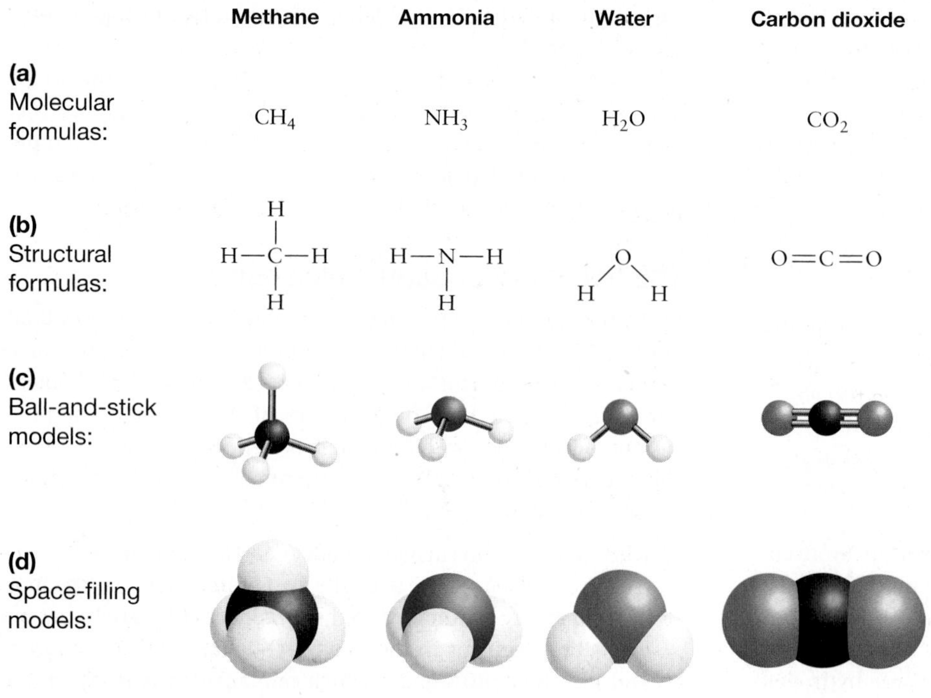

Figure 2.10 Molecules Can Be Represented Several Ways. Each method of representing a molecule has particular advantages.

Representing Molecules

Molecules can be represented in a variety of increasingly complex ways—only some of which reflect their actual shape. Each method has advantages and disadvantages (see **BioSkills 14**).

- **Molecular formulas** are compact, but don't contain a great deal of information—they indicate only the numbers and types of atoms in a molecule (**Figure 2.10a**).
- **Structural formulas** indicate which atoms in a molecule are bonded together. Single, double, and triple bonds are represented by single, double, and triple dashes, respectively. Structural formulas also indicate geometry in two dimensions (**Figure 2.10b**). This method is useful for planar molecules such as water and CO_2.
- **Ball-and-stick models** take up more space than structural formulas, but provide information on the three-dimensional shape of molecules and often indicate the relative sizes of the atoms involved (**Figure 2.10c**).
- **Space-filling models** are more difficult to read than ball-and-stick models but more accurately depict the relative sizes of atoms and their spatial relationships (**Figure 2.10d**).

In both ball-and-stick and space-filling models, biologists use certain colors by convention to represent certain atoms. A black ball, for example, always symbolizes carbon.

Some of the small molecules you've just learned about are found in volcanic gases, the atmospheres of nearby planets, and in deep-sea hydrothermal vents like those shown in the photograph at the start of this chapter. Based on these observations, researchers propose that they were important components of Earth's ancient atmosphere and oceans. If so, then they could have provided the building blocks for chemical evolution. The question is: How did these simple building blocks combine to form more complex products early in Earth's history?

Researchers postulate that most of the critical reactions in chemical evolution occurred in an **aqueous**, or water-based, environment. To understand what happened and why, let's delve into the properties of water and then turn to analyzing the reactions that triggered chemical evolution.

CHECK YOUR UNDERSTANDING

✔ If you understood this section, you should be able to . . .

1. Explain what changes when the radioactive isotope tritium (^3H) decays and one of its neutrons becomes a proton.
2. **MODEL** Draw the structural formula of the substance known as formaldehyde (CH_2O), using dots to indicate where the electrons are being shared in each covalent bond. Note any partial charges that may be associated with each atom based on the relative electronegativities of C, H, and O.
3. Predict whether the bonds that chlorine forms with sodium (to form NaCl) and with carbon (to form CCl_4) would be the same in both compounds.

Answers are available in Appendix A.

2.2 Properties of Water and the Early Oceans

Life is based on water. It arose in an aqueous environment and remains dependent on water today. In fact, 75 percent of the volume in a typical cell is water; water is the most abundant molecule in organisms. You can survive for weeks without eating, but you aren't likely to live more than 3 or 4 days without drinking.

After you complete this section, you should be able to . . .

- Summarize the properties of water and explain how they are necessary for life.

Water is vital for a simple reason: It is an excellent **solvent**—that is, an agent for dissolving, or getting substances into **solution**. Consider what happens when you add a packet of sugar to a cup of coffee. The sugar "disappears" as the sugar molecules disperse in the aqueous solution. The sugar molecules become separated from one another and interact with water's partial charges instead. Why is this so important?

The reactions that were responsible for chemical evolution some 3.5 billion years ago, like those occurring inside your body right now, depended on direct, physical interaction between molecules. Substances are most likely to come into contact with one another and react when they are **solutes**—meaning, when they are dissolved in a solvent like water. The formation of Earth's first ocean, about 3.8 billion years ago, was a turning point in chemical evolution because it gave the process a place to happen.

What Properties Are Correlated with Water's Structure?

Water's small size, highly polar covalent bonds, and bent shape resulting in overall polarity are unique among molecules. Because the structure of molecules routinely correlates with their function, it's not surprising that water has some remarkable properties.

Water Is an Efficient Solvent To understand why water is such an effective solvent, recall that each of the covalent O—H bonds in a water molecule is polar, owing to the difference in the electronegativities of hydrogen and oxygen. As a result, the oxygen atom has a partial negative charge and each hydrogen atom has a partial positive charge. Also recall that water molecules have a bent geometry. This shape makes the partial negative charge on the oxygen atom stick out, away from the partial positive charges on the hydrogen atoms, and gives a water molecule an overall **polar** nature (Figure 2.11a). Like polar covalent bonds, when molecules are polar, they carry a partial positive charge on one side and a partial negative charge on the other.

Figure 2.11b illustrates how water's polarity affects its interactions with other water molecules. When two water molecules approach each other, the partial positive charge on hydrogen attracts the partial negative charge on oxygen. This weak electrical interaction is an example of a **hydrogen bond**—an attraction between a hydrogen atom with a partial positive charge and another atom, usually oxygen or nitrogen, with a partial negative charge.

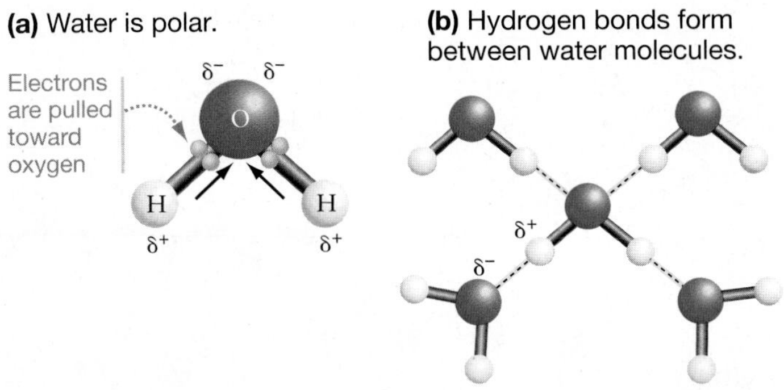

Figure 2.11 Water Is Polar and Participates in Hydrogen Bonds. **(a)** The polar covalent bonds in water give the oxygen a partial negative charge and each hydrogen atom a partial positive charge. **(b)** The partial charges on water molecules can form up to four hydrogen bonds. The oxygen can form two; each hydrogen can form one.

✔ **If you understand how water's structure makes hydrogen bonding possible, you should be able to (1) compare and contrast CO_2 and H_2O in terms of electronegativity, types of covalent bonds, and overall geometry; and (2) explain why electrical attractions between water molecules would be much weaker if their structure resembled CO_2.**

In an aqueous solution, hydrogen bonds also form between water molecules and polar solutes, such as the sugar glucose (Figure 2.12a). Similar interactions occur between water and ions, such as Na^+ and Cl^- from dissolved table salt (Figure 2.12b). Ions and polar molecules stay in solution because of their interactions with water's partial charges. Substances that interact with water in this way are said to be **hydrophilic** ("water-loving"). These interactions make it possible for almost any ionic compound and polar molecule to dissolve in water.

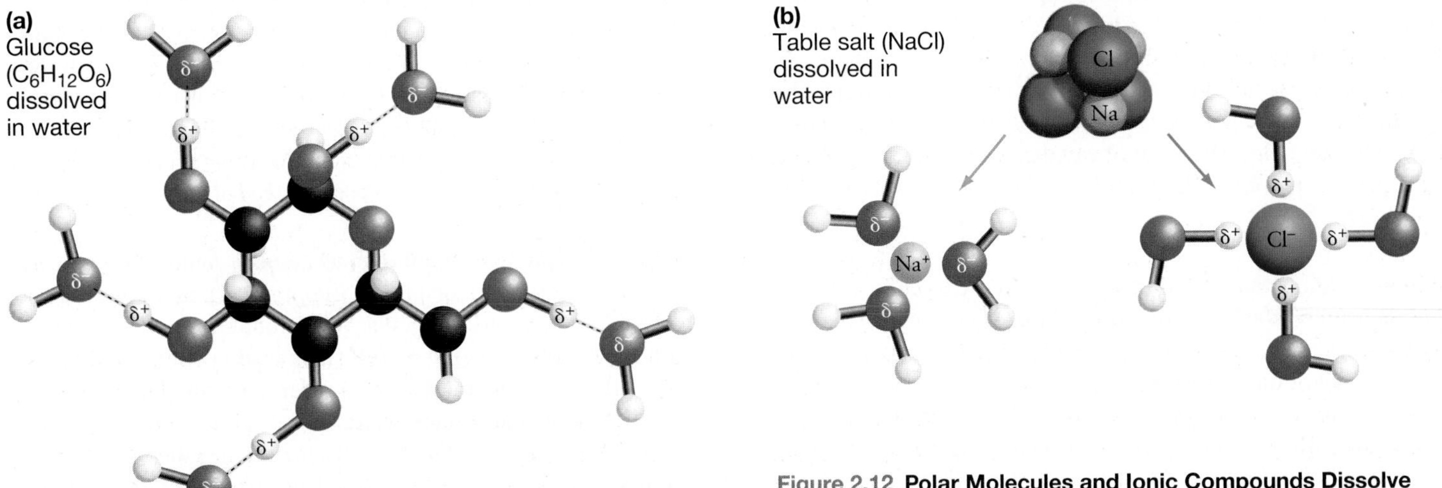

Figure 2.12 Polar Molecules and Ionic Compounds Dissolve Readily in Water. Water's polarity makes it a superb solvent.

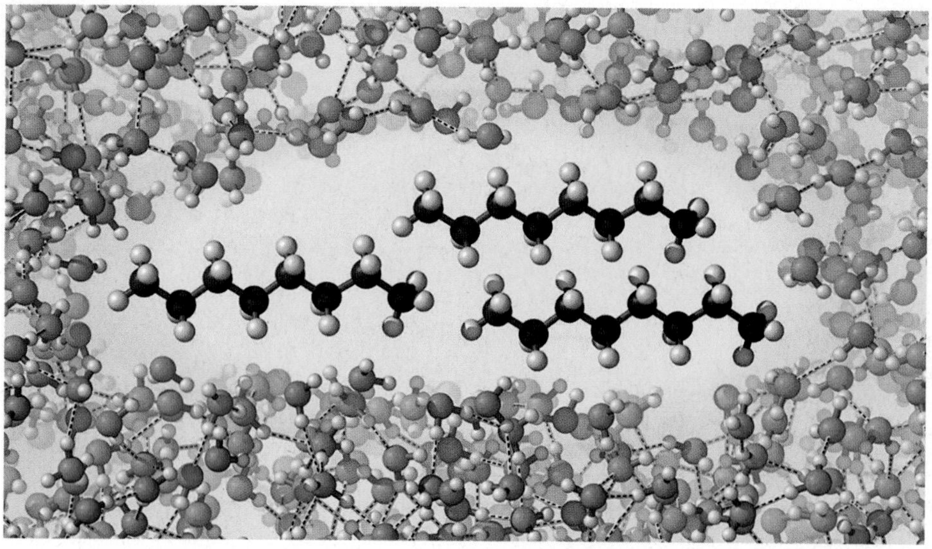

Figure 2.13 Nonpolar Molecules Do Not Dissolve in Water. In aqueous solution, nonpolar molecules such as octane (C_8H_{18})—a component of gasoline—are forced to interact with themselves. This occurs because water is much more stable when it interacts with itself rather than with the nonpolar molecules.

✔ Explain why when an oil is added to water, it forms a separate layer.

Although individual hydrogen bonds are not as strong as covalent or ionic bonds, many of them occur in a solution. Hydrogen bonds, and other similar interactions between water and hydrophilic solutes, are extremely important in biology owing to their sheer number and their role in dissolving substances in an aqueous solution.

In contrast, nonpolar molecules do not readily dissolve in aqueous solutions. Substances that do not interact with water are said to be **hydrophobic** ("water-fearing"). Because the interactions of hydrophobic molecules with water are minimal or nonexistent, the surrounding water molecules are forced to form more hydrogen bonds with each other. As a result, the hydrophobic molecules are drawn close together to minimize the disruption of the normal hydrogen bonding between water molecules (Figure 2.13). The interactions between nonpolar molecules that result from being brought together in an aqueous solution are called **hydrophobic interactions**.

Once hydrophobic molecules are close to one another, their association is further stabilized by weak electrical attractions known as **van der Waals interactions**. These weak attractions occur because the constant motion of electrons gives molecules a tiny asymmetry in charge that changes with time. If nonpolar molecules get extremely close to each other, the minute partial charge on one molecule induces an opposite partial charge in the nearby molecule and causes an attraction. Although the interaction is very weak relative to covalent bonds or even hydrogen bonds, a large number of van der Waals interactions can significantly increase the stability of clustered hydrophobic molecules.

Cohesion, Adhesion, and Surface Tension Attraction between like molecules is called **cohesion**. Water is cohesive—meaning that it stays together—because of the hydrogen bonds that form between individual molecules.

Attraction between unlike molecules, in contrast, is called **adhesion**. Adhesion is usually analyzed in regard to interactions between a liquid and a solid surface. Water adheres to surfaces that have any polar or charged components.

Cohesion and adhesion are important in explaining how water can move from the roots of plants to their leaves against the force of gravity (Ch. 35, Section 35.2). But you can also see them in action in the concave surface, or meniscus, that forms in a glass tube (Figure 2.14a). A meniscus forms as a result of

1. ***Adhesion*** At the perimeter of the surface, partial positive charges on water molecules adhere to the negative charges on glass, resulting in an upward pull.
2. ***Cohesion*** Along the surface, water molecules hydrogen-bond to those next to them and below them, resulting in a net lateral and downward pull that resists the upward pull of adhesion.

Cohesion is also instrumental in the phenomenon known as surface tension. **Surface tension** is the cohesive force caused by attraction between the molecules at the surface of a liquid. When water molecules are at the surface, there are no other molecules above them for hydrogen bonding. But hydrogen bonds do form between surface molecules and their nearest neighboring water molecules—next to and below them—resulting in tension that minimizes total surface area.

This fact has an important consequence: Water resists any force that increases its surface area. More specifically, any force that depresses a water surface meets with resistance. This resistance makes a water surface act like an elastic membrane (Figure 2.14b). Hydrogen bonds cause the "elastic membrane" of water to be stronger than that of other liquids. Water's extraordinarily high surface tension explains why it is better to cut the water's surface with your fingertips when you dive into a pool, instead of doing a belly flop.

Water Is Denser as a Liquid than as a Solid When factory workers pour liquid metal or plastic into a mold and allow it to cool to a solid, the material shrinks. When molten lava pours out of a volcano and cools to solid rock, it shrinks. But when you fill an ice tray with water and put it in the freezer, the water expands as ice.

Unlike most substances, water is denser as a liquid than it is as a solid. In other words, there are more molecules of water in a given volume of liquid water than there are in the same volume of solid water, or ice. Figure 2.15 illustrates why this is so.

(a) A meniscus forms where water meets a solid surface, as a result of two forces.

(b) Water has high surface tension.

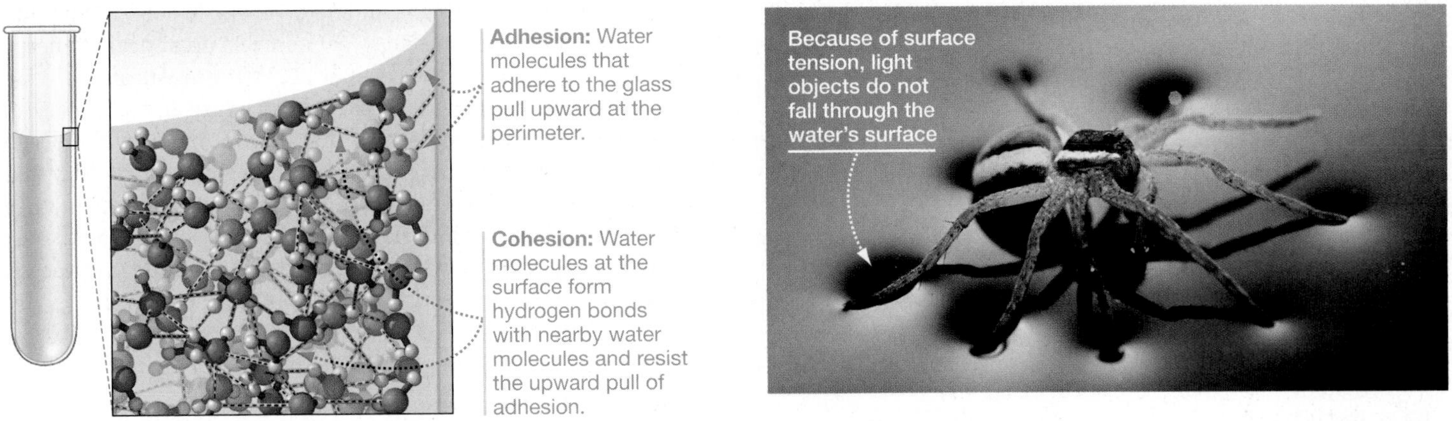

Figure 2.14 Cohesion, Adhesion, and Surface Tension. (a) Meniscus formation is based on hydrogen bonding and other interactions with glass that are represented here by highlighted dashed lines. **(b)** Water resists forces—like the weight of a spider—that increase its surface area. The resistance is great enough that light objects do not break the surface.

Note that in ice, each water molecule participates in four hydrogen bonds. These hydrogen bonds cause the water molecules to form a regular and repeating lattice structure, or crystal (Figure 2.15a). The crystal structure of ice is fairly open, meaning that there is a relatively large amount of space between molecules.

Normally, heating a substance causes it to expand because molecules begin moving faster and colliding more often and with greater force. But heating ice causes hydrogen bonds to break and the open crystal structure to collapse. Compare the arrangement of water molecules in ice with that of liquid water, illustrated in Figure 2.15b. In liquid water, hydrogen bonds are constantly being formed and broken, so the extent of hydrogen bonding in liquid water is much less than that found in ice. As a result, molecules in the liquid phase are packed much more closely together than in the solid phase, making liquid water denser than ice.

This property of water has an important result: Ice floats (Figure 2.15c). If it didn't, ice would sink to the bottom of lakes, ponds, and oceans soon after it formed. The ice would stay frozen in the cold depths. Instead, ice serves as a blanket, insulating the liquid below from the cold air above. If water weren't so unusual, Earth's oceans almost certainly would have frozen solid before life had a chance to start.

(a) In ice, water molecules form a crystal lattice.

(b) In liquid water, no crystal lattice forms.

(c) Liquid water is denser than ice. As a result, ice floats.

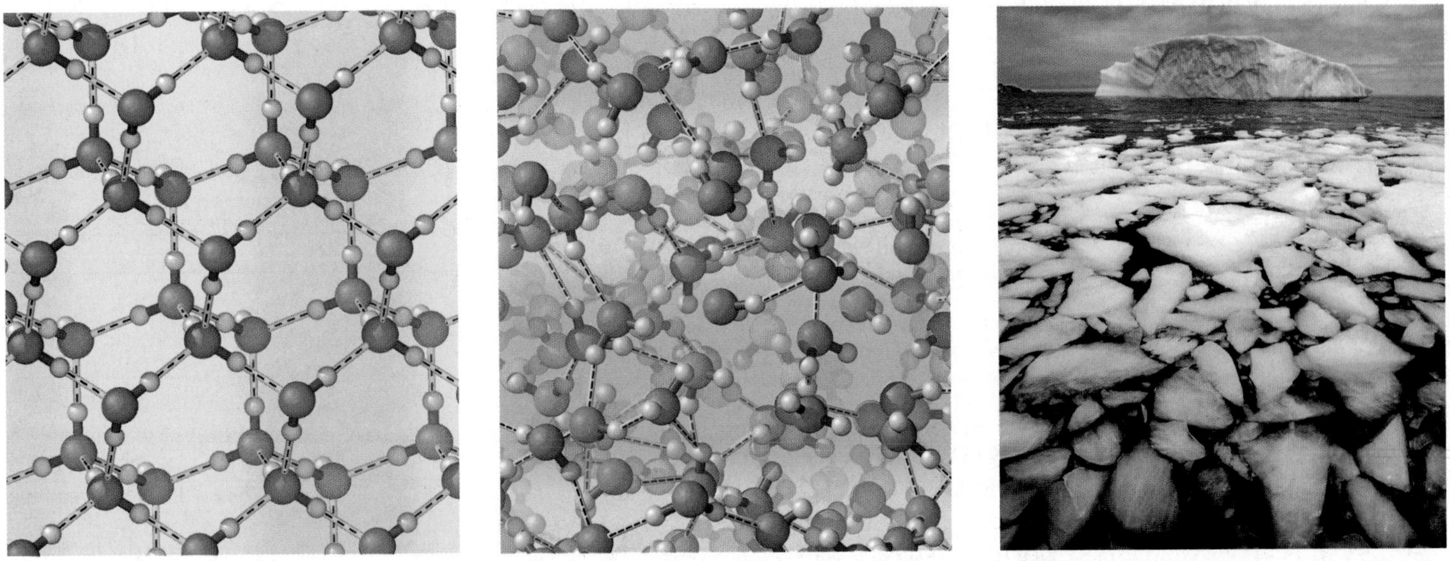

Figure 2.15 Hydrogen Bonding in Ice and Water. (a) In ice, each molecule forms four hydrogen bonds (yellow dashed lines) at one time. **(b)** In a liquid, bonds are continually broken and formed, so no lattice develops. **(c)** As a result, ice is less dense than water.

Table 2.1 Specific Heats of Some Liquids

The specific heats reported in this table were measured at 25°C and are given in units of joules per gram of substance per degree Celsius. (The joule is a unit of energy.)

With extensive hydrogen bonding	Specific Heat
Water (H_2O)	4.18
With some hydrogen bonding	
Ethanol (C_2H_6O)	2.44
Glycerol ($C_3H_8O_3$)	2.38
With little or no hydrogen bonding	
Benzene (C_6H_6)	1.74
Xylene (C_8H_{10})	1.72

DATA: D. R. Lide (ed.). 2008. Standard thermodynamic properties of chemical substances. In *CRC Handbook of Chemistry and Physics*. 89th ed. Boca Raton, FL: CRC Press.

Water Has a High Capacity for Absorbing Energy Hydrogen bonding is also responsible for another of water's remarkable physical properties: Water has a high capacity for absorbing energy.

Specific heat is the amount of energy required to raise the temperature of 1 gram of a substance by 1°C. Water has a high specific heat because when a source of energy hits it, hydrogen bonds must be broken before heat can be transferred and the water molecules begin moving faster. As **Table 2.1** indicates, as molecules increase in overall polarity and thus in their ability to form hydrogen bonds, it takes more energy to change their temperature.

Similarly, it takes a large amount of energy to break the hydrogen bonds in liquid water and change the molecules from the liquid phase to the gas phase. Water's **heat of vaporization**—the energy required to change 1 gram of water from a liquid to gas—is higher than that of most molecules that are liquid at room temperature. To put it another way, water has to absorb a great deal of energy to evaporate. Water's high heat of vaporization is the reason that sweating is an effective way to cool off on a hot day. The water molecules in sweat absorb a great deal of energy from your body before they evaporate, so you lose heat.

Water's ability to absorb energy is critical to the theory of chemical evolution. Molecules that were formed in the ocean were well protected from excessive heat or sources of energy that could break them apart, such as intense sunlight. As a result, they would have persisted and slowly increased in concentration over time, making them more likely to react and continue the process. **Table 2.2** summarizes some of the key properties of water.

The Role of Water in Acid–Base Chemical Reactions

You've seen that water's high specific heat and heat of vaporization tend to resist temperature and phase changes. One other aspect of water's chemistry is important for understanding chemical evolution and how organisms work: Water is not a completely stable molecule. In reality, water molecules continually undergo a chemical reaction with themselves.

When a **chemical reaction** occurs, one substance is combined with others or broken down into another substance. Atoms may also be rearranged; in most cases, chemical bonds are broken and new bonds form. Chemical reactions are written in a format similar to mathematical equations: The initial, or **reactant**, molecules are shown on the left and the resulting, or **product**, molecules are shown on the right. Arrows in between are like equal signs—the number of atoms of each element in the reactants must equal (be "balanced" with) the atoms of each element in the products. **Making Models 2.1** introduces some important conventions used to represent chemical reactions.

The chemical reaction that takes place between water molecules is called a "dissociation" reaction. It can be written as follows:

$$H_2O \rightleftharpoons H^+ + OH^-$$

SUMMARY Table 2.2 Properties of Water

Property	Cause	Biological Consequences
Solvent for charged or polar compounds	Electrostatic attractions occur between partial charges on water and opposite charges on other polar molecules or ions.	
Denser as a liquid than a solid	As water freezes, each molecule maintains four hydrogen bonds, leading to the formation of the low-density crystal structure called ice.	
High specific heat	Water molecules must absorb lots of heat energy to break hydrogen bonds and experience increased movement (and thus temperature).	
High heat of vaporization	Water molecules must absorb lots of heat energy to break hydrogen bonds and change water from liquid to gas.	

✔ You should be able to fill in the biological consequences column in this table.

Making Models 2.1 Tips on Writing Chemical Equations

In chemical equations, arrows represent the direction of the reaction. Single arrows indicate a nonreversible reaction, where reactants are shown on the left and products on the right. Double arrows are used to show that a reaction is reversible—the reactants of the forward reaction can become the products in the reverse.

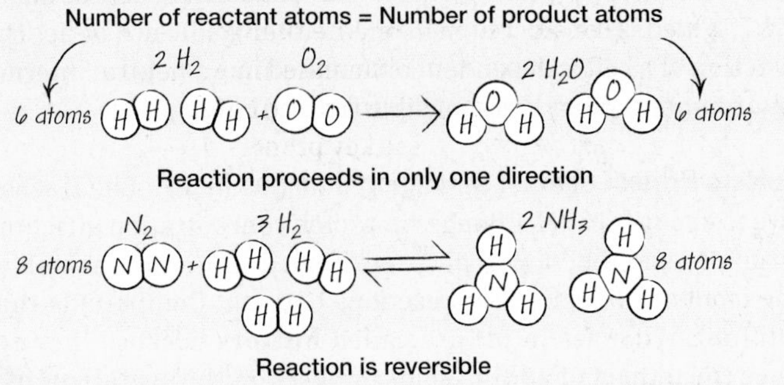

MODEL When CH_4 and O_2 react, CO_2 and H_2O are produced. This reaction is not normally reversible. Draw the chemical equation of this reaction. Make sure your equation is balanced.

To see this model in action, go to the Study Area of **Mastering Biology**

The products on the right-hand side of the expression are the **hydrogen ion** (H^+) and the **hydroxide ion** (OH^-). Note that this reaction is reversible. When the forward and reverse reactions proceed at the same rate, the quantities of reactants and products remain constant, although not necessarily equal. A dynamic but stable state such as this is termed a **chemical equilibrium**.

In a sample of pure water, the concentration of hydrogen ions is always equal to the concentration of hydroxide ions. Recall that a hydrogen ion is simply a proton (see Figure 2.1a). In reality, however, protons do not exist by themselves. In water, for example, protons associate with water molecules to form hydronium ions (H_3O^+). Thus, the dissociation of water is more accurately written as

$$2\,H_2O \rightleftharpoons H_3O^+ + OH^-$$

One of the water molecules on the left-hand side of the expression has given up a proton, while the other water molecule has accepted a proton.

Substances that give up protons during chemical reactions and raise the hydronium ion concentration of water are called **acids**; molecules or ions that acquire protons during chemical reactions and lower the hydronium ion concentration of water are called **bases**. Most acids act only as acids, and most bases act only as bases; but water can act as both an acid and a base.

A chemical reaction that involves a transfer of protons is called an acid–base reaction. Every acid–base reaction requires a proton donor and a proton acceptor—an acid and a base, respectively.

Water is an extremely weak acid and base—at any given moment, very few water molecules dissociate to form hydronium ions and hydroxide ions. In contrast, strong acids like the hydrochloric acid (HCl) in your stomach readily give up a proton to form hydronium ions when they react with water.

$$HCl + H_2O \rightleftharpoons H_3O^+ + Cl^-$$

Strong bases readily acquire protons when they react with water. For example, sodium hydroxide (NaOH, commonly called lye) dissociates completely in water to form Na^+ and OH^-.

$$NaOH(aq) \longrightarrow Na^+ + OH^-$$

(The "*aq*" indicates that NaOH is in aqueous solution.[1]).

The hydroxide ion produced by that reaction then accepts a proton from a hydronium ion in the water, forming two water molecules.

$$OH^- + H_3O^+ \rightleftharpoons 2\,H_2O$$

To summarize, adding an acid to a solution increases the concentration of protons; adding a base to a solution lowers the concentration of protons. The stronger the acid or base, the more likely it is to donate or accept protons.

Determining the Concentration of Protons In a solution, the tendency for acid–base reactions to occur is largely a function of the number of protons present. The problem is, there's no simple way to count the actual number of protons present in a sample. Researchers solve this problem using the mole concept.

A **mole** refers to the number 6.022×10^{23}—just as the unit dozen refers to the number 12 or the unit million refers to the number 1×10^6. The mole is a useful unit because the mass of one mole of an atom is the same as its atomic weight expressed in grams. The mass of one mole of a molecule, called its **molecular weight**, is the sum of the atomic weights of all the atoms in the molecule.

For example, to get the molecular weight of H_2O, you add the atomic weights of two atoms of hydrogen and one atom of oxygen. Since the atomic weights of hydrogen and oxygen are very close to their mass numbers (see Figure 2.2), the molecular weight of water would be $1 + 1 + 16$, or a total of 18. Thus, if you weighed a sample of 18 grams of water, it would contain around 6×10^{23} water molecules, or about 1 mole of water molecules.

When substances are dissolved in water, their concentration is expressed in terms of molarity (symbolized by "M"). **Molarity** is the number of moles of the solute present per liter of solution. A 1-molar solution of hydrogen ions in water, for example, means that 1 mole of protons is contained in 1 liter of solution.

The pH of a Solution Reveals Whether It Is Acidic or Basic The concentration of protons in water is very low. In a sample of pure water at 25°C, the concentration of H^+ is 1.0×10^{-7} M, or 0.0000001 molar. Because exponential notation is cumbersome, scientists prefer to express the concentration of protons in a solution, and thus whether it is acidic or basic, with a logarithmic notation called **pH**. The term "pH" is derived from the French *puissance d'hydrogéne*, or "power of hydrogen."

[1] The physical state of each reactant and product in a chemical expression is indicated as gas (*g*), liquid (*l*), solid (*s*), or in aqueous solution (*aq*).

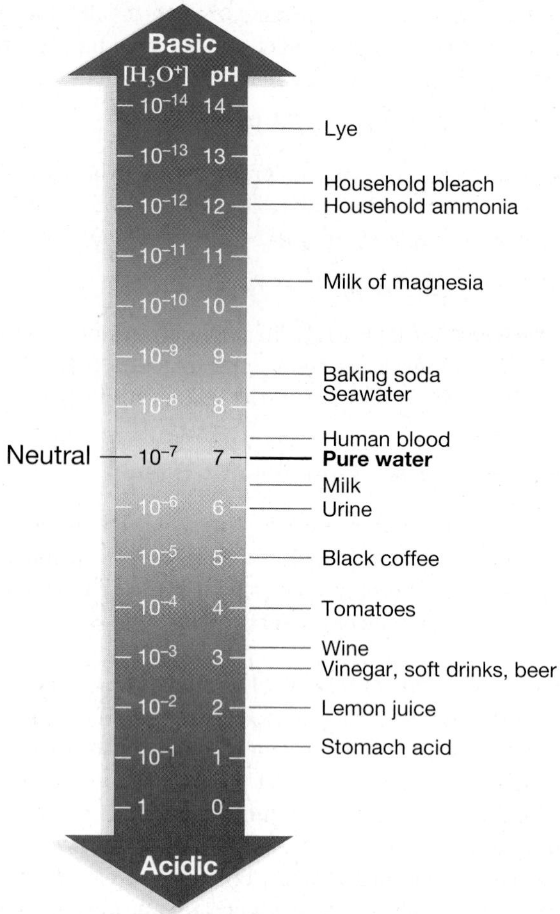

Figure 2.16 The pH Scale. Because the pH scale is logarithmic, a change in one unit of pH represents a change in the concentration of hydrogen ions equal to a factor of 10. Coffee has a hundred times more H^+ than pure water has.

✔ What happens to the concentration of protons in black coffee after you add milk?

By definition, pH is the negative of the base-10 logarithm, or −log, of the molar concentration of hydrogen ions in a solution:

$$pH = -\log [H^+]$$

(The square brackets are a standard notation for indicating "concentration" of a substance in solution.) Chemists can measure the concentration of protons in a solution directly using an instrument called a pH meter. Recall that the concentration of H^+ in a sample of pure water is 1.0×10^{-7} M. Using the formula just considered, the pH of pure water is therefore 7. (To review logarithms, see **BioSkills 5**.)

Taking antilogs gives:

$$[H^+] = \text{antilog}(-pH) = 10^{-pH}$$

Solutions that contain acids have a proton concentration larger than 1×10^{-7} M and thus a $pH < 7$. This is because acidic molecules tend to release protons into solution. In contrast, solutions that contain bases have a proton concentration less than 1×10^{-7} M and thus a $pH > 7$. This is because basic molecules tend to accept protons from solution.

Although pH is a convenient way to indicate the concentration of protons in a solution, take note of what the number represents. For example, a pH change from 7 to 5 might not seem like a big deal, but it means that the sample contains 100 times more protons and has become 100 times more acidic.

The pH scale in **Figure 2.16** gives the pH of some selected solutions. Pure water is used as a standard, or point of reference, for pH 7 on the pH scale. The solution inside living cells is about pH 7, which is considered neutral—neither acidic nor basic. The function of a cell is dependent on maintaining a neutral internal environment. What is responsible for regulating pH?

Buffers Protect against Damaging Changes in pH Life is sensitive to changes in pH. Changes in proton concentration affect the structure and function of polar or charged substances as well as the tendency of acid–base reactions to occur. Compounds that minimize changes in pH are called **buffers** because they reduce the impact of adding acids or bases on the overall pH of a solution. Buffers are important in maintaining relatively constant conditions, or **homeostasis**, in organisms.

Most buffers are weak acids, meaning they are somewhat likely to give up a proton in solution, but once the proton concentration rises, the acid is regenerated. In cells and bodily fluids, a wide array of substances act as buffers. Your blood, for example, contains carbonic acid. To see how buffers work, consider the dissociation of carbonic acid in an aqueous solution to form bicarbonate ions and protons:

$$\underset{\text{carbonic acid}}{CH_2O_3(aq)} \rightleftharpoons \underset{\text{bicarbonate}}{CHO_3^-} + H^+$$

When carbonic acid and bicarbonate are present in about equal concentrations in a solution, such as blood, they function as a buffering system. If the concentration of protons increases slightly, the protons react with bicarbonate ions to form carbonic acid and pH does not change. If the concentration of protons decreases slightly, carbonic acid gives up protons and pH does not change.

✔ **If you understand this concept, you should be able to predict what would happen to the concentration of bicarbonate ions if a strong base like sodium hydroxide (NaOH) were added to a solution of carbonic acid.**

As chemical evolution began, then, water provided the physical environment for key chemical reactions to take place. Now let's consider what happened in solution, some 3.5 billion years ago.

CHECK YOUR UNDERSTANDING

✔ **If you understood this section, you should be able to . . .**

1. Contrast the effect of water on dissolving NaCl versus glucose.
2. Explain which property of water underlies the observation that pigs often cover themselves in mud on hot days, and indicate how this activity benefits the pig.
3. Calculate the pH of a 0.5-M solution of HCl. If you wanted to determine the number of protons in the solution, what additional information would you need?

Answers are available in Appendix A.

2.3 Chemical Reactions, Energy, and Chemical Evolution

Proponents of the theory of chemical evolution contend that simple molecules present in the atmosphere and oceans of early Earth participated in chemical reactions that eventually produced larger, more complex organic (carbon-containing) molecules—such as the proteins, nucleic acids, sugars, and lipids introduced in the next four chapters. Currently, researchers are investigating two environments where these reactions may have occurred:

After you complete this section, you should be able to . . .

- Explain the changes in bonds, energy, and entropy that occur in a chemical reaction.

1. *The atmosphere,* which in early Earth was probably dominated by gases ejected from volcanoes. Water vapor, carbon dioxide (CO_2), and nitrogen (N_2) are the dominant gases ejected by volcanoes today; a small amount of molecular hydrogen (H_2) and carbon monoxide (CO) may also be present.
2. *Deep-sea hydrothermal vents,* where extremely hot rocks contact deep cracks in the seafloor. In addition to gases such as CO_2 and H_2, certain deep-sea vents are rich in minerals containing reactive metals such as nickel and iron.

When gases like CO_2, N_2, H_2, CO, and water vapor are allowed to interact on their own, however, very little happens. The simple molecules do not suddenly link together to create large, complex substances like those found in living cells. Instead, their bonds remain intact. To understand why the bonds of these molecules remain unchanged, you must first learn more about how chemical reactions proceed.

How Do Chemical Reactions Happen?

The most common reaction in the mix of gases that emerges from volcanoes results in the production of carbonic acid, which can be precipitated in rainwater:

$$CO_2(g) + H_2O(g) \rightleftharpoons \underset{\text{carbonic acid}}{CH_2O_3\ (aq)}$$

Note that the expression above is balanced; that is, there are 1 carbon, 3 oxygen, and 2 hydrogen atoms present on each side of the reaction arrows. The interacting molecules in this equation constitute what a chemist would call a system, while everything else would be called the surroundings. A **system** is defined as just those substances you want to focus attention on—the surroundings are ignored, except for their effect on the system. In this example, the system is defined as being "closed" relative to its surroundings (**Figure 2.17**), so the mass of reactants is always equal to the mass of products. Chemical equations such as this one illustrate the *law of conservation of mass*—mass cannot be created or destroyed, but it may be rearranged through chemical reactions.

Changing the concentration of reactants or products can disturb a chemical equilibrium. For example, adding more CO_2 to the system would drive the reaction to the right, creating more CH_2O_3 until the equilibrium proportions of reactants and products are reestablished. Removing CO_2 or adding more CH_2O_3 would drive the reaction to the left.

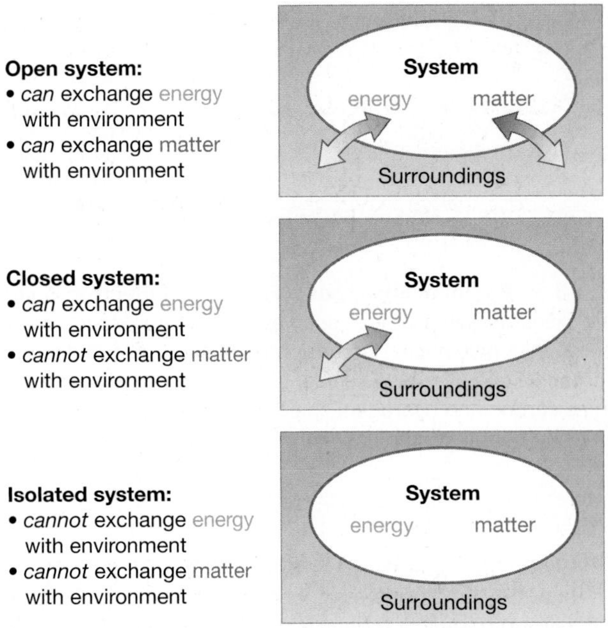

Figure 2.17 Types of Systems Used to Evaluate Chemical Reactions.

A chemical equilibrium can also be altered by changes in temperature. For example, consider a closed system comprising a combination of liquid water and water vapor:

$$H_2O(l) \rightleftharpoons H_2O(g)$$

If this system absorbs enough thermal energy from the surrounding environment, the liquid water molecules will overcome the heat of vaporization and undergo a physical state change from liquid to gas. (Keep in mind that this would not represent a chemical reaction because the molecules are the same on both sides of the equation.) Such a change is termed **endothermic** ("within heating") because thermal energy is absorbed by the system during the process. In contrast, the transformation of water vapor to liquid water releases thermal energy to the environment and is **exothermic** ("outside heating"). This release of energy increases disorder of the surroundings at the same time that order is increased in the system by the formation of liquid water. Raising the temperature of this system drives the equilibrium to the right; cooling the system drives it to the left. This same type of energy transfer occurs in the phase change between liquid water and ice.

Besides driving physical state changes, transfer of energy is also required for chemical reactions to occur. To understand how life could have evolved from chemical reactions, you will first need a brief introduction to energy.

What Is Energy?

Energy can be defined as the capacity to do work or supply heat. This capacity exists in one of two ways—as a stored potential or as an active motion.

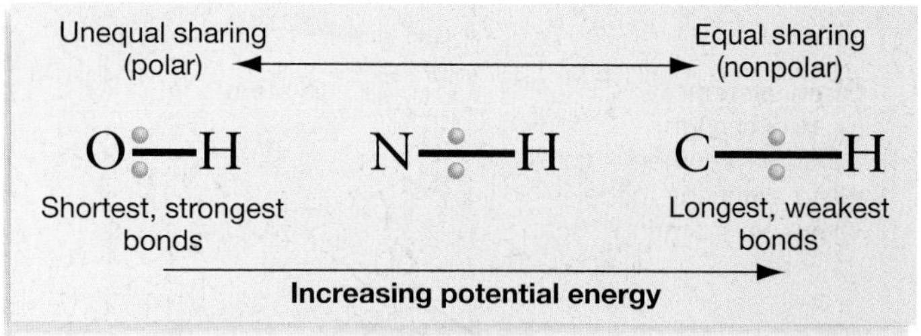

Figure 2.18 **Potential Energy as a Function of Electron Sharing.** Highly electronegative atoms, such as oxygen, pull shared electrons closer to their own nuclei, increasing bond strength and decreasing the potential energy of a molecule. Less electronegative atoms, such as carbon and hydrogen, share electrons more equally, decreasing bond strength and increasing potential energy.

Stored energy is called **potential energy**. An object gains or loses its ability to store energy because of its position. In molecules, potential energy is related to the position of shared electrons in covalent bonds. When the shared electrons are far from the atomic nuclei, the bond is long and weak. If the electrons are shifted closer to one or both of the atoms, the bond becomes shorter and stronger. Atoms bound together with weak bonds have a greater capacity to be broken apart to reform into new, stronger bonds during a reaction than do atoms held together by strong bonds. A molecule's potential to form stronger bonds is a type of potential energy called **chemical energy**.

What factors contribute to bond strength? Recall that the electronegativities of atoms affect the position of shared electrons. For example, the shared electrons in polar covalent bonds are pulled closer to the more electronegative atom, resulting in shorter, stronger bonds compared to what would be expected if the bond were nonpolar (Figure 2.18). Increased bond strength also occurs as more electron pairs are shared when double and triple bonds are formed.

Energy of motion is called **kinetic energy**. All molecules have some degree of kinetic energy because they are constantly in motion. The kinetic energy of molecular motion is called **thermal energy**.

- The **temperature** of an object is a measure of how much thermal energy its molecules possess. A cold object has a low temperature and consists of molecules that are moving slower than if the temperature were raised to make it hot.
- When two objects with different temperatures come into contact, thermal energy is transferred between them. This transferred energy is called **heat**.

There are many forms of potential energy and kinetic energy, and energy can change from one form into another. However, according to the **first law of thermodynamics**, energy is conserved—it cannot be created or destroyed, but only transferred and transformed. (A more thorough explanation of energy transformation is provided in Ch. 8, Section 8.1.)

Energy transformation was the heart of chemical evolution. According to the best data available, molecules that were part of the early Earth were exposed to massive inputs of energy. Kinetic energy, in the form of heat, was present in the gradually cooling molten mass that initially formed the planet. The atmosphere and surface of the early Earth were also bombarded with electricity from lightening and radiation from the Sun. Energy stored in the chemical bonds of molecules was also abundant.

Now that you understand the different forms of energy that can be transferred in chemical reactions, a big question remains: What determines if a reaction will take place?

What Makes a Chemical Reaction Spontaneous?

When chemists say that a reaction is spontaneous, they have a precise meaning in mind: Chemical reactions are spontaneous if they are able to proceed on their own, without any continuous external influence such as added energy. Two factors determine if a reaction will proceed spontaneously:

1. Reactions tend to be spontaneous if *the products have lower potential energy than the reactants*—that is, when the shared electrons in the reaction products are held more tightly than those in the reactants. For example, when hydrogen and oxygen gases react, water is produced spontaneously:

$$2H_2(g) + O_2(g) \longrightarrow 2H_2O(g)$$

The electrons involved in the O—H bonds of water are held much more tightly by the more electronegative oxygen atom than when they were shared equally in the H—H and O=O bonds of H_2 and O_2 (Figure 2.19a). As a result, the products have much lower potential energy than the reactants.

2. Reactions tend to be spontaneous when *the product molecules are less ordered than the reactant molecules*. For example, glucose is a single, highly ordered molecule. But when glucose burns in air, it breaks up into gaseous carbon dioxide and water vapor (Figure 2.20). These product molecules are much less ordered than the reactant glucose molecules. The amount of disorder in a system (or the surrounding environment) is called **entropy**. Entropy increases in the system when the products of a chemical reaction are less ordered than the reactant molecules.

It is important to note that spontaneous reactions can occur without appearing to increase entropy in the system. For example, the reaction in Figure 2.19a converts three molecules of gas into just two molecules of water vapor, an apparent decrease in entropy. However, the difference in chemical energy between reactants and products is given off as kinetic energy—heat, light, or sound, to name just a few forms. This released energy of motion results in an overall increase in entropy when you include its effect on the surrounding environment.

The **second law of thermodynamics** states that in all spontaneous reactions, entropy always increases when both the system and its environment are taken into account. The *Hindenburg* disaster of 1937 illustrates the large and terrifying amount of kinetic energy that can be given off by this relatively simple reaction (Figure 2.19b).

To summarize: Physical and chemical changes tend to proceed in the direction that results in lower potential energy,

(a) When hydrogen and oxygen gas react, the products have much lower potential energy than the reactants.

(b) The difference in potential energy is released as **heat and light**, which vaporizes the water produced.

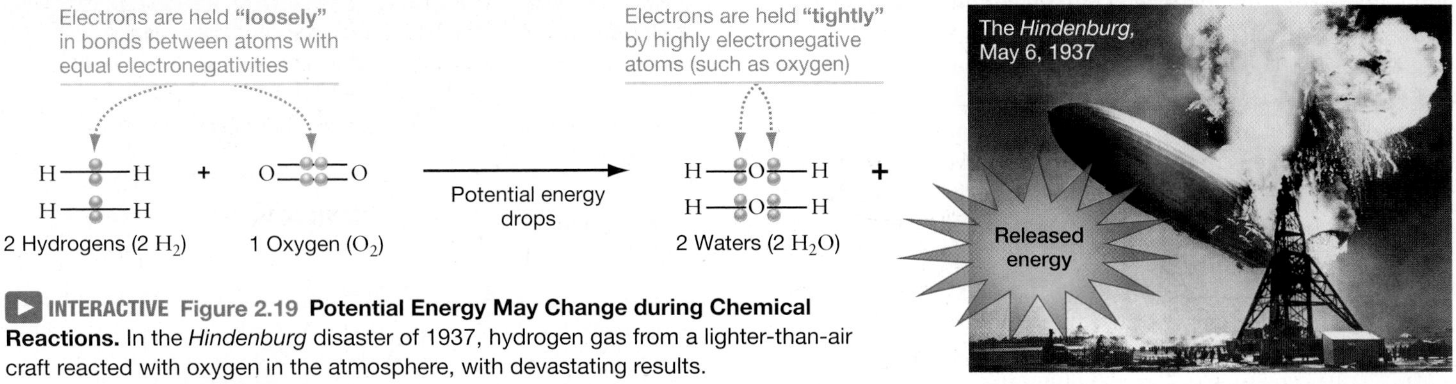

INTERACTIVE Figure 2.19 Potential Energy May Change during Chemical Reactions. In the *Hindenburg* disaster of 1937, hydrogen gas from a lighter-than-air craft reacted with oxygen in the atmosphere, with devastating results.

✔ In part (a), which bonds have relatively low potential energy and which bonds have relatively high potential energy? How does potential energy relate to the strength of the bonds?

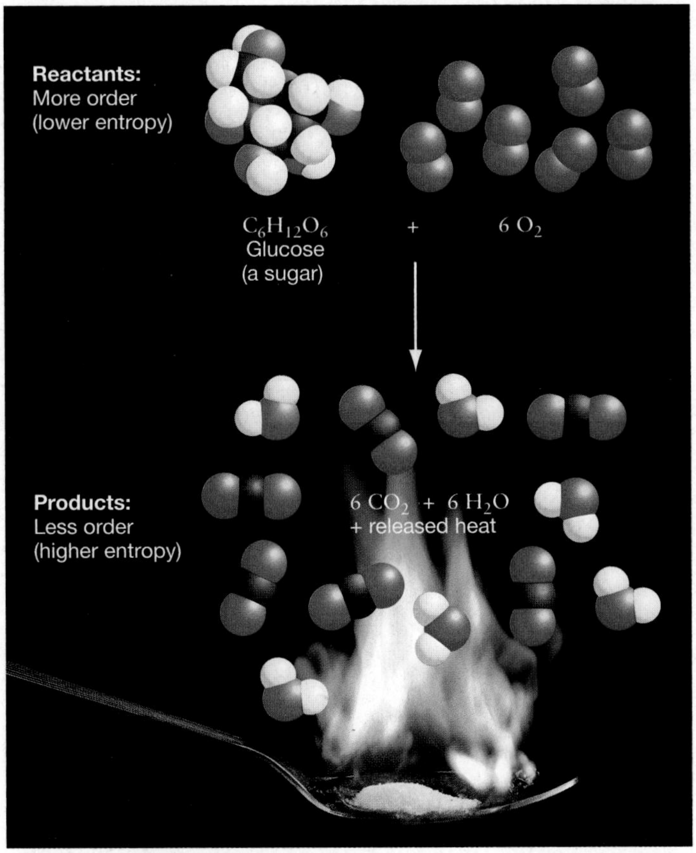

Figure 2.20 Spontaneous Processes Result in Increased Disorder—That Is, Increased Entropy. Combustion of glucose illustrates this principle. (As you will see in Chapter 8, your cells perform a similar task through a series of reactions.)

increased entropy, or both. Potential energy and entropy are used to figure out whether a reaction is spontaneous or not. Were the reactions that led to chemical evolution spontaneous? Next you will learn how researchers have tried to address this question (Section 2.4).

CHECK YOUR UNDERSTANDING

✔ If you understood this section, you should be able to . . .

1. Explain how the positions of the valence electrons in carbon and hydrogen change as methane (CH_4) reacts with oxygen (O_2) to produce carbon dioxide and water in the following reaction:

$$CH_4 + 2\ O_2 \longrightarrow CO_2 + 2\ H_2O$$

2. Determine if the reaction above is spontaneous or not, addressing both potential energy and entropy.
3. **THINK CAREFULLY** Explain what happens to the potential energy removed from chemical bonds in reactants after they are converted to products with lower potential energy.

Answers are available in Appendix A.

2.4 Investigating Chemical Evolution

After you complete this section, you should be able to . . .

- Describe how experiments are used to investigate chemical evolution.

Chemical evolution was first proposed by Alexander I. Oparin in 1924. The hypothesis was published again—independently and five years later—by J. B. S. Haldane. Today, the Oparin–Haldane proposal is considered a formal scientific theory (Ch. 1, Section 1.1). At the time, however, since Oparin and Haldane were unable to conduct definitive experiments, their proposal remained an untested hypothesis.

Chemical evolution was first taken seriously in 1953 when a graduate student named Stanley Miller performed a breakthrough experiment. Miller wanted to answer a simple question: Can complex organic (carbon-containing) compounds be synthesized from the simple molecules present in Earth's early atmosphere? In other words, is it possible to re-create the first steps in chemical evolution by simulating early-Earth conditions in the laboratory?

Miller's experimental setup (Figure 2.21) was designed to produce a microcosm of early Earth. The large glass flask represented the atmosphere and contained the gases methane (CH_4), ammonia (NH_3), and hydrogen (H_2), all of which have high potential energy. This large flask was connected to a smaller flask by glass tubing. The small flask held a tiny ocean—200 milliliters (mL) of liquid water.

To connect the mini-atmosphere with the mini-ocean, Miller boiled the water constantly. This added water vapor (H_2O) to the mix of gases in the large flask. As the vapor cooled and condensed, it flowed back into the smaller flask, where it boiled again. In this way, water vapor circulated continuously through the system. This was important: If the molecules in the simulated atmosphere reacted with one another, the "rain" would carry them into the mini-ocean, where additional reactions might occur.

Had Miller stopped at merely boiling the molecules, little or nothing would have happened. Even at the boiling point of water (100°C), the starting molecules used in the experiment are stable and do not undergo spontaneous chemical reactions. A different form of energy would be required for these substances to react.

Something started to happen in the apparatus when Miller sent electrical discharges across the electrodes he'd inserted into the atmosphere. These miniature "lightning bolts" added a crucial component to the reaction mix—pulses of intense electrical energy. After a day of continuous boiling and sparking, the solution in the boiling flask began to turn pink. After a week, it was deep red and cloudy.

When Miller analyzed samples from the mini-ocean, he found large quantities of hydrogen cyanide and formaldehyde. Although these chemicals are poisonous, they are also highly reactive and can promote the synthesis of larger, more complex compounds. Even more exciting, the samples also contained newly synthesized amino acids, which are the building blocks of proteins (Ch. 3, Section 3.1).

RESEARCH

QUESTION: Can simple molecules and kinetic energy lead to chemical evolution?

HYPOTHESIS: This chemical evolution of organic molecules will occur in environments simulating early-Earth conditions.

NULL HYPOTHESIS: Chemical evolution will not occur in early-Earth simulations.

EXPERIMENTAL SETUP:

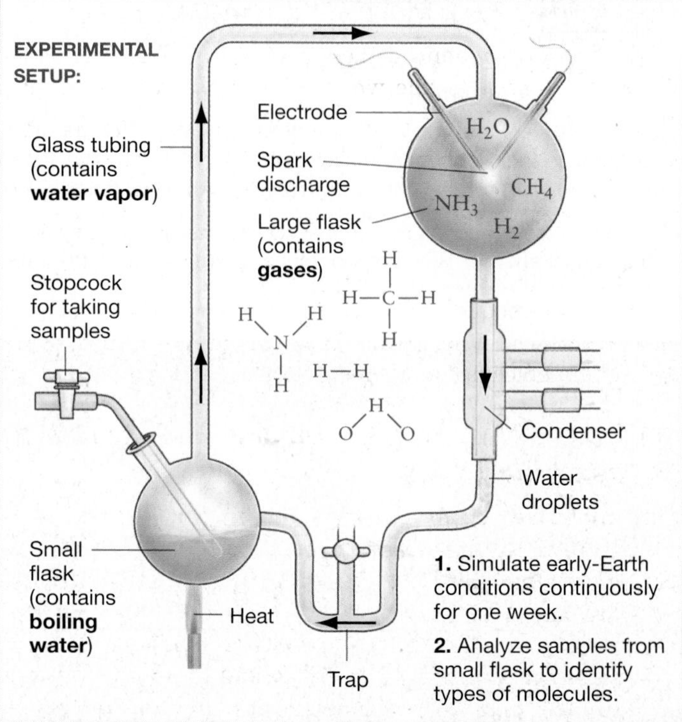

PREDICTION OF HYPOTHESIS: If kinetic energy is added to a mix of simple molecules, complex organic compounds will be produced.

PREDICTION OF NULL HYPOTHESIS: No complex organic compounds will be produced.

RESULTS

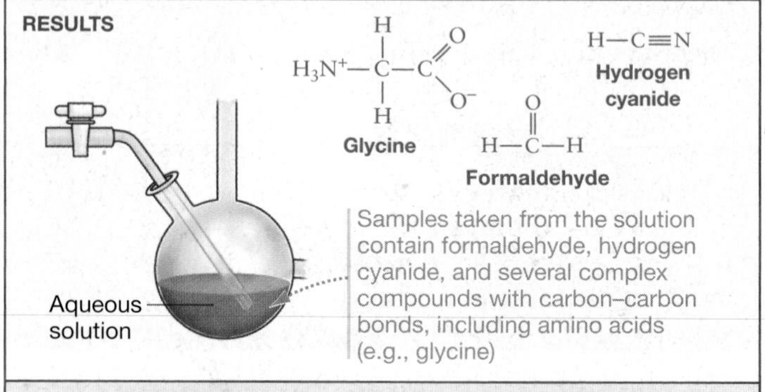

CONCLUSION: Chemical evolution occurs readily if simple molecules with high free energy are exposed to a source of kinetic energy.

Figure 2.21 **Miller's Spark-Discharge Experiment.** The arrows in the "Experimental Setup" diagram indicate the flow of water vapor or liquid. The condenser is a jacket with cold water flowing through it.

SOURCE: S. L. Miller. 1953. A production of amino acids under possible primitive Earth conditions. *Science* 117: 528–529.

✔ **PROCESS OF SCIENCE** Which parts of the apparatus were intended to mimic the early ocean, atmosphere, rain, and lightning?

CHECK YOUR UNDERSTANDING

✔ **If you understood this section, you should be able to . . .**

PROCESS OF SCIENCE Describe what would be necessary to conclude that the results from Miller's experiment support the theory of chemical evolution before the existence of life on Earth.

Answers are available in Appendix A.

2.5 Life Is Carbon Based

After you complete this section, you should be able to . . .

- Differentiate the structural and functional characteristics of organic molecules.

Life has been called a carbon-based phenomenon, and with good reason. Except for water, almost all of the molecules found in organisms contain this atom. Many molecules that contain carbon bonded to other elements, such as hydrogen, are called **organic compounds**. (Other types of molecules are referred to as being *inorganic*.)

Carbon has great importance in biology because it is the most versatile atom on Earth. Because of its four valence electrons, it will form four covalent bonds. The formation of carbon–carbon bonds was an important event in chemical evolution: It represented a crucial step toward the production of organic molecules found in living organisms. As a result, organic molecules come in an almost limitless array of molecular shapes, made possible by different combinations of single and double bonds.

Carbon Provides a Molecular Skeleton

You have already examined the shapes of single-carbon molecules such as carbon dioxide, which is linear, and methane, which is tetrahedral (see Figure 2.8 and Figure 2.10). In molecules that contain more than one carbon atom, the shapes can become much more complex. For example, several carbon and hydrogen atoms can bond to one another to form long hydrocarbon chains, as in octane (C_8H_{18}; **Figure 2.22a**). In glucose ($C_6H_{12}O_6$; **Figure 2.22b**), a ring structure is formed from carbon, hydrogen, and oxygen atoms that are bonded together. Octane is one of the primary components in gasoline, and the sugar glucose is the primary energy storage molecule for organisms (Ch. 5, Section 5.1).

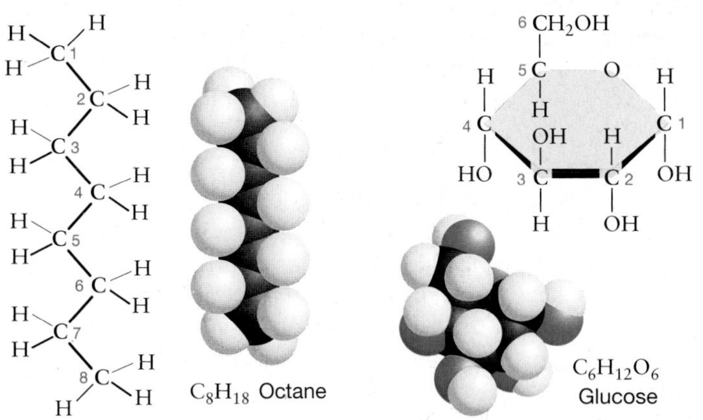

Figure 2.22 The Shapes of Carbon-Containing Molecules.
(a) Octane is a hydrocarbon chain present in gasoline. **(b)** Glucose is a sugar that can form a ring-like structure.

Functional Groups Define the Chemical Behavior of Organic Molecules

In general, the carbon atoms in an organic molecule furnish a skeleton that gives the molecule its overall shape. But the chemical behavior of the compound—meaning the types of reactions that it participates in—is dictated by groups of H, N, O, P, or S atoms that are bonded to one of the carbon atoms in a specific way.

The critically important H-, N-, O-, P-, and S-containing groups found in organic compounds are called **functional groups**. The composition and properties of six prominent functional groups that are commonly found in organic molecules and recognized by organic chemists are summarized in **Table 2.3** on page 76. To understand the role that organic compounds play in organisms, it is important to analyze how these functional groups behave.

- *Amino and carboxyl functional groups* tend to attract or release a hydrogen ion (proton), respectively, when in solution. Amino groups function as bases; carboxyl groups act as acids. During chemical evolution and in organisms today, the most important types of amino- and carboxyl-containing molecules are the amino acids (analyzed in detail in Ch. 3, Section 3.1). Amino acids contain both an amino group and a carboxyl group. Amino acids can be linked together by covalent bonds that form between amino and carboxyl groups. In addition, both of these functional groups participate in hydrogen bonding.
- *Carbonyl* groups are found on molecules such as acetaldehyde and acetone. This functional group is the site of reactions that link these molecules into larger, more complex organic compounds.
- *Hydroxyl groups* are important because they act as weak acids. In many cases, the protons involved in acid–base reactions that occur in cells come from hydroxyl groups on organic compounds. Because hydroxyl groups are polar, molecules containing hydroxyl groups will form hydrogen bonds and tend to be soluble in water.
- *Phosphate groups* carry negative charges on two of their oxygen atoms. When phosphate groups are transferred from one organic compound to another, the change in charge often dramatically affects the structure of the recipient molecule. In addition, phosphate groups that are bonded together store chemical energy that can be used in chemical reactions (some of these are discussed in Chapter 4).
- *Sulfhydryl groups* consist of a sulfur atom bonded to a hydrogen atom. They are important because sulfhydryl groups can link to one another via disulfide (S—S) bonds.

To summarize, functional groups make things happen. The number and types of functional groups attached to a framework of carbon atoms imply a great deal about how that molecule is going to behave.

When you encounter an organic molecule that is new to you, it's important to do the following three things:

1. Examine the overall size and shape provided by the carbon framework.
2. Identify the types of covalent bonds present based on the electronegativities of the atoms. Use this information to estimate the polarity of the molecule and the amount of potential energy stored in its chemical bonds.
3. Locate any functional groups and note the properties these groups give to the molecule.

Understanding these three features will help you predict the molecule's role in the chemistry of life.

Small Organic Molecules Can Assemble into Large Molecules

Once carbon-containing molecules with functional groups had appeared early in Earth's history, what happened next? For chemical evolution to continue, small organic molecules had to

SUMMARY Table 2.3 Six Functional Groups Commonly Attached to Carbon Atoms

Functional Group	Formula*	Family of Molecules	Properties of Functional Group	Example
Amino	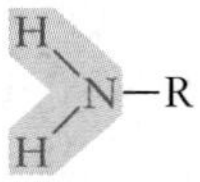	Amines	Acts as a base—tends to attract a proton to form: 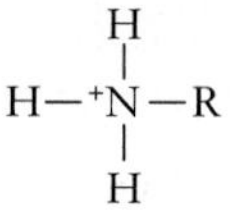	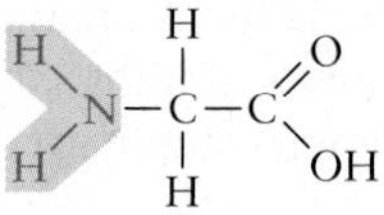 Glycine (an amino acid)
Carboxyl	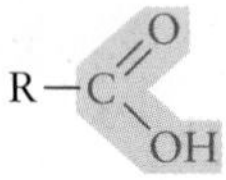	Carboxylic acids	Acts as an acid—tends to lose a proton in solution to form: 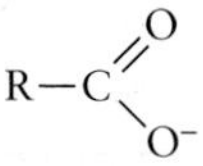	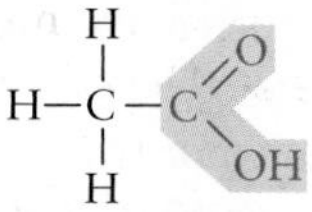 Acetic acid
Carbonyl	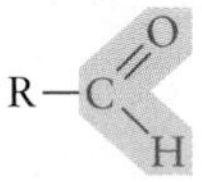	Aldehydes	Aldehydes, especially, react with certain compounds to produce larger molecules to form: 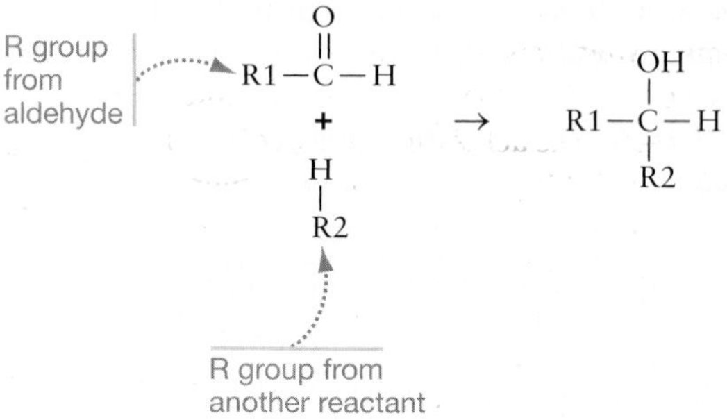	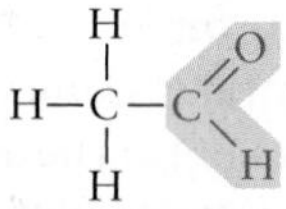 Acetaldehyde
	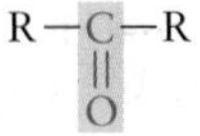	Ketones		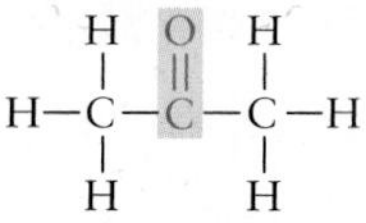 Acetone
Hydroxyl	R—OH	Alcohols	Highly polar, so makes compounds more soluble through hydrogen bonding with water; may also act as a weak acid and drop a proton	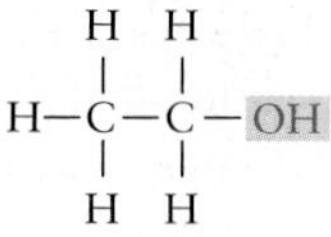 Ethanol
Phosphate	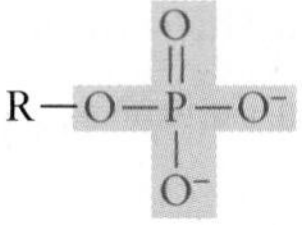	Organic phosphates	Molecules with more than one phosphate linked together store large amounts of chemical energy	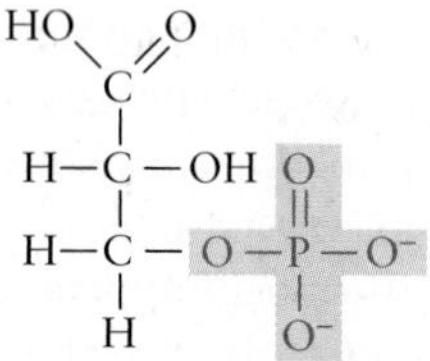 3–Phosphoglyceric acid
Sulfhydryl	R—SH	Thiols	When present in proteins, can form disulfide (S–S) bonds that contribute to protein structure	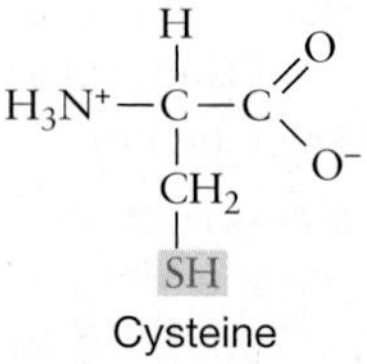 Cysteine

*In these structural formulas, "R" stands for the rest of the molecule.

✔ Based on the relative electronegativities of the atoms involved, predict whether each functional group is polar or nonpolar.

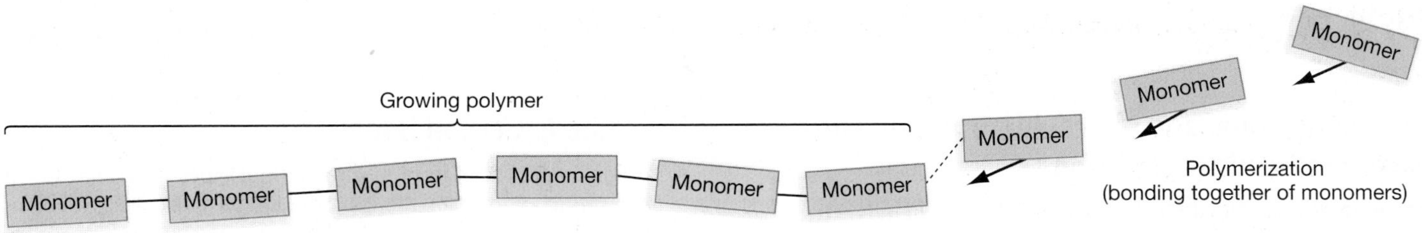

Figure 2.23 Monomers Are the Building Blocks of Polymers.

form still larger, more complex molecules like those found in living cells. These large molecules, made up of smaller molecular subunits joined together, are called **macromolecules**. In general, a molecular subunit used to build a macromolecule is called a **monomer** ("one-part").

When a large number of monomers are bonded together, the resulting structure is called a **polymer** ("many-parts"). The process of linking monomers together is called **polymerization** (Figure 2.23). The amino acids produced from Miller's experiment are an example of monomers that polymerize to form proteins (Ch. 3, Section 3.1). And as you will also learn (Chapters 4 and 5), other macromolecules of life—nucleic acids and carbohydrates—are also polymers formed by the linking together of monomers.

The theory of chemical evolution states that monomers polymerized to form the larger and more complex macromolecules that are now found in organisms. But according to the second law of thermodynamics (see Section 2.3), a pool of free monomers would not be expected to spontaneously self-assemble into a polymer. The polymerization reaction organizes multiple simpler monomers into a single more complex and ordered structure. Stated another way, polymerization decreases the disorder, or entropy, of the molecules involved. What is necessary for monomers to be linked together?

Monomers polymerize through **condensation reactions**, also known as **dehydration reactions**. These reactions are aptly named because the newly formed bond results in the loss of a water molecule (Figure 2.24a). The reverse reaction, called **hydrolysis**, breaks polymers apart by adding a water molecule (Figure 2.24b). The water molecule reacts with the bond linking the monomers, separating one monomer from the polymer chain.

In a solution, condensation and hydrolysis represent the forward and reverse reactions of a chemical equilibrium:

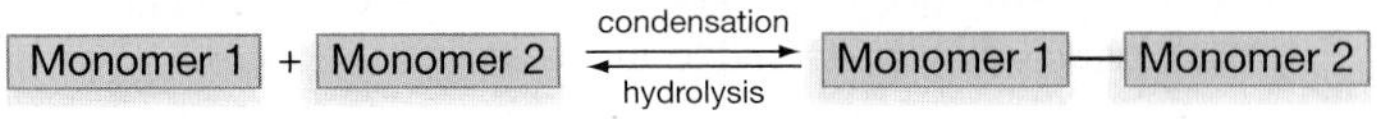

Based on what you know about chemical reactions, it should be clear that hydrolysis dominates because it both increases entropy and is favorable energetically. This means that polymerization would occur only if there were a very high concentration of monomers to push the reaction toward condensation.

Since the equilibrium favors free monomers over polymers even under concentrated conditions, a polymer is unlikely to have grown much beyond a short chain. However, as you will learn in later chapters, there are several ways the macromolecules of life—proteins, nucleic acids, and carbohydrates—may have polymerized early in chemical evolution. The rest of this unit explores the structure and function of these macromolecules, culminating in the formation of the first living cell.

CHECK YOUR UNDERSTANDING

✔ If you understood this section, you should be able to . . .

1. Explain what property of the element silicon has led some biologists to propose that silicon-based (rather than carbon-based) life-forms may have arisen on other planets. (Hint: Look back at Figure 2.8.)
2. Determine what type of functional group is present on formaldehyde (CH_2O). What property is associated with this group?

Answers are available in Appendix A.

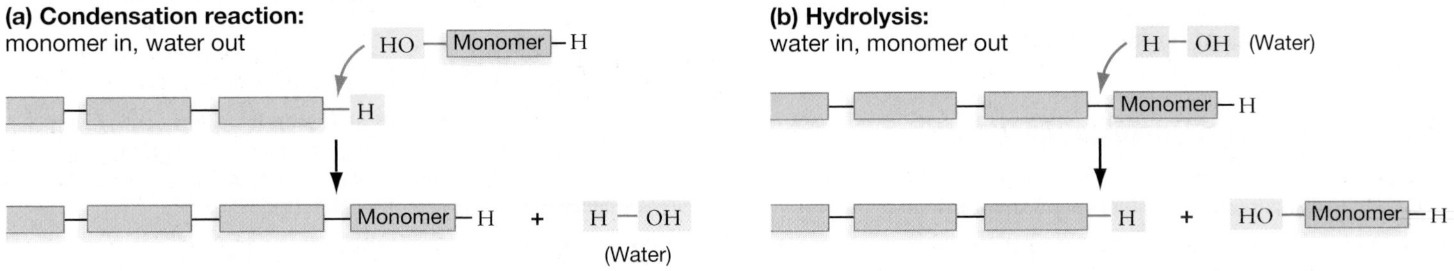

Figure 2.24 Polymers Can Be Extended or Broken Apart.

2.1 Atoms, Ions, and Molecules: The Building Blocks of Chemical Evolution

- When atoms participate in chemical bonds, the shared or transferred electrons often give the atoms full valence shells and thus contribute to the atoms' stability.
- The electrons in a covalent bond may be shared equally or unequally, depending on the relative electronegativities of the two atoms involved.
- Nonpolar covalent bonds result from equal sharing; polar covalent bonds are due to unequal sharing. Ionic bonds form when an electron is completely transferred from one atom to another.

2.2 Properties of Water and the Early Oceans

- The chemical reactions required for life take place in water.
- Water is polar—meaning that it has partial positive and negative charges—because it is bent and has two polar covalent bonds.
- Solutes dissolve in water. Water interacts with polar molecules via hydrogen bonding and ions via similar electrical attractions.
- Water's ability to participate in hydrogen bonding also gives it an extraordinarily high capacity to absorb heat and cohere to other water molecules.
- Water spontaneously dissociates into hydrogen ions (or protons, H^+) and hydroxide ions (OH^-). The concentration of protons in a solution determines the pH, which can be altered by acids and bases or stabilized by buffers.

2.3 Chemical Reactions, Energy, and Chemical Evolution

- The first step in chemical evolution was the formation of small organic compounds from molecules such as molecular hydrogen (H_2) and carbon dioxide (CO_2).
- Chemical reactions involve bonds being broken, atoms being rearranged, and new bonds being formed. This process involves energy from either the potential energy within bonds of the reactants or kinetic energy of external sources (e.g., thermal energy).
- Energy comes in different forms. Although energy cannot be created or destroyed, one form of energy can be transformed into another.

2.4 Investigating Chemical Evolution

- Experiments that attempt to mimic conditions of early Earth are used to investigate the process of chemical evolution.
- Stanley Miller demonstrated that external sources of energy, such as lightning, could have driven chemical reactions to form highly reactive molecules and amino acids from simple molecules that were likely present in early Earth.

2.5 Life Is Carbon Based

- Carbon is the foundation of organic molecules based on its valence, which allows for the construction of molecules with complex shapes.
- Organic molecules are critical to life because they possess versatility of chemical behavior due to the presence of functional groups.
- Functional groups promote further interactions between organic molecules to form macromolecules.

Answers are available in Appendix A.

✓ TEST YOUR KNOWLEDGE

1. Which of the following occurs when a covalent bond forms?
   - **a.** Electrons in valence shells are transferred from one atom to another.
   - **b.** Electrons in valence shells are shared between atoms.
   - **c.** Partial charges on polar molecules interact.
   - **d.** Nonpolar molecules are pushed together.by surrounding water molecules.
2. What are the defining characteristics of a condensation reaction?
   - **a.** Two monomers are covalently bonded together and a water molecule is produced.
   - **b.** Two monomers are covalently bonded together and a water molecule is used up.
   - **c.** A polymer is broken down into monomers and a water molecule is produced.
   - **d.** A polymer is broken down into monomers and a water molecule is used up.
3. Which of these functional groups is known to be used for storing large amounts of chemical energy?
   - **a.** amino group
   - **b.** carbonyl group
   - **c.** phosphate group
   - **d.** sulfhydryl group
4. What factors determine whether a chemical reaction is spontaneous or not?

✓ TEST YOUR UNDERSTANDING

5. Which of these molecules would you predict to have the largest number of polar covalent bonds based on their molecular formulas?
   - **a.** C_2H_6O (ethanol)
   - **b.** C_2H_6 (ethane)
   - **c.** $C_2H_4O_2$ (acetic acid)
   - **d.** C_3H_8O (propanol)
6. Locate fluorine (F) on the partial periodic table provided in Figure 2.2. Predict its relative electronegativity compared to hydrogen, sodium, and oxygen. State the number and type of bond(s) you expect it would form if it reacted with sodium (Na).
7. **QUANTITATIVE** If you were given a solution that has a pH of 8.5, what would be its concentration of protons? What is the difference in proton concentration between this solution and one that has a pH of 7?

8. Consider the reaction between carbon dioxide and water to form carbonic acid:

$$CO_2(g) + H_2O(l) \rightleftharpoons CH_2O_3(aq)$$

In the ocean, carbonic acid immediately dissociates to form a proton and bicarbonate ion, as follows:

$$CH_2O_3(aq) + H^+(aq) \rightleftharpoons CHO_3^-(aq)$$

If an underwater volcano bubbled additional CO_2 into the ocean, would this sequence of reactions be driven to the left or the right? How would this affect the pH of the ocean?

✔ TEST YOUR PROBLEM-SOLVING SKILLS

9. When H_2 and CO_2 react, acetic acid can be formed spontaneously while the production of formaldehyde requires an input of energy. Which of these conclusions can be drawn from this observation?

a. More heat is released when formaldehyde is produced compared to the production of acetic acid.

b. Compared to the reactants that it is formed from, formaldehyde has more potential energy than does acetic acid.

c. Entropy decreases when acetic acid is produced and increases when formaldehyde is produced.

d. Only acetic acid could be produced under conditions that existed in early Earth.

10. From what you have learned about water, why do coastal regions tend to have milder climates with cooler summers and warmer winters than do inland areas at the same latitude?

✔ PUT IT ALL TOGETHER: Case Study

How do organisms survive below freezing temperatures?

The winter flounder (*Pseudopleuronectes americanus*) lives in the Mid-Atlantic Ocean and along the New England coast, where frigid water temperatures would normally turn the water inside the flounder's cells into ice. Does the flounder produce some type of antifreeze compound that prevents it from freezing solid?

11. The flounder is able to survive in very cold water, but only when water is in its liquid state. What property of water prevents the ocean from freezing solid when the temperature in the air is well below water's freezing point?

12. The salty ocean has a higher level of entropy compared to fresh water due to the dissolved ions it contains, which interact with and disperse water molecules. For ice to form in the ocean, this entropy must be reduced to allow the crystalline structure shown in Figure 2.15a to form. If you were to break off a piece of this ice and melt it, would the water taste fresh or salty? Explain your answer.

13. THINK CAREFULLY Evaluate the following statements related to the process of freezing water. Select True or False for each.

T/F It does not follow the second law of thermodynamics.

T/F It is exothermic.

T/F It results in an overall increase in entropy.

T/F It requires an input of energy.

14. PROCESS OF SCIENCE *P. americanus* produces a small antifreeze protein (AFP) that binds to ice crystals as they form. The structure of AFP has polar groups on one side and nonpolar groups on the other. Propose a hypothesis to explain how AFP prevents cells from freezing solid.

15. PROCESS OF SCIENCE Flounder AFP has been used to genetically modify plants to reduce tissue damage under freezing temperatures. Data from an early experiment using a potato plant are provided below. What additional information is needed to determine if AFP is useful in protecting against frost damage?

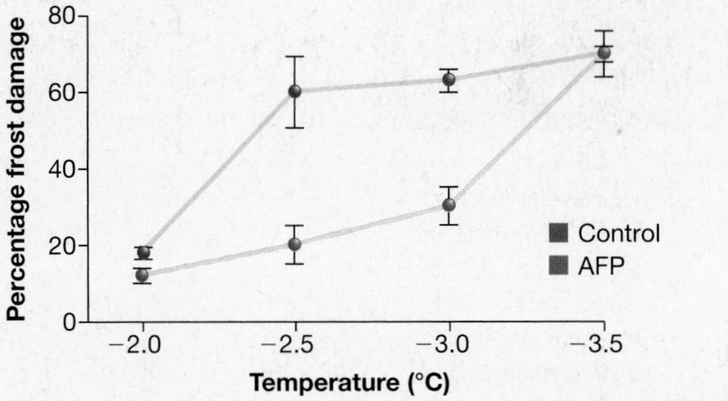

Source: J. G. Wallis, H. Wang, D. J. Guerra. 1997. *Plant Molecular Biology* 35: 323–330.

16. SOCIETY Tomatoes were also genetically modified to produce flounder AFP to protect the fruit from frost damage, but were met with considerable resistance and never made it to market. More recently, tomatoes have been modified to express a similar AFP that is normally produced by carrots. Would you be more comfortable eating genetically modified tomatoes expressing AFP from carrot instead of flounder? Explain your response, and then state what you think are likely pros and cons of producing these genetically modified organisms.

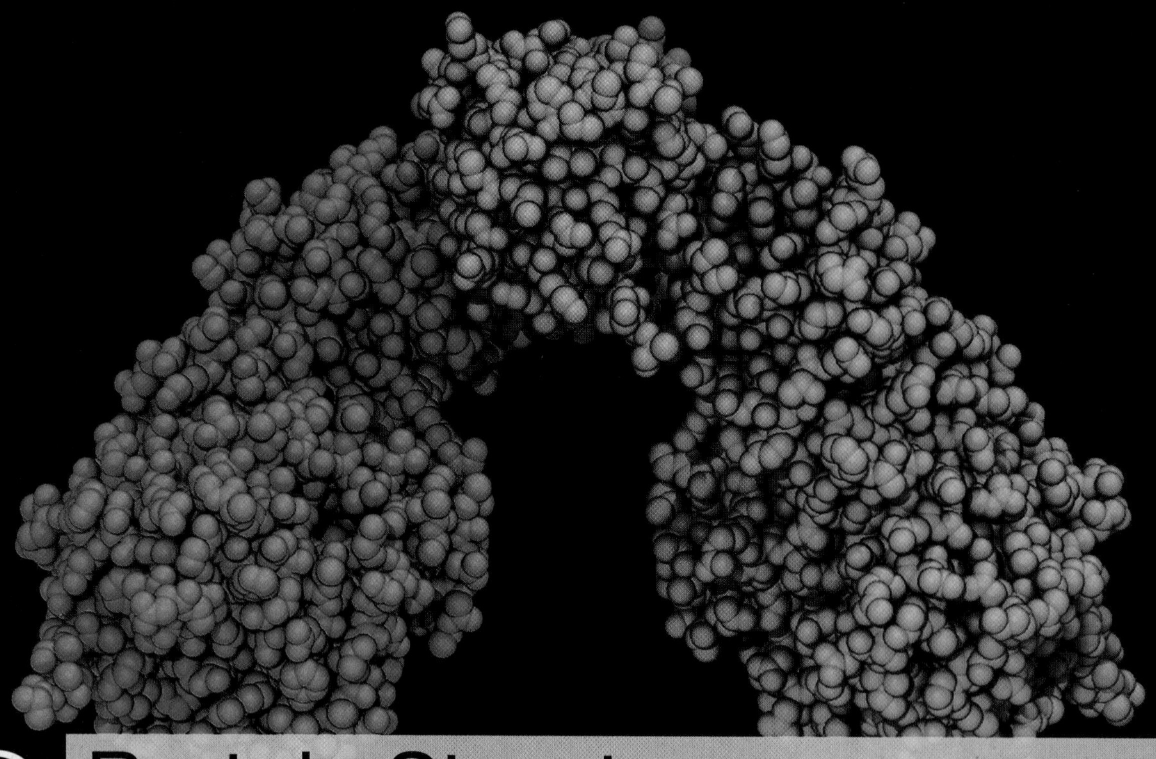

3 Protein Structure and Function

A molecular model of Hsp90 (heat shock protein 90)—a molecular chaperone that helps other proteins fold into their functional shape.

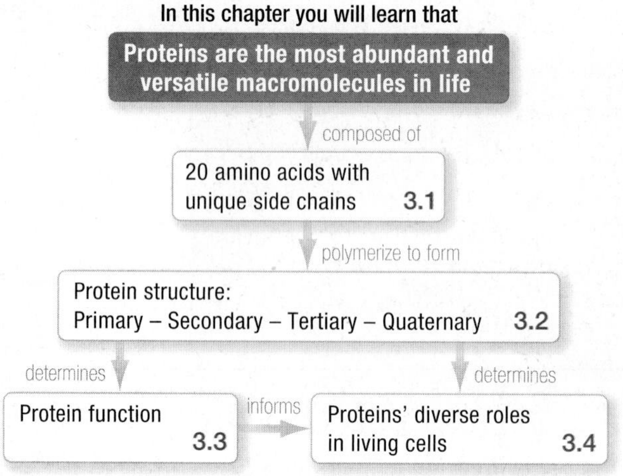

What type of molecule was responsible for the origin of life? Answering this question is a recurring theme in this and the next three chapters.

To address the question of life's origins, researchers designed experiments to identify the types of molecules that could have been produced in the waters of prebiotic Earth (Ch. 2, Section 2.4). Early Earth simulations designed by Stanley Miller and others who followed up on his work sparked particular excitement for origin-of-life researchers, because the same molecules were repeatedly discovered among their products—amino acids.

BIG PICTURE

This chapter is part of the Big Picture. See how on pages 144–145.

Amino acids have also been found in meteorites and produced in experiments that approximate the environment of interstellar space. Taken together, these observations have led researchers to conclude that amino acids were present and probably abundant during chemical evolution. Because amino acids are the building blocks of proteins, and proteins are vital, tremendously versatile components of today's cells, many researchers have asked, Could a protein have been the initial spark of life? For the answer to be yes, or even maybe, proteins would need to possess three of the fundamental attributes of life, namely, information, replication, and evolution.

To determine if proteins do have these attributes, let's begin by looking at their basic structural unit, the amino acid, and at how amino acids link to form proteins.

3.1 Amino Acids and Their Polymerization

Modern cells, such as those that make up your body, produce tens of thousands of distinct proteins. Most of these molecules are composed of just 20 different building blocks, called **amino acids**. All 20 of these building blocks share a common core structure.

After you complete this section, you should be able to . . .

- Analyze the characteristics of amino acids and the bonds that link them together in polypeptides

The Structure of Amino Acids

To understand how amino acids are put together, recall that carbon atoms have a valence of four—they can form up to four covalent bonds (Ch. 2, Section 2.1). In all 20 amino acids, a central carbon atom (referred to as the α-carbon) bonds covalently to four different atoms or groups of atoms (see also Figure 3.1a):

1. H—a hydrogen atom
2. NH_2—an amino functional group
3. COOH—a carboxyl functional group
4. a distinctive "R-group" (often referred to as a "side chain")

The combination of amino and carboxyl functional groups is key to how these molecules behave. In water, which has a pH of 7, amino acids ionize. The concentration of protons at this pH causes the amino group to act as a base, and it attracts a proton to

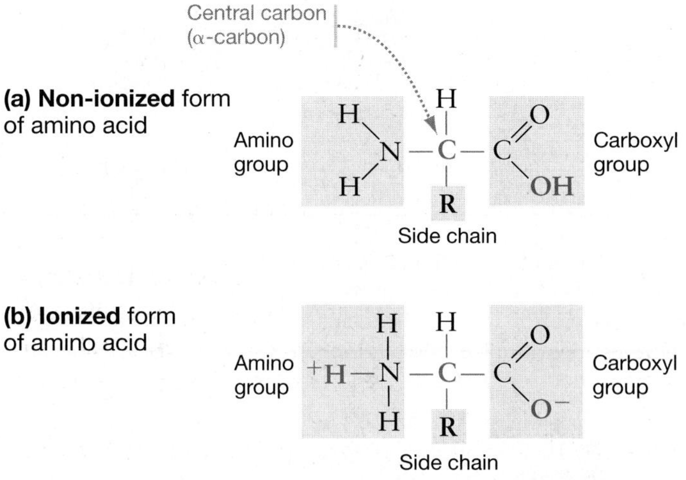

Figure 3.1 **All Amino Acids Have the Same Core Structure.** The presence of amino (basic) and carboxyl (acidic) functional groups inspired the name *amino acid*.

form $-NH_3^+$ (Figure 3.1b, left). The carboxyl group, in contrast, acts as an acid. The two highly electronegative oxygen atoms in this group pull the electron away from its hydrogen atom, which means that it is relatively easy for this group to lose a proton to form $-COO^+$ (Figure 3.1b, right).

The charges on these functional groups are important for two reasons: **(1)** They help amino acids stay in solution, where they can interact with one another and with other solutes, and **(2)** they affect the amino acid's chemical reactivity.

The Nature of Side Chains

What about the R-group? The **R-group**, or **side chain**, represents the part of the amino acid core structure that makes each of the 20 different amino acids unique. R-groups vary from a single hydrogen atom to large structures containing carbon atoms linked into rings. The properties of amino acids vary because their R-groups vary. Figure 3.2 on page 82 highlights the R-groups on the 20 most common amino acids found in cells.[1]

Functional Groups Affect Reactivity Several of the side chains found in amino acids contain carboxyl, sulfhydryl, hydroxyl, or amino functional groups. Under the right conditions, these functional groups can participate in chemical reactions. For example, amino acids with a sulfhydryl group (—SH) in their side chains can form disulfide (S—S) bonds that help link different parts of large proteins. Such bonds form naturally between proteins in hair; curly hair contains many of these cross-links while straight hair has far fewer.

In contrast, some amino acids contain side chains that have no functional groups—they consist solely of carbon and hydrogen atoms. These R-groups rarely participate in chemical reactions. Thus, the influence of these amino acids on protein function depends primarily on their size and shape rather than reactivity.

Polarity and Charge of R-Groups Affect Solubility The nature of its R-group affects the solubility of an amino acid in its natural environment—the aqueous interior of the cell.

- Both polar and electrically charged R-groups interact readily with water and are **hydrophilic**. Hydrophilic R-groups dissolve easily in water.
- Nonpolar R-groups lack charged or highly electronegative atoms capable of forming hydrogen bonds with water. These R-groups are **hydrophobic**, meaning that they do not interact with water. Instead of dissolving, hydrophobic R-groups tend to coalesce in aqueous solution.

Amino acid R-groups can be divided into three general types: charged, which includes acidic and basic; uncharged polar; and nonpolar. If given the structural formula for an amino acid at cellular pH, as in Figure 3.2, you can determine which type of amino acid it is by asking three questions:

1. Does the R-group have a negative charge? If so, it is acidic and has lost a proton, like aspartate.
2. Does the R-group have a positive charge? If so, it is basic and has picked up a proton, like lysine.

[1]There are actually 22 amino acids found in proteins that occur in organisms, but two are very rare.

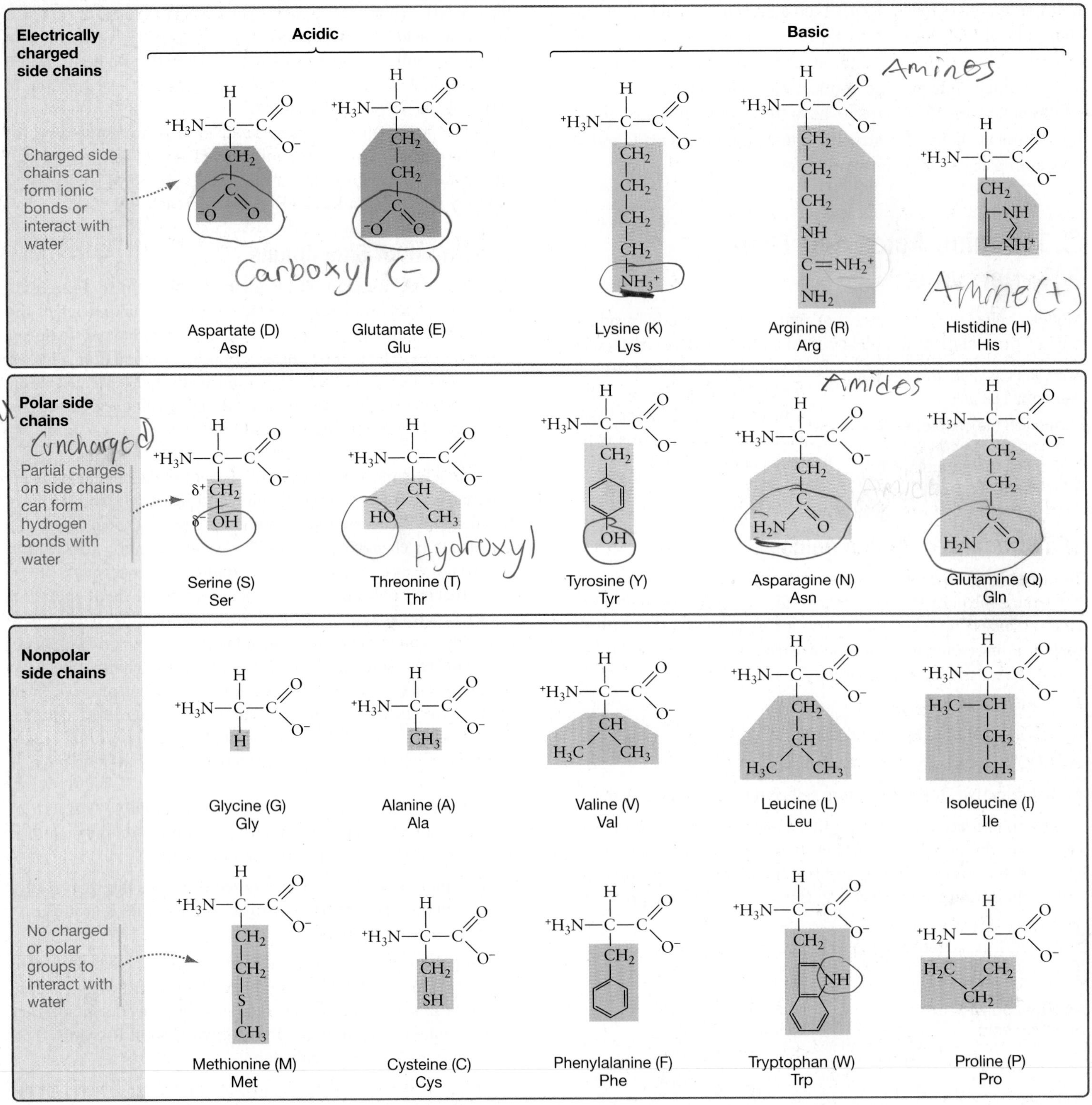

Figure 3.2 The 20 Major Amino Acids Found in Organisms. At the pH of most cells (about pH 7), the 20 major amino acids found in organisms are ionized and have the structural formulas shown here. Their R-groups (side chains) are highlighted, and standard one- and three-letter abbreviations for each amino acid are given. For clarity, carbon and hydrogen atoms are not shown in the ring structures of histidine, tyrosine, penylalanine, and tryptophan; each bend in a ring is the site of a carbon atom. A double line inside a ring indicates a double bond.

✔ Based on the relative electronegativities of O, N, C, S, and H (see Ch. 2, Section 2.1), explain why the R-groups highlighted in green are nonpolar and why R-groups highlighted in pink are polar.

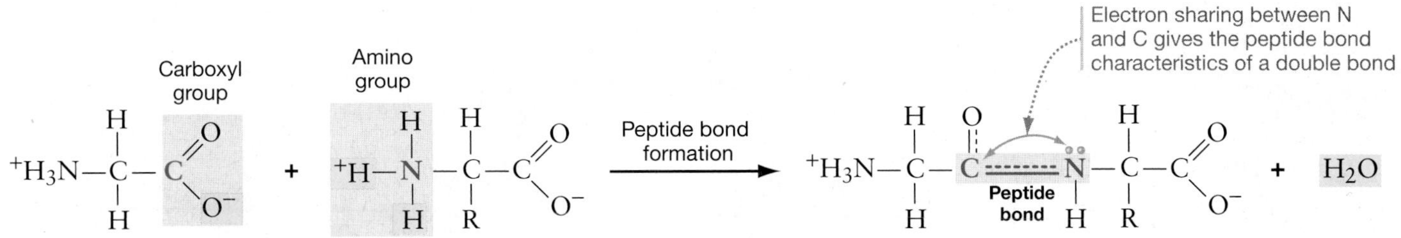

Figure 3.3 Peptide Bonds Form When the Carboxyl Group of One Amino Acid Reacts with the Amino Group of a Second Amino Acid.

3. If the R-group is uncharged, does it have an oxygen atom? If so, then the highly electronegative oxygen will form a polar covalent bond in the R-group, thus making it uncharged polar like serine. (Although somewhat less electronegative, nitrogen atoms can also contribute polarity.) The overall polarity of an R-group is based on the number of highly polar covalent bonds relative to nonpolar bonds.

If the R-group in your amino acid does not have a negative charge, a positive charge, or an oxygen atom, then you are looking at a nonpolar amino acid, such as methionine.

Now that you have seen the diversity of structures in amino acids, let's put them together to make a protein.

How Do Amino Acids Link to Form Proteins?

As Figure 3.3 shows, amino acids polymerize when a bond forms between the carboxyl group of one amino acid and the amino group of another. (See Ch. 2, Section 2.5 to review the general principles of polymerization.) The C—N covalent bond that results from this condensation reaction is called a **peptide bond.** When a water molecule is removed in the condensation reaction, the carboxyl group is converted to a carbonyl functional group (C═O) and the amino group becomes simply N—H in the resulting polymer.

The Peptide Bond Compared to monomer linkages in other types of macromolecules, peptide bonds are unusually stable. This is because the nitrogen can intermitantly donate its pair of unshared valence electrons to the carbon in the C—N bond, forming a C═N double bond (see Figure 3.3). When this occurs, a pair of electrons are pushed from the carbonyl (C═O) to the oxygen atom, forming and a single bond with an oxygen anion ($C—O^-$). These different bond configurations oscillate, but the degree of electron sharing is great enough that peptide bonds have some of the characteristics of a double bond. For example, the peptide bond is planar, limiting the movement of the atoms participating in the peptide bond.

When amino acids are linked by peptide bonds into a chain, they are referred to as residues to distinguish them from free amino acid monomers. Figure 3.4a shows how the chain of peptide bonds in a short polymer gives the molecule a structural framework, or a "backbone." There are three key points to note about the peptide-bonded backbone:

1. ***R-group orientation*** The side chains of each residue extend out from the backbone, making it possible for them to interact with each other and with water.

2. ***Directionality*** There is an amino group ($—NH_3^+$) on one end of the backbone and a carboxyl group ($—COO^-$) on the

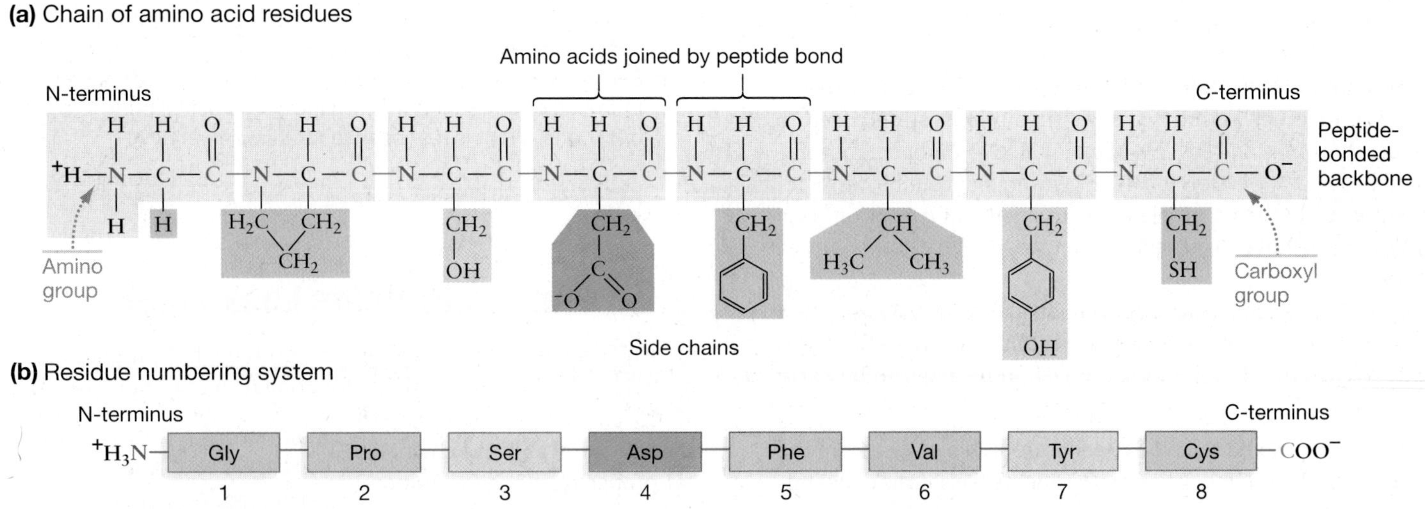

Figure 3.4 Amino Acids Polymerize to Form Chains. Note that the sequences in both parts are identical.

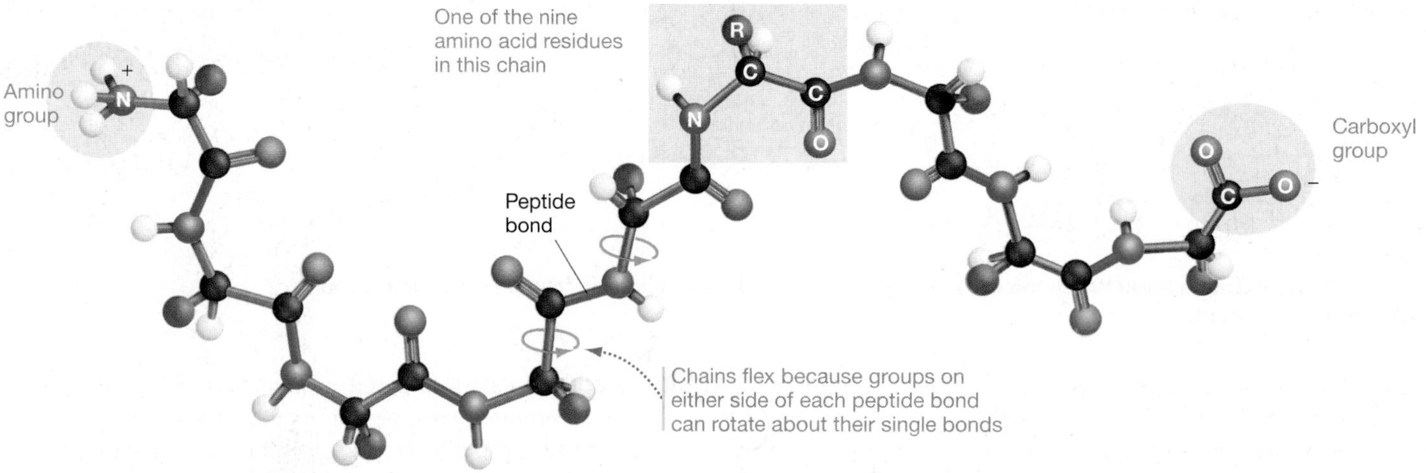

Figure 3.5 **Amino Acid Chains Are Flexible.**

other. The end of the residue sequence that has the free amino group is called the N-terminus, or amino terminus, and the end with the free carboxyl group is called the C-terminus, or carboxy terminus. By convention, biologists always write amino acid residue sequences from the N-terminus to the C-terminus (Figure 3.4b), because the N-terminus is the start of the chain when proteins are synthesized in cells.

3. ***Flexibility*** Although the peptide bond itself cannot rotate because of its double-bond nature, the single bonds on either side of the peptide bond can rotate. As a result, the structure as a whole is flexible (Figure 3.5).

Generally, when fewer than 50 amino acids are linked together in this way, the resulting polymer is called an **oligopeptide** ("few-peptides") or simply a **peptide**. Polymers that contain 50 or more amino acids are called **polypeptides** ("many-peptides").

The term **protein** is often used to describe any chain of amino acid residues. But in formal use, "protein" refers to the complete, often functional form of the molecule. Most proteins are large enough to be considered polypeptides, some consist of a single polypeptide, and others are functional only when multiple polypeptide subunits interact with one another.

Polymerization of Proteins in Early Earth In cells, the process of linking together amino acids requires other cellular factors such as ribosomes and several additional RNA and protein components (Ch. 17, Section 17.3). However, if proteins were responsible for the start of life early in chemical evolution, this reaction must have occurred in the absence of these cellular factors.

So a key question is: Can amino acids spontaneously assemble into proteins? The current consensus is that several mechanisms could have led to self-polymerization of amino acids under early Earth conditions.

- Researchers have been able to generate stable polymers in the laboratory by mixing free amino acids with a source of chemical energy and tiny mineral particles. Apparently, growing macromolecules are protected from hydrolysis if they cling, or adsorb, to a mineral surface.
- In conditions that simulate the hot, metal-rich environments of undersea volcanoes, researchers have observed amino acids being formed and even polymerized.
- In other laboratory experiments, amino acids have also joined into polymers in cooler water if an energy-rich carbon- and sulfur-containing gas—one that is commonly ejected from undersea volcanoes—is present.

While it remains unknown if chemical evolution gave rise to proteins, they clearly are the stuff of modern life. Let's look at how these strings of amino acid residues become proteins and then see what they do.

CHECK YOUR UNDERSTANDING

✔ If you understood this section, you should be able to . . .

1. **MODEL** Draw the structural formula for two glycine residues (glycine's R-group is an H) linked by a peptide bond, and label the amino and carboxy termini.
2. Use Figure 3.2 to order the following amino acids from most hydrophylic to most hydrophobic: valine, aspartate, asparagine, and tyrosine. Explain how you determined the order of these amino acids.
3. Explain what is unusual about the peptide bond compared to the other covalent bonds that form the polypeptide backbone. What is responsible for this difference?

Answers are available in Appendix A.

3.2 What Do Proteins Look Like?

Structure gives rise to function, and proteins have unparalleled diversity when it comes to their functional roles in life. The variability in protein size and shape, as well as in the chemical properties of their amino acid residues, is responsible for the diverse functions that proteins perform in cells.

After you complete this section, you should be able to . . .

- Describe the four different levels of protein structure.

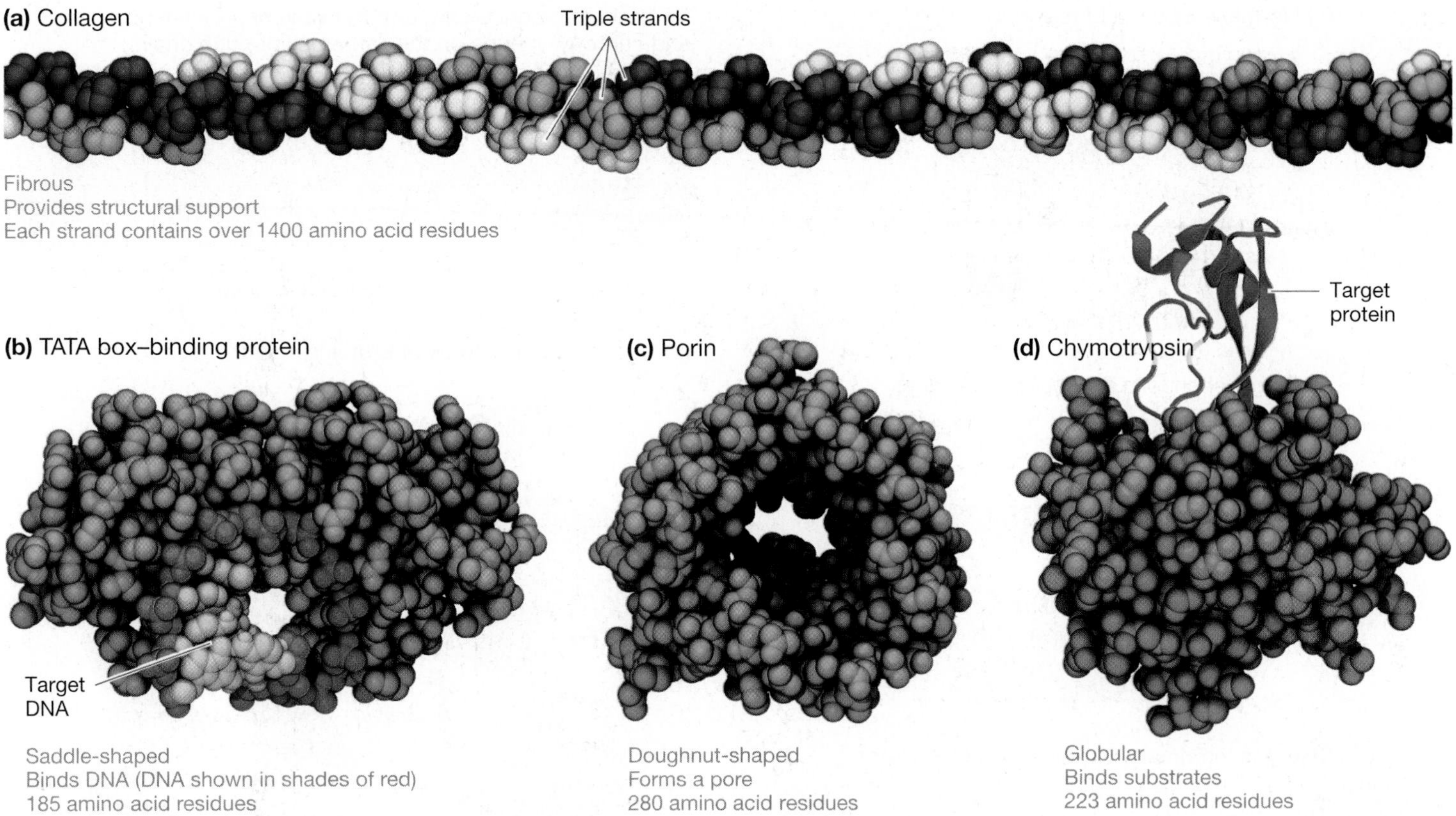

Figure 3.6 In Overall Shape, Proteins Are the Most Diverse Class of Molecules Known. These space-filling models illustrate the three-dimensional appearance of the proteins. Each atom is represented by a sphere with a diameter proportional to its size.

Figure 3.6 illustrates some of the variety in the sizes (number of amino acid residues) and shapes observed in proteins. Proteins that provide structural support for cells or tissues, such as the collagen triple helix in Figure 3.6a, often form long, cable-like fibers. In the case of the TATA box–binding protein in Figure 3.6b and the porin protein in Figure 3.6c, the shape of the molecule has a clear correlation with its function. The TATA box–binding protein has a groove where a molecule of DNA (a type of nucleic acid) fits; porin has a hole that forms a pore. The groove in the TATA box–binding protein interacts with specific regions of a DNA molecule to regulate gene activity while porin fits in cell membranes and allows certain hydrophilic molecules to pass through.

But many of the proteins found in cells do not have overall shapes that are noticeably correlated with their functions. For example, the chymotrypsin protein in Figure 3.6d has a globular shape that tells little about its function, which is binding to and cleaving the peptide bonds of other proteins. How can biologists make sense of this diversity of protein size and shape? Initially, the amount of variation seems overwhelming. Fortunately, it is not. No matter how large or complex a protein may be, its underlying structure can be categorized into just four basic levels of organization.

Primary Structure

Each protein has a unique sequence of amino acids. That simple conclusion was the culmination of 12 years of study by Frederick Sanger and co-workers during the 1940s and 1950s. Sanger's group worked out the first techniques for determining the amino acid sequence of insulin, a hormone that helps regulate sugar concentrations in the blood of humans and other mammals. When other proteins were analyzed, it rapidly became clear that each protein has a definite and distinct amino acid sequence.

Biochemists refer to the unique sequence of amino acids in a protein as its **primary structure**. With 20 types of amino acids available and chain lengths of up to tens of thousands of amino acid residues, the number of primary structures that are possible is practically limitless. There may, in fact, be 20^n different combinations of amino acid residues for a polymer with a given length of n. For example, a chain of just 10 amino acids has 20^{10} possible sequences. This is over 10,000 billion variations.

Why is the order and type of residues in the primary structure of a protein important? Recall that the R-groups present on each amino acid affect its size, shape, chemical reactivity, and solubility. It's therefore reasonable to predict that the order of the R-groups present in a polypeptide will affect that molecule's properties and function. This prediction is correct. In some cases, even a single change in the sequence of amino acids can cause striking changes in the way the protein as a whole behaves.

As an example, consider hemoglobin, an oxygen-binding protein in human red blood cells. In some individuals, one of the two different polypeptide sequences that make up hemoglobin (see section on quaternary structure) has a valine instead of a glutamate at

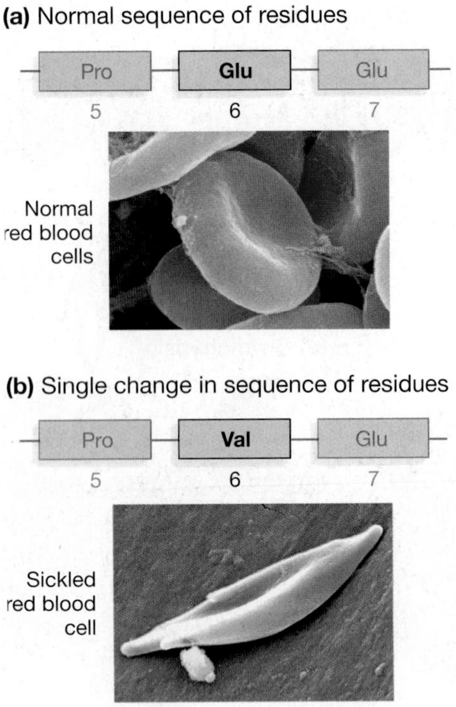

Figure 3.7 Changes in Primary Structure Affect Protein Function. Compare the primary structure of normal hemoglobin **(a)** with that of hemoglobin molecules in people with sickle cell disease **(b)**. The single amino acid change at residue 6 causes red blood cells to alter from their normal disc shape (a) to a sickled shape (b) when oxygen concentrations are low.

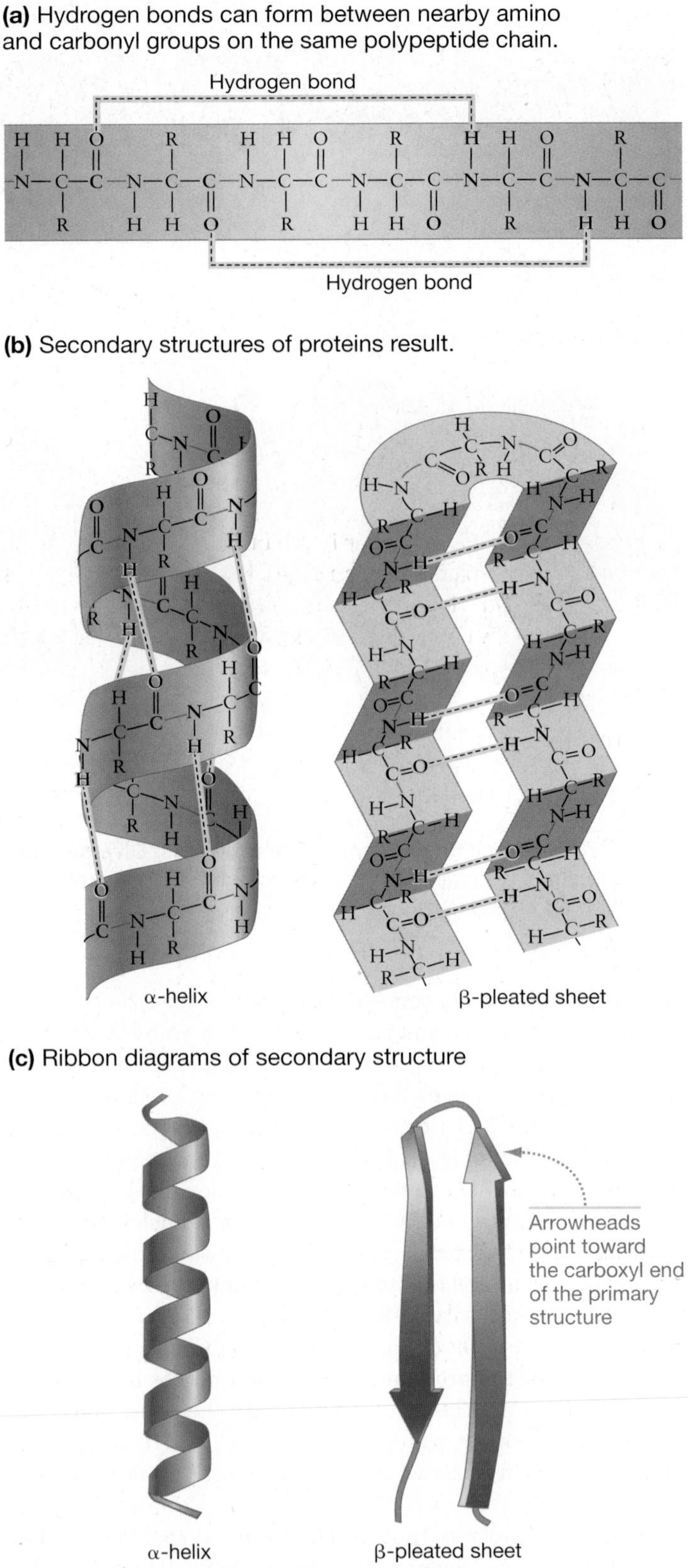

Figure 3.8 Secondary Structures of Proteins. (a) Neighboring regions in a polypeptide chain can form hydrogen bonds between N—H and C═O groups along the peptide-bonded backbone. **(b)** These interactions can result in helical coils or folds that form pleated-sheet structures in the polypeptide. **(c)** Ribbon diagrams represent secondary structures using coils for helices and arrows for pleated sheets.

the 6th position (Figure 3.7a and 3.7b). Valine and glutamate have radically different side chains. The change in R-group produces hemoglobin molecules that stick to one another and form fibers when oxygen concentrations in the blood are low. Red blood cells that carry these fibers adopt a sickle-like shape (Figure 3.7b inset). Sickled red blood cells get stuck in small blood vessels called capillaries, thereby starving downstream cells of oxygen. A debilitating illness called sickle cell disease results.

A protein's primary structure is fundamental to its function. Primary structure is also fundamental to the higher levels of protein structure: secondary, tertiary, and quaternary.

Secondary Structure

Even though variation in the amino acid sequence of a protein is virtually limitless, it is only the tip of the iceberg in terms of generating structural diversity.

The next level of organization in proteins—**secondary structure**—is generated in part by interactions between functional groups in the peptide-bonded backbone. Secondary structures are distinctively shaped sections of the linear sequence that are stabilized largely by hydrogen bonding that occurs between the oxygen on the carbonyl (C═O) group of one amino acid residue and the hydrogen on the amino (N—H) group of another (Figure 3.8a). The oxygen atom in the C═O group has a partial negative charge due to its high electronegativity, while the hydrogen atom in the N—H group has a partial positive charge because it is bonded to nitrogen, which also has high electronegativity.

Here's a key point: Hydrogen bonding between sections of the same backbone is possible only when a polypeptide bends in a way that brings C=O and N—H groups close together. In most proteins, these polar groups are aligned and form hydrogen bonds with one another when the backbone bends to form one of two possible structures:

1. an **α-helix (alpha-helix)**, in which the polypeptide's backbone is coiled (Figure 3.8b, left); or

2. a **β-pleated sheet (beta-pleated sheet)**, in which segments of a peptide chain bend 180° and then fold in the same plane (Figure 3.8b, right).

In both structures, the residues that hydrogen-bond to one another are often close together in the linear sequence of a polypeptide's primary structure. In an α-helix, H-bonds form between residues that are just four linear positions apart (see Figure 3.8a). The linear distance between residues that form a β-pleated sheet may be larger because 180° bends in the polypeptide chain can bring them close enough together to hydrogen bond.

Biologists use illustrations called ribbon diagrams to portray the secondary structures within the overall shape of a protein. Ribbon diagrams represent α-helices as coils and β-pleated sheets as groups of arrows side by side in a plane (Figure 3.8c). Unlike space-filling models, ribbon diagrams don't show the presence of each atom and its volume—only the underlying contours of the protein backbone.

Which secondary structures form, if either, depend on the molecule's primary structure—specifically, the identities of the amino acids in the sequence. Certain amino acids are more likely to be involved in α-helices than in β-pleated sheets, and vice versa, due to the specific geometry of their side chains. Proline, for example, is rarely found in α-helices due to its unusual R-group, which bonds not only to the central carbon of the residue but also to the nitrogen of the residue's core amino group (see residue 2 in Figure 3.4a). Proline often introduces kinks in the peptide-bonded backbone that do not conform to the shape of an α-helix.

Each of the hydrogen bonds in an α-helix or a β-pleated sheet is weak relative to a covalent bond, but the large number of hydrogen bonds in these secondary structures makes them highly stable. As a result, they increase the stability of the molecule as a whole and help define its shape. For overall shape and stability, though, the tertiary structure of a protein is even more important.

Tertiary Structure

A protein's distinctive overall three-dimensional shape, or **tertiary structure**, results from interactions between residues that are brought together as the backbone bends and folds in space. The residues that interact with one another are often far apart in the linear sequence. In contrast to secondary structures, which involve only hydrogen bonds between backbone amino and carbonyl groups, tertiary structures form using a variety of bonds and interactions between R-groups or between R-groups and the backbone.

Five types of interactions involving R-groups are particularly important (Figure 3.9):

1. ***Hydrogen bonding*** Hydrogen bonds form between polar side chains and opposite partial charges either on the peptide backbone or other R-groups.

2. ***Hydrophobic interactions*** In an aqueous solution, water molecules interact with the hydrophilic polar side chains of a polypeptide, forcing the hydrophobic nonpolar side chains to coalesce in the interior of the resulting globular mass. Water molecules surrounding the mass form more hydrogen bonds with each other and the polar residues on the surface of the protein, increasing the stability of their own interactions and the disorder of the rest of the aqueous solution.

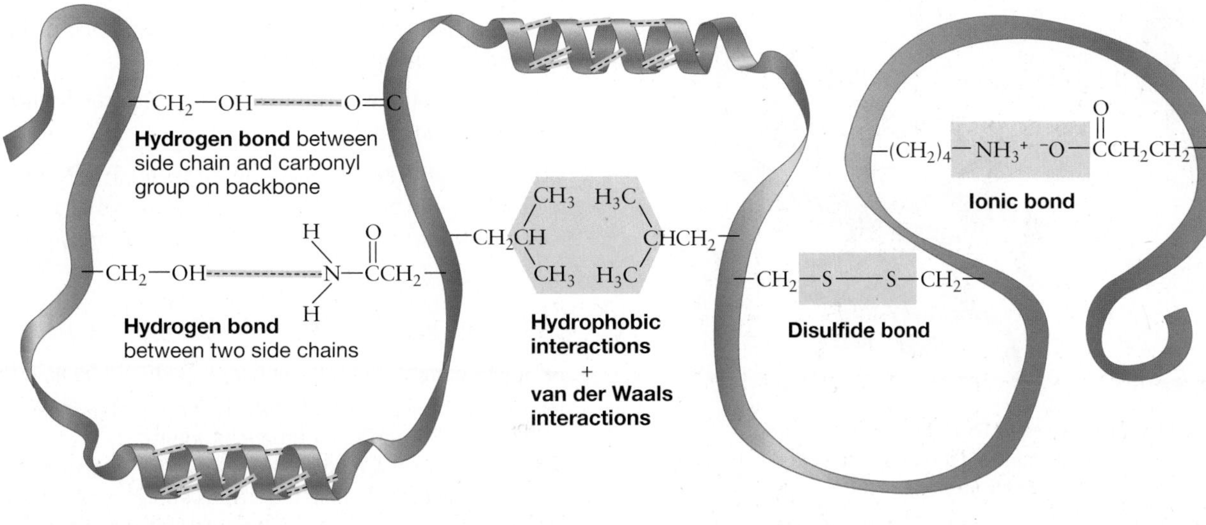

Figure 3.9 **Tertiary Structure of Proteins Results from Interactions Involving R-Groups.** Tertiary structure is stabilized by bonds and other interactions.

3. ***van der Waals interactions*** Once nonpolar side chains are forced close to one another by hydrophobic interactions, their association is further stabilized by van der Waals interactions. A large number of these weak electrical attractions can significantly increase the stability of the protein.
4. ***Covalent bonding*** Covalent bonds can form between the side chains of two cysteines through a reaction between the sulfhydryl groups. These **disulfide** ("two-sulfur") **bonds** are frequently referred to as bridges, because they create strong links between distinct regions of the same polypeptide or two separate polypeptides.
5. ***Ionic bonding*** Ionic bonds may form between groups that have full and opposing charges, such as the ionized acidic and basic side chains highlighted on the right in Figure 3.9.[2]

In addition, the tertiary structure of many proteins depends in part on the presence of secondary structures like α-helices and β-pleated sheets (Figure 3.10). Thus, tertiary structure depends on both primary and secondary structures.

With so many interactions possible between side chains and peptide-bonded backbones, it's not surprising that proteins vary in shape from rod-like filaments to ball-like masses.

[2]Ionic bonding is rare in proteins unless ionized R-groups are located in the interior where there is little water. Ionized groups on the exterior are normally exposed to the aqueous environment and enveloped by a shell of water, which prevents them from interacting with one another.

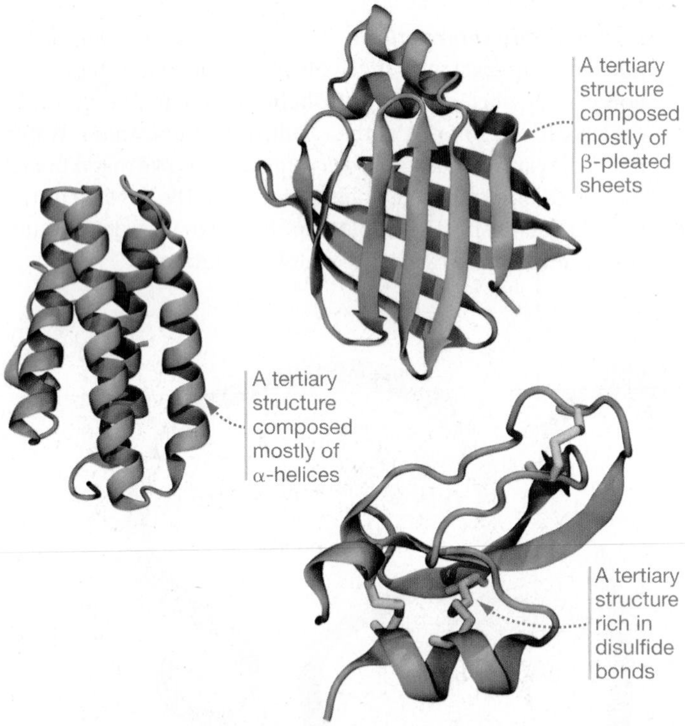

Figure 3.10 Tertiary Structures of Proteins. These different examples of tertiary structure include interactions between α-helices and β-pleated sheets (secondary structures).

✔ If you understand how tertiary structure is formed, you should be able to predict which of these structures would be most stable. Explain your answer.

Quaternary Structure

The first three levels of protein structure involve single polypeptides. But some proteins contain multiple polypeptides that interact to form a single functional structure. The combination of polypeptides, referred to as subunits, gives some proteins **quaternary structure**. The individual polypeptides are held together by many of the same types of bonds and interactions found in the tertiary level of structure.

In the simplest case, a protein with quaternary structure can consist of just two subunits that are identical. The Cro protein found in a virus called bacteriophage λ (pronounced *LAMB-da*) is an example (Figure 3.11a). Proteins with two polypeptide subunits are called dimers ("two-parts")—when the two polypeptide subunits are identical, they are called homodimers; heterodimers when they are non-identical. The quaternary structure of a protein may also include polypeptides that are distinct in primary, secondary, and tertiary structures. For example, hemoglobin consists of four polypeptides: two identical copies of an α subunit and two identical copies of a β subunit (Figure 3.11b). Hemoglobin is an example of a tetramer ("four-parts").

In addition, cells contain **macromolecular machines**: complexes of multiple proteins that assemble to carry out a particular function. Some protein complexes also include other types of macromolecules. The ribosome (introduced in Ch. 7, Section 7.1) is an example; it consists of several nucleic acid molecules as well as over 50 different proteins.

Table 3.1 summarizes the four levels of protein structure, using hemoglobin as an example. The key thing to note is that protein structure is hierarchical. The order and type of amino acids in the primary structure is responsible for the secondary structures, which then fold up to form tertiary structure. Quaternary structure (if present) is based on interactions between the tertiary structures of the polypeptide subunits.

The summary table and preceding discussion convey three important messages:

1. The combination of primary, secondary, tertiary, and quaternary levels of structure is responsible for the fantastic diversity of sizes and shapes observed in proteins.
2. Protein folding is directed by the sequence of amino acids present in the primary structure.
3. Most elements of protein structure are the result of folding polypeptide chains.

Does protein folding occur spontaneously? What happens to the function of a protein if normal folding is disrupted? Let's use these questions as a guide to dig deeper into how proteins fold.

CHECK YOUR UNDERSTANDING

✔ If you understood this section, you should be able to . . .

1. Explain how secondary, tertiary, and quaternary levels of structure depend on primary structure.
2. Predict where amino acid residues with nonpolar R-groups would be found within the overall structure of a folded globular protein such as chymotrypsin, shown in Figure 3.6d.

Answers are available in Appendix A.

(a) Cro protein, a dimer

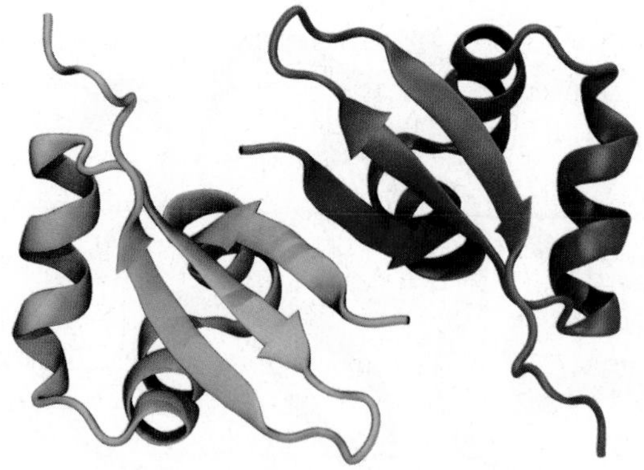

(b) Hemoglobin, a tetramer

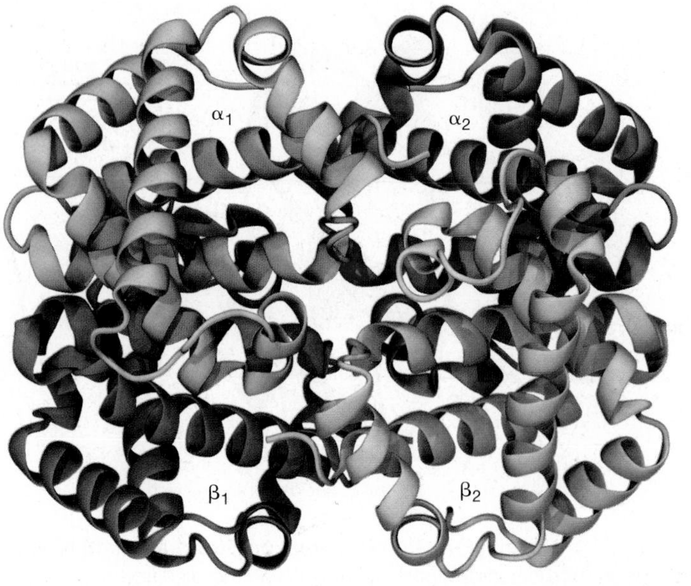

Figure 3.11 **Proteins with Quaternary Structure Have Multiple Polypeptides. (a)** The Cro protein is a homodimer—it consists of two identical polypeptide subunits, colored light and dark green in this figure. **(b)** Hemoglobin is a tetramer—it consists of four polypeptide subunits: two identical α subunits (light and dark green) and two identical β subunits (light and dark blue).

SUMMARY Table 3.1 **Protein Structure**

Level	Description	Stabilized by	Example: Hemoglobin
Primary	The sequence of amino acids in a polypeptide	Peptide bonds	–Gly–Ser–Asp–Cys–
Secondary	Formation of α-helices and β-pleated sheets in a polypeptide (depends on primary structure)	Hydrogen bonding between groups along the peptide-bonded backbone	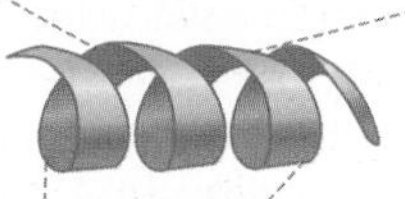 One α-helix
Tertiary	Overall three-dimensional shape of a polypeptide (includes contribution from secondary structures)	Bonds and other interactions between R-groups, or between R-groups and the peptide-bonded backbone	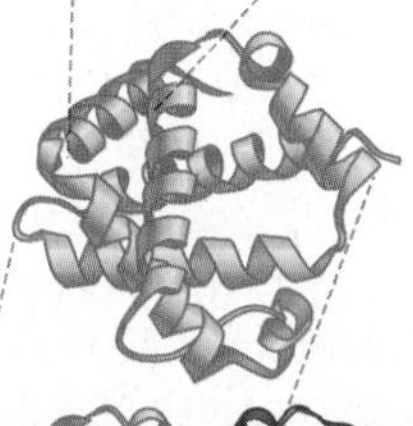 One of hemoglobin's subunits
Quaternary	Shape produced by combinations of polypeptides (thus, combinations of tertiary structures)	Bonds and other interactions between R-groups, and between peptide backbones of different polypeptides	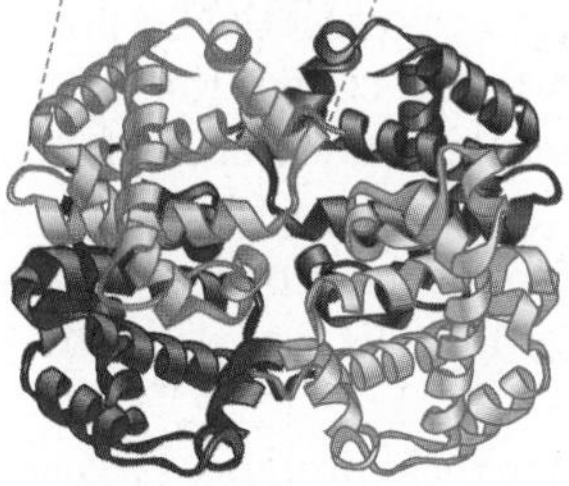 Hemoglobin consists of four polypeptide subunits

3.3 Folding and Function

If you could synthesize one of the polypeptides in hemoglobin from individual amino acids, and you then placed the resulting chain in an aqueous solution, it would spontaneously fold into the shape of the tertiary structure shown in Table 3.1.

After you complete this section, you should be able to …

- Explain the relationship between protein folding and function.

In terms of entropy, this result may seem to conflict with the second law of thermodynamics (Ch. 2, Section 2.3). Because an unfolded protein has many more ways to move about, it has much higher entropy than the folded version. But folding *does* tend to be spontaneous because the chemical bonds and interactions that occur release enough energy to overcome this decrease in entropy and will also increase entropy in the surrounding environment. As a result, the folded molecule has less potential energy and is thus more stable than the unfolded molecule.

Folding is also crucial to the function of a completed protein. This relationship between protein structure and function was hammered home in a set of classic experiments by Christian Anfinsen and colleagues during the 1950s.

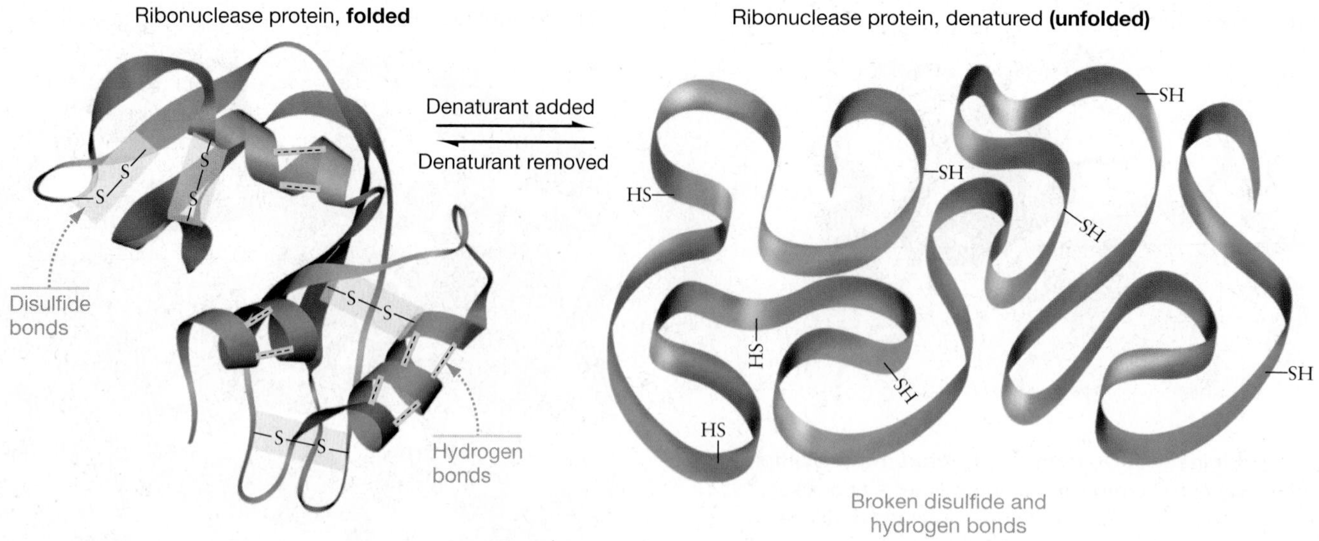

Figure 3.12 Protein Function Depends on Structure. (left) Ribonuclease is functional when properly folded via hydrogen and disulfide bonds. **(right)** When the disulfide and various noncovalent bonds are broken, ribonuclease is no longer able to function. The double arrow indicates that in this case, the process is reversible.

Normal Folding Is Crucial to Function

Anfinsen studied a protein called ribonuclease that cleaves ribonucleic acid (RNA) polymers. He found that ribonuclease could be unfolded, or **denatured**, by treating it with compounds that break hydrogen bonds and disulfide bonds. The denatured ribonuclease was unable to function normally—it could no longer degrade nucleic acids.

When the chemical denaturing agents were removed, however, ribonuclease refolded spontaneously and began to function normally again (Figure 3.12). These experiments confirmed that the primary sequence contained all the information required for folding and that folding is essential for protein function.

More recent work has shown that cells contain special proteins called **molecular chaperones** that can facilitate protein folding. Many molecular chaperones belong to a family of molecules called the heat-shock proteins, because they are produced in large quantities after cells experience the denaturing effects of high temperatures.

The name given to these proteins is a good indicator of their activity. The chaperones at a high school dance are responsible for blocking inappropriate interactions, and molecular chaperones do the same with unfolded proteins. As shown in Figure 3.13, the nonpolar side chains of unfolded polypeptides can clump together and disrupt the normal folding process. Molecular chaperones, such as the heat shock protein (Hsp90) shown at the start of this chapter, step in and attach to these hydrophobic patches before aggregates can form, then release them to fold properly. In this way, chaperones help new proteins, and in some cases denatured proteins, fold into the shape specified by their primary sequence.

So what is the "normal shape" of a protein? Is only one shape possible for each protein, or could there be several different folded shapes?

Protein Shape Is Flexible

Although each protein has a characteristic folded shape that is necessary for its function, most proteins are flexible and dynamic when they are not actively performing that function. Over half of the proteins that have been analyzed to date have disordered regions lacking any apparent structure when they are in an inactive state. Each of these proteins will exist in an assortment of shapes until they are prompted to adopt a single folded and functional form. This step is often accomplished when the proteins interact with particular ions or molecules, or when they are chemically modified.

Protein Folding Is Often Regulated Since the function of a protein is dependent on its shape, controlling when or where it is folded into its active shape will regulate its activity. Proteins involved in cell signaling, for example, are often regulated in this way. Many of these proteins are disordered and do not complete their folding until after binding to other molecules or ions that are present only during a signaling event. This interaction induces the protein to fold into an ordered, active conformation. Such regulated folding plays a major role in controlling and coordinating cellular activities (Ch. 11, Section 11.3).

Folding Can Be "Infectious" In 1982, Stanley Prusiner published what may be the most surprising result to emerge from research on protein folding: Certain normal proteins can be induced to fold into infectious, disease-causing agents. These proteins are called **prions** (pronounced *PREE-ons*), or proteinaceous infectious particles.

Infectious prions are alternately folded forms of normal proteins that are present in healthy individuals. The two versions of the protein have the same primary structure, but

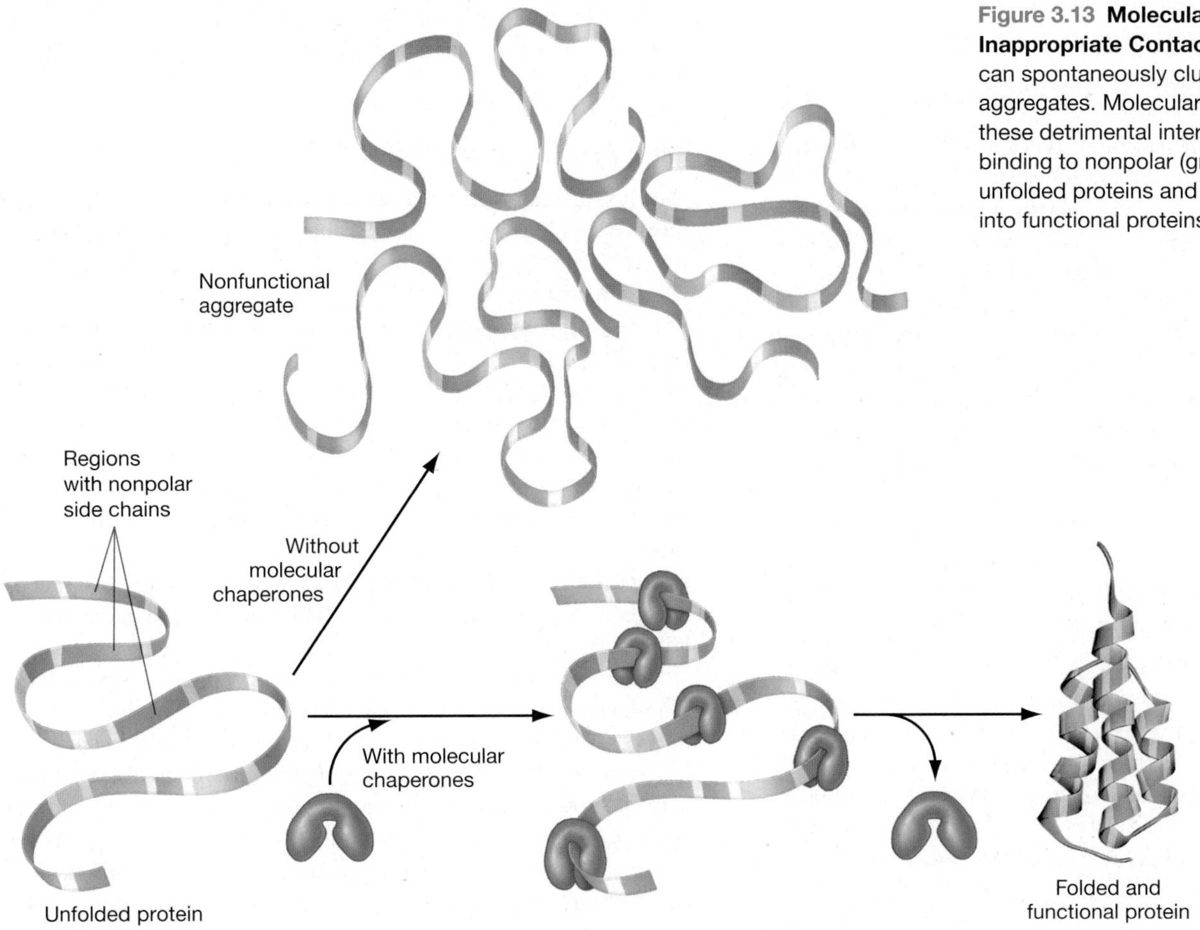

Figure 3.13 Molecular Chaperones Block Inappropriate Contact. Unfolded proteins can spontaneously clump into nonfunctional aggregates. Molecular chaperones prevent these detrimental interactions by reversibly binding to nonpolar (green) regions of unfolded proteins and allowing them to fold into functional proteins.

their *shapes* are radically different. Figure 3.14 illustrates the differences in shape observed between the normal and infectious forms of the prion protein (PrP) responsible for "mad cow disease"—a disease in cattle that destroys the central nervous system. Figure 3.14a shows the normal folded form seen in healthy cattle cells. The infectious version of this protein is shown in Figure 3.14b.

Mad cow disease is one of a family of diseases caused by prions known as the spongiform encephalopathies—literally, "sponge-brain-illnesses." Cows, sheep, goats, and humans afflicted with these diseases undergo massive degeneration of the brain. Although some spongiform encephalopathies can be inherited, in many cases the disease is transmitted when individuals eat tissues containing the infectious form of PrP.

Infectious prions propagate by binding to normal prions and inducing conformational changes that cause the normal versions to adopt the alternate, infectious shape. This shape change stabilizes the interactions between prion proteins, resulting in the assembly of long fibrils that often leads to cell death.

Prions are a particularly dramatic example of how a protein's function depends on its shape as well as how the final shape of a protein depends on folding.

CHECK YOUR UNDERSTANDING

✔ **If you understood this section, you should be able to . . .**

Explain how ions, such as calcium (Ca^{2+}) released during a signaling event, are involved in controlling cellular activities. How is this different from the changes that occur in prion proteins?

Answers are available in Appendix A.

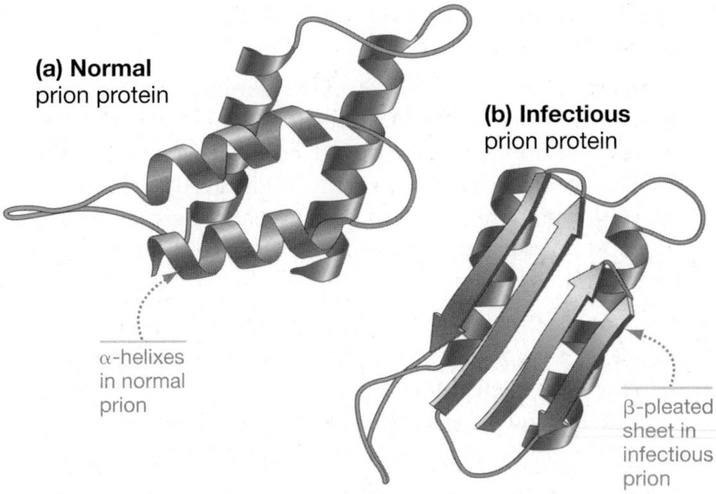

Figure 3.14 Prion Infectivity Is Linked to Structure. Ribbon diagrams of **(a)** normal, noninfectious prion protein with α-helices; and **(b)** infectious prion protein with β-pleated sheets, which causes mad cow disease in cattle.

3.4 Protein Functions Are as Diverse as Protein Structures

As a group, proteins perform more types of cell functions than any other type of molecule. It makes sense to hypothesize that life began with proteins, simply because proteins are so vital to the life of today's cells.

After you complete this section, you should be able to ...

- Describe the different functions of proteins in living systems.

Consider the red blood cells that are moving through your veins and arteries right now. Each of these cells contains about 300 million copies of hemoglobin. Hemoglobin carries oxygen from your lungs to cells throughout the body. But every red blood cell also has thousands of copies of a protein called carbonic anhydrase, which is important for moving carbon dioxide from cells back to the lungs, where it can be breathed out. These are just two examples of the incredible variety of proteins in your body. Proteins are crucial to most tasks required for cells and organisms to exist:

- ***Catalysis*** Many proteins are specialized to **catalyze**, or speed up, chemical reactions. A protein that functions as a catalyst is called an **enzyme**. The carbonic anhydrase molecules in red blood cells are enzymes. So is the salivary amylase protein in your mouth. Salivary amylase begins the digestion of starch into simple sugars. Most chemical reactions that make life possible depend on enzymes (Chapters 8, 9, and 10).
- ***Structure*** Structural proteins make up body components such as fingernails and hair, and form the internal "skeleton" of individual cells (Ch. 7, Section 7.6). Structural proteins keep red blood cells flexible and in their normal disc-like shape.
- ***Movement*** Motor proteins and contractile proteins are responsible for moving the cell itself, or for moving large molecules and other types of cargo inside the cell. As you turn this page, for example, specialized proteins called actin and myosin will slide past one another to flex or extend muscle cells in your fingers and arm (Ch. 7, Section 7.6 and Ch. 45, Section 45.1).
- ***Signaling*** Proteins are involved in carrying and receiving signals from cell to cell inside the body. Many of them reside on the cell's membrane to interact with neighboring cells. If sugar levels in your blood are low, a small peptide called glucagon will bind to receptor proteins on your liver cells, triggering enzymes inside to release sugar into your bloodstream (Chapters 11, 37, and 46).
- ***Transport*** Proteins allow particular molecules to enter and exit cells or carry them throughout the body. Hemoglobin is a particularly well-studied transport protein, but virtually every cell is studded with membrane proteins that control the passage of specific molecules and ions (Ch. 6, Section 6.4).
- ***Defense*** Proteins called antibodies attack and destroy viruses and bacteria that cause disease (Ch. 48, Section 48.2).

Of all the functions that proteins perform in cells, catalysis may be the most important. The reason is speed. Life, at its most basic level, consists of chemical reactions. But most don't occur fast enough to support life unless a catalyst is present. Enzymes are the most effective catalysts on Earth. Why is this so?

Why Are Enzymes Good Catalysts?

Catalyzed reactions involve one or more reactants, called **substrates**. Part of the reason enzymes are such effective catalysts is that they hold substrates in a precise orientation so they can react.

The initial hypothesis for how enzymes work was proposed by Emil Fischer in 1894. According to Fischer's "lock-and-key" model, enzymes are analogous to a lock and the keys are substrates that fit into the lock and then react. Several important ideas in this model have stood the test of time. For example, Fischer was correct in proposing that enzymes bring substrates into a precise orientation that makes reactions more likely. His model also accurately explained why most enzymes effectively catalyze one specific reaction. Enzyme specificity is a product of the geometry and types of functional groups in the sites where substrates bind.

As researchers began to test Fischer's model, the location where substrates bind and react became known as the enzyme's **active site**. The active site is where catalysis actually occurs.

When techniques for determining the three-dimensional structure of enzymes became available (see x-ray crystallography in **BioSkills 6**), the active sites were identified as clefts or cavities in the overall shapes. The digestive enzyme chymotrypsin, at work in your body now, is a good example. The active site in chymotrypsin contains three key amino acid residues, called a catalytic triad, with functional groups that catalyze the cleavage of peptide bonds in other proteins (**Figure 3.15**).

No other class of macromolecules can match proteins for their catalytic potential. The variety of reactive functional groups present in amino acids is much better suited for this activity than those found in nucleotides or sugars.

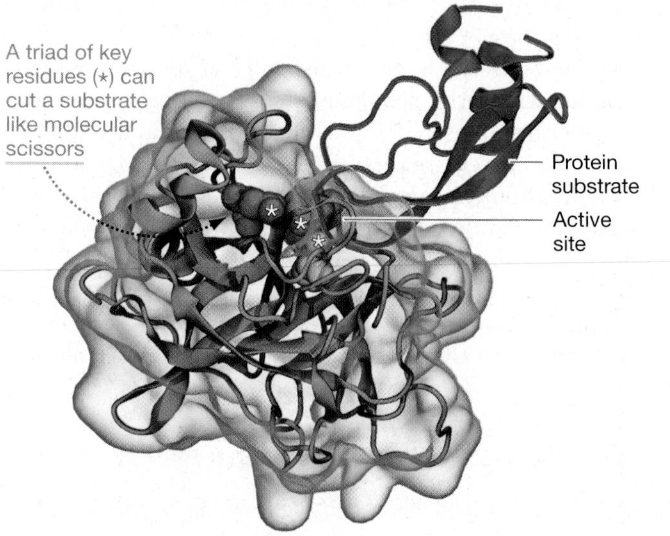

Figure 3.15 Substrates Bind to a Specific Location in an Enzyme Called the Active Site. The active site in chymotrypsin, as in many enzymes, contains three key amino acid residues that bind substrates and catalyze a reaction.

The role of enzymes in catalyzing reactions is discussed in more detail in the next unit (Ch. 8, Section 8.3). There you will see that Fischer's model had to be refined as research on enzyme action progressed.

Did Life Arise from a Self-Replicating Enzyme?

Based on several observations in the preceding sections, it could be argued that a protein catalyst was the first molecule capable of replication. Experimental studies have shown that amino acids were likely abundant during chemical evolution (see, for example, Ch. 2, Figure 2.20) and that they could have polymerized to form small proteins. To date, however, attempts to simulate the origin of life with proteins alone have not been successful.

Although it is too early to arrive at definitive conclusions, most origin-of-life researchers are skeptical that life began with a protein. To achieve the attributes of life, proteins would need to possess information, replicate, and evolve. The information carried in proteins is necessary for their function, but it cannot be used as a template or mold for their own replication. If they cannot replicate, then they cannot evolve on their own. Nucleic acids, in contrast, do carry this type of information. How they use it is the subject of the next chapter.

CHECK YOUR UNDERSTANDING

✔ **If you understood this section, you should be able to . . .**

1. List the six different tasks performed by proteins in your body.
2. Explain how the primary and tertiary levels of protein structure relate to enzyme–substrate specificity.

Answers are available in Appendix A.

CHAPTER 3 Review

For media, go to **Mastering Biology**

3.1 Amino Acids and Their Polymerization

- Amino acids have a central carbon bonded to an amino group, a hydrogen atom, a carboxyl group, and an R-group.
- The structure of the R-group affects the chemical reactivity and solubility of the amino acid.
- In proteins, amino acids are joined by a peptide bond between the carboxyl group of one amino acid and the amino group of another amino acid.

3.2 What Do Proteins Look Like?

- A protein's primary structure, or sequence of amino acids, is responsible for most of its chemical properties.
- Interactions that take place between C═O and N—H groups in the same peptide-bonded backbone create secondary structures, which are stabilized by hydrogen bonding.
- Tertiary structure results from interactions between R-groups—or R-groups and the peptide-bonded backbone—that stabilize a complete polypeptide into an overall three-dimensional shape.
- In many cases, a complete protein consists of more than one polypeptide bonded together. The combination of polypeptides represents the protein's quaternary structure.

3.3 Folding and Function

- Protein folding is a spontaneous process.
- A protein's overall folded shape is essential to its function.
- Many proteins must first bind to other molecules or ions before they can adopt their active conformation.
- Improperly folded proteins can be detrimental to life, and certain folded variations of proteins can cause deadly infectious diseases.

3.4 Protein Functions Are as Diverse as Protein Structures

- In organisms, proteins function in catalysis, defense, movement, signaling, structural support, and transport of materials.
- Proteins can have diverse functions in cells because they have such diverse structures and chemical properties.
- Catalysis takes place at the enzyme's active site, which has unique chemical properties and a distinctive size and shape that is specific to its substrates.

Answers are available in Appendix A.

✔ TEST YOUR KNOWLEDGE

1. What two functional groups are bound to the central carbon of every free amino acid monomer?
   a. an R-group and a hydroxyl group
   b. an N—H group and a C═O group
   c. an amino group and a hydroxyl group
   d. an amino group and a carboxyl group
2. What type of bond is directly involved in the formation of an α-helix?
   a. peptide bonds between amino acid residues
   b. hydrogen bonds between amino acid residues
   c. van der Waals interactions between nonpolar residues
   d. disulfide bonds between cysteine residues
3. What type of information is used to direct different polypeptides to fold into different shapes?
4. Which of the following *correctly* describe an active site? Select True or False for each statement.
   T/F It is the location in an enzyme where substrates bind.
   T/F It is the place where a molecule or ion binds to an inactive enzyme to induce a shape change to make it active.
   T/F It is the portion of an enzyme where chaperones bind to help enzymes fold.
   T/F It is the site on an enzyme where catalysis occurs.

✔ TEST YOUR UNDERSTANDING

5. **QUANTITATIVE** If a cell were to use only 10 of the 20 possible amino acids, how much effect would you expect this to have on protein diversity? Calculate and compare the number of different sequences that can be generated by randomly assembling either 10 or 20 amino acids into peptides that are five residues long.
6. Explain how molecular chaperones facilitate protein folding in many different polypeptides, each with their own specific shape.
7. Why are proteins not considered to be a good candidate for the first living molecule?
   a. Their catalytic capability is not sufficient for most biological reactions.
   b. Their amino acid monomers were not likely present in the prebiotic soup.
   c. They cannot serve as a template for replication.
   d. They could not have polymerized from amino acid monomers under early Earth conditions.
8. Predict the effect on protein function if each polypeptide adopted only a single, inflexible shape based on its primary structure.

✔ TEST YOUR PROBLEM-SOLVING SKILLS

9. Based on what you know of the peptide bonds that link together amino acid residues, why would proline's side chain reduce the flexibility of the backbone?
10. Make a concept map (see **BioSkills 12**) that relates the four levels of protein structure and shows how they can contribute to the formation of hemoglobin. Your map should include the following boxed terms: Primary structure, Secondary structure, Tertiary structure, Quaternary structure, Amino acid sequence, R-groups, α-helices, and β-pleated sheets.

✔ PUT IT ALL TOGETHER: Case Study

Why do some people need to avoid eating gluten?

Gluten is a mixture of proteins abundant in wheat, barley, and rye. Although these grains are staples in the Western diet, almost 1 percent of Americans (over 3 million people) have celiac disease—a disorder of the intestines caused by an abnormal immune response after eating gluten. This immune response damages the fingerlike villi of the small intestine, which can lead to intestinal upset and malnourishment. Currently, the only treatment is to maintain a strict gluten-free diet.

11. The typical college student diet is rich in gluten-containing foods. List six items that you would not be able to eat if you were diagnosed with celiac disease.
12. When you eat gluten, enzymes present in your stomach and intestines digest all but a few short peptides. How do enzymes accomplish this activity?
    a. The active site of an enzyme binds to a gluten protein and facilitates the hydrolysis reaction that breaks apart peptide bonds.
    b. An enzyme binds to the active site of a gluten protein and speeds up the condensation reaction that breaks apart peptide bonds.
    c. The active site of an enzyme binds to a gluten protein and reacts with it to break apart hydrogen bonds.
    d. An enzyme binds to the active site of a gluten protein and catalyzes the reactions that break apart hydrogen bonds.
13. **QUANTITATIVE** One of the peptides that can be recovered after gluten digestion is 33 residues long; 13 of the 33 residues are proline. How many times would you expect proline to appear in this peptide if it were made up of a completely random assortment of the 20 most common amino acids?
14. Recall that proline often introduces kinks in the backbone of a polypeptide. These kinks make it difficult for enzymes in your gut to fully digest gluten. In people with celiac disease, certain proline-rich peptides left over after gluten digestion will trigger an abnormal immune response. Researchers have identified a mold enzyme called AN-PEP that effectively digests proline-rich peptides. Predict where the structural differences would occur between AN-PEP and other enzymes that do not digest the peptides.
15. **PROCESS OF SCIENCE** AN-PEP was tested for its ability to digest gluten peptides in a system that mimics the human stomach. White bread was loaded into the artificial stomach with or without AN-PEP. The following graph shows some of the experiment's results. Interpret the effect of AN-PEP on the accumulation of proline-rich gluten peptides. Explain why the peptide concentration increases when AN-PEP is absent (the negative experimental control).

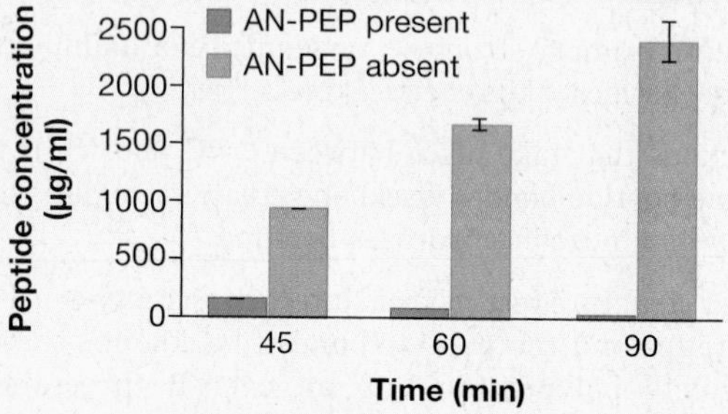

Source: C. Mitea et al. 2008. *Gut* 57:25–32.

16. **SOCIETY** Based on the experiment in question 15, how might AN-PEP be used to treat celiac disease?

Mastering Biology

Students Go to Mastering™ Biology for assignments, the eText, and the Study Area with animations, practice tests, and activities.

Professors Go to Mastering™ Biology for automatically graded tutorials and questions that you can assign to your students, plus Instructor Resources.

4 Nucleic Acids and an RNA World

This is part of the sheet-metal-and-wire model that James Watson and Francis Crick used to figure out the secondary structure of DNA. The large "T" stands for the nitrogen-containing base thymine.

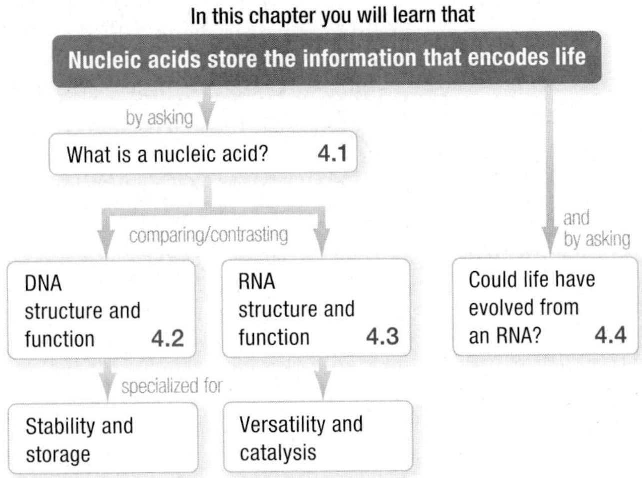

Life began when chemical evolution led to the production of a molecule that could promote its own replication. In today's cells, **deoxyribonucleic acid (DNA)** stores genetic information and is replicated using proteins. Relatively few researchers favor the hypothesis that life began as a DNA or protein molecule, however. Instead, the **RNA world hypothesis** proposes that there was a stage in the evolution of life when **ribonucleic acid (RNA)** both stored the genetic information and catalyzed its own replication.

There is compelling evidence that such an RNA world existed on early Earth. The nature of the first "living molecule" has, however, been the subject of many investigations and heated debates. Did life on Earth begin with RNA first, or did some other replicating, evolving molecule come before RNA, just as RNA likely preceded DNA and proteins?

BIG PICTURE

This chapter is part of the Big Picture. See how on pages 144–145.

Regardless of the answer to this question, once the first self-replicating molecules evolved, chance errors in the copying process created variations that would undergo natural selection—the evolutionary process by which individuals, in this case molecules, with certain characteristics reproduce more frequently than others (Ch. 1, Section 1.1). At this point, chemical evolution was over and biological evolution was off and running.

This chapter focuses on the structure and function of nucleic acids in today's cells, but also explores how they could have triggered the evolution of life on Earth. Let's begin with an analysis of nucleic acid monomers and how they are linked together into polymers.

4.1 What Is a Nucleic Acid?

Nucleic acids are polymers, just as proteins are polymers. But instead of being assembled from amino acids, **nucleic acids** are made up of monomers called **nucleotides**.

> **After you complete this section, you should be able to ...**
>
> - Analyze the characteristics of nucleotides and the bonds that link them together in nucleic acids.

Figure 4.1a diagrams the three components of a nucleotide: **(1)** a phosphate group, **(2)** a five-carbon sugar, and **(3)** a nitrogenous (nitrogen-containing) base. The phosphate group is bonded to the sugar molecule, which in turn is bonded to the base.

The sugar is the central component of the nucleotide, much like the α-carbon in amino acids (Ch. 3, Section 3.1). The five carbons in this sugar are labeled with numbers and prime (′) symbols to provide a frame of reference. For example, the base is attached to the 1′ carbon and the phosphate group is attached to the 5′ carbon.

The monomers of ribonucleic acid (RNA) are referred to as **ribonucleotides**, and the monomers of deoxyribonucleic acid (DNA) are called **deoxyribonucleotides**. In ribonucleotides, the sugar is ribose; in deoxyribonucleotides, it is deoxyribose (*deoxy* means "lacking oxygen"). As Figure 4.1b shows, both of these sugars have an —OH group bonded to the 3′ carbon, but ribose has an —OH group bonded to the 2′ carbon while deoxyribose has an H at the same location—a difference of just a single oxygen atom.

Ribonucleotides and deoxyribonucleotides also differ in one of their nitrogenous bases. These bases, diagrammed in Figure 4.1c, belong to structural groups called **purines** and **pyrimidines**. The purines are adenine (A) and guanine (G); the pyrimidines are cytosine (C), uracil (U), and thymine (T). Ribonucleotides use uracil (U), while deoxyribonucleotides use thymine (T).

Note that purines consist of double rings formed from nine atoms, compared to the six atoms that make up the single ring in each pyrimidine. This makes identifying the structure of purines easy, since both adenine and guanine include "nine" in their names.

To summarize: After the different sugars and bases are taken into account, eight different nucleotides are used to build nucleic acids—four ribonucleotides (A, G, C, and U) and four

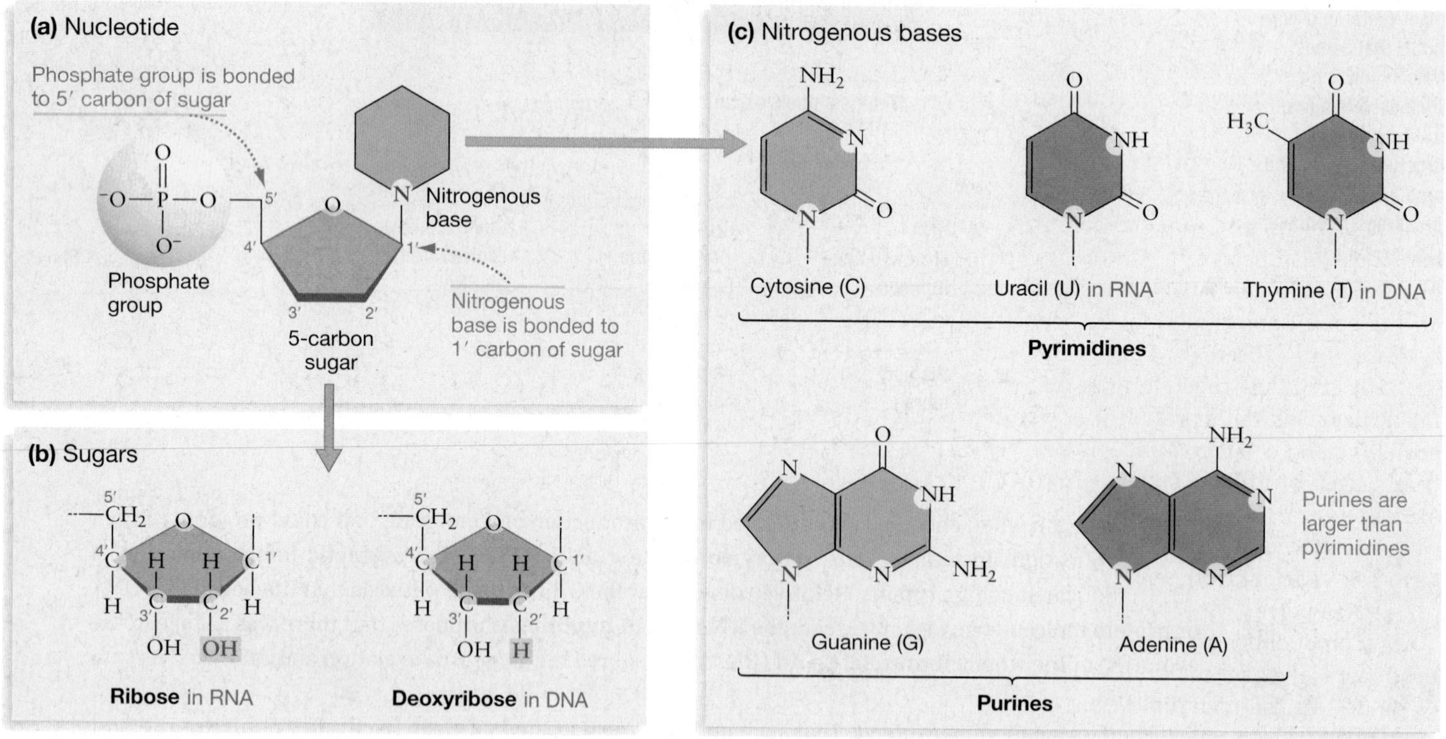

Figure 4.1 **The General Structure of a Nucleotide.** Note that in the bases, the nitrogen (N) that bonds to the sugar is colored blue.

Making Models 4.1 Tips on Drawing Nucleotides

Models of complex chemical structures may be drawn by using simple shapes for the different components. Nucleotides are often drawn using a circle for the phosphate group, a pentagon with an upright tail for the sugar, and a hexagon for the base.

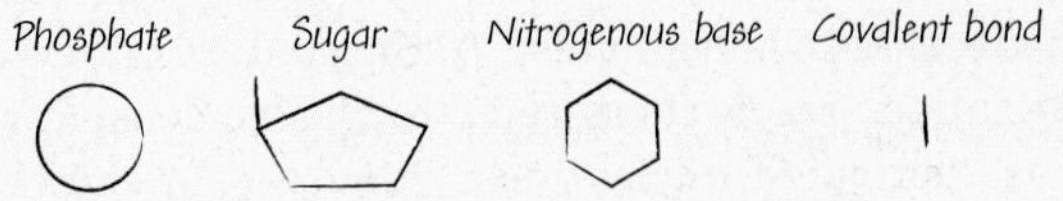

MODEL Using the shapes above, draw two nucleotide models—one ribonucleotide and one deoxyribonucleotide. Use lines to represent covalent bonds connecting the different components. Label the 2′, 3′, and 5′ carbons on the sugars and show what is bonded to each carbon.

To see this model in action, go to the Study Area of **Mastering Biology**

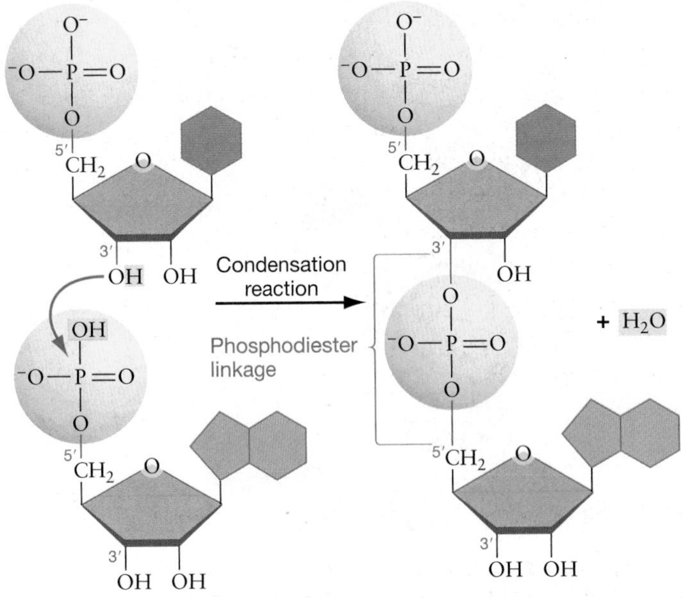

Figure 4.2 **Nucleotides Polymerize via Condensation Reactions.** The resulting phosphodiester linkage connects the 3′ carbon of one nucleotide and the 5′ carbon of another nucleotide. The two ester bonds connecting the nucleotides are highlighted in red. (Ribonucleotides are shown here, but the same reaction occurs between deoxyribonucleotides.)

deoxyribonucleotides (A, G, C, and T). See **Making Models 4.1** to learn how simple models are commonly used to represent nucleotides.

If nucleic acids played any role in the chemical evolution of life, then at least some of these nucleotides must have been present in the prebiotic oceans. Is there any evidence to suggest that this was possible?

Could Chemical Evolution Result in the Production of Nucleotides?

Based on data from Stanley Miller (see Ch. 2, Figure 2.20) and researchers who followed, most biologists accept the idea that amino acids could have been synthesized early in Earth's history. The reactions behind the prebiotic synthesis of nucleotides, however, have been more difficult to identify.

Miller-like laboratory simulations have shown that nitrogenous bases and many different types of sugars, including ribose, can be synthesized readily under conditions that mimic those in early Earth oceans. Recent work has focused on the conditions that exist in deep-sea hydrothermal vent systems (introduced in Chapter 2). What researchers have found is striking—reactive minerals on the surface of walls inside deep-sea vents preferentially bind to ribose, effectively enriching and concentrating ribose (the sugar found in RNA) from a pool of diverse sugars. Did this likewise occur in the ancient vents? If yes, the implications are exciting: A high concentration of ribose would have been present in the same deep-sea vent environment where the evolution of life may have taken place.

The production of nucleotides has been a challenge for the theory of chemical evolution, but research on this issue continues. In the meantime, let's consider the next question: Once nucleotides formed, how would they polymerize to form nucleic acids? This question has a definitive answer.

How Do Nucleotides Polymerize to Form Nucleic Acids?

As Figure 4.2 shows, nucleotides polymerize via condensation reactions between the hydroxyl on the sugar component of one nucleotide and the phosphate group of another nucleotide. The reaction forms a new covalent bond between the nucleotides, and a molecule of water is released. The bridge formed by the phosphate group is called a **phosphodiester linkage**—also frequently referred to as a phosophodiester bond.

When phosphodiester linkages join ribonucleotides together, the polymer that is produced is RNA. Phosphodiester linkages between deoxyribonucleotides produce DNA.

DNA and RNA Strands Are Directional Figure 4.3 on page 98 shows how the chain of linked sugars and phosphates in a nucleic acid acts as a backbone, analogous to the peptide-bonded backbone found in proteins.

Like the peptide-bonded backbone of a polypeptide, the sugar–phosphate backbone of a nucleic acid is directional. In a strand of RNA or DNA, one end has an unlinked 5′ phosphate while the other end has an unlinked 3′ hydroxyl—meaning the groups are not bonded to another nucleotide.

The order of the different nucleotides forms the primary structure of the nucleic acid. When biologists write the primary structure of a stretch of DNA or RNA, they use shorthand and simply list the sequence of bases by their single-letter

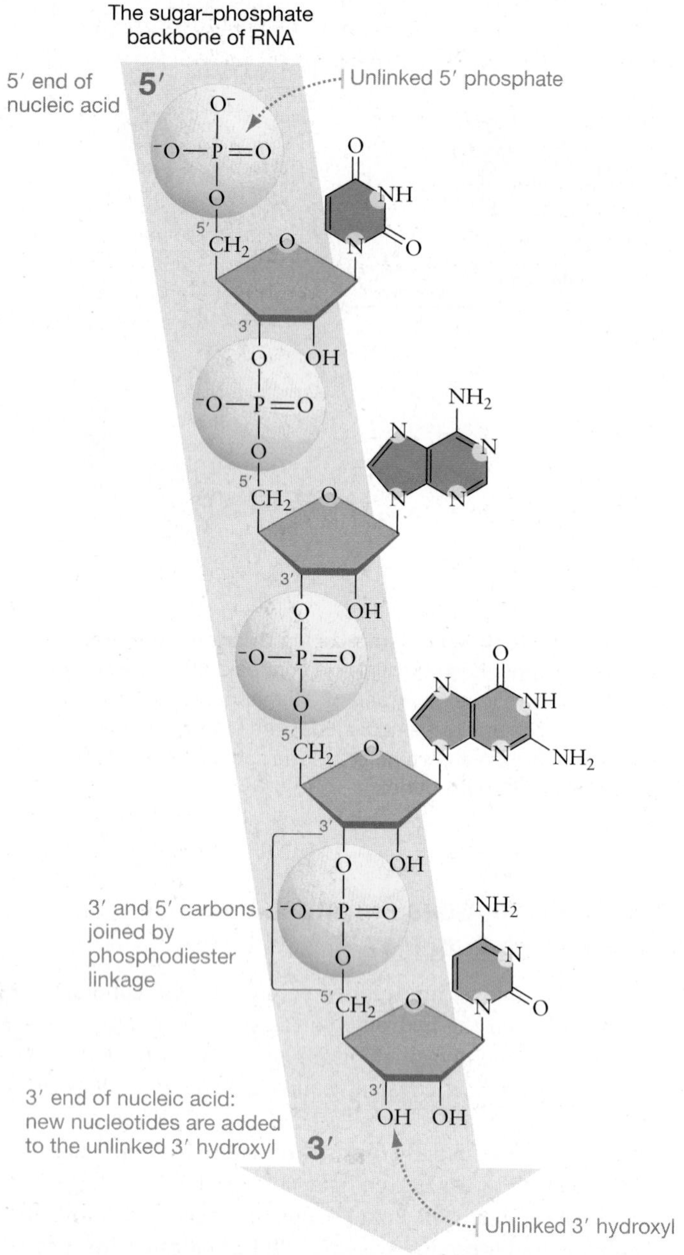

Figure 4.3 Nucleic Acids Have a Sugar–Phosphate Backbone.

✔ Identify the four bases in this RNA strand, using Figure 4.1c as a key. Then write down the base sequence, starting at the 5′ end.

abbreviations. For example, a DNA sequence consisting of six nucleotides might be ATTAGC. It would take roughly 6 billion of these letters to write the primary structure of the DNA in most of your cells.

By convention, the sequence of bases found in an RNA or DNA strand is always written in the 5′ → 3′ direction. This system is logical because in cells, RNA and DNA are always synthesized in this direction. Nucleotides are added only at the 3′ end of a growing nucleic acid molecule.

Polymerization Requires an Energy Source As with other polymerization reactions, the joining of nucleotides into nucleic acids dramatically decreases entropy and is thus not spontaneous. An input of energy is needed to tip the energy balance in favor of polymerization.

Nucleic acid polymerization can take place in cells (assisted by enzymes; Ch. 3, Section 3.4) because the potential energy of the nucleotide monomers is first raised by reactions that add two additional phosphate groups to the 5′ phosphates of ribonucleoside or deoxyribonucleoside monophosphates, creating nucleoside triphosphates.[1] In the context of nucleic acid polymerization, researchers refer to the nucleoside triphosphates as "activated nucleotides." Figure 4.4a shows an example of an activated ribonucleotide; this molecule is called **adenosine triphosphate**, or **ATP**. The equivalent nucleotide used for DNA synthesis would be deoxyadenosine triphosphate (dATP).

How does adding phosphate groups raise the potential energy of a nucleotide? Recall that phosphates are negatively charged and that like charges repel (Ch. 2, Section 2.5). Linking two or more phosphates together generates covalent bonds that carry a large amount of potential energy due to strong repulsive forces. The energy is released when the phosphates form new, more stable bonds with other atoms (Figure 4.4b).

When activated nucleotides polymerize, the energy released from the condensation reaction compensates for the decrease in entropy, making the reaction spontaneous. You will see in later chapters that potential energy stored in ATP is also used to drive other cellular activities, independent of nucleotide polymerization. (Ch. 8, Section 8.2 explains how this happens in more detail.)

CHECK YOUR UNDERSTANDING

✔ **If you understood this section, you should be able to ...**

1. **MODEL** Draw a simplified diagram of two nucleotides connected by a phosphodiester linkage. Indicate the 5′ → 3′ polarity, and mark where the next nucleotide would be added to the growing chain.
2. Identify the changes that you would need to make to Figure 4.3 if the strand were DNA instead of RNA.
3. Describe how nucleotides are activated for incorporation into a polymer. Why is this activation required?

Answers are available in Appendix A.

4.2 DNA Structure and Function

After you complete this section, you should be able to ...

- Analyze the different levels of DNA structure and how they are related to DNA function.

The primary structure of DNA is somewhat similar to the primary structure of proteins. Proteins have a peptide-bonded backbone with a series of R-groups that

[1] A note about nomenclature: A molecule consisting of just a sugar and one of the bases in Figure 4.1c is called a nucleoside. A molecule consisting of a sugar and a base plus one or more phosphate groups is called a nucleotide. By convention, the phosphorylation state of a nucleotide is conveyed by using nucleoside, then appending mono-, di-, or tri- to the word "phosphate." For example, a sugar attached to a base and one phosphate group is called a nucleoside monophosphate; with three attached phosphates, you have a nucleoside triphosphate.

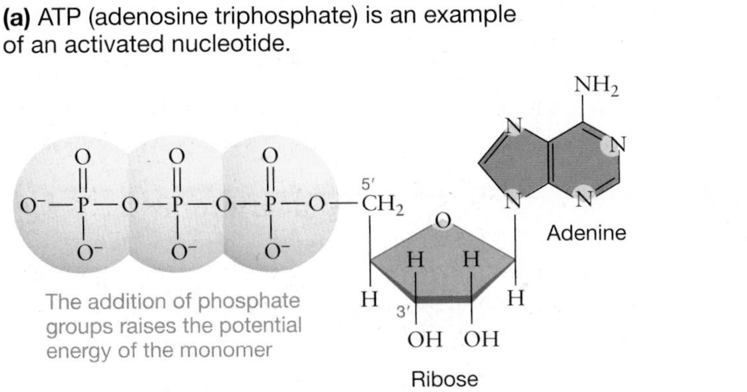

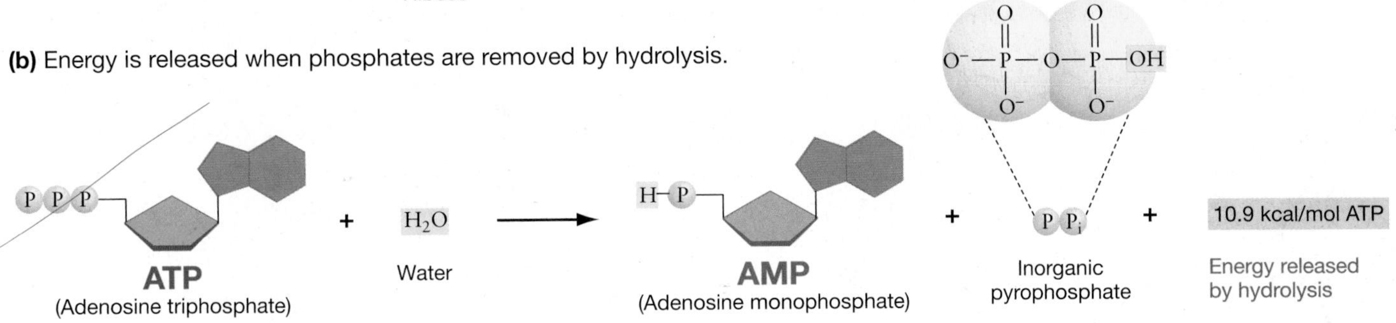

INTERACTIVE Figure 4.4 Activated Monomers Drive Polymerization Reactions. The potential energy in activated nucleotides, such as ATP, is primarily stored in the bonds between the phosphates. When ATP reacts with water, one of the bonds between two phosphates is replaced with a lower potential energy bond, resulting in the release of energy, and either a monophosphate (P_i) or a diphosphate (PP_i), usually called, for historical reasons, pyrophosphate. (The subscript "i" signifies that the groups are inorganic; i.e., not bonded to a carbon.) A similar release of potential energy occurs when activated nucleotides are used as substrates for polymerization of nucleic acids.

extend from it. DNA molecules have a sugar–phosphate backbone, created by phosphodiester linkages, and a sequence of any of four nitrogenous bases that extend from it.

Like proteins, DNA also has secondary structure. But while the α-helices and β-pleated sheets of proteins are formed by hydrogen bonding between groups in the backbone, the secondary structure of DNA is formed in a very different way. Let's look at details of this structure and how it relates to DNA's function as an information-carrying molecule.

What Is the Nature of DNA's Secondary Structure?

The discovery of DNA's secondary structure, announced in 1953, ranks among the great scientific breakthroughs of the twentieth century. James Watson and Francis Crick presented their celebrated model for the secondary structure of DNA in a single page that was published in the scientific journal *Nature*.

Early Data Provided Clues Watson and Crick's model was a hypothesis based on a series of results from other laboratories. They were trying to propose a secondary structure that could explain several important observations about the DNA found in cells:

- Chemists had worked out the structure of nucleotides and knew that DNA polymerized through the formation of phosphodiester linkages. Thus, Watson and Crick knew that the molecule had a sugar–phosphate backbone.
- By analyzing the nitrogenous bases in DNA samples from different organisms, Erwin Chargaff had established two empirical rules: **(1)** The number of purines in a given DNA molecule is equal to the number of pyrimidines, and **(2)** the DNA molecule has an equal number of T's and A's, and it has an equal number of C's and G's.
- By bombarding DNA with X-rays and analyzing how it scattered the radiation, Rosalind Franklin and Maurice Wilkins had calculated the distances between groups of atoms in the molecule. The technique they used is called **X-ray crystallography** (see **BioSkills 6** for an introduction to this technique). The scattering patterns showed that three distances were repeated many times: 0.34 nanometer (nm), 2.0 nm, and 3.4 nm. Because the measurements repeated, the researchers inferred that DNA molecules had a regular and repeating structure. The pattern of X-ray scattering suggested that the molecule was helical, or spiral, in nature.

Based on this work, understanding DNA's structure boiled down to understanding the nature of the helix involved. What type of helix would have a sugar–phosphate backbone and explain both Chargaff's rules and the Franklin–Wilkins measurements?

DNA Strands Form an Antiparallel Double Helix Watson and Crick began by analyzing the size and geometry of the three nucleotide components: deoxyribose, phosphate, and base. The bond angles and measurements suggested that the distance of 2.0 nm represented the width of the helix and that 0.34 nm was likely to be the distance between bases stacked in a spiral.

How could they make sense of Chargaff's rules and the 3.4-nm distance, which appeared to be exactly 10 times the distance between a single pair of bases?

To solve this problem, Watson and Crick constructed a series of physical models that allowed them to tinker with different types of helical configurations. After many false starts, certain things started to click:

- They arranged two strands of DNA side by side with the sugar–phosphate backbones on the outside and the bases on the inside. If the bases extending from each backbone are to fit

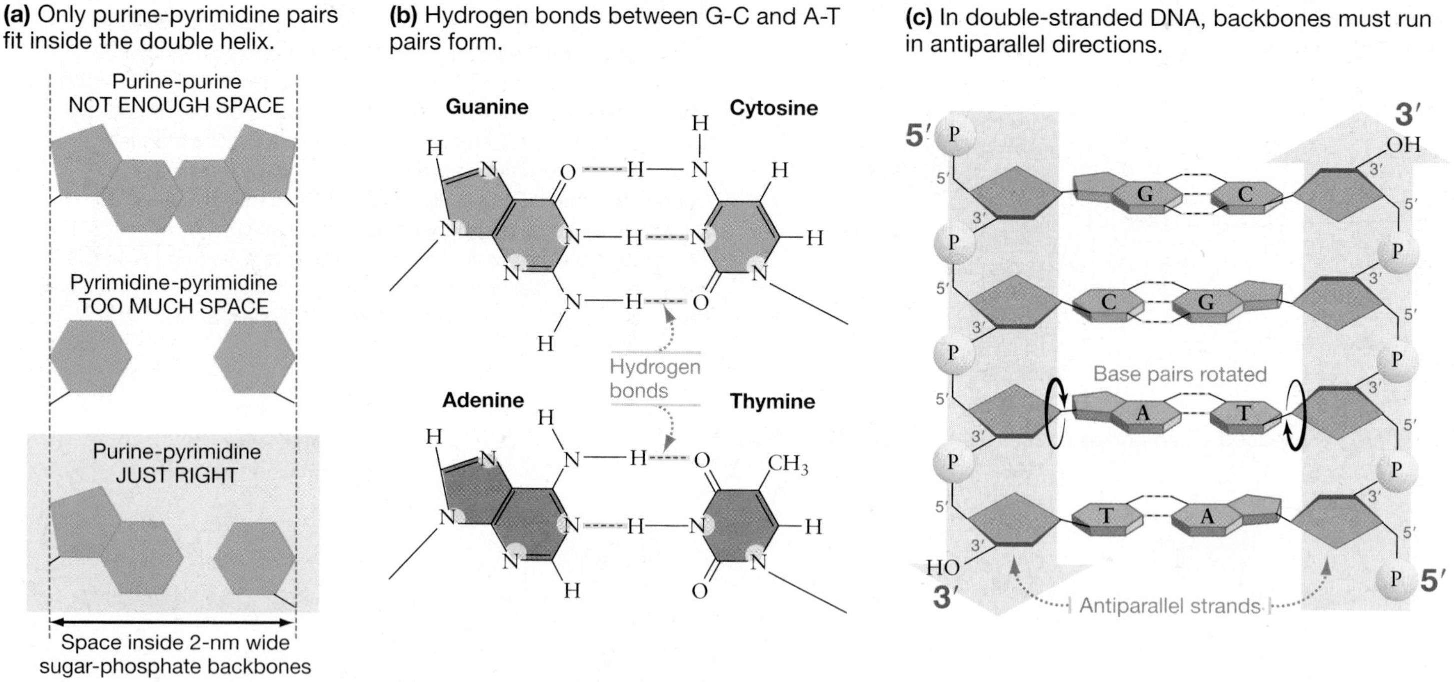

Figure 4.5 Complementary Base Pairing Is Based on Hydrogen Bonding.

within the interior of a 2.0-nm-wide structure, then they have to form purine-pyrimidine pairs (see Figure 4.5a).

- Purine-pyrimidine pairing allows hydrogen bonds to form only between certain bases, said to be complementary. Adenine will form two hydrogen bonds with thymine, and guanine will form three hydrogen bonds with cytosine (see Figure 4.5b). The third hydrogen bond in G-C pairs makes them slightly stronger than A-T pairs.
- The patterns of hydrogen bonding shown in Figure 4.5b could form only if the bases on opposite strands were flipped 180 degrees relative to one another. For this to happen, the two parallel strands of DNA must be oriented in opposite directions—meaning that one strand runs in the $5' \rightarrow 3'$ direction while the other strand runs $3' \rightarrow 5'$ (Figure 4.5c). Strands with this orientation are said to be **antiparallel**.
- After these parameters were in place, the antiparallel strands were predicted to be twisted together to form a **double helix**.

By creating this model, Watson and Crick had discovered **complementary base pairing** between the A-T and G-C bases. In fact, the term **Watson–Crick pairing** is now used interchangeably with the phrase "complementary base pairing." This discovery explains the purine-pyrimidine ratios that Chargaff observed.

As Figure 4.5c shows, DNA is put together like a ladder. The antiparallel sugar–phosphate backbones form the ladder side rails. The bases attached to the sugars are rotated and pair up via hydrogen bonding to form the ladder rungs.

Although each base has polar groups involved in the hydrogen bonds, the bases' carbon-nitrogen rings are mostly nonpolar. This is a key point, because in aqueous solution (the environment inside a cell) hydrophobic interactions cause double-stranded DNA to twist into a helix to minimize contact between the hydrophobic bases and surrounding water molecules (Figure 4.6a). The physical restraints posed by these interactions result in a full helical turn every 10 bases—the 3.4-nm distance observed by Franklin and Wilkins (Figure 4.6b).

The two strands are further stabilized by **base stacking**, which results from van der Waals interactions between the tightly packed adjacent bases. The rotated orientation of the interior base pairs allows the rings of adjacent bases to stack on top of one another like coins (see Figure 4.6b). This nonpolar interior is sandwiched between the negatively charged phosphate groups of the outward-facing backbone, which make the double helix hydrophilic overall and thus soluble in aqueous solutions.

Additional features of DNA's secondary structure are highlighted in Figure 4.6b. It's important to note that the outside of the helical DNA molecule forms two types of grooves. The wider of the two is known as the major groove, and the narrower one is known as the minor groove. This groove asymmetry is vital for granting access to proteins that bind to particular base sequences in DNA (see Ch. 19, Section 19.3, to learn more about the roles of these proteins).

As Figure 4.6 shows, the secondary structure of DNA can be illustrated at different levels of detail. In Making Models 4.2, you will learn how to draw highly simplified versions of DNA that can be used to model key concepts of its secondary structure.

Since Watson and Crick's model of the double helix was published, experimental tests have shown that the hypothesis is correct in almost every detail.

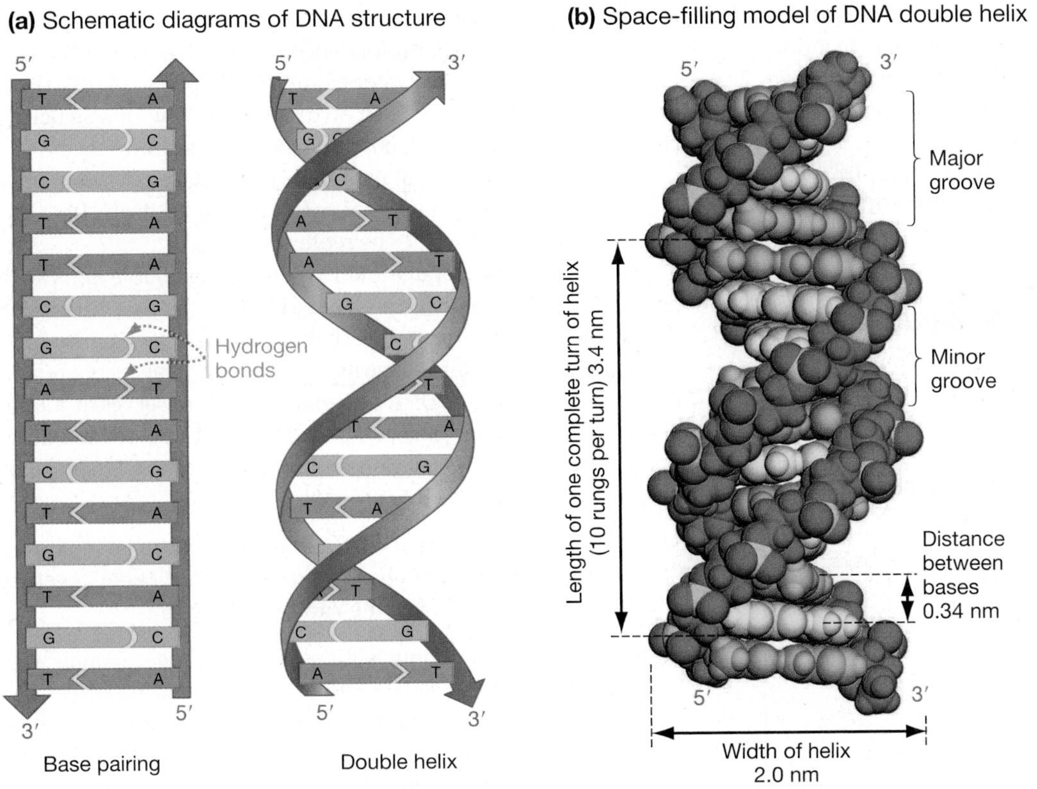

Figure 4.6 The Secondary Structure of DNA Is a Double Helix. (a) The schematic diagrams illustrate complementary base pairing (left) and how strands are twisted into a double helix (right). (Yellow bands represent hydrogen bonding.) **(b)** The space-filling model shows tight packing of the bases inside the double helix. The double-helix structure explains the measurements inferred from X-ray analysis of DNA molecules.

To summarize:

- DNA's secondary structure consists of two antiparallel strands twisted into a double helix.
- The double helix is shaped and stabilized by hydrogen bonding between the complementary base pairs, hydrophobic interactions, and van der Waals interactions.

So far, the focus of this chapter has been on DNA's secondary structure. Does the DNA double helix also form tertiary structures?

Making Models 4.2 Tips on Drawing Nucleic Acids

When you're drawing models of DNA, the molecular details of the structure are not necessary. The sugar–phosphate backbones can be simplified to single lines with arrowheads to identify the 3′ ends. Base pairing is drawn as short lines between the backbones and, if the sequence is part of the model, nucleotides are represented by letters.

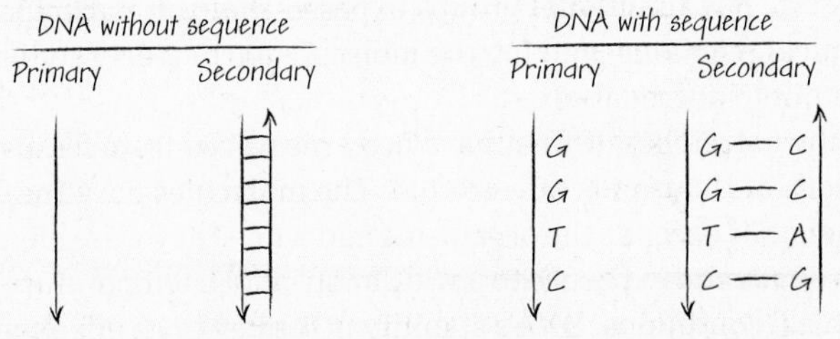

MODEL Draw a double-stranded DNA molecule with the sequence A-G-C-T. Label the 5′ and 3′ ends, one of the sugar–phosphate backbones, and the hydrogen bonds involved in base pairing.

To see this model in action, go to the Study Area of **Mastering Biology**

The Tertiary Structure of DNA

Recall that the secondary structure of a protein often leads to a more compact tertiary structure when the polypeptide folds on itself (Ch. 3, Section 3.2). It turns out that the DNA in cells is also normally found in more compact three-dimensional structures. The need for this compaction is evident, when you consider that the total length of DNA in each of your cells is roughly six feet long.

Compared to proteins, tertiary structure in DNA is less dependent on primary structure, and so it is far less variable between different sequences. Two forms of DNA tertiary structure are commonly found in cells. When DNA becomes wound too tightly or loosely with respect to the number of base pairs per helical turn, it can twist on itself to form compact, three-dimensional structures called supercoils (see supercoiled circular DNA in Ch. 7, Figure 7.2). In addition, DNA in the cells of eukaryotes and certain archaea will form tertiary structures by wrapping around specialized DNA-binding proteins called histones. The resulting DNA–protein complexes compact the DNA into discrete, movable units during cell division (i.e., condensed chromosomes; Ch. 12, Section 12.1), and they also contribute to DNA's ability to store and transmit information (Ch. 19, Section 19.2).

Now let's take a closer look at how the structure of DNA is involved in storing biological information.

DNA Functions as an Information-Containing Molecule

Watson and Crick's model created a sensation for a simple reason: It revealed the role of DNA as a biological reservoir of information. In literature, information consists of letters on a page. In music, information is composed of the notes on a staff. But inside cells, information consists of a sequence of nucleotides in a nucleic acid. The four nitrogenous bases function like letters of the alphabet. A sequence of bases is like the sequence of letters in a word—it has meaning.

In all organisms that have been examined to date, from tiny bacteria to gigantic redwood trees, DNA stores the information

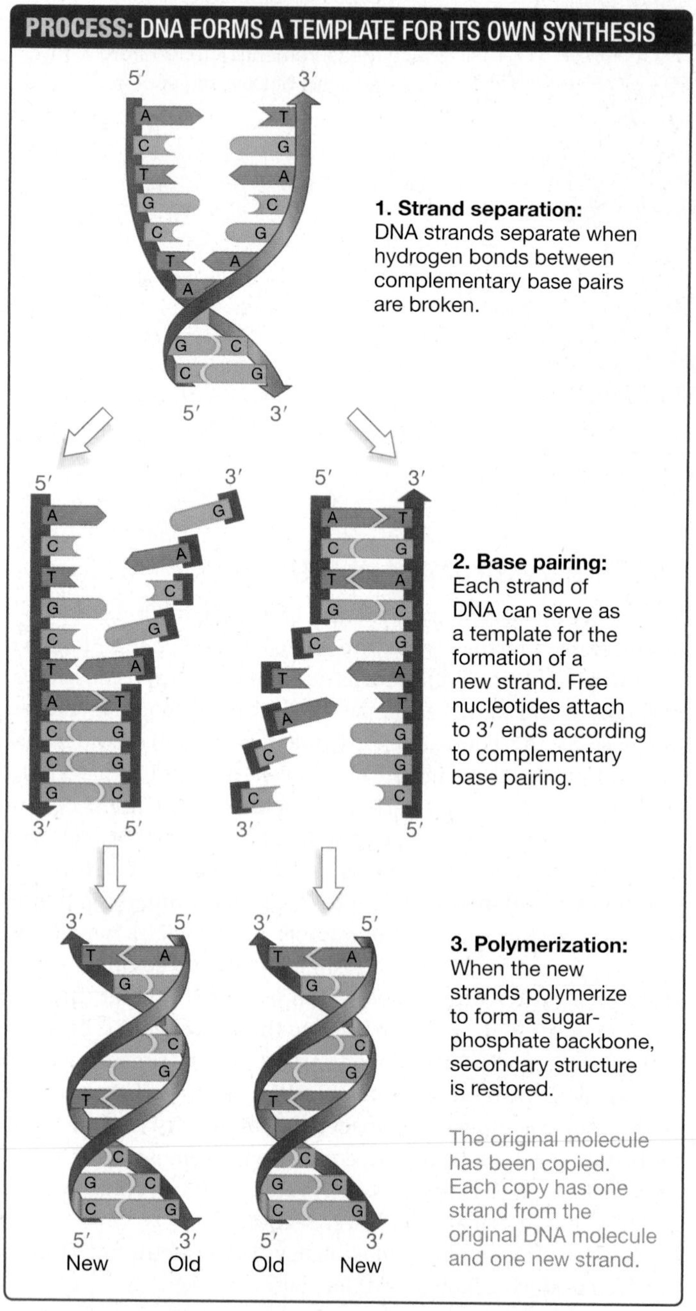

Figure 4.7 **Making a Copy of DNA.** If new bases are added to each of the two strands of DNA via complementary base pairing, a copy of the DNA molecule can be produced.

✔ When double-stranded DNA is heated to 95°C, the bonds between complementary base pairs break and single-stranded DNA results. Considering this observation, is the reaction shown in step 1 spontaneous?

required for the organism's growth and reproduction. Exploring how hereditary information is encoded and translated into action is the heart of several later chapters (Chapters 16–19). Here, however, our focus is on DNA structure and how this structure relates to its function and, possibly, how life began.

The theory of chemical evolution holds that life began once a molecule emerged that could make a copy of itself. Does the structure of DNA allow it to be replicated?

Watson and Crick ended their paper on the double helix with one of the classic understatements in the scientific literature: "It has not escaped our notice that the specific pairing we have postulated immediately suggests a possible copying mechanism." Here's the key insight: DNA's primary structure serves as a template for the synthesis of a complementary strand, meaning that DNA contains the information required for a copy of itself to be made. Figure 4.7 shows this process.

Step 1 The two strands of a DNA double helix can be separated by breaking the hydrogen bonds that hold them together using either heat or enzyme-catalyzed reactions.

Step 2 Free deoxyribonucleotides form hydrogen bonds with complementary bases on the original strand of DNA—also called a **template strand**. As they do, their sugar–phosphate groups form phosphodiester linkages to create a new strand—also called a **complementary strand**. Note that the 5′ → 3′ directionality of the complementary strand is the opposite to that of the template strand.

Step 3 Complementary base pairing allows each strand of a DNA double helix to be copied exactly, producing two identical daughter molecules.

Making copies of DNA represents one of the five characteristics of life (introduced in Ch. 1, Section 1.1): replication. But can DNA catalyze the reactions needed to *self*-replicate? In today's cells and in laboratory experiments, the answer is no. Instead, the molecule is copied through a complicated series of reactions that are catalyzed by enzymes (Ch. 15, Section 15.3). Why can't DNA catalyze these reactions itself?

The DNA Double Helix Is a Stable Structure

The DNA double helix is highly structured. It is regular, symmetric, and held together by phosphodiester linkages, hydrogen bonding, and hydrophobic interactions. In addition, the double helix has few functional groups exposed that can participate in chemical reactions, making the molecule particularly stable and resistant to degradation.

Intact stretches of DNA have been recovered from fossils that are tens of thousands of years old. The molecules have the same sequence of bases as the organisms had when they were alive, despite death and exposure to a wide array of pH, temperature, and chemical conditions. DNA's stability is the key to its effectiveness as a reliable information-storage molecule. DNA's structure is consistent with its function in cells.

The orderliness and stability that make DNA such a dependable information repository also make it inept at catalysis. Recall that enzymes function by forming a structure that will specifically bind to a substrate and catalyze a reaction

(Ch. 3, Section 3.4). A wide variety of catalytic activities can be generated in protein enzymes thanks to variation in the reactivity among R-groups in amino acids and to the enormous diversity of shapes found in proteins. In comparison, the structure of DNA is simple and nonreactive. It's not surprising, then, that DNA has never been observed to catalyze any reaction in any organism.

In short, DNA furnishes an extraordinarily stable template for storing information encoded in a sequence of bases. But owing to DNA's inability to act as an effective catalyst and therefore to self-replicate, there is virtually no support for the hypothesis that the first life-form consisted of DNA alone. Instead, most biologists who are working on the origin of life support the hypothesis that life began with RNA.

CHECK YOUR UNDERSTANDING

✔ If you understood this section, you should be able to ...

1. Differentiate between primary, secondary, and tertiary levels of DNA structure.
2. **MODEL** Use Figure 4.5b to evaluate the possibility of hydrogen bonding between the nucleotides G-T and A-C. Explain why these pairs of bases are not complementary.
3. Explain how the structure of DNA allows it to be copied.

Answers are available in Appendix A.

4.3 RNA Structure and Function

The first living molecule would have needed to perform two key functions: carry information and catalyze reactions that promoted its own replication. At first glance, these two functions appear to conflict. Information storage requires regularity and stability; catalysis requires variation in chemical composition and flexibility in shape. How is it possible for a molecule to do both? The answer lies in structure.

After you complete this section, you should be able to ...

- Analyze the different levels of RNA structure and how they are related to RNA function.

Structurally, RNA Differs from DNA

How does the structure of RNA differ from DNA? To answer this question, let's take a look at the primary, secondary, and tertiary structures of RNA.

Primary Structure Like DNA, RNA has a primary structure consisting of four types of nitrogenous bases extending from a sugar–phosphate backbone. But it's important to recall two significant differences between these nucleic acids:

1. The sugar in the sugar–phosphate backbone of RNA is ribose, not deoxyribose as in DNA.
2. The pyrimidine base thymine does not exist in RNA. Instead, RNA contains the closely related pyrimidine base uracil.

The first point is critical. Look back at Figure 4.1b and compare the functional groups attached to ribose and deoxyribose. Notice the hydroxyl (—OH) group on the 2′ carbon of ribose. This additional hydroxyl is much more reactive than the hydrogen atom on the 2′ carbon of deoxyribose. When RNA molecules fold in certain ways, the hydroxyl group can attack the phosphate linkage between nucleotides, which would generate a break in the sugar–phosphate backbone. This —OH group makes RNA much less stable than DNA, but as you will see later, it can also support other catalytic activities.

Secondary Structure Like DNA, most RNA molecules have secondary structure that results from complementary base pairing between purine and pyrimidine bases. In RNA, adenine forms two hydrogen bonds with uracil, and guanine again forms three hydrogen bonds with cytosine. (Other, non-Watson–Crick base pairs can occur, although less frequently.)

This hydrogen bonding should seem familiar, since DNA bonds in a similar manner—so how do the secondary structures of RNA and DNA differ? In the vast majority of cases, the purine and pyrimidine bases in RNA undergo hydrogen bonding with complementary bases on the *same strand*, rather than forming hydrogen bonds with complementary bases on a different strand, as in DNA.

Figure 4.8 shows how within-strand base pairing works. The key is that when bases on one part of an RNA strand fold over and align with bases on another part of the same strand, the two sugar–phosphate strands are antiparallel. In this orientation, hydrogen bonding between complementary bases results in a helical structure that resembles the double helix of DNA—but unlike DNA, the RNA structure forms from a single nucleic acid strand.

If the section where the fold occurs includes unpaired bases, then the stem-and-loop configuration shown in Figure 4.8

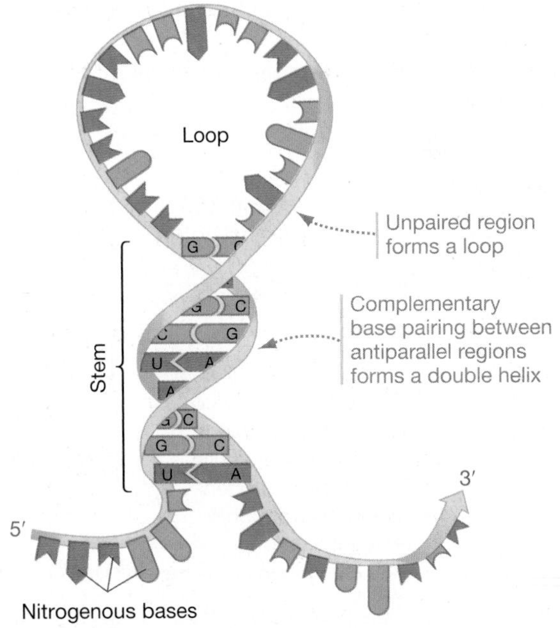

Figure 4.8 Complementary Base Pairing Directs Secondary Structure in RNA. This RNA molecule has secondary structures in the form of a double helical "stem" and unpaired "loop." Note that in secondary structures, the bases participating in hydrogen bonding are antiparallel.

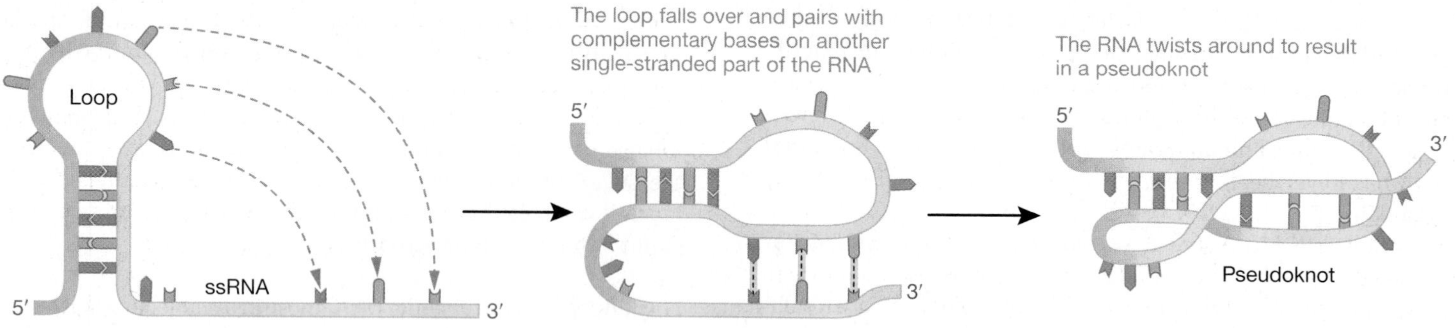

Figure 4.9 Complementary Base Pairing Directs Tertiary Structure in RNA. Base pairing between different regions of an RNA molecule causes it to fold into a more complex tertiary structure. Note that in tertiary structures, as in secondary structures, the bases participating in hydrogen bonding are antiparallel.

results. Several other types of RNA secondary structures are possible, each involving a different length and arrangement of base-paired segments.

Like the α-helices and β-pleated sheets observed in many proteins, RNA secondary structures will form spontaneously. The bases are brought together by hydrophobic interactions and stabilized by hydrogen bonding and base stacking interactions.

Tertiary Structure RNA molecules can also have tertiary structure, which arises when secondary structures fold into more complex shapes. The pseudoknot structure in **Figure 4.9** is an example of how three-dimensional shapes can be formed by base pairing between distant regions of folded RNA molecules. As a result, RNA molecules with different base sequences can have very different overall shapes and chemical properties. RNA molecules are much more diverse in size, shape, and reactivity than DNA molecules.

Table 4.1 summarizes the similarities and differences in the structures of RNA and DNA.

RNA's Versatility

In cells, RNA molecules are highly versatile, like a pocket tool with an array of functions. In terms of structure, you've seen that RNA is a nucleic acid like DNA, but RNA folds into complex three-dimensional shapes much like proteins.

The structural flexibility of RNA molecules allows them to perform many different tasks. The central dogma introduced RNA as an intermediate between DNA and protein (see Ch. 1, Section 1.4). This intermediate, called messenger RNA, transmits information needed to synthesize polypeptides. Further research has brought new insights into the diversity of roles that RNAs play in cells. RNA molecules also help regulate the production of messenger RNA from DNA, process and edit information stored in these messages, and even catalyze the synthesis of proteins, among other things. (Functions of RNA are explored in more depth in Chapters 17, 18, and 19.)

For this chapter, let's focus on the roles that RNA may have played in the origin of life—as a catalyst and as an information-containing entity.

SUMMARY **Table 4.1 DNA and RNA Structure**

Level of Structure	DNA		RNA	
Primary	Sequence of deoxyribonucleotides; bases are A, T, G, C		Sequence of ribonucleotides; bases are A, U, G, C	
Secondary	Two antiparallel strands twist into a double helix, stabilized by hydrogen bonding, hydrophobic interactions, and van der Waals interactions		Most commonly, a single strand folds back on itself to form a double-helical "stem" and an unpaired "loop"	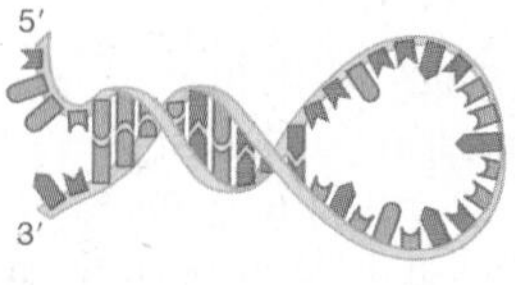
Tertiary	Double helical DNA forms compact structures by wrapping around histone proteins (shown) or twisting into supercoils (not shown)	 DNA (red) + histones (green)	Secondary structures fold to form a wide variety of distinctive three-dimensional shapes Example: pseudoknot	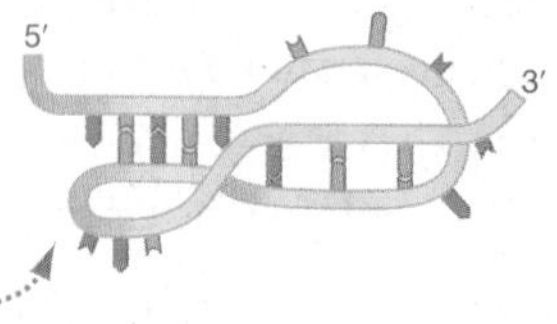

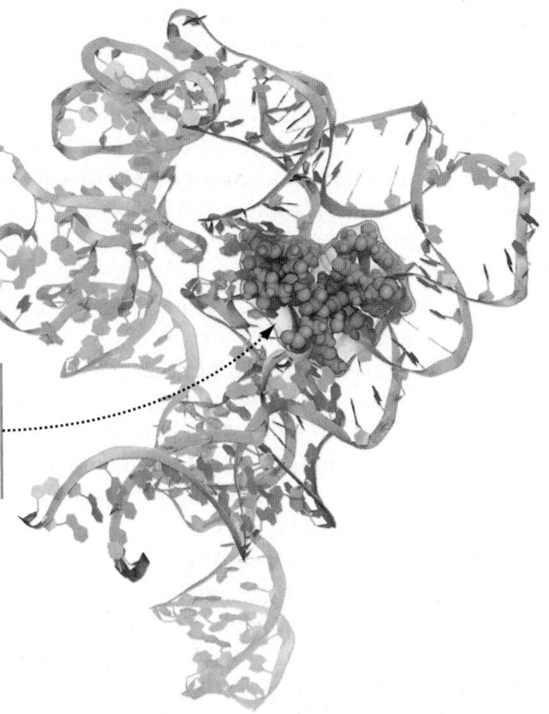

Figure 4.10 Tertiary Structure of the *Tetrahymena* Ribozyme. The folded structure brings together bases from distant locations in the primary structure to form the active site (the highlighted residues shown in spacefill) where catalysis occurs.

RNA Can Function as a Catalytic Molecule

In terms of diversity in shape and chemical reactivity, the four types of nucleotides in RNA molecules are no match for the 20 different amino acid residues in proteins. Nevertheless, because RNA has a degree of structural and chemical complexity, it's capable of forming structures that catalyze a number of chemical reactions. Sidney Altman and Thomas Cech shared the 1989 Nobel Prize in chemistry for showing that organisms have catalytic RNAs. These RNAs are called **ribozymes**, or RNA enzymes, because they catalyze reactions similar to protein enzymes.

Figure 4.10 shows the structure of a ribozyme Cech isolated from a single-celled eukaryote called *Tetrahymena*. This ribozyme catalyzes both the hydrolysis and the condensation of phosphodiester linkages in RNA. Researchers have since discovered a variety of ribozymes that catalyze several important reactions in cells. For example, ribozymes are responsible for the catalytic activity of the ribosomes that polymerize amino acids to form polypeptides. Ribozymes are at work in your cells right now.

The three-dimensional nature of ribozymes is vital to their catalytic activity. To catalyze a chemical reaction, substrates must be brought together in an environment that will promote the reaction. As with protein enzymes, the region of the ribozyme that is responsible for this activity is called the active site. When the *Tetrahymena* ribozyme was compared to enzymes that catalyze similar reactions, their active sites were found to be similar in structure. This observation about two very different molecules demonstrates the critical relationship between structure and function.

The discovery of ribozymes was a watershed event in origin-of-life research. Before Altman and Cech published their results, most biologists thought that the only molecules capable of catalyzing reactions in cells were proteins. The fact that a ribozyme could catalyze the formation of a phosphodiester bond raised the possibility that an RNA molecule could polymerize a copy of itself. Such a molecule could qualify as the first living entity. Is there any experimental evidence to support this hypothesis?

CHECK YOUR UNDERSTANDING

✔ **If you understood this section, you should be able to ...**

1. Contrast the effect of different sequences on the secondary and tertiary levels of structure in RNA versus DNA.
2. Explain what is primarily responsible for the functional versatility observed in RNA.

Answers are available in Appendix A.

4.4 In Search of the First Life-Form

After you complete this section, you should be able to ...

- Evaluate the hypothesis that life began as an RNA molecule.

The theory of chemical evolution maintains that life began as a naked self-replicator—a molecule that existed by itself in solution, without being enclosed in a membrane. To make a copy of itself, that first living molecule had to **(1)** provide a template that could be copied, and **(2)** catalyze polymerization reactions that would link monomers into a copy of that template. Because RNA is capable of both processes, most origin-of-life researchers propose that the first life-form was an RNA.

RNA contains a sequence of bases analogous to the letters in a sentence, so it can function as an information-containing molecule. Like DNA, the information stored in RNA can also be used to make copies of itself via complementary base pairing. Figure 4.11 on page 106 illustrates how the information stored in an RNA molecule could have been used to direct its own replication on early Earth. This process still occurs today in some viruses that store their hereditary information in RNA (Ch. 33, Section 33.2).

Step 1 To replicate a single-stranded RNA, first a complementary copy of the RNA is made. Using the original strand as a template, free ribonucleotides form hydrogen bonds with complementary bases on the template.

Step 2 A new strand is polymerized when 3′ hydroxyls and 5′ phosphates on adjacent nucleotides are linked together via condensation reactions. The product is a double-stranded RNA molecule.

Step 3 To arrive at a duplicate of the original single-stranded RNA in step 1 (the template), the hydrogen bonds between the double-stranded product in step 2 must first be broken by heating or by a catalyzed reaction. The newly made complementary RNA molecule now exists independently of the original template strand.

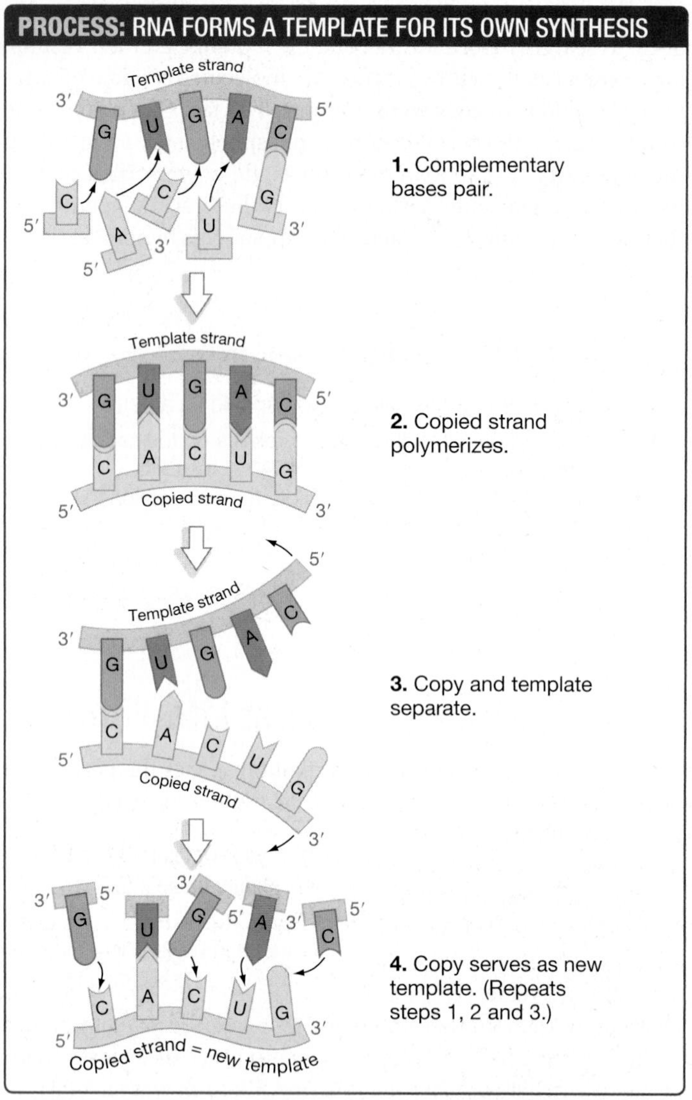

Figure 4.11 **RNA Molecules Contain Information That Allows Them to Be Replicated.** For a single-stranded RNA to be copied, it must make a complementary strand and pass through double-stranded RNA intermediates.

If steps 1–3 were repeated with the new strand serving as a template (step 4 in Figure 4.11), the resulting molecule would be an exact copy of the original strand. This series of steps can repeat to make more copies of the template and complementary strands.

How Biologists Study the RNA World

Researchers test the RNA world hypothesis by establishing an environment in the laboratory that selects for ribozymes that catalyze key steps required for an RNA world. To understand how this is done, consider two experiments performed by researchers in David Bartel's laboratory.

In one study, the team attempted to generate an RNA molecule that could catalyze the kind of template-directed polymerization needed for RNA replication—an RNA "replicase." The experimental protocol was designed to mimic the process of natural selection (introduced in Ch. 1, Section 1.3), but targeting the characteristics of molecules instead of organisms. The team succeeded in isolating a ribozyme that could add 14 nucleotides to an existing RNA strand.

The results from this research created considerable excitement among biologists interested in the origin of life. However, an RNA replicase capable of self-replication was still far from being discovered.

In another study, Bartel's group asked a different question: Would it be possible to select for a ribozyme that could make RNA nucleotides? This type of ribozyme has not been observed in cells, but it would have been a key player in an RNA world.

Recall that the direction of a chemical reaction and how much product it makes is influenced by the amount of reactants present (Ch. 2, Section 2.3). Nucleotides may have been a scarce resource in early Earth conditions, so ribozymes that catalyze the production of RNA nucleotides would be more likely to be copied due to local accumulation of monomers.

Starting with a large pool of randomly generated RNA sequences, the researchers selected for RNAs that could catalyze the addition of a uracil base (the base unique to RNA) to a ribose sugar. By again mimicking the process of natural selection, they purified a ribozyme that could perform this task 1 million times more efficient than the uncatalyzed reaction. In effect, molecular evolution had occurred in the reaction tubes.

Thanks to similar efforts at other laboratories around the world, biologists have produced an increasingly impressive set of catalytic activities from RNA molecules. The results from each of these studies help clarify our view of what might have occurred in an RNA world. If a living ribozyme ever existed, then each round of simulated molecular evolution brings us closer to resurrecting it.

An RNA World May Have Sparked the Evolution of Life

Most of the discovered ribozymes that continue to exist in modern cells play key roles in the synthesis of proteins. If these ribozymes were removed from cells, proteins could no longer be made. This relationship suggests an order of events in the evolution of life—an RNA world preceded proteins.

The evolution of protein enzymes would have marked the end of an RNA world—providing the means for catalyzing reactions necessary for life to emerge in a cellular form. After this milestone, three of the five fundamental characteristics of life (Ch. 1, Section 1.1) would have been solidly in place:

1. ***Information*** Proteins and ribozymes would be processing information stored in nucleic acids for the synthesis of more proteins.
2. ***Replication*** Enzymes, and possibly ribozymes, would be replicating the nucleic acids that stored the hereditary information.
3. ***Evolution*** Random changes in the nucleic acids would lead to the synthesis of different proteins and ribozymes. Selective advantages resulting from some of these changes would allow for the evolution of new functions.

If these events occurred in a hydrothermal vent, the molecular assemblages of nucleic acids and proteins would have been constantly fed with thermal and chemical energy. To gain independence from their undersea hatchery, enzymes would have evolved to store energy as something more portable—carbohydrates. The structure and function of carbohydrates will be the focus of the next chapter.

CHECK YOUR UNDERSTANDING

✔ **If you understood this section, you should be able to . . .**

1. Describe the key attributes of RNA that make it a candidate for the first living molecule.
2. Provide two examples of catalytic activities that would have been necessary for ribozymes to replicate in an RNA world.

Answers are available in Appendix A.

CHAPTER 4 Review

For media, go to Mastering Biology

4.1 What Is a Nucleic Acid?

- Nucleic acids are polymers of nucleotide monomers, which consist of a sugar, a phosphate group, and a nitrogenous base. Ribonucleotide monomers polymerize to form RNA. Deoxyribonucleotide monomers polymerize to form DNA.
- Ribonucleotides have a hydroxyl (—OH) group on their 2′ carbon; deoxyribonucleotides have a hydrogen (—H) instead.
- Nucleic acids polymerize when condensation reactions join nucleotides together via phosphodiester linkages.
- Nucleic acids are directional: they have a 5′ end and a 3′ end. During polymerization, new nucleotides are added only to the 3′ end.

4.2 DNA Structure and Function

- DNA's primary structure consists of a sequence of linked deoxyribonucleotides.
- The secondary structure of DNA consists of two DNA strands running in opposite directions that are twisted into a double helix.
- The tertiary structure of DNA forms compact structures by twisting the double helix into supercoils or wrapping around histone proteins.
- DNA is an extremely stable molecule that serves as a superb archive for information in the form of base sequences. Its secondary structure is stabilized by hydrogen bonds and base-stacking interactions between bases inside of the helix.
- DNA is readily copied via complementary base pairing. Complementary base pairing in DNA occurs between A-T and G-C bases.

4.3 RNA Structure and Function

- Like DNA, RNA's primary structure consists of a sequence of linked ribonucleotides.
- Compared to DNA, in RNA, thymine (T) is replaced by uracil (U) so that complementary base pairing occurs between A-U and C-G bases.
- RNA's secondary structure includes a variety of configurations including short regions of complementary base pairing within a single strand to form double-helical stems and unpaired loops.
- The secondary structures in a strand of RNA can further fold into more complex shapes via complementary base pairing to give the molecule tertiary structure.
- RNA is versatile and can function as an information-carrying molecule and a catalyst.

4.4 In Search of the First Life-Form

- To test the RNA world hypothesis, researchers have attempted to synthesize new ribozymes in the laboratory. Using artificial selection strategies, they have succeeded in identifying RNAs that catalyze several different reactions.
- One origin-of-life theory holds that ribozymes that catalyzed reactions necessary for the production of ribonucleotides may have preceded the evolution of RNA replicases.

Answers are available in Appendix A.

✔ TEST YOUR KNOWLEDGE

1. What are the four nitrogenous bases found in RNA?
   a. cytosine, guanine, thymine, uracil (C, G, T, U)
   b. adenine, cytosine, guanine, thymine (A, C, G, T)
   c. adenine, cytosine, guanine, uracil (A, C, G, U)
   d. alanine, cysteine, glycine, threonine (A, C, G, T)
2. What determines the primary structure of a DNA molecule?
   a. stem-and-loop configuration
   b. complementary base pairing
   c. deoxyribonucleotide sequence
   d. hydrophobic interactions and hydrogen bonding
3. Evaluate the following statements related to the synthesis of nucleic acids. Select True or False for each statement.
   T/F Ribonucleotides are added to the 3′ end of a DNA strand.
   T/F Polymerization of nucleic acids occurs by the formation of phosphodiester bonds.
   T/F Complementary pairing between sugars is required for copying nucleic acids.
   T/F Strands in a double helix are synthesized in an antiparallel orientation.
4. Single strands of nucleic acids are directional, meaning that there are two different ends. What functional groups define the two different ends of a strand?

✔ TEST YOUR UNDERSTANDING

5. What is responsible for the increased stability of DNA compared to RNA?
6. **QUANTITATIVE** If nucleotides from the DNA of a human were quantified and 30 percent of them contained the base adenine, what percentage of them would contain the base guanine?
   a. 20 percent **b.** 30 percent **c.** 40 percent **d.** 70 percent

7. What would be the sequence of the strand of DNA that is made from the following template: 5′-GATATCGAT-3′? (Your answer must be written 5′ → 3′.) How would the sequence be different if RNA were made from this DNA template?

8. According to the RNA world model, a ribozyme would replicate by creating a double-stranded RNA intermediate. Would you expect the intermediate to have the same catalytic activity as the original ribozyme? Justify your answer with an explanation.

✔ TEST YOUR PROBLEM-SOLVING SKILLS

9. Make a concept map (see **BioSkills 12**) that relates DNA's primary structure to its secondary structure. Your diagram should include deoxyribonucleotides, base-stacking interactions, purines, pyrimidines, phosphodiester linkages, DNA's primary structure, DNA's secondary structure, complementary base pairing, and antiparallel double helix.

10. **MODEL** In the field of nanotechnology, DNA is used like Velcro to assemble tiny particles into structures that are < 0.0001 mm in size. Draw a model to illustrate how two particles (a circle and a square) could be brought together by linking them to short single-stranded DNA molecules. If the DNA sequence linked to the circle is GGATC, then provide the sequence linked to the square and identify the 5′ and 3′ ends of each strand.

✔ PUT IT ALL TOGETHER: Case Study

Who deserves credit for discovering the structure of the double helix?

The famous model featured in the photo above was built by Watson and Crick in 1953 to demonstrate the secondary structure of DNA. This was not the first attempt at modeling DNA's structure, however. Before 1953, there were several failed attempts. The accumulation of data from different research groups was key to arriving at the correct model. What were the incorrect versions of DNA structure, and how did Watson and Crick get it right?

11. Phoebus Levene was the first to describe the structure of nucleotides and how they were bonded together with phosphodiester linkages. In 1919, he incorrectly proposed the tetranucleotide hypothesis, which stated that nucleic acids were polymers consisting of GCTA repeated over and over. If his model had been correct, then how would it affect the information that could be stored in the DNA?

12. In the 1950s, the race to solve the secondary structure of DNA became intense. In an uncharacteristic rush to publish, Linus Pauling erroneously proposed a triple-stranded structure in February 1953. This model had the nitrogenous bases on the exterior and the sugar–phosphate backbones clustered in the middle. How does the orientation of the sugar–phosphate backbone in this model compare with the one proposed by Watson and Crick? Do you think Pauling's structure could exist in cells? Why or why not?

13. Rosalind Franklin was the first person to obtain X-ray crystallographic data on the form of DNA that is most commonly found in cells. Other researchers, including Pauling, used data from DNA samples that were more concentrated (contained less water) than Franklin's samples. Why would you expect the amount of water to affect the helical structure of DNA?

14. **QUANTITATIVE** In 1951, Erwin Chargaff was accumulating data on the molar ratios of nucleotides using DNA obtained from a variety of sources. Some of these data are provided in the following table:

Molar Ratios in DNA Preparations

Source	A:G	T:C	A:T	G:C	purines:pyrimidines
Ox	1.29	1.43	1.04	1.00	1.10
Human	1.56	1.75	1.00	1.00	1.00
Chicken	1.45	1.29	1.06	0.91	0.99
Salmon	1.43	1.43	1.02	1.02	1.02
Wheat	1.22	1.18	1.00	0.97	0.99
Yeast	1.67	1.92	1.03	1.20	1.00

SOURCE: E. Chargaff. 1951. *Federal Proceedings* 10: 654–659.

Compare the molar ratios presented from each organism tested and between different organisms. Explain how these data could be used to show that Levene's tetranucleotide model is incorrect. What do they imply about the primary structure of DNA in different organisms?

15. Watson and Crick met with Chargaff to discuss his work in 1952. Explain how Chargaff's observations helped Watson and Crick to propose the complementary base pairing in their model of double-stranded DNA. Would you expect similar ratios of nucleotides if Chargaff had used RNA instead? Explain why or why not.

16. **PROCESS OF SCIENCE** Now that you have learned a little more of the history behind the elucidation of the secondary structure of DNA, what does it tell you about the role of the scientific community in advancing science? Were Watson and Crick solely responsible?

Mastering Biology

Students Go to Mastering™ Biology for assignments, the eText, and the Study Area with animations, practice tests, and activities.

Professors Go to Mastering™ Biology for automatically graded tutorials and questions that you can assign to your students, plus Instructor Resources.

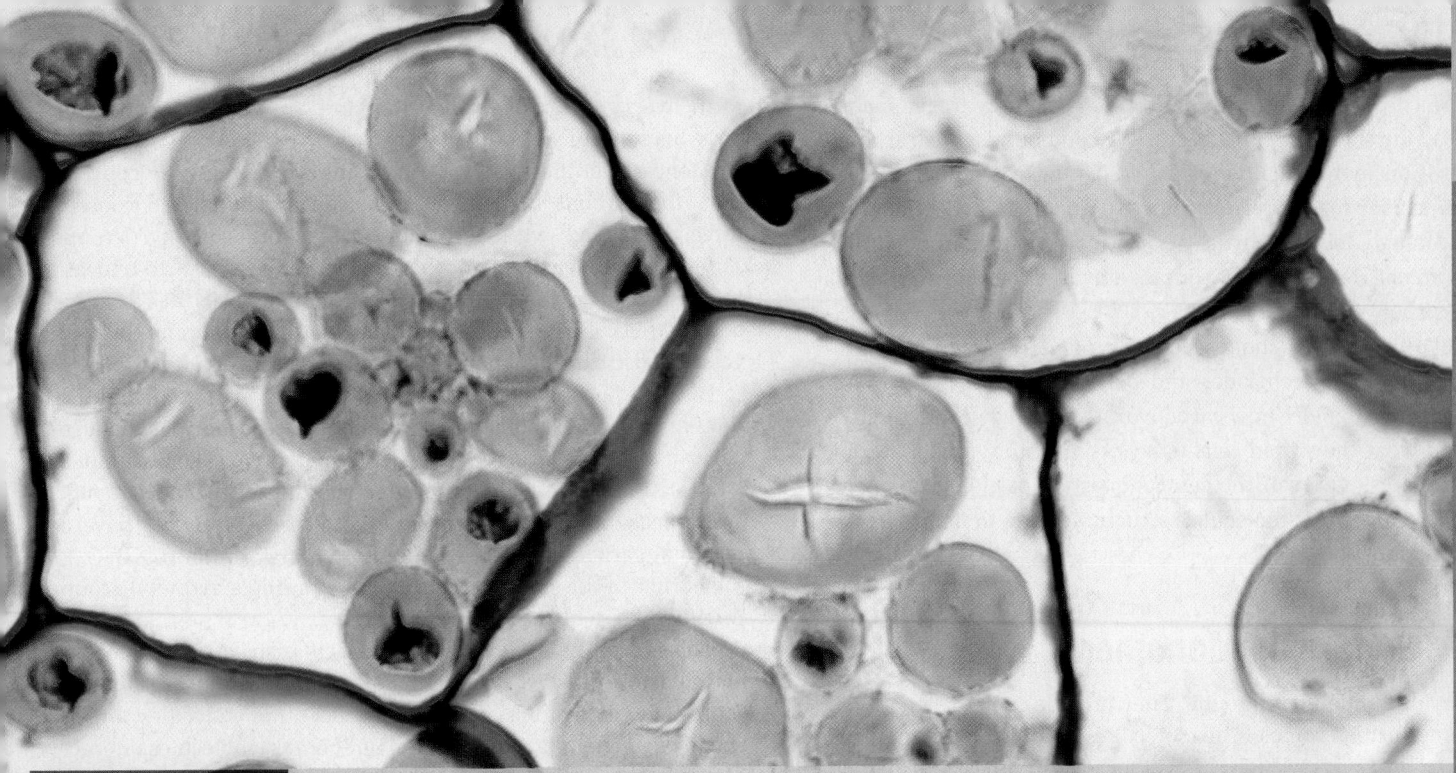

5 An Introduction to Carbohydrates

A cross section through a potato tuber. Starch-filled structures are stained red; cellulose-rich cell walls are stained green. Starch is an energy-storage carbohydrate; cellulose is a structural carbohydrate.

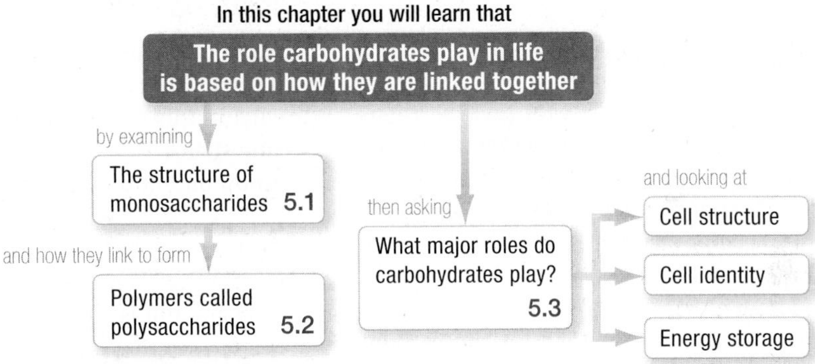

This unit highlights the four types of macromolecules that were key to the evolution of the cell: proteins, nucleic acids, carbohydrates, and lipids. Understanding the structure and function of macromolecules is a basic requirement for exploring how life began and how organisms work. Recall that nucleic acids could satisfy three of the five fundamental characteristics of life: information, replication, and evolution (Ch. 1, Section 1.1). Carbohydrates, the subject of this chapter, play an important role in a fourth characteristic—energy.

The term **carbohydrate**, or **sugar**, encompasses the monomers called **monosaccharides** (literally, "one-sugar"), small polymers called **oligosaccharides** ("few-sugars"), and the large polymers called **polysaccharides** ("many-sugars"). The name "carbohydrate" is logical because the molecular formula of many of these molecules is $(CH_2O)_n$, where the n indicates the number of "carbon-hydrate" groups (the word "hydrate" refers to water). The value of n can vary from 3, for the smallest sugar, to well over a thousand for some of the large polymers.

The name "carbohydrate" is a little misleading, though, because carbohydrates do not consist of carbon atoms bonded to water molecules. Instead, they are made up of a carbonyl group (C═O), several hydroxyl

BIG PICTURE

This chapter is part of the Big Picture. See how on pages 144–145.

groups (—OH), along with multiple carbon–hydrogen bonds (C—H). Furthermore, not all CH_2O compounds are carbohydrates. Consider formaldehyde, a carbon-based compound with the molecular formula CH_2O (introduced in Ch. 2, Section 2.5). Even though formaldehyde fits the pattern of carbohydrates (with an *n* of 1), it does not contain a hydroxyl group and so it is not a carbohydrate.

Let's begin with monosaccharides and then put them together into oligosaccharides and polysaccharides. From there, you will then explore how carbohydrates figured in the origin of life and what they do in cells today. As you study this material, be sure to ask yourself the central question of biological chemistry: How does this molecule's structure relate to its properties and function?

5.1 Sugars as Monomers

Sugars are fundamental to life. They provide chemical energy in cells and furnish some of the molecular building blocks required for the synthesis of larger, more complex compounds. Monosaccharides were important during chemical evolution, early in Earth's history, too. For example, as you've seen, the sugar called ribose is required for the formation of the nucleotides that make up nucleic acids (Ch. 4, Section 4.1).

After you complete this section, you should be able to ...

- Describe the structural variations that exist among monosaccharides.

What Distinguishes One Monosaccharide from Another?

Monosaccharides, or simple sugars, are the monomers of carbohydrates. **Figure 5.1** illustrates two of the smallest monosaccharides. Although these two sugars share the same molecular formula ($C_3H_6O_3$), their molecular structures are different. The carbonyl group that serves as one of monosaccharides' distinguishing features can be found either at the end of the molecule, forming an aldehyde sugar (an aldose), or within the carbon chain, forming a ketone sugar (a ketose).

The presence of a carbonyl group along with multiple polar hydroxyl groups means that even the simplest sugars have many reactive and hydrophilic functional groups. Based on this observation, it's not surprising that sugars are polar molecules that form hydrogen bonds with water and are easily dissolved in aqueous solutions.

Monosaccharides also vary in their carbon count. By convention, the carbons in a monosaccharide are numbered consecutively, starting with the end nearest the carbonyl group. Three-carbon sugars, such as those in Figure 5.1, are called **trioses**. Ribose, which acts as a building block for nucleotides, has five carbons and is called a **pentose**; the glucose that's coursing through your bloodstream right now is a six-carbon sugar, or a **hexose**.

Besides varying in the location of the carbonyl group and the total number of carbon atoms present, monosaccharides can also vary in the spatial arrangement of their atoms. For example, **Figure 5.2** illustrates the different structures of two hexose sugars: glucose and galactose. While both of these sugars have a terminal carbonyl and the same molecular formula ($C_6H_{12}O_6$), they differ in the spatial arrangement of a single hydroxyl group (highlighted in blue in Figure 5.2).

This is a key point: Because the molecular structures of glucose and galactose differ, their functions differ. In cells, glucose is both a source of carbon atoms, used to construct other molecules, and of chemical energy that sustains life. But because molecules interact in precise ways based on their shape, galactose must first be converted to glucose via an enzyme-catalyzed reaction to be used in these same ways. This example underscores a general theme: Even seemingly simple changes in structure—like the spatial orientation of a single hydroxyl group—will have functional consequences.

Can the Same Monosaccharide Exist in More Than One Form?

Although Figure 5.2 shows monosaccharides as linear chains, it's actually rare for sugars consisting of five or more carbons to exist in this form. In aqueous solution, they spontaneously form ring structures when the carbonyl group reacts with a hydroxyl group on another carbon. An example of this process is shown in **Figure 5.3**. When glucose (a six-carbon aldose) forms a ring, the C-1 carbon (the first carbon in the linear chain) forms a bond with the oxygen atom of the C-5 hydroxyl. In this reaction, a hydrogen atom is removed from the C-5 hydroxyl and a hydrogen is added to the C-1 carbonyl to generate a new hydroxyl group.[1]

[1] The hydrogen atom transferred to the C-1 glucose carbonyl does not necessarily originate directly from the C-5 hydroxyl. The new C-1 hydroxyl may be formed from a proton (H^+) present in the surrounding aqueous solution, balancing the loss of the hydrogen released into solution from the C-5 hydroxyl.

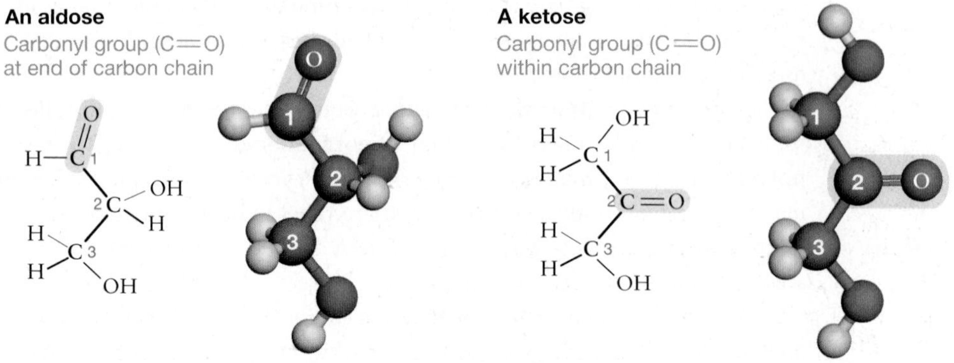

Figure 5.1 The Carbonyl Group in a Sugar Occurs in One of Two Configurations.

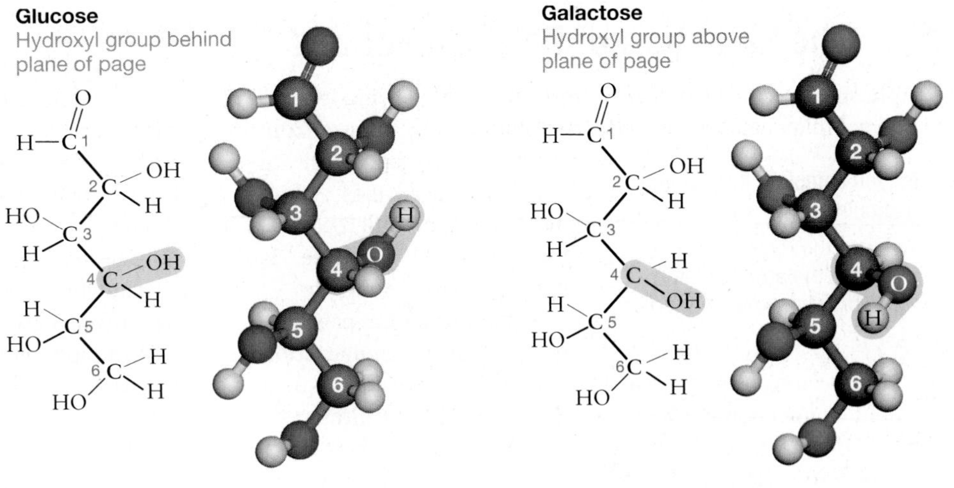

Figure 5.2 The Hydroxyl Groups in Sugars May Vary in Their Configuration. The two six-carbon sugars shown here vary only in the spatial orientation of their hydroxyl groups on carbon number 4 (highlighted in blue).

✔ **MODEL** Mannose is a six-carbon sugar that is similar to glucose, except for the orientation of the hydroxyl (—OH) group on the second carbon. Draw the structural formula of mannose, and circle carbon number 2 and the hydroxyl group.

This balanced exchange preserves the number of atoms and hydroxyls between the ring and linear forms.

When sugars form a ring structure, the position of the newly formed hydroxyl group (e.g., at C-1 in glucose) will be fixed in one of two possible orientations: below or above the plane of the ring. The arrangement of the other hydroxyl groups remains the same, so there are two possible forms of glucose: α-glucose and β-glucose. The two forms exist in equilibrium, but β-glucose is more common because it is slightly more stable than α-glucose. The significance of these two possible forms becomes apparent when they are linked together, which is the subject of the next section.

To summarize, many distinct monosaccharides exist because so many aspects of their structure are variable: aldose or ketose placement of the carbonyl group, the number of carbons, and the different arrangements of hydroxyl groups in space. Ring forms of the same molecule also have alternative shapes. These variations give each monosaccharide a unique structure and function.

How do these varied monomers join together to form polymers? Is the process similar to how amino acids link together to form proteins and how nucleotides join together to form nucleic acids? Let's next explore how the functional groups in monosaccharides influence the polymerization of carbohydrates.

CHECK YOUR UNDERSTANDING

✔ **If you understood this section, you should be able to ...**

MODEL Draw the structural formula of a three-carbon sugar ($C_3H_6O_3$) in linear form and then draw three different sugars that illustrate three ways monosaccharides differ from one another.

Answers are available in Appendix A.

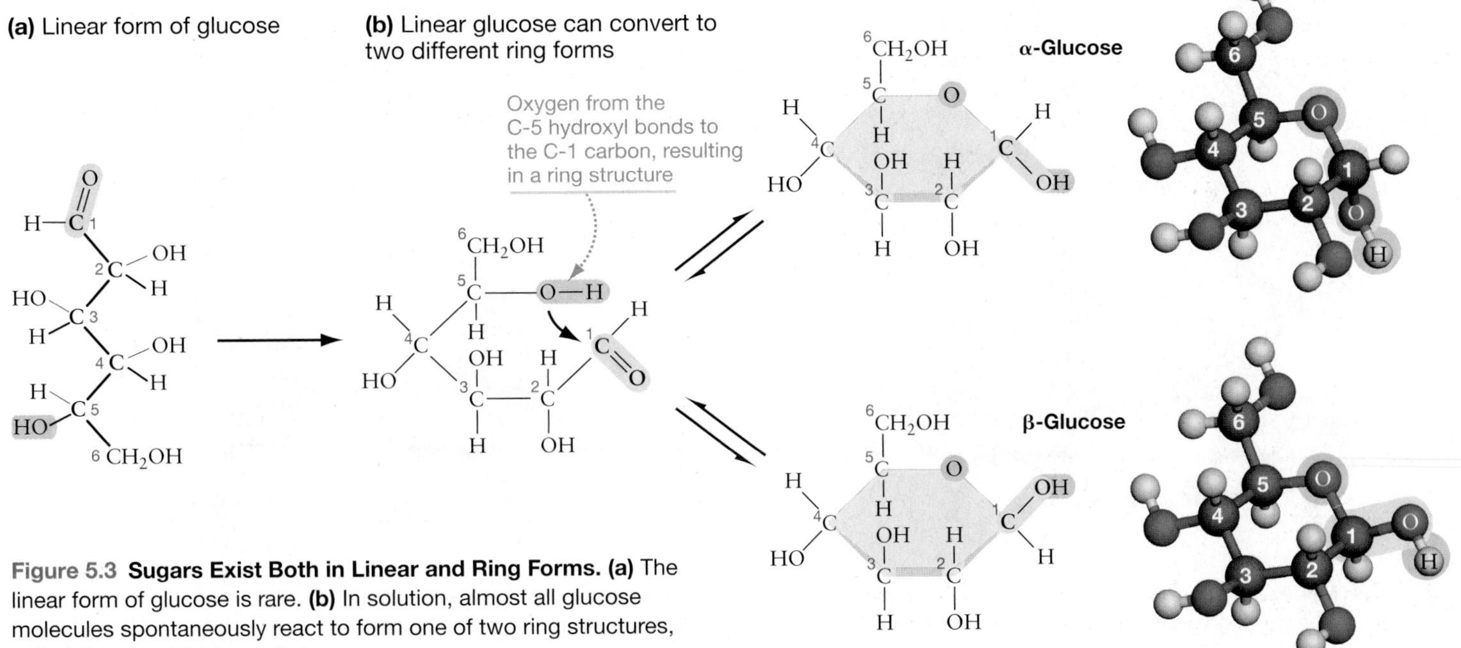

Figure 5.3 Sugars Exist Both in Linear and Ring Forms. (a) The linear form of glucose is rare. **(b)** In solution, almost all glucose molecules spontaneously react to form one of two ring structures, called the α and β forms of glucose.

5.2 The Structure of Polysaccharides

Simple sugars covalently link to form chains of varying lengths called complex carbohydrates. These chains range in size from short oligosaccharides to long polysaccharides. When just two sugars link together, the resulting molecule is known as a **disaccharide**.

> **After you complete this section, you should be able to ...**
>
> - Analyze the relationship between the structure and function of polysaccharides.

Monosaccharides polymerize when a condensation reaction occurs between two hydroxyl groups, resulting in a covalent connection (—O—) called a **glycosidic linkage**, or glycosidic bond. The inverse reaction, hydrolysis, cleaves these linkages. (To review condensation and hydrolysis reactions, see Ch. 2, Section 2.5.)

Glycosidic linkages connect monomers together, just like peptide bonds and phosphodiester linkages do for proteins and nucleic acids. There is an important difference, however. Peptide bonds and phosphodiester linkages form between the same locations in their monomers, giving proteins and nucleic acids a standard backbone structure, but this is not the case for carbohydrates. Because glycosidic linkages form between hydroxyl groups, and because every monosaccharide contains at least two hydroxyls, the location and geometry of glycosidic linkages can vary widely among different oligosaccharides and polysaccharides.

The structure and function of larger carbohydrates depend not only on how the monomers are linked together but also on the type of monomers involved. In this way, carbohydrates are similar to proteins and nucleic acids.

The same principles that dictate the overall structure and function in polysaccharides hold true for small oligosaccharides as well. For example, maltose, or malt sugar, is a disaccharide made up of two glucose molecules and is abundant in the starter liquid used to brew beer (**Figure 5.4a**). Lactose, an important sugar in milk, is a disaccharide of glucose and galactose (**Figure 5.4b**).

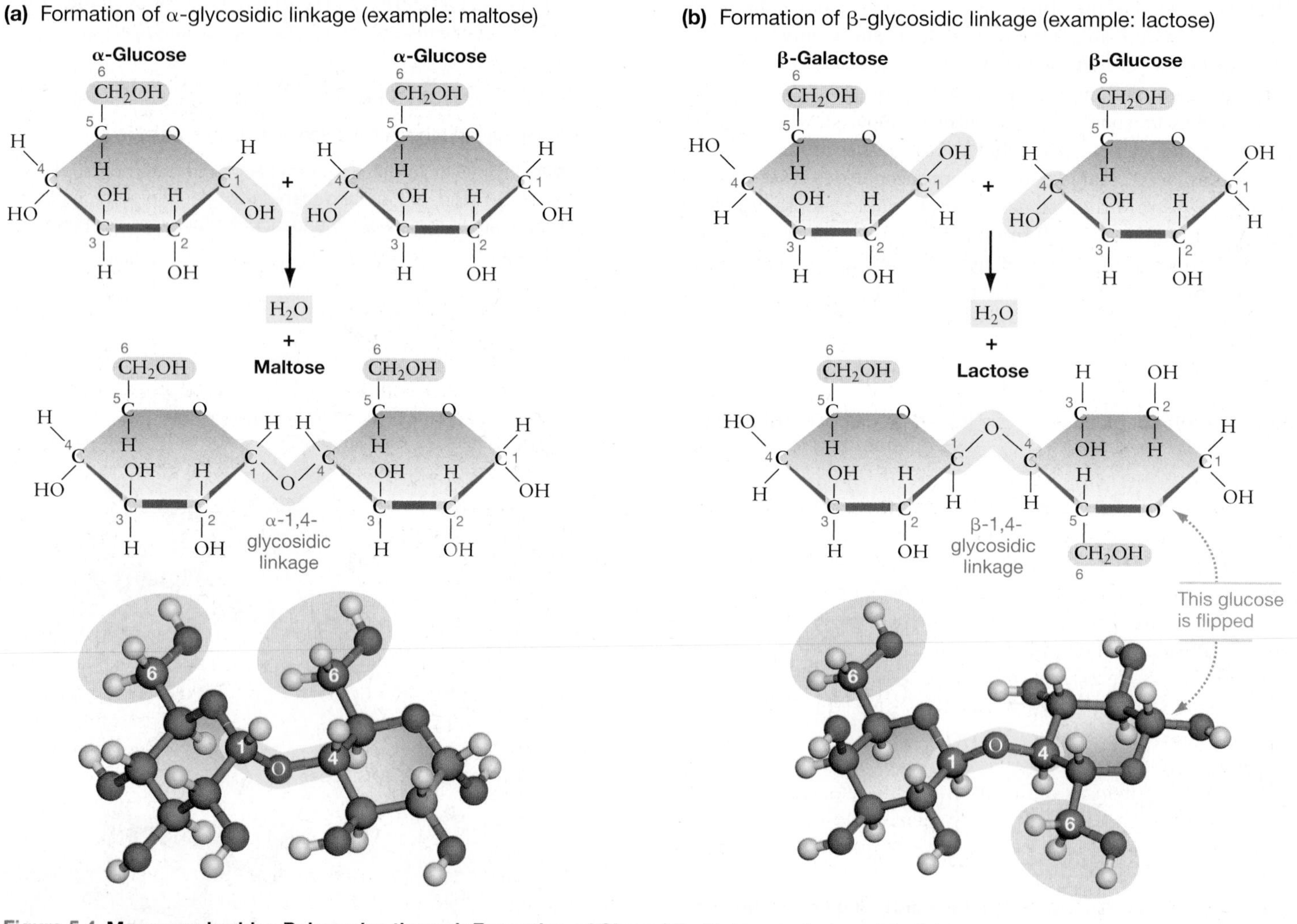

Figure 5.4 Monosaccharides Polymerize through Formation of Glycosidic Linkages. A glycosidic linkage occurs when hydroxyl groups on two monosaccharides undergo a condensation reaction. Maltose and lactose are disaccharides.

Making Models 5.1 Tips on Drawing Carbohydrates

Drawing simple models is the best way to understand the structures of monosaccharides and glycosidic linkages. In these models, focus on the overall shape of each monomer and how the monomers' carbons are numbered. You can keep the drawings simple by showing only the hydroxyl groups on the carbons being linked together, as in these examples based on α-glucose:

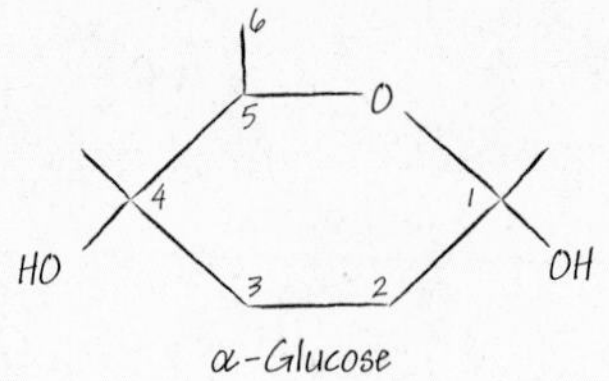

α-Glucose

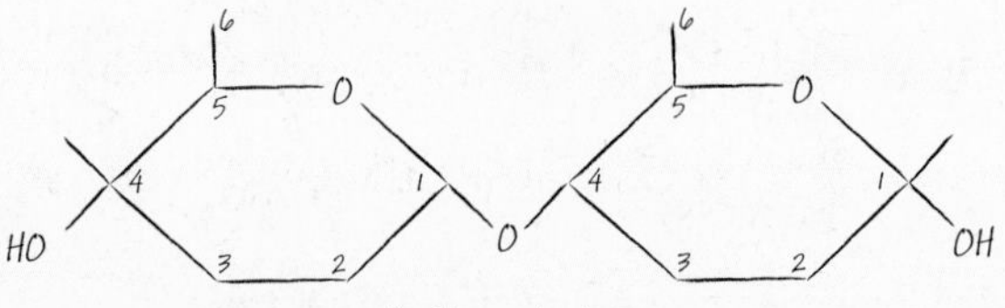

α-1,4-Glycosidic linkage

MODEL Use the examples above and Figure 5.4b to draw simplified models of a β-glucose monosaccharide and a β-glucose disaccharide with a β-1,4-glycosidic linkage.

To see this model in action, go to the Study Area of **Mastering Biology**

(a) Starch—used for energy storage in plant cells

(b) Chitin—used for structural support in insect exoskeletons

Figure 5.5 Polysaccharides Are Used in Many Different Ways. The variation in carbohydrate monomers and the glycosidic linkages that bond them together result in very different structures, including **(a)** the starch in potatoes and **(b)** the exoskeleton of a cicada.

Maltose and lactose illustrate two of the most common glycosidic linkages, called the α-1,4-glycosidic linkage and the β-1,4-glycosidic linkage. The numbers refer to the carbons on either side of the linkage, indicating that the linkages are between the C-1 and C-4 carbons. Their geometry, however, is different: α and β refer to the contrasting orientations of the C-1 hydroxyls—on opposite sides of the plane of the glucose rings (i.e., "below" versus "above" the plane). See **Making Models 5.1** to learn how simple drawings may be used to make sense of the names given to glycosidic linkages.

A functional consequence of the structural differences between maltose and lactose is that the enzymes used to hydrolyze maltose will not cleave lactose. Instead, lactose is digested by lactase—an enzyme that many humans stop secreting after childhood. Without lactase, adults may become lactose intolerant and suffer intestinal discomfort if they consume dairy products. This is just one example of how the orientation of glycosidic linkages affects the structure, function, and durability of the carbohydrates.

The variation in how polysaccharides are formed allows organisms to use them in radically different ways. For example, polysaccharides may be used to store chemical energy in the cells of plants or provide structural support in the exoskeletons of insects (**Figure 5.5**).

Let's now consider the structures of the most common polysaccharides found in organisms today: starch, glycogen, cellulose, and chitin, along with a modified polysaccharide called peptidoglycan. As you will shortly see, each of these macromolecules is joined by particular α-1,4 or β-1,4-glycosidic linkages. These polysaccharides can consist of a few hundred to many thousands of residues.

As you learn more about each of these five carbohydrate polymers, pay special attention to how the different glycosidic linkages affect the structure of the molecule and its interaction with other molecules. The differences in their functions result directly from differences in their structures, which generally comes down to a simple twist of a link.

Starch: A Storage Polysaccharide in Plants

In plant cells, some monosaccharides are polymerized and stored for later use in the form of starch. **Starch** consists entirely of α-glucose joined by glycosidic linkages. Most of these linkages are between C-1 and C-4 carbons, and the angle of these bonds causes the chain of glucose residues to coil into a helix.

SUMMARY TABLE 5.1 Polysaccharides Differ in Structure

Polysaccharide | **Chemical Structure** | **Three-dimensional Structure**

Starch

Used for energy storage in plant cells (such as in potatoes)

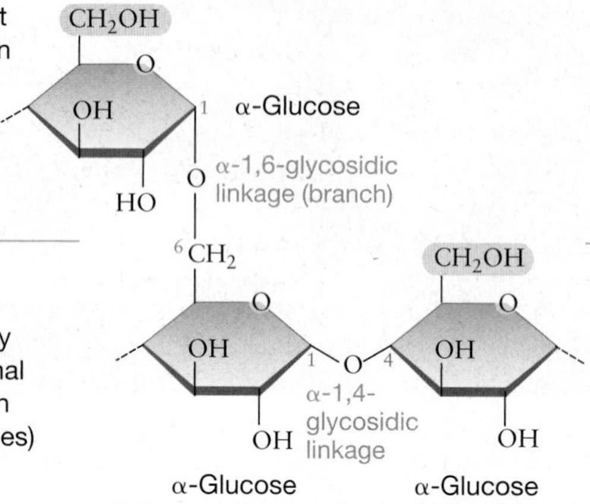

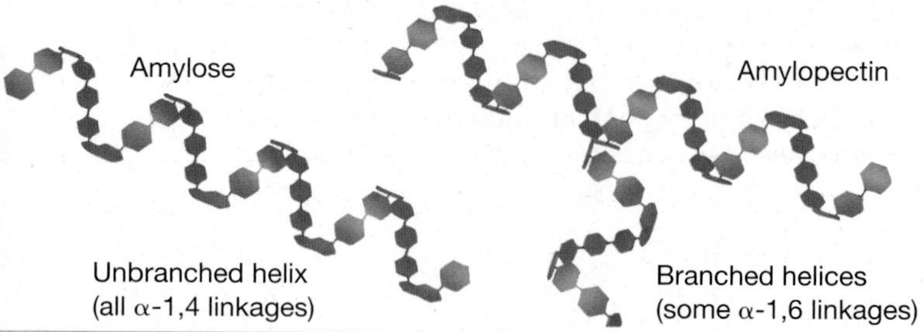

Glycogen

Used for energy storage in animal cells (such as in liver and muscles)

Highly branched helices (many α-1,6 linkages)

Cellulose

Used for structural support in cell walls of plants and many algae

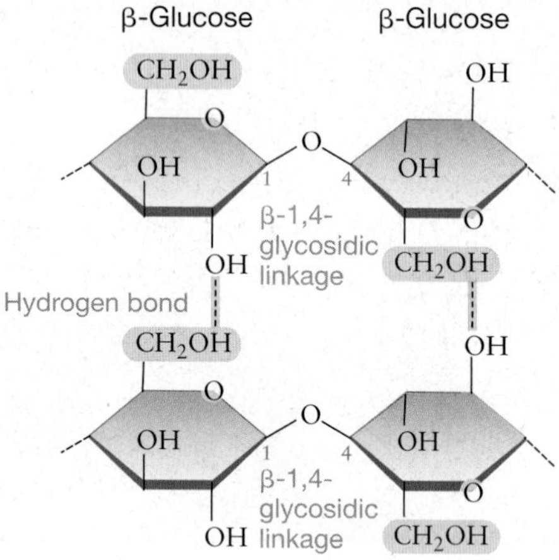

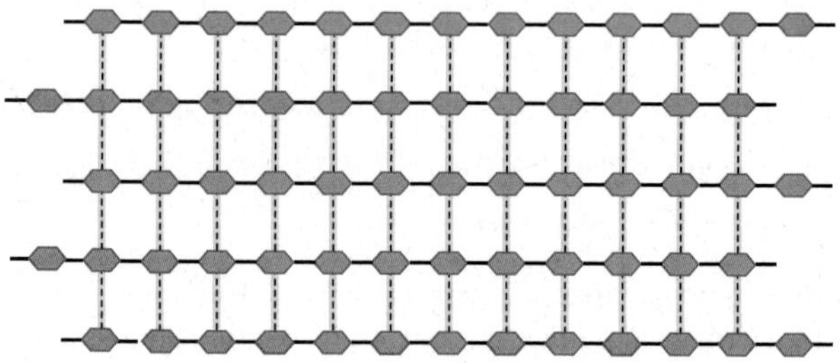

Parallel strands joined by hydrogen bonds

Chitin

Used for structural support in the cell walls of fungi and the external skeletons of insects and crustaceans

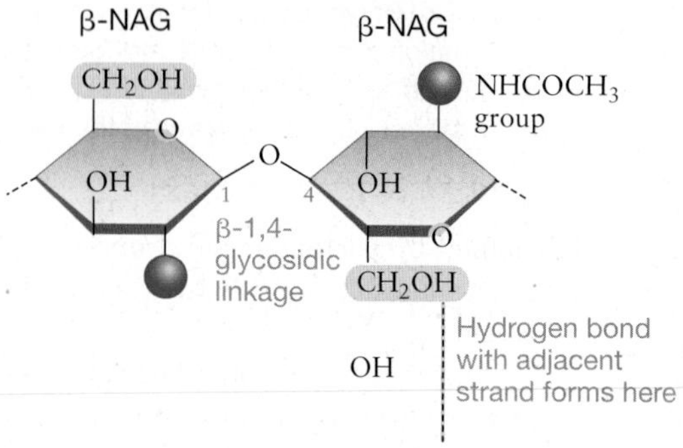

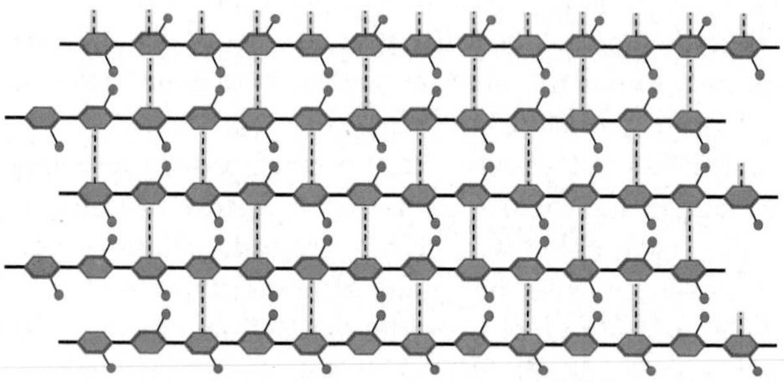

Parallel strands joined by hydrogen bonds

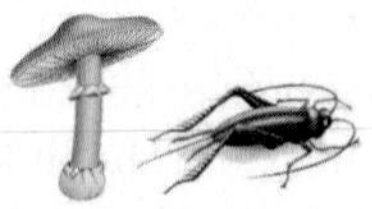

Peptidoglycan

Used for structural support in bacterial cell walls

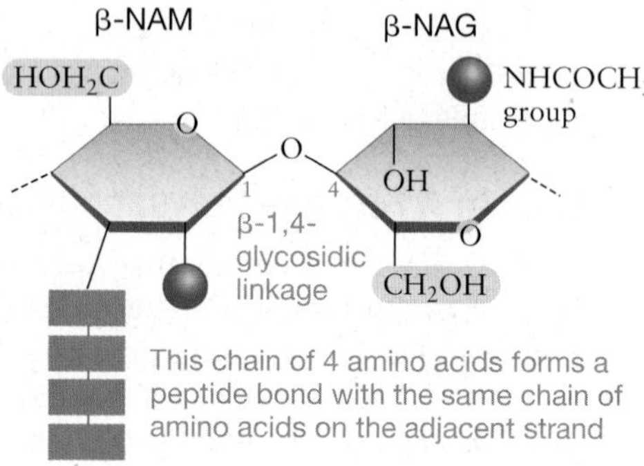

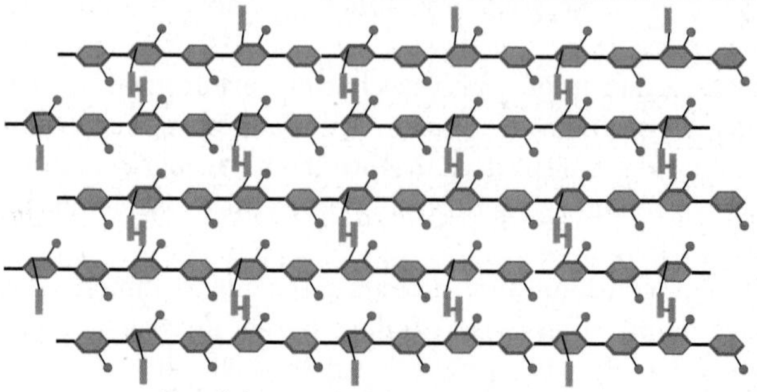

Parallel strands joined by peptide bonds

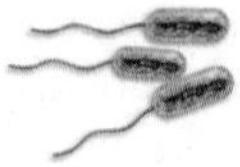

As shown in the top panel in **Table 5.1**, starch is made up of two types of polymers. One is an unbranched molecule called amylose, which contains only α-1,4-glycosidic linkages. The other is a branched molecule called amylopectin. Branching occurs when a glycosidic linkage forms between a C-1 carbon and a C-6 carbon (an α-1,6 linkage). In amylopectin, branching occurs at about one out of every 30 glucose residues.

Glycogen: A Highly Branched Storage Polysaccharide in Animals

Glycogen performs the same storage role in animals as starch does in plants. In humans, for example, glycogen is stored in the cells of liver and muscle tissues. When you start exercising, enzymes begin breaking glycogen into glucose monomers, which are then processed in muscle cells to supply energy.

Glycogen is a helical polymer of α-glucose and is nearly identical to the branched form of starch. However, instead of an α-1,6-glycosidic linkage occurring in about 1 out of every 30 residues in amylopectin, a branch occurs in about 1 out of every 10 glucose subunits (see Table 5.1). The branches provide more ends for enzymes to release glucose when your body needs it.

Cellulose: A Structural Polysaccharide in Plants

All cells are enclosed by a membrane (Ch. 1, Section 1.2), and the cells of many organisms are also surrounded by a protective layer of material called a **cell wall**. Polysaccharides are the primary building materials for most of these cell walls, including those in plants, fungi, and bacteria.

In plants, the major component of the cell wall is cellulose. **Cellulose** is a polymer made from β-glucose monomers joined by β-1,4-glycosidic linkages. As Table 5.1 shows, the geometry of the linkage is such that each glucose residue in the chain is flipped in relation to the adjacent residue. The flipped orientation is important because **(1)** it generates a linear molecule, rather than the helix seen in starch; and **(2)** it permits multiple hydrogen bonds to form between adjacent, parallel strands of cellulose. As a result, cellulose forms strong fibers consisting of multiple parallel strands joined by hydrogen bonds. Interacting cellulose fibers give plant cells structural support.

Chitin: A Structural Polysaccharide in Fungi and Animals

Chitin is a polysaccharide that stiffens the cell walls of fungi. It's also found in a few types of protists and in many animals. It is, for example, the most important component of the external skeletons of insects and crustaceans.

Chitin is similar to cellulose, but instead of consisting of glucose residues, the monosaccharide involved is one called *N*-acetylglucosamine (abbreviated as NAG). These NAG monomers are joined by β-1,4-glycosidic linkages (see Table 5.1). As in cellulose, the geometry of these bonds results in every other residue being flipped in orientation. The NAG subunits in chitin also form hydrogen bonds between adjacent strands to produce a stiff protective armor.

Peptidoglycan: A Structural Polysaccharide in Bacteria

Most bacteria, like all plants and fungi, have cell walls. The primary structural component of bacterial cell walls consists of a polysaccharide called **peptidoglycan**.

Peptidoglycan is the most complex of the polysaccharides discussed so far. It has a long backbone formed by NAG and *N*-acetylmuramic acid (NAM) that alternate with each other and are linked by β-1,4-glycosidic linkages (see Table 5.1). In addition, a short chain of amino acids is attached at the C-3 carbon of NAM. When molecules of peptidoglycan align, peptide bonds link the amino acid chains on adjacent strands. These links serve the same purpose as the hydrogen bonds between the parallel strands of cellulose and chitin in the cell walls of other organisms.

While there is no evidence to suggest polysaccharides played a significant role in chemical evolution, they became enormously important once cellular life evolved. So in the next section, let's look closely at how they function in today's cells.

CHECK YOUR UNDERSTANDING

✔ **If you understood this section, you should be able to ...**

1. Describe four structural differences that could result in different oligosaccharides consisting of two glucose residues and two galactose residues.
2. Predict how the structure of cellulose would change if all of the β-1,4-glycosidic linkages were changed to α-1,4-glycosidic linkages.

Answers are available in Appendix A.

5.3 What Do Carbohydrates Do?

After you complete this section, you should be able to ...

- Analyze how structure influences the different roles of carbohydrates in cells.

One of the basic functions that carbohydrates perform in organisms is to serve as a building block for more-complex molecules. For example, the nucleotides that make up RNA and DNA polymers consist of a ribose ($C_5H_{10}O_5$) or a deoxyribose ($C_5H_{10}O_4$) sugar, a phosphate group, and a nitrogenous base (Ch. 4, Section 4.1). The sugar itself acts as a subunit of each of these monomers.

In addition, sugars frequently furnish the raw "carbon skeletons" that are broken down for the synthesis of important molecules. Your cells are using sugars right now, for example, as a starting point for the synthesis of amino acids.

Carbohydrates have diverse functions in cells: Besides serving as precursors to other molecules, they **(1)** provide fibrous structural materials, **(2)** mark cell identity, and **(3)** store chemical energy. Let's look at each of these functions in turn.

Carbohydrates Can Provide Structural Support

Cellulose and chitin, along with the modified polysaccharide peptidoglycan, are key structural compounds. They form fibers that give cells and organisms strength and elasticity.

To appreciate why cellulose, chitin, and peptidoglycan are effective structural molecules, recall that they form long strands and that bonds can form between adjacent strands. In the cell walls of plants, for example, a collection of about 80 cellulose molecules are cross-linked by hydrogen bonding to produce a fiber. These cellulose fibers, in turn, crisscross to form a tough sheet that is able to withstand pulling and pushing forces—what an engineer would call tension and compression (see Ch. 11, Section 11.1).

Structural carbohydrates are not just tough, but durable. Almost all organisms produce enzymes that cleave the α-glycosidic linkages in starch and glycogen molecules, but only a few organisms have enzymes capable of digesting cellulose, chitin, or peptidoglycan. These fibers tend to be insoluble thanks to the strong interactions between strands consisting of β-1,4-glycosidic linkages. The exclusion of water within these fibers makes them more difficult to hydrolyze, so they are resistant to degradation and decay.

Ironically, its indigestibile structure makes cellulose extremely important for your digestive health. The cellulose that you ingest when you eat plants—what biologists call dietary fiber—forms a porous mass that absorbs and retains water. This sponge-like mass adds moisture and bulk that helps fecal material move through the intestinal tract more quickly, preventing constipation and other problems.

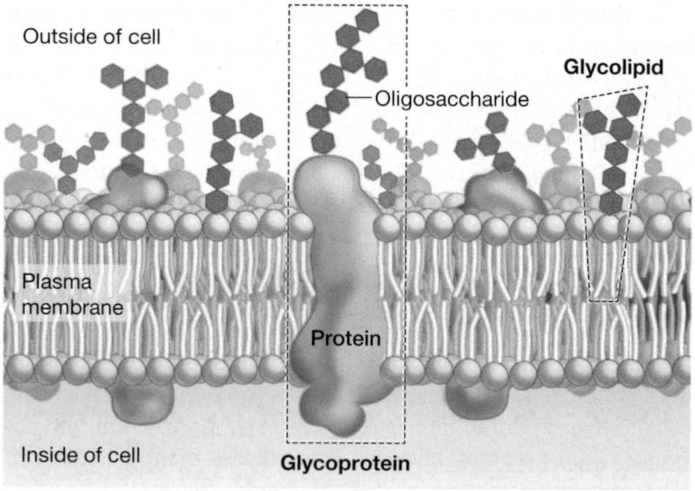

Figure 5.6 Carbohydrates Are an Identification Badge for Cells. Glycolipids and glycoproteins bear carbohydrates that project outside the cell from the surface of the cell-enclosing plasma membrane. These sugar groups have distinctive structures that identify the type or species of the cell.

The Role of Carbohydrates in Cell Identity

Structural polymers tend to be repetitive, made up of only one or two types of monosaccharides. The same is not true for all complex carbohydrates. Some polysaccharides exhibit enormous structural diversity, because their component monomers—and the linkages between them—vary a lot. As a result, they are capable of displaying information to other cells through their structure. More specifically, polysaccharides act as an identification badge on the outer surface of the plasma membrane that surrounds a cell. (Ch. 6, Section 6.2, describes plasma membranes and their components in detail.)

Figure 5.6 shows how this information about cell identity is displayed. Carbohydrates attached to membrane components, such as lipids and proteins, project outward from the cell surface into the surrounding environment. A **glycolipid** is a lipid that has been glycosylated, meaning it has one or more covalently attached carbohydrates. A **glycoprotein** is a protein that is similarly linked to carbohydrates. The carbohydrates attached to glycolipids and glycoproteins are usually short, branched oligosaccharides.

Glycolipids and glycoproteins are key molecules in what biologists call cell–cell recognition and cell–cell signaling. Each cell in your body has carbohydrates on its surface that identify it as part of your body. For example, your blood type (A, B, AB, or O) is determined by the type of marker oligosaccharides presented on the surface of your blood cells. The A, B, and O markers arise from different modifications of the carbohydrates in glycolipids (Ch. 14, Figure 14.6). In addition, each distinct type of cell in a multicellular organism—for example, the nerve cells and muscle cells in your body—displays a different set of glycoproteins on its surface. This identification information helps cells recognize and communicate with each other.

The key point here is to recognize that the variety in the types of monosaccharides and how they can be linked together makes it possible for an enormous number of unique complex carbohydrates—and therefore glycolipids and glycoproteins—to exist. As a result, each cell type and each species can display a unique identity.

Carbohydrates are also at the root of one of the most important events in the life cycle of many multicellular organisms—sexual reproduction. During the 1980s, Paul Wassarman and colleagues investigated the role of glycoproteins in cell–cell recognition between sperm and egg during fertilization. This step guarantees specificity—sperm normally recognize and bind only to eggs of their own species.

In one experiment, the researchers mixed sperm with purified egg-surface glycoproteins and discovered that most of the sperm subsequently lost their ability to attach to eggs. Such loss of function is an example of what researchers call competitive inhibition. The glycoproteins had bound to—and thus blocked—the same structure on the sperm that it uses to bind to eggs. This result showed that sperm attach to eggs via egg glycoproteins.

But which part of the egg glycoproteins is essential for recognition and attachment—the protein or the carbohydrate? In follow-up experiments, Wassarman's group used the same type

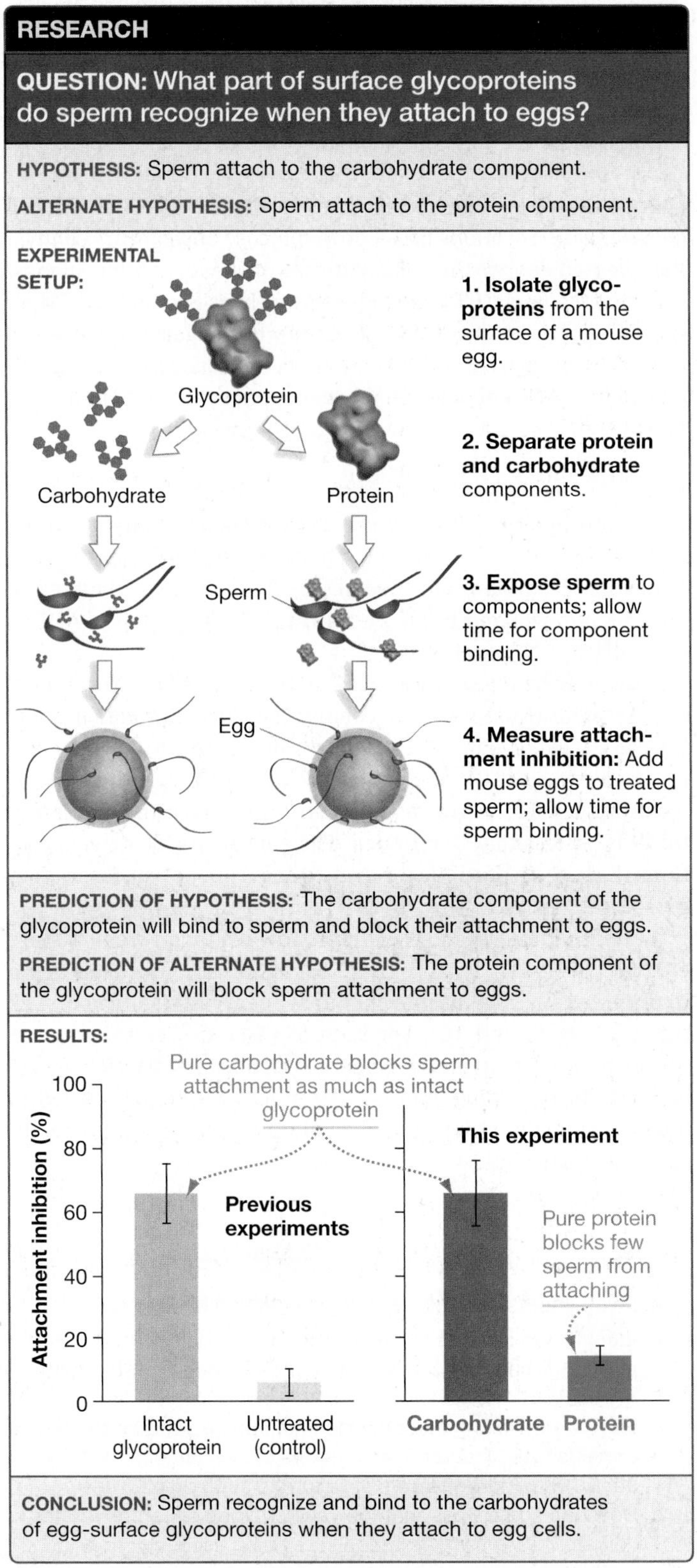

Figure 5.7 Carbohydrates Are Required for Cellular Recognition and Attachment during Fertilization.

SOURCES: H. M. Florman, K. B. Bechtol, and P. M. Wassarman. 1984. Enzymatic dissection of the functions of the mouse egg's receptor for sperm. *Developmental Biology* 106: 243–255. Also H. M. Florman and P. M. Wassarman. 1985. O-linked oligosaccharides of mouse egg ZP3 account for its sperm receptor activity. *Cell* 41: 313–324.

✔ **PROCESS OF SCIENCE** How would the graph bars change if sperm attachment required only the protein portion of egg glycoproteins?

of competitive-binding assay to answer this question (**Figure 5.7**). When sperm were mixed with purified carbohydrates alone, most were unable to attach to eggs. In contrast, most sperm treated with purified protein alone were not inhibited and still attached to eggs. Both results show that the carbohydrate component plays a fundamental role in the process of egg-cell recognition.

Carbohydrates and Energy Storage

Candy-bar wrappers promise a quick energy boost, and ads for sports drinks claim that their products provide the "carbs" needed for peak activity. If you were to ask friends or family members what carbohydrates do in your body, they would probably say something like "They give you energy." And you'd agree, but only after pointing out that carbohydrates are also used in cell identity, as a structural material, and as a source of carbon atoms for the synthesis of other complex molecules.

Carbohydrates store and provide chemical energy in cells. What aspect of carbohydrate structure makes this function possible?

Carbohydrates Store Energy from Sunlight as Chemical Energy

The essence of chemical evolution was the transformation of energy. Recall Stanley Miller's early Earth experiment (see Ch. 2, Figure 2.20), where he showed that the energy in electrical discharges could be converted into chemical energy and stored in bonds of molecules such as formaldehyde (CH_2O).

In cells today, the energy in light is likewise transformed into chemical energy, but instead of making formaldehyde, the energy is stored in sugars. For example, plants harvest the energy in sunlight and store it in the bonds of carbohydrates by the process known as **photosynthesis**. (Photosynthesis is the focus of Chapter 10.)

Photosynthesis entails a complex set of reactions that can be summarized most simply as follows:

$$CO_2 + H_2O + \text{sunlight} \longrightarrow (CH_2O)_n + O_2$$

where $(CH_2O)_n$ represents a carbohydrate. The key to understanding the energy conversion that takes place during photosynthesis is to compare the positions of the electrons associated with carbon in the reactants to those in the products.

1. The electrons in the C═O bonds of carbon dioxide (CO_2) and the C—O bonds of carbohydrates are held tightly because of oxygen's high electronegativity. Thus, they have relatively low potential energy.
2. The electrons involved in the C—H bonds of carbohydrates are shared equally because the electronegativity of carbon and hydrogen is about the same. Thus, bonds are weaker and these electrons have relatively high potential energy.
3. Electrons are also shared equally in the carbon–carbon (C—C) bonds of carbohydrates—meaning that they, too, have relatively high potential energy.

So, because C—C and C—H bonds have much higher potential energy than C—O bonds have, carbohydrates store much more chemical energy than carbon dioxide does.

(a) Carbon dioxide

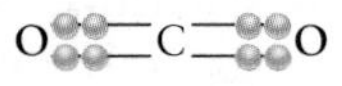

(b) A carbohydrate

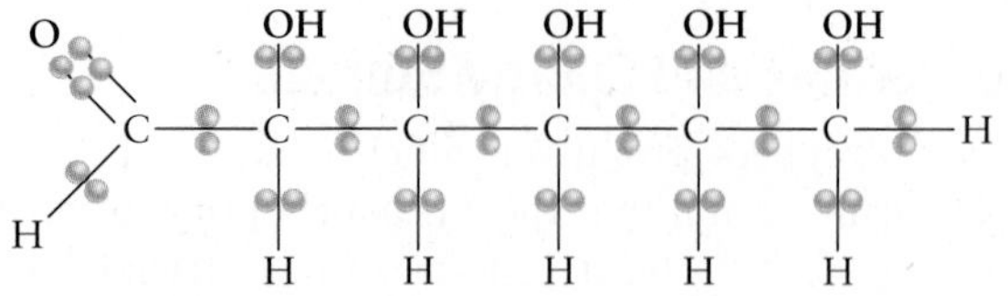

Figure 5.8 **In Organisms, Potential Energy Is Stored in the Bonds of Molecules. (a)** Carbon dioxide has low potential energy because the electrons involved in covalent bonds are held tightly by oxygen atoms. **(b)** Carbohydrates, such as the sugar shown here, have high potential energy because many of the covalent bonds are weak and the electrons are held equally between C and H atoms.

✔ Circle the bonds in Figure 5.8 that have high potential energy.

Figure 5.8 summarizes these points. Compare the structure of carbon dioxide in Figure 5.8a with the carbohydrate in Figure 5.8b. The main difference is the presence of C—C and C—H bonds in the carbohydrate.

This point is important. C—C and C—H bonds have high potential energy because the participating atoms have low electronegativities and share the electrons equally. C—O bonds, in contrast, have low potential energy because the highly electronegative oxygen atom holds the electrons so tightly, resulting in a stronger bond. Recall that the potential energy in bonds is released when they are broken and new, stronger bonds are formed (Ch. 2, Section 2.3). Thus carbohydrates are valuable as fuel in cells.

Enzymes Hydrolyze Energy-Storage Polysaccharides to Release Glucose Starch and glycogen are efficient energy-storage molecules because they polymerize via α-glycosidic linkages instead of the β-glycosidic linkages in the structural polysaccharides. The α-linkages in storage polysaccharides are readily hydrolyzed to release glucose, while the structural polysaccharides resist enzymatic degradation.

The most important enzyme involved in catalyzing the hydrolysis of α-glycosidic linkages in glycogen molecules is a protein called **phosphorylase**. Many of your cells contain phosphorylase, so they can break down glycogen to provide glucose on demand. The enzymes involved in breaking the α-glycosidic linkages in starch are called **amylases**. Your salivary glands and pancreas produce amylases that are secreted into your mouth and small intestine, respectively. These amylases are responsible for digesting the starch that you eat.

The glucose subunits that are hydrolyzed from glycogen and starch are processed in reactions that result in the production of chemical energy that can be used in the cell. Glycogen and starch are like a candy bar that has segments, so you can break off chunks whenever you need a boost.

Energy Stored in Glucose is Used to Make ATP When a cell needs energy, reactions break down glucose and capture some of the released energy through synthesis of the nucleotide adenosine triphosphate (ATP) (introduced in Ch. 4, Section 4.1). More specifically, the energy that's released when sugars are processed is used to synthesize ATP from a precursor called adenosine diphosphate (ADP) plus a free inorganic phosphate (P_i) molecule. The overall reaction can be written as follows:

$$(CH_2O)_n + O_2 + ADP + P_i \longrightarrow CO_2 + H_2O + ATP$$

To put this in terms of bonds, some of the chemical energy stored in the C—H and C—C bonds of carbohydrate [$(CH_2O)_n$] is released as new C=O bonds are formed in CO_2. This energy is then transferred to a new bond linking a third phosphate group to ADP to form ATP.

How much sugar does it take to form ATP? Not much! Consider this example: A cell can use the sugar stored in a single LifeSavers candy (about 15 Calories of energy) to produce approximately 3×10^{23} molecules of ATP. Although this sounds like a lot of ATP, an average person would burn through all of this ATP in less than 2 minutes! The energy in ATP drives reactions that are responsible for everything from polymerization to muscle movement.

Later chapters analyze in detail how cells capture and store energy in sugars and how these sugars are then broken down to provide cells with usable chemical energy in the form of ATP (Chapters 8, 9, and 10). For both of these processes to occur, however, a selectively permeable membrane barrier is required. The following chapter introduces the lipids needed to build these membranes and the role they played in the evolution of the first cell.

CHECK YOUR UNDERSTANDING

✔ If you understood this section, you should be able to ...

1. Identify two aspects of the structures of cellulose, chitin, and peptidoglycan that correlate with their function as structural molecules.
2. Evaluate the claim that polysaccharides are superior to proteins (see Chapter 3) as markers of cellular identity. Cite attributes of carbohydrate structure to support your answer.
3. Describe some ways that the various types of carbohydrates you ate during breakfast today are being used in your body right now.

Answers are available in Appendix A.

5.1 Sugars as Monomers

- Monosaccharides are organic compounds that have a carbonyl group and several hydroxyl groups. The molecular formula for a sugar is typically $(CH_2O)_n$, but the number of "carbon-hydrate" groups may vary between sugars, as indicated by the subscript *n*.
- Monosaccharides have either an aldose or ketose configuration, depending on whether the carbonyl group is located at the end or within the carbon chain.
- Although some monosaccharides may have the same molecular formula, the spatial arrangement of their functional groups leads to differences in their molecular structures and therefore functions.
- In solution, monosaccharides can form ring structures that may differ from one another in the orientation of a hydroxyl group, even among molecules of the same monosaccharide.

5.2 The Structure of Polysaccharides

- Monosaccharides can be covalently bonded to one another via glycosidic linkages, which join hydroxyl groups on adjacent molecules.
- In contrast to proteins and nucleic acids, polysaccharides do not always form a single uniform backbone structure. The numerous hydroxyls found in each monosaccharide allow glycosidic linkages to form at different sites and new strands to branch from existing chains.
- The types of monomers involved and the geometries of the glycosidic linkages between monomers distinguish different polysaccharides from one another.
- The most common polysaccharides in organisms today are starch, glycogen, cellulose, and chitin; peptidoglycan is an abundant polysaccharide that consists of sugar monosaccharides and short chains of attached amino acids.

5.3 What Do Carbohydrates Do?

- In carbohydrates, as in proteins and nucleic acids, structure correlates with function.
- Cellulose, chitin, and peptidoglycan are polysaccharides that function in support. They are made up of monosaccharide monomers joined by β-1,4-glycosidic linkages. When individual molecules of these polysaccharides align side by side, bonds form between them—resulting in strong, flexible fibers or sheets that resist hydrolysis.
- The oligosaccharides on cell-surface glycoproteins and glycolipids can function as specific signposts or identity tags because their constituent sugar residues are so diverse in geometry and composition.
- Both starch and glycogen function as energy-storage molecules. They are made up of glucose molecules that are joined by α-glycosidic linkages. These linkages are readily hydrolyzed to release glucose for the production of ATP and raw materials for building new molecules.

Answers are available in Appendix A.

✔ TEST YOUR KNOWLEDGE

1. What are three ways monosaccharides differ from one another?
2. What type of bond is formed between two sugars in a disaccharide?
   a. glycosidic linkage
   b. phosphodiester bond
   c. peptide bond
   d. hydrogen bond
3. What holds cellulose molecules together in bundles large enough to form fibers?
   a. the cell wall
   b. peptide bonds
   c. hydrogen bonds
   d. hydrophobic interactions
4. What are the primary functions of carbohydrates in cells?
   a. cell identity, energy storage, raw material source for synthesis, and structure
   b. catalysis, energy storage, metabolism, and structure
   c. catalysis, digestion, energy storage, and information storage
   d. energy storage, information storage, polymerization, and raw material source for synthesis

✔ TEST YOUR UNDERSTANDING

5. Which of the differences listed here could be found among molecules of the same monosaccharide? Select True or False for each statement.
   T/F There is a difference in the orientation of a hydroxyl group in the ring form.
   T/F There is a difference in the number of carbons.
   T/F There is a difference in the position of the carbonyl group in the linear form.
   T/F There is a difference in the overall shape of the molecule—one is a ring and the other is linear.
6. Although cellulose and starch are identical in terms of stored chemical energy, our ability to harvest the energy from these two polysaccharides differs considerably. What is responsible for this difference?
7. Contrast the structure of glycogen and chitin in terms of their monosaccharides, glycosidic linkages, and interactions between polysaccharide chains.
8. Lysozyme, an enzyme found in human saliva, tears, and other secretions, catalyzes the hydrolysis of the β-1,4-glycosidic linkages in peptidoglycan. Predict the effect of this enzyme on bacteria and how it may be involved in human health.

✔ TEST YOUR PROBLEM-SOLVING SKILLS

9. **SOCIETY** Galactosemia is a potentially fatal disease that occurs in humans who lack the enzyme that converts galactose to glucose. If you were a physician treating a person with this disease, which of the following would you have them exclude from their diet?
   a. maltose **b.** starch **c.** mannose **d.** lactose
10. If you hold a salty cracker in your mouth long enough, it will begin to taste sweet. What is responsible for this change in taste?

✔ PUT IT ALL TOGETHER: Case Study

Is the sugar in your soda affecting your mind?

A 20-ounce soda typically contains 15 teaspoons of either sucrose or high-fructose corn syrup. These sugary beverages are a leading source of calories in the American diet and have been correlated with obesity and diabetes. There may be even more to this sweet dilemma, however. Could the large amount of sugars found in soda (and other processed foods and beverages) be altering your brain function? Consider the following questions.

11. Sucrose is a disaccharide consisting of α-glucose and β-fructose. Review the disaccharides shown in Figure 5.4 and state what type of glycosidic linkage connects these monosaccharides in sucrose.

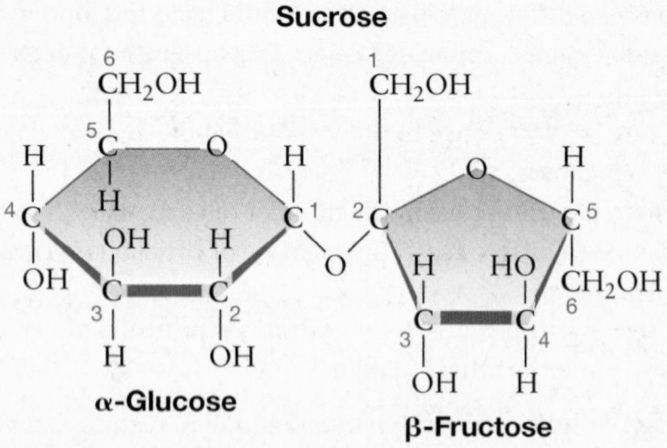

12. Sucrose is cleaved in your saliva by the enzyme sucrase to release glucose and fructose. Use the structural formula of sucrose to describe fructose using the terms that define its carbon number and placement of the carbonyl group.

13. You perceive the sweetness of sucrose based on a specific interaction between fructose and proteins on your tongue's taste buds. What structural difference between glucose and fructose would you predict to be responsible for the fact that fructose tastes sweeter?

14. High-fructose corn syrup is produced by converting starch from corn into a mixture of glucose and fructose monosaccharides. What two events must occur in this process in order to turn starch into these simple sugars?

15. QUANTITATIVE The effect of diet on human health is often evaluated using animal models. Researchers tested the impact of drinking fructose solutions on memory by using rats trained to find an escape chamber in a maze. The rats were fed diets consisting of either a control chow (deficient in omega-3 fatty acids, to simulate popular unhealthy human diets) or a chow that was supplemented with omega-3 fatty acids, plus drinking water with or without fructose (15 percent solution). After six weeks on the diets, the rats were tested again on the same maze with the results shown in the bar graph (* means $P < 0.05$ and ** means $P < 0.01$; P values and statistical significance are discussed in **BioSkills 3**). What can you conclude from these results about learning retention in rats whose diet included fructose versus those without fructose? How did the presence or absence of dietary omega-3 fatty acids (an essential nutrient) affect memory in this test?

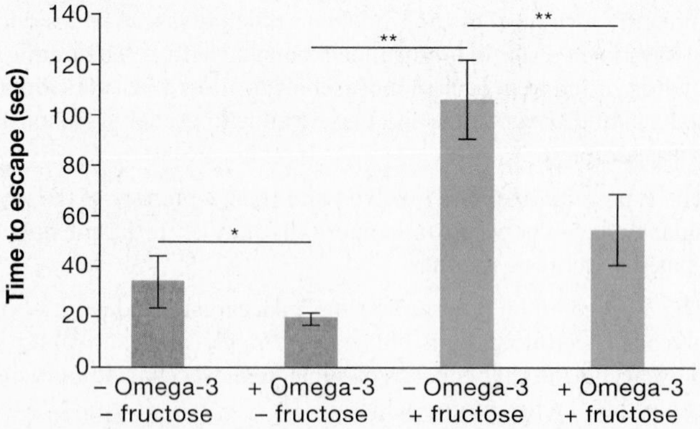

Source: R. Agrawal and F. G. Pinilla. 2012. *Journal of Physiology* 590: 2485–2499.

16. SOCIETY How might the results from this study influence your diet while studying biology?

Mastering Biology

Students Go to Mastering™ Biology for assignments, the eText, and the Study Area with animations, practice tests, and activities.

Professors Go to Mastering™ Biology for automatically graded tutorials and questions that you can assign to your students, plus Instructor Resources.

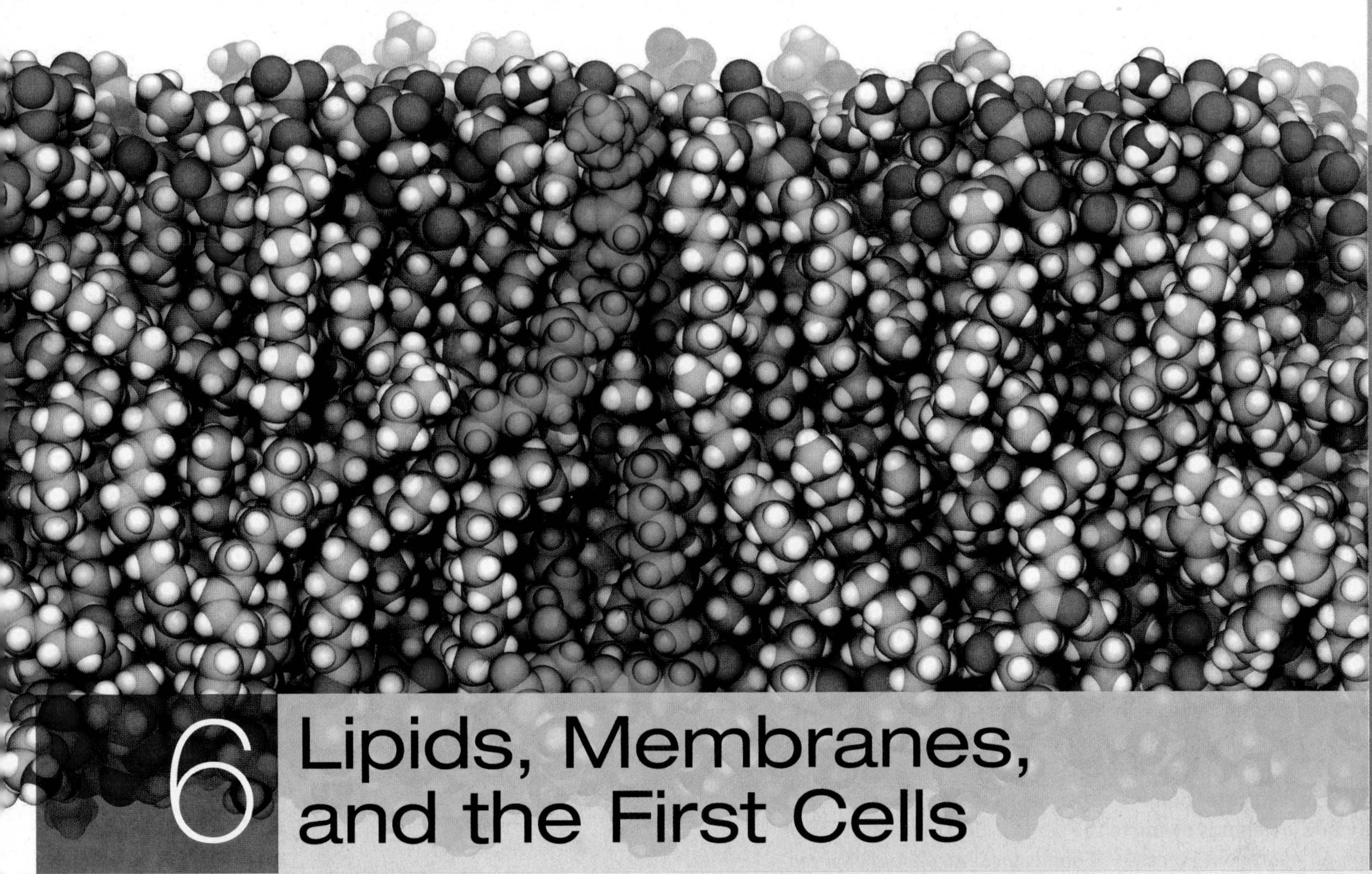

6 Lipids, Membranes, and the First Cells

A space-filling model of a phospholipid bilayer. Within this ordered cluster of molecules, two phospholipids (one in the upper lipid sheet and one in the lower) are highlighted in blue. A phospholipid bilayer, also known as the cell membrane, forms the boundary between life (inside the cell) and nonlife (outside the cell).

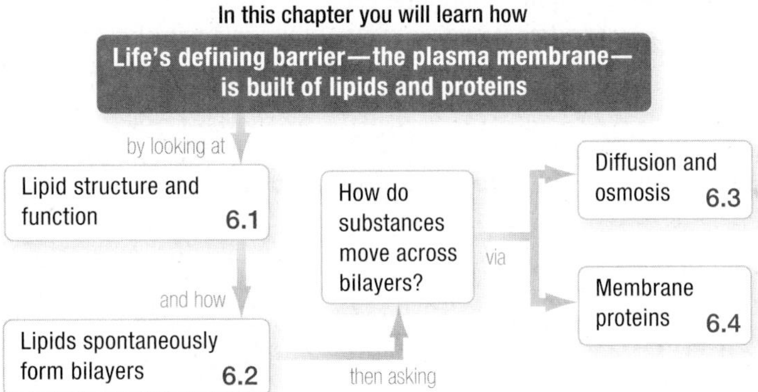

Currently, most biologists support the hypothesis that biological evolution began with a molecule that could replicate itself. As the offspring of this molecule multiplied, natural selection would have favored the versions that reproduced most frequently. A second great milestone in the history of life occurred when descendants of these replicators became enclosed within a membrane.

Why was the emergence of a membrane so important? The **plasma membrane**, often called the **cell membrane**, separates life from nonlife. It is a layer of molecules that surrounds and separates the cell interior from the environment.

- The plasma membrane serves as a selective barrier: It can keep damaging substances out of the cell and allow entry of substances needed by the cell.
- Because the plasma membrane sequesters the appropriate chemicals in an enclosed area, reactants collide more frequently, allowing the chemical reactions necessary for life to occur much more efficiently.

BIG PICTURE

This chapter is part of the Big Picture. See how on pages 144–145.

How do membranes form? Which ions and molecules can pass through a membrane and which cannot, and why? These are some of the most fundamental questions in all of biological science. Let's delve into them, beginning with the membrane's foundation—lipids.

6.1 Lipid Structure and Function

Lipid is a catchall term for carbon-containing compounds that are characterized by a physical property—their insolubility in water. This insolubility results from a high proportion of nonpolar C—C and C—H bonds relative to polar functional groups. Lipids do dissolve, however, in organic solvents consisting of nonpolar compounds like benzene (C_6H_6).

> **After you complete this section, you should be able to ...**
>
> - Analyze the relationship between the molecular structure and the properties of lipids.

To understand why lipids are insoluble in water, examine the five-carbon compound, called isoprene, illustrated in Figure 6.1a. Note that isoprene consists entirely of carbon atoms bonded to hydrogen atoms. Molecules that contain only carbon and hydrogen are known as **hydrocarbons**.

Hydrocarbons are nonpolar because electrons are shared equally in C—H bonds owing to the similar electronegativities of carbon and hydrogen. Since C—H bonds form no partial charges, hydrocarbons do not dissolve in water. (Recall from Ch. 2, Section 2.2, that water is a polar solvent.) Lipids, therefore, are mostly hydrophobic because they have a significant hydrocarbon component.

Figure 6.1a also shows the structural formula of a chain of linked isoprenes, called an isoprenoid. Isoprenoids serve a wide range of functions in organism—from pigments and scents to vitamins and precursors of sex hormones. As you will see, they are also important building blocks for other, more complex lipids.

How Does Bond Saturation Affect Hydrocarbon Structure?

Figure 6.1b gives the structural formula of a **fatty acid**, a simple lipid consisting of a hydrocarbon chain bonded to a polar carboxyl functional group (—COOH). Fatty acids typically contain a total of 14–20 carbon atoms, most found in their long nonpolar hydrocarbon "tails." Like isoprenoids, fatty acids are also key building blocks of important lipids found in organisms. Let's look at how these molecules are put together.

Just as subtle differences in the orientation of hydroxyl (—OH) groups can lead to dramatic effects in the structure and function of sugars, the type of bond between carbons in hydrocarbon chains is a key factor in lipid structure and function.

When two carbon atoms form a double bond, the attached atoms are found in a plane (for example, ethylene, $H_2C{=}CH_2$):

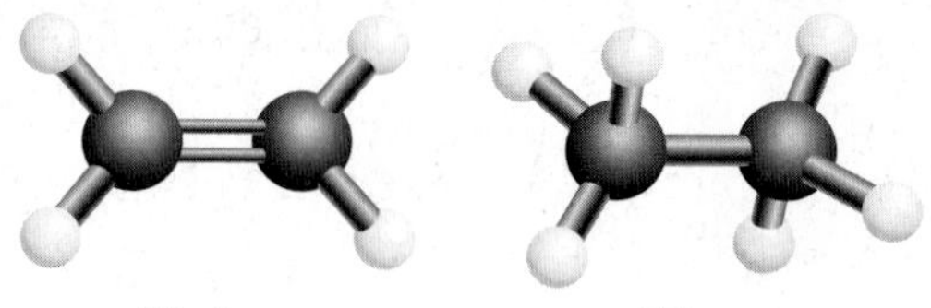

instead of three-dimensional tetrahedra (for example, ethane, $H_3C{—}CH_3$) (Ch. 2, Section 2.1).

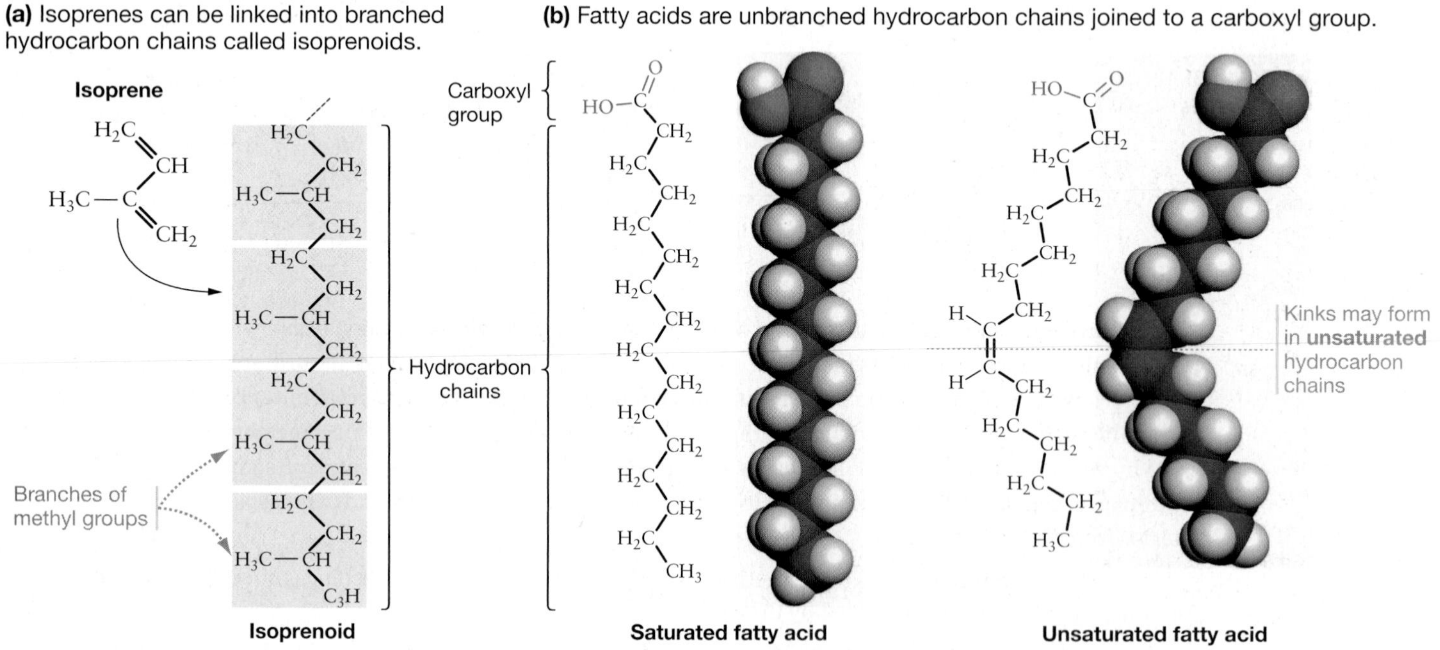

Figure 6.1 Hydrocarbon Structure. (a) Isoprene subunits, like the one shown to the left, can link to each other, end to end, to form long branched hydrocarbon chains called isoprenoids. **(b)** Fatty acids have unbranched hydrocarbon chains. Unsaturated hydrocarbons contain carbon–carbon double bonds; saturated hydrocarbons do not.

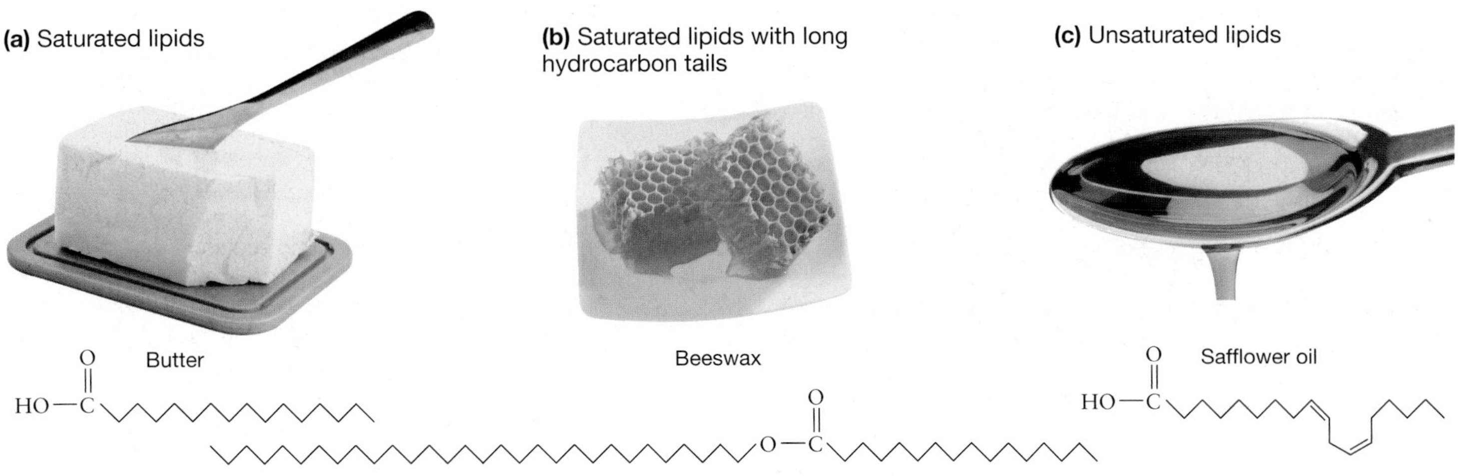

(d) Hydrogenation

Figure 6.2 The Fluidity of Lipids Depends on the Length and Saturation of Their Hydrocarbon Chains. (a) Butter consists primarily of saturated lipids. **(b)** Waxes are lipids with extremely long saturated hydrocarbon chains. **(c)** Oils are dominated by "polyunsaturates"—lipids with hydrocarbon chains that contain multiple C=C double bonds. **(d)** The product Crisco is made by converting polyunsaturates into saturated lipids by hydrogenation.

The carbon atoms involved in double bonds are also locked into place. They cannot rotate freely, as can carbons in C—C single bonds. As a result, certain double bonds between carbon atoms (called *cis* bonds) produce a "kink" in an otherwise straight hydrocarbon chain (compare the two structures in Figure 6.1b).

Hydrocarbon chains that consist of only single bonds between the carbons are called **saturated**. If one or more double bonds exist in the hydrocarbon chains, then they are **unsaturated**. The choice of terms is logical. If a hydrocarbon chain does not contain a double bond, it is saturated with the maximum number of hydrogen atoms that can attach to the carbon skeleton. If it is unsaturated, then a C—H bond is removed to form a C=C double bond, resulting in fewer than the maximum number of attached hydrogen atoms.

Foods that contain lipids with many double bonds are said to be polyunsaturated and are advertised as healthier than foods with saturated lipids. Some recent research suggests that polyunsaturated lipids may help protect the heart from disease. Exactly how this occurs is under investigation.

Besides affecting the molecular structure of lipids, saturation also profoundly changes their physical state. Recall that van der Waals interactions allow nonpolar molecules to stick together (see Ch. 2, Section 2.2). If the lipids are composed of straight chains, many of these interactions will form along the chain and allow the lipids to pack together tightly to form a solid. If the hydrocarbons are bent, like the unsaturated fatty acid on the right side of Figure 6.1b, they will have fewer interactions, move freely, and form a liquid.

Highly saturated lipids, such as in butter, have relatively high melting points and are solid at room temperature (20–22°C) (Figure 6.2a). Saturated lipids that have extremely long hydrocarbon tails, like **waxes**, form particularly stiff solids at room temperature (Figure 6.2b). Highly unsaturated lipids are liquid at room temperature and called **oils** (Figure 6.2c). Unsaturated lipids may be converted to saturated lipids by breaking double bonds and adding hydrogen atoms via the process of hydrogenation (Figure 6.2d).

A Look at Three Types of Lipids Found in Cells

Unlike amino acids, nucleotides, and monosaccharides, lipids do not possess a shared chemical structure. The structure of lipids varies widely because their hydrocarbon skeletons can be put together in many different ways. For example, consider three of the most important types of lipids found in cells: steroids, fats, and phospholipids.

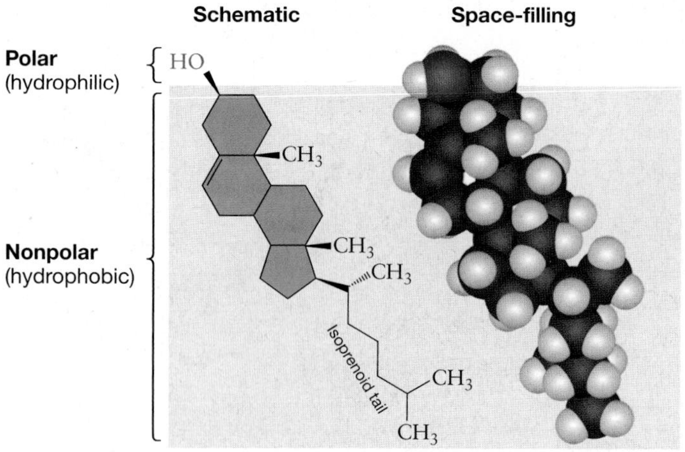

Figure 6.3 Steroids Are Defined by a Common Structure. All steroids have a distinctive four-ring structure, shown in orange. Cholesterol, the steroid shown here, has a polar hydroxyl group and a nonpolar isoprenoid chain attached to these rings.

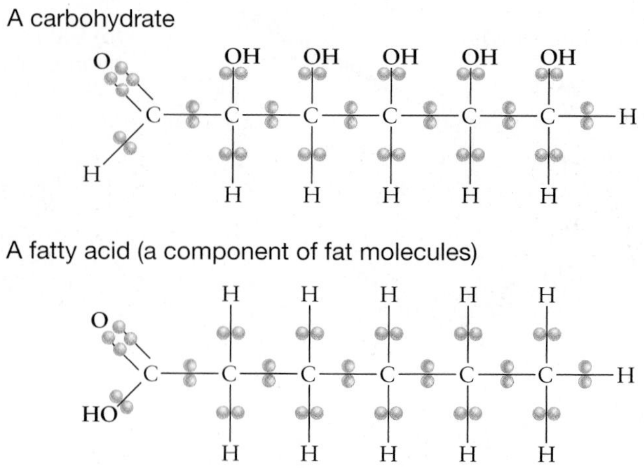

Figure 6.4 Fats Store More Energy than Carbohydrates. When compared to carbohydrates, the fatty acids of fats have a higher ratio of bonds with high potential energy to bonds with low potential energy.

Steroids **Steroids** are a family of lipids distinguished by the bulky, four-ring structure highlighted in orange in Figure 6.3. The various steroids differ from one another by the functional groups or side groups attached to different carbons in those hydrophobic rings. Steroids such as estrogens and testosterone are known for their role as hormones in cell signaling (Ch. 11, Section 11.3). The steroid shown in the figure is cholesterol, which has a polar hydrophilic hydroxyl group attached to the top ring and nonpolar isoprenoid "tail" attached at the bottom. Cholesterol is an important component of plasma membranes in many organisms.

Fats **Fats** are nonpolar molecules composed of three fatty acids that are linked to a three-carbon molecule called **glycerol**. Because of this structure, fats are also called triacylglycerols or simply triglycerides. (If the glycerol-linked fatty acids are polyunsaturated, the resulting triglycerides are liquid at room temperature.)

In organisms, energy storage is the primary role of fats. This makes sense when you note the large number of high-energy bonds in the fatty acid chains. The ratio of C—C and C—H bonds (high potential energy) to C—O bonds (low potential energy) is much greater in fats than in energy storage carbohydrates (Figure 6.4). Thanks to all of these bonds, fats can store about twice as much chemical energy per gram as carbohydrates.

As Figure 6.5a shows, fats form when a dehydration reaction occurs between a hydroxyl group of glycerol and the carboxyl group of a free fatty acid (when fatty acids are not attached to other molecules, they are referred to as free fatty acids). The glycerol and fatty acid molecules become joined by what is called an **ester linkage**. An ester linkage occurs when two atoms (one of them carrying a double-bonded oxygen, often a carbonyl group) are linked together by an oxygen. But notice that since fatty acids are not linked into chains, they are not considered monomers, and thus fats are not polymers. In this way, the structure of fats differs from the polymers that are formed when amino acids, nucleotides, and monosaccharides link together (Chapters 3–5).

Phospholipids **Phospholipids** consist of a glycerol that is linked to a phosphate group and two hydrocarbon chains of either isoprenoids or fatty acids. The phosphate group is also bonded to a small organic molecule that is charged or polar (Figure 6.5b).

Phospholipids with fatty acid tails are found in the domains Bacteria and Eukarya; phospholipids with isoprenoid tails are found in the domain Archaea.(The domains of life were introduced in Ch. 1, Section 1.5.) In all three domains, phospholipids are crucial components of the plasma membrane. The branched isoprenoid chains in archaeal phospholipids provide greater membrane stability and protection in the extreme environments inhabited by certain archaea (Ch. 26, Section 26.1).

How Membrane Lipids Interact with Water

The lipids found in organisms have a wide array of functions. In addition to storing chemical energy, lipids act as pigments that capture or respond to sunlight, serve as signals between cells, form waterproof coatings on leaves and skin, and act as vitamins used in many cellular processes. The most prominent function of lipids is their role in cell membranes, even though not all lipids can form membranes.

Besides the nonpolar, hydrophobic region that defines lipids, membrane-forming lipids have a polar, hydrophilic region. To better understand this structure, take another look at the phospholipid illustrated in Figure 6.5b. Notice that the molecule has a "head" region containing a negatively charged phosphate group attached to a polar group. The charges and polar covalent bonds in the head region interact with water molecules when a phospholipid is placed in solution. In contrast, the long hydrocarbon tails of a phospholipid are nonpolar and hydrophobic. Water molecules cannot form hydrogen bonds with the hydrocarbon tail, so they do not interact extensively with this part of the molecule.

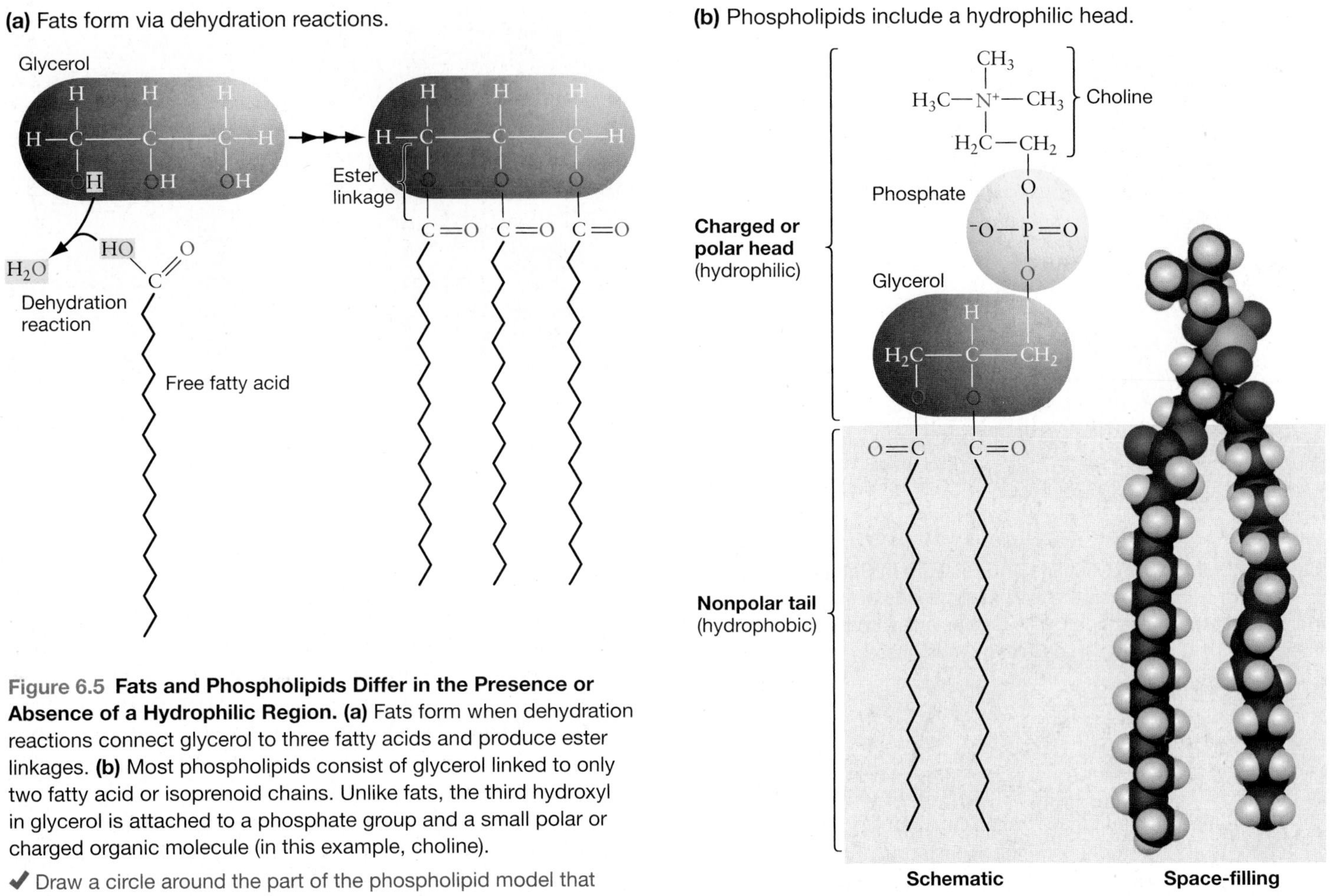

Figure 6.5 **Fats and Phospholipids Differ in the Presence or Absence of a Hydrophilic Region. (a)** Fats form when dehydration reactions connect glycerol to three fatty acids and produce ester linkages. **(b)** Most phospholipids consist of glycerol linked to only two fatty acid or isoprenoid chains. Unlike fats, the third hydroxyl in glycerol is attached to a phosphate group and a small polar or charged organic molecule (in this example, choline).

✔ Draw a circle around the part of the phospholipid model that would interact with water if it were placed in an aqueous solution.

Substances that contain both hydrophilic and hydrophobic regions are **amphipathic** (literally, "dual-sympathy"). Phospholipids are amphipathic. As Figure 6.3 shows, cholesterol is also amphipathic because it has a hydrophilic hydroxyl functional group attached to its hydrophobic rings.

The amphipathic nature of many lipids is by far their most important biological feature. It is responsible for life's defining barrier—the plasma membrane. If mixed with an aqueous solution, amphipathic lipids will spontaneously form hollow "bubbles" that resemble small cells. Why do they do this?

CHECK YOUR UNDERSTANDING

✔ If you understood this section, you should be able to ...

1. Explain how the saturation status of hydrocarbon chains affects the physical characteristics of lipids.
2. Compare and contrast the structure of a steroid, a fat, and a phospholipid.
3. Determine if free fatty acids (see Figure 6.1b) and fats (see Figure 6.5b) are amphipathic, and explain why or why not.

Answers are available in Appendix A.

6.2 Phospholipid Bilayers

After you complete this section, you should be able to ...

- Analyze the relationship between the structure and function of phospholipid bilayers.

Amphipathic lipids do not dissolve when they are placed in water. Their hydrophilic heads interact with water, but their hydrophobic tails do not. Instead of dissolving in water, amphipathic lipids assume one of two types of structures: micelles or lipid bilayers.

- Micelles (**Figure 6.6a**, on page 126) are tiny spherical aggregates created when the hydrophilic heads of a set of lipids face outward and interact with the water, while the hydrophobic tails interact with each other in the interior, away from the water.
- A **lipid bilayer** is created when lipid molecules align in paired sheets. As **Figure 6.6b** on page 126 shows, the hydrophilic heads in each layer face the surrounding solution while the hydrophobic tails face one another inside the bilayer. In this way, the hydrophilic heads interact with water while the hydrophobic tails interact with one another.

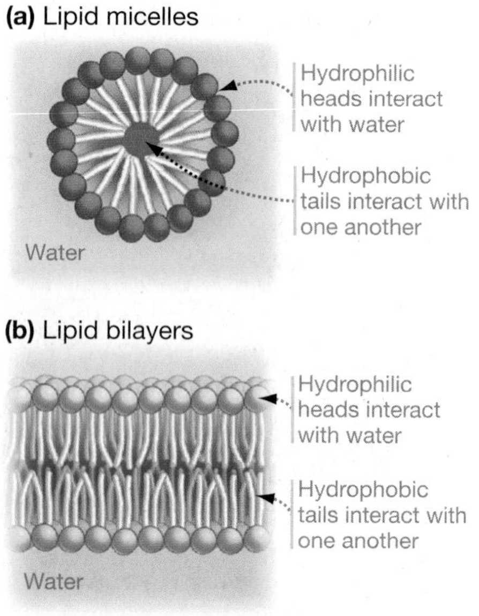

Figure 6.6 Lipids Form Micelles and Bilayers in Solution. In **(a)** a micelle or **(b)** a lipid bilayer, the hydrophilic heads of lipids face out, toward water; the hydrophobic tails face in, away from water. Lipid bilayers are the foundation of cellular membranes.

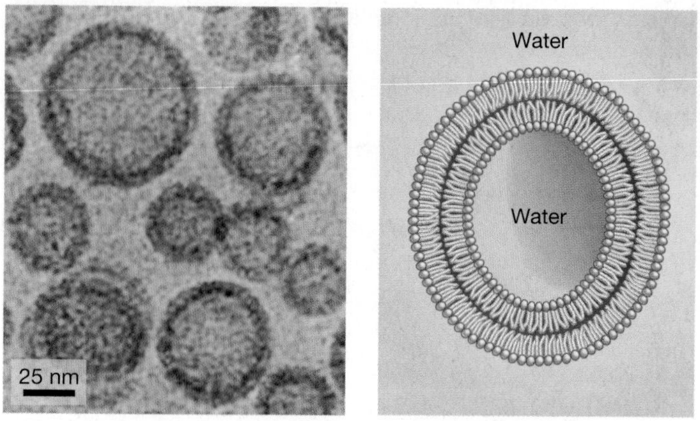

Figure 6.7 Liposomes Are Artificial Membrane-Bound Vesicles. Electron micrograph of liposomes in cross section (left) and a cross-sectional diagram of the lipid bilayer in a liposome (right).

Micelles tend to form from free fatty acids or other simple amphipathic lipids with single hydrocarbon chains. Phospholipids, which have bulkier nonpolar regions consisting of two hydrocarbon tails, tend to form bilayers. Lipid bilayers consisting of phospholipids are often called *phospho*lipid bilayers.

Micelles and phospholipid bilayers form spontaneously in water—no input of energy is required. This fact may come as a surprise because, at the level of lipid organization, entropy seemingly decreases—the lipids become less disordered as micelles and phospholipid bilayers form. (Recall that entropy is a measure of the randomness or disorder in a system; see Ch. 2, Section 2.3.) How can this be explained if spontaneous processes tend to increase entropy? The answer involves understanding how amphipathic lipids aggregate.

First you must consider the organization of water molecules. Recall that hydrophobic interactions occur when nonpolar structures become surrounded by a "cage" of highly organized water molecules (Ch. 2, Section 2.2). When amphipathic lipids are dispersed in an aqueous solution, cages of water form around each of the nonpolar tails. If the tails aggregate to form micelles and bilayers, then only the hydrophilic regions of the lipids are exposed and the water cages will melt. This decrease in water molecule organization results in an overall increase in the entropy of the system.

Artificial Membranes as an Experimental System

To explore how membranes work, researchers began by creating and experimenting with artificial membranes in the lab. In one experimental method, phospholipids are added to an aqueous solution and agitated, causing them to form **vesicles**—small bubble-like structures consisting of lipid bilayers surrounding a small amount of aqueous solution. In this configuration, the hydrophobic tails are shielded from water, and the hydrophilic heads remain in contact with water on the inside or outside of the vesicle. Artificially generated membrane-bound vesicles like these are called **liposomes** (**Figure 6.7**). Liposomes provide a three-dimensional model that mimics a membrane-bound cell.

Another artificial membrane often used in experiments is the planar bilayer, which provides a model unlike a vesicle. In this case, the lipid bilayer is constructed across a hole in a glass or plastic wall separating two aqueous solutions (**Figure 6.8a**).

Some of the first questions scientists posed about membranes concerned the permeability of lipid bilayers. The **permeability** of a structure is its tendency to allow a given substance to pass through it. Using liposomes and planar bilayers, researchers can study what happens when a known ion or molecule is added to one side of a lipid bilayer. **Figure 6.8b** shows how a planar bilayer could be used to answer experimental questions like these:

- Does the substance cross the membrane and show up on the other side?
- If so, how rapidly does the movement take place?
- What happens when a different type of phospholipid is used to make the artificial membrane?
- Does the membrane's permeability change when proteins or other types of molecules become part of it?

Biologists describe such an experimental system as elegant and powerful because it lets them precisely control which factor changes from one experimental treatment to the next. Control, in turn, is why experiments are such an effective way to explore scientific questions. Recall that good experimental design allows researchers to alter one factor at a time and determine what effect, if any, each has on the process being studied (Ch. 1, Section 1.6).

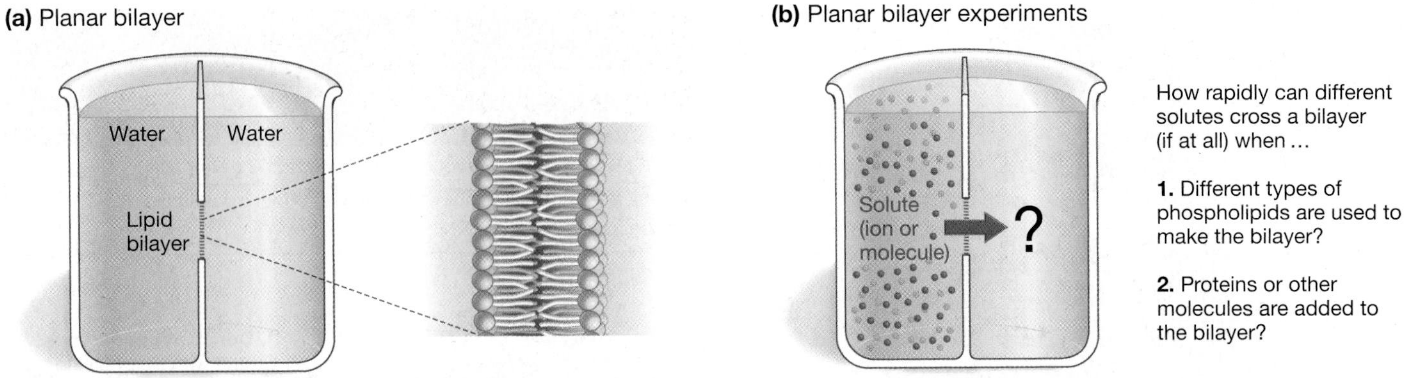

Figure 6.8 Use of Planar Bilayers in Experiments. (a) The construction of a planar bilayer across a hole in a wall separating two water-filled compartments. **(b)** A wide variety of experiments are possible with planar bilayers; just a few experimental questions are suggested here.

Selective Permeability of Lipid Bilayers

When researchers put molecules or ions in a solution containing liposomes, or on one side of a planar bilayer, and measure the rate at which the particles cross the membrane barrier, a clear pattern emerges: Lipid bilayers are highly selective.

Selective permeability means that some substances cross a membrane more easily than other substances do. This difference in membrane permeability is a critical issue because controlling what passes between the exterior and interior environments is a key characteristic of cells.

According to the data in Figure 6.9, small nonpolar molecules such as oxygen (O_2) move across bilayers quickly. If the small molecules are polar but uncharged, such as water (H_2O), the rate of transport decreases. Larger polar molecules cross the membrane even slower. For example, glucose moves across a lipid bilayer more than 10,000 times slower than do water molecules.

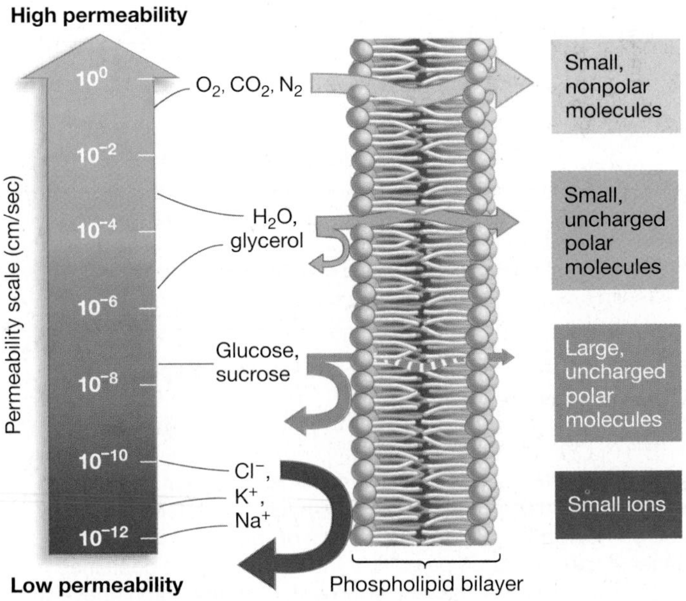

Figure 6.9 Lipid Bilayers Show Selective Permeability. Only certain substances cross lipid bilayers readily. The polarity, size, and charge of solutes affect their rate of diffusion across a membrane.

What about the permeability of very small substances, such as ions? It turns out that charged solutes—even tiny ions like sodium (Na^+)—do not effectively cross lipid bilayers without "help" from membrane proteins (introduced later in the chapter). Without these proteins, sodium ions would cross the membrane a billion times slower than water.

So if membranes don't simply filter solutes based on size, what is responsible for the differences in permeability? The leading hypothesis to explain the pattern of permeability described earlier is that charged substances and polar molecules above a certain size are more stable dissolved in water—a polar environment—than they would be in the nonpolar interior of membranes. ✔ **If you understand this hypothesis, you should be able to predict where amino acids and nucleotides would be placed in Figure 6.9 and explain your reasoning.**

How Does Lipid Structure Affect Membrane Permeability?

The amphipathic nature of phospholipids causes them to spontaneously form into bilayers consisting of two lipid sheets held together by hydrophobic interactions. But not all phospholipid bilayers are the same. The length and saturation state of the hydrocarbon tails, in addition to the presence of cholesterol molecules, profoundly influences the physical properties of a membrane and its permeability.

Bond Saturation and Hydrocarbon Chain Length Affect Membrane Permeability A phospholipid's degree of saturation—along with the length of its hydrocarbon tails—affects key aspects of its behavior in a membrane.

- When unsaturated hydrocarbon tails are packed into a lipid bilayer, kinks created by double bonds produce spaces among the tails. These spaces reduce the number of van der Waals interactions that help hold the hydrophobic tails together, weakening the barrier to solutes.

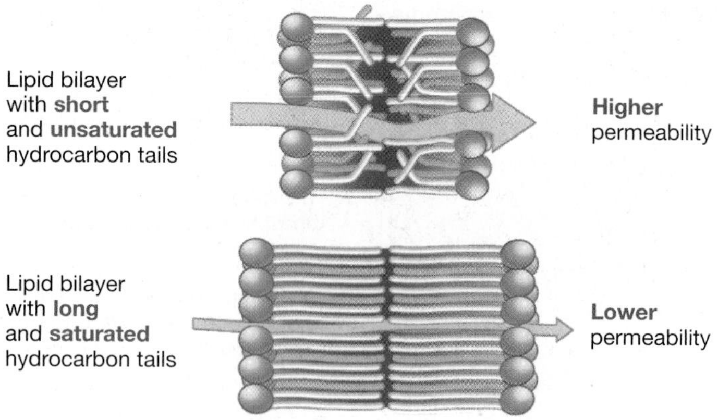

Figure 6.10 Degree of Hydrocarbon Saturation Affects the Permeability of Membranes. In general, phospholipids containing unsaturated hydrocarbon tails form bilayers that have more gaps and are more permeable than bilayers formed from phospholipids with saturated hydrocarbon tails.

- Packed saturated hydrocarbon tails have fewer spaces and more van der Waals interactions. As the length of saturated hydrocarbon tails increases, the forces that hold them together also increase, making the membrane even denser.

These factors profoundly affect membrane permeability. As **Figure 6.10** shows, lipid bilayers are more permeable when they contain many short, kinked, unsaturated hydrocarbon tails. A largely unsaturated membrane allows more materials to pass because its interior is held together less tightly. Bilayers containing mostly long, straight, saturated hydrocarbon tails are much less permeable.

Cholesterol Affects Membrane Permeability Cholesterol molecules are present, to varying extents, in the membranes of every cell in your body. What effect does adding cholesterol have on a membrane? Researchers have found that adding cholesterol molecules to artificial membranes dramatically reduces their permeability.

Some data behind this conclusion are presented in **Figure 6.11**. To read the graph in the "Results" section of the figure, put your finger on the x-axis at the point marked 20°C, and note that permeability to glycerol is much higher at this temperature in membranes that contain no cholesterol versus membranes with 20 percent or 50 percent cholesterol. Repeating this procedure at other temperature points should convince you that membranes lacking cholesterol are more permeable than the other two membranes at every temperature tested in the experiment.

What explains this result? Cholesterol orients in the membrane with its hydrophobic steroid rings buried deeply in the hydrocarbon tails of the phospholipids. The bulky cholesterol rings force the phospholipid tails closer to each other, increasing their packing density. When cholesterol was added to the experimental membranes, the closer packing of the tails caused the membranes to become less permeable.

RESEARCH

QUESTION: Does adding cholesterol to a membrane affect its permeability?

HYPOTHESIS: Cholesterol reduces membrane permeability.

NULL HYPOTHESIS: Cholesterol has no effect on membrane permeability.

EXPERIMENTAL SETUP:

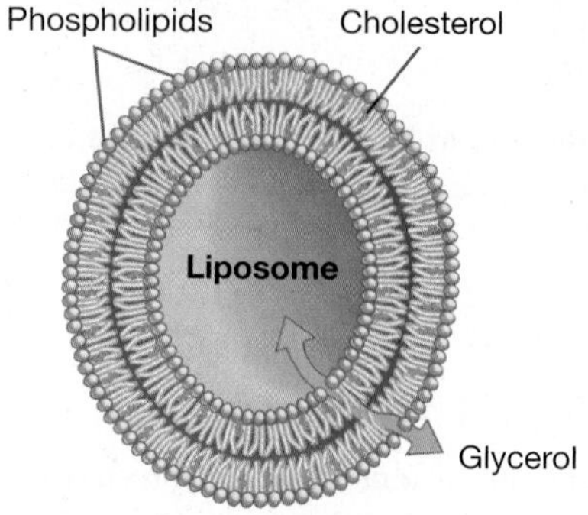

1. Construct liposomes: Create with no cholesterol, 20% cholesterol, and 50% cholesterol.

2. Measure glycerol movement: Record how quickly glycerol moves across each type of membrane at different temperatures.

PREDICTION OF HYPOTHESIS: Liposomes with higher cholesterol levels will have reduced permeability to glycerol.

PREDICTION OF NULL HYPOTHESIS: All liposomes will have reduced permeability to glycerol.

RESULTS:

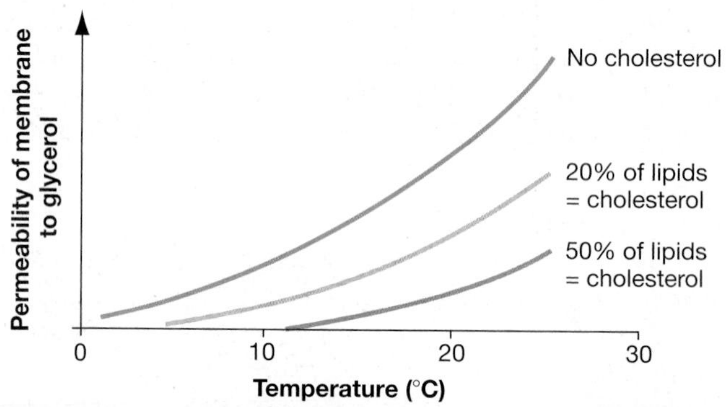

CONCLUSION: Adding cholesterol to membranes decreases their permeability to glycerol. The permeability of all membranes analyzed in this experiment increases with increasing temperature.

Figure 6.11 The Permeability of a Membrane Depends on Its Lipid Composition.

SOURCE: J. de Gier, J. G. Mandersloot, L. L. M. Van Deenen. 1968. Lipid composition and permeability of liposomes. *Biochimica et Biophysica Acta* 150: 666–675.

✔QUANTITATIVE Suppose the investigators had created a set of liposomes using phospholipids with fully saturated tails and compared them to two other sets of liposomes where either 20 percent or 50 percent of the phospholipids contained polyunsaturated tails. Label the three lines on the graph above with your prediction for the three different sets of liposomes in this new experiment.

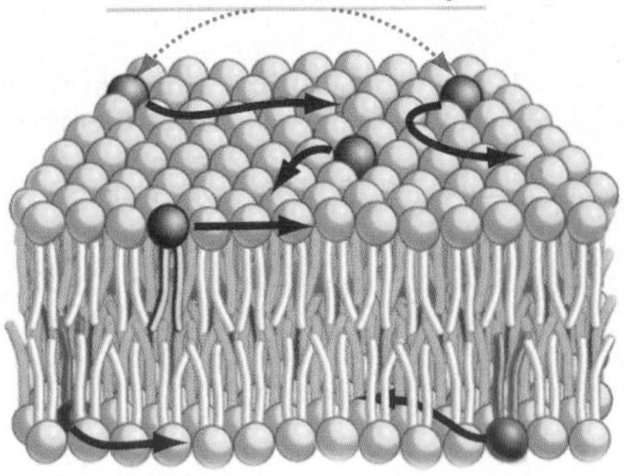

Figure 6.12 Phospholipids Move within Membranes. Membranes are dynamic—in part because phospholipid molecules randomly move laterally within each layer in the structure.

How Does Temperature Affect the Fluidity and Permeability of Membranes?

The phospholipids in the plasma membrane of a cell have a consistency resembling olive oil. This fluid physical state allows individual lipid molecules to move laterally within each layer, a little like a person moving about in a dense crowd (**Figure 6.12**). By tagging individual phospholipids and following their movement, researchers have clocked average speeds of 2 micrometers (μm) per second. At these speeds, a phospholipid could travel the length of a small bacterial cell in a second.

A membrane's permeability is closely related to its level of fluidity, which is a measure of molecular mobility. As temperature drops, molecules in a bilayer move more slowly and become less fluid. As a result, the hydrophobic tails in the interior of membranes pack together more tightly. At very low temperatures, lipid bilayers even begin to solidify. As the "Results" in Figure 6.11 indicated, low temperatures can make membranes impervious to molecules that would cross them readily at more moderate temperatures. Put your finger at 0°C on the *x*-axis of that graph (just about the freezing point of water), and note that membranes that lack cholesterol are almost completely impermeable to glycerol. But if you trace any of the three data lines in the same figure to the right (increasing temperature), you will see that permeability increases.

CHECK YOUR UNDERSTANDING

✔ **If you understood this section, you should be able to . . .**

1. Describe the characteristics of phospholipids that allow them to spontaneously form lipid bilayers in water.
2. Fill in a chart that has columns labeled Factor, Effect on permeability, and Reason and rows under the Factor column labeled Temperature, Cholesterol, Length of hydrocarbon tails, and Saturation of hydrocarbon tails.

Answers are available in Appendix A.

6.3 How Substances Move across Lipid Bilayers: Diffusion and Osmosis

Small uncharged polar and nonpolar molecules can cross membranes readily and spontaneously—that is, without an input of energy. The question now is: How is this possible? What process is responsible for movement of molecules across lipid bilayers?

After you complete this section, you should be able to . . .

- Predict how gradients can affect the movement of water and solutes across a membrane.

Diffusion

A thought experiment can help explain how substances can cross membranes spontaneously. Suppose you rack up a set of billiard balls in the middle of a pool table and then begin to vibrate the table.

1. Because of the vibration, the billiard balls will move about randomly. They will also bump into one another.
2. After these collisions, some balls will move outward—away from their original position.
3. As movement and collisions continue, the overall or net movement of balls will be outward. This occurs because the random motion of the balls disrupts their original, nonrandom position. As the balls move at random, they are more likely to move away from one another than to stay together.
4. Eventually, the balls will be distributed randomly across the table. The entropy of the billiard balls has increased. The second law of thermodynamics states that in an isolated system, entropy always increases (Ch. 2, Section 2.3).

This hypothetical example illustrates how vibrating billiard balls move at random. More to the point, it also explains how substances located on one side of a lipid bilayer can move to the other side spontaneously—because like the billiard balls, dissolved solutes are in constant random motion due to their thermal energy. Spontaneous movement of molecules and ions is known as **diffusion**.

A difference in solute concentrations creates what is called a **concentration gradient**. Solutes move randomly in all directions, but when a concentration gradient exists, there is a net movement from regions of high concentration to regions of low concentration. Diffusion down a concentration gradient, or away from the higher concentration, is a spontaneous process because it results in an increase in entropy.

Once the molecules or ions are randomly distributed throughout a solution, an equilibrium is established. For example, consider two aqueous solutions separated by a lipid—like the situation shown in step 1 of **Figure 6.13** on page 130. Steps 2 and 3 of the figure show how solutes that can pass through the bilayer diffuse to the other side. When substances diffuse across a membrane in the absence of an outside energy source, it is known as **passive transport**.

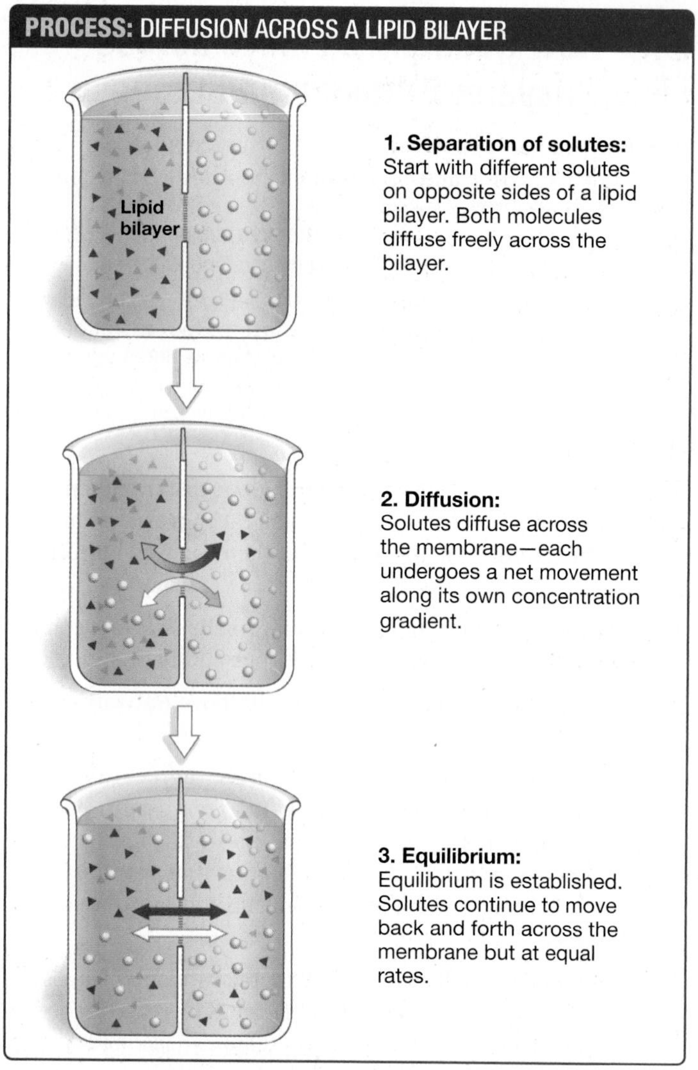

Figure 6.13 Diffusion across a Selectively Permeable Membrane Establishes an Equilibrium.

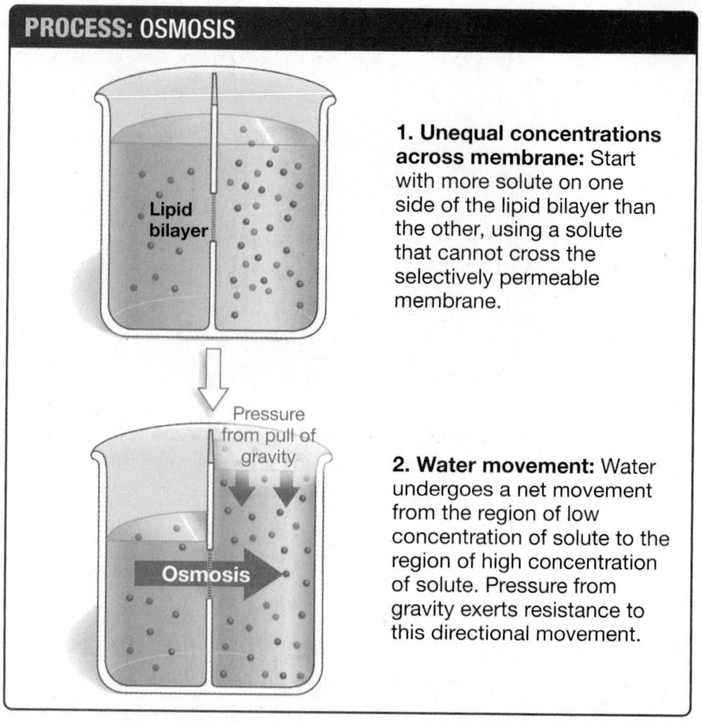

Figure 6.14 Osmosis Is the Diffusion of Water across a Selectively Permeable Membrane.

At equilibrium, movement across the membrane does not stop. Instead, these solutes continue to move back and forth across the membrane due to their constant random motion. At this point, there is no longer a net movement of solutes across the membrane because they are equally likely to move in any direction. ✔ **If you understand passive transport, you should be able to predict how increasing the temperature would affect the rate of achieving equilibrium in Figure 6.13.**

Osmosis

What about water? As the data in Figure 6.9 show, water moves across lipid bilayers relatively quickly. The movement of water is a special case of diffusion that is given its own name: **osmosis**. Osmosis occurs only when solutions are separated by a membrane that permits water to cross, but holds back some or all of the solutes—that is, a selectively permeable membrane. To drive this point home, let's suppose the concentration of a particular solute is higher on one side of a selectively permeable membrane than it is on the other side (Figure 6.14, step 1). Also suppose that this solute cannot diffuse through the membrane to establish equilibrium. What happens? Water will move from the side with a lower concentration of solute to the side with a higher concentration of solute (Figure 6.14, step 2).

It's also important to note that the solute affects the movement of water across a membrane. Recall that water molecules interact with charged particles and form hydrogen bonds with polar molecules (see Ch. 2, Section 2.2). If a solute can't cross the membrane, then any associated water molecules are also prevented from crossing. Thus, only unbound water molecules are able to diffuse across the membrane during osmosis.

The overall result is that osmosis dilutes the higher concentration of solute as water diffuses across the membrane. This directional movement is spontaneous because entropy will increase as the difference in solute concentrations decreases.

When water moves by osmosis, the solutions on both sides of the membrane experience a change in volume as well as a change in solute concentration. The greater the initial difference in solute concentration, the greater the volume change will be. However, opposing forces, such as the pressure resulting from the downward pull of gravity, exert resistance to the directional movement of water. ✔ **THINK CAREFULLY If you understand what forces contribute to osmosis, you should be able to predict whether or not the concentration of the solute in Figure 6.14 would be the same across the membrane after reaching equilibrium. Explain your answer.**

How is life affected by osmosis? When water moves across the membranes of cells and vesicles, the volume and concentration of

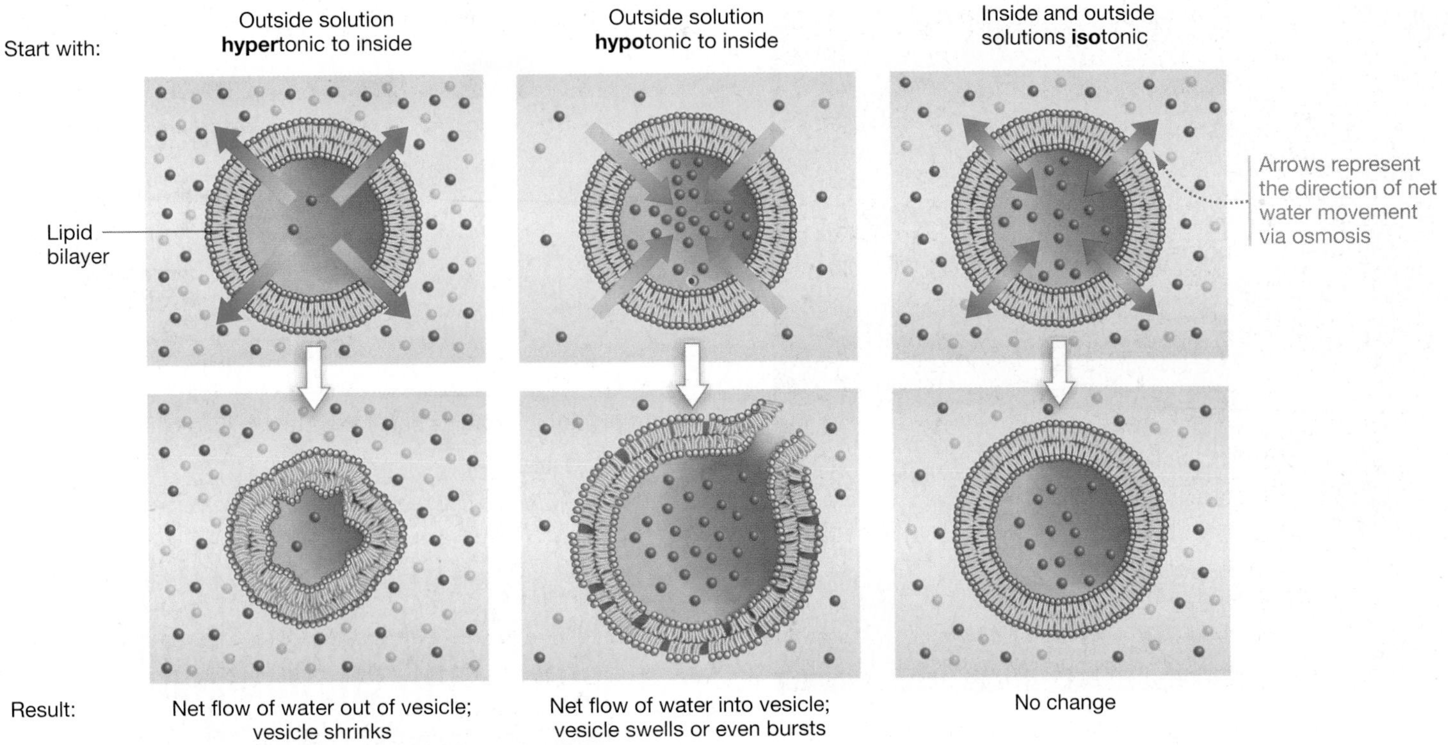

INTERACTIVE Figure 6.15 Osmosis Can Shrink or Burst Membrane-Bound Vesicles.

solutes enclosed within the membrane will change. In cells, a rapid change in amount of water can be catastrophic. To understand the effect of osmosis on the volume of a cell, consider the liposomes illustrated in Figure 6.15. (Remember that osmosis occurs only when a solute cannot pass through a separating membrane.)

- ***Left*** If the solution outside the vesicle has a higher concentration of solutes than the interior has, the solution outside is said to be **hypertonic** relative to the inside of the vesicle. Water moves out of the vesicle into the solution outside. As water leaves, the vesicle shrinks and the membrane shrivels.
- ***Middle*** If the solution outside the vesicle has a lower concentration of solutes than the interior has, the outside solution is said to be **hypotonic** relative to the inside of the vesicle. Water moves into the vesicle via osmosis. The incoming water causes the vesicle to swell, or even burst.
- ***Right*** If solute concentrations are equal on both sides of the membrane, the outside is said to be **isotonic**. There is no net movement of water, and the vesicle maintains its size and shape.

Note that the terms "hypertonic," "hypotonic," and "isotonic" are used only in referring to the effect of water movement into or out of a membrane-enclosed structure, such as a vesicle or cell.[1]

[1] The language of tonicity derives from how structures like vesicles or cells are affected by water transport, where "tone" (from the Greek root *tonos*, meaning "to stretch") refers to the structure's firmness. For example, if a vesicle is placed in a hypertonic solution, then the inside solution would be hypotonic ("lower-tone") because the vesicle's firmness is reduced as water leaves. The opposite is true for a vesicle placed in a hypotonic solution, where the inside solution would be hypertonic ("excess-tone") and cause the vesicle to swell as water enters.

Membranes and Chemical Evolution

What do diffusion and osmosis have to do with the first membranes floating in the prebiotic oceans of early Earth? Both processes tend to *reduce* differences in chemical composition between the inside and outside of membrane-bound compartments. But if liposome-like structures first arose in early Earth oceans, their interiors probably didn't offer a radically different environment from the surrounding solution.

In all likelihood, the primary importance of the first lipid bilayers was simply to provide a container for replicating the first "living" molecule—thought by many researchers to have been RNA (Ch. 4, Section 4.4). But ribonucleotide monomers would need to be available for these RNAs to replicate. Can negatively charged ribonucleotides get across lipid bilayers and inside lipid-bounded vesicles?

For lipid bilayers consisting of phospholipids, the answer is no. However, Jack Szostak and colleagues considered this question using alternative membranes that consist of fatty acids and other simple amphipathic lipids thought to be present in the early oceans. Like phospholipids, fatty acids are amphipathic and able to assemble into lipid bilayers and form water-filled vesicles. Their experiments showed that ions, and even ribonucleotides, can diffuse across fatty acid bilayers—meaning that monomers could have been available for RNA synthesis inside early Earth vesicles.

Simple vesicle-like structures that harbor nucleic acids are referred to as **protocells** (Figure 6.16 on page 132). Most origin-of-life researchers view protocells as possible intermediates in the evolution of the cell.

For a cell to emerge from these protocells, its membrane must be modified to import solutes necessary for life while excluding

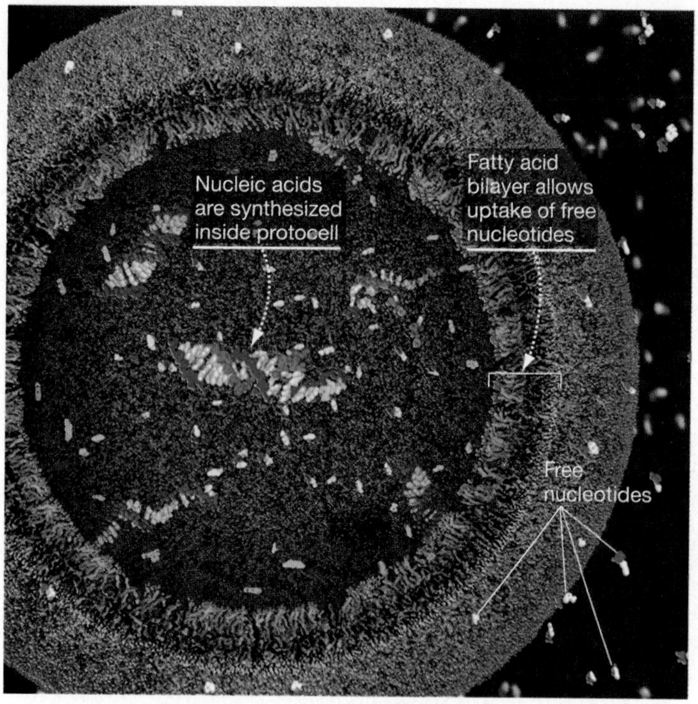

Figure 6.16 Protocells May Have Had Simple, Permeable Membranes. This image shows a cross section of a computer model of a protocell. Like this model, the membranes of early cells may have been built of fatty acids. Passive transport of nucleotides across these membranes, as well as replication of nucleic acids inside, has been observed in the laboratory.

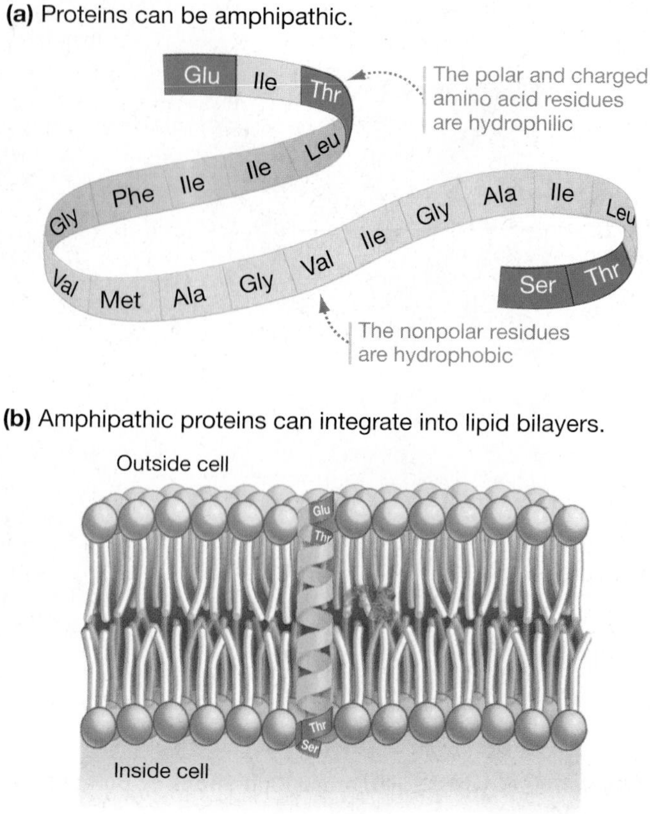

Figure 6.17 The Hydrophobic Region of an Amphipathic Protein Can Be Anchored into a Lipid Bilayer.

those that might damage it. What sort of molecule could become incorporated into a lipid bilayer and influence the bilayer's permeability? The title of the next section gives away the answer—proteins.

CHECK YOUR UNDERSTANDING

✔ **If you understood this section, you should be able to ...**

Make a concept map (see **BioSkills 12**) that includes these boxed terms: water molecules, solute molecules, osmosis, diffusion, areas of high-to-low concentration, selectively permeable membranes, concentration gradients, hypertonic solutions, hypotonic solutions, and isotonic solutions.

Answers are available in Appendix A.

6.4 Proteins Alter Membrane Structure and Function

After you complete this section, you should be able to ...

- Analyze the different roles of membrane proteins in regulating the transport of solutes.

If amphipathic lipids are responsible for the lipid bilayer of a cell membrane, how do proteins fit in? Can a protein be amphipathic too? Recall that proteins consist of amino acids, which have side chains that range from highly nonpolar to highly polar or charged (see Ch. 3, Figure 3.2). It's conceivable, then, that a protein could have a series of nonpolar amino acid residues in the middle of its primary structure flanked by polar or charged amino acid residues (**Figure 6.17a**). The nonpolar residues would be stable in the interior of a lipid bilayer, while the polar or charged residues would be stable alongside the polar lipid heads and surrounding water (**Figure 6.17b**).

Further, because the secondary and tertiary structures of proteins are almost limitlessly variable, it is possible for proteins to form openings and function as a selective passageway across a lipid bilayer.

From these considerations, it's not surprising that when researchers began analyzing the chemical composition of plasma membranes, they found that proteins were often just as common, in terms of mass, as phospholipids. How are these two types of molecules arranged in a membrane?

Development of the Fluid-Mosaic Model

In 1935 Hugh Davson and James Danielli proposed that cell membranes were structured like a sandwich in which hydrophilic proteins coat both sides of a pure lipid bilayer (**Figure 6.18a**). Early electron micrographs of plasma membranes seemed to be consistent with the sandwich model, and for decades it was widely accepted.

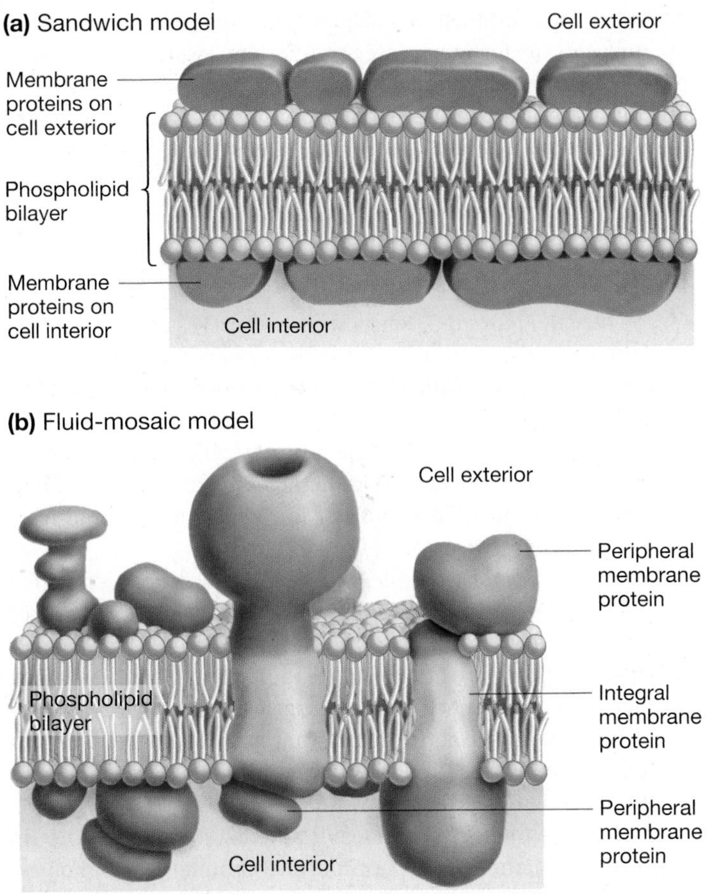

Figure 6.18 Past and Current Models of Membrane Structure Differ in Where Membrane Proteins Reside. (a) The protein-lipid-lipid-protein sandwich model was the first hypothesis for the arrangement of lipids and proteins in cell membranes. **(b)** The fluid-mosaic model was a radical departure from the sandwich hypothesis.

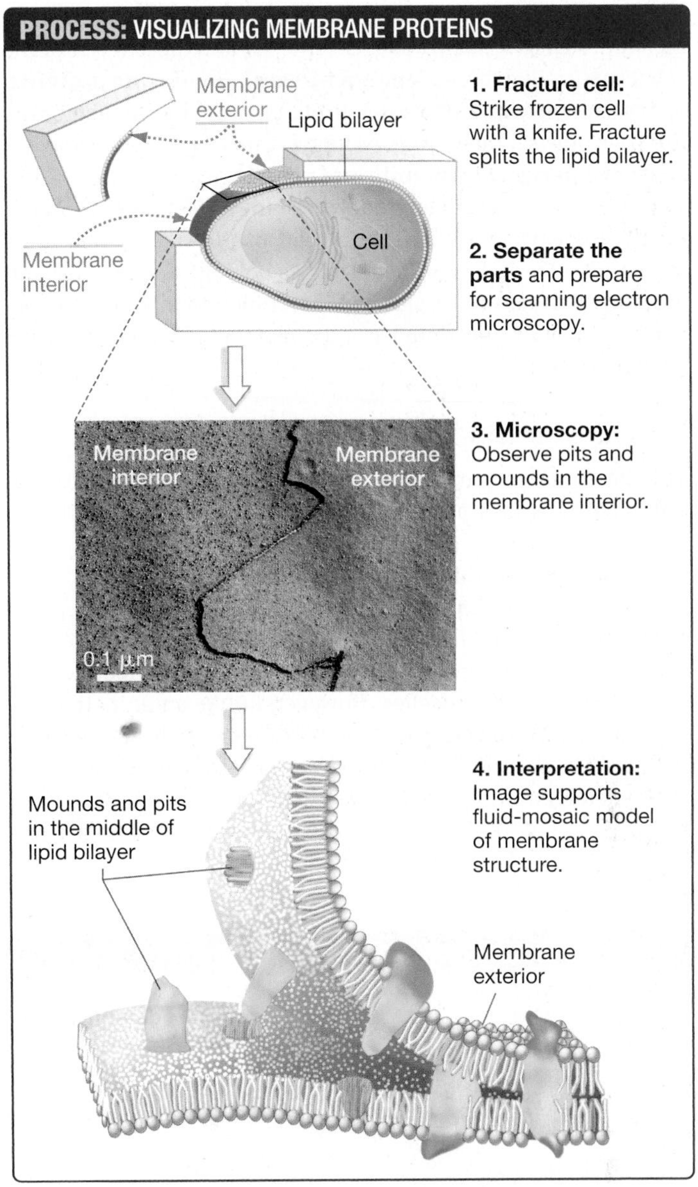

Figure 6.19 Freeze-Fracture Preparations Allow Biologists to View Membrane Proteins.

✔ **PROCESS OF SCIENCE** What would be an appropriate control to show that the pits and mounds were not simply irregularities in the lipid bilayer caused by the freeze-fracture process?

The realization that membrane proteins could be amphipathic, however, led S. Jon Singer and Garth Nicolson to suggest an alternative hypothesis. In 1972, they proposed that at least some proteins span the membrane instead of being found only outside the lipid bilayer. Their hypothesis was called the **fluid-mosaic model** (Figure 6.18b). Singer and Nicolson suggested that membranes are a dynamic and fluid mosaic of phospholipids and different types of proteins.

The debate over the nature of the cell membrane was resolved in the early 1970s with the development of an innovative technique for visualizing the surfaces of plasma membranes. The method is called freeze-fracture electron microscopy because the steps involve freezing and fracturing the membrane before examining it with a **scanning electron microscope (SEM)**, which produces images of an object's surface (see BioSkills 9).

As Figure 6.19 shows, the freeze-fracture technique allows researchers to split cell membranes and view the middle of the structure. The scanning electron micrographs that result show pits and mounds studding the inner surfaces of the lipid bilayer. Researchers interpreted these structures as the locations of membrane proteins. As step 4 in the figure shows, the mounds represent proteins that remained attached to one side of the split lipid bilayer and the pits are the holes they left behind.

These observations conflicted with the sandwich model but were consistent with the fluid-mosaic model. Based on these and subsequent observations, the fluid-mosaic model is now widely accepted. This basic model of membrane structure has been expanded over the past 40 years as researchers have learned more about the organization and function of membrane lipids, proteins, and the carbohydrates that may be bound to them (Ch. 5, Section 5.3, and Ch. 7, Section 7.5).

Notice in Figure 6.18b that some proteins span the membrane and have segments facing both the interior and the exterior of the cell. Proteins like these are called **integral membrane proteins**, or **transmembrane proteins**. Proteins that bind to membrane lipids or integral membrane proteins without passing through it are called **peripheral membrane proteins**.

Certain peripheral membrane proteins are found only on the membrane surface facing the interior of the cell, while others are found only on the cell exterior. As a result, the interior and exterior surfaces of the plasma membrane are distinct—the peripheral membrane proteins and the ends of transmembrane proteins differ.

Systems for Studying Membrane Proteins

The discovery of transmembrane proteins was consistent with the hypothesis that proteins affect membrane permeability. To test this hypothesis, researchers needed some way to isolate and purify membrane proteins.

Figure 6.20 outlines one method that researchers developed to separate proteins from membranes. The key to the technique is the use of detergents. A detergent is a small amphipathic molecule that can form micelles. However, unlike amphipathic lipids, detergents are water soluble. When detergents are added to the solution surrounding a lipid bilayer, the hydrophobic tails of the detergent molecule interact with the hydrophobic tails of the lipids and with the hydrophobic portions of transmembrane proteins. These interactions displace the membrane phospholipids and end up forming water-soluble detergent–protein complexes that can be isolated.

Since intensive experimentation on membrane proteins began, researchers have identified three broad classes of proteins that affect membrane permeability: channels, carriers, and pumps. Let's consider each class in turn.

Channel Proteins Facilitate Diffusion

As you saw in Figure 6.9, ions almost never cross pure phospholipid bilayers on their own. But in cells, ions routinely cross membranes by way of specialized transmembrane proteins called **ion channels**.

Ion channels form pores, or openings, in a membrane. Ions diffuse through these pores in a predictable direction: from regions of high concentration to regions of low concentration and from areas of like charge to areas of unlike charge.

In Figure 6.21, for example, a large sodium ion concentration gradient across the membrane favors the diffusion of these ions from the region of higher sodium concentration to the region of lower sodium concentration. But in addition, the solution above the membrane has a net positive charge while the solution below the membrane has a net negative charge. When considered together, concentration and electrical gradients are called an **electrochemical gradient**.

In response to electrochemical gradients, ions will diffuse in a directional manner if an appropriate channel exists. For example, if a sodium ion channel were inserted into the membrane in Figure 6.21, the net movement of sodium ions would occur down the electrochemical gradient. At equilibrium, sodium ions would continue to move back and forth through the channel, but at equal rates.

✔ **If you understand the basis of electrochemical gradients, you should be able to add another arrow to Figure 6.21 indicating the electrochemical gradient for chloride ions.**

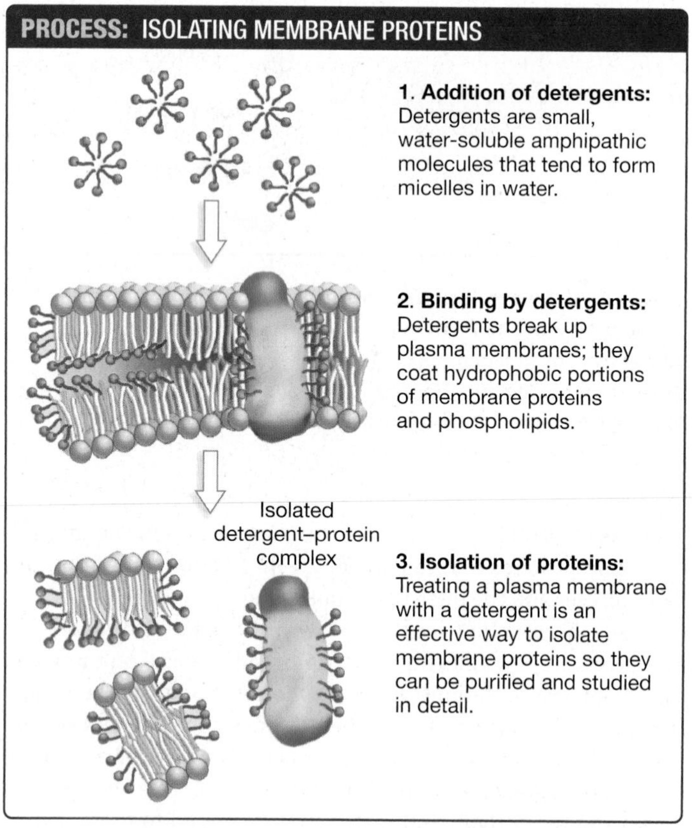

Figure 6.20 Detergents Can Be Used to Isolate Proteins from Membranes.

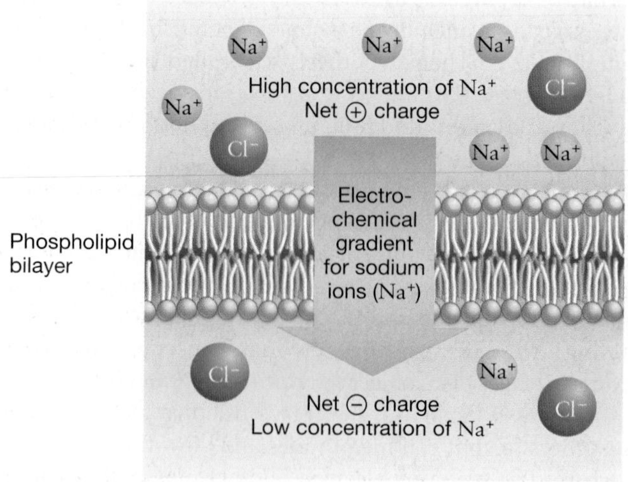

Figure 6.21 An Electrochemical Gradient Is a Combined Concentration and Electrical Gradient. Electrochemical gradients are established when ions build up on one side of a membrane.

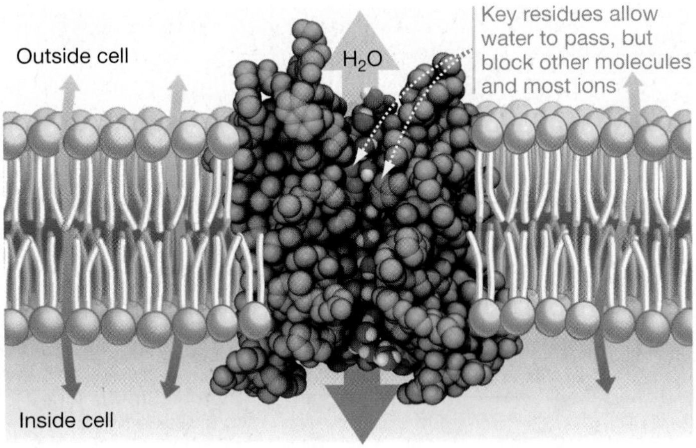

Figure 6.22 **Membrane Channels Are Highly Selective.** A cutaway view looking at the inside of a membrane channel, aquaporin. The key residues identified in the space-filling model selectively filter other molecules and most ions, allowing only water (red and white structures) to pass through.

Protein Structure Determines Channel Selectivity Cells have many different types of pore-like **channel proteins** in their membranes. Some of these channels are for ions, and others are for small polar molecules. Each channel protein has a structure that permits only a particular type of ion or small molecule to pass through it. What is responsible for this selectivity?

To understand what makes channels selective, first look at the structure of the pore formed across the membrane. The amino acid residues that line a channel's pore are hydrophilic relative to those facing the hydrocarbon tails of the membrane. For example, Figure 6.22 shows a cutaway side view of a channel called **aquaporin** ("water-pore"). Peter Agre and co-workers discovered that aquaporins allow water to cross the plasma membrane but exclude other molecules and most ions. Although water can move across lipid bilayers without aquaporins, they are transported over 10 times faster when these channels are present. This increased rate of transport is particularly important for the absorption of water in your gastrointestinal tract.

But how can aquaporin be selective for water and yet exclude other substances that might be associated with water? Researchers found the answer when examining the structure of aquaporin. Key side chains in the interior of the pore function as a filter. The position of these groups across the channel allows only water molecules, which are capable of interacting with all of the functional groups in a precise manner, to pass through to the other side.

Movement through Many Membrane Channels Is Regulated Recent research has shown that aquaporins and many other ion channels are **gated channels**—meaning that they open or close in response to a signal, such as the binding of a particular substance or a change in the electrical voltage across the membrane.

As an example of how voltage-gated channels work, Figure 6.23 shows a potassium channel in closed and open configurations. The electrical charge on the membrane is normally negative on the inside relative to the outside, which causes the channel to adopt a closed shape that prevents potassium ions (K^+) from passing through. When this charge asymmetry is reversed, the shape changes in a way that opens the channel and allows potassium ions to cross. The key point here is that in almost all cases, the flow of ions and small molecules through membrane channels is carefully controlled.

The movement of substances through channel proteins is passive—meaning it does not require an input of energy. Channel proteins simply enable ions (e.g., K^+) or small polar molecules (e.g., water) to diffuse across lipid bilayers efficiently in response to an existing gradient. When transmembrane proteins assist the passive transport of substances that otherwise would not cross a membrane readily, the process is called **facilitated diffusion**.

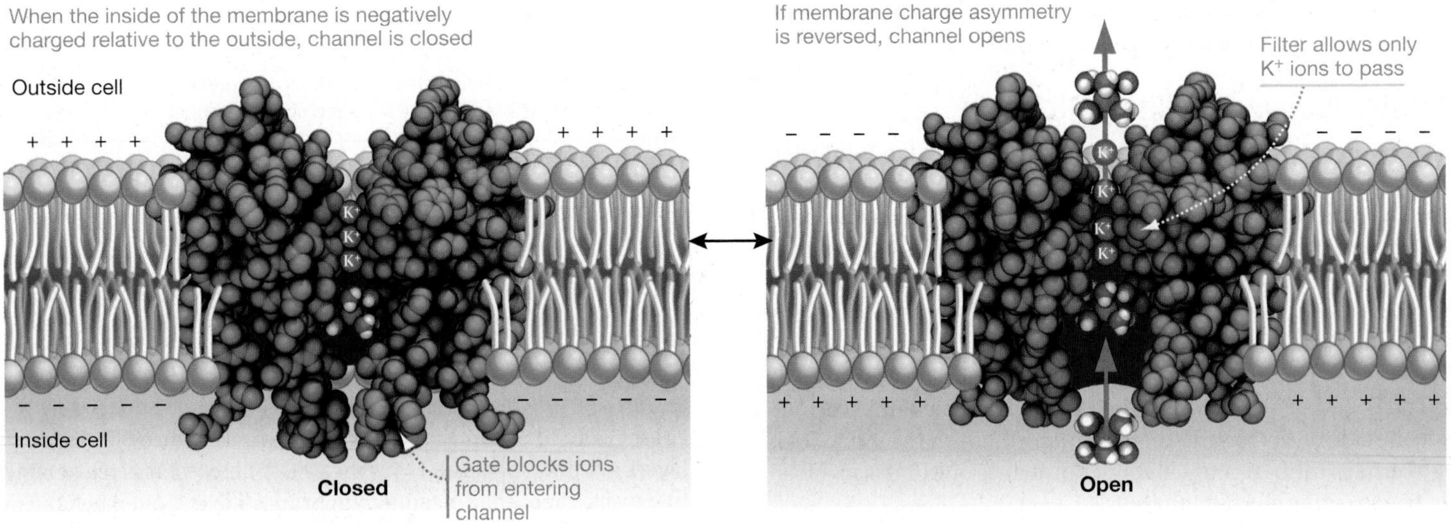

Figure 6.23 **Some Membrane Channels Are Highly Regulated.** A model of a voltage-gated K^+ channel in the closed and open configurations. The channel gate (at the bottom of the pore) changes shape based on the voltage across the membrane. The channel filter (blue residues at the top of the pore displaces water molecules that normally surround the K^+ ions in an aqueous solution before ions pass through the pore.

Is an Ion Channel Involved in Cystic Fibrosis? To drive home the importance of ion channels, let's consider work on the cause of cystic fibrosis (CF).

Cystic fibrosis is the most common genetic disease in humans of Northern European descent. It affects cells that produce mucus, sweat, and digestive juices. Normally these secretions are thin and slippery and act as lubricants. In individuals with CF, however, the secretions become abnormally concentrated and sticky, which can cause them to clog passageways in organs like the lungs.

Experiments published in 1983 suggested that cystic fibrosis is caused by defects in a transmembrane protein—the cystic fibrosis conductance transmembrane regulator, CFTR—which allows passage of chloride ions (Cl^-). It was proposed that a reduced rate of chloride ion transport was responsible for the thick mucus observed in the airway passages of affected individuals.

How is the transport of chloride ions involved in mucus consistency? If a defective CFTR channel prevents chloride ions from leaving cells surrounding airway passageways, a hypertonic state can't be established on airway surfaces. As a result, water isn't pulled from the cells by osmosis to maintain the proper mucus consistency inside the airways. In effect, the disease results from the mismanagement of osmosis.

Using molecular techniques introduced in Unit 3 (Ch. 20, Section 20.5), biologists were able to **(1)** find the gene that is defective in people suffering from CF and **(2)** isolate the gene from a healthy individual and use it to produce copies of the normal CFTR protein.

How did researchers confirm that CFTR is in fact a chloride channel? To do so, they inserted purified CFTR into planar bilayers and measured the flow of electric current across the membrane. Because ions carry a charge, ion movement across a membrane produces an electric current.

The graphs in Figure 6.24 show the results of this experiment by plotting the amount of current flowing across the membrane over time. Notice that when CFTR was absent, no electric current passed through the membrane. But when CFTR was inserted into the membrane, current began to flow. This was strong evidence that CFTR was indeed a chloride ion channel.

RESEARCH

QUESTION: Is CFTR a chloride channel?

HYPOTHESIS: CFTR increases the flow of chloride ions across a membrane.

NULL HYPOTHESIS: CFTR has no effect on membrane permeability.

EXPERIMENTAL SETUP:

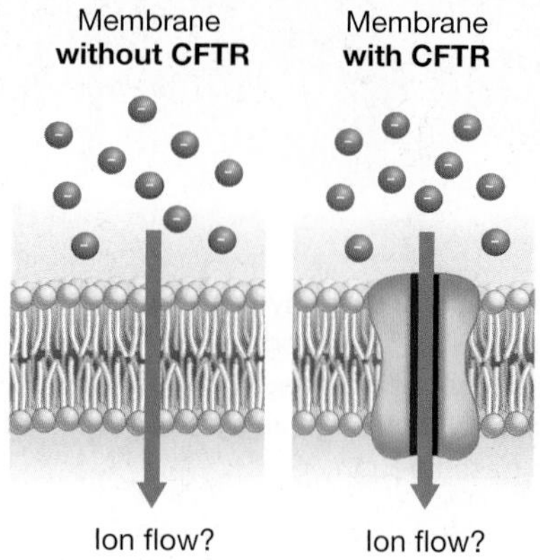

1. **Create planar bilayers** with and without CFTR.
2. **Add chloride ions** to one side of the planar bilayer to create an electrochemical gradient.
3. **Record electrical currents** to measure ion flow across the planar bilayers.

PREDICTION OF HYPOTHESIS: Ion flow (current) will be higher in the membrane with CFTR.

PREDICTION OF NULL HYPOTHESIS: Ion flow will be the same in both membranes.

RESULTS:

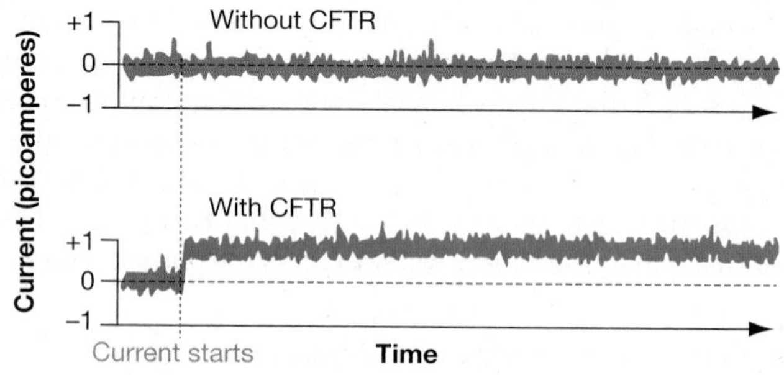

CONCLUSION: CFTR facilitates diffusion of chloride ions along an electrochemical gradient. CFTR is a chloride channel.

Figure 6.24 Electric Current Measurements Indicate that Chloride Ions Flow through CFTR.

SOURCE: C. A. Bear, C. Li, N. Kartner, et al. (1992). Purification and functional reconstitution of the cystic fibrosis transmembrane conductance regulator (CFTR). *Cell* 68: 809–818.

✔ If the researchers had repeated the "with CFTR" treatment, but reversed the orientation of channel protein in the membrane, would the results be different? Explain why or why not.

Carrier Proteins Facilitate Diffusion

The movement of water and K^+ are examples of facilitated diffusion through channel proteins, but facilitated diffusion can also occur through specialized membrane proteins called **carrier proteins**.

The primary difference between channels and carrier proteins is the mechanism of transport. Channels allow movement through a selective pore, much like bridges allow people to cross back and forth over a river. In contrast, carrier proteins selectively pick up a solute on one side of the membrane, then drop it off on the other side. This would be like a ferry picking up people on one side of a river and then dropping them on the other side.

Perhaps the best-studied carrier protein is one that is involved in transporting glucose into cells.

The Search for a Glucose Carrier Next to ribose, the six-carbon sugar glucose is the most prevalent sugar found in organisms. Virtually all living organisms use glucose as a building block for important macromolecules and as a source of stored chemical energy (Ch. 5, Section 5.3). But as shown in Figure 6.9, lipid bilayers are only moderately permeable to glucose. It is reasonable to expect, then, that plasma membranes have some mechanism for increasing their permeability to this sugar.

This prediction was supported in experiments on pure preparations of plasma membranes from human red blood cells. These plasma membranes turned out to be much more permeable to glucose than are pure lipid bilayers. Why?

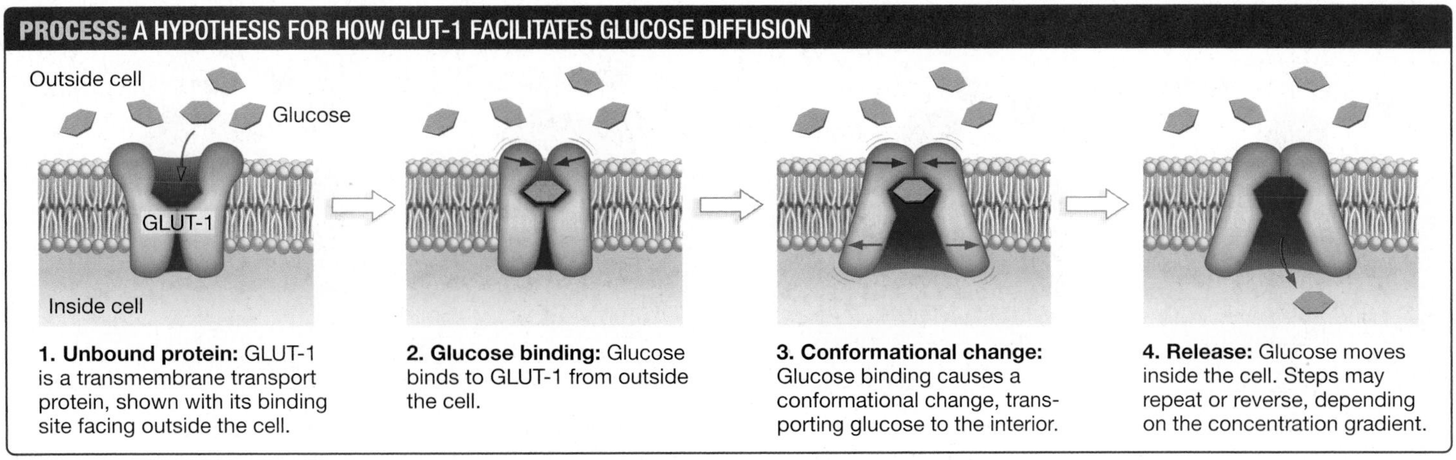

Figure 6.25 Carrier Proteins Move Substances via Structural Changes. This model shows that when GLUT-1 binds a glucose molecule, it undergoes a conformational change to move glucose across the membrane.

After isolating and analyzing many proteins from red blood cell membranes, researchers found one protein that specifically increases membrane permeability to glucose. When they added this purified protein to liposomes, the artificial membrane transported glucose at the same rate as a membrane from a living cell. This experiment convinced biologists that the membrane protein—now called GLUT-1 (short for glucose transporter 1)—was indeed responsible for transporting glucose across plasma membranes.

How Does GLUT-1 Work? Recall that proteins frequently change shape when they bind to other molecules and that such conformational changes are often a critical step in their function (Ch. 3, Section 3.3).

Figure 6.25 illustrates the current hypothesis for how GLUT-1 works to facilitate the movement of glucose. The idea is that when glucose binds to GLUT-1, it changes the shape of the protein in a way that moves the sugar through the hydrophobic region of the membrane and releases it on the other side.

What drives the movement of molecules through carriers? The answer is diffusion. GLUT-1 facilitates diffusion by allowing glucose to enter the carrier from either side of the membrane. Glucose will pass through the carrier in the direction dictated by its concentration gradient. A large variety of molecules move across plasma membranes via specific carrier proteins. To practice modeling the process of transport across a membrane, see Making Models 6.1.

Pumps Perform Active Transport

Diffusion—whether it is facilitated by proteins or not—is a passive process that moves substances in either direction across a membrane to make the cell interior and exterior environments more similar. But it is also possible for cells to move molecules or ions in a directed manner, often against an existing gradient. Accomplishing this task requires an input of energy to counteract the decrease in entropy that occurs when molecules or ions are concentrated. It makes sense, then, that transport against a gradient is called **active transport**.

Making Models 6.1 Tips on Drawing Membranes

Drawing models of membranes can help you understand membrane structure and transport, but the amount of detail you include depends on your goal. For example:

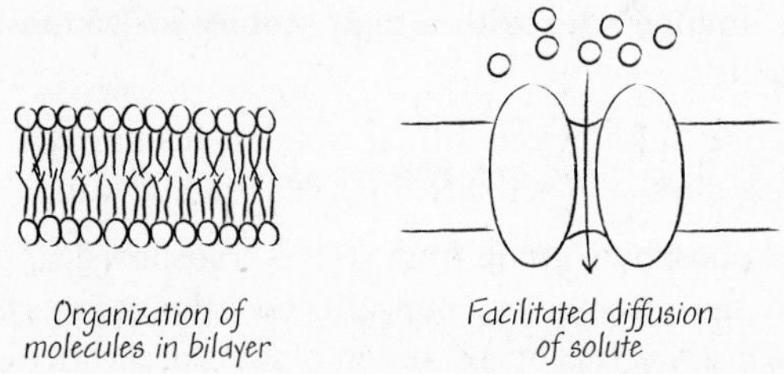

MODEL Draw what happens when there is a concentration gradient of glucose across a membrane that contains the GLUT-1 carrier protein. Use an arrow to indicate the gradient.

To see this model in action, go to the Study Area of **Mastering Biology**

In cells, ATP (the nucleotide adenosine triphosphate) often provides the energy for active transport by transferring a phosphate group (HPO_4^{2-}) to an active transport protein called a **pump**. Recall that ATP contains three phosphate groups (see Ch. 4, Section 4.1), and that phosphate groups carry two negative charges. When a phosphate group is transferred from ATP to a pump, its negative charges interact with charged amino acid residues in the protein. As a result, the pump's potential energy increases and its shape changes.

The Sodium–Potassium Pump A classic example of how structural changes can lead to active transport is provided in the **sodium–potassium pump**, or more formally, Na^+/K^+-ATPase. The Na^+/K^+ part of the name refers to the ions that are transported, ATP indicates that adenosine triphosphate is used, and *–ase* identifies the molecule as an enzyme.

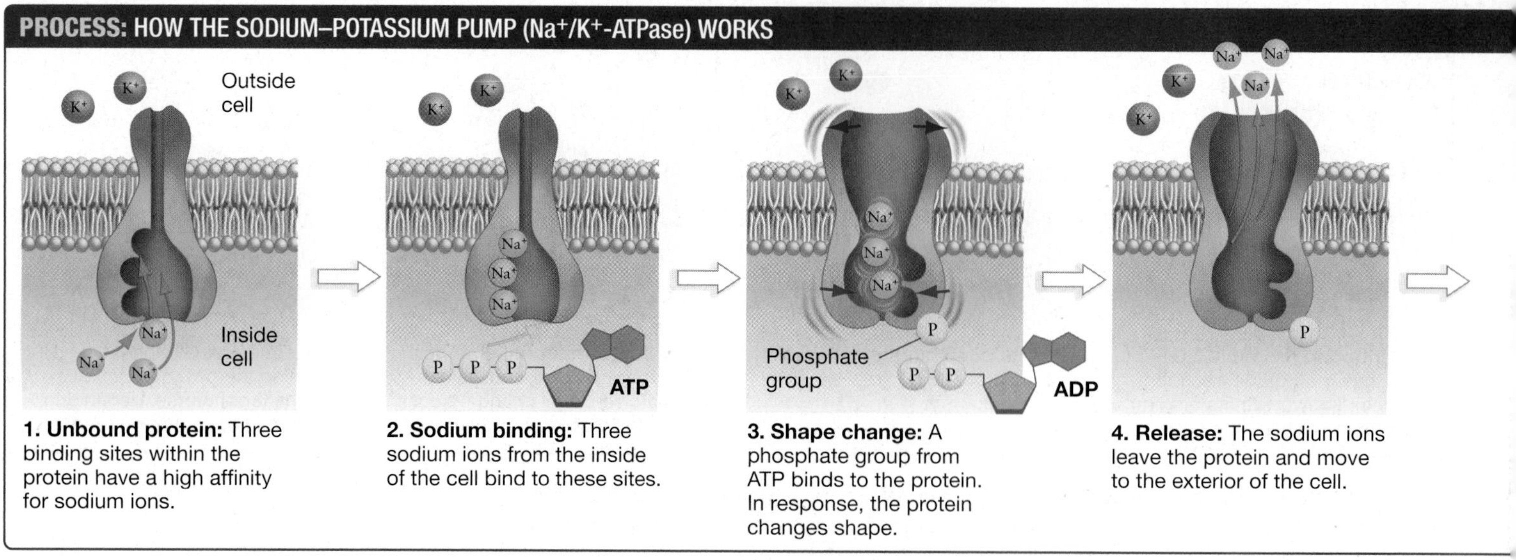

Figure 6.26 The Sodium–Potassium Pump Depends on an Input of Chemical Energy Stored in ATP.

As shown in **Figure 6.26**, sodium and potassium ions move in a multistep process:

Step 1 When Na^+/K^+-ATPase is in the conformation shown here, binding sites with a high affinity for sodium ions are available.

Step 2 Three sodium ions diffuse from the inside of the cell, bind to these sites, and activate the ATPase activity in the pump.

Step 3 A phosphate group from ATP is transferred to the pump. When the phosphate group attaches, the pump changes its shape in a way that opens the ion-binding pocket to the external environment and reduces the pump's affinity for sodium ions.

Step 4 The sodium ions exit the protein and diffuse to the exterior of the cell.

Step 5 In this conformation, the pump has binding sites with a high affinity for potassium ions facing the external environment.

Step 6 Two potassium ions from outside the cell bind to the pump.

Step 7 When the potassium is bound, the phosphate group is cleaved from the protein and its structure changes in response—back to the original shape with the ion-binding pocket facing the interior of the cell.

Step 8 In this conformation, the pump has low affinity for potassium ions. The potassium ions exit the protein and diffuse into the interior of the cell. The cycle then repeats.

Other types of pumps move protons (H^+), calcium ions (Ca^{2+}), or other ions or molecules across membranes in a directed manner, regardless of the existing gradient. This is an important point. If the gradient were to reverse, pumps would continue to use the same energy source to transport the solutes in the same direction, even if it is not against the existing gradient.

As a result, cells can import and concentrate valuable nutrients and ions inside the cell despite their relatively low external concentration. They can also expel molecules or ions, even when a gradient favors diffusion of these substances into the cell.

Secondary Active Transport Approximately 30 percent of all the ATP generated in your body is used to drive the Na^+/K^+-ATPase cycle. With each cycle, three Na^+ ions are exported for every two K^+ ions imported. In this way, the outside of the membrane becomes positively charged relative to the inside. In other words, the sodium–potassium pump converts energy from ATP to an electrochemical gradient across the membrane that favors a flow of anions (negative ions) out of the cell and a flow of cations (positive ions) into the cell.

The electrochemical gradient established by Na^+/K^+-ATPase represents a form of stored energy, much like the electrical energy stored in a battery. Do cells use this energy?

Gradients are crucial to the function of the cell, in part because they make it possible for cells to engage in **secondary active transport**—also known as cotransport. When cotransport occurs, ATP is not directly used to power transport, but instead an ATP pump provides the energy in the form of a gradient that is used to power the movement of a different solute in a directed manner, often against its particular gradient.

Recall that GLUT-1 facilitates the movement of glucose into or out of cells in the direction of its gradient. Can glucose be moved against its gradient? The answer is yes—a cotransport protein in your gut cells uses the Na^+ gradient created by Na^+/K^+-ATPases to import glucose against its chemical gradient. When Na^+ ions bind to this cotransporter, its shape changes in a way that allows glucose to bind. Once glucose binds, the cotransporter changes shape and transports both Na^+ ions and glucose to the inside of the cell. Note that in this case, Na^+ is moving down its gradient and the glucose is moving against its gradient. After dropping off Na^+ ions and glucose, the protein's original shape returns to repeat the cycle.

In this way, glucose present in the food you digest is actively transported into your body. The glucose molecules eventually

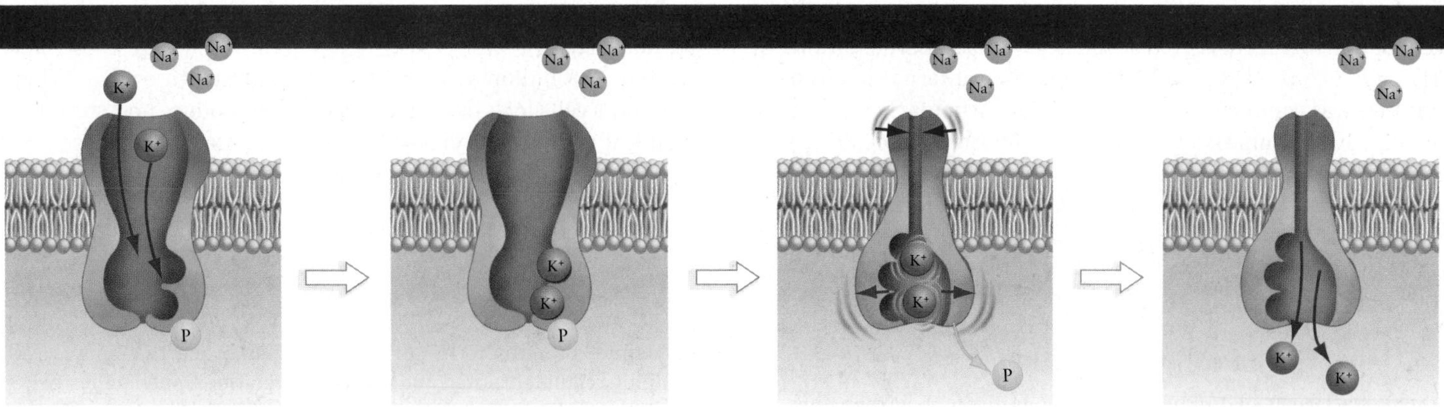

5. Unbound protein: In this conformation, the protein has binding sites with a high affinity for potassium ions.

6. Potassium binding: Two potassium ions bind to the pump.

7. Shape change: The phosphate group is cleaved from the protein, allowing the pump to return to its original shape.

8. Release: The potassium ions leave the protein and diffuse to the interior of the cell. *These 8 steps repeat.*

diffuse into your bloodstream and are transported to your brain, where they provide the chemical energy you need to stay awake and learn some biology. (You will learn more about secondary active transport in Units 6 and 7.)

Plasma Membranes Define the Intracellular Environment

Taken together, biological membranes combine the selective permeability of the lipid bilayer and the specificity of proteins involved in passive and active transport. These characteristics enable cells to create an internal environment that is much different from the external one (**Figure 6.27**).

CHECK YOUR UNDERSTANDING

✔ **If you understood this section, you should be able to ...**

1. Differentiate between peripheral and integral membrane proteins.
2. Explain what is passive about passive transport and active about active transport.
3. **MODEL** Draw a model to explain the process of secondary active transport of glucose. In this drawing, include the Na^+/K^+-ATPase and a Na^+/glucose cotransporter, along with the relevant gradients across the membrane.

Answers are available in Appendix A.

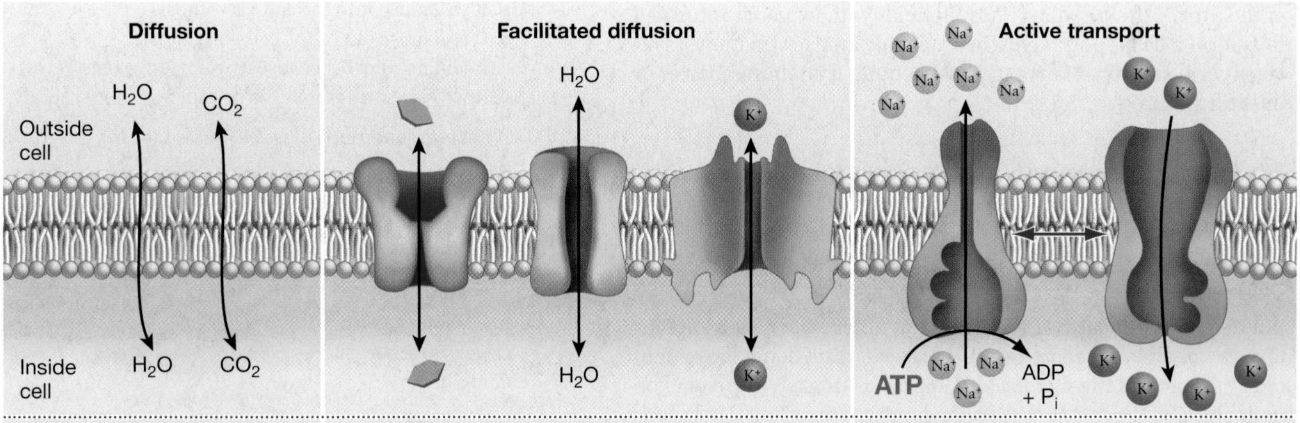

	Diffusion	Facilitated diffusion	Active transport
Description:	Passive movement of small, uncharged molecules along an electrochemical gradient, through a membrane	Passive movement of ...	Active movement of ...
Protein(s) involved:	None		

Figure 6.27 **Summary of the Passive and Active Mechanisms of Membrane Transport.**

✔ Complete the chart.

With the evolution of membrane proteins that either passively or actively transported substances across the membrane, early cells acquired the ability to create an internal environment that was conducive to life. Cells with particularly efficient and selective membrane proteins that imported molecules required for survival and replication would be favored by natural selection and would come to dominate the population. Cellular life had begun.

Some 3.5 billion years later, cells continue to evolve. What do today's cells look like, and how do they produce and store the chemical energy that makes life possible? Answering these and related questions is the focus of the following unit.

CHAPTER 6 Review

For media, go to **Mastering Biology**

6.1 Lipid Structure and Function

- Lipids are largely hydrophobic compounds due to their high number of nonpolar C—H bonds.
- The three main types of lipids found in cells are fats, steroids, and phospholipids. These molecules vary considerably in structure and function. Fats store chemical energy; certain steroids and phospholipids are key components of cell membranes.
- In hydrocarbon chains, the length and degree of saturation profoundly affect their physical properties.
- All lipids have a hydrophobic region, but amphipathic lipids also possess a distinct hydrophilic region containing polar or charged groups. Phospholipids have a polar or charged head and a nonpolar tail. The nonpolar tail usually consists of fatty acids or isoprenoids.

6.2 Phospholipid Bilayers

- In solution, phospholipids spontaneously assemble into bilayers that can serve as a physical barrier between an internal and external environment.
- Small nonpolar molecules tend to move directly across lipid bilayers readily; ions cross rarely, if at all.
- The permeability and fluidity of lipid bilayers depend on the temperature, the concentration of cholesterol, and the chemical structure of the lipids present, such as the saturation status and length of the hydrocarbon chains. Phospholipids with longer or saturated tails form a dense and highly hydrophobic interior that lowers bilayer permeability, relative to phospholipids containing shorter or unsaturated tails.

6.3 How Substances Move across Lipid Bilayers: Diffusion and Osmosis

- Diffusion is the random movement of ions or molecules owing to their thermal energy.
- If a membrane separates solutions that differ in concentration or charge, passive transport results in the net directional movement of solutes across the membrane that makes the environments on both sides of the membrane more similar. This is a spontaneous process driven by an increase in entropy.
- The diffusion of water across a membrane in response to a concentration gradient is called osmosis.

6.4 Proteins Alter Membrane Structure and Function

- The permeability of lipid bilayers can be altered significantly by membrane proteins.
- Channel proteins form pores in the membrane that may have highly regulated closed and open conformations, and facilitate the diffusion of specific solutes into and out of the cell.
- Carrier proteins undergo conformational changes that facilitate the diffusion of specific molecules into and out of the cell.
- Pumps use energy to actively move ions or molecules in a single direction, often against the electrical or chemical gradient.
- In combination, the selective permeability of phospholipid bilayers and the specificity of transport proteins make it possible to create an environment inside a cell that is radically different from the exterior environment.

Answers are available in Appendix A.

✔ TEST YOUR KNOWLEDGE

1. How do the phospholipids in archaea differ from those in other cells?
   a. They have tails made of unsaturated fatty acids instead of saturated fatty acids.
   b. They do not contain hydrocarbon chains.
   c. They have isoprenoid tails instead of fatty acid tails.
   d. They have two hydrocarbon chains instead of three hydrocarbon chains.
2. If a solution surrounding a cell is hypertonic relative to the inside of the cell, how will water move?
   a. It will move into the cell via osmosis.
   b. It will move out of the cell via osmosis.
   c. It will not move, because equilibrium exists.
   d. It will evaporate from the cell surface more rapidly.
3. What two conditions must be present for osmosis to occur?
4. Integral membrane proteins are anchored in lipid bilayers. Which of the following groups of amino acid residues (see Ch. 3, Figure 3.2) would likely be found in the portion that crosses the lipid bilayer?
   a. acidic
   b. basic
   c. polar uncharged
   d. nonpolar

✔ TEST YOUR UNDERSTANDING

5. Cooking oil lipids consist of long, unsaturated hydrocarbon chains. Would you expect these molecules to form membranes spontaneously? Why or why not? Describe, on a molecular level, how you would expect these lipids to behave in water.
6. **MODEL** Draw and label the plasma membrane of a cell that is placed in a solution with concentrations of calcium ions and lactose that are greater than those on the inside of the cell. Use arrows to show the relevant gradients and the activity of the following membrane

proteins: (1) a pump that exports protons; (2) a calcium channel; and (3) a lactose carrier.

7. In terms of structure, how do channel proteins differ from carrier proteins?

8. **THINK CAREFULLY** Suppose a cell is placed in a solution with a high concentration of potassium and no sodium. How would the cellular sodium–potassium pump function in this environment?
   a. It would stop moving ions across the membrane.
   b. It would continue using ATP to pump sodium out of the cell and potassium into the cell.
   c. It would move sodium and potassium ions across the membrane, but no ATP would be used.
   d. It would reverse the direction of sodium and potassium ions to move them against their gradients.

✔ TEST YOUR PROBLEM-SOLVING SKILLS

9. **PROCESS OF SCIENCE** In an experiment, you create two groups of liposomes in a solution containing 0.1 M NaCl—one made from red blood cell membranes and the other from frog egg cell membranes. When the liposomes are placed in water, those with red blood cell membranes burst more rapidly than those made from egg membranes. What could explain these results? Select True or False for each of the following statements.
   T/F The red blood cell liposomes are more hypertonic relative to water than the frog egg liposomes.
   T/F The red blood cell liposomes are more hypotonic relative to water than the frog egg liposomes.
   T/F The red blood cell liposomes contain more aquaporins than the frog egg liposomes.
   T/F The frog egg liposomes contain ion channels, which are not present in the red blood cell liposomes.

10. **QUANTITATIVE** Examine the experimental chamber in Figure 6.8a. Explain what would occur by osmosis if you added a 1-M solution of sodium chloride on the left side and an equal volume of a 1.5 M solution of potassium ions on the right. How might the addition of the CFTR protein to the lipid bilayer impact the direction of water movement?

✔ PUT IT ALL TOGETHER: Case Study

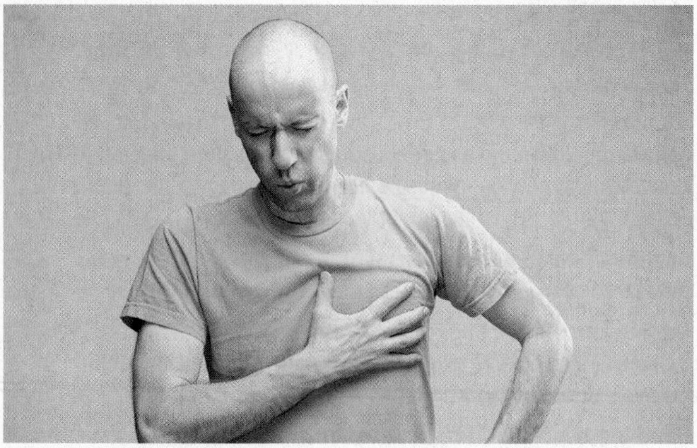

Is lipid structure in foods you eat linked to heart disease?

The media is full of advice on what you should and should not eat, but these recommendations often change from year to year. You may have heard that polyunsaturated fats are heart healthy and saturated fats are not, but is it really that simple?

11. How you prepare food can affect the amount of saturated fats that you eat. For example, if you cook meat on a grill, the product will have less saturated fat than if you were to eat the same meat raw. What occurs during the process of cooking the meat on a grill that would explain this result?

12. How is the chemical structure of saturated fats different from that of unsaturated fats? What physical property is often associated with these chemical differences?

13. An industrial process called hydrogenation is used to convert vegetable oils into semisolid compounds such as margarine. Explain why hydrogenation is an appropriate term for this process.

14. A side effect of hydrogenation is that a small percentage of oil lipids retain their double bonds, but instead of causing kinks, they straighten out the hydrocarbon chain. These are called trans fats based on the configuration of the double bond (*cis* bonds generate kinks—see Figure 6.1b, right—but *trans* bonds do not). How do trans fats complicate the previous description of saturated and unsaturated fats?

15. **PROCESS OF SCIENCE** To study the effect of food on heart disease, researchers fed a variety of diets to mice and then examined them for atherosclerosis—the narrowing of arteries that is a leading cause of heart attacks. Data observed from mice fed with trans fats are provided below. Do these data identify the cause of atherosclerosis in the mice? Explain. (Note that "causation" means that one event is responsible for the occurrence of the other, while "correlation" means that the two events appear to occur together).

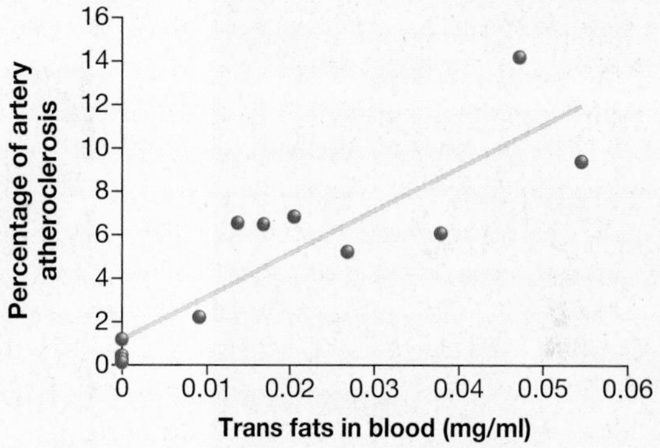

Source: C. M. C. Bassett, R. S. McCullough, A. L. Edel, et al. 2009. *Metabolism: Clinical and Experimental* 58: 1802–1808.

16. **SOCIETY** Recent studies on humans have shown that trans fats are correlated with heart disease while saturated fats are not. In addition to diet, what other factors should be evaluated before drawing conclusions on the health risk of trans fats?

END-OF-UNIT
CASE STUDY

UNIT 1

PUT IT ALL TOGETHER

For an introduction to the Mystery of the Newt case study, see page 17.

Now that you've learned about the molecules that are required for life, let's investigate the properties of a molecule that can cause death. Recall from the introduction to the Mystery of the Newt (page 17) that a single newt killed three hunters after it got into their coffee pot. As the newt was boiled over the campfire, a substance on its skin—called tetrodotoxin (TTX)—contaminated the water to make a lethal brew.

✔ If you understand Unit 1, you should be able to apply your learning to this case study:

What's So Toxic about Tetrodotoxin?

TTX is found in a variety of animals and is often used as a defense against predation. Puffer fish—a delicacy in Japanese cuisine—is a well-known cause of accidental, sometimes fatal, TTX food poisoning in humans. In our story, the newt's body was not eaten—the campers simply drank water in which it was boiled.

1. If the poisoning resulted from drinking contaminated water alone, what does this tell you about the solubility and stability of the TTX present on the newt's skin?
2. Consider the structure of the TTX secreted by the newt (**Figure 1**). Which of the following functional groups is most common on this molecule? (See Section 2.4)
   a. Amino **b.** Carbonyl **c.** Hydroxyl **d.** Carboxyl
3. Explain how the abundant functional group contributed to the solubility of TTX in the water surrounding the boiled newt. (See Section 2.2)

Takifugu rubripes (puffer fish, fugu)

Figure 1

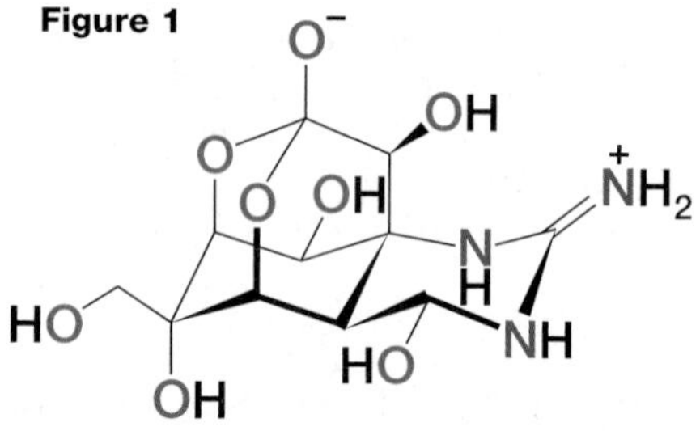

Newt TTX (tetrodotoxin)

How did TTX kill the campers? In the early 1960s, researchers determined that TTX interferes with the activity of nerve and muscle tissue by binding to a sodium ion channel (see Section 6.4) in the plasma membranes of nerve cells.

4. Describe the functional role of sodium ion channels. How do these channels differ from carrier proteins and active transport pumps?
5. Cells use the sodium electrical gradient to perform many activities. What membrane protein is used to establish the sodium electrochemical gradient and how does it work?
6. After a sodium gradient is established, which side of the plasma membrane would you expect to be more positively charged?

Ion channels allow passive diffusion of particular ions along their electrochemical gradient. Ion movement through the TTX-sensitive channel is regulated by the electrical gradient, or voltage, across the plasma membrane. Voltage in resting cells is normally a negative value (–90 mV). Researchers used artificial lipid bilayers loaded with the TTX-sensitive sodium channel to determine how its activity is affected by variations in membrane voltage (**Figure 2**).

7. What relationship did researchers find between the electrical gradient and the activity of the sodium ion channel?

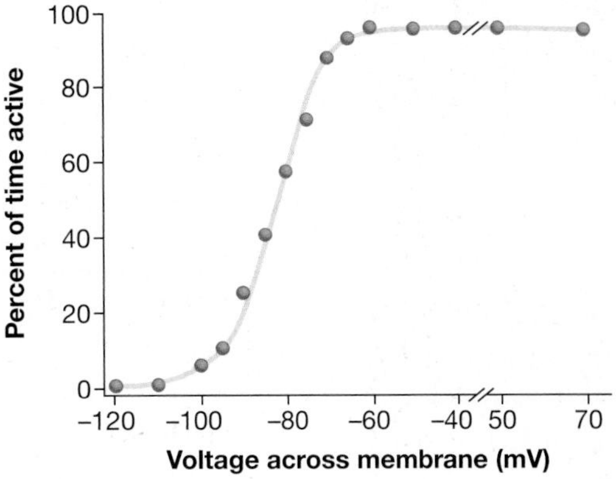

Figure 2

DATA: R. P. Hartshorne et al. 1985. *PNAS* | *Neurobiology* 82: 240–244, Figure 1C.

Figure 3 shows the activated and deactivated conformations of the voltage-gated sodium channel. Changes in conformation are based on a regulatory alpha helix that slides closer to the cytoplasm in the deactivated state and closer to the exterior in the activated state.

8. In Figure 3, add plus and minus symbols to represent the electrical gradient across the membrane in the two different conformations.

9. What type of amino acids would you expect to be included in the regulatory alpha helix based on its change in position? (See Figure 3.2)

Figure 3

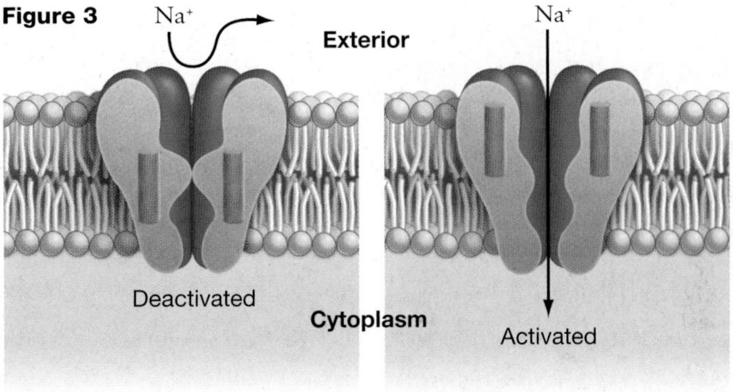

Figure 4

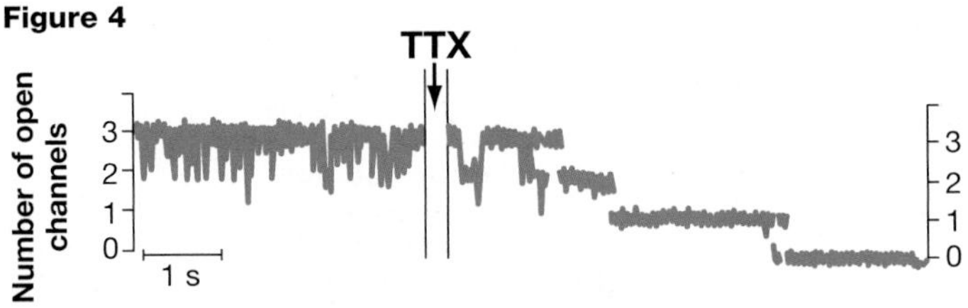

DATA: R. P. Hartshorne et al. 1985. *PNAS* | *Neurobiology* 82: 240–244, Figure 2B.

Toxins that alter the activity of voltage-gated ion channels are common. Some of these toxins keep the channels continuously active while others inactivate them. Using the same artificial bilayer system, the researchers tested the effect of TTX on the activity of the voltage-gated sodium channels.

10. Based on the data in **Figure 4**, how does TTX impact the transport of ions through the voltage-gated sodium channels?

11. MODEL Use Figure 3 to draw a simple model to predict two different ways that TTX could affect sodium transport by the voltage-gated sodium channel.

Research on the structure of voltage-gated sodium channels revealed that TTX binds to certain amino acid residues in the pore of the channel, interfering with its normal function (**Figure 5**). In the absence of functional voltage-gated sodium channels, signals between the brain and muscles are disrupted and the campers likely died of either respiratory or heart failure.

Figure 5

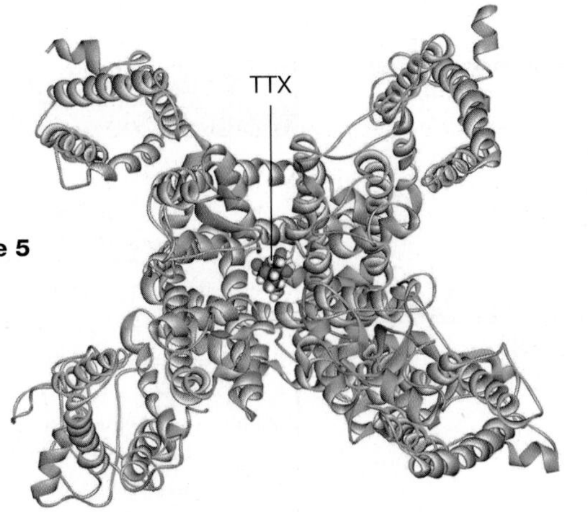

Top down view of channel pore blocked by TTX

Each unit ends with a continuation of this story. ■

BIG PICTURE

The first spark of life ignited when simple chemical reactions began to convert small molecules into larger, more complex molecules with novel 3-D structures and activities. According to the theory of chemical evolution, these reactions eventually led to the formation of the four types of macromolecules characteristic of life—proteins, nucleic acids, carbohydrates, and lipids.

As you look through this concept map, consider how the functions of the four types of macromolecules are determined by their structures, and how these structures stem from the chemical properties of the atoms and bonds used to build them.

Note that most boxes in the concept map indicate the chapters and sections where you can go for review. Also, be sure to do the blue exercises in the Check Your Understanding box below.

THE BIG PICTURE

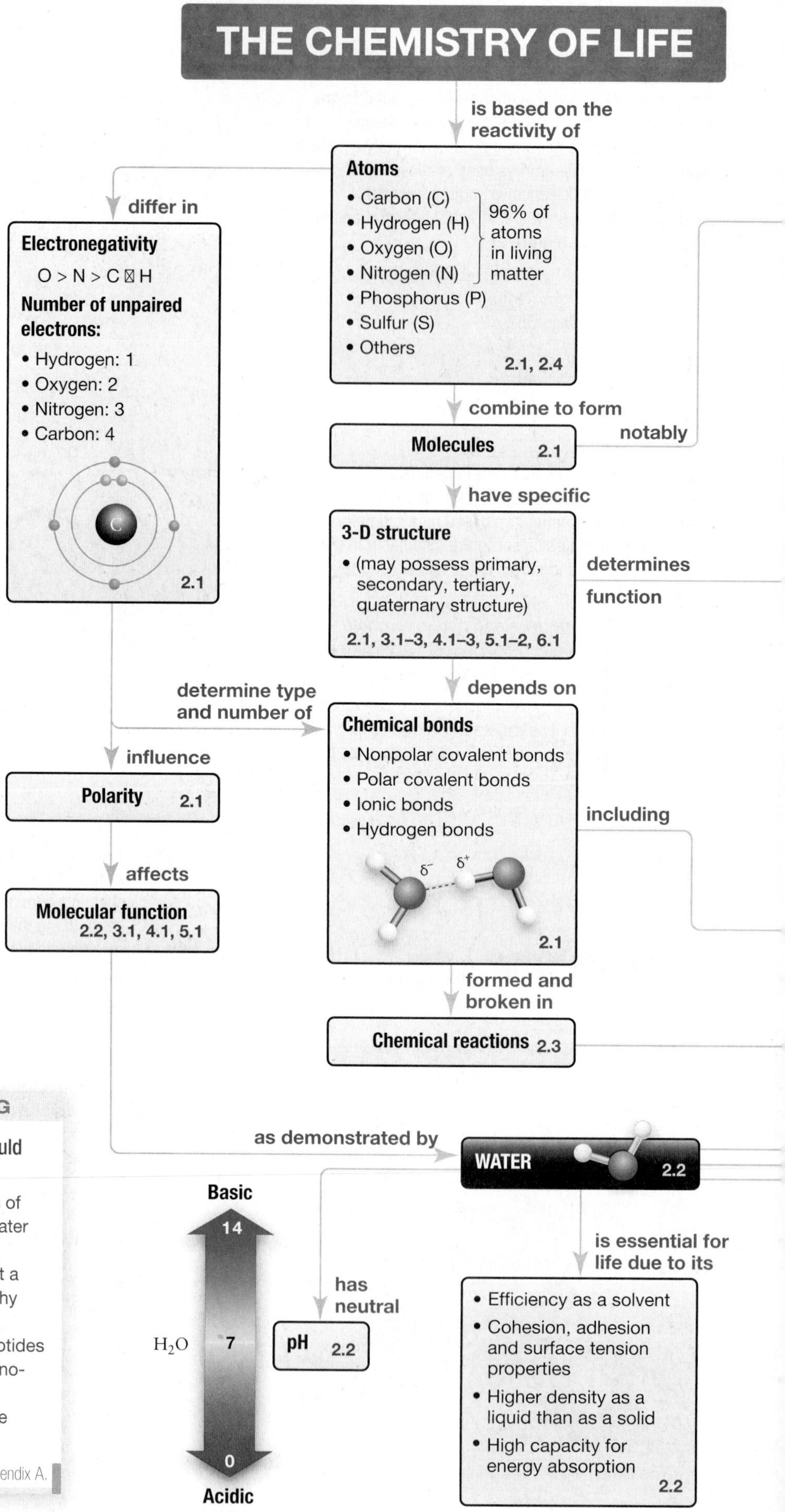

CHECK YOUR UNDERSTANDING

✔ If you understand the big picture, you should be able to . . .

1. Explain how the relative electronegativities of atoms affect bonding within and among water molecules.
2. Describe the attributes of RNA that make it a candidate for the origin of life molecule. Why isn't DNA considered a viable candidate?
3. Circle the atoms in amino acids and nucleotides that engage in creating bonds that link monomers to form polymers.
4. Draw a protein in the lipid bilayer. What role might it play?

Answers are available in Appendix A.

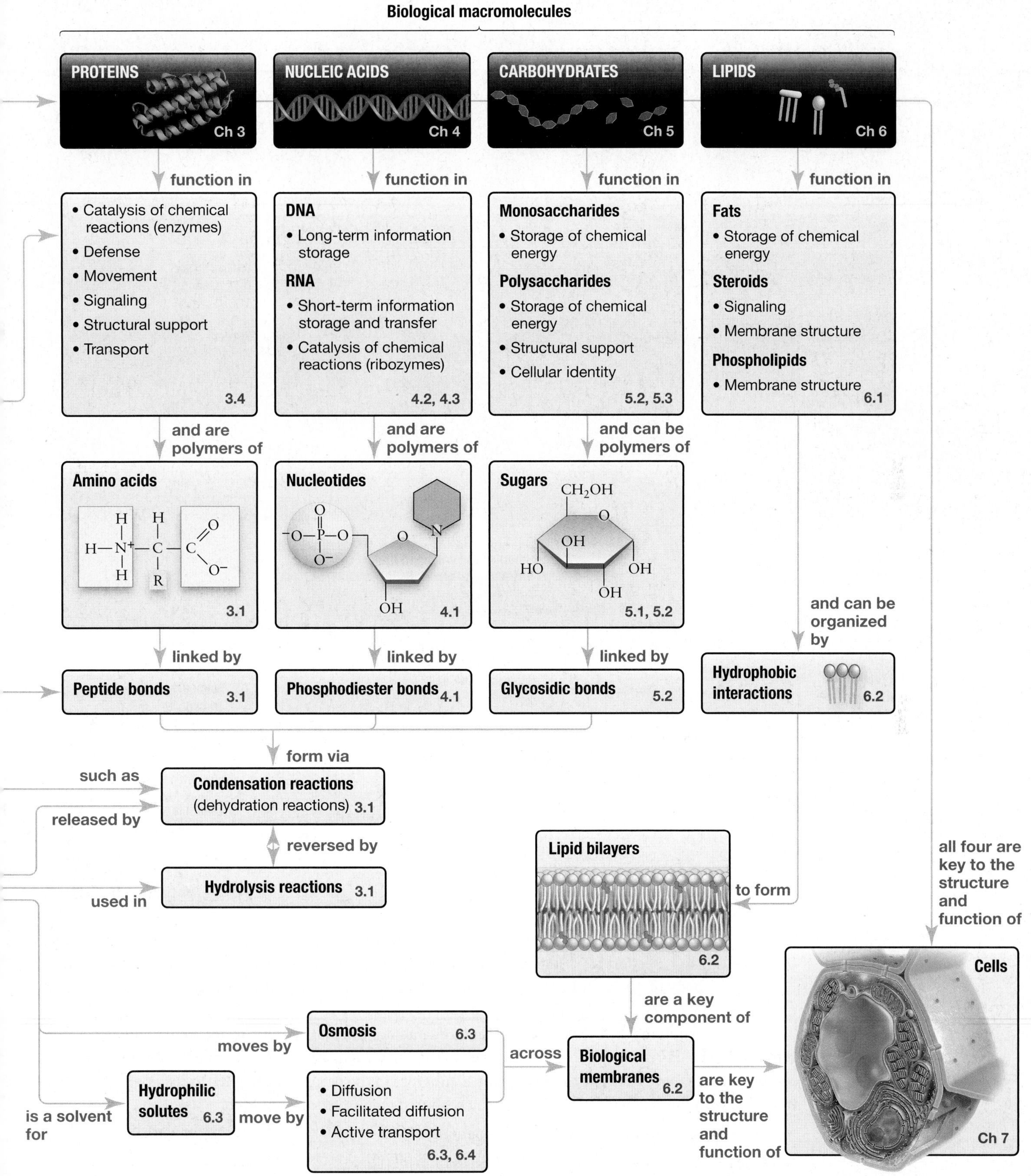

Biological macromolecules
PROTEINS Ch 3
NUCLEIC ACIDS Ch 4
CARBOHYDRATES Ch 5
LIPIDS Ch 6
function in
• Catalysis of chemical reactions (enzymes)
• Defense
• Movement
• Signaling
• Structural support
• Transport
3.4
DNA
• Long-term information storage
RNA
• Short-term information storage and transfer
• Catalysis of chemical reactions (ribozymes)
4.2, 4.3
Monosaccharides
• Storage of chemical energy
Polysaccharides
• Storage of chemical energy
• Structural support
• Cellular identity
5.2, 5.3
Fats
• Storage of chemical energy
Steroids
• Signaling
• Membrane structure
Phospholipids
• Membrane structure
6.1
and are polymers of
and can be polymers of
Amino acids 3.1
Nucleotides 4.1
Sugars 5.1, 5.2
and can be organized by
linked by
Peptide bonds 3.1
Phosphodiester bonds 4.1
Glycosidic bonds 5.2
Hydrophobic interactions 6.2
form via
such as
Condensation reactions (dehydration reactions) 3.1
released by
reversed by
Hydrolysis reactions 3.1
used in
Lipid bilayers 6.2
to form
all four are key to the structure and function of
are a key component of
Cells Ch 7
Osmosis 6.3
moves by
across
Biological membranes 6.2
are key to the structure and function of
is a solvent for
Hydrophilic solutes 6.3
move by
• Diffusion
• Facilitated diffusion
• Active transport
6.3, 6.4

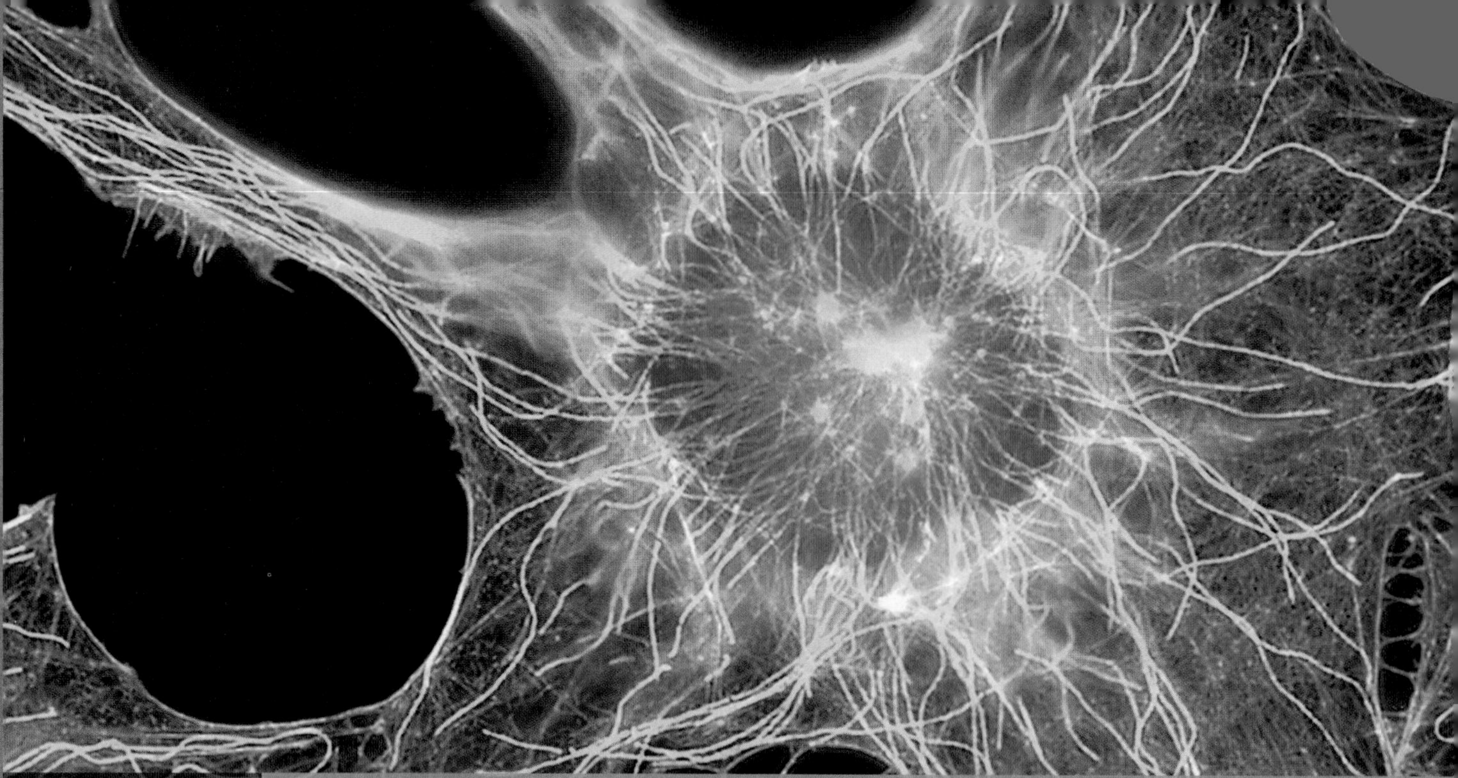

7 Inside the Cell

This cell has been treated with fluorescing molecules that bind to its fibrous cytoskeleton. Microtubules (large protein fibers) are yellow; actin filaments (smaller fibers) are blue. The cell's nucleus has been stained green.

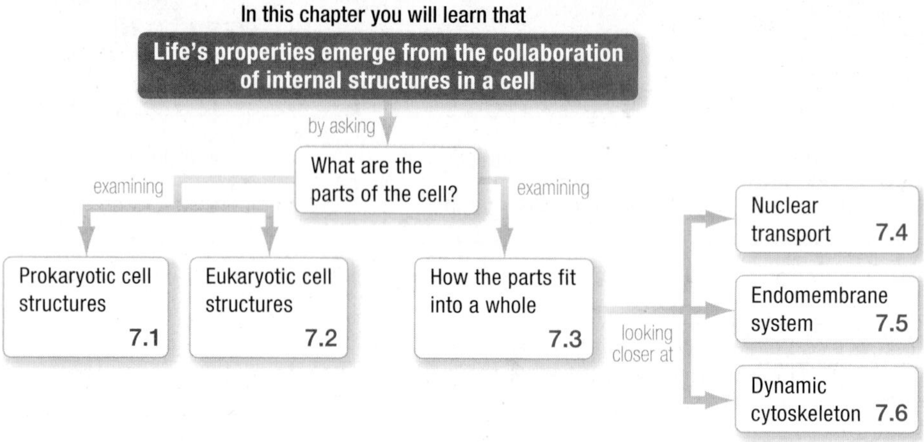

The cell theory states that all organisms consist of cells, and all cells are derived from preexisting cells (Ch. 1, Section 1.2). Since this theory was initially developed and tested in the 1850s, an enormous body of research has confirmed that the cell is the fundamental structural and functional unit of life. Life on Earth is cellular.

Previous chapters (Unit 1) delved into the fundamental attributes of life by looking at biologists' current understanding of how the cell evolved—from the early chemistry to the assembly and replication of a protocell. As the first cells left their deep-sea incubators in the oceans of early Earth (Ch. 6, Section 6.3), they took with them characteristics that are now shared among all known life-forms:

- *Proteins* that perform most of the cell's functions
- *Nucleic acids* that store, transmit, and process information

- *Carbohydrates* that provide chemical energy, carbon, support, and identity
- *A plasma membrane* that serves as a selectively permeable barrier

Thanks to the selective permeability of phospholipid bilayers and the activity of membrane transport proteins, the plasma membrane creates an internal environment that differs from conditions outside the cell. Our task now is to explore the structures contained within this membrane to understand how the properties of life emerged from their collaboration.

Cells are divided into two fundamental types: eukaryotes and prokaryotes (Ch. 1, Section 1.5). This division is mostly based on cell **morphology** (literally, "form-science")—eukaryotic cells have a membrane-bound compartment called a nucleus, and prokaryotic cells do not.

But according to **phylogeny** ("tribe-source"), or evolutionary history, organisms are divided into three broad domains called **(1)** Bacteria, **(2)** Archaea, and **(3)** Eukarya. Members of the Bacteria and Archaea are prokaryotic; members of the Eukarya—including algae, fungi, plants, and animals—are eukaryotic.

Let's begin by analyzing how the parts inside a cell function individually and then exploring how they work as a unit. This approach is analogous to studying individual organs in the body and then learning about how they work together to form the nervous system or digestive system. As you study this material, keep asking yourself this key question: How does the structure of this part or group of parts correlate with its function?

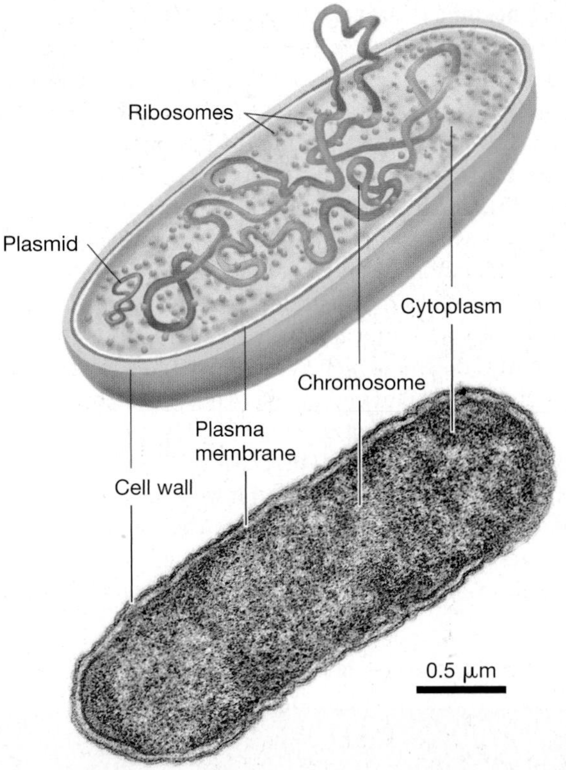

Figure 7.1 Overview of a Prokaryotic Cell. Prokaryotic cells are identified by a negative trait—the absence of a membrane-bound nucleus. Common characteristics among all bacterial and archaeal cells include a plasma membrane, a chromosome, and protein-synthesizing ribosomes.

7.1 Bacterial and Archaeal Cell Structures and Their Functions

Until the mid-twentieth century, biologists thought that prokaryotic cells had a simple morphology and that little structural diversity existed among species. This conclusion was valid at the time, given the resolution of the microscopes that were available and the number of species that had been studied. But our view of prokaryotes changed dramatically in 1931 with the invention of the **transmission electron microscope**, which passes a beam of electrons through extremely thin sections of cells to visualize their internal structure (see BioSkills 9). Recent improvements in microscopy and other research tools are changing our view even more.

After you complete this section, you should be able to ...

- Differentiate the structure and function of prokaryotic cell components.

A Revolutionary New View

Biologists are now convinced that prokaryotic cells, among which bacteria are the best understood, have an array of distinctive structures and functions found among millions of species. Discovering this diversity among prokaryotes was one of the most exciting developments in cell biology.

To start with the basics, Figure 7.1 offers a low-magnification electron micrograph and a stripped-down diagram of a prokaryotic cell. This electron micrograph, along with others in this chapter, was generated using a transmission electron microscope.

Prokaryotic Cell Structures: A Parts List

Figure 7.1 highlights the components common to most prokaryotes studied to date—they all have at least one chromosome and many protein-synthesizing ribosomes enclosed within a plasma membrane (for more about membranes, see Ch. 6, Sections 6.1 and 6.2). The phospholipid components of archaeal and bacterial membranes, however, have different structures.

Bacterial phospholipids consist of fatty acids bound to glycerol while archaeal phospholipids use highly branched isoprenoid chains (Ch. 6, Figure 6.1) bound to glycerol. These molecules also vary in the structure of their hydrocarbon chains and the types of linkages used to join the hydrocarbon tails to glycerol heads. Because of these variations, the archaeal membrane is more stable in the extreme environments that are inhabited by certain species in this domain (Ch. 26, Section 26.1).

All the contents of a cell inside the membrane (excluding the nucleus in eukaryotes) are collectively termed the **cytoplasm** ("cell-formed"). Let's explore these parts one by one, starting from the inside and working out, and then look at more specialized structures found in particular species.

The Chromosome Is Organized into a Nucleoid The most prominent structure inside a prokaryotic cell is the **chromosome**. Most bacterial and archaeal species have a single, circular chromosome that consists of a large DNA molecule associated with proteins. The DNA molecule contains information, and the proteins give the DNA structural support.

Recall that the information in DNA is encoded in its sequence of nitrogenous bases (Ch. 4, Section 4.1). Segments of DNA that contain information for building functional RNAs, some of which may be used to make polypeptides, are called **genes**. The synthesis of RNA based on information stored in DNA is the first step in the central dogma, which describes the flow of genetic information in cells (Ch. 1, Section 1.4). Thus, chromosomes contain DNA, which contains genes that code for RNA.

In the well-studied bacterium *Escherichia coli*, the circular chromosome would be over 1 mm long if it were linear—500 times longer than the cell itself (see Figure 7.2a). This situation is typical in prokaryotes. To fit into the cell, the DNA double helix coils on itself with the aid of enzymes to form a compact, "supercoiled" structure.

(a) Compared to the cell, chromosomal DNA is very long.

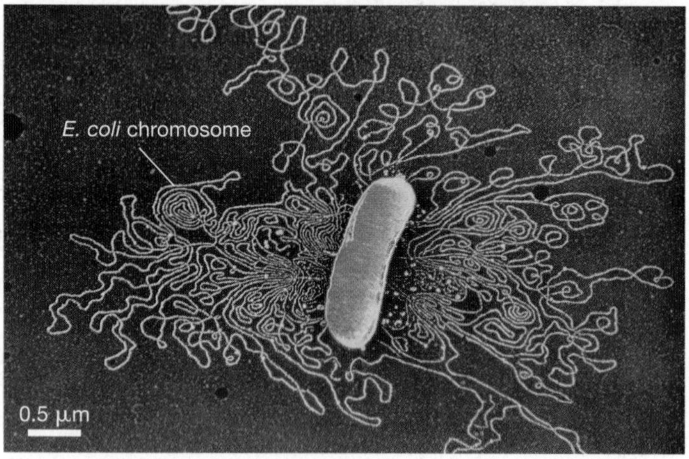

(b) DNA is packaged by supercoiling.

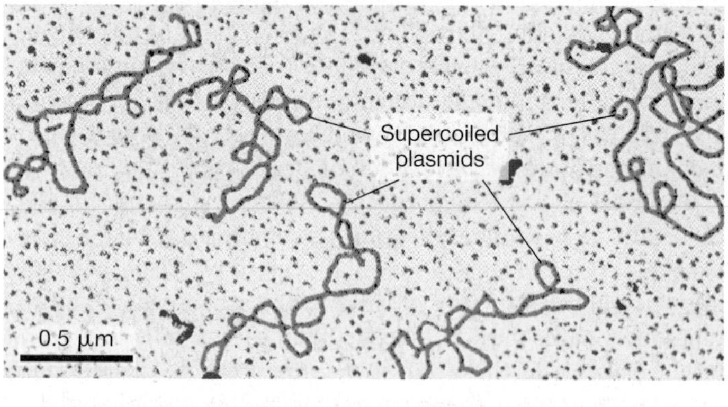

Figure 7.2 **Prokaryotic DNA is Supercoiled. (a)** Chromosomes of bacteria and archaea are often over 500 times the length of the cell, as shown in this micrograph of *E. coli* that has been treated to release its DNA. To fit inside cells, this DNA must be highly compacted by supercoiling. **(b)** A colorized electron micrograph showing the effect of supercoiling on the DNA of isolated plasmids (colored green).

Supercoiled regions of DNA resemble a rubber band that has been held at either end and then twisted until it coils back onto itself.

The region of the cell where the circular chromosome is located is called the **nucleoid** (pronounced *NEW-klee-oyd*). The genetic material in the nucleoid is often organized by clustering loops of DNA into distinct domains, but it is not separated from the rest of the cell interior by a membrane. There is currently intense research into the functional role of these domains and how it changes over time.

Besides their chromosomes, prokaryotic cells may contain from one to about 100 small, usually circular, supercoiled DNA molecules called **plasmids** (Figure 7.2b). Plasmids contain genes but are physically independent of the cellular chromosome. In many cases the genes carried by plasmids are not required under normal conditions; instead, they help cells adapt to unusual circumstances, such as the sudden presence of a poison in the environment. As a result, plasmids can be considered auxiliary genetic elements.

Ribosomes Manufacture Proteins **Ribosomes** are observed in all prokaryotic cells and are found throughout the cell interior. It is not unusual for a single cell to contain 10,000 ribosomes, each functioning as a protein-manufacturing center.

Ribosomes are complex structures composed of large and small subunits, each of which contains RNA and protein molecules. Biologists often refer to ribosomes, along with other multicomponent complexes that perform specialized tasks, as "macromolecular machines." These tiny machines are responsible for the second step of the central dogma, where the information stored in RNA may be used to direct the synthesis of protein. (Ch. 17, Section 17.5 analyzes the structure and function of ribosomes in detail.)

While the ribosomes in bacteria and archaea are similar in size and function, the primary structures of their RNA and protein components are different.

The Cytoskeleton Structures the Cell's Interior and Its Shape Researchers have also observed long, thin protein filaments in the cytoplasm of bacteria and archaea that serve a variety of roles. All bacterial species, for example, contain cytoplasmic protein filaments that are essential for cell division to take place. Some species also have internal filaments that help maintain cell shape. Protein filaments such as these form the basis of the **cytoskeleton** ("cell skeleton").

Recent research has revealed a much more complex cytoskeletal network in prokaryotes than previously thought. Researchers are working to identify how these different filaments participate in cell morphology, growth, and division. (For more about how elements of the cytoskeleton participate in bacterial cell division, see Ch. 12, Section 12.2.)

Photosynthetic Species Have Internal Membrane Complexes Photosynthesis is the set of chemical reactions responsible for converting the energy in sunlight into chemical energy stored in sugars (Chapter 10). In bacteria that perform photosynthesis, it is common to observe multiple membranes passing through the internal region of the cells. The photosynthetic membranes observed in bacteria develop as infoldings of the plasma membrane and contain the enzymes and pigment molecules required for these reactions to occur.

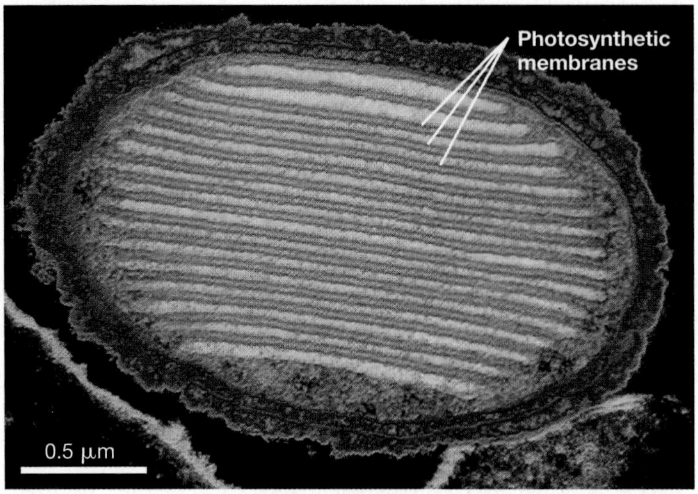

Figure 7.3 **Photosynthetic Membranes in Bacteria.** In this electron micrograph of a photosynthetic bacterium, the areas shown in green are membranes containing the pigments and enzymes required for photosynthesis. (The image has been colorized to identify different structures.)

In some cases, membrane-bound vesicles pinch off as the plasma membrane folds in. In other cases, flattened stacks of photosynthetic membrane remain connected to the plasma membrane, like those shown in Figure 7.3. These internal membranes provide an extensive surface area that allows more photosynthetic reactions to occur and thus increases the cell's ability to make food.

Certain Species Have Organelles for Specialized Functions Several bacterial species have internal compartments classified as **organelles** ("little organs"). An organelle is a compartment inside the cell—often bounded by a membrane—that contains enzymes or structures specialized for a particular function.

Bacterial organelles perform specialized tasks, including

- storing calcium ions;
- holding crystals of the mineral magnetite, which function like compass needles to help cells swim in a directed way; and
- concentrating enzymes responsible for synthesizing complex carbon compounds from carbon dioxide.

The Cell Wall Forms a Protective "Exoskeleton" Because the cytoplasm contains a high concentration of solutes, it is hypertonic relative to the surrounding environment in most habitats. Under these conditions, water enters the cell via osmosis and makes the cell's volume expand. In most bacteria and archaea, this pressure is resisted by a stiff **cell wall**.

Bacterial and archaeal cell walls are a tough, fibrous layer that surrounds the plasma membrane. This structure protects the organisms and gives them shape and rigidity, much like the exoskeleton (external skeleton) of a crab or insect. In prokaryotes, the osmotic pressure that pushes the plasma membrane against the cell wall has a force similar to the pressure in an automobile tire.

The molecular structure of cell walls differs between bacteria and archaea. In most bacteria, the primary structural component of the cell wall is the modified polysaccharide peptidoglycan

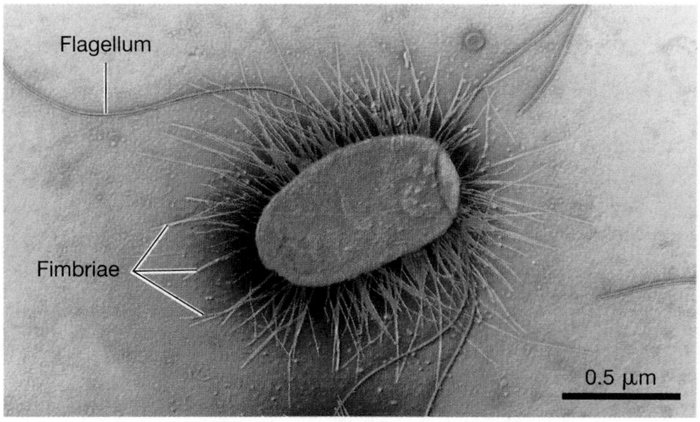

Figure 7.4 **Extracellular Appendages Found on Prokaryotes.** Some species of bacteria and archaea, such as the *E. coli* bacterial cell shown here, assemble large protein structures used for swimming through liquid (flagella) or adhering to surfaces (fimbriae).

(Ch. 5, Section 5.2). Some bacterial cell walls are also surrounded by an outer membrane consisting of glycolipids. The cell walls of archaea are highly variable among the different species, but peptidoglycan is markedly absent among those studied to date.

Certain Species Have External Structures for Movement or Attachment Many prokaryotes interact with their environment via structures that grow from the plasma membrane. The flagella and fimbriae shown in Figure 7.4 are structures commonly found on bacterial surfaces. Archaea also have flagella and appendages similar to fimbriae, but they are structurally distinct from those found on bacteria.

A prokaryotic **flagellum** (plural: **flagella**) is assembled from many different proteins at the cell surface of certain species. The feature that is common to both archaeal and bacterial flagella is a molecular motor embedded in the plasma membrane. The proteins that make up these motors and filaments differ between the two groups, but their functions are the same—to rotate a long rigid filament that propels the cell through water. At top speed, flagella can drive a bacterial cell at 60 cell lengths per second. In contrast, the fastest animal in the ocean—the sailfish—can swim at a mere 10 body lengths per second.

A **fimbria** (plural: **fimbriae**) is a needlelike projection that extends from the plasma membrane of some bacteria and promotes attachment to other cells or surfaces. Similar but unrelated structures are also found in archaea. These structures tend to be more numerous than flagella and are often distributed over the cell's entire surface. Fimbriae are not involved in cell motility, but their ability to glue bacteria to the surface of tissues makes them crucial in establishing many infections.

CHECK YOUR UNDERSTANDING

✔ **If you understood this section, you should be able to . . .**

Describe the structure and function of the (1) ribosomes, (2) photosynthetic membranes, (3) flagella, and (4) cell wall.

Answers are available in Appendix A.

7.2 Eukaryotic Cell Structures and Their Functions

The domain Eukarya includes species that range from microscopic algae to 100-meter-tall redwood trees. Protists, fungi, plants, and animals are all eukaryotic. Although multicellularity has evolved several times among eukaryotes (Ch. 27, Section 27.3), many species are unicellular.

After you complete this section, you should be able to ...

- Differentiate the structure and function of eukaryotic cell components.

The first thing that strikes biologists about eukaryotic cells is how much larger they are on average than bacteria and archaea. Most prokaryotic cells measure 1 to 10 μm in diameter, while most eukaryotic cells range from about 5 to 100 μm in diameter. A micrograph of an average eukaryotic cell, at the same scale as the bacterial cell in Figure 7.1, would fill this page. For many species of unicellular eukaryotes, this size difference allows them to make a living by ingesting bacteria and archaea whole.

Large size has a downside, however. As a cell increases in diameter, its volume increases more than its surface area. In other words, the relationship between them—the surface-area-to-volume ratio—changes. (To see how plants and animals are affected by the surface-area-to-volume ratio, check out Ch. 34, Section 34.1, and Ch. 39, Section 39.3.) Since the surface is where the cell exchanges substances with its environment, the reduction in this ratio decreases the rate of exchange: Diffusion only allows for rapid movement across very small distances.

Prokaryotic cells tend to be small enough so that ions and small molecules arrive where they are needed via diffusion. The random movement of diffusion alone, however, is insufficient for this type of transport as the cell's diameter increases.

How do eukaryotic cells overcome the problems associated with a low surface-area-to-volume ratio? The answer lies in their many organelles.

The Benefits of Organelles

Organelles compartmentalize the volume inside a eukaryotic cell into many small bins. Because eukaryotic cells are subdivided, the **cytosol**—the fluid portion between the plasma membrane and these organelles—is only a fraction of the total cell volume. This relatively small volume of cytosol offsets the effects of a low cell surface-area-to-volume ratio with respect to the exchange of nutrients and waste products.

Figure 7.5 **Overview of Eukaryotic Cells.** Generalized images of **(a)** an animal and **(b)** a plant cell that illustrate the cellular structures in the "typical" eukaryote. The structures have been color-coded for clarity. Compare with the prokaryotic cell, shown at true relative size.

(a) Generalized animal cell

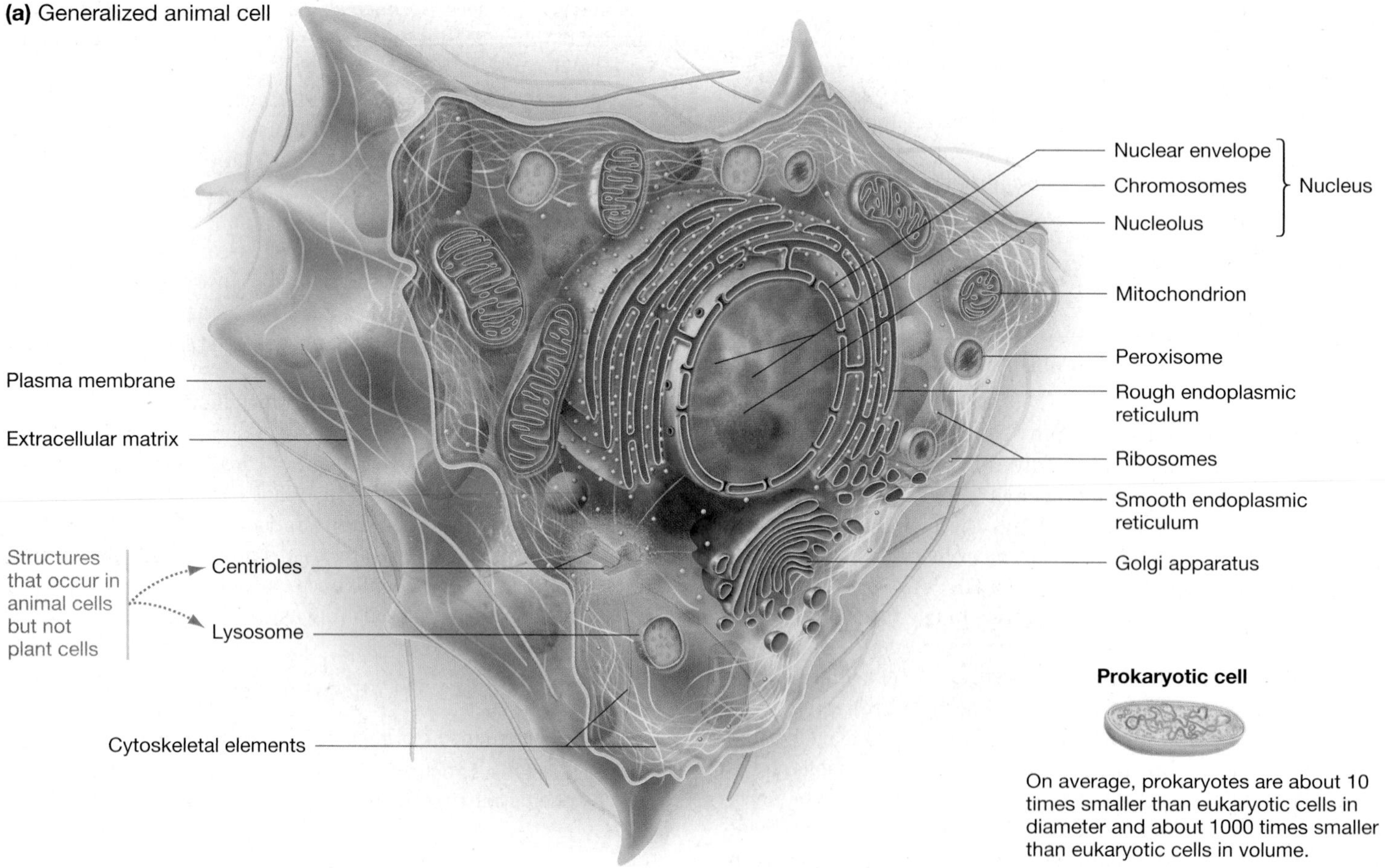

Compartmentalization also offers two key advantages:

1. Incompatible chemical reactions can be separated. For example, new fatty acids can be synthesized in one organelle while excess or damaged fatty acids are degraded and recycled in a different organelle.
2. Chemical reactions become more efficient. First, the substrates required for particular reactions can be localized and maintained at high concentrations within organelles. When substrates are used up in a particular part of the organelle, they can be replaced by substrates that have only a short distance to diffuse. Second, groups of enzymes that work together can be clustered within or on the membranes of organelles instead of floating free in the cytosol. When the product of one reaction is the substrate for a second reaction, clustering the two enzymes increases the speed and efficiency of both reactions.

The complex internal organization of eukaryotic cells resembles a sprawling industrial park. The organelles and other structures found in eukaryotes are like highly specialized buildings that act as administrative centers, factories, transportation corridors, waste and recycling facilities, warehouses, and power stations. Let's now dig a little deeper into eukaryotic cells to investigate how their internal structures perform these different roles.

Eukaryotic Cell Structures: A Parts List

Figure 7.5 provides simplified views of a typical animal cell and a typical plant cell. The artist has removed most of the cytoskeletal elements to make the organelles and other cellular parts easier to see. As you read about each cell component in the pages that follow, focus on identifying how its structure correlates with its function. Then use **Table 7.1** at the end of this section (page 157) as a study guide. As with bacterial cells, let's start from the inside and move to the outside.

The Nucleus While prokaryotic chromosomes are in a loosely defined nucleoid region, eukaryotic chromosomes are enclosed within a membrane-bound compartment called the **nucleus**. Among the largest and most highly organized of all organelles (**Figure 7.6** on page 152), the nucleus functions as an administrative center for information storage and processing. It is enclosed by a unique double-membrane structure called the **nuclear envelope**. As Section 7.4 will detail, the nuclear envelope is studded with pore-like openings, and the inside surface is linked to fibrous proteins that form a lattice-like sheet called the **nuclear lamina**. The nuclear lamina stiffens the double membrane and maintains organelle shape.

Chromosomes do not float freely inside the nucleus—instead, each chromosome occupies a distinct area, which may vary in

Figure 7.5 Overview of Eukaryotic Cells (continued).

(b) Generalized plant cell

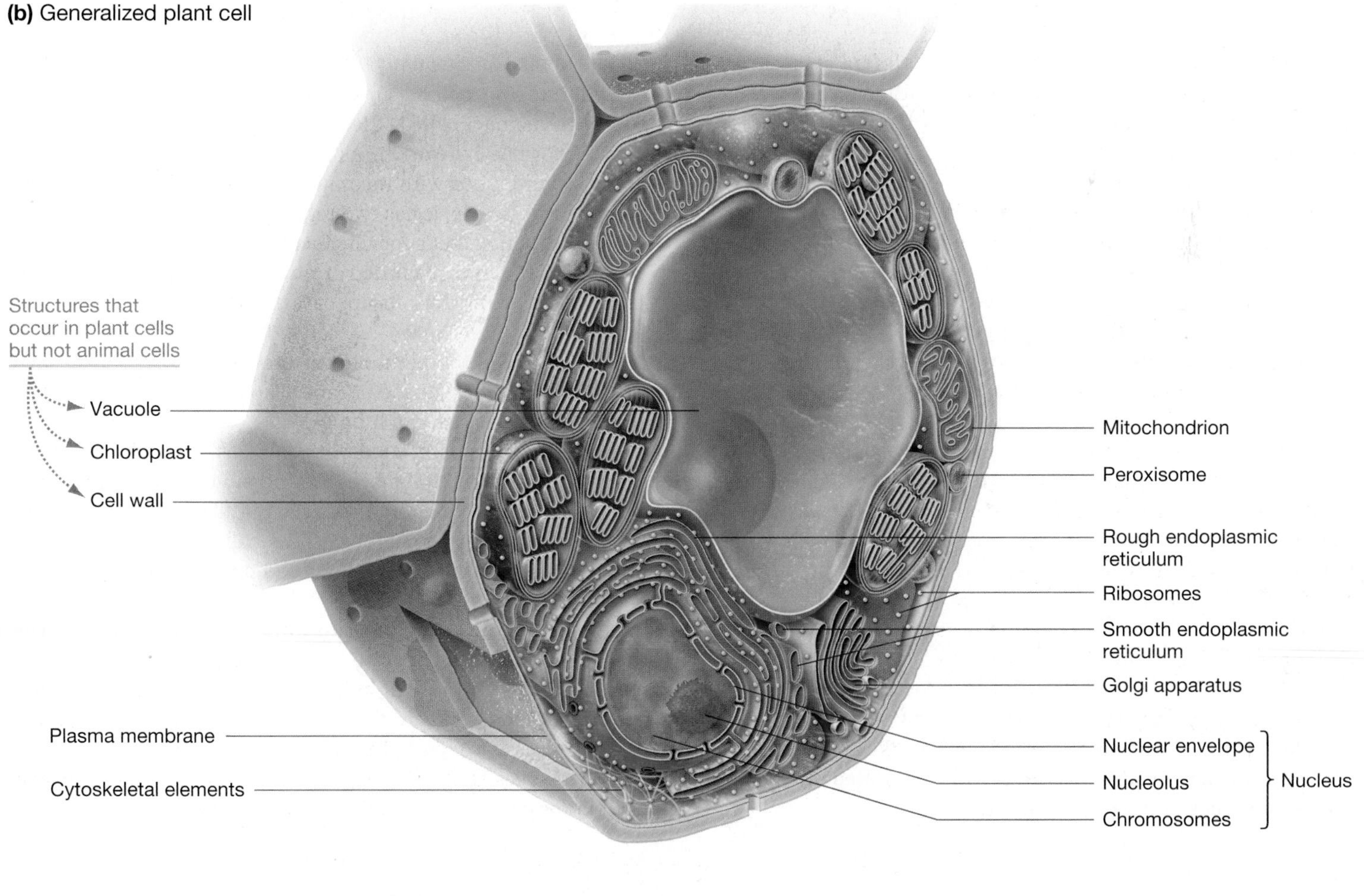

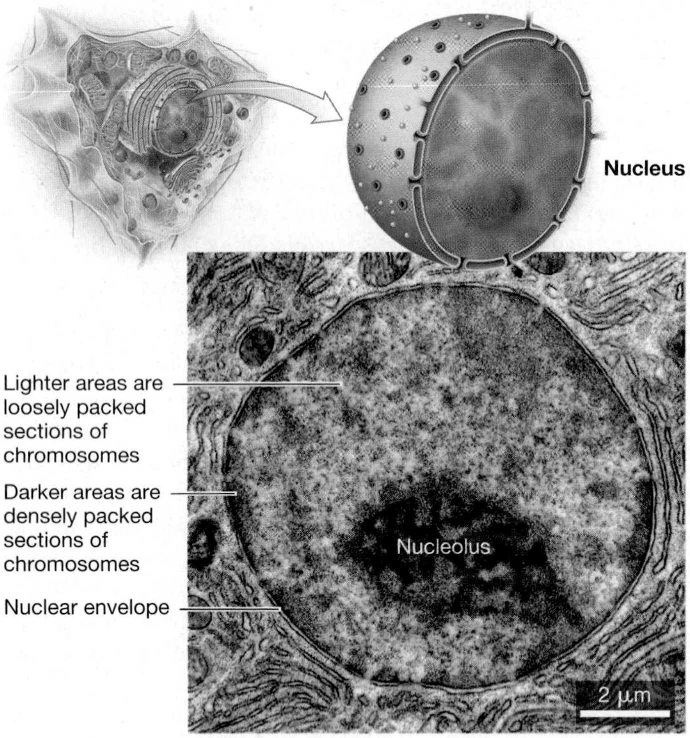

Figure 7.6 The Nucleus Stores and Transmits Information. The nucleus houses genetic, or hereditary, information encoded in DNA, which is a component of the chromosomes inside the nucleus.

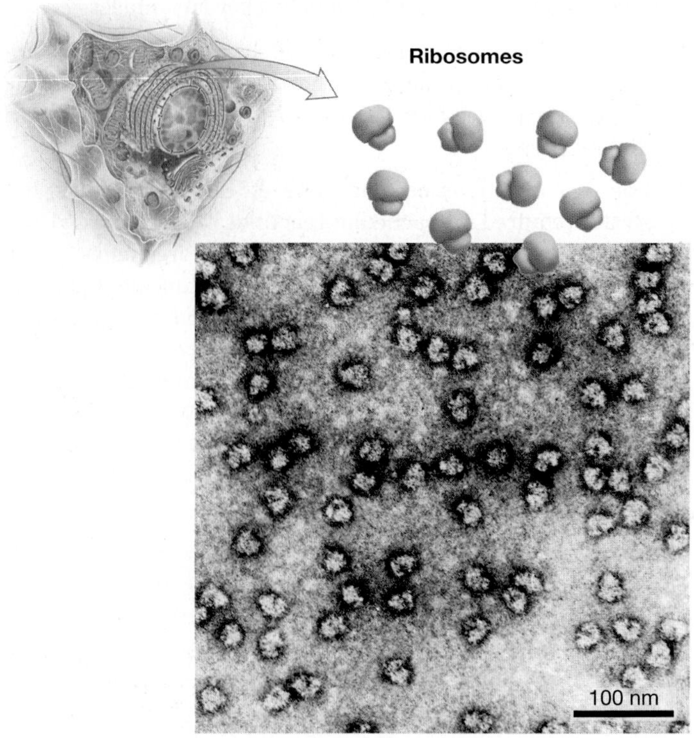

Figure 7.7 Ribosomes Are the Site of Protein Synthesis. Eukaryotic ribosomes are larger than bacterial and archaeal ribosomes, but similar in overall structure and function.

different cell types and over the course of cell replication. The chromosomes are arranged in the nucleus with densely packed sections concentrated at the periphery and loosely packed sections toward the interior.

The nucleus is the site where the first step of the central dogma takes place—the synthesis of RNA from information encoded in DNA. The nucleus also contains specific sites where RNA molecules are processed into their functional form (for more about RNA processing, see Ch. 17, Section 17.2). One of these regions, called the **nucleolus**, is responsible for manufacturing and processing the RNA molecules that assemble into large and small ribosomal subunits.

Ribosomes Scattered throughout the cytosol of eukaryotic cells are millions of ribosomes (Figure 7.7). Like bacterial ribosomes, eukaryotic ribosomes are complex macromolecular machines that use the information in RNA to manufacture proteins. Note that ribosomes are not compartments inside a cell, so they are not classified as organelles.

Eukaryotic ribosomes are not only scattered free in the cytosol, but are also associated with the surface of an organelle called the endoplasmic reticulum. Proteins manufactured by free ribosomes either remain in the cytosol or are imported into other organelles, such as the nucleus (see Section 7.4). Those made at the surface of the endoplasmic reticulum have a different fate. Let's take a closer look at this organelle to learn more.

Endoplasmic Reticulum Portions of the nuclear envelope extend into the cytoplasm to form an extensive membrane-enclosed factory called the **endoplasmic reticulum (ER)** ("inside-formed-network"). As Figure 7.5 shows, the ER membrane is continuous with the nuclear envelope. Although the ER is a single organelle, it has two regions that are distinct in structure and function. Let's consider each region in turn.

The **rough endoplasmic reticulum (RER)**, or more simply **rough ER**, is named for its appearance in transmission electron micrographs (see Figure 7.8, left). The dark, knobby-looking structures lining the perimeter of the rough ER are ribosomes that are attached to its membrane. The membrane forms a network of flattened sacs and tubules.

The ribosomes associated with the surface of the rough ER synthesize proteins that move into the interior of the organelle. The inside compartment of the rough ER, like the interior of any sac-like structure in a cell or body, is called the **lumen**. In the rough ER lumen, newly manufactured proteins undergo folding and other types of processing.

After the proteins are processed in the rough ER, they will either function in the ER or be packaged into vesicles and transported as cargo to other destinations, such as a different organelle, the plasma membrane, or the cell exterior (see Section 7.5). The proteins that are exported play various roles for the cell. Some carry messages to other cells; some act as membrane transport proteins or pumps; others catalyze reactions.

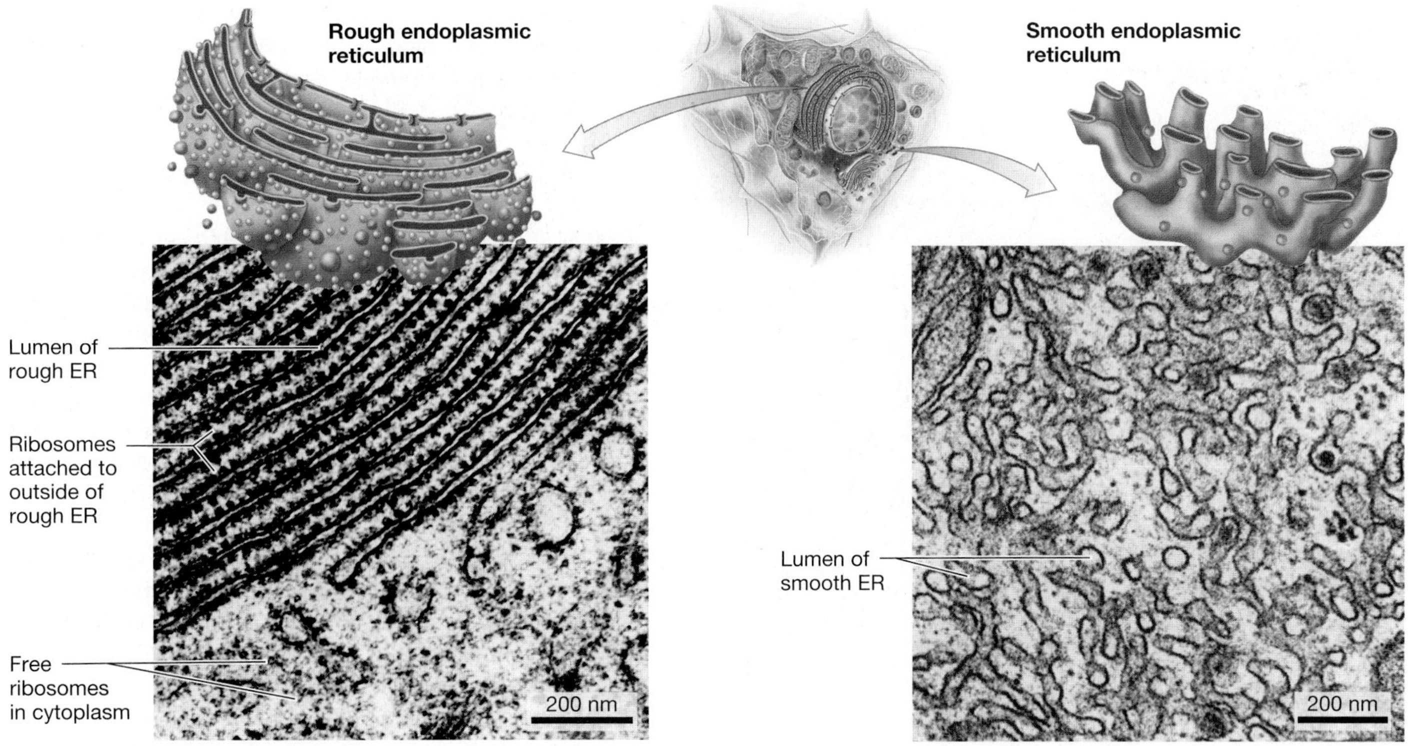

Figure 7.8 The Endoplasmic Reticulum Is a Site of Synthesis, Processing, and Storage. The ER is continuous with the nuclear envelope and possesses two distinct regions: The rough ER is a system of membrane-bound sacs and tubules with ribosomes attached; the smooth ER is a system of membrane-bound sacs and tubules that lacks ribosomes.

In electron micrographs, parts of the ER that are free of ribosomes appear smooth and even. Appropriately, these regions of the ER are called the **smooth endoplasmic reticulum (SER)**, or **smooth ER** (see Figure 7.8, right).

The smooth ER contains enzymes that catalyze reactions involving lipids. Depending on the type of cell, these enzymes may synthesize lipids needed by the organism (such as phospholipids used in membranes) or modify lipids and other molecules that are toxic. In addition, the smooth ER functions as a reservoir for calcium ions (Ca^{2+}) that can be released to trigger a wide array of activities inside the cell.

The structure of the endoplasmic reticulum correlates closely with its function. The rough ER has ribosomes and functions primarily as a protein-manufacturing center; the smooth ER lacks ribosomes and functions primarily as a lipid-processing center.

Golgi Apparatus Most of the proteins that leave the rough ER must pass through the **Golgi apparatus** before reaching their final destination. The Golgi apparatus consists of a set of membranous compartments called **cisternae** (singular: **cisterna**). Each cisterna varies in function based on the distribution of different enzymes across the set of compartments. In most eukaryotes, these compartments are flattened sacs that are stacked on top of each other like pancakes (Figure 7.9). When the cisternae are arranged in this way, the Golgi apparatus has a distinct polarity or sidedness. The *cis* ("on this side") surface is closest to the nucleus, and the *trans* ("across") surface is oriented toward the plasma membrane.

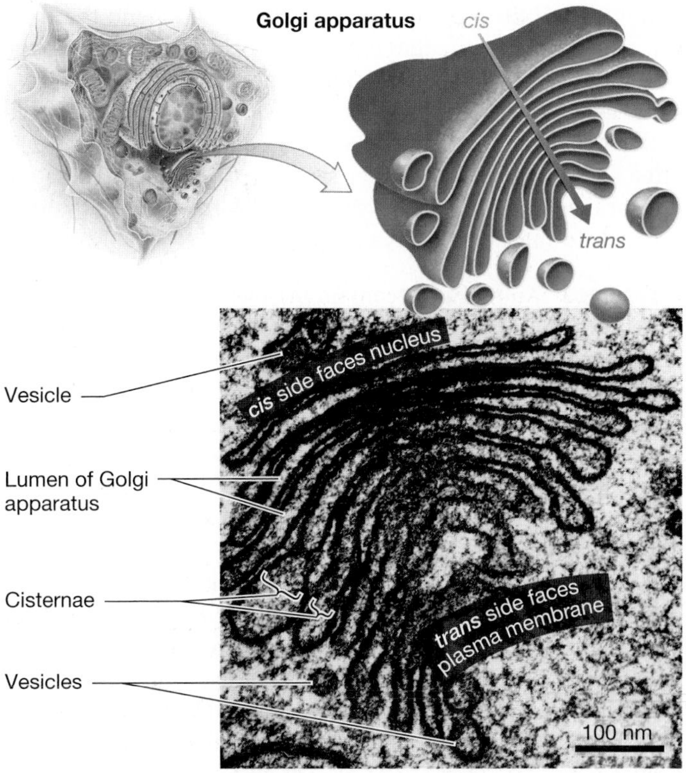

Figure 7.9 The Golgi Apparatus Is a Site of Protein Processing, Sorting, and Shipping. The Golgi apparatus is a collection of flattened sacs called cisternae.

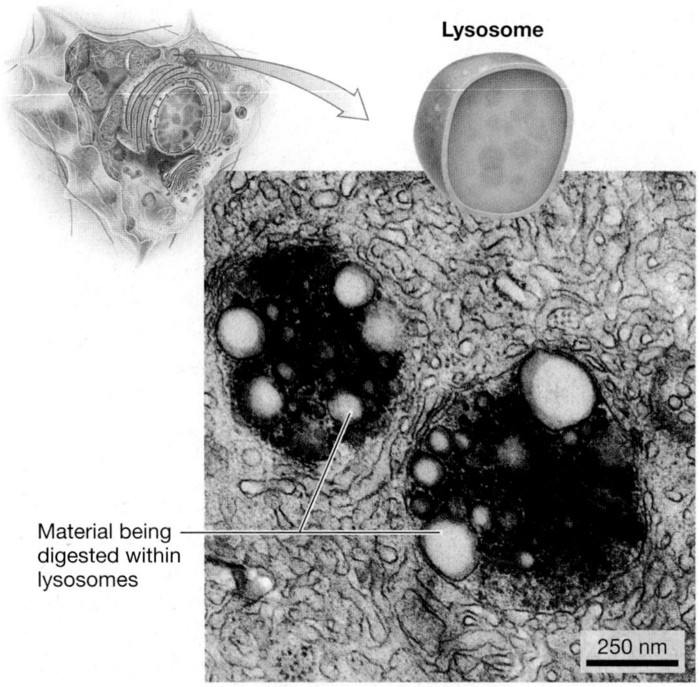

Figure 7.10 **Lysosomes Are Recycling Centers.** Lysosomes are oval or globular organelles that contain enzymes to digest macromolecules.

The *cis* side of a Golgi apparatus receives the vesicles containing rough ER products, referred to as cargo, and the *trans* side ships them out to other organelles or the cell surface. As the cargo moves through the Golgi apparatus from the *cis* to *trans* surfaces, it is sequentially processed by the enzymes resident in each compartment and then packaged for delivery. Micrographs often show "bubbles" on either side of a Golgi stack. These are membrane-bound transport vesicles that carry proteins or other products to and from the organelle.

Lysosomes Animal cells contain organelles called **lysosomes** that function as recycling centers (Figure 7.10). Lysosomes contain about 40 different enzymes, each specialized for hydrolyzing different types of macromolecules—proteins, nucleic acids, lipids, or carbohydrates. The amino acids, nucleotides, sugars, and other molecules that result from hydrolysis are exported from the lysosome via transport proteins in the organelle's membrane. Once in the cytosol, they can be used as sources of energy or building blocks for new molecules.

The digestive enzymes inside lysosomes are collectively called acid hydrolases because under acidic conditions (a pH of 5.0), they use water to break monomers from macromolecules. In the cytosol, where the pH is about 7.2, acid hydrolases would be less active. Proton pumps in the lysosomal membrane maintain an acidic pH in the lumen of the lysosome by importing hydrogen ions.

Even though lysosomes are physically separated from the Golgi apparatus and the endoplasmic reticulum, these various organelles jointly form a key functional grouping referred to as the **endomembrane system**. The endomembrane ("inner-membrane") system is a center for producing, processing, and transporting proteins, carbohydrates, and lipids in eukaryotic cells. For example, acid hydrolases are synthesized in the ER, processed in the Golgi apparatus, and then shipped to the lysosome. Section 7.5 analyzes the intracellular movement of molecules through the endomembrane system in more detail.

Vacuoles The cells of plants, fungi, and certain other eukaryotes contain a prominent organelle called a **vacuole**. Compared with the organelles of animal cells, the vacuoles of plant and fungal cells are large—in plants, sometimes taking up as much as 80 percent of a cell's volume (Figure 7.11).

The cellular role of a vacuole can vary between organisms and even between cells of the same multicellular organism. Some vacuoles are known to contain hydrolases to digest and recycle macromolecules, which has led to their being classified as functionally equivalent to the lysosomes of animal cells.

A more common role of vacuoles is to serve as storage depots. In many cases, ions such as potassium (K^+) and chloride (Cl^-), among other solutes, are stored at such high concentrations that they draw water in from the environment. As the vacuole expands in volume, it pushes the cytoplasm and plasma membrane against the cell wall. The effect of this change in volume is observed when wilted green plants regain their rigid structure after water is added to the soil.

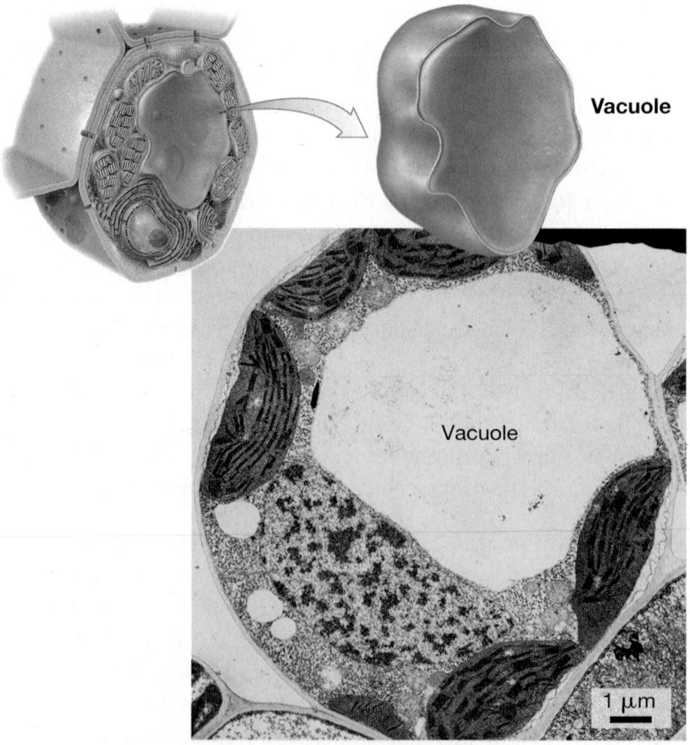

Figure 7.11 **Vacuoles Are Generally Storage Centers in Plant and Fungal Cells.** Vacuoles vary in size and function. Some contain digestive enzymes and serve as recycling centers; most are large storage containers.

✔ Propose a hypothesis to explain why toxins like nicotine, cocaine, and caffeine are stored in vacuoles instead of the cytosol.

In certain plant cells, vacuoles can include more specialized storage functions:

- In seeds, cells may contain a large vacuole filled with proteins. When the embryonic plant inside the seed begins to grow, enzymes begin digesting these proteins to provide amino acids for the growing individual.
- In flower petals or fruits, cells may contain vacuoles that are filled with colorful pigments.
- Elsewhere, vacuoles may be packed with noxious compounds that protect leaves and stems from being eaten by predators. The type of chemical involved varies by species, ranging from bitter-tasting tannins to toxins such as nicotine, morphine, caffeine, or cocaine.

Peroxisomes Virtually all eukaryotic cells contain globular organelles called **peroxisomes** (Figure 7.12). These organelles originate when empty vesicles from the ER are loaded with peroxisome-specific enzymes from the cytosol. Once peroxisomes are formed, they do not exchange materials with other organelles, so they are not considered part of the endomembrane system.

Although different types of cells from the same individual may have distinct types of peroxisomes, these organelles all share a common function: Peroxisomes are centers for reduction–oxidation (redox) reactions. (Ch. 8, Section 8.2, details how redox reactions transfer electrons between atoms and molecules.) For example, the peroxisomes in your liver cells contain enzymes that oxidize the ethanol in alcoholic beverages. In the leaves of plants, specialized peroxisomes called **glyoxysomes** are packed with enzymes that oxidize fatty acids to form a compound that can be used to store energy for the cell.

In animals and plants, these reactions often include hydrogen peroxide (H_2O_2), which is highly reactive. If hydrogen peroxide escaped from the peroxisome, it would quickly react with and damage DNA, proteins, and cellular membranes. This event is rare, however, because inside the peroxisome, the enzyme catalase quickly "detoxifies" hydrogen peroxide to form water and oxygen. The enzymes found inside the peroxisome make a specialized set of oxidation reactions possible and safe for the cell.

Mitochondria Most of the work required to maintain the structure and function of a cell depends on the chemical energy stored in adenosine triphosphate (ATP). The organelle primarily responsible for supplying ATP in animals, plants, and virtually all other eukaryotic cells is the **mitochondrion** (plural: **mitochondria**).

As Figure 7.13 shows, each mitochondrion has two membranes. The outer membrane defines the organelle's surface, while the inner membrane forms a series of sac-like **cristae** (singular: **crista**). The solution enclosed within the inner membrane is called the **mitochondrial matrix**. In eukaryotes, the chemical energy in carbohydrates and fats is used to produce ATP. Most of the enzymes and molecular machines responsible

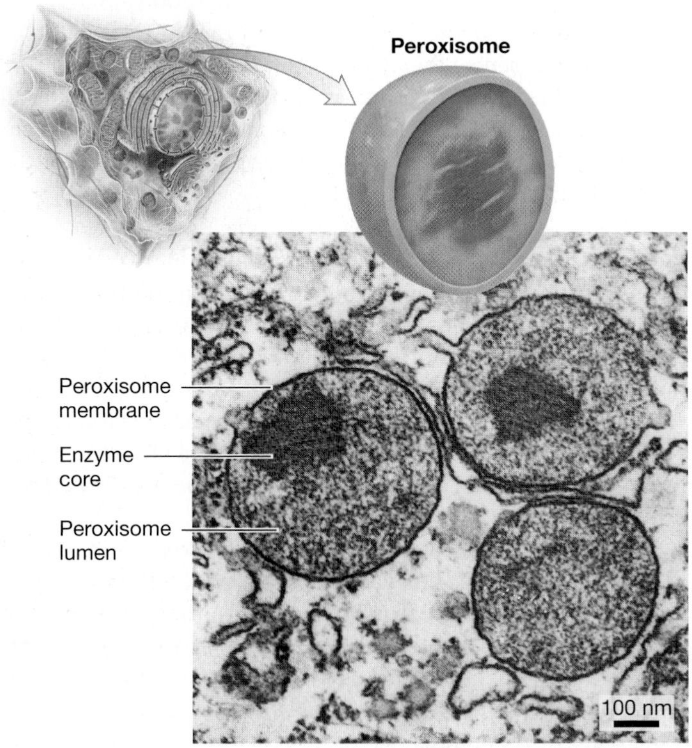

Figure 7.12 Peroxisomes Are the Site of Oxidation Reactions. Peroxisomes are globular organelles that contain enzymes involved in detoxifying reactive molecules, such as hydrogen peroxide.

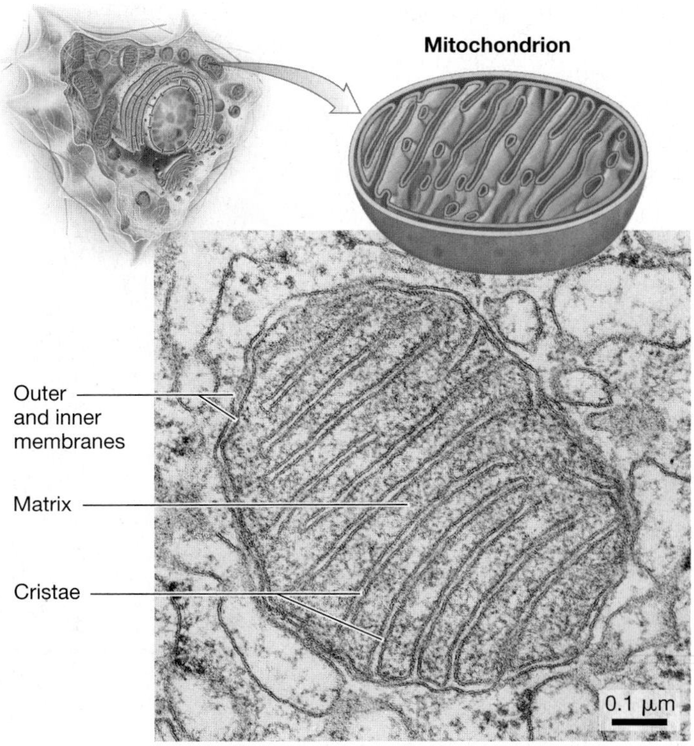

Figure 7.13 Mitochondria Are Power-Generating Stations. Mitochondria vary in size and shape, but all have two membranes with sac-like cristae formed from the inner membrane that are involved in producing ATP.

for synthesizing ATP are embedded in the inner membrane or suspended in the matrix (Ch. 9, Sections 9.3, 9.4, and 9.5).

Mitochondria are typically drawn as small bean-shaped organelles, but in cells, their morphology is often much more dynamic. These organelles are prone to fusion and fission, resulting in either elongated and branched structures, called mitochondrial networks, or multiple individual organelles. Depending on the type of cell, the number of mitochondria can range from one to more than a million.

Each mitochondrion has many copies of a small, circular—or, in some species, linear—chromosome called **mitochondrial DNA (mtDNA)** that is independent of the nuclear chromosomes. The mitochondrial DNA contains only a tiny fraction of the genes responsible for the function of the organelle—the other genes reside in the nuclear DNA.

Among the genes present in mitochondrial DNA are those that encode RNAs for mitochondrial ribosomes. These ribosomes are smaller than those found in the cytosol, yet they still function to produce some of the mitochondrial proteins. (Most of the proteins found in mitochondria are produced from ribosomes in the cytosol and imported into the organelle.)

Chloroplasts Most algal and plant cells possess an organelle called the **chloroplast**, in which sunlight is converted to chemical energy during photosynthesis (Figure 7.14). The number of chloroplasts per cell varies from one to several dozen.

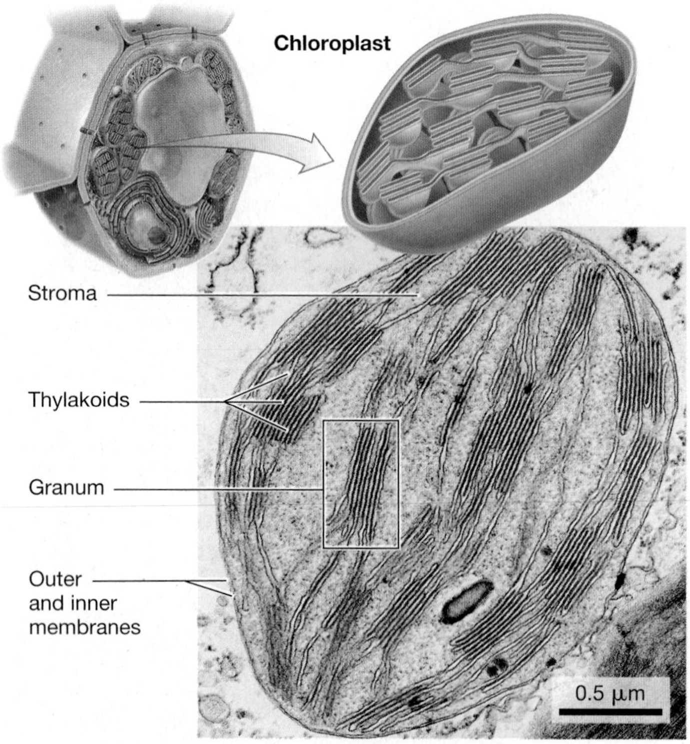

Figure 7.14 Chloroplasts Are Sugar-Manufacturing Centers in Plants and Algae. Many of the enzymes and other molecules required for photosynthesis are located in membranes inside the chloroplast. These membranes form thylakoids that consist of discs stacked into grana.

Like the mitochondrion, the chloroplast is surrounded by a double membrane. Unlike mitochondria, however, chloroplasts have no cristae extending from the inner membrane into the interior. Instead, a third membrane forms an independent network of hundreds of flattened, sac-like structures called **thylakoids** throughout the interior. Most thylakoids are arranged in interconnected stacks called **grana** (singular: **granum**).

Many of the pigments, enzymes, and macromolecular machines responsible for converting light energy into chemical energy are embedded in the thylakoid membranes (Ch. 10, Section 10.3). The fluid-filled space surrounding grana, called the **stroma**, contains enzymes that use this chemical energy to produce sugars.

Like mitochondria, each chloroplast contains copies of its own circular chromosome and small ribosomes that manufacture some, but not all, of the organelle's proteins. And like mitochondria, chloroplasts also grow and divide independently of cell division.

Because these attributes are odd compared with those of other organelles, biologists propose that mitochondria and chloroplasts were once free-living bacteria. According to the **endosymbiosis theory**, these bacteria were engulfed by the ancestors of modern eukaryotes, but were not destroyed—instead, a mutually beneficial relationship evolved. (In Ch. 27, Section 27.3, you will learn more about the origins of these unique eukaryotic organelles.)

Cytoskeleton The final major structural feature common to all eukaryotes is the cytoskeleton. This extensive system of protein fibers gives the cell its shape and structural stability. It is also involved in moving materials within the cell as well as the cell itself. In essence, the cytoskeleton organizes all the organelles and other cellular structures into a cohesive whole. Recall that prokaryotes also have a cytoskeleton, but it is far less extensive. Section 7.6 will analyze the structure and functions of the eukaryotic cytoskeleton in detail.

The Eukaryotic Cell Wall and Extracellular Matrix In fungi, algae, and plants, cells have an outer cell wall in addition to their plasma membrane. Like the cell wall in bacteria and archaea (Section 7.1), a eukaryotic cell wall is located outside the plasma membrane and furnishes a durable outer layer that gives structural support to the cell. The cells of animals lack a cell wall, but are often supported by a more diffuse mixture of secreted proteins and polysaccharides that form the **extracellular matrix**, or **ECM**.

Although the composition of the eukaryotic cell wall and ECM varies among species and even among types of cells in the same individual, the general plan is similar: Rods or fibers run through a stiff matrix made of polysaccharides and proteins (see Ch. 11, Section 11.1 for details). This organization of extracellular molecules provides cells with structural support and may be used to attach cells to one another.

To summarize: Within a cell, the structure of each component correlates with its function. As you will see in the next section, the overall size, shape, and composition of a cell similarly correlate with its function.

SUMMARY Table 7.1 Eukaryotic Cell Components

Icons Not to Scale		Structure		
		Membrane	**Components**	**Function**
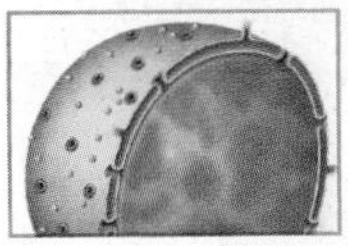	**Nucleus**	Double ("envelope"); openings called nuclear pores	Chromosomes Nucleolus Nuclear lamina	Information storage and transmission Ribosome subunit assembly Structural support
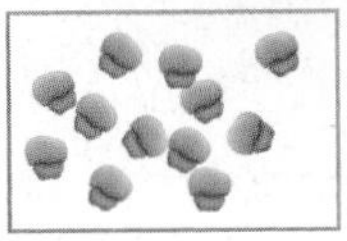	**Ribosomes**	None	Complex of RNA and proteins	Protein synthesis
	Endomembrane system			
	*Endoplasmic reticulum: rough*	Single; contains receptors for entry of selected proteins	Network of branching sacs Ribosomes associated	Protein synthesis and processing
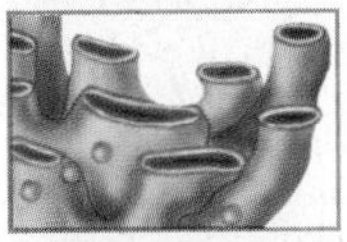	*Endoplasmic reticulum: smooth*	Single; contains enzymes for synthesizing phospholipids	Network of branching sacs Enzymes for synthesizing or modifying lipids	Lipid synthesis and processing
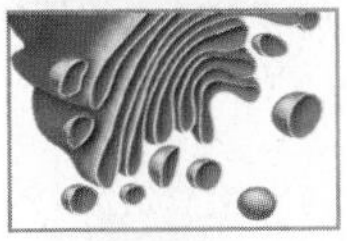	*Golgi apparatus*	Single; contains receptors for products of rough ER	Distinct cisternae, often as stack of flattened vesicles	Protein, lipid, and carbohydrate processing
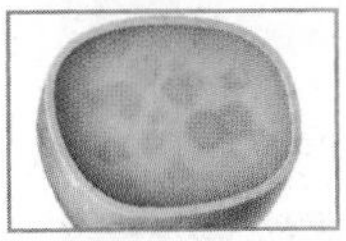	*Lysosomes*	Single; contains transporters for selected molecules and proton pumps	Acid hydrolases (catalyze hydrolysis reactions)	Digestion and recycling
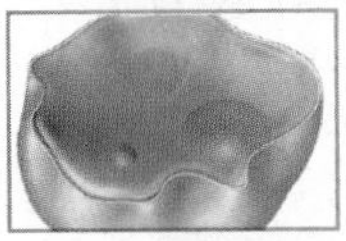	**Vacuoles**	Single; contains transporters for selected molecules	Varies—carbohydrates, water, pigments, oils, toxins, or hydrolases	Storage, digestion, and recycling
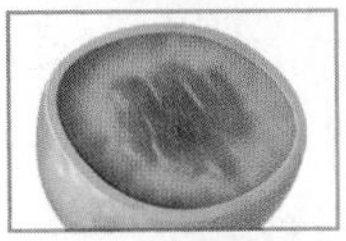	**Peroxisomes**	Single; contains transporters for selected macromolecules	Enzymes that catalyze oxidation reactions Catalase (processes peroxide)	Oxidation of fatty acids, ethanol, or other compounds
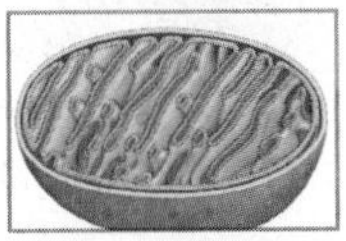	**Mitochondria**	Double; inner contains enzymes for ATP production	Enzymes that harvest energy from molecules to make ATP	ATP production
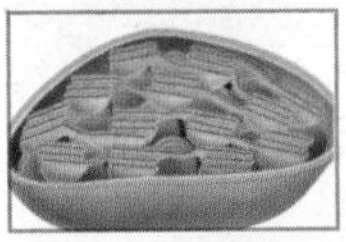	**Chloroplasts**	Double; plus membrane-bound sacs in interior	Pigments Enzymes that use light energy to make sugars	Production of sugars via photosynthesis
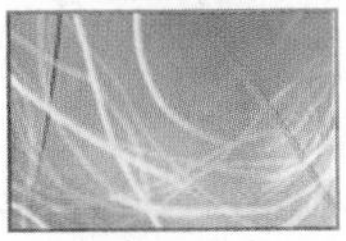	**Cytoskeleton**	None	Actin filaments Intermediate filaments Microtubules	Structural support; movement of materials; in some species, movement of whole cell
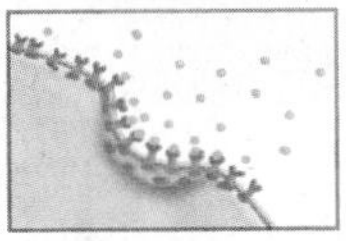	**Plasma membrane**	Single; contains transport and receptor proteins	Phospholipid bilayer with transport and receptor proteins	Selective permeability—maintains intracellular environment
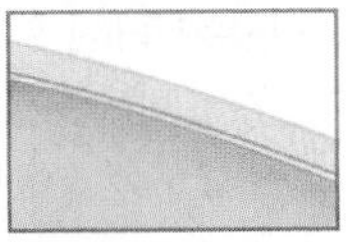	**Cell wall and extracellular matrix**	None	Fibers running through carbohydrate or protein matrix	Protection, structural support

CHECK YOUR UNDERSTANDING

✔ **If you understood this section, you should be able to . . .**

1. Explain how the structures of different organelles correlate with their different functions; use lysosomes and peroxisomes as examples.
2. In Table 7.1, label each component with one of the following analogies: administrative/information hub, power station, warehouse, large molecule manufacturing and shipping facility (with subtitles for lipid factory, protein finishing and shipping line, protein synthesis and folding center, waste processing and recycling center), support beams, perimeter fencing, protein factory, food-manufacturing facility, and fatty-acid processing and detox center.

Answers are available in Appendix A.

7.3 Putting the Parts into a Whole

If the industrial complex serves as an analogy for a eukaryotic cell, then a city that consists of many different industries could represent a multicellular organism. Just as a clothing manufacturing center has a very different layout and composition from an airplane production facility, cells in your muscles differ from nearby fat cells. How does the physical and chemical makeup of an individual cell correlate with its function?

After you complete this section, you should be able to . . .

- Explain how diverse cellular activities are correlated with the function of organelles.

Structure and Function at the Whole-Cell Level

An individual plant or animal is made up of cells specialized for certain tasks. These different cells have structures that correlate with their functions. For example, the cardiac muscle cells in your heart are long and tapered. They are filled with protein fibers that slide past one another as the muscle flexes or relaxes to generate the heartbeat. Muscle cells are also packed with mitochondria, which produce the ATP required for the sliding motion to occur.

In contrast, fat cells are rounded, globular structures that store lipids. They consist of little more than a plasma membrane, a nucleus, and a fat droplet. Neither cell bears a close resemblance to the generalized animal cell pictured in Figure 7.5a. To drive home the correlation between the overall structure and function of a cell, examine the transmission electron micrographs in **Figure 7.15**. Each one shows key details from a different cell type.

- The animal cell in Figure 7.15a is from the pancreas. It is packed with rough ER that manufactures and secretes digestive enzymes.
- The animal cell in Figure 7.15b is from the testis. It is dominated by smooth ER that synthesizes lipids like testosterone—a steroid hormone.

(a) Pancreatic cell: : Exports digestive enzymes.

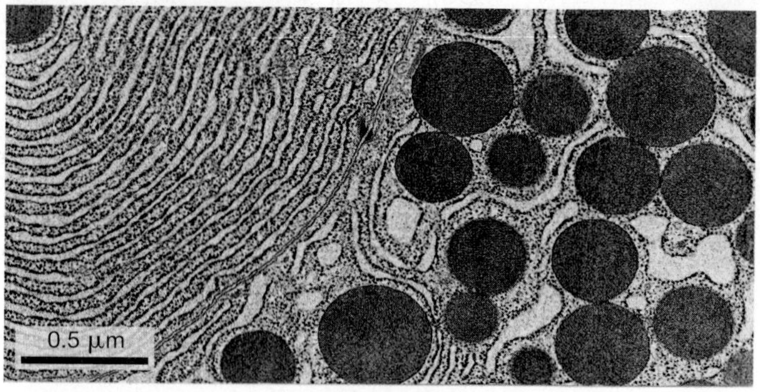

(b) Testis cell: Exports lipid-soluble signals.

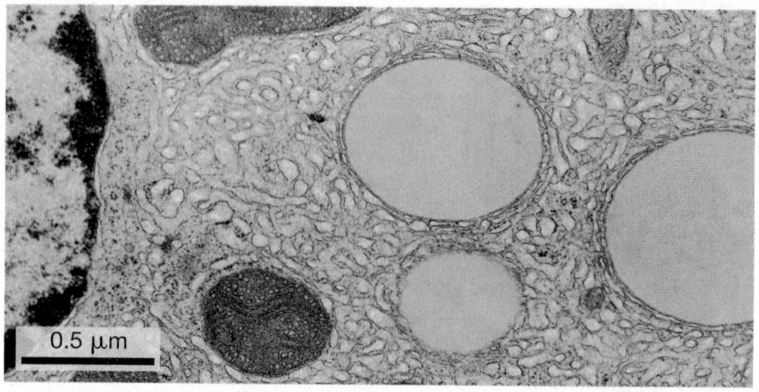

(c) Cardiac muscle cell: Uses ATP to generate the heartbeat.

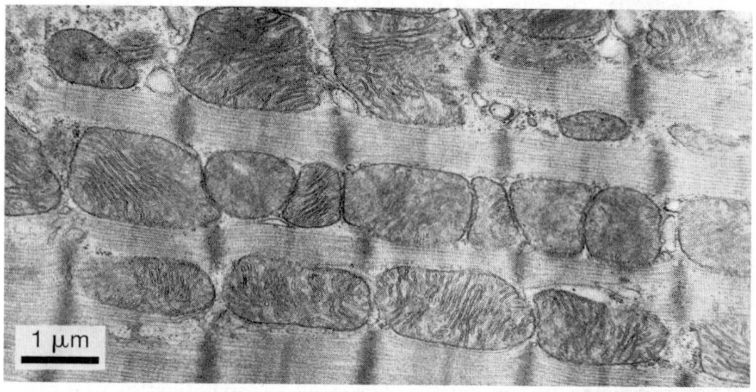

(d) Leaf cell: Manufactures ATP and sugar.

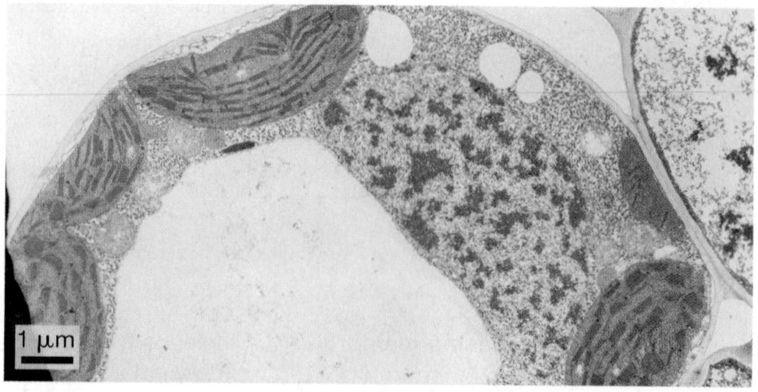

Figure 7.15 **Cell Structure Correlates with Function.**

✔ In part (a), label the rough ER and the dark, round secretory vesicles. In (b), label the smooth ER. In (c), label the mitochondria. In (d), label the chloroplasts, vacuole, and nucleus.

- The animal cells in Figure 7.15c come from cardiac muscle. The cells have numerous mitochondria that use the energy stored in sugars to produce ATP.
- The plant cell in Figure 7.15d is from the leaf of a potato. It has many chloroplasts that are specialized for absorbing light and manufacturing sugar.

In each case, the size and number of different types of organelles in each cell correlate with the cell's specialized function. In addition to the variation in organelle content, cells also differ in size. For example, the cells that make up your sciatic nerve stretch from the base of your spine to your big toe.

The Dynamic Cell

Biologists study the structure and function of organelles and cells with a combination of tools and approaches. For several decades, a technique called **differential centrifugation** was particularly important because it allowed researchers to isolate particular cell components and analyze their chemical composition. Differential centrifugation is based on breaking cells apart to release cell components and separating components in a centrifuge (see **BioSkills 7**). The individual parts of the cell can then be purified and studied in detail, in isolation from other parts of the cell.

Historically and currently, however, the most important research in cell biology is based on imaging—simply looking at cells. Recent innovations allow biologists to put fluorescing tags or other types of markers on particular cell components and then look at them with increasingly sophisticated light microscopes and electron microscopes. Advances in microscopy provide increasingly high magnification and better resolution (some of the most important imaging techniques are featured in **BioSkills 9**).

It's important to recognize, though, that a technique can have limitations. For example, differential centrifugation splits cells into parts that are analyzed independently, and electron microscopy gives a fixed "snapshot" of the cell or organisms being observed. Neither of these techniques allows investigators to explore directly how things move from place to place in the cell or how parts interact. The information gleaned from these techniques can make cells seem static. In reality, however, cells are dynamic—as current live-cell imaging techniques (using markers like green fluorescent protein, GFP) now make abundantly clear (**BioSkills 9**).

The amount of chemical activity and the speed of molecular movement inside cells are nothing short of fantastic. Here are just a few remarkable cellular feats:

- In an average second, a typical cell in your body uses about 10 million ATP molecules and synthesizes just as many.
- It's not unusual for a cellular enzyme to catalyze 25,000 or more reactions per second; most cells contain hundreds or thousands of different enzymes.
- A minute is more than enough time for each membrane phospholipid in your body to travel the breadth of the organelle or cell where it resides.
- The hundreds of trillions of mitochondria inside you are completely replaced about every 10 days, for as long as you live.

Within a cell, events take nanoseconds, which is one billionth of a second (10^{-9} or 1/1,000,000,000 s). Movement within a cell is measured in micrometers per second (10^{-6} m/s or 1/1,000,000 m/s). This is the speed and scale of life.

In the rest of this chapter, you will focus on this theme of cellular dynamism and movement. The goal is to put some of the individual pieces of a cell together and ask how they work as systems to accomplish key tasks.

To begin, let's first look at how molecules move into and out of the cell's control center—the nucleus. To understand how organelles are integrated in the endomembrane system, you will investigate how proteins move from ribosomes into the lumen of the rough ER and then to the Golgi apparatus and beyond. In the final section of this chapter, you'll read about the cytoskeletal elements and how their associated motor proteins are used to transport cargo inside the cell or move the cell itself.

CHECK YOUR UNDERSTANDING

✔ **If you understood this section, you should be able to . . .**

1. Predict what type of organelle would be dominant in cells of your immune system that consume and digest bacterial cells.
2. Explain why you should expect an increased Golgi apparatus in addition to copious rough ER in a secretory cell, like the one from the pancreas shown in Figure 7.15a.

Answers are available in Appendix A.

7.4 Cell Systems I: Nuclear Transport

After you complete this section, you should be able to . . .

- Analyze the process of transporting molecules across the nuclear envelope.

The nucleus is the information center of eukaryotic cells—a corporate headquarters, design center, and library all rolled into one. Appropriately enough, its interior is highly organized. Specific centers exist where the genetic information in DNA is decoded and processed. At these locations, large suites of enzymes interact to produce RNA messages from specific genes at specific times. Meanwhile, the nucleolus functions as the site of ribosome assembly.

Structure and Function of the Nuclear Envelope

The nuclear envelope separates the nucleus from the rest of the cell. Starting in the 1950s, transmission electron micrographs of cross sections through the nuclear envelope showed that the structure is supported by an internal fibrous nuclear lamina and bounded by two membranes. How does this administrative center communicate with the rest of the cell across the double-membrane barrier?

Micrographs like the one on the left in **Figure 7.16** on page 160 show that the nuclear envelope is perforated with openings that are approximately 60 nanometers (nm) in diameter. Follow-up

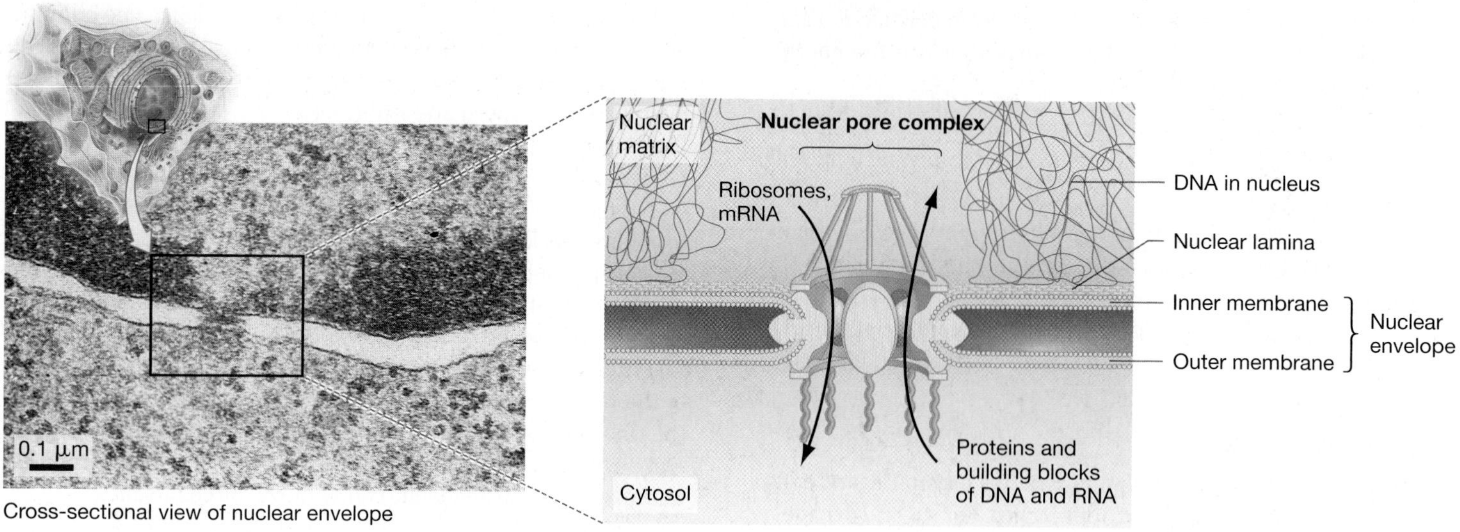

Figure 7.16 Structure of the Nuclear Envelope and Nuclear Pore Complex.

research showed that these openings are formed by an elaborate structure called the **nuclear pore complex**. As shown in the diagram on the right in Figure 7.16, the nuclear pore complex extends through both the inner and outer nuclear membranes, connecting the inside of the nucleus with the cytosol. Each nuclear pore complex consists of approximately 30 different proteins.

What substances travel through nuclear pore complexes, and how are they transported? Chromosomal DNA clearly does not travel—it remains in the nucleus as long as the nuclear envelope is intact. But most of the RNA that is synthesized from DNA is exported through nuclear pore complexes to the cytosol.

Several types of RNA molecules are produced in the nucleus, each distinguished by size and function. For example, **ribosomal RNAs** are manufactured in the nucleolus, where they bind to proteins to form the large and small subunits of ribosomes. Molecules called **messenger RNAs (mRNA)** carry the information required to manufacture proteins. Both the newly assembled ribosome subunits and the mRNAs must be transported from the nucleus to the cytosol, where protein synthesis takes place.

Inbound traffic is also impressive. Nucleoside triphosphates that act as building blocks for DNA and RNA enter the nucleus, as do a variety of proteins responsible for copying DNA, synthesizing RNAs, or assembling ribosomes.

To summarize, a typical cell imports or exports over 500 molecules through each of 2000–5000 nuclear pores every second. The scale of traffic through nuclear pore complexes is mind-boggling. How is it regulated and directed?

How Do Molecules Enter the Nucleus?

Small molecules, like nucleotides, can freely diffuse through nuclear pores into the nucleus along their concentration gradients. The import of larger molecules, such as proteins and RNA, appears to be more selective: Some are allowed to enter, others are excluded. What role do the nuclear pores play in this selective transport across the nuclear envelope?

In experiments evaluating the import of proteins, biologists observed that certain proteins would rapidly enter the nucleus, while others were completely excluded—even if they were similar in size. Based on these observations, researchers concluded that nuclear pores do not function as a static filter that selects proteins based on their size alone. They hypothesized that the nuclear pore complex instead serves as a dynamic gate to control passage through the envelope. If this is the case, then what is required to open and close these gates so that only certain proteins may pass?

A series of experiments in the 1980s on a protein called nucleoplasmin helped researchers understand the nature of nuclear import. Nucleoplasmin was chosen because it is strictly found in the nucleus, where it plays an important role in the assembly of chromatin. When labeled nucleoplasmin was injected into the cytoplasm of living cells, the protein was found to be quickly concentrated inside the nucleus. What is special about nucleoplasmin? Does its protein sequence have a "send-to-nucleus" signal for nuclear import?

To address these questions, researchers used enzymes called proteases to cleave nucleoplasmin into two structurally distinct pieces—one referred to as the core and the other as the tail. After separating the two types of fragments, they labeled each type with radioactive atoms and injected them into the cytoplasm of different cells. At various times after the injections, researchers examined the nuclei and cytoplasm of the cells to track down the radioactive label.

The results were striking. They found that tail fragments were rapidly transported from the cytosol into the nucleus. Core fragments, in contrast, were not allowed to pass through the nuclear envelope and remained in the cytosol.

These data led to a key hypothesis: Nuclear proteins contain a kind of "zip code"—a molecular address tag—that marks them for transport through the nuclear pore complex. These proteins are synthesized by ribosomes that are free in the cytosol, and the zip code allows them to pass into the nucleus.

By analyzing different stretches of the tail portion of nucleoplasm, the biologists found a 17-amino-acid-residue-long section of the polypeptide that had to be present to direct nucleoplasmin to the nucleus. Follow-up work confirmed that other proteins transported into the nucleus, even those expressed by some viruses, have similar amino acid sequences directing their transport. This common sequence came to be called the **nuclear localization signal (NLS)**. (Proteins that leave the nucleus have a different signal, required for nuclear export.)

Once a nuclear localization signal had been identified, researchers could evaluate its activity by adding it to a cytosolic protein that is normally excluded from the nucleus. For example, Figure 7.17 shows what happens to pyruvate kinase, a cytosolic protein involved in glycolysis, when it is modified to include a nuclear localization signal. (To learn more about the function of pyruvate kinase in glycolysis, see Ch. 9, Section 9.2).

More recent research has shown that the movement of proteins and other large molecules into and out of the nucleus is an energy-demanding process that involves special transport proteins. These nuclear transport proteins function like trucks that haul cargo into or out of the nucleus through the nuclear pore complex, depending on whether they have an import or export zip code. Biologists are now trying to unravel how all this traffic in and out of the nucleus is regulated to avoid backups and head-on collisions.

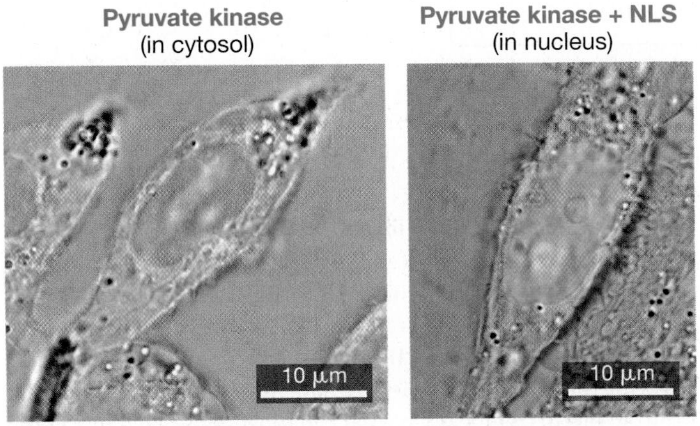

Figure 7.17 Nuclear Localization Signals Direct Cytosolic Proteins into the Nucleus. The cytosolic protein pyruvate kinase (green) is imported from the cytosol (left) into the nucleus (right) when the protein sequence is modified to include a nuclear localization signal.

✔ **PROCESS OF SCIENCE** Use BioSkills 9 to determine the technique that was most likely used to identify the location of pyruvate kinase by live-cell imaging. How was this protein modified to make it fluoresce green in the micrographs?

CHECK YOUR UNDERSTANDING

✔ **If you understood this section, you should be able to ...**

Compare and contrast the movement of (1) nucleotides and (2) large proteins through the nuclear pore complex. Which process would you expect to require an input of energy?

Answers are available in Appendix A.

7.5 Cell Systems II: The Endomembrane System Manufactures, Ships, and Recycles Cargo

The nuclear membrane is not the only place in cells where cargo moves in a regulated and energy-demanding fashion. Most of the proteins found in peroxisomes, mitochondria, and chloroplasts are also actively imported after being manufactured by ribosomes that are free in the cytosol.

After you complete this section, you should be able to ...

- Analyze the process of transporting molecules into and through the endomembrane system.

If you think about it for a moment, the need to sort proteins and ship them to specific destinations should be clear. Proteins are produced by ribosomes that are either free in the cytosol or on the surface of the ER. Many of these proteins must be transported to a compartment inside the eukaryotic cell. Acid hydrolases must be shipped to lysosomes and catalase to peroxisomes. To get to the right location, each protein must have a specific zip code and a delivery system.

To get a better understanding of protein sorting and transport in eukaryotic cells, let's consider perhaps the most intricate of all manufacturing and shipping complexes: the endomembrane system. In this system, proteins that are synthesized on the rough ER move to the Golgi apparatus for processing, and from there they travel to the cell surface or other destinations.

Studying the Pathway through the Endomembrane System

The idea that materials move through the endomembrane system in an orderly way was inspired by a simple observation. According to electron micrographs, cells that secrete digestive enzymes or other proteins have particularly large amounts of rough ER and Golgi (the extensive rough ER is shown in Figure 7.15a). This correlation led to the idea that these organelles may participate in a "secretory pathway" that starts in the rough ER and ends with products leaving the cell (Figure 7.18 on page 162). How does this hypothesized pathway work?

Tracking Protein Movement via Pulse–Chase Assay George Palade and colleagues did pioneering research on the secretory pathway using a **pulse–chase experiment** to track protein movement. This strategy is based on two steps:

1. ***The "Pulse"*** Expose experimental cells to a high concentration of a modified amino acid for a short time. For example, if a cell is briefly exposed to a large amount of radioactively labeled amino acid, virtually all the proteins synthesized during that interval will be radiolabeled.
2. ***The "Chase"*** End the pulse by washing away the modified amino acid and replacing it with the normal version of the same molecule. The time following the end of the pulse is referred to as the chase. The proteins synthesized during the chase period will *not* be radiolabeled.

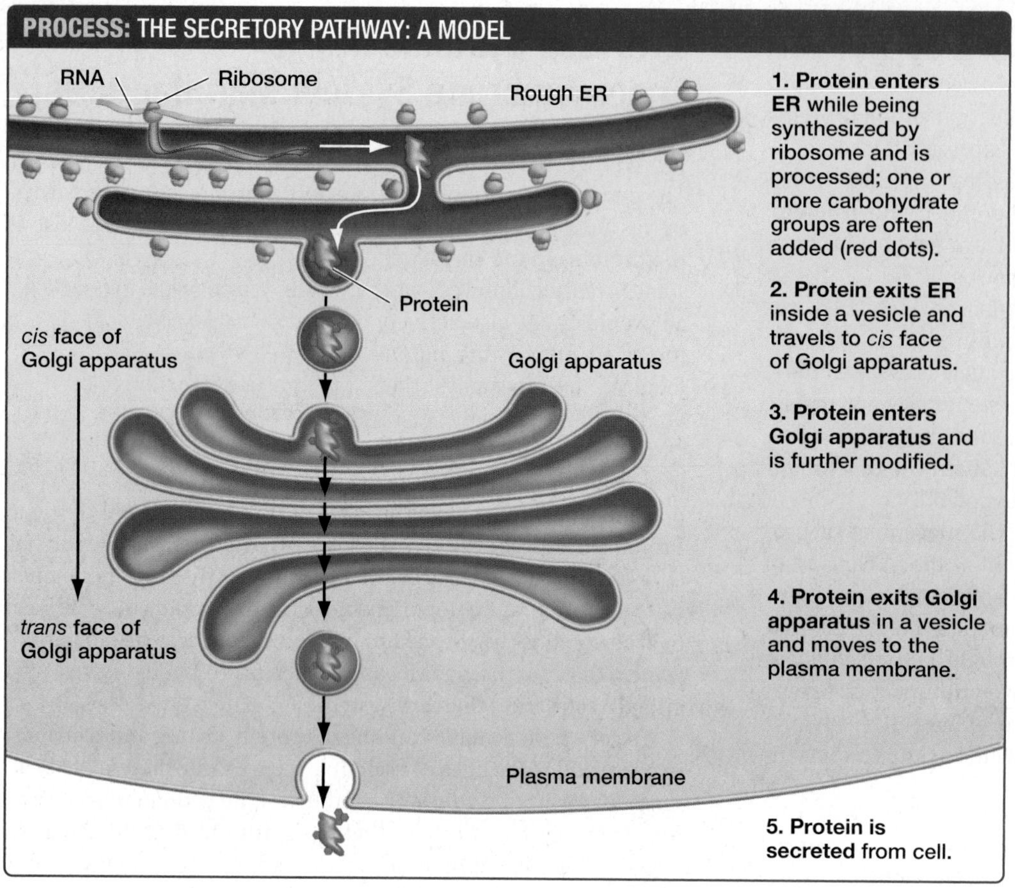

Figure 7.18 The Secretory Pathway Hypothesis: An Overview. This hypothesis proposes that proteins intended for secretion from the cell are synthesized and processed in a highly prescribed series of steps. Note that proteins are packaged into vesicles in order to move them from the rough ER to the Golgi and from the Golgi to the cell surface.

The idea is to mark a population of molecules at a particular interval (the pulse) and then follow their fate over time (the chase). This approach is analogous to adding a small amount of dye to a stream and then following the movement of the dye to track the pattern of water flow.

To understand why the chase requires unlabeled amino acids in these experiments, imagine what would happen if you added dye to a stream continuously. Soon the entire stream would be dyed—you could no longer track a specific population of dye molecules.

In testing the secretory pathway hypothesis, Palade's team focused on pancreatic cells. Pancreatic cells are specialized for secreting digestive enzymes into the small intestine and are packed with rough ER and Golgi. The cells for the experiment were grown in **culture**, or in vitro.[1]

The basic experimental approach was to pulse the cells for 3 minutes with a radiolabeled version of the amino acid leucine, followed by a long chase period with nonradioactive leucine (**Figure 7.19**). The pulse produced a population of proteins that were related to one another by the timing of their synthesis. At different points during the chase, the researchers tracked the movement of these proteins by preparing samples of the cells for autoradiography and electron microscopy (see **BioSkills 6** and **9**). The drawings in Figure 7.19a illustrate what the researchers observed from electron micrographs taken at the end of the pulse and at different times during the chase.

Results of the Pulse–Chase Experiment The graph in Figure 7.19b was based on the electron microscopy results, which revealed that proteins are trafficked through the secretory pathway in a highly organized and directed manner. Track the movement of proteins through the cell during the chase by covering the graph with a piece of paper and then slowly sliding it off from left to right. Notice what is happening to each line at the following time points:

0 minutes Immediately after the pulse, most of the newly synthesized proteins are inside this cell's rough ER.

37 minutes During the chase, the situation changes. At this time, most of the labeled proteins have left the rough ER and entered the Golgi apparatus, and some of them have accumulated inside structures called secretory vesicles.

117 minutes By the end of the chase, most of the labeled proteins have left the Golgi and are either in secretory vesicles or were secreted from the cells.

Over a period of two hours, the labeled population of proteins moved along a defined trail through the rough ER, Golgi apparatus, and secretory vesicles to reach the exterior of the cell.

Jamieson and Palade's results support the hypotheses that a secretory pathway exists and that the rough ER and Golgi apparatus function together as parts of an integrated endomembrane system.

[1] The term "in vitro" is Latin for "in glass." Experiments that are performed outside living organisms are done in vitro. The term "in vivo," in contrast, is Latin for "in life." Experiments performed within living organisms are done in vivo.

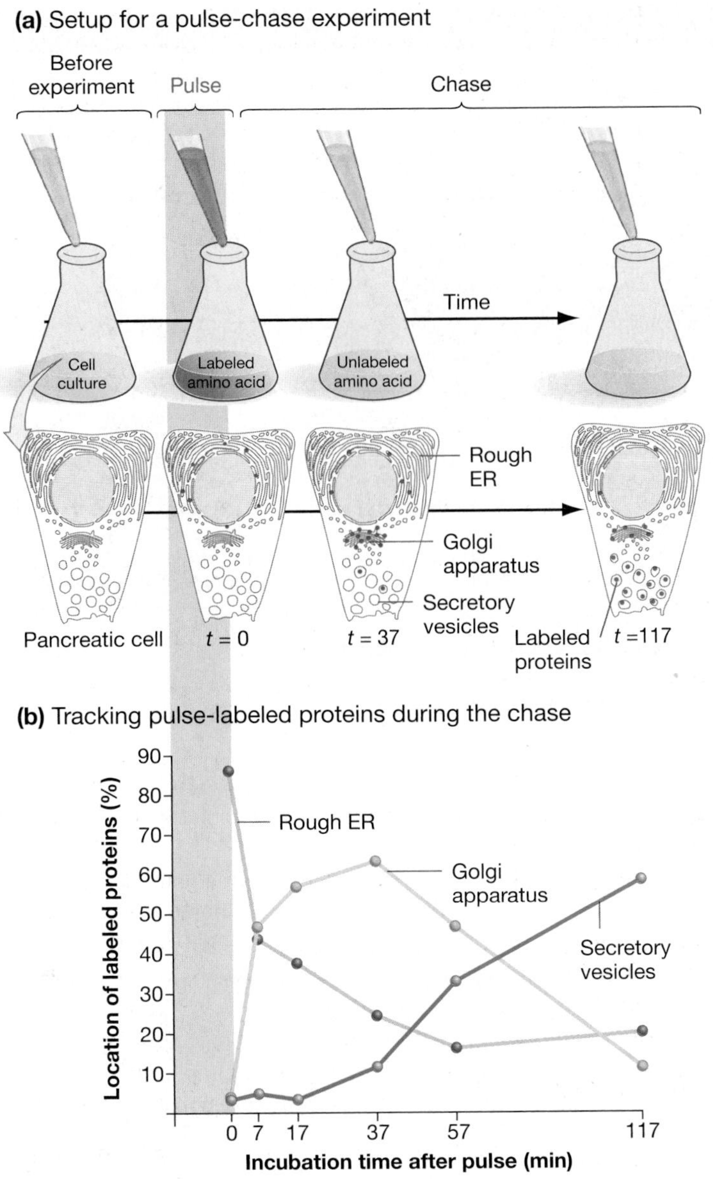

INTERACTIVE Figure 7.19 A Pulse-Chase Experiment. Part **(a)** shows how newly synthesized proteins are labeled during the pulse when exposed to medium containing radioactive amino acids (red). At the start of the chase, this medium is replaced with medium containing non-radioactive amino acids (yellow) so only proteins labeled in the pulse will be tracked (red dots). **(b)** A graph plots the relative abundance of radiolabeled proteins in three different organelles during the chase.

SOURCE: J. D. Jamieson and G. E. Palade. 1967. Intracellular transport of secretory proteins in the pancreatic exocrine cell. *Journal of Cell Biology* 34: 597–615.

✔ **QUANTITATIVE** Use the graph in part (b) to estimate the shortest time it would take for a protein to pass through the Golgi apparatus.

Next, let's examine four of this pathway's steps in more detail:

1. How do proteins enter the lumen of the rough ER?
2. How do the proteins move from the ER to the Golgi apparatus?
3. Once proteins are inside the Golgi apparatus, what happens to them?
4. And finally, how does the Golgi apparatus sort out the proteins so each one goes to the appropriate place?

How Do Proteins Enter the Endomembrane System?

The synthesis of proteins destined to be secreted or embedded in membranes begins in ribosomes free in the cytosol. Günter Blobel and colleagues proposed that at some point these ribosomes become attached to the outside of the ER. But what directs ribosomes to the ER surface? The "signal hypothesis" predicts that proteins bound for the endomembrane system have a molecular zip code that serves a similar role to the nuclear localization signal in nucleoplasmin. Blobel proposed that the first amino acid residues of the growing protein act as a signal that marks the ribosome for transport to the ER membrane.

Blobel's group went on to produce convincing data that supported the hypothesis: They identified a "send-to-ER" signal, or **ER signal sequence**, that guides the growing protein and associated ribosome to the rough ER. The ER signal sequence typically is present in the first 20 amino acid residues and is removed when protein synthesis is complete.

More recent work has documented the mechanisms responsible for receiving this send-to-ER signal and inserting the protein into the rough ER. Figure 7.20 on page 164 illustrates the key steps involved for a protein that will eventually be shipped to the inside of an organelle or secreted from the cell.

Step 1 Protein synthesis begins on a free ribosome in the cytosol. The ribosome synthesizes the ER signal sequence, using information carried in an mRNA.

Step 2 The signal sequence binds to a **signal recognition particle (SRP)**—a complex of RNA and protein. The attached SRP causes protein synthesis to stop.

Step 3 The ribosome + signal sequence + SRP complex moves to the rough ER membrane, where it attaches to the SRP receptor. Think of the SRP as a key that is activated by an ER signal sequence. The SRP receptor in the ER membrane is the lock.

Step 4 Once the lock (the receptor) and key (the SRP) connect, the SRP is released and protein synthesis continues through a channel called the translocon.

Step 5 The growing protein is fed into the ER lumen, and the ER signal sequence is removed.

After cleavage of the signal sequence, the protein may be completely released into the ER lumen. Some proteins, however, remain associated with the membrane as integral membrane proteins. How do such proteins get inserted into the ER membrane? Current models propose that the translocon has molecular "gates" that divert stretches of nonpolar amino acids in the growing protein into the phospholipid bilayer. Recall that transmembrane portions of integral membrane proteins consist of nonpolar amino acids (Ch. 6, Section 6.4). After pushing these regions into the membrane, the rest of the protein continues to be made in the cytosol or the ER lumen.

Once proteins are inside the rough ER or inserted into its membrane, they fold into their three-dimensional shape with the help of chaperone proteins (Ch. 3, Section 3.3). In addition, proteins that enter the ER lumen interact with enzymes that catalyze the addition of carbohydrate side chains (see Figure 7.18).

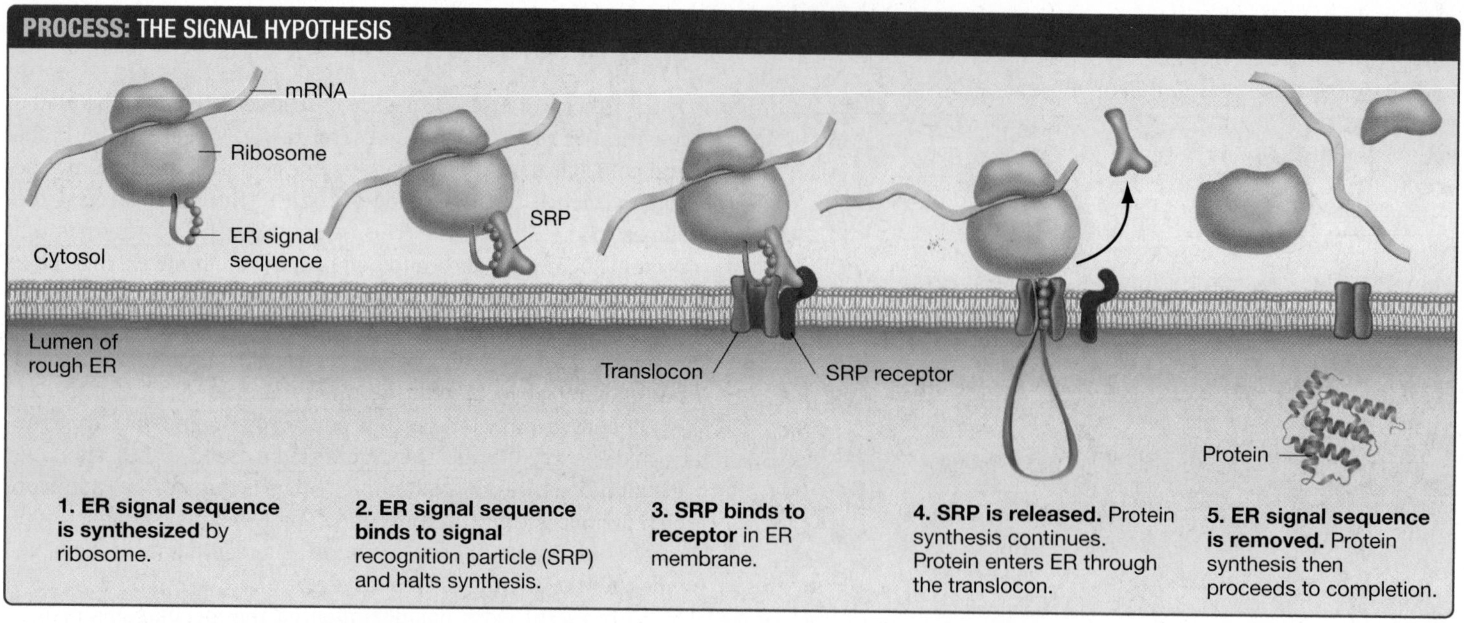

Figure 7.20 **The Signal Hypothesis Explains How Proteins Destined for Secretion Enter the Endomembrane System.** According to the signal hypothesis, proteins destined for secretion contain a short stretch of amino acids that interact with a signal recognition particle (SRP) in the cytosol. This interaction directs the synthesis of the remaining protein into the rough ER lumen.

Because carbohydrates are polymers of sugar monomers, the addition of one or more carbohydrate groups is called **glycosylation** ("sugar-together"). The resulting molecule is a **glycoprotein** ("sugar-protein"; Ch. 5, Section 5.3). The structure of these carbohydrates changes as the proteins are folded, serving as an indicator for shipment to the next destination—the Golgi apparatus.

Moving from the ER to the Golgi Apparatus

How do proteins travel from the ER to the Golgi apparatus? In Palade's pulse–chase experiment, labeled proteins were observed in small membrane-bound structures between the rough ER and the Golgi apparatus. Based on these observations, Palade's group suggested that proteins are transported in vesicles that bud off from the ER and move to the *cis* face of the Golgi apparatus.

This hypothesis was supported when other researchers used differential centrifugation to isolate and characterize the vesicles that contained the pulse-labeled proteins. They found that a distinctive type of vesicle carries proteins from the rough ER to the Golgi apparatus. Ensuring that only appropriate cargo is loaded into these vesicles and that the vesicles dock and fuse only with the *cis* face of the Golgi apparatus involves a complex series of events and is an area of active research.

What Happens Inside the Golgi Apparatus?

Section 7.2 described this organelle as a stack of flattened compartments called cisternae, with cargo entering one side of the organelle and exiting the other. Recent research has shown that the composition of the Golgi apparatus is dynamic. New cisternae constantly form at the *cis* face of the Golgi apparatus, while old cisternae break apart at the *trans* face, to be replaced by the cisternae behind it.

By separating individual cisternae and analyzing their contents, researchers found that cisternae at various stages of maturation contain different suites of enzymes. As a result, the cargo gets modified in a stepwise manner as it moves within compartments at different stages of maturation.

If the rough ER is like a foundry and stamping plant where rough parts are manufactured, then the Golgi apparatus can be considered a finishing area where products are polished, painted, and readied for shipping.

How Do Proteins Reach Their Proper Destinations?

The rough ER and Golgi apparatus constitute an impressive assembly line. Certain proteins manufactured by this process remain in these organelles, replacing worn-out resident molecules. But those proteins that are simply passing through as cargo must be sorted and sent to their intended destination as the *trans* cisterna breaks up into vesicles. How is cargo put into the right shipping containers, and how are the containers addressed for transport to different locations?

Studies on hydrolytic enzymes that are shipped to lysosomes have provided some answers to both questions. A key finding was that lysosome-bound proteins have a phosphate group attached to a specific sugar on their surface, forming the compound mannose-6-phosphate. If this phosphorylated sugar is removed from these proteins, they are not transported to a lysosome.

The mannose-6-phosphate tag serves as a zip code, like the nuclear localization and ER signal sequences discussed earlier. Data indicate that a receptor protein in the membrane of the *trans*-Golgi cisterna binds to this tag. Regions that are enriched with these receptor–cargo complexes will form cargo-filled vesicles. Besides these receptors, the vesicles also include specific membrane proteins that direct their transport to pre-lysosomal compartments. In this way, the presence of mannose-6-phosphate targets proteins to organelles that eventually become lysosomes.

Figure 7.21 presents a simplified model of how cargo is sorted and loaded into specific vesicles that are shipped to different destinations. Each cargo protein has a molecular tag that directs it to particular vesicle budding sites by interacting with receptors in the *trans* cisterna. These receptors, along with other membrane and cytosolic proteins that are not shown, direct the transport vesicles to the correct destinations. Take a moment to observe how the vesicles shown in the middle of Figure 7.21 fuse with the pre-lysosomal compartment membrane and deliver their contents.

Next, notice that the transport vesicle shown on the right of Figure 7.21 is bound for the plasma membrane, where it will secrete its contents to the outside. This process is called **exocytosis** ("outside-cell-act"). When exocytosis occurs, the vesicle membrane and plasma membrane make contact. As the two membranes fuse, the interior of the vesicle is exposed to the outside of the cell. The vesicle's contents then diffuse into the space outside the cell. This is how cells in your pancreas deliver digestive enzymes to the duct that leads to your small intestine, where food is digested.

When illustrating the process of cargo transport through the endomembrane system, biologists often use arrows to represent different processes. To learn more about how to model the secretory pathway using arrows and simplified structures, see **Making Models 7.1** on page 166.

While the movement of protein cargo has been the focus of this section, another key role of the endomembrane system is the distribution of membrane. For example, when vesicles fuse with the cell surface during exocytosis, the membrane of the vesicle becomes part of the plasma membrane. The transported membrane proteins will profoundly impact the function of the plasma membrane. In addition, the resulting increase in the surface area of a cell is particularly important as cells grow and divide.

Recycling Material in the Lysosome

Now that you have seen how the endomembrane system transports cargo to different destinations in the cell, let's look at how cargo transport contributes to protein recycling. Recall that large molecules, such as proteins, do not readily pass through membranes (Ch. 6, Section 6.2). This means that to get amino acid monomers back into the cytosol for new protein synthesis, existing proteins must first be digested in the lysosome—but how do they get there?

Figure 7.21 In the Golgi Apparatus, Proteins Are Sorted into Vesicles That Are Targeted to a Destination.

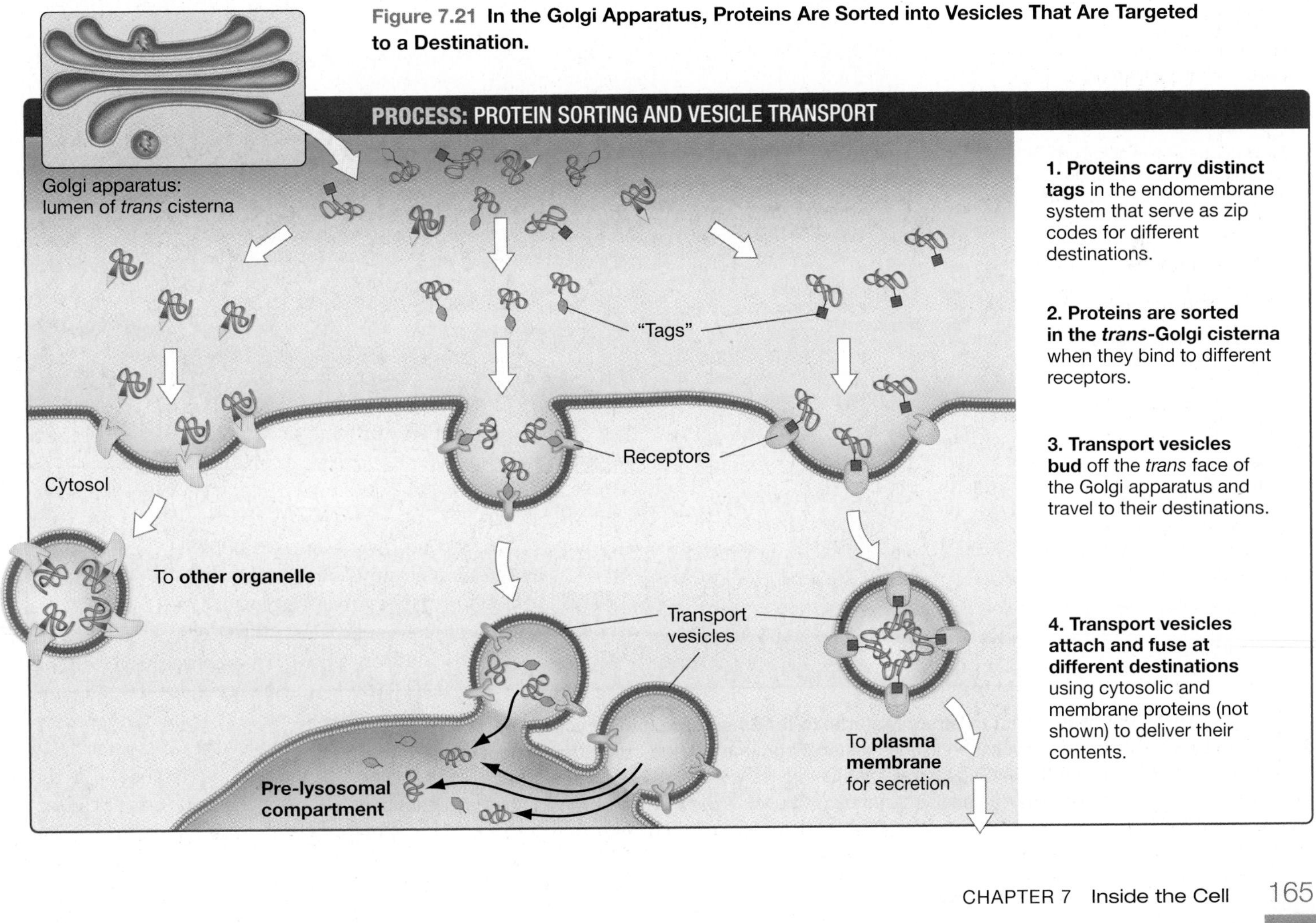

Making Models 7.1 Tips on Drawing Cell Structures and Processes

In models of a cell, scientists often draw single lines to represent membranes and use different shapes to indicate structures within the cell. Arrows can indicate a variety of processes, such as the transport of cargo, the fusion of a vesicle with another membrane, or the maturation of an organelle.

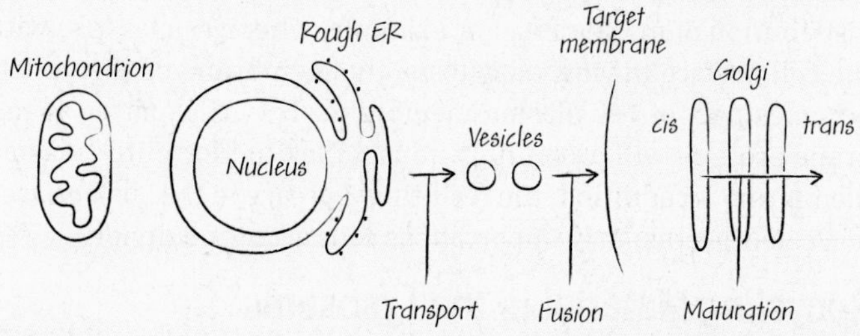

MODEL Draw the path of a secretory protein within a cell from the site of synthesis to exocytosis. Identify all the relevant cell structures and add notes indicating what happens to the protein at each step.

To see this model in action, go to the Study Area of **Mastering Biology**

There are three pathways that animal cells commonly use to recycle material in the lysosome (see Figure 7.22). Two of these three pathways involve pinching off the plasma membrane to take up material from outside the cell—a process called **endocytosis** ("inside-cell-act"). The third pathway targets materials already inside the cell.

Receptor-Mediated Endocytosis As its name implies, **receptor-mediated endocytosis** is a sequence of events that begins when particles outside the cell bind to receptors on the plasma membrane. More than 25 distinct cargo receptors have now been characterized, each specialized for binding to different molecules.

Once receptor binding occurs, the plasma membrane folds in and pinches off to form an endocytic vesicle. These vesicles then drop off their cargo in an organelle called the **early endosome** ("inside-body"). The activity of proton pumps in the membrane of this organelle acidifies its lumen, which causes the cargo to be released from their receptors. Many of these emptied cargo receptors are then repackaged into vesicles and returned to the plasma membrane.

As proton pumps continue to lower the early endosome's pH, it undergoes a series of processing steps that cause it to mature into a **late endosome**. The late endosome is the pre-lysosomal

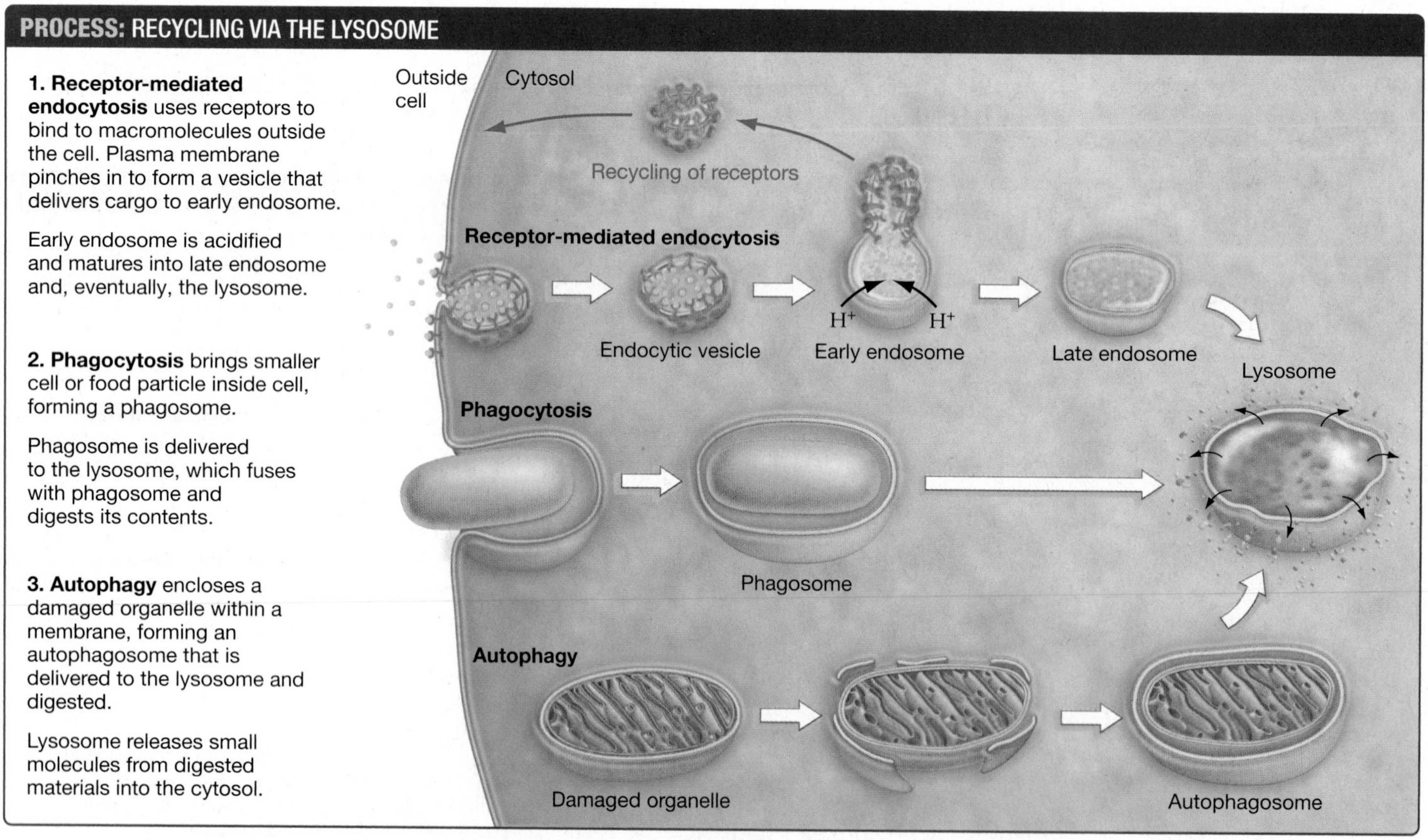

Figure 7.22 Three Pathways Exist to Direct Material to the Lysosome for Recycling. *Receptor-mediated endocytosis* and *phagocytosis* involve bringing in material from the outside and surrounding it with a lipid bilayer from the plasma membrane. Endosomes mature into lysosomes or, like phagosomes, will fuse with existing lysosomes. In *autophagy*, material within the cytoplasm is encapsulated with an internal membrane before fusing with the lysosome.

compartment introduced earlier (Figure 7.21), where acid hydrolases from the Golgi apparatus are dropped off. As before, the emptied cargo receptors transported from the Golgi apparatus are removed from the late endosome as it matures into a fully active lysosome.

Phagocytosis A second pathway that involves recycling material brought in from the outside of the cell is called **phagocytosis** ("eat-cell-act"). In phagocytosis, the plasma membrane of a cell surrounds a smaller cell or food particle and engulfs it, forming a structure called a phagosome. This structure is delivered to a lysosome, where the phagosome and lysosome membranes fuse and the contents of the phagosome are digested.

Autophagy Cells are also involved in recycling large structures and organelles that exist within the cytoplasm through a process called **autophagy** ("same-eating"). During autophagy, portions of the cytoplasm, including damaged organelles marked for destruction, are enclosed within an internal membrane to form an autophagosome. Like the phagosome, the membrane of the autophagosome fuses with the lysosome and the contents are digested.

Regardless of whether the materials digested by lysosomes originate via autophagy, phagocytosis, or receptor-mediated endocytosis, the result is similar: Molecules are hydrolyzed and the products are transported across the lysosomal membrane into the cytosol for recycling.

Throughout this section, vesicles have been key to the transport of cargo. If these transport steps depended on the random movement of diffusion alone, however, then the vesicles and their cargo might never reach their intended destinations. Instead, there are defined tracks that direct the movement of these shipping containers. What are these tracks, and what molecule or molecules function to transport the vesicles along them? Let's delve into these questions in the next section.

CHECK YOUR UNDERSTANDING

✔ If you understood this section, you should be able to ...

1. Compare and contrast the movement of proteins into the nucleus versus the ER lumen.
2. Predict the final location of a protein that was experimentally modified to include an ER signal sequence, mannose-6-phosphate tag, and a nuclear localization signal. Justify your answer by addressing the impact of each signal on its transport.

Answers are available in Appendix A.

7.6 Cell Systems III: The Dynamic Cytoskeleton

After you complete this section, you should be able to ...

- Differentiate the structure and function of the three major components of the eukaryotic cytoskeleton.
- Differentiate the roles of the three types of cytoskeletal motor proteins.

The endomembrane system may be the best-studied example of how individual organelles work together in a dynamic, highly integrated way. This integration depends in part on the physical relationship of organelles, which is organized by the cytoskeletal system.

The cytoskeleton is a dense and complex network of fibers that helps maintain cell shape by providing structural support. However, the cytoskeleton is not a static structure like the bones that make up your skeleton. Like the rest of the cell, the cytoskeleton is dynamic. Its fibrous proteins move and change to alter the cell's shape, shift its contents, and even move the cell itself.

As **Table 7.2** shows, there are three major types of cytoskeletal elements in eukaryotic cells: actin filaments, intermediate

SUMMARY Table 7.2 Cytoskeletal Filaments

The three types of filaments that make up the cytoskeleton are distinguished by their size, structure, and type of protein subunit.

Filament	Structure	Subunits	Functions
Actin filaments (microfilaments)	Two coiled strands 	Actin 	• maintain cell shape by resisting tension (pull) • move cells via muscle contraction or cell crawling • divide animal cells in two • move organelles and cytoplasm in plants, fungi, and animals
Intermediate filaments	Fibers wound into thicker cables 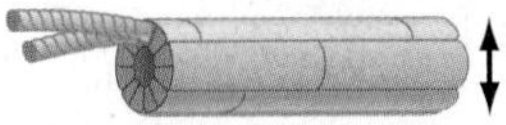	Keratins, lamins, or others 	• maintain cell shape by resisting tension (pull) • anchor nucleus and some other organelles
Microtubules	Hollow tube 	α- and β-tubulin dimers 	• maintain cell shape by resisting compression (push) • move cells via flagella or cilia • move chromosomes during cell division • assist formation of cell plate during plant cell division • provide tracks for intracellular transport

filaments, and microtubules. These three eukaryotic filaments are structurally and functionally related to cytoskeletal elements in bacteria and archaea.

Each of the three cytoskeletal filaments found in eukaryotes has a distinct size, structure, and function. Let's look at each one in turn, starting at the smallest and working up to the largest.

Actin Filaments

Sometimes called **microfilaments** because they are the cytoskeletal element with the smallest diameter, **actin filaments** are fibrous structures made of globular protein subunits called actin (Table 7.2). In animal cells, actin is often the most abundant of all proteins—typically it represents 5–10 percent of the total protein in the cell. For example, a single liver cell might contain about half a billion of these molecules.

Actin Filament Structure A completed actin filament resembles two long strands that coil around each other. Actin filaments form when individual actin protein subunits assemble, or polymerize, from head to tail through the formation of noncovalent bonds.

Because actin proteins are not symmetrical, this head-to-tail arrangement of subunits results in filaments that have distinct polarity, or two different ends. The two distinct ends of an actin filament are referred to as plus and minus ends. These names do not reflect the electrical charge of the filaments, but instead the differences between rates of assembly—the plus end grows faster than the minus end.

In animal cells, actin filaments are particularly abundant just under the plasma membrane. They are organized into long, parallel bundles or dense, crisscrossing networks in which actin filaments are linked to one another by other proteins. The reinforced bundles and networks of actin filaments help stiffen the cell and define its shape.

Actin Filament Function In addition to providing structural support, actin filaments are involved in movement. In some cases, actin's role in movement depends on the protein myosin. Myosin is a **motor protein**—a protein that converts the chemical energy in ATP into the kinetic energy of mechanical work, just as a car's motor converts the chemical energy in gasoline into spinning wheels.

The interaction between actin and myosin is frequently presented in the context of how it produces muscle contraction and movement (Ch. 45, Section 45.1). For now, it's enough to recognize that when myosin binds and hydrolyzes ATP to ADP, it undergoes a series of shape changes that extends the "head" region, attaches it to actin, and then contracts to pull itself along the actin filament. These changes cause the actin and myosin to slide past each other. After repeated rounds of this attachment and contraction cycle, the myosin gradually moves toward the plus end of the actin filament (Figure 7.23a). This type of movement resembles the arms of a sailor hauling in a rope.

As Figure 7.23b shows, the ATP-powered interaction between actin and myosin is the basis for an array of cell movements:

- **Cytokinesis** ("cell-moving") is the final stage in cell division when the cytoplasm is divided to form two cells. In animals, this accomplished by actin filaments that are connected to the plasma membrane and arranged in a ring. Myosin causes the filaments to slide past one another, reducing the diameter of the ring and pulling in the membrane that eventually fuses to produce two cells.
- **Cytoplasmic streaming** is the directed flow of cytosol and organelles that is often seen within plant and fungal cells. The movement occurs along actin filaments and is powered by myosin. It is especially common in large cells, where the circulation of cytoplasm facilitates material transport.

In addition, the movement called **cell crawling** occurs when groups of actin filaments grow, causing bulges in the plasma membrane that extend and move the cell. Cell crawling occurs in a wide range of organisms and cell types, including amoebae, slime molds, and certain animal cells.

Intermediate Filaments

There are many types of **intermediate filament**, each consisting of a different—though similar in size and structure—type of protein subunit (Table 7.2). Humans, for example, have 70 genes

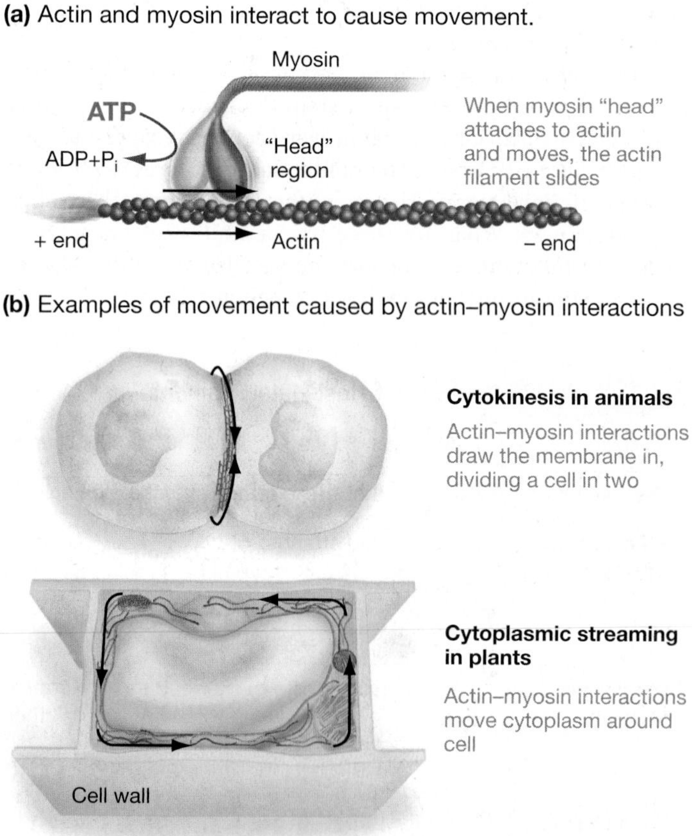

Figure 7.23 Many Cellular Movements Are Based on Actin–Myosin Interactions. (a) ATP hydrolysis in the "head" region of myosin causes the protein to attach to actin and change shape. The movement slides the myosin toward the plus end of actin. **(b)** Actin–myosin interactions can divide cells and move organelles and cytoplasm.

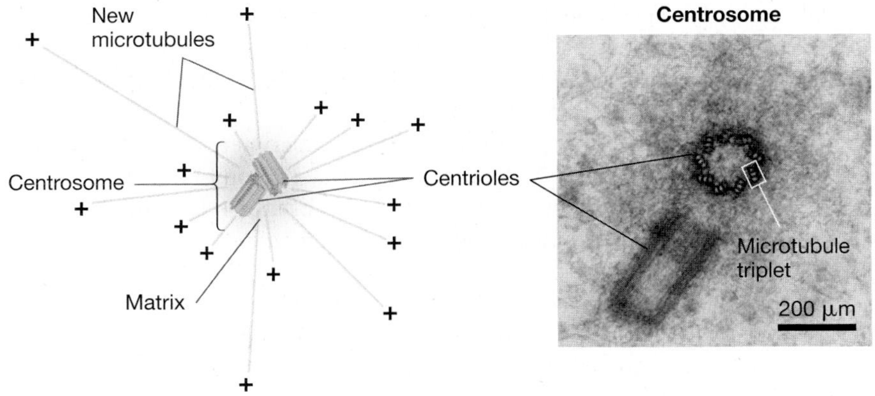

Figure 7.24 Centrosomes Are a Type of Microtubule-Organizing Center. Microtubule-organizing centers, such as the centrosomes of animal cells, are the sites where new microtubules are made. Microtubules grow from a matrix surrounding the centrioles, and their positive ends point away from the centrosomes. The centrosome's two centrioles each consist of nine triplets of microtubules arranged in a circle.

that code for different intermediate filament proteins. This is in stark contrast to actin filaments and microtubules, which are made from the same protein subunits in all eukaryotic cells.

Also unlike actin filaments, intermediate filaments do not exhibit filament polarity, nor do they serve as tracks for the directed movement of motor proteins. Instead, these filaments have identical ends and appear to serve only a structural role in eukaryotic cells.

The most familiar intermediate filaments belong to a family of molecules called the keratins. The cells that make up your skin and line surfaces inside your body contain about 20 types of keratin. These intermediate filaments provide the mechanical strength required for these cells to resist pressure and abrasion. Certain cells in the skin can also produce secreted forms of keratin. Depending on the location of the cell and keratins involved, the secreted filaments form fingernails, toenails, or hair.

Nuclear lamins, which make up the nuclear lamina layer introduced in Section 7.4, also are intermediate filaments. Nuclear lamins form a dense mesh inside the nuclear envelope that anchors chromosomes, defines the shape of the nucleus, and stabilizes the envelope. By controlling the interactions between these lamins, many eukaryotic cells will break down and reform the nuclear envelope during cell division.

Various types of intermediate filaments are found throughout the cytoplasm of the cell. Some intermediate filaments project from the nucleus to hold it in place. Others run parallel to the cell surface and interact with proteins embedded in the plasma membrane. To summarize, intermediate filaments function like a flexible internal scaffolding to help secure the shape and stability of the cell.

Microtubules

Microtubules have the largest diameter of the three cytoskeletal filaments. As Table 7.2 shows, they are assembled from subunits consisting of two closely related proteins, α-tubulin and β-tubulin, that exist under normal conditions as stable protein **dimers** ("two-parts").

Tubulin dimers polymerize in a polar head-to-tail fashion via noncovalent bonds to form thin chains called protofilaments that, in turn, interact with one another to form hollow tubes. Because protofilaments always align in the same orientation, microtubules also exhibit polarity—they have α-tubulins at one end (the minus end) and β-tubulins at the other end (the plus end). Similar to actin filaments, microtubules are dynamic and their plus ends grow faster than their minus ends.

Microtubules originate from a structure called the **microtubule-organizing center (MTOC)**. The plus ends of microtubules grow outward from the MTOC, radiating throughout the cell. Although plant cells typically have hundreds of sites where microtubules start growing, most animal and fungal cells have just one MTOC that is near the nucleus. In animal cells this site is called the **centrosome**. As Figure 7.24 shows, animal centrosomes consist of two bundles of microtubules, called **centrioles**, surrounded by an amorphous matrix of proteins that help initiate the growth of new microtubules.

In function, microtubules are similar to actin filaments: They provide stability and are involved in movement. Like the ribs that emanate from your spine, microtubules that radiate from an organizing center stiffen the cell by resisting compression forces. Microtubules also provide a structural framework for organelles. If microtubules are prevented from forming, the network-like configuration of the ER collapses and the Golgi apparatus breaks up into vesicles.

Microtubules are best known for their role in separating chromosomes during mitosis and meiosis (Ch. 12, Section 12.2 and Ch. 13, Section 13.1), but they are involved in many other types of cellular movement as well. Let's first consider their role in moving materials inside cells and then explore how microtubules can help cells to swim.

Microtubules Serve as Tracks for Vesicle Transport Recall from Section 7.5 that vesicles are used to transport materials to a wide array of destinations inside cells. In the 1980s, to study how this movement happens, Ronald Vale and colleagues focused on an extremely large nerve cell in squid called the giant axon. This cell runs the length of the animal's body. If a squid is disturbed, the giant axon signals muscles to contract so the squid can jet away to safety.

The squid giant axon provided a system that could be observed and manipulated efficiently in the lab. The researchers found that if they gently squeezed the cytoplasm out of the giant axon, vesicle transport still occurred in the released cytoplasmic material. This

(a) Electron micrograph

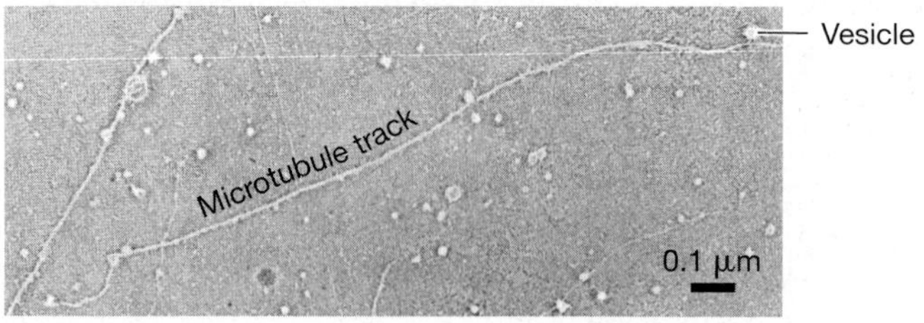

(b) Video image

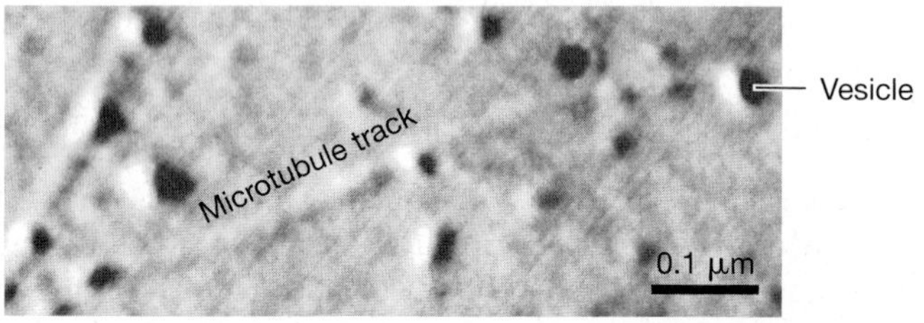

Figure 7.25 **Transport Vesicles Move along Microtubule Tracks.** The two images show extruded cytoplasm from a squid giant axon. **(a)** An electron micrograph that allowed researchers to identify the filaments as microtubules. In the upper part of this image, you can see a vesicle on a "track." **(b)** An image of a video, made using enhanced contrast microscopy, that allowed researchers to watch vesicles move in real time.

allowed them to do experiments on vesicle transport without the cell's plasma membrane being in the way. To watch vesicle transport in action, the researchers mounted a video camera to a microscope. As **Figure 7.25** shows, this technique allowed them to document that vesicle transport occurred along filamentous tracks.

To identify the filament involved, the biologists measured the diameter of the tracks and analyzed their chemical composition. Both types of data indicated that the tracks consist of microtubules. Microtubules also appear to be required for movement of materials elsewhere in the cell. For instance, the movement of vesicles from the rough ER to the Golgi apparatus requires microtubule tracks.

The general message of these experiments is that transport vesicles move through the cell along microtubules. How? Do the tracks themselves move, like a conveyer belt, or are vesicles carried along on some sort of molecular vehicle?

Motor Proteins Pull Vesicles Along the Tracks To study the way vesicles move along microtubules, Vale's group took the squid axon's transport system apart and then determined what components were required to put it back together. A simple experiment convinced the group that vesicle movement is an energy-dependent process: If they depleted the amount of ATP in the axonal cytoplasm, vesicle transport stopped.

To examine this process further, they mixed purified microtubules and vesicles with ATP, but no transport occurred. Something had been left out—but what? To find the missing component or components, the researchers purified one subcellular part after another and added it to the microtubule + vesicle + ATP system.

Through trial and error, and further purification steps, the researchers finally succeeded in isolating a protein that generated vesicle movement. They named the molecule **kinesin**, from the Greek word *kine* ("movement").

Like myosin, kinesin is a motor protein. Kinesin converts the chemical energy in ATP into mechanical energy in the form of movement. More specifically, when ATP is hydrolyzed by kinesin, the protein moves along microtubules in a directional manner: toward the plus end.

Biologists began to understand how kinesin works by first focusing on its structure. Kinesin consists of multiple protein subunits. Each of the two large subunits have three major functional regions: a head section, a tail associated with small polypeptides, and a stalk that connects the head and tail (**Figure 7.26a**). The head

(a) Structure of kinesin

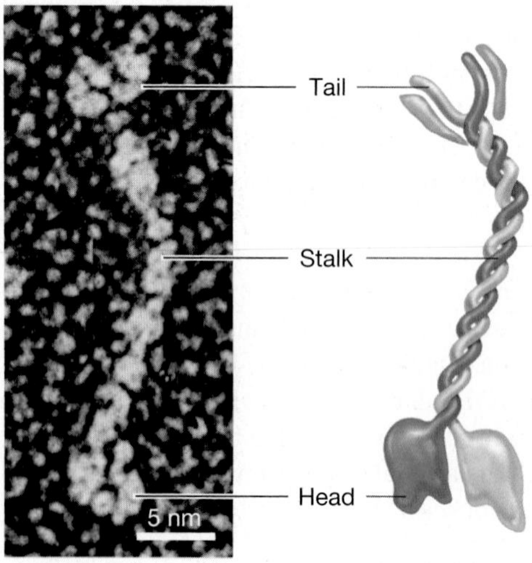

(b) Kinesin "walks" along a microtubule track.

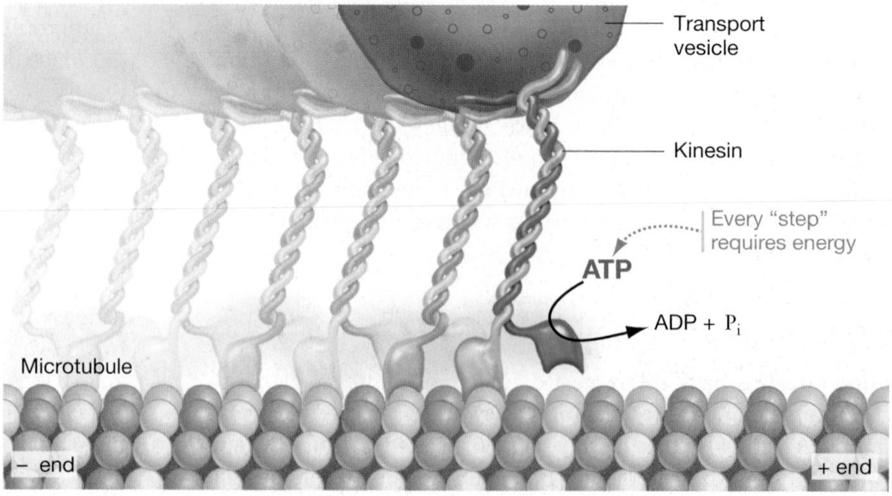

Figure 7.26 **Motor Proteins Move Vesicles Along Microtubules. (a)** Kinesin has three distinct regions. **(b)** The current model to explain how kinesin "walks" along a microtubule track to transport vesicles. The two head segments act like feet that alternately attach, pivot, and release in response to the attachment and hydrolysis of ATP.

region binds to the microtubule while the tail region binds to the transport vesicle. Recent work has shown that kinesin uses these domains to "walk" along the microtubule through a series of conformational changes as it hydrolyzes ATP (Figure 7.26b). In this way, kinesin races along at an amazing 375 "steps" per second.

The discovery of kinesin explained how secretory vesicles could be moved toward the plus end of microtubules, during their transport from the *trans*-Golgi cisterna to the plasma membrane. What about movement in the opposite direction, toward the minus end of microtubules located at the centrosome? By studying whole-cell locomotion, researchers discovered a second type of motor protein that fulfills this role.

Flagella and Cilia: Moving the Entire Cell

Some eukaryotic cells use long, whiplike flagella that project from the cell surface for locomotion. While many prokaryotes also have flagella, the structure is completely different in the two groups.

- A prokaryotic flagellum consists of a single helical rod made of flagellin (in bacteria) or other types of proteins (in archaea); a eukaryotic flagellum consists of several microtubules constructed from tubulin dimers.
- Prokaryotic flagella move the cell by rotating like a ship's propeller; eukaryotic flagella move the cell by undulating—they whip back and forth.
- Eukaryotic flagella are surrounded by the plasma membrane; prokaryotic flagella are not.

Based on these observations, biologists conclude that the prokaryotic and eukaryotic flagella evolved independently, even though their purpose is similar.

The eukaryotic flagellum is closely related to a structure called the **cilium** (plural: **cilia**), which is a short, hairlike projection that is also found in some eukaryotic cells (Figure 7.27, left). Eukaryotic flagella (Figure 7.27, right) are generally much longer than cilia, and the two structures differ in their abundance and pattern of movement. But when researchers examined the two structures with an electron microscope, they found that their underlying organization is identical.

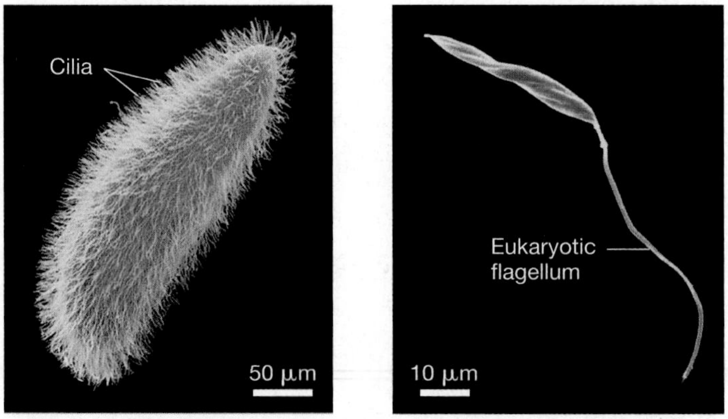

Figure 7.27 Cilia and Eukaryotic Flagella Differ in Length and Number. Cilia range in length from 1 to 10 μm, while eukaryotic flagella are typically longer and can exceed 1 mm. Flagellated cells typically possess only 1–4 flagella. Cilia tend to occur in larger numbers, and certain ciliated cells have up to 14,000 cilia. (The cells in these scanning electron micrographs have been colorized.)

How are Cilia and Flagella Constructed? In the 1950s, anatomical studies established that most cilia and flagella have a characteristic "9 + 2" arrangement of microtubules called the **axoneme** ("axle-thread"). As Figure 7.28a shows, the axoneme consists of nine microtubule pairs, or doublets, surrounding two central microtubules.

The nine doublets of the axoneme originate from a structure called the **basal body**. The basal body is identical in structure to a centriole (it has nine microtubule triplets arranged in a circle) and serves as an MTOC for growth of the axoneme doublets.

Through further study, biologists gained a more detailed view of the axoneme's structure. A series of spoke-like proteins connect each doublet to the central pair of microtubules, and molecular links connect adjacent doublets to one another (Figure 7.28b). Each doublet also has pairs of "arms" along its length that project toward an adjacent doublet.

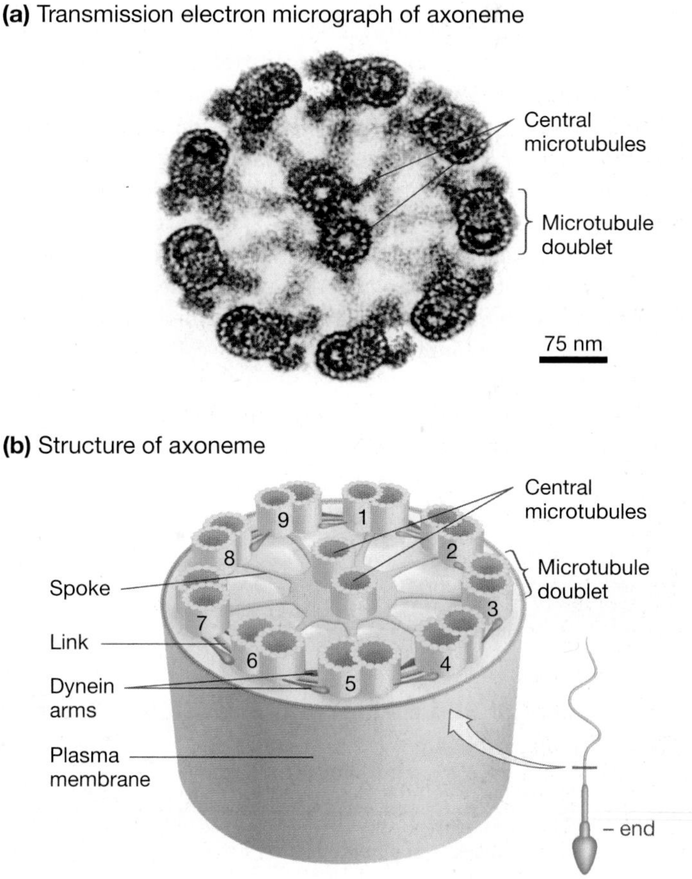

Figure 7.28 Cilia and Flagella Share a Structure Called an Axoneme. (a) Transmission electron micrograph of a cross section through an axoneme. **(b)** The microtubules in cilia and flagella are connected by links and spokes, and the entire structure is surrounded by the plasma membrane.

Axonemes are complex. How do their components interact to generate motion?

What Provides the Force Required for Movement? In the 1960s Ian Gibbons began studying the cilia of a common unicellular eukaryote called *Tetrahymena*. He isolated the axonemes from cilia and found that they would beat only if he supplied them with ATP, confirming that the beating of cilia requires energy.

In another experiment, Gibbons treated the isolated axonemes with a molecule that disrupts the arms from the doublets. The resulting axonemes could not beat even after being supplied with ATP. This result suggested that the arms are required for movement. Follow-up work showed that the arms are made of a large protein that Gibbons named **dynein** (from the Greek root *dyne*, meaning "force").

Like myosin and kinesin, dynein is a motor protein that uses ATP to undergo conformational changes. These shape changes move dynein along microtubules toward the minus end. In the cytoplasm, dynein motors are known to play various roles similar to those of other motor proteins, including the transport of vesicles. In the context of the axoneme, however, the outcome of dynein walking is very different.

So what is special about the axoneme? Remember that each of the nine doublets in the axoneme is connected to the central pair of microtubules by radial spokes, and linking proteins connect the adjacent doublets (Figure 7.28b). As a result, the sliding motion produced by dynein walking is constrained—if one doublet slides, it transmits force to the rest of the axoneme via the links and spokes (Figure 7.29). If the dynein arms on just one side of the axoneme are activated, then the localized movement results in bending. The bending of cilia or flagella results in a swimming motion.

Scaled for size, axoneme-powered swimming can be rapid. In terms of the number of body or cell lengths traveled per second, a sperm cell from a bull moves faster than a human world-record-holder does when swimming freestyle. At the cellular level, life is fast paced.

Taken together, the data reviewed in this chapter can be summed up in six words: Cells are dynamic, highly integrated structures. To maintain the level of organization that is required for life, chemical reactions must take place at mind-boggling speeds. How cells accomplish this feat is taken up elsewhere (Chapter 8).

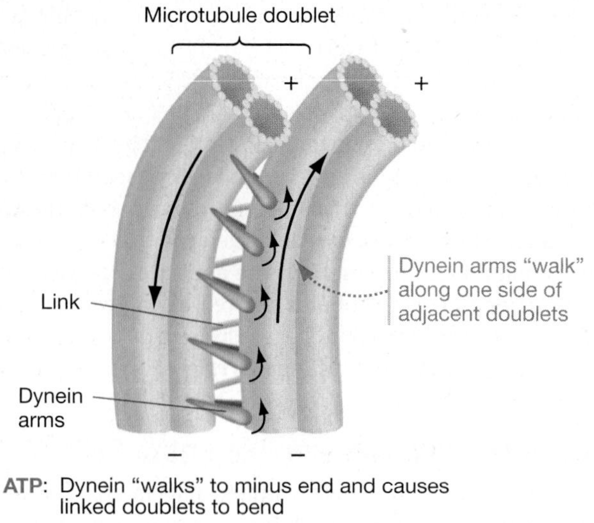

Figure 7.29 Mechanism of Axoneme Bending. When dynein arms "walk" along the microtubule doublets on one side of a flagellum, force is transmitted to these links and spokes, causing the entire axoneme to bend.

✔ **MODEL** If the links were removed from the pair of doublets, draw how the structure would appear after adding ATP. Would link removal affect bending or not?

CHECK YOUR UNDERSTANDING

✔ **If you understood this section, you should be able to ...**

1. Differentiate the structures of actin filaments, intermediate filaments, and microtubules.
2. List the three cytoskeletal motor proteins and provide a distinct cellular role for each.

Answers are available in Appendix A.

CHAPTER 7 Review

For media, go to Mastering Biology

7.1 Bacterial and Archaeal Cell Structures and Their Functions

- There are two basic cellular designs: prokaryotic and eukaryotic. A defining characteristic that differentiates prokaryotes from eukaryotes is the absence of a nucleus.
- Structures common to most, if not all, prokaryotes are ribosomes, a cell wall, a plasma membrane, an interior cytoskeleton, and a nucleoid.
- Many prokaryotes also possess flagella, fimbriae, and internal compartments, some of which are considered organelles.

7.2 Eukaryotic Cell Structures and Their Functions

- Eukaryotic cells are usually much larger and more structurally complex than prokaryotic cells.
- Eukaryotic cells contain numerous specialized organelles that compartmentalize the cytoplasm and enable the cells to grow to a large size. Common eukaryotic organelles are as follows:
  1. The *nucleus*, which contains the cell's chromosomes and serves as its control center.

2. The *endomembrane system*, which consists of a diverse group of interrelated organelles, including the endoplasmic reticulum, Golgi apparatus, lysosomes or vacuoles, and endosomes. These organelles work together to synthesize, process, sort, transport, and recycle material.
3. *Peroxisomes* are organelles where key reactions take place that often result in the generation of toxic by-products. Specialized enzymes within the peroxisomes safely disarm these byproducts soon after they are generated.
4. *Mitochondria and chloroplasts*, which have extensive internal membrane systems where the enzymes responsible for ATP generation and photosynthesis reside.

7.3 Putting the Parts into a Whole

- Cells have a tightly organized interior, where the presence and quantity of organelles often reflect the function of the cell.
- The activity in a cell illustrates the dynamic nature of life. Organelles and cytosolic proteins continually bustle about like a seemingly nonstop rush hour.
- Much of what is known about cellular activity has come from advances in cell imaging and techniques for isolating cellular components.

7.4 Cell Systems I: Nuclear Transport

- Traffic across the nuclear envelope occurs through nuclear pore complexes that serve as gatekeepers.
- Small molecules passively diffuse through nuclear pore complexes while larger molecules require a nuclear localization signal to direct them through the nuclear pore complex via nuclear transport proteins.

7.5 Cell Systems II: The Endomembrane System Manufactures, Ships, and Recycles Cargo

- Molecules synthesized in the ER may be transported as cargo to the Golgi apparatus and then to other organelles or outside the cell.
- Before products leave the Golgi apparatus, they are sorted by their molecular "zip codes" that help package them into specific vesicles. Other membrane and cytosolic proteins deliver the vesicles to their target locations.
- Lysosomes consist of enzymes and membranes that are made and processed through the endomembrane system. Lysosomes are involved in recycling products via receptor-mediated endocytosis, phagocytosis, and autophagy.

7.6 Cell Systems III: The Dynamic Cytoskeleton

- The cytoskeleton is an extensive system of fibers that provides (1) structural support for organizing organelles and other cell components; (2) paths for moving intracellular structures and organelles; and (3) cellular locomotion via cell crawling or propulsion by flagella or cilia.
- The cytoskeleton is dynamic. Actin filaments and microtubules are polarized, meaning they have different ends designated as plus or minus. The plus ends have a higher growth rate than the minus ends.
- Motor proteins move along actin filaments and microtubules using chemical energy stored in ATP. Myosin motor proteins move toward the plus ends of actin filaments. Kinesin and dynein motor proteins move along microtubules toward the plus and minus ends, respectively.
- In the axonemes of eukaryotic cilia and flagella, dynein motors move microtubules to generate forces that bend the structures and enable cells to swim.

Answers are available in Appendix A.

✔ TEST YOUR KNOWLEDGE

1. What are three attributes of mitochondria and chloroplasts that suggest they were once free-living bacteria?
2. **PROCESS OF SCIENCE** Which of the following results provided evidence of a discrete nuclear localization signal somewhere on the nucleoplasmin protein?
   a. The nucleoplasmin protein was small and easily slipped through the nuclear pore complex.
   b. After cleavage of the nucleoplasmin protein, only the tail segments appeared in the nucleus.
   c. Removing the tail from the nucleoplasmin protein allowed the core segment to enter the nucleus.
   d. The SRP bound only to the tail of the nucleoplasmin protein, not the core segment.
3. Molecular zip codes direct molecules to particular destinations in the cell. How are these signals read?
   a. They bind to receptor proteins.
   b. They enter transport vesicles.
   c. They bind to motor proteins.
   d. They are glycosylated by enzymes.
4. How does the hydrolysis of ATP result in the movement of a motor protein along a cytoskeletal filament?

✔ TEST YOUR UNDERSTANDING

5. Which of the following cell structures would you expect to be most important in the growth of bacteria on the surface of your teeth?
   a. cell wall
   b. fimbriae
   c. flagella
   d. cilia
6. Cells that line your intestines are known to possess a large number of membrane proteins that transport small molecules and ions across the plasma membrane. Which of the following cell structures would you expect to be required for this characteristic of the cells?
   a. the endoplasmic reticulum
   b. peroxisomes
   c. lysosomes
   d. the cell wall
7. Most of the proteins that enter the nucleus possess a nuclear localization signal (NLS), even if they are small enough to pass through the nuclear pore complex unhindered. Why would a small protein have an NLS, when it naturally diffuses across the nuclear pore complex without one?

8. The eukaryotic cytoskeleton is a highly dynamic network of filaments and motor proteins. Which of the following correctly describe activities of these cytoskeletal components? Select True or False for each statement.
   T/F Myosin motors walk toward the plus ends of intermediate filaments.
   T/F Dynein motors are responsible for the whip-like movement of eukaryotic flagella.
   T/F Kinesin motors move vesicles along tracks toward the microtubule-organizing center.
   T/F Actin filaments are required for cytoplasmic streaming.

✔ TEST YOUR PROBLEM-SOLVING SKILLS

9. When analyzing a sample of cells from a patient, you find the lysosomes are filled with undigested material. This observation makes you think that the lysosomes are not functioning properly. What are three different defects that could be responsible for malfunctioning lysosomes?

10. George Palade's research group used the pulse–chase assay to elucidate the secretory pathway in pancreatic cells. If they had instead performed this assay on muscle cells, where would you expect the labeled proteins to end up during the chase? (Muscle cells consist primarily of actin and myosin filaments and have high energy demands for muscle contraction.)

PUT IT ALL TOGETHER: Case Study

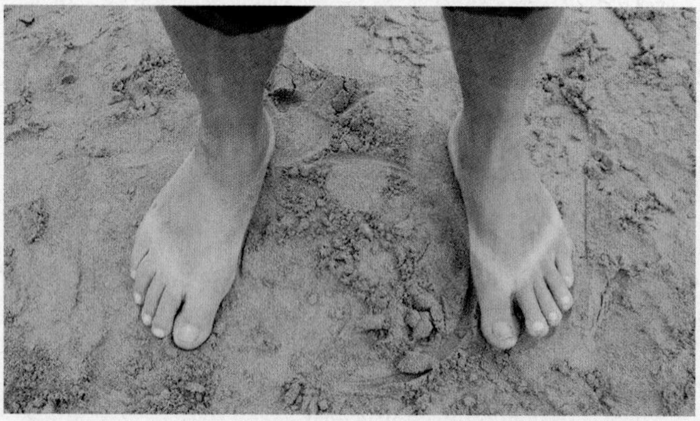

What organelles are required to color your cells?

The color of your eyes, skin, and hair is a product of cellular activity. In some animals, pigmentation is dynamic. You may be familiar with scenes of an octopus blending in with its environment or chameleons turning bright red to ward off threats. Although less dynamic, the color of human skin can also change—or tan—when exposed to sunlight. How do cells pull off this remarkable feat?

11. Colorful animal cells are rich in melanosomes, a specialized organelle that contains pigments, like melanin, and the enzymes that produce them. Melanosomes are found only in certain cells and are called lysosome-related organelles based on their origin, but they do not contain hydrolases or have a low pH. How would you expect melanosomes to form?

12. **MODEL** The distribution of melanosomes in cells is tightly regulated in animals that rapidly change color. Dark-colored cells have melanosomes scattered throughout the cytoplasm, while light-colored cells have them aggregated near the nucleus. Draw a model of a cell representing each state, and predict how cells could use microtubules and motor proteins to change the distribution of melanosomes.

13. The color of human skin is determined by the abundance of melanosomes in keratinocytes—the dominant cell type in skin. But keratinocytes do not produce melanosomes. Instead, less abundant melanosome-producing skin cells called melanocytes make and secrete the organelles. How could keratinocytes take up these organelles? What step in the process must be altered to ensure that the organelles remain in the keratinocytes?

14. **PROCESS OF SCIENCE** Human skin can darken, or tan, in response to the damaging effects of UV radiation from the sun. Propose a hypothesis to explain this phenomenon, and describe how you could set up an experiment to test your hypothesis using microscopy.

15. Even without being exposed to sunlight, humans exhibit a wide range of skin colors due to differences in the abundance of melanosomes. A recent hypothesis to address this difference is that autophagy plays a role in these differences. What is autophagy, and how might you expect this process to differ between individuals with darker skin and those with lighter skin?

16. **QUANTITATIVE** One research group investigated the role of autophagy on skin color using drugs that either induce autophagy (rapamycin) or inhibit autophagy (HCQ) in cells. The graph below compares the level of melanin in the treated samples to that of an untreated control. Explain these results, and use the P values from the graph to determine if the differences are significant or not (* means $P < 0.05$, ** means $P < 0.01$; see **BioSkills 3**). What conclusion, if any, can be drawn from this study?

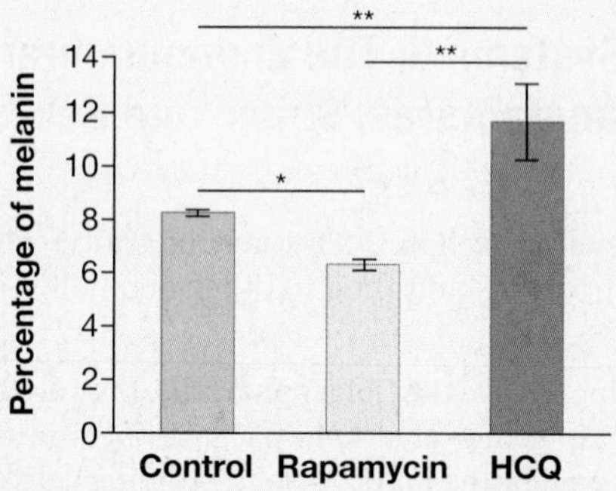

Source: D. Murase et al. 2013. *Journal of Investigative Dermatology* 133: 2416–2424.

Mastering Biology

Students Go to Mastering™ Biology for assignments, the eText, and the Study Area with animations, practice tests, and activities.

Professors Go to Mastering™ Biology for automatically graded tutorials and questions that you can assign to your students, plus Instructor Resources.

8 Energy and Enzymes: An Introduction to Metabolism

When provoked, the bombardier beetle mixes reactants with enzymes in a special chamber near the tip of its abdomen. An enzyme-catalyzed reaction releases heat energy and oxygen gas. This enables the beetle to project boiling hot liquid at a predator, or in this case, a human finger.

In this chapter you will learn how

Enzymes use energy to drive the chemistry of life

looking at energy, asking

- What happens to energy in chemical reactions? 8.1
- Can chemical energy drive nonspontaneous reactions? 8.2

looking at enzymes, asking

- How do enzymes help speed chemical reaction rates? 8.3
- What factors affect enzyme function? 8.4
- How do enzymes work together in metabolic pathways? 8.5

Cells are dynamic. In each of your cells, millions of macromolecules are continually assembling and disassembling, proteins transport substances across cellular membranes, vesicles ferry cargo between organelles and to other destinations. These activities change constantly in response to signals from other cells or the environment.

What is responsible for all this activity? The answer is twofold—energy and enzymes. Because staying alive takes work, there is no life without energy. Life, at its most basic level, consists of chemical reactions catalyzed by enzymes. By using enzymes to direct which reactions occur and when, life possesses the distinguishing feature of creating order from a naturally disordered environment.

This chapter is about how enzymes work to help cells acquire and use energy. It is also your introduction to metabolic pathways—the ordered series of chemical reactions that build up or break down a particular molecule.

Let's begin by reviewing some fundamental concepts about energy and how it is used in cells.

BIG PICTURE

This chapter is part of the Big Picture. See how on pages 236–237.

8.1 What Happens to Energy in Chemical Reactions?

When biologists consider the amount of energy in chemical reactions that is available to do work, they evaluate two types of energy: kinetic energy and potential energy (Ch. 2, Section 2.3). **Kinetic energy** is energy of motion. All moving objects have kinetic energy. Sound, thermal energy, electricity, and electromagnetic radiation (e.g., light) are other forms of kinetic energy. **Potential energy** is energy that is stored in position or configuration. Different forms of potential energy include gravitational, electrical, or chemical gradients, as well as energy in chemical bonds.

After you complete this section, you should be able to . . .

- Explain how energy is transformed in a chemical reaction.
- Describe the factors that determine the spontaneity and rate of a chemical reaction.

The existence of two types of energy—kinetic and potential—does not mean that energy is locked into either type. Rather, it is often transformed from one type to the other. To drive this point home, consider a drop of water sitting at the top of a waterfall, as in Figure 8.1.

Step 1 The water drop has potential energy (E_p) when it is at the top of the waterfall—this is gravitational potential energy.

Step 2 As the drop passes over the waterfall, some of its potential energy is converted to kinetic energy (E_k), in the form of motion.

Step 3 As the drop approaches the rocks below, most of the potential energy has been converted to kinetic energy. When the drop reaches the bottom, some of the kinetic energy generates a force that breaks up rocks (mechanical energy). The rest of the energy is transformed into different forms of kinetic energy, such as thermal energy that raises the temperature of the water, and sound.

PROCESS: ENERGY TRANSFORMATION IN A WATERFALL

1. Potential energy
A water drop sitting at the top of a waterfall has a defined amount of potential energy, $E_{p\ (top)}$.

2. Kinetic energy
As the drop of water falls, some of this potential energy is converted to kinetic energy, E_k.

3. Other forms of kinetic energy
When the water drop strikes the rocks below, its potential energy is now much lower. The change in potential energy has been transformed into an equal amount of kinetic energy.

Result: $E_{p\ (top)} = E_{p\ (bottom)} + E_{k\ (total)}$

Conclusion: Energy is neither created nor destroyed; it simply changes form.

Figure 8.1 **Energy Transformations.** The total amount of energy in the system remains constant during energy transformation.

Chemical Reactions Involve Energy Transformations

At the molecular level, kinetic energy typically exists in the form of thermal energy while potential energy is stored in chemical bonds. The amount of potential energy in a covalent bond reflects the position of the shared electrons relative to the nuclei of the bonded atoms (see Figure 8.2). In general, longer, weaker bonds with equally shared electrons have high potential energy, and shorter, stronger bonds with unequally shared electrons have low potential energy. The potential energy in bonds is analogous to the gravitational energy present in the water drop at the top of the waterfall. If the shared negatively charged electrons are far from the positive charges in both nuclei, then the bond has high potential energy.

In chemical reactions, the products formed often have shorter, stronger covalent bonds than the reactants, thus the potential energy stored within the bonds decreases. This would be like the water drop in Figure 8.1 falling down the waterfall. The difference in potential energy between the reactants and products is

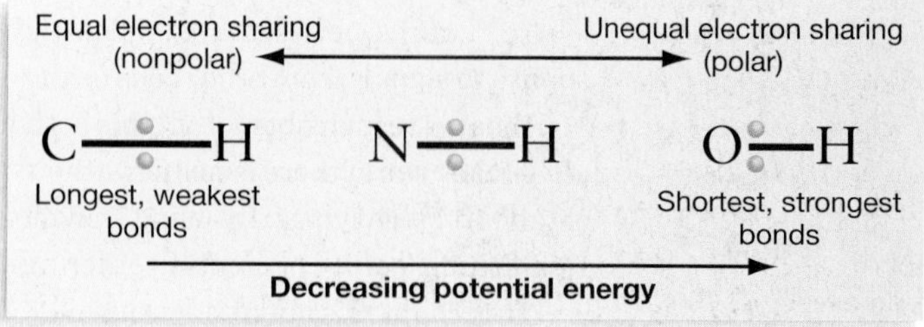

Figure 8.2 **Potential Energy in Molecules Is Based on Bonds.** Potential energy stored in the covalent bonds is directly related to electron position between atomic nuclei.

transformed into an equal amount of kinetic energy. In chemical reactions, this kinetic energy usually takes the form of thermal energy, but sometimes it takes the form of light.

These examples illustrate the **first law of thermodynamics**, which states that energy is conserved. Put another way, the first law is that energy cannot be created or destroyed, it can only be transferred and transformed.

The total energy in a molecule is referred to as its **enthalpy** (represented by H). Enthalpy includes the potential energy in the bonds of the molecule, plus the effect of the molecule's kinetic energy (movement) on the pressure and volume of its surroundings.

The contributions of potential energy, pressure, and volume to the enthalpy of a molecule are best understood by observing the change in enthalpy in a chemical reaction. For example, let's examine the reaction responsible for the explosive bursts of scalding hot liquid a bombardier beetle can produce when provoked, as seen in the photograph featured on the first page of this chapter.

$$2\,H_2O_2\,(aq) \rightarrow 2\,H_2O(l) + O_2(g)$$

In this reaction, a large change in enthalpy occurs as aqueous hydrogen peroxide (H_2O_2) is broken down into liquid water and O_2 gas. Heat given off from the reaction also increases the temperature of the products dramatically. The products expand to over 500 times the original volume of the H_2O_2. These massive increases in temperature and volume generate the pressure that propels boiling liquid and heated oxygen out of an opening at the tip of the beetle's abdomen.

Changes in enthalpy in chemical reactions can be measured and are represented by ΔH. (The uppercase Greek letter delta, Δ, is often used in chemical and mathematical notation to represent change.) The value of ΔH is primarily based on *the difference in potential energy*, since—unlike the reaction in the bombardier beetle—most biological reactions do not result in substantial changes in pressure and volume. When a reaction releases heat (products have less potential energy than the reactants), it is **exothermic** and the ΔH is negative. If heat is taken up during the reaction, generating products that have higher potential energy than the reactants, the reaction is **endothermic** and ΔH is positive.

Another factor that changes during a chemical reaction is the amount of disorder or **entropy** (symbolized by S). When the products of a chemical reaction become less ordered than the reactant molecules were, entropy increases and ΔS is positive (see Figure 8.3). The **second law of thermodynamics**, in fact, states that total entropy always increases in a system that includes the surroundings as well as the products of the reaction (see Ch. 2, Section 2.3, to review the first and second laws of thermodynamics).

To determine whether a chemical reaction is spontaneous, it's necessary to assess the amount of energy in the reaction available to do work—what chemists call **Gibbs free energy** (symbolized by G). This is accomplished by determining the change in Gibbs free energy (ΔG) for the reaction, which is based on changes in enthalpy and entropy as shown in the equation for **standard free-energy change**:

$$\Delta G = \Delta H - T\Delta S$$

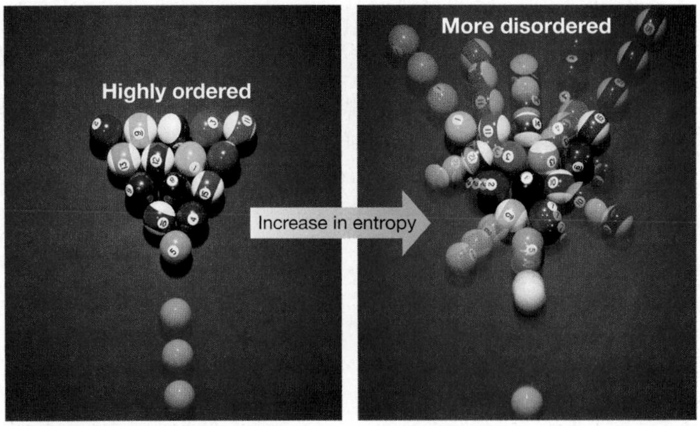

Figure 8.3 When a Reaction Generates More Disorder, Entropy Increases.

Here, T stands for temperature measured on the Kelvin scale (see **BioSkills 1**).

Putting the equation into words: The change in Gibbs free energy (ΔG) in a reaction is equal to the change in enthalpy (ΔH) minus the product of the temperature and the change in entropy ($T\Delta S$). Since thermal energy (T) increases the amount of disorder in the system (S), multiplying $T \times \Delta S$ simply means that the faster molecules are moving, the more important the entropy value becomes in determining the change in Gibbs free energy.

Chemical reactions are spontaneous when ΔG is less than zero. Such reactions are said to be **exergonic**. Reactions are nonspontaneous when ΔG is greater than zero. Such reactions are termed **endergonic**. When ΔG is equal to zero, reactions are at equilibrium.

Spontaneous chemical reactions run in the direction that lowers the free energy of the system. Exergonic reactions are spontaneous and release energy; endergonic reactions are nonspontaneous and require an input of energy to proceed. Be careful not to confuse these terms with the exothermic and endothermic reactions introduced earlier, which address only the change in enthalpy and do not include the effect of changes in entropy.

✔ If you understand the concepts introduced in this section, you should be able to explain (1) why the same reaction can be nonspontaneous at low temperature but spontaneous at high temperature, and (2) why some exothermic reactions are nonspontaneous.

Temperature and Concentration Affect Reaction Rates

Even if a chemical reaction occurs spontaneously, it may not happen quickly. For example, the reactions that convert iron to rust, or sugar molecules to carbon dioxide and water, are spontaneous, but at room temperature they occur very slowly, if at all.

For most reactions to proceed, one or more chemical bonds have to break and others have to form. These two events occur together. For this to happen, the substances involved must collide at a specific orientation that brings the electrons involved near each other. (See Ch. 2, Section 2.1, to review the forces involved in bond formation.)

RESEARCH

QUESTION: Can the timing of the iodine clock reaction be changed?

HYPOTHESIS: Chemical reaction rates increase as the number of collisions between reactants increase.

ALTERNATIVE HYPOTHESIS: Chemical reaction rates do not change based on the frequency of reactant collisions.

EXPERIMENTAL SETUP:

1. Set up reactants that vary in reactant concentration or temperature. Prepare three replicates at each condition.

2. Mix reactants and starch and begin timing reaction. Starch will serve as an indicator for the production of iodine (I_2).

	Reactant concentration constant, temperature varies	Treatment 1	Treatment 2	Treatment 3
Temp (°C)	6° → 55°	23°	23°	23°
[IO_3^-] (M)	0.02 → 0.02	0.02	0.04	0.04
[5 HSO_3^-] (M)	0.02 → 0.02	0.02	0.02	0.04

3. Record time when the reaction is complete, which is indicated by the iodine reacting with starch to produce a blue-black color. Record as reaction rate (1/time to completion).

Iodine clock reaction:

$$2\,IO_3^-(aq) + 5\,HSO_3^-(aq) + 2\,H^+(aq) \rightarrow I_2(aq) + 5\,HSO_4^-(aq) + H_2O(l)$$

PREDICTION OF HYPOTHESIS: The rate of the iodine clock reaction will be faster with increased concentration and temperature.

PREDICTION OF ALTERNATIVE HYPOTHESIS: There will be no difference in reaction rates among the different treatments.

RESULTS:

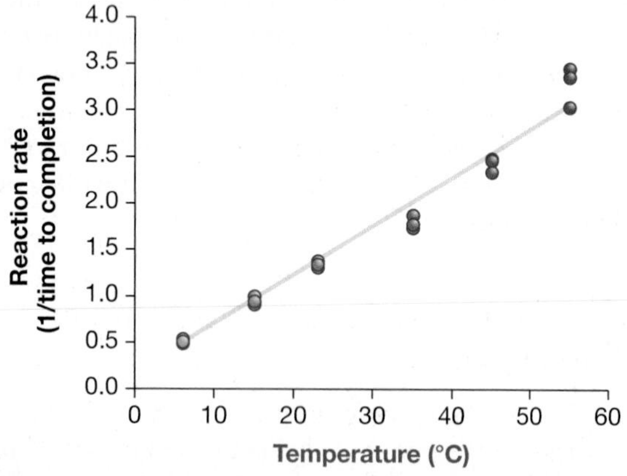

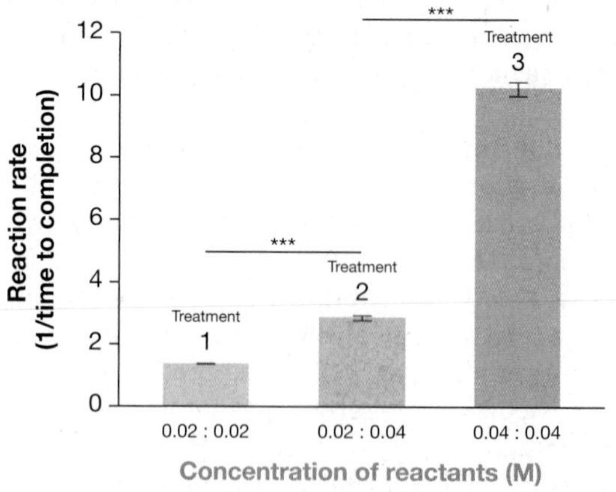

CONCLUSION: The rates of chemical reactions increase as the number of collisions between reactants increase, such as when the concentration or temperature of the reactants increase.

Figure 8.4 **Testing the Hypothesis that Reaction Rates Are Sensitive to Changes in Temperature and Concentration.**

✔ **QUANTITATIVE** Estimate the slope of the trend line shown in the graph on the left side of the "Results" section. What does this value mean with respect to the magnitude of the effect temperature has on the reaction rate?

The number of collisions occurring between the substances in a mixture depends on their temperature and concentration:

- When the concentration of reactants is high, more collisions will occur and reactions should proceed more quickly.
- When their temperature is high, reactants will move faster and should collide more frequently.

Higher concentrations and higher temperatures should therefore speed up chemical reactions. To test this hypothesis, experiments to measure the rate of what's popularly known as the "iodine clock" reaction can be performed under different conditions, as shown in Figure 8.4. Pay special attention to the two graphs in the "Results" section:

- ***Temperature versus reaction rate*** The results graph on the left is based on experiments where the concentration of the reactants was the same, but the temperature varied. Each data point represents one experiment. Notice that the trend line based on these points rises from left to right—meaning, in this case, that the reaction rate increased when the temperature of the reaction mixture was higher.
- ***Concentration versus reaction rate*** The results graph on the right is based on experiments where the temperature was constant, but the concentration of reactants varied. Each bar represents the average reaction rate over many replicates of each treatment, or set of concentrations. The thin lines at the top of each bar indicate the standard error of the mean—a measure of variability (see **BioSkills 3**). The take-home message of this graph is that reaction rates are higher when reactant concentrations are higher.

The reactions shown in Figure 8.4 were spontaneous (exergonic), meaning that the products had lower free energy than the reactants, so no input of energy was required to drive the reactions. But, what drives nonspontaneous, endergonic reactions? Let's take a closer look.

CHECK YOUR UNDERSTANDING

✔ If you understood this section, you should be able to ...

1. Predict how energy is transformed in a chemical reaction where the products have much lower enthalpy than the reactants.
2. Explain how the reactant concentration affects the rate of an exergonic reaction.

Answers are available in Appendix A.

8.2 Nonspontaneous Reactions May Be Driven Using Chemical Energy

After you complete this section, you should be able to ...

- Analyze the process of energetic coupling to drive nonspontaneous reactions.

By definition, nonspontaneous (endergonic) reactions require an input of energy to proceed. Recall that radiation from the Sun and electricity from lightning may have supplied the energy necessary to drive key endergonic reactions during chemical evolution (Ch. 2, Sections 2.3 and 2.4). What source of energy drives these reactions inside cells?

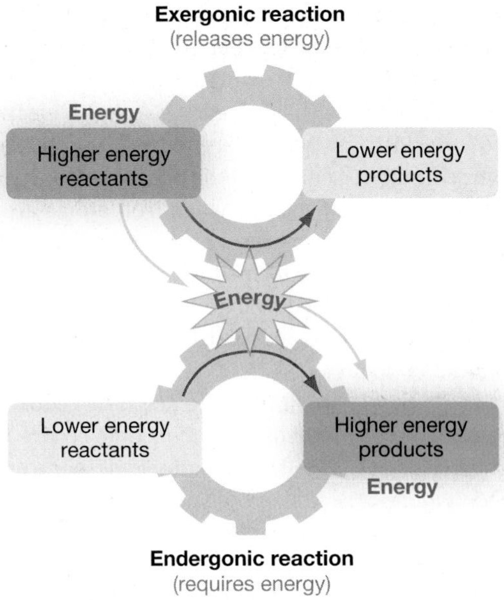

Figure 8.5 Energetic Coupling Allows Endergonic Reactions to Proceed Using the Free Energy from Exergonic Reactions.

Figure 8.5 shows how **energetic coupling** between exergonic and endergonic reactions transfers free energy from one reaction to drive another. In cells, this process generally occurs in one of two ways, either through the transfer of electrons or the transfer of a phosphate group.

Redox Reactions Transfer Energy via Electrons

Chemical reactions that involve the loss or gain of one or more electrons are called **reduction–oxidation reactions**, or **redox reactions**. When an atom or molecule loses one or more electrons, it is said to be oxidized. This terminology makes sense if you notice that the word "oxidized" sounds as if oxygen has done something to an atom or molecule. Recall that oxygen is highly electronegative and often pulls shared electrons away from the nuclei of other atoms in covalent bonds (Ch. 2, Section 2.3). On the other hand, when an atom or molecule gains one or more electrons, it is said to be reduced. To keep these terms straight, students often use the mnemonic "OIL RIG"—**Oxidation** *Is Loss* of electrons; **Reduction** *Is Gain* of electrons.

Oxidation (electron loss) events are always paired with a reduction (electron gain) event: If one atom loses an electron, another has to gain it, and vice versa. Recall, too, that electron position within a covalent bond is related to the potential energy of a bond (see Figure 8.2).

Redox reactions represent the energetic coupling of two half-reactions, one exergonic and one endergonic. Oxidation is the spontaneous, exergonic half-reaction; it lowers the potential energy of a reactant. Reduction is the nonspontaneous, endergonic half-reaction; it increases the potential energy of a reactant. Some of the potential energy from the oxidation step is used to increase the potential energy of a molecule in the reduction step. In cases where more free energy is lost by oxidation than is necessary for reduction, the overall reaction is exergonic.

It is important to note that in redox reactions, electrons may be "gained" or "lost" in two different ways. First, transferred electrons may result in a change in the number of electrons in the valence shell of an atom, such as when iron atoms are reduced from Fe^{3+} to Fe^{2+}. In other cases, electrons are transferred as new covalent bonds that are formed with other atoms. In the latter cases, an atom is reduced or oxidized based on the change in the position of shared electrons relative to the atom's nucleus.

An Example of Redox in Action To see how redox reactions work, consider the spontaneous reaction that occurs when the sugar glucose ($C_6H_{12}O_6$) is burned (combusted) in the presence of oxygen (O_2) (**Figure 8.6**). The orange dots in the illustration represent the positions of the electrons involved in covalent bonds.

Compare the position of the shared electrons in the first reactant, glucose, with the position of the shared electrons in the first product, carbon dioxide (CO_2). Notice that the *number* of electrons around carbon has not changed, but after the reaction, the shared electrons in CO_2 have moved farther from the carbon nucleus. This means that carbon has been oxidized: it has "lost" electrons. The change occurred because the carbon and hydrogen atoms in glucose share electrons equally, while the carbon and oxygen atoms in CO_2 don't. In CO_2, the high electronegativity of the oxygen atoms pull electrons away from the carbon atom.

Now compare the position of the shared electrons in the reactant O_2 molecules with their position in the product water molecules. In water, these electrons have moved closer to the oxygen nuclei than they were in the O_2 molecules, meaning that the oxygen atoms have been reduced. Oxygen has "gained" electrons. Thus, when glucose burns, carbon atoms are oxidized while oxygen atoms are reduced.

These changes in electron position correspond to changes in chemical energy levels. When glucose reacts with oxygen, the new bonds formed in the products are stronger and the electrons are held much tighter than in the reactant molecules. This means the enthalpy (*H*), including the potential energy, is lower in the products than the reactants. Conversely, the entropy (*S*) of the products is much higher than that of the reactants, as indicated by the increase in the number of molecules. Because of the negative ΔH and positive ΔS, this reaction is exergonic. The difference in potential energy between the reactants and products is transformed into kinetic energy, which is released in the form of heat and light.

Another Approach to Understanding Redox During many redox reactions that occur in cells, electrons (e^-) may be transferred between molecules—accompanied by changes to structure that are minor compared to those seen in the combustion of glucose. When this occurs, the transferred electron may be accompanied by a proton (H^+), which would result in the addition of a neutral hydrogen (H) atom to the molecule receiving the electron. In such a reaction, the molecule giving up an electron is called the **electron donor**, while the recipient molecule is called the **electron acceptor**.

Most electron acceptors gain potential energy as they are reduced because the new bonds, often formed with hydrogen atoms, are weaker than the original bonds. This observation should sound familiar, from what you have learned about carbohydrates and fats (Ch. 5, Section 5.3, and Ch. 6, Section 6.1). Molecules that have a large number of C—H bonds, such as carbohydrates and fats, store a great deal of potential energy.

Conversely, molecules that are oxidized in cells often lose potential energy. To understand why, review Figure 8.6 and note that the oxidized carbons in CO_2 have a greater number of C—O bonds compared to the reactants. Because oxygen atoms have extremely high electronegativity, the electrons involved in bonds with oxygen atoms have low potential energy.

In many redox reactions in biology, understanding where oxidation and reduction have occurred is a matter of following hydrogen atoms—reduction often "adds H's" and oxidation often "removes H's." For example, **flavin adenine dinucleotide (FAD)** is a cellular electron acceptor that is reduced by two electrons accompanied by two protons to form **$FADH_2$** (**Figure 8.7a**). $FADH_2$ can readily donate these electrons to other molecules. As a result, it is called an **electron carrier** and is said to have "reducing power."

Another common electron acceptor is **nicotinamide adenine dinucleotide (NAD^+)**, which is reduced to form **NADH**. As with FAD, two electrons reduce NAD^+. These reductions differ, however, in the number of protons transferred. NAD^+ acquires only one of the two H's to form NADH; the second is released into the environment as H^+ (**Figure 8.7b**). These two examples illustrate an

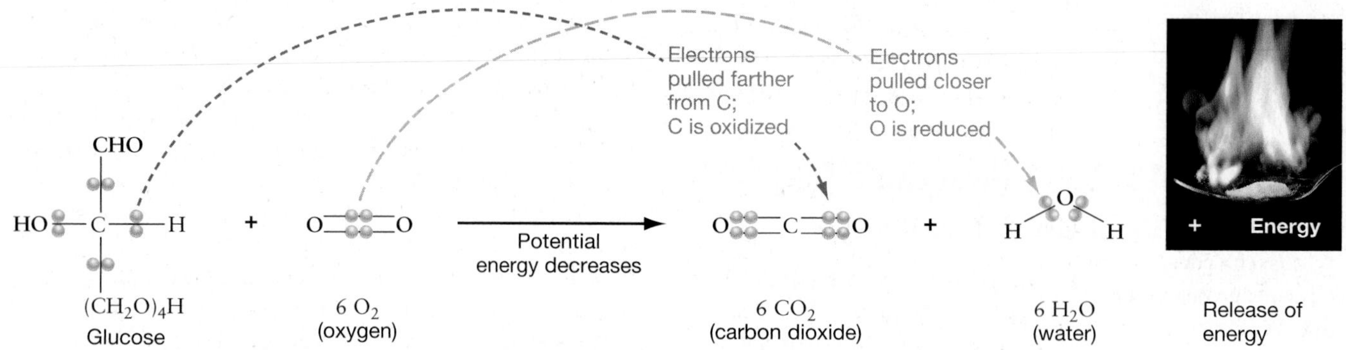

Figure 8.6 Redox Reactions Involve the Gain or Loss of One or More Electrons. This diagram shows how the position of shared electrons changes when glucose reacts with oxygen. The carbons of glucose become oxidized while the oxygen atoms of O_2 become reduced.

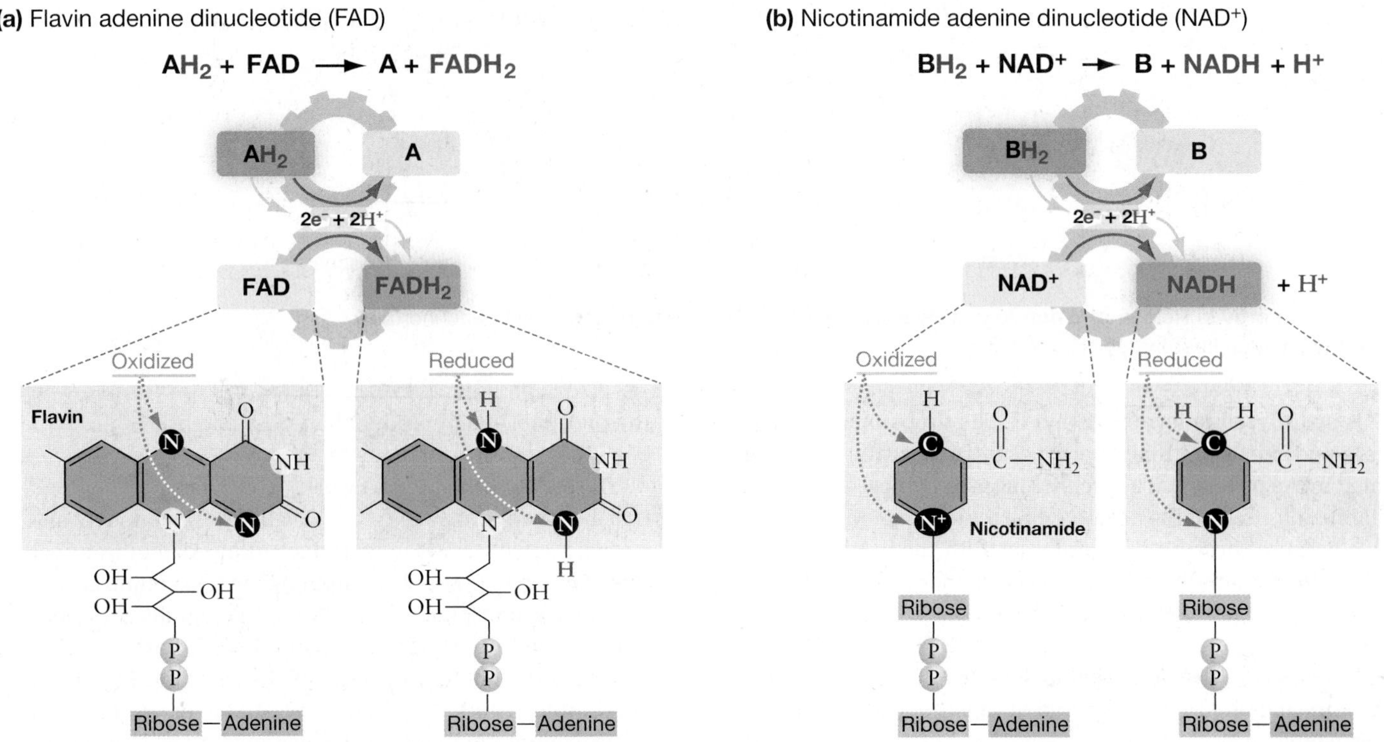

Figure 8.7 Redox Reactions May Transfer Electrons Alone or Accompanied by Protons. $FADH_2$ and NADH are important electron carriers that are formed by transferring electrons (e^-) from other molecules (symbolized as AH_2 or BH_2) to FAD and NAD^+. Some of these transferred electrons are accompanied by protons (H^+), which result in the formation of new covalent bonds with hydrogen atoms.

important point—all redox reactions involve the transfer of electrons, but not all of the transferred electrons combine with protons to form hydrogen atoms.

✔ **If you understand how electrons are transferred in redox reactions, you should be able to identify the electron donors and electron acceptors in the two reactions shown in Figure 8.7.**

Redox reactions are central in biology—they transfer energy via electrons. The potential energy in $FADH_2$ and NADH drives a set of key redox reactions used to produce ATP from ADP and P_i (Chapter 9). By now you have learned that ATP is a major player in meeting life's energy demands, but how do cells actually use the energy stored in ATP?

ATP Transfers Energy via Phosphate Groups

Adenosine triphosphate (ATP) (introduced in Ch. 4, Section 4.1) is a ribonucleotide used for RNA synthesis, but it is more commonly known for its ability to make things happen in cells. As Figure 8.8 shows, ATP has a great deal of potential energy based on the presence of four negative charges confined to a small area in its three phosphate groups. In part because these negative charges repel each other, the covalent bonds between the phosphate groups are very weak so the potential energy in the molecule is extraordinarily high.

ATP Hydrolysis Releases Free Energy When ATP reacts with water during a hydrolysis reaction, the bond between ATP's outermost phosphate group and its neighbor is broken, resulting in the formation of ADP and an inorganic phosphate, abbreviated P_i, which has the formula HPO_4^{2-} (Figure 8.9 on page 182). This reaction is highly exergonic. Under standard conditions of temperature and pressure in the laboratory, a total of 7.3 kilocalories of energy per mole of ATP (or 7.3 kcal/mol) is released during the reaction. A **kilocalorie (kcal)** of energy raises 1 kilogram (kg) of water 1°C.

ATP hydrolysis is exergonic because the entropy of the product molecules is higher than that of the reactants (positive ΔS), and because there is a large drop in enthalpy (negative ΔH) when

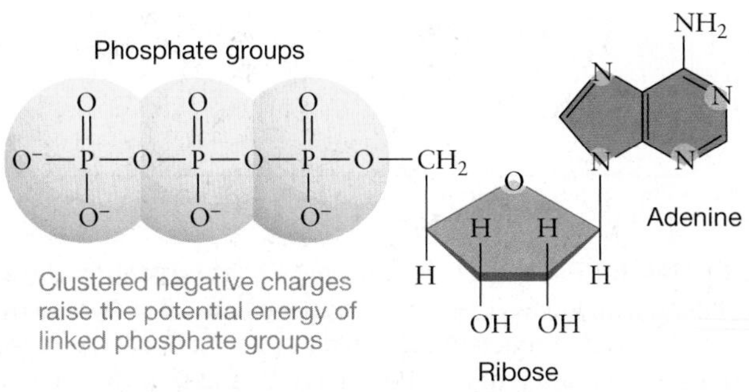

Figure 8.8 Adenosine Triphosphate (ATP) Stores a Large Amount of Potential Energy. ATP's high potential energy results, in part, from the four negative charges clustered in its three phosphate groups. The negative charges repel each other, raising the potential energy of the bonds between groups.

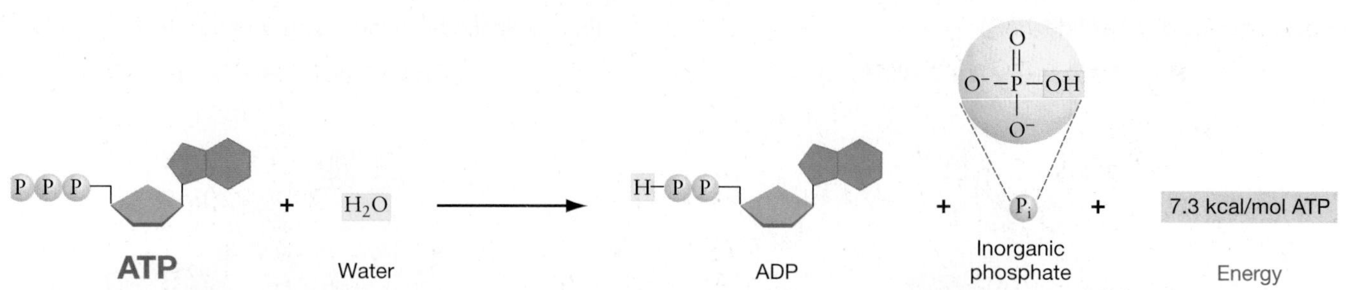

Figure 8.9 Energy Is Released When ATP Is Hydrolyzed. When ATP is hydrolyzed to ADP and inorganic phosphate, a large free-energy change occurs.

ATP is hydrolyzed to form ADP and P_i. The drop in potential energy included in the enthalpy term occurs because the new bonds formed in the products are much stronger than those in ATP. The destabilizing effect of the negative charges is also reduced in ADP and P_i because these products interact with the partial positive charges on surrounding water molecules more efficiently than the clustered negative charges on ATP did.

How Does ATP Drive Endergonic Reactions? If the reaction diagrammed in Figure 8.9 occurred in a test tube, the energy released would be lost as heat. But cells don't lose that 7.3 kcal/mol as heat. Instead, they use it to make things happen. Specifically, instead of hydrolyzing ATP and releasing P_i into solution, the phosphate group is covalently bonded to a target molecule.

The addition of a phosphate group to a molecule is called **phosphorylation**. When ATP is the phosphate donor, phosphorylation is exergonic because the newly formed bonds in ADP and the phosphorylated product have less potential energy than the bond that was broken in the reactant.

To see how this process works, consider an endergonic reaction between two reactant molecules—compound A and compound B—that results in a product AB needed by your cells. For this reaction to proceed, an input of energy is required.

When a phosphate group from ATP is added to one of the reactant molecules, the new bond formed with the phosphate increases the potential energy of the reactant. The phosphorylated intermediate is referred to as being activated. This is the critical point: Activated intermediates have high enough potential energy that the reaction between compound A and, for example, the activated form of compound B is now exergonic. The two compounds then go on to react and form the product molecule AB.

Figure 8.10 graphs how phosphorylation can couple exergonic and endergonic reactions. After the exergonic transfer of a

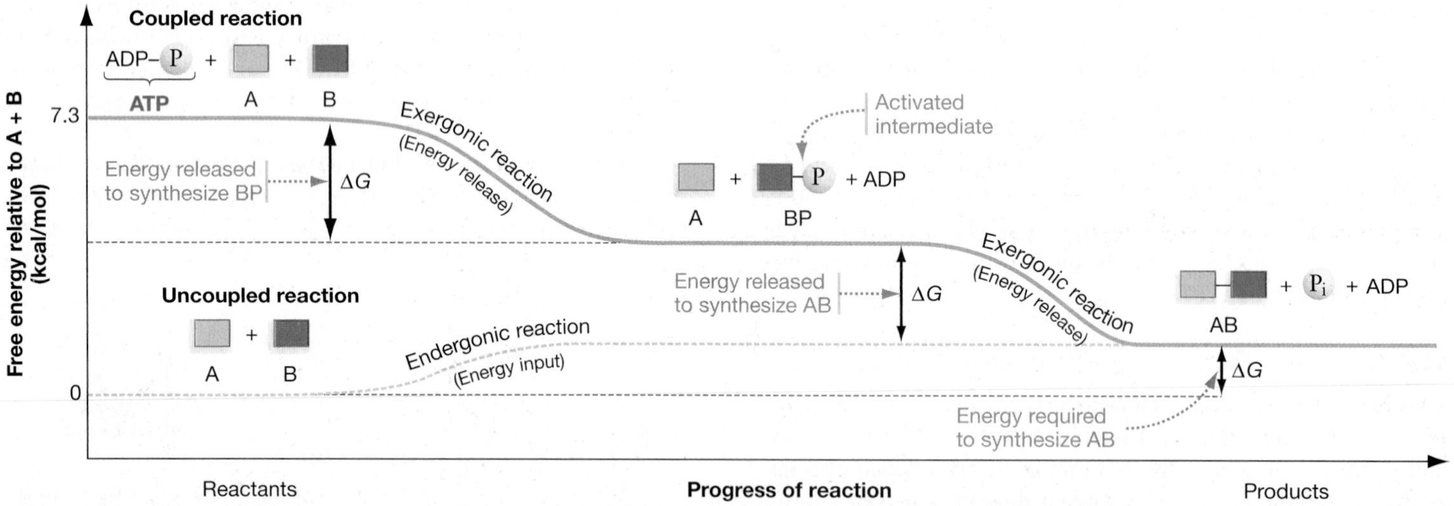

INTERACTIVE Figure 8.10 Exergonic Phosphorylation Reactions Are Coupled to Endergonic Reactions. In cells, many reactions occur only if one reactant is activated by phosphorylation. The increased free energy in the phosphorylated reactant molecule makes the subsequent reaction exergonic. In this graph, the free energy being tracked on the *y*-axis represents A, B, and the 7.3 kcal/mol that is released when ATP is hydrolyzed. For simplicity, the free energy in ADP and P_i is not shown. ΔG represents the change in free energy between the reactants and products for each indicated step.

✔ Indicate if the ΔG is positive (> 0) or negative (< 0) in the uncoupled reaction and the two steps of the coupled reaction.

phosphate group from ATP to B occurs, the free energy of the reactants A and BP is high enough to make the reaction that forms AB exergonic. When a reactant molecule in an endergonic reaction is phosphorylated, the increase in potential energy in that molecule makes the combined overall reaction exergonic.

✔ **If you understand the principles of energetic coupling, you should be able to compare and contrast how energy is transferred via redox reactions and ATP hydrolysis.**

It is hard to overstate the importance of energetic coupling. In the time it takes to read this sentence, ATP has been used to convert millions of endergonic reactions into exergonic reactions. If the cells in your body could no longer couple these reactions, you would die within minutes.

Now the question is, What role do enzymes play in biological reactions?

CHECK YOUR UNDERSTANDING

✔ **If you understood this section, you should be able to . . .**

1. Explain why reduced molecules with many C—H bonds store more potential energy than oxidized molecules with many C—O bonds.
2. Explain why ATP has such high potential energy.

Answers are available in Appendix A.

8.3 How Enzymes Work

Regardless of whether reactions in cells are spontaneous or not, none would occur at the speed required for life without the support of enzymes. Before looking at enzymes' role in these events, let's consider why unassisted reactions are so slow.

After you complete this section, you should be able to . . .

- Explain how enzyme structure affects the rate of a chemical reaction.

In all reactions, even spontaneous ones, a certain minimum amount of kinetic energy—called the **activation energy**—is required to sufficiently strain the chemical bonds in molecules so they can react to form products. In the absence of enzymes, how do reactants clear this energy barrier? Where do they get the needed kinetic energy? Recall from the "iodine clock" experiment that the kinetic energy of molecules is a function of temperature; higher temperatures result in more frequent and forceful collisions. (This is why reactions tend to proceed faster at higher temperatures.)

Figure 8.11 graphs the changes in free energy that take place during the course of a chemical reaction. As you read along the x-axis from left to right, note that free energy rises dramatically when the reactants combine to form an intermediate configuration called the **transition state**. The transition state is the intermediate point between breaking old bonds and forming new ones. The free energy of the transition state is high because the bonds that existed in the original reactants must be destabilized. Once products form, the free energy drops sharply.

The ΔG label on the graph indicates the overall change in free energy in the reaction—that is, the energy of the products minus the energy of the reactants. In this case, the products have lower free energy than the reactants, meaning that the reaction is exergonic. But because the activation energy for this reaction, symbolized by E_a, is high, the reaction would proceed slowly—even at high temperature.

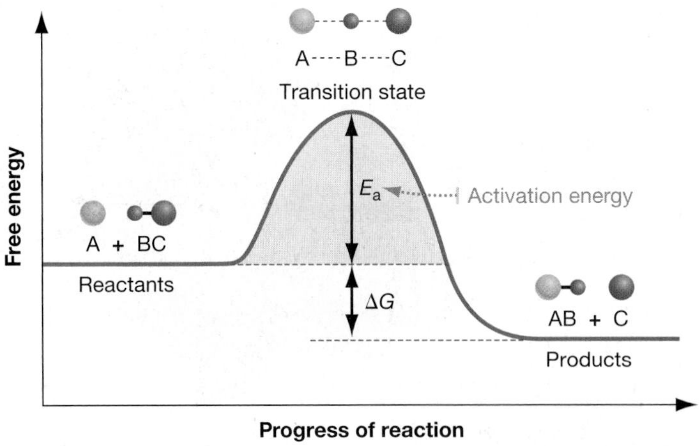

Figure 8.11 Changes in Free Energy during a Chemical Reaction. The energy profile shows changes in free energy that occur during a hypothetical reaction between a molecule A and a molecule containing parts B and C. The overall reaction would be written as A + BC → AB + C. E_a is the activation energy of the reaction.

Here is an important point: The more unstable the transition state, the higher the activation energy and the less likely a reaction is to proceed quickly.

Reaction rates, then, depend on both the kinetic energy of the reactants and the activation energy of the particular reaction—meaning the free energy of the transition state. If the kinetic energy of the reactants is high, such as at high temperatures, then molecular collisions are more likely to overcome the activation energy barrier. At this point, the transition state is formed and the reaction takes place.

In the presence of enzymes, however, rates of reaction can proceed faster without increasing the temperature. Let's next look at why this happens.

Enzymes Help Reactions Clear Two Hurdles

Before any chemical reaction can take place, two hurdles must be cleared: Reactants need to **(1)** collide in a precise orientation and **(2)** have enough kinetic energy to overcome the activation energy barrier and achieve the transition state. To appreciate how enzymes speed up this process, let's consider each hurdle in turn.

Enzymes Bring Substrates Together Recall the initial hypothesis for how enzymes speed up reactions—the "lock-and-key" model, first proposed in 1894 by Emil Fischer (see Ch. 3, Section 3.4). In Fischer's model, reactants would fit into enzymes like a key being inserted into a lock. When reactants undergo a chemical reaction by binding to an enzyme, they are referred to as **substrates**. Enzymes are **catalysts**—they bring substrates together in a precise orientation that makes reactions more likely to occur.

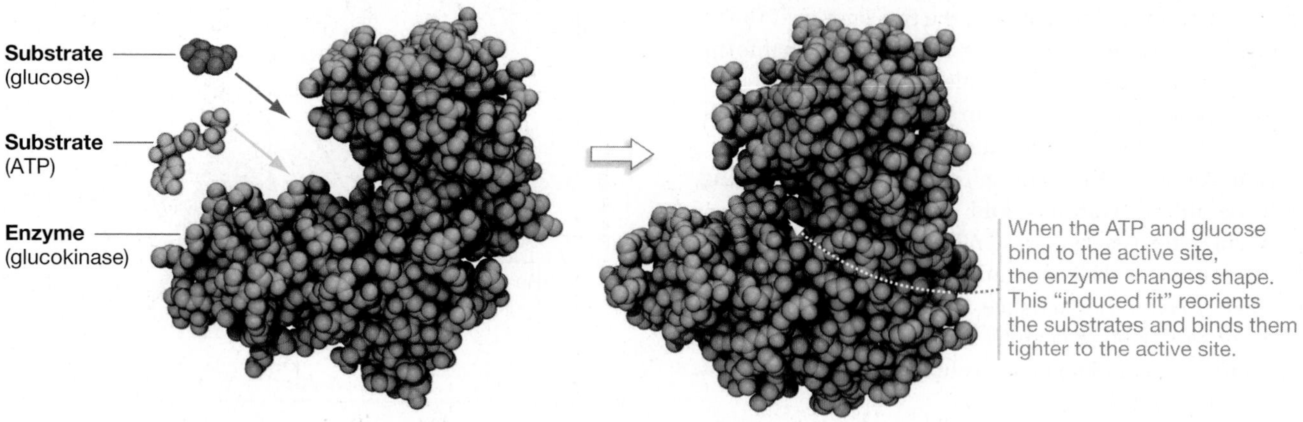

Figure 8.12 Substrate Molecules Bind to Specific Locations in an Enzyme. The substrate (reactant) molecules, shown in red and orange, fit into a precise location, called the active site, in the green enzyme. In this enzyme and in many others, the binding event causes the protein to change shape.

Part of the reason enzymes are such effective catalysts is that they bring substrates together in the enzyme's **active site** (Ch. 3, Section 3.4). Active-site binding helps substrates to collide in a precise orientation so that particular bonds can break and new bonds can form to generate products. To understand how this takes place, let's look at the structure of enzymes.

Enzymes are macromolecules that are generally much larger than their substrates. Most of the enzymes in your cells are proteins, but some consist of RNA and are called ribozymes (Ch. 4, Section 4.3, and Ch. 17, Section 17.5). The active site is often present within a cleft or cavity of the enzyme. A good example is the protein glucokinase, which is an enzyme that catalyzes the phosphorylation of the sugar glucose. (Many enzymes have names that hint at the identity of the substrate and end with *–ase*.) As the left side of Figure 8.12 shows, the active site in glucokinase is a small notch in an otherwise large, crescent-shaped enzyme.

In Fischer's original lock-and-key model, enzymes were conceived of as being rigid—almost literally as rigid as a lock. As research on enzyme action progressed, however, Fischer's model had to be modified. Perhaps the most important realization was that enzymes are not rigid and static, but flexible and dynamic. In fact, many enzymes undergo a significant change in shape, or conformation, when substrates bind to the active site. You can see this conformational change, called an **induced fit**, in the glucokinase molecule on the right side of Figure 8.12. Once glucokinase binds its substrates—ATP and glucose—the enzyme rocks forward over the active site to bring the two substrates together.

Modeling can help drive home how enzymes and substrates interact (Making Models 8.1). There is more to achieving the transition state than simply an enzyme binding to its substrates, however.

Making Models 8.1 Tips on Drawing Enzymes and Substrates

Drawing simple models of enzymes and their substrates often helps you understand how they interact with one another. Enzymes may be drawn as circles with small clefts to show the active site. Draw substrates as shapes that fit within this cleft.

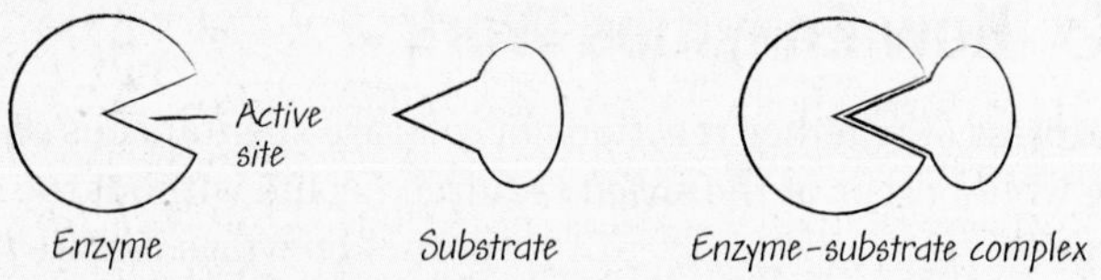

MODEL Draw a model to show how an enzyme will interact with only its substrate and not with other molecules. Explain what is responsible for the specificity of this interaction.

To see this model in action, go to the Study area of **Mastering Biology**

Enzymes Lower the Activation Energy

When substrate molecules enter the active site, they are initially held in place through hydrogen bonding or other weak interactions with amino acid residues in the active site. Once the substrate is bound, one or more R-groups in the active site come into play. The degree of interaction between the substrate and enzyme increases and reaches a maximum strength, at which point bonds in the substrates are destabilized to form the transition state. Thus, by establishing an environment in the active site that promotes the transition state, enzymes decrease the activation energy required for the reaction to proceed.

Figure 8.13 diagrams how enzymes lower the activation energy for a reaction. Note that the presence of an enzyme does not affect the overall energy change, ΔG, or change the energy of the reactants or the products. In other words, an enzyme will not make an endergonic reaction exergonic, it only lowers the amount of activation energy required for the substrates to achieve the transition state.

As suggested in Fischer's lock-and-key model, most enzymes are highly specific in their activity—they catalyze just a single reaction by lowering the required activation energy. Enzymes are also astonishingly efficient. Most of the important reactions in biology would occur at imperceptible rates without a catalyst.

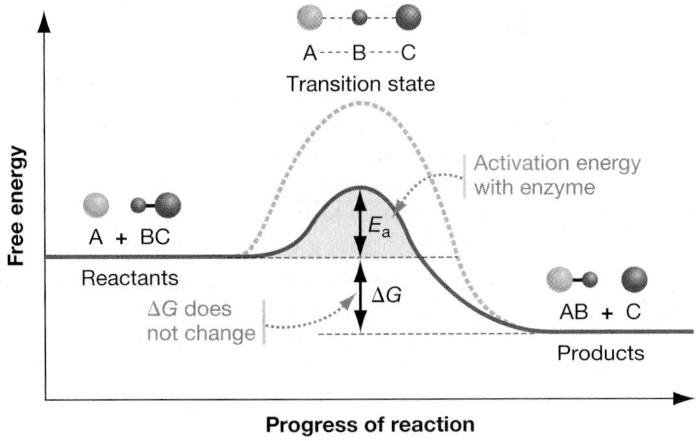

Figure 8.13 An Enzyme Changes the Activation Energy of a Reaction. The energy profile for the same reaction diagrammed in Figure 8.11, but now with a catalyst (enzyme) present. Even though the energy barrier to the reaction, E_a, is much lower, ΔG does not change.

✔ **CAUTION** Can a catalyst alone make a nonspontaneous reaction occur spontaneously? Explain why or why not.

It's not unusual for enzymes to speed up reactions by a factor of a million; some enzymes make reactions go many *trillions* of times faster than they would without a catalyst.

It's also important to note that an enzyme is not consumed in a chemical reaction, even though it participates in the reaction. The composition of an enzyme is exactly the same after the reaction as it was before. Figure 8.14 summarizes how enzymes catalyze reactions.

Step 1 Initiation Instead of substrates occasionally colliding in a random fashion, enzymes precisely orient substrates as they bind at specific locations within the active site.

Step 2 Transition state facilitation Inside a catalyst's active site, substrate molecules are more likely to reach their transition state. In many cases the transition state is stabilized by a change in the enzyme's shape. Interactions between the substrate and R-groups in the enzyme's active site lower the activation energy required for the reaction. Thus, the catalyzed reaction proceeds much more rapidly than the uncatalyzed reaction.

Step 3 Termination The reaction products have less affinity for the active site than the transition state does. Binding ends, the products are released, and the enzyme returns to its original conformation.

✔ **If you understand the basic principles of enzyme catalysis, you should be able to complete the following sentences: (1) Enzymes speed reaction rates by ______ and lowering activation energy. (2) Enzyme specificity is a function of the active site's shape and the chemical properties of the ______ within the active site. (3) In enzymes, as in many molecules, function follows from ______.**

What Limits the Rate of Catalysis?

For several decades after Fischer's model was published, most research on enzymes focused on rates of enzyme action, or what biologists call enzyme kinetics. Researchers observed that, when the amount of product produced per second—indicating the speed of the reaction—is plotted as a function of substrate concentration, a graph like that shown in Figure 8.15 on page 186 results.

In this graph, each data point represents an experiment where reaction rate was measured when substrates were present in various concentrations. The two lines represent two series of experiments: one with the reactions catalyzed by an enzyme and the

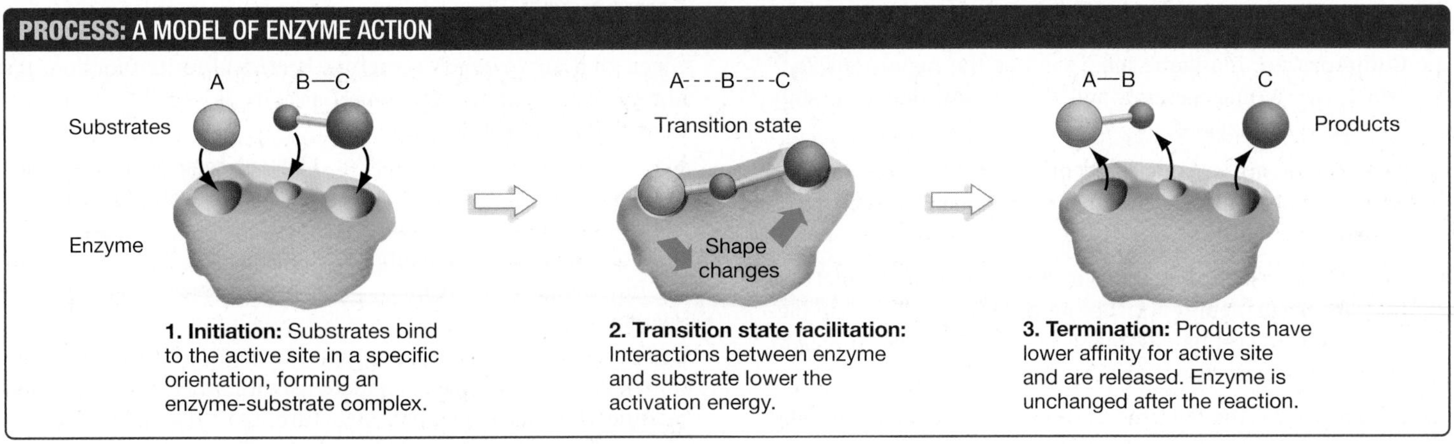

Figure 8.14 A Three-Step Process to Model Enzyme Action.

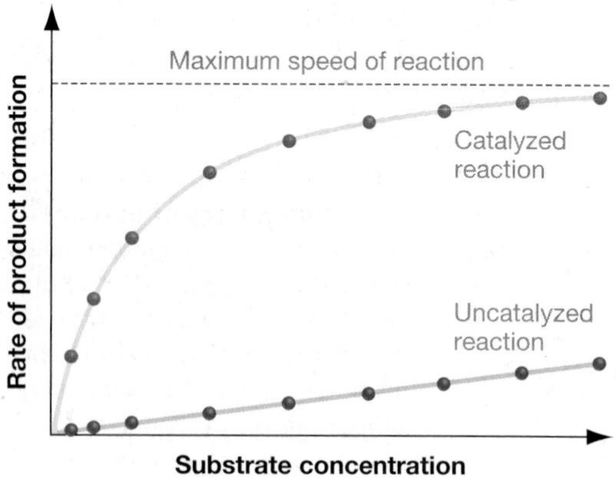

Figure 8.15 Enzyme-Catalyzed Reactions Can Be Saturated. When the enzyme concentration is held constant, adding more substrate will markedly increase the rate of enzyme-catalyzed reactions until they reach a maximum.

✔ After reading the rest of this section, predict what would need to be added to the reaction to increase the rate of product formation above the maximum shown in the graph.

other uncatalyzed. As you examine the curve for the catalyzed reaction in Figure 8.15, note that it has three sections:

1. When substrate concentrations are low, the speed of an enzyme-catalyzed reaction increases in a steep, linear fashion.
2. At intermediate substrate concentrations, the increase in speed begins to slow.
3. At high substrate concentration, the reaction rate plateaus.

This leveling-off pattern is in striking contrast to the situation for the uncatalyzed reactions, where the reaction speed is far slower, but tends to show a continuing linear increase with substrate concentration.

What's going on in the catalyzed reaction? At some point, active sites cannot accept substrates any faster, no matter how large the concentration of substrates gets. Stated another way, reaction rates level off because all available enzyme molecules are being used—the enzyme is said to be saturated. In fact, experiments revealing such "saturation kinetics" in enzyme-catalyzed reactions were taken as strong evidence that the enzyme–substrate complex proposed by Fischer actually exists.

Do Enzymes Work Alone?

The answer to this question, in many cases, is no. Ions or molecules that are not part of an enzyme's primary structure are often required for an enzyme to function normally. These enzyme "helpers" can be divided into three different types:

1. **Cofactors** are inorganic ions, such as the metal ions Zn^{2+} (zinc), Mg^{2+} (magnesium), and Fe^{2+} (iron), that reversibly interact with enzymes.
2. **Coenzymes** are organic molecules that reversibly interact with enzymes, such as the earlier mentioned electron carriers NADH or $FADH_2$.
3. **Prosthetic groups** are atoms or non–amino acid molecules that are permanently attached to proteins, such as the pigment molecule retinal. Retinal is involved in converting light energy into nerve impulses (Ch. 44, Section 44.3).

In many cases, the "helpers" are part of the active site and play a key role in stabilizing the transition state. Their presence is therefore essential for the catalytic activity of many enzymes.

To appreciate why this is important, consider that many of the vitamins in your diet are actually coenzymes or used for the production of coenzymes. Vitamin deficiencies thus can disrupt enzyme function and cause disease. For example, vitamin C (ascorbic acid) is required as a coenzyme for two different enzymes involved in synthesizing collagen, an important component of the extracellular matrix (Ch. 11, Section 11.1). Lack of dietary vitamin C dramatically reduces the activity of these enzymes and causes abnormalities in collagen formation, which leads to bleeding, anemia, tooth loss, and an array of other disorders collectively known as scurvy.

CHECK YOUR UNDERSTANDING

✔ If you understood this section, you should be able to . . .

1. Explain how enzymes lower the activation energy of a chemical reaction.
2. Propose a hypothesis to explain why Mg^{2+} is a common cofactor in enzymes that use ATP to phosphorylate substrates.

Answers are available in Appendix A.

8.4 What Factors Affect Enzyme Function?

Given that an enzyme's structure is critical to its function, it's not surprising that an enzyme's activity is sensitive to conditions that alter protein shape. Recall that protein structure is dependent on the chemical bonds and interactions that fold the polypeptide into its functional form (Ch. 3, Section 3.3).

After you complete this section, you should be able to . . .

- Contrast the different mechanisms used by cells to regulate enzyme activity.

The activity of an enzyme often changes drastically as a function of temperature, pH, interactions with other molecules, and modifications of its primary structure. Let's take a look at how enzyme function is affected by, and sometimes even regulated by, each of these factors.

Enzymes Are Optimized for Particular Environments

Temperature affects the folding and movement of an enzyme as well as the kinetic energy of its substrates. The concentration of protons in a solution, as measured by pH, also affects enzyme structure and function. pH affects the charge on acidic and basic groups in residue side chains, and also the active site's ability to participate in reactions that involve the transfer of protons or electrons.

The effect of temperature and pH on enzymatic activity has been evaluated using an enzyme called chitinase, which digests the cell walls of fungi. This enzyme was purified from two different species of bacteria—one that lives in a cool neutral environment and another that lives in hot acidic conditions.

While the two versions of chitinase have the same catalytic activity, the primary structures of the proteins are different. The difference in protein sequence between the two enzymes is a consequence of natural selection (introduced in Ch. 1, Section 1.3), resulting in each species having a version of the enzyme best suited for its environment. Do data support this assertion?

Figure 8.16a shows how the activities of the two enzymes, plotted on the *y*-axis, change as a function of temperature, plotted on the *x*-axis. In this graph, each data point represents the enzyme's relative activity—meaning that individual reaction rates were scaled relative to the highest rate observed. Note that, in both bacterial species, the enzymes have distinct optimums or peaks—temperatures at which they function best. The temperature optimums for the different enzymes reflect the environments these bacteria normally inhabit.

Figure 8.16b makes the same point for pH. The effect of pH on enzyme activity was tested on the same chitinases used in Figure 8.16a, but this time applying conditions that varied only in pH. As with temperature, the pH optimum for each bacterial chitinase reflects the environment where the bacteria live.

To summarize, the rate of an enzyme-catalyzed reaction depends not only on substrate concentration and the enzyme's intrinsic affinity for the substrate, but also on temperature and pH (among other factors). Temperature affects the kinetic energy; both temperature and pH affect enzyme shape and reactivity.

Most Enzymes Are Regulated

Controlling when and where enzymes will function is vital to the work of a cell. But while temperature and pH affect the activity of enzymes, they are not often used as a means of regulating enzyme function in a cellular setting. Instead, other molecules, in some cases other enzymes, regulate most of the cell's enzymatic activity. These regulatory molecules often change the enzyme's structure or its ability to bind its substrate in ways that either activate or inactivate the function of the enzyme.

Regulating Enzymes via Noncovalent Interactions Many molecules that regulate enzyme activity (i.e., alter reaction rate) bind to enzymes noncovalently. Since the interaction does not permanently affect the enzyme's primary structure, it is often referred to as being "reversible."

Reversible regulatory interactions work in one of two ways:

1. The regulatory molecule is similar in size and shape to the enzyme's natural substrate and inhibits catalysis by binding to the enzyme's active site. This tactic is called **competitive inhibition** because the molecule involved competes with the substrate for access to the enzyme's active site (Figure 8.17a on page 188).

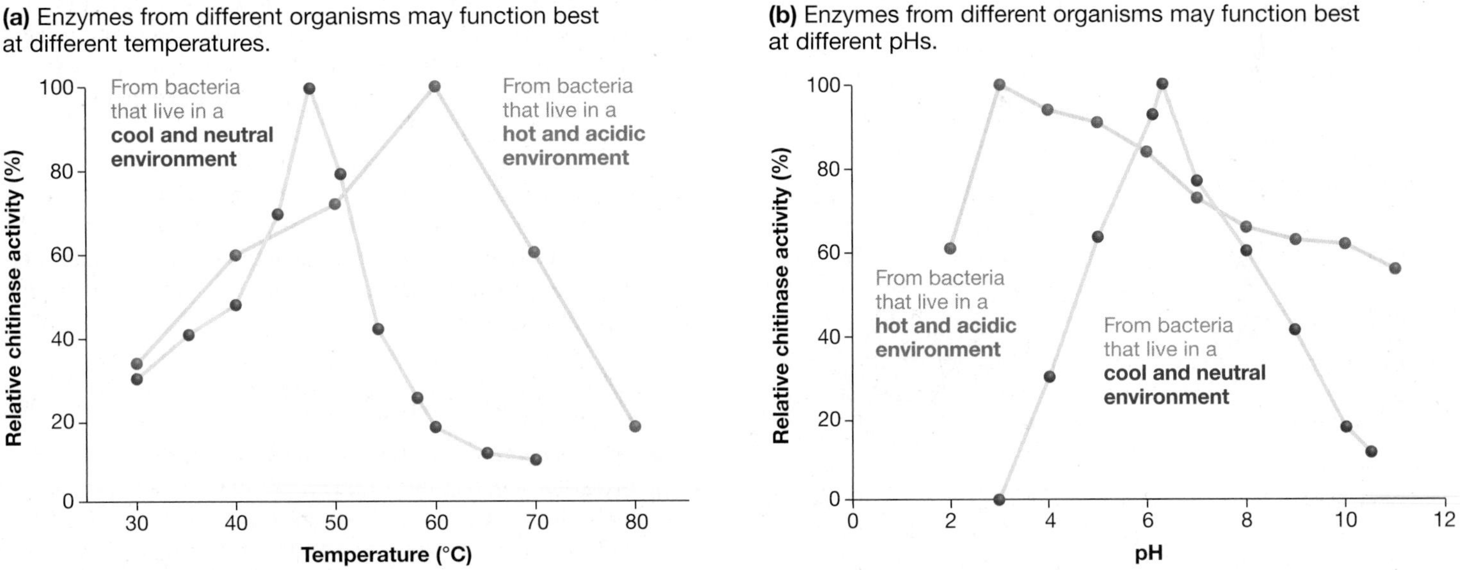

Figure 8.16 Enzymes Have an Optimal Temperature and pH. The activity of enzymes is sensitive to changes in temperature **(a)** and pH **(b)**.

DATA: N. Nawani, B. P. Kapadnis, A. D. Das, et al. 2002. *Journal of Applied Microbiology* 93: 865–975. Also N. Nawani and B. P. Kapadnis. 2001. *Journal of Applied Microbiology* 90: 803–808.

2. The regulatory molecule binds at a location other than the active site and changes the shape of the enzyme. This type of interaction is called **allosteric** ("different-structure") **regulation** because the binding event changes the shape of the enzyme in a way that makes the active site available or unavailable (Figure 8.17b and 8.17c).

Both strategies depend on the concentration of the regulatory molecule—the more regulatory molecule present, the more likely it will be to bind to the enzyme and affect its activity. These regulatory molecules may be acquired from the environment or produced by enzymes within the cell itself. The amount of regulatory molecules produced by cells is often tightly controlled and, as you'll see in Section 8.5, the enzymes that produce regulatory molecules are often themselves managed by their products.

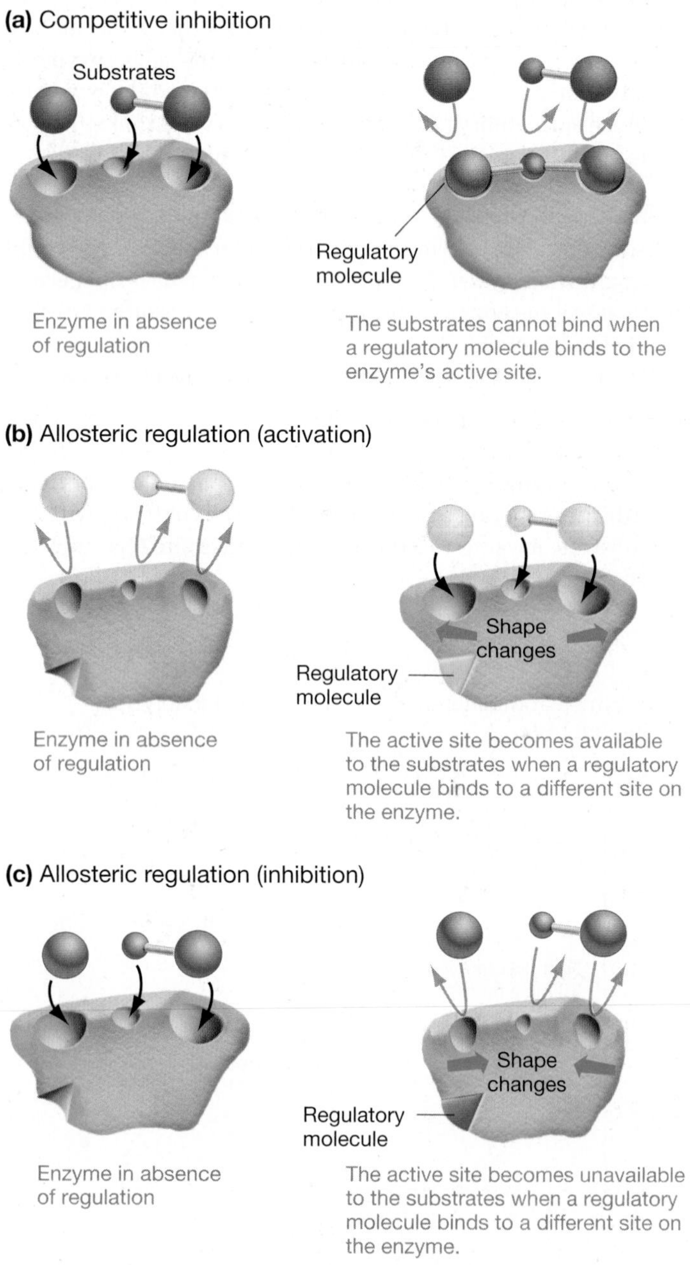

Figure 8.17 An Enzyme's Activity Is Precisely Regulated. Enzymes are turned on or off when specific regulatory molecules bind to them.

Regulating Enzymes via Covalent Modifications In some cases, the function of an enzyme is altered by a chemical change in its primary structure. This change may be reversible or irreversible, depending on the type of modification.

Irreversible changes often result from the cleavage of peptide bonds that make up the primary structure of the enzyme. The enzyme trypsin, for example, is not functional until a small section of the protein is removed by a specific protease.

The most common modification of enzymes is the addition of one or more phosphate groups, similar to what was described for activated substrates in Section 8.2. In this case, however, the enzyme is phosphorylated instead of the substrate molecule. The transfer of a phosphate from ATP to the enzyme may be catalyzed by the enzyme itself or by a different enzyme.

When phosphorylation adds a negative charge to one or more of an enzyme's amino acid residues, the chemical bonds that are responsible for the enzyme's structure change configuration. This change in conformation may activate or inactivate the function of the active site.

Note that the term "activated" means something different when used to describe the effect of phosphorylation on an enzyme versus a substrate. When a substrate is activated, its potential energy has been increased, and this extra energy is used to convert an endergonic reaction to one that is exergonic (see Figure 8.10). When an enzyme is activated, its catalytic function has been turned on—but any change in the potential energy of the enzyme is not directly used to drive the catalytic reaction.

To see how phosphorylation can regulate the activity of an enzyme, let's look at an example called mitogen-activated protein (MAP) kinase, which is involved in cell signaling (Ch. 11, Section 11.3). As shown in Figure 8.18, when key amino acid residues in a particular loop of the folded protein are phosphorylated, the loop changes its shape and acts like a switch to activate the enzyme.

Phosphorylation of an enzyme is a reversible modification to the protein's structure. Dephosphorylation—removal of phosphates—can quickly return the protein to its previous shape. The relative abundance or activity of enzymes that catalyze phosphorylation and dephosphorylation, then, regulates the function of the protein.

CHECK YOUR UNDERSTANDING

✔ **If you understood this section, you should be able to . . .**

1. Explain why the relative activity appears to drop off in Figure 8.16a, when it has been shown that reaction rates tend to increase at higher temperatures (Figure 8.4).
2. Compare and contrast competitive inhibition and allosteric regulation.

Answers are available in Appendix A.

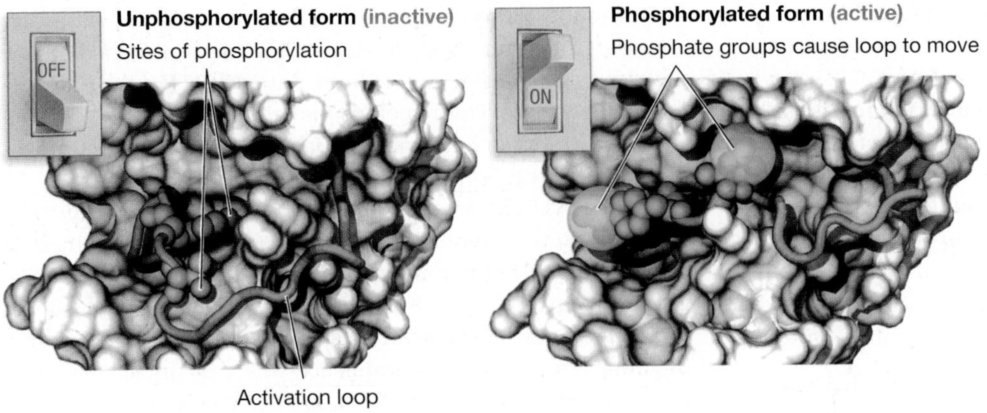

Figure 8.18 Phosphorylation Changes the Shape and Activity of Proteins. When proteins are phosphorylated, they often change shape in a way that alters their activity. The figure shows the structural change that occurs when the activation loop of MAP kinase is phosphorylated. This small change in the protein's structure can have a big effect on its function, like flipping a switch to turn on a light.

8.5 Enzymes Can Work Together in Metabolic Pathways

After you complete this section, you should be able to ...

- Describe the organization, regulation, and evolution of metabolic pathways.

The eukaryotic cell has been compared to an industrial complex, where distinct organelles are functionally integrated into a cooperative network with a common goal—life (Ch. 7, Sections 7.2 and 7.5). Similarly, enzymes often work together in a manner resembling an assembly line in a factory. Each of the molecules presented in this book is built by a series of reactions, each catalyzed by a different enzyme. These multistep processes are referred to as **metabolic pathways**.

The following equation is an example of this type of teamwork, where an initial substrate A is sequentially modified by enzymes 1–3 to produce product D:

$$A \xrightarrow{\text{enzyme 1}} B \xrightarrow{\text{enzyme 2}} C \xrightarrow{\text{enzyme 3}} D$$

The B and C molecules are referred to as intermediates in the pathway—they serve as both a product and a reactant. For example, molecule B is the product of reaction 1 and the reactant for reaction 2.

Although these reactions have been written in a single direction, from left to right, the directionality often depends on the relative concentrations of the reactants and products, and the change in free energy (ΔG) for each reaction. For each step in the pathway, however, the concentration of the product will generally be higher than the concentration of its respective reactant at equilibrium. Since D is the overall product for this pathway, it would be expected to have the highest concentration relative to A, B, and C.

Metabolic Pathways Are Regulated

Since enzymes catalyze the reactions in metabolic pathways, the mechanisms that regulate enzyme function introduced in Section 8.4 also apply to the individual steps in a pathway. A convenient way to regulate metabolic pathways is to use the final product of the reaction sequence to inactivate one of the pathway's own enzymes. This type of regulation is called **feedback inhibition**. As the concentration of the product molecule becomes abundant, it "feeds back" to stop the reaction sequence (Figure 8.19). By inhibiting a step early in the pathway, the amount of the initial substrate is not depleted unnecessarily, allowing it to be stored or used for other reactions.

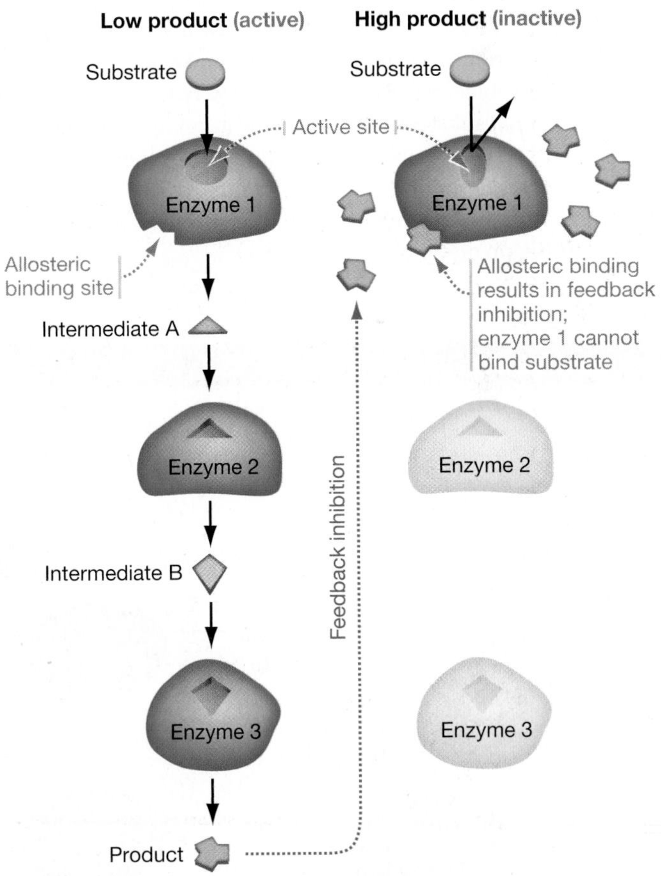

Figure 8.19 Feedback Inhibition Regulates Some Metabolic Pathways. Feedback inhibition occurs when the product of a metabolic pathway inhibits an enzyme that functions early in the pathway.

For example, to think about how blocking an individual reaction can affect an entire pathway, go back to the equation showing the reaction sequence for production of product D and "inactivate" enzyme 2 by crossing it out. ✔ **MODEL If you understand the assembly-line behavior of enzymes in a metabolic pathway, you should be able to predict how inactivating enzyme 2 would affect the concentration of molecules A, B, C, and D relative to what they would be if the pathway were fully functional.**

Metabolic Pathways Evolve

While many enzymes are extraordinarily specific, some have active sites capable of interacting with groups of related substrates, which enables them to catalyze a range of different reactions. Research suggests that such flexibility allowed new enzymes to evolve and that enzymes specialized for catalyzing key reactions provided cells with a selective advantage. Could the same flexibility also help explain the evolution of the stepwise series of reactions seen in metabolic pathways?

In 1945, Norman Horowitz proposed a simple, stepwise process that could have directed pathway evolution on early Earth. In Horowitz's model, enzymes first would have evolved to make the building blocks of life from readily available substrates, such as small organic compounds (Ch. 2, Section 2.4).

If an original substrate were depleted, natural selection would favor the evolution of a new enzyme to make more of it from other existing molecules. By evolving a new reaction step to produce the original substrate—now serving as an intermediate in a two-step pathway—the original enzyme would have been able to continue its work. Figure 8.20 illustrates this model—referred to as retro-evolution—in which repeated backward evolution produces a multistep metabolic pathway.

Researchers also speculate that as early metabolic pathways emerged, the enzymes involved may have been recruited to new pathways, where they evolved new catalytic activities that performed new tasks. This hypothesis is called patchwork evolution, since the new reaction series would consist of enzymes brought together from different pathways.

Evidence of patchwork evolution has been observed in modern organisms, where new metabolic activities have emerged in response to human-made chemicals. For example, a novel pathway has recently evolved in one bacterial species to use the pesticide pentachlorophenol as a source of energy and carbon building blocks. Pentachlorophenol was first introduced into the environment in the 1930s as a timber preservative. The new pathway evolved by using enzymes from two preexisting pathways in a novel series of reactions. The metabolic activity of other microbes is now being scrutinized and engineered to clean up a variety of human-made pollutants—giving rise to a new technology called **bioremediation** (Ch. 26, Section 26.1).

Regardless of how they evolved, metabolic pathways are now vital to the function of all cells. Those that break down molecules for sources of energy and carbon building blocks are called **catabolic pathways**; those that use energy and carbon building blocks to synthesize molecules are called **anabolic pathways**.

CHECK YOUR UNDERSTANDING

✔ **If you understood this section, you should be able to . . .**

1. Explain how regulating the activity of a single enzyme can affect an entire metabolic pathway.
2. Differentiate the retro-evolution and patchwork evolution hypotheses for the origin of metabolic pathways.

Answers are available in Appendix A.

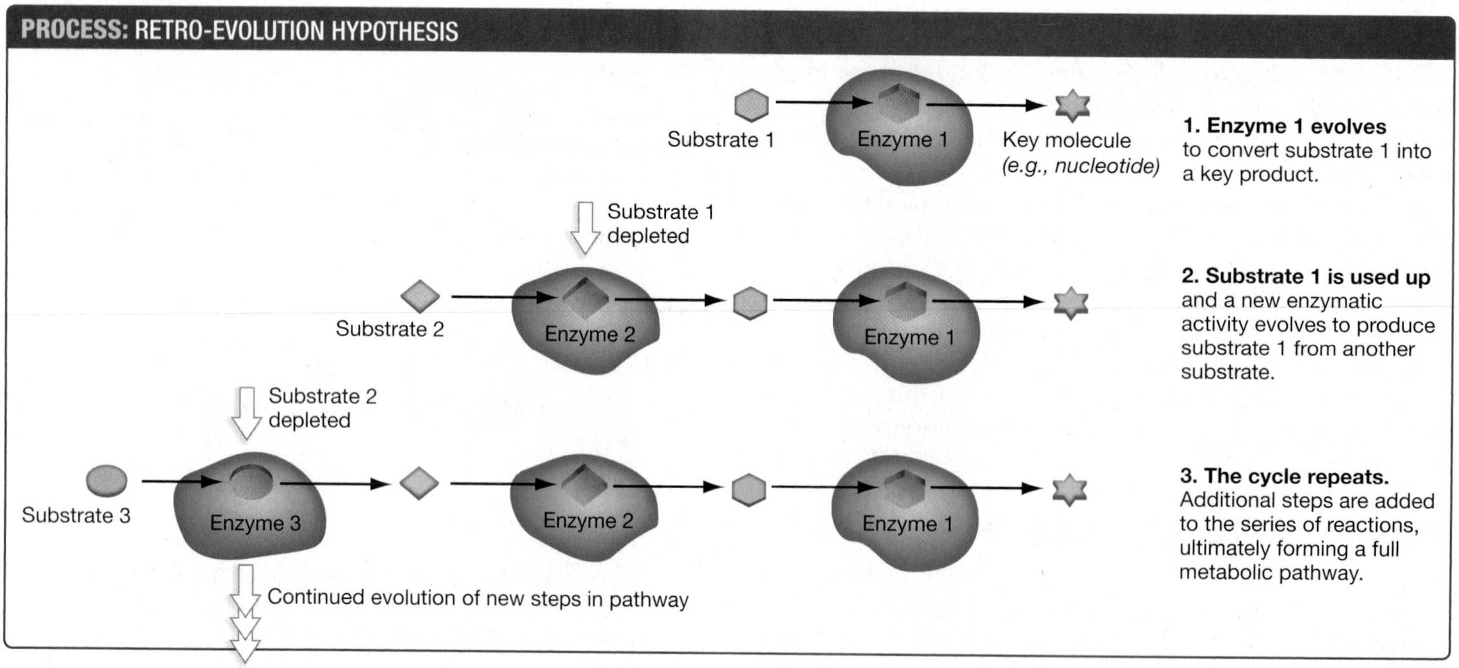

Figure 8.20 A Hypothetical Model for Metabolic Pathway Evolution.

You are being kept alive by key catabolic and anabolic pathways. For example, the catabolic pathways of cellular respiration (introduced in Chapter 9) break down energy-rich molecules like glucose, harvest electrons from their reduced carbons, and pass the electrons through redox reactions to generate ATP. Reduced carbons are in turn produced by anabolic pathways of photosynthesis that are driven by light energy (described in Chapter 10). Reduced involved in cellular respiration and photosynthesis perform the most important energy transformations to life on Earth.

CHAPTER 8 Review

For media, go to **Mastering Biology**

8.1 What Happens to Energy in Chemical Reactions?

- The standard free-energy change for a chemical reaction is an equation used to measure the overall change in Gibbs free energy (ΔG)—that is, to calculate the combined effects of changes in enthalpy (ΔH) and entropy (ΔS).
- Reactions with a negative ΔG are exergonic and will occur spontaneously.
- Reactions with a positive ΔG are endergonic and will not occur without an input of energy.

8.2 Nonspontaneous Reactions May Be Driven Using Chemical Energy

- Redox reactions transfer energy by coupling exergonic oxidation reactions to endergonic reduction reactions.
- Bonds with high potential energy may be formed during the reduction step of a redox reaction. Many of these bonds form when an electron is transferred along with a proton (H^+).
- The potential energy in ATP is used to drive a variety of cellular processes.
- When a phosphate group from ATP is added to a substrate, it increases the potential energy of the substrate and may be used to convert endergonic reactions into exergonic reactions.

8.3 How Enzymes Work

- Activation energy is the amount of kinetic energy required to reach the transition state of a reaction.
- Enzymes are catalysts. They speed reaction rates but do not affect the change in free energy of the reaction.
- Enzymes have active sites that bring substrates together and may change shape to stabilize the transition state.
- Enzymes speed up a reaction by lowering the activation energy, often with the help of cofactors, coenzymes, or prosthetic groups.

8.4 What Factors Affect Enzyme Function?

- Most enzymes are proteins, and thus their activity can be directly influenced by environmental factors, such as temperature and pH, or by modifications that alter their three-dimensional structure.
- Enzyme activity may be regulated by molecules that compete with substrates to occupy the active site or whose binding alters the enzyme shape.
- Protein cleavage and phosphorylation may regulate enzyme activity by modifying the primary structure of the enzyme.

8.5 Enzymes Can Work Together in Metabolic Pathways

- In cells, enzymes often work together in metabolic pathways that sequentially modify a substrate to make a product.
- Feedback inhibition may be used to regulate a pathway, often by controlling the activity of the first enzyme in the pathway.
- Metabolic pathways were vital to the evolution of life, and new pathways continue to evolve in cells.

Answers are available in Appendix A.

✔ TEST YOUR KNOWLEDGE

1. Which of the following *correctly* describe an exergonic reaction? Select True or False for each statement.
T/F The products have lower Gibbs free energy than the reactants.
T/F Activation energy is required for the reaction to proceed.
T/F The products always have lower entropy than the reactants.
T/F The reaction always occurs quickly.

2. What is a transition state?
a. the shape adopted by an enzyme that has an inhibitory molecule bound at its active site
b. the amount of kinetic energy required for a reaction to proceed
c. the intermediate complex formed as covalent bonds in the reactants are being broken and re-formed during a reaction
d. the enzyme shape after binding an allosteric regulatory molecule

3. How does pH affect enzyme-catalyzed reactions?
a. Protons serve as substrates for most reactions.
b. Energy stored in protons is used to drive endergonic reactions.
c. Proton concentration increases the kinetic energy of the reactants, enabling them to reach their transition state.
d. The concentration of protons affects an enzyme's folded structure and reactivity.

4. Explain how feedback inhibition regulates metabolic pathways.

✔ TEST YOUR UNDERSTANDING

5. Explain the lock-and-key model of enzyme activity. What is incorrect about this model?

6. If you were to expose glucose to oxygen on your lab bench, why would you not expect to see it burn as described by the reaction in Figure 8.6?
a. The reaction is endergonic and requires an input of energy.
b. The reaction is not spontaneous unless an enzyme is added.
c. The sugar must first be phosphorylated to increase its potential energy.
d. Activation energy is required for the sugar and oxygen to reach their transition state.

7. **QUANTITATIVE** In Figure 8.10, the energetic coupling of substrate phosphorylation and an endergonic reaction are shown. If the hydrolysis of ATP releases 7.3 kcal of free energy, use the graph in this figure to estimate what you would expect the ΔG values to be for the uncoupled reaction and the two steps in the coupled reaction.

8. Using what you have learned about changes in Gibbs free energy, would you predict the ΔG value of catabolic reactions to be positive or negative? What about anabolic reactions? Justify your answers using the terms "enthalpy" and "entropy."

✔ TEST YOUR PROBLEM-SOLVING SKILLS

9. **MODEL** Draw a chemical equation to represent the redox reaction that occurs when methane (CH_4) burns in the presence of oxygen (O_2). Identify the reactant that is reduced and the reactant that is oxidized. Of the four molecules that should be in your equation, point out the one that has bonds with the highest potential energy.

10. You have discovered an enzyme that appears to function only when a particular sugar accumulates. Which of the following scenarios would you predict to be responsible for activating this enzyme?
    a. The sugar cleaves the enzyme to form the active conformation.
    b. The sugar is an allosteric regulatory molecule for the enzyme.
    c. The sugar is a competitive inhibitor for the enzyme.
    d. The sugar phosphorylates the enzyme to form the active conformation

✔ PUT IT ALL TOGETHER: Case Study

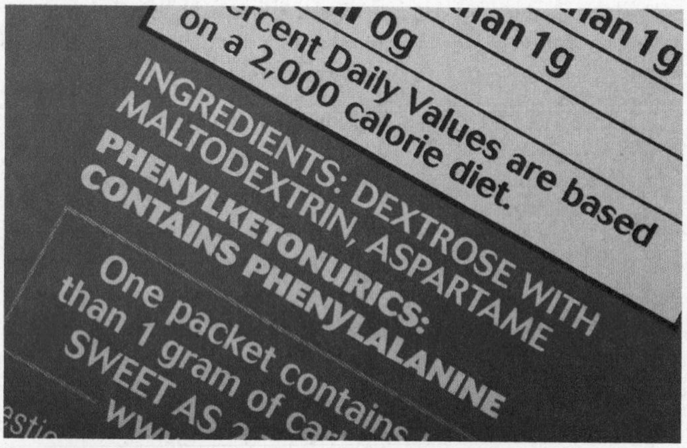

How can an essential nutrient—the amino acid phenylalanine—be toxic?

The amino acid phenylalanine is present in many foods and commonly used in artificial sweeteners, but it is toxic when high levels accumulate in cells. People born with a metabolic disorder called phenylketonuria (PKU) cannot clear excess phenylalanine from their bodies. As a result, they will suffer irreversible brain damage and mental impairment if they ingest high levels of the amino acid. What type of metabolic defect causes this problem, and how might knowledge about the cause lead to a cure?

11. Phenylalanine is required for the production of many bodily proteins, but none of your cells are capable of synthesizing this essential amino acid. What type of metabolic pathway must be responsible for providing phenylalanine to your cells? How would your diet affect the amount of phenylalanine that is available?

12. **MODEL** PKU can result from defects in phenylalanine hydroxylase (PAH), an enzyme that converts phenylalanine into the amino acid tyrosine. It has been hypothesized that phenylalanine is both the substrate and an allosteric regulatory molecule for PAH. If this hypothesis is true, draw two models to show how PAH would appear if phenylalanine concentration is low (inactive enzyme) or high (active enzyme). Label phenylalanine, the allosteric binding site, and the active site.

13. **PROCESS OF SCIENCE** To evaluate the effect of phenylalanine on the structure of PAH, enzymes like trypsin can be used to cut the protein at specific amino acid residues exposed on the surface. Electrophoretic data from such an experiment in the presence (+) or absence (–) of phenylalanine are shown below (see **BioSkills 6** to review gel electrophoresis). Do these results support the hypothesis that phenylalanine is an allosteric regulator of PAH? Explain your conclusion.

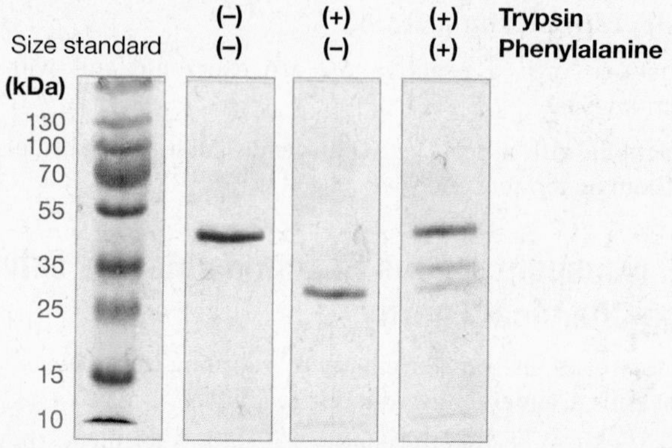

Source: E. K. Jaffe et al. 2013. *Archives of Biochemistry and Biophysics* 530: 73–82.

14. The catalytic activity of PAH requires a coenzyme that is oxidized in the reaction. This coenzyme is subsequently reduced by NADH to regenerate it for additional PAH reactions. If a person were diagnosed with PKU, but their PAH enzymes were fully functional, what metabolic defects would you look for to explain accumulation of phenylalanine in the patient?

15. The functional form of PAH contains four identical active sites, but based on its amino acid sequence, only one active site can be formed. What does this imply about the structure of the functional enzyme?

16. Starting in the 1960s, newborns identified as having PKU were placed on a strict low-protein diet. This treatment mitigates or prevents mental impairment. What avenues of treatments, beyond a restrictive diet, might be researched?

Mastering Biology

Students Go to Mastering™ Biology for assignments, the eText, and the Study Area with animations, practice tests, and activities.

Professors Go to Mastering™ Biology for automatically graded tutorials and questions that you can assign to your students, plus Instructor Resources.

9 Cellular Respiration and Fermentation

This hydroelectric dam on the Duero, a river between Spain and Portugal, uses pumps to move water from the lower reservoir to the upper reservoir. When energy demand is high, the potential energy stored as a result of this activity is used to generate electricity. Cells use an analogous process to produce ATP during cellular respiration.

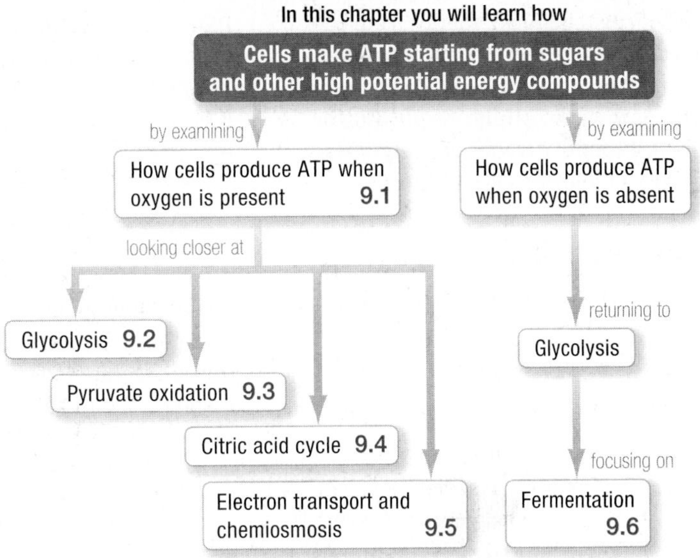

BIG PICTURE

This chapter is part of the Big Picture. See how on pages 236–237.

Life requires energy. Even before life arose on early Earth, chemical evolution was driven by energy from chemicals, radiation, heat, or other sources (Ch. 2, Section 2.3). Harnessing energy and controlling its flow has been the single most important event in the evolution of life.

What fuels life in cells? In many cases, the answer will be the nucleotide adenosine triphosphate (ATP). ATP has high potential energy and allows cells to overcome a lot of life's energy barriers (Ch. 8, Section 8.2).

In this chapter you will see how cells make ATP, starting with an introduction to the metabolic pathways that harvest energy from high-energy molecules like the sugar **glucose**—the most common source of chemical energy used by organisms. As cells break down, or catabolize, sugar, the energy that is released is used to transfer a phosphate group to adenosine diphosphate (ADP), generating ATP. (You can see the Big Picture of how the production of glucose in photosynthesis is related to its catabolism in cellular respiration on pages 236–237.)

9.1 An Overview of Cellular Respiration

> **After you complete this section, you should be able to ...**
>
> - Describe the four interconnected processes in cellular respiration and explain their central role in catabolism and anabolism.

In general, a cell contains only enough ATP to sustain from 30 seconds to a few minutes of normal activity. Because it has such high potential energy, ATP is unstable and is not stored. As a result, most cells are making ATP all the time.

Much of the ATP your cells produce is made using the chemical energy from glucose. How do cells obtain glucose? Photosynthetic organisms can produce glucose from the products of photosynthesis, where the energy in sunlight is used to reduce carbon dioxide (CO_2). These organisms will either use the glucose to make ATP or store it in other energy-rich molecules like starch. When photosynthetic organisms are eaten or decompose, their glucose molecules are obtained by animals, fungi, and many bacteria and archaea.

Storage carbohydrates, such as starch and glycogen, act like savings accounts for chemical energy (Ch. 5, Section 5.3). ATP, in contrast, is like cash. To withdraw chemical energy from the accounts to get cash, storage carbohydrates are first hydrolyzed into their glucose monomers. The glucose is then used to produce ATP through one of two general pathways: cellular respiration or fermentation (Figure 9.1). The primary difference between these two pathways lies in the degree to which glucose is oxidized.

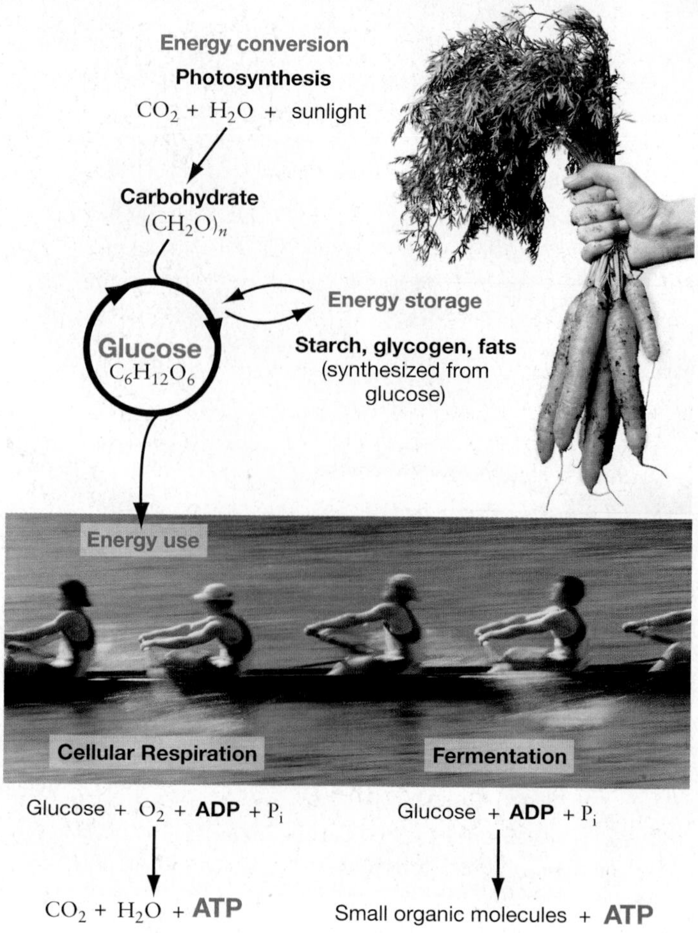

Figure 9.1 Glucose Is the Hub of Energy Processing in Cells. Glucose is a product of photosynthesis. Both plants and animals store glucose and oxidize it to provide chemical energy in the form of ATP. [$(CH_2O)_n$ is the fundamental molecular formula for all carbohydrates (Chapter 5). The molecular formula for glucose is $(CH_2O)_6$, also written as $C_6H_{12}O_6$.]

What Happens When Glucose Is Oxidized?

When glucose undergoes the uncontrolled oxidation reaction called burning, much of the potential energy stored in its chemical bonds is converted to kinetic energy in the form of heat and light:

$$\underset{\text{glucose}}{C_6H_{12}O_6} + \underset{\text{oxygen}}{6\,O_2} \longrightarrow \underset{\text{carbon dioxide}}{6\,CO_2} + \underset{\text{water}}{6\,H_2O} + \underset{\text{energy}}{\text{Heat and light}}$$

More specifically, a total of about 685 kilocalories (kcal) of heat is released when one mole of glucose is oxidized. To put this in perspective, if you burned one mole of glucose (~180 grams), it would give off enough heat to bring almost 2.5 gallons of room-temperature water to a boil.

Glucose does not burn in cells, however. Instead, it is oxidized through a long series of carefully controlled redox reactions (Ch. 8, Section 8.2). These reactions are occurring, millions of times per minute, in your cells right now. Instead of releasing all the energy stored in glucose as heat, much of it is used to synthesize ATP from ADP and P_i. Cellular respiration is responsible for making most of the ATP that you use to read, think, move, and stay alive.

Fermentation is another pathway that oxidizes glucose and synthesizes ATP. So how does fermentation differ from cellular respiration? Cellular respiration, like burning, results in the complete oxidation of the carbons in glucose to CO_2. Fermentation, on the other hand, does not fully oxidize glucose. Instead, small, reduced organic molecules are produced as waste. As a result, cellular respiration harvests more energy from glucose than fermentation.

You can think of the oxidation of glucose via cellular respiration as a set of four interconnected processes that together convert most of the chemical energy in glucose to chemical energy in ATP. Each of the four processes consists of a distinctive starting molecule, a series of chemical reactions, and a characteristic set of products.

1. ***Glycolysis*** In glycolysis, one six-carbon molecule of glucose is broken into two molecules of the three-carbon compound pyruvate. During this process, ATP is produced from ADP and P_i, and nicotinamide adenine dinucleotide (NAD^+) is reduced to form NADH.

2. ***Pyruvate processing*** Each pyruvate produced by glycolysis is processed to release one molecule of CO_2, and the remaining two carbons are used to form the compound acetyl CoA. The oxidation of pyruvate results in more NAD^+ being reduced to NADH.

3. ***Citric acid cycle*** The two carbons from each acetyl CoA produced by pyruvate processing are oxidized to two molecules of CO_2. During this sequence of reactions, more ATP and NADH are produced, and flavin adenine dinucleotide (FAD) is reduced to form $FADH_2$.
4. ***Electron transport and oxidative phosphorylation*** Electrons from the NADH and $FADH_2$ produced by pyruvate processing and the citric acid cycle move through a series of electron carriers that together are called an electron transport chain (ETC). The energy obtained from this chain of redox reactions is used to create a proton gradient across a membrane; the ensuing flow of protons back across the membrane is used to make ATP. Because this mode of ATP production links oxidation of NADH and $FADH_2$ with phosphorylation of ADP, it is called **oxidative phosphorylation**.

Figure 9.2 summarizes the four stages of cellular respiration. Formally, **cellular respiration** is defined as any set of reactions that uses electrons harvested from high-energy molecules to produce ATP via an electron transport chain. Making Models 9.1 provides some tips for how you can use models like the one shown in Figure 9.2 as references to draw your own models of cellular respiration. Such models are essential in biology to distill complex topics into understandable narratives.

The enzymes, products, and intermediates involved in cellular respiration do not exist in isolation. Instead, they are part of a huge and dynamic inventory of chemicals inside the cell.

Making Models 9.1 Tips on Drawing Flowcharts (I)

Cellular respiration is complex. By drawing your own simple models, you can practice keeping track of the main events. The details you choose to include depend on the focus of your model. For example, the flowchart in Figure 9.2 summarizes the four stages of cellular respiration. The flowchart below uses "balls" to represent carbons to track the fate of carbon during cellular respiration.

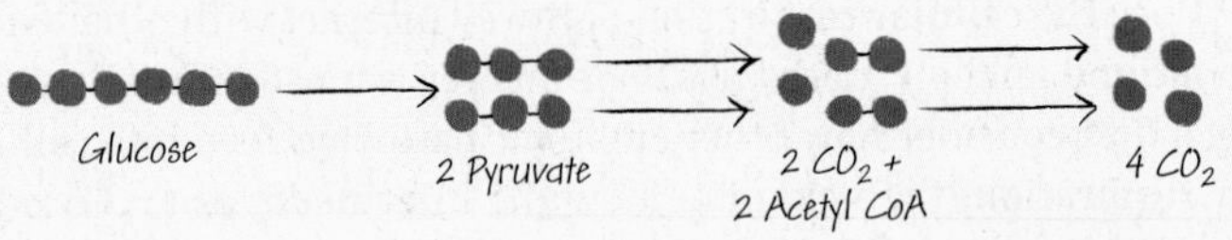

MODEL Where do all the carbons of glucose end up when glucose is completely oxidized? Add detail to the model by labeling the processes represented by the arrows.

To see this model in action, go to the Study Area of **Mastering Biology**

This complexity can be boiled down to a simple idea, however: Two of the most fundamental requirements of a cell are energy and carbon. They need a source of energy for generating ATP and a source of carbon that can be used as raw material to synthesize DNA, RNA, proteins, fatty acids, and other molecules. With these requirements in mind, let's take a closer look at the central role cellular respiration plays in cellular metabolism.

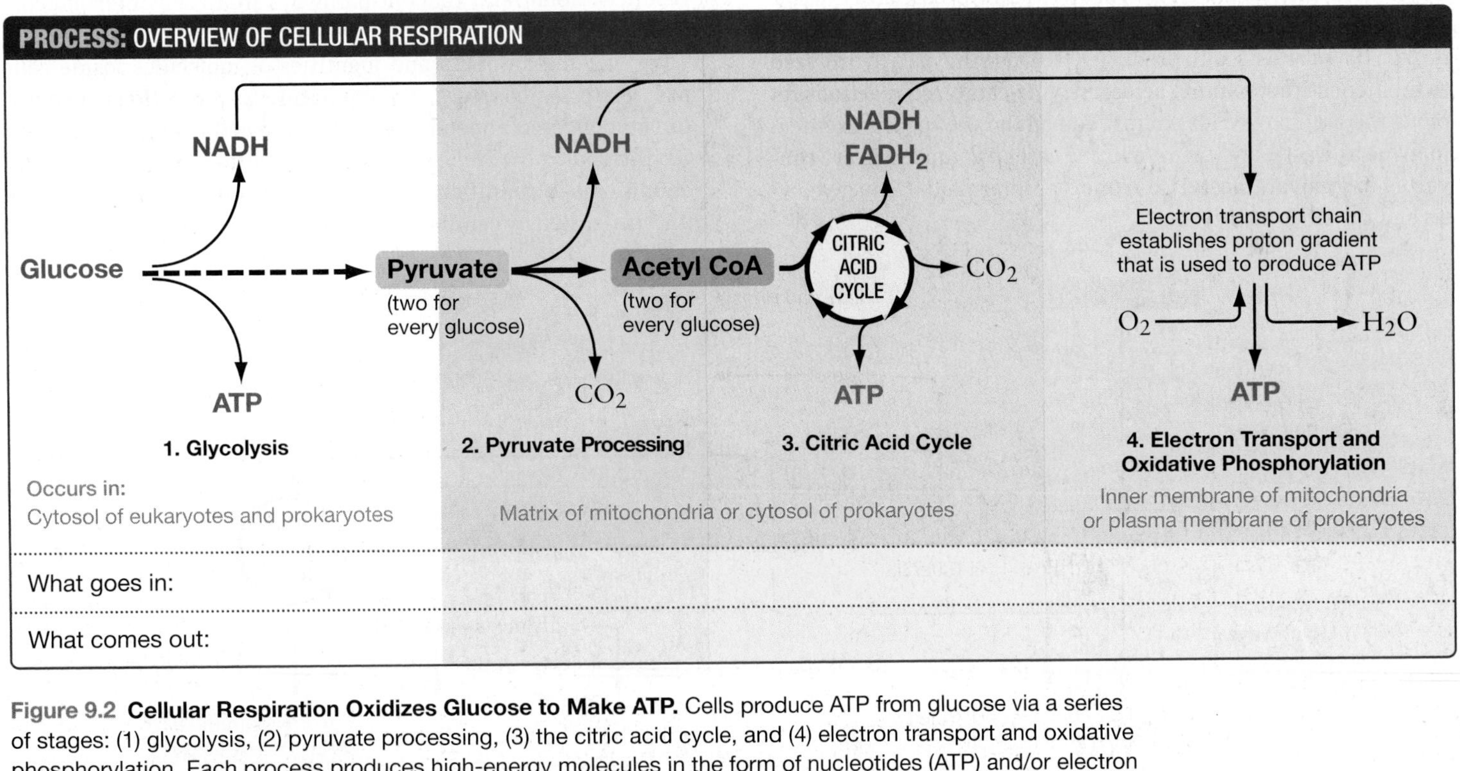

Figure 9.2 Cellular Respiration Oxidizes Glucose to Make ATP. Cells produce ATP from glucose via a series of stages: (1) glycolysis, (2) pyruvate processing, (3) the citric acid cycle, and (4) electron transport and oxidative phosphorylation. Each process produces high-energy molecules in the form of nucleotides (ATP) and/or electron carriers (NADH or $FADH_2$). Because the four processes are connected, cellular respiration is an integrated metabolic pathway. The first three stages oxidize glucose and intermediate products to produce NADH and $FADH_2$, which then feed the electron transport chain.

✔ Use what you have learned in the text to fill in the chart along the bottom of the figure.

Cellular Respiration Plays a Central Role in Metabolism

Recall that sets of reactions that break down molecules are called catabolic pathways (Ch. 8, Section 8.5). These reactions often harvest stored chemical energy to produce ATP. Anabolic pathways, on the other hand, are sets of reactions that synthesize larger molecules from smaller components. Anabolic reactions often use energy in the form of ATP.

Does the cellular respiration pathway interact with other catabolic and anabolic pathways? The answer is most definitely yes! Let's first consider how other catabolic pathways feed into cellular respiration, then examine how the intermediates and products of glycolysis, pyruvate processing, and the citric acid cycle feed into anabolic pathways.

Catabolic Pathways Break Down a Variety of Molecules Most organisms ingest, absorb, or synthesize many different carbohydrates—not just glucose. These molecules range from sucrose, maltose, and other simple sugars to large polymers such as glycogen and starch (Ch. 5, Section 5.2). Using enzyme-catalyzed reactions, cells can break down and transform these other carbohydrates to produce glucose or intermediates in cellular respiration.

Carbohydrates are not the only important source of carbon compounds used in catabolic pathways, however. Fats are highly reduced macromolecules consisting of glycerol bonded to chains of fatty acids (Ch. 6, Section 6.1). In cells, enzymes routinely break down fats to release the glycerol and convert the fatty acids into acetyl CoA molecules. Glycerol can be further processed and enter glycolysis as an intermediate. Acetyl CoA enters the citric acid cycle.

Proteins can also be catabolized, meaning that they can be broken down and used to produce ATP. Once they are hydrolyzed to their constituent amino acids, enzyme-catalyzed reactions remove the amino (—NH_2) groups. The amino groups are excreted in urine as waste, and the remaining carbon compounds are converted to pyruvate, acetyl CoA, or other intermediates in glycolysis and the citric acid cycle.

The top half of **Figure 9.3** summarizes the catabolic pathways of carbohydrates, fats, and proteins and shows how their breakdown products feed an array of steps in cellular respiration. When all three types of molecules are available in the cell to generate ATP, carbohydrates are used up first, then fats, and finally proteins.

Catabolic Intermediates Are Used in Anabolic Pathways Where do cells get the precursor molecules required to synthesize amino acids, RNA, DNA, phospholipids, and other cell components? Not surprisingly, the answer often involves intermediates in cellular respiration. For example,

- Intermediates in glycolysis can be used in the synthesis of ribonucleotides and deoxyribonucleotides. Nucleotides, in turn, are building blocks used in RNA and DNA synthesis.
- Acetyl CoA is the starting point for anabolic pathways that result in the synthesis of fatty acids. Fatty acids can then be used to build phospholipids and fats.
- In humans, about half the required amino acids can be synthesized from molecules siphoned from the citric acid cycle.
- If ATP is abundant, pyruvate can be used in the synthesis of glucose. Excess glucose may be converted to glycogen or starch and stored.

The bottom half of Figure 9.3 summarizes how intermediates in carbohydrate metabolism are drawn off to synthesize macromolecules. The take-home message is that the same molecule can be processed to serve many different functions in the cell. As a result, catabolic and anabolic pathways are closely intertwined.

Metabolism comprises thousands of different chemical reactions, yet the amounts and identities of molecules inside cells are relatively constant. By regulating key reactions involved in catabolic and anabolic pathways, the cell is able to maintain its internal environment even under different environmental conditions—a condition referred to as **homeostasis**. While the ATP generated by cellular respiration and fermentation is crucial

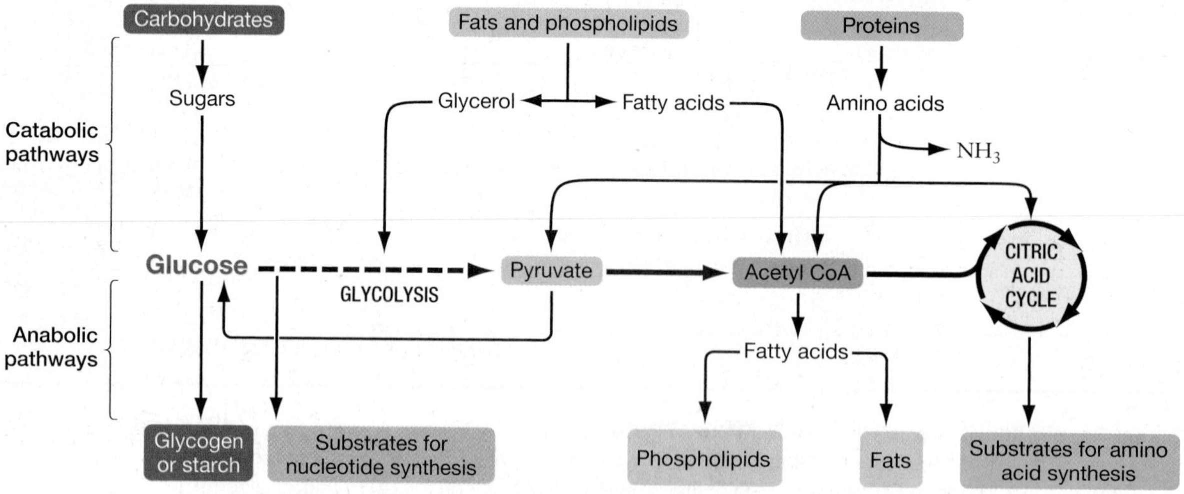

Figure 9.3 Cellular Respiration Interacts with Other Catabolic and Anabolic Pathways. A variety of high-energy compounds from carbohydrates, fats, or proteins can be broken down in catabolic reactions and used by cellular respiration for ATP production. Several of the intermediates in cellular respiration serve as precursor molecules in anabolic reactions leading to the synthesis of carbohydrates, nucleotides, lipids, and amino acids.

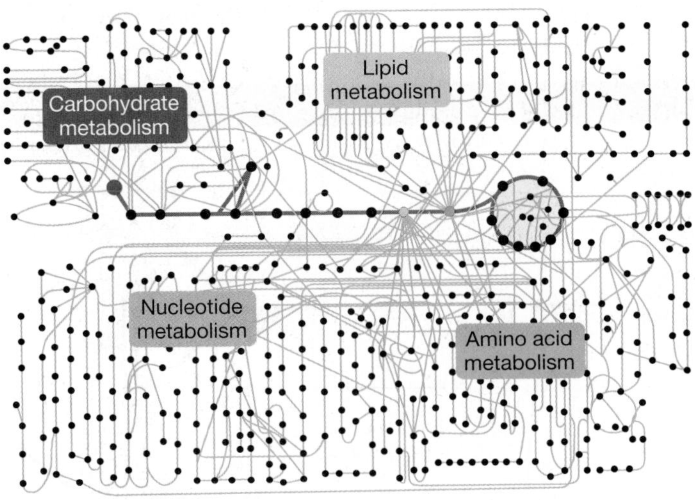

Figure 9.4 Cellular Respiration Plays a Central Role in the Metabolic Activity of Cells. Cellular respiration is connected to a multitude of different chemical reactions. In this schematic diagram, dots represent a few of the many thousands of molecules involved in metabolism, and green lines represent enzyme-catalyzed reactions. At the center of all this, the first three stages involved in cellular respiration (see Figure 9.3) are emphasized by bold dots along a thick black line. For reference, the dots along this line representing glucose, pyruvate, and acetyl CoA are identified by the same distinctive colors used in Figure 9.3.

for survival, the intermediates in these pathways also are central parts of a highly integrated metabolism (**Figure 9.4**).

Once you've filled in the chart at the bottom of Figure 9.2, you'll be ready to analyze each of the four stages that make up cellular respiration in detail. As you delve in, keep asking yourself the same key questions: What goes in and what comes out? What happens to the chemical energy? Where does each process occur, and how is it regulated? Then take a look in the mirror. All these processes are occurring right now, in virtually all your cells.

CHECK YOUR UNDERSTANDING

✔ **If you understood this section, you should be able to ...**

THINK CAREFULLY Explain why many different molecules—including lipids, amino acids, and CO_2—end up radiolabeled when cells are fed glucose with radioactive carbons (^{14}C).

Answers are available in Appendix A.

9.2 Glycolysis: Oxidizing Glucose to Pyruvate

After you complete this section, you should be able to ...

- Describe the process cells use to turn glucose into pyruvate.

Because the enzymes responsible for the breakdown of glucose into pyruvate—the process we call **glycolysis**—have been observed in nearly every prokaryote and eukaryote, it is logical to infer that the ancestor of all organisms living today made ATP by glycolysis. It's ironic, then, that the process was discovered by accident.

In the 1890s, Hans and Edward Buchner were working out techniques for breaking open baker's yeast cells and extracting the contents for commercial and medicinal use. (Yeast extracts are still added to some foods as a flavor enhancer or nutritional supplement today.) In one set of experiments, the Buchners added sucrose (common table sugar) to their extracts. At the time, sucrose was commonly used as a preservative—a substance used to prevent food from decaying.

Instead of preserving the yeast extracts, though, the sucrose was quickly broken down and alcohol appeared as a by-product. This was a key finding: It showed that metabolic pathways could be studied in vitro—outside the organism. Until then, researchers thought that metabolism could take place only in intact organisms.

When researchers studied how the sugar was being processed, they found that the reactions could go on much longer than normal if inorganic phosphate (P_i) were added to the mixture. This result implied that some of the compounds involved were being phosphorylated. Soon after, a molecule called fructose bisphosphate was isolated. (The prefix *bis–* means that single phosphate groups are attached to the fructose molecule at two different locations.) Subsequent work showed that all but the starting and ending molecules in glycolysis—glucose and pyruvate—are phosphorylated.

In 1905, researchers found that the processing of sugar by yeast extracts stopped if they boiled the reaction mix. Because it was known that enzymes could be inactivated by heat, this discovery suggested that enzymes were involved in at least some of the processing steps. After decades of research, investigators identified the 10 enzyme-catalyzed reactions of the most common type of glycolysis[1], referred to as the Embden–Meyerhof–Parnas (EMP) pathway to credit those who led this work. Let's take a closer look at these reactions to see how they extract energy from the reduced bonds in glucose.

Glycolysis Is a Sequence of 10 Reactions

In both eukaryotes and prokaryotes, glycolysis occurs in the cytosol. **Figure 9.5** on page 198 details the 10 reactions that make up the pathway. As you study the sequence of reactions (steps 1–10), consider the following three key points:

1. Glycolysis starts by *using* ATP, not producing it. In the initial step, glucose is phosphorylated to form glucose-6-phosphate. After the second reaction rearranges the sugar to form fructose-6-phosphate, the third reaction adds a second phosphate group, forming the compound fructose-1,6-bisphosphate observed by early researchers. Thus, in steps 1–5, two ATP

[1] Other metabolic pathways exist that use the term "glycolysis" and convert glucose to pyruvate, such as the Entner–Doudoroff pathway often used in prokaryotes. Enzymes different from those used in Embden–Meyerhof–Parnas glycolysis may be employed.

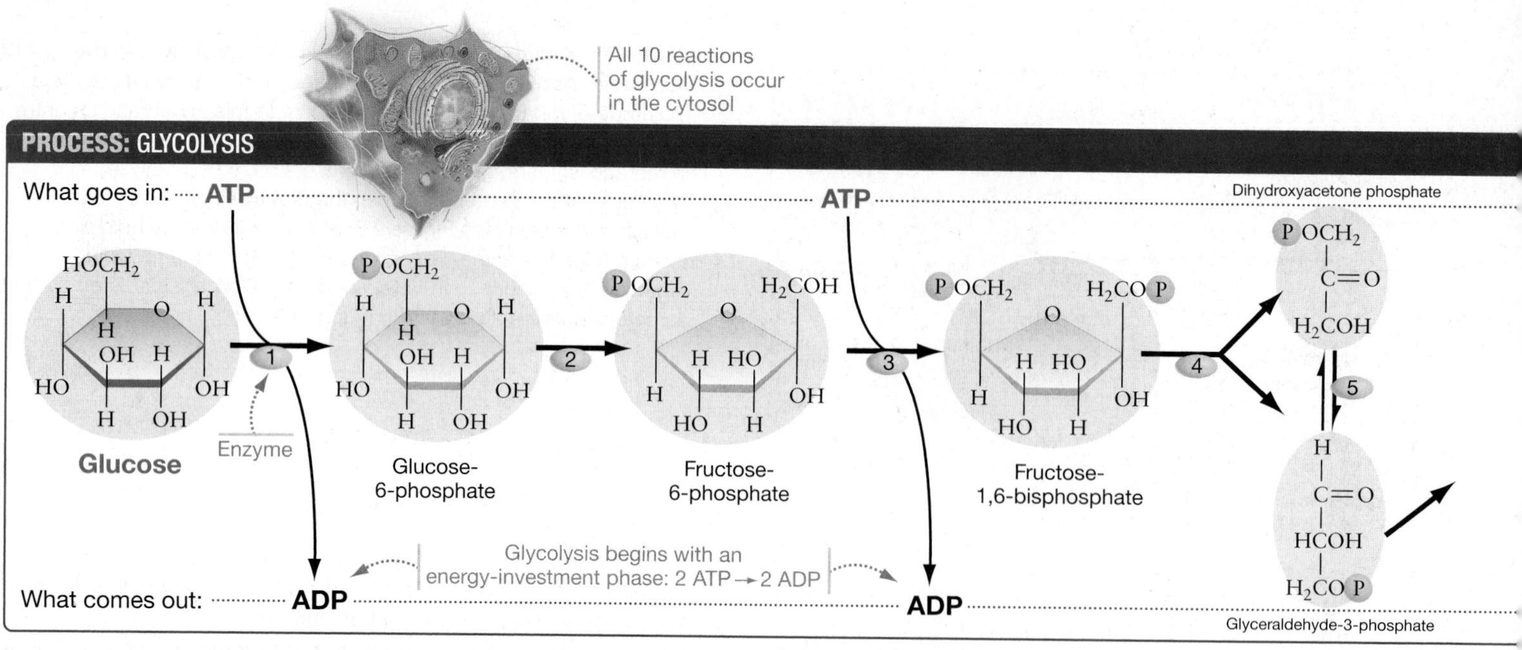

Figure 9.5 **Glycolysis Pathway.** This sequence of 10 cytosolic reactions oxidizes glucose to pyruvate. Each reaction step is catalyzed by a different enzyme (represented by numbered green ovals) to produce two net ATP (4 ATP are produced, but 2 are invested), two molecules of NADH, and two molecules of pyruvate. In step 4, fructose-1,6-bisphosphate is divided into two products that both proceed through steps 6–10. (The amounts for "What goes in" and "What goes out" in 6–10 are the combined totals for both reactant molecules.)

molecules are used up before any ATP is produced. Without ATP, these reactions would be endergonic, so this part of glycolysis is referred to as the energy-investment phase.

2. The energy-payoff phase of glycolysis (steps 6–10) consists of exergonic reactions that do not require an input of energy. The first high-energy molecules are produced in the sixth reaction, where two molecules of NAD^+ are reduced to form two NADH. In reactions 7 and 10, enzymes catalyze the transfer of a phosphate group from a phosphorylated substrate to ADP, forming ATP. When ATP is produced in this manner, it is termed **substrate-level phosphorylation** (Figure 9.6).
3. For each molecule of glucose processed by glycolysis, the net yield is two molecules of NADH, two of ATP, and two of pyruvate.

The discovery and elucidation of the glycolytic pathway ranks as one of the great achievements in the history of biochemistry.

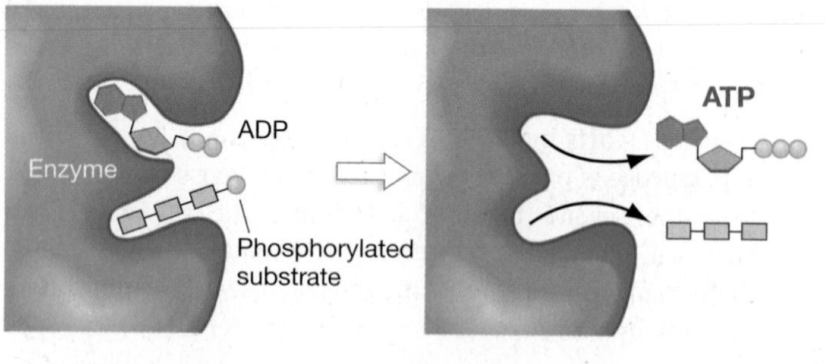

Figure 9.6 **Substrate-Level Phosphorylation Involves an Enzyme and a Phosphorylated Substrate.** Substrate-level phosphorylation occurs when an enzyme catalyzes the transfer of a phosphate group from a phosphorylated substrate to ADP, forming ATP.

For more detail about the enzymes that catalyze each step, see **Table 9.1**. While the catabolism of glucose can occur via other pathways, this set of reactions is among the most ancient and fundamental of all life processes.

How Is Glycolysis Regulated?

An important advance in understanding how glycolysis is regulated occurred when biologists observed that high levels of ATP inhibit a key glycolytic enzyme called phosphofructokinase. **Phosphofructokinase** catalyzes reaction 3 in Figure 9.5—the synthesis of fructose-1,6-bisphosphate from fructose-6-phosphate. This is a key step in the sequence.

The products of steps 1 and 2 can be easily converted back to glucose by an array of enzymes. Before step 3, then, the sequence is not committed to glycolysis and glucose can be used in other pathways. But once fructose-1,6-bisphosphate is synthesized, it will not be converted back to glucose. Based on these observations, it makes sense that the pathway is regulated at the first committed step—reaction 3. How do cells do it?

As shown in Figure 9.5, ATP serves as a substrate for the addition of a phosphate to fructose-6-phosphate. In the vast majority of cases, increasing the concentration of a substrate would *speed* the rate of a chemical reaction, but in this case, it inhibits it. Why would ATP—a substrate that is required for the reaction—also serve as an inhibitor of the reaction? The answer lies in knowing that ATP is also the end product of the overall catabolic pathway.

Recall that when an enzyme in a pathway is inhibited by the product of the reaction sequence, feedback inhibition occurs (Ch. 8, Section 8.5). When the product molecule is abundant, it

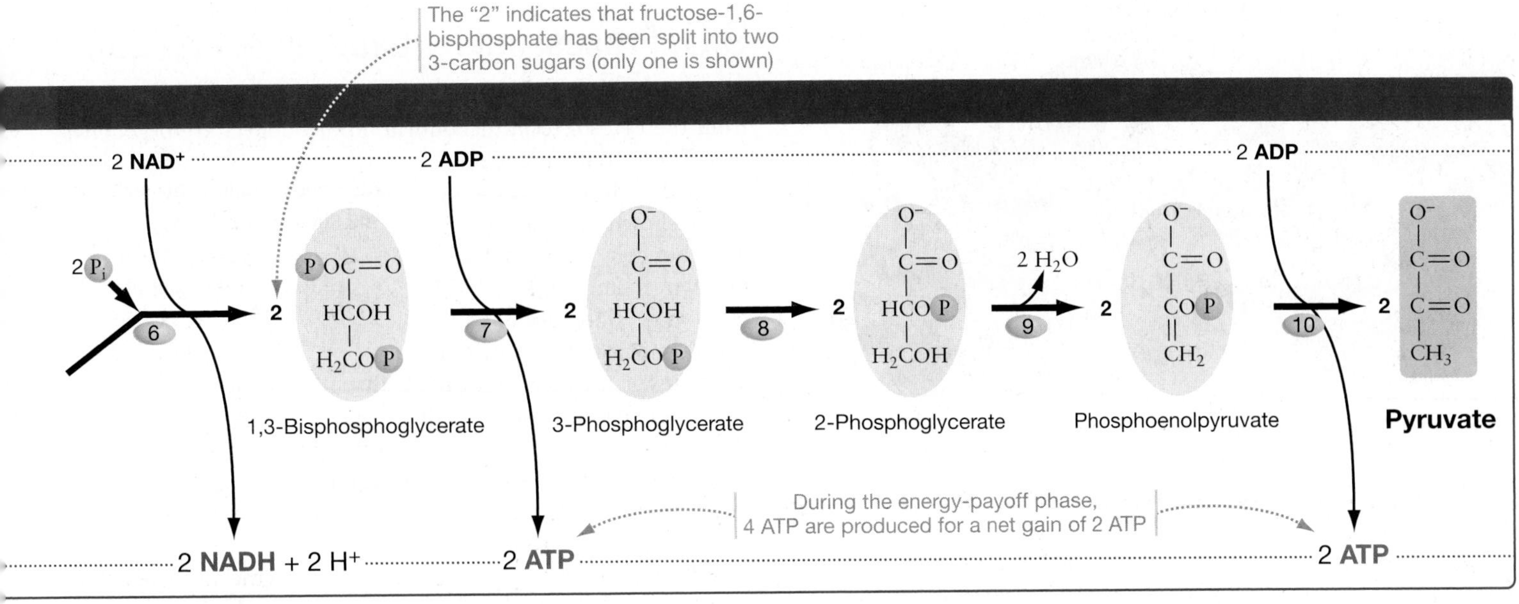

SUMMARY **Table 9.1** **The Reactions of Glycolysis**

Step	Enzyme	Reaction
1	Hexokinase	Uses **ATP** to phosphorylate glucose, increasing its potential energy.
2	Phosphoglucose isomerase	Converts glucose-6-phosphate to fructose-6-phosphate; referred to as an isomer of glucose-6-phosphate.
3	Phosphofructokinase	Uses **ATP** to phosphorylate the opposite end of fructose-6-phosphate, increasing its potential energy.
4	Fructose-bis-phosphate aldolase	Cleaves fructose-1,6-bisphosphate into two different 3-carbon sugars.
5	Triose phosphate isomerase	Converts dihydroxyacetone phosphate (DAP) to glyceraldehyde-3-phosphate (G3P). Although the reaction is fully reversible, the DAP-to-G3P reaction is favored because G3P is immediately used as a substrate for step 6.
6	Glyceraldehyde-3-phosphate dehydrogenase	In a two-step reaction, G3P is first oxidized using the **NAD^+** coenzyme to produce **NADH**. Energy from the first reaction is used to attach a P_i to the oxidized product to form 1,3-bisphosphoglycerate.
7	Phosphoglycerate kinase	Transfers a phosphate from 1,3-bisphosphoglycerate to **ADP** to make 3-phosphoglycerate and **ATP**.
8	Phosphoglycerate mutase	Rearranges the phosphate in 3-phosphoglycerate to make 2-phosphoglycerate.
9	Enolase	Removes a water molecule from 2-phosphoglycerate to form a C=C double bond and produce phosphoenolpyruvate.
10	Pyruvate kinase	Transfers a phosphate from phosphoenolpyruvate to **ADP** to make pyruvate and **ATP**.

can inhibit its own production by interfering with one of the reactions used to create it. Cells that are able to stop glycolytic reactions when ATP is abundant can conserve their stores of glucose for times when ATP is scarce. As a result, feedback inhibition in glycolysis contributes to cellular homeostasis.

How do high levels of the substrate inhibit the enzyme? As **Figure 9.7** on page 200 shows, phosphofructokinase has two distinct binding sites for ATP. ATP can bind at the enzyme's active site, where it is used to phosphorylate fructose-6-phosphate, or at a regulatory site, where it turns off the enzyme's activity.

The key to feedback inhibition lies in the ability of the two sites to bind to ATP. When concentrations are low, ATP binds only to the active site, which has a greater affinity for ATP than does the regulatory site. As ATP concentrations increase, however, it also binds at the regulatory site on phosphofructokinase. When ATP binds at this second location, the enzyme's conformation changes in a way that dramatically lowers the reaction rate at the active site. In phosphofructokinase, ATP serves as both a substrate and an allosteric regulator (Ch. 8, Section 8.4).

To summarize, glycolysis starts with one 6-carbon glucose molecule and ends with two 3-carbon pyruvate molecules. The reactions occur in the cytosol, and the energy that is released is used to produce a net total of two ATP and two NADH. Now the question is, what happens to the pyruvate?

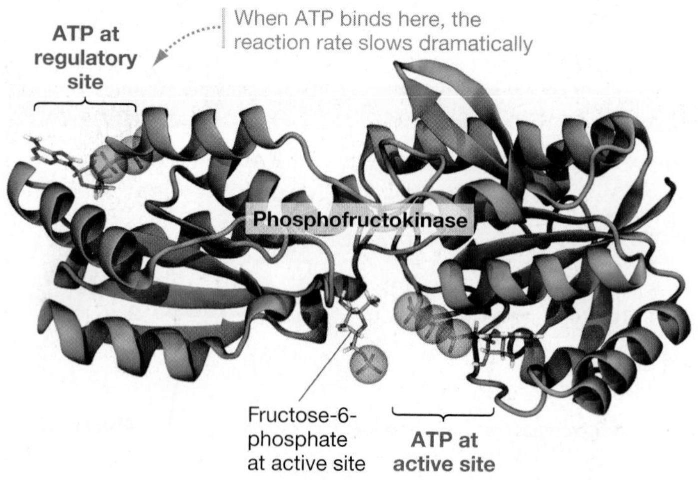

Figure 9.7 Phosphofructokinase Has Two Binding Sites for ATP. In the active site (lower right), ATP is used as a substrate to transfer one of its phosphate groups to fructose-6-phosphate. In the regulatory site (upper left), ATP binding inhibits the reaction by changing the shape of the enzyme.

CHECK YOUR UNDERSTANDING

✔ **If you understood this section, you should be able to . . .**

1. Explain the role of ATP in the energy-investment phase. Why is it not necessary for the energy-payoff phase?
2. **QUANTITATIVE** Draw a graph showing the rate of ATP production as a function of ATP concentration. Predict how the rate would change if the regulatory site in phosphofructokinase had higher affinity for ATP than the active site did.

Answers are available in Appendix A.

9.3 Processing Pyruvate to Acetyl CoA

After you complete this section, you should be able to . . .

- Describe the process cells use to turn pyruvate into acetyl CoA.

In eukaryotes, the pyruvate produced by glycolysis is transported from the cytosol to mitochondria. Mitochondria are organelles found in virtually all eukaryotes (Ch. 7, Section 7.2).

As shown in Figure 9.8, mitochondria are enclosed by two membranes, called the outer membrane and inner membrane. Portions of the inner membrane protrude into the interior of the organelle and expand to form sac-like compartments called **cristae**. The regions between the outer and inner membranes, including the space within the cristae, make up the intermembrane space. The compartment enclosed within the inner membrane is the **mitochondrial matrix**.

Pyruvate moves from the cytosol across the mitochondrial outer membrane through small pores and is then transported into the matrix through a carrier protein in the inner membrane. Once pyruvate is inside the matrix, it is processed by an enormous and intricate enzyme complex called **pyruvate dehydrogenase**. Mitochondria do not exist in bacteria and archaea, so these organisms process pyruvate using a similar enzyme complex located in the cytosol.

Pyruvate dehydrogenase couples the oxidation of one carbon in pyruvate to the reduction of NAD^+, resulting in the release of a CO_2 molecule and production of NADH. The same enzyme complex then takes the two-carbon acetyl unit ($-COCH_3$) and covalently bonds it to a compound called **coenzyme A (CoA)**. Coenzyme A is sometimes abbreviated as CoA—SH to call attention to its key sulfhydryl (—SH) functional group. The product

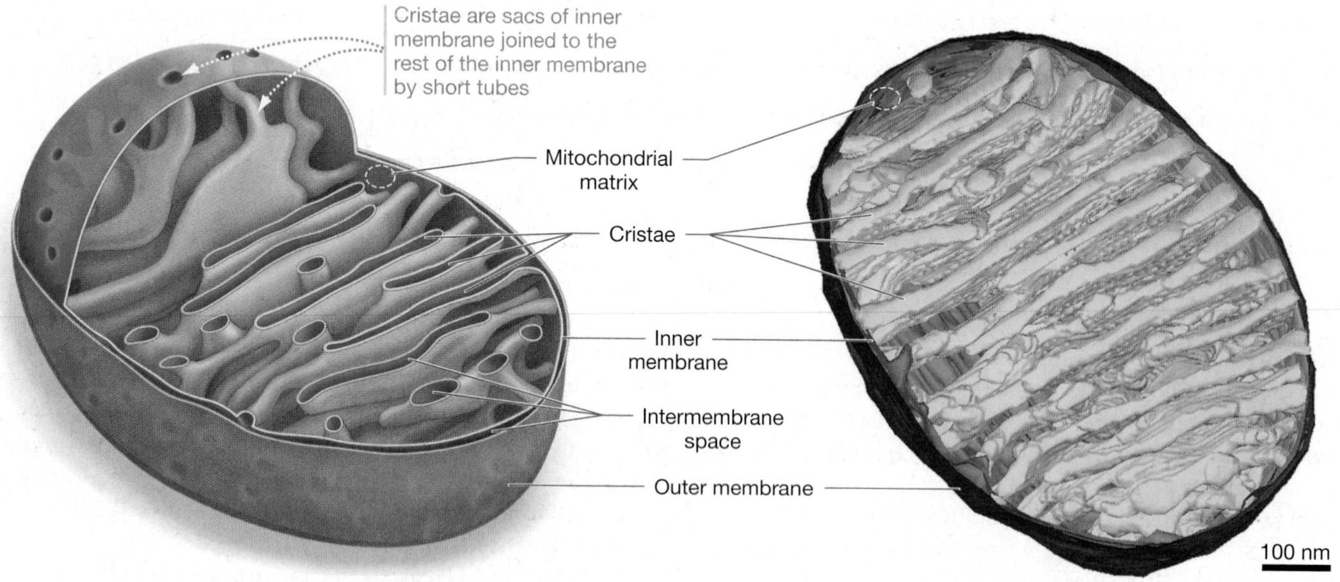

Figure 9.8 The Structure of the Mitochondrion. Mitochondria have outer and inner membranes that define the intermembrane space and matrix. Recent research using cryo-electron tomography (the colorized image on the right) shows that the sac-like cristae are expansions of short tubes formed from the inner membrane. Pyruvate processing occurs within the mitochondrial matrix.

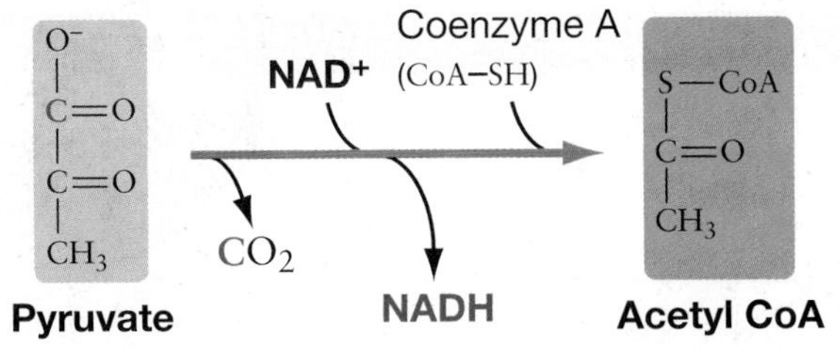

Figure 9.9 Pyruvate Is Oxidized to Acetyl CoA. The reaction shown here is catalyzed by pyruvate dehydrogenase. In the process, one carbon (red in the diagram) is fully oxidized to CO_2 and released. The remaining two carbons are bonded to coenzyme A, forming acetyl CoA.

of this final reaction is called acetyl CoA (**Figure 9.9**). In this and many other reactions, CoA acts as a coenzyme by accepting and then later transferring an acetyl group to another substrate ("A" stands for acetylation).

Acetyl CoA is the final product of the pyruvate-processing step in glucose oxidation. Pyruvate, NAD^+, and CoA go in; CO_2, NADH, and acetyl CoA come out.

Like the regulated phosphofructokinase reaction in glycolysis, pyruvate processing is a key commitment step that is not reversible. Feedback inhibition shuts down pyruvate dehydrogenase when the products of pyruvate processing (NADH and acetyl CoA) or glycolysis (ATP) are in abundant supply. But instead of binding to these products and undergoing allosteric regulation, pyruvate dehydrogenase is shut down when it is phosphorylated by other enzymes. The rate of phosphorylation increases when one or more of the feedback products are at high concentration.

In contrast, a high concentration of one or more of the substrates in glycolysis and pyruvate processing (ADP, pyruvate, CoA, or NAD^+) results in more dephosphorylated and active forms of the pyruvate dehydrogenase complex.

Pyruvate processing is thus under both positive and negative regulation. Large supplies of products inhibit the enzyme complex; large supplies of reactants and low supplies of products stimulate it.

To summarize, pyruvate processing starts with a three-carbon pyruvate molecule and ends with one carbon released as CO_2 and the remaining two carbons in the form of acetyl CoA. The reactions occur in the mitochondrial matrix, and the released free energy is used to produce one NADH for each pyruvate that is processed. Now the question is, what happens to the acetyl CoA?

CHECK YOUR UNDERSTANDING

✔ If you understood this section, you should be able to ...

Circle the three molecules in Figure 9.9 that would speed up the reaction when in abundance and the two molecules that would slow it down when in abundance. State how the structure of pyruvate dehydrogenase is affected by each set of molecules.

Answers are available in Appendix A.

9.4 The Citric Acid Cycle: Oxidizing Acetyl CoA to CO_2

While researchers were working out the sequence of reactions in glycolysis, biologists in other laboratories were focusing on redox reactions that oxidize small organic acids called **carboxylic acids**. Note that carboxylic acids all have carboxyl functional groups (R-COOH), hence the name.

After you complete this section, you should be able to ...

- Describe the pathway used by cells to fully oxidize the remaining two carbons from glucose that are present in acetyl CoA.

Early researchers identified eight small carboxylic acids that are rapidly oxidized in sequence, from most reduced to most oxidized. Redox reactions that involve carboxylic acids often produce carbon dioxide, which is the endpoint of glucose oxidation via cellular respiration. When the researchers added one of the eight carboxylic acids to cells, the rate of glucose oxidation increased, suggesting that the reactions were somehow connected to pathways involved in glucose catabolism. What they found next was puzzling. Whichever carboxylic acid they added, it did not appear to be used up. Instead, virtually all the acids could be recovered later. How is this possible?

Hans Krebs solved the mystery when he proposed that the reaction sequence occurs in a cycle instead of a linear pathway. Krebs had another crucial insight when he suggested that the reaction sequence was tied to pyruvate, which we now know is first oxidized to acetyl CoA in the pyruvate processing stage.

To test these hypotheses, Krebs and a colleague set out to determine if adding pyruvate could link the two ends of the sequence of eight carboxylic acids. (At the time, the pyruvate processing stage, where pyruvate is oxidized to acetyl CoA, had not been discovered.) When Krebs added pyruvate, the most oxidized of the eight carboxylic acids (oxaloacetate) was converted to the most reduced carboxylic acid (citrate). The conclusion? The sequence of eight carboxylic acid reactions is indeed arranged in a cycle (see **Figure 9.10** on page 202).

Many biologists now refer to the cycle as the **citric acid cycle** because it starts with citrate, which is a deprotonated form of citric acid. The citric acid cycle is also known as the tricarboxylic acid (TCA) cycle, because citrate has three carboxyl groups, or as the Krebs cycle, after its discoverer.

In each turn of the cycle, the energy harvested by oxidizing the acetyl group ($-COCH_3$) from acetyl CoA (product of the pyruvate processing stage) generates three molecules of NADH, one of $FADH_2$, and one of ATP or guanosine triphosphate (GTP). As in glycolysis, the ATPs and GTPs are produced by substrate-level phosphorylation. Whether ATP or GTP is produced depends on the version of the enzyme used in the fifth reaction.[2] For

[2] Traditionally it was thought that the citric acid cycle produced GTP, which was later converted to ATP in the same cell. Recent work suggests that ATP is produced directly in some cell types, while GTP is produced in other cells. See J. D. Johnson et al., Genetic evidence for the expression of ATP- and GTP-specific succinyl-CoA synthetases in multicellular eukaryotes. *Journal of Biological Chemistry* 42 (1998): 27580–27586.

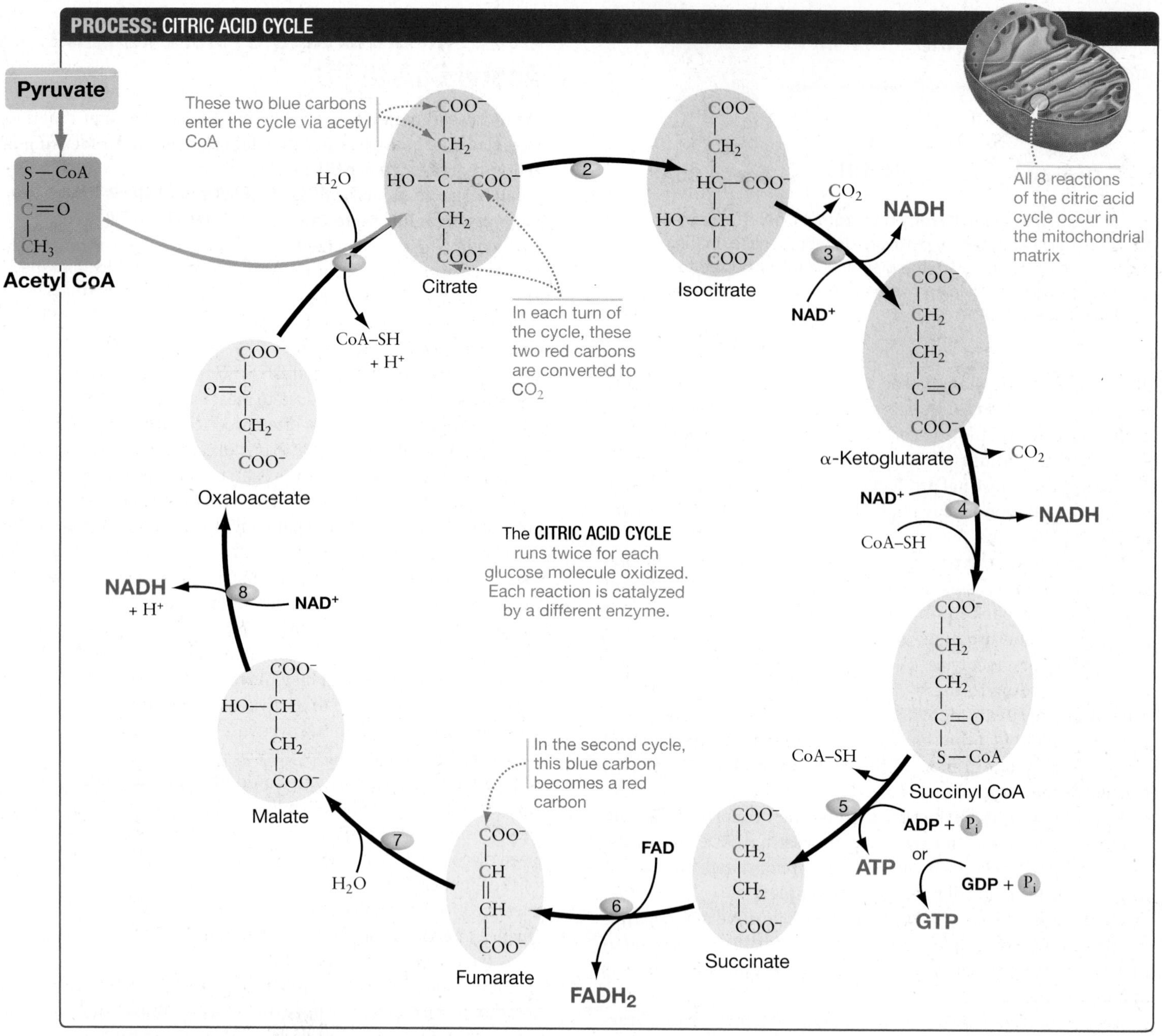

Figure 9.10 The Citric Acid Cycle Completes the Oxidation of Glucose. Acetyl CoA (from pyruvate) goes into the citric acid cycle, and carbon dioxide, NADH, $FADH_2$, and ATP or GTP come out. ATP or GTP is produced by substrate-level phosphorylation. If you follow individual carbon atoms around the cycle several times, you'll come to an important conclusion: Each of the carbons in the cycle is eventually a "red carbon" that is released as CO_2.

example, the enzyme used in muscle cells of mammals produces ATP, while the enzyme used in liver cells produces GTP. For simplicity, ATP has been used as the product of the citric acid cycle throughout this chapter.

In prokaryotes (bacteria and archaea), the enzymes responsible for the citric acid cycle are located in the cytosol. In eukaryotes, these enzymes are located in the mitochondrial matrix. Because glycolysis produces two molecules of pyruvate, each being converted into an acetyl CoA, the cycle turns twice for each molecule of glucose processed in cellular respiration.

How Is the Citric Acid Cycle Regulated?

By now, it shouldn't surprise you to learn that the citric acid cycle is also carefully regulated. The citric acid cycle can be turned off at multiple points via feedback inhibition. Reaction rates are high when ATP and NADH are scarce; the rates are low when ATP or NADH is abundant.

Figure 9.11 highlights the major control points. In step 1, the enzyme that combines acetyl CoA and oxaloacetate to form citrate is shut down when ATP binds at an allosteric regulatory site. In step 3, NADH interferes with the reaction by binding to the enzyme's active

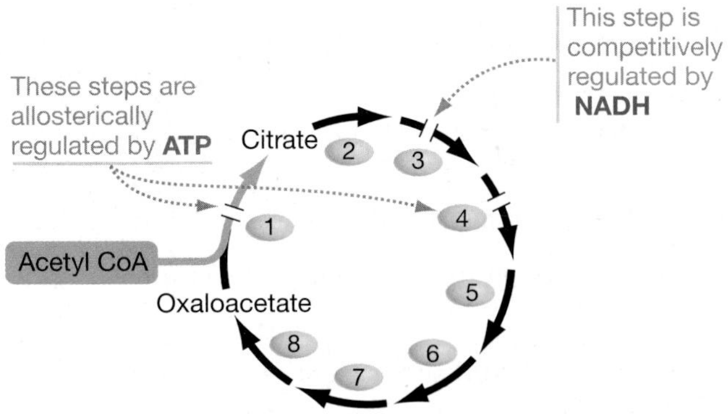

Figure 9.11 The Citric Acid Cycle Is Regulated by Feedback Inhibition. The citric acid cycle slows down when ATP and NADH are plentiful. ATP acts as an allosteric regulator, while NADH acts as a competitive inhibitor.

site. This is an example of competitive inhibition (Ch. 8, Section 8.4). In step 4, ATP again functions as an allosteric regulator.

To summarize, each round of the citric acid cycle starts with a two-carbon acetyl molecule in the form of acetyl CoA and ends with the release of two CO_2. The completion of this cycle marks the point when all of the carbons from glucose have been fully oxidized to CO_2. For more detail about the enzymes that catalyze each step of the citric acid cycle, see **Table 9.2** on page 204.

In addition to producing CO_2, the free energy obtained from these reactions is used to produce three NADH, one $FADH_2$, and one ATP for each acetyl oxidized. But a major question remains.

What Happens to the NADH and $FADH_2$?

Figure 9.12 reviews the relationships of glycolysis, pyruvate processing, and the citric acid cycle and identifies where each process takes place in eukaryotic cells. As you study the figure, note that for each molecule of glucose that is fully oxidized to carbon dioxide (2 CO_2 molecules during pyruvate processing, 4 during the citric acid cycle), the cell also produces 10 molecules of NADH, 2 of $FADH_2$, and 4 of ATP. The ATP molecules are produced by substrate-level phosphorylation during glycolysis and the citric acid cycle, and can be used to drive endergonic reactions. The CO_2 molecules are a gas that is disposed of as you exhale.

What happens to the product NADH and $FADH_2$? To approach this question, take a step back and recall the overall reaction for glucose oxidation (introduced in Section 9.1):

$$C_6H_{12}O_6 + 6O_2 \longrightarrow 6CO_2 + 6H_2O + \text{Energy}$$

Glycolysis, pyruvate processing, and the citric acid cycle account for the glucose, the CO_2, and—because ATP is produced—also some of the chemical energy that results from the overall reaction. But the O_2 and the H_2O components of the equation are still unaccounted for. As it turns out, so is much of the chemical energy. Where do these components come from; where do they go? Let's start by analyzing the energy.

The relative changes in free energy that occur as the carbons in glucose are being oxidized are shown in **Figure 9.13** on page 204.

In glycolysis, pyruvate processing, and the citric acid cycle, redox reactions transfer electrons to NAD^+ and FAD to form NADH and $FADH_2$. However, at this point in cellular respiration, the following reaction has yet to occur:

$$10NADH + 2FADH_2 + 6O_2 + 10H^+ \longrightarrow 10NAD^+ + 2FAD + 12H_2O + \text{Energy}$$

In the above reaction, another set of redox reactions transfers electrons from NADH and $FADH_2$ to oxygen, oxidizing NADH and $FADH_2$ back to NAD^+ and FAD and reducing O_2 to form water. These events occur in steps that will be described in Section 9.5.

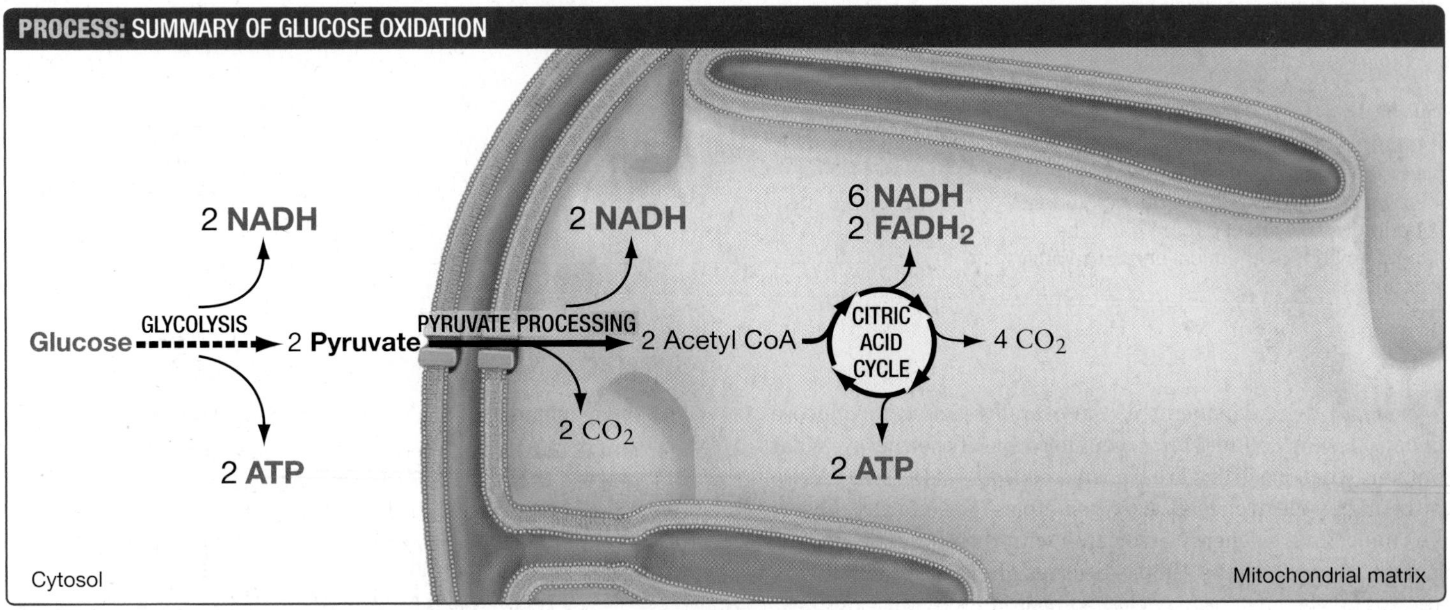

Figure 9.12 Glucose Oxidation Produces ATP, NADH, $FADH_2$, and CO_2. The carbons in glucose are completely oxidized to carbon dioxide via glycolysis, pyruvate processing, and the citric acid cycle. In eukaryotes, glycolysis occurs in the cytosol; pyruvate oxidation and the citric acid cycle take place in the mitochondrial matrix.

SUMMARY Table 9.2 The Reactions of the Citric Acid Cycle

Step	Enzyme	Reaction
1	Citrate synthase	Transfers the 2-carbon acetyl group from acetyl CoA to the 4-carbon molecule oxaloacetate to produce the 6-carbon molecule citrate.
2	Aconitase	Converts citrate to isocitrate by the removal of one water molecule and the addition of another water molecule.
3	Isocitrate dehydrogenase	Oxidizes isocitrate using the **NAD**$^+$ coenzyme to produce **NADH** and release one CO_2, resulting in the formation of the 5-carbon molecule α-ketoglutarate.
4	α-Ketoglutarate dehydrogenase	Oxidizes α-ketoglutarate using the **NAD**$^+$ coenzyme to produce **NADH** and release one CO_2. The remaining 4-carbon molecule is added to coenzyme A (CoA) to form succinyl CoA.
5	Succinyl-CoA synthetase	Replaces CoA with an inorganic phosphate (P_i), converting succinyl CoA to succinyl phosphate. This phosphate is then transferred to **ADP** to form **ATP**, or to **GDP** to form **GTP,** depending on the version of the enzyme used. What remains after the transfer is succinate.
6	Succinate dehydrogenase	Oxidizes succinate by transferring two hydrogens to the coenzyme **FAD** to produce **FADH**$_2$, resulting in the formation of fumarate.
7	Fumarase	Converts fumarate to malate by the addition of one water molecule.
8	Malate dehydrogenase	Oxidizes malate by using the **NAD**$^+$ coenzyme to produce **NADH**, resulting in the regeneration of the oxaloacetate that will be used in step 1 of the cycle.

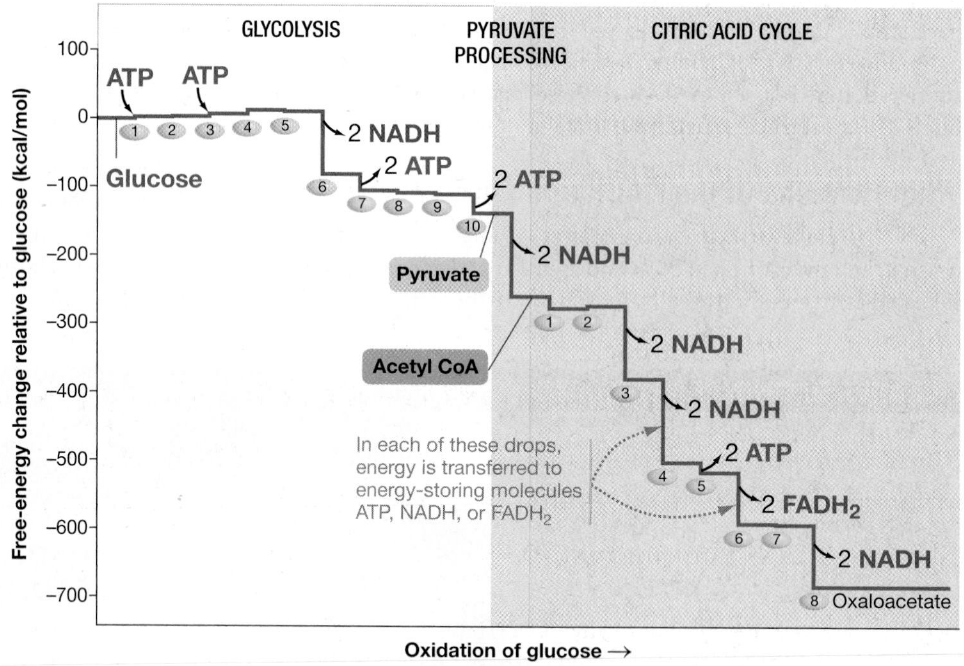

Figure 9.13 Free Energy Changes as Glucose Is Oxidized. If you read the vertical axis of this graph carefully, it should convince you that the change in free energy from the oxidation of glucose is about 685 kcal/mol. Some of this energy is harnessed in the form of ATP, and some in NADH, and $FADH_2$. The numbered green ovals identify the reaction steps in glycolysis and the citric acid cycle (see Tables 9.1 and 9.2).

DATA: X. Li, R. K. Dash, R. K. Pradhan, et al. 2010. *Journal of Physical Chemistry B*. 114: 16068–16082.

✔ **QUANTITATIVE** Based on the data in this graph, which one of the three high-energy molecules produced during glucose oxidation would you expect to carry the highest amount of chemical energy? Justify your answer.

Now, all the components of the overall reaction for glucose oxidation are accounted for, except for some of the energy. What happens when electrons are transferred from NADH and $FADH_2$ to the highly electronegative oxygen atoms? Specifically, how is the change in free energy associated with these redox reactions used by the cell? In the 1960s—decades after the details of glycolysis and the citric acid cycle had been worked out—an unexpected answer to this question emerged. The answer ultimately includes the production of a lot more ATP.

CHECK YOUR UNDERSTANDING

✔ **If you understood this section, you should be able to . . .**

1. **MODEL** Draw a flowchart to track the flow of electrons from glucose to NADH or $FADH_2$ as the carbons are oxidized to CO_2. (Hint: Rather than showing balls for carbons as in Making Models 9.1, use triangles to represent pairs of electrons, starting with 12 triangles for glucose.)
2. Describe the three regulation points in the citric acid cycle in terms of the type of regulatory molecule and how each enzyme is affected.

Answers are available in Appendix A.

9.5 Electron Transport and Chemiosmosis: Building a Proton Gradient to Produce ATP

After you complete this section, you should be able to …

- Analyze the relationship between the electron transport chain and oxidative phosphorylation.

A first step toward answering the question about energy and the oxidation of NADH and $FADH_2$ turned out to be relatively straightforward. By isolating different parts of mitochondria, researchers determined that NADH is oxidized when combined with the inner membrane of the mitochondria, including the cristae. In prokaryotes, which lack mitochondria, the oxidation of NADH requires the plasma membrane. It was hypothesized that these membranes contain the components responsible for oxidizing NADH and $FADH_2$.

Biologists made a key discovery when they isolated the membrane components after exposing them to NADH and $FADH_2$—the components were found to cycle between oxidized and reduced states. What are these molecules, and how do they work?

The Electron Transport Chain

Collectively, the molecules responsible for the oxidation of NADH and $FADH_2$ are designated the **electron transport chain (ETC)**. Several points are fundamental to understanding how the ETC works:

- Most of the molecules are proteins that contain distinctive cofactors and coenzymes where the redox events take place (Ch. 8, Section 8.3). The latter include inorganic iron–sulfur complexes, organic ring-containing structures called flavins, or organic iron-containing heme groups called cytochromes. Each of these "helper" groups is readily reduced or oxidized.
- The inner membrane of the mitochondrion also contains a pool of nonprotein molecules called **ubiquinone**. Ubiquinone got its name because it is nearly ubiquitous in organisms and belongs to a family of compounds called quinones. Also called **coenzyme Q**, or simply Q, ubiquinone is lipid soluble and moves efficiently throughout the hydrophobic interior of the inner mitochondrial membrane.
- The molecules involved in oxidizing NADH and $FADH_2$ differ in their ability to accept electrons in a redox reaction. This ability is referred to as the **redox potential** of the electron acceptors. In addition, some of the molecules pick up a proton with each electron, forming bonds to uncharged hydrogen atoms, while others obtain only electrons.

Because Q and the ETC proteins differ in redox potential, investigators realized that it should be possible to arrange their redox reactions into a logical sequence. The idea was that electrons would pass from a molecule with a lower redox potential to one with a higher redox potential. As electrons moved through the chain, they would be held more and more tightly in the newly formed bonds, and the potential energy in each successive bond would lessen.

Organization of the Electron Transport Chain Researchers worked out the sequence of the redox reactions in the ETC by experimenting with poisons that inhibit particular proteins in the inner membrane. It was expected that if part of the chain were inhibited, then the components upstream of the block would become reduced and those downstream would remain oxidized.

Experiments with various poisons showed that NADH donates an electron to a flavin-containing protein (FMN) at the top of the chain, while $FADH_2$ donates electrons to an iron- and sulfur-containing protein (Fe • S). Each of these reduced proteins then passes their electrons directly to a Q molecule. After being transferred through each of the remaining components in the chain, the electrons are finally accepted by oxygen.

Figure 9.14 shows how the potential energy in shared electrons steps down from the electron carriers NADH and $FADH_2$ to O_2. The x-axis plots the sequence of redox reactions in the ETC;

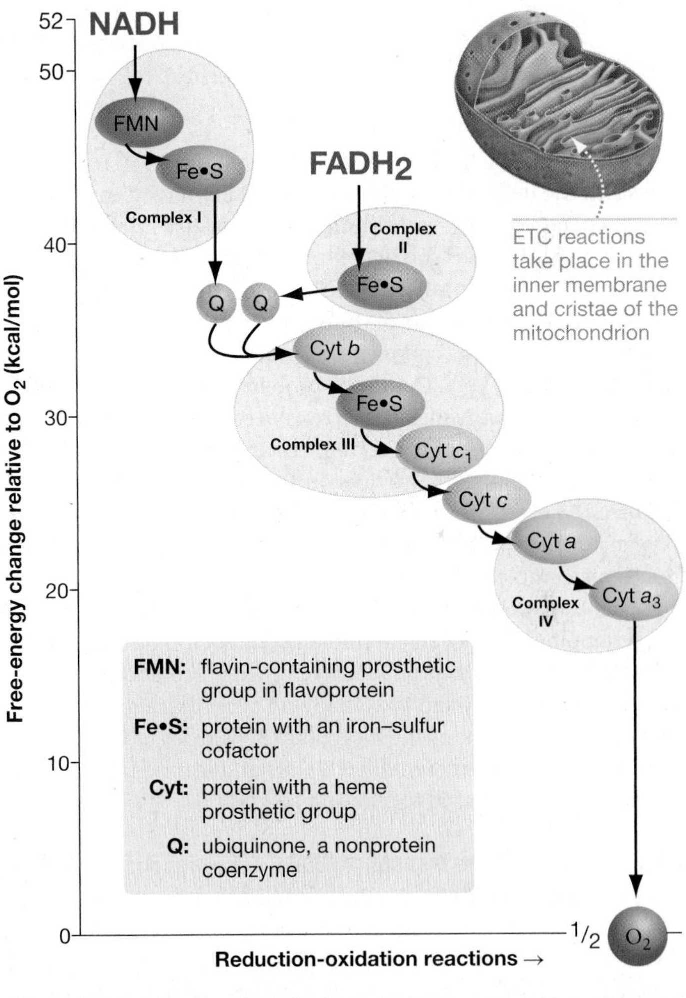

Figure 9.14 A Series of Reduction–Oxidation Reactions Occur in an Electron Transport Chain. The potential energy in shared electrons steps down from the electron carriers NADH and $FADH_2$ through an electron transport chain to a final electron acceptor. In this electron transport chain, oxygen is the final electron acceptor and it forms water as a by-product. The overall free-energy change of 52 kcal/mol (from NADH to oxygen) is broken into small steps.

DATA: D. F. Wilson, M. Erecinska, and P. L. Dutton. 1974. *Annual Review of Biophysics and Bioengineering* 3: 203–230. Also V. D. Sled, N. I. Rudnitzky, Y. Hatefi, et al. 1994. *Biochemistry* 33: 10069–10075.

SUMMARY **Table 9.3 Components and Reactions of the Electron Transport Chain**

ETC Component	Descriptive Name	Reaction
Complex I	NADH dehydrogenase	Oxidizes **NADH** and transfers the two electrons through proteins containing FMN prosthetic groups and (Fe · S) cofactors to reduce an oxidized form of ubiquinone (Q). Four $\mathbf{H^+}$ are pumped out of the matrix to the intermembrane space.
Complex II	Succinate dehydrogenase	Oxidizes $\mathbf{FADH_2}$ and transfers the two electrons through proteins containing (Fe · S) cofactors to reduce an oxidized form of Q. This complex is also used in step 6 of the citric acid cycle.
Q	Ubiquinone	Reduced by two electrons from complexes I or II and moves throughout the hydrophobic interior of the ETC membrane, where it is oxidized by complex III.
Complex III	Cytochrome *c* reductase	Oxidizes Q and transfers one electron at a time through proteins containing heme prosthetic groups and (Fe · S) cofactors to reduce an oxidized form of cytochrome *c* (cyt *c*). A total of four $\mathbf{H^+}$ for each pair of electrons is transported from the matrix to the intermembrane space.
Cyt *c*	Cytochrome *c*	Reduced by accepting a single electron from complex III and moves within the intermembrane space, where it is oxidized by complex IV.
Complex IV	Cytochrome *c* oxidase	Oxidizes cyt *c* and transfers each electron through proteins containing heme groups. After two cyt *c*, will pick up two $\mathbf{H^+}$ from the matrix to produce water from oxygen gas (O_2) and pump two additional $\mathbf{H^+}$ from the matrix to the intermembrane space.

the *y*-axis plots the free-energy changes that occur. ✔ **If you understand the logic of how electrons are transferred in the electron transport chain, you should be able to use Figure 9.14 to identify the ETC electron acceptor with the highest redox potential and the acceptor with the lowest redox potential.**

Most of the components of the electron transport chain are organized into four large complexes of proteins, often referred to as simply complexes I–IV. Q and the protein **cytochrome *c* (cyt *c*)** act as shuttles that transfer electrons between these complexes. Details on the names of the complexes and their role in the electron transport chain are provided in **Table 9.3**.

Once the electrons at the bottom of the ETC are accepted by oxygen to form water, the extraction of chemical energy from glucose is complete. Under controlled conditions in the laboratory, the total potential energy difference from NADH to oxygen is a whopping 52 kilocalories/mole (kcal/mol). Oxidation of the 10 molecules of NADH produced from each glucose by glycolysis, pyruvate processing, and the citric acid cycle therefore accounts for almost 80 percent of the total 685 kilocalories of energy harvested from the sugar (Section 9.1). What does the ETC do with all this energy? Is it used to make more ATP?

Role of the Electron Transport Chain Throughout the 1950s, most biologists working on cellular respiration assumed that the ETC included enzymes that would use energy from redox reactions to produce ATP via substrate-level phosphorylation. Recall that when substrate-level phosphorylation occurs, a phosphate group is transferred from a phosphorylated substrate to ADP, forming ATP (see Figure 9.6). Despite intense efforts, however, no one was able to find an enzyme that could perform this task among the components of the ETC.

But researchers did find that the energy from redox reactions would drive the active transport of protons across the inner membrane from the matrix into the intermembrane space (**Figure 9.15**). The exact route and mechanism used by ETC complexes to pump protons is still being worked out. In some cases, it is not clear how the complex uses redox reactions to transport protons.

The best-understood connection between electron transport and proton transport takes place in complex III. Research has shown that when Q accepts electrons from complex I or complex II, it picks up protons from the matrix side of the inner membrane. The reduced form of Q then diffuses through the inner membrane, where its electrons are used to reduce a component of complex III near the intermembrane space. The protons held by Q are then released to the intermembrane space.

In this way, through redox reactions alone, Q shuttles protons from one side of the membrane to the other. The transported protons contribute to an electrochemical gradient.

Once the nature of the electron transport chain became clear, biologists understood the fate of the electrons and the energy carried by NADH and $FADH_2$. Much of the chemical energy that was originally present in glucose is now accounted for in the proton electrochemical gradient. This is satisfying, except for a key question: Is the proton gradient used to make ATP?

The Discovery of ATP Synthase

In 1960 Efraim Racker and associates made several key observations about the structure and function of the inner mitochondrial membrane. When they used high-frequency vibration to disrupt mitochondrial membranes (see **BioSkills 7**), Racker's group noticed that the inner membrane often formed vesicles inside out—the side that normally faced the mitochondrial matrix faced the outside of the vesicles instead of the inside. Electron microscopy revealed that the inside-out membranes had many large proteins studded along their surfaces. Each protein appeared to have a base in the membrane, from which a lollipop-shaped stalk and knob projected (**Figure 9.16**).

Racker's group seized on this technique to isolate the inside-out vesicles and do experiments with them. For example, they

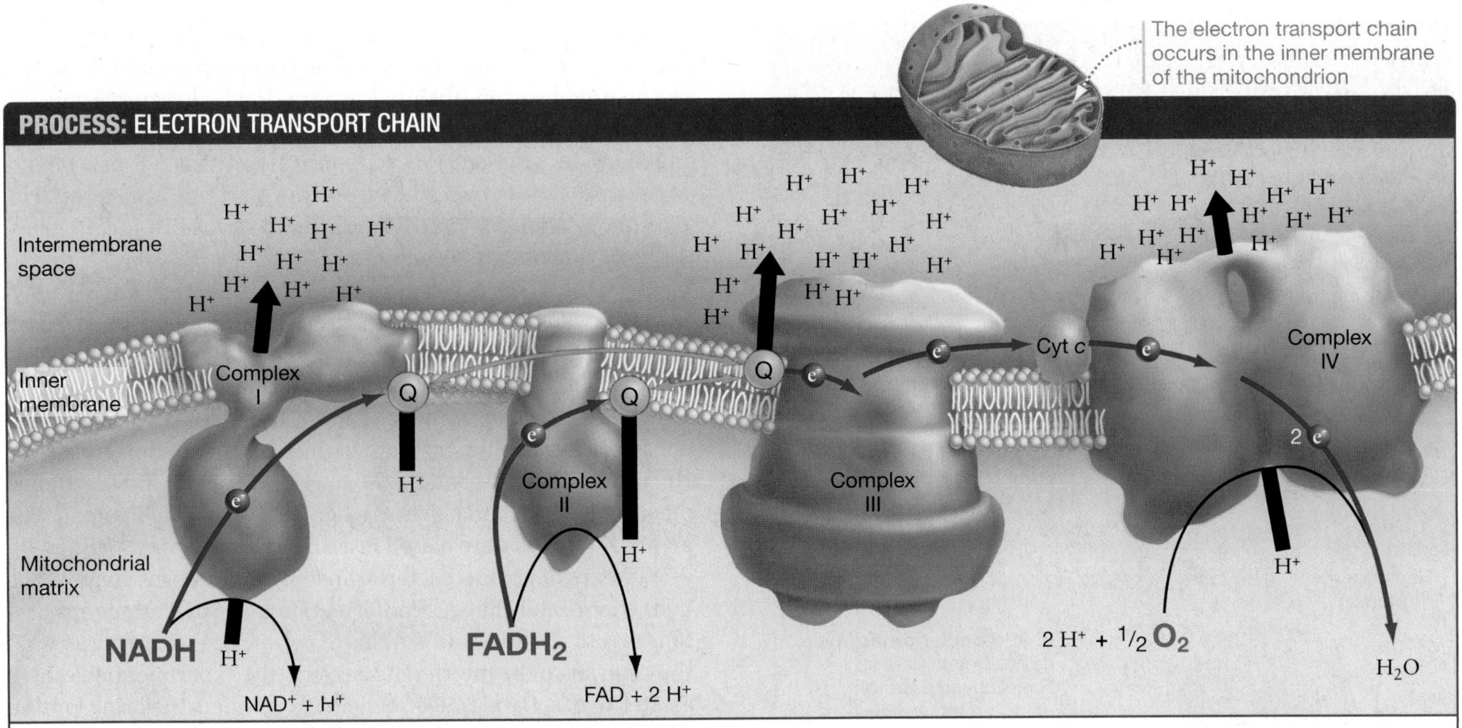

	Complex I	Complex II	Complex III	Complex IV
What goes in:				
What comes out:				

Figure 9.15 How Does the Electron Transport Chain Work? The individual components of the electron transport chain that were diagrammed abstractly in Figure 9.14 are found in the inner membrane of mitochondria. Electrons are carried from one complex to another by Q or by cytochrome *c*; Q also shuttles protons across the membrane (black half-arrows). The orange arrows indicate Q moving back and forth between complexes I, II, and III. Complexes I and IV use redox reactions to pump protons directly from the mitochondrial matrix to the intermembrane space.

✔ Draw an arrow across the membrane from low to high proton concentration and label it "Proton gradient." In the boxes at the bottom, list "What goes in" and "What comes out" for each complex.

found that the stalks and knobs on these vesicles could both synthesize ATP and also hydrolyze it to form ADP and inorganic phosphate. If the vesicles were further vibrated or treated with a compound called urea, the stalks and knobs fell off. Vesicles that contained just the base component, without the stalks and knobs, could not make ATP. The base components were, however, capable of transporting protons across the membrane.

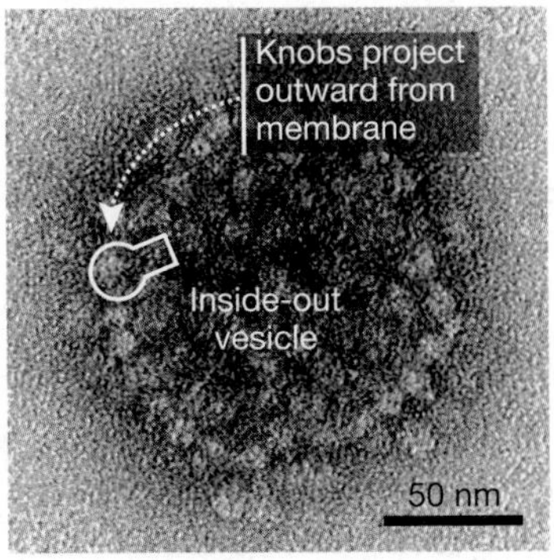

Figure 9.16 The Discovery of ATP Synthase. Vesicles formed from disrupted inner mitochondrial membranes contain lollipop-shaped stalk-and-knob structures extending from their surface. These structures are part of the ATP synthase proteins and normally face inward, toward the mitochondrial matrix.

Based on these observations, Racker proposed that the stalks and knobs were the component of the protein responsible for ATP hydrolysis and synthesis. To test his idea, the researchers added the stalks and knobs back to vesicles that had been stripped of them and confirmed that the vesicles regained the ability to synthesize ATP. The entire protein complex is now known as **ATP synthase**. Follow-up work also confirmed the hypothesis that the membrane-bound base component of the complex is a proton channel. Is there a connection between proton transport and ATP synthesis?

The Chemiosmosis Hypothesis

In 1961 Peter Mitchell broke with prevailing ideas that electron transport produces ATP via substrate phosphorylation. Instead, he proposed something completely new—an indirect connection between electron transport and ATP production. Mitchell proposed that the only job of the electron transport chain is to pump protons across the inner membrane of mitochondria from the matrix to the intermembrane space. After a proton gradient is established, an enzyme in the inner membrane, like Racker's ATP synthase, would synthesize ATP from ADP and P_i.

RESEARCH

QUESTION: How are the electron transport chain and ATP production linked?

CHEMIOSMOTIC HYPOTHESIS: The linkage is indirect. The ETC creates a proton gradient and ATP synthase uses the gradient to synthesize ATP.

ALTERNATIVE HYPOTHESIS: The linkage is direct. Specific ETC proteins are required for ATP synthesis by ATP synthase.

EXPERIMENTAL SETUP:

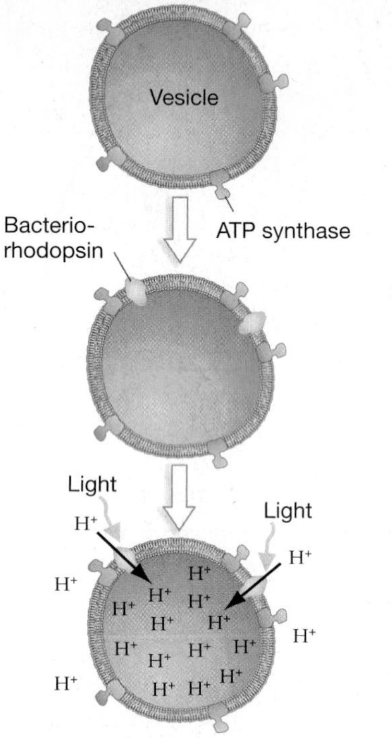

1. Produce vesicles from artificial membranes; add ATP synthase, an enzyme found in mitochondria.

2. Add bacteriorhodopsin, a protein that acts as a light-activated proton pump, to the vesicles.

3. Illuminate vesicle so that bacteriorhodopsin pumps protons into the vesicle, creating a proton gradient.

4. Determine if ATP is produced after adding ADP and inorganic phosphate (P_i) to the outside solution.

PREDICTION OF CHEMIOSMOTIC HYPOTHESIS: ATP will be produced outside the vesicle when bacteriorhodopsin forms a proton gradient.

PREDICTION OF ALTERNATIVE HYPOTHESIS: No ATP will be produced in the presence of a proton gradient without components of the ETC.

RESULTS:

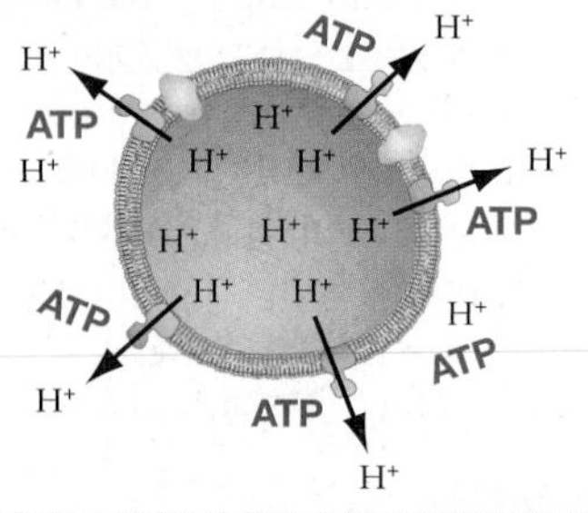

ATP is produced outside the vesicle using a proton gradient alone, in the absence of the electron transport chain.

CONCLUSION: The linkage between electron transport and ATP production by ATP synthase is indirect; the synthesis of ATP only requires a proton gradient.

Figure 9.17 Evidence for the Chemiosmotic Hypothesis.

SOURCE: E. Racker and W. Stoeckenius. 1974. Reconstitution of purple membrane vesicles catalyzing light-driven proton uptake and adenosine triphosphate formation. *Journal of Biological Chemistry* 249: 662–663.

✔ **PROCESS OF SCIENCE** Propose an alternative approach that could be used to generate a proton gradient for this study in the absence of bacteriorhodopsin.

Mitchell introduced the term **chemiosmosis** to describe the use of a proton gradient to drive energy-requiring processes, like the production of ATP. Similar to osmosis, chemiosmosis involves diffusion across a membrane, but in this case, protons are diffusing along its gradient rather than water. Although proponents of a direct link between electron transport and substrate-level phosphorylation objected vigorously to Mitchell's idea, several key experiments supported it.

Figure 9.17 illustrates how one key element of Mitchell's hypothesis was confirmed—that a proton gradient alone can be used to synthesize ATP via ATP synthase. The Racker team made vesicles from artificial membranes that contained the ATP synthase they had isolated from mitochondria. To generate a proton gradient across the membrane, they also included membrane fragments from bacteria that contained bacteriorhodopsin, a well-studied membrane protein that acts as a light-activated proton pump. The artificial vesicles were placed in a solution containing ADP and P_i.

When light strikes bacteriorhodopsin, it absorbs some of the light energy and changes conformation in a way that pumps protons across a membrane—in this experiment, pumping was from the exterior to the interior. As a result, the experimental vesicles established a strong electrochemical gradient favoring proton movement to the exterior. When the vesicles were illuminated to initiate proton pumping, ATP began to be produced from ADP and P_i present outside the vesicles.

Mitchell's prediction was correct: In this situation, ATP production depended solely on the existence of a **proton-motive force**, which is based on a proton electrochemical gradient across a membrane. It could occur in the *absence* of an electron transport chain. This result, along with many others, has provided strong support for the hypothesis of chemiosmosis.

✔ **If you understand chemiosmosis, you should be able to explain the relationships among glucose oxidation, the proton gradient, and ATP synthase.**

Organisms throughout the tree of life use electron transport chains and ATP synthases. These systems are humming away in your cells now and produce most of the ATP that keeps you alive. Let's look in more detail at how they function.

The Proton-Motive Force Couples Electron Transport to ATP Synthesis As Figure 9.18 shows, the structure of ATP synthase is now well understood. The ATP synthase "knob" component is called the F_1 unit; the membrane-anchored, proton-transporting base component is the F_o unit. The F_1 and F_o units are connected by a shaft, as well as by a stator, which holds the two units in place.

ATP synthase is a molecular rotary motor. The F_o unit serves as a rotor whose turning is conveyed to the F_1 unit via the shaft. A flow of protons through the F_o unit causes the rotor and shaft to spin. The motor's rotation can reach speeds of 350 revolutions per second. To put this into perspective, if the wheels on a typical car were spinning at this rate, the car would be traveling 1,500 miles per hour.

As the shaft spins within the F_1 unit, it is thought to change the conformation of the F_1 subunits in a way that catalyzes the phosphorylation of ADP to ATP. This mode of chemosmotically driven ATP production was introduced in Section 9.1 as *oxidative phosphorylation*—a process powered indirectly by the electron transport chain and executed by ATP synthase.

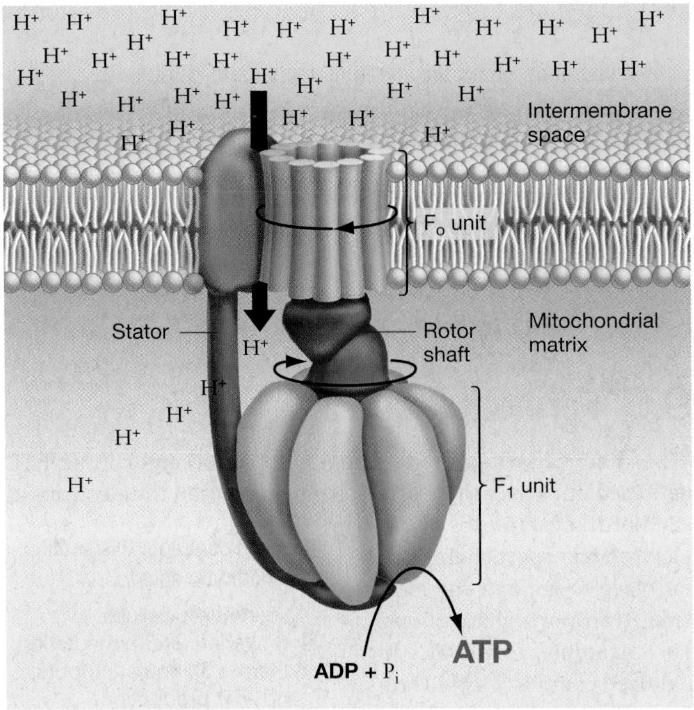

Figure 9.18 ATP Synthase Includes a Proton-Driven Rotor and an ATP-Generating Enzyme. ATP synthase has two major components, designated F_o and F_1, connected by a shaft. The F_o unit spins as protons pass through. The shaft transmits the rotation to the F_1 unit, causing it to make ATP from ADP and P_i.

Oxidative phosphorylation resembles the process of generating electricity in a hydroelectric dam like the one pictured on the first page of this chapter. The ETC pumps protons across the inner membrane, similar to the way a series of gigantic pumps force water up and behind a dam. When protons pass through the ATP synthase, it spins and releases energy used to synthesize ATP. This is analogous to how water passing through the turbines of hydroelectric dams causes them to spin and generate electricity.

The key idea to note here is that the energy to produce ATP in oxidative phosphorylation comes from an established proton gradient, not phosphorylated substrates as used in substrate-level phosphorylation.

It has been determined that the ETC transports enough protons to produce approximately three ATP for each NADH and two for each $FADH_2$, depending on the type of ATP synthase used. These yields, however, are not observed in cells, since the proton-motive force is also used to drive other activities, such as the import of phosphates into the mitochondrial matrix.

In some cells, such as those in brown adipose tissue of mammals, the ATP yield is even lower due to the presence of special channel proteins—called uncoupling proteins—that allow protons to freely pass through the inner membrane of mitochondria. In these cells, much of the energy stored in the proton gradient is used to produce heat instead of ATP. Brown adipose tissue is particularly important for maintaining the body temperature in neonates.

Figure 9.19 summarizes cellular respiration by tracing the fate of the carbon atoms and electrons that originated in glucose. Notice that the electrons harvested from glucose are first carried by NADH and $FADH_2$, and then passed through the ETC where they reduce oxygen to form water. The proton pumping activity of the ETC creates the proton-motive force that drives ATP synthesis.

The diagram also indicates the approximate yield of ATP from each stage of the pathway. Recent research shows that about 29 ATP molecules are produced from each molecule of

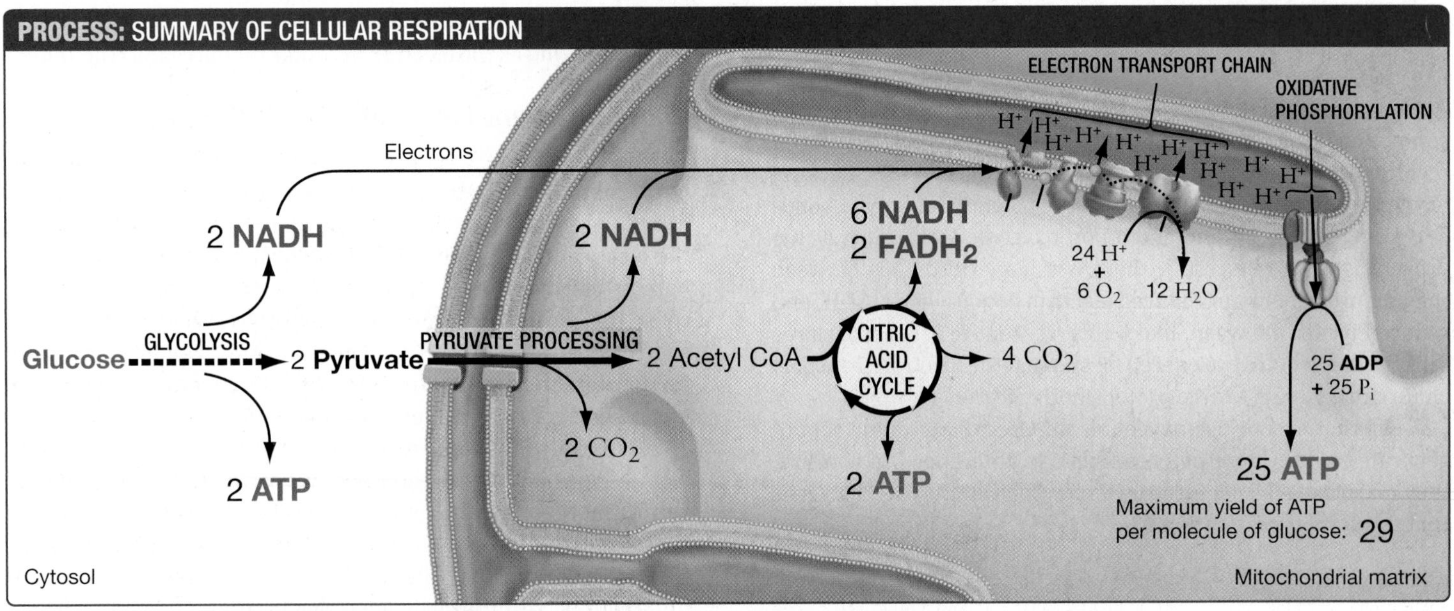

Figure 9.19 ATP Yield during Cellular Respiration. The actual yield of ATP per glucose (29 ATP) is lower than the theoretical calculation (38 ATP) because the proton motive force is used to drive other mitochondrial activities, such as the active transport of P_i into the mitochondrial matrix.

glucose.[3] Of these, 25 ATP molecules are produced by ATP synthase. What is the fundamental message here? The vast majority of the "payoff" from the oxidation of glucose occurs via oxidative phosphorylation.

The chemical equation that represents the overall process involved in cellular respiration is

$$C_6H_{12}O_6 + 6H_2O + 6O_2 + 29ADP + 29P_i \longrightarrow 6CO_2 + 12H_2O + 29ATP$$

The reactants include six water molecules that are used in glycolysis and the citric acid cycle (some of these are depicted in Figure 9.10). For simplicity, the equation does not include the NADH and $FADH_2$ electron carriers.

Organisms Use a Diversity of Electron Acceptors

During cellular respiration, oxygen is the electron acceptor used by all eukaryotes and a wide diversity of prokaryotes. Species that depend on oxygen as an electron acceptor for the ETC use **aerobic** respiration and are called aerobic organisms. (The Latin root *aero* means "air.")

Variations of Cellular Respiration It is important to recognize, though, that cellular respiration can also occur without oxygen. Many thousands of bacterial and archaeal species rely on electron acceptors other than oxygen, and electron donors other than glucose. For example, nitrate (NO_3^-) and sulfate (SO_4^{2-}) are particularly common electron acceptors in species that live in oxygen-poor environments (Ch. 26, Section 26.3). In addition, many bacteria and archaea use H_2, H_2S, CH_4, or other inorganic compounds as electron donors—not glucose.

Cells that depend on electron transport chains with electron acceptors other than oxygen are said to use **anaerobic** ("no air") respiration. Even though the starting and ending points of cellular respiration differ, aerobic and anaerobic respiring cells still use an ETC to create a proton-motive force that drives the synthesis of ATP. In bacteria and archaea, the ETC and ATP synthase are located in the plasma membrane.

Aerobic Respiration Is Most Efficient Even though an array of compounds can serve as the final electron acceptor in cellular respiration, oxygen provides the greatest energy yield. Because oxygen is so highly electronegative, the potential energy in bonds between an oxygen atom and a non-oxygenic atom, such as hydrogen, is low. As a result, there is a large difference between the potential energy of reduced electron donors, like NADH, and reduced forms of oxygen, like water (see Figure 9.14). This large differential in potential energy means that the electron transport chain can generate a large proton-motive force.

Cells that do not use oxygen as an electron acceptor cannot generate such a large potential energy difference. As a result, they make less ATP per electron donor, such as glucose, than cells that use aerobic respiration.

[3] Traditionally, biologists thought that up to 38 ATP would be synthesized for every molecule of glucose oxidized in cells. More recent work has shown that actual yield is only about 29 ATP (see P. R. Rich, The molecular machinery of Keilin's respiratory chain. *Biochemical Society Transactions* 6 [2003]: 1095–1105). Also, it's important to note that yield varies with conditions in the cell.

CHECK YOUR UNDERSTANDING

✔ **If you understood this section, you should be able to ...**

1. Explain the relationship between the structure of ATP synthase and its function. What are the different roles of the F_1 and F_o units of ATP synthase?
2. Differentiate the electron transport chains in aerobic versus anaerobic respiration. How does this difference affect the amount of ATP produced by oxidative phosphorylation?

Answers are available in Appendix A.

9.6 Fermentation

After you complete this section, you should be able to ...

- Differentiate cellular respiration and fermentation in terms of inputs, outputs, and ATP production.

What happens when the terminal electron acceptor in an ETC gets used up? When there is no terminal acceptor, the electrons in each of the complexes of the electron transport chain have no place to go, and the electron transport chain stops. For example, without an oxidized complex I, NADH remains reduced. The concentration of NAD^+ drops rapidly as cells continue to convert NAD^+ to NADH.

This situation is life threatening. When there is no longer any NAD^+ to drive glycolysis, pyruvate processing, and the citric acid cycle, then no ATP can be produced. If NAD^+ cannot be regenerated somehow, the cell will die. How do cells cope?

Fermentation is a metabolic pathway that includes glycolysis and an additional set of reactions that oxidize stockpiles of NADH to regenerate NAD^+. In respiring cells, fermentation serves as an emergency backup so glycolysis can continue to produce ATP even when the ETC and oxidative phosphorylation is shut down (Figure 9.20).

How does fermentation regenerate the NAD^+, and what molecule is serving as the electron acceptor in this redox reaction?

Many Different Fermentation Pathways Exist

When you run up a long flight of stairs, the cells of your muscles begin metabolizing glucose so fast that the supply of oxygen is rapidly used up by their mitochondria. When oxygen runs out, the electron transport chains shut down and NADH cannot donate its electrons there. When fermentation takes place in your cells, the pyruvate produced by glycolysis then begins to accept electrons from NADH. This pathway, called **lactic acid fermentation**, regenerates NAD^+ by reducing pyruvate to form lactate—a deprotonated form of lactic acid (Figure 9.21a).

When your muscles are deprived of oxygen, your body reacts by making you breathe faster and increasing your heart rate. By getting more oxygen to your muscle cells, the electron transport chain is revived. The lactic acid produced by fermentation can be converted back to pyruvate and used as a source of energy to continue cellular respiration when oxygen is present.

In many cases, however, the cell cannot use the molecule that is formed when pyruvate (or another electron acceptor) accepts electrons from NADH. This by-product may even be toxic and

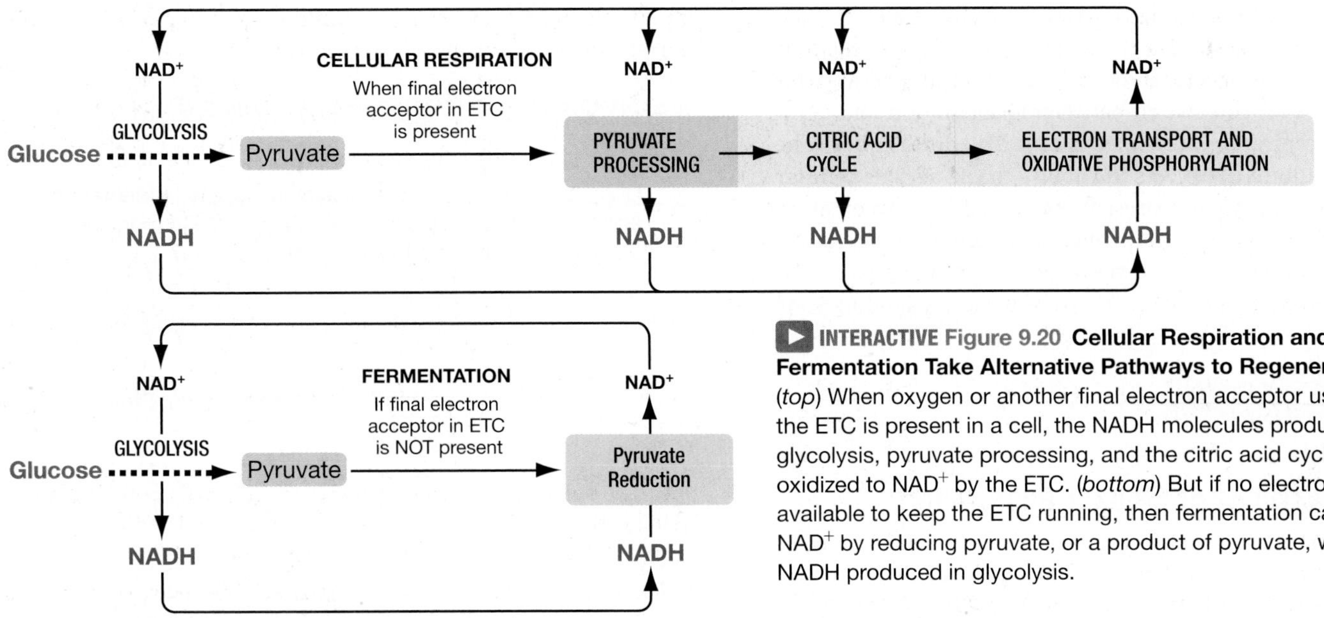

INTERACTIVE Figure 9.20 Cellular Respiration and Fermentation Take Alternative Pathways to Regenerate NAD$^+$. (*top*) When oxygen or another final electron acceptor used by the ETC is present in a cell, the NADH molecules produced by glycolysis, pyruvate processing, and the citric acid cycle are oxidized to NAD$^+$ by the ETC. (*bottom*) But if no electron acceptor is available to keep the ETC running, then fermentation can regenerate NAD$^+$ by reducing pyruvate, or a product of pyruvate, with the NADH produced in glycolysis.

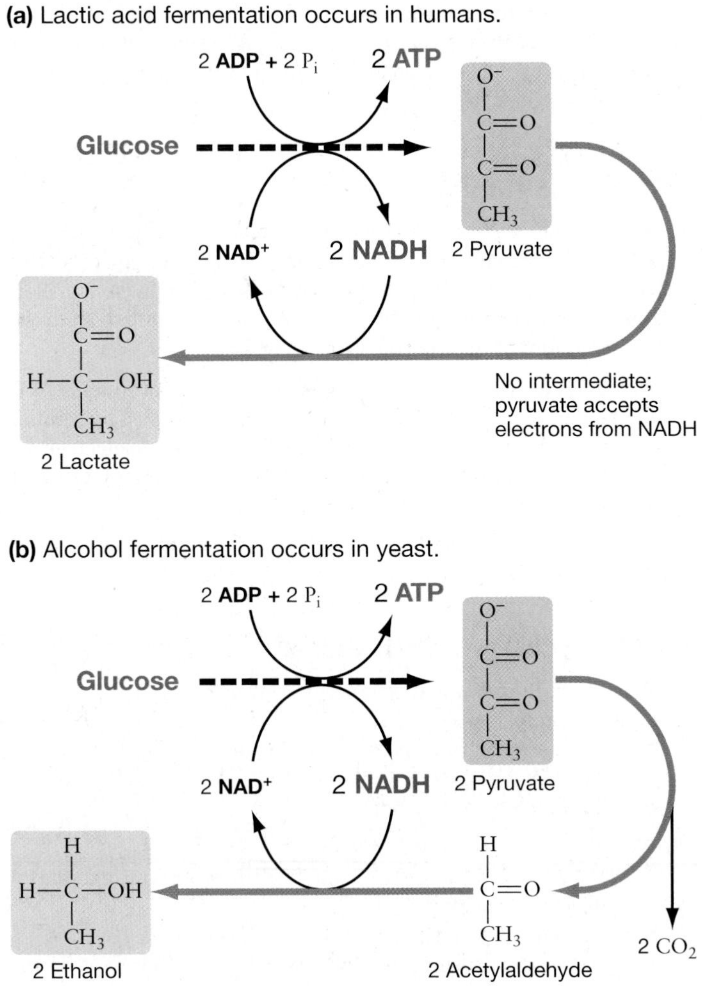

Figure 9.21 Fermentation Regenerates NAD$^+$ So that Glycolysis Can Continue. Shown here are just two examples of the many types of fermentation that occur in prokaryotes and eukaryotes.

excreted from the cell as waste even though it has not been fully oxidized.

Figure 9.21b illustrates **alcohol fermentation**, which occurs in the eukaryote *Saccharomyces cerevisiae*, strains of which are used to make baker's and brewer's yeast. When yeast cells grow in bread dough or a bottle of grape juice, they quickly use up the available oxygen. Instead of using NADH to reduce pyruvate, yeast first convert pyruvate to the two-carbon compound acetaldehyde. This reaction gives off carbon dioxide, which causes bread to rise and produces the bubbles in champagne and beer.

Acetaldehyde then accepts electrons from NADH, forming the NAD$^+$ required to keep glycolysis going. The addition of electrons to acetaldehyde forms ethanol as a waste product. The yeast cells excrete ethanol as waste. In essence, the active ingredient in alcoholic beverages is like yeast urine.

Cells that employ other types of fermentation are used commercially in the production of soy sauce, tofu, yogurt, cheese, vinegar, and other products. The products of these reactions are responsible for many of the complex flavors in these foods.

Bacteria and archaea that rely exclusively on fermentation are called obligate anaerobes. These organisms are present in phenomenal numbers in your intestines and in the first compartment of a cow's stomach, called the rumen. The rumen is a specialized digestive organ that contains over 10^{10} (10 billion) bacterial and archaeal cells per *milliliter* of fluid. The fermentations that occur in these cells produce an array of high-energy products, like fatty acids. Cattle don't actually live off grass directly—they eat it to feed their bacteria and archaea and then use the fermentation by-products for energy.

Fermentation as an Alternative to Cellular Respiration

Even though fermentation is a widespread type of metabolism, it is extremely inefficient compared with cellular respiration. Fermentation produces just 2 molecules of ATP per molecule of glucose metabolized, while aerobic cellular respiration produces about 29—almost 15 times more ATP per glucose molecule than fermentation. The reason for the disparity is that the reactions that follow glycolysis in fermentation are not used to generate ATP.

Organisms that can switch between fermentation and aerobic cellular respiration are called **facultative anaerobes**. The adjective "facultative" reflects the ability to use aerobic cellular respiration when oxygen is present and anaerobic fermentation when it is absent. Many of your cells can function as facultative anaerobes to a certain extent; however, you cannot survive for long without oxygen. To make this point clear, try holding your breath—it should take only a minute for you to realize how important electron transport is to your cells.

CHECK YOUR UNDERSTANDING

✔ If you understood this section, you should be able to ...

Compare and contrast the roles of lactic acid fermentation and the electron transport chain. Why would fermentation alone not be sufficient to keep your cells alive?

Answers are available in Appendix A.

CHAPTER 9 Review

For media, go to **Mastering Biology**

9.1 An Overview of Cellular Respiration

- Cellular respiration is based on redox reactions that oxidize a compound with high potential energy, such as glucose, and produce molecules with low potential energy, such as CO_2 and water.
- In eukaryotes, cellular respiration consists of four stages: (1) glycolysis, (2) pyruvate processing, (3) the citric acid cycle, and (4) electron transport coupled to oxidative phosphorylation.
- Glycolysis, pyruvate processing, and the citric acid cycle are central to the metabolism of most cells. Other catabolic pathways feed into them, and the intermediates of these components of cellular respiration are used in the synthesis of many key molecules.

9.2 Glycolysis: Oxidizing Glucose to Pyruvate

- The glycolytic pathway is a 10-step reaction sequence in which glucose is broken down into two molecules of pyruvate. It takes place in the cytosol and produces ATP and NADH.
- Glycolysis slows when ATP binds to a regulatory site in phosphofructokinase.

9.3 Processing Pyruvate to Acetyl CoA

- Pyruvate processing is a series of reactions that convert pyruvate to acetyl CoA in the mitochondrial matrix in eukaryotes or the cytosol of prokaryotes. NADH and CO_2 are produced.
- The pyruvate dehydrogenase complex is inhibited when it is phosphorylated by ATP. It speeds up in the presence of reactants and slows down in the presence of products.

9.4 The Citric Acid Cycle: Oxidizing Acetyl CoA to CO_2

- The citric acid cycle is an eight-step reaction cycle in the matrix of mitochondria or cytosol of prokaryotes. It begins with acetyl CoA and produces $FADH_2$, NADH, and ATP or GTP. By the end of the citric acid cycle, all of the carbons from glucose are completely oxidized to CO_2.
- Certain enzymes in the citric acid cycle are inhibited when NADH or ATP binds to them.

9.5 Electron Transport and Chemiosmosis: Building a Proton Gradient to Produce ATP

- The electron transport chain resides in the inner membrane of mitochondria or plasma membrane of prokaryotes. It consists of a series of electron acceptors that vary in their redox potential starting with the oxidation of NADH and $FADH_2$ and ending with the reduction of a terminal electron acceptor, like O_2.
- The change in energy that accompanies the redox reactions in the electron transport chain is used to transport protons across the inner mitochondrial membrane, creating an electrochemical gradient.
- ATP production is coupled to the ETC by oxidative phosphorylation. The potential energy stored in the proton gradient built up by the ETC is used to spin components of the ATP synthase to produce ATP. This process is responsible for most of the ATP made by cellular respiration.

9.6 Fermentation

- Fermentation occurs in the cytosol of many cells and consists of glycolysis followed by reactions that regenerate NAD^+ from NADH. The oxidation of NADH via fermentation is required when an electron transport chain is not present or it is inactive due to an insufficient amount of the final electron acceptor.
- Production of NAD^+ enables glycolysis to continue producing ATP, albeit significantly less ATP per glucose than produced by cellular respiration. Depending on the molecule that acts as an electron acceptor, fermentation pathways produce lactate, ethanol, or other reduced organic compounds as a by-product.

Answers are available in Appendix A.

✔ TEST YOUR KNOWLEDGE

1. Where does the citric acid cycle occur in eukaryotes?
   a. in the cytosol of cells
   b. in the intermembrane space of mitochondria
   c. in the inner membrane of mitochondria
   d. in the matrix of mitochondria
2. What does the chemiosmotic hypothesis claim?
   a. ATP is generated using phosphates taken from intermediates in the electron transport chain.
   b. ATP is generated using a phosphate gradient produced by glycolysis and the citric acid cycle.
   c. ATP is generated using a proton-motive force that is produced by the electron transport chain.
   d. Water is generated using electrons taken from NADH and $FADH_2$ and transported through the electron transport chain.

3. After glucose is fully oxidized by glycolysis, pyruvate processing, and the citric acid cycle, where is most of its energy stored?

4. Which of the following correctly describe the fermentation pathway? Select True or False for each statement.
   T/F It includes a reaction that oxidizes NADH to NAD^+.
   T/F It synthesizes ATP by substrate-level phosphorylation.
   T/F It includes a reaction that reduces NAD^+ to NADH.
   T/F It synthesizes electron acceptors, so that cellular respiration can continue.

✔ TEST YOUR UNDERSTANDING

5. Compare and contrast substrate-level phosphorylation and oxidative phosphorylation.

6. If you were to expose cells that are undergoing aerobic respiration to a radioactive oxygen isotope in the form of O_2, which of the following molecules would you expect to be radiolabeled?
   a. pyruvate **b.** water **c.** NADH **d.** CO_2

7. In step 3 of the citric acid cycle, the enzyme isocitrate dehydrogenase is regulated by NADH. Compare and contrast the regulation of this enzyme with the regulation of phosphofructokinase in glycolysis.

8. Explain the relationship between electron transport and oxidative phosphorylation. How do uncoupling proteins "uncouple" this relationship in brown adipose tissue?

✔ TEST YOUR PROBLEM-SOLVING SKILLS

9. Cyanide ($C{\equiv}N^-$) blocks complex IV of the electron transport chain. Suggest a hypothesis for what happens to the ETC when complex IV stops working. Your hypothesis should explain why cyanide poisoning in humans is fatal.

10. **QUANTITATIVE** Early estimates suggested that the oxidation of glucose via aerobic respiration would produce 38 ATP. Based on what you know of the theoretical yields of ATP from cellular respiration, show how this total was determined. Why do biologists now think this amount of ATP per molecule of glucose is not achieved in cells?

✔ PUT IT ALL TOGETHER: Case Study

Does the "Fountain of Youth" spring from the mitochondrial proton gradient?

For thousands of years, explorers sought mythical waters that promote a long life. In modern times, the quest for extending the human life span continues. Current research points to changes in the mitochondrial electron transport chain and the proton gradient as a cause of aging. How is the ETC involved in aging? Can it be manipulated to increase longevity?

11. Research has shown that cellular damage associated with aging occurs via the formation of chemically reactive molecules derived from oxygen radicals called reactive oxygen species (ROS). ROS are produced when an excessive proton gradient across the inner mitochondrial membrane slows down the rate of electron transport in the ETC. Propose a hypothesis to explain how a proton gradient can reduce the rate of electron transport.

12. **QUANTITATIVE** The production of ROS can be reduced using drugs that allow protons to freely pass through the inner membrane, simulating the uncoupling of electron transport and oxidative phosphorylation that occurs in brown adipose tissue. The effect of a drug of this type, called DNP, on the life span of mice is shown in the graph below. Each point represents the death of a single mouse. At what age is the difference in survival between the DNP-treated and the untreated control the greatest?

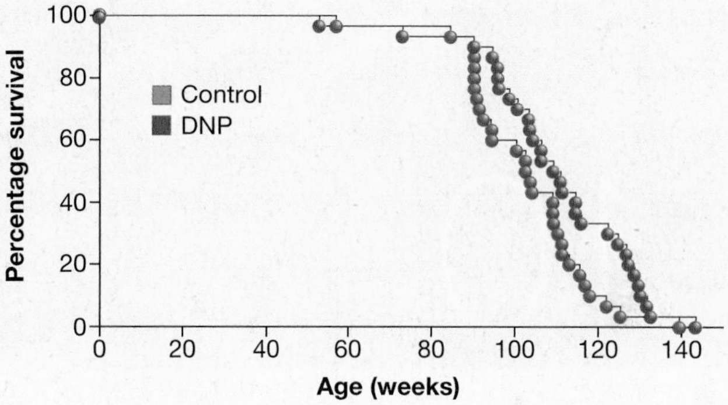

Source: C. C. Caldeira da Silva et al. 2008. *Aging Cell* 7: 552–560.

13. **QUANTITATIVE** In the above study, the investigators determined that a *low* concentration of DNP increased the average life span from 719 days (Control) to 770 days (DNP). If the U.S. population has an average life span of 79 years, then how many years would be added if the same percentage increase were observed?

14. **PROCESS OF SCIENCE** How could you determine if the proton motive force is affected in DNP-treated mice? Propose an experiment to determine if there is a correlation between life span and oxidative phosphorylation in mitochondria isolated from mice used in the experiment described in Question 12.

15. Besides an increased life span, mice treated with *low* concentrations of DNP also showed a significantly lower weight gain compared to the control group despite ingesting the same amount or type of food. Propose an explanation for why DNP would have this effect.

16. **SOCIETY** In the 1930s, DNP was introduced as a diet drug until it was banned from human use because of adverse side effects when *high* concentrations of the drug were used. These included rapid breathing, fever, and even death. Propose an explanation for the side effects based on the effect DNP has on the proton gradient.

Mastering Biology

Students Go to Mastering™ Biology for assignments, the eText, and the Study Area with animations, practice tests, and activities.

Professors Go to Mastering™ Biology for automatically graded tutorials and questions that you can assign to your students, plus Instructor Resources.

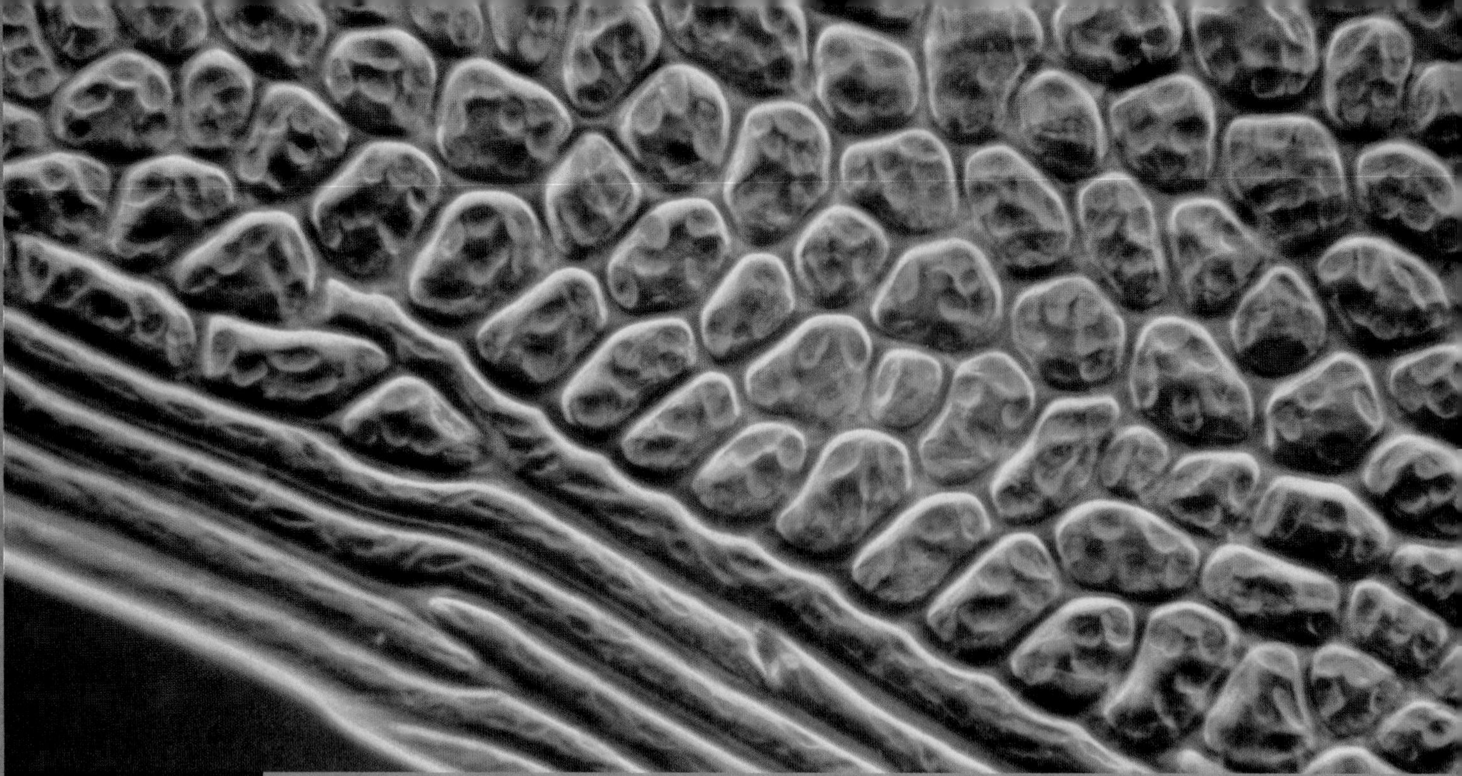

10 Photosynthesis

A close-up of moss cells filled with chloroplasts, where photosynthesis converts the energy in sunlight to chemical energy in the bonds of sugar. Directly or indirectly, most organisms on Earth get their energy from photosynthesis.

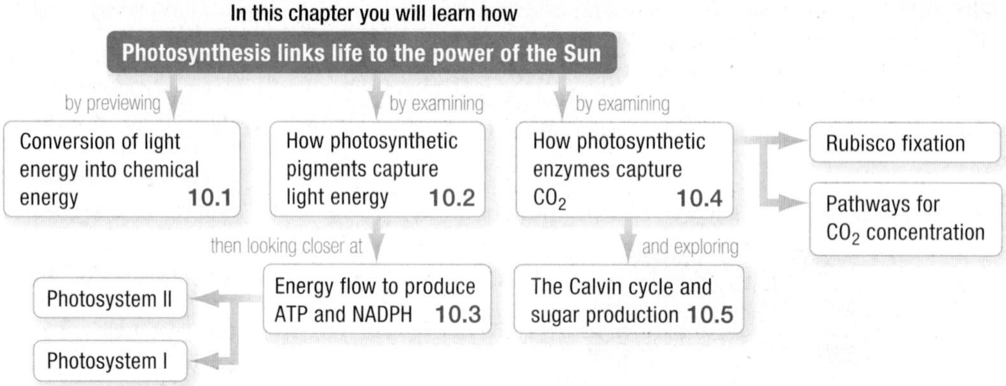

Some 3 billion years ago, a novel combination of light-absorbing molecules and enzymes gave a bacterial cell the capacity to capture energy in light and then convert it to chemical energy in C—C and C—H bonds. The origin of **photosynthesis**—the use of sunlight to manufacture carbohydrate—ranks as one of the great events in the history of life.

The vast majority of organisms alive today rely on photosynthesis, either directly or indirectly, to stay alive. Maples, mosses, and other photosynthetic organisms are termed **autotrophs** (literally, "self-feeders") because they make all their own food from ions and simple molecules. Humans, houseflies, and other non-photosynthetic organisms are called **heterotrophs** ("different-feeders") because they have to obtain the sugars and many of the other molecules they need from other organisms.

Because heterotrophs depend on the molecules produced by autotrophs, photosynthesis is fundamental to almost all life. From an ecological viewpoint, meaning how organisms interact with one another, photosynthesis is easily the most important energy-related pathway.

How does photosynthesis work? Let's begin with an overview and then delve into a step-by-step analysis of some of the most remarkable chemistry on Earth.

BIG PICTURE

This chapter is part of the Big Picture. See how on pages 236–237.

10.1 Photosynthesis Harnesses Sunlight to Make Carbohydrate from CO_2

After you complete this section, you should be able to ...

- Summarize the process of photosynthesis.

Photosynthetic organisms cannot store the electromagnetic energy of light unless it is first converted into another form. In photosynthesis, energy in sunlight is transformed to chemical energy by converting the C═O bonds in CO_2 to the C—C and C—H bonds of carbohydrate. The overall reaction—the sum of many independent reactions—can be simplified and written as

$$CO_2 + H_2O + \text{Light energy} \rightarrow (CH_2O)_n + O_2$$

In this reaction, the carbohydrate product is represented as a ratio of C, H, and O; n represents a number of three or more (Ch. 5, Section 5.1). As you will see in Section 10.5, the actual carbohydrate produced by photosynthesis is a phosphorylated three-carbon sugar.

Now read the reaction again, and note how it contrasts with the overall reaction for cellular respiration (Chapter 9):

$$(CH_2O)_6 + 6\,O_2 \rightarrow 6\,CO_2 + 6\,H_2O + \text{Energy}$$

Photosynthesis is an energy-demanding series of redox reactions that produce sugar and oxygen (O_2) from carbon dioxide (CO_2) and water (H_2O). Cellular respiration is an energy-releasing series of redox reactions that produces CO_2 and water from a sugar (glucose) and O_2.

Figure 10.1 is an incomplete electron-sharing diagram that shows the formation of a three-carbon sugar from CO_2 and H_2O. ✔ **If you understand the fundamental principles of reduction–oxidation (see Ch. 8, Section 8.2), you should be able to complete the right side of Figure 10.1 (following the instructions in the caption exercise) and then explain why the reaction requires the input of energy.**

So how does photosynthesis produce O_2 and carbohydrate? Research on this process began early in the history of biological science. In the 1770s, Joseph Priestley performed a series of experiments showing that the green parts of plants would "restore air" that had been consumed by animals or fire. This work led to the discovery of the element oxygen and the finding that plants produce O_2 in the presence of sunlight, CO_2, and water.

Early investigators assumed that CO_2 and H_2O molecules react directly to produce the $(CH_2O)_n$ found in carbohydrates and release O_2 as a by-product. This idea, however, turned out to be incorrect. Instead, CO_2 and H_2O participated in entirely different reactions, and the oxygen atoms in O_2 came from water. How was this discovered?

Photosynthesis: Two Linked Sets of Reactions

Starting in the 1930s, two independent lines of research on photosynthesis converged, leading to a major advance in biologists' understanding of how oxygen gas (O_2) and carbohydrate are produced.

The first research program, led by Cornelius van Niel, focused on photosynthesis in organisms called purple sulfur bacteria. Van Niel and his group found that these bacteria are autotrophs that manufacture their own carbohydrates from CO_2, sunlight, and hydrogen sulfide (H_2S, the equivalent of H_2O in plant reactions).

Van Niel also showed that these bacteria did not produce oxygen as a by-product of photosynthesis. Instead, they released elemental sulfur (S). In these organisms, a simplified version of the overall reaction for photosynthesis is

$$CO_2 + 2\,H_2S + \text{Light energy} \rightarrow (CH_2O)_n + H_2O + 2\,S$$

Van Niel's work was crucial for two reasons:

1. It showed that in these bacteria, H_2S and CO_2 do *not* combine directly during photosynthesis.
2. It showed that the oxygen atoms in CO_2 are *not* released as O_2. The purple sulfur bacteria produced no oxygen, even though carbon dioxide participated in the reaction—just as it did in plants.

Based on these findings, biologists hypothesized that the oxygen atoms released during plant photosynthesis must come from H_2O. The hypothesis was confirmed when heavy isotopes of oxygen—^{18}O in contrast to the normal isotope, ^{16}O—became available to researchers. They observed the ^{18}O in oxygen gas only when algae or plants were exposed to ^{18}O-labeled H_2O, not to ^{18}O-labeled CO_2.

In addition, the reactions responsible for producing O_2 occurred only in the presence of sunlight, whether or not CO_2 was present. These data suggested that there were two distinct

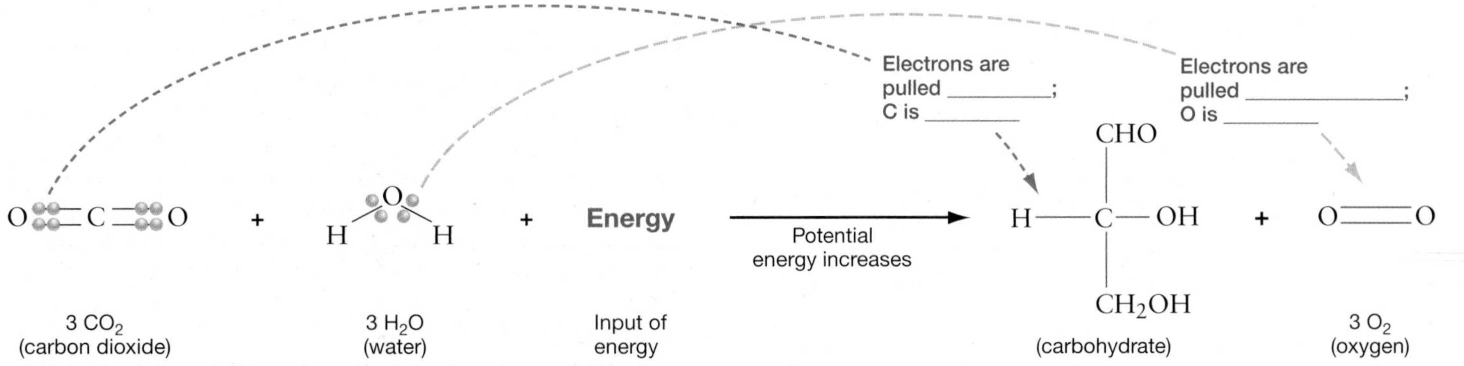

Figure 10.1 During Photosynthesis, Electron Transfer Reduces and Oxidizes Products.

✔ Fill in the electron positions for each bond in the reaction products, and complete the labels explaining which product is reduced and which is oxidized.

sets of reactions: one that uses light to produce O_2 from H_2O and one that converts CO_2 into sugars.

A second major line of research supported the idea of two sets of reactions. Between 1945 and 1955, a team led by Melvin Calvin began introducing radioactively labeled carbon dioxide ($^{14}CO_2$) to algae and identifying the molecules that subsequently became labeled with the radioisotope. These experiments allowed researchers to identify the sequence of reactions involved in reducing CO_2 to sugars. Because Calvin played an important role in this research, the reactions that reduce carbon dioxide and produce sugar came to be known as the **Calvin cycle**.

Like the reactions that produced O_2 from water, those involved in producing sugar from CO_2 occurred only in the presence of light. Later research showed that while the Calvin cycle does not directly require light, it functions only if the light-capturing reactions that produce O_2 are also occurring. The inverse is also true—the light-capturing reactions eventually stop producing O_2 if the Calvin cycle is not occurring.

What is responsible for the apparent interdependence of these two sets of reactions? Let's look at what's happening.

The light-capturing reactions and the Calvin cycle are linked by a coordinated series of redox reactions. During the light-capturing reactions, electrons are promoted to a high-energy state by light. This energy elevation ignites a chain of electron transport steps that starts with the oxidation of water to form O_2 and ends with the reduction of **$NADP^+$ (nicotinamide adenine dinucleotide phosphate)** to form **NADPH**. $NADP^+$ and NADPH are phosphorylated versions of the NAD^+ and NADH used in cellular respiration (Ch. 9, Section 9.1). Some of the energy released from the redox reactions is also used to produce ATP (**Figure 10.2**). During the Calvin cycle, the electrons in NADPH and the potential energy in ATP are used to reduce CO_2 to carbohydrate. Thus, the light-capturing reactions provide high-energy molecules that drive the Calvin cycle, which in turn regenerates the ADP, P_i, and $NADP^+$ used by the light-capturing reactions. (See **Making Models 10.1** to learn how to track the process of photosynthesis using flowcharts.)

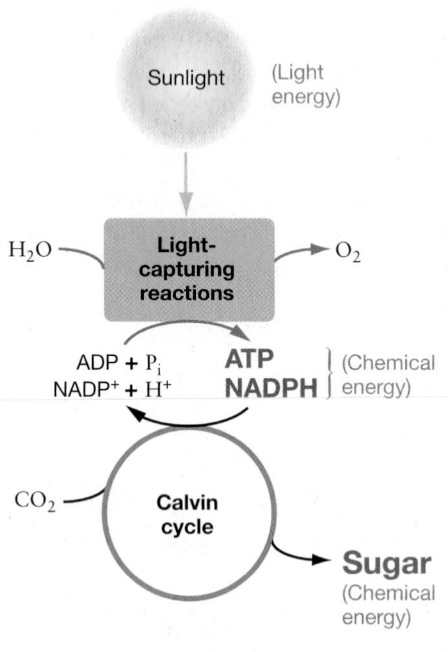

Figure 10.2 Photosynthesis Has Two Linked Components. In the light-capturing reactions of photosynthesis, light energy is transformed to chemical energy in the form of ATP and NADPH. The Calvin cycle uses the ATP and NADPH to reduce carbon dioxide to sugar and regenerates ADP, P_i, and $NADP^+$ for the light-capturing reactions.

Making Models 10.1 Tips on Drawing Flowcharts (II)

By drawing your own flowcharts, you can practice keeping track of the main events in complex processes like photosynthesis. Use boxes or ovals to represent sets of reactions, inputs, and outputs. Use arrows to represent the movement of molecules and/or energy. Be sure to label your flowchart clearly. Remember, models are flexible, so there are many correct ways to show the same idea.

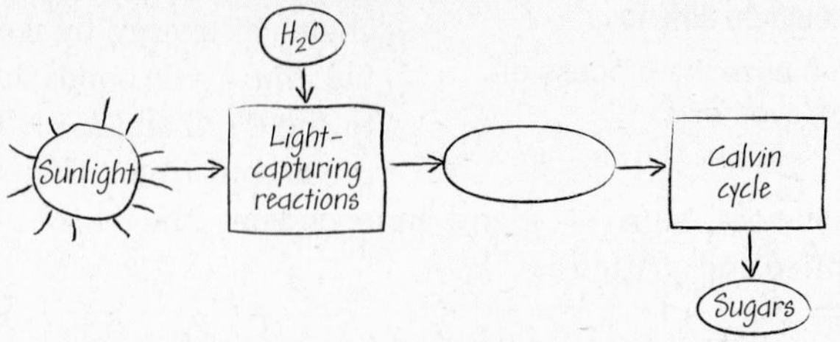

MODEL Add detail to the model by showing what links the light-capturing reactions and the Calvin cycle. Also, add ovals and arrows to show where O_2 and CO_2 fit in.

To see this model in action, go to the Study Area of **Mastering Biology**

Photosynthesis Occurs in Chloroplasts

Once later experiments established that photosynthesis takes place only in the green portions of plants, biologists focused on the bright green organelles called **chloroplasts** ("green-formed elements"). One leaf cell typically contains 40 to 50 chloroplasts, and a square millimeter of leaf tissue averages about 500,000 (**Figure 10.3**).

When membranes derived from chloroplasts were found to release oxygen after exposure to sunlight, the hypothesis that chloroplasts are the site of photosynthesis became widely accepted.

As Figure 10.3 shows, a chloroplast is enclosed by an outer membrane and an inner membrane. The interior of the organelle is dominated by flattened, membranous sac-like structures called **thylakoids**, which often occur in interconnected stacks called **grana** (singular: granum). The space inside a thylakoid is its **lumen**. (Lumen is a general term for the interior of any sac-like structure. Your stomach and intestines have a lumen.) The fluid-filled space between the thylakoids and the inner membrane is the **stroma**.

When researchers analyzed the chemical composition of thylakoid membranes, they found huge quantities of pigments. **Pigments** are molecules that absorb only certain wavelengths of light—other wavelengths are either reflected or transmitted (pass through). Pigments appear colored because people see only the wavelengths that are *not* absorbed.

The most abundant pigment in the thylakoid membranes of green plants is chlorophyll ("green-leaf"), which reflects or transmits green light. As a result, chlorophyll is responsible for the green color of plants, some algae, and many photosynthetic bacteria.

Before plunging into the details of how photosynthesis occurs, take a moment to consider just how astonishing the process is. Using some CO_2 gas, water, and sunlight, photosynthesis produces the raw materials for every organic molecule in the organism. Chemical energy in the products of the Calvin cycle may also be harvested to produce ATP. In virtually every eukaryote and many prokaryotes, these sugars are oxidized to produce ATP via aerobic cellular respiration, which is also dependent on the O_2 liberated from light-capturing reactions. If photosynthesis is not *the* most important biochemistry on Earth, it is certainly a contender.

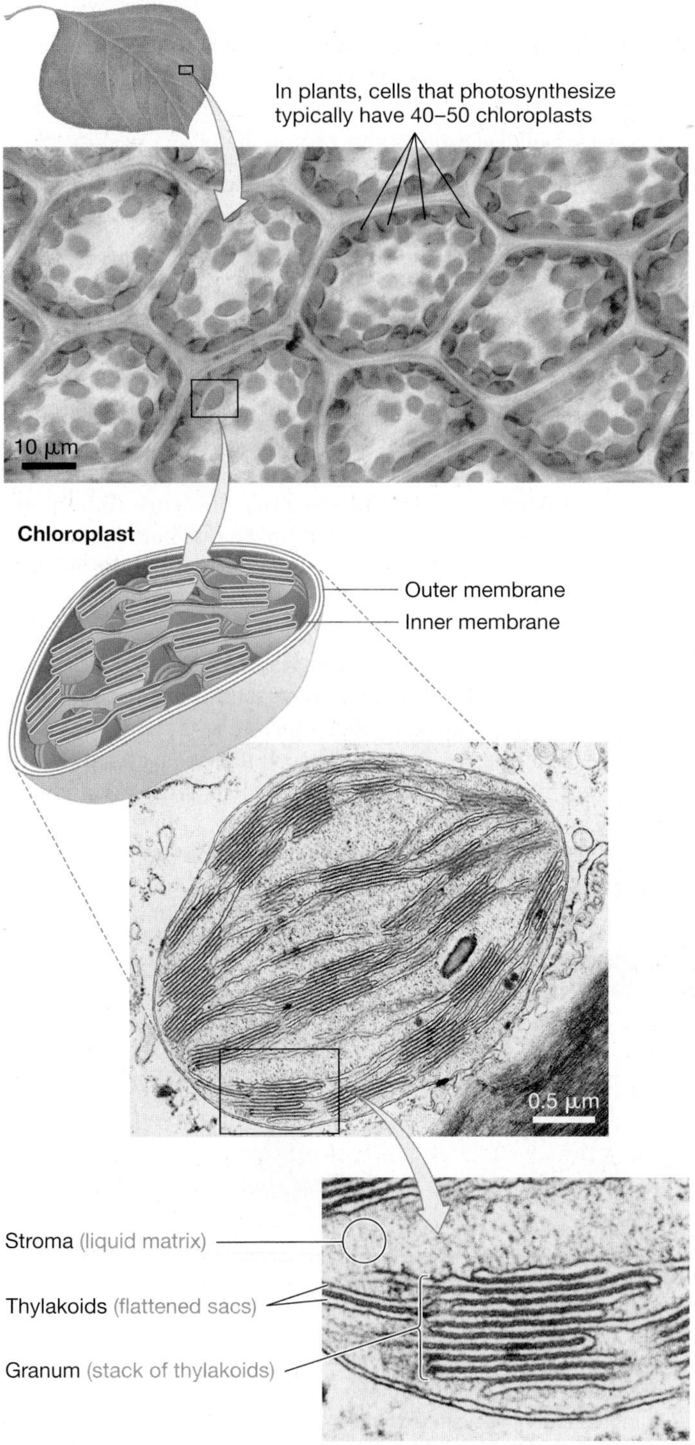

Figure 10.3 Photosynthesis Takes Place in Chloroplasts.

CHECK YOUR UNDERSTANDING

✔ **If you understood this section, you should be able to ...**

Explain the interdependence observed between the light-capturing reactions and the Calvin cycle.

Answers are available in Appendix A.

10.2 How Do Pigments Capture Light Energy?

After you complete this section, you should be able to ...

- Analyze how pigment structure and organization allow cells to acquire energy from light.

The light-capturing reactions of photosynthesis begin with the simple act of sunlight striking chlorophyll. To understand the consequences of this event, it's helpful to review the nature of light.

Light is a type of electromagnetic radiation—a form of energy. Physicists describe light's behavior as both wavelike and particle-like. Like water waves or sound waves, electromagnetic radiation is characterized by its **wavelength**—the distance between two successive wave crests or troughs. The wavelength determines the type of electromagnetic radiation.

Figure 10.4 illustrates the range of wavelengths of electromagnetic radiation—the **electromagnetic spectrum**. The

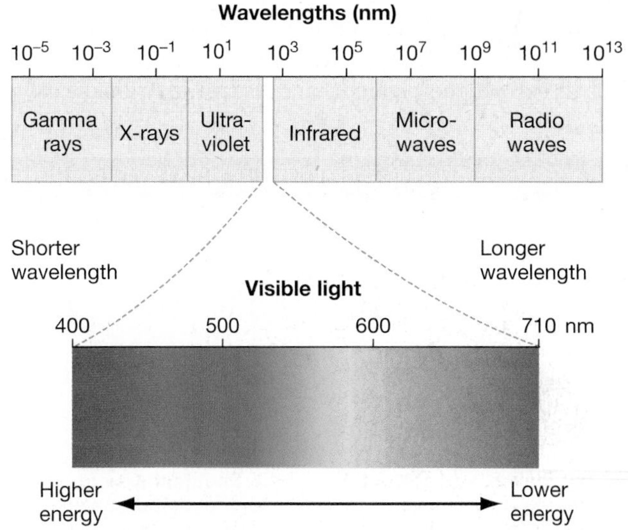

Figure 10.4 The Electromagnetic Spectrum. Electromagnetic energy radiates through space in the form of waves. The shorter the wavelength of electromagnetic radiation, the higher its energy. Humans can see radiation at wavelengths between about 400 nm to 710 nm.

electromagnetic radiation that humans can see, **visible light**, ranges in wavelength from about 400 to about 710 nanometers (nm, or 10^{-9} m). Shorter wavelengths of electromagnetic radiation contain more energy than longer wavelengths do. Thus, there is more energy in blue light than in red light.

To emphasize the particle-like nature of light, physicists point out that it exists in discrete packets called **photons**. Each photon of light has a characteristic wavelength and energy level. Pigment molecules absorb the energy of some of these photons. How?

Photosynthetic Pigments Absorb Light

Sunlight includes white light, which consists of all wavelengths in the visible portion of the electromagnetic spectrum at once. When a photon strikes an object, the photon may be absorbed, transmitted, or reflected. A pigment molecule absorbs photons of particular wavelengths.

If a pigment absorbs all the visible wavelengths, the pigment appears black because no visible wavelength of light is reflected back to your eye. If a pigment absorbs many or most of the wavelengths in the blue and green parts of the spectrum but transmits or reflects longer wavelengths, it appears red.

What wavelengths do various plant pigments absorb? In one approach to answering this question, researchers grind up leaves in a liquid that acts as a solvent to extract pigment molecules from the leaf mixture. A technique called thin layer chromatography (see BioSkills 6) separates the pigments in the extract (Figure 10.5a).

To begin, spots of raw leaf extract are placed near the bottom of a stiff support that is coated with a thin layer of silica gel, cellulose, or similar porous material. The coated support is then placed in a solvent solution. As the solvent wicks upward through the porous coating, it carries the pigment molecules in the extract with it. Because the pigment molecules vary in size, solubility, or both, they are carried at different rates.

Figure 10.5b shows a chromatograph from a grass-leaf extract. Notice that this leaf contains an array of pigments. To find out which wavelengths are absorbed by each of these molecules, researchers cut out a single region (color band) of the porous material, extract the pigment, and use an instrument called a spectrophotometer to record the wavelengths absorbed. (For more information on spectrophotometry, see BioSkills 8).

Different Pigments Absorb Different Wavelengths of Light Research based on the techniques shown in Figure 10.5 has confirmed that there are two major pigment classes in plant leaves: chlorophylls and carotenoids.

1. **Chlorophylls**, designated chlorophyll *a* and chlorophyll *b*, absorb strongly in the blue and red regions of the visible spectrum. The presence of chlorophylls makes plants look green because they mostly reflect green light, which is not absorbed well.
2. **Carotenoids** absorb wavelengths in the blue and green parts of the visible spectrum. Thus, carotenoids appear yellow, orange, or red. The carotenoids found in plants belong to two classes, called carotenes and xanthophylls.

Which of these wavelengths drive photosynthesis?

In 1882, T. W. Engelmann answered this question by studying the photosynthetic activity of a filamentous alga that was spread across a microscope slide and illuminated with a spectrum of colors (Figure 10.6). The idea was that the algal cells would begin performing photosynthesis in response to the various wavelengths of light and produce oxygen as a by-product. To determine exactly where oxygen was being produced, Engelmann added bacterial cells from a species that is attracted to oxygen.

As the drawing in the "Results" section of Figure 10.6 shows, most of the bacteria congregated in the violet-to-blue and red regions of the slide. Because wavelengths in these parts of the spectrum were associated with high oxygen concentrations, Engelmann concluded that they defined the **action spectrum** for photosynthesis—the wavelengths that drive the light-capturing reactions. Engelmann's data indicate that *violet-to-blue* and *red* light are the most effective at driving photosynthesis. Because the chlorophylls absorb these wavelengths, this early experiment showed that chlorophylls are the main photosynthetic pigments.

Using thin layer chromatography, and more advanced techniques to evaluate photosynthetic activity, biologists have

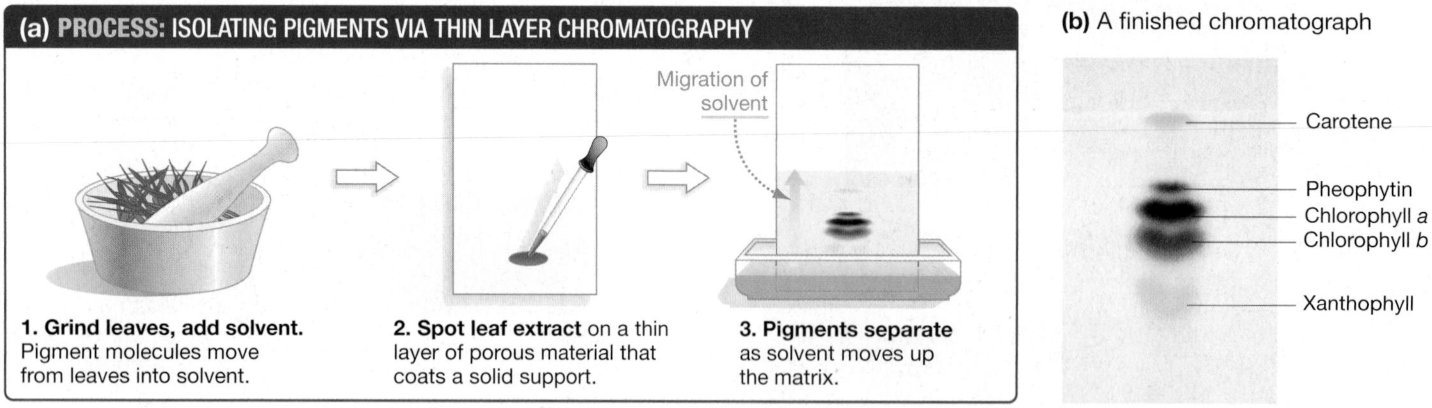

Figure 10.5 Chromatography Is a Technique for Separating Molecules. Different species of photosynthetic organisms may contain different types and quantities of pigments. The pigments in this example were extracted from grass leaves.

RESEARCH

QUESTION: Which of the wavelengths found in white light are responsible for driving photosynthesis?

HYPOTHESIS: Certain wavelengths of light are more useful for driving photosynthesis than others.

NULL HYPOTHESIS: The wavelength of light does not affect the rate of photosynthetic activity.

EXPERIMENTAL SETUP:

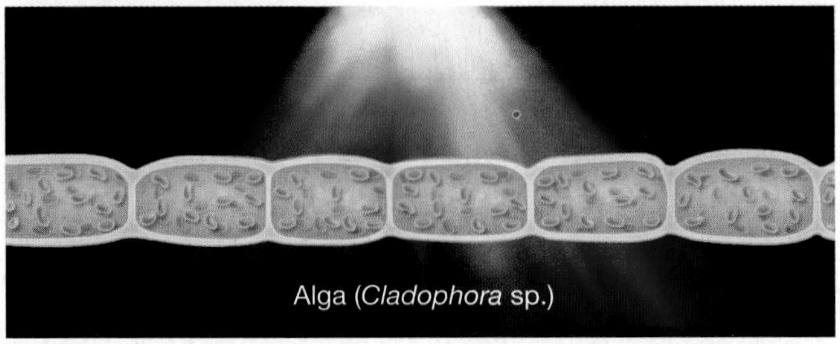

1. Expose algal cells to all the wavelengths of visible light in the presence of oxygen-seeking bacteria.

2. Record the number of bacteria swarming near algal cells exposed to different wavelengths.

PREDICTION OF HYPOTHESIS: Oxygen-seeking bacteria will only swarm around algal cells exposed to particular wavelengths of light.

PREDICTION OF NULL HYPOTHESIS: There will be no difference in where the oxygen-seeking bacteria will swarm around the algal cells.

RESULTS:

Swarming bacteria, observed and sketched by Engelmann (1882)

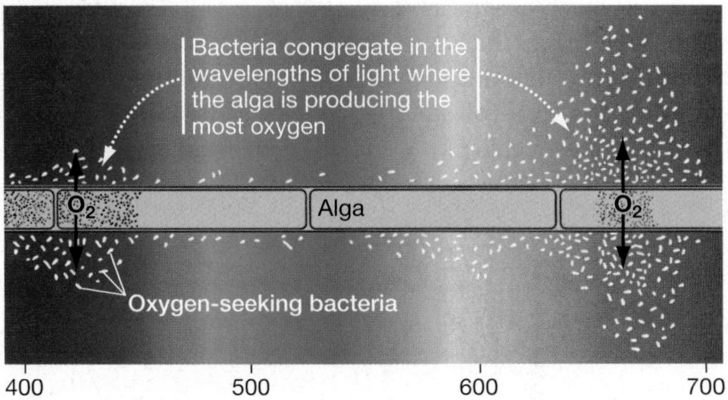

CONCLUSION: Pigments that absorb violet-to-blue and red wavelengths are most effective at triggering photosynthesis in algae.

Figure 10.6 **Certain Wavelengths of Light Are Used to Drive Photosynthesis.**

SOURCE: T. W. Engelmann. 1882. Oxygen excretion from plant cells in a microspectrum. *Botanische Zeitung* 40: 419–426.

✔ Draw what you expect the results of this experiment would look like if the pigments that drive photosynthesis in the algae were to absorb most strongly at 500 nm and 560 nm.

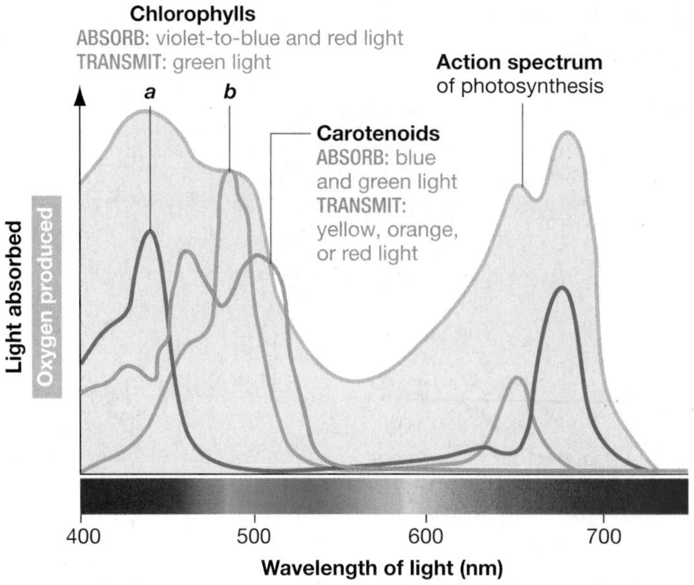

Figure 10.7 **There Is a Strong Correlation between the Absorption Spectra of Pigments and the Action Spectrum for Photosynthesis.**

DATA: G. S. Singhal et al. 1999. *Concepts in Photobiology: Photosynthesis and Photomorphogenesis*. Dordrecht: Kluwer Academic; co-published with Narosa Publishing House (New Delhi), 11–51.

produced data like those shown in **Figure 10.7**. This graph shows the action spectrum (production of O_2) of photosynthesis and the absorption spectra for three different pigments found in chloroplasts. An **absorption spectrum** measures how the wavelength of photons influences the amount of light absorbed by a pigment. In the combined graph, peaks indicate wavelengths where absorbance or photosynthetic activity is high; troughs indicate wavelengths where absorbance or photosynthetic activity is low.

Which Part of a Pigment Absorbs Light? As **Figure 10.8a** on page 220 shows, chlorophyll *a* and chlorophyll *b* are similar in structure. Both have two fundamental parts: a long isoprenoid "tail" (introduced in Ch. 6, Section 6.1) and a "head" consisting of a large ring structure with a magnesium atom in the middle. The tail interacts with proteins embedded in the thylakoid membrane; the head is where light is absorbed.

The structure of β-carotene, shown in **Figure 10.8b**, consists of an isoprenoid chain connecting two rings that are responsible for absorbing light. This pigment is what gives carrots their orange color. A xanthophyll called zeaxanthin, which gives corn kernels their bright yellow color, is nearly identical to β-carotene, except that the ring structures on either end of the molecule contain a hydroxyl (—OH) group.

Researchers had shown that chlorophylls are the main photosynthetic pigments, but carotenoids also absorb light. What do they do? Before analyzing what happens when chlorophyll pigments absorb light, let's first look at the function of the carotenoids.

What Is the Role of Carotenoids and Other Accessory Pigments? Carotenoids are called accessory pigments because they absorb light and pass the energy on to chlorophyll. Both xanthophylls

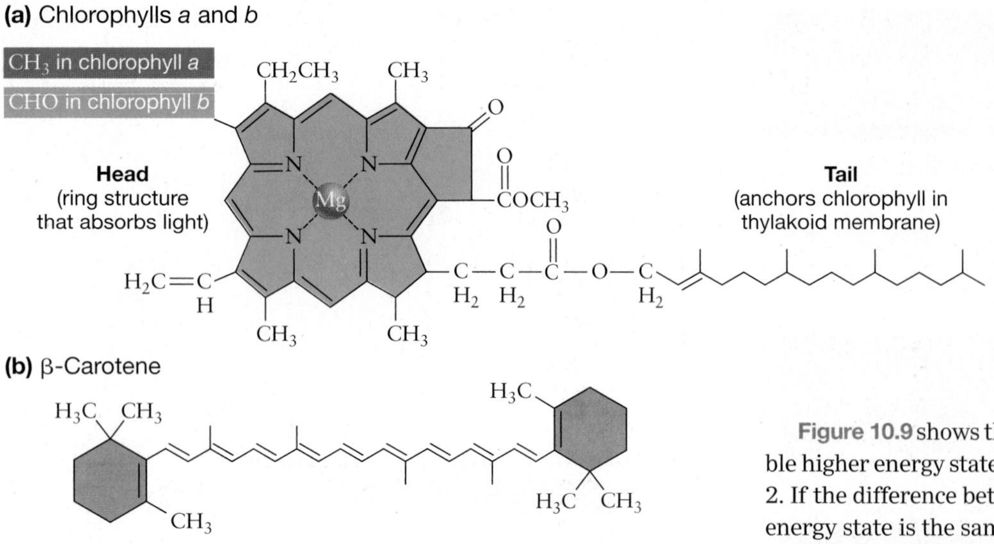

Figure 10.8 Photosynthetic Pigments Contain Ring Structures. (a) Although chlorophylls *a* and *b* are very similar structurally, they have the distinct absorption spectra shown in Figure 10.7. **(b)** β-Carotene is an orange carotenoid found in carrot roots and other plant tissues.

and carotenes are found in chloroplasts. In the autumn, when the leaves of deciduous trees start to die, their chlorophyll degrades first, leaving the carotenoids and other pigments that turn forests into spectacular displays of yellow, orange, and red. As the variation in color suggests, these accessory pigments can extend the range of wavelengths that drive photosynthesis by absorbing some of the photons not readily absorbed by chlorophyll.

Researchers discovered an even more important function for carotenoids, though, by analyzing what happens to leaves when these pigments are destroyed. Many herbicides, for example, work by inhibiting enzymes that are involved in carotenoid synthesis. Plants lacking carotenoids rapidly lose their chlorophyll, turn white, and die. Based on these results, researchers have concluded that the primary function of carotenoids is to protect the plant.

To understand why carotenoids are protective, recall that photons—especially the high-energy, short-wavelength photons in the ultraviolet part of the electromagnetic spectrum—contain enough energy to knock electrons out of atoms and create free radicals. Free radicals, in turn, trigger reactions that can disrupt and degrade molecules.

Carotenoids "quench" free radicals by accepting or stabilizing unpaired electrons. As a result, they protect chlorophyll molecules from harm. When carotenoids are absent, chlorophyll molecules are destroyed by free radicals and photosynthesis stops. Starvation and death follow.

When Light Is Absorbed, Electrons Enter an Excited State

Just what is absorption? What happens when a photon of a particular wavelength—say, red light with a wavelength of 680 nm—strikes a chlorophyll molecule? When a chlorophyll molecule absorbs a photon, the photon's energy is transferred to bonds in the chlorophyll molecule's head region. In response, an electron becomes "excited," or bumped up to a higher energy state.

The excited electron states that are possible in a particular pigment are discrete—meaning incremental rather than continuous—and can be represented as lines on an energy scale. These discrete energy levels are a property of the electron configurations in a particular pigment.

Figure 10.9 shows the ground, or unexcited, state as 0; two possible higher energy states in a chlorophyll molecule are labeled 1 and 2. If the difference between the ground state and a possible higher energy state is the same as the energy in a photon, the photon can be absorbed and an electron excited to a higher energy state.

In chlorophyll, for example, the energy difference between the ground state and state 1 is equal to the energy in a red photon, while the energy difference between state 0 and state 2 is equal to the energy in a blue photon. Thus, chlorophyll can readily absorb red photons and blue photons.

Chlorophyll does not absorb green light well, because there is no discrete step—no difference in possible energy states for its electrons—that corresponds to the amount of energy in a green photon.

Wavelengths in the ultraviolet part of the spectrum have so much energy that they may actually eject electrons from a pigment molecule and create a free radical. In contrast, wavelengths in the infrared regions have so little energy that in most cases they merely increase the movement of atoms in the pigment, generating heat rather than exciting electrons.

But if a pigment absorbs a photon with the right amount of energy in the form of electromagnetic radiation, an electron is excited. The electron now has high potential energy, but this

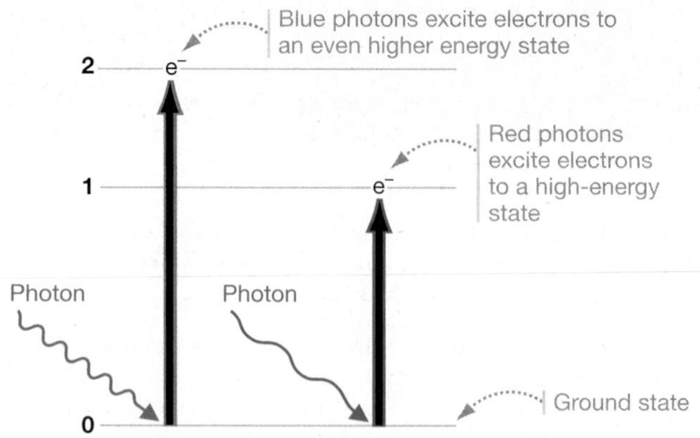

Possible energy states of electrons in chlorophyll

Figure 10.9 Electrons in Chlorophyll Are Promoted to High-Energy States When Photons Are Absorbed. The unexcited, or ground state of the electron, is labeled 0; discrete elevated energy states are labeled 1 and 2.

✔ Where you would predict the energy states to be on this diagram for a carotenoid pigment that absorbs blue and green photons? (Hint: Review Figure 10.4.)

excited state is unstable and it lasts for only a short time—on the order of picoseconds (1×10^{-12} seconds). What happens next?

If the excited electron simply falls back to its ground state, the absorbed energy is released as heat or a combination of heat and electromagnetic radiation (light). When the electron energy produces light, it is called **fluorescence**. Because some of the original photon's energy is transformed to heat, the electromagnetic radiation that is given off during fluorescence has lower energy and a longer wavelength than the original photon did.

When pigments in chloroplasts absorb photons, about 2 percent of the excited electrons produce fluorescence. The other 98 percent of the energized pigments use their excited electrons to drive photosynthesis.

Photosystems

To understand what happens to these excited electrons during photosynthesis, it's important to recognize that chlorophyll molecules work in groups—not individually. In the chloroplast thylakoid membrane, 200–300 chlorophyll molecules and accessory pigments are organized by associated proteins to form large complexes called **photosystems**. Most pigment molecules in photosystems serve as light-gathering **antenna pigments** that guide energy toward a central **reaction center**.

Antenna Pigments When antenna pigments absorb photons, the energy—but not the electron itself—is passed to a nearby pigment molecule, where another electron is excited in response. This phenomenon is known as **resonance energy transfer**.

Resonance energy transfer is possible only between pigments that are able to absorb different wavelengths of photons—from those absorbing higher-energy photons to those absorbing lower-energy photons. Proteins organize and tune the absorption potential of antenna pigments so that resonance energy is efficiently moved between pigments, as the potential energy drops at each step.

Once the energy is transferred, the original excited electron falls back to its ground state. In this way, energy is transferred inside the photosystem in a manner that may be likened to the transfer of sound between tuning forks, or excitement between fans at a sports event during the "wave." But unlike the stadium wave, most of this resonance energy is directed to a particular location in a photosystem, called the reaction center.

The Reaction Center When a photon or resonance energy from an antenna pigment reaches the reaction center, the energy is absorbed by one of two specialized chlorophyll molecules (together called the *special pair*). When this pigment is energized, an excited electron is transferred from the special pair pigment to an electron acceptor. As the acceptor becomes reduced, its potential energy increases. This is a key step in the transformation of light energy: Electromagnetic energy from sunlight has now been transformed to chemical energy.

Note that in the absence of light, the electron acceptor does not accept electrons. It remains in an oxidized state because the redox reaction that transfers an electron to the electron acceptor is endergonic. But when light excites electrons in chlorophyll to a high-energy state, the reaction becomes exergonic. In this way, the energy in light transforms an endergonic reaction to an exergonic one.

Figure 10.10 summarizes the four possible fates of electrons in photosynthetic pigments that are excited by photons. The energy released from these electrons can

1. be emitted in the form of light via fluorescence, and/or
2. be given off as heat, or
3. excite an electron in a nearby pigment and induce resonance, or
4. be transferred to an electron acceptor in a redox reaction.

Fluorescence is typical of isolated pigments and is less frequent in chloroplasts, resonance energy transfer occurs in antenna pigments, and redox occurs in reaction center pigments.

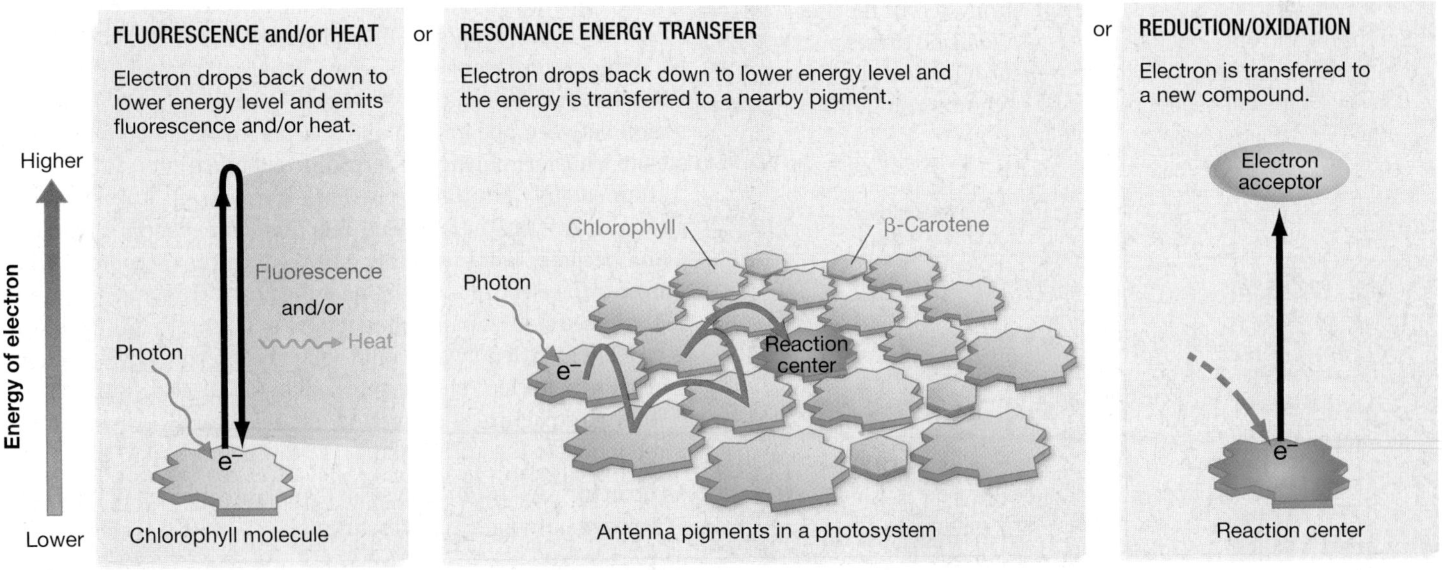

Figure 10.10 Four Fates for Excited Electrons in Photosynthetic Pigments. When sunlight promotes electrons in pigments to a high-energy state, four things can happen: They can fluoresce, release heat, pass energy to a nearby pigment via resonance, or transfer the electron to an electron acceptor.

Now the question is, what happens to the reduced electron acceptor in the reaction center? Specifically, how is the energy stored in this molecule used to make sugar?

CHECK YOUR UNDERSTANDING

✔ **If you understood this section, you should be able to . . .**

Predict how the pigments of a photosystem would be organized, with regard to the wavelength of photons absorbed, to allow directional transport of energy from the outer pigments to the reaction center.

Answers are available in Appendix A.

10.3 The Discovery of Photosystems I and II

After you complete this section, you should be able to . . .

- Explain how photosystems use light energy to produce chemical energy.

During the 1950s, the fate of excited electrons in photosystems was the central issue facing biologists interested in photosynthesis. A key breakthrough began with simple experiments by Robert Emerson on how green algae responded to various wavelengths of light. The algal cells being studied responded to wavelengths in the red and far-red regions of the visible spectrum.

Emerson found that if the algal cells were illuminated with either red or far-red wavelengths of light, the photosynthetic response was moderate. But if cells were exposed to a combination of both wavelengths, the rate of photosynthesis increased more than the sum of the rates produced by each wavelength independently. This phenomenon was called the enhancement effect, and it is not limited to algal cells. In follow-up work by other researchers, it was also observed in isolated chloroplasts from plants (**Figure 10.11**). Why the enhancement effect occurred was a complete mystery at the time.

Robin Hill and Faye Bendall proposed a solution to this puzzle. They hypothesized that this enhancement effect resulted from two distinct types of reaction centers, each absorbing different wavelengths of light. According to the two-photosystem hypothesis, the enhancement effect occurs because photosynthesis is much more efficient when both photosystems operate together.

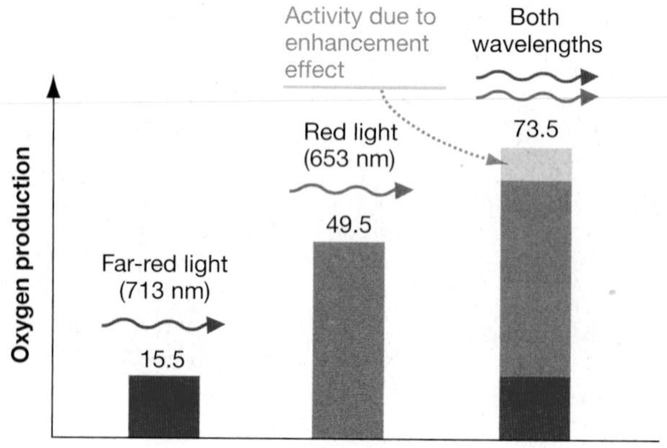

Figure 10.11 The "Enhancement Effect" of Two Different Wavelengths in Isolated Chloroplasts.

DATA: R. Govindjee, Govindjee, and G. Hoch. 1964. *Plant Physiology* 39: 10–14.

Subsequent work has shown that the two-photosystem hypothesis is correct for cyanobacteria ("blue-green bacteria") and the chloroplasts of eukaryotes, such as algae and plants. These two photosystems differ in structure and function, but work together in the light-capturing reactions.

To figure out how the two photosystems work, investigators focused on species of photosynthetic bacteria that possess one or the other of the two photosystems, but not both. Once each type of photosystem was understood in isolation, researchers explored how they work in combination. Let's do the same—by analyzing first **photosystem II**, then **photosystem I** (so named because it was discovered first), and then how the two interact.

How Does Photosystem II Work?

To study photosystem II, researchers first focused on the single photosystem of purple photosynthetic bacteria, including the purple sulfur bacteria that were studied by van Niel (see Section 10.1). The bacteria's photosystem shares many components with the photosystem II of cyanobacteria and eukaryotic chloroplasts of algae and plants, but was easier to study in the lab. For simplicity, the eukaryotic chloroplast—and its now well-characterized photosystems—will serve as the model system for the remainder of the chapter.

Converting Light Energy into Chemical Energy In photosystem II, the action often begins when a mobile accessory structure called the **light-harvesting complex** transmits resonance energy to an antenna pigment inside the photosystem. From there, resonance is relayed by other antenna pigments to transfer energy to the special pair pigments in the reaction center. At this point, another type of pigment molecule, called pheophytin, comes into play. Structurally, **pheophytin** is identical to chlorophyll except that pheophytin lacks a magnesium atom in its head region. Functionally, the two molecules are very different.

Unlike other pigments, pheophytin does not become excited by photons or resonance energy—it accepts excited electrons from the reaction center special pair chlorophylls. The redox reaction between pheophytin and special pair pigments is the key step in transforming light energy into chemical energy.

Immediately after an excited electron is transferred to pheophytin, however, the oxidized reaction center special pair pigment becomes an incredibly strong electron acceptor. What prevents the electron from being pulled back from pheophytin to the oxidized special pair pigment? The answer is that the electron is quickly shuttled away from the reaction center by a quinone molecule to an electron transport chain (ETC) (**Figure 10.12**).

In both structure and function, the thylakoid ETC is similar to components in the mitochondrial ETC (Ch. 9, Section 9.5).

- Structurally, the photosystem II and mitochondrial ETCs both contain quinones and cytochromes.
- Functionally, the redox reactions that occur in both ETCs result in protons being actively transported from one side of an internal membrane to the other. The resulting proton-motive force drives ATP production via ATP synthase.

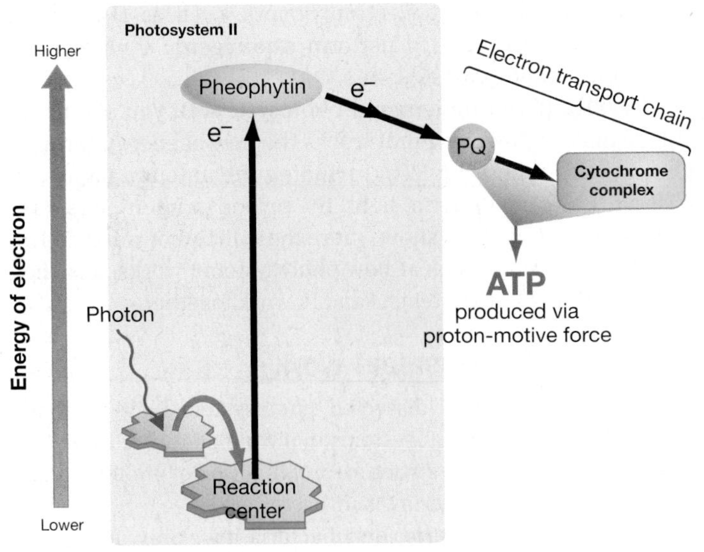

Figure 10.12 Photosystem II Feeds Excited Electrons to an Electron Transport Chain. When the energy from a single photon excites a special pair chlorophyll in the reaction center of photosystem II, the electron is accepted by pheophytin, transferred to a quinone molecule (identified as PQ), and then stepped down in energy along an electron transport chain.

Figure 10.13 explains how the electron transport chain associated with photosystem II works in more detail. Start by focusing on the molecule called **plastoquinone (PQ)**—a quinone similar to ubiquinone in the ETC of cellular respiration. Recall that ubiquinone is a small hydrophobic molecule that can transport two electrons between molecules (Ch. 9, Section 9.5). Because PQ is lipid soluble, it is free to move within the thylakoid membrane.

When PQ is reduced by picking up two electrons from photosystem II, it carries them through the membrane to the lumen side of the thylakoid and delivers them to molecules with a higher redox potential in the cytochrome complex. In this way, PQ shuttles electrons from photosystem II to the cytochrome complex much like ubiquinone shuttled electrons between complexes I or II and complex III in mitochondria.

Similar to the mitochondrial ubiquinone, PQ also picks up protons when it becomes reduced. These protons are pulled from the chloroplast stroma. After being oxidized by the cytochrome complex, PQ drops off the protons in the thylakoid lumen.

The protons transported by PQ result in a high concentration of protons in the thylakoid lumen. The pH in the thylakoid reaches 5 while the pH of the stroma hovers around 8. Because the pH scale is logarithmic (see **BioSkills 5**), the difference of 3 units means that the concentration of H^+ is $10 \times 10 \times 10 = 1000$ times higher

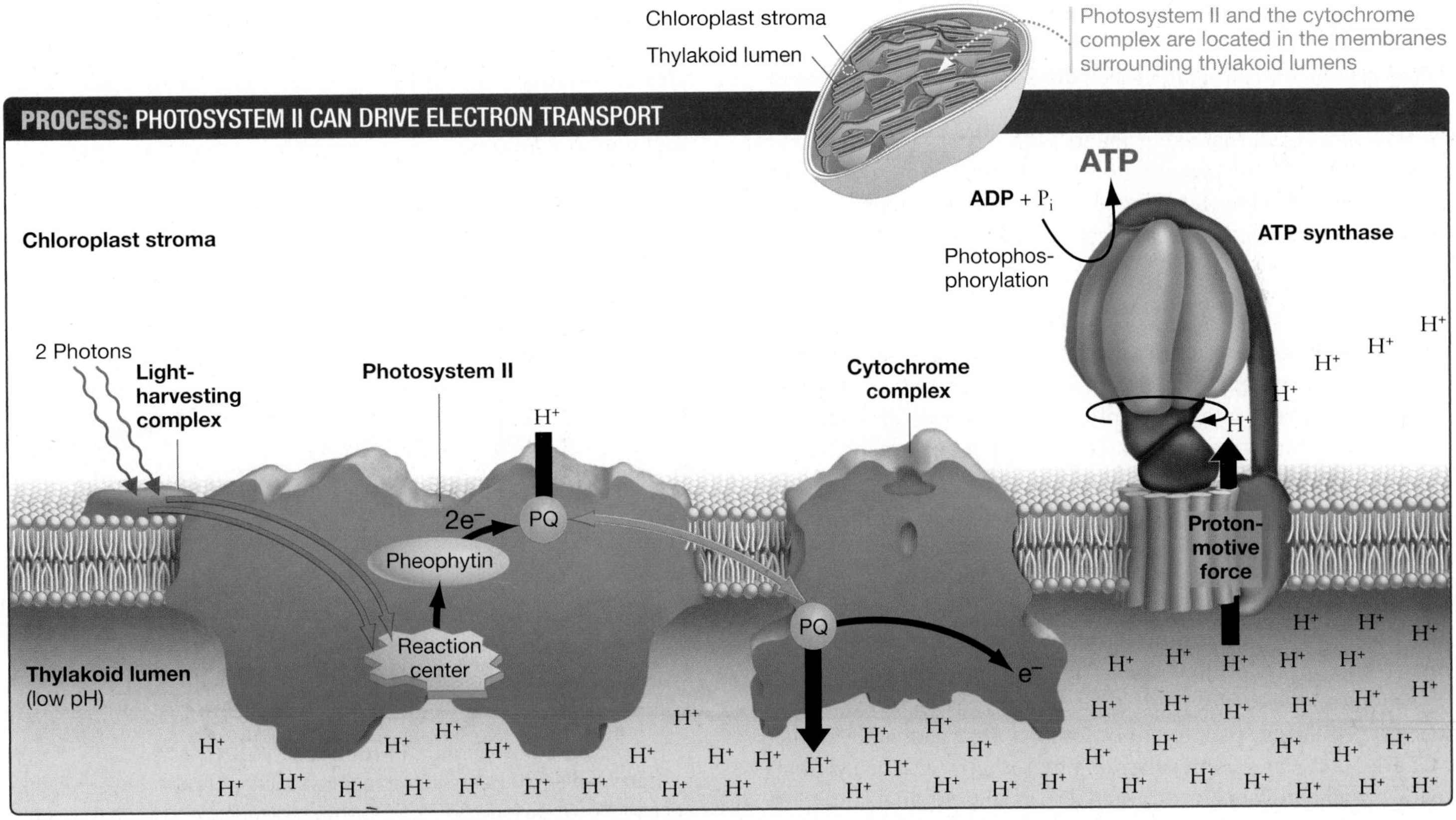

Figure 10.13 Electron Transport between Photosystem II and the Cytochrome Complex. Two photons excite two electrons to reduce one plastoquinone (PQ), which carries the electrons from photosystem II along with protons from the stroma. The cytochrome complex oxidizes plastoquinone, releasing the protons into the thylakoid lumen that drive ATP synthesis.

in the lumen than in the stroma. In addition, the stroma becomes negatively charged relative to the thylakoid lumen.

The net effect of electron transport, then, is a large proton electrochemical gradient. This gradient results in a proton-motive force that, in turn, drives H^+ out of the thylakoid lumen and into the stroma. Proton flow down the electrochemical gradient is an exergonic process that is coupled to the endergonic synthesis of ATP from ADP and P_i. The stream of protons flows through ATP synthase, causing conformational changes in the enzyme that drive production of ATP.

Since the synthesis of ATP in chloroplasts is initiated by the energy from light, it is called **photophosphorylation**. Although photophosphorylation is similar to the oxidative phosphorylation that occurs in plant and animal mitochondria, there is a key difference in how this ATP is used. In mitochondria, ATP is exported and fuels many different cellular processes. In chloroplasts, however, the ATP remains within the organelle and is used for the production of carbohydrate.

The photosystem II story is not yet complete, however. The electrons carried by PQ are passed through the cytochrome complex, but what about the oxidized photosystem II reaction center that PQ left behind? For the ETC to continue functioning, the electrons removed from the reaction center special pair pigments need to be replaced. Where do the electrons required to regenerate photosystem II come from?

Photosystem II Obtains Electrons by Oxidizing Water Think back to the simplified overall reaction for photosynthesis:

$$CO_2 + H_2O + \text{Light energy} \longrightarrow (CH_2O)_n + O_2$$

In the presence of sunlight, carbon dioxide and water are used to produce carbohydrate and oxygen gas.

Now also recall that experiments with heavy isotopes of oxygen showed that the oxygen atoms in O_2 come from water, not from carbon dioxide (Section 10.1). For this to happen, water must be oxidized (i.e., release electrons). The oxygen-generating reaction can be written as

$$2\,H_2O \rightarrow 4\,H^+ + 4\,e^- + O_2$$

This reaction is referred to as "splitting" water. It supplies electrons for photosystem II and is catalyzed by a complex of enzymes that are physically integrated into photosystem II. Since oxygen is very electronegative, this reaction is highly endergonic. What supplies the energy necessary to oxidize water?

As it turns out, the light energy harvested by photosystem II is responsible for splitting water. When excited electrons are removed from the photosystem II reaction center pigments, the redox potential of the oxidized pigments becomes so strong that enzymes can pull electrons from water, pass them to the pigments, and release protons and oxygen.

Among all life-forms, photosystem II in cyanobacteria and eukaryotic chloroplasts is the only complex that can catalyze the oxidation of water molecules. Photosynthetic organisms that oxidize water will generate oxygen (O_2) as a by-product, and thus perform **oxygenic** ("oxygen-producing") photosynthesis.

Other organisms that have only a single photosystem do not oxidize water, and thus do not produce O_2. Instead, these organisms use different electron donors, such as H_2S in the purple sulfur bacteria, to perform **anoxygenic** ("no oxygen-producing") photosynthesis.

✔ **MODEL If you understand photosystem II, you should be able to make a flowchart that tracks the flow of energy through photosystem II and the ETC, including the antenna pigments, ATP synthase, pheophytin, light, the proton gradient, and a reaction center. Be sure to show where the splitting of water fits in.**

Let's turn next to look at how photosystem I works, and then at how the two photosystems, I and II, work together.

How Does Photosystem I Work?

Recall that researchers dissected photosystem II by studying similar, but simpler, photosystems in purple photosynthetic bacteria. To understand the structure and function of photosystem I, they turned to heliobacteria ("sun-bacteria").

Like the purple bacteria, heliobacteria have only one photosystem that uses the energy in sunlight to promote electrons to a high-energy state. But instead of being passed to an electron transport chain that pumps protons across a membrane, the excited electrons in heliobacteria are used to reduce NAD^+. When NAD^+ gains two electrons and a proton, NADH is produced.

In cyanobacteria and the chloroplasts of eukaryotes, a similar set of light-capturing reactions reduces a phosphorylated version of NAD^+, symbolized $NADP^+$, to yield NADPH. Both NADH and NADPH function as strong reducing agents—that is, because they have been reduced, they become carriers of electrons that can readily be transferred to reduce other molecules.

Figure 10.14 explains how photosystem I works in chloroplasts. To reduce one $NADP^+$, two electrons must be excited by photons—put your finger on the "2 Photons" arrows and trace the steps that follow.

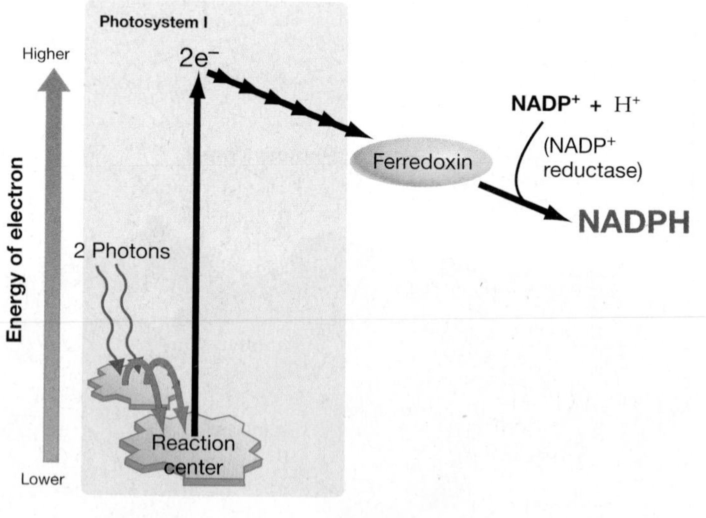

Figure 10.14 Photosystem I Produces NADPH. When two photons excite pigments in the reaction center of photosystem I, the excited electrons leave the chlorophyll molecules and pass through a series of iron- and sulfur-containing proteins until they are accepted by ferredoxin. In an enzyme-catalyzed reaction, the two electrons are transferred from ferredoxin to $NADP^+$ to produce NADPH.

1. Antenna pigments absorb photons and pass the energy to the photosystem I reaction center.
2. Two electrons (one for each of the photons absorbed in step 1) are excited in reaction center chlorophyll molecules.
3. The reaction center pigments are oxidized, and the excited electrons are passed through a series of carriers inside the photosystem, then to a molecule called **ferredoxin**, and then to the enzyme called $NADP^+$ reductase.
4. $NADP^+$ reductase transfers the two electrons and a proton to reduce $NADP^+$ and form NADPH.

To summarize: Electrons from photosystem I are used to produce NADPH, which is a reducing agent similar in function to the NADH and $FADH_2$ produced by the citric acid cycle (see Ch. 9, Section 9.4). Electrons from photosystem II, in contrast, are used to produce a proton-motive force that drives the synthesis of ATP. The ATP and the reducing power of NADPH will ultimately be used in the manufacture of sugar.

In combination, then, photosystems II and I produce chemical energy stored in ATP and NADPH. But there are still gaps in the flow of electrons through these two photosystems. Where do the electrons from photosystem II end up? How does the oxidized reaction center of photosystem I obtain electrons so NADPH will continue to be made?

The Z Scheme: Photosystems II and I Work Together

Figure 10.15 illustrates the **Z-scheme** model for how photosystems II and I interact. The name was inspired by the changes occurring in electron potential energy as plotted on a vertical axis, which takes on the shape of a Z that has fallen over.

To drive home how energy flows through the light-capturing reactions, trace the route of electrons through Figure 10.15 with your finger. Start on the lower left. The process starts when photons excite electrons in photosystem II's antenna pigments. When the energy in the excited electrons is transferred to the reaction center, a special pair of chlorophyll molecules, each called P680, passes excited electrons to pheophytin. (These are the same special pair pigments described previously, and the name P680 represents the optimal wavelength absorbed by the pigments—680 nm.)

Before moving on, direct your attention back to the equation at the bottom of photosystem II. Note that once the electrons pass from photosystem II's reaction center, they are be replaced by electrons that are stripped away from water. Taking electrons from water produces O_2 as a by-product and H^+ ions that contribute to the electrochemical gradient. In this figure, the effect of four photons is shown to illustrate the events that would be required to produce one O_2 molecule from water.

Returning to the reduced pheophytin, follow the transfer of the four electrons to an electron transport chain that follows photosystem II. There the electrons are gradually stepped down in potential energy through redox reactions among a series of quinones and cytochromes. Each reduced plastoquinone (PQ) picks up protons from the stroma and transfers them to the lumen after being oxidized by the cytochrome complex. ATP synthase uses the resulting proton-motive force to phosphorylate ADP, creating ATP.

When electrons reach the end of the cytochrome complex, they are passed to a small diffusible protein called **plastocyanin (PC)**. Each reduced plastocyanin diffuses through the lumen of the thylakoid and donates one electron to an oxidized reaction center pigment in photosystem I.

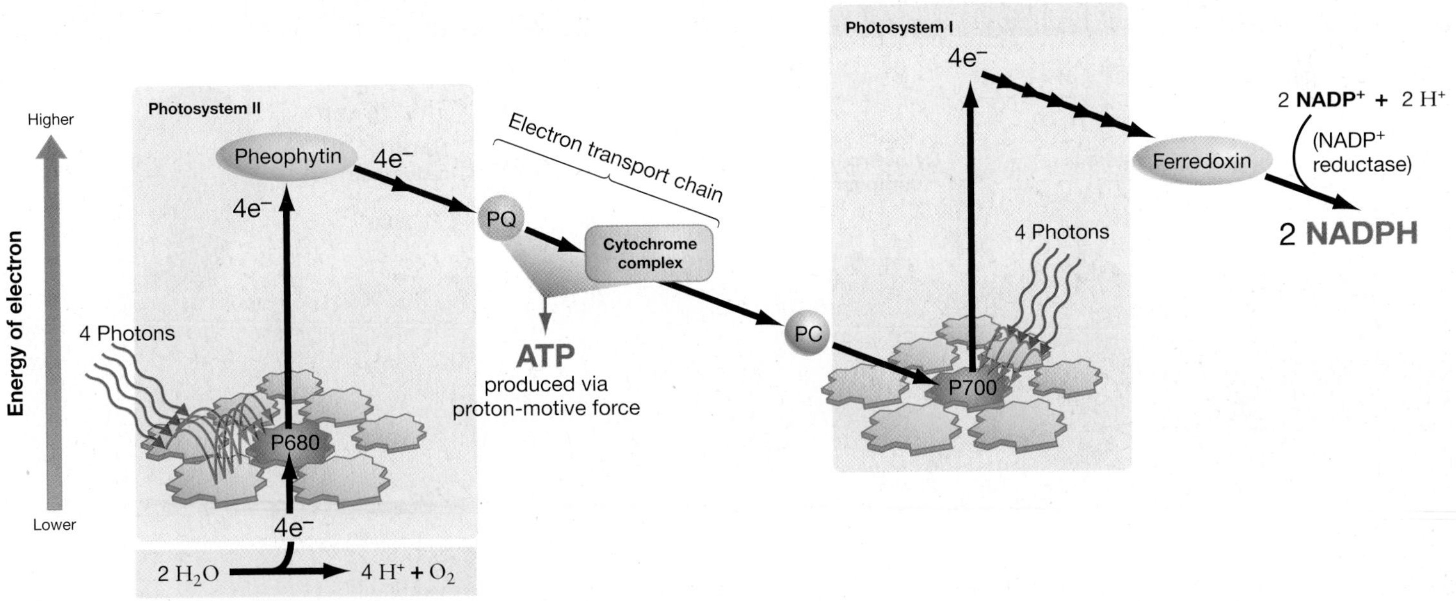

Figure 10.15 **The Z-Scheme Model Links Photosystems II and I.** The Z scheme explains the linear pathway that connects photosystems II and I to generate ATP and NADPH. For each O_2 molecule produced, four electrons are transferred through the Z scheme.

Stop tracing for a moment, and consider the following:

- Plastocyanin is critical—it forms a physical link between the ETC following photosystem II and photosystem I.
- A single plastocyanin can shuttle over 1000 electrons per second between the cytochrome complex and photosystem I.
- The flow of electrons between photosystems, by means of plastocyanin, is important because it replaces electrons that are carried away from the special pair of pigments in the photosystem I reaction center. These special pair chlorophyll molecules are called P700 (optimal absorption of 700-nm wavelength).

Now keep going. The electrons that flow into P700 are eventually excited and transferred to the protein ferredoxin, which passes electrons to the enzyme that catalyzes reduction of $NADP^+$ to NADPH. Note that for each O_2 molecule produced by photosystem II, four electrons have been transferred along the Z scheme to make two molecules of NADPH.

Understanding the Enhancement Effect The Z-scheme model helps explain the enhancement effect shown in Figure 10.11. When chloroplasts are illuminated with wavelengths in the red portion of the spectrum, only photosystem II can run at a maximum rate. The overall rate of electron flow through the Z scheme is moderate because photosystem I's efficiency is reduced.

Similarly, when chloroplasts are illuminated with wavelengths in the far-red portion of the spectrum, only photosystem I is capable of peak efficiency; photosystem II is working at a below-maximum rate, so the overall rate of electron flow is reduced.

But when both wavelengths are available at the same time, both photosystems are activated and work at a maximum rate, leading to enhanced efficiency.

Noncyclic Electron Flow between Water and $NADP^+$ The complete path that electrons follow from photosystem II to photosystem I and how it is oriented in the thylakoid membrane is shown in Figure 10.16. Note that the electrons that pass from water to $NADP^+$ move through a chain of redox reactions in a linear fashion, referred to as **noncyclic electron flow**.

Compare the movement of electrons and protons in Figure 10.16 with what you have learned about electron transport chains in mitochondria (see Ch. 9, Figure 9.15). In both these organelles, redox reactions are used to build a proton gradient for ATP production. At the end of the chains, electrons are donated to terminal electron acceptors.

Chloroplasts and mitochondria differ, however, in how electron potential energy changes between the primary electron donor and the terminal electron acceptor. In the mitochondrial ETC, electron potential energy starts high and then steadily drops as the electrons are transferred to the terminal electron acceptor, which has the lowest potential energy. In chloroplasts, the electron donor (H_2O) has much lower potential energy than the reduced terminal electron acceptor (NADPH; see Figure 10.15).

✔ **MODEL If you understand how the photosystems work, you should be able to draw a model of a section of thylakoid membrane showing how the following elements are related: the two photosystems, the cytochrome complex, ATP synthase, photons, electrons, and protons.**

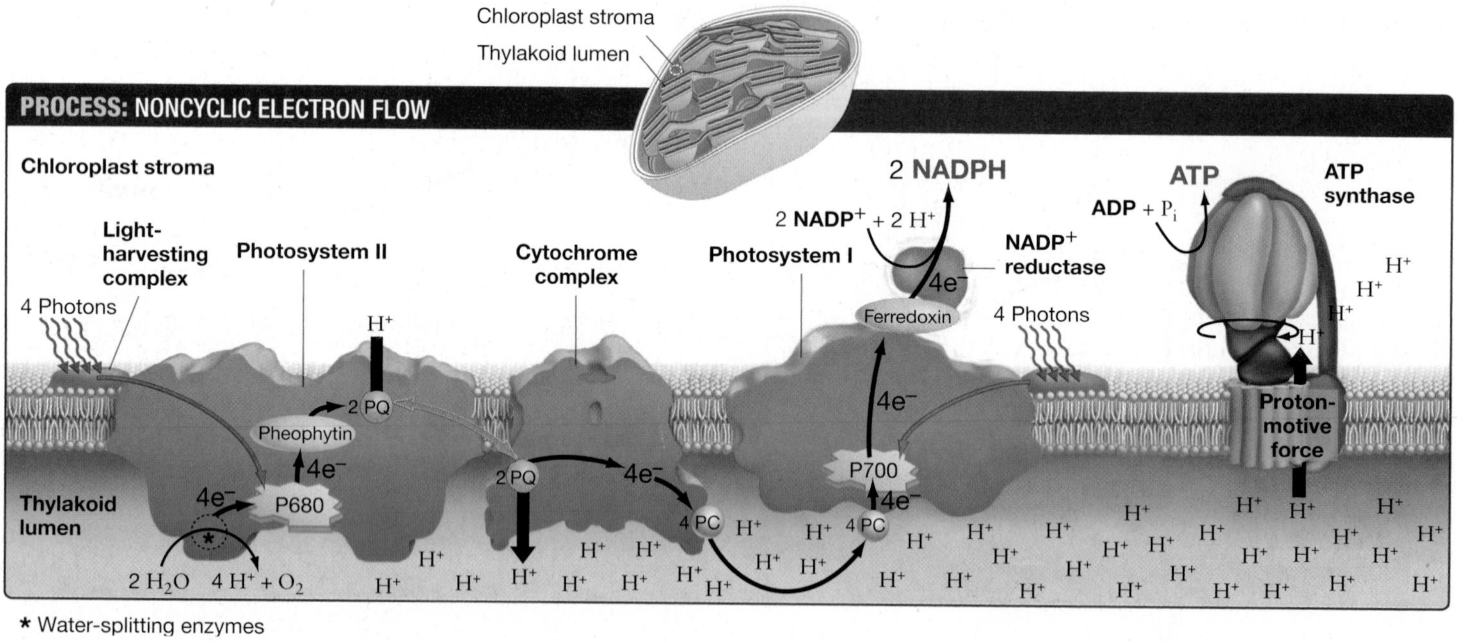

Figure 10.16 Electrons Are Passed from Water to $NADP^+$ in a Linear Pathway. In the thylakoid membrane, photosystem II uses light to excite electrons taken from water and pass them through an ETC including plastoquinone (PQ), the cytochrome complex, and plastocyanin (PC). The ETC produces a proton-motive force that is used to make ATP. Photosystem I uses light to excite electrons from PC and passes them on to ferredoxin to reduce $NADP^+$ to NADPH.

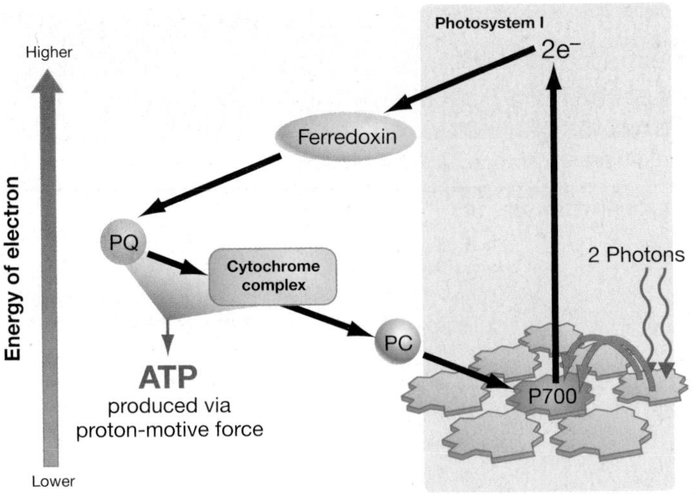

Figure 10.17 Cyclic Electron Flow Leads to ATP Production. Cyclic electron flow is an alternative to the Z scheme. Instead of being donated to $NADP^+$, electrons are returned to plastoquinone (PQ) and cycle between photosystem I and the ETC, resulting in the production of additional ATP via cyclic photophosphorylation.

Cyclic Electron Flow Recycles Electrons and Drives Photophosphorylation More recent evidence indicates that an alternative electron path, called **cyclic electron flow**, also occurs in green algae and land plants (Figure 10.17). In these organisms, ATP is produced via cyclic as well as noncyclic photophosphorylation.

During cyclic electron flow, electrons excited in photosystem I are transferred back to the electron transport chain, generating ATP through photophosphorylation instead of reducing $NADP^+$. Cyclic electron flow coexists with noncyclic electron flow and produces additional ATP to meet the energy demand for manufacturing sugars from carbon dioxide (CO_2).

Oxygenic Photosynthesis and the Evolution of Earth Although oxygen is a by-product of oxygenic photosynthesis, the impact of producing this molecule on the environment of early Earth cannot be overstated. Photosynthesis produces the oxygen that is keeping you alive right now. Biologists rank the evolution of Earth's oxygen-rich atmosphere as one of the most important events in the history of life. Why?

According to the geologic record, oxygen levels in the atmosphere and oceans began to rise only about 2 billion years ago, as organisms that performed oxygenic photosynthesis increased in abundance. O_2 was, in fact, almost nonexistent on Earth before enzymes evolved that could catalyze the splitting of water.

Since ozone (O_3) is formed from O_2 gas, a protective layer of ozone could have arisen in our atmosphere only after the evolution of oxygenic photosynthesis. Without the ozone layer, Earth's surface would have been bombarded continually by the searing intensity of ultraviolet radiation—making the evolution of life on land nearly impossible.

As oxygen became more abundant, bacterial cells that evolved the ability to use it as an electron acceptor via cellular respiration flourished. O_2 is so electronegative that it creates a huge potential energy drop for the electron transport chains involved in cellular respiration. As a result, organisms that use O_2 as an electron acceptor in cellular respiration can produce ATP more efficiently than can organisms that use other electron acceptors (Ch. 9, Section 9.5).

Despite the importance of oxygen in the evolution and maintenance of life, in photosynthesis, oxygen is simply waste. The useful products of the light-capturing reactions are ATP and NADPH, which are required to reduce carbon dioxide to sugar. Your life, and the lives of most organisms, depends on the production of sugar from CO_2. How does it happen?

CHECK YOUR UNDERSTANDING

✔ **If you understood this section, you should be able to ...**

Compare and contrast the flow of electrons in mitochondria (see Ch. 9, Section 9.5) and chloroplasts. What are the primary electron donors and terminal electron acceptors, and how do they differ in terms of initial and final energy states?

Answers are available in Appendix A.

10.4 How Do Cells Capture Carbon Dioxide?

After you complete this section, you should be able to ...

- Compare the different pathways used for carbon fixation and explain why they are necessary.

Atmospheric carbon dioxide is a key reactant in photosynthesizing cells. It would seem straightforward, then, for CO_2 to diffuse directly into plants along a concentration gradient. But the situation is not this simple, because plants are covered with a waxy coating called a cuticle (Figure 10.18). This lipid layer prevents water from evaporating out of tissues, but it also prevents gases like CO_2 and O_2 from crossing the plasma membrane.

Figure 10.18 Waxy Cuticles Block Passage of O_2, H_2O, and CO_2.

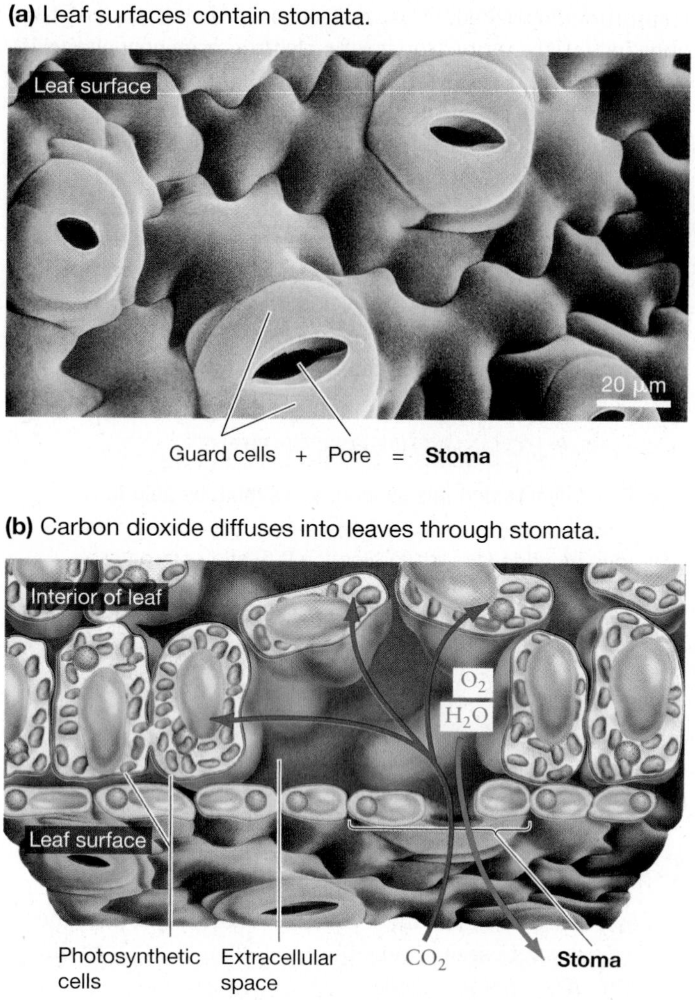

Figure 10.19 Leaf Cells Obtain Carbon Dioxide through Stomata.

Instead, CO_2 gets into photosynthesizing tissues through specialized pores. The surface of a leaf is dotted with openings bordered by two distinctively shaped cells called **guard cells** (Figure 10.19a). The opening between these paired cells is called a pore, and the entire structure is a **stoma** (plural: **stomata**).

An open stoma allows CO_2 from the atmosphere to diffuse into uncoated air-filled spaces inside the leaf and excess O_2 to diffuse out (Figure 10.19b). Eventually the CO_2 diffuses along a concentration gradient into the chloroplasts of photosynthesizing cells. A strong concentration gradient favoring entry of CO_2 is maintained by the Calvin cycle, which constantly uses up the CO_2 in chloroplasts. How does this happen?

The Calvin Cycle Fixes Carbon

Carbon fixation is the addition of carbon atoms from inorganic carbon dioxide to an organic compound. The word "fix" is appropriate because the process converts—or fixes—CO_2 gas to a biologically useful form. Once carbon atoms are added to an organic compound, they can be used as sources of energy and as building blocks to construct the molecules found in cells.

RESEARCH

QUESTION: What intermediates are produced as carbon dioxide is reduced to sugar?

HYPOTHESIS: No specific hypothesis.

EXPERIMENTAL SETUP:

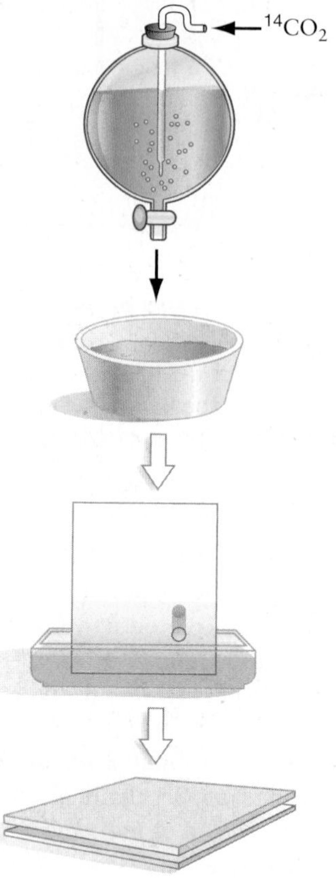

1. Add $^{14}CO_2$ to actively photosynthesizing algae.

2. Stop reaction after varying amounts of time and homogenize cells by immersing in hot alcohol.

3. Separate molecules via chromatography.

4. Lay X-ray film on chromatograph to locate radioactive label.

PREDICTION: No specific prediction.

RESULTS:

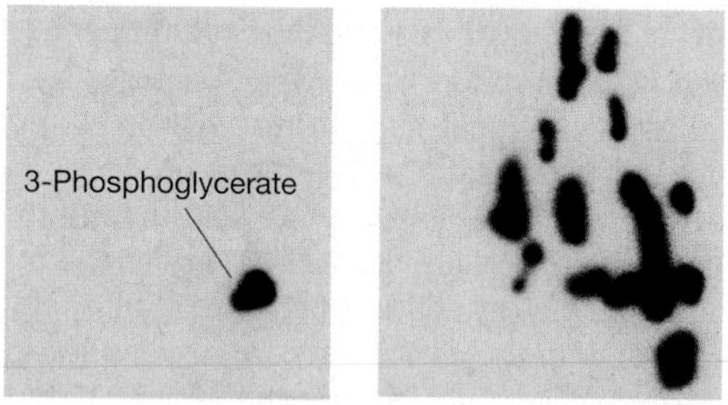

Compounds produced after 5 seconds

Compounds produced after 60 seconds

CONCLUSION: 3-Phosphoglycerate is the first intermediate product. Other intermediates appear later.

Figure 10.20 Experiments Revealed the Reaction Pathway Leading to Reduction of CO_2.

SOURCE: A. A. Benson, J. A. Bassham, M. Calvin, et al. 1950. The path of carbon in photosynthesis. V. Paper chromatography and radioautography of the products. *Journal of the American Chemistry Society* 72: 1710–1718.

✔ **PROCESS OF SCIENCE** Why wasn't this experiment based on a specific hypothesis and set of predictions?

Carbon fixation is a redox reaction—the carbon atom in CO_2 is reduced by attaching it to another carbon. Research on how this happens in chloroplasts gained momentum just after World War II, when radioactive isotopes of carbon became available for research purposes.

Melvin Calvin's group (Section 10.1) made great strides early in this effort by tracking the incorporation of $^{14}CO_2$ into molecules during photosynthesis (Figure 10.20). After injecting $^{14}CO_2$ into a culture of algae that were undergoing photosynthesis, they stopped the reaction at different periods of time by killing the cells in hot alcohol. This treatment immediately denatured the enzymes involved in the reactions, effectively halting any further change in the radiolabeled intermediates.

The molecules labeled with the ^{14}C in this extract were separated by chromatography and detected using X-ray film. If radioactively labeled molecules were present in the chromatograph, the energy they emitted would expose the film and create a dark spot as shown in the "Results" section of Figure 10.20. The labeled compounds could then be isolated and identified.

By varying the amount of time the algae were exposed to labeled $^{14}CO_2$, Calvin and co-workers pieced together the sequence in which various intermediates formed. For example, when the team analyzed cells almost immediately after adding the $^{14}CO_2$, they found that the ^{14}C was predominantly in a three-carbon organic acid called 3-phosphoglycerate (3PGA). This result suggested that 3PGA was the first product of carbon fixation. Stated another way, it appeared that carbon dioxide reacted with some unknown molecule to produce 3PGA. Which compound reacted with CO_2 to produce 3PGA?

Since 3PGA has three carbons, Calvin's group searched for a two-carbon compound that might serve as the initial carbon dioxide acceptor. They were unsuccessful. Then, while Calvin was running errands one day, it occurred to him that the molecule reacting with carbon dioxide might contain five carbons, not two. Adding CO_2 to a five-carbon molecule would produce a six-carbon compound, which could then split in half to form 2 three-carbon molecules.

Experiments to test this hypothesis confirmed that the five-carbon compound **ribulose bisphosphate (RuBP)** is the initial reactant. Ecologically, the fixation of CO_2 to RuBP is the most important chemical reaction on Earth. The enzyme that catalyzes it is fundamental to life. How does this protein work?

The Discovery of Rubisco

To find the enzyme that fixes CO_2 to RuBP, Arthur Weissbach and colleagues ground up spinach leaves, purified a large series of proteins from the resulting cell extracts, and tested each protein to see if it could catalyze this step. Eventually they isolated the catalyst, which happens to be the most abundant enzyme in leaf tissue. The researchers' data suggested that it constituted almost 50 percent of the total protein in spinach leaves.

All photosynthetic organisms that use the Calvin cycle to fix carbon require the CO_2-fixing enzyme ribulose-1,5 bisphosphate carboxylase/oxygenase (commonly referred to as **rubisco**). As shown in Figure 10.21, the rubisco enzyme is roughly cube-shaped and consists of 16 polypeptides that form eight active sites where CO_2 is fixed. Some of these polypeptide subunits are made in the chloroplast while others are made in the cytoplasm and then imported into the organelle.

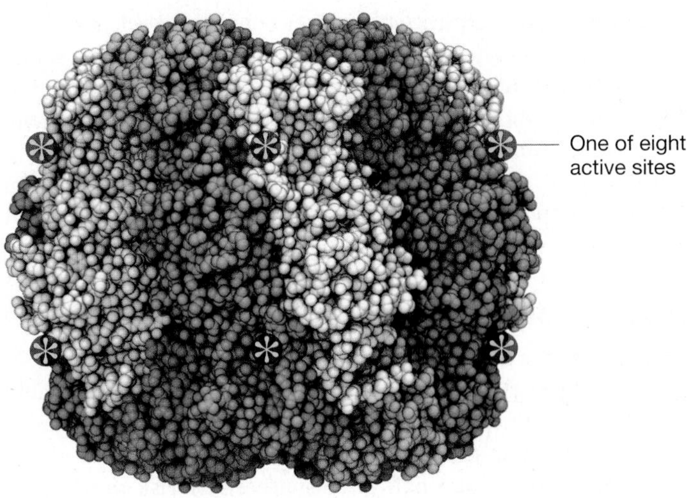

Figure 10.21 Rubisco Is a Large Enzyme Complex. Rubisco consists of 16 polypeptides that form 8 catalytic active sites.

Despite its large number of active sites, rubisco is a slow enzyme. Each active site catalyzes just three reactions per second; other enzymes typically catalyze thousands of reactions per second. Plants synthesize huge amounts of rubisco, possibly as an adaptation compensating for its lack of speed. As a result, rubisco is thought to be the most abundant enzyme on Earth.

Besides being slow, rubisco is also inefficient because it will catalyze the addition of either O_2 or CO_2 to RuBP. This is a key point: Oxygen and carbon dioxide compete at the enzyme's active sites, which slows the rate of CO_2 reduction.

The reaction of O_2 with RuBP actually does more than just compete with the reaction of CO_2 at the same active site. One of the molecules produced from the addition of oxygen to RuBP (2-phosphoglycolate) is processed in reactions that require ATP and release CO_2, regenerating 3PGA. Part of this pathway occurs in chloroplasts, and part occurs in peroxisomes and mitochondria. The reaction sequence resembles respiration, because it consumes oxygen and produces carbon dioxide. As a result, it is called **photorespiration** (Figure 10.22).

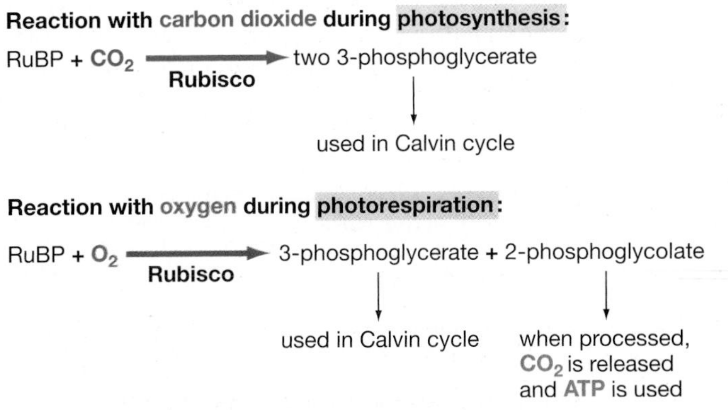

Figure 10.22 Rubisco Can React with CO_2 or O_2. In addition to fixing CO_2 in photosynthesis, rubisco catalyzes a competing reaction with O_2 with a very different outcome.

Because photorespiration requires energy and releases fixed CO_2, it "undoes" photosynthesis. When photorespiration occurs, the overall rate of CO_2 fixation declines. This does not mean that the plant does not benefit, however. Some of the products from photorespiration are known to be involved in plant signaling and development. In addition, a protective role for photorespiration has been proposed when plants are in conditions with high light and low CO_2.

Mechanisms for Increasing CO_2 Concentration

Stomata are normally open during the day, when photosynthesis is occurring, and closed at night. But if the daytime is extremely hot and dry, leaf cells may lose a great deal of water to evaporation through their stomata. When this occurs, they must either close the openings and halt photosynthesis or risk death from dehydration.

When conditions are hot and dry, then, stomata must close and CO_2 and O_2 transport stops—meaning that photosynthesis slows and photorespiration increases. Even with the stomata open, the atmosphere is 21 percent oxygen and only 0.04 percent carbon dioxide. How can photosynthesizing cells raise CO_2 concentrations to make photosynthesis more efficient, especially in hot, dry environments? An answer emerged in a surprising experimental result.

The C_4 Pathway After Calvin's work observing carbon fixation in algae (Figure 10.20), researchers in a variety of labs used the same radioactive carbon dioxide tracking approach to investigate how carbon fixation occurs in other species. Researchers in the labs of Hugo Kortschak and Y. S. Karpilov exposed leaves of sugarcane and maize (corn) to $^{14}CO_2$ and sunlight; then they isolated and identified the intermediates.

Both research teams expected to find the first of the radioactive carbon atoms in 3-phosphoglycerate—the normal product of carbon fixation by rubisco. Instead, they found that in their species, the radioactive carbon atom ended up in four-carbon compounds such as oxaloacetate. (Note that oxaloacetate is also used as a substrate in the first step of the citric acid cycle—see Figure 9.10.)

Instead of creating a three-carbon molecule as in the Calvin cycle, it appeared that these species were able to fix CO_2 to produce four-carbon molecules. This newly identified set of reactions became known as the **C_4 pathway** to distinguish it from Calvin's CO_2 fixation via what is now termed the **C_3 pathway** (Figure 10.23).

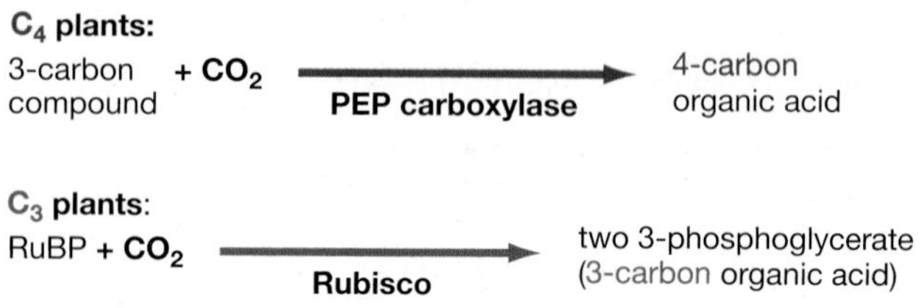

Figure 10.23 Initial Carbon Fixation in C_4 Plants Is Different from That in C_3 Plants.

Researchers who followed up on the initial reports found that the C_4 pathway does not replace the Calvin cycle, but serves as an additional fixation step. C_4 plants actually fix carbon dioxide using both pathways—to a three-carbon compound by an enzyme called **PEP carboxylase** (C_4) and to RuBP by rubisco (C_3). They also showed that the two pathways are found in distinct cell types within the same leaf. PEP carboxylase is common in **mesophyll cells** near the surface of leaves, while rubisco is found in **bundle-sheath cells** that surround the vascular tissue in the interior of the leaf (Figure 10.24a). Vascular tissue conducts water and nutrients in plants (Ch. 35, Section 35.4).

Based on the observations about C_4 plants, Hal Hatch and Roger Slack proposed a four-step model to explain how the CO_2 that is fixed to form a four-carbon molecule can be released to feed the Calvin cycle (Figure 10.24b):

Step 1 PEP carboxylase fixes CO_2 to phosphoenolpyruvate, or PEP (a three-carbon molecule), in mesophyll cells.

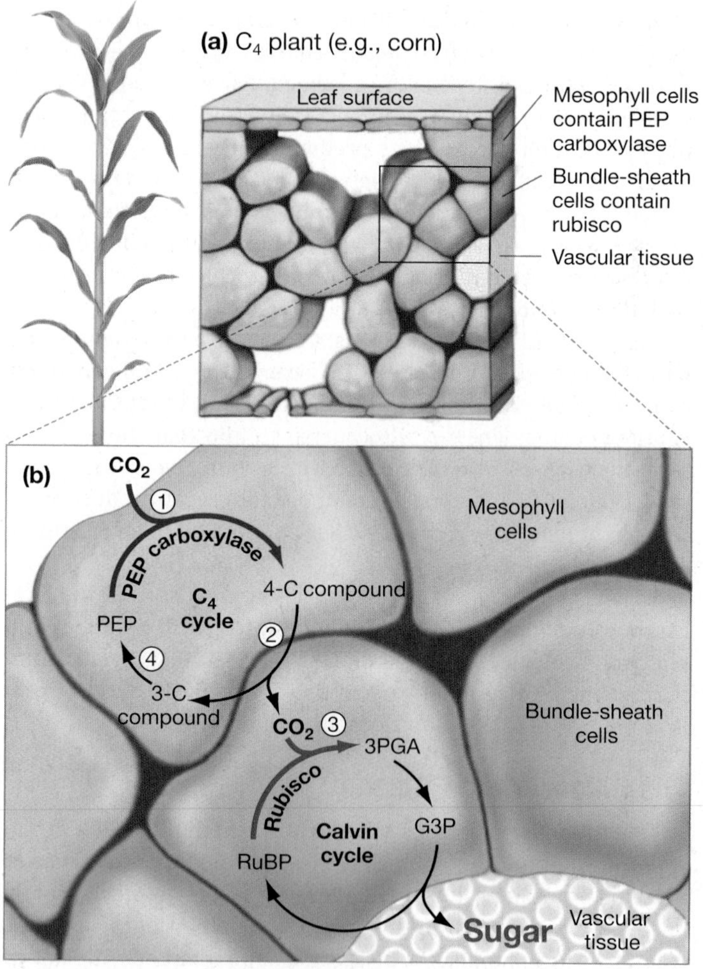

Figure 10.24 In C_4 Plants, Carbon Fixation and the Calvin Cycle Occur in Different Cell Types. (a) The carbon-fixing enzyme PEP carboxylase is located in mesophyll cells, while rubisco is in bundle-sheath cells. **(b)** CO_2 is fixed to the three-carbon compound PEP by PEP carboxylase, forming a four-carbon organic acid. A CO_2 molecule can be released from the organic acid to feed the Calvin cycle.

Step 2 The four-carbon molecules that result are transported to bundle-sheath cells via channels called plasmodesmata (Ch. 11, Section 11.2).

Step 3 The four-carbon molecules are broken down to release CO_2 molecules. Bundle-sheath cells are less permeable to gases, so CO_2 concentrations rise and promote carbon fixation by rubisco to form 3PGA. This step initiates the Calvin cycle.

Step 4 The three-carbon compound remaining after CO_2 is released is returned to the mesophyll cell to regenerate PEP.

In effect, then, the C_4 pathway acts as a CO_2 concentrator. The reactions that take place in mesophyll cells require energy in the form of ATP, but they increase CO_2 concentrations in bundle-sheath cells where rubisco is active. Because it increases the ratio of carbon dioxide to oxygen in photosynthesizing cells, less O_2 binds to rubisco's active sites. As a result, the C_4 pathway improves the efficiency of the Calvin cycle.

Although the C_4 pathway requires additional ATP to fix CO_2 for the Calvin cycle, the energy expenditure is justified by the increased efficiency of photosynthesis in conditions where stomata are mostly closed to prevent dehydration. The affinity for CO_2 by PEP carboxylase is also much higher than that of rubisco, which means that stomata can be open for shorter periods in C_4 plants.

This strategy is not the only mechanism that plants use to continue growth in hot, dry climates, however. Some environments are so arid that even C_4 plants are unable to avoid dehydration. Nevertheless, certain plants use the C_4 pathway in a unique way that allows them to thrive in these deserts. How do they do it?

CAM Plants Researchers studying a group of flowering plants called the Crassulaceae discovered a second mechanism for limiting the effects of dehydration and photorespiration. This photosynthetic pathway, **crassulacean acid metabolism**, or **CAM**, resembles the C_4 pathway in some ways. It is a CO_2 concentrator that acts as an additional, preparatory step to the Calvin cycle. It also generates a four-carbon organic acid in its first CO_2 fixation step. But unlike the C_4 pathway, CAM occurs at a different time than the Calvin cycle does—not in a different place.

CAM occurs in cacti and other species that routinely keep their stomata closed on hot, dry days. At night, when conditions are cooler and more humid, CAM plants open their stomata and take in huge quantities of CO_2. The CO_2 is temporarily fixed to organic acids and stored in the central vacuoles of photosynthetic cells. During the day, when stomata are closed, these acids are processed in reactions that release the CO_2 and feed the Calvin cycle (**Figure 10.25**).

The C_4 and CAM pathways function as CO_2 pumps. They minimize photorespiration when stomata are closed and CO_2 cannot diffuse in directly from the atmosphere. Both are found in species that live in hot, dry environments.

But while C_4 plants stockpile CO_2 by fixing and storing organic acids in cells *where* rubisco is not active, CAM plants store CO_2 *when* rubisco is inactive. In C_4 plants, the reactions catalyzed by PEP carboxylase and rubisco are separated in space; in CAM plants, the reactions are separated in time.

While the process of carbon fixation is a significant event in photosynthesis, still more needs to be done to convert 3PGA into a high-energy carbohydrate. To accomplish this feat, the ATP and NADPH from the light-capturing reactions will be required. Let's now return to the Calvin cycle and see how this is done.

CHECK YOUR UNDERSTANDING

✔ If you understood this section, you should be able to . . .

1. Explain why photorespiration is generally considered to be detrimental to the productivity of photosynthesis.
2. Describe how CO_2 is delivered to rubisco in (a) C_3 plants, (b) C_4 plants, and (c) CAM plants.

Answers are available in Appendix A.

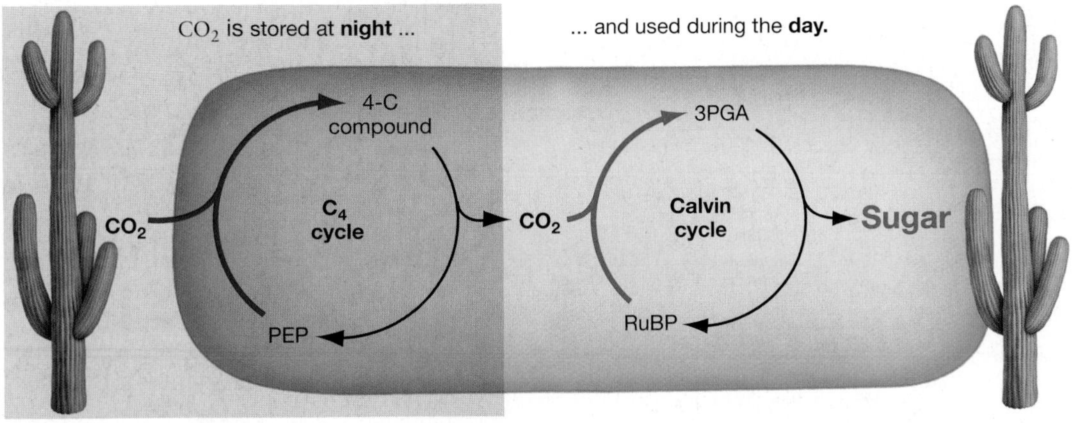

Figure 10.25 In CAM Plants, Carbon Fixation Occurs at Night and the Calvin Cycle Occurs during the Day.

✔ At what part of the day would there be the highest concentration of four-carbon organic acids in the vacuoles of CAM plants?

10.5 Captured Carbon Dioxide Is Reduced to Make Sugar

After you complete this section, you should be able to . . .

- Analyze the process used to reduce carbon dioxide to sugars.

The reactions analyzed in Section 10.3 are triggered by light. This is logical, because their entire function is focused on transforming electromagnetic energy in the form of sunlight into chemical energy in the bonds of ATP and NADPH. The reactions that produce sugar from carbon dioxide, in contrast, are not triggered directly by light. Instead, they depend on the ATP and NADPH produced by the light-capturing reactions.

The Calvin Cycle Is a Three-Step Process

The discovery of 3-phosphoglycerate (3PGA) as the first stable product formed after carbon fixation was an intriguing result, because 3PGA is also one of the 10 intermediates in glycolysis. The Calvin cycle manufactures carbohydrate; glycolysis breaks it down. Because the two processes are related in this way, it was logical that at least some intermediates in glycolysis and the Calvin cycle are the same. The complete Calvin cycle has three phases (Figure 10.26):

1. ***Fixation phase*** The Calvin cycle begins when CO_2 reacts with RuBP. This phase fixes carbon and produces two molecules of 3PGA, which is a phosphorylated three-carbon organic acid.
2. ***Reduction phase*** The 3PGA is phosphorylated by ATP and then reduced by accepting electrons from NADPH as the phosphate is removed. The product is the phosphorylated three-carbon sugar **glyceraldehyde-3-phosphate (G3P)**. Some of the G3P that is synthesized is drawn off to produce other organic molecules, like the six-carbon sugar glucose.
3. ***Regeneration phase*** The rest of the G3P keeps the cycle going by serving as the substrate for the third phase in the cycle: reactions that use additional ATP in the regeneration of RuBP.

All three phases take place in the stroma of chloroplasts. One turn of the Calvin cycle fixes one molecule of CO_2. Three turns of the cycle fix three molecules of CO_2, yielding one molecule of G3P and three fully regenerated RuBP (Figure 10.26).

The discovery of the Calvin cycle clarified how the ATP and NADPH produced by light-capturing reactions allow cells to fix and reduce CO_2 gas to carbohydrate $(CH_2O)_n$. Because sugars store a great deal of potential energy, producing them takes a great deal of chemical energy. In the Calvin cycle, each mole of CO_2 requires the energy from 3 moles of ATP and 2 moles of NADPH to fix it and reduce it to sugar.

✔ **QUANTITATIVE If you understand the Calvin cycle, you should be able to provide the *minimum number* of RuBP, ATP, and NADPH molecules that would be required to run through six complete cycles. Explain why you would not need six RuBP molecules to fix and reduce six CO_2.**

The conversion of CO_2 gas into carbohydrate is, without doubt, worthy of this energy investment. Plants use sugars to fuel cellular respiration and build all of the organic molecules in leaves

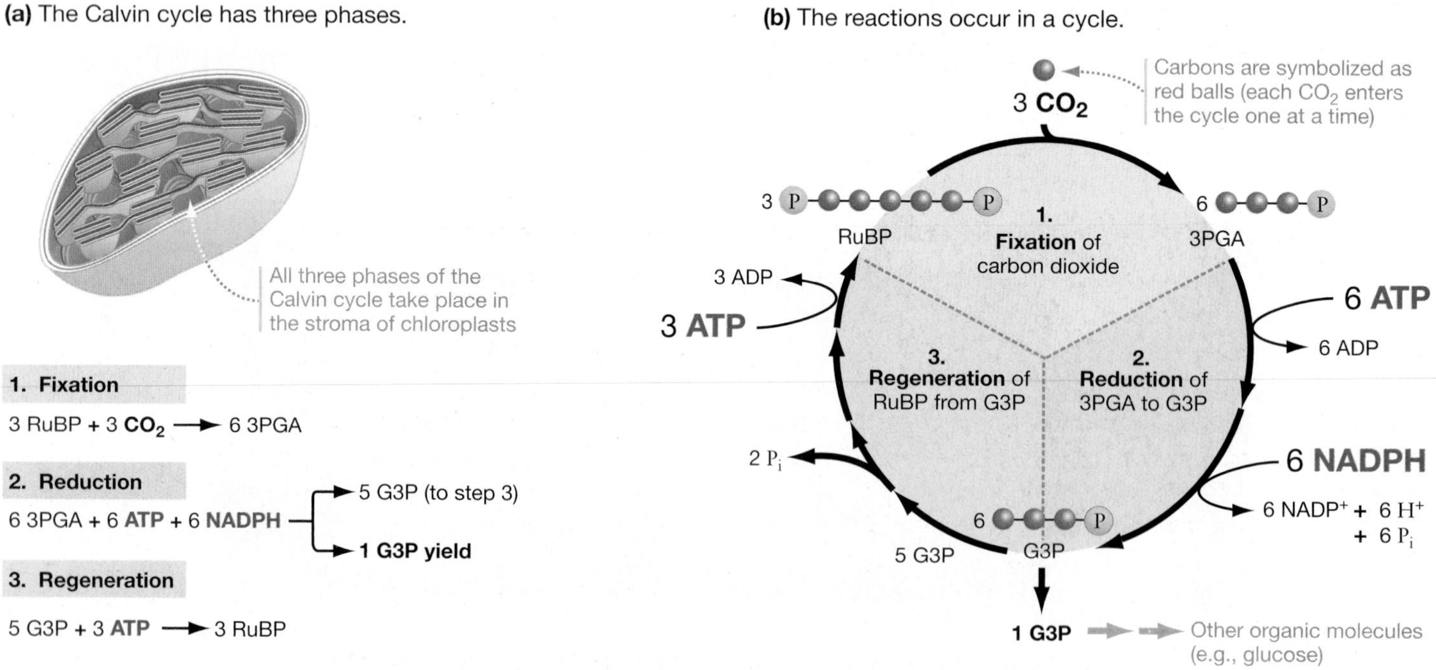

Figure 10.26 The Calvin Cycle Fixes and Reduces the Carbon in CO_2 and Regenerates RuBP. The number of reactants and products resulting from three turns of the cycle are shown. Of the six G3Ps that are generated during the reduction phase, one is used in the synthesis of other molecules, such as glucose, and the other five are used to regenerate RuBP. The three RuBPs that are regenerated participate in fixation reactions for additional turns of the cycle.

and other structures. Millions of non-photosynthetic organisms also depend on this reaction to provide the sugars they need for cellular respiration and raw material for synthesis of their own organic molecules (Ch. 9, Section 9.1).

How Is Photosynthesis Regulated?

Like cellular respiration, regulation of photosynthesis is based on the presence of inputs and outputs. Although the mechanisms responsible for turning photosynthesis on or off are still under investigation, several patterns have emerged:

- The presence of light triggers the production of proteins required for photosynthesis.
- When sugar supplies are high, the production of proteins required for photosynthesis is inhibited, but the production of proteins required to process and store sugars is stimulated.
- Rubisco is activated by regulatory molecules that are produced when light is available, but inhibited in conditions of low CO_2 availability—when photorespiration is favored.

The central message here is that the rate of photosynthesis is finely tuned to use resources efficiently in response to changes in environmental conditions.

What Happens to the Sugar That Is Produced by Photosynthesis?

The products of the Calvin cycle enter one of several reaction pathways that result in the production of every organic molecule in the photosynthetic organism. The most important of these reaction sequences uses G3P to produce the monosaccharide glucose, a process called **gluconeogenesis**. This glucose is often combined with fructose, which is also made from G3P, to form the disaccharide ("two-sugar") **sucrose**.

When photosynthesis is taking place slowly, almost all the G3P that is produced is used to make sucrose. Sucrose is water soluble and readily transported to other parts of the plant. If sucrose is delivered to rapidly growing parts of the plant, it is broken down to fuel cellular respiration and growth.

An alternative pathway occurs when photosynthesis is proceeding rapidly and sucrose is abundant. Under these conditions, the glucose molecules are polymerized to form **starch**, which is stored in the cells of leaves and roots. Starch production occurs inside the chloroplast; sucrose synthesis takes place in the cytosol.

In photosynthesizing cells, starch acts as a temporary sugar-storage product. At night, the starch that is stored in leaf cells is broken down to glucose molecules. The glucose is then fed into cellular respiration or used to manufacture sucrose for transport to other parts of the plant. In this way, chloroplasts provide sugars for cells throughout the plant by day and by night.

The role of photosynthesis in producing organic molecules extends well beyond the autotrophs, however. For example, when a mouse eats the starch that is stored in the leaves or seeds of a plant, the reduced carbons of starch supply the energy and carbon for the mouse's growth and reproduction. If an owl eats the mouse, the reduced carbons in the mouse's tissues fuel the predator's growth and reproduction. (You can see how energy is processed via photosynthesis and cellular respiration in the Big Picture on pages 236–237.)

In this way, virtually every carbon present in organic molecules, and the energy stored within their bonds, can be traced back to photosynthesis. Photosynthesis is the staff of life.

CHECK YOUR UNDERSTANDING

✔ If you understood this section, you should be able to . . .

1. Explain how light, sugar, and CO_2 levels affect the production and/or activity of proteins involved in photosynthesis.
2. Predict the relative concentration of starch in leaves at the start of the day versus the end of the day.

Answers are available in Appendix A.

CHAPTER 10 Review

For media, go to **Mastering Biology**

10.1 Photosynthesis Harnesses Sunlight to Make Carbohydrate from CO_2

- The light-capturing reactions of photosynthesis occur in internal membranes of the chloroplast that are organized into structures called thylakoids that stack to form grana.
- The Calvin cycle takes place in a fluid portion of the chloroplast called the stroma.
- The CO_2-reduction reactions of photosynthesis depend on the products of the light-capturing reactions: ATP and NADPH.

10.2 How Do Pigments Capture Light Energy?

- Pigment molecules capture light energy by the excitement of their electrons when a photon is absorbed. Each pigment absorbs photons of particular wavelengths.
- After a pigment molecule absorbs a photon, the excitation energy is quickly released as fluorescence and heat, heat alone, or resonance energy that excites another pigment; or it is transferred as an excited electron to reduce an electron acceptor.
- Antenna pigments in the thylakoid membrane transfer absorbed light energy via resonance to the reaction center, where an excited electron is transferred to an electron acceptor.

10.3 The Discovery of Photosystems I and II

- In photosystem II, excited electrons are transferred to plastoquinone (PQ) at the start of an electron transport chain. The redox reactions in the ETC are used to generate a proton-motive force that drives the synthesis of ATP. Electrons taken from photosystem II are replaced by splitting water, releasing oxygen and protons.

- In photosystem I, excited electrons are passed to ferredoxin. In an enzyme-catalyzed reaction, the reduced form of ferredoxin passes electrons to $NADP^+$, forming NADPH.
- The Z scheme connects photosystems II and I. Electrons excited by light in photosystem II are passed through the ETC, picked up by plastocyanin (PC), and transferred to oxidized pigments in the photosystem I reaction center. These electrons are again excited by light in photosystem I and subsequently used to reduce $NADP^+$ to NADPH.
- Electrons from photosystem I may occasionally be passed back to plastoquinone instead of $NADP^+$. This cyclic flow of electrons between the photosystem I and the ETC boosts ATP supplies.

10.4 How Do Cells Capture Carbon Dioxide?

- Stomata regulate the exchange of CO_2 and O_2 between the leaf tissue of a plant and its environment.
- Rubisco catalyzes the fixation of CO_2 to a five-carbon compound called ribulose bisphosphate (RuBP) as the first step in the Calvin cycle. The six-carbon compound that results immediately splits to form two molecules of 3-phosphoglycerate (3PGA).
- Rubisco catalyzes the addition of oxygen as well as carbon dioxide to RuBP. The reaction with oxygen leads to a loss of fixed CO_2 and ATP through a process called photorespiration.
- In C_4 plants and CAM plants, CO_2 is initially fixed to four-carbon compounds, then released to be used by rubisco. This increases CO_2 levels in plant tissues and reduces the effect of photorespiration when stomata are closed.

10.5 Captured Carbon Dioxide Is Reduced to Make Sugar

- The 3PGA produced after CO_2 fixation by rubisco is phosphorylated by ATP and reduced by NADPH to produce glyceraldehyde-3-phosphate (G3P).
- Some G3P is used to synthesize other organic molecules, like glucose; ATP phosphorylates the rest in a series of reactions to regenerate RuBP so the cycle can continue.
- Photosynthesis is stimulated by the presence of light and reduced when sugar supplies are high or when CO_2 availability is low.
- G3P is mostly converted into sucrose for distribution throughout the plant when photosynthetic activity is low, or into starch for storage when the activity is high.

Answers are available in Appendix A.

✔ TEST YOUR KNOWLEDGE

1. How is energy transferred among antenna pigment molecules?
   a. by heat
   b. by redox reactions
   c. by fluorescence
   d. by resonance
2. Why is chlorophyll green?
   a. It absorbs all wavelengths in the visible spectrum.
   b. It absorbs wavelengths only in the red portions of the spectrum (680–700 nm).
   c. It absorbs wavelengths only in the blue part of the visible spectrum (450–480 nm).
   d. It absorbs wavelengths in only the blue and red parts of the visible spectrum.
3. What is the role of PEP carboxylase in C_4 and CAM plants?
   a. It fixes CO_2 into an organic acid.
   b. It produces ATP for the Calvin cycle.
   c. It replaces rubisco in the Calvin cycle.
   d. It releases CO_2 from organic acids.
4. At what point in photosynthesis is the electromagnetic energy of light first converted into chemical energy?

✔ TEST YOUR UNDERSTANDING

5. Why is the chlorophyll in chloroplasts less likely to produce fluorescence compared to extracted chlorophyll molecules?
6. Describe the three phases of the Calvin cycle and how the products of the light-capturing reactions participate in this process.
7. **QUANTITATIVE** Apply what you know of the relationship between the light-capturing reactions and the Calvin cycle to calculate the number of photons used to produce a new G3P and regenerate RuBP. (Assume 1 ATP is produced for each pair of electrons used to form NADPH.)
8. **THINK CAREFULLY** Which of the following correctly identify a role of the ATP produced in the light-capturing reactions? Select True or False for each statement.
   T/F It is used by rubisco to fix CO_2 to RuBP.
   T/F It serves the same role as ATP produced by mitochondria.
   T/F It is used to regenerate RuBP from G3P molecules.
   T/F It is used to produce G3P molecules

✔ TEST YOUR PROBLEM-SOLVING SKILLS

9. Predict how the following conditions would affect the production of O_2, ATP, and NADPH and state whether noncyclic or cyclic electron flow would occur in each: (1) Only blue photons hit a chloroplast; (2) blue and red photons hit a chloroplast, but no $NADP^+$ is available; (3) blue and red photons hit a chloroplast, but a proton channel has been introduced into the thylakoid membrane, so it is fully permeable to protons.
10. An investigator exposes chloroplasts to 700-nm photons and observes low O_2 production, but high ATP production. Which of the following best explains this observation?
    a. The electrons from water are directly transferred to $NADP^+$, which is used to generate ATP.
    b. Photosystem II is not splitting water, and the ATP is being produced by cycling electrons via photosystem I.
    c. The O_2 is being converted to water as a terminal electron acceptor in the production of ATP.
    d. Electron transport has stopped and ATP is being produced by the Calvin cycle.

✔ PUT IT ALL TOGETHER: Case Study

Can green algae replace "black gold"?

Over the past decade, there has been great interest in developing a sustainable and renewable energy source that would displace our dependence on fossil fuels. Photosynthetic organisms, like the microalgae being cultured in the tanks shown above, are currently being used to produce bioethanol, biodiesel, and bio-H_2. How are photosynthetic organisms being engineered to promote the production of these biofuels?

11. Aquatic plants and algae carry out photosynthesis using the same light reactions and Calvin cycle as in land plants. Would you expect any of these organisms to also require the C_4 pathway? Describe the role of the C_4 pathway in land plants and speculate if this could ever be necessary in an aquatic environment.

12. Fossil fuel consists of coal, oil, and natural gas that have been formed from the remains of plants and animals that died millions of years ago. What was the source of the energy used to produce the high-energy molecules in these fossil fuels?

13. One of the contributing factors to environmental pollution and climate change is the production of CO_2 via the combustion of fossil fuels. What is a benefit of using biofuels, produced by living organisms, if they will release similar amounts of CO_2 when used to meet our energy demand?

14. A promising group of organisms for biofuel production are green algae like *Chlamydomonas reinhardtii*. What type of carbohydrate is most likely being produced from G3P when these single-celled organisms are grown under conditions that maximize photosynthetic activity?

15. **PROCESS OF SCIENCE** To increase the photosynthetic efficiency of *C. reinhardtii*, researchers engineered the organism by decreasing the size of light-harvesting complexes used in the two photosystems. Comparisons of the photosynthetic activity and growth rate between this engineered strain and a control under high-light conditions are shown below. Summarize the results, and propose a hypothesis to explain how reducing the size of the light-harvesting complexes could benefit cultures shown in the photograph that are grown in full sunlight.

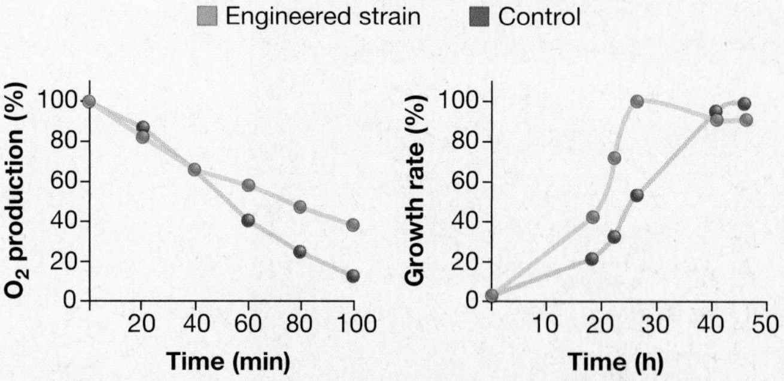

Source: J. H. Mussgnug et al. 2007. *Plant Biotechnology Journal* 5: 802–814.

16. Propose an alternative method to enhance the rate of photosynthesis in microalgae cultures exposed to sunlight, but without genetic modification.

Mastering Biology

Students Go to Mastering™ Biology for assignments, the eText, and the Study Area with animations, practice tests, and activities.

Professors Go to Mastering™ Biology for automatically graded tutorials and questions that you can assign to your students, plus Instructor Resources.

BIG PICTURE

THE BIG PICTURE

It takes energy to stay alive. Use this concept map to review how energy is harvested and used by cells to drive the reactions required to sustain life.

As you read the map, remember that chemical energy is a form of potential energy. Potential energy is based on the position of matter in space, and chemical energy is all about the position of electrons in covalent bonds. When hydrogen gas reacts explosively with oxygen, all that's happening is that electrons are moving from high-energy positions to lower-energy positions.

In essence, organisms transform energy from the Sun into chemical energy in the C—C and C—H bonds of sugars, such as glucose, and then into chemical energy in the bonds between phosphates in ATP.

The potential energy in ATP allows cells to do work: pump ions, synthesize molecules, move cargo, and transfer information.

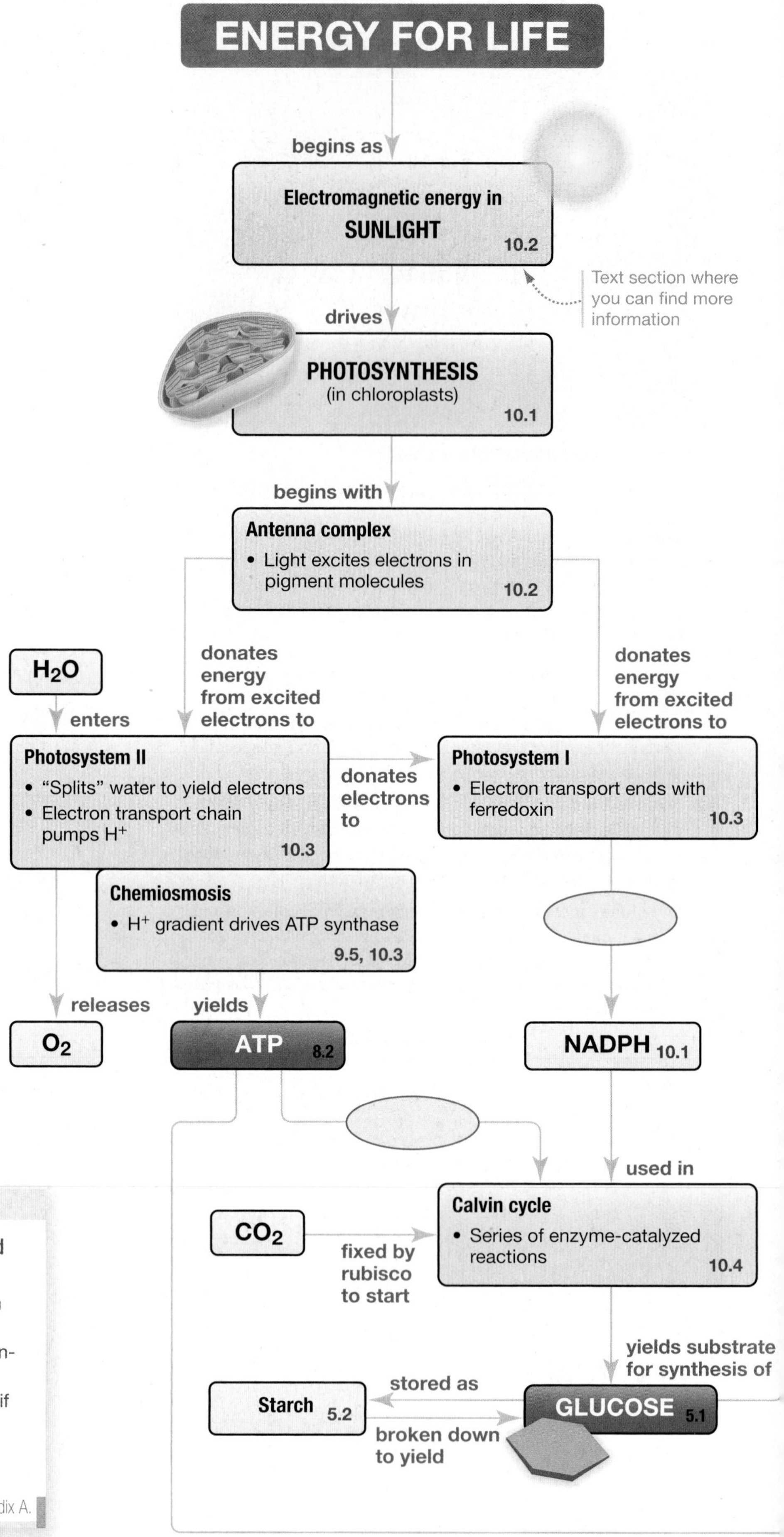

CHECK YOUR UNDERSTANDING

✔ If you understand the big picture, you should be able to . . .

1. Explain how H_2O and O_2 are cycled between photosynthesis and cellular respiration.
2. Explain how CO_2 is cycled between photosynthesis and cellular respiration.
3. Describe what might happen to life on Earth if rubisco were suddenly unable to fix CO_2.
4. Fill in the blue ovals with appropriate linking verbs or phrases.

Answers are available in Appendix A.

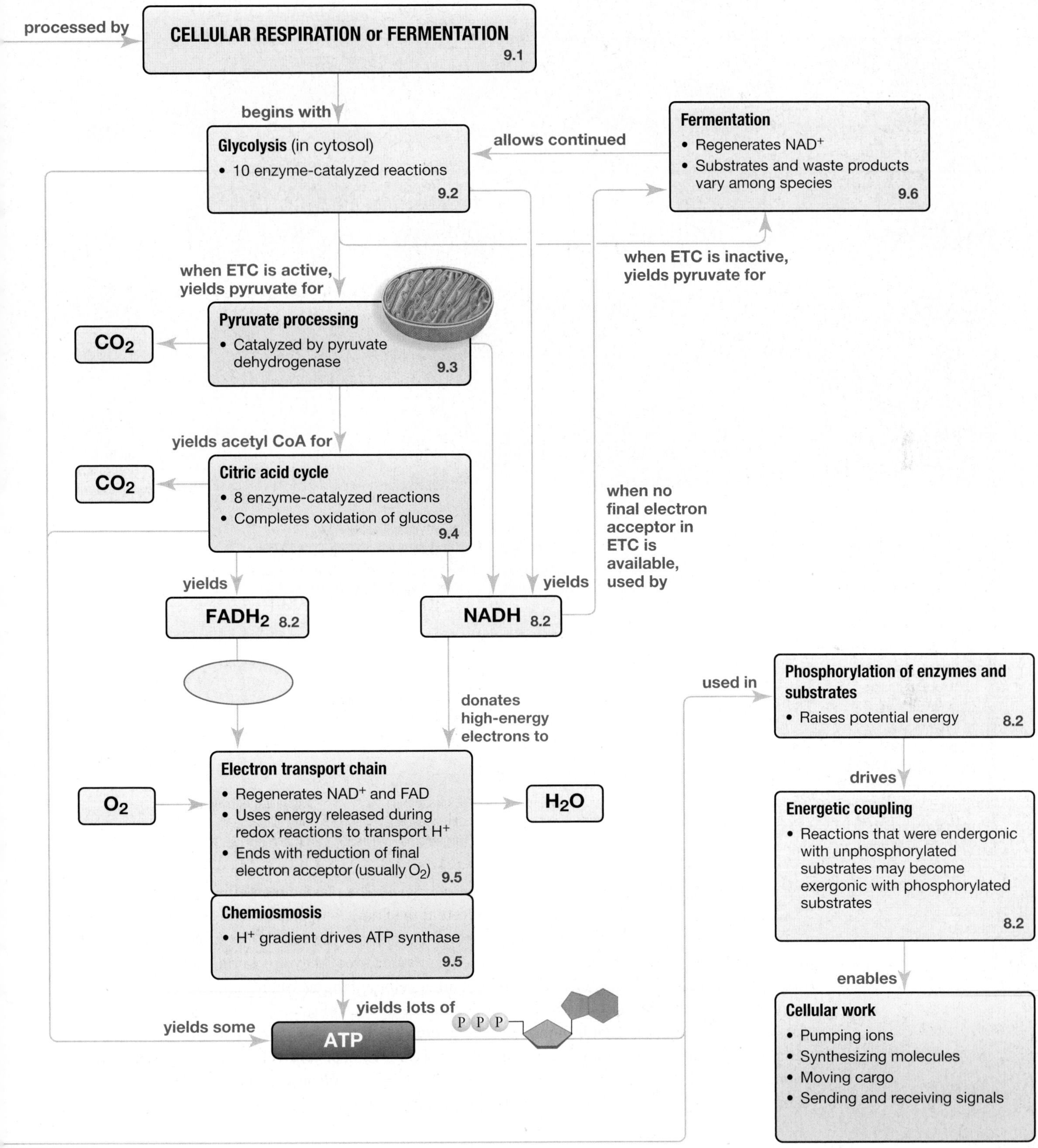
processed by
CELLULAR RESPIRATION or FERMENTATION 9.1
begins with
Glycolysis (in cytosol)
• 10 enzyme-catalyzed reactions 9.2
allows continued
Fermentation
• Regenerates NAD^+
• Substrates and waste products vary among species 9.6
when ETC is active, yields pyruvate for
when ETC is inactive, yields pyruvate for
Pyruvate processing
• Catalyzed by pyruvate dehydrogenase 9.3
CO_2
yields acetyl CoA for
Citric acid cycle
• 8 enzyme-catalyzed reactions
• Completes oxidation of glucose 9.4
CO_2
when no final electron acceptor in ETC is available, used by
yields
yields
$FADH_2$ 8.2
NADH 8.2
donates high-energy electrons to
Electron transport chain
• Regenerates NAD^+ and FAD
• Uses energy released during redox reactions to transport H^+
• Ends with reduction of final electron acceptor (usually O_2) 9.5
O_2
H_2O
Chemiosmosis
• H^+ gradient drives ATP synthase 9.5
yields lots of
yields some
ATP
P P P
used in
Phosphorylation of enzymes and substrates
• Raises potential energy 8.2
drives
Energetic coupling
• Reactions that were endergonic with unphosphorylated substrates may become exergonic with phosphorylated substrates 8.2
enables
Cellular work
• Pumping ions
• Synthesizing molecules
• Moving cargo
• Sending and receiving signals

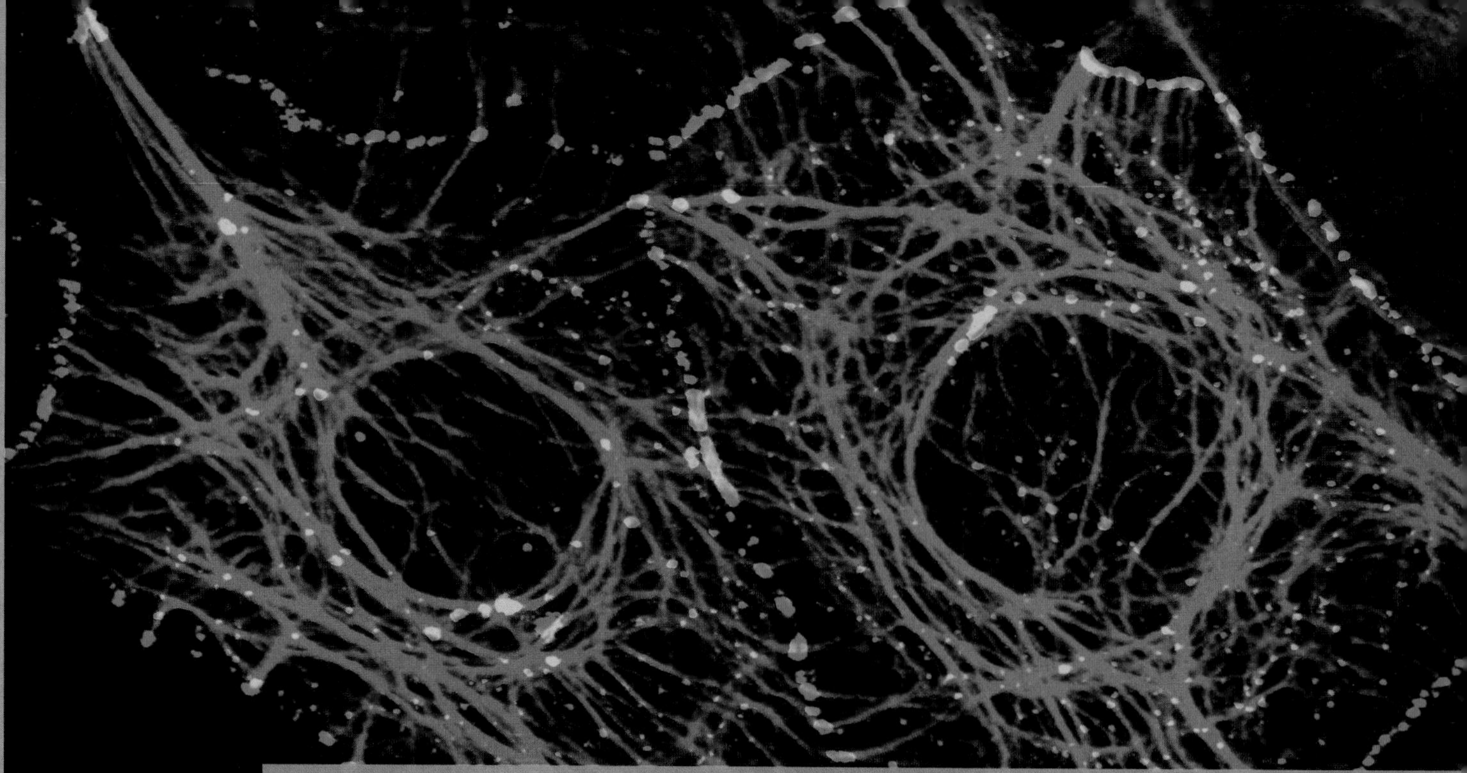

11 Cell–Cell Interactions

In this confocal micrograph of cultured human skin cells, intermediate filaments are stained red and desmosomes are stained green. The desmosomes anchor adhesion proteins at the membrane to intermediate filaments in the cytosol. Without these strong intercellular connections, a simple handshake could rip your skin cells apart.

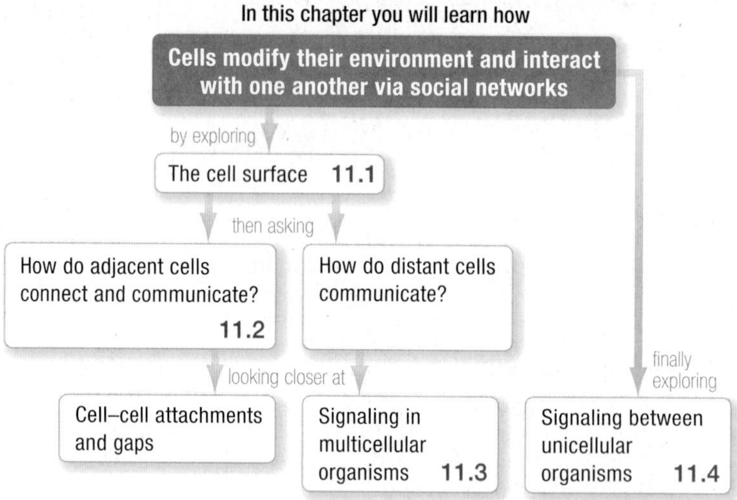

A diversity of events takes place at the cellular level. The plasma membrane surrounds a bustling enterprise consisting of organelles, molecular machines, and cytoskeletal elements (see Chapters 6 and 7). Molecular motors transport cargo throughout the cell at breathtaking speed. It would be a mistake, however, to think that cells are self-contained—that they are worlds in and of themselves. Instead, cells are dependent on interactions with other cells and the surrounding environment.

For most unicellular species, the outside environment is teeming with other organisms. Inside your gut, for example, hundreds of billions of bacterial cells are jostling for space and resources. Besides interacting with other individuals, every unicellular organism must contend with constant shifts in environmental conditions, such as heat, light, ion concentrations, and food supplies. If unicellular organisms cannot sense these conditions and respond appropriately, they die.

In multicellular species, the environment outside the cell is made up of other cells, both neighboring and distant. The cells that make up a redwood tree, a mushroom, or your body are intensely social. Although biologists often study cells in isolation, an individual tree, fungus, or person is actually an interdependent community of cells. If those cells do not communicate and cooperate, the whole will break into dysfunctional parts and die.

To understand the life of a cell, then, it is critical to analyze how the cell interacts with the world outside its membrane. How do cells obtain information about the world and respond to that information? In particular, how do cells interact with other cells? To answer these questions, let's begin at the cell surface—with the molecules that connect the cell with its environment.

11.1 The Cell Surface

The line between life and nonlife is drawn by the plasma membrane that surrounds every cell. Recall that the structure of this membrane consists of a phospholipid bilayer studded with membrane proteins. These proteins are integral, meaning embedded in the bilayer, or peripheral, meaning attached to one surface. Some membrane proteins regulate the transport of substances as part of the primary function of the plasma membrane: to create an environment inside the cell that is different from the conditions outside (Ch. 6, Section 6.4).

After you complete this section, you should be able to...

- Analyze the extracellular material produced by cells.

The plasma membrane does not exist in isolation, however. Many membrane proteins attach to cytoskeletal elements on the interior surface of the bilayer (Ch. 7, Section 7.6) or to a complex array of extracellular structures, including those attached to the membranes of neighboring cells. Let's consider the nature of the material outside the cell and then analyze how the cell interacts with it and other cells.

The Structure and Function of an Extracellular Layer

It is extremely rare for cells to be bounded simply by a plasma membrane. Most cells secrete products that are assembled into a layer or wall just beyond the membrane. This extracellular material helps define the cell's shape and either attaches it to other cells or acts as a first line of defense against the outside world.

The structure of cell walls surrounding prokaryotic cells is remarkably different between bacteria and archaea. In bacteria, cell walls mostly consist of polymers of the polysaccharide peptidoglycan that are connected to one another by peptide bonds (Ch. 5, Section 5.2). Archaea do not share any unifying characteristics in their cell walls apart from the absence of peptidoglycan. Often the cell walls of these organisms are formed as a dense coat of proteins on the surface of the cell, called an S-layer.

In contrast, virtually all types of extracellular layers in eukaryotes—from the cell walls of algae, fungi, and plants to the extracellular material that surrounds most animal cells—have the same fundamental organization. Like reinforced concrete, they are fiber composites: They consist of a cross-linked network of long filaments embedded in a stiff surrounding material called the ground substance (**Figure 11.1**). The molecules that make up the filaments and ground substance vary among organisms, but the engineering principle is the same. Why?

Figure 11.1 Fiber Composites Resist Tension and Compression. Reinforced concrete is an example of a fiber composite consisting of a ground substance (concrete) that fills spaces between cross-linked fibers (steel rods).

- The rods or filaments in a fiber composite are extremely effective at withstanding stretching and straining forces, or tension. The filaments in the extracellular material of most cells are functionally similar to the steel rods in reinforced concrete—they resist being pulled or pushed lengthwise.
- The stiff ground substance is effective at withstanding pressing forces, called compression. Concrete performs this function in highways, and a gel-forming mixture of polysaccharides plays the same role in extracellular material.

Thanks to the combination of tension- and compression-resisting elements, fiber composites are particularly rugged. In many living cells, fiber composites are flexible as well as strong.

While the overall organization of these composites is shared among many organisms, the molecular components vary considerably. Let's focus our attention on the extracellular layers produced by multicellular eukaryotes. What molecules make up the filaments and ground substance found on the surface of plant and animal cells? Where are these extracellular materials synthesized, and what do they do?

The Cell Wall in Plants

Virtually all plant cells are surrounded by a cell wall—a fiber composite that is the basis of major industries. The paper in this book, the threads in your cotton clothing, and the wood in your neighborhood's houses are made up primarily of plant cell walls.

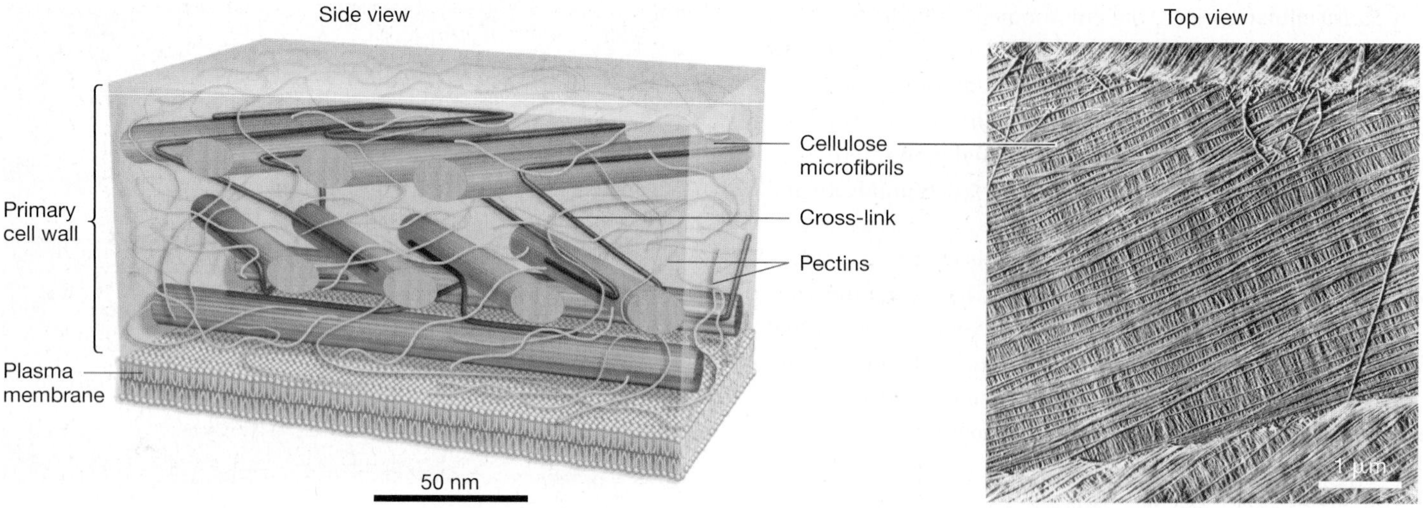

Figure 11.2 Primary Cell Walls of Plants Are Fiber Composites. In a plant's primary cell wall, cellulose microfibrils are connected by polysaccharide cross-links. The spaces between the microfibrils and cross-links are filled with molecules such as pectin, which form a gelatinous solid.

Primary Cell Walls When plant cells first form, they secrete an initial fiber composite called a **primary cell wall**.

- The fibrous component of the primary cell wall consists of long strands of the polysaccharide cellulose. These strands are bundled into stout structures termed **microfibrils**, which are cross-linked via hydrogen bonds to other polysaccharide filaments. The microfibrils are synthesized directly into the extracellular space by a complex of enzymes in the plasma membrane, where they form a crisscrossed network (Figure 11.2).
- The spaces between microfibrils are filled with gelatinous polysaccharides such as **pectins**—the molecules that are used to thicken jams and jellies. Because these polysaccharides are hydrophilic, they attract and hold large amounts of water, keeping the cell wall moist. The gelatinous components of the cell wall are synthesized in the rough endoplasmic reticulum (ER) and Golgi apparatus and secreted into the extracellular space (Ch. 7, Section 7.5).

The primary cell wall helps shape a plant cell. Under normal conditions, the concentration of solutes is higher inside the cell than outside, causing water to enter the cell via osmosis. The incoming water increases the cell's volume, pushing the plasma membrane up against the wall. The force exerted by the cell against the wall is known as **turgor pressure**.

Although plant cells exert turgor pressure throughout their lives, it is particularly important in young cells that are actively growing. Young plant cells secrete proteins named expansins into their cell wall. **Expansins** disrupt the hydrogen bonds that cross-link microfibrils to other polymers in the wall, loosening the structure and allowing the microfibrils to slide past one another. Turgor pressure then forces the wall to elongate and expand, allowing for cell growth (Ch. 37, Section 37.2).

✔ **If you understand the structure and function of the primary cell wall, you should be able to predict what would happen to a plant cell if it were treated with an enzyme that digests cellulose microfibrils.**

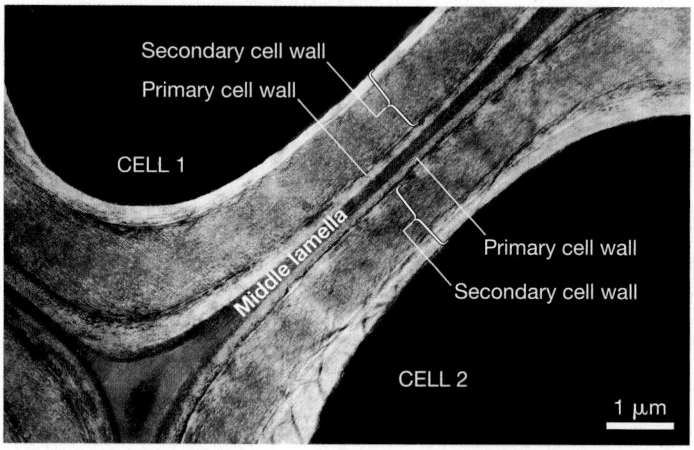

Figure 11.3 Plant Cells Often Secrete Two Cell Wall Layers.

Secondary Cell Walls As plant cells mature and stop growing, they may secrete an additional layer of material—a **secondary cell wall**—between the plasma membrane and the primary cell wall (Figure 11.3). The makeup of the secondary cell wall varies from cell to cell in the plant and correlates with each cell's function. Cells on the surface of a leaf have secondary cell walls containing waxes that form a waterproof coating; cells that support a plant's stem have stiff secondary cell walls that contain a great deal of cellulose.

In cells that form wood, the secondary cell wall also contains **lignin**, a complex polymer that forms an exceptionally rigid network. Thick secondary cell walls of cellulose and lignin help woody plants withstand the forces of gravity and wind.

The Extracellular Matrix in Animals

Most animal cells secrete a fiber composite called the **extracellular matrix**, or simply **ECM**. Like the extracellular materials found in other organisms, the ECM provides structural support.

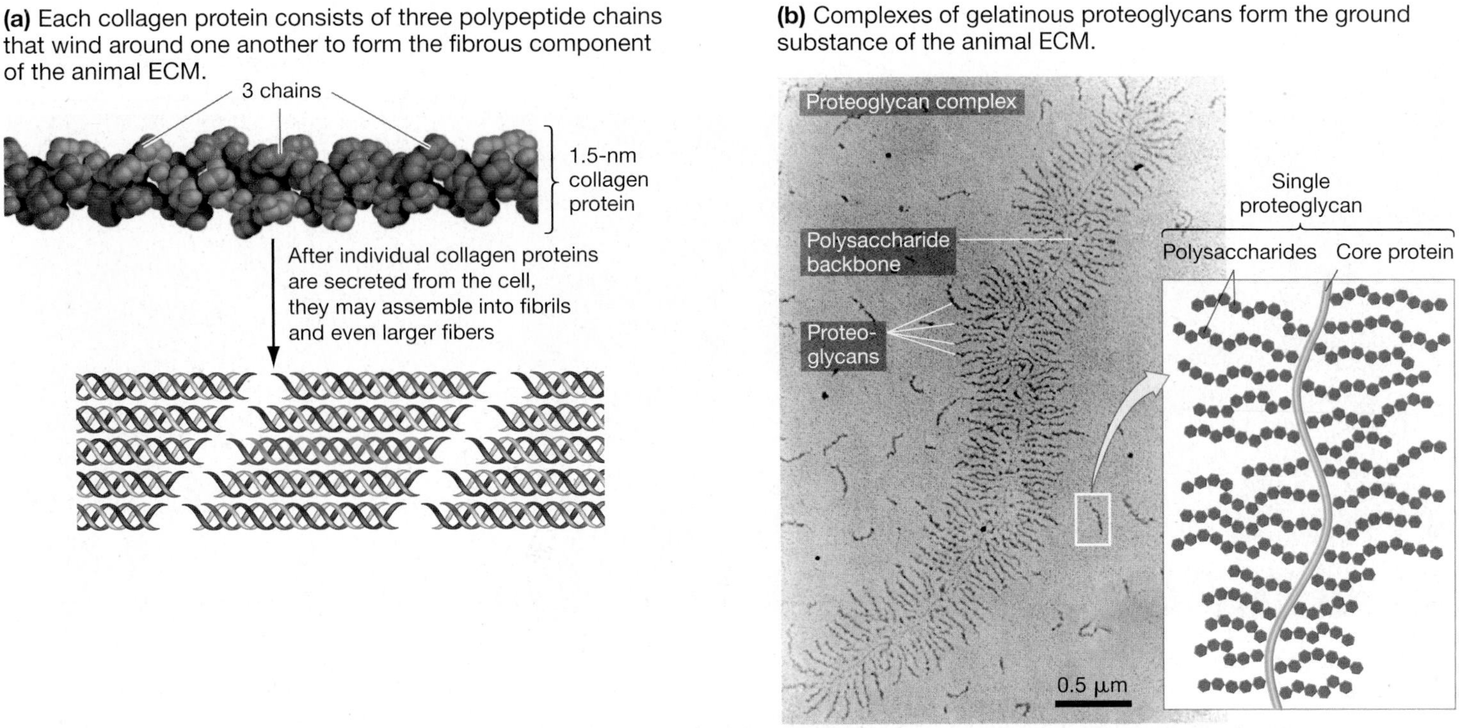

Figure 11.4 The Extracellular Matrix of Animals Is a Fiber Composite. **(a)** Although several types of fibrous proteins are found in the ECM, the most abundant is collagen. After collagen is secreted from the cell, the triple-helix proteins can assemble into fibrils and even larger cable-like fibers. **(b)** The spaces between collagens are filled with a ground substance consisting of proteoglycans. Each individual proteoglycan consists of a core protein attached to many polysaccharides (inset). In some tissues, proteoglycans are assembled into even larger complexes.

ECM organization follows the same principles observed in the cell walls of algae, fungi, and plants. There is a key difference, however: The animal ECM contains much more protein relative to carbohydrate than does a cell wall.

The fibrous component of animal ECM is dominated by glycoproteins named **collagen** (Figure 11.4a, top). About a quarter to a third of all the protein in your body is collagen.

Most ECM proteins are synthesized in the rough ER, processed in the Golgi apparatus, and secreted from the cell via exocytosis. After secretion, the individual proteins may assemble into larger structures. For example, groups of collagen triple helixes may coalesce to form collagen fibrils (Figure 11.4a, bottom), and bundles of fibrils may link to form even larger fibrous complexes.

The ground substance that surrounds collagen and other fibrous components of the ECM contains highly glycosylated, gel-forming proteins called **proteoglycans** (Figure 11.4b, inset). In addition, secreted proteoglycans may be attached to long polysaccharides synthesized by cellular enzymes in the extracellular space. The resulting huge complexes, such as the one shown in the photo in Figure 11.4b, are responsible for the rubber-like consistency of cartilage.

Even in the same organism, the amount of ECM varies among different types of tissues. A **tissue** consists of similar cells that function as a unit. Bone and cartilage, for example, have relatively few cells surrounded by a large amount of ECM. Skin cells, in contrast, are packed together with a minimal amount of ECM.

The composition of the ECM also varies among tissue types. For example, the ECM surrounding cells in your lungs contains large amounts of a rubber-like protein called elastin, which allows the ECM to expand and contract when you breathe. The structure of a tissue's ECM correlates with the function of the tissue.

Although collagen and the other common ECM proteins are much more elastic and bendable than the stiff cell walls of plants, they support cell structure via their attachments to the cell surface. As Figure 11.5 on page 242 shows, membrane proteins called **integrins** bind to extracellular cross-linking proteins, including **laminins**, which in turn bind to other components of the ECM. (Don't confuse laminins with lamins, which are intermediate filaments found in the nucleus; see Ch. 7, Section 7.6.)

The intracellular portions of the integrins bind to proteins that are connected to the cytoskeleton, effectively linking the cytoskeleton and ECM. This linkage is critical. Besides keeping individual cells in place, it helps adjacent cells adhere to each other via their common connection to the ECM.

Cells monitor the cytoskeleton–ECM linkage via signaling pathways that will be introduced in Section 11.3. When integrins bind to the ECM, they transmit signals that inform the cell it is in the right place and properly anchored. If this linkage breaks down, the signals are not transmitted and cells normally die as a result. For most of the cells in your body, anchorage to the ECM is a matter of life and death.

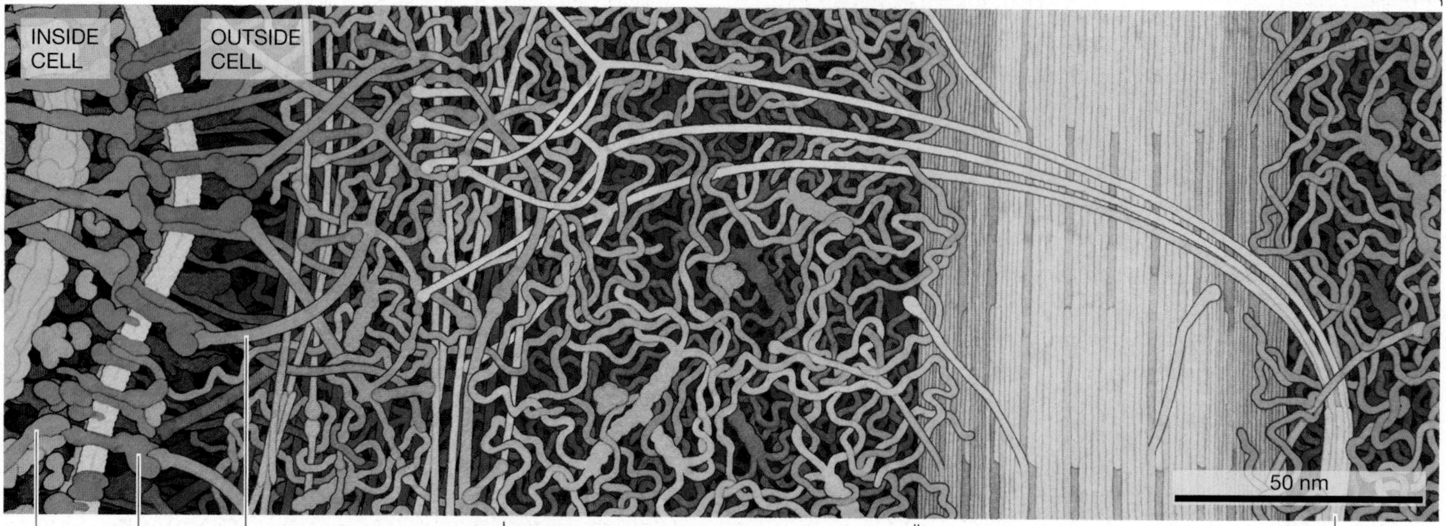

Figure 11.5 Integrins Connect the Cytoskeleton to the Extracellular Matrix.

Much like the ECM of animal cells, the walls surrounding plant cells can also be used to connect adjacent cells. Let's now turn to intercellular connections and learn more about their structure and function.

> **CHECK YOUR UNDERSTANDING**
>
> ✔ **If you understood this section, you should be able to...**
>
> Compare and contrast the molecular composition of the animal cell ECM and cell walls of plant cells.
>
> Answers are available in Appendix A.

11.2 How Do Adjacent Cells Connect and Communicate?

> **After you complete this section, you should be able to...**
>
> - Analyze the interactions between adjacent cells.

Direct physical connections between cells are the basis of **multicellularity**. These cell–cell connections—along with indirect connections via the ECM—maintain the structure and function of tissues. The muscle tissue in your heart, for example, depends on cell–cell connections to maintain its integrity and synchronize the individual cells as they contract and relax with each heartbeat.

Let's look first at the structures that attach cells to each other and then examine how adjacent cells are able to exchange substances and information.

Cell–Cell Attachments in Multicellular Organisms

Materials and structures that bind cells together are particularly important in **epithelia** (singular: **epithelium**)—tissues that form external and internal surfaces. Epithelia function as barriers between the external and internal environments of plants and animals. In animals, epithelia also serve as gatekeepers that regulate the transport of substances, such as the absorption of water and nutrients across the epithelia of the intestines (Ch. 39, Section 39.2).

The adhesive structures that hold cells together vary among organisms. Let's examine this variation by considering the intercellular connections in plants and animals.

Indirect Cell–Cell Attachments The extracellular space between the walls of adjacent plant cells sandwich a central layer, the middle lamella, which consists primarily of gelatinous pectins. Because the middle lamella is continuous with the primary cell walls of adjacent cells, it serves to glue them together (**Figure 11.6**). The two cell walls are like slices of bread, and the middle lamella is like a layer of peanut butter. If enzymes degrade the middle lamella, as they do when flower petals and leaves detach and fall, the adjacent cells separate.

Recall that in many animal tissues, integrins in the plasma membranes of cells will form connections between their cytoskeletal structures and the extracellular matrix (see Figure 11.5). By interacting with the same network of ECM components, multiple cells both within and between different tissues can be indirectly attached to one another. The large collagen fibrils that run throughout the ECM are particularly important in reinforcing these interactions.

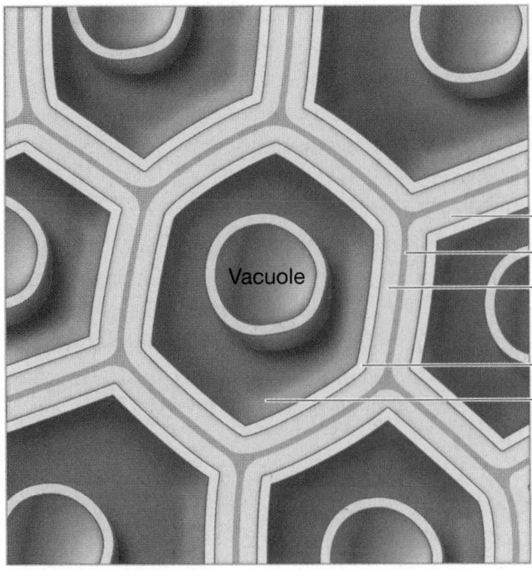

Figure 11.6 The Middle Lamella Connects Adjacent Plant Cells.

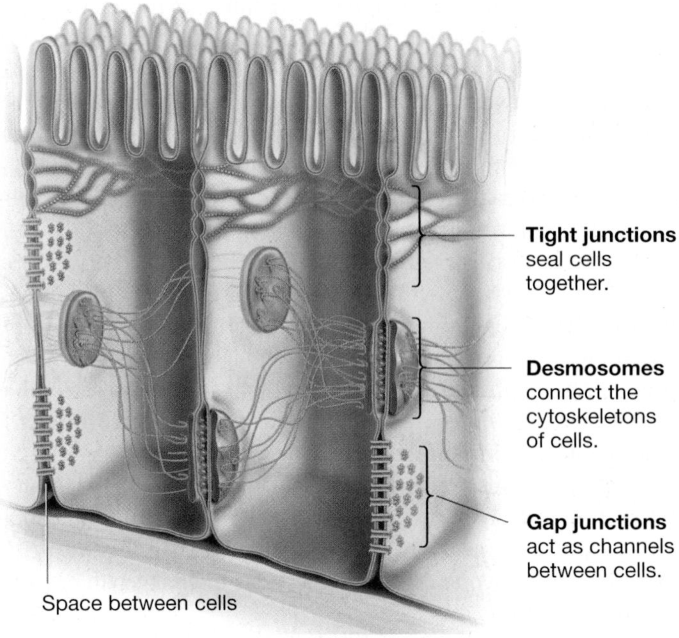

Figure 11.7 An Array of Structures Are Involved in Cell–Cell Adhesion and Communication between Animal Cells: An Overview. Shown here are idealized intestinal epithelial cells. (Cell–cell attachments are not drawn to scale.)

In contrast to such indirect intercellular connections, in animals, where cell walls do not exist, a variety of membrane proteins allow for *direct* cell–cell attachments in epithelia and other tissues (Figure 11.7). Let's start by looking at tight junctions and desmosomes.

Tight Junctions Form Waterproof Seals A **tight junction** is a cell–cell attachment composed of specialized proteins in the plasma membranes of adjacent animal cells (Figure 11.8a). As the drawing in Figure 11.8b indicates, long chains of these proteins form on the surface of a cell and attach to the same proteins on adjacent cells. The tight interactions between these proteins will pull the membranes of the two cells very close together. The resulting structure resembles a quilt, where the proteins "stitch" the membranes of two cells together. In cells, the structure forms a watertight seal that prevents solutions from flowing through the space between two cells.

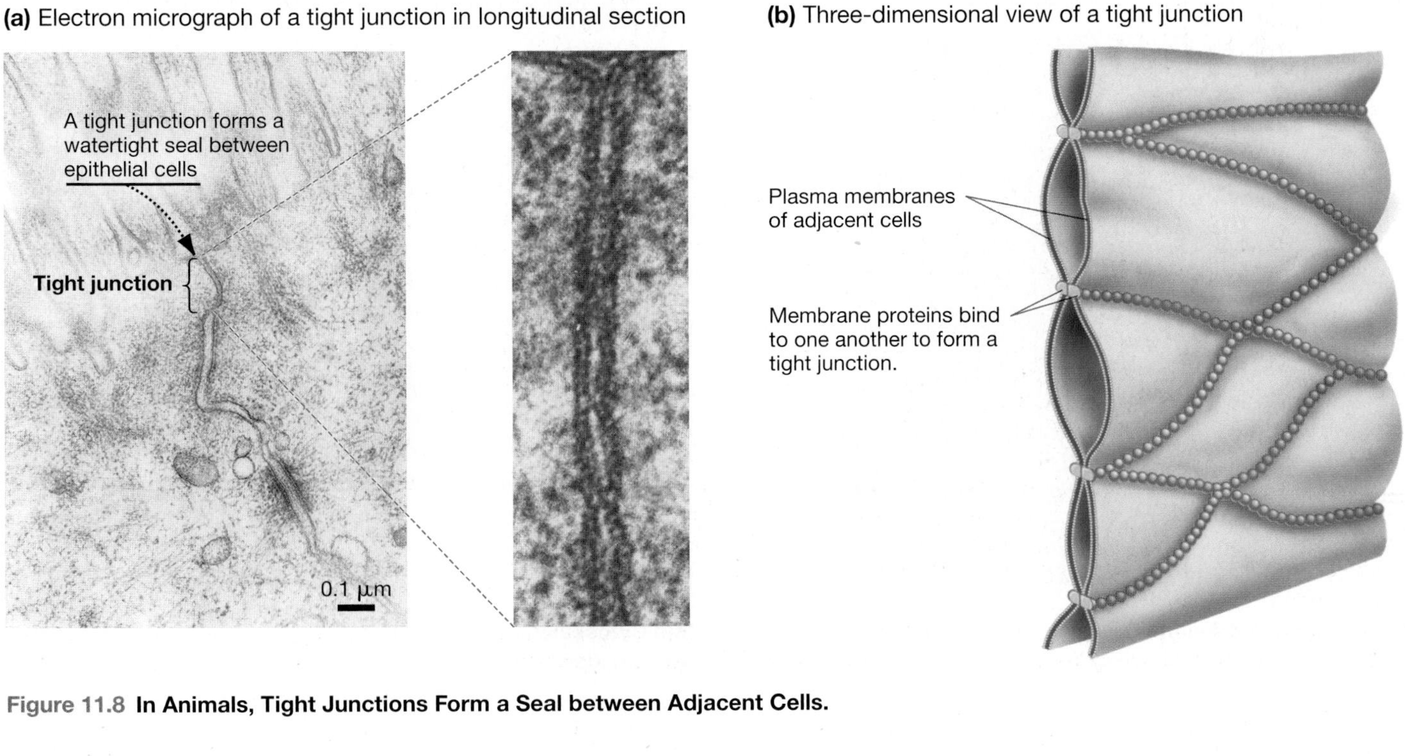

Figure 11.8 In Animals, Tight Junctions Form a Seal between Adjacent Cells.

Tight junctions are commonly found between cells that form a barrier, such as the epithelial cells lining your stomach and intestines. There, tight junctions restrict the passive movement of substances between the inside of your gut and the rest of your body. Selected nutrients may cross the epithelia via specialized transport proteins in the plasma membrane (Ch. 6, Section 6.4).

Although all tight junctions stitch together adjacent cells, their ability to restrict the movement of substances will vary in different tissues. For example, the tight junctions between the cells lining your bladder draw the cells closer together than those between the cells lining your small intestine, because they consist of different proteins. As a result, small ions can pass between the cells lining the surface of the small intestine more easily than between those lining the bladder—helping you absorb ions in your food and eliminate them in your urine.

Tight junctions are also dynamic. For example, they loosen to permit more transport between epithelial cells lining the small intestine after a meal and then retighten later. In this way, tight junctions can open and close in response to changes in environmental conditions.

Although tight junctions are very good at holding cells close together, they are weak adhesions that can be easily broken. Since epithelial cells often experience pulling and shearing forces, other intercellular adhesions are required to help hold cells together in such a tissue. What are these other adhesions, and how do they resist being pulled apart?

Desmosomes Form Secure Adhesions Figure 11.9a illustrates a **desmosome**, a strong cell–cell attachment particularly common in animal epithelial cells and certain muscle cells. In their structure and function, desmosomes are analogous to the rivets that hold pieces of sheet metal together.

As Figure 11.9b indicates, desmosomes comprise linking proteins and cytosolic anchoring proteins. The linking proteins span the membrane and directly connect adjacent cells and their anchoring proteins located on the inner faces of each cell membrane. Cytoskeletal intermediate filaments (Ch. 7, Section 7.6) help reinforce desmosomes by attaching to the intracellular anchoring proteins. In this way, desmosomes help form a continuous structural support system between all the cells in the tissue (see Figure 11.7 and the micrograph on the opening page of this chapter).

What are the linking proteins that serve this cell connection function in desmosomes? The answer to this question traces back to some of the first experiments conducted on cell–cell interactions.

Cell–Cell Adhesions Are Selective Long before electron micrographs revealed the presence of desmosomes, biologists realized that some sort of molecule must bind animal cells to one another. This insight grew out of experiments conducted by H. V. Wilson on sponges in the early 1900s.

Sponges are aquatic animals, and the sponge species used in these experiments consists of just two basic types of tissues. When Wilson treated adult sponges with chemicals that made the cells separate from one another, the result was a jumbled mass of individual, unconnected cells. But when normal chemical conditions were restored, the cells gradually began to move and stick together.

As the experiment continued, cells began to aggregate based on their origin—adhering to other cells of the same tissue type. This phenomenon is now called **selective adhesion**. Eventually, the cells re-formed functional adult sponges with two distinct tissues. How could this happen?

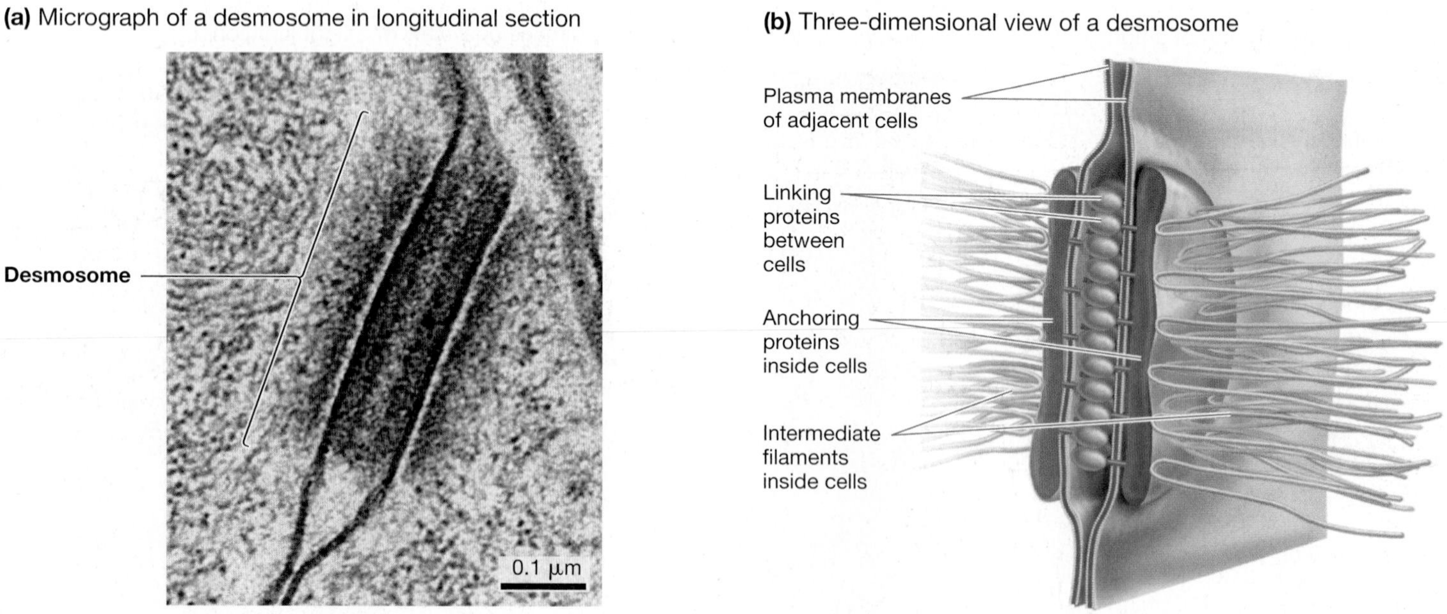

Figure 11.9 Adjacent Animal Cells Are Linked by Desmosomes, which Bind Cytoskeletons Together.
(Figure 11.10 shows how the linking components of desmosomes were isolated and identified.)

The Discovery of Cell–Cell Adhesion Proteins What is the molecular basis of selective adhesion? The initial hypothesis, proposed in the 1970s, was that specialized membrane proteins were involved. The idea was that different types of cells have different types of adhesion proteins in their membranes, and only cells with the same or complementary adhesion proteins can attach to one another.

This hypothesis was tested through experiments that relied on molecules called antibodies. An **antibody** is a protein produced by an immune response that binds specifically to a unique type of molecule, often another protein (Ch. 48, Section 48.2). When an antibody binds to a protein, it can interfere with the protein's ability to interact with other molecules. This property of antibodies was crucial to these experiments.

Figure 11.10 shows how researchers tested the hypothesis that cell–cell adhesion takes place via interactions between membrane proteins:

Step 1 Isolate the membrane proteins from cells that adhere to one another in a certain tissue. Produce pure preparations of each protein.

Step 2 Inject one of the purified membrane proteins into a rabbit. The rabbit's immune system cells recognize the protein as foreign and respond by producing antibodies to it. Purify those antibodies and then repeat this procedure for the other membrane proteins that were isolated. In this way, obtain a large collection of different antibodies—each of which binds specifically to only one of the membrane proteins.

Step 3 Dissociate cells from the original tissue. Take identical samples of the cells and add antibodies directed against a different membrane protein to each sample. Observe each sample, looking for signs of cell adhesion.

If treatment with a particular antibody prevents the cells from adhering to one another, the antibody is probably bound to an adhesion protein. The logic is that if the antibody binds to the adhesion protein, it blocks the adhesion protein's ability to bind to other adhesion proteins and attach the cells to one another.

This approach allowed biologists to identify several major classes of cell adhesion proteins, including **cadherins**—the linking proteins in desmosomes (Figure 11.9). There are various types of cadherins, and cells from different tissues have different types of cadherins in their plasma membranes. Each cadherin can bind only to cadherins of the same type. In this way, cells of the same tissue type attach specifically to one another.

To summarize: Animal cells attach to one another in a selective manner because different types of cell adhesion proteins can bind and rivet certain cells together. Cadherins provide the physical basis for selective adhesion in cells that form tissues and are a critical component of desmosomes.

✔ **If you understand cell–cell adhesion, you should be able to predict what would happen if all of the cells in a developing frog embryo expressed the same type of cadherin on their surfaces.**

Besides giving structural support to tissues, intercellular connections can direct cell–cell communication. But how can cellular connections pass information between cells?

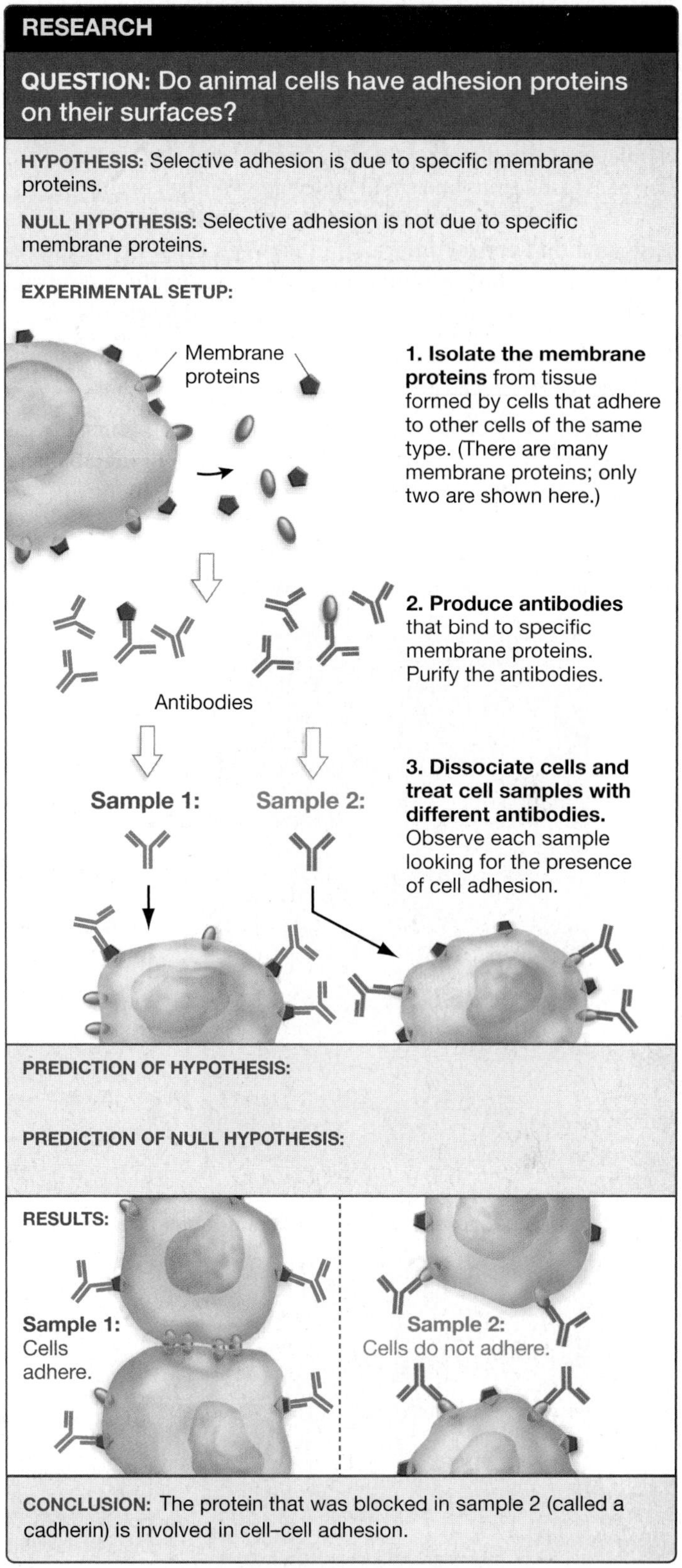

Figure 11.10 Evidence for Adhesion Proteins on Animal Cells.

Sources: K. Hatta and M. Takeichi. 1986. Expression of N-cadherin adhesion molecules associated with early morphogenetic events in chick development. *Nature* 320: 447–449; M. Takeichi. 1988. The cadherins: Cell–cell adhesion molecules controlling animal morphogenesis. *Development* 102: 639–655.

✔ **PROCESS OF SCIENCE** Fill in the prediction made by each hypothesis.

Cells Communicate via Cell–Cell Gaps

In both plants and animals, direct connections between cells in the same tissue help the cells to work in a coordinated fashion. One way of accomplishing this is to have channels in the membranes of adjacent cells, allowing the cells to communicate via the diffusion of cytosolic ions and small molecules from cell to cell.

Ions and small molecules are just two of many different forms of signals that convey information between cells. How cells respond to this exchange of information depends on the type of cell and the type of signal, but there are two general mechanisms:

1. Signals may regulate gene expression, altering which proteins are produced and which are not; or
2. Signals may activate or inactivate particular proteins that already exist in the cell—often those involved in metabolism, membrane transport, secretion, and the cytoskeleton.

Whatever the mechanism, the cell's activity often changes dramatically after the signal arrives. Let's now take a look at how gap junctions and plasmodesmata transmit these signals.

Gap Junctions Connect Animal Cells via Protein Channels In many animal tissues, structures called **gap junctions** connect adjacent cells. In a gap junction, specialized proteins assemble in the membranes of adjacent cells, creating interconnected channels that allow water, ions, and small molecules such as amino acids, sugars, and nucleotides to move between the cells (**Figure 11.11a**).

Gap junctions are communication portals. They can help adjacent cells coordinate their activities by allowing the rapid passage of regulatory ions or small molecules. In the muscle cells of your heart, for example, a flow of ions through gap junctions acts as a signal that coordinates contractions. Without this cell–cell communication, a normal heartbeat would be impossible.

(a) Gap junctions create gaps that connect animal cells.

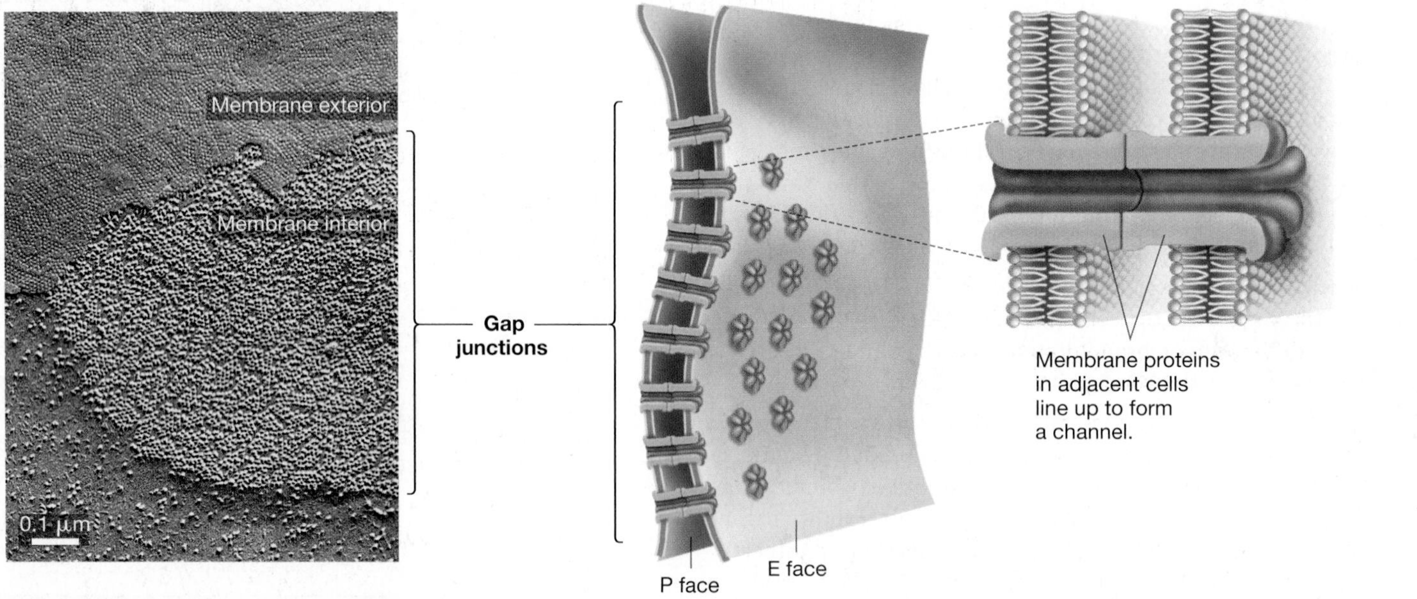

(b) Plasmodesmata create gaps that connect plant cells.

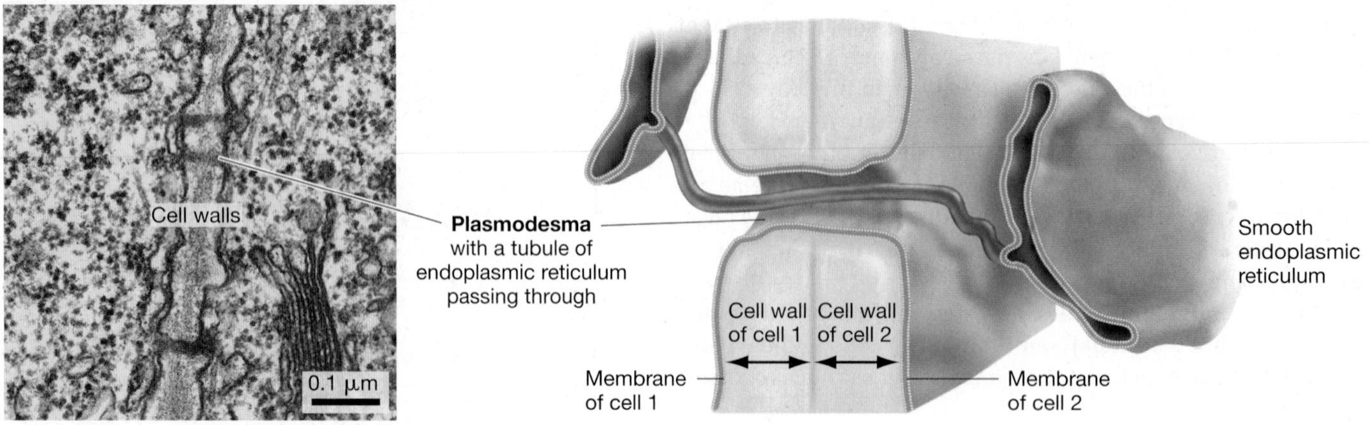

Figure 11.11 Adjacent Animal Cells and Adjacent Plant Cells Communicate Directly. (a) In certain animal tissues, gap junction proteins form channels between adjacent cells. In freeze fracture electron micrographs like the one shown here, they appear as clusters of small dots. (See Ch. 6, Figure 6.18 to review the freeze-fracture technique.) **(b)** In plant cells, plasmodesmata connect the cytoplasm of adjacent cells by forming membrane-lined channels through the cell walls.

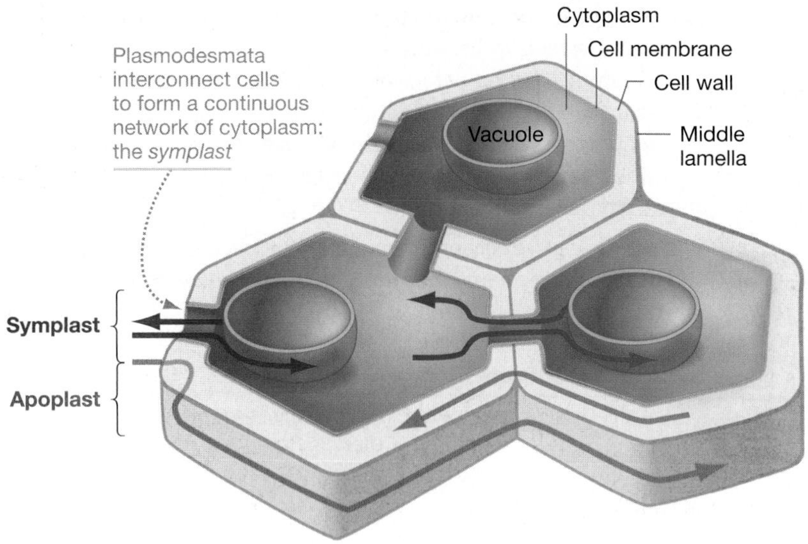

Figure 11.12 Most Plant Tissues Are Divided into Two Corridors: Symplast and Apoplast. Small molecules may travel through plant tissues either within the shared cytoplasm (symplast pathway, shown in purple) or through the extracellular space (apoplast pathway, shown in orange).

Plasmodesmata Connect Plant Cells via Membrane-Lined Channels In plants, direct interactions between cells via membrane proteins are impossible due to the presence of cell walls. How do adjacent plant cells communicate?

In plants, gaps through cell walls allow direct connections between the cytoplasm of adjacent cells. At these connections, named **plasmodesmata** (singular: **plasmodesma**), the plasma membrane and cytoplasm of the two cells are continuous. Tubular extensions from the smooth ER run through these membrane-lined channels (Figure 11.11b).

Like gap junctions, plasmodesmata are communication portals through the plasma membrane. In plants, the plasma membrane separates most tissues into two independent corridors: **(1)** the **symplast**, which is a continuous network of cytoplasm connected by plasmodesmata, and **(2)** the **apoplast**, which is the region outside the plasma membrane (Figure 11.12). The apoplast consists of cell walls, the middle lamella, and air spaces. Small molecules can move through plant tissues in either of these compartments without ever crossing a membrane (Ch. 35, Section 35.2).

Gap junctions and plasmodesmata allow adjacent cells to transmit information, like a conversation between neighbors. But how do multicellular organisms send messages between different tissues, where in most cases there is no direct contact? For example, suppose that you become dehydrated while exercising or that you are startled by a loud noise. How do cells that sense these threats signal tissues or organs elsewhere in your body to promote rehydration or prepare for fleeing from danger? Distant cell communication is the subject of Section 11.3.

CHECK YOUR UNDERSTANDING

✔ If you understood this section, you should be able to...

1. Compare and contrast the structure and function of the middle lamella of plants and the tight junctions and desmosomes of animals.
2. Describe the structure and function of plasmodesmata and gap junctions.

Answers are available in Appendix A.

11.3 How Do Distant Cells Communicate?

After you complete this section, you should be able to...

- Analyze how information is exchanged between distant cells in multicellular organisms.

Cells that are not in physical contact can communicate with one another. This is true for unicellular organisms, where hundreds or thousands of cells may live in close proximity, as well as for multicellular organisms like humans and maple trees, which typically contain trillions of cells and dozens of tissue types.

Cell–cell communication is one of the most dynamic research areas in biology. Let's begin by analyzing how distant cells in humans and other multicellular organisms exchange information, and then—in Section 11.4—explore how unicellular organisms communicate.

Cell–Cell Signaling in Multicellular Organisms

Suppose that cells in your brain sense that you are becoming dehydrated. Brain cells can't do much about the water you lose during urination, but kidney cells can. In response to dehydration, certain brain cells release a signaling molecule that travels to the kidneys and activates the expression of water channels called aquaporins (Ch. 6, Section 6.4). As a result, water moves out of the urine and back into the blood, preventing further dehydration.

Biologists have classified many types of signaling molecules that keep distant tissues in touch. One type, neurotransmitters, may open or close ion channels in the plasma membrane of distant cells, changing the electrical properties of the membrane. This type of signal is responsible for the transmission of information through the nervous system, allowing your brain to control the movements of the rest of your body (see Chapters 43 and 45).

The best-studied means of distant signaling, however, may be via **hormones**—information-carrying molecules that are secreted by plant and animal cells into bodily fluids and act on distant target cells. Hormones are usually small molecules and include certain peptides, steroids, and even gases. Although hormones are typically present in minute concentrations, they have a large impact on the activity of target cells.

The most important point about a hormone or other signaling molecule is how a cell interprets the information it conveys. How do cells receive and process signals from distant cells? The basic steps are common to all cell signaling systems. Let's consider each step in turn.

Signal Reception

Hormones and other types of cell–cell signaling molecules deliver their message by binding to receptor molecules. The key characteristic of this interaction is that it changes the shape, or conformation, of the receptor. A **signal receptor**, then, is a protein that changes its shape and activity after binding to a signaling molecule. This change in shape is how a message is passed from the signaling molecule to its receptor.

The presence of an appropriate signal receptor dictates which cells will respond to a particular signaling molecule. For example, even though the molecule that carries the message "We're getting dehydrated—conserve water" is broadcast throughout the body, only certain kidney cells respond because only they have the receptor that binds to that molecule.

Cells in a wide array of tissues may respond to the same signaling molecule, though, if they have the appropriate receptor. If you are startled by a loud noise, cells in your adrenal glands secrete the hormone adrenaline (also called epinephrine), which carries the message "Get ready to fight or run." In response, your heart rate increases, your breathing rate increases, and cells in your liver release glucose, which is used in your muscles to power rapid movement. This response is the basis of an "adrenaline rush." Cells in your heart, lung, and liver respond to adrenaline because they all have the receptor that binds to it. Identical receptors in diverse cells and tissues allow long-distance signals to coordinate the activities of cells throughout a multicellular organism.

Where does the interaction between a signaling molecule and its receptor occur—inside the target cell or outside? The answer depends on the signaling molecule's ability to pass through plasma membranes.

- Most lipid-soluble signaling molecules can diffuse across the hydrophobic region of the membrane and enter the cytosol of their target cells. The receptors for these molecules exist inside the cell.
- Large or hydrophilic signaling molecules are lipid insoluble, and most cannot cross the plasma membrane. To affect a target cell, they have to be recognized at the cell surface. Their receptors are usually located in the plasma membrane.

It's critical to note two additional points about signal receptors, no matter where they're located:

1. ***Receptors are dynamic.*** The number of receptors in a particular cell may decline if hormonal stimulation occurs at high levels over a long time. The ability of a receptor to bind tightly to a signaling molecule may also decline in response to intensive stimulation. As a result, the sensitivity of a cell to a particular hormone may change over time.
2. ***Receptors can be blocked.*** Many drugs are used to block the interaction between hormones and their receptors. For example, certain beta-blocker drugs will prevent adrenaline from binding to its receptor on heart cells. So if a physician wants to reduce the strength of a patient's heart cell contractions as a way to lower blood pressure, she is likely to prescribe a beta-blocker.

The change in receptor structure that occurs after a signaling molecule binds means that the signal has been received. It's like throwing an "on" switch. What happens next?

Signal Processing

Once a cell receives a signal, it has to process the signal to initiate a response. This step happens in one of two ways, depending on whether the receptors are located in the cytosol or at the membrane surface.

Processing Lipid-Soluble Signaling Molecules Steroid hormones such as estrogens and cortisol are examples of lipid-soluble signaling molecules. Because they are hydrophobic, most lipid-soluble signaling molecules must be carried through the bloodstream by hydrophilic proteins. After reaching their target cells, these signaling molecules are released from the carrier proteins, diffuse through the plasma membrane, and enter the cytosol. Often, a hormone–receptor complex is formed in the cytosol and then transported to the nucleus, where it triggers changes in gene expression (Figure 11.13). By altering the expression of genes (Chapter 17), the cell produces different proteins that will directly affect the function or shape of the cell.

Processing Lipid-Insoluble Signaling Molecules Hormones that *cannot* diffuse across the plasma membrane and enter the cytosol do not *directly* participate in intracellular activities, like changing gene expression. Instead, the signal that arrives at the surface of the cell has to produce an intracellular signal—the processing step is indirect.

When a signaling molecule binds at the cell surface, it triggers **signal transduction**—the conversion of a signal from one form to another. A long and often complex series of events ensues, collectively called a signal transduction pathway.

Signal transduction is a common occurrence in everyday life. For example, the text messages you receive are sent from one phone to another as radio wave transmissions. These electronic signals can be transmitted efficiently over long distances but would be meaningless to you. Software in your phone has to transduce, or convert, the signals into a form that you can understand and respond to, such as words in a text message.

Signal transduction pathways in cells work in a similar way. In a cell, signal transduction converts an extracellular signal to an intracellular signal (Figure 11.14). As in a text message, a signal that is easy to transmit is converted to a signal that is easily understood and that triggers a response.

Intracellular Signals May Be Amplified and Diversified Recall that hormones are present in minuscule concentrations but trigger a large response from cells. Signal amplification is one reason this is possible. When a hormone arrives at the cell

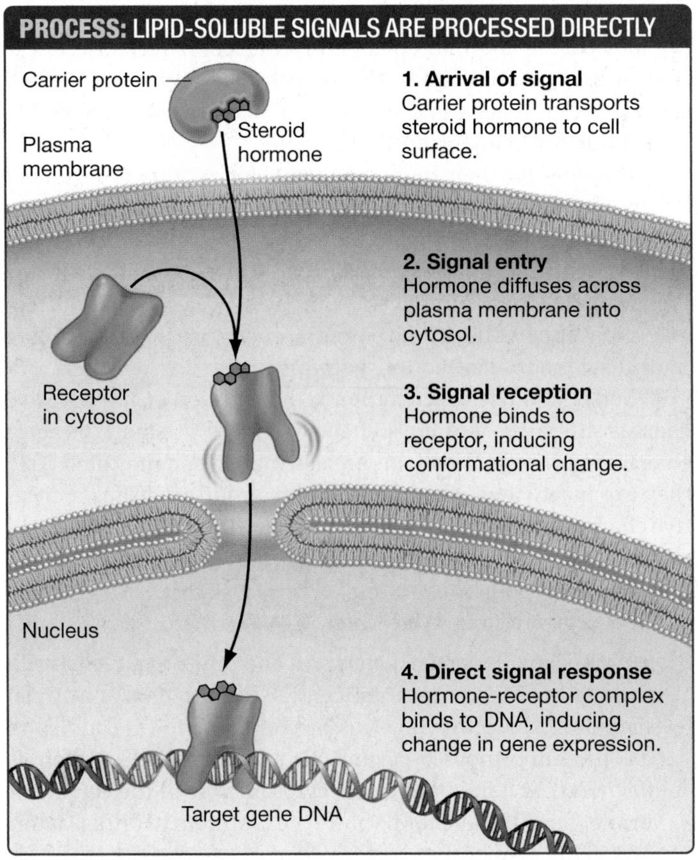

Figure 11.13 Some Cell–Cell Signaling Molecules Enter the Cell and Bind to Receptors in the Cytosol. Because they are lipids, steroid hormones can diffuse across cell membranes and bind to signal receptors located in the cytosol. The hormone–receptor complex may then be transported to the nucleus, where it changes the activity of genes.

✔ Based on what you learned about nuclear transport (see Ch. 7, Section 7.4), what type of signal would you expect to be exposed on the cytosolic receptor after the steroid hormone changes the receptor's conformation?

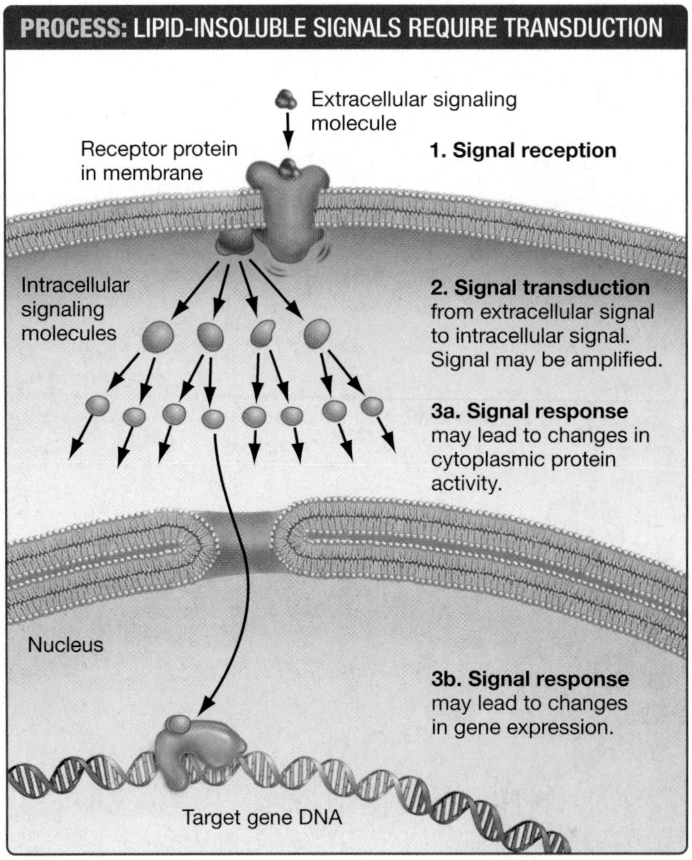

Figure 11.14 Signal Transduction Converts an Extracellular Signal to an Intracellular Signal. A lipid-insoluble signaling molecule will not pass through the membrane to direct a cellular response. Instead, the molecule activates a surface receptor that directs a multistep process to generate intracellular signals. One or more of the intracellular signaling molecules may alter the activity of proteins in the cytoplasm or enter the nucleus to change the activity of genes.

surface, the message it transmits may be amplified as the signal changes form. An increased number of intracellular signals also makes it possible for hormones to affect different molecules in the cell. Your portable music player performs an analogous function: By amplifying a tiny sound signal, it enables the sound to affect a whole roomful of people (amplification)—some may dance while others may plug their ears (diversification).

In cells, signal transduction begins at the plasma membrane; amplification and diversification of the signal takes place inside the cell. This may occur in a variety of ways, depending on the mechanism of signal transduction. In general, the arrival of a single signaling molecule results in a secondary signal that involves many ions or molecules that can affect several different cellular activities.

For example, when liver cells are stimulated by adrenaline to release glucose into the bloodstream, the signal is amplified by the production of numerous small molecules called second messengers. The signal then diversifies to activate enzymes that break down glycogen into glucose, inhibit enzymes that synthesize glycogen, and produce new enzymes that make glucose.

Now let's focus on two major types of signal transduction systems that are distinguished based on how they are initiated:

1. G-protein-coupled receptors initiate the production of intracellular second messengers, which then amplify and diversify the signal.
2. Enzyme-linked receptors activate a series of proteins inside the cell, through the addition of phosphate groups. The number and type of proteins activated lead to the amplification and diversification of the signal.

Although there are many variations in the signaling pathways that fall within these two categories, the common features are emphasized here.

Signal Transduction via G-Protein–Coupled Receptors Many signal receptors span the plasma membrane and are closely associated with peripheral membrane proteins inside the cell called **G proteins**. When G proteins are activated by a signal receptor, they often trigger a key step in signal transduction: the

PROCESS: HOW DO G-PROTEIN-COUPLED RECEPTORS WORK?

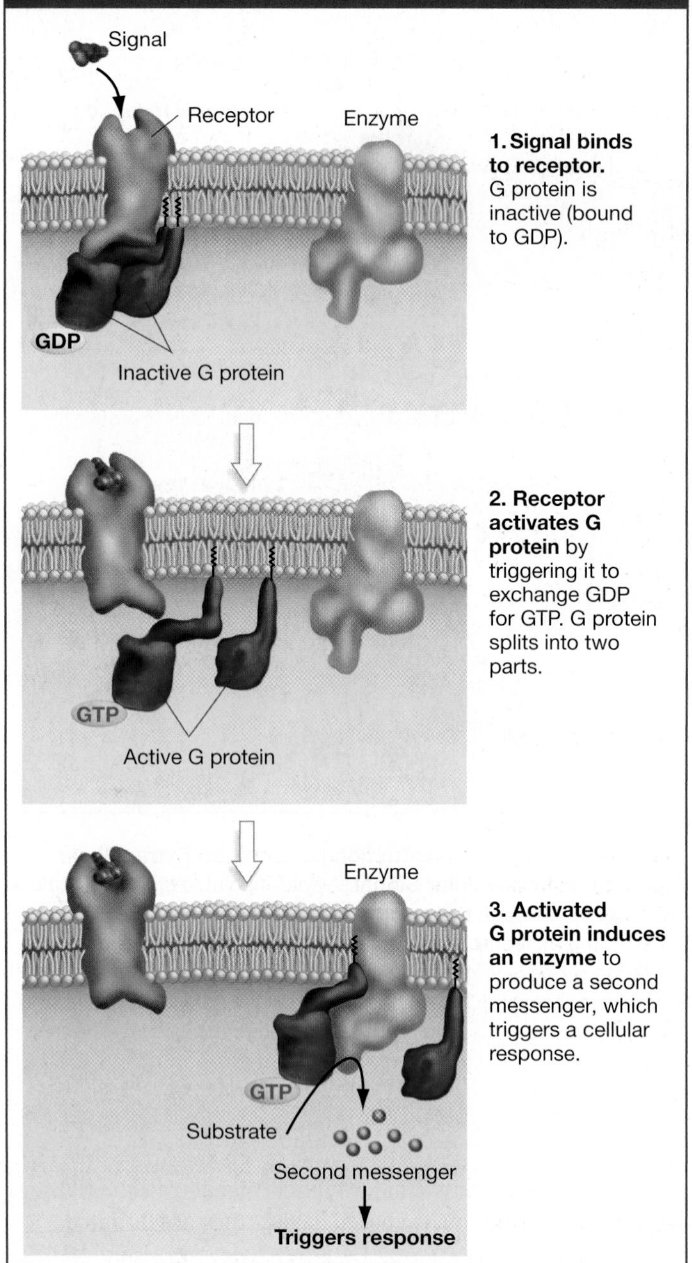

Figure 11.15 G-Protein-Coupled Receptors Trigger the Production of a Second Messenger.

production of a **second messenger**—a small, nonprotein signaling molecule or ion that elicits an intracellular response to the first messenger (the signaling molecule that arrived at the cell surface). G proteins link the receipt of an extracellular signal to the production of an intracellular signal.

G proteins got their name because their activity is regulated by the type of nucleotide they are bound to: either **guanosine triphosphate (GTP)** or guanosine diphosphate (GDP). GTP is a nucleoside triphosphate that is similar in structure to adenosine triphosphate (ATP; introduced in Chapter 4). Recall that nucleoside triphosphates have high potential energy because their three phosphate groups have four negative charges close together.

When GTP binds to a G protein, the addition of the negative charges alters the protein's shape. Changes in shape produce changes in activity. G proteins are activated when they bind GTP; they are inactivated when a phosphate group, and thus a negative charge, is removed from GTP to form GDP. The G protein will remain inactive until the GDP is replaced with a new GTP.

To understand how G proteins fit into an overall signal transduction pathway, follow the events in Figure 11.15.

Step 1 A signaling molecule arrives and binds to a receptor in the plasma membrane. Notice that the receptor is a transmembrane protein whose intracellular portion is coupled to a G protein composed of multiple subunits. The G protein is anchored by a lipid tail to the cytosolic side of the cell membrane. The lipid anchor permits the G protein to diffuse laterally in the membrane.

Step 2 In response to binding of the signaling molecule, the receptor changes shape and activates its G protein. Specifically, the receptor kicks out the GDP from the inactive G protein, allowing GTP to bind to the protein. When GTP is bound, the G protein will change shape radically: The active GTP-binding subunit splits off.

Step 3 The active G protein subunit interacts with a nearby enzyme that is embedded in the plasma membrane. This interaction stimulates the enzyme to catalyze production of a second messenger.

Second messengers are effective because they are small and therefore can diffuse rapidly to spread the signal throughout the cell. In addition, they can be produced quickly in large quantities. This characteristic is important. Because the arrival of a single signaling molecule can stimulate the production of many second messengers, the signal transduction event amplifies the original signal.

Several types of small molecules and ions act as second messengers in cells. **Table 11.1** lists some of the best-studied second

Table 11.1 Examples of Second Messengers

Name	Type of Response
Calcium ion (Ca^{2+})	Binds to a protein called calmodulin; Ca^{2+}/calmodulin complex then activates proteins
Cyclic adenosine monophosphate (cAMP)	Activates certain protein kinases
Cyclic guanosine monophosphate (cGMP)	Opens ion channels; activates certain protein kinases
Diacylglycerol (DAG)	Activates certain protein kinases
Inositol trisphosphate (IP_3)	Opens calcium channels, allowing stored calcium ions to enter cytosol

messengers and provides an example of how cells respond to each of them. Note that several second messengers activate **protein kinases**—enzymes that activate or inactivate other proteins by adding a phosphate group to them. For example, the signal diversification observed in adrenaline-stimulated liver cells is due to protein kinases activated by the cAMP second messenger.

It's also important to note two things:

1. Second messengers aren't restricted to a single role—the same second messenger can initiate dramatically different events in the same cell or in different cell types receiving the same signaling molecule.
2. More than one type of second messenger may be involved in triggering a cell's response to the same extracellular signaling molecule.

To make sure that you understand how G proteins and second messengers work, imagine the following movie scene: A spy arrives at a castle gate. The castle guard receives a note from the spy, but he cannot read the coded message on the note. Instead, the guard brings the note to the queen. She reads the message and summons the commander of the guard, who sends soldiers throughout the castle to warn everyone of approaching danger. ✔ **MODEL You should be able to identify which characters in the scene correspond to the second messenger, G protein, signaling molecule, receptor, and enzyme activated by the G protein.**

It's difficult to overstate the importance of signal transduction by G-protein-coupled receptors. Biomedical researchers estimate that a third of all human drugs target these signal receptors. These drugs include the beta-blockers that regulate heart rate. In 2012, Brian Kobilka and Robert Lefkowitz won the Nobel Prize in Chemistry for their work on the structure and function of G-protein-coupled receptors.

Signal Transduction via Enzyme-Linked Receptors Enzyme-linked receptors transduce hormonal signals by directly catalyzing a reaction inside the cell. Figure 11.16 focuses on the best-studied group of enzyme-linked receptors: the **receptor tyrosine kinases (RTKs)**.

Step 1 A hormone binds to two subunits of an RTK and causes them to form a dimer.

Step 2 The conformational change in the RTK turns on its catalytic activity, allowing the RTK to phosphorylate itself at tyrosine residues using ATP inside the cell.

Step 3 Proteins inside the cell bind to the phosphorylated RTK, forming a bridge between the receptor and a lipid-anchored peripheral membrane protein called **Ras**, which is a single subunit G protein. Bridge formation activates Ras by causing it to exchange its bound GDP for a GTP.

Step 4 When Ras is activated, it triggers the phosphorylation and activation of a protein kinase.

Step 5 The Ras-activated kinase catalyzes the phosphorylation and activation of a second kinase, which then phosphorylates and activates a third kinase. The third kinase triggers the cell response by phosphorylating additional proteins.

This sequence of protein modifications that culminates in a cell response is termed a **phosphorylation cascade**. Since these

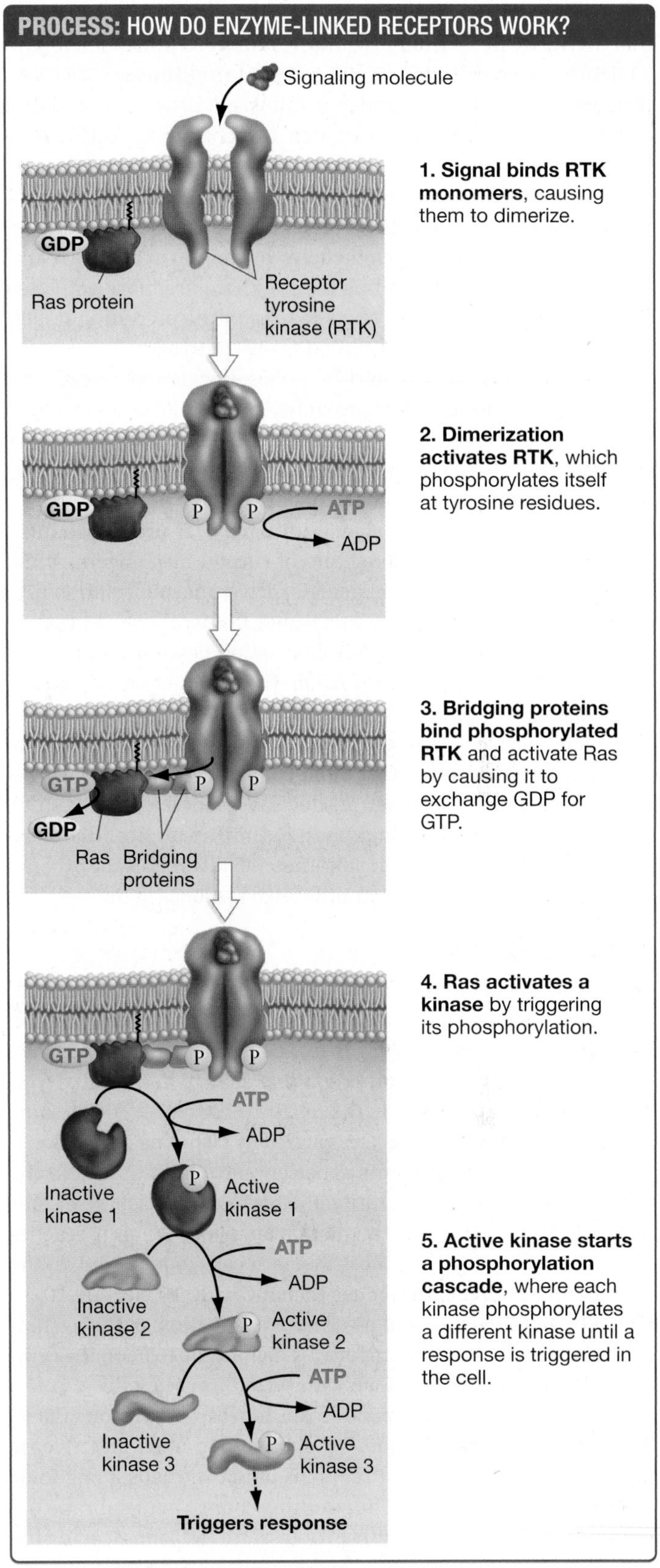

Figure 11.16 Enzyme-Linked Receptors Trigger a Phosphorylation Cascade.

cascades are often initiated by mitogens—signaling molecules that activate cell division—the three numbered kinases in Figure 11.16 are called **mitogen-activated protein kinases (MAPKs)**. (The *mito–* in mitogen stands for mitosis, a process involved in eukaryotic cell division.) Although the change in MAPK conformation after it is phosphorylated is very subtle (see Ch. 8, Figure 8.17), it has a dramatic effect in MAPK catalytic activity.

In some cases, each copy of an enzyme in a cascade catalyzes the phosphorylation of many copies of the next "downstream" enzyme, and so on. When this occurs, there are more activated enzymes at each stage in a cascade than in the preceding stages, and the original signal is amplified many times over.

The proteins that take part in a phosphorylation cascade are often held in close physical proximity by scaffold proteins. This arrangement increases the speed and efficiency of the reaction sequence, but it decreases the amplification of the response by limiting the number of proteins that are phosphorylated.

In general, intracellular signals initiated by G-protein-coupled receptors result in the production of second messengers, while enzyme-linked receptors, like RTKs, drive phosphorylation cascades. Variations, however, sometimes occur: Some G-protein-coupled receptors trigger phosphorylation cascades, and some enzyme-linked receptors result in the production of second messengers.

To summarize: Many of the key signal transduction events observed in cells occur via G-protein-coupled receptors or enzyme-linked receptors. A signal transduction event has two results: **(1)** It converts an extracellular message into an intracellular message, and **(2)** in some cases it amplifies and diversifies the original message to elicit a large and multifaceted response in the cell.

Signal Response

What is the ultimate response to the messages carried by signaling molecules? Recall that when adjacent cells share information through cell–cell gaps, two general categories of response may occur: a change in gene expression or a change in the activity of proteins that already exist in the cell (see Section 11.2). The same holds for responses to messages carried by signaling molecules.

For example, when plants experience drought, tissues in the root system respond by secreting the hormone abscisic acid. This hormone travels through the plant's apoplast. When it reaches the leaves, it binds to receptors in guard cells, which control the stomatal pores that allow for gas exchange (Ch. 10, Section 10.4). Receptor binding initiates a signal transduction pathway that opens ion channels in the smooth ER membrane to flood the cytosol with the calcium ions (Ca^{2+}) stored in this organelle.

How do guard cells respond to an increase in cytosolic Ca^{2+}? The Ca^{2+} activates chloride ion channels to open and release large amounts of chloride (Cl^-) ions out of the cells along their electrochemical gradient. The resulting change in membrane potential from Cl^- export results in potassium (K^+) ions leaving the cell. The accumulation of Cl^- and K^+ ions outside the cell leads to the movement of water out of the guard cells via osmosis. The guard cells deflate and close the pores, which prevents water loss from the plant (for more information about this process, see Ch. 37, Figure 37.22b).

At this point, you've analyzed the first three steps of cell–cell communication: signal reception, signal processing, and signal response. ✔ **If you understand how cells receive and process signals, you should be able to explain how adrenaline can bind to the same receptor in cells of the heart and liver but trigger different responses (increasing contraction rate in heart cells and releasing glucose by liver cells).**

Now the question is, how is a signal turned off? Consider the response plants have to abscisic acid when they experience drought. If this response continued indefinitely, gas exchange through the closed stomata would be insufficient to maintain photosynthetic activity. What limits the response to a cell–cell signal?

Signal Deactivation

Cells have built-in systems for turning off intracellular signals. Although many different mechanisms may be used, most signal transduction systems are exquisitely sensitive to small changes in the concentration of signaling molecules or the number and activity of signal receptors. As a result, they trigger a rapid response and can be shut down quickly.

For example, once an activated G protein turns on a downstream enzyme, the bound GTP is hydrolyzed by the G protein to GDP and P_i. This reaction changes the G protein's conformation and returns the protein to its inactive state. Activation of its downstream target stops, and production of the second messenger ceases. To produce a high concentration of second messengers, the pool of inactive G proteins must be continuously reactivated by the signal receptor to keep the process going. Otherwise, the signal transduction system quickly shuts down.

Phosphorylation cascades are also sensitive to the continuing presence of external signaling molecules. If stimulation of a receptor tyrosine kinase ends, enzymes called **phosphatases** will remove phosphate groups from components of the phosphorylation cascade, causing the signal transduction to cease.

The presence of second messengers in the cytosol is also short lived. For example, pumps in the membrane of the smooth ER return cytosolic calcium ions to storage in the ER lumen, and enzymes called phosphodiesterases convert active cAMP and cGMP (see Table 11.1) to inactive AMP and GMP, respectively. If the production of second messengers is halted, then they are quickly cleared from the cytosol and the signal transduction stops.

To appreciate what happens when a signal transduction system does not shut down properly, let's return to the phosphorylation cascade illustrated in Figure 11.16. Recall that Ras is active when it is bound to GTP, but it is deactivated when it hydrolyzes GTP to GDP and P_i. If this hydrolysis activity were defective, however, Ras would remain active and continue stimulating the cascade even when the external signal is no longer present.

Why is continuously active Ras a problem? Many of the mitogen-activated protein kinases that are activated by Ras induce cells to divide. Cells expressing such a defective Ras would receive a never-ending "divide now" signal that could lead to the development of cancer. In fact, an estimated 25–30 percent of all human cancers involve cells that express this type of defective Ras. (To learn more about the family of diseases called cancer, see Chapters 12 and 19.)

Crosstalk: Synthesizing Input from Many Signals

In this section you have learned how cells respond to individual signals, but it's crucial to realize that cells receive and process an almost constant stream of different signals. Just as you receive information about your environment via text messages, e-mails, phone calls, and talking to friends, cells process an array of chemical signals about changes in their environment.

The signal transduction pathways that are triggered by different signals and receptors often intersect. In reality, they are not strictly linear pathways as illustrated in Figures 11.14 through 11.16. Instead, signal transduction pathways form a network. This complexity is important: It allows cells to respond to many different signals in an integrated way.

The diverse signals that a cell receives are integrated by **crosstalk**—interactions among different signaling pathways that modify the cell response (Figure 11.17). Crosstalk is like getting advice from multiple people before making a decision.

Here are three key things to note:

1. One pathway may inhibit steps in a second pathway, reducing the cell's response to the second pathway even though the appropriate signal is present.
2. One pathway may stimulate steps in a second pathway, leading to the production of two different responses to a single signaling molecule.
3. The presence of multiple steps in a signaling pathway provides a series of points where crosstalk can regulate the flow of information. This regulation is vital because it allows the cell to respond appropriately to many signals at the same time.

CHECK YOUR UNDERSTANDING

✔ **If you understood this section, you should be able to...**

1. Explain why only certain cells respond to particular signaling molecules that are sent throughout the body.
2. Explain what it means for a signal to be "amplified" and "diversified" by cells.
3. Predict how plants could deactivate the signal transduction event initiated by abscisic acid that causes their stomata to close.

Answers are available in Appendix A.

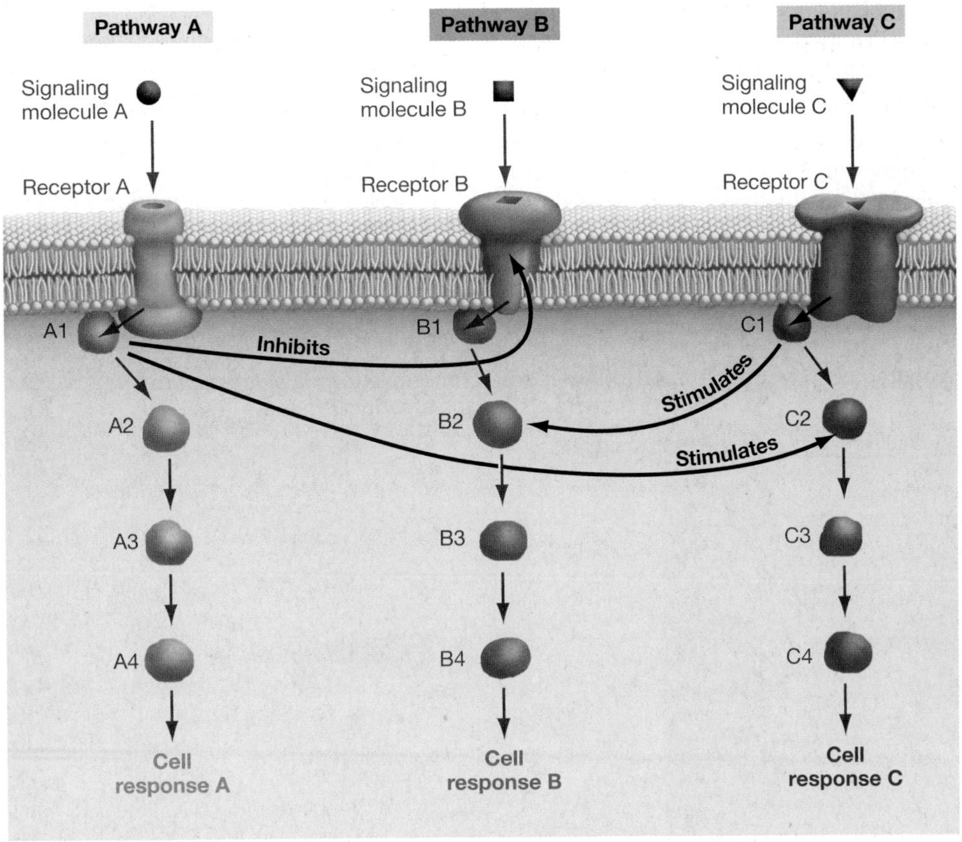

Figure 11.17 Signaling Pathways Interact via Crosstalk.

✔ Predict which responses would occur in cells exposed to the following signaling molecules: (1) A + B; (2) A + C; (3) C alone.

11.4 Signaling between Unicellular Organisms

Surprisingly, much of what we know about signal transduction in multicellular organisms has come from the study of unicellular organisms. While the signal reception and processing events in signal transduction pathways are similar in unicellular and multicellular organisms, the topic of conversation often differs. Rather than involving calls for help, as when a dehydrated brain asks the kidney to conserve water, the conversations between unicellular microbes are often about changes in the environment.

After you complete this section, you should be able to...

- Analyze the role of intercellular signaling between unicellular organisms.

One environmental factor that is closely monitored by populations of unicellular organisms is the density of the population. The use of signaling pathways to respond to population density in prokaryotic and eukaryotic microbes is referred to as **quorum sensing**. The name was inspired by the observation that cells of the same species may undergo dramatic changes in activity when their numbers reach a threshold, or quorum.

Quorum sensing is based on signaling molecules that are secreted by cells and diffuse through the environment. The response to these molecules depends on the species. In bacteria, quorum sensing is often used to help glue a community of microbes to a surface in a biofilm (Ch. 26, Section 26.1), such as the plaque that forms on your teeth. Quorum sensing is also involved in light emission (bioluminescence) by certain bacteria. For example, bacterial species including *Vibrio fischeri* are actively cultured in the light organs of the bobtail squid; after reaching a certain density, they express enzymes that catalyze a light-producing reaction (see the Chapter 18 Case Study).

Quorum sensing allows unicellular organisms to communicate and coordinate their activities. When it occurs, these cells take on some of the characteristics of multicellular organisms. For example, quorum sensing via a G-protein-coupled receptor causes the free-living cells (amoebae) of the slime mold *Dictyostelium* to aggregate into multicellular mounds (**Figure 11.18**). Amazingly, the slug-like body that is formed from one of these aggregates can crawl across a surface and eventually organize itself into a fruiting body that releases spores into the air (see **BioSkills 11**).

Cell signaling has been one of the hottest research areas in biological science over the past two decades. It has taken a great deal of painstaking work to identify each step in individual signaling pathways. Biologists are now investigating cell signaling at a whole-system level—examining how the major pathways interact and how they are integrated at the tissue level within multicellular organisms.

The number of molecules involved in cell signaling and the complexity of their interactions can seem overwhelming, but the punch line is simple: Cell signaling helps organisms ranging from bacteria to blue whales receive information about their environment and respond appropriately to changing conditions.

CHECK YOUR UNDERSTANDING

✔ If you understood this section, you should be able to...

1. Compare and contrast intercellular signaling in unicellular and multicellular organisms.
2. Explain how quorum sensing occurs, and give an example of how this system coordinates the activities of unicellular organisms.

Answers are available in Appendix A.

(a) Slime mold amoebae aggregate in response to sensing a quorum.

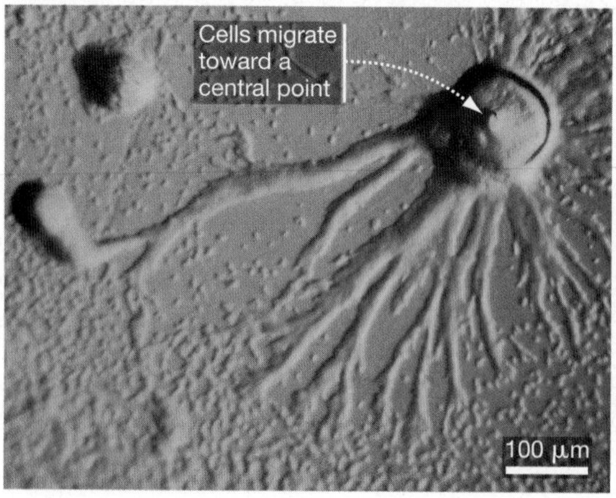

(b) Aggregated amoebae form a sluglike body that crawls across a surface.

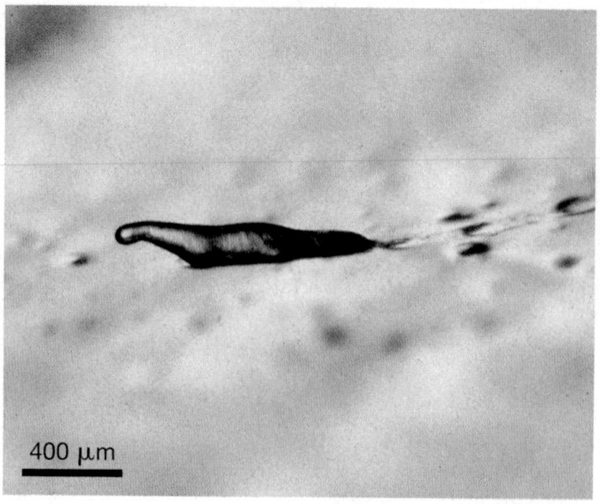

Figure 11.18 Unicellular Organisms Use Quorum Sensing to Coordinate Activities.

11.1 The Cell Surface

- The vast majority of cells produce material that forms a layer outside the plasma membrane.
- In bacteria, archaea, algae, fungi, and plants, the extracellular material is stiff and forms a cell wall. In animals, the material is flexible and is called the extracellular matrix (ECM).
- In eukaryotes, extracellular layers are fiber composites. They consist of cross-linked filaments that provide tensile strength and a ground substance that fills space and resists compression.
- In plants, the extracellular filaments are cellulose microfibrils; in animals, the most abundant filaments are made of the protein collagen. In both plants and animals, the ground substance is composed primarily of gel-forming polysaccharides.

11.2 How Do Adjacent Cells Connect and Communicate?

- In multicellular organisms, molecules in the extracellular layer and plasma membrane mediate interactions between adjacent cells.
- Adjacent cells may be physically bound to one another by a glue-like middle lamella in plants or by tight junctions and desmosomes in animals.
- The cytoplasm of adjacent cells may be in direct contact through openings called plasmodesmata in plants and gap junctions in animals. These openings allow adjacent cells to communicate via cytosolic signals.
- Cells may respond to signals by altering the expression of their genes or by changing the activity of existing proteins. These responses enable cells within tissues to coordinate their activities.

11.3 How Do Distant Cells Communicate?

- Distant cells in multicellular organisms communicate by secreting signaling molecules that bind to receptors located in the cytosol or on the surface of specific target cells.
- Lipid-soluble signaling molecules often pass through the plasma membrane and bind to cytosolic signal receptors. Signaling molecules that are not lipid soluble often bind to signal receptors in the plasma membrane.
- Signal receptors in the plasma membrane change their conformation on binding to the signal, which triggers production of a second messenger or activates a phosphorylation cascade.
- The cell's response to signals is tightly regulated. Intracellular signals are quickly deactivated without constant signaling from the receptors, and signaling pathways often interact.

11.4 Signaling between Unicellular Organisms

- Unicellular organisms use chemical signals to sense aspects of their environment, such as their population density. Quorum sensing allows populations and communities of cells to coordinate changes in their activities when population density is high.

Answers are available in Appendix A.

✔ TEST YOUR KNOWLEDGE

1. What is a fiber composite? How do cellular fiber composites resemble reinforced concrete?
2. **THINK CAREFULLY** Where are protein components of the extracellular matrix synthesized?
   a. in the rough ER
   b. in the Golgi apparatus
   c. in the plasma membrane
   d. in the extracellular layer itself
3. Which of the following actions correctly describe a role performed by at least one type of intercellular connection? Select True or False for each statement.
   T/F Allows communication between adjacent cells.
   T/F Forms a watertight barrier between the cells.
   T/F Uses components of the extracellular matrix to indirectly connect adjacent cells.
   T/F Associates with cytoskeletal components to resist pulling forces.
4. What does it mean to say that a signal is transduced?
   a. The signaling molecule enters the cell directly and elicits a cellular response.
   b. The signal is generated by the production of proteins.
   c. The physical form of the signal changes between the outside of the cell and the inside.
   d. The signal is amplified.

✔ TEST YOUR UNDERSTANDING

5. How do the extracellular filaments in plants differ from those in animals?
   a. Plant filaments resist compression forces; animal filaments resist pulling forces.
   b. Animal filaments consist of proteins; plant filaments consist of polysaccharides.
   c. Plant extracellular filaments never move; animal filaments can slide past one another.
   d. Plant filaments run parallel to one another; animal filaments crisscross.
6. Summarize the experimental evidence in sponges showing that animal cells adhere to each other selectively. Explain the molecular basis of selective adhesion.
7. **MODEL** Suppose you were to model amplification by the phosphorylation cascade in Figure 11.16, using a penny for each kinase 1, a nickel for each kinase 2, and a dime for each kinase 3. Also suppose that Ras and each of the kinases can activate 10 proteins. How much money would you need to construct your model?
8. What is the significance of the observation that many signal transduction pathways create a network, where they intersect or overlap?

✔ TEST YOUR PROBLEM-SOLVING SKILLS

9. Steroid hormones, like most lipid-soluble signaling molecules, are processed directly. How does the absence of a signal transduction cascade in the processing of steroid hormones affect (a) signal amplification, (b) signal regulation, and (c) the cellular response that is possible?

10. Suppose you have an antibody that binds to the receptor tyrosine kinase illustrated in Figure 11.16. When you add this antibody to the cell, you find that it activates the cell response, even when no signaling molecule is present. Explain this result.

✔ PUT IT ALL TOGETHER: Case Study

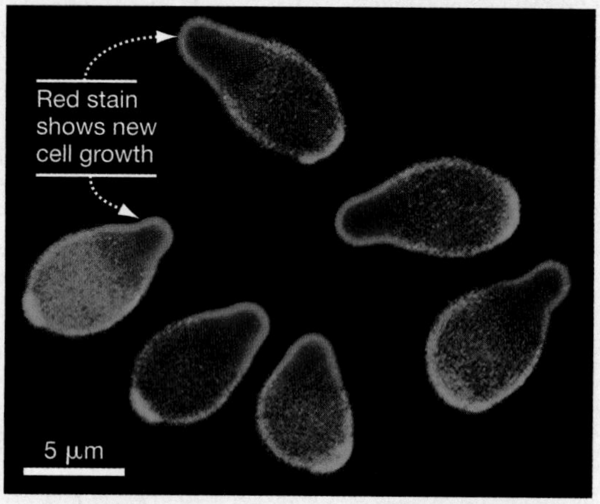

What type of foreplay is required for sexual reproduction in yeast?

Some unicellular eukaryotes, including the yeast *Saccharomyces cerevisiae*, can reproduce sexually (Ch. 13, Section, 13.4). At the most basic level, sexual reproduction involves the fusion of two cells and the combining of genetic material from each cell into one nucleus. Yeast cells are not motile, so in order to attach and fuse, the opposite mating types—referred to as "**a**" cells and "alpha (**α**)" cells—must alter their growth (indicated by red staining in the accompanying photograph). How does a yeast cell arouse this cellular response in the opposite mating type?

11. Yeast cells secrete signaling molecules called pheromones that are specific for their mating type. Type **α** cells produce the pheromone α-factor, which binds to a G-protein-coupled receptor on type **a** cells, and vice versa. Explain what happens to the associated G protein when one of these receptors binds to a pheromone.

12. Instead of producing second messengers, the pheromone-activated response in yeast involves a peripheral membrane protein related to Ras. Based on what you've learned about signal processing in multicellular organisms, predict how a pheromone signal is further processed in yeast cells to trigger a cellular response.

13. Some of the intracellular proteins involved in the pheromone-activated response in yeast are organized by a scaffold protein called Ste5. After pheromone binding occurs at the surface, Ste5 is phosphorylated and one of these scaffolded proteins, a mitogen-activated protein kinase (MAPK) called Fus3, is released from Ste5 and triggers the intracellular response. Explain how a scaffolded organization affects the speed of the response and amplification of the signal.

14. **QUANTITATIVE** Phosphorylated Ste5 can be dephosphorylated by a phosphatase named Ptc1. To examine the role of Ste5 phosphorylation on the release of Fus3, researchers created a yeast strain with the *ptc1* gene deleted (*ptc1*Δ). They then measured the amount of binding between Fus3 and Ste5 in cells of this strain and in wild-type cells as they added increasing concentrations of α-factor. Their results are shown in the graph that follows. Use the graph to evaluate how the phosphorylation status of Ste5 affects its ability to release Fus3, and predict how the *ptc1*Δ mutation would affect the mating response.

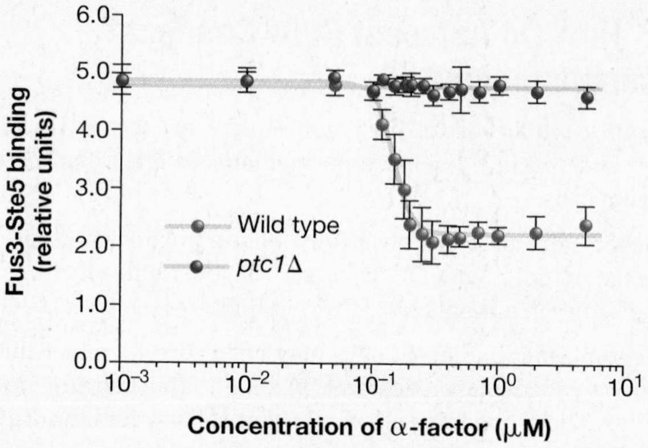

Source: M. Malleshaiah et al. 2010. *Nature* 456: 101–105.

15. **PROCESS OF SCIENCE** A yeast cell's response to pheromone signaling includes assembly of microfilaments that direct its growth toward the signal source. Propose a hypothesis that explains how growth might be limited to the region of the cell where signal receptors are activated.

16. **SCIENCE AND SOCIETY** The development of novel drug treatments that target signal transduction pathways is often unsuccessful or leads to unexpected side effects. Based on what you have learned about signaling pathways in a cell, propose an explanation for the limited success of these approaches.

Mastering Biology

Students Go to Mastering™ Biology for assignments, the eText, and the Study Area with animations, practice tests, and activities.

Professors Go to Mastering™ Biology for automatically graded tutorials and questions that you can assign to your students, plus Instructor Resources.

12 The Cell Cycle

This cell, from a hyacinth plant, is undergoing a type of nuclear division called mitosis. Understanding how mitosis occurs is a major focus of this chapter.

In this chapter you will learn how

The life cycle of a cell culminates in division

- starting with: The four phases of the cell cycle 12.1
- asking: How does cell division take place? 12.2
  - via: Mitosis
  - and: Cytokinesis
- by examining: Control of the cell cycle 12.3
  - and applying: Cancer: out-of-control cell division 12.4

The cell theory maintains that all organisms are made of cells and that all cells arise from preexisting cells (Ch. 1, Section 1.2). Although the cell theory was widely accepted among biologists by the 1860s, most thought that new cells arose within preexisting cells by a process that resembled the growth of mineral crystals. But Rudolf Virchow proposed that new cells are formed by the splitting of preexisting cells—that is, by **cell division**.

In the late 1800s, microscopic observations of newly developing organisms, or **embryos**, confirmed Virchow's hypothesis. Plants and animals start life as single-celled embryos and grow through a series of cell divisions.

Early studies revealed two fundamentally different ways that nuclei divide before cell division: meiosis and mitosis. In animals, **meiosis** leads to the production of sperm and eggs, which are the male and female reproductive cells termed **gametes**. **Mitosis** leads to the production of all other cell types, referred to as **somatic** (literally, "body-belonging") **cells**. (You can see how meiosis and mitosis are related to each other and to the transmission of genetic information in the Big Picture on pages 404–405.)

BIG PICTURE

This chapter is part of the Big Picture. See how on pages 404–405.

Mitosis and meiosis are usually accompanied by **cytokinesis**—the division of the cytoplasm into two distinct cells. When cytokinesis is complete, a so-called parent cell has given rise to two daughter cells.

Mitotic and meiotic cell divisions are responsible for one of the five fundamental attributes of life: reproduction (see Ch. 1, Section 1.1). But even though mitosis and meiosis share many characteristics, they are fundamentally different. During mitotic division, the genetic material is copied and then divided equally between two cells. This is referred to as cellular *replication*, as the daughter cells are genetically identical to the parent cell. In contrast, meiosis results in daughter cells that are genetically different from each other and that have half the amount of hereditary material as the parent cell.

In this chapter the focus is on mitotic cell division; meiotic cell division is the subject of another chapter (Chapter 13). Let's begin with a look at the key events in a cell's life cycle, continue with an in-depth analysis of mitosis and the regulation of the cell cycle, and end by examining how uncontrolled cell division can lead to cancer.

12.1 How Do Cells Replicate?

For life on Earth to exist, cells must replicate. The basic steps in cellular replication are **(1)** copying the DNA (deoxyribonucleic acid), **(2)** separating the copies, and **(3)** dividing the cytoplasm to create two complete cells. In this chapter, you'll focus on a process that has been studied for well over a century: how eukaryotic cells replicate. Like much work in biology, the research on eukaryotic cell replication began with simple observations of the process.

After you complete this section, you should be able to…

- Analyze the eukaryotic cell cycle.

What Is a Chromosome?

As studies of cell division in eukaryotes began, biologists found that certain chemical dyes made threadlike structures visible within nuclei. In 1879, Walther Flemming used a dye made from a coal tar to observe these structures and watch them change in the dividing cells of salamander embryos. The threads first appeared in pairs just before cell division and then split to produce single, unpaired threads in the daughter cells. Flemming introduced the term "mitosis," from the Greek *mitos* ("thread"), to describe this process.

Others studied the roundworm *Ascaris* and noted that the number of threads in a cell was the same before and after mitotic division. All of these cells had the same number of threads.

In 1888, Wilhelm Waldeyer coined the term **chromosome** ("colored-body") to refer to these threadlike structures (visible in the chapter-opening photo). Research carried out since then has shown that a chromosome consists of a single long DNA double helix that is wrapped around proteins, called **histones**, in a highly organized manner (Ch. 19, Section 19.2). DNA encodes the cell's hereditary information, or genetic material. A **gene** is a region of DNA in a chromosome that codes for a particular protein or ribonucleic acid (RNA).

Before mitosis, each chromosome is replicated (Ch. 15, Section 15.3). As mitosis starts, the chromosomes condense into compact structures that can be moved around the cell efficiently. Then one copy of each chromosome is distributed to each of two daughter cells.

Figure 12.1 illustrates an unreplicated chromosome, the same chromosome after it has been replicated, and the replicated chromosome that has condensed at the start of mitosis. Each of the double-stranded DNA copies in a replicated chromosome is called

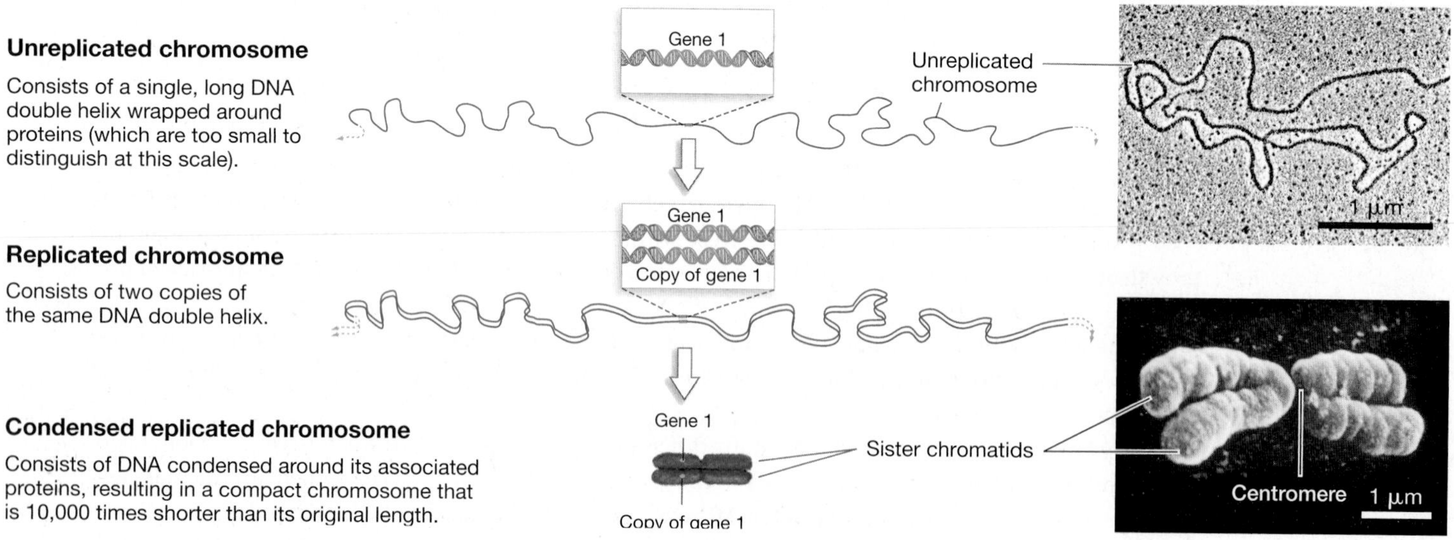

Figure 12.1 Changes in Chromosome Morphology. After chromosomes replicate, the two identical copies of the double-stranded DNA are attached to each other along their entire length. Early in mitosis, replicated chromosomes condense and sister chromatids remain attached at a region called the centromere.

a **chromatid**. Before mitosis, the two chromatids are joined along their entire length by proteins called cohesins. Once mitosis begins, however, these connections are removed except for those at a specialized region of the chromosome called the **centromere**. Chromatid copies that remain attached at their centromere are referred to as **sister chromatids**. Even though a replicated chromosome consists of two chromatids, it is still considered a single chromosome. In Making Models 12.1, you will practice drawing chromosomes to represent the difference between unreplicated and replicated versions.

Cells Alternate between M Phase and Interphase

The division of eukaryotic cells is like a well-choreographed stage performance. The most visually stimulating part of the show, called **M** (*m*itotic or *m*eiotic) **phase**, occurs when cells are in the process of separating their chromosomes. Stained chromosomes can be observed with a light microscope when they condense into compact structures during M phase.

The rest of the time, the cell is in **interphase** ("between-phase"). No dramatic changes in the nucleus are visible by light microscopy during interphase. The chromosomes uncoil into the extremely long, thin structures shown in Figure 12.1 and no longer appear as individual threads. However, this does not mean that the cell is idle. Interphase is an active time: The cell is either growing and preparing to divide or fulfilling its specialized function in a multicellular individual. Cells actually spend most of their time in interphase.

The Discovery of S Phase

Once M phase and interphase were identified by microscopy, researchers could start assigning roles to these distinct phases. They could see that the separation of chromosomes and cytokinesis take place during M phase, but when are the chromosomes replicated?

To answer this question, researchers needed to distinguish cells that were making copies of their DNA from those that were not. They were able to do this by adding radioactive phosphorus, in the form of phosphates, to cells. Those cells that were synthesizing DNA would incorporate the radioactive isotope into nucleotides. (See Ch. 4, Section 4.1, to review where phosphates are in DNA.) There were three steps in this procedure:

1. Label DNA as chromosomes were being replicated.
2. Wash away any radioactive phosphorus that hadn't been incorporated and remove RNA, which would also incorporate phosphorus.
3. Visualize the labeled, newly synthesized DNA by exposing the treated cells to X-ray film. Emissions from radioactive phosphorus create black dots in the film. This technique is called autoradiography (see BioSkills 6).

In 1951, Alma Howard and Stephen Pelc performed this procedure and found black dots—indicating active DNA synthesis—in some interphase cells, but not in M-phase cells. This result showed that DNA replication occurs during a period in interphase. Several years later, this result was verified using radioactive thymidine, which is incorporated into DNA but not RNA.

Thus, biologists had identified a new stage in the life of a cell. They called it **S** (or **synthesis**) **phase**. S phase is part of interphase. The process of copying the genetic material is separated, in time, from the partitioning of replicated chromosomes during M phase.

Howard and Pelc coined the term **cell cycle** to describe the orderly sequence of events that leads a eukaryotic cell through the duplication of its chromosomes to the time it divides.

Making Models 12.1 Tips on Drawing Chromosomes (I)

Drawing models of chromosomes can help you understand how chromosomes change during cell division. There are many ways to make simple drawings of chromosomes. Here are three examples:

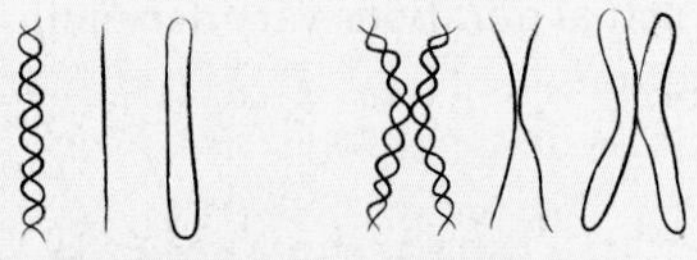

MODEL Draw a model to represent a cell with two different chromosomes before and after the chromosomes are replicated. Use a circle to represent the cell and one of the models shown here to represent the chromosomes. Label the chromatids.

To see this model in action, go to the Study Area of **Mastering Biology**

The Discovery of the Gap Phases

In addition to discovering S phase, Howard and Pelc made another key observation—not all interphase cells were labeled. This meant that there was at least one "gap" in interphase when DNA was not being replicated.

Howard and Pelc, along with researchers in other labs, followed up on these early results by asking where S phase was positioned in interphase. There were three possible scenarios:

1. The S phase is immediately before M phase, with a single gap between the end of M phase and the start of S phase.
2. The S phase is immediately after M phase, with a single gap between the end of S phase and the start of M phase.
3. Two gaps exist: one before and one after S phase.

To address which of these scenarios, if any, was correct, many experiments were done on cells in culture. Cultured cells are powerful experimental tools because they can be manipulated much more easily than cells in an intact organism (see **BioSkills 11**). In most of these experiments, researchers used cultures that were *asynchronous*, meaning that the cells were randomly distributed in various stages of the cell cycle.

To understand the value of asynchronous cultures, imagine the cell cycle as a clock. Every complete rotation of the second hand around the clock would represent one cell division, and each tick would represent a different point in the cycle. At any given time, an asynchronous culture would have at least one cell at each of the ticks on the clock. As time passed, these cells would move around this cell-cycle clock at the same rate and in the same direction.

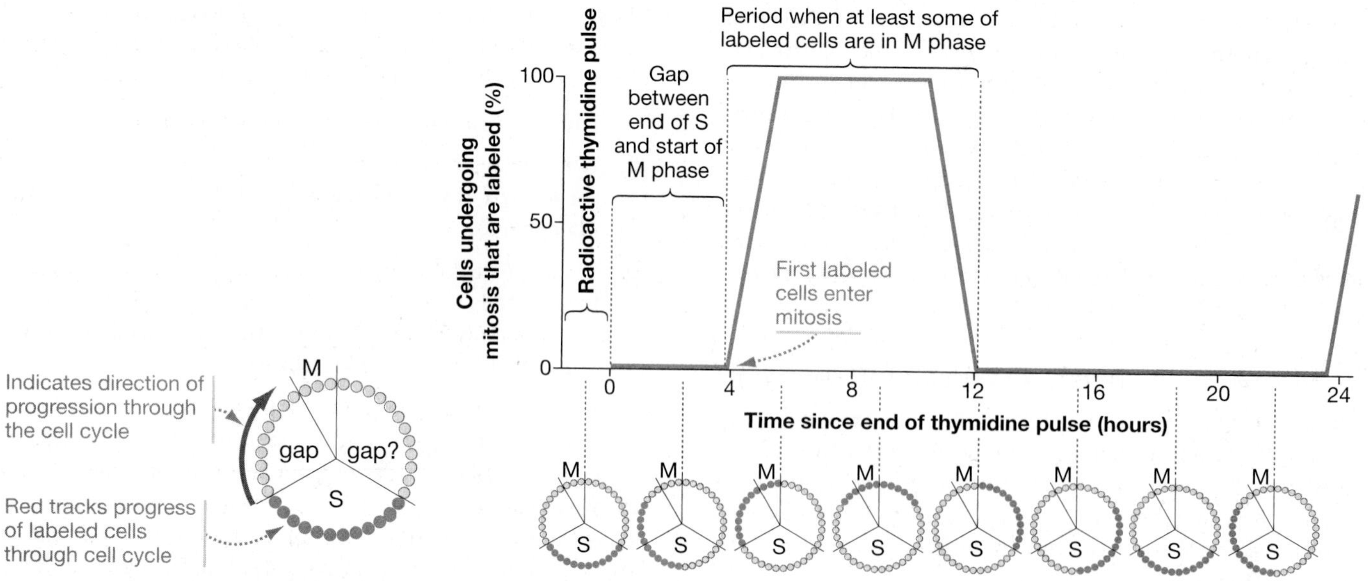

Figure 12.2 A Pulse–Chase Experiment Reveals a Gap Phase. Cells labeled with radioactive thymidine during the pulse were tracked during the chase. The period between the end of the pulse and the appearance of the first labeled mitotic cells represents a gap between the end of S phase and start of M phase.

In one experiment, researchers added radioactively labeled thymidine to the cells in a human cell culture. A short time later, they stopped the labeling by flooding the solution surrounding the cultured cells with nonradioactive thymidine, which washed away any labeled thymidine that had not already been incorporated into DNA. This pulse–chase approach (introduced in Ch. 7, Section 7.5) labeled only those cells that were in S phase during the radioactive pulse. Imagine these labeled cells moving together through the cell cycle like the second hand moving around a clock.

Once the pulse ended, the researchers took samples of cells from the culture at different times during the chase. In each sample, they recorded how many labeled cells were undergoing mitosis, meaning how many cells that were in S phase during the pulse had entered M phase. Figure 12.2 summarizes the results of this experiment.

One striking result emerged early on: None of the labeled cells started mitosis immediately. Because the cultures were asynchronous, at least some of the cells must have been at the very end of their S phase when they were exposed to the pulse. If S phase were immediately followed by M phase, then some of these labeled cells would have entered M phase just as the chase began. Instead, it took several hours before any of the labeled cells began mitosis.

The time between the end of the pulse and the appearance of the first labeled mitotic nuclei corresponds to a gap between the end of S phase and the beginning of M phase. This gap is a period when chromosome replication is complete but mitosis has not yet begun. The graph in Figure 12.2 shows how cells labeled with radioactive thymidine can be tracked as they progress through M phase.

✔ **If you understand how the pulse–chase approach was used in Figure 12.2, you should be able to predict how the graph would have appeared if there had been no gap between the S and M phases.**

This result narrowed the possible scenarios for the organization of the cell cycle: There could be either one gap between the end of S phase and the start of M phase, or two gaps flanking S phase. Which scenario represents the eukaryotic cell cycle? Once researchers determined the lengths of the S and M phases, they found that the combined time, including the gap between them, was shorter than the length of the cell cycle. This discrepancy indicated that there must be an additional gap between the end of M phase and the start of S phase.

The cell cycle was thus finally mapped out. The gap between the end of M phase and the start of S phase is called **G_1 phase.** The second gap, between the end of S phase and the start of M phase, is called **G_2 phase.**

The Cell Cycle

Figure 12.3 pulls these results together into a comprehensive view of the cell cycle. The cell cycle involves four phases: M phase and an interphase consisting of the G_1, S, and G_2 phases. In the cycle diagrammed here, G_1 phase is about twice as long as G_2 phase, but their actual durations vary depending on the cell type and growth conditions.

CHECK YOUR UNDERSTANDING

✔ **If you understood this section, you should be able to ...**

1. Compare and contrast the roles of the two gap phases in the eukaryotic cell cycle.
2. **QUANTITATIVE** Use the graph in Figure 12.2 to determine the length of the G_2 phase in the cells being evaluated.

Answers are available in Appendix A.

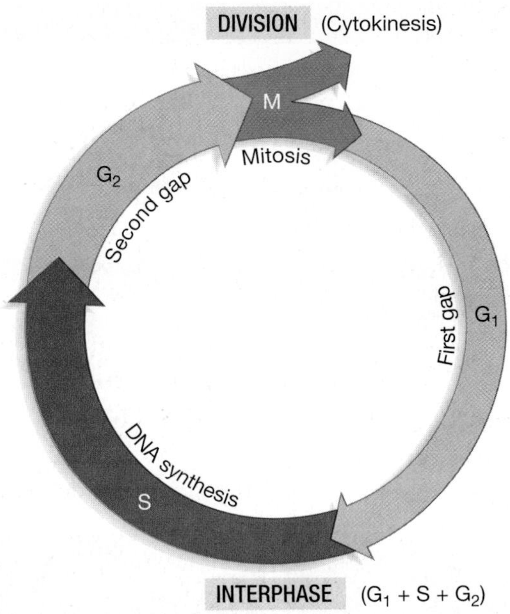

Figure 12.3 The Cell Cycle Has Four Phases. The duration of the four phases varies dramatically among cells and organisms.

Why do the gap phases exist? In multicellular organisms, cells perform their functional roles mostly during G_1 phase. G_1 is also the period when the cell "decides" to begin replication and transitions to S phase (explained in Section 12.4). Before mitosis can take place, a cell uses G_2 phase to prepare for M phase. The time spent in both G_1 and G_2 allows the cell to grow and replicate organelles so it will be able to divide into two cells that can function normally.

Now let's turn to M phase. Once the genetic material has been copied in S phase, how is it divided between daughter cells?

12.2 What Happens during M Phase?

M phase typically consists of two distinct events: the division of the nucleus and the division of the cytoplasm. Mitosis divides the replicated chromosomes to form two daughter nuclei with identical chromosomes and genes. Cytokinesis usually follows mitosis and divides the cytoplasm of the parent cell to form two daughter cells.

After you complete this section, you should be able to . . .

- Describe how the nuclear and cytoplasmic components of a cell are divided during M phase.

Figure 12.4 is an overview of how chromosomes change before, during, and after mitosis and cytokinesis, beginning with a hypothetical plant or animal cell in G_1 phase. The first drawing shows a total of four chromosomes in the cell, but chromosome number varies widely among species—potato plants have 48 chromosomes in each cell, dogs have 78, fruit flies have 8, and you have 46.

Eukaryotic chromosomes consist of DNA wrapped around globular histone proteins. This DNA–histone complex is called **chromatin**. During interphase, the chromatin of each chromosome is in a "relaxed" or less condensed state, forming long, thin strands (see Figure 12.1, top).

The second drawing in Figure 12.4 shows G_2 phase, where the cell contains replicated chromosomes before mitosis. Each chromosome now consists of two sister chromatids. Each chromatid contains one long DNA double helix, and sister chromatids represent exact copies of the same genetic information.

At the start of mitosis, then, each chromosome consists of two sister chromatids that are attached to each other at the centromere.

✔ You should be able to explain the relationship between chromosomes and (1) DNA, (2) chromatin, and (3) sister chromatids.

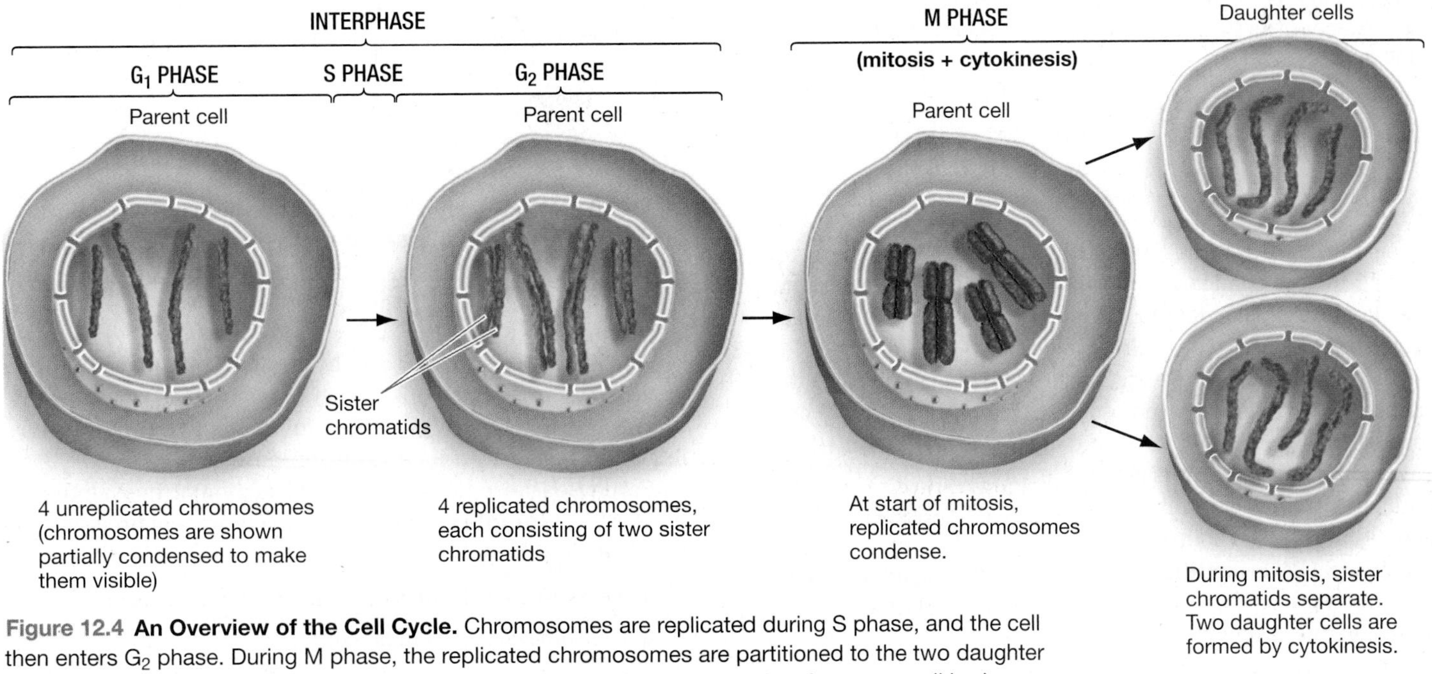

Figure 12.4 An Overview of the Cell Cycle. Chromosomes are replicated during S phase, and the cell then enters G_2 phase. During M phase, the replicated chromosomes are partitioned to the two daughter cells. Each daughter cell contains the same complement of chromosomes that the parent cell had.

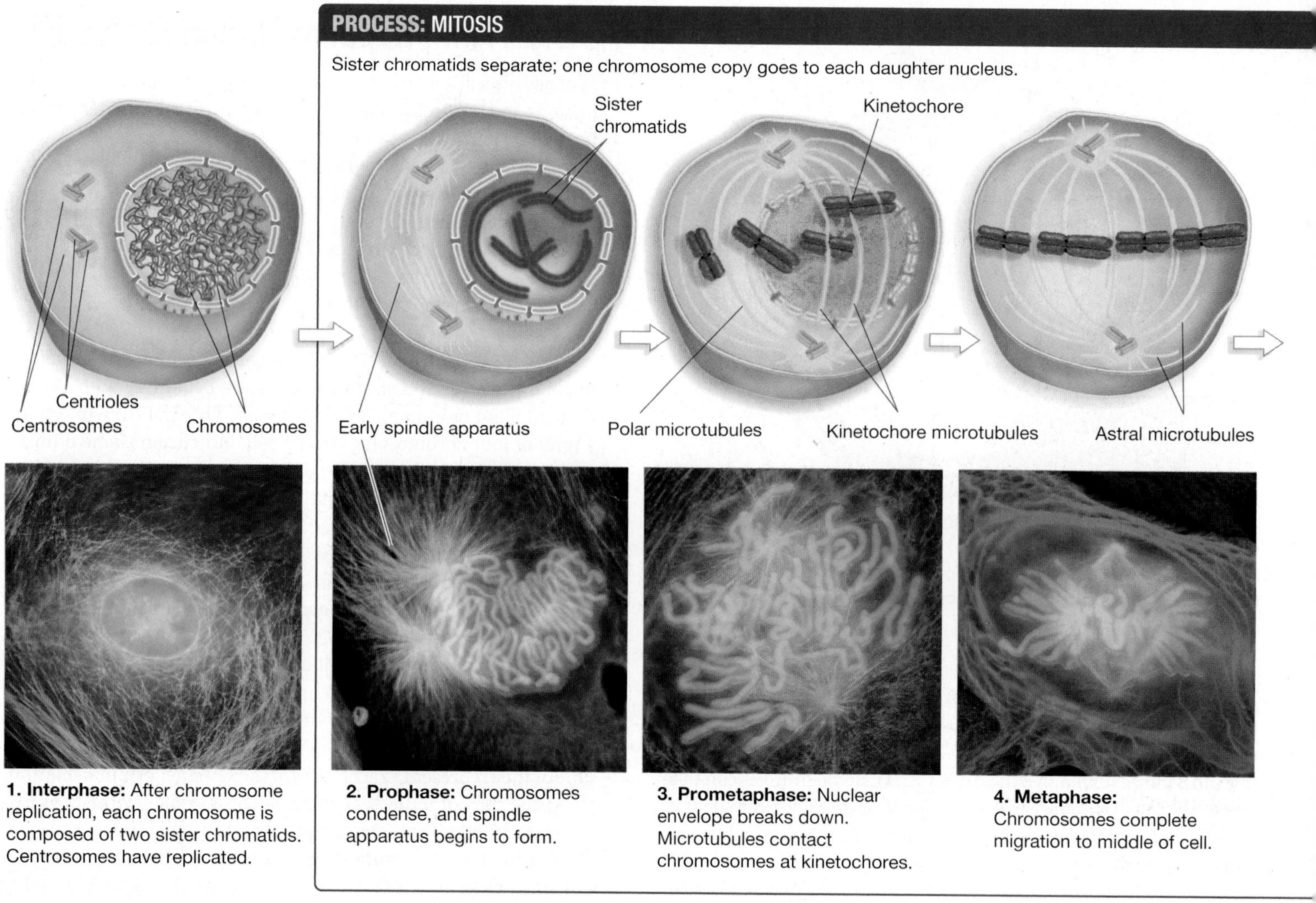

Figure 12.5 M Phase: Mitosis and Cytokinesis. In the micrographs of newt lung cells under the drawings, chromosomes are stained blue, microtubules are yellow/green, and intermediate filaments are red.

✔ **THINK CAREFULLY** If the cell shown in the micrographs has 60 picograms of DNA (6×10^{-11}g) and 22 chromosomes in its G_1 phase, how much DNA and how many chromosomes are in (1) the prophase cell, (2) the anaphase cell, and (3) each daughter cell?

Events in Mitosis

As the third drawing in Figure 12.4 indicates, mitosis begins when chromatin condenses to form a much more compact structure. Replicated, condensed chromosomes correspond to the paired threads observed by early biologists.

During mitosis, the two sister chromatids separate to form independent daughter chromosomes. One copy of each chromosome goes to each of the two daughter cells. (See the final drawing in Figure 12.4.) As a result, each cell receives the same complement of chromosomes (identical copies of each chromosome) as the parent cell had.

Biologists have identified five subphases within mitosis based on distinctive events that occur: prophase, prometaphase, metaphase, anaphase, and telophase.

Recall that before mitosis begins, chromosomes are replicated during S phase of interphase. Now let's investigate each subphase of mitosis to look at how cells separate the chromatids of these replicated chromosomes (Figure 12.5).

Prophase Mitosis begins with the events of **prophase** ("before-phase," Figure 12.5, step 2), when chromosomes condense into compact structures. Individual chromosomes first become visible in the light microscope during prophase.

Prophase is also marked by the formation of the spindle apparatus. The **spindle apparatus** is a structure that produces mechanical forces that

- move replicated chromosomes during early mitosis and
- pull chromatids apart in late mitosis.

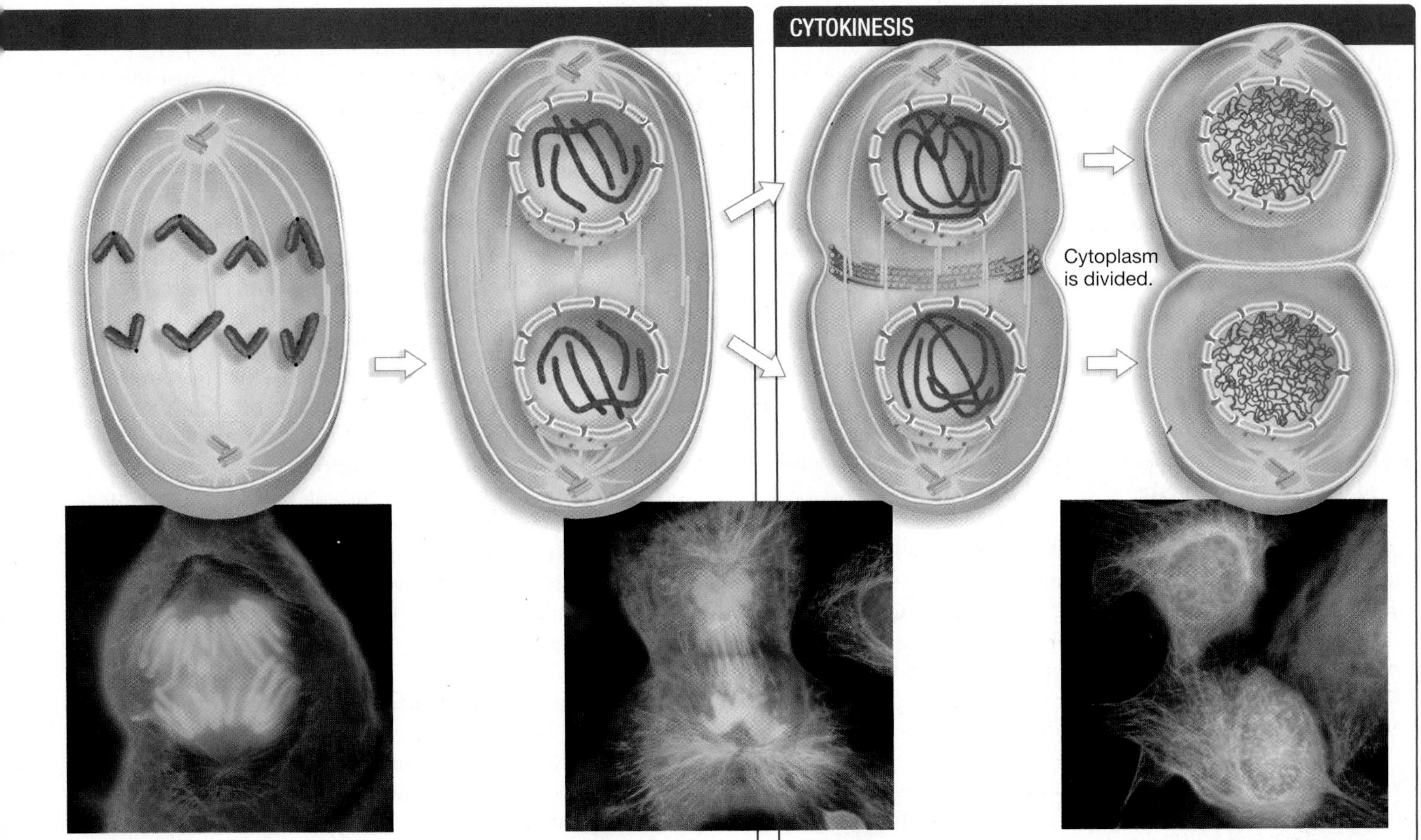

5. Anaphase: Sister chromatids separate into daughter chromosomes, which are pulled to opposite poles of spindle apparatus.

6. Telophase: Nuclear envelope re-forms, and chromosomes de-condense.

7. Cell division begins: Actin–myosin ring causes plasma membrane to begin pinching in.

8. Cell division is complete: Two daughter cells form.

The spindle apparatus consists of microtubules—components of the cytoskeleton. Microtubules have the following characteristics (see Ch. 7, Section 7.6):

- They are composed of α-tubulin and β-tubulin dimers.
- They have a plus end and a minus end—meaning they are asymmetric.
- The plus end is the site where microtubule growth normally occurs. Microtubule disassembly is more frequent at the minus end.

Microtubules originate from microtubule-organizing centers (MTOCs). MTOCs define the two poles of the spindle apparatus and produce large numbers of microtubules, whose plus ends grow outward through the cytoplasm. Although the nature of the MTOC varies among plants, animals, fungi, and other eukaryotic groups, the spindle apparatus has the same function.

Figure 12.5 illustrates mitosis in newt lung cells, where the MTOC is a **centrosome**—a structure that contains a pair of **centrioles** (Ch. 7, Section 7.6). During the S phase, the single centrosome replicates along with the DNA. At the start of prophase, the two centrosomes move to opposite sides of the nucleus to begin forming the spindle apparatus. Some of these microtubules extend from each spindle pole and overlap with one another—these are called **polar microtubules**.

Prometaphase In many eukaryotes, once chromosomes have condensed, the nuclear envelope disintegrates. Removal of the envelope allows the cytoplasmic microtubules to attach to chromosomes at specialized structures called **kinetochores**. These events define the start of **prometaphase** ("before middle-phase"; see Figure 12.5, step 3).

Each sister chromatid has its own kinetochore, which is assembled at the centromere. Because the centromere is also the attachment site for chromatids, the result is two kinetochores on opposite sides of each replicated chromosome. The microtubules attached to these structures are called **kinetochore microtubules**.

Early in prometaphase, kinesin and dynein motors attached to the kinetochores "walk" the chromosomes up and down microtubules. This process is similar to the way the same motors transport vesicles and organelles along microtubules (Ch. 7, Section 7.6). When the chromosomes reach the plus ends of the microtubules, the kinetochore proteins secure their attachment.

Eventually, each chromosome will have its two kinetochores attached to microtubules that originate from opposite sides of the spindle apparatus. The chromosomes are then pushed and pulled by microtubules and motor proteins until they reach the middle of the spindle.

Metaphase Once all the chromosomes have migrated to the middle of the spindle (Figure 12.5, step 4), the mitotic cell enters **metaphase** ("middle-phase"). At this point, the chromosomes are lined up on an imaginary plane between the two spindle poles called the **metaphase plate**.

Formation of the spindle apparatus is now complete. The polar microtubules that extend from each spindle pole overlap in the middle of the cell, thereby forming a pole-to-pole connection. Each chromosome is held by kinetochore microtubules reaching out from opposite poles and exerting the same amount of tension, or pull. The spindle poles are held in place partly because of **astral microtubules** that extend from the MTOCs and interact with proteins on the plasma membrane.

The polarized growth and disassembly of the kinetochore microtubules contributes to the alignment of the chromosomes at the metaphase plate. The slow disassembly of the minus ends at the MTOCs is balanced by the slow growth of the plus ends at the kinetochores. Because the sister chromatids of each chromosome are connected to opposite poles, a tug of war between the poles begins during metaphase.

Anaphase At the start of **anaphase** ("against-phase"), the cohesins that hold sister chromatids together at the centromeres are cleaved by an enzyme (Figure 12.5, step 5). Because the chromatids are under tension, each replicated chromosome is pulled apart, creating two independent daughter chromosomes. By definition, this separation of chromatids instantly doubles the number of chromosomes in the cell.

Two types of movement occur during anaphase. First, the daughter chromosomes move to opposite poles via the attachment of kinetochore proteins to the shrinking kinetochore microtubules. Second, the two poles of the spindle are pushed and pulled farther apart. The push comes from motor proteins in overlapping polar microtubules, which force the poles away from each other. The pull comes from different motors on the plasma membrane, which walk along on the astral microtubules and drag the poles to opposite sides of the cell.

The separation of replicated chromosomes to opposite poles is a critical step in mitosis because it ensures that each daughter cell receives the same complement of chromosomes. When anaphase is complete, two complete sets of chromosomes are fully separated, each set identical to that of the parent cell before chromosome replication.

Telophase During **telophase** ("end-phase"), the nuclear envelope re-forms around each set of chromosomes, and the chromosomes begin to de-condense (Figure 12.5, step 6). Once two independent nuclei have formed, mitosis is complete.

Table 12.1 summarizes the key structures involved in mitosis. ✔ **After you've studied Table 12.1 and reviewed Figure 12.5, you should be able to make a new table that summarizes what happens to (1) the spindle apparatus, (2) the nuclear envelope, and (3) the chromosomes in each of the five phases of mitosis.**

SUMMARY Table 12.1 Structures Involved in Mitosis

Structure	Definition
Chromosome	A structure containing genetic information in the form of genes
Chromatin	The material that makes up eukaryotic chromosomes; consists of a DNA molecule complexed with histone proteins
Sister chromatids	The two attached, double-stranded DNA copies of a replicated chromosome. When chromosomes are replicated, they consist of two genetically identical sister chromatids. When sister chromatids separate during mitosis, they become independent chromosomes.
Centromeres	Specialized regions of chromosomes where sister chromatids are most closely joined to each other
Kinetochores	The structures on sister chromatids where microtubules attach
Centrosome	The microtubule-organizing center in animals and certain plants and fungi. Each pole in the spindle apparatus is a centrosome.
Microtubules	Cytoskeletal filaments that form the spindle apparatus, which consists of polar microtubules, kinetochore microtubules, and astral microtubules
Microtubule motor proteins	The dynein and kinesin motors that participate in moving chromosomes and the poles of the spindle apparatus

RESEARCH

QUESTION: How do kinetochore microtubules pull chromatids apart during anaphase?

HYPOTHESIS: Microtubules shorten at the spindle pole.

ALTERNATIVE HYPOTHESIS: Microtubules shorten at the kinetochore.

EXPERIMENTAL SETUP:

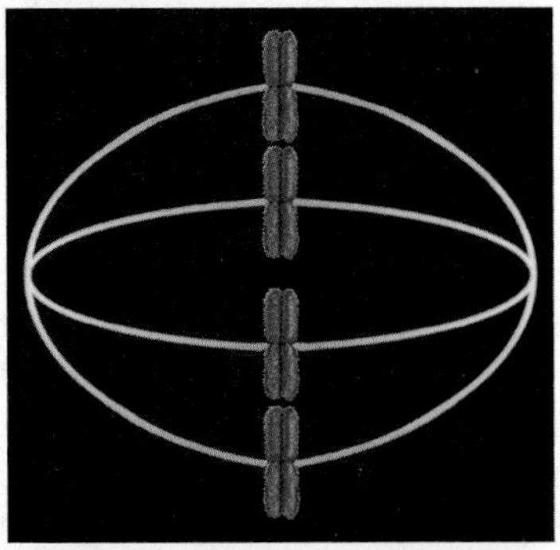

1. Label targets: Use fluorescent labels to make the metaphase chromosomes fluoresce blue and the microtubules fluoresce yellow.

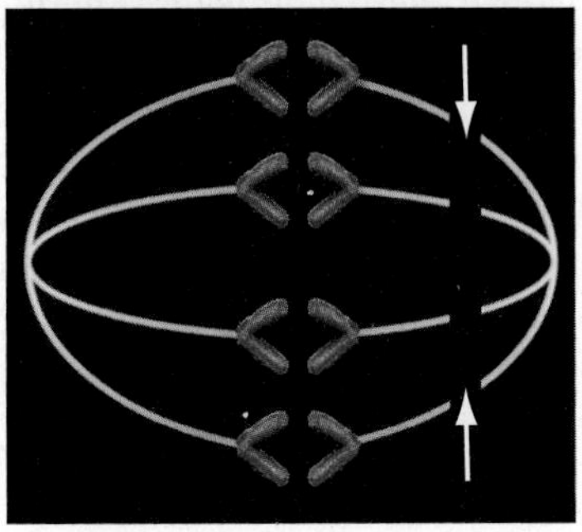

2. Mark microtubules: At the start of anaphase, use laser to darken sections of microtubules to mark them without changing their function.

3. Using microscopy, note the movement of chromosomes relative to the movement of the darkened section of the microtubules.

PREDICTION OF HYPOTHESIS:

PREDICTION OF ALTERNATIVE HYPOTHESIS: Darkened sections will remain stationary while the daughter chromosomes move toward the pole.

RESULTS:

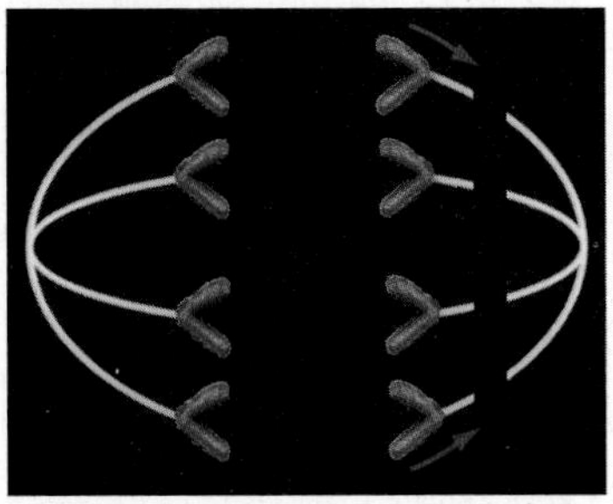

The darkened sections of the microtubules remained stationary as the chromosomes moved through them toward the pole.

CONCLUSION: Kinetochore microtubules shorten at the kinetochore to pull daughter chromosomes apart during anaphase.

Figure 12.6 During Anaphase, Microtubules Shorten at the Kinetochore.

SOURCE: G. J. Gorbsky, P. J. Sammak, and G. G. Borisy. 1987. Chromosomes move poleward in anaphase along stationary microtubules that coordinately disassemble from their kinetochore ends. *Journal of Cell Biology* 104: 9–18.

✔ Complete the prediction for the hypothesis that microtubules shorten at the spindle pole.

How Do Chromosomes Move during Anaphase?

The exact and equal partitioning of genetic material to the two daughter nuclei is the most fundamental aspect of mitosis. To understand how sister chromatids separate and move to opposite sides of the spindle, biologists have focused on the role of kinetochore microtubules. How do these microtubules pull chromatids apart?

Mitotic Spindle Forces During mitosis, the microtubules originating from the spindle poles are highly dynamic. Rapid growth and disassembly ensures that some of the microtubules will be able to attach to kinetochores with their plus ends. Others will be stabilized by different proteins in the cytoplasm and become polar or astral microtubules.

These observations suggest two hypotheses for the movement of chromosomes during anaphase. The simpler hypothesis is that kinetochore microtubules stop growing at their plus ends but remain attached to the kinetochores. As the minus ends disassemble at the spindle poles, the chromosomes would be reeled in like hooked fish. An alternative hypothesis is that the chromosomes move along microtubules that are being disassembled at their plus ends at the kinetochores. In this case, each chromosome would be like a yo-yo running up a string into your hand.

To test these hypotheses, biologists introduced fluorescently labeled tubulin subunits into prophase or metaphase cells. This treatment made the kinetochore microtubules visible (Figure 12.6, step 1). Once anaphase began, the researchers marked a bar-shaped region of these microtubules with a beam of laser light. The laser permanently bleached sections of the fluorescently labeled microtubules, darkening them—although they were still functional (Figure 12.6, step 2).

As anaphase progressed, two things happened: **(1)** The darkened sections of the microtubules appeared to remain stationary, and **(2)** the chromosomes moved closer to the darkened sections, eventually overtaking them.

This result was a complete surprise. It suggested that when cells transition from metaphase to anaphase, the plus ends of the kinetochore microtubules switch from adding tubulin dimers to removing them. As these plus ends shrink back to the spindle poles, the chromosomes are pulled along. But if the microtubule is disassembling at the kinetochore, how does the chromosome remain attached?

Kinetochores Are Linked to Retreating Microtubule Ends The kinetochore is a complex of many proteins that attaches the centromere region of the chromosome to one or more microtubules. Figure 12.7 on page 266 shows a current model of kinetochore structure and function during chromosome movement in anaphase. For simplicity, a yeast kinetochore is shown, which attaches to only one microtubule. (Other eukaryotes can have as many as 30 microtubules attached to each kinetochore.)

Fibers that extend from the yeast kinetochore are tethered to a ring that surrounds the kinetochore microtubule (Figure 12.7, top). Biologists have found that as anaphase gets under way, the plus end of the kinetochore microtubule begins to fray and disassemble. As the fraying end widens, its expansion forces the ring,

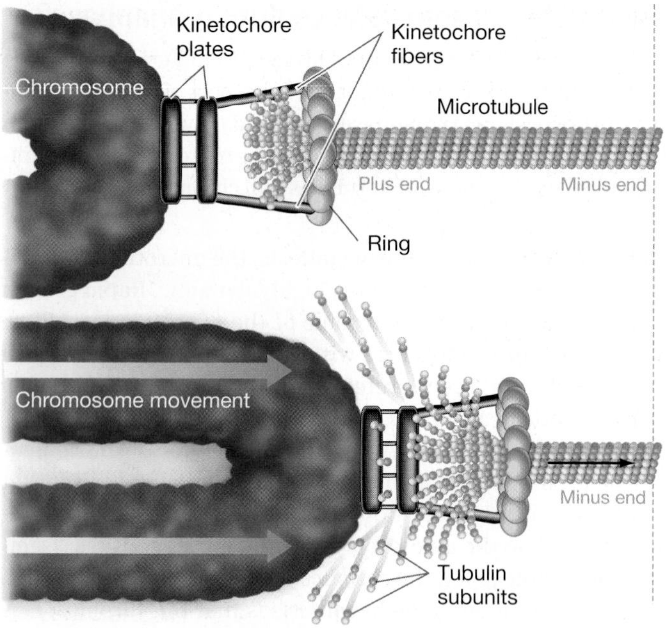

Figure 12.7 How Do Microtubules Move Chromosomes during Anaphase? Microtubules are disassembled at the kinetochore during anaphase. In yeast, kinetochore plates and fibers tether the chromosome to a ring that is pushed toward the spindle pole by the fraying plus end of the microtubule.

and the attached chromosome, toward the minus end of the microtubule (see Figure 12.7, bottom). The result is that the chromosome is pulled to the spindle pole by the force generated when kinetochore microtubules depolymerize.

Cytokinesis Results in Two Daughter Cells

At this point, the chromosomes have been replicated in S phase and distributed to opposite sides of the spindle via mitosis. Now it's time to divide the cell into two daughter cells that contain identical copies of each chromosome. If these cells are to survive, however, the parent cell must also ensure that more than just chromosomes make it into each daughter cell.

While the cell was in interphase, the cytoplasmic contents, including the organelles, increased in number or volume. During cytokinesis (Figure 12.5, steps 7 and 8), the cytoplasm divides to form two daughter cells, each with its own nucleus and complete set of organelles. In most types of cells, cytokinesis directly follows mitosis.

In plant cells, polar microtubules left over from the spindle apparatus help define and organize the region where the new plasma membranes and cell walls will form. Vesicles from the Golgi apparatus carry components for a new cell wall to the middle of the dividing cell. These vesicles are moved along the polar microtubules via motor proteins. In the middle of what was the spindle, the vesicles start to fuse and form a flattened, sac-like structure called the **cell plate** (Figure 12.8a). The cell plate continues to grow as new vesicles fuse with it. Eventually, the cell plate contacts and fuses with the existing plasma membrane, dividing the cell into two daughter cells.

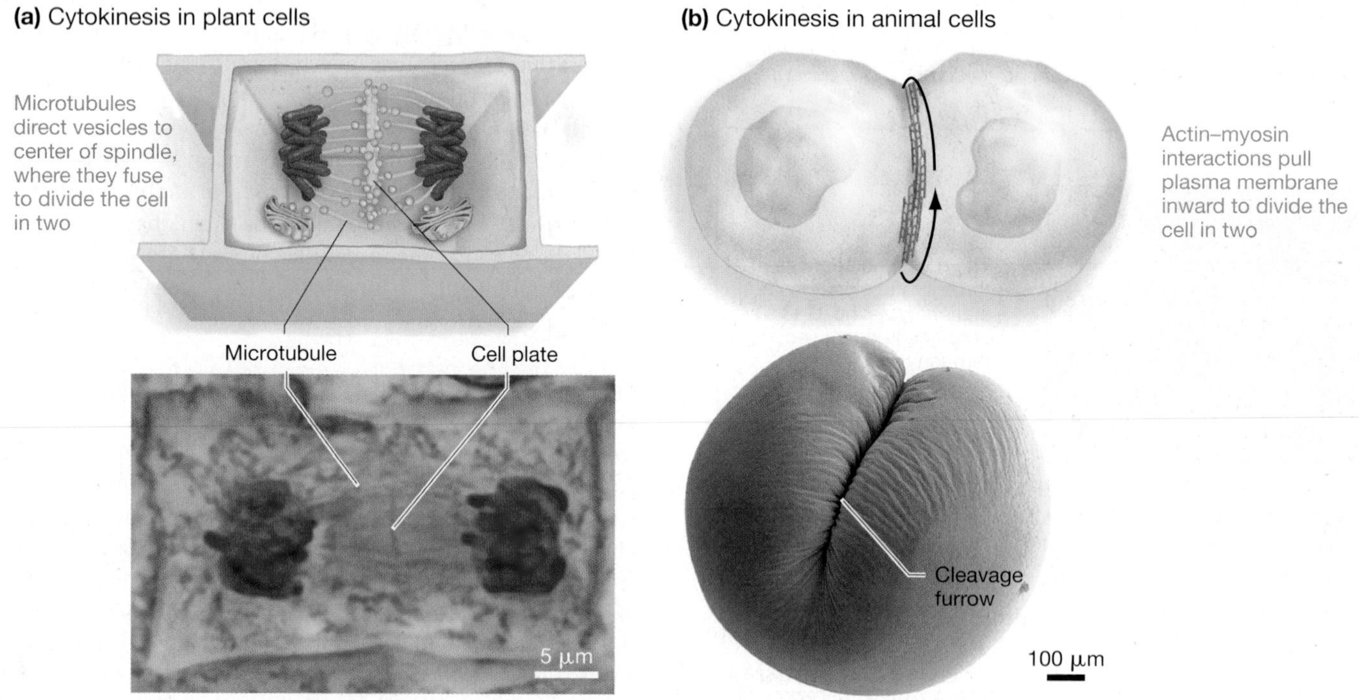

Figure 12.8 The Mechanism of Cytokinesis Varies among Eukaryotes. (a) In plant cells, the cytoplasm is divided by a cell plate that forms in the middle of the parent cell. **(b)** In animal cells, the cytoplasm is divided by a cleavage furrow. (The cells in both micrographs have been stained or colorized.)

In animals and many other eukaryotes, cytokinesis begins with the formation of a **cleavage furrow** (Figure 12.8b). The furrow appears when a ring of overlapping actin filaments starts to contract just inside the plasma membrane, in the middle of what used to be the spindle. This contraction is caused by myosin motor proteins that bind to the actin filaments and use adenosine triphosphate (ATP) to slide the filaments past one another (Ch. 7, Section 7.6).

As myosin moves the actin filaments, the ring shrinks and tightens. Because the ring is attached to the inside of the plasma membrane, the contracting ring pulls the membrane with it. As a result, the plasma membrane is drawn inward. Myosin continues to slide the actin filaments past each other, tightening the ring further, until the plasma membrane fuses and cell division is complete.

Chromosome separation and cytoplasmic division are common requirements for all organisms, not just eukaryotes. What is known about cell division in prokaryotes? Is the process of cell division in your cells similar to that in bacteria?

Bacterial Cell Replication

Many bacteria divide using a process called **binary fission**. Although binary fission does not involve mitosis, recent research has shown that chromosome segregation and cytokinesis in bacteria are strikingly similar to what occurs in the eukaryotic M phase (Figure 12.9). Protein filaments attach to the replicating bacterial chromosomes, then the filaments pull the chromosomes apart.

Once the chromosome copies have been moved to opposite sides of the cell, other filaments attach to the plasma membrane and form a ring between the chromosome copies. A signal from the cell causes the filaments to draw in the membrane, eventually cleaving the parent cell into two genetically identical cells.

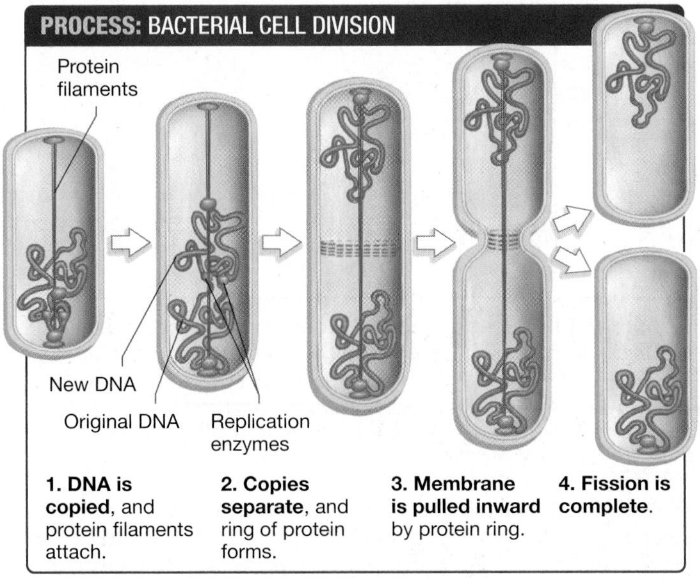

Figure 12.9 Bacterial Cells Divide but Do Not Undergo Mitosis.

Having explored what occurs during cell division, let's focus on how it is controlled in eukaryotes. When does a eukaryotic cell divide, and when does it stop dividing?

CHECK YOUR UNDERSTANDING

✔ **If you understood this section, you should be able to ...**

1. **MODEL** Draw the mitotic spindle for an animal cell that has two chromosomes in metaphase. Label the sister chromatids, kinetochores, centrosomes, and three types of microtubules.
2. Use your drawing to explain the two types of movement that are responsible for separating daughter chromosomes during anaphase.
3. Compare and contrast cytokinesis in plant and animal cells.

Answers are available in Appendix A.

12.3 Control of the Cell Cycle

After you complete this section, you should be able to ...

- Explain how cells monitor and regulate the progression of the cell cycle.

Although the events of mitosis are similar in all eukaryotes, control of the cell cycle often varies—even among cells in the same organism. In humans, for example, intestinal cells routinely divide twice a day to replace tissue that is lost during digestion, whereas mature nerve and muscle cells do not divide at all.

Most of these differences are due to variation in the length of the G_1 phase. In rapidly dividing cells, G_1 is essentially eliminated. Most nondividing cells, in contrast, are permanently stuck in G_1. Researchers refer to this arrested state as the G_0 state, or simply "G zero." Nerve cells, muscle cells, and many other cell types enter G_0 once they have matured.

A cell's division rate can also vary in response to changing conditions. For example, human liver cells normally divide about once per year. But if part of the liver is damaged or lost, the remaining cells divide every one or two days until repair is accomplished. Cells of unicellular eukaryotes, such as yeasts and some protists, divide rapidly only if the environment is rich in nutrients; otherwise, they enter G_0.

To explain these differences, biologists hypothesized that the cell cycle must be regulated in some way. Cell-cycle control is now the most prominent issue in research on cell division—partly because defects in control can lead to uncontrolled cell growth and cancer.

The Discovery of Cell-Cycle Regulatory Molecules

The first solid evidence for cell-cycle control molecules came to light in 1970. Researchers found that when they fused cells that were in different stages of the cell cycle, forming a single cell with two nuclei, one of the nuclei often changed phases. For example, when a cell in M phase was fused with one in interphase, the nucleus of the interphase cell immediately initiated mitosis, even if its chromosomes had not been replicated.

To explain these results, the researchers hypothesized that the cytoplasm of M-phase cells contains a regulatory molecule that induces interphase cells to enter M phase. But cell-fusion experiments were difficult to control and didn't explain whether the nucleus or the cytoplasm was responsible for the induction. To address this issue, they turned to the South African clawed frog, *Xenopus laevis*.

As an egg of these frogs matures, it changes from a cell called an immature oocyte, which is arrested in G_2 of interphase, to a mature egg that is arrested in M phase. The large size of these cells—more than 1 mm in diameter—makes them relatively easy to manipulate. For example, using extremely fine pipets, researchers could remove cytoplasm from one cell and inject it into another.

When biologists injected cytoplasm from M-phase frog eggs into the cytoplasm of frog oocytes arrested in G_2, the oocytes immediately entered M phase. But when the same experiment was done using the cytoplasm from immature oocytes, the injected cells remained in the G_2 phase. The researchers concluded that the cytoplasm of M-phase cells—but not the cytoplasm of G_2 interphase cells—contains a substance that drives immature oocytes into M phase to complete their maturation.

The substance that initiates M-phase in oocytes is now called the **M-phase-promoting factor**, or **MPF**. Subsequent experiments showed that virtually all eukaryotes use a similar MPF to induce M phase. As a result, G2-arrested frog oocytes have been used extensively to measure MPF activity from a variety of sources. For example, when frog oocytes are injected with M-phase cytoplasm from human cells, the oocytes immediately entered M phase.

MPF appears to be a general signal that says "Start M phase," but several questions still remain.

What Proteins Are Responsible for MPF Activity? After discovering that MPF caused cells to enter M phase, researchers began looking for proteins that might be responsible for this activity. One candidate belonged to a family of proteins called **cyclins**, based on the cyclic nature of their expression during the cell cycle. The pattern of cyclin expression was similar to MPF—its concentration increased just before M phase, and then plummeted after cells divided and reentered interphase. How could researchers distinguish between cyclin being responsible for MPF versus it being just a product of the MPF activity?

Differentiating between correlation and causation is a common problem faced in biology. Researchers addressed the relationship between cyclin and MPF by again using the frog eggs just described. In this experiment, the eggs were stimulated to divide, then the cytoplasm was extracted and all of the mRNA destroyed. Only cyclin mRNA was reintroduced to the extracts, and samples were tested for the presence of cyclin and MPF activity (Figure 12.10). The results showed that the MPF increased only after cyclin protein was translated from the added cyclin mRNA.

After further research, it was determined that MPF is made up of two distinct polypeptide subunits. One subunit is the cyclin protein, and the other is a protein kinase that is more or less constant in its concentration throughout the cell cycle. Protein

RESEARCH

QUESTION: Is the expression of cyclin required for MPF activity?

HYPOTHESIS: Cyclin protein is a required component of MPF activity.

NULL HYPOTHESIS: Production of cyclin protein is only correlated with MPF: The cyclin protein is not a requirement for MPF activity.

EXPERIMENTAL SETUP:

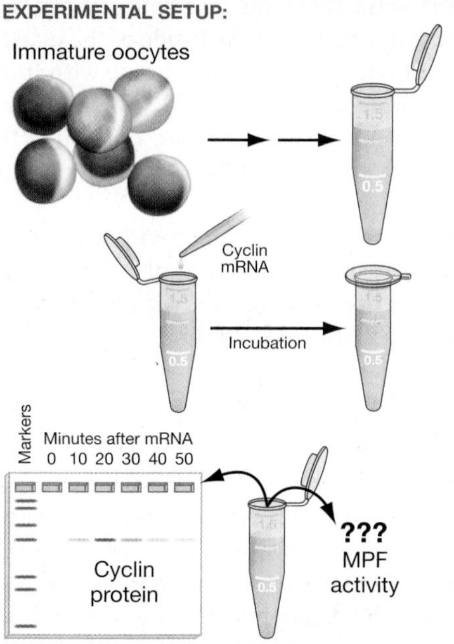

1. Prepare a cell extract from frog eggs that have been activated to undergo the cell cycle, but with all mRNAs destroyed. In the absence of mRNA, the cell cycle is halted.

2. Add only cyclin mRNA to the treated egg extract and pull samples at regular intervals.

3. Evaluate samples for the presence of cyclin protein (see gel electrophoresis in BioSkills 6) and for MPF activity.

PREDICTION OF HYPOTHESIS: MPF activity will only be present when cyclin protein is produced.

PREDICTION OF NULL HYPOTHESIS: The production of cyclin protein will not result in MPF activity.

RESULTS:

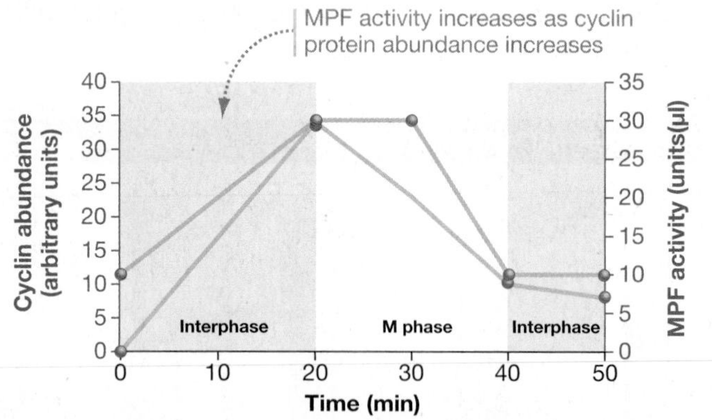

CONCLUSION: The activity of MPF is dependent on the presence of cyclin protein, suggesting that cyclin is a required part of MPF.

Figure 12.10 **Is Cyclin a Key Component of MPF Activity?**

SOURCE: A. W. Murray and M. W. Kirschner. 1989. Cyclin synthesis drives the early embryonic cell cycle. *Nature* 339: 275–280.

✔ **PROCESS OF SCIENCE** Based on what you have learned about the discovery of MPF activity, propose an approach that could be used in the Experimental Setup to measure MPF activity in the extract after cyclin mRNA was added.

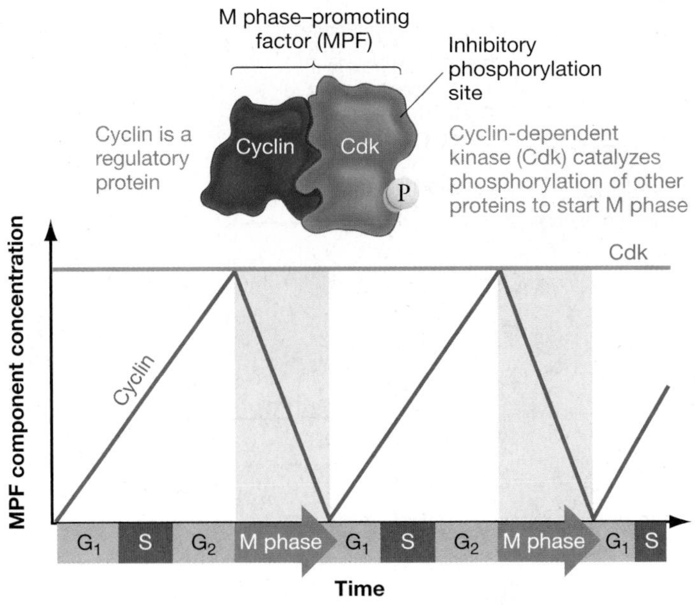

Figure 12.11 Cyclin Concentration Regulates the Concentration of the MPF Dimer. The concentration of cyclin fluctuates in dividing cells, reaching a peak in M phase. The activity of MPF, shown in the blue shaded areas, requires both cyclin and Cdk components.

✔ In this figure, the cyclin concentration declines rapidly during M phase. Why do you think this decline is important?

kinases are enzymes that catalyze the transfer of a phosphate group from ATP to a target protein. Recall that phosphorylation may activate or inactivate the function of proteins by changing their shape (Ch. 8, Section 8.4). This observation suggested that MPF phosphorylates proteins that trigger the onset of M phase.

As Figure 12.11 shows, the concentration of the cyclin associated with MPF builds during interphase and peaks in M phase. The timing of this increase is important because the protein kinase subunit in MPF is functional only when it is bound to the cyclin subunit. As a result, the protein kinase subunit of MPF is called a **cyclin-dependent kinase**, or **Cdk**.

To summarize, MPF is a dimer consisting of a cyclin and a cyclin-dependent kinase. The cyclin subunit regulates the formation of the MPF dimer; the kinase subunit catalyzes the phosphorylation of other proteins to start M phase.

How Is MPF Turned On? According to Figure 12.11, the concentration of cyclin builds up steadily during interphase. Why doesn't the resulting increase in the concentration of MPF trigger the onset of M phase earlier in the cell cycle?

The answer is that the activity of MPF's Cdk subunit is further regulated by two phosphorylation sites on the subunit. Phosphorylation of one site activates the kinase, but phosphorylation of the second site inhibits the kinase. Both sites are phosphorylated after cyclin binds to the Cdk subunit. This allows the concentration of the dimer to increase without prematurely starting M phase. Late in G_2 phase, however, a phosphatase removes the inhibitory phosphate. This dephosphorylation reaction, coupled with the addition of the activating phosphate, changes the Cdk's shape in a way that turns on its kinase activity.

Once MPF is active, it triggers a chain of events. Although the exact mechanisms involved are still under investigation, the result is that chromosomes begin to condense and the spindle apparatus starts to form. In this way, MPF triggers the onset of M phase.

How Is MPF Turned Off? During anaphase, an enzyme complex begins degrading MPF's cyclin subunit, triggering a chain of events that leads to the deactivation of MPF.

MPF deactivation illustrates two key concepts about regulatory systems in cells:

1. **Negative feedback** occurs when a process is slowed or shut down by one of its products. Thermostats shut down furnaces when temperatures are high; enzymes in glycolysis are inhibited by ATP (Ch. 9, Section 9.2); MPF is turned off by an enzyme complex that is activated by events in mitosis.
2. Destroying specific proteins is a common way to control cell processes. In the case of MPF, the enzyme complex that is activated in anaphase attaches small proteins called ubiquitins to MPF's cyclin subunit. This marks the subunit for destruction by a protein complex known as the proteasome.

In response to MPF activity, then, the concentration of cyclin declines rapidly. It slowly builds up again during interphase.

✔ **If you understand this aspect of cell-cycle regulation, you should be able to explain the relationship between MPF activity and the presence of cyclin, Cdk, Cdk kinases, Cdk phosphatase, and the enzymes that degrade cyclin.**

Cell-Cycle Checkpoints Can Arrest the Cell Cycle

The dramatic changes in cyclin concentration and Cdk activity drive the ordered events of the cell cycle. These events are occurring in your body right now. Over a 24-hour period, you swallow millions of cheek cells and lose millions of cells from your intestinal lining as waste. To replace them, other cells in your cheek and intestinal tissue are making and degrading cyclin to push themselves through the cell cycle.

MPF is only one of many protein complexes involved in regulating the cell cycle, however. A different cyclin–Cdk complex triggers the passage from G_1 phase into S phase, and several regulatory molecules hold cells in particular stages.

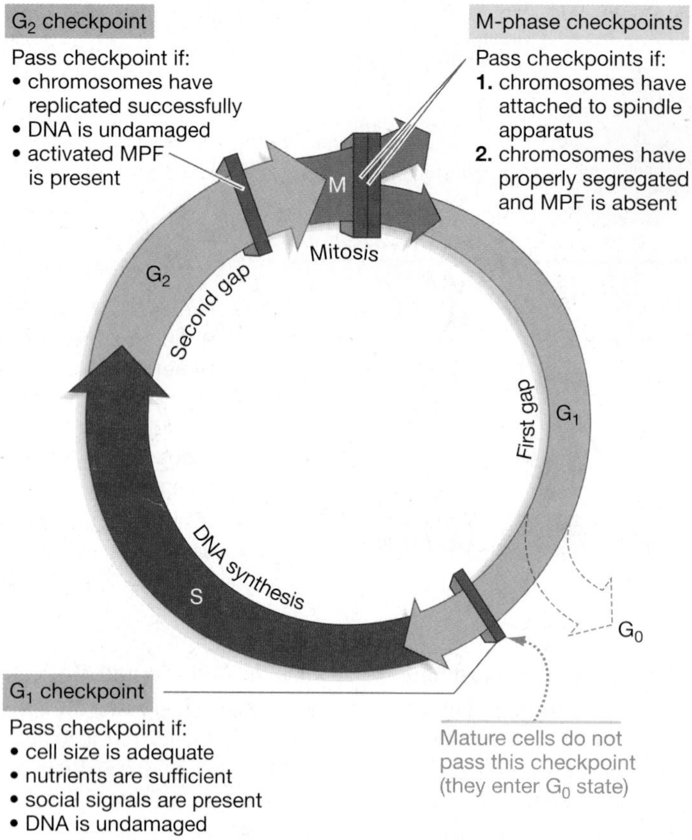

Figure 12.12
The Four Cell-Cycle Checkpoints.

To make sense of these observations, Leland Hartwell and Ted Weinert introduced the concept of **cell-cycle checkpoints**. A cell-cycle checkpoint is a critical point in the cell cycle that is regulated.

Hartwell and Weinert identified checkpoints by analyzing yeast cells with defects in the cell cycle. The defective cells kept dividing under culture conditions that caused normal cells to stop dividing, because the defective cells lacked a specific checkpoint. In multicellular organisms, cells that keep dividing in this way may die or form a mass of cells called a **tumor**.

There are distinct checkpoints in three of the four phases of the cell cycle (Figure 12.12). In effect, interactions among regulatory molecules at each checkpoint allow a cell to "decide" whether to proceed with division or not. If these regulatory molecules are defective, the checkpoint may fail and cells may start dividing in an uncontrolled fashion.

G$_1$ Checkpoint The first cell-cycle checkpoint occurs late in G$_1$ phase. For most cells, this checkpoint is the most important in establishing whether the cell will continue through the cycle and divide, or exit the cycle and enter G$_0$. What factors are important in determining whether a cell passes the G$_1$ checkpoint?

- ***Size*** Because a cell must reach a certain size before its daughter cells will be large enough to function normally, biologists hypothesize that some mechanism exists to arrest the cell cycle if the cell is too small.
- ***Availability of nutrients*** Unicellular organisms arrest at the G$_1$ checkpoint if nutrient conditions are poor.
- ***Social signals*** Cells in multicellular organisms pass (or do not pass) the G$_1$ checkpoint in response to signaling molecules from other cells, which are termed social signals.
- ***Damage to DNA*** If DNA is physically damaged, the **p53 protein** activates genes that either stop the cell cycle until the damage can be repaired or cause the cell's programmed, controlled destruction—a phenomenon known as **apoptosis** (pronounced *a-pop-TOH-sis*). In this way, p53 acts as a brake on the cell cycle.

If "brake" molecules such as p53 are defective, damaged DNA remains unrepaired. Damage in genes that regulate cell growth can lead to uncontrolled cell division. Consequently, regulatory proteins such as p53 are called **tumor suppressors**.

G$_2$ Checkpoint The second checkpoint occurs after S phase, at the boundary between the G$_2$ and M phases. Because MPF is the key signal triggering the onset of M phase, investigators were not surprised to find that it is involved in the G$_2$ checkpoint.

Data suggest that if DNA is damaged or if chromosomes are not replicated correctly, the phosphatase that removes the inhibitory phosphate on MPF's Cdk subunit is not active. As a result, MPF is not turned on, and cells remain in G$_2$ phase. Cells at the G$_2$ checkpoint may also respond to signals from other cells and to internal signals relating to cell size.

M-Phase Checkpoints The final two checkpoints occur during mitosis. The first regulates the transition from metaphase to anaphase. This checkpoint ensures that the sister chromatids do not split until all kinetochores are attached properly to the spindle apparatus. If the metaphase checkpoint does not function properly, chromosomes may not separate correctly, and daughter cells could receive either too many or too few chromosomes.

The second checkpoint regulates the transition from anaphase to telophase. To exit M phase and progress into G_1 phase, cells must degrade all of their cyclins and thus turn off MPF activity. The enzymes responsible for degrading cyclins are activated only when all the chromosomes have been properly separated. If chromosomes do not fully separate during anaphase, the remaining MPF activity will prevent the cell from entering telophase and undergoing cytokinesis. If cells are arrested by either of these two checkpoints, they will remain in M phase.

To summarize, the four cell-cycle checkpoints have the same purpose: They prevent the division of cells that are damaged or that have other problems. The G_1 checkpoint also prevents mature cells that are in the G_0 state from dividing.

Understanding cell-cycle regulation is fundamental. If one of the checkpoints fails, the affected cells may begin dividing in an uncontrolled fashion. For a multicellular organism as a whole, the consequences of uncontrolled cell division may be dire: cancer.

CHECK YOUR UNDERSTANDING

✔ If you understood this section, you should be able to ...

List where the four cell-cycle checkpoints occur in the cell cycle, and explain why they are important.

Answers are available in Appendix A.

12.4 Cancer: Out-of-Control Cell Division

After you complete this section, you should be able to ...

- Explain how the properties of cancer are related to the breakdown of social control that regulates cell division.

Forty percent of American men and women will develop cancer during their lifetime. In the United States, one in four of all deaths is from cancer. It is the second leading cause of death, exceeded only by heart disease.

Cancer is a general term for disease caused by cells that divide in an uncontrolled fashion, invade nearby tissues, and spread to other sites in the body. Cancerous cells cause disease because they use nutrients and space needed by normal cells and disrupt the function of normal tissues.

Humans suffer from at least 200 types of cancer. Stated another way, cancer is not a single illness but a complex family of diseases that affect an array of organs, including the breast, colon, brain, lung, and skin (**Figure 12.13**). In addition, several types of cancer can affect the same organ. Skin cancers, for example, come in multiple forms (see melanoma in Figure 12.13).

Although cancers vary in time of onset, growth rate, lethality, and cause, they have a unifying feature: Cancers arise from cells in which cell-cycle checkpoints have failed.

Cancerous cells have two types of defects related to cell division: **(1)** defects that activate the proteins required for cell growth when they shouldn't be active, and **(2)** defects that prevent tumor suppressor genes from shutting down the cell cycle.

For example, the protein Ras is a key component in signal transduction systems—including phosphorylation cascades that trigger cell growth (Ch. 11, Section 11.3). Many cancers have defective forms of Ras that do not become inactivated. Instead, the defective Ras constantly sends signals that trigger mitosis and cell division.

Likewise, a large percentage of cancers have defective forms of the tumor suppressor protein p53. Instead of being arrested or destroyed, cells with damaged DNA are allowed to continue dividing.

Let's first review the general characteristics of cancer and then explore the regulatory mechanisms that are most often disabled in the disease.

Properties of Cancer Cells

When even a single cell in a multicellular organism begins to divide in an uncontrolled fashion, a mass of cells called a tumor may result. Some tumors can be surgically removed without damage to the affected organ. Often, though, tumor removal doesn't cure cancer. Why?

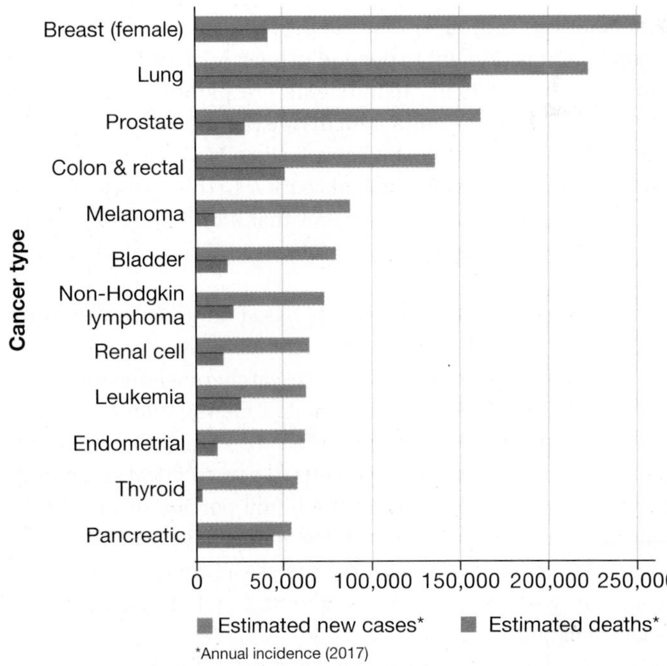

Figure 12.13 Cancers Vary in Type and Lethality.

DATA: The website of the National Cancer Institute (http://www.cancer.gov/types/common-cancers), Common Cancer Statistics, February 2017.

In addition to uncontrolled replication, cancer cells are invasive—meaning that they are able to spread to adjacent tissues and throughout the body via the bloodstream or the lymphatic vessels (introduced in Ch. 42, Section 42.5), which collect excess fluid from tissues and return it to the bloodstream.

Invasiveness is a defining feature of a **malignant tumor**—one that is cancerous. Masses of noninvasive cells are noncancerous and form **benign tumors**. Some benign tumors are largely harmless. Others grow quickly and can cause problems if they are located in the brain or other sensitive parts of the body.

Cells in a tumor become cancerous if they gain the ability to detach from the tumor and invade other tissues. By spreading from the primary tumor site, cancer cells can establish secondary tumors elsewhere in the body (Figure 12.14). This process is called **metastasis**.

If metastasis has occurred by the time the original tumor is detected, secondary tumors may have formed already, and surgically removing the primary tumor will not lead to a cure. This is why early detection is the key to treating cancer most effectively.

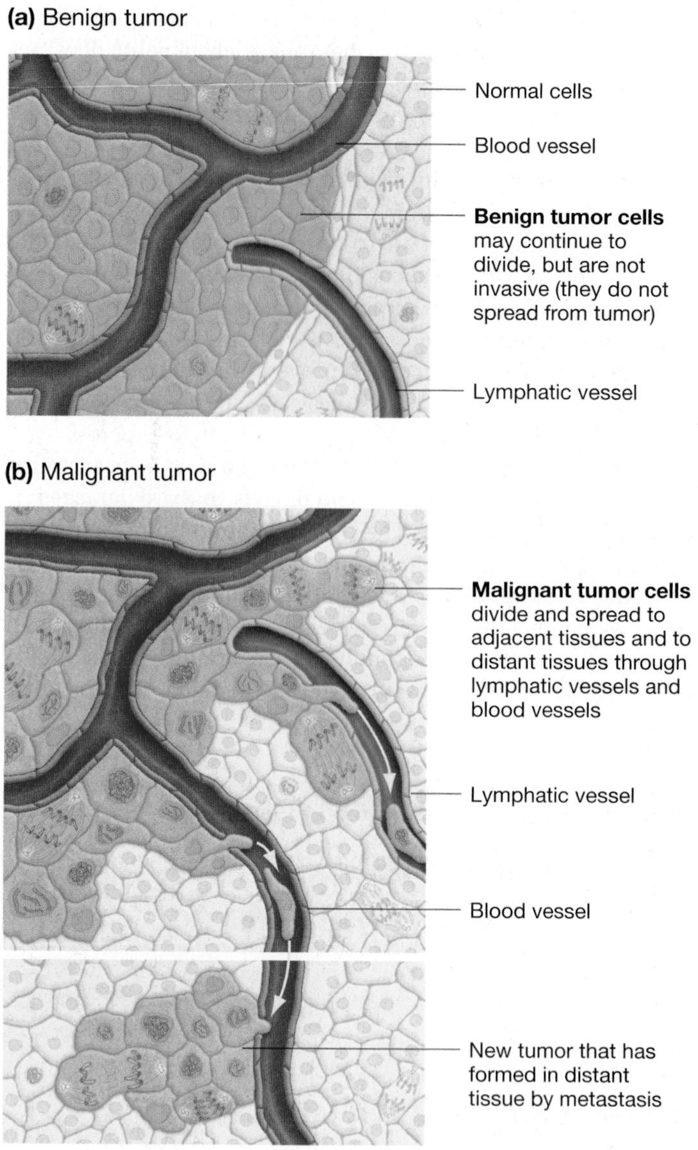

Figure 12.14 **Cancers Spread to New Locations in the Body.** **(a)** Benign tumors grow in a single location. **(b)** Malignant tumors are invasive and may be metastatic—meaning that their cells can spread to distant parts of the body and initiate new tumors. Malignant tumors cause cancer.

Cancer Involves Loss of Cell-Cycle Control

What causes cancer at the molecular level? Recall that when many cells mature, they enter the G_0 state—meaning their cell cycle is arrested at the G_1 checkpoint. In contrast, cells that do pass through the G_1 checkpoint are irreversibly committed to replicating their DNA and entering G_2.

Based on this observation, biologists hypothesize that many types of cancer involve defects in the G_1 checkpoint. To understand the molecular nature of the disease, then, researchers have focused on understanding the normal mechanisms that operate at that checkpoint. Cancer research and research on cell-cycle regulation have become two sides of the same coin.

Social Control In unicellular eukaryotes, passage through the G_1 checkpoint is thought to depend primarily on cell size and the availability of nutrients. If nutrients are plentiful, cells grow, pass through the checkpoint, and divide rapidly.

In multicellular organisms, however, cells divide in response to signals from other cells. Biologists refer to this as *social control* over cell division. The general idea is that individual cells are allowed to divide only when it is in the best interests of the organism as a whole.

Social control of the cell cycle is based on **growth factors**—polypeptides or small proteins that stimulate cell division. Many growth factors were discovered by researchers who were trying to grow cells in culture. When isolated mammalian cells were placed in a culture flask and provided with adequate nutrients, they arrested in G_1 phase. The cells began to grow again only when biologists added **serum**—the liquid portion of blood that remains after blood cells and cell fragments have been removed. Researchers identified growth factors as the components in the serum that were responsible for allowing cells to pass through the G_1 checkpoint.

Cancer cells are an exception. They can often be grown successfully in culture without externally supplied growth factors. This observation suggests that the normal social controls on the G_1 checkpoint have broken down in cancer cells.

How Does the G_1 Checkpoint Work? In G_0 cells, the arrival of growth factors stimulates the production of key regulatory proteins. One of these proteins, called E2F, triggers the expression of genes required for S phase.

When E2F is first produced, however, its activity is blocked by a tumor suppressor protein called Rb. The **Rb protein** serves as a gatekeeper that enforces the G_1 checkpoint. (Rb is short for retinoblastoma, which is a cancer in the retina of the eye that was first discovered in children with a nonfunctional version of Rb.)

PROCESS: THE G_1 CHECKPOINT IS SUBJECT TO SOCIAL CONTROL

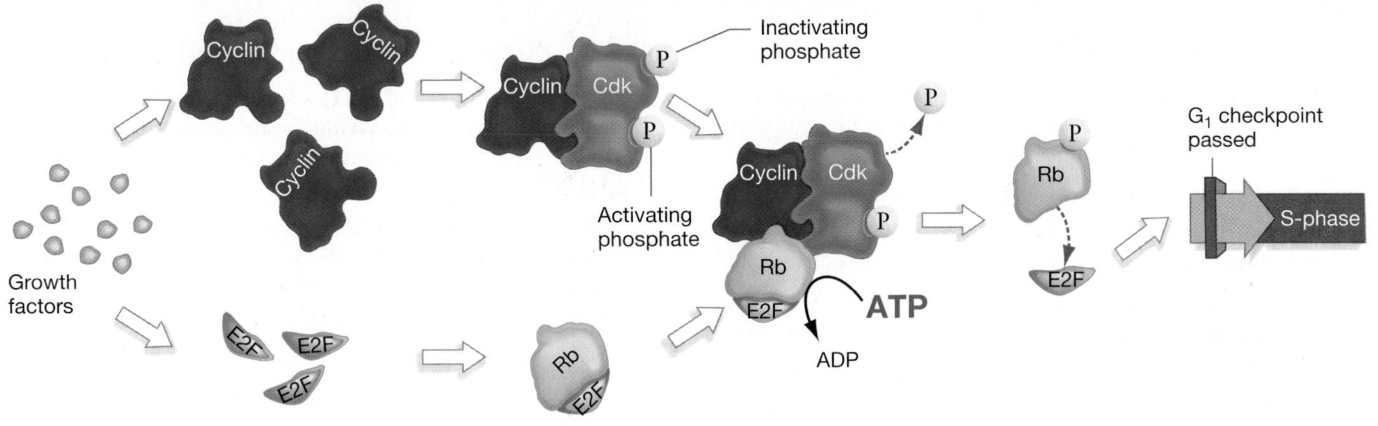

1. Growth factors arrive from other cells.

2. Cells produce cyclin and E2F after stimulation by growth factors.

3. Cyclin binds to Cdk; Cdk is phosphorylated. Rb inactivates E2F by binding to it.

4. Cdk phosphorylates Rb after inactivating phosphate is removed.

5. Phosphorylated Rb releases E2F.

6. E2F triggers production of S-phase proteins.

Figure 12.15 Growth Factors Move Cells through the G_1 Checkpoint.

When Rb is bound to E2F, it switches E2F to the "off" position—it can't activate the genes required for S phase. As long as Rb stays bound to E2F, the cell remains in G_1.

But as **Figure 12.15** shows, the growth factors also stimulate the production of a cyclin. To understand how this cyclin affects E2F activity, think back to how cells progress from G_2 to M phase. As in passage from G_2 to M phase, an activated cyclin–Cdk dimer permits passage from G_1 to S by phosphorylating certain regulatory proteins—in this case, the Rb protein. The phosphorylated Rb releases its hold on E2F, which then activates genes involved in the transition to S phase. In this way, growth factors function as a social signal that says, "It's OK to pass the G_1 checkpoint and divide."

How Do Social Controls and Cell-Cycle Checkpoints Fail? Cells can become cancerous when social controls fail—meaning, when cells begin dividing in the absence of the go-ahead signal from growth factors. One of two things can go wrong: The G_1 cyclin is overproduced, or Rb is defective.

When G_1 cyclins are overproduced and stay at high concentrations, they bind to Cdk and help activate it so that Rb is continuously phosphorylated and unable to bind to E2F. The pool of free E2F sends the cell into S phase.

What happens if Rb is defective? When Rb is missing or mutated so it does not bind normally to E2F, any E2F that is present will activate the genes that push the cell into S phase. The loss of Rb activity, either by mutation or keeping it phosphorylated by overproduction of G_1 cyclins, leads to uncontrolled cell division.

Because cancer is a family of diseases with a complex and highly variable molecular basis, there will be no "magic bullet," or single therapy, that cures all forms of the illness. Still, recent progress in understanding the cell cycle and the molecular basis of cancer has been dramatic, and cancer prevention and early detection programs are increasingly effective. The prognosis for many cancer patients is remarkably better now than it was even a few years ago. Thanks to research, almost all of us know someone who is a cancer survivor.

CHECK YOUR UNDERSTANDING

✔ **If you understood this section, you should be able to . . .**

1. Compare and contrast benign and malignant tumors.
2. Explain how growth factors allow cells to bypass the G_1 checkpoint and enter S phase.

Answers are available in Appendix A.

12.1 How Do Cells Replicate?

- For a cell to replicate, it must copy its chromosomes, separate the copies, and divide the cytoplasm to generate daughter cells that have the same chromosomal complement as the parent cell.
- Eukaryotic cells divide by cycling between interphase and M phase.
- Interphase consists of S phase, when chromosomes replicate, and the G_1 and G_2 phases, when cells grow and prepare for division.
- M phase consists of mitosis or meiosis, when chromosomes separate, and cytokinesis, when the parent cell divides into two daughter cells.

12.2 What Happens during M Phase?

- Mitosis can be described as a sequence of five phases:
  1. ***Prophase*** Chromosomes condense. The spindle apparatus begins to form, and polar microtubules overlap each other.
  2. ***Prometaphase*** In cells of many organisms, the nuclear envelope disintegrates. Microtubules attach to the kinetochores of chromosomes, which begin moving to the middle of the spindle.
  3. ***Metaphase*** All the chromosomes are positioned in the middle of the spindle. The spindle is anchored to the plasma membrane by astral microtubules.
  4. ***Anaphase*** Sister chromatids are pulled apart by the disassembly of kinetochore microtubules at the kinetochore. The separated chromatids are now daughter chromosomes. The spindle poles are moved farther apart to fully separate the replicated chromosomes.
  5. ***Telophase*** Daughter chromosomes are fully separated and are clustered at opposite poles of the spindle. A nuclear envelope forms around each set, and the chromosomes decondense.
- In most cells, mitosis is followed by cytokinesis—division of the cytoplasm to form two daughter cells.

12.3 Control of the Cell Cycle

- The onset of the S and M phases is primarily determined by the activity of protein complexes consisting of a cyclin and a cyclin-dependent kinase (Cdk).
- Cyclin concentrations oscillate during the cell cycle, regulating the formation of the complexes. The activity of Cdk is further regulated by addition of a phosphate in its activating site and removal of one from its inhibitory site.
- Progression through the cell cycle is controlled by checkpoints in three phases.
  1. The G_1 checkpoint regulates progress based on nutrient availability, cell size, DNA damage, and social signals.
  2. The G_2 checkpoint delays progress until chromosome replication is complete and any damaged DNA present is repaired.
  3. The two M-phase checkpoints **(1)** delay anaphase until all chromosomes are correctly attached to the spindle apparatus and **(2)** delay the onset of cytokinesis and G_1 until all chromosomes have been properly partitioned.

12.4 Cancer: Out-of-Control Cell Division

- Cancer is characterized by **(1)** loss of control at the G_1 checkpoint, resulting in cells that divide in an uncontrolled fashion; and **(2)** metastasis, or the ability of tumor cells to spread throughout the body.
- The G_1 checkpoint depends in part on Rb protein, which prevents progression to S phase, and G_1 cyclin–Cdk complexes, which trigger progression to S phase. Alterations in expression of Rb and G_1 cyclin are common in cancer.

Answers are available in Appendix A.

✔ TEST YOUR KNOWLEDGE

1. Which statement about the daughter cells following mitosis and cytokinesis is correct?
   a. They are genetically different from each other and from the parent cell.
   b. They are genetically identical to each other and to the parent cell.
   c. They are genetically identical to each other but different from the parent cell.
   d. Only one of the two daughter cells is genetically identical to the parent cell.
2. After S phase, what makes up a single chromosome?
   a. two daughter chromosomes
   b. a double-stranded DNA molecule
   c. two single-stranded DNA molecules
   d. two sister chromatids
3. Progression through the cell cycle is regulated by oscillations in the concentration of which type of molecule?
   a. p53, Rb, and other tumor suppressors
   b. receptor tyrosine kinases
   c. cyclins
   d. cyclin-dependent kinases
4. What major events occur during anaphase of mitosis?

✔ TEST YOUR UNDERSTANDING

5. Identify at least two events in the cell cycle that must be completed successfully for daughter cells to share an identical complement of chromosomes.
6. What evidence suggests that during anaphase, kinetochore microtubules shorten at the kinetochore?
7. Evaluate each of the following defects. Which could lead to uncontrolled growth in cancer? Select True or False for each statement.
   T/F The overexpression of MPF activity.
   T/F A nonfunctional Rb protein.
   T/F The overexpression of G_1 cyclin.
   T/F A nonfunctional E2F protein.
8. Compare and contrast the effects of removing growth factors from asynchronous cultures of human cells that are normal and those that are cancerous.

✔ TEST YOUR PROBLEM-SOLVING SKILLS

9. **QUANTITATIVE** A particular cell type spends 4 hours in G_1 phase, 2 hours in S phase, 2 hours in G_2 phase, and 30 minutes in M phase. If a pulse–chase experiment were performed with radioactive thymidine on an asynchronous culture of such cells, what percentage of mitotic cells would be radiolabeled 9 hours after the pulse?
   a. 0 percent
   b. 50 percent
   c. 75 percent
   d. 100 percent

10. When a fruit fly embryo first begins to develop, a large cell is generated that contains over 8000 genetically identical nuclei. What is most likely responsible for this result?

✔ PUT IT ALL TOGETHER: Case Study

What are the molecular targets of anticancer drugs?

The bark of the Pacific yew tree (*Taxus brevifolia*) was the original source of one of the most effective drugs for treating tumors of the breast, lung, and other sites. Taxol, a chemical extracted from this bark, kills actively replicating cells by inhibiting the depolymerization of microtubules. Why are microtubules good targets for killing cancerous cells?

11. During what phases in the cell cycle would you expect there to be large changes in the polymerization or depolymerization of microtubules? Why are these changes necessary?

12. When actively growing cells are treated with Taxol, they often are unable to complete the cell cycle. Based on what you have learned about cell-cycle checkpoints, which checkpoint likely causes these cells to arrest? Explain your reasoning.

13. **QUANTITATIVE** Suppose you performed the pulse–chase experiment illustrated in Figure 12.2 but included Taxol in the medium during the chase. Draw a new line on the graph to show the results you would expect, and explain why you would expect them.

14. **PROCESS OF SCIENCE** Aggressive forms of breast cancer are resistant to Taxol chemotherapy. In these cancers, the gene encoding a protein called stathmin is overexpressed. To investigate the mechanism of action of stathmin, researchers measured tumor volume over time in mice with aggressive cancers under three conditions: no treatment (control), Taxol treatment, and Taxol treatment with stathmin gene expression turned off (Taxol + Δ stathmin). Their results are shown in the graph here. Use these results to hypothesize how the stathmin protein affects microtubule stability.

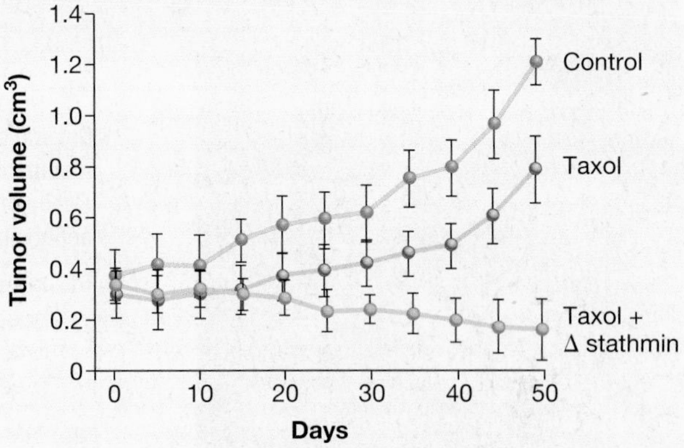

Source: C. Miceli et al. 2013. *Cancer Gene Therapy* 20: 298–307.

15. In normal cells, stathmin is inactivated by phosphorylation at the start of M phase. Phosphatases remove these phosphates as the cell transitions from M phase to G_1. What enzyme is likely to be responsible for phosphorylating stathmin during M phase?

16. **SCIENCE AND SOCIETY** Inhibiting expression of the stathmin gene arrests cells in M phase and is being investigated as an alternative therapy for treating cancer. What additional genes could be therapeutic targets that, when inactivated, would arrest cancerous cells in G_1 phase?

Mastering Biology

Students Go to Mastering™ Biology for assignments, the eText, and the Study Area with animations, practice tests, and activities.

Professors Go to Mastering™ Biology for automatically graded tutorials and questions that you can assign to your students, plus Instructor Resources.

END-OF-UNIT
CASE STUDY

UNIT 2

PUT IT ALL TOGETHER

▌For an introduction to the Mystery of the Newt case study, see page 17.

The use of tetrodotoxin (TTX) as a defense against predation is not unique to newts. While many animals in both marine and terrestrial environments coat themselves with TTX, the origin of this deadly toxin remains largely unknown. Three hypotheses for the source of TTX in animals are being actively investigated:

1. ***Symbiotic*** Microbes that live within the animal's body produce TTX.
2. ***Exogenous*** Environmental TTX is taken up by the animal and concentrated.
3. ***Endogenous*** The animal produces enzymes that manufacture TTX.

✔ If you understand Unit 2, you should be able to apply your learning to this case study:

How Did the Newt Become So Toxic?

Several different species of bacteria (such as *Bacillus* sp. 1839) have been identified that manufacture TTX and have been isolated from the tissues of the toxic marine ribbon worm *Cephalothrix simula*. When the bacteria were cultured in the laboratory and stained with antibodies against TTX, many of the cells were found to contain the toxin.

1. In the symbiotic hypothesis, bacteria may be stimulated to produce TTX inside a host animal once a "quorum" has been reached. Explain how bacteria sense the presence of a quorum and how this could lead to the induction of TTX synthesis. (See Section 11.4)

Most of the research on the origin of TTX in animals has been done on puffer fish. It is known that when the TTX-producing puffer fish *Takifugu rubripes* is held in captivity, it quickly loses its toxicity. To test the effect of diet on TTX accumulation, researchers fed a TTX-containing diet to captive bred *T. rubripes* and control fish that do not normally harbor TTX (**Figure 1**). The amount of TTX present in the fish's livers was measured at different times after introducing the new diet.

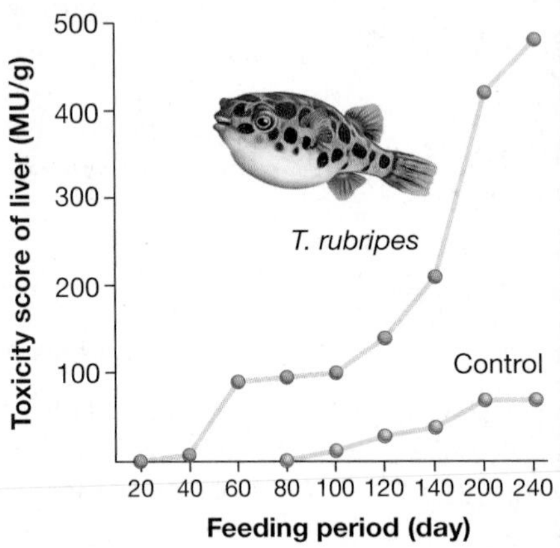

Figure 1

DATA: T. Noguchi et al. 2006. *Comparative Biochemistry and Physiology Part D: Genomics and Proteomics* 1: 145–152, Figure 2.

2. What conclusion may be drawn from the experimental results? Which of the three hypotheses is supported by these data?
3. **PROCESS OF SCIENCE** Propose an experiment to demonstrate that the TTX in the liver of puffer fish was the same as what was present in their diet? (See Figure 7.20)
4. In addition to the liver, researchers found that puffer fish ingesting TTX accumulate the toxin in their skin, while there was little TTX distributed in other areas of the body. Speculate as to how puffer fish collect TTX only in the liver and skin. What kind of protein would you expect to be present in the plasma membranes of cells in these tissues? (See Section 11.3)

Tetrodotoxin is also stored in the skin on the backs of *T. granulosa*. This region of the newt's skin is rough in texture because it is filled with several kinds of secretory glands. Mucus glands are responsible for covering the newt with a slimy coating. Granular glands have a more ominous role—they harbor the deadly TTX.

5. Researchers discovered TTX within numerous secretory vesicles that appear inside cells of granular glands. Describe the general pathway and organelles involved in the production of secretory vesicles. (See Section 7.5)

6. Since TTX is not a protein, carbohydrate, or lipid (see Unit 1 case study), it would not follow the normal pathway used by such molecules to enter secretory vesicles. Propose two possible processes that could result in TTX being present inside secretory vesicles. (See Sections 6.4 and 7.5)

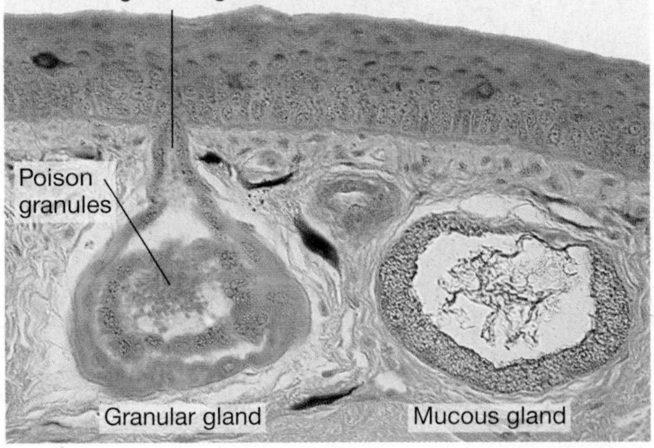

Glands in cross section under surface of amphibian skin

How do newts coat themselves with TTX? To address this question, researchers measured TTX levels in skin samples taken from newts before and after they were stimulated by electric shock, then again nine months later. It was hypothesized that electric shock would elicit a panic response in newts much like when they are attacked by a predator. Results of the experiment are shown in **Figure 2**.

7. Interpret the results. Does electric shock induce secretion of TTX? What additional information was gained by sampling the skin nine months after stimulation?

8. Although the signaling pathway controlling TTX secretion from granular gland cells is unknown, adrenaline-induced secretion has been observed in other cell types. Based on what you learned about G-protein-coupled receptors, describe how they transduce signals to generate a cellular response. (See Section 11.3)

9. When newts are stimulated to secrete TTX, what type of microtubule motor would be used to transport the secretory vesicles to the plasma membrane? (See Section 7.6)

a. Myosin **b.** Kinesin **c.** Dynein **d.** Actin

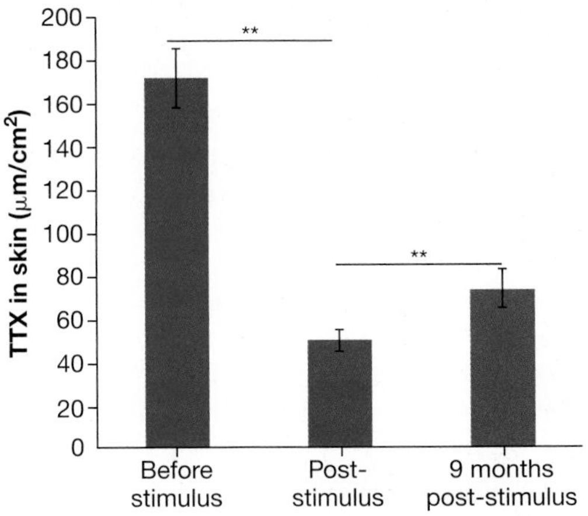

Figure 2

DATA: B. L. Cardall et al. 2004. *Toxicon* 44: 933–938, Table 1.

Contrary to what was observed in puffer fish, toxic *T. granulosa* bred in captivity and fed a TTX-free diet maintained, or even increased, the level of TTX present in their skin.

10. What does the toxicity maintenance in captive newts suggest about the source of their toxin compared to the toxin in puffer fish?

Although the origin of newt TTX remains a mystery, its role as a defense against predation is clear. Are there any predators that can get around this toxic coating? As you will see in later case studies, evolution has a way of evening the odds in ongoing battles between predator and prey. To understand how certain snakes have evolved to survive while dining on toxic newts, you first need to learn more about genes and how they work—the subject of Unit 3.

Each unit ends with a continuation of this story.

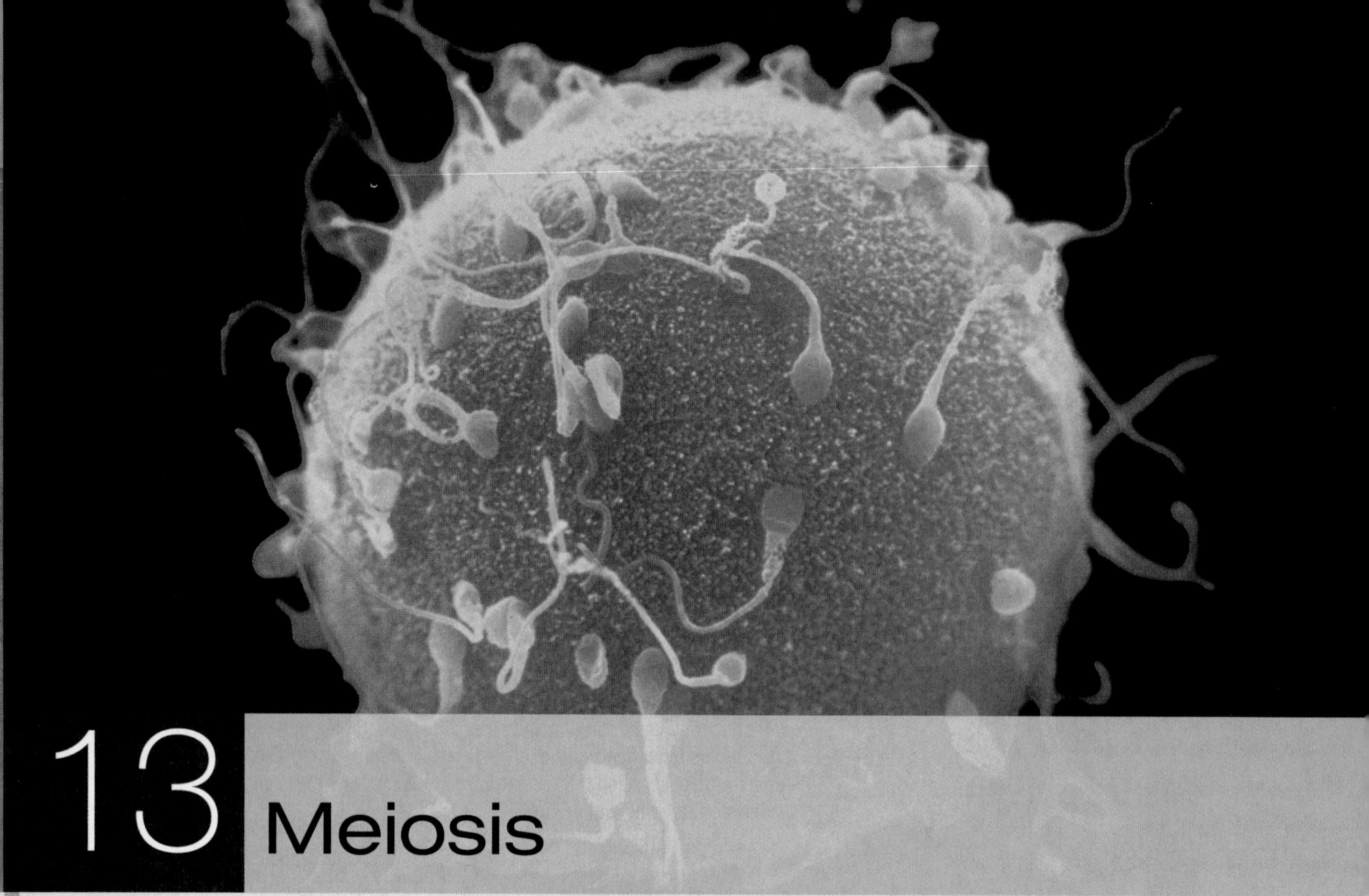

13 Meiosis

Scanning electron micrograph (with color added) showing human sperm attempting to enter a human egg. This chapter introduces the type of nuclear division called meiosis, which in animals occurs during the formation of sperm and eggs.

In this chapter you will learn how

Meiosis promotes genetic diversity and allows for the benefits of sex

- starting with: How does meiosis occur? **13.1**
  - and comparing it to: Mitosis **Ch. 12**
- by examining: How meiosis promotes genetic variation
  - looking at: Independent assortment, crossing over, and fertilization **13.2**
  - then asking: What happens when things go wrong? **13.3**
- by asking: What are the benefits of meiosis? **13.4**
  - Purifying selection hypothesis
  - Changing-environment hypothesis

BIG PICTURE

This chapter is part of the Big Picture. See how on pages 404–405.

Why sex?

Simple questions—such as why sexual reproduction exists—are sometimes the best for getting to the heart of things. In this chapter you will consider what sexual reproduction is and why some organisms employ it. The focus here is on how organisms reproduce, or replicate—one of the five fundamental attributes of life introduced in Ch. 1, Sections 1.1 and 1.2.

For centuries people have known that during sexual reproduction, a male reproductive cell—a **sperm**—and a female reproductive cell—an **egg**—unite in a process called **fertilization** to form a new individual. Biologists wanted to know how the chromosomes from a sperm and an egg could combine, but form an offspring that has the same chromosome number as in each of its parents. A hint at the answer came in 1883 with the observation that cells in the body of roundworms of the genus *Ascaris* have four chromosomes, while their sperm and egg nuclei each have only two chromosomes.

Four years later, German biologist August Weismann proposed a hypothesis to explain the riddle: During the formation of **gametes**—reproductive cells such as sperm and eggs—some type of cell division must take place that reduces chromosome number. If the sperm and egg contribute an equal number of chromosomes to the fertilized egg, Weismann reasoned that they must each contain half of the usual number of chromosomes. In this way, when sperm and egg combine, the resulting cell has the same chromosome number as its mother's cells and its father's cells.

In the decades that followed, biologists confirmed this hypothesis by observing gamete formation in a variety of plant and animal species. Eventually this form of cell division came to be called meiosis (literally, "lessening-act"). **Meiosis** is nuclear division that leads to a halving of chromosome number and ultimately to the production of sperm and egg. (Meiosis is an important part of The Big Picture of Genetic Information on pages 404–405.) To a biologist, asking "Why sex?" is another way of asking "Why meiosis?" Let's delve into this question by first looking at how meiosis happens.

13.1 How Does Meiosis Occur?

To understand meiosis, it is critical to grasp some key ideas about chromosomes. For example, when cell biologists began to study the cell divisions that lead to gamete formation, they made an important observation: Each organism has a characteristic number of chromosomes. For example, your cells have 46 chromosomes, those of dogs have 78, and those of some ferns have over 1000.

After you complete this section, you should be able to . . .

- Describe the transmission of chromosomes during meiosis.

Consider the drawings in Figure 13.1, based on research begun by the American cell biologist Nettie Maria Stevens in 1906. Stevens was the first person to study chromosomes during the cell divisions leading to the formation of egg and sperm. Using the fruit fly *Drosophila melanogaster*, a major model organism (see **BioSkills 13**), she identified a total of eight chromosomes in *Drosophila* cells.

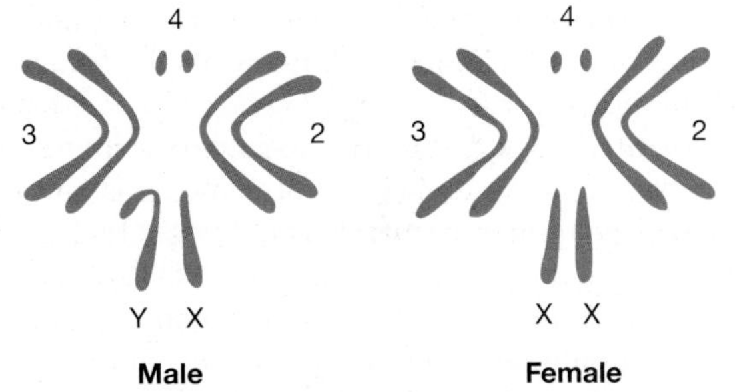

Figure 13.1 **Cells Contain Different Types of Chromosomes That Often Come in Pairs.** Numbers and letters designate the types of *Drosophila* chromosomes.

Chromosomes Come in Distinct Sizes and Shapes

Stevens found that although each *Drosophila* cell has eight chromosomes, there were just five distinct types, distinguished by their size and shape. Three of these chromosome types always occurred in pairs and are labeled chromosomes 2–4 in Figure 13.1. In males, Stevens observed that two chromosomes, now known as the X and the Y chromosomes, were unpaired. In females, Stevens found that there was a pair of X chromosomes and no Y chromosome. This is the same configuration found in mammals. The X and Y chromosomes are called **sex chromosomes** and are associated with an individual's sex. Any chromosomes other than sex chromosomes, such as chromosomes 2–4 in *Drosophila*, are **autosomes**.

The terminology for chromosomes and chromosome number is some of the trickiest in all of biology. You'll need to pay careful attention. Chromosomes that are the same size and shape are called **homologous chromosomes**, or **homologs** ("same-proportion"), and the pair is called a **homologous pair**. Importantly, homologous pairs contain the same genes in the same positions along the chromosomes. Typically, however, the two chromosomes are *not* identical. To understand how this could be, you need to know the difference between a gene and an allele.

A **gene** is a segment of DNA, found at a specific place on a chromosome, that influences a trait. For example, each copy of chromosome 2 in *Drosophila* carries many different genes, including one that influences eye color. However, this gene comes in two different forms—one that contributes to normal-colored red eyes and another that can result in purple eyes. Biologists use the term **allele** to denote a particular version of a gene. In the case of the eye-color gene, one homolog of chromosome 2 may carry the allele associated with red eyes, whereas the other homolog may carry the allele associated with purple eyes (Figure 13.2). To sum up, homologous chromosomes carry the same genes in the same positions but are seldom identical, because each homolog may contain different alleles of any particular gene.

Another thing to watch for as you develop the vocabulary of meiosis and genetics is that sometimes both biologists and the media use terms imprecisely. This is especially true of the words "gene" and "allele." For instance, you might read about the discovery of a new breast cancer gene. What's really meant is that a defective allele of a gene has been discovered. The normal function of the gene isn't to cause breast cancer, but this newly discovered defective form contributes to breast cancer. You'll often need to pay attention to know whether the discussion really is about a gene or an allele.

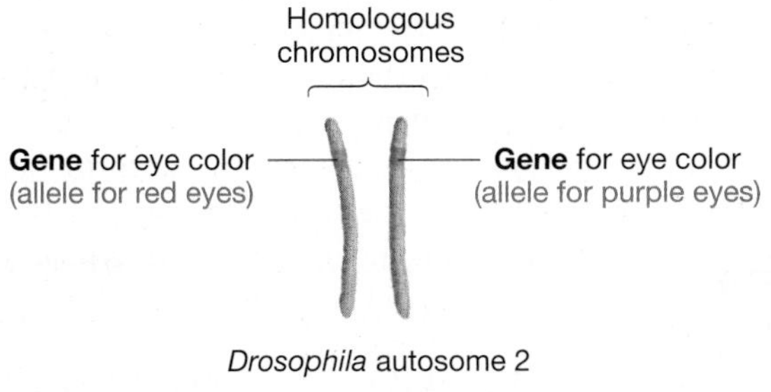

Figure 13.2 **Homologous Chromosomes May Contain Different Alleles of the Same Gene.** The homologous pair of *Drosophila* chromosome 2 is shown with the location of only one of many genes indicated.

The Concept of Ploidy

By identifying the number and types of chromosomes present in a *Drosophila* cell, Stevens had determined the *Drosophila* **karyotype**. As cell biologists studied the chromosomes of more organisms, they realized that, like *Drosophila*, the vast majority of plants and animals have more than one of each type of chromosome.

Insects, humans, trees, and other organisms that have two versions (homologs) of each type of chromosome are called **diploid** ("double-form"). Diploid organisms have two alleles of each gene. One allele is carried on each of the homologs. Recall that these alleles may be identical or they may be different. Although a diploid individual can carry at most two different alleles of a gene, there can be many different alleles of a gene in a population.

Bacteria, archaea, and many algae and fungi have cells that contain just one of each type of chromosome. These cells and organisms are called **haploid** ("single-form"). Because haploid organisms have only one copy of each chromosome, an individual has just one allele of each gene. Biologists use a compact notation to indicate the number of chromosomes, and chromosome sets, in a particular organism or type of cell:

- The letter n stands for the number of *distinct types* of chromosomes in a given cell and is called the **haploid number**. If different sex chromosomes exist, they are counted as a single type in the haploid number. In humans, n is 23.
- To indicate the number of complete chromosome sets when it is greater than one, a number is placed before the n: $2n$, or $3n$, and so on.

The number of chromosome sets is termed the cell's **ploidy**. Diploid cells or species are designated $2n$ because two chromosomes of each type are present—one from each parent. A **maternal chromosome** comes from the mother, and a **paternal chromosome** comes from the father. Humans are diploid, so for us $2n = 46$. Haploid cells or species are labeled simply n, because they have just one set of chromosomes—no homologs are present. In haploid cells, the number 1 in front of n is implied and is not written out.

To summarize, the haploid number n indicates the number of distinct types of chromosomes present. In contrast, a cell's ploidy (n, $2n$, $3n$, etc.) indicates the number of each type of chromosome present. Stating a cell's ploidy is the same as stating the number of haploid chromosome sets present. ✔ **If you understand how these terms relate, you should be able to state the haploid number, ploidy, and total number of chromosomes present in a male *Drosophila*.**

Later work revealed that it is common for species—particularly many plants—to contain more than two of each type of chromosome. Instead of having two homologous chromosomes per cell, these species are **polyploid** ("many-form"), meaning that they have three or more of each type of chromosome in each cell. Depending on the number of homologs present, polyploid species are called triploid ($3n$), tetraploid ($4n$), hexaploid ($6n$), octoploid ($8n$), and so on.

Stevens and other early cell biologists who tracked how chromosome number changes during meiosis confirmed Weismann's idea that a special type of cell division occurs during gamete formation.

Making Models 13.1 shows how drawing chromosomes can help you understand the meaning of ploidy and chromosome number.

Making Models 13.1 Tips on Drawing Chromosomes (II)

Drawing models of cells with different numbers and types of chromosomes can help you understand the meaning of ploidy. The drawing below provides an example of the unreplicated chromosomes in both a $2n$ cell in which $n = 2$, and a $3n$ cell in which $n = 1$.

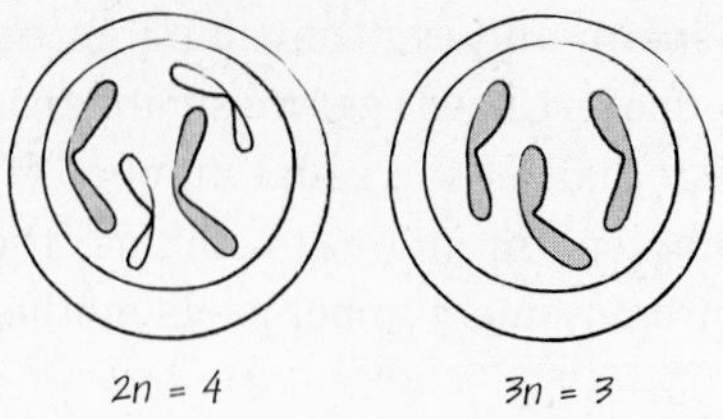

MODEL Draw unreplicated chromosomes in both a $2n$ cell in which $n = 3$ and a $3n = 6$ cell.

To see this model in action, go to the Study Area of **Mastering Biology**

An Overview of Meiosis

Cells replicate each of their chromosomes before starting meiosis. At the beginning of meiosis, chromosomes are in the same state they are in at the start of mitosis.

Recall that an unreplicated eukaryotic chromosome consists of a single, long DNA double helix organized around proteins called histones (see Ch. 12, Section 12.1). During S phase of the cell cycle the DNA is replicated, and therefore each chromosome is also replicated. A replicated chromosome consists of two **sister chromatids**. **Table 13.1** summarizes the terms used to describe chromosomes and illustrates the relationship between chromosomes and chromatids. Each sister chromatid contains an identical copy of the DNA double helix that was present in the unreplicated chromosome before S phase. Therefore, each sister chromatid contains the same genetic information. Sister chromatids remain physically joined during much of meiosis.

To understand meiosis, it is critical to understand the relationship between chromosomes and sister chromatids. The trick is to recognize that an unreplicated chromosome and a replicated chromosome are both *single* chromosomes, even though the replicated chromosome contains *two* sister chromatids. This makes sense if you consider that a chromosome carries a particular set of genetic information in its DNA and that the amount of *unique* information is the same whether there is one copy of it present or two. It's similar to thinking about the amount of information present in one copy of a book or two copies—two copies have more pages, but there's no more new information.

By convention, an unreplicated chromosome is never called a chromatid; the term "chromatid" is used only to describe the structures in a replicated chromosome. ✔ **THINK CAREFULLY If you understand the relationship between chromosomes and chromatids, you should be able to draw one chromosome in both an unreplicated and replicated state, label the sister chromatids, indicate the number of double-helical molecules of DNA present in each drawing, and explain why both of your drawings represent single chromosomes.**

SUMMARY **Table 13.1 Terms for Describing Chromosomes**

Term	Definition	Example or Comment
Chromosome	Structure made up of DNA and proteins; carries the cell's hereditary information (genes)	Eukaryotes have linear chromosomes; most bacteria and archaea have just one circular chromosome
• **Sex chromosome**	Chromosome associated with an individual's sex	X and Y chromosomes of humans (males are XY, females XX); Z and W chromosomes of birds and butterflies (males are ZZ, females ZW)
• **Autosome**	Any chromosome other than a sex chromosome	Chromosomes 1–22 in humans
Unreplicated chromosome	A chromosome that consists of one double-helical molecule of DNA packaged with proteins	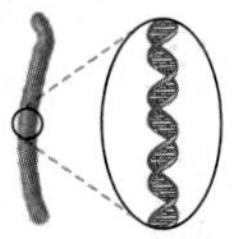
Replicated chromosome	A chromosome after DNA replication. Consists of two identical chromatids, each containing one double-helical DNA molecule packaged with proteins	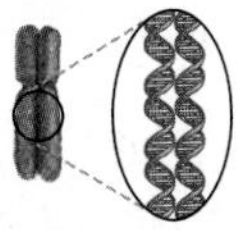
Sister chromatids	The two identical chromatids in a replicated chromosome	
Homologous chromosomes (homologs)	Chromosomes that have the same genes in the same position and are the same size and shape. (Because the alleles of particular genes are often different between the homologs, homologs are not called identical chromosomes.)	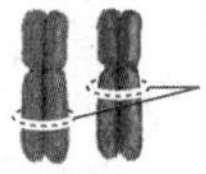
Non-sister chromatids	Chromatids on different members of a homologous chromosome pair. (To be non-sister chromatids, one of the chromatids is on one homolog and the other chromatid is on the other homolog.)	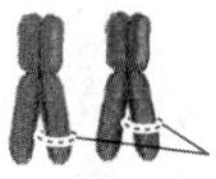
Bivalent	Paired, replicated homologous chromosomes that exist during prophase I and metaphase I of meiosis	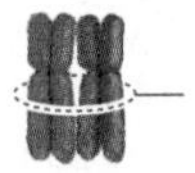
Haploid number	The number of different types of chromosomes in a cell; symbolized n	Humans have 23 different types of chromosomes ($n = 23$)
Diploid number	The number of chromosomes present in a diploid cell (see below); symbolized $2n$	All human cells except gametes are diploid and contain 46 chromosomes ($2n = 46$)
Ploidy	The number of each type of chromosome	The number of haploid chromosome sets present; shown by the number in front of n (for example, $2n$)
• **Haploid**	Having one of each type of chromosome (n)	Bacteria and archaea are haploid, as are many algae; most plant and animal gametes are haploid
• **Diploid**	Having two of each type of chromosome ($2n$)	Most familiar plants and animals are diploid
• **Polyploid**	Having more than two of each type of chromosome; may be triploid ($3n$), tetraploid ($4n$), hexaploid ($6n$), and so on	Seedless bananas are triploid; many ferns are tetraploid; bread wheat is hexaploid

Meiosis Consists of Two Cell Divisions Two back-to-back cell divisions occur in meiosis, **meiosis I** and **meiosis II**. As Figure 13.3 on page 282 shows, these divisions differ sharply.

The homologs of each chromosome pair are separated in meiosis I, when one homolog goes to one daughter cell and the other homolog goes to the other. At the end of meiosis I, each of the two daughter cells has one of each type of chromosome instead of two, and thus half as many chromosomes as the parent cell. During meiosis I, the diploid ($2n$) parent cell produces two haploid (n) daughter cells. Notice, however, that at the end of meiosis I, each chromosome still consists of *two sister chromatids*—chromosomes remain in their replicated form.

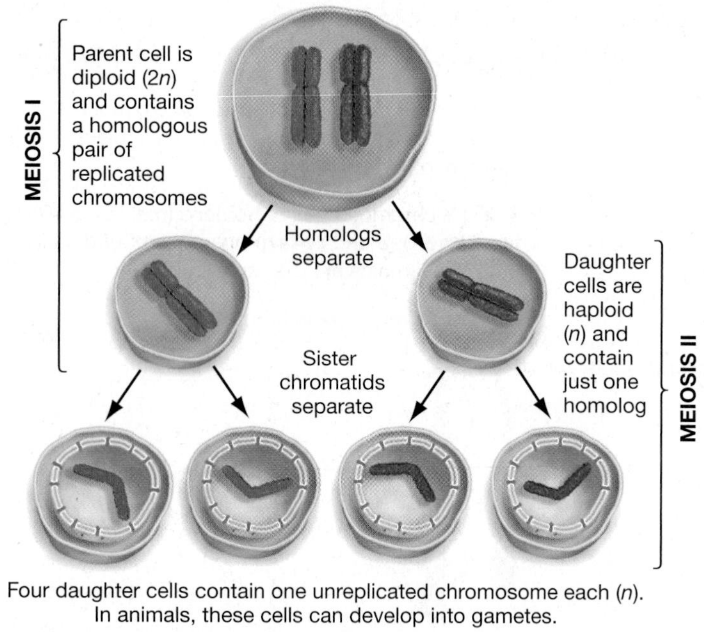

Figure 13.3 Meiosis First Separates Homologs and Then Separates Sister Chromatids. Before undergoing meiosis, DNA and chromosomes are replicated, so there are two chromatids per chromosome. Meiosis halves chromosome number by separating the homologous pairs. In diploid organisms, the cells produced by meiosis are haploid. The color scheme of showing the maternal chromosome in red and the paternal chromosome in blue is used throughout this chapter.

During meiosis II, the sister chromatids of each chromosome separate, with each sister chromatid becoming an individual chromosome called a daughter chromosome. At the end of meiosis II, each of the daughter cells possesses one copy of each daughter chromosome. Remember that each haploid cell that started meiosis II also had only one of each type of chromosome, but each chromosome consisted of two sister chromatids. The cells produced by meiosis II are haploid, but contain unreplicated daughter chromosomes that each consist of just one double-helical molecule of DNA.

To reiterate, sister chromatids separate into daughter chromosomes during meiosis II. This is just what happens during mitosis. Meiosis II is equivalent to mitosis in a haploid cell. In meiosis I, however, sister chromatids stay together. This sets meiosis I apart from both meiosis II and mitosis.

As in mitosis, chromosome movement during meiosis I and II depends on microtubules of the **spindle apparatus** that attach to **kinetochores** located at the **centromere** of each chromosome (Ch. 12, Section 12.2).

Meiosis I Is a Reduction Division The reduction of chromosome number during meiosis I makes this division fundamentally different from meiosis II or mitosis. In most plants and animals, the original cell entering meiosis is diploid and the four final daughter cells are haploid. In animals, some or all of the haploid daughter cells go on to form egg cells or sperm cells in the process of **gametogenesis** ("gamete-origin"; see Ch. 47, Section 47.1). In many organisms, the haploid cells may divide by mitosis and even go on to produce a haploid multicellular organism.

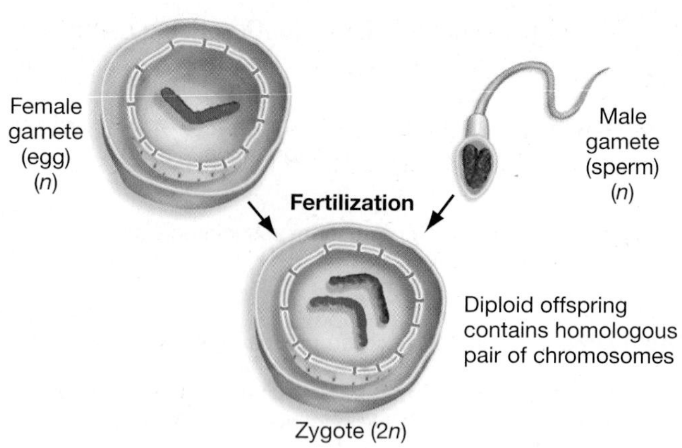

Figure 13.4 Fertilization Restores a Diploid Set of Chromosomes.

When two haploid gametes fuse during fertilization, the full diploid complement of chromosomes is restored (**Figure 13.4**). The diploid cell that results from fertilization is called a **zygote**, and it is the first cell of a new individual. In this way, each diploid individual receives a haploid chromosome set from its mother and a haploid set from its father.

Figure 13.5 puts these events into the context of a **life cycle**—the sequence of events that occurs over the life of an individual, from fertilization to the production of offspring. As you study the figure, note how ploidy changes as the result of meiosis and fertilization. In the case of the diploid adult dog illustrated here, meiosis in specialized cells of the reproductive system results in the formation of haploid gametes, which combine at fertilization to form a diploid zygote. The zygote marks the start of a new generation. Mitotic divisions that begin with the zygote and extend through development produce an adult. The dog life cycle is typical of familiar animals. However, life cycles of other organisms can be wildly different (see Chapters 29 and 38 for some examples).

✔ **If you understand the events of meiosis, you should be able to predict how many double-helical DNA molecules will**

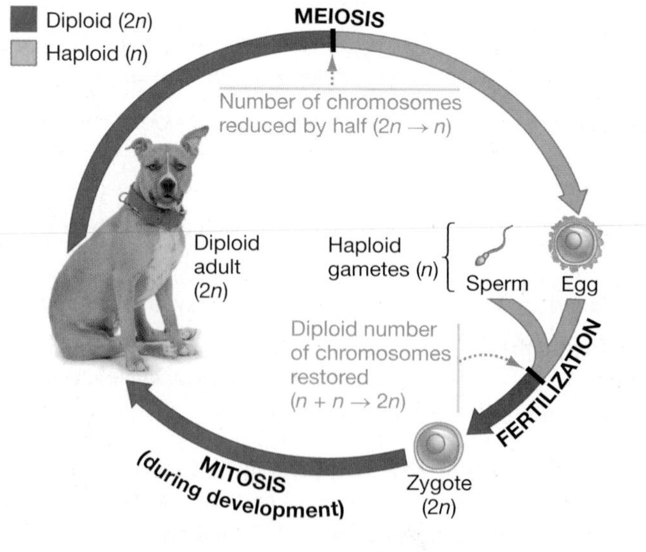

Figure 13.5 Ploidy Changes during the Life Cycle of an Animal. The dog life cycle, which is typical of most familiar animals, is spent mostly in the diploid state.

be present in the sperm and egg cells of dogs, diploid animals that contain 78 replicated chromosomes in each cell that enters meiosis.

Once early research had filled in the details of meiosis and the accompanying changes in ploidy, the mystery of fertilization was solved. Let's analyze the phases of meiosis in detail.

The Phases of Meiosis I

Like mitosis, meiosis is a set of continuous events. However, just as for mitosis, to understand the process, it helps to divide meiosis into a set of distinct phases. Note that the names of meiotic phases parallel those of mitosis, with the addition of a Roman numeral I or II to indicate whether the phase is occurring in meiosis I or II. The major steps are shown in Figure 13.6 on pages 284–285. You can see in step 1 that before meiosis, the cell is in interphase.

Early Prophase I During early prophase I, the nuclear envelope begins to break down, chromosomes condense, and the spindle apparatus begins to form. At this point, sister chromatids are held together along their full length by proteins known as *cohesins* (Ch. 12, Section 12.1). Then, a crucial event that is unique to meiosis I occurs: Homologous chromosome pairs come together. The end result is **synapsis**, a tight side-by-side pairing of homologous chromosomes along their corresponding regions. This is illustrated in step 2 of Figure 13.6. The homologs are held together by a network of proteins called the **synaptonemal complex**. In most organisms, synapsis requires breaking and then connecting DNA of the two homologs at many spots along their length.

The structure that results from synapsis is called a **bivalent** (*bi* means "two" and *valens* means "strong" in Latin). A bivalent consists of paired homologous, replicated chromosomes, where each homolog consists of two sister chromatids. Chromatids from different homologs are referred to as **non-sister chromatids**. In Figure 13.6, the red-colored (maternal) chromatids are non-sister chromatids with respect to the blue-colored (paternal) chromatids.

Late Prophase I During late prophase I, the nuclear envelope is completely broken down, and each homolog in the bivalent is attached to microtubule fibers that come from a single spindle pole. One homolog is attached to one pole, and the other homolog is attached to the other. This form of attachment, which is unique to meiosis I, is essential for separating the homologous pairs.

The synaptonemal complex disassembles in late prophase I, and the homologs that were so closely paired in synapsis now begin to separate at many points along their length. They stay joined, however, by X-shaped structures called **chiasmata** (singular: **chiasma**). (In the Greek alphabet, the letter "X" is called chi, pronounced *kye*.) Normally, at least one chiasma forms in every pair of homologous chromosomes. In humans, for example, an average of 1.5 chiasmata form on each chromosome. The chiasmata mark particular sites of DNA breakage and rejoining between homologs.

As step 3 of Figure 13.6 shows, the chromatids that form a chiasma are non-sister chromatids. At each chiasma there is an exchange of parts of chromosomes between paternal and maternal homologs. These reciprocal exchanges between different homologs lead to chromatids that have both paternal and maternal segments. This process of chromosome exchange is called **crossing over**. Crossing over is a fundamental part of meiosis.

Metaphase I In metaphase I, the kinetochore microtubules move the pairs of homologous chromosomes (bivalents) to a region called the **metaphase plate** (step 4). The metaphase plate is not a physical structure but an imaginary plane midway between the poles of the spindle apparatus. The bivalents are still connected by at least one chiasma.

Here are two key points about chromosome movement: **(1)** In metaphase I, each bivalent straddles the metaphase plate with one homolog on one side and the other homolog on the other; and **(2)** the alignment of each bivalent is independent of any other bivalent. This means that if one bivalent has a maternal homolog above the metaphase plate and the paternal homolog below the metaphase plate, the alignment of this pair of homologs has no influence on how the maternal and paternal chromosomes of any other bivalent will align. The independent alignment of bivalents may seem a trivial detail, but it accounts for independent assortment, a fundamental principle of genetics (see Section 13.2 and Ch. 14, Section 14.3 for more information).

Anaphase I and Telophase I Anaphase I starts as each of the homologs begins moving to a different pole of the spindle apparatus. The unique attachment of the kinetochores of each homolog to spindle fibers that come from one spindle pole means that each homolog is attached to a different spindle pole. This allows the homologous chromosomes in each bivalent to separate from each other during anaphase I.

Chiasmata are broken as the cell enters anaphase I. This is made possible by removal of the cohesin proteins that link sister chromatids together from all parts of the chromosome except those around the centromere. Notice how the separating homologs (step 5 of Figure 13.6) are a combination of maternal (red) and paternal (blue) segments. This is due to the crossing over that occurred between non-sister chromatids during prophase I. Consequently, the chromosomes have a mixture of maternal and paternal alleles.

Meiosis I concludes with telophase I, when the homologs finish moving to opposite sides of the spindle (step 6). When meiosis I is complete, **cytokinesis** (division of cytoplasm) occurs (Ch. 12, Section 12.2) and two haploid daughter cells form.

Meiosis I: A Recap Chromosome movement during meiosis I occurs through the dynamic assembly and disassembly of the microtubules attached to the kinetochore. When meiosis I is complete, the cell divides and two haploid daughter cells are produced. A reduction in chromosome number has occurred. The daughter cells produced by meiosis I are haploid, having only one copy of each homologous pair. The sister chromatids remain attached in each chromosome, however, meaning that the haploid daughter cells at the end of meiosis I still contain replicated chromosomes.

An important point is that the chromosomes in each daughter cell are a random assortment of maternal and paternal chromosomes as a result of **(1)** crossing over and **(2)** the independent alignment of maternal and paternal copies of different chromosomes at metaphase I.

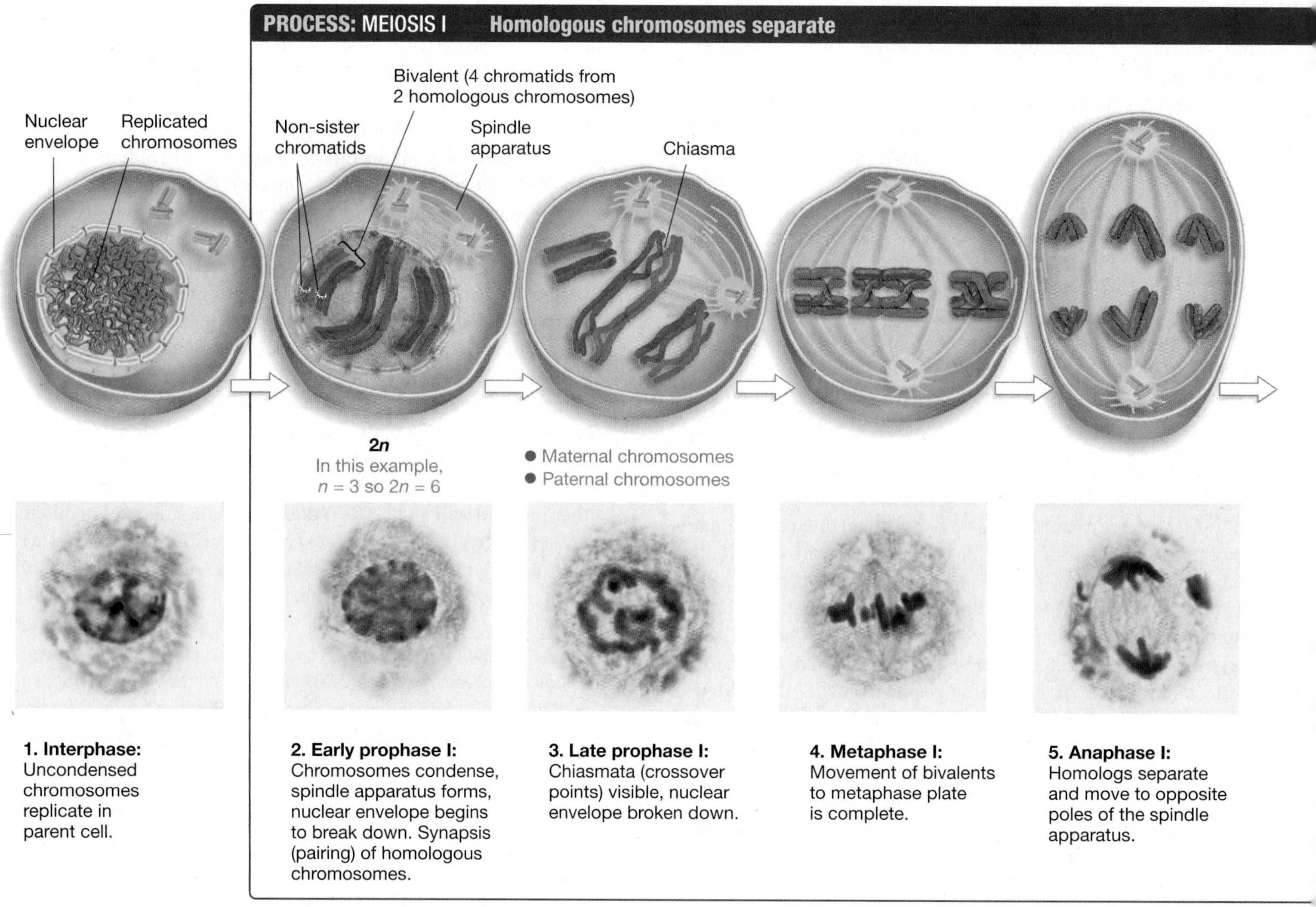

Figure 13.6 Meiosis Can Be Divided into Phases. The micrographs of each phase shown in the lower panel are from a species of salamander. See pages 283–284 for a full description of all the phases.

The Phases of Meiosis II

Recall that chromosome replication occurred before meiosis I. A key feature of the period between meiosis I and II is that there is no DNA replication and therefore no chromosome replication. Meiosis II separates the sister chromatids of the replicated chromosomes into individual cells. Each of these cells will contain unreplicated daughter chromosomes.

During prophase II, a spindle apparatus forms in both daughter cells (step 7 of Figure 13.6). Microtubules that polymerize from the two spindle poles attach to kinetochores on opposite sides of every chromosome and begin moving the chromosomes toward the middle of each cell. This attachment is the same as observed in mitosis (Ch. 12, Section 12.2).

In metaphase II, the chromosomes are lined up at the metaphase plate (step 8). In contrast to metaphase I, each chromosome is attached by spindle fibers to both of the spindle poles. The sister chromatids of each chromosome are separated during anaphase II (step 9) and move to different daughter cells during telophase II (step 10). Once they are separated, each chromatid is considered an independent daughter chromosome. At the end of meiosis II, there are four haploid cells, each with one daughter chromosome of each type in the chromosome set. Because meiosis II occurs in both daughter cells produced by meiosis I, the overall process results in four daughter cells from each parent cell. To summarize meiosis, one diploid cell with replicated chromosomes gives rise to four haploid cells with unreplicated chromosomes.

Mitosis versus Meiosis

How do mitosis and meiosis compare? **Table 13.2** summarizes some important similarities and differences. A key difference between the two processes is that homologous chromosomes pair early in meiosis but do not pair at all during mitosis. Because homologs in prophase of meiosis I are connected by chiasmata, they can migrate to the metaphase plate together. In meiotic prophase I, the unique attachment of each homolog to spindle fibers coming from only one spindle pole allows for the separation of homologs during anaphase of meiosis I. This results in a reduction division. ✔ **If you understand key differences between meiosis and mitosis, you should be able to explain why mitosis in a triploid (3*n*) cell can occur easily but meiosis is difficult.**

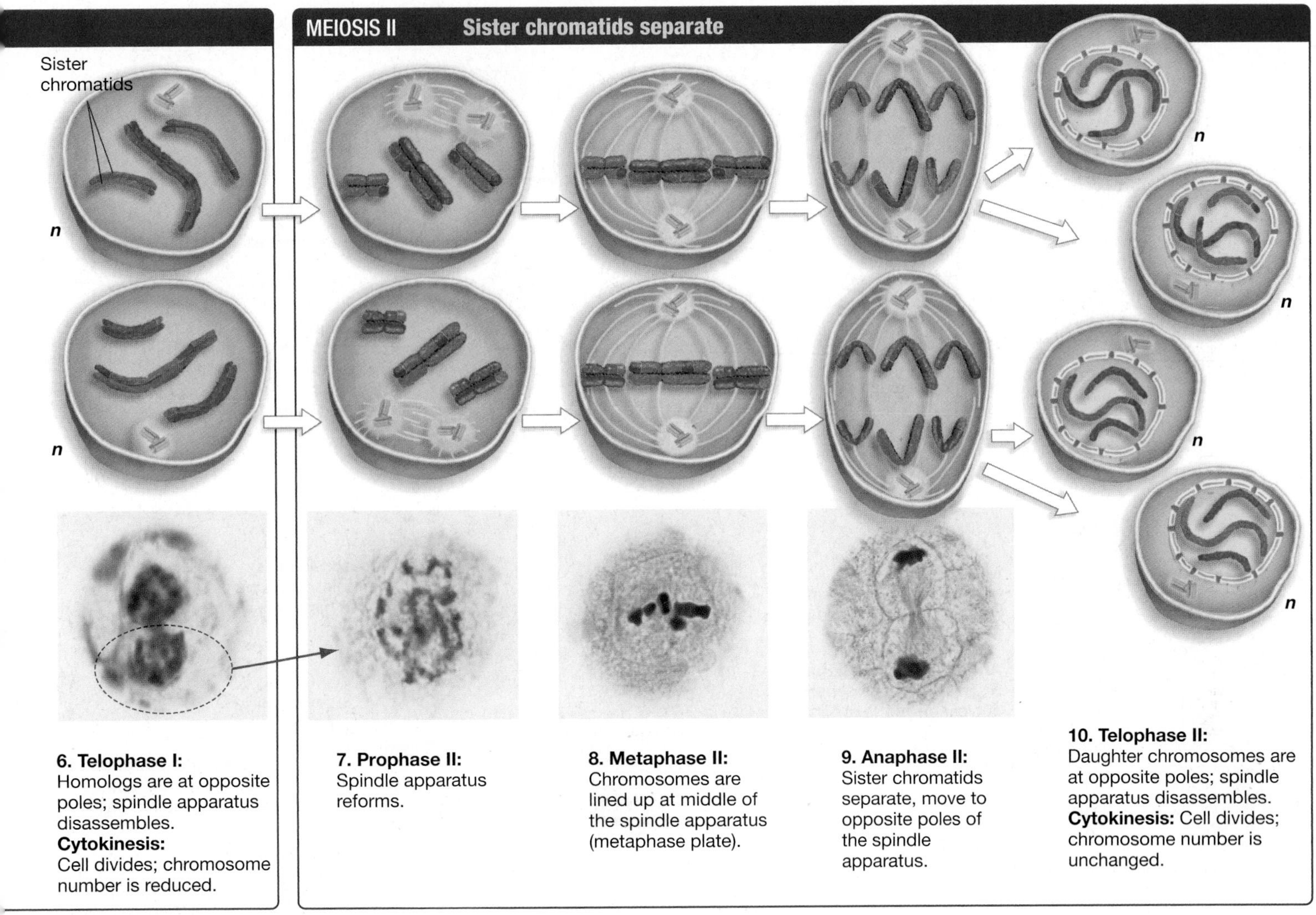

SUMMARY Table 13.2 Key Differences between Mitosis and Meiosis

Feature	Mitosis	Meiosis
Number of cell divisions	One	Two
Number of chromosomes in daughter cells compared with parent cell	Same	Half
DNA content of daughter cells compared with parent cell	Reduced to 1/2 as chromosomes go from replicated → unreplicated	Reduced to 1/4 as chromosomes go from replicated diploid sets → replicated haploid sets (meiosis I) → unreplicated haploid sets (meiosis II)
Synapsis of homologs	No	Yes
Spindle fiber attachment	Individual chromatids in each chromosome attach to spindle fibers from different spindle poles.	Both chromatids in each chromosome attach to spindle fibers from the same spindle pole.
Number of chiasmata	None	One or more per pair of homologous chromosomes
Makeup of chromosomes in daughter cells	Identical	Different—various combinations of maternal and paternal chromosomes, paternal and maternal segments mixed within chromosomes
Role in organism life cycle	Asexual reproduction in some eukaryotes; cell division for growth, replacement of cells, and wound healing	Halving of chromosome number in cells that will produce gametes

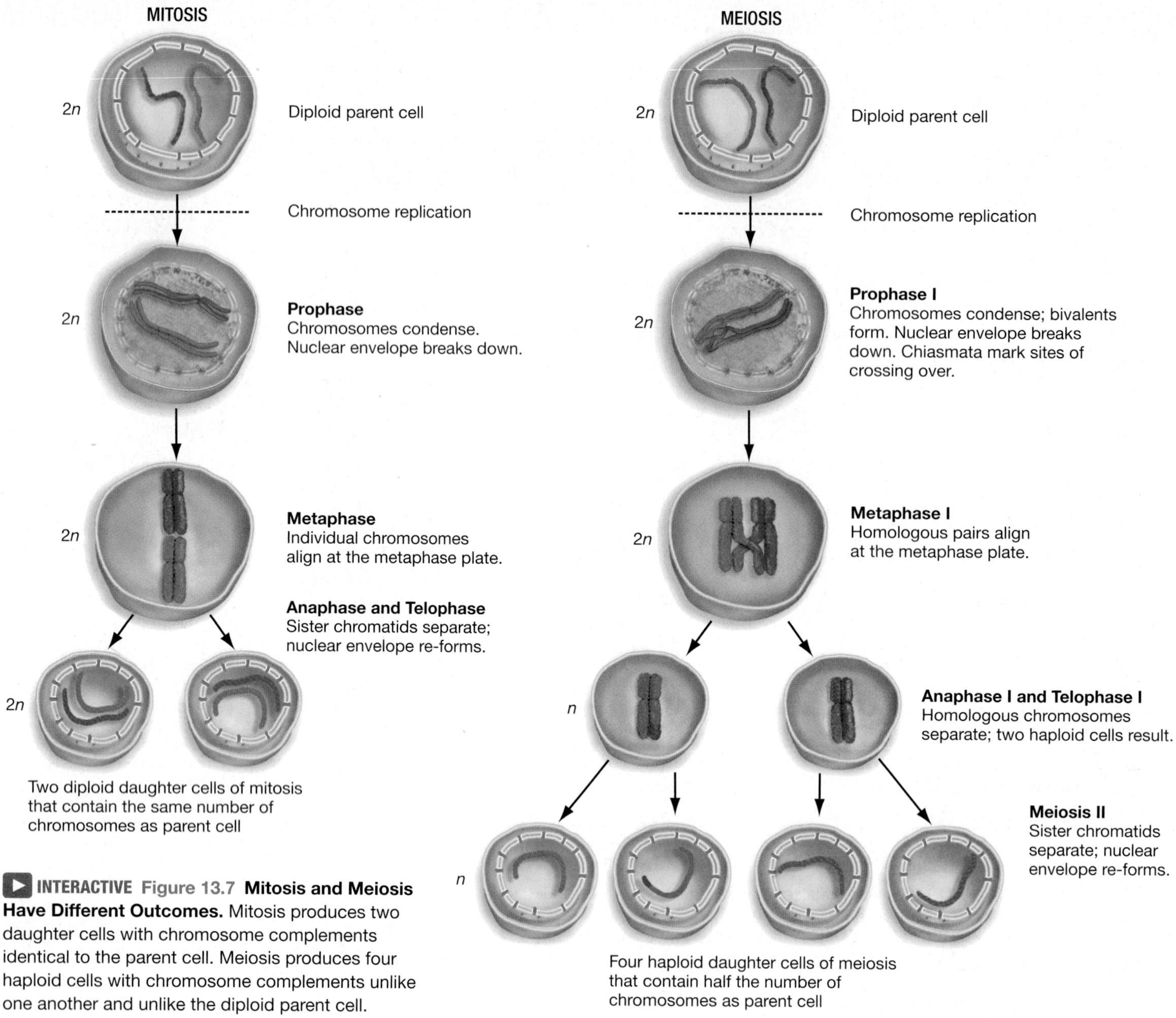

INTERACTIVE Figure 13.7 Mitosis and Meiosis Have Different Outcomes. Mitosis produces two daughter cells with chromosome complements identical to the parent cell. Meiosis produces four haploid cells with chromosome complements unlike one another and unlike the diploid parent cell.

See **Figure 13.7** for a visual summary of these two processes. Given that mitosis works perfectly well for cell division, why does meiosis exist at all? What are its consequences?

CHECK YOUR UNDERSTANDING

✔ **If you understood this section, you should be able to . . .**

1. Explain how DNA content is reduced by half during both meiosis I and mitosis, yet chromosome number is reduced only during meiosis I.
2. **QUANTITATIVE** Predict the amount of DNA in each daughter cell at the end of meiosis if the parent cell that entered meiosis contained 20 units of DNA.

Answers are available in Appendix A.

13.2 Meiosis Promotes Genetic Variation

Thanks to crossing over during meiosis I and the random shuffling of maternal and paternal chromosomes into haploid cells, the chromosomes in one gamete are different from the chromosomes in every other gamete and different from those in the parent cell. Amplifying this variation, fertilization creates genetically varied diploid offspring by bringing together different haploid sets of parental chromosomes.

After you complete this section, you should be able to . . .

- Explain how independent assortment, crossing over, and random fertilization work to promote genetic variation in offspring.

The changes in chromosome sets that result from meiosis and fertilization are significant because chromosomes are the repositories of genes. The critical idea is that changes in chromosome sets and varied combinations of paternal and maternal alleles occur only during sexual reproduction—*not* during asexual reproduction.

- **Asexual reproduction** is any mechanism of producing offspring that does not involve the production and fusion of gametes. Asexual reproduction in eukaryotes is based on mitosis. The chromosomes in cells produced by mitosis are identical to the chromosomes in the parental cell. Because mitosis is the only cell division involved in asexual reproduction, asexually produced offspring are **clones**—or exact genetic copies—of their parent. A familiar example of asexual reproduction is growing a new plant from a cutting.
- **Sexual reproduction** is the production of offspring through the generation and fusion of gametes. Sexual reproduction results in offspring that have chromosome complements unlike those of their siblings or their parents.

To understand the consequences of sexual reproduction, let's take a closer look at the two aspects of meiosis that promote variation: **(1)** separation and distribution of homologous chromosomes and **(2)** crossing over.

Independent Assortment

Each somatic cell in your body contains 23 homologous pairs of chromosomes, or 46 chromosomes in total. Half of these chromosomes came from your mother, half from your father. Each chromosome contains genes, and genes influence particular traits. For example, one gene that affects the type of hemoglobin in your red blood cells is located on chromosome 11, while a gene that affects the form of an ion channel protein is located on chromosome 7 (**Figure 13.8a**). Why are the type of hemoglobin and the specific form of ion channel important? It's because abnormal forms of these two proteins can lead to sickle cell disease and cystic fibrosis, two relatively common and serious genetic diseases.

Suppose that the chromosomes you inherited from your mother contain alleles associated with normal hemoglobin (called Hb-β) and an abnormal channel protein that causes cystic fibrosis. In contrast, suppose the chromosomes you inherited from your father include the alleles for the abnormal hemoglobin that causes sickle cell disease and a normal channel protein, called cystic fibrosis transmembrane conductance regulator (CFTR), that does not cause cystic fibrosis. Everyone who has either sickle cell disease or cystic fibrosis possesses two copies of the disease-associated allele.

Will some gametes that you produce contain the instructions you inherited from your mother while others contain the instructions you inherited from your father? **Figure 13.8b** shows that when pairs of homologous chromosomes line up during meiosis I and the homologs separate, different combinations of maternal and paternal chromosomes can result. Each daughter cell gets a random assortment of maternal and paternal chromosomes (which will be reduced from $2n$ to n during meiosis).

This phenomenon is known as **independent assortment**. In the example shown in Figure 13.8, meiosis will result in some gametes with only alleles for normal hemoglobin (no sickle cell disease) and for cystic fibrosis, the traits from your mother, as well as some gametes with alleles for sickle cell disease and normal CFTR (no cystic fibrosis), the traits from your father. But two additional combinations will also occur: gametes having alleles for no sickle cell disease and no cystic fibrosis, and gametes having alleles for both sickle cell disease and cystic fibrosis. The creation of new combinations of alleles is called **genetic recombination**. Four different combinations of paternal and maternal chromosomes are possible for two chromosomes distributed to daughter cells during meiosis I. **Making Models 13.2** on page 288 suggests one strategy for predicting the possible outcomes of independent assortment.

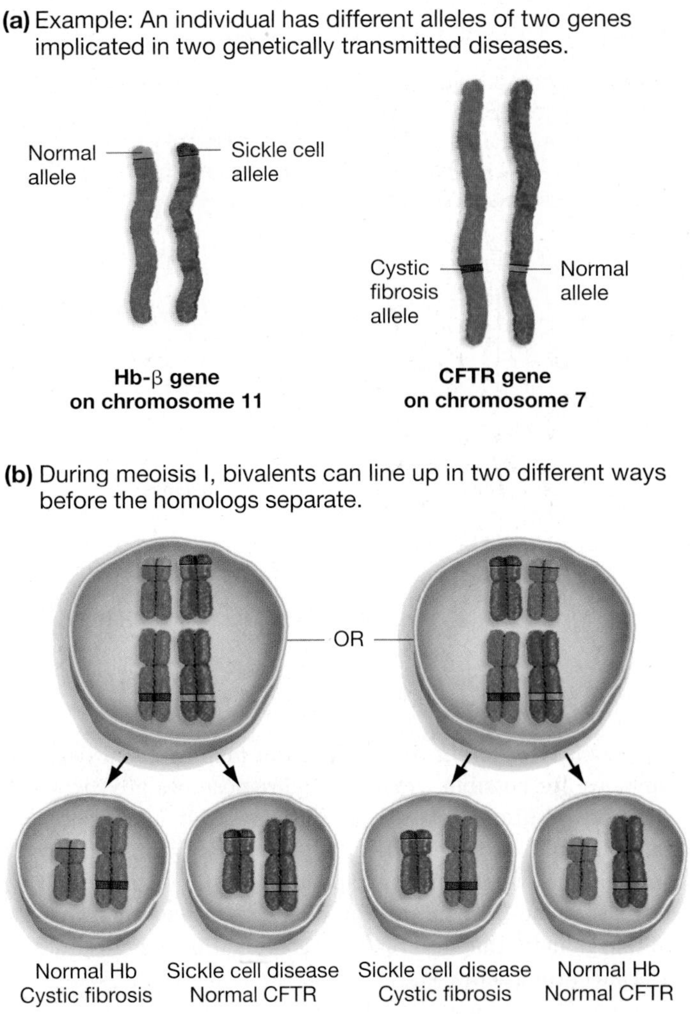

Figure 13.8 Independent Assortment of Homologous Chromosomes Results in Varied Combinations of Maternal and Paternal Alleles. CFTR stands for the ion channel protein that causes cystic fibrosis when cells possess two copies of the defective allele; Hb-β is an abbreviation for a gene that determines the type of hemoglobin. Sickle cell disease occurs when there are two copies of the defective Hb-β allele.

✔ **If you understand how independent assortment leads to genetic variation in the daughter cells produced by meiosis, you should be able to explain how genetic variation would be affected if maternal chromosomes always lined up on one side of the metaphase plate during meiosis I and paternal chromosomes always lined up on the other side.**

Making Models 13.2 Tips on Drawing Chromosomes (III)

Drawing models of chromosomes is a useful way to see how genetic diversity can be created during independent assortment. It's important in these models to distinguish different types of chromosomes as well as the maternal and paternal homologs. Two of many possible approaches are shown here.

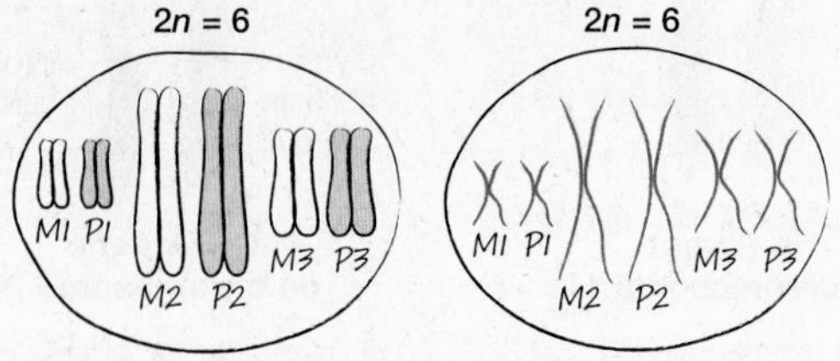

M = maternal, P = paternal
1, 2, and 3 = different types of chromosomes

MODEL Draw the different types of gametes that could be produced by independent assortment from a $2n = 6$ cell.

To see this model in action, go to the Study Area of **Mastering Biology**

How many different combinations of maternal and paternal homologs are possible when more chromosomes are involved? With each additional pair of chromosomes, the number of combinations doubles. In general, a diploid organism can produce 2^n combinations, where n is the haploid chromosome number. This means that you ($n = 23$) can produce 2^{23}, or an impressive 8.4 million, gametes that differ in their combination of maternal and paternal chromosome sets.

Crossing Over

Recall from Section 13.1 that segments of paternal and maternal chromatids exchange when crossing over occurs during meiosis I (for example, see the mixed blue and red segments of chromosomes shown in Figure 13.6 and Figure 13.7). Therefore, crossing over produces new combinations of alleles *within* a chromosome—combinations that did not exist in either parent.

Genetic recombination from both crossing over and independent assortment is important because it creates genetically diverse gametes. Independent assortment of homologous chromosomes generates varied combinations of chromosomes in gametes; crossing over produces new combinations of alleles along each chromosome. With crossing over, the number of genetically different gametes that you can produce is much more than 8.4 million—it is virtually limitless.

How Does Fertilization Affect Genetic Variation?

As long as some alleles vary in a diploid individual, crossing over and the independent assortment of maternal and paternal chromosomes ensure that each gamete is genetically unique. These differences between gametes lead to another important source of genetic variation—random fertilization. Random fertilization means that sperm and egg from different individuals come together without regard for any particular alleles they carry.

How many genetically distinct offspring can be produced by random fertilization? Let's answer this question using humans as an example. Recall that a single human can produce about 8.4 million different gametes by independent assortment alone. When a sperm and egg come together at random, the number of possible genetic combinations that can result is equal to the product of the numbers of different gametes produced by each parent. (To understand this logic, see **BioSkills 4**.) In humans, this means that through independent assortment and random fertilization, two parents can potentially produce 8.4 million $\times$ 8.4 million $= 70.6 \times 10^{12}$ genetically distinct offspring. This number is far greater than the total number of people who have ever lived. Adding in crossing over further amplifies this number. Unless you have an identical twin, you really are genetically unique.

In any complicated process such as meiosis, things can and do go wrong. What happens if there is a mistake?

CHECK YOUR UNDERSTANDING

✔ **If you understood this section, you should be able to . . .**

1. **QUANTITATIVE** Choose true or false and explain your thinking: One pair of homologous chromosomes is the minimum number needed to observe independent assortment.
2. Choose true or false and explain your thinking: Crossing over creates new combinations of alleles along single types of chromosomes, and independent assortment creates many different combinations of alleles that are carried on different chromosomes.
3. **QUANTITATIVE** Choose true or false and explain your thinking: Considering only independent assortment and random fertilization, in a mating between two $2n = 4$ individuals, there are 16 different possible combinations of chromosomes in offspring.

Answers are available in Appendix A.

13.3 What Happens When Things Go Wrong in Meiosis?

Errors in meiosis are disturbingly common. If meiotic errors were like a spelling mistake, it might be only an annoyance. But in humans, as many as half of conceptions are spontaneously terminated because of problems that occurred in meiosis. What are the consequences for offspring if gametes contain an abnormal set of chromosomes?

After you complete this section, you should be able to . . .

- Explain how problems in separating chromosomes during meiosis can lead to problems in human health.

In 1866 the physician Langdon Down described a distinctive set of conditions that included mental retardation, a high risk for heart problems and leukemia, and a degenerative brain disorder similar to Alzheimer's disease. **Down syndrome**, as the disorder came to be called, is observed in about one infant in every 700 live births in the United States.

For over 80 years the cause of the syndrome was unknown. Then, in the late 1950s, a study of the chromosome sets of nine children with Down syndrome suggested that the condition is associated with an extra copy of chromosome 21. This situation is called **trisomy** ("three-bodies"), in this case trisomy-21, because each cell has three copies of that chromosome (Figure 13.9). The explanation proposed for the trisomy was that a mistake had occurred during meiosis in either the mother or the father.

How Do Mistakes Occur?

For a gamete to get one complete set of chromosomes, two steps in meiosis must be perfectly executed.

1. The chromosomes in each homologous pair must separate from each other during the first meiotic division, so that only one homolog ends up in each daughter cell.
2. Sister chromatids must separate from each other and move to opposite poles of the dividing cell during meiosis II.

If both homologs in meiosis I or both sister chromatids in meiosis II move to the same daughter cell, the products of meiosis will be abnormal. This sort of meiotic error is referred to as **nondisjunction**, because the homologs or sister chromatids fail to separate, or disjoin.

Figure 13.10 shows what happens when homologs do not separate correctly during meiosis I. Note that at the end of meiosis, two daughter cells have two copies of the same chromosome, while the other two lack that chromosome entirely. Gametes that contain an extra chromosome are symbolized as $n + 1$ gametes that lack one chromosome are symbolized as $n - 1$.

If an $n + 1$ gamete is fertilized by a normal n gamete, the resulting zygote will be $2n + 1$. This situation is trisomy because there are three copies of one type of chromosome. If the $n - 1$ gamete is fertilized by a normal n gamete, the resulting zygote will be $2n - 1$. This situation is called **monosomy** because there is only one copy of one of the chromosomes. Cells that have too

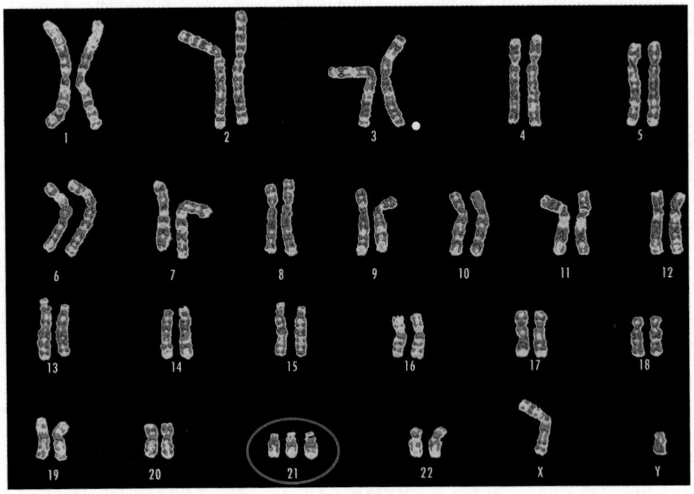

Figure 13.9 Karyotype of an Individual with Down Syndrome. The three copies of chromosome 21 are circled.

many or too few chromosomes of a particular type are said to be **aneuploid** ("without-form").

Researchers estimate that 20–40 percent of all human conceptions produce a zygote that is aneuploid. Most of these errors result from the failure of a homologous pair to separate in anaphase of meiosis I; less often, sister chromatids stay together during anaphase of meiosis II.

Aneuploidy is the leading cause of miscarriage (spontaneous abortion) in humans. Trisomy can occur for all human chromosomes and, in almost all cases, it leads to miscarriage because the embryo or fetus is not viable. Only trisomies for chromosomes 13, 18, 21, and the sex chromosomes are seen in live births. Monosomy is also common at conception, but with the exception of the X chromosome, monosomy prevents completion of development.

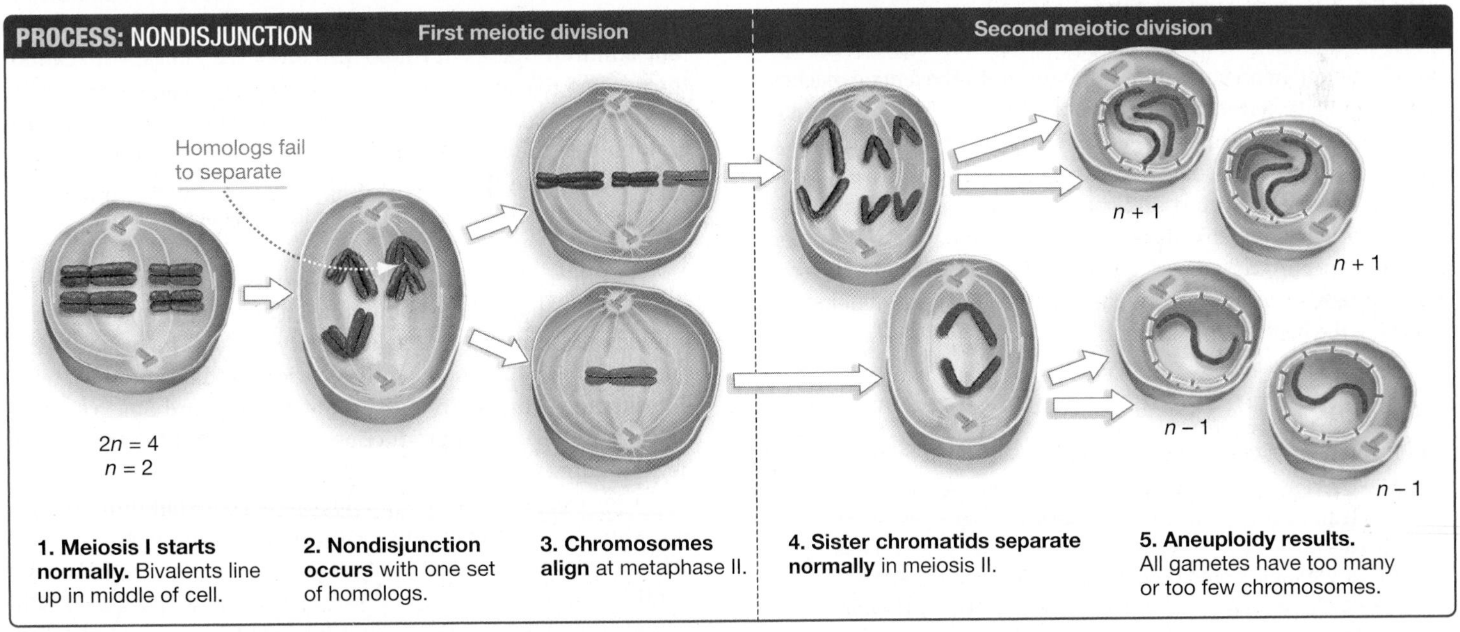

Figure 13.10 Nondisjunction Leads to Gametes with Abnormal Chromosome Numbers. If homologous chromosomes fail to separate during meiosis I, the gametes that result will have an extra chromosome or will lack a chromosome. Nondisjunction can also occur during meiosis II if sister chromatids fail to separate.

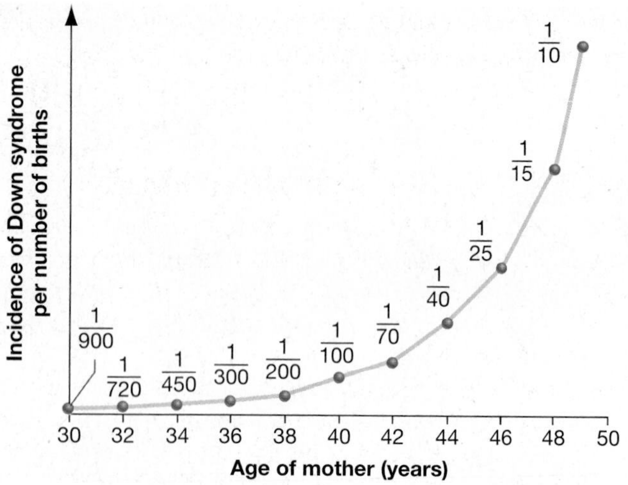

Figure 13.11 The Frequency of Down Syndrome Increases as a Function of a Mother's Age.

DATA: www.ndss.org, National Down Syndrome Society (2012).

Why Do Mistakes Occur?

Trisomy and other meiotic mistakes result from random errors. Errors are especially common in **(1)** microtubule attachment to kinetochores early in meiosis I, **(2)** separation of chromosomes that have a single chiasma near their ends or their centromeres, **(3)** the failure to form a chiasma, and **(4)** the premature separation of sister chromatids.

A striking pattern is the steep increase in trisomic offspring born to older mothers. For example, as Figure 13.11 shows, in the case of Down syndrome, the incidence increases dramatically in babies born to mothers over 35 years old. A father's age has a much weaker influence on mistakes in meiosis.

Why is there such a strong correlation between advanced maternal age and trisomy-21? In part, the answers involve an unusual feature of human egg development, or **oogenesis**. **Primary oocytes**, which are diploid precursors to eggs, enter meiosis I during embryonic development and arrest in prophase I until the female reaches sexual maturity. After this point, once each month, a small group of primary oocytes reenter meiosis, with only one of them typically producing an egg. For some oocytes, this means a wait as long as 50 years between the start and completion of meiosis. The maintenance of chiasma, spindle apparatus function, and the proper separation of chromosomes decline after this long period. Much remains to be discovered, but one thing is clear: Successful meiosis is critical to the health of offspring.

CHECK YOUR UNDERSTANDING

✔ If you understood this section, you should be able to . . .

1. Make a drawing that shows how nondisjunction in meiosis II can lead to aneuploid daughter cells.
2. Explain how the following can be true: Down syndrome is by far the most common autosomal (non–sex chromosome) aneuploid condition in human populations, but when gametes are examined, the frequency of aneuploidy for chromosome 21 is similar to that of many other chromosomes.

Answers are available in Appendix A.

13.4 What Are the Benefits of Meiosis?

After you complete this section, you should be able to . . .

- Make predictions about when sexual reproduction is favored over asexual reproduction.

Meiosis and sexual reproduction occur in only a small fraction of the lineages on the tree of life. Many organisms—in fact, most of them when considering the vast number of bacterial and archaeal species—reproduce asexually. Given all the mishaps in meiosis, why do any organisms use it in their reproduction? Of course, without meiosis sexual reproduction would not be possible, but then, why sex?

Although sexual reproduction plays a central role in the life of many familiar organisms—including us—until recently scientists had no clear idea of why it occurs. In fact, on the basis of evolutionary theory, biologists had good reason to think that sexual reproduction should *not* exist.

The Paradox of Sex

In 1978 evolutionary biologist John Maynard Smith pointed out that the existence of sexual reproduction presents a paradox. Maynard Smith developed a mathematical model showing that because asexually reproducing individuals have no need for males, their progeny on average can produce twice as many offspring as individuals that reproduce sexually. Figure 13.12 diagrams this model by showing the number of females (♀), males (♂), and asexually reproducing organisms (○) produced over three generations by asexual versus sexual reproduction.

In this example of the model, each asexually reproducing individual and each sexually reproducing couple produces four offspring over the course of their lifetimes. Two out of every four children that each couple produces are males, who cannot themselves give birth to children. As a result, at generation 2 the asexual population has twice the number of child-producing individuals as the sexual population. Maynard Smith referred to this result as the "two-fold cost of males." Asexual reproduction is twice as efficient as sexual reproduction because no males are produced.

What does the model predict when asexual and sexual individuals exist in the same population? If all other things are equal, individuals that reproduce asexually should increase in frequency in the population, while individuals that reproduce sexually should decline in frequency. In fact, Maynard Smith's model predicts that sexual reproduction is so inefficient that it should be eliminated.

To resolve this paradox, biologists began examining the assumption "If all other things are equal." Were there ways that meiosis and random fertilization could lead to the production of offspring that reproduce more than asexually produced individuals do? What kinds of conditions could favor sexual reproduction in spite of its obvious disadvantages? After decades of debate and analysis, two solid hypotheses to explain the paradox of sex have emerged.

The Purifying Selection Hypothesis

The first clue to unraveling the paradox of sex is a simple observation: If a gene is altered in a way that causes it to function poorly, the alteration will be inherited by *all* of the individual's offspring when asexual reproduction occurs. Suppose the altered gene arose in generation 1 of Figure 13.12. If this gene is important for reproductive success, its alteration would likely cause the four asexual individuals present in generation 2 to produce fewer than four offspring apiece. If so, then generation 3 will not have twice as many individuals in the asexual lineage compared with the sexual lineage.

An allele that functions poorly and lowers the fitness of an individual is said to be deleterious. Natural selection (Ch. 22, Section 22.4) against deleterious alleles is called purifying selection. Because asexual individuals are doomed to transmit all their deleterious alleles to all of their offspring, purifying selection will reduce the numerical advantage of asexual reproduction.

In contrast, suppose that the same deleterious allele arose in the sexually reproducing female in generation 1 of Figure 13.12. If the female also has a normal copy of the gene and she mates with a male that has two normal copies of the gene, then on average half her offspring will have this deleterious allele and half will lack it. This is important: Once a deleterious allele appears in a sexually reproducing individual, not all of the offspring from that point forward are doomed to carry it.

These ideas predict that there will be more deleterious alleles in asexually reproducing species. To test this notion, researchers compared the same genes in two closely related species of *Daphnia*, a tiny crustacean that is a common inhabitant of ponds and lakes (see Ch. 47, Section 47.1). One species reproduces asexually and the other reproduces sexually. As predicted, the scientists found that individuals in the asexual species contained many more deleterious alleles than individuals in the sexual species. Results like these have convinced biologists that purifying selection is an important factor promoting sexual reproduction.

The Changing-Environment Hypothesis

The second hypothesis to explain sexual reproduction focuses on the benefits of producing genetically diverse offspring. Here's the key idea: If parents are adapted to a particular environment, their offspring that are genetic clones are unlikely to thrive if the environment changes.

What type of environmental change might favor genetically diverse offspring? One ever-present change is the rapid evolution of pathogens and parasites—bacteria, viruses, fungi, and other entities that cause disease. These disease agents exert tremendous selective pressure on the organisms they infect. In your own lifetime, for example, several new disease-causing agents have emerged that afflict humans. These include the Zika virus, multiple drug-resistant strains of the tuberculosis bacterium, methicillin-resistant *Staphylococcus aureus* (MRSA), and tick-borne diseases such as Lyme disease. Hundreds of genes help defend you against these types of invaders, with some alleles being much more effective than others. In what amounts to an evolutionary arms race, pathogens and parasites constantly evolve new ways to infect the most common types of host, and in turn, host populations are driven to evolve resistance to the new forms of infectious agents.

What happens if all the offspring produced by a well-adapted individual are genetically identical? If a new strain of pathogen evolves that is more dangerous for that individual, then all the asexually produced offspring will be equally susceptible to that new strain. But if the offspring are genetically varied, then at

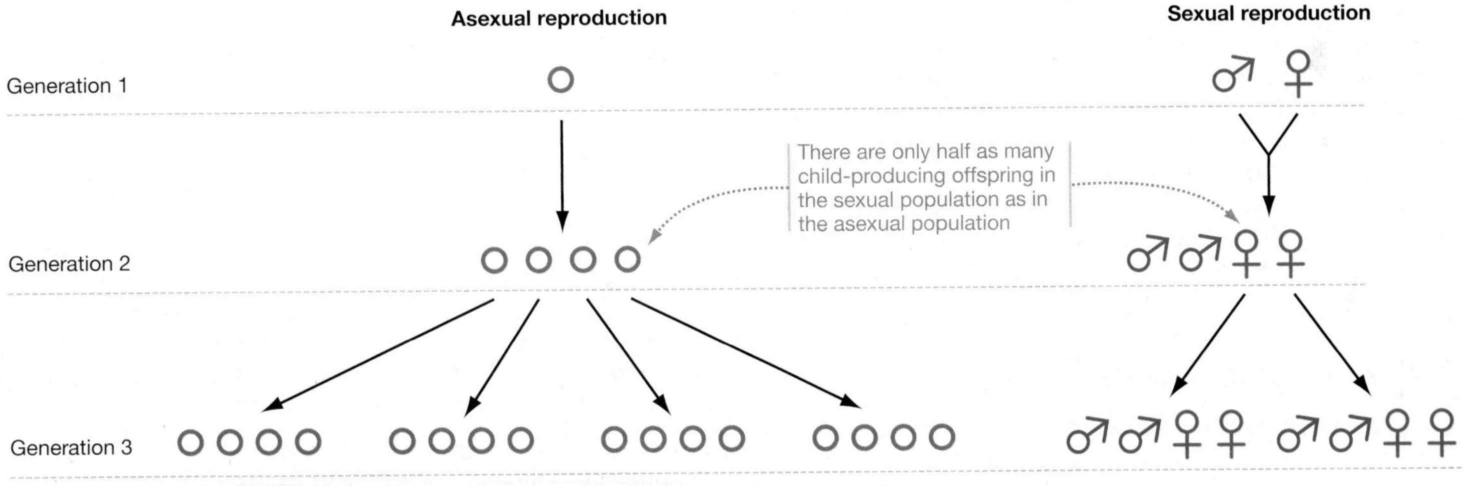

Figure 13.12 Asexual Reproduction Produces More Offspring. Each female (♀), male (♂), and asexual (○) symbol represents an individual. In this model, (1) every asexual individual or sexually reproducing couple produces four offspring over the course of a lifetime, (2) sexually reproducing individuals produce half males and half females, and (3) all offspring survive to breed.

✔ **QUANTITATIVE** In generation 2, there are only two more child-producing individuals in the asexual population than there are in the sexual population. If the same twofold difference in reproductive rates were continued, how many more child-producing individuals would there be in the asexual compared to the sexual population of generation 5?

least some of them are likely to have combinations of alleles that enable them to better resist the new strain of pathogen and produce offspring of their own.

A logical question is, does exposure to evolving pathogens favor genetic diversity in sexually reproducing organisms? To address this question, Levi Morran, Curtis Lively, and colleagues carried out a pivotal study on a tiny 1-mm-long roundworm named *Caenorhabditis elegans*.

C. elegans is an important model organism (see **BioSkills 11**) that was chosen for this study because it leads an unusual sex life. There are no females, only males and hermaphrodites. Because hermaphrodites have both male and female sex organs, *C. elegans* can reproduce either by self-fertilization or by mating with males. In contrast to mating with unrelated individuals, (termed *outcrossing*), self-fertilization reduces the genetic variation in offspring. The proportion of worms that reproduce by self-fertilization versus outcrossing can vary between strains or between different environments. Knowing this, the research team could test whether the rate of outcrossing increased in response to selection by a pathogen.

The setup of Morran, Lively, and colleagues' experiment is shown in **Figure 13.13**. The team began with a population of worms that had not been exposed to the pathogen and that reproduced predominantly by self-fertilization. The researchers then split the starting population into different groups. Half the groups were grown in the presence of a pathogen—a deadly bacterium—and the other half were grown without it. Once ingested by a worm, the bacterial pathogen could kill a susceptible individual within 24 hours.

At each generation, bacteria were collected from the carcasses of killed worms. Companion experiments showed that the pathogen evolved to become even more infectious over the course of the study. The graph in Figure 13.13 displays the results. The rate of outcrossing stayed low over 32 generations in populations that did not encounter the pathogen. In contrast, populations that were exposed to the evolving pathogen showed a rapid increase in the rate of outcrossing. The interpretation was that genetic variation promoted by sexual reproduction between unrelated individuals was favored in the changing environment created by the evolving pathogen.

At the end of the experiment, the worms in the pathogen-exposed population were significantly more resistant to the evolved pathogen than their ancestors were. This means that the worms in the predominantly outcrossing population had evolved

RESEARCH

QUESTION: Does exposure to evolving pathogens favor outcrossing?

HYPOTHESIS: In environments where evolving pathogens are present, sexual reproduction by outcrossing will be favored.

NULL HYPOTHESIS: The presence of evolving pathogens will not favor outcrossing.

EXPERIMENTAL SETUP:

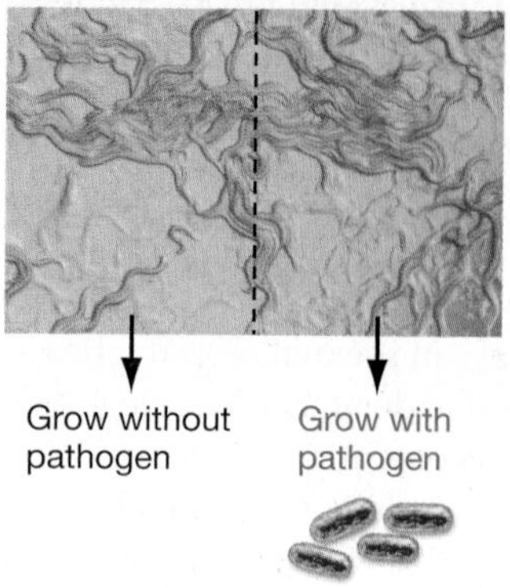

1. Start with a pathogen-free population of *C. elegans* roundworms with a 20% rate of outcrossing.

2. Divide the population; grow one subgroup in the absence of a pathogen and another subgroup in the presence of an evolving pathogen.

3. Assess the rate of outcrossing over many generations.

PREDICTION OF HYPOTHESIS: The rate of outcrossing will increase in response to exposure to a pathogen.

PREDICTION OF NULL HYPOTHESIS: The rate of outcrossing will not be influenced by a pathogen.

RESULTS:

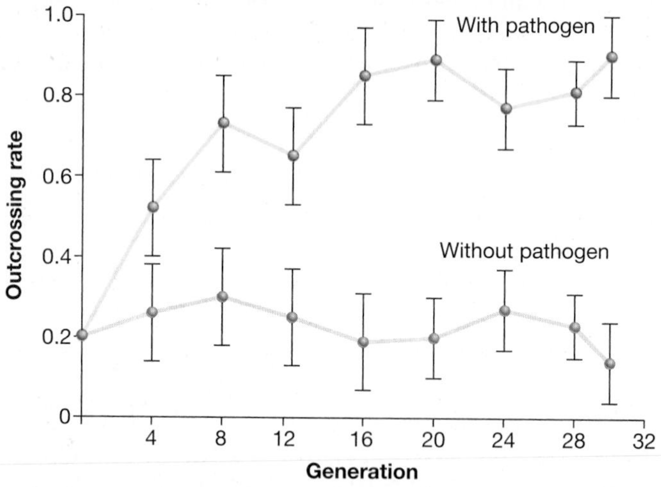

CONCLUSION: Exposure to evolving pathogens favors outcrossing.

Figure 13.13 Does Exposure to Pathogens Favor Sexual Reproduction through Outcrossing? Each point in the graph shows the average percentage of reproduction by outcrossing for five populations. The bars indicate the degree of variation in the data (see **BioSkills 3** for a description of error bars).

SOURCE: L. T. Morran et al. 2011. Running with the Red Queen: Host–parasite coevolution selects for biparental sex. *Science* 333: 216–218.

✔ **PROCESS OF SCIENCE** What would you predict if a non-evolving pathogen were used?

CHECK YOUR UNDERSTANDING

✔ **If you understood this section, you should be able to . . .**

1. Use the purifying selection hypothesis to predict how changes in the rate of mutation would influence whether sexual or asexual reproduction is favored.
2. Use the changing-environment hypothesis to predict whether there would be an advantage of sexual reproduction over asexual reproduction in a population of genetically identical individuals.

Answers are available in Appendix A.

along with the pathogen. In striking contrast, when a parallel experiment was done with a strain of worms that could reproduce only by self-fertilization, those populations were unable to evolve resistance to the pathogen. In fact, they became extinct.

These results and many others support the changing-environment hypothesis. Although the advantages of sexual reproduction remain an active area of research, more biologists are becoming convinced that sexual reproduction is helpful for two reasons: **(1)** Offspring are not doomed to inherit harmful alleles, and **(2)** the production of genetically varied offspring means that at least some may be able to resist rapidly evolving pathogens and parasites.

CHAPTER 13 Review

For media, go to **Mastering Biology**

13.1 How Does Meiosis Occur?

- Meiosis is a nuclear division resulting in cells that have only one of each type of chromosome and half as many chromosomes as the parent cell. In animals, meiosis leads to the formation of eggs and sperm.
- In diploid ($2n$) organisms, individuals have two versions of each type of chromosome. The two versions are called homologs. One homolog is inherited from the mother and one from the father. Haploid organisms (n) have just one version of each type of chromosome.
- Each chromosome is replicated during the S phase of the cell cycle, well before meiosis begins. At the start of meiosis I, each chromosome consists of a pair of sister chromatids, each with a double-helical DNA molecule identical to that of the other sister chromatid.
- Homologous pairs of replicated chromosomes synapse early in meiosis I, forming a bivalent—two closely paired homologous chromosomes. Non-sister chromatids undergo crossing over.
- The pair of homologous chromosomes, connected by at least one chiasma, is moved to the metaphase plate.
- At the end of meiosis I, the homologous pairs of replicated chromosomes are separated and distributed to two daughter cells. The daughter cells are haploid—each contains one of each type of replicated chromosome.
- During meiosis II, sister chromatids of the replicated chromosomes separate and are distributed to two daughter cells.
- From one diploid cell with replicated chromosomes, meiosis produces four haploid daughter cells with unreplicated chromosomes.

13.2 Meiosis Promotes Genetic Variation

- Each cell produced by meiosis receives a different combination of maternal and paternal versions of each chromosome. Because genes are located on chromosomes, and often come in different forms (alleles), each cell produced by meiosis receives a different complement of alleles for its genes. The resulting offspring are genetically distinct from one another and from their parents.
- When meiosis and mating between unrelated individuals occur, the chromosome complements of offspring differ from one another and from their parents for these three reasons:
  1. Gametes receive a random assortment of maternal and paternal chromosomes when homologs separate in meiosis I. This is independent assortment.
  2. Because of crossing over, each chromosome contains a random assortment of paternal and maternal alleles.
  3. Random fertilization results in a combination of different chromosome sets from each parent.

13.3 What Happens When Things Go Wrong in Meiosis?

- Mistakes during meiosis often result in egg and sperm cells that contain the wrong number of chromosomes. Human embryos formed from such gametes usually do not complete development. Children with Down syndrome, who have an extra copy of chromosome 21, are one notable exception.
- Most of the mistakes in meiosis are failures of either homologous chromosomes or sister chromatids to separate.

13.4 What Are the Benefits of Meiosis?

- In asexual reproduction, all individuals can bear offspring. From this standpoint, asexual reproduction is more efficient than sexual reproduction, in which on average half the offspring (males) cannot bear offspring.
- The leading hypotheses to explain the existence of sexual reproduction and therefore the benefits of meiosis are that:
  1. a parent with a harmful allele can produce offspring without that allele; and
  2. production of genetically diverse offspring leads to populations in which some individuals are better able to resist evolving pathogens and parasites.

Answers are available in Appendix A.

✓ TEST YOUR KNOWLEDGE

1. **THINK CAREFULLY** What are homologous chromosomes?
   a. chromosomes that are similar in their size, shape, and gene content
   b. similar chromosomes that are found in different individuals of the same species
   c. the two chromatids in a replicated chromosome
   d. any chromosome in a diploid cell
2. What is an outcome of genetic recombination?
   a. the synapsing of homologs during prophase of meiosis I
   b. the new combination of maternal and paternal chromosome segments that results when homologs cross over
   c. the new combinations of chromosome segments that result when self-fertilization occurs
   d. the combination of a haploid phase *and* a diploid phase in a life cycle
3. What proportion of chromosomes in a man's skin cell are maternal chromosomes?
4. Nondisjunction that leads to problems in offspring can occur in:
   a. mitosis
   b. meiosis I only
   c. meiosis I and II
   d. mitosis, meiosis I, and meiosis II

✔ TEST YOUR UNDERSTANDING

5. Select True or False for each statement.
   T/F Sister chromatids are homologous chromosomes.
   T/F Non-sister chromatids are found on two different homologs.
   T/F Crossing over occurs between sister chromatids.
   T/F A chiasma forms between two of the four molecules of double-stranded DNA on duplicated homologous chromosomes.

6. If you followed a woman's cells through meiosis, at what stage of meiosis would the amount of DNA in one of these cells be equal to the amount of DNA in one of her G_1 phase (before DNA replication) kidney cells?

7. Norway rats have 42 chromosomes in their diploid cells. If such a cell enters meiosis, how many chromosomes and double-helical molecules of DNA will be present in each daughter cell at the end of meiosis II?
   a. 21 chromosomes and 21 double-helical DNA molecules
   b. 21 chromosomes and 42 double-helical DNA molecules
   c. 42 chromosomes and 42 double-helical DNA molecules
   d. 42 chromosomes and 84 double-helical DNA molecules

8. Triploid ($3n$) watermelons, which are seedless, are produced by crossing a tetraploid ($4n$) strain with a diploid ($2n$) plant. Explain why this mating produces a triploid individual.

✔ TEST YOUR PROBLEM-SOLVING SKILLS

9. **QUANTITATIVE** Meiosis results in independent assortment of maternal and paternal chromosomes. If $2n = 6$ for a given organism, and there is no crossing over, what is the chance that a gamete produced by this diploid organism will receive *only* paternal chromosomes?
   a. 0 **b.** 1/16 **c.** 1/8 **d.** 1/3

10. **PROCESS OF SCIENCE** A species of rotifer, a small freshwater invertebrate, lost the ability to reproduce sexually millions of years ago. A remarkable feature of its life cycle is the ability to withstand dry conditions. When the rotifer's environment dries out, so does the rotifer, and it can be blown to a new area. Rotifers that land in water will rehydrate and resume an active life. A major pathogen of these rotifers is a species of fungus that cannot survive drying. Some scientists hypothesize that drying rids the rotifers of this pathogen.
    a. Design an experimental study to test this hypothesis.
    b. Why might the ability to withstand drying reduce any potential advantage of sexual reproduction in this rotifer species?

✔ PUT IT ALL TOGETHER: Case Study

What links maternal age to aneuploidy?

In analyzing the events of meiosis for answers to why older women produce more aneuploid eggs, biologists have hypothesized that binding between sister chromatids becomes weaker in oocytes of older women. If so, does it lead to problems in chromosome segregation that cause aneuploid conditions such as Down syndrome?

11. **QUANTITATIVE** Researchers used female mice of different ages as a model for aging in women. They measured the percentage of aneuploid eggs, analyzed chromosomes in meiosis for the amount of cohesin (a protein complex that connects sister chromatids), and measured the distance between sister kinetochores. Does the evidence shown in the graphs support the hypothesis that weakened binding between sister chromatids leads to aneuploidy? Explain.

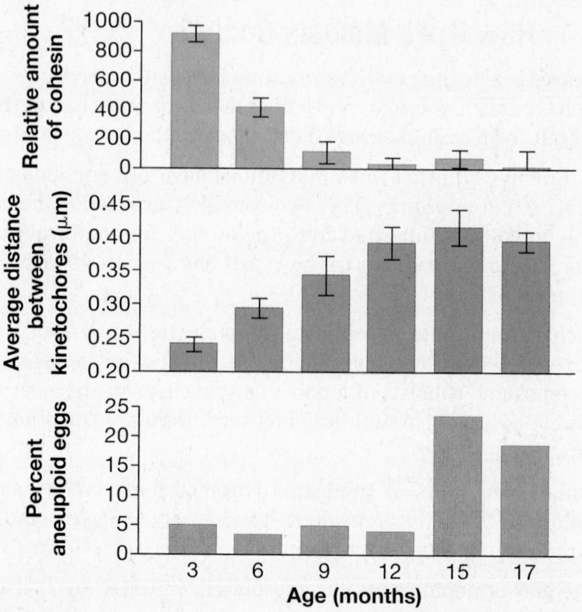

Source: T. Chiang, F. E. Duncan, K. Schindler, et al. 2010. *Current Biology* 20: 1522–1528.

12. Data in the graphs show an inverse relationship between the amount of cohesin on each chromosome and the average distance between kinetochores. Why is this relationship logical?

13. A unique feature of meiosis I is the attachment of both kinetochores of a replicated chromosome to spindle fibers that come from a single spindle pole. Why is this unusual attachment important?

14. In meiosis I in older females, chromosomes often incorrectly attach to spindle fibers coming from both spindle poles. Based on data shown in the graphs, what is a possible explanation for this increase in attachment error?

15. If the results showing increased aneuploidy in older mice can be related to humans, roughly what age in women would correspond to 15-month-old female mice (see Figure 13.11)?

16. **SOCIETY** Imagine that you are a genetic counselor and are consulting with a 42-year-old woman who has just learned she's pregnant. She knows that Down syndrome births increase sharply with a mother's age. You want to explain her chances of *not* having a child with Down syndrome. What will you say?

Mastering Biology

Students Go to Mastering™ Biology for assignments, the eText, and the Study Area with animations, practice tests, and activities.

Professors Go to Mastering™ Biology for automatically graded tutorials and questions that you can assign to your students, plus Instructor Resources.

14 Mendel and the Gene

Experiments on pea plants helped launch the science of genetics.

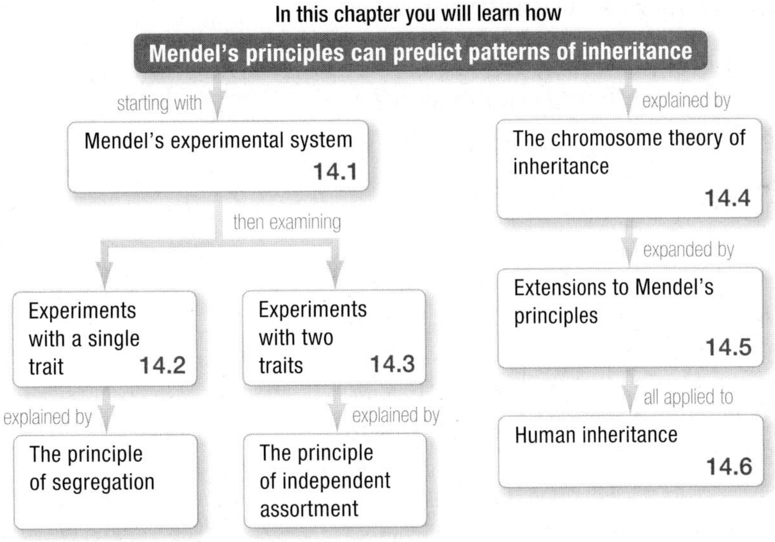

The science of biology is built on a series of great ideas. In this chapter you'll take a closer look at one of these concepts, the chromosome theory of inheritance, which explains how genetic information is transmitted from one generation to the next.

An Austrian monk named Gregor Mendel laid the groundwork for the chromosome theory of inheritance in 1865 when he announced that he had worked out the rules of inheritance through a series of experiments on garden peas. Another key insight emerged during the final decades of the nineteenth century, when biologists described meiosis (Ch. 13, Section 13.1).

BIG PICTURE

This chapter is part of the Big Picture. See how on pages 404–405.

Walter Sutton and Theodor Boveri linked these two discoveries in 1902 to form the **chromosome theory of inheritance**. This theory states that genes are located on chromosomes and that the transmission of chromosomes to daughter cells at meiosis accounts for the patterns of inheritance that Mendel observed. (Genes determine inherited traits and are center stage in the Genetic Information Big Picture on pages 404–405.)

The chromosome theory of inheritance launched the study of **genetics**, the branch of biology that focuses on the inheritance of traits. Let's start at the beginning: What are the rules of inheritance that Mendel discovered?

14.1 Mendel's Experimental System

When biological science began to emerge, questions about **heredity**—meaning inheritance, or the transmission of traits from parents to offspring—were primarily the concern of animal and plant breeders. A **trait** is any observable characteristic of an individual, ranging from outward appearance such as height to molecular characteristics such as the primary structure of a particular membrane protein.

After you complete this section, you should be able to ...

- Explain Mendel's rationale for setting up his studies of inheritance.

Mendel's community was keenly interested in how selective breeding could result in hardier and more productive varieties of sheep, fruit trees, and grape vines. An agricultural association had formed to promote research into making selective breeding more efficient, and Mendel was an active member. The rich intellectual climate extended into Mendel's monastery, which was devoted to scientific teaching and research.

What Questions Was Mendel Trying to Answer?

Mendel set out to address a fundamental issue of heredity: What are the patterns of the transmission of traits from parents to offspring? At the time, there were two hypotheses:

1. *The blending inheritance hypothesis* proposed that the traits of a mother and father blend together to form traits in their offspring. As a result, an offspring has characteristics that are intermediate between its mother and father.
2. *The inheritance of acquired characters hypothesis* proposed that traits are modified through use and then passed on from parents to their offspring in the modified form.

Each of these hypotheses made predictions. Blending inheritance predicted that when black sheep and white sheep mate, their hereditary determinants blend to give offspring with gray wool. Inheritance of acquired characters predicted that if giraffes extend their necks by straining to reach leaves high in the tops of trees, they transmit this acquired trait to produce longer-necked offspring.

Both of these hypotheses seem sensible. Are they correct?

The Garden Pea Served as the First Model Organism in Genetics

After investigating and discarding several candidate species to study, Mendel chose the garden pea, *Pisum sativum*. His reasons were practical: Peas are inexpensive and easy to grow from seed, have a relatively short generation time, and produce large numbers of seeds. These features made it possible for him to continue experiments over several generations and collect data from many individuals.

Peas served as a **model organism**: a species that is used for research because it is easy to work with and conclusions drawn from studying it may apply to many other species. **BioSkills 11** introduces some of the important model organisms used in biological science today.

Two additional features of the pea made it possible for Mendel to design his experiments: He could control which parents were involved in a mating, and individuals were available that differed in easily recognizable traits, such as flower color or seed shape. Whenever a trait appears commonly in two or more different forms, for example purple or white flowers, it is called a **polymorphic trait**.

How Did Mendel Control Matings? Each garden pea flower contains both male and female reproductive structures (**Figure 14.1a**). Sperm cells are produced in pollen grains, which are small sacs that mature in the male reproductive organ of the plant. Eggs are produced in the female reproductive organ.

Normally, garden peas undergo **self-fertilization**: that is, a flower's pollen falls on the female reproductive organ of that same flower. As **Figure 14.1b** shows, however, Mendel could prevent self-fertilization by removing the male reproductive organs from a flower before any pollen formed. Later he could use a brush to transfer pollen from another pea plant to the target flower's female reproductive organ. This type of mating is referred to as *cross-fertilization*, or simply a **cross**. Using this technique, Mendel could control the matings of his model organism.

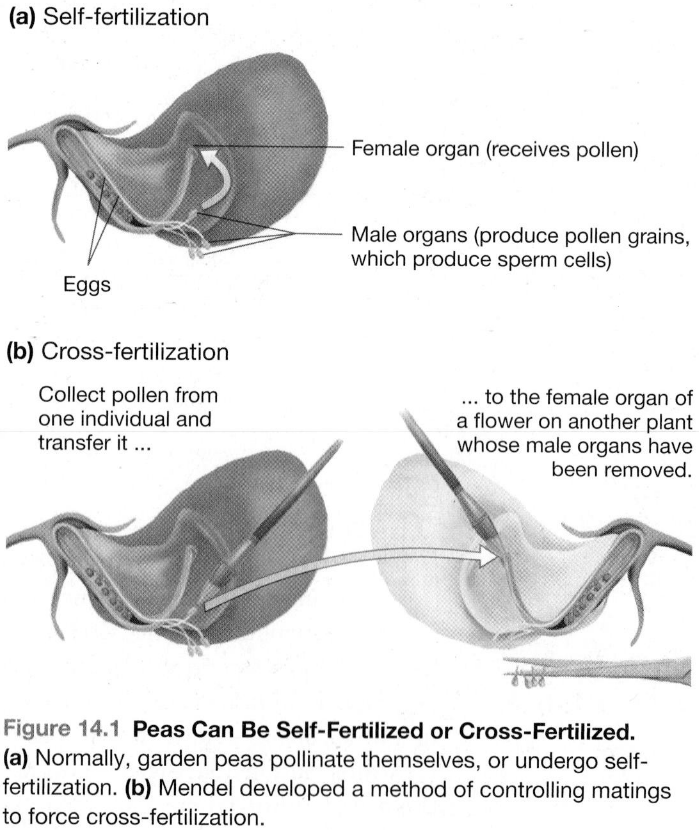

Figure 14.1 Peas Can Be Self-Fertilized or Cross-Fertilized. **(a)** Normally, garden peas pollinate themselves, or undergo self-fertilization. **(b)** Mendel developed a method of controlling matings to force cross-fertilization.

What Traits Did Mendel Study? Mendel used varieties of peas that differed in seven traits: seed shape, seed color, pod shape, pod color, flower color, flower and pod position, and stem length. Biologists refer to the observable traits of an individual, such as the shape of a pea seed or the eye color of a person, as its **phenotype**. Phenotype can be any observable characteristic, from molecular-level traits such as protein shape to the familiar traits of entire organisms that Mendel studied. Phenotype is just one term in the rich vocabulary of genetics. You can review many of these terms in **Table 14.1**. In the peas that Mendel studied, two distinct phenotypes existed for each of the seven traits.

Mendel began his work by obtaining individuals from pure lines, or true-breeding lines. A **pure line** consists of individuals that produce offspring identical to the parents when they are crossed to another member of the same pure-line population or are self-fertilized. For example, breeders had developed pure lines for wrinkled seeds and round seeds. During two years of trial experiments, Mendel confirmed that individuals that germinated from his wrinkled seeds produced only wrinkle-seed offspring when they were mated to themselves or to another pure-line individual that germinated from a wrinkled seed.

Why is this important? Remember that Mendel wanted to find out how traits are transmitted from parents to offspring. Was it by blending inheritance, inheritance of acquired characteristics, or some other process? By starting with predictable pure lines, he could be certain that any new phenotype that appeared after a cross was due to the cross rather than some random event.

Suppose that Mendel arranged matings between a pure-line individual with round seeds and a pure-line individual with

SUMMARY **Table 14.1** Terms Used in Mendelian Genetics

Term	Definition	Example or Comment
Autosomal inheritance	The patterns of inheritance of any genes not on a sex chromosome. These are the "standard" patterns of inheritance.	Mendel studied only autosomal inheritance.
Gene	A hereditary factor that influences a particular trait.	This is the earliest definition. A more modern definition of a gene is a region of DNA in a chromosome that codes for a particular protein or RNA.
Allele	A particular form of a gene.	The two alleles in a diploid may be the same or different.
Genotype	A listing of the alleles of particular genes in an individual.	In diploids, the genotype lists two alleles of each gene; in haploid gametes, the genotype lists one allele of each gene.
Phenotype	An individual's observable traits.	Phenotype can be observed at levels from molecules to the whole organism; influenced to different degrees by the genotype.
Homozygous	Having two of the same allele.	Refers to a particular gene.
Heterozygous	Having two different alleles.	Refers to a particular gene.
Dominant allele	An allele that produces the same phenotype in heterozygous and homozygous genotypes.	Dominance does not imply high frequency or high fitness.
Recessive allele	An allele that produces its phenotype only in homozygous genotypes.	Phenotype "recedes" or disappears in heterozygous individuals; recessive does not imply low frequency or low fitness.
Pure line	Individuals identical in phenotype that, when crossed, produce offspring that all have the same phenotype.	Pure-line individuals are homozygous for the gene in question.
Hybrid	Offspring from crosses between homozygous parents with different genotypes.	Hybrids are heterozygous.
Polymorphic trait	A trait that appears commonly in two or more different forms	Flower color in pea plants, which is commonly purple or white, is a polymorphic trait.
Reciprocal cross	A cross in which the phenotypes of the male and female are reversed compared with a prior cross.	Used to test if the sex of the parent influences transmission of the trait.
Testcross	A cross of a homozygous recessive individual and an individual with the dominant phenotype but unknown genotype.	Usually used to determine whether a parent with a dominant phenotype is homozygous or heterozygous.
X-linkage	Referring to a gene located on the X chromosome or a trait associated with a gene on the X chromosome. Also known as X-linked.	X-linked genes and traits show different patterns of inheritance in males and females.
Y-linked	Referring to a gene located on the Y chromosome or a trait associated with a gene on the Y chromosome. Also known as Y-linked.	In humans, most Y-linked genes determine male-specific development.

wrinkled seeds. He knew that one parent carried a hereditary determinant for round seeds, while the other carried a hereditary determinant for wrinkled seeds, and that offspring from this mating must contain both types of hereditary determinants. They would be **hybrids**—offspring from matings between true-breeding parents that differ in one or more traits.

Would these hybrid offspring have wrinkled seeds, round seeds, something between wrinkled and round, or perhaps some new trait? What would the seed shape in subsequent generations be when hybrid individuals self-pollinated?

CHECK YOUR UNDERSTANDING

✔ **If you understood this section, you should be able to . . .**

1. Predict the outcome of a mating between a longer-necked giraffe and a shorter-necked giraffe that both live in an environment with tall trees, if (a) the blending inheritance hypothesis was correct, and (b) the inheritance of acquired characters hypothesis was correct.
2. **PROCESS OF SCIENCE** Justify why, when investigating fundamental genetic principles, pea plants are likely to be a better model organism than elephants.

Answers are available in Appendix A.

14.2 Mendel's Experiments with a Single Trait

In his first set of experiments, Mendel crossed pure lines of garden peas that differed in just one trait. Starting with the simplest possible situation as Mendel did is a key research strategy. Once you understand what's going on in the simplest case, you can extend this knowledge by considering more complex questions.

After you complete this section, you should be able to . . .

- Analyze the results of monohybrid crosses using the principle of segregation.

Mendel crossed individuals from round-seed and wrinkle-seed pure lines. The individuals used in the initial cross are the **parental generation**. Their progeny (offspring) are the $\mathbf{F_1}$ **generation**. F_1 stands for "first filial"; the Latin roots *filius* and *filia* mean "son" and "daughter," respectively. Offspring from a mating between two F_1 individuals are called the F_2 generation; an F_2 mating leads to an F_3 generation, and so on.

The Monohybrid Cross

In his first set of crosses, Mendel took pollen from plants in the round-seed line and placed it on the female reproductive organs of plants from the wrinkle-seed line. As Figure 14.2a shows, all the seeds produced by progeny from this cross were round. This was a remarkable result:

- The traits did not blend together to form an intermediate phenotype. Instead, only the round-seed trait appeared. This result was in stark contrast to the prediction of the blending-inheritance hypothesis shown in Figure 14.2b.
- The wrinkle-seed trait disappeared.

Dominant and Recessive Traits To learn what was going on, Mendel did something brilliantly simple: He planted the F_1 seeds and allowed the pea plants to self-fertilize when they matured.

Remember, Mendel knew that each of the F_1 plants must have inherited a genetic determinant for round seeds and a genetic determinant for wrinkled seeds. A mating like this—between parents that each carry two different genetic determinants for the same trait—is called a **monohybrid cross**. The Greek root *mono* means "one" or "single," so the name tells you that the cross produces a hybrid for a single trait.

The seeds produced by the F_1 plants were the next, or F_2, generation. When Mendel examined these F_2 seeds, he found that 5474 were round and 1850 were wrinkled (see Figure 14.2a). He was astounded to find the wrinkled seeds. The wrinkle-seed

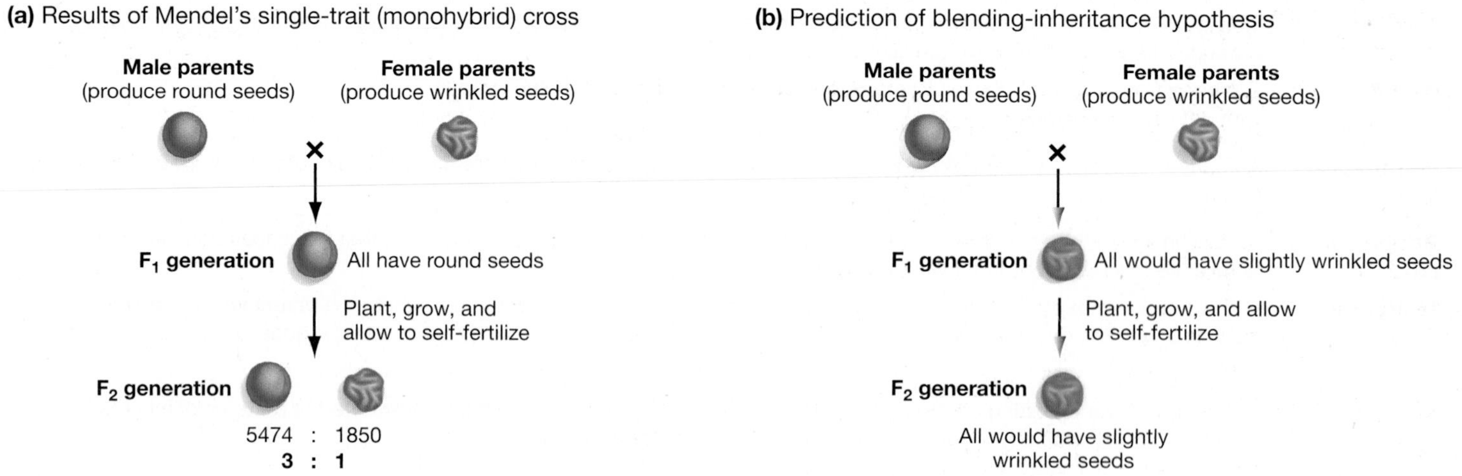

Figure 14.2 Mendel Performed a Monohybrid Cross. The results of Mendel's crosses involving a single trait **(a)** contrasted with the predictions of the blending-inheritance hypothesis **(b)**. Did the genetic determinant for the trait also disappear? If not, where was it hiding?

shape had reappeared in the F_2 generation after disappearing completely in the F_1 generation. No one had observed this phenomenon before.

- Mendel called wrinkled shape a **recessive** trait relative to round-seed trait. This term was appropriate because none of the F_1 individuals had wrinkled seeds, even though they must have inherited the hereditary determinant from one of their parents. The wrinkle-seed phenotype appeared to recede or be hidden in the hybrid F_1.
- Mendel called round shape a **dominant** trait relative to wrinkled shape. This term was apt because the round-seed phenotype appeared to dominate over the wrinkle-seed determinant when both were present in the hybrid F_1.

Note, though, that in genetics the term "dominant" has nothing to do with its everyday English usage of being powerful or superior. Individuals with a dominant phenotype are not necessarily more fit than individuals with the recessive phenotype. Nor are the genetic determinants associated with a dominant phenotype necessarily more common than recessive ones. For example, a rare fatal disorder called Huntington disease is caused by an equally rare, dominant genetic determinant. In genetics, the terms "dominant" and "recessive" identify *only* which phenotype is observed and which is masked in individuals carrying two different genetic determinants for a given trait.

Mendel also noticed that the round and wrinkled seeds of the F_2 generation were in a ratio of 2.96 : 1, or essentially 3 : 1. This means that for every four individuals, on average three have the dominant phenotype and one has the recessive phenotype. The results can also be stated in terms of proportions: In this case, about ¾ of the F_2 seeds were round and ¼ were wrinkled.

✔ **If you understand Mendel's method for doing a monohybrid cross, you should be able to explain why self-fertilization of a hybrid is equivalent to crossing two hybrid individuals of the same type together.**

A Reciprocal Cross Mendel wondered if it mattered whether the male or female parent had a particular genetic determinant. To test this, he performed a second set of crosses between two pure-line populations—this time "in reverse," using the pollen of an individual from a pure line of wrinkle-seed peas (**Figure 14.3**).

RESEARCH

QUESTION: Is the inheritance of seed shape in peas affected by whether the genetic determinant comes from a male or female gamete?

HYPOTHESIS: The type of gamete *does* affect the inheritance of seed shape.

NULL HYPOTHESIS: The type of gamete does *not* affect the inheritance of seed shape.

EXPERIMENTAL SETUP:

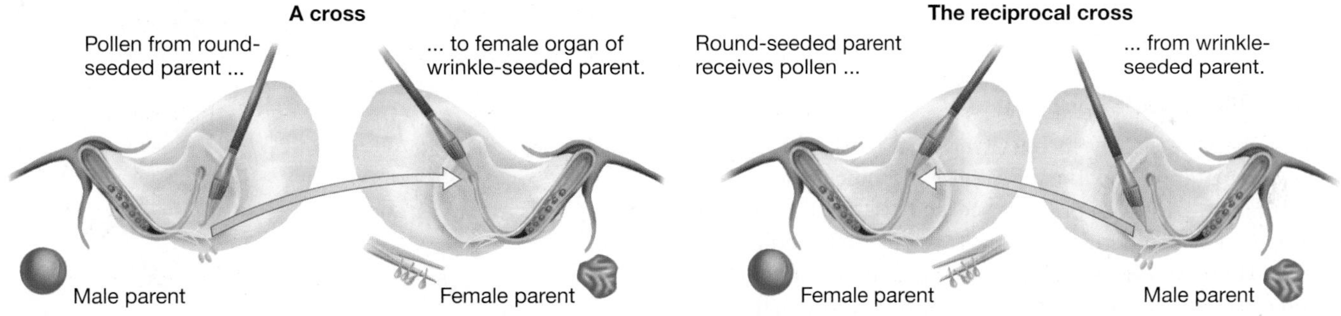

PREDICTION OF "SEX MATTERS" HYPOTHESIS: Offspring phenotypes will be different in the two crosses.

PREDICTION OF NULL HYPOTHESIS: Offspring phenotypes will be identical in the two crosses.

RESULTS:

CONCLUSION: It makes no difference whether the genetic determinant for seed shape comes from the male gamete or from the female gamete.

Figure 14.3 Mendel Also Performed a Reciprocal Cross.

SOURCE: G. Mendel. 1866. Versuche über Pflanzen-hybriden. Verhandlungen des naturforschenden Vereines in Brünn 4: 3–47. English translation available from ESP: Electronic Scholarly Publishing (www.esp.org).

✔ **PROCESS OF SCIENCE** Given that the "sex matters" hypothesis was not supported, was this experiment a failure?

These experiments used a **reciprocal cross**—a set of matings where the mother's phenotype in the initial cross is the father's phenotype in a subsequent cross, and the father's phenotype in the initial cross is the mother's phenotype in a subsequent cross.

The results of the two crosses were identical: All the F_1 progeny from the reciprocal cross had round seeds, just as in the initial cross. The reciprocal cross established that it does not matter if the genetic determinants for seed shape come from the male or female parent. As you'll learn later, reciprocal crosses of other traits sometimes do produce different results. When this happens, it reveals important insights about the underlying genes.

Do Mendel's Results Hold for Other Traits? Before drawing broad conclusions, Mendel established that the results of crosses with round and wrinkled seeds were general by repeating the experiments with six other traits.

Table 14.2 shows the F_2 data Mendel collected from thousands of offspring. Some important patterns emerged:

- The F_1 progeny showed only the dominant trait and did not exhibit an intermediate or new phenotype.
- Reciprocal crosses produced the same results.
- The ratio of dominant to recessive phenotypes in F_2-generation individuals was about 3 to 1.

How could these patterns be explained? Mendel answered this question with a series of propositions about the nature and behavior of the hereditary determinants. His model was built on some of the most brilliant insights in the history of biology.

Table 14.2 The F_2 Produced from Mendel's Monohybrid Crosses

Trait	Dominant Phenotype	Recessive Phenotype	Ratio
Seed shape	5474 round 	1850 wrinkled 	2.96 : 1
Seed color	6022 yellow 	2001 green 	3.01 : 1
Pod shape	882 inflated 	299 constricted 	2.95 : 1
Pod color	428 green 	152 yellow 	2.82 : 1
Flower color	705 purple 	224 white 	3.15 : 1
Flower and pod position	651 axial 	207 terminal 	3.14 : 1
Stem length	787 tall 	266 dwarf 	2.96 : 1

DATA: G. Mendel. 1866. *Verhandlungen des naturforschenden Vereines in Brünn* 4: 3–47.

Particulate Inheritance

Mendel's results were clearly inconsistent with the hypothesis of blending inheritance and the hypothesis of inheritance of acquired characters. To explain the patterns that he observed, Mendel proposed a competing hypothesis called **particulate inheritance**. He maintained that the hereditary determinants for traits do not blend together or become modified through use. Instead, hereditary determinants maintain their integrity from generation to generation. Rather than blending together, they act as discrete, unchanging particles.

Mendel's hypothesis was the only way to explain the observation that phenotypes disappeared in one generation and reappeared unchanged in the next. It also represented a break with ideas that had prevailed for hundreds of years.

Genes, Alleles, and Genotypes Today, geneticists use the word **gene** for the hereditary determinant of a trait. For example, the hereditary determinant that controls whether pea seeds are round or wrinkled is referred to as the gene for seed shape.

Mendel also proposed that each individual can have two versions of any gene. Today different versions of the same gene are called **alleles**. In the case of the gene for seed shape, one allele of this gene is responsible for the round form of the seed while another allele is responsible for the wrinkled form. The combination of alleles found in an individual is the **genotype**.

The hypothesis that pea plants have two copies of each gene—either two of the same allele or two different alleles—sprang from the need to explain how a trait could disappear in one generation only to resurface in another. Mendel reasoned that if each individual carries two copies of a gene instead of one, and one allele is dominant over another, then in a hybrid, the recessive trait will be hidden. That's why, in a cross between plants from a round-seed pure line and a wrinkle-seed pure line, the hybrid F_1 offspring are all round-seed. The round-seed allele is dominant because it is the only one with an observable phenotype in the hybrid, and the wrinkle-seed allele is recessive because its phenotype is hidden in the hybrid.

Mendel's postulate that there are two copies of each gene opened the possibility that when F_1 hybrids are crossed, some of the F_2 offspring may inherit two copies of the recessive allele, one from each of the hybrid parents. These F_2 offspring with only the recessive allele would be expected to show the recessive phenotype that was hidden in the F_1 hybrids.

This idea explains why the wrinkled-seed phenotype disappeared in the F_1 generation and could reappear in the F_2 generation. But why were the F_2 generation seeds in a 3 : 1 ratio of round and wrinkled?

The Principle of Segregation To account for the 3 : 1 ratio of phenotypes in F_2 individuals, Mendel reasoned that the two members of each gene pair must segregate—that is, separate—into different gamete cells during the formation of eggs and sperm. As a result, each gamete contains one allele of each gene. This idea is called the **principle of segregation**.

To model the segregation of alleles, Mendel used letters to indicate the gene for a particular trait. For example, for seed shape he used an uppercase italic *R* to symbolize a dominant allele and a lowercase italic *r* to symbolize a recessive allele. (Using uppercase versus lowercase letters to distinguish dominant versus recessive alleles has become a common, though not universal, convention. Note, too, that letters representing genes are italicized.)

Using this notation, Mendel described the genotype of the individuals in the round-seed pure line as *RR* (having two of the dominant allele). The genotype of the wrinkle-seed pure line is *rr* (two of the recessive allele). Because *RR* and *rr* individuals have two copies of the same allele, they are said to be **homozygous** for the seed-shape gene (*homo* is the Greek root for "same," while *zygo* means "yoked"). Crosses of individuals from the same pure line always produce offspring with the same phenotype because they are homozygous—no other allele is present.

Figure 14.4a uses a diagram called a Punnett square to show what happened to these alleles when Mendel crossed the *RR* and *rr* pure lines. R. C. Punnett invented this method for predicting the genotypes and phenotypes of different crosses years after Mendel published his work. According to Mendel's hypothesis, *RR* parents produce gametes that all carry the *R* allele, while *rr* parents produce gametes with the *r* allele only. This is because the two alleles segregate in the formation of gametes.

An important extension of the principle of segregation is that segregated alleles present in the gametes from each parent come together at fertilization to produce a zygote, the first cell of an offspring. When the gametes from each parent of this cross come together at fertilization, they create offspring with the *Rr* genotype. Such individuals, with two different alleles for the same gene, are said to be **heterozygous** (*hetero* is the Greek root for "different"). Heterozygous individuals, or heterozygotes, show that the *R* allele is dominant because only the round phenotype is seen even though the wrinkled allele is present.

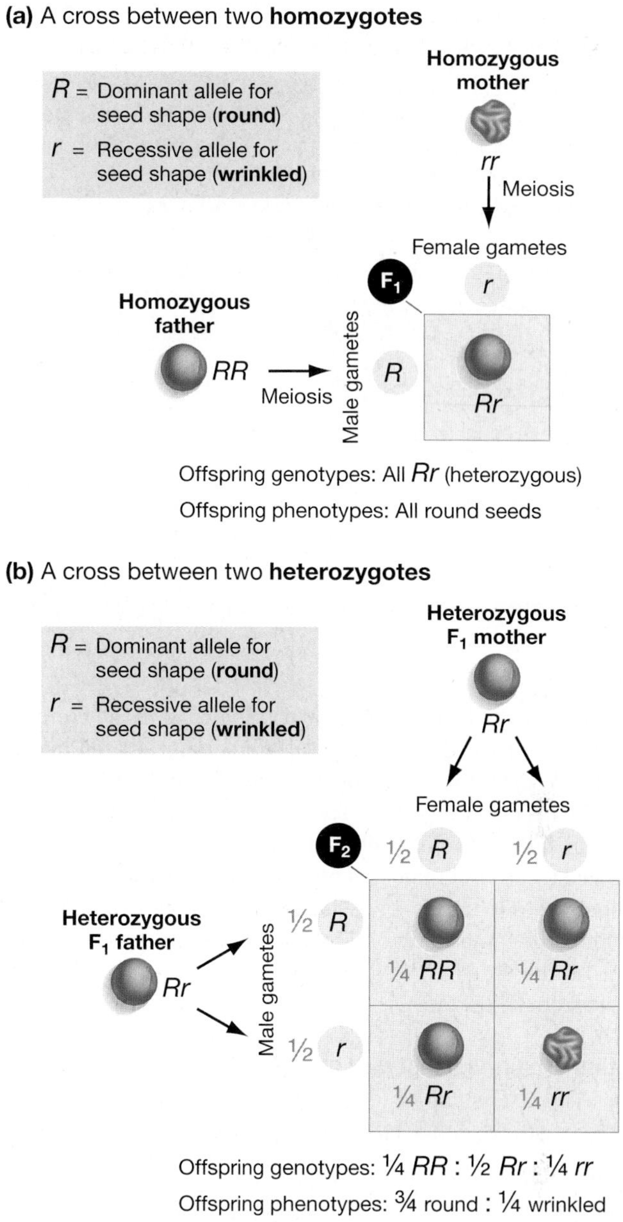

Figure 14.4 Mendel Analyzed the Offspring of a Monohybrid Cross. See Making Models 14.1 on page 302 for tips on how Punnett squares are constructed.

Why do the two phenotypes appear in a 3 : 1 ratio in the F_2 generation? Mendel proposed that during gamete formation in the F_1 heterozygotes, the paired *Rr* alleles separate into different gamete cells. As a result, and as shown in the Punnett square of Figure 14.4b, half the gametes should carry the *R* allele and half should carry the *r* allele. Importantly, a given sperm has an equal chance of fertilizing either an *R*-bearing egg or an *r*-bearing egg. This is random fertilization. The end result? The 3 : 1 ratio in the F_2 generation.

Making Models 14.1 Tips on Drawing Punnett Squares (I)

When you draw Punnett squares: **(1)** You get the same results whether you put the female or male gametes at the top or the side. **(2)** You only need to list each unique gamete type once at the head of a row or a column (but if you list it twice, you will still get the same result). Consider this cross: *Rr* female × *RR* male.

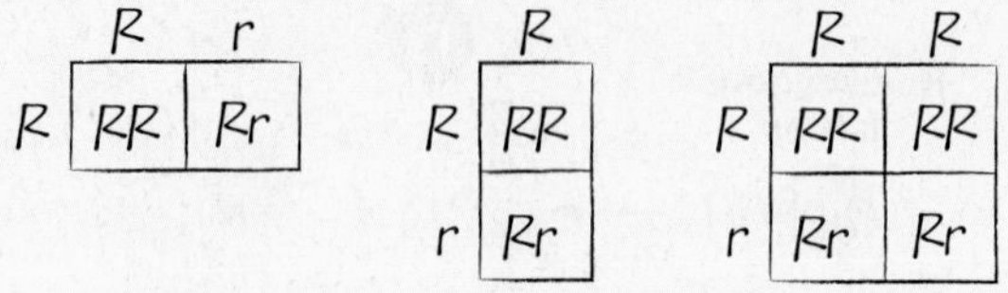

These Punnett squares produce the same results

MODEL Label the male and female gametes in these figures and then determine the frequencies of each offspring genotype to show that the results are the same.

To see this model in action, go to the Study Area of **Mastering Biology**

The outcome of a 3 : 1 ratio is shown graphically in the Punnett square in Figure 14.4b, but it can also be demonstrated mathematically. As described in **BioSkills 4**, the chance that two independent events occur together—such as a sperm from an *Rr* male carrying an *r* allele (event 1) fertilizing an egg with an *r* allele that is produced by an *Rr* female (event 2)—is given by the product of the probabilities (chances) of each event. In this case, it is ½ × ½, or ¼. This is the same result obtained using a Punnett square and agrees with the observed frequency of *rr* wrinkle-seed offspring produced from the *Rr* × *Rr* cross.

Predicting Offspring Genotypes and Phenotypes with a Punnett Square The box you've just studied in Figure 14.4b is a simple Punnett square. To produce a Punnett square:

1. Write each of the *unique* gamete genotypes produced by one parent in a horizontal row along the top of the diagram.
2. Write each of the *unique* gamete genotypes produced by the other parent in a vertical column down the left side of the diagram.
3. Create a table under the horizontal row of gametes and to the right of the vertical column of gametes.
4. In each box of the table, enter the parental gamete genotypes that are written at the top of the box's column and at the left side of the box's row. This step represents fertilization and produces the offspring genotypes.
5. Finally, calculate the proportions or ratios of each offspring genotype and phenotype.

✔ **If you understand these concepts, you should be able to state how filling in the top and side of a Punnett square is related to the principle of segregation, and predict the offspring phenotype and genotype ratios for a cross between Rr and rr peas.**

As an example of the final step in analyzing a cross, the Punnett square in Figure 14.4b predicts that ¼ of the F_2 offspring will be *RR*, ½ will be *Rr*, and ¼ will be *rr*. Because the *R* allele is dominant to the *r* allele, ¾ of the offspring should be round-seed (the sum of the *RR* and the *Rr* offspring) and ¼ should be wrinkle-seed (the *rr* offspring). These predictions are based on three key principles: **(1)** segregation during meiosis (see Section 14.4), **(2)** a dominant and a recessive allele, and **(3)** random fertilization. Mendel obtained these predicted results, and his interpretation explains the 3 : 1 ratio of round to wrinkled seeds in the F_2 offspring and the reappearance of the wrinkled seeds.

The term "genetic model" refers to a set of hypotheses that explains how a particular trait is inherited. **Table 14.3** summarizes Mendel's model for explaining the patterns in the transmission of traits from parents to offspring; these hypotheses are sometimes referred to as Mendel's rules.

SUMMARY **Table 14.3 Mendel's Model to Explain the Results of a Monohybrid Cross***

Mendel's Claims	Comments
1. Peas have two copies of each gene and thus may have two different alleles of the gene.	This is true for many other organisms.
2. Genes are particles of inheritance that do not blend together.	Genes do not change when being transmitted between generations.
3. Each gamete contains one copy of each gene (one allele).	This is due to the principle of segregation—the members of each gene pair segregate during the formation of gametes.
4. Males and females contribute equally to the genotype of their offspring.	When gametes fuse, offspring acquire two of each gene—one from each parent.
5. Some alleles are dominant to other alleles.	When a dominant and a recessive allele for a gene are paired in a heterozygote, that individual has the dominant phenotype.

* Mendel did not use these terms, but he expressed these ideas in different words.

CHECK YOUR UNDERSTANDING

✔ **If you understood this section, you should be able to . . .**

1. Explain how the disappearance of one of the two parental traits in the F_1 generation and the reappearance of that trait in the F_2 generation led Mendel to conclude that there were two copies of a gene for each trait.
2. Explain how the 3 : 1 ratio of dominant to recessive offspring in the F_2 generation led Mendel to conclude that the F_1 hybrids produced gametes with a 1 : 1 ratio of dominant to recessive alleles.

Answers are available in Appendix A.

14.3 Mendel's Experiments with Two Traits

Working with one trait at a time allowed Mendel to infer that each pea plant had two copies of each gene, to recognize the principle of segregation, and to learn that one allele could be dominant when paired with a different allele.

After you complete this section, you should be able to ...

- Analyze the results of dihybrid crosses using the principle of independent assortment.

Mendel's next question was, Do alleles of different genes segregate together, or do they segregate independently?

The Dihybrid Cross

Mendel crossed a pure-line parent that produced round yellow seeds with a pure-line parent that produced wrinkled green seeds. According to his model, the F_1 offspring of this cross should be heterozygous for both genes. These F_1 individuals are called dihybrids, and a mating between dihybrids is a **dihybrid cross**.

Mendel's earlier experiments had established that the allele for yellow seeds was dominant to the allele for green seeds; these alleles were designated *Y* for yellow and *y* for green. As **Figure 14.5** on page 304 indicates, there are two possibilities for how the alleles of these two different genes—the gene for seed shape and the gene for seed color—would be transmitted to offspring.

1. The first possibility was that the alleles for seed shape and the alleles for seed color originally present in each parent would be transmitted independently. This hypothesis is called independent assortment because the two alleles of each gene would be sorted into gametes independently of each other (see Figure 14.5a).
2. The second possibility was that the allele for seed shape and the allele for seed color originally present in each parent would be transmitted to gametes together. This hypothesis can be called dependent assortment because the transmission of one particular allele would be tied to or depend on the transmission of another (see Figure 14.5b).

As Figure 14.5 shows, the F_1 offspring of Mendel's mating are expected to have the dominant round and yellow phenotypes whether the different genes are transmitted together or independently. This is exactly what Mendel found.

The two hypotheses make radically different predictions, however, about what will be observed when the F_1 individuals self-fertilize and produce an F_2 generation. If the alleles of different genes assort independently to form gametes, then each heterozygous parent should produce four different gamete genotypes present in equal amounts, as shown in Figure 14.5a. This Punnett square predicts that there should be 9 different offspring genotypes and 4 phenotypes. Further, the yellow-round, green-round, yellow-wrinkled, and green-wrinkled phenotypes should be present in frequencies of $\frac{9}{16}$, $\frac{3}{16}$, $\frac{3}{16}$, and $\frac{1}{16}$, respectively. This is a ratio of 9:3:3:1.

But if the alleles from each parent stay together, then the prediction is for only three offspring genotypes and a 3:1 ratio of two phenotypes—yellow-round or green-wrinkled—in the F_2, as Figure 14.5b shows.

When Mendel examined the phenotypes of the F_2 offspring, he found that they conformed to the predictions of the independent assortment hypothesis. Four phenotypes were present in proportions that closely approximated the predicted ratio of 9:3:3:1 (see Figure 14.5c). Based on these data, Mendel accepted the hypothesis that alleles of different genes are transmitted independently of one another. This hypothesis became known as the **principle of independent assortment**.

Seeing what kinds of gametes can be produced by the combination of segregation and independent assortment can get tricky as the number of traits under consideration increases. A helpful tip is to remember that each gamete has to receive one allele for each trait. This idea is illustrated in **Making Models 14.2**. An additional approach to track proportions of different gametes is a method called a forked-line diagram (**Making Models 14.3**).

Making Models 14.2 Tips on Drawing Punnett Squares (II)

When you set up Punnett squares for crosses involving two or more traits, make sure that each gamete receives one allele for *each* trait. For example, consider the cross *RRYy* × *rrYy*.

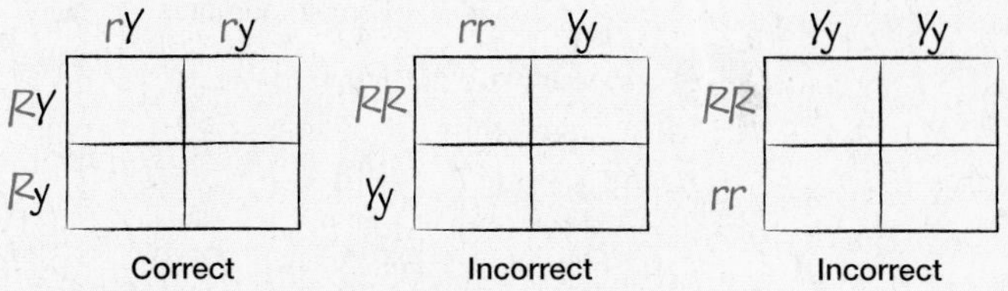

MODEL Explain why the square on the right is incorrect; then draw a Punnett square showing a cross between two *PPRr* individuals.

To see this model in action, go to the Study Area of **Mastering Biology**

Making Models 14.3 Tips on Drawing Forked-Line Diagrams

You can use a forked-line diagram to determine what types of gametes an individual can produce. This method helps you focus on the alleles of one gene at a time. By multiplying the fractions (or frequencies) of each allele, you can calculate the proportion of gametes that are predicted to have each combination of alleles (see **BioSkills 4** to learn why probabilities are multiplied).

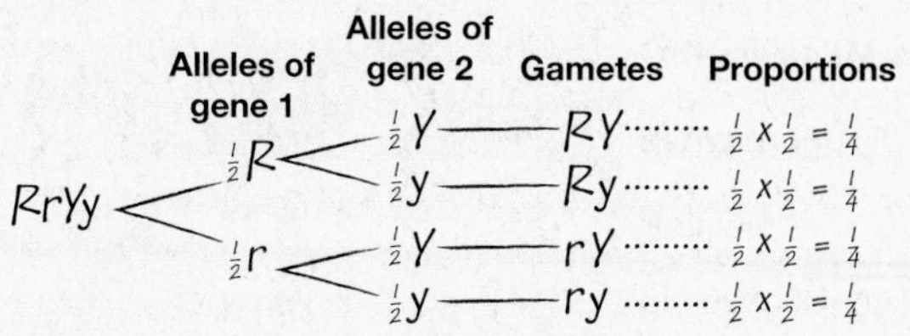

MODEL Use the forked-line method to predict the types and proportions of gametes produced by an *RrPPTt* pea plant.

To see this model in action, go to the Study Area of **Mastering Biology**

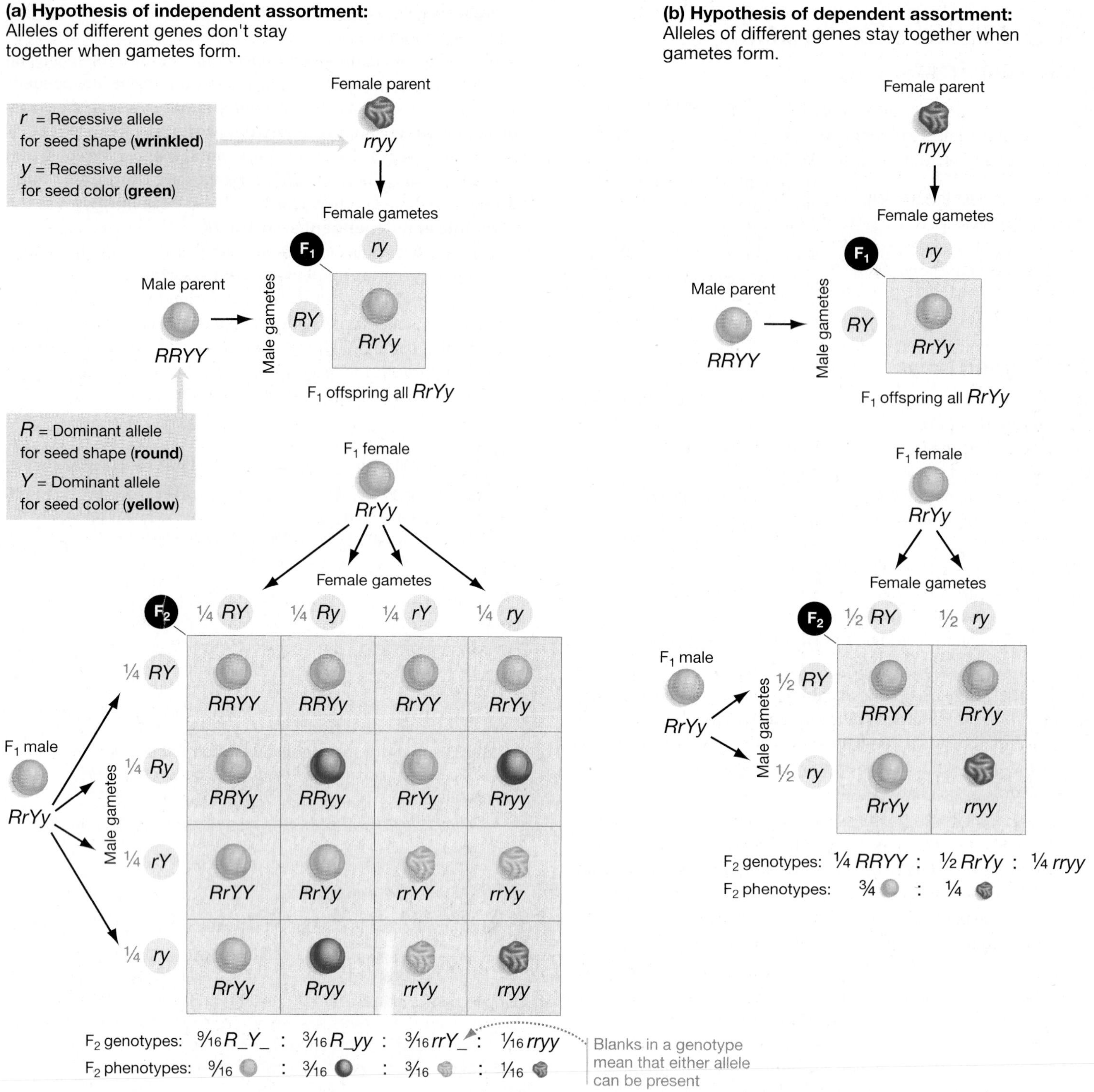

(c) Mendel's results

F_2 phenotypes	(round, yellow)	(round, green)	(wrinkled, yellow)	(wrinkled, green)
Number	315	108	101	32
Fraction of offspring	9/16	3/16	3/16	1/16

556 total

Data are consistent with the predictions of independent assortment.

Figure 14.5 Mendel Tested Two Hypotheses by Analyzing the Offspring of a Dihybrid Cross. Pure-line parents differing in two traits were crossed to produce a dihybrid F_1 generation. These F_1 then were allowed to self-fertilize to produce an F_2. Parts **(a)** and **(b)** depict the two hypotheses: **(a)** The alleles are sorted into gametes independently of each other, and **(b)** particular alleles are always transmitted together. Mendel's results **(c)** supported independent assortment.

Using a Testcross to Confirm Predictions

Mendel did experiments with combinations of traits other than seed shape and color and obtained results similar to those in Figure 14.5c. Each dihybrid cross produced a 9:3:3:1 ratio of progeny phenotypes in the F_2 generation. He even did a limited set of crosses examining three traits at a time. Although all these data were consistent with the principle of independent assortment, Mendel looked for an alternative test of his hypothesis to provide more evidence that it was correct.

Mendel's goal was to test the prediction that an *RrYy* plant produces four different types of gametes in equal proportions. To accomplish this, he used a technique that is now called a testcross. In a traditional **testcross**, a parent with a dominant phenotype but unknown genotype is crossed with a parent that contributes only recessive alleles. By analyzing the phenotypes of the offspring, the unknown parental genotype can be determined.

The genius of the testcross is that the genetic contribution of the homozygous recessive parent is known, which allows researchers to infer whether the other parent is homozygous or heterozygous for the dominant allele. The reasoning is shown in Figure 14.6 for a cross of parents with the dominant traits of yellow and round seeds, but two different genotypes, either *RrYy* or *RRYY*.

The types and proportions of offspring that could result from a testcross of these parents with a plant grown from a wrinkled and green seed (*rryy*) can be predicted with a Punnett square (see Figure 14.6). If the principle of independent assortment is valid, the testcross should produce four types of offspring in equal proportions if the tested parent is *RrYy*, and only one type of offspring if the tested parent is *RRYY*. When used in a testcross, plants with other genotypes that produce round and yellow seeds will produce different results that can be used to deduce the genotype of the parent with the dominant phenotypes.

What did Mendel find when he testcrossed an *RrYy* pea plant? Among the 110 seeds produced, 31 were round and yellow, 26 were round and green, 27 were wrinkled and yellow, and 26 were wrinkled and green. Each type was almost exactly ¼ of the total, which matched the proportions predicted for offspring of an *RrYy* parent with independently assorting alleles. The testcross had confirmed the principle of independent assortment.

Notice that Mendel analyzed large numbers of offspring to support his hypotheses of segregation and independent assortment. His results came very close to matching the predicted ratios of offspring. What if he had studied smaller numbers of offspring? In that case, the chance of coming close to the predicted ratios would have been lower. Because of the random chance of which sperm fertilizes which egg, analyzing a small number of offspring is more likely to produce ratios that deviate from the expected outcomes. These ideas of randomness and the need to analyze sufficient numbers of outcomes—in this case, offspring—are cornerstones of experimental design and interpretation. Even without formal statistics, Mendel had an intuitive feel for the importance of these principles.

Mendel's work provided a powerful conceptual framework for thinking about transmission genetics—the patterns that occur as alleles pass from one generation to the next. This framework was based on **(1)** the segregation of alleles for the same trait into separate gametes, and **(2)** the independent assortment of alleles of different genes. The experiments were brilliant in design, execution, and interpretation, but they were ignored for 34 years.

CHECK YOUR UNDERSTANDING

✔ If you understood this section, you should be able to . . .

1. Explain why Mendel needed to study two different genes rather than two different alleles of one gene in order to derive the principle of independent assortment.
2. Explain why using an *RrYy* pea plant in a testcross produces a 1:1:1:1 ratio of offspring phenotypes, but self-fertilization produces a 9:3:3:1 ratio of offspring phenotypes.

Answers are available in Appendix A.

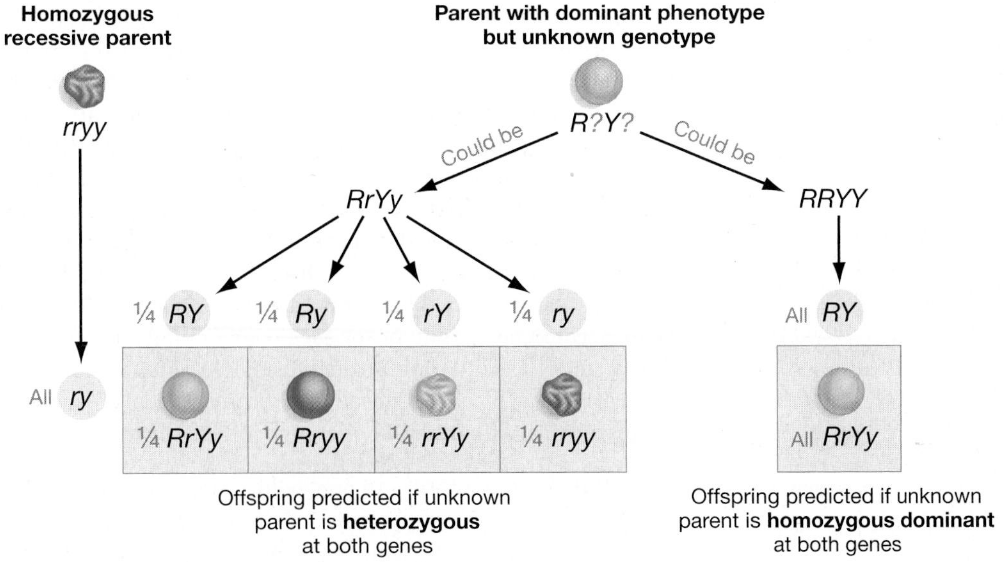

Figure 14.6 Predictions Made by the Principle of Independent Assortment Can Be Evaluated in a Testcross. If the principle of independent assortment is correct, then crossing *RrYy* parents with *rryy* parents will produce four types of gametes in equal proportions, as shown on the left. Testcrosses can also reveal the genotype of any parent with dominant phenotypes, as seen by comparing the different results obtained from parents with *RrYy* versus *RRYY* genotypes.

14.4 The Chromosome Theory of Inheritance

Historians debate why Mendel's work was overlooked for so long. Maybe his reliance on ratios and proportions made his work difficult for biologists of that time to understand or accept. Or maybe the hypotheses of blending inheritance and the inheritance of acquired characteristics were so entrenched that his results were dismissed.

After you complete this section, you should be able to . . .

- Analyze how the transmission of chromosomes to daughter cells during meiosis explains the principles of segregation and independent assortment.

Whatever the reason, Mendel's work was not appreciated until 1900. In that year, three biologists independently discovered Mendel's findings and, through their own work with a variety of plants and animals, reached the same conclusions.

The discovery of Mendel's work more than three decades after its publication ignited the field of genetics. His experiments established the basic patterns of inheritance, but what process is responsible for these patterns? Two other biologists, also working independently, came up with the answer. Walter Sutton and Theodor Boveri each realized that meiosis could account for Mendel's rules. (For a review of meiosis, see Chapter 13.) When this hypothesis was published in 1902, research in genetics exploded.

Meiosis Explains Mendel's Principles

What Sutton and Boveri grasped is that meiosis explains the principle of segregation and the principle of independent assortment. To appreciate how, look closely at **Figure 14.7**. The cell at the top illustrates Sutton and Boveri's central insight: Mendel's hereditary determinants, or genes, are located on chromosomes. In this example, alleles of the gene for seed shape are shown at a particular position—known as a **locus** ("place"; plural, **loci**)—along a certain chromosome.

The paternal and maternal chromosomes of the parent cell shown in Figure 14.7 happen to possess different alleles at the seed-shape gene locus: One allele specifies round seeds (*R*) and the other specifies wrinkled seeds (*r*). Figure 14.7 shows how these alleles segregate into different daughter cells during meiosis I, when homologous chromosomes separate. The physical separation of homologous chromosomes during anaphase of meiosis I is responsible for Mendel's principle of segregation.

Figure 14.8 follows the segregation of two different genes—in this case, for seed shape and seed color—as meiosis proceeds. If the alleles for different genes are located on different chromosomes, they will assort independently of one another at meiosis I. The figure illustrates a key point: there are two equally likely ways for the homologous pairs to line up. This is the physical basis of Mendel's principle of independent assortment. Over many meiotic divisions, four types of gametes will be produced in equal proportions.

Recall that Sutton and Boveri formalized these observations in the chromosome theory of inheritance: Mendel's rules can be explained by postulating that genes are located on chromosomes and that chromosomes line up independently before being separated at meiosis I.

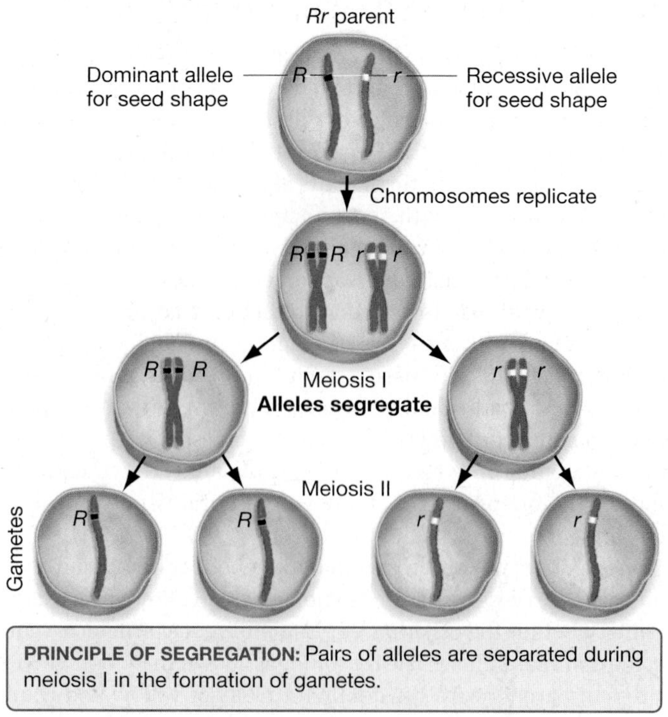

Figure 14.7 Meiosis Explains the Principle of Segregation. Alleles of genes come in pairs because they are located on chromosomes that come in pairs. Alleles segregate into different gametes because the homologous chromosomes that carry them separate during meiosis I.

When Sutton and Boveri published their theory, however, the idea that genes are located on chromosomes was simply based on the observed association between movement of chromosomes at meiosis and Mendel's principles. What experiments confirmed that chromosomes carry genes?

Testing the Chromosome Theory of Inheritance

During the first decade of the twentieth century, an unassuming insect rose to prominence as a model organism for genetics. This organism—the fruit fly *Drosophila melanogaster*—has been at the center of genetic studies ever since (see **BioSkills 11**).

Drosophila melanogaster has all the attributes of a useful model organism for studies in genetics: small size, ease of rearing in the lab, a short generation time (about 10 days), and abundant offspring (up to a few hundred per mating). The elaborate external anatomy of this insect also makes it possible to identify the phenotypic variations that are essential for genetics research.

Drosophila was adopted as a model organism by Thomas Hunt Morgan and his students. But because *Drosophila* is not a domesticated species, common variants such as Mendel's round and wrinkled seeds were not available. Morgan had access only to flies with the most common phenotype for each trait. Common phenotypes are referred to as **wild type**. To do informative genetic crosses, genetic variants are needed. Consequently, an early goal of Morgan's was to find flies with different phenotypes to use in crosses.

The White-Eyed Mutant At one point, Morgan discovered a male fly that had white eyes rather than the wild-type red eyes. Morgan inferred that the white-eyed phenotype resulted from a

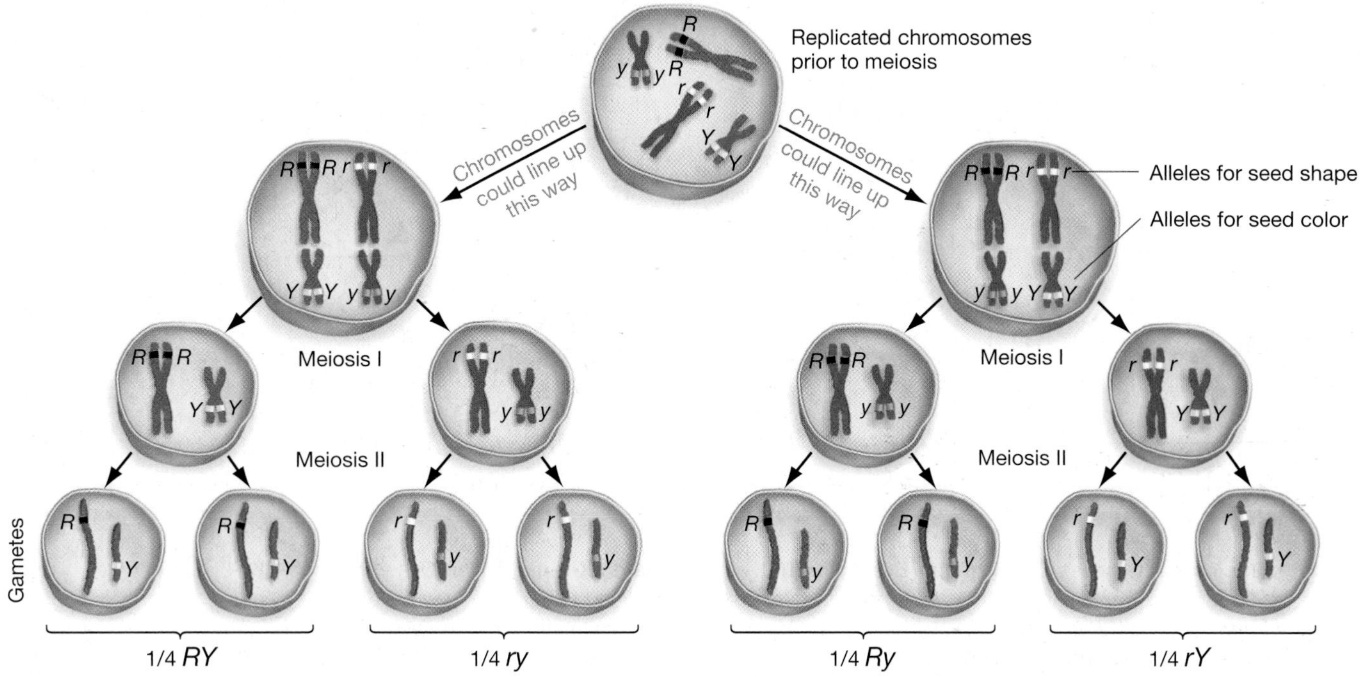

INTERACTIVE Figure 14.8 **Meiosis Is Responsible for the Principle of Independent Assortment.**

mutation—a heritable change in a gene. An individual with an unusual phenotype due to a mutation is referred to as a **mutant**.

With his first mutant in hand, Morgan set out to explore how the eye-color trait was inherited. He mated a red-eyed female fly with the mutant white-eyed male fly. All the F_1 progeny had red eyes. By continued crosses, Morgan obtained white-eyed female flies. However, when he tried a reciprocal cross between a white-eyed female and a red-eyed male, he found something puzzling: All the F_1 females had red eyes, but all F_1 males had white eyes.

Recall that Mendel's reciprocal crosses had always given results that were essentially the same. But Morgan's reciprocal crosses suggested a relationship between the sex of the parent and the inheritance of eye color. What was going on?

The Discovery of Sex Chromosomes Nettie Stevens began studying the chromosomes of insects about the time that Morgan began working with *Drosophila*. First in the beetle *Tenebrio molitor*, and later in other insects, including *Drosophila*, she noticed a striking difference in the chromosome complements of males and females.

Recall that Stevens and others discovered that there were sex chromosomes (the X and the Y) and autosomes (Ch. 13, Section 13.1). Female flies have a pair of X chromosomes, and male flies have an X and a Y chromosome. Morgan's knowledge of Stevens' findings was the key to explaining his puzzling results and to providing support for the chromosome theory of inheritance.

Sex-Linkage and the Chromosome Theory of Inheritance Morgan realized that the transmission pattern of the X chromosome in males and females could explain the results of his reciprocal crosses. He reasoned that half the gametes produced by males would contain an X chromosome and half a Y chromosome (**Figure 14.9**). Morgan proposed that the gene for eye color in fruit flies is located on the X chromosome and that the Y chromosome does not carry this gene.

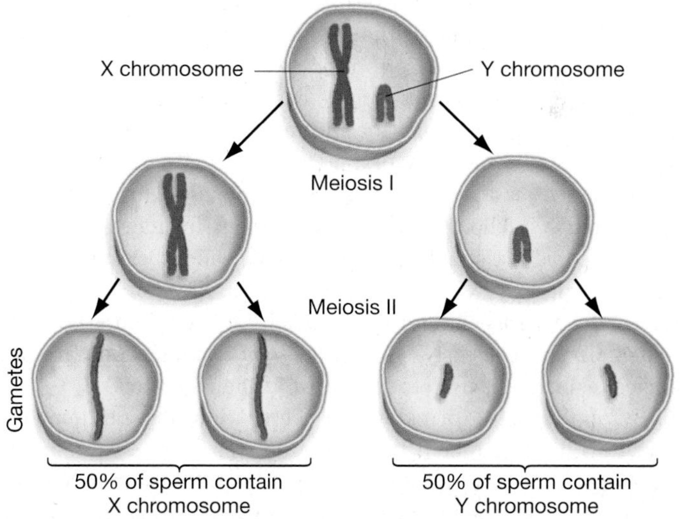

Figure 14.9 **Sex Chromosomes Segregate in Males to Form X-Bearing and Y-Bearing Gametes.** Sex chromosomes pair at meiosis I in male fruit flies because of a small, gene-free region shared by the X and Y chromosomes. This allows normal segregation, so half the sperm cells bear an X chromosome and half have a Y chromosome. Only the X and Y chromosomes are shown, but other chromosomes are also present in a *Drosophila* cell.

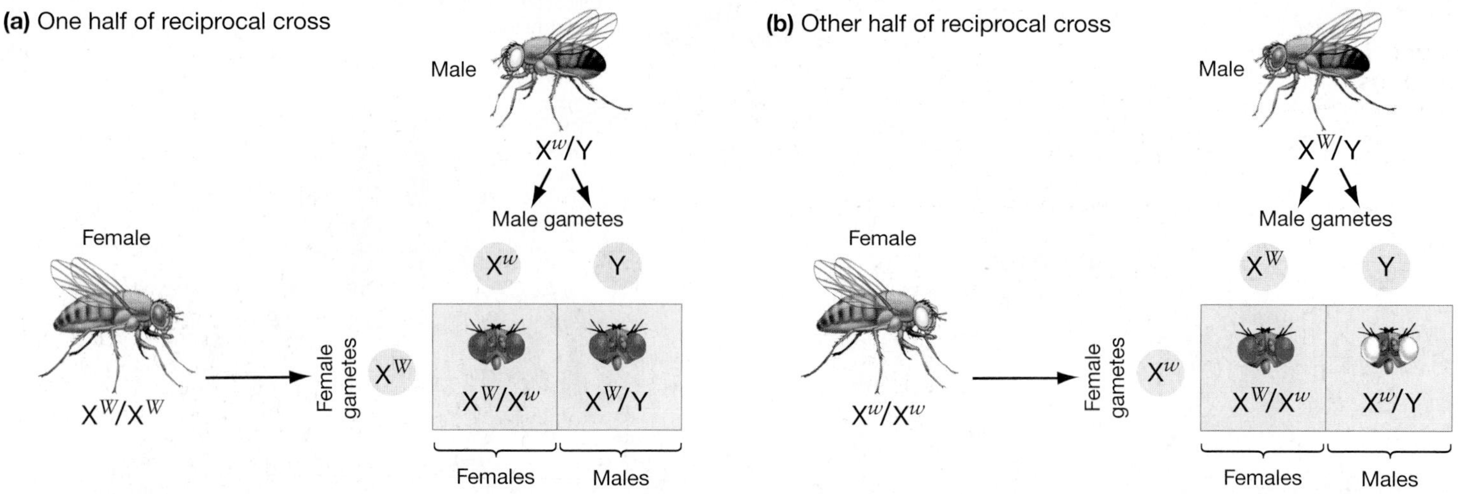

Figure 14.10 Reciprocal Crosses Confirm that Eye Color in *Drosophila* Is an X-Linked Trait. When Morgan crossed red-eyed females with white-eyed males **(a)** and then crossed white-eyed females with red-eyed males **(b)**, he observed strikingly different results. These results can be explained by postulating that the eye-color gene resides on the X chromosome—it is X-linked.

Morgan went on to postulate that a female fruit fly has two copies of the gene that specifies eye color because she has two X chromosomes. One of these chromosomes came from her mother, the other from her father. A male, in contrast, has only one copy of the eye-color gene because he has only one X chromosome, inherited from his mother.

The Punnett squares in **Figure 14.10** show that Morgan's unusual results can be explained if the gene for eye color is located on the X chromosome, and if the allele for red color is dominant to the allele for white color. A gene on the X chromosome is now described as an **X-linked gene**; a gene on the Y chromosome is a **Y-linked gene**, and the general term for genes being located on either sex chromosome is **sex-linked genes**.

Notice in Figure 14.10 that a different system of symbols is used for *Drosophila* alleles to represent the mutant phenotype. Instead of showing alleles for the eye-color gene with an *R* for wild-type red eyes and an *r* for mutant white eyes, the alleles are shown with a *W* (red eyes) or a *w* (white eyes). In this figure, the allele for red eyes is denoted X^W while the allele for white eyes is denoted X^w. The Y chromosome present in males is simply designated as Y. Using this notation,[1] the genotypes are written as X^W/X^W and X^W/X^w for red-eyed females; X^w/Y for white-eyed males; X^w/X^w for white-eyed females; and X^W/Y for red-eyed males. To more clearly show alleles on homologous chromosomes, a slash (/) is used to separate the allele symbols.

By applying the principles of segregation and random fertilization to an X-linked trait, Morgan was able to explain his puzzling results. At the same time, his proposal that the X chromosome carried the eye-color gene provided strong support for the chromosome theory of inheritance.

[1]Scientific papers on fruit fly genetics use a different notation. The wild-type allele is designated with a superscript +, and no X is used for X-linked traits. The red-eyed allele, for example, is denoted w^+; the white-eyed allele is w. The notation used here is simplified and conforms to conventions used in human genetics.

When reciprocal crosses give different results, such as those illustrated in Figure 14.10, the gene in question is likely to be sex-linked. The patterns of inheritance of these genes are said to show **sex-linked inheritance**. If the gene is on the X chromosome, it is **X-linked inheritance**, if on the Y chromosome, it's **Y-linked inheritance**. Non-sex chromosomes are called autosomes (Ch. 13, Section 13.1). Genes on non-sex chromosomes are said to be autosomal, and their patterns of inheritance—which are the patterns seen by Mendel—are called **autosomal inheritance**.

The discovery of X-linked inheritance convinced most biologists that the chromosome theory of inheritance was correct.

CHECK YOUR UNDERSTANDING

✔ **If you understood this section, you should be able to . . .**

1. Explain why the simplest drawing to model the principle of independent assortment requires four chromosomes, but the simplest drawing to model the principle of segregation requires only two chromosomes.
2. **MODEL** Create a drawing that shows all possibilities for the independent assortment of three different genes, *A*, *B*, and *C*, each with two different alleles, on three different homologous pairs of chromosomes.

Answers are available in Appendix A.

14.5 Extending Mendel's Rules

After you complete this section, you should be able to . . .

- Explain the extensions of Mendel's rules that account for common patterns of inheritance.

Mendel analyzed the simplest possible genetic system. The traits he studied were not sex-linked, they were influenced by just two alleles of each gene, and each allele was fully dominant or recessive.

With this model system, Mendel was able to discover the fundamental rules of inheritance. Mendel probably would have failed, as so many others did before him, had he tried to analyze more complex patterns of inheritance.

Once Mendel's work was discovered, researchers began to analyze traits and alleles whose inheritance was more complicated. If experimental crosses produced F_2 progeny that did not conform to the expected 3:1 or 9:3:3:1 ratios, researchers had a hint that something interesting was going on. The discovery of sex-linkage is a prominent example. How can other traits that don't appear to follow Mendel's rules contribute to a more complete understanding of heredity?

Linkage: What Happens When Genes Are Located on the Same Chromosome?

Once the chromosome theory of inheritance gained support, biologists began to reevaluate Mendel's principle of independent assortment. It seemed unlikely that genes on the same chromosome would assort independently.

Linkage is the tendency of alleles of particular genes to be inherited together. Linkage is seen when genes are on the same chromosome. Notice that the overlapping terms "linkage" and "sex-linkage" have different meanings. Linkage refers to two or more genes that are located on the same chromosome. Sex-linkage can refer to a single gene, and it means that this gene is located on a sex chromosome. Sex-linkage says nothing about the location of one gene relative to another.

The first examples of linked genes happened to be those on the X chromosome of fruit flies. After Morgan established that the *white eye* gene was located on *Drosophila*'s X chromosome, he and colleagues established that a gene that affects body color is also located on the X chromosome. Red eyes and gray body are the wild-type phenotypes; white eyes and yellow body occur as mutant phenotypes. The alleles for red eyes (X^W) and gray body (X^Y) are dominant to the alleles for white eyes (X^w) and yellow body (X^y). Be sure not to confuse the notation for the Y chromosome in males, Y, with the gray-body allele, X^Y.

Does Linked Mean Inseparable?

Because linked genes are located on the same chromosome, it would be logical to predict that they should always be transmitted together in meiosis. The problem with this prediction is that it's wrong.

Figure 14.11 shows a cell of a female fruit fly with one X chromosome carrying the *white eye* and *gray body* alleles, written X^{wY}, and the homologous X chromosome carrying the *red eye* and *yellow body* alleles, written X^{Wy}. In this case, the genotype is written X^{wY}/X^{Wy}.

When the chromosomes of an X^{wY}/X^{Wy} female segregate during meiosis I, this female might be expected to generate haploid cells with just two genotypes in equal proportions, not the four genotypes that are predicted by the principle of independent assortment. What actually occurs?

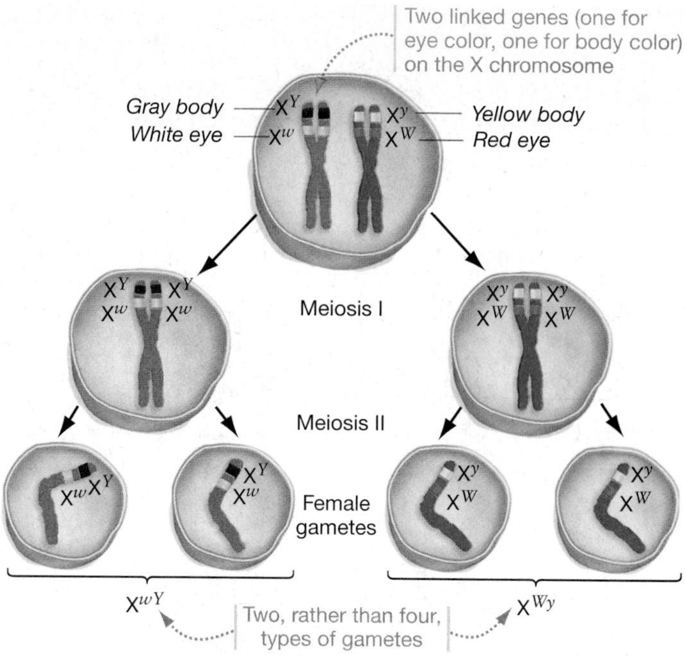

Figure 14.11 Linked Genes Are Often Inherited Together.

✔ List the genotypes and their proportions that would be generated if the *white eye* and *yellow body* genes were not linked but were on different chromosomes.

The Role of Crossing Over

To determine whether alleles on the same chromosome always stay linked, Morgan performed crosses like the one depicted in the "Experimental Setup" section of **Figure 14.12** on page 310. In this case, X^{wY}/X^{Wy} females were crossed with X^{wY}/Y males.

The "Results" table in Figure 14.12 summarizes the phenotypes and genotypes of male offspring produced in this cross. Examining only males was a clever experimental approach that allowed Morgan to learn which alleles were present on the different X chromosomes produced during meiosis in the mother. This is because there is only one X chromosome in each male, so the phenotype associated with any X-linked allele is expressed:

- Most of these males carried an X chromosome with one of the combinations of alleles found in the chromosomes of their mothers: X^{wY} or X^{Wy}.
- A small percentage of males carried an X chromosome with new combinations of alleles: X^{wy} and X^{WY}. Morgan referred to these individuals as **recombinant** because the alleles on their X chromosome were different (recombined) from the combinations present in their mother.

Morgan concluded that alleles on the same chromosome often stay together, but not always. To explain the recombinant chromosomes, Morgan proposed that gametes with new combinations of alleles were generated when crossing over occurred during prophase of meiosis I in the females.

Recall that crossing over involves an exchange of corresponding segments of non-sister chromatids between homologous chromosomes (Ch. 13, Section 13.1). Crossing over typically occurs at least once in every synapsed pair of homologous chromosomes.

RESEARCH

QUESTION: Will genes undergo independent assortment if they are on the same chromosome?

LINKAGE HYPOTHESIS: Linked genes will always stay linked, violating the principle of independent assortment.

NULL HYPOTHESIS: Linked genes will adhere to the principle of independent assortment.

EXPERIMENTAL SETUP:

Red-eyed gray-bodied female X^{wY}/X^{Wy}

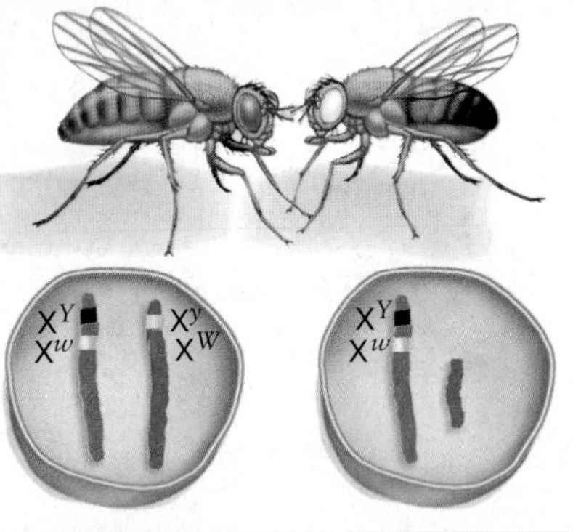

White-eyed gray-bodied male X^{wY}/Y

PREDICTION OF LINKAGE HYPOTHESIS: Because these two genes are X-linked, male offspring will have only one copy of each gene, inherited from their mother; the two possible male offspring genotypes are X^{wY}/Y and X^{Wy}/Y.

PREDICTION OF NULL HYPOTHESIS: Four male genotypes will be produced ($X^{wY}/Y : X^{Wy}/Y : X^{wy}/Y : X^{WY}/Y$), each at equal frequency.

RESULTS:

Male offspring

	Phenotype	Genotype	Number
		X^{wY}/Y	4292
		X^{Wy}/Y	4605
Recombinant genotypes		X^{wy}/Y	86
		X^{WY}/Y	44

Four male genotypes were observed, not two, but not at the equal frequencies predicted by independent assortment

CONCLUSION: ______________________________

Figure 14.12 Do Linked Genes Assort Independently?

SOURCE: T. H. Morgan. 1911. An attempt to analyze the constitution of the chromosomes on the basis of sex-limited inheritance in *Drosophila*. *Journal of Experimental Zoology* 11: 365–414.

✔ **PROCESS OF SCIENCE** Without looking at the text, fill in the conclusion of this experiment.

(Male fruit flies are an exception to this rule. For unknown reasons, crossing over does not occur in male fruit flies.)

Figure 14.13 shows how a crossover between the eye-color and body-color genes in the X^{wY}/X^{Wy} females can explain the recombinant gametes. Male progeny produced from fertilization with these gametes are predicted to have either yellow bodies and white eyes or gray bodies and red eyes, just as Morgan observed.

Notice that the predictions of Figure 14.13 don't fit either the model of independent assortment or complete linkage. Independent assortment predicts a 1:1:1:1 ratio of all four combinations of phenotypes, while complete linkage would give only the two phenotypes associated with the nonrecombinant, or parental, chromosomes. Instead, most flies have parental phenotypes and a smaller number have recombinant phenotypes.

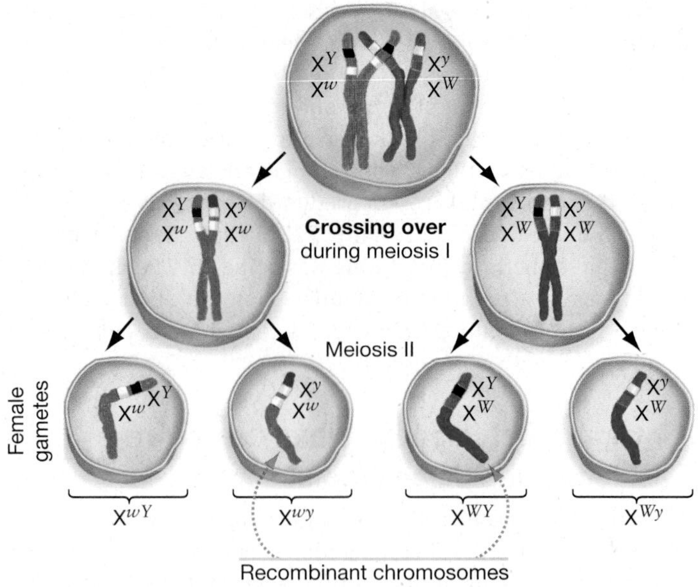

Figure 14.13 Genetic Recombination Results from Crossing Over. To explain the results in Figure 14.12, Morgan proposed that crossing over occurred between the body-color (*y*) and eye-color (*w*) genes in a small percentage of meiotic divisions in the female parent. The recombinant chromosomes would produce the recombinant phenotypes observed in the male offspring.

✔ **PROCESS OF SCIENCE** Given Morgan's predictions, why did he observe *unequal* numbers of white-eyed, yellow-bodied males and red-eyed, gray-bodied males, as shown in Figure 14.12?

As **Quantitative Methods 14.1** explains, the percentage of recombinant offspring that occur in crosses like the one diagrammed in Figure 14.13 can be used to estimate the relative distance between genes. The reasoning is that the farther apart genes on the same chromosome are, the more likely a crossover is to occur someplace between these genes. Data on the frequency of crossing over between different genes on the same chromosome can be used to create a **genetic map**—a diagram showing the relative positions (loci) of genes along a particular chromosome.

Knowing the locus of a gene can be important. For example, genetic mapping approaches have led to the discovery of many human genes involved with disease. First, researchers map the position of a disease-associated gene on a chromosome. Once they know the location, they can investigate the DNA sequence of the gene using molecular biology tools and techniques (Chapter 20 and **BioSkills 10**). Study of the gene often gives insights into what goes wrong in the disease and leads to potential treatments and cures.

Morgan's investigations showed that genes on the same chromosome are linked and are inherited together unless crossing over occurs between them. When crossing over occurs, the result is genetic recombination. Linkage is an important exception to Mendel's principle of independent assortment.

Quantitative Methods 14.1 Linkage and Genetic Mapping

In experiments like the one shown in Figure 14.12, researchers can establish the relative distance between genes by calculating how often genes recombine due to crossover at meiosis. This is the *recombination frequency*, and it is found by dividing the number of offspring with recombinant chromosomes by the total number of offspring. For crosses involving the X-linked traits of white eyes and yellow bodies, the recombination frequency is 1.4 percent: 130 of the male offspring were recombinant among 9027 total male offspring [(130/9027) × 100 = 1.4%]. (Recall that the phenotype of male offspring reflects the type of X chromosome each male received from its mother.)

For autosomal genes, an experimental approach to determining the number of recombinant chromosomes is to do a testcross between a parent heterozygous for all genes being studied and a parent homozygous for the recessive alleles of these genes. From the homozygous recessive parent, every offspring gets a recessive allele for each gene. From the heterozygous parent, offspring get either recessive or dominant alleles for each gene. Whether this allele is dominant or recessive can be determined by the offspring's phenotype. This is because each offspring's phenotype is determined by the alleles it receives from the heterozygous parent.

In crosses with different pairs of linked genes, the fraction of recombinant offspring differs. For example, in crosses of fruit flies involving the *white eye* gene and another X-linked gene (*singed bristles*) for the shape of bristles, recombinant chromosomes are seen 19.6% of the time.

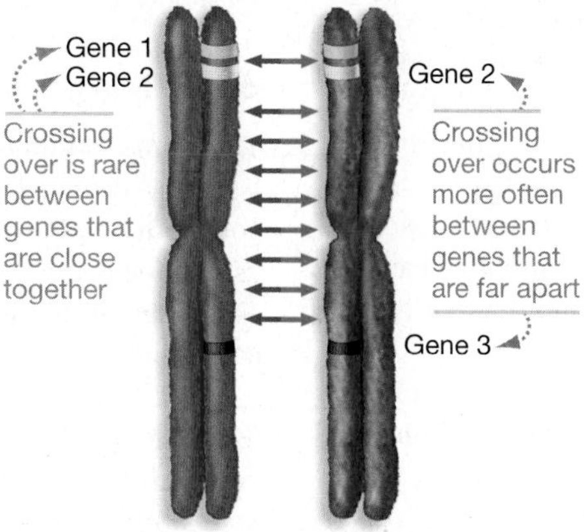

Figure 14.14 The Distance between Genes Determines the Frequency of Crossing Over. The arrows show that crossing over is possible at any point between the genes. Given that crossing over occurs a small number of times between a homologous pair of chromosomes, the chance of a crossover between a pair of genes increases when the distance between the genes is large.

Alfred Sturtevant, an undergraduate student working with Morgan, proposed that the physical distance between genes determines how frequently crossing over occurs between them. His idea was that crossing over occurs randomly and can take place at any location along a chromosome. The shorter the distance between a pair of genes, the lower the probability that crossing over will take place between them (**Figure 14.14**).

Sturtevant realized that this principle could be used to make genetic maps. He defined one unit of his genetic map as the distance between genes that produces 1 percent recombinant chromosomes. Then he termed this distance one map unit, a unit also known as 1 centiMorgan (cM).

The eye-color and the bristle-shape genes are 19.6 cM apart on the X chromosome, because recombination between these genes results in 19.6 percent of males with recombinant chromosomes. The genes for body color and eye color, in contrast, are just 1.4 cM apart, once again because recombination between these genes produces 1.4 percent of males with recombinant chromosomes.

Where is the *yellow body* gene relative to the *singed bristles* gene? In Sturtevant's studies, 21 percent of the offspring had recombinant chromosomes for these traits, meaning that the *yellow body* and *singed bristles* genes are 21.0 cM apart. Sturtevant inferred that the *white eye* gene must be located *between* the *yellow body* and *singed bristles* genes, as shown in **Figure 14.15a**.

Mapping genes relative to one another is like fitting pieces into a puzzle: Placing *white eye* between *yellow body* and *singed bristles* is the only way to make the distances between each pair sum correctly.

Figure 14.15b provides a partial map of the X chromosome in *Drosophila melanogaster*, along with the data used by Sturtevant to establish this first genetic map. Maps have been an important tool in genetics ever since.

(a) Mapping genes

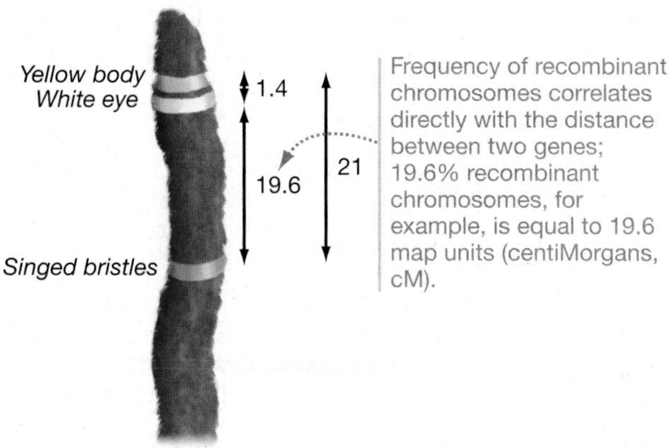

(b) Constructing a genetic map

% Frequency of recombinant chromosomes for some genes on the X chromosome of fruit flies

	Miniature wings	*Ruby eye*
Yellow body	36.1	7.5
White eye	34.7	6.1
Singed bristles	15.1	13.5
Miniature wings	—	28.6

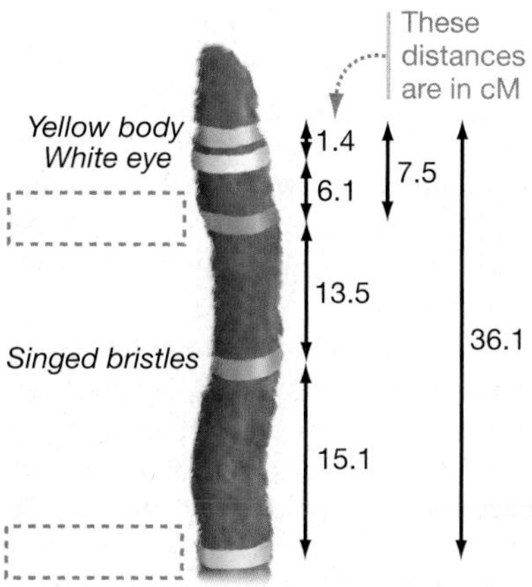

Figure 14.15 Gene Locations Can Be Mapped by Analyzing the Frequency of Recombination. (a) The *yellow body* gene is on the end of the fruit fly X chromosome. To explain the recombination frequencies observed in experimental crosses, the *yellow body, white eye,* and *singed bristles* genes must be spaced as shown here. **(b)** A partial genetic map of the X chromosome in fruit flies.

✔ In part (b), label the orange and blue genes. (Which is *ruby eyes* and which is *miniature wings*?)

How Many Alleles Can a Gene Have?

Mendel worked with genes that had two alleles. In most populations, however, it's not unusual to find many alleles of a single gene. The existence of more than two common alleles of the same gene is **multiple allelism**. Considering that genes are made of DNA sequences that can change over time, the idea of multiple alleles isn't surprising.

The ABO blood group phenotypes in humans are determined by a gene with three common alleles. The gene is symbolized as *I* (shorthand for the historical name of isoagglutinogen) and it has I^A, I^B, and *i* alleles.

Each *I* gene allele controls the production of a different polysaccharide attached to a glycoprotein found in the plasma membrane of red blood cells. (To review glycoproteins, see Ch. 5, Section 5.3.) The I^A and I^B alleles code for different forms of an enzyme that adds a different sugar to the end of a core polysaccharide. The *i* allele codes for a nonfunctional form of this enzyme, so no sugar is added. ABO blood group phenotypes are important in blood transfusions. Some mismatches of blood type between a donor and recipient are tolerated, but others can cause fatal reactions.

The type of polysaccharide associated with each allele is shown in **Figure 14.16** along with all genotypes and phenotypes. Six genotypes are possible for the three alleles of the *I* gene. The I^A allele codes for the type A polysaccharide, and the I^B allele codes for the type B polysaccharide. What happens in a person with both the I^A and I^B alleles?

(a) Three products of the *I* gene alleles

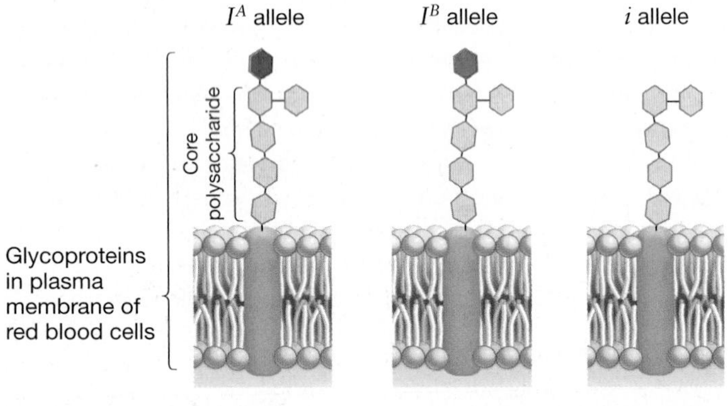

(b) Six genotypes and four phenotypes

Genotype	I^AI^A	I^Ai	I^BI^B	I^Bi	I^AI^B	ii
Blood group phenotype	A	A	B	B	AB	O

Figure 14.16 Multiple Alleles of the *I* Gene Lead to Different Products and ABO Blood Group Phenotypes. **(a)** The types of polysaccharides associated with each *I* gene allele. These polysaccharides form part of a glycoprotein on the surface of red blood cells. The different-colored hexagons on the ends of the first two core polysaccharides represent distinct sugars that are added by different forms of an enzyme encoded by the I^A and I^B alleles. The *i* allele polysaccharide lacks this terminal sugar group. **(b)** The six possible genotypes created by combinations of the three alleles and the associated ABO blood group phenotypes. The colored boxes below each genotype represent the type of polysaccharide that is expressed.

Are Alleles Always Dominant or Recessive?

The terms "dominant" and "recessive" describe which phenotype is observed when two different alleles of a gene occur in a heterozygous individual. In the traits Mendel studied, only the phenotype associated with the dominant allele appeared in heterozygotes. This form of dominance is called **complete dominance**. Not all combinations of alleles work this way.

Codominance Many alleles show a relationship called **codominance**. The type AB phenotype shown in Figure 14.16 is an example of codominance. In this case, an AB heterozygote expresses both the A and the B polysaccharides together on the surface of red blood cells. This is the essence of codominance—the simultaneous expression of the phenotype associated with each of the alleles in a heterozygote. An AB individual exhibits both the A *and* the B phenotypes.

The three alleles of the ABO blood group system illustrate another important point—a single allele can show more than one form of dominance. Notice in Figure 14.16 that the I^A and I^B alleles are both completely dominant to the *i* allele while the I^A and I^B alleles are codominant with each other.

Incomplete Dominance Complete dominance and codominance are not the end of the story. Consider flowers called four-o'clocks, pictured in **Figure 14.17a**. Plant breeders have developed a pure (*RR*) line that has red flowers and a pure (*rr*) line that has white flowers. When individuals from these strains are mated, all their F_1 offspring are pink (*Rr*) (**Figure 14.17b**). In Mendel's peas, crosses between parents with different traits produced offspring that expressed only one of these traits—the dominant one. What's happening in four-o'clocks?

Biologists answered this question by examining the phenotypes of F_2 four-o'clocks that were produced after self-fertilization of F_1 plants. Of the F_2 plants, ¼ have red flowers, ½ have pink flowers, and ¼ have white flowers. This 1 : 2 : 1 ratio of phenotypes exactly matches the 1 : 2 : 1 ratio of genotypes hypothesized in Figure 14.17. It occurs because in this case, heterozygotes have a phenotype that is between the phenotypes of the two different homozygous parents.

This is the hallmark of a form of dominance that is called **incomplete dominance**. In this example, the pink *Rr* heterozygotes have a color between the red of *RR* homozygotes and the white of *rr* homozygotes.

Clearly, alleles are not always dominant or recessive. Instead, there are three possible dominance relationships between different alleles: complete, incomplete, and codominance.

Does Each Gene Affect Just One Trait?

Mendel's results led him to hypothesize that one gene influences one trait. The gene for seed color in garden peas, for example, did not appear to affect other phenotypes. In reality, however, almost all genes influence more than one trait.

(a) Flower color is variable in four-o'clocks.

(b) Incomplete dominance in flower color

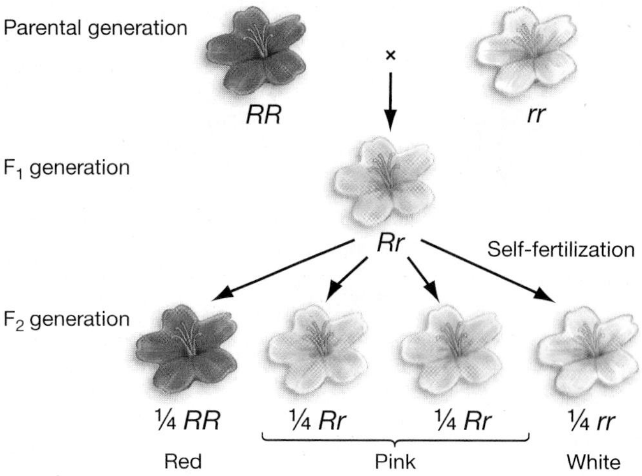

Figure 14.17 When Incomplete Dominance Occurs, Heterozygotes Have Intermediate Phenotypes. The cross in part (b) is explained by hypothesizing that a single gene influences flower color and that alleles *R* and *r* exhibit incomplete dominance.

A gene that influences many traits is said to be **pleiotropic** ("more-turning"). For example, mutations in one gene, *FBN1*, cause Marfan syndrome in humans. If you guessed that this rare mutation would be a recessive trait, you'd be wrong. Like many other genes, the mutant *FBN1* allele is dominant to the wild-type allele. Dominance and recessiveness have nothing to do with whether an allele is wild type or mutant, or beneficial or harmful. There isn't a single phenotype associated with the mutant *FBN1* allele—individuals with Marfan syndrome are tall, have disproportionately long limbs and fingers, an abnormally shaped chest, and often have heart and vascular problems. Therefore, the gene associated with Marfan syndrome is pleiotropic. Based on this set of phenotypes, many scientists believe Abraham Lincoln had Marfan syndrome.

Is a Particular Trait Determined by One Gene?

Each trait Mendel studied seemed to be controlled by one gene. What's more, there was a simple correspondence between a genotype and a trait. Pea seeds were always round if they had at least one dominant allele and wrinkled if they had two recessive alleles.

In many cases, however, two or more genes work together to control a single trait, and the effect of a particular genotype depends on alleles of other genes. Consider a classic study published in 1905 on comb shape in chickens. William Bateson and R. C. Punnett crossed parents from pure lines with comb shapes called rose and pea and found that the F_1 offspring had a new phenotype not seen in either parent: walnut combs. When these walnut-comb individuals bred, their offspring had walnut, rose, pea, and a fourth phenotype called single combs in a 9:3:3:1 ratio (**Figure 14.18a** on page 314).

The genetic model in **Figure 14.18b** shows how these results can be explained if two different genes work together to control one trait (comb shape). When two or more genes influence a single trait, it is called **gene interaction**. Bateson and Punnett's model proposed that comb shape results from interaction between two genes (symbolized *R* and *P*), that there is a dominant and a recessive allele for each gene, and that alleles of these genes work together as indicated at the bottom of the figure to produce the four comb phenotypes.

When gene interaction occurs, one trait is influenced by the alleles of two or more different genes. If a chicken has an *R* allele, its phenotype depends not only on the *R* allele but also on alleles of the *P* gene. For example, note in Figure 14.18b how chickens with a dominant allele of the *R* gene (*R_*) can have a walnut comb or a rose comb, depending on whether the genotype of the *P* gene is *P_* or *pp*. (A blank in a genotype means that either allele can be present in the marked position.)

There are many forms of gene interaction. One important form is epistasis. **Epistasis** occurs when the expression of a phenotype associated with a particular genotype of one gene can be completely masked by a particular genotype of a different gene. One example of epistasis comes from the ABO blood group phenotypes. Researchers had observed that some individuals with I^AI^A and I^Ai genotypes have type O blood instead of the expected type A (see Figure 14.16). How can this be explained?

It turns out that another gene, called *H*, is required for the synthesis of the core polysaccharide to which the "A type" sugar is added. A recessive allele of the *H* gene, *h*, prevents the synthesis of the core polysaccharide. Therefore, individuals with *hh* I^AI^A or *hh* I^Ai genotypes will have type O blood, because there is no core polysaccharide to add the A-type sugar to when the *H* gene genotype is *hh*. The same masking effect also occurs for the I^B allele. This example illustrates epistasis because the *hh* genotype masks the expression of alleles of the *I* gene.

Gene interaction is common and has important implications in human genetics. Imagine that two people have the same genotype for one gene that increases risk for a heart disease. If there is gene interaction, then the risk of developing heart disease also depends on the genotype at other genes. Even if they experience

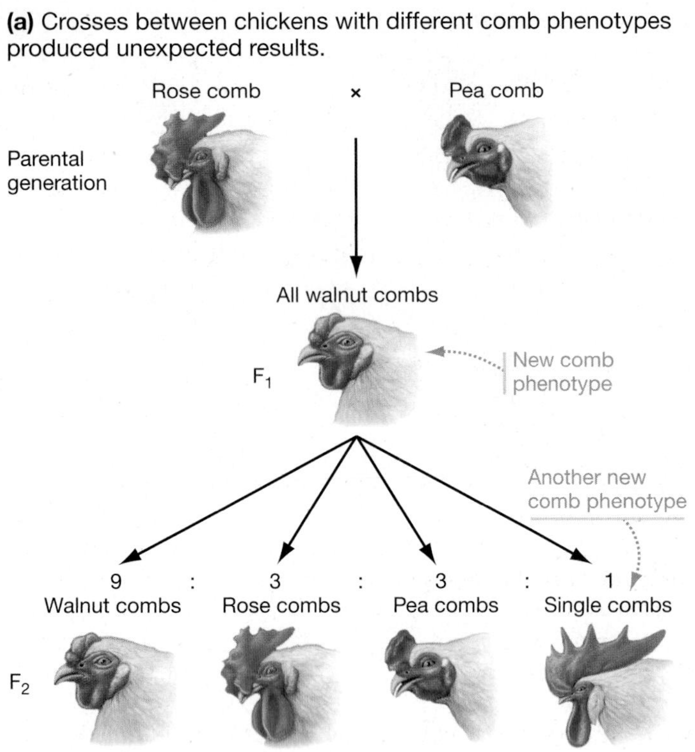

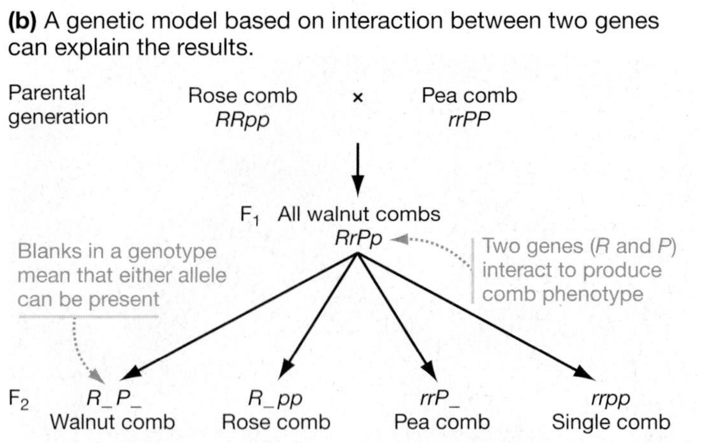

Figure 14.18 Genes at Different Loci Can Interact to Influence One Trait. (a) Notice how new phenotypes appeared in the F_1 and the F_2 generations in Bateson and Punnett's breeding experiment. **(b)** To explain the results, the researchers hypothesized that comb shape depends on two interacting genes. In this case, the phenotype associated with the genotype of one gene depends on the particular alleles of a second gene.

✔ What is different about the 9:3:3:1 ratio in the F_2 of this cross compared with the ratio observed in a standard dihybrid cross?

identical environments, two people with the same genotype for one gene may have very different outcomes.

Are Traits Determined Exclusively by Genes?

After analyzing Mendel's results, it would be tempting to conclude that only genes control phenotypes. But in Mendel's studies, each pea plant received a similar amount of sunlight and water, and was grown in similar soil. Keeping these conditions the same was critical because if plants are deprived of sunlight, water, or nutrients, even plants with alleles that promote growth will be stunted and resemble plants with alleles for dwarfing.

The phenotypes produced by most genes are strongly affected by the environment experienced by an individual. Thus an individual's phenotype is often as much a product of the environment as it is a product of the genotype. In the context of genetics, **environmental effects** are anything that influences phenotypes other than the genotype, including temperature, sunlight, nutrient availability, competition, and even a mother's hormone levels during development of an embryo. Genes and environment work together to shape phenotype.

Controlling environmental influences often is the basis for the treatment of a genetic disease. One example is phenylketonuria (PKU). Individuals with PKU are homozygous for a recessive allele of an enzyme-coding gene. The enzyme helps convert the amino acid phenylalanine, which is present in many types of food, to the amino acid tyrosine. In PKU, this enzyme is absent and, as a result, phenylalanine and a related molecule accumulate. These compounds interfere with the development of the nervous system and produce severe mental retardation.

But are people who cannot metabolize phenylalanine genetically fated to mental retardation? In many countries, newborns are routinely tested for the defect. If identified at birth, individuals who are homozygous for the recessive allele can be placed on a low-phenylalanine diet. The change in environment—reduced phenylalanine in the diet—has a dramatic influence on phenotype. Treated individuals develop normally. PKU is a genetic disease phenotype, but controlling the environment ensures it is neither inevitable nor invariant.

Can Mendel's Principles Explain Traits That Don't Fall into Distinct Categories?

Mendel worked with **discrete traits**—traits that are clearly different from each other. In peas, seed color is either yellow or green—no intermediate phenotypes exist. But many traits don't fall into such neat categories. In humans, for example, height, weight, and skin color vary continuously. People are not limited to being tall or short—countless heights are possible.

For height and many other characteristics, individuals differ by degree. These types of continuously varying traits that don't fall into distinct categories are called **quantitative traits**. Like many discrete traits, quantitative traits are often greatly influenced by the environment. The effects of nutrition on human height and intelligence, for example, are well known.

For many quantitative traits, when the frequencies of different trait values observed in a population are plotted as a histogram, or frequency distribution (see **BioSkills 2**), they form a bell-shaped curve, or normal distribution (**Figure 14.19**).

At the turn of the 20th century, geneticists debated whether Mendel's principles could account for the inheritance of quantitative traits, or if an entirely different set of genetic principles was at work. In 1909, Herman Nilsson-Ehle had an important insight: If many genes each contribute a small amount to the value of a trait, then a normal distribution results for the population as a whole.

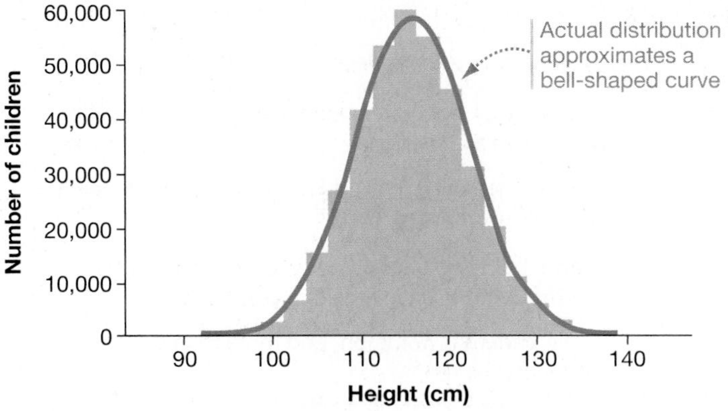

Figure 14.19 Quantitative Traits Often Have a Normal Distribution. A histogram showing the heights (measured in whole centimeters) of first-grade schoolchildren in Guatemala in 2001.

DATA: Pan American Health Organization/WHO. 2004. *Epidemiological Bulletin* 25: 9–13, Graph 1.

Nilsson-Ehle established this finding using strains of wheat that differed in the intensity of kernel color, ranging from no color (white) to darkest red (**Figure 14.20a**). Nilsson-Ehle's genetic model proposed the following:

- The parental strains differ with respect to three genes that control kernel color: *AABBCC* produces dark-red kernels, and *aabbcc* produces white kernels.
- The three genes assort independently.
- The *a*, *b*, and *c* alleles do not contribute to pigment production, but the *A*, *B*, and *C* alleles do contribute to pigment production in an equal and additive way. This is a form of incomplete dominance. As a result, the degree of red pigmentation is determined by the number of *A*, *B*, or *C* alleles present. Each dominant allele that is present makes a wheat kernel slightly darker by the same amount, regardless of the gene.

For example, an individual of genotype *AaBbCc* and an individual of genotype *AABbcc* would be the same shade of red because both of them have three alleles that contribute to coloration in each genotype, and this shade would be between the extremes of the *aabbcc* (white; zero contributing alleles) and *AABBCC* (dark red; six contributing alleles) genotypes.

As **Figure 14.20b** shows, a cross of white and dark-red parents is predicted to produce offspring that are heterozygous for all three genes and intermediate in coloration relative to the parents. What is predicted when the heterozygous F_1 plants self-fertilize? Here, things get interesting. The F_1 plants are expected to produce eight different gamete genotypes. If a Punnett square is used to predict the F_2 genotypes, it would require 64 cells. Tallying the genotypes that have different numbers of alleles that contribute to coloration—ranging from zero to six—produces the result summarized in Figure 14.20b. As you can see, the extreme phenotypes of coloration are rare, the mid-range phenotypes of coloration are common, and the entire frequency distribution approaches a normal distribution, or bell-shaped curve.

The outcome predicted in this model closely matched what Nilsson-Ehle observed in the cross (see Figure 14.20a). This

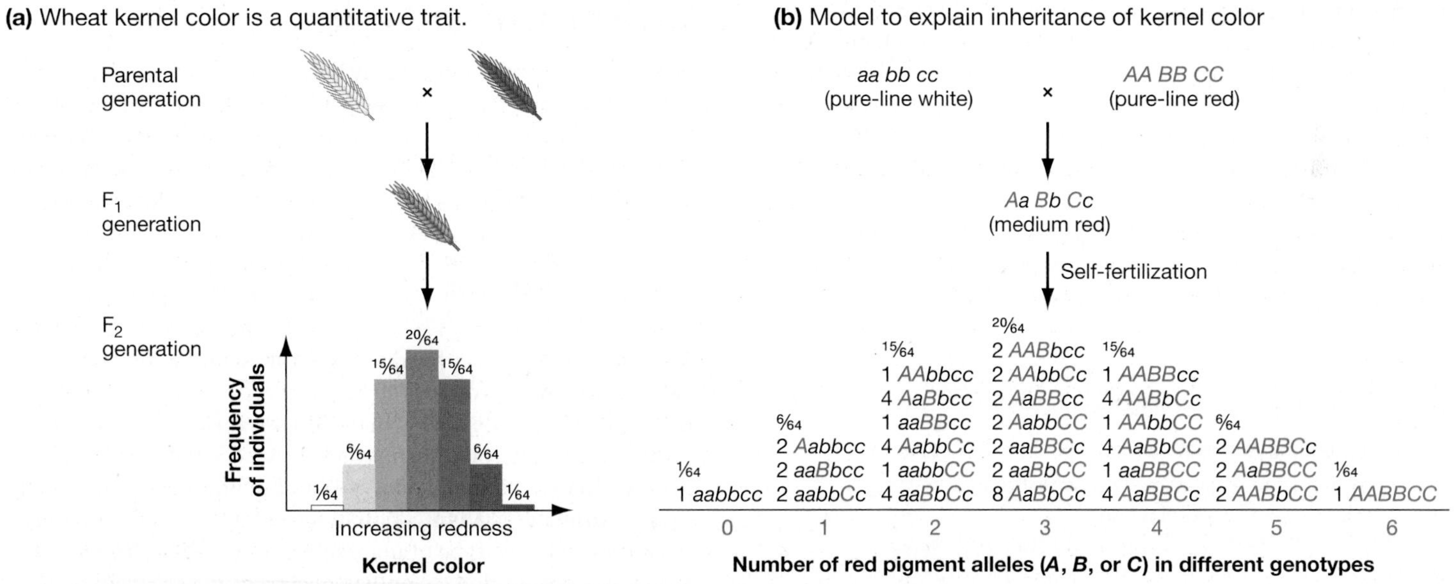

Figure 14.20 Quantitative Traits Result from the Effects of Many Genes. (a) When F_1 wheat plants produced from a cross between parents with white kernels and with dark-red kernels were self-fertilized, the F_2 offspring showed a wide range of kernel colors. The fractions show the frequency of each phenotype, which approximates a normal distribution. **(b)** This model of Mendel's principles and incomplete dominance of alleles of three genes explains the results shown in part (a). The predicted frequency of each phenotype is shown above the predicted genotypes, which have from zero to six alleles that contribute to coloration.

✔ Why are there fewer dark-red or very light wheat kernels compared with kernels of intermediate coloration?

SUMMARY **Table 14.4** Some Exceptions and Extensions to Mendel's Rules

Type of Inheritance	Definition	Consequences or Comments
Sex-linkage	A gene is located on a sex chromosome.	Patterns of inheritance in males and females differ. Compare with *linkage*.
Linkage	Two or more genes are on the same chromosome.	Linked genes typically do not assort independently. Compare with *sex-linkage*.
Codominance	Heterozygotes have phenotypes of both alleles.	More than two phenotypes are seen because heterozygotes have a unique phenotype.
Incomplete dominance	Heterozygotes have an intermediate phenotype.	This pattern is common, and more than two phenotypes are seen.
Multiple allelism	In a population, there are more than two common alleles for a locus.	More than two phenotypes are often seen.
Pleiotropy	A single gene affects many traits.	Pleiotropy is common.
Polygenic (quantitative) trait	A trait that is influenced by many genes and often exhibits continuous variation rather than distinct phenotypes.	Unlike genes that determine discrete traits, each gene adds a small amount to phenotype.
Gene interaction	The phenotype associated with an allele depends on which alleles of another gene are present.	This pattern leads to one allele being associated with different phenotypes.
Epistasis	A particular form of gene interaction in which particular alleles of one gene mask the expression of alleles at another gene.	Epistasis shows that there are many forms of gene interaction and leads to surprising results in which an expected phenotype is not seen.
Environmental effects	Phenotype is influenced by the environment experienced by individuals with the same genotype.	Same genotypes can lead to different phenotypes.

meant that inheritance of a quantitative trait could be predicted using Mendel's principles of segregation and independent assortment—no new genetic principles needed to be invoked.

Quantitative traits are produced by the independent effects of many genes. This idea is captured in the term **polygenic trait** ("many-genes" trait), in which each of many different genes adds a small amount to the value of the trait.

The contributions of Morgan, Nilsson-Ehle, and many others in the decades immediately after the discovery of Mendel's work helped answer why offspring resemble their parents. **Table 14.4** summarizes some of the key exceptions and extensions to Mendel's rules discovered by these researchers.

CHECK YOUR UNDERSTANDING

✔ If you understood this section, you should be able to . . .

1. Explain how a single allele can exhibit both complete dominance and codominance.
2. **PROCESS OF SCIENCE** Outline a study to test whether individuals who share the same genotype of one gene, say *aa*, but express different phenotypes influenced by this genotype do so because of gene interaction or because of environmental effects.
3. **QUANTITATIVE** Explain why in Nilsson-Ehle's genetic model, 1/64 of the F_2 offspring were predicted to be white and 1/64 of these offspring were expected to be dark red.

Answers are available in Appendix A.

14.6 Applying Mendel's Rules to Human Inheritance

When researchers set out to study how a particular gene is transmitted in wheat, fruit flies, or garden peas, they begin by making a series of controlled crosses. For obvious reasons, this is not possible with humans. But suppose you are concerned about an illness that runs in your family and go to a genetic counselor to find out how likely your children are to have the disease. To advise you, the counselor needs to know whether the gene involved is autosomal or sex-linked and what type of dominance is associated with the disease allele.

After you complete this section, you should be able to . . .

- Analyze simple pedigrees to deduce the mode of transmission of a trait.

To understand the transmission of human traits, investigators have to analyze human genotypes and phenotypes that already exist. A **mode of transmission** describes a trait as autosomal or sex-linked and the type of dominance of the allele. To learn the mode of transmission, scientists construct a **pedigree**, or family tree, of affected and unaffected individuals. By analyzing pedigrees, biomedical researchers have been able to discover how more than 3000 human genetic diseases are inherited.

A pedigree records the relationships between the individuals in a family along with each person's sex and phenotype with respect to a trait. If the trait is governed by a single gene, then analyzing the pedigree may reveal whether a given phenotype is due to a dominant or recessive allele and whether the gene responsible is located on a sex chromosome or on an autosome. Let's look at some case histories to see how this work is done.

Identifying Alleles as Recessive or Dominant

To analyze the inheritance of a discrete trait, biologists begin with the simplest possible model by assuming that a single autosomal gene is responsible and that the alleles present in the population have a simple dominant–recessive relationship. If the pattern of inheritance fits this model, then the assumptions of inheritance by a single autosomal gene and simple dominance are supported. Let's first analyze the pattern of inheritance that is typical of autosomal recessive traits and then examine patterns that emerge in pedigrees for autosomal dominant traits.

Autosomal Recessive Traits If a phenotype is due to an autosomal recessive allele, then

- Individuals with the trait must be homozygous.
- If the parents of an affected individual do not have the trait, then both parents are heterozygous for the trait.

Heterozygous individuals who do not have an inherited disease but carry a recessive allele for it are called **carriers** of the disease. When two carriers have children, about ¼ of their offspring are expected to exhibit the recessive disease phenotype.

Figure 14.21a is a pedigree from a family in which an autosomal recessive trait, such as sickle cell disease, occurs. The key feature is that both boys and girls can exhibit the trait even though their parents do not. This is the expected pattern when the parents of an individual who shows the trait are heterozygous. It is also logical to observe that when an affected (homozygous) individual has children, often none of those children exhibit the trait. This pattern is predicted if affected people marry individuals who are homozygous for the dominant allele. This is likely to occur if the recessive allele is rare in the population.

A recessive phenotype shows up in offspring when both parents have that recessive allele and pass it on to their offspring. By definition, a recessive allele produces a given phenotype only when the individual is homozygous for that allele.

Autosomal Dominant Traits When a trait is autosomal dominant, individuals who are homozygous or heterozygous for it have the dominant phenotype. If one parent is heterozygous and the other is homozygous recessive, on average half their children should show the dominant phenotype. And unless a new mutation has occurred in a gamete, any child with the trait must have a parent with this trait. The latter observation contrasts with the pattern seen in autosomal recessive traits.

Figure 14.21b shows the inheritance of the degenerative brain disorder Huntington disease. This pedigree has two features that indicate Huntington disease is due to an autosomal dominant allele. First, if a child shows the trait, then one of the parents shows the trait. Second, if families have a large number of children, the trait usually shows up in every generation—owing to the ½ chance that any child born to a heterozygous parent will be affected.

Identifying Traits as Autosomal or Sex-Linked

When it is not possible to arrange reciprocal crosses, can data in a pedigree indicate whether a trait is autosomal or sex-linked? The answer is based on a simple premise. If a trait appears about equally often in males and females, then it is likely to be autosomal. But if males express the trait more often than females, then the allele is likely to be recessive and found on the X chromosome.

(a) Autosomal recessive trait (e.g., sickle-cell disease)

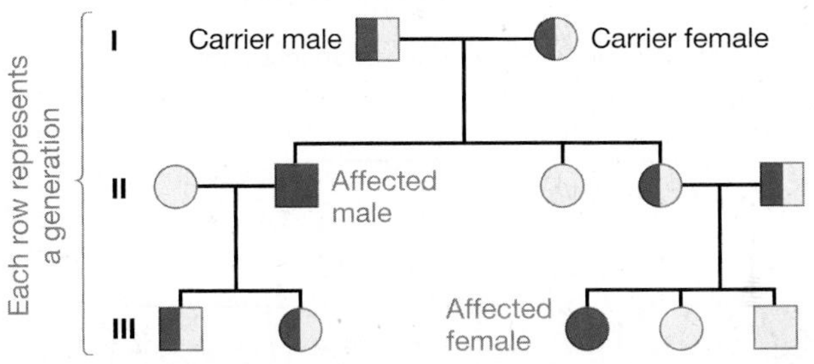

CHARACTERISTICS:
- Males and females are equally likely to be affected
- Affected offspring often have unaffected parents
- Unaffected parents of affected offspring are heterozygous (carriers)
- Affected offspring are homozygous
- If both parents are heterozygous, about ¼ of the offspring will be affected
- Trait often skips generations

(b) Autosomal dominant trait (e.g., Huntington disease)

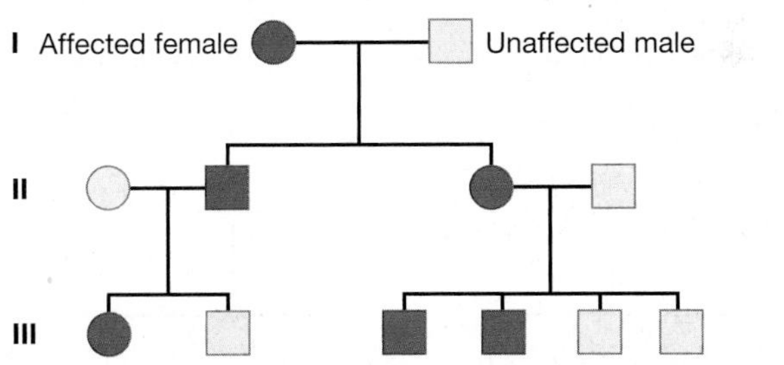

CHARACTERISTICS:
- Males and females are equally likely to be affected
- Affected offspring have at least one affected parent
- Affected offspring are heterozygous if only one parent is affected
- Unaffected offspring are homozygous recessive
- If one parent is heterozygous, about ½ of the offspring will be affected
- Trait does not skip generations

Figure 14.21 **Pedigrees Can Differentiate Autosomal Recessive and Autosomal Dominant Traits.** Pedigrees use standard symbols: squares = males, circles = females; unfilled symbols = unaffected individuals (those without the trait), filled symbols = affected individuals, half-filled symbols = known heterozygotes for a recessive trait (carriers); horizontal lines connect parents, vertical lines connect parents to children.

X-linked Recessive Traits X-linked recessive traits are common. They include the form of red–green color blindness that affects about 10 percent of males and the devastating blood clotting disorder hemophilia. A pedigree for red–green color blindness is shown in **Figure 14.22a**.

Why do males exhibit X-linked recessive traits more often than females? It's because males have only one copy of the X chromosome. Therefore, any X-linked allele will determine the phenotype in a male. This is exactly why Morgan examined only male fruit flies when he studied recombination of X-linked genes (see Figure 14.12). Because women have two X chromosomes, they have two copies of each X-linked gene, and a recessive phenotype will only be seen in homozygotes. For rare alleles, such as disease alleles, homozygotes are very uncommon. In the pedigree of Figure 14.22a, seeing that only males are affected almost certainly means this is an X-linked recessive trait.

Another characteristic of an X-linked recessive trait is that it usually skips a generation. Notice how the trait appears in generations I and III but not in generation II. This pattern occurs for two reasons. An affected male passes his only X chromosome on to his daughters, but because they usually receive a wild-type allele from their unaffected mother, the daughters don't show the trait. These heterozygous daughters will, however, pass the recessive allele on to about half their sons. This pattern is seen on the left side of Figure 14.22a. In contrast, sons receive their only X chromosome from their mother. If the unaffected mother does not carry the recessive allele (see the female in generation I), none of her sons will express the recessive trait.

X-linked Dominant Traits X-linked dominant traits are rare. One example is a bone disorder known as hypophosphatemia, or vitamin D–resistant rickets. As the pedigree in **Figure 14.22b** shows, the most telling feature of an X-linked dominant trait is that an affected male passes the trait to all his daughters and none of his sons. This is because daughters receive their father's only X chromosome. In contrast, a heterozygous female will pass the trait to half her daughters and half her sons. This is because there is an equal chance that a heterozygous mother will transmit an X chromosome with either the dominant or the recessive trait.

What about Y-linked traits? Although the patterns for Y-linked inheritance can easily be predicted, in reality the Y chromosome carries few genes, and these are involved with male-specific sexual development. Except for maleness, there are no other known human Y-linked traits.

Within a few decades of the discovery of Mendel's work, the burning question in genetics was no longer the nature of inheritance but the nature of the gene. What are genes made of? How are they copied so that parents pass their alleles on to their offspring? These questions are the focus of the next chapter (Chapter 15).

CHECK YOUR UNDERSTANDING

✔ If you understood this section, you should be able to . . .

1. Predict what mode of transmission is most likely to account for the following pedigree:

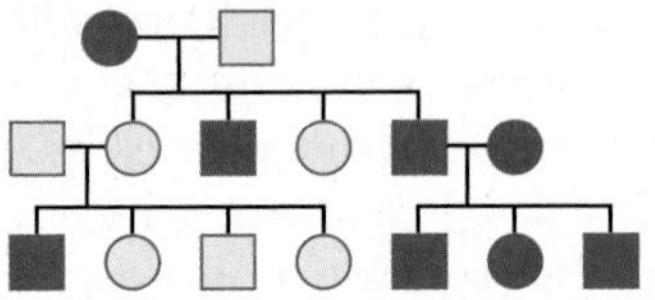

2. Predict the chance that the first daughter of generation III in the family on the left is a carrier (do not assume that the lack of a half-filled symbol indicates she is homozygous). Also predict the chance that, if the parents of this family have another child, the new child will be a carrier.

Answers are available in Appendix A.

(a) X-linked recessive trait (e.g., red–green color blindness)

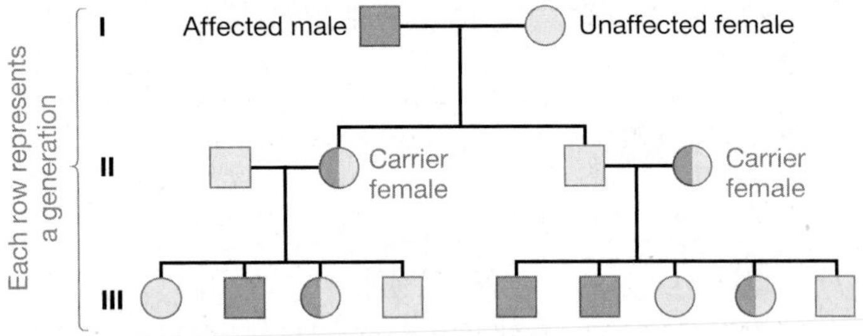

CHARACTERISTICS:
- Males are affected more frequently than females
- Trait is never passed from father to son
- Affected sons are usually born to carrier mothers
- About ½ of the sons of a carrier mother will be affected
- All daughters of an affected male and an unaffected non-carrier female are carriers
- Trait often skips generations

(b) X-linked dominant trait (e.g., hypophosphatemia)

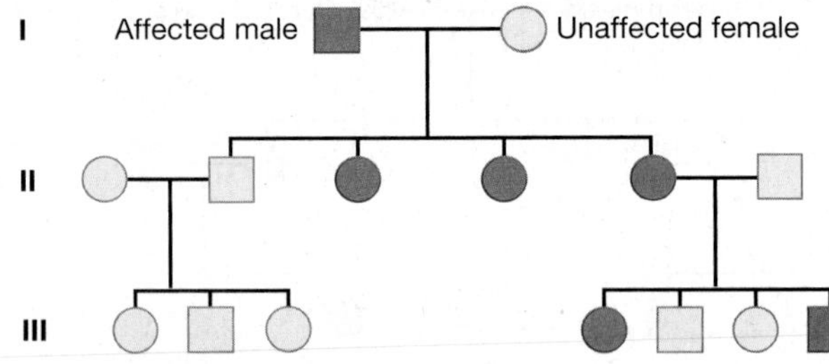

CHARACTERISTICS:
- Males and females are equally likely to be affected
- All daughters of an affected father are affected, but no sons
- Affected sons always have affected mothers
- About ½ of the offspring of an affected mother will be affected
- Affected daughters can have an affected mother or father
- Trait does not skip generations

Figure 14.22 Pedigrees Can Differentiate X-Linked Recessive and X-Linked Dominant Traits.

✔ What genotype in a mother and a father would be predicted to produce a 1 : 1 ratio of normal : color-blind offspring? From this mating, what would be the ratio of color-blind male : color-blind female?

14.1 Mendel's Experimental System

- When Mendel began his work, there were two leading hypotheses of inheritance: blending inheritance and the inheritance of acquired characteristics.
- Mendel chose pea plants as a model organism and started his studies with pure lines that he crossed to produce hybrids.
- Mendel sought to discover rules of heredity that explained how phenotypes were transmitted from parents to offspring.

14.2 Mendel's Experiments with a Single Trait

- From analyzing traits expressed in hybrids, Mendel concluded that one trait was dominant and the other recessive.
- From the results of monohybrid crosses, Mendel concluded that inheritance is particulate—genes do not blend together.
- The traits that Mendel studied are specified by pairs of alleles that separate from each other during meiosis.
- Analysis of monohybrid crosses led to the principle of segregation: During the formation of gametes by meiosis, the pairs of alleles separate so that each egg or sperm cell receives only one of them.

14.3 Mendel's Experiments with Two Traits

- Analysis of dihybrid crosses led to the principle of independent assortment: Alleles of different genes are transmitted to gametes independently of each other.
- Testcrosses allow an investigator to determine whether an individual of dominant phenotype is homozygous or heterozygous.

14.4 The Chromosome Theory of Inheritance

- The chromosome theory of inheritance states that chromosomes contain genes and that Mendel's rules can be explained by the segregation and independent assortment of homologous chromosomes at meiosis I.
- The chromosome theory of inheritance was supported by the discovery of sex-linkage. Puzzling results of crosses could be explained by postulating that the trait was determined by a gene on the X chromosome.
- X-linked traits give different results in reciprocal crosses.

14.5 Extending Mendel's Rules

- There are important exceptions and extensions to the patterns of inheritance that Mendel discovered.
- All genes follow the principle of segregation, but genes close together on the same chromosome do not follow the principle of independent assortment.
- Crossing over between homologous chromosomes creates new combinations of alleles along each chromosome.
- The frequency of crossing over can be used to create genetic maps that show the order and spacing of genes along chromosomes.
- Many genes have more than two alleles. This is called multiple allelism.
- Not all heterozygotes show a dominant phenotype. In addition to complete dominance, incomplete dominance and codominance are also common.
- Many genes are pleiotropic, meaning that they influence more than one trait.
- The phenotype associated with an allele may be influenced by the effects of alleles of other genes and by the environment that the individual experiences.
- One trait is often influenced by many genes. These polygenic traits show quantitative instead of discrete variation. The frequency of different phenotypes often is distributed normally (a plot produces a "bell curve," or normal distribution).

14.6 Applying Mendel's Rules to Human Inheritance

- Pedigrees are used to map the transmission of human traits.
- Applying Mendel's rules to pedigrees can reveal the mode of transmission—whether a trait is dominant or recessive, autosomal or X-linked.

Answers are available in Appendix A.

✔ TEST YOUR KNOWLEDGE

1. The genes for the traits that Mendel worked with are either all located on different chromosomes or behave as if they were. How did this help Mendel recognize the principle of independent assortment?
   a. Otherwise, his dihybrid crosses would not have produced a 9 : 3 : 3 : 1 ratio of F_2 phenotypes.
   b. The occurrence of individuals with unexpected phenotypes led him to the discovery of recombination.
   c. It led him to the realization that the behavior of chromosomes during meiosis explained his results.
   d. It meant that the alleles involved were either dominant or recessive, which gave 3 : 1 ratios in the F_1 generation.
2. **THINK CAREFULLY** Why is the pea wrinkle-seed allele a recessive allele?
   a. It "recedes" in the F_2 generation when homozygous parents are crossed.
   b. The trait associated with the allele is not exhibited in heterozygotes.
   c. Individuals with the allele have lower fitness than that of individuals with the dominant allele.
   d. The allele is less common than the dominant allele. (The wrinkled allele is a rare mutant.)
3. **THINK CAREFULLY** The alleles found in haploid organisms cannot be dominant or recessive. Why?
   a. Dominance and recessiveness describe which of two possible phenotypes are exhibited when two different alleles occur in the same individual.
   b. Because only one allele is present, alleles in haploid organisms are always dominant.
   c. Alleles in haploid individuals are transmitted like mitochondrial DNA or chloroplast DNA.
   d. Most haploid individuals are bacteria, and bacterial genetics is completely different from eukaryotic genetics.

✔ APPLY PROBLEM-SOLVING STRATEGIES

The best way to test and extend your knowledge of genetics is to work problems. In most genetics problems, you are given some information about the genotypes or phenotypes of individuals. Your task is to generate a set of hypotheses—a genetic model—to explain the results. Typically, your hypotheses must address questions like these:

- Is the trait discrete or quantitative?
- Is the phenotype a product of one gene or many genes?
- For each gene involved, how many alleles are at work?
- Do the alleles show complete dominance, incomplete dominance, or codominance?
- Are the genes sex-linked or autosomal?
- If there is more than one gene, are they linked or unlinked? If they are linked, does crossing over occur frequently?
- Are there gene interactions? What about pleiotropy? Environmental effects?

In working the problem, start with the simplest possible explanation. For example, if you are dealing with a discrete trait, you might hypothesize that the cross involves a single autosomal gene with two alleles that show complete dominance. Get fancier only if you need to. Your next step often is to infer what the parental genotypes are and then make a Punnett square to predict what the offspring phenotypes and their frequencies should be based on your hypothesis. Next, check whether these predictions match the observed results given in the problem. If the answer is no, you need to go back and change one of your hypotheses, redo the Punnett square, and see if the predictions and observations match. Repeat until you have a model that fits the data.

4. **Example Problem** Two black female mice are crossed with a brown male. In several litters, female I produced 9 black offspring and 7 brown; female II produced 57 black offspring. What deductions can you make about the inheritance of black and brown coat color in mice? What are the genotypes of the parents?

   A worked solution is available in Appendix A.

5. **Example Problem** A plant with orange, spotted flowers was grown in the greenhouse from a seed collected in the wild. The plant was self-pollinated and gave rise to the following progeny: 88 orange with spots, 34 yellow with spots, 32 orange with no spots, and 8 yellow with no spots. What can you conclude about the dominance relationships of the alleles responsible for the spotted and unspotted phenotypes? What can you conclude about the genotype of the original plant that had orange, spotted flowers?

   A worked solution is available in Appendix A.

✔ TEST YOUR PROBLEM-SOLVING SKILLS

6. In peas, purple flowers are dominant to white. If a purple-flowered, heterozygous plant were crossed with a white-flowered plant, what is the expected ratio of genotypes and phenotypes among the F_1 offspring? If two of the purple-flowered F_1 offspring were randomly selected and crossed, what is the expected ratio of genotypes and phenotypes among the F_2 offspring?

7. In flies, small wings are recessive to normal wings. If a cross between two flies produces 8 small-wing offspring and 28 normal-wing offspring, what are the most likely genotypes of the parents? (Use *S* to represent the normal-wing allele and *s* to represent the short-wing allele.)

8. In garden peas, yellow seeds (*Y*) are dominant to green seeds (*y*), and inflated pods (*I*) are dominant to constricted pods (*i*). Suppose you have crossed *YYII* parents with *yyii* parents.
   - Draw the F_1 Punnett square and predict the expected F_1 phenotype(s).
   - List the genotype(s) of gametes produced by F_1 individuals.
   - Draw the F_2 Punnett square. Based on this Punnett square, predict the expected phenotype(s) in the F_2 generation and the expected frequency of each phenotype.

9. In parakeets, two autosomal genes that are located on different chromosomes control the production of feather pigment. Gene *B* codes for an enzyme that is required for the synthesis of a blue pigment, and gene *Y* codes for an enzyme required for the synthesis of a yellow pigment. Green results from a mixture of yellow and blue pigments, and recessive mutations that prevent production of either pigment are known for both genes. Suppose that a breeder has two green parakeets and mates them. The offspring are green, blue, yellow, and albino (unpigmented). Based on this observation, what are the genotypes of the green parents? What genotypes produce each color in the offspring? What fraction of the progeny should exhibit each type of color?

10. The smooth feathers on the back of the neck in pigeons can be reversed by a mutation to produce a "crested" appearance in which feathers form a distinctive spike at the back of the head. A pigeon breeder examined offspring produced by a single pair of non-crested birds and recorded the following: 22 non-crested and 7 crested. She then made a series of crosses using offspring from the first cross. When she crossed two of the crested birds, all 20 of the offspring were crested. When she crossed a non-crested bird with a crested bird, 7 offspring were non-crested and 6 were crested.
    - For these three crosses, provide genotypes for parents and offspring that are consistent with these results.
    - Which allele is dominant?

11. **SOCIETY** As a genetic counselor, you advise couples about the possibility of genetic disease in their offspring. Today you met with an engaged couple, both of whom are phenotypically normal. The man, however, has a brother who died of Duchenne-type muscular dystrophy, an X-linked recessive condition. His fiancée, whose family has no history of the disease, is worried that the couple's sons or daughters might be afflicted.
    - How would you advise this couple?
    - The sister of this man is planning to marry his fiancée's brother. How would you advise this second couple?

12. Suppose you are heterozygous for two genes that are located on different chromosomes. You carry alleles *A* and *a* for one gene and alleles *B* and *b* for the other. Draw a diagram illustrating what happens to these genes and alleles when meiosis occurs in your reproductive tissues. Label the stages of meiosis, the homologous chromosomes, sister chromatids, nonhomologous chromosomes, genes, and alleles. Be sure to list all the genetically different gametes that could form and indicate how frequently each type should be observed. On the diagram, identify the events responsible for the principle of segregation and the principle of independent assortment.

13. Select True or False for each statement.
    T/F Linked genes are always inherited together.
    T/F Genetic map distances measure the number of nucleotides between a pair of genes.
    T/F The farther apart genes are on a chromosome, the more likely there is to be a crossover between these genes during meiosis.
    T/F Crossing over occurs between genes on different homologs of a homologous chromosome pair.

14. **PROCESS OF SCIENCE** The blending inheritance hypothesis proposed that the genetic material from parents is mixed in the offspring. As a result, traits of offspring and later descendants should lie between the phenotypes of parents. Mendel, in contrast, proposed that genes are discrete and that their integrity is maintained in the offspring and in subsequent generations. Suppose the year is 1890. You are a horse breeder who has just read Mendel's paper. You don't believe his results, however, because you often work with cremello (very light-colored) and chestnut (reddish-brown) horses. You know that when you breed a cremello individual from a pure-breeding line with a chestnut individual from a pure-breeding line, the offspring are palomino—meaning they have an intermediate (golden-yellow) body color. What additional cross would you do to test whether Mendel's model is valid in the case of genes for horse color? According to his model, what offspring phenotype frequencies would you get from your experimental cross? Explain why your cross would test Mendel's model versus blending inheritance.

15. **SOCIETY** Two mothers give birth to sons at the same time in a busy hospital. The son of couple 1 is afflicted with hemophilia A, which is a recessive X-linked disease. Neither parent has the disease. Couple 2 has a normal son even though the father has hemophilia A. The two couples sue the hospital in court, claiming that a careless staff member swapped their babies at birth. You appear in court as an expert witness. What do you tell the jury? Make a diagram that you can submit to the jury.

16. You have crossed two *Drosophila melanogaster* individuals that have long wings and red eyes—the wild-type phenotype. In the progeny, curved wings and lozenge eyes mutant phenotypes appear as follows:

Females	Males
600 long wings, red eyes	300 long wings, red eyes
200 curved wings, red eyes	100 curved wings, red eyes
	300 long wings, lozenge eyes
	100 curved wings, lozenge eyes

- According to these data, is the curved-wing allele autosomal recessive, autosomal dominant, sex-linked recessive, or sex-linked dominant?
- Is the lozenge-eyed allele autosomal recessive, autosomal dominant, sex-linked recessive, or sex-linked dominant?
- What is the genotype of the female parent?
- What is the genotype of the male parent?

17. **QUANTITATIVE** Recall that hemophilia is an X-linked recessive disease. If a woman with hemophilia had children with a man without hemophilia, what is the chance that their first child will have the disease? What is the chance that their first child will be a carrier?

Mastering Biology

Students Go to Mastering™ Biology for assignments, the eText, and the Study Area with animations, practice tests, and activities.

Professors Go to Mastering™ Biology for automatically graded tutorials and questions that you can assign to your students, plus Instructor Resources.

✔ PUT IT ALL TOGETHER: Case Study

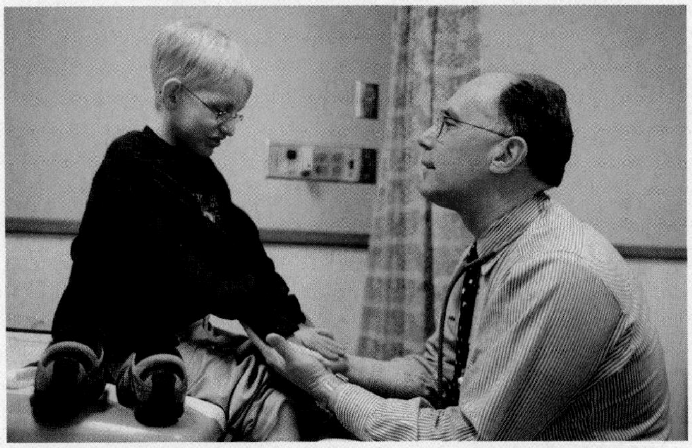

How can genetic principles help advise prospective parents?

The boy in this photo is terminally ill with a rare genetic disorder called adrenoleukodystrophy (ALD). ALD is caused by a mutation that leads to progressive brain damage, failure of the adrenal glands, and eventual death. A pedigree of a family with ALD follows.

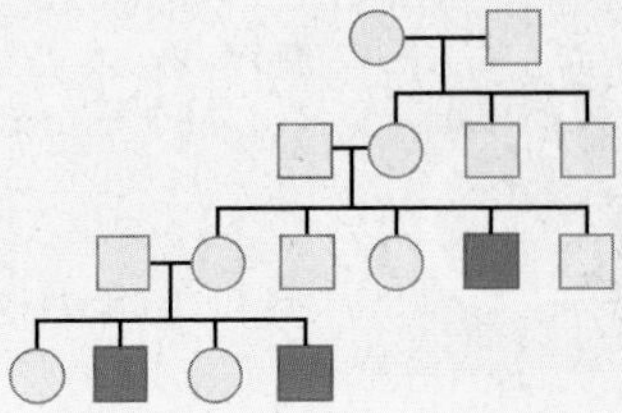

18. ALD is caused by mutations in one gene. Given the symptoms of ALD, which of the following terms describes the disease-associated allele?
a. pleiotropic
b. dominant
c. recessive
d. polygenic

19. Given the pedigree, what is the most likely mode of transmission of ALD?
a. autosomal recessive
b. autosomal dominant
c. X-linked recessive
d. X-linked dominant

20. Which phrase describes the female in the first generation?
a. likely to be homozygous for the recessive allele
b. likely to be homozygous for the dominant allele
c. likely to be heterozygous
d. equally likely to be homozygous dominant or heterozygous

21. **QUANTITATIVE** If the son with ALD in the third generation survives to have children with a woman without any history of ALD in her family, what is the chance that a newborn son of this couple will *not* develop ALD?

22. Imagine that a woman is heterozygous for a color blindness allele. At a site on the chromosome with the color blindness allele, a new mutation occurs that causes ALD, creating one chromosome with an allele for color blindness and an allele for ALD. A son of this woman is color-blind but does not have ALD. Assuming that no new mutations have occurred, what could account for this color-blind son without ALD?

23. **SOCIETY** Imagine that you are a genetic counselor working with the first daughter in the last generation of the pedigree shown. She is married to a man with no history of ALD in his family. The couple plans to have children and they want to know the chance of their first child having ALD. How will you advise them?

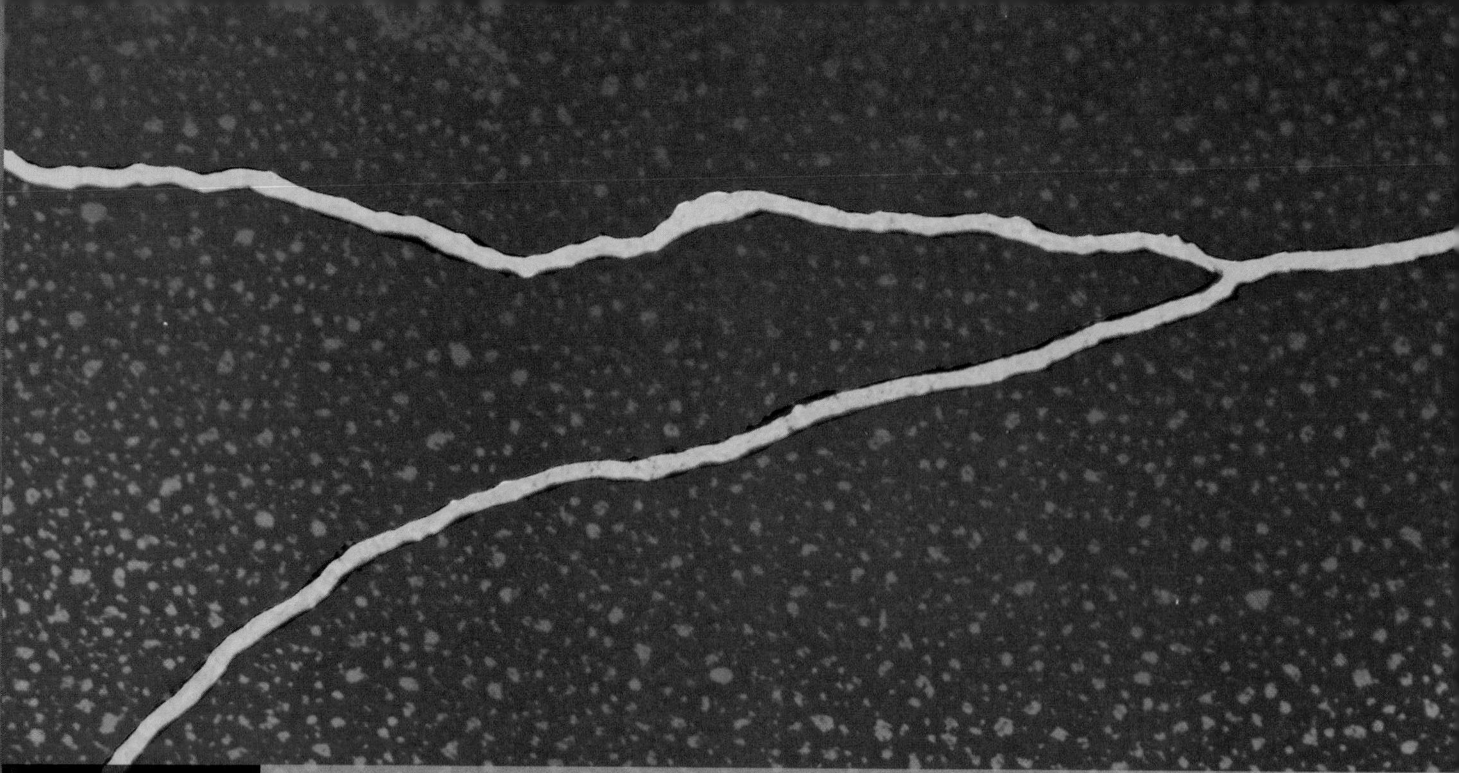

15 DNA and the Gene: Synthesis and Repair

Electron micrograph (with color added) showing DNA replicating. The original DNA double helix (far right) is being replicated into two DNA double helices (on the left). The helices diverge at the replication fork, which is where DNA synthesis is taking place.

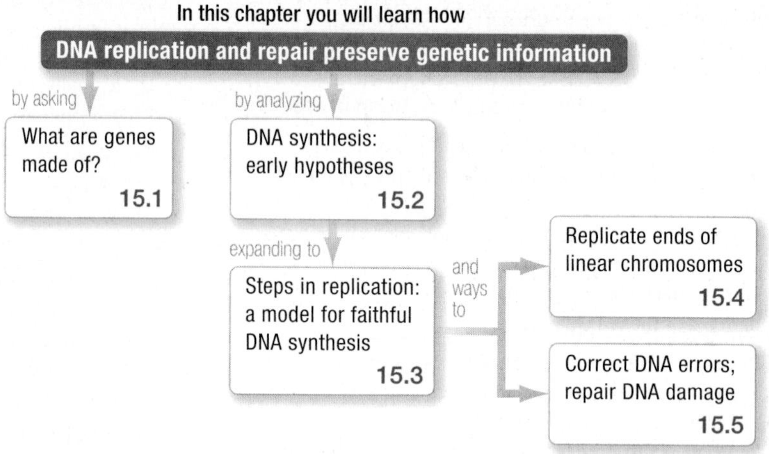

What are genes made of, and how are they copied so that they are faithfully passed on to offspring? These questions dominated biology during the middle of the twentieth century. The investigations that answered these questions are among the most inspiring studies in biology and are a focus of this chapter.

Since Mendel's time, the predominant research strategy in genetics had been to conduct a series of experimental crosses, create a genetic model to explain the types and proportions of phenotypes that resulted, and then test the model's predictions through other crosses. This strategy led to virtually all the discoveries of classical genetics, including Mendel's rules, sex linkage, linked genes, and quantitative inheritance (Chapter 14).

Although biologists knew that genes and chromosomes were replicated during the cell cycle—with copies distributed to daughter cells during mitosis and meiosis (Chapters 12 and 13)—the molecular nature of

BIG PICTURE

This chapter is part of the Big Picture. See how on pages 404–405.

genes and how they were copied remained mysteries for the first half of the twentieth century.

This chapter explores how researchers solved these mysteries. The knowledge they gained links two of the five attributes of life (introduced in Chapter 1): processing of genetic information and replication. (You can see how DNA synthesis and repair fits into the Big Picture of Genetic Information on pages 404–405.)

Let's begin with studies that identified the nature of the genetic material, then explore how genes are copied during the synthesis (S) phase of the cell cycle, and conclude by analyzing how incorrectly copied or damaged genes are repaired. Once the molecular nature of the gene was known, the nature of biological science changed forever.

15.1 What Are Genes Made Of?

After you complete this section, you should be able to ...

- Explain the experimental evidence that led to the acceptance of DNA as the hereditary material.

The chromosome theory of inheritance (Ch. 14, Section 14.4) proposed that chromosomes contain genes, and biologists had known since the late 1800s that chromosomes in living cells are a complex of DNA and proteins. So the question, What are genes made of? came down to a simple choice: DNA or protein?

Initially, most biologists favored the hypothesis that genes are made of proteins. The arguments for this hypothesis were compelling. Because of their almost limitless variation in structure and function, proteins seemed perfectly suited to contain the wealth of information needed to program a cell.

In contrast, DNA was known to contain just four types of deoxyribonucleotide building blocks (see Ch. 4, Section 4.2). What's more, an early but incorrect model of DNA suggested that it was a monotonous, repetitive molecule. It seemed impossible that something as simple as DNA could hold complex information.

DNA or protein? The experiment that finally settled the question is a classic in biological science.

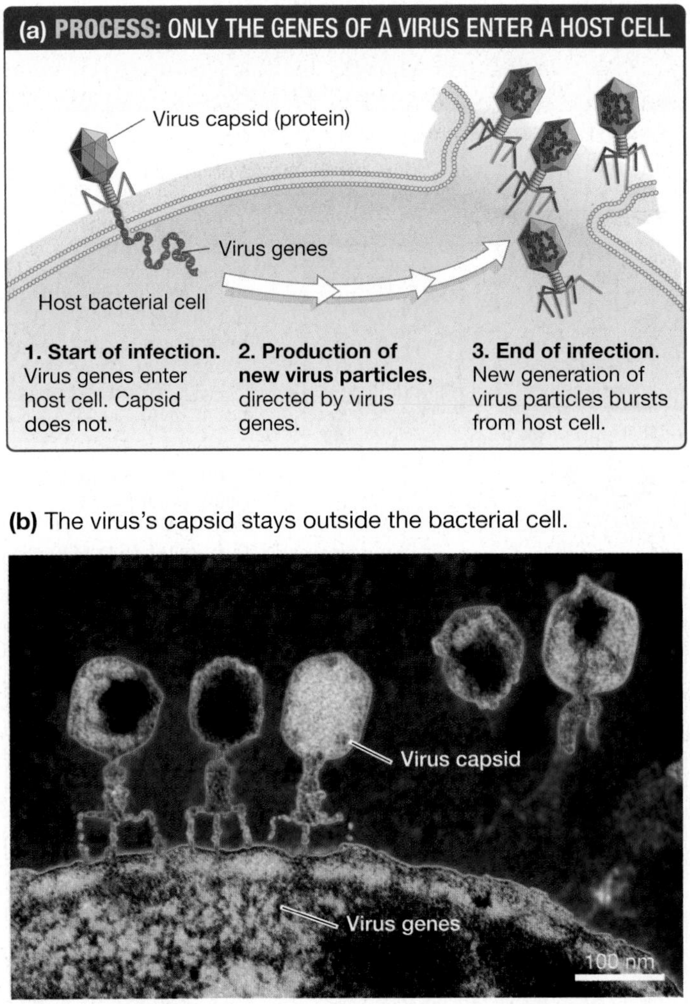

Figure 15.1 The T2 Virus Injects Genes into Bacterial Cells and Leaves a Capsid Behind. The transmission electron micrograph in (b) is colorized to make key structures more visible.

The Hershey–Chase Experiment

In 1952 Alfred Hershey and Martha Chase took up the question of whether genes were made of protein or DNA. Eight years before Hershey and Chase began their study, Oswald Avery and colleagues had used bacterial cells to show that DNA could serve as genetic material. But because these researchers drew their conclusion from a specialized aspect of bacterial genetics, coupled with an entrenched genes-as-proteins hypothesis, many scientists remained unconvinced of their finding or its generality.

Hershey and Chase took a different approach. They began by studying how a virus called T2 infects and replicates within the bacterium *Escherichia coli*, an inhabitant of the human gut. Because *E. coli* is tiny and grows quickly and readily in the laboratory, it quickly became a favored model organism in studies of biochemistry and molecular genetics. (See **BioSkills 11** for more on *E. coli*.) Choosing a virus that infects bacterial cells to answer a central biological question may seem odd, but investigations using bacterial viruses—and the cells they infect—laid the foundations of molecular biology.

When beginning their study, Hershey and Chase already knew that T2 contained only DNA and proteins, and that somehow genes from the virus made their way into infected bacterial cells. Once inside the cells, the viral genes then direct the production of new virus particles. Were these genes protein or DNA? After conducting many experiments, Hershey and Chase came to a model, shown in **Figure 15.1a**, of how T2 infects bacterial cells and replicates within them. During the infection, the exterior protein coat, or **capsid**, of the virus is left behind on the exterior of the host cell (**Figure 15.1b**). The biologists realized that by studying the T2 infection cycle, they could learn the molecular nature of genes.

Hershey and Chase's strategy for determining the composition of the viral substance that enters the cell and acts as the hereditary material was based on two biochemical facts: **(1)** T2 proteins contain sulfur but not phosphorus, and **(2)** DNA contains phosphorus but not sulfur.

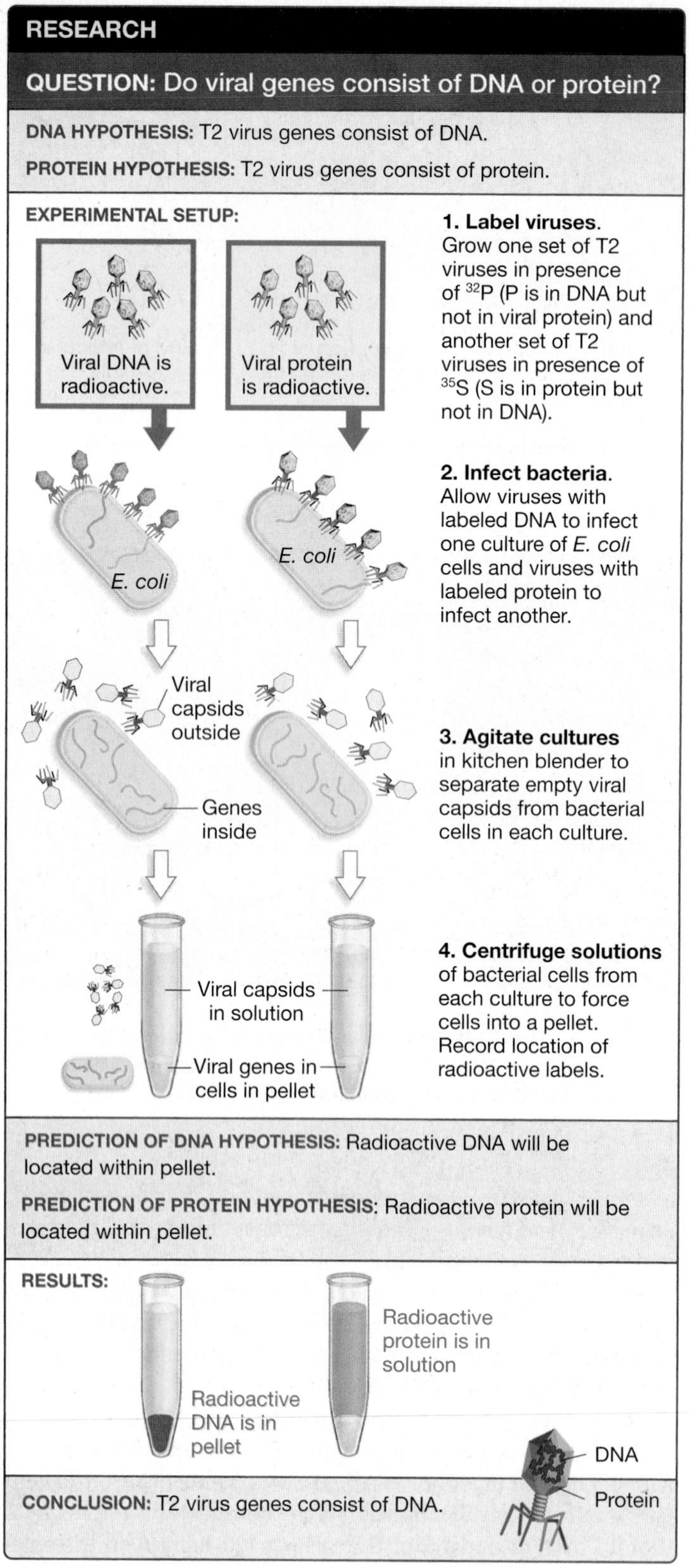

Figure 15.2 Experimental Evidence that DNA Is the Hereditary Material.

SOURCE: A. D. Hershey and M. Chase. 1952. Independent functions of viral protein and nucleic acid in growth of bacteriophage. *Journal of General Physiology* 36: 39–56.

As **Figure 15.2** shows, the researchers grew viruses in the presence of either a radioactive isotope of sulfur (^{35}S) or a radioactive isotope of phosphorus (^{32}P). Because these isotopes were incorporated into newly synthesized proteins and DNA, this step produced a population of viruses with radioactive proteins and a population with radioactive DNA. (See **BioSkills 6** for more on the uses of radioactive isotopes.)

Hershey and Chase allowed each set of radioactive viruses to infect *E. coli* cells. If genes consist of DNA, and only genes are injected into the infected cell, then radioactive protein should be found only in the capsids outside the infected host cells, while radioactive DNA should be located inside the cells. But if genes consist of proteins, then radioactive protein should be inside the cells.

To test these predictions, Hershey and Chase sheared the capsids off the cells by vigorously agitating each of the cultures in kitchen blenders. When the researchers spun the samples in a centrifuge, the small virus capsids remained in the solution while the much larger cells formed a pellet at the bottom of the centrifuge tube (**BioSkills 7** reviews how centrifugation works).

As predicted by the DNA hypothesis, the biologists found that virtually all the radioactive protein was outside cells in the emptied capsids, while virtually all the radioactive DNA was inside the host cells. What's more, the newly created virus particles contained radioactive DNA but not radioactive protein. Because the injected component of the virus directs the production of a new generation of virus particles, this component must represent the virus's genes. **✔PROCESS OF SCIENCE If you understand the work of Hershey and Chase, you should be able to predict what the investigators would have concluded if they had found both radioactive DNA and protein in the pellet.**

After these results were published, other biologists came to accept the hypothesis that DNA, not protein, must be the hereditary material. (It is now known that many viruses contain RNA instead of DNA as their hereditary material; see Ch. 33, Section 33.2). A seemingly implausible claim—that DNA contained all the information for life's complexity—was correct.

The Structure of DNA

In 1953, just one year after Hershey and Chase's landmark results were published, Watson and Crick proposed a model for the structure of DNA. Recall that DNA is typically double-stranded, and each strand consists of a long polymer made up of monomers called deoxyribonucleotides (Ch. 4, Section 4.2).

Each deoxyribonucleotide consists of a deoxyribose sugar, a phosphate group, and a nitrogenous base (**Figure 15.3a**). Deoxyribonucleotides link together into a polymer when a hydroxyl group on the 3′ carbon of one deoxyribose and the phosphate group attached to the 5′ carbon of another deoxyribose are joined by a covalent bond called a **phosphodiester linkage**, or phosphodiester bond (**Figure 15.3b**). The two strands together make up one DNA molecule that functions as the genetic information storage molecule of cells.

As Figure 15.3b shows, a strand of DNA has two major features: **(1)** a "backbone" made up of the sugar and phosphate groups of covalently linked deoxyribonucleotide monomers, and **(2)** a series of bases that project from each sugar in the backbone.

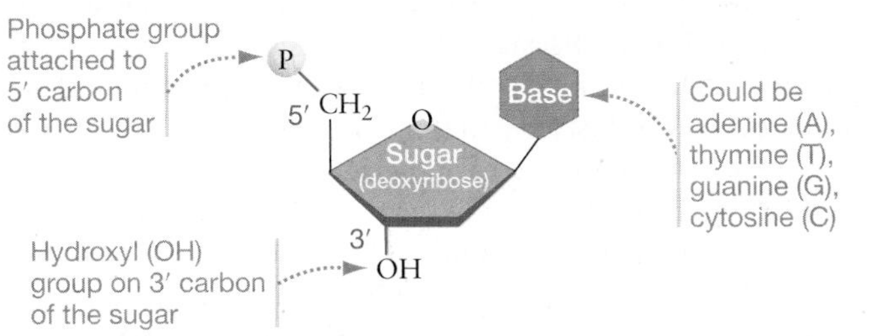

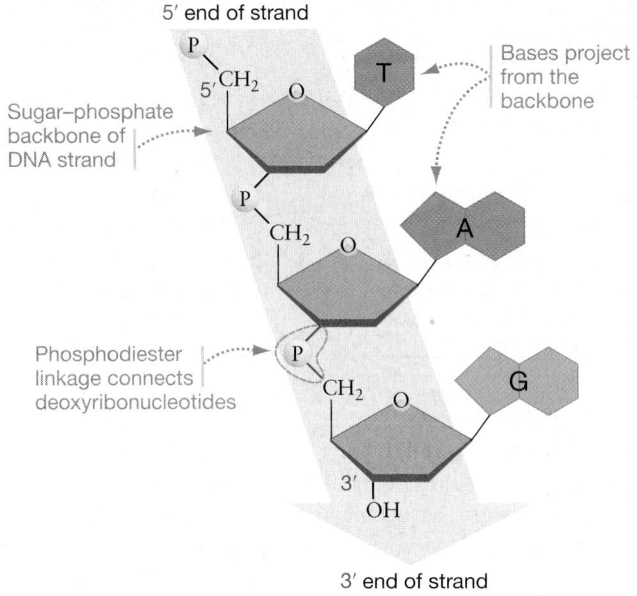

Figure 15.3 DNA Is a Polymer (a) Deoxyribonucleotides are monomers (building blocks) that polymerize to form DNA. **(b)** A strand of DNA is made of deoxyribonucleotides joined by phosphodiester linkages. Notice that DNA has a sugar–phosphate "backbone" with nitrogen-containing bases attached to each sugar. Most DNA molecules contain two strands, but only one strand is shown here for simplicity.

Each strand of DNA has a directionality, or polarity: One end has an exposed hydroxyl group on the 3′ carbon of a deoxyribose, while the other has an exposed phosphate group on a 5′ carbon. So the molecule has distinctly different 3′ and 5′ ends, a feature that turns out to be very important for many functions of DNA.

As they explored different models for the structure of DNA, Watson and Crick hit on the idea of lining up two of these long strands in opposite directions in what is called antiparallel fashion (Figure 15.4a). The key structural features then fell into place: working with molecular models, Watson and Crick realized that these antiparallel strands will twist around each other into a double-stranded spiral—or **double helix**—and that only certain bases fit together snugly in pairs within the helix by **complementary base pairing**. Hydrogen bonds form inside the spiral between the bases adenine (A) and thymine (T) and between the bases guanine (G) and cytosine (C) (Figure 15.4b).

A DNA double helix is considered one molecule even though it's made of two strands. The double-helical DNA molecule is stabilized in two ways: by **(1)** complementary base pairing, and **(2)** interactions between the stacked base pairs inside the helix.

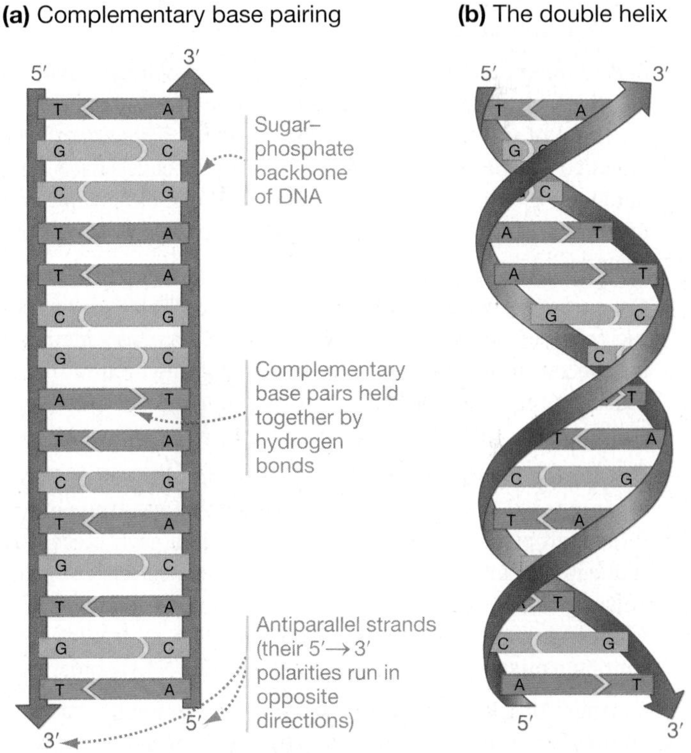

Figure 15.4 Complementary Base Pairing and the DNA Double Helix. (a) DNA normally consists of two strands, each with a sugar–phosphate backbone. Bases project from each strand and form hydrogen bonds between the strands. Only A-T and G-C pairs fit together in a way that allows hydrogen bonding to occur. Together, the two complementary strands constitute a single DNA molecule. **(b)** Bonding between complementary bases twists DNA into a double helix.

CHECK YOUR UNDERSTANDING

✔ If you understood this section, you should be able to ...

1. **PROCESS OF SCIENCE** In Hershey and Chase's actual study, there was a trace amount of radioactive sulfur in the pellet of cells that formed after centrifugation. Explain how this might have occurred, although their hypothesis was correct.
2. Justify why two complementary single strands of DNA are considered two separate molecules, yet when these strands come together and form a double-stranded DNA, the double-stranded DNA is considered a single molecule of DNA.
3. Describe the differences at the molecular level between the 5′ and 3′ ends of a DNA strand.

Answers are available in Appendix A.

15.2 Testing Early Hypotheses about DNA Synthesis

Watson and Crick realized that the A-T and G-C pairing rules suggested a way for DNA to be copied when chromosomes are replicated. They proposed that the existing strands of DNA served as a template (pattern) for the production of new strands and that deoxyribonucleotides were added to the new strands according to the rules of complementary base pairing. For example, if the template strand contained a T, then an A would be added to the new strand to pair with that T. Similarly, a G on the template strand would dictate the addition of a C on the new strand.

After you complete this section, you should be able to ...

- Interpret the results of the Meselson-Stahl experiment to argue that DNA is replicated semiconservatively.

Complementary base pairing provided a basis for copying DNA, but it did not reveal a mechanism. How was DNA replicated?

Three Alternative Hypotheses

Biologists at the time proposed these alternative hypotheses about DNA replication:

- ***Semiconservative replication*** If the **parental strands** of DNA separated, each one could then be used as a template for the synthesis of a new **daughter strand**. This hypothesis is called **semiconservative replication** because each new daughter DNA molecule would consist of one old strand and one new strand. This would keep, or conserve, only one of the strands, leading to the name "semiconservative."
- ***Conservative replication*** If the bases of both strands temporarily turned out from the helix, they could serve as a template for the synthesis of an entirely new double helix all at once. This hypothesis is called conservative replication because both strands of the initial DNA molecule would be kept, or conserved.
- ***Dispersive replication*** This more complicated hypothesis proposed that the parental double helix was fragmented into small pieces before replication and then each piece was replicated by either a conservative or semiconservative mechanism. Following replication of these small fragments, the fragments would be rejoined into two DNA molecules that contain a mixture of parental and daughter DNA strands.

Matthew Meselson and Franklin Stahl, both graduate students when they tested the replication hypotheses, realized that if they could somehow distinguish parental and daughter strands of DNA, they could determine whether replication was conservative, semiconservative, or dispersive. They devised an elegant experimental approach that led to a classic study in molecular biology.

The Meselson–Stahl Experiment

Meselson and Stahl decided to work with *Escherichia coli*—the same bacterium that Hershey and Chase used. Like all organisms, bacterial cells copy their entire complement of DNA, their **genome**, before every cell division. To distinguish parental strands of DNA from daughter strands, Meselson and Stahl came up with a plan to grow *E. coli* cells for successive generations in the presence of different isotopes of nitrogen: ^{15}N first, followed by ^{14}N in the next generation. Because ^{15}N contains an extra neutron, it is heavier than the normal isotope, ^{14}N. The nitrogen isotopes were supplied in a chemical form that the *E. coli* cells used to synthesize the nitrogen-containing bases of their DNA.

The difference in the mass of the two isotopes was the key to the experiment summarized in **Figure 15.5**. The logic ran as follows:

- If a different nitrogen isotope were available in the growth medium as new DNA was synthesized, then the parental and daughter strands would have different densities.
- A technique called density-gradient centrifugation separates molecules based on their density (**BioSkills 7**). Lower-density molecules cluster in bands high in the centrifuge tube; higher-density molecules cluster in bands lower in the centrifuge tube.
- When intact, double-stranded DNA molecules are subjected to density-gradient centrifugation, DNA that contains ^{14}N should form a band higher in the centrifuge tube; DNA that contains ^{15}N should form a band lower in the centrifuge tube.

In short, DNA containing ^{14}N and DNA containing ^{15}N could be separated into two distinct bands. How could this tagging system be used?

Meselson and Stahl began by growing *E. coli* cells with nutrients that contained only ^{15}N. They purified DNA from a sample of these cells (generation 0) and transferred the rest of the culture to a growth medium containing only the ^{14}N isotope. After enough time had elapsed for these cells to divide once (generation 1)—meaning that the DNA had been copied once—they removed a sample and isolated the DNA. After the remainder of the culture had divided again (generation 2), they removed another sample and isolated its DNA. As Figure 15.5 shows, the conservative, semiconservative, and dispersive models make distinct predictions about the makeup of the DNA molecules after replication occurs in the first and second generations. Examine the figure carefully to understand these distinct predictions.

The photographs at the bottom of Figure 15.5 show the experiment's results. After one generation, DNA molecules had an intermediate density. This finding suggested that the hypothesis of conservative replication was wrong, because it predicted two different densities in the first generation. After two generations, a lower-density band appeared in addition to the intermediate-density band. Had dispersive replication occurred, the second generation would have produced only a single, intermediate density band. Meselson and Stahl's results offered strong support for the semiconservative replication hypothesis. Each newly made DNA molecule is made up of one old strand and one new strand.

CHECK YOUR UNDERSTANDING

✔ **If you understood this section, you should be able to ...**

1. Explain the differences between the conservative and semiconservative DNA replication hypotheses.
2. Explain why the mass difference between ^{14}N and ^{15}N was important in Meselson and Stahl's investigation.

Answers are available in Appendix A.

RESEARCH

QUESTION: Is replication semiconservative, conservative, or dispersive?

HYPOTHESIS 1: Replication is semiconservative.	HYPOTHESIS 2: Replication is conservative.	HYPOTHESIS 3: Replication is dispersive.

EXPERIMENTAL SETUP:

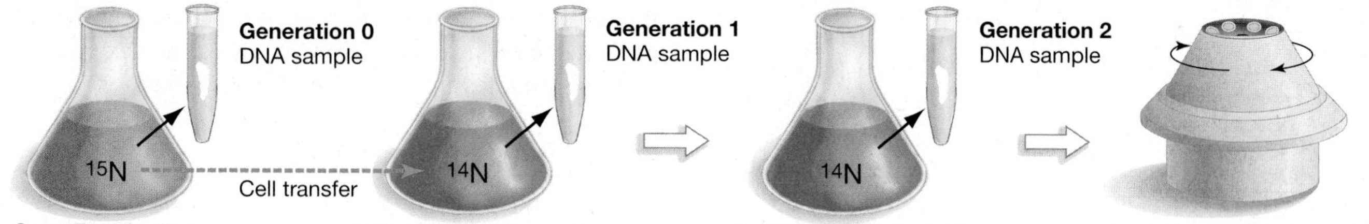

1. Grow *E. coli* cells in medium with ^{15}N as sole source of nitrogen for many generations. Collect sample and purify DNA.

2. Transfer cells to medium containing ^{14}N. After cells divide once, collect sample and purify DNA.

3. After cells have divided a second time in ^{14}N medium, collect sample and purify DNA.

4. Centrifuge the three samples separately. Compare the locations of the DNA bands in each sample to determine density.

PREDICTIONS:

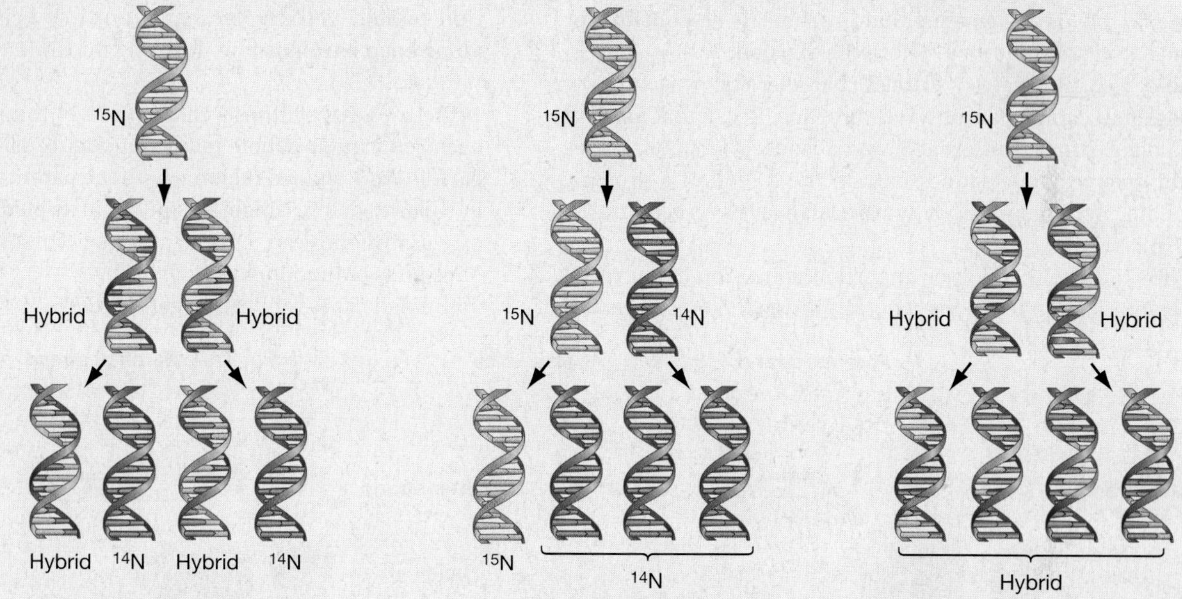

Semiconservative replication	Conservative replication	Dispersive replication
After 2 generations: **1/2 low-density DNA** (^{14}N) **1/2 intermediate-density DNA** (hybrid)	After 2 generations: **1/4 high-density DNA** (^{15}N) **3/4 low-density DNA** (^{14}N)	After 2 generations: **All intermediate-density DNA** (hybrid)

RESULTS:

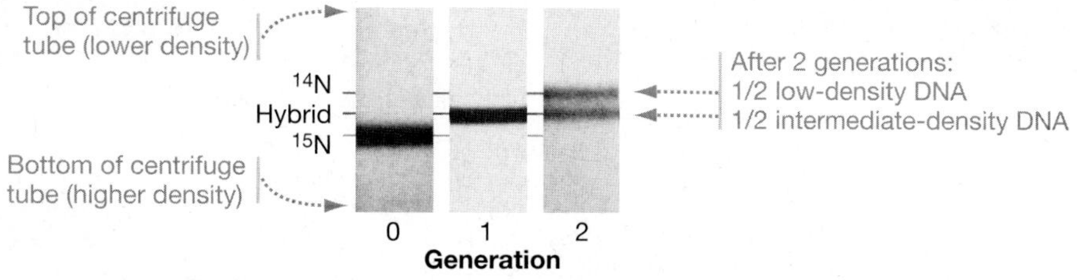

CONCLUSION: Data from generation 1 conflict with conservative replication hypothesis. Data from generation 2 conflict with dispersive replication hypothesis. Replication is semiconservative.

Figure 15.5 **The Meselson–Stahl Experiment Settled a Key Question about Replication.**

SOURCE: M. Meselson and F. W. Stahl. 1958. The replication of DNA in *Escherichia coli*. *Proceedings of the National Academy of Sciences USA* 44: 671–682.

✔ **PROCESS OF SCIENCE** Meselson and Stahl let their experiment run for four generations (two more generations than shown in Figure 15.5), with cultures growing in the presence of ^{14}N. Explain what data from third- and fourth-generation DNA should look like—that is, where the DNA band(s) should be—assuming semiconservative replication.

15.3 A Model for DNA Synthesis

The DNA inside a cell is like an ancient text that has been painstakingly copied and handed down, generation after generation. But while the most ancient of texts contain messages that are thousands of years old, the DNA in cells today has been copied and passed down for billions of years, reaching back at least as far as the last universal common ancestor (LUCA; see Ch. 1, Section 1.5) of all life on Earth. What molecules are responsible for copying DNA, and how do they work?

After you complete this section, you should be able to …

- Analyze synthesis of the leading and lagging strands of DNA.

The initial breakthrough on DNA replication came with the discovery of an enzyme called **DNA polymerase**, so named because it polymerizes deoxyribonucleotide monomers into DNA. This protein catalyzes DNA synthesis. Subsequent work showed that there are several types of DNA polymerase. DNA polymerase III, for example, is the enzyme that is primarily responsible for copying *E. coli*'s chromosome before cell division.

Figure 15.6 illustrates a critical characteristic of DNA polymerases: They can work in only one direction. Recall that a strand of DNA has distinctly different 3′ and 5′ ends. DNA polymerases can add deoxyribonucleotides only to the 3′ end of a growing DNA chain. As a result, DNA synthesis always proceeds in the 5′ → 3′ direction.

Figure 15.6 makes another important point. You might recall that polymerization reactions generally are endergonic, meaning they require an input of energy (Ch. 8, Section 8.1). DNA synthesis also requires an input of energy, but with a twist. The potential energy of the deoxyribonucleotide monomers is first raised by reactions that add two phosphate groups to form **deoxyribonucleoside triphosphates (dNTPs)**. (The *N* in dNTP stands for any of the four bases found in DNA.) Because they have three closely spaced phosphate groups, dNTPs have high potential energy—high enough to make the subsequent formation of phosphodiester bonds in a growing DNA strand exergonic as two of the phosphates are cleaved off (Ch. 4, Section 4.1).

Where Does Replication Start?

Another major insight into the mechanism of DNA synthesis emerged when biologists used electron microscopy to catch DNA replication in action. As Figure 15.7a shows, a replication bubble forms as DNA is synthesized. Initially, the replication bubble forms at a specific sequence of bases called the **origin of replication** (Figure 15.7b). Bacterial chromosomes have only one origin of replication, and they form a single replication bubble. Eukaryotes have multiple origins of replication along each chromosome, forming multiple replication bubbles (Figure 15.7c).

Active DNA synthesis takes place at the replication forks of each replication bubble (shown in Figure 15.7c). A **replication fork** is the Y-shaped region where the parental DNA double helix is separated into single strands and copied. Starting from an origin of replication, DNA synthesis occurs in both directions at once—it is bidirectional.

But how does replication get started?

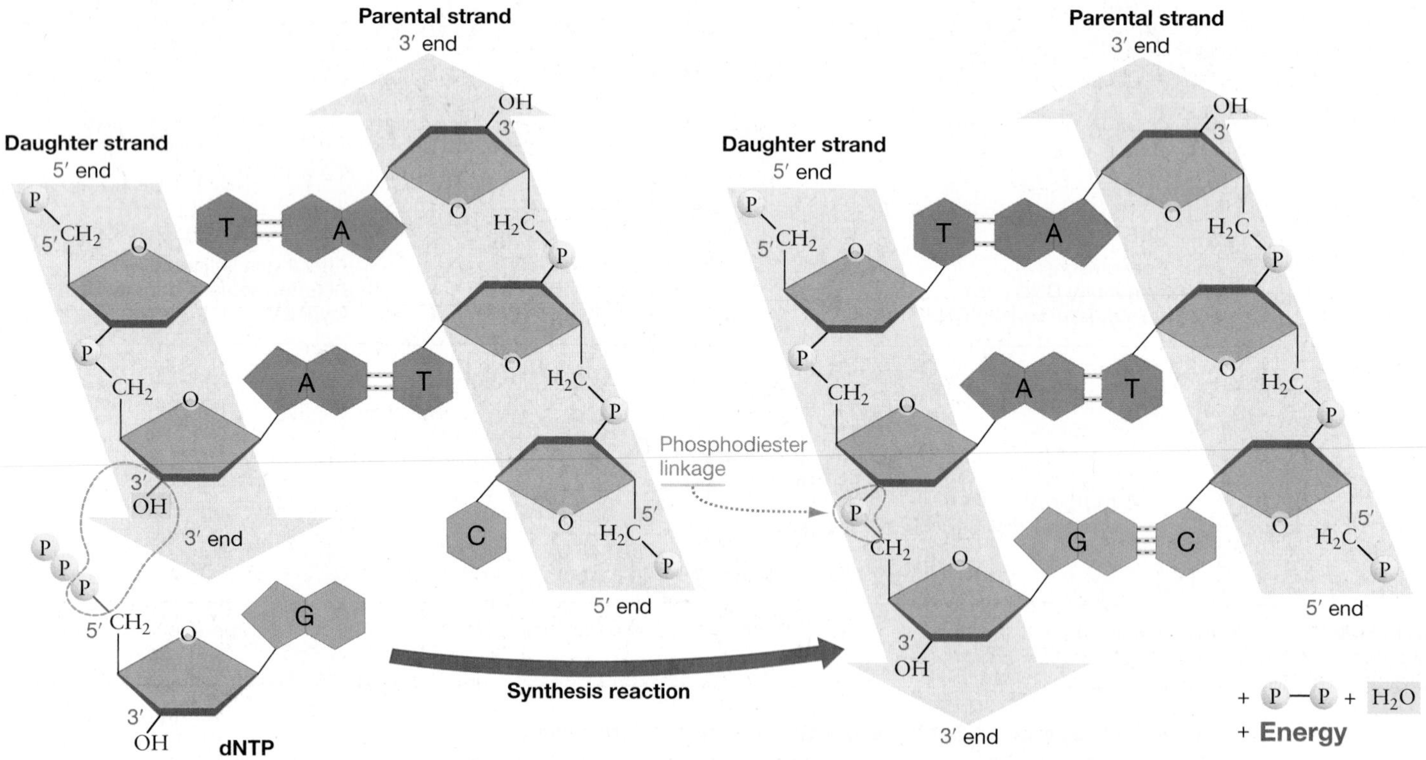

Figure 15.6 DNA Synthesis Proceeds in Only One Direction. DNA synthesis involves a condensation reaction that forms a phosphodiester linkage attaching the incoming deoxynucleotide to the 3′ end of an existing DNA strand. Note that a deoxyribonucleoside triphosphate (dNTP) is used in the reaction, but two phosphates are lost as the monomer is added.

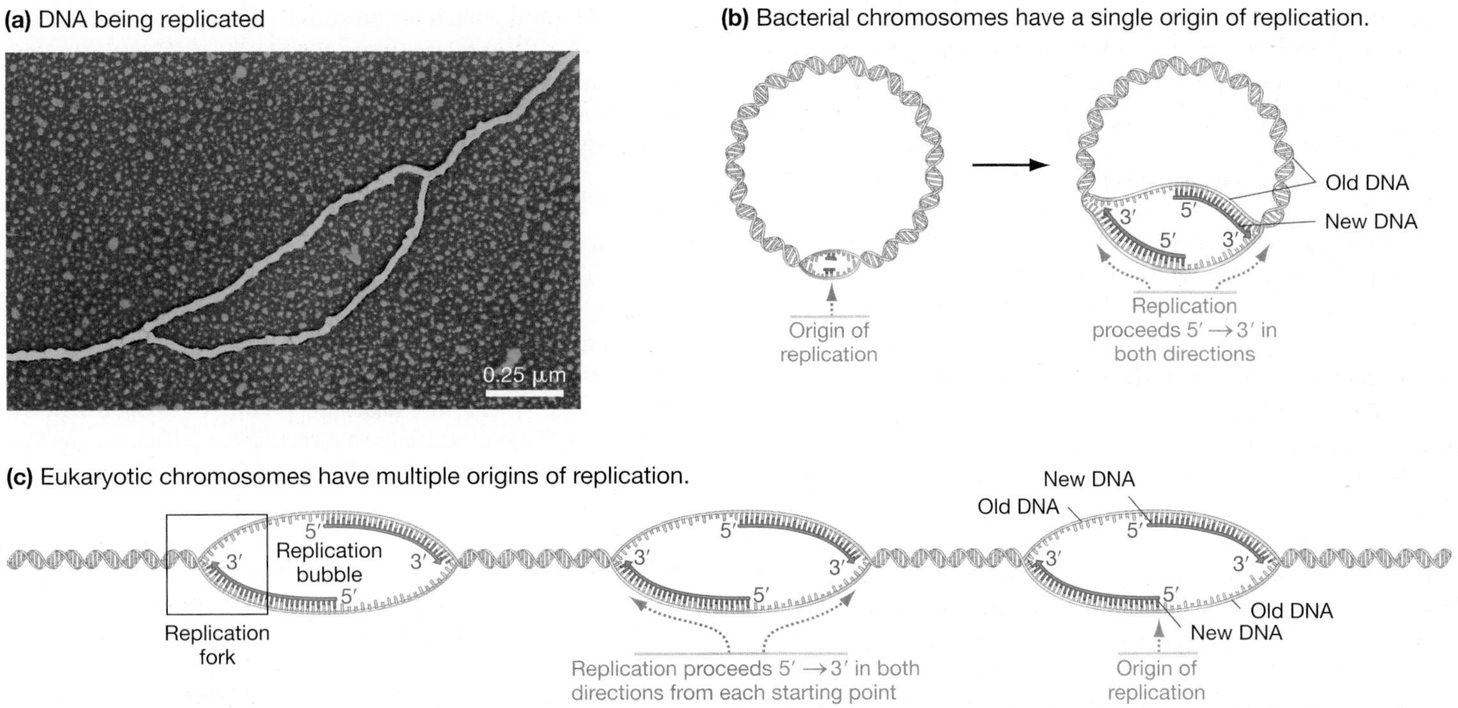

Figure 15.7 DNA Replication Forks Move in Two Directions from an Origin of Replication. Color has been added to the micrograph in part (a).

How Is the Helix Opened and Stabilized?

Although many proteins converge at an origin of replication in both bacteria and eukaryotes, these types of organisms use different mechanisms to control the initiation of DNA synthesis. In bacteria, once a specific set of proteins recognizes the origin on a chromosome, strands near the origin are separated. This is followed by binding of the enzyme **DNA helicase** to one of the single strands of DNA near each of the forming replication forks. In eukaryotes, the proteins that initiate DNA replication are under tight control by cell-cycle regulatory proteins (Ch. 12, Section 12.3). DNA helicase is loaded onto double-stranded DNA at the origin. In all organisms, DNA helicase uses the energy of ATP hydrolysis (breakdown) to separate the two strands of DNA at the replication forks. **Single-strand DNA–binding proteins (SSBPs)** attach to the separated strands to prevent them from snapping back into a double helix (Figure 15.8, step 1).

The unwinding of DNA strands at the replication fork creates twists farther down the helix. To understand why, imagine what happens if you start to pull apart the twisted strands of a

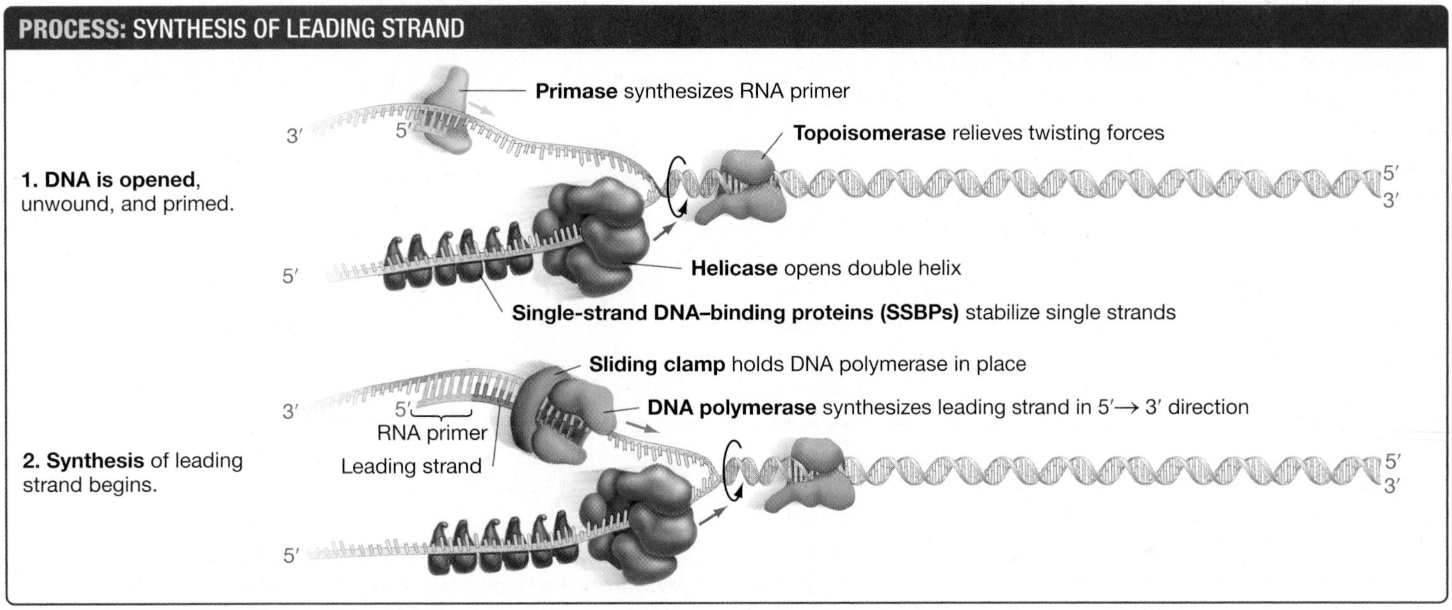

Figure 15.8 Leading-Strand Synthesis. This process is essentially the same in both bacteria and eukaryotes.

rope—the intact section rotates in response. If the end of the rope is fixed in place, the rope coils on itself as the untwisting continues from the other end. The same thing happens to DNA. Without a solution to the DNA twisting problem, replication would soon halt. However, there is a way out, led by a class of enzymes called topoisomerases. A **topoisomerase** is an enzyme that cuts DNA, allows it to unwind, and rejoins it. Topoisomerases that act in DNA replication work ahead of the advancing replication fork to relieve the twists generated by DNA helicase.

Once single strands of DNA are open, what's next?

How Is the Leading Strand Synthesized?

The keys to understanding what happens in DNA synthesis are to recognize the consequences of the antiparallel nature of DNA strands and the limitations of DNA polymerases. These limitations include **(1)** DNA polymerases can synthesize DNA only in the $5' \rightarrow 3'$ direction, and **(2)** DNA polymerases cannot start synthesis from scratch on a template strand. Instead, DNA polymerases can only *extend from* the 3′ end of an existing strand that is hydrogen-bonded by complementary base pairing to the template.

As shown in the figure here, the 3′ end is supplied by a short strand of RNA called a **primer** that is base-paired to the DNA template (actual RNA primers are about a dozen nucleotides long). Both DNA and RNA strands can serve as primers, but RNA is used exclusively at the start of synthesis.

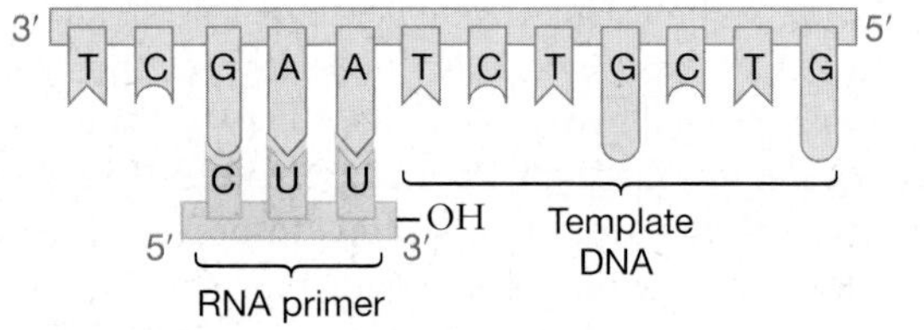

How is the primer added? An enzyme called **primase** does this (see Figure 15.8, step 1). Primase is one type of **RNA polymerase**—a class of enzymes that catalyze the polymerization of ribonucleotides into RNA (see Ch. 4, Section 4.3 to review RNA's structure). Critically, unlike DNA polymerases, RNA polymerases can start synthesis from scratch on a template strand—no primer is needed.

Once a primer has been laid down on a single-stranded template, DNA polymerase can go to work adding deoxyribonucleotides in the $5' \rightarrow 3'$ direction. As Figure 15.8, step 2, shows, a protein ring encircles the DNA and binds to DNA polymerase. This ring—the sliding clamp—helps keeps the DNA polymerase from falling off the DNA during synthesis.

What's so important about the antiparallel orientation of DNA strands? Because they run in opposite directions and because DNA polymerases can extend DNA only in a $5' \rightarrow 3'$ direction, there are important differences in how DNA synthesis proceeds on each template strand at the replication fork. The strand of DNA that is synthesized toward the replication fork is called the **leading strand**, or **continuous strand**, because its synthesis can proceed continuously in the direction of the moving replication fork. Synthesis of the leading strand is straightforward: After an RNA primer is in place, DNA polymerase moves along reading the template, adding deoxyribonucleotides one by one to the extending 3′ end of that strand. The enzyme moves into the replication fork, which is unwound ahead of it. Events on the opposite strand are more involved. ✔ **If you understand leading-strand synthesis, you should be able to list the enzymes involved and predict whether each of them is absolutely required for DNA synthesis.**

How Is the Lagging Strand Synthesized?

If you put together the facts that DNA is an antiparallel DNA double helix and DNA polymerase only synthesizes DNA in a $5' \rightarrow 3'$ direction, you're faced with a paradox: Only one of the strands at the replication fork, the leading strand, can be synthesized in a direction that follows the moving replication fork. The other strand, appropriately called the **lagging strand**, or **discontinuous strand**, must be synthesized in a direction *away* from the moving replication fork. This process is illustrated in **Figure 15.9**.

The Discontinuous Replication Hypothesis The discontinuous replication hypothesis was proposed to explain how the lagging strand is synthesized. It held that primase synthesizes new RNA primers for lagging strands as the moving replication fork opens single-stranded regions of DNA, and that DNA polymerase synthesizes short DNA fragments from these primers. The fragments eventually are linked together into a continuous strand.

In the 1960s, Reiji and Tsuneko Okazaki and their colleagues tested a central prediction of the discontinuous replication hypothesis—the existence of many short DNA fragments on one of the two newly synthesized strands of DNA. Their study was based on the pulse–chase strategy (Ch. 7, Section 7.5). They briefly exposed a culture of growing *E. coli* cells to radioactive deoxyribonucleotides (the pulse) and then transferred the cells to a growth medium with nonradioactive deoxyribonucleotides (the chase). According to the discontinuous replication model, some of these radioactive deoxyribonucleotides should first appear in short fragments of DNA.

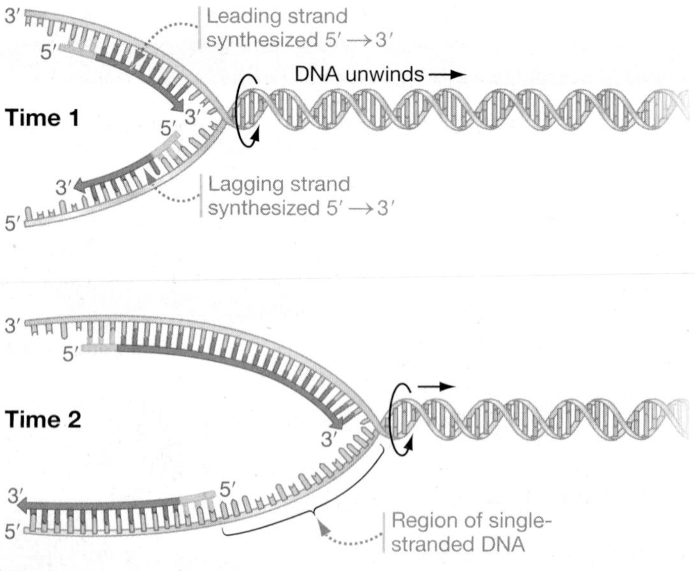

INTERACTIVE Figure 15.9 The Lagging Strand Is Synthesized Away from the Replication Fork. This occurs because the DNA strands are antiparallel, and DNA polymerase works only in the $5' \rightarrow 3'$ direction.

The Discovery of Okazaki Fragments The researchers found short radioactively labeled DNA fragments averaging about 1000 deoxyribonucleotides long when they separated DNA strands obtained from the cells early in the chase period. These short DNAs, which were attached to RNA primers (added in step 1 of Figure 15.10), came to be known as **Okazaki fragments** (Figure 15.10, steps 2 and 3). The short DNAs gradually became longer during the chase. This is because Okazaki fragments are linked together soon after they're formed.

How are Okazaki fragments connected? In bacteria, one way works like this: DNA polymerase III dissociates from the 3′ end of an Okazaki fragment when the polymerase encounters the RNA primer that begins the next Okazaki fragment. Next, a different polymerase called DNA polymerase I attaches to the 3′ end of the Okazaki fragment. DNA polymerase I then moves in the 5′ → 3′ direction, simultaneously removing the RNA primer ahead of it and replacing the ribonucleotides with deoxyribonucleotides, as shown in step 4 of Figure 15.10.

Once the RNA primer is removed and replaced by DNA, an enzyme called **DNA ligase** catalyzes the formation of a phosphodiester bond between the 3′ and 5′ ends of adjacent Okazaki fragments, closing up the backbone (see Figure 15.10, step 5). If you think about what happened in the Okazakis' pulse–chase study, the action of DNA polymerase I followed by DNA ligase accounts for the lengthening of the labeled DNA during the chase.

✔ **If you understand lagging-strand synthesis during DNA replication, you should be able to draw what the two newly synthesized molecules of DNA at a single replication fork would look like if (a) DNA polymerase I was defective or (b) DNA ligase was defective.**

PROCESS: SYNTHESIS OF LAGGING STRAND

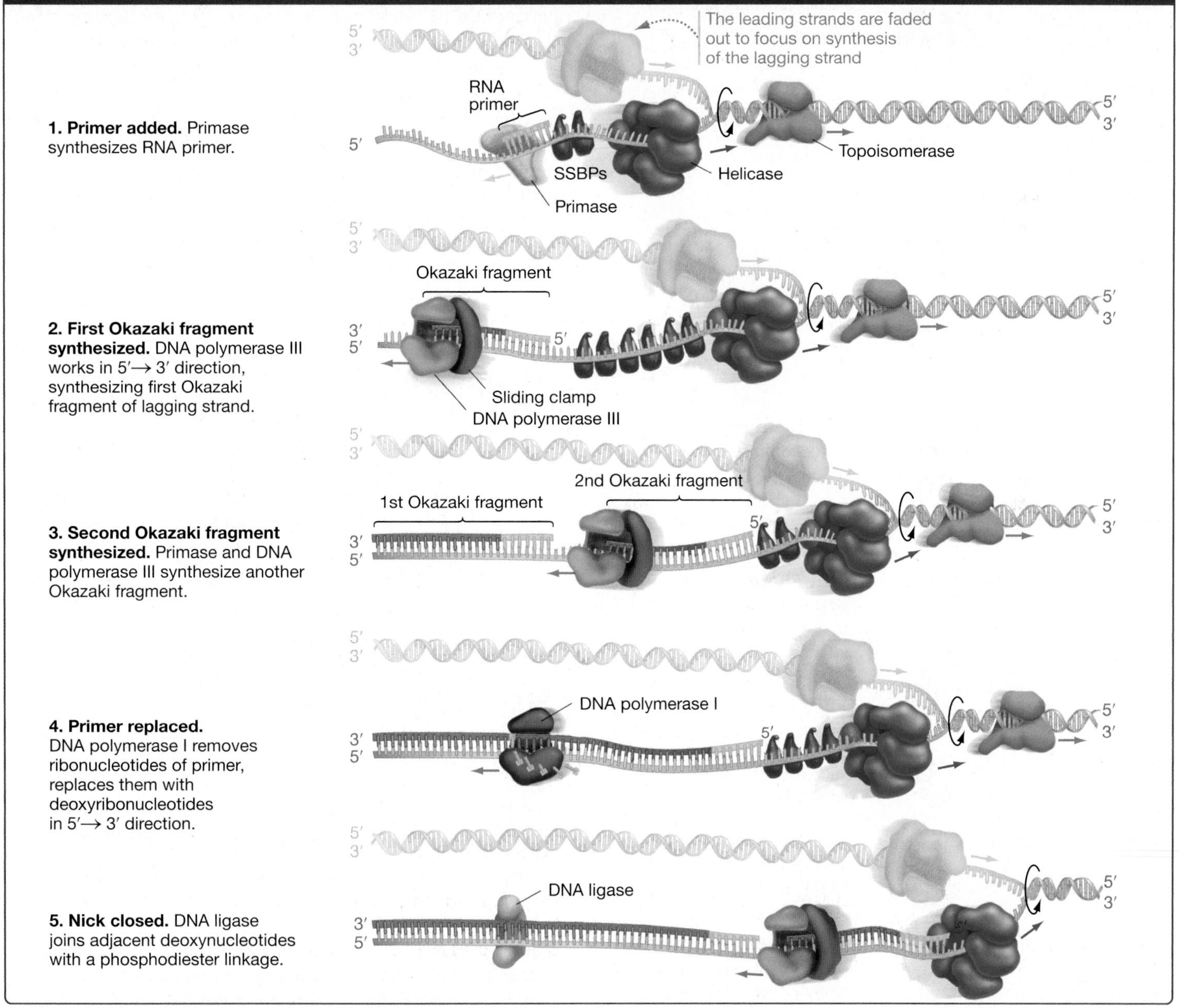

Figure 15.10 **Lagging-Strand Synthesis in Bacteria.**

Note that Figure 15.10 shows how lagging-strand synthesis occurs in *E. coli*. The overall process, however, applies to all groups of organisms—bacteria, archaea, and eukaryotes. The differences lie mostly in details. For example, Okazaki fragments in eukaryotes are shorter (100–200 deoxyribonucleotides) than in *E. coli*, they're synthesized by a DNA polymerase dedicated to lagging-strand synthesis, and the primers are removed in a different way. However, the essential elements are the same: One of the two template strands of DNA is used to synthesize Okazaki fragments that are joined into an unbroken chain of DNA. Working together, the enzymes involved in DNA synthesis (**Table 15.1**) copy the DNA with extraordinary speed and accuracy.

SUMMARY Table 15.1
Proteins Required for DNA Synthesis in Bacteria

Name	Structure	Function
Opening the helix		
Helicase	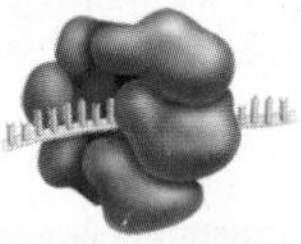	Catalyzes the separation of DNA strands to open the double helix
Single-strand DNA–binding proteins (SSBPs)		Stabilizes single-stranded DNA
Topoisomerase	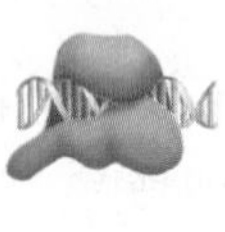	Breaks and rejoins the DNA double helix to relieve twisting forces caused by the opening of the helix
Leading-strand synthesis		
Primase	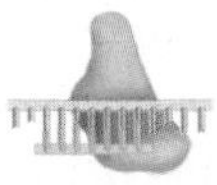	Catalyzes the synthesis of the RNA primer
DNA polymerase III		Extends the leading strand
Sliding clamp		Holds DNA polymerase in place during strand extension
Lagging-strand synthesis		
Primase	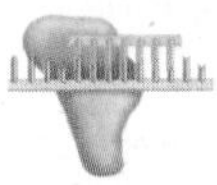	Catalyzes the synthesis of the RNA primer on an Okazaki fragment
DNA polymerase III	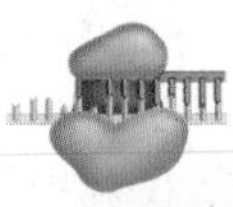	Extends an Okazaki fragment
Sliding clamp		Holds DNA polymerase in place during strand extension
DNA polymerase I	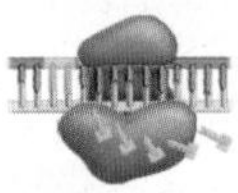	Removes the RNA primer and replaces it with DNA
DNA ligase		Catalyzes the joining of Okazaki fragments into a continuous strand

New Discoveries in DNA Synthesis

Although DNA synthesis has been studied for more than half a century, new insights into this remarkable process continue to emerge. For example, although synthesis of both the leading and lagging strands have been presented as independent processes, the reality is much more interesting and complex. The proteins and enzymes you've learned about work together in a large macromolecular machine called the **replisome** (**Figure 15.11**).

In *E. coli*, the replisome may contain up to three copies of DNA polymerase III. As shown in Figure 15.11, the lagging strand forms a loop, an arrangement that allows the replisome to move as a single unit that follows the replication fork opened by DNA helicase. Through some enzymatic gymnastics, once the DNA polymerase on the lagging strand completes synthesis of an Okazaki fragment, it releases the DNA. Some models propose that a third, previously disengaged DNA polymerase jumps into the act to begin synthesis from the most recently created primer.

The replisome model of DNA replication has been around in various forms since the mid-1970s, so it is hardly a recent development.

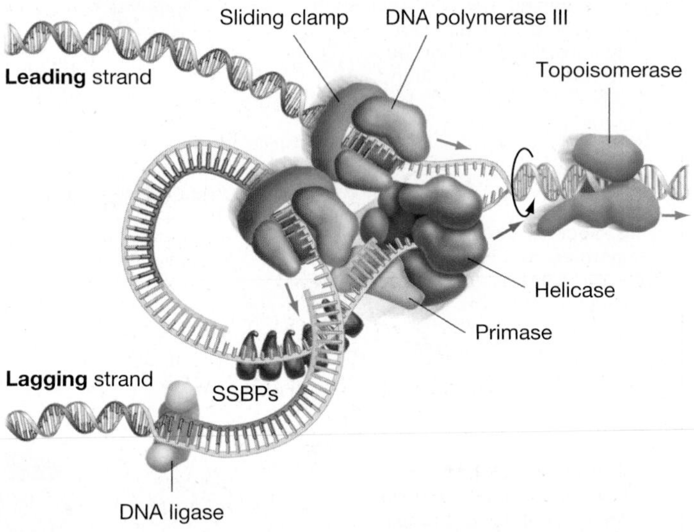

Figure 15.11 The Replisome Is a DNA-Synthesizing Machine. Many of the proteins required for DNA synthesis (for example, DNA polymerases, sliding clamp, helicase, primase, and SSBPs) are organized into a macromolecular machine called the replisome. Other proteins, such as topoisomerase and DNA ligase, work with the replisome during DNA synthesis.

✔**QUANTITATIVE** Assuming that the replisome at each replication fork works at a rate of 1000 base pairs per second, how long would it take to replicate the *E. coli* chromosome (with 4.6 million base pairs) from a single origin of replication?

What is recent are findings made possible by newly developed microscopy techniques that allow tracking of single molecules in real time (BioSkills 9). Visualizing the behavior of individual proteins in the replisome suggests that **(1)** the replisome is a dynamic structure that frequently exchanges proteins between the replisome and the bacterial cell cytoplasm, **(2)** the continuous strand may not be synthesized as continuously as once believed, and **(3)** the rates of DNA synthesis of both the leading and lagging DNA strands may vary widely during the course of DNA synthesis. The bottom line is that the mechanisms of molecular genetics, even for a long-studied process like DNA synthesis, continue to surprise and amaze.

CHECK YOUR UNDERSTANDING

✔ **If you understood this section, you should be able to ...**

1. Predict which of these events would most rapidly arrest (stop) a replication fork: loss of topoisomerase or DNA helicase.
2. Explain the reason that distinct leading and lagging strands appear at each replication fork.
3. Explain why primase works more often on the lagging strand than on the leading strand.

Answers are available in Appendix A.

15.4 Replicating the Ends of Linear Chromosomes

After you complete this section, you should be able to ...

- Analyze the replication of chromosome ends.

The circular DNA molecules in bacteria and archaea can be synthesized by the enzymes introduced in Section 15.3, and so can almost all of the DNA along the linear chromosomes of eukaryotes. But replication of the very ends of eukaryotic chromosomes is another story.

The End-Replication Problem

The region at the end of a eukaryotic chromosome is called a **telomere** (literally, "end-part"). Figure 15.12 on page 334 illustrates why replicating the ends of linear chromosomes presents such a problem.

Steps 1 and 2 (leading strand) When a replication fork reaches the end of a linear chromosome, a DNA polymerase synthesizes the leading strand (top) all the way to the end of the parent DNA template. As a result, leading-strand synthesis produces a double-stranded copy of the DNA molecule.

Steps 2 and 3 (lagging strand) Meanwhile on the lagging strand (bottom), primase adds an RNA primer close to the end of the chromosome. After a DNA polymerase synthesizes the final Okazaki fragment of the lagging strand, an enzyme that degrades ribonucleotides removes the primer.

Step 4 Now here's the problem: Without a primer, DNA polymerase is unable to add new DNA at the very end of the lagging-strand template. As a result, the single-stranded DNA that is left stays single stranded.

Although not shown in Figure 15.12, the same process occurs on the other end of the chromosome.

The single-stranded DNA at the end of the lagging strand is eventually degraded, which results in the shortening of the replicated chromosome. In a cell, this process would result in the shortening of each end of a chromosome by about 50 to 100 deoxyribonucleotides each time DNA replication occurred. Over time, linear chromosomes would vanish. How do cells protect their chromosomal ends from degradation?

Telomerase Solves the End-Replication Problem

One answer to the end-replication problem emerged after Elizabeth Blackburn, Carol Greider, and Jack Szostak reported these striking discoveries:

1. Telomeres are made of short stretches of bases that are repeated over and over. In human telomeres, for example, the base sequence TTAGGG is repeated thousands of times.
2. An enzyme called **telomerase** replicates telomeric DNA. Telomerase is extraordinary because it catalyzes the synthesis of DNA using an RNA template that is an integral part of the enzyme.

These researchers were awarded the 2009 Nobel Prize in Physiology or Medicine.

Figure 15.13 on page 335 shows a simplified model of telomerase activity:

Step 1 Lagging-strand synthesis leaves telomeric DNA with an unreplicated single-stranded "overhang" with a 3′ end.

Step 2 Telomerase binds to the 3′ end of the overhanging single-stranded DNA. Telomerase then catalyzes extension of the overhang to the end of the template region of its RNA molecule.

Step 3 Telomerase shifts down the newly synthesized DNA and catalyzes the addition of another copy of the same short DNA sequence to the end of the single strand. *This step is repeated over and over.*

Step 4 Once the overhang of parental DNA is lengthened sufficiently, the standard complement of DNA synthesizing enzymes can use it as a template to produce a complementary strand. The resulting double-stranded addition counteracts the shortening of the lagging strand caused by standard DNA synthesis.

Effect of Telomere Length on Cell Division and Aging

Telomerase works in only a limited number of cell types. In humans, for example, telomerase is found primarily in the cells that produce gametes and in stem cells (undifferentiated cells that retain the potential to divide throughout an individual's life; see Ch. 21, Section 21.3), but not in most other cells. Any cell not involved in gamete formation is a **somatic cell**. Without telomerase, the chromosomes of somatic cells—the vast majority of cells in an organism—should

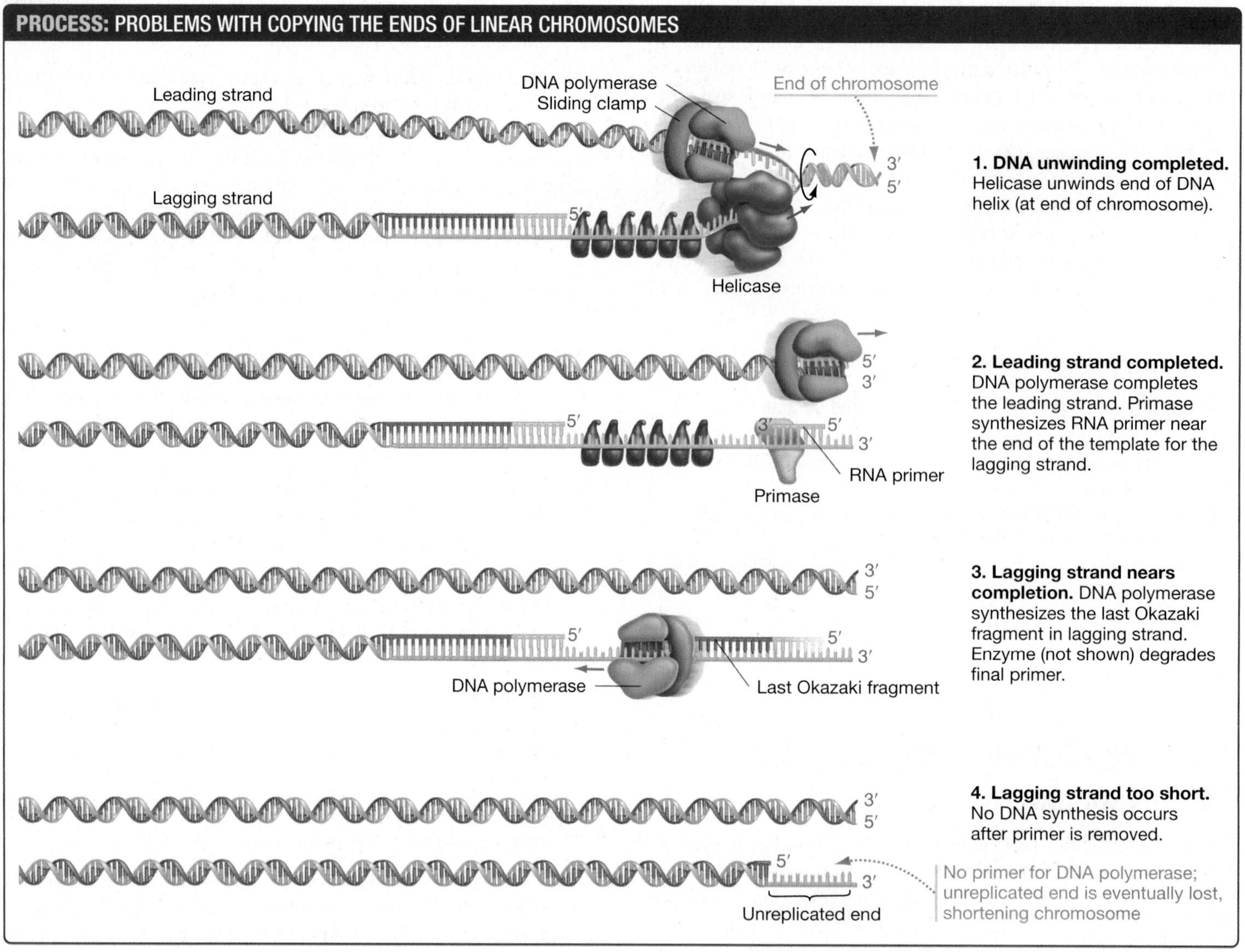

Figure 15.12 Chromosomes Shorten during Normal DNA Replication. When the final primer is removed from the 3′ end of the lagging strand (step 3), that end of the template strand cannot be replicated. The result is a single-stranded section of parental DNA (step 4) that is eventually degraded, leading to chromosome shortening. The same process occurs on the other end of the chromosome but is not shown here. See page 333 for a discussion of these steps.

✔ The upper double-stranded DNA molecule in step 4 has been replicated all the way to its right-hand end. If this DNA represents all the DNA of a chromosome, predict what the left-hand end of this same, newly replicated molecule would look like.

gradually shorten with each mitotic division and become shorter on average as an individual ages. This is exactly what's observed. Although reduced telomere length isn't the full explanation of aging, many biologists think that gradual shortening of telomeres contributes to the declines associated with old age.

The idea that chromosomes get shorter and shorter in cells without telomerase led to the hypothesis that the number of cell divisions possible for a somatic cell is limited by the initial length of its telomeres. The hypothesis proposed that once a chromosome shortened to a threshold length, an alarm signal would go off to shut down further divisions and prevent the loss of critical DNA sequences. Carol Greider and colleagues tested this hypothesis by obtaining cells with a variety of telomere lengths from donors aged newborn to 90 years old and growing these cells in culture. (For an introduction to cell culture, see **BioSkills 11**.)

If telomere shortening controls the final number of cell divisions, then another prediction is that restoring telomerase activity in somatic cells should free those cells from growth limitations. Researchers tested this hypothesis by introducing telomerase into human cells growing in culture. As predicted, these cells continued dividing long past the age when otherwise identical cells stopped growing. Biologists are convinced that telomere shortening has a role in limiting the number of somatic cell divisions.

There's also a dark side to telomerase. Unlike the somatic cells they derive from, most cancer cells have an active telomerase. Many cancer biologists have proposed that telomerase activity is one change that allows unlimited division of cancer cells. A simple prediction is that by inhibiting telomerase, the progression of cancer can be slowed or stopped. Could drugs that knock out telomerase be an effective way to fight cancer? The answer is unclear.

PROCESS: TELOMERE REPLICATION

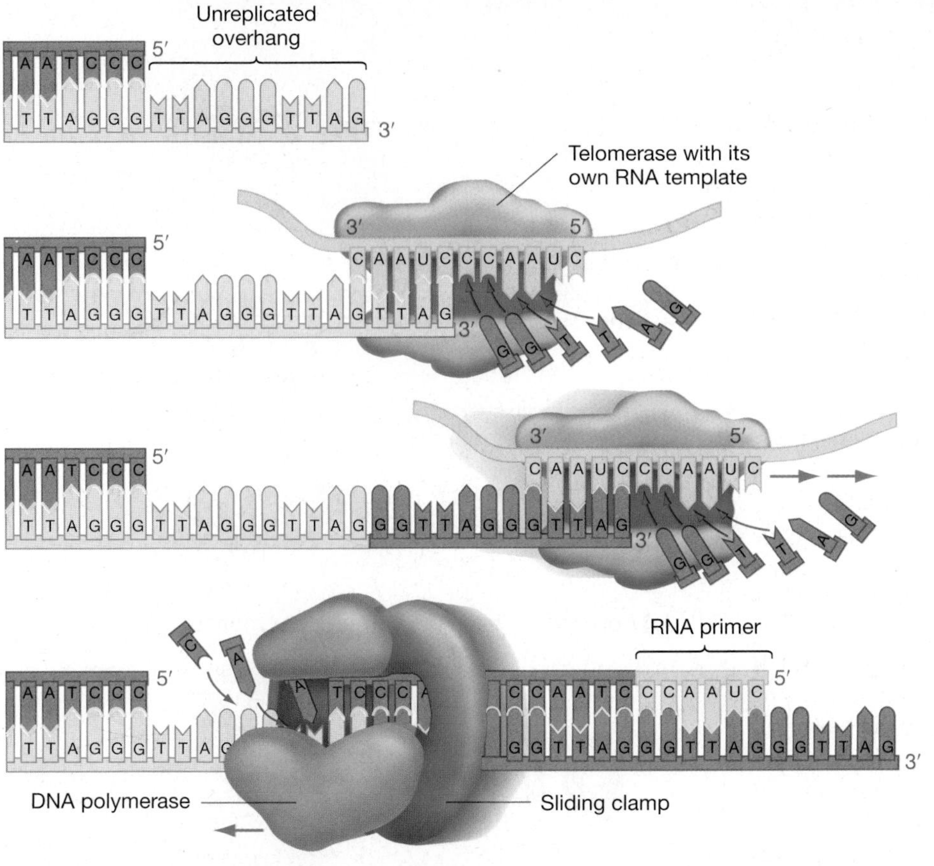

1. A strand of parental DNA remains unreplicated after the RNA primer is removed from the end of the lagging strand (see Figure 15.12, steps 3 and 4).

2. Telomerase extends unreplicated end. Telomerase binds to 3′ end of the overhanging strand of parental DNA and, using its own internal RNA template, extends the strand.

3. Telomerase shifts and repeats activity. Telomerase extends the DNA strand by shifting down the newly synthesized DNA and adding additional repeats, multiple times.

4. Extended single-stranded DNA acts as template. Standard DNA synthesis on this template creates double-stranded DNA to prevent chromosome shortening.

Figure 15.13 **Telomerase Prevents Shortening of Telomeres during Replication.** DNA extension by telomerase makes enough room for primase to add an RNA primer to the unreplicated end of lagging-strand DNA. Addition of a primer allows other replicative enzymes to fill in the missing section of the lagging strand. In this example, a portion of human telomerase RNA is used as a template for synthesizing telomeric DNA. See page 333 for a discussion of these steps.

CHECK YOUR UNDERSTANDING

✔ If you understood this section, you should be able to ...

1. Explain why the DNA template strand with a 3′ end cannot be replicated to its end, but the template strand with a 5′ end can be.
2. Explain how the short repetitive DNA sequences of telomeres arise during replication of chromosome ends.

Answers are available in Appendix A.

15.5 Repairing Mistakes and DNA Damage

After you complete this section, you should be able to ...

- Explain some of the ways that damage to DNA is repaired.

DNA polymerases work fast. In *E. coli*, for example, each replication fork advances at the breakneck speed of 600–1,000 deoxyribonucleotides per second. But it's not enough simply to be fast—DNA replication also has to be accurate. This is a tall order. But in organisms ranging from *E. coli* to humans, the overall error rate during DNA replication is about one mistake per *billion* deoxyribonucleotides. This level of accuracy can be compared to copying *War and Peace*, one of the longest novels, by hand more than 300 times without making a single spelling or punctuation error.

Being this accurate is vital. Consider that humans develop from a fertilized egg that has roughly 12 billion deoxyribonucleotides in its DNA. This DNA is replicated over and over to create the trillions of cells that eventually make up the adult body. If more than a few mutations occurred during each of the cell divisions required in a person's development, genes would be riddled with errors. Sadly, even though errors in DNA replication are exceedingly rare, over the course of a person's lifetime these mistakes appear to be major contributors to the development of cancers. It is no exaggeration to say that the accurate replication of DNA is a matter of life and death.

How can the enzymes involved in DNA replication be so precise?

Correcting Mistakes in DNA Synthesis

As DNA polymerase speeds along a DNA template, hydrogen bonding occurs between the complementary bases on the incoming deoxyribonucleotides and the deoxyribonucleotides on the template strand. Researchers estimate that DNA polymerase inserts an incorrect deoxyribonucleotide only about once in every 100,000 deoxyribonucleotides added. This impressively low rate still produces far too many mistakes. How is the actual error rate of one in a billion achieved during DNA replication? What happens when DNA polymerase makes a mistake?

DNA Polymerase Proofreads Biologists learned more about how DNA synthesis could be so accurate when they discovered *E. coli* mutants with DNA synthesis error rates 100 times higher than normal. Recall that a mutant is an individual with a new trait caused by a mutation—a permanent change in DNA sequence (Ch. 14, Section 14.4). In the case of these *E. coli* mutants, biologists found a defect in a part of DNA polymerase III that acts as a DNA-specific exonuclease—an enzyme that removes deoxyribonucleotides from the ends of DNA strands.

The ability of DNA polymerases to select the correct deoxyribonucleotide to add to a growing strand comes from two sources: **(1)** correct base pairs (A-T and G-C) are the most energetically favorable, and **(2)** the shape of an incorrect base pair differs from that of the correct A-T and G-C pairs. DNA polymerase's active site for DNA synthesis can discriminate between these shapes and will add a new deoxyribonucleotide only when the previous base pair is correct.

When an incorrect deoxyribonucleotide is added, its 3′ OH does not align properly in the polymerase active site, causing the enzyme to pause (**Figure 15.14a**). The mismatched deoxyribonucleotide moves to another part of the enzyme where it does fit, the *exonuclease active site*. This site catalyzes the removal of the incorrect deoxyribonucleotide (**Figure 15.14b**). Now, using the remaining correctly matched base pair, DNA polymerase restarts synthesis (**Figure 15.14c**). The ability of DNA polymerase to recognize and remove an incorrect deoxyribonucleotide is called **proofreading**. Proofreading is an important mechanism for achieving accuracy in DNA synthesis.

Eukaryotic DNA polymerases have the same type of proofreading ability. Typically, proofreading reduces the overall error rate of DNA synthesis to about one mistake in 10 million deoxyribonucleotides added. Is this accurate enough? The answer remains no.

Mismatch Repair Even proofreading is not foolproof. Occasionally (about 1 percent of the time), DNA polymerase leaves a mismatched base pair behind in the newly synthesized strand. Once DNA synthesis moves beyond a mismatch, proofreading is no longer possible. Luckily, a battery of proteins can spring into action to correct the problem by **mismatch repair**, a form of error correction that cleans up errors introduced during DNA synthesis.

Just as for proofreading, analysis of *E. coli* mutants was the key to discovering proteins involved in mismatch repair. In this case, the mutants had abnormally high mutation rates but without any alterations in DNA polymerase III. The first mismatch repair mutation was identified in the late 1960s and was called *mutS* (*mut* is short for "mutator").

Researchers eventually identified 10 *E. coli* proteins involved in different aspects of mismatch repair. Recall that mismatched base "pairs" have a different shape from complementary base pairs. Mismatch repair proteins scan the DNA, and when they detect a mismatch, other proteins are brought in to remove a section of the DNA strand that includes the incorrect base. Finally,

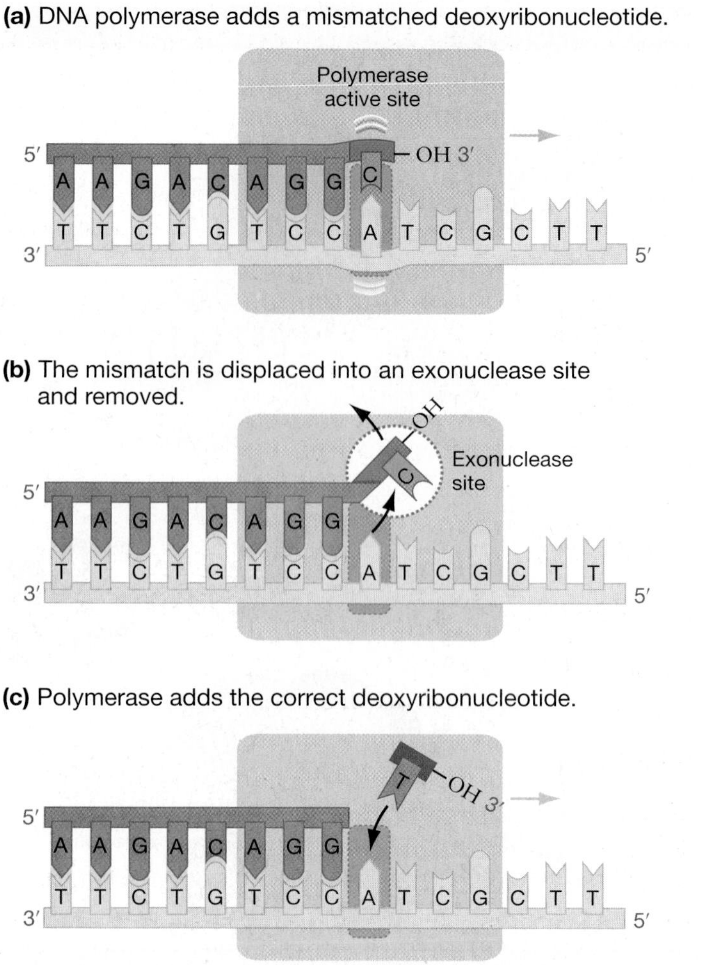

Figure 15.14 DNA Polymerase Can Proofread. If a mismatch occurs, such as the pairing of C with A **(a)**, DNA polymerase can remove the mismatched deoxyribonucleotide in its exonuclease active site **(b)** and start over to add the correct deoxyribonucleotide in its polymerase active site **(c)**.

DNA polymerase resynthesizes the missing DNA using the parental (original) strand as a template, and DNA ligase seals the remaining nick in the backbone.

How do mismatch repair proteins determine which strand has the incorrect base and needs repair? In *E. coli*, chemical marks on the parental strand allow these proteins to distinguish between the parental and daughter strands. Eukaryotes also must distinguish older from newly synthesized DNA, but they use a different system.

Mismatch repair is the final layer of error detection and correction for errors introduced during DNA synthesis. The mismatch repair proteins are like a copyeditor who corrects the errors that a writer—DNA polymerase—did not catch. Coupled with the abilities of DNA polymerase to discriminate between correct and incorrect bases and to proofread its occasional errors, mismatch repair brings the overall error rate of DNA synthesis down to roughly one mistake per billion deoxyribonucleotides incorporated into a DNA strand. The importance of mismatch repair is revealed by a grim fact: Mutations in many genes of the mismatch repair system are observed frequently in cancers, and they play a major role in cancer development and progression.

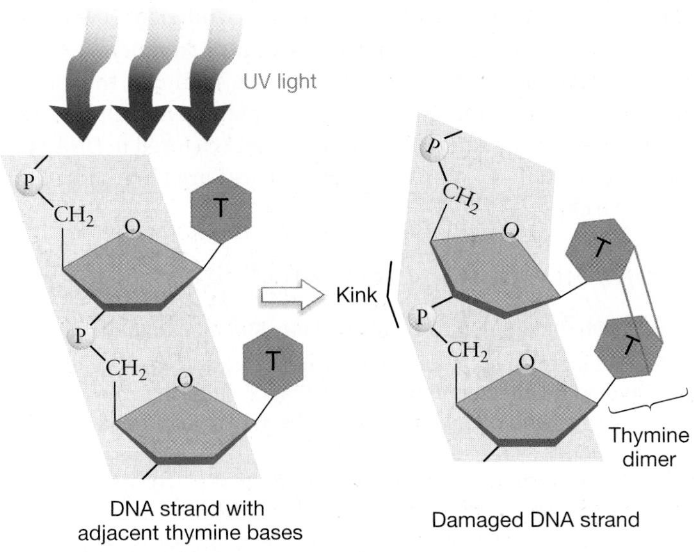

Figure 15.15 UV Light Damages DNA. When UV light is absorbed by a section of DNA that has adjacent thymine bases, the energy can lead to the formation of bonds between them. The resulting thymine dimer distorts the sugar–phosphate backbone of DNA and is not recognized as a template by standard DNA polymerases.

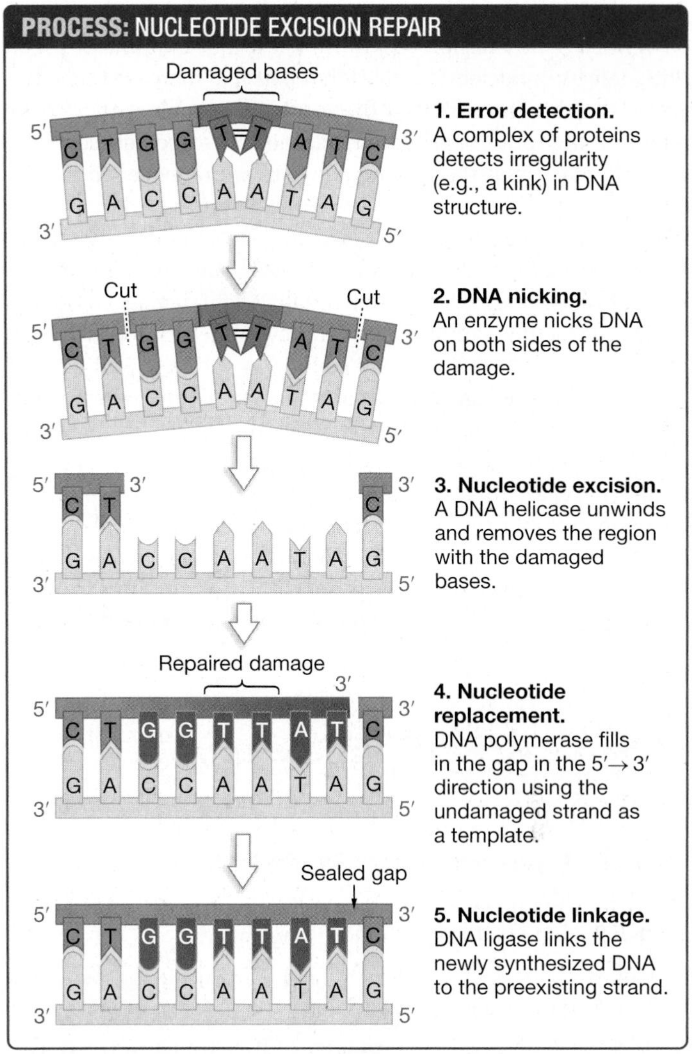

Figure 15.16 Nucleotide Excision Repair Removes and Replaces Defective Bases. Only the changes to DNA are shown here. Each step requires different proteins.

Repairing Damaged DNA

Even after DNA is synthesized, proofread, and mismatches are corrected, the job of ensuring accuracy continues. Genes are under constant assault. DNA is damaged by sunlight, X-rays, and countless chemicals—such as the hydroxyl (OH) radicals produced during aerobic metabolism, and benzopyrenes in cigarette smoke—and even by chemical instabilities in DNA itself. If this damage were ignored, mutations would quickly accumulate to a lethal level. To fix damaged DNA, organisms have evolved a wide array of DNA damage-repair systems. As one example, consider the **nucleotide excision repair** system that works on DNA damage caused by ultraviolet light and many different chemicals.

Ultraviolet (UV) light is found in a range of wavelengths too small for detection by most visual systems (about 180–380 nm; see **BioSkills 1** for metric system units). UV light in sunlight—and tanning booths—can cause a covalent bond to form between adjacent pyrimidine bases within the same DNA strand. The thymine–thymine pair illustrated in Figure 15.15 is a prominent example. This defect, called a thymine dimer, cannot be recognized as a template for DNA synthesis by standard DNA polymerases, and it creates a kink in the sugar–phosphate backbone of DNA. Most DNA polymerases stall at pyrimidine dimers. If the damage is not dealt with, the cell will die.

Nucleotide excision repair removes thymine dimers and many other types of damage that distort the DNA helix, and replaces the damaged DNA strand with correct newly synthesized DNA. Figure 15.16 illustrates the process in bacteria.

First, a protein complex recognizes an irregularity in the DNA helix (step 1 in Figure 15.16). Then an enzyme makes nicks on both sides of the defect (step 2) and a DNA helicase removes the strand of DNA that contains the damage, leaving behind a single-stranded gap (step 3). The intact DNA strand provides a template for synthesis of a new strand by DNA polymerase (step 4). Finally, DNA ligase links the newly synthesized DNA to the original strand (step 5).

As with mismatch repair, excision repair requires multiple proteins to work together and involves DNA synthesis. These are hallmarks of DNA repair—of which there are many types, each specialized to deal with different types of DNA damage.

How important is nucleotide excision repair? This question can be answered by seeing what happens when this repair pathway is defective.

A DNA Repair Defect in Humans

Xeroderma pigmentosum (XP) is a rare autosomal recessive disease in humans. Individuals with this condition are extremely sensitive to UV light. Their skin sunburns and develops rough, scaly patches and irregular dark spots after even slight exposure to sunlight.

In 1968 James Cleaver proposed a connection between XP and DNA nucleotide excision repair. He knew that mutants of *E. coli* with defects in nucleotide excision repair had increased sensitivity to UV light. Cleaver's hypothesis was that people with XP have similar mutations. He proposed that they are extremely sensitive to sunlight because they are unable to repair damage induced by UV light.

Cleaver's ideas proved to be correct. Genetic analyses of XP patients showed that the condition can result from mutations in any of eight genes involved in the repair of UV light damage to DNA.

The devasting symptoms of XP have much to tell us about the frequency of UV light damage to DNA and how well most of us repair this damage. XP patients avoid sunlight and therefore have much less UV damage in their DNA than the vast majority of people with normal sensitivity to sunlight. Even so, without the ability to repair this relatively small amount of damaged DNA, XP patients suffer serious effects, including a staggering 1000- to 2000-fold increase in the risk of skin cancer. In fact, few people with XP *fail* to develop skin cancer. The symptoms of XP imply that even average sunlight exposure creates potentially deadly damage to DNA, and that it's only because of efficient DNA repair mechanisms that most of can tolerate a sunny summer day.

This chapter has presented the evidence leading to the conclusion that genes are made of DNA, and explained how DNA is copied and repaired. But how can information be stored in DNA, and how can this information be decoded? These are the topics of the next two chapters.

CHECK YOUR UNDERSTANDING

✔ **If you understood this section, you should be able to ...**

1. Predict how the mutation rate associated with DNA synthesis would be affected if the system that distinguishes the parental and daughter DNA strands in mismatch repair were defective.
2. Predict how mutation rate would be affected if all base pairs had the same shape and degree of stability.
3. Explain the evidence for a role of functioning DNA repair systems in cancer protection.

Answers are available in Appendix A.

CHAPTER 15 Review

For media, go to **Mastering Biology**

15.1 What Are Genes Made Of?

- Experiments using viruses that had labeled proteins or labeled DNA showed that DNA is the hereditary material.
- Each strand of DNA consists of a sequence of nitrogenous bases held on a sugar–phosphate backbone.
- DNA is a double helix that consists of two strands of deoxyribonucleotides that run in opposite directions (antiparallel orientation). The strands twist into a double helix and are held together by hydrogen bonding between complementary bases and stacking of base pairs within the helix.

15.2 Testing Early Hypotheses about DNA Synthesis

- By labeling DNA with ^{15}N or ^{14}N, researchers validated the semiconservative hypothesis of DNA replication.
- In semiconservative replication, each strand of a parental DNA molecule provides a template for the synthesis of a daughter strand, resulting in two complete DNA double helices.

15.3 A Model for DNA Synthesis

- DNA synthesis requires many different proteins, and it occurs in only one direction ($5' \rightarrow 3'$).
- DNA synthesis requires both a template and a short RNA primer. It takes place at the replication fork where the double helix is opened.
- Synthesis of the leading strand is continuous, but synthesis of the lagging strand is discontinuous because on that strand, the DNA polymerase moves away from the replication fork.
- On the lagging strand, short DNA fragments called Okazaki fragments form and are joined together.
- Synthesis of the leading and lagging strands is coordinated by the replisome—a large group of enzymes and proteins involved in DNA synthesis that are held together in a group (a complex) that synchronously produces both the leading and lagging strands of DNA.

15.4 Replicating the Ends of Linear Chromosomes

- To prevent shortening at the ends of linear chromosomes in eukaryotes, the enzyme telomerase adds short, repeated sequences of single-stranded DNA that is made double-stranded by enzymes used for standard DNA replication.
- Telomerase is active in reproductive cells that eventually undergo meiosis. As a result, the length of chromosomes in gametes is maintained.
- Chromosomes in cells without telomerase shorten with continued cell division until their telomeres reach a critical length at which cell division no longer occurs.

15.5 Repairing Mistakes and DNA Damage

- DNA replication is remarkably accurate because **(1)** DNA polymerase selectively adds a deoxyribonucleotide that correctly base-pairs with the template strand; **(2)** DNA polymerase proofreads each added deoxyribonucleotide; and **(3)** mismatch repair proteins remove deoxyribonucleotides with incorrect bases that escape proofreading and replace them with the correct deoxyribonucleotide.
- Other types of DNA repair occur after DNA has been damaged by chemicals or radiation.

- Nucleotide excision repair cuts out damaged portions of DNA and replaces them with correct sequences.
- If DNA repair proteins are defective, mutation rate increases. Defects in the genes responsible for DNA repair lead to several types of cancer.

Answers are available in Appendix A.

✔ TEST YOUR KNOWLEDGE

1. Which of the following is not a property of DNA polymerase?
   a. It adds dNTPs only in the $5' \rightarrow 3'$ direction.
   b. It requires a primer to begin synthesis.
   c. It opens the two strands of DNA at the replication fork.
   d. Its exonuclease activity is involved in proofreading.
2. What is the function of primase?
   a. synthesis of the repetitive, short double-stranded DNA sequences required by DNA polymerase
   b. synthesis of a short RNA strand that is complementary to single-stranded DNA
   c. closing the gap at the $3'$ end of DNA after excision repair
   d. removing primers and synthesizing a short section of DNA to replace them
3. How are Okazaki fragments synthesized?
   a. by using the leading-strand template, and synthesizing $5' \rightarrow 3'$
   b. by using the leading-strand template, and synthesizing $3' \rightarrow 5'$
   c. by using the lagging-strand template, and synthesizing $5' \rightarrow 3'$
   d. by using the lagging-strand template, and synthesizing $3' \rightarrow 5'$
4. Telomerase synthesizes DNA in the ____________ direction.

✔ TEST YOUR UNDERSTANDING

5. **PROCESS OF SCIENCE** Researchers design experiments so that only one thing is different between the treatments that are being compared. In the Hershey–Chase experiment, what was this single difference?
6. Analyze the following statements about DNA synthesis. Select True or False for each statement.
   T/F An RNA polymerase is essential for DNA synthesis.
   T/F Okazaki fragments would be unnecessary if DNA polymerase could synthesize DNA in both the $3' \rightarrow 5'$ and $5' \rightarrow 3'$ directions.
   T/F DNA ligase is used more frequently on the lagging strand than on the leading strand.
   T/F Toposiomerase is required to separate the two strands of DNA at the replication fork.
7. How does telomerase prevent linear chromosomes from shortening during replication?
8. What aspect of DNA structure makes it possible for the proteins of nucleotide excision repair to recognize many different types of DNA damage?
   a. the polarity of each DNA strand
   b. the antiparallel orientation of strands in the double helix
   c. the energy differences between correct and incorrect base pairs
   d. the regularity of DNA's structure

✔ TEST YOUR PROBLEM-SOLVING SKILLS

9. **MODEL** In the late 1950s, Herbert Taylor grew bean root-tip cells in a solution of radioactive thymidine (a precursor to one of the deoxyribonucleotides in DNA) and allowed them to undergo one round of DNA replication. He then transferred the cells to a solution without radioactive thymidine, allowed them to replicate again, and examined their chromosomes for the presence of radioactivity. His results are shown in the following figure, where red indicates a radioactive chromatid.

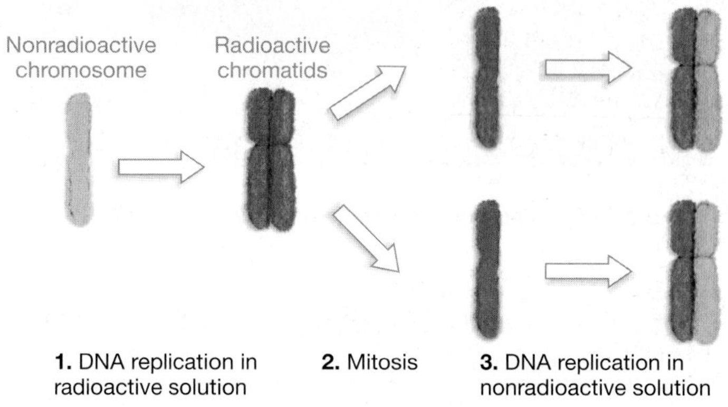

   a. Draw labeled diagrams of double-stranded DNA molecules that explain the pattern of radioactivity observed in the sister chromatids after the first and second rounds of replication.
   b. What would the results of Taylor's experiment be if eukaryotes used a conservative mode of DNA replication?
10. **QUANTITATIVE** The graph that follows shows the survival of four different *E. coli* strains after exposure to increasing doses of ultraviolet light. The wild-type strain is normal, but the other strains have a mutation in either a gene called *uvrA*, a gene called *recA*, or both.

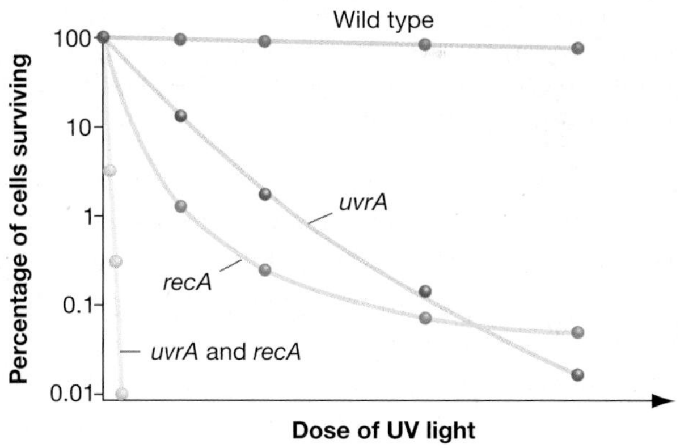

DATA: P. Howard-Flanders and R. P. Boyce. 1966. *Radiation Research Supplement* 6: 156–184, Fig. 8.

   a. Which strains are most sensitive to UV light? Which strains are least sensitive?
   b. What are the relative contributions of these genes to the repair of UV damage?

✔ PUT IT ALL TOGETHER: Case Study

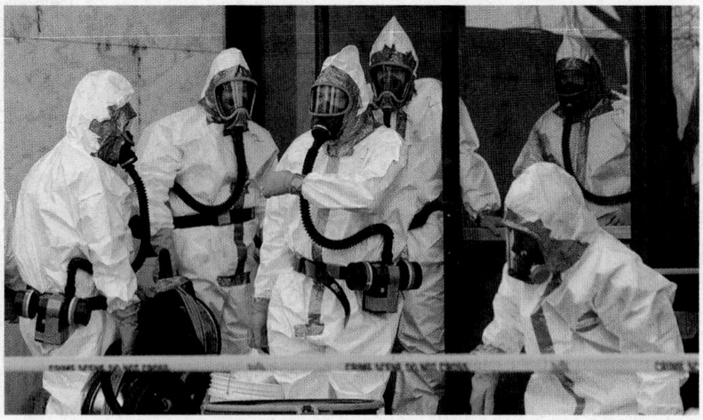

How are DNA synthesis inhibitors used to fight bioterrorism?

The bacterium *Bacillus anthracis*, commonly known as anthrax, causes an often fatal disease that is also called anthrax. In this photo, a biohazard team waits to enter the Hart U.S. Senate Office Building on November 7, 2001, after the Senate Majority Leader's office received an anthrax-tainted letter. More than 30 Senate workers showed signs of exposure.

11. The U.S. Centers for Disease Control and Prevention (CDC) recommends that anyone who might have been exposed to anthrax begin treatment with the antibiotic ciprofloxacin, a DNA-synthesis inhibitor. For ciprofloxacin to be useful, it must work
- **a.** exclusively against telomerase
- **b.** exclusively against bacterial DNA synthesis proteins
- **c.** exclusively against eukaryotic DNA synthesis proteins
- **d.** against both bacterial and eukaryotic DNA synthesis proteins

12. MODEL To understand how ciprofloxacin works, it's important to be able to visualize the events of DNA synthesis. To help with this, draw a diagram of a replication bubble that shows (1) the $5' \rightarrow 3'$ polarity of the two parental DNA strands, (2) the leading and lagging daughter strands at each replication fork, (3) helicase, and (4) topoisomerase.

13. Ciprofloxacin inhibits DNA gyrase, a bacterial topoisomerase that cuts DNA ahead of the replication fork, relieves the twists added during DNA synthesis, and then reseals the DNA. In ciprofloxacin-treated bacteria, newly synthesized DNA is found in fragments. Based on this evidence, what activity of DNA gyrase is likely to be inhibited by ciprofloxacin?

14. QUANTITATIVE Ciprofloxacin belongs to a family of antibiotics called quinolones. Researchers studying three quinolones added them separately to growing bacteria and then examined DNA synthesis rates over time compared to a culture not exposed to any drug. The results are shown in the figure here. Which antibiotic inhibited DNA synthesis most rapidly?

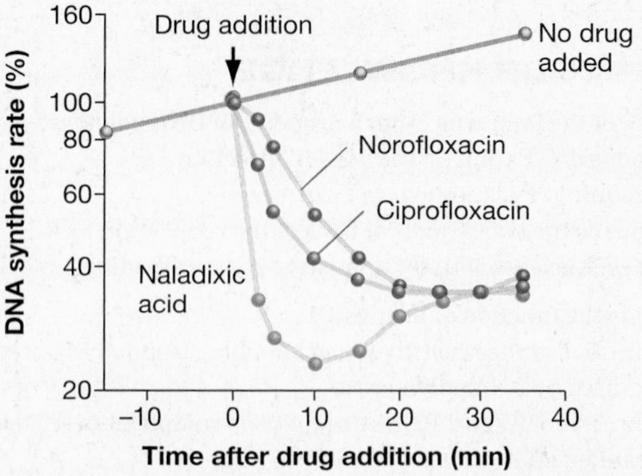

Source: B. Fournier et al. 2000. *Antimicrobial Agents and Chemotherapy* 44: 2160–2165.

15. QUANTITATIVE Based on results shown in the previous figure, how much more effective was naladixic acid than norofloxacin at inhibiting DNA synthesis 10 minutes after drug addition?

16. SOCIETY You are a reporter working in 2001 who's just been assigned a story on how ciprofloxacin may save the lives of Senate staff exposed to anthrax. Write a short paragraph that links knowing how DNA is replicated to defense against this act of bioterrorism.

Mastering Biology

Students Go to Mastering™ Biology for assignments, the eText, and the Study Area with animations, practice tests, and activities.

Professors Go to Mastering™ Biology for automatically graded tutorials and questions that you can assign to your students, plus Instructor Resources.

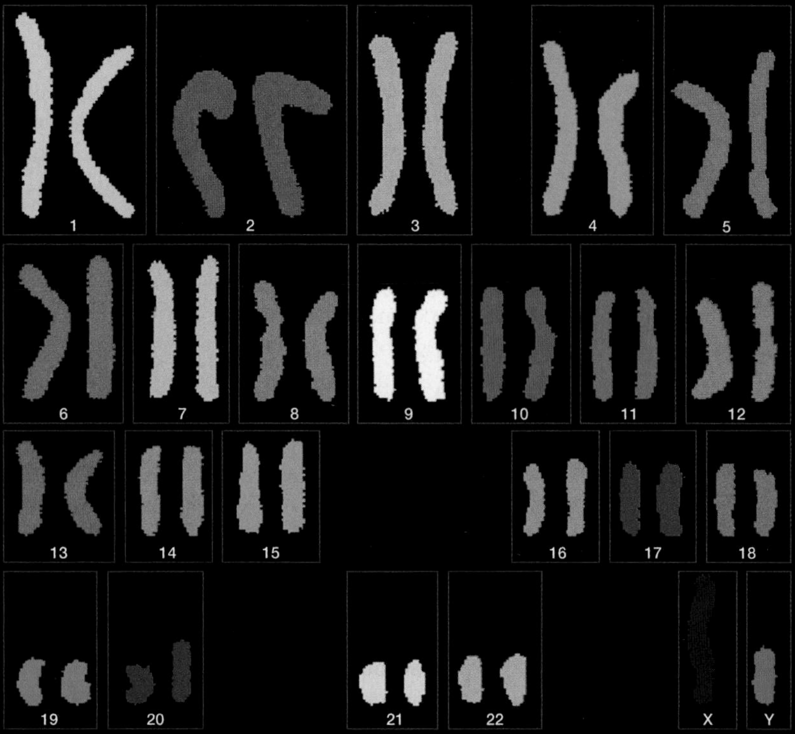

16 How Genes Work

This image shows a normal human male spectral karyotype—a micrograph of metaphase chromosomes stained and arranged to show different homologous pairs.

In this chapter you will learn how

Genetic information flows from DNA → RNA → proteins

by asking — What do genes do? **16.1**

then examining — The central dogma of molecular biology **16.2**

and analyzing — The genetic code, with its 3-letter "words" **16.3**

which together explain — How mutations can modify genes and genomes **16.4**

DNA has been called the blueprint of life. But how is this blueprint brought to life to specify the materials needed to build and maintain the cell, and remodel it when conditions change?

The science of **molecular biology**—the branch of biology that seeks to understand life by studying the molecules that create cells—began with the discovery that DNA is the hereditary material and that DNA is a double-helical structure containing sequences of four bases. But biologists had no idea how the information in DNA is translated into action. How does **gene expression**—the process of converting information coded in DNA into molecules that actually do things—occur?

This chapter introduces some of the most pivotal ideas in biology—ideas that connect genotypes to phenotypes by revealing how genes work at the molecular level. They also are at the heart of a key attribute of life: processing genetic information to produce a living organism.

Understanding how genes work triggered a major transition in biological science. Instead of thinking about genes as abstract hereditary determinants that somehow influence eye color in fruit flies or seed shape in garden peas, biologists could begin analyzing the molecular composition of genes, their products, and how these lead to phenotypes. The molecular revolution in biology took flight.

BIG PICTURE

This chapter is part of the Big Picture. See how on pages 404–405.

16.1 What Do Genes Do?

Although biologists of the early twentieth century made tremendous progress in understanding how genes are inherited, an explicit hypothesis explaining what genes do did not appear until 1941. That year George Beadle and Edward Tatum published a series of breakthrough experiments.

After you complete this section, you should be able to ...

- Explain the function of genes.

Beadle and Tatum's research was inspired by an idea that was brilliant in its simplicity. As Beadle said, "One ought to be able to discover what genes do by making them defective." The idea was to knock out a gene by damaging it and then infer what the gene does by observing the phenotype of the mutant individual. Recall that a mutation is a permanent change in a gene and that a mutant is an individual that carries a mutation (Ch. 14, Section 14.4).

Alleles that do not function at all are called **null alleles**, or **loss-of-function alleles**. The approach pioneered by Beadle and Tatum of creating loss-of-function mutant alleles and analyzing their effects remains one of the most common research strategies in studies of gene function.

The One-Gene, One-Enzyme Hypothesis

Beadle and Tatum used a model organism in their work—the bread mold *Neurospora crassa*. To start their work, they exposed a large number of *N. crassa* cells to radiation. Radiation can damage DNA—often in a way that makes the affected gene nonfunctional. Their next step was to examine the *N. crassa* cells, looking for mutants that failed to make specific compounds needed for normal growth.

For example, one of the mutants could not make a compound called pyridoxine, also known as vitamin B_6, even though normal cells can. Analysis of these mutants showed that the inability to synthesize pyridoxine was due to a defect in one gene, and that the failure to synthesize other molecules was due to defects in other genes.

These results inspired the **one-gene, one-enzyme hypothesis**. Beadle and Tatum proposed that each of the mutants could not make a particular compound because it lacked an enzyme required to synthesize the compound. They further proposed that the lack of the enzyme was due to a genetic defect. Therefore, Beadle and Tatum proposed that each gene contains the information needed to make an enzyme.

An Experimental Test of the Hypothesis

Three years later, Adrian Srb and Norman Horowitz published a rigorous test of the one-gene, one-enzyme hypothesis. These biologists focused on the ability of *N. crassa* to synthesize the amino acid arginine. In the lab, normal cells of this bread mold grow well on a medium without arginine. This is possible because wild-type cells produce their own arginine.

Previous work had shown that arginine is synthesized in a series of steps called a **metabolic pathway**. As Figure 16.1 shows, compounds called ornithine and citrulline are intermediate products in the metabolic pathway leading to arginine. Specific enzymes are required to convert the precursor into ornithine, alter ornithine to form citrulline, and change citrulline to arginine. Srb and Horowitz hypothesized that particular *N. crassa* genes are responsible for producing each of the three enzymes.

To test this idea, Srb and Horowitz used radiation to create a large number of mutant cells. To find the mutants they were looking for, the researchers performed a genetic screen. A **genetic screen** is any technique for picking particular types of mutants out of many randomly generated mutants—not unlike finding the proverbial needle in a haystack.

Srb and Horowitz began their genetic screen by growing colonies of irradiated cells on a medium that included arginine. Then they transferred a sample of each colony to a medium that *lacked* arginine. If a cell could grow with arginine but failed to grow without it, they concluded that it couldn't make its own arginine. Therefore, it must be a mutant with a defect somewhere in the arginine metabolic pathway shown in Figure 16.1.

To test the one-gene, one-enzyme hypothesis, Srb and Horowitz grew each mutant that couldn't synthesize arginine under four different conditions: **(1)** on a medium without added arginine, and on the same medium but supplemented with either **(2)** ornithine, **(3)** citrulline, or **(4)** arginine. As Figure 16.2 shows, each mutant was able to grow on one or more of these media but not on one or more of the others. The mutants fell into three distinct classes, called *arg1, arg2*, and *arg3*.

As the "Interpretation" section of the figure shows, the data make sense if each mutant was unable to carry out a different, specific step in a metabolic pathway. Because the metabolic defects of each mutant were due to a missing enzyme, Srb and Horowitz concluded that specific genetic defects that produced each of the mutants resulted in a specific enzymatic defect. Therefore, genes must encode enzymes. This experiment convinced most investigators that the one-gene, one-enzyme hypothesis was correct.

Later work showed that genes contain the information for all the proteins produced by an organism—not just enzymes. Biologists finally had a handle on what most genes do: They contain the instructions for making proteins. In many cases, though, a protein is made up of several different polypeptides (Ch. 3, Section 3.2), each of which is a product of a different gene. Consequently, the one-gene, one-enzyme hypothesis was modified to the one-gene, one-polypeptide hypothesis.

Ideas about the nature of genes continue to change as biologists probe deeper into genomes and how they work. The one-gene, one-polypeptide hypothesis has been broadened to account for genes that have RNA as a final product and (as you'll see in Ch. 17, Section 17.2) for genes that code for two or more related products.

CHECK YOUR UNDERSTANDING

✔ If you understood this section, you should be able to ...

1. **PROCESS OF SCIENCE** Predict how Srb and Horowitz's results and interpretations might have differed if they had studied a metabolic pathway with enzymes that were each composed of two different polypeptide chains.
2. Propose a definition of a gene based on current understanding described in this section.

Answers are available in Appendix A.

Figure 16.1 Different Enzymes Catalyze Each Step in the Metabolic Pathway for Arginine.

✔ If a cell lacked enzyme 2 but was placed in growth medium with only (1) ornithine or (2) citrulline, could it grow?

RESEARCH

QUESTION: What do genes do?

HYPOTHESIS: Each gene contains the information required to make one enzyme.

NULL HYPOTHESIS: Genes do not have a one-to-one correspondence with enzymes.

EXPERIMENTAL STRATEGY: Produce mutants unable to synthesize arginine. Test to see if each mutant also lacks one of the enzymes required for different steps in the pathway for synthesizing arginine.

EXPERIMENTAL SETUP: Isolate mutant *Neurospora crassa* that cannot synthesize arginine. Grow each type of mutant on growth medium that is:

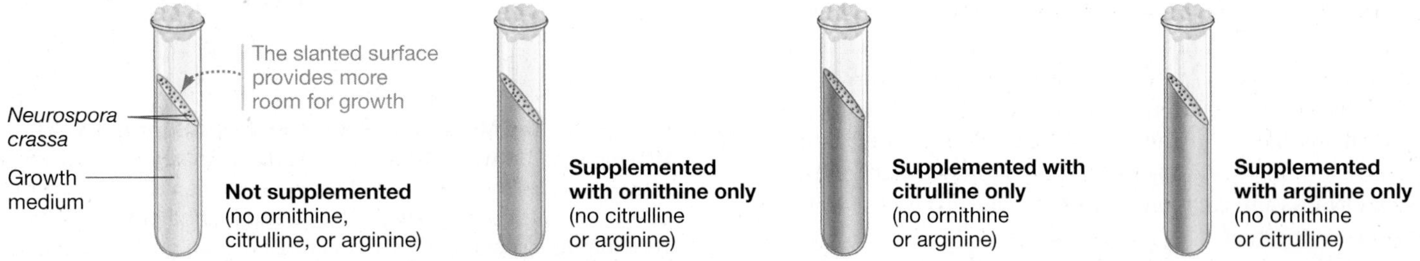

PREDICTION OF HYPOTHESIS: There will be three distinct types of mutants, corresponding to defects in enzyme 1, enzyme 2, and enzyme 3 in the pathway for synthesizing arginine. Each type of mutant will be able to grow on different combinations of the four kinds of media.

PREDICTION OF NULL HYPOTHESIS: There will not be a simple correspondence between a particular mutation and a particular enzyme.

RESULTS: There are three distinct types of mutants, called *arg1*, *arg2*, and *arg3*.

Mutant type	Supplement type: None	Ornithine only	Citrulline only	Arginine only
arg1	no growth	GROWTH	GROWTH	GROWTH
arg2	no growth	no growth	GROWTH	GROWTH
arg3	no growth	no growth	no growth	GROWTH

INTERPRETATION:

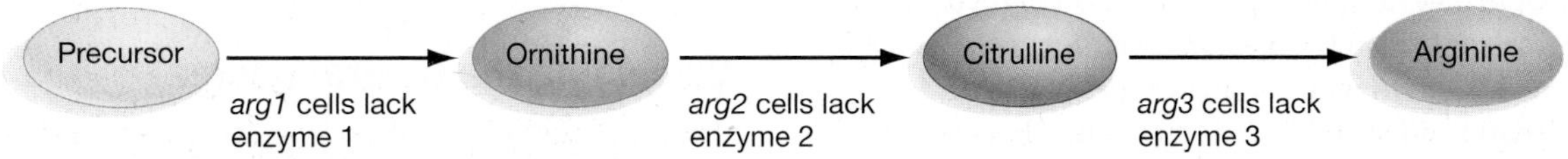

CONCLUSION: The one-gene, one-enzyme hypothesis is supported.

Figure 16.2 Evidence for the One-Gene, One-Enzyme Hypothesis. The one-to-one correspondence between specific genetic defects in different *N. crassa* mutants and specific defects in the metabolic pathway for arginine synthesis supported the one-gene, one-enzyme hypothesis.

SOURCE: A. M. Srb and N. H. Horowitz. 1944. The ornithine cycle in *Neurospora* and its genetic control. *Journal of Biological Chemistry* 154: 129–139.

16.2 The Central Dogma of Molecular Biology

How *does* a gene specify the production of a protein? As soon as Beadle and Tatum's hypothesis gained support, this question became a central issue in biology.

Part of the answer lay in the structure of DNA. Watson and Crick's model (Ch. 4, Section 4.2) offered little hope that DNA could directly catalyze reactions that produce proteins. DNA's structural regularity would not allow it to

After you complete this section, you should be able to . . .

- Explain the link between genotype and phenotype.

bind to the wide variety of substrates needed for protein synthesis. So what, then, did DNA do?

The Genetic Code Hypothesis

Francis Crick proposed that the sequence of bases in DNA acted as a code. His idea was that DNA was an information-storage molecule only. The instructions it contained would somehow have to be read and then used to somehow produce proteins.

Crick offered Morse code as an analogy. He proposed that different combinations of bases could specify the 20 amino acids, just as different combinations of dots and dashes in Morse code specify the 26 letters of the alphabet. A stretch of DNA, then, could contain the information needed to produce the amino acid sequence of a particular polypeptide.

The information required to build and operate a cell could be stored in a sequence of A's, T's, G's, and C's. This information could also be copied through complementary base pairing and transmitted faithfully from one generation to the next.

It soon became apparent, however, that the information encoded in the base sequence of DNA is not translated into the amino acid sequence of proteins directly. Instead, the link between DNA as information repository and proteins as cellular machines is indirect.

RNA as the Intermediary between Genes and Proteins

The first clue that the biological information in DNA must go through an intermediary to produce proteins came from knowledge of cell structure. In eukaryotic cells, DNA is enclosed within the nucleus (Ch. 7, Section 7.2), but the ribosomes, where protein synthesis takes place, lie outside the nucleus.

To make sense of this, François Jacob and Jacques Monod suggested that RNA molecules link genes in the nucleus and the protein-manufacturing centers that work in the cytoplasm. Jacob and Monod's hypothesis is illustrated in Figure 16.3. They predicted that single-stranded molecules of RNA, which they called **messenger RNA**, or **mRNA** for short, carry information out of the nucleus from DNA to the site of protein synthesis in the cytoplasm. Follow-up research confirmed the messenger RNA hypothesis and showed that messenger RNA is one of many types of RNA.

Biochemists searched for and found an enzyme that could use DNA to synthesize a complementary RNA molecule. This enzyme was called **RNA polymerase** because it polymerized ribonucleotides into strands of RNA. Like DNA polymerase, RNA polymerase uses a DNA strand as a template to specify which complementary nucleotide to add to the growing strand. However, unlike DNA polymerase (Ch. 15, Section 15.3), RNA polymerase does not require a primer to begin connecting ribonucleotides together to produce a strand of RNA.

An important element of the mRNA hypothesis was that a strand of DNA is used as a template for synthesis of a complementary strand of RNA. To test this, researchers created a reaction mix containing **(1)** the enzyme RNA polymerase; **(2)** ribonucleotides containing the bases adenine (A), guanine (G), cytosine (C), and uracil (U) (recall from Ch. 4, Section 4.3 that RNA contains the base U in place of the T found in DNA); and **(3)** single strands

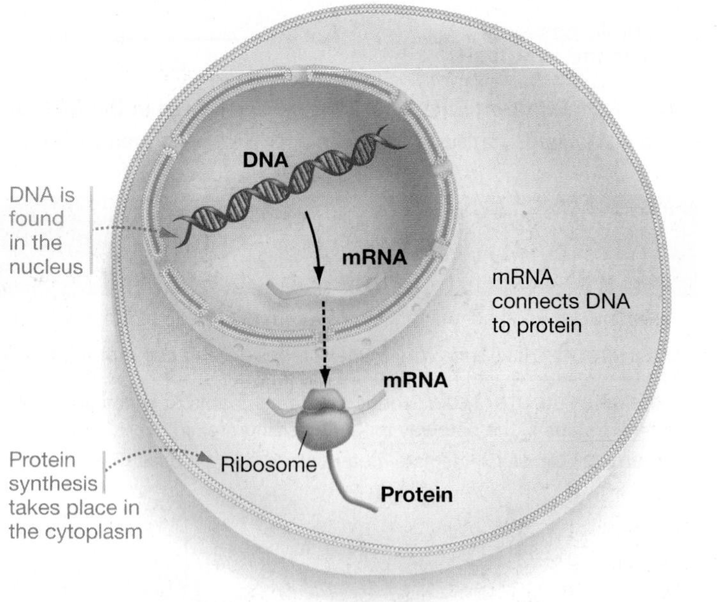

Figure 16.3 The Messenger RNA Hypothesis Supplied a Missing Link between DNA and Protein. In eukaryotic cells, most DNA is found in the nucleus, but proteins are manufactured using ribosomes outside the nucleus. Jacob and Monod proposed that the information coded in DNA is carried from the nucleus to the ribosomes by messenger RNA (mRNA).

✔ **PROCESS OF SCIENCE** Explain why the mRNA hypothesis was much more likely to be proposed based on observations of eukaryotic cells rather than bacterial cells.

of DNA in which the only base was thymine (T). After allowing the reaction to proceed, the biologists isolated RNA molecules that contained only the base adenine.

Because adenine pairs with thymine, this result supported the hypothesis that RNA polymerase synthesizes RNA using a strand of DNA as a template. Similar experiments showed that DNAs containing only cytosine result in the production of RNA molecules containing only guanine.

Dissecting the Central Dogma

Once the mRNA hypothesis was accepted, Francis Crick articulated a set of core principles that became known as the **central dogma** of molecular biology. The central dogma summarizes the flow of information from DNA to proteins. It states that DNA codes for RNA, which codes for proteins.

Crick devised what became iconic shorthand for this process:

$$\text{DNA} \longrightarrow \text{RNA} \longrightarrow \text{Proteins}$$

In this model, arrows represent the flow of information between molecules, not the conversion of one molecule into another.

The central dogma is elegant in its simplicity. DNA is the hereditary material. Genes consist of specific stretches of DNA that code for products used in the cell. The sequence of bases in DNA specifies the sequence of bases in an RNA molecule, which, in turn, specifies the sequence of amino acids in a protein. In this way, genes ultimately code for proteins. Proteins are the workers of cells, functioning as enzymes, motors, structural elements, transporters, signals, and regulators.

The Roles of Transcription and Translation Biologists use the terms "transcription" and "translation" to describe crucial steps in the flow of information from DNA to protein. These terms are similar, but they describe very different processes:

- **Transcription** is the process of using a DNA template to make an RNA molecule that has a base sequence complementary to the DNA. DNA is transcribed to RNA by RNA polymerase.
- **Translation** is the process of using the information in the base sequence of mRNA to synthesize proteins. Information in the messenger RNA is translated into proteins by ribosomes.

The terms transcription and translation make sense if you think about what each step accomplishes. Transcription means copying information. In the first step of the central dogma, information in DNA is copied (transcribed) into information in RNA. Compare this to translation, which in general refers to converting information in one language to another. In biology, translation is the conversion of information held in the nucleotide-based language of mRNA into the amino acid–based language of proteins. Whenever you hear "translation" in the context of biology, you should think "protein synthesis."

The following diagram shows where transcription and translation fit into the central dogma:

DNA (information storage) —Transcription→ **mRNA** (information carrier) —Translation→ **Proteins** (active cell machinery)

Linking Genotypes and Phenotypes An organism's genotype is determined by the sequence of bases in its DNA, while its phenotype is largely a product of the proteins it produces. Stated another way, the collection of proteins in a particular cell at a particular time plays the preeminent role in connecting genotype and phenotype. Here is a useful extension of Crick's idea:

$$\text{DNA (genotype)} \longrightarrow \text{mRNA} \longrightarrow \text{Proteins} \longrightarrow \text{Phenotype}$$

Transcription and translation, then, link genotype and phenotype.

To appreciate the connection between genotype and phenotype, let's consider coat color in mice (*Peromyscus polionotus*) native to southeastern North America. The mice have a gene for a protein called the melanocortin receptor. Melanocortin is a hormone that works through the melanocortin receptor to influence how much dark pigment is deposited in fur. Coat color is an important phenotype, and it is determined partly by the DNA sequence of the melanocortin receptor gene (**Figure 16.4**).

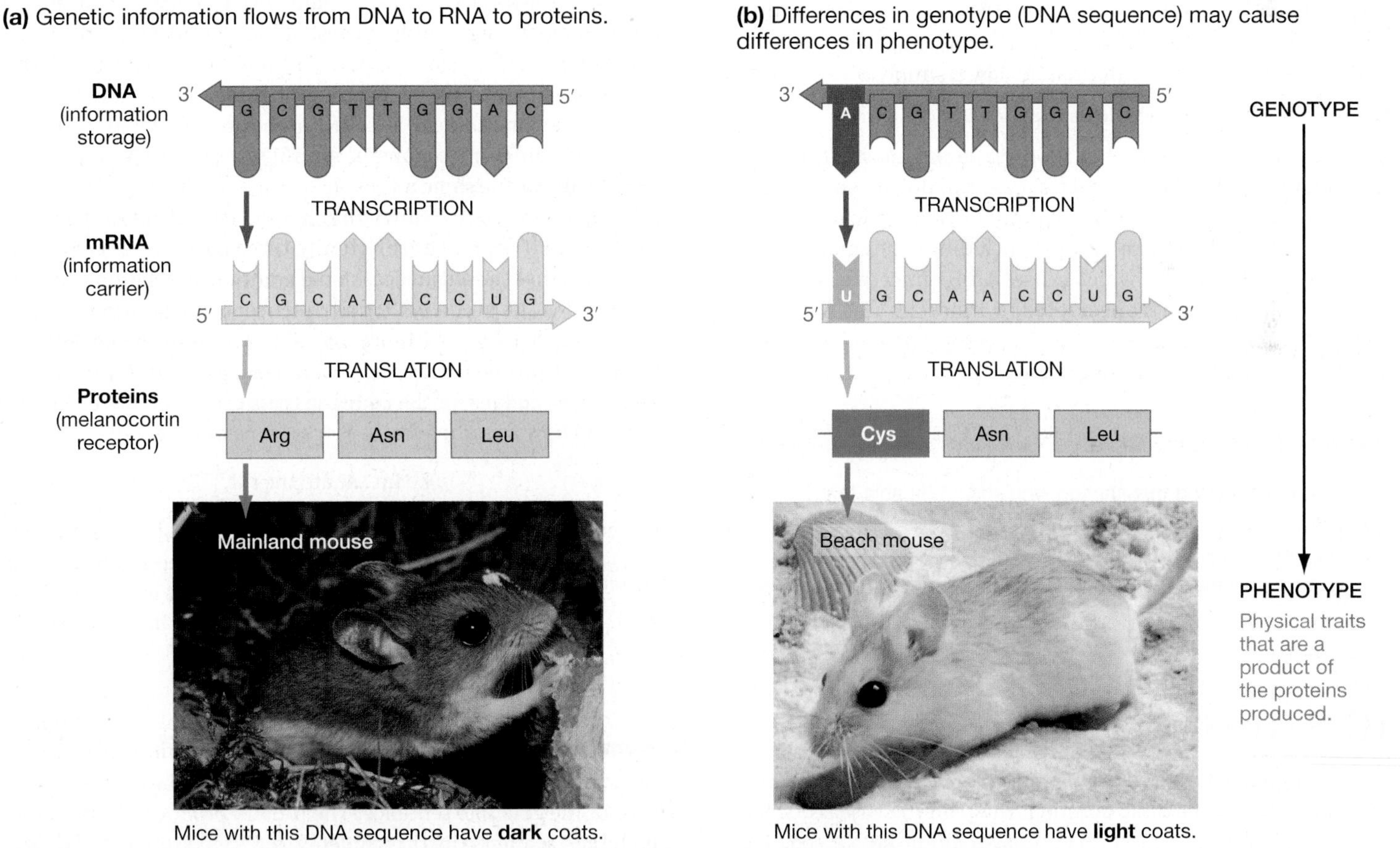

INTERACTIVE Figure 16.4 The Central Dogma Explains the Relationship between Genotype and Phenotype. The DNA sequences given in parts **(a)** and **(b)** are from different genotypes that influence mouse coat-color phenotypes. Mainland mice are camouflaged in abandoned fields, and beach mice are camouflaged in sandy habitats. Only the DNA strand used as a template during transcription is shown.

Figure 16.4 shows two related but distinct DNA sequences of a small part of the melanocortin receptor gene (the DNA is double stranded, but only the strand used as a template during transcription is shown). Each DNA sequence constitutes a particular allele (form) of the gene. The result of these slightly different DNA sequences is the production of proteins that differ in their amino acid sequence. One allele specifies an arginine residue where the other specifies a cysteine residue. If the amino acid sequences of proteins vary, their functions are likely to vary as well. Is this the case for the proteins produced from the two melanocortin receptor alleles?

Melanocortin receptors that have arginine in the location shown in Figure 16.4a result in a large amount of coat pigment, but receptors that have cysteine in this location result in small amounts of coat pigment. Whether a mouse is dark or light depends largely on a single base change in its DNA sequence. In this case, a tiny difference in DNA sequence produces a large change in phenotype. This is how the central dogma links genotype to phenotype.

Modifications of the Central Dogma The core of the central dogma has stood the test of time, but new discoveries have required some modifications to this simple scheme. These exceptions will be explored in detail in later chapters. Here, let's simply take a look at some highlights:

- Many genes code for RNA molecules that do not function as mRNAs—they are transcribed from DNA but never translated into proteins. These RNAs control many important phenotypes. For these genes, information flow is simply DNA → RNA.
- Information sometimes flows from RNA back to DNA.

The discovery in the early 1970s of "reverse" information flow from RNA to DNA seemed to turn the central dogma on its head. The first evidence came from one group of viruses whose genes consist of RNA. When these viruses—including the human immunodeficiency virus (HIV), which causes AIDS—infect a cell, a viral enzyme called **reverse transcriptase** synthesizes a DNA version of the RNA genes. The conclusion? The central dogma remains a core principle, but life isn't so dogmatic.

CHECK YOUR UNDERSTANDING

✔ **If you understood this section, you should be able to ...**

1. Explain why mRNA is aptly named.
2. Describe the action of each arrow in the central dogma.
3. Explain how the central dogma links genotype and phenotype.

Answers are available in Appendix A.

16.3 The Genetic Code

After you complete this section, you should be able to ...

- Describe the properties of the genetic code.

Once biologists understood the pattern of information flow in the cell, they faced the next big question: How can the base sequence in a strand of mRNA code for the sequence of amino acids in a protein? At the heart of this question is the **genetic code**—the rules that specify the relationship between a sequence of nucleotides in DNA or RNA and the sequence of amino acids in a protein. Researchers from all over the world took up the challenge. A race was on.

How Long Is a "Word" in the Genetic Code?

The first step in cracking the genetic code was to determine how many bases make up a "word." In this genetic language, words specify amino acids. In a sequence of mRNA, how long is a message that designates one amino acid?

Based on some simple logic, George Gamow suggested that each genetic word contains three bases. His proposal rested on the observation that 20 amino acids are used in cells, and on the hypothesis that each amino acid must be specified by a particular sequence of mRNA. His reasoning was as follows:

- There are only four different bases in ribonucleotides (A, U, G, and C), so a one-base code could specify only four different amino acids.
- A two-base code could represent just 4×4, or 16, different amino acids.
- A three-base code could specify $4 \times 4 \times 4$, or 64, different amino acids.

A three-base code, known as a **triplet code**, is the shortest genetic word to code for at least 20 amino acids.

Because a three-base code could specify 64 different amino acids, Gamow's hypothesis suggested that the genetic code might be redundant. That is, an amino acid could be specified by more than one triplet of bases. A group of three bases that specifies a particular amino acid is called a **codon**. Redundancy would mean that different codons in an mRNA—say, AAA and AAG—might code for the same amino acid.

Francis Crick and Sydney Brenner devised an elegant experiment to see if codons were actually three bases long. They used chemicals that caused an occasional addition or deletion of a base pair in DNA. As predicted for a triplet code, an addition or deletion of one base pair led to a loss of function in the gene being studied. This is because a single addition or deletion throws the sequence of codons, or the **reading frame**, out of register. To understand how a reading frame works, consider the sentence

"The fat cat ate the rat."

The reading frame of this sentence is a set of sequential three-letter words. If the fourth letter in this sentence—the *f* in "fat"—were deleted, the reading frame, which is set by the first three-letter word, would transform the sentence into gibberish:

"The atc ata tet her at."

When the reading frame in a DNA sequence is thrown out of register by the addition or deletion of a base pair, the composition of each subsequent codon changes, just like the letters in each word of the example sentence. The protein produced from the altered DNA sequence has a completely different sequence of amino acids. In terms of its normal function, this protein is gibberish.

Crick and Brenner were also able to produce DNA sequences that had deletions or additions of two or three base pairs. The only time functional proteins were produced was when three base pairs were added or removed. As an example, removing the three letters that spell "fat" from the original sentence results in

"The cat ate the rat."

Just as the altered sentence still conveys meaning, the genes with deletions of three base pairs in Crick and Brenner's study produced a functional protein. The researchers interpreted these results as evidence for a triplet genetic code. The question now became, which amino acid is specified by each of the 64 codons?

How Did Researchers Crack the Code?

The initial advance in deciphering the genetic code came in 1961, when Marshall Nirenberg and Heinrich Matthaei developed a method for synthesizing RNAs composed of a single type of ribonucleotide. They began by creating a long polymer of uracil-containing ribonucleotides. These synthetic RNAs were added to an in vitro (literally, "in glass," and used to mean "in a test tube" or "outside the cell") system for synthesizing proteins. The researchers analyzed the resulting amino acid chain and determined that it was a polymer of the amino acid phenylalanine.

This result showed that the RNA triplet UUU codes for phenylalanine. The researchers then synthesized the RNA triplet AAA, which produced a polypeptide composed entirely of lysine, and the RNA triplet CCC, which produced a polypeptide composed entirely of proline.

Extending this work, Nirenberg and Matthaei created RNAs from mixtures of different ribonucleotides. The RNA synthesis method they used did not require a template. This meant that by knowing the proportions of the ribonucleotides used to synthesize RNA, the researchers could predict how often a particular triplet would occur on average in the RNA molecules. Based on the types and proportions of amino acids incorporated into polypeptides synthesized using the RNA templates, they were able to infer the meanings of many more codons—but far from all of them.

A breakthrough came when Nirenberg and another colleague, Philip Leder, devised a way to synthesize single codons with specific base sequences. They then added these synthetic codons individually to a cell extract containing ribosomes and the 20 different amino acids in a form suitable for protein synthesis. The researchers then determined which amino acid became bound to ribosomes when a particular codon was present. For example, when the codon CAC was used in the reaction mix, the amino acid histidine would bind to the ribosomes. This indicated that CAC codes for histidine. These ribosome-binding experiments enabled Nirenberg and Leder to fill in the meanings of all the remaining unknown codons.

Later, researchers discovered that some codons did not specify amino acids. Instead, one particular codon signaled an end to the reading frame, much like a period in a sentence. Another codon marked the beginning of a reading frame, essentially opening each "sentence" (reading frame) with the same word.

- There is one **start codon**. Working with nearby ribonucleotide sequences, the AUG start codon signals that protein synthesis should begin at that point on the mRNA molecule. The start codon specifies the amino acid methionine. It sets the reading frame of the message, locking in which set of three-base triplets constitute "words."
- There are three **stop codons**, also called termination codons, with the sequences UAA, UAG, and UGA. Stop codons do not code for any amino acid but signal the end of the reading frame and, therefore, the end of the polypeptide.

The complete genetic code is given in Figure 16.5. Deciphering it was a landmark achievement.

FIRST BASE	SECOND BASE: U	SECOND BASE: C	SECOND BASE: A	SECOND BASE: G	THIRD BASE
U	UUU, UUC – Phenylalanine (Phe)	UCU, UCC, UCA, UCG – Serine (Ser)	UAU, UAC – Tyrosine (Tyr)	UGU, UGC – Cysteine (Cys)	U, C
U	UUA, UUG – Leucine (Leu)		UAA – **Stop codon**	UGA – **Stop codon**	A
U			UAG – **Stop codon**	UGG – Tryptophan (Trp)	G
C	CUU, CUC, CUA, CUG – Leucine (Leu)	CCU, CCC, CCA, CCG – Proline (Pro)	CAU, CAC – Histidine (His)	CGU, CGC, CGA, CGG – Arginine (Arg)	U, C
C			CAA, CAG – Glutamine (Gln)		A, G
A	AUU, AUC, AUA – Isoleucine (Ile)	ACU, ACC, ACA, ACG – Threonine (Thr)	AAU, AAC – Asparagine (Asn)	AGU, AGC – Serine (Ser)	U, C
A	AUG – Methionine (Met) **Start codon**		AAA, AAG – Lysine (Lys)	AGA, AGG – Arginine (Arg)	A, G
G	GUU, GUC, GUA, GUG – Valine (Val)	GCU, GCC, GCA, GCG – Alanine (Ala)	GAU, GAC – Aspartic acid (Asp)	GGU, GGC, GGA, GGG – Glycine (Gly)	U, C
G			GAA, GAG – Glutamic acid (Glu)		A, G

Figure 16.5 **The Genetic Code Consists of Three-Letter "Words."** To read a codon in mRNA, locate its first base in the red band on the left; then move rightward at that level to the box under the codon's second base in the blue band along the top. Finally, in that box, locate the codon's third base according to the green band on the right side to learn the amino acid. By convention, codons are written in the $5' \rightarrow 3'$ direction.

Analyzing the Code Once biologists had cracked the genetic code, they saw a set of important properties:

- ***The code is redundant.*** All amino acids except methionine and tryptophan are coded for by more than one codon. This is referred to as degeneracy, or redundancy, of the code. Codons specifying the same amino acid are synonymous codons.
- ***The code is unambiguous.*** A given codon never codes for more than one amino acid.
- ***The code is non-overlapping.*** Once the ribosome locks onto the first codon, the reading frame is established, and the ribosome then reads each separate codon one after another.
- ***The code is (nearly) universal.*** With a few minor exceptions, all codons specify the same amino acids in all organisms.
- ***The code is conservative.*** When several codons specify the same amino acid, the first two bases in those codons are usually identical.

The last point is subtle, but important. Here's the key: If a change in DNA sequence leads to a change in the third position of a codon, it is less likely to alter the amino acid in the protein. In this way, the genetic code minimizes the phenotypic effects of small alterations in DNA sequence. Modeling studies indicate that the genetic code was not assembled randomly, but has been honed by natural selection and is remarkably efficient.

The Value of Knowing the Code Knowing the genetic code and the central dogma allows biologists to

- Predict the amino acid sequence encoded by a particular DNA sequence (**Figure 16.6**).
- Determine mRNA and DNA sequences that could code for a particular sequence of amino acids.

Typically, a set of mRNA or DNA sequences, not just one, can code for a particular amino acid sequence. This is because the code is redundant. For example, if a polypeptide contains phenylalanine, you don't know if the codon responsible is UUU or UUC.

When scientists analyze genomes, they use the genetic code to learn what stretches of DNA code for proteins and what kinds of proteins they are. Conversely, when biologists have studied a protein and then want to find the corresponding gene, they use the amino acid sequence of the polypeptide to deduce DNA sequences that could encode the protein. These DNA sequences are then searched for in genome sequences by using methods described in Chapter 20 (Section 20.3) to find the corresponding gene. None of this work would be possible without knowing the genetic code.

✔ **If you understand how to read the genetic code, you should be able to: (1) Identify the codons in Figure 16.4 and decide whether they are translated correctly. (2) Write an mRNA that codes for the amino acid sequence Ala-Asn-Asp-Phe-Gln but is different from the one given in Figure 16.6a. Indicate the 5′ → 3′ polarity of the mRNA. Then write the double-stranded DNA that corresponds to this mRNA, indicating the 5′ → 3′ polarity of both DNA strands.**

Once biologists understood the central dogma and genetic code, they were able to explore and eventually understand the molecular basis of mutation. Questions like how do novel traits such as dwarfing in garden peas and white eye color in fruit flies come to be could finally be answered.

CHECK YOUR UNDERSTANDING

✔ **If you understood this section, you should be able to ...**

1. **PROCESS OF SCIENCE** Predict what Crick and Brenner would have found if codons were two bases long.
2. Underline the start and stop codons in the mRNA sequence 5′-UAUCCAUGGCACUUUAAAC-3′.
3. **QUANTITATIVE** Predict how many different mRNA sequences could code for the following amino acid sequence plus a stop codon: Met-Trp-Cys-(Stop).

Answers are available in Appendix A.

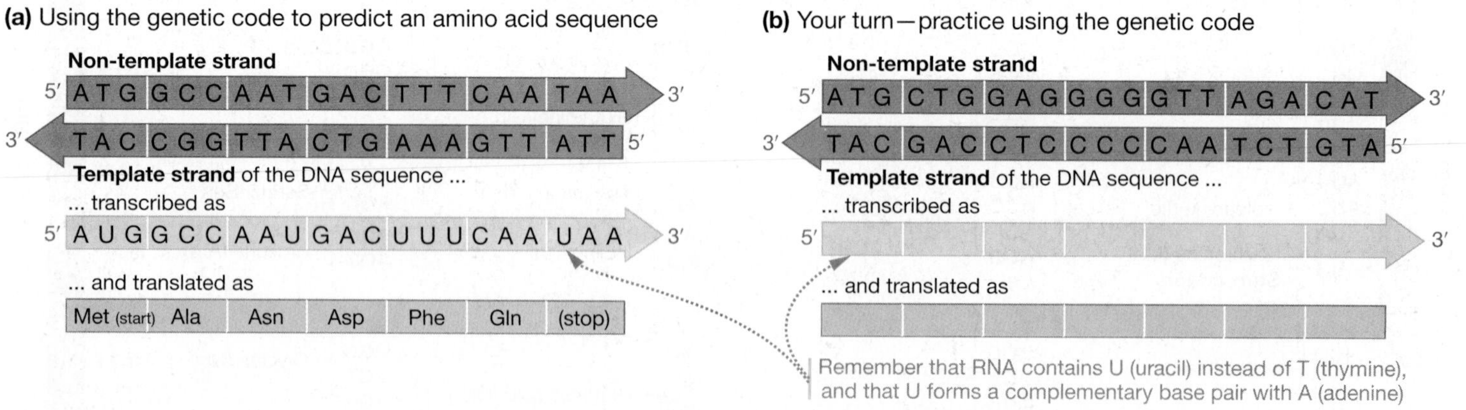

Figure 16.6 The Genetic Code Can Predict Amino Acid Sequences. The strand of DNA that is transcribed is the template strand, and the strand of DNA that is not transcribed is the non-template strand. The non-template strand of a protein-coding DNA sequence has the same polarity and sequence as the RNA except that where a T occurs in DNA, a U is found in RNA.

✔ Fill in the mRNA and amino acid sequences in part (b).

16.4 What Are the Types and Consequences of Mutation?

Mutations change genetic information. Earlier, a mutation was defined as a heritable change in a gene. Heritability can be from mother cell to daughter cells or between generations of multicellular organisms. Now that you know more about the nature of genes, it's possible to provide a molecular view of **mutation**: a mutation is any permanent change in an organism's DNA. It is a modification in a cell's information archive—a change in its genotype. In this way, mutations create new alleles.

> **After you complete this section, you should be able to ...**
>
> - Explain the ways mutations can change genetic information.

Mutations can alter DNA sequences that range in size from a single base pair in DNA to whole sets of chromosomes. They can occur in DNA sequences anywhere in the genome—in sequences that code for amino acids, in regions important for regulating gene expression, and in regions of DNA that have no known function. Later chapters will cover many additional aspects of mutation. Here, let's get a first look at some different types of mutations and their consequences.

Point Mutations

Figure 16.7 shows how a common type of mutation occurs. If a mistake is made during DNA synthesis or repair, a change in the sequence of bases in the DNA results that can lead to mutation. A mutation that alters the sequence of one or a small number of base pairs is called a **point mutation**.

What happens when a point mutation occurs in a region of DNA that codes for proteins? To answer this question, look back at Figure 16.4 and recall that a change in a single base pair in DNA is associated with a difference in amino acid sequence in a protein that controls coat color in mice. The DNA sequence in Figure 16.4a is found in dark-colored mice that live in mainland habitats with dark backgrounds; the sequence in Figure 16.4b is found in light-colored mice that live on beaches.

Which of the melanocortin receptor alleles was the original, or ancestral, allele and which allele was changed by mutation? Because beach-dwelling populations are evolutionarily younger than the mainland populations, researchers hypothesize the following sequence of events:

1. Dark-coated mainland mice colonized beach habitats.
2. Either before or after the colonization event, a random point mutation occurred in a mouse that altered the melanocortin receptor gene and resulted in some offspring with light coats.
3. Light-colored mice are camouflaged in beach habitats; in these sandy environments, they experience less predation than do dark-colored mice.
4. Over time, the mutated allele increased in frequency in beach-dwelling populations.

Point mutations like this one that change the identity of an amino acid in a protein are called **missense mutations**.

Note that if the same G-to-A change had occurred in the third base of the DNA sequence coding for arginine, instead of the first, there would have been no change in the protein produced. This mutation would change the codon CGC to CGU, and both of these code for arginine. A point mutation like this that does not change the amino acid sequence of the gene product is called a **silent mutation**.

Some point mutations alter many amino acids of a protein, not just one. How can this be? Recall that addition or deletion of a single base pair within a protein-coding region throws the sequence of codons out of register and alters the meaning of all subsequent codons. Such mutations shift the reading frame and are aptly called **frameshift mutations**. These almost always destroy the function of the protein.

Another type of point mutation with a large effect is a **nonsense mutation**. Nonsense mutations occur when a codon that specifies an amino acid is changed by mutation to one that specifies a stop codon. This causes early termination of the polypeptide chain and, like frameshift mutations, results in a nonfunctional protein.

The same mutation can be viewed many different ways. An important perspective is the effect a mutation has on the fitness

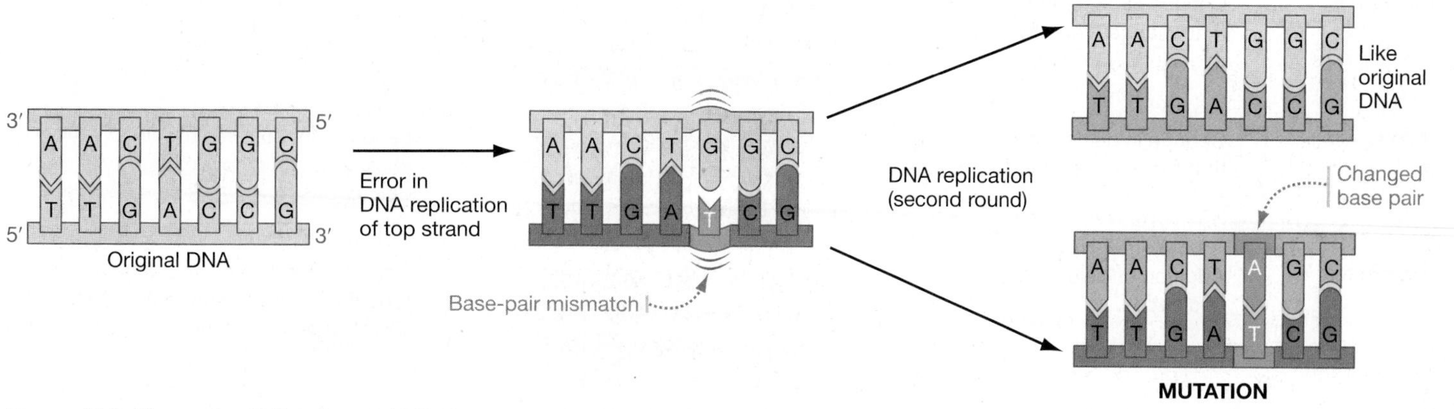

Figure 16.7 Unrepaired Mistakes in DNA Synthesis Lead to Point Mutations.

of an organism. Biologists taking this perspective divide mutations into three categories:

1. **Beneficial** Some mutations increase the fitness of the organism—its ability to survive and reproduce in a particular environment. For example, the G-to-A missense mutation is beneficial in beach habitats because it camouflages mice.
2. **Neutral** If a mutation has no effect on fitness, it is termed neutral. Silent mutations are usually neutral.
3. **Deleterious** Because most individuals are well adapted to their current habitat, and because mutations are random changes in genotype, many mutations lower fitness. These mutations are termed harmful, or deleterious. The same G-to-A mutation in mice that was beneficial in a beach habitat is deleterious in a mainland habitat. Note that whether a mutation is harmful or beneficial is not a property of the mutation alone—it depends on the environment.

Most point mutations are slightly deleterious or neutral. **Table 16.1** summarizes the types of point mutations that occur in protein-coding sequences of a gene and reviews their consequences for the amino acid sequences of proteins and for fitness.

Point mutations can and do occur in DNA sequences that do not code for proteins. These mutations are not referred to as missense, silent, frameshift, or nonsense mutations, because those terms apply only to mutations that can change the protein-coding portions of a gene. However, if point mutations alter DNA sequences that are important for *gene expression*, they can have important effects on phenotype even though they do not change the amino acid sequence of a protein. As discussed in Chapter. 21 (Section 21.6), many mutations important for evolution are exactly of this type.

✔ If you understand point mutations, you should be able to predict the kind of mutation—missense, silent, nonsense, or frameshift—that would be created by changing the codon-specifying sequence 5′-ATA-3′ in the non-template strand of DNA to ATC, TTA, or ACA, respectively.

Chromosome Mutations

Besides point mutations, there are much larger-scale mutations that change either the structure or number of chromosomes. Like point mutations, these types of mutations are permanent changes in the DNA. You might recall, for example, that polyploidy is the state of having more than two of each type of chromosome, while aneuploidy results from the addition or deletion of individual chromosomes (Ch. 13, Section 13.3).

Changes in chromosome number result from mistakes in moving chromosomes into daughter cells during meiosis or mitosis. Mutations producing polyploidy and aneuploidy don't change DNA sequences but do cause a permanent change in an organism's DNA by altering the number of copies of each sequence. Not all mutations that change chromosome number are deleterious: The formation of polyploids plays a major role in the evolution of many species, particularly plants (Ch. 24, Section 24.3).

In addition to changes in chromosome number, the structure of individual chromosomes can change in significant ways. These

SUMMARY Table 16.1 Consequences of Point Mutations That Alter Codons

Name	Definition	Example	Consequence
	Original sequence of non-template DNA	TAT TGG CTA GTA CAT	
	Original mRNA transcript	UAU UGG CUA GUA CAU	
		Tyr–Trp–Leu–Val–His	Original polypeptide
Silent	Change in nucleotide sequence that does not change the amino acid specified by a codon	TAC TGG CTA GTA CAT UAC UGG CUA GUA CAU Tyr–Trp–Leu–Val–His	No change in phenotype; neutral with respect to fitness
Missense	Change in nucleotide sequence that changes the amino acid specified by codon	TAT TGT CTA GTA CAT UAU UGU CUA GUA CAU Tyr–Cys–Leu–Val–His	Change in primary structure of protein; may be beneficial, neutral, or deleterious
Nonsense	Change in nucleotide sequence that results in an early stop codon	TAT TGA CTA GTA CAT UAU UGA CUA GUA CAU Tyr STOP	Leads to mRNA breakdown or a shortened polypeptide; usually deleterious
Frameshift	Addition or deletion of a nucleotide	TAT TCG GCT AGT ACA T UAU UCG GCU AGU ACA U Tyr–Ser–Ala–Ser–Thr	Reading frame is shifted, altering the meaning of all subsequent codons; almost always deleterious

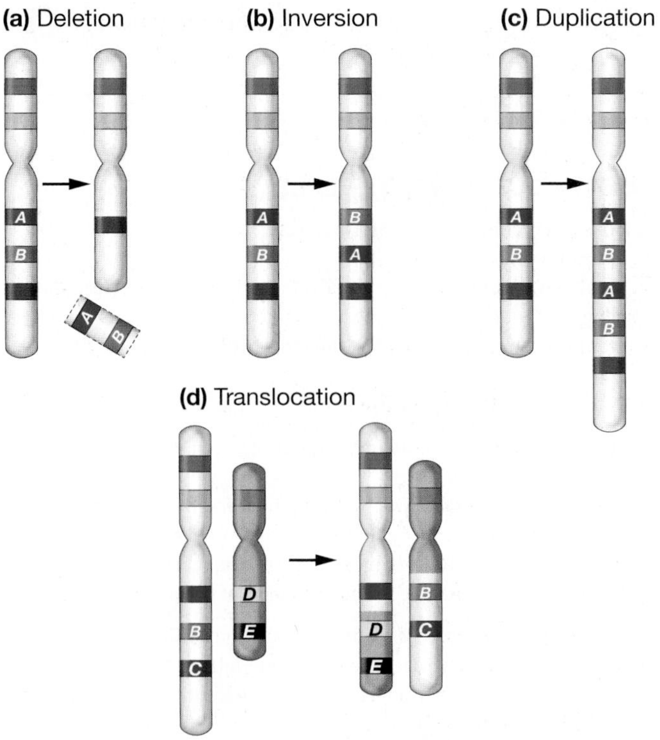

Figure 16.8 **There Are Four Types of Chromosome Structural Mutations.**

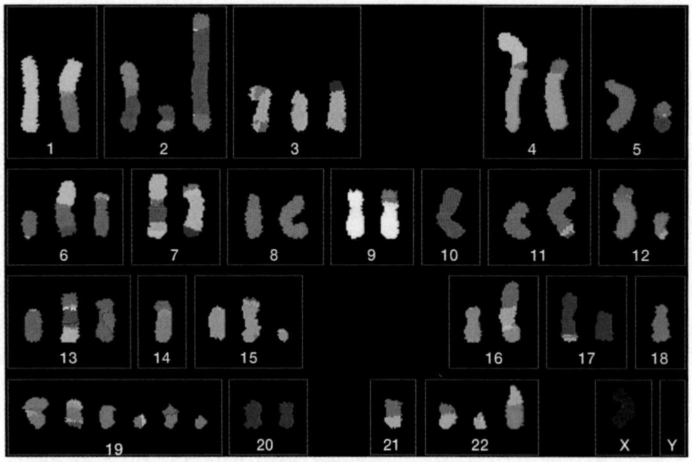

Figure 16.9 **Karyotypes Can Reveal Chromosome-Level Mutations.** This spectral karyotype of a breast cancer cell from a female shows chromosome rearrangements and aneuploidy typical of cancer. In a normal spectral karyotype, as seen in the chapter-opening image, each chromosome is stained a single, solid color, which varies for each chromosome pair.

✔ Compare this karyotype to the one shown in the chapter-opening image. Remember that females normally have two X chromosomes and males normally have one X chromosome. Which chromosomes show evidence of aneuploidy? Which chromosomes show evidence of rearrangements?

structural changes are an important type of chromosome mutation that profoundly affects evolution and health. There are four major types:

1. Breaks in chromosomes are common and are the source of many structural mutations in chromosomes. A broken segment of a chromosome can be lost, causing a **deletion** (Figure 16.8a).
2. Segments of a broken chromosome may be flipped and rejoined, creating a chromosome **inversion** (Figure 16.8b).
3. Errors in crossing over or in DNA synthesis can lead to the presence of one or more additional copies of a segment—a **duplication** (Figure 16.8c). Duplication of chromosome segments duplicates genes, opening the door to a central process of evolution—duplication and divergence (Ch. 20, Section 20.6)—that creates families of related genes.
4. Finally, a broken piece of a chromosome can become attached to a different chromosome, an event called chromosome **translocation**. Relatively often, two different chromosomes experience breaks in their DNA. The broken DNA is usually rejoined, but sometimes to the incorrect chromosomes. This creates a reciprocal translocation, the type of translocation shown in Figure 16.8d.

Like point mutations, chromosome mutations can be beneficial, neutral, or deleterious. For example, more than 200 different inverted sections of chromosomes were found in comparisons of the DNA from eight phenotypically normal people. These mutations appear to be neutral. Not all chromosome mutations are so harmless, however. Chromosomes of cancer cells exhibit deleterious chromosome mutations that include aneuploidy, inversions, translocations, deletions, and duplications. Figure 16.9 drives this point home by showing the **karyotype**—the complete set of chromosomes in a cell—of a cancerous human cell. Chromosome mutations abound.

To summarize, point mutations and chromosome mutations are random changes in DNA that can produce new genes, alleles, and traits. At the level of individuals, mutations can cause disease or death or lead to increases in fitness. At the level of populations, mutations furnish the heritable variation that makes evolution possible. The central role of mutation in evolution is explored in depth in Unit 5.

CHECK YOUR UNDERSTANDING

✔ If you understood this section, you should be able to . . .

1. Predict whether the position of a mutation within a reading frame, for example near the beginning of the reading frame or near its end, is likely to play a bigger role in protein function in the case of a missense or a nonsense mutation.
2. Describe two ways that a mutation in a coding sequence can be a neutral mutation.
3. Explain why changes in both chromosome structure and number are considered mutations.

Answers are available in Appendix A.

16.1 What Do Genes Do?

- Experiments with mutants of the bread mold *N. crassa* led to the one-gene, one-enzyme hypothesis.
- The original one-gene, one-enzyme hypothesis has been broadened to account for genes that code for proteins other than enzymes, for genes that have RNA as a final product, and for genes that have more than one product.

16.2 The Central Dogma of Molecular Biology

- DNA is transcribed to messenger RNA (mRNA) by RNA polymerase, and then mRNA is translated to proteins by ribosomes. In this way, genetic information is converted from DNA to RNA to protein.
- The flow of information from DNA to RNA to protein is called the central dogma of molecular biology.
- The central dogma has been extended to account for many RNAs that do not code for proteins. Instead, these RNAs perform other important functions in the cell.

16.3 The Genetic Code

- Each amino acid in a protein is specified by a codon—a group of three bases in mRNA that is determined by a complementary sequence in DNA.
- By synthesizing RNAs with different base sequences and observing the results of translation, researchers were able to decipher the genetic code.
- The genetic code is redundant—meaning that most of the 20 amino acids are specified by more than one codon.
- Certain codons signal where translation starts and stops.

16.4 What are the Types and Consequences of Mutation?

- Mutations are random, permanent changes in DNA that range from alterations in a single base pair to changes in the structure or number of chromosomes.
- Point mutations in protein-coding regions may have no effect on the protein (silent mutation), may change a single amino acid (missense mutation), may shorten the protein (nonsense mutation), or may shift the reading frame and alter many amino acids (frameshift mutation).
- Mutations can occur anywhere in the genome, including regions that do not code for proteins.
- Mutations can have beneficial, neutral, or harmful effects on the fitness of organisms.

Answers are available in Appendix A.

✔ TEST YOUR KNOWLEDGE

1. If a base-pair change occurs in DNA, this
  a. is a mutation.
  b. would be a mutation only if it falls in a protein-coding part of a gene.
  c. would be a mutation only if it falls in a transcribed part of the genome.
  d. is not a mutation, because only one base pair has been altered.

2. Which of the following is an important exception to the central dogma of molecular biology?
  a. Many genes code for RNAs that function directly in the cell.
  b. DNA is the repository of genetic information in all cells.
  c. Messenger RNA is a short-lived "information carrier."
  d. Proteins are responsible for most aspects of the phenotype.

3. DNA's primary structure is made up of just four different bases, and its secondary structure is regular and highly stable. How can a molecule with these characteristics hold the information required to build and maintain a cell?

4. Which of the following describes the experimental strategy that was used to decipher the genetic code?
  a. comparing the amino acid sequences of proteins with the base sequence of their genes
  b. analyzing the sequence of RNAs produced from known DNA sequences
  c. analyzing mutants that changed the code
  d. examining the polypeptides produced when RNAs with particular sequences were translated

✔ TEST YOUR UNDERSTANDING

5. THINK CAREFULLY A friend says, "Geneticists spend all their time talking about DNA, but that's silly because DNA really isn't that important in the functions of a cell." In what ways is she right, and in what ways might she be wrong?

6. A minimal genetic code requires only 21 codons—one for each amino acid, and one for a stop signal. Given this, what advantage might be offered by having a code with 64 codons?

7. Which of the following describes mutations? Select True or False for each statement.
  T/F Point mutations can occur in any DNA sequence.
  T/F Frameshift mutations can occur in any DNA sequence.
  T/F Neutral mutations depend on the degeneracy of the genetic code.
  T/F Deleterious mutations occur only in protein-coding sequences of DNA.

8. THINK CAREFULLY Explain what's wrong with this statement: All point mutations change the genotype and the phenotype.

✔ TEST YOUR PROBLEM-SOLVING SKILLS

9. **MODEL** Draw a hypothetical metabolic pathway in *Neurospora crassa* composed of five substrates, five enzymes, and a product called nirvana. Number the substrates 1–5, and label the enzymes A–E, in order. (For instance, enzyme A catalyzes the reaction between substrates 1 and 2.)
   - Suppose a mutation made the gene for enzyme C nonfunctional. What molecule would accumulate in the affected cells?
   - Suppose a mutant strain can survive if substrate 5 is added to the growth medium, but it cannot grow if substrates 1, 2, 3, or 4 are added. Which enzyme in the pathway is affected in this mutant?

10. **QUANTITATIVE** One of the possibilities considered about the genetic code was that the code was overlapping, meaning that a single base could be part of up to three codons. How many amino acids would be encoded in the sequence 5′-AUGUUACGGAAU-3′ by a non-overlapping and a maximally overlapping triplet code?
    a. 4 (non-overlapping) and 16 (overlapping)
    b. 4 and 12
    c. 4 and 10
    d. 12 and 4

✔ PUT IT ALL TOGETHER: Case Study

What can fish color tell us about human pigmentation?

Skin color is often one of the first traits people notice in each other. Studies in zebrafish uncovered a mutation that altered a transport protein and resulted in light-colored fish. This discovery led to the finding that the same gene in humans has a strong influence on skin pigmentation in many populations.

11. The zebrafish mutation that reduced coloration created a null allele of the transport protein gene. Which of the following types of mutation would be most likely to create this null allele?
    a. a missense mutation
    b. a frameshift mutation
    c. a neutral mutation
    d. a silent mutation

12. **PROCESS OF SCIENCE** Investigators examined the expression of transporter mRNA and protein produced in zebrafish homozygous for each of the alleles and obtained the results summarized here (+ = present, − = absent). Does the allele associated with light color appear to be altering transcription or translation? Why?

	mRNA	Protein
Dark-colored zebrafish allele	+	+
Light-colored zebrafish allele	−	−

Source: R. L. Lamason et al. 2005. *Science* 310: 1782–1786.

13. A small portion of the human transport protein amino acid sequence is shown here. The upper sequence is associated with darker skin, and the lower sequence is associated with lighter skin. What DNA base-pair change created the light-skin form of the human protein from the gene that coded for the dark-skin form?

Ala–Gly–Ala–Thr–Phe

Ala–Gly–Thr–Thr–Phe

14. Researchers compared the amino acid sequences of the transport protein in zebrafish, puffer fish, mice, and humans. They found many stretches with identical sequences in all four species. Does this mean that the corresponding mRNA base sequences are also the same in these four species? Explain why or why not.

15. The allele of the human transport protein associated with lighter skin is found almost exclusively in people with European ancestry. The other common allele for darker skin, which appears to be the ancestral allele, is found in people with African ancestry. What is a plausible explanation for how the lighter-skin allele came to be so common in those with European ancestry?

16. **SOCIETY** Some elected officials have argued that research dollars intended to answer questions about human biology should be spent exclusively on studies of humans. Imagine that your congresswoman took this position. Using the information in this case study, write a paragraph to her to argue that studies in animal models can lead to important discoveries about human biology.

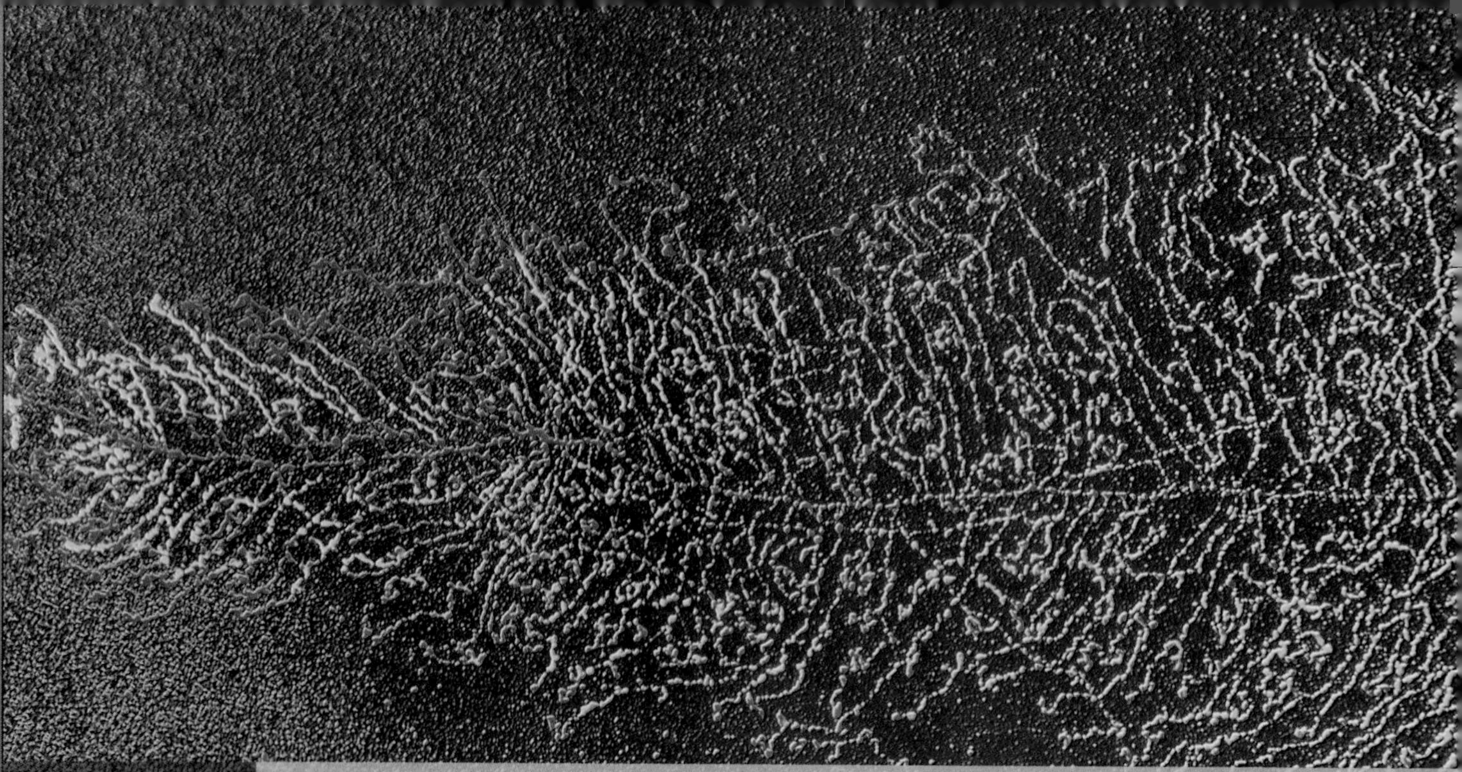

17 Transcription, RNA Processing, and Translation

Extensive transcription is occurring along this gene within a frog cell. The horizontal strand (colored red) in the middle of this micrograph is DNA; the strands (yellow and red) coming off above and below the DNA are RNA molecules.

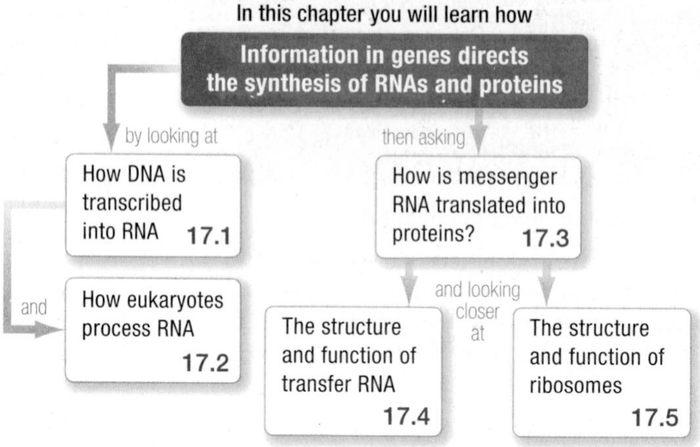

Proteins are the stuff of life. They give shape to our cells, control the chemical reactions that go on inside them, and regulate how materials move into, out of, and through them. Some proteins may not be produced at all in some types of cells; others may be present in quantities ranging from fewer than a dozen to millions of copies.

A cell builds the proteins it needs from instructions encoded in its DNA. The central dogma of molecular biology states that the flow of information in cells is from DNA to messenger RNA (mRNA) to protein (Ch. 16, Section 16.2). (Although the focus of this chapter will be on genes that code for proteins, recall that many genes code for important RNA molecules that are not mRNAs.) Once this pattern of information flow had been established, biologists puzzled over how cells actually accomplish the two major steps of the central dogma: transcription and translation.

This chapter answers these and other questions by delving into the mechanisms of gene expression—the heart of the central dogma, and of life itself.

BIG PICTURE

This chapter is part of the Big Picture. See how on pages 404–405.

17.1 An Overview of Transcription

The first step in using genetic information is transcription—synthesizing an RNA version of the instructions archived in DNA. Controlling which genes are transcribed, and when, where and to what level this transcription occurs, is the most important element in regulating how information in DNA is used by a cell. This makes understanding the mechanisms of transcription fundamentally important in knowing how genetic information is used to build and maintain organisms.

> **After you complete this section, you should be able to ...**
>
> - Describe the steps of transcription.

Many of the things you've learned about DNA synthesis and DNA polymerases (Ch. 15, Section 15.3) apply to understanding transcription. Enzymes called **RNA polymerases** are front and center in transcription (**Figure 17.1**). RNA polymerases use monomers called ribonucleoside triphosphates, or NTPs. NTPs are like the deoxyribonucleoside triphosphates (dNTPs) used for DNA synthesis, except that they have a hydroxyl (—OH) group on the 2′ carbon. This makes the sugar in an NTP a ribose instead of the deoxyribose sugar of DNA.

Once an NTP with a base that matches a base on the DNA template is in place, RNA polymerase catalyzes a reaction that cleaves off two phosphates and forms a phosphodiester linkage between the 3′ end of the growing mRNA chain and the new ribonucleoside monophosphate. As this matching-and-linking process continues, an RNA that is complementary to one of the DNA strands of the gene is synthesized in the 5′ → 3′ direction. This is the essence of transcription.

Notice in Figure 17.1 that only one of the two DNA strands is used as a template and transcribed, or "read," by RNA polymerase.

- The strand that is read by RNA polymerase is the **template strand**.
- The other strand is the **non-template strand**, or **coding strand**. Coding strand is an appropriate name, because, with one exception, its sequence matches the sequence of the RNA that is transcribed from the template strand.

What is the exception to a perfect match between the sequences of the transcribed RNA and DNA of the coding strand? The only difference is that RNA contains the base uracil (U) rather than the thymine (T) found in DNA. Adenine (A) in the DNA template strand specifies a U rather than a T in the complementary RNA strand. Like DNA polymerases, an RNA polymerase performs template-directed synthesis in the 5′ → 3′ direction. But unlike DNA polymerases, RNA polymerases do not require a primer to begin transcription—they can start "from scratch."

Bacteria have a single RNA polymerase. In contrast, eukaryotes have at least three distinct types. Let's first take a look at general principles of transcription using bacteria as a model, and then examine the ways transcription differs in eukaryotes.

Initiation: How Does Transcription Begin in Bacteria?

One way of thinking about what constitutes a gene is that it's a stretch of DNA that is transcribed to produce a functional product for the cell. Given this, how does RNA polymerase "know" which sequences of DNA to transcribe? In short, how does RNA polymerase find genes?

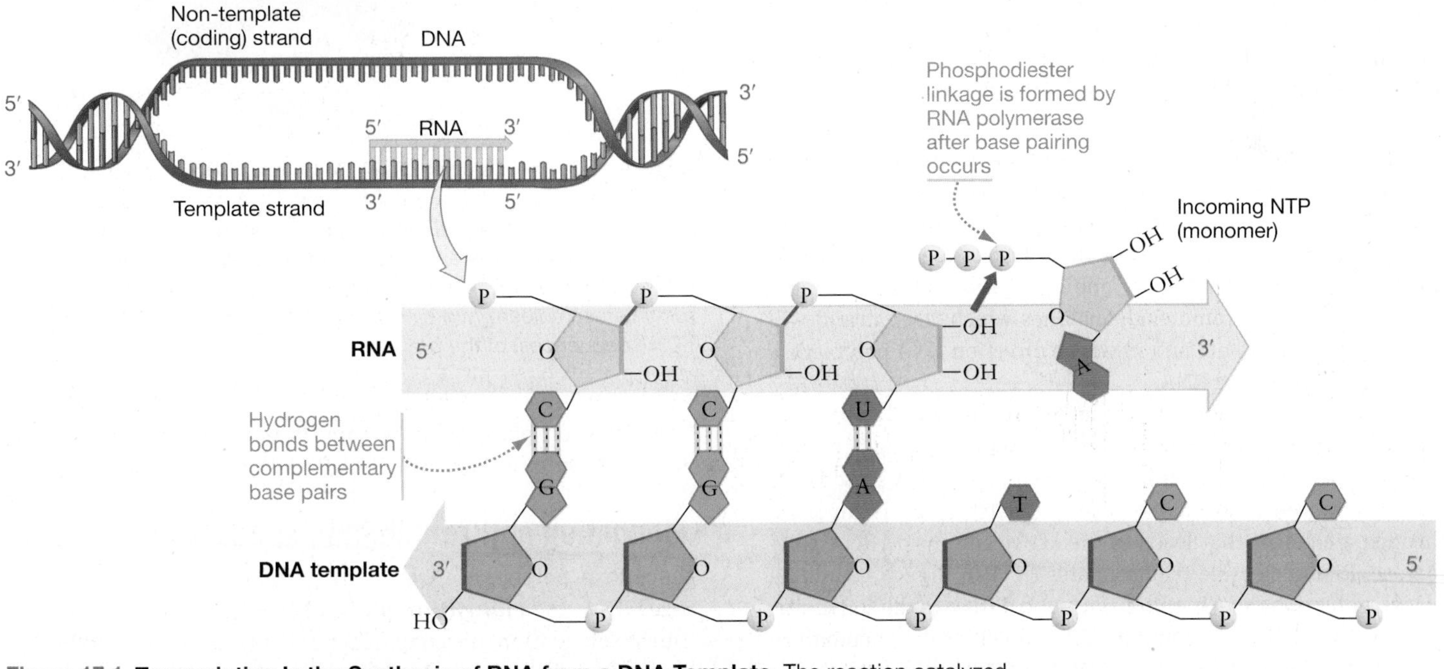

Figure 17.1 Transcription Is the Synthesis of RNA from a DNA Template. The reaction catalyzed by RNA polymerase (not shown) results in the formation of a phosphodiester linkage between ribonucleotides. RNA polymerase produces an RNA strand whose sequence is complementary to the bases in the DNA template.

✔ Are the RNA and DNA strands parallel or antiparallel? In which direction is the DNA template "read"?

If transcription enables the use of genetic information, then the ability to begin transcription—**initiation**—is the key to its control. (You'll learn more about this control in Chapters 18 and 19.) Researchers soon discovered that bacterial RNA polymerase, the enzyme that transcribes genes—the **core enzyme**—wasn't sufficient to recognize the start of genes. Instead, another protein, called **sigma**, must first bind to the core enzyme to recognize sites where transcription should begin. These sites were named **promoters** because they are regions of DNA that promote the start of transcription. Together, the bacterial RNA polymerase core enzyme and sigma form a **holoenzyme** (literally, "whole enzyme"; Figure 17.2a). The discovery of promoters suggested that sigma was responsible for guiding RNA polymerase to specific locations where transcription should begin.

What do promoters look like, and how do they work?

Bacterial Promoters David Pribnow offered an initial answer to these questions in the mid-1970s. Analyzing the base sequence of promoters from various bacteria and from viruses that infect bacteria, he found that promoters were 40–50 base pairs long and had a particular section in common: a series of bases on one strand of DNA identical or similar to TATAAT. This six-base-pair sequence is now known as the −10 box, because it is centered about 10 bases from the point where transcription starts (Figure 17.2b).

DNA that is located in the direction RNA polymerase moves during transcription is said to be **downstream** from a point of reference; DNA located in the opposite direction is said to be **upstream** from the same point of reference. Thus, the −10 box is centered about 10 bases upstream from the transcription start site. The place where transcription begins is numbered +1.

Soon after the discovery of the −10 box, researchers recognized that the sequence TTGACA also commonly occurred in promoters. This additional sequence is centered about 35 bases upstream from the +1 site and is called the −35 box. The sequences within the promoter but outside the −10 and −35 boxes vary.

Events inside the Holoenzyme In bacteria, transcription can be initiated only when sigma, as part of the holoenzyme complex, binds to the −35 and −10 boxes in the DNA. Because of the asymmetry and distinct sequences of these boxes, the sigma protein can bind the promoter in only one orientation. In this way, the orientation of the promoter determines which DNA strand will be used as the template and in which direction RNA polymerase will start synthesizing RNA.

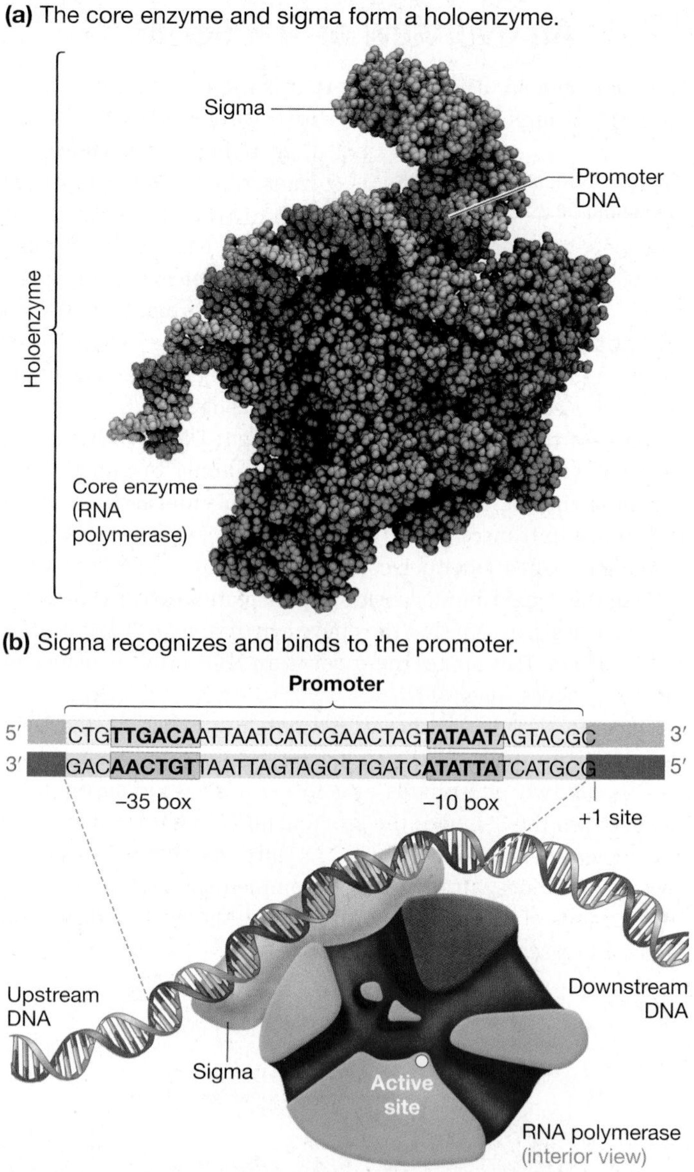

Figure 17.2 **The Bacterial RNA Polymerase Holoenzyme.** The holoenzyme is made up of the core RNA polymerase (core enzyme) and sigma protein. **(a)** A space-filling model of bacterial RNA polymerase holoenzyme bound to the DNA of the promoter. **(b)** Sigma functions within the bacterial RNA polymerase holoenzyme by binding to the −35 and −10 boxes (DNA sequences) of the promoter.

Once the holoenzyme is bound to a promoter (Figure 17.3, step 1), RNA polymerase opens the DNA helix, creating two separated strands of DNA (step 2). These separated DNA strands are called a transcription bubble. The template strand is threaded through a channel that leads to the active site inside RNA polymerase. Ribonucleoside triphosphates (NTPs), the RNA building blocks, enter a channel in the enzyme and diffuse to the active site.

When the first two incoming NTPs pair with a complementary base on the template strand of DNA, RNA polymerization begins. The reaction catalyzed by RNA polymerase is exergonic and spontaneous because NTPs have significant potential energy, due to their three phosphate groups. Step 3 of Figure 17.3 shows that the initiation phase of transcription is complete as RNA polymerase extends a short stretch of RNA from the +1 site.

Elongation and Termination in Bacteria

Once RNA polymerase leaves the promoter region as it synthesizes RNA, the **elongation** phase of transcription is under way. During elongation, the enzyme reads the DNA template as it catalyzes the addition of nucleotides to the 3′ end of the growing RNA at the rate of about 50 nucleotides per second. Like DNA polymerase (Ch. 15, Section 15.3), RNA polymerase proofreads and corrects errors during synthesis.

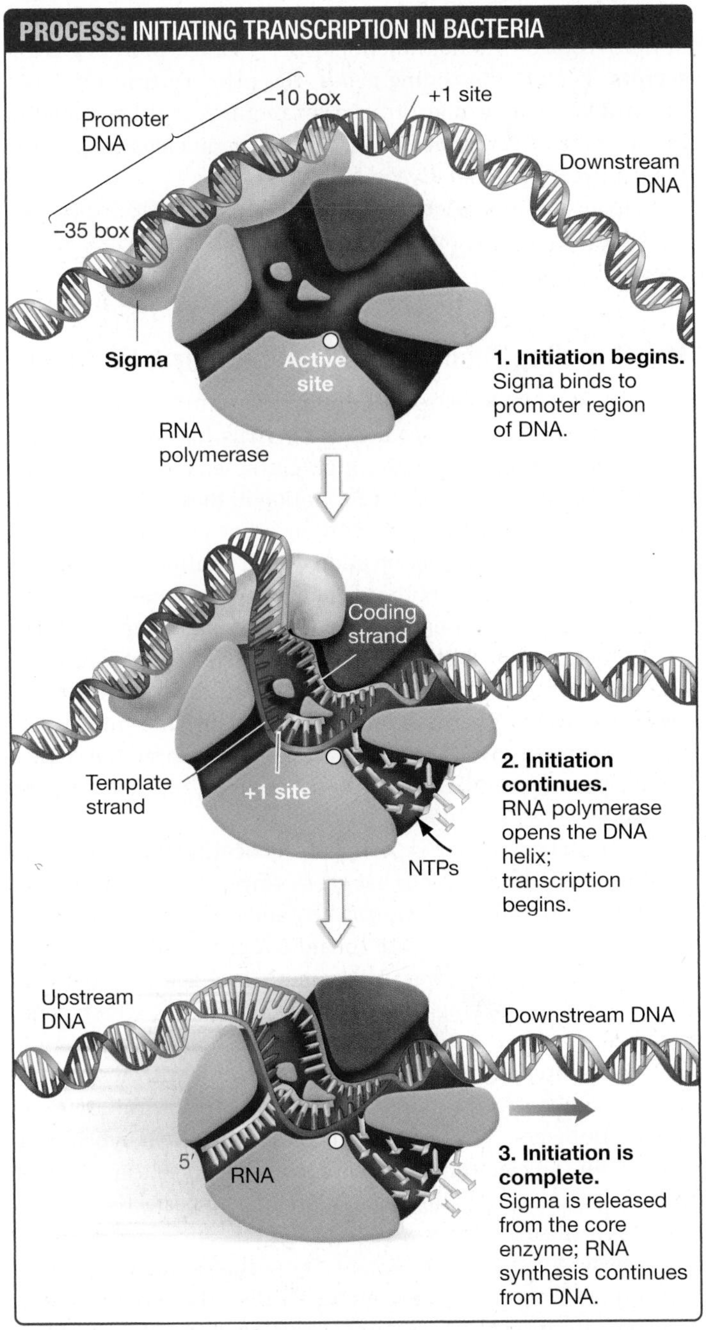

Figure 17.3 **Sigma Orients the RNA Polymerase on DNA During the Initiation of Transcription.** Sigma binds to the promoter, and RNA polymerase opens the DNA helix and threads the template strand through the active site.

RNA polymerase is a macromolecular machine with a structure that is critical for its function. Different parts of the enzyme help to steer the template and coding strands through channels inside the enzyme and to separate the newly synthesized RNA from the DNA template (see Figure 17.3, step 3). Double-stranded DNA goes into and out of one groove, NTPs enter another, and the growing RNA strand exits to the rear.

Termination ends transcription. In bacteria, transcription stops when RNA polymerase transcribes a DNA sequence called a transcription-termination signal.

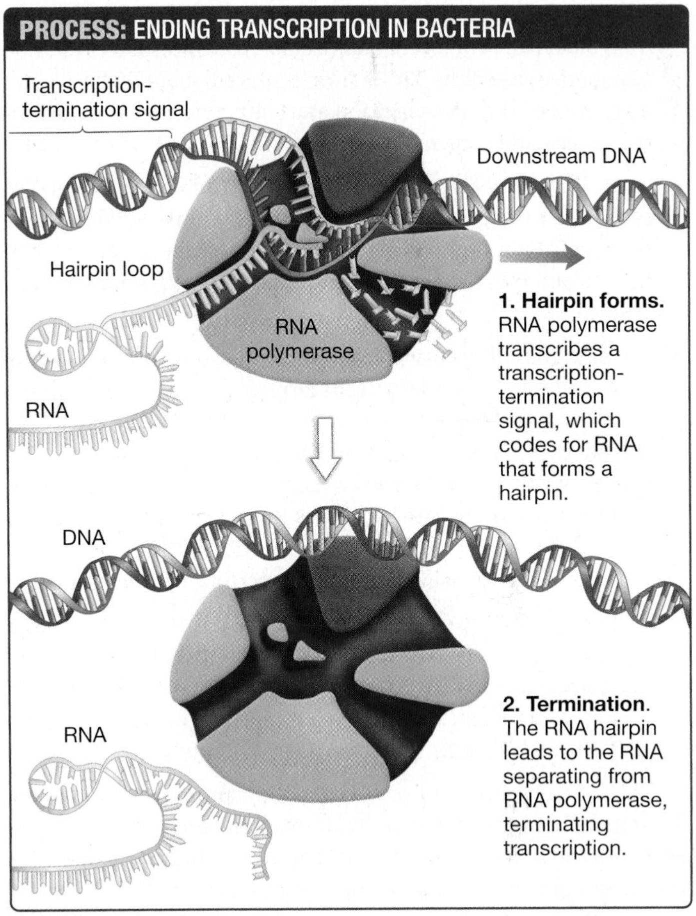

Figure 17.4 **Transcription in Bacteria Terminates When an RNA Hairpin Forms.**

The bases in bacterial DNA that make up the termination signal are transcribed into a stretch of RNA that folds back on itself to form a short double helix held together by complementary base pairing (Figure 17.4). Recall that this type of RNA secondary structure is called a hairpin, or stem-and-loop, structure (Ch. 4, Section 4.3). The hairpin weakens the interaction between RNA polymerase and the RNA transcript, resulting in the separation of the enzyme and its product. In some genes, the RNA hairpin is sufficient to end transcription, but in other genes, an additional protein is needed that works with the RNA hairpin to end transcription.

Transcription in Eukaryotes

Fundamental features of transcription are similar in bacteria and eukaryotes. In fact, these similarities provide compelling evidence for a common ancestor of all cells. There are, however, some important differences.

- Eukaryotes have three major polymerases—RNA polymerase I, II, and III—that are often referred to as pol I, pol II, and pol III. Each polymerase produces only certain types of RNA. For example, mRNAs are derived from genes transcribed by pol II. The eukaryotic RNA polymerases share structural features with the single bacterial RNA polymerase.

- Promoters in eukaryotic DNA are larger and more diverse than bacterial promoters. Many eukaryotic promoters include a sequence called the **TATA box**, centered about 30 base pairs upstream of the transcription start site, and other important sequences that vary more widely.
- Instead of using a sigma protein, eukaryotic RNA polymerases recognize promoters using a group of proteins called **general transcription factors**. General transcription factors are the first to assemble at the promoter, and then RNA polymerase follows.
- Termination of transcription for eukaryotic protein-coding genes differs substantially from termination in bacteria. In eukaryotes, a DNA sequence near the end of each gene called the polyadenylation signal, or **poly(A) signal**, is transcribed. The RNA downstream of the poly(A) signal sequence is cut by an enzyme as the polymerase continues to transcribe the DNA template. Eventually, and at varying locations, RNA polymerase comes off the DNA template, and transcription is terminated.
- In eukaryotes, transcription and translation are separated in time and space—transcription occurs in the nucleus, and translation occurs by using ribosomes located in the cytoplasm. In contrast, bacteria sometimes begin translating an mRNA even before its transcription is complete.

Table 17.1, at the end of this chapter, summarizes some key differences between transcription in bacteria and eukaryotes (see page 369). For one thing, the molecular machinery required for transcription is much more complex in eukaryotes than in bacteria. But these differences pale compared with what's required to turn eukaryotic RNA transcripts into functional RNAs. In bacteria, when mRNA transcription terminates, the result is a mature mRNA that's ready to be translated into a protein. Eukaryotic transcripts cannot be used for protein synthesis—or for any other purposes—until they are extensively modified. Let's take a closer look.

CHECK YOUR UNDERSTANDING

✔ If you understood this section, you should be able to ...

1. Explain why bacterial terminator sequences are expected to vary more than bacterial promoter sequences.
2. Explain how the function of eukaryotic general transcription factors is more similar to the bacterial sigma protein than it is to the bacterial core RNA polymerase.

Answers are available in Appendix A.

17.2 RNA Processing in Eukaryotes

In the late 1970s, researchers came to a startling realization: unlike bacterial genes that are transcribed into RNAs ready for action, newly transcribed eukaryotic RNAs are nonfunctional and often many times larger than the corresponding RNAs that worked in the cell. How could this be?

After you complete this section, you should be able to ...

- Describe the steps needed to process primary transcripts in eukaryotic gene expression.

Work from many teams confirmed that eukaryotic genes are initially copied into nonfunctional RNAs called **primary transcripts**. For protein-coding genes, the primary transcript is a **pre-mRNA**. The primary transcript requires multistep modification, called **RNA processing**, within the nucleus to generate the mature, functional RNA.

RNA processing is critically important for gene expression in eukaryotes. How does processing occur, and what are its consequences?

The Startling Discovery of Split Eukaryotic Genes

In 1977, Richard Roberts and Phillip Sharp independently discovered that a common-cold-causing virus had protein-coding genes that were studded with intervening sequences of noncoding DNA. In essence, the key information in these genes was split into pieces. Roberts, Sharp, and many other workers went on to show that this puzzling observation wasn't confined to a particular virus. Instead, it is a hallmark of eukaryotic genes.

What were the findings that led to such a startling claim? The first evidence came when the researchers tried to determine the location of genes within the DNA of the virus. As you've already seen, the study of viruses often provides insights into fundamental processes of the cells they infect, so much so that Arthur Kornberg, a pioneer of DNA synthesis research, advised, "Depend on viruses to open windows."

The team began their experiments by heating the virus's DNA to break the hydrogen bonds between complementary bases. This treatment separated the two DNA strands. The single-stranded DNA was then incubated with the mRNA produced by the virus. The team's intention was to promote base pairing between the mRNA and the single-stranded DNA to reveal where genes occurred.

The researchers expected that each viral mRNA sequence would align perfectly along a complementary viral DNA sequence. But when the team examined the DNA–RNA hybrid molecules using an electron microscope, they observed the structure shown in **Figure 17.5a**. Instead of matching up perfectly, parts of the DNA formed loops.

What was going on? As **Figure 17.5b** shows, the researchers interpreted these loops as stretches of DNA that are present in the template strand but are *not* represented in the corresponding mRNA.

Roberts, Sharp, and their colleagues went on to perform similar studies on eukaryotic genes. The results were the same as for the viral genes. They proposed that between the full nucleotide sequence of a eukaryotic gene and its mRNA, there is not a one-to-one correspondence. As an analogy, eukaryotic genes do not carry messages such as "Biology is my favorite course of all time." Instead, they carry messages that read something like this:

BIOLτηεπροτεινχοδινγρεγιονσοφγενεσOGY IS MY
FAVORαρειντερρθπτεϑβυνονψοϑινγϑITE COURSE
OF ανϑηαωετοβεσπλιχεϑτογετηερ ALL TIME

Here the sections of noncoding sequence are represented with Greek letters. They must be removed from the RNA before it can carry an intelligible message to the ribosome.

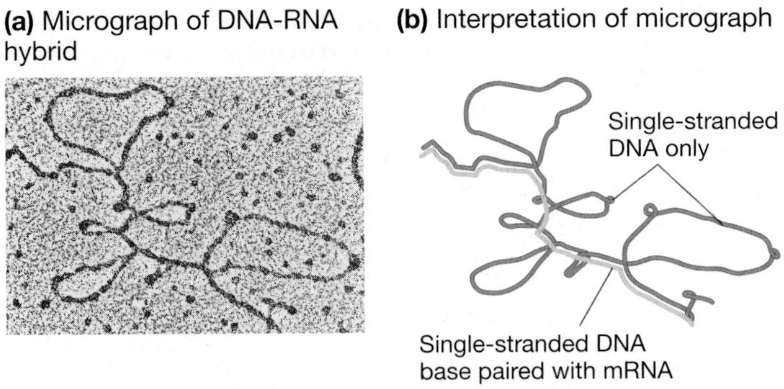

Figure 17.5 The Discovery of Introns. The loops in the micrograph and drawing represent regions of DNA that are transcribed but are not found in the final mRNA. These regions are called introns.

✔ **PROCESS OF SCIENCE** If introns were not present, what would the micrograph in part (a) look like?

Regions of a gene that are transcribed but not represented in the final RNA came to be called **introns** (because they are *int*ervening), and regions that are transcribed and represented in the final mature RNA were called **exons** (because they are *ex*pressed in the mature RNA). The terms "intron" and "exon" are also used for the regions of the transcribed RNA that are cut out (introns) or retained (exons) in the fully processed RNA molecule.

A common misconception is that exons are protein-coding regions. Many of them are, but exons also consist of RNA sequences that do not code for protein. Because of introns, eukaryotic genes are much larger than their corresponding mature RNAs. Although introns were first discovered in genes that produce mRNA, they were later found in genes for many other types of RNA.

RNA Splicing

Transcription of eukaryotic genes generates a primary transcript that contains exons and introns (Figure 17.6a). As transcription proceeds, the introns are removed from the growing RNA strand by a remarkable process known as **splicing**. In this phase of RNA processing, pieces of the primary transcript are removed and the remaining segments are joined together. Splicing, which occurs within the nucleus while transcription is still under way, results in an RNA that contains an uninterrupted genetic message. Given that many eukaryotic genes have dozens of introns that must be spliced out of the transcribed RNA, and an equally large number of exons that must be joined precisely, the mechanism for splicing has to be incredibly accurate to ensure the creation of a functioning mRNA, or any other type of RNA.

Figure 17.6b is an overview of how introns are removed. Splicing of primary transcripts is catalyzed by RNAs called small nuclear RNAs (snRNAs) working with an extraordinarily dynamic complex of proteins. These protein-plus-RNA macromolecular machines are known as **small nuclear ribonucleoproteins** (or **snRNPs**, pronounced *snurps*). The snRNAs of the snRNPs recognize RNA sequences critical for splicing.

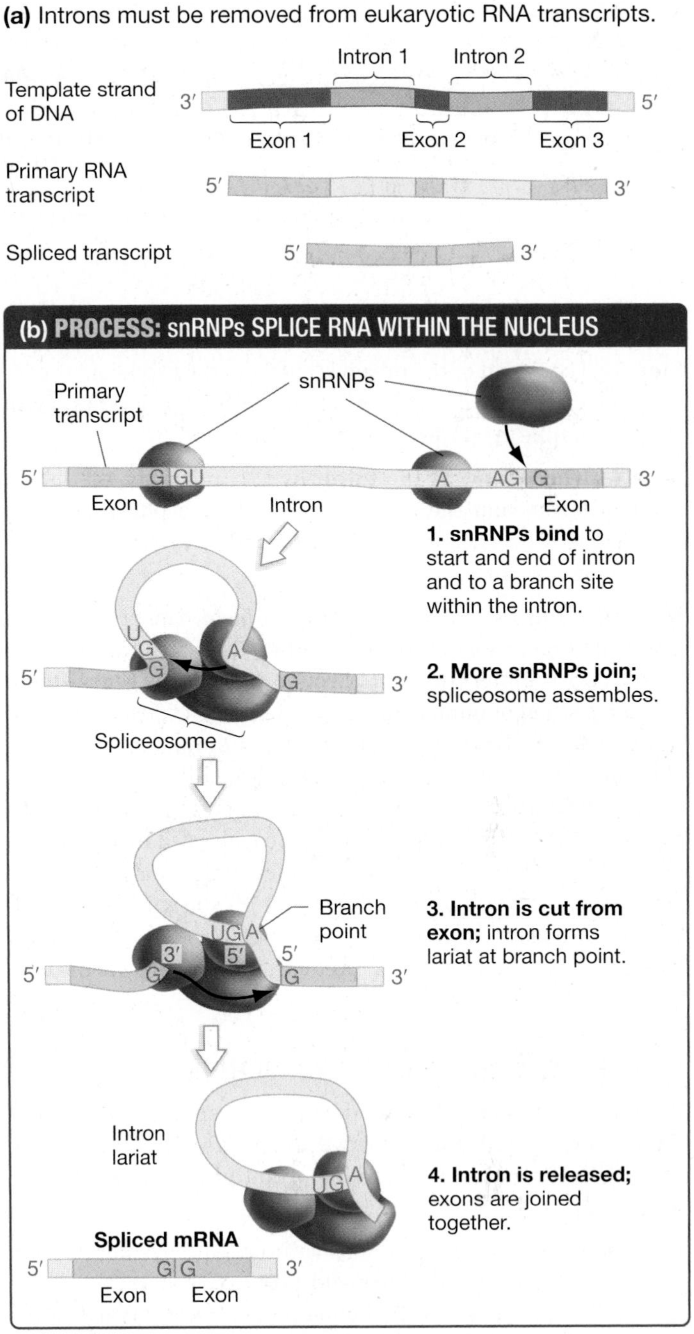

Figure 17.6 During Splicing, Introns Are Cut Out of the Primary Transcript. Characteristic sequences at the junctions between exons and introns are used by the snRNPs to recognize where exons and introns begin and end.

A simplified view of splicing can be broken into four steps:

Step 1 The snRNPs bind to the 5′ exon–intron and 3′ intron–exon boundaries, and also to a region near the end of the intron that contains an adenine (A) ribonucleotide.

Step 2 The **spliceosome** assembles as more snRNPs join the complex. Mammalian spliceosomes contain 5 different snRNAs and more than 300 different proteins, making spliceosomes the largest macromolecular machine.

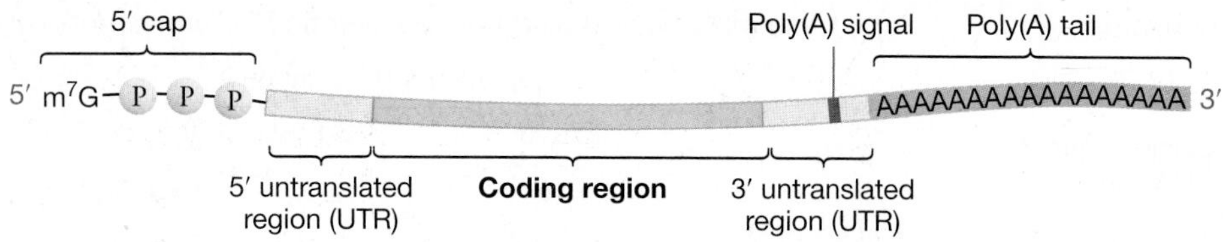

Figure 17.7 In Eukaryotes, a Cap and a Tail Are Added to mRNAs. A 5′ cap consisting of a modified guanine (G) nucleotide (symbolized as m^7G) is added through an unusual linkage to the end of the mRNA. At the 3′ end, a poly(A) tail made up of a long series of adenine (A) residues is synthesized after cleavage of the primary transcript downstream from a sequence called the poly(A) signal.

Step 3 The 5′ end of the intron is cut from the exon, and the intron forms a single-stranded stem plus a loop—a lariat—with the adenine at its connecting (branch) point.

Step 4 The 3′ end of the intron is cut, releasing the intron as a lariat, and the exons are joined by a phosphodiester linkage. The excised intron is degraded to ribonucleoside monophosphates.

Splicing of this pair of exons is now complete. The process is repeated for other exons and introns within the primary transcript.

As you'll see later (in Ch. 19, Section 19.4), many genes—more than 95 percent of human genes, for example—code for RNA that can be spliced in more than one way. This allows the production of different mRNAs and proteins from one gene.

Recent findings indicate that both the cutting and rejoining reactions that occur during splicing are catalyzed by the snRNA molecules in the spliceosome. This means that these reactions are catalyzed by a ribozyme. In Section 17.5 you will see that a ribozyme also plays a central role in translation. As the RNA world hypothesis (Ch. 4, Section 4.4) predicts, proteins are not the only important catalysts in cells.

Adding Caps and Tails to Transcripts

Considering everything that's involved in splicing, you might think splicing is all that's needed to change a pre-mRNA into an mRNA. However, two more important processing events are required.

1. As soon as the 5′ end of a eukaryotic pre-mRNA emerges from RNA polymerase, enzymes add a **5′ cap** (Figure 17.7). The cap consists of a modified guanine nucleotide linked to the transcript in an unusual way. The cap enables ribosomes to bind to the mRNA, and it also protects the 5′ end of the mRNA from enzymes that degrade RNA (ribonucleases).

2. An enzyme cleaves the 3′ end of the pre-mRNA after a sequence called the poly(A) signal (introduced in Section 17.1) that is located downstream of the protein-coding portion of the mRNA. A specialized RNA polymerase that does not require a template adds about 100–250 adenine (A) nucleotides. This string of adenines is known as the **poly(A) tail**. The poly(A) tail is not encoded by the DNA template strand. Like the 5′ cap, the poly(A) tail is required for ribosomes to start translation and to protect the end of the mRNA from attack by enzymes.

With the addition of the cap and tail and completion of splicing, processing of the pre-mRNA finally is complete. The product is a mature mRNA. Figure 17.7 also shows another important point about the mature mRNA—only some of it codes for a polypeptide. In addition to the noncoding poly(A) tail, a pair of noncoding sequences at the ends of the mRNA surround the protein-coding region. These 5′ and 3′ untranslated regions (or UTRs) were part of the primary transcript and therefore are exons or parts of exons. The 5′ and 3′ UTRs are essential for the mRNA to function in protein synthesis and also help stabilize the mature RNA and regulate its translation. mRNAs in bacteria also possess 5′ and 3′ UTRs.

CHECK YOUR UNDERSTANDING

✔ **If you understood this section, you should be able to ...**

1. Give an example of an exon or part of an exon that does not code for amino acids.
2. Explain how the RNA of spliceosomes is able to recognize which areas of a primary transcript to splice.
3. Predict the effect of removing the 5′ cap from an mRNA.

Answers are available in Appendix A.

17.3 An Introduction to Translation

After you complete this section, you should be able to ...

- Describe the roles of ribosomes, mRNA, and tRNAs in translation.

To synthesize a protein, the sequence of bases in a messenger RNA molecule must be translated into a sequence of amino acids in a polypeptide. This is easier said than done.

Ribosomes Are the Site of Protein Synthesis

One of the first questions about translation was, Where does it occur? The answer grew from the observation that there is a strong correlation between the number of **ribosomes** in a cell and the rate of protein synthesis. Based on this observation, investigators proposed that ribosomes are the site of protein synthesis.

To test this hypothesis, Roy Britten and collaborators did a pulse–chase experiment, which labels a population of molecules as they are being produced (see Ch. 7, Section 7.5). The location of the tagged molecules is then followed over time.

In Britten's study, the tagging was done by supplying a pulse of radioactive sulfur atoms that would be incorporated into the amino acids methionine and cysteine, followed by a chase of unlabeled sulfur atoms. If the ribosome hypothesis were correct, the radioactive signal should be associated with ribosomes for a short period of time—when the amino acids were being polymerized into proteins. Later, when translation was completed, the radioactivity should be in proteins no longer associated with ribosomes.

This is exactly what the researchers found. Based on these data, biologists concluded that proteins are synthesized at ribosomes and then released.

An Overview of Translation

About a decade after the ribosome hypothesis was confirmed, electron micrographs showed bacterial ribosomes in action. The images indicated that in bacteria, ribosomes can attach to mRNAs and begin synthesizing proteins even before transcription is complete, coupling transcription and translation (**Figure 17.8a**). In fact, multiple ribosomes attach to each mRNA. When two or more ribosomes simultaneously translate one mRNA, the structure is called a **polyribosome**, as seen in **Figure 17.8b**. Polyribosomes greatly increase the number of copies of a protein that can be produced from a single mRNA.

Transcription and translation can be coupled in bacteria because there is no nuclear envelope to separate the two processes. Simultaneous transcription and translation does not always occur, but when it does, gene expression is extremely rapid. In eukaryotes, transcription and processing of the primary transcripts occur only in the nucleus. The mature mRNA is then exported to the cytoplasm for translation. This means that in eukaryotes, transcription and translation are separated in time and space. Once mRNAs are outside the nucleus, ribosomes can attach and begin translation. As in bacteria, polyribosomes form.

Making Models 17.1 Tips on Showing Polarity Relationships between DNA, RNA, and Polypeptides

Drawing simple pictures of DNA, RNA, and polypeptides can help you understand how transcription and translation fit together. The examples here show how the polarities of DNA and RNA are related during transcription (top) and the polarity relationship between an mRNA and the polypeptide chain produced by translation (bottom).

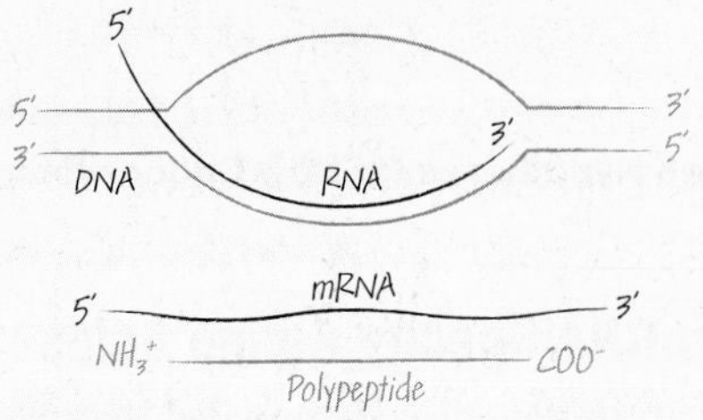

MODEL Draw coupled transcription and translation with a polyribosome containing two ribosomes. Show both strands of DNA and the relative lengths of the growing polypeptide chains. Label the polarities of the DNA strands, the mRNA, and the polypeptide chains. [Hint: Polypeptides are always synthesized in an amino (NH_3^+) → carboxyl (COO^-) terminal direction.]

To see this model in action, go to the Study Area of **Mastering Biology**

Understanding the relationships between DNA, RNA, and protein during transcription and translation is important and can be challenging. Drawing the processes is a good way to see how these molecules are related (see **Making Models 17.1**).

(a) Electron micrograph of coupled transcription and translation in bacteria

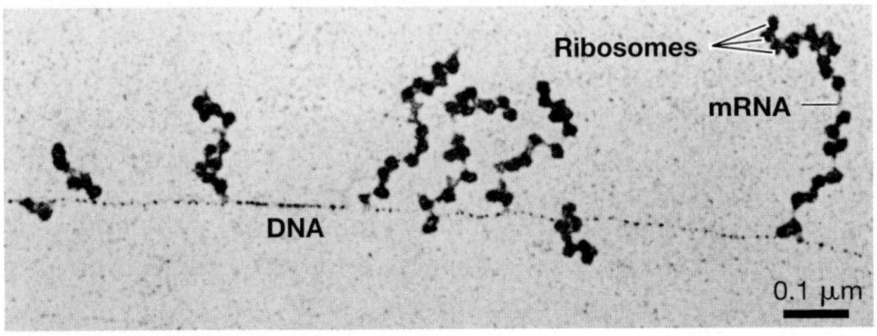

(b) Schematic depiction of coupled transcription and translation in bacteria

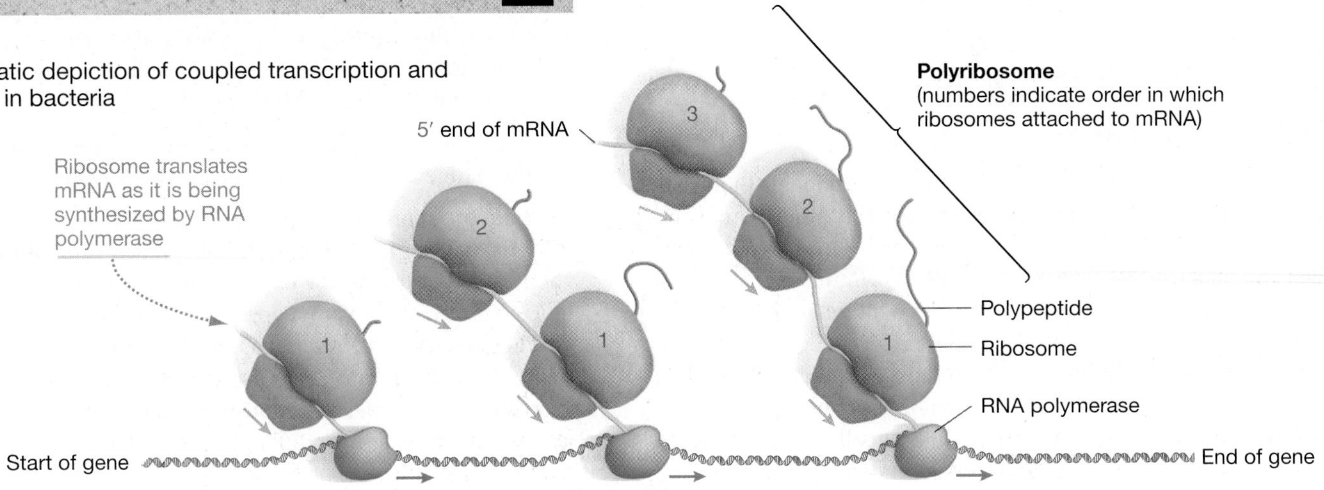

Figure 17.8 Transcription and Translation Can Be Coupled, or Occur Simultaneously, in Bacteria. In bacteria, ribosomes sometimes attach to mRNA transcripts and begin translation while RNA polymerase is still transcribing the DNA template strand.

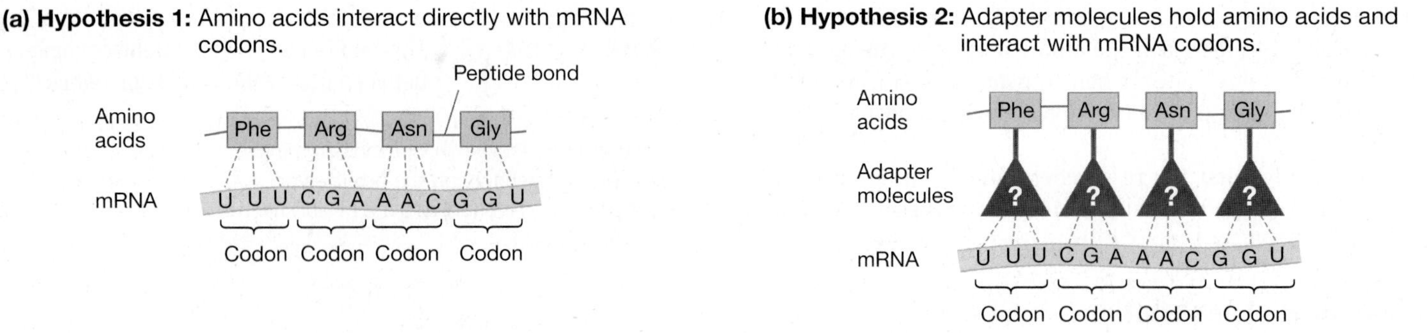

Figure 17.9 Two Hypotheses for How Codons Interact with Amino Acids.

How Does mRNA Specify Amino Acids?

The discovery of the genetic code revealed that triplet codons in mRNA specify particular amino acids in a protein. How does this conversion occur?

One early hypothesis was that mRNA codons and amino acids interact directly. This hypothesis proposed that the bases in a particular codon were complementary in shape or charge to the side group of a particular amino acid (Figure 17.9a). But Francis Crick pointed out problems with this idea. For example, how could the nucleic acid bases interact with a nonpolar amino acid side group, which does not form hydrogen bonds?

Crick proposed an alternative hypothesis. As Figure 17.9b shows, he suggested that some sort of adapter molecule holds amino acids in place while interacting directly and specifically by hydrogen bonding with a codon in mRNA. In essence, Crick predicted the existence of a chemical go-between that produced a physical connection between the two types of molecules. Crick was right.

CHECK YOUR UNDERSTANDING

✔ If you understood this section, you should be able to ...

1. Explain why polyribosomes can be found in both bacteria and eukaryotes, but only bacteria can make use of linked transcription and translation.
2. Explain why base pairing between an adapter molecule and codons in mRNA seemed a much more likely way to read the genetic code than matching amino acids directly to codons.

Answers are available in Appendix A.

17.4 The Structure and Function of Transfer RNA

After you complete this section, you should be able to ...

- Analyze the structure and function of transfer RNA.

Crick's adapter molecule was discovered by accident. Biologists were trying to work out an in vitro protein-synthesis system and reasoned that ribosomes, mRNA, amino acids, and a source of energy would be needed. The logic was that ribosomes provide the catalytic machinery, mRNAs contribute the message to be translated, amino acids are the building blocks of proteins, and ATP or other related molecules would supply potential energy to drive the endergonic polymerization reactions required to form proteins.

These molecules weren't enough: A cellular component that contained a previously unknown type of RNA turned out to be indispensable. If this type of RNA was missing, protein synthesis did not occur. What was this mysterious RNA, and why was it needed?

The novel class of RNA eventually became known as **transfer RNA (tRNA)**. The role of tRNA was a mystery until some researchers happened to add a radioactively labeled amino acid—leucine—to an in vitro protein-synthesis system. The treatment was actually done as a control for an unrelated experiment. To the researchers' amazement, some of the radioactive leucine attached to tRNA molecules. When a tRNA has an amino acid attached, it is known as an **aminoacyl tRNA**.

What happens to the amino acids bound to tRNAs? To answer this question, Paul Zamecnik and colleagues tracked the fate of radioactive leucine molecules attached to tRNAs. They found that the amino acids are transferred from tRNAs to proteins.

The data supporting this conclusion are shown in the "Results" section of Figure 17.10. The graph shows that radioactive amino acids are lost from tRNAs and incorporated into polypeptides synthesized by ribosomes. To understand this conclusion, follow the events in the graph:

1. Put your finger on a position along the *x*-axis that indicates one minute has passed since the start of the experiment.
2. Read up until you hit the green line and then the gray line. The green line represents data from polypeptides; the gray line represents data from tRNAs.
3. Check the values on the *y*-axis that correspond to the 1-minute point for the amount of radioactive leucine present in polypeptides (green line) or tRNAs (gray line).
4. You'll see that at this early stage in the experiment, almost all the radioactive leucine is attached to tRNA, not polypeptides.

Next, do the same four steps at the point on the *x*-axis labeled 10 minutes (since the start of the experiment). Your conclusion now should be that late in the experiment, almost all the radioactive leucine is attached to polypeptides, not tRNA.

These results inspired the use of the word "transfer" in tRNA's name, because amino acids are transferred from the RNA to a growing polypeptide. The experiment also confirmed that tRNAs act as the interpreter during translation: tRNAs are Crick's adapter molecules.

What Is the Structure of tRNAs?

Transfer RNAs serve as chemical go-betweens that allow amino acids to interact with an mRNA template. But precisely how is the connection made? This question was answered by research on tRNA's structure.

The initial studies established the sequence of nucleotides in various tRNAs. Transfer RNA sequences are relatively short, ranging from 75 to 95 nucleotides in length. Biologists noticed that certain parts of each tRNA could form hydrogen bonds with complementary base sequences elsewhere in the same molecule. As a result, portions of the molecule form stem-and-loop structures (**Figure 17.11a**).

RESEARCH

QUESTION: What happens to amino acids attached to tRNAs?

HYPOTHESIS: Aminoacyl tRNAs transfer amino acids to growing polypeptides.

NULL HYPOTHESIS: Aminoacyl tRNAs do not transfer amino acids to growing polypeptides.

EXPERIMENTAL SETUP:

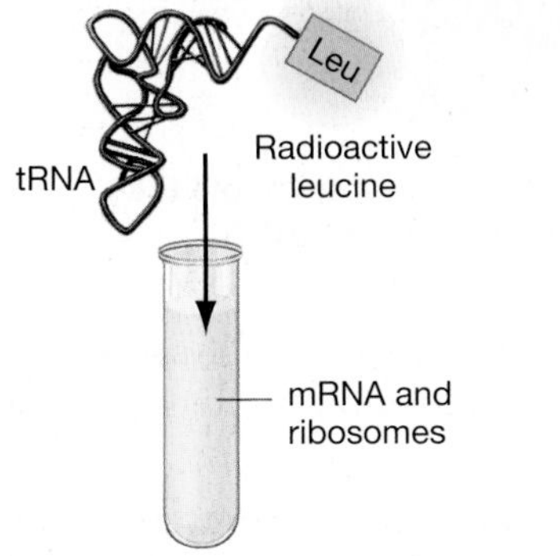

1. Attach radioactive leucine molecules to tRNAs.

2. Add these aminoacyl tRNAs to an in vitro system that allows protein synthesis. Follow fate of the radioactive amino acids.

PREDICTION OF HYPOTHESIS: Radioactive amino acids will be found in polypeptides.

PREDICTION OF NULL HYPOTHESIS: Radioactive amino acids will not be found in polypeptides.

RESULTS:

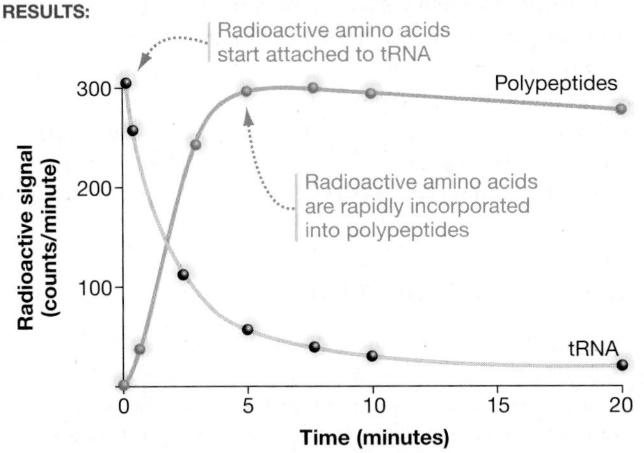

CONCLUSION: Aminoacyl tRNAs transfer amino acids to growing polypeptides.

Figure 17.10 **Amino Acids Are Transferred from tRNAs to Polypeptides.**

SOURCE: M. B. Hoagland, M. L. Stephenson, J. F. Scott, et al. 1958. A soluble ribonucleic acid intermediate in protein synthesis. *Journal of Biological Chemistry* 231: 241–257.

✔ **PROCESS OF SCIENCE** What would the graphed results look like if the null hypothesis were correct?

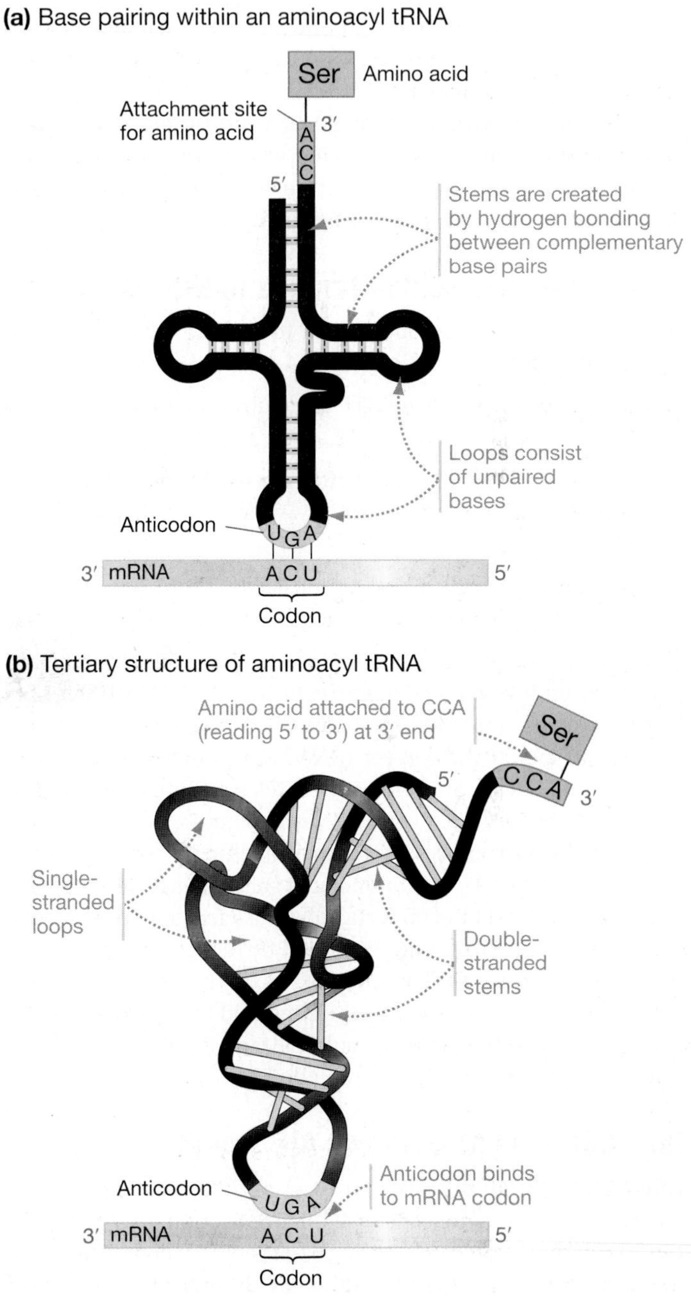

Figure 17.11 **The Structure of an Aminoacyl Transfer RNA.** **(a)** A schematic view that flattens the tRNA in two dimensions to emphasize internal complementary base pairing. In this view, the tRNA resembles a cloverleaf. **(b)** The tertiary (3-D) structure of tRNA is L-shaped.

Two parts of tRNA proved especially important. A CCA sequence at the 3′ end of each tRNA molecule was the site for amino acid attachment. The second critical part of each tRNA was a loop opposite the amino acid attachment site. Within this loop, scientists found that each tRNA contained three ribonucleotides that served as an **anticodon**—a triplet of ribonucleotides able to form base pairs with the codon for the amino acid in mRNA. Just as for strands in double-stranded DNA or the strands in an RNA–DNA hybrid, the codon and anticodon pair in an antiparallel orientation.

Later, X-ray crystallography studies revealed the tertiary—or three-dimensional—structure of tRNAs (Ch. 4, Section 4.3). As Figure 17.11b shows, tRNAs fold into an L-shaped molecule with the anticodon at one end and the CCA sequence and attached amino acid at the other end.

✔ **If you understand the structure and function of tRNAs, you should be able to (1) describe where the amino acid attaches, and (2) explain the relationship between the anticodon of a tRNA and a codon in an mRNA.**

How Are Amino Acids Attached to tRNAs?

How do amino acids become linked to tRNAs? What allows the right amino acid for a particular tRNA to be attached?

- An input of energy from ATP is required to attach an amino acid to a tRNA.
- Enzymes called **aminoacyl-tRNA synthetases** catalyze the addition of amino acids to tRNAs—what biologists call "charging" a tRNA.
- For each of the 20 major amino acids, there is a different aminoacyl-tRNA synthetase and one or more tRNAs.

Each aminoacyl-tRNA synthetase has a binding site for a particular amino acid and a particular tRNA. Subtle differences in tRNA shape and base sequence allow the enzymes to recognize and match the correct tRNA to the correct amino acid. Figure 17.12 shows an aminoacyl-tRNA synthetase bound to a tRNA that has just been charged with an amino acid. Note how tightly the two structures fit together—making it possible for the enzyme and its tRNA and amino acid substrates to interact precisely. This precision is important because frequent mistakes in choosing the matching amino acid and tRNA would be disastrous for the cell. Errors would produce tRNAs that recognize an appropriate codon but carry the wrong amino acid into a growing polypeptide. In turn, the protein would have no function, an altered function, or even possibly kill the cell.

How Many Types of tRNAs Are There?

After cataloging all the different types of tRNAs, biologists faced a paradox. The genetic code (see Ch. 16, Section 16.3) specifies the 20 common amino acids using 61 different codons. This leads to a prediction that there should be 61 different tRNAs to read these codons. Instead, most cells contain only about 40. How can 61 codons be translated with only 40 tRNAs?

To resolve this paradox, Francis Crick proposed what is known as the wobble hypothesis. Recall that:

- Many amino acids are specified by more than one codon.

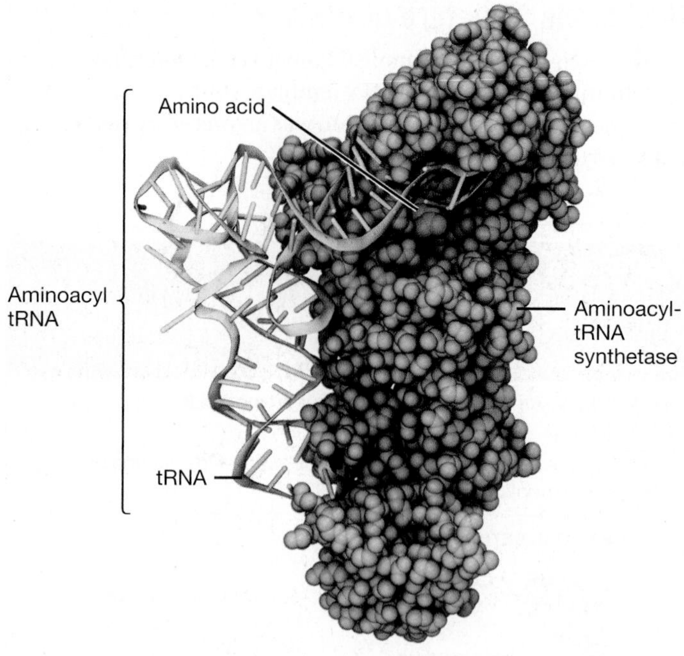

Figure 17.12 Aminoacyl-tRNA Synthetases Couple the Appropriate Amino Acid to the Appropriate tRNA.

- Codons for the same amino acid often have the same nucleotides at the first and second positions but a different nucleotide at the third position.

For example, both of the codons CAA and CAG code for the amino acid glutamine. Surprisingly, biologists found that a tRNA with an anticodon of GUU can base-pair with both CAA and CAG in mRNA. The GUU anticodon matches the CAA codon perfectly but only the first two bases in the CAG codon. What allows the tRNA to read this codon? It turns out that the U in the anticodon's third position[1] can form a nonstandard base pair with a G in the CAG codon.

Crick proposed that certain bases in the third position of tRNA anticodons can bind to bases in the third position of a codon in ways that do not match Watson–Crick base pairing. This allows some flexibility, or "wobble," in base pairing. **Wobble pairing** allows one tRNA to read more than one codon. Wobble in the third position of a codon is the reason that 40 or so tRNAs can translate all 61 codons.

CHECK YOUR UNDERSTANDING

✔ **If you understood this section, you should be able to . . .**

1. Explain how tRNA fulfills the adapter role proposed by Crick.
2. Predict the average number of different tRNA molecules that can be recognized by a single type of aminoacyl tRNA synthetase.
3. Predict the number of different tRNAs that would be required if wobble pairing did not occur.

Answers are available in Appendix A.

[1]Because anticodons and codons are antiparallel and nucleic acids are numbered in a 5′ → 3′ direction, the first position in the anticodon contains the U that forms a nonstandard base pair with the G in the third position of the codon. For simplicity, this formal numbering rule will be ignored, and the same numbering will be used for positions in codons and anticodons.

Some might conclude that wobble pairing accounts for the redundancy of the genetic code (Ch. 16, Section 16.3)—but don't fall into that trap. Wobble pairing explains only how one tRNA can read more than one codon, not how one amino acid can be specified by more than one codon.

17.5 Ribosome Structure and Function in Translation

After you complete this section, you should be able to ...

- Explain the events of translation initiation, elongation, and termination.

The translation of each codon in mRNA into the next amino acid in a polypeptide chain begins when the anticodon of an aminoacyl tRNA binds to the codon. Translation of a codon is complete when a peptide bond forms between the amino acid originally carried by the tRNA and the growing polypeptide.

Both of these events take place inside a ribosome. Biologists have known since the 1930s that ribosomes contain many proteins and **ribosomal RNAs (rRNAs)**. Later work showed that ribosomes can separate into two parts, called the large subunit and small subunit. Each ribosome subunit consists of a complex of rRNA molecules and proteins. During translation, the small subunit holds the mRNA and the large subunit is where peptide bonds are formed.

Figure 17.13 shows two views of how the molecules required for translation fit together. Note that the ribosome can accommodate three tRNAs bound to codons in mRNA.

- The tRNA on the right in the figure (colored red) carries an amino acid. The site in the ribosome that holds this tRNA is the A site—"A" for acceptor, or aminoacyl.
- The tRNA that is in the middle (green) holds the growing polypeptide chain and occupies the P (for peptidyl) site inside the ribosome. (Also think of "P" for peptide-bond formation.)
- The left-hand (blue) tRNA no longer has an amino acid attached and is about to leave the ribosome. It occupies the ribosome's E site—"E" for exit.

The ribosome is a remarkably sophisticated macromolecular machine that synthesizes proteins in a three-step sequence:

1. An aminoacyl tRNA diffuses into the A site; if its anticodon matches a codon in mRNA, it stays in the ribosome.
2. A peptide bond forms between the amino acid held by the aminoacyl tRNA in the A site and the growing polypeptide, which was held by a tRNA in the P site.
3. The ribosome moves relative to the mRNA by one codon, and all three tRNAs are shifted one position within the ribosome. The tRNA in the E site exits; the tRNA in the P site moves to the E site; the tRNA in the A site switches to the P site; the A site is now empty and ready to accept another aminoacyl tRNA.

The protein that is being synthesized grows by one amino acid each time this three-step sequence repeats. The process occurs up to 20 times per second in bacterial ribosomes and at half this rate or less in eukaryotic ribosomes. Protein synthesis starts at the amino end (N-terminus) of a polypeptide and proceeds to the carboxy end (C-terminus; see Ch. 3, Section 3.1).

This introduction to how tRNAs, mRNAs, and ribosomes interact during protein synthesis leaves several key questions unanswered. How do mRNAs and ribosomes get together to start the process? Once protein synthesis is under way, how is peptide-bond formation catalyzed inside the ribosome? And how does protein synthesis conclude when the ribosome reaches the end of the protein-coding portion of the mRNA? Let's consider each question in turn.

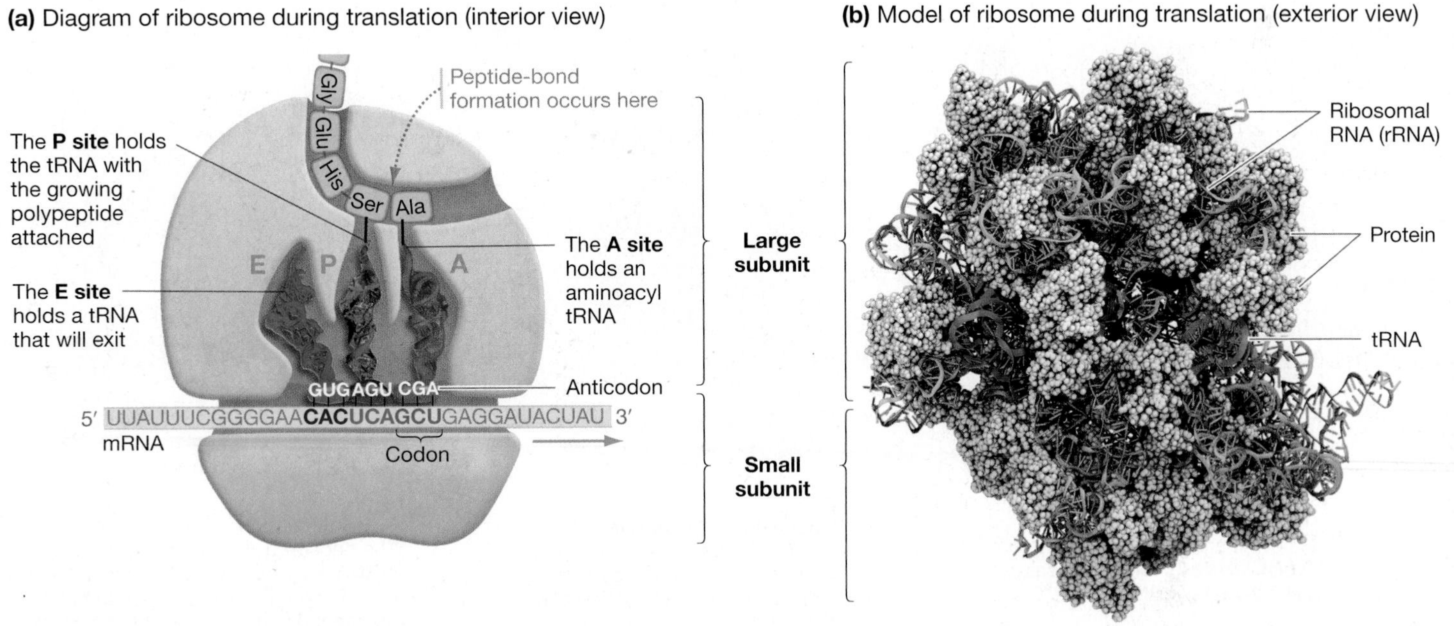

Figure 17.13 **Ribosomes Contain Three tRNA Binding Sites.** The large and small subunits assemble on the mRNA and allow tRNAs to decode codons.

Initiating Translation

To translate an mRNA, a ribosome must begin at the first codon in a message, translate the mRNA up to the message's termination codon, and then stop. Using the same terminology that they apply to transcription, biologists call these three phases of protein synthesis initiation, elongation, and termination.

One key to understanding translation initiation is to recall that a start codon (usually AUG) is found near, but never at, the 5′ end of mRNAs, and that it codes for the amino acid methionine (Ch. 16, Section 16.3). Make sure to distinguish initiation of translation at a start codon from initiation of transcription at a promoter—these are separate processes that are easily confused.

Figure 17.14 shows how translation gets under way in bacteria. Translation begins when a section of rRNA in a small ribosomal subunit binds to a complementary sequence on an mRNA. This mRNA region is called the **ribosome binding site**, or

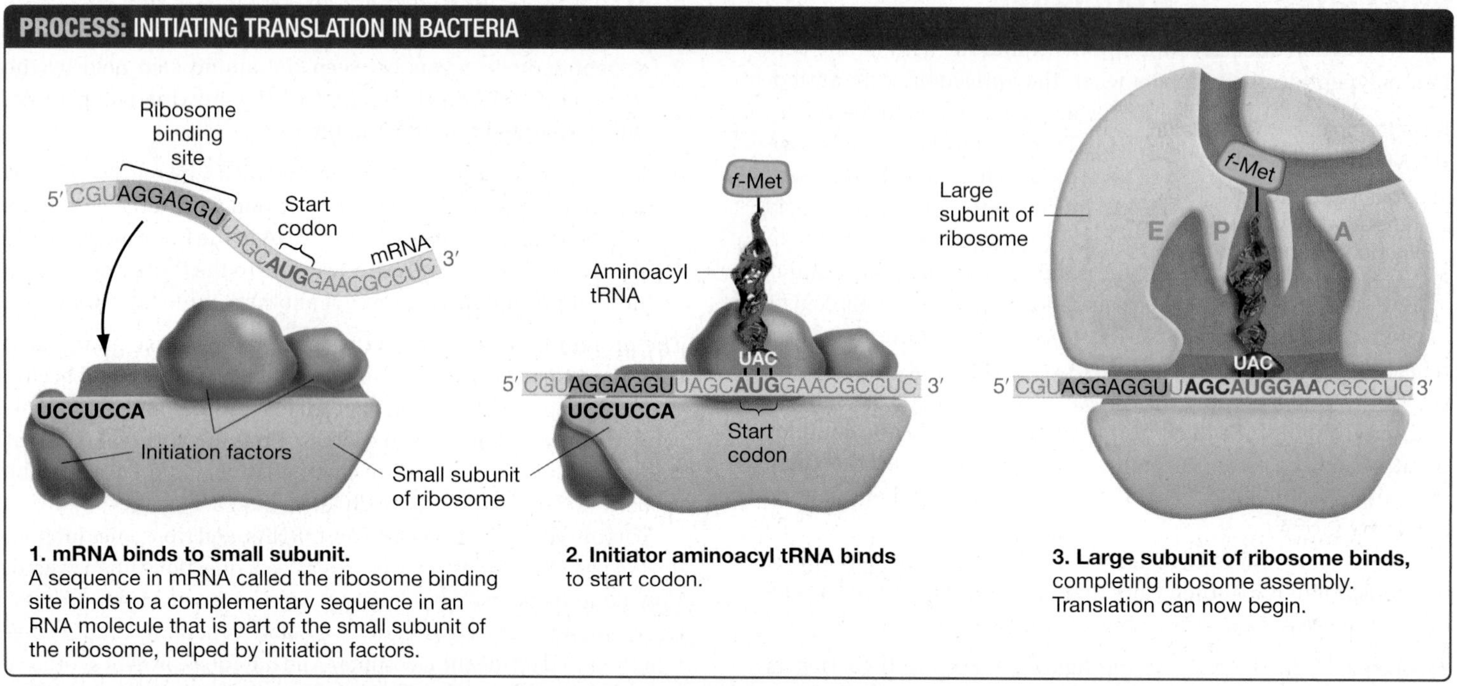

Figure 17.14 Initiation Requires Binding of an Initiator tRNA to the mRNA, and Assembly of the Ribosome.

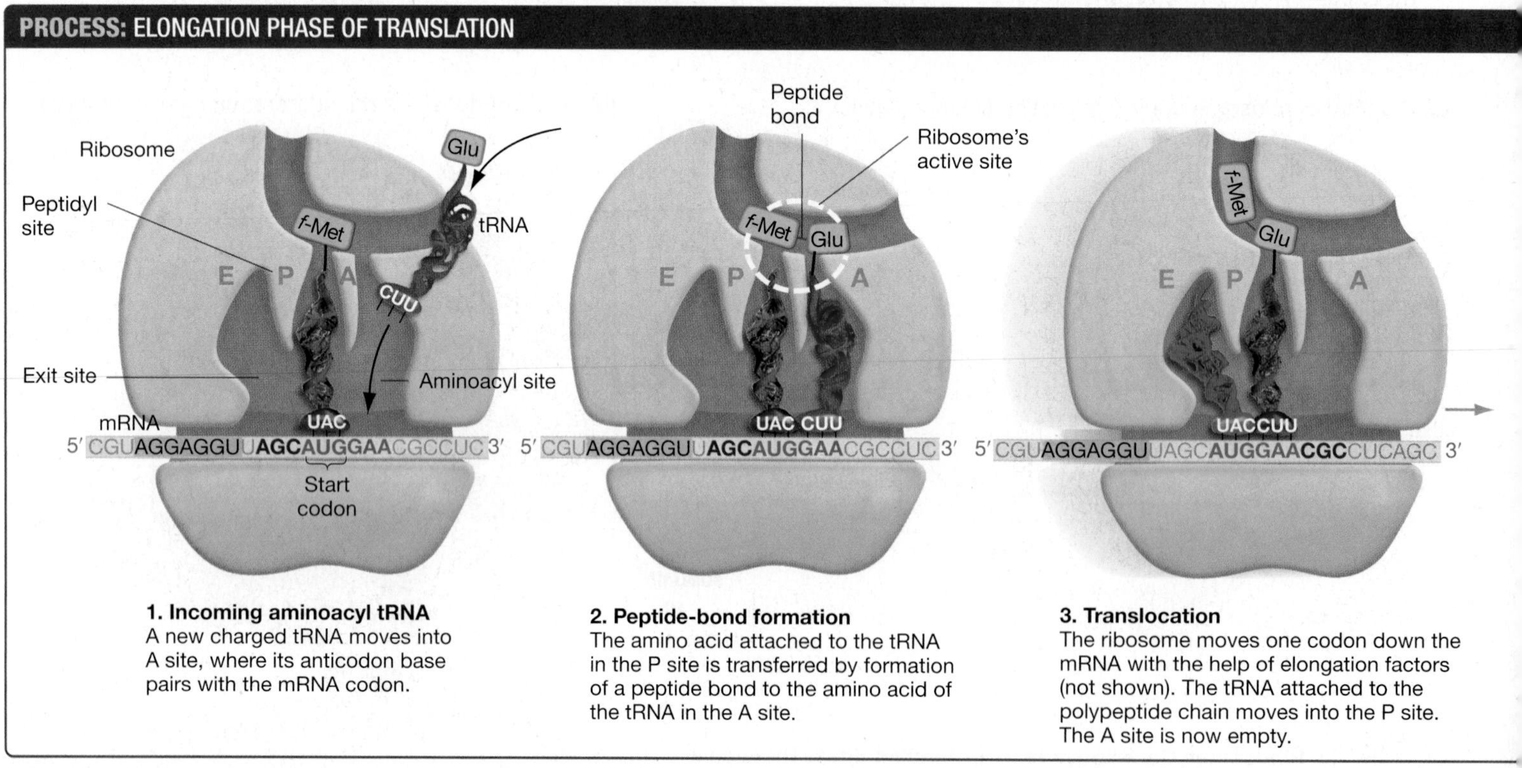

Figure 17.15 Elongation Extends the Polypeptide Chain.

Shine–Dalgarno sequence, after the biologists who discovered it. This site is about six nucleotides upstream from the start codon.

The interactions between the small subunit, the message, and the tRNA are mediated by proteins called **initiation factors** (see Figure 17.14, step 1). Initiation factors help in preparing the ribosome for translation and in binding the first aminoacyl tRNA to the ribosome. In bacteria this is a specialized tRNA, called the initiator tRNA, that bears a modified form of methionine—N-formylmethionine (abbreviated *f*-Met; step 2). Initiation in eukaryotes differs in a number of ways. For example, many more initiation factors are needed, the ribosome first associates with the cap at the 5′ end of the mRNA, and the initiator tRNA carries a normal methionine.

Initiation is complete when the large subunit joins the complex (step 3). At this point, the initiator tRNA occupies the P site of the assembled ribosome. This is the only time a tRNA carrying a single amino acid occupies the P site.

To summarize, translation initiation in bacteria is a three-step process: **(1)** The mRNA binds to a small ribosomal subunit, **(2)** the initiator tRNA bearing *f*-Met binds to the start codon, and **(3)** the large ribosomal subunit binds, completing the complex.

Although many aspects of initiation are different in eukaryotes, it still involves recognition of a start codon, assembly of the ribosome, assistance from initiation factors, and the positioning of a methionine-carrying initiator tRNA in the P site.

Elongation: Extending the Polypeptide

At the start of elongation, the E and A sites in the ribosome are empty of tRNAs. As a result, an mRNA codon is exposed in the A site. As step 1 in Figure 17.15 illustrates, the elongation phase begins when an aminoacyl tRNA binds to the codon in the A site by complementary base pairing between the anticodon and codon.

When both the P and A sites are occupied by tRNAs, the amino acids on the tRNAs are in the ribosome's active site. This is where peptide-bond formation—the essence of protein synthesis—occurs. Peptide-bond formation is one of the most important reactions that takes place in cells because manufacturing proteins is central to all cell processes.

Is the Ribosome an Enzyme or a Ribozyme? Because ribosomes contain about equal amounts of protein and RNA, and because both molecules can serve as catalysts, researchers argued over whether peptide bond formation was catalyzed by protein or RNA. The debate was finally resolved in the year 2000 when researchers completed three-dimensional models detailed enough to reveal the structure of the active site. The active site consists entirely of RNA. The ribosome is a ribozyme—not a protein-based enzyme.

The finding that protein synthesis is catalyzed by RNA supports the RNA world hypothesis (Ch. 4, Section 4.4): RNA emerged before DNA and proteins in the early history of life.

Moving Down the mRNA What happens after a peptide bond forms? Step 2 in Figure 17.15 shows that peptide-bond formation involves the transfer of the amino acid linked to the tRNA in the P site to the amino acid held by the tRNA in the A site.

Step 3 shows **translocation**, the process in which the ribosome moves one codon down the mRNA once a new peptide bond is formed. In reality, it is often the mRNA that is ratcheted through a stationary ribosome, but the important point is that translocation involves the codon-by-codon movement of the

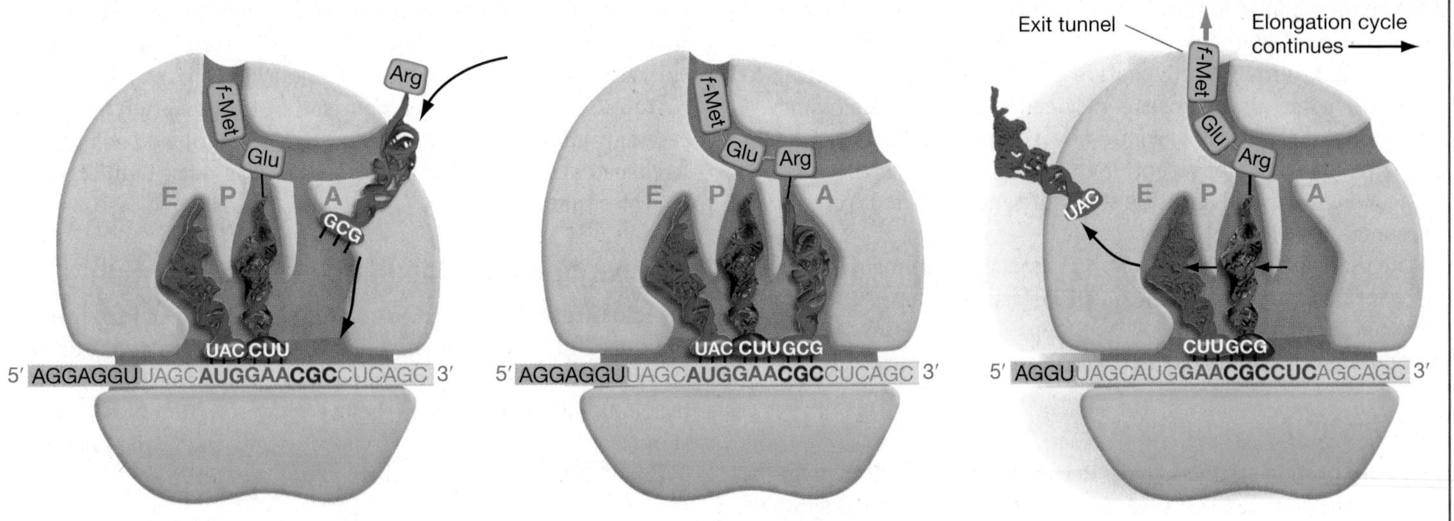

4. Incoming aminoacyl tRNA
A new charged tRNA moves into A site, where its anticodon base-pairs with the mRNA codon.

5. Peptide-bond formation
The polypeptide chain attached to the tRNA in the P site is transferred by peptide bond formation to the aminoacyl tRNA in the A site.

6. Translocation
The ribosome once again moves one codon down the mRNA. The tRNA attached to polypeptide chain moves into P site. Uncharged tRNA from P site moves to E site, where tRNA is ejected. The A site is empty again.

PROCESS: TERMINATING TRANSLATION

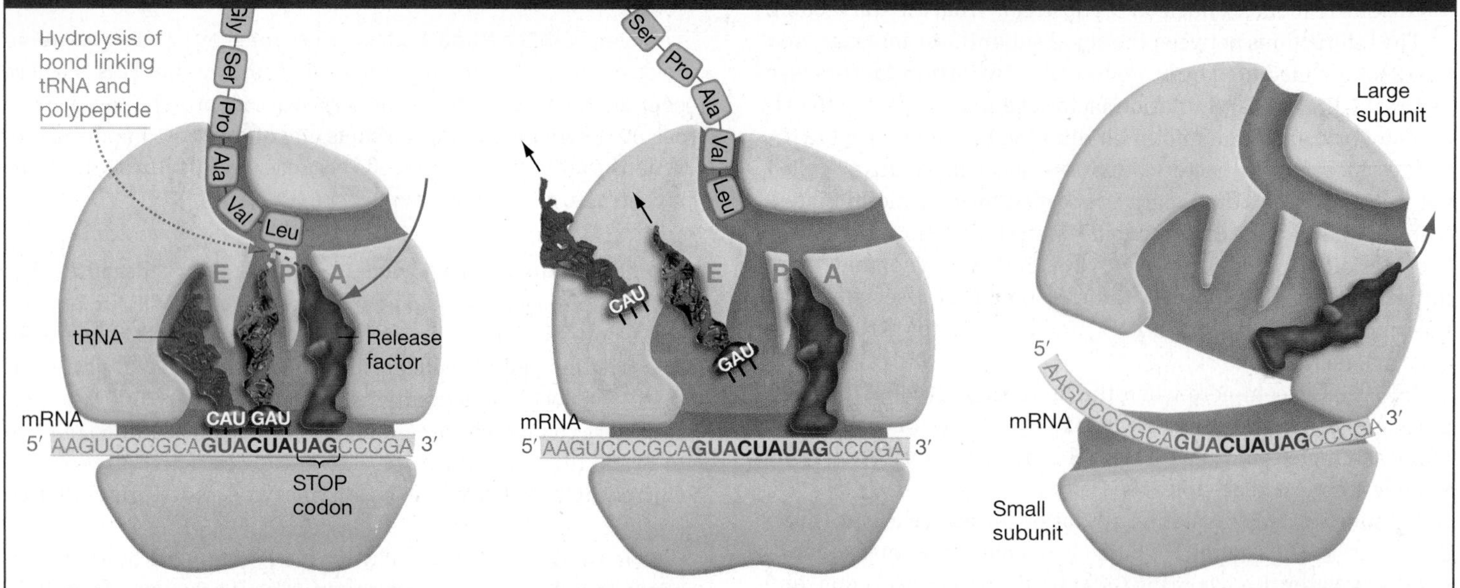

1. Release factor binds to stop codon. When the translocating ribosome reaches a stop codon, a protein release factor fills the A site. The release factor breaks the bond linking the tRNA in the P site to the polypeptide chain.

2. Polypeptide and uncharged tRNAs are released.

3. Ribosome subunits separate. The subunits are ready to attach to the start codon of another message.

Figure 17.16 Termination Occurs When a Release Factor Binds to a Stop Codon Encountered by the Ribosome.

mRNA relative to the ribosome. Translocation requires a type of protein called an **elongation factor**. (Other elongation factors that are involved in steps beside translocation are not shown in Figure 17.15.) Translocation requires energy, and this energy is obtained as one of the elongation factors binds to the ribosome and breaks down the energy-rich molecule GTP.

Translocation does several things: Because the anticodons of tRNA remain bound to the codons in mRNA, movement of the ribosome brings the uncharged tRNA into the E site and the tRNA containing the growing polypeptide into the P site. Translocation of the ribosome also exposes a new codon in the A site, which is now open and free to accept a new charged tRNA. The tRNA that transferred the polypeptide to the amino acid of the tRNA in the A site now finds itself in the E site and is ejected from the ribosome.

The three steps in elongation—**(1)** arrival of aminoacyl tRNA, **(2)** peptide-bond formation, and **(3)** translocation—repeat at each codon along the mRNA (see steps 4–6 of Figure 17.15). Elongation occurs similarly in bacteria and eukaryotes.

Terminating Translation

How does protein synthesis end? Recall that the genetic code includes three stop codons: UAA, UAG, and UGA (Ch. 16, Section 16.3). Instead of tRNAs working to terminate translation, translation is brought to an end when the translocating ribosome reaches one of the stop codons and a protein called a **release factor** recognizes the stop codon and fills the A site (Figure 17.16, step 1). Stop codons are found in the 3′ region of an mRNA, but never at the very end of an mRNA.

Release factors fit tightly into the A site because they have the size and shape of an aminoacyl tRNA coming into the ribosome. Once in the A site, the release factor triggers the hydrolysis of the bond that links the tRNA in the P site to the polypeptide chain. This reaction frees the polypeptide.

With the help of other proteins, the newly synthesized polypeptide and uncharged tRNAs are released from the ribosome (step 2). Then the ribosome separates from the mRNA, and the two ribosomal subunits dissociate (step 3). Termination occurs in very similar ways in bacteria and eukaryotes.

Table 17.1 provides a comparison of transcription, RNA processing, and translation as they occur in bacteria and eukaryotes. Archaea carry out these processes in ways more similar to eukaryotes than to bacteria.

Polypeptides Are Modified after Translation

Proteins are not fully formed and functional at the end of translation. Recall from earlier chapters that most proteins go through an extensive series of processing steps, collectively called post-translational modification, before they are functional. These steps require a wide array of molecules and events and take place in many different locations throughout the cell.

Polypeptide Folding A fundamental principle of biology is that a protein's function depends on its shape, and in turn, a protein's shape depends on how it folds (Chapter 3). Folding is determined by the amino acid sequence of a polypeptide chain. Although folding can occur spontaneously, it is frequently guided and accelerated by proteins called **molecular chaperones**.

SUMMARY **Table 17.1 Transcription, RNA Processing, and Translation in Bacteria and Eukaryotes**

Process	Bacteria	Eukaryotes
Transcription		
• RNA polymerase(s)	One	Three; each produces a different class of RNA
• Promoter structure	Typically contains a −35 box and a −10 box	More variable and larger; often includes a TATA box about −30 from the transcription start site
• Proteins that associate with promoter	Sigma; different versions of sigma bind to different promoters	Many general transcription factors
RNA processing	Rare	Extensive; several processing steps occur in the nucleus before RNA is exported to the cytoplasm: **(1)** enzyme-catalyzed addition of 5′ cap on mRNAs, **(2)** splicing (intron removal) by the spliceosome to produce mRNA, and **(3)** enzyme-catalyzed addition of 3′ poly(A) tail on mRNAs
Translation (initiation, elongation, and termination)	Initiation and termination less complex; elongation similar to eukaryotes	Initiation and termination more complex; elongation similar to bacteria

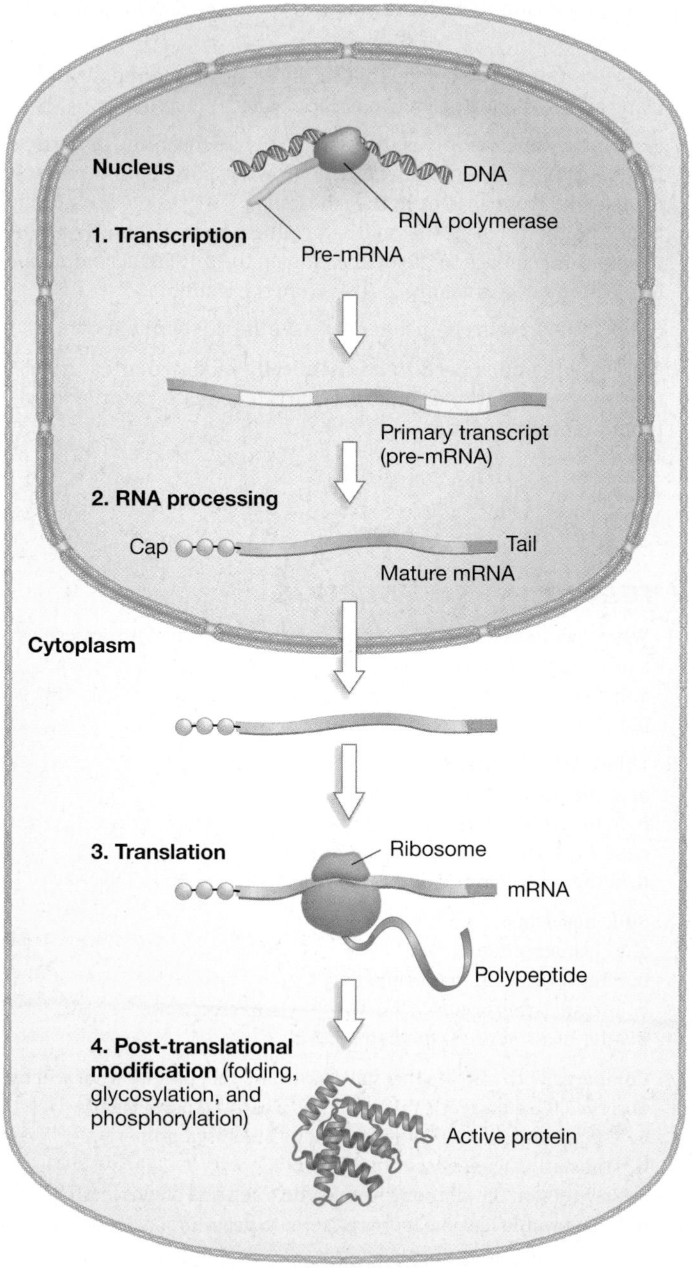

Figure 17.17 The Major Steps of Gene Expression in a Eukaryotic Cell.

Chemical Modifications An earlier chapter described how eukaryotic proteins are often extensively modified after they are synthesized. For example, in the organelles called the rough endoplasmic reticulum and the Golgi apparatus, small chemical groups may be added to proteins—often sugars (in the process of glycosylation; Ch. 7, Section 7.5) or lipid groups that are critical for normal functioning or to target them to specific locations. Another common post-translational modification is the addition of a phosphate group (in the process of phosphorylation) by enzymes called protein kinases. Adding a phosphate group—and removing it later—often dramatically affects the protein's activity (Ch. 11, Section 11.3).

Figure 17.17 reviews how gene expression works in a eukaryotic cell. Take a close look to see how all these steps fit together. The take-home message is that gene expression is a multistep process that begins with transcription. What's critical to remember is that the RNAs and proteins produced during gene expression give the cell and organism its characteristics. These molecules truly are the basis of life.

But genes can't be turned on all at the same time. How does a cell "decide" which of its many genes should be expressed and when to express them? These fundamental questions are the focus of the next two chapters.

CHECK YOUR UNDERSTANDING

✔ **If you understood this section, you should be able to ...**

1. Explain why it's important that the initiator tRNA be placed in the P site instead of the A site.
2. Explain how a tRNA that was once in the A site comes to be in the P site during the elongation phase of translation.
3. Explain why it's logical that a release factor has the same structure as an aminoacyl tRNA.

Answers are available in Appendix A.

17.1 An Overview of Transcription

- In transcription, RNA polymerase produces an RNA molecule with a base sequence complementary to the base sequence of the DNA template strand.
- RNA polymerase begins transcription by binding to promoter sequences in DNA with the help of other proteins.
- In bacteria, this binding is accomplished through a protein called sigma. Sigma associates with RNA polymerase and then recognizes sequences within promoters that are centered 10 bases (−10 box) and 35 bases (−35 box) upstream from where transcription begins.
- Eukaryotic promoters vary more than bacterial promoters.
- In eukaryotes, transcription begins when a large array of proteins called general transcription factors bind to a promoter. In response, RNA polymerase binds at the promoter.
- In bacteria and eukaryotes, RNA is elongated in a $5' \rightarrow 3'$ direction.
- Transcription in bacteria ends when a hairpin structure forms in the transcribed RNA; in eukaryotes, transcription terminates after the RNA is cleaved downstream of the poly(A) signal.

17.2 RNA Processing in Eukaryotes

- In eukaryotes, the primary (initial) transcript must be processed to produce a mature RNA.
- Splicing of primary transcripts removes stretches of RNA called introns and joins together regions called exons.
- Complex macromolecular machines called spliceosomes cut introns out of pre-mRNA.
- A "cap" is added to the 5′ end of pre-mRNAs, and a poly(A) tail is added to their 3′ end.
- The cap and tail serve as recognition signals for starting translation, and also protect the mRNA from degradation.
- RNA processing occurs in the nucleus.

17.3 An Introduction to Translation

- Ribosomes translate mRNAs into proteins with the help of adapter molecules called transfer RNAs (tRNAs) that act as a chemical bridge between codons in mRNA and amino acids added to the synthesized polypeptide.
- In bacteria, an RNA can sometimes be transcribed and translated at the same time.
- In eukaryotes, transcription and translation of an RNA are physically separated processes—transcription occurs in the nucleus and translation occurs in the cytoplasm.

17.4 The Structure and Function of Transfer RNA

- Each transfer RNA carries an amino acid that is specified by the tRNA's three-base-long anticodon.
- tRNAs have an L-shaped tertiary structure. One "end" of the L contains the anticodon, which forms complementary base pairs with the mRNA codon. The other end holds the amino acid specified by that codon.
- Enzymes called aminoacyl-tRNA synthetases link the correct amino acid to the correct tRNA.
- Imprecise pairing—"wobble pairing"—in the third position of the codon and anticodon allows the approximately 40 types of tRNA to translate all 61 codons that code for amino acids.

17.5 Ribosome Structure and Function in Translation

- Ribosomes are large macromolecular machines made of many proteins and RNAs.
- In the ribosome, the anticodon of the tRNA binds to a three-base-long codon in mRNA to hold the correct amino acid in the ribosome.
- The ribosome is a ribozyme that catalyzes peptide-bond formation using an RNA, not a protein-based enzyme.
- Protein synthesis involves three steps: **(1)** an incoming aminoacyl tRNA occupies the A site; **(2)** the growing polypeptide chain is transferred from a tRNA in the ribosome's P site to the amino acid bound to the tRNA in the A site, forming a peptide bond; and **(3)** the ribosome moves to the next codon on the mRNA, accompanied by ejection of the uncharged RNA from the E site.
- Chaperone proteins help fold newly synthesized proteins.
- Most polypeptides need to be chemically modified after translation (post-translational modification) to activate them or target them to specific locations.

Answers are available in Appendix A

✔ TEST YOUR KNOWLEDGE

1. What does a bacterial RNA polymerase produce when it transcribes a protein-coding gene?
   a. rRNA
   b. tRNA
   c. mRNA
   d. snRNA
2. **THINK CAREFULLY** Where is the start codon located?
   a. at the start (5′ end) of the mRNA
   b. in the DNA just upstream of where transcription starts
   c. at the downstream end of the 5′ untranslated region (UTR)
   d. at the upstream end of the 3′ untranslated region (UTR)
3. Splicing begins:
   a. as transcription occurs.
   b. after transcription is complete.
   c. as translation occurs.
   d. after translation is complete.
4. Compared with mRNAs that have a cap and tail, predict what will be observed if a eukaryotic mRNA lacked a cap and poly(A) tail.
   a. The primary transcript would not be processed properly.
   b. Translation would occur inefficiently.
   c. Enzymes on the ribosome would add a cap and poly(A) tail.
   d. tRNAs would become more resistant to degradation.

✔ TEST YOUR UNDERSTANDING

5. RNases and proteases are enzymes that destroy RNAs and proteins, respectively. Which of the following enzymes, if added to a spliceosome, would be predicted to prevent recognition of pre-mRNA regions critical for splicing?
   a. an RNase specific for tRNAs
   b. an RNase specific for snRNAs
   c. a protease specific for initiation factors
   d. a protease specific for a release factor

6. **THINK CAREFULLY** For each of these statements about the genetic code, select True or False.
   T/F Wobble pairing accounts for the redundancy of the genetic code.
   T/F There are 64 different tRNAs that read the 64 possible codons.
   T/F All possible codons are used, but not all codons specify an amino acid.
   T/F Some codons are recognized by proteins, not by tRNAs.

7. In a particular bacterial species, temperature-sensitive conditional mutations cause expression of a wild-type phenotype at one growth temperature and a mutant phenotype at another—typically higher—temperature. Imagine that when a bacterial cell carrying such a mutation is shifted from low to high growth temperatures, RNA polymerases in the process of elongation complete transcription normally, but no new transcripts can be started. The mutation in this strain most likely affects:
   a. the terminator sequence
   b. the start codon
   c. sigma
   d. one of the polypeptides of the core RNA polymerase

8. **THINK CAREFULLY** In what ways are a promoter and a start codon similar? In what ways are they different?

✔ TEST YOUR PROBLEM-SOLVING SKILLS

9. The nucleotide shown here is called cordycepin triphosphate. It is a natural product of a fungus that is used in traditional medicines.

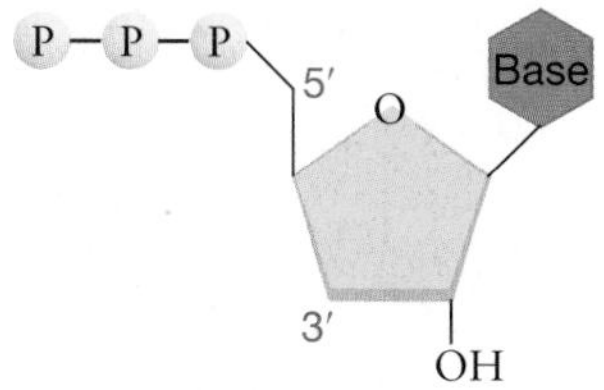

   If cordycepin triphosphate is added to a cell-free transcription reaction, the nucleotide is added onto the growing RNA chain but then no more nucleotides can be added. Examine the structure of cordycepin and explain why it ends transcription.

10. **QUANTITATIVE** Controlling the rates of transcription and translation is important in bacteria to avoid collisions between ribosomes and RNA polymerases. Calculate what the maximum rate of translation by a ribosome in a bacterial cell would have to be, in units of amino acids per second, so as not to overtake an RNA polymerase that is synthesizing mRNA at a rate of 60 nucleotides per second. How long would it take for this bacterial cell to translate an mRNA containing 1800 codons?

✔ PUT IT ALL TOGETHER: Case Study

What better *not* be for dinner?

Eating even a single death cap mushroom (*Amanita phalloides*) can be fatal due to a compound called α-amanitin, a toxin that inhibits transcription.

11. What would you predict to be the immediate outcome of adding α-amanitin to a cell?
    a. reduced DNA synthesis
    b. reduced production of one or more types of RNA
    c. reduced binding of tRNAs to anticodons
    d. reduced rate of translocation of ribosomes translating mRNA

12. α-Amanitin inhibits transcription by binding inside an RNA polymerase to a region other than the active site that catalyzes addition of a nucleotide to the RNA chain. Based on the model of RNA polymerase shown in Figure 17.3, predict how the toxin might function to inhibit transcription.

13. Toxins like α-amanitin are used for research in much the same way as null mutants (Chapter 16)—to disrupt a process and see what happens when it no longer works. Researchers examined the ability of α-amanitin to inhibit different RNA polymerases. They purified RNA polymerases I, II, and III from rat liver, incubated the enzymes with different concentrations of α-amanitin, and then tested their activity. The results of this experiment are shown here. These findings suggest that cells treated with α-amanitin will have a reduced level of:
    a. tRNAs **c.** snRNAs
    b. rRNAs **d.** mRNAs

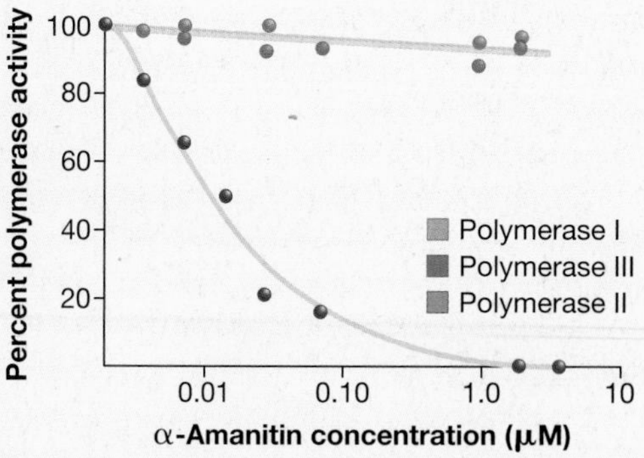

Source: T. J. Lindell, F. Weinberg, P. W. Morris, et al. (1970). *Science* 170: 447–449.

14. **QUANTITATIVE** If you wanted to use α-amanitin to shut down 95 percent of transcription by RNA polymerase II, roughly what concentration of α-amanitin would you use? Note that the scale on the *x*-axis of the graph in Question 13 is logarithmic rather than linear, so that each tick mark shows a tenfold higher concentration. (See **BioSkills 5** for tips on working with logarithms.)

15. **PROCESS OF SCIENCE** Biologists have investigated how fast pre-mRNA splicing occurs by treating cells with a toxin that blocks the production of new pre-mRNAs, then following the rate of splicing of the pre-mRNAs that were transcribed before adding the toxin. Why is addition of a toxin important in this study?

16. **PROCESS OF SCIENCE** The primary cause of death from α-amanitin poisoning is liver failure. Suppose a physician informs you that liver cells die because their rate of protein production falls below a level needed to maintain active metabolism. Given that α-amanitin is an inhibitor of transcription, you wonder if this information is correct. Propose an experiment to determine whether the toxin also has an effect on protein synthesis.

Mastering Biology

Students Go to Mastering™ Biology for assignments, the eText, and the Study Area with animations, practice tests, and activities.

Professors Go to Mastering™ Biology for automatically graded tutorials and questions that you can assign to your students, plus Instructor Resources.

UNIT 3

GENE STRUCTURE AND EXPRESSION

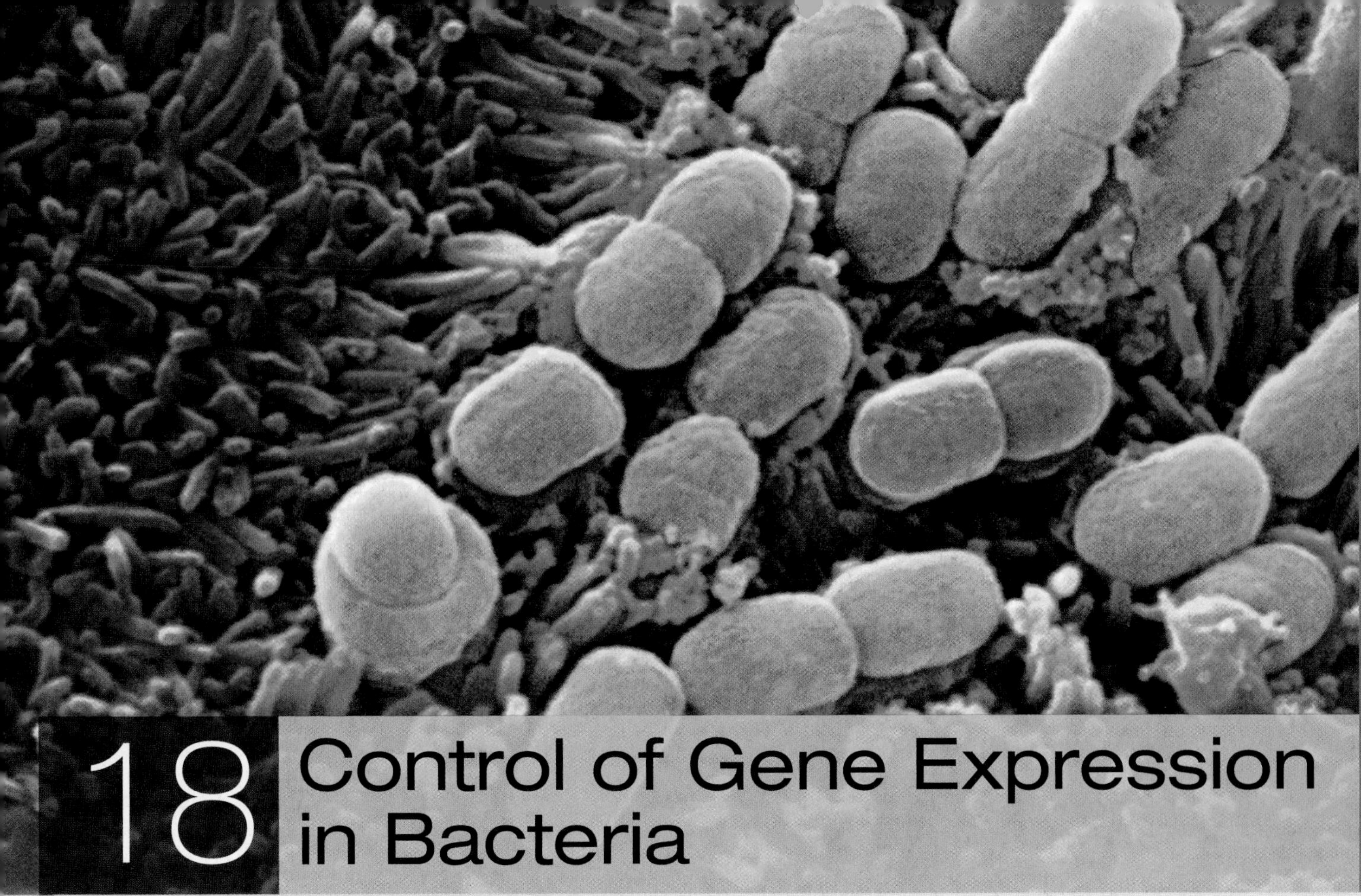

18 Control of Gene Expression in Bacteria

This false-color micrograph shows *E. coli* bacteria (yellow) and projections from human intestinal cells (blue). In the intestine, the nutrients available to bacteria constantly vary. This chapter explores how changes in gene expression help bacteria respond to environmental changes.

In this chapter you will learn how

Bacteria turn their genes on and off to adapt to changing environments

surveying

Different ways genes can be regulated **18.1**

looking closer at

Negative and positive control of gene expression **18.2**

Ways bacteria regulate many genes together **18.3**

Imagine waiting to hear the opening lines of a melodic symphony played by a renowned orchestra. The crowd applauds as the conductor comes onstage and then hushes as he takes the podium. He cocks the baton; the musicians raise their instruments. As the baton comes down, every instrument begins blaring a different tune at full volume. A tuba plays "Dixie," a violinist renders "Hey Jude," and a vocalist starts to rap Twenty-One Pilots' "Stressed Out" as the bass drum simulates cannons in the "1812 Overture." Instead of music, there is pandemonium. The conductor staggers offstage, clutching his heart.

Cacophony like this would result if a bacterial cell "played" all its genes at full volume all the time. For instance, the *Escherichia coli* cells living in your gut right now have over 4300 genes. If all those genes were turned on full blast at all times, these cells would stagger off the stage too. But this doesn't happen. Cells must be exquisitely selective about which genes are turned on, when and where these genes are activated, and how much of the gene products to make.

In this chapter, you will explore how bacterial cells control the activity, or expression, of their genes. This is important not only for the bacterial cell, but also for us. Changes in gene expression can control

BIG PICTURE

This chapter is part of the Big Picture. See how on pages 404–405.

whether a bacterial cell can infect people. **Gene expression** is the multistep process of converting information that is archived in DNA into molecules that actually do things in the cell. It occurs when a gene product—an RNA or a protein—is synthesized and becomes active. (You can see on pages 404–405 how gene expression fits into the Big Picture of Genetic Information.)

In previous chapters you saw how genetic information is processed in cells; this chapter is the first of two that focus on ways to control *when* and *to what level* genetic information is used. Here, you'll learn how bacteria control gene expression, and in the following chapter you'll come to understand some different strategies used by eukaryotes to regulate expression of their genes. Let's begin by reviewing some of the environmental challenges that bacterial cells face and then explore how they regulate gene expression to meet them.

18.1 An Overview of Gene Regulation and Information Flow

To understand why control over gene expression is so important, consider that cells of thousands of bacterial species are packed along your intestinal walls. All of these organisms are competing for space and nutrients. In an environment like this, a cell has to use resources efficiently so it can survive and reproduce. An individual that synthesizes proteins it doesn't need has fewer resources for making essential proteins. Such cells compete less successfully for the resources required to produce offspring.

After you complete this section, you should be able to ...

- Describe how gene expression can be regulated.

Realizing this need for gene regulation, biologists predicted that most bacterial gene expression is triggered by specific signals from the environment, such as the presence of specific nutrients. Like a conductor who needs to regulate the orchestra's musicians, cells need to regulate which proteins they produce.

Mechanisms of Regulation

In bacteria, the flow of information from DNA to a final active protein product occurs in three steps, represented by arrows in the following pathway:

$$\text{DNA} \longrightarrow \text{mRNA} \longrightarrow \text{protein} \longrightarrow \text{activated protein}$$

Gene expression can be controlled at any of these steps. The arrow from DNA to messenger RNA (mRNA) represents transcription—producing an RNA complementary to the template strand of DNA. The arrow from mRNA to protein represents translation, in which ribosomes read and use the information in mRNA to synthesize a protein. The arrow from protein to activated protein represents post-translational (literally, "after-translation") modifications that can lead to changes in shape and activity. (Eukaryotes add another step called RNA processing, discussed in Chapter 19.)

How can a bacterial cell produce only the proteins that are needed at a particular time? A look at the flow of information from DNA to active protein suggests three possible mechanisms:

1. **Transcriptional control**: The cell could make mRNAs only for proteins it needs. If genes for unneeded proteins are *not* transcribed into mRNA (indicated by the symbol $\not\longrightarrow$ in the next pathway), then ribosomes cannot make these proteins. This form of control occurs when regulatory proteins affect RNA polymerase's ability to bind to a promoter and initiate transcription:

$$\text{DNA} \not\longrightarrow \text{mRNA} \longrightarrow \text{protein} \longrightarrow \text{activated protein}$$

2. **Translational control**: The cell could prevent the mRNAs for unneeded proteins from being translated. This form of control occurs either through regulation of the mRNA's life span or ability to be translated:

$$\text{DNA} \longrightarrow \text{mRNA} \not\longrightarrow \text{protein} \longrightarrow \text{activated protein}$$

3. **Post-translational control**: After translation, many proteins have to be activated by chemical modification, such as the addition of a phosphate group, in order to function. These modifications can be controlled to regulate gene expression.

$$\text{DNA} \longrightarrow \text{mRNA} \longrightarrow \text{protein} \not\longrightarrow \text{activated protein}$$

Which of these forms of control occur in bacteria? The short answer is, they all do. **Figure 18.1** shows how these types of regulation can control the production of active protein from a particular gene.

- *Transcriptional control* is particularly important due to its efficiency—it saves the most energy for the cell because it controls gene expression before the cell expends many resources.
- *Translational control* allows more rapid changes than transcriptional control in the amounts of different proteins because the mRNA has already been made and is available for translation.
- *Post-translational control* provides the most rapid response of all three mechanisms because only one step is needed to activate or inactivate an existing protein.

Among these mechanisms of gene regulation, there is a clear trade-off between the speed of response and the conservation of nucleotides, amino acids, energy, and other resources. Transcriptional control provides a slower response, but conserves resources; post-translational control is fast but energetically expensive.

In this chapter, you'll focus largely on mechanisms of transcriptional control, but remember that bacteria also use translational and post-translational forms of control. (These means of control are covered in Chapter 19.) Just as important, some genes—such as those that code for the enzymes required for glycolysis—are transcribed all the time, or **constitutively**. Finally, it is critical to realize that gene expression is not an all-or-none proposition. Genes are not just fully "on" or "off"—instead, the level of expression can vary between these extremes.

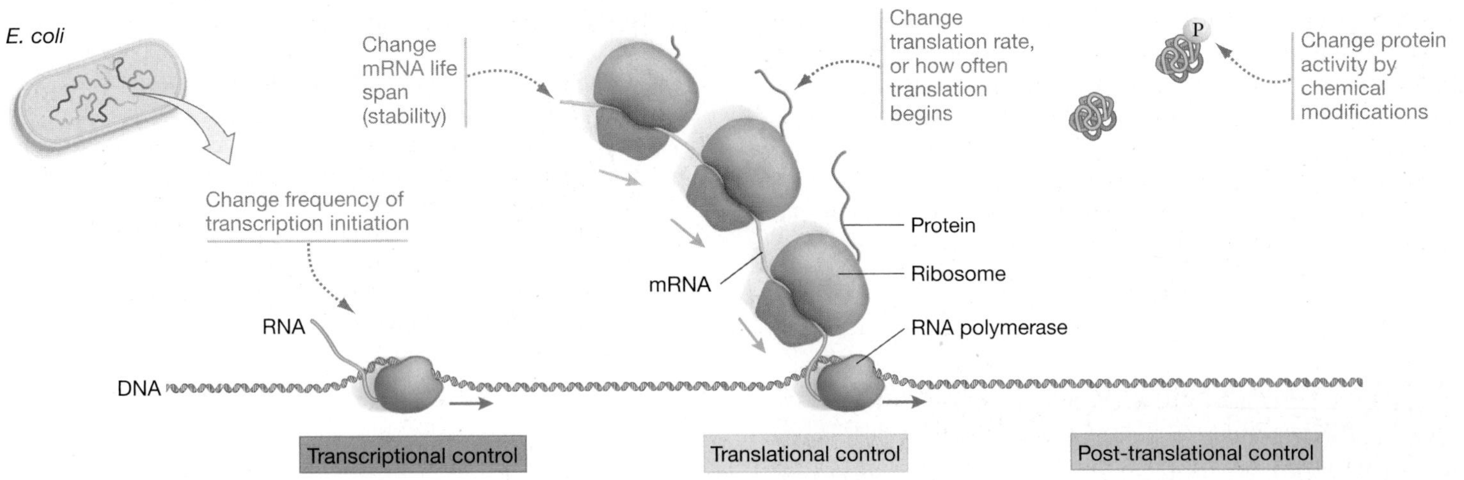

Figure 18.1 Gene Expression in Bacteria Can Be Regulated at Three Levels.

Metabolizing Lactose—A Model System

Many great advances in genetics have been made by analyzing model systems (see **BioSkills 11**). During the 1950s and 1960s, Jacques Monod, François Jacob, and many colleagues used the bacterium *E. coli* to study the regulation of lactose metabolism. Although they worked with just one species and a small set of genes, their results profoundly affected thinking about gene regulation in all organisms. In this chapter you will see how the insights these researchers gained from studying the regulation of genes for lactose metabolism in *E. coli* have provided a foundation for understanding general principles of gene regulation.

E. coli can use a wide variety of sugars for ATP production via cellular respiration or fermentation. These sugars also serve as raw material in the synthesis of amino acids, vitamins, and other complex compounds. Glucose, however, is *E. coli*'s preferred carbon source—meaning it is the source of energy and carbon atoms that the organism uses first.

A preference for glucose makes sense, because glycolysis begins with glucose and is the main pathway for the production of ATP. Lactose, the predominant sugar found in milk, can also be used by *E. coli*, but it is left untouched until glucose supplies are depleted. Lactose is a disaccharide made up of a molecule of glucose and a molecule of galactose (Ch. 5, Figure 5.5).

To use lactose, *E. coli* must first transport the sugar into the cell using a membrane protein called galactoside permease. Then the enzyme β-galactosidase catalyzes a reaction within the cell that breaks down lactose into glucose and galactose. These steps in lactose metabolism are summarized in **Figure 18.2**. The glucose released by this reaction is used directly for glycolysis; other enzymes convert the galactose to a substance that can also be processed in the glycolytic pathway.

In the early 1950s, biologists learned that *E. coli* produces a high level of β-galactosidase only when lactose is present in the environment. Based on this observation, researchers proposed that lactose itself regulates the gene for β-galactosidase—meaning that lactose acts as an inducer. An **inducer** is a small molecule that triggers transcription of a specific gene.

Jacques Monod wondered what effect the presence of glucose, the preferred carbon source, might have on the regulation of the β-galactosidase gene. Would *E. coli* produce a high level of

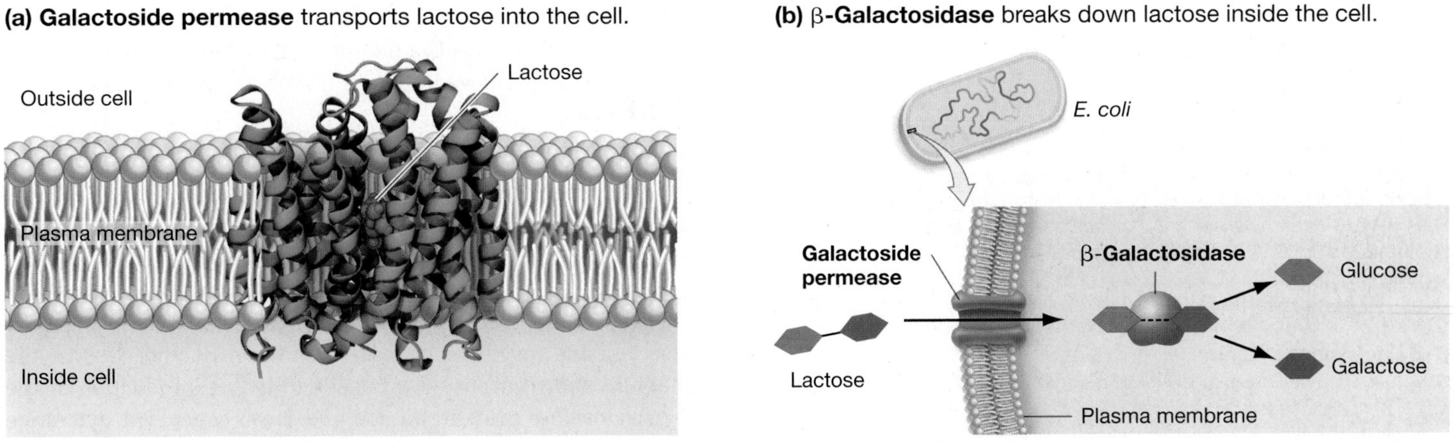

Figure 18.2 Two Proteins Needed to Use Lactose Are Galactoside Permease and β-Galactosidase.

β-galactosidase when both glucose and lactose were present in the surrounding environment? As the experiment summarized in **Figure 18.3** shows, the answer is no. Significant amounts of β-galactosidase are produced only when lactose is present and glucose is absent.

How does lactose induce expression of the β-galactosidase gene? And how does glucose prevent the expression of the β-galactosidase gene? The answers to these specific questions illuminated the general ways that genes in all organisms are controlled.

RESEARCH

QUESTION: Will *E. coli* produce β-galactosidase when both glucose and lactose are present?

HYPOTHESIS: *E. coli* will not produce β-galactosidase when glucose is present, even if lactose is present. (Glucose is the preferred food source.)

NULL HYPOTHESIS: *E. coli* will produce β-galactosidase whenever lactose is present, regardless of the presence or absence of glucose.

EXPERIMENTAL SETUP:

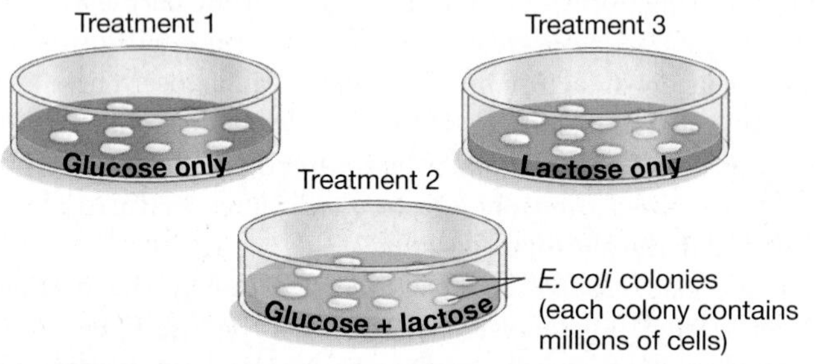

PREDICTION OF HYPOTHESIS: β-Galactosidase will be produced only in treatment 3.

PREDICTION OF NULL HYPOTHESIS: β-Galactosidase will be produced in treatments 2 and 3.

RESULTS:

Treatment 1	Treatment 2	Treatment 3
No β-galactosidase	No β-galactosidase	Production of β-galactosidase

E. coli does not produce β-galactosidase if glucose is present

CONCLUSION: Glucose prevents expression of the gene for β-galactosidase. The presence of lactose without glucose stimulates expression of the β-galactosidase gene.

Figure 18.3 Glucose Affects the Regulation of the β-Galactosidase Gene.

SOURCE: A. B. Pardee, F. Jacob, and J. Monod. 1959. The genetic control and cytoplasmic expression of "inducibility" in the synthesis of β-galactosidase by *E. coli*. *Journal of Molecular Biology* 1: 165–178.

✔ Predict how the results would be different if instead of glucose, another sugar that *E. coli* can use—say table sugar (sucrose)—were provided instead.

CHECK YOUR UNDER STANDING

✔ **If you understood this section, you should be able to ...**

1. Explain why transcriptional control provides a slower response than post-translational control, but why cells often regulate genes at the level of transcription despite the relatively sluggish response.
2. Explain why an *E. coli* cell that always expressed galactoside permease and β-galactosidase would be at a disadvantage when competing with a wild-type *E. coli* cell that regulated expression of these proteins.

Answers are available in Appendix A.

18.2 Negative and Positive Control of Transcription

There are two general ways that the transcription of any gene can be regulated: by negative control or positive control.

After you complete this section, you should be able to ...

- Explain how negative and positive control of transcription controls the expression of the *lac* operon.

1. **Negative control** occurs when a regulatory protein called a **repressor** binds to DNA and shuts down transcription (symbolized by a red "X"; **Figure 18.4a**, bottom).
2. **Positive control** occurs when a regulatory protein called an **activator** binds to DNA and triggers transcription (symbolized by a green arrow; **Figure 18.4b**, bottom).

Drawing figures like the ones in Figure 18.4 can help you understand how transcription is controlled. **Making Models 18.1** gives some tips on how to do this.

Imagine a car that's in gear and moving forward at a constant speed. In this case, negative control of the speed would be using the brake, while positive control would be using the gas pedal. It turns out that the "speed" of initiating transcription of the genes involved in lactose metabolism are controlled both ways—with a brake and a gas pedal. These genes are under both negative and positive control.

Before getting into the details of how negative and positive regulation are achieved, let's first look at the genes for lactose metabolism.

A Gene Needed to Regulate Lactose Metabolism

To understand how *E. coli* controls the utilization of lactose, Monod and Jacob isolated and analyzed mutants unable to use lactose.

They found three types of mutants. In one class, the cells were unable to cleave lactose—even when lactose was in the medium and transported into cells to induce production of the β-galactosidase protein. Monod and Jacob concluded that these mutants must lack a functioning version of the β-galactosidase protein. Therefore, the gene that encodes β-galactosidase is defective. This gene was designated *lacZ*, and the mutant allele $lacZ^-$.

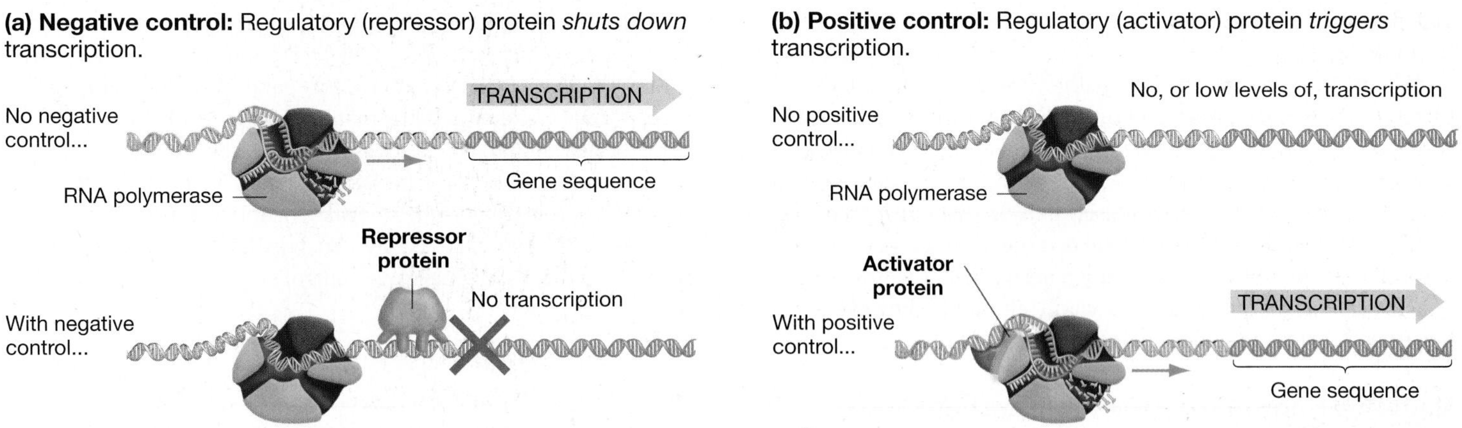

Figure 18.4 Gene Transcription Is Regulated by Negative Control, Positive Control, or Both. (To review transcription initiation, see Ch. 17, Figure 17.3.)

Making Models 18.1 Tips on Drawing How Transcription Is Controlled

There are many ways to diagram how transcription is controlled. Here's one simple approach for drawing positive control of a gene. It's most important to show how the gene, regulatory protein (here an activator), and regulatory DNA sequences work together—other details can be omitted.

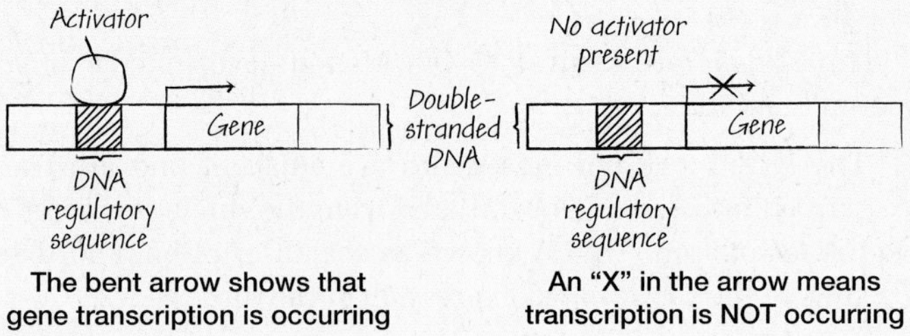

MODEL Draw the same arrangement of the gene and DNA regulatory sequence as shown here, but this time show what happens if the regulatory protein exerts negative control.

To see this model in action, go to the Study Area of **Mastering Biology**

In the second class of mutants, the cells failed to accumulate lactose inside the cell. Monod and Jacob hypothesized that these cells had a defective membrane protein responsible for transporting lactose into the cell. This protein was identified and, as shown in Figure 18.2, named galactoside permease; the gene that encodes it was designated *lacY*.

A third class of mutants could produce both β-galactosidase and galactoside permease, but had defects in *regulating* the expression of these proteins. In wild-type cells, β-galactosidase and galactoside permease are turned on (induced) only when lactose is present. In contrast, these regulatory mutants made the proteins all the time (they expressed the proteins constitutively), even without lactose in the cell. **Table 18.1** summarizes these three types of mutants.

Cells that are abnormal because they produce a product at all times instead of regulating expression of the product are called **constitutive mutants**. The gene that was mutated to produce constitutive β-galactosidase and galactoside permease expression was named *lacI*. The letter "I" signified that these mutants did not need an inducer—lactose—to express β-galactosidase or galactoside permease. This means that $lacI^-$ mutants have a

Table 18.1 Three Types of Lactose Metabolism Mutants

Observed Phenotype	Interpretation	Genotype
Cells cannot cleave lactose, even in the presence of inducer (lactose).	No β-galactosidase; gene for β-galactosidase is defective. This gene is named *lacZ*.	$lacZ^-$
Cells cannot accumulate lactose.	No membrane protein (galactoside permease) to import lactose; gene for galactoside permease is defective. This gene is named *lacY*.	$lacY^-$
Cells can cleave lactose even if lactose is absent as an inducer.	Constitutive (constant) expression of *lacZ* and *lacY*; gene for regulatory protein that shuts down *lacZ* and *lacY* is defective. This gene is named *lacI*.	$lacI^-$

defect in gene regulation. In these mutants, the gene is on when it should be off.

To pull these observations together, the researchers hypothesized that the normal product of the *lacI* gene *prevents* the transcription of *lacZ* and *lacY* when lactose is absent. In short, *lacI* was proposed to act as a negative regulator. Because lactose triggers production of β-galactosidase, the scientists proposed that lactose interacts with the *lacI* gene or gene product. (Later work showed that the inducer is actually a derivative of lactose called *allolactose*. For historical accuracy and simplicity, however, this discussion refers to lactose itself as the inducer.)

Negative Control of Lactose Utilization Genes

In the late 1950s, Leo Szilard, a physicist turned biologist, suggested that the *lacI* gene codes for a protein product that represses transcription of the *lacZ* and *lacY* genes. Genetic mapping experiments had found that the *lacZ, lacY*, and *lacI* genes are clustered together in the positions shown in Figure 18.5. This finding suggested that the *lacZ* and *lacY* genes might be transcribed together under the control of the *lacI* gene.

The model proposed by Szilard, Monod, and Jacob for how the *lacI* gene works is shown in Figure 18.6. The *lacI* gene produces a repressor protein that binds directly to DNA and overlaps the promoter for the *lacZ* and *lacY* genes (see Figure 18.6a). Binding of the repressor blocks RNA polymerase, either by preventing initiation (as shown in the figure) or by interfering with the binding of RNA polymerase to the promoter.

To explain how lactose triggers transcription, Szilard and Monod proposed that lactose binds to the repressor, changes its shape, and causes the repressor to release from its binding site in DNA (see Figure 18.6b). In negative control, genes are normally transcribed, with the repressor acting as a brake. In the case of genes for lactose metabolism, lactose works to release this brake.

What about the constitutive mutants? Figure 18.6c shows that constitutive transcription is observed in *lacI*$^-$ mutants because a functional repressor is absent—the brake is gone.

To test the hypothesis of negative control by a repressor, Monod, Jacob, and co-workers added a functioning copy of the *lacI* repressor gene to the *lacI*$^-$ mutants that made β-galactosidase all the time (constitutively). When these cells were grown without lactose, β-galactosidase production declined and then stopped.

This result supported the hypothesis that the repressor codes for a protein that shuts down transcription. Significantly, if the mutant cells with an introduced *lacI* gene were grown using lactose, β-galactosidase activity resumed. This result supported the hypothesis that lactose removes the repressor.

What's the take-home message? The *lacI* gene codes for a repressor protein that exerts negative control on *lacZ* and *lacY*. Lactose acts as an inducer by causing the repressor to release from DNA and ending negative control.

The Operon Model

One of Jacob and Monod's key conclusions was that the genes for β-galactosidase and galactoside permease are controlled together and transcribed into a single mRNA. An mRNA like this codes for two or more different polypeptides and is known as a polycistronic mRNA. Many bacterial mRNAs are polycistronic, but almost all eukaryotic mRNAs code for only a single polypeptide.

Jacob and Monod coined the term **operon** to describe a set of coordinately regulated bacterial genes that are transcribed together into one polycistronic mRNA. Many bacterial genes are organized into operons. The group of genes involved in lactose metabolism was termed the ***lac* operon**.

Later, another gene called *lacA* was found to be within the *lac* operon. The *lacA* gene codes for the enzyme transacetylase. This enzyme catalyzes reactions that allow certain types of sugars to be exported from the cell when they are too abundant and could harm the cell. The components of the *lac* operon are summarized in Figure 18.7.

Three ideas are central to the Monod–Jacob model of *lac* operon regulation:

1. The *lacZ, lacY*, and *lacA* genes are adjacent and are transcribed into one mRNA initiated from the single promoter of the *lac* operon. This is known as cotranscription, and it results in the coordinated expression of the three genes.
2. The repressor protein encoded by *lacI* binds to a specific sequence in DNA and prevents transcription of the *lac* operon genes (*lacZ, lacY*, and *lacA*). Jacob and Monod proposed that *lacI* is expressed constitutively; that the repressor binds to a DNA sequence in the *lac* operon called the **operator**; and that repressor bound to the operator prevents the RNA polymerase holoenzyme from binding to the promoter of the *lac* operon.

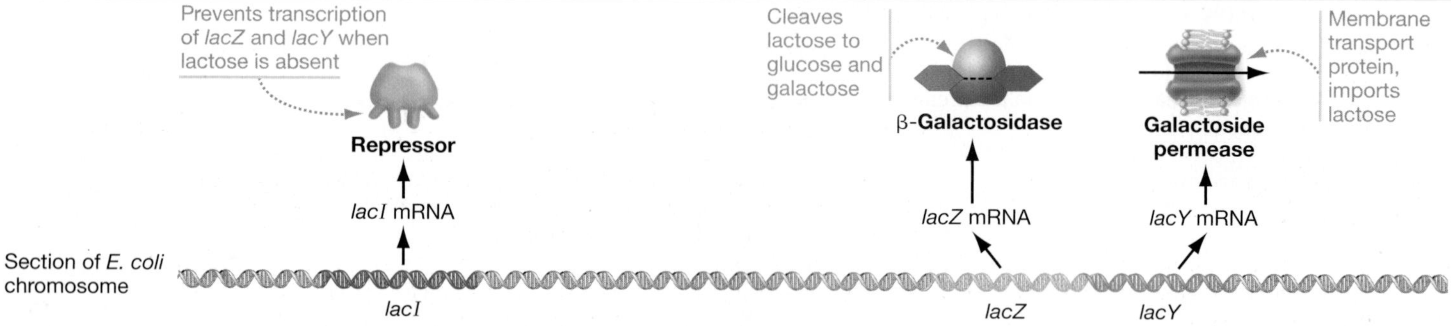

Figure 18.5 The *lac* Genes Are Clustered Together. The associated proteins and their functions are shown above each gene.

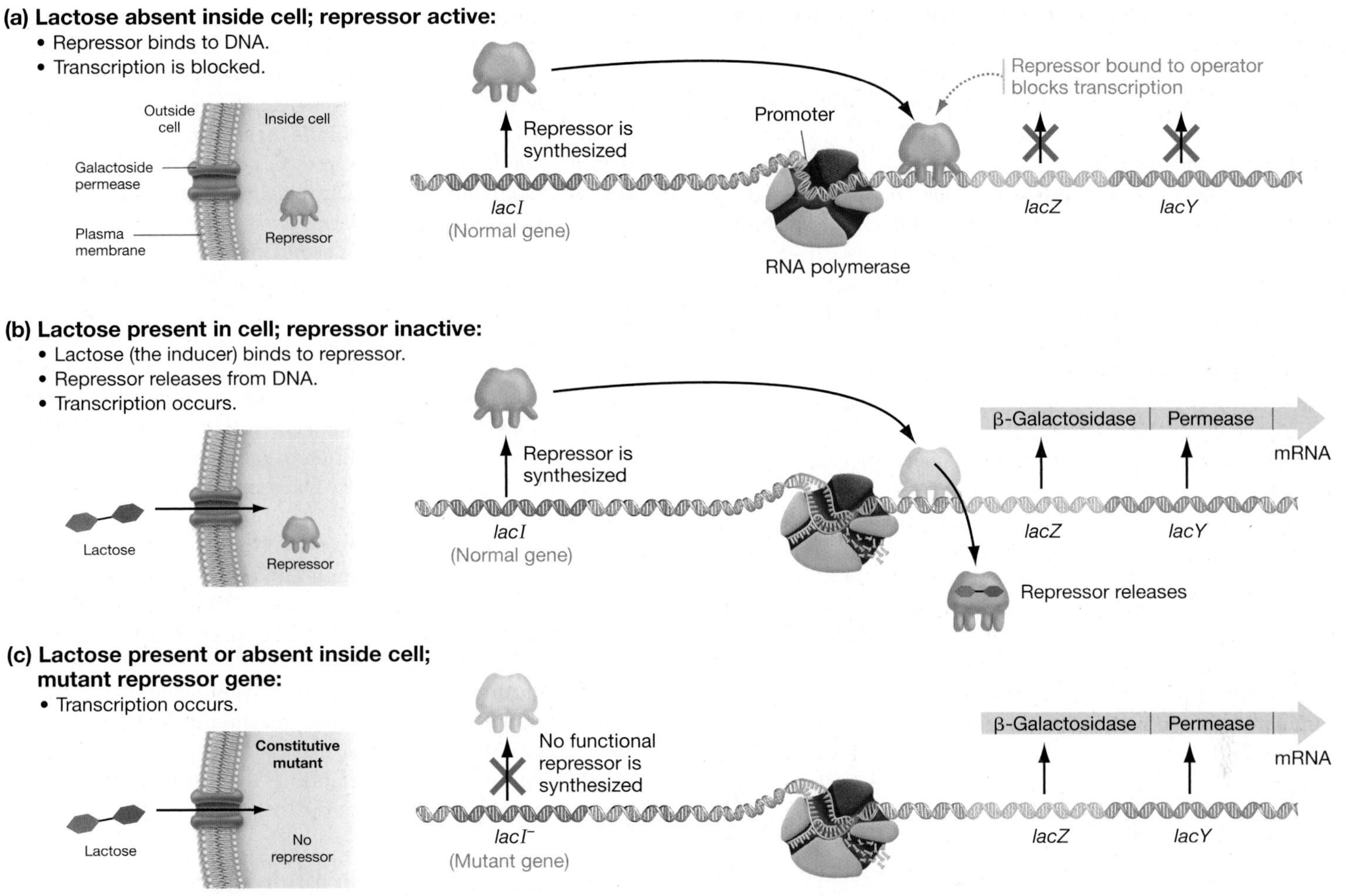

Figure 18.6 Genes Involved in Lactose Metabolism Are under Negative Control. The plasma membrane and galactoside permease are shown as a reminder that lactose comes from outside the cell. Repression of *lac* genes is never complete, so there is always some galactoside permease to transport lactose into the cell and begin induction of gene expression.

3. The inducer (lactose) binds to the repressor. When it does, the repressor changes shape. The shape change causes the repressor to come off the DNA. Recall that this form of control over protein function is **allosteric regulation** (Ch. 8, Section 8.4). In allosteric regulation, a small molecule binds to a protein and causes it to change its shape and activity. When the inducer binds to the repressor, the repressor can no longer bind to DNA, and transcription can proceed.

Positive Control of Lactose Utilization Genes

Transcription of the *lac* operon is drastically reduced when glucose is present in the environment—even when lactose is available to induce β-galactosidase expression (see Figure 18.3). This makes sense, given that glucose is *E. coli*'s preferred carbon source. When glucose is present, there is no need for the cell to cleave lactose as a way of acquiring glucose.

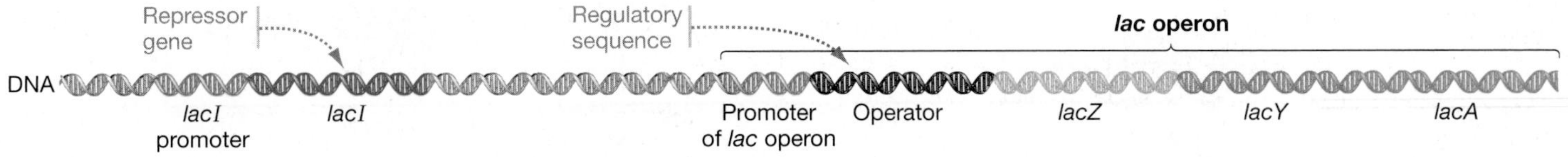

Figure 18.7 The *lac* Operon Is a Set of Coordinately Regulated Genes. This view emphasizes the arrangement of genes and regulatory sequences and is not drawn to scale. (The *lacI* gene is not part of the *lac* operon, but it is close to the operon genes and is included in this figure because of its importance in *lac* operon control.)

✔ **MODEL** Using small, colored bits of candy or paper, add the repressor protein to the figure. Next, add RNA polymerase; then add lactose. At each step, explain what happens after each molecule is added.

There are two different ways that glucose acts to prevent *lac* operon expression. Let's start with the first of these ideas, because it was the initial model for how glucose exerts control, and it offers a classic illustration of positive control of transcription.

Positive Control by CAP Regulation A protein known as the **catabolite activator protein (CAP)** has long been viewed as exerting positive control of many operons in *E. coli*, including the *lac* operon. Like the repressor protein, CAP is transcribed and translated constitutively and is always available to regulate transcription. As shown in **Figure 18.8a**, when CAP binds to a regulatory sequence of DNA just upstream of the promoter, the CAP binding site, the frequency of initiating transcription is increased. CAP bound to the regulatory sequence in DNA promotes the association of the RNA polymerase holoenzyme with the promoter through a stabilizing interaction between CAP and the holoenzyme.

There's a twist, however: CAP must be bound to cyclic AMP (cAMP) in order to bind to DNA. (You learned about cAMP and its role as a second messenger in Ch. 11, Section 11.3.)

How does glucose in the environment influence CAP binding to DNA? When the glucose level outside the cell is high, cAMP synthesis is inhibited—and without cAMP, CAP does not bind to the CAP binding site in DNA. In this case, transcription is not stimulated. In contrast, when the glucose level outside the cell is low, cAMP synthesis within the cell ramps up, the cAMP level increases, and CAP forms a complex with cAMP. CAP associated with cAMP then binds to the CAP binding site upstream of the promoter and stabilizes the interaction of RNA polymerase with the *lac* operon promoter. Transcription now begins much more frequently. These ideas are summarized in Figure 18.8a.

The CAP binding site is found in other operons that function in the use of sugars other than glucose. This allows glucose to exert control over many other operons in addition to the *lac* operon.

Control by Inducer Exclusion There is another way that glucose prevents activation of the *lac* operon and other operons that allow the cell to use other sugars, and it has nothing to do with positive control. The mechanism is known as inducer exclusion, and it works when glucose inhibits the transport of sugars other than glucose into the cell. In the case of the *lac* operon, when glucose is abundant in the environment, the transport of lactose into the cell by galactoside permease is inhibited (**Figure 18.8b**, top). Because lactose does not accumulate in the cytoplasm, the repressor remains bound to the operator. Negative control (as in Figure 18.6a) is in place.

In contrast, when the glucose level outside the cell is low, galactoside permease is active. If lactose is present, it is transported into the cell (see Figure 18.8a, bottom) and induces *lac* operon expression. Inducer exclusion affects the activity of many different sugar transporters in addition to galactoside permease. It allows *E. coli* to preferentially use glucose, even when other sugars are also present outside the cell.

Although the exact balance between positive control of the *lac* operon by the CAP–cAMP complex and maintenance of negative control by inducer exclusion is debated, what is clear is that the CAP protein when bound to cAMP acts as a positive regulator—an activator—of *lac* operon transcription.

✔ **If you understand the alternative mechanisms for how glucose regulates the *lac* operon, you should be able to explain how CAP–cAMP and inducer exclusion can achieve the same outcome (preventing transcription when glucose and lactose are both present) by different means.**

Why Has the *lac* Operon Model Been So Important Scientifically?

The *lac* operon has been an immensely important model system for understanding how genes are regulated. Here are some of the most significant insights:

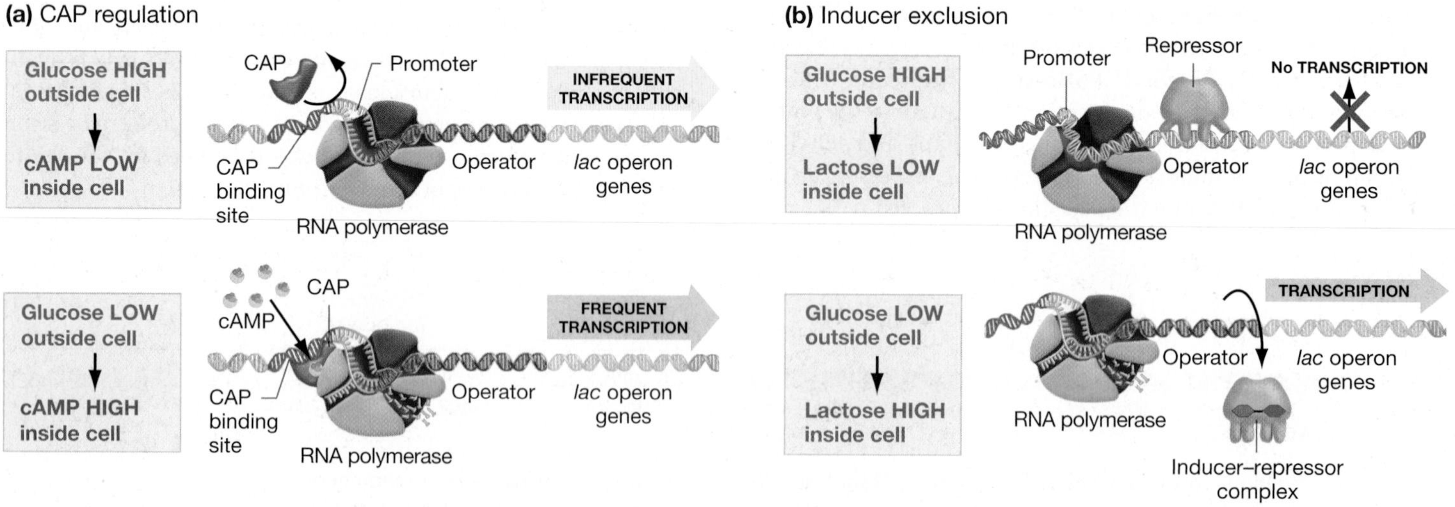

Figure 18.8 Two Models to Explain How Glucose Regulates the *lac* Operon. Panels (a) and (b) both show the situation when lactose is present outside of the cell.

- Many bacterial genes and operons are under negative control by repressor proteins.
- Activator proteins often work by enhancing the binding of RNA polymerase to the promoter.
- Gene transcription is regulated by physical contact between regulatory proteins and specific regulatory sequences in DNA.
- Post-translational control allows a rapid response to changing environments.

On this last point, remember that both the repressor and CAP proteins are transcribed and translated constitutively and are always present. When a rapid change in *lac* operon activity is called for, there's no need to synthesize either of these regulatory proteins. Instead, the activity of *existing* repressor or CAP proteins is altered allosterically. This is a common type of post-translational control.

In most cases, the activity of key regulatory proteins is controlled primarily by post-translational modifications.

The *trp* Operon: A Twist on Negative Control

The *lac* operon works to break down something—lactose—into its component sugars. Can the operons that create compounds, called biosynthetic operons, be controlled by a repressor protein? The simple answer is yes. An important example is the *trp* operon—an operon involved in synthesizing the amino acid tryptophan.

Tryptophan synthesis follows a multistep biochemical pathway. Starting with a precursor compound, each reaction step converts one intermediate to another until tryptophan is produced. Each reaction in the pathway is catalyzed by an enzyme encoded by a particular *trp* operon gene. Five enzymes are needed, and there are five corresponding *trp* operon genes. The logic of bacterial operons is evident in the *trp* operon: All the genes needed to work together in a process, tryptophan synthesis in this case, can be expressed together by transcribing the genes into one polycistronic mRNA. Like the *lac* operon, the *trp* operon contains an operator that overlaps the promoter, and control of transcription is mediated by a repressor protein. **Figure 18.9** is an overview of the *trp* operon and its regulation.

Think about the problem the *trp* operon has evolved to solve: Genes for tryptophan synthesis should be expressed only when tryptophan is required. In short, the genes need to be turned on when something—tryptophan—is absent. This challenge is the opposite of that faced by the *lac* operon, in which genes should be transcribed only when something—lactose—is present. How can regulation using a repressor and operator solve these different problems?

The short answer is by controlling the activity of the repressor proteins in opposing ways. The *trp* repressor binds to its operator only when it is bound by its regulator, tryptophan. The *lac* repressor only binds to the *lac* operator when it is *not* bound to its regulator, lactose. The logic is simple and elegant. When the tryptophan level drops, the repressor no longer binds the operator. The *trp* operon genes are now transcribed, the mRNA is translated, and the tryptophan level is restored. This regulation is a form of **negative feedback** control, a form of control in which the final product of a pathway inhibits the production of the product. You'll see many instances of negative feedback control when you study physiological processes in later chapters.

Note that in both the *lac* and *trp* operons, a small molecule binding to the repressor changes the repressor's shape and activity. This is allosteric regulation. The only difference is that for the *trp* operon the small molecule regulator is called a **co-repressor**, because it works with the repressor to make it active.

The regulatory logic of the *lac* and *trp* operons is wonderful in its simplicity and efficiency, but it is not unique. Countless operons in millions of species of bacteria use the same regulatory scheme.

CHECK YOUR UNDERSTANDING

✓ If you understood this section, you should be able to . . .

1. Explain why it makes sense for the cell to use lactose to induce transcription of the *lac* operon.
2. **MODEL** Diagram the *lac* operon, showing the relative positions of the operator, the promoter, and the three protein-coding genes; indicate what would happen if there were a mutant repressor protein that could not bind lactose.
3. Explain how, under the CAP–cAMP model, glucose could prevent the transcription of operons for the use of many different sugars.
4. Explain why it's logical for tryptophan to act as a co-repressor in the *trp* operon.

Answers are available in Appendix A.

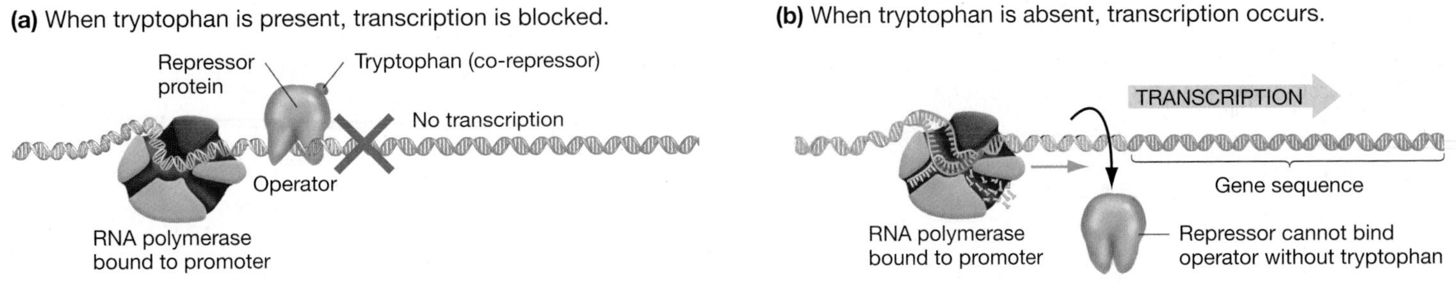

Figure 18.9 Negative Control of the *trp* Operon.

18.3 Global Gene Regulation

Sometimes it's not enough to turn on genes or even operons individually. Under some conditions, such as a sudden rise in temperature, a loss of nutrients, a shift to a new environment, or extensive DNA damage, bacteria must coordinately change the expression of large sets of genes to meet these and many other challenges. How can bacterial cells manage responses that require the expression of dozens or even hundreds of genes?

After you complete this section, you should be able to ...

- Compare global gene regulation to the regulation of single operons.

Global gene regulation is the coordinated regulation of many genes. One means of global gene regulation is grouping genes into a **regulon**—a set of separate genes and operons that contain the same regulatory sequences and are controlled by a single type of regulatory protein.

Like operons, regulons can be under negative control by a repressor protein or under positive control by an activator protein. For example, the SOS response regulon, which allows bacterial cells to survive and repair extensive damage to DNA, is under negative control. As shown in Figure 18.10, damage to DNA sets off an SOS signal that induces the transcription of more than 40 genes that code for enzymes needed for DNA repair, recombination, and specialized DNA polymerases that can use damaged DNA as a template. These proteins work to allow the cell first to survive the DNA damage and ultimately to repair it. Without the SOS response, bacteria whose DNA is extensively damaged face almost certain death. Most bacterial species have an SOS response and an SOS regulon.

A key part of the regulatory logic of the SOS system is that it is under the control of one of its own genes. The gene, called *lexA*, codes for the LexA protein that represses transcription of SOS regulon genes while the cell is healthy. DNA damage sets off a signal for LexA to be cleaved and therefore inactivated. Inactivation of LexA releases the brake on transcription of all SOS genes—not only those required for DNA repair, but *lexA* itself. While repair proteins are being produced and going to work, the DNA damage that signals LexA cleavage diminishes and levels of active LexA are replenished. By the time damaged DNA is repaired, the SOS regulon is restored to its starting point by active LexA. The entire system is ready to spring into action again should DNA be signficantly damaged.

What are the general messages of this chapter? Interactions between protein regulators and the DNA sequences they bind to produce finely tuned control over gene expression, regulating individual genes, operons, or large sets of genes. With these exquisite controls over gene expression, bacteria have been able to grow, reproduce, and compete for resources for more than 3 billion years.

Do eukaryotes control their genes the same way as bacteria? If not, what are the differences? These questions are the focus of the next chapter.

PROCESS: REGULON CONTROL OF GENE EXPRESSION

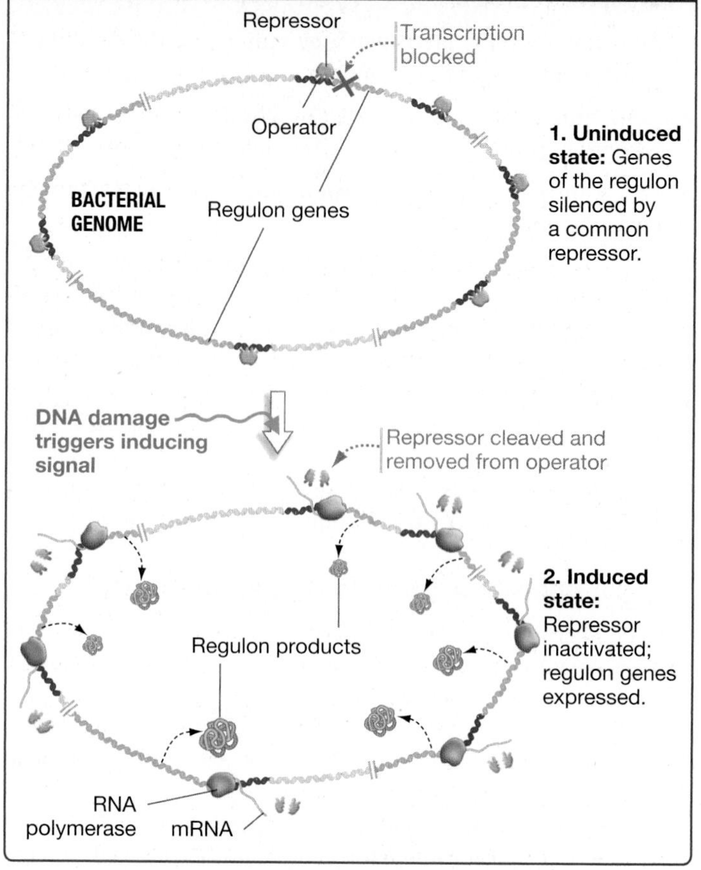

Figure 18.10 The Genes of the SOS Regulon Are Expressed Together. The "||" symbols indicate regions of the bacterial chromosome not shown between regulon genes.

CHECK YOUR UNDERSTANDING

✔ If you understood this section, you should be able to ...

1. Explain why a mutation that altered a repressor protein would affect the expression of all genes in a regulon under negative control, but why a mutation of an operator would affect the expression of only one gene in the regulon.
2. Discuss the benefit of having the *lexA* gene be part of the regulon that it controls.

Answers are available in Appendix A.

18.1 An Overview of Gene Regulation and Information Flow

- Changes in gene expression allow bacterial cells to respond to changes in the environment.
- Most gene products are produced or activated only when needed.
- Gene expression can be controlled at three levels: transcription, translation, or post-translation (protein activation or inactivation).
- Transcriptional control can be negative or positive. Negative control occurs when a regulatory protein prevents transcription. Positive control occurs when a regulatory protein increases the frequency of initiating transcription.

18.2 Negative and Positive Control of Transcription

- Mutants that failed to cleave lactose, to transport lactose into the cell, or to regulate transcription of *lac* operon genes were the basis for understanding how the *lac* operon worked.
- Transcription may be constitutive or regulated. Constitutive expression occurs in genes whose products are required at all times, such as genes that encode glycolytic enzymes.
- The *lac* operon is transcribed efficiently when lactose is present and glucose is absent.
- The *lac* operon is under negative and positive control.
- In negative control of the *lac* operon, a repressor protein binds to an operator sequence in DNA near the *lac* operon promoter to prevent transcription of *lac* operon genes.
- When lactose is present, it binds to the repressor and causes it to release from the operator, allowing transcription to occur.
- Glucose inhibits transcription of the *lac* operon by a combination of preventing an activator protein from binding near the promoter and inhibiting lactose transport into the cell.
- Positive control of transcription occurs when a protein called an activator binds to a regulatory sequence in DNA.
- Activator proteins bind to RNA polymerase in addition to DNA. Binding between the activator and RNA polymerase increases the rate of transcription initiation.
- The *trp* operon codes for genes required in the synthesis of tryptophan. The operon is under negative control by a repressor protein that binds to an operator. But unlike the *lac* repressor, the *trp* repressor binds to the operator only when it binds to a regulatory molecule, tryptophan, which acts as a co-repressor.
- The *trp* operon exhibits negative feedback control, an important form of regulation in which the end product of a pathway inhibits the activity of the pathway.

18.3 Global Gene Regulation

- Bacterial cells often need to coordinate the expression of large sets of genes in response to changing environments.
- Regulons coordinate the expression of different genes by using a shared regulator that acts on a regulatory sequence found in all genes and operons of the regulon. Regulons can work through negative control using repressors, or through positive control using activators.
- The SOS regulon contains more than 40 genes needed to repair DNA damage. The SOS regulon is controlled by a repressor protein coded for by the *lexA* gene. DNA damage leads to cleavage of the LexA repressor protein, activating expression of regulon genes. The LexA repressor level returns to normal when DNA damage is repaired, turning off SOS regulon genes.

Answers are available in Appendix A.

✓ TEST YOUR KNOWLEDGE

1. Why are the genes involved in lactose metabolism considered to be an operon?
   a. They occupy adjacent locations on the *E. coli* chromosome.
   b. They have a similar function.
   c. They are all required for normal cell function.
   d. They are all controlled by the same promoter.
2. In the *lac* operon, the repressor inhibits transcription when
   a. the repressor is bound to the inducer.
   b. the repressor is not bound to the inducer.
   c. the repressor is bound to glucose.
   d. the repressor is not bound to the operator.
3. Activators bind to regulatory sequences in ________ and to ________ polymerase.
4. A regulon is a set of genes controlled by
   a. one type of regulator of transcription.
   b. two or more different alternative sigma proteins.
   c. many different types of promoters.
   d. glucose.

✓ TEST YOUR UNDERSTANDING

5. Evaluate these statements about regulation of the *lac* operon. Select True or False for each statement.
   T/F The *lac* operon is transcribed at the highest rate when extracellular glucose and lactose are abundant.
   T/F The repressor protein is bound to DNA of the operator when lactose is present.
   T/F A mutation in the operator is likely to prevent transcription of the *lac* operon under any condition.
   T/F A mutation that alters the catabolite activator protein is predicted to alter the regulation of many different operons.
6. Predict what would happen to regulation of the *lac* operon if the *lacI* gene were moved 50,000 nucleotides upstream of its normal location.
7. Explain why it makes sense for the *lexA* regulatory gene of the SOS regulon to be expressed constitutively.
8. IPTG is a molecule with a structure much like lactose. IPTG can be transported into cells by galactoside permease and can bind to the *lac* repressor protein. However, unlike lactose, IPTG is not broken down by β-galactosidase. Predict what would occur to *lac* operon regulation if IPTG were added to *E. coli* growth medium containing no glucose or lactose.

✔ TEST YOUR PROBLEM-SOLVING SKILLS

9. In a mutant that lacks adenylyl cyclase, the enzyme that synthesizes cAMP, predict which of the following conditions of extracellular lactose and glucose would cause regulation of the *lac* operon to differ from that of wild-type cells.
a. no lactose, no glucose
b. no lactose, abundant glucose
c. abundant lactose, no glucose
d. abundant lactose, abundant glucose

10. PROCESS OF SCIENCE X-gal is a colorless, lactose-like molecule that can be split into two fragments by β-galactosidase. One of these product molecules creates a blue color. The photograph here shows *E. coli* colonies growing in a medium that contains X-gal.

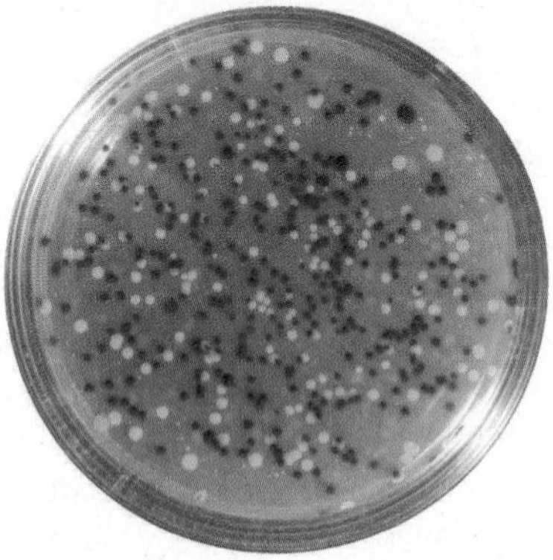

Find three colonies whose cells have functioning copies of β-galactosidase. Find three colonies whose cells might have mutations in the *lacZ* or the *lacY* genes. Suppose you analyze the protein-coding sequence of the *lacZ* and *lacY* genes of cells from the three mutant colonies and find that these sequences are wild type (normal). What other region of the *lac* operon might be altered to account for the mutant phenotype of these colonies?

✔ PUT IT ALL TOGETHER: Case Study

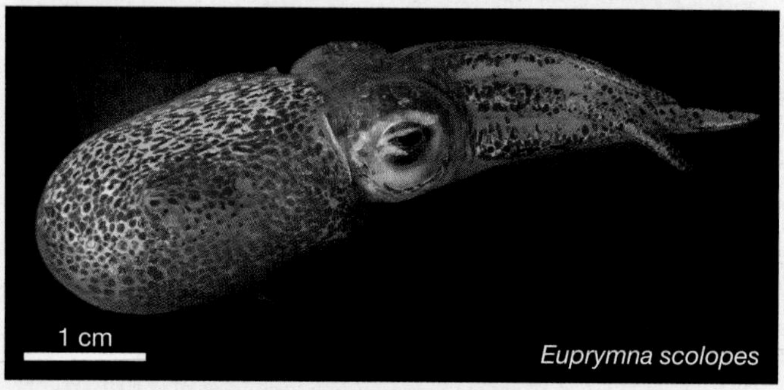

How does bacterial gene regulation light up a squid like a starry sky?

The Hawaiian bobtail squid (*Euprymna scolopes*) is able to glow from luminescent *Vibrio fischeri* bacteria held in its light organs. As it swims at night near the ocean surface, it adjusts the amount of light visible to predators below to match the light from the stars and moon. Predators have difficulty seeing the illuminated squid against the night sky.

11. The bacteria glow in response to a molecule that regulates expression of genes involved in light-producing chemical reactions. The regulator controls production of the genes' mRNA. Therefore, the light-producing genes are under
a. transcriptional control.
b. translational control.
c. post-translational control.
d. negative control.

12. The light-producing genes of *V. fischeri* are organized in an operon that is under positive control by an activator protein called LuxR. Would you expect the genes of this operon to be transcribed when LuxR is bound or not bound to a DNA regulatory sequence? Explain.

13. MODEL The diagram shown here is a model of the gene regulatory circuit for light production by *V. fischeri* cells. The *lux* operon contains genes for luminescence (*luxCDABE*) and a gene, *luxI*, that encodes an enzyme that catalyzes the production of an inducer. This inducer easily moves back and forth across the plasma membrane and acts as a signaling molecule. The *lux* operon is never completely turned off. The *luxR* gene codes for the activator LuxR. The inducer can bind to LuxR, and when it does, the LuxR–inducer complex can bind to a regulatory site to activate transcription of the *lux* operon *and* inhibit transcription of *luxR*. Explain how this gene regulatory circuit accounts for bacteria emitting light only when they reach a high cell density.

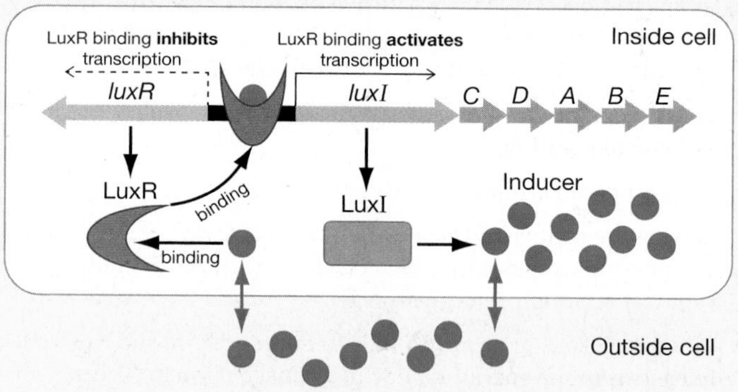

14. LuxR is allosterically regulated by the inducer molecule secreted by *V. fischeri*. What does it mean that LuxR is allosterically regulated?

15. What characteristic of the light-producing regulatory circuit is consistent with the idea that it may be a regulon? What characteristic of this circuit stretches the definition for a regulon?

16. Quorum sensing (introduced in Ch. 11, Section 11.4) allows bacteria to detect the number of neighboring cells and to trigger a response only when this number reaches a critical level. Quorum sensing is used by *V. fischeri* in light production and by many pathogenic bacteria, including *Vibrio cholerae*, to turn on genes for toxin production only when a critical cell density is reached. Why might quorum sensing be beneficial to pathogenic bacteria?

Mastering Biology

Students Go to Mastering™ Biology for assignments, the eText, and the Study Area with animations, practice tests, and activities.

Professors Go to Mastering™ Biology for automatically graded tutorials and questions that you can assign to your students, plus Instructor Resources.

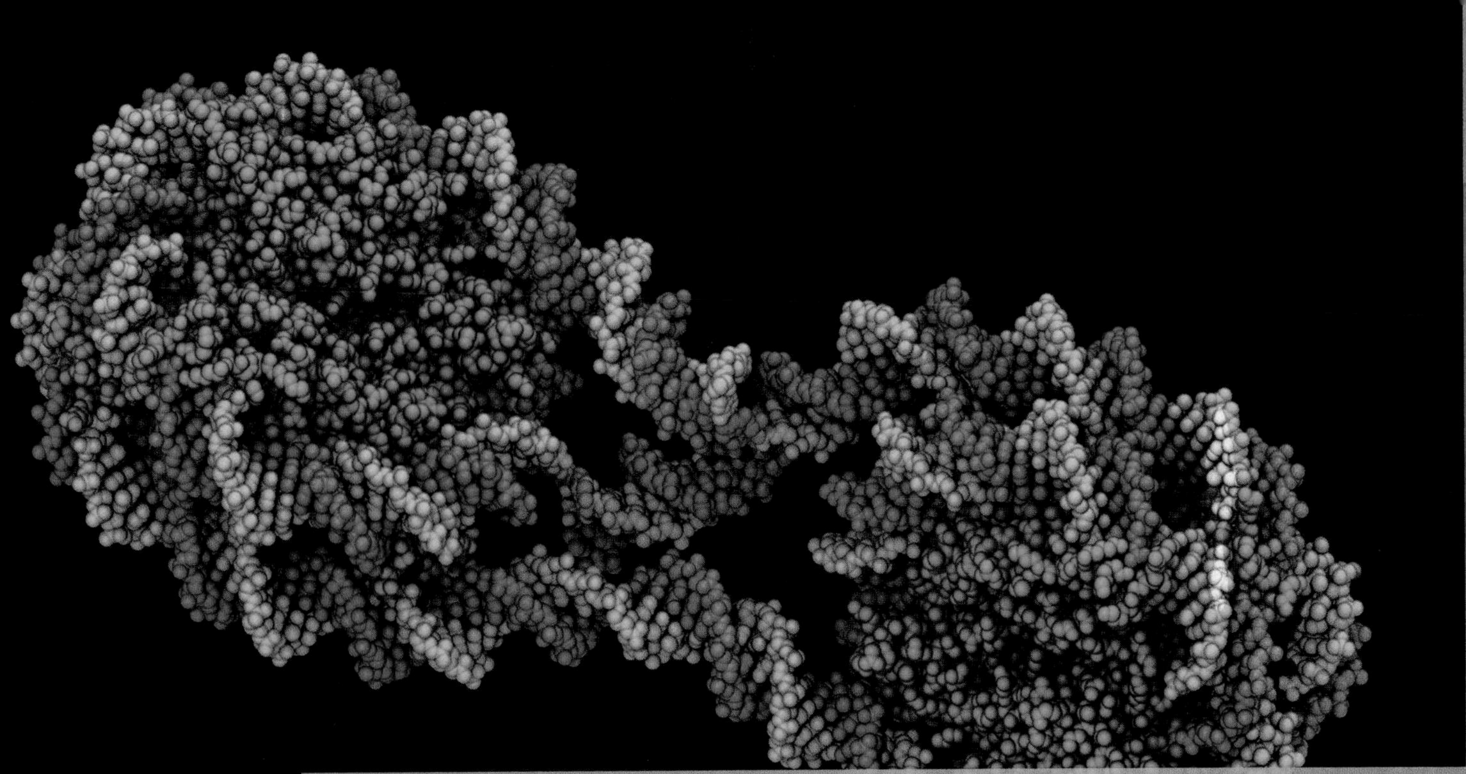

19 Control of Gene Expression in Eukaryotes

A model of a small region of eukaryotic chromatin. The DNA (red and pink) is wrapped around proteins (green). The DNA has to be uncoiled from proteins before gene expression can begin.

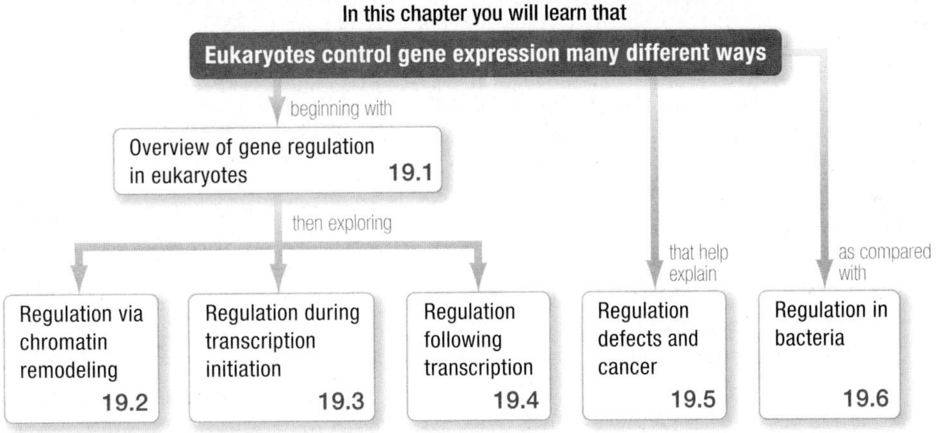

Bacteria regulate gene expression to respond to changes in their environment. For example, *Escherichia coli* express the genes required to import and cleave lactose only when the cells have to rely on lactose as a source of energy (Chapter 18).

Unicellular eukaryotes face similar challenges. And for cells that make up multicellular eukaryotes, there is a new demand—how to create and maintain different types of cells using the same set of genes. Consider your body, which contains trillions of cells, each with a specialized structure and function. You have muscle cells, lung cells, nerve cells, skin cells, and many others. Although cells differ from each other, they all contain the same genes. Your muscle cells have liver-cell genes and your liver cells have muscle-cell genes—they just don't *express* them.

BIG PICTURE

This chapter is part of the Big Picture. See how on pages 404–405.

Why not? The answer is that besides responding to the environment outside the organism, as bacteria and unicellular eukaryotes do, cells in a multicellular eukaryote also respond to changes in the organism's *internal* environment—specifically, to signals from other cells. As a human or an oak tree develops, cells at different times and in different locations within the organism are exposed to different signals. As a result, cells with the same genome express different sets of genes. This process, called **differential gene expression**, is responsible for forming specialized cell types, arranging them into tissues, and coordinating their activity in the multicellular society that is the individual.

How does all of this regulation and specialization of cell types happen? In later chapters, you'll explore the signals that trigger the formation of muscle, bone, leaf, and flower cells (Chapters 21, 38, and 47). In this chapter, you'll focus on how gene expression in a eukaryotic cell is controlled in response to these signals. Let's start with an overview of gene expression, take a look at how defects in the process can trigger cancer, and close with a comparison of eukaryotic and bacterial gene expression.

19.1 Gene Regulation in Eukaryotes—An Overview

Like bacteria, eukaryotes can control gene expression at the levels of transcription, translation, and post-translation. But as **Figure 19.1** shows, additional levels of control occur in eukaryotes as genetic information flows from DNA to proteins.

After you complete this section, you should be able to ...

- Explain the steps of gene regulation in eukaryotes.

The first additional level of control involves the DNA–protein complex at the top of the figure. In eukaryotes, DNA is wrapped around proteins to create a structure called **chromatin**. Eukaryotic genes have promoters, just as bacterial genes do, but before transcription can begin in eukaryotes, the stretch of DNA containing the promoter must be released from tight interactions with proteins so that RNA polymerase can make contact with the promoter. To capture this idea, biologists say that **chromatin remodeling** must occur before transcription, transitioning from a condensed or "closed" state to a decondensed or "open" state.

Another level of regulation that is unique to eukaryotes is **RNA processing**—the steps required to produce a mature, processed mRNA from a primary RNA transcript. Recall that introns have to be spliced out of primary transcripts (Chapter 17). In many cases, carefully orchestrated alternative splicing occurs—meaning that different combinations of exons are included in the mRNA. If different cells use different splicing patterns, different gene products result.

In addition, mRNA stability is regulated in eukaryotes. Those mRNAs that remain in the cell for a long time tend to be translated more than mRNAs that have a shorter life span.

In this chapter you'll explore all the control points shown in Figure 19.1. Let's begin with the series of events that occur as a cell responds to an external signal.

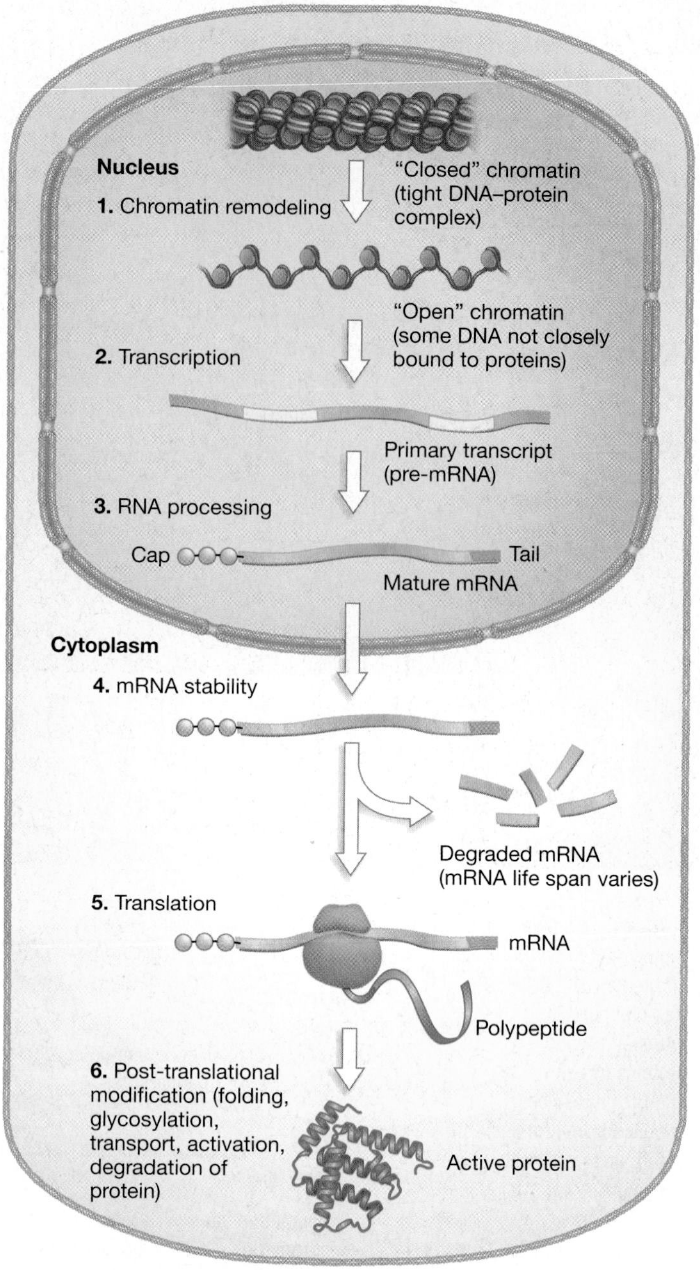

Figure 19.1 In Eukaryotes, Gene Expression Can Be Controlled at Many Different Levels.

CHECK YOUR UNDERSTANDING

✔ **If you understood this section, you should be able to ...**

1. Explain how eukaryotic chromatin poses both challenges and opportunities for regulating gene expression.
2. Describe how RNA splicing opens possibilities for controlling gene expression that are unavailable to bacteria.

Answers are available in Appendix A.

19.2 Chromatin Remodeling

Suppose a cell detects a signal that tells it to produce a specific protein. What happens next? For a molecular signal to trigger the transcription of a specific gene, the chromatin around the target gene must be remodeled. To appreciate why, consider that a typical cell in your body contains about 6 billion base pairs of DNA. Lined up end to end, these nucleotide pairs would form a double helix about 2 m (6.5 feet) long. But the nucleus that holds this DNA is only about 5 μm in diameter—far less than the thickness of this page. To fit inside the nucleus, the DNA must be packed tightly—so tightly that RNA polymerase can't access it. How is DNA packaged? And how can it be unpacked at particular genes so RNA polymerase can transcribe it?

After you complete this section, you should be able to ...

- Describe the ways that chromatin is remodeled.

Chromatin Structure

The first studies on the composition of chromatin were published in the early 1900s, when researchers established that eukaryotic DNA is intimately associated with proteins. Later work documented that a group of proteins called **histones** are the most abundant DNA-associated proteins. Chromatin consists of DNA complexed with histones and other proteins.

In the 1970s electron micrographs like the one in Figure 19.2a revealed that chromatin has a characteristic structure. In some preparations for electron microscopy, chromatin looked like beads on a string. The "beads" came to be called **nucleosomes**.

More information emerged in 1984 when researchers determined the three-dimensional structure of one form of eukaryotic chromatin by using X-ray crystallography (see BioSkills 6). These studies indicated that each nucleosome consists of about 200 base pairs of DNA wrapped almost twice around a core of eight histone proteins. As Figure 19.2b indicates, a histone called H1 "seals" DNA to each set of histone proteins. Between each pair of nucleosomes there is a stretch of linker DNA with no histones.

The intimate association between DNA and histones occurs in part because DNA is negatively charged and histones are positively charged. DNA has a negative charge because of its phosphate groups; histones are positively charged because they contain many lysines and arginines, two positively charged amino acids.

Individual nucleosomes are linked to each other in increasingly complex structures to form chromatin. First, H1 histones interact with one another and with histones in other nucleosomes to produce a tightly packed structure like that shown in Figure 19.2b. Based on its width, this structure is called the 30-nanometer fiber. (Recall that a nanometer is one-billionth of a meter and is abbreviated nm.)

Moving up the organizational ladder, 30-nm fibers are attached at intervals along their length to proteins that form a scaffold or framework inside the nucleus. In this way, the entire chromosome is organized and held in place. Finally, when chromosomes condense before mitosis or meiosis, the scaffold proteins and 30-nm fibers are folded into even more tightly packed structures that ultimately lead to the chromosomes that are visible during cell division.

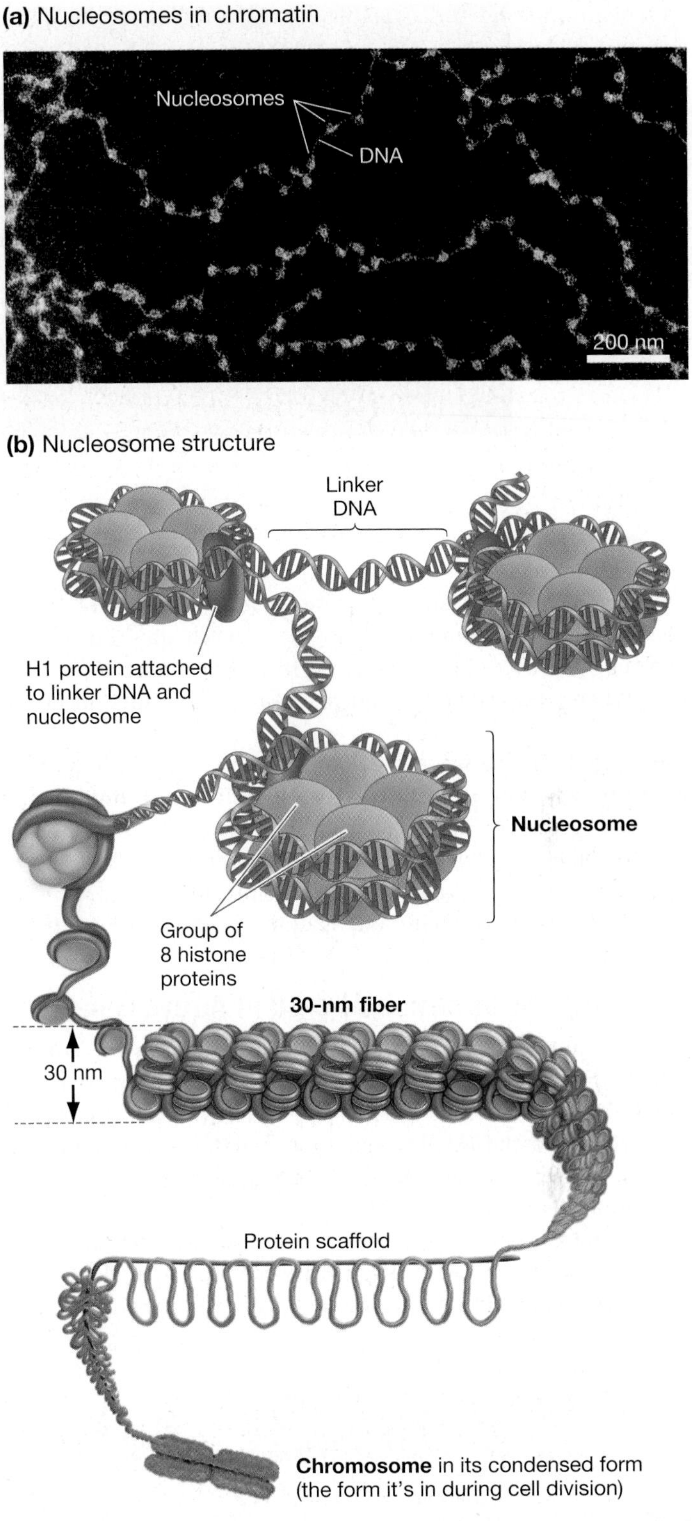

Figure 19.2 Chromatin and Chromosomes Have Several Levels of Organization.

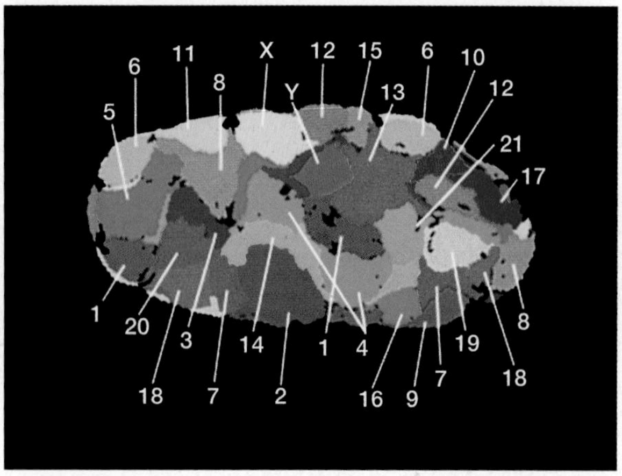

Figure 19.3 Chromosomes Reside in Distinct Territories. This image represents a cross section through the chromosomes (numbered) in the nucleus of one human cell during interphase.

Recent studies have found that each chromosome lies in its own distinct region, or territory, within the interphase nucleus. This arrangement is shown in **Figure 19.3**. Portions of these segregated chromosomes can fold out to reach sites within the nucleus where genes are actively transcribed and then fold back when transcription is finished. As in real estate, location matters.

Similar to the situation in eukaryotes, bacterial DNA interacts with proteins, but 30-nm fibers or higher-order arrangements have not been observed in bacterial chromosomes.

The elaborate structure of eukaryotic chromatin does more than just package DNA so that it fits into the nucleus. Chromatin structure also has profound implications for the control of gene expression.

Chromatin Structure Is Altered in Active Genes

Once the nucleosome-based structure of chromatin was established, scientists realized that the close physical interaction between DNA and histones must be altered for RNA polymerase to make contact with DNA. They hypothesized that a gene could not be transcribed until the condensed chromatin near its promoter was remodeled.

The central idea is that chromatin must be decondensed to expose the promoter so RNA polymerase can bind to it. If so, then chromatin remodeling would be the first step in the control of eukaryotic gene expression. Studies that examined the accessibility of DNA to DNases have provided strong support for this hypothesis. DNases are enzymes that cut DNA, but they cannot cut efficiently if DNA is tightly wrapped with proteins. As **Figure 19.4** shows, DNase works effectively only if DNA is in a decondensed, or open, configuration.

Harold Weintraub and Mark Groudine used this observation to test the hypothesis that the DNA of actively transcribed genes is in an open configuration. Using chicken blood cells, they compared chromatin structure in two genes: β-globin and ovalbumin.

β-globin is a protein that is part of the hemoglobin found in red blood cells; ovalbumin is a protein found in egg white. In blood cells, the β-globin gene is transcribed at high levels, but the ovalbumin gene is not transcribed at all.

After treating blood cells with DNase and then comparing the state of the β-globin and ovalbumin genes, the researchers found that DNase cut the β-globin gene DNA much more readily than DNA of the ovalbumin gene did. They interpreted this finding as evidence that in blood cells, chromatin of the actively transcribed β-globin gene was decondensed and that the chromatin of the non-transcribed ovalbumin gene was condensed. Similar studies of many different genes in various cell types led to similar conclusions—chromatin is decondensed in genes that are being transcribed.

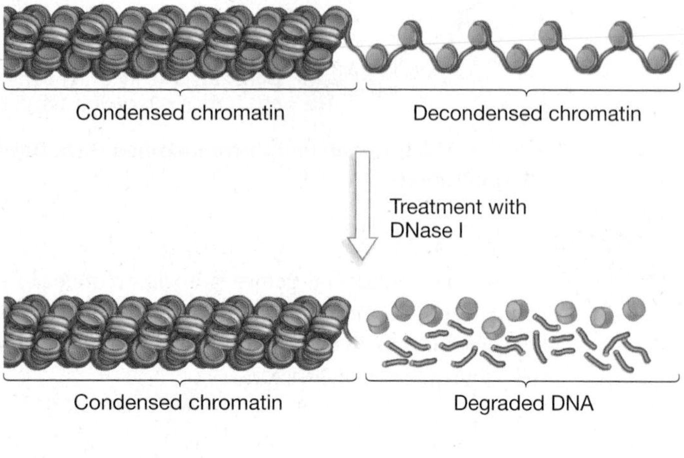

Figure 19.4 DNase Treatment Can Reveal Chromatin Structure. DNase is an enzyme that cuts DNA at random locations. However, it cannot cut DNA in condensed chromatin.

These results suggest that in their normal, or default, state, eukaryotic genes are turned off because DNA is normally wrapped tightly in chromatin. Gene expression then depends on opening up chromatin in the promoter region.

How Is Chromatin Altered?

Research on chromatin remodeling has advanced at a furious pace, and biologists have succeeded in identifying key players that work to change the state of chromatin condensation. There are three major ways to remodel chromatin: DNA methylation, histone modification, and the use of chromatin-remodeling complexes.

DNA Methylation A group of enzymes known as **DNA methyltransferases** add methyl groups (—CH_3) to cytosine residues in DNA, by a process called **DNA methylation**. In mammals, the sequence recognized by these enzymes is a C next to a G in one strand of the DNA. This sequence is abbreviated CpG and is shown here in its methylated form:

```
         CH3
          |
5'-ATATCGTA-3'
3'-TATAGCAT-5'
           |
          CH3
```

Why is DNA methylation important? Methylated CpG sequences are recognized by proteins that trigger chromatin condensation. Actively transcribed genes usually have relatively few methylated CpG sequences near their promoters, while non-transcribed genes usually have many methylated CpG sequences.

Histone Modification A large set of enzymes adds a variety of chemical groups to specific amino acids of histone proteins. These include acetyl groups (—$COCH_3$), methyl groups, phosphate groups, and short polypeptide chains. Modifying histones with these and other chemical groups alters the association of DNA with histone proteins and promotes condensed or decondensed chromatin, depending on the specific set of modifications made to particular histones.

Researchers have proposed that particular combinations of histone modifications on specific amino acids of histone proteins set the state of chromatin condensation for a particular gene. This idea is the **histone code** hypothesis. Let's take a look at one way histone modifications can control chromatin structure and the accessibility of DNA sequences needed to initiate transcription.

Figure 19.5 shows two different types of enzymes that add or remove acetyl groups from histones. Shown to the right, **histone acetyltransferases (HATs)** *add* acetyl groups to the positively charged lysine residues in histones, and to the left, **histone deacetylases (HDACs)** *remove* them. Adding an acetyl group to lysines in histones neutralizes the positive charge on this amino acid. Addition of an acetyl group reduces the electrostatic interactions between the negatively charged phosphates in the DNA backbone with histones, and it interferes with the assembly of nucleosomes into condensed chromatin. For these reasons, histone **acetylation** usually promotes decondensed chromatin, a state associated with active transcription. The addition of acetyl groups also creates a binding site for other proteins that help open the chromatin.

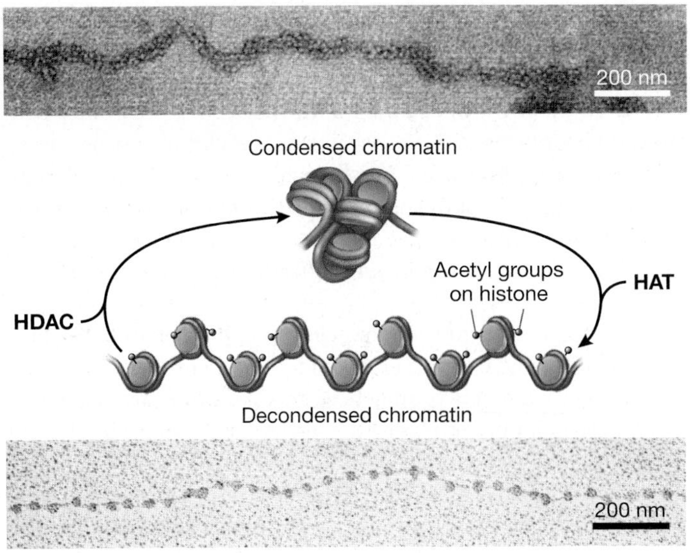

Figure 19.5 Adding or Removing Acetyl Groups on Histones Changes Chromatin Condensation. Histone acetyltransferases (HATs) cause chromatin to decondense; histone deacetylases (HDACs) cause it to condense.

✔ Do HATs and HDACs work in positive control or negative control? Explain your reasoning.

In contrast, when HDACs remove acetyl groups from histones, this reverses the events promoted by histone acetylation, reverting chromatin to a condensed state associated with no transcription. HATs are an on switch for transcription, while HDACs are an off switch.

Chromatin-Remodeling Complexes Another major player in chromatin alteration and gene regulation are proteins that form macromolecular machines called **chromatin-remodeling complexes**. These complexes harness the energy in ATP to reshape chromatin. Chromatin-remodeling complexes either cause nucleosomes to slide along the DNA or to knock the histones completely off the DNA to open up stretches of chromatin for transcription.

DNA methylation, histone modifications, and chromatin-remodeling complexes work together to fine-tune chromatin condensation at specific genes. The take-home message is that the condensation state of chromatin is critical in determining whether transcription can occur.

DNA and Chromatin Modifications Vary and Can Be Inherited

The pattern of chromatin modifications varies from one cell type to another. For example, suppose you analyzed the same gene in a skin cell and a liver cell from one individual. This gene and others are likely to have different patterns of DNA methylation and histone modifications in the two cell types, as well as different patterns of chromatin condensation.

If the skin or liver cells divide, the patterns of DNA methylation and histone modifications such as acetylation can be passed on to their daughter cells. This provides a way for each daughter cell to inherit patterns of gene expression by **epigenetic inheritance**—the collective term for any mechanism of inheritance that is due to something *other* than differences in DNA sequences.

With epigenetic inheritance, when a cell receives a "become a skin cell" signal early in development, the chromatin is modified in distinctive ways, and those modifications are passed on to its descendants. Skin cells are different from liver cells, for example, not because they contain different genes, but largely because they have inherited different patterns of DNA methylation and histone modifications during their development. The shared set of genes in these cells are differentially accessible for expression because of distinct and stable patterns of chromatin condensation.

But the story of epigenetic inheritance involves more than just formation of specialized cell types during development. Emerging evidence indicates that epigenetic mechanisms can record life events and that this archive is sometimes difficult to erase. This is the case when prenatal conditions make long-lasting alterations in the chromatin of embryos that ultimately influence phenotypes in later life.

Long-Lasting Epigenetic Marks Imposed during Development

One example of how events in early development can make lifelong differences comes from a study of rats. Investigators examined what happened to rats born to mothers fed low-protein diets during pregnancy and while nursing. Even when provided a normal diet after these early deprivations, these animals have a greatly increased risk of developing disorders in later life that are similar to type 2 diabetes. Type 2 diabetes is a serious and increasingly common disease that alters the cellular uptake of glucose (Chapter 41). Both genetic factors and environmental factors, such as diet, play important roles in diabetes development.

One significant gene associated with diabetes is *Hnf4a*. The *Hnf4a* gene codes for a regulator of genes involved in glucose uptake. The diabetic rats born to protein-deprived mothers express the *Hnf4a* gene at lower levels than normal rats. Could epigenetic inheritance be at work in silencing *Hnf4a* expression?

One team's approach to probing this question is shown in Figure 19.6. The researchers measured the types of histone modifications found at a key regulatory region of the *Hnf4a* gene in control and treatment groups. A regulatory region is a section of DNA that, like operators in bacteria (Chapter 18), is involved in controlling the activity of a gene. The chromatin at this *Hnf4a* gene regulatory region has to be opened up for transcription to occur. What did the team learn?

As the graph on the left in the "Results" section of Figure 19.6 shows, they found that histone modifications of the *Hnf4a* gene that led to condensed chromatin were *elevated* in rats born to malnourished mothers as compared to control offspring.

RESEARCH

QUESTION: Does poor nutrition in a mother produce epigenetic effects in offspring?

HYPOTHESIS: Protein-deprived mothers will produce offspring with abnormal histone modifications.

NULL HYPOTHESIS: Protein-deprived mothers will produce offspring with normal histone modifications.

EXPERIMENTAL SETUP:

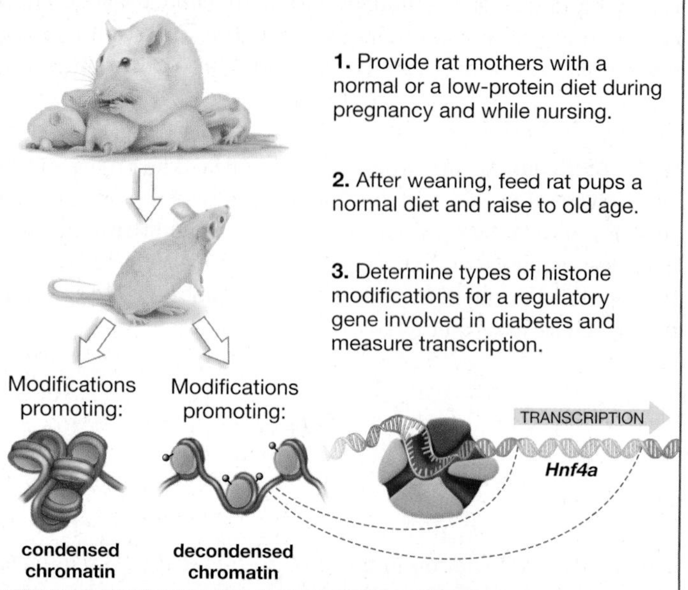

1. Provide rat mothers with a normal or a low-protein diet during pregnancy and while nursing.

2. After weaning, feed rat pups a normal diet and raise to old age.

3. Determine types of histone modifications for a regulatory gene involved in diabetes and measure transcription.

PREDICTION OF HYPOTHESIS: Adult offspring of mothers fed a low-protein diet will have abnormal histone modifications.

PREDICTION OF NULL HYPOTHESIS: Adult offspring of mothers fed a low-protein diet will have normal histone modifications.

RESULTS:

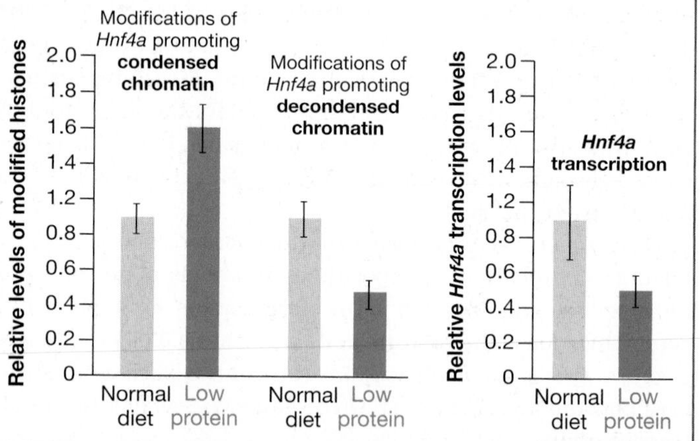

CONCLUSION: A mother's diet influences chromatin modifications and gene expression patterns throughout her offspring's life.

Figure 19.6 **Events in Early Life Can Be Recorded through Epigenetic Mechanisms.**

SOURCE: I. Sandovici, N. H. Smith, and M. D. Nitert. 2011. Maternal diet and aging alter the epigenetic control of a promoter–enhancer interaction at the *Hnf4a* gene in rat pancreatic islets. *Proceedings of the National Academy of Sciences USA* 108: 5449–5454.

✔ **PROCESS OF SCIENCE** What could researchers do to provide evidence that the histone modifications are causing reduced regulatory gene transcription?

Conversely, histone modifications associated with decondensed chromatin were significantly *reduced* in the treatment group. The graph on the right confirms that transcription of *Hnf4a* was much lower in the treatment group than the control group. Together these results demonstrate correlations between maternal diet, altered histone modifications, and decreased levels of *Hnf4a* gene expression in *adult* offspring.

Remember that in this study, all offspring were provided a healthy diet after weaning. This finding implies that a mother's nutritional status during pregnancy and nursing is responsible for the types of chromatin modifications seen in offspring that develop diabetes in later life.

Epigenetic Inheritance Is Widespread The finding that environmental influences early in a rat's life can be recorded in patterns of chromatin condensation is remarkable, but by no means unique. The floodgates have opened for stunning discoveries about epigenetic inheritance across a range of species. As just one example, there is a hereditary form of human mental retardation called Rett syndrome. Symptoms include stunted brain growth, problems with muscles and coordination, and limited or no language development.

Rett syndrome occurs when a mutation leads to the absence of a protein that binds to methylated DNA to induce chromatin condensation. One of the remarkable unexplained features of Rett syndrome is how the lack of a protein that can bind to many methylated DNA sequences found across the entire genome can cause the specific phenotypes of Rett syndrome. Whatever the explanation, the failure to normally condense chromatin leads to this human developmental disorder. Epigenetics is at work again.

CHECK YOUR UNDERSTANDING

✔ **If you understood this section, you should be able to ...**

1. Predict how gene expression will be affected if a cell is grown with compounds that prevent DNA methylation.
2. Explain how certain patterns of histone acetylation or DNA methylation could influence whether a cell became a muscle cell or a liver cell.
3. Explain how genetic inheritance and epigenetic inheritance differ.

Answers are available in Appendix A.

19.3 Initiating Transcription

As in bacteria, the promoter in eukaryotes is a site in DNA where RNA polymerase binds to initiate transcription. However, eukaryotic promoters are significantly more complex.

After you complete this section, you should be able to ...

- Describe how the initiation of transcription is regulated.

In eukaryotes the term **core promoter** is often used to indicate the specific sequence where RNA polymerase binds, as opposed to the other sequences needed for regulation of transcription. When you hear the word "promoter" in the context of a eukaryote, it is important to think about whether this refers narrowly to the core promoter or if it is meant more broadly to include other DNA sequences involved in controlling the initiation of transcription. The most intensively studied core promoter sequence is a short stretch of DNA known as the **TATA box**.

Once a core promoter that contains a TATA box has been exposed by chromatin remodeling, the first step in initiating transcription is binding of the **TATA-binding protein (TBP)**. The TBP works in the context of a large, multi-subunit protein complex. There are also proteins, related in function to the TBP, that work on promoters with other conserved sequences. But the binding of TBP or any of its relatives does not guarantee that a gene will be transcribed. A wide array of other DNA sequences and proteins must work together with RNA polymerase to allow transcription to begin.

Promoter-Proximal Elements Are Regulatory Sequences Near the Core Promoter

Besides the core promoter, other DNA sequences—called **regulatory sequences**—allow the binding of proteins that control the initiation of transcription. Regulatory sequences in eukaryotes are analogous to the operator and CAP binding site that you learned about in Ch. 18, Section 18.2.

The first eukaryotic regulatory sequences were discovered in the late 1970s, when Yasuji Oshima and co-workers set out to understand how yeast cells control the metabolism of the sugar galactose. When galactose is absent, *S. cerevisiae* cells produce only tiny quantities of the enzymes required to metabolize it. But when galactose is present, transcription of the genes encoding these enzymes increases by a factor of 1000.

The team isolated mutant cells that failed to produce any of the five enzymes required for galactose metabolism, even if galactose was present. To interpret this phenotype, they developed these hypotheses:

- The five enzyme-coding genes are regulated together, even though they are not on the same chromosome.
- Normal cells have an activator protein that exerts positive control over the five genes.
- The mutant cells have a mutation that disables the activator protein.

Other researchers isolated the regulatory protein and found that it binds to a short stretch of DNA—a regulatory sequence—located just upstream from the core promoter for each of the five genes required for galactose use.

In bacteria, genes that need to be regulated together are often clustered into a single operon and transcribed into a single mRNA. In contrast, eukaryotes use the strategy uncovered by Oshima for galactose-metabolizing genes in yeast—instead of being clustered together, each co-regulated gene has the same regulatory DNA sequence that binds the same type of regulatory protein. Regulatory DNA sequences exist in all eukaryotic genes.

Regulatory sequences such as the ones discovered in yeast that are close to the promoter are termed **promoter-proximal elements**. Unlike the core promoter, different types of promoter-proximal elements are associated with each gene. In this way, they allow eukaryotic cells to express certain genes but not others.

The discovery of promoter-proximal elements and a mechanism of positive control suggested a satisfying parallel between gene regulation in bacteria and in eukaryotes. This simple picture changed when researchers discovered a new class of DNA regulatory sequences that were unlike anything known in bacteria.

Enhancers Are Regulatory Sequences Far from the Core Promoter

Susumu Tonegawa and colleagues made a startling discovery while exploring how human cells regulate gene expression. The gene studied by Tonegawa's group was broken into many introns and exons. Recall that introns are transcribed sequences that are spliced out of the primary transcript; exons are transcribed regions that are included in the mature RNA once splicing is complete (Ch. 17, Section 17.2). The researchers found a regulatory sequence required for enhanced transcription *within* one of the introns.

This finding was remarkable for two reasons: **(1)** The regulatory sequence was thousands of bases away from the promoter, and **(2)** it was downstream of the promoter. Regulatory sequences that are far from the promoter and activate transcription are termed **enhancers**. Later work showed that enhancers occur in all eukaryotes.

Enhancers have several key characteristics:

- Enhancers can be vast distances (sometimes more than 100,000 base pairs) away from the promoter. They can be located in introns or in nontranscribed sequences and can be upstream, downstream, or even within a gene (**Figure 19.7**).
- Like promoter-proximal elements, there are many types of enhancers.
- Most genes have more than one enhancer.
- An enhancer is composed of many short regulatory sequences that each bind a different specific regulatory protein.
- Enhancers can work even if they are flipped from their normal $5' \rightarrow 3'$ orientation or moved to new locations near the gene.

Enhancers are regulatory DNA sequences primarily found in eukaryotes. When regulatory proteins called **transcriptional activators**, or **activators** for short, bind to enhancers, transcription begins. Thus, enhancers and activators are like a gas pedal—an element of positive control. Eukaryotes also possess regulatory sequences that are similar in structure and share key characteristics with enhancers but work to inhibit transcription. These DNA sequences are called **silencers**. When regulatory proteins called **repressors** bind to silencers, transcription is shut down. Silencers and repressors are like a brake—an element of negative control.

The Role of Transcription Factors in Differential Gene Expression

Follow-up work supported the hypothesis that enhancers and silencers are binding sites for activators and repressors that regulate transcription. Collectively, these proteins are termed **regulatory transcription factors**, or often **transcription factors** for short. There are hundreds of transcription factors that bind to enhancers, silencers, and promoter-proximal elements.

These findings support one of the most important statements researchers can make about gene regulation in eukaryotes: Different types of cells express different genes primarily because they have different transcription factors. In multicellular species, the genes encoding transcription factors, in turn, are expressed largely in response to signals that arrive from other cells, especially during embryonic development.

For example, if a signal that says "become a muscle cell" reaches a cell in the early embryo, it triggers a signal transduction cascade (Chapter 11) that leads to the production of transcription factors specific to muscle cells. Because different transcription factors bind to specific regulatory sequences, they turn on the production of muscle-specific proteins. But if a become-a-muscle-cell signal isn't present, then no active muscle-specific transcription factors are produced and no muscle-specific gene expression takes place in that embryonic cell.

Differential gene expression results largely from the production or activation of specific transcription factors. Eukaryotic genes are turned on when transcription factors bind to enhancers and promoter-proximal elements; the genes are turned off when transcription factors bind to silencers, when a particular transcription factor is not present, or when chromatin is condensed.

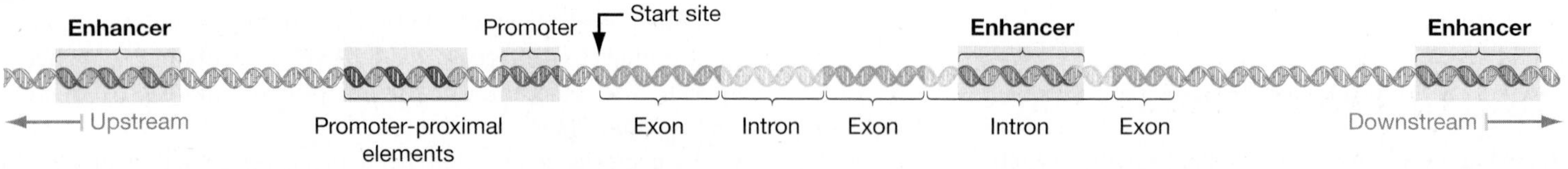

Figure 19.7 Enhancers and Promoter-Proximal Elements Regulate the Expression of Eukaryotic Genes. Promoter-proximal elements are near the core promoter. Enhancers are located farther away, may be upstream or downstream from the promoter, and may even be within introns. Each of the colored areas in the enhancers or the promoter-proximal elements represents a regulatory sequence that is recognized by a protein. This drawing gives you an idea of the arrangement of introns, exons, and regulatory elements, but not their actual sizes, which vary widely.

✔ Compare and contrast the structure of this typical eukaryotic gene and the structure of a bacterial operon.

Distinctive sets of transcription factors are what make a muscle cell a muscle cell and a liver cell a liver cell.

How Do Transcription Factors Recognize Specific DNA Sequences?

Each transcription factor must be able to recognize and bind to a specific DNA sequence. How can it do this? Recall that DNA bases are partially exposed in the grooves of the DNA double helix, and that one of these grooves, the major groove, is wide, and the other groove, the minor groove, is narrow. The edges of an A-T base pair and a C-G base pair that project into the major and minor grooves contain different sets of atoms and have different surface shapes (Figure 19.8a). These differences in chemical composition and shape can be recognized by transcription factors.

Just as base pairs come together by complementary molecular interactions, so too can proteins and specific DNA sequences. An example is shown in Figure 19.8b. In this case, a transcription factor that is essential for the development of muscle cells inserts amino acid side chains into two major grooves of DNA. This particular transcription factor binds to a specific regulatory sequence because of complementary interactions between its amino acids and a particular sequence of base pairs in DNA. Without such specific interactions between transcription factors and DNA, the development of muscle cells—or any other cell—would not be possible.

A Model for Transcription Initiation

Although gene expression can be controlled at many levels, regulating the start of transcription is at center stage. For a process so important, many questions remain.

Besides the regulatory transcription factors you've learned about that bind to enhancers, silencers, and promoter-proximal elements, there is another type: **general transcription factors**. These proteins interact with the core promoter and are not restricted to particular genes or cell types. The term "general" implies that these proteins are necessary for transcription to occur, but they do not provide much in the way of regulation. The TATA-binding protein (TBP) that you learned about is part of a large general transcription factor that is used for many genes. ✔ **If you understand this concept, you should be able to compare and contrast the regulatory and general transcription factors expected to be found in muscle cells versus nerve cells.**

In addition to transcription factors, a large complex of proteins called the **Mediator** acts as a bridge between regulatory transcription factors, general transcription factors, and RNA polymerase II. The Mediator plays a critical role in integrating the input of many regulatory transcription factors and delivers a signal to RNA polymerase to initiate transcription. Figure 19.9 on page 394 summarizes a model for how transcription is initiated in eukaryotes.

Step 1 Activators bind to DNA and recruit chromatin-remodeling complexes and histone acetyltransferases (HATs). Chromatin decondenses.

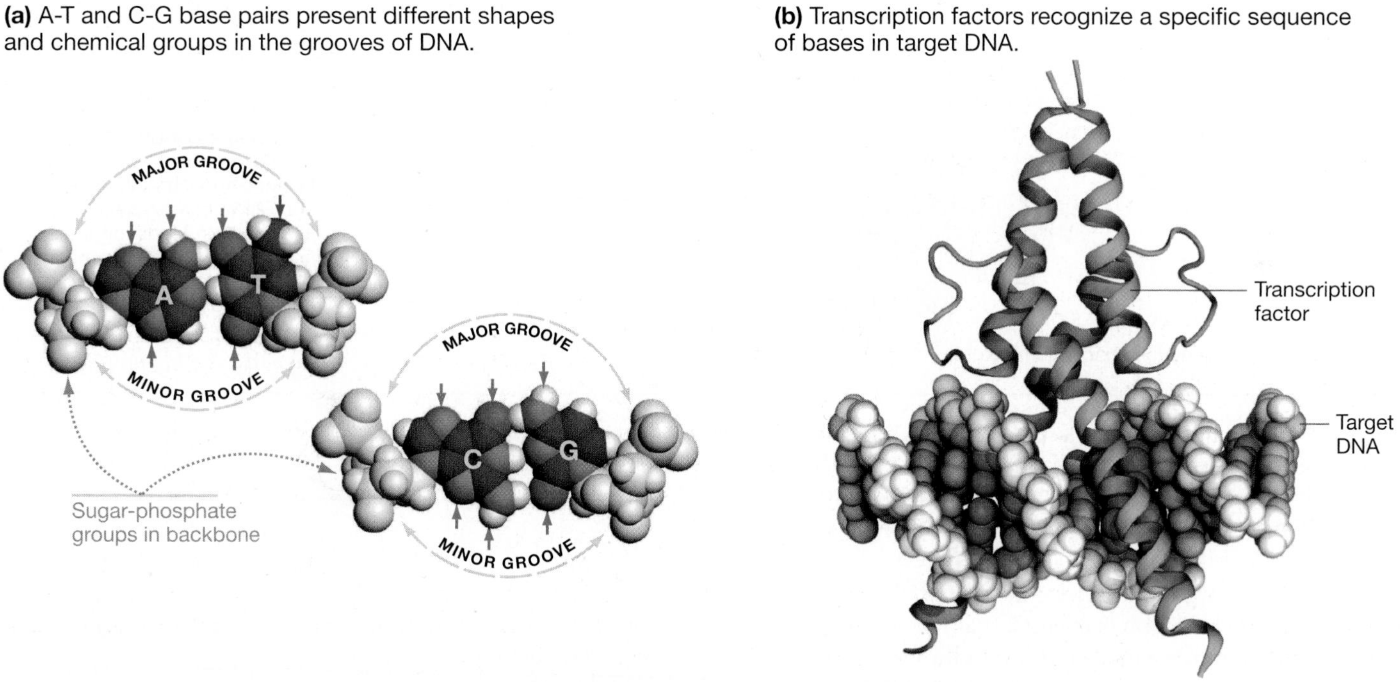

Figure 19.8 **Transcription Factors Bind to Regulatory Sequences by Recognizing Base Pairs. (a)** The edges of different base pairs projecting into the major and minor grooves of DNA present different shapes and chemical groups. Atoms of these chemical groups that participate in hydrogen bonding with amino acids of transcription factors are indicated by green arrows. The methyl group on thymine (T), indicated with a blue arrow, is also important in recognition. **(b)** A transcription factor (green) binding to a regulatory sequence in DNA. This transcription factor recognizes edges of base pairs that project into the major groove. The bases within the regulatory sequence recognized by the protein are highlighted in red.

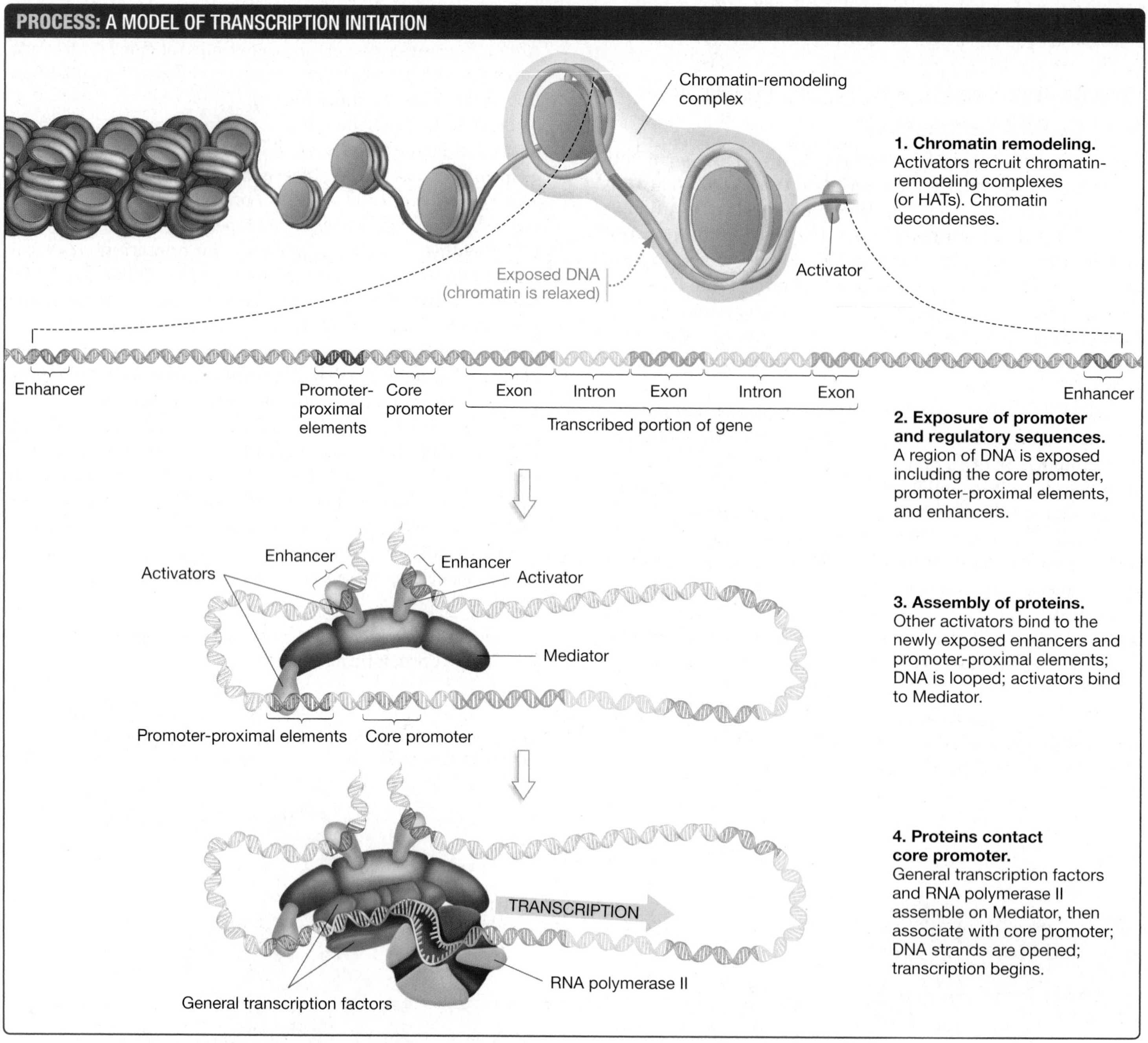

Figure 19.9 Transcription Initiation in Eukaryotes Is a Multistep Process. For simplicity, the figure shows enhancers and promoter-proximal elements as containing a single binding site for an activator, and only one activator binding to each enhancer and set of promoter-proximal elements. In reality, many different activators bind to each enhancer and set of promoter-proximal elements.

Step 2 A swath of chromatin is exposed that includes the core promoter, promoter-proximal elements (only one is shown in the figure), and enhancers.

Step 3 Other activators bind to the exposed enhancers and promoter-proximal elements; DNA loops, allowing DNA-bound activators to bind to the Mediator.

Step 4 General transcription factors and RNA polymerase II assemble on the Mediator, then associate with the core promoter. DNA strands are opened, and RNA polymerase II can now begin transcription.

✔ **If you understand this model, you should be able to explain why DNA forms loops near the core promoter in order for transcription to begin.**

An important point in this model of transcription initiation is the dual role of transcriptional activators. Activators work not only to stimulate transcription but also to bring chromatin-remodeling proteins to the right place at the right time. None

SUMMARY **Table 19.1 Elements of Transcriptional Regulation**

Type of Element	Element	Function
DNA sequence	**Core promoter**	Allows RNA polymerase to initiate transcription. Unlike bacterial promoters, RNA polymerase associates with core promoter only after other proteins have been bound.
	Promoter-proximal element	A DNA sequence near the core promoter that is a binding site for an activator or a repressor protein. Moving a promoter-proximal element alters or eliminates its function. Promoter-proximal elements are typically found in clusters.
	Enhancer	A set of clustered DNA sequences that each are bound by a regulatory transcription factor. Found at a distance from the core promoter, either upstream or downstream from the transcription start site. May be moved to other sites near the gene without loss of function.
	Silencer	Sequence that binds repressor proteins to inhibit transcription initiation.
Protein	**Activator**	A regulatory transcription factor that binds to promoter-proximal elements and enhancers to promote transcription initiation or to recruit chromatin-remodeling proteins.
	Repressor	A regulatory transcription factor that binds to promoter-proximal elements and silencers to inhibit initiation of transcription.
	General transcription factor	A protein or protein complex that associates with the core promoter to recruit RNA polymerase or to aid RNA polymerase in initiating transcription. Most general transcription factors are used at almost all core promoters and do not regulate transcription.
	DNA methylase	Enzyme that adds methyl groups to specific sequences of DNA (methyl added to C in CpG sequences in vertebrates). Methylation recruits methyl DNA-binding proteins that condense chromatin and inhibit initiation of transcription.
	Histone acetyl transferases (HATs)	Enzymes that add acetyl (CH_3COO^-) groups to specific amino acids of particular histones. Addition of acetyl groups usually promotes an open state of chromatin and transcription.
	Chromatin-remodeling complex	Group of proteins that shifts the position and alters the density of nucleosomes along DNA. This event either exposes or covers DNA regulatory sequences that control transcription.

of the proteins that remodel chromatin can recognize specific DNA sequences. Transcriptional activators bind to regulatory sequences of particular genes to recruit the proteins needed to alter chromatin structure.

The role of transcriptional activators in bringing in proteins that decondense chromatin leads to a chicken-and-egg paradox: How can an activator bind to DNA in the first place if chromatin is condensed? There appear to be two answers. Some transcription factors can bind to DNA that is associated with histones. It also turns out that most chromatin is dynamic. DNA occasionally dissociates from the histone proteins in nucleosomes, particularly DNA sequences near the linker DNA that connects nucleosomes, exposing regulatory sequences to activators.

CHECK YOUR UNDERSTANDING

✔ If you understood this section, you should be able to ...

1. Compare and contrast the nature of regulatory sequences and regulatory proteins in bacteria versus eukaryotes.
2. **THINK CAREFULLY** Predict whether regulatory sequences or transcription factors—or both—would be the same in muscle cells and brain cells within an individual.
3. Explain how changes in a promoter-proximal element DNA sequence could lead to changes in the rate of transcription initiation.

Answers are available in Appendix A.

Getting gene transcription started requires an elaborately choreographed dance between many interacting proteins and DNA sequences. The cast is reviewed in **Table 19.1**. The result is a large, macromolecular machine positioned at a gene's start site and capable of initiating transcription. Compared with what happens in bacteria, where just a few proteins interact at the promoter, the process in eukaryotes is amazingly complex.

19.4 Post-Transcriptional Control

Chromatin remodeling and transcription are just the opening to the story of gene regulation. Once a gene is transcribed, a series of events has to occur before a final product appears (see Figure 19.1). Each of these events offers an opportunity to regulate gene expression. Any regulation that occurs after transcription is a form of post-transcriptional control. These regulatory mechanisms include **(1)** different ways of splicing the same primary transcript, **(2)** altering the ability to translate particular mRNAs, or destroying them, and **(3)** altering the activity of proteins after translation has occurred. Let's consider each mechanism in turn.

After you complete this section, you should be able to ...

- Describe ways to achieve post-transcriptional control of gene expression.

Many Primary Transcripts Are Alternatively Spliced

Introns are spliced out in the nucleus as the primary RNA is transcribed. Recall that the mRNA that results from splicing consists of sequences encoded by exons, and that it is protected by a cap on the 5′ end and a long poly(A) tail on the 3′ end (Ch. 17, Section 17.2). You may also recall that splicing is accomplished by macromolecular machines called spliceosomes, and that many primary transcripts can be spliced in more than one way. This is a major way of regulating eukaryotic gene expression.

During splicing, gene expression is regulated by differential processing of introns and exons—introns can be retained and exons can be skipped. As a result, the same primary RNA transcript can be spliced together in different ways, yielding more than one kind of mature, processed mRNA. Because these mature mRNAs contain differences in their sequences, the polypeptides translated from them will likewise differ. Splicing the same primary RNA transcript in different ways is **alternative splicing**.

To see how alternative splicing works, consider the protein tropomyosin. The tropomyosin gene is expressed in at least five different types of cells, including two kinds of muscle cells that make up skeletal muscle responsible for voluntary movement and smooth muscle that lines the gut and certain blood vessels.

As Figure 19.10a shows, the rat α-tropomyosin gene contains 13 exons. However, in each cell, a different subset of the 13 exons present in the primary transcript is spliced together to produce distinctly different mRNAs (Figure 19.10b). Each of these mRNAs is referred to as an isoform. As a result of alternative splicing, the same tropomyosin gene is expressed to produce six distinct proteins. Tropomyosin is important in muscle function, and one reason skeletal muscle and smooth muscle are different is that they contain different types of tropomyosin protein.

Think of what alternative splicing of tropomyosin means—one gene, one pre-mRNA, six different mRNAs, six different polypeptides. Across the entire set of protein-coding genes, this principle of "from one, many" made possible by alternative splicing amplifies the number of proteins that can be specified by a much smaller set of genes.

Alternative splicing is controlled by proteins that bind to RNAs in the nucleus and interact with spliceosomes to influence which sequences are used for splicing. Embryonic cells that develop into skeletal muscle or smooth muscle receive signals leading to the production or activation of proteins that regulate splicing.

Before the discovery of alternative splicing, a gene was considered to be a nucleotide sequence that encodes one protein or RNA, along with the necessary regulatory sequences. When biologists realized that primary transcripts could be alternatively spliced, the definition of a gene had to be modified yet again. A gene is now viewed as a nucleotide sequence that allows the production of one *or more* related RNAs or polypeptides. Over 90 percent of human genes undergo alternative splicing. Alternative splicing is a major mechanism in the control of gene expression in multicellular eukaryotes.

✔ MODEL If you understand alternative splicing, you should be able to draw a model showing some possible mRNAs that could be spliced from a primary transcript containing two exons that surround an intron.

mRNA Stability and Translation Are Important Mechanisms of Post-Transcriptional Control

Even after an mRNA is produced and exported to the cytoplasm, other crucial levels of gene regulation come into play (see Figure 19.1). Let's begin with a look at how the translation

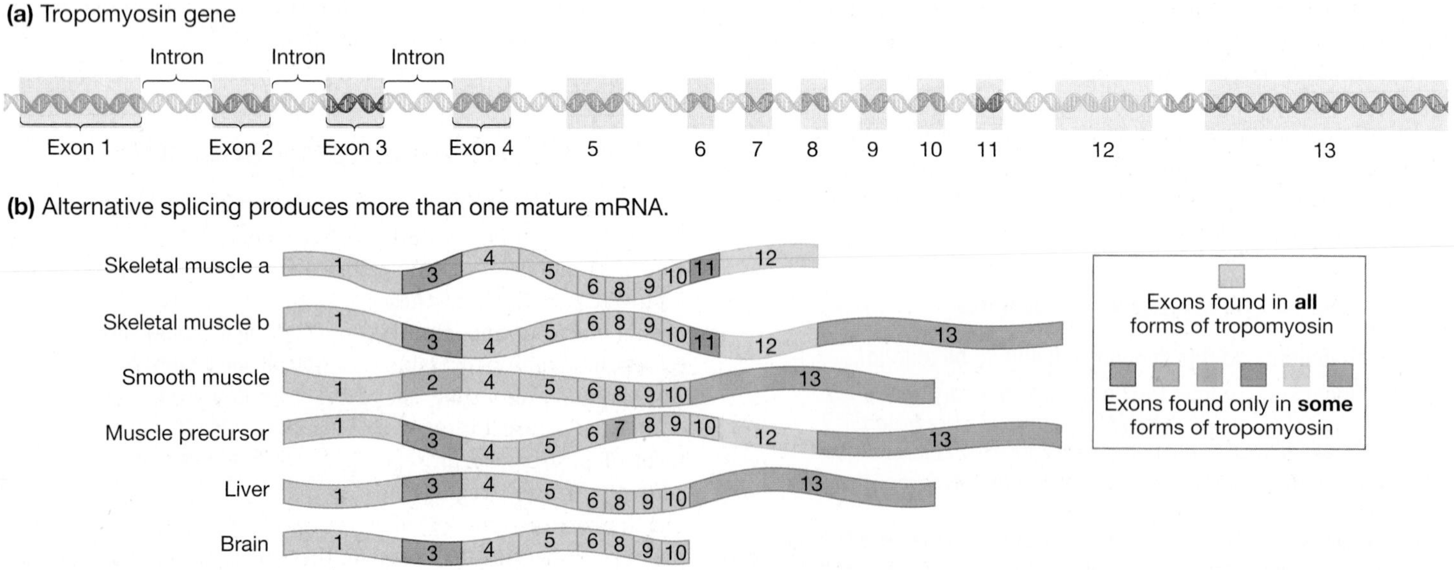

Figure 19.10 Alternative Splicing Produces More than One Mature mRNA from the Same Gene. There are six isoforms (types) of tropomyosin mRNA, including two different types in skeletal muscle cells.

of particular mRNAs can be regulated by a set of small RNAs that have a long reach in controlling gene expression.

RNA Interference Controls the Expression of RNAs For many years, biologists knew of only a handful of instances where the expression of specific mRNAs was controlled, and these seemed to be special cases. The discovery of RNA interference turned this view on its head.

RNA interference (RNAi) occurs when a tiny, single-stranded RNA held by a protein complex binds to a complementary sequence in another RNA. In the case of mRNAs, depending on how well the small RNA matches its mRNA target, this binding of complementary RNAs either leads to the destruction of the mRNA or blocks the mRNA's translation. How does it work?

There are several types of RNA interference. As Figure 19.11 shows, one form of RNA interference works through a small RNA called a **microRNA (miRNA)** that is derived from transcription of cellular genes. This mechanism proceeds as:

Step 1 RNA polymerase transcribes genes coding for miRNAs; the newly transcribed RNAs double back on themselves to form hairpins (only one hairpin is shown in the figure). Hairpins form because sets of bases within the RNA are complementary.

Step 2 Hairpin-containing precursor miRNA is bound to proteins in the nucleus that form an RNA-processing complex. The single-stranded 5′ and 3′ ends are trimmed off.

Step 3 The partially processed miRNA is exported to the cytoplasm and bound by another RNA-processing complex called Dicer. Dicer trims off the loop, leaving a small (~21 base pair), double-stranded RNA with short single-stranded overhangs at each end.

Step 4 The double-stranded miRNA fragments are bound by the RNA-induced silencing complex (RISC) proteins.

Step 5 The RNA strands are unwound. One strand of RNA is expelled and ultimately degraded, and the other strand, called the guide RNA, is retained by RISC. The guide RNA is the mature miRNA.

Step 6 As part of RISC, the miRNA binds to its complementary sequences in a target mRNA.

Step 7 If the match between an miRNA and an mRNA is perfect, an enzyme in RISC cuts the mRNA in two. If the match is imperfect, the guide RNA bound to the mRNA with RISC inhibits translation without destroying the mRNA. Either way, miRNAs interfere with gene expression.

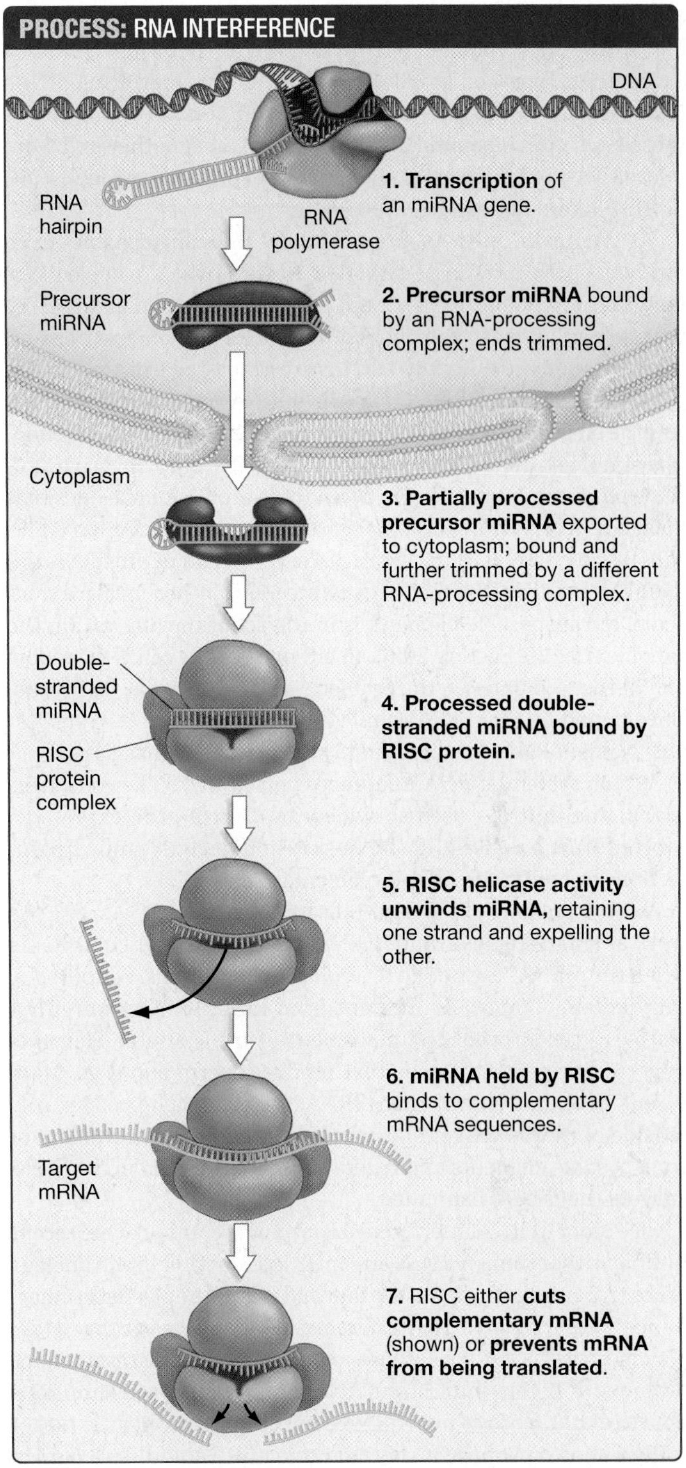

Figure 19.11 **MicroRNAs Either Target mRNAs for Destruction or Prevent Their Translation.** MicroRNAs are held by RISC and bind to target mRNAs by complementary base pairing.

What kind of genes code for miRNAs? There are a lot of different types, and only a few of these genes are primarily devoted to producing miRNAs. In most cases, the double-stranded RNA precursors for miRNAs are coded for in both protein-coding and noncoding exons of mRNAs, in the introns of pre-mRNAs, and in the introns of non-protein-coding RNAs. In short, almost any region of almost any type of RNA transcript can encode an miRNA.

An important feature of miRNA-based control of gene expression is that a single type of miRNA can regulate many different mRNAs. Conversely, one mRNA can be targeted by many different miRNAs. Considering the large number of miRNAs found in most species—for example, there are at least 1100 different human miRNAs—this suggests a huge network of gene regulation. Many, if not most, genes of multicellular eukaryotes are controlled by miRNAs. miRNAs are critical for development, and mutations in miRNA genes are associated with many diseases. These tiny RNAs are big elements of gene regulation.

siRNAs and piRNAs Protect against Viruses and Transposable Elements RNA interference involves more than just miRNAs. Two other types of small RNAs are short interfering RNAs (siRNAs) and PIWI-interacting RNAs (piRNAs). The origins of these two types of small RNAs differ from one another and from microRNAs, but all work to silence gene expression through RISC or RISC-like complexes.

At maturity, siRNAs are virtually indistinguishable from miRNAs—both are single-stranded RNAs about 21 nucleotides long that are bound to RISC and that target particular RNAs by complementary base pairing. But the stories of their creation and their function are different. siRNAs are generated from much longer double-stranded RNAs than the miRNA precursors—and this sets the stage for differences in how miRNAs and siRNAs function. The double-stranded RNA precursors for siRNAs are invasive molecules that can harm the cell. Often these are RNA molecules that make up viral genomes or that form as part of viral infection cycles (Ch. 33, Section 33.3). In other cases, the origin of an siRNA is a double-stranded RNA that is generated as another invasive molecule, a transposable element, is in the act of moving within the genome (Ch. 20, Section 20.6). In either case, the cell is in trouble and needs to shut down the problem—and this is what siRNAs do. Once bound by the siRNA coupled to RISC, the viral or transposable element RNAs are chopped to bits. The danger has passed.

Which system of RNA inference evolved first? The consensus is that the miRNAs used so widely to control gene expression evolved from an ancestral siRNA system that could stop viral infections or hold transposable elements in check.

What is the role of piRNAs? Like miRNAs and siRNAs, piRNAs work as small single-stranded RNAs that target particular RNAs when the piRNAs are bound to a complex of proteins. For piRNAs, this protein complex is different from RISC. piRNAs were first discovered to prevent the movement of transposable elements in germ-line cells—the cells that produce sperm and eggs. More recent work has shown that piRNAs play a much broader role outside of reproductive cells, where they block the movement of transposable elements and regulate gene expression in every eukaryote that's been examined.

The story of RNA interference continues to unfold. One recent finding is that some piRNAs and miRNAs function in the nucleus to control chromatin condensation and regulate epigenetic inheritance. Research on RNA interference has made it clear that RNAs play many regulatory roles previously thought to be reserved for proteins. ✔ **If you understand RNA interference, you should be able to explain how a microRNA can recognize a specific target mRNA and to propose a way that one miRNA could recognize more than one target mRNA.**

Post-Translational Control

In bacteria, controlling the activity of proteins after they are made—post-translational regulation—is important in allowing cells to respond rapidly to new conditions (Chapter 18). The same is true for eukaryotes. Instead of waiting for transcription, RNA processing, and translation to occur, the cell can keep an existing but inactive protein handy and then quickly activate it in response to altered conditions. Similarly, an active protein can rapidly be inactivated by removing a modification. This is the essence of post-translational control. There is a trade-off, however: Speed comes at the expense of efficiency. Transcription, RNA processing, and translation use energy and materials.

You have already encountered several important mechanisms of post-translational control of gene expression.

- Proteins may be activated when a protein kinase adds a phosphate group. One example is activation of cyclin–Cdk complexes by phosphorylation, which triggers entry into M phase of the cell cycle (Ch. 12, Section 12.3).
- Proteins may be targeted for destruction. When a protein such as a cyclin needs to be destroyed, enzymes mark it by adding copies of a small polypeptide called ubiquitin. Ubiquitin got its name because it is ubiquitous (widespread and common) in cells. A macromolecular machine called the **proteasome** recognizes proteins that have a ubiquitin tag and cuts them into short segments.

As you can see, the regulation of gene expression in eukaryotes includes everything from opening chromatin in the nucleus to controlling the life span of proteins.

Normal regulation of gene expression results in the orderly development of an embryo and appropriate responses to environmental change in adults. What happens when gene expression goes awry? Unfortunately, one answer is uncontrolled cell growth and the set of diseases called cancer. Understanding how changes in gene expression can lead to cancer is one of today's great research frontiers

CHECK YOUR UNDERSTANDING

✔ **If you understood this section, you should be able to ...**

1. Explain how alternative splicing could allow two different species with the same number of genes to produce vastly different numbers of proteins.
2. Explain why RNA interference is aptly named.
3. Explain the costs and benefits to a cell of regulating gene expression at the post-transcriptional level rather than the transcriptional level.

Answers are available in Appendix A.

19.5 Linking Cancer to Defects in Gene Regulation

All cancers involve uncontrolled cell division. What allows this unbridled increase in cell number? Each type of cancer is caused by a different set of mutations that lead to cancer when they alter two classes of genes: **(1)** genes that stop or slow the cell cycle, and **(2)** genes that trigger cell growth and division. It turns out that many of the genes that are mutated in cancer control gene expression, either directly as transcription factors or indirectly as proteins involved in cell communication. Let's take a closer look at how altered gene regulation can lead to the uncontrolled cell growth that is a hallmark of cancer.

After you complete this section, you should be able to ...

- Explain the relationship between cancer and defects in gene regulation.

The Genetic Basis of Uncontrolled Cell Growth

As you learned in the chapter on the cell cycle (Ch. 12, Section 12.3), proteins that stop or slow the cell cycle when conditions are unfavorable for cell division are called **tumor suppressors**. The genes that code for these proteins are called tumor suppressor genes. If the function of a tumor suppressor gene is lost because of mutation, then a brake on the cell cycle is eliminated.

On the flip side of the coin are genes that *stimulate* cell division. These genes are called **proto-oncogenes** (literally, "first cancer genes"). In normal cells, the proteins produced from proto-oncogenes are active only when conditions are appropriate for division. In cancerous cells, defects in the regulation of proto-oncogenes or their protein products spur cells to divide all the time. In such cases, a mutation has converted the proto-oncogene into an **oncogene**—a mutant allele that promotes cancer.

For cancers to develop, many mutations are required within a single cell, and these alter both tumor suppressor genes and proto-oncogenes.

The p53 Tumor Suppressor

To understand how defects in gene expression can lead to cancer, let's consider research on the gene that is most often defective in human cancers. The gene is called *p53* because when it was first discovered, researchers knew only that the protein it codes for has a molecular weight of about 53 kilodaltons. Sequencing studies have revealed that mutant, nonfunctional forms of the *p53* gene are found in well over half of all human cancers. This gene codes for a regulatory transcription factor.

Researchers began to understand one function of *p53* when they exposed noncancerous human cells to UV radiation and noticed that the level of active p53 protein increased markedly. Recall that UV radiation damages DNA (Ch. 15, Section 15.5). Follow-up studies confirmed that there is a close correlation between DNA damage and the activity of p53 in a cell. Analyses of the p53 protein's primary structure suggested that it might contain a DNA-binding region similar to the one shown for the muscle-specific transcription factor in Figure 19.8b.

These observations inspired the hypothesis that p53 is a regulatory transcription factor that works as a master brake on the cell cycle, making it a "guardian of the genome." In this model, shown in Figure 19.12a, p53 activity is induced by DNA damage. Activated p53 binds to the enhancers of genes that arrest the cell cycle, repair DNA damage, and when all else fails, trigger apoptosis (cell death). Expression of these genes allows the cell to halt the cell cycle and repair its DNA, if this is possible, or undergo apoptosis if the DNA damage is too severe.

In mutant cells with an altered form of p53 unable to bind to enhancers, DNA damage cannot trigger either arrest of the cell cycle or apoptosis, and damaged DNA is replicated (Figure 19.12b). This situation leads to chromosome breaks and mutations that move the cell farther down the road to cancer. The p53 protein is like a quality control officer—if it is missing, errors are made and things go downhill.

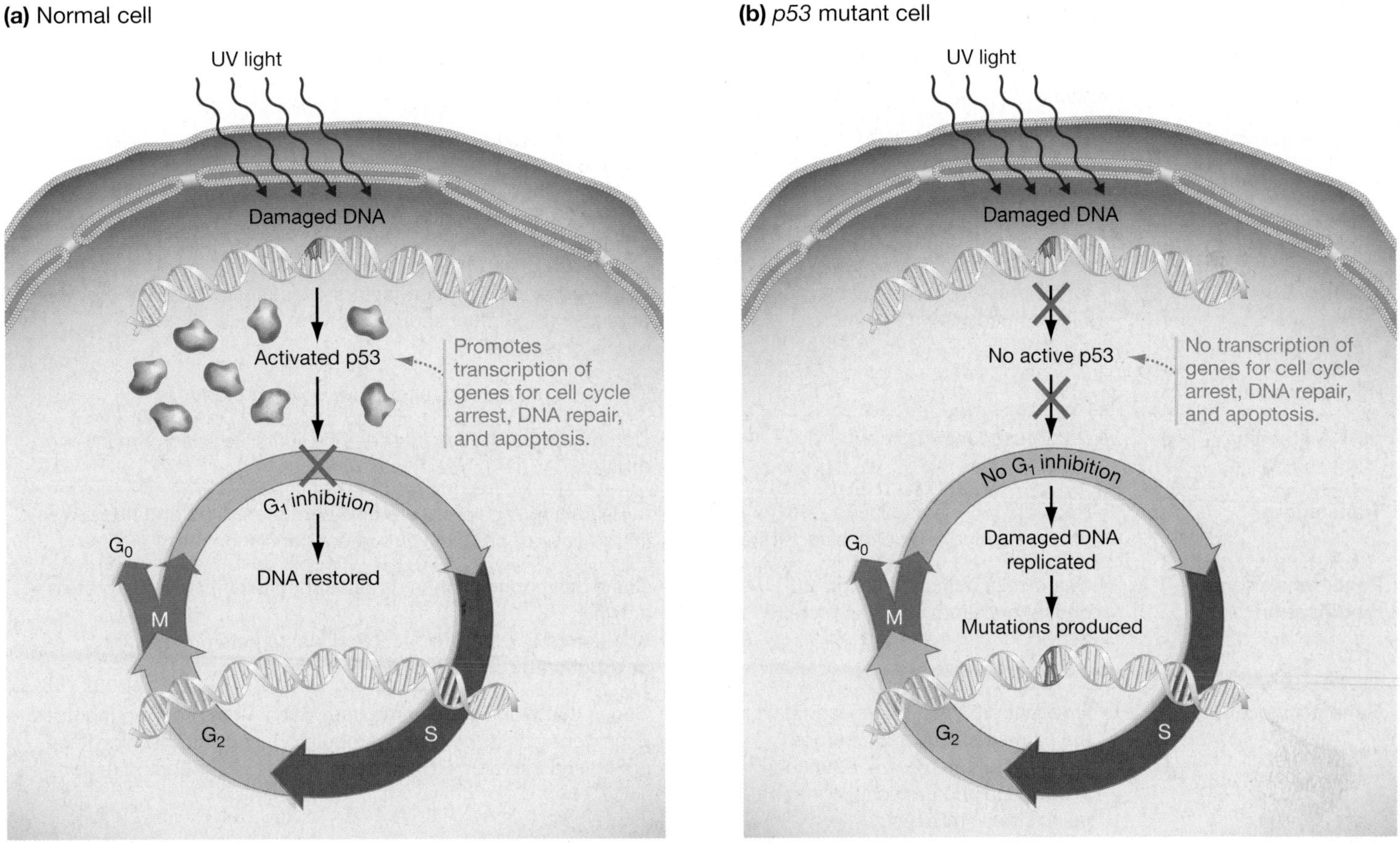

Figure 19.12 ***p53* Gene Mutations Allow Cells to Replicate DNA and Divide Even When DNA Is Damaged.**

CHECK YOUR UNDERSTANDING

✔ **If you understood this section, you should be able to ...**

1. Explain why cancer has a common pattern of uncontrolled cell growth, but not a common cause.
2. Explain how cancer development requires inactivating mutations in tumor suppressor genes and activating mutations in proto-oncogenes.
3. Explain why loss-of-function mutations in *p53* often lead to mutations in other genes.

Answers are available in Appendix A.

19.6 A Comparison of Gene Expression in Bacteria and Eukaryotes

After you complete this section, you should be able to ...

- Compare and contrast how gene expression is controlled in bacteria and eukaryotes.

How does gene expression compare in bacteria and eukaryotes? These two major groups of organisms share a very distant common ancestor, and they often experience different selective pressures.

Look at **Table 19.2** for a quick comparison of gene expression and its control in these two domains of life.

The control of gene expression in bacteria and eukaryotes shares many similarities, but also shows significant differences.

- ***DNA packaging*** The chromatin of eukaryotic DNA must be decondensed for general and regulatory transcription factors to gain access to genes and for RNA polymerase to initiate transcription. The tight packaging of eukaryotic DNA means that the default state of transcription in eukaryotes is "off." In contrast, the default state of transcription in bacteria, which lack the condensed chromatin of eukaryotes and have more accessible promoters, is "on." Condensed chromatin prevents transcription and provides a mechanism of negative control that does not exist in bacteria.
- ***Complexity of transcription*** Transcription initiation is much more elaborate in eukaryotes. The sheer number of eukaryotic proteins needed to regulate the start of transcription dwarfs that in bacteria, as does the number of their interactions.
- ***Coordinated transcription*** In bacteria, genes that take part in the same cellular process are often organized into operons and transcribed together from a single promoter. In contrast, operons are rare in eukaryotes. Instead, for coordinated gene expression, eukaryotes rely on the strategy used in bacterial regulons—physically scattered genes are expressed together when the same regulatory transcription factors trigger the transcription of genes with the same DNA regulatory sequences.

SUMMARY Table 19.2 Regulating Gene Expression in Bacteria and Eukaryotes

Level of Regulation	Bacteria	Eukaryotes
Chromatin remodeling	• Less packaging of DNA • Remodeling not a major issue in regulating gene expression.	• Extensive packaging of DNA • Chromatin must be decondensed for transcription to begin.
Transcription	• Positive and negative control by regulatory proteins that act at sites close to the promoter • Sigma interacts with promoter.	• Positive and negative control by regulatory proteins that act at sites close to and far from promoter • Large set of transcription factors interact with the core promoter and many regulatory sequences. • Mediator required.
RNA processing	• Rare	• Extensive processing: alternative splicing of introns
mRNA stability	• Rarely used for control	• Commonly used: RNA interference limits life span of many mRNAs.
Translation	• Regulatory proteins bind to mRNAs and ribosomes and affect translation rate.	• miRNAs and regulatory proteins bind to mRNAs and ribosomes (in the case of proteins) and affect translation rate.
Post-translational modification	• Chemical modification (e.g., phosphorylation) changes protein activity.	• Chemical modification (e.g., phosphorylation) changes protein activity. • Addition of short proteins (ubiquitins) targets proteins for destruction by proteasome.
Gene organization	• Frequent grouping of genes that work in the same pathway into operons • Genes in an operon are regulated in common and transcribed into polycistronic mRNAs.	• Genes that work in the same pathway are not grouped together. • Each gene has its own set of regulators and is transcribed and processed into an mRNA with only one protein-coding region.

- ***Reliance on post-transcriptional control*** Eukaryotes make much greater use of post-transcriptional control, such as alternative splicing and RNA interference. Bacteria use other forms of post-transcriptional control, but use these to a lesser extent.

Why is gene regulation in multicellular eukaryotes so much more complex than in bacteria? One hypothesis is that the demands of differential gene expression during development have driven the evolution of both new and more elaborate ways of controlling genes.

Now that you know how gene expression is controlled, let's see in the next chapter how scientists have exploited this knowledge in technological applications, and how, in turn, these technologies have advanced our understanding of life.

CHECK YOUR UNDERSTANDING

✔ **If you understood this section, you should be able to ...**

1. Predict how the number of regulatory elements would change if eukaryotes made extensive use of operons.
2. Knowing that mRNAs in bacteria have very short half-lives (they don't last long), explain why it is reasonable to expect bacteria *not* to use a system analogous to miRNA-mediated control of gene expression.

Answers are available in Appendix A.

CHAPTER 19 Review

For media, go to **Mastering Biology**

19.1 Gene Regulation in Eukaryotes—An Overview

- Changes in gene expression allow eukaryotic cells to respond to changes in the environment and make it possible for distinct cell types to develop.
- In a multicellular eukaryote, cells are different because they express different genes, not because they have different genes.
- Gene expression is regulated at six levels: Chromatin has to be remodeled, the transcription of specific genes may be initiated or repressed, pre-mRNAs may be spliced in different ways to produce different mRNAs, the life span of specific mRNAs may be extended or shortened, the translation rate may be controlled, and the life span or activity of particular proteins may be altered.

19.2 Chromatin Remodeling

- Eukaryotic DNA is packaged with proteins into chromatin that must be opened before transcription can occur.
- Eukaryotic DNA is wrapped around histone proteins to form nucleosomes connected by linker DNA that are then coiled into 30-nm fibers and higher-order chromatin structures.
- Transcription cannot be initiated until the chromatin around regulatory regions is decondensed.
- The state of chromatin condensation depends on the methylation of cytosines in DNA, on acetylation and many other modifications of histones, and on chromatin-remodeling complexes that remove nucleosomes from stretches of DNA.
- Patterns of DNA methylation and histone modifications can be passed from mother cells to daughter cells.
- Epigenetic inheritance is the inheritance of different phenotypes due to anything other than differences in alleles; transmitting patterns of chromatin condensation from mother to daughter cells or from parent to offspring is the primary mechanism of epigenetic inheritance.

19.3 Initiating Transcription

- In eukaryotes, transcription is triggered by regulatory proteins called transcription factors that bind to sequences both close to and far from the core promoter.
- Regulatory transcription factors can be activators or repressors; these bind to regulatory sequences—either (1) promoter-proximal sequences that are near promoters or (2) enhancers and silencers that are often located far from core promoters.
- Amino acids on regulatory transcription factors interact with the surfaces of base pairs that project in the grooves of the DNA helix to allow binding to specific regulatory sequences.
- The first regulatory transcription factors that bind to DNA recruit proteins that loosen the interaction between nucleosomes and DNA, making the core promoter, promoter-proximal elements, and enhancers accessible to other transcription factors.
- Interactions between regulatory and general transcription factors occur through a complex of proteins called the Mediator. This leads to the positioning of RNA polymerase at the core promoter and the start of transcription.

19.4 Post-Transcriptional Control

- Once transcription is complete, gene expression is controlled by (1) alternative splicing, (2) RNA interference, and (3) activation or inactivation of protein products.
- Alternative splicing allows a single gene to produce more than one version of an mRNA and more than one kind of protein.
- RNA interference occurs when a tiny strand of RNA held by the protein complex RISC (or a related complex for piRNAs) binds to a complementary RNA. In the case of mRNA, this marks the mRNA for destruction or prevents its translation. MicroRNAs (miRNAs) derive from processed transcripts of cellular genes and play an important role in controlling gene expression.

- Once translation occurs, proteins may be activated or inactivated by the addition or removal of chemical groups such as phosphates, or marked for destruction in the proteasome by the addition of copies of a polypeptide known as ubiquitin.

19.5 Linking Cancer to Defects in Gene Regulation

- If mutations alter regulatory proteins that promote or inhibit progression through the cell cycle, then uncontrolled cell growth and tumor formation may result. These regulatory proteins are encoded by proto-oncogenes (promote) and tumor suppressor genes (inhibit), respectively.
- *p53* is a major tumor suppressor gene that is often mutated in cancers, leading to unrestrained cell growth under conditions that normally halt cell division.

19.6 A Comparison of Gene Expression in Bacteria and Eukaryotes

- Review Table 19.2, Regulating Gene Expression in Bacteria and Eukaryotes.

Answers are available in Appendix A.

✔ TEST YOUR KNOWLEDGE

1. What is chromatin?
   a. the histone-containing protein core of the nucleosome
   b. the 30-nm fiber
   c. the complex of DNA and proteins found in the nucleus
   d. the histone *and* non-histone proteins in eukaryotic nuclei
2. Which of these statements about enhancers is correct?
   a. They contain a unique base sequence called a TATA box.
   b. They are located only in 5′ untranslated regions.
   c. They are located only in introns.
   d. They are found both upstream and downstream from the transcription start site and are functional in any orientation.
3. **THINK CAREFULLY**In eukaryotes, what allows only certain genes to be expressed in different types of cells?
4. What is alternative splicing?
   a. phosphorylation that leads to different types of post-translational regulation
   b. mRNA processing that leads to different combinations of exons being spliced together
   c. folding that leads to proteins with alternative conformations
   d. the outcome of regulatory proteins that leads to changes in the life span of an mRNA

✔ TEST YOUR UNDERSTANDING

5. Compare and contrast the items in each pair: (a) enhancers and the *E. coli* CAP binding site (Chapter 18); (b) promoter-proximal elements and the operator of the *lac* operon; (c) general transcription factors and sigma.
6. **THINK CAREFULLY** Imagine discovering a loss-of-function mutation in a eukaryotic gene. You determine the gene's nucleotide sequence from the start site for transcription to the termination point of transcription and find no differences from the wild-type sequence. Explain where you think the mutation might be and how the mutation might be acting.
7. The following statements are about the control of chromatin condensation. Select True or False for each.
   T/F Reducing histone acetylase activity is likely to decrease gene transcription.
   T/F Mutations that reduce the number of positively charged amino acids on histones should promote open chromatin.
   T/F Chromatin remodeling complexes add chemical groups to histones.
   T/F Adding an inhibitor of DNA methylation is likely to reduce gene transcription.
8. Predict how a mutation that caused continuous production of active p53 would affect the cell.

✔ TEST YOUR PROBLEM-SOLVING SKILLS

9. In the follow-up work to the experiment shown in Figure 19.6, the researchers used a technique that allowed them to see if two DNA sequences are in close physical proximity (association). They applied this method to examine how often an enhancer and the core promoter of the *Hnf4a* regulatory gene were near each other. A logical prediction is that compared with rats born to mothers fed a healthy diet, the *Hnf4a* gene in rats born to mothers fed a protein-poor diet would
   a. show no difference in how often the promoter and enhancer associated.
   b. never show any promoter–enhancer association.
   c. show a lower frequency of promoter–enhancer association.
   d. show a higher frequency of promoter–enhancer association.
10. **QUANTITATIVE** Imagine repeating the experiment on epigenetic inheritance that is shown in Figure 19.6. You measure the amount of radioactive uridine (U) incorporated into *Hnf4a* mRNA in counts per minute (cpm) to determine the level of *Hnf4a* gene transcription in rats born to mothers fed either a normal diet or a low-protein diet. The results are 11,478 cpm for the normal diet and 7368 cpm for the low-protein diet. For this problem, your task is to prepare a graph similar to the one at the bottom of Figure 19.6 that shows the normalized results for the low-protein diet relative to the normal diet. Normalizing values means that the value obtained from one condition is expressed as 1.0 (the norm; the normal diet in this case) and the values obtained from any other conditions (low-protein diet in this case) are expressed as decimal values relative to the norm.

✔ PUT IT ALL TOGETHER: Case Study

What does gene regulation have to do with being blond?

Scientists discovered that the difference between blond and dark hair comes down in part to a single nucleotide difference in the DNA sequence of an enhancer that lies more than 350,000 base pairs away from the gene it controls. On average, blonds transcribe this gene less efficiently than people with dark hair. How can a tiny difference in a distant enhancer make a blond?

11. Which of the following statements about this enhancer is correct?
   a. the enhancer codes for a large protein.
   b. the enhancer codes for a small protein.
   c. the enhancer is associated with a protein when the gene is being transcribed.
   d. the enhancer is within the sequence that codes for the mRNA's 5′ untranslated region.

12. How is it possible for this enhancer to regulate transcription of a gene that is so far away?

13. Scientists discovered that this hair-color-associated enhancer has a binding site for a particular transcription factor. One form of the binding site has the sequence CACTAAAG and is associated with dark hair, and the other form of the binding site has the nearly identical sequence CGCTAAG and is associated with blond hair. How could these two nearly identical enhancer binding sites lead to different rates of initiating transcription of the regulated gene?

14. QUANTITATIVE The human gene controlled by the enhancer is called *KITLG*, and it codes for a signaling protein that binds to a cell-surface receptor. The signaling pathway activated by KITLG controls many important phenotypes, including hair color. Using genetic engineering methods (described in Chapter 20 and **BioSkills 10**), scientists fused a copy of *Kitl*—the mouse equivalent of human *KITLG*—and its core promoter to forms of the human enhancer associated with either dark hair or blond hair. These engineered DNAs were introduced into mouse embryos, and the amount of *Kitl* mRNA present in embryonic mouse skin cells was measured and compared to untreated controls. The results are shown here. What can be concluded about the effectiveness of the enhancers associated with either dark hair or blond hair in driving expression of *Kitl* mRNA? (* means $P < 0.05$, *** means $P < 0.001$; box-and-whisker plots and P values are explained in **BioSkills 2** and **BioSkills 3**.)

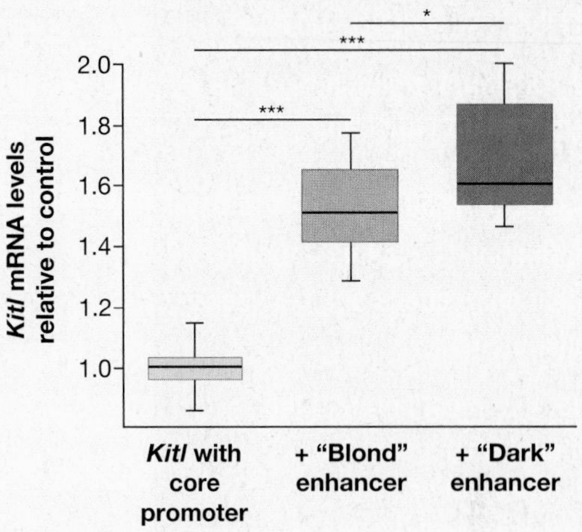

Source: C. A. Guenther et al. 2014. *Nature Genetics* 46: 748–752.

15. PROCESS OF SCIENCE Why can't the results shown here be used as proof that the two different forms of the enhancer are important in determining dark versus blond hair? Propose a study that would test this idea more directly (more than one approach is possible).

16. PROCESS OF SCIENCE Imagine working with the scientists who did this research. One of them proposes testing the hypothesis that *Kitl* is important to coat coloration. To do this, she plans to observe what happens when an engineered double-stranded miRNA precursor is expressed in mouse embryos. One strand of the miRNA precursor would be complementary to *Kitl* mRNA. How could this approach work to test the hypothesis?

BIG PICTURE

THE BIG PICTURE

Copying, using, and transmitting genetic information is fundamental to life. Cells use the genetic information archived in their DNA to respond to changes in the environment and, in multicellular organisms, to develop into specific cell types.

Hereditary information is transmitted to offspring with random changes called mutations. Thus, genetic information is dynamic—both within generations and between generations.

Note that most boxes in the concept map indicate the chapters and sections where you can go for review. Also, be sure to do the blue exercises in the Check Your Understanding box below.

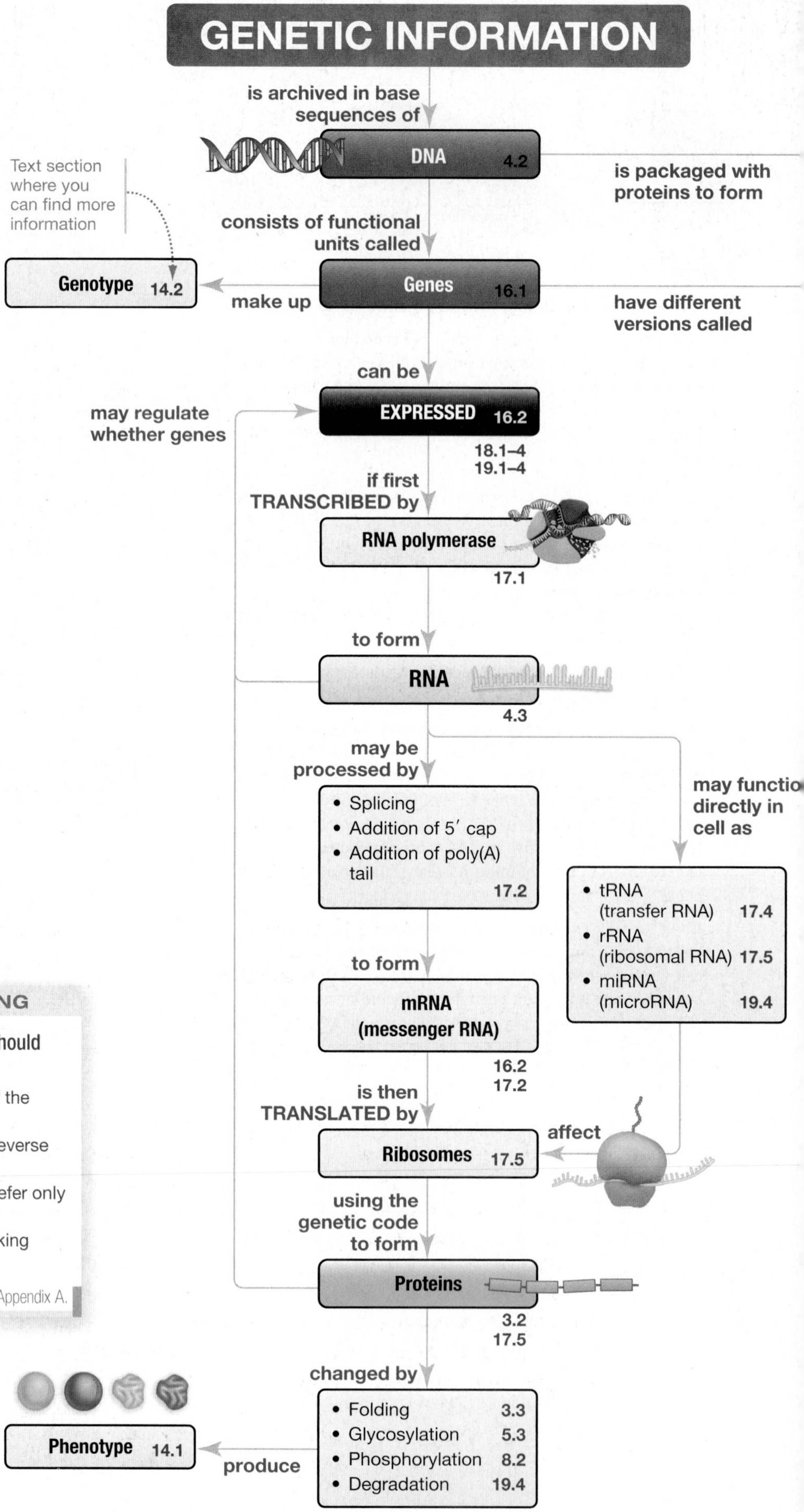

CHECK YOUR UNDERSTANDING

✔ If you understand the big picture, you should be able to . . .

1. Draw stars next to the three elements of the central dogma of molecular biology.
2. Add arrows and labels indicating what reverse transcriptase does.
3. Draw an E in the corners of boxes that refer only to eukaryotes, not prokaryotes.
4. Fill in the blue ovals with appropriate linking verbs or phrases.

Answers are available in Appendix A.

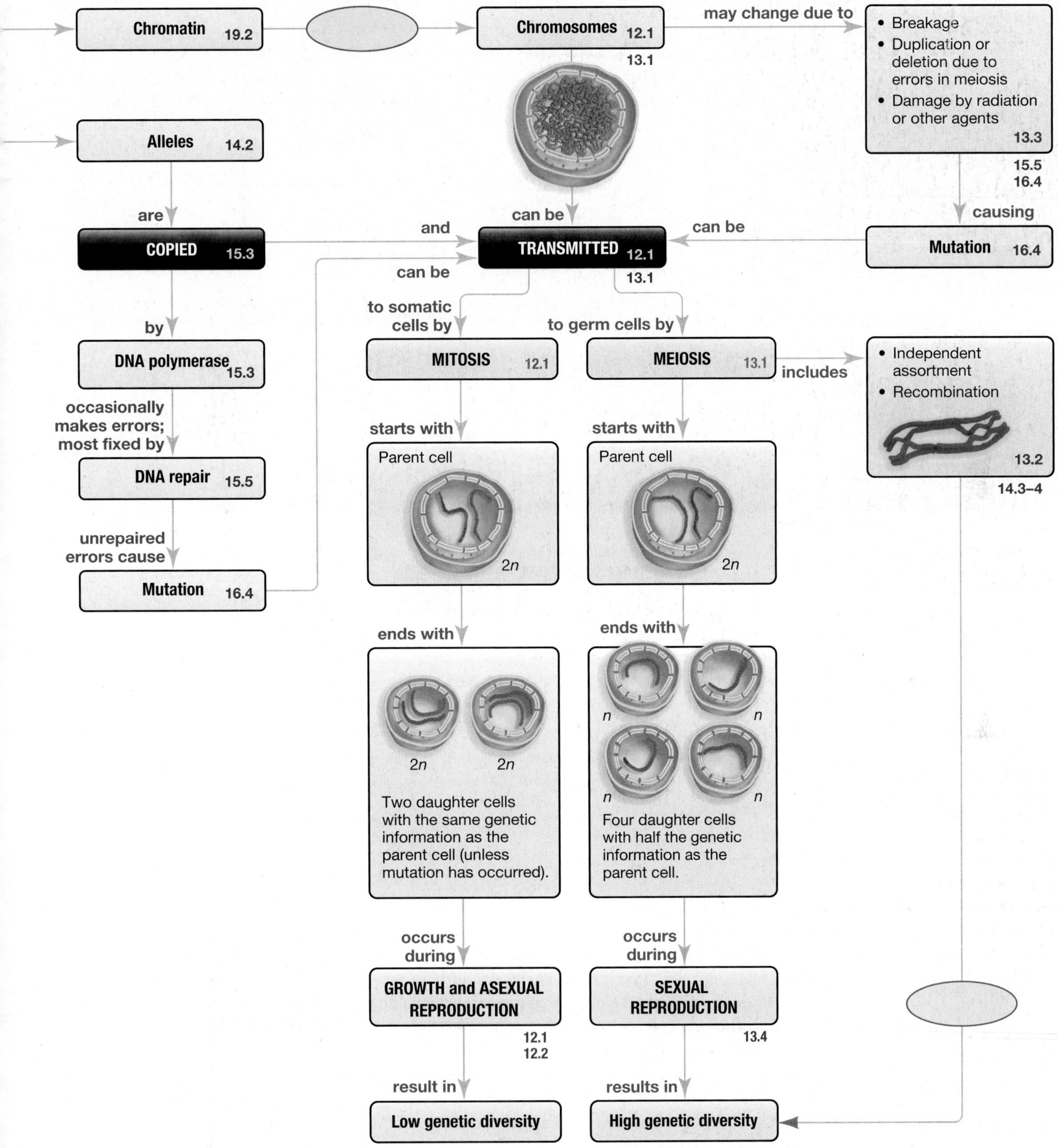
Chromatin 19.2
Chromosomes 12.1
13.1
may change due to
• Breakage
• Duplication or deletion due to errors in meiosis
• Damage by radiation or other agents
13.3
15.5
16.4
causing
Mutation 16.4
can be
Alleles 14.2
are
COPIED 15.3
and
can be
TRANSMITTED 12.1
13.1
can be
by
DNA polymerase 15.3
occasionally makes errors; most fixed by
DNA repair 15.5
unrepaired errors cause
Mutation 16.4
to somatic cells by
MITOSIS 12.1
to germ cells by
MEIOSIS 13.1
includes
• Independent assortment
• Recombination
13.2
14.3–4
starts with
Parent cell
2n
starts with
Parent cell
2n
ends with
2n
2n
Two daughter cells with the same genetic information as the parent cell (unless mutation has occurred).
ends with
n
n
n
n
Four daughter cells with half the genetic information as the parent cell.
occurs during
GROWTH and ASEXUAL REPRODUCTION
12.1
12.2
occurs during
SEXUAL REPRODUCTION
13.4
result in
Low genetic diversity
results in
High genetic diversity

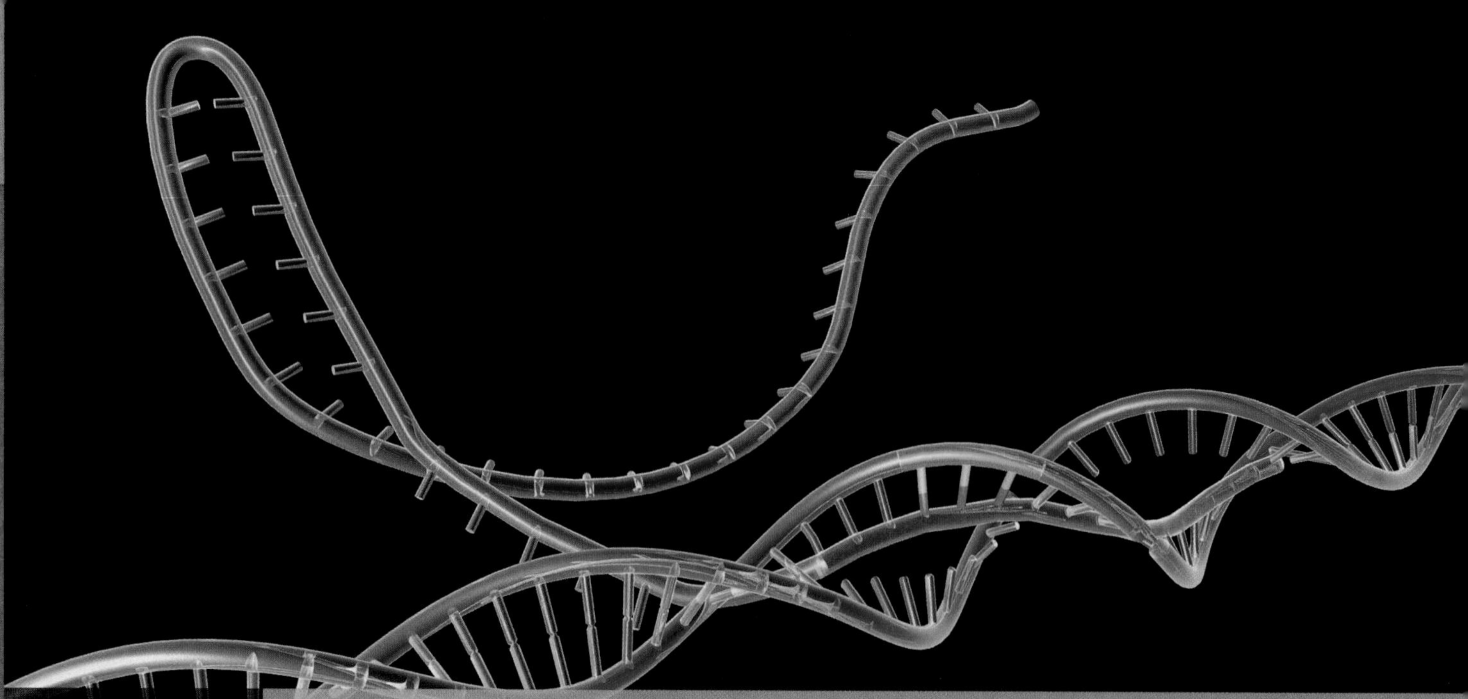

20 The Molecular Revolution: Biotechnology, Genomics, and New Frontiers

Genetic modifications are now possible to create in almost any organism by using the CRISPR-Cas9 editing system. The system uses a guide RNA and an enzyme to cut DNA at a site complementary to the RNA. This image shows an artist's depiction of the guide RNA (red) base paired to target DNA (blue).

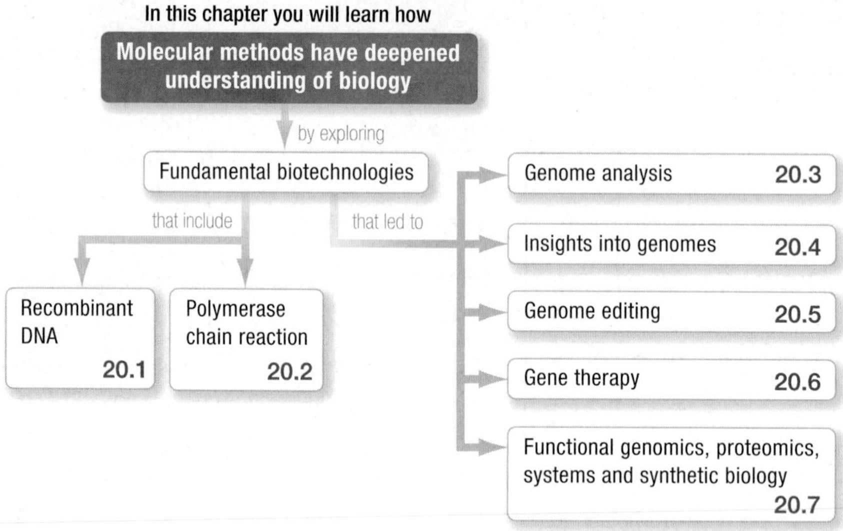

Advances in technology and a deeper understanding of biology go hand in hand. Nowhere is this more apparent than in our knowledge of how genes work. Transformative technologies have opened windows to biology that were unimaginable just a few years ago. In turn, the questions spurred by this new knowledge have driven the development of even more powerful experimental tools. We truly are living through a molecular revolution.

How so? Consider this: DNA's role in information storage was first established in 1952 following the work of Alfred Hershey and Martha Chase (Ch. 15, Section 15.1), but biologists had no idea of how DNA worked. Fast-forward less than 50 years later to 2001. Not only had the nature of the gene been established, but the first catalog of what DNA sequences it takes to be human were brought to light after all the DNA of a person—the human **genome**—was sequenced.

Jump ahead to today, and the news is about newfound abilities to precisely engineer the human—or any other—genome, to create designer organisms, to genetically alter wild populations, or to create synthetic cells. Of course, with this technological power comes difficult ethical questions. Just because we can do something, should we?

In this chapter you'll delve into selected topics to explore some of the technologies that have revolutionized biology and the insights they've provided. A common theme is that discoveries in *basic* biology lie at the heart of every one of these technologies. (Use **BioSkills 10** as a companion to this chapter to dig deeper into the details of some of these molecular tools.) Let's begin with the ability to mix and match DNA sequences from any organisms.

20.1 Recombinant DNA Technology

An essential tool in the molecular biology toolkit is **recombinant DNA technology**. This set of methods allows researchers to mix and match ("recombine") specific DNA sequences from any organism to create DNA molecules not found in nature. With recombinant DNA methods, biologists no longer have to rely solely on random mutations and controlled breeding to change the genetic characteristics of plants, animals, or other organisms. Recombinant DNA is a cornerstone of **biotechnology**: the engineering of genes, cells, and organisms for basic research as well as practical purposes—for example, in agriculture.

After you complete this section, you should be able to ...

- Describe how a gene can be cloned.

Recombinant DNA technology became a realty in the 1970s and has since touched all our lives. It's responsible for the production of many medicines—for example, insulin used by diabetics—and for the genetic modification of many of our most widely grown crops. Unless you work hard to avoid consuming **genetically modified organisms (GMOs)**, in the last few days you've almost certainly eaten a food modified by recombinant DNA technology. Some applications of recombinant DNA technology, such as GMOs, are controversial, and a recurrent theme of this chapter is the societal and ethical issues arising from these advances.

Using Plasmids in Cloning

One of the most basic requirements of recombinant DNA technology is the ability to produce many copies of a gene or other DNA sequence of interest, a process referred to as **DNA cloning**. If a researcher says that she has cloned a gene, it means that she has isolated it and then produced many identical copies.

In many cases, researchers clone a sequence of DNA by inserting it into a small, circular DNA molecule called a **plasmid**. You might recall that plasmids are common in bacterial cells (Ch. 7, Section 7.1). They are physically separate from the bacterial chromosome, carry genes that are not required for normal growth and reproduction, and can replicate independently of the chromosome. Plasmids used in cloning carry genes for antibiotic resistance that allow a cell to grow in the presence of an antibiotic that kills cells without the plasmid.

Researchers realized that if they could add a piece of foreign DNA to a plasmid and then insert the recombinant molecule into a bacterial cell, the engineered plasmid would be replicated and passed on to daughter cells. If this bacterium was grown in a nutrient broth, then billions of copies of the original cell, each containing many identical versions of the recombinant plasmid DNA, would result. When a plasmid is used to make copies of a foreign DNA sequence, it is called a **cloning vector**, or simply a **vector**. It's important to realize that the DNA added to the plasmid can be from any organism. Most often in cloning, the DNA fragment added to the plasmid comes from a vastly different species—say a human or a corn plant—than the bacterial cell that replicates it.

Biologists harvest the recombinant plasmids by breaking the bacteria open, isolating all the DNA, and then separating the plasmids from the main chromosomes. But how do they insert a gene into a plasmid in the first place?

Using Restriction Endonucleases and DNA Ligase to Cut and Paste DNA

To cut out a gene, or any DNA sequence, for insertion into a cloning vector, researchers use enzymes from bacteria called restriction endonucleases. A **restriction endonuclease** cuts DNA molecules at specific base sequences called recognition sites. There are many types of restriction endonucleases, each cutting DNA at a different recognition site.

It's easy to think that restriction endonucleases were discovered in a search for enzymes that could be used for DNA cloning. Not so. Scientists found these enzymes while investigating bacterial defense systems. In nature, these enzymes are used by bacteria to cut up DNA from viruses and other invaders to prevent a fatal infection.

One way of inserting foreign DNA into plasmids is shown in the sequence of steps outlined in **Figure 20.1** on page 408.

Step 1 The left side of the figure shows a plasmid that contains a gene for antibiotic resistance and a recognition site in the double-stranded DNA that is cut by a specific restriction endonuclease. As the right side of the figure shows, the same recognition sequences are present on a piece of foreign DNA.

Step 2 A restriction endonuclease cuts the recognition sites in the plasmid (left) and in the foreign DNA (right).

Step 3 The cuts made in the recognition site by most restriction enzymes are staggered (offset on opposing strands) as shown here, rather than blunt (straight across both strands). With staggered cuts, the resulting DNA fragments are described as having **sticky ends**, because the single-stranded bases exposed on the plasmid are complementary to the single-stranded bases exposed on the foreign DNA. As a result, the complementary ends hydrogen-bond the two molecules to each other. DNAs from any sources that are cut with the same restriction endonuclease can be joined together in this way.

Step 4 Next, **DNA ligase**—the enzyme that connects Okazaki fragments during DNA replication (Ch. 15, Section 15.3)—is added to seal the pieces of DNA together. This is the essence of recombinant DNA technology—forming new combinations of DNA by cutting specific sequences and linking them together in new ways.

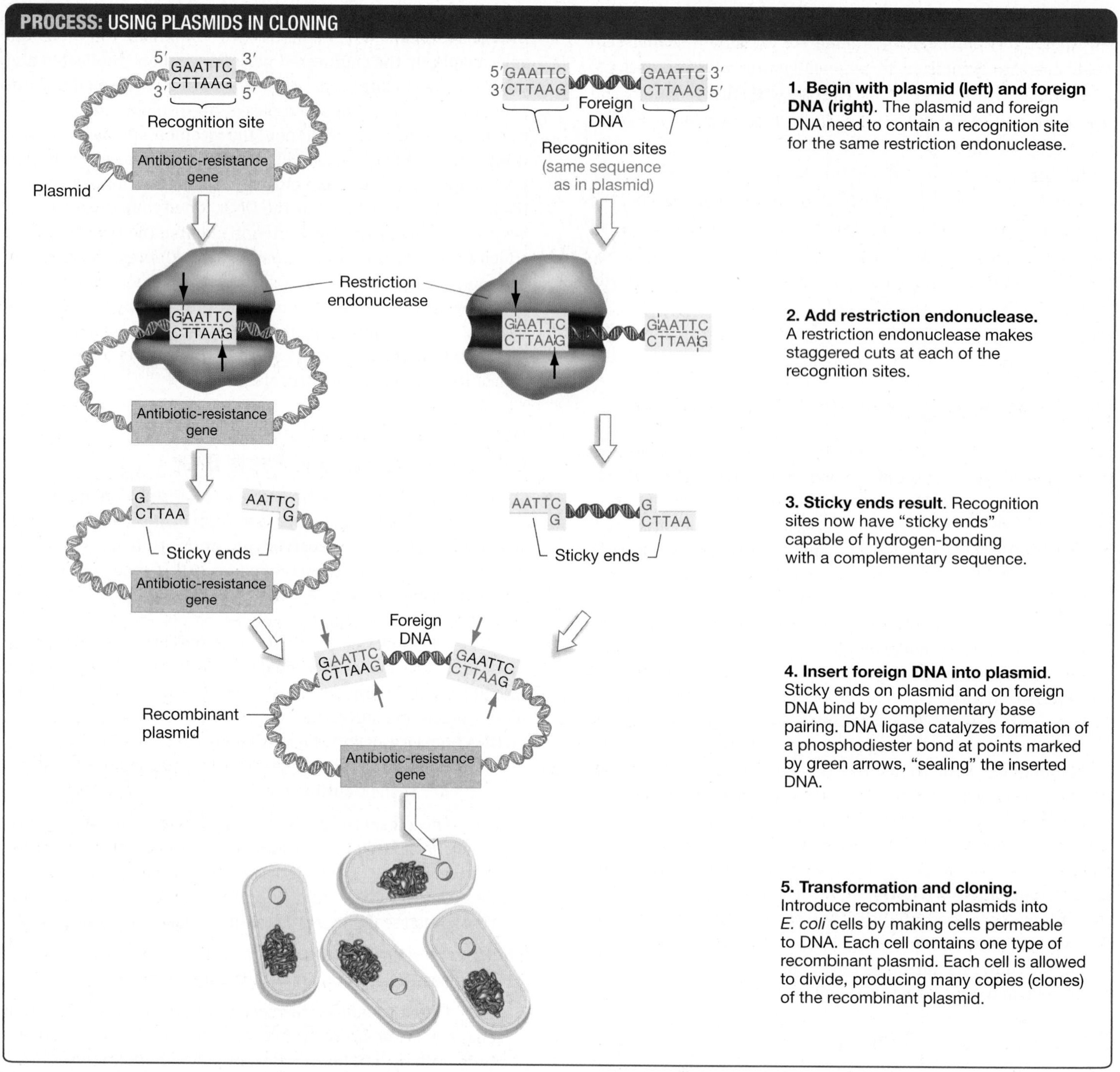

Figure 20.1 Foreign DNA Can Be Inserted into Plasmids for Cloning. Once foreign DNA has been inserted into a plasmid, the recombinant plasmid can be introduced into bacterial cells that grow and divide to produce many identical copies of the recombinant plasmid. The inserted DNA can be any DNA sequence, including a gene or part of a gene.

Step 5 Finally, recombinant plasmids are introduced into *E. coli* by making the cells permeable to DNA. The resulting "transformed" bacterial cells are then allowed to grow and divide to produce many identical copies of the recombinant DNA.

How can a bacterial cell take up DNA?

Transformation: Introducing Recombinant Plasmids into Bacterial Cells

Cells that take up DNA from the environment and incorporate it into their genomes are said to undergo **transformation**. Bacterial cells do not routinely take up DNA on their own, even under laboratory conditions. So, to get plasmid DNA into *E. coli* or other species, researchers treat the cells to increase their permeability to DNA. One common method mixes cold, chemically treated cells and the recombinant plasmid and then heats the mixture briefly to allow the uptake of DNA. Transformation is inefficient, with few bacteria in a population ever taking up plasmid DNA. Typically, just a single plasmid enters any one cell during this treatment.

The cells are then spread out on plates under conditions that allow only cells with the plasmids to grow into colonies. This is where the antibiotic-resistance gene on the plasmid comes into play. Cells are grown in the presence of the antibiotic, so only those with the plasmid can grow into a colony; cells without the plasmid do not multiply. Each colony contains millions of identical cells, each with many identical copies of the recombinant plasmid.

Complementary DNA (cDNA)

What kinds of DNA are cloned? Often, it's DNA taken directly from an organism—called **genomic DNA**—but another important approach is to make a DNA copy of an mRNA.

As you learned earlier (Ch. 16, Section 16.2), a viral enzyme called reverse transcriptase catalyzes the synthesis of DNA from an RNA template. DNA that is transcribed from RNA is called **complementary DNA**, or **cDNA**. Reverse transcriptase initially produces a single-stranded cDNA, but once this is done, it can use this cDNA strand to synthesize the complementary DNA strand to yield double-stranded DNA (see Ch. 33, Figure 33.12). (**BioSkills 10** explains how geneticists clone cDNAs in the lab.)

Researchers use cDNAs in many ways. For instance, they can isolate mRNAs from specialized cells—muscle cells, for example—and convert those mRNAs to double-stranded cDNAs. These cDNAs, which correspond to each gene that is actively expressed into mRNA in the source cells, can be sequenced to learn which genes are expressed in the given cell type.

Another important use of cDNAs is in deducing the amino acid sequence of the polypeptide encoded by a gene. Because cDNAs are copies of mRNAs, they do not contain introns. This makes it a simple matter to identify the uninterrupted set of codons that specify the polypeptide and then use the genetic code to deduce its amino acid sequence.

Biotechnology in Agriculture

A major focus in biotechnology is to improve crop varieties to reduce losses from herbivore and pathogen damage, make crops herbicide resistant so weeds can be suppressed without killing the crop, and improve food quality. To achieve these goals, scientists create **transgenic** (literally, "across-genes") plants using the bacterium *Agrobacterium tumefaciens*. This bacterium infects many different plants and does so by transferring *part* of a plasmid it carries—the T-DNA region of the tumor-inducing plasmid, or Ti plasmid for short—into plant cells (**Figure 20.2**).

The T-DNA genes that are moved into the plant cell are incorporated into the plant genome, where they promote uncontrolled plant cell growth and formation of tumors.

When researchers want to make a transgenic plant, they remove the tumor-inducing genes from the T-DNA region of the Ti plasmid and replace these with genes they want to introduce into the plant. *Agrobacterium* cells containing this engineered Ti plasmid are used

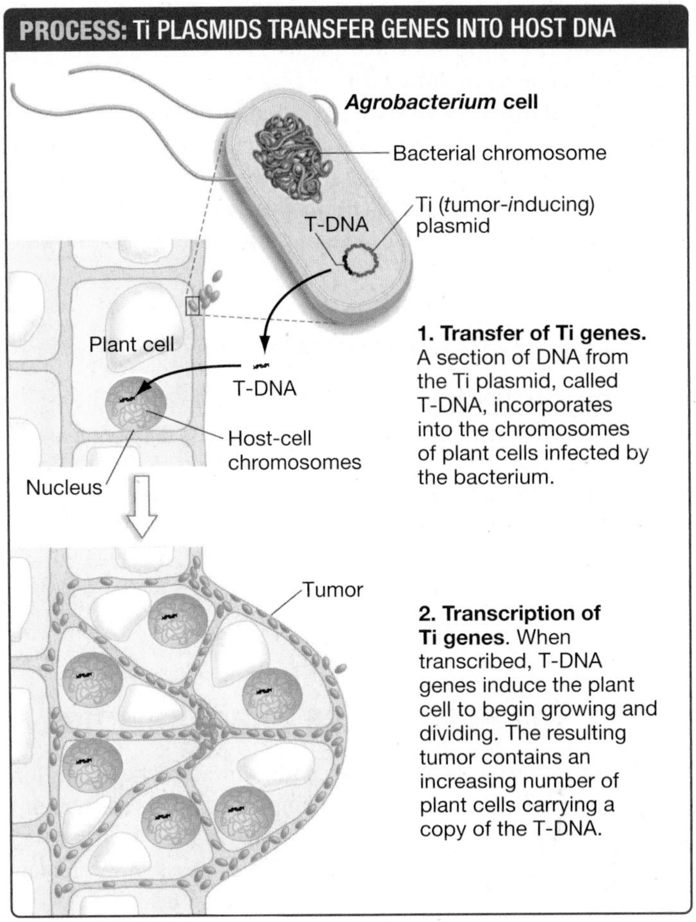

Figure 20.2 The Bacterium *Agrobacterium tumefaciens* Transfers DNA into Plant Cell Chromosomes. Genes on part of this plasmid called the T-DNA region are transferred into a plant cell and are then incorporated into the plant genome. For genetic engineering, some of the T-DNA genes are replaced with genes that researchers aim to incorporate into the plant cell.

to transfer the recombinant plasmid DNA into the plant. The introduced genes now become part of the plant cell's genome.

An entire genetically modified plant can be made from a single cell using cell and tissue culture techniques (described in **BioSkills 11**). In brief, a cell that contains a foreign gene is placed on a nutrient-rich medium and allowed to grow into a mass of cells called a callus. The callus then is transferred to another medium that promotes development into a tiny plant that can be grown in soil. This genetically modified plant is propagated, and seeds are produced for future plantings.

GMO crops are a mainstay of U.S. agriculture. For example, more than 90 percent of corn, cotton, and soybeans are genetically modified. But although widespread, GMO crops continue to stir controversy because of lingering concerns some people have about negative health or environmental effects.

No negative health effects have yet been documented from the consumption of GMO crops. Some early studies suggested that pollen from corn genetically modified to resist insect pests could spread to milkweed plants, the preferred food source for Monarch butterfly caterpillars, and therefore harm butterfly populations. Later studies concluded that under field conditions, there were no significant consequences for butterflies from the spread of pollen from GMO corn.

Another frequently voiced concern about GMOs is the environmental effects of herbicides and herbicide runoff when herbicides are applied infrequently at high dose to herbicide-resistant GMO crops compared to more frequent, lower-dose applications to non-GMO crops. Although long used and widespread, GMOs continue to stir debate.

CHECK YOUR UNDERSTANDING

✔ **If you understood this section, you should be able to ...**

1. Predict what aspect of cloning DNA would be different if an antibiotic gene were not present on a plasmid cloning vector.
2. Contrast how laboratory-based genetic engineering of a plant cell using *Agrobacterium tumefaciens* differs from the natural genetic modification of plant cells by this bacterium.

Answers are available in Appendix A.

20.2 The Polymerase Chain Reaction

After you complete this section, you should be able to ...

- Explain how PCR can create many copies of a DNA sequence.

Revolutions often advance in waves, and so it was for the revolution in molecular biological methods. Cloning genes using recombinant DNA technology transformed what was possible in biology, but it's a slow business. A technique developed in the 1980s called the **polymerase chain reaction (PCR)** provided a quicker way to clone DNAs. Starting with minute quantities of DNA, millions of copies of any DNA sequence could be obtained in hours using purely chemical means, instead of the days or weeks it might take using plasmids. PCR opened unimagined new possibilities in biotechnology. When you hear about DNA fingerprinting, paternity testing, testing for many bacterial or viral pathogens, or DNA-based genealogies, PCR is what makes all these methods possible.

Requirements of PCR

PCR is fast and simple, but there's a catch: To amplify a particular DNA sequence that's of interest, a researcher needs some information about the surrounding DNA to design primers—short lengths of single-stranded DNA that match sequences on either side of the region to be amplified. These primers allow the DNA synthesis reaction at the core of PCR to begin (Ch. 15, Section 15.3). For the wide array of organisms with sequenced genomes, researchers can easily find primer sequences to use in cloning almost any target gene by PCR. Even for organisms without sequenced genomes, PCR is often possible if a genome sequence or target gene sequence from a related organism is known.

In outline, PCR involves repeating three steps over and over in an automated PCR machine. Each repeat is called a cycle and, because a round of DNA replication occurs in each cycle, the number of target DNA sequences doubles. During PCR, each strand of DNA synthesized in a previous cycle serves as a template in the next cycle. In just 20 cycles of PCR, a single starting DNA sequence can be amplified to over a million copies. (In **BioSkills 10**, you can see how PCR works in greater detail.)

PCR depends on separating the complementary strands of DNA at each amplification cycle. This is done by heating the solution of DNA to near boiling, something no conventional DNA polymerase could withstand. When PCR was first developed, researchers had to add fresh DNA polymerase for each cycle. A technological breakthrough came when scientists discovered microbes that thrive in the near boiling waters of Yellowstone National Park's hot springs. It's the DNA polymerase from one of these heat-resistant microbes that made automated PCR possible.

Even when starting from tiny quantities of DNA, sometimes just a single molecule, PCR can produce enough copies of a particular DNA fragment for countless research opportunities and applications. For example:

- Forensic scientists amplify DNA from tiny samples of blood or hair and then analyze the copied DNA to determine paternity, identify victims of crime or natural disasters, implicate perpetrators, or exonerate the falsely accused.
- Genetic counselors use PCR to learn if prospective parents have alleles associated with serious illness.
- Researchers investigate patterns of gene expression in cells or tissues by analyzing PCR-amplified copies of RNAs.
- Physicians use PCR to identify bacterial and viral pathogens, monitor the course of infections such as HIV-1, and diagnose cancers.
- Genealogists use PCR to learn a person's ancestry.

The polymerase chain reaction has become one of the most widely used techniques in biology. Let's look at an example of PCR in action.

PCR In Action: DNA Fingerprinting

DNA fingerprinting (also known as DNA profiling or DNA typing) refers to any technique for identifying individuals based on the unique features of their genomes. Many eukaryotic genomes have thousands of sites that contain short DNA sequences that are repeated one after another along part of a chromosome and that vary in repeat number between individuals. These repetitive sequences, called tandem repeats, are at the core of today's DNA fingerprinting.

The type of tandem repeats used in most current DNA fingerprinting approaches is **short tandem repeats (STRs)** (also known as **microsatellites** or **simple sequence repeats [SSRs]**). They consist of simple repeating units from two to about eight nucleotides long. STRs occur outside of protein-coding regions of genes, and usually between genes. This means that differences in the number of STRs and the overall length of the repeat region have no effect on phenotypes. STRs are thought to originate when DNA polymerase skips or mistakenly adds extra bases during replication.

Soon after STRs were identified, the number of sequence repeats of different STRs was found to vary widely among individuals. This makes them ideal for DNA fingerprinting. Each variant of an STR, known as an STR allele, is transmitted from parent to offspring just like any other DNA sequence. For example, different alleles of a four-nucleotide STR used in the FBI's DNA fingerprinting system have from 6 to 15 tandem repeats of the sequence GATA.

How can STRs be used for DNA fingerprinting? Investigators obtain a DNA sample and then perform PCR (see **BioSkills 10**) using primers that flank a region containing an STR (**Figure 20.3a**). Once an STR region is amplified, gel electrophoresis (see **BioSkills 6**) is used to determine the number of repeats it contains—the fewer the number, the shorter the PCR product, and the farther it will travel in a gel during electrophoresis. **Figure 20.3b** shows the principle of using DNA fingerprinting in paternity tests.

Sets of primers are available that allow the analysis of many different STRs in a single PCR sample. For example, today's FBI DNA fingerprinting system amplifies 20 different STRs found at various sites in the human genome.

(a) Use PCR to amplify an STR.

(b) Compare bands on a gel to test paternity.

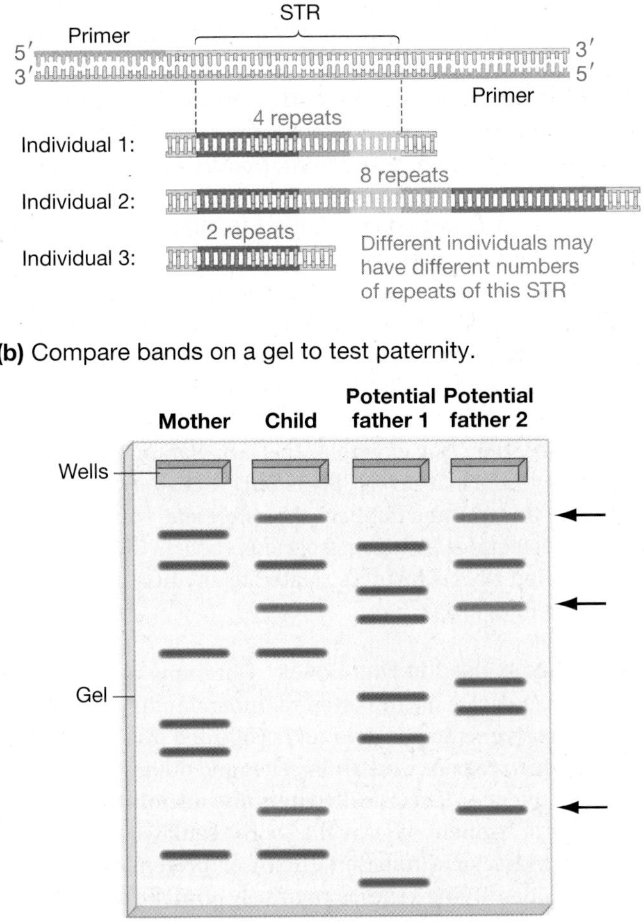

Figure 20.3 DNA Fingerprinting Can Be Used to Identify Parents. (a) Here, only one STR allele is shown for each individual. Individuals are often heterozygous, so the repeat number for a given STR varies within and between individuals. **(b)** A hypothetical gel analysis of three different STRs for four individuals. In this example, each individual is heterozygous and produces two bands for each STR. Note how the mother transmits half of her STR alleles to the child, as does one of the possible fathers (father 2). Arrows indicate the bands that match between the likely father and child. The child shares no STR alleles with possible father 1, eliminating him as a parent. Typically, from 6 to 16 different STRs are examined to establish paternity.

CHECK YOUR UNDERSTANDING

✔ If you understood this section, you should be able to ...

1. Explain why DNA fingerprinting for paternity analysis is better for excluding the possibility of paternity than it is for proving paternity.
2. Explain why the sequences targeted by the PCR primers used in DNA fingerprinting (Figure 20.3) need to lie on either side of the region containing the STRs, not within this region.

Answers are available in Appendix A.

20.3 Analyzing Genomes

After you complete this section, you should be able to ...

- Describe how genomes are studied.

At the same time that researchers were developing recombinant DNA technology, a breakthrough on another front unfolded in the 1970s: a method to sequence DNA. About 40 years later, in the first decade of the new millennium, a technological revolution in DNA sequencing dawned, and this revolution continues today. The amount of DNA sequence that can now be generated by a researcher in a day is greater than the entire DNA sequence of the first human genome that took hundreds of scientists working across the world 5 years to produce.[1]

These technological leaps are remarkable, but what's the value of learning a gene or genome nucleotide sequence?

- When the nucleotide sequence of a gene is known, the genetic code can be used to infer the amino acid sequence of the gene's protein product. The amino acid sequence often provides clues to the protein's function.
- Comparing sequences often reveals why alleles vary in function—for example, why one allele causes disease and another doesn't.
- Evolutionary relationships can be inferred by comparing genome sequences in different species or between individuals of the same species.

The first method for sequencing DNA that came into widespread use is called **dideoxy sequencing**, a method developed by Frederick Sanger in 1977. Dideoxy sequencing could not have developed without recombinant DNA technology, because in its original form many identical copies of a cloned fragment of DNA

[1]The Human Genome Project (Section 20.6) spanned 15 years, but the first 10 years were devoted largely to developing tools and approaches. The bulk of the actual DNA sequencing occurred in the last 5 years of the project.

were required. A modification of the original Sanger dideoxy sequencing approach that employed PCR (see **BioSkills 10**) was used extensively in obtaining the first human genome sequence.

Since the first reports of the human genome sequence in 2001, many innovative ways to sequence DNA have followed, each vastly increasing the sequence output and reducing overall cost. Collectively, these methods are now referred to as **next-generation sequencing (NGS)**. Most of these methods are based on amplification steps that create many copies of a template DNA molecule. This allows sequencing minute quantities of DNA—even the DNA from a single cell may be sufficient. Just as important, in contrast to dideoxy sequencing, next-generation methods are massively parallel, meaning that they can simultaneously sequence millions of DNA fragments in a single run. Next-generation sequencing technologies have made it possible to rapidly determine the sequence of entire genomes, and they are changing the face of biology.

Whole-Genome Sequencing

Genomes range in size from about a half million base pairs to hundreds of billions. But even under the best conditions, a single dideoxy sequencing reaction can analyze only about 1000 nucleotides, and most next-generation methods give much shorter read lengths. How do investigators piece together a whole genome from these short sequences?

When researchers first sequence the genome of a species, they rely on an approach known as **shotgun sequencing** (see **BioSkills 10**). In shotgun sequencing, many copies of a genome are broken up randomly into a set of fragments of various sizes. Since the breaks are random and there are many copies of the genomic DNA, these fragments are expected to overlap in many regions. The DNA fragments are sequenced, and regions of overlap between different fragments are used as a guide to put the whole genome back together again in a process called genome assembly.

In many current applications of shotgun sequencing, genome assembly depends entirely on having a computer algorithm piece together the sequence reads produced by next-generation sequencing. This is called de novo (loosely translated, "from scratch") genome assembly. The first high-quality genome sequence obtained of a given species is called a **reference genome**. It becomes the standard for comparing and analyzing the genomes from other individuals of that species (Section 20.7).

Genome sequencing is accelerating at a dizzying pace. The first cellular genome to be sequenced was from the bacterium *Haemophilus influenzae*. It was chosen because it is a human pathogen and its genome was believed to be small, a necessity with the technology available in the early 1990s. The *H. influenzae* genome sequence was published in 1995, followed quickly by reports of complete genome sequences from an assortment of bacteria and archaea.

Sequencing of the first eukaryotic genome, from the yeast *Saccharomyces cerevisiae*, was a multiyear international effort and was finished in 1996. The human genome sequence, announced in 2001 and published in more complete form in 2003, was a milestone. In the years after the release of the human genome sequence, next-generation sequencing methods changed the landscape of sequencing possibilities.

Today, just 15 years after completion of the human genome sequence, the genome sequences of more than 24,000 species from all domains of life are known, including nearly 900 animal and 300 plant genomes. As of this writing, the primary international repository for DNA sequence data contained almost three *trillion* nucleotides. An almost mind-boggling number of sequences of genes and whole genomes are now being generated. This wealth of sequence data has opened windows to a deeper understanding of biology.

Analyzing Genomes Through Bioinformatics

What tools do researchers use to align the millions of fragments produced by sequencing into a complete genome sequence? Once a complete genome is assembled, how are the raw sequence data and information about genes and their products made available to the international community of researchers? How can researchers compare the genomes of different species to learn about evolutionary relationships?

The broad answer is **bioinformatics**—a field that fuses mathematics, computer science, and biology to manage and analyze sequence data. Researchers in bioinformatics have created vast searchable databases of sequence information that allow investigators anywhere in the world to evaluate the similarities between newly discovered genes and genes that have been studied in any species. The immense quantity of data generated by genome sequencing centers makes bioinformatics an indispensable element of **genomics**, the branch of biology that obtains and analyzes genome sequences to gain insights into life.

Sequence databases are at the fingertips of anyone with an Internet connection. For example, the U.S. National Center for Biotechnology Information (NCBI) is only a click away on your computer. At this free and publicly accessible site, you can search billions of nucleotides by using programs such as BLAST, which can quickly find DNA sequences related to any new gene uncovered in a genomics project.

Bioinformatics Is Used to Find Genes Obtaining raw sequence data is just the necessary first step to understanding a genome. Once this massive sequence data set is collected, the first task is to identify which regions constitute genes and other functionally important sequences. This is called **genome annotation**. Recall that a gene is a segment of DNA that both regulates the production of and codes for a functional RNA or protein product. In prokaryotes, identifying genes is relatively straightforward. The task is much more difficult in eukaryotes.

To annotate genomes, biologists turn to bioinformatics and use an array of computer programs. When looking for protein-coding sequences, they use a program that predicts the RNA sequence corresponding to each strand of DNA, then searches all possible reading frames in each of the predicted RNA strands. Recall that a reading frame is a sequence of codons (Ch. 16, Section 16.3).

Codons consist of three nucleotides, so there are three potential reading frames on an RNA strand, and a total of six possible reading frames on the two hypothetical RNAs derived from double-stranded DNA (**Figure 20.4**). Because random sequences are predicted to contain a stop codon about 1 in every 20 codons on average (3 stop codons among 64 total codons), the presence of

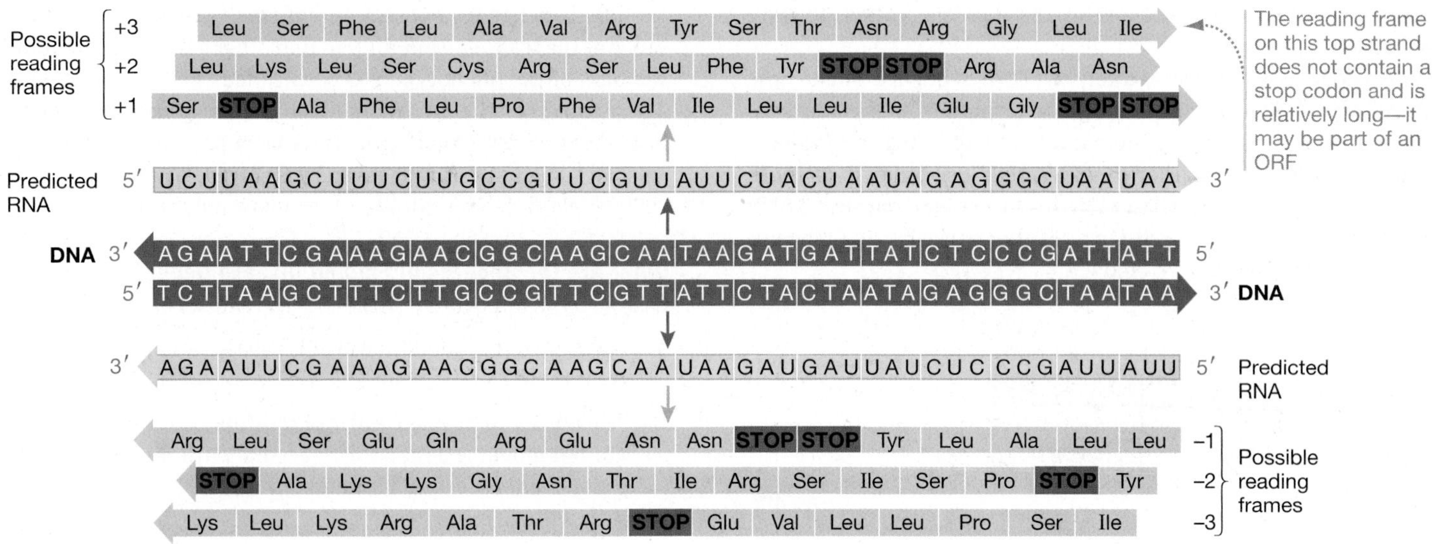

Figure 20.4 Open Reading Frames Can Identify Genes. Computer programs use the genetic code to translate the three possible reading frames encoded on each strand of DNA. This figure shows the RNAs that would correspond to each DNA strand and the amino acids that would be encoded if each of the RNAs were translated.

✔ Why would an ORF identified in a eukaryotic gene usually lack a stop or a start codon?

a long stretch of codons without a stop codon—an **open reading frame**, or **ORF**—is a good indication of a protein-coding sequence.

For bacteria, gene-finding programs draw attention to any "gene-sized" open reading frames that are flanked by a stop codon and a start codon. Because polypeptides range in size from a few dozen amino acids to many hundreds of amino acids, gene-sized stretches of sequence range from a hundred or so bases to thousands of bases. In addition, the computer programs look for consensus sequences typical of promoters, ribosome binding sites, or other regulatory sites in the proper position relative to the potential coding sequence.

Finding ORFs in eukaryotes is not so simple, because coding sequences are divided into exons and introns. Algorithms are designed to identify potential coding exons (ORFs) that usually lack start and stop codons, and are separated by splice site consensus sequences from interspersed introns that lack long ORFs. Sequences with these characteristics are likely to be protein-coding genes in a eukaryote.

Once an ORF is found, another computer program compares its sequence with the sequences of known genes from other species. If the ORF is unlike any gene that has so far been described in any species, further research is required before it can be considered a gene. In contrast, if an ORF is similar in sequence to a known gene from another species, then it is very likely to be part of a gene.

Similarities in sequence and in function between genes in different species are usually due to **homology**. If genes are homologous, it means they are similar (show homology) because they are related by descent from a common ancestor. For example, consider the genes that code for enzymes involved in repairing mismatches in DNA (Ch. 15, Section 15.5). The mismatch-repair genes in *E. coli*, yeast, and humans are similar in DNA sequence and function. To explain this similarity, biologists hypothesize that the common ancestor of all cells living today had mismatch-repair genes and that the descendants of this ancestral species have retained this gene with modifications specific to each lineage.

Given the challenges of identifying eukaryotic genes solely through bioinformatics approaches, additional, biologically based methods are often used. One common method makes use of **expressed sequence tags**, or **ESTs**. The rationale behind the use of ESTs is based on the fact that genes are transcribed into RNAs. If a cDNA copy can be made of even a part of the RNA, particularly for mRNAs, and the sequence of that cDNA is determined, then sequence-matching algorithms can be used to search for regions of the genome that correspond to the cDNA. These regions are likely to be part of genes. In this method, the cDNA is the expressed sequence tag, because it tags or marks parts of the eukaryotic genome that have been expressed as an RNA.

Genome-Wide Association Studies—An Alternative Approach to Identifying Genes

Another method of leveraging the power of genomics to locate genes begins as traditional genetic approaches do, starting with a set of mutants with interesting phenotypes. But in contrast to classical approaches where geneticists might generate mutants in an experimental organism, this method, called a **genome-wide association study** (or **GWAS**), begins with large populations that show variation in a phenotype.

In studies of humans, these phenotypic variations are often disease versus non-disease phenotypes, for example, having type 2 (adult-onset) diabetes or not. GWAS uses an idea you learned about earlier (Ch. 14, Section 14.5): Genes that are close together on the same chromosome will be inherited together, with the frequency of co-inheritance depending on the distance between the genes. This is a simple principle. A GWAS is more complicated. The idea is to look for associations (co-inheritance) between a phenotype of interest, such as adult-onset diabetes, and DNA sequences at known locations in the genome, called **genetic markers**.

To work as a genetic marker, the DNA sequence must come in at least two common sequence variants. Genetic markers are established through genomic analysis of many individuals. The type of genetic marker in widespread use today is a **single nucleotide polymorphism (SNP)**. As the name implies, a SNP (pronounced "snip") is a site in DNA that varies at a single base pair. An example is shown below.

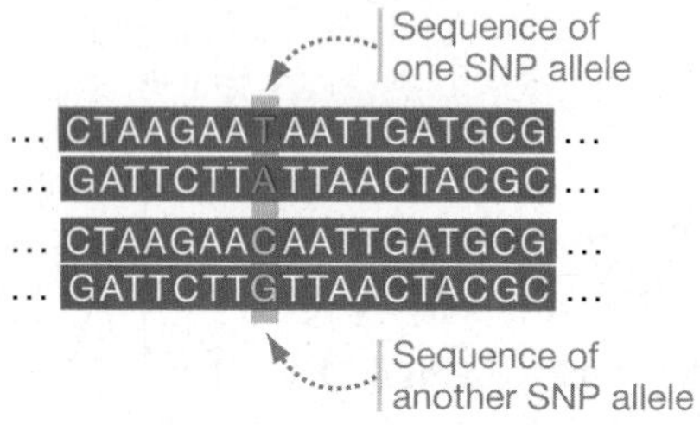

Researchers have compiled huge catalogs of SNPs in the genomes of many organisms, including roughly 10 million SNPs in the human genome. Most SNPs fall outside of coding or regulatory regions and have no direct effect on phenotype. Importantly, there are rapid, automated methods for establishing SNP genotypes in many individuals. The scale of analysis is critical to a GWAS investigation. SNP genotypes comprising thousands to hundreds of thousands of SNP loci are established in hundreds to thousands of people with and without the trait of interest.

Once this baseline is established, investigators use sophisticated statistical methods to look for the co-inheritance (association) of a particular SNP allele and a particular trait. If the trait and the SNP allele almost always occur together, then the gene for the trait must lie near the known location of the SNP.

What Have GWAS Investigations Revealed? Thousands of genome-wide association studies have been performed in many species. Studies have looked at agriculturally and evolutionarily important traits, as well as human health traits. In humans, GWAS investigations have searched for genes associated with macular degeneration, type 1 and type 2 diabetes, breast and ovarian cancer, obesity, coronary heart disease, bipolar disorder, Crohn's disease, rheumatoid arthritis, and many others. What are the conclusions of these studies?

One take-home message is that many of the genes uncovered by GWAS are complete surprises—biologists wouldn't have inferred them based on what's known about a particular trait. For example, in a study of genes associated with type 2 diabetes, two genes with the strongest influence on disease development are involved in protein degradation and in controlling cell shape. The upshot is that a lot of work needs to be done to understand how the genes revealed by GWAS influence a specific trait.

Another important finding of GWAS investigations is that many genes contribute to the development of most traits. This discovery is consistent with what you've learned about polygenic traits (Ch. 14, Section 14.5). But there's a persistent mystery.

Missing Heritability Thousands of genes for many different traits have been discovered by GWAS, but in virtually all cases, the identified genes explain only a small portion of the inheritance of the trait. The inability to find other genes that account for the bulk of the phenotype has been called "missing heritability." There must be other genes that lead to phenotypes like diabetes, but what are they? Many ideas have been proposed to account for this paradox, but until the mystery is solved, GWAS can only partially explain the inheritance of complex, polygenic traits.

What Are the Benefits of Finding Disease Genes? Finding disease genes is important. Three of the greatest benefits are:

1. An improved understanding of diseases and possible ways to treat them.
2. The ability to produce **animal models** of human disease. An animal model is a laboratory animal with disease symptoms that parallel those of a human disease.
3. The development of genetic tests to screen for individuals who may be at risk of disease development.

On the last point, genetic testing often creates serious ethical and legal dilemmas. The U.S. Congress has enacted legislation prohibiting the discriminatory use of genetic information by employers or insurers. But what about harrowing personal choices? Should people be tested for diseases that have no cure? Would you choose to be tested for a defective allele and risk finding out that you were almost certain to develop an incurable disease?

Stepping back, what has genomics provided to date? It's served up complete surprises, such as the case of missing heritability, but has also offered insights, such as the identity of genes at the root of many diseases. Taking an even longer view, what's been learned about the biology of genomes?

CHECK YOUR UNDERSTANDING

✔ **If you understood this section, you should be able to ...**

1. Explain how de novo genome assembly is similar to putting Humpty Dumpty back together again.
2. **SOCIETY** Propose reasons for and against being tested for an allele that causes a serious genetic disease with no current cure.

Answers are available in Appendix A.

20.4 Insights into Genomes

After you complete this section, you should be able to ...

- Compare and contrast bacterial and eukaryotic genomes.

In a sense, biologists who first worked in genomics were like naturalists of the eighteenth and nineteenth centuries. These naturalists explored the globe, observing the plants and animals they encountered. Their goals were to describe what existed and identify patterns in nature. Similarly, the goal of early genomics investigations was to catalog what is in a genome—specifically, the number, type, and organization of genes and other DNA sequences—and then look for patterns within and between different genomes. What observations have biologists made about the nature of genomes?

Prokaryotic Genomes

As of this writing, biologists have obtained the genome sequences of nearly 140,000 distinct prokaryotic species and strains. For example, researchers have sequenced the genome of a harmless laboratory population of *E. coli* bacteria as well as the genome of forms that cause disease. This allows researchers to identify genes that differ between these strains and begin experiments to learn what accounts for infectious properties.

Here are some principles that have emerged from the study of prokaryotic genomes:

- Prokaryotic genomes are compact. Introns are rare, the vast majority of their coding sequences are uninterrupted, they have little space between genes, make extensive use of operons, and compared to eukaryotes, have few regulatory sequences. Look at the graph of bacterial and archaeal genomes in Figure 20.5a, and notice how genome size and the number of genes increase together in a nearly straight line. In contrast, this simple relationship does not hold for eukaryotic genomes (Figure 20.5b).
- Bacterial species that live in a variety of habitats and use a wide array of molecules for food have large genomes; parasitic species that make use of a host's biochemical machinery rather than synthesizing their own molecules have small genomes.
- Most of the genes found in one species are not shared widely with others. Only a relatively small set of genes involved in fundamental processes such as DNA replication, transcription, and translation are similar across a wide range of prokaryotes.
- Genome size and content vary extensively even within species.
- Prokaryotic genomes are frequently rearranged during evolution. Even closely related species show little similarity in gene order.
- A significant proportion of many prokaryotic genomes has been acquired from other, often distantly related species by **lateral gene transfer**. This observation has caused scientists to rethink what it means to be a species.

Lateral Gene Transfer There are many ways to define a species (Ch. 24, Section 24.1), but in all traditional definitions, members of one species cannot exchange genes with members of another species. Lateral gene transfer counters this view: Instead of moving vertically from generation to generation within a species, genes move "laterally" between different species. Lateral gene transfer is a major force in the evolution of prokaryotes, and its extent has been revealed through comparative genome analysis.

How are laterally transferred genes identified? Biologists primarily use two criteria:

1. The gene of interest is more similar to genes in distantly related species than to those in closely related species. In the bacterium *Thermotoga maritima* that thrives near deep-sea vents, almost 25 percent of its genes are closely related to genes found in archaea that live in the same habitats. The archaea-like genes occur in well-defined clusters within the *T. maritima* genome, supporting the hypothesis that these sequences were transferred in large pieces between these two domains of life.
2. The proportion of G-C base pairs to A-T base pairs in a particular gene or series of genes is markedly different from the base composition of the rest of the genome. This second criterion works because the proportion of G-C base pairs in a genome is characteristic of the particular genus or species.

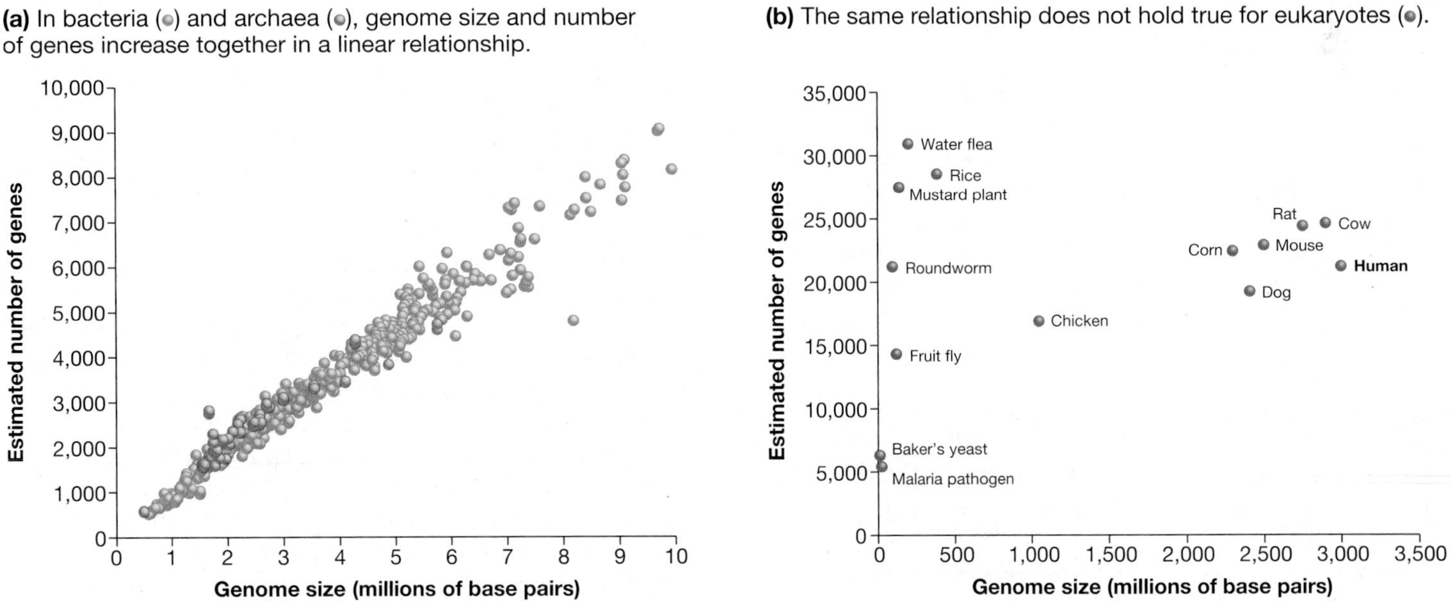

Figure 20.5 Eukaryotes and Prokaryotes Differ in the Relationship between Genome Size and Gene Number.

DATA: (a) Y. Hou and S. Lin. 2009. *PLoS ONE* 4(9): e6978, Supplemental Table S1. (b) KEGG: Kyoto Encyclopedia of Genes and Genomes, KEGG Organisms: Complete Genomes. www.genome.jp/kegg/.

How can genes move from one species to another?

In some cases, plasmids are responsible. For example, researchers have documented the transfer of plasmid-borne antibiotic-resistance genes between distantly related bacteria. Genes from plasmids then can become integrated into the chromosome of a bacterium through genetic recombination. Lateral gene transfer may also occur by transformation—when bacteria and archaea take up raw pieces of DNA from the environment—and by infection with viruses that pick up DNA from one cell and transfer it to another cell.

Does lateral gene transfer occur in eukaryotes? Although it's less common than in prokaryotes, more and more instances are being discovered in eukaryote evolution. One key example is the capture of bacterial cells that were predecessors of today's mitochondria and chloroplasts and the subsequent transfer of many bacterial genes to the host genome (Ch. 27, Section 27.3). Another more recently discovered case is the pigment-producing genes found in aphids—common insects that feed on the sap of plants—that allow aphids to be colored and help avoid predation. These pigment genes were transferred from fungal species into an ancestral species of modern aphids.

Lateral gene transfer shakes the tree of life. If lateral gene transfer were rare, then evolutionary paths in the bacteria and archaea would form a set of branches that began at common ancestors and split out to descendants. If lateral gene transfer is as widespread as most biologists now believe, then evolutionary paths form an interconnected network that links species in a web of vertical and lateral gene transfers, not a branching tree. These alternative views of life's evolution, revealed largely by genomics, are shown in Figure 20.6.

Metagenomics Sequencing the genomes of individual species and strains continues as the central approach to genomics, but recently, many researchers have taken a different tack: cataloging all the genes present in a complex community of organisms. This type of research, typically done with microbial communities, is called **metagenomics** or **environmental sequencing** (Ch. 26, Section 26.2).

Metagenomics is the only way to get a good handle on the types of microbes present in an environment. This is because many bacteria, archaea, and single-celled eukaryotes cannot be grown outside their native environments. Traditional methods of species identification, which rely on growing cells in the lab, can detect only a tiny fraction of these organisms. With advances in the ability to analyze DNA sequences, it's often possible to learn not just which genes are present, but what constitutes a species and how abundant it is in complex microbial communities. Metagenomics investigations have uncovered many new microbial species in almost any environment that's studied, and they have revealed an incredible, previously unimagined diversity of the living world.

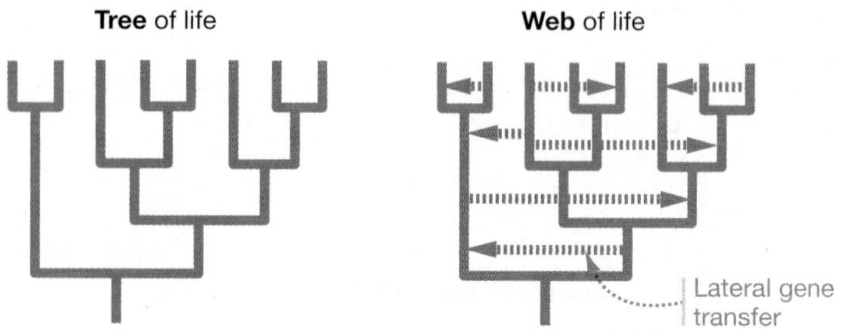

Figure 20.6 Tree of Life and Web of Life Present Different Views of Evolution.

Eukaryotic Genomes

DNA sequence analysis has revealed extraordinary features of eukaryotic genomes. One of these characteristics is a wide variation in genome size with relatively little variation in the number of genes. For example, the eukaryotic genomes shown in Figure 20.6b span a roughly 250-fold range of size but there is less than a sixfold difference in the number of genes. How can gene number be so similar among species with vastly different genome sizes and in organisms that range from single-celled parasites to large plants and animals?

Repetitive DNA and Transposable Elements Repeated DNA sequences partially explain the paradox of the immense variation in eukaryotic genome size. Many eukaryotic genomes contain vast amounts of repeated DNA sequences. For example, about 50 percent of the human genome and other mammalian genomes is composed of repetitive DNA, and more than 90 percent of the corn genome is made of repetitive DNA. Most of this repetitive DNA doesn't code for products essential for the cell, and it was initially considered unimportant and uninteresting "junk DNA."

Subsequent work showed that many of these repeated sequences are derived from **transposable elements**, DNA segments that can move from one place to another within the genome. Transposable elements were discovered in the late 1940s in corn by Barbara McClintock and later shown to be present in organisms from every domain of life. Transposable elements come in a wide variety of types and spread through genomes many different ways. They are anything but unimportant or uninteresting.

Some transposable elements behave similarly to retroviruses that insert into the genome and are clearly related to this group of viruses. In contrast to retroviruses, however, these transposable elements do not leave their host cell. Instead, they often make copies of themselves that become inserted in new locations, increasing the number of transposable elements. Because they are part of the genome, transposable elements are passed from mother to daughter cell and from parents to offspring.

As an example, consider a well-studied type of transposable element called a **long interspersed nuclear element (LINE)**, found in humans and other mammals. Like retroviruses (Ch. 33, Section 33.2), LINEs code for a reverse transcriptase and have a similar way of inserting into and excising from the genome. Therefore, biologists hypothesize that LINEs evolved from retroviruses. Your genome contains nearly 1 million LINEs, each from about 1000 to 5000 base pairs long, which together make up about 20 percent of your DNA. As LINES, or any other transposable element, move from place to place, insertion into a new site creates a mutation that can have negative, neutral, or positive effects on fitness (Ch. 16, Section 16.4). They can disrupt the coding sequence of genes, change patterns of gene regulation, or promote gene duplication and loss. In this way, these mobile DNAs shape the structure, function, and evolution of genomes.

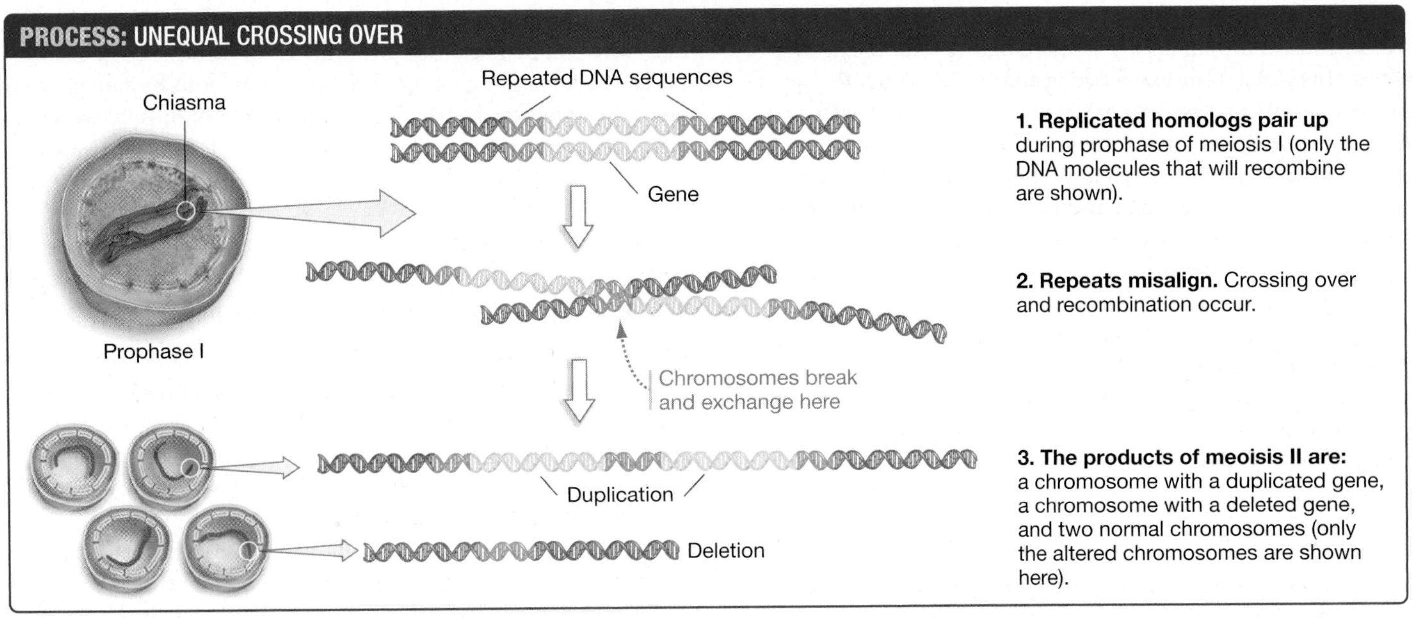

Figure 20.7 Gene Duplication Can Occur through Unequal Crossing Over at Repeated DNA Sequences. If a gene lies between the repeats, a crossover between misaligned repeats will result in two normal chromosomes, a chromosome with a deleted gene, and a chromosome with a duplicated gene once meiosis is complete.

Many transposable elements are no longer able to move under their own power, because they have acquired mutations in a critical gene or regulatory element. This is the case for the human genome. Genomics has shown that these inactive transposable elements, which are sometimes referred to as molecular fossils, litter the genome.

Gene Families How do new genes arise? In eukaryotes, one important source of new genes is the duplication of existing genes. Biologists infer that **gene duplication** has occurred when they find groups of genes that are similar in sequence and structure, such as the arrangement of exons, introns, and regulatory sequences. Within a species, genes that are similar to each other in structure and function are considered to be part of a **gene family** and to have arisen from a shared ancestral sequence through gene duplication.

A common way that genes are duplicated is through a process called **unequal crossing over** during meiosis. Unequal crossing over often involves repeated DNA sequences such as transposable elements. As Figure 20.7 shows, if the same type of transposable element or any other repeated DNA sequence surrounds a gene, in rare instances a crossover in meiosis I may occur between misaligned copies of the sequence repeat. When this happens, it results in one of the chromatids having a deletion of the DNA segment that was between the sequence repeats, and the homologous chromatid having a duplication of the segment. Like STRs, the duplicated segments are arranged in tandem—one after the other.

Gene duplication is important because the original gene is still functional and produces a normal product. As a result, the new, duplicated stretches of sequence are redundant. If mutations occur in the duplicated sequence, then a new gene may be added to the genome. These mutations can change the function of the gene product, changing the phenotype of the cell or organism. Together, these processes that create new genes are **duplication and divergence**.

Thousands of gene families exist in every known mammalian genome. One example of a gene family is the subset of human globin genes diagrammed in Figure 20.8. Collectively, this group of genes is known as the β-globin gene family, and they code for proteins that form part of hemoglobin, the oxygen-carrying molecule in your red blood cells. Each coding gene in the family serves a slightly different function and is expressed in different tissues and times of

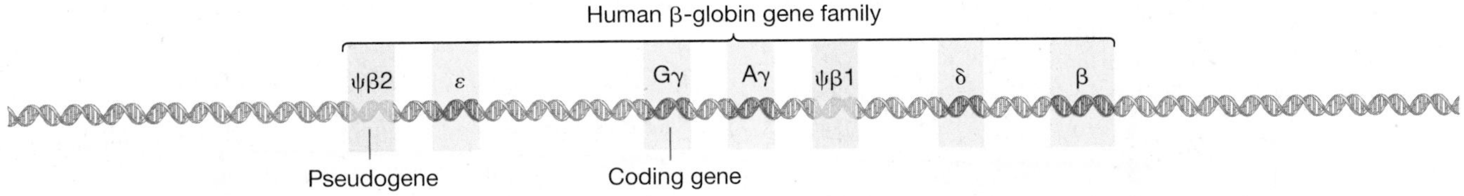

Figure 20.8 Gene Families Are Closely Related Genes Derived by Gene Duplication and Divergence. The β-globin gene family is shown with coding genes in red and pseudogenes in yellow. Pseudogenes are always indicated by preceding a gene name with the Greek letter psi (ψ).

✔ Suppose that during prophase of meiosis I, the β locus of the β-globin gene family on one chromosome aligned with the ψβ2 locus on another chromosome, and crossing over occurred in the noncoding sequences just to the left (as oriented in the figure) of this β–ψβ2 pairing. List the order of the β-globin-family genes that would result on each chromosome.

development. For example, the protein product of the globin gene expressed in the fetus binds oxygen more tightly than the protein made from the globin gene expressed in adults. Consequently, oxygen moves readily from the mother's blood, where it is not as tightly bound to hemoglobin, to the fetus's blood (Ch. 42, Section 42.4).

Although mutations in duplicated regions can create genes with new functions, more often, mutations lead to nonfunctional genes, called **pseudogenes**. For example, a mutation could produce a stop codon in the middle of an exon. Note in Figure 20.8 that the β-globin gene family contains two pseudogenes along with several functional genes. In the human genome there are roughly as many pseudogenes as functional genes. Genomics has revealed that just as nonfunctional transposable elements litter the genome as molecular fossils, so too do pseudogenes that are the outcomes of gene duplication and divergence.

Insights from the Human Genome Project

The multinational **Human Genome Project** required more than 15 years and about $3 billion to assemble the first human genome sequence. When it was completed in 2001, then President Clinton announced the human genome sequence as "the most wondrous map ever produced by humankind." Today, sequencing a human genome is becoming a routine procedure with costs as low as $1000 per genome. Very possibly, you will someday have your own genome sequenced.

What has analysis of human genomes revealed? Given biologists' long-standing focus on protein-coding genes, the composition of the human genome was astounding. As Figure 20.9 reveals, less than 2 percent of the genome consists of protein-coding exons, and nearly half is made of transposable elements. Introns make up over one-quarter of the genome and are 17 times more abundant than protein-coding exons.

Another striking discovery came from comparisons of the human genome to genomes of other species. Humans, whose morphology, biochemistry, and behavior are complex, do not have particularly large numbers of genes. Look back at Figure 20.5b and notice that the total number of human genes is only 50 percent more than the number in fruit flies, is about the same as in roundworms, and is substantially lower than the number of genes in water fleas, rice, and the tiny plant *Arabidopsis thaliana*.

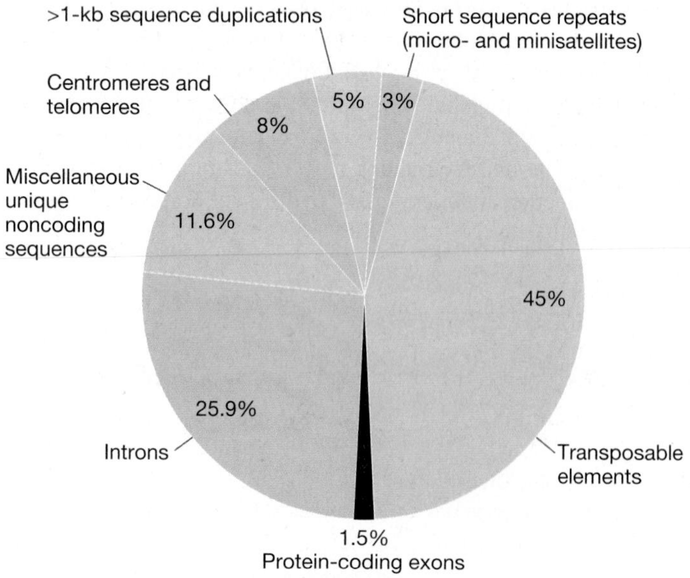

Figure 20.9 Composition of the Human Genome.
DATA: T. R. Gregory. 2005. *Nature Reviews Genetics* 6: 699–708.

Before the human genome was sequenced, many biologists expected that humans would have at least 100,000 genes. But we may have only a fifth of that number. How can this be?

Alternative Splicing One hypothesis to explain the discrepancy between genome size and organismal complexity is based on **alternative splicing**. Recall that the exons of a particular gene can be spliced in ways that produce distinct mature mRNAs (Ch. 19, Section 19.4). Consequently, a single eukaryotic gene can code for multiple transcripts and thus multiple proteins.

Researchers estimate that more than 95 percent of human genes produce transcripts that are alternatively spliced, with an average of more than three distinct mRNAs per protein-coding gene. This means that the number of different proteins that can be produced is more than triple the gene number. Alternative splicing makes it possible for a relatively small number of genes to produce a much larger set of proteins.

Noncoding RNAs Roughly 90 percent of the human genome is transcribed—far more than imagined. Some of these transcripts code for regulatory RNAs with known roles, such as microRNAs (Ch. 19, Section 19.4), but what many others do is currently mysterious. A new class of RNA, **long noncoding RNA (lncRNA)**, has caught the attention of biologists. Thousands of genes encoding these transcribed, but non-protein-coding, RNAs have been identified. Some lncRNAs play known and important roles in regulating gene expression. An open question is whether most lncRNAs are regulatory, or whether many of these lncRNAs are the result of background transcriptional "noise."

If many of the newly discovered RNAs have a function, then the human genome and the genomes of other complex organisms may not be so gene poor after all. If so, this casts the genome in a new light: Instead of coding for proteins, perhaps most genes produce regulatory RNAs that are never translated. As you read this, biologists are working hard on this central question that has been opened by functional studies of the human genome.

Once again, some of the most interesting outcomes of genome analysis are what it's revealed about what we don't know. The years ahead will be exciting as biologists tackle fundamental questions about the function and regulation of genomes.

CHECK YOUR UNDERSTANDING

✔ **If you understood this section, you should be able to ...**

1. Explain why it is logical to observe that parasitic bacteria have small genomes.
2. Predict, assuming the alternative splicing hypothesis is correct, whether there will be a difference in the extent of alternative splicing between humans and roundworms, organisms that have a similar number of genes.

Answers are available in Appendix A.

20.5 Genome Editing

You've seen how essential tools in biotechnology often come from the least expected places. Serendipity coupled with insight and perseverance put another invaluable tool into the hands of biologists: precise genome editing using the **CRISPR-Cas** system. This section is about the biology, techniques, and applications of a tool discovered in prokaryotic cells and modified by researchers to become an indispensable part of biology's molecular toolkit.

After you complete this section, you should be able to ...

- Describe how the CRISPR-Cas system is used to modify genomes.

The Biology of the CRISPR-Cas System

The story of CRISPR-Cas-based genome editing begins in 1993. While studying a salt-tolerant archaeon, investigators uncovered a set of short repeating DNA sequences separated by curious spacer sequences that had no similarity to each other or to any other sequences in the genome. As more genome sequences became available, it became clear that this repeat-separated-by-spacer sequence was widespread in both bacteria and archaea. What was it doing?

In 2005, a set of research teams converged on an answer: The spacer sequences between the DNA sequence repeats weren't random bits of DNA—instead, each of these spacers came from the DNA of different viruses that could infect the cell. Follow-up work showed that the viral spacer DNAs provided immunity against future infections by the corresponding virus. But how?

Answers came quickly. The DNA sequences that protected against viral infections came to be known as the *C*lustered *R*egularly *I*nterspersed *S*hort *P*alindromic *R*epeat locus, or, the **CRISPR** locus. The CRISPR locus, which is shown in Figure 20.10a, is transcribed into a long pre-crRNA ("crRNA" is shorthand for CRISPR RNA). This pre-crRNA is then processed into a set of shorter RNA fragments, with each fragment, called crRNA, containing a spacer region that is an RNA copy of part of a viral genome flanked by short stretches of bacterially encoded repeat sequences.

A crRNA binds to the complementary DNA sequence of an invading virus, but it doesn't act alone. Figure 20.10b shows how crRNAs are base-paired at one end with another short, noncoding RNA called the tracrRNA (short for *tra*ns-activating *CR*ISPR RNA). The crRNA and tracrRNA in turn are bound by a protein called a *C*RISPR-*as*sociated protein, or **Cas protein**. Cas proteins are enzymes (more precisely, endonucleases) that serve to cut DNA that is complementary to a crRNA.

The details of which Cas protein is involved and whether one protein or a complex of proteins binds the cr- and tracrRNAs vary between four distinct CRISPR systems. The Cas protein shown in Figure 20.10 is Cas9, and it is part of the simplest and most widely used of the CRISPR-Cas systems for genome editing.

Once biologists understood that a DNA-cleaving enzyme—a Cas protein—could be guided to specific genome sequences by a complementary crRNA, they saw the possibilities for editing the genome by cutting it in specific places.

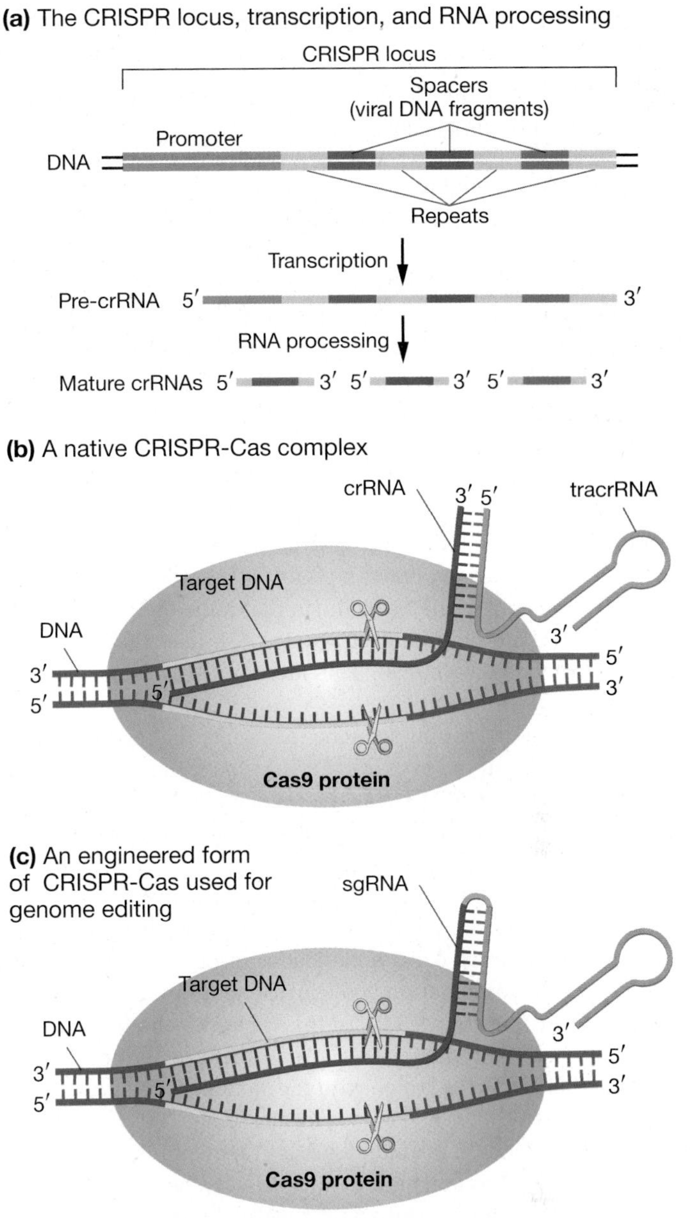

Figure 20.10 The CRISPR-Cas Genome Editing System
(a) Repeats (bacterial DNA) range from about 20 to 50 base pairs, and spacers (viral DNA fragments) range from about 30 to 70 base pairs. The mature crRNAs are processed from a pre-crRNA transcribed from the entire CRISPR locus. Each mature crRNA contains an RNA copy of a spacer flanked by parts of the repeat sequence. **(b)** The spacer region of a processed crRNA base-pairs with a complementary strand of viral DNA (the crRNA's 5′ repeat sequence was removed in a prior step) and a portion of the crRNA's 3′ repeat sequence base-pairs with a small RNA called tracrRNA. Both RNAs are held within a Cas protein (Cas9 here) that cuts the DNA in the region bound by the crRNA. **(c)** In the engineered form, the crRNA and tracrRNA are joined into a single guide RNA (sgRNA).

Using the CRISPR-Cas System for Genome Editing

The next step in making the leap from basic biology to biotechnology was to simplify the CRISPR-Cas system for practical use. This was done by designing a chemically synthesized DNA that could be transcribed in vitro to create a fusion of cr- and tracrRNAs, producing what is called a *single guide RNA*, or sgRNA. This tool makes many approaches to genome editing possible.

A common method is to mix the sgRNA with a plasmid DNA that contains the gene for Cas9 and then introduce this mixture of nucleic acids into a cell. Once the Cas9 gene is expressed, the Cas9 protein associates with the sgRNA, and the sgRNA guides Cas9 to a complementary target sequence in the genome. Here, the Cas9 endonuclease makes double-stranded cuts in the DNA. This process is shown in Figure 20.10c.

How does cutting DNA lead to genome editing? There are two ways, and both depend on DNA repair mechanisms (Ch. 15, Section 15.5) activated in the cell by double-strand breaks in DNA. As Figure 20.11a shows, a double-strand break can be repaired by an error-prone mechanism that joins the broken DNA back together, although almost always with insertions or deletions (abbreviated as indels). This mechanism is called nonhomologous end joining (NHEJ).

If DNA is cut in a coding sequence of a gene, the reading frame will likely be disrupted after repair. If instead, the cut occurs in a regulatory sequence of DNA, the regulatory sequence will be altered. When the goal is to disrupt gene function, this is how CRISPR-Cas genome editing is used.

What about modifying or replacing genes rather than destroying them? As shown in Figure 20.11b, cells also can repair a double-strand break using homologous recombination, similar to the type of recombination you learned about in the context of meiosis (Ch. 13, Section 13.2). In this repair mechanism, called homology-directed repair (HDR), if an intact segment of DNA is available that spans the region with the double-strand break, this segment of DNA can be exchanged for the broken DNA. The trick is for the researcher to introduce a DNA fragment into cells along with the CRISPR-Cas machinery. This DNA can be a modification of the original sequence, say a wild-type allele when the original allele is a defective mutant allele, or a sequence that would give a protein-coding gene a new property, such as production of an enzyme with enhanced activity.

CRISPR-Cas-based genome editing is the next revolution in biology. Like recombinant DNA technology, PCR, and DNA sequencing that came before it, CRISPR-Cas-based genome editing will change the faces of science and society. A host of ethical issues come with genome editing. You have the privilege of watching this revolution unfold as you study biology, and perhaps even shaping it in your career.

What's Been Achieved With CRISPR-Cas Genome Editing?

There are too many answers to this question for anything but an overview. Let's look briefly at two areas actively being explored—agriculture and ecosystems.

CRISPR-Cas Genome Editing In Agriculture Animal and crop scientists have embraced CRISPR-Cas genome editing because of its ease, precision, ability to introduce multiple genome changes at once (multiplex editing), and the "invisibility" of the edits. On this final point, the genome-modifying methods that produce conventional genetic modifications in crops or animals leave bits and pieces of foreign DNA in the genome. With CRISPR-Cas editing, unless the investigator chooses to introduce a foreign DNA sequence, nothing foreign is inserted into the genome. A rice genome remains solely a rice genome, and a pig genome remains purely a pig genome.

Three genome-edited agricultural products that will soon hit the market are a mushroom that resists browning (Figure 20.12); an extra-oil-producing crop plant, *Camelina sativa*, commonly known as false flax; and soybeans altered to improve drought and salt tolerance. In all cases, the crop genomes were edited by disrupting genes using the method shown in Figure 20.11a. Only one gene was modified in mushrooms, but three were altered in false flax and two in soybeans. It's certain that future CRISPR-Cas-edited crops will have many more genes modified within a single variety.

Genomes of domestic animals have also been altered using CRISPR-Cas, but bringing these animals into production has been a slower process than with plants. One notable development is the generation of pigs that are resistant to the devastating porcine

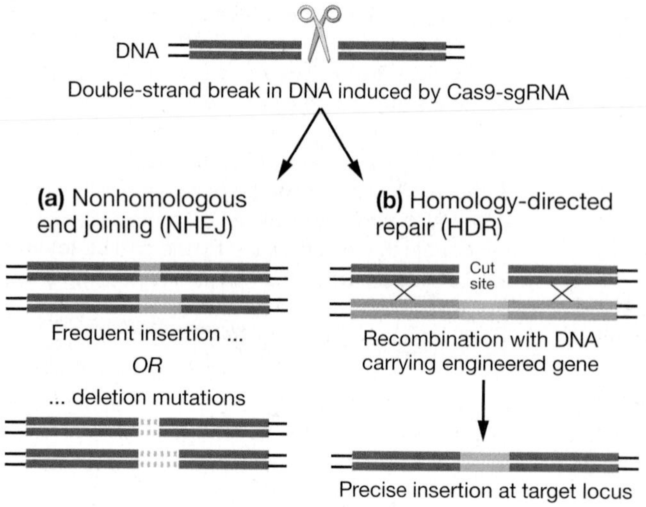

Figure 20.11 **Genome Editing Depends on Two Ways of Repairing Double-Strand Breaks in DNA.**
(a) Nonhomologous end joining creates insertion mutations (blue) or deletion mutations (dots) that typically inactivate the target gene.
(b) Homology-directed repair between a target locus (break in red strands) and a DNA fragment (green) added to the cell can precisely modify the target locus without inactivating it or can replace it.

Figure 20.12 **Using CRISPR-CAS9 Genome Editing, the White Button Mushroom (*Agaricus bisporus*) Has Been Modified to Resist Browning.**

respiratory and reproductive syndrome virus (PRRSV). PRRSV attaches to cells via a specific cell-surface protein, CD163, that regulates the immune response. Using CRISPR-Cas genome editing, scientists removed a small portion of the *CD163* gene that encodes the part of the protein the virus recognizes. Importantly, this edit still allowed CD163 to regulate the immune system. The result? Pigs homozygous for this deletion are significantly protected from PRRSV infection.

How are CRISPR-Cas-edited organisms regulated? The short answer is that they're not, at least not currently in the United States. As of this writing the U.S. Department of Agriculture (USDA), the regulatory agency for crops and livestock, has determined that genomes that do not contain foreign DNA are not substantially different from genomes produced by standard breeding. Therefore, the USDA has ruled that in contrast to conventional GMOs, CRISR-Cas-produced organisms do not need to be regulated. This ruling may change, and regulations of other nations are likely to be different once they're in place, but for now, U.S. companies have the green light to move quickly to develop and introduce genome-edited crops and domestic animals to the market. An open question is how consumers will react to these new varieties.

CRISPR-Cas Genome Editing In Ecosystems—Technology That Alters Populations Thoughts of CRISPR-Cas genome editing typically turn to agriculture or medicine. But there's another side of CRISPR-Cas, one that's fraught with possibility and peril—using it in a genetic engineering process called a "gene drive." This process uses CRISPR-Cas to spread ("drive") a particular form of a gene throughout a population in just a few generations.

The story of gene drives could be confused with science fiction. What if an organism genetically engineered in the lab was released into the wild with the intent of spreading its engineered gene through the entire population? What if this gene caused sterility? Could it wipe out entire populations or species? This is the stuff of gene drives, a development that's been tested in the lab and in carefully controlled field trials, but not yet in the wild.

A CRISPR-Cas-based gene drive converts organisms that begin development as heterozygotes for an engineered allele into homozygotes for this altered allele. How does this work? Figure 20.13 shows how.

Modifying a gene in one individual by the method shown in Figure 20.13 is only the first step of a gene drive. The next step is to spread the altered gene across a population. How is this possible? If the engineered gene is modfied to contain sequences coding for Cas9 and an sgRNA—that is, if it is a gene drive locus (Figure 20.13)—then it can spread rapidly though a population, as shown for the mosquitos in Figure 20.14. A mating between a mosquito homozygous for the engineered allele and a wild-type mosquito will produce offspring that are initially heterozygous for the gene drive allele but quickly become homozygous for this altered allele. Soon, the entire population contains only the modified form of the gene. In most cases, the modified gene is engineered to lack the gene's normal activity.

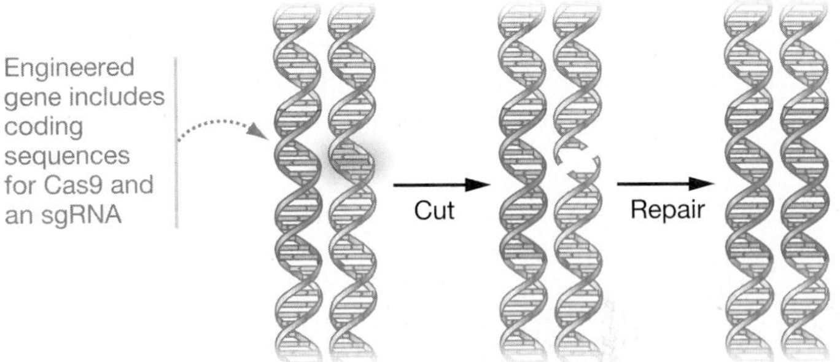

Figure 20.13 **A CRISPR-Cas9-based Gene Drive System Can Convert a Heterozygote into a Homozygote.** A gene (green) on one chromosome is engineered to include coding sequences for Cas9 protein and an sgRNA. This altered gene is called a gene drive locus. The sgRNA expressed from the gene drive locus directs cutting at the corresponding position on the homologous wild-type chromosome. Once the wild-type chromosome is cut, the gene drive locus on the engineered chromosome serves as a template for homology-directed repair (see Figure 20.11b), converting the heterozygote into a homozygote for the engineered allele. If the cell is a zygote, then all the cells of the organism that develops will be homozygous for the engineered allele.

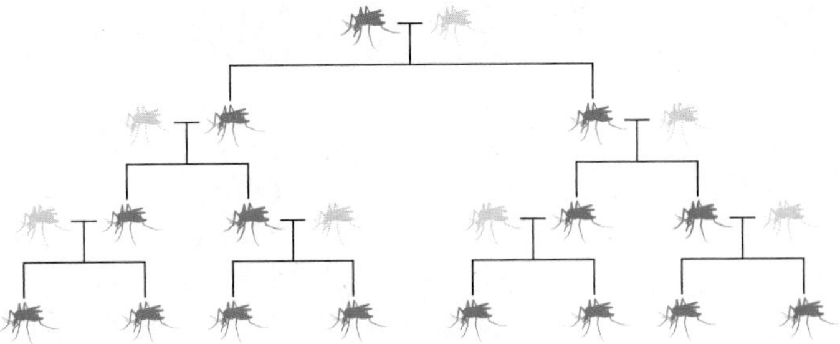

Figure 20.14 **A CRISPR-Cas9-based Gene Drive Can Genetically Alter a Mosquito Population.** All the offspring of a mosquito containing the engineered allele (green) become homozygous for the engineered allele. This results in the spread of the modified gene throughout the population in just a few generations.

✔ Predict the effect of creating a gene drive allele by inserting the sgRNA/Cas9 gene into a mosquito gene that is required to support replication of the malaria parasite in mosquitos.

What could be wrong with modifying, or even wiping out, populations of disease-carrying organisms like mosquitos that transmit malaria and sicken over 200 million people a year? Maybe nothing, but there are many unknowns. What if the gene drive locus spreads to nontarget species? Do we know enough about the effect of species removal on ecosystems? Most scientists and regulatory agencies believe there are too many uncertainties and risks to move ahead. At a minimum, great caution is needed as oversights are put into place.

Let's now shift focus and move from the field to the bedside. One of the great hopes for CRISPR-Cas genome editing is its ability to precisely modify the human genome to cure disease. What's been accomplished using molecular methods for therapy?

CHECK YOUR UNDERSTANDING

✔ **If you understood this section, you should be able to ...**

1. Explain what accounts for the precision of CRISPR-Cas9 genome editing.
2. **SOCIETY** Explain why U.S. defense and security agencies have expressed concern about gene drive technology.

Answers are available in Appendix A.

20.6 Gene Therapy

After you complete this section, you should be able to ...

- Explain the methods used in gene therapy.

Not long after the development of recombinant DNA technology, physicians began to dream about using it for **gene therapy** to cure genetic diseases. Broadly speaking, gene therapy is any approach to treating or curing disease that involves modifications of the genome.

More than 20 years ago, and with great optimism, the first gene therapy trials began. The ensuing years made it painfully clear that gene therapy was easier in principle than in practice and that much more had been promised than delivered in gene therapy. Confidence dropped to even lower levels when tragedy struck in two gene therapy trials. Patients died, including young children.

The pessimism about gene therapy's future began to lift in 2009, when new tools for delivering genes to cells led to some notable successes. Improved gene delivery systems have been used successfully to treat two forms of blindness, a brain disorder, an immune system disorder, and most recently, infants who faced certain death from a neurological disease. Today, particularly with the CRISPR-Cas system as part of the molecular toolkit, biologists and physicians are again feeling positive about gene therapy's potential.

How Does Gene Therapy Work? There are some stringent requirements for successful gene therapy.

- The disease must be due to defects in a single gene.
- The sequence of the wild-type allele must be known.
- There must be a way to introduce this therapeutic allele into affected individuals and have it expressed in the correct tissues, in the correct amount, and at the correct time.

Adding to the list, in the case of a dominant disease allele, there must be a way to replace the defective allele with one that functions normally, not simply to add a functioning allele. The CRISPR-Cas system of genome editing holds great promise in this area.

To deliver therapeutic genes, researchers have focused on packaging these genes into genetically engineered viruses for transport into human cells. These modified viruses are referred to as **vectors**, and their genomes have been altered to allow the incorporation of therapeutic genes and to disable replication of the virus in target cells. Critically, the virus vector can still gain access to the cell to deliver the gene. (Chapter 33 details how viruses enter cells.) Note that the term vector has multiple meanings. The use of the term here for a gene delivery vector is different from that of a cloning vector you learned about earlier.

How Has Gene Therapy Been Done? There are two primary approaches to gene therapy. In the first, known as *ex vivo* ("outside the body") gene therapy (shown in **Figure 20.15**), cells that

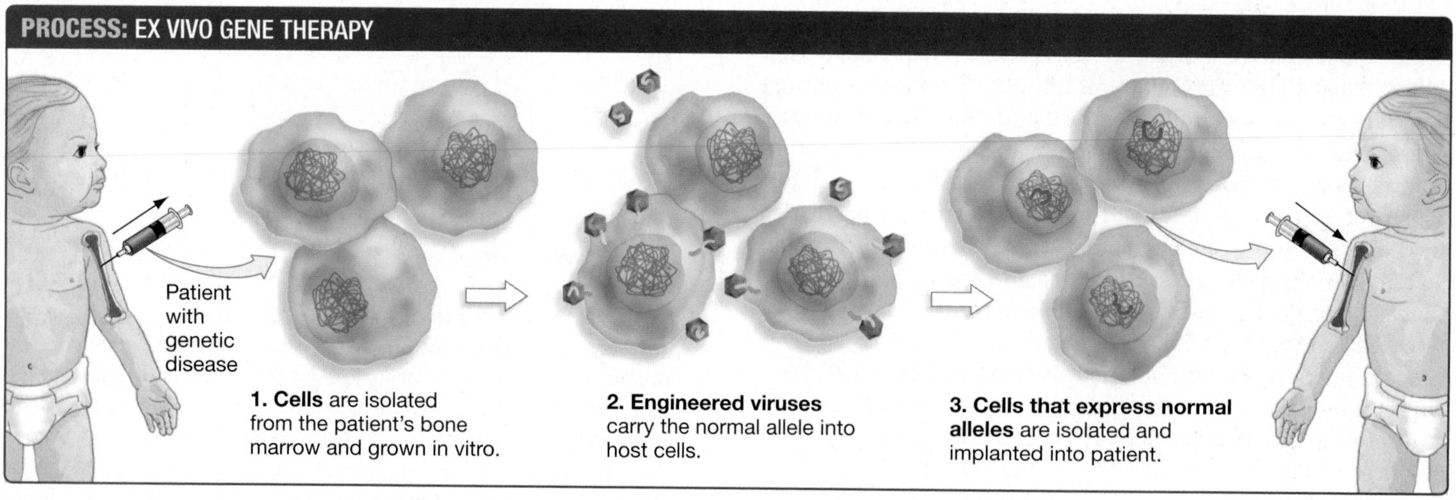

Figure 20.15 One Approach to Gene Therapy.

require the therapeutic gene are removed from the patient and then infected with the viral gene therapy vector. Vectors for ex vivo gene therapy typically integrate the therapeutic gene into the patient's genome.

The example in Figure 20.15 shows a case of using ex vivo gene therapy to treat an immune deficiency in which particular types of bone marrow cells fail to express an essential gene. Cells removed from the patient are infected with the gene therapy vector and then examined for expression of the therapeutic gene. Those with sufficient expression are grown in tissue culture (see **BioSkills 11**) to increase their number, and this population of cells is then introduced into the patient at a site where the genetically modified cells will be most effective in curing the disease.

Another approach to gene therapy is *in vivo* ("in the body") gene therapy. In this method, the viral gene delivery vector is injected into the bloodstream and the virus is transported throughout the patient's body.

More Recent Work and Looking Ahead A breakthrough in gene therapy came in 2017 when physician-scientists used in vivo gene therapy to treat spinal muscular atrophy type 1 (SMA1), a degenerative genetic disease that causes spinal motor neurons to die in early life. Children with severe forms of this disorder lose motor control, cannot lift their heads, and typically die by 2 years of age because they can no longer activate the muscles needed to breathe (**Figure 20.16**).

A group of 15 children with SMA1 were treated with a virus vector that carried a functional cDNA copy of the defective gene. This virus had the unusual property of being able to cross the blood-brain barrier, allowing it to access and infect cells of the central nervous system. The symptoms of SMA1 reversed in many of the children.

Looking ahead, biomedical scientists are excited about the possibilities of CRISPR-Cas genome editing applied to gene therapy. In addition to its specificity and relative ease of use, CRISPR-Cas genome editing allows for altering many genes at once—a property called multiplex editing. In diseases caused by more than one genetic defect, this approach offers a huge advantage. The possibilities of CRISPR-Cas genome editing have been tested in mouse models of the common and debilitating disease Duchenne muscular dystrophy, as well as a liver disease. In these cases, genome editing by CRISPR-Cas reversed disease symptoms. As of this writing, a handful of human clinical trials are under way. Will the potential of CRISPR-Cas genome editing in gene therapy translate into reality? It's too early to tell, but scientists and clinicians are hopeful.

With or without CRISPR-Cas genome editing, the good news is that gene therapy may finally be poised to deliver on some of its promises.

Figure 20.16 A Family with a Child Afflcted with Spinal Muscular Atrophy.

CHECK YOUR UNDERSTANDING

✔ If you understood this section, you should be able to ...

1. Explain why an ex vivo approach to gene therapy is better suited to correct diseases due to deficiencies in secreted proteins than diseases due to defective intracellular proteins.
2. Explain why it is easier to use gene therapy for a disease caused by a nonfunctional allele than a disease caused by a misfunctioning allele.

Answers are available in Appendix A.

20.7 New Frontiers: Functional Genomics, Proteomics, Systems and Synthetic Biology

After you complete this section, you should be able to ...

- Compare and contrast disciplines in the new frontiers of molecular biology.

A genome sequence is essentially a parts list. Once that list is assembled, scientists delve deeper to understand how genes interact to produce an organism. While whole-genome sequencing and annotation supply the list of parts, functional genomics answers questions about what particular genes do, how they're expressed, and how they work together to produce phenotypes.

Genomics has spawned a host of related fields. These are often referred to as the *–omics*, owing to names like proteomics, metabolomics, and transcriptomics, but they also include emerging areas like systems biology. All these fields take a holistic approach to learning about the entire set of proteins, metabolites, RNA transcripts, or interactions between genes in a given cell at a particular time.

Functional Genomics Seeks to Understand How Genes and Genomes Work

For decades, biologists have worked at understanding how and when individual genes are expressed. Research on the *lac* operon exemplifies this type of study (Ch. 18, Section 18.1). But now, starting with the data provided by genome sequencing projects, researchers in **functional genomics** can ask how, when, and where all the genes in an organism are expressed and what they do.

An important focus of functional genomics is genome-wide patterns of gene expression. This work is motivated by the realization that the biology of an organism ultimately comes down to which genes it expresses. Knowing only about the expression of single genes or small sets of genes—the only thing possible for biologists in the pre-genomics era—is not enough to understand how organisms work.

There are many approaches to learning genome-wide patterns of gene expression. One of these uses a **DNA microarray** to assess the expression of thousands of genes at a time. Researchers use microarrays in many ways, such as to establish which genes are transcribed in different organs and tissues, in cancers, or in response to signals. (The technology behind microarrays and their use is described in **BioSkills 10**.)

Today, however, the most common approach for discovering genome-wide patterns of gene expression is a method called **deep sequencing** or **RNA-seq** (pronounced "RNA-seek"). This method leverages next-generation sequencing technologies and knowledge of whole-genome sequences. In outline, RNA-seq involves the following steps:

1. Extract RNA from a cell or tissue type.
2. Enrich the sample for the type of RNA being studied (for example, mRNA).
3. Create cDNA from the RNA using reverse transcriptase.
4. Use next-generation sequencing to sequence millions of the cDNAs.
5. Map the cDNA sequences back to a reference genome sequence.
6. Analyze which regions of the genome were transcribed and at what frequency.

Single types of cells growing under one condition can be analyzed by RNA-seq. But more often, researchers compare patterns of gene expression between different conditions, or between different cell types or states (for example, cancerous cells and noncancerous cells from the same organ).

One study of breast cancer used RNA-seq to analyze gene expression patterns in *individual* breast cancer cells and noncancerous cells within breast tumors. The researchers found many of the expected changes in gene expression in the breast cancer cells, but with the single-cell analysis, they also discovered striking cell-to-cell differences in gene expression among the cancer cells. These differences are likely to be important in how breast cancers respond to treatment. A particular treatment may stop most of the cancer cells from proliferating, but with the extensive variation in gene expression revealed by single-cell RNA-seq, some of the cancer cells may be untouched by a generally effective therapy. Knowing the existence and extent of this heterogeneity within a single tumor may help shape more effective cancer therapies.

Proteomics Seeks to Identify the Entire Set of Proteins Expressed in a Cell

The Greek root *–ome*, meaning all, inspired the term "genome." Similarly, biologists use the term **transcriptome** in referring to the complete set of RNA molecules that have been transcribed in a particular cell, and **proteome** in referring to the complete set of proteins that are produced. It follows that **proteomics** is the large-scale study of all the proteins in a cell or organism.

The methods used to identify proteins are distinct from those used in working with DNA. In proteomics, scientists use a technique called mass spectroscopy, coupled with computational methods, to learn the precise size—and ultimately the amino acid sequence—of each protein in a sample. The ultimate goal is to identify every protein and its abundance in a cell.

Just as genomics moves from a catalog of parts—genes—to understanding how genes work together to create phenotypes, once individual proteins are identified in proteomics, researchers seek to understand how all the proteins—the proteome—change over time, vary between different cells, and interact.

Scientists across the world are collaborating on the Human Proteome Project, a natural extension of the Human Genome Project. Two early aims of this project are to confirm that annotated protein-coding genes produce protein products and to learn where these proteins are made. To date, researchers have searched for the proteins produced in 30 different tissues. Proteins encoded by nearly 20,000 human genes have been discovered. The researchers also turned up nearly 600 proteins that are encoded in regions of the genome previously thought to be noncoding. The genome continues to offer surprises.

Systems and Synthetic Biology—New Approaches to Understanding Life

For much of the history of molecular biology and genetics, biologists have pulled complex systems apart, one piece at a time. The aim has been to understand the function of an individual molecule or gene with the hope that knowing about the part will offer insights into the function of the entire system—for example, how a particular protein adds to the phenotypes of a cell.

Systems Biology—The Whole Is Greater Than the Sum of Its Parts With the explosion of parts lists from the *–omics*, advances in computational methods, and improved theoretical frameworks to understand networks, some scientists are taking a different path to explore how life works—**systems biology**. Systems biology seeks knowledge of how the network of interactions between the individual parts of a biological system lead to the properties of life. It is a holistic approach based on the notion that the whole is greater than the sum of its parts.

A goal of systems biology is understanding **emergent properties**, properties that arise at one level of organization from the interaction of simpler elements at a lower organizational level. For example, the phenotypes of a cell emerge from the interaction of the molecules that compose it, and consciousness emerges from the activity of individual neurons in the brain. Proteins do not have the complex behaviors of a cell, nor are neurons conscious, but these simpler parts working together give rise to the emergent properties of cell behaviors and consciousness. **✔ If you understand this concept, you should be able to explain why homeostasis (the ability to maintain relatively constant internal conditions in the face of fluctuating external conditions) is an emergent property.**

Many systems biology investigations focus on mapping the network of interactions between genes or proteins in particular types of cells. Genomics and proteomics provide the set of parts needed to start these systems biology studies. It's the job of a systems biologist to learn how these parts are linked together into networks and how new properties emerge from these interactions.

Synthetic Biology—Learning through Engineering Life In contrast to systems biology, which seeks understanding from observations of the natural world, **synthetic biology** pursues knowledge by building biological systems and discovering what properties they demonstrate. Synthetic biology also tinkers with living systems to engineer them for useful purposes.

One example of synthetic biology seems lifted from science fiction. Biologists have long wondered, what is the minimum set of genes needed for life? A team of synthetic biologists set out to find the answer. Starting with a bacterium with one of the smallest genomes known, the biologists mutated genes one by one to identify the genes required for life.

This analysis provided a rough idea of the minimum set of essential genes. With that information in hand, the team began an audacious project—to do a complete chemical synthesis of a minimal genome and introduce it into the cell of another bacterial species whose DNA was removed. The researchers produced a type of life never seen before—a hybrid cell derived from the cytoplasm of one bacterial species directed by a purely synthetic genome that mirrored genes from a different bacterial species. In the laboratory, this hybrid cell divided rapidly and produced colonies with millions of cells. In short, a new form of life had been synthesized.

How many genes did this synthetic genome have? There were 473 genes housed in a genome that spanned slightly more than 500,000 base pairs, a size far smaller than any found in nature. Revisiting the theme that genomics often opens our eyes to what we *don't* know, although all the genes were essential for viability, more than 30 percent of them had unknown biological functions.

Advances in genomics are revealing how cells work in health and disease, but they're also laying a foundation for understanding one of the most extraordinary aspects of life—how a single cell develops into a complete organism. This is the topic of the next chapter.

CHECK YOUR UNDERSTANDING

✔ If you understood this section, you should be able to . . .

1. Explain why a genome sequence is needed for RNA-seq analysis.
2. Develop an argument to answer this question: Did the synthetic biologists who produced a cell with a minimal genome create life?

Answers are available in Appendix A.

CHAPTER 20 Review

For media, go to **Mastering Biology**

20.1 Recombinant DNA Technology

- In recombinant DNA technology, DNA is added to a cell either to modify the cell's properties or to clone (obtain many identical copies of) the DNA.
- Restriction endonucleases cut DNA at specific locations. The resulting DNA fragments can be inserted into plasmids or other vectors with the help of DNA ligase.
- Many crop plants are genetically engineered for traits that include pest and herbicide resistance as well as improved food quality.

20.2 The Polymerase Chain Reaction

- The polymerase chain reaction (PCR) produces many identical copies of a gene by repeated rounds of DNA synthesis without using cells for cloning.
- PCR has many applications, such as DNA fingerprinting.

20.3 Analyzing Genomes

- Advances in DNA sequencing technologies have allowed investigators to sequence DNA more rapidly and cheaply.
- Thousands of genomes have been sequenced to date for many different purposes.
- Bioinformatics is the application of computer science to genome analysis, and it is essential for genome research.
- Researchers annotate genome sequences by using sequence data to find genes and to provide clues about their function.
- Two common approaches to finding genes in a genome sequence are to search for open reading frames (ORFs) or to match the sequences of expressed sequence tag (EST) cDNAs to regions of the genome.
- Genome-wide association studies (GWAS) find genes by association between the phenotypes they cause and genetic markers called SNPs that reside at known positions in the genome.

20.4 Insights into Genomes

- There is a linear relationship between prokaryotic gene number and genome size, and a similar relationship between the complexity of metabolism and size of the genome.
- There is a huge amount of genetic diversity in prokaryotic genomes, even among different strains of the same species.
- Lateral gene transfer is common in prokaryotes and is an important source of new genes in many species.
- Metagenomics allows the analysis of all the genes in a community of prokaryotes and can provide information on the diversity and abundance of species.

- There is no correlation between morphological complexity and gene number in eukaryotes.
- Because of alternative splicing, the number of distinct transcripts produced in many eukaryotes is larger than the gene number.
- Gene duplication has been an important source of new genes in eukaryotes.
- Much of the eukaryotic genome is transcribed, but the function of most noncoding transcripts is unknown.

20.5 Genome Editing

- The CRISPR-Cas system for genome editing allows for rapid and precise modification of genomes without the introduction of foreign DNA.
- The CRISPR-Cas system evolved in bacteria and archaea as a mechanism for immunity against viruses and other invaders.
- CRISPR-Cas genome editing has applications in agriculture, ecology (in the form of gene drives), and medicine, though some of those uses are controversial.

20.6 Gene Therapy

- Gene therapy is the modification of human genomes to treat or cure genetically based diseases.
- Gene therapy uses viral vectors to carry therapeutic genes into cells.
- Gene therapy has had limited success and notable failures, but appears poised for a brighter future.

20.7 New Frontiers: Functional Genomics, Proteomics, Systems and Synthetic Biology

- Functional genomics uses tools such as bioinformatics, DNA microarrays, and deep sequencing to learn the function of genes and patterns of gene expression.
- Proteomics is similar to genomics but works to identify the complete set of proteins expressed in a cell, how this set changes under different conditions, and how it relates to phenotype.
- Systems biology starts with genomics and proteomics data and studies the set of interactions between different genes or proteins to understand how biological systems work.
- Synthetic biology seeks to both understand life and to engineer it by creating artificial biological systems.

Answers are available in Appendix A.

✔ TEST YOUR KNOWLEDGE

1. What do restriction endonucleases do?
2. What is a plasmid?
   a. an organelle found in many bacteria and certain eukaryotes
   b. a circular DNA molecule that replicates independently of the main chromosome(s)
   c. a type of virus that has a DNA genome and infects certain types of human cells, including lung and respiratory tract tissue
   d. a type of virus that has an RNA genome, codes for reverse transcriptase, and inserts a cDNA copy of its genome into cells
3. After finding a gene that causes a disease, researchers often introduce the defective allele into mice to create an animal model of the disease. Why are these models valuable?
   a. They allow the testing of potential drug therapies without endangering human patients.
   b. They allow the sequencing of the mutant allele.
   c. They allow the production of large quantities of the defective gene product, usually a protein.
   d. They allow the study of how the gene was transmitted from parents to offspring.
4. **THINK CAREFULLY** The human genome size is 3 billion base pairs, and the size of the baker's yeast genome, a single-celled organism, is 12 million base pairs. Therefore, the predicted genome size for another single-celled organism, an amoeba,
   a. is about the size of the human genome
   b. is about the size of the yeast genome
   c. is somewhere between the sizes of the yeast and human genomes
   d. cannot be predicted with any certainty

✔ TEST YOUR UNDERSTANDING

5. Explain how RNA-seq can be used to analyze patterns of gene expression.
6. Consider the validity of the following statements about genome editing. Select True or False for each statement.
   T/F Cas proteins work as endonucleases.
   T/F sgRNA is used by bacterial cells to detect which DNA to cut.
   T/F Homologous recombination is always used to join pieces of broken DNA.
   T/F It is possible to modify genes as well as disrupt them by genome editing.
7. **QUANTITATIVE** Gene density is the number of genes per million base pairs (Mbp). Using Figure 20.5b, find the approximate number of genes estimated in water fleas and in humans, and note the size of each genome. Calculate the gene density in water fleas relative to that in humans.
8. **THINK CAREFULLY** A friend who works in a research lab performed a GWAS and discovered a tight association between a SNP allele and the disease she is studying. She concluded that the SNP allele must be the mutation that causes the disease. Explain why she is likely to be wrong.

✔ TEST YOUR PROBLEM-SOLVING SKILLS

9. Revolutionaries executed Nicholas II, the last czar of Russia, along with his wife and five children, the family physician, and about a dozen servants. Many decades later, a grave said to hold the remains of the royal family was discovered. Biologists were asked to analyze DNA from the bodies. If the remains of the family were in this grave, predict how similar the DNA fingerprints would be between the parents, the children, and the unrelated individuals in the grave.
10. One hypothesis for differences between humans and chimpanzees involves differences in gene regulation. A study using RNA-seq showed that the overall patterns of gene expression were similar in the liver and blood of the two species, but the expression patterns were strikingly different in the brain. How do these results relate to the hypothesis?

✔ PUT IT ALL TOGETHER: Case Study

Could genetic modification protect potatoes from disease?

Potato blight causes potato plants to shrivel and rot. The disease is caused by the pathogen *Phytophthora infestans*, infamous for its role in Ireland's Great Potato Famine in the mid-1840s. The disease can devastate crops during wet weather, sometimes leading to total crop loss. Researchers aim to use recombinant DNA methods to transfer blight resistance genes from resistant varieties into susceptible varieties of potato.

11. Explain how restriction endonucleases and DNA ligase could be used to insert a potato blight resistance gene into a plasmid.

12. Transgenic plants usually contain genes of bacterial plasmid origin. In a recent study, researchers designed a strategy that avoided using any plasmid genes. They transformed cells from a susceptible potato variety with a potato blight resistance gene cloned from a resistant variety. Next, to determine which plants from this group were also free of plasmid DNA (cloning vector) sequences, they performed PCR using primers specific for the plasmid. The positive control lane shows PCR amplification of plasmid DNA only, and the negative control lane shows an attempted PCR amplification of no added DNA. Based on the gel analysis of PCR products shown below, which plants contain only the potato gene? Explain your answer. (See **BioSkills 10** and **BioSkills 6** for an overview of PCR and for help in reading a gel).

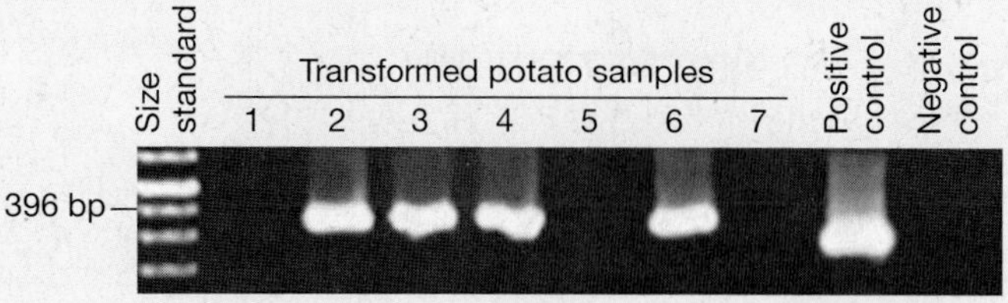

Source: K-R. Jo et al. 2014. *BMC Biotechnology* 14: 50.

13. QUANTITATIVE If the sequence of DNA in Question 12 were amplified using 25 PCR cycles, then the amount of this DNA would be predicted to increase by ____-fold.

14. PROCESS OF SCIENCE Why was it important to include a positive control and a negative control in the PCR analysis?

15. How could the research group determine whether a homologous gene for blight resistance exists in the human genome?

16. SOCIETY Many environmental groups are strongly opposed to GM crops that contain foreign (non-crop plant) genes. What do you think are some of the reasons for concern? How might the study described in Question 12 get around some of these concerns?

Mastering Biology

Students Go to Mastering™ Biology for assignments, the eText, and the Study Area with animations, practice tests, and activities.

Professors Go to Mastering™ Biology for automatically graded tutorials and questions that you can assign to your students, plus Instructor Resources.

21 Genes, Development, and Evolution

Developing frogs. In this chapter the focus is on the processes responsible for transforming a fertilized egg into an individual with specialized cells, tissues, and organs.

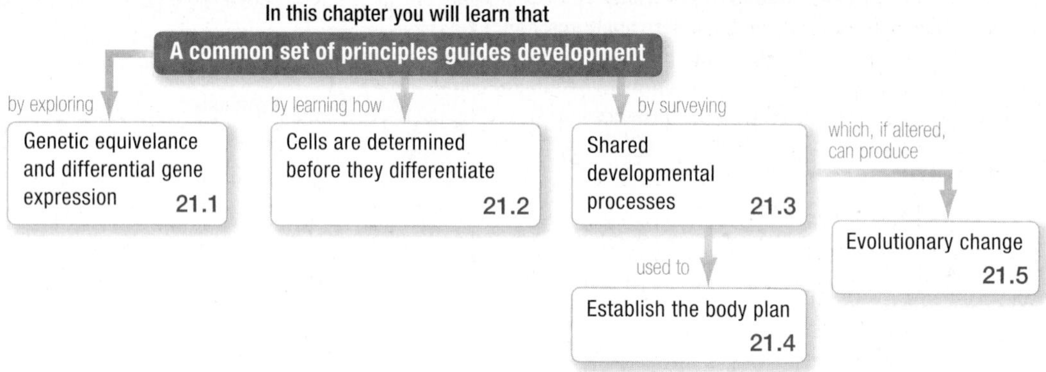

What is today's greatest challenge in biological science? Although there are many candidates, one of the most compelling is understanding **development**—the processes that allow a multicellular individual to form from a single cell.

It's important to pause for a moment and think about the incredible nature of development. Consider that once you were a single cell—a fertilized egg, or **zygote**. If you could have watched your own development, you would have seen that single cell divide rapidly and form a ball of tiny, identical-looking cells. At that point, the fertilized egg had become an **embryo**—a young, developing organism. After continued cell division, large groups of cells suddenly began moving into the embryo's interior. Cell division continued at a dizzying pace. After a week or two the embryo elongated, and a recognizable head and tail portion appeared. Tiny precursors of vertebrae became visible, along with rudimentary eyes. Eventually buds emerged that went on to form your limbs. As development continued, the embryo became you. Biologists who have watched this process in humans or other organisms never cease to marvel at it. How does a body form and grow?

To answer this question, you'll need to draw on what you've already learned about gene expression, how cells interact, and a host of other topics. Part of the excitement surrounding **developmental biology** is that it weaves together discoveries from genetics, biochemistry, cell biology, and evolution. It is one of the most interdisciplinary fields in all of biology. This chapter is about the basic genetic and cellular processes of development that apply to almost all multicellular organisms. Later you will learn how these principles apply to development in plants (Ch. 38, Sections 38.2 and 38.5) and animals (Ch. 47, Sections 47.1, 47.3, and 47.4). Let's delve in.

21.1 Genetic Equivalence and Differential Gene Expression in Development

A fundamental aspect of development is **differentiation**—the process in which cells acquire specialized properties. An organism like you has hundreds of distinct types of cells. For example, there are at least three different kinds of muscle cells, dozens of different nerve cells, and an equally large number of blood cell types. Yet all of your cells originated from one cell, the zygote. How did this amazing array of cells types arise?

After you complete this section, you should be able to...

- Explain the importance of differential gene expression in development.

It's obvious that cells have different structures and functions because they contain different molecules. What wasn't so clear to early developmental biologists was whether cells are different because they contain different genes, or whether different cells in an individual contain the same genes but express only a specialized subset.

It turns out that the differentiation of a cell occurs through **differential gene expression**, when different cells use the same set of genes in different ways. The muscle cells in your body are different from your nerve cells because they express different genes and therefore produce different proteins, and the water-transport cells in an oak tree are different from its leaf-surface cells for the same reason. The first compelling evidence that differential gene expression accounts for cell differentiation came from studies of plants.

Evidence that Differentiated Plant Cells Are Genetically Equivalent

The idea that all the different cells in an organism contain the same genes is known as **genetic equivalence**. Support for this hypothesis came from classical agricultural practices as much as it did from scientific investigations. Gardeners and farmers have long propagated many plants from cuttings. When a branch or stem is removed from a plant, cells in the cut region can **de-differentiate** and begin dividing to form a mass of undifferentiated cells. These cells can then re-differentiate to form roots, as shown for a coleus plant in Figure 21.1.

If cells from a cut branch or stem can de-differentiate to form roots, these differentiated cells must contain the genes required by root cells.

The notion of genetically equivalent cells was confirmed in the 1950s when biologists were able to grow entire tobacco plants or carrots from a single, differentiated cell taken from an adult. These carrots plants are **clones**—a genetically identical copy—of the plant from which the single cell was obtained. As you learned earlier (Ch. 20, Section 20.1), the ability to produce entire genetically identical plants from a single, undifferentiated cell is a core cloning technique in plant biotechnology. (For more information on how plants are cloned and grown in culture, see BioSkills 11.)

Evidence that Differentiated Animal Cells Are Genetically Equivalent

The issue of genetic equivalence proved more difficult to resolve in animals. Experiments by Robert Briggs, Thomas King, and John Gurdon that began in the 1950s involved transferring

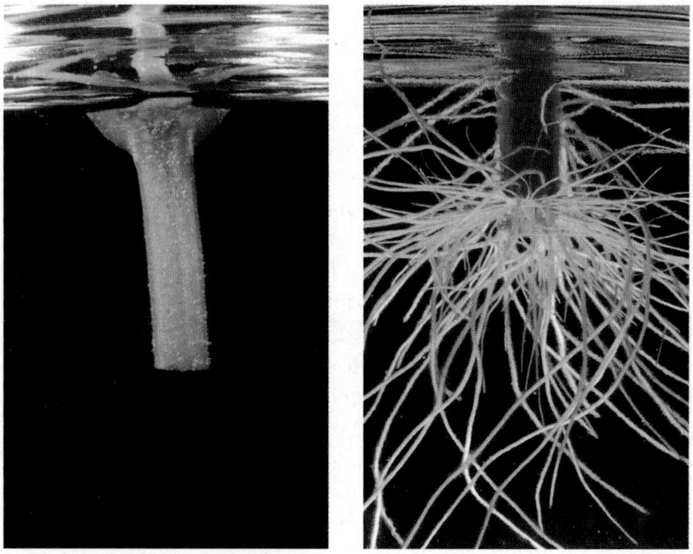

Figure 21.1 Genetic Equivalence Is Demonstrated by Plant Cells. The cells in a cut coleus stem (left) de-differentiate and divide to form a mass of undifferentiated cells, which then re-differentiate into root cells that form functional roots (right). This is possible only if shoot cells contain all the genes needed in root cells.

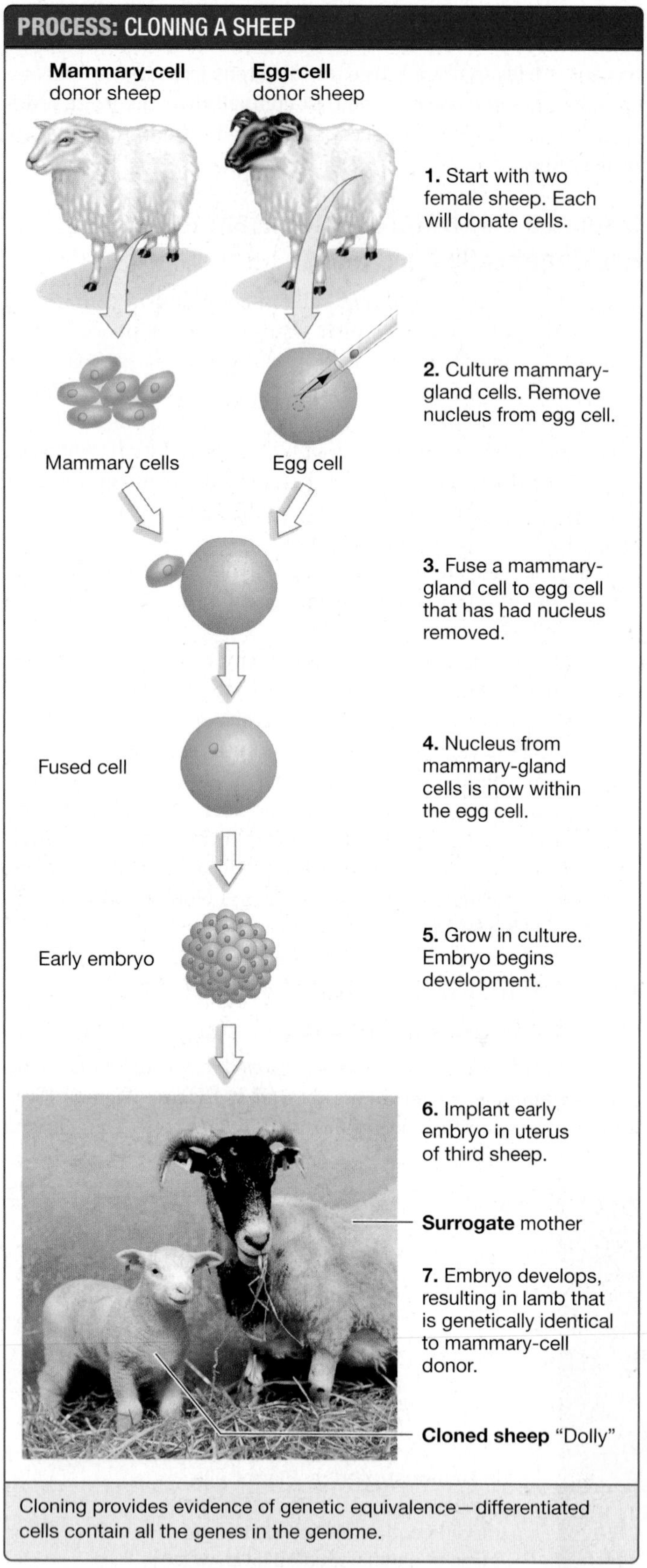

Figure 21.2 Genetic Equivalence Is Demonstrated by the Ability to Clone Mammals. The lamb that resulted from this experiment was genetically identical to the white-faced individual that donated the nucleated cell, not the black-faced egg donor or surrogate mother.

nuclei from differentiated frog cells into unfertilized egg cells whose nuclei had been removed. Some of these transferred nuclei were able to direct the development of tadpoles. These results provided evidence that all cells in the same frog are genetically equivalent. But what about cells in mammals?

Genetic equivalence in mammals was demonstrated by Ian Wilmut and colleagues with their stunning 1997 report of nuclear transfer experiments in sheep. As **Figure 21.2** shows, the researchers removed mammary-gland cells from a female, grew them in culture, and fused them with eggs whose nuclei had been removed. The eggs came from a black-faced breed of sheep, while the nucleated donor cells came from a white-faced breed. After developing in culture, the resulting embryos were implanted into surrogate mothers. In one of several hundred nuclear transfers, an apparently normal white-faced lamb, Dolly, was born.

Genetic tests showed that Dolly was a clone of the white-faced donor of the mammary-gland cell. Dolly grew into a fertile adult. Soon after, other research groups reported similar results in mice and cows. A menagerie of animals, including horses, monkeys, dogs, and cats, have since been cloned. But the success rate of cloning remains low; the vast majority of such attempts fail to produce healthy newborns.

Recent molecular evidence suggests that there may be more to the story of genetic equivalence. Random mutations during the course of development often set up genetic differences between cells of an individual. However, even with these occasional genetic differences between cells, the central conclusion drawn from cloning plants and animals is unshaken: Cellular differentiation typically does not involve changes in the genetic makeup of cells. Instead, it results from differential gene expression.

How Does Differential Gene Expression Occur?

Eukaryotic cells regulate gene expression at several different levels: control of chromatin condensation, transcriptional regulation, alternative splicing and selective destruction of mRNAs, translation rate, and activation and deactivation of proteins after they are translated. These processes (all explored in Ch. 19, Sections 19.2, 19.3, and 19.4) play vital roles during development. Which is most important in differentiation?

The answer is the control of transcription via chromatin condensation. Genes in condensed regions of chromatin are not transcribed while genes in regions of relaxed chromatin have the possibility of being transcribed. Regulating how tightly chromatin is packaged is an essential element of differential gene expression. Differences in the patterns of chromatin condensation are the basis of **epigenetic inheritance**, any form of inheritance that is not based on nucleotide sequence differences in alleles (Ch. 19, Section 19.2). Epigenetic inheritance—epigenetics for short—has become a focal point of genetics research over the past decade. A large part of this interest stems from something you'll see in the following section: Epigenetic inheritance is the cornerstone of how an embryonic cell differentiates into a particular cell type and remains that type of cell.

CHECK YOUR UNDERSTANDING

✔ **If you understood this section, you should be able to ...**

1. Explain how the cloning of mammals provides evidence for genetic equivalence.
2. Explain why knowing how gene expression occurs is important in understanding development.

Answers are available in Appendix A.

21.2 Cells Are Determined Before They Differentiate

After you complete this section, you should be able to ...

- Contrast determination and differentiation in development.

An analogy for following embryonic cells through development is following a group of children in their development to adult professional lives. As infants, almost any possibility is open to them. But as children grow, their experiences, aptitudes, circumstances, and chance set them on different career paths. When does a child decide on what to be? How irrevocable is their decision? Do the range of possibilities for any child gradually narrow with age? The same questions can be asked of differentiating cells.

For us, it can be difficult but still possible to change between radically different careers. For cells, switching between adult "professions" is harder. Let's first see how an embryonic animal cell commits to its adult state and then look at ways of unlocking this commitment to open new possibilities for development and medicine.

Commitment and Determination

Commitment occurs when an embryonic animal cell becomes dedicated to follow a particular path of cell specialization. Commitment doesn't work like an on-off switch; rather, cells become committed gradually. Initially, commitment is weak and can easily be reversed. Over time, a cell becomes more and more deeply committed until—under normal conditions—the path it follows does not change.

Once a cell has locked into becoming a particular type of cell, it is said to be **determined**. Importantly, during and shortly after determination, a cell does not appear outwardly different from its embryonic state. The cell has not yet acquired the specialized features of a differentiated cell, even though its developmental options have become restricted.

To learn if cells are determined, biologists transplant them from one location to another within a developing embryo. The reasoning is that if a cell has irreversibly committed to becoming a particular type of cell, then moving it to a different part of the embryo where it will receive a new set of signals that control development will not alter its path of differentiation to that cell type. This cell is determined. On the other hand, if a cell is not determined, then placing it in a different location will cause it to differentiate into another type of cell—a type appropriate to the new location.

What changes inside a cell commit it to a particular pathway of development? There is strong evidence that epigenetic changes (Ch. 19, Section 19.2) play a central role. Signals in the embryo induce a cascade of events that alter patterns of chromatin condensation that in turn control which genes can be expressed and which are silenced. Once these patterns are established, they are maintained, locking the cell into its determined state.

Determination is a feature that distinguishes plant and animal development. While animal cells play by the rule of "once determined, (nearly) always determined," plant cells don't seem to be locked into a particular pathway. If plant cells were irreversibly determined, cloning plants from many different types of cells, tissues, or cuttings would not be so simple.

Not all animal cells play by the rules, however. Many cancer biologists hypothesize that tumor cells are at least partially dedifferentiated. In tumors, differentiated cells that once divided slowly, if at all, now divide rapidly, become migratory, and lose their specialized properties, coming to have properties of early embryonic cells. In short, it appears that determination and differentiation have been at least partially reversed. As you'll see next, the determined state of differentiated animal cells can also be reversed using experimental techniques, and this is the basis for many forms of stem cell therapy.

Master Regulators of Differentiation and Development

If transcription factors regulate much of development, are there master regulators that, once expressed, can change the fate of developing cells? Harold Weintraub and colleagues were the first to answer this question in the late 1980s. Weintraub worked with embryonic cells called myoblasts. These cells look nothing like muscle cells but are determined to become muscle cells.

Weintraub and co-investigators hypothesized that myoblasts must express at least one crucial transcription factor that can trigger their differentiation to muscle cells. The researchers set out to find a transcription factor in myoblasts that dictates, "Become a muscle cell."

Figure 21.3 on page 432 outlines how the biologists went about searching for this hypothetical regulatory protein via the following steps:

Step 1 They began by isolating mRNAs from myoblasts.

Step 2 They used reverse transcriptase to convert the mRNAs to cDNAs (see **BioSkills 10**). Because myoblasts were predicted to transcribe genes that control muscle-cell differentiation, they reasoned that the cDNAs must include copies of this key transcript or transcripts.

Step 3 They attached a type of promoter to the cDNAs that would ensure expression of each cDNA in any type of cell (a "general" promoter).

Step 4 They introduced the modified cDNAs into non-muscle cells called fibroblasts and monitored the development of the cells.

Result One of the myoblast-derived cDNAs converted fibroblasts into muscle-like cells.

RESEARCH

QUESTION: What causes some cells in a developing embryo to produce muscle?

HYPOTHESIS: Production of an mRNA for a regulatory protein (or proteins) by myoblasts determines their developmental fate.

NULL HYPOTHESIS: Myoblasts do not produce an mRNA for a regulatory protein that determines their developmental fate.

EXPERIMENTAL SETUP:

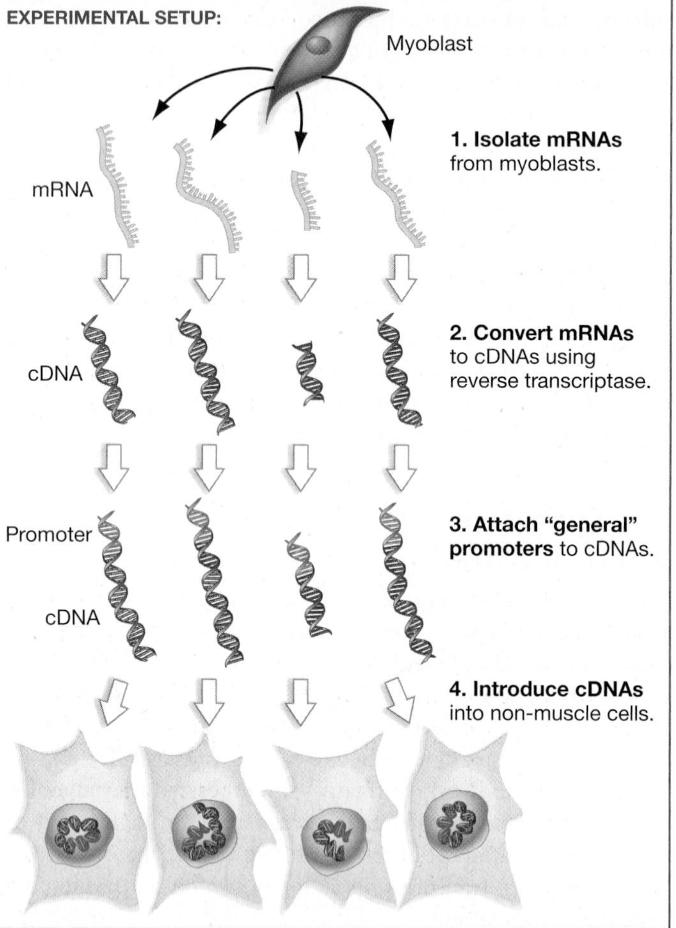

PREDICTION OF HYPOTHESIS: One of the myoblast-derived cDNAs will convert non-muscle cells into cells that produce muscle-specific proteins.

PREDICTION OF NULL HYPOTHESIS: None of the myoblast-derived cDNAs will convert non-muscle cells into muscle-like cells.

RESULTS:

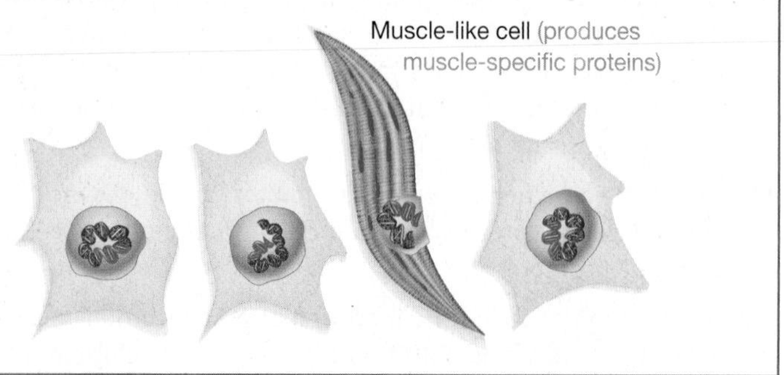

CONCLUSION: There is a master regulator of muscle cell differentiation.

Figure 21.3 The Search for a Protein That Causes Muscle-Cell Differentiation.

SOURCE: H. Weintraub, S. J. Tapscott, R. L. Davis, et al. 1989. Activation of muscle-specific genes in pigment, nerve, fat, liver, and fibroblast cell lines by forced expression of *MyoD*. *Proceedings of the National Academy of Sciences, USA* 86: 5434–5438.

✔ **PROCESS OF SCIENCE** Why did the researchers have to attach a "general" promoter to the cDNAs?

Follow-up experiments showed that the same myoblast mRNA, once expressed as a protein, could convert pigment cells, nerve cells, fat cells, and liver cells into cells that produced muscle-specific proteins and exhibit many properties of muscle cells.

Weintraub's group called the protein produced by this gene MyoD, for *myo*blast *d*etermination. Consistent with the hypothesis that regulatory transcription factors control determination and differentiation, subsequent work showed that the *MyoD* gene encodes a regulatory transcription factor and that the protein, MyoD, binds to enhancer elements located upstream of muscle-specific genes. MyoD is a **master regulator** of muscle differentiation. A master regulator is a gene product that can unleash a series of events that produce a specialized cell type, tissue, or body structure. Master regulators are used in the development of both animals and plants.

If master regulators can drive cells forward in development, might there be regulators that can push individual cells backwards, to an undifferentiated state? Biologists began testing this possibility with the goal of using de-differentiated cells for treating disease. Before you get to this story, you'll need to learn about a type of cell that is critical in embryonic development and for maintaining and repairing tissue in adults.

Stem Cells and Stem Cell Therapy

As development proceeds in both plants and animals, most cells differentiate and then no longer divide. But in later stages of development and in the adult, some populations of undifferentiated cells are maintained that preserve the potential to divide throughout the individual's life. These are **stem cells**.

As Figure 21.4 shows, stem cells divide to produce a cell that remains a stem cell and another cell that ultimately differentiates into a specialized cell type. Producing one daughter that stays a stem cell ensures that populations of stem cells are maintained. This is known as **self-renewal**.

Stem cell populations are found in specialized locations in the organism, called stem cell niches, that provide unique environments to keep stem cells in an undifferentiated state and to control their division to meet the needs of the embryo or adult.

- In plants, stem cell niches are found in **meristems**. Meristems exist in embryos and adults and contain the stem cells that produce all the cells needed to construct stems, roots, leaves, flowers, and other structures that develop throughout a plant's life.
- In animals, stem cell niches are found in tissues where there is a demand to generate new cells for growth or to replace worn-out or damaged cells. In adults, stem cells proliferate

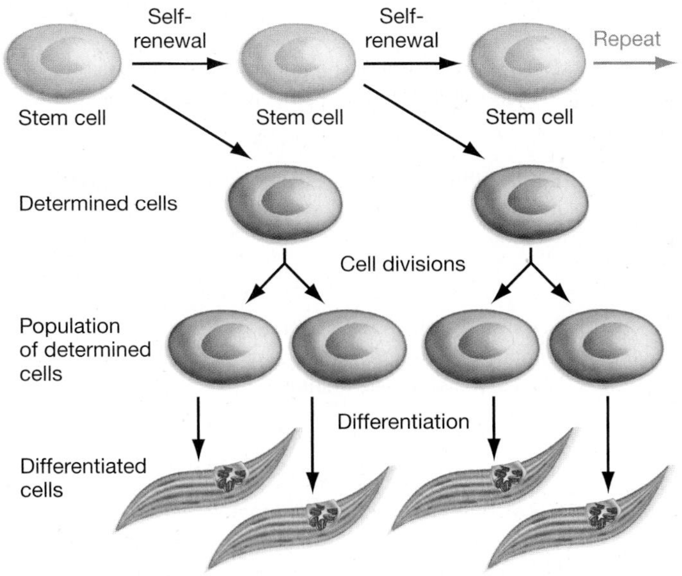

Figure 21.4 **Stem Cells.** Stem cells are undifferentiated cells that divide to produce another stem cell and a cell that becomes determined to follow a particular path of specialization. The stem cell maintains the population of stem cells, and the determined cell divides to produce a population of cells, each of which differentiate into a specific cell type, such as the muscle cells shown here.

to replace skin, blood, and gut cells that die; repair wounds; replace some types of damaged muscle; and create a constant supply of disease-fighting cells in the immune system.

Knowledge of the role that stem cells play in cell replacement and tissue repair spurred researchers to look for a way to create stem cells in the laboratory that might be used for therapeutic purposes.

Shinya Yamanaka and colleagues began with fully differentiated adult fibroblast cells found in the connective tissue of mammalian skin and looked for ways to "reprogram" the skin cells into the type of stem cells found in an early embryo. These cells, called **embryonic stem cells**, can form any cell in the body, a property known as **pluripotency**.

The researchers systematically introduced and expressed within the adult skin cells different combinations of genes that encode transcription factors active in the earliest stages of embryonic development. They eventually hit on a combination of four transcription factors that caused some of the differentiated adult cells to de-differentiate to an embryonic stem-cell-like state. The resulting cells were called **induced pluripotent stem cells**, or **iPS cells** for short.

The creation of iPS cells has ignited a massive research effort. iPS cells can be coaxed into differentiating to form particular specialized cell types by controlling the culture conditions (see **BioSkills 10** for a description of cell culture methods). The hope is that these differentiated cells derived from iPS cells will be a ready source of patient-specific cells that can replace those destroyed in diseases such as diabetes, Parkinson's disease, or heart disease. Medical researchers are making progress in turning this promise into a reality.

CHECK YOUR UNDERSTANDING

✔ If you understood this section, you should be able to ...

1. Interpret the following results in light of determination: Cells from a region of an early embryo that normally forms part of the eye are transplanted to a skin-forming region of another early embryo. These cells form skin. In another study, cells from the same region of an embryo later in development are transplanted to a skin-forming region of an early embryo. These cells form eye tissue.
2. Explain the meaning of "self-renewal" in the context of stem cells.
3. Explain why transcription factors are added to differentiated cells when creating pluripotent cells.

Answers are available in Appendix A.

21.3 Shared Developmental Processes

After you complete this section, you should be able to ...

- Explain the developmental processes that are shared by multicellular organisms.

Over a hundred years of developmental biology research culminated in one of the great insights of contemporary biology: A few fundamental principles are common to development in every multicellular organism. This discovery has brought a unified understanding to how all embryos grow and develop from a single cell into complex, multicellular individuals.

So far, you've learned how single cells become determined and differentiate during development, and how this process can be controlled in the laboratory. These are important ideas, but to fully understand development, it's necessary to know how groups of cells coordinate their behaviors to form an adult.

An individual develops as

1. cells divide;
2. signal what they are doing, and what type of cell they are becoming;
3. begin to express certain genes rather than others;
4. move, expand, or contract in specific directions; and
5. in the case of some cells, die (**Table 21.1** on page 434).

Let's consider each process in turn, while keeping in mind that in the embryo they are interdependent and occur together.

Cell Division

For an embryo to grow and develop, its cells have to divide to make more cells. This is obvious, but what's less apparent—and equally important—is that the location, timing, and extent of cell division have to be tightly controlled.

How is cell division controlled? You might recall that mitosis and cytokinesis are responsible for cell division in eukaryotes (Ch. 12, Section 12.2) and that cells initiate mitosis in response to a regulatory protein complex called M-phase-promoting factor

SUMMARY Table 21.1 Essential Developmental Processes

Cell division	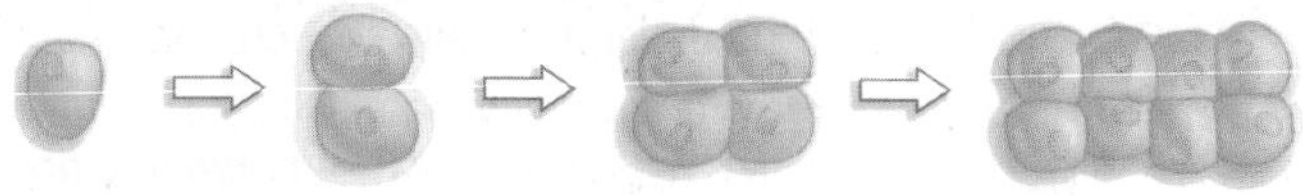	Cells divide by mitosis and cytokinesis. The timing, location, and amount of cell division are regulated.
Cell–cell interactions	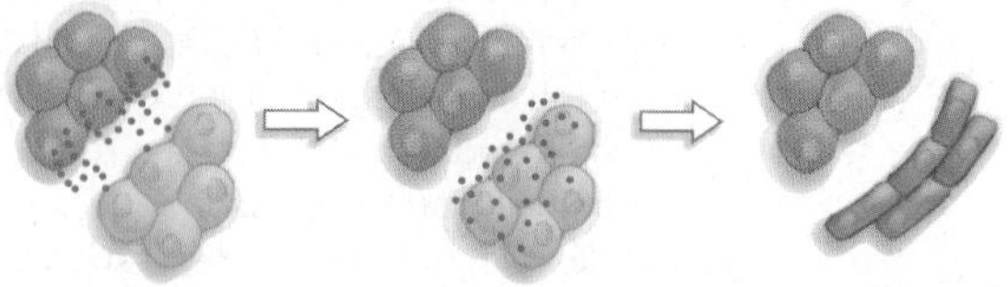	Signals that are produced by cells influence their neighbors to divide, differentiate, move, expand, or die.
Cell differentiation	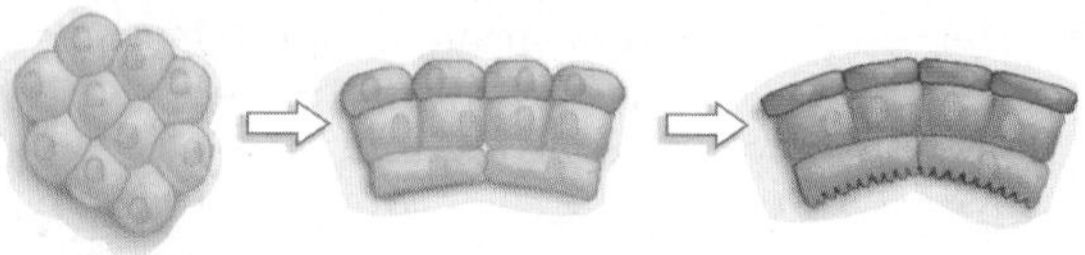	Undifferentiated cells specialize at specific times and places in a stepwise fashion.
Cell movement and coordinated cell-shape changes	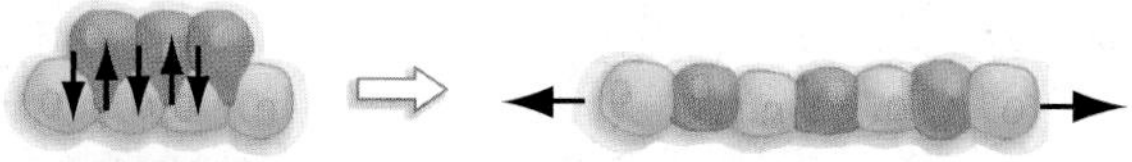	Cells can move past one another within a block of animal cells, causing drastic shape changes in the embryo.
	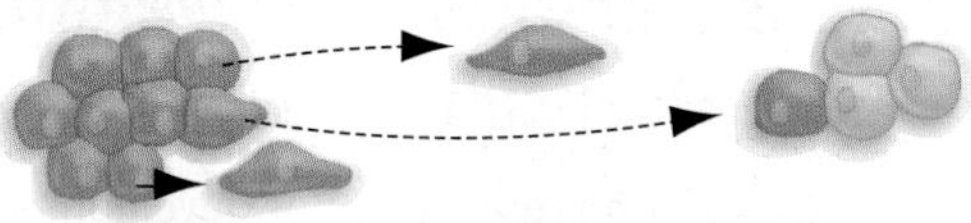	Cells can break away from a block of animal cells and migrate to new locations.
	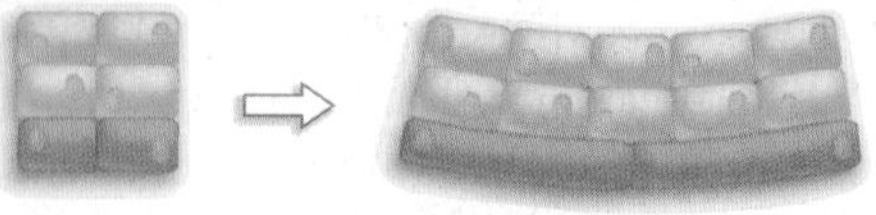	Cells can regulate the plane of cell division and expand in specific directions, causing dramatic changes in shape.
Programmed cell death	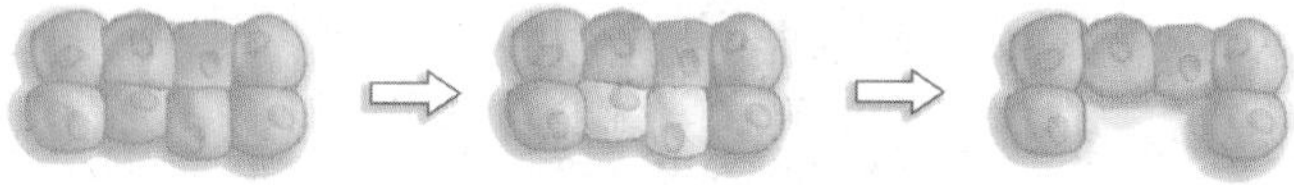	The timing, location, and amount of cell death are regulated.

(MPF). You also may remember that there are checkpoints at different stages of the cell cycle that regulate progression through the cycle, and that cells control their division in response to what biologists call "social controls"—signals from other cells (Ch. 12, Section 12.3).

All of these controls are used throughout development to generate the right number of cells in the appropriate places and at the correct times.

Cell–Cell Interactions

Cells interact constantly during development. They engage in "conversations" with their neighbors through a diverse set of signaling molecules. This ongoing communication drives most of the cellular behaviors that are the foundation of development.

Signaling molecules used in development may diffuse in the watery environment that surrounds cells, but they also can be present on the surface of other cells or bound to the extracellular matrix. These nondiffusible signals are detected by developing cells as they expand or move within the embryo.

You might recall that when a signaling molecule is received by a cell with an appropriate receptor, the message is processed and relayed through a signal transduction cascade (Ch. 11, Section 11.3). In a developing embryo, some signals are received inside the cell by intracellular receptors, but most signals are received by receptors on the cell surface. Once signal transduction cascades are stimulated, transcription factors are activated (Ch. 19, Section 19.3), patterns of gene expression change, and the embryonic cell's activity is altered.

In response to a signaling molecule, embryonic cells may divide, differentiate, move, change shape, or die. In this way, the fate of a cell inside an embryo hinges on the signals it receives from other cells.

Cell Differentiation

Most embryonic cells differentiate and acquire the specialized properties of adult cells throughout development. At crucial points, the signals received by cells specify their **fate**—the type of cell they will ultimately form. These signals direct embryonic cells to follow particular pathways of differentiation to become, for example, particular types of bone, skin, muscle and blood cells. As development proceeds, the embryonic cells produced by a fertilized egg ultimately give rise to hundreds of distinct adult cell types.

But just how is the fate of a cell specified? The answer is by signals arising inside the cell (cytoplasmic determinants) or outside the cell (induction):

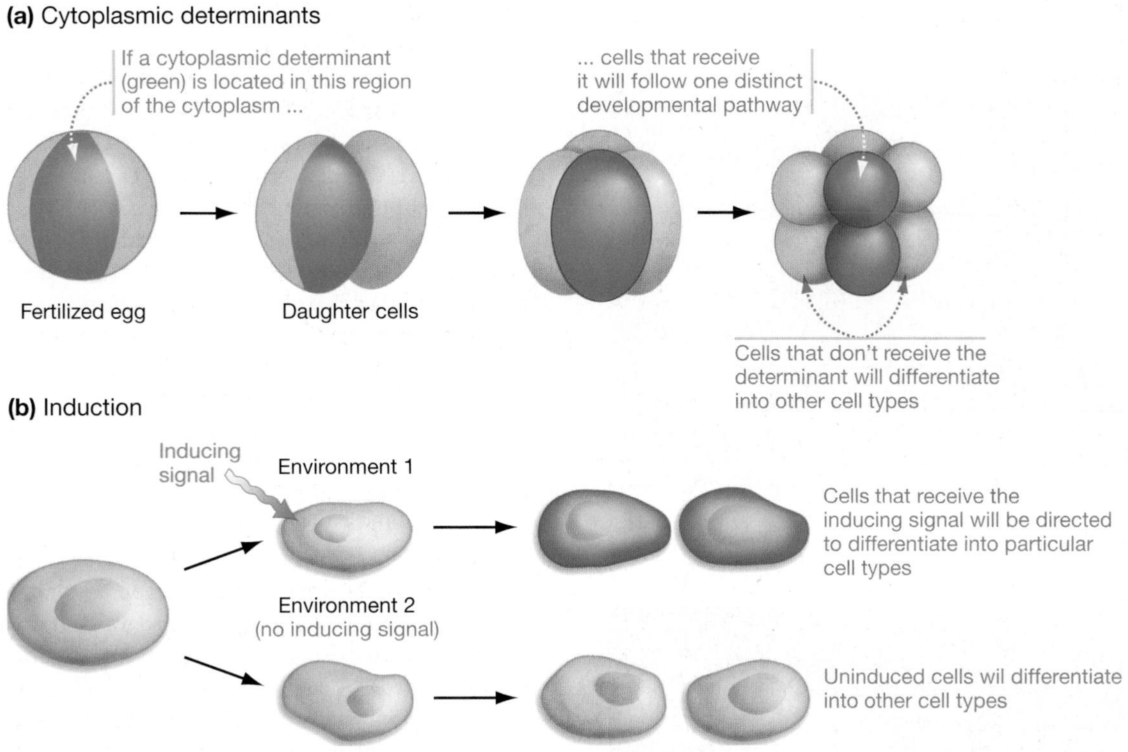

Figure 21.5 Cytoplasmic Determinants and Induction Specify the Fate of Cells. Unequal distribution of a cytoplasmic determinant depends on the plane of cell division.

✔ Predict what might occur if *all* the cell divisions shown in panel (a) occurred in the horizontal plane.

- **Cytoplasmic determinants** (Figure 21.5a) are gene regulatory molecules that are present in the cytoplasm of a dividing cell. Because they can be localized asymmetrically, they can become unequally distributed among daughter cells after cell division. If a cytoplasmic determinant is segregated to one region of a cell, cell division may result in one of the daughter cells receiving this cytoplasmic determinant and the other not receiving any. In this case, the two daughter cells will be specified to have different fates.
- **Induction** (Figure 21.5b) works through external signals. With induction, a cell that receives the signal is prompted to follow a different pathway of development than it would without the signal. The developmental pathway followed by a cell that does not receive the inducing signal is often called the default pathway. Inducing signals may be diffusible signals secreted by nearby cells; they may present on the surface of one cell that contacts and thereby signals another; or they may be anchored in the extracellular matrix.

All organisms use induction frequently during the course of development. On the other hand, the use of cytoplasmic determinants varies across species. For example, cytoplasmic determinants are used often in insects, rarely in mammals, and almost never in plants.

Cytoplasmic determinants and induction both lead to the distinct patterns of gene expression characteristic of each type of differentiated cell.

Cell Movement and Changes in Shape

Development in animals requires many cells to move relative to their neighbors or migrate to more distant locations. Some of the most dramatic rearrangements of animal cells occur once rapid cell divisions have produced a mass of cells. At this point, in a stage of development called **gastrulation**, cells in different parts of the mass rearrange themselves into three distinctive layers, which then give rise to the skin, gut, and other basic parts of the body. (Gastrulation is described in more detail in Ch. 47, Section 47.4.)

Later in development, some animal cells break away from their original sites and migrate to far-flung locations across the embryo. Depending on the species, these migratory cells give rise to germ cells (cells that produce sperm or eggs), pigment-containing cells, precursors of blood cells, certain nerve cells, and many others.

In short, animal development depends on the ability of groups of cells to coordinate their behavior in several ways: to change their shape, to make new arrangements with their neighbors, to break away from neighboring cells to begin migrations, and to come together in new locations.

Plant cells, in contrast, are encased in stiff cell walls and do not move. However, plants are masters at controlling both changes in cell shape and the orientation of the plane of cell division to determine where cells are added. The directions in which cells divide and expand are meticulously regulated to form stems, leaves, roots, and all other parts of the plant.

(a) Chicken embryo with normal (left) and defective (right) cell-death genes.

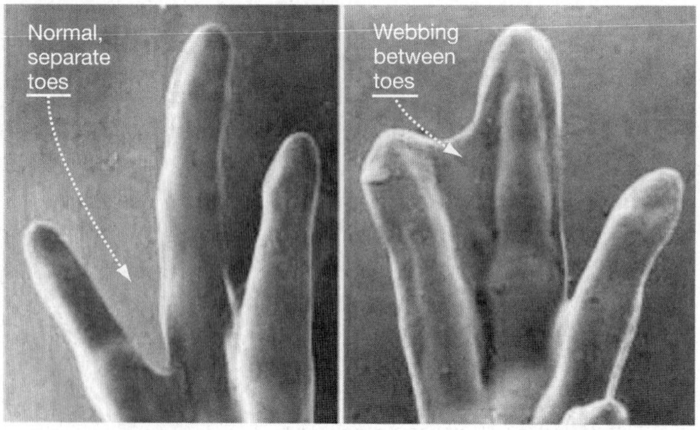

(b) Mouse embryo with normal (left) and defective (right) cell-death genes.

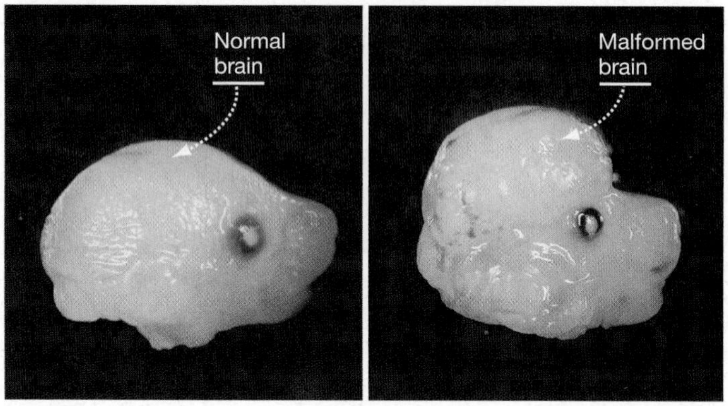

Figure 21.6 Programmed Cell Death Is a Normal Part of Development.

Programmed Cell Death

Often, signals trigger a cell to divide or differentiate, but sometimes they convey a message of death. **Programmed cell death** is a highly regulated and essential aspect of plant and animal development that occurs as tissues and organs take shape.

There are a few different forms of programmed cell death. In animals, the predominant one used in development is **apoptosis** (literally, "falling away"). Although biologists debate whether to call the programmed cell death that occurs in plants apoptosis, the important role of cell death in plant development is undisputed.

Apoptosis is widespread in animal development. For example, the feet of a chicken embryo initially develop with webs. Cells that are present between the toes must die to form separate toes (Figure 21.6a). The same is true for you—without apoptosis, your hands and feet would be webbed like a duck's feet. The webbing of a duck foot comes from the cells between the toes that do *not* undergo apoptosis.

Another important role for apoptosis is in the development of the vertebrate nervous system. Nerve cells are overproduced during development. As a normal part of "wiring" an effective nervous system, at least half of embryonic nerve cells are destroyed by apoptosis. A similar situation occurs in the mammalian immune system, where some differentiated immune system cells produce proteins that could harm the individual. These dangerous cells are eliminated by apoptosis.

Apoptosis was discovered in studies using the model organism *Caenorhabditis* (pronounced *see-no-rab-DIE-tiss*) *elegans*—a tiny roundworm with a name longer than the animal. The adult contains only about a thousand cells that develop according to reproducible and well-known pathways, and the embryonic worm is transparent. This last point makes it possible for biologists to follow individual cells throughout development. (For more on *C. elegans*, see **BioSkills 11**.)

As a *C. elegans* worm matures, exactly 131 of its cells undergo apoptosis. To explore how this happens, Hilary Ellis and Robert Horvitz studied mutant embryos that do not exhibit the normal pattern of cell death and identified the mutated genes responsible for the defect. Their early work uncovered two genes required for apoptosis. The researchers proposed that these were cell-death genes involved in a genetic program for apoptosis.

In follow-up work, investigators used bioinformatics to search databases of DNA sequences (Ch. 20, Section 20.3) and discovered that mice have similar genes. When a research team used genetic engineering techniques to reduce cell-death gene expression in mice, the embryos had severe malformation of the head caused by an abnormally large brain (Figure 21.6b). The defect occurred because nerve cells that normally die during development survived.

The same cell-death genes exist in humans and other mammals. Abnormal apoptosis—either too much or too little—occurs in dozens of diseases. For example, inappropriate activation of programmed cell death is involved in some devastating neurodegenerative diseases such as ALS (Lou Gehrig's disease), and a failure to undergo apoptosis contributes to many cancers.

The developmental processes you've learned about so far are the most important events that establish a multicellular organism from a single cell. But how are the different parts of an embryo specified? What determines the head and the tail, the back and the belly, and everything in between? These fundamental questions were first answered in studies of the fruit fly *Drosophila melanogaster*.

CHECK YOUR UNDERSTANDING

✔ If you understood this section, you should be able to ...

1. Use the following information to make a prediction: Cells removed from one region of a frog embryo differentiate into skin cells when they are isolated. In contrast, when these same cells are placed next to cells from a different region of the embryo, they differentiate into muscle cells. Is the specification of muscle cells predicted to occur using cytoplasmic determinants or induction?
2. Predict whether a mutation that causes the loss of a cell-surface receptor is more likely to alter cell specification by cytoplasmic determinants or by induction.
3. Explain why researchers concluded that there are normal gene products that kill some cells during development.

Answers are available in Appendix A.

21.4 Establishing the Body Plan

An early event in development is establishing the body plan—the position of different body parts in three dimensions. Position within an animal embryo and adult (Figure 21.7) is determined by where cells are located along three axes that, like an *x, y, z* coordinate system, specify points within three-dimensional space.

After you complete this section, you should be able to...

- Explain how an animal body plan is established.

1. One axis runs **anterior** (toward the head) to **posterior** (toward the tail).
2. One axis runs **dorsal** (toward the back) to **ventral** (toward the belly).
3. One axis runs left to right.

How do cells "learn" where they are within the developing embryo? In essence, different signal molecules are present at different concentrations along each of the axes. Cells read the types and amounts of these signals to establish where they are, and the cells develop according to their position.

Recall that information from signaling molecules is transmitted through receptors and signal transduction cascades that often activate transcription factors (Ch. 11, Section 11.3). Different signals—or different concentrations of the same signal molecule—are found at different positions within the embryo and at different times during development. As a result of signals that change with time and position, distinct transcription factors turn specific genes on or off, and the patterns of gene expression at successive stages of development determine the fate of each cell.

Let's consider some important ways transcription factors come to vary among embryonic cells. Although you'll be analyzing what happens as a fruit fly embryo develops, keep an important point in mind: Principles that were discovered in fruit flies are relevant to all multicellular organisms—from cedar trees to humans.

(a) The three body axes observed in humans and other animals ...

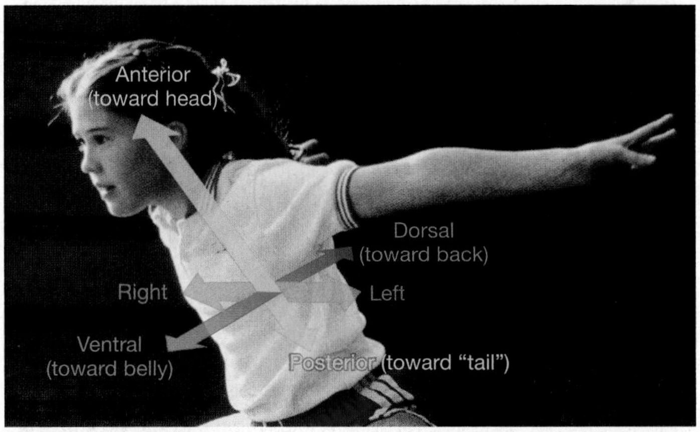

(b) ... are initially established in embryos.

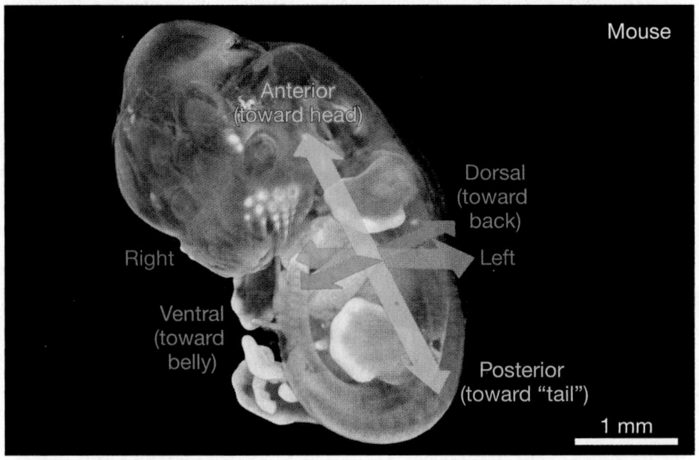

Figure 21.7 Most Animals Have Three Body Axes.

Morphogens Set Up the Body Axes in *Drosophila*

Biologists use the term **pattern formation** to describe the events that determine the spatial organization of cells in an embryo. If a molecule signals that a target cell is in the anterior or posterior, or dorsal or ventral side of an embryo, then that molecule is involved in pattern formation.

Many molecules that work in pattern formation exist in a concentration gradient, with high concentrations near the source of the molecule and lower concentrations farther away. If embryonic cells detect these different concentrations and use this information to determine their location, then this molecule is a **morphogen**. In flies and most other organisms, morphogens play central roles in setting up the body axes.

Complex patterns don't arise all at once. Instead, pattern formation is progressive. The major elements of the body—the anterior–posterior, dorsal–ventral, and left–right axes—are set up first by morphogens. Some of the genes activated by these morphogens turn on other genes that generate new signal molecules that provide more specific information about parts of the body. As development continues, the process repeats: New signals arrive and activate genes that specify finer and finer regions within the embryo. Development unfolds in a cascade of events unleashed by a few key initial signals.

Discovery of the Bicoid Morphogen Morphogens had long been proposed on theoretical grounds, but confirmation of their existence emerged from work on the fruit fly *Drosophila melanogaster* (BioSkills 11)—a model organism whose developmental stages have been studied intensively.

Christiane Nüsslein-Volhard and Hans Fröhnhofer started work in the mid-1980s on the early stages of pattern formation in *Drosophila* embryos and larvae. A **larva** (plural: **larvae**) is a juvenile form between the hatched embryo and the adult (Figure 21.8 on page 438). Nüsslein-Volhard and Fröhnhofer took a genetic approach and searched for mutants that exhibited altered patterns along the anterior–posterior and dorsal–ventral axes of early stage larvae.

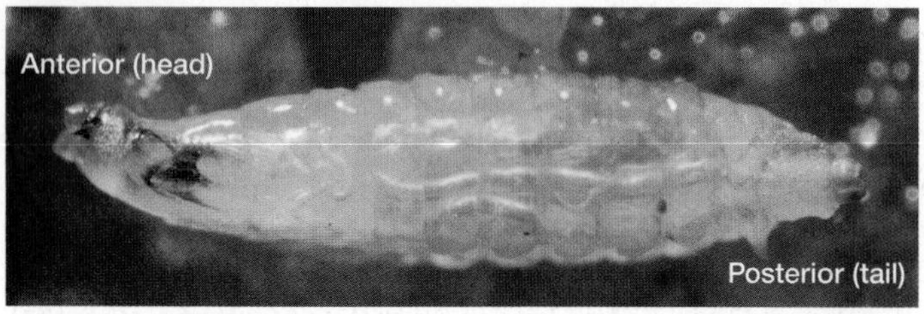

Figure 21.8 A *Drosophila* larva.

There was a twist, however, in their method that was demanded by the developmental biology of *Drosophila* and most other insects. In *Drosophila*, the body axes are laid down in the egg before fertilization. The egg has distinct front-and-back and up-and-down poles that foreshadow the anterior–posterior and dorsal–ventral axes of the future embryo.

The significance of establishing these axes in the egg is that genes in the mother, not the embryo, are responsible for setting the embryo's axes. In other words, the mutations that altered patterning sought by Nüsslein-Volhard and Fröhnhofer turned out to be mutations that alter gene expression in the mother but affect the phenotype of her offspring. These types of mutations are called **maternal effect mutations**.

To understand the types of alterations the researchers searched for, you have to understand body **segments**. Insects such as *Drosophila*, and many other animals, are built of a set of repeating units called segments (**Figure 21.9a**). Segments arise early in development and produce characteristic body structures such as wings, legs, and antennae.

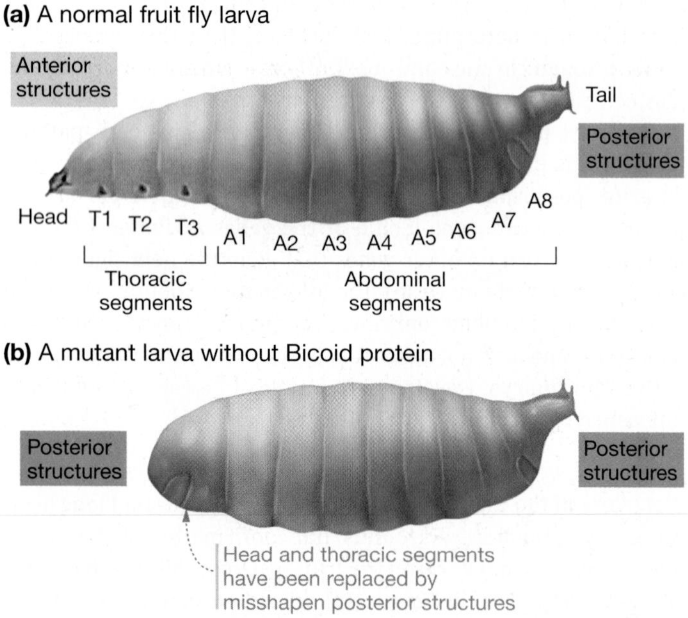

Figure 21.9 The *bicoid* Pattern-Formation Mutant Lacks Anterior Portions of the Body. (a) A normal *Drosophila* larva. The blue and yellow colors are added here to suggest a gradation of segment types moving from the anterior (yellow) and posterior (blue) ends of the larva. **(b)** Larvae produced by mothers who are homozygous for *bicoid* loss-of-function mutations do not develop anterior segments. Instead, the segments that normally develop into the head and thorax of the adult are replaced with misshapen segments found at the posterior-most part of the normal larva.

In their search for genes that control the body plan, Nüsslein-Volhard and Fröhnhofer discovered many mutants with bizarre alterations of the normal pattern of segments in embryos and larvae. A standout was a headless mutant they called *bicoid* (two-tailed) that replaced the anterior segments that form the head and thorax with misshapen segments that are normally found at the posterior end of the larvae (**Figure 21.9b**).

Based on this phenotype, Nüsslein-Volhard and Fröhnhofer were convinced that the product of the normal *bicoid* gene must provide positional information needed to specify the anterior region of *Drosophila*. They hypothesized that the *bicoid* gene coded for a morphogen that tells cells where they are along the anterior–posterior body axis.

The Importance of Morphogen Concentration Gradients Nüsslein-Volhard and her colleagues set out to test their hypothesis that *bicoid* encodes a morphogen. One way they did this began with DNA cloning and sequencing (see Ch. 20, Sections 20.1 and 20.3, and **BioSkills 10**). Working with the cloned *bicoid* gene, they used **in situ** (literally, "in place") **hybridization** to find where *bicoid* mRNAs are located in whole embryos. In situ hybridization works by adding a label to single-stranded DNA or RNA molecules to create a probe that is complementary in sequence to the mRNA of interest in an embryo or any other biological specimen (see **BioSkills 6**).

Figure 21.10 shows a form of this method—fluorescence in situ hybridization, or FISH for short—that makes use of a probe labeled with a fluorescent molecule. Recall that a fluorescent molecule absorbs light at one wavelength and emits the light at a different wavelength (Ch. 10, Section 10.2). In the example shown in Figure 21.10, the probe sequence was designed to bind to *bicoid* mRNA. As a result, when viewed by fluorescence microscopy (see **BioSkills 9**), the probes that mark the location of *bicoid* mRNAs can be detected by the glow of their emitted light.

Nüsslein-Volhard's group found that the *bicoid* mRNA was located in the anterior region of early embryos, as shown in step 5 of Figure 21.10. The clustering of *bicoid* mRNA in the anterior tip is achieved gradually, beginning even before fertilization occurs. The egg does not produce *bicoid* mRNA; instead, maternal cells connected to the developing egg transfer *bicoid* mRNA into the egg, where the *bicoid* mRNA is then actively concentrated at the future anterior region. The mRNA is not translated until fertilization.

Later work by Nüsslein-Volhard and Wolfgang Driever indicated that when these mRNAs are translated, the protein product diffuses to form a concentration gradient. More recent findings show that the *bicoid* mRNA itself is distributed in a gradient and that the Bicoid protein gradient is dictated by the distribution of the mRNA. Whatever the details, Bicoid protein is abundant in the anterior and decreases to progressively lower concentrations in the posterior.

Recall that forming a concentration gradient is a key criterion of a morphogen. But, does Bicoid meet the other requirement—does it provide spatial information? Would altering the Bicoid protein gradient alter the body pattern along the anterior–posterior axis?

To answer this question, Nüsslein-Volhard and her colleagues prepared *bicoid* mRNAs and injected them into the anterior end

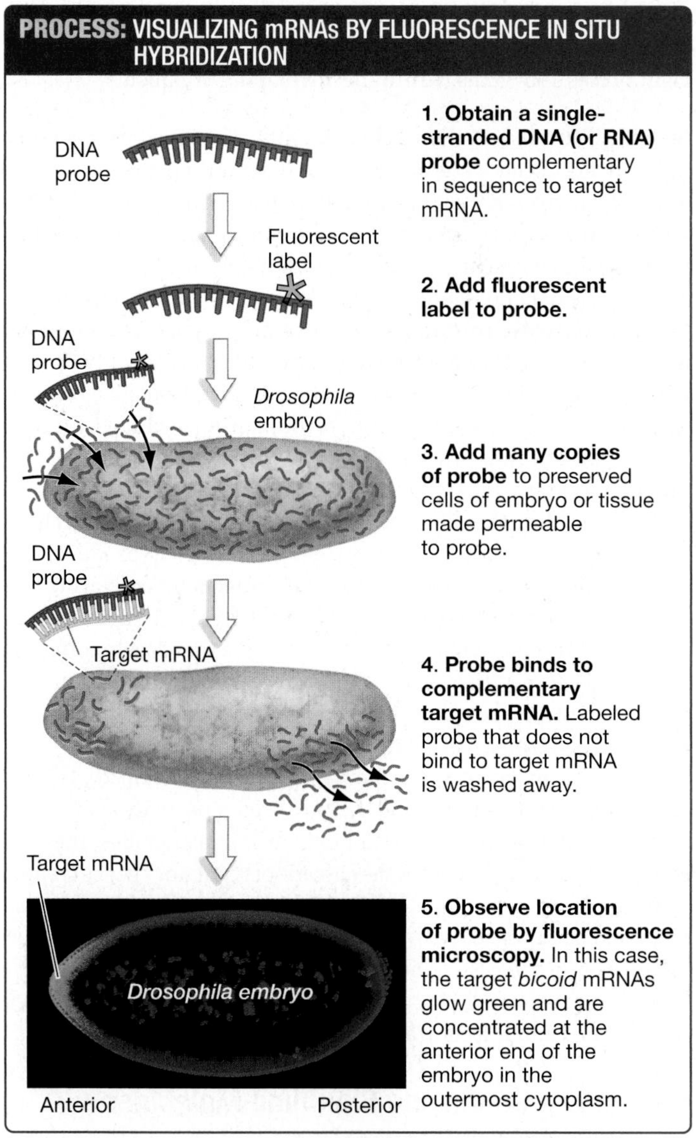

Figure 21.10 Fluorescence In Situ Hybridization (FISH) Allows Researchers to Pinpoint the Location of Specific mRNAs. The micrograph in step 5 shows the location of the *bicoid* mRNA (green) in an early fruit fly embryo. (Blue stain indicates DNA.)

of a normal embryo. This altered the *bicoid* mRNA and protein gradients. In turn, the body pattern was altered, with anterior body parts developing farther than normal toward the posterior. This crucial piece of evidence supports the hypothesis that Bicoid is a morphogen.

In fact, follow-up work showed that no matter where *bicoid* mRNAs were injected, anterior structures were produced. Heads followed by anterior segments could be made in the middle of embryos and even on what amounts to the "tail" of an embryo.

How does Bicoid work? Sequencing the *bicoid* gene provided the first evidence that Bicoid is a regulatory transcription factor. Later work confirmed that the Bicoid protein binds to enhancers in DNA and activates genes required for the formation of anterior structures. In effect, cells learn their position along much of the anterior–posterior axis through the concentration of Bicoid.

Higher levels of Bicoid result in higher levels of transcription of genes with Bicoid binding sites. In this way, high Bicoid concentrations tell a cell that it's in the anterior, and lower concentrations tell the cell that it's more posterior. When Bicoid is absent due to a mutation, cells throughout the embryo get a "you're in the posterior" message from a different morphogen that has its source in the posterior of the embryo—leading to the mutant phenotype you saw in Figure 21.10. ✔ **If you understand what causes the *bicoid* phenotype, you should be able to predict the phenotype of a mutant that produces twice the normal levels of Bicoid.**

Genetic Regulatory Cascades Provide Increasingly Specific Positional Information

Work on *bicoid* revealed the importance of morphogens in development. Another fundamental developmental principle—one common to both plants and animals—soon emerged: Genetic regulatory cascades progressively provide more and more detailed information about where cells are located and what they are to become.

Genetic Regulatory Cascades A **genetic regulatory cascade** is a set of regulatory genes that are linked in such a way that one initially activated gene turns on the expression of other regulatory genes, which in turn trigger the expression of yet more regulatory genes. Each level in a genetic regulatory cascade provides more specific information about where a cell is located in the embryo and what it should become. The cascade controlling early *Drosophila* development involves the following genes and gene products:

1. ***Maternal effect genes*** Morphogens encoded by maternal effect genes, such as *bicoid*, define the anterior–posterior axis of the early embryo.
2. ***Gap genes*** Gap gene products, expressed under the control of maternal-effect morphogens, mark out large, coarse subdivisions along the anterior–posterior axis.
3. ***Pair-rule genes*** Pair-rule genes are expressed next, primarily in response to gap gene products. Pair-rule genes are expressed in alternating bands. Each band roughly corresponds to a body segment that is forming along the embryo.
4. ***Segment polarity genes*** Pair-rule genes turn on segment polarity genes that are expressed in a portion of each segment. Segment polarity gene products define front, middle, and rear subregions within the segments laid down by the pair-rule genes.
5. **Hox *genes*** Even though body segments have been established, the unique identity of each segment has not yet been specified. *Hox* genes, which are activated by gene products at many different levels of the regulatory cascade, specify what type of adult structure will develop from each segment in the mature organism.
6. ***Effector genes*** Expression of effector genes is activated by *Hox* gene products working with segment polarity and pair-rule gene products. All the genes upstream of effector genes in the cascade are involved in regulation. They are

managers In contrast, the effector genes are like workers that actually get the job of development done by executing processes such as cell proliferation, cell movement, cell–cell interactions, programmed cell death (see Table 21.1), and ultimately differentiation.

Figure 21.11 summarizes the *Drosophila* genetic regulatory cascade. The take-home message? Developmental genes are linked and work within genetic regulatory cascades. Morphogens at the top of the cascade trigger the production of other regulatory signals and transcription factors, which initiate the production of yet another set of signals and regulatory proteins, and so on down the chain. Ultimately, the genes needed to form adult body parts like antennae, wings, and legs are expressed in the right places and times within the developing organism.

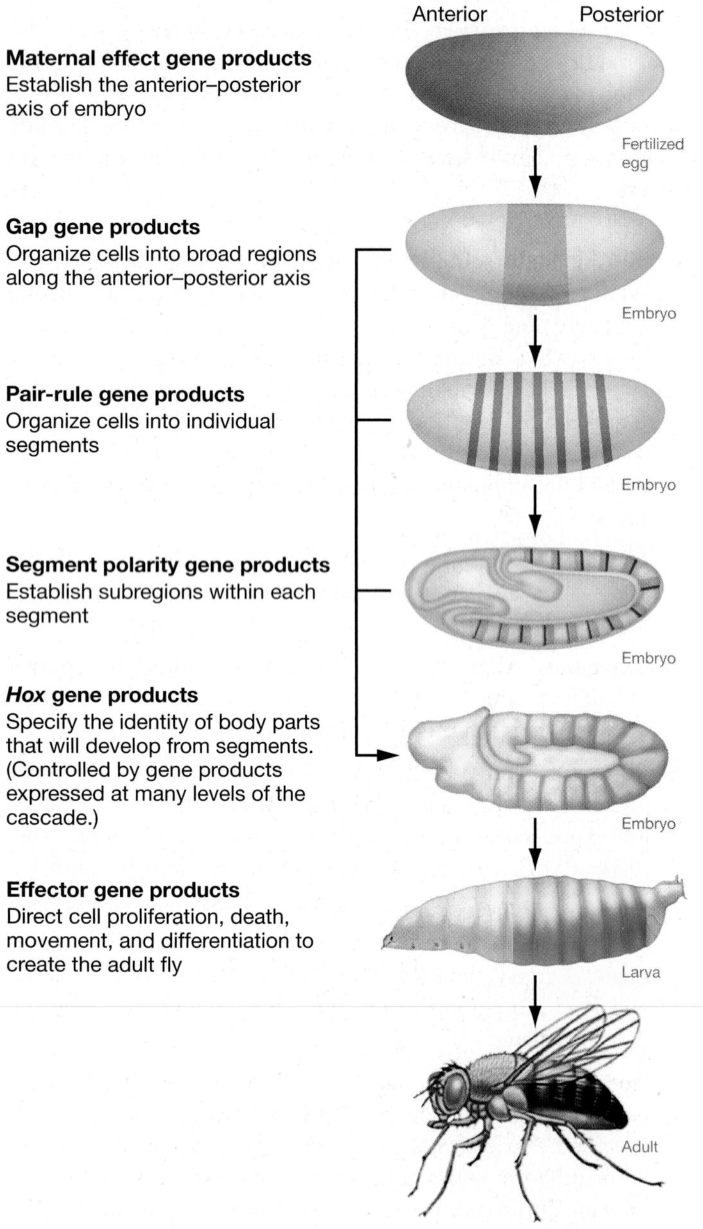

Figure 21.11 **A Genetic Regulatory Cascade in *Drosophila*.** This cascade establishes the positions and identities of future body parts along the anterior-to-posterior axis. Different colors mark where each gene product is expressed. There are other interactions between classes of gene products that are not shown here.

Insights Provided by Homeotic Mutants Some of the strangest mutants ever seen were the key to understanding how particular parts of the fly—and virtually all animals—acquire their identity. More than 100 years ago, researchers came across insects, including fruit flies, with body parts in the wrong place. This bizarre outcome is called homeosis, and the types of mutations that cause it are termed **homeotic mutations**. For example, homeotic mutations in *Drosophila* can turn a segment in the middle part of the body into a segment just like the one that lies in front of it. Instead of bearing a pair of small balancing structures, the transformed segment now bears a pair of wings—as does the normal segment in front of it. The mutant has four wings instead of the normal two (Figure 21.12).

Decades after the discovery of homeotic mutants, researchers identified the **homeotic**, or ***Hox*, genes**. *Hox* gene products are transcription factors that achieve their effects by activating the expression of effector genes (see Figure 21.11) that ultimately lead to the production of particular proteins that create differentiated cells. Every *Hox* gene contains a conserved DNA sequence called the **homeobox** that codes for the DNA-binding domain of a *Hox*-encoded transcription factor.

Hox genes make up a gene family (see Ch. 20, Section 20.4) in all organisms that have them, and they are used repeatedly at different stages in the development of all familiar animals. In flies, the *Hox* genes are first activated after the cascade of regulatory genes creates body segments (see Figure 21.11). Particular combinations of transcription factors encoded by *Hox* genes give each segment its unique identity or, in the case of homeotic mutations, the wrong identity. Later in development, *Hox* genes specify the identity of many other body parts. Their effects are spectacular and far-reaching.

Regulatory Genes and Signaling Molecules Are Evolutionarily Conserved

Early genetic mapping studies revealed that the *Drosophila Hox* genes were present in clusters along a single chromosome, as shown in Figure 21.13a. Once methods to isolate and sequence DNA were available (Ch. 20, Section 20.3, and **BioSkills 10**), researchers applied them to investigate the *Hox* genes. The results of these molecular studies were stunning. One of the first surprises was that the order of *Hox* genes along the chromosome corresponds to the order of where the genes are expressed along the anterior-to-posterior axis.

Conservation of *Hox* Gene Organization With information about fruit fly *Hox* genes in hand, a few biologists thought it was worth looking for similar genes in other animals. This was a long shot, because it seemed unlikely that creatures as different as flies, sea urchins, mice, and humans would use the same genes to build such vastly different bodies. The gamble paid off. Virtually all animals were found to possess related sets of *Hox* genes. Incredibly, these *Hox* genes were organized into clusters and in an order that aligned with the *Drosophila* genes. Even more amazing, the curious relationship first seen in *Drosophila* between *Hox* gene order along the chromosome and where a particular *Hox*

Figure 21.12 Homeotic Mutants Have Structures in the Wrong Locations. These colorized scanning electron micrographs show a normal fruit fly (left) and homeotic mutants (middle and right). In *Ultrabithorax*, wings grow where balancing structures called halteres should be. In *Antennapedia*, legs grow from the head instead of antennae.

gene is expressed along the anterior–posterior axis was shared by all these animals.

A comparison of *Hox* genes in flies and mice (**Figure 21.13b**) reveals that there are more *Hox* genes in the mouse, but their chromosomal organization is similar. Notice how genes at the "left-hand" side of the *Hox* clusters are expressed in the anterior end of both embryos, while genes at the opposite end of the cluster are expressed in posterior regions. The discovery that similar regulatory genes exist in and are organized so similarly in such a diverse array of animals was extraordinary. Could the nearly unimaginable prove true—that *Hox* genes play similar roles in specifying the body parts of vastly different animals?

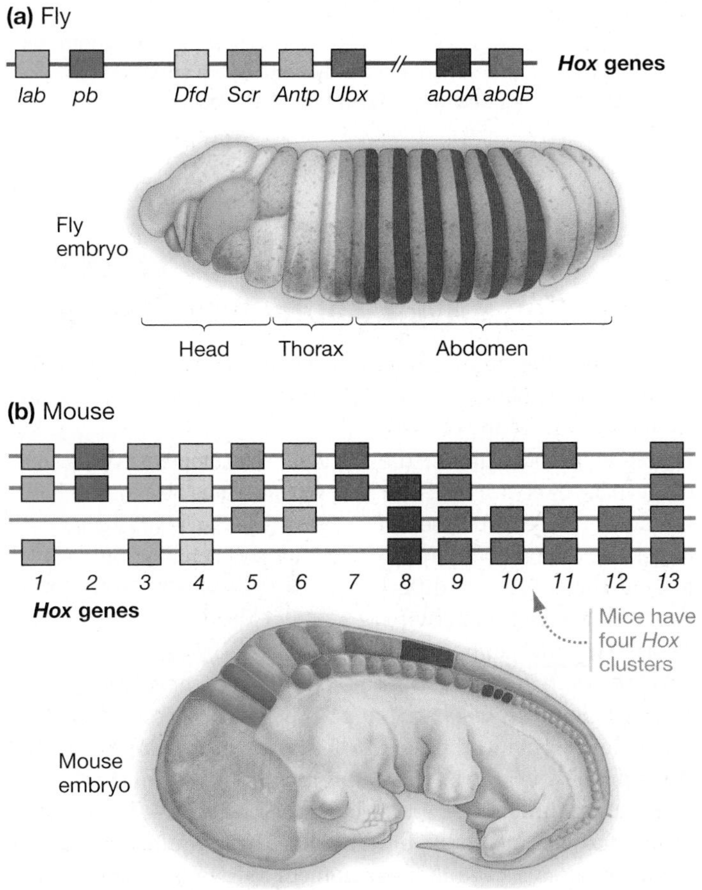

Figure 21.13 *Hox* Genes in Vastly Different Species Are Similar in Organization and Expression. The order of *Hox* genes on **(a)** fly chromosome 3 and **(b)** mouse chromosomes 6, 11, 15, and 2 corresponds with their pattern of expression in fly and mouse embryos. Matching colors of genes and regions in the embryo indicate where each gene is expressed at its highest levels. The genes represented by same-color boxes in the fly and mouse are evolutionarily conserved. The four *Hox* clusters in mouse arose during evolution through gene duplication divergence (see Ch. 25, Section 25.3).[1]

✔ What evidence would support the claim that the fly and mouse genes are related?

[1]For simplicity, two features are not shown in Figure 21.13: (1) The fly genes are present in two separate groups that were created by a relatively recent chromosome breakage in the fly lineage. (2) There is overlap in the regions of the body in which *Hox* genes are expressed.

Conservation of *Hox* Gene Function Evidence that *Hox* genes in animals other than *Drosophila* specify body parts was soon found. Mutations discovered in many animals, including humans, showed that alteration of *Hox* genes could produce defects in pattern formation, including switching one body part for another. Biologists concluded that for most animals, *Hox* genes play a central role in specifying which body structures to build.

This conclusion was supported dramatically when researchers in William McGinnis's lab introduced the mouse *HoxB6* gene, stripped of its normal regulatory sequences, into fruit fly eggs. The mouse *HoxB6* gene sequence is most similar to the *Antennapedia* (*Antp*) gene of flies (Figure 21.13a), the gene that when expressed in the wrong location causes legs to sprout on a fly's head instead of antennae (Figure 21.12). What did the mouse gene do when expressed in a fly? The mouse gene not only altered the development of the fly, it mimicked the effect of the fly *Antp* gene mutation—it replaced the fly's antennae with legs.

To interpret these observations, biologists hypothesize that the genes in *Hox* clusters of animals are **homologous**—meaning that they are similar because they are descended from genes in a common ancestor. This hypothesis implies that the first *Hox* genes arose very early in animal evolution, when the shared ancestor of flies, mice, and many other creatures lived. For roughly the past 600 million years, *Hox* gene products have been instrumental in directing the development of animals.

The take-home message from these studies is that key molecular mechanisms of pattern formation have been highly conserved during animal evolution. The discovery of these shared mechanisms is one of the most significant results to have emerged from studies of animal development. Although animal bodies are spectacularly diverse in size and shape, the underlying mechanisms responsible for their development are similar.

One Regulator Can Be Used in Different Ways

Genetic regulatory cascades and the evolutionary conservation of key developmental genes are general features of development. But there's another central organizing principle: During development, the same regulatory genes and signaling molecules are often used over and over in a variety of contexts.

The Wnt signal protein offers an example of repeated uses of the same molecule to achieve different ends. *Drosophila* has a single *Wnt* gene known as *wingless*. From the name, you can guess that *wingless* is involved in wing development—flies that have a particular mutated version of the *wingless* gene lack wings. However, the initial role of *wingless* during development is not in controlling wing formation, but in creating boundaries between segments. The list goes on—*wingless* is also a regulator of many developmental decisions. The *wingless* gene produces a signal protein, called Wingless, that binds to specific receptors on different target cells to control development of the leg, head, and digestive system as well as specifying particular nerve cells. In development, as in human communication, the context in which a signal such as Wingless is sent and received—its location, timing, and intensity—has a major effect on the signal's meaning and consequences.

In contrast to genes such as *wingless* that are used in many different ways during the development of one organism, recall the examples described earlier in which different organisms have related genes that control similar events in development. These are **tool-kit genes**—conserved sets of genes that code for signal proteins, signal-transduction pathway components, and transcription factors that direct related aspects of development in many different species.

Hox genes are an important example of tool-kit genes because they are present in a wide array of animals and are used to specify body structures. Similar tool-kit genes can, however, produce dramatically different structures if they are used at different times, in different places, and in different developmental contexts. For example, human arms, bat wings, and whale flippers are forelimbs whose development is controlled by similar tool-kit genes, yet the final structures are vastly different.

What controls how a tool kit gene works? The answer lies largely in the regulatory sequences of DNA. Mutation of these sequences allows conserved tool-kit genes to achieve different outcomes in different species. As you'll see next, changes in regulatory sequences have profound effects on evolution.

CHECK YOUR UNDERSTANDING

✔ **If you understood this section, you should be able to . . .**

1. Explain why a mating between a normal male and a female homozygous for a loss-of-function *bicoid* mutation produces misshapen embryos, but the reciprocal cross produces normally patterned embryos.
2. Propose the type of evidence needed to distinguish between Bicoid protein being an inducer of the anterior region and Bicoid protein being a morphogen.
3. Propose the type of evidence needed to support the hypothesis that individual segments of the fly embryo are specified only after larger regions of the fly embryo are specified.

Answers are available in Appendix A.

21.5 Changes in Developmental Gene Expression Drive Evolutionary Change

After you complete this section, you should be able to . . .

- Explain the relationship between changes in gene expression and evolutionary changes.

You've seen that for an embryo to develop, cells have to proliferate, move or expand, differentiate, interact in specific ways, and sometimes die. If any of these processes are disrupted, the embryo is likely to die. But if one of these processes is *modified*, the effect may be a structure with a different size, shape, or activity. As a result, the embryo will develop new features, and the adult will have a novel phenotype.

Once biologists began to understand the processes that bring about development, they realized that genetic changes altering these processes must be the foundation of evolutionary change. The increase in body size that has occurred during human evolution, for example, must have resulted from mutations that altered the signals, regulatory sequences in DNA, or transcription factors that are involved in the amount and timing of cell proliferation throughout the body.

An emerging research field called evolutionary developmental biology, or **evo-devo**, focuses on understanding how changes in developmentally important genes have led to the evolution of new forms such as the flower, the leaf, and animal limbs. As an example of this exciting work, let's consider how snakes came to be limbless. The fossil record shows that the ancestor of all snakes had four functional legs. This means that snake evolution proceeded by a reduction in limbs that existed in ancestral organisms.

How did limbs come to disappear in snakes? Researchers were able to answer this question through an understanding of the genes and signals involved in limb development.

Vertebrate limbs form from protrusions—called, logically enough, limb buds—that grow out from the sides of embryos. Scientists had long known that a small group of cells in each limb bud was a source of a signaling protein called Sonic hedgehog that is essential for limb development. The *sonic hedgehog* gene had been intensively studied and was found to contain a distant enhancer that controlled the expression of the Sonic hedgehog protein in developing limbs. Mutations in this enhancer region were known to cause limb malformations, including severe shortening of limbs in humans. Could modifications of this enhancer sequence be involved in the loss of limbs during the evolution of snakes?

Researchers discovered that key elements of the *sonic hedgehog* limb enhancer were conserved in all limbed vertebrates, but were missing in snakes. This suggested that degradation of the *sonic hedgehog* enhancer in the evolutionary lines leading to snakes might account for the limb loss. The loss of enhancer elements and lack of limbs in snakes is an interesting correlation, but is there stronger evidence that changes in the enhancer *caused* limb loss?

Using the CRISPR-Cas technology you learned about in an earlier chapter (Ch. 20, Section 20.5), biologists knocked out the *sonic hedgehog* limb enhancer in mice. They found that the knockout mice developed severely shortened limbs, a phenotype they called "serpentized." The researchers then used the serpentized mice as a test bed for the activity of *sonic hedgehog* limb enhancers from mice, humans, and the coelacanth, a species of fish relatively closely related to limbed vertebrates (see Ch. 32, Section 32.5). Could *sonic hedgehog* limb enhancers from these limbed (and finned) species rescue limb development in serpentized mice? What about the corresponding enhancer from snakes?

Remarkably, when the researchers used CRISPR-Cas genome editing to add back fish, mouse, or human *sonic hedgehog* limb enhancers to serpentized mice, normal mouse limb development was restored. In sharp contrast, *sonic hedgehog* limb enhancers from two divergent species of snakes—python and cobra—failed to rescue limb development in the test mice. Intriguingly, the enhancer from python promoted a tiny amount of leg development and the enhancer from cobra promoted no leg development at all. It turns out that pythons have tiny, stunted vestiges of hind limbs and cobras completely lack hind limbs.

✔ **If you understand these concepts, you should be able to explain why researchers first created a knockout line of mice lacking the *sonic hedgehog* limb enhancer before testing the activity of *sonic hedgehog* limb enhancers from different species.**

What does all this mean? It appears that snakes lost their limbs, at least in part, because of changes to a conserved gene enhancer sequence. This enhancer normally drives expression of the Sonic hedgehog signal molecule, which is required for limb development. In snakes, however, the enhancer barely functions or doesn't function at all. No enhancer activity, no Sonic hedgehog signal, no limbs. A profound evolutionary change—limblessness—has been propelled by a change in the regulation of a tool-kit gene for development.

CHECK YOUR UNDERSTANDING

✔ **If you understood this section, you should be able to . . .**

1. Propose a way that a mutation that falls outside of a gene's coding sequence can alter the body plan of an organism.
2. Explain how knowing about development can help with understanding evolution.

Answers are available in Appendix A.

CHAPTER 21 Review

For media, go to **Mastering Biology**

21.1 Genetic Equivalence and Differential Gene Expression in Development

- Cloning of plants and animals from differentiated adult cells demonstrates genetic equivalence—that is, all cells of an individual contain the same genetic information.
- Genetic equivalence implies that cells are different because they use a common set of genes in different ways. The expression of different genes from the same genome is differential gene expression.
- Transcriptional control through the production of different sets of transcription factors in different types of cells is the most important form of gene control during differentiation.

21.2 Cells Are Determined Before They Differentiate

- Commitment is the gradual process by which the developmental fate of an embryonic cell is specified, directing it to form a particular type of specialized cell.
- Determination is the final phase of commitment. A cell that reaches determination is essentially locked into a particular path of differentiation.
- Stem cells are undifferentiated cells that divide to produce a cell that differentiates and a stem cell that maintains the stem cell population.

- In the adult, stem cells provide a source of cells needed to replace differentiated cells that are lost or damaged.
- Induced pluripotent stem cells (iPS cells) are embryonic-like cells made by reversing determination and differentiation of specialized adult cells.
- iPS cells can be differentiated into diverse cell types.

21.3 Shared Developmental Processes

- During development, cells divide, interact, differentiate, move, expand or contract in a directed manner, and sometimes die.
- Cells have to divide in a regulated manner to promote growth of the body.
- Signaling molecules sent between cells provide a constant flow of information about where cells are in space and time.
- Cells undergo a step-by-step process of differentiation that leads to specialized cell types.
- Daughter cells can become different from one another through the effects of cytoplasmic determinants or induction.
- Development requires the controlled movement and expansion of cells. In animals, many cells move in a directed way throughout development. Plant cells cannot move because of the cell wall, but plant cells precisely control the directions of cell division and cell expansion to shape the embryo and the adult.
- Programmed cell death of some cells is an essential part of normal development.

21.4 Establishing the Body Plan

- Cells establish where they are in the body and how far along development has progressed because they receive and process a steady stream of signals.
- Morphogens are signals present in a gradient. Cells can establish where they are along a body axis by sensing the concentration of morphogen.
- Cells gradually acquire positional information using signals produced from genetic regulatory cascades, first establishing where they are in broad regions and later in smaller parts of each region.
- *Hox* genes specify the identity of body parts in many animals.
- Many signaling pathways are used over and over in development in different contexts to achieve different outcomes.
- A set of conserved tool-kit genes is used to control core developmental processes in a wide array of species.

21.5 Changes in Developmental Gene Expression Drive Evolutionary Change

- Mutations that alter the regulation of genes responsible for development lead to the evolution of new body sizes, shapes, and structures. These changes are important components of evolution.

Answers are available in Appendix A.

✔ TEST YOUR KNOWLEDGE

1. What is apoptosis?
   a. an experimental technique used to kill specific cells
   b. programmed cell death that is required for normal development
   c. a pathological condition observed only in damaged or diseased organisms
   d. a developmental mechanism unique to the roundworm *C. elegans*
2. In adult animals, ____________ are a source of undifferentiated cells that can divide to produce cells that can specialize.
3. What is a homeotic mutant?
   a. an individual with a structure located in the wrong place
   b. an individual with an abnormal head-to-tail axis
   c. an individual that is missing segments
   d. an individual with double the normal number of structures
4. A tool-kit gene is ____________.

✔ TEST YOUR UNDERSTANDING

5. **THINK CAREFULLY** A friend is interested in isolating genes that are expressed solely in liver cells but only has access to skin cells. She asks you for advice on whether to start her studies. What will you say?
6. The following predictions ask you to consider how genetic regulatory cascades provide positional information. Select True or False for each statement.
   T/F Mutation of a gene at one level of a regulatory cascade will affect the expression of genes at all levels of the cascade.
   T/F Mutation of a gene that is expressed later in a regulatory cascade will affect a smaller region of the body than mutation of gene that is expressed early in the cascade.
   T/F In the regulatory cascade used by *Drosophila,* a gene at one level of the cascade will be controlled only by genes at the level immediately above it.
   T/F Genes that control the largest regions of the *Drosophila* embryo are not transcribed in the embryo.
7. What is the connection between genetic regulatory cascades and the observation that differentiation is a step-by-step process?
8. Which of the following provides the strongest evidence for the conservation of tool-kit genes?
   a. Bicoid moved from one fly embryo into the posterior of another fly embryo causes the formation of two head regions.
   b. Mutation of an unrelated gene in another species of fly has a similar effect to mutation of *bicoid* in *Drosophila.*
   c. A mouse *Hox* gene can be used to take over the function of a mutated *Drosophila Hox* gene.
   d. Sheep can be cloned by fusing a differentiated adult cell with an enucleated egg.

✔ TEST YOUR PROBLEM-SOLVING SKILLS

9. **QUANTITATIVE** Imagine a situation in which a morphogen has its source at the posterior end of a *Drosophila* embryo. Every 100 μm from the posterior pole, the morphogen concentration decreases by half. If a cell required 1/16th the amount of morphogen found at the posterior pole to form part of a leg, how far from the posterior pole would the leg form?
   a. 100 μm
   b. 160 μm
   c. 400 μm
   d. 1600 μm
10. **PROCESS OF SCIENCE** Some stickleback fish develop protective spines, and other stickleback fish are spineless. Spine development is controlled by the expression of a gene known as *Pitx1*. The spineless phenotype is due to a mutation in *Pitx1* that results in no expression of *Pitx1* during development in regions where spines would otherwise form. When scientists compared the *Pitx1* coding

sequence in spined and spineless fish, they found this sequence was the same in both types of fish. Propose plausible hypotheses for the location of this mutation and for how it alters spine development.

✔ PUT IT ALL TOGETHER: Case Study

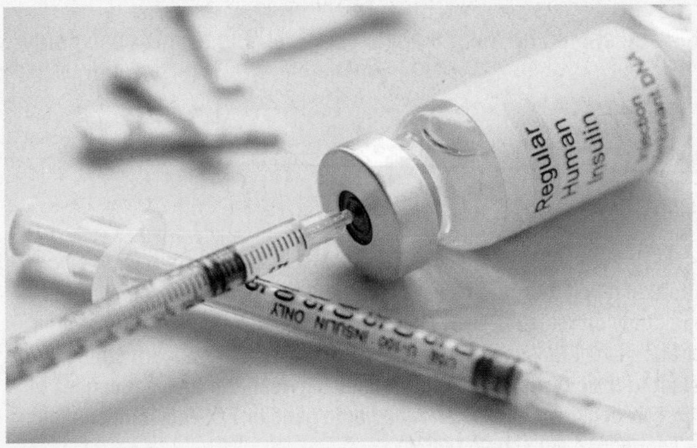

Will iPS cells be the cure for diabetes?

Type I diabetes is a form of diabetes that is due to the loss of insulin-producing cells of the pancreas. The potential of stem cells—in particular, induced pluripotent stem (iPS) cells—for therapy has gotten a lot of press. Can iPS cells be coaxed to differentiate into insulin-producing cells? Is there any evidence that they might actually work to cure type I diabetes?

11. What are iPS cells?
   a. cells taken from early human embryos
   b. cells taken from the pancreas of people without diabetes
   c. cells derived by de-differentiating specialized adult cells
   d. cells derived by differentiating pancreas precursor cells

12. If researchers were attempting to stimulate the differentiation of iPS cells, which of the following would they most likely add to the cell-culture medium (the liquid surrounding the cells)?
   a. activin A, an extracellular signal protein
   b. Sox-2, a transcription factor active in early development
   c. Grb-2, an intracellular signal transduction protein
   d. lactase, an enzyme that catalyzes the breakdown of lactose

13. When researchers are trying to differentiate iPS cells, they often monitor the differentiation of the cells by looking for the expression of particular proteins. What is the logic behind this method?

14. **PROCESS OF SCIENCE** iPS cells are typically generated by inserting genes into the genome of differentiated cells. This method alters the genome and raises some concern about unintended side effects. A team of scientists recently tested a new way to generate iPS cells without permanently modifying the genome. They succeeded in generating cells that looked like iPS cells when grown in culture. To more rigorously examine the characteristics of these cells, the researchers determined the patterns of expression of key genes that mark undifferentiated cells in the two potential iPS cell lines and compared these to the pattern seen in conventionally produced iPS cells. They used a technique called RT-PCR, which involves the isolation of mRNA, production of a cDNA copy, amplification of the cDNA by PCR, and visualization of the PCR products by gel electrophoresis (see **BioSkills 10** for a review of cDNAs and PCR; see **BioSkills 6** for how to interpret bands on a gel). The results are shown in the figure below. Do the results support the conclusion that the new method works to produce iPS cells? What further studies would you propose to test if the cells produced by the new procedure behave like conventionally produced iPS cells?

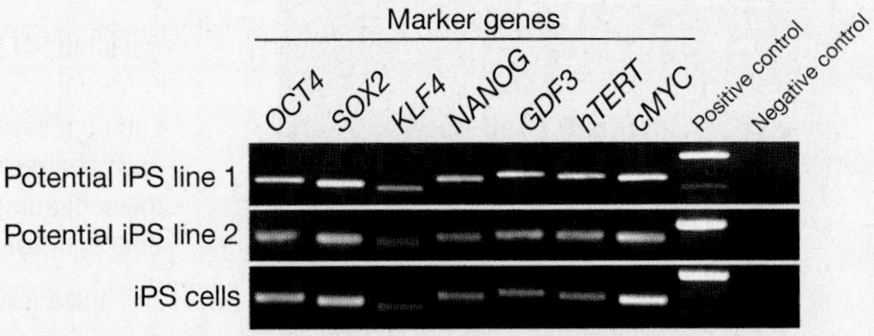

Source: C. B. Driscoll, J. M. Tonne, M. El Katib, et al. 2015. *Stem Cell Research and Therapy* 6: 48–61.

15. Many of the proteins used to induce differentiation in human iPS cells also work to differentiate mouse iPS cells. Why is this ability of a human protein to work in mouse cells *not* a surprise?

16. **SOCIETY** A research team has explored the possibility of differentiating human embryonic stem (hES) cells into pancreatic cells that might be used to treat diabetes. In contrast to iPS cells, hES cells are taken from an early human embryo, and obtaining them destroys the embryo. The team showed that cells differentiated from hES cells could cure diabetes in a mouse model of the disease. If the same outcome of curing diabetes could be achieved in humans, would there be any advantage to starting with hES or iPS cells to produce pancreatic cells in order to gain widespread acceptance of this therapeutic approach? Why?

Mastering Biology

Students Go to Mastering™ Biology for assignments, the eText, and the Study Area with animations, practice tests, and activities.

Professors Go to Mastering™ Biology for automatically graded tutorials and questions that you can assign to your students, plus Instructor Resources.

END-OF-UNIT
CASE STUDY

UNIT 3

PUT IT ALL TOGETHER

For an introduction to the Mystery of the Newt case study, see page 17.

The tetrodotoxin (TTX) found in rough-skinned newts is not only deadly to campers, it can also kill predators of newts, such as snakes. What's surprising is that while some snakes are very sensitive to the toxin, others are extraordinarily resistant. In fact, some snakes can survive amounts of tetrodotoxin that would kill 900 people! How can some snakes be so resistant? In large part, it's in the genes.

✔ If you understand Unit 3, you should be able to apply your learning to this case study:

How Can Mutations Save a Snake?

Recall that TTX plugs the pore of a voltage-gated sodium channel needed for muscle contraction (see Unit 2 case study on pages 276–277). No sodium flow, no muscle contraction, and no movement or respiration in a TTX-sensitive snake. Scientists hoping to learn what makes some snakes resistant to toxic newts found differences between the DNA sequences of the sodium channel gene in TTX-sensitive snakes and the same gene in TTX-resistant snakes. Small portions of the gene sequences are shown below:

Tetrodotoxin-sensitive:	5′-GGT TTG GAC GGC TTG-3′	(template strand)
	3′-CCA AAC CTG CCG AAC-5′	(non-template strand)
Tetrodotoxin-resistant:	5′-GGT TTG AAC GGC TTG-3′	(template strand)
	3′-CCA AAC TTG CCG AAC-5′	(non-template strand)

1. What are the predicted deoxyribonucleotide sequences of the corresponding mRNAs? (See Section 16.3)
2. Assuming that each block of three deoxynucleotides in the sequences above specifies a codon in mRNA, what are the predicted amino acid sequences that would be translated from each mRNA? (See Section 16.3)
3. Assuming that a mutation occurred to convert the TTX-sensitive form of the channel gene into the TTX-resistant form, then this is a ________ mutation. (See Section 16.4)
   a. nonsense **b.** silent **c.** missense **d.** neutral
4. The TTX-sensitive and TTX-resistant sequences represent:
   a. different alleles **b.** different genes **c.** different mutants

The gene sequence for the TTX-resistant form of the sodium channel ultimately results in weaker binding of TTX to the channel pore.

5. A snake heterozygous for the TTX-sensitive and TTX-resistant forms of the gene expresses both alleles. In this case, the TTX-resistant allele is likely to act as which type of allele for the phenotype of sodium ions moving across nerve and muscle cell membranes? (Sections 14.2 and 14.5)
   a. dominant
   b. codominant
   c. incompletely dominant
   d. recessive

If this snake was TTX-resistant, it would be impervious to a toxic newt.

Biologists soon discovered that many different amino acids in the pore region of the sodium channel protein influence TTX sensitivity. A two-dimensional diagram of the protein's snaking path across the plasma membrane is shown in **Figure 1**. Four regions of the protein come together to form the pore, with pore-forming regions 3 and 4 frequently containing altered amino acids in resistant snakes.

6. Explain in broad terms how mutations in the coding sequences for pore-forming regions 3 and 4 could influence tetrodotoxin sensitivity of the sodium channel protein. (See Section 16.2)

Figure 1

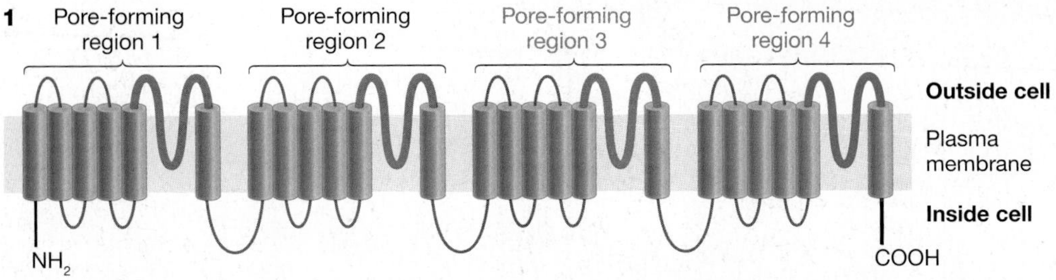

The results of many studies examining TTX sensitivity of snake populations and the amino acid sequences of pore-forming regions 3 and 4 of the sodium channel are shown in **Figure 2**. The lineage tree shows evolutionary relationships among the snakes; the table shows the amino acid sequences of pore-forming regions 3 and 4 for each snake population (see Ch 3., Figure 3.2 for a key to single-letter amino acid abbreviations). The topmost row of sequence is from a TTX-sensitive species; below, amino acid substitutions are indicated for snakes from other populations. Colored dots indicate TTX sensitivity levels—sensitive (•), modestly resistant (••), and highly resistant (•••), respectively.

7. Based on the data in Figure 2, select True or False for each of the following statements. (See Section 16.4)

T/F Mutations that result in amino acid changes in the pore-forming portions of the sodium channel protein account for TTX resistance in all of the snakes considered in this study.

T/F The data suggest that either another region of the sodium channel gene or a different gene can influence TTX sensitivity.

T/F The data suggest that the greater the number of amino acid substitutions in pore forming regions, the higher the TTX resistance.

T/F Many different mutations can lead to TTX resistance.

Mutations with an important role in evolution can alter either coding or non-coding regions of genes. (See Section 21.4)

8. Predict whether it is more likely for TTX resistance to evolve by mutations in the coding region of the sodium channel gene or by mutations in regulatory regions of the gene.

In the next unit, you'll extend the concepts of genetics to discover how populations evolve, including populations of toxic newts and resistant snakes.

Figure 2

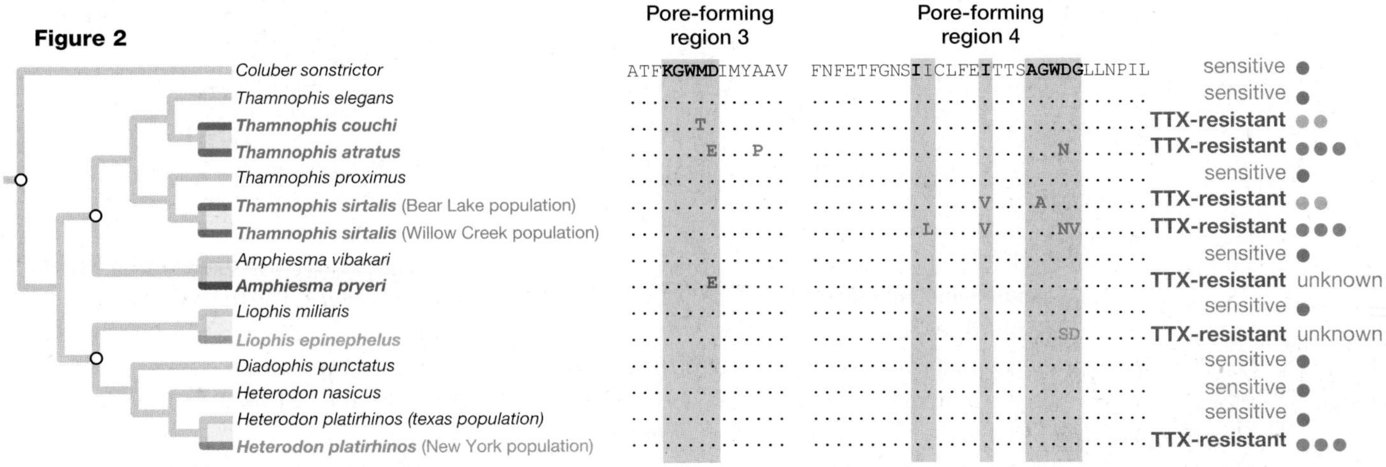

DATA: C. R. Feldman et al. 2008. *Heredity* 116: 84–91, Figure 3.

Each unit ends with a continuation of this story. ∎

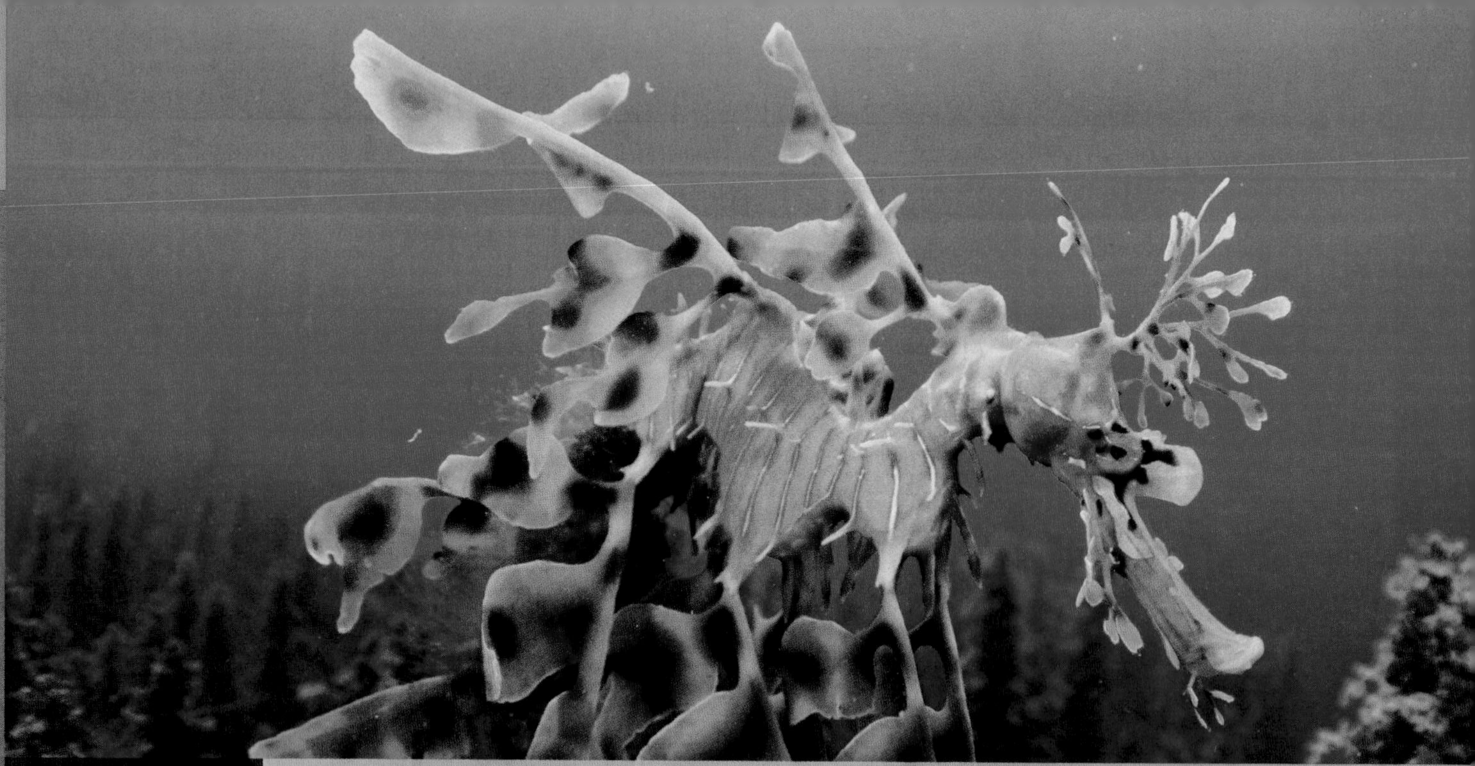

22 Evolution by Natural Selection

Natural selection explains how populations become well suited to their environments over time. The shape and coloration of leafy sea dragons (fishes closely related to seahorses) are heritable traits that help them to hide from predators.

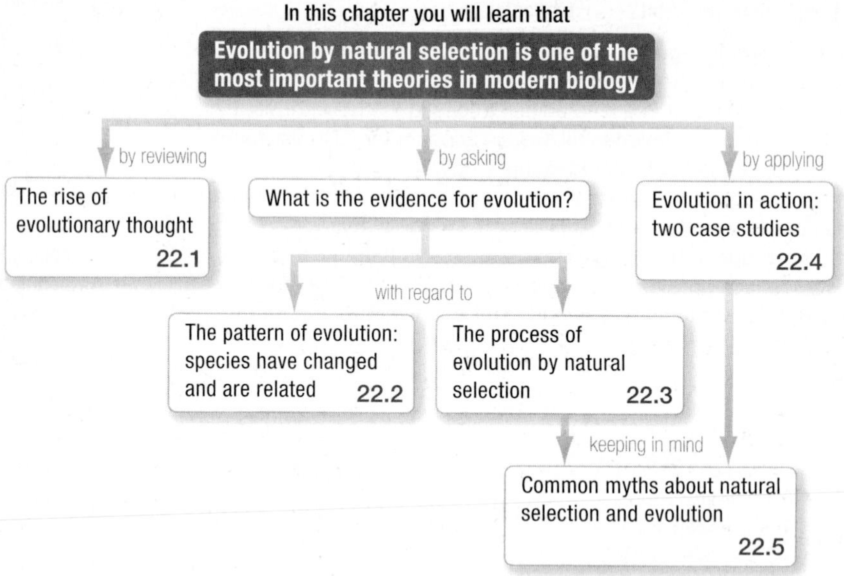

BIG PICTURE

This chapter is part of the Big Picture. See how on pages 532–533.

This chapter is about one of the great ideas in science: the theory of *evolution by natural selection*, formulated independently by Charles Darwin and Alfred Russel Wallace. The theory explains how **populations**—individuals of the same species that live in the same area at the same time—have come to be adapted to environments ranging from arctic tundra to tropical wet forest. It revealed one of the five key attributes of life: Populations of organisms evolve (Ch. 1, Section 1.1). In other words, the heritable characteristics of populations change over time.

Evolution by natural selection is one of the best supported and most important theories in the history of scientific research. But like most scientific breakthroughs, this one did not come easily. When Darwin published his theory in 1859 in a book called *On the Origin of Species by Means of Natural Selection*, it unleashed a firestorm of protest throughout Europe.

In everyday English, the word "theory" suggests a thoughtful guess, but a scientific **theory** is an explanation for a broad class of observations that is widely supported by overwhelming evidence. (For help with problematic words, see **BioSkills 17**.) Scientific theories usually have two components: a pattern and a process.

1. The *pattern component* is a statement that summarizes a series of observations about the natural world. The pattern component is about facts—about how things *are* in nature.
2. The *process component* is a mechanism that produces that pattern or set of observations.

This chapter begins by giving you an overview of the development of evolutionary thought. Then you'll examine the pattern and process components of the theory of evolution by natural selection. This chapter also clarifies several common misconceptions about evolution. For example, did you know there are other processes of evolution besides natural selection?

22.1 The Rise of Evolutionary Thought

People often describe the theory of evolution by natural selection as revolutionary. A scientific revolution overturns an existing idea about how nature works and replaces it with a radically different idea. Revolutionary scientific theories include Copernicus's theory of the Sun as the center of our solar system, Newton's laws of motion and theory of gravitation, the germ theory of disease, the theory of plate tectonics, and Einstein's general theory of relativity. These theories are the foundation of modern science.

After you complete this section, you should be able to ...

- Contrast typological thinking, Lamarckian evolution, and Darwinian evolution.

The advance of the theory of evolution by natural section represented a profound scientific revolution. The idea that Darwin and Wallace overturned—that species were supernaturally, not naturally, created—had dominated thinking about the nature of organisms in Western civilization for over 2000 years.

To help you understand the conceptual change that occurred, let's walk through a time sequence of visual models of the diversity of life, summarized in Figure 22.1.

Plato and Typological Thinking

The Greek philosopher Plato, who lived over 2400 years ago in ancient Greece, claimed that every organism was an example of a perfect essence, or type, created by God, and that these types were unchanging (Figure 22.1a). Plato acknowledged that individuals of a species sometimes varied slightly from one another, but that these were just trivial deviations around a "perfect essence."

Today, philosophers and biologists refer to ideas like this as typological thinking. Typological thinking also occurs in the Bible's book of Genesis and in the creation stories of many other religions, where a divine being creates each type of organism.

Aristotle and the Scale of Nature

One of Plato's students, Aristotle, organized typological thinking into a linear scheme called the great chain of being, or scale of nature, where "scale" means a ladder or stairway (Figure 22.1b). Aristotle proposed that species were fixed types organized into a sequence based on increased size and complexity. The scale started with minerals and lower plants at the bottom, then rose through higher plants, lower and higher invertebrates (animals without backbones), and lower and higher vertebrates (animals with backbones) to humans at the top, surpassed only by angels and God.

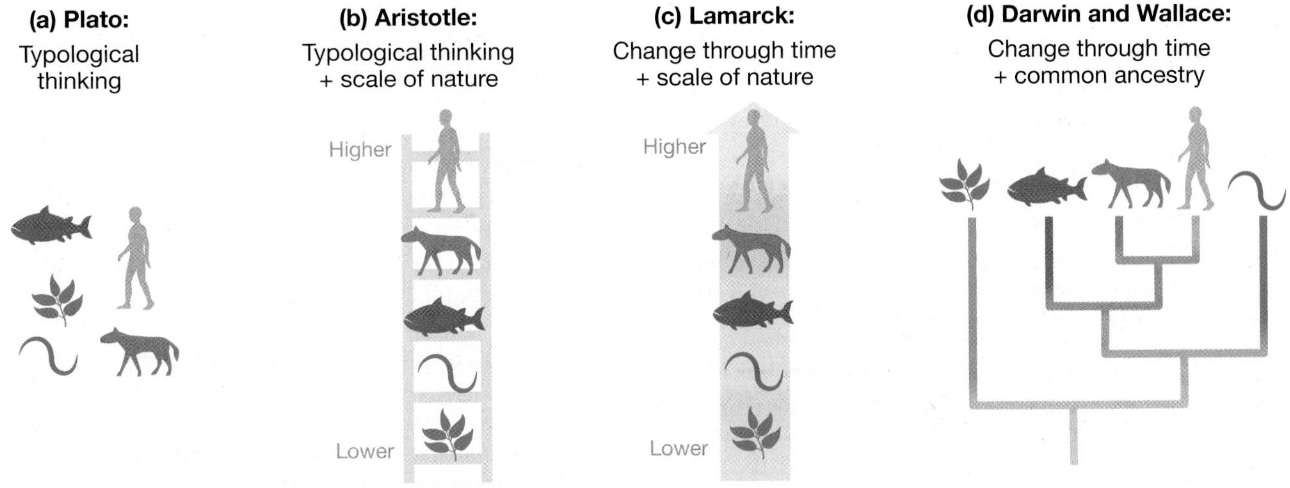

Figure 22.1 Models of the Diversity of Life Have Changed through Time. Visual models are helpful for comparing ideas. The models shown here include only five living species, for simplicity. Each model is explained in the text (see also **BioSkills 12**).

Aristotle's ideas remained popular in scientific and religious circles well into the 1700s. Since many cultures had embraced Aristotle's model for so long—over 20 centuries—the notion of "lower" and "higher" species, including the separation of humans from other animals, lingers as a cultural habit even today (discussed in Section 22.5).

Lamarck and the Idea of Evolution as Change through Time

Typological thinking eventually began to break down. In 1809 the biologist Jean-Baptiste de Lamarck proposed the first formal theory of **evolution**—that species are not static but change through time. However, the pattern component of Lamarck's theory was initially based on the scale of nature.

When he started his work on evolution, Lamarck claimed that simple organisms originate at the base of the scale by spontaneous generation (Ch. 1, Section 1.2) and then evolve by moving up the scale over time (Figure 22.1c). Thus, Lamarckian evolution is progressive in the sense of always producing larger and more complex, or "better," species.

Lamarck also contended that species change through time via the inheritance of acquired characters. The idea here is that as an individual develops, its phenotype changes in response to challenges posed by the environment, and it passes on these phenotypic changes to offspring. A classic Lamarckian scenario is that giraffes develop long necks as they stretch to reach leaves high in treetops, and they then produce offspring with elongated necks.

Darwin and Wallace and Evolution by Natural Selection

As his thinking matured, Lamarck eventually abandoned his linear and progressive view of life. Darwin and Wallace agreed. But more important, they emphasized that the process responsible for change through time—natural selection—occurs because traits vary among the individuals in a population, and because individuals with certain traits leave more offspring than others do.

Darwin claimed that instead of being unimportant, variation among individuals in a population was the key to understanding the nature of species. Biologists refer to this view as **population thinking**.

The theory of evolution by natural selection was revolutionary for several reasons:

- It overturned the idea that species are static and unchanging. Instead, it suggested that species change through time and are related by common ancestry (Figure 22.1d).
- It replaced typological thinking with population thinking.
- It was scientific. It proposed a mechanism that could account for change through time and made predictions that could be tested through observation and experimentation.

Plato and his followers emphasized the existence of fixed types, whereas evolution by natural selection is all about change and diversity. With the advent of evolutionary thought, new questions arose: What evidence backs the claim that species are not fixed types? What data support the theory of evolution by natural selection?

CHECK YOUR UNDERSTANDING

✔ **If you understood this section, you should be able to ...**

Imagine you are standing in a meadow of sunflowers. Compare your view of sunflowers from the perspective of typological thinking and from that of population thinking.

Answers are available in Appendix A.

22.2 The Pattern of Evolution: Have Species Changed, and Are They Related?

After you complete this section, you should be able to ...

- Summarize evidence that supports the theory of evolution.

In *On the Origin of Species*, Darwin repeatedly described evolution as **descent with modification**. In essence, the pattern component of the theory of evolution by natural selection makes two predictions about the nature of species:

1. Species change through time.
2. Species are related by common ancestry.

Let's consider the evidence for each prediction in turn. The different types of evidence are numbered to help you keep track as you read along: 1–5 for *species change through time*, 6–8 for *species are related by common ancestry*.

Evidence for Change through Time

When Darwin began his work, biologists and geologists had just begun to assemble and interpret the fossil record. A **fossil** is any trace of an organism that lived in the past. These traces range from bones, branches, shells, and dung to tracks or impressions left by organisms in soft sediments such as sand and clay (Ch. 25, Section 25.2). The **fossil record** consists of all the fossils that have been found on Earth and described in the scientific literature.

Why did data in the fossil record support the hypothesis that species have changed through time? And what data from **extant species**—those living today—support the claim that they are modified forms of ancestral species?

Evidence 1: The Vastness of Geologic Time Evidence for the Earth's vast age began to mount in the late 1700s with James Hutton's proposal of the principle of uniformitarianism—the idea that geological processes occurring today are similar to what occurred in the past. Hutton reached his insight by evaluating evidence—he traveled around Europe and measured geological patterns and rates of rock formation and erosion.

Figure 22.2 Sedimentary Rocks Reveal the Vastness of Geologic Time. The relative ages of sedimentary rocks are used to determine the relative ages of fossil organisms because younger layers are deposited on top of older ones. The deepest rock layer in the Grand Canyon is over a billion years old, and the top layer is 270 million years old. (mya = million years ago)

Sedimentary rocks form from sand or mud or other materials deposited in layers at locations such as beaches or river mouths. Hutton calculated that sedimentary rocks form at quite a slow rate. When he extrapolated this rate to determine how long it would take for massive rock formations to form, it was clear that Earth was very old. It was much, much older than the 6000 years claimed by many proponents of special creation.

Hutton's ideas were popularized by Darwin's close friend, the geologist Charles Lyell. Sedimentary rocks, along with rocks derived from episodic lava flows, form with younger layers deposited on top of older layers. Lyell and others used this information to place fossils in a younger-to-older sequence, based on the fossils' relative position in layers of sedimentary rock (Figure 22.2).

As the scientists observed similarities in rocks and fossils at different sites, they began to create a **geologic time scale**: a sequence of named intervals called eons, eras, and periods that represented the major events in Earth history (see Ch. 25, Section 25.2). This geologic time scale was a relative one, however. The absolute age of Earth was still unknown.

After Marie Curie's discovery of radioactivity in the late 1800s, researchers realized that radioactive decay—the steady rate at which unstable "parent" atoms are converted into more stable "daughter" atoms—furnished a way to assign *absolute* ages, in years, to the relative ages in the geologic time scale.

Radioactive decay functions as a "natural clock" in which the decay of different elements can be used to measure different time scales. For example, the half-life of uranium-238 is about 4.5 billion years, which means that 50 percent of uranium-238 atoms will decay to lead-206 atoms during this time. Knowing the half-life, geologists can use the ratio of uranium to lead in a rock sample to infer the age of the sample.

According to data from radiometric dating, Earth is about 4.6 billion years old, and the earliest signs of life appear in rocks that formed 3.4–3.8 billion years ago. Data from relative and absolute dating techniques agree: Life on Earth is ancient. A great deal of time has gone by for change to occur.

Evidence 2: Extinction Changes the Species Present over Time In the early nineteenth century, researchers began discovering fossil bones, leaves, and shells that were unlike structures from any known animal or plant. At first, many scientists insisted that living examples of these species would be found in unexplored regions of the globe. But as research continued and the number and diversity of fossil collections grew, the argument became less and less plausible.

The issue was finally settled in 1812 when Baron Georges Cuvier published a detailed analysis of several **extinct species**—that is, species that no longer exist. Cuvier intentionally focused

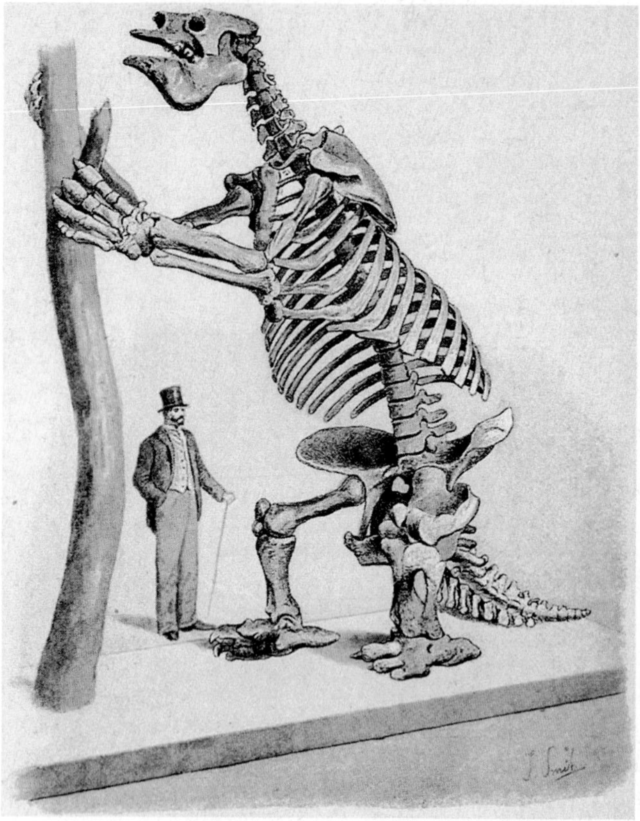

Figure 22.3 **Evidence of Extinction.** This 19th-century drawing depicts Cuvier's fossil evidence. Scientists agreed that the sloth, like other giant fossil vertebrates, was too large and unique to be overlooked if it were alive; it must have gone extinct.

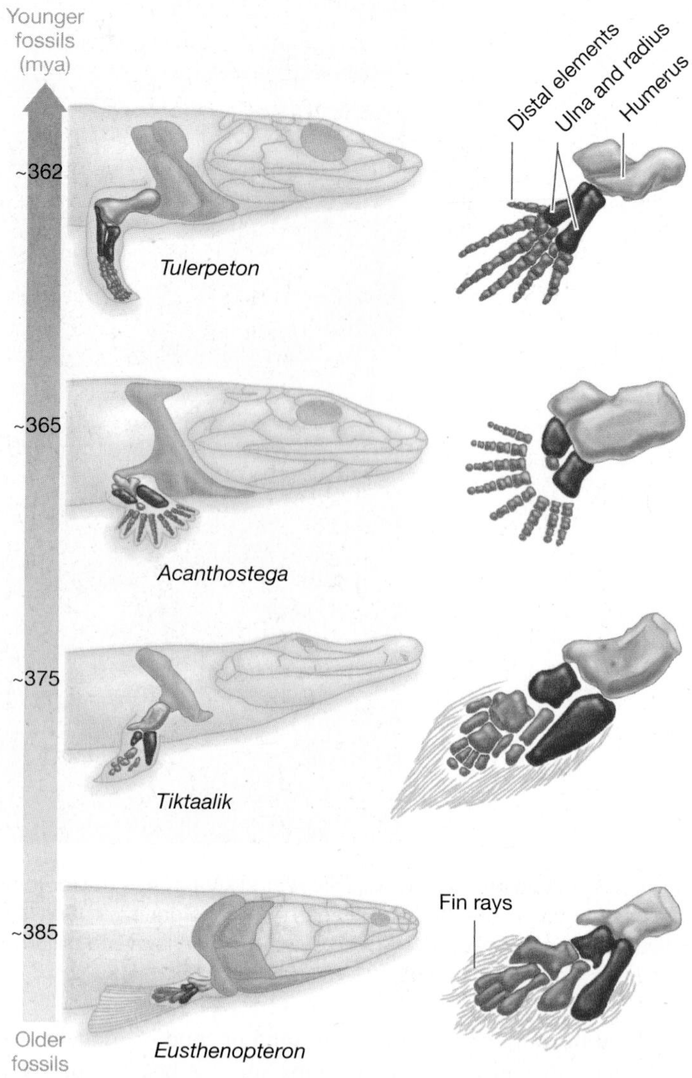

Figure 22.4 **Transitional Features during the Evolution of the Tetrapod Limb.** Fossil species similar to today's lungfish and tetrapods have fin and limb bones that are transitional features. *Eusthenopteron* was aquatic; *Tulerpeton* was probably semiaquatic.

✔ Contrast how the transitions shown here would fit into Lamarck's early model of evolution (see Figure 22.1c) versus Darwin and Wallace's model of evolution (see Figure 22.1d).

on the fossils of large terrestrial animals such as mammoths, mastodons, giant armadillos, giant deer, and giant sloths. He reasoned that unlike many plants or marine animals, those large, distinctive terrestrial animals probably would have been discovered if they were still alive (**Figure 22.3**). Cuvier's overwhelming evidence convinced scientists of the fact of extinction.

Darwin interpreted extinct forms as evidence that species are not static, immutable entities, unchanged since the moment of special creation. He reasoned that if species have gone extinct, then the array of species living on Earth has changed through time.

Recent analyses of the fossil record suggest that over 99 percent of all the species that have ever lived are now extinct. The data also indicate that species have gone extinct continuously throughout Earth's history—not just in one or even a few catastrophic events (Ch. 25, Section 25.4).

Evidence 3: Transitional Features Link Older and Younger Species Long before Darwin published his theory, researchers reported striking resemblances between the fossils found in the rocks underlying certain regions and the living species succeeding them in the same geographic areas, such as the extinct giant sloths of South America and the sloths that occur there today. The pattern was so widespread that it became known as the "law of succession."

Darwin pointed out that this pattern provided strong evidence in favor of the hypothesis that species had changed through time. He proposed that the extinct forms and living forms were related—that they represented ancestors and descendants.

As the fossil record expanded, researchers discovered species with characteristics that broadened the scope of the law of succession. A **transitional feature** is a trait in a fossil species that is intermediate between those of ancestral (older) and derived (younger) species. For example, intensive work over the past several decades has yielded fossils that document a gradual change over time from aquatic animals that had fins to terrestrial animals that had limbs (**Figure 22.4**). Over a period of about

25 million years, the fins of species similar to today's lungfish transitioned into limbs similar to those found in today's amphibians, reptiles, and mammals—a group called the tetrapods (literally, "four-footed").

Note that such evolutionary transitions are not goal oriented or purposeful. Rather, some individuals with favorable traits managed to survive and reproduce in the new environment, resulting in change in the population over time (explained further in Section 22.3).

Similar sequences of transitional features document changes that led to the evolution of feathers and flight in birds; stomata and vascular tissue in plants; upright posture, flattened faces, and large brains in humans; jaws in vertebrates; the loss of limbs in snakes; and other traits. Data like these are consistent with predictions from the theory of evolution: If the traits observed in more recent species evolved from traits in more ancient species, then transitional forms are expected to occur in the appropriate time sequence. Note, however, that individual fossils of transitional forms are not necessarily *direct* ancestors of later species—they may be relatives of the direct ancestor.

Evidence 4: Vestigial Traits Are Evidence of Change through Time Darwin was the first to provide a widely accepted interpretation of vestigial traits. A **vestigial trait** is a reduced or incompletely developed structure that has no function, or reduced function, but is clearly similar to functioning organs or structures in ancestral or closely related species.

Biologists have documented thousands of examples of vestigial traits.

- Some whales and snakes have tiny hip and leg bones that do not help them swim or slither.
- Ostriches and kiwis have reduced wings and cannot fly.
- Blind cave-dwelling fish have eye sockets but no eyes.
- Monkeys and many other primates have long tails; but our tiny tailbone, or coccyx, illustrated in Figure 22.5, is too small to help us maintain balance or grab tree limbs for support.

The change in species over time is logical because the environments of species change dramatically over time. Change can occur due to abiotic factors, such as a rise in temperature or a drop in sea level, and biotic factors, such as arrival of a new parasite or the extinction of a predator (Ch. 49, Section 49.2). For example, a tail can be a beneficial, heritable trait for life in the forest canopy, but may become less beneficial if changes in rainfall patterns cause the forest to transition to grassland.

Evidence 5: Species Can Be Observed Changing Today Biologists have documented hundreds of contemporary populations that are evolving in response to changes in their environment. In Section 22.4 you will find a detailed analysis of research on two examples of evolution in action: the evolution of antibiotic resistance in bacteria and the evolution of beak shape in finches. One key insight from these examples is that evolution can be observed on the scale of days, weeks, and months, not just thousands or millions of years.

To summarize, evidence from the fossil record and living species indicates that life is ancient and that species have changed through the course of Earth's history. Further, change through time continues today and can be measured directly.

Evidence of Descent from a Common Ancestor

Data from the fossil record and contemporary species refute the idea that species are static, unchanging types. What about the claim that species were created independently—meaning that they are unrelated to each other?

Evidence 6: Similar Species Are Found in the Same Geographic Area Charles Darwin began to realize that species are related by common ancestry during a five-year voyage he took aboard the English naval ship HMS *Beagle*. While fulfilling its mission to explore and map the coast of South America, the *Beagle* spent a few weeks in the Galápagos Islands off the coast of present-day Ecuador (see Figure 22.6a on page 454).

Darwin had taken over the role of ship's naturalist and, as the first scientist to study the area, gathered extensive collections of the plants and animals found in these islands. Most famous among the birds he collected were the Galápagos mockingbirds and finches.

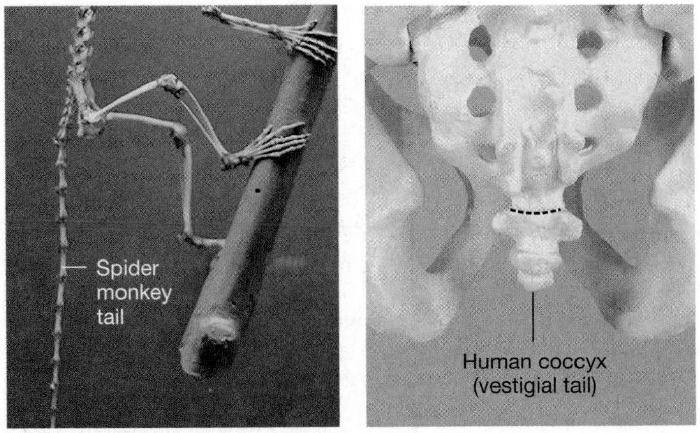

Figure 22.5 **Vestigial Traits Are Reduced Versions of Traits in Other Species.** The tailbone is a human trait that has reduced function. It is no longer useful for balance and locomotion.

✔ How could vestigial traits such as the tiny human tailbone be explained via inheritance of acquired characters?

(a) Mainland birds colonized the Galápagos.

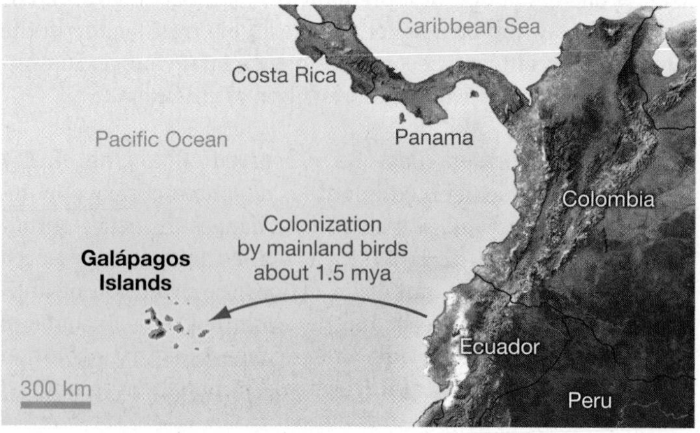

(b) Galápagos finches share a common ancestor.

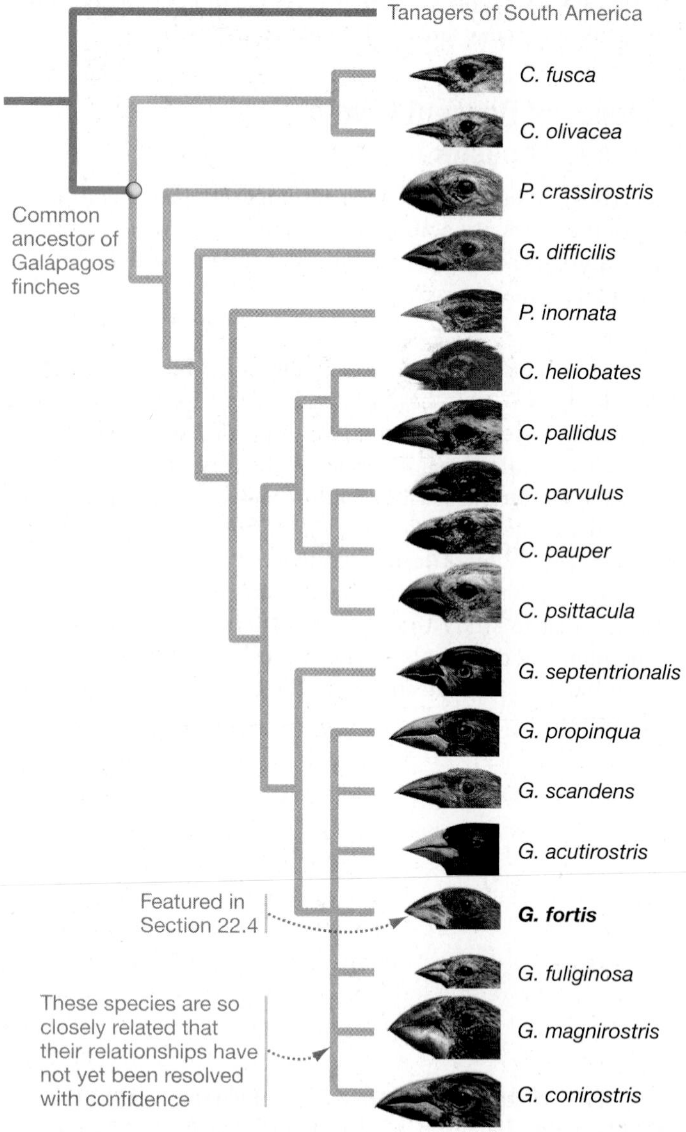

Figure 22.6 The Galápagos Finches Share a Common Ancestor.
DATA: S. Lamichhaney et al. 2018. *Science* 359: 224–228; C. E. Wagner. 2018. *Science* 359: 157–159.

Several years after Darwin returned to England, a colleague pointed out that Darwin's finch collection contained several distinct species, based on differences in coloration and in beak size and shape. This struck Darwin as remarkable. Why would so many species that were so similar occur in such a small geographic area?

Darwin realized that this pattern—puzzling when examined as a product of special creation—made perfect sense when interpreted in the context of evolution, or descent with modification. The finches were similar, he proposed, because they had descended from the same common ancestor from the mainland. That is, instead of being created independently, finch populations that colonized different islands had changed through time and formed new species.

Recent analyses of DNA sequences in these finches support Darwin's hypothesis. DNA sequence comparisons have enabled researchers to place the 18 currently recognized species on a **phylogenetic tree**—a branching diagram that depicts the ancestor–descendant relationships among species or other taxa (Ch. 25, Section 25.1). As **Figure 22.6b** shows, the Galápagos finches share a single common ancestor that is estimated to have arrived from the mainland about 1.5 million years ago.(For help with reading phylogenetic trees, see **BioSkills 13**.)

Evidence 7: Related Species Share Homologies Translated literally, *homology* means "the study of likeness." When biologists first began to study the anatomy of humans and other vertebrates, they were struck by the remarkable similarity of their skeletons, muscles, and organs. But because the biologists who did these early studies were advocates of special creation, they could not explain why striking similarities existed among certain organisms but not others.

Today biologists recognize that **homology** is a similarity that exists in species due to common ancestry. Human hair and dog fur are homologous. Humans have hair and dogs have hair (fur) because they share a common ancestor—an early mammal species—that had hair (see **Making Models 22.1**).

Making Models 22.1 Tips on Drawing Phylogenetic Trees (I)

Phylogenetic trees can help you understand the concept of homology (see **BioSkills 13** and Ch. 25, Section 25.1). You can draw a bar on a branch to symbolize the origin (or loss) of a trait. All the organisms to the right of that bar are assumed to possess that trait. Compare the meaning of two trees shown here:

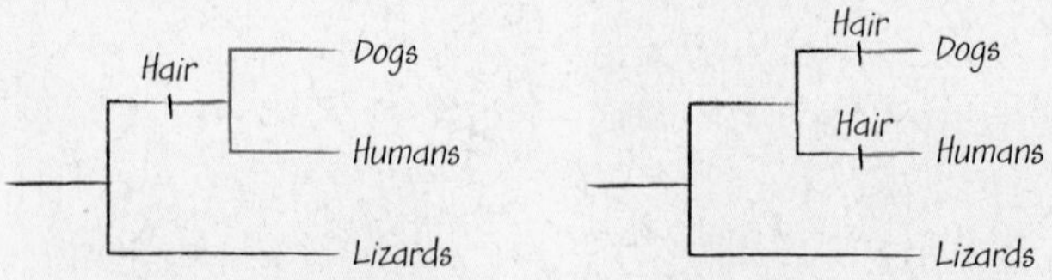

"Hair is homologous in dogs and humans"

"Hair is not homologous in dogs and humans"

MODEL Draw a dot on the node (branching point) of each tree that represents the most recent common ancestor of dogs and humans, and indicate if it had hair. Which tree is correct?

To see this model in action, go to the Study Area of **Mastering Biology**

SUMMARY **Table 22.1 Three Levels of Homology**

Level | **Example**

Genetic homology

Similarity in DNA, RNA, or amino acid sequences due to inheritance from a common ancestor

Amino acid sequences from a portion of the *Aniridia* gene product found in humans are 90 percent identical to those found in the *Drosophila eyeless* gene product.

Gene:	Amino acid sequence (single-letter abbreviations):
Aniridia (Human)	LQRNRTSFTQEQIEALEKEFERTHYPDVFARERLAAKIDLPEARIQVWFSNRRAKWRREE
eyeless (Fruit fly)	LQRNRTSFTNDQIDSLEKEFERTHYPDVFARERLAGKIGLPEARIQVWFSNRRAKWRREE

(For a key to the single-letter abbreviations used for the amino acids, see Figure 3.2.)

Developmental homology

Similarity in embryonic form or developmental processes due to inheritance from a common ancestor

The early embryonic stages of a chick, a human, and a cat show a strong resemblance and are the product of similar developmental processes. (These embryos are not shown at the same scale.)

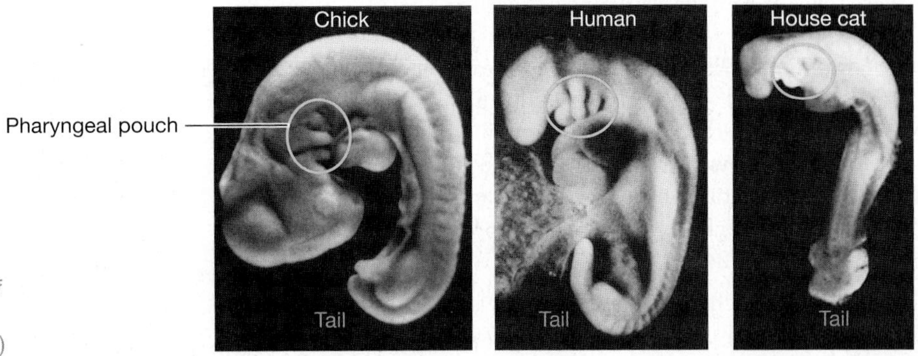

Structural homology

Similarity in adult organismal structures due to inheritance from a common ancestor

Even though their function varies, all vertebrate limbs are modifications of the same number and arrangement of bones. (These limbs are not drawn to scale.)

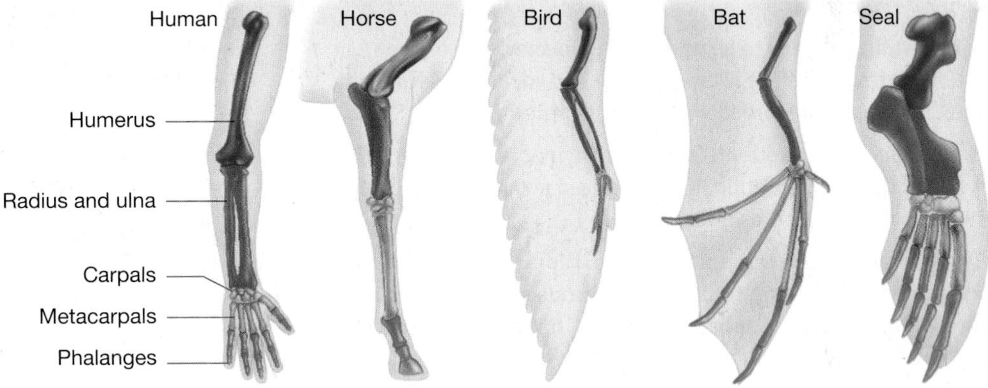

Homology can be recognized and studied at three levels (summarized in **Table 22.1**):

1. **Genetic homology** is similarity that occurs in DNA nucleotide sequences, RNA nucleotide sequences, or amino acid sequences. Perhaps the most fundamental of all homologies is the genetic code. With a few minor exceptions, all organisms use the same rules for transferring the information coded in DNA into proteins (Ch.16, Section 16.3).

2. **Developmental homology** is similarity in developmental structures or processes. For example, early chick, human, and cat embryos have tails and structures called pharyngeal pouches that are a product of similar developmental processes inherited from a common ancestor. Later in development, these pouches are lost in all three species, and tails are lost in humans. But in fish, the pharyngeal pouches stay intact and give rise to functioning gills in adults. To explain this observation, biologists hypothesize that pharyngeal pouches and tails exist in chicks, humans, and cats because they existed in the fishlike species that was the common ancestor of today's vertebrates.

3. **Structural homology** is a similarity in adult **morphology**, or form. A classic example is the common structural plan observed in the limbs of vertebrates. In Darwin's own words, "What could be more curious than that the hand of a man, formed for grasping, that of a mole for digging, the leg of the horse, the paddle of the porpoise, and the wing of the bat, should all be constructed on the same pattern, and should include the same bones, in the same relative positions?" The structural homology exists because mammals evolved from a lungfish-like ancestor that had the same general arrangement of bones in its fins.

The three levels of homology interact. Genetic homologies cause the developmental homologies observed in embryos, which then lead to the structural homologies recognized in adults.

Homology is a key concept in contemporary biology:

- The revolutionary genome-editing system called CRISPR-Cas was discovered in bacteria but can be used to edit the genome of any type of organism due to the homology of the structure and function of DNA in all of life (Ch. 20, Section 20.4).
- Model organisms such as fruit flies, zebrafish, and mice can be used to understand human illnesses such as cancer due to the homology of many of the molecular structures and processes involved (see **BioSkills 11**).

The theory of evolution by natural selection predicts that homologies will occur. If species were created independently of one another, these types of similarities would not occur.

Evidence 8: Formation of New Species from Preexisting Species Can Be Observed Today Biologists have documented dozens of contemporary populations that are undergoing *speciation*—a process that results in new species being formed from preexisting species. Some populations have served as particularly well-studied examples of speciation in action (explored in detail in Ch. 24, Sections 24.2 and 24.3), such as:

- Speciation among finches in the Galápagos is ongoing and can be observed directly. Most recently, a new species of finch (dubbed Big Birds) was formed via hybridization in just three generations.
- Killer whales appear to be undergoing speciation. Some populations are so different in prey choice and social behavior that they no longer interbreed with other populations.
- A new species of maidenhair fern arose in a single generation when chromosome duplication produced a viable species with twice as many chromosomes.

The contemporary examples of speciation are powerful evidence that species living today are descendants of species that lived in the past. They support the claim that all organisms are related by descent from a common ancestor.

SUMMARY **Table 22.2 Evidence for Evolution**

Type of Evidence	✔ Your Turn: Cite an Example
Prediction 1: Species Change through Time	
1. Life on Earth is ancient.	1. Layers of sedimentary rock in the Grand Canyon
2. The fossil record shows a change in life over time, including extinctions.	2. Fossils discovered of extinct mammoths and giant sloths
3. Transitional features document change in traits through time.	3.
4. Vestigial traits are common.	4.
5. The characteristics of populations vary within species and can be observed changing today.	5.
Prediction 2: Species Are Related by Common Ancestry	
6. Similar species are found in the same geographic area.	6. Similar species of finches found on Galápagos Islands
7. Related species share genetic, developmental, and structural homologies.	7.
8. The formation of new species from preexisting species can be observed today.	8.

✔ Complete the table using examples from your reading.

Evolution's "Internal Consistency"—The Importance of Independent Data Sets

Biologists draw on data from several sources to challenge the hypothesis that species are immutable and were created independently. The data support the idea that species have descended, with modification, from a common ancestor. **Table 22.2** summarizes this evidence.

Perhaps the most powerful evidence for any scientific theory, including evolution by natural selection, is what scientists call internal consistency. This is the observation that data from independent sources agree in supporting predictions made by a theory.

As an example, consider the evolution of whales and dolphins—a group called the cetaceans. At first glance, whales have the outward appearance of fish. Yet on closer inspection, whales have a lot in common with terrestrial mammals. Did whales evolve from terrestrial mammals?

- ***The fossil record*** contains a series of species that are clearly identified as cetaceans based on the unusual ear bones found only in this group. Some of the species have the long legs and compact bodies typical of mammals that live primarily on land; some are limbless and have the streamlined bodies typical of aquatic mammals; some have intermediate features.
- ***A phylogeny of the fossil cetaceans***, estimated on the basis of similarities and differences in morphological traits other than limbs and overall body shape, indicates that a gradual transition occurred between terrestrial forms and aquatic, whale-like forms (**Figure 22.7**).
- ***Relative dating,*** based on the positions of sedimentary rocks where the fossils were found, agrees with the order of species indicated in the phylogeny.
- ***Absolute dating,*** based on analyses of radioactive atoms in rocks in or near the layers where the fossils were found, also agrees with the order of species indicated in the phylogeny.
- ***A phylogeny of living whales and dolphins,*** estimated from similarities and differences in DNA sequences, indicates that hippos—which spend much of their time in shallow

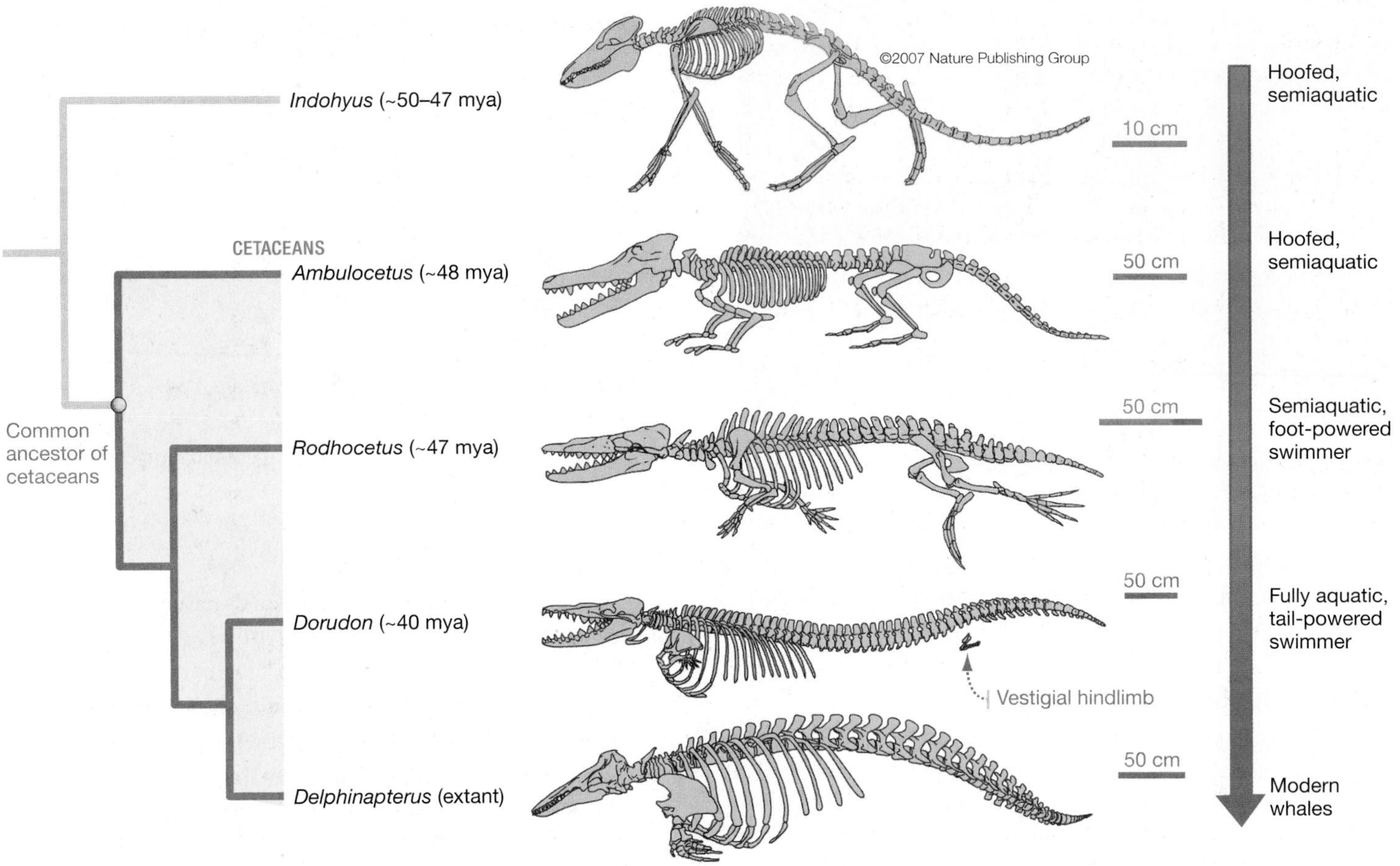

Figure 22.7 Data on Evolution from Independent Sources Are Consistent. This phylogeny of fossil cetaceans is consistent with data from relative dating, absolute dating, and phylogenies estimated from molecular traits in living species—all agree that whales evolved from terrestrial ancestors that also gave rise to hippos.

water—are the closest living relative of cetaceans (shown in Ch. 25, Figure 25.4). This observation supports the hypothesis that cetaceans and hippos shared a common ancestor that was semiaquatic.

- ***Vestigial hip and hindlimb bones*** occur in some whales as adults, and vestigial hindlimb buds—outgrowths where legs form in other mammals—occur in some dolphin embryos.

The general message here is that many independent lines of evidence converge on the same conclusion: Whales gradually evolved from a terrestrial ancestor about 50 million years ago.

As you evaluate the evidence supporting the pattern component of the theory of evolution, though, it's important to recognize that no single observation or experiment instantly "proved" the fact of evolution. Rather, data from many different sources are much more consistent with evolution than with special creation. Descent with modification is a successful and powerful scientific theory because it explains observations—such as vestigial traits and the close similarities among species on neighboring islands—that special creation does not.

What about the process component of the theory of evolution by natural selection? If hippos and whales were not created independently, how did they come to be?

CHECK YOUR UNDERSTANDING

✔ If you understood this section, you should be able to . . .

1. Determine what kind of evidence would support the hypothesis that birds evolved from dinosaurs.
2. Explain why the DNA sequences of chimpanzees and humans are about 96 percent similar.

Answers are available in Appendix A.

22.3 The Process of Evolution: How Does Natural Selection Work?

After you complete this section, you should be able to . . .

- Explain how natural selection causes evolution.

Darwin's greatest contribution was not just recognizing the fact of evolution. Lamarck and other researchers, including Darwin's own grandfather Erasmus Darwin, had proposed evolution long before Charles Darwin began his work. Instead, Darwin's crucial insight was to recognize a process, called natural selection, that could explain the pattern of descent with modification.

Figure 22.8 **Diversity of Pigeon Breeds in Captivity.** Darwin used the breeding of pigeons as a model system to study how the characteristics of populations can change over time.

Darwin's Inspiration

How did Darwin arrive at his insight? First, he had spent decades exploring and documenting the diversity of plants and animals, both around the globe and in his native England. So he had a wealth of data on variation within and among species, and he viewed this variation in the context of the ancient and changing Earth as popularized by his geologist friend, Charles Lyell.

All of this careful work gave Darwin an especially strong foundation in the pattern of evolution. To make sense of the process of evolution, Darwin turned in part to pigeon breeding—a model system that would be easier to study and manipulate than populations in the wild. Pigeon breeding was popular in England at the time, and in Darwin's words, "The diversity of the breeds is something astonishing" (**Figure 22.8**).

Darwin crossbred pigeons and observed how characteristics were passed on to offspring. He chose certain individuals with desirable traits to reproduce, thus manipulating the composition of the population by a process called **artificial selection**. It was clear to Darwin and other breeders that the diverse varieties were all descended from the wild rock pigeons.

Another influence on Darwin was the publication of a book by Thomas Robert Malthus, *An Essay on the Principle of Population*, which inspired heated discussion in England at the time. Malthus's studies of human populations in England and elsewhere led him to a startling conclusion: Since many more individuals are born than can survive, a "struggle for existence" occurs as people compete for food and places to live.

Darwin combined his observations of artificial selection with this notion of "struggle for existence" in natural populations, which he knew—from his countless studies—contained variation. From this synthesis arose his concept of natural selection.

Although both Darwin and Wallace arrived at the same idea, Darwin's name is more closely associated with natural selection because of his extensive evidence for it in *On the Origin of Species*.

Making Models 22.2 Tips on Drawing Darwin's Postulates

Drawing models can help you understand and remember abstract ideas such as Darwin's postulates. Simple drawings can be very effective, so focus on the ideas rather than trying to make lifelike illustrations. Consider this example:

MODEL Using the flower example shown here, make a model of Darwin's first postulate.

To see this model in action, go to the Study Area of **Mastering Biology**

Darwin's Four Postulates

Darwin distilled the process of evolution by natural selection into four simple postulates (criteria) that form a logical sequence (see **Making Models 22.2**):

1. ***Variation exists among individual organisms that make up a population,*** such as variation in size and shape.
2. ***Some of the trait differences are heritable,*** meaning that they are passed on to offspring. For example, parent flowers with long petals tend to have offspring with long petals.
3. ***Survival and reproductive success are highly variable.*** Many more offspring are produced than can possibly survive. Thus, only some individuals in each generation survive long enough to produce offspring, and among the individuals that produce offspring, some will produce more than others.
4. ***The subset of individuals that survive best and produce the most offspring is not a random sample of the population.*** Instead, individuals with certain heritable traits are more likely to survive and reproduce.

Altogether, **natural selection** occurs when individuals with certain heritable traits produce more surviving offspring than do individuals without those traits. Thus, the frequency of the selected traits increases from one generation to the next. Biologists now know that traits are determined by alleles, particular versions of genes (Chapter 14). So the outcome of evolution by natural selection is a change in allele frequencies in a population over time (Chapter 23).

In studying these criteria, you should realize that variation among individuals in a population is essential for evolution to occur. Darwin had to introduce population thinking into biology because it is populations that change over time. To come up with these postulates and understand their consequences, Darwin had to think in a revolutionary way.

Today, biologists usually condense Darwin's four postulates into a two-part statement that more forcefully communicates the essence of evolution by natural selection: Evolution by natural selection occurs when **(1)** heritable variation leads to **(2)** differential reproductive success.

Biological Definitions of Fitness, Adaptation, and Selection

To explain the process of natural selection, Darwin referred to successful individuals as "more fit" than other individuals. In doing so, he gave the word "fitness" a definition different from its everyday English usage. Biological **fitness** is the ability of an individual to produce surviving, fertile offspring relative to that ability in other individuals in the population.

Note that fitness can be measured. When researchers study a population in the lab or in the field, they can measure the relative fitness of individuals by comparing the number of surviving offspring each individual produces.

The concept of fitness, in turn, provides a compact way of formally defining adaptation. The biological meaning of adaptation, like the biological meaning of fitness, is different from its everyday English usage (see **BioSkills 17**). In evolutionary biology, an **adaptation** is a heritable trait that increases the fitness of an individual in a particular environment relative to individuals lacking the trait. Adaptations increase fitness—the ability to produce viable, fertile offspring. (You can see the Big Picture of how adaptation and fitness relate to natural selection on pages 532–533.)

Lastly, the term "selection" has a commonsense meaning in the context of artificial selection. Breeders *choose* which characteristics they want to keep or get rid of in their plant and animal breeds. However, selection has a very different meaning in the biological context of natural selection. Here, it refers to a passive process—differential reproduction as a result of heritable variation—not a purposeful choice.

CHECK YOUR UNDERSTANDING

✔ If you understood this section, you should be able to . . .

1. **THINK CAREFULLY** Long petals increase the fitness of certain flowers by increasing their ability to attract pollinators. What does fitness mean in this context?
2. **MODEL** Using a population of flowers with different petal lengths as the example, draw a model of Darwin's fourth postulate.

Answers are available in Appendix A.

22.4 Evolution in Action: Measuring Natural Selection in Populations Today

The theory of evolution by natural selection is testable, meaning biologists can test the validity of each of Darwin's postulates. Indeed, researchers have documented heritable variation and differential reproductive success in a wide array of natural populations.

After you complete this section, you should be able to . . .

- Apply Darwin's postulates of natural selection to a real-world scenario.

This section summarizes two examples of evolution by natural selection being observed in nature. Hundreds of other case studies are available, involving a wide variety of traits and organisms. Let's start by exploring the evolution of drug resistance, one of the great challenges of our time.

Case Study 1: How Did *Mycobacterium tuberculosis* Become Resistant to Antibiotics?

Mycobacterium tuberculosis, the bacterium that causes the infectious disease **tuberculosis**, or **TB**, has long been a scourge of humankind. It usually infects the lungs and causes fever, coughing, sweats, weight loss, and often death. In Europe and the United States, TB was once as great a public health issue as cancer is now. It receded in importance during the early 1900s, though, for two reasons:

1. Advances in nutrition made people better able to fight off most *M. tuberculosis* infections quickly.
2. Development of antibiotics allowed physicians to stop infections.

In the late 1980s, however, rates of *M. tuberculosis* infection surged in many countries, and in 1993 the World Health Organization declared TB a global health emergency. Physicians were particularly alarmed because the strains of *M. tuberculosis* responsible for the increase were largely or completely resistant to antibiotics that were once extremely effective.

How and why did the evolution of drug resistance occur? The case of a single patient—a young man who lived in Baltimore—illustrates what is happening all over the world.

A Patient History The story begins when the man was admitted to the hospital with fever and coughing. Chest X-rays, followed by bacterial cultures of fluid coughed up from the lungs, showed that he had an active TB infection. He was given several antibiotics for 6 weeks, followed by twice-weekly doses of the antibiotic rifampin for an additional 33 weeks. Ten months after therapy started, bacterial cultures from the patient's chest fluid indicated no *M. tuberculosis* cells. His chest X-rays were also normal. The antibiotics seemed to have cleared the infection.

Just two months after the TB tests proved normal, however, the young man was readmitted to the hospital with a fever, severe cough, and labored breathing. Despite being treated with a variety of antibiotics, including rifampin, he died of respiratory failure 10 days later. Samples of fluid from his lungs showed that *M. tuberculosis* was again growing actively there. But this time the bacterial cells were completely resistant to rifampin.

Drug-resistant *M. tuberculosis* cells had killed this patient. Where did they come from? Could a strain that was resistant to antibiotic treatment have evolved *within* him? To answer this question, a research team analyzed DNA from the drug-resistant strain and compared it with stored DNA from *M. tuberculosis* cells that had been isolated a year earlier from the same patient. After examining extensive stretches from each genome, the biologists were able to find only one difference: a point mutation in a gene called *rpoB*.

A Mutation in a Bacterial Gene Confers Resistance The *rpoB* gene codes for a component of RNA polymerase. This enzyme transcribes DNA to mRNA, so it is essential to the survival and reproduction of bacterial cells (Ch. 17, Section 17.1). In this case, the point mutation in the *rpoB* gene changed a cytosine to a thymine, forming a new allele for the *rpoB* gene (see Ch. 16, Section 16.4). This missense mutation caused a change in the amino acid sequence of the RNA polymerase (from a serine to a leucine at the 153rd amino acid)—and a change in its shape.

This shape change proved critical. Rifampin, the antibiotic that was being used to treat the patient, works by binding to the RNA polymerase of *M. tuberculosis* and interfering with transcription. Bacterial cells with the C → T mutation continue to produce offspring efficiently even in the presence of the drug, because the drug cannot bind efficiently to the mutant RNA polymerase. In summary, the *rpoB* mutation, the shape of RNA polymerase, and drug resistance are related in this way:

DNA ——→ **Protein** ——→ **Phenotype**

Mutation in *rpoB* gene | Shape change in RNA polymerase inhibits binding of antibiotic | Drug resistance

These results suggest the steps that led to this patient's death (**Figure 22.9**):

1. Completely by chance, one or a few of the large population of *M. tuberculosis* bacterial cells present in the patient *before* the onset of drug therapy had the *rpoB* allele with the C→T mutation. Under normal conditions, mutant forms of RNA polymerase do not work as well as the more common form, so cells with the C→T mutation would not produce many offspring and would stay at low frequency—even while the overall population grew to the point of inducing symptoms that sent the young man to the hospital.
2. Rifampin drug therapy began, killing most *M. tuberculosis* cells with normal RNA polymerase. As a result, the overall bacterial population declined in size so drastically that the patient seemed cured and drug therapy ended.
3. Mutant cells increased in number after therapy ended. Eventually the *M. tuberculosis* population regained its former abundance, and the patient became sick again.
4. A second round of drug therapy began but had no effect on the drug-resistant *M. tuberculosis* cells that then dominated the population, so the patient died.

Testing Darwin's Postulates Does the sequence of events illustrated in Figure 22.9 indicate that evolution by natural selection occurred? One way of answering this question is to review Darwin's four postulates and test whether each one was verified:

1. ***Did variation exist in the population?*** Yes. Due to mutation, both resistant and nonresistant strains of TB were present before administration of the drug. Most *M. tuberculosis* populations, in fact, exhibit variation for the trait; studies on cultured *M. tuberculosis* show that a mutation conferring resistance to rifampin is present in one out of every 10^7 to 10^8 cells.

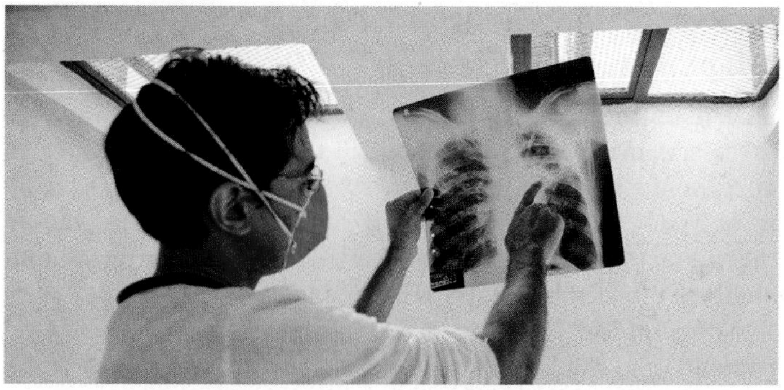

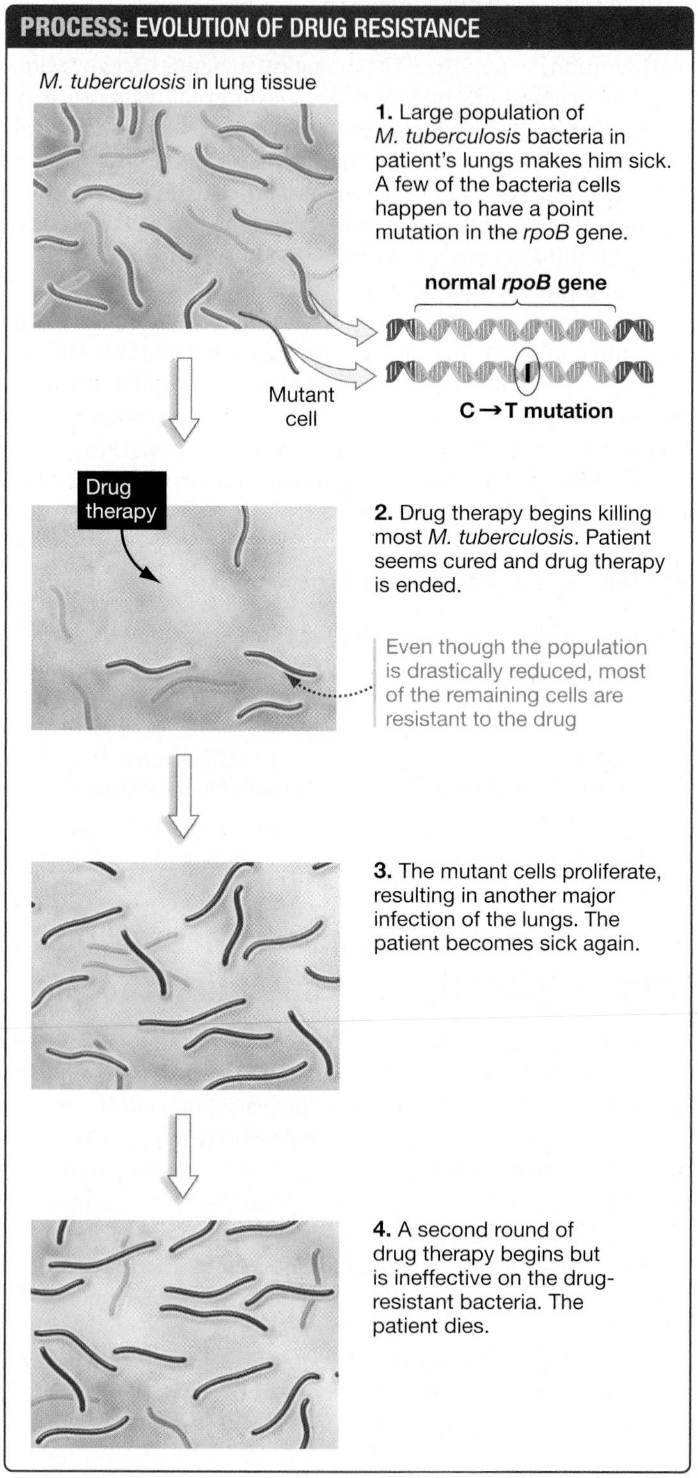

INTERACTIVE Figure 22.9 Alleles That Confer Drug Resistance Increase in Frequency When Drugs Are Used.

✔ Explain (1) why the relapse in step 3 occurred, and (2) whether a family member or health-care worker who got TB from this patient at step 3 or step 4 would respond to drug therapy.

2. ***Was this variation heritable?*** Yes. The researchers showed that the variation in the phenotypes of the two strains—from drug susceptibility to drug resistance—was due to variation in their genotypes. Because the mutant *rpoB* gene is passed on to daughter cells when an *M. tuberculosis* cell replicates, the allele and the phenotype it produces—drug resistance—are passed on to offspring.
3. ***Was there variation in survival and reproductive success?*** Yes. Only a tiny fraction of *M. tuberculosis* cells in the patient survived the first round of antibiotics long enough to reproduce. Most cells died and left no or almost no offspring.
4. ***Were survival and reproduction nonrandom?*** Yes. When rifampin was present, certain cells—those with the drug-resistant allele—had higher reproductive success than cells with the normal allele.

M. tuberculosis individuals with the mutant *rpoB* allele had higher fitness in an environment where rifampin was present. The mutant allele produces a protein that is an adaptation when the cell's environment contains the antibiotic.

This study verified all four postulates and confirmed that evolution by natural selection had occurred. The *M. tuberculosis* population evolved because the mutant *rpoB* allele increased in frequency.

Note, however, that *the individual cells themselves did not evolve*—they could not mutate their genes in order to survive the antibiotics. When natural selection occurred, the individual bacterial cells simply survived or died, or produced more or fewer offspring. This is a fundamental point: Natural selection acts on individuals, because individuals experience differential reproductive success. But only populations evolve. Allele frequencies change in populations, not in individuals. Understanding evolution by natural selection requires population thinking.

The Rise of Drug Resistance The events reviewed for this single patient have occurred many times in other patients. Resistance to a wide variety of antibiotics is skyrocketing globally due to the escalating use of antibiotics in applications ranging from toothpaste to livestock feed. In some cases, physicians have no effective antibiotics available to treat infections. The incidence of antibiotic resistance has risen so quickly that the World Health Organization recently announced that the world could soon enter a "post-antibiotic era." Drug-resistant strains of bacteria kill about 700,000 people each year worldwide. This number could jump to 10 million per year by 2050 if innovative solutions are not found. Biologists like you could really make a difference.

Further, resistance to a wide variety of insecticides, fungicides, antiviral drugs, and herbicides has evolved in hundreds of insects, fungi, bacteria, viruses, and plants (**Figure 22.10**). In every case, evolution has occurred because the original population included individuals with the heritable ability to resist a chemical compound. As the susceptible individuals died from the pesticide, herbicide, or drug, the alleles that confer resistance increased in frequency.

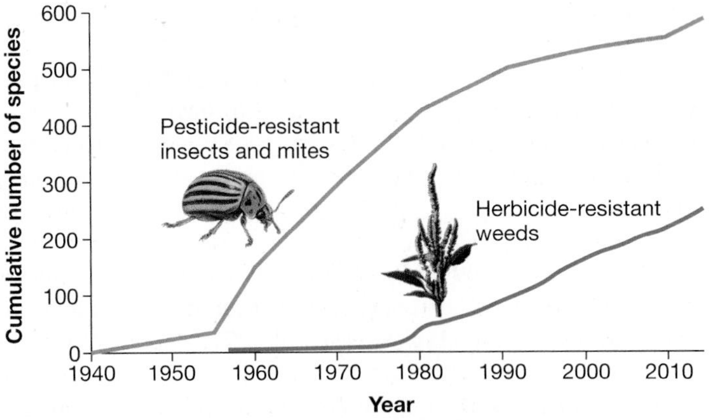

Figure 22.10 Like Antibiotic Resistance, Pesticide and Herbicide Resistance Are on the Rise. Chemical sprays are not as effective at killing crop pests and weeds as they used to be, because so many species are becoming resistant to these treatments.

DATA: B. Borel 2017. *Nature* 543: 302–304.

Case Study 2: Why Do Beak Sizes and Shapes Vary in Galápagos Finches?

Human use of antibiotics, insecticides, and herbicides has clearly influenced the evolution of many species. Have biologists also observed evolution caused by natural environmental changes? Most certainly yes. As an example, consider research led by Peter and Rosemary Grant. For over four decades, these biologists have been investigating changes in beak size, beak shape, body size, and other traits in Galápagos finches (see Figure 22.6).

Because the island of Daphne Major of the Galápagos is so small—about the size of 80 football fields—the Grants' team has been able to catch, weigh, and measure all the medium ground finches in the island's population (**Figure 22.11**) and mark each one with a unique combination of colored leg bands. The medium

Figure 22.11 The Grants' Field Site in the Galápagos.

ground finch makes its living by eating seeds. Finches crack seeds with their beaks.

Early studies of the finch population established that parents with particularly deep beaks tend to have offspring with deep beaks. Large parents also tend to have large offspring. That is, beak size and shape and body size are traits with heritable variation.

Selection during Drought Conditions Not long after the Grants began to study the finch population, a dramatic selection event occurred. In the annual wet season of 1977, Daphne Major received just 24 millimeters (mm) of rain instead of the 130 mm that normally falls. During the drought, few plants were able to produce seeds, and 84 percent (about 660 individuals) of the medium ground finch population disappeared.

Two observations support the hypothesis that most or all of these individuals died of starvation:

1. The researchers found 38 dead birds, and all were emaciated.
2. None of the missing individuals were spotted on nearby islands, and none reappeared once the drought had ended and food supplies returned to normal.

The research team realized that the die-off was a **natural experiment**. Instead of comparing groups created by direct manipulation under controlled conditions, natural experiments allow researchers to compare treatment groups created by an unplanned change in conditions. In this case, the Grants' team could test whether natural selection occurred by comparing the population before and after the drought.

Were the survivors different from nonsurvivors? The histograms in Figure 22.12 show the distribution of beak sizes in the population before and after the drought. Notice the different scales of the *y*-axes of the two graphs. (For more on how histograms are constructed, see BioSkills 2.) On average, survivors tended to have much deeper beaks than did the birds that died.

Were deeper beaks adaptive? At the peak of the drought, most seed sources were absent and the tough fruits of a plant called *Tribulus cistoides* served as the finches' primary food source. These fruits are so difficult to crack that finches ignore them in years when food supplies are abundant. The Grants hypothesized that individuals with particularly large and deep beaks were more likely to crack these fruits efficiently enough to survive.

At this point, the Grants had shown that natural selection led to an increase in average beak depth in the population. When breeding resumed in 1978, the offspring produced had beaks that were half a millimeter deeper, on average, than those in the population that existed before the drought. This result confirmed that evolution had occurred. In only one generation, natural selection led to a measurable change in the heritable characteristics of the population. Large, deep beaks were an adaptation for cracking large fruits and seeds.

Continued Changes in the Environment, Continued Selection, Continued Evolution In 1983, the environment on the Galápagos Islands changed again. Over a seven-month period, 1359 mm of rain fell. Plant growth was luxuriant, and finches fed primarily

RESEARCH

QUESTION: Did natural selection on ground finches occur when the environment changed?

HYPOTHESIS: Beak characteristics changed in response to a drought.

NULL HYPOTHESIS: No changes in beak characteristics occurred in response to a drought.

EXPERIMENTAL SETUP:

Weigh and measure all birds in the population before and after the drought.

PREDICTION OF HYPOTHESIS:

PREDICTION OF NULL HYPOTHESIS:

RESULTS:

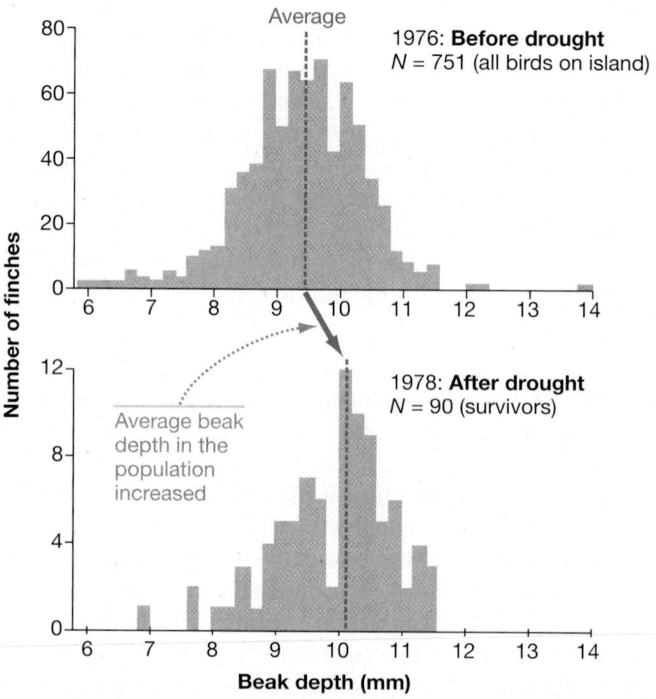

CONCLUSION: Natural selection occurred. Heritable characteristics of the population have changed.

Figure 22.12 A Natural Experiment: Changes in a Medium Ground Finch Population in Response to a Change in the Environment (a Drought). The results show the distribution of beak depth in the population of medium ground finches on Daphne Major before and after the drought of 1977. *N* is the population size.

SOURCE: P. T. Boag and P. R. Grant. 1981. Intense natural selection in a population of Darwin's finches (Geospizinae) in the Galápagos. *Science* 214: 82–85.

✔ **PROCESS OF SCIENCE** Fill in both predictions.

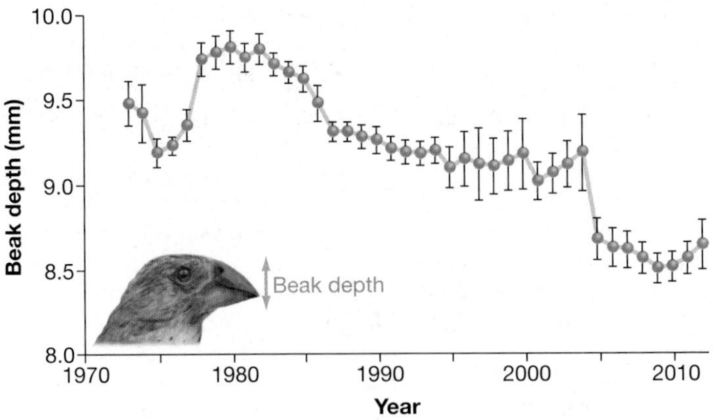

Figure 22.13 **Beak Depth in Medium Ground Finches Fluctuated over a 40-Year Interval.**

DATA: P. R. Grant and B. R. Grant. 2014. *40 Years of Evolution.* Princeton UP.

✔ **MODEL** Label the graph to show where the drought of 1977 occurred.

on the small, soft seeds that were being produced in abundance. During this interval, individuals with small, shallow beaks had exceptionally high reproductive success—meaning that they had higher fitness than those with large, deep beaks—because they were better able to harvest the small seeds. As a result, the characteristics of the population changed again.

The Grants documented changes in finch beak depth over a 40-year period and observed fluctuation in this trait over time (**Figure 22.13**). The take-home message from the long-term data is that most traits are not inherently "good" or "bad"—the adaptive value of a trait depends on context, and context can change over time. ✔ **THINK CAREFULLY Evaluate this statement in the context of the finch data: "Bigger is always better." How do the error bars (see BioSkills 3) affect your evaluation?**

Which Genes Are under Selection? Characteristics such as beak shape and body size are polygenic, meaning that many genes—each one exerting a relatively small effect—influence the trait (Ch. 14, Section 14.5). Because many genes are involved, it can be difficult for researchers to know exactly which alleles are changing in frequency when polygenic traits evolve.

The Grants have recently collaborated with a team of geneticists to sequence the genomes of 120 birds representing all of the finch species on the Galápagos. Among other insights, this research has pinpointed two transcription factor genes that appear especially important:

- The *ALX1* gene regulates beak *shape*.
- The *HMGA2* gene regulates beak *size*.

Both genes influence the development of facial structures in humans and other animals (a genetic homology), so their role in finch beak development makes sense. Further, a number of genes coding for signaling molecules, such as *BMP4, CAM*, and *β-catenin*, also appear to play a role because they have different expression patterns in different finches. This area of research is exciting because it integrates evolution with genetics, development, and ecology.

CHECK YOUR UNDERSTANDING

✔ **If you understood this section, you should be able to ...**

1. Apply Darwin's four postulates to the Galápagos finch case study to verify that natural selection has occurred.
2. **MODEL** Use simple sketches of finches with different beak depths to illustrate how Darwin's four postulates apply in the finch case study (see Making Models 22.2).

Answers are available in Appendix A.

22.5 Debunking Common Misconceptions about Natural Selection and Evolution

After you complete this section, you should be able to ...

- Correct common misconceptions about natural selection and evolution.

Evolution by natural selection is a simple process—just the logical outcome of some straightforward postulates. However, natural selection and population thinking are relatively new concepts in the scope of human culture. After more than two thousand years of belief in typological thinking, the scale of nature, and goal-oriented evolution (see Section 22.1), cultural inertia has fostered some misconceptions about evolution by natural selection. Let's clarify three of the most common types of confusion, summarized in **Table 22.3** on page 464 (also see **BioSkills 17**).

Natural Selection Does *Not* Change Individuals

Perhaps the most important point to clarify about natural selection is that in the process, individuals do not change—only the population does. During the drought, the beaks of individual finches did not become deeper. Rather, the average beak depth in the population increased over time because more deep-beaked individuals survived and produced offspring than shallow-beaked individuals did.

In the same way, individual *M. tuberculosis* cells did not change when rifampin was introduced to their environment. Each of these bacterial cells had the same RNA polymerase allele throughout its life. But because the mutant allele increased in frequency in the population over time, the average characteristics of the bacterial population changed.

Natural Selection Is *Not* "Lamarckian" Inheritance There is a sharp contrast between evolution by natural selection and evolution by the inheritance of acquired characters. If you recall (see Section 22.1), Lamarck proposed that **(1)** individuals change in response to challenges posed by the environment (such as giraffes stretching their necks to reach leaves high in the treetops), and **(2)** the changed traits are then passed on to offspring. The key claim is that the important evolutionary changes occur in individuals.

In contrast, Darwin realized that individuals do not change when they are selected. Instead, they simply produce more

SUMMARY **Table 22.3** Common Misconceptions about Natural Selection, Clarified

Examples of Common Confusing/Confused Statements in Biology Classes		Clarification	✔ Your Turn: Revise the Statements
1. "Giraffe necks grew longer over evolutionary time because giraffes stretched their necks to reach leaves higher in the treetops." **2. "Since the cave fish did not use their eyes any more, their offspring didn't inherit eyes."** **3. "Each finch adapted to eating tougher seeds by growing a larger beak."** **4. "I adapted to the cold temperature in the classroom."**		**Natural selection does not change individuals** • Natural selection is not Lamarckian inheritance; organisms do not gain traits through use, nor lose traits through disuse • Evolutionary adaptations arise in populations, not in individuals; note that "adaptation" has a different meaning in biology than in everyday English	*1. Average neck length increased in the population over time because giraffes born with longer necks had more surviving offspring, on average, than giraffes born with shorter necks.* 2. 3. 4.
5. "Rose plants needed to protect themselves from being eaten so evolved sharp thorns." **6. "*M. tuberculosis* cells mutated their *rpoB* gene so that they could survive the antibiotics."** **7. "Antibiotics caused a mutation in *M. tuberculosis* cells that enabled them to survive."** **8. "Tapeworms are a lowly animal because they do not even have a digestive system."**		**Natural selection is not goal directed** • Adaptations do not occur because organisms want or need them • Mutation, the source of new alleles, occurs by chance • Evolution is not progressive; for example, loss of traits can be adaptive • There is no such thing as a "higher" or "lower" organism	*5. Rose plants that happened to have sharp "thorns" (prickles) were more protected so had more surviving offspring than roses without them.* 6. 7. 8.
9. "Since humans have tailbones, the bones must be an adaptation for something." **10. "The beaks of finches are perfectly adapted for opening seeds."** **11. "The ear bones of humans are ideal for transmitting sound."**		**Natural selection does not lead to perfection** • Not all traits are adaptive • Some traits cannot be optimized due to fitness trade-offs • Some traits are limited by genetic, historical, or environmental constraints • Natural selection is not the only process of evolution	*9. Tailbones are a vestigial trait in humans; they are no longer useful for balance and locomotion compared to their function in primate ancestors.* 10. 11.

✔ Use the examples in the text to help you rewrite the statements to align them with current evolutionary theory.

or fewer offspring than other individuals do. When this happens, alleles found in the selected individuals become more or less frequent in the population. Natural selection sorts existing variants—it doesn't change them.

Individuals Do *Not* Adapt The issue of change in individuals is tricky because individuals often *do* change in response to changes in the environment. For example, if you were to travel to the Tibetan Plateau in Asia, your body would experience oxygen deprivation due to the low partial pressure of oxygen at high elevations (Ch. 42, Section 42.2). As a result, your body would produce more of the oxygen-carrying pigment hemoglobin and more hemoglobin-carrying red blood cells. Your body does not normally produce excessive red blood cells, because viscous (thick) blood can cause a disease—chronic mountain sickness—that can lead to heart failure.

Figure 22.14 Adaptation Has a Different Meaning in Science than in Everyday Life. In 1953, New Zealand mountaineer Edmund Hillary (left) and Tibetan sherpa Tenzing Norgay (right) were the first humans to reach the highest mountain peak in the world, the summit of Mount Everest. Climbers must *acclimatize* slowly to extremely high altitudes to avoid "mountain sickness," but sherpas like Norgay also have *adaptations* to life at high altitudes—genetically determined characteristics that improve their ability to survive in the mountains.

The increase in red blood cells is an example of what biologists call **acclimatization**—a change in an individual's phenotype that occurs in response to a change in natural environmental conditions. (When this process occurs in study organisms in a laboratory, it is called **acclimation**.) Phenotypic changes due to acclimatization are not passed on to offspring, because no alleles have changed. As a result, acclimatization does not cause evolution.

In contrast, populations that have lived in Tibet for many generations are adapted to this environment through genetic changes. Among native Tibetans, for example, alleles that increase the ability of hemoglobin to hold oxygen have increased to high frequency. In populations that do not live at high elevations, these alleles are rare or nonexistent (Figure 22.14).

Natural Selection Is *Not* Goal Directed

It is tempting to think that evolution by natural selection is goal directed. But adaptations do not occur because organisms want or need them.

Mutations Occur by Chance, *Not* on Purpose Mutations are the source of new alleles, but mutations occur by chance and are random with respect to fitness—they might be beneficial, neutral, or deleterious (see Ch. 16, Section 16.4). For example, you might hear a fellow student say that Tibetans "needed" the new hemoglobin allele so that they could survive at high altitudes, or that *M. tuberculosis* cells "wanted" the drug-resistant allele so that they could survive and continue to reproduce in an environment that included rifampin. This purposeful change does not happen.

The mutations that created the mutant alleles in both examples occurred randomly, due to errors in DNA synthesis, and they just happened to be advantageous when the environments changed. There is no mechanism that enables the environment to direct which mistakes DNA polymerase makes when copying genes.

Evolution Is *Not* Progressive It is often appealing to think that evolution by natural selection is progressive—meaning organisms have improved over time. (In this context, "improved" usually suggests that they have become bigger, stronger, or more complex.) And it is true that groups appearing later in the fossil record are often more morphologically complex than closely related groups that appeared earlier: For example, flowering plants are considered more complex than mosses. But there is nothing predetermined or absolute about this tendency.

In fact, complex traits are routinely lost or simplified over time as a result of evolution by natural selection. You've already analyzed evidence documenting limb loss in whales (see Section 22.2). Populations that become parasitic are particularly prone to loss of complex traits. For example, the tapeworm parasites of humans and other mammals have lost their sophisticated digestive tracts and mouths due to natural selection—tapeworms are well adapted to absorb nutrients directly from their environment.

There Is *No Such Thing* as a Higher or Lower Organism The nonprogressive nature of evolution by natural selection contrasts sharply with Lamarck's conception of the evolutionary process, in which organisms progress over time to higher and higher levels on a scale of nature (see Figure 22.1c). Under evolution by natural selection, there is no such thing as a higher or lower organism (see Figure 22.1d). For example, mosses may be a more ancient group than flowering plants, but neither group is higher or lower than the other (see **Making Models 22.3** on page 466). Mosses simply have a different suite of adaptations than do flowering plants, so they thrive in different types of environments.

Natural Selection Does *Not* Lead to Perfection

Although organisms are often exquisitely adapted to their environment, adaptation is far from perfect. A long list of circumstances limits the effectiveness of natural selection; only a few of the most important ones are discussed here.

Traits Are *Not* Always Adaptive Vestigial traits such as the human coccyx (tailbone) and the tiny hindlimbs of early whales do not increase the fitness of individuals with those traits. The structures are not adaptive. They exist because they were present in the ancestral population.

Perhaps the best example of nonadaptive traits involves changes in DNA sequences. A mutation may change a base in

Making Models 22.3 Tips on Drawing Phylogenetic Trees (II)

Phylogenetic trees can be challenging to interpret and draw (see **BioSkills 13** and Ch. 25, Section 25.1). Note that the labels on the branch tips are not arranged in a "lower to higher" order. Rather, branches can be rotated around the nodes (marked by dots) without changing their relationships.

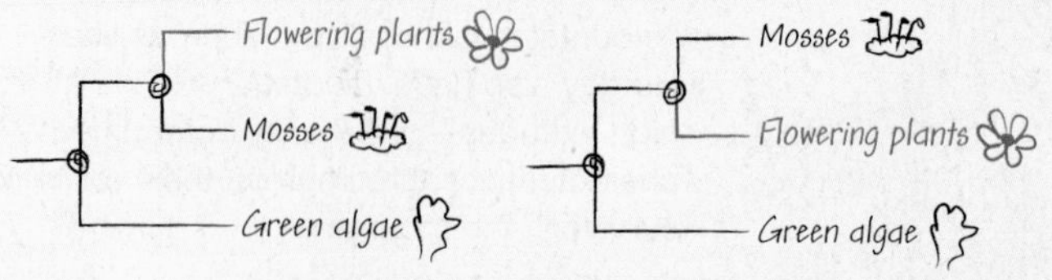

These trees have the same meaning

MODEL Draw a third tree with the same meaning as the two shown here by rotating the green algae branch to the top.

To see this model in action, go to the Study Area of **Mastering Biology**

the third position of a codon without changing the amino acid sequence of the protein encoded by that gene. Changes such as these are said to be silent. They occur due to the redundancy of the genetic code (Ch. 16, Section 16.4). Silent changes in DNA sequences are extremely common. But because they don't change the phenotype, they can't be acted on by natural selection and are not adaptive.

Fitness Trade-offs Exist In everyday English, the term "trade-off" refers to a compromise between competing goals. It is difficult to design a car that is both large and fuel efficient, a bicycle that is both rugged and light, or a plane that is both steady and maneuverable.

In nature, selection occurs in the context of fitness trade-offs. A **fitness trade-off** (or simply **trade-off**) is a compromise between two traits that cannot be optimized simultaneously. In medium ground finches, for example, there is a trade-off between having a beak that is large and deep enough to crack open tough seeds and one that is small and shallow enough to extract tiny seeds. Similarly, there is a trade-off between bright coloration to attract mates and cryptic coloration to hide from predators.

Traits Are Genetically Constrained There are a number of genetic constraints on natural selection, such as **genetic correlation**, when selection on one trait causes a change in another trait. One mechanism of genetic correlation is pleiotropy, which occurs when a single gene affects multiple traits (Ch. 14, Section 14.5). For example, the *MC1R* gene influences fur color in mammals but also plays a role in pain perception. Thus the evolution of one of these traits is constrained by the other.

Lack of genetic variation is also important. Consider that salamanders have the ability to regrow severed limbs. Some eels and sharks can sense electric fields. Birds can see ultraviolet light. Even though these traits would possibly confer increased reproductive success in humans, they do not exist—because humans lack the necessary genes.

Traits Are Historically Constrained All traits evolve from previous traits, so are constrained by evolutionary history. For example, the tiny incus, malleus, and stapes bones found in your middle ear evolved from bones that were either part of the jaw or supported the jaw in the ancestors of mammals. Are the bones a "perfect" solution to the problem of transmitting sound from the outside of the ear to the inside? No. Other vertebrates have different structures involved in transmitting sound to the ear, and some of them may be more efficient than our incus, malleus, and stapes in some conditions.

Traits Are Environmentally Constrained Natural selection often occurs in the context of a changing environment. Recall that the amount of rain fluctuates on Daphne Major in the Galápagos Islands, causing dramatic changes in vegetation and seed availability over time. Thus, a beak shape that is adaptive in one season may not be adaptive in the next season, preventing the "perfection" of beak shape in the finch population (see Figure 22.13).

Besides fluctuating over time, the abiotic and biotic features of the environment can change over the geographic range of the population. For example, one of the most notorious invasive plants in North America, purple loosestrife, has spread from the southern United States into Canada over a distance of 1000 kilometers (km) in less than a century. In the south, where the growing season is long, it is adaptive for purple loosestrife to grow large and then flower late. But in the north, it is more adaptive for purple loosestrife to grow less and flower early. No growth rate or flowering time is optimal for all environments.

Natural Selection Is *Not* the Only Process of Evolution Some students confuse evolution (the pattern of change through time) with natural selection (a process that causes adaptive evolution). This is understandable because Darwinian evolution is focused on natural selection.

But in the century and a half since the *Origin of Species*, the Modern Synthesis of Darwinian ideas with modern genetics has revealed that three other processes—*genetic drift, gene flow*, and *mutation*—also change allele frequencies over time. Unlike natural selection, these processes are random with respect to fitness, and so they *do not* lead to adaptation. The four processes of evolution are compared in detail in Chapter 23. (You can see the Big Picture of how natural selection relates to other evolutionary processes on pages 532–533.)

CHECK YOUR UNDERSTANDING

✔ If you understood this section, you should be able to ...

1. Explain the difference between the evolutionary definition of adaptation and its use in everyday English.
2. Predict how prickles ("thorns") originated in roses.

Answers are available in Appendix A.

22.1 The Rise of Evolutionary Thought

- Plato, Aristotle, and several religious texts present species as unchanging types. This view is called typological thinking.
- Lamarck proposed a theory of evolution—that species are not static but change through time. He proposed that evolution occurs by the inheritance of acquired characteristics.
- Darwin and Wallace proposed that evolution occurs by natural selection. This was the beginning of population thinking, whereby variation among individuals is the key to understanding evolution.

22.2 The Pattern of Evolution: Have Species Changed, and Are They Related?

- Evidence that species change through time includes data on **(1)** Earth's age; **(2)** extinction in the fossil record; **(3)** transitional features in the fossil record; **(4)** the presence of vestigial traits; and **(5)** change in populations that can be observed today.
- Evidence that species are related by common ancestry includes data on **(6)** the geographic proximity of closely related species; **(7)** the existence of genetic, developmental, and structural homologies; and **(8)** the formation of new species from preexisting species that can be observed today.
- Evidence for evolution is internally consistent, meaning that data from several independent sources are mutually reinforcing.

22.3 The Process of Evolution: How Does Natural Selection Work?

- Darwin developed four postulates that outline the process of evolution by natural selection. These postulates can be summarized by the following statement: Heritable variation leads to differential reproductive success.
- Alleles or traits that increase the reproductive success of an individual are said to increase the individual's fitness. A trait that leads to higher fitness, relative to individuals without the trait, is an adaptation. If a particular allele increases fitness and leads to adaptation, the allele will increase in frequency in the population.

22.4 Evolution in Action: Measuring Natural Selection in Populations Today

- The evolution of antibiotic resistance in the bacterium responsible for tuberculosis is an example of natural selection that can be observed and measured in populations today. The incidence of drug resistance, herbicide resistance, and pesticide resistance are sharply rising.
- The evolution of size and shape of finch beaks in the Galápagos is a result of natural selection for beaks that can harvest available seeds. Because seed availability fluctuates over time, the average dimensions of finch beaks fluctuate over time as well, as has been documented over the last four decades.
- Evolution by natural selection has been confirmed by a wide variety of studies and is considered to be a central organizing principle of biology.

22.5 Debunking Common Misconceptions about Natural Selection and Evolution

- Natural selection acts on individuals, but evolutionary change occurs in populations. Individual organisms do not evolve.
- Evolution is not goal directed and does not lead to perfection. Mutations occur by chance, not because organisms "want" or "need" them to survive. There is no such thing as a higher or lower organism.
- Not all traits are adaptive, and even adaptive traits are limited by fitness trade-offs and genetic, historical, and environmental constraints.
- Natural selection is not the only process of evolution. Genetic drift, gene flow, and mutation also cause evolutionary change but do not lead to adaptation, because they are random with respect to fitness.

Answers are available in Appendix A.

✔ TEST YOUR KNOWLEDGE

1. True or false? Some traits are considered vestigial because they existed long ago.
2. **THINK CAREFULLY** Why does the presence of extinct forms and transitional features in the fossil record support the pattern component of the theory of evolution by natural selection? Select True or False for each statement.
   T/F It supports the hypothesis that individuals change over time.
   T/F It supports the hypothesis that weaker species are eliminated by natural selection.
   T/F It supports the hypothesis that species evolve to become more complex and better adapted over time.
   T/F It supports the hypothesis that species change over time.
3. Traits that are derived from a common ancestor, like the bones of human arms and bird wings, are said to be ________.
4. **THINK CAREFULLY** How can evolutionary fitness be measured? Select True or False for each statement.
   T/F Document how long individuals survive.
   T/F Count the number of healthy, fertile offspring produced.
   T/F Determine which individuals are strongest.
   T/F Determine which phenotype is the most common.

✔ TEST YOUR UNDERSTANDING

5. **THINK CAREFULLY** According to data presented in this chapter, which of the following are correct? Select True or False for each statement.
   T/F When individuals change in response to challenges from the environment, their altered traits are passed on to offspring.
   T/F Species are created independently of each other and do not change over time.
   T/F Populations—not individuals—change when natural selection occurs.
   T/F The traits of populations become more perfect over time.
6. Some biologists summarize evolution by natural selection with the phrase "mutation proposes, selection disposes." Mutation is a process that creates heritable variation. Explain what the phrase means.

7. **THINK CAREFULLY** Why don't the biggest and strongest individuals in a population always produce the most offspring?
   a. The biggest and strongest individuals always have higher fitness.
   b. In some environments, being big and strong lowers fitness.
   c. Sometimes the biggest and strongest individuals may choose to have fewer offspring.
   d. Sometimes the number of offspring is not related to fitness.
8. **SOCIETY** Explain why the overprescription of antibiotics by doctors, or the overuse of everyday soaps containing antibiotics, can be a health risk.

✔ TEST YOUR PROBLEM-SOLVING SKILLS

9. **SOCIETY** Coral reefs are biodiverse ecosystems that provide food, income, coastal protection, and many other services to millions of people. Yet coral reefs are under threat from human impacts such as climate change—many corals cannot tolerate the warming ocean water and have suffered massive "bleaching" events, which can be fatal. Researchers have discovered that some corals have the capacity to acclimatize to warmer water, while other corals do not. The researchers fear that because corals are long-lived colonial animals and thus evolve slowly, they may not be able to adapt to global warming fast enough to avoid extinction. Explain how the difference between *acclimatize* and *adapt* is important to the fate of corals.
10. **PROCESS OF SCIENCE** The geneticist James Crow wrote that successful scientific theories have the following characteristics: (1) They explain otherwise puzzling observations; (2) they provide connections between otherwise disparate observations; (3) they make predictions that can be tested; and (4) they are heuristic, meaning that they open up new avenues of theory and experimentation. Crow added two other elements of scientific theories that he considered important on a personal, emotional level: (5) They should be elegant, in the sense of being simple and powerful; and (6) they should have an element of surprise. How well does the theory of evolution by natural selection fulfill these six criteria?

✔ PUT IT ALL TOGETHER: Case Study

Is fur color in mice a matter of life and death?

Most mice living on the mainland of Florida are brown, but the mice that live on the sand dunes of the barrier islands have white fur. The differences in color are heritable, determined by genes such as *MC1R* (Ch. 16, Section 16.2). It is intuitive that the light color of beach-dwelling mice is an adaptation for blending into their environment—and thus evading predators. How can this hypothesis be tested?

11. Compare how evolution by inheritance of acquired characters and the theory of evolution by natural selection would explain the observation of white mice living primarily on light soil and brown mice living primarily on dark soil.
12. **THINK CAREFULLY** What is an evolutionary adaptation?
    a. a trait that improves the fitness of its bearer, compared with individuals without the trait
    b. a trait that changes in response to environmental influences within the individual's lifetime
    c. the ability of an individual to adjust to its environment
    d. a trait that an individual wants so that it can survive
13. Apply Darwin's four postulates to a population of mice living on sand dunes in coastal Florida.
14. **PROCESS OF SCIENCE** A team led by evolutionary biologist Hopi Hoekstra set out to test the hypothesis that predators are an agent of natural selection on mouse color. They made 250 plasticine models of mice that were alike in every way except that half were painted white and half were painted brown. Suggest one advantage and one disadvantage of using model mice instead of real mice in this experiment.
15. **QUANTITATIVE** The researchers placed white and brown mouse models both in abandoned fields on the mainland (dark soil) and on sand dunes on the islands (light soil) and then measured the percentage of models that were attacked by predators. What is the take-home message of the data? Do the data support or reject the hypothesis that mouse color is adaptive?

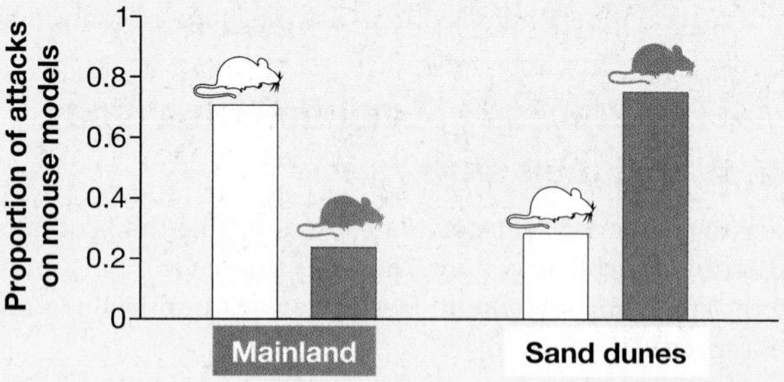

SOURCE: S. N. Vignieri, J. G. Larson, and H. E. Hoekstra. 2010. *Evolution* 64: 2153–2158.

16. **PROCESS OF SCIENCE** When a statistical test was used to compare the results for white models versus brown models, the *P* value (see **BioSkills 3**) was 0.01 for each habitat. Does this result increase or decrease your confidence in your take-home message? Explain.

Mastering Biology

Students Go to Mastering™ Biology for assignments, the eText, and the Study Area with animations, practice tests, and activities.

Professors Go to Mastering™ Biology for automatically graded tutorials and questions that you can assign to your students, plus Instructor Resources.

UNIT 4

EVOLUTIONARY PATTERNS AND PROCESSES

23 Evolutionary Processes

This albino sea turtle carries rare alleles that prevent pigment formation. The frequencies of these alleles change in populations of sea turtles over time due to the evolutionary processes discussed in this chapter.

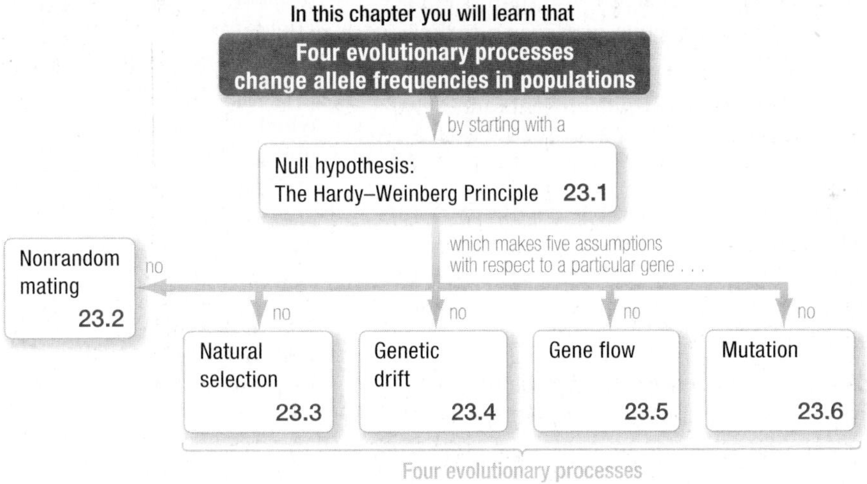

Understanding *evolution*—a change in heritable traits, and thus allele frequencies, in a population over time—is essential to understanding ourselves and the diversity of organisms around us. What causes biological evolution and how can evolution be measured?

BIG PICTURE

This chapter is part of the Big Picture. See how on pages 532–533.

Charles Darwin made a major contribution to the study of evolution by proposing a mechanism of descent with modification called *natural selection* (discussed in detail in Chapter 22). But Darwin knew that a clear understanding of inheritance would be needed to complete his theory of evolution.

Although Darwin was a contemporary of Gregor Mendel, the Austrian monk who demonstrated particulate inheritance (Chapter 14), both died in the 1880s without knowledge of the other's major contributions. It wasn't until the first few decades of the 1900s that the "Modern Synthesis" occurred:

the application of Mendelian genetics to Darwinian evolution. During that time, evolutionary biologists, mathematicians, and geneticists collaborated to make enormous leaps in *quantifying* evolution.

One of the fruits of the Modern Synthesis was **population genetics**, the study of processes that change the allele and genotype frequencies in populations. Recall that a **population** is a group of individuals of the same species that live in the same area at the same time and can interbreed. Darwin embraced "population thinking" by emphasizing the importance of heritable *variations* in phenotypes among individuals in populations (Ch. 22, Section 22.1). The Modern Synthesis provided a clear link between alleles, genotypes, and phenotypes. You can see these connections in this example using sea turtles:

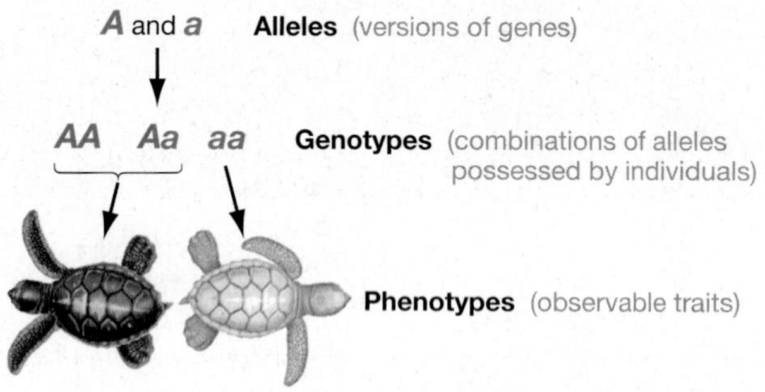

Evolution is now understood to be driven by four processes:

1. ***Natural selection*** increases the frequency of certain alleles—the ones that contribute to reproductive success in a particular environment (Ch. 22, Section 22.3).
2. ***Genetic drift*** causes allele frequencies to change randomly. In some cases, drift may cause alleles that decrease fitness to increase in frequency.
3. ***Gene flow*** occurs when individuals leave one population, join another, and breed. Allele frequencies may change when gene flow occurs, because arriving individuals introduce alleles to their new population and departing individuals remove alleles from their old population.
4. ***Mutation*** modifies allele frequencies by continually introducing new alleles. The alleles created by mutation may be beneficial, deleterious (detrimental), or neutral in their effects on fitness.

As you read the chapter, keep in mind that your everyday experience with the topic of evolution may have set you up for some common misconceptions (see Ch. 22, Section 22.5). Try to focus on the following fundamental ideas as you read ahead:

- Natural selection is not the only process responsible for evolution.
- Each of the four evolutionary processes has different consequences for genetic variation and fitness.

Let's examine what happens to allele frequencies over time in the absence of evolution, then consider each of the four evolutionary processes in turn.

23.1 Null Hypothesis: The Hardy–Weinberg Principle

After you complete this section, you should be able to . . .

- Apply the Hardy–Weinberg principle to determine if a population is evolving.

In 1908, a British mathematician, G. H. Hardy, and a German doctor, Wilhelm Weinberg, each published a major result independently. At the time, it was commonly believed that changes in allele frequency occur simply as a result of sexual reproduction—meiosis followed by the random fusion of gametes (eggs and sperm) to form offspring. Some biologists claimed that dominant alleles inevitably increase in frequency when gametes combine at random. Others predicted that two alleles of the same gene inevitably reach a frequency of 0.5.

To test these hypotheses, Hardy and Weinberg analyzed the frequencies of alleles when many individuals in a population mate and produce offspring. Instead of thinking about the consequences of a mating between two parents with a specific pair of genotypes, as Mendel did in his pea experiments (**Figure 23.1**), Hardy and Weinberg wanted to know what happened in an entire population when *all* of the individuals—and thus all possible genotypes—bred.

The Gene Pool Concept

To analyze the consequences of matings among all of the individuals in a population, Hardy and Weinberg invented a novel approach: They imagined a hypothetical scenario in which all of the alleles from all the gametes produced in each generation go into a single group called the **gene pool**. Something similar to this happens in actual populations of species such as clams and sea stars, which release their gametes into the water where the gametes mix randomly before combining to form zygotes.

To determine which genotypes would be present in the next generation and in what frequencies, Hardy and Weinberg calculated what would happen if pairs of gametes are plucked at random out of the gene pool, many times, and each of these pairs then combines to form offspring. These calculations predict the genotypes of the offspring that would be produced, as well as the frequency of each genotype.

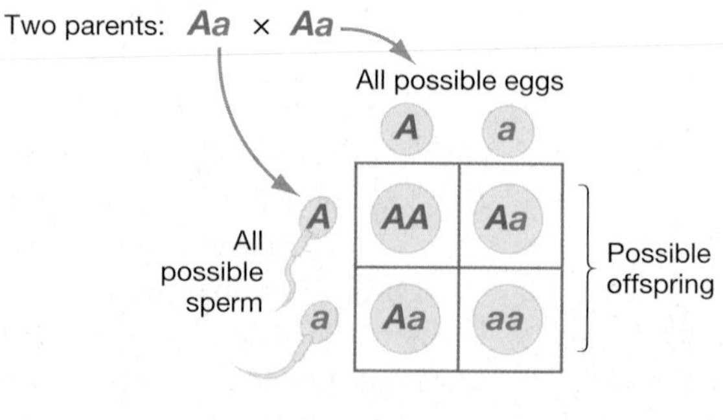

Figure 23.1 Mendelian Genetics Focuses on Matings between Two Parents. Recall that you can use Punnett squares to help you figure out the likely genotypes of the offspring produced when two parents mate. (Gamete icons are not drawn to the same scale.)

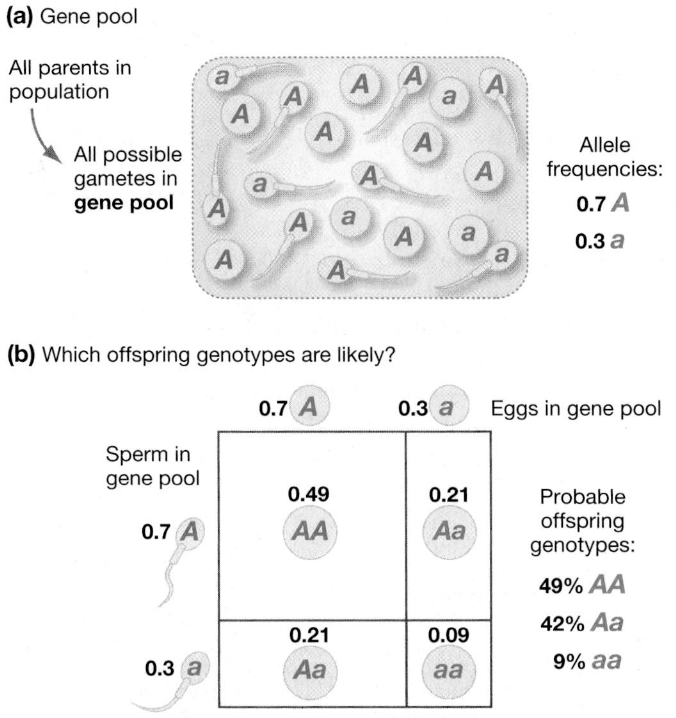

INTERACTIVE Figure 23.2 Population Genetics Focuses on Matings among Individuals in Entire Populations. Punnett squares (Ch. 14, Section 14.2) can help you to figure out the likely offspring genotypes of a whole population of parents. In this example, sizes of boxes inside the square reflect differences in offspring probabilities.

For example, Figure 23.2a shows a scenario with just one gene having two alleles, *A* and *a*. The frequency of the dominant *A* allele in the gene pool happens to be 0.7, or 70 percent, and the frequency of the recessive *a* allele happens to be 0.3, or 30 percent. When the haploid gametes randomly combine to form diploid offspring, how many of the offspring are likely to have the genotypes *AA*, *Aa*, and *aa*?

You can use a Punnett square to calculate the probable offspring in this population, as shown in Figure 23.2b.

- There is a $0.7 \times 0.7 = 0.49$ chance that offspring will have the homozygous dominant genotype *AA* (meaning that about half of the population is likely to be *AA*).
- There is a $0.3 \times 0.3 = 0.09$ chance that offspring will have the homozygous recessive genotype *aa*.
- Since there are two ways to produce a heterozygote, there is a $0.3 \times 0.7 = 0.21 \times 2 = 0.42$ chance that the offspring will have the heterozygote genotype, typically written as *Aa* with the dominant allele first, regardless of which alleles came from which gametes.

(To review tips for calculating probabilities, see **BioSkills 4**.)

The allele frequencies happen to be 0.7 and 0.3 in this example, but other allele frequencies are possible. Since Hardy and Weinberg wanted to be able to write a general formula that would work with all allele frequencies, they used letters to represent these variables. They used the letter p to symbolize the frequency of *A* alleles in the gene pool, and the letter q to symbolize the frequency of *a* alleles in the same gene pool. (They assumed that there are only two alleles of this gene.) This logic enabled them to articulate the **Hardy–Weinberg principle**, which serves as a mathematical null hypothesis for the study of evolutionary processes.

The Hardy–Weinberg principle makes two fundamental claims:

1. If the frequencies of alleles *A* and *a* in a population are given by p and q, then the frequencies of genotypes *AA*, *Aa*, and *aa* will be given by p^2, $2pq$, and q^2, respectively, for generation after generation. That is:

$$\textbf{Allele frequencies: } p + q = 1$$

$$\textbf{Genotype frequencies: } p^2 + 2pq + q^2 = 1$$

2. When alleles are merely transmitted via meiosis and random combinations of gametes, their frequencies do not change over time, meaning that evolution does not occur.

This last statement is very important: *allele frequencies do not change over time*. In the example in Figure 23.2, the Hardy–Weinberg principle predicts that if allele frequencies are $p = 0.7$ and $q = 0.3$ in the parent generation, then the allele frequencies will be the same in the offspring generation. If you figure out the allele frequencies in the offspring population, this turns out to be true: $p = 0.7$ and $q = 0.3$. **Quantitative Methods 23.1** on page 472 walks through Hardy's and Weinberg's calculations to clarify the relationships between allele frequencies and genotype frequencies in different generations when no evolution is occurring.

Given a set of parental allele frequencies, the Hardy–Weinberg principle predicts what genotype frequencies and allele frequencies will occur in the next generation. If biologists observe frequencies that conform to the Hardy–Weinberg predictions, the frequencies are said to be in **Hardy–Weinberg equilibrium**. In real populations, however, allele frequencies often *do* change over time. How can this be explained? The answer lies in the assumptions of the Hardy–Weinberg principle.

The Hardy–Weinberg Principle Makes Important Assumptions

The mathematical model underlying the Hardy–Weinberg principle is based on five important assumptions about how populations and alleles behave:

1. ***Random mating*** The model assumes that gametes from the gene pool combine at random. Individuals are not allowed to *choose* a mate.
2. ***No natural selection*** The model assumes that *all* members of the parental generation survive and contribute equal numbers of gametes to the gene pool, no matter what their genotype.
3. ***No genetic drift (random allele frequency changes)*** The model assumes that alleles are picked in their exact frequencies p and q, and not at different frequencies caused by chance—that is, the model behaves as though the population is infinitely large. In the example in Figures 23.2 and 23.3, allele *a* did not "get lucky" and get drawn more than 30 percent of the time.

Quantitative Methods 23.1 Logic of the Hardy–Weinberg Principle

Hardy and Weinberg began by analyzing the simplest situation possible—just two alleles of a particular gene. The letter p is used to symbolize the frequency of A alleles in the gene pool and q is used to symbolize the frequency of a alleles in the same gene pool. Because there are only two alleles, the two frequencies must add up to 1:

$$p + q = 1$$

Now follow the steps in **Figure 23.3**:

Parental Allele Frequencies

Although p and q can have any value between 0 and 1, let's suppose that the initial frequency of A is 0.7 and that of a is 0.3. In this gene pool, 70 percent of the gametes carry A and 30 percent carry a.

Offspring Genotype Frequencies

In this example, each time a gamete is involved in forming an offspring, there is a 70 percent chance that it carries A and a 30 percent chance that it carries a. In general, there is a p chance that it carries A and a q chance that it carries a.

Because only two alleles are present, three genotypes are possible in the offspring generation: AA, Aa, and aa. What will the frequency of these three genotypes be? According to the logic of Hardy's and Weinberg's model:

- Frequency of the AA genotype: p^2
- Frequency of the Aa genotype: $2pq$
- Frequency of the aa genotype: q^2

The genotype frequencies in the offspring generation must add up to 1, which means that

$$p^2 + 2pq + q^2 = 1$$

In this numerical example:

$$0.49 + 0.42 + 0.09 = 1$$

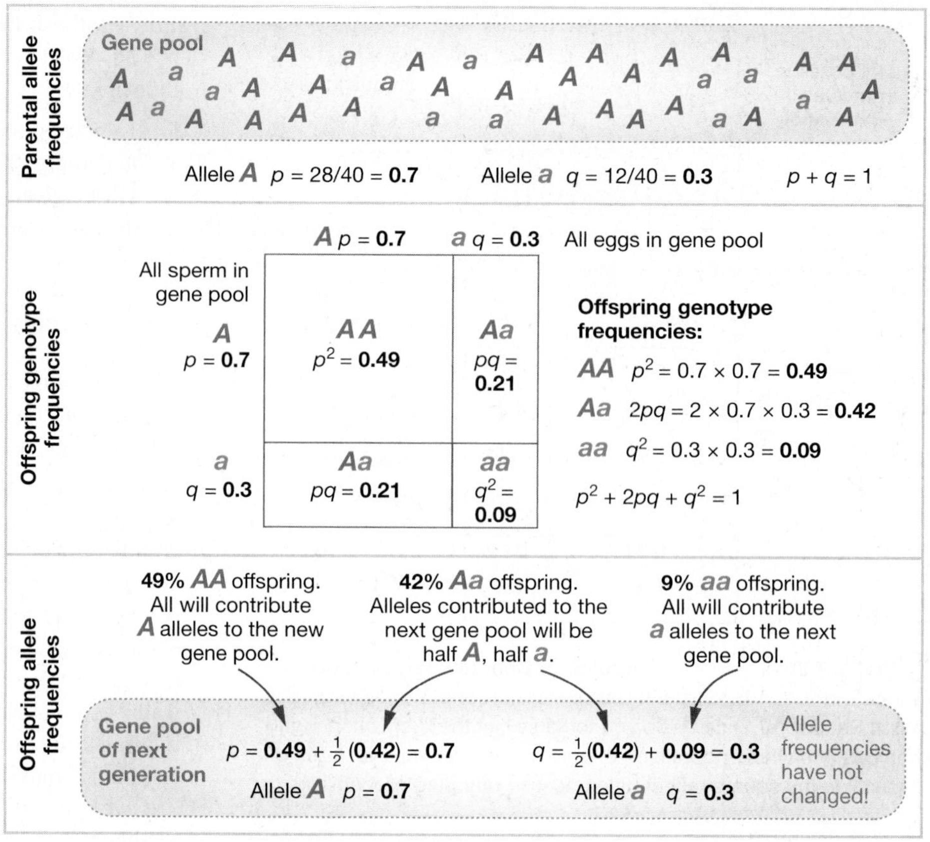

Figure 23.3 Logic of the Hardy–Weinberg Principle.

Offspring Allele Frequencies

The easiest way to calculate the allele frequencies in the offspring is to imagine that they form gametes that go into a gene pool. All of the gametes from AA individuals carry A, so the frequency of A gametes contributed to the gene pool by these individuals will be 0.49 (that is, p^2). But half of the gametes from Aa will also carry A, so an additional $\frac{1}{2}(0.42) = 0.21$ gametes in the gene pool will carry A (that is, $\frac{1}{2} \times 2pq = pq$). The allele frequency for A in the offspring is thus $0.49 + 0.21 = 0.70$ (that is, $p^2 + pq$). Using the same logic, the frequency of a in the offspring is $pq + q^2$, or $0.21 + 0.09 = 0.3$.

The frequency of allele A in the offspring generation is still 0.7 and the frequency of allele a is still 0.3. The values of p and q have not changed from one generation to the next. This result is called the Hardy–Weinberg principle.

✔ **QUANTITATIVE** If you understand these calculations, you should be able to **(1)** calculate the frequencies of the three offspring genotypes if $p = 0.6$ and $q = 0.4$ in the parental gene pool; **(2)** use your answer to calculate the allele frequencies in the offspring gene pool (show your math).

4. ***No gene flow*** The model assumes that no new alleles are added by immigration or lost through emigration. As a result, all of the alleles in the offspring population come from the original population's gene pool.

5. ***No mutation*** The model assumes that no new alleles, such as A_2, are introduced into the gene pool.

Note that *four* of these factors—natural selection, genetic drift, gene flow, and mutation—are the processes of evolution mentioned in this chapter's introduction. Each of these processes, alone or in combination, can cause allele frequencies in a population to change over time. In contrast, nonrandom mating can cause only genotype frequencies to change, not allele frequencies, unless it occurs in concert with other processes such as selection (as will be explained in Section 23.2). Biologists often want to test a hypothesis that nonrandom mating is occurring, or that natural selection or one of the other evolutionary processes is affecting a particular trait in a population. In addressing questions like these, the Hardy–Weinberg principle functions as a null hypothesis. Recall that a null hypothesis specifies what should be observed if the hypothesis being tested is *not* correct (Ch. 1, Section 1.6).

Let's consider an example to illustrate how the Hardy–Weinberg principle is used as a null hypothesis in a study of a trait in real populations: MN blood types in humans.

Case Study: Are MN Blood-Type Alleles in Humans in Hardy–Weinberg Equilibrium?

One of the first genes that geneticists could analyze in natural populations was the MN blood group of humans. Most human populations have two alleles, designated *M* and *N*, at this gene.

Because the *MN* gene codes for a protein found on the surface of red blood cells—the *M* allele codes for the M version, and the *N* allele codes for the N version—researchers could determine whether individuals are *MM, MN*, or *NN* by treating blood samples with antibodies to each protein (Ch. 11, Section 11.2). The *M* and *N* alleles are codominant—heterozygotes have both M and N versions of the protein on their red blood cells (Ch. 14, Section 14.5). Are the *M* and *N* alleles in Hardy–Weinberg equilibrium?

Table 23.1 shows *MN* genotype frequencies for three populations and compares observed genotype frequencies with the frequencies expected if they are in Hardy–Weinberg equilibrium. The analysis is based on the following steps:

Step 1 Estimate genotype frequencies by observation—in this case, by testing blood samples from a large number of individuals for the *M* and *N* alleles, then dividing the number of individuals with each genotype by the total number of individuals in the sample. These frequencies are given in the rows labeled "Observed" in Table 23.1.

Step 2 Calculate observed allele frequencies from the observed genotype frequencies. In this case, the frequency of the *M* allele is the frequency of *MM* homozygotes plus half the frequency of *MN* heterozygotes; the frequency of the *N* allele is the frequency of *NN* homozygotes plus half the frequency of *MN* heterozygotes. (You can review the logic behind this calculation in Quantitative Methods 23.1.)

Step 3 Use the observed allele frequencies to calculate the genotypes expected according to the Hardy–Weinberg principle (see the "Expected" rows in Table 23.1). Under the null hypothesis of random mating and no evolution, the expected genotype frequencies are $p^2 : 2pq : q^2$.

Step 4 Compare the observed and expected values. Researchers use statistical tests to determine whether the differences between the observed and expected genotype frequencies are small enough to be due to chance (in which case the null hypothesis is accepted) or large enough to reject the null hypothesis.

Although using statistical testing is beyond the scope of this text (see **BioSkills 3** for a brief introduction), you should be able to inspect the numbers and comment on them. In these populations, for example, the observed and expected *MN* genotype frequencies are almost identical. (A statistical test shows that the small differences observed are probably due to chance.) For every population surveyed, genotypes at the *MN* gene are in Hardy–Weinberg equilibrium. As a result, biologists conclude that the assumptions of the Hardy–Weinberg principle are valid for this trait.

The results imply that when these data were collected, the *M* and *N* alleles in these populations were not being affected by the four evolutionary processes and that mating was random with respect to this gene—meaning that humans were not choosing mates based on their *MN* genotype. The null hypothesis is accepted.

Before moving on, however, note that a study such as this does not mean that the *MN* gene has never been subject to nonrandom mating, selection, or genetic drift. Even if selection has been very strong for many generations, one generation of random mating and no evolution will result in genotype and allele frequencies that conform to Hardy–Weinberg expectations.

If biologists observe frequencies that are not in Hardy–Weinberg equilibrium, it means that something interesting is going on. Further research is then needed to determine which of the Hardy–Weinberg assumptions is being violated. Let's consider each of the processes that can violate the Hardy–Weinberg assumptions and that therefore influence evolution.

Table 23.1 The MN Blood Group of Humans: Observed and Expected Genotype Frequencies

You can calculate the expected genotype frequencies from the observed allele frequencies using the Hardy–Weinberg principle.

		Genotype Frequencies			Allele Frequencies	
Population and Location	**Data Type**	***MM***	***MN***	***NN***	***M***	***N***
Caucasians (U.S.)	Observed	0.292	0.494	0.213	0.540	0.460
	Expected	0.290	0.497	0.212		
Aborigines (Australia)	Observed	0.025	0.304	0.672	0.176	0.824
	Expected	0.031	0.290	0.679		
Ainu (Japan)	Observed (Step 1)	0.179	0.502	0.319	(Step 2)	
	Expected (Step 4)	(Step 3)				

DATA: W. C. Boyd. 1950. Boston: Little, Brown and Company.

✔ **QUANTITATIVE** Fill in the values for observed allele frequencies (*Step 2*) and expected genotype frequencies (*Step 3*) for the Ainu people of Japan, and compare the expected values to the observed values (*Step 4*).

CHECK YOUR UNDERSTANDING

✔ **If you understood this section, you should be able to ...**

QUANTITATIVE Analyze whether a gene suspected of causing hypertension in humans is in Hardy–Weinberg equilibrium. In one study, the observed genotype frequencies were *HH* 0.574; *Hh* 0.339; *hh* 0.087. (Note: The sample size in this study was so large that a difference of 3 percent or more in any of the observed versus expected frequencies indicated a statistically significant difference—meaning a difference that was not due to chance; see **BioSkills 3**.)

Answers are available in Appendix A.

23.2 Nonrandom Mating

After you complete this section, you should be able to ...

- Explain the effect of nonrandom mating on allele and genotype frequencies.

In the Hardy–Weinberg model, gametes were picked from the gene pool at random and paired to create offspring genotypes. In nature, however, matings between individuals may not be random with respect to the gene in question. Even in species such as clams that broadcast their gametes into the surrounding water, gametes from individuals living close to each other are more likely to combine than gametes from individuals living farther apart.

The most intensively studied form of nonrandom mating is called **inbreeding**, the mating between relatives.

How Does Inbreeding Affect Allele Frequencies and Genotype Frequencies?

To understand how inbreeding affects populations, let's follow the fate of alleles and genotypes when inbreeding occurs. As before, focus on a single locus with two alleles, *A* and *a*, and suppose that these alleles initially have equal frequencies of 0.5.

Now imagine that the gametes produced by individuals in the population don't go into a gene pool. Instead, individuals self-fertilize. Many flowering plants, for example, contain both male and female organs and routinely self-pollinate. Self-fertilization, or selfing, is the most extreme form of inbreeding.

As **Figure 23.4a** shows, homozygous parents that self-fertilize produce all homozygous offspring. Heterozygous parents, in contrast, produce homozygous and heterozygous offspring in a 1 : 2 : 1 ratio. **Figure 23.4b** shows the outcome for the population as a whole. In this figure, the width of the boxes represents the frequency of the three genotypes, which start out in Hardy–Weinberg equilibrium at a ratio of $p^2 : 2pq : q^2$. Notice that the homozygous proportion of the population increases with each generation, while the heterozygous proportion is halved. At the end of the three generations illustrated, heterozygotes are rare. The same outcomes occur more slowly with less extreme forms of inbreeding.

This simple exercise demonstrates two fundamental points about inbreeding:

- Inbreeding increases homozygosity.
- Inbreeding itself does not cause evolution, because, although genotype frequencies change, allele frequencies do not change in the population as a whole.

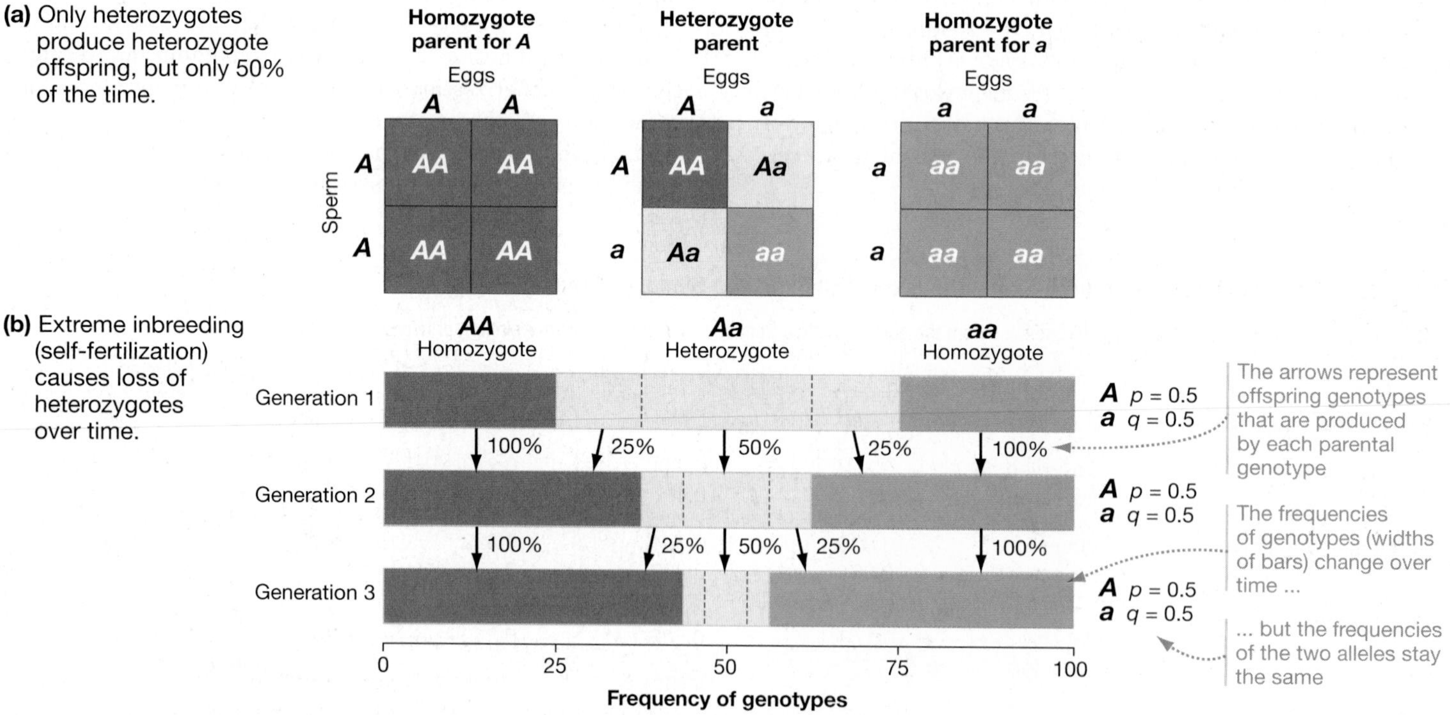

Figure 23.4 Inbreeding Increases Homozygosity and Decreases Heterozygosity. (a) Heterozygous parents produce homozygous and heterozygous offspring in a 1 : 2 : 1 ratio. **(b)** The width of the boxes corresponds to the frequency of each genotype

How Does Inbreeding Influence Evolution?

The trickiest point to grasp about inbreeding is that even though it does not cause evolution directly, it can speed the rate of evolutionary change. More specifically, it increases the rate at which natural selection eliminates recessive **deleterious alleles**—alleles that lower fitness—from a population.

Many recessive deleterious alleles are rare and exist in heterozygous individuals where they have little or no effect—one normal allele usually produces enough functional protein to support a normal phenotype. But inbreeding increases the frequency of homozygous recessive individuals, exposing the rare allele to selection.

Consider **inbreeding depression**, the decline in average fitness that may take place when homozygosity increases and heterozygosity decreases in a population. Inbreeding depression is a common problem in small populations, such as the dwindling population of endangered Florida panthers. Figure 23.5 shows results from recent efforts to reduce illness in this population by introducing females from Texas. Note that the two sets of data points compare the fitnesses of offspring from non-inbred Texas × Florida matings and inbred Florida × Florida matings. Inbreeding depression is represented by the vertical distance between data points at each age. (For help interpreting error bars, see **BioSkills 3**).

Several studies show similar results on inbreeding depression in human populations around the world. Children of first-cousin marriages consistently have a higher mortality rate than children of marriages between nonrelatives. Because inbreeding has such deleterious consequences in humans, it is not surprising that many human societies have laws forbidding marriages between individuals who are related as first cousins or closer.

Nonrandom Mating via Sexual Selection

Sometimes nonrandom mating occurs not due to inbreeding but because an organism actively choses a certain mate based on its physical and/or behavioral traits. This form of nonrandom mating—called **sexual selection**—is fundamentally different from inbreeding because it *does* lead to changes in allele frequencies in the population and increases fitness, and thus it is a form of natural selection. Let's take a closer look at this type of selection in the next section.

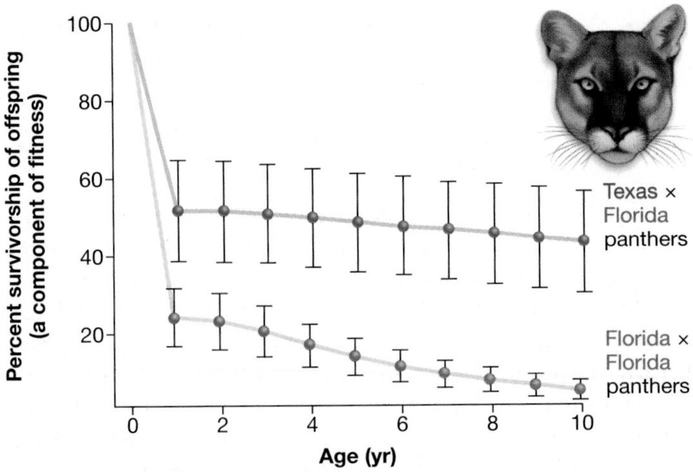

Figure 23.5 Inbreeding Depression Occurs in Florida Panthers. Inbreeding depression is the fitness difference between non-inbred and inbred individuals.

DATA: W. E. Johnson, E. P. Onorato, M. E. Roelke, et al. 2010. *Science* 329: 1641–1645.

CHECK YOUR UNDERSTANDING

✔ **If you understood this section, you should be able to ...**

Predict the change of allele frequencies in a small population of sea turtles over time due to inbreeding alone versus inbreeding plus natural selection. Assume the gene that causes albinism has the starting allele frequencies $A = 0.99$ and $a = 0.01$, where only *aa* individuals are white (see chapter opener photo).

Answers are available in Appendix A.

23.3 Natural Selection

After you complete this section, you should be able to ...

- Explain why natural selection violates the Hardy–Weinberg assumptions.
- Compare directional, stabilizing, and disruptive selection.
- Compare intersexual and intrasexual selection.

Evolution by natural selection occurs when heritable variation leads to differential success in survival and reproduction (Ch. 22, Section 22.3). Stated another way, natural selection occurs when individuals with certain phenotypes produce more surviving offspring than do individuals with other phenotypes. If certain alleles are associated with the favored phenotypes, they increase in frequency while other alleles decrease in frequency. Natural selection thus violates the assumptions of the Hardy–Weinberg principle.

Note that Darwin originally conceived of sexual selection as a second category of selection *in addition to* natural selection. He did so because he wanted to compare sexual selection to selection for traits that enable organisms to survive in their physical and biological environments, sometimes called **ecological** (or **environmental**) **selection**. Other researchers use this two-category framework for selection as well. However, in the context of the Hardy–Weinberg principle, sexual selection is considered to be a subset of natural selection because both produce adaptation, increasing fitness. The one-category framework (natural selection = ecological selection + sexual selection) is used throughout this book.

Also note that different agents of natural selection can act on organisms simultaneously, sometimes favoring the same traits, other times resulting in fitness trade-offs (Ch. 22, Section 22.5). For example, sexual selection may favor long display feathers, called trains, in peacocks, while ecological selection may favor shorter trains that make the peacocks better able to maneuver and less vulnerable to predators. The relative importance of different agents of selection can change over time and over the geographic range of the species.

Let's begin by comparing the different ways that selection can affect variation in any kind of trait, then consider the specific case of sexual selection. (Recall that two case studies on ecological selection are explored in Ch. 22, Section 22.4: antibiotic resistance in *M. tuberculosis* and the evolution of beak depth in Galápagos finches.)

How Does Selection Affect Genetic Variation?

When biologists analyze the consequences of selection, they often focus on **genetic variation**—the number and relative frequency of alleles that are present in a particular population. The reason is simple: Lack of genetic variation in a population is usually a bad thing (Ch. 13, Section 13.4).

To understand why this is so, recall that selection can occur only if heritable variation exists in a population (Ch. 22, Section 22.3). If genetic variation is low and the environment changes—perhaps due to the emergence of a new disease-causing virus or a rapid change in climate—there is a good chance that none of the available alleles will enable individuals to survive and reproduce under the new conditions. As a result, the average fitness of the population will decline. If the environmental change is severe enough, the population may even be faced with extinction.

Natural selection occurs in four main patterns, or modes, each affecting genetic variation in different ways:

1. *Directional selection* changes the average value of a trait.
2. *Stabilizing selection* reduces variation in a trait.
3. *Disruptive selection* increases variation in a trait.
4. *Balancing selection* maintains variation in a trait.

Let's take a closer look at each of these four modes in turn.

Mode 1: Directional Selection When **directional selection** occurs, the average phenotype of a population changes in one direction. To get a sense of how directional selection affects genetic variation, look at the top graph in Figure 23.6a, which plots the value of a trait on the x-axis and the number of individuals with a particular value of that trait on the y-axis (for help reading histograms, see BioSkills 2). Note that the trait in question is determined by multiple genes, so it has a bell-shaped, normal distribution in this hypothetical population (Ch. 14, Section 14.5).

The second graph in Figure 23.6a shows what happens when directional selection acts on this trait. Directional selection tends to reduce the genetic diversity of populations. If directional selection continues over time, the favored alleles will eventually approach a frequency of 1.0 while deleterious alleles will approach a frequency of 0.0, a clear violation of the assumptions of the Hardy–Weinberg principle. Alleles that reach a frequency of 1.0 are said to be fixed; those that reach a frequency of 0.0 are said to be lost. When deleterious alleles decline in frequency, **purifying selection** is said to occur.

Now consider an example. Recall that Peter and Rosemary Grant happened to catch, weigh, and measure all the medium ground finches on Daphne Major Island in the Galápagos before a drought. Then in the annual wet season of 1977, Daphne Major received just 24 millimeters (mm) of rain instead of the 130 mm that normally falls. During the drought, few plants were able to produce seeds, and 84 percent (about 660 individuals) of the medium ground finch population disappeared or were found dead.

Were the survivors different from nonsurvivors? The histogram in Figure 23.6b shows the distribution of beak sizes in the population before and after the drought. On average, survivors tended to have much deeper beaks than did the birds that died—the deeper beaks enabled the finches to open the tough seeds that were available during the drought. Directional selection had occurred.

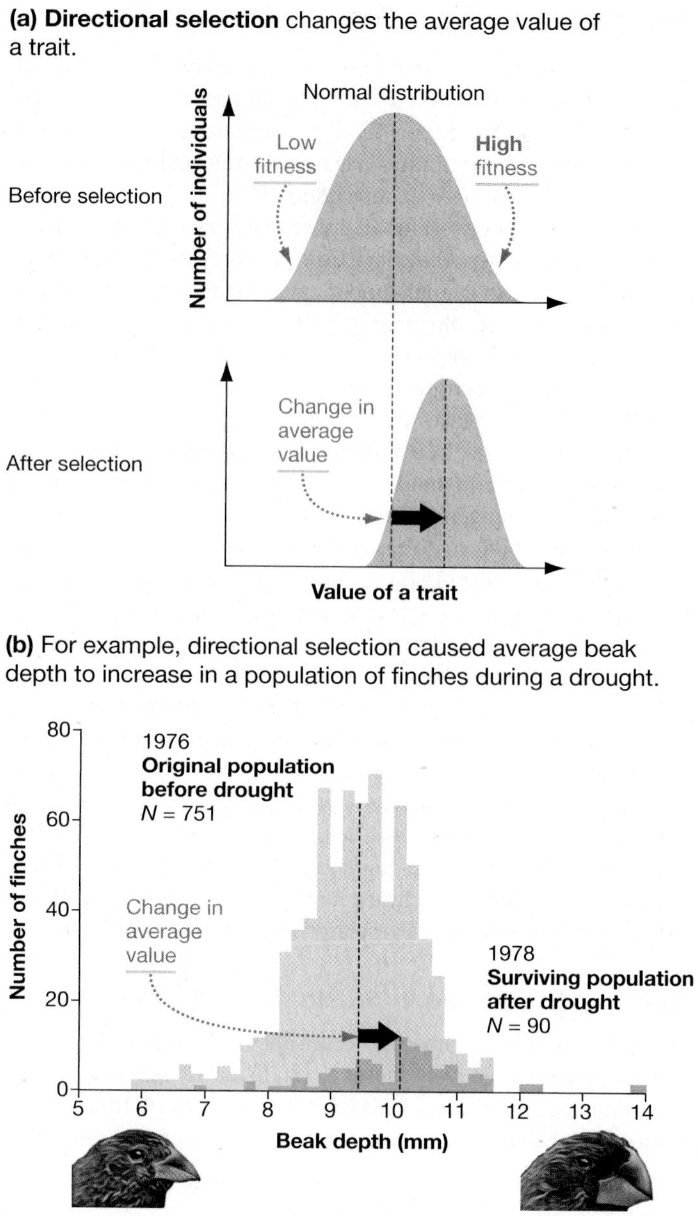

Figure 23.6 Directional Selection. When directional selection acts on traits that have a normal distribution, individuals at one end of the distribution have poor reproductive success.

SOURCE: P. T. Boag and P. R. Grant. 1981. *Science* 214: 82–85.

Since beak depth is a heritable trait governed by several genes such as *ALX1, HMGA2*, and *BMP4* (Ch. 22, Section 22.4), some alleles increased in frequency and some alleles decreased in frequency during the drought. Evolution had occurred in just one year.

As it turned out in this case, directional selection was short-lived because the severe weather event was temporary. Since the amount of rainfall, and thus the availability of different types of seeds, varies each year on the Galápagos Islands, the direction of selection can change year to year (look back at Figure 22.13). In other cases, directional selection can be more persistent in one direction, such as the directional selection that occurs due to global warming.

✔ **QUANTITATIVE If you understand this concept, you should be able to predict how the allele frequencies of a sea turtle**

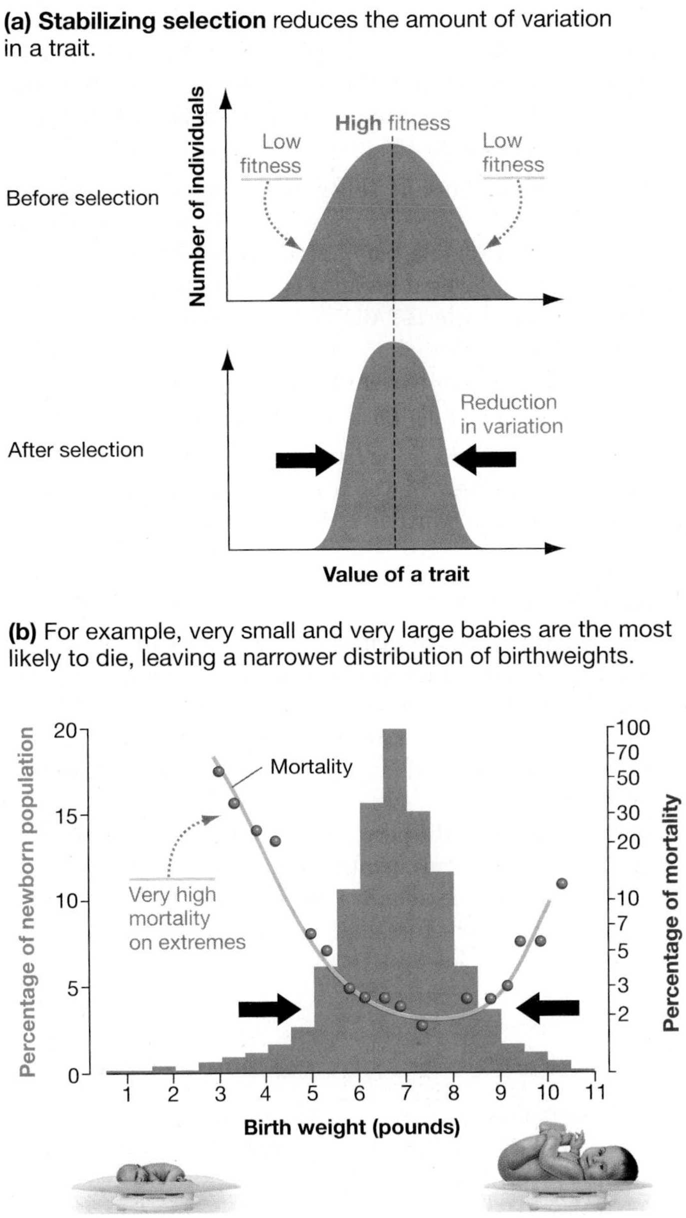

Figure 23.7 **Stabilizing Selection.** When stabilizing selection acts on normally distributed traits, individuals with extreme phenotypes have poor reproductive success.

DATA: M. N. Karn, H. Lang-Brown, J. J. MacKenzie, et al. 1951. *Annals of Eugenics* 15: 306–322.

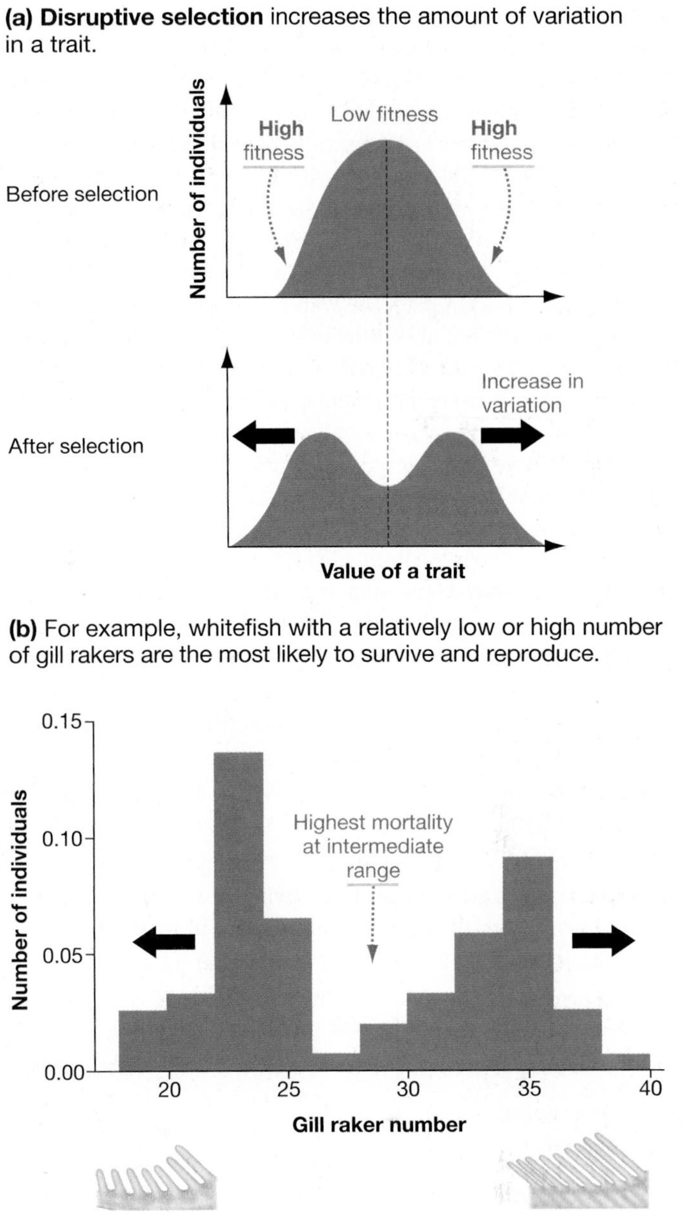

Figure 23.8 **Disruptive Selection.** When disruptive selection occurs on traits with a normal distribution, individuals with extreme phenotypes have high reproductive success.

DATA: S. Bhat. 2014. *PLOS ONE* 9: e91208.

population are likely to change if 50 percent of the population starts with normal pigmentation (*AA* and *Aa*) and 50 percent are albino (*aa*). Assume albino turtles are more visible to predators.

Mode 2: Stabilizing Selection Selection can also reduce both extremes in a population, as illustrated in Figure 23.7a. This mode of selection, called **stabilizing selection**, has two important consequences: Genetic variation in the population is reduced, but there is no change in the average value of a trait over time.

Figure 23.7b shows a classical data set in humans that illustrates stabilizing selection. Biologists who analyzed birth weights and mortality in 13,730 babies born in British hospitals in the 1940s found that babies of average size (slightly over 7 pounds) survived best. Mortality was high for very small babies and very large babies. This is persuasive evidence that birth weight was under strong stabilizing selection in this population. Alleles associated with high birth weight or low birth weight were subject to purifying selection, and alleles associated with intermediate birth weight increased in frequency.

Mode 3: Disruptive Selection **Disruptive selection** has the opposite effect of stabilizing selection. Instead of favoring phenotypes near the average value and eliminating extreme phenotypes, it eliminates phenotypes near the average value and favors extreme phenotypes (Figure 23.8a). When disruptive selection occurs, the overall amount of genetic variation in the population is increased.

Recent research has shown that disruptive selection is responsible for the striking distribution of gill raker numbers in whitefish in Norwegian lakes (**Figure 23.8b**). Gill rakers are comb-like projections of the gill arches that can function in feeding in some fish. The data plotted in the graph show that individuals with either long, dense gill rakers or with few, sparse gill rakers survived and that fish with intermediate phenotypes perished.

What is the cause of this pattern? Food availability. Researchers observed that the whitefish with dense gill rakers are efficient at filtering tiny plankton from the water. The whitefish with sparse gill rakers are efficient at capturing small animals without getting their rakers clogged up with mud. Whitefish with an intermediate number of gill rakers are inefficient at gathering both types of food. Since gill raker number is a genetically determined trait, disruptive selection maintains high genetic variation in this population.

Disruptive selection is important because it can play a role in *speciation*, the formation of new species (detailed in Chapter 24). If whitefish tend to mate only with fish with similar raker number in the same part of the lake, reproductive isolation develops between the two populations. If this isolation continues over time, the populations may eventually form distinct species.

Mode 4: Balancing Selection **Balancing selection** occurs when no phenotype has a distinct advantage. Instead, there is a balance among several phenotypes in terms of their fitness and frequency. Of the various mechanisms of balancing selection, the following are the most common:

- **Heterozygote advantage** occurs when heterozygous individuals have higher fitness than homozygous individuals do. This pattern of selection is commonly observed in genes involved in the ability of immune systems to recognize and destroy invading bacteria and viruses. The consequence is that selection on different alleles is balanced and genetic variation in the population is maintained.
- **Frequency-dependent selection** occurs when certain alleles are favored when they are rare, but not when they are common. For example, rare alleles responsible for coloration in guppies are favored because predators learn to recognize common color patterns. Alleles for common colors get eliminated; alleles for rare colors increase in frequency. As a result, allele frequencies shift over the short term, but overall genetic variation in the population is maintained in the long term.

Table 23.2 summarizes the four modes of natural selection. No matter how natural selection occurs, though, its most fundamental attribute is the same: It increases fitness and leads to adaptation.

Sexual Selection

Darwin was the first to recognize that selection based on success in courtship is a mechanism of evolutionary change. Although he introduced the idea of sexual selection in *On the Origin of Species* in 1859, it was not until 12 years later that he detailed his views on sexual selection in *The Descent of Man, and Selection in Relation to Sex*.

Darwin was initially perplexed by seemingly nonadaptive traits such as the extravagantly long and iridescent trains of peacocks. Why was it adaptive for the peacock to spend energy producing a structure that would increase the peacock's

SUMMARY **Table 23.2 Modes of Selection**

Mode of Selection		Effect on Phenotype	Example	Effect on Genetic Variation
Directional selection	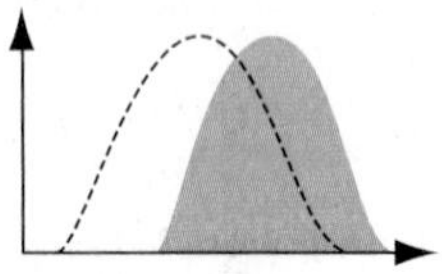	Favors one extreme phenotype, causing the average phenotype in the population to change in one direction.	Average beak depth increased in ground finches during the 1977 drought.	Genetic variation is reduced.
Stabilizing selection	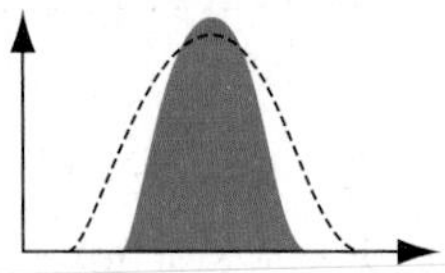	Favors phenotypes near the middle of the range of phenotypic variation, maintaining average phenotype.	Human babies of average size are most likely to survive.	Genetic variation is reduced.
Disruptive selection	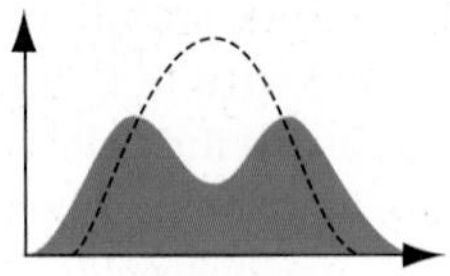	Favors extreme phenotypes at both ends of the range of phenotypic variation.	Whitefish with low or high numbers of gill rakers are most likely to survive.	Genetic variation is increased.
Balancing selection	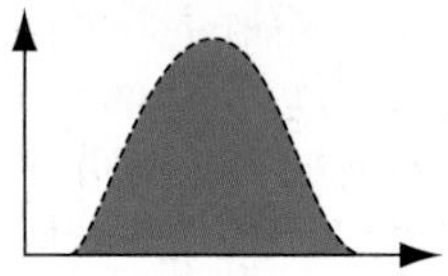	No single phenotype is favored in all populations of a species at all times.	Guppies with rare color patterns are favored.	Genetic variation is maintained.

vulnerability to predation? Darwin concluded that the peacock's train is adaptive because it helps the peacock to attract mates, thereby increasing its reproductive success.

The peacock example demonstrates mate choice, sometimes referred to as ***inter*sexual selection**—the selection of an individual of one sex for mating by an individual of the other sex. Darwin also recognized that individuals of the same sex sometimes compete with one another to obtain mates. This form of selection *within* a sex is referred to as ***intra*sexual selection**.

The Fundamental Asymmetry of Sex Why would the extravagant trains of peacocks occur in the males but not the peahens? A. J. Bateman addressed this question in the 1940s, followed by Robert Trivers three decades later. The Bateman–Trivers hypothesis contains an observed pattern and a hypothesized process:

- ***Pattern*** Traits that attract members of the opposite sex tend to be more highly elaborated in males.
- ***Process*** The energetic cost of creating a large egg is enormous, whereas sperm contain few energetic resources (**Figure 23.9a**). That is, "eggs are expensive, but sperm are cheap." As a result, females tend to invest more in their offspring than do males (**Figure 23.9b**).

The Bateman–Trivers hypothesis, sometimes referred to as the *fundamental asymmetry of sex*, makes several predictions:

- If females invest a lot in few eggs, then they should protect that investment by being choosy about their mates. Conversely, if males invest little in each sperm, then they should be willing to mate with almost any female.
- If there are an equal number of males and females in the population, and if males are trying to mate with any female possible, then males will compete with each other for mates.
- If male fitness is limited by access to mates, then any allele that increases a male's attractiveness to females or success in male–male competition should increase in the population. Thus, sexual selection should act more strongly on males than on females.

Do data from experimental or observational studies agree with these predictions? Yes, but only in some cases. Trivers himself recognized that some males invest more in parental care for offspring than females and/or there can be more females than males in a population, resulting in stronger sexual selection on females rather than males. Also:

- Sexual selection varies in the context of different mating strategies; some females benefit from being promiscuous (having multiple mates) while some males benefit from being monogamous (having one mate; see Ch. 50, Table 50.1).
- Sexual selection can vary in different environmental contexts, depending on the availability of mates, food resources, and/or other factors at a given time and place.

In sum, the Bateman–Trivers hypothesis holds up in some cases, but the *fundamental asymmetry of sex* is not valid across all species all the time.

***Inter*sexual Selection: Mate Choice** If mating is nonrandom, how does an organism choose a mate? Research has revealed a spectacular variety of physiological, physical, and behavioral traits that are subject to mate choice.

One long-standing hypothesis for mate choice is that organisms select traits that are "honest signals" of health and genetic quality. Consider the following observations:

- In some bird species, the existence of colorful feathers or a colorful beak is due to the presence of the red and yellow pigments called carotenoids.

Figure 23.9 In Many Species, Females Make a Large Investment in Each Offspring. (a) The human egg (red) dwarfs the sperm (blue) in this colorized scanning micrograph. Human females usually produce one mature egg per month, while human males can produce a billion sperm per month. **(b)** This female dusky salamander is protecting her small clutch of enormous eggs.

- Carotenoids protect tissues and stimulate the immune system to fight disease more effectively.
- To obtain carotenoids, animals have to find and eat carotenoid-rich plant tissues.

These observations suggest that the healthiest and best-nourished birds in a population have the most colorful beaks and feathers. Conversely, dull coloration can occur because birds are sick and using their carotenoids to stimulate their immune system, or because the birds are poorly fed and have few carotenoids available. The assumption is that the ability to fight off disease and feed efficiently is based at least in part on "good alleles" that can be passed on to offspring.

To test the hypothesis that bright colors are an honest signal of mate quality, a team of researchers experimented with zebra finches (**Figure 23.10**). They identified pairs of brothers and randomly assigned one brother to the treatment group and one brother to the control group. They fed all the brothers the same diet, but supplemented the treatment group with carotenoids.

As predicted, the brothers eating the carotenoid-enriched diet developed more colorful beaks and larger cell-mediated immune responses than did those fed the unsupplemented diet. When given a choice of perching with either of the two brothers, 9 out of the 10 females tested preferred the more-colorful male (brightly colored bars in Figure 23.10). These results support the hypothesis that females of this species are choosy about their mates and that they prefer to mate with healthy, well-fed males.

Note that "honest signals" of mate quality can indicate not only good health but also desirable behaviors or the ability to provide resources or care for offspring. For example, some female fish prefer to mate with males that protect a nest site and care for the eggs until they hatch. In humans and many species of birds, mates provide food, protection, and other resources for rearing young.

Is there an alternate hypothesis to the "honest signals" hypothesis? In some cases, the trait under selection appears to be chosen merely because it is *attractive*. The logic here is that the signal (such as a red beak) and the sensitivity to the signal (such as the ability to see red) can reinforce one another over time because the offspring inherit these traits and preferences.

Recall that sexually-selected traits, like all traits, are subject to fitness trade-offs: Red beaks and feathers may be attractive to mates, but they also draw predators. In some species, ecological selection favors females and/or males that have "cryptic" coloration that helps them to blend in with their environments for part or all of their life cycle.

RESEARCH

QUESTION: Do females zebra finches choose mates with brighter colors?

HYPOTHESIS: Females choose mates that are healthy and well fed as indicated by the brighter color of their beaks.

NULL HYPOTHESIS: Females do not prefer mates with brighter beaks.

EXPERIMENTAL SETUP:

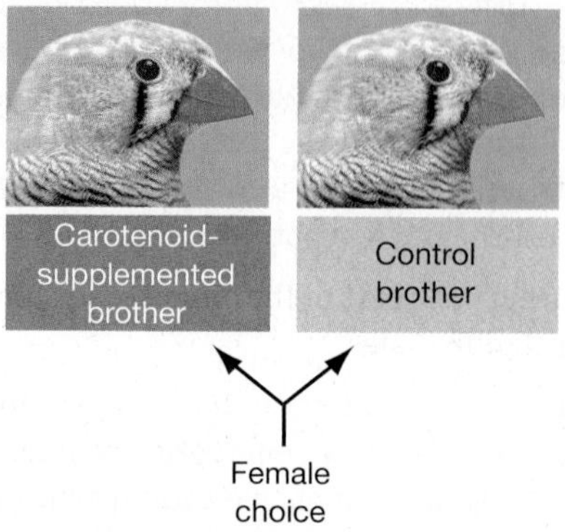

1. Find 10 brother pairs of zebra finches. Select one brother from each pair at random to receive food with extra carotenoids.

2. Present a female to each pair of brothers and record how much time she spends perched next to each brother.

PREDICTION OF HYPOTHESIS: Females will spend more time with brothers with brighter beaks.

PREDICTION OF NULL HYPOTHESIS: Females will spend the same amount of time, on average, with both types of brothers.

RESULTS:

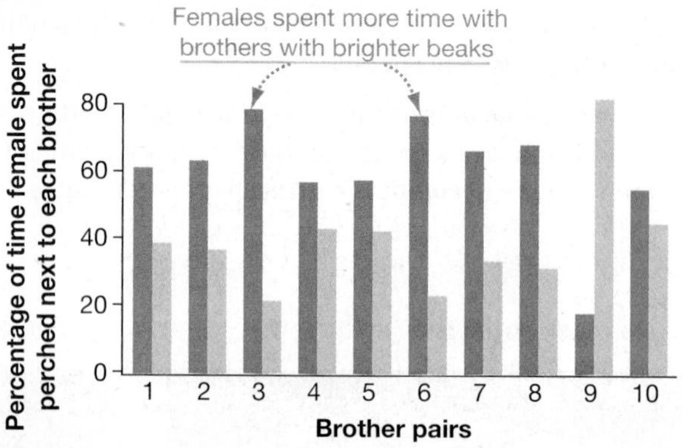

CONCLUSION: On average, females chose males with more colorful beaks.

Figure 23.10 An Experimental Test of Mate Choice in Zebra Finches.

SOURCE: J. D. Blount, N. B. Metcalf, T. R. Birkhead, et al. 2003. Carotenoid modulation of immune function and sexual attractiveness in zebra finches. *Science* 300: 125–127.

✔ **PROCESS OF SCIENCE** Why do you think the researchers chose brother pairs for the study rather than unrelated males?

***Intra*sexual Selection: Competition for Mates** In some species, individuals compete for access to mates. As an example, consider data from a long-term study of a northern elephant seal population breeding on Año Nuevo Island, off the coast of California.

Elephant seals feed mostly on marine fish, squid, and octopus, and spend most of the year in the water. But when females are ready to give birth, nurse their pups, then mate again, they haul themselves out of the water and congregate in large numbers on the few island beaches in their habitat that have gentle slopes.

Male elephant seals establish territories on breeding beaches by fighting. A **territory** is an actively defended area where the owner has exclusive or semi-exclusive use. Males that win battles with other males protect the females residing in their territories, then monopolize matings with them. Males that lose battles are relegated to territories with fewer females or are

(a) Males compete for the opportunity to mate with females.

(b) Variation in reproductive success is high in males.

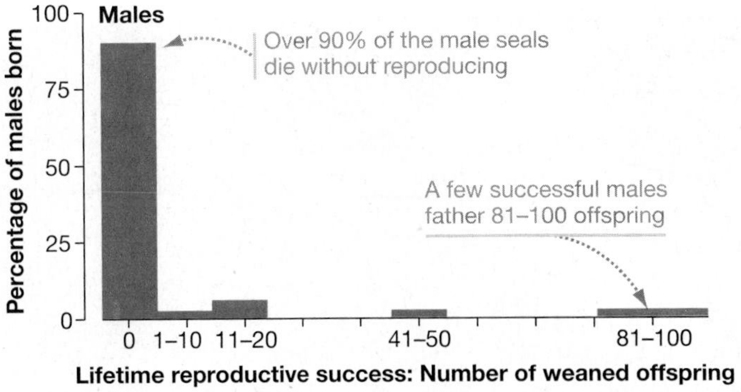

(c) Reproductive success varies less in females.

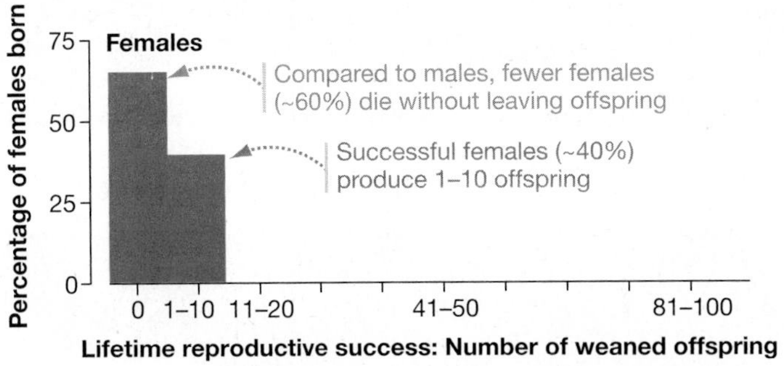

Figure 23.11 Sexual Selection Is Intense in Male Elephant Seals. The histograms show that variation in the lifetime reproductive success of northern elephant seals is higher in **(b)** males than it is in **(c)** females.

DATA: B. J. Le Boeuf and R. S. Peterson. 1969. *Science* 163: 91–93.

✔ **QUANTITATIVE** Compare the number of offspring produced by the most reproductively successful males and females.

excluded from the beach. Fights are essentially slugging contests and are usually won by the larger male. The males, which frequently weigh 2700 kg (~6000 lbs), stand face to face, bite each other, and land blows with their heads (Figure 23.11a).

Male northern elephant seals are over four times more massive, on average, than females. This dramatic size difference, a dimorphic trait, suggests that strong intrasexual selection is occurring. Do data on reproductive success in males support this hypothesis?

Biologists have marked many individuals in the seal population on Año Nuevo to track their lifetime reproductive success. The data for males provides evidence of intense sexual selection (Figure 23.11b). The *x*-axis indicates fitness, plotted as the number of offspring produced and weaned over a lifetime. The *y*-axis indicates the percentage of males born in the population that achieved each category of offspring production (0, 1–10, and so on). In this population a few large males father most of the offspring, while most males father few or none, an example of directional selection for large body size. Among females, variation in reproductive success is much lower than in males (Figure 23.11c).

If you look closely at any example of sexual selection, you will discover some interesting nuances. For example, in some elephant seal populations, females sometimes mate with small males at the margins of the beach. These "sneaked" matings have trade-offs but can be adaptive for both individuals and increase genetic diversity in the population. Just as Darwin's "population thinking" predicts, organisms within populations vary in their mating behavior.

Also keep in mind that competition for an opportunity to mate is not always as obvious as bloody elephant seal fights on the beach. For example, competition among sperm is a hot area of research (as is female choice via physiology, anatomy, and behavior to accept or reject sperm).

Sexual Dimorphism Can Result from Sexual Selection Both intersexual and intrasexual selection help to explain why some traits differ between the sexes of the same species, a difference called **sexual dimorphism** (literally, "two-forms"). Such traits range from elaborate ornamentation and courtship displays that attract mates, as observed in peacocks and many other birds, to weapons that can be used to deter competing suitors, such as the antlers possessed by red deer males but not females (see Figure 23.12 on page 482). For elephant seals, body size is dimorphic.

Some displays and behaviors can be involved in both intersexual and intrasexual selection, as is the case with push-up displays in Anolis lizards (Ch. 50, Section 50.3).

(a) Male raggiana birds of paradise display for females.

(b) Red deer males use weaponry to compete for females.

Figure 23.12 **Sexually Dimorphic Traits Can Help to Attract Mates and/or Deter Competitors for Mates.**

Sexual Selection in Humans Humans have a lot in common with other animals due to common ancestry, but it's important to avoid jumping to conclusions about humans based on anecdotes of finches and elephant seals. In humans, as in some other highly social animals, sexual selection is especially complex because courtship and mating are behaviors that can serve functions other than producing offspring—such as creating and maintaining social bonds, both between sexes and within sexes. Thus, sexual selection for reproductive success and ecological selection for social behavior are closely intertwined.

Take-Home Messages Mate choice and competition for mates illustrate how sexual selection can favor certain phenotypes in a population. Many fascinating twists on sexual selection occur in diverse species, including plants. The key insights in every case, however, are as follows:

- Multiple forms of sexual and/or ecological selection can occur at the same time in the same population, either favoring the same phenotypes or resulting in fitness trade-offs (Ch. 22, Section 22.5).
- Overall, the alleles responsible for adaptive morphological, physiological, and/or behavioral phenotypes increase in frequency over time. Thus, evolution occurs and the assumptions of the Hardy–Weinberg principle are violated.
- Selection can change over time and space because it occurs in the context of ecology, which is highly complex and dynamic.

Natural selection is the only evolutionary process that results in adaptation, but it is not the only evolutionary process that violates the Hardy–Weinberg assumptions. Next let's consider what happens when random changes in allele frequency occur.

CHECK YOUR UNDERSTANDING

✔ **If you understood this section, you should be able to ...**

Predict which genotype is likely to be most frequent in a population of Soay sheep that has a single gene, *RXFP2*, that determines horn size. Incomplete dominance occurs for this trait such that homozygotes for the Ho^+ allele have large horns, homozygotes for the Ho^P allele have small horns, and heterozygotes have intermediate horns. Note that rams with large horns have the highest mating success per season, but rams with small horns survive through more mating seasons by avoiding combat with other rams.

Answers are available in Appendix A.

23.4 Genetic Drift

After you complete this section, you should be able to ...

- Compare genetic drift to the other processes of evolution.

Genetic drift is defined as a change in allele frequencies in a population that is due to chance. The process is aptly named, because allele frequencies drift up and down over time, randomly with respect to fitness. Genetic drift occurs in every population, in every generation, but it especially affects small populations.

The cause of genetic drift is known in statistics as sampling error. **Sampling error** occurs when the allele frequencies of a chosen subset of a population (the sample) are different from those in the total population, by chance (see **BioSkills 3**). Sampling error is not a "mistake" in the everyday sense—it is just the statistical reality that if you have, for example, a 50 percent chance of flipping a coin and getting heads, you will not necessarily get heads every other time you flip a coin.

Simulation Studies of Genetic Drift

To understand why genetic drift occurs, imagine a married couple marooned on a deserted island with enough resources to support them. Suppose that at gene *A*, the wife's genotype is *Aa* and

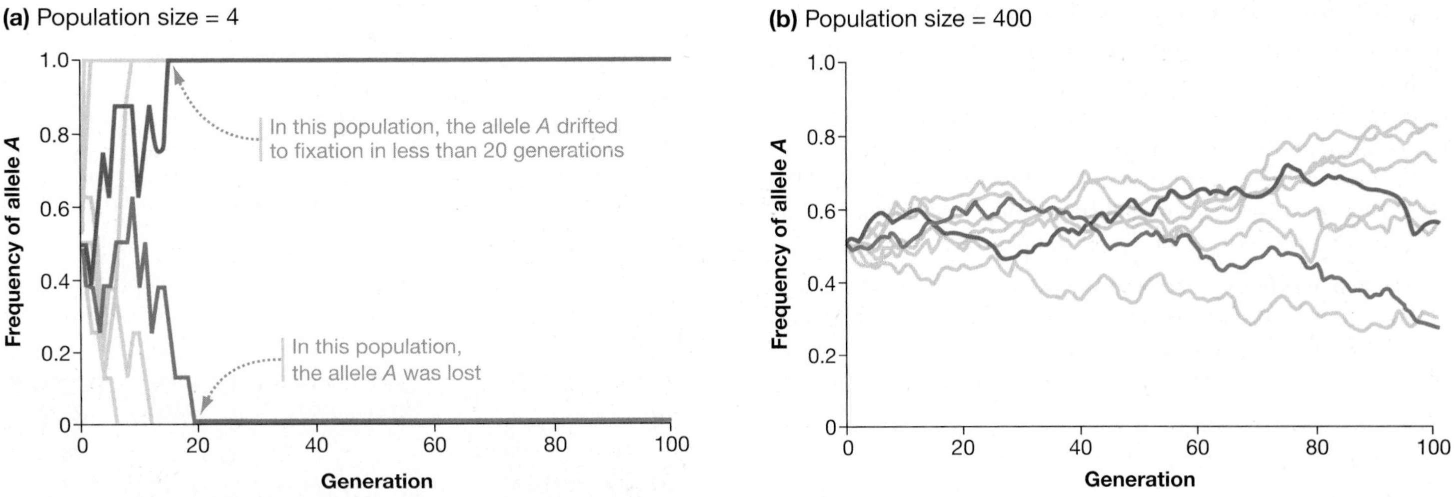

Figure 23.13 Genetic Drift Is More Pronounced in Small Populations than in Large Populations. In each computer simulation, one of the eight replicates is colored blue and another is colored red to ease comparison.

✔ **MODEL** Draw a graph predicting what these simulation results would look like for a population size of 4000.

the husband's is also *Aa*. In this tiny population, the two alleles are each at a frequency of 0.5.

Now suppose that the couple produce five children over their lifetime. Half of the eggs produced by the wife carry allele *A* and half carry allele *a*. Likewise, half of the sperm produced by the husband carry allele *A* and half allele *a*. To simulate which sperm and which egg happen to combine to produce each of the five offspring, you can flip a coin for each sperm and egg, where heads represents allele *A* and tails represents allele *a*.

The following coin flips were done by a pair of students in a recent biology class:

	Sperm	Egg	Genotype
First offspring	*A*	*A*	*AA*
Second offspring	*a*	*a*	*aa*
Third offspring	*a*	*A*	*Aa*
Fourth offspring	*A*	*A*	*AA*
Fifth offspring	*a*	*A*	*Aa*

When the parents die, just 10 alleles are left in the population. But note that the allele frequencies have changed. In this generation, six of the 10 alleles (60 percent) are *A*; four of the 10 alleles (40 percent) are *a*. Evolution—a change in allele frequencies in a population—occurred due to genetic drift.

Instead of each allele being sampled in exactly its original frequency (0.5) when offspring formed, as the Hardy–Weinberg principle assumes, a chance sampling error occurred. Allele *A* "got lucky," and allele *a* was "unlucky."

Computer Simulations Figure 23.13 shows what happens when a computer simulates the same process of random combinations in gametes over time. The program that generated the graphs combines the alleles in a gene pool at random to create an offspring generation, calculates the allele frequencies in the offspring, and uses those frequencies to create a new gene pool.

In this example, the process was continued for 100 generations. The *x*-axis on each graph plots time in generations; the *y*-axis plots the frequency of one of the two alleles present at the *A* gene in a hypothetical population.

The graph in Figure 23.13a shows eight replicates of this process with a population size of 4; the graph in Figure 23.13b shows eight replicates with a population of 400. Notice the striking differences between the effects of drift in the small versus large population and the consequences for genetic variation when alleles drift to fixation (frequency of 1.0) or loss (frequency of 0.0). Although drift is more dramatic in small populations, it still occurs in large populations and can be an important factor in some cases.

✔ **If you understand genetic drift, you should be able to examine the MN blood group data in Table 23.1 on page 473 and describe how drift could explain differences in genotype frequencies among populations.**

Key Points about Genetic Drift The data from the simulations illustrate three important points about genetic drift:

1. ***Genetic drift is random with respect to fitness.*** The changes in allele frequency that it produces are not adaptive.
2. ***Genetic drift is most pronounced in small populations.*** In the computer simulation, allele frequencies changed much less in the large population than in the small population. And if the couple on the deserted island had hypothetically produced 50 children instead of five, allele frequencies in the next generation almost certainly would have been much closer to 0.5.
3. ***Over time, genetic drift can lead to the random loss or fixation of alleles.*** In the computer simulation with a population of 4, it took at most 20 generations for one allele to be fixed or lost. When random loss or fixation occurs, genetic variation in the population declines.

Let's see how these simulations apply to real organisms.

Experimental Studies of Genetic Drift

If you were asked to test the predictions of the simulations in Figure 23.13 in real organisms, what organism would you choose to study? Which heritable trait would you measure? How would you control for other processes such as natural selection?

In the mid-1950s, Warwick Kerr and Sewall Wright were the first to do an experiment to determine how drift works in practice. Here are some of the choices they made:

- ***Study organism?*** Kerr and Wright set up a laboratory population of fruit flies, *Drosophila melanogaster*. These animals are easy and inexpensive to maintain in controlled conditions in the lab and have a short generation time (see **BioSkills 11**).
- ***Heritable trait?*** The researchers focused on the shape of body bristles. Some fruit flies had normal, straight bristles and some had forked bristles. The variations were easy to recognize and were known to depend on a single gene with just two alleles—let's call them *B* for normal and *b* for forked.
- ***Control for natural selection?*** Kerr and Wright knew from previous studies that the two alleles do not affect the fitness of flies. They could be confident that if changes in the frequency of alleles occurred, they would not be due to natural selection.

The researchers set up 96 small populations of eight flies (four females, four males) that had starting allele frequencies 0.5 *B* and 0.5 *b*, just like the simulations in Figure 23.13. Of the many offspring produced, the researchers selected just eight (four males and four females) at random to form the next generation. They did this for 16 generations. Did the experimental results agree with the simulation results?

After 16 generations, both bristle alleles were still present in only 26 of the 96 populations—significantly less than the full 96 expected by Hardy–Weinberg equilibrium. The message of the study is startling: In 73 percent of the experimental populations (70 out of the 96), genetic drift had reduced the diversity of alleles at this gene to zero—very similar to the results shown in Figure 23.13a. As predicted, genetic drift decreased genetic variation within populations and increased genetic differences among populations that were random with respect to fitness.

What Causes Genetic Drift in Natural Populations?

The random sampling process that takes place during fertilization occurs in every population in every generation in every species that reproduces sexually. Similarly, accidents that remove individuals at random occur in every population in every generation.

It is important to realize, though, that because drift is caused by sampling error, it can occur by *any* process or event that involves sampling—not just the sampling of gametes that occurs during fertilization or the loss of unlucky individuals due to accidents. Let's consider two special cases of genetic drift, called the *founder effect* and the *bottleneck effect*.

Founder Effects on the Green Iguanas of Anguilla When a group of individuals immigrates to a new geographic area and establishes a new population, a *founder event* is said to occur. If the new population is small enough, the allele frequencies in the new population are almost guaranteed to be different from those in the source population—due to sampling error. A change in allele frequencies that occurs when a new population is established is called a **founder effect** (see model in **Figure 23.14a**).

(a) Founder effect

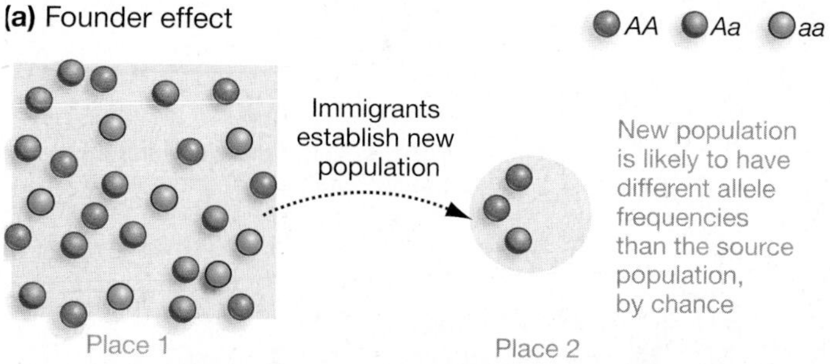

(b) Green iguanas founded a new island population.

Figure 23.14 The Founder Effect Is a Type of Genetic Drift. The smaller the new population, the higher the likelihood that genetic drift will result in differences in allele frequencies as well as in loss of alleles.

✔ **QUANTITATIVE** The original population in **(a)** consists of 9 *AA*, 11 *Aa*, and 7 *aa* individuals (this hypothetical population is very small for simplicity). Compare the frequencies of the *A* allele in the original and new populations.

In 1995, fishermen on the island of Anguilla in the Caribbean witnessed a founder event involving green iguanas—a type of large, tree-dwelling lizard (**Figure 23.14b**). A few weeks after two major hurricanes swept through the region, a large raft composed of downed logs tangled with other debris floated onto a beach on Anguilla. The fishermen noticed green iguanas on the raft and several onshore.

Because green iguanas had not previously been found on Anguilla, the fishermen notified biologists. The researchers determined that at least 15 individuals had arrived; two years later, they confirmed that at least some of the individuals were breeding. A new population had formed. During this founder event, it was extremely unlikely that allele frequencies in the new Anguilla population of green iguanas exactly matched those of the source population (see **Making Models 23.1**).

Colonization events like these have been the major source of populations that occupy islands all over the world, as well as island-like habitats such as mountain meadows, caves, and ponds. Each time a founder event occurs, a founder effect is likely to accompany it, changing allele frequencies through genetic drift.

Making Models 23.1 Tips on Modeling Evolution

Drawing models can help you to understand and remember how evolutionary processes work. Use simple icons (even circles) to represent individuals in a population, and add shading, color, or other marks to show variation among individuals.

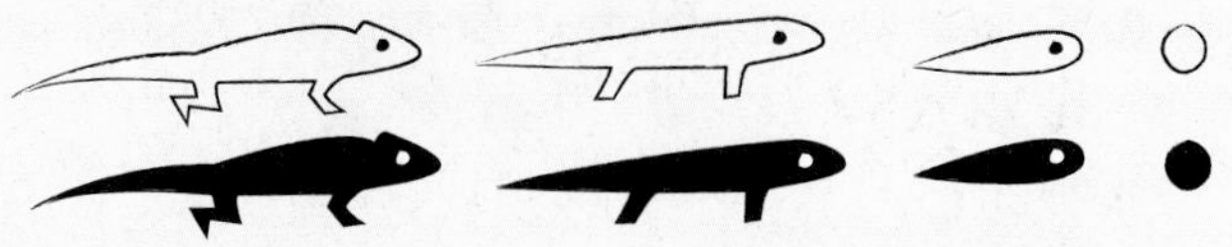

Different styles of lizard symbols (showing variation in color)

MODEL Using simple symbols for iguanas, draw a cartoon of a source population and a new population to illustrate the founder effect.

To see this model in action, go to the Study Area of **Mastering Biology**

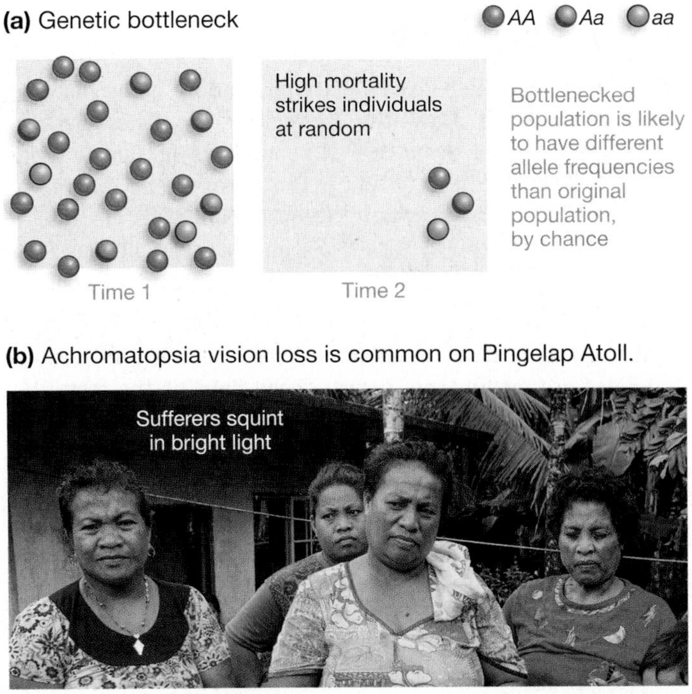

Figure 23.15 The Bottleneck Effect Is a Type of Genetic Drift. The smaller the surviving population, the higher the likelihood that genetic drift will result in differences in allele frequencies as well as in loss of alleles.

✔ **QUANTITATIVE** The original population in **(a)** consists of 20 *AA*, 5 *Aa*, and 2 *aa* individuals (this hypothetical population is very small for simplicity). Compare the frequencies of the *a* allele at Time 1 and Time 2.

Genetic Bottleneck on Pingelap Atoll If a large population experiences a sudden reduction in size, a *population bottleneck* is said to occur. The term comes from the metaphor of pouring colored beads or candies out of a bottle with a narrow neck—when you tilt the bottle and pour, only a few beads of random color will pass through the neck. Population bottlenecks can be caused by disease outbreaks; natural catastrophes such as floods, fires, or storms; and human-caused catastrophes such as deforestation. Afterward, only a few individuals are left to establish the next generation. A **genetic bottleneck** is a sudden reduction in the diversity of alleles in a population (**Figure 23.15a**). Genetic drift occurs during genetic bottlenecks and causes a change in allele frequencies.

For example, consider the humans who occupy Pingelap Atoll in the South Pacific. On this island, only about 20 people out of a population of several thousand managed to survive the effects of a typhoon that occurred around 1775. The survivors apparently included at least one individual who carried a loss-of-function allele at a gene called *CNGB3*, which codes for a protein involved in color vision. The loss-of-function *CNGB3* allele is recessive, and when it is homozygous it causes a serious vision deficit called achromatopsia. People with this condition have poor vision and are nearly or totally color-blind. The condition is extremely rare in most populations, and the frequency of the recessive *CNGB3* allele is estimated to be under 1.0 percent.

In today's population on Pingelap Atoll, however, over 1 in 20 people are afflicted with achromatopsia, and the allele that causes the affliction is at a frequency of well over 20 percent (**Figure 23.15b**). Because the loss-of-function allele is extremely unlikely to be favored by directional selection or heterozygote advantage, researchers hypothesize that the frequency of the allele in this small population increased during the extreme population bottleneck of 1775, then continued to increase over the past 240 years due to genetic drift.

Take-Home Messages

Even if a population formed by a founder event or a bottleneck event happens to have the same allele frequencies as the original population by chance, the smaller population is more likely to experience a loss or fixation of alleles in subsequent generations due to genetic drift, as demonstrated by the simulations in Figure 23.13. Random fluctuations in allele frequency happen in all populations, but most dramatically in small ones.

What's more, even though genetic drift is neutral with respect to fitness (meaning that fitness may increase, decrease, or stay the same by chance), the loss of genetic variation tends to be a problem, especially in small populations. Conservation biologists are especially concerned about genetic drift because many natural populations are being drastically reduced in size by habitat destruction and other human activities (Ch. 54, Section 54.2). The loss of genetic diversity due to genetic drift reduces a population's ability to adapt via natural selection to changes in the environment in the future.

In any species severely affected by genetic drift—whether they're cheetahs or soybeans—the only short-term prospect for restoring genetic diversity is gene flow.

CHECK YOUR UNDERSTANDING

✔ **If you understood this section, you should be able to ...**

Predict whether evolution due to genetic drift is likely to occur in the Brazilian Atlantic rain forest, where juçara palms are becoming isolated in smaller "islands" of forest because of human activities. Explain your reasoning.

Answers are available in Appendix A.

23.5 Gene Flow

When an individual leaves one population, joins another, and breeds, **gene flow**—the movement of alleles between populations—is said to occur. The term "gene flow" can cause confusion because it sounds very ambiguous. Take a few moments to consider what the term really means:

After you complete this section, you should be able to ...

- Compare gene flow to the other processes of evolution.

- While "allele flow" might be a better description—because alleles are moving, not genes—"gene flow" is the term that evolutionary biologists traditionally use.
- While *organisms* can emigrate from a source population or immigrate to a new population, the movement of their *alleles* is called gene flow.
- Despite the similar-sounding terms, genetic drift (a random fluctuation in allele frequency) is very different from gene flow (the movement of alleles between populations). Try this mnemonic: Genetic drift drops and lifts; gene flow goes to and fro.

As an evolutionary process, gene flow usually has one outcome: It equalizes allele frequencies between the source population and the recipient population. When alleles move from one population to another, the populations tend to become more alike. To capture this point, biologists say that gene flow *homogenizes* allele frequencies among populations (Figure 23.16).

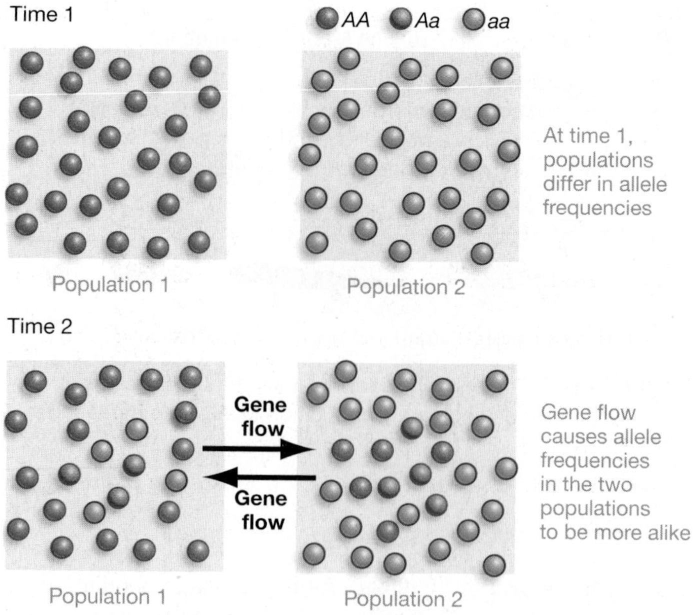

Figure 23.16 Gene Flow Makes Allele Frequencies More Similar between Populations. Gene flow can occur in one or both directions. If gene flow continues and no selection occurs, allele frequencies will eventually be identical.

Measuring Gene Flow between Populations

The presence or absence of gene flow has particularly important implications for the conservation of threatened and endangered species. Numerous studies have documented the decline of gene flow between wild populations that have been isolated from one another—for example, by habitat fragmentation.

Many studies have also documented the effects of gene flow between wild populations and captive populations. For example, the captive breeding of fish is increasing around the world as wild populations are being depleted. In some cases, wild fish are used to start captive populations, which are then kept isolated in "farms" until they go to market—except when they escape. In other cases, captive-bred fish are purposefully released into the wild in an effort to supplement the size of wild populations. What are the effects of accidental or purposeful gene flow between captive-bred populations and wild populations?

A team of biologists recently studied steelhead trout (*Oncorhynchus mykiss*) in the Hood River of Oregon to try to answer this question (Figure 23.17a). Some of the trout in the Hood River are wild, while others were raised in a hatchery and released to supplement the diminishing wild population. The researchers were able to catch, sample, and release all of the steelhead trout in the Hood River each year for three years by trapping them in a dam near the mouth of the river as the fish traveled upstream to their wintering streams.

The researchers extracted DNA from tissue samples and calculated genotypes using eight microsatellite alleles (Ch. 20, Section 20.2) to determine whether the parents of each fish were wild or captive-bred. The team then compared the reproductive fitness—the number of surviving adult offspring—of the trout in three groups:

- Individuals with two wild parents (Wild $\times$ Wild)
- Individuals with one wild parent and one captive-bred parent (Wild $\times$ Captive)
- Individuals with two captive-bred parents (Captive $\times$ Captive)

The graph in Figure 23.17b shows the fitness results for female steelhead trout. Individuals with one or two captive-bred parents are assigned a value from 0 to 1 relative to the fitness of trout with two wild parents (red bar, fitness = 1). Individuals with one captive-bred parent (purple bar) have 16 percent lower fitness than wild trout on average, while individuals with two captive-bred parents (blue bar) have significantly lower fitness—38 percent lower than wild-bred trout, on average. The researchers observed similar results for males.

These data demonstrate that gene flow is occurring from the captive-bred population to the wild population and is causing a reduction in fitness in that population. Related research on farmed Atlantic salmon in Ireland demonstrated that farmed salmon produce smaller eggs, spawn at different times, and have different predator-avoidance behaviors than wild salmon. When farmed salmon were introduced into wild populations, the reproductive success of the wild population decreased.

The conclusion of these studies on steelhead trout and Atlantic salmon is that gene flow between captive-bred and wild populations reduces fitness in the wild populations. As a result, efforts to augment wild populations with captive-bred fish may ultimately result in a *decline* in population size—contrary to the intentions of fisheries biologists.

(a) Wild steelhead trout populations are declining.

(b) Captive-bred trout reduce the fitness of wild populations.

Figure 23.17 Gene Flow Reduces Fitness in a Population of Steelhead Trout. In this study, the reproductive fitness of female steelhead trout with one or two captive-bred parents is shown relative to females with two wild parents. Data for males are comparable (not shown).

DATA: H. Araki, B. Cooper, and M. S. Blouin. 2009. *Biology Letters* 5(5): 621–624.

Gene Flow Is Random with Respect to Fitness

Gene flow doesn't always reduce fitness in the receiving population. If a population has lost alleles due to genetic drift, then the arrival of new alleles via gene flow should increase genetic diversity. If increased genetic diversity results in better resistance to infections by bacteria or viruses or other parasites, for example, gene flow would increase the average fitness of individuals. This was the result when gene flow was used to aid an endangered population of the Florida panther (see Figure 23.5).

The arrival or departure of alleles can increase, decrease, or have no effect on average fitness, depending on the situation. But in every case, movement of alleles between populations tends to reduce their genetic differences.

This latter generalization is particularly important in our own species right now. Large numbers of people from Africa, the Middle East, Mexico, Central America, and Asia are immigrating to the countries of the European Union and the United States. Now, more than ever before, individuals from different cultural and ethnic groups are intermarrying and having children. Therefore, allele frequencies are becoming more similar across human populations.

Gene flow moves alleles among populations. How do these alleles originate in the first place?

CHECK YOUR UNDERSTANDING

✔ If you understood this section, you should be able to ...

SOCIETY Predict the outcome of gene flow in this agricultural scenario: One farmer is growing organic corn (not using pesticides or herbicides), while the neighboring farmer is growing genetically modified corn with an allele for herbicide resistance (so that the farmer can spray the crop with herbicides and kill the weeds, not the corn).

Answers are available in Appendix A.

23.6 Mutation

A central question that biologists ask about an evolutionary process is: How does it affect genetic variation in a population? Recall that:

After you complete this section, you should be able to ...

- Compare mutation to the other processes of evolution.

- Natural selection often favors certain alleles and leads to a decrease in overall genetic variation in a population.
- Genetic drift tends to decrease genetic diversity over time, as alleles are randomly lost or fixed in a population.
- Gene flow increases genetic diversity in a recipient population if new alleles arrive with immigrating individuals. But gene flow may decrease genetic variation in the source population if alleles leave with emigrating individuals.

If these evolutionary processes often lead to a loss of genetic diversity over time, what processes increase genetic diversity? In sexually reproducing organisms, independent assortment and recombination (crossing over) during meiosis are major sources of genetic diversity (Ch. 13, Section 13.2). But these processes only create new combinations of existing alleles. Where do entirely new alleles come from? The answer to this question is **mutation**.

Mutations can occur in a number of ways:

- **Point mutation**: A change in a single base pair in DNA. If a change in nucleotide sequence occurs in a stretch of DNA that codes for a protein, the new allele may result in a polypeptide with a novel amino acid sequence (Ch. 16, Section 16.4). If the mutation occurs in a stretch of DNA that codes for regulatory RNA, the new allele may result in a change in regulation of the expression of other genes.
- **Chromosome-level mutation**: A change in the number or composition of chromosomes. One result of chromosome

mutation is gene duplication, which increases the number of copies of a gene. If duplicated genes diversify via point mutations, they can lose their function or create new alleles (Ch. 20, Section 20.4).

- **Lateral gene transfer** (also known as *horizontal gene transfer*): The transfer of genetic information from one species to another, rather than from parent to offspring. New studies suggest that lateral gene transfer might be a more important source of genetic variation than previously thought (Ch. 20, Section 20.4).

Because DNA synthesis errors and chromosome damage are inevitable, mutation constantly introduces new alleles into populations in every generation. However, mutation is random with respect to the affected allele's impact on the fitness of the individual. Changes in the makeup of chromosomes or in specific DNA sequences do not occur in ways that tend to increase or decrease fitness. Mutation just happens.

On rare occasions, mutation in coding sequences produces a **beneficial** allele—an allele that allows individuals to produce more surviving offspring, increasing fitness. Beneficial alleles tend to increase in frequency in the population due to natural selection. A mutation can also produce a **neutral** allele—an allele with no effect on fitness, as occurs when a point mutation is silent (a change in the DNA sequence does not cause a change in the amino acid sequence; Ch. 16, Section 16.4).

But because most organisms are well adapted to their current habitat, random changes in genes usually result in products that do not work as well as the products of alleles that currently exist. Stated another way, most mutations in sequences that code for a functional protein or RNA result in deleterious alleles, which lower fitness. Deleterious alleles tend to be eliminated by purifying selection (introduced in Section 23.2).

Because mutation produces new alleles, it can in principle change the frequencies of alleles through time. But does mutation *alone* occur often enough to make it an important factor in changing allele frequencies? The short answer is no.

Mutation as an Evolutionary Process

To understand why mutation is not a significant mechanism of evolutionary change by itself, consider that the highest mutation rates that have been recorded at individual genes in humans are on the order of 1 mutation in every 10,000 gametes produced. Now suppose that 195,000 humans live in a population and that in one year, no individuals die and 5000 offspring are born—the population now numbers 200,000. Over the course of a year, the allele frequency change introduced by mutation is 0.0000023. At this rate, it would take 4000 years for mutation to produce a change in allele frequency at a specific gene of just 1 percent.

Although mutation can be a significant evolutionary process in bacteria and archaea, which have extremely short generation times, mutation in eukaryotes rarely causes a change from the genotype frequencies expected under the Hardy–Weinberg principle. As an evolutionary process, mutation is slow compared with selection, genetic drift, and gene flow. However, mutation can have a very large effect on evolution when *combined* with genetic drift, gene flow, and selection. Let's explore two examples.

Experimental Studies of Mutation

Consider an ongoing long-term lab experiment designed by Richard Lenski and colleagues to evaluate evolution in action over many generations. Lenski's group studies *Escherichia coli*, a bacterium that is a common resident of the human intestine and a model organism in biology (see **BioSkills 11**). The scientists set up 12 populations of *E. coli* in 1988, each founded with a single cell from a clone of identical cells, and allowed them to replicate in identical culture conditions for three decades—over 50,000 generations so far! The strain of *E. coli* used in the experiment is completely asexual and reproduces by cell division. Thus, mutation is the only source of genetic variation in these populations.

The biologists saved a sample of cells from each population at regular intervals and stored them in a freezer for later analysis, such as genome sequencing. The resulting evolutionary archive is a powerful tool for studying many different kinds of evolutionary questions. For example, do different populations experience the same mutation rates?

Figure 23.18 shows the number of mutations that accumulated in each of the 12 populations over the 50,000 generations since 1988. (For context, this number of generations would represent a span of about 1,250,000 years for human populations.) Note that six of the *E. coli* populations are overlapping in the bottom-most line—the differences between these lines are too small to see at this scale, but these populations ended up with 60–120 mutations.

All 12 populations accumulated mutations over time, but at very different rates, varying from 0.0012 to 0.05 mutations per generation—some changed about 100 times faster than others. When the researchers examined the types of mutations that occurred, they observed that:

- *Silent mutations* (neutral with respect to fitness) accumulated at a constant rate.
- *Beneficial mutations* (increasing fitness) occurred at a higher rate at first, then at a slower rate later, probably because the bacteria were living in an unchanging lab environment.

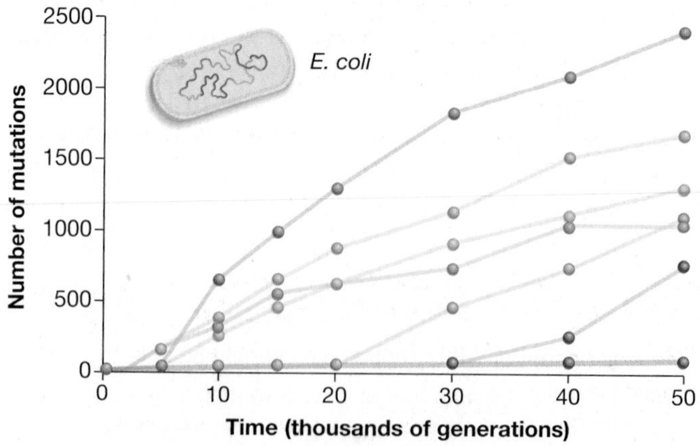

Figure 23.18 Evidence of Mutation in 12 Lab Populations of *E. coli* during Experimental Evolution. Each line represents the average number of mutations in one of 12 populations of *E. coli* (six lines overlap at the bottom).

DATA: O. Tenaillon et al. 2016. *Nature* 536: 165–170.

Figure 23.19 Evidence of Mutation in Pea Aphids. Red and green pea aphids occur together in populations of the same species. Both morphs of aphid feed on plant sap, but they do not derive their color from their food.

- *Deleterious mutations* (decreasing fitness) did not accumulate in the populations because, as you would expect, individuals that happened to inherit a deleterious mutation died or reproduced at a slower rate than other bacteria.

Lenski's experiment demonstrates the random patterns of cumulative mutations in laboratory populations. Now consider a recent study showing how several forms of mutation have combined with natural selection to create color variations, or polymorphisms, in wild populations of pea aphids.

Studies of Mutation in Natural Populations

Pea aphids (*Acyrthosiphon pisum*) are small insects that feed on plant sap. They occur in two colors in the wild, red and green, and both colors coexist within populations (Figure 23.19a). The two phenotypes are maintained in the population due to balancing selection—ladybird beetles are more likely to prey on red aphids, whereas parasitoid wasps are more likely to lay their eggs in green aphids.

The color of aphids is determined by carotenoid pigments, the same group of pigments that give zebra finches their bright orange beaks. Finches and other animals get their carotenoid pigments from the food they eat, mostly from plants.

But there's a twist. Pea aphids do *not* acquire their carotenoids from the plant sap they eat. So where do the pigments come from?

Lateral Gene Transfer Researchers recently examined several bacteria associated with pea aphids and concluded that the carotenoids did not originate from these bacteria, either. Instead, other researchers found that the aphids produce enzymes that allow them to synthesize their *own* carotenoids—the first animals ever discovered to have this ability.

How did the pea aphids obtain this biosynthetic pathway? The answer was surprising—researchers discovered that DNA had moved from the genome of a fungus to the genome of a recent ancestor of the pea aphids. They made this inference based on the close similarity of DNA sequences in the aphid and fungus genes. This finding was clear evidence of lateral gene transfer from one eukaryote to another.

Gene Duplication, Diversification, and Deletion Lateral gene transfer is only part of the mutation story in pea aphids. Carotenoids are a family of molecules with similar pigments and a number of intermediate forms. Two types of enzymes are primarily responsible for the biosynthesis of these molecules in both pea aphids and their fungal ancestor.

A comparison of aphid and fungal genome sequences suggests that after the fungal genes for these two types of enzymes were transferred to aphids, they underwent further mutations. First, duplications and sequence diversification by point mutations produced red pea aphids (Figure 23.19b), which have the enzymes necessary to synthesize yellow, green, and red carotenoid pigments. Then, a deletion produced green pea aphids, which have the enzymes necessary to synthesize only the yellow and green pigments. The deletion of a gene in the green aphid genome illustrates an important point: *Sometimes a loss-of-function mutation can be adaptive.* In this case, being green helps aphids to avoid predation by ladybird beetles.

Take-Home Messages

The research on mutations in *E. coli* populations in the lab and the origin of color polymorphism in pea aphids reinforces four fundamental messages about the role of mutation in evolution.

1. ***Mutation is the ultimate source of genetic variation.*** Crossing over and independent assortment of chromosomes shuffle existing alleles into new combinations, but only mutation creates new alleles.
2. ***Mutations are random with respect to fitness.*** When a mistake is made in copying genetic information, the resulting mutation can be beneficial, neutral, or deleterious with respect to fitness. Mutations just happen—organisms cannot create mutations because they "want" or "need" them (Ch. 22, Section 22.5).

SUMMARY Table 23.3 Evolutionary Processes

	Definition/Description	Effect on Genetic Variation	Effect on Average Fitness
Natural selection	Certain alleles are favored and increase in frequency, while others are disfavored and decrease in frequency 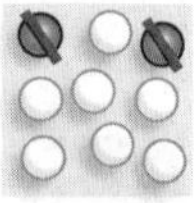	Can maintain, increase, or reduce genetic variation	Can produce adaptation, increasing fitness
Genetic drift	Random changes in allele frequencies; most important in small populations 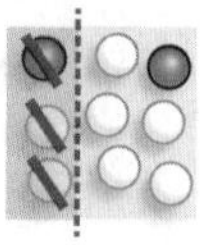	Tends to reduce genetic variation via loss or fixation of alleles	Random with respect to fitness; usually reduces average fitness
Gene flow	Movement of alleles between populations; reduces differences between populations 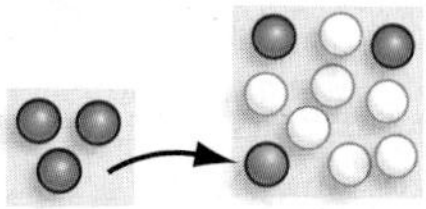	May increase genetic variation by introducing new alleles; may decrease it by removing alleles	Random with respect to fitness; may increase, decrease, or have no effect on average fitness, depending on the alleles
Mutation	Random production of new alleles 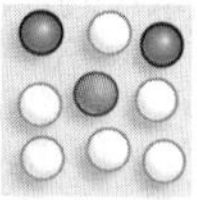	Increases genetic variation by producing new alleles	Random with respect to fitness; most mutations in coding sequences lower fitness

3. ***If mutation did not occur, evolution would eventually stop.*** Recall that natural selection and genetic drift tend to eliminate alleles. Without mutation, eventually there would be no variation for selection to act on, and alleles that were not lost would become fixed.

4. ***Mutation alone is usually inconsequential in changing allele frequencies at a particular gene.*** However, when considered across the genome and when combined with natural selection, genetic drift, and gene flow, mutation becomes an important evolutionary process.

Table 23.3 summarizes the take-home messages for mutation and the other three evolutionary processes. Each has different consequences for allele frequencies, all of them violating Hardy–Weinberg assumptions. Gene flow reduces differences between populations, while natural selection, genetic drift, and mutation cause populations to diverge, sometimes leading to the formation of new species—that is, new branches on the tree of life (detailed in Chapter 24). The ultimate result is biological diversity.

You can see how nonrandom mating and the four evolutionary processes fit into the Big Picture of Evolution on pages 532–533.

CHECK YOUR UNDERSTANDING

✔ **If you understood this section, you should be able to ...**

1. **THINK CAREFULLY** Evaluate this statement: Red aphids mutated their genes so that they could be green and avoid predation by ladybird beetles.
2. **THINK CAREFULLY** Evaluate this statement: Mutations are bad.

Answers are available in Appendix A.

CHAPTER 23 Review

For media, go to **Mastering Biology**

23.1 Null Hypothesis: The Hardy–Weinberg Principle

- The Hardy–Weinberg principle played an important role in the synthesis of Mendelian genetics and Darwinian evolution.
- The Hardy–Weinberg principle can serve as a null hypothesis in evolutionary studies because it predicts what genotype and allele frequencies are expected if mating is random with respect to the gene in question and if none of the four evolutionary processes is operating on that gene.

23.2 Nonrandom Mating

- Nonrandom mating changes only genotype frequencies, not allele frequencies, so is not an evolutionary process itself.
- Inbreeding—mating among relatives (or, in some species, self-fertilization), is a form of nonrandom mating. It leads to an increase in homozygosity and a decrease in heterozygosity.
- Inbreeding can expose alleles to natural selection and can cause inbreeding depression.

23.3 Natural Selection

- Natural selection is the only evolutionary process that produces adaptation.
- Directional selection favors phenotypes at one end of a distribution, causing the average phenotype to change. It tends to decrease the diversity of alleles in populations.
- Stabilizing selection eliminates phenotypes with extreme characteristics. It decreases the diversity of alleles in populations.
- Disruptive selection favors extreme phenotypes and thus increases genetic variation in populations. Disruptive selection sometimes leads to the formation of new species.
- Balancing selection occurs when no single phenotype is favored; there is a balance among alleles in terms of fitness and frequency. Balancing selection preserves genetic variation.
- Sexual selection is a type of natural selection that leads to the evolution of traits that increase reproductive success.
- *Inter*sexual selection occurs when an individual of one sex, often a female, choses an individual of another sex, often a male, to mate with.
- *Intra*sexual selection occurs when suitors, often males, compete with each other to gain access to a mate, often a female.
- Sexual selection in humans is especially complex because courtship and mating are social behaviors that serve functions other than producing offspring.

23.4 Genetic Drift

- Genetic drift causes random changes in allele frequencies.
- Genetic drift is particularly important in small populations, and it tends to reduce overall genetic diversity.
- Genetic drift can result from random fusion of gametes at fertilization, founder events, and population bottlenecks.

23.5 Gene Flow

- Gene flow tends to equalize allele frequencies among populations.
- Gene flow can introduce alleles from one population to another when individuals move among populations.
- The introduced alleles may have a beneficial, neutral, or deleterious effect.

23.6 Mutation

- Mutation is the only evolutionary process that creates new alleles, which may be beneficial, neutral, or deleterious. Sometimes a loss-of-function allele can be adaptive.
- Mutation occurs by chance, not because an organism wants or needs a new allele.
- Mutation occurs too infrequently to be a major cause of change in allele frequency alone, but it is important when combined with natural selection, genetic drift, and gene flow.

Answers are available in Appendix A.

✔ TEST YOUR KNOWLEDGE

1. **PROCESS OF SCIENCE** In what sense is the Hardy–Weinberg principle a null hypothesis?
2. Why isn't inbreeding considered an evolutionary process?
   a. It does not change genotype frequencies.
   b. It does not change allele frequencies.
   c. It does not occur often enough to be important in evolution.
   d. It does not violate the assumptions of the Hardy–Weinberg principle.
3. **THINK CAREFULLY** Why is genetic drift aptly named? Select True or False for each statement.
   T/F It causes allele frequencies to drift up or down randomly.
   T/F It occurs when alleles from one population drift into another.
   T/F It occurs when mutations drift into a genome.
4. True or false? Gene flow can either increase or decrease the average fitness of a population. Explain.

✔ TEST YOUR UNDERSTANDING

5. **QUANTITATIVE** In a population of 2500, how many babies would you expect to have cystic fibrosis, a homozygous recessive condition, if the frequency of the dominant allele is 0.9 and the population is at Hardy–Weinberg equilibrium?
   a. $0.9 \times 2500 = 2250$
   b. $2 \times 0.9 \times 0.1 \times 2500 = 450$
   c. $0.9 \times 0.1 \times 2500 = 225$
   d. $0.1 \times 0.1 \times 2500 = 25$
6. In the 1700s and 1800s, royalty in Europe often married their close relatives; furthermore, recessive genetic diseases such as hemophilia showed up much more often among royals than in the general population. Explain the likely connection.
7. **THINK CAREFULLY** Determine what is incorrect in the following statement: Red deer mutated their genes so that they could have larger antlers to compete for mates.
8. **MODEL** Draw a small concept map (**BioSkills 12**) showing how selection, genetic drift, gene flow, and mutation relate to genetic variation.

✔ TEST YOUR PROBLEM-SOLVING SKILLS

9. **QUANTITATIVE** In humans, albinism is caused by loss-of-function mutations in genes involved in the synthesis of melanin, the dark pigment in skin. Only people homozygous for a loss-of-function allele (genotype *aa*) have the albino phenotype. In Americans of northern European ancestry, albino individuals are present at a frequency of about 1 in 10,000 (or 0.0001). Assuming that genotypes are in Hardy–Weinberg equilibrium, what is the predicted frequency of Caucasians in the United States who carry a single allele for albinism?
10. Suppose you were studying several species of monkeys. In one, males never helped females raise offspring. In another, males provided just as much parental care as females except for actually carrying the fetus during pregnancy. How does the fundamental asymmetry of sex compare in the two species? How would you expect sexual dimorphism to compare between the two species?

✔ PUT IT ALL TOGETHER: Case Study

Are toucans important to tropical forests?

Human activities are causing the fragmentation of the Brazilian Atlantic rain forest. One result is that toucans have become extinct or nearly extinct in some of the forest fragments. Does the absence of toucans affect the forest?

11. Toucans disperse seeds of key forest species such as juçara palms by eating the fruit and defecating the seeds in new locations, sometimes more than a kilometer away. If there are no toucans, is the genetic diversity of palms likely to increase or decrease within forest fragments? Why?
a. increase (due to increased genetic drift)
b. decrease (due to decreased gene flow)
c. decrease (due to decreased mutation rate)
d. decrease (due to decreased natural selection)

12. QUANTITATIVE Toucans can eat fruits with large seeds because their large bills can open wide. Most other birds in the same forest can only eat small seeds. Ecologist Mauro Galetti and his colleagues measured the seed sizes of palms in forest fragments with and without toucans. The graphs show two of the forest populations they studied. What is the take-home message of the data?

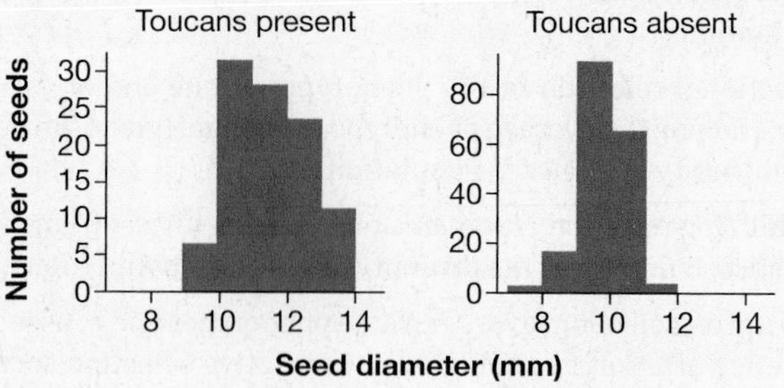

Source: M. Galetti, R. Guevara, and M. C. Côrtes, et al. 2013. *Science* 340: 1086–1090.

13. Do these data illustrate directional, stabilizing, disruptive, or balancing selection? Justify your answer in terms of fitness.

14. Large seeds carry more resources than small seeds and tend to have a higher rate of survival, especially after being dispersed by a bird. Predict how the local extinction of toucans will affect the palm population over time.

15. PROCESS OF SCIENCE The data in the graphs are from two of the 22 forest fragments studied by the researchers: 7 with toucans present, 15 with toucans absent. Why do you think the researchers bothered to study so many forest fragments?

16. SOCIETY If you were a journalist covering this story, how could you use data from this study to respond to the following social media post? "Evolution is a slow process. Humans do not cause evolution in other organisms."

Mastering Biology

Students Go to Mastering™ Biology for assignments, the eText, and the Study Area with animations, practice tests, and activities.

Professors Go to Mastering™ Biology for automatically graded tutorials and questions that you can assign to your students, plus Instructor Resources.

24 Speciation

New species of killer whales are forming around the world. This chapter explains how.

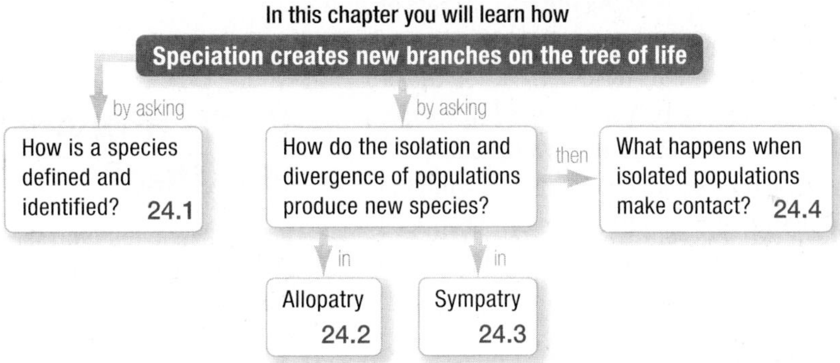

Think for a second about killer whales. How did they originate? How many species are there now? How can you tell?

Charles Darwin described the origin of species as the "mystery of mysteries." The title of his masterwork is *On the Origin of Species by Means of Natural Selection*, but Darwin wrote little on how new species arise. What he did do, though, was advance the study of species tremendously by proposing that species change through time via natural selection and are related by common ancestry (Chapter 22).

Then during the modern synthesis of Darwinian evolution and Mendelian genetics in the early 1900s, three other processes of evolution emerged in addition to natural selection: genetic drift, mutation, and gene flow (Chapter 23). How are the four processes of evolution involved in the origin of new species? In most cases, **speciation** is a splitting event that results from a two-step process:

1. ***Genetic isolation*** occurs when some sort of barrier to gene flow isolates two populations within a species. Recall that gene flow makes allele frequencies more similar among populations—so the absence of gene flow means that alleles are no longer exchanged.
2. ***Genetic divergence*** occurs when mutation, natural selection, and genetic drift occur in each of the isolated populations—meaning that the populations begin to evolve independently of each other. If isolated populations diverge sufficiently, distinct species form.

BIG PICTURE

This chapter is part of the Big Picture. See how on pages 532–533.

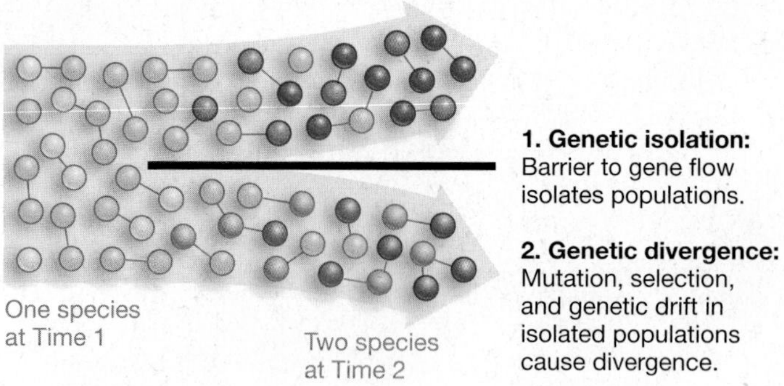

Figure 24.1 General Model of Speciation. The dots represent individuals, the lines connecting them represent mating, and the colors represent genetic variation.

This general model of speciation is summarized in **Figure 24.1**, showing genetic isolation followed by genetic divergence.

The process of speciation can be rapid or gradual. For example, speciation occurred in a single generation in maidenhair ferns when a chromosomal mutation took place. By contrast, a single population of chicken-sized birds called trumpeters (*Psophia* spp.) diverged into eight species over the course of 3 million years (**Figure 24.2**). Whether speciation is rapid or gradual, it adds branches to the tree of life, increasing biodiversity (Ch. 54, Section 54.1). **Making Models 24.1** provides tips on how to draw these branches.

Scientists have named about 1.5 million species on Earth but estimate that 5 to 100 million species exist. As the current rate of extinction rises due to human-related threats, the study of speciation has urgent practical implications for conservation biologists attempting to preserve biodiversity. You can see how speciation fits into the Big Picture of Evolution on pages 532–533.

Let's begin by considering the *pattern* component of studying speciation—how can we define and identify species? Then let's explore the *process* of speciation—how new species form.

Making Models 24.1 Tips on Drawing Phylogenetic Trees (III)

Phylogenetic trees show relationships among species (see **BioSkills 13** and Ch. 25, Section 25.1). However, the shape of a fork in a tree—where one branch becomes two—usually does *not* contain information about the rate of speciation. For example, speciation in the tree on the left is not more abrupt than speciation in the tree on the right; both trees have the same meaning.

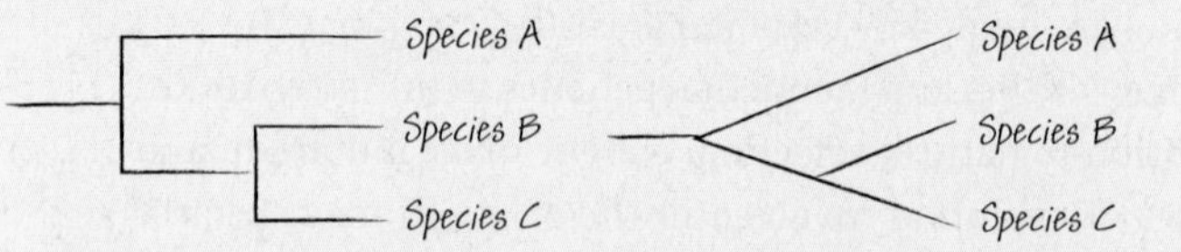

Neither tree contains information about the rate of speciation

MODEL Draw a third tree with the same meaning as the two above; for example, using U-shaped curved lines instead of straight ones.

To see this model in action, go to the Study Area of **Mastering Biology**

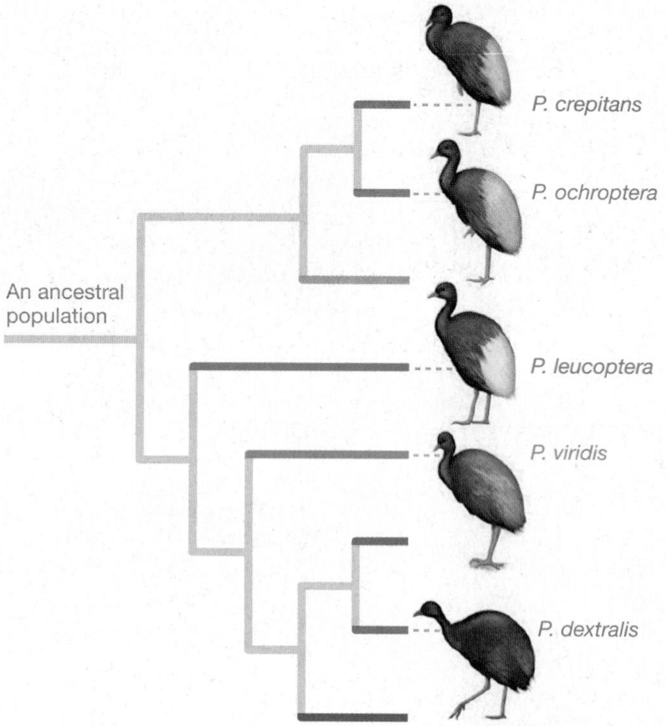

Figure 24.2 Speciation Forms Evolutionarily Independent Populations. Many species of trumpeters have diverged from one ancestral population in the Amazon basin in the last 3 million years, adding branches to the tree of life.

DATA: C. Ribas et al. 2012. *Proceedings of the Royal Society B* 279: 681–689.

24.1 How Are Species Defined and Identified?

If your friend tells you she's planning to study polar bears and grizzly bears for her summer research project, you'd likely know that these animals are distinct species. But what if your friend is going to compare forest elephants and savanna elephants of Africa? Are they the same species or two different species?

After you complete this section, you should be able to . . .

- Compare mechanisms of reproductive isolation.
- Compare the advantages and disadvantages of different species concepts.

Evolutionary biologists have been wrangling with the definition of species for decades—how can you reliably distinguish two or more species of bears, elephants, or bacteria in the field or fossil record? Although there is no single, universal answer, scientists do agree there is a distinction between the *general definition* of a species and the criteria used in the *practical identification* of species in particular cases.

In general, a **species** is defined as an evolutionarily independent population or group of populations. Of the many criteria that can be used to identify species, the three most common ones are **(1)** the biological species concept, **(2)** the morphospecies concept, and **(3)** the phylogenetic species concept.

Table 24.1 Mechanisms of Reproductive Isolation

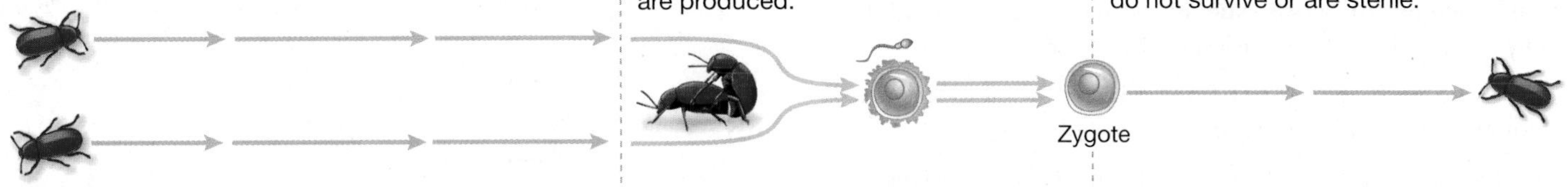

Prezygotic Isolation					Postzygotic Isolation	
Mating does not occur.			Mating occurs but no zygotes are produced.		Zygotes are produced but offspring do not survive or are sterile.	
Temporal	**Habitat**	**Behavioral**	**Mechanical**	**Gametic barrier**	**Hybrid inviability**	**Hybrid sterility**
Populations are isolated because they breed at different times.	Populations are isolated because they breed in different habitats.	Populations do not interbreed because they have different courtship displays.	Mating fails because male and female reproductive structures are incompatible.	Mating fails because eggs and sperm are incompatible.	Hybrid offspring do not develop normally and die at some point during early development.	Hybrid offspring mature but are sterile as adults.
Examples						
Bishop pines and Monterey pines release their pollen at different times of year.	Mainland mice and beach mice mate in different habitats.	To attract females, male songbirds sing species-specific songs.	The genital structures of some male and female insects fit together with "lock and key" specificity.	Differences in the bindin protein determine whether sea urchin sperm will penetrate eggs.	When ring-necked doves mate with rock doves, less than 6 percent of eggs hatch.	Horses and donkeys can mate but their hybrid offspring (mules) are sterile.

The Biological Species Concept

According to the **biological species concept**, the main criterion for identifying species is reproductive isolation. This is a logical yardstick because no gene flow occurs between populations that are reproductively isolated from each other. If two different populations do not interbreed in nature, or if they fail to produce viable and fertile offspring when matings take place, then they are considered distinct species. The influential evolutionary biologist Ernst Mayr strongly promoted the biological species concept because the criterion of reproductive isolation offers clear evidence of evolutionary independence.

Reproductive isolation can result from a wide variety of events and processes. To organize the various mechanisms that stop gene flow between populations, biologists distinguish between

- **prezygotic** ("before-zygote") **isolation**, which prevents individuals of different species from mating successfully; and
- **postzygotic** ("after-zygote") **isolation**, in which the hybrid offspring of matings between members of different species either do not survive or do not reproduce.

(See **BioSkills 15** for a review of common Latin and Greek roots.)

Table 24.1 outlines some of the more important mechanisms of prezygotic and postzygotic isolation and gives examples of each.

✔ If you understand Table 24.1, you should be able to identify what type(s) of reproductive isolation occur in these two examples: (1) In the Sierra Nevada mountains of California, the monkeyflower *Mimulus lewisii* lives at high elevation and attracts bumblebee pollinators, whereas the closely related monkeyflower *M. cardinalis* lives at lower elevation and attracts hummingbird pollinators. (2) When fruit flies *Drosophila melanogaster* and *D. simulans* mate, female larvae survive but male larvae die.

Although the biological species concept has a strong theoretical foundation, it also has disadvantages:

- The criterion of reproductive isolation cannot be evaluated in fossils or in species that reproduce asexually.
- The concept is difficult to apply when closely related populations don't overlap with each other geographically. In this case, biologists are left to guess whether interbreeding would occur if the populations happened to come into contact.
- Reproductive isolation can sometimes be a complex gradient rather than an all-or-nothing scenario. For example, if a grizzly bear and polar bear mate and produce a rare "pizzly bear" or "grolar bear" that survives and reproduces, are grizzly bears and polar bears necessarily the same species?

The Morphospecies Concept

How do biologists identify species when the criterion of reproductive isolation cannot be applied? Under the **morphospecies** ("form-species") **concept** (also known as the *morphological species concept*), researchers identify evolutionarily independent lineages by differences in size, shape, or other morphological features. The logic behind the morphospecies concept is that distinguishing features are most likely to arise if populations are independent and isolated from gene flow.

The morphospecies concept is compelling simply because it is so widely applicable. It is a useful criterion when biologists have no data on the extent of gene flow, and it applies equally to sexual, asexual, or fossil species. However, it also has disadvantages:

- It may lead to the naming of two or more species when there is only one **polymorphic species** with differing phenotypes, such as the spotted and black morphs of jaguars.
- It cannot identify **cryptic species**, which differ in traits other than morphology, such as the songs in some songbirds.
- The morphological features used to distinguish species are subjective—different researchers working on the same populations may disagree on the characters that distinguish species. For example, paleontologists sometimes disagree on the identity of an extinct human species based on measurements of fossil bone fragments.

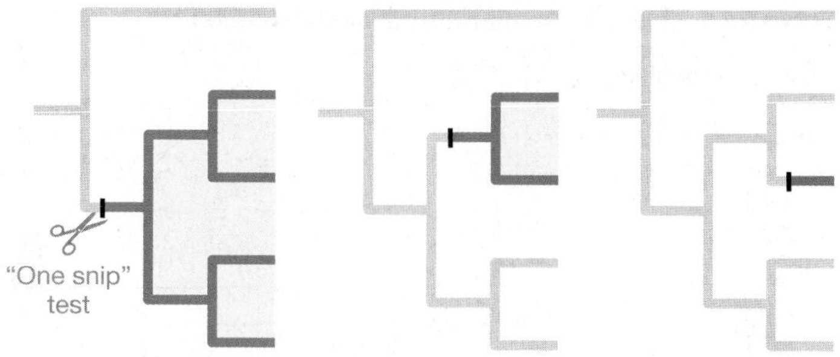

Figure 24.3 The Phylogenetic Species Concept Is Based on Monophyletic Groups. All three phylogenetic trees show examples of monophyletic groups in blue, as you could verify using the "one snip" test. The phylogenetic species concept considers the smallest monophyletic groups on the tree of life (branch tips such as the one highlighted in the tree on the right) as species.

The Phylogenetic Species Concept

The **phylogenetic species concept** identifies species based on the evolutionary history of populations. The reasoning behind this concept begins with Darwin's theory that all species are related by common ancestry. In modern terms, all species form a monophyletic ("one-tribe") group—the tree of life.

A **monophyletic group**, also called a **clade** or **lineage**, consists of an ancestral population, all of its descendants, and *only* those descendants. Phylogenetic trees are useful tools for visualizing monophyletic groups (**BioSkills 13** reviews how to read phylogenetic trees). On any given evolutionary tree, there are many monophyletic groups nested within one another (**Figure 24.3**). One strategy you can use to identify monophyletic groups is called the "one snip" test: If you imagine cutting one branch with scissors, everything that would fall off is a monophyletic group. Under the phylogenetic species concept, species are defined as the smallest monophyletic groups on the tree of life.

Monophyletic groups, in turn, are identified by traits called synapomorphies ("unique-forms"). A **synapomorphy** is a trait that is found in certain groups of organisms and their common ancestor but is missing in more distant ancestors. White fur is a synapomorphy that identifies polar bears as a monophyletic group within bears. In many cases, researchers use DNA sequence data to identify synapomorphies, which in turn help to identify phylogenetic species.

The phylogenetic species concept is versatile because it can be applied to any population (fossil, asexual, or sexual). It is also logical because different species have different synapomorphies only if they are isolated from gene flow and have evolved independently. The approach has disadvantages, however:

- Carefully estimated phylogenies are available only for a tiny (though growing) subset of populations on the tree of life.
- Because it makes slight differences among groups easier to detect, the phylogenetic species concept tends to recognize many more species than the morphospecies or biological species concepts. While this might better reflect the extent of life's diversity, it complicates the naming of species.

The three major species concepts are summarized in **Table 24.2**. In practice, researchers use a combination of species concepts to identify evolutionarily independent populations in nature. Let's use elephants as a case study to explore the application of species concepts to everyday life.

SUMMARY Table 24.2 The Three Most Common Species Concepts

		Criterion for Identifying Populations as Species	Advantages	Disadvantages
Biological species		Reproductive isolation between populations (they don't breed and don't produce viable, fertile offspring)	Reproductive isolation = evolutionary independence	Not applicable to asexual or fossil species; difficult to assess if populations do not overlap geographically.
Morphospecies	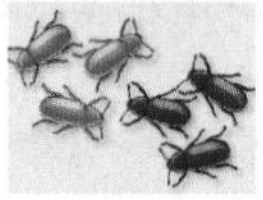	Morphologically distinct populations	Widely applicable	Misidentifies polymorphic species; misses cryptic species; subjective (researchers often disagree about how much or what kinds of morphological distinction indicate speciation).
Phylogenetic species	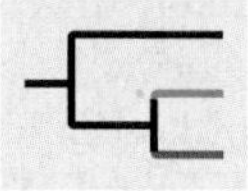	Smallest monophyletic group on phylogenetic tree	Widely applicable; based on testable criteria	Relatively few well-estimated phylogenies are currently available.

Species Definitions in Action: How Many Species of Elephants Are There?

Systematics is the discipline of biology that characterizes and classifies the relationships among all organisms on Earth. Within systematics, **taxonomy** is the practice of describing, naming, and classifying individual species. It is the job of a taxonomist, then, to answer the question: How many species of elephants are there?

What do *you* think? The most common answer is *two*: the larger, floppy-eared elephants in Africa and the smaller-eared elephants of southeast Asia. Indeed, elephants have traditionally been classified into two species: African elephants (*Loxodonta africana*) and Asian elephants (*Elephas maximus*).

However, detailed morphological comparisons of the African populations have recently raised the question of whether the smaller forest elephants in central Africa and the larger savanna elephants in eastern and southern Africa are distinct enough to be considered different species. How would the different species concepts address this question?

Applying the Biological Species Concept Do the forest and savanna elephants interbreed and, if so, are the offspring viable and fertile? The most important form of reproductive isolation in this case is habitat isolation, because the forest elephants and savanna elephants live in different habitats and rarely interact. There is evidence that hybrid offspring have been born and survived (unlike the hybrid offspring of Asian and African elephants, when tested in captivity). However, this evidence does not convincingly reject the three-species hypothesis.

Applying the Morphospecies Concept Careful anatomical comparisons of forest and savanna elephants support the three-species hypothesis. Characteristics such as body size, skull dimensions, tusk size and shape, and number of toenails are consistently different between forest and savanna elephants. Just how different do these variations have to be to reject the two-species hypothesis?

Applying the Phylogenetic Species Concept Geneticist Alfred Roca and his colleagues employed the phylogenetic species concept to answer the elephant species question. They used dart guns to retrieve small tissue samples from 195 elephants from 4 populations in Asia, 3 populations in the forest of central Africa, and 18 populations in the savanna of central, eastern, and southern Africa (**Figure 24.4**). After sequencing four nuclear genes in all the samples, they discovered 1732 nucleotide differences they could use as unique traits (synapomorphies) to build a phylogenetic tree. How many monophyletic groups emerged?

The phylogenetic tree in Figure 24.4 shows the results. You can see that the populations cluster into three distinct monophyletic groups, not two. Statistical analysis supports these findings:

- The DNA sequences in elephant populations within monophyletic groups (such as Dzanga-Sangha, Lopé, and Odzala in central Africa) are not distinguishable from each other.
- The DNA sequences of pairs of elephants in different groups are consistently distinguishable, even when comparing forest and savanna elephants that live in close proximity.

RESEARCH

QUESTION: How many species of elephants are there?

THREE-SPECIES HYPOTHESIS: There are three species of elephants.

TWO-SPECIES HYPOTHESIS: There are two species of elephants.

STUDY DESIGN:

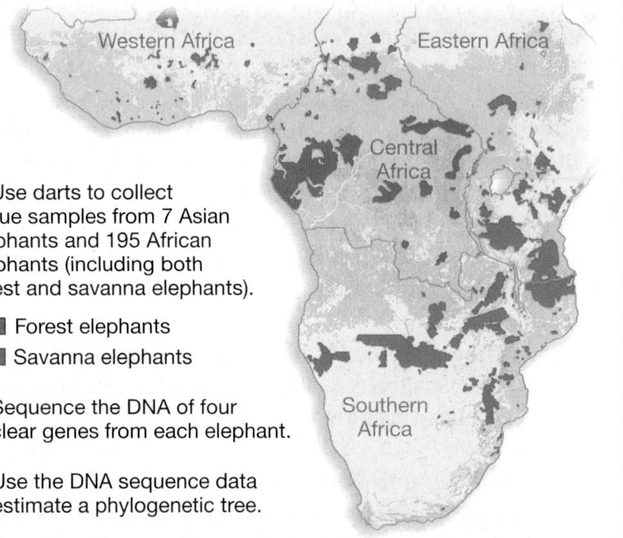

1. Use darts to collect tissue samples from 7 Asian elephants and 195 African elephants (including both forest and savanna elephants).

- Forest elephants
- Savanna elephants

2. Sequence the DNA of four nuclear genes from each elephant.

3. Use the DNA sequence data to estimate a phylogenetic tree.

PREDICTION OF THREE-SPECIES HYPOTHESIS: Three monophyletic groups of elephants will be apparent in the phylogenetic tree.

PREDICTION OF TWO-SPECIES HYPOTHESIS: Two monophyletic groups of elephants will be apparent in the phylogenetic tree.

RESULTS:

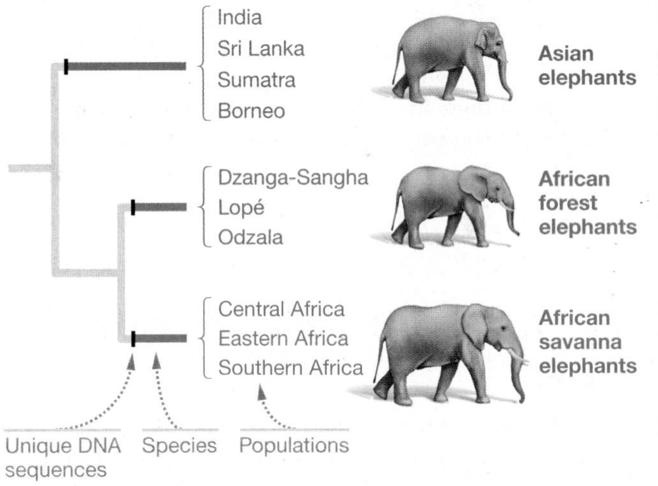

Three monophyletic groups are apparent in the phylogenetic tree.

CONCLUSION: There are three species of elephants.

Figure 24.4 **The Phylogenetic Species Concept Tests the Two-Species Hypothesis for Elephants.** Morphological and habitat differences distinguish forest and savanna elephant populations in Africa. The phylogenetic species concept supports giving forest and savanna elephants species status.

DATA: Map, IUCN Elephant Specialist Group, 2017; A. L. Roca et al. 2001. *Science* 293: 1473–1477.

✔ **MODEL** Sketch what the tree would look like if it supported the two-species hypothesis.

These results are strong evidence that African forest elephants are sufficiently distinct from African savanna elephants and Asian elephants to justify the recognition of three elephant species: *Elephas maximus* in Asia and *Loxodonta africana* (savanna elephants) and *Loxodonta cyclotis* (forest elephants) in Africa.

So is the change official? Not yet. It takes time for international networks of researchers to agree and formalize a major change such as naming a new species of elephant—and new data continue to emerge.

Why Does It Matter? It's important to know how many species of elephants there are, because the elephant population is plummeting and becoming more fragmented—largely due to the illegal trade of ivory and changes in land use such as habitat destruction and fencing. Species designations help conservation biologists and communities to make important decisions about where to collect data, spend resources, prioritize habitat preservation, prevent local extinction, and decide where to move individuals when relocation is necessary. Similar reevaluations of species status are under way for many other types of organisms.

This case study introduces you to the *pattern* component of studying speciation—defining and identifying species. Now let's explore the *process* of speciation—how do new species form?

CHECK YOUR UNDERSTANDING

✔ **If you understood this section, you should be able to . . .**

1. Predict which mechanism of reproductive isolation played a role in trumpeter speciation in the Amazon basin. Note: Trumpeters cannot fly across large rivers.
2. Determine which species concept(s) could be used to identify the number of trumpeter species in the Amazon.

Answers are available in Appendix A.

24.2 Isolation and Divergence in Allopatry

After you complete this section, you should be able to . . .

- Use examples to explain the process of allopatric speciation.

Recall that speciation begins when gene flow between populations is reduced or eliminated, causing genetic isolation (see Figure 24.1). Genetic isolation happens routinely when populations become geographically separated. Populations that are geographically separated are said to be in **allopatry** ("different-homeland"). Thus, speciation that begins with geographic isolation is known as **allopatric speciation**.

Allopatric speciation starts in one of two ways: **(1) dispersal**, the movement of individuals from one place to another, or **(2) vicariance**, the physical splitting of a habitat. As the general model in **Figure 24.5a** illustrates, a population can disperse to a new habitat, colonize it, and found a new population. Alternatively, a new geographic barrier can split a population into two or more subgroups that are physically isolated from each other (**Figure 24.5b**). These processes take us into the field of **biogeography**—the study of how species and populations are distributed geographically (Ch. 49, Section 49.2).

Models are useful to biologists because they summarize concepts without the complicated details of specific examples, and they point to testable hypotheses. Let's apply the dispersal and vicariance models in Figure 24.5 to a couple of specific case studies to answer two questions. First, how do dispersal and vicariance events occur? And second, once populations are physically isolated, how do mutation, genetic drift, and selection produce divergence?

Allopatric Speciation by Dispersal

Peter Grant and Rosemary Grant had the good fortune to witness a colonization event while working in the Galápagos Islands off the coast of South America. The Grants have been studying finches on the island of Daphne Major since 1971 (Ch. 22, Section 22.4).

In 1973 the Grants began to observe a yearly migration of large ground finches from their home island to Daphne Major for a few months between breeding seasons. Finally, in 1982, thirteen colonists stayed on Daphne Major throughout the breeding season, and five of them produced offspring. These colonists represented a new population, allopatric with their migrant source population, because the two populations bred and nested on different islands.

Could this dispersal event lead to speciation? To evaluate this question, the Grants caught, weighed, and measured most of the parents and offspring produced on Daphne Major over the succeeding 12 years. When they compared these data with measurements of large ground finches in the migrant population, they discovered that the average beak size in the colonist population was much larger. Two evolutionary processes could be responsible for the change in beak size:

1. The founder event produced a colonizing population that had particularly large beaks relative to the migrant population. This is an example of genetic drift (Ch. 23, Section 23.4).
2. Natural selection in the new environment favored alleles associated with large beaks.

The Grants concluded from their detailed observations that both genetic drift and natural selection were at play.

The new population of large ground finches is not yet considered a separate species from the migrant population, because some gene flow continues to occur—such as when a new migrant joins the colonist population. Given enough time, however, the populations may continue to diverge. Dispersal and colonization, accompanied by genetic drift and natural selection, are thought to be responsible for speciation in Galápagos finches and many other island groups.

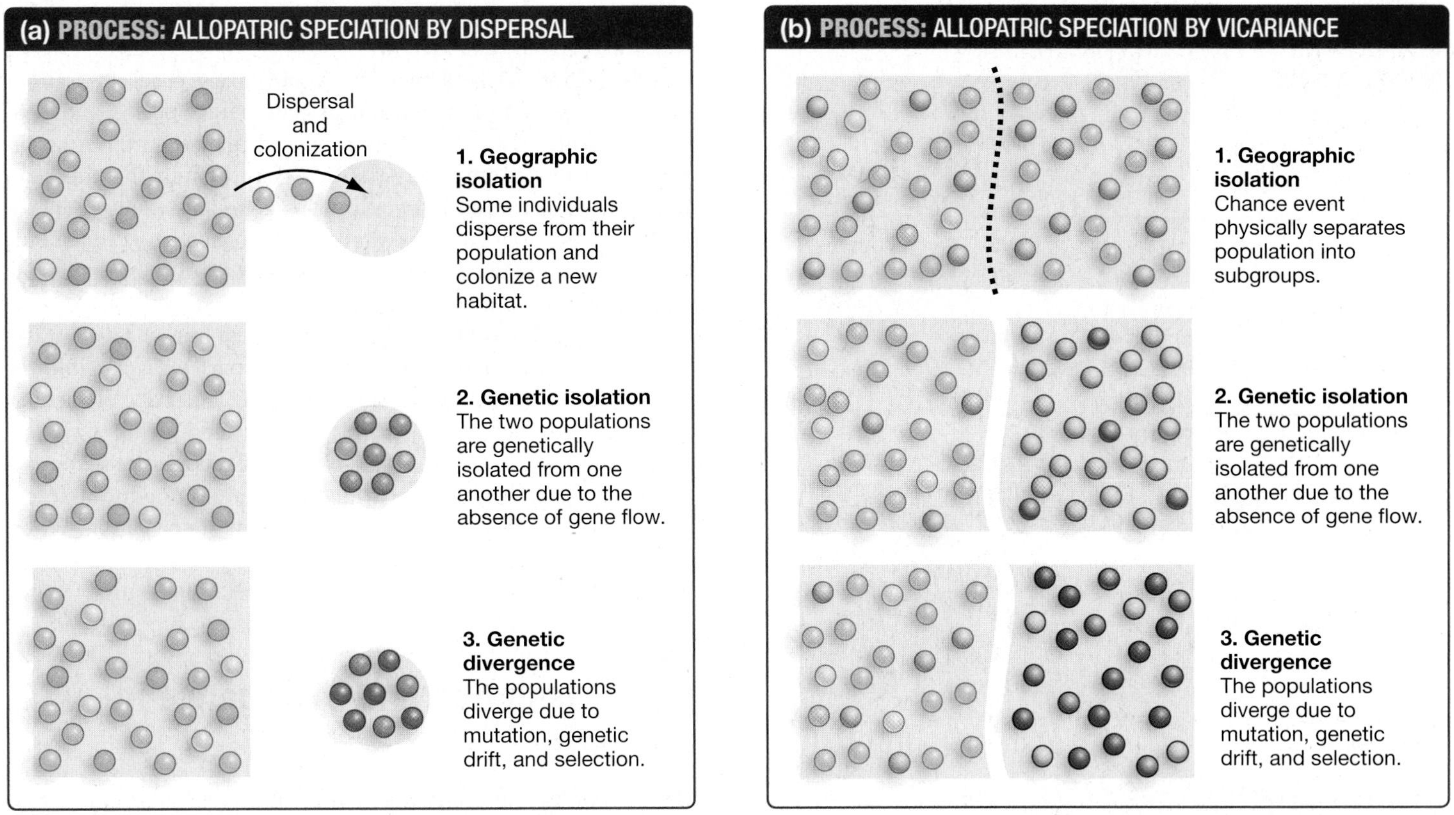

Figure 24.5 **Allopatric Speciation Begins via Dispersal or Vicariance. (a)** When dispersal occurs, colonists establish a new population in a novel location. **(b)** In vicariance, a widespread population becomes fragmented into isolated subgroups. Different colors represent genetic variation.

✔ MODEL If you understand ongoing allopatric speciation in finches, you should be able to apply the model in Figure 24.5a to this example. What does the small population on the right side of the model represent? What does the red color represent? How could you modify the model to show that gene flow is still occurring?

Allopatric Speciation by Vicariance

If a new physical barrier such as a mountain range or river splits the geographic range of a species, vicariance has taken place. Such changes are common on Earth due to continental drift, climate fluctuations, and many other factors (Ch. 49, Section 49.2). However, it is often unclear which event or combination of events in the past triggered the formation of species seen today.

The Amazonian rain forest in South America is famous as a hotspot of biodiversity (Ch. 54, Section 54.1). Why has speciation been so common in this tropical ecosystem? Evolutionary biologists Camila Ribas, Joel Cracraft, and their colleagues took a biogeographical approach to the question. They set out to test the hypothesis that the history of the Amazon River system itself was key to the diversification of birds called trumpeters, introduced in Figure 24.2. Trumpeters are able to fly and swim short distances, but they spend most of their time on the ground.

The researchers collected tissue samples from trumpeters distributed throughout the Amazon basin and compared the DNA sequences of four genes to estimate a phylogeny (see Figure 24.2). Using molecular dating techniques, they were also able to estimate the timing of each speciation event on the tree.

The researchers then compared these phylogenetic data to a biogeographical analysis of the formation of the Amazon River system. Although the continents seem static to us today, there was no Amazon River 3 million years ago. The Andes Mountains had not yet risen on the west side of the continent, and coastal

CHECK YOUR UNDERSTANDING

✔ If you understood this section, you should be able to . . .

1. Use the map of current elephant populations in Africa in Figure 24.4 to determine whether this is an example of allopatric speciation by dispersal or vicariance. Explain your reasoning. Note: These are remnant populations that represent about 1 percent of the 30 million elephants that once occupied much of the continent of Africa.
2. Predict whether allopatric speciation would be likely to occur again if the current isolated populations of elephants were to remain isolated over time (and avoid extinction). Explain.

Answers are available in Appendix A.

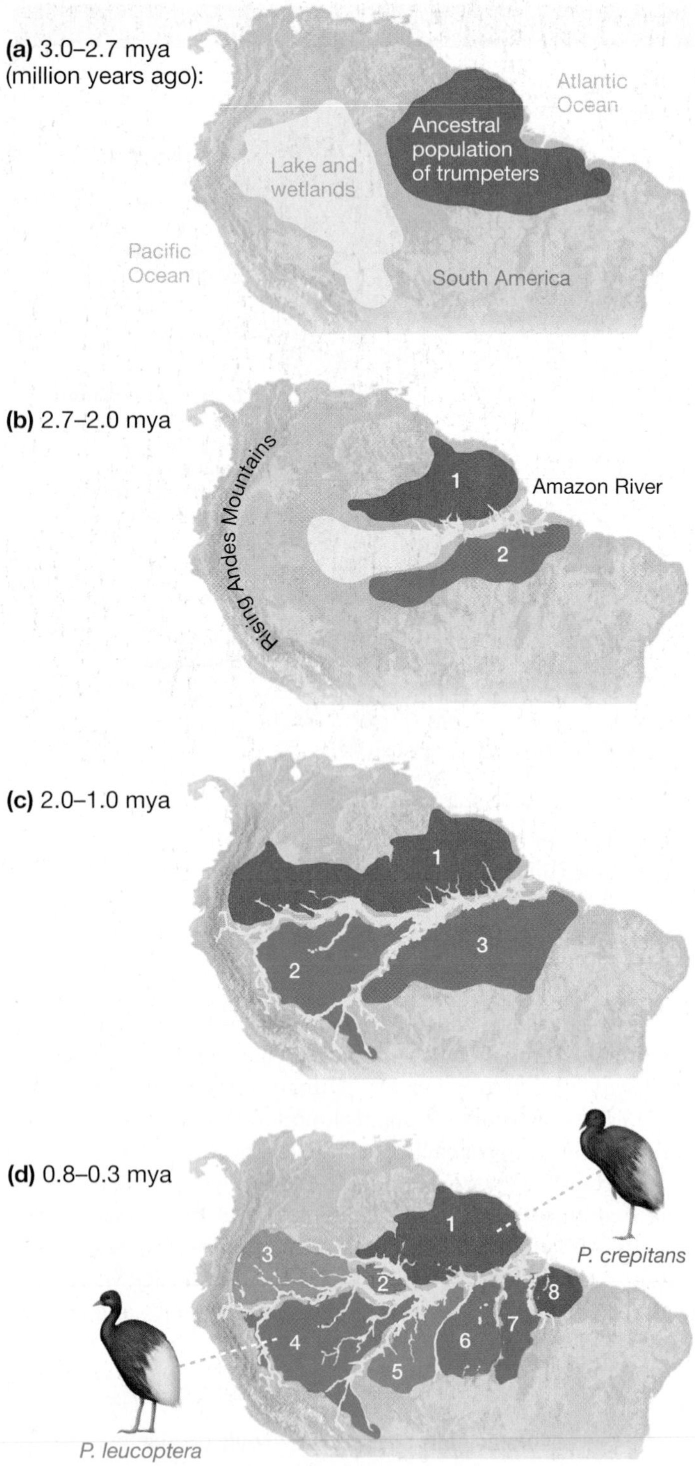

Figure 24.6 Evidence Supports Speciation by Vicariance in Trumpeters. (a) The Amazon River did not exist about 3 million years ago. **(b–d)** As the rising Andes Mountains caused the Amazon River and its tributaries to form, the rivers isolated populations of trumpeters (birds in the genus *Psophia*), leading to speciation. The colors in **(d)** correspond to the branches in Figure 24.2.

DATA: C. Ribas et al. 2012. *Proceedings of the Royal Society B* 279: 681–689.

✔ Compare the data in this series of maps to the data in Figure 24.2. Propose a hypothesis to explain why there are two clusters of species in the phylogenetic tree.

hills on the east side prevented the drainage of a massive lake and wetland system (**Figure 24.6a**).

Ribas, Cracraft, and their colleagues discovered close alignment between the distribution of trumpeter species and geological events. The researchers proposed that when the Amazon River first flowed into the Atlantic Ocean, it split the ancestral trumpeter population in two, limiting gene flow (**Figure 24.6b**). As the other large rivers of the system formed, they too subdivided the trumpeter populations (**Figure 24.6c–d**). The result was eight species isolated by vicariance.

These data provide strong support for the hypothesis that the formation of the river system itself was partly responsible for creating biodiversity in the Amazon basin. The scope of this hypothesis is now being tested using other species.

To summarize, geographic isolation of populations via dispersal or vicariance produces genetic isolation due to the interruption of gene flow—the first requirement of speciation. When genetic isolation is accompanied by genetic divergence—due to genetic drift, mutation, and selection—speciation results.

24.3 Isolation and Divergence in Sympatry

When populations or species live in the same geographic area, or close enough to one another to make interbreeding possible, biologists say that they live in **sympatry** ("together-homeland"). Traditionally, evolutionary biologists such as Ernst Mayr predicted that speciation could not occur among sympatric populations because gene flow would easily overwhelm any differences caused by genetic drift and natural selection.

After you complete this section, you should be able to . . .

- Use examples to explain the process of sympatric speciation.

Recently, a number of studies have countered this prediction. **Sympatric speciation**—speciation that occurs even though populations live within the same geographical area (**Figure 24.7**)—can indeed occur, and serves as a source of new branches on the tree of life. How does it work?

Two types of factors can initiate the process of sympatric speciation: **(1)** extrinsic (external) factors, such as disruptive selection for extreme phenotypes based on different ecological niches and/or mate preferences; and **(2)** intrinsic (internal) factors, such as chromosomal mutations. Let's consider an example of each of these factors in turn.

Sympatric Speciation by Disruptive Selection

Recall that *disruptive selection* favors extreme phenotypes (Ch. 23, Section 23.3). For example, whitefish in some northern lakes have higher fitness if they have either a small or large number of gill rakers (part of their feeding apparatus), but lower fitness if they have an intermediate number of gill rakers (see Ch. 23, Figure 23.8). If disruptive selection persists over time, the two subsets of the population can become genetically isolated and diverge from each other, leading to speciation.

Disruptive selection (sometimes called divergent selection once populations have separated) can act on diverse types of traits ranging from morphology to behaviors. Further, disruptive selection can involve the survival of organisms in their environment (ecological selection) and/or their ability to attract mates (sexual selection).

Let's consider the case of killer whales, where all of these factors are in play.

Disruptive Selection for Ecological Niche in Killer Whales

Despite their global distribution in the world's oceans, killer whales are currently described as a single species (*Orcinus orca*). Recent estimates put their number at about 50,000.

Michael Bigg was one of the first marine biologists to make careful observations of killer whales. In the early 1970s, he realized that individual killer whales in the northeast Pacific Ocean could be recognized by the shapes of their dorsal fins, the saddle patches on their backs, and scrapes and scars. Using hundreds of volunteers and photographic records, Bigg created a system to track the behavior, survival, and relationships of specific killer whales.

One of the first patterns Bigg noticed was that killer whales tend to occur in two kinds of groups in the coastal waters of British Columbia: larger groups of 10–25 individuals that he called resident groups, and smaller, traveling groups of 2–6 individuals that he called transient groups. Bigg used these names because he thought the transients were moving from one resident group to another. But he soon discovered that he was wrong: the transient killer whales looked different from the resident killer whales. For example, the transients were larger, on average, and had pointier dorsal fins than the residents (**Table 24.3**). The two so-called *ecotypes* were not mixing.

This observation was perplexing: How could two types of killer whales coexist in the same waters without intermingling? Wouldn't gene flow erase any differences between them?

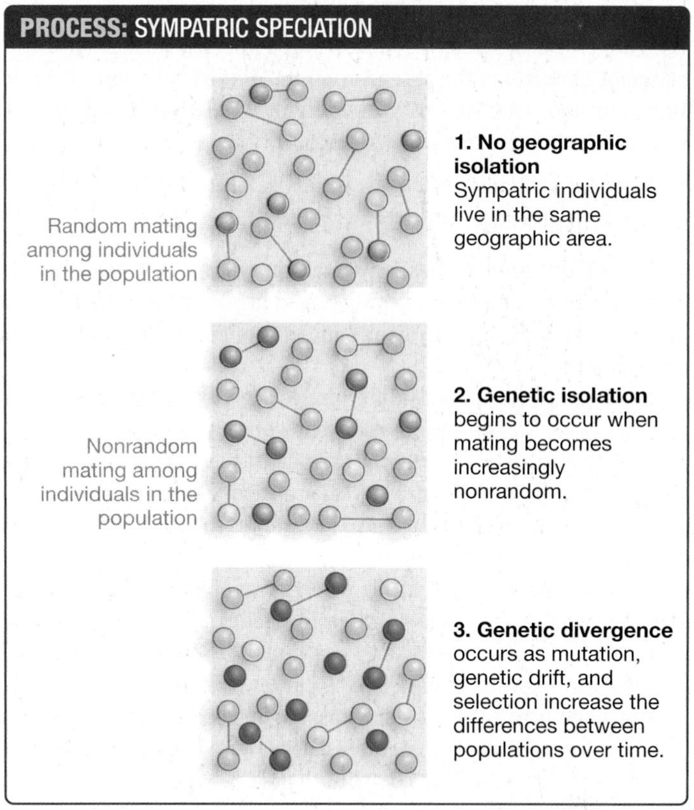

Figure 24.7 Sympatric Speciation Does Not Require Geographic Isolation. In this model, the dots represent individuals, the colors represent variation in a trait among individuals, and lines connecting the dots represent mating.

As more researchers began to observe killer whale behavior and to inspect the gut contents of killer whales that had died, a theme began to emerge: The killer whale groups were separated by their feeding cultures.

Table 24.3 Two Killer Whale Ecotypes in the Northeast Pacific

Ecotype	Appearance	Group Size	Preferred Prey	Hunting Behavior	Sample of Call
Resident		Large groups, typically 10–25 individuals	Fish, especially Chinook salmon, the largest of the salmon species in the area	Very chatty while hunting: use calls to communicate with group and clicks for echolocation to locate prey; individuals in group fan out and hunt solo while staying in contact with others	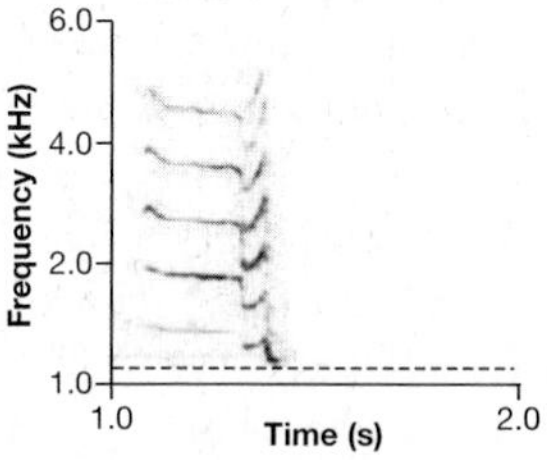
Transient		Small groups, typically 2–6 individuals	Marine mammals such as harbor seals, harbor porpoises, and Steller sea lions	Use stealth to hunt: travel in near silence (no calls or clicks) to avoid detection by prey; often work together in surprise attacks, or to exhaust or drown prey; share the catch; resume calls after detected by prey	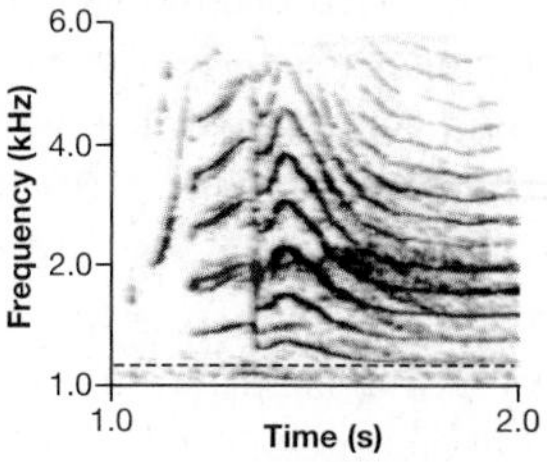

DATA: Artwork by Uko Gorter in R. Riesch et al. 2011. *Biological Journal of the Linnean Society* 106: 1–17; call data from A. D. Foote and J. A. Nystuen. 2007. *Journal of the Acoustic Society of America* 123: 1747–1752.

Biologists use the term **niche** (pronounced *nitch*) to describe the range of ecological resources that a species can use and the range of conditions that it can tolerate (Ch. 52, Section 52.1). It turns out that the two ecotypes of killer whales occupy different niches within the same area of ocean:

- *Resident ecotype* eats fish, especially salmon (never mammals).
- *Transient ecotype* eats marine mammals (never fish).

Killer whales' preferences for certain prey appear to be learned. The whales live in small, family-based groups where they learn specific methods of hunting and dialects of communication from their elders.

For example, resident killer whales are very chatty underwater; they stay in communication with others as they hunt individually for salmon. These killer whales use a vocabulary of calls that is unique to their group. They also make numerous click sounds for echolocation to find prey underwater (fish do not detect these clicks).

In contrast, transient killer whales travel longer distances in smaller groups and do not use calls or clicks while hunting because the porpoises, seals, and sea lions they seek are sensitive to these underwater frequencies. Rather, transients use stealth to approach prey—they are experts at surprise attacks. Once the prey is cornered, the transients use their own dialect of calls (see example in Table 24.3) to coordinate their team effort to trap, exhaust, and/or drown their savvy prey, then share the feast.

Consider these additional observations:

- When a transient killer whale was taken into captivity, it died of starvation rather than eat fish.
- A different transient killer whale learned to eat fish in captivity when taught by a resident killer whale.
- In captivity, transient and resident killer whales can mate and produce viable offspring.

These observations tell us that resident and transient killer whales have the capacity to mate and to learn each other's cultural practices, yet don't appear to do so in the wild. Why not?

One hypothesis is that disruptive selection for hunting behavior selects against killer whales that have a poor match between their learned hunting methods and food type. This has the effect of enforcing their fidelity to ecotype. Because the unique dialects of calls that are used for hunting are also used for mate choice, divergence in hunting culture could cause reproductive isolation. Natural selection and genetic drift would begin to occur separately in the different ecotypes, causing other traits like body size to diverge too.

Researchers can't yet rule out the alternate hypothesis that the ecotypes originally diverged in allopatry and then moved back together to sympatry. However, ecotypes have also emerged among sympatric killer whales in waters other than the northeast Pacific, such as in the northern Atlantic and the Antarctic.

The upshot is that although killer whales live in the same geographic area, pre-mating behavioral isolation is causing reproductive isolation due to natural selection for adaptations to two different niches. Resident killer whales and transient killer whales, as well as ecotypes in other oceans, are currently in the process of becoming distinct species. ✔ **MODEL If you understand the ongoing sympatric speciation in killer whales, you should be able to apply the model in Figure 24.7 to this example. What do the colored balls represent?**

Are Speciation Genes under Selection in Sympatric Speciation?

In the case of killer whales, many behavioral, morphological, and physiological traits are under selection in association with prey choice. Are there examples where a specific gene is pivotal to the formation of new species?

Molecular techniques are revealing a number of cases where certain genes play a critical role in speciation. For example, evolutionary biologist Sean Carroll and colleagues recently identified a single gene, *mFAS*, that affects ecological divergence as well as mate choice in two species of fruit flies on the east coast of Australia. The *mFAS* gene codes for fatty acid synthase, an enzyme that catalyzes the production of waxy hydrocarbons that make the external cuticles of flies more resistant to drying out. Carroll and his team made these observations:

- *Drosophila serrata* expresses *mFAS* and is a generalist that lives in a variety of habitats, both moist and dry.
- *D. birchii* does not express *mFAS* and survives only within the moist part of the *D. serrata* range.

Besides protecting flies from drying out, the waxy hydrocarbons act as pheromones—chemical signals used by mates to communicate (Ch. 44, Section 44.4). Carroll and his colleagues propose that *D. birchii* arose by sympatric speciation from the generalist *D. serrata* when it lost the ability to express *mFAS*. The mutant flies were able to persist in the moist part of the range, but females in the generalist population no longer mated with the mutant males because their chemically altered pheromones were not attractive—a form of prezygotic isolation.

If single genes can play an important role in speciation, what about changes in entire chromosomes?

Sympatric Speciation by Polyploidization

Even though mutation is the ultimate source of genetic variation in populations, it tends to occur at a low rate. Therefore, mutation is an inefficient mechanism of evolutionary change on its own (Ch. 23, Section 23.6). Natural selection and genetic drift are usually responsible for amplifying the effects of mutations.

One particular type of mutation, though, turns out to be extremely important in speciation—especially in plants, but also in animal groups such as flatworms, fishes, frogs, and salamanders. **Polyploidy**, the condition of possessing more than two complete sets of chromosomes, is caused by a massive error in mitosis or meiosis. There are two types of polyploids:

1. **Autopolyploid** ("same-many-form") individuals are produced when a mutation results in a doubling of chromosome number and the chromosomes all come from the same species (**Figure 24.8a**).
2. **Allopolyploid** ("different-many-form") individuals are created when parents of different species mate and then an error in mitosis occurs, resulting in viable, nonsterile offspring with two full sets of chromosomes (**Figure 24.8b**).

How can these two types of polyploidy lead to speciation?

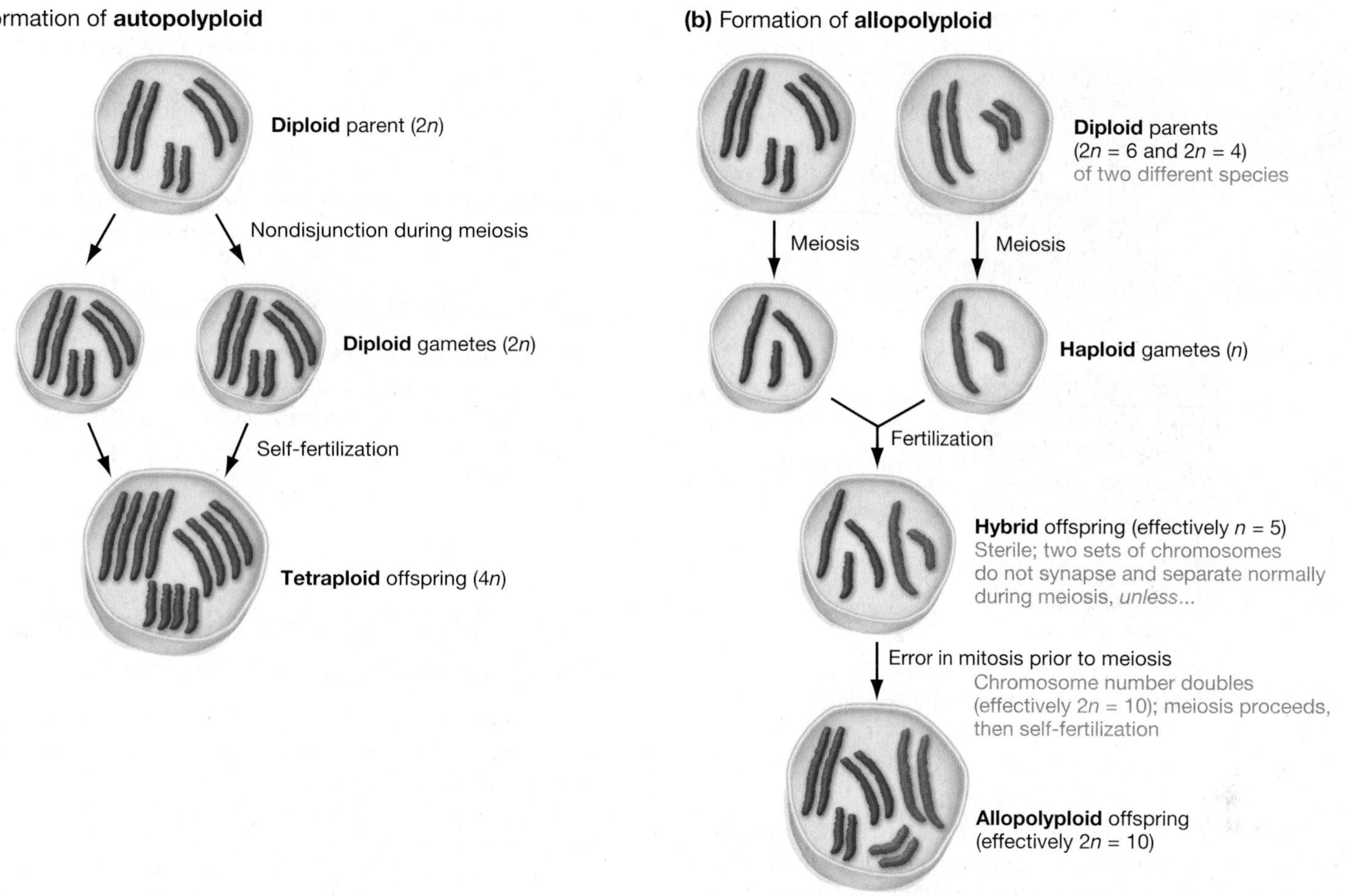

Figure 24.8 **Autopolyploids and Allopolyploids. (a)** The key event leading to autopolyploidy is nondisjunction during meiosis resulting in diploid, rather than haploid, gametes. **(b)** Allopolyploidy can occur by different methods; one example is shown. In this case an error during mitosis leads to the duplication of chromosomes, enabling the pairing of homologs. In both auto- and allopolyploidy, the polyploids are able to produce gametes that can self-fertilize.

Autopolyploidy Is the Duplication of the Same Genome Autopolyploidy is thought to be much less common than allopolyploidy. One example is found in the maidenhair fern, native to woodlands across North America.

Biologists initially set out to do a routine survey of the diversity of alleles in a population of these ferns. By chance, they stumbled upon individuals that could produce offspring that were tetraploid ($4n$) rather than diploid—that is, polyploid mutants within a normal population.

To follow up on the observation, researchers studied the parents of the mutant individuals and found they have a defect in meiosis that caused nondisjunction of chromosomes (Ch. 13, Section 13.1). Instead of producing normal, haploid cells as a result of meiosis, the mutant individuals produced diploid cells. These diploid cells eventually led to the production of diploid gametes (see Figure 24.8a). Because maidenhair ferns can self-fertilize, the diploid gametes combined to form tetraploid offspring.

How could this event lead to speciation? The tetraploid offspring can self-fertilize or mate with their tetraploid parent or each other, but not with other diploids in the population. Tetraploid individuals produce diploid gametes, but diploid individuals produce haploid gametes. When diploid and haploid gametes unite, a triploid ($3n$) offspring forms. Triploids usually cannot produce normal gametes because three chromosomes cannot synapse and separate correctly during meiosis. That is, triploids are sterile. Thus polyploid individuals are reproductively isolated from the original diploid population. Speciation (as defined by the biological species concept) occurs in a single generation—instantaneously, in evolutionary terms. ✔ **If you understand how autopolyploidy works, you should be able to propose a scenario explaining how a tetraploid grape plant with extra-large fruit could come from a diploid population with smaller fruit. (You've probably seen both types of fruit in the supermarket.)**

Allopolyploidy Is the Combination of Two Genomes As Figure 24.8b illustrates, new polyploid species may be created when two species hybridize. Notice that the allopolyploid offspring in Figure 24.8b appears diploid in the sense that it has two copies of each chromosome ($2n = 10$). However, it is functionally polyploid because it has two copies each of *two sets* of chromosomes (indicated by blue and red). In terms of genetic information, this offspring has four copies of genes that are common to both parental species, instead of just two.

Figure 24.9 Allopolyploids Can Form New Species.

Exactly this chain of events occurred after a few European species of weedy plants in the genus *Tragopogon* were introduced to western North America in the early 1900s. In 1950 a botanist described the first new allopolyploid species (**Figure 24.9**). An analysis of chromosomes showed the new species to be the result of mating between introduced diploid species.

Follow-up work by plant geneticists Pam and Doug Soltis has shown that the polyploidization has occurred multiple times between the same two diploid *Tragopogon* species. The Soltises raise an interesting question: Should distinct polyploid populations that result from matings between the same two parental species be considered the same or different daughter species?

✔ If you understand how allopolyploidy works, you should be able to explain how a cross between Emmer wheat, a tetraploid, and a wild, diploid wheat gave rise to the hexaploid bread wheat grown throughout the world today.

Why Is Speciation by Polyploidy So Common in Plants? The claim that speciation by polyploidization has been particularly important in plants is backed by the observation that many diploid species have close relatives that are polyploid. Why have polyploids been so successful? The Soltises have addressed this question by examining the genetic implications of polyploidy:

- Polyploids have higher levels of heterozygosity than their diploid relatives.
- Polyploids can tolerate higher levels of self-fertilization because they are not as affected by inbreeding depression as their diploid relatives (Ch. 23, Section 23.2).
- Genes on duplicated chromosomes can diverge independently, increasing genetic variation in the population.

Genetic variation is a prerequisite for evolution. Thus, the high genetic diversity of polyploids has enabled a rapid diversification of plant species.

To summarize, speciation by polyploidization is driven by chromosome-level mutations and occurs in sympatry. Compared to the gradual process of speciation by geographic isolation or by disruptive selection in sympatry, speciation by polyploidization is virtually instantaneous. It is fast, sympatric, and common.

CHECK YOUR UNDERSTANDING

✔ If you understood this section, you should be able to . . .

Compare speciation in trumpeters in the Amazon rain forest and killer whales in the northeast Pacific Ocean. What do these processes have in common, and how are they different?

Answers are available in Appendix A.

24.4 What Happens When Isolated Populations Come into Contact?

Suppose two populations that have been isolated come into contact again. If divergence has taken place and prezygotic isolation exists, then mating between the populations is rare, gene flow is minimal, and the populations continue to diverge. But what if there is no prezygotic isolation, and the populations begin interbreeding?

After you complete this section, you should be able to . . .

- Compare the possible outcomes of secondary contact between populations that have been isolated.

Fusion or Extinction

The simplest outcome of two populations coming into contact is that the populations fuse over time, as gene flow erases any distinctions between them.

For example, this fusion process appears to be taking place in a Norwegian lake after an invasive species of fish called vendace (*Coregonus albula*) was introduced in 1993. The vendace is outcompeting one of the two sympatric populations of European whitefish there (*Coregonus lavaretus*)—it's pushing the open-water population (with more gill rakers) out of its plankton-feeding niche to the edges of the lake, where it is now interbreeding with the benthic population (with fewer gill rakers). That is, competition with vendace is causing gene flow between whitefish populations. Virtually all of the whitefish survivors are hybrids and the populations are fusing (**Figure 24.10**).

What will happen to whitefish if the invasive vendace persists in the lake for generations? Whitefish with intermediate or high gill raker numbers will probably continue to have lower fitness living in the benthic niche, because their gill rakers are still likely to clog with mud. Thus, after several generations of directional selection for low gill raker number, the fused population of whitefish will probably have an average gill raker number that is similar to the original benthic population. That is, the open-water whitefish population effectively will have gone extinct.

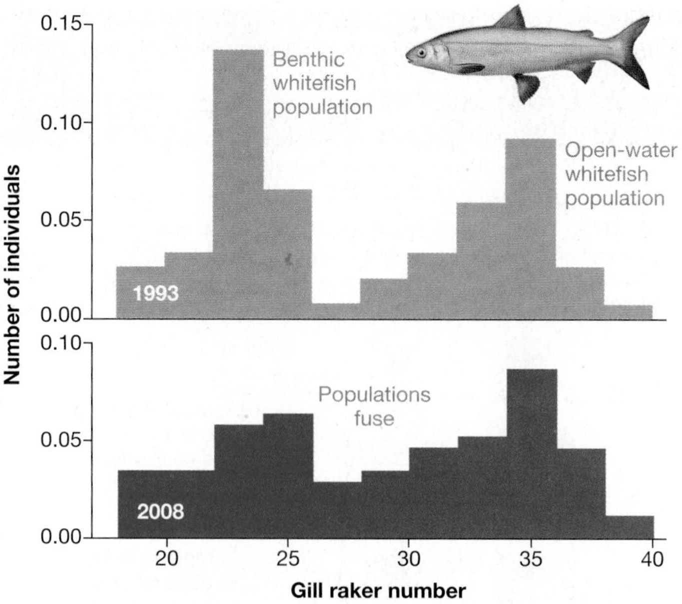

Figure 24.10 **Fusion of Sympatric Populations of Whitefish.** Just 15 years after the introduction of an open-water competitor, two sympatric populations of whitefish are fusing back into one.

DATA: S. Bhat et al. 2014. *PLOS ONE* 9: e91208.

✔ **MODEL** Sketch a graph showing the distribution of gill raker number you would expect to see within many generations if the invasive fish persists in the lake and if whitefish with high gill raker number have low fitness in the benthic environment.

Reinforcement of Divergence

If two populations have diverged extensively and are distinct genetically, it's reasonable to expect that their hybrid offspring will have lower fitness than their parents because they will not be well adapted to either habitat. The hybrid offspring also may fail to develop normally or may be infertile (see postzygotic isolation in Table 24.1).

When postzygotic isolation occurs, there should be strong selection against interbreeding because hybrid offspring represent a wasted effort by the parents, reducing their fitness. Natural selection for traits that prevent interbreeding among populations is called **reinforcement**. The name is descriptive because the selected traits reinforce differences that evolved while the populations were isolated from one another.

Some of the best data on reinforcement come from laboratory studies of closely related fruit fly species in the genus *Drosophila*. Evolutionary biologists Jerry Coyne and Allen Orr conducted a series of experiments that tested whether members of closely related fly species will mate with one another. They found an interesting pattern:

- If closely related species are *sympatric*—meaning that they live in the same area—individuals from the two species will seldom mate with one another in the lab.
- If the species are *allopatric*—meaning that they live in different areas—then individuals from the two species will often mate with one another in the lab.

The pattern is logical because natural selection to reduce mating between species (prezygotic isolation) will exist only where their ranges overlap. Thus it is reasonable to find that sympatric fruit fly species exhibit prezygotic isolation, but allopatric species do not. Although the concept of reinforcement was controversial at first, several new examples of reinforcement are emerging.

Hybrid Zone Formation

Hybrid offspring are not always dysfunctional. In some cases two diverged populations are capable of mating and producing viable and fertile offspring. When this happens, hybrid zones can form. A **hybrid zone** is a geographic area where interbreeding occurs and hybrid offspring are common. Depending on the fitness of hybrid offspring and the extent of breeding between parental species, hybrid zones can be narrow or wide, can be long- or short-lived, and can move over time or remain stable in one place.

As an example, consider recent work on two bird species. Townsend's warblers and hermit warblers live in the coniferous forests of North America's Pacific Northwest. In western Washington State, where their ranges overlap, the two species hybridize extensively. As the drawings in Figure 24.11 show, hybrid offspring have characteristics that are intermediate relative to the two parental species.

To explore the dynamics of this hybrid zone, ornithologist Sievert Rohwer teamed up with population geneticist Eldredge Bermingham and others to examine the mitochondrial gene

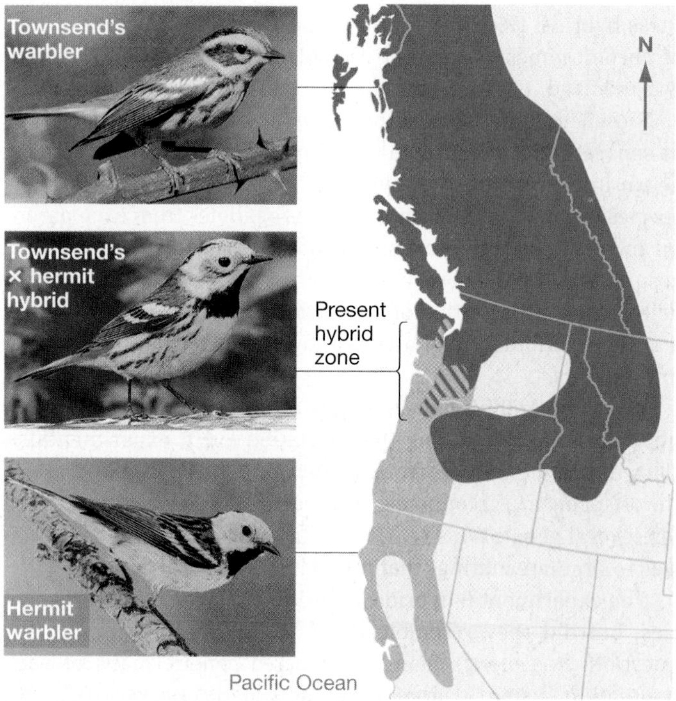

Figure 24.11 **Analyzing a Hybrid Zone.** The current ranges of Townsend's warblers (in red) and hermit warblers (in orange) are shown on the map, with the hybrid zone where they meet.

DATA: S. Rohwer, E. Bermingham, and C. Wood. 2001. *Evolution* 55: 405–422.

(mtDNA) sequences of a large number of warblers collected from forests throughout the region. The mtDNA is maternally inherited in most animals and plants—offspring receive their mitochondria from their mother's egg during fertilization. Thus, identifying mtDNA types allowed the research team to infer the parentage of each hybrid individual. For example, if a hybrid individual has hermit mtDNA, its mother is a hermit warbler, while its father is a Townsend's.

The mtDNA and behavioral data suggest that Townsend's males invade hermit territories, drive off the hermit males, and mate with hermit females. The researchers hypothesized that hermit warblers were once found as far north as Alaska and that Townsend's warblers have gradually taken over their range, causing the hybrid zone to move south over time. If this hypothesis is correct, hermit warblers may eventually become extinct. In some cases, however, hybridization does not lead to extinction, but rather the opposite—the formation of new species.

New Species through Hybridization

If two species interbreed and produce hybrid offspring that can not only survive and reproduce but also possess a unique combination of traits that are adaptive in their particular environment, a new species may result. How would you identify such a species in nature?

Evolutionary biologist Loren Rieseberg and colleagues were among the first to provide convincing evidence of speciation by hybridization. They examined the relationships of three sunflower species native to the Southwest: Recently diverged *Helianthus annuus* and *H. petiolaris* are known to hybridize in regions where their ranges overlap, and *H. anomalus* resembles these hybrids. However, all three species have the same number of chromosomes, so neither allopolyploidy nor autopolyploidy was involved.

Rieseberg hypothesized that *H. annuus* and *H. petiolaris* arose as separate species through isolation and divergence, then later began interbreeding. The hybrid offspring became a third, new species that had unique combinations of alleles from each parental species and therefore distinct characteristics. This hypothesis is supported by the observation that *H. anomalus* grows in much drier habitats than either of the parental species—suggesting that a unique combination of alleles allowed *H. anomalus* to thrive in dry habitats.

Rieseberg and his colleagues set out to test the hybridization hypothesis by re-creating the speciation event experimentally. They mated individuals from the two parental species and raised the offspring in a greenhouse, then crossbred the hybrid offspring for several generations (**Figure 24.12**). The goal of these crosses was to simulate matings that might have occurred naturally.

The experimental hybrids looked like the natural hybrid species, but did they resemble them genetically? To answer this question, Rieseberg's team constructed genetic maps of each population, using a large series of genetic markers (Ch. 20, Section 20.3). Because both parental populations had many unique markers in their genomes, the research team hoped to identify which genes found in the hybrids came from which parental species.

RESEARCH

QUESTION: Can new species arise by hybridization between existing species?

HYPOTHESIS: *Helianthus anomalus* originated by hybridization between *H. annuus* and *H. petiolaris*.

NULL HYPOTHESIS: *Helianthus anomalus* did not originate by hybridization between *H. annuus* and *H. petiolaris*.

EXPERIMENTAL SETUP:

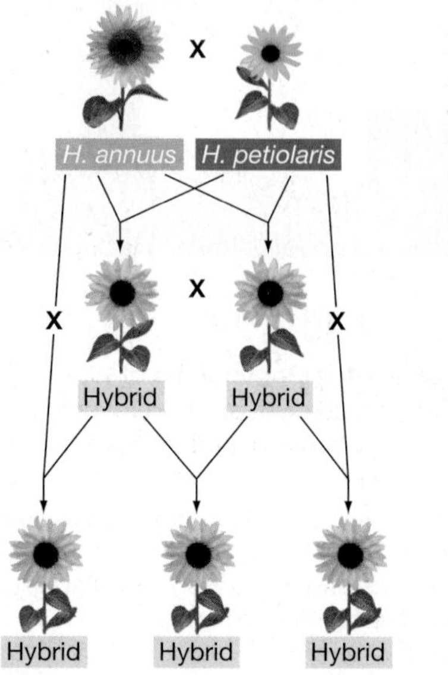

1. Mate *H. annuus* and *H. petiolaris* and raise offspring.

2. Mate F_1 hybrids to each other or backcross F_1s to parental species; raise offspring.

3. Repeat crosses for four more generations, then compare genetic information of experimental hybrids versus *H. anomalus*.

PREDICTION OF HYPOTHESIS: Experimental hybrids will have a similar mix of *H. annuus* and *H. petiolaris* genes as *H. anomalus*.

PREDICTION OF NULL HYPOTHESIS: Experimental hybrids will not have a similar mix of *H. annuus* and *H. petiolaris* genes as *H. anomalus*.

RESULTS:

DNA comparison of a chromosomal region called *S*:

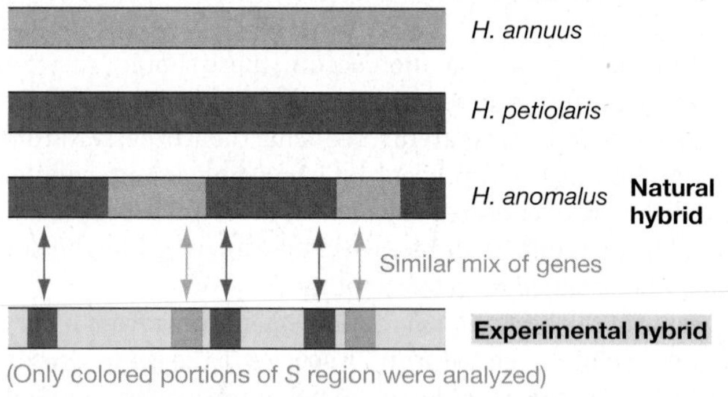

CONCLUSION: New species may arise via hybridization between existing species.

Figure 24.12 Experimental Evidence that New Species Can Originate in Hybridization Events.

DATA: L. H. Rieseberg, B. Sinervo, C. R. Linder, et al. 1996. Role of gene interactions in hybrid speciation: Evidence from ancient and experimental hybrids. *Science* 272: 741–745.

✔ **PROCESS OF SCIENCE** Suggest one result that would have caused the researchers to reject the hybridization hypothesis.

SUMMARY **Table 24.4 Possible Outcomes of Secondary Contact between Populations.**

		Process	Example
Fusion of the populations	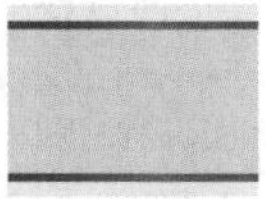	The two populations freely interbreed.	Open-water whitefish population fused with sympatric benthic whitefish population after introduction of invasive open-water competitor.
Extinction of one population	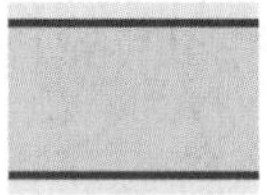	If one population or species is a better competitor for shared resources, then the poorer competitor may be driven to extinction.	Open-water whitefish population may go extinct after fusion with benthic population. Townsend's warblers may be driving hermit warblers to extinction.
Reinforcement of divergence		If hybrid offspring have low fitness, natural selection favors the evolution of traits that prevent interbreeding between the populations.	Appears to be common for distinct populations of fruit fly species that occupy the same geographic areas.
Hybrid zone formation	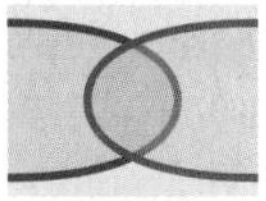	Hybridization occurs in a well-defined geographic area. This area may move over time or be stable.	The hybrid zone between hermit and Townsend's warblers appears to have moved over time.
Formation of new species	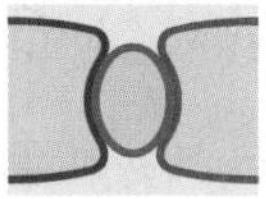	If the combination of genes in hybrid offspring allows them to occupy distinct habitats or use novel resources, they may form a new species.	Hybridization between sunflowers gave rise to a new species with unique characteristics.

Some of the data are diagrammed in the "Results" section of Figure 24.12. The bars in the illustration show the genetic composition of a chromosomal region called S in the genome of *H. annuus* (color coded orange), *H. petiolaris* (color coded red), the natural hybrid *H. anomalus*, and the experimental hybrid. The genetic markers revealed that the genetic makeups of the natural hybrids and experimental hybrids were very similar. In effect, the researchers succeeded in re-enacting a speciation event. Their results provide strong support for the hybridization hypothesis for the origin of *H. anomalus*.

Rieseberg and his colleagues observed a pattern that has now become common in studies of speciation by hybridization:

- The genome of the hybrid species is a mosaic of the genomes of the two parental species—a diagnostic characteristic that can be convincingly demonstrated in the lab.
- The hybrid species is ecologically isolated from the parental species. *H. anomalus* grows in much drier habitats than either of the parental species—suggesting that a unique combination of alleles allowed *H. anomalus* to thrive in dry habitats.

This pattern has now emerged in diverse species such as butterflies in the Sierra Nevada, bats in the Caribbean, and finches in the Galápagos. The mosaic genomes contain combinations of alleles that are adaptive in certain environments where parental species don't thrive—and geographic isolation promotes genetic isolation by limiting gene flow between species.

Secondary contact of two populations can produce a dynamic range of possible outcomes: fusion of the populations, extinction of one population, reinforcement of divergence, founding of hybrid zones, or the formation of new species. **Table 24.4** summarizes the outcomes of secondary contact.

CHECK YOUR UNDERSTANDING

✔ **If you understood this section, you should be able to . . .**

Compare the fitness of hybrids in cases where reinforcement occurs versus cases where speciation by hybridization occurs.

Answers are available in Appendix A.

Take-Home Messages

Speciation is a two-step process involving genetic isolation and genetic divergence. This process typically involves the splitting of an ancestral species into two or more distinct species. However, there is increasing evidence that new species also form when species merge via allopolyploidy or hybridization. That is, new branches on the tree of life can form not only from splitting events, but also form from merging events.

Speciation is an ongoing, dynamic, and sometimes complex process. Fortunately, new genetic and phylogenetic tools, coupled with careful field observations, are enabling scientists to more easily identify species and to discover the genetic mechanisms involved in reproductive isolation and speciation. These efforts help to show how the processes of evolution (Chapters 22 and 23) in the context of ecology (Unit 8) create branches on the tree of life (Chapter 25 and Unit 5). Many of the concepts and methods discussed in this chapter are especially important to conservation biology (Chapter 54).

CHAPTER 24 Review

For media, go to Mastering Biology

24.1 How Are Species Defined and Identified?

- A species is defined as an evolutionarily independent population or group of populations. Researchers use several criteria to identify whether populations represent distinct species.
- The biological species concept uses reproductive isolation as a criterion to identify species.
- The morphospecies concept identifies species using distinctive morphological traits.
- The phylogenetic species concept identifies species as the smallest monophyletic groups on the tree of life.

24.2 Isolation and Divergence in Allopatry

- Speciation is typically a splitting event in which one lineage gives rise to two or more independent descendant lineages.
- Speciation occurs when populations of the same species become genetically isolated by lack of gene flow and then diverge from each other due to selection, genetic drift, and mutation.
- Allopatric speciation occurs when populations diverge in geographic isolation.
- Geographic isolation occurs through dispersal, when small groups of individuals colonize a new habitat, or through vicariance, when a large, continuous population becomes fragmented into isolated habitats.

24.3 Isolation and Divergence in Sympatry

- Sympatric speciation occurs when populations diverge genetically despite living in the same geographic area.
- Sympatric speciation can occur when disruptive selection favors individuals that breed in different ecological niches and/or that choose different mates.
- Mutations that produce polyploidy can trigger rapid speciation in sympatry because they lead to reproductive isolation between diploid and polyploid populations.

24.4 What Happens When Isolated Populations Come into Contact?

- If prezygotic isolation exists, populations that come back into contact will probably continue to diverge.
- If gene flow occurs, populations that have diverged may fuse into a single species.
- Secondary contact can lead to reinforcement—the evolution of mechanisms that prevent hybridization.
- Gene flow between different species can lead to the formation of hybrid zones that move over time or are stable.
- In some cases, hybridization between species can result in new species with unique combinations of traits.
- When speciation by hybridization occurs, a new branch is added to the tree of life by the merging of parts of preexisting branches.

Answers are available in Appendix A.

✔ TEST YOUR KNOWLEDGE

1. What distinguishes a morphospecies?
   a. It has distinctive characteristics, such as size, shape, or coloration.
   b. It represents a distinct branch in a phylogeny of populations.
   c. It is reproductively isolated from other species.
   d. It is a fossil from a distinct time in Earth history.
2. Which of the following describes vicariance?
   a. Small populations coalesce into one large population.
   b. A population is fragmented into isolated subpopulations.
   c. Individuals colonize a novel habitat.
   d. Individuals disperse and found a new population.
3. Select True or False to indicate which of the following groups could be identified using the biological species concept.
   T/F lizard species living today
   T/F sunflower species living today
   T/F extinct dinosaurs
   T/F bacteria living today
4. Evaluate this statement: Gene flow increases the genetic divergence of populations.

✔ TEST YOUR UNDERSTANDING

5. When the ranges of two different species meet, a stable "hybrid zone" occupied by hybrid individuals may form. How is this possible?
   a. Two diverged populations are capable of mating and producing viable and fertile offspring.
   b. Hybrid individuals are always allopolyploid and are thus unable to mate with either of the original species.
   c. Hybrid individuals may have reduced fitness and thus be strongly selected against.
   d. One species has a selective advantage, so as hybridization continues, the other species will go extinct.
6. Sexual selection favors individuals with traits that increase their ability to obtain mates, such as mating calls in crickets. Using this example, propose a scenario where sexual selection could contribute to divergence in sympatric speciation.
7. **QUANTITATIVE** If one species ($2n = 10$) crosses with another species ($2n = 18$), producing an allopolyploid offspring, what is the ploidy of the offspring?
   a. $2n = 10$
   b. $2n = 18$
   c. $2n = 10 + 18 = 28$
   d. $4n = 36 + 20 = 56$

8. **THINK CAREFULLY** True or false? Speciation is a slow process. Justify your answer.

✓ TEST YOUR PROBLEM-SOLVING SKILLS

9. Three-spine sticklebacks are small fish that originated in the ocean and continue to exist there, but then some took up residence in hundreds of streams and freshwater lakes in the Northern Hemisphere. Predict whether you would expect to find different species in lakes today.

10. **SOCIETY** All over the world, natural habitats are being fragmented into tiny islands by suburbs, ranches, farms, and roads. If asked to join a science panel to speak to Congress, could you explain how this fragmentation process could lead to extinction? How it could lead to speciation?

✓ PUT IT ALL TOGETHER: Case Study

Have human species hybridized in the past?

A revolution in the study of human evolution is under way thanks to the invention of techniques that enable DNA sequencing of not only modern humans, but also ancient humans (Ch. 32, Section 32.5).

11. **SOCIETY** Human populations today are *not* considered separate species under the biological species, morphospecies, and phylogenetic concepts. Explain what kind of evidence supports this statement.

12. Before the application of DNA sequencing to fossils, which species concept was most useful for distinguishing human fossils? What is one disadvantage of this approach?

13. Modern humans (*Homo sapiens*) and archaic humans called Neanderthals (*H. neanderthalensis*) shared a common ancestor in Africa but diverged in different geographic areas. When modern humans migrated out of Africa and dispersed around the world starting about 100,000 years ago, they overlapped with Neanderthals in Europe. This scenario is best described as
    a. species living in sympatry following allopatric speciation
    b. species living in sympatry following sympatric speciation
    c. species living in allopatry following allopatric speciation
    d. species living in allopatry following sympatric speciation

14. Svante Pääbo and colleagues were the first to sequence the Neanderthal genome from fossils and compare the sequences to modern humans. According to the data shown here, did the two species interbreed when they overlapped in Europe? Explain.

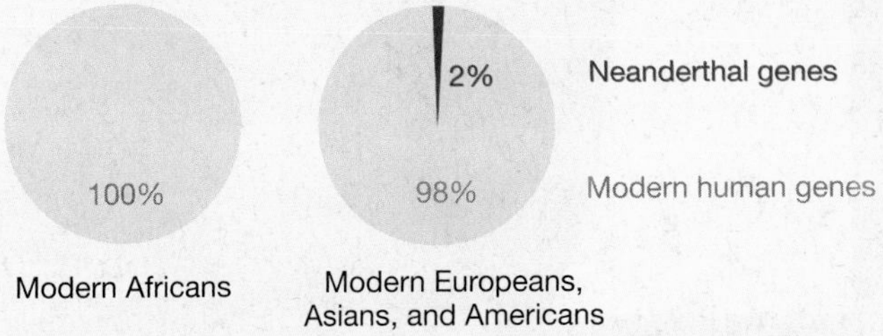

Source: K. Prüfer et al. 2014. *Nature* 505: 43–49.

15. **PROCESS OF SCIENCE** If you sequenced the DNA of all your classmates, you would probably observe that the percentage of Neanderthal genes in their genomes varies from 0–4%. How would you know whether there is a significant difference in percentage between students with blue eyes and students with brown eyes (see **BioSkills 3**)?

16. Neanderthals disappeared about 40,000 years ago when the modern human population increased. This is an example of what outcome(s) of secondary contact (see Table 24.4)? Predict one way this result might have come about.

Mastering Biology

Students Go to Mastering™ Biology for assignments, the eText, and the Study Area with animations, practice tests, and activities.

Professors Go to Mastering™ Biology for automatically graded tutorials and questions

25 Phylogenies and the History of Life

A fossilized trilobite. The last trilobites disappeared during a mass extinction event analyzed in Section 25.4.

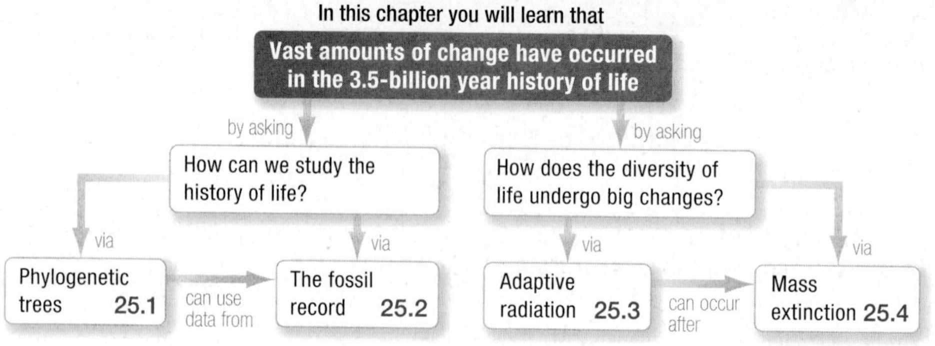

This chapter is about time and change. More specifically, it's about vast amounts of time and profound change in life on Earth.

Evolution observed at the population level, such as the evolution of antibiotic resistance (Ch. 22, Section 22.4), is referred to by some biologists as *microevolution*. Microevolution is a relatively easy concept to understand because it can be observed and measured directly within a human lifetime. But it takes practice to get comfortable with studying evolution at the species level (Chapter 24) or higher. At this scale, sometimes called *macroevolution*, events often occur over millions of years. How can biologists study the history of life over such vast amounts of time?

The first two sections of this chapter introduce two major analytical tools that biologists use to reconstruct the history of life. The remaining two sections explore the major patterns that emerge from these reconstructions. Visit the Big Picture of Evolution on pages 532–533 to see how these topics are connected.

BIG PICTURE

This chapter is part of the Big Picture. See how on pages 532–533.

25.1 Tools for Studying Life's History: Phylogenetic Trees

The branching evolutionary history of species or other groups of organisms is called a **phylogeny**. A **phylogenetic tree** is a simplified diagram of this history. **Systematics** is the discipline of biology that characterizes and classifies the relationships among all organisms on Earth, piecing together the **tree of life**, the most universal of all phylogenetic trees (see Chapter 1 and the Big Picture of Diversity on pages 722–723).

After you complete this section, you should be able to ...

- Use phylogenetic trees to answer biological questions.

Phylogenetic trees can be used to answer a wide range of questions in biology, for example:

- in **taxonomy**, to help describe, name, and classify individual species and higher-level groups, called **taxa** (Ch. 24, Section 24.1);
- in medicine, to track the spread of disease, such as AIDS and Ebola (Ch. 33, Section 33.3);
- in conservation, to identify species that are a priority for preservation (Ch. 54, Section 54.1).

It takes time and practice to interpret and draw phylogenetic trees correctly. Let's start with a warm-up to make sure that you are comfortable with the basics, then consider how phylogenetic trees can be used as a tool to study the history of life.

Fish Fraud Case Study: How Do Biologists Interpret Phylogenetic Trees?

If you have not had much exposure to phylogenetic trees and "tree thinking," now would be a good time to turn to **BioSkills 12** for help with visual models in general and **BioSkills 13** for an introduction to phylogenetic trees in particular. When you are ready, use the following example to review what you have learned.

The Question Biology students in Seattle wondered if salmon sold at markets and restaurants was really the species of salmon that merchants claimed. Some species of salmon are more expensive than others, so there is an incentive for merchants to distort the truth to increase profits. Is fish fraud taking place?

The Method First the students collected several fish samples from markets and restaurants. Next, they sequenced a certain gene (*cytochrome* c *oxidase*) from the samples and downloaded the sequence data from known salmon species from the DNA Barcode of Life project (Ch. 54, Section 54.1). The students then used the DNA sequence data to estimate a phylogenetic tree using inference methods described in the next section. **Table 25.1** shows a sample of the results. Take a moment to review the parts of the tree to make sure that you understand the conventions used.

SUMMARY Table 25.1 Anatomy of a Phylogenetic Tree

Graphical Symbols and Definitions Applied to Fish Fraud Case Study

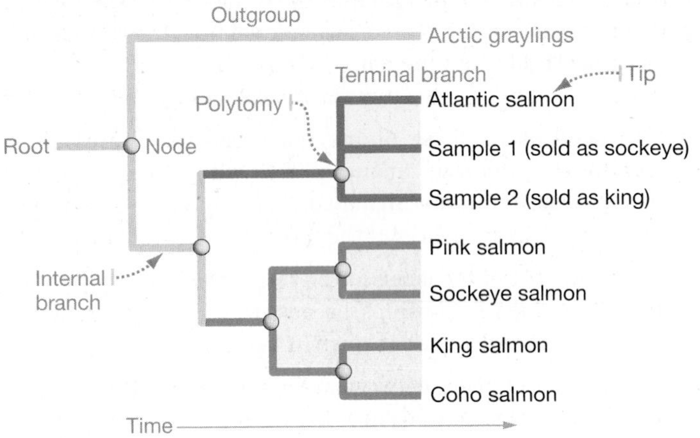

Branch A line representing a species or other taxon through time

Root The most ancestral branch in the tree

Tip Endpoint of a branch; represents a living or extinct species or other taxon

Outgroup A taxon that diverged before the taxa that are the focus of the study; helps to root the tree

Node (fork) A point within the tree where a branch splits into two or more branches; the node represents the most recent common ancestor of the descendant groups

Polytomy A node that depicts an ancestral branch dividing into three or more (rather than two) descendant branches; usually indicates that insufficient data were available to resolve which taxa are more closely related

DATA: E. Cline. 2012. *Food Research International* 45: 388–393.

The Interpretation Notice that the tree in this table is "rooted" on the left, which means that the **branches** on the left are more ancient, or ancestral, while the ones on the right are most recent, or derived. Notice also that the **nodes** (forks) represent *hypothetical* common ancestors before splitting events—the taxa themselves are always located on the tips of terminal branches, never on internal branches or nodes. That is, none of the taxa are presumed to be ancestors of others, even if some of the taxa are extinct.

This is an important point that is easily misunderstood. For example, in Table 25.1, fish called arctic graylings are not the ancestors of salmon. Rather, graylings and salmon are **sister groups**—the two descendants that split from the same node and therefore are each other's closest relatives. Once the two lineages were genetically isolated, they diverged due to mutation, natural selection, and genetic drift, producing today's graylings and salmon (Chapter 24). ✔ **MODEL If you understand how ancestor–descendant relationships are shown in trees, you should be able to draw a small tree that shows the relationship between humans and our closest living relatives, chimpanzees, and label and explain what the node means.**

What does the tree say about fish fraud? To find out, look at the relationships between Sample 1 (sold in a restaurant as sockeye salmon) and Sample 2 (sold in a restaurant as king salmon) and the reference salmon species. The restaurant samples and the Atlantic salmon share a more recent common ancestor than the samples do with any of the Pacific salmon species (green branches), as indicated by the node at the base of the blue branches. Because the fish samples were sold as sockeye salmon and king salmon rather than as Atlantic salmon, the restaurants were not truthful about the source of the fish.

There are a few more features of Table 25.1 to note:

- The node at the base of the blue branches represents a **polytomy** (literally, "many-split"), with three branches emerging from it rather than two. This polytomy occurred because the DNA sequences of the three samples were identical.
- The proximity of the labels at the tips of the tree is not useful in deciphering relationships, because the branches of the tree can rotate at each node as shown in **Making Models 25.1**.
- The number of nodes between species in the tree is not useful for deciphering relationships, because the number of nodes depends on the number of taxa that are included in a tree, which is flexible as shown in **Making Models 25.2**.

Now that you have reviewed the basics of phylogenetic trees, let's examine how biologists use data to estimate relationships in phylogenies, focusing on trees that reveal the history of life.

How Do Biologists Estimate Phylogenies?

Humans have not been around to observe most evolutionary events directly. Thus, the relationships depicted in phylogenetic trees must be *estimated* from the best available data. That is, phylogenetic trees are hypotheses that can be tested, and they are often revised as new data become available.

Creating the Data Matrix The first step in inferring evolutionary relationships is to decide which populations, species, or higher taxa to compare and which characteristics to use. Consider the simple case of estimating a phylogenetic tree for lizards, dogs, and humans using a data set of four characters. A useful **character**, or **trait**, is any heritable genetic, morphological, physiological, developmental, or behavioral characteristic that varies among the taxa to be studied.

Most modern phylogenetic analyses (such as the fish fraud example in Table 25.1) use molecular characters, such as DNA nucleotide positions, because they are abundant and relatively easy to analyze. However, morphological characters are typically the only kind of characters available for phylogenetic analyses of fossil taxa.

Once you have selected the taxa and characters that you want to include in your study, the next step is to set up a data matrix—a table that enables you to score the character states of each taxon. The simple example in **Figure 25.1a** uses four characters to build a data matrix: skull, limbs, hair, and lactation (milk production by females with offspring). In this case, each character has two possible character states: present (1) or absent (0).

Notice that a fourth taxon, a lungfish, is included in the data matrix as an **outgroup**—a taxon that is not part of the taxonomic group being studied but is closely related. Researchers add outgroups to their analyses to help establish the direction of change of each character. An **ancestral trait** is a character that existed in an ancestor; a **derived trait** is one that is a modified form of the ancestral trait, found in a descendant. Derived traits originate via mutation, selection, and genetic drift (Chapter 23).

It's important to recognize that ancestral and derived traits are relative. If you are comparing mammals with lungfish, then hair and lactation are derived traits. But if you are comparing dogs and humans, then hair and lactation are ancestral traits. In the example in Figure 25.1, the common ancestor of lungfish, lizards, dogs, and humans is inferred to have had a skull, but no limbs, hair, or lactation based on the outgroup; any traits that differ from those in this inferred ancestor are considered derived.

Note that outgroup lineages evolve too, so they do not represent ancestors of the other taxa in the tree. For this reason, biologists often use multiple outgroups to estimate phylogenetic relationships. Additional outgroups in this example could include ray-finned fishes and sharks, which also possess a skull but lack limbs, hair, and lactation (Chapter 32)—just like lungfish.

Making Models 25.1 Tips on Drawing Phylogenetic Trees (IV)

The closeness of taxon labels cannot be used to determine relationships among taxa. To understand why, you must view and draw trees as flexible models that can rotate at each node (like mobiles hanging from a ceiling) rather than as static structures.

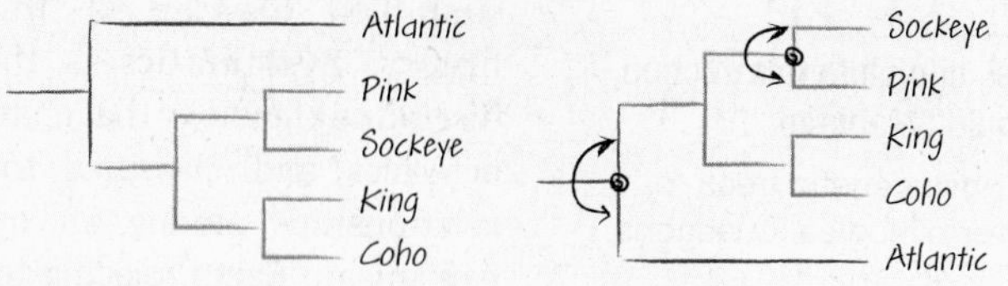

These trees have the same meaning.

MODEL Draw one more "equivalent" tree with the same meaning as the two shown here, rotating one or more of the nodes.

To see this model in action, go to the Study Area of **Mastering Biology**

Making Models 25.2 Tips on Drawing Phylogenetic Trees (V)

The number of nodes between species cannot be used to determine relationships among taxa. To understand why, you can practice adding and subtracting branches from trees, which does not change the relationships of the taxa.

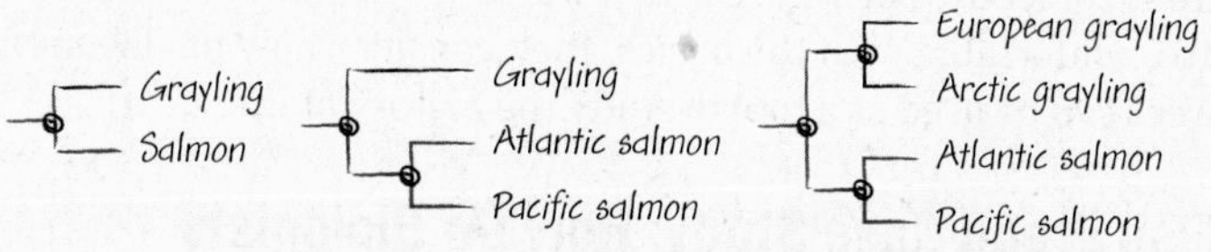

MODEL Draw a tree showing the relationships among four taxa: Arctic grayling, Atlantic salmon, and two species of Pacific salmon, king and coho. Draw a dot on the node that represents the most recent common ancestor of all the salmon. Is the Arctic grayling more closely related to Atlantic salmon or coho? Explain.

To see this model in action, go to the Study Area of **Mastering Biology**

(a) Data matrix

	Characters				
	Skull	Limbs	Hair	Lactation	
	1	0	0	0	Lungfish (outgroup)
	1	1	0	0	Lizard
	1	1	1	1	Dog
	1	1	1	1	Human

(b) Phylogenetic tree inferred from the data

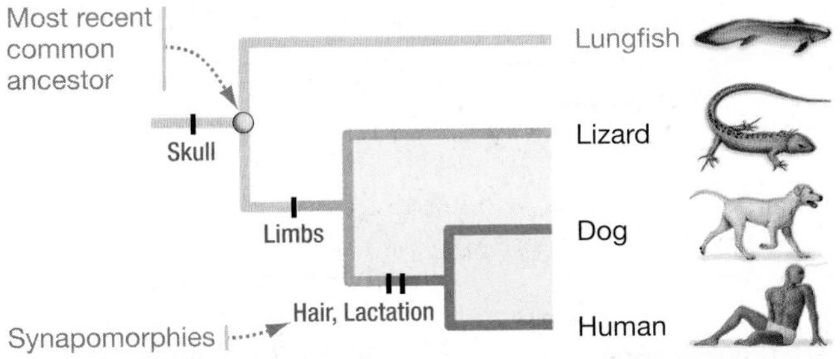

INTERACTIVE Figure 25.1 Using Data to Construct a Phylogenetic Tree. Phylogenetic analysis **(a)** begins with a data matrix containing the scored character states of selected characters in selected taxa (in this case 1 = present, 0 = absent) and **(b)** uses these data to infer evolutionary relationships among the taxa.

✔ **MODEL** Draw an equivalent tree with the same meaning as the one shown here but with different branch rotations.

SUMMARY Table 25.2 Monophyletic vs. Non-Monophyletic Groups

Graphical Symbol	Definition
Monophyletic group	
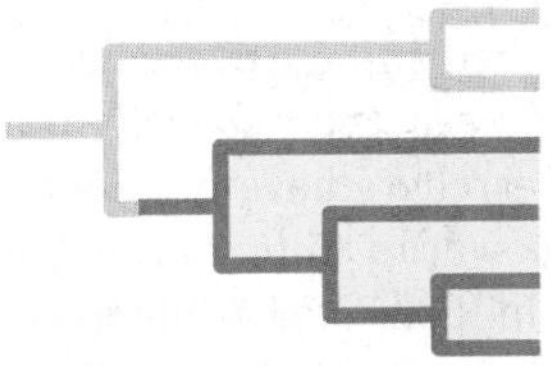	Any group that forms an evolutionary unit including an ancestral population and all of its descendants but no others (also called a *lineage*, or *clade*)
Non-monophletic groups	
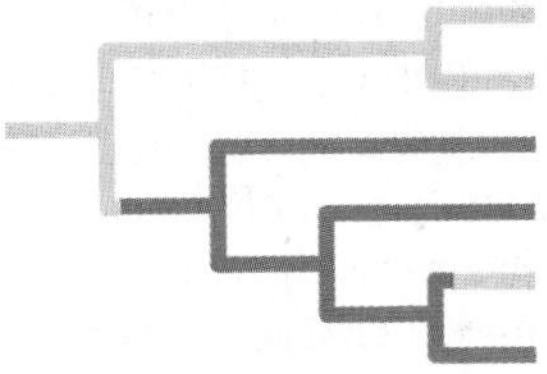	This "group" in blue is *paraphyletic*, not monophyletic, because it includes an ancestral population and some of its descendents, but not all (see dicots in Chapter 28 and fish in Chapter 32)
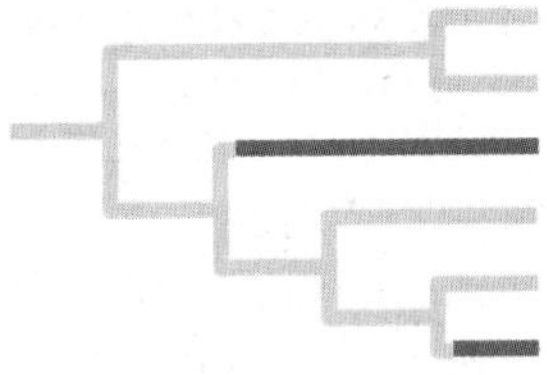	This "group" in purple is *polyphyletic*, not monophyletic, because it shares similar traits but does not include the most recent common ancestor (see segmented animals in Chapter 31)

Using the Data Matrix to Estimate a Tree In 1950, German biologist Willi Hennig was the first to come up with a clear, objective method for using traits to infer evolutionary relationships. Hennig based his method on the principle that relationships among species can be reconstructed by identifying shared derived traits, called synapomorphies (introduced in Ch. 24, Section 24.1).

A **synapomorphy** is a trait found in two or more taxa that is present in their most recent common ancestor but is missing in more distant ancestors. Synapomorphies include the origin of novel characteristics, such as the origin of limbs in tetrapod vertebrates. Synapomorphies are important because they allow biologists to recognize monophyletic groups—also called **clades** or **lineages**. A **monophyletic group** is an evolutionary unit that includes an ancestral species and all of its descendants, but no others. **Table 25.2** compares monophyletic groups and two types of non-monophyletic groups.

In the character set in Figure 25.1a, lizards, dogs, and humans are recognized as a monophyletic group (tetrapods) based on their shared, derived trait—limbs (circled in orange). Similarly, dogs and humans can be recognized as a different monophyletic group (mammals) based on their shared, derived traits—hair and lactation (circled in red). The logic here is that smaller monophyletic groups are nested within larger ones, as summarized by the tree in **Figure 25.1b**.

In this simple example, it is relatively easy to infer a tree that represents the data set in a logical way. In common practice, however, there are several complications. For example:

- Hennig's method assumes that derived character states evolve only once—if two organisms share this trait, they must have inherited it from a common ancestor. That is, Hennig's method assumes **homology** ("same-source"), or similarity due to common ancestry. However, sometimes **homoplasy** ("same-form") occurs when traits evolved independently in two or more different lineages and thus are similar for reasons other than common ancestry.
- Hennig's method assumes that reversals in a character state do not occur. However, such reversals do occur, such as the gain of a trait in one branch followed by the loss of the same trait in a descendant branch, creating the appearance that no change occurred (such as the loss of limbs in snakes).

These complications pose a problem for Hennig's method, because a species can be part of one monophyletic group according to one trait in the matrix, but part of a different monophyletic group according to a different trait in the matrix. Hennig's solution to this challenge was to reevaluate his data matrix and, if necessary, remove instances of homoplasy and reversals by hand. This solution is not feasible with large data sets and especially not for molecular data.

In general, the more data the matrix includes, the more accurate the resulting tree will be. However, the larger the number of species and traits added to the matrix, the greater the number of possible trees and character combinations. There are 15 possible trees for a phylogeny of 4 taxa but over 100,000 trees for a phylogeny of 8 taxa!

Given the challenges to Hennig's method and the overwhelming number of possible trees, how can researchers identify the tree with the best fit to the data? Researchers have developed several methods to harness computer power to "filter" probable trees out of a pool of thousands or even millions of possible trees. These methods make different assumptions. The following short descriptions will give you a sense of their approaches:

- ***Parsimony*** The first "filtering" method to emerge invoked the principle of **parsimony**, which assumes that the most likely explanation or pattern is the one that requires the fewest steps. To implement a parsimony analysis, biologists compare the branching patterns that are theoretically possible in a phylogenetic tree and count the number of character changes required to produce each pattern. The tree that requires the fewest overall character changes is hypothesized to be the one that most likely reflects what really happened during evolution. Today, parsimony analysis is still applied to morphological character sets but less often to molecular data sets.
- ***Evolutionary distance*** Unlike parsimony analysis, distance methods acknowledge that more rapid evolutionary change may occur in some branches than others. Evolutionary distance methods begin by quantifying the average frequency of character changes between individual *pairs* of taxa, then they search for the tree that best aligns with the total data set.
- ***Maximum likelihood and Bayesian analysis*** These mathematical models are more complex than the other methods because they require more information than is provided in the data matrix, such as the probability that in a DNA sequence, a G would be substituted by an A rather than a C or T. These methods are popular with molecular data sets because they can produce relatively precise estimates of evolutionary history with more statistical power than the other methods.

Note that the meaning of branch lengths varies depending on the methods used to estimate the tree, as shown in **Table 25.3**. Many trees focus on branching *patterns* to emphasize relationships—the branch lengths themselves are arbitrary. However, other analyses produce trees whose branch lengths represent genetic distance or time since divergence. Such trees will always include a scale.

Evaluating Confidence in Estimated Trees Once you have estimated a phylogenetic tree, how confident can you be that this tree represents true evolutionary history? Do you have enough information to accept or reject a hypothesis based on the tree?

Researchers use a number of strategies to determine their confidence in the result. For example, they may use more than one data set to estimate trees (such as nuclear DNA and mitochondrial DNA), and they can use more than one method to estimate the trees and compare the results. Several statistical tests are available to quantify the levels of support that a particular set of characters provide for nodes in an estimated phylogeny.

SUMMARY Table 25.3
Branch Lengths in Phylogenetic Trees

Graphical Symbol	Meaning of Branch Lengths
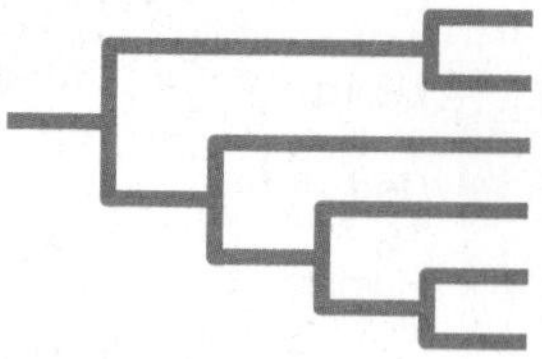	Branch lengths are arbitrary. Emphasis is on the branching pattern, which estimates evolutionary relationships among populations.
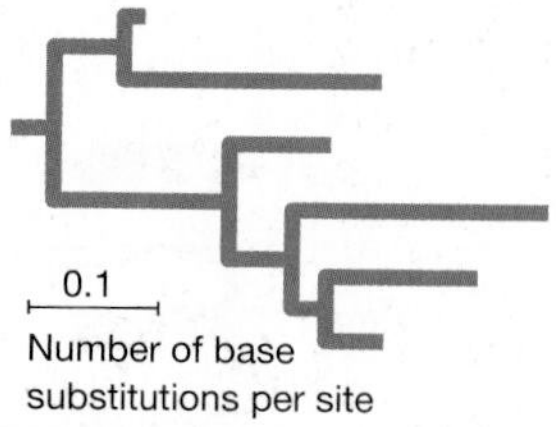	Branch lengths show the extent of genetic difference among populations. A scale bar is included.
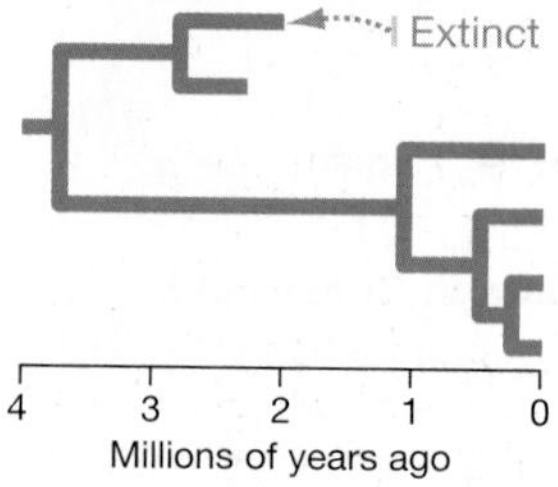	Branch lengths show the extent of evolutionary time between nodes. A scale bar is included.

How Can Biologists Use Phylogenetic Trees to Study the History of Life?

Regardless of the method that is used to estimate phylogenetic trees, the concepts of homology and homoplasy are central to using trees to study the history of life. A common cause of homoplasy is **convergent evolution**. Convergent evolution is the independent evolution of similar traits in distantly related organisms due to adaptation to similar environments and lifestyles. Because species have many traits, some may be homologous with traits in other species, while others may be convergent. Let's consider two examples to contrast homology and convergent evolution, then put it all together in a case study about whales.

Are the Flowers of Water Lilies and Wild Roses Homologous or Convergent? Flowers are extremely diverse in color, size, and shape but serve the same general function—to house the male and female gametes and to facilitate fertilization (Ch. 38, Section 38.2). Did flowers arise in the common ancestor of water lilies and roses (meaning that flowers are homologous), or did they arise independently in different lineages (meaning that flowers are convergent)? Several lines of evidence support homology:

- ***Phylogenetic evidence*** A molecular phylogeny of plants (based on DNA sequences) shows that water lilies and roses fall into a monophyletic group, called angiosperms (**Figure 25.2**). Note that all the other lineages in this group also have flowers. This result suggests that water lilies and

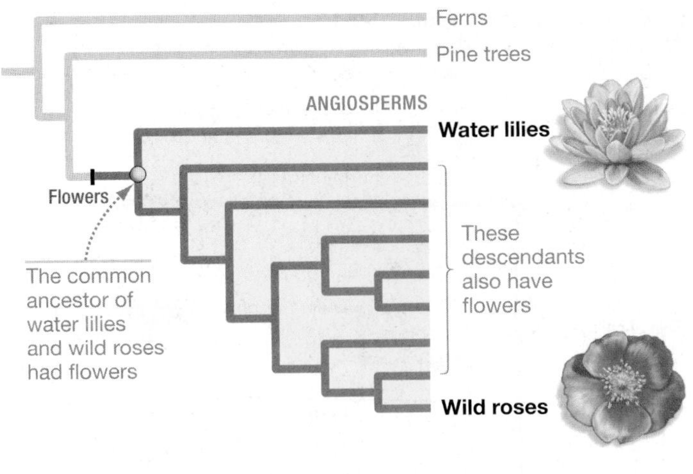

Figure 25.2 Homology: Similarities Are Inherited from a Common Ancestor. Flowers in diverse plants, such as water lilies and roses, are homologous.

wild roses have flowers because they inherited these structures from a common ancestor (indicated by the node in the tree) that also had flowers.

- ***Structural evidence*** Although flowers are diverse, most are built on the same general template: four concentric whorls of cells that form sepals, petals, stamen, and carpels.
- ***Genetic and developmental evidence*** The genetic toolkit and developmental patterns responsible for the development of floral structures are similar in different plants, including the involvement of *MADS-box* genes (Ch. 38, Section 38.6).

The agreement of the phylogenetic, structural, genetic, and developmental data provide strong evidence that flowers in diverse types of plants are homologous. The different lineages diverged over time, sometimes gaining and losing different traits, but they share a common ancestor that had flowers.

Are Streamlined Bodies in Dolphins and Ichthyosaurs Homologous or Convergent? The extinct aquatic reptiles called ichthyosaurs were strikingly similar to modern dolphins (**Figure 25.3**). Like dolphins, they were large marine predators with streamlined bodies and large dorsal fins. Are these traits homologous in dolphins and ichthyosaurs?

Figure 25.3 shows a phylogenetic tree based on skeletal characters, such as the number and placement of openings in the skull (because ichthyosaurs are known only from their fossilized skeletons). This analysis shows that dolphins occur within the mammal clade, whereas ichthyosaurs are most closely related to lizards. Notice that the sister groups to dolphins and ichthyosaurs—which are the branches on the tree most closely related to them—do not have streamlined bodies, flippers, and fins.

Based on these data, it is logical to infer that the similarities between ichthyosaurs and dolphins result from convergent evolution. The convergent traits do not occur in the common ancestor of the similar species. Streamlined bodies, flippers, and dorsal fins are adaptations that help *any* aquatic species—whether it's a reptile or a mammal—chase down fish through water.

Case Study: Where Do Whales Belong on the Tree of Life?

Let's apply what you've learned about phylogenetic trees and the concepts of homology to an interesting evolutionary case study that has puzzled scientists since Darwin: Where do whales belong on the tree of life?

Data Set 1: A Phylogeny Based on Morphological Traits Hippos, cows, deer, pigs, and camels are artiodactyls. Members of this mammal group have hooves and an even number of toes. They also share another trait: the unusual pulley shape of an ankle bone called the astragalus. Their unique foot morphology and shape of the astragalus are synapomorphies that identify the artiodactyls as a monophyletic group. Members of the perissodactyls (including horses and rhinos) do not possess a pulley-shaped astragalus and have an odd number of toes.

Whales (including dolphins, which are classified as a type of toothed whale) share a number of other morphological characteristics with artiodactyls, but they do not have an astragalus (indeed, no ankles at all). These morphological data support the tree shown in **Figure 25.4a** on page 516, where whales are

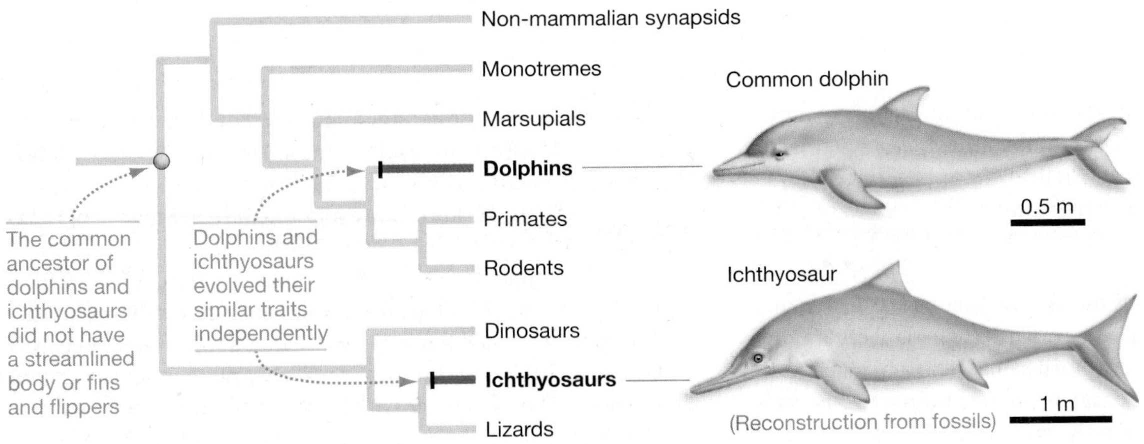

Figure 25.3 Homoplasy: Traits Are Similar but Not Inherited from a Common Ancestor. Dolphins and ichthyosaurs look similar but are not closely related.

(a) Data set 1 (morphological traits):
Whales diverged before the origin of artiodactyls.

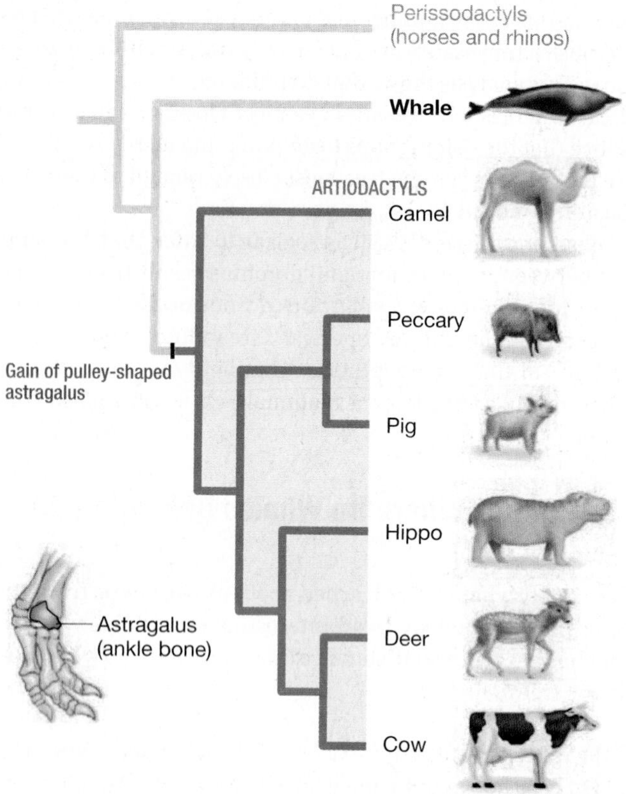

(b) Data set 2 (DNA sequences):
Whales and hippos share a recent common ancestor.

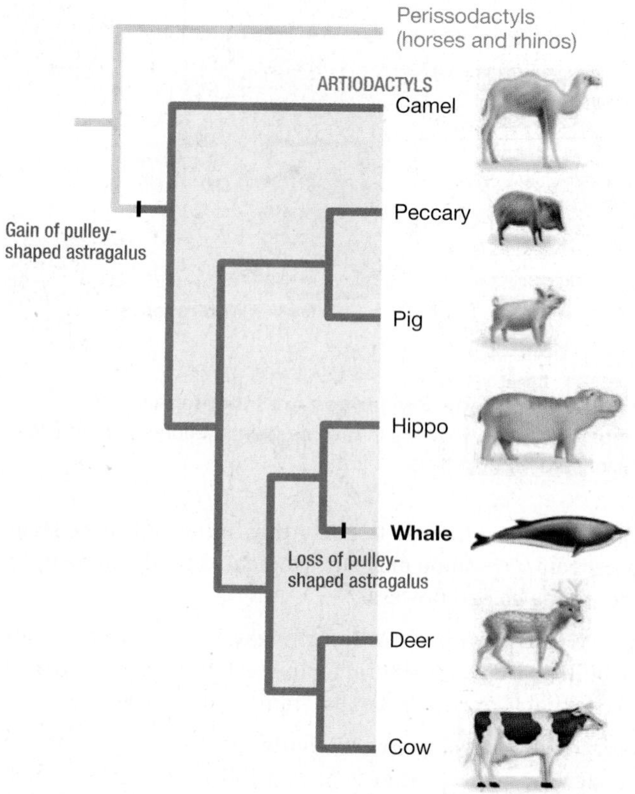

(c) Data set 3 (presence and absence of SINEs): Supports the close relationship between whales and hippos.

Locus	1	2	3	4	5	6	7	8	9	10	11	12	13	14	15	16	17	18	19	20
Cow	0	0	0	0	0	0	0	**1**	**1**	**1**	**1**	**1**	**1**	**1**	**1**	**1**	**1**	**1**	0	0
Deer	0	0	0	0	0	0	0	**1**	?	**1**	**1**	**1**	**1**	**1**	**1**	?	**1**	**1**	0	0
Whale	**1**	**1**	**1**	**1**	**1**	**1**	**1**	0	?	**1**	0	**1**	**1**	0	0	0	?	**1**	0	0
Hippo	0	?	0	**1**	**1**	**1**	**1**	0	**1**	**1**	0	**1**	**1**	0	0	0	?	**1**	0	0
Pig	0	0	0	?	0	0	0	0	?	0	0	0	?	?	0	0	?	**1**	**1**	**1**
Peccary	?	?	?	?	?	?	?	?	?	?	?	?	?	?	?	?	?	?	**1**	**1**
Camel	0	0	0	0	0	0	0	0	0	0	0	0	0	0	0	0	?	0	0	0

1 = SINE present
0 = SINE absent
? = still undetermined

Whales and hippos share four *unique* SINEs (4, 5, 6, and 7)

Figure 25.4 Where Do Whales Belong on the Tree of Life?

DATA for (c): M. Nikaido, A. P. Rooney, and N. Okada. 1999. *Proceedings of the National Academy of Sciences USA* 96: 10261–10266.

✔ Map the origin of the following groups of SINEs on the tree in part (b): Group (4, 5, 6, 7) identifies hippos and whales as part of a monophyletic group; (8, 11, 14, 15, 17) identifies deer and cows as part of a monophyletic group; (10, 12, 13) identifies hippos, whales, deer, and cows as part of a monophyletic group. What two species does SINEs group (19, 20) identify as part of a monophyletic group?

an outgroup to the artiodactyls. It is most parsimonious to map the gain of the pulley-shaped astragalus in the ancestral population, marked by a black bar in Figure 25.4a, because all descendants of that ancestor have the trait, but members of the outgroup do not.

Data Set 2: A Phylogeny Based on DNA Sequence Data When researchers began comparing DNA sequences of artiodactyls and whales, however, the data were surprising. The phylogenetic trees showed that whales were a sister group to hippos *within* the artiodactyl clade, as illustrated in **Figure 25.4b**.

Here, it is still most parsimonious to map the evolution of the pulley-shaped astragalus at the same ancestor as in Figure 25.4a. But this placement assumes that the trait was lost in an ancestor of today's whales. To show this event, the tree in Figure 25.4b includes a second black bar mapping the trait loss.

The tree supported by the DNA data conflicts with the tree supported by morphological data because it implies that the pulley-shaped astragalus evolved in artiodactyls and then was lost during whale evolution. These two character changes (a gain *and* a loss of the astragalus) are less parsimonious than just one change (a gain only).

Data Set 3: Transposable Elements The conflict between the two phylogenies was resolved when researchers analyzed the distribution of DNA sequences called SINEs (short interspersed nuclear elements), which occasionally insert themselves into the genomes of mammals. SINEs provide convincing evidence of relationships because the chance of the same SINE inserting itself in exactly the same place in the genomes of two species is astronomically small.

As the data in **Figure 25.4c** show, whales and hippos share several types of SINEs (4, 5, 6, and 7) that are not found in other groups. Other SINEs are present in some artiodactyls but not in others. No SINEs were discovered in camels.

To explain these data, biologists hypothesize that none of the SINEs were present in the population that is ancestral to all of the species in the study. Then, after the branching event that led to the split between the camels and all the other artiodactyls, different SINEs inserted themselves into the genomes of descendant populations. If this hypothesis is correct, then the presence of a particular SINE represents a derived character.

Conclusion: Whales Are Closely Related to Hippos A phylogenetic tree estimated based on the SINE data had the same branching pattern as the one shown in Figure 25.4b. (To convince yourself, try the exercise in the figure caption.) Thus, most biologists accept this phylogeny as the most accurate estimate of whale evolutionary history. According to this phylogeny, whales are artiodactyls and share a relatively recent common ancestor with hippos. This common ancestor may have been a population of land-dwelling artiodactyls that spent most of their time feeding in shallow water, much as hippos do today.

In 2001, the discovery of fossil artiodactyls supported this hypothesis in spectacular fashion. These fossil species were clearly related to whales—they have an unusual ear bone found only in whales—and yet had a pulley-shaped astragalus as part of still-functional hindlimbs.

Scaling Up to the Tree of Life The whale case study demonstrates how systematists can sort out relationships within a group like artiodactyls. Other specialists use the same process for their taxonomic area of expertise, whether studying humans, water lilies, or butterflies. The beauty of phylogenetic trees is that monophyletic groups "nest" within one other. For example, the 90 or so living species of whales nest within the artiodactyls, which nest within the mammals, and so on. This nesting pattern enables biologists to link numerous independently estimated phylogenetic trees to form the tree of life (see the Big Picture of Diversity on pages 722–723).

CHECK YOUR UNDERSTANDING

✔ **If you understood this section, you should be able to ...**

1. **MODEL** Draw what the tree in Table 25.1 would look like if the fish samples had been labeled accurately by the restaurants—that is, if Sample 1 was actually sockeye salmon and Sample 2 was actually king salmon.
2. **THINK CAREFULLY** Develop an argument to agree or disagree that whales evolved from hippos.
3. Hypothesize whether the following traits represent homology or homoplasy: hair in humans and whales; extensive hair loss in humans and whales; limbs in humans and whales; social behavior in humans and certain whales (e.g., dolphins). Explain your reasoning.

Answers are available in Appendix A.

25.2 Tools for Studying Life's History: The Fossil Record

The fossil record is a powerful tool for studying evolutionary history because it provides *direct* evidence about what organisms that lived in the past looked like, where they lived, and when they existed. Fossil evidence can also be used in phylogenetic analyses to study relationships (such as in the ichthyosaur example in Figure 25.3) or to help validate estimated times since divergence (see the third example in Table 25.3).

After you complete this section, you should be able to ...

- Discuss the major strengths and limitations of the fossil record.
- Describe the three major eras of the Phanerozoic eon, including the major evolutionary events and dominant taxa.

Paleontologists are scientists who study **fossils**, pieces of physical evidence from an organism that lived in the past. The **fossil record** is the total collection of fossils that paleontologists and amateur collectors have found throughout the world and archived in thousands of private and public collections.

Let's review the kinds of fossil evidence that are available, analyze the strengths and weaknesses of the fossil record, and then summarize major events that have taken place in life's approximately 3.5-billion-year history.

Biologists Study Many Types of Fossils

Different types of fossils form depending on the geological processes involved. Most of the processes that form fossils begin when part or all of an organism is buried in ash, sand, mud, or some other type of sediment. Once burial occurs, several things can happen depending on the rate and timing of decomposition, as summarized in **Table 25.4** on page 518. For example, the trilobite in the chapter-opening photo decomposed completely after it was buried. Dissolved minerals filled up the cavity that remained, leaving a lifelike cast fossil of an animal that went extinct 250 million years ago.

Buried fossils can be exposed at the surface by erosion, a road cut, quarrying, or other processes. If researchers find a fossil, they can prepare it for study by painstakingly clearing away the surrounding rock. If the species represented is new, researchers describe its morphology in a scientific publication, name the species, estimate the fossil's age based on dates assigned to nearby rock layers, and add the specimen to a collection so that it is available for study by other researchers. It is now part of the fossil record.

✔ **If you understand the types of fossils, you should be able to predict what type of fossil is most likely to yield DNA evidence.**

Table 25.4 Types of Fossils

Type of Fossil	Example
Intact fossil Forms when decomposition does not occur and the organic remains are preserved intact.	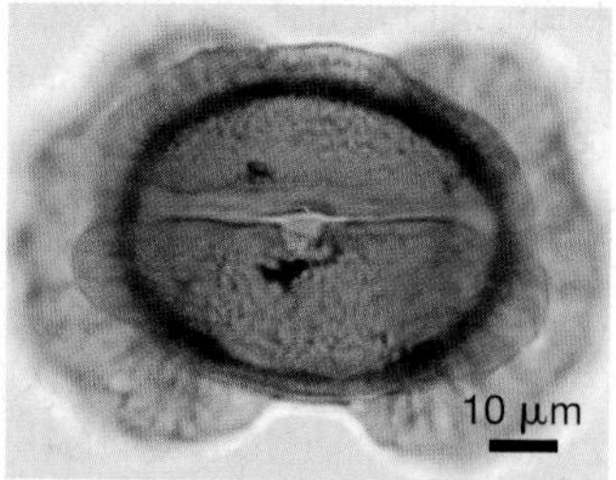 Pollen
Compression fossil Forms when sediments accumulate on top of the organism and become cemented into rocks such as mudstone or shale. The sediments' weight compresses the organic material below into a thin, carbonaceous film.	 Leaf
Cast fossil Forms when a buried organism decomposes, leaving an empty cavity in the sediments that fills with dissolved minerals and hardens into an accurate cast of the remains.	 Ammonite
Permineralized fossil Forms when organisms decompose extremely slowly. Dissolved minerals gradually infiltrate the interior of cells and harden into stone.	 Petrified wood
Trace fossil Forms when sedimentation and mineralization preserve indirect evidence of an organism in the environment, including footprints, tracks, burrows, feeding marks, and feces.	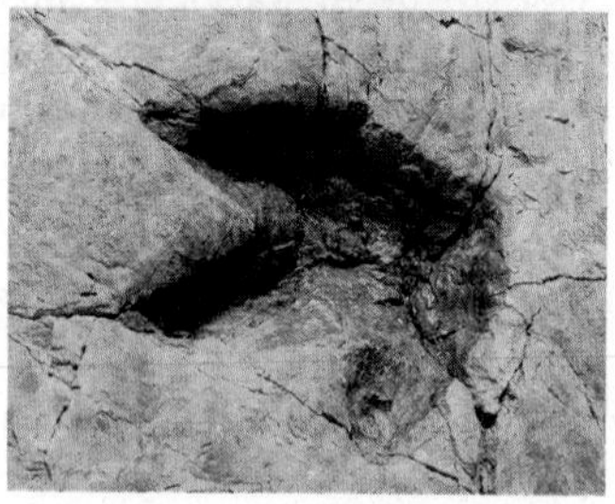 Dinosaur footprint

Fossils form only under particular conditions: Organisms must perish in an environment where they are buried rapidly and then decompose slowly. In most habitats the opposite situation occurs—burial is slow and decomposition is rapid. In reality, then, fossilization is an extremely rare event.

The chance of discovery of individual fossils is also rare. Consider that there are currently 12 specimens of *Archaeopteryx*, the first birdlike dinosaur to appear in the fossil record. If you accept an estimate that crow-sized birds native to wetland habitats would have a population size of around 10,000 and a life span of 10 years, and if you accept the current estimate that the species existed for about 2 million years, then you can calculate that about 2 billion *Archaeopteryx* lived. But as far as researchers currently know, only 1 out of every 200,000,000 individuals were fossilized and discovered. For this species, the odds of becoming part of the fossil record were almost 40 times worse than your odds of winning the state lottery.

What Are the Opportunities and Limitations of the Fossil Record?

Before looking at how researchers use the fossil record to answer questions about the history of life, it is essential to review the opportunities and limitations of this archive.

Habitat Bias Because burial in sediments is so crucial to fossilization, there is a strong habitat bias in the database. Organisms that live in areas where sediments are actively being deposited—including beaches, mudflats, and swamps—are much more likely to form fossils than are organisms that live in other habitats. Within these habitats, burrowing organisms such as clams are already underground at death and are therefore much more likely to fossilize. Organisms that live aboveground in dry forests, grasslands, and deserts are much less likely to fossilize.

Taxonomic and Tissue Bias Slow decay is almost always essential to fossilization, so organisms with hard parts such as bones or shells are most likely to leave fossil evidence. This requirement introduces a strong taxonomic bias into the record. Clams, snails, and other organisms with hard parts have a much higher tendency to be preserved than do soft-bodied organisms such as worms.

A similar bias exists for tissues within organisms. For instance, pollen grains are encased in a tough outer coat that resists decay, so they fossilize much more readily than do flowers. Shark teeth are abundant in the fossil record, but shark skeletal elements, which are made of cartilage, are almost nonexistent.

Temporal Bias Recent fossils are much more common than ancient fossils. This causes a temporal bias in the fossil record. The reasons are straightforward—the older a fossil is, the longer it has been exposed to potentially destructive forces in deep layers of rock and the less likely it is to be exposed to the surface by erosion (such as in the Grand Canyon) or by uplift.

Abundance Bias The fossil record has an abundance bias; it is weighted toward common species. Organisms that are abundant, widespread, and/or present on Earth for long periods of time leave evidence much more often than do species that are rare, local, or ephemeral.

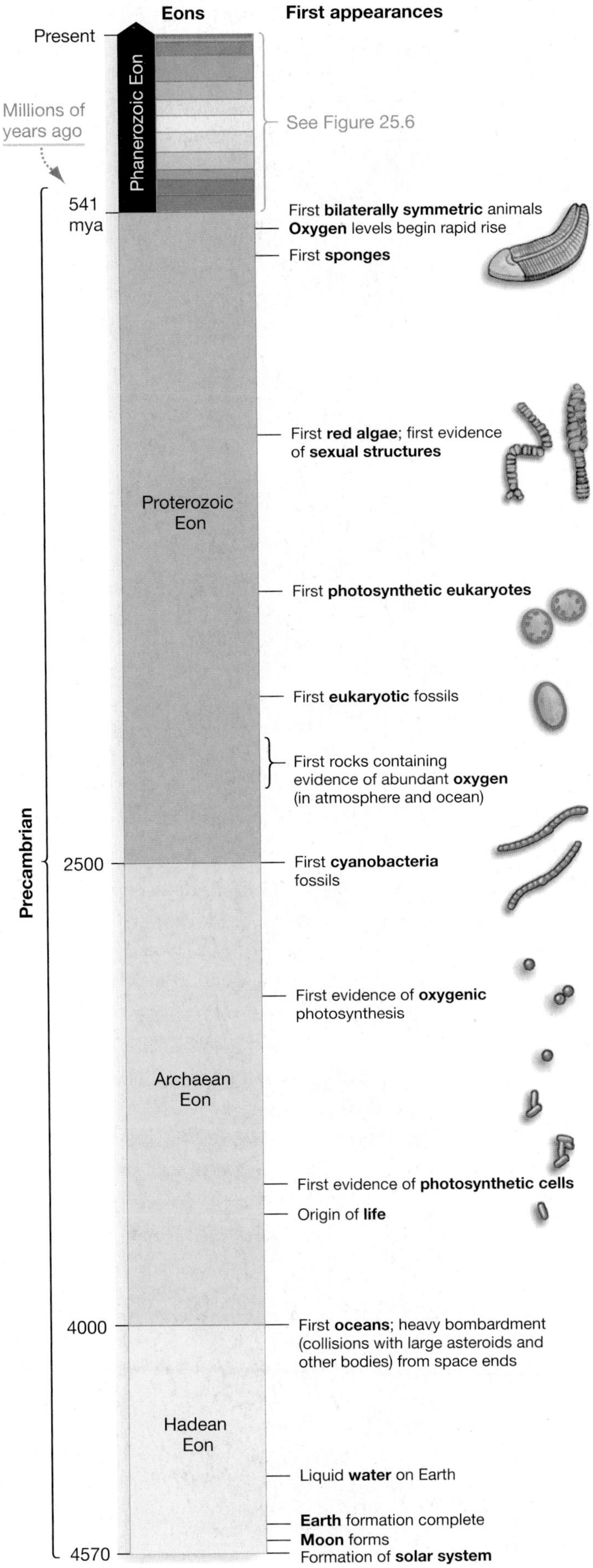

To summarize, the fossil database represents a highly nonrandom sample of the past. Paleontologists recognize that they are limited to asking questions about tiny and scattered segments on the tree of life. And yet, as this chapter shows, the fossil record is a scientific treasure trove. Analyzing fossils is the only way scientists can directly examine the physical appearance of extinct forms and infer how they lived.

✔ PROCESS OF SCIENCE If you understand the opportunities and limitations of the fossil record, you should be able to determine whether the discovery of feathers on *Tyrannosaurus* dinosaur species is consistent with the hypothesis that feathers are a synapomorphy (review Section 25.1) for birds. Explain.

How Are Fossils Used to Estimate Life's Time Line?

It is very difficult to date the origin of life precisely. Our best estimate is that the Earth started to form about 4.6 billion years ago, and that life began before 3.5 billion years ago—probably in warm, deep-sea hydrothermal vents (Ch. 2, Section 2.4). To organize the tremendous sweep of time between then and now, researchers divide Earth history into segments called eons, eras, periods, and epochs.

Originally, geologists used distinctive rock formations or fossilized organisms to identify the boundaries between named time intervals in the geologic time scale, which is generally organized with oldest rock layers buried under newer layers (see Ch. 22, Figure 22.2). Later, researchers were able to use radiometric dating to assign absolute dates—expressed as years before the present—to events in the fossil record. Radiometric dating is based on the well-studied decay rates of certain radioactive isotopes (Ch. 22, Section 22.1). By dating rocks that contain fossils, researchers can also assign an absolute age to many of the species in the fossil record.

To summarize the history of life, researchers create time lines that record key "evolutionary firsts"—the appearance of important new lineages or innovations—organized from oldest at bottom to most recent at top. It's important to recognize, though, that the times assigned to these first appearances underestimate when lineages appeared and events occurred. The reason is simple: A particular species or lineage can exist for millions of years before leaving fossil evidence. The fossil record and efforts to date fossils are constantly improving, so time lines are always a work in progress.

Precambrian The interval between the formation of Earth, about 4.6 billion years ago, and the appearance of most animal groups about 541 million years ago (abbreviated mya) is called the **Precambrian** (Figure 25.5); it is divided into the Hadean, Archaean, and Proterozoic eons.

Figure 25.5 **Major Events of the Precambrian.** Life, photosynthesis, and the oxygen atmosphere all originated in the Precambrian.

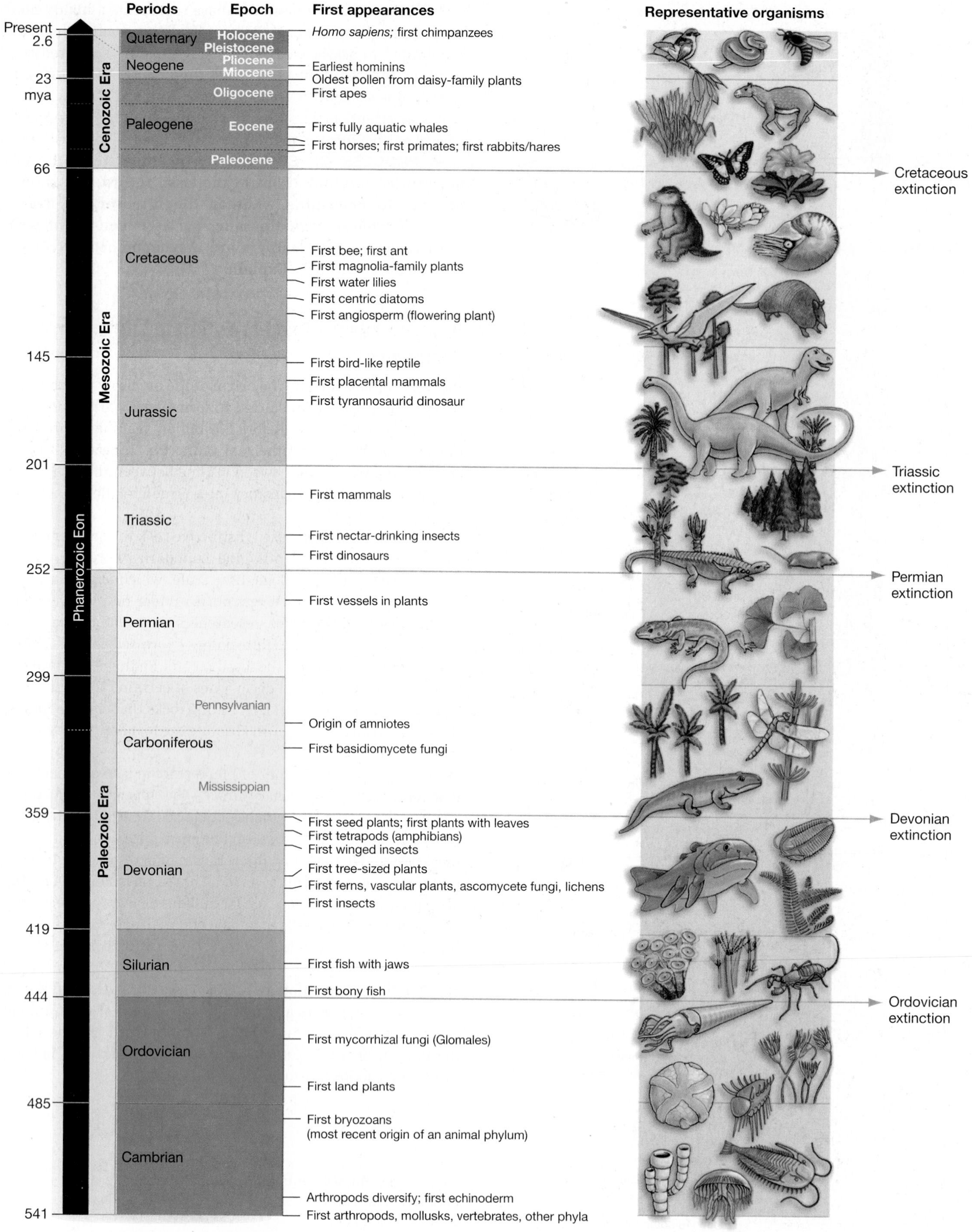

Periods
Epoch
First appearances
Representative organisms
Present
2.6
23
mya
66
145
201
252
299
359
419
444
485
541
Phanerozoic Eon
Cenozoic Era
Mesozoic Era
Paleozoic Era
Quaternary
Holocene
Pleistocene
Neogene
Pliocene
Miocene
Oligocene
Paleogene
Eocene
Paleocene
Cretaceous
Jurassic
Triassic
Permian
Pennsylvanian
Carboniferous
Mississippian
Devonian
Silurian
Ordovician
Cambrian
Homo sapiens; first chimpanzees
Earliest hominins
Oldest pollen from daisy-family plants
First apes
First fully aquatic whales
First horses; first primates; first rabbits/hares
First bee; first ant
First magnolia-family plants
First water lilies
First centric diatoms
First angiosperm (flowering plant)
First bird-like reptile
First placental mammals
First tyrannosaurid dinosaur
First mammals
First nectar-drinking insects
First dinosaurs
First vessels in plants
Origin of amniotes
First basidiomycete fungi
First seed plants; first plants with leaves
First tetrapods (amphibians)
First winged insects
First tree-sized plants
First ferns, vascular plants, ascomycete fungi, lichens
First insects
First fish with jaws
First bony fish
First mycorrhizal fungi (Glomales)
First land plants
First bryozoans
(most recent origin of an animal phylum)
Arthropods diversify; first echinoderm
First arthropods, mollusks, vertebrates, other phyla
Cretaceous extinction
Triassic extinction
Permian extinction
Devonian extinction
Ordovician extinction

Here are the important things to know about the Precambrian:

- Life was exclusively unicellular for most of Earth's history.
- Oxygen was virtually absent from the oceans and atmosphere for almost 2 billion years after the origin of life. Photosynthetic bacteria called cyanobacteria were responsible for the creation of the oxygen-rich atmosphere (Ch. 26, Section 26.3).

Phanerozoic Eon The interval between 541 mya and the present is called the Phanerozoic eon. This eon is divided into three eras (Figure 25.6), each of them further divided into intervals called periods, which are divided into epochs.

1. The **Paleozoic** ("ancient life") **era** begins with the appearance of most major animal lineages and ends with the obliteration of almost all multicellular life-forms at the end of the Permian period. The Paleozoic includes initial diversification of the animals, land plants, and fungi, as well as the appearance of land animals.
2. The **Mesozoic** ("middle life") **era** begins with the end-Permian extinction events and ends with the extinction of the dinosaurs (except birds) and other groups at the boundary between the Cretaceous period and Paleogene period. In terrestrial environments of the Mesozoic, gymnosperms were the dominant plants and dinosaurs were the dominant vertebrates.
3. The **Cenozoic** ("recent life") **era** is divided into the Paleogene, Neogene, and Quaternary periods. On land, angiosperms were the dominant plants, and birds and mammals were the dominant vertebrates.

At the top of Figure 25.6, you can see that the periods of the Cenozoic are divided into epochs and that we are presently in the Holocene epoch. However, many scientists have called for the recognition of a new epoch—the **Anthropocene** ("human epoch")—to reflect the dramatic physical, chemical, and biological changes that humans are causing on Earth, especially since the industrial revolution. Intense discussion about this proposal continues, including debates about the evidence that would define the boundary between the Holocene and the Anthropocene.

Taken together, the fundamental message of the time-line data is one of constant change of life on Earth. The water chemistry, atmospheric composition, climate, and continental positions also changed dramatically during this time (see Ch. 49, Figure 49.3). The changes are well documented, but the expanse of time is still difficult to comprehend. If Earth's history were scaled to fit in one calendar year, our species, *Homo sapiens*, would not appear until one hour before the stroke of midnight on December 31st.

CHECK YOUR UNDERSTANDING

✔ If you understood this section, you should be able to ...

1. Use Table 25.4 to decide what type of fossil is represented by drill holes in a Jurassic clam shell. (Drill holes are evidence of ancient predation.)
2. Use the time line in Figure 25.6 to estimate how long mollusks, such as clams and snails, have lived on Earth.
3. **PROCESS OF SCIENCE** Propose two reasons that researchers might prefer to use drill holes in clamshells as a measure of historic predation rather than another measure.

Answers are available in Appendix A.

25.3 Large-Scale Pattern in Life's History: Adaptive Radiation

After you complete this section, you should be able to ...

- Use examples to explain how adaptive radiations occur.

When biologists study the history of life using phylogenetic trees and the fossil record, two of the most compelling types of evolutionary events they discover are **(1)** periods when species originate and diversify rapidly, and **(2)** periods when species go extinct rapidly. Let's focus on dramatic events that generate biological diversity first; the chapter's concluding section analyzes how that diversity gets wiped out.

When a single lineage rapidly produces *many* descendant species with a wide range of adaptive forms, biologists say that an **adaptive radiation** has occurred. An adaptive radiation may be inferred by phylogenetic analysis, or it may be observed as the sudden appearance of related, diverse species in the fossil record.

Figure 25.7 on page 522 provides an example: the Hawaiian silverswords. The 30 species in this lineage evolved from a species of tarweed, native to present-day California, that colonized the Hawaiian Islands about 5 million years ago. This is an example of allopatric speciation by dispersal (Ch. 24, Section 24.2). Compared to speciation rates in other groups, this rate is extremely rapid.

The Hawaiian silverswords fulfill the three hallmarks of an adaptive radiation:

1. They are a monophyletic group.
2. They speciated rapidly.
3. They diversified ecologically into many niches.

Figure 25.6 **Major Events of the Phanerozoic Eon (page 520)** The Phanerozoic began with the initial diversification of animals, continued with the evolution and early diversification of land plants and fungi, and includes the subsequent movement of animals to land. A total of five mass extinctions occurred during the eon.

DATA: International Commission on Stratigraphy, 2017.

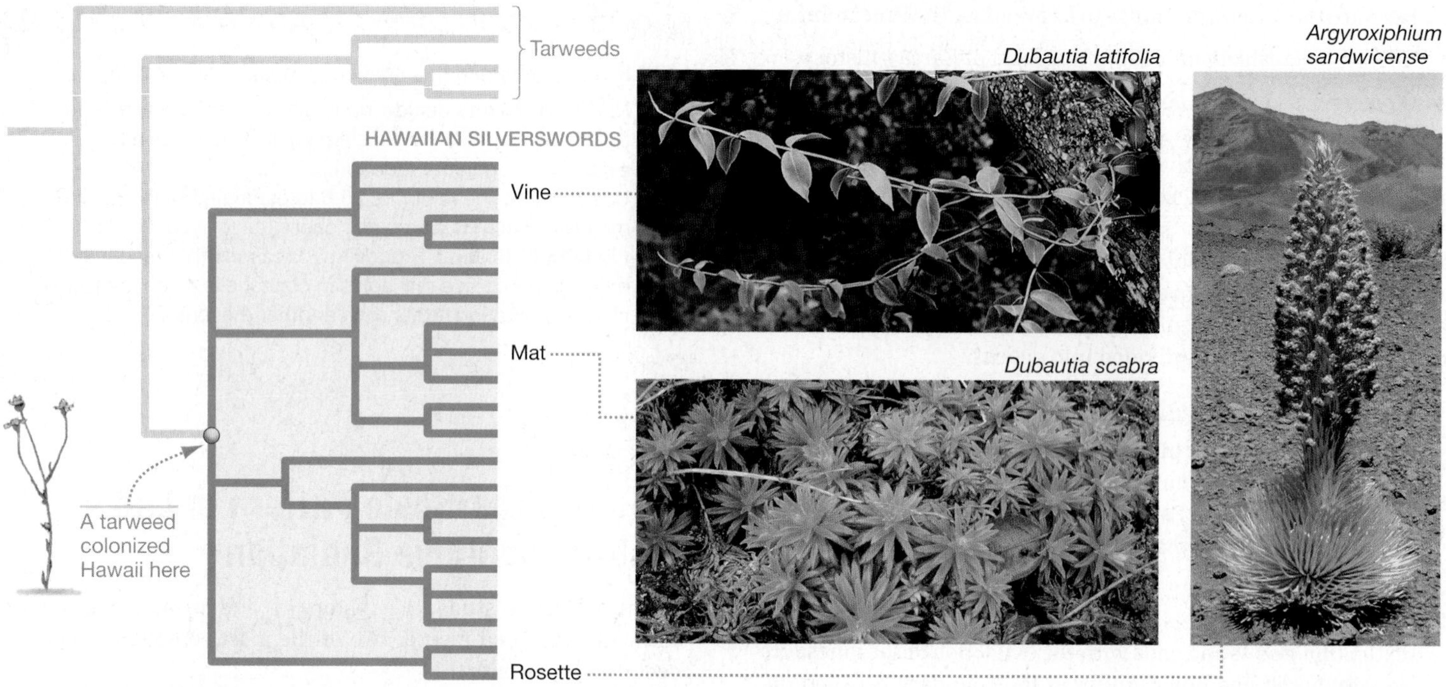

Figure 25.7 **An Adaptive Radiation.** This tree shows a subset of the Hawaiian silverswords, illustrating the extent of morphological divergence into a wide array of niches.

Biologists use the term "niche" to describe the range of conditions that a species can tolerate and the range of resources that it can use (Ch. 24, Section 24.3). Today's silverswords vary from mosslike mats to vines and rosettes. They live in habitats ranging from lush forests to barren lava flows.

Why Do Adaptive Radiations Occur?

Adaptive radiations are a major pattern in the history of life. But why do some lineages diversify rapidly, while others do not? Two general mechanisms can trigger adaptive radiations: factors extrinsic (external) to organisms, such as favorable new conditions in the environment, and factors intrinsic to (within) organisms, such as the evolution of key morphological, physiological, or behavioral traits. Let's consider each in turn.

Ecological Opportunity Evolution occurs in an ecological context. Ecological opportunity—marked by the availability of more resources or new types of resources—has driven a wide array of adaptive radiations. For example, biologists hypothesize that few other flowering plant species were present on the Hawaiian Islands 5 million years ago, so silverswords were able to diversify into many vacant niches. With few competitors, the descendants of the original colonizing tarweed began to grow in a wide range of habitats. Over time—by the processes of mutation, genetic drift, and natural selection (Chapter 23)—some became specialized for growth in different niches. ✔ **If you understand this concept, you should be able to generate a hypothesis to explain why tarweeds in California are not particularly species rich or ecologically diverse.**

Morphological, Physiological, or Behavioral Innovation The evolution of a key trait—such as one that allowed descendants to live in new areas, exploit new sources of food, or select mates in a new way—may have triggered many of the important diversification events in the history of life. For example:

- *Flowers* are a unique reproductive structure that helped trigger the diversification of angiosperms (Ch. 28, Section 28.3). Today angiosperms are the most species-rich lineage of land plants: Over 250,000 species are known.
- *Feathers and wings* gave some dinosaurs the ability to fly. Feathers originally functioned in mating display or as insulation; later they were used in gliding and in powered flight. Today the lineage called birds contains about 10,000 species, which live in virtually every habitat on the planet.

The genetic and developmental mechanisms underlying the origins of these traits are the subject of a thriving new area of research. Note that ecological opportunity and trait innovation can occur together, as they did during the Cambrian explosion.

The Cambrian Explosion

Almost all life-forms were unicellular for almost 3 *billion* years after the origin of life (see Figure 25.5). The exceptions were several lineages of small multicellular algae, which show up in the fossil record about 1 billion years ago. The first animals—early sponges—appear about 700 million years ago.

Then life-forms became much larger and more complex in just 50 million years—1/60th of the total time life had existed. These life-forms included creatures with shells, exoskeletons, internal skeletons, legs, heads, tails, eyes, antennae, jaw-like mandibles, segmented bodies, muscles, and brains. It was arguably the most spectacular period of evolutionary change in the history of life. Because much of it occurred during the Cambrian period, this adaptive radiation of animals is called the **Cambrian explosion**.

Early Animal Fossils The Cambrian explosion is documented by fossil assemblages—primarily in Canada, China, and Australia—that record the state of animal life near the beginning of the Cambrian. Each fossil assemblage records a distinctive **fauna**—or collection of animal species:

- ***Microfossils*** The first fauna to appear were tiny microfossils of sponges and corals (less than 1 mm across) from the Ediacaran period before the Cambrian. These creatures probably made their living by filtering organic debris from the water.
- ***First macroscopic fossils*** The first macroscopic fauna appeared during the Ediacaran and included sponges, jellyfish, and comb jellies as well as fossilized burrows, tracks, and other traces from unidentified animals. None of these organisms have shells, limbs, heads, or feeding appendages (**Figure 25.8**, bottom). They likely filtered or absorbed organic material as they burrowed in sediments, floated in the water, or were immobile on the seafloor.
- ***Cambrian macroscopic fossils*** The most sensational additions ever made to the fossil record were the macroscopic Cambrian fauna. Sponges, jellyfish, and comb jellies are abundant in these rocks, but so are several wormlike creatures and arthropods (see Figure 25.8, top), in addition to mollusks, echinoderms, and all the other major animal phyla (Chapter 30). This list includes our own phylum—the chordates—represented by early vertebrate fossils with fishlike bodies and a skull made of cartilage. Species in the Cambrian fauna slithered, swam, burrowed, crawled, and grasped, many using skeletons and appendages (**Figure 25.9**); there were filter feeders, predators, scavengers, and grazers, many with specialized mouthparts. The diversification filled many of the ecological niches found in marine habitats today.

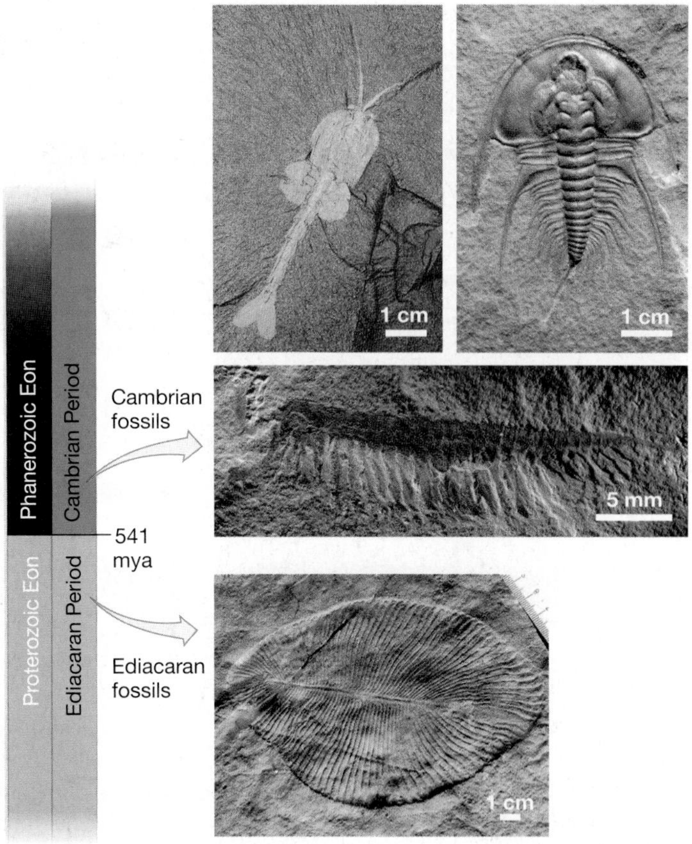

Figure 25.8 **Fossils Document the Cambrian Explosion.** Early macroscopic animals were soft-bodied, but they diversified to include hard-bodied animals during the Cambrian explosion.

What Triggered the Cambrian Explosion? The fossils document what happened during the Cambrian explosion. Now the question is: *How* did all this speciation, morphological change, and ecological diversification come about?

Recent molecular phylogenetic analyses suggest that the most recent common ancestor to all living animals may have arisen about 800 million years ago—long before diversification. Further, key developmental toolkit genes probably evolved before the diversification. These data suggest that the explosion had a "long fuse." So what triggered the onset of the adaptive radiation?

To answer this question, biologists point to an array of hypotheses. It's important to recognize that most or all of these hypotheses could be correct—they are not mutually exclusive.

- ***Higher oxygen levels*** By analyzing the composition of rocks formed during the Proterozoic, geologists have established that oxygen levels gradually rose in the atmosphere and ocean. Increased oxygen levels make aerobic respiration more efficient (Chapter 9); increased aerobic respiration supported larger bodies and more active movements. Some biologists suggest that oxygen levels reached a critical threshold at the start of the Cambrian explosion, making the evolution of big, mobile animals possible.

Figure 25.9 **Life in the Cambrian.** This artist's interpretation of animal diversity during the Cambrian is based on fossils found in the Chengjiang formation of China.

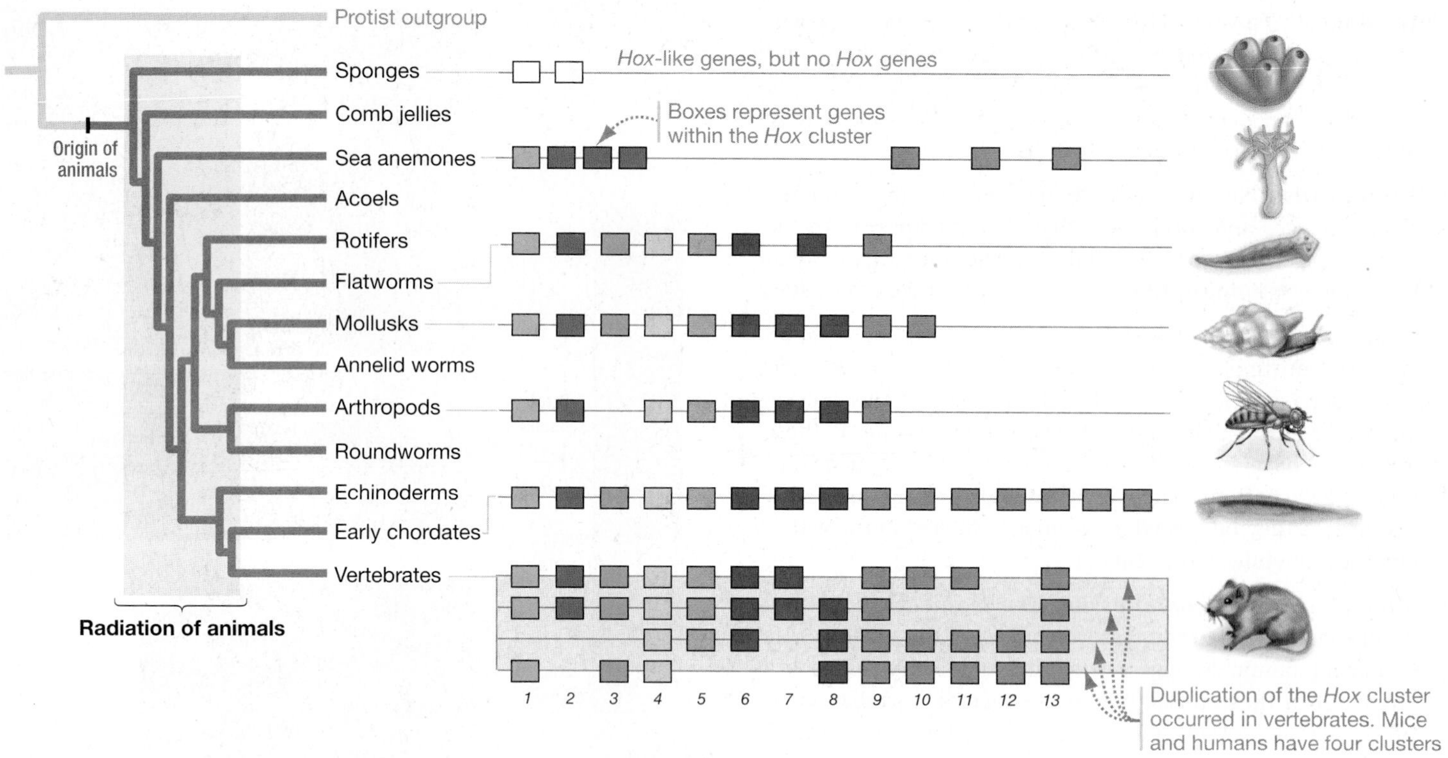

Figure 25.10 ***Hox* Genes and Animal Diversification.** The duplication and diversification of *Hox* genes may have enabled larger, more complex bodies to evolve. Each color represents a set of homologous genes with a similar gene expression pattern (see Ch. 21, Figure 21.12).

- ***Rise of algae*** Cyanobacteria were the primary photosynthetic organism during the Precambrian, but around 650 million years ago, an increase in the availability of phosphorous enabled the rise of algae. These larger, more nutrient-rich photosynthetic organisms not only produced oxygen but also served as a higher-quality food source for early animals—making more energy and mass available for growth, reproduction, and diversification.
- ***The evolution of predation*** Before the Cambrian explosion, animals made their living by eating organic material that settled on the seafloor or by filtering cells and debris from the water. But Cambrian fossils include shelled animals with holes in the shells—evidence that a predator drilled through and ate the animal inside. Once predation evolved, natural selection for shells, hard exoskeletons, rapid movement, and other defensive adaptations drove morphological divergence among prey animals.
- ***New niches beget more new niches*** Once animals could move off the seafloor, they could exploit other resources. The presence of animals at an array of depths created ecological opportunity for the evolution of species that could live on them and/or eat them. In this way, the ability to exploit new niches created more new niches, driving speciation and ecological diversification.
- ***New genes, new bodies*** *Hox* genes play a key role in organizing the development of the body in diverse animals by signaling where cells are in the embryo (Ch. 21, Section 21.3). **Figure 25.10** summarizes the *Hox* genes found in some major animal groups. The key observation is that the earliest animals in the fossil record had few or no *Hox* genes; most groups that appeared later have more. The hypothesis here is that gene duplication and diversification increased the number of *Hox* genes in animals and made it possible for larger, more complex bodies to evolve.

How might some of these hypotheses be interrelated? If increased levels of oxygen and algae in the ocean made larger bodies and more rapid movement possible, then animals could move into new habitats off the seafloor and become large enough to eat smaller animals. Selection would favor individuals in these new niches with mutations in *Hox* genes and other DNA sequences that make the development of a large, complex body possible.

About 100 million years after the Cambrian explosion, a similar adaptive radiation occurred after the first land plants—the descendants of green algae—adapted to life on land. In the span of about 56 million years, an array of growth forms and most major lineages of plants appear in the fossil record. This "Devonian explosion" resulted from the morphological innovations that allowed green plants to thrive on land, coupled with the ecological opportunity provided by an environment awash in sunlight (Ch. 28, Section 28.3).

Adaptive radiations occur when species originate and diversify rapidly. Now let's consider how dramatic events can *wipe out* diversity.

> **CHECK YOUR UNDERSTANDING**
>
> ✔ **If you understood this section, you should be able to ...**
>
> Explain how **(1)** ecological opportunity and **(2)** morphological innovation might have contributed to the Cambrian explosion.
>
> Answers are available in Appendix A.

25.4 Large-Scale Pattern in Life's History: Mass Extinction

> **After you complete this section, you should be able to ...**
>
> - Compare and contrast the causes of mass extinction and background extinction.

A **mass extinction** refers to the rapid extinction of a large number of diverse species around the world. More specifically, a mass extinction occurs when at least 60 percent of the species present are wiped out within 1 million years.

Mass extinction events are evolutionary hurricanes. They buffet the tree of life, snapping twigs and breaking branches. They are catastrophic episodes that wipe out huge numbers of species and lineages in a short time, giving the tree of life a drastic pruning. They are the opposite of adaptive radiation.

Before analyzing two of the best-studied mass extinctions in the fossil record, let's step back and ask how they differ from normal extinction events.

How Do Mass Extinctions Differ from Background Extinctions?

Background extinction refers to the lower, average rate of extinction observed when a mass extinction is not occurring. Although there is no hard-and-fast rule for distinguishing between background and mass extinction rates, paleontologists traditionally recognize and study five historic mass extinction events. Figure 25.11, for example, plots the percentage of plant and animal lineages called families that died out during each stage in the geologic time scale since the Cambrian explosion. Five spikes in the graph—denoting a large number of extinctions within a short time—are drawn in red. These are referred to as "The Big Five."

Biologists are interested in distinguishing between background and mass extinctions because these events have contrasting causes and effects.

- *Background extinctions* occur when normal environmental change, emerging disease, predation pressure, or competition with other species reduces certain populations to zero.
- *Mass extinctions* result from extraordinary, sudden, and temporary changes in the environment. During a mass extinction, species do not die out because individuals are poorly adapted to normal or gradually changing environmental conditions. Rather, species die out from exposure to exceptionally harsh, short-term conditions—such as huge volcanic eruptions or catastrophic sea-level changes.

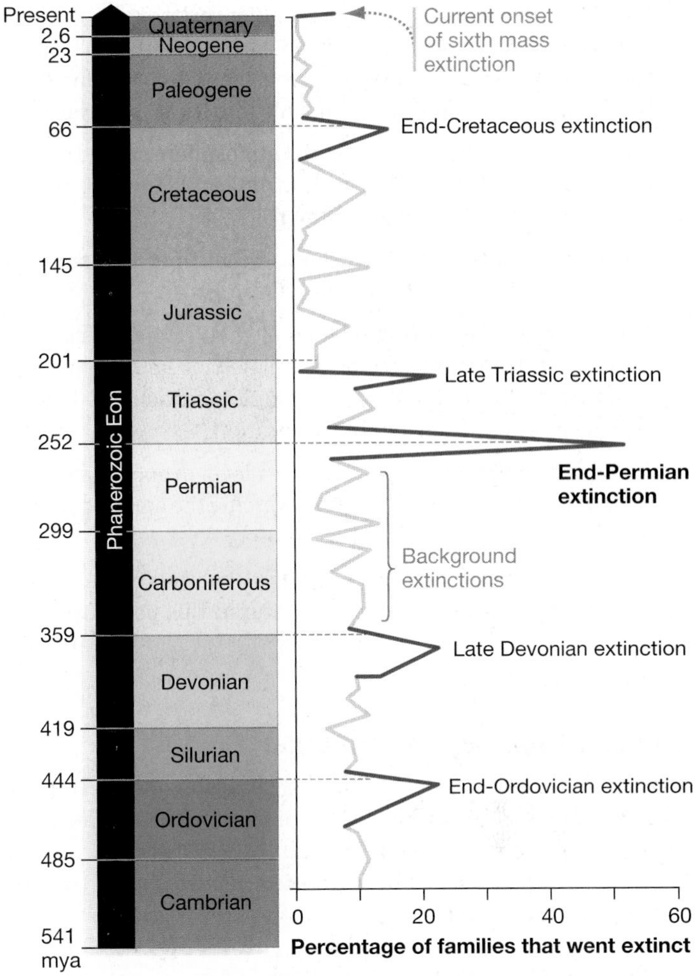

Figure 25.11 The Big Five Mass Extinction Events. This graph shows the percentage of plant and animal lineages called families that went extinct over each period in the fossil record since the Cambrian explosion. Over 50 percent of families and 90 percent of species went extinct during the end-Permian extinction.

DATA: M. J. Benton. 1995. *Science* 268: 52–58.

✔ **QUANTITATIVE** Which extinction event ended the era of the dinosaurs 66 million years ago? About what percentage of families went extinct in this event?

To drive these points home, let's consider a mass extinction event that nearly uprooted the tree of life entirely. The end-Permian extinction, which occurred about 252 million years ago, came close to ending multicellular life on Earth.

The End-Permian Extinction

The end-Permian has been called the "Mother of Mass Extinctions." To appreciate the scale of the event, imagine that while walking through a dense forest, you snapped your fingers and nine out of ten species disappeared. An area that was teeming with diverse forms of life would look barren. This is what happened, all over the world, during the end-Permian extinction.

Although biologists have long appreciated the scale of the end-Permian extinction, research on its causes is ongoing. Consider the following:

- Flood basalts are outpourings of molten rock that flow across Earth's surface. The largest flood basalts on Earth, called the Siberian traps, occurred during the end-Permian. They belched heat, CO_2, and sulfur dioxide into the atmosphere.
- A high level of sulfur dioxide in the atmosphere caused severe acid rain, which would have devastated plants and all the other organisms dependent on them.
- The flood basalts ignited widespread coal fires that pumped toxic ash, including mercury, into the air.
- Rocks that formed during the end of the Permian period indicate that the oceans became completely or largely anoxic—meaning that they lacked oxygen. These conditions are fatal to organisms that rely on aerobic respiration.
- There is convincing evidence that sea level dropped dramatically during the extinction event, reducing the amount of habitat available for marine organisms.

In short, both marine and terrestrial habitats were devastated. A prominent researcher has captured this point by naming the suite of killing mechanisms the "world-went-to-hell hypothesis."

The End-Cretaceous Extinction

What killed the dinosaurs 66 million years ago? The **impact hypothesis** for the extinction of the dinosaurs, first put forth in 1980 by father-and-son team Luis and Walter Alvarez, proposed that an asteroid struck Earth and snuffed out an estimated 60–80 percent of multicellular species.

Evidence for the Impact Hypothesis The impact hypothesis was controversial at first. As researchers set out to test its predictions, however, support began to grow. For example, researchers studying events at the Cretaceous–Paleogene (K–Pg) boundary found high quantities of iridium indicative of asteroids and a giant crater off the coast of Mexico that dates to 66 million years ago (**Figure 25.12**). Taken together, these data provided conclusive evidence in favor of the impact hypothesis.

Astronomers and paleontologists estimate that the asteroid was about 10 km across—a massive rock about the size of Mt. Everest. It would have struck with the force of several million nuclear bombs. According to both computer models and geologic data, the consequences of the K–Pg asteroid strike were far-reaching and devastating.

- A tremendous fireball of hot gas would have spread from the impact site; large soot and ash deposits in sediments dated to 66 mya testify to catastrophic wildfires, worldwide.
- The largest tsunami in the last 3.5 billion years would have disrupted ocean sediments and circulation patterns.
- The impact site itself is underlain by a sulfate-containing rock that would have reacted with water in the atmosphere to form sulfuric acid, triggering extensive acid rain.
- Enormous quantities of dust, ash, and soot would have blocked the Sun for long periods, leading to rapid global cooling and a crash in plant productivity.

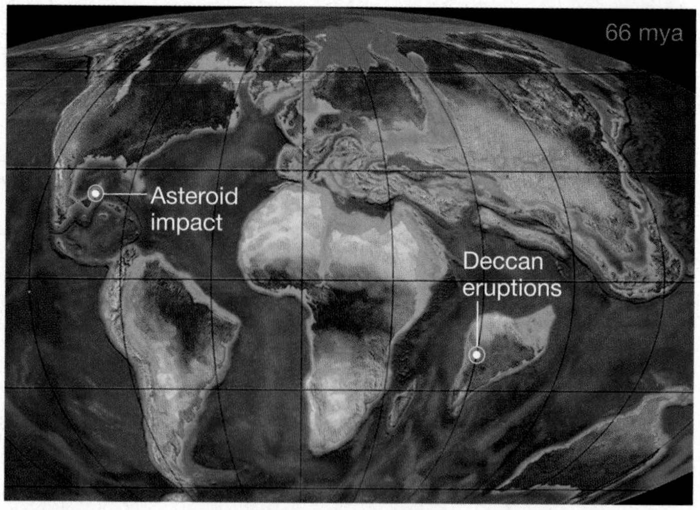

Figure 25.12 Cataclysmic Events at the End of the Cretaceous Period.
DATA: R. Blakey. Colorado Plateau Geosystems, Arizona, USA.

Evidence for Cataclysmic Volcanism The asteroid may not have doomed life on its own. Increasing evidence suggests that before the asteroid struck, cataclysmic volcanic flood basalts called the Deccan Traps were already spewing molten rock in vast areas of what is now India (see Figure 25.12), destabilizing ecosystems and causing decline in some taxonomic groups. It appears that the jolt of the asteroid impact accelerated these eruptions, compounding the devastation. Researchers are working to clarify the time frame and complex interactions between these deadly events.

Recovery from the Extinction Both terrestrial and marine ecosystems around the world were radically simplified by the extinction event, and recovery of diversity was slow—taking millions of years. Certain lineages disappeared forever while others survived and flourished. For example:

- After the K–Pg extinction, more than half of plant species may have disappeared, especially plants that depended on animals for pollination. Extensive stands of ferns replaced diverse assemblages of cone-bearing and flowering plants after the impact. Eventually, stress-tolerant flowering plants began to flourish and diversify.
- Not all dinosaurs disappeared during the K–Pg extinction—birds survived. Why? One hypothesis is that they ate seeds. During the nuclear winter that followed the impact, seeds would have been a more persistent food source than other plant materials or animals. Birds subsequently experienced an adaptive radiation, spreading from South America around the globe and into diverse ecological niches.
- The lineage called Mammalia had begun to diversify before the K–Pg extinction, during a period of rapid diversification of flowering plants. But it wasn't until after the dinosaur extinction that placental mammals diversified into the orders we recognize today—from bats to primates—and average body size increased exponentially. Why did mammals survive? One hypothesis is that their small size during the K–Pg extinction,

nocturnal habits (good sight in the dark, good sense of smell), and flexible diet contributed to their survival.

- About 15 percent of marine groups disappeared. While large species suffered higher losses on land, this size selection is not apparent in marine records. However, organisms in the open ocean suffered higher extinction rates than those in coastal waters, possibly due to changes in ocean stratification and nutrient cycling.

Although the K–Pg extinction was devastating to some lineages, others survived and benefited from the sudden availability of new ecological niches. Extinctions are often followed by adaptive radiations due to ecological opportunity.

The Sixth Mass Extinction

Are mass extinctions events of the distant past? No, life on Earth is currently engaged in a sixth mass extinction event, this time precipitated by human impacts such as habitat loss, pollution, overexploitation, invasive species, and climate change (Chapter 54).

Some researchers estimate that the current extinction rate is 100 or even 1000 times higher than the background extinction rate—the highest it's been since the asteroid impact 66 million years ago. **Figure 25.13a** is a glimpse of the rising extinction rate in vertebrates over the past century using conservative estimates. The trend is similar for other organisms. But researchers emphasize that current extinction rates vastly underrepresent the problem, for a couple of reasons:

- Current extinction rates show only the percentage of species that are already extinct. But surviving species are in serious decline. For example, about 45 percent of the land mammals sampled have lost 81–100 percent of their range during the past century—the blink of an eye in geologic time (**Figure 25.13b**). Larger species are disappearing more rapidly than smaller species.
- Biologists realize that the species count itself is not the most important concern—it's ecosystem function. Even the *rarity* of species, such as wolves in North America, tigers in Asia, and sharks in the oceans, can have dramatic and cascading effects on community structure and ecosystem function. Further, these effects diminish the ecosystem services that humans rely on, such as the abundance of food and clean water.

In Unit 6, you will explore the diversity of organisms alive on Earth today. Then, in Unit 9, you will examine how these organisms interact in complicated networks that are increasingly impacted by humans.

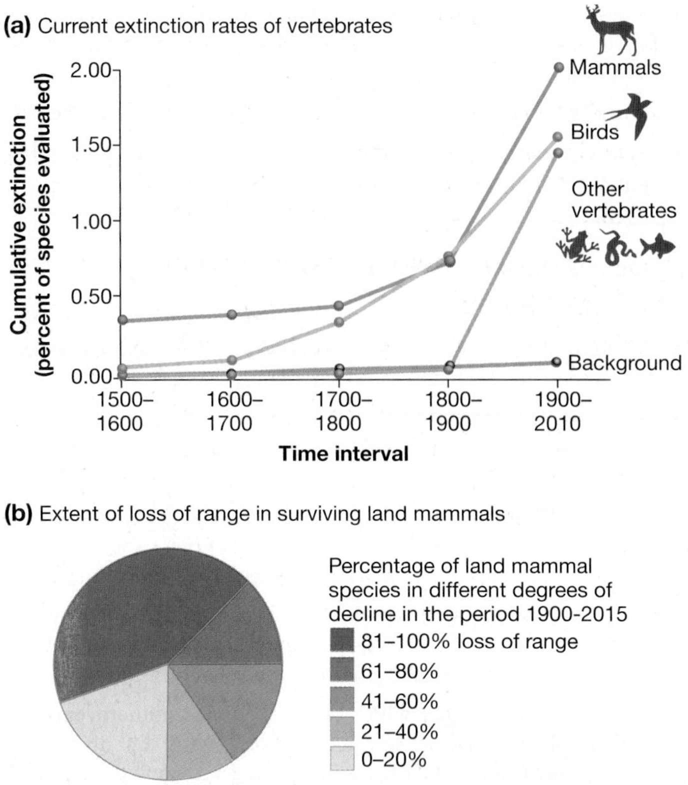

Figure 25.13 Current Extinction and Decline in Vertebrates.

DATA: G. Ceballos et al. 2015. *Science Advances* 1(5): e1400253; G. Ceballos et al. 2017. *Proceedings of the National Academy of Sciences, USA* 114(30): e6089–e6096.

✔ **PROCESS OF SCIENCE** The null hypothesis for the sixth mass extinction is that current extinction rates represent background extinctions. How would the graph in (a) look if the data supported the null hypothesis?

CHECK YOUR UNDERSTANDING

✔ **If you understood this section, you should be able to . . .**

1. Predict how flying reptiles called pterosaurs went extinct 66 million years ago. Were they poorly adapted to their environment?
2. **SOCIETY** Assess whether the current dramatic decline in sharks is more likely the effect of background extinction or a mass extinction. (Some shark populations have dropped to just 5 percent of their former abundance in just a few decades due to overfishing.)

Answers are available in Appendix A.

CHAPTER 25 Review

For media, go to **Mastering Biology**

25.1 Tools for Studying Life's History: Phylogenetic Trees

- Phylogenetic trees show relationships among taxa. They have revolutionized the study of evolution and many related fields.
- The correct interpretation of phylogenetic trees requires a focus on most recent common ancestors (rather than the position of tips or the number of nodes).
- Phylogenetic trees are often estimated by analyzing shared, derived characters (synapomorphies) that identify monophyletic groups.
- Estimates of phylogeny from morphological characters are sometimes based on parsimony, which assumes that the most likely explanation is the one that implies the fewest character changes.

- Estimates of phylogeny from molecular characters are most often based on statistical methods that involve a mathematical model of DNA evolution.
- Homology occurs when traits are shared due to common ancestry.
- Homoplasy occurs when traits are similar due to reasons other than common ancestry, such as convergent evolution.

25.2 Tools for Studying Life's History: The Fossil Record

- The fossil record is the only direct source of data about what extinct organisms looked like and where they lived.
- The fossil record is biased: Common, recent, and abundant species that burrow and that have hard parts are most likely to be present in the record.
- During the Precambrian, which started with Earth's formation 4.6 billion years ago and ended with the origin of most animals about 541 million years ago, life was almost exclusively unicellular.
- Oxygen was virtually absent from the oceans and atmosphere for almost 2 billion years after the origin of life. The oxygen-rich atmosphere was generated by photosynthetic bacteria.
- During the Phanerozoic, which began about 541 million years ago and continues to the present, animals originated and diversified first in the ocean and then on land. Land plants and fungi also originated and diversified.
- If Earth's history were scaled to fit in one calendar year, our species, *Homo sapiens*, would not appear until one hour before the stroke of midnight on December 31st.

25.3 Large-Scale Pattern in Life's History: Adaptive Radiation

- Adaptive radiations occur when a single lineage rapidly produces many descendant species with a wide range of forms.
- Adaptive radiations can be triggered by ecological opportunity, such as the availability of unoccupied niches, and/or by morphological, physiological, or behavioral innovations.
- The Cambrian explosion was one of the most spectacular adaptive radiations of all time, producing all of the major animal phyla existing today. During this event, the size and morphological complexity of animals increased tremendously, and animals diversified in how they made a living.

25.4 Large-Scale Pattern in Life's History: Mass Extinction

- Unlike background extinctions, mass extinctions are short-term environmental catastrophes that eliminate most of the species alive. Mass extinctions have altered the course of evolutionary history at least five times.
- Mass extinctions such as the end-Permian and end-Cretaceous have marked the end of several major branches on the tree of life and the subsequent rise of new branches.
- A sixth mass extinction is presently under way due to the drastic effects of humans on the Earth in recent history.

Answers are available in Appendix A.

✓ TEST YOUR KNOWLEDGE

1. Choose the best definition of a fossil.
   a. a rock that contains information about an organism
   b. a bone, tooth, shell, or other hard part of an organism that has been preserved
   c. any trace of an organism that lived in the past
   d. any part of a dead organism
2. **THINK CAREFULLY** Critique the following statement: The absence of a trait cannot be used as a synapomorphy in phylogenetic analysis; only shared derived traits that are present in the clade can be used.
3. Which of the following best characterizes an adaptive radiation?
   a. Descendant species occupy a large geographic area.
   b. A single lineage diversifies rapidly, and descendant species occupy many habitats and ecological roles.
   c. Natural selection is particularly intense, because disruptive selection occurs.
   d. Species recover after a mass extinction.
4. Which of the following is an example of homoplasy?
   a. hair in humans and fur in mice
   b. astragalus ankle bones in hippos and deer
   c. *Hox* genes in humans and flies
   d. streamlined bodies in dolphins and ichthyosaurs

✓ TEST YOUR UNDERSTANDING

5. What important assumption does parsimony make when assessing which phylogenetic tree is most accurate? Why was parsimony misleading in the case of the astragalus during the evolution of artiodactyls?
6. You can use a "one-snip test" to identify monophyletic groups—meaning that if you "cut" any branch on a tree, everything that "falls off" is a monophyletic group (**BioSkills 13**). Why is this valid?
7. Fossilization continues even today. If you wanted to increase the probability that your deceased favorite pet would become fossilized, what actions would you take?
   a. Place the corpse in an environment where decomposition is rapid (such as a forest).
   b. Place the corpse in an environment where decomposition is slow (such as a swamp or bog).
   c. Place the corpse in an environment where there are plenty of rocks (such as a boulder field).
   d. Place the corpse in an environment where plenty of oxygen is available (such as in a mountain stream).
8. Use the fossil evidence shown in Figure 25.6 to determine whether flight evolved earlier in insects or in birds. Is flight an example of homology or convergent evolution? Explain.

✔ TEST YOUR PROBLEM-SOLVING SKILLS

9. MODEL Use the data matrix shown here to draw a phylogenetic tree with the minimum number of character changes. Label the origin of each *Hox* gene on the tree. How could you change your analysis to increase your confidence in the relationships shown in the tree?

Presence or Absence of *Hox* Genes*

	Hox1	***Hox2***	***Hox3***	***Hox4***	***Hox12***
Mouse	1	1	1	1	1
Snail	1	1	1	1	0
Flatworm	1	1	1	1	0
Sea anemone	1	1	0	0	0
Sponge	0	0	0	0	0

*1 = gene present; 0 = gene absent.

10. SOCIETY Malaria is a tropical, mosquito-borne disease caused by a parasitic protist, *Plasmodium*. The species *P. falciparum* is particularly deadly, killing about 1 million people each year. Use this phylogenetic evidence to determine if *P. falciparum* originated from another strain of *Plasmodium* in humans, chimpanzees, or gorillas. Explain your reasoning. How might this information be useful in developing medical treatments for malaria?

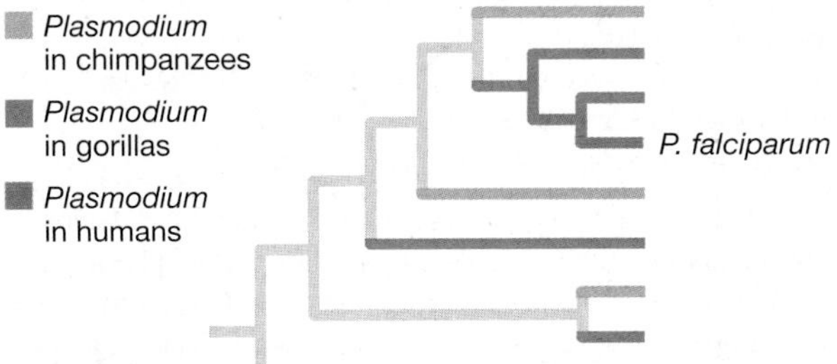

DATA: W. Liu et al. 2013. *Nature* 467: 420–425.

✔ PUT IT ALL TOGETHER: Case Study

Are dinosaurs extinct?

Theropod dinosaurs are the bipedal ("two-footed") carnivores of *Tyrannosaurus rex* fame that many people imagine when they think of dinosaurs. Did these dinosaurs go extinct?

11. The colored bars in the phylogenetic tree that follows represent the time range of a selection of theropod dinosaurs. Use Figure 25.6 and the time scale beneath the tree as references to draw vertical lines through the tree to label the geologic eras and periods represented.

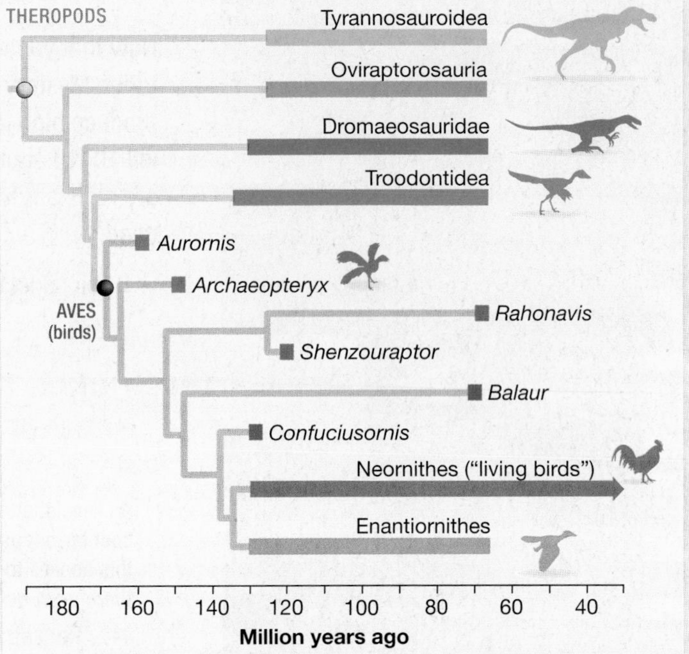

12. Are all the theropod dinosaurs extinct?

a. Yes, because all theropod dinosaurs died 66 million years ago.
b. No, because theropod dinosaurs are a monophyletic group that includes both extinct and living birds.
c. Yes, because theropod dinosaurs are a monophyletic group that includes extinct birds but not living birds.
d. No, because birds are a monophyletic group that includes theropod dinosaurs.

13. QUANTITATIVE The tree shows evidence of what event 66 million years ago? If the 12 hours of a clock represented the duration of the Phanerozoic eon (starting at noon and ending at midnight), at what time would this event have occurred?

14. THINK CAREFULLY Referring to the tree in Question 11, select True or False for each of the following statements. Correct any statements that are false.

T/F *Archaeopteryx* is more closely related to *Aurornis* than to living birds.
T/F Living birds are more closely related to Enantiornithes than to *Confuciusornis*.
T/F *Archaeopteryx* is an ancestor of living birds.
T/F *Tyrannosaurus rex* (Tyrannosauroidea) is an ancestor of living birds.

15. Redraw the tree in Question 11 to show the relationships among only the branches with animal icons (for simplicity) and using arbitrary branch lengths (not scaled to time). Then (1) add an outgroup to the theropods called the Sauropoda (which includes the enormous *Apatosaurus*); (2) mark the origin of downy feathers, which occur in all theropods; and (3) mark the origin of shafted feathers and wings, which occur in all the groups except Tyrannosauroidea. Can the presence of feathers and wings be considered a synapomorphy that triggered the adaptive radiation of birds? Why or why not?

16. PROCESS OF SCIENCE Consider the fossil data and phylogenetic analyses that were required to create the illustration in Question 11. What aspects of the data are researchers likely to be confident about? What aspects of the data are more likely to change as more data are collected?

END-OF-UNIT
CASE STUDY

UNIT 4

PUT IT ALL TOGETHER

■ For an introduction to the Mystery of the Newt case study, see page 17.

Now that you've learned about evolutionary processes and patterns, it's time to return to the Mystery of the Newt. Recall that rough-skinned newts (*Taricha granulosa*) produce high levels of tetrodotoxin (TTX) in their skin—so high that one newt could kill 10–20 adult humans (if they shared the newt for dinner). Yet some garter snakes are resistant to this powerful toxin, allowing them to eat newts without dropping dead.

✔ If you understand Unit 4, you should be able to apply your learning to this case study:

Are Garter Snakes and Newts Engaged in an Arms Race?

An "evolutionary arms race" occurs when two or more species evolve adaptations and counter-adaptations in response to interactions with each other. To begin thinking about how this applies to toxic newts and resistant snakes, recall that for evolution to occur in a trait, there must be heritable variation.

1. **PROCESS OF SCIENCE** Which of the following types of evidence would support the hypothesis that there is heritable variation in TTX levels in rough-skinned newts? Select True or False for each statement. (See Section 22.3)

   T/F All newts have the same TTX levels.
   T/F Newts with high TTX levels tend to have offspring with high TTX levels.
   T/F Different newts have different levels of TTX.
   T/F Young newts raised on a diet lacking TTX grow up to be toxic.

2. Recall that TTX is lethal because it blocks sodium channels in skeletal muscles, leading to paralysis and asphyxiation. Garter snakes have different levels of resistance to TTX depending on the amino acid sequences of their sodium channels. For example, compare the two amino acid sequences below, both found in *Thamnophis sirtalis* snakes. (For a key to single-letter amino acid abbreviations, see Ch 3., Figure 3.2.) How many alleles are represented? By what process of evolution do new alleles arise in the garter snake population? (See Section 23.6)

   ...KGWMEIMYP...ILCLFE**V**TTSAGW**NVL**... → 100 units of TTX resistance
   ...KGWMEIMYP...IICLFE**I**TTSAGW**DGL**... → 5 units of TTX resistance

Rough-skinned newts have higher fitness when they are not eaten by snakes, but producing TTX is metabolically expensive and newts can be endangered by their own toxin. Likewise, garter snakes have higher fitness if they are not killed or paralyzed by eating toxic newts, but having TTX-resistant sodium channels makes them slower than non-resistant snakes and thus more vulnerable to predators.

3. Complete the matrix (**Figure 1**) to summarize the fitness trade-offs of TTX production in newts and TTX resistance in garter snakes. (See Section 22.5)

4. **THINK CAREFULLY** Does evolution make species perfect? Use the matrix to explain your answer. (See Section 22.5)

Figure 1

	Benefits of trait	Costs of trait
Newts		
Snakes		

Now that you have considered heritable variation and fitness trade-offs, the next step is to consider whether natural selection is occurring (see Section 22.3). Researchers measured and mapped the toxicity of newts within their range on the West Coast of North America. They also mapped the distribution of TTX-resistant snakes in this same area. (Maps in Figure 2 show data for Washington and Oregon.)

5. What conclusions can you draw by comparing the two maps? Predict where a group of hunters would be most likely to die from newt-infused coffee.

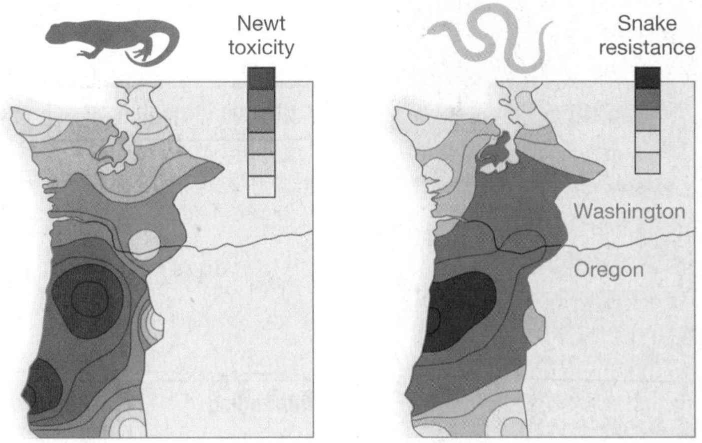

Figure 2

DATA: C. T. Hanifin, E. D. Brodie Jr., E. D. Brodie III. 2008. *PLoS Biology* 6 (3): e60.

Does a similar pattern appear between other species pairs of garter snakes and toxic newts? Figure 3 shows phylogenetic relationships among different species of garter snakes. The colored branches show which snakes have TTX-resistant sodium channels. Different colors represent different amino acid sequences in the sodium channels. (See Section 25.1)

6. Predict which snakes live in areas with toxic newts.

7. MODEL Draw a dot on the tree that represents the most recent common ancestor of all the TTX-resistant snakes.

8. CAUTION Is there a way that you can rotate the branches of the tree to move all the TTX-resistant snakes into a monophyletic group containing no TTX-sensitive snakes? Explain.

9. Is it more parsimonious to hypothesize that the most recent common ancestor of all the TTX-resistant snakes was also TTX-resistant, or that resistance arose independently in multiple lineages?

10. Is TTX-resistance in garter snakes an example of homology (shared ancestry) or homoplasy (convergent evolution)?

Figure 3

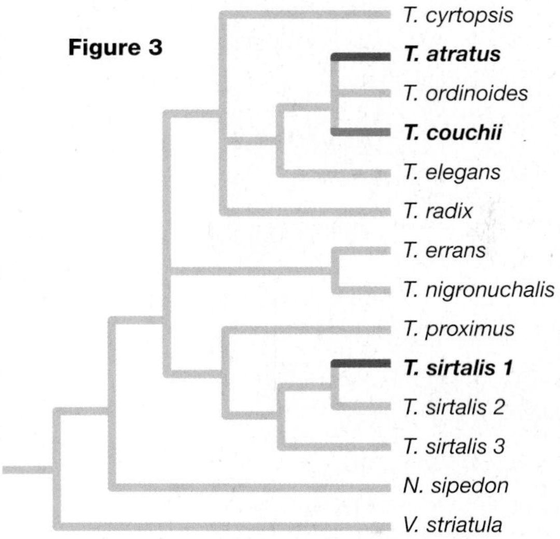

DATA: C. R. Feldman et al. 2009. *PNAS* 106: 13415–13420.

11. Now that you have evaluated a number of types of data, explain whether you think an arms race may be occurring between toxic newts and garter snakes.

It turns out that newts aren't the only organisms to use TTX or other toxins for defense. A number of prey species throughout the tree of life, including protists and plants, produce toxins while a number of predator species have evolved resistance to those toxins, a theme which will be revisited in Unit 5: The Diversification of Life.

Toxic newts and TTX-resistant snakes coincide in West-Coast forests.

Each unit ends with a continuation of this story.

BIG PICTURE

Geneticist and evolutionary biologist Theodosius Dobzhansky said that "Nothing in biology makes sense except in the light of evolution." Use this concept map to study how ideas introduced in Unit 4 fit together.

The key is to connect the four evolutionary processes that work at the level of populations—natural selection, genetic drift, mutation, and gene flow—to processes, events, and outcomes at higher levels of organization: speciation, adaptive radiation, mass extinction, and the tree of life.

It's all about changes in allele frequencies. Over time, small changes that occur between populations lead to large changes that distinguish major lineages on the tree of life.

Note that each box in the concept map indicates the chapter and section where you can go for review. Also, be sure to do the blue exercises in the Check Your Understanding box below.

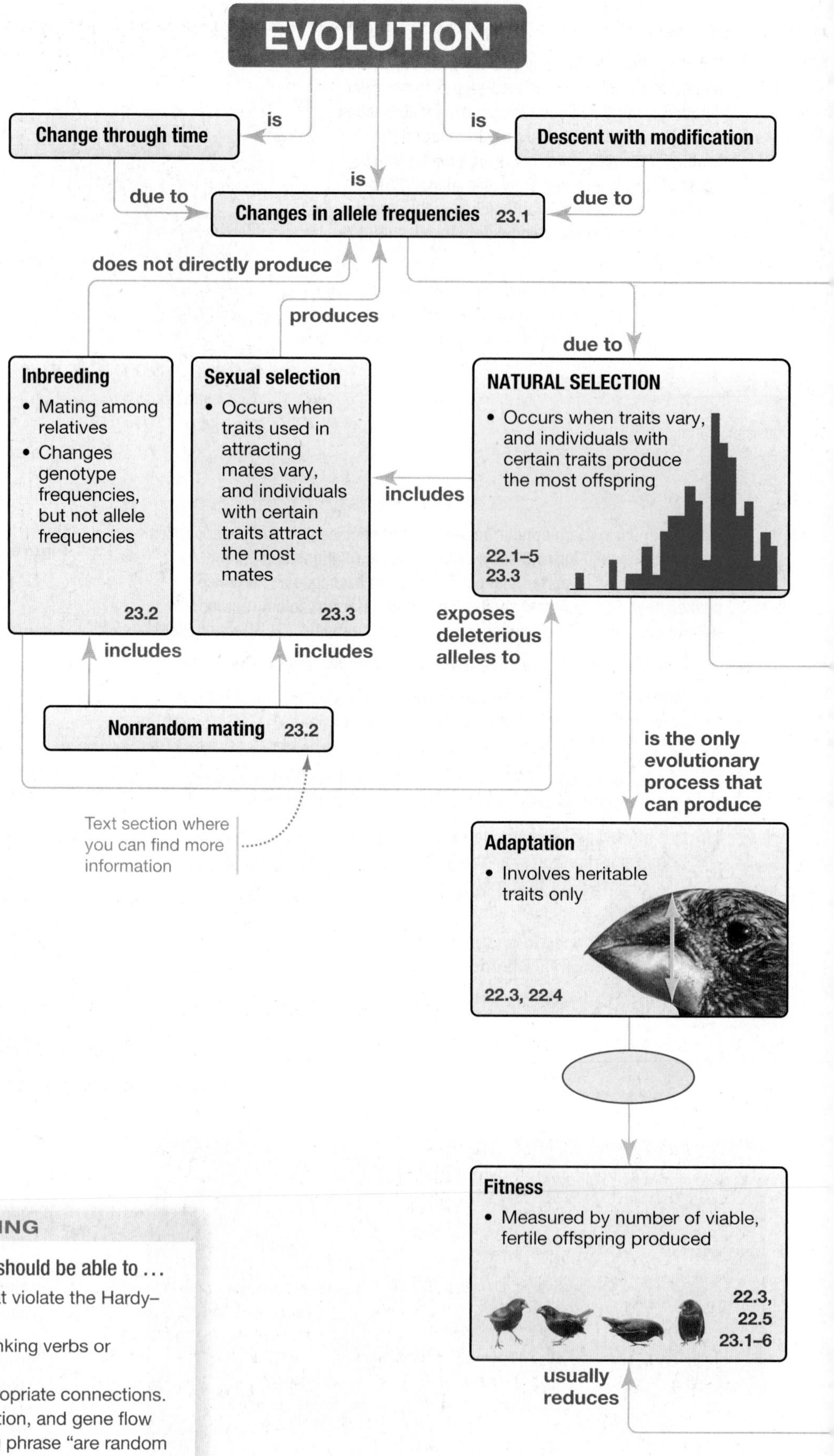

CHECK YOUR UNDERSTANDING

✔ **If you understand the big picture, you should be able to . . .**

1. Draw a circle around the processes that violate the Hardy–Weinberg assumptions.
2. Fill in the blue ovals with appropriate linking verbs or phrases.
3. Add a box for "Fossil record" with appropriate connections.
4. Draw arrows linking genetic drift, mutation, and gene flow to the appropriate box using the linking phrase "are random with respect to."

Answers are available in Appendix A.

THE BIG PICTURE

due to

GENETIC DRIFT
- Changes in allele frequencies due entirely to chance
- Especially important in small populations

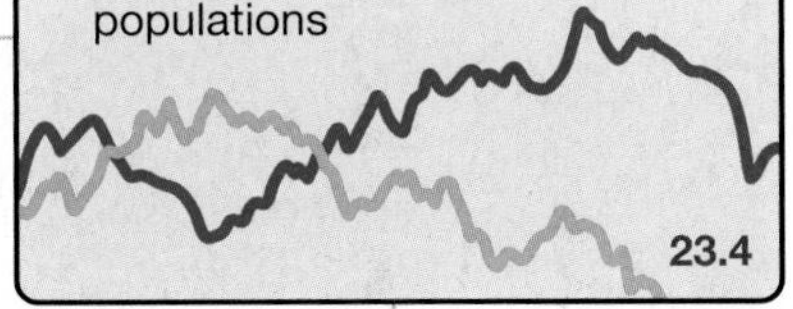

23.4

due to

MUTATION
- Random changes in DNA
- Creates new alleles
- Occurs in every individual in every generation, at low frequency

AT TAT TGT CTA GTA CCC CA

16.4, 23.6

due to

GENE FLOW
- Occurs when individuals move between populations
- Homogenizes allele frequencies between populations

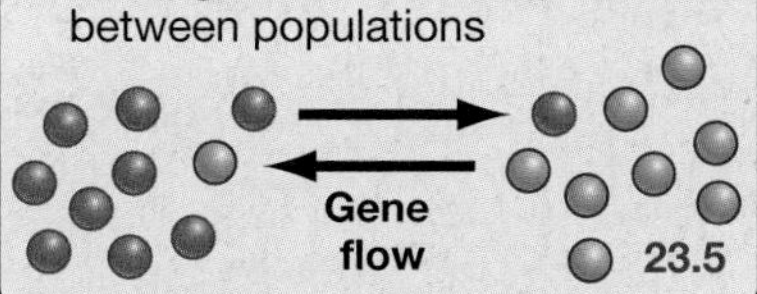

23.5

due to lack of

produces divergence required for

produces divergence required for

provides raw material for

Speciation
Results from:
1. Genetic isolation, followed by
2. Genetic divergence

24.2–4

creates new branches on

form smallest possible tips on

The Tree of Life
- Describes the evolutionary relationships among all species

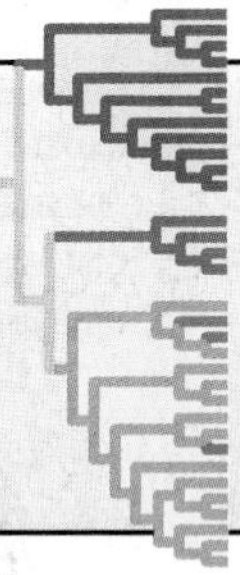

1.5, 25.1

forms new

Species
Evolutionarily independent units in nature, identified by:
1. Reproductive isolation, and/or
2. Phylogenetic analysis, and/or
3. Morphological differences

24.1

"prune"

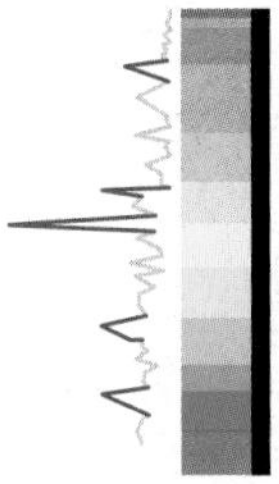

Mass extinctions
- 60% of species are lost in less than 1 million years
- 5 events in the past 541 million years. A sixth mass extinction is now underway

25.4

may occur after

with

Synapomorphies
- Shared, derived traits that are unique to a single lineage
- Arise in a common ancestor

24.1
25.1

that may be

Key innovations
- Traits that allow species to exploit resources in a new way or use new habitats

25.4

may result in

Adaptive radiations
- Rapid and extensive speciation in a single lineage
- Dramatic divergence in physiology, morphology, or behavior (species use a wide array of resources/habitats)

25.3

26 Bacteria and Archaea

Although this hot spring looks devoid of life, it is actually teeming with billions of bacterial and archaeal cells.

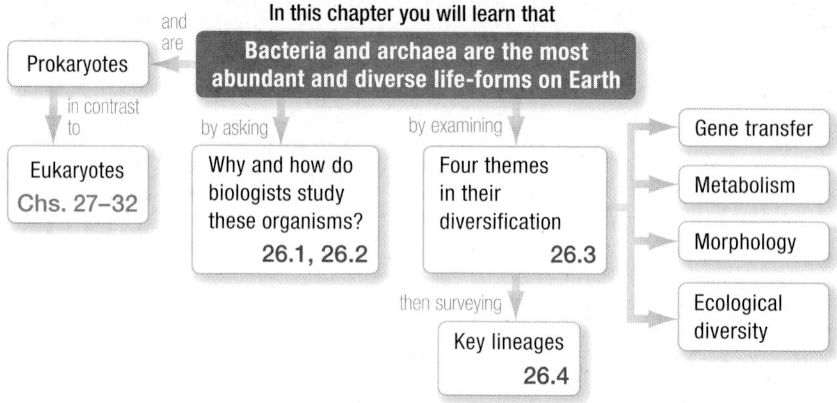

By the early 1950s biologists had reached a staggering conclusion: At least one-half of the "living protoplasm" on Earth is invisible to the naked eye. Most of these microscopic organisms are members of the **Bacteria** and **Archaea** (*ar-KEE-ah*), two of the three largest branches on the tree of life. Recall that virtually all bacteria and archaea are unicellular, and all are **prokaryotic**—meaning that they lack a membrane-bound nucleus (Ch. 7, Section 7.1). The third major branch consists of eukaryotes and is called the **Eukarya**. **Eukaryotic** organisms have cells with membrane-bound nuclei.

Biologists estimate the total number of individual bacteria and archaea at over 3×10^{30}. Virtually all are unnamed and undescribed. If you want to discover and name new species, then study prokaryotes. You share your body with a dynamic population—as many as ten bacterial or archaeal cells for every one of your own cells. You are a walking, talking habitat for prokaryotes. Those prokaryotes that live on your body are part of a human microbiome. A **microbiome** is a community of microbes that naturally inhabits a particular area and encompasses all the genetic material contained within it. The human microbiome can refer to the microbes that inhabit the whole body or parts of it, such as the mouth, gut, and other tissues.

BIG PICTURE

This chapter is part of the Big Picture. See how on pages 722–723.

Table 26.1 Distinguishing Features of Bacteria and Archaea

Molecular Feature	Bacteria	Archaea
RNA polymerase	Just one type (consists of five subunits)	Just one type (consists of 13 subunits; similar to RNA polymerase II in eukaryotes)
Peptidoglycan in cell wall	Present	Absent
First amino acid incorporated during translation	Formylmethionine	Methionine
Histones associated with DNA	No	Yes

Although both bacteria and archaea have a relatively simple morphology, they are strikingly different at the molecular level. If you were shrunk to their size, bacteria and archaea would seem to you as different from each other as mammals and fish.

What criteria are used to distinguish bacteria from archaea? For one thing, there are key differences in the types of molecules that make up their plasma membranes and cell walls (Ch. 7, Section 7.1). In addition, the machinery that bacteria and archaea use to process genetic information is strikingly different. Specifically, the DNA polymerases, RNA polymerases, transcription-initiation proteins, and ribosomes found in archaea are more similar to those found in eukaryotes than to those found in bacteria (see Chapters 17 and 18 for more detail on these molecular machines). **Table 26.1** lists some of the key molecular differences between the domains Bacteria and Archaea.

These differences have practical consequences: Antibiotics that poison bacterial ribosomes do not affect the ribosomes of archaea or eukaryotes. If all ribosomes were identical, these antibiotics would kill your cells along with the bacterial species that was supposed to be targeted. In addition, as you'll see later in the chapter, studying these shared features has helped biologists understand the early evolution of life on Earth.

Although prokaryotes are tiny, they have an enormous impact on you and the planet in general. By the time you finish reading the chapter, you should understand why a researcher summed up the bacteria and archaea by claiming, "They run this joint."

26.1 Why Do Biologists Study Bacteria and Archaea?

Biologists study bacteria and archaea for the same reasons they study any organisms. First, prokaryotes are intrinsically fascinating. Novel discoveries—such as finding prokaryotic cells living over 2 kilometers underground or in 95°C hot springs—keep biologists engaged in figuring out how those cells stay alive. Second, there are important benefits to studying these species that share the planet with us—in terms of both understanding life on Earth and improving human health and welfare.

After you complete this section, you should be able to ...

- Explain why bacteria and archaea are considered the most important, diverse, and abundant organisms on Earth.

Biological Impact

The lineages in Bacteria and Archaea are ancient, diverse, abundant, and ubiquitous. The oldest fossils of any type found to date are 3.7-billion-year-old carbon-rich deposits derived from bacteria. Because eukaryotes do not appear in the fossil record until 1.75 billion years ago, biologists infer that prokaryotes were the only form of life on Earth for at least 1.9 billion years.

Just how many species of bacteria and archaea are alive today? Although only about 10,000 species have been formally named and described to date, hundreds of thousands are virtually certain to exist. Consider that over 1000 species of prokaryotes are living in your large intestine right now, and another 700 species are living in your mouth. Microbiologist Norman Pace points out that there may be tens of millions of different insect species but notes, "If we squeeze out any one of these insects and examine its contents under the microscope, we find hundreds or thousands of distinct microbial species." Most of these **microbes** (microscopic organisms) are bacteria or archaea.

Abundance Besides recognizing how diverse bacteria and archaea are in terms of numbers of species, it's also critical to appreciate their abundance. For example, trillions of prokaryotic cells can live on or within eukaryotes. The following examples further illustrate the extent of microbe abundance:

- A mere teaspoon of good-quality soil contains *billions* of microbial cells, most of which are bacteria and archaea.
- Overall, the average human body is estimated to contain roughly 100 trillion bacteria and archaea cells.
- In sheer numbers, species in a lineage of marine archaea may be the most successful organisms on Earth. Biologists routinely find these cells at concentrations of over 10,000 individuals per milliliter in most of the world's oceans. At these concentrations, one liter of seawater contains a population equivalent to that of the largest human cities. Yet this lineage was not described until the early 1990s.
- Enormous numbers of prokaryotes thrive inside rocks and sediments as much as 1600 meters underneath the world's oceans. Although recently discovered, the bacteria and archaea living under the ocean may make up 10 percent of the world's total mass of living material.

In terms of the total volume of living material on our planet, bacteria and archaea are the dominant life-forms.

Habitat Diversity Prokaryotes are found almost everywhere. They live in environments as unusual as oxygen-free mud, hot springs, and salt flats. In seawater they are found from the surface to depths of 10,000 meters (m), and at temperatures over 120°C (well above water's boiling point of 100°C) near submarine volcanoes.

Although there are far more prokaryotes than eukaryotes, much more is known about eukaryotic diversity than about prokaryotic diversity. Researchers who study prokaryotic diversity are exploring one of the most wide-open frontiers in all of science. In fact, recent collecting expeditions have turned up entirely new **phyla**—major lineages—of prokaryotes (see Ch. 1,

Section 1.5 to review taxonomic levels). This achievement is like suddenly discovering a new lineage of eukaryotes as distinctive as flowering plants or animals with backbones.

The physical world has been explored and mapped, and many of the larger plants and animals are named. But in **microbiology**—the study of organisms that can be seen only with the aid of a microscope—this is an age of exploration and discovery.

Some Prokaryotes Thrive in Extreme Environments

Bacteria or archaea that live in high-salt, high-temperature, low-temperature, or high-pressure habitats are **extremophiles** (literally, "extreme-lovers"). Studying them has been extraordinarily fruitful for understanding the tree of life, developing industrial applications, and exploring the structure and function of enzymes.

As an example of these habitats, consider hydrothermal vents at the bottom of the ocean, where water as hot as 300°C emerges and mixes with 4°C seawater. At locations like these, archaea are abundant forms of life.

Researchers recently discovered a species of archaea that grows so close to these hydrothermal vents that its surroundings are at 121°C—a record for life at high temperature. This organism can live and grow in water that is heated past its boiling point and at pressures that would instantly destroy a human. Because high temperature breaks non-covalent bonds holding macromolecules together, extreme heat usually denatures proteins, makes membranes leaky, and separates the strands of the DNA double helix. Biologists are intrigued by how these cells can thrive under such extreme conditions.

Other bacteria and archaea can grow:

- at a pH less than 1.0;
- at temperatures of 0°C under Antarctic ice;
- in water that is 5–10 times saltier than seawater (Figure 26.1).

Extremophiles have become a hot area of research. The genomes of a wide array of extremophiles have been sequenced, and expeditions regularly seek to characterize new species. Why?

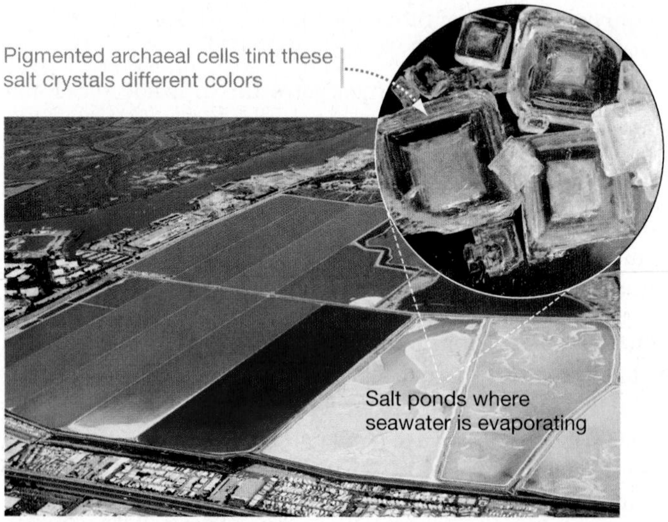

Figure 26.1 **Some Archaea Live in High-Salt Habitats.** Industrial salt is harvested after evaporating water molecules from seawater held in large pools. The remains of archaea that live in this habitat give the salt its color.

- ***Origin of life*** Based on models of conditions that prevailed early in Earth's history, the first forms of life probably lived at high temperature and pressure in environments that lacked oxygen—conditions that humans would call extreme. Therefore, understanding extremophiles may help explain how life on Earth began.
- ***Extraterrestrial life?*** In a similar vein, many astrobiologists ("space-biologists") use extremophiles as model organisms in the search for extraterrestrial life. The idea is that if certain prokaryotic cells can thrive in extreme habitats on Earth, cell-based life might possibly be found in similar environments on other planets or moons of planets.
- ***Commercial applications*** Because enzymes that function at extreme temperatures and pressures are useful in many industrial processes, extremophiles are of commercial interest as well. For example, *Taq* polymerase—a DNA polymerase that is stable up to 95°C—is used to run the polymerase chain reaction (PCR) in research and commercial settings (Ch. 20, Section 20.2, and **BioSkills 10**). This enzyme was isolated from a bacterium called *Thermus aquaticus* ("hot water"), which was discovered in a hot spring in Yellowstone National Park.

Medical Importance

The first paper documenting that an archaeal species was associated with a human disease—a dental condition called periodontitis—was published in 2004. But biologists have been studying disease-causing bacteria for over a century.

Of the thousands of bacterial species living in and on your body, only a tiny fraction can disrupt body functions enough to cause illness. Bacteria that cause disease are said to be **pathogens** ("disease-producers"). Pathogenic bacteria have been responsible for some of the most devastating epidemics in human history.

Table 26.2 lists some of the bacteria that cause illness in humans. Here are the important things to note:

- Pathogenic bacteria come from several different phyla.
- Pathogenic bacteria tend to infect tissues at the entry points to the body, such as wounds or pores in the skin, the respiratory and gastrointestinal tracts, and the urogenital canal.

Koch's Postulates Robert Koch was the first person to establish a link between a particular species of bacterium and a specific disease. When Koch began his work on the nature of disease in the late 1800s, microscopists had confirmed the existence of the particle-like organisms people now call bacteria, and Louis Pasteur had shown that bacteria and other microbes are responsible for spoiling milk, wine, broth, and other foods. Koch hypothesized that bacteria might also be responsible for causing diseases that were spread by being passed from an infected individual to an uninfected individual.

Koch set out to test this hypothesis by identifying the organism that causes anthrax, a disease of cattle and other grazing mammals, common in agricultural regions, that can result in fatal blood poisoning in humans. People can get sick with anthrax if they come in contact with infected animals or contaminated animal products.

To establish a causative link between a specific microbe and a specific disease, Koch proposed that four criteria had to be met:

Table 26.2 Some Bacteria That Cause Illness in Humans

Lineage	Species	Tissues Affected	Disease
Firmicutes	*Clostridium tetani*	Wounds, nervous system	Tetanus
	Staphylococcus aureus	Skin, urogenital canal	Acne, boils, impetigo, toxic shock syndrome
	Streptococcus pneumoniae	Respiratory tract	Pneumonia
	Streptococcus pyogenes	Respiratory tract	Strep throat, scarlet fever
Spirochaetes	*Borrelia burgdorferi*	Skin and nerves	Lyme disease
	Treponema pallidum	Urogenital canal	Syphilis
Actinobacteria	*Mycobacterium tuberculosis*	Respiratory tract	Tuberculosis
	Mycobacterium leprae	Skin and nerves	Leprosy
	Propionibacterium acnes	Skin	Acne
Chlamydiales	*Chlamydia trachomatis*	Urogenital canal	Genital tract infection
ε-Proteobacteria	*Helicobacter pylori*	Stomach	Ulcer
β-Proteobacteria	*Neisseria gonorrhoeae*	Urogenital canal	Gonorrhea
γ-Proteobacteria	*Haemophilus influenzae*	Ear canal, nervous system	Ear infections, meningitis
	Pseudomonas aeruginosa	Urogenital canal, eyes, ear canal, lungs	Infections of eye, ear, urinary tract, lungs
	Salmonella enterica	Gastrointestinal tract	Food poisoning
	Yersinia pestis	Lymph and blood	Plague

1. The microbe must be present in individuals suffering from the disease and absent from healthy individuals. By careful microscopy, Koch was able to show that the bacterium *Bacillus anthracis* was always present in the blood of cattle suffering from anthrax, but absent from healthy individuals. Note that Koch later abandoned this absolute requirement when he discovered that, in some cases, infectious microbes could be present in an individual who was nearly or completely free of signs of disease.
2. The organism must be isolated and grown in a pure culture away from the host organism. Koch was able to grow pure colonies of *B. anthracis* in glass dishes on a nutrient medium, using gelatin as a substrate.
3. If organisms from the pure culture are injected into a healthy experimental animal, the disease symptoms should appear. Koch demonstrated this crucial causative link in mice injected with *B. anthracis*. The symptoms of anthrax infection appeared, and then the infected mice died. "Should" is the key word here—not all organisms will become infected (see Chapter 48 for more on immunity).
4. The organism should be isolated from the diseased experimental animal, again grown in pure culture, and demonstrated by its size, shape, and color to be the same as the original organism. Koch did this by purifying *B. anthracis* from the blood of diseased experimental mice.

These criteria, now called **Koch's postulates**, are still used in modified form to confirm a causative link between new diseases and a suspected infectious agent. Microbiologists now recognize that many bacteria cannot be grown in culture, so they use other means of detection for those organisms, such as DNA sequence-based identification. As such, new molecular criteria are being established as a standard for judging disease causation.

The Germ Theory Koch's experimental results were the first test of the **germ theory of disease**. The germ theory of disease states that infectious diseases are caused by specific microbes in the body—such as bacteria, archaea, and viruses. Viruses are acellular particles that propagate by hijacking the genetic machinery of cells (Chapter 33). **Infectious diseases** are spread in three main ways: **(1)** some are passed from person to person; **(2)** some are transmitted by bites from insects or animals; and **(3)** some are acquired by ingesting contaminated food or water, or being exposed to microbes in the surrounding environment.

The germ theory of disease laid the foundation for modern medicine. Initially its greatest impact was on sanitation—efforts to prevent transmission of pathogenic bacteria. For example, during the American Civil War more soldiers died of bacterial infections contracted from drinking water contaminated with human feces than from wounds in battle.

What Makes Some Bacterial Cells Pathogenic? Virulence, or the ability to cause disease, is a heritable trait that varies among individuals in a population. Most *Escherichia coli*, for example, are harmless inhabitants of the gastrointestinal tract of humans and other mammals. But some *E. coli* cells cause potentially fatal food poisoning.

What makes some cells of the same species pathogenic while others are harmless? Biologists have answered this question for *E. coli* by sequencing the entire genome of a harmless lab strain and a strain called O157:H7, which is harmful to humans. The genome of the pathogenic strain is slightly larger because it has acquired virulence genes, including one coding for a protein **toxin** (a "poison"). After entering a host cell, this toxin binds to ribosomes and inhibits protein synthesis, killing the host cells. Because of key differences between the ribosomes of bacteria and eukaryotic cells, only the host's—not the bacterium's—protein synthesis is blocked by the toxin. Cells lining the blood vessels near the host's intestinal

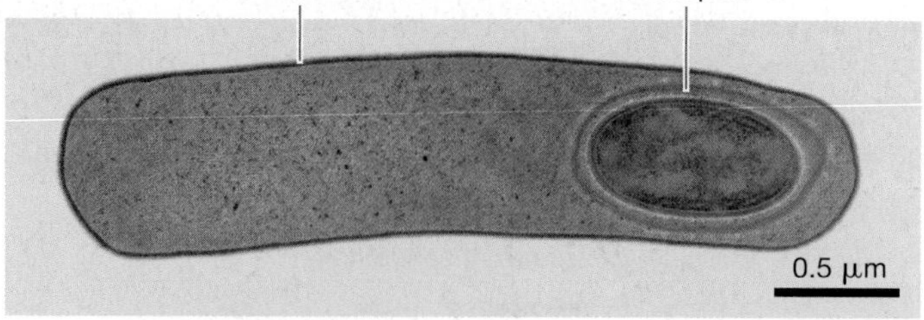

Figure 26.2 Endospore Formation in *Clostridium difficile*. *C. difficile* is a common bacterium in the human gut. Large numbers of this bacterium can irritate the large intestine and cause diarrhea.

epithelium are most affected by the toxin, and the resulting damage leads to bloody diarrhea and possible death. If sanitation is poor, the pathogenic bacteria are likely to infect many new hosts.

Similar types of studies are identifying the genes responsible for virulence in a wide array of pathogenic bacteria.

Some Pathogenic Bacteria Produce Resistant Endospores **Endospores** are tough, thick-walled, dormant structures formed during times of environmental stress, often in response to a lack of nutrients. Endospores contain a copy of the cell's DNA, RNA, ribosomes, and essential enzymes that become surrounded by a tough, resistant wall (Figure 26.2). Once walled off from the surrounding cell, metabolic activity within the endospore stops and the rest of the original cell breaks down. Endospores are resistant to high temperatures, UV radiation, and even antibiotics. When conditions become favorable, endospores resume normal growth, becoming actively dividing cells.

Some bacterial endospores are involved in transmitting disease to humans. Examples include tetanus, gangrene, botulism, and anthrax. In the case of anthrax, endospores of *Bacillus anthracis* can survive in soil for decades. Cattle, herbivores, and humans can develop anthrax if they consume or come in contact with endospores. In 2001, anthrax was used in several incidents of bioterrorism. Envelopes containing a mysterious white powder were sent to various government offices—the powder in some of those envelopes was purified endospores of *B. anthracis*.

The Past, Present, and Future of Antibiotics **Antibiotics** are molecules that kill bacteria or stop them from growing. They are produced naturally by a wide array of soil-dwelling bacteria and fungi. In these environments, antibiotics are hypothesized to help cells reduce competition for nutrients and other resources.

The discovery of antibiotics in 1928, their development over subsequent decades, and widespread use starting in the late 1940s gave physicians effective tools to combat many bacterial infections. Unfortunately, extensive use of antibiotics in the late twentieth century in clinics and animal feed led to the evolution of drug-resistant strains of pathogenic bacteria (Ch. 22, Section 22.4). One study found that there are now soil-dwelling bacteria in natural environments that—far from being killed by antibiotics—actually use them as food.

New research indicates that bacteria have another advantage: They usually grow as **biofilms**, dense bacterial colonies enmeshed in a polysaccharide-rich matrix that helps shield the bacteria from antibiotics. Antibiotic-resistant biofilms on medical devices such as catheters are a growing problem in hospitals. Biofilms also grow on surfaces such as phone touchscreens, plastic food containers, and teeth.

Role in Bioremediation

Bacteria are often in the news because of their dire medical effects. However, only a tiny proportion of prokaryotes actually cause disease in humans or other organisms. In the vast majority of cases, bacteria and archaea either have no direct impact on humans or are beneficial. In fact, microbes play an important role in wastewater treatment efforts, and researchers are using them to clean up sites polluted with organic solvents—an effort called **bioremediation**.

Some of the most serious pollutants in soils, rivers, and ponds consist of organic compounds that were originally used as solvents or fuels but leaked or were spilled into the environment. Most of these compounds are highly hydrophobic and because they do not dissolve in water, they tend to accumulate in sediments. If these compounds are subsequently ingested by burrowing worms, clams, or other organisms, they can be passed along to fish, insects, humans, birds, and other consumers. At moderate to high concentrations, these pollutants are toxic to eukaryotes. Petroleum from oil spills and compounds that contain carbon ring structures and chlorine atoms, such as the family of compounds called dioxins, are particularly notorious because of their toxicity to humans. Fortunately, naturally existing populations of bacteria and archaea can grow in spills and degrade the toxins. This growth can be enhanced using two complementary bioremediation strategies:

1. ***Fertilizing contaminated sites to encourage the growth of existing prokaryotes that degrade toxic compounds ("biostimulation")*** After several oil spills, researchers added nitrogen to affected sites as a fertilizer. Dramatic increases occurred in the growth of bacteria and archaea that use hydrocarbons in cellular respiration, probably because the cells used the added nitrogen to synthesize enzymes and other key compounds. In at least some cases, the fertilized shorelines cleaned up much faster than unfertilized sites.
2. ***"Seeding"—or adding—specific species of prokaryotes to contaminated sites*** Seeding shows promise of alleviating pollution in some situations. For example, researchers have recently discovered bacteria that can render harmless certain chlorinated compounds, such as polychlorobiphenyls (PCBs). Instead of being poisoned by the pollutants, these bacteria use the chlorinated compounds as electron acceptors during cellular respiration. In at least some cases, the by-product is dechlorinated and nontoxic to humans and other eukaryotes.

Researchers and civil engineers have used bioremediation successfully in many cases. For example:

- ***London's Olympic Park*** The site of the 2012 Olympics in London had been polluted due to years of industrial activity. Bioremediation helped to clean soil and groundwater by using archaea to convert ammonia into nitrogen gas.
- ***Deepwater Horizon oil spill*** Microbes such as *Alcanivorax borkumensis* were used to break down oil that had spilled into the Gulf of Mexico in 2010.

CHECK YOUR UNDERSTANDING

✔ **If you understood this section, you should be able to ...**

1. Describe the environmental conditions that extremophiles are able to live in.
2. Explain the role of endospores in the life cycle of certain prokaryotes.
3. Describe how bacteria and archaea can be used to help clean up polluted environments.

Answers are available in Appendix A.

26.2 How Do Biologists Study Bacteria and Archaea?

After you complete this section, you should be able to ...

- Describe the methods used to study bacteria and archaea.

Biologists' understanding of the domains Bacteria and Archaea is advancing more rapidly right now than at any time during the past 100 years—and perhaps faster than their understanding of any other lineages on the tree of life. As an introduction to the domains Bacteria and Archaea, let's examine a few of the techniques that biologists use to answer questions about them. Some of these research strategies have been used since bacteria were first discovered; some were invented in the past decade.

Using Enrichment Cultures

Which species of prokaryotes are present at a particular location, and what do they use as food? To answer questions like these, biologists rely heavily on their ability to culture organisms in the lab. Of the 10,000 species of bacteria and archaea that have been formally described to date, many were discovered when they were isolated from natural habitats and grown under controlled conditions in the laboratory.

One classical technique for isolating new types of bacteria and archaea is called **enrichment culture**. Enrichment cultures are based on establishing a specified set of growing conditions—temperature, lighting, substrate, types of available food, and so on. Cells that thrive under the specified conditions increase in numbers enough to be isolated and studied in detail.

To appreciate how this strategy works in practice, consider research on bacteria that live deep below Earth's surface. One study began with samples of rock and fluid from drilling operations in Virginia and Colorado. The samples came from sedimentary rocks at depths ranging from 860 to 2800 meters below the surface, where temperatures are between 42°C and 85°C. The questions posed in the study were simple: Is anything alive down there? If so, what do the organisms use to fuel cellular respiration?

The research team hypothesized that if organisms were living deep below Earth's surface, the cells might use hydrogen molecules (H_2) as electron donors and ferric ions (Fe^{3+}) as electron acceptors (**Figure 26.3**). Ferric ions—the oxidized form of iron—are abundant in the rocks the biologists collected from great depths in the form of ferric oxyhydroxide, $Fe(OH)_3$. The researchers

RESEARCH

QUESTION: Can bacteria live a mile below Earth's surface?

HYPOTHESIS: Bacteria are capable of cellular respiration deep below Earth's surface by using H_2 as an electron donor and Fe^{3+} as an electron acceptor.

NULL HYPOTHESIS: Bacteria from this environment are not capable of using H_2 as an electron donor and Fe^{3+} as an electron acceptor.

EXPERIMENTAL SETUP:

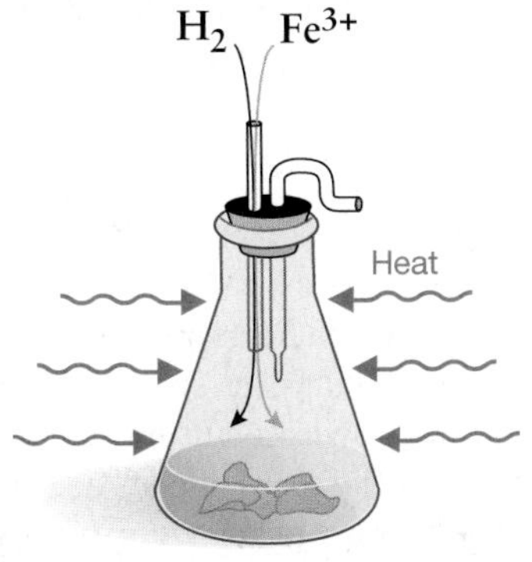

1. Prepare enrichment culture abundant in H_2 and Fe^{3+}; raise temperatures above 45°C.

2. Add rock and fluid samples extracted from drilling operations at depths of about 1000 m below Earth's surface.

PREDICTION OF HYPOTHESIS: Black, magnetic grains of magnetite (Fe_3O_4) will accumulate because Fe^{3+} is reduced by growing cells and shed as a waste product. Cells will be visible.

PREDICTION OF NULL HYPOTHESIS: No magnetite will appear. No cells will grow.

RESULTS: Cells are visible, and magnetite is detectable.

CONCLUSION: At least one bacterial species that can live deep below Earth's surface grew in this enrichment culture. Different culture conditions might result in the enrichment of different species present in the same sample.

Figure 26.3 Enrichment Cultures Isolate Large Populations of Cells That Grow under Specific Conditions.

SOURCE: S. V. Liu, J. Zhou, C. Zhang, et al. 1997. Thermophilic Fe(III)-reducing bacteria from the deep subsurface: The evolutionary implications. *Science* 277: 1106–1109.

✔ **PROCESS OF SCIENCE** Suppose no organisms had grown in this culture. Explain why the lack of growth would be either strong or weak evidence in determining whether organisms live a mile below Earth's surface.

predicted that if an organism in the samples reduced the ferric ions during cellular respiration, then a black, oxidized, and magnetic mineral called magnetite (Fe_3O_4) would start appearing in the cultures as a by-product of cellular respiration.

What did the team's enrichment cultures produce? In some cultures, a black compound began to appear within a week. An array of tests confirmed that the black substance was indeed magnetite. As the "Results" section of Figure 26.3 shows, microscopy revealed the organisms themselves—bacteria previously unknown to science. Because they grow only when incubated at 45°C–75°C, these organisms are considered **thermophiles** ("heat-lovers"). The discovery was spectacular—one of the first studies to demonstrate that Earth's crust is teeming with organisms to depths of over a mile below the surface.

Using Metagenomics

While enrichment culture continues to be a productive way to isolate and characterize new species of bacteria and archaea, biologists estimate that of all the prokaryotic species living today, less than 1 percent have been grown in culture. To augment research based on enrichment cultures, biologists are employing a technique called **metagenomics**, or **environmental sequencing**.

Metagenomics is being used to catalog all the genes present in a mixed community of prokaryotes. It is based on extracting and sequencing much of the DNA from an environmental sample that contains numerous unknown species. Genes present in the community are identified by comparing the DNA sequences with those of known genes. In this way, previously undiscovered species, genes, and biochemical pathways can be characterized.

Here is a summary of the steps performed in a metagenomics study:

1. Collect samples from an environment containing a mixed community of unknown organisms and extract DNA.
2. Generate small DNA fragments and sequence as many as possible.
3. Compare these sequences with those of known genes. This information can be used to identify species and investigate biological processes.

In combination with **direct sequencing**—a technique based on isolating and sequencing a specific gene from organisms found in a particular habitat—metagenomics is revolutionizing biologists' understanding of prokaryotic diversity.

Investigating the Human Microbiome

With the advent of next-generation sequencing techniques (Ch. 20, Section 20.3), the pace of metagenomic analysis has accelerated. Biologists are rapidly acquiring information about the bacteria and archaea found in all sorts of environments, including those that make up the human microbiome (Figure 26.4). Recall that the human microbiome encompasses all the microbes that inhabit the body or parts of the body, such as the mouth, gut, and other tissues. Some biologists are even exploring the ancient microbiome preserved on fossilized human teeth.

In one microbiome study that focused on bacteria, biologists extracted DNA from 125 human fecal samples and identified over 500 billion base pairs of bacterial DNA sequence, over 150 times more than the entire human genome. The results they obtained are fascinating:

- In total, the samples contained about 1000 different species of bacteria. Some species were found in most of the samples while others were found in only a few of the humans sampled.
- The identified bacterial genes that were shared by all of the human subjects suggest that bacteria play important roles in human physiology, including digestion of complex carbohydrates and synthesis of essential amino acids and vitamins.

The take-home message? Humans harbor a diverse ecosystem of symbiotic prokaryotes. The human gut microbiome alone is composed of 100 trillion bacteria and archaea that play a key role in human health. Some microbes may make us sick, but we depend on many others to stay healthy. Changes in the gut microbiome, for example, have been linked to both inflammatory bowel disease and obesity. Results like these have led some biologists to conclude, "You are what you host."

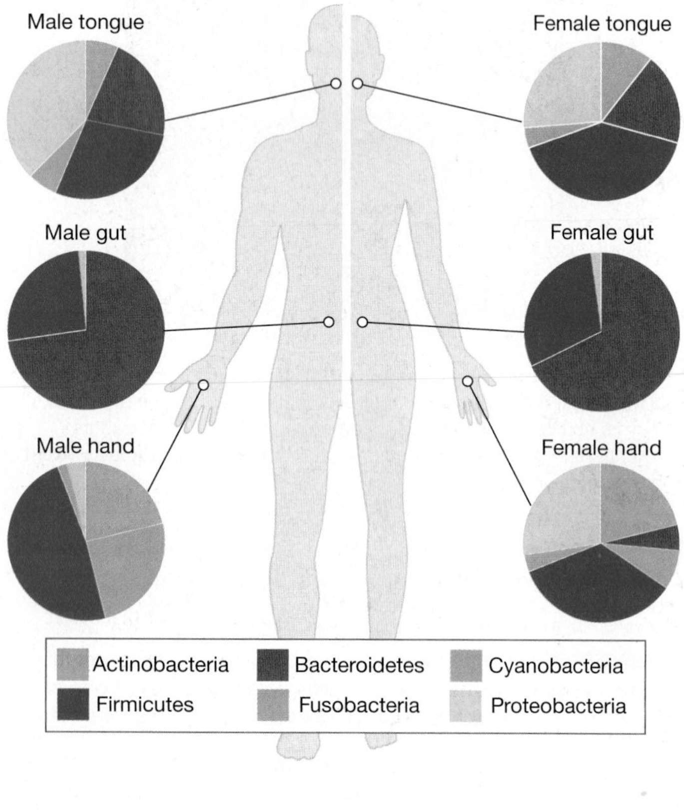

Figure 26.4 The Human Body Is Home to Diverse Microbiomes. Metagenomic analysis is used to identify the major groups of prokaryotes found in different parts of the body.

DATA: Caporaso, J.G., et al., 2011. *Genome Biology,* 12:R50

Evaluating Molecular Phylogenies

Biologists use data from enrichment culture and metagenomic studies to build more accurate phylogenetic trees. Recall that phylogenetic trees illustrate the evolutionary relationships among species and lineages (Chapters 1 and 25; **BioSkills 13**).

Some of the most useful phylogenetic trees for the Bacteria and the Archaea have been based on studies of the RNA molecules found in the small subunit of ribosomes, or what biologists call 16S ribosomal RNA (rRNA) in prokaryotes and 18S rRNA in eukaryotes. (See Ch. 17, Section 17.5 for more information on the structure and function of ribosomes.) In the late 1960s Carl Woese and colleagues began a massive effort to determine and compare the base sequences of 16S and 18S rRNA molecules from a wide array of species. The result of their decade-long analysis was the three-domain **tree of life**, which forms the framework for the phylogeny of bacteria and archaea shown in Figure 26.5.

Woese's tree is now considered a classic result. Before its publication, biologists thought that the major division among organisms was between prokaryotes and eukaryotes. But based on data from ribosomal RNA molecules, the major divisions of life-forms were acknowledged to be the Bacteria, Archaea, and Eukarya.

Tracing the early evolutionary history of these domains is extremely difficult, however, because the events distinguishing the lineages took place so long ago. In addition, lateral gene transfer (Ch. 20, Section 20.4) has blurred the boundaries of the domains.

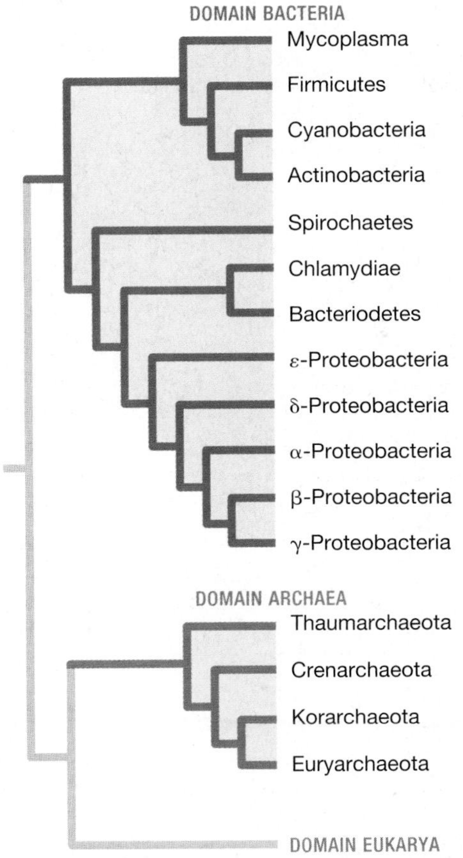

Figure 26.5 **Phylogeny of Some Major Lineages in Bacteria and Archaea.** Note that Bacteria and Archaea represent two separate monophyletic groups.

As new experiments are conducted and new data are collected, our understanding of the relationships among ancient lineages continues to become clearer.

For example, one hypothesis, first proposed in the mid-1980s, suggests that the tree of life is better represented by just two domains. In the two-domain scheme, Bacteria is one domain and the rest of life is the other; so all Eukarya are nested within—that is, descended from—Archaea. This hypothesis is gaining support based on results from whole-genome sequencing (Ch. 20, Section 20.3). But in 2014, another group of researchers put forth an entirely different model: Archaea form the root of the tree, with Bacteria and Eukarya more closely related to each other than either is to Archaea. The debate surrounding the relationships among these ancient lineages continues.

Although scientists continue to puzzle over the relationships among Bacteria, Archaea, and Eukarya, analyses of morphological and molecular characteristics have succeeded in identifying a large series of **monophyletic groups** (clades or lineages) within Bacteria and Archaea. The phylogenetic tree in Figure 26.5 summarizes recent results, but it is still considered highly provisional.

CHECK YOUR UNDERSTANDING

✔ **If you understood this section, you should be able to . . .**

1. **PROCESS OF SCIENCE** Design an enrichment culture that would isolate species that could be used to clean up oil spills.
2. **PROCESS OF SCIENCE** Outline a study designed to identify the bacteria and archaea present in a soil sample near the biology building on your campus.

Answers are available in Appendix A.

26.3 What Themes Occur in the Diversification of Bacteria and Archaea?

After you complete this section, you should be able to . . .

- Describe reproductive, morphological, and metabolic diversity among prokaryotes.

At first, the diversity of prokaryotes can seem almost overwhelming. To make sense of the variation among lineages and species, biologists focus on three themes in diversification: (1) genetic variation through gene transfer, (2) morphology, and (3) metabolism. Let's explore each in turn.

Genetic Variation through Gene Transfer

Lateral gene transfer (introduced in Ch. 20, Section 20.4) is central to the evolution of bacteria and archaea. Through this process, prokaryotes can acquire diverse traits—such as antibiotic resistance—that would otherwise not be available when cells divide by **binary fission** (Ch. 12, Section 12.2). Gene transfer occurs in three ways:

1. **Transformation**—when bacteria or archaea naturally take up DNA from the environment that has been released by cell lysis or secreted (Figure 26.6a, page 542);
2. **Transduction**—when viruses pick up DNA from one prokaryotic cell and transfer it to another cell (Figure 26.6b); and

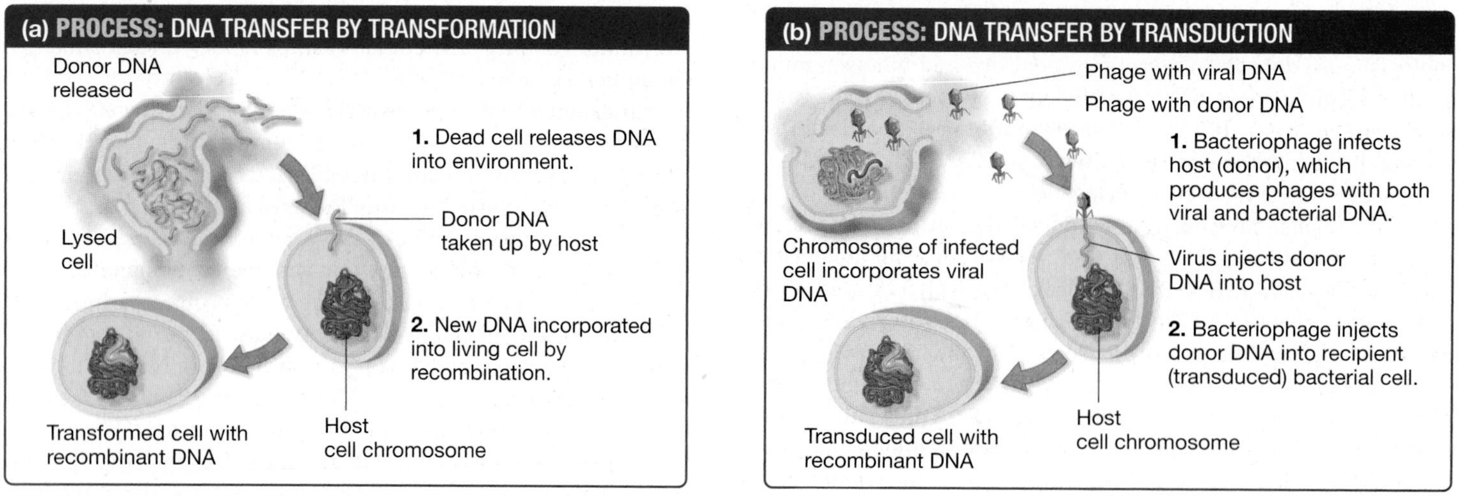

Figure 26.6 Prokaryotes Can Acquire New DNA via Transformation or Transduction.

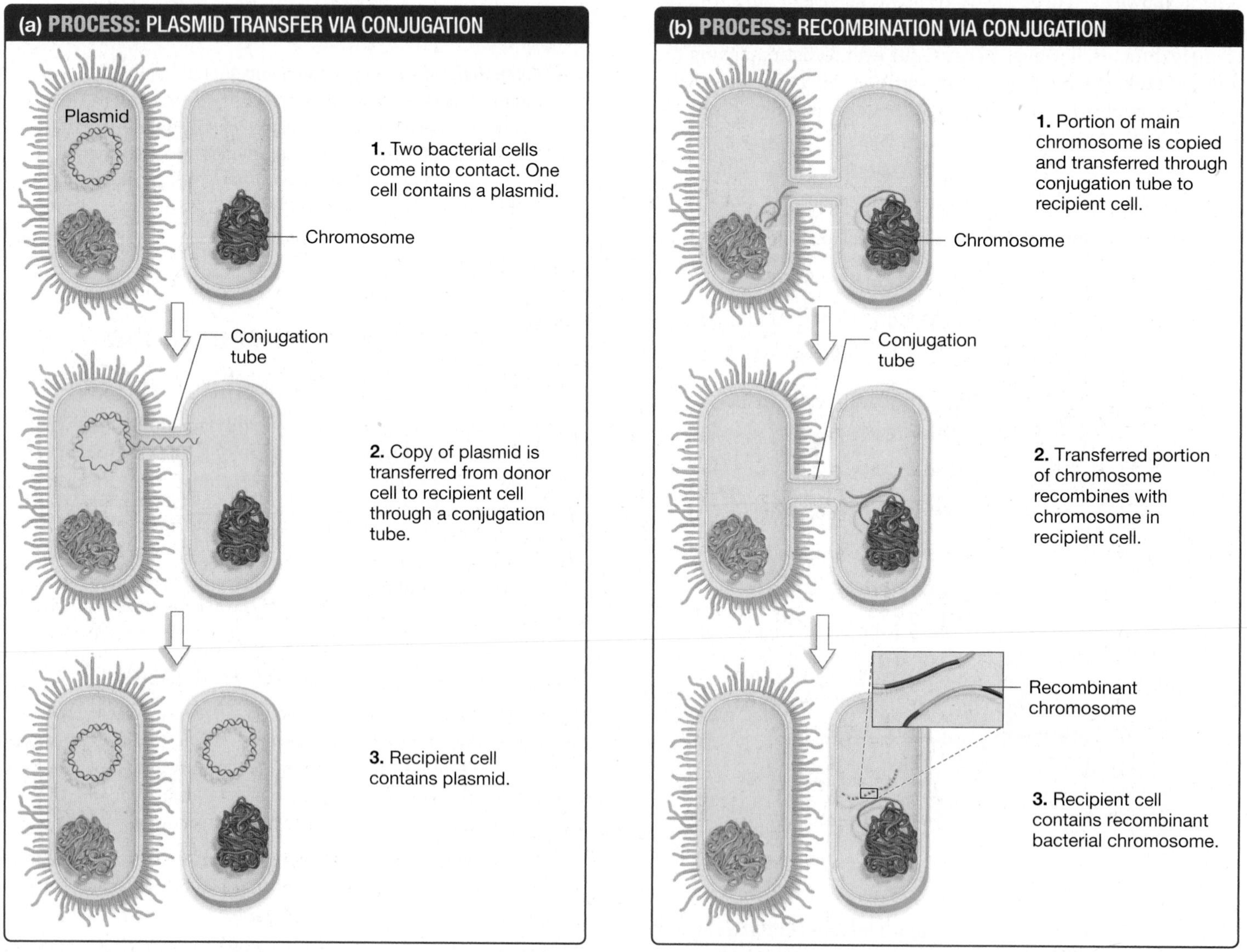

Figure 26.7 Prokaryotic "Sex," or Conjugation, Is One-Way and Involves Only a Few Genes.

3. **Conjugation**—when genetic information is transferred by direct cell-to-cell contact (**Figure 26.7**).

Let's take a closer look at conjugation. Recall that in many eukaryotes, sexual reproduction occurs when haploid cells that are formed by meiosis combine to form a new diploid individual (Ch. 13, Section 13.2). Sexual reproduction does not exist in prokaryotes, though. Bacteria and archaea are haploid throughout their lives and do not undergo meiosis. Genes can move from one individual to another, however, via conjugation.

Conjugation takes place when a physical connection, called a conjugation tube, forms between two prokaryotic cells. In many cases conjugation is followed by an event called plasmid transfer. As Figure 26.7a shows, a small circular piece of DNA called a **plasmid** is copied in one cell and transferred to the other cell. (Plasmids were introduced in Ch. 7, Section 7.1.) Plasmids often contain accessory genes that aren't required for normal growth, but that can be advantageous to the bacterium. For example, genes responsible for antibiotic resistance are found on plasmids.

Conjugation can also result in genetic recombination. This occurs when a plasmid that has become integrated into the main bacterial chromosome is copied and transferred, along with genes from the main bacterial chromosome, through the conjugation tube to a recipient cell (see Figure 26.7b). In some cases the transferred portion of chromosome replaces the homologous portion in the recipient cell's original chromosome, resulting in a new combination of alleles on that chromosome—meaning, recombination.

Figure 26.7 highlights two key points about bacterial "sex": **(1)** It is a one-way transfer of genetic material instead of an exchange between individuals, and **(2)** instead of involving all the genes present, the transfer is limited to a plasmid or a small portion of the genes in the main chromosome.

Morphological Diversity

Different species of bacteria and archaea can be fairly diverse in morphology. For example, prokaryotes are found in a wide variety of sizes and shapes. Some are motile while others are not. Most bacteria have cell walls, but the chemical composition of cell walls can be significantly different from one species to the next. Let's explore some of the morphological variation found among prokaryotes.

Bacterial Size, Shape, and Motility To appreciate how diverse these organisms are in their morphology, consider bacteria alone:

- ***Size*** Bacterial cells range in size from the smallest of all free-living cells—bacteria called mycoplasmas with volumes as small as 0.15 μm^3—to the largest bacterium known, *Thiomargarita namibiensis*, which has volumes as large as 200×10^6 μm^3. Over a billion *Mycoplasma* cells could fit inside an individual *T. namibiensis* (**Figure 26.8a**).
- ***Shape and Arrangement*** Bacterial cells range in shape from filaments, spheres (cocci), and rods (bacilli) to spirals (**Figure 26.8b**). They are often arranged in chains (the strep arrangement) or clusters (the staph arrangement).

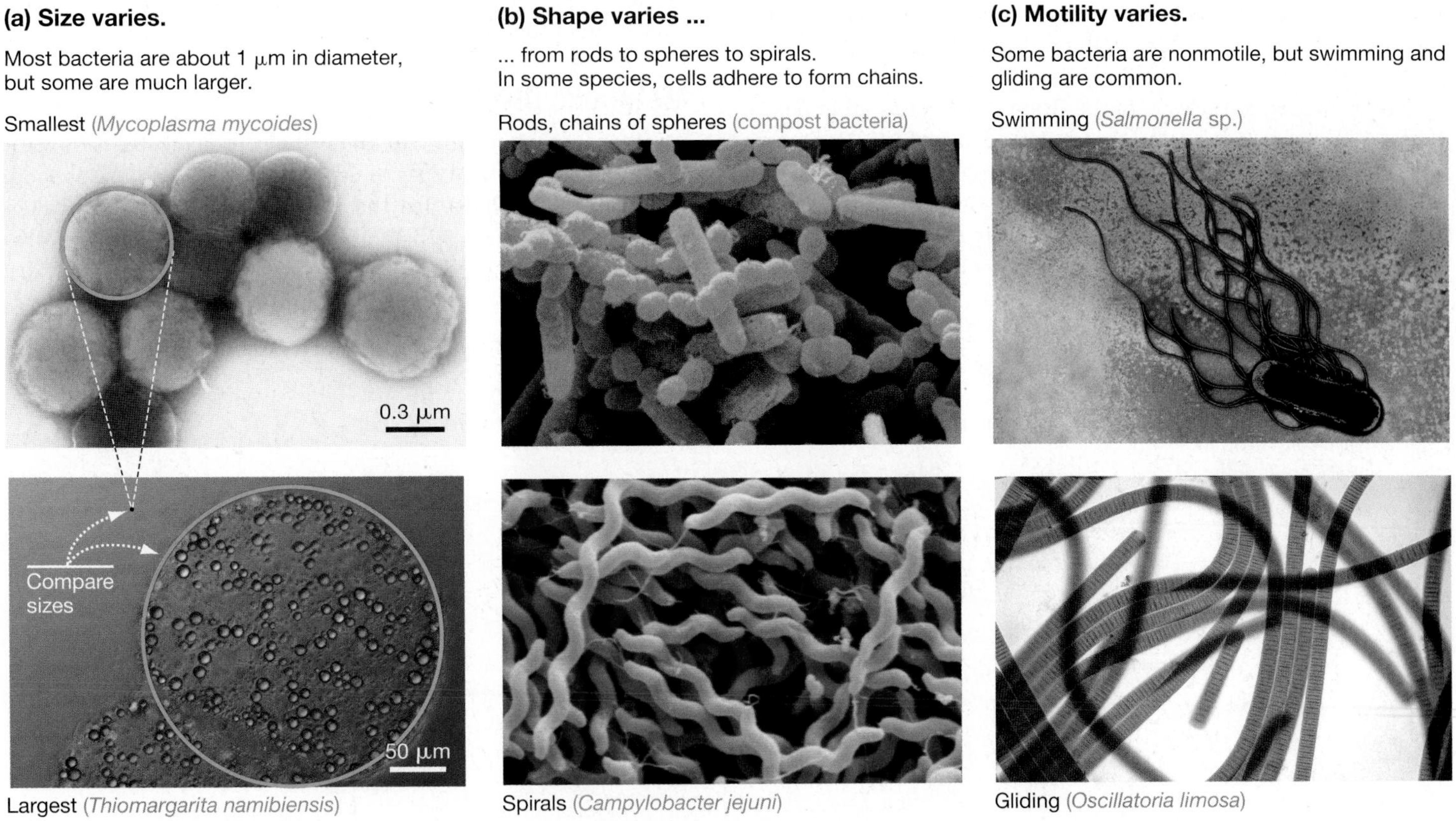

Figure 26.8 Morphological Diversity among Bacteria Is Extensive. Some of the cells in these micrographs have been colorized to make them more visible.

✔ **QUANTITATIVE** The diameter of *Thiomargarita* is roughly how many times larger than that of *Mycoplasma*?

- ***Motility*** Many bacterial cells are motile; their swimming movements are powered by rotating flagella. Instead of undulating back and forth, as is the case in eukaryotic flagella, bacterial flagella spin like a propeller. Depending on the direction of rotation, cells either swim ahead or tumble, which allows them to change direction. (Bacterial flagella may represent the only biological wheel found in organisms; see Ch. 7, Figure 7.5.) Gliding movement, which enables cells to creep along a surface, also occurs in several groups, though the molecular mechanism responsible for this form of motility is still unknown (Figure 26.8c).

Cell-Wall Composition For single-celled organisms, the composition of the plasma membrane and cell wall are particularly important. There are two key differences between the plasma membranes and cell walls of bacteria versus archaea. Archaea have unique phospholipids in their plasma membranes—the hydrocarbon tails of the phospholipids are made from isoprene (Chapter 6), while bacteria have a unique compound called peptidoglycan in their cell walls.

Within bacteria, biologists distinguish between two general types of cell wall by using a dyeing system called the **Gram stain**. As Figure 26.9 shows, Gram-positive cells look purple but Gram-negative cells look pink.

At the molecular level, most cells that are **Gram-positive** have a plasma membrane surrounded by a cell wall with extensive peptidoglycan (Figure 26.10a). You might recall that peptidoglycan is a complex substance composed of carbohydrate strands that are cross-linked by short chains of amino acids (Ch. 5, Section 5.2). Most cells that are **Gram-negative**, in contrast, have a plasma membrane surrounded by a cell wall that has two components—a thin gelatinous layer containing peptidoglycan and an outer phospholipid bilayer (Figure 26.10b).

Analyzing cell cultures with the Gram stain can be an important preliminary step in treating bacterial infections. Because they contain so much peptidoglycan, Gram-positive cells may respond to treatment by penicillin-like antibiotics that disrupt peptidoglycan synthesis. Gram-negative cells, in contrast, are more likely to be affected by erythromycin or other types of antibiotics that poison bacterial ribosomes.

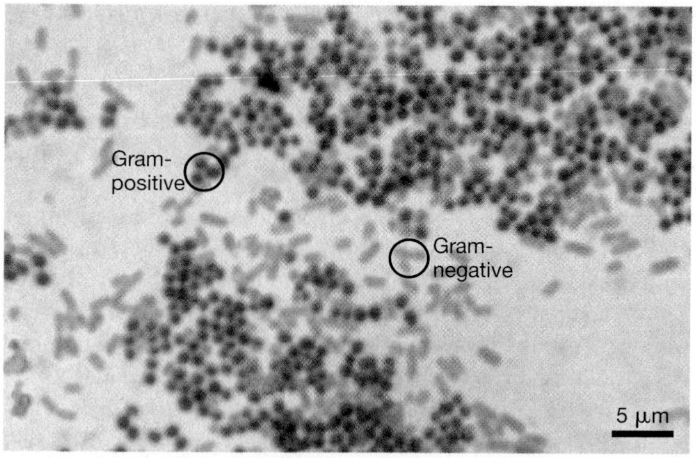

Figure 26.9 Gram Staining Distinguishes Two Types of Cell Walls in Bacteria. As seen in this micrograph, cells with extensive peptidoglycan retain a large amount of stain and look purple. These are referred to as Gram-positive. Others retain little stain and look pink. These are referred to as Gram-negative.

To summarize, members of the Bacteria and the Archaea are remarkably diverse in their overall size, shape, and motility as well as in the composition of their cell walls and plasma membranes. But when asked to name the innovations that were most responsible for the diversification of these two domains, biologists do not point to their morphological diversity. Instead, they point to metabolic diversity—variation in the chemical reactions that go on inside these cells.

Metabolic Diversity

Another important thing to know about prokaryotes is how diverse they are in the types of compounds they can use as food. Bacteria and archaea are the masters of metabolism. Taken together, they can subsist on almost anything—from hydrogen molecules to crude oil. Their biochemical capabilities are dazzling.

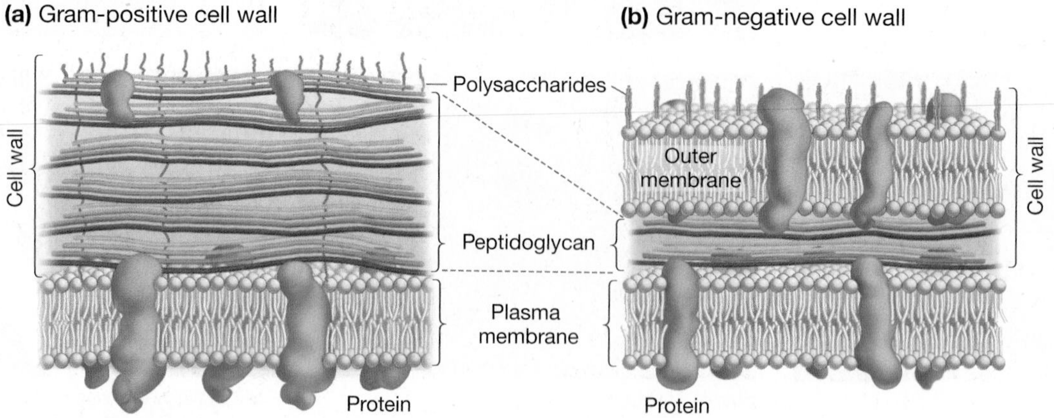

Figure 26.10 Two Types of Cell Walls in Bacteria. (a) Gram-positive bacterial cells have a thick layer of peptidoglycan lining the cell wall. **(b)** Gram-negative bacterial cells have very little of this carbohydrate and have an outer membrane. Thus Gram-positive cells stain more than Gram-negative cells.

Just how varied are bacteria and archaea when it comes to making a living? To appreciate the answer, recall that all organisms have two fundamental nutritional needs—acquiring chemical energy that can be used to make adenosine triphosphate (ATP) and obtaining molecules with carbon–carbon bonds that can be used as building blocks for the synthesis of lipids, proteins, DNA, RNA, and other large, complex compounds required by the cell. Bacteria and archaea acquire energy to produce ATP in three ways:

1. **Phototrophs** ("light-feeders") use light energy to excite electrons. ATP is produced by photophosphorylation.
2. **Chemoorganotrophs** ("chemical–carbon-feeders") oxidize organic molecules with high potential energy, such as sugars. ATP may be produced by cellular respiration—with sugars serving as electron donors—or via fermentation pathways.
3. **Chemolithotrophs** ("chemical–rock-feeders") oxidize inorganic molecules with high potential energy, such as ammonia (NH_3) or hydrogen sulfide (H_2S). ATP is produced by cellular respiration, and inorganic compounds serve as the electron donor.

Bacteria and archaea fulfill their second nutritional need—obtaining building-block compounds with carbon–carbon bonds—in two ways:

1. **Autotrophs** ("self-feeders") synthesize their own compounds from simple starting materials such as CO_2 and methane (CH_4).
2. **Heterotrophs** ("other-feeders") absorb ready-to-use organic compounds—called building-block compounds—produced by other organisms in their environment.

Because there are three distinct ways of producing ATP and two general mechanisms for obtaining carbon, there are a total of six methods for producing ATP and obtaining carbon. The names that biologists use for organisms that employ these six "feeding strategies" are given in **Table 26.3** on page 546.

Of the six possible ways to produce ATP and obtain carbon, just two are observed among eukaryotes (photoautotrophs and chemoorganoheterotrophs). But bacteria and archaea do them all. In addition, certain species of prokaryotes can switch among modes of living, depending on environmental conditions. In their metabolism, eukaryotes are simple compared to bacteria and archaea.

The story of prokaryotic metabolism can be boiled down to two sentences: The basic chemistry required for photosynthesis, cellular respiration, and fermentation originated in Bacteria and Archaea. Then the evolution of variations on each of these processes allowed prokaryotes to diversify into millions of species that occupy diverse habitats. Let's take a closer look.

Producing ATP through Cellular Respiration: Variation in Electron Donors and Acceptors Millions of prokaryotic and eukaryotic species—including animals and some plants—are chemoorganotrophs. These organisms obtain the energy required to make ATP by the process of cellular respiration, breaking down organic compounds such as sugars, starch, or fatty acids.

During cellular respiration in both prokaryotes and eukaryotes, enzymes strip electrons from organic molecules that have high potential energy and then transfer these high-energy electrons to the electron carriers NADH and $FADH_2$ (Chapter 9). These compounds feed electrons to an electron transport chain (ETC), where electrons are stepped down from a high-energy state to a low-energy state (**Figure 26.11a**). In eukaryotic cells the ETC is located in the highly folded inner mitochondrial membrane. However, in bacteria and archaea, the ETC is located in the plasma membrane.

The energy that is released allows components of the ETC to generate a proton gradient across the plasma membrane (**Figure 26.11b**). The resulting flow of protons back through the enzyme ATP synthase results in the production of ATP, via the process called chemiosmosis.

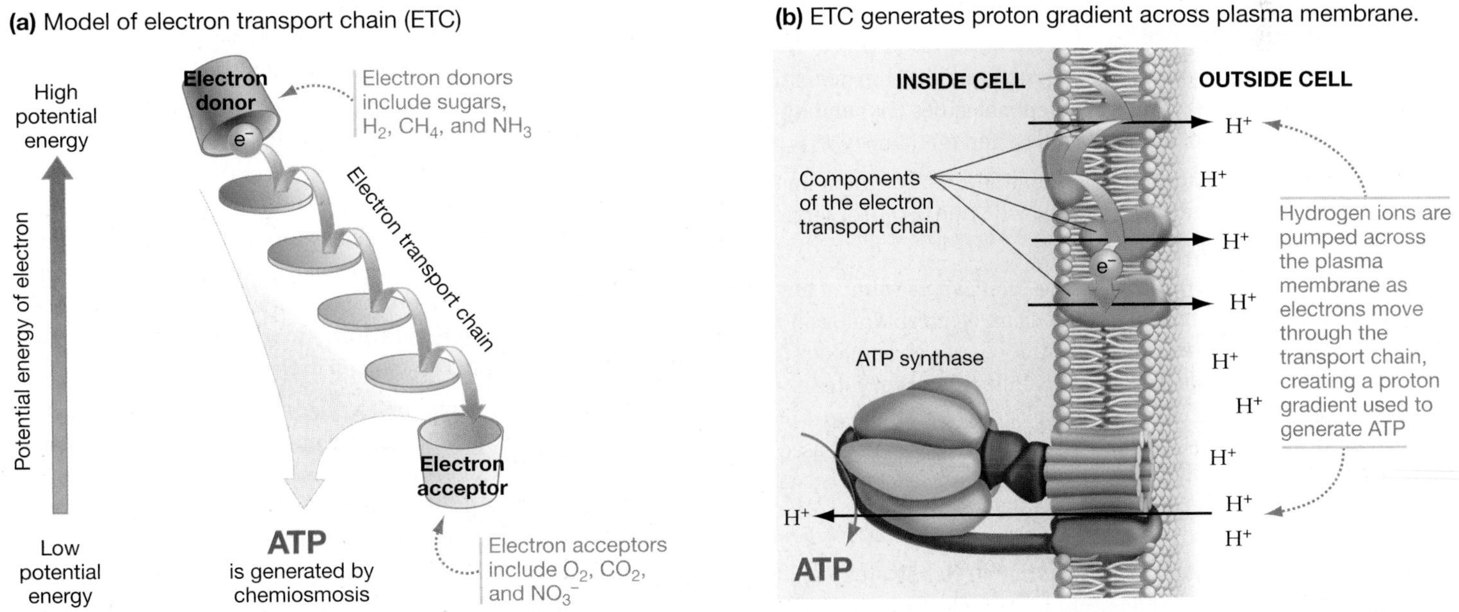

Figure 26.11 Cellular Respiration Is Based on Electron Transport Chains. Protons are pumped across the cell membrane during electron transport. Some protons may diffuse away from the cell, but a gradient will still form.

SUMMARY **Table 26.3 Six General Methods for Obtaining Energy and Carbon–Carbon Bonds**

	Source of C–C Bonds (for synthesis of complex organic compounds)	
Source of Energy (for synthesis of ATP)	**Autotrophs:** self-synthesized from CO_2, CH_4, or other simple molecules	**Heterotrophs:** from molecules produced by other organisms
Phototrophs: from sunlight	Photoautotrophs Example: Cyanobacteria photosynthesizing in a pond	Photoheterotrophs Example: Heliobacteria in waterlogged soil, both photosynthesizing and absorbing organic molecules
Chemoorganotrophs: from organic molecules	Chemoorganoautotrophs Example: *Methylomonas* bacteria that use methane as a source of carbon and energy	Chemoorganoheterotrophs Example: *Escherichia coli* bacteria using organic matter in animal intestines as a source of carbon and energy
Chemolithotrophs: from inorganic molecules	Chemolithoautotrophs Example: Ocean-dwelling *Nitrosopumilus maritimus* archaea, using ammonia waste excreted by fish as an electron donor and CO_2 as a source of carbon	Chemolithoheterotrophs Example: *Beggiatoa* bacteria in a hydrothermal vent, using hydrogen sulfide for energy and absorbing organic molecules as a source of carbon

The list that follows summarizes key points regarding cellular respiration carried out by most eukaryotes and many prokaryotes.

- Organic compounds with high potential energy—often glucose—serve as the original electron donor. When cellular respiration is complete, glucose is completely oxidized to CO_2, which is given off as a by-product.
- Oxygen is the final electron acceptor, and water is also produced as a by-product.

This kind of cellular respiration is called aerobic, because oxygen is used as the final electron acceptor.

However, many prokaryotes employ an electron donor other than sugars and an electron acceptor other than oxygen during cellular respiration. These species produce by-products other than carbon dioxide and water.

- Molecules with high potential energy serve as electron donors. Electron donors range from hydrogen molecules (H_2) and hydrogen sulfide (H_2S) to ammonia (NH_3) and methane (CH_4).
- Compounds with relatively low potential energy—including sulfate (SO_4^{2-}), nitrate (NO_3^-), carbon dioxide (CO_2), or ferric ions (Fe^{3+})—act as electron acceptors.

Researchers have identified prokaryotes that can use almost any compound with high potential energy as an electron donor and almost any compound with low potential energy as an electron acceptor. The remarkable metabolic diversity of prokaryotes explains why they play such a key role in cleaning up some types of pollution. Species that use organic solvents or petroleum-based fuels as electron donors or electron acceptors may excrete waste products that are less toxic than the original compounds.

Producing ATP via Fermentation: Variation in Substrates One strategy for making ATP that does not involve electron transport chains is called **fermentation** (Ch. 9, Section 9.6). In fermentation, no outside electron acceptor is used.

Because fermentation is a much less efficient way to make ATP compared with cellular respiration, in many species it occurs as an alternative metabolic strategy when no electron acceptors are available to make cellular respiration possible. In other species, fermentation does not occur at all. In still other prokaryotes, fermentation is the *only* way that cells make ATP.

Prokaryotes that produce ATP via fermentation use a wide variety of substrates. For example:

- Species of *Clostridium* ferment complex carbohydrates (including cellulose or starch), proteins, purines, or amino acids. Species that ferment amino acids produce by-products with names such as cadaverine and putrescine. These molecules are responsible for the odor of rotting flesh.
- Other bacteria can ferment lactose, a prominent component of milk. In some species this fermentation has two end products: propionic acid and CO_2. Propionic acid is responsible for the taste of Swiss cheese, and the CO_2 produced during fermentation creates the holes in Swiss cheese.

Producing ATP via Photophosphorylation: Variation in Electron Sources and Pigments Instead of using molecules as a source of high-energy electrons, phototrophs pursue a radically different strategy: **photophosphorylation**. Among bacteria and archaea, photophosphorylation can happen in three different ways:

1. Light activates a pigment called bacteriorhodopsin or archaerhodopsin, which uses the absorbed energy to transport protons across the plasma membrane and out of the cell. The resulting flow of protons back into the cell drives the synthesis of ATP via chemiosmosis (Ch. 9, Section 9.5).
2. A recently discovered bacterium that lives near hydrothermal vents on the ocean floor performs photophosphorylation not by absorbing sunlight, but by absorbing faint light from geothermal radiation.

3. Pigments that absorb light raise electrons to high-energy states. As these electrons are stepped down to lower energy states by electron transport chains, the energy released is used to generate ATP. Researchers have isolated nine different chlorophyll pigments from different lineages of bacterial phototrophs. Each lineage has one or more of these distinctive chlorophylls.

An important feature of this last mode of photophosphorylation is that the process requires a source of electrons (Ch. 10, Section 10.3). Recall that during **photosynthesis** in cyanobacteria and plants, the required electrons come from water. When these organisms "split" water molecules apart to obtain electrons, they generate oxygen as a by-product. Species that use water as a source of electrons are said to complete **oxygenic** ("oxygen-producing") photosynthesis.

In contrast, many phototrophic bacteria use a molecule other than water as the source of electrons. In some cases, the electron donor is hydrogen sulfide (H_2S); a few species can use the ion known as ferrous iron (Fe^{2+}). Instead of producing oxygen as a by-product of photosynthesis, these cells produce elemental sulfur (S) or the ferric ion (Fe^{3+}). This type of photosynthesis is said to be **anoxygenic** ("no oxygen-producing").

Obtaining Building-Block Compounds: Variation in Pathways for Fixing Carbon In many autotrophs, including plants, the enzymes of the Calvin cycle transform carbon dioxide (CO_2) into organic molecules that can be used in synthesizing cell material (Ch. 10, Section 10.4). The carbon atom in CO_2 is reduced during the process and is said to be "fixed." Animals and fungi, in contrast, are heterotrophs. They obtain carbon from living plants or animals, or by absorbing the organic compounds released as dead tissues decay.

Prokaryotes pursue these same two strategies, autotrophy and heterotrophy. However, not all autotrophic bacteria and archaea use the Calvin cycle to make building-block molecules, and not all start with CO_2 as a source of carbon atoms. For example, consider these biochemical pathways:

- Some bacteria are called **methanotrophs** ("methane-eaters") because they use methane (CH_4) as their carbon source. (They also use CH_4 as an electron donor in cellular respiration.) Methanotrophs process CH_4 into more complex organic compounds via one of two enzymatic pathways, depending on the species.
- Some archaea are considered **methanogens**. These organisms produce methane as a by-product of cellular respiration. The formation of methane in the guts of humans and other mammals is largely due to the activity of methanogens.

These observations highlight an important message from this chapter: Compared with eukaryotes, the metabolic capabilities of prokaryotes are remarkably complex and diverse.

Ecological Diversity and Global Impacts

The metabolic diversity observed among bacteria and archaea explains why these organisms can thrive in such a wide array of habitats.

- The array of electron donors, electron acceptors, and fermentation substrates they exploit allows heterotrophic prokaryotes to live just about anywhere.
- The evolution of three types of photophosphorylation—based on bacteriorhodopsin or archaerhodopsin, geothermal energy, or pigments that donate high-energy electrons to ETCs—extends the types of habitats that can support phototrophic prokaryotes.

The complex chemistry that these cells carry out, combined with their abundance, has made them potent forces for global change throughout Earth's history. Bacteria and archaea have altered the chemical composition of the oceans, atmosphere, and terrestrial environments for billions of years. They continue to do so today.

The Oxygen Revolution Today, oxygen represents almost 21 percent of the molecules in Earth's atmosphere. But researchers who study the composition of the atmosphere are virtually certain that no free molecular oxygen (O_2) existed for the first 2.3 billion years of Earth's existence. This conclusion is based on two observations:

1. There was no plausible source of oxygen at the time the planet formed.
2. Chemical analysis of the oldest Earth rocks suggests that they formed in the absence of atmospheric oxygen.

Early in Earth's history, the atmosphere was dominated by nitrogen and carbon dioxide. Where did the oxygen we breathe come from? The answer is cyanobacteria.

Cyanobacteria is a lineage of photosynthetic bacteria. According to the fossil record, species of cyanobacteria first became numerous in the oceans about 2.7–2.5 billion years ago. Their appearance was momentous because cyanobacteria were the first organisms to perform oxygenic ("oxygen-producing") photosynthesis (Figure 26.12). The fossil record and geologic record indicate that oxygen concentrations in the oceans and atmosphere began to increase 2.3–2.1 billion years ago. Once oxygen

Figure 26.12 **Cyanobacteria Were the First Organisms to Produce Oxygen during Photosynthesis.** This image shows a mat of cyanobacteria that are releasing oxygen.

was common in the oceans, cells could begin to use it as the final electron acceptor during cellular respiration. *Aerobic* respiration was now a possibility. Before that, organisms had to use compounds other than oxygen as a final electron acceptor—only *anaerobic* respiration was possible. (Aerobic and anaerobic respiration were introduced in Chapter 9.)

The evolution of aerobic respiration was a crucial event in the history of life. Because oxygen is extremely electronegative, it is an efficient electron acceptor. Much more energy is released as electrons move through electron transport chains with oxygen as the ultimate electron acceptor than is released with other substances as the electron acceptor.

To drive home this point, study the graph in Figure 26.13. Notice that the vertical axis plots free energy changes; the graph shows the energy released when glucose is oxidized with iron, nitrate, or oxygen as the final electron acceptor. Once oxygen was available, cells could produce much more ATP for each electron donated by NADH or $FADH_2$. As a result, the rate of energy production rose dramatically.

To summarize, data indicate that cyanobacteria were responsible for a fundamental change in Earth's atmosphere—a high concentration of oxygen. The presence of oxygen allowed formation of the ozone layer, which protected Earth's surface from ultraviolet radiation, making evolution of life on land possible. Never before, or since, have organisms done so much to alter the nature of our planet.

Nitrogen Fixation and the Nitrogen Cycle Organisms must have nitrogen to synthesize proteins and nucleic acids. Although molecular nitrogen (N_2) is extremely abundant in the atmosphere, most organisms cannot use it because of the strong triple bond linking the nitrogen atoms. To incorporate nitrogen atoms into amino acids and nucleotides, all eukaryotes and many prokaryotes have to obtain N in a form such as ammonia (NH_3) or nitrate (NO_3^-).

Certain bacteria and archaea are the only organisms capable of converting molecular nitrogen to ammonia. The steps in the process, referred to as **nitrogen fixation**, are highly endergonic reduction-oxidation (redox) reactions (Ch. 8, Section 8.2). The key enzyme that catalyzes the reaction—nitrogenase—is found only in a few phyla of bacteria and archaea. Many of these organisms are free living, but some form important relationships with plants:

- Some species of cyanobacteria live in association with a water fern that grows in rice paddies and help fertilize the rice plants.
- In terrestrial environments, nitrogen-fixing bacteria live in close association with certain plants (e.g., legumes)—often taking up residence in special root structures called nodules (Ch. 36, Section 36.4).

Why is nitrogenase not found in all organisms? The answer lies in an interesting property of the enzyme. When exposed to O_2, nitrogenase is irreversibly poisoned and is degraded. The only organisms with the nitrogenase gene are those that live in anaerobic habitats or are able to protect the enzyme from O_2.

Nitrogen fixation is only the beginning of the story, however. Bacteria and archaea use a wide array of nitrogen-containing compounds as electron donors and electron acceptors during cellular respiration. To understand why this is important, consider that the nitrite (NO_2^-) produced by some bacteria as a by-product of respiration does not build up in the environment. Instead, other species of bacteria and archaea use it as an electron donor, and it is oxidized to molecular nitrate (NO_3^-). Nitrate, in turn, is reduced to molecular nitrogen (N_2) by yet another suite of prokaryotic species. In this way, bacteria and archaea are responsible for driving the movement of nitrogen atoms through ecosystems around the globe in a process called the **nitrogen cycle** (Figure 26.14).

Similar types of interactions occur with molecules that contain phosphorus, sulfur, and carbon. In this way, bacteria and archaea play a key role in the cycling of these nutrients, too. **✔ If you understand the role of bacteria and archaea in the nitrogen cycle, you should be able to offer a plausible explanation of what the composition of the atmosphere and what the nitrogen cycle might be like if prokaryotes did not exist.**

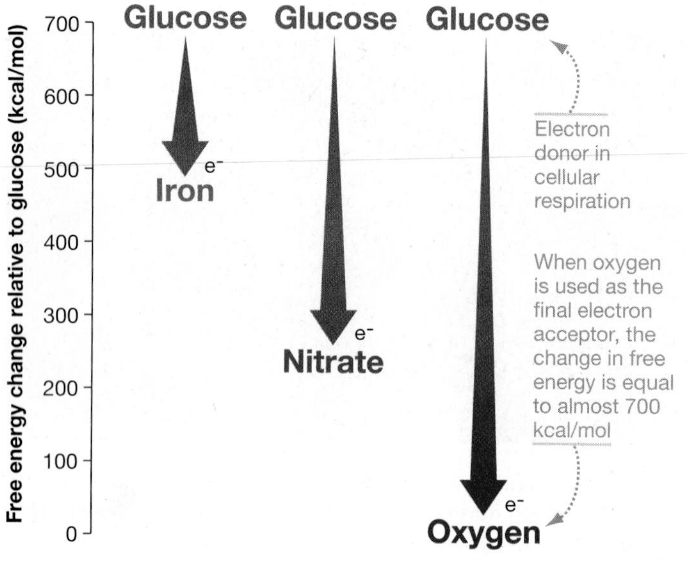

Figure 26.13 Cellular Respiration Can Produce More Energy When Oxygen Is the Final Electron Acceptor. More potential energy in glucose can be released when oxygen is the final acceptor compared to other molecules or ions.

DATA: D. F. Wilson, M. Erecińska, and P. L. Dutton. 1974. *Annual Review of Biophysics and Bioengineering* 3: 203–230; Tables 1 and 3.

✔ QUANTITATIVE If two species of bacteria use glucose as an electron donor, but species A uses iron as the final electron acceptor and species B uses oxygen, then approximately how many times more glucose will species A need to maintain the same metabolic rate as species B?

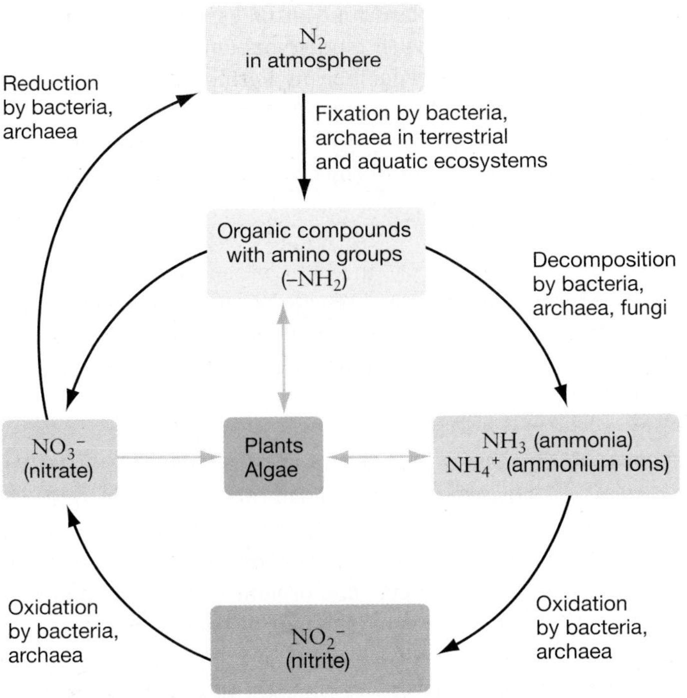

Figure 26.14 The Nitrogen Cycle: Bacteria and Archaea Drive the Movement of Nitrogen Atoms through Ecosystems. Nitrogen atoms cycle in different molecular forms.

✔ **MODEL** Add arrows and labels to indicate that animals ingest amino groups from plants or other animals and release amino groups or ammonia.

Nitrate Pollution Most crop plants—including corn, rice, and wheat—do not live in association with nitrogen-fixing bacteria. To increase yields of these crops, farmers use fertilizers that are high in nitrogen. In parts of the world, massive additions of nitrogen in the form of ammonia are causing serious pollution problems.

To understand why nitrates can pollute the oceans, consider what happens in the Gulf of Mexico (Figure 26.15):

1. When ammonia is added to a cornfield—in midwestern North America, for example—much of it is not absorbed by the growing corn plants.
2. Instead, prokaryotes in the soil use a significant fraction of the ammonia as food. Microbes that use ammonia as an electron donor to fuel cellular respiration release nitrite (NO_2^-) as a waste product. Other microbes use nitrite as an electron donor and release nitrate (NO_3^-).
3. Nitrate molecules are extremely soluble in water and tend to be washed out of soils into groundwater or streams, ultimately ending up the Mississippi River.
4. Nitrates carried by the river are used as a nutrient by cyanobacteria and algae that live in the Gulf. The population numbers of these organisms "bloom" in response.
5. When these organisms die and sink to the bottom of the Gulf, bacteria and archaea and other decomposers use them as food. The decomposers use oxygen as an electron acceptor in cellular respiration.

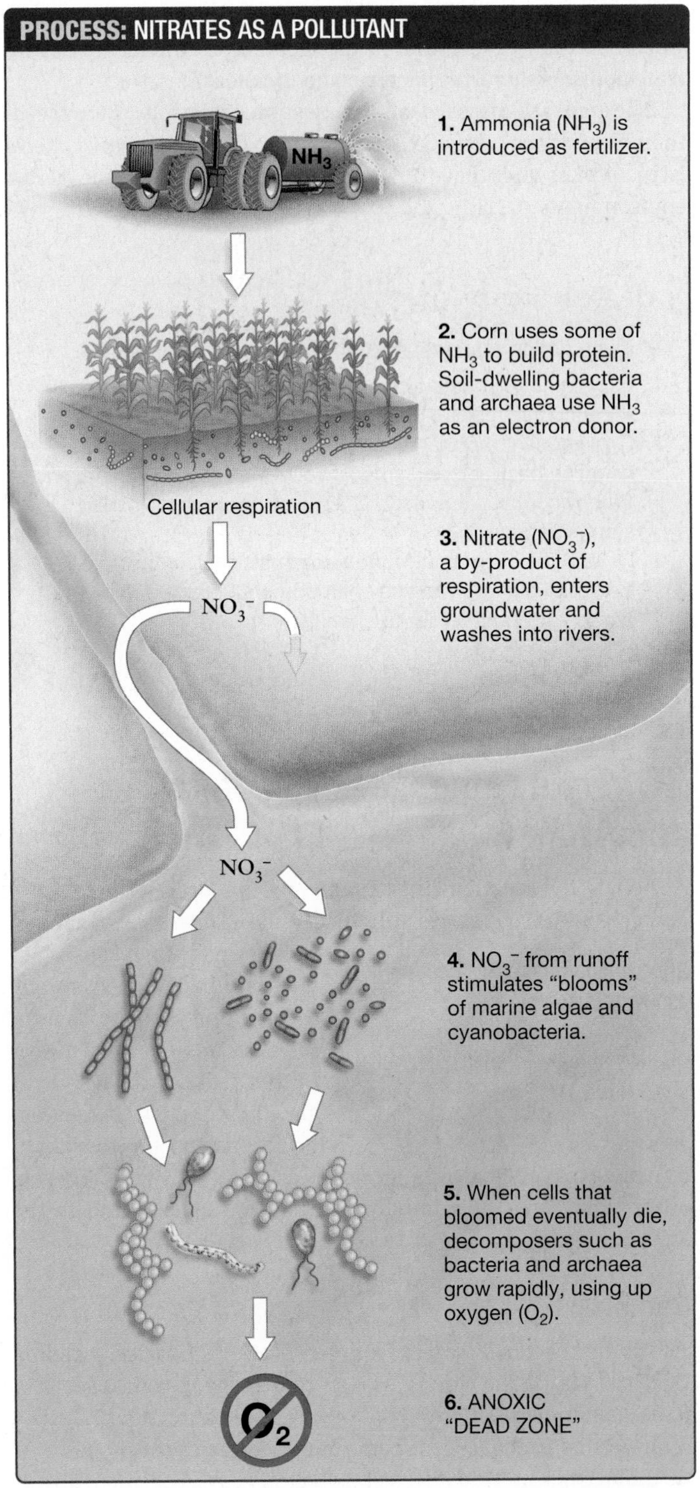

Figure 26.15 Nitrates Act as Pollutants in Aquatic Ecosystems.

6. Oxygen levels in Gulf waters decline significantly, producing an anoxic (lacking in oxygen) "dead zone."

Nitrate pollution has been so severe that large areas in the Gulf of Mexico are anoxic. Lately the dead zone has encompassed, on average, about 14,350 square kilometers (km^2)—roughly the size of the state of Connecticut. The dead zone in the Gulf of Mexico is devoid of fish and other organisms that require oxygen.

Similar problem spots are cropping up in other parts of the world. Virtually every link in the chain of events leading to nitrate pollution involves bacteria and archaea.

The general message of this section is simple: Because of their abundance, ubiquity, and ability to do sophisticated chemistry, prokaryotes have an enormous influence on the global environment.

CHECK YOUR UNDERSTANDING

✔ If you understood this section, you should be able to ...

1. Explain how a bacterium can acquire a plasmid with genes for antibiotic resistance from another bacterium.
2. Explain the structural differences between *E. coli* (a Gram-negative bacillus sometimes involved in food poisoning) and *Staphylococcus aureus* (a Gram-positive coccus involved in staph infections).
3. Defend the claim that photoautotrophic prokaryotes are similar to plants, whereas chemoorganoheterotrophic prokaryotes are similar to animals.

Answers are available in Appendix A.

26.4 Key Lineages of Bacteria and Archaea

After you complete this section, you should be able to ...

- Describe key features of the major lineages of Bacteria and Archaea.

Although the relationships among the major lineages within Bacteria and Archaea are still uncertain in some cases, many of the lineages themselves are well studied. Let's survey the attributes of species from selected major lineages of Bacteria and Archaea, with an emphasis on themes explored earlier in the chapter: their morphological and metabolic diversity, their impacts on humans, and their importance to other species and to the environment.

Bacteria

The name "bacteria" comes from the Greek root *bacter*, meaning "rod" or "staff." The name was inspired because the first bacteria to be seen under a microscope were rod shaped. Later biologists realized that bacterial cells come in a wide variety of shapes.

Biologists who study bacterial diversity currently recognize at least 29 lineages, or phyla, within the domain. Some species are classified into these lineages by distinctive morphological or metabolic characteristics, others by phylogenetic analyses of gene sequence data. The lineages described here and summarized in **Table 26.4** are just a sampling of bacterial diversity.

Actinobacteria Many species in this lineage are filamentous, forming branching chains. Species in the genus *Streptomyces* and *Arthrobacter* are abundant in soil and are important decomposers of dead plant and animal material. Some species live in association with plant roots and fix nitrogen; others can break down toxins such as herbicides, nicotine, and caffeine. Over 3000 distinct antimicrobial compounds have been isolated from species in the genus *Streptomyces*, many of which are now actively prescribed to treat diseases in humans or livestock.

Chlamydiae In terms of numbers of species living today, Chlamydiae may be the least diverse of all major bacterial lineages. Only 13 species are known, all of them spherical and very small. They live as parasitic **endosymbionts** (i.e., they live inside of living host cells).

Cyanobacteria These photosynthetic bacteria are found as independent cells, in chains that form filaments, or in loose aggregations of individual cells called colonies. Cyanobacteria are among the most abundant organisms on Earth. They produce much of the oxygen, nitrogen, and organic compounds that feed organisms living in freshwater and marine environments.

Firmicutes Members of this lineage are extremely common in animal intestines, including yours, where they live in symbiotic mutualism, aiding the digestive process. Several species play useful roles in agriculture and food processing; others cause a variety of human diseases.

Proteobacteria Because its members are so diverse in their morphology and metabolism, this lineage is named after the Greek god Proteus, who could assume many shapes. Some species form stalked cells, while others form aggregates of cells organized as spore-forming **fruiting bodies**. While several species cause disease, others play key roles in the cycling of nitrogen atoms through terrestrial and aquatic ecosystems.

Spirochaetes (Spirochetes) Members of this lineage are distinguished by their unique corkscrew shape and flagella. Instead of extending into the water surrounding the cell, a spirochete's flagella are contained within a structure called the outer sheath, which surrounds the cell. As a flagellum beats, the cell lashes back and forth, moving forward. Parasitic, disease-causing species are propelled by this motion into the tissues of their host. Other spirochete species are extremely common in freshwater and marine habitats. And one species is found only in the hindgut of termites, where it fixes nitrogen.

Archaea

The name "archaea" comes from the Greek root *archae*, for "ancient." The name was inspired by the hypothesis that this is a particularly ancient group. Initially biologists thought that archaea were restricted to hot springs, salt ponds, and other extreme habitats, but we now know they live in virtually all types of habitats.

Recent phylogenies based on DNA sequence data indicate that the domain is composed of several major phyla. A summary

DIVERSITY Table 26.4 Key Lineages of Bacteria

		Morphology	Metabolism	Relevance
	**Actinobacteria** Common in soil and freshwater habitats ◀ This *Streptomyces griseus* produces the antibiotic streptomycin.	Cells found as rods or filaments. Some form chains or branching chains (called mycelia).	Chemoheterotrophs. Use a variety of organic electron donors and oxygen as an electron acceptor. Several are parasites. Some fix nitrogen.	Members of this group cause tuberculosis and leprosy. Produce hundreds of antibiotics (e.g., streptomycin, neomycin, tetracycline, and erythromycin).
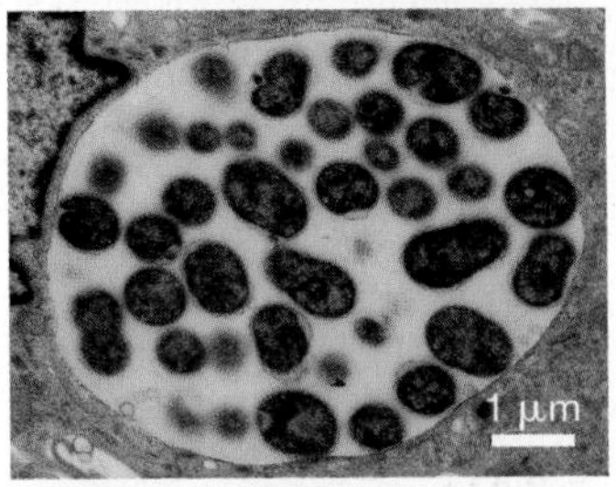	**Chlamydiae** Common in host cells of many vertebrates ◀ *Chlamydia trachomatis* cells infect a human cell.	Spherical cells. Found in clusters. Very small (as small as some viruses).	Chemoheterotrophs. All species live as parasites inside host cells. Can produce ATP by electron transport.	The sexually transmitted disease caused by *Chlamydia trachomatis* can lead to ectopic pregnancy and infertility.
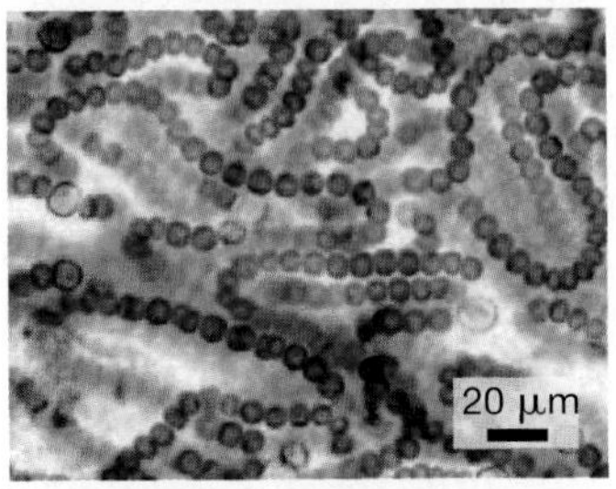	**Cyanobacteria** Common in lakes, rivers, oceans ◀ Cells of this *Nostoc* sp., like all cyanobacteria, contain chlorophyll.	Filaments, spheres, spirals. Individual cells, chains, or colonies. Some contain heterocysts where nitrogen fixation occurs (brown structures in photo).	Photoautotrophs. Involved in nitrogen fixation. Form symbiotic relationships with fungi (called lichens) and with protists, sponges, and legume plants.	Responsible for the origin of Earth's oxygen-rich atmosphere. Some involved in harmful algal blooms. Provide much of the nitrogen used by other organisms.
	**Firmicutes** Common in the human gut ◀ Firmicutes (false-colored yellow and pink) are found among the milk solids (white) in yogurt.	Most are rods or spheres. Many form chains or clusters of four cells. One group produces a cell wall made of cellulose. Some produce a durable resting stage called an endospore.	Chemoheterotrophs. Some fix nitrogen. Some perform anoxygenic photosynthesis. Some can use hydrogen gas as an electron donor.	Members of this group cause anthrax, botulism, tetanus, gangrene, strep throat. *Bacillus thuringiensis* produces BT toxin, an important insecticide. Some are used in yogurt and cheese production.
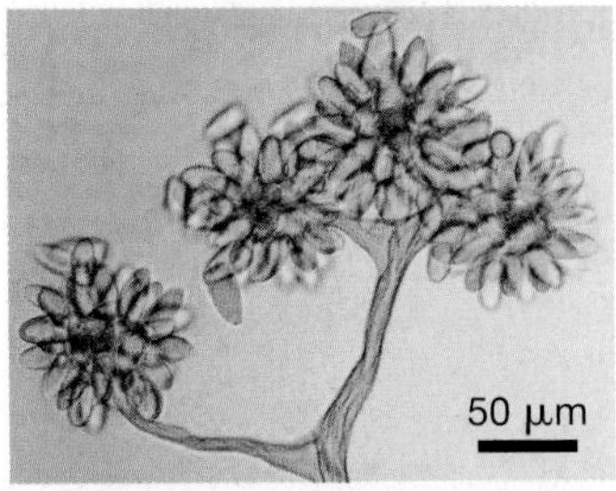	**Proteobacteria** Common in aquatic environments and as pathogens ◀ This *Chondromyces* sp. forms fruiting bodies.	Diverse morphology. Rods, spheres, spirals. Some form colonies that aggregate into a fruiting body and produce reproductive spores at their tips.	Most are heterotrophs or chemoautotrophs. Some (e.g., "purple bacteria") contain bacteriochlorophyll and obtain energy through photosynthesis.	*Escherichia coli* and *Agrobacterium* are often used in biotechnology. Members of this group cause cholera, food poisoning, plague, dysentery, typhus.
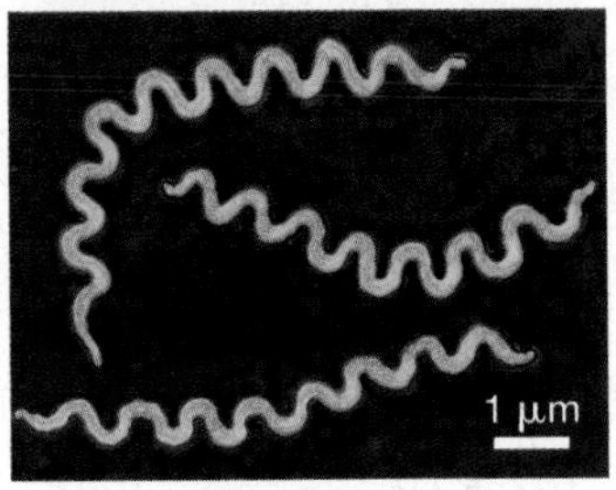	**Spirochaetes (Spirochetes)** Common in the guts of animals and as pathogens ◀ *Treponema pallidum* causes syphilis.	Corkscrew-shaped. Found as individual cells. Flagella are found inside cells and cause cells to move in a spiral fashion.	Chemoheterotrophs. Produce ATP via fermentation. Can thrive in anaerobic conditions.	Members of this group are responsible for leptospirosis, syphilis, and Lyme disease. Corkscrew-like movement enables cells to burrow into host tissue.

For more go to https://goo.gl/B3gha2

DIVERSITY **Table 26.5 Key Lineages of Archaea**

		Morphology	Metabolism	Relevance
	**Crenarchaeota (Eocytes)** Common in sulfur-rich hot springs, acidic environments, and deep-ocean sediments ◀ *Sulfolobus* sp.	Rods, spheres, filaments, and discs. Flagella are common. Some produce protein fibers that help attach to sulfur granules. One species produces a glycoprotein cell wall.	Chemoheterotrophs and chemolithoautotrophs. Use sulfur, hydrogen gas, or Fe^{2+} as electron donors. Some make ATP only by fermentation.	May be the only life-forms in extremely hot, high-pressure, acidic environments.
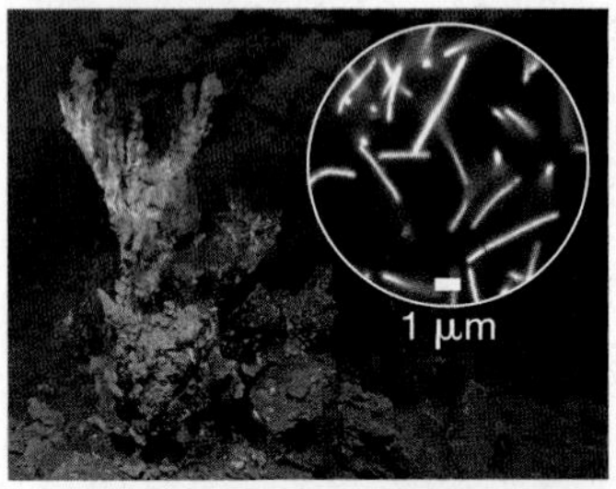	**Euryarchaeota** Diverse habitats (human gut, highly acidic and alkaline environments, deep-ocean sediments) ◀ *Methanopyrus kandleri*, on the wall of hydrothermal vent	Rods, spheres, filaments, spirals, and discs. Found as chains or clusters. Flagella are common. Some lack cell wall.	Chemoheterotrophs and chemolithoautotrophs. Many produce methane as a by-product of respiration.	Some members found near abandoned mines and produce acids that pollute streams. Methanogens (found in guts of mammals and swamps) add billions of tons of methane to the atmosphere each year.
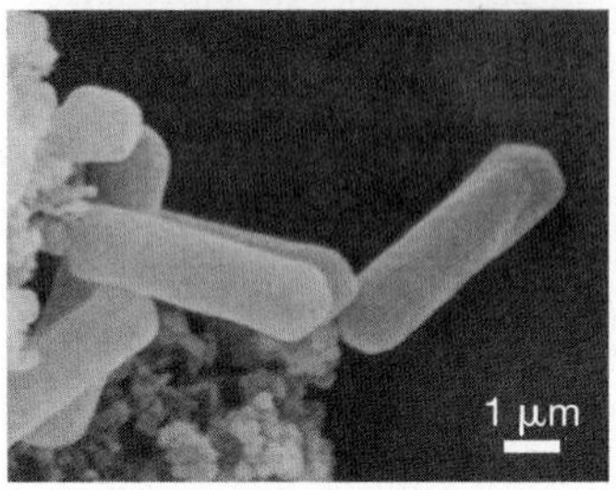	**Thaumarchaeota** Common in fresh and saltwater habitats and soil ◀ *Nitrosopumilus maritimus*	Rod shaped. Found as individual cells.	Chemolithoautotrophs. Use ammonia as a source of energy and produce nitrite as a by-product.	Only a few members of the group have been observed. Very abundant in oceans. One member lives as an endosymbiont in marine sponges.

For more go to https://goo.gl/hOpOCF

of the major attributes of three of those phyla are shown in **Table 26.5**. In addition there are the Korarchaeota, known only from direct-sequencing studies and very difficult to grow in culture; and the Nanoarchaeota, represented by only one species.

Crenarchaeota Members of this phylum are also called eocytes. They are found in harsh environments, including the hot springs of Yellowstone National Park. They thrive in hot, acidic, and even high-pressure environments.

Euryarchaeota This phylum is aptly named, because the word root *eury–* means "broad." Members of this phylum live in nearly every conceivable habitat. Some species are adapted to high-salt habitats with pH 11.5—almost as basic as ammonia. Other species are adapted to acidic conditions with a pH as low as 0. Species in the genus *Methanopyrus* live near hot springs called black smokers that are 2000 m (over 1 mile) below sea level.

Thaumarchaeota Members of this recently recognized, ancient lineage are extremely abundant in oceans, estuaries, and terrestrial soils. Unlike the extremophiles, species in this lineage are considered mesophilic ("middle-loving") because they grow best at moderate temperatures.

Collectively, bacteria and archaea play several pivotal roles in nature. Research on documenting new species, understanding their role in disease, developing new biotechnology techniques by exploiting their unique genetics and biochemistry, and clarifying their relationships to other organisms continues at a brisk pace. Prokaryotes and their microbiomes are hot areas of research for biologists across the globe.

CHECK YOUR UNDERSTANDING

✔ If you understood this section, you should be able to . . .

1. Explain the relationship between cyanobacteria and other types of organisms that rely on oxygen, nitrogen, and organic compounds for their survival.
2. Describe the environmental conditions where you might expect to find *Sulfolobus* (a member of the Crenarchaeota).

Answers are available in Appendix A.

26.1 Why Do Biologists Study Bacteria and Archaea?

- Bacteria and archaea are the most abundant organisms on Earth and are found in every habitat that has been sampled.
- Bacteria and archaea are very small, prokaryotic cells, and most are unicellular.
- Bacteria and archaea can be distinguished by their different kinds of membrane lipids and cell walls as well as by their different transcription machinery.
- Bacteria play many beneficial roles in animal digestion, bioremediation, and the production of antibiotics.
- Bacteria cause some of the most dangerous human diseases, including plague, syphilis, botulism, cholera, and tuberculosis.

26.2 How Do Biologists Study Bacteria and Archaea?

- Enrichment cultures are used to grow large numbers of bacterial or archaeal cells that thrive under specified conditions.
- Using metagenomic analysis, biologists can study bacteria and archaea that cannot be cultured by extracting DNA directly from an environment and then sequencing and characterizing DNA fragments. The information obtained is used to identify biochemical processes and novel organisms that are then placed on the tree of life.
- Biologists have analyzed DNA from samples of different parts of the human body (e.g., mouth, intestine) and have discovered microbiomes consisting of large numbers of diverse species of prokaryotes.

26.3 What Themes Occur in the Diversification of Bacteria and Archaea?

- Genetic variation in prokaryotes can be achieved through transformation (uptake of DNA from the environment), transduction (transfer of DNA from one cell to another by a virus), and conjugation (direct transfer of DNA from one cell to another).
- Metabolic diversity and complexity are the hallmarks of the bacteria and archaea, just as morphological diversity and complexity are the hallmarks of the eukaryotes.
- Among bacteria and archaea, a wide array of inorganic or organic compounds with high potential energy may serve as electron donors in cellular respiration, and a wide variety of inorganic or organic molecules with low potential energy may serve as electron acceptors. Dozens of distinct organic compounds are fermented.
- Many bacteria and archaea are phototrophs, meaning that they use the energy in light to produce high-energy electrons that are ultimately used to generate ATP. In cyanobacteria, water is used as a source of electrons and oxygen gas is generated as a by-product. In other species, the electrons come from a source other than water, and no oxygen is produced.
- To acquire building-block molecules containing carbon–carbon bonds, some bacteria and archaea species use the enzymes of the Calvin cycle to reduce CO_2. But several other biochemical pathways found in bacteria and archaea can also reduce simple organic compounds to sugars or carbohydrates.
- Because of their metabolic diversity, bacteria and archaea play a large role in carbon and nitrogen cycling and alter the global atmosphere, oceans, and terrestrial environments.
- Nitrogen-fixing species provide nitrogen in forms that can be used by many other species, including plants and animals.

26.4 Key Lineages of Bacteria and Archaea

- Prokaryotes can be divided into two lineages, the Bacteria and the Archaea, based on a wide variety of morphological, biochemical, and molecular characters.
- Bacteria are divided into at least 29 major lineages including organisms that play major roles in ecosystems as primary producers, decomposers, and parasites.
- Archaea are divided into at least five major lineages and were thought to exist only in extreme environments; they are now recognized to be widespread.

Answers are available in Appendix A.

✓ TEST YOUR KNOWLEDGE

1. What do some photosynthetic bacteria use as a source of electrons instead of water?
   a. oxygen (O_2)
   b. hydrogen sulfide (H_2S)
   c. organic compounds (e.g., CH_3COO^-)
   d. nitrate (NO_3^-)
2. What are organisms called that use inorganic compounds as electron donors in cellular respiration?
   a. phototrophs
   b. heterotrophs
   c. organotrophs
   d. lithotrophs
3. Unlike plant cell walls that contain cellulose, bacterial cell walls are composed of ________.
4. Evaluate these statements about Koch's postulates, which are used to establish a causative link between a specific microbe and a specific disease. Select True or False for each statement.
   T/F The microbe must be present in individuals suffering from the disease and absent from healthy individuals.
   T/F The microbe must be isolated and grown in pure culture.
   T/F If organisms from the pure culture are injected into a healthy experimental animal, the disease symptoms should appear.
   T/F The microbe does not have to be isolated from the experimental animal as long as the disease is present.

✔ TEST YOUR UNDERSTANDING

5. What has metagenomic analysis allowed researchers to do for the first time?
   a. sample organisms from an environment and grow them under defined conditions in the lab
   b. isolate organisms from an environment and sequence their entire genome
   c. study organisms that cannot be cultured (grown in the lab)
   d. identify important morphological differences among species
6. Biologists often use the term "energy source" as a synonym for "electron donor." Why?
7. The text claims that the evolution of an oxygen-rich atmosphere paved the way for increasingly efficient cellular respiration and higher growth rates in organisms. Explain.
8. **PROCESS OF SCIENCE** The researchers who observed that magnetite was produced by bacterial cultures from the deep subsurface carried out a follow-up experiment. These biologists treated some of the cultures with a drug that poisons the enzymes involved in electron transport chains. In cultures where the drug was present, no more magnetite was produced. Does this result support or undermine their hypothesis that the bacteria in the cultures perform cellular respiration? Explain your reasoning.

✔ TEST YOUR PROBLEM-SOLVING SKILLS

9. **PROCESS OF SCIENCE** *Streptococcus mutans* obtains energy by oxidizing sucrose. This bacterium is abundant in the mouths of Western European and North American children and is a prominent cause of cavities. The organism is virtually absent in children from East Africa, where tooth decay is rare. Propose a hypothesis to explain this observation. Outline the design of a study that would test your hypothesis.
10. **PROCESS OF SCIENCE** Suppose that you've been hired by a firm interested in using bacteria to clean up organic solvents found in toxic waste dumps. Your new employer is particularly interested in finding cells that are capable of breaking a molecule called benzene into less-toxic compounds. Where would you go to look for bacteria that can metabolize benzene as an energy or carbon source? How would you design an enrichment culture capable of isolating benzene-metabolizing species?

✔ PUT IT ALL TOGETHER: Case Study

Three-Domain Hypothesis

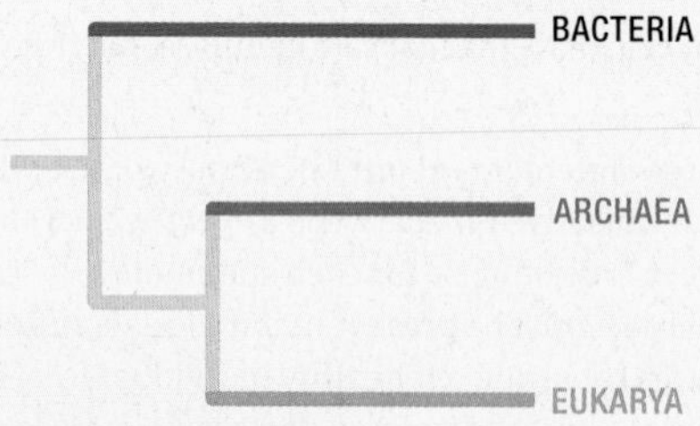

Have we really figured out the tree of life?

The traditional tree of life (shown above) presents the three domains as distinct, monophyletic lineages. However, other hypotheses propose different views on the relationships among the Archaea, Bacteria, and Eukarya. In particular, the two-domain hypothesis—or eocyte hypothesis—is emerging as a well-supported alternative to the three-domain hypothesis. The eocyte hypothesis, illustrated below, suggests that eukaryotes evolved from eocytes (also known as the Crenarchaeota—a major lineage of the Archaea). Resolving the relationships among these ancient lineages is difficult, but it has profound implications on our understanding of the origin of eukaryotic cells.

Two-Domain Hypothesis

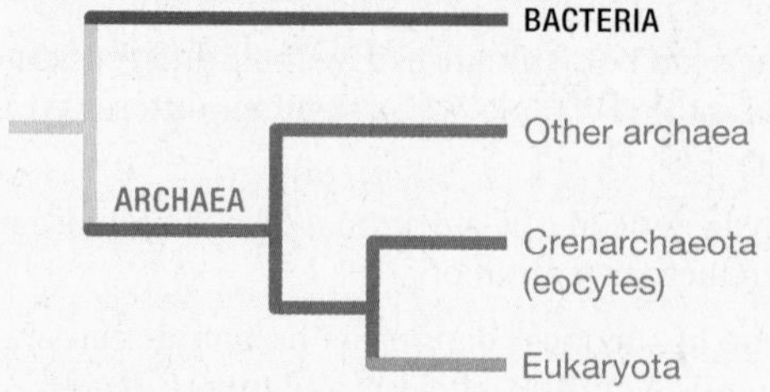

11. Why are Archaea considered a monophyletic group according to the three-domain hypothesis?
    a. Because this group includes all organisms except eukaryotes.
    b. Because this group includes an ancestral population and all of its descendants.
    c. Because all members of this group lack membrane-bound organelles.
    d. Because this group evolved after the origin of bacteria.
12. The Bacteria and Archaea both include microscopic prokaryotes that lack membrane-bound nuclei. What criteria have led to the classification of these two groups as separate domains?
13. Early ideas on the classification of life recognized all organisms as belonging to one of two fundamental lineages—prokaryotes or eukaryotes. Is this view compatible with either of the hypotheses illustrated here? Explain.
14. **MODEL** Other hypotheses for the tree of life present the Archaea as the ancestors to all other organisms. Sketch a phylogenetic tree that presents Bacteria and Eukarya as more closely related to each other than to Archaea, and that has Archaea as a sister group to Bacteria and Eukarya.
15. Evaluate this statement: According to the two-domain hypothesis, all members of the domain Archaea are prokaryotes and therefore lack membrane-bound nuclei.
16. **PROCESS OF SCIENCE** What other types of evidence or features might be used to ascertain whether the tree of life is best represented according to the three-domain or the eocyte hypothesis?

Mastering Biology

Students Go to Mastering™ Biology for assignments, the eText, and the Study Area with animations, practice tests, and activities.

Professors Go to Mastering™ Biology for automatically graded tutorials and questions that you can assign to your students, plus Instructor Resources.

27 Diversification of Eukaryotes

Organisms that are referred to as protists, including the *Paramecia* pictured here, are abundant in aquatic environments and range in size from single cells to giant seaweeds.

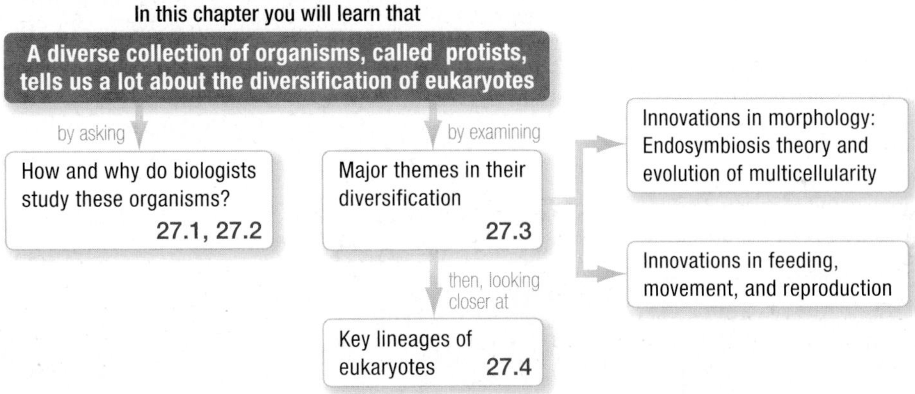

This chapter introduces the third domain on the tree of life: the **Eukarya**, which range from single-celled organisms to giant sequoia trees and blue whales. Although eukaryotes are astonishingly diverse, they share fundamental features that distinguish them from bacteria and archaea. For example:

- Compared to bacteria and archaea, most eukaryotic cells are larger, have many more organelles (e.g., chloroplasts, mitochondria), and have a much more extensive cytoskeleton.
- The nuclear envelope is a unique trait that defines the eukaryotes.
- Multicellularity is rare in bacteria and unknown in archaea, but has evolved multiple times in eukaryotes.
- Bacteria and archaea reproduce asexually by binary fission; eukaryotes reproduce asexually via mitotic cell division, or sexually by producing gametes or spores via meiotic cell division.

This chapter explores how the morphological innovations of eukaryotes—features like organelles and the nuclear envelope—evolved. It then analyzes how these structures enabled eukaryotes to achieve novel ways of performing basic life tasks such as feeding, moving, and reproducing.

BIG PICTURE

This chapter is part of the Big Picture. See how on pages 722–723.

In introducing the Eukarya, this chapter focuses on a diverse collection of lineages known as "protists" (**Figure 27.1**). The term **protist** (literally, "the very first") refers to all eukaryotes that are not land plants, fungi, or animals (these groups are the focus of Chapters 28 through 32). Protists have no defining feature that is found only in protists and in no other organisms. As a result, they do not form a naturally occurring unified group. Historically, protists were thought to represent their own separate kingdom, but this viewpoint is now obsolete. Nevertheless, "protist" is still used as a collective term for practical purposes.

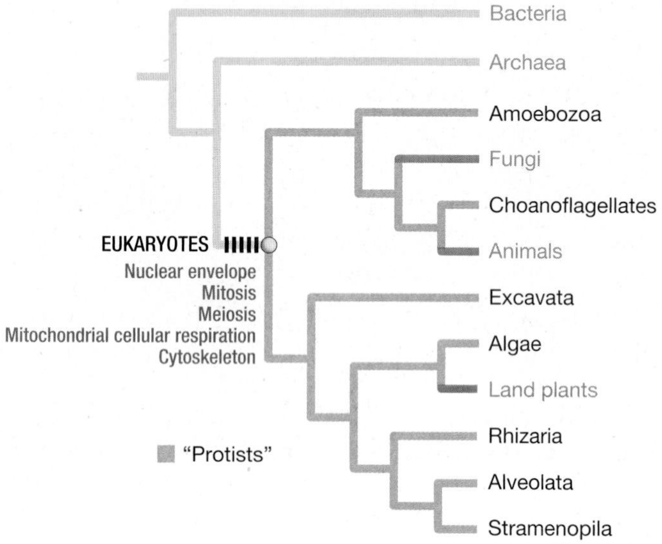

Figure 27.1 The Term "Protist" Refers to All Eukaryotes Except Land Plants, Fungi, and Animals. This figure indicates just a few of the defining features of eukaryotes.

One common feature of protists is the tendency to live in environments where they are surrounded by water most of the time (**Figure 27.2**). Most plants and fungi, and many animals, are terrestrial. But protists are found in wet soils and aquatic habitats or inside the bodies of other organisms—including, perhaps, you. ✔ **A monophyletic group includes all of the descendants of a common ancestor. Why are protists not considered a monophyletic group?**

27.1 Why Do Biologists Study Protists?

Biologists study protists for three reasons, in addition to their intrinsic interest: **(1)** they are important medically, **(2)** they are important ecologically, and **(3)** they are critical to understanding the evolution of plants, fungi, and animals. The rest of the chapter focuses on why protists are interesting in their own right and how they evolved. But first, let's consider their impact on human welfare.

After you complete this section, you should be able to . . .

- Summarize the medical and ecological importance of protists.

Impacts on Human Health and Welfare

The most spectacular crop failure in history, the Irish potato famine, was caused by a protist. In 1845 most of the 3 million acres that had been planted to grow potatoes in Ireland became infested with *Phytophthora infestans*—a parasite that belongs to a group of protists called water molds. Potato tubers that were infected with *P. infestans* rotted in the fields or in storage.

Open ocean: Surface waters teem with microscopic protists, such as these diatoms.

Shallow coastal waters: Gigantic protists, such as these kelp, form underwater forests.

Intertidal habitats: Protists such as these brown algae are particularly abundant in tidal habitats.

Figure 27.2 Protists Are Particularly Abundant in Aquatic Environments.

As a result of crop failures in Ireland for two consecutive years, an estimated 1 million people out of a population of less than 9 million died of starvation or starvation-related illnesses. Several million others emigrated. Many people of Irish heritage living in North America, New Zealand, and Australia trace their ancestry to relatives who left Ireland to evade the famine. As devastating as the potato famine was, however, it does not begin to approach the misery caused by the protist *Plasmodium*.

Malaria Physicians and public health officials point to three major infectious diseases that are currently afflicting large numbers of people worldwide: tuberculosis, HIV, and malaria. Tuberculosis is caused by a bacterium (Ch. 22, Section 22.4); HIV is caused by a virus (Ch. 33, Section 33.1). Malaria is caused by five species of *Plasmodium*—a protist in a group called the apicomplexans.

Malaria ranks as one of the world's worst infectious diseases. Approximately 200 million people worldwide are sickened by it each year, and roughly 600,000 people die from the disease annually. The toll is equivalent to four 747s, loaded with passengers, crashing every day. Most of the dead are children of preschool age.

As Figure 27.3 shows, *Plasmodium* has a complex life cycle involving both mosquito and human hosts. Infections start when a *Plasmodium*-infected mosquito takes a blood meal from a human. As the mosquito feeds, *Plasmodium* cells enter the human bloodstream. Inside their human host, the protists travel to the liver, where they infect a small number of liver cells and begin replicating asexually. At this point, the infected person has no symptoms. What happens next causes the illness.

After increasing in number up to 40,000-fold, the *Plasmodium* cells differentiate into a specialized form. In this new form they are released from the liver and invade red blood cells, where they feast on stored hemoglobin and continue to replicate asexually. When the daughter cells mature, the blood cells burst open, releasing cellular products and the next wave of invaders that go on to infect and destroy other red blood cells (see Figure 27.3). The bursting of the red blood cells correlates with the onset of malarial symptoms, including headaches, fever, and fatigue.

Eventually some of the asexual blood-stage protists differentiate into male and female cells that will become gametes. The infection cycle continues when mosquitoes feed on the blood of infected humans. Large numbers of *Plasmodium* cells in a human host's blood increase the chance that a mosquito's blood meal will contain some of them. Inside the mosquito, *Plasmodium* completes its life cycle by reproducing sexually in the gut, then migrating to the salivary glands, where it awaits transfer to its next human victim.

Although *Plasmodium* is arguably the best studied of all protists, researchers have still not been able to devise effective and sustainable measures to control it.

- Efforts to develop a vaccine against *Plasmodium* have been difficult, in part because the parasite evolves so quickly. Flu virus and HIV pose similar problems (Chapter 48).
- Natural selection has favored mosquito strains that are resistant to the insecticides that have been sprayed in their breeding habitats in attempts to control malaria's spread.

A novel approach to controlling malaria's spread, however, shows promise. Researchers are using CRISPR-cas9 technology (Ch. 20, Section 20.5) to genetically modify mosquitoes so that the females do not produce eggs and therefore do not reproduce.

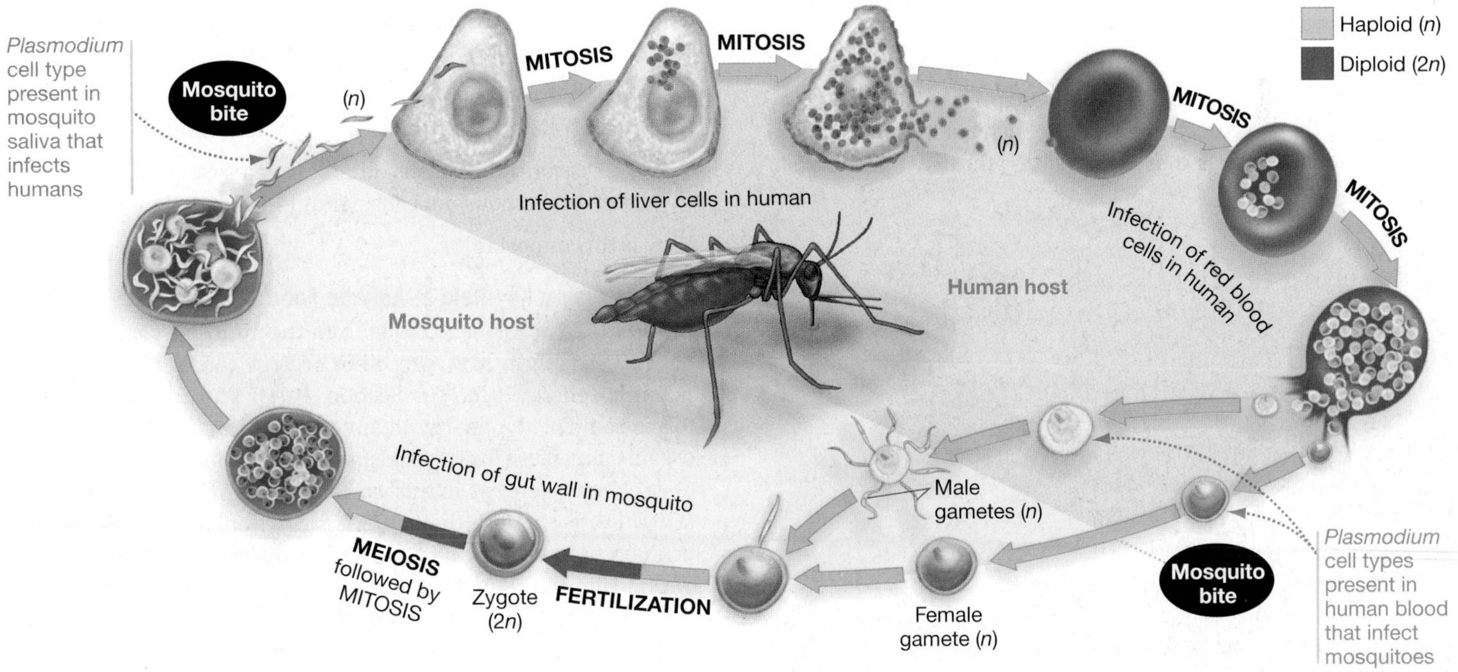

Figure 27.3 *Plasmodium* Lives in Mosquitoes and in Humans, Where It Causes Malaria. Over the course of its life cycle, *Plasmodium falciparum* alternates between a mosquito host and a human host.

Table 27.1 Some Human Health Problems Caused by Protists

Species	Disease
Five species of *Plasmodium,* primarily *P. falciparum* and *P. vivax*	These organisms cause malaria. Over 3.4 billion people (roughly one-half of the world's population) live in areas at risk of malaria transmission.
Naegleria fowleri	Brain-eating amoeba causes meningoencephalitis; infections are almost always fatal.
Toxoplasma gondii	Toxoplasmosis may cause eye and brain damage in infants and in AIDS patients.
Many species of dinoflagellates	Toxins released during harmful algal blooms accumulate in shellfish and poison people if eaten.
Trypanosoma gambiense and *T. rhodesiense*	Trypanosomiasis ("sleeping sickness") is a potentially fatal disease transmitted through bites from tsetse flies. Occurs in Africa.
Trypanosoma cruzi	Chagas disease affects 6–7 million people and causes 50,000 deaths annually, primarily in South and Central America.
Entamoeba histolytica	Amoebic dysentery results in severe diarrhea and subsequent dehydration.

Unfortunately, malaria is not the only important human disease caused by protists. **Table 27.1** lists a few of the protists that cause human suffering and economic losses.

Harmful Algal Blooms When a unicellular species experiences rapid population growth and reaches high densities in an aquatic environment, it is said to "bloom." A handful of the many protist species involved in blooms can be harmful to other organisms.

Harmful algal blooms are usually due to photosynthetic protists called dinoflagellates. Certain dinoflagellates synthesize toxins that give protection from predation by small animals called copepods. Because dinoflagellates sometimes contain high concentrations of accessory pigments called xanthophylls (Ch. 10, Section 10.2), their blooms can discolor seawater (**Figure 27.4**). These blooms are often referred to as red tides. Certain dinoflagellates can also display bioluminescence in the ocean. When water is disturbed, they glow in the dark (for more see Section 27.4).

Algal blooms can be harmful to people because clams and other shellfish feed on the photosynthetic protists after filtering them out of the water. During a bloom, high levels of toxins can build up in the flesh of these shellfish. Typically, the shellfish themselves are not harmed. But if a person eats contaminated shellfish, several types of poisoning can result.

Figure 27.4 Harmful Algal Blooms Caused by Dinoflagellates.

Paralytic shellfish poisoning, for example, occurs when people eat shellfish that have fed heavily on protists that synthesize poisons called saxitoxins. Saxitoxins block ion channels that have to open for electrical signals to travel through nerve cells (Ch. 43, Section 43.2). In humans, a high dose of saxitoxins causes unpleasant symptoms such as prickling sensations in the mouth or even life-threatening symptoms such as muscle weakness and paralysis. Most species of protists, however, are not harmful. Let's have a look at some protists that are essential to healthy ecosystems.

Ecological Importance of Protists

As a whole, protists represent just 10 percent of the total number of named eukaryote species. Although this number of named species is relatively low, the abundance of protists is extraordinarily high. The number of individual protists found in some habitats is astonishing:

- One milliliter of pond water can contain well over 500 single-celled protists that swim with the aid of flagella.
- Under certain conditions, dinoflagellates can reach a concentration of 60 thousand cells per milliliter of seawater.

Why is this important?

Protists Play a Key Role in Aquatic Food Chains Photosynthetic protists take in carbon dioxide from the atmosphere and reduce, or "fix," it to form sugars or other organic compounds with high potential energy (Ch. 10, Section 10.4). Photosynthesis transforms some of the energy in sunlight into chemical energy that organisms can use to grow and produce offspring. Species that produce chemical energy in this way are called **primary producers**. Diatoms are photosynthetic protists that rank among the leading primary producers in the oceans simply because they are so abundant. Primary production in the ocean represents almost half of the total carbon dioxide that is fixed on Earth.

Diatoms and other organisms that drift in the open oceans or lakes are called **plankton**. The sugars and other organic compounds produced by photosynthetic plankton are the basis of food chains in freshwater and marine environments.

A **food chain** describes nutritional relationships among organisms, and thus how chemical energy flows within ecosystems. In this case, photosynthetic protists and other primary producers are eaten by primary consumers, many of which are protists. Primary consumers are eaten by fish, shellfish, and other secondary consumers, which in turn are eaten by tertiary consumers—whales, squid, and large fish (such as tuna). Many of the species at the base of food chains in aquatic environments are protists.

Could Protists Help Limit Global Climate Change? The carbon dioxide level in the atmosphere is increasing rapidly because humans are burning fossil fuels and forests (Ch. 53, Section 53.3). Carbon dioxide traps heat that is radiating from Earth, so a high CO_2 level in the atmosphere contributes to the rise in temperatures associated with global climate change—an issue that many observers consider today's most pressing environmental problem.

The carbon atoms in carbon dioxide molecules move to organisms on land or in the oceans and then back to the atmosphere, in what researchers call the **global carbon cycle**. To restrict global climate change, researchers are trying to figure out ways to decrease the carbon dioxide concentration in the atmosphere and increase the amount of carbon stored in terrestrial and marine environments.

To understand how this might be done, consider the marine carbon cycle diagrammed in Figure 27.5. The cycle starts when CO_2 from the atmosphere dissolves in water and is taken up by primary producers—photosynthetic plankton called phytoplankton—and converted to organic matter. The plankton are eaten by primary consumers, die, and are consumed by decomposers or scavengers; or they die and sink to the bottom of the ocean, where they become part of a **carbon sink**—a long-term reservoir of carbon. Once at the bottom of the ocean, they may enter one of two carbon sinks:

1. ***Sedimentary rocks*** Several lineages of protists have shells made of calcium carbonate ($CaCO_3$). When these shells rain down from the ocean surface and settle in layers at the bottom, the deposits that result are compacted by the weight of the water and by sediments accumulating above them. Eventually the deposits turn into rock. The limestone used to build the pyramids of Egypt consists largely of protist shells.
2. ***Petroleum*** Although the process of petroleum (oil) formation is not well understood, it begins with accumulations of dead bacteria, archaea, and protists at the bottom of the ocean.

Recent experiments have shown that the carbon cycle speeds up when habitats in the middle of the ocean are fertilized with iron. Iron is a critical component of the electron transport chains responsible for photosynthesis and respiration, but it is in particularly short supply in the open ocean. After iron is added to ocean waters, populations of protists and other primary producers often increase by a factor of 10. In theory, the potential exists to use iron fertilization to promote protist blooms and thereby reduce CO_2 concentrations in the atmosphere.

The effectiveness of iron fertilization is hotly debated, however. Fertilizing the ocean with iron might lead to large accumulations of dead organic matter and the formation of anaerobic dead zones (Ch. 26, Section 26.3). But if further research shows that iron fertilization is safe and effective, it could be added to the list of approaches to limit global climate change.

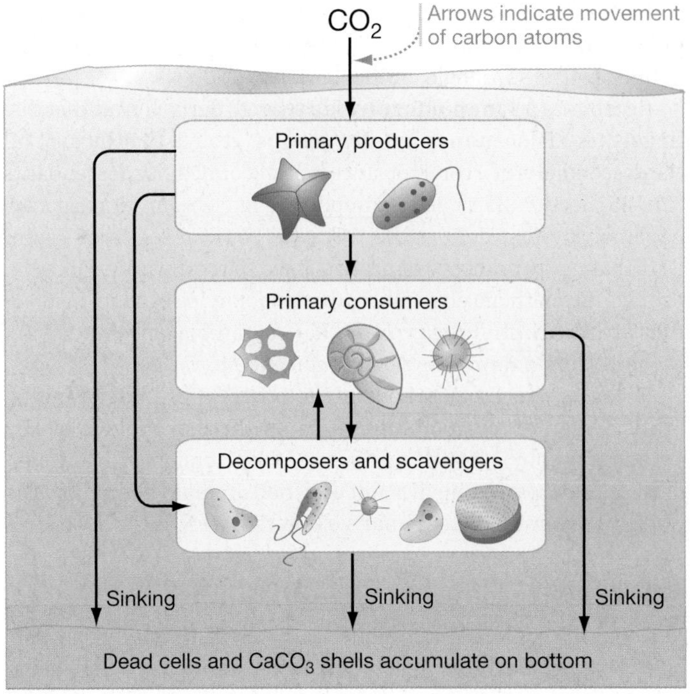

Figure 27.5 Protists Play a Key Role in the Marine Carbon Cycle. At the surface, carbon atoms tend to shuttle quickly among organisms. But if carbon atoms sink to the bottom of the ocean in the form of dead cells or shells, they may be locked up for long periods in carbon sinks miles below the surface.

CHECK YOUR UNDERSTANDING

✔ If you understood this section, you should be able to . . .

1. **SOCIETY** Explain why public health workers promote the use of insecticide-treated sleeping nets to protect against malaria.
2. Make a flowchart showing the chain of events that would start with massive iron fertilization of the ocean and end with large deposits of carbon-containing compounds on the ocean floor.

Answers are available in Appendix A.

27.2 How Do Biologists Study Protists?

Although biologists have made great strides in understanding pathogenic protists and the role that protists play in the global carbon cycle, they have found it extremely difficult to gain any sort of solid insight into how the group as a whole diversified over time. The problem is that the eukaryotic lineages split over a billion years ago

After you complete this section, you should be able to . . .

- Evaluate evolutionary relationships among protists, and explain how this group is paraphyletic.

and diverged so much that it's not easy to find any overall patterns in the evolution of the group. There are no synapomorphies that define protists as a single, unified group.

Recall that a **synapomorphy** is a shared, derived trait that distinguishes major **monophyletic groups**, groups that include all the descendants of a common ancestor and only those descendants (Ch. 25, Section 25.1). For example, the nuclear envelope is a synapomorphy that defines the Eukarya. You can see from Figure 27.1 that many branches of protists have split off within Eukarya. Protists do not themselves make up a monophyletic group. Instead, they are a **paraphyletic group**—they represent some, but not all, of the descendants of a single common ancestor.

Recently, researchers have made progress in understanding protist diversity by combining data on the morphology of key groups with phylogenetic analyses of DNA sequences. However, despite the progress, significant questions remain about how the Eukarya diversified. Let's analyze how this work has been done.

Microscopy: Studying Cell Structure

Various microscopic techniques are used to study protists (**BioSkills 9**). Early biologists were able to identify and name many of the protist species known today using traditional light microscopy. For example, the early microscopist Anton van Leeuwenhoek identified the parasite *Giardia intestinalis* by examining samples of his own feces.

When transmission electron microscopes became available, a major breakthrough in understanding protist diversity occurred: Detailed studies revealed that protists could be grouped according to overall cell structure, according to organelles with distinctive features, or both.

For example, both light and electron microscopy confirmed that the species that caused the Irish potato famine has reproductive cells with an unusual type of **flagellum**. Flagella are organelles that project from the cell and whip back and forth to produce swimming movements. In reproductive cells of this species, one of the two flagella present has tiny, hollow, hairlike projections. Biologists noted that kelp and other species of brown algae also have cells with this type of flagellum.

To make sense of these results, researchers interpreted these types of distinctive morphological features as synapomorphies. Species that have a flagellum with hollow, hairlike projections became known as stramenopiles ("straw-hairs"); the hairs typically have branches at the tip (**Figure 27.6**).

In recognizing the Stramenopila lineage, investigators hypothesized that because an ancestor had evolved a distinctive flagellum, all or most of its descendants also had this trait. The qualifier "most" is important, because it is not unusual for certain subgroups within a lineage to lose particular traits over the course of evolution, much as humans gradually lost fur and tailbones.

Eventually, seven major groups of eukaryotes came to be identified on the basis of diagnostic morphological characteristics. These groups and some of the key synapomorphies that identify them are listed in **Table 27.2**. Notice that in almost every case, the synapomorphies that define eukaryotic lineages are changes in structures that protect or support the cell or that influence how the organism moves, feeds, or reproduces. The land plants, fungi, and animals represent subgroups within two of the seven major eukaryotic lineages. How are the major groups of eukaryotes related to each other?

Evaluating Molecular Phylogenies

When researchers began using DNA sequence data to estimate the evolutionary relationships among eukaryotes, the analyses suggested that each of the seven groups identified on the basis of distinctive morphological characteristics were indeed monophyletic. This was important support for the hypothesis that the distinctive morphological features were synapomorphies that existed in a common ancestor of each lineage.

The phylogenetic tree in **Figure 27.7** is the current best estimate of the eukaryotes' evolutionary history (see **BioSkills 13** for more on phylogenetic trees). The tree continues to be refined as databases expand, and as techniques for analyzing data improve.

SUMMARY **Table 27.2** Distinguishing Features (Synapomorphies) of Major Lineages of Eukaryotes

Lineage	Distinguishing Morphological Features (Synapomorphies)
Amoebozoa	Cells lack cell walls. When portions of the cell extend outward to move the cell, they form large lobes.
Opisthokonta	Reproductive cells have a single flagellum at their base. The cristae inside mitochondria are flat, not tube shaped as in other eukaryotes. This lineage includes protists as well as the fungi and the animals. (Fungi and animals are discussed separately in Chapters 29–32.)
Excavata	Most cells have a pronounced "feeding groove" where prey or organic debris is ingested. Most species lack typical mitochondria, although genes derived from mitochondria are found in the nucleus.
Plantae	Cells have chloroplasts with a double membrane. (The diversity of green algae and land plants is covered in Chapter 28.)
Rhizaria	Cells lack cell walls, although some produce an elaborate shell-like covering. When portions of the cell extend outward to move the cell, they are threadlike in shape.
Alveolata	Cells have sac-like structures called alveoli that form a continuous layer just under the plasma membrane. Alveoli are thought to provide support.
Stramenopila	If flagella are present, cells usually have two—one of which is covered with hairlike projections.

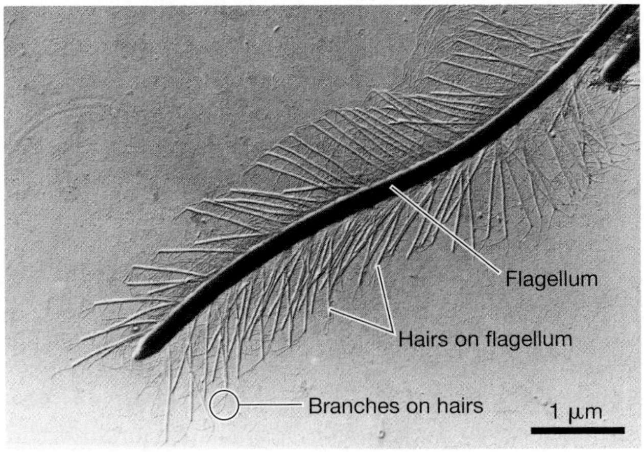

Figure 27.6 Species in the Lineage Called Stramenopila Have a Distinctive Flagellum. The unusual, hollow "hairs" on the flagella of stramenopiles often have three branches at the tip.

For example, some recent studies suggest that Excavata may be paraphyletic. As you read this tree, note that:

- The Amoebozoa and the Opisthokonta—which include fungi and animals—form a monophyletic group called the Unikonta, meaning "one-tailed." Any cells among this group that are flagellated have only one flagellum.
- The other five major eukaryotic lineages form a monophyletic group called the Bikonta, meaning "two-tailed." Any flagellated cells among this group have two flagella.

Understanding where the root or base of the tree of Eukarya lies has been more problematic. Finding it will help researchers understand what the common ancestor looked like and how eukaryotic lineages evolved over time. One hypothesis suggests that the first eukaryotic split was between unikonts and bikonts. This hypothesis is tentative and controversial, however, given the data analyzed to date. Researchers continue to work on the issue of placing the root of the Eukarya.

Discovering New Lineages via Direct Sequencing

As the effort to refine the phylogeny of the Eukarya continues, one of the most exciting research frontiers involves a technique called direct sequencing. **Direct sequencing** is based on sampling soil or water, analyzing the DNA sequence of specific genes in the sample, and using the data to place the organisms from the sample on a phylogenetic tree. Direct sequencing and metagenomics have led to the discovery of previously unknown but major lineages of Archaea (Ch. 26, Section 26.2). To the amazement of biologists all over the world, the same thing happened when researchers used direct sequencing to survey eukaryotes.

The first direct sequencing studies that focused on eukaryotes were published in 2001. One study sampled organisms at depths from 250 to 3000 meters (m) below the surface in waters off Antarctica; another focused on cells at depths of 75 m in the Pacific Ocean, near the equator. Both studies detected a wide array of species that were new to science.

Investigators who examined the samples microscopically were astonished to find that many of the newly discovered eukaryotes

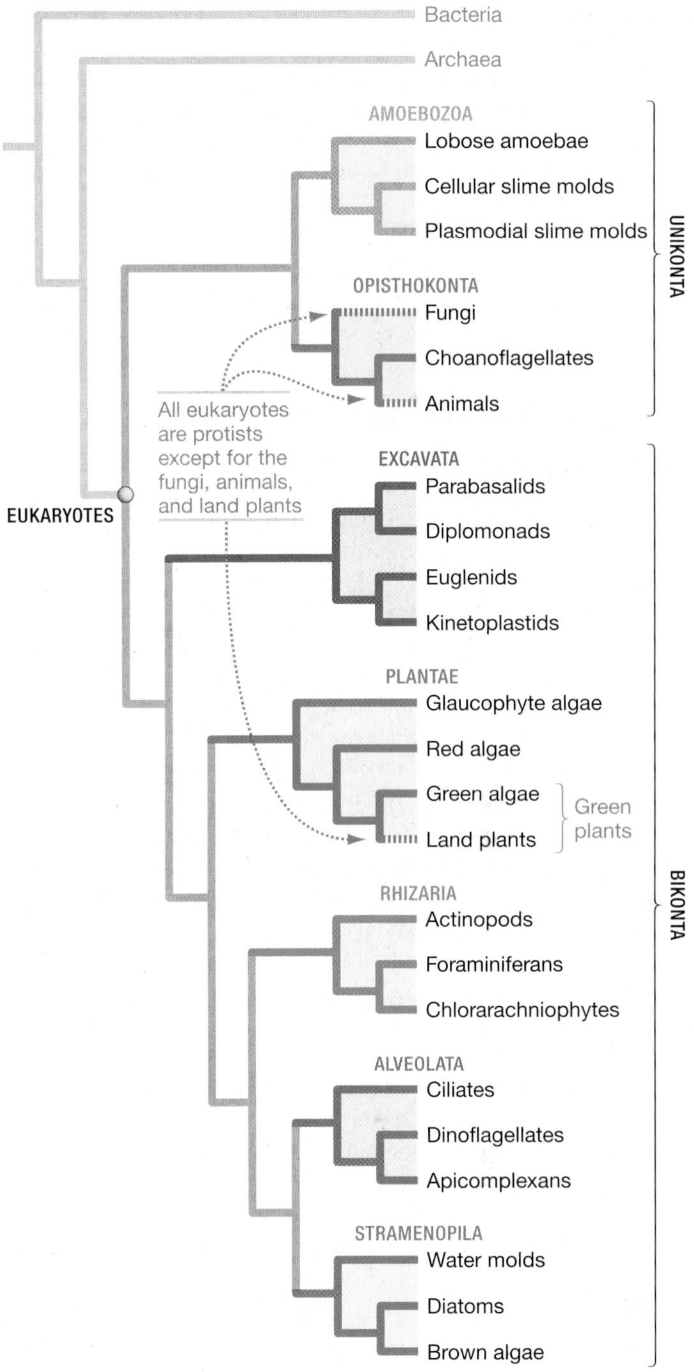

Figure 27.7 Phylogenetic Analyses Have Identified Seven Major Lineages of Eukaryotes. This tree shows selected subgroups for each major lineage. (Kinetoplastids are not discussed elsewhere in the chapter, but they include the pathogen *Trypanosoma* described in Table 27.1.)

were tiny—from 0.2 micrometers (μm) to 2 μm in diameter. Subsequent work has confirmed the existence of many, diverse species of these so-called **picoplankton**. These cells overlap in size with bacteria, which typically range from 0.5 μm to 2 μm in diameter. Hence, eukaryotic cells are much more variable in size than previously imagined. A whole new world of tiny protists awaits further exploration.

CHECK YOUR UNDERSTANDING

✔ **If you understood this section, you should be able to . . .**

1. Explain why the terms "opisthokonts" and "stramenopiles" are appropriate names for these lineages. (Hint: The root *opistho* refers to the backside or base of something.)
2. Explain how direct sequencing studies allow researchers to characterize species that are new to science.

Answers are available in Appendix A.

27.3 What Themes Occur in the Diversification of Protists?

Although land plants, fungi, and animals are large and diverse lineages, much of eukaryote diversity is found among the protists. They range in size from single cells to giant kelp and live in habitats from the open oceans to the guts of termites. They are almost bewildering in their morphological and ecological diversity. Because they are a paraphyletic group, they do not share derived characteristics that set them apart from all other lineages on the tree of life. Once an important new innovation arose in protists, it triggered the evolution of species that live in a wide array of habitats and make a living in diverse ways.

After you complete this section, you should be able to . . .

- Describe the life history innovations that evolved in protists.

What Morphological Innovations Evolved in Protists?

Virtually all bacteria and all archaea are unicellular. Given that most subgroups of the major lineages of eukaryotes are also unicellular, it is logical to conclude that the first eukaryote was a single-celled organism.

Further, all eukaryotes alive today have **(1)** either mitochondria or genes that are normally found in mitochondria, **(2)** a nucleus and endomembrane system, and **(3)** a cytoskeleton. Based on the distribution of cell walls in living eukaryotes—some have cell walls and some don't—the first eukaryotes likely lacked this feature.

Using these observations, biologists hypothesize that the earliest eukaryotes were probably single-celled organisms with mitochondria, a nucleus and endomembrane system, and a cytoskeleton, but no cell wall.

The first eukaryotic cells probably also swam using a novel type of flagellum. Eukaryotic flagella are completely different structures from bacterial flagella and evolved independently. The eukaryotic flagellum is made up of microtubules, and dynein is the major motor protein. An undulating motion occurs as dynein molecules "walk" along microtubules (see Ch. 7, Figure 7.29). The flagella of bacteria and archaea, in contrast, are composed primarily of a protein called flagellin. Instead of undulating, prokaryotic flagella rotate to produce movement. Recent phylogenetic studies indicate that the earliest eukaryotic cells likely had two flagella. ✔ **If you understand the synapomorphies that identify the eukaryotes as a monophyletic group, you should be able to map the origin of the eukaryotic flagellum on Figure 27.7.**

Now let's consider how several new morphological features arose and influenced the subsequent diversification of eukaryotes, beginning with one of the key organelles that define the Eukarya, the mitochondrion.

Endosymbiosis and the Origin of Mitochondria Mitochondria are organelles that generate adenosine triphosphate (ATP) using pyruvate as an electron donor and oxygen as the ultimate electron acceptor.

In the nineteenth century, a radical explanation for the origin of mitochondria was suggested. The hypothesis—which today is part of an overarching **endosymbiosis theory**—proposes that mitochondria originated when a bacterial cell took up residence inside another cell about 2 billion years ago (**Figure 27.8**).

The theory's name was inspired by the Greek word roots *endo*, *sym*, and *bio* ("inside-together-living"). **Symbiosis** is said to occur when individuals of two different species live in close and prolonged physical contact; **endosymbiosis** occurs when an organism of one species lives inside the cells of an organism of another species. Like any theory in science, the endosymbiosis hypothesis was not recognized as a *theory* until a large body of supporting evidence was amassed.

Interest in the endosymbiosis hypothesis was rekindled in the early 1970s by evolutionary biologist Lynn Margulis. At first, her efforts were met with a storm of criticism—largely because the idea seemed slightly preposterous. But gradually biologists began to examine it rigorously. For example, endosymbiotic relationships between protists and bacteria exist today. Among the α-proteobacteria alone (Ch. 26, Section 26.4), three major groups are found *only* inside eukaryotic cells. In many cases, the bacterial cells are transmitted to offspring in eggs or sperm and are required for survival.

Several observations about the structure of mitochondria are also consistent with the endosymbiosis theory:

- Mitochondria are similar in size to α-proteobacteria.
- Mitochondria replicate by fission, as do bacterial cells. The duplication of mitochondria takes place independently of division by the host cell. When eukaryotic cells divide, each daughter cell receives some of the many mitochondria present.
- Mitochondria have their own ribosomes and manufacture some of their own proteins. Mitochondrial ribosomes closely resemble bacterial ribosomes in size and composition and are poisoned by antibiotics such as streptomycin that inhibit bacterial, but not eukaryotic, ribosomes.
- Mitochondria have double membranes, consistent with the engulfing mechanism of origin illustrated in Figure 27.8.
- Mitochondria have their own genomes, which are organized as circular molecules—much like a bacterial chromosome. Mitochondrial genes code for a few of the proteins needed to conduct electron transport and RNAs needed to translate the mitochondrial genome.

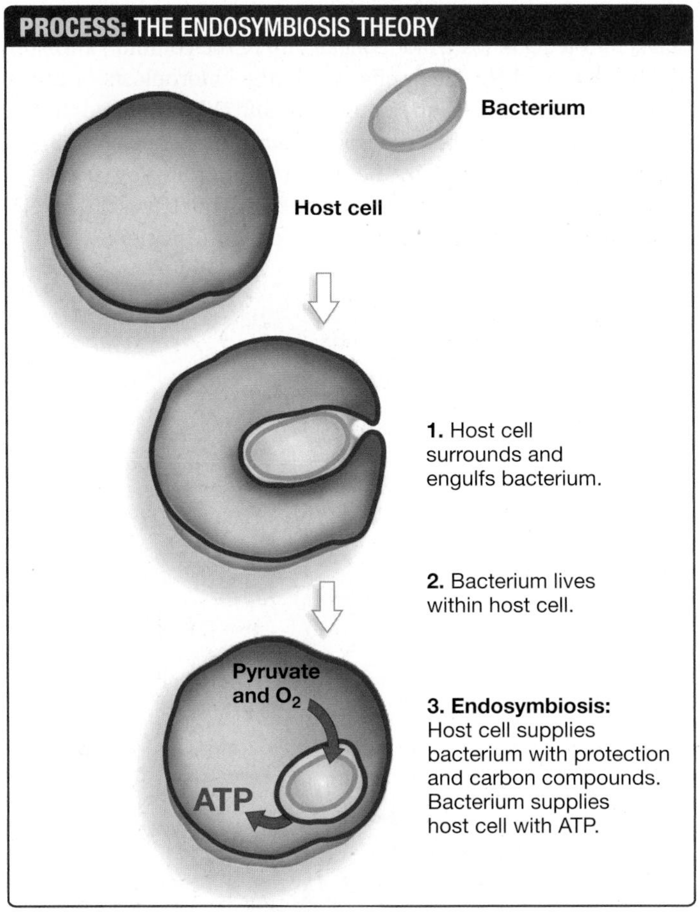

INTERACTIVE Figure 27.8 Proposed Initial Steps in the Evolution of Mitochondria.

✔ According to this hypothesis, mitochondria should consist of how many membranes? Explain your logic.

Although these data are impressive and are consistent with the endosymbiosis theory, they do not exclude other explanations. This is a general principle in science: Evidence is considered strong when it cannot be explained by reasonable alternative hypotheses. In this case, one reasonable alternative is that mitochondria evolved within eukaryotic cells, separately from bacteria.

A breakthrough occurred when researchers realized that according to the "within-eukaryotes" hypothesis, the genes found in mitochondria would be derived from nuclear genes in ancestral eukaryotes. In contrast, according to endosymbiosis, genes found in mitochondria would be bacterial in origin.

Researchers tested these alternative hypotheses by studying the phylogenetic relationships of mitochondrial genes. Specifically, they compared gene sequences isolated from eukaryotic mitochondrial DNA with sequences of similar genes isolated from eukaryotic nuclear DNA and with DNA from several species of bacteria. They found that mitochondrial gene sequences much more closely match sequences from α-proteobacteria DNA than sequences from the nuclear DNA of eukaryotes. The researchers concluded that the within-eukaryotes hypothesis could be rejected. Combined with other observations about mitochondria, the once controversial endosymbiosis hypothesis became a solidly supported theory.

One intriguing feature of mitochondrial genomes is that they typically encode fewer than 50 genes, whereas the genomes of their bacterial cousins encode about 1500 genes. Most of the genes from the endosymbiotic bacterium moved into the nuclear genome in what was one of the most spectacular **lateral gene transfer** events in the history of life (Ch. 20, Section 20.4).

How and when the momentous symbiosis that led to the mitochondrion occurred is hotly debated. Some scientists hypothesize that a primitive amoeba-like eukaryote engulfed a bacterium and simply failed to digest it in its lysosome, allowing it to take up residence. Evidence used to support this theory includes the similar feeding habits of today's amoebas and the existence of some protists that lack typical mitochondria.

Recent evidence, however, indicates that the original protists all had mitochondria when they first evolved (and any modern protists lacking mitochondria simply lost them over the course of evolution). A new hypothesis gaining support is that the first eukaryote (and its mitochondria) may have been formed at the same time, as a result of an endosymbiosis between two prokaryotes—an archaeal host and a bacterium.

Endosymbiosis accounts for the origin of mitochondria, but can it explain the origin of other energy-processing organelles such as chloroplasts?

Endosymbiosis and the Origin of Chloroplasts You might recall that photosystems I and II evolved in bacteria, and that both photosystems occur in cyanobacteria (Ch. 10, Section 10.3). In eukaryotes, photosynthesis takes place in chloroplasts—organelles with multiple membranes that contain photosystems I and II. None of the basic machinery required for photosynthesis evolved in eukaryotes. Instead, they likely "captured" it via endosymbiosis.

All photosynthetic protists have chloroplasts. An extension of the endosymbiosis theory contends that the eukaryotic chloroplast originated when a protist engulfed a cyanobacterium. Once inside the protist, the photosynthetic bacterium provided its eukaryotic host with oxygen and glucose, while the host provided the bacterium protection and access to light. The evidence for an endosymbiotic origin for the chloroplast is persuasive:

- Chloroplasts have the same list of bacteria-like characteristics presented earlier for mitochondria.
- Chloroplasts contain a circular DNA molecule containing genes that are extremely similar to genes found in various species of cyanobacteria.
- The photosynthetic organelle of one group of protists, the glaucophyte algae, has an outer layer containing the same constituent (peptidoglycan) found in the cell walls of cyanobacteria (Ch. 5, Section 5.2).
- There are many examples of endosymbiotic cyanobacteria living inside the cells of protists or animals today.

Which eukaryote originally obtained a photosynthetic organelle? Because all species in the Plantae have chloroplasts with two membranes, biologists infer that the original, or primary, endosymbiosis occurred in these species' common ancestor. That ancestor eventually gave rise to all subgroups in the Plantae lineage—the glaucophyte algae, red algae, and green plants (green algae and land plants).

But chloroplasts also occur in four of the other major lineages of protists—the Excavata, Rhizaria, Alveolata, and Stramenopila. In these species, the chloroplast is surrounded by more than two membranes—usually four.

To explain this observation, researchers hypothesize that the ancestors of these groups acquired their chloroplasts by ingesting photosynthetic protists that already had chloroplasts. This process, called secondary endosymbiosis, occurs when an organism engulfs a photosynthetic eukaryotic cell and retains the chloroplasts as intracellular symbionts. Secondary endosymbiosis is illustrated in **Figure 27.9**.

Figure 27.10 shows where primary and secondary endosymbiosis occurred on the phylogenetic tree of eukaryotes. Once certain protists obtained the chloroplast, it was "swapped around" to new lineages via secondary endosymbiosis. For example, the chloroplasts in red algae were likely picked up in the common ancestor of Alveolata and Stramenopila, while the chloroplasts of green algae were likely picked up by euglenids and chlorarachniophytes.

Although endosymbiosis accounts for mitochondria and chloroplasts, it doesn't explain the origin of another key organelle found in most eukaryotic cells—namely, the nucleus. What steps were likely involved during the evolution of the nucleus?

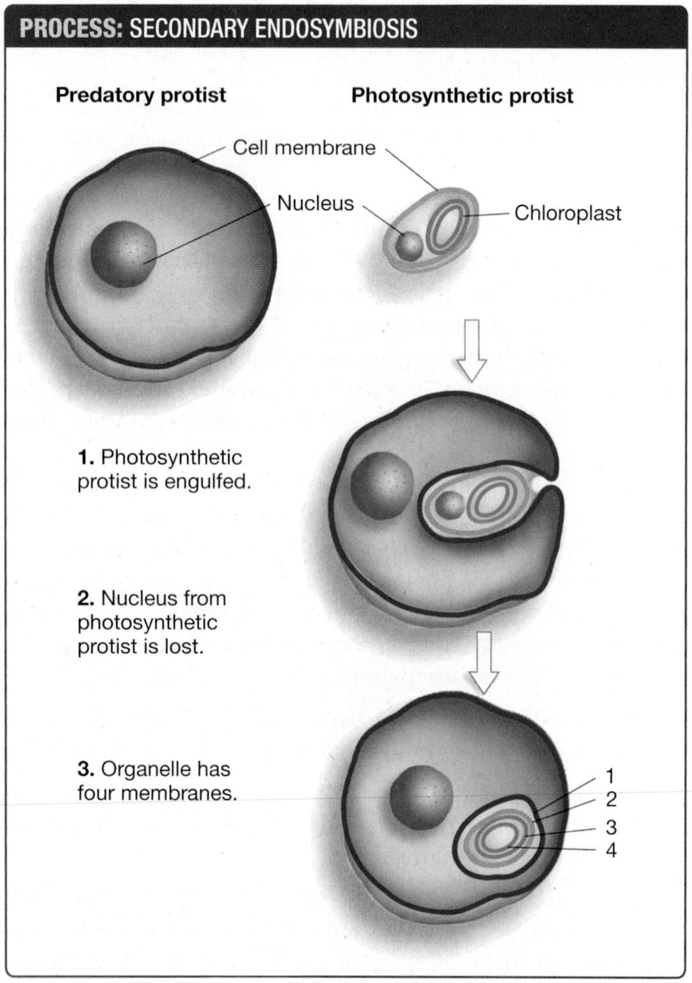

INTERACTIVE Figure 27.9 Secondary Endosymbiosis Leads to Organelles with Four Membranes. The chloroplasts found in some protists have four membranes and are hypothesized to be derived by secondary endosymbiosis. In species where chloroplasts have three membranes, biologists hypothesize that secondary endosymbiosis was followed by the loss of one membrane.

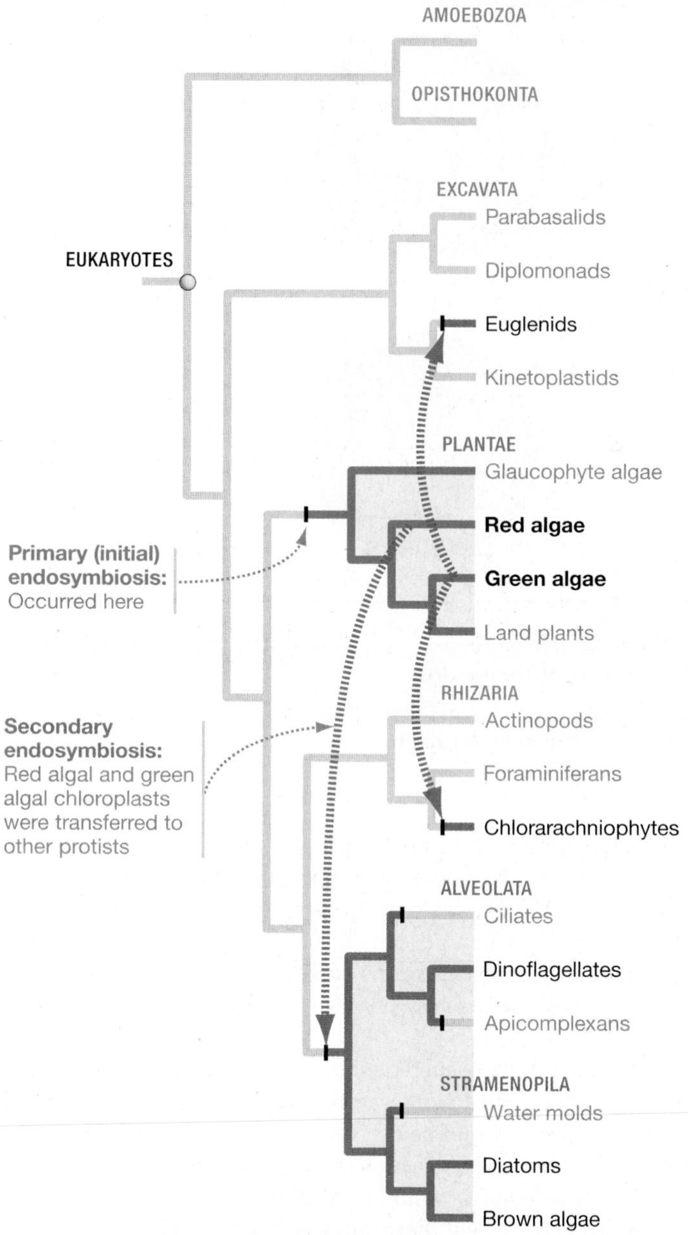

Figure 27.10 Photosynthesis Arose in Eukaryotes by Primary Endosymbiosis and Then Spread among Lineages via Secondary Endosymbiosis. Biochemical similarities link the chloroplasts found in alveolates and stramenopiles with red algae, and the chloroplasts found in euglenids and chlorarachniophytes with green algae. Notice that among the alveolates and stramenopiles, only the dinoflagellates, diatoms, and brown algae have photosynthetic species. In ciliates, apicomplexans, and water molds, the chloroplast has been lost or has changed function.

The Nuclear Envelope The leading hypothesis to explain the origin of the nuclear envelope is based on infoldings of the plasma membrane. As the drawings in Figure 27.11 show, a stepwise process could have given rise to small infoldings that were elaborated by mutation and natural selection over time, and the infolding eventually could have been detached from the plasma membrane. Note that the infoldings would have given rise to the nuclear envelope and the endoplasmic reticulum (ER) together.

Two lines of evidence support this hypothesis: Infoldings of the plasma membrane occur in some bacteria living today, and the nuclear envelope and ER of today's eukaryotes are continuous. ✔ **If you understand the infolding hypothesis, you should be able to explain why these observations support it.**

According to current thinking, the evolution of the nuclear envelope was advantageous because it separated transcription and translation. Recall that in bacteria and archaea, transcription and translation occur together, while in eukaryotes RNA transcripts are processed inside the nucleus but translated outside the nucleus. With a simple nuclear envelope in place, alternative splicing and other forms of RNA processing could occur (Chapters 17 and 19). The take-home message here is that an important morphological innovation gave the early eukaryotes a novel way to control gene expression.

Structures for Support and Protection Many protists have cell walls outside their plasma membrane; others have the hard external structure of a **shell**; others have rigid structures inside the plasma membrane. In many cases, these novel structures represent synapomorphies that identify monophyletic groups among protists. For example:

- Diatoms (Stramenopila) are surrounded by a glass-like cell wall of silicon dioxide. The cell wall is made up of two pieces that fit together in a box-and-lid arrangement (Figure 27.12a), like the petri plates you may have seen in lab.
- Dinoflagellates (Alveolata) have a cell wall made up of cellulose plates (Figure 27.12b).
- Within the foraminiferans (Rhizaria), some subgroups secrete an intricate, chambered calcium carbonate *shell* (Figure 27.12c).
- The euglenids (Excavata) have a collection of protein strips, called a pellicle (from Latin *pellicula*, "thin skin"), located just under the plasma membrane (Figure 27.12d). The strips are supported by microtubules and stiffen the cell.

As this overview demonstrates, the diversification of protists has frequently been associated with the evolution of innovative structures for support and protection.

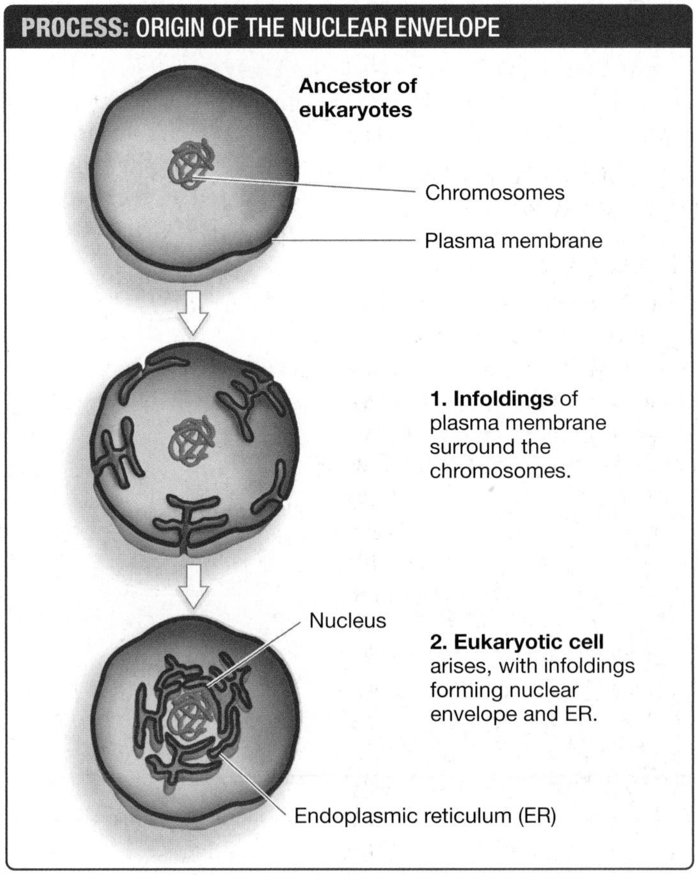

Figure 27.11 **A Hypothesis for the Origin of the Nuclear Envelope.** Infoldings of the plasma membrane, analogous to those shown here, have been observed in bacteria living today.

Figure 27.12 **Hard Outer Coverings in Protists Vary in Chemical Composition and Structure.**

Multicellularity One of the most significant changes in the history of life on Earth occurred when organisms containing more than one cell evolved. The mutations leading to **multicellularity** probably first caused cells to simply stick together after cell division. Selection pressures could then act on these larger, colonial organisms, allowing them to evolve and diversify.

Eventually cells became specialized for different functions. In the simplest multicellular species, certain cells are specialized for producing or obtaining food, while other cells are specialized for reproduction. The key point about multicellularity is that not all cells express the same genes.

Multicellularity arose independently in a wide array of eukaryotic lineages. Even when defined in a strict sense—requiring that cells be connected, communicate, and cooperate with each other—multicellularity appears to have evolved once in animals, three times in fungi, and six times in algae. Once a new type of eukaryotic cell or multicellular individual existed, subsequent diversification was often triggered by novel ways of finding food, moving, or reproducing. Let's consider each of these life processes in turn.

How Do Protists Obtain Food?

Bacteria and archaea can use a wide array of molecules as electron donors and electron acceptors during cellular respiration (Ch. 26, Section 26.3). Some get these molecules by absorbing them directly from the environment. Other bacteria make their own food via photosynthesis. Many groups of protists are similar to bacteria in the way they obtain food: They either absorb nutrients or perform photosynthesis.

But one of the most important stories in the diversification of protists was the evolution of a novel method for securing nutrients—ingesting food. Many protists eat bacteria, archaea, or even other protists whole. This process is called **phagocytosis**. When phagocytosis occurs, an individual takes in packets of food much larger than individual molecules.

Some protists ingest food as well as perform photosynthesis—meaning that they use a combination of feeding strategies. Note that all three lifestyles—ingestive, absorptive, and photosynthetic—can occur within a single lineage.

To drive this last point home, consider the monophyletic group called the alveolates. This lineage has three major subgroups—dinoflagellates, apicomplexans, and ciliates. About half of the dinoflagellates are photosynthetic, but many others are parasitic and absorb nutrients from their hosts. Apicomplexans are all parasitic. Ciliates include many species that ingest prey, but some ciliates live in the guts of cattle or the gills of fish, where they absorb nutrients. Other ciliate species make a living by holding algae or other types of photosynthetic endosymbionts inside their cells.

The key point? Alveolates aren't the only group that uses such a variety of feeding methods. The seven major lineages of eukaryotes all show similar diversification. Protists can feed by either **(1)** ingesting packets of food, **(2)** absorbing organic molecules directly from the environment, or **(3)** performing photosynthesis.

Ingestive Feeding Ingestive lifestyles are based on eating live or dead organisms or on scavenging bits of organic debris. Protists such as the slime mold *Dictyostelium discoideum* are large enough to engulf bacteria; many other protists are large enough to surround and ingest other protists or even microscopic animals.

Feeding by phagocytosis is possible in protists that lack a cell wall. A flexible membrane and dynamic cytoskeleton give these species the ability to surround and "swallow" prey using long, finger-like projections called **pseudopodia** ("false-feet") (Figure 27.13a).

Although many ingestive feeders actively hunt down prey and engulf them, others do not. Instead of moving toward food, these species attach themselves to a surface and let food come to them. Unicellular protists that feed in this way typically have cilia that surround a "mouth" and beat in a coordinated way. The motion creates water currents that sweep food particles into the cell (Figure 27.13b).

(a) Pseudopodia engulf food.

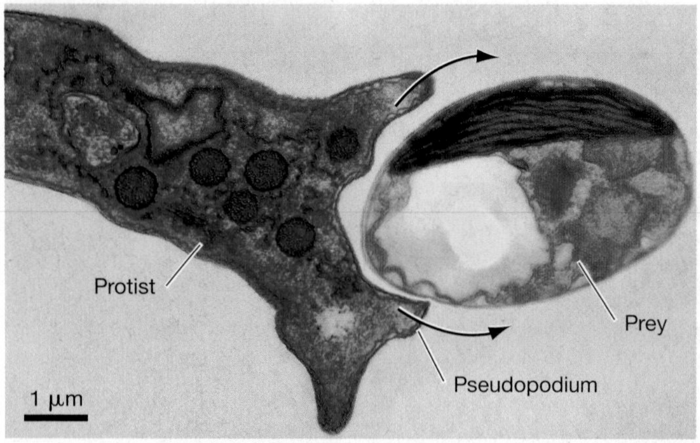

(b) Ciliary currents sweep food into gullet.

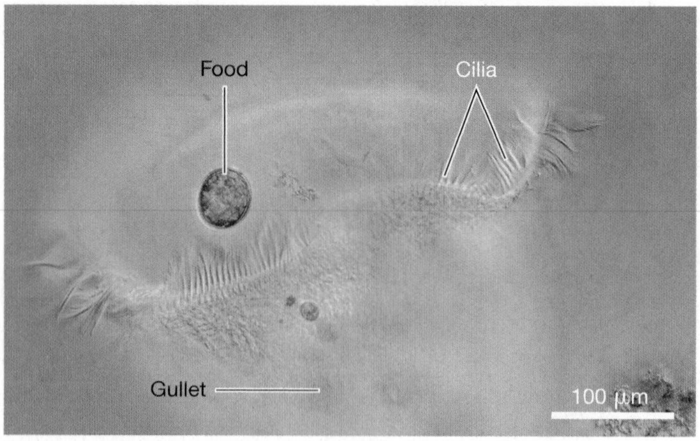

Figure 27.13 **Ingestive Feeding.** Methods of prey capture vary among ingestive protists. **(a)** Some predators engulf prey with pseudopodia; **(b)** other predators sweep them into their gullets with water currents set up by the beating of cilia. Note that the cells in part (a) have been colorized.

✔ Explain how the observation of ingestive feeding in euglenids relates to the hypothesis that this lineage gained chloroplasts via secondary endosymbiosis.

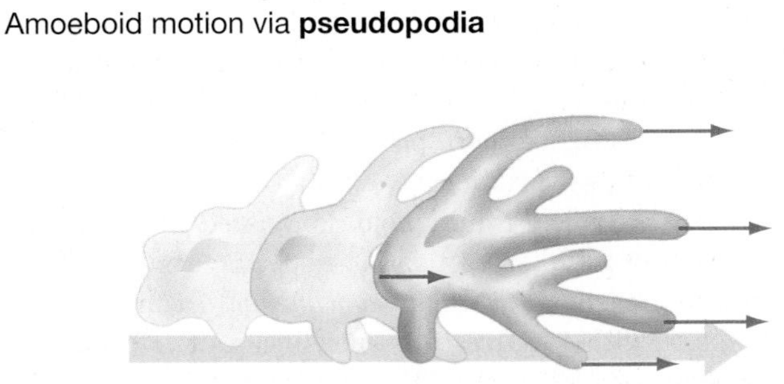

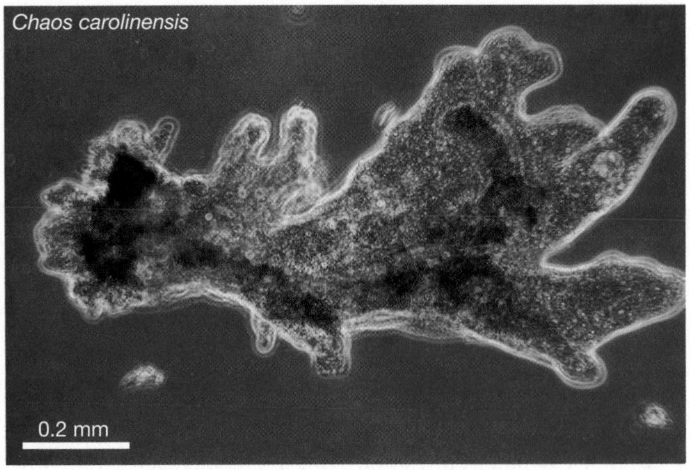

Figure 27.14 Amoeboid Motion Is Possible in Species That Lack Cell Walls. In amoeboid motion, long pseudopodia stream out from the cell. The rest of the cytosol, organelles, and external membrane follow.

Absorptive Feeding Absorptive feeding occurs when nutrients are taken up directly from the environment, across the plasma membrane, usually through transport proteins. Absorptive feeding is common among protists.

Some protists that live by absorptive feeding are **decomposers**, meaning that they feed on dead organic matter, or **detritus**. But many of the protists that absorb their nutrition directly from the environment live inside other organisms. If an absorptive species damages its host, that species is called a **parasite**.

Photosynthesis Protists that engage in ingestive and absorptive feeding are heterotrophs—they acquire organic compounds by feeding on other organisms. Other protists are able to produce their own organic compounds through photosynthesis and are therefore autotrophs. Photosynthetic protists range in size from single-celled organisms to giant seaweeds. These plant-like protists use carbon dioxide as their primary source of carbon.

How Do Protists Move?

Many protists actively move to find food. Members of the Amoebozoa (amoebas and slime molds) appear to crawl over a substrate in search of food. Most of the unicellular, photosynthetic species are capable of swimming to sunny locations, though some just drift passively in water currents. How are crawling and swimming movements possible?

Amoeboid motion is a sliding movement observed in some protists. In the classic mode illustrated in Figure 27.14, pseudopodia stream forward over a substrate, and the rest of the cytosol, organelles, and plasma membrane follow. The motion requires ATP and involves interactions between proteins called actin and myosin inside the cytoplasm. The mechanism is related to muscle movement in animals. But at the level of the whole cell, the precise sequence of events during amoeboid movement is still uncertain. The topic is attracting researchers' attention, because key immune system cells in humans use amoeboid motion as they hunt down and destroy disease-causing agents.

The other major mode of locomotion in protists involves flagella or cilia (Figure 27.15). Recall that flagella and cilia both consist of nine sets of doublet (paired) microtubules arranged around two central, single microtubules. Flagella are long and are usually found alone or in pairs; cilia are short and usually occur in large numbers on any one cell.

Even closely related protists can use radically different forms of locomotion. For example, within the Alveolata, the ciliates swim by beating their cilia, the dinoflagellates swim by whipping their flagella, and mature apicomplexan cells move by amoeboid motion (though their gametes swim via flagella). In addition, many types of protists do not exhibit active movement, but instead float passively in water currents.

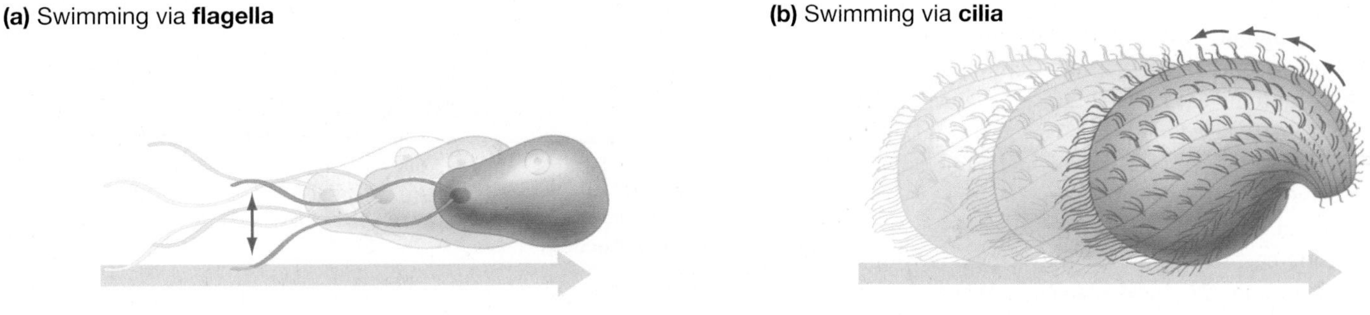

Figure 27.15 Many Protists Swim Using Flagella or Cilia. (a) Flagella are long and few in number. **(b)** Cilia are short and numerous.

How Do Protists Reproduce?

Some species of protists reproduce sexually, others asexually, and some by either method depending on environmental conditions. Sexual reproduction is a uniquely eukaryotic trait that originated in protists.

Sexual reproduction (Chapter 13) can best be understood in contrast to asexual reproduction. The key issues are the type of cell division involved and the consequences for genetic diversity in the offspring produced.

- In eukaryotes, asexual reproduction is based on mitotic cell division, whereas in prokaryotes (bacteria and archaea) it's based on binary fission. Both types of cell division result in daughter cells that are genetically identical to the parent.
- Sexual reproduction (found only in eukaryotes) is based on meiotic cell division and fusion of gametes. Meiotic cell division results in daughter cells that are genetically different from their parents and from each other.

Most protists undergo asexual reproduction routinely. But sexual reproduction occurs only intermittently in many protists—often at one particular time of year, or when individuals are stressed by overcrowding or scarce food supplies.

The evolution of sexual reproduction ranks among the most significant evolutionary innovations observed in eukaryotes.

Sexual versus Asexual Reproduction The leading hypothesis to explain why sexual reproduction evolved states that genetically variable offspring may have a better chance at thriving if the environment changes (the "changing-environment" hypothesis). For example, offspring with genotypes different from those of their parents may be better able to withstand attacks by parasites that successfully attacked their parents. This could be a key advantage because many types of parasites, including bacteria and viruses, have short generation times and evolve quickly.

Because the genotypes and phenotypes of parasites are constantly changing, natural selection favors host individuals with new genotypes. The idea is that novel genotypes in offspring generated by meiosis may contain combinations of alleles that allow hosts to withstand attack by new strains of parasites. In short, many biologists view sexual reproduction as an adaptation to fight disease. ✔ **If you understand the changing-environment hypothesis for the evolution of sex, you should be able to evaluate and explain whether it is consistent with the observation that many protists undergo meiosis when food is scarce or the population density is high.**

Life Cycles—Haploid versus Diploid Dominated A **life cycle** describes the sequence of events that occur as individuals grow, mature, and reproduce. The evolution of meiosis introduced a new event in the life cycle of protist species; what's more, it established a distinction between haploid and diploid phases in the life of an individual.

Recall that diploid individuals have two of each type of chromosome inside each cell. Haploid individuals have just one of each type of chromosome inside each cell. When meiosis occurs in diploid cells, it results in the production of haploid cells.

Virtually every aspect of the life cycle is variable among protists—whether meiosis occurs, whether asexual reproduction occurs, and whether the haploid or the diploid phase of the life cycle is the longer and/or more prominent phase. **Figure 27.16** illustrates some of this variation.

Figure 27.16a depicts the haploid-dominated life cycle observed in many unicellular protists. The specific example given here is the dinoflagellate *Gyrodinium uncatenum*. To analyze this life cycle, start with **fertilization**—the fusion of two haploid gametes to form a diploid zygote. Then trace what happens to the zygote. In this case, the diploid zygote undergoes meiosis. The haploid products of meiosis then grow into mature cells that can take one of two pathways. They can undergo asexual reproduction by producing new cells by mitosis. Or, they can form new cells by mitosis that differentiate into gametes. The gametes can then participate in fertilization, thus completing sexual reproduction. They key point is that the life cycle is dominated by haploid cells.

Now turn to the diploid-dominated life cycle in Figure 27.16b. The specific organism shown here is the diatom *Thalassiosira punctigera*. Notice that after fertilization, the diploid zygote grows and develops into a mature, diploid adult cell that can reproduce asexually by dividing mitotically. After several rounds of mitosis, in which the daughter cells become successively smaller, the cells cannot further divide and meiosis takes place. This leads the cells back into sexual reproduction and the formation of haploid gametes, which then fuse to form a diploid zygote. The important contrasts with the haploid-dominated life cycle are that **(1)** meiosis in diploid-dominated life cycle occurs in the mitotically produced offspring rather than in the zygote, and **(2)** gametes are the only haploid cells in the diploid-dominated life cycle. **Making Models 27.1** offers another aid to understanding sexual and asexual life cycle phases.

Making Models 27.1 Tips on Drawing Life Cycles

Drawing life cycles can help you understand complex concepts. Life cycles have two components: **(1)** life stages arranged in sequential order and marked as haploid or diploid, and **(2)** arrows representing processes that connect the stages.

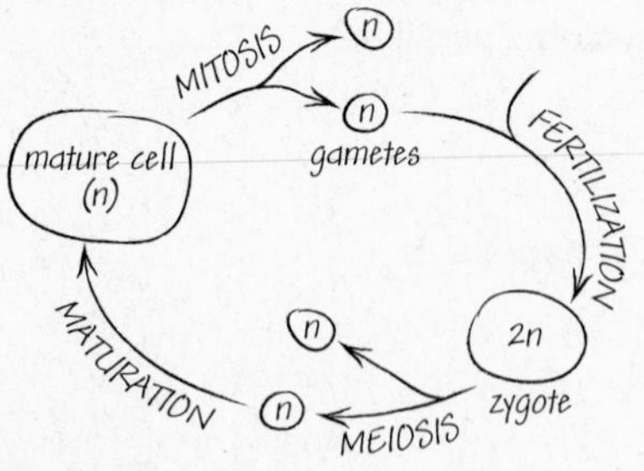

MODEL The life cycle shown here models sexual reproduction in a haploid-dominated protist. Expand this model to include asexual reproduction.

To see this model in action, go to the Study Area of **Mastering Biology**

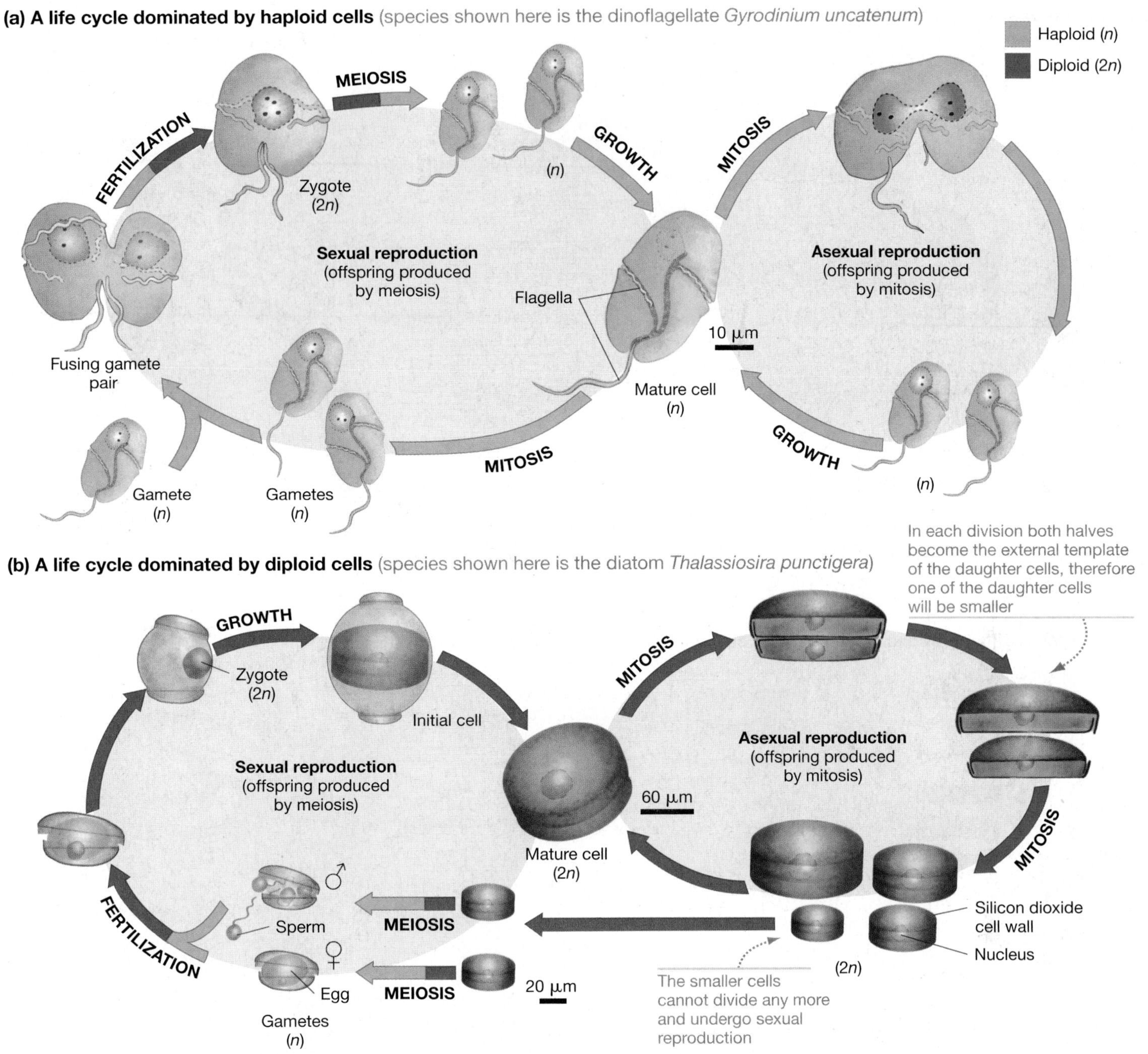

Figure 27.16 Life Cycles Vary Widely among Unicellular Protists. Many unicellular protists can reproduce by both sexual reproduction and asexual reproduction. The cell may be haploid for most of its life **(a)** or diploid for most of its life **(b)**.

Life Cycles—Alternation of Generations In contrast to the relatively simple life cycles of *single-celled* protists shown in Figure 27.16, many *multicellular* protists have a multicellular haploid form in one phase of their life cycle alternating with a multicellular diploid form in another phase. This alternation of multicellular haploid and diploid forms is known as **alternation of generations**.

- The multicellular haploid form is called a **gametophyte**, because specialized cells in this individual produce gametes, also haploid, by *mitotic* cell division.

- The multicellular diploid form is called a **sporophyte** because it has specialized cells that undergo *meiotic* cell division to produce haploid cells called spores.

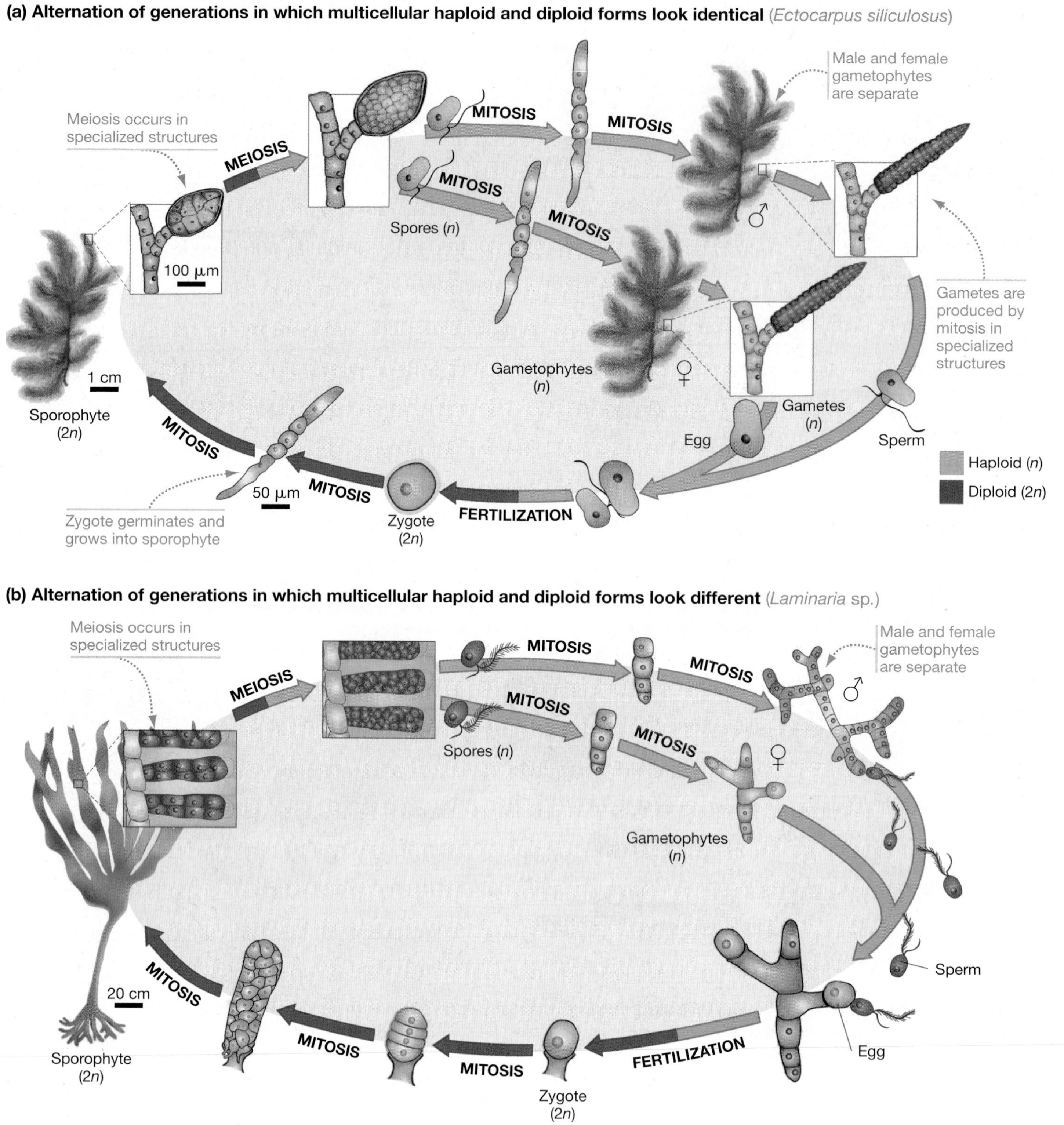

Figure 27.17 Alternation of Generations Occurs in Many Multicellular Protists. Two species of brown algae are compared here.

✔ **THINK CAREFULLY** Are the gametes in brown algae formed by mitosis or meiosis?

A **spore** is a single haploid cell that divides mitotically to form a multicellular, haploid gametophyte. A gamete is also a single haploid cell, but its role is to fuse with another gamete to form a diploid zygote.

Gametophytes and sporophytes may be identical in appearance, as in the brown alga *Ectocarpus siliculosus* (**Figure 27.17a**). In many cases, however, the gametophyte and sporophyte look different, as in the brown algae in the genus *Laminaria* (**Figure 27.17b**).

Take a moment to go back and reread the bulleted list on page 569. If it strikes you as odd, you might be thinking of how humans and other animals produce gametes (egg and sperm) by *meiotic* cell division. To understand alternation of generations, some students find it helpful to memorize this sentence: Sporophytes produce spores by meiosis; gametophytes produce gametes by mitosis.

Be aware that alternation of generations represents the life cycle not only of some protists, including brown and green algae, but also of land plants. If you understand this fundamental pattern and the relationship between sporophytes, spores, gametophytes, and gametes, then you can explain the life cycle of numerous protists as well as any land plants you observe.

Why does so much variation occur in the types of life cycles observed among protists? The answer is not known; developing and testing hypotheses to explain this variation is an active area of research and is discussed further in the chapter on green algae and land plants (Chapter 28).

CHECK YOUR UNDERSTANDING

✔ If you understood this section, you should be able to . . .

1. Explain why the primary and secondary endosymbiosis events introduced in this section represent the most massive lateral gene transfers in the history of life, in terms of the number of genes moved at once. (To review lateral gene transfer, see Chapter 20.)
2. Evaluate whether the life cycle of humans is more comparable to that of *Gyrodinium* or *Thallassiosira* (see Figure 27.16) and whether it is more comparable to *Ectocarpus* or *Laminaria* (see Figure 27.17). Explain your reasoning.

Answers are available in Appendix A.

27.4 Key Lineages of Eukaryotes

After you complete this section, you should be able to . . .

- Describe the key features of the major lineages of eukaryotes.

Each of the seven major Eukarya lineages has at least one distinctive morphological characteristic. But once an ancestor evolved a distinctive cell structure, its descendants diversified into a wide array of lifestyles. Despite this diversification, some features evolved separately in different lineages and thus represent examples of convergent evolution. For example, parasitic species evolved independently in all seven major lineages. Photosynthetic species exist in most of the seven, and multicellularity evolved independently in at least four. Similar statements could be made about the evolution of life cycles and modes of locomotion.

Let's examine these major lineages, starting with an overview of each one. The lineages described here and summarized in **Table 27.3** on pages 572–573 are just a sampling of eukaryote diversity.

Amoebozoa

Species in the Amoebozoa lack cell walls and take in food by engulfing it. They move via amoeboid motion and produce large, lobe-like pseudopodia. They are abundant in freshwater habitats and in wet soils; some are parasites of humans and other animals.

Major subgroups in the lineage include amoebas, cellular slime molds—including the model organism *Dictyostelium discoideum* (**BioSkills 11**)—and plasmodial slime molds. *Dictyostelium* consists of numerous amoeboid cells that live independently part of the time, but join together to form a cohesive multicellular reproductive structure under certain conditions.

The plasmodial slime mold *Physarum polycephalum* (featured in the end-of-chapter case study) can grow over 10 cm in diameter but is just one cell that contains thousands of nuclei. The large "supercell," called a plasmodium, moves by streaming its cytoplasm in a regular back-and-forth motion. **✔ You should be able to mark the origin of the supercell and the origin of this lineage's amoeboid form on Figure 27.7.**

Opisthokonta

This is a large and diverse group of eukaryotes that includes fungi as well as animals, and their closest living relatives—the group of protists called choanoflagellates.

Fungi play a large role as decomposers in many ecosystems due largely to their ability to produce enzymes that break down the cell walls of plants and other organisms. (Fungi are explored more fully in Chapter 29.)

Animals include many diverse lineages, all of which are multicellular heterotrophs. All animals (except for sponges) also contain neurons and muscle cells. Unlike choanoflagellates and some fungi, all animals display true multicellularity with differential cell development and function. (In Chapters 30, 31, and 32, you will look at the tremendous amount of diversity found among animals, including their close evolutionary relationship with choanoflagellates.)

The choanoflagellates are a group of protists that can live as single cells or as colonies. Individual choanoflagellate cells are almost identical in structure and function to the feeding cells found in sponges.

Excavata

The unicellular species that form the Excavata are named for their distinguishing morphological feature—an "excavated" feeding groove found on one side of the cell (see Table 27.2). Because some lack recognizable mitochondria, excavates were once thought to trace their ancestry to eukaryotes that existed before the origin of mitochondria. But researchers have found that excavates either have **(1)** genes in their nuclear genomes that are of mitochondrial origin, or **(2)** unusual organelles that appear to be vestigial mitochondria. These observations support the hypothesis that the ancestors of excavates had mitochondria, but that these organelles were lost or reduced over time in this lineage. **✔ You should be able to mark the loss of mitochondria on Figure 27.7.**

DIVERSITY **Table 27.3** **Key Lineages of Eukaryotes**

		Morphology and Movement	Reproduction	Relevance
	Amoebozoa **Lobose amoebas, cellular slime molds, plasmodial slime molds** ◀ *Dictyostelium discoideum* (a cellular slime mold)	Lack cell walls; flexible and dynamic cell membranes. Display amoeboid movement with cytoplasmic streaming.	Both asexual and sexual reproduction. Cellular slime molds are haploid dominant. Plasmodial slime molds are diploid dominant.	Slime molds influence nutrient cycling by feeding on microorganisms. *Dictyostelium discoideum* is used as a model organism for cell biology.
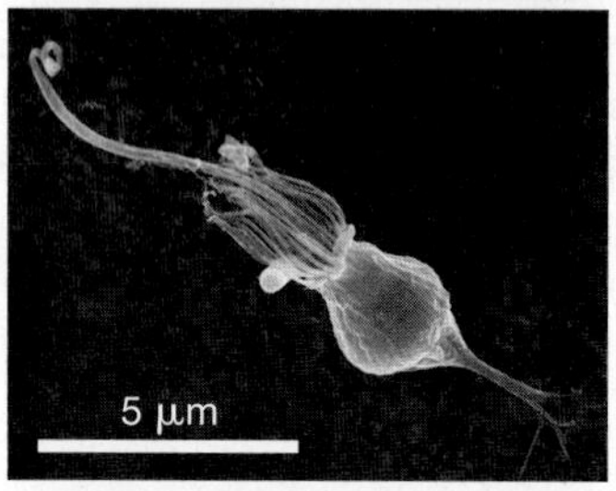	Opisthokonta **Choanoflagellates, animals, and fungi** ◀ *Salpingoeca* sp. (a choanoflagellate)	Single celled to large, multicellular. Some with cell walls, some without. Most produce motile cells as part of their life cycle.	Both sexual and asexual reproduction.	Choanoflagellates are the closest living relatives to animals. Fungi play a large role as decomposers. Animals play a large role as consumers.
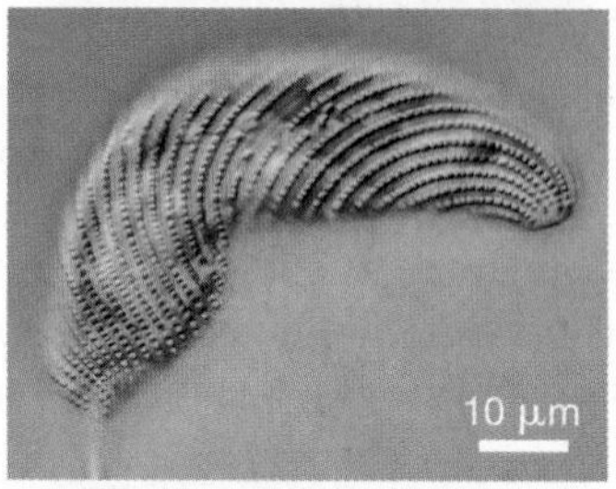	Excavata **Parabasalids, diplomonads, euglenids** ◀ *Euglena spirogyra* (a euglenid)	Excavated feeding groove on side of cell. Most swim via flagella. Diplomonads have two nuclei. Many euglenids are photosynthetic.	Primarily asexual reproduction. Sexual reproduction has been observed in only a few members.	*Trichomonas* causes trichomoniasis. *Giardia* is responsible for giardiasis. Euglenids are abundant in freshwater plankton.
	Plantae **Glaucophyte algae, red algae, green algae, land plants** ◀ *Palmaria palmata*, or dulse (a red alga)	Algae can be unicellular, colonial, or multicellular. Contain chloroplasts. Cell walls composed of cellulose. Algae produce flagellated cells.	Both asexual and sexual reproduction. Some algae and all land plants exhibit alternation of generations.	Primary producers in most ecosystems. Agar is derived from cell walls of red algae. Some red algae aid in development of coral reefs.

(Continued)

Major groups in the lineage include parabasalids, diplomonads, and euglenids. All 300 known species of parabasalids live inside animals; some live only in the guts of termites, where they aid in the digestion of cellulose. About 100 species of diplomonads have been named. Many live in the guts of animal species without causing harm to their host; other species live in stagnant-water habitats. There are about 1000 known species of euglenids, many of which are photosynthetic. Although most live in freshwater, a few are found in marine habitats.

Plantae

Biologists use the name **Plantae** to refer not just to the commonly named land plants, but to the entire monophyletic group that includes glaucophyte algae, red algae, and green algae as well. All subgroups within Plantae are thought to have descended from a common ancestor that engulfed a cyanobacterium, beginning the endosymbiosis that led to the evolution of the chloroplast—their distinguishing morphological feature. Various lines of evidence suggest that this endosymbiosis event probably occurred in an ancestor of today's glaucophyte algae. Recall from Figure 27.10 that the chloroplast later spread to other lineages via secondary endosymbiosis. ✔ **You should be able to mark the origin of the chloroplast on Figure 27.7.**

The glaucophyte algae are unicellular or colonial. They live as plankton or attached to substrates in freshwater environments—particularly in bogs or swamps. Some glaucophyte species have flagella or produce flagellated spores, but sexual reproduction has never been observed in the group. The chloroplasts of glaucophytes have a distinct bright blue-green color.

DIVERSITY Table 27.3 Key Lineages of Eukaryotes (continued)

		Morphology and Movement	Reproduction	Relevance
	Rhizaria **Foraminiferans, actinopods, chlorarachniophytes** ◀ *Acanthometra sp.* (an actinopod)	Lack cell walls, but many have a shell-like covering. Move by amoeboid motion via slender pseudopodia.	Asexual reproduction is common, but sexual reproduction is possible. Meiosis does produce haploid gametes that fuse to form a diploid individual. Many display alternation of generations.	Shells of dead foraminiferans form extensive sediments on the ocean bottom and can form limestone. Presence of certain shells can be used to date rocks during petroleum exploration.
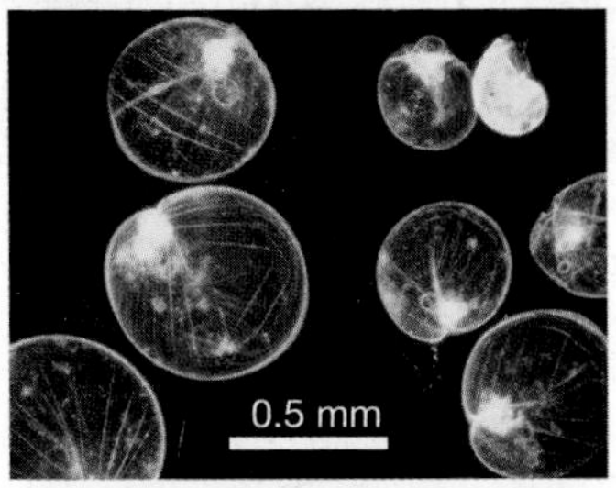	Alveolata **Ciliates, dinoflagellates, apicomplexans** ◀ *Noctiluca scintillans* (a bioluminescent dinoflagellate)	Contain vesicles called alveoli that support the plasma membrane. Move by cilia or flagella. Some apicomplexans move by amoeboid motion.	Both asexual and sexual reproduction.	Dinoflagellates are photosynthetic, and some are bioluminescent. Apicomplexans are parasites responsible for several diseases.
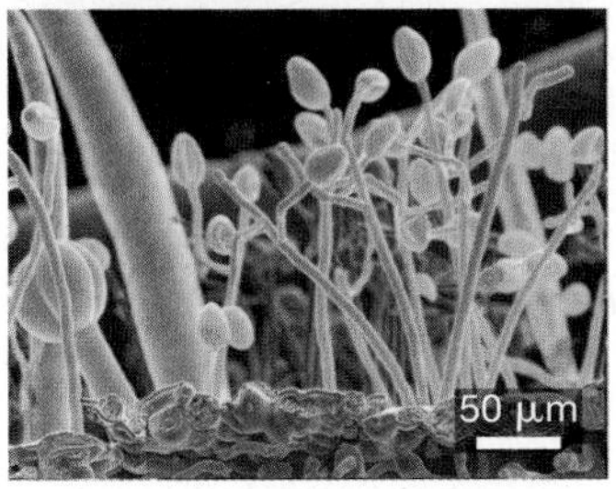	Stramenopila **Water molds, diatoms, brown algae** ◀ *Phytopthora infestans* (a water mold)	Contain flagella with hollow projections. Diatoms and brown algae contain chloroplasts. Diatoms form ornate glass-like shells.	Both asexual and sexual reproduction. Many exhibit diploid-dominant life cycles. Brown algae (e.g., kelp) display alternation of generations.	*Phytopthora infestans* caused the Irish potato famine. Diatoms are important primary producers. Giant kelp forests form habitats for a wide variety of animals.

For more go to https://goo.gl/PdUYps

The 6000 or so species of red algae live primarily in marine habitats. Although their color varies, many species are red because their chloroplasts contain large amounts of the accessory pigment phycoerythrin, which absorbs strongly in the blue and green portions of the visible spectrum. Because blue light penetrates water better than other wavelengths, red algae are able to live at considerable depth in the oceans.

The evolution and diversity of the largest subgroups of Plantae (the green algae and land plants) are covered separately (Chapter 28).

Rhizaria

The rhizarians are single-celled organisms that lack cell walls, though some species produce elaborate shell-like coverings. They move by amoeboid motion and produce long, slender pseudopodia. Over 11 major subgroups in this lineage have been identified and named, including the planktonic organisms called actinopods—which synthesize glassy, silicon-rich shells—and the chlorarachniophytes, which most likely obtained their chloroplasts via secondary endosymbiosis and are photosynthetic. ✔ **You should be able to mark the origin of this lineage's amoeboid form on Figure 27.7.**

The best-studied and most abundant group is the foraminiferans, or forams, which got their name from the Latin *foramen*, meaning "hole." Forams produce calcium carbonate shells with holes through which pseudopodia protrude (see Figure 27.12). Fossil shells of foraminifera are abundant in marine sediments—a continuous record of fossilized forams dates back 530 million years. One species was recently found living in sediments at a depth of 11,000 m below sea level. They are abundant in marine plankton as well as bottom habitats.

Alveolata

Alveolates are distinguished by systems of small membrane-bound sacs, called alveoli, that are located just under the plasma membrane. The three major subgroups within the lineage include ciliates, apicomplexans, and dinoflagellates. ✔ **You should be able to mark the origin of alveoli on Figure 27.7.**

Ciliates are named for the cilia that cover them. Some 12,000 species are known from freshwater, marine, and wet soil environments. All of the 5000 known species of apicomplexans are parasitic. For example, several species of *Plasmodium* cause malaria (see Figure 27.3), and *Toxoplasma* is an important opportunistic pathogen in people infected with HIV.

Most of the 4000 known species of dinoflagellates are marine or freshwater plankton. Dinoflagellates are responsible for many harmful algal blooms (see Figure 27.4). Some species are capable of **bioluminescence**, meaning they emit light via an enzyme-catalyzed reaction. If you've seen ocean waves glowing at night, then you've probably experienced this remarkable phenomenon. Bioluminescent dinoflagellates typically glow when they are disturbed, such as in cresting waves (**Figure 27.18a**) or when boat propellers churn the water. What selective advantage might bioluminescence provide to dinoflagellates?

The leading hypothesis suggests that bioluminescence serves as a defense mechanism in dinoflagellates. The light produced may help prevent them from being eaten by startling potential predators (e.g., copepods, which are tiny crustaceans) and allowing the dinoflagellate time to escape. Alternatively, the bioluminescence could attract small fish that feed on copepods which would indirectly reduce the number of dinoflagellates eaten.

In one experiment to test the "startle" hypothesis, researchers first prepared "bright" and "dim" dinoflagellates. When grown in bright sunlight for several days and then placed in the dark, dinoflagellates are able to bioluminesce brightly for about 12 hours; after 12 hours, they lose this ability. The "bright" and "dim" dinoflagellates were then used to test the impact of bioluminescence on the likelihood of being eaten by copepods. The results provide strong support for the "startle" hypothesis: the ability to produce light significantly reduced the number of dinoflagellates eaten by copepods (**Figure 27.18b**).

Stramenopila (Heterokonta)

Stramenopiles are sometimes called heterokonts ("different-hairs"). At some stage of their life cycle, all stramenopiles have flagella that are covered with distinctive hollow "hairs." The structure of these flagella is the synapomorphy of this lineage. **✔ You should be able to mark the origin of the "hairy" flagellum on Figure 27.7.**

The stramenopiles are a large and morphologically diverse group of eukaryotes. They include parasitic, saprophytic, and photosynthetic species, and they range in size from tiny unicellular organisms to large multicellular seaweeds. Major subgroups within the lineage include water molds, diatoms, and brown algae.

Water molds are extremely important decomposers in freshwater ecosystems. Most species feed on decaying organic material; a few are parasitic. Mature individuals are **sessile**—that is, permanently fixed to a substrate. Parasitic forms that affect humans indirectly include the organism that caused the Irish potato famine.

Diatoms are considered the most important primary producer in fresh and salt water. Their glassy cell walls settle into massive accumulations that are mined and sold as diatomaceous earth, which is used in filtering applications and as an ingredient in polishes, paint, cosmetics, and other products.

Brown algae are multicellular. They include many of the seaweeds found in intertidal habitats and also the kelp forests that grow in many coastal areas with cool water.

(a) Observation: Dinoflagellates bioluminesce when physically disturbed.

(b) "Startle" hypothesis: Ingestion rates by predators suggest bioluminescence serves as a defense mechanism.

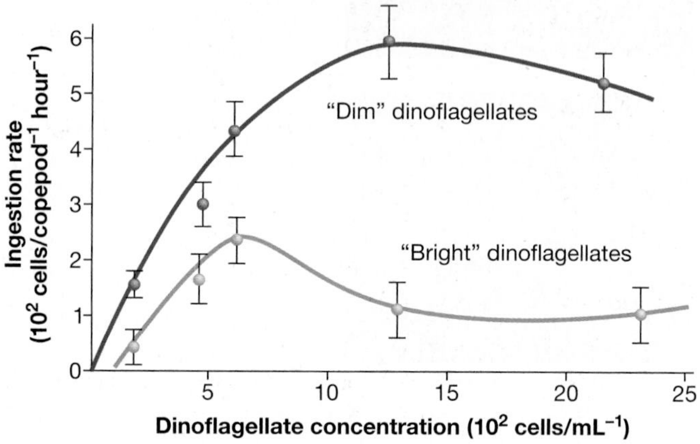

Figure 27.18 Bioluminescent Dinoflagellates Light Up Ocean Waves at Night. (a) Glowing plankton on beach. **(b)** The results of a simple experiment to test the "startle" hypothesis provide evidence that bioluminescence protects dinoflagellates by reducing predation.

DATA: W. E. Esaias and H. C. Curl Jr. 1972. *Limnology and Oceanography* 17: 901–906.

✔ PROCESS OF SCIENCE What would the results of the experiment in part (b) look like if the startle hypothesis were not supported?

CHECK YOUR UNDERSTANDING

✔ If you understood this section, you should be able to . . .

1. Explain whether the amoeboid form evolved just once or independently in separate eukaryote lineages. (Hint: Use the information in this section as well as Figure 27.7.)
2. Explain whether bioluminescence in eukaryotes, which is found not only in dinoflagellates but also in certain members of the Opisthokonta, evolved just once or independently in separate lineages. (Hint: Use Figure 27.7 to help answer this question).

Answers are available in Appendix A.

27.1 Why Do Biologists Study Protists?

- Protists include all eukaryotes except the land plants, fungi, and animals.
- Protists are often tremendously abundant in marine and freshwater plankton and other aquatic habitats. Protists provide food for many organisms in aquatic ecosystems and fix so much carbon that they have a large impact on the global carbon budget.
- Parasitic protists cause several important diseases in humans, including malaria.
- Toxin-producing protists that grow to high densities can result in a harmful algal bloom.

27.2 How Do Biologists Study Protists?

- Microscopic investigation of protists led to the understanding that many are single celled. Unlike bacteria, they contain mitochondria, a nucleus and endomembrane system, and a cytoskeleton.
- Direct sequencing has revealed important new groups of protists.
- Phylogenies based on molecular data confirmed that there are seven major lineages of protists and that the group is paraphyletic.

27.3 What Themes Occur in the Diversification of Protists?

- The first eukaryote probably resulted from an ancient endosymbiosis that occurred when a prokaryotic host cell (likely an archaeal cell) engulfed an α-proteobacterium. This bacterium evolved into today's mitochondria.
- The chloroplast's size and its circular chromosome, ribosomes, and double membrane are consistent with the hypothesis that this organelle originated as an endosymbiotic cyanobacterium.
- After primary endosymbiosis occurred, chloroplasts were "passed around" to new lineages of protists via secondary endosymbiosis.
- The evolution of the nuclear envelope allowed for spatial separation of the processes of transcription and translation.
- Multicellularity evolved in several different protist groups independently.
- Protists vary widely in the way they obtain food. They exhibit predatory, parasitic, or photosynthetic lifestyles, which evolved in many groups independently.
- The evolution of ingestive feeding was important for two reasons: **(1)** It allowed eukaryotes to obtain resources in a new way—by eating bacteria, archaea, and other eukaryotes; and **(2)** it made endosymbiosis and the evolution of mitochondria and chloroplasts possible.
- Protists vary widely in the way they reproduce. They undergo cell division based on mitosis and reproduce asexually. Many protists also undergo meiosis and sexual reproduction at some phase in their life cycle.
- Alternation of generations is common in multicellular protist species—meaning they have separate haploid and diploid forms.
- When alternation of generations occurs, haploid gametophytes produce gametes by mitosis; diploid sporophytes produce spores by meiosis.

27.4 Key Lineages of Eukaryotes

- Protists are a highly diverse group of eukaryotic species organized into seven lineages.
- Protist lineages are defined by DNA sequence evidence and morphological traits such as the presence of chloroplasts or unique flagella.
- A wide range of traits including parasitism, autotrophy, and multicellularity evolved independently in several different lineages of protists.

Answers are available in Appendix A.

✔ TEST YOUR KNOWLEDGE

1. Why are protists considered paraphyletic?
 a. They include many extinct forms, including lineages that no longer have any living representatives.
 b. They include some but not all descendants of their most recent common ancestor.
 c. They represent all of the descendants of a single common ancestor.
 d. Not all protists have all of the synapomorphies that define the Eukarya, such as a nucleus.

2. The most important primary producers in marine ecosystems are ________________.

3. Evaluate the following statements regarding motility in protists. Select True or False for each statement.
 T/F Amoeboid motion is common in species with cell walls.
 T/F Some protists use flagella to swim.
 T/F Some protists use cilia to swim, which are shorter and more numerous than flagella.
 T/F Amoeboid motion requires ATP and interactions between the proteins actin and myosin.

4. How can dinoflagellates be harmful to humans?
 a. They are transmitted by mosquitoes and cause malaria.
 b. They produce toxins that can be absorbed by clams and other shellfish which, when eaten by people, can lead to paralytic shellfish poisoning.
 c. They cause amoebic dysentery which leads to severe diarrhea and dehydration.
 d. They are transmitted by tsetse flies and cause "sleeping sickness."

✔ TEST YOUR UNDERSTANDING

5. Explain the logic behind the claim that the nuclear envelope is a synapomorphy that defines eukaryotes as a monophyletic group.

6. Consider the endosymbiosis theory for the origin of the mitochondrion. How did each endosymbiotic partner benefit from the relationship?

7. **PROCESS OF SCIENCE** Why was finding a close relationship between mitochondrial DNA and bacterial DNA considered particularly strong evidence in favor of the endosymbiosis theory?

8. The text claims that the evolutionary history of protists can be understood as a series of morphological innovations that established seven distinct lineages, each of which subsequently diversified based on innovative ways of feeding, moving, and reproducing. Explain how the Alveolata support this claim.

✔ TEST YOUR PROBLEM-SOLVING SKILLS

9. **SOCIETY** Consider the following:
   - *Plasmodium* has an unusual organelle called an apicoplast. Recent research has shown that apicoplasts are derived from chloroplasts via secondary endosymbiosis and have a large number of genes related to chloroplast DNA.
   - Glyphosate is one of the most widely used herbicides. It works by poisoning an enzyme located in chloroplasts.
   - Biologists are testing the hypothesis that glyphosate could be used as an antimalarial drug in humans.

   How are these observations connected?

10. **SOCIETY** Suppose a friend says that we don't need to worry about the rising temperatures associated with global climate change. She claims that increased temperatures will make planktonic algae grow faster and that carbon dioxide (CO_2) will be removed from the atmosphere faster. According to her, this carbon will be buried at the bottom of the ocean in calcium carbonate shells. As a result, the amount of carbon dioxide in the atmosphere will decrease and global warming will decline. Comment.

✔ PUT IT ALL TOGETHER: Case Study

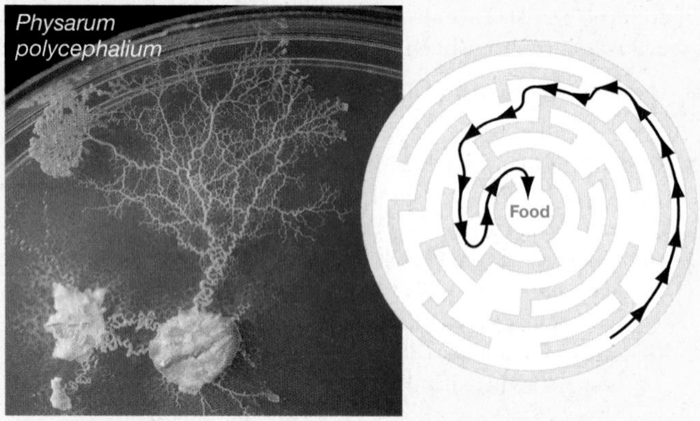

Does a slime mold "know" where it's going?

When placed at the perimeter of a maze with food in the center, the plasmodial slime mold *Physarum polycephalum* explores the maze, retracts branches from dead-end corridors, and then grows exclusively along the shortest path possible to the food. How does *Physarum* do this? One theory is that it leaves behind slime deposits—an externalized "memory" that "reminds" it not to retry dead ends.

11. Which of the following best describes movement in *Physarum*?
    a. Cilia propel the slime mold.
    b. Flagella propel the slime mold.
    c. The slime mold moves by amoeboid motion.
    d. The slime mold moves by gliding motility.

12. *Physarum* is a plasmodial slime mold, whereas *Dictyostelum* is a cellular slime mold (**BioSkills 11**). Compare and contrast movement by the migrating slug stage of *Dictyostelium* to the plasmodial stage of *Physarum*.

13. **QUANTITATIVE** Does an organism without a brain have the ability to use an externalized "memory"—a spatial "slime map" that the organism uses to avoid moving to regions where it has been before? Researchers addressed this question by placing a U-shaped trap between *Physarum* and its food (see diagram that follows). Twenty-three out of 24 slime molds reached the food when plain agar was used as the growth substrate. However, when the agar was coated with extracellular slime, only 8 of 24 found the food. The mean time in hours that it took the successful slime molds to reach the food when placed on plain agar or agar pre-coated with extracellular slime was compared ($P = 0.012$). Use the P value provided to determine if the difference is significant or not (hint: See **BioSkills 3**). What conclusion can be drawn from the graph shown here?

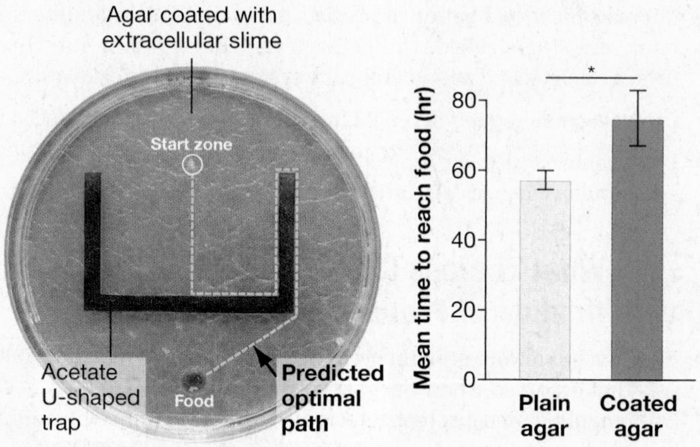

Source: C. R. Reid, T. Latty, A. Dussoutour, et al. 2012. *Proceedings of the National Academy of Science USA* 109: 17490–17494.

14. **PROCESS OF SCIENCE** Propose an experiment that would test whether the coating of extracellular slime changed the speed at which the slime mold moved across the substrate.

15. **PROCESS OF SCIENCE** Develop simple experiments to test whether *Physarum* prefers (1) brightly lit or dark environments; (2) dry or moist conditions; (3) oats or sugar as a food source.

16. **SOCIETY** Researchers have proposed that slime molds could be used to help to plan the paths of future roadways and railways. Justify this statement.

Mastering Biology

Students Go to Mastering™ Biology for assignments, the eText, and the Study Area with animations, practice tests, and activities.

Professors Go to Mastering™ Biology for automatically graded tutorials and questions that you can assign to your students, plus Instructor Resources.

28 Green Algae and Land Plants

Earth has been called the Blue Planet, but it could also be called the Green Planet.

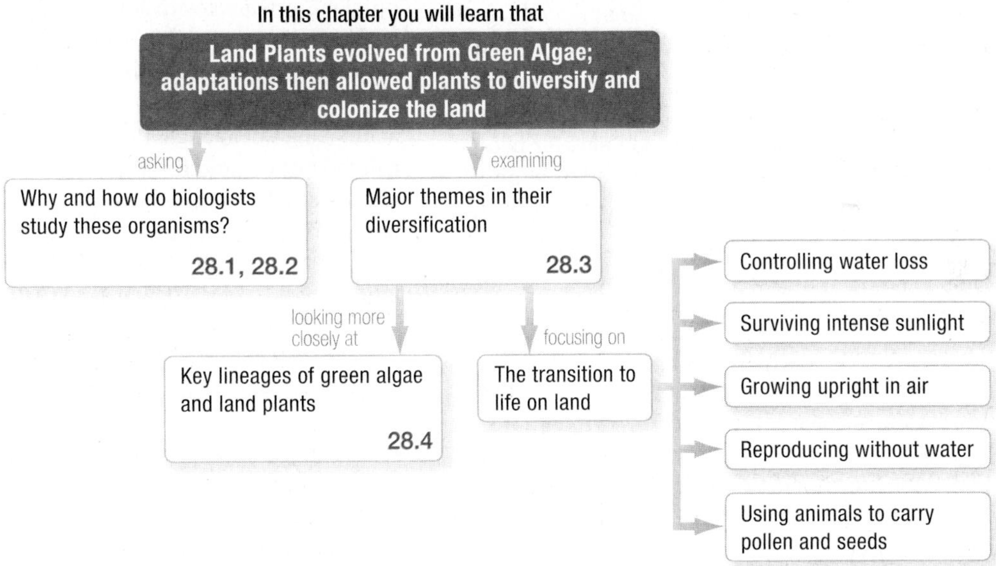

In total mass and importance to other organisms, the green plants dominate terrestrial and freshwater habitats. When you walk through a forest or meadow, you are surrounded by plants. If you look at pond or lake water under a microscope, green algae are everywhere.

The Viridiplantae, or green plants, are made up of two major and astonishingly diverse types of organisms: the green algae and the land plants. **Green algae** are important photosynthetic organisms in aquatic habitats—particularly lakes, ponds, and other freshwater settings. **Land plants** are the key photosynthesizers in terrestrial environments. While green algae have traditionally been considered protists, studying them alongside land plants is logical for two reasons: **(1)** They are the closest living relatives to land plants and form a monophyletic group with them, and **(2)** the transition from aquatic to terrestrial life occurred when land plants evolved from green algae.

BIG PICTURE

This chapter is part of the Big Picture. See how on pages 722–723.

Land plants were the first organisms that could thrive with their tissues completely exposed to the air instead of being partially or completely submerged. Before they evolved, likely the only life on the continents consisted of bacteria, archaea, and single-celled eukaryotes that thrive in wet soils. By colonizing the continents, plants transformed the nature of life on Earth. It was, in the words of biologist Karl Niklas, "One of the greatest adaptive events in the history of life." Plants made the Earth green.

28.1 Why Do Biologists Study Green Algae and Land Plants?

Biologists study the green plants because they are fascinating, and because we couldn't live without them. Along with most other animals and fungi, humans depend almost completely on plants for food. People rely on plants for other necessities of life as well—fuel, the fibers used to make clothing and other products, building materials, and pharmaceuticals. Agriculture, forestry, and horticulture are among the most important industries based on plants. To drive this point home, consider that the sale of cut flowers generates over $7 billion each year in the United States alone.

After you complete this section, you should be able to . . .

- Summarize the ecosystem services and foods, medicines, and materials plants provide humans.

Tens of thousands of biologists are employed in research designed to better understand plants, increase their productivity, and find new ways of using them to benefit people. Biologists also study two categories of plants that cause problems for people: weeds that decrease the productivity of crop plants, and newly introduced species that invade and then degrade natural areas.

Plants Provide Ecosystem Services

An **ecosystem** consists of all the organisms in a particular area, along with physical components of the environment such as the atmosphere, precipitation, surface water, sunlight, soil, and nutrients. Green algae and land plants provide **ecosystem services** because they enhance the life-supporting attributes of the atmosphere, soil, surface water, and other physical components of an ecosystem.

Plants Produce Oxygen Recall that plants perform oxygenic (literally, "oxygen-producing") photosynthesis (Chapter 10). In this process, electrons that are removed from water molecules are used to reduce carbon dioxide (CO_2). In the process of stripping electrons from water, plants release oxygen molecules (O_2) as a by-product.

Oxygenic photosynthesis evolved in cyanobacteria and was responsible for the origin of an oxygen-rich atmosphere (Ch. 26, Section 26.3). The evolutionary success of plants continued this trend because plants add huge amounts of oxygen to the atmosphere.

Figure 28.1 **Devegetation Leads to Soil Erosion.**

Plants Build and Hold Soil Fallen leaves and decaying roots and stems provide food for worms, fungi, bacteria, archaea, single-celled eukaryotes, and other decomposers in the soil. These organisms add organic matter to the soil, which changes the soil structure and the ability of soils to hold nutrients and water.

The extensive network of fine roots produced by trees, grasses, and other land plants helps hold soil particles in place. And by taking up nutrients in the soil, plants prevent the nutrients from being blown or washed away. When vegetation is removed by grazing, farming, logging, or suburbanization, large quantities of soil and nutrients are lost to erosion by wind and water (Figure 28.1).

Plants Hold Water and Moderate Climate Because plant tissues take up and retain water, intact forests, prairies, and wetlands act like a giant sponge. Plant leaves soften the physical impact of rainfall on soil, and the bulk of plant organic matter enhances the soil's water-holding capacity. In addition, plants reduce local temperatures by shading the area beneath them, and they increase relative humidity by buffering the impact of winds that dry out landscapes.

When humans clear vegetation to make way for houses or other development, the land becomes much drier and is subject to more extreme temperature swings.

Plants Are Primary Producers Land plants are the dominant primary producers in terrestrial ecosystems. Primary producers convert energy in sunlight into chemical energy (Chapter 10). The sugars that plants produce by photosynthesis support virtually all of the other organisms present in terrestrial habitats through different levels of the food chain (Ch. 53, Section 53.1).

Because of their role in primary production, land plants are the key to the carbon cycle on the continents. Plants take CO_2 from the atmosphere and reduce it to make sugars. Although green algae and land plants also produce a great deal of CO_2 as a result of cellular respiration, they fix much more CO_2 than they release.

The loss of plant-rich prairies and forests due to suburbanization and other development has contributed to an increased concentration of CO_2 in the atmosphere. Consider that each acre of deciduous forest can absorb roughly 3,500 pounds of carbon each year. A higher atmospheric carbon dioxide level, in turn, is partly responsible for the rising temperatures associated with global climate change (Ch. 53, Section 53.3).

Plants Provide Humans with Food, Fuel, Fiber, Building Materials, and Medicines

If you ate an apple for lunch, or are wearing a cotton T-shirt, you have plants to thank. It is difficult to overstate the importance of plant research to the well-being of human societies. Let's take a closer look.

Food Agricultural research began with the initial domestication of crop plants, which occurred independently at several locations around the world starting about 12,000 years ago (Figure 28.2a). By actively selecting individuals with the largest and most nutritious seeds, leaves, or other plant parts year after year, our ancestors gradually changed the characteristics of certain wild species. This process is called **artificial selection** (Ch. 22, Section 22.3).

Figure 28.2b compares kernel size in modern maize and its wild ancestor—a grass called teosinte that is native to Central America. Artificial selection has been responsible for dramatic increases in the protein content of maize kernels over the past 100 years. A current focus in agricultural research is the improvement of crop varieties through genetic engineering (Ch. 20, Section 20.1).

Fuel For perhaps 100,000 years, wood burning was the primary source of energy used by all humans. Wood has now been replaced in the industrialized countries by fossil fuels, including coal and petroleum or natural gas. Each of these forms of energy, however, is derived from plants that either grew during our lifetime or millions of years ago. The major deposits of fossil fuels formed when the remains of plants and animals from the Carboniferous period some 359–299 million years ago were buried and subjected to intense pressure.

Coal still supplies about 20 percent of the energy used in Japan and Western Europe and 70 percent of the energy used in China. In the United States, coal accounts for roughly 30 percent of electricity production. Current research is focused on finding cleaner and more efficient ways of mining and burning coal.

Other exciting areas of research on plant-based fuels include the production of ethanol, as well as the production of biodiesel. Plants such as corn are typically used as the raw material for ethanol production. Biodiesel, however, consists of oils from various plants such as soybean (but can also include animal fats). Biodiesel and other biofuels represent alternative energy sources for cars, heating, and other energy needs.

Fiber and Building Materials Nylon and polyester derived from petroleum are increasingly important in manufacturing, but cotton, hemp, and other types of plant fibers are still important sources of raw material for clothing, rope, and household articles such as towels and bedding.

Woody plants also provide lumber for houses and furniture. Relative to its density, wood is a stiffer and stronger building material than concrete, cast iron, aluminum alloys, or steel.

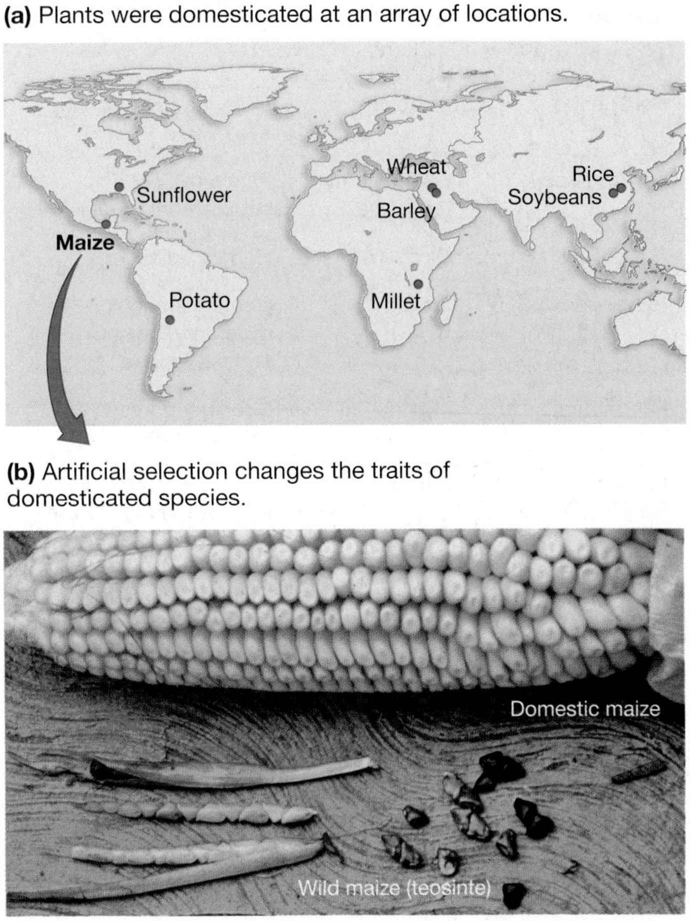

Figure 28.2 Crop Plants Are Derived from Wild Species via Artificial Selection. **(a)** The domestication of crop plants has occurred on nearly every continent. **(b)** In some cases, artificial selection has changed domesticated species so radically that they bear little resemblance to their wild ancestors.

Most of the fibers used in papermaking come from woody plants. The cellulose fibers obtained from trees or bamboo and then used in paper manufacturing are stronger under tension (pulling) than nylon, silk, chitin, collagen, tendon, or bone. One line of research is focused on bioengineering woody plants to reduce the tough, resistant compound called lignin that is typically found along with cellulose. Reducing wood lignins results in the generation of fewer pollutants during the papermaking process.

Medicines In both traditional and modern medicine, plants are a key source of drugs. Table 28.1 on page 580 lists some familiar medicines derived from land plants; overall, an estimated 25 percent of the prescriptions written in the United States each year include at least one molecule derived from plants.

In most cases, plants synthesize these compounds in order to repel insects, deer, or other types of herbivores. For example, experiments have confirmed that morphine, cocaine, nicotine, caffeine, and other toxic compounds found in plants are effective deterrents to herbivores. Researchers continue to isolate and test new plant compounds for medicinal use in humans and domesticated animals.

Table 28.1 Some Drugs Derived from Land Plants

Source Plant	Compound	Use
Belladonna	Atropine	Dilating pupils during eye exams
Opium poppy	Codeine, morphine	Pain relief, cough suppressant (codeine)
Foxglove	Digitalin	Heart medication
Ipecacuanha	Ipecac	Treating amoebic dysentery, poison control, emetic
Peppermint	Menthol	Cough suppressant, relief of stuffy nose
Papaya	Papain	Reducing inflammation, treating wounds
Cinchona	Quinine	Malaria prevention
	Quinidine	Heart medication
Aspen, willow	Salicin	Pain relief (aspirin)
Wild yam	Steroids	Precursor compounds for manufacture of birth control pills and cortisone (to treat inflammation)
Pacific yew	Taxol	Treating ovarian cancer
Curare vine	Tubocurarine	Muscle relaxant used in surgery
Rosy periwinkle	Vinblastine, vincristine	Treating leukemia (cancer of blood)

CHECK YOUR UNDERSTANDING

✔ **If you understood this section, you should be able to . . .**

1. **SOCIETY** Explain the potential impacts of deforestation on ecosystem services provided by plants.
2. Explain how most of what you have eaten within the past week with caloric value can be traced to a photosynthetic plant or alga.

Answers are available in Appendix A.

28.2 How Do Biologists Study Green Algae and Land Plants?

After you complete this section, you should be able to . . .

- Analyze research strategies for studying key features of green plants and the relationships among the major groups.

Given the importance of plants to the planet in general and humans in particular, it is not surprising that knowing as much as possible about plants, including how they evolved, is a key component of biological science. To understand how green plants originated and diversified, biologists analyze **(1)** morphological traits, **(2)** the fossil record, and **(3)** phylogenetic trees estimated from similarities and differences in DNA sequences from homologous genes and whole genomes.

These approaches are complementary and have produced a relatively clear picture of how land plants evolved from green algae and then diversified. Let's consider each of these research strategies.

Analyzing Morphological Traits

The green algae include species that are unicellular, colonial, or multicellular and that live in marine, freshwater, or moist terrestrial habitats. Most green algae are aquatic. Although some land plants, such as water lilies and the waterweed Elodea, live in ponds or lakes or rivers, the vast majority live on land.

Similarities between Green Algae and Land Plants The freshwater green algae have long been hypothesized to be closely related to land plants, because key traits are similar in the two groups:

- Their chloroplasts contain the photosynthetic pigments chlorophyll *a* and *b* and the accessory pigment β-carotene.
- They have similar arrangements of the internal, membrane-bound sacs called thylakoids (Ch. 10, Section 10.1).
- Their cell walls, sperm, and peroxisomes are similar in structure and composition. (Recall that peroxisomes are organelles in which specialized oxidation reactions take place; see Ch. 7, Section 7.2.) The earliest land plants even have motile sperm—a reflection of their algal ancestry.
- Their chloroplasts synthesize starch as a storage product.

Of all the green algal groups, the three most similar to land plants—based on DNA sequence analysis (Ch. 20, Section 20.3)—are the Zygnematophyceae (conjugating algae), Coleochaetophyceae (coleochaetes), and Charophyceae (stoneworts), shown in **Figure 28.3**. Because the species that make up these groups are largely multicellular (the Zygnematophyceae also have unicellular forms) and live in ponds and other types of freshwater environments, biologists hypothesize that land plants evolved from green algae that lived in freshwater habitats. Study continues, but evidence is accumulating that the Zygnematophyceae are the closest relatives to land plants.

Major Morphological Differences among Land Plants Based on morphology, the land plants were traditionally clustered into three broad categories:

1. **Nonvascular plants** lack **vascular tissue**—specialized groups of cells that conduct water and nutrients from one part of the plant body to another. Members of this group also lack seeds. Instead, they rely on **spores** for reproduction and dispersal. Recall that a spore is a single haploid cell that divides mitotically and grows into a new organism (Ch. 27, Section 27.3). Mosses are one group of nonvascular plants (**Figure 28.4a**).
2. Seedless vascular plants have vascular tissue but do not make seeds. Like the nonvascular plants, they rely on spores for dispersal. Ferns are an example of seedless vascular plants (**Figure 28.4b**).

Zygnematophyceae (conjugating algae)

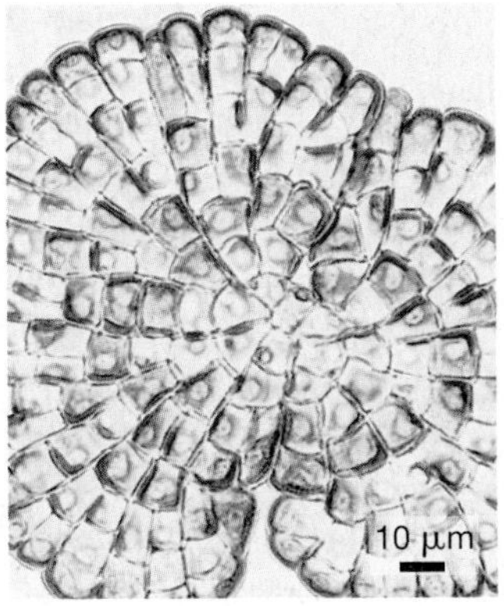

Coleochaetophyceae (coleochaetes)

Charophyceae (stoneworts)

Figure 28.3 Most Green Algae Are Aquatic. Examples of species from green algal lineages that are closely related to land plants.

(a) Nonvascular plants

Do not have vascular tissue; produce spores, not seeds (for example, mosses)

(b) Seedless vascular plants

Have vascular tissue; produce spores, not seeds (for example, ferns)

(c) Seed plants: Gymnosperms

Have vascular tissue; produce seeds but no flowers or fruits (for example, ginkgos)

(d) Seed plants: Angiosperms

Have vascular tissue; produce flowers, . . .

. . . seeds, and fruits (for example, viburnums)

Figure 28.4 Morphological Diversity in Land Plants.

3. Seed plants have vascular tissue, and they produce seeds. A **seed** consists of an embryo and a store of nutritive tissue, surrounded by a tough protective layer. The first seed plants to appear were **gymnosperms** ("naked-seeds," Figure 28.4c), some of which produce their seeds in cones. The flowering plants, or **angiosperms** ("encased-seeds"), followed with flowers and seeds contained in fruits (Figure 28.4d), which can aid in seed dispersal.

How are nonvascular plants, seedless vascular plants, and seed plants related to each other, and to green algae?

Using the Fossil Record

The first green plants in the fossil record are green algae, found in rocks from 700 to 725 million years old. The first land plant fossils are found in rocks that are about 475 million years old. Because green algae appear long before land plants, the data support the hypothesis that land plants are derived from green algae.

At roughly the same time that green algae evolved and began to diversify, the oceans and atmosphere were starting to become oxygen rich—as never before in Earth's history. Based on this time correlation, it is reasonable to hypothesize that the evolution of green algae contributed to the rise of the oxygen level on Earth. The origin of an oxygen-rich atmosphere occurred not long before the appearance of animals in the fossil record and may have played a role in their origin and early diversification.

Origin of Land Plants The plant fossil record is massive. To organize and synthesize the database, Figure 28.5 breaks it into five time intervals—each encompassing a major event in the diversification of plants.

The oldest interval begins about 475 million years ago (mya), spans some 60 million years, and documents the origin of the group. Most of the fossils dating from this period are fragments of plants and microscopic spores. Despite not having complete organisms to study, biologists hypothesize that these fossils represent the first land plants. Here's why:

- Some of the fossils are of thin sheets of waxy material that resembles cuticle. **Cuticle** is a watertight barrier that coats the aboveground parts of today's land plants and helps them resist drying.
- Fossilized spores from this period are surrounded by a material that, under the electron microscope, appears almost identical in structure to sporopollenin. **Sporopollenin** is a waxy substance (similar to cuticle) that encases the spores and pollen of modern land plants and helps them resist drying.
- The fossilized spores have been found in association with spore-producing structures, called **sporangia** (singular: **sporangium**), that are similar to sporangia observed in some modern nonvascular plants.

Silurian–Devonian Explosion The second major interval in the fossil record of land plants is called the "Silurian–Devonian explosion." In rocks dated 416–359 mya, biologists find fossils from most of the major plant lineages. Virtually all of the adaptations that allow plants to occupy dry, terrestrial habitats are present, including water-conducting tissue and roots.

According to the fossil record, plants colonized the land in conjunction with fungi that grew in a mutually beneficial symbiosis. Fungi that grew belowground channeled nutrients from the soil to plants; at the same time, plants produced sugars and other products of photosynthesis that were useful to fungi (Chapter 29).

The Carboniferous Period The third interval in the fossil history of plants spans the aptly named Carboniferous period. In sediments dated from about 359 to 299 mya, biologists find extensive deposits of coal, a type of carbon-rich rock packed with fossil spores, branches, leaves, and tree trunks. Because coal is formed only in the presence of water, the Carboniferous fossils indicate the presence of extensive forested swamps.

Most of these fossils are derived from seedless vascular plants that were ancestors of today's club mosses, horsetails, and ferns. During the Carboniferous these groups were species rich and included a wide array of tree-sized forms (significantly larger than the species living today).

Diversification of Gymnosperms The fourth interval in plant history is characterized by seed plants called gymnosperms. Examples of gymnosperms alive today are ginkgos, redwoods, junipers, yews, pines, spruces, and firs.

Because gymnosperms grow readily in dry habitats, biologists infer that both wet and dry environments on the continents became blanketed with green plants for the first time during this interval. Gymnosperms are prominent in the fossil record from 299 mya to 145 mya.

Diversification of Angiosperms The fifth interval in the history of land plants is still under way. This is the age of flowering plants—the angiosperms. The first flowering plants in the fossil record appear about 125 mya (although a recent discovery of angiosperm-like pollen may push back the origin of angiosperms to 240 mya). The plants that produced the first flowers

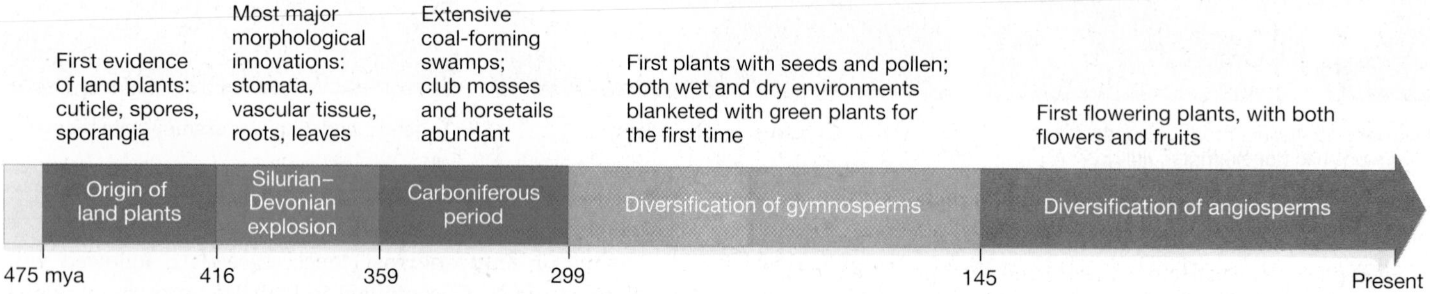

Figure 28.5 The Fossil Record of Land Plants Can Be Divided into Five Major Intervals. Note that the first insects are found in the fossil record at about 400 mya, the first terrestrial vertebrate animals at 365 mya, and the first mammals at 210 mya.

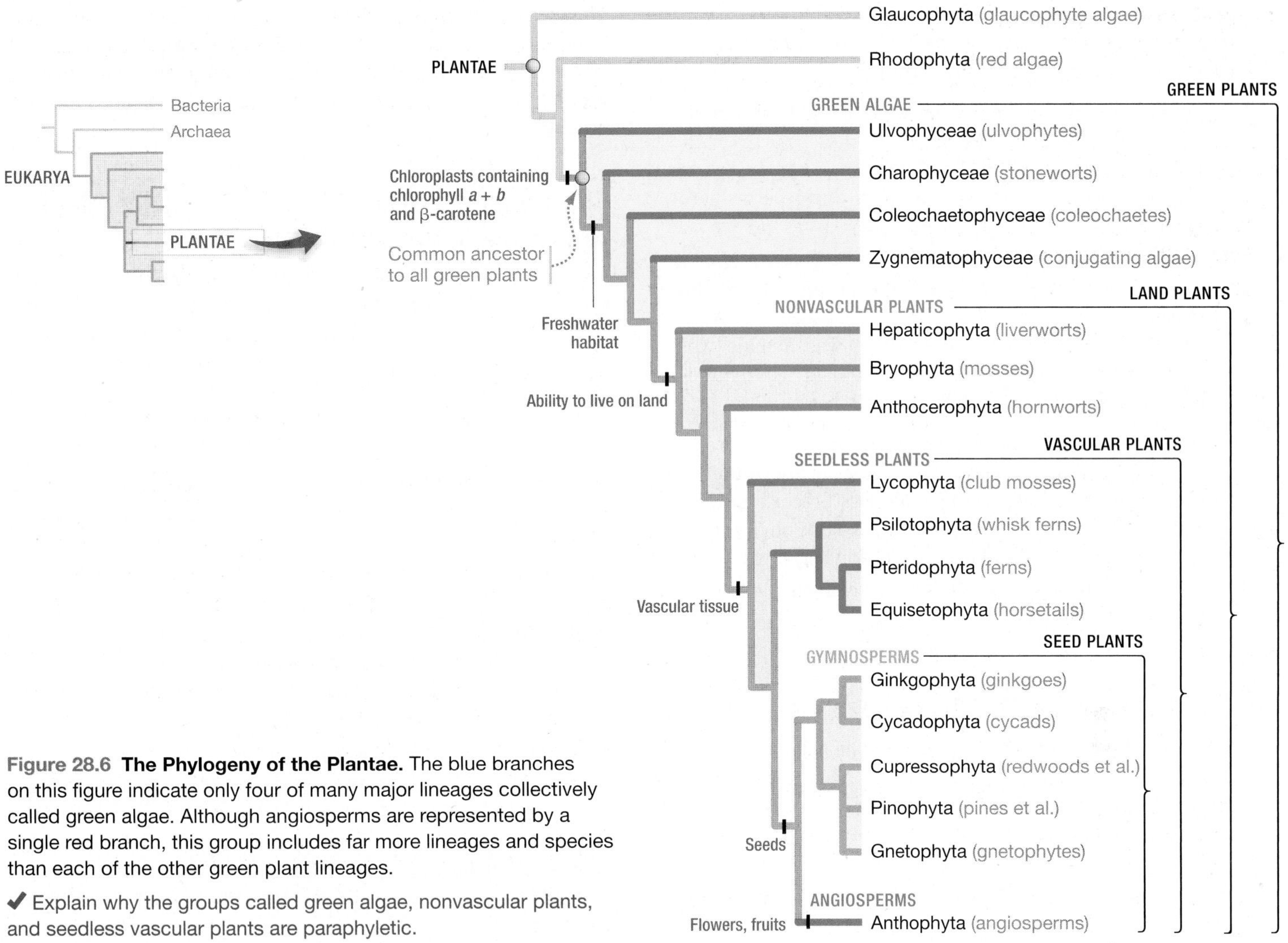

Figure 28.6 The Phylogeny of the Plantae. The blue branches on this figure indicate only four of many major lineages collectively called green algae. Although angiosperms are represented by a single red branch, this group includes far more lineages and species than each of the other green plant lineages.

✔ Explain why the groups called green algae, nonvascular plants, and seedless vascular plants are paraphyletic.

are the ancestors of today's grasses, orchids, daisies, oaks, maples, and roses.

According to the fossil record, then, the green algae appear first, followed by the nonvascular plants, seedless vascular plants, and seed plants. Organisms that appear late in the fossil record are often less dependent on moist habitats than are groups that appear earlier. For example, the sperm cells of mosses and ferns swim to accomplish fertilization, while gymnosperms and angiosperms produce pollen grains that are transported via wind or insects and that carry the cells that will later develop into sperm.

To test the validity of the hypothesis that land plants evolved from green algae, biologists analyze data sets that are independent of the fossil record. Foremost among these are DNA sequences used to infer phylogenetic trees. Does the phylogeny of land plants confirm or contradict the patterns in the fossil record?

Evaluating Molecular Phylogenies

The phylogenetic relationships among the major groups of plants continue to be debated. However, the phylogenetic tree in Figure 28.6 is a recent version of results emerging from laboratories around the world. Each black bar marked across a branch indicates a **synapomorphy**, or a shared, derived trait that distinguishes monophyletic groups. (For more information on how to read phylogenetic trees, see **BioSkills 13**.)

Each of the following bulleted points makes a key observation about this tree, followed by an interpretation—often, a hypothesis that is supported by the observation.

- The green plants are monophyletic.
  Interpretation: A single common ancestor gave rise to all of the green algae and land plants.
- Green algae are paraphyletic.
  Interpretation: The green algae include some but not all of the descendants of a single common ancestor.
- Zygnematophyceae is the closest living relative to land plants.
  Interpretation: Land plants evolved from a green alga that lived in freshwater habitats.
- Land plants are monophyletic.
  Interpretation: There was only one successful transition from freshwater environments to land.
- The nonvascular plants are the earliest-branching groups among land plants.
  Interpretation: The nonvascular plants are the most ancient living group of land plants.

- The nonvascular plants are paraphyletic.
  Interpretation: The nonvascular plants include some but not all descendants of a single common ancestor.
- The seedless vascular plants are paraphyletic, but the vascular plants as a whole are monophyletic.
  Interpretation: Vascular tissue evolved once.
- The seed plants—the gymnosperms plus angiosperms—are monophyletic.
  Interpretation: The seed evolved once.
- The gymnosperms are a monophyletic group, as are the angiosperms.
  Interpretation: Among seed plants, there was a major divergence in how seeds develop—either "naked" (in gymnosperms) or protected inside a fruit (in angiosperms).

Although the tree in Figure 28.6 will undoubtedly change and improve as additional data accumulate, biologists are confident about its most fundamental message: The fossil record and the phylogenetic tree based on DNA sequence data agree on the order in which groups appeared. Land plant evolution began with nonvascular plants, proceeded to seedless vascular plants, and continued with the evolution of seed plants.

CHECK YOUR UNDERSTANDING

✔ If you understood this section, you should be able to . . .

Explain why (1) morphological data, (2) the fossil record, and (3) molecular phylogenies all support the hypothesis that land plants evolved from green algae.

Answers are available in Appendix A.

28.3 What Themes Occur in the Diversification of Land Plants?

After you complete this section, you should be able to . . .

- Compare and contrast the life cycles and key traits of green algae, nonvascular plants, seedless vascular plants, gymnosperms, and angiosperms.

Land plants have evolved from green algae that grew on the muddy shores of ponds 475 million years ago to organisms that enrich the soil and produce much of the oxygen you breathe and most of the food you eat. How did this happen?

Answering this question begins with recognizing the most striking trend in the phylogeny and fossil record: The most ancient groups in the lineage are dependent on wet habitats, while the more recently evolved groups can live and reproduce in dry—or even desert—conditions. The story of land plants is the story of adaptations that allowed photosynthetic organisms to move from aquatic to terrestrial environments.

Let's first consider adaptations that allowed plants to grow in dry conditions, and then analyze the evolution of traits that allowed plants to reproduce efficiently on land where water may not always be readily available. This section closes with a brief look at the radiation of flowering plants, which are by far the most important plants in many of today's terrestrial environments.

The Transition to Land I: How Did Plants Adapt to Dry Conditions with Intense Sunlight?

For aquatic green algae, terrestrial environments are deadly. Compared with a habitat in which the entire organism is bathed in water, in terrestrial environments only a portion, if any, of the plant's tissues are in direct contact with water. Tissues that are exposed to air tend to dry out and die.

Once the first green plants made the transition to survive out of water, though, growth on land offered bountiful resources.

- ***Light*** On land, the amount of light reaching plant leaves is plentiful. In contrast, the water in ponds, lakes, and oceans absorbs and reflects light. As a result, the amount of light available to drive photosynthesis is drastically reduced even a meter or two below the water's surface.
- ***Carbon dioxide*** CO_2—the most important molecule required by photosynthetic organisms—is more readily available to plants growing in air rather than water. In water, CO_2 diffuses into leaves more slowly than on land.

Natural selection favored early land plants with three main adaptations that solved the drying problem by **(1)** preventing water loss, which kept cells from drying out and dying; **(2)** providing protection from harmful ultraviolet (UV) radiation; and **(3)** moving water from tissues with direct access to water to tissues without direct access. Let's examine each adaptation in turn.

Preventing Water Loss: Cuticle and Stomata If biologists had to point to one innovation that made the transition to land possible, it would be the random mutations that led to the production of cuticle. Recall that cuticle is a waxy, watertight sealant that covers the aboveground parts of plants and gives them the ability to survive in dry environments (**Figure 28.7a**). In fact, the presence of cuticle in fossils is used as a diagnostic indicator for land plants.

Covering surfaces with wax creates a problem, however, regarding the exchange of gases across those surfaces. Plants need to take in carbon dioxide (CO_2) from the atmosphere in order to perform photosynthesis. But cuticle is almost as impervious to CO_2 as it is to water.

Most plants solve this problem with a structure called a **stoma** ("mouth"; plural: **stomata**). A stoma consists of an opening surrounded by specialized **guard cells** (**Figure 28.7b**). The opening, or pore, opens or closes as the guard cells change shape. When guard cells lose water, they become flaccid, or "limp," and the stomata close. Pores are normally closed at night to limit water loss from the plant when CO_2 uptake is not needed. But when guard cells absorb water, they become turgid, or "taut," opening the pore. Open stomata allow CO_2 to diffuse into the interior of leaves and

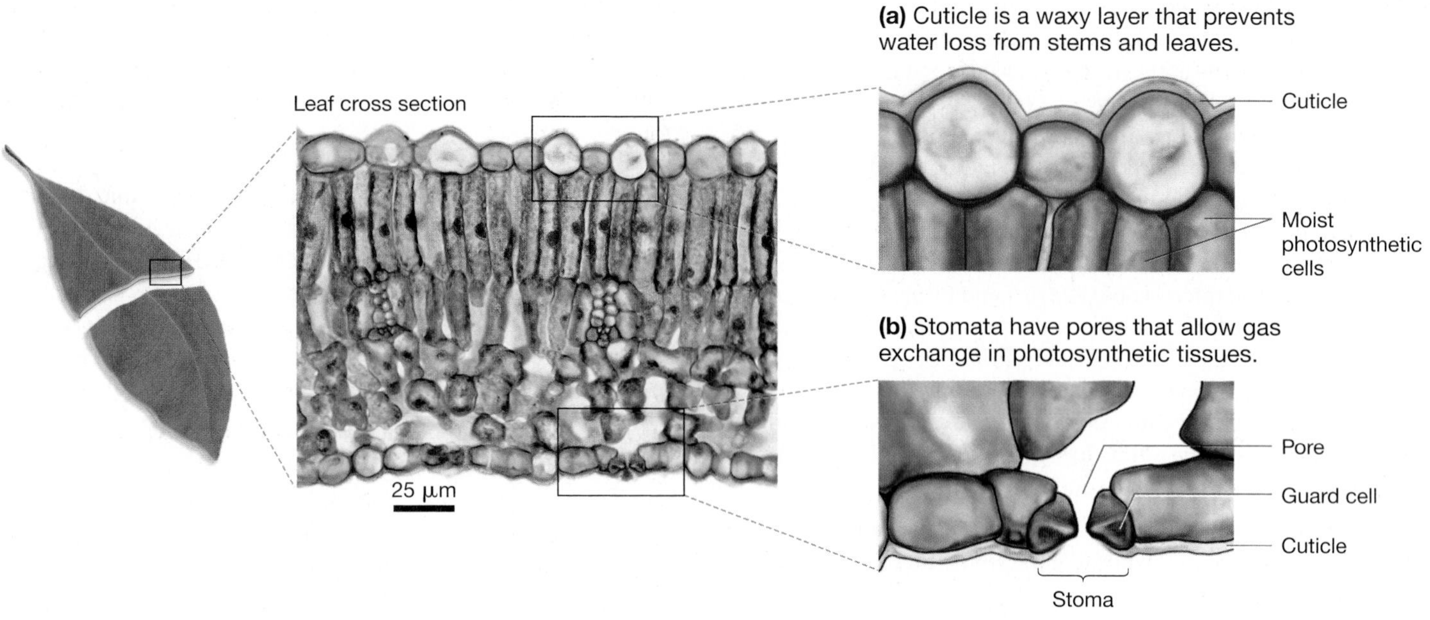

Figure 28.7 Cuticle and Stomata Are the Most Fundamental Plant Adaptations to Life on Land. In these micrographs, leaf cells have been stained blue to make their structure more visible. **(a)** The interior of plant leaves and stems is extremely moist; a waxy cuticle (on outer surface) prevents water loss by evaporation. **(b)** Stomata, which are opened and closed by guard cells, create pores that allow CO_2 to diffuse into the interior of leaves and stems where cells are actively photosynthesizing. Stomata are often on the underside of leaves.

stems where cells are actively photosynthesizing. (The mechanism behind guard-cell movement is explored in Ch. 37, Section 37.6.)

Stomata are present in all land plants except the liverworts, which have pores but no guard cells. These data suggest that the earliest land plants had pores that allowed gas exchange to occur at breaks in the cuticle-covered surface. Later, the evolution of guard cells gave land plants the ability to regulate gas exchange—and control water loss—by opening and closing their pores.

Providing Protection from UV Irradiation Life out of water gave land plants a distinct advantage. Because they were exposed to higher light intensities, the plants could carry out photosynthesis faster and more efficiently. However, out of water they were also exposed to the harmful UV rays of the sun. UV light is known to damage DNA by causing thymine dimers (Ch. 15, Section 15.5). Water absorbs UV light, so algae did not face this problem to the same extent.

Researchers think that many species of algae probably colonized the wet soil near their home ponds, but only some were able to survive the harsh sunlight. The plants that survived were those that by chance made compounds that absorb UV light. Most plants today accumulate UV-absorbing compounds, called flavonoids, that protect their DNA from damage. These flavonoids are plant pigments that function as a sunscreen for leaves and stems. Humans use similar compounds to manufacture sunscreen.

The first plants had a cuticle that kept them from drying out, and UV-absorbing compounds that acted as sunscreen, allowing them to grow on the water-saturated soils of lake or pond edges. The next challenge? Defying gravity.

The Importance of Upright Growth Multicellular green algae are able to maintain their shape and sometimes even grow quite tall in water because the density of their cells is similar to that of water. But outside of water, the body of a multicellular green alga collapses. The water that fills its cells is roughly 800 times denser than air. Although the cell walls of green algae are strengthened by the presence of cellulose, their bodies lack the structural support to withstand the force of gravity and to keep an entire algal organism upright in air.

Based on these observations, biologists hypothesize that the first land plants were small or had a low, sprawling growth habit. Besides lacking rigidity, early land plants would have had to obtain water through pores or through cells that lacked the waterproof cuticle that surrounds the aboveground portions of many plant organs—meaning they would have had to grow in a way that kept many or most of their tissues in direct contact with moist soil. If this hypothesis is correct, then competition for space and light would have become intense soon after the first plants began growing on land.

In a terrestrial environment, individuals that can grow upright have much better access to sunlight than individuals that cannot. But two problems have to be overcome first:

1. land plants must transport water from tissues that are in contact with wet soil to tissues that are in contact with dry air, against the force of gravity; and
2. land plants must be rigid enough to avoid falling over in response to gravity and wind.

As it turns out, vascular tissue solved both problems.

The Origin of Vascular Tissue Paul Kenrick and Peter Crane explored the origin of water-conducting cells and upright growth in plants by examining the extraordinary fossils found in a rock formation in Scotland called the Rhynie Chert. These rocks formed about 400 million years ago and contain some of the first large plant specimens in the fossil record—much larger than the microscopic spores and cuticle found in older rocks.

The Rhynie Chert contains numerous plants that fossilized in an upright position. How did they stay vertical? By examining fossils with the electron microscope, Kenrick and Crane established that species from the Rhynie Chert contained elongated cells that were organized into tissues along the length of the plant. Based on these data, the biologists hypothesized that the elongated cells were part of water-conducting tissue and that water could move from the base of the plants upward to the rest of the plant through these specialized water-conducting cells.

- Some of the fossilized water-conducting cells had simple, cellulose-containing cell walls like the water-conducting cells found in today's mosses (Figure 28.8a).
- Some of the water-conducting cells present in the early fossils had cell walls with thickened rings containing a molecule called lignin (Figure 28.8b).

Lignin is a complex polymer built from six-carbon rings. It is extraordinarily strong for its weight and is particularly effective in resisting compressing forces such as gravity. Lignin is the compound mentioned earlier in the chapter that makes wood tough and durable.

These observations inspired the following hypothesis: The evolution of lignified vascular tissue allowed early plants to support upright stems in the face of wind and gravity and to transport water from roots to aboveground tissues. Today, the presence of lignin in the cell walls of water-conducting cells is considered the defining feature of vascular tissue. (See Ch. 34, Section 34.2 for additional details on the structure and function of plant vascular tissue.)

The Elaboration of Vascular Tissue: Tracheids and Vessels Once simple water-conducting tissues evolved, evolution by natural selection favored more complex tissues that were more efficient in providing support and transport.

In fossils that are about 380 million years old, biologists find the advanced water-conducting cells called tracheids. **Tracheids** are long, thin, tapering cells that have the following components:

- A thickened, lignin-containing **secondary cell wall** in addition to a cellulose-based **primary cell wall**
- Pits in the sides and ends of the cell where the secondary cell wall is absent, where water can flow efficiently from one tracheid to the next (Figure 28.8c)

Thanks to the secondary cell wall, tracheids provided better structural support, but water could still move easily through the cells because of the pits. Today, all vascular plants contain tracheids.

In fossils dated to 250–270 million years ago, biologists have documented the most recently evolved type of water-conducting cells observed in plants. **Vessel elements** are shorter and wider than tracheids, and their upper and lower ends have perforations where both the primary and secondary cell wall are missing. The width of vessels and the presence of perforations reduce resistance and make water movement more efficient (Figure 28.8d). In vascular tissue, vessel elements are lined up end to end to form a continuous pipe-like structure.

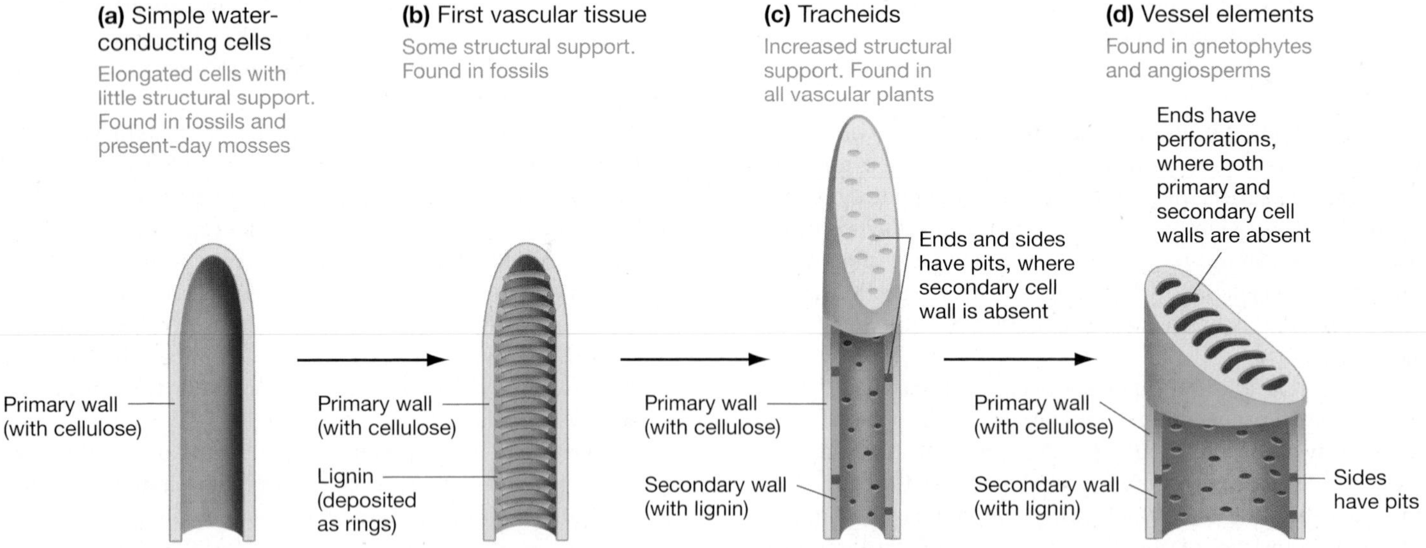

Figure 28.8 **Evolutionary Sequence Observed in Water-Conducting Cells.** According to the plant fossil record and phylogenetic data, water-conducting cells became stronger over time due to the evolution of lignin and secondary cell walls. Efficient water transport was maintained through pits where the secondary cell wall is missing or through perforations where both the secondary and primary cell walls are absent.

✔ Biologists claim that vessels are more efficient than tracheids at transporting water. Why does this claim make sense?

In the stems and branches of some vascular plant species, tracheids or a combination of tracheids and vessels form the extremely strong support material called **wood**. (The anatomy of wood is explained in detail in Ch. 34, Section 34.4.) The ability of plants to make lignified vascular tissues allowed some species such as the giant redwood to grow to 115 m tall and transport water from soil all the way to the top—a remarkable engineering feat.

Mapping Evolutionary Changes on the Phylogenetic Tree

Cuticle, stomata, and vascular tissue were key adaptations that allowed early plants to colonize and thrive on land. Figure 28.9 summarizes how land plants adapted to dry conditions by mapping on the phylogenetic tree where major innovations occurred.

As you study the tree in Figure 28.9, note that fundamentally important adaptations to dry conditions—such as cuticle, pores, stomata, vascular tissue, and tracheids—evolved just once. However, convergent evolution also occurred. When convergence occurs, similar traits evolve independently in two or more distinct lineages (Ch. 25, Section 25.1). For example, vessels evolved independently in angiosperms, gnetophytes, and several species of seedless vascular plants.

These key adaptations equipped land plants with basic survival skills. Now the question is, how did their reproductive strategies evolve for success on land?

The Transition to Land II: How Do Plants Reproduce in Dry Conditions?

Life cycles of plants, and other sexually reproducing eukaryotes, serve several different functions:

- increase genetic variability as a result of meiosis and fertilization;
- increase the number of individuals; and
- disperse individuals to new habitats.

To understand the details of any organism's life cycle, biologists first consider the context in which it lives—for example, does the organism live in water or on land, and is it mobile or sessile? Biologists are particularly interested in how the massive change in context during land plant evolution—from water to dry land—affected the dispersal stage of plant life cycles. Whereas water currents move gametes, spores, and even whole organisms and keep them moist, air poses many challenges as a dispersal medium, and it is relatively dry. Organisms living on land have to protect themselves, including their reproductive structures, from desiccation.

Three innovations that occurred early in land plant evolution were instrumental for efficient reproduction in a dry environment: **(1)** spores were produced that resist drying, **(2)** gametes were produced in complex, multicellular structures; and **(3)** the embryo was retained on the parent (mother) plant and was nourished by it. Let's examine each of them in turn.

Desiccation-Resistant Spores Section 28.2 introduced the first of several key adaptations for reproducing on land: spores that resist drying because they are encased in a tough coat of sporopollenin. Sporopollenin is a very durable polymer—if you wanted to remove it from spores, you'd have to boil them in acid.

The tough outer coating of spores helps them survive for fairly long periods of time. For example, studies on fern spores demonstrate that spores can remain viable for several months when kept at cool temperatures. Although spores are tiny, single-celled structures, they can be dispersed hundreds of miles by wind currents. The resistant, lightweight nature of spores helps increase their chances of being dispersed to a suitable environment for growth.

Protective, Complex Reproductive Organs The fossilized structures of early plants contain specialized reproductive organs called **gametangia** (singular: **gametangium**). Although members of the green algae group Charophyceae (stoneworts) also develop gametangia, the gametangia found in land plants are larger and more complex.

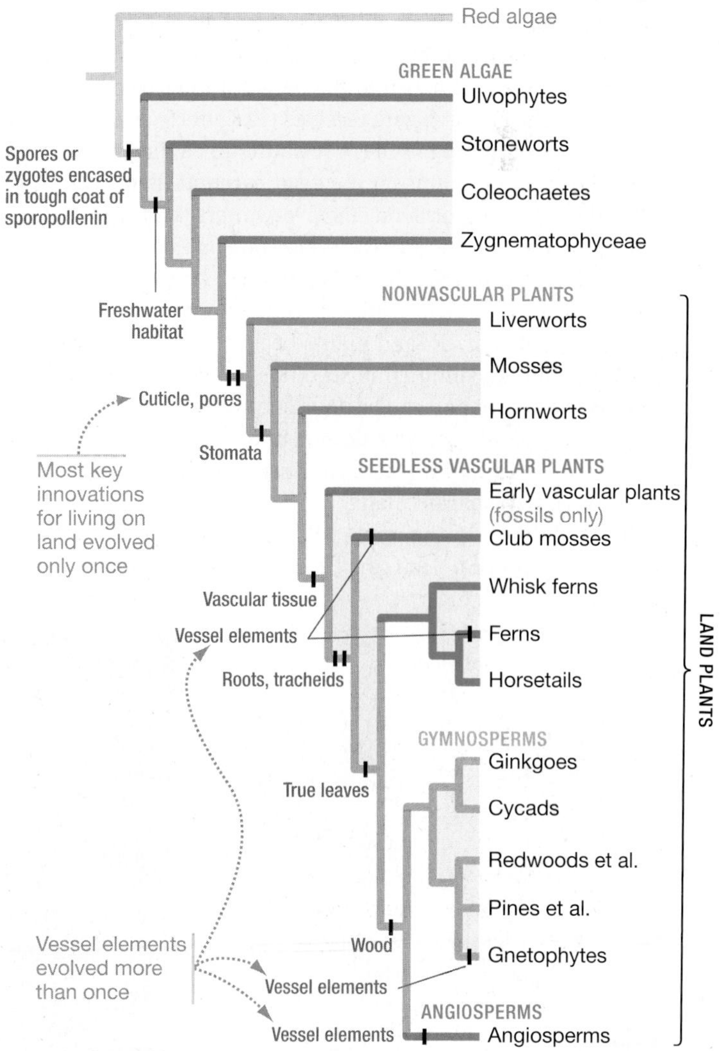

Figure 28.9 **Plant Phylogeny Emphasizing the Evolutionary Innovations That Allowed Plants to Adapt to Life on Land.** Several features of plants help prevent desiccation and facilitate the transport of water and nutrients.

The evolution of an elaborate gametangium was important because it protected gametes from drying and from physical damage. Gametangia are present in all groups of land plants living today except angiosperms, where structures inside the flower perform the same functions.

In both the Charophyceae and the land plants, individuals produce distinctive male and female gametangia (Figure 28.10).

- The sperm-producing gametangium is called an **antheridium** (plural: **antheridia**).
- The egg-producing gametangium is called an **archegonium** (plural: **archegonia**).

The antheridia and archegonia in plants have functions comparable to the testes and ovaries of animals.

Embryos Nourished by Parental Tissues The third innovation that occurred early in land plant evolution involved the eggs that formed inside archegonia. Instead of shedding their eggs into the water or soil, land plants retain them.

Eggs are also retained in the green algal lineages that are closely related to land plants: In Charophyceae and other closely related groups, sperm swim to the egg, fertilization occurs, and the resulting zygote stays attached to the parent. Either before or after fertilization, the egg or zygote receives nutrients from the mother plant. But the parent plant dies each autumn as the temperature drops. The zygote remains on the dead parental tissue, settles to the bottom of the lake or pond, and overwinters. In spring, meiosis occurs, and the resulting spores develop into haploid adult plants.

In land plants, the zygote is also retained on the parent plant after fertilization. But in contrast to the zygotes of most green algae, the zygotes of all land plants begin to develop on the living parent plant, forming a multicellular embryo that remains attached to the parent and can be nourished by it. This is important because land plant embryos do not have to manufacture their own food early in life. Instead, they receive most or all of their nutrients from the parent plant.

The retention of the embryo was such a key event in land plant evolution that the formal name of the group is **Embryophyta**—literally, the "embryo-plants." The retention of the fertilized egg in **embryophytes** is analogous to pregnancy in mammals, where offspring are retained by the mother and nourished through the initial stages of growth.

Alternation of Generations The innovation that integrates all the other reproductive adaptations of plants isn't a structure, but a phenomenon—the alternation of generations. When **alternation of generations** occurs, individuals exist in either a multicellular haploid or diploid phase. The multicellular haploid stage is called the **gametophyte**; the multicellular diploid stage is called the **sporophyte**. The two phases of the life cycle are connected by distinct types of reproductive cells—spores and gametes.

Although alternation of generations is observed in a wide array of eukaryotic lineages and in some groups of green algae, it does not occur in algal groups closely related to land plants. In the coleochaetes, stoneworts, and conjugating algae, the multicellular form is haploid. Only the zygote is diploid. As Figure 28.11 shows, the zygote undergoes meiosis to form haploid spores. After dispersing with the aid of flagella, the spores begin dividing by mitosis and eventually grow into adult, haploid individuals. You might recall that this haploid-dominant life cycle is also common in other eukaryote lineages (Ch. 27, Section 27.3).

These data suggest that alternation of generations originated in land plants independently of its evolution in other groups of eukaryotes, and that it originated early in their history—soon after they evolved from green algae. Why was this strategy successful?

Diploid, multicellular sporophytes can make many more spores via meiosis than a single diploid zygote can. In contrast to the aquatic spores of green algae, spores produced on land are easily dispersed through wind currents and can travel long distances from the parent plant, increasing the likelihood of reaching environmental conditions favorable to the multicellular phase. In early lineages of land plants, male gametes still require moist soil or water droplets in order to swim to female gametes. In contrast, later lineages (i.e., most seed plants) no longer need a watery environment for fertilization—male gametes are carried to female gametes in pollen grains.

Figure 28.10 In Many Land Plant Groups, Gametes Are Produced in Gametangia. Gametangia such as in this moss *(Polytrichium)* are complex, multicellular structures that protect developing gametes from drying and physical damage. (Angiosperms and some gymnosperms are exceptions. Among most gymnosperms, for example, only eggs are produced in gametangia; sperm are produced in pollen tubes, see Figure 28.18.)

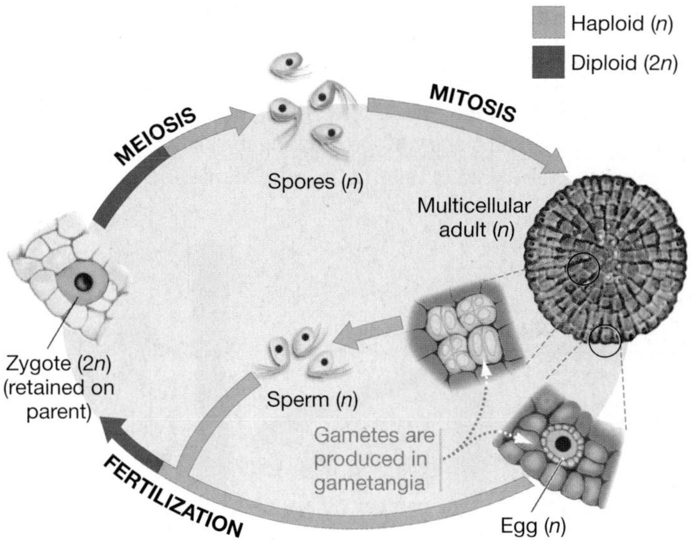

Figure 28.11 **In Green Algae That Are Closely Related to Land Plants, Only the Zygote Is Diploid.** The coleochaetes (shown here), stoneworts, and conjugating algae do not have alternation of generations. The multicellular stage is haploid.

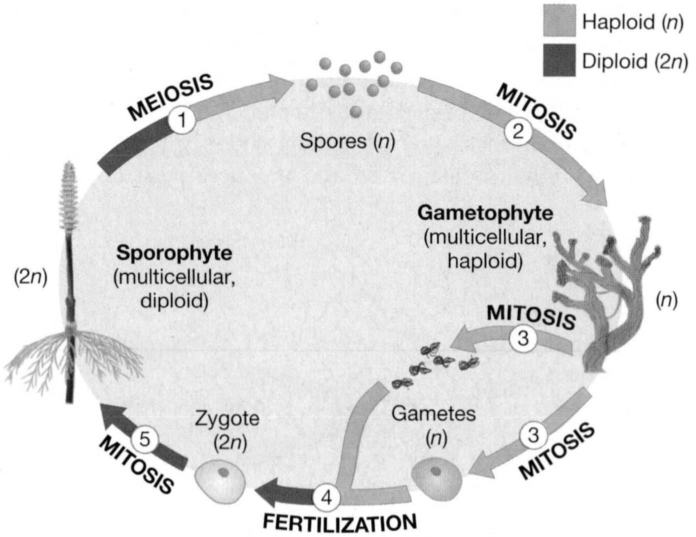

Figure 28.12 **All Land Plants Undergo Alternation of Generations.** Alternation of generations always involves the same sequence of five events, even though the details of the reproductive structures vary among different groups of plants.

✔ **THINK CAREFULLY** Are plant sperm and eggs produced by mitosis or meiosis? Explain.

Alternation of generations always involves the same basic sequence of events, illustrated in Figure 28.12. To review how this type of life cycle works, put your finger on the sporophyte in the figure and trace the cycle clockwise to find the following five key events in a land plant:

1. The sporophyte produces spores by meiosis. Spores are haploid.
2. Spores germinate and divide by mitosis to develop into multicellular, haploid gametophytes.
3. Gametophytes produce gametes by mitosis. Both the gametophyte and the gametes are haploid, but gametophytes are multicellular while gametes are unicellular.
4. Two gametes unite during fertilization to form a diploid zygote.
5. The zygote divides by mitosis and develops into a multicellular, diploid sporophyte.

Once you've traced the cycle successfully, take a moment to compare and contrast zygotes, spores, and gametes. Here are the main points:

- Zygotes and spores are both single cells that divide by mitosis to form a multicellular individual. Zygotes develop into sporophytes; spores develop into gametophytes.
- Zygotes are diploid, while spores and gametes are haploid.
- Zygotes result from the fusion of two haploid cells, such as a sperm and an egg, but spores are not formed by the fusion of gametes.
- Spores are produced by meiosis inside structures called sporangia; gametes are produced by mitosis inside gametangia.

Making Models 28.1 gives you an opportunity to compare and contrast the life cycle of land plants (alternation of generations) with that of coleochaetes (green algae).

Alternation of generations can be a difficult topic to master, for two reasons: **(1)** It is unfamiliar because it does not occur in humans or other animals, and **(2)** gamete formation results from mitosis—not meiosis, as it does in animals (Chapter 13). ✔ **If you understand the basic principles of alternation of generations, you should be able to summarize the sequence of events in a life cycle with alternation of generations, starting with a sporophyte.**

The five steps illustrated in Figure 28.12 occur in all organisms with alternation of generations. But in land plants, the relationship between the gametophyte and sporophyte is highly variable. Let's explore how some of the details differ across the major groups of land plants.

Making Models 28.1 Tips on Comparing Life Cycles

Life cycles often have stages arranged in sequential order and marked as haploid (n) or diploid (2n), with arrows representing processes that connect the stages. The life cycle shown here represents a generic alga with a life cycle similar to that of *Coleochaete,* in which the only multicellular stage is haploid.

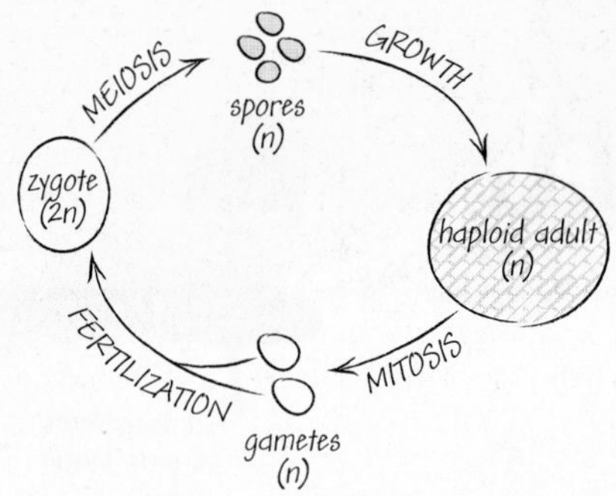

MODEL Draw a life cycle for a generic land plant. How is it different from the life cycle modeled here?

To see this model in action, go to the Study Area of **Mastering Biology**

From Gametophyte Dominant to Sporophyte Dominant In non-vascular plants such as mosses, the sporophyte is small and short lived and is largely dependent on the gametophyte for nutrition (**Figure 28.13**). When you see a green mat of mosses growing on a tree trunk or on rocks, you are looking at gametophytes. Because the gametophyte is long lived and produces most of the food required by the individual, it is considered the dominant part of the life cycle.

In contrast, in ferns and other vascular plants the sporophyte is much larger and longer lived than the gametophyte (**Figure 28.14**). The ferns you see growing in gardens or forests are sporophytes. You'd have to hunt on your hands and knees to find

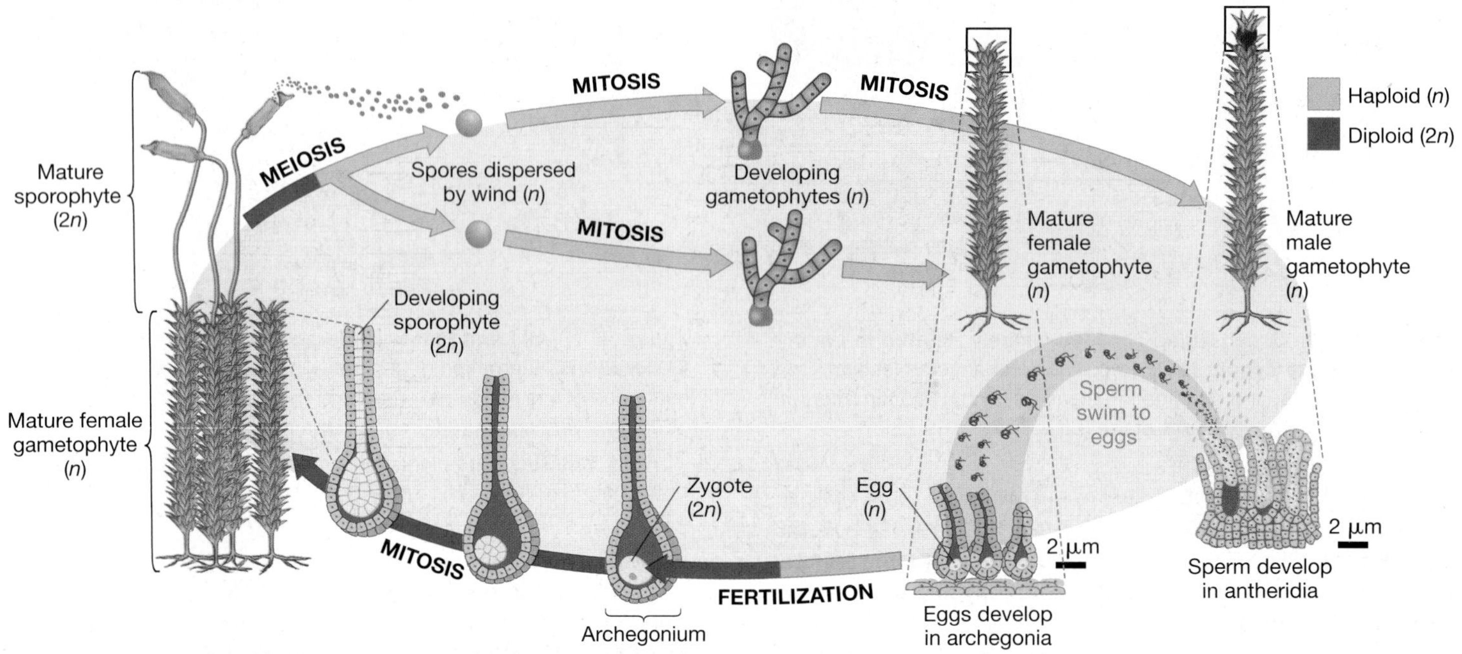

Figure 28.13 Gametophyte-Dominant Life Cycles Evolved Early. Like today's mosses, the earliest land plants in the fossil record have gametophytes that are much larger and longer lived than the sporophyte. The sporophyte depends on the gametophyte for nutrition.

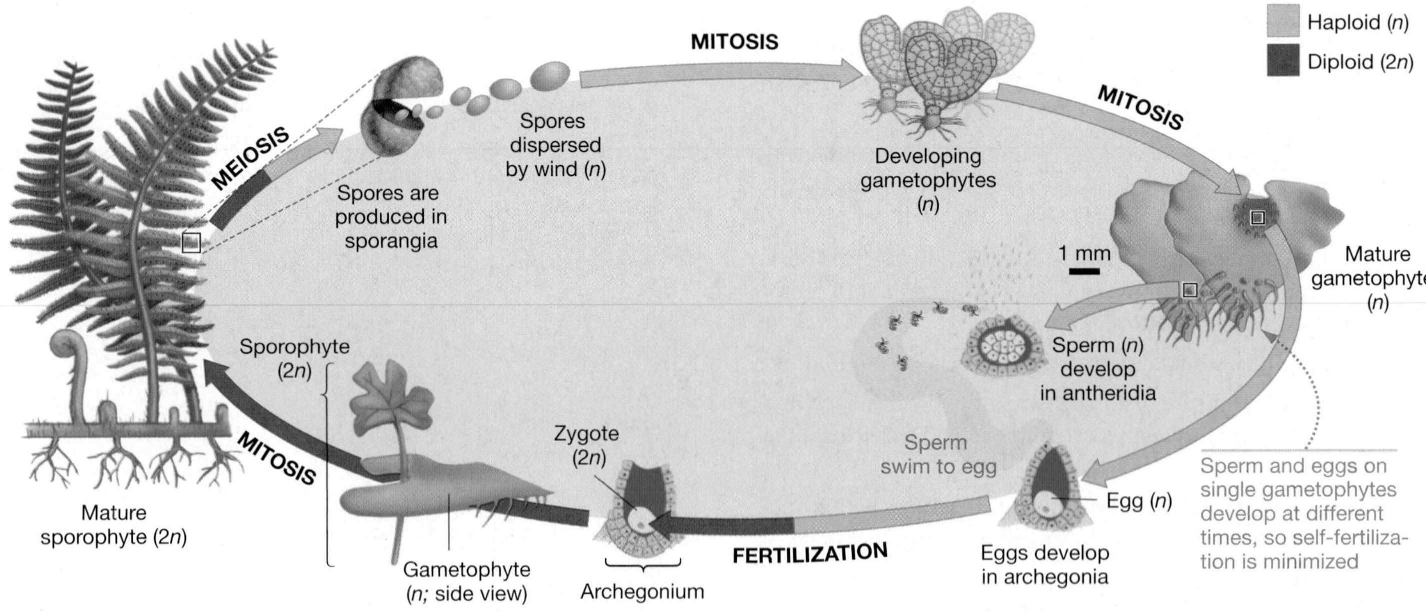

Figure 28.14 Sporophyte-Dominant Life Cycles Evolved Later. In lineages that evolved later, such as ferns, the sporophyte is much larger and longer lived than the gametophyte. However, when young, the sporophyte also depends on the gametophyte for nutrition.

✔ How can you tell that alternation of generations occurs in mosses (shown in Figure 28.13) and ferns (shown in Figure 28.14)?

their free-living gametophytes, which are typically just a few millimeters in diameter and grow flat against the ground. As you'll learn later in the chapter, the gametophytes of gymnosperms and angiosperms are even smaller—they are microscopic. Ferns and other vascular plants are said to have a sporophyte-dominant life cycle. ✔ **If you understand the difference between life cycles of gametophyte-dominant and sporophyte-dominant organisms, you should be able to examine the photos of hornworts (a nonvascular plant) and horsetails (a seedless vascular plant) in Figure 28.15 and identify which is the gametophyte and which is the sporophyte.**

The transition from gametophyte-dominant life cycles to sporophyte-dominant life cycles is one of the most striking of all trends in land plant evolution. To explain why it occurred, biologists hypothesize that sporophyte-dominant life cycles were advantageous because diploid cells can respond to varying environmental conditions more efficiently than haploid cells can—particularly if the individual is heterozygous at many genes. This idea has yet to be tested rigorously, however.

Heterospory In addition to sporophyte-dominant life cycles, another important innovation found in seed plants is called **heterospory**—the production of two distinct types of spores by different structures.

In contrast, **homospory** is the production of a single type of spore. All of the nonvascular plants and most of the seedless vascular plants are homosporous, although some club mosses and a few ferns are heterosporous. Homosporous species produce spores that develop into bisexual gametophytes that produce both eggs and sperm (Figure 28.16a). If these gametophytes are isolated in nature, they can self-fertilize and produce offspring. However, if two young gametophytes are close enough for the exchange of sperm, outcrossing is favored because it increases genetic variation (see Chapter 13). This tends to occur by developing sperm and eggs at different times on the same gametophyte. The gametophyte that begins to grow first will form eggs and will also release a hormone that induces the formation of sperm in the nearby gametophyte.

The two types of spore-producing structures in heterosporous species are often found on the same individual (Figure 28.16b).

1. **Microsporangia** (singular: **microsporangium**) are spore-producing structures that produce microspores. **Microspores** develop into male gametophytes, which produce sperm by mitosis.
2. **Megasporangia** (singular: **megasporangium**) are spore-producing structures that produce megaspores. **Megaspores** develop into female gametophytes, which produce eggs by mitosis.

Thus, the gametophytes of seed plants are either male or female, but never both.

The evolution of heterospory was a key event in land plant evolution because it made possible two of the most important adaptations for life in dry environments—pollen, and the subsequent formation of seeds.

(a) Hornwort gametophytes and sporophytes

(b) Horsetail gametophyte and sporophyte

Figure 28.15 **Reduction of the Gametophyte Is One of the Strongest Trends in Land Plant Evolution. (a)** A hornwort. **(b)** A horsetail species.

(a) Nonvascular plants and most seedless vascular plants are **homosporous.**

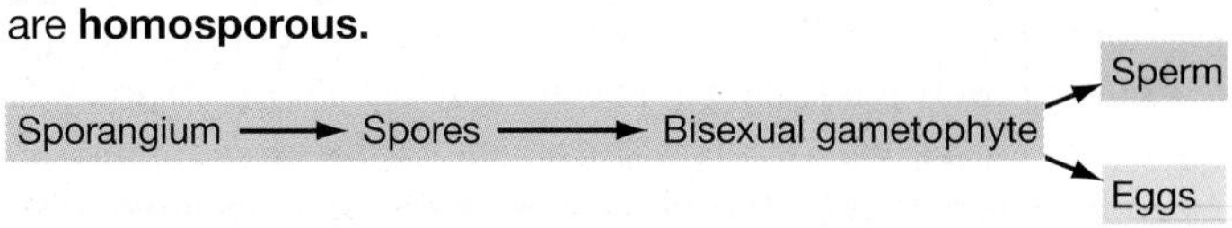

(b) Seed plants are **heterosporous.**

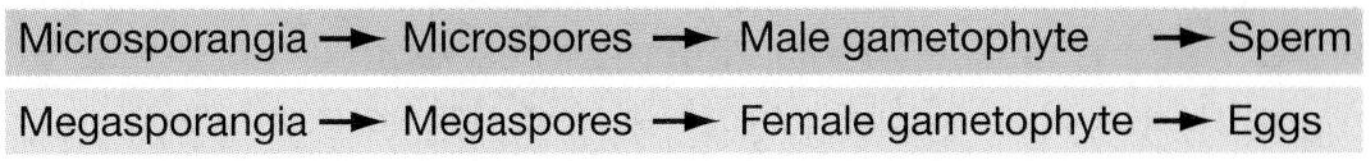

Figure 28.16 **Homospory versus Heterospory.** Unlike homosporous plants **(a)**, heterosporous plants **(b)** produce male and female spores that are morphologically distinct.

Pollen The nonvascular plants and the seedless vascular plants have sperm that swim to the egg to perform fertilization. For a sperm cell to swim, there has to be a continuous sheet of water between the male and female gametophytes, or a raindrop has to splash sperm onto a female gametophyte.

In species that live in dry environments, these conditions are rare. The land plants made their final break with their aquatic origins and were able to reproduce efficiently in dry habitats when a structure evolved that could move their gametes without the aid of water.

In heterosporous seed plants, the microspore germinates to form a tiny male gametophyte that is surrounded by a tough coat of sporopollenin, resulting in a **pollen grain**.

Pollen grains can be exposed to the air for long periods of time without dying from dehydration. They are also tiny enough to be carried to female gametophytes by wind or animals. Upon landing near the egg, the male gametophyte releases the sperm cells that accomplish fertilization.

When pollen evolved, then, heterosporous plants lost their dependence on water to accomplish fertilization. Instead of swimming to the egg as a naked sperm cell, their tiny gametophytes took to the skies.

Seeds The evolution of large gametangia protected the eggs and sperm of land plants from drying. Embryo retention allowed offspring to be nourished directly by their parent, and pollen enabled fertilization to occur in the absence of water.

Retaining embryos has a downside, however: In ferns and horsetails, for example, sporophytes have to live in the same place as their parent gametophyte. Seed plants overcome this limitation via well-packaged embryos that can be dispersed to new locations. The dispersal stage of the life cycle thus shifted from the haploid spore to the young diploid sporophyte.

Seeds develop from an **ovule**. Ovules contain a megasporangium with one or more eggs. Similar to a bird's egg, seeds include an embryo and a food supply surrounded by a tough coat that allows for effective dispersal (Figure 28.17). Spores are an effective dispersal stage for nonvascular plants and seedless vascular plants, but they lack the stored nutrients found in seeds.

The evolution of heterospory, pollen, and seeds triggered a dramatic radiation of seed plants starting about 365 million years ago. To make sure that you understand these key processes and structures, study the life cycle of the pine tree pictured in Figure 28.18.

1. Starting with the sporophyte on the left, note that this and many other gymnosperm species have separate structures, called cones, where microsporangia and megasporangia develop. In this case, the two types of spores associated with heterospory develop in separate cones.
2. The microsporangia contain microspore mother cells that divide by meiosis to form microspores, which then divide by mitosis to form pollen grains—tiny male gametophytes.
3. Megasporangia are found inside protective structures called ovules. Each megasporangium contains a megaspore mother cell that divides by meiosis to form a megaspore.

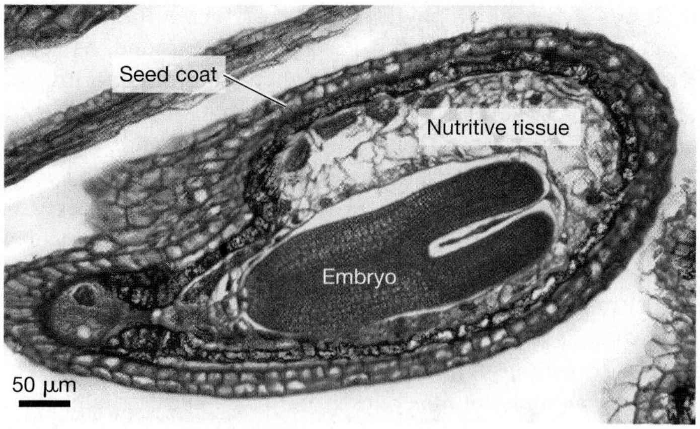

Figure 28.17 Seeds Contain an Embryo and a Food Supply with a Protective Coat That Allows Them to Be Dispersed. This *Capsella* (shepherd's purse) seed has been sectioned and stained.

4. The megaspore undergoes mitosis to form the female gametophyte, which then produces egg cells through further rounds of mitosis.
5. The female gametophyte stays attached to the sporophyte as pollen grains arrive and produce sperm that fertilize the eggs. Sperm reach the egg through a structure called a pollen tube.
6. Seeds mature as the embryo develops. Inside the seed, cells derived from the female gametophyte become packed with nutrients provided by the sporophyte. If you have ever eaten pine nuts, you consumed nutrient-rich female gametophytes, each with an embryo located inside.

When the seed disperses and germinates, the embryo grows into a young sporophyte, and the cycle of life begins anew. ✔ **Compare the life cycle of the pine tree in Figure 28.18 with that of the fern in Figure 28.14. How are gametophytes similar in these two groups? How are they different?**

Flowers Flowering plants, or angiosperms, are the most diverse land plants living today. About 300,000 species have been described, and more are discovered each year. Their success in terms of geographical distribution, number of individuals, and number of species revolves around a reproductive organ—the **flower**.

Most flowers contain two key reproductive structures: the stamens and carpels illustrated on the left-hand side of Figure 28.19. Stamens and carpels are responsible for heterospory in angiosperms. Note the structures and events below as you study Figure 28.19.

1. A **stamen** includes a structure called an anther, where microsporangia develop. Meiosis occurs inside the microsporangia, forming microspores. Microspores then divide by mitosis to form pollen grains (male gametophytes).
2. A **carpel** contains a protective structure called an **ovary**, where the ovules are found. Meiosis occurs inside ovules, forming megaspores within megasporangia. Megaspores then divide by mitosis, forming female gametophytes.

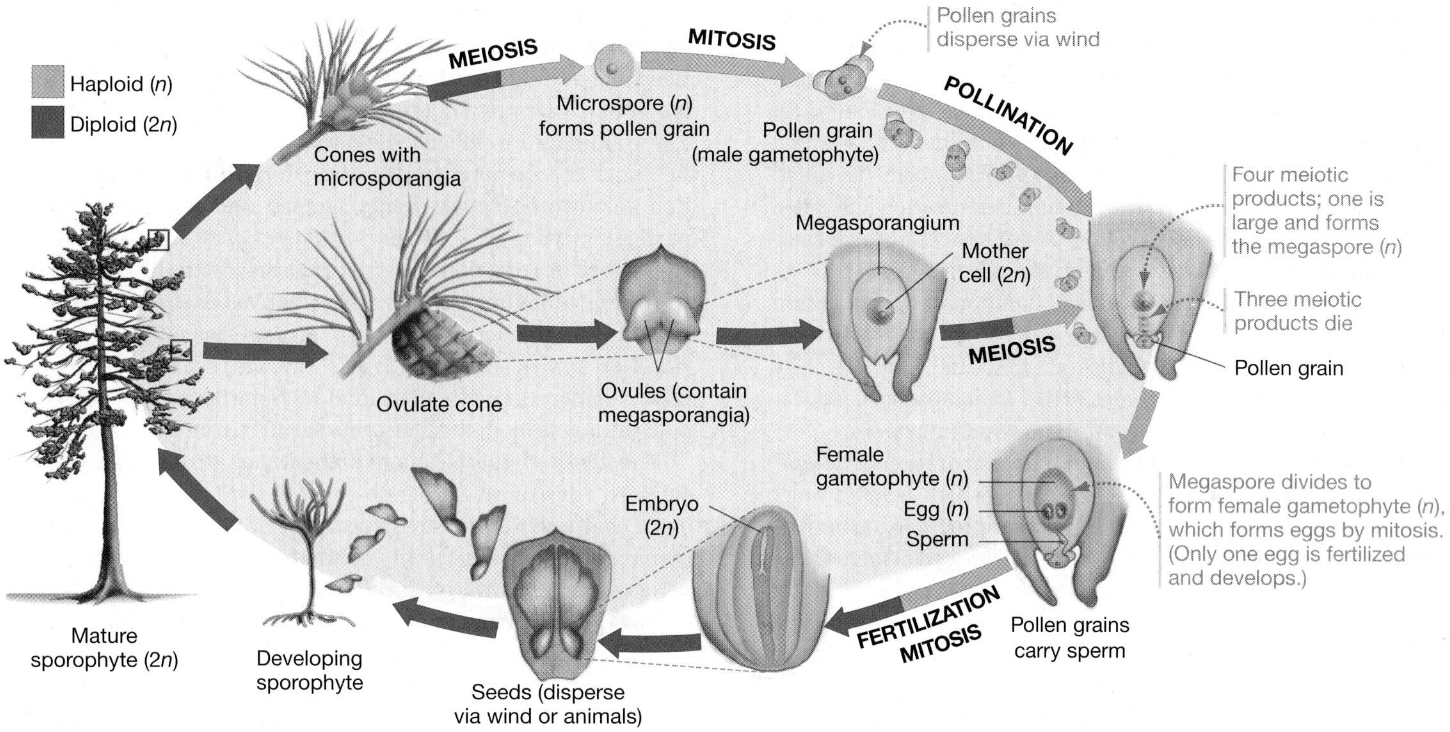

Figure 28.18 Heterospory in Gymnosperms: Microspores Produce Pollen Grains; Megaspores Produce Female Gametophytes.

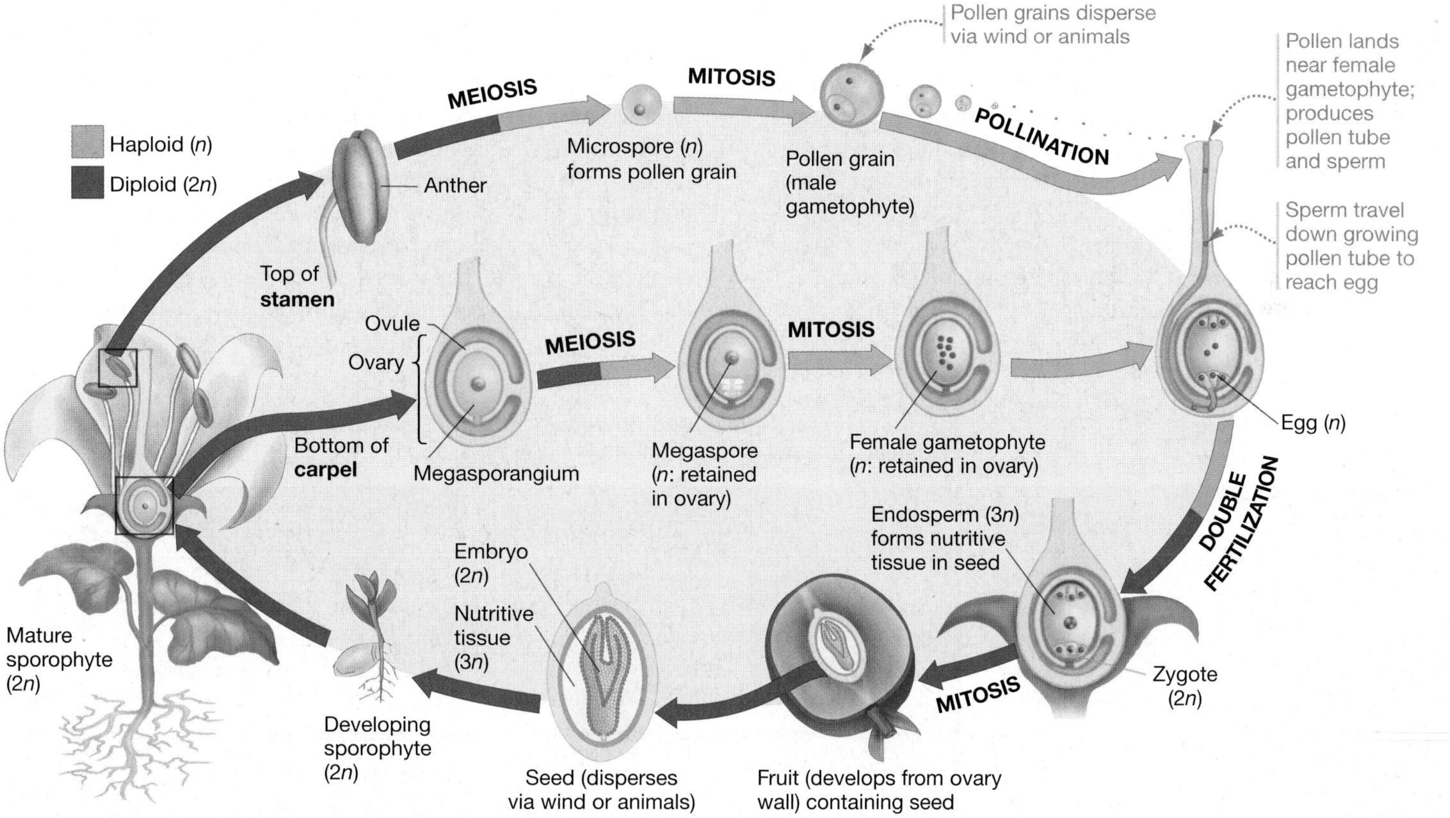

INTERACTIVE Figure 28.19 Heterospory in Angiosperms: Flowers Contain Microspores and Megaspores.

✔ Gymnosperm pollen grains typically contain from four to 40 cells; mature angiosperm pollen grains contain three cells. Gymnosperm female gametophytes typically contain hundreds of cells; angiosperm female gametophytes typically contain seven. Do these observations conflict with the trend of reduced gametophytes during land plant evolution, or are they consistent with it? Explain your logic.

After a pollen grain lands on a carpel and produces sperm, fertilization takes place. Angiosperm fertilization involves two sperm cells. One sperm fuses with the egg to form the diploid ($2n$) zygote, while a second sperm fuses with two nuclei in the female gametophyte to form a triploid ($3n$) nutritive tissue called **endosperm**. The involvement of two sperm nuclei is called **double fertilization**. A form of double fertilization has been shown to occur in a few gymnosperms, but results in the formation of multiple embryos—endosperm is not formed.

Ovaries develop into fruits after pollination and fertilization take place. Each fruit contains one or more seeds, each with a developing embryo. The seeds are dispersed by wind, animals, or by some other means. If a seed comes to rest in an appropriate location, it will germinate and grow into a new sporophyte.

The evolution of the flower, then, is an elaboration of heterospory. The key innovation was the ovary, which helps protect female gametophytes from insects and other predators. After pollination, ovaries develop into fruits that aid in seed dispersal.

Double fertilization with subsequent formation of endosperm is another striking innovation associated with the flower, but its adaptive significance is still not well understood. Explaining the significance of double fertilization is a major challenge for biologists interested in understanding how seed plants diversified.

The rest of this section briefly describes the role of flowers in pollination, fruit development, and seed dispersal. (These topics are explored in more depth in Chapter 38.)

Pollination by Insects and Other Animals Once stamens and carpels evolved, they became enclosed by modified leaves called **sepals** and **petals**. The four structures then diversified to produce a fantastic array of sizes, shapes, and colors—from red roses to blue violets. Specialized cells inside flowers also began producing a wide range of scents.

To explain these observations, biologists hypothesize that flowers are adaptations that increase an animal's likelihood of performing **pollination**—the transfer of pollen from one plant's stamen to another plant's carpel. Instead of leaving pollination to an undirected agent such as wind, the hypothesis is that natural selection favored structures that reward an animal—usually an insect—for carrying pollen directly from one flower to another. These animals are called *pollinators*.

Under the **directed-pollination hypothesis**, natural selection has favored flower scents, shapes, and colors that attract particular types of pollinators (Figure 28.20). In general, the characteristics of a flower correlate closely with the characteristics of its pollinator. Flowers are attractive because they provide pollinators with food in the form of protein-rich pollen or a sugar-rich fluid known as **nectar**. In this way, the relationship between flowering plants and their pollinators is mutually beneficial. The pollinator gets food; the plant gets sex (fertilization).

The directed-pollination hypothesis has strong experimental support. For example, field observations led researchers to conclude that a white-flowered species of South American petunia (*Petunia axillaris*) was mainly pollinated at night by hawk moths, while a purple-flowered species (*P. integrifolia*) was pollinated during the day by bumblebees. (Note that white is highly visible at night, and purple is the color most easily seen by bees.)

Was flower color the key to flower choice by these insects? This is difficult to determine because of other differences in flower characteristics between two plant species, such as flower shape and scent. In this experiment, researchers identified a gene named *AN2* to be responsible for the pigmentation in the purple petunia and introduced it into the genome of the white petunia, producing new plants that now had purple flowers. The researchers then documented that other than color, the wild-type and genetically altered flowers were identical.

In a controlled greenhouse experiment, hawk moths and bumblebees were allowed to choose between the two *P. axillaris* plants, and as predicted by the directed-pollination hypothesis, flower color was key to pollinator preference (Figure 28.21). These data strongly support the hypothesis that flower color is an adaptation that increases the frequency of pollination by specific pollinators.

(a) Fly-pollinated flowers: Carrion flower smells like rotting flesh

(b) Hummingbird-pollinated flowers: Red, long tubes with nectar at the base

(c) Bee-pollinated flowers: Often bright purple

Figure 28.20 Flowers with Different Scents, Shapes, and Colors Attract Different Pollinators.

RESEARCH

QUESTION: Does flower color influence pollinator preference?

HYPOTHESIS: Hawk moths prefer white petunias while bumblebees prefer purple, even when the flowers are otherwise identical.

NULL HYPOTHESIS: Neither insect has a clear preference for white- or purple-flowered petunias.

EXPERIMENTAL SETUP:

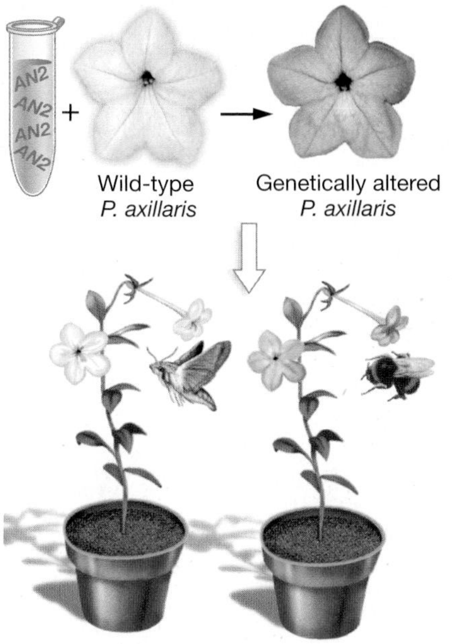

1. Isolate *AN2* gene for purple flower color from a related species (*P. integrifolia*) and insert it into *P. axillaris.*

2. Grow wild-type (white) and genetically altered (purple) *P. axillaris* in controlled greenhouse conditions.

3. Count number of visits by hawk moths and bumblebees to white and purple flowers.

PREDICTION OF HYPOTHESIS: Hawk moths will visit white-flowered *P. axillaris* more frequently, while bumblebees will visit purple-flowered *P. axillaris* more frequently.

PREDICTION OF NULL HYPOTHESIS: Both insects will visit white and purple flowers equally.

RESULTS:

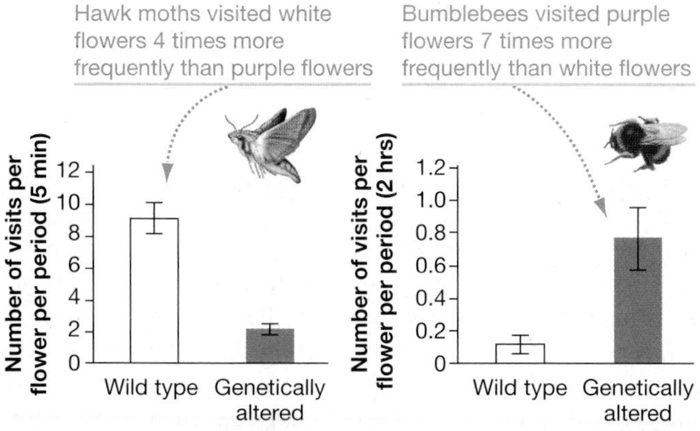

CONCLUSION: Petunia flower preference among bumblebees and hawk moths is significantly influenced by flower color.

Figure 28.21 **The Adaptive Significance of Flower Color.**

SOURCE: M. E. Hoballah, T. Gübitz, J. Stuurman, et al. 2007. Single gene-mediated shift in pollinator attraction in *Petunia. Plant Cell* 19: 779–790.

✔ **PROCESS OF SCIENCE** Why did the researchers measure pollination in a greenhouse? Why did they plot flower visits on different scales?

Based on results like this, biologists contend that the spectacular diversity of angiosperms resulted, at least in part, from **coevolution** with animal pollinators (Chapter 38). Evolutionary changes in angiosperms and corresponding changes in their pollinators were highly dependent on each other. The outcome? Many animals and the flowering plants they pollinate depend on each other for their survival.

Fruits The evolution of the ovary was an important event in land plant diversification, but not only because it protected the female gametophytes of angiosperms. It also made the evolution of fruit possible.

A **fruit** is a structure that is derived from the ovary and encloses one or more seeds (Figure 28.22a). Tissues derived from the ovary are often nutritious and brightly colored (Figure 28.22b). Animals eat these types of fruits, digest the nutritious tissue around the seeds, and disperse seeds in their feces. In other cases, the tissues derived from the ovary help fruits disperse via wind or water. Coconuts, for example, can float and drift for thousands of miles before coming to rest on a shore and growing into a new

(a) Fruits are derived from ovaries and contain seeds.

(b) Many fruits are dispersed by animals.

Figure 28.22 **Angiosperms Produce Fruits Derived from Ovaries That Contain Seeds. (a)** A pea pod is one of the simplest types of fruit. **(b)** The ovary wall often becomes thick, fleshy, and nutritious enough to attract animals that disperse the seeds inside.

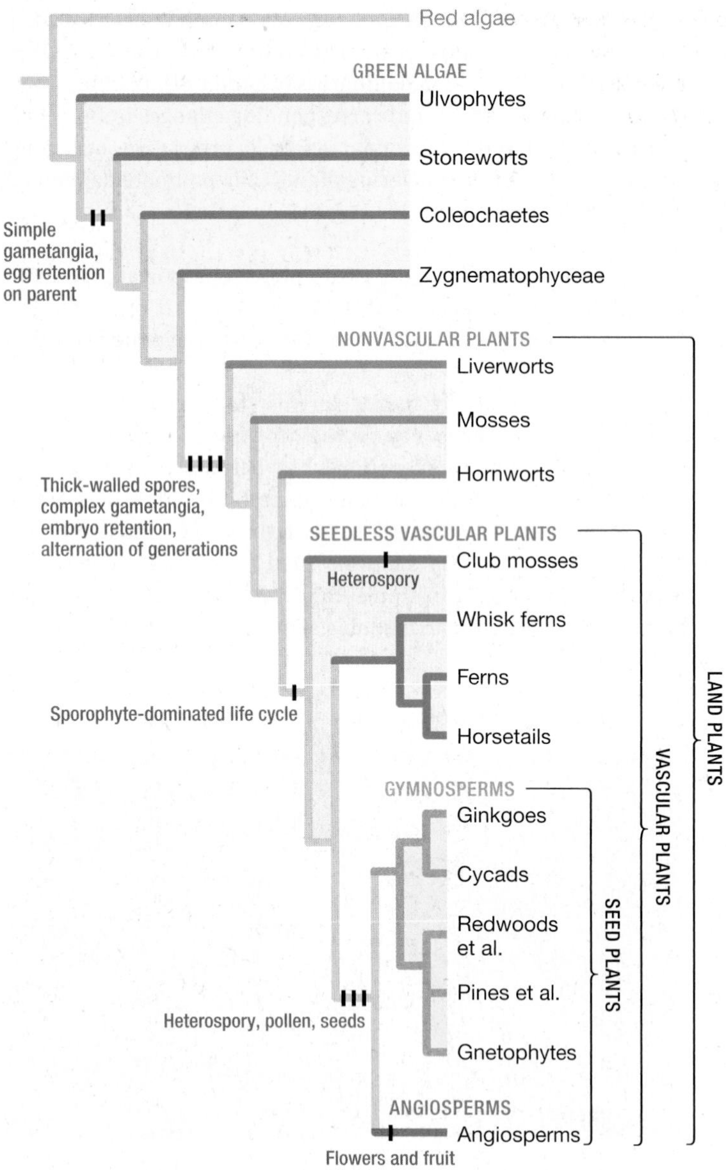

Figure 28.23 Plant Phylogeny Emphasizing the Evolutionary Innovations That Allowed Plants to Reproduce Efficiently on Land.

✔ Redwoods, pines, gnetophytes, and angiosperms are land plants that do not have flagellated sperm that swim to the egg. Describe where the loss of flagellated sperm occurred on the phylogenetic tree shown in Figure 28.6.

palm tree. The evolution of flowers made efficient pollination possible; the evolution of fruits made efficient seed dispersal possible. (See Chapter 38, Section 38.4, for a more in-depth look at fruit development and seed dispersal.)

The list of adaptations that allow land plants to reproduce in dry environments is impressive; **Figure 28.23** summarizes them. Once land plants had vascular tissue and could grow efficiently in dry habitats, the story of their diversification revolved around traits that allowed sperm cells to reach eggs efficiently and helped seeds disperse to new locations.

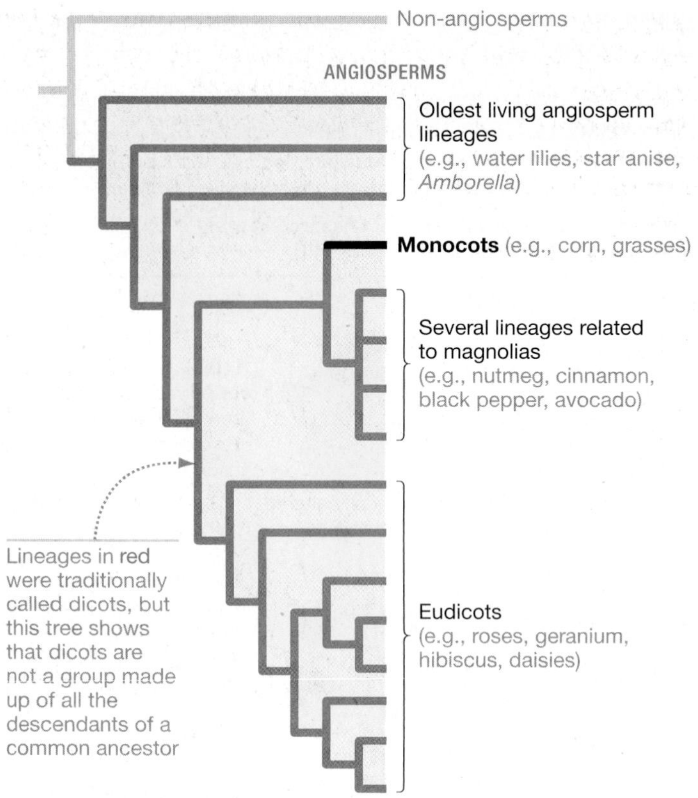

Figure 28.24 Monocots Are Monophyletic, Dicots Are Paraphyletic.

The Angiosperm Radiation

For the past 125 million years, land plant diversification has really been about angiosperms. The Anthophyta, or angiosperms, represent one of the great adaptive radiations in the history of life. An **adaptive radiation** occurs when a single lineage produces a large number of descendant species that are adapted to a wide variety of habitats (Chapter 25).

The diversification of angiosperms is associated with three key adaptations: **(1)** water-conducting vessels, **(2)** flowers, and **(3)** fruits. In combination, these traits allow angiosperms to transport water, pollen, and seeds efficiently. Based on these observations, it is not surprising that most land plants living today are angiosperms.

On the basis of morphological traits, the 300,000 species of angiosperms identified to date have historically been classified into two major groups: the monocotyledons, or **monocots**, and the dicotyledons, or **dicots**. Some familiar monocots include the grasses (such as corn and wheat), orchids, palms, and lilies; familiar dicots include beans, roses, buttercups, daisies, oaks, and maples. The names of the two groups were inspired by differences in the embryonic leaf that develops in seeds, a structure called the cotyledon. A **cotyledon** ("seed-leaf") stores nutrients and supplies them to the developing embryonic plant.

Recent work has shown that dicots do not form a monophyletic group consisting of a common ancestor and all of its descendants. To drive this point home, consider the phylogeny illustrated in **Figure 28.24**. These relationships were estimated

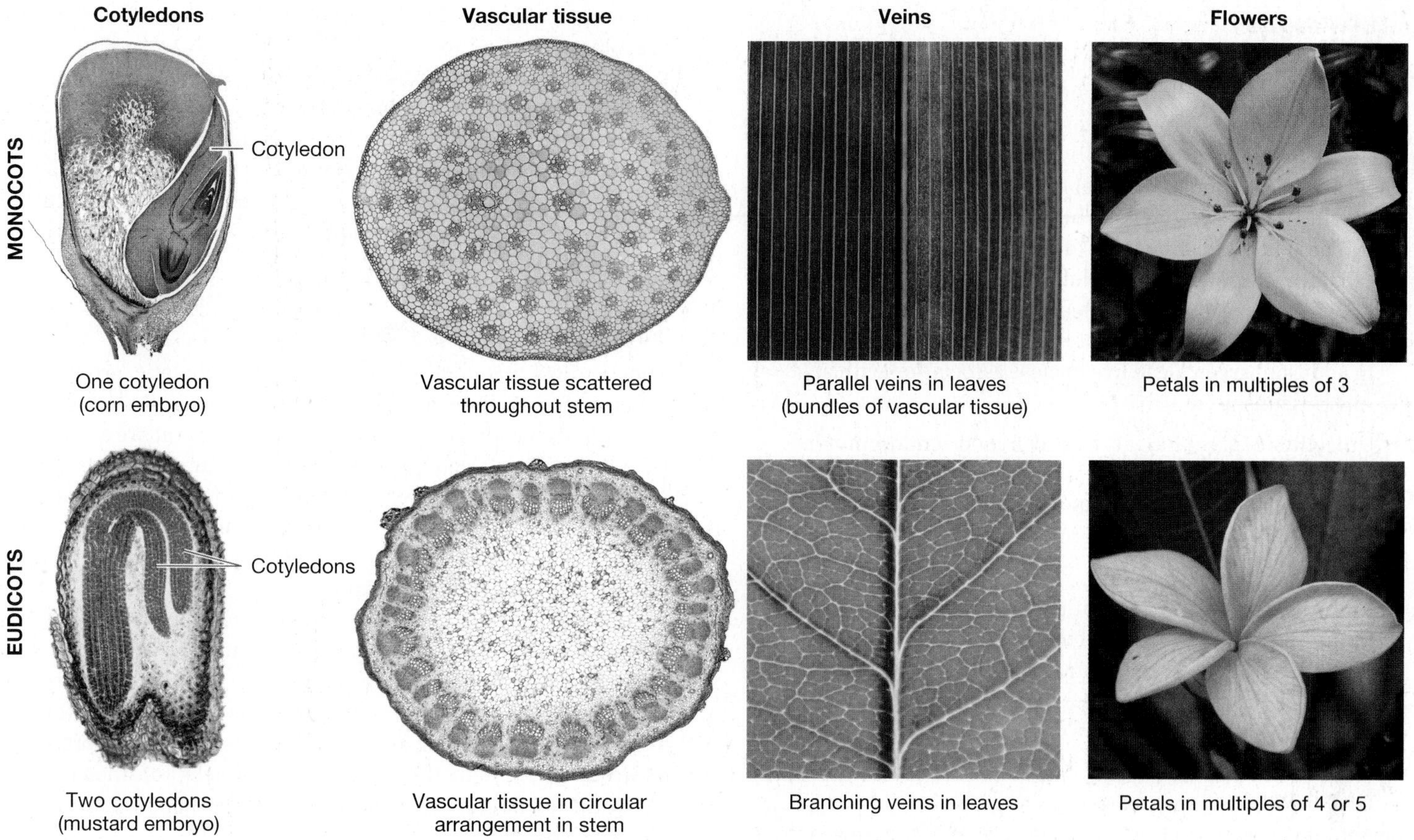

Figure 28.25 **Four Morphological Differences between Monocots and Eudicots.** The seed and stem sections, showing cotyledons and vascular tissue, have been stained.

by comparing the sequences of several genes that are shared by all angiosperms. Notice that species with dicot-like characters are found in multiple lineages of the angiosperm phylogenetic tree. Based on this analysis, biologists have concluded that although monocots are monophyletic, dicots are not. Dicots are paraphyletic.

Biologists have adjusted the names assigned to angiosperm lineages to reflect this new knowledge of phylogeny. The most important of these changes was identifying the **eudicots** ("true-dicots") as a monophyletic lineage that includes most of the plants once considered dicots.

As Figure 28.25 shows, monocots have a single cotyledon (hence the root *mono*) while eudicots have two cotyledons (hence the *di*). The figure also highlights other major morphological differences observed in monocots and eudicots, including the arrangement of vascular tissue and leaf veins and the characteristics of flowers.

One flowering plant that is neither a monocot nor a eudicot is the magnolia tree. Many species of magnolias are common in the southern United States and produce large, fragrant flowers.

Plant systematists continue to work toward understanding relationships throughout the angiosperm phylogenetic tree. There will undoubtedly be more name changes as our knowledge grows.

CHECK YOUR UNDERSTANDING

✔ **If you understood this section, you should be able to ...**

1. Explain why the evolution of cuticle, UV-absorbing compounds, and vascular tissue were important in plant survival or reproduction.
2. Using Figure 28.6, describe where the origins of pollen, flowers, and fruits occurred.

Answers are available in Appendix A.

28.4 Key Lineages of Green Algae and Land Plants

The adaptations introduced in Section 28.3 allowed the land plants to make one of the most important water-to-land transitions in the history of life. To explore green plant diversity in more detail, let's take a closer look at a few representatives from the major groups of green algae and land plants.

After you complete this section, you should be able to . . .

- Describe the key features of the major lineages of green algae and land plants.

Green Algae

The green algae are a paraphyletic group that totals about 8000 species. **Table 28.2** illustrates four of this group's major lineages.

Green algae are important primary producers in nearshore ocean environments and in all types of freshwater habitats. They are also found in several types of more exotic environments, including snowfields at high elevations, pack ice, and ice floes. These habitats are often splashed with bright colors due to large concentrations of unicellular green algae (**Figure 28.26**). Although these cells live at near-freezing temperatures, they make all their own food via photosynthesis.

In addition, green algae live in close association with an array of other organisms.

- Unicellular green algae are common endosymbionts in planktonic eukaryotes that live in lakes and ponds. The association is considered mutually beneficial: The algae supply the protists with food; the protists provide protection to the algae.
- **Lichens** are stable associations between green algae and fungi, or between cyanobacteria and fungi, and are often found in terrestrial environments that lack soil, such as on tree bark or bare rock. The algae or cyanobacteria in a lichen are protected from drying by the fungus; the fungus receives sugars from the algae produced by photosynthesis.

Of the 17,000 species of lichens described to date, about 85 percent involve green algae. The green algae that are involved are unicellular or grow in long filaments. (Lichens are explored in more detail in Chapter 29.)

Nonvascular Plants

The initial lineages to branch off early during the evolution of land plants are collectively called the nonvascular plants. **Table 28.3** on page 600 shows a small sampling of this group's diversity, which includes mosses as well as the less familiar hornworts and liverworts.

All of the nonvascular plant species present today have a low, sprawling growth habit. In fact, it is unusual to find species that are more than 5 to 10 centimeters tall. The gametophyte is the dominant and longer-lived phase of the life cycle in all of the nonvascular plants.

Individuals are anchored to soil, rocks, or tree bark by structures called **rhizoids**. Although simple water-conducting cells and tissues are found in some mosses, nonvascular plants lack vascular tissue with lignin-reinforced cell walls.

All nonvascular plants have flagellated sperm that swim to eggs through raindrops or small puddles on the plant surface. Spores are dispersed by wind.

Seedless Vascular Plants

The seedless vascular plants are a paraphyletic group that evolved after the nonvascular plants. All species of seedless vascular plants have conducting tissues with cells that are reinforced with lignin. Tree-sized club mosses and horsetails are abundant in the fossil record, and tree ferns are still common inhabitants of certain habitats, such as in New Zealand and on mountain slopes in the tropics. **Table 28.4** on page 601 summarizes the major groups of seedless vascular plants.

The sporophyte is the dominant and longer-lived phase of the life cycle in all of the seedless vascular plants. The gametophyte is physically independent of the sporophyte, however. Eggs are retained on the gametophyte, and sperm swim to the egg with the aid of flagella. Thus, seedless vascular plants depend on the presence of water for reproduction. Sporophytes develop on the gametophyte and are nourished by the gametophyte when they are small.

Seed Plants: Gymnosperms and Angiosperms

The seed plants are a monophyletic group consisting of the gymnosperms—ginkgos, cycads, redwoods, pines, and gnetophytes—and the angiosperms. The group as a whole is defined by two key synapomorphies, traits common to a lineage and found in no other lineages: the production of seeds and the production of pollen grains. Whereas angiosperms produce their seeds in ovaries, gymnosperms do not.

Table 28.5 on page 602 summarizes the major groups of gymnosperms, while **Table 28.6** on page 603 summarizes some of the major groups of angiosperms. The groups presented in Table 28.6 are just a small sampling of angiosperm diversity. Angiosperms are by far the most diverse group of plants on Earth.

Seed plants are found in virtually every type of habitat, and they adopt every growth habit known among land plants. Their forms range from tiny floating duckweed plants, barely larger than a grain of salt, to shrubs and vines to 100-meter-tall trees. Gymnosperms live for many years. The oldest non-clonal

Figure 28.26 Some Green Algae Live in Unusual Environments. Green algae are responsible for green snow.

example is a bristlecone pine that has been alive for approximately 5000 years. Angiosperms can be **annual** (have a single growing season) or **perennial** (live for many years), with life spans ranging from a few weeks to several thousand years. (See Chapters 34 through 38 to further explore the structure and function of seed plants.)

CHECK YOUR UNDERSTANDING

✔ **If you understood this section, you should be able to . . .**

1. Describe where spores are produced on ferns.
2. Differentiate between gymnosperms and angiosperms.

Answers are available in Appendix A.

DIVERSITY Table 28.2 **Key Lineages of Green Algae**

		Known Species	Reproduction	Life Cycle	Relevance
	Ulvophyceae **Ulvophytes** ◂ *Ulva lactuca* (sea lettuce)	**4000** Marine Freshwater Unicellular/ multicellular	Sexual and asexual. Spores and gametes are motile. Male and female gametes look identical. External fertilization.	Zygotes are the only diploid stage in unicellular species. Alternation of generations only in multicellular species. Gametophytes and sporophytes may look identical or different.	Important primary producers in freshwater environments and in coastal areas of oceans.
	Charophyceae **Stoneworts** ◂ *Chara vulgaris* (common stonewort) showing red antheridia	**6000** Fresh water (lakes) Multicellular (some species over a meter in length)	Sexual and asexual. Spores and sperm are motile. Eggs are retained on the parent and nourished after fertilization.	Zygotes are the only diploid stage (no alternation of generations).	Form extensive beds on lake bottoms or ponds and provide food for waterfowl and shelter for fish. Heavily studied given their close relationship to land plants.
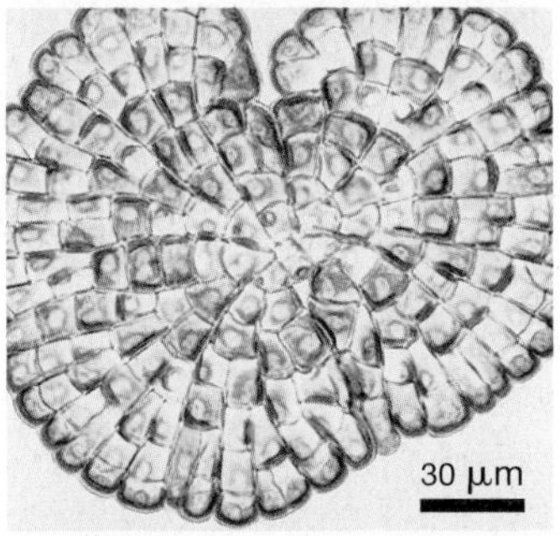	Coleochaetophyceae **Coleochaetes** ◂ *Coleochaete orbicularis*	**19** Fresh water Multicellular	Sexual and asexual. Spores and sperm are motile. Eggs are retained on the parent and nourished after fertilization.	Zygotes are the only diploid stage (no alternation of generations).	Heavily studied given their close relationship to land plants.
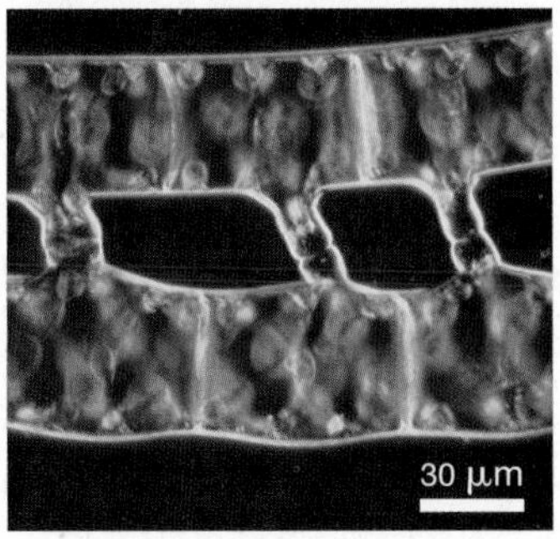	Zygnematophyceae **Conjugating algae** ◂ *Spirogyra* sp. showing conjugation tubes	**2700** Fresh water Unicellular/ multicellular (filamentous)	Sexual and asexual. Conjugation occurs between cells of adjacent filaments. Gametes are of equal size and pass from one cell to another through conjugation tubes.	Zygotes are the only diploid stage (no alternation of generations).	Heavily studied given recent evidence indicating their close relationship to land plants.

For more go to https://goo.gl/h873sW

DIVERSITY **Table 28.3 Key Lineages of Nonvascular Plants**

		Known Species	Reproduction	Life Cycle	Relevance
	**Hepaticophyta** **Liverworts** ◂ *Marchantia polymorpha*	**8500** Some have liver-shaped leaves. Many have pores similar to stomata.	Sexual and asexual by dispersal of small fragments called gemmae. Motile sperm. Spores dispersed by wind or rain.	Gametophyte dominant. Sporophyte is small and depends on gametophyte for nutrition.	Thought to resemble first land plants. Their decaying tissues contribute to initial stages of soil formation.
	**Bryophyta** **Mosses** ◂ *Sphagnum sp.* showing spore capsules	**12,000** Some able to withstand extreme drying. Some have rudimentary conducting tissues.	Sexual and asexual by clonal propagation of gametophyte. Motile sperm. A sporophyte can produce 50 million spores. Spores usually disperse by wind.	Gametophyte dominant. Sporophyte is small and depends on gametophyte for nutrition.	Partially decayed mosses form large patches of peat, which is used as fuel for cooking, heating. Sphagnum can hold up to 20 times its weight in water and is a common soil additive.
	**Anthocerophyta** **Hornworts** ◂ *Anthoceros* sp.	**19** Sporophytes have stomata.	Sexual and asexual by fragmentation of gametophyte. Motile sperm. Gametophytes are either unisexual or bisexual. Spores disperse by wind or rain.	Gametophyte dominant. Sporophyte is small and obtains some nutrition from gametophyte, but also performs photosynthesis.	Some have symbiotic cyanobacteria that fix nitrogen.

For more go to https://goo.gl/9oZ9eX

DIVERSITY Table 28.4 Key Lineages of Seedless Vascular Plants

		Known Species	Reproduction	Life Cycle	Relevance
	Lycophyta **Lycophytes (also called club mosses or ground cedar)** ◄ *Lycopodium sp.*	**1000** Most ancient lineage with roots.	Sexual and asexual; in some species by dispersal of small fragments called gemmae. Motile sperm. Some species show heterospory.	Sporophyte dominant. Gametophyte of some lives with symbiotic fungi. Gametophyte may live up to 15 years.	Tree-sized lycophytes were abundant during the Carboniferous period. Spores of some are flammable and were used as flash powder for early photography.
	Psilophyta **Whisk ferns** ◄ *Psilotum nudum*	**6** Body consists of only branching stems.	Sexual and asexual by extension of underground stems. Spores disperse by wind. Bisexual gametophytes.	Sporophyte dominant. Gametophyte is only about 2 cm long. Sporophyte may be up to 30 cm tall; grows directly on the gametophyte.	Some derive nutrition from symbiotic fungi. Used in landscaping, especially in Japan. Often grows as a weed in greenhouses.
	Pteridophyta **Ferns** ◄ *Polypodium virginianum* showing orange clusters of sporangia	**12,000** Common in moist, humid habitats. Have large leaves called fronds.	In bisexual gametophytes, sperm and eggs mature at different times, minimizing self-fertilization. Motile sperm.	Sporophyte dominant. Gametophyte is photosynthetic. Sporangium pops open in dry conditions, releasing spores.	Young fronds, called "fiddleheads," are a culinary delicacy. Widely used as ornamental plants.
	Equisetophyta **Horsetails** ◄ *Equisetum sp.* showing sterile (vegetative) and fertile (reproductive) stems	**25** Hollow stems allow oxygen to move down to roots that often grow in oxygen-poor soil.	Sexual and asexual. Reproductive stems produce clusters of sporangia. Spore dispersal in wind is facilitated by wing-like structures called elaters.	Sporophyte dominant. Gametophytes are small and short lived. Sperm and eggs mature at different times, minimizing self-fertilization.	Stems are rich in abrasive silica granules. Used to scour pots and pans. (The plants are sometimes called scouring rushes.)

For more go to https://goo.gl/7EdH4N

DIVERSITY **Table 28.5** Key Lineages of Seed Plants: Gymnosperms

		Known Species	General Features	Life Cycle	Relevance
	Ginkgophyta **Ginkgos (also called maidenhair tree)** ◂ *Ginkgo biloba*	**1**	Loses leaves in winter (is deciduous). Leaves are virtually identical to 150-million-year-old fossils, so *Ginkgo* is considered a "living fossil."	Separate male and female plants. Pollen carried by wind; motile sperm released into female gametophyte swim to egg.	Widely planted as ornamentals; resistant to air pollution. Leaf extracts used as dietary supplement thought to improve memory, vision, and help maintain overall health.
	Cycadophyta **Cycads (also called sago palms)** ◂ *Cycas revoluta*, male with cone	**140**	Have been around for approximately 300 million years. Resemble palms. Unique among gymnosperms in having compound leaves.	Separate male and female plants. Pollen carried by wind; motile sperm released into female gametophyte swim to egg.	Roots contain symbiotic cyanobacteria that fix nitrogen. Popular landscaping plants.
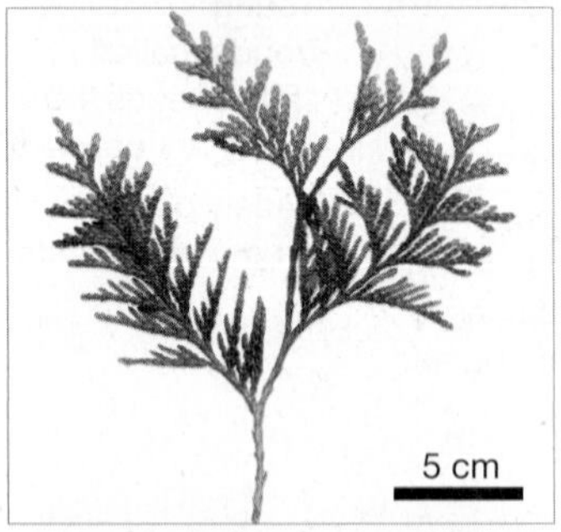	Cupressophyta **Redwoods, junipers, yews** ◂ *Thuja plicata*	**150**	Size ranges from small shrubs to giant redwood trees. Small leaf surface area reduces water loss.	Wind pollinated. Separate male and female cones. Seeds disperse by wind, birds, or mammals.	Cedar wood is commonly used for furniture, decks, or roofing. Juniper "berries" are used to flavor gin. Chemotherapy drug taxol was originally obtained from bark of yew trees.
	Pinophyta **Pines, spruces, firs** ◂ *Picea abies*	**240**	Needle-like leaves have small surface area. Often found growing in dry areas.	Wind pollinated. Separate male and female cones. Female cones may take two years to mature.	Dominant plants at high latitudes and high elevations. Seeds are source of food for mice, squirrels, and other animals. Wood is used for building and paper industry.
	Gnetophyta **Gnetophytes** ◂ *Welwitschia mirabilis*	**70**	Grow as vines, trees, or shrubs. Closely related to angiosperms. Display several angiosperm features, including wood that contains vessel elements.	Pollen is transferred by wind or insects. Double fertilization occurs, but results in the formation of two embryos (endosperm is not formed).	The drug ephedrine was originally isolated from *Ephedra* (also called Mormon's tea). *Welwitschia* lives only in the Namib desert and may live up to 1500 years.

DIVERSITY **Table 28.6 Key Lineages of Seed Plants: Angiosperms**

		Known Species	General Features	Phylogenetic Relationships	Relevance
	Basal angiosperms ***Amborella*, water lilies, Austrobaileyales (e.g., star anise)** **Oldest living angiosperm lineages** ◀ *Amborella trichopoda*	**200**	Share some features with monocots (e.g., pollen with one groove) and eudicots (e.g., two seed leaves). Numerous flattened stamens. Some lack vessel elements.	A paraphyletic assemblage of species at the base of the angiosperm lineage tree. Includes many species traditionally categorized as dicots (but they are separate from the eudicots).	Evolved before the vast majority of other angiosperms. *Amborella* is considered the sister taxon to all other flowering plants.
	Monocotyledons (monocots) **Orchids, grasses, palms, bananas, ginger, and others** ◀ *Iris sibirica* (Siberian iris)	**60,000**	Embryo has a single seed leaf (cotyledon). Flower parts typically in multiples of three. Lack secondary (woody) growth. Pollen grains with a single groove. Leaf veins run parallel.	A monophyletic group closely related to the magnoliids.	Some of the major crops worldwide are monocots (e.g., corn, wheat, rice, sugarcane).
	Magnoliids **Magnolias, black pepper, tulip tree, and others** ◀ *Magnolia grandiflora* (Southern magnolia)	**9000**	Large trees, shrubs, or vines. Large flowers with numerous petals and sepals. Large net-veined leaves.	A monophyletic group that includes many species traditionally categorized as dicots (but they are separate from the eudicots).	Many economically important species used for food, medicines, perfumes, and timber.
	Eudicotyledons (eudicots) **Roses, daisies, oak, maple, beans, peas, and many others** ◀ *Rudbeckia hirta* (black-eyed Susan)	**200,000**	Embryo has two seed leaves (cotyledons). Distinguished from other groups by producing pollen grains with three grooves.	A monophyletic group that is sister to a group including the monocots and magnoliids. By far the largest group of angiosperms.	Many economically important species; used for food, medicines, perfumes, and timber. Most trees in deciduous forests are eudicots.

For more go to https://goo.gl/QftyNm

28.1 Why Do Biologists Study the Green Algae and Land Plants?

- Green algae are hypothesized to be the closest relatives of land plants.
- Plants are important primary producers and a source of oxygen, and they hold soil and water in place.
- Plants improve the quality of the environment for other organisms.
- Humans depend on plants for food, fuel, fiber, building materials, and medicines.

28.2 How Do Biologists Study Green Algae and Land Plants?

- Scientists analyze morphological traits, examine the fossil record, and evaluate molecular phylogenies to study and compare green algae and land plants.
- Recent DNA sequence analyses suggest that land plants evolved from freshwater green algae from the Zygnematophyceae (conjugating algae) family.
- The fossil record reveals that the first land plants evolved about 475 million years ago and then diversified into plant lineages with vascular systems, seeds, and flowers.

28.3 What Themes Occur in the Diversification of Land Plants?

- The evolution of cuticle allowed plant tissues to be exposed to air without dying.
- UV-absorbing compounds allowed the first plants growing on land to tolerate the harsh UV rays from the sun.
- The evolution of stomata, breaks in the cuticle consisting of pores controlled by guard cells, allowed plants to maximize gas exchange and minimize water loss.
- Vascular tissue conducts water and has secondary cell walls reinforced with lignin, which provides structural support, allowing plants to grow upright. This adaptation reduces competition for light.
- Tracheids are water-conducting cells found in all vascular plants; in addition, angiosperms, gnetophytes, and a few species of seedless vascular plants have water-conducting cells called vessels.
- All land plants are embryophytes, meaning that eggs and embryos are retained on the parent plant. Consequently, the developing embryo can be nourished by the mother plant.
- All land plants have alternation of generations. Over the course of land plant evolution, the gametophyte phase decreased in size and life span, and the sporophyte phase became more prominent. Heterospory also became more common.
- The evolution of pollen was an important breakthrough in the history of life, because sperm no longer needed to swim to the egg—instead, tiny male gametophytes could be transported through the air via wind or animals.
- Seed plant embryos are dispersed from the parent plant to a new location, encased in a protective housing, and supplied with a store of nutrients.
- Flowers and fruits are defining features of angiosperms that have played a role in the diversification of this large group.

28.4 Key Lineages of Green Algae and Land Plants

- Green algae include single-celled and multicellular marine and freshwater species. Some freshwater species share numerous characteristics with land plants. Some single-celled species are symbionts with protists, fungi, and animals.
- Nonvascular plants include mosses and liverworts that are small and have a gametophyte-dominant life cycle.
- Seedless vascular plants include club mosses and ferns, which have small, free-living gametophytes and large, longer-lived sporophytes. Dispersal is carried out using spores.
- Seed plants include gymnosperms and angiosperms that dominate the world's floras. Animals are often used to disperse seeds and, in angiosperms, to transport pollen between flowers.

Answers are available in Appendix A

✔ TEST YOUR KNOWLEDGE

1. What important role does lignin play in vascular plants?
   a. It is the major component of the cuticle, which prevents plants from drying out.
   b. It forms the outer layer of pollen grains.
   c. It forms the tough, rigid cell walls of water-conducting cells.
   d. It nourishes the embryo in seed plants.
2. The appearance of cuticle and stomata correlated with what event in the evolution of green plants?
   a. the first upright growth forms
   b. the first woody tissues
   c. growth on land
   d. the evolution of the first water-conducting tissues
3. What is a pollen grain?
   a. male gametophyte
   b. female gametophyte
   c. male sporophyte
   d. sperm
4. Evaluate the following statements regarding seeds. Select True or False for each statement.
   T/F They contain an embryo.
   T/F They contain nutritive tissue for the embryo.
   T/F They consist entirely of diploid cells.
   T/F They are also referred to as spores.

✔ TEST YOUR UNDERSTANDING

5. What does it mean to say that a life cycle is gametophyte dominant versus sporophyte dominant?
6. **SOCIETY** Soils, water, and the atmosphere are major components of the abiotic (nonliving) environment. Describe how green plants affect the abiotic environment in ways that are advantageous to humans.

7. The evolution of cuticle presented plants with a challenge that threatened their ability to live on land. Describe this challenge, and explain why stomata represent a solution. Compare and contrast stomata with the pores found in liverworts.

8. Explain the difference between homosporous and heterosporous plants. Where are the microsporangium and megasporangium found in a tulip? What happens to the spores that are produced by these structures?

✔ TEST YOUR PROBLEM-SOLVING SKILLS

9. Angiosperms such as grasses, oaks, and maples are wind pollinated. The ancestors of these subgroups were probably pollinated by insects, however. As an adaptive advantage, why might a species "revert" to wind pollination? (Hint: Think about the costs and benefits of being pollinated by insects versus wind.) Why is it logical to observe that wind-pollinated species usually grow in dense stands containing many individuals of the same species? Why is it logical to observe that in wind-pollinated deciduous trees, flowers form very early in spring—before leaves form?

10. **PROCESS OF SCIENCE** You have been hired as a field assistant for a researcher interested in the evolution of flower characteristics in orchids. Design an experiment to determine whether color, size, shape, scent, or amount of nectar is the most important factor in attracting pollinators to a particular species. Assume that you can change any flower's color with a dye and that you can remove petals or nectar stores, add particular scents, add nectar by injection, or switch parts among species by cutting and gluing.

✔ PUT IT ALL TOGETHER: Case Study

Is long-distance sex possible in mosses?

Moss sperm are motile and capable of swimming short distances to fertilize an egg. However, it was unknown until recently how sperm make their way from male to female gametophytes that may be separated by a distance of several centimeters or more.

11. **QUANTITATIVE** Researchers hypothesized that small animals called springtails (a type of arthropod) facilitate sperm transfer in mosses. To test this hypothesis, the researchers grew patches of male and female mosses at varying distances from each other and in the presence or absence of springtails. They later checked to see how many sporophytes were found in each female patch (as an indicator that fertilization had occurred). Their results are shown in the graph here (** means $P < 0.01$, * means $P > 0.05$; see **BioSkills 3** for help interpreting P values). What conclusions can be drawn about the role of the springtails during the reproductive cycle of the mosses?

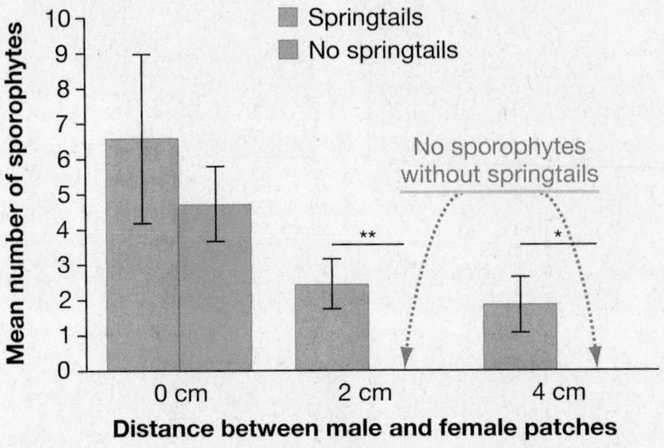

Source: N. Cronberg, R. Natcheva, and K. Hedlund. 2006. *Science* 313: 1255.

12. What structure do moss sperm use for motility, and how does that reflect the evolution of land plants from green algae?

13. **PROCESS OF SCIENCE** Beyond the results presented here, what other evidence would be useful when drawing conclusions about the role of springtails in moss reproduction?

14. **THINK CAREFULLY** A friend in your biology study group claims that eggs in mosses are fertilized by spores. What is wrong with that statement?

15. Besides springtails, are there other ways moss sperm may be transferred from male to female gametophytes?

16. Animals are responsible for pollinating many flowering plants. Did animal-mediated fertilization originate with flowering plants, or was this mode of sperm dispersal likely in place before the origin of flowering plants? Explain.

Mastering Biology

Students Go to Mastering™ Biology for assignments, the eText, and the Study Area with animations, practice tests, and activities.

Professors Go to Mastering™ Biology for automatically graded tutorials and questions that you can assign to your students, plus Instructor Resources.

29 Fungi

Netted stinkhorn (*Phallus indusiatus*) is an edible fungus related to mushrooms. This species forms a greenish, slimy, foul-smelling cap that produces spores. Insects attracted by the foul smell help disperse the spores.

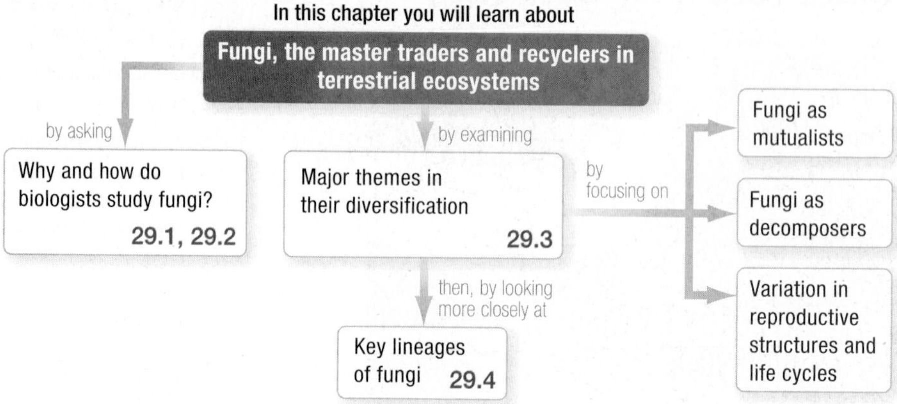

Fungi are eukaryotes that grow as single cells or as large, branching networks of multicellular filaments. Familiar fungi include the mushrooms you've encountered in the woods or a grocery store, the molds and mildews that grow in homes, the organism that causes athlete's foot, and the yeasts used in baking and brewing.

Along with the land plants and the animals, the **fungi** are one of three major lineages of large, multicellular eukaryotes that occupy terrestrial environments. When it comes to making a living, the species in these three groups use radically different strategies. Land plants make their own food through photosynthesis. Animals and fungi both feed on plants, protists, or each other by releasing digestive enzymes and absorbing small molecules. But while many animals carry out digestion in a dedicated digestive tract, fungi release their digestive enzymes to the external environment.

BIG PICTURE

This chapter is part of the Big Picture. See how on pages 722–723.

Fungi that absorb nutrients from dead organisms are the world's most important decomposers. Although a few types of organisms are capable of digesting the cellulose in plant cell walls, certain fungi and a handful of bacterial species are the only organisms capable of digesting both the lignin and cellulose that make up wood. Without fungi, Earth's surface would be piled so high with dead tree trunks and branches that there would be almost no room for animals to move or plants to grow.

Other fungi specialize in absorbing nutrients from living organisms. When fungi absorb these nutrients without benefiting their host, they lower the fitness of their host organism and act as parasites. If you've ever had athlete's foot or a vaginal yeast infection, you've hosted a parasitic fungus.

Most of the fungi that live in association with other organisms benefit their hosts, however. In these cases, fungi are not parasites but **mutualists**.

- The roots of most land plants are colonized by an array of mutualistic fungi that provide water and key nutrients such as nitrogen and phosphorus to the host plant. The soil around you is alive with an enormous network of fungi that are fertilizing the plants you see aboveground.
- Fungi living inside the shoots of certain plants help ward off herbivores by making toxic compounds.
- Many insects harbor single-celled fungi in their guts that aid their digestion of plant material.
- Some insects grow gardens of fungi that they feed with pieces of leaves. The insects then feed off the fungi, which they maintain and continue to cultivate.

In short, fungi are the master traders and recyclers in terrestrial ecosystems. Some fungi release nutrients from dead plants and animals into the soil; others obtain nutrients and then transfer them directly to living plants and animals. Because they recycle key elements such as carbon, nitrogen, and phosphorus and because they transfer key nutrients to plants and animals, fungi profoundly influence ecosystem productivity and biodiversity. In terms of nutrient cycling, fungi make the world go around.

29.1 Why Do Biologists Study Fungi?

Given their importance to life on land and their intricate relationships to other organisms, it's no surprise that fungi are fascinating to biologists. But there are important practical reasons to study them as well. Fungi can cause debilitating diseases in humans and crop plants; they nourish the plants that nourish us; they affect global climate change, because they are critical to the carbon cycle on land. Let's take a closer look at some of the ways that fungi affect land-dwelling organisms.

After you complete this section, you should be able to ...

- Summarize the important roles fungi play in the environment.

Fungi Have Important Economic and Ecological Impacts

In humans, parasitic fungi cause athlete's foot, vaginitis, diaper rash, ringworm, pneumonia, and thrush, among other miseries. But even though these maladies can be serious, in reality no more than approximately 300 species of fungi—out of some hundreds of thousands—regularly cause illness in humans. Compared with the frequency of diseases caused by bacteria, viruses, and protists, the incidence of fungal infections in humans is low.

In fact, fungi have arguably done more to promote human health than degrade it. The first antibiotic that was widely used, penicillin, was isolated from a fungus, and soil-dwelling fungi continue to be the source of many of the most important antibiotics prescribed against bacterial infections.

The major destructive impact that fungi have on people is through the food supply. Fungi known as rusts, smuts, mildews, wilts, and blights cause annual crop losses estimated in the billions of dollars. These fungi are particularly troublesome in wheat, corn, barley, and other grain crops (**Figure 29.1a**). Saprophytic fungi are also responsible for enormous losses due to spoilage—particularly for fruit and vegetable growers (**Figure 29.1b**).

(a) Parasitic fungi infect corn and other crops.

(b) Saprophytic fungi rot fruits and vegetables.

Figure 29.1 Fungi Cause Problems with Crop Production and Storage. (a) A wide variety of grain crops are parasitized by fungi. Corn smut is a serious disease in sweet corn, although in Mexico the smut fungus is eaten as a delicacy. **(b)** Fungi decompose fruits and vegetables as well as leaves and tree trunks.

However, fungi also play many important *positive* roles in the human food supply:

- Mushrooms are consumed in many cultures; in the industrialized nations they are used in sauces, salads, pizza, and many other diverse types of foods.
- The yeast *Saccharomyces cerevisiae* was domesticated thousands of years ago; today it and other fungi are essential to the manufacture of bread, soy sauce, cheese, beer, wine, whiskey, and other products. In most cases, domesticated fungi are used in conditions where the cells grow via fermentation, creating ethanol and by-products such as the CO_2 that causes bread to rise and beer and champagne to fizz.
- Chocolate is made from the seeds of *Theobroma cacao*, but it is edible only after the seeds are fermented by several species of fungi.
- Enzymes derived from different types of fungi are used to improve the characteristics of foods ranging from fruit juice and candy to meat.

In ecosystems, the impact of fungi can be widespread. Epidemics caused by fungi have killed 4 billion American chestnut trees and tens of millions of American elm trees. The fungal species responsible for these epidemics were accidentally imported on species of chestnut and elm native to other regions of the world. When the fungi arrived in North America and began growing in native chestnuts and elms, the results were catastrophic. The local chestnut and elm populations had virtually no genetic resistance to the pathogens and quickly succumbed. Living chestnut root systems continue to sprout shoots, but these trees rarely live long enough to reproduce.

The chestnut and elm epidemics radically altered the composition of upland and floodplain forests in the eastern United States. Before these fungal epidemics occurred, chestnuts and elms dominated these habitats, providing abundant food and shelter for many animals, including humans. With the loss of so many trees from disease came new opportunities for other tree species to dominate the landscape. In fact, despite the widespread harmful impact fungi can have on an ecosystem, the role of fungi in the healthy functioning of terrestrial ecosystems is critical.

Let's look at two important benefits fungi provide.

Mycorrhizal Fungi Provide Nutrients for Land Plants

Fungi that live in close association with plant roots are said to be **mycorrhizal** (literally, "fungal-root"; Figure 29.2a). The fungi along with the roots they are associated with are referred to as **mycorrhizae** (pronounced *my-koh-RY-zee*). When biologists first discovered how extensive these fungi–plant associations are, they asked an obvious question: Does plant growth suffer if mycorrhizal fungi are absent?

Figure 29.2b shows a result typical of many experiments. In this case, seedlings were grown in the presence and absence of the mycorrhizal fungi normally found on their roots. The photograph documents that this species grows three to four times faster in the presence of its normal fungal associates than it does without them.

For farmers, foresters, and ranchers, the presence of normal mycorrhizal fungi can mean the difference between profit and loss. Fungi are critical to the productivity of forests, croplands, and rangelands. Many commercially available products improve crop yield by supplying plants with mycorrhizal fungi.

Figure 29.2 Plants Grow Better in the Presence of Mycorrhizal Fungi. (a) Root system of a larch tree seedling; the white threads of a mycorrhizal fungus are visible. **(b)** Typical experimental results when plants are grown with and without their normal mycorrhizal fungi.

Saprophytic Fungi Accelerate the Carbon Cycle on Land

Fungi that make their living by digesting dead plant material are called **saprophytes**. To understand how saprophytic fungi play a key role in today's terrestrial environments, recall that cells in the vascular tissues of land plants have secondary cell walls containing both lignin and cellulose (Ch. 28, Section 28.3). Wood forms when stems grow in girth by adding layers of lignin-rich vascular tissue.

When trees die, certain fungi are among the organisms that break down wood into sugars and other small organic compounds (Figure 29.3). Fungi use these molecules as food, as do many microorganisms. In addition, when fungi die or are eaten, the molecules are passed along to a wide array of other organisms.

Fungi play an important role in cycling carbon atoms through terrestrial ecosystems. Note that there are two basic components of the **carbon cycle** on land:

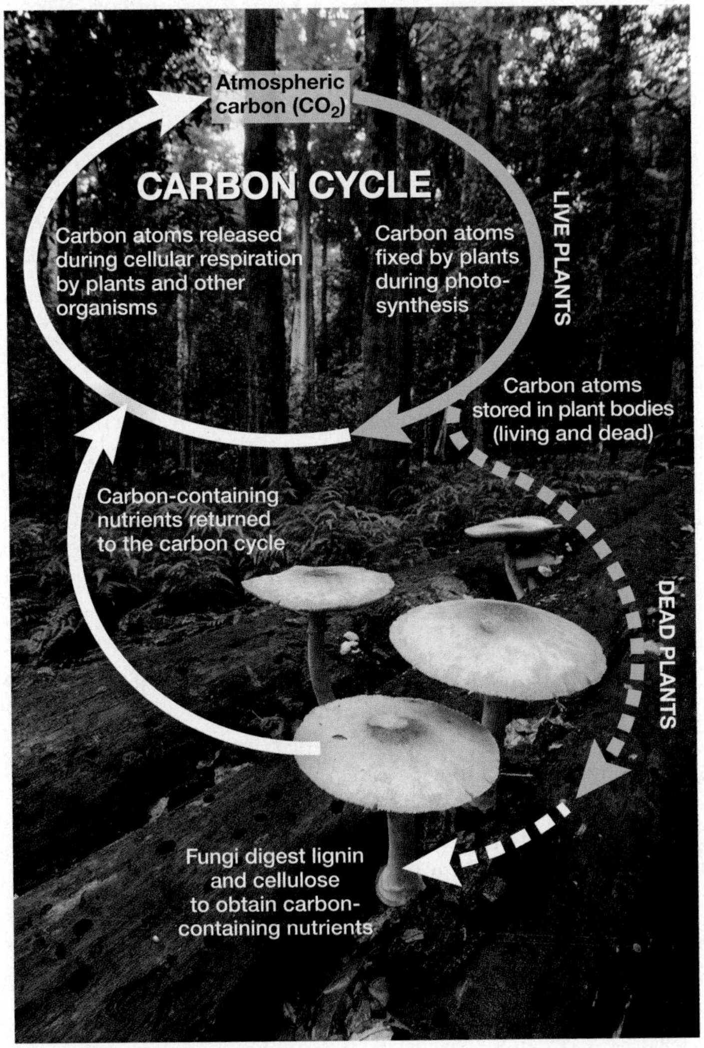

Figure 29.3 Fungi Speed Up the Carbon Cycle as They Break Down Dead Trees in Terrestrial Ecosystems. Mushrooms often grow on, digest, and absorb nutrients from dead trees and other organisms.

1. The fixation of carbon by land plants—meaning that carbon in atmospheric CO_2 is reduced to form sugar, which is then used to synthesize cellulose, lignin, and other complex organic compounds in the bodies of plants. (See Ch. 10, Section 10.4, to review carbon fixation.)
2. The release of CO_2 from nearly all organisms as the result of cellular respiration—meaning the oxidation of glucose and production of the ATP that sustains life. (See Chapter 9 to review cellular respiration.)

The fundamental point is that, for many carbon atoms, saprophytic fungi connect the two parts of the carbon cycle.

If fungi had not evolved the ability to digest lignin and cellulose soon after land plants evolved the ability to make these compounds, carbon atoms would have been sequestered in wood for millennia instead of being rapidly recycled into glucose molecules and CO_2. Terrestrial environments would be radically different than they are today, and probably much less productive. On land, fungi make the carbon cycle turn much more rapidly than it would without these organisms.

To summarize, biologists study fungi because they affect a wide range of species in nature, including humans. What tools are helping researchers understand the diversity of fungi?

CHECK YOUR UNDERSTANDING

✔ **If you understood this section, you should be able to ...**

1. **SOCIETY** Identify at least three human diseases caused by fungi.
2. Explain why it can be beneficial to inoculate seeds of crop plants with mycorrhizal fungi.
3. Explain how fungi help connect two major parts of the cycle: carbon fixation and carbon release.

Answers are available in Appendix A.

29.2 How Do Biologists Study Fungi?

After you complete this section, you should be able to ...

- Analyze key features used to study fungi.

About 110,000 species of fungi have been described and named to date, and several hundreds more are discovered each year. But the fungi are so poorly studied that the known species are widely regarded as a tiny fraction of the actual total.

Consider what researchers found when they analyzed fungi growing on Barro Colorado Island, Panama: Living on the healthy leaves of just two tropical tree species were a total of 418 distinct morphospecies of fungi. (Recall from Ch. 24, Section 24.1, that morphospecies are distinguished from each other by some aspect of morphology.) Because over 310 species of trees and shrubs grow on Barro Colorado, the data suggest that tens of thousands of fungi may be native to this island alone. If further work on fungal diversity in the tropics supports these conclusions, there may turn out to be many millions of fungal species. Indeed, a similar and more recent study on soil fungi suggests that as many as 6 million species of fungi may be found worldwide.

This viewpoint of fungal diversity was reinforced by an analysis of the fungi living in the guts of beetles. In this study, researchers isolated fungi from 27 species of beetles and characterized them by observing various morphological and physiological traits. They also used the direct sequencing approach to analyze the gene that codes for the RNA molecule in the small subunit of fungal ribosomes (Ch. 17, Section 17.3). The data showed that in the beetle guts there were over 650 fungal species, 200 of which had never been described. Biologists are only beginning to realize the extent of species diversity in fungi.

Let's consider how biologists are working to make sense of all this diversity, beginning with an overview of fungal morphology.

Analyzing Morphological Traits

Compared with animals and land plants, fungi have simple bodies. Only two growth forms occur among them: **(1)** single-celled forms called **yeasts** (Figure 29.4a), and **(2)** multicellular, filamentous structures called **mycelia** (singular: **mycelium**; Figure 29.4b). Many species of fungus grow either as yeasts or as mycelia, but some regularly adopt both growth forms.

The fossil record for land plants and animals has been extremely useful to understanding their early evolution (covered in Chapters 22 and 28). In contrast, fossils of fungi are very rare, so biologists rely on morphological and molecular data to explore their past. Because most fungi form mycelia and because this body type is so fundamental to the absorptive mode of life, most studies of fungal morphology have focused on mycelia.

The Nature of the Fungal Mycelium If food sources are plentiful, mycelia can be long lived and grow to be extremely large. Researchers in Oregon discovered a mycelium of the fungus *Armillaria* growing beneath 2100 acres (8.8 km^2) of soil—an area substantially larger than most college campuses. The biologists estimated the individual's weight at hundreds of tons and its age at thousands of years, making it one of the largest and oldest organisms known.

Although most mycelia are much smaller and shorter lived than the individual in Oregon, all mycelia are dynamic. Mycelia constantly grow in the direction of food sources and die back in areas where food is running out. The body shape of a fungus can change almost continuously throughout its life.

The Nature of Hyphae The filaments making up a mycelium are called **hyphae** (singular: **hypha**). As Figure 29.5a shows, hyphae are long and narrow.

In most terrestrial fungi, each filament is divided into cells by cross-walls called **septa** (singular: **septum**; Figure 29.5b). Septa do not close off the cells along hyphae completely. Instead, gaps called pores enable a wide variety of materials, even nuclei and other organelles, to flow from one cell-like compartment to the next.

Some fungal lineages have hyphae that are **coenocytic** ("common-celled"; pronounced *see-no-SIT-ik*)—meaning that they are not divided into separate cells; they lack septa entirely (Figure 29.5c). Coenocytic fungi have hundreds or thousands of nuclei scattered throughout the mycelium.

Fungi lack complex, long-distance transport systems like those found in plants and animals. One advantage of having pores in septa (or no septa) is that nutrients can move rapidly from regions of uptake to regions of mycelial growth. Because nutrients and some organelles can flow through the entire mycelium—at least to a degree—the fungal mycelium is intermediate between a multicellular land plant or animal and an enormous single-celled organism.

Some fungi produce hyphae with unique adaptations for prey capture and nutrient uptake. For example, some are able to capture microscopic nematodes (roundworms) by means of sticky substances on their cell walls, or in snares consisting of looped hyphae (Figure 29.6). Once a worm is captured, hyphae invade its body, digest it, and absorb the nutrients that are released.

(a) Single-celled fungi are called yeasts.

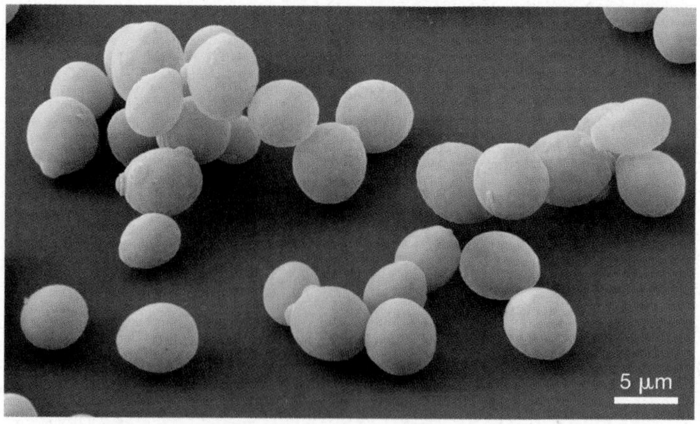

(b) Multicellular fungi have weblike bodies called mycelia.

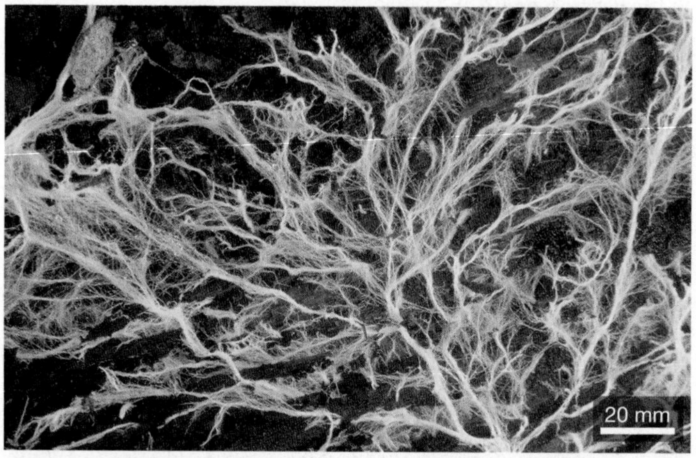

Figure 29.4 **Fungi Have Just Two Growth Forms.** Fungi grow as **(a)** single-celled yeasts and/or **(b)** multicellular mycelia made up of long, thin, highly branched filaments. The scanning electron micrograph in (a) is colorized.

✓ **THINK CAREFULLY** People sometimes mistakenly refer to yeasts as bacteria. What is a fundamental difference between yeasts and bacteria?

(a) Both the reproductive structure and mycelium are composed of hyphae.

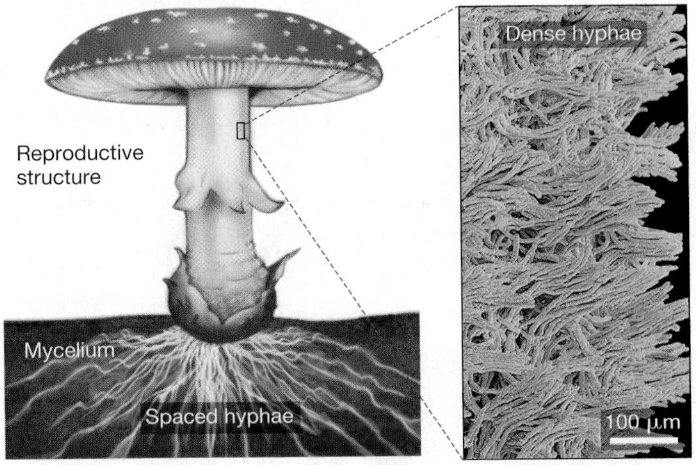

(b) Most hyphae are divided into compartments by septa.

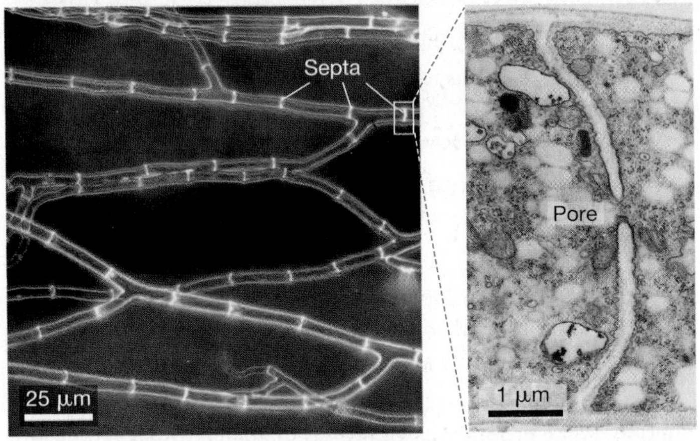

(c) Coenocytic hyphae consist of multinucleate cells.

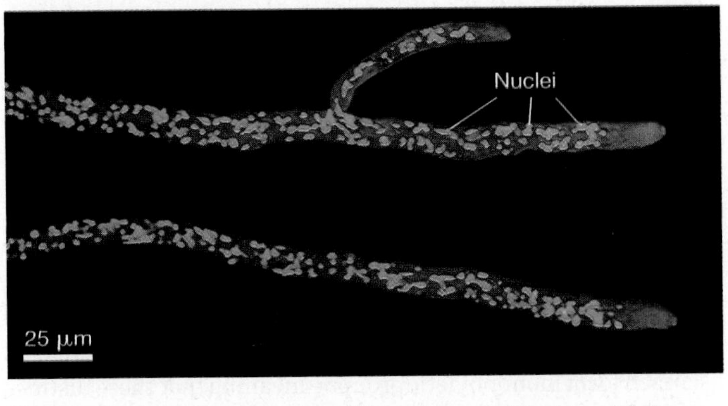

Figure 29.5 Multicellular Fungi Have Unusual Bodies. (a) The feeding portion of a fungus is a mycelium, which is made up of hyphae. In some species, hyphae are densely clustered, forming multicellular, fleshy reproductive structures such as mushrooms, brackets, or morels that emerge from the ground. **(b)** Hyphae of some terrestrial fungi are divided into cell-like compartments by partitions called septa that contain pores. As a result, the cytoplasm of different compartments is continuous. **(c)** Hyphae of some coenocytic fungi, which lack septa, are composed of giant, multinucleate cells.

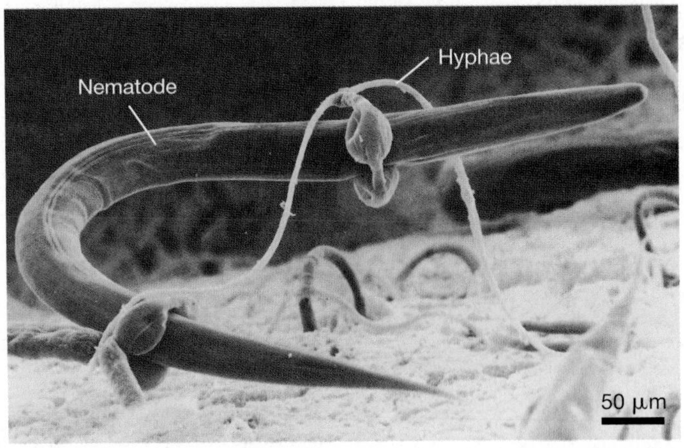

Figure 29.6 Some Fungi Are Predators. This nematode-trapping fungus (*Arthrobotrys anchonia*) forms specialized hyphae with lasso-like traps.

Mycelia Have a Large Surface Area It's important to appreciate just how thin hyphae are. Plant root tips are typically about 1 mm in diameter, but fungal hyphae are typically less than 10 μm in diameter, or 1/100th the width of a root tip. Fungal mycelia can penetrate tiny fissures in soil and absorb nutrients that are inaccessible to plant roots.

Perhaps the most important aspect of mycelia and hyphae, however, is their overall shape. Because mycelia are composed of complex, branching networks of extremely thin hyphae, fungi have the highest surface-area-to-volume ratios observed in multicellular organisms. To drive this point home, consider that the hyphae found in any fist-sized ball of rich soil typically have a surface area equivalent to half a page of this book. Thanks to their large surface area, fungi are extremely efficient at absorbing nutrients.

The extraordinarily high surface area of a mycelium has a downside, however. The amount of water that evaporates from an organism is a function of its surface area—meaning that fungi are prone to drying out. As a result, fungi are most abundant in moist habitats.

When conditions dry, the fungal mycelium may die back partially or completely. However, fungi produce reproductive cells called **spores** by sexual or asexual reproduction that are resistant to drying. As a result, spores can endure dry periods and then germinate to form a new mycelium when conditions improve. Mycelial growth is dynamic, changing with moisture availability and food supply.

Reproductive Structures Mycelia are an adaptation that supports external digestion and the absorptive lifestyle of fungi. Many fungi also produce dense, fleshy, multicellular structures—such as mushrooms and puffballs—that do not absorb food. Instead, they function in reproduction. Typically they are the only part of a fungus that is exposed to air, where drying is a potential problem. The mass of hyphal filaments on the inside of mushrooms is protected from drying by the densely packed hyphae forming the surface (see the micrograph in Figure 29.5a).

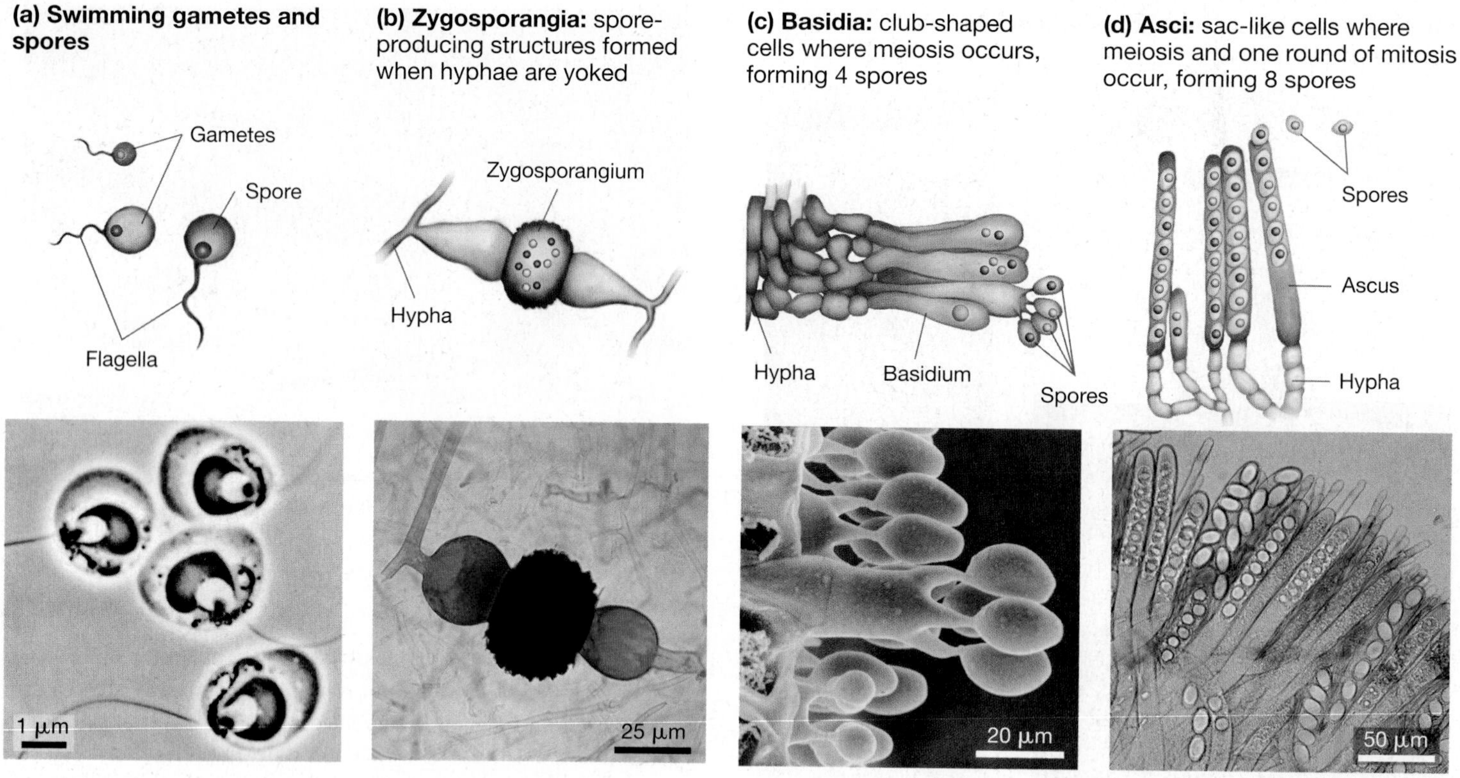

Figure 29.7 Four Types of Sexual Reproductive Structures Are Observed in Fungi. The light and dark dots in the illustrations represent nuclei.

In many fungi, including some entire lineages, sexual reproduction has never been observed. (Recall from Chapter 13 that asexual reproduction is based on mitosis, while sexual reproduction is based on meiosis.) Among those lineages of fungi that do reproduce sexually, important morphological differences are seen. Most fungal species that undergo sexual reproduction produce one of four types of distinctive reproductive structures.

1. ***Swimming gametes and spores*** In certain species that live primarily in water or wet soils, the gametes produced during sexual reproduction have flagella, as do the spores produced during asexual reproduction (Figure 29.7a). These are the only motile cells known in fungi. Species with swimming gametes are traditionally known as chytrids (pronounced *KIE-tridz*).
2. ***Zygosporangia*** In some species, haploid hyphae from two individuals meet and become joined, like oxen with a yoke, as shown in Figure 29.7b. Cells from yoked hyphae fuse to form a distinctive spore-producing structure called a **zygosporangium** (plural: **zygosporangia**; the Greek root *zygo* means "to be yoked"; BioSkills 15). Species with a zygosporangium are traditionally known as zygomycetes.
3. ***Basidia*** Mushrooms, brackets, and puffballs form specialized club-shaped cells at the ends of hyphae called **basidia** (singular: **basidium**; "little-club"). Each basidium produces four spores by meiosis (Figure 29.7c). Species with basidia are traditionally called basidiomycetes, or "club fungi."
4. ***Asci*** Cups, morels, and some other types of fungi form specialized sac-like cells called **asci** (singular: **ascus**) at the tips of hyphae. Within each ascus, meiosis and one round of mitosis produces eight spores (Figure 29.7d). Species with asci are traditionally known as ascomycetes, or "sac fungi."

Many fungi use asexual reproduction as well as sexual reproduction. Some produce asexual spores called **conidia** by the trillions (Figure 29.8). Conidia can be dispersed by water or wind currents and grow into new hyphae when conditions are right. Mold, for example, spreads rapidly across decaying food and other items by producing lots and lots of asexual spores.

In sum, morphological studies allowed biologists to describe and interpret the mycelial growth habit as an adaptation that makes external digestion and nutrient absorption extremely efficient. Careful analyses of morphological features also allowed researchers to identify four major types of sexual reproductive structures. We return to these structures in Section 29.3, where we discuss their function as part of fungi life cycles.

Now the question is, within the Fungi, do species that produce either swimming gametes and spores, zygosporangia, basidia, or asci each form monophyletic groups—meaning that these distinctive reproductive structures evolved just once? Also, which eukaryotes are most closely related to fungi? These questions and more are explored in the next section.

Evaluating Molecular Phylogenies

Researchers have sequenced and analyzed an array of genes and genomes to establish where fungi fit on the tree of life. In fact, the first complete genome sequence of a eukaryotic species was that

Figure 29.8 **Asexual Spores.** *Penicillium roquefortii* is the dark-colored fungus used in the production of blue cheese. A close-up of the fungus reveals asexual spores (conidia).

✔ Are conidia produced from a single fungus genetically identical or different? Explain.

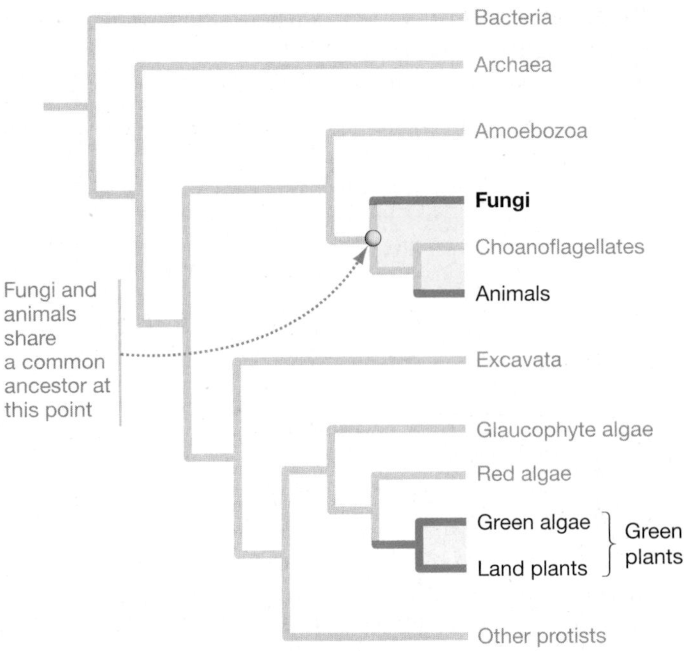

Figure 29.9 **Fungi Are More Closely Related to Animals than to Land Plants.** Phylogenetic tree showing the evolutionary relationship of the Fungi to other major lineages. (Recall from Chapter 27 that choanoflagellates are solitary or colonial protists found in fresh water.)

of a fungus, the model organism baker's yeast, *Saccharomyces cerevisiae* (**BioSkills 11**). The phylogenetic position of the Fungi is now well established; the position of lineages within the Fungi is still the subject of intense research.

Fungi Are Closely Related to Animals Figure 29.9 shows that fungi are more closely related to animals than they are to land plants. The close evolutionary relationship between fungi and animals explains why fungal infections in humans are much more difficult to treat than bacterial infections. Fungi and humans shared a more recent common ancestor than bacteria and humans. As a result, fungal and human enzymes and cell components are relatively similar in structure and function. Drugs that disrupt fungal enzymes and cells are also likely to damage humans.

In addition to DNA sequence data, three key morphological traits link animals and fungi:

1. Many animals and most fungi synthesize the tough structural material called chitin (Ch. 5, Section 5.2). Chitin is a prominent component of the cell walls of fungi.
2. The flagella that develop in chytrid spores and gametes are similar to those observed in animals. As in animals, the flagella in chytrids are single, are located at the back of reproductive cells, and move in a whiplike manner.
3. Both animals and fungi store food by synthesizing the polysaccharide glycogen. (Green plants, in contrast, synthesize starch as their storage product.)

What Are the Relationships among the Major Fungal Groups?

To understand the relationships among species with swimming gametes and spores, zygosporangia, basidia, or asci, biologists have sequenced a series of genes from an array of fungal species and used the data to estimate the phylogeny of the group. The results, shown in Figure 29.10 on page 614, support several important conclusions:

- The single-celled, parasitic eukaryotes called microsporidians are fungi.
  Interpretation: They are not a distantly related sister group to fungi, as initially thought. This point is important. Researchers are now testing the hypothesis that **fungicides**—substances that can kill fungi or slow their growth—can cure microsporidian infections in bee colonies, silkworm colonies, and AIDS patients.
- The chytrids and zygomycetes are poorly resolved. The order of branching events among lineages with these reproductive structures is still not known (in Figure 29.10, they are collapsed into a **polytomy**; see Ch. 25, Section 25.1).
  Interpretation: Either swimming gametes and the zygosporangium evolved more than once, or both structures were present in a common ancestor but then were lost in certain lineages.
- The group called the Glomeromycota is monophyletic.
  Interpretation: The adaptations that allow glomeromycete species to live in association with plant roots as mycorrhizae (discussed in Section 29.3) evolved once.

- The basidiomycetes are monophyletic—they form a lineage called Basidiomycota, or club fungi.
  Interpretation: The basidium evolved once.
- The ascomycetes are monophyletic—they form a lineage called Ascomycota, or sac fungi.
  Interpretation: The ascus evolved once.
- Together, the Basidiomycota and Ascomycota form a monophyletic group.
  Interpretation: Because basidiomycetes and ascomycetes both form septate hyphae and large "fruiting" structures, this growth habit evolved once.
- The sister group to fungi is made up of animals plus protists called choanoflagellates.
  Interpretation: Because choanoflagellates and the most ancient groups of animals are aquatic, and because chytrids are aquatic, it is reasonable to hypothesize that the earliest fungi were aquatic and that the switch to terrestrial life occurred early in the evolution of the Fungi.

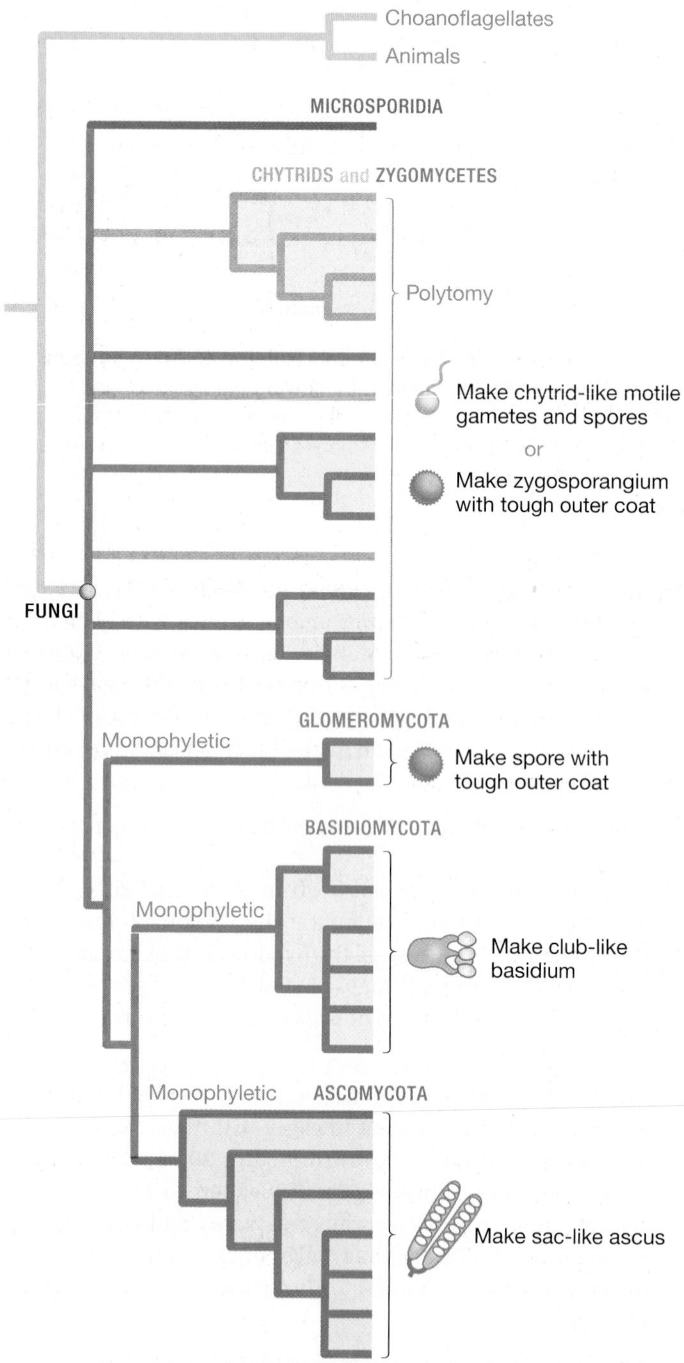

Figure 29.10 Phylogeny of the Fungi. A phylogenetic tree based on analyses of DNA sequence data. The icons represent the types of sexual reproductive structures observed in each major lineage.

Although recent advances have been made in understanding the evolutionary history of fungi, the phylogenetic tree in Figure 29.10 is still a work in progress. For example, it is not yet clear where microsporidians should be placed relative to several lineages of chytrids and zygomycetes. (Microsporidians lack both swimming gametes and zygosporangia.) Future work should clarify exactly how fungi diversified.

CHECK YOUR UNDERSTANDING

✔ If you understood this section, you should be able to . . .

1. Identify the four types of sexual reproductive structures observed in fungi.
2. Explain how mold on a single piece of fruit can easily spread to nearby fruit.
3. Evaluate whether morels (ascomycetes) or bread molds (zygomycetes) are more closely related to mushrooms (basidiomycetes).

Answers are available in Appendix A.

29.3 What Themes Occur in the Diversification of Fungi?

After you complete this section, you should be able to . . .

- Summarize the types of symbiotic relationships that fungi can have with other organisms.
- Compare and contrast the life cycles and modes of nutrition among fungi, animals, and plants.

Why are there so many different species of fungi? This question is particularly puzzling given that fungi share a common attribute: They all make their living by absorbing food directly from their surroundings. In contrast to the diversity of food-obtaining strategies observed in bacteria, archaea, and protists, all fungi make their living in the same basic way. In this respect, fungi are like plants—virtually all of which make their own food via photosynthesis.

The diversification of land plants was driven not by novel ways of obtaining food, but by adaptations that allowed plants to grow and reproduce in a diverse array of terrestrial habitats (Ch. 28, Section 28.3). Recall that associations with animal pollinators were especially important in the diversification of flowering plants. What drove the diversification of fungi? The answer is the evolution of novel methods for absorbing nutrients from a diverse array of food sources.

In this section you'll look at a few of the ways that fungi go about absorbing nutrients from different food sources, as well as how they produce offspring. Let's explore the diverse ways that fungi do what they do.

Fungi Often Participate in Symbioses

The first land plants in the fossil record are closely associated with fungal fossils; the ability to absorb nutrients from fungi may have been crucial in the early evolution of land plants. Close associations between land plants and fungi continue today. Researchers estimate that over 90 percent of land plants live in physical contact with fungi. Stated another way, fungi and land plants often have a **symbiotic** ("together-living") relationship.

Although some species of fungi live in association with an array of different land plant species, some fungi–plant associations are very specific. It is common for one fungal species to live in only a particular type of tissue, in one plant species.

Scientists categorize these symbiotic relationships as **mutualistic** if they benefit both species, **parasitic** if one species benefits at the other's expense, or **commensal** if one species benefits while the other is unaffected. However, it is sometimes difficult to know if a species really benefits from the relationship.

To understand the nature of the association, biologists have turned to experimental approaches. In these experiments, treatments lacking fungal symbionts are created by sterilizing soils with heat or by treating soils and seeds with fungicides. Presence–absence experiments have generally shown that plants grow much larger with their normal symbiotic fungi than they do without (see Figure 29.2b). Similarly, fungi that are typically symbiotic are usually unable to grow and reproduce if their regular host plant is absent.

To explore the nature of fungi–plant symbioses in detail, researchers have used isotopes as tracers for specific elements. For example, to test the hypothesis that fungi obtain food in the form of carbon-containing compounds from their plant associates, biologists introduced radioactively labeled carbon dioxide into the air surrounding plants that did or did not contain symbiotic fungi (see **Figure 29.11** on page 616). The plants incorporated the labeled CO_2 molecules into the sugars they produced during photosynthesis. The researchers then followed the location of the radioactive atoms over time by using a device that detects radioactivity. If plants feed their fungal symbionts, then labeled carbon compounds should be transferred from the plant to the fungi.

To further test the hypothesis that plants are receiving nutrients from their symbiotic fungi, other researchers added radioactive phosphorus atoms or the heavy isotope of nitrogen (^{15}N) to potted plants that did or did not contain symbiotic fungi (see Figure 29.11). If fungi facilitate the transfer of nutrients from soil to plants, then plants grown in the presence of their symbiotic fungi should receive much more of the radioactive phosphorus or heavy nitrogen than do plants grown in the absence of fungi.

Other experiments with isotopes used as tracers have shown that sugars and other carbon-containing compounds produced by plants via photosynthesis are transferred to their fungal symbionts. In some cases, as much as 20 percent of the sugars produced by a plant end up in their symbiotic fungi. In addition, the symbiotic fungi facilitate the transfer of phosphorus or nitrogen—or both—from soil to the plant.

Phosphorus and nitrogen are in short supply in many environments. Thus, the role of symbiotic fungi in transferring ions and other nutrients that contain these atoms has likely been critical to the evolutionary success of plants.

Not long after associations between fungi and the roots of land plants were discovered, researchers found that two types of plant–mycorrhizal interactions are particularly common, involving **ectomycorrhizal fungi (EMF)** and **arbuscular mycorrhizal fungi (AMF)**. The two major types of mycorrhizae have distinctive morphologies, as shown in **Figure 29.12** on page 617, as well as distinctive functions and geographic distributions.

Ectomycorrhizal Fungi (EMF) EMF like the one shown in Figure 29.12a are found on many tree species in the temperate regions of the world, where warm summers alternate with cold winters (Ch. 49, Section 49.4). In this type of association, hyphae form a dense network that covers a plant's root tips. Individual hyphae penetrate between cells in the outer layer of the root, but hyphae do not enter the root cells. The Greek root *ecto*, which means "outer," describes this association accurately: The fungi form an outer sheath on root tips that is often 0.1 mm thick. Hyphae also extend out from the sheath-like portion of the mycelium into the soil.

How do these trees and fungi interact, once they start living together? In the habitats where EMF are abundant, nitrogen atoms—for example, in the form of amino acids—tend to remain in dead tissues instead of dispersing into the soil. The hyphae of EMF penetrate decaying material and release enzymes called peptidases that cleave the peptide bonds between amino acids in dead tissues. The amino acids released by this reaction are absorbed by the hyphae and transported to spaces between the root cells of trees, where they can be absorbed by the plant.

EMF also can acquire phosphate ions that are bound to soil particles and transfer the ions to host plants. The fungi receive sugars and other complex carbon compounds from the tree.

Researchers have found that when birch tree seedlings are grown with and without their normal EMF in pots filled with forest soil, only the seedlings with EMF are able to acquire significant quantities of nitrogen and phosphorus. Inspired by such data, one biologist has referred to ectomycorrhizae as the "dominant nutrient-gathering organs in most temperate forest ecosystems."

The hyphae of EMF are like an army of miners that discover, excavate, and deliver precious nuggets of nitrogen and phosphorus to trees. The productivity of the world's most important commercial forests depends on EMF.

RESEARCH

QUESTION: Is the relationship between mycorrhizal fungi and plants mutualistic?

HYPOTHESIS: Host plants provide mycorrhizal fungi with sugars and other photosynthetic products. Mycorrhizal fungi provide host plants with phosphorus and/or nitrogen from the soil.

NULL HYPOTHESIS: No exchange of food or nutrients occurs between plants and mycorrhizal fungi. The relationship is not mutualistic.

EXPERIMENTAL SETUP:

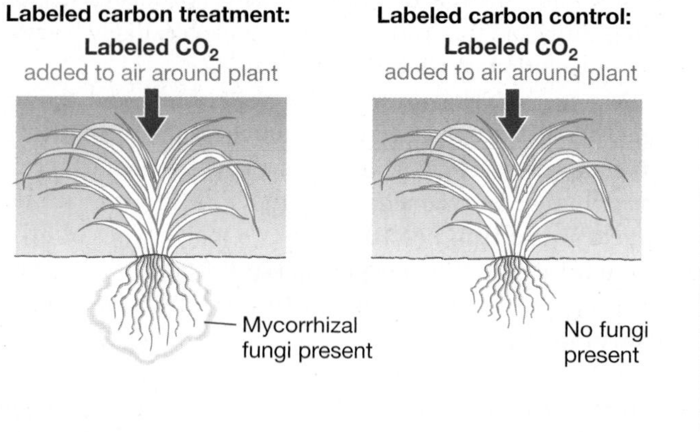

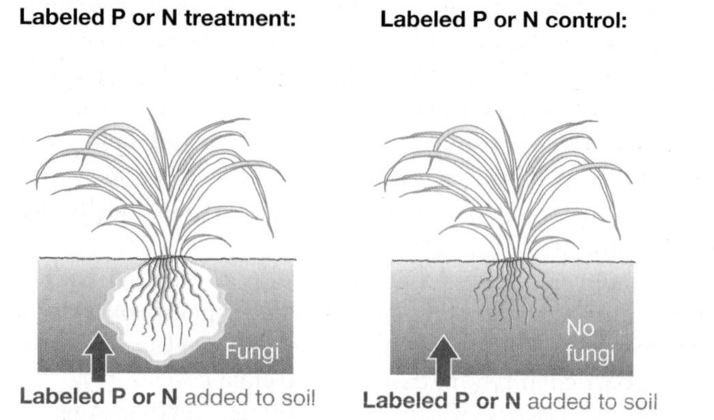

PREDICTION OF HYPOTHESIS, LABELED CARBON: A large percentage of the labeled carbon taken up by the plant will be transferred to mycorrhizal fungi. In the control, little labeled carbon will be present in the soil surrounding the roots.

PREDICTION OF NULL HYPOTHESIS, LABELED CARBON: There will be no difference in the localization of carbon in the two treatments.

PREDICTION OF HYPOTHESIS, LABELED P OR N: A large percentage of the labeled P or N taken up by the fungi will be transferred to the plant. In the control, little or no labeled P or N will be taken up by the plant.

PREDICTION OF NULL HYPOTHESIS, LABELED P OR N: There will be no difference between amounts of labeled P or N found in the plant in the presence or absence of fungi.

RESULTS:

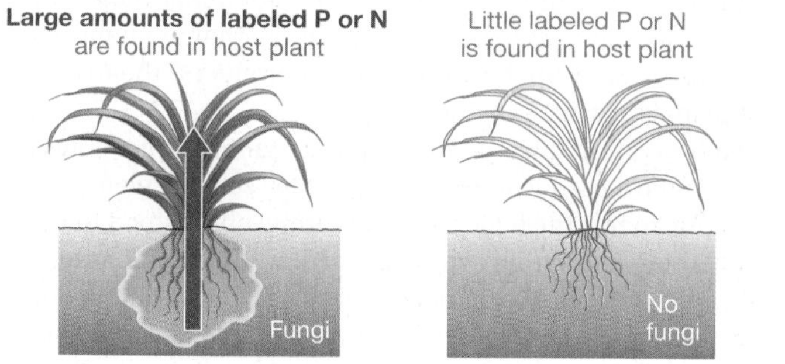

CONCLUSION: The relationship between plants and mycorrhizal fungi is mutualistic. Plants provide mycorrhizal fungi with carbohydrates. Mycorrhizal fungi supply host plants with nutrients.

Figure 29.11 Experimental Evidence that Mycorrhizal Fungi and Plants Are Mutualistic. Sugars flow from plants to mycorrhizal fungi; key nutrients flow from mycorrhizal fungi to plants.

SOURCES: R. D. Finlay and B. Söderström. 1992. Mycorrhiza and carbon flow to the soil. In *Mycorrhizal Functioning. An Integrative Plant-Fungal Process.* Ed. M. J. Allen. New York: Chapman and Hall, 134–160. H. Bücking and W. Heyser. 2001. Microautoradiographic localization of phosphate and carbohydrates in mycorrhizal roots of *Populus tremula* × *Populus alba* and the implications for transfer processes in ectomycorrhizal associations. *Tree Physiology* 21: 101–107.

✔ **PROCESS OF SCIENCE** What do these "labeled nutrient" experiments tell you that you didn't already know from experiments like that shown in Figure 29.2, where plants are grown with and without mycorrhizae?

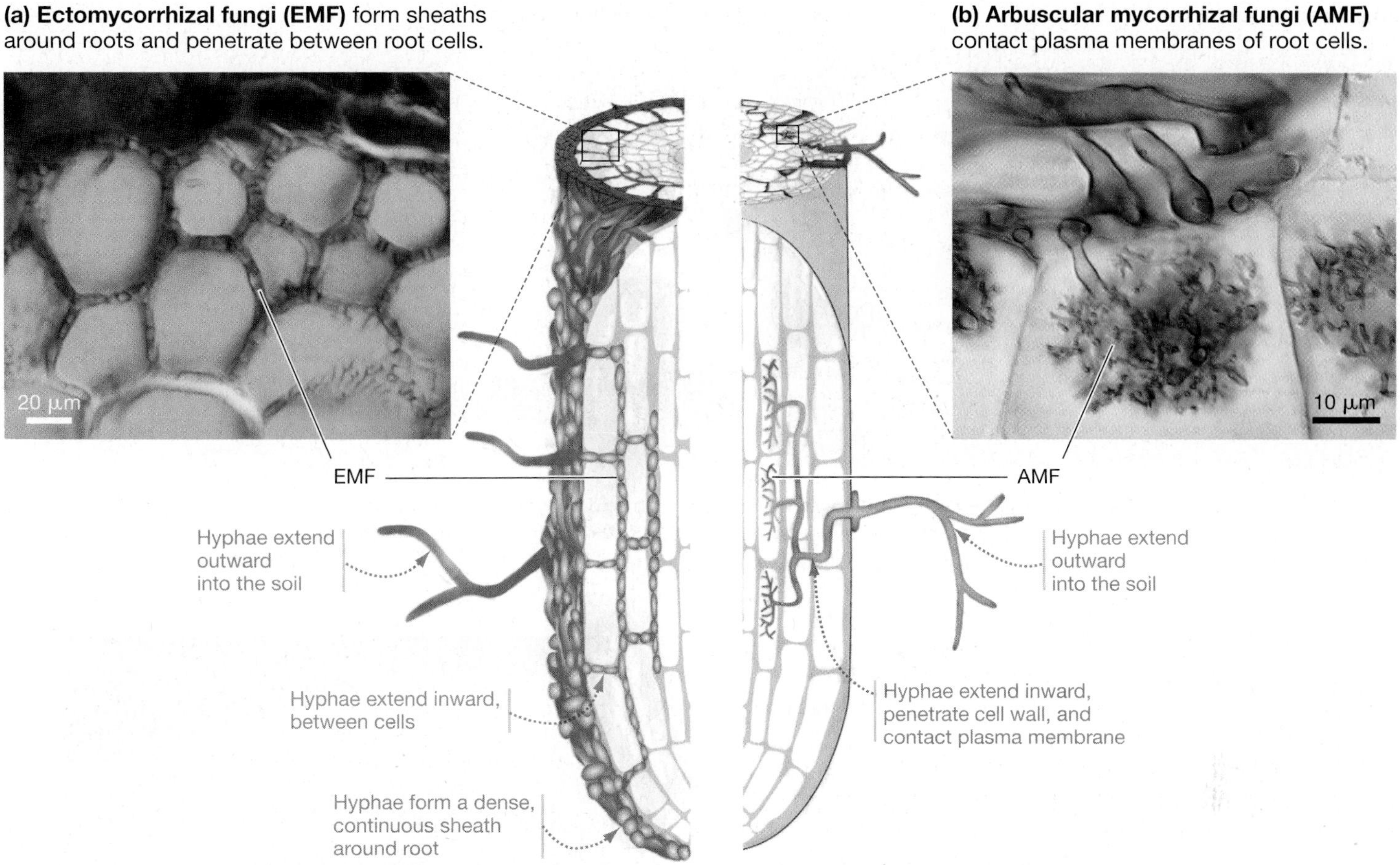

Figure 29.12 **Mutualistic Fungi Interact with the Roots of Plants in Two Distinct Ways.** **(a)** Ectomycorrhizal fungi (EMF) form a dense network around the roots of plants. Their hyphae penetrate the intercellular spaces of the root but do not enter the root cells. **(b)** The hyphae of arbuscular mycorrhizal fungi (AMF) penetrate the walls of root cells, where they branch into bushy structures or balloon-like vesicles that are in close contact with the root cell's plasma membrane.

Arbuscular Mycorrhizal Fungi (AMF) In contrast to the hyphae of EMF, the hyphae of arbuscular mycorrhizal fungi (AMF) grow *into* the cells of root tissue without penetrating cell membranes. The name *arbuscular* ("little-tree") was inspired by the bushy, highly branched hyphae, shown in Figure 29.12b, that form between the cell walls and the plasma membrane of root cells. AMF are also called **endomycorrhizal fungi**, because they grow inside root cell walls.

The key point is that the hyphae of AMF penetrate the cell wall and make direct contact with the plasma membrane of root cells. The highly branched hyphae inside the plant cell wall of AMF are thought to be an adaptation that increases the surface area available for exchange of molecules between the fungus and its host. However, AMF do not form a tight sheath around roots, as do EMF. Instead, they form a pipeline extending from inside plant roots into the soil well beyond the root.

AMF are found in a whopping 80 percent of all land plant species. They are particularly common in grasslands and in tropical forests. They are also widespread in temperate climates.

What do AMF do? Plant tissues decompose quickly in the grasslands and tropical forests where AMF flourish, because the growing season is long and warm. As a result, nitrogen is often readily available to plants. Phosphorus is usually in short supply, though, because it tends to leach out of soils that experience high rainfall.

Endophytes Mycorrhizae aren't the only type of symbiotic fungi found in plants. Researchers have also become interested in fungi that are **endophytes** ("inside-plants")—organisms that live between and within plant cells. Endophytic fungi live in close association with the roots or the aboveground tissues (leaves and stems) of land plants.

Although endophytic fungi were unknown before the 1940s, they are turning out to be both extremely common and highly diverse.

- Biologists in Brazil are examining tree leaves for the presence of fungi; each time they do, they are discovering several new species of endophytes.
- Recall from Section 29.2 that a study in Panama found hundreds of fungal species living in the leaves of just two tree species. These fungi are all endophytes.

Recent research has shown that some endophytes increase the drought tolerance of their host plants. Other endophytes, found

Figure 29.13 Lichens Are Symbiotic Associations between an Ascomycete and a Cyanobacterium or Green Alga. (a) Lichens often colonize surfaces, such as tree bark or bare rock, where other organisms are rare. **(b)** In a lichen, cyanobacteria or green algae live within a dense network of fungal hyphae.

in some grasses and morning glories, also produce compounds that benefit plants by deterring or even killing herbivores. Endophytes have been shown to benefit from these symbioses, too, by absorbing sugars from the plants.

Based on these results, biologists have concluded that the relationship between endophytes and some plants is mutualistic. In other types of plants, however, researchers have not been able to document benefits for the plant host. The current consensus is that at least some endophytic fungi may be commensals—meaning the fungi and the plants simply coexist with no observable effect, either bad or good, on the host plant.

The take-home message about mycorrhizae and other endophytes is that most plants are covered with fungi, inside and out, from the ends of their branches to the tips of their roots. Throughout their lives, many or even most plants are involved in several distinct types of commensal or mutualistic relationships with fungi.

Other Symbioses Fungi also take up residence with species other than land plants. The many unique properties of fungi have made them interesting partners in a wide array of symbiotic relationships.

- Many plant-eating insects harbor diverse arrays of yeast species in their guts, which may aid digestion or detoxification of plant compounds.
- **Lichens** are a mutualistic partnership usually between a species of ascomycete and either a cyanobacterium or an alga (Figure 29.13).
- Some fungi that parasitize insects dramatically change their hosts' behavior. For example, the fungus *Ophiocordyceps* causes ants to follow a zombie-like sequence of motions: An infected ant climbs to the top of a plant, bites down hard on the stem or leaf, and eventually dies. The fungus then produces a reproductive structure that sprouts from the head of the "zombie ant" (Figure 29.14). From the high perch atop the plant, the fungal spores are readily dispersed. Behavioral control by "zombie ant" fungi remains an active area of research. Recent studies indicate that *Ophiocordyceps* releases a brain-controlling chemical cocktail that includes compounds such as guanidinobutyric acid and sphingosine—both of which appear to play a role in certain neurological disorders in humans.

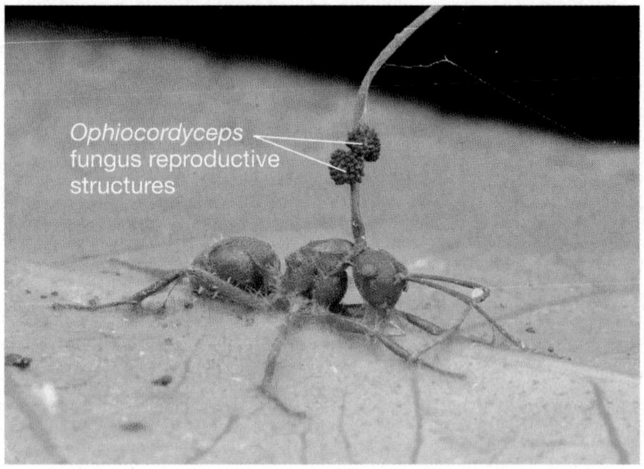

Figure 29.14 "Zombie Ant" Fungus. After being infected by the *Ophiocordyceps* fungus, an ant will climb a plant, bite down hard, and soon die. The fungus sprouts from the ant's head and releases spores.

What Adaptations Make Fungi Such Effective Decomposers?

The saprophytic fungi are master recyclers. Although bacteria and archaea are also important decomposers in terrestrial environments, some fungi and a few bacterial species are the only organisms that can digest wood completely. Given enough time, fungi can turn even the hardest, most massive trees into soft soils.

How do fungi do it? You've already been introduced to two key adaptations: **(1)** The large surface area of a mycelium makes nutrient absorption exceptionally efficient. **(2)** Saprophytic fungi can grow toward the dead tissues that supply their food. What other adaptations help fungi decompose plant tissues?

Extracellular Digestion Large molecules such as starch, lignin, cellulose, proteins, and RNA cannot diffuse across plasma membranes. Only sugars, amino acids, and other small molecules can enter the cytoplasm of cells through membrane carrier proteins or transporters (Ch. 6, Section 6.4). As a result, fungi, like most animals, have to digest their food before they can absorb it.

Fungi perform **extracellular digestion**—digestion that takes place outside the organism. The simple compounds that result from enzymatic action are then absorbed by the hyphae. Most animals also perform extracellular digestion, but it takes place in a digestive tract (Ch. 41, Section 41.3).

The two most abundant organic molecules on Earth—lignin and cellulose—are digested by fungi.

- Lignin is an extremely strong, complex polymer built from monomers that contain six-carbon rings. Recall that most lignin is found in the secondary cell walls of plant vascular tissues, where it provides structural support (Ch. 28, Section 28.3).
- Cellulose is a polymer of glucose and is found in the primary and secondary cell walls of all plant cells.

Some basidiomycetes can degrade lignin and digest cellulose completely—to CO_2 and H_2O. Let's look at how they do it.

Lignin Degradation Biologists have been keenly interested in understanding how basidiomycetes digest lignin. Paper manufacturers are also interested in this process because they need safe, efficient ways to degrade lignin in order to make paper products soft and absorbent.

To find out how lignin-digesting fungi do it, biologists began analyzing the proteins that these species secrete into extracellular spaces. After purifying these molecules, the investigators tested each protein for the ability to degrade lignin. Using this approach, investigators from two labs independently discovered an enzyme called lignin peroxidase.

Lignin peroxidase catalyzes the removal of a single electron from an atom in the ring structures of lignin. This oxidation step creates a free radical—an atom with an unpaired electron (Ch. 2, Section 2.1). This is an extremely unstable electron configuration, and it leads to a series of uncontrolled and unpredictable reactions that split the polymer into smaller units.

The unpredictable nature of the reaction is remarkable, because virtually all of the other reactions catalyzed by enzymes are extremely predictable. The lack of predictability here makes sense, however. Unlike proteins, nucleic acids, and most other polymers with a regular and predictable structure, lignin is extremely heterogeneous. Over 10 types of covalent linkages are routinely found between the monomers that make up lignin. But once lignin peroxidase has created a free radical in the ring structure, any of these linkages can be broken.

However, fungi cannot grow with lignin as their sole source of food. The six-carbon rings in lignin are extremely difficult to metabolize, and as a result they accumulate in soil. If wood-rotting fungi don't use lignin as food, why do they produce enzymes to digest it? The answer is simple. In wood, lignin forms a dense matrix around long strands of cellulose.

Degrading the lignin matrix gives hyphae access to huge supplies of energy-rich cellulose. Similar to the mycorrhizal fungi discussed earlier, saprophytic fungi are like miners. But instead of seeking out rare, gem-like nitrogen or phosphorus atoms, the saprophytes use lignin peroxidase to blast away enormous lignin molecules, exposing rich veins of cellulose that can fuel growth and reproduction.

Cellulose Digestion Once lignin peroxidase has softened wood by stripping away its lignin matrix, the long strands of cellulose that remain can be attacked by enzymes called cellulases.

Like lignin peroxidase, cellulases are secreted into extracellular spaces by fungi. But unlike the uncontrolled chain reaction catalyzed by lignin peroxidase, degradation of cellulose by cellulases is extremely predictable.

Biologists have purified seven different cellulases from the fungus *Trichoderma reesei*. Two of these enzymes catalyze a critical early step in digestion—they cleave long strands of cellulose into a disaccharide called cellobiose. The other cellulases are equally specific and also catalyze hydrolysis reactions. In combination, the suite of seven enzymes in *T. reesei* transforms long strands of cellulose into a simple monomer—glucose—that the fungus can absorb and use as a source of food.

Variation in Reproduction

Recall from Section 29.2 that fungi may produce swimming gametes and spores, yoked hyphae in which nuclei from different individuals fuse to form a zygosporangium inside a protective structure, or specialized spore-producing cells called basidia and asci. As fungi diversified, besides adapting to a variety of food sources, they evolved an array of ways to reproduce.

Spores as Key Reproductive Cells The spore is the most fundamental reproductive cell in fungi. Spores are the dispersal stage in the fungal life cycle and are produced during both asexual and sexual reproduction. Fungi produce spores in such prodigious quantities that it is not unusual for fungal spores to outnumber pollen grains in air samples.

If a spore falls on a food source and is able to germinate, a mycelium begins to form. As the fungus expands, hyphae grow in the direction in which food is most abundant. But if food begins to run out, mycelia respond by making spores, which are dispersed by wind or animals.

Why would mycelia reproduce when food is low? The leading hypothesis is that spore production allows starving mycelia to disperse offspring to new habitats where more food might be available. Thus, spore production is favored by natural selection when individuals are under nutritional stress.

Multiple Mating Types In some fungal lineages, hyphae come in many different mating types. Instead of having morphologically distinct males and females that produce sperm and eggs, hyphae of different mating types appear identical.

Hyphae of the same mating type will not combine during sexual reproduction. When zygomycete hyphae grow close to each other, for example, they will not fuse unless the individuals have different alleles of one or more genes involved in mating. If chemical messengers released by two hyphae indicate that they are of different mating types, then fusion and zygosporangium formation follows.

In this way, mating types function as sexes. Instead of having just two sexes, a single fungal species may have tens of thousands of sexes. The basidiomycete *Schizophyllum commune*, for example, is estimated to have 28,000 mating types.

Why so many? The leading hypothesis to explain the existence of mating types is that it promotes genetic diversity in offspring by increasing the probability of outcrossing. Genetic diversity, in turn, is known to be helpful in fighting off infections and responding to changes in the environment (Ch. 13, Section 13.4).

How Does Fertilization Occur? Compared with fertilization in land plants, protists, and animals, fertilization in fungi has important unique features:

- Among all the lineages of fungi, only some chytrids produce gametes, which are not considered eggs and sperm. Both male and female gametes are motile, and female gametes are only slightly larger than male gametes.
- In all other lineages of fungi, fertilization occurs in two distinct steps: **(1)** fusion of hyphae and **(2)** fusion of nuclei from the fused hyphae. These two steps can be separated by long time spans and even long distances.

In many fungi, the process of sexual reproduction begins when hyphae from two different mating types grow near each other and fuse to form a hybrid cell. When the cytoplasm of two individuals fuses in this way, **plasmogamy** (pronounced *plaz-MAH-ga-mee*) is said to occur (**Figure 29.15**). Often, plasmogamy does not immediately lead to nuclear fusion. Rather, the genetically distinct, haploid nuclei persist in the same cell or mycelium. When genetically distinct nuclei from the two different mating types exist within a single mycelium, it is considered **heterokaryotic** ("different-kernel"). Most hyphae in heterokaryotic mycelia are **dikaryotic** ("two-kernel")—their cells are divided by septa, and each cell contains two nuclei, one from each mating type.

The distinct nuclei in dikaryotic and heterokaryotic mycelia function independently, even though gene expression must be coordinated for growth and development to occur. For example, in septate dikaryotic fungi the two types of nuclei divide as the hyphae expand, so each compartment that is divided by a septum contains only one of the two types of nuclei. Biologists are currently investigating how the activities of the two nuclei are coordinated.

In a dikaryotic or heterokaryotic mycelium, pairs of unlike nuclei may eventually fuse to form diploid zygotes. The fusion of nuclei is called **karyogamy** (pronounced *ka-ree-AH-ga-mee*). The diploid nuclei that are produced by karyogamy then divide by meiosis to form haploid spores. Because of meiosis, the spores are genetically distinct and different mating types are produced.

✔ **If you understand the relationship between plasmogamy and karyogamy, you should be able to compare and contrast**

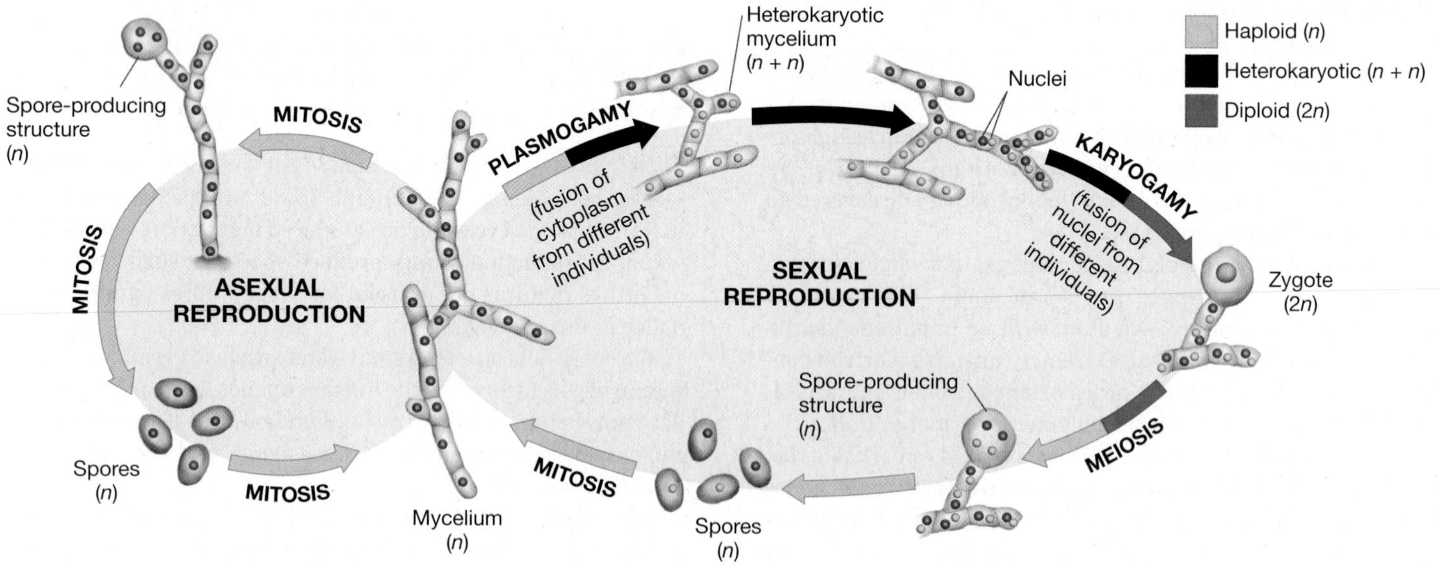

Figure 29.15 Fungi Have Unusual Life Cycles. A generalized fungal life cycle, showing both asexual and sexual reproduction.

✔ Fungi spend most of their lives feeding. Which is the longest-lived component of this life cycle?

these events with the life cycles of other eukaryotes. For example, which human cells undergo plasmogamy and karyogamy? Are any human cells similar to a heterokaryotic fungal cell? Explain.

Asexual Reproduction As the left side of Figure 29.15 indicates, most fungal species can reproduce asexually as well as sexually. In fact, large numbers of ascomycetes have never been observed to reproduce sexually. Researchers don't yet know if these fungi are unable to reproduce sexually, or if it simply hasn't happened while they have been growing under observation.

During asexual reproduction, spore-forming structures are produced by a haploid mycelium, and spores are generated by mitosis. As a result, offspring are clones—meaning that they are genetically identical to their parent.

Producing microscopic spores that are carried by wind was an adaptation to life on land that allowed terrestrial fungi to disperse widely and to colonize new habitats. The production of large numbers of asexual spores has proven to be an effective strategy for many fungi.

Four Major Types of Life Cycles

Among the sexually reproducing species of fungi, the presence of a heterokaryotic stage and the morphology of the spore-producing structure vary. Morphologies of reproductive structures were introduced in Section 29.2. Let's now take a closer look at the sexual reproduction portion of each of the four major types of life cycles that have been observed in fungi. Each life cycle is summarized in **Figure 29.16** on pages 622–623.

Chytrid Life Cycle The chytrids are the only types of fungi with species that exhibit alternation of generations. Recall that in organisms with alternation of generations, there are multicellular haploid forms in some phases of reproduction and diploid forms in others (Ch. 27, Section 27.3). Figure 29.16a shows how alternation of generations occurs in the well-studied genus *Allomyces*. As you read through the stages of one full life cycle, pay attention to changes in ploidy:

- Haploid adults form structures called gametangia, in which motile male and female haploid gametes are produced by mitosis.
- Gametes from the same individual or different individuals fuse to form a diploid zygote.
- The zygote grows by mitosis into a diploid sporophytic mycelium.
- Meiosis occurs inside the sporophytic mycelium, in structures called sporangia.
- The haploid spores produced by meiosis disperse by swimming; then each grows by mitosis into a haploid gametophytic mycelium.

Zygomycete Life Cycle In zygomycetes, sexual reproduction starts when hyphae from different mating types fuse, as shown in Figure 29.16b. Plasmogamy forms a spore-forming structure called a zygosporangium that develops a tough, resistant coat. Inside the zygosporangium, nuclei from the mating partners fuse—meaning that karyogamy occurs.

The zygosporangium can persist if conditions become too cold or dry to support growth. When temperature and moisture conditions are again favorable, however, meiosis occurs and the zygosporangium germinates and forms a sporangium. The meiotic products (haploid spores) within the sporangium are released and germinate, growing into new mycelia.

Zygomycete mycelia frequently also reproduce asexually by making sporangia whose haploid spores (conidia) are produced by mitosis and dispersed by wind.

Basidiomycete Life Cycle Mushrooms, bracket fungi, boletes, stinkhorns, and puffballs are sexual reproductive structures produced by members of the Basidiomycota (see Figure 29.16c). Even though their size, shape, and color vary enormously from species to species, all basidiomycete reproductive structures originate from the dikaryotic hyphae of mated individuals.

Inside a mushroom or bracket fungus or puffball, the club-shaped, spore-producing cells called basidia form at the ends of dikaryotic hyphae. Karyogamy occurs within the basidia. The diploid nucleus that results undergoes meiosis, yielding four haploid spores.

Spores are eventually ejected from the end of the basidia and are dispersed by the wind. It is not unusual for a single puffball or mushroom to produce a billion spores. If they land in a suitable location, the spores germinate and grow by mitosis to form new hyphae.

Ascomycete Life Cycle As in basidiomycetes, the mature sexual reproductive structure in ascomycetes is also produced by a dikaryotic hypha.

Figure 29.16d illustrates the ascomycete life cycle. The process usually begins when hyphae or specialized structures from different mating types fuse, forming a heterokaryotic cell containing many independent nuclei. A short dikaryotic hypha with cells containing two nuclei, one from each parent, emerges and eventually grows into a complex reproductive structure whose hyphae have the sac-like, spore-producing structures called asci at their tips.

After karyogamy occurs inside each ascus, meiosis and one round of mitosis takes place, resulting in the production of eight haploid spores. When the ascus matures, the spores inside are forcibly ejected.

CHECK YOUR UNDERSTANDING

✔ **If you understood this section, you should be able to ...**

1. Describe evidence that mutualism occurs in EMF.
2. Explain the role of spores in fungal reproduction.
3. Compare and contrast what happens inside a zygosporangium, a basidium, and an ascus.

Answers are available in Appendix A.

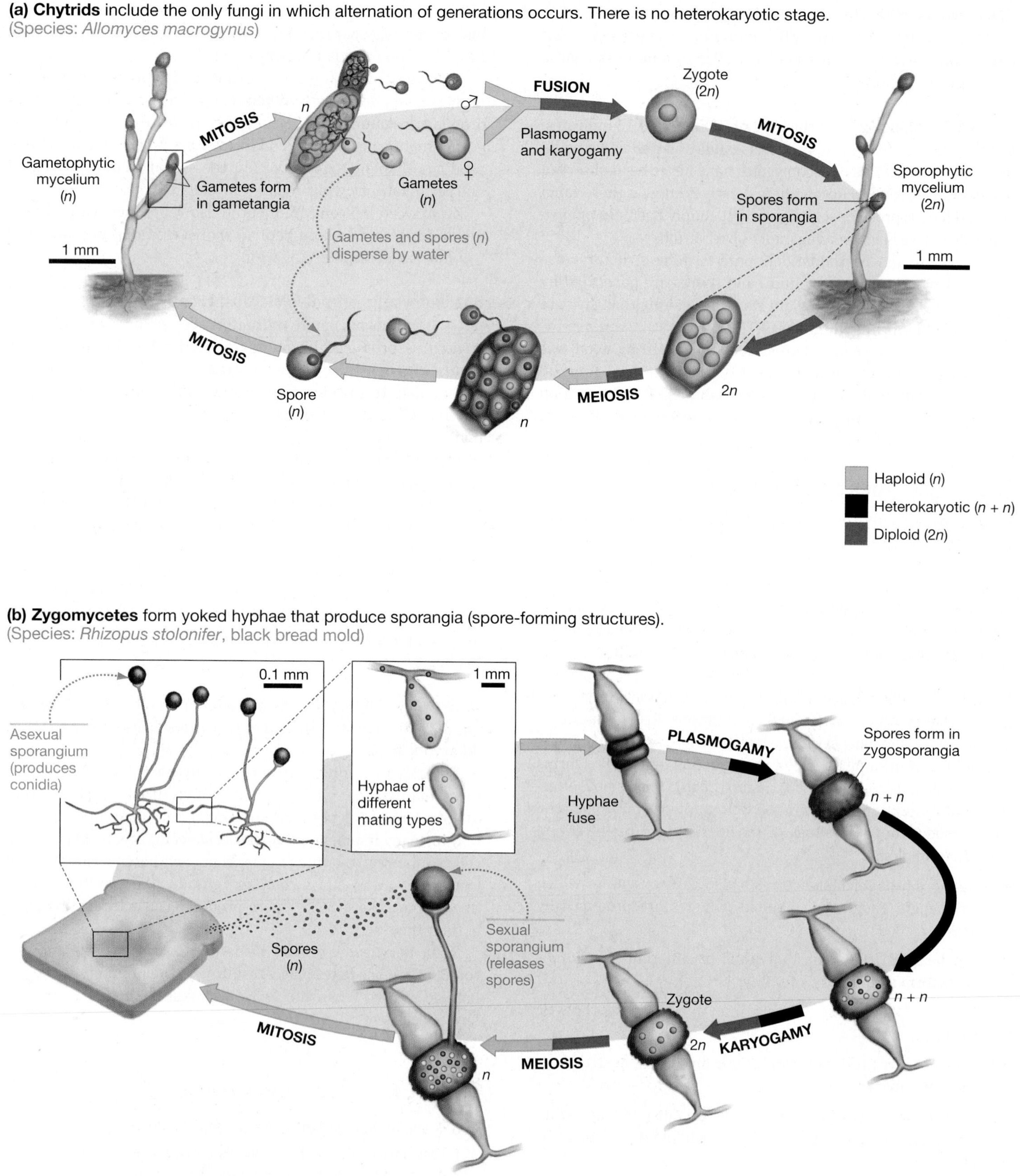

Figure 29.16 **Variation in Sexual Reproduction in Fungi.** The sexual part of the life cycle in four major groups of fungi: **(a)** chytrids, **(b)** zygomycetes, **(c)** basidiomycetes, and **(d)** ascomycetes.

(c) Basidiomycota have reproductive structures with many spore-producing basidia.
(Species: *Amanita muscaria,* fly agaric)

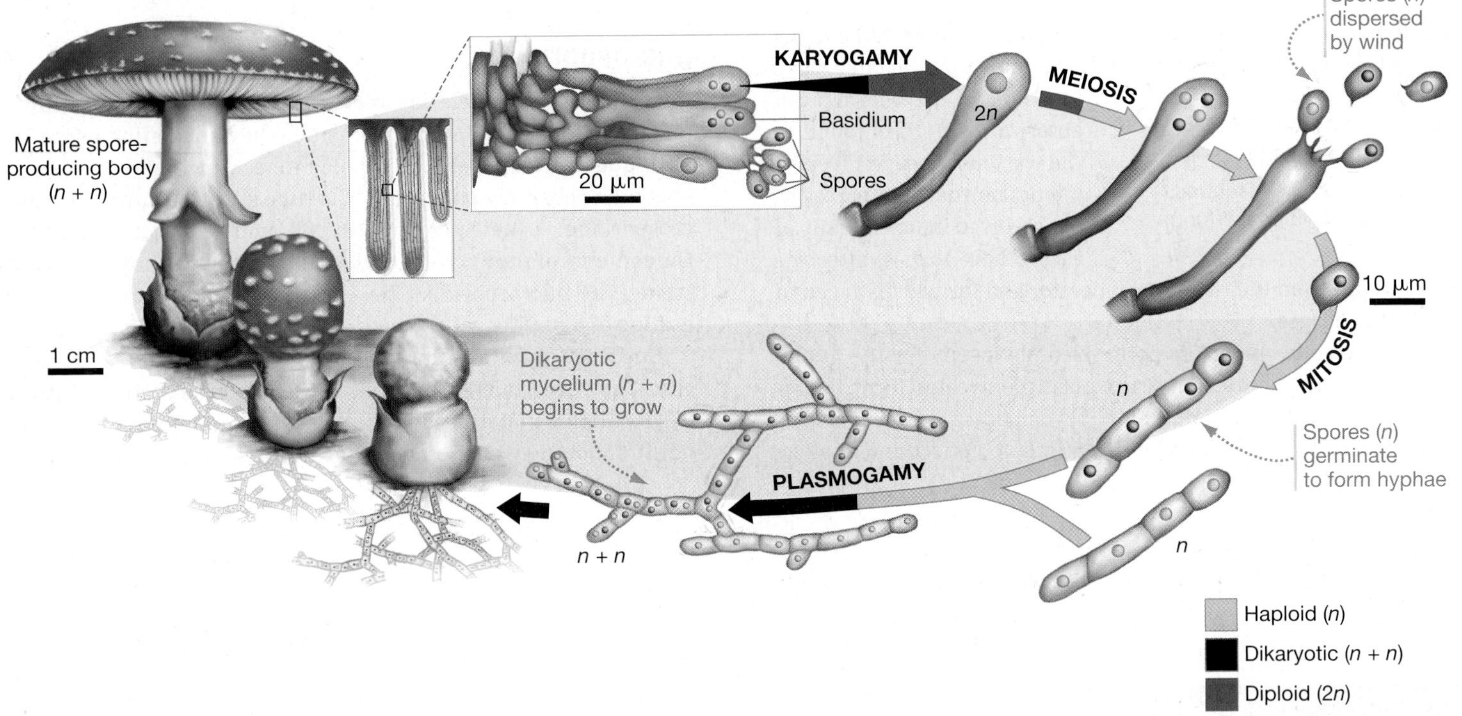

(d) Ascomycota have reproductive structures with many spore-producing asci.
(Species: *Cookeina speciosa,* Venus wine cup fungus)

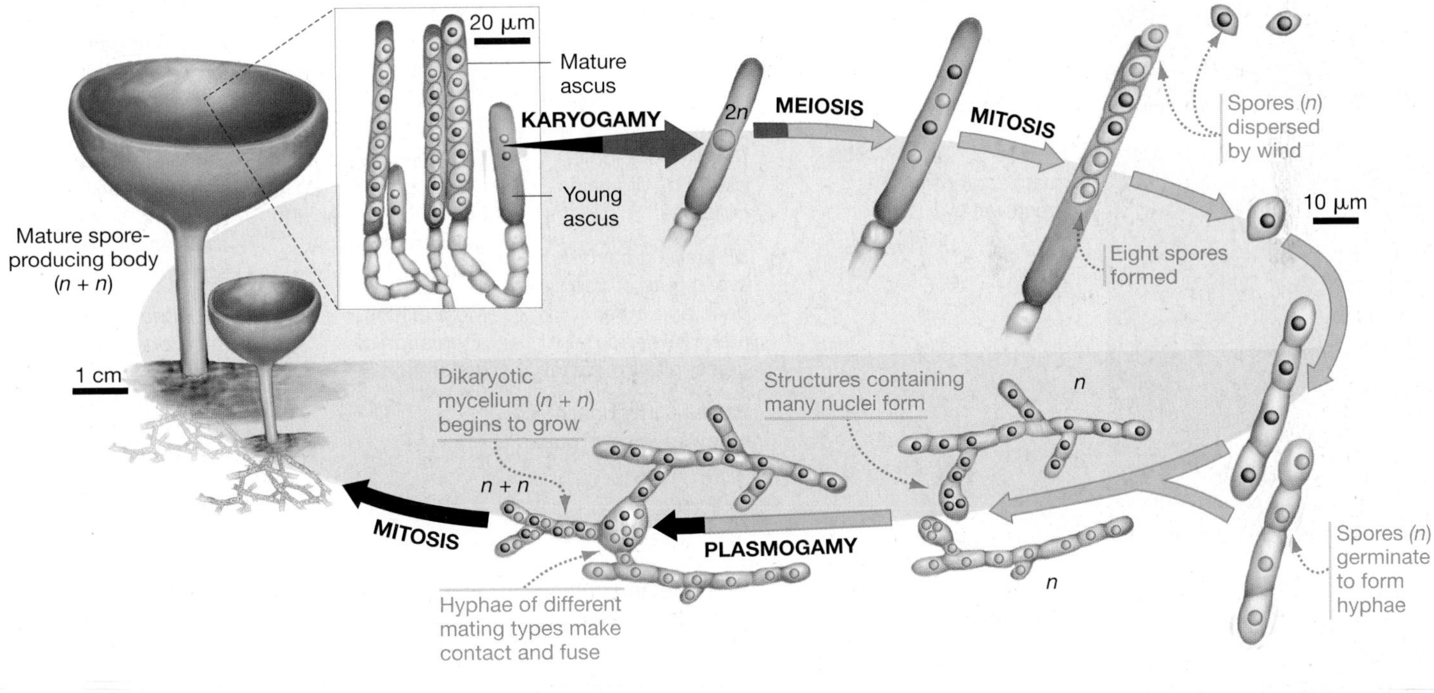

Figure 29.16 Variation in Sexual Reproduction in Fungi (continued).

29.4 Key Lineages of Fungi

The lineages described here and summarized in **Table 29.1** are just a small sampling of fungal diversity. As you study the table, keep in mind that the wide array of absorptive lifestyles found in the various lineages (saprophytic, mutualistic, parasitic, etc.) evolved independently of each other—that is, after the groups split from their common ancestor and formed distinct and separate lineages.

> **After you complete this section, you should be able to ...**
> - Describe the key features of the major lineages of fungi.

Based on the current hypothesized phylogeny for the Fungi, it is clear that neither chytrids nor zygomycetes form monophyletic groups and should not be named as single lineages (see Figure 29.10). Researchers continue to collect and analyze additional data—mostly DNA sequences—to identify monophyletic groups and explain the early evolution of fungi.

Microsporidia

More than 1200 species of microsporidia have been identified, and all are single celled and parasitic. They are distinguished by a unique structure called the polar tube, which allows them to enter the cells they parasitize. The tube shoots out from the microsporidian, penetrates the host cell, and acts as a conduit for the contents of the microsporidian cell to enter the host cell. Once inside, the microsporidian DNA replicates and produces a generation of daughter cells, which go on to infect other host cells.

Species from nine different genera can infect humans. In most cases, however, microsporidians cause serious infections only in AIDS patients and other individuals whose immune systems are not functioning well. Some microsporidians are serious pests in

DIVERSITY Table 29.1 **Key Lineages of Fungi**

		Known Species	Absorptive Life Style	Life Cycle	Relevance
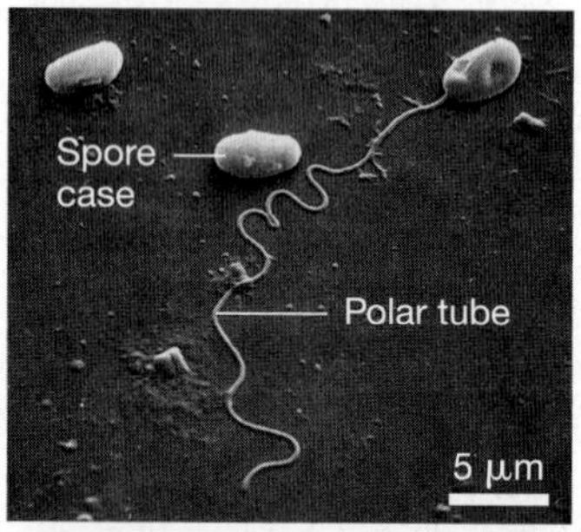	**Microsporidia** ◀ *Nosema tractabile*	1300	All are parasites of animal cells, especially in insects or fish. Enter host cell through polar tube (which microsporidian shoots into host cell).	Some appear to reproduce only asexually. Others produce various types of sexual spores. Some infect several hosts to complete life cycle.	Many species can infect humans, but cause serious infections only in patients with a compromised immune system. Some are pests in honeybee and silkworm colonies.
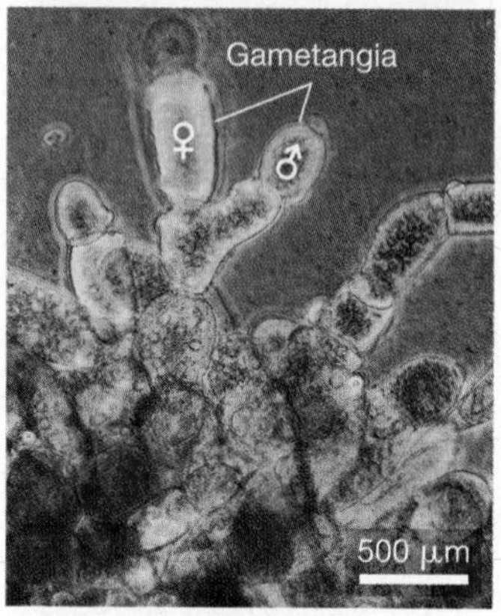	**Chytrids** Aquatic; common in fresh water ◀ *Allomyces* sp. showing male and female gametangia	750	Many decompose plants by digesting cellulose. Mutualistic chytrids live in guts of cows, deer, and other mammals and help digest plant material. Parasitic chytrids infect many species of plants and animals.	The only fungi that produce motile cells—both their spores and gametes swim via flagella. Most exhibit alternation of generations.	Some are parasites of mosquitoes and are being studied as a biological control agent. *Batrachochytrium dendrobatidis* is largely responsible for declines in amphibian populations worldwide.
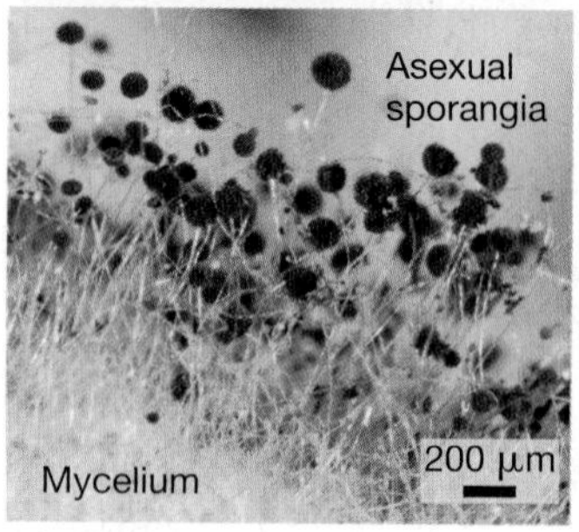	**Zygomycetes** **Food molds** ◀ *Rhizopus nigricans* (a bread mold)	1050	Many are saprophytes and live on plant debris. Some parasitize other fungi, insects, or spiders.	Asexual reproduction is very common; asexual sporangia produce spores. Sexual reproduction involves fusion of hyphae and formation of zygosporangium and sexual sporangia.	Several species often seen growing on bread and soft fruits. Some are used in the production of steroids, pigments, alcohols, and fermented foods.

honeybee and silkworm colonies. However, other species that infect grasshoppers are used as a biological control agent to reduce crop damage caused by these insects.

Chytrids

Chytrids are largely aquatic and are common in freshwater environments. Some species live in wet soils, and a few have been found in desert soils that are wet only during a rainy season. Species found in dry soils have tough spores that endure harsh conditions. Spores from chytrids have been shown to germinate after a resting period of as long as 31 years.

Many species of chytrids have enzymes that allow them to digest cellulose. As a result, these species are important decomposers of plant material in wet soils, ponds, and lakes. However, many of the freshwater species are parasitic; some of these species cause disease epidemics in algae or aquatic insects. Other species parasitize mosses, ferns, flowering plants, or animals. Chytrids are the likely culprit behind the decline of amphibian populations worldwide.

Zygomycetes

The zygomycetes ("yoked-fungi") are primarily soil-dwellers. When hyphae of opposite mating types grow near each other, they fuse during sexual reproduction and then form a thick-walled zygosporangium. Asexual reproduction is also common.

The common black bread mold *Rhizopus stolonifer* is a frequent household pest—it's probably the zygomycete that is most familiar to you. Saprophytic and parasitic members of this lineage are also responsible for rotting strawberries and other fruits and vegetables and causing large losses in these food industries. On the positive side, other species are used beneficially in commercial production of various products, including enzymes used in the production of various detergents.

DIVERSITY **Table 29.1 Key Lineages of Fungi (continued)**

		Known Species	Absorptive Life Style	Life Cycle	Relevance
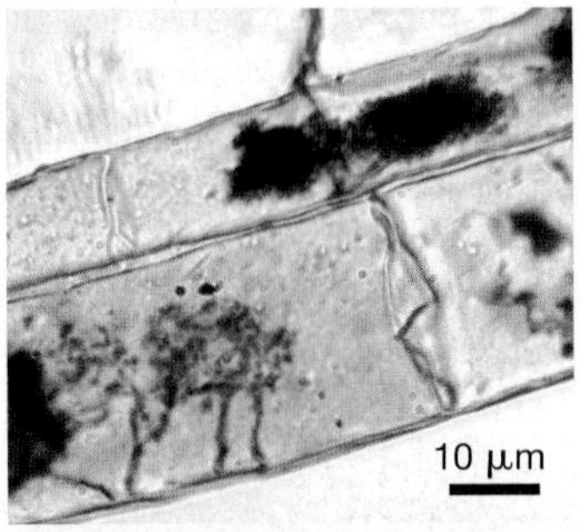 	**Glomeromycota** ◀ An arbuscular mycorrhizal fungus penetrating a root	**200**	Form mutualistic associations with plant roots in the form of arbuscular mycorrhizal fungi.	Most produce large spores. Difficult to grow and study. Sexual reproduction has not been documented.	Play large roles in the ecology of prairies and tropical forests.
 	**Basidiomycota** ("club fungi") **Mushrooms, boletes, stinkhorns, puffballs, brackets** ◀ *Lycoperdon perlatu* (a puffball) releasing spores	**32,000**	Decompose wood by producing lignin peroxidase. Some form ectomycorrhizal associations with tree roots. Smut and rust fungi are plant parasites.	Produce haploid spores in club-shaped basidia. Basidia of mushroom-forming species line gills found under the cap.	Mushrooms are used as a source of food. Some mushrooms are toxic or hallucinogenic. Ectomycorrhizal basidiomycetes enhance growth of hardwoods and softwoods. Smuts and rusts cause serious economic damage to crop plants.
 	**Ascomycota** ("sac fungi") **Morels, truffles, yeasts, cup fungi, lichen-forming fungi** ◀ *Cookeina speciosa* (a cup fungus)	**64,000**	Many form ectomycorrhizal associations with tree roots. Some form mutualistic associations with photosynthetic algae or bacteria in lichens. Some are predatory on protists or nematodes.	Produce haploid spores in sac-like asci. Asci of many species found on fleshy, cup-shaped structures. Asexual reproduction in many.	Lichens are often sensitive to air pollution and are used as bioindicators.

Glomeromycota

All members of this group (except for a single species) comprise the arbuscular mycorrhizal fungi (AMF). Recall from Section 29.3 that AMF absorb phosphorus-containing ions or molecules in the soil and transfer them, along with nitrogen and other nutrients and water, into the roots of most of the plants living in grasslands and tropical forests. The host plant provides the symbiotic fungi with sugars and other organic compounds.

Basidiomycota

Almost all basidiomycetes form mycelia and produce multicellular reproductive structures that include basidia, the club-shaped cells where meiosis and spore formation occur. About 31,000 species of basidiomycetes have already been described, and more are discovered each year.

Throughout the world, mushrooms are cultivated or collected from the wild as a source of food. The white button, crimini, and portabella mushrooms you may have seen in grocery stores are all varieties of the same species, *Agaricus bisporus*. Some of the toxins found in poisonous mushrooms are used in biological research; others are hallucinogens.

Basidiomycetes are important saprophytes. Along with a few soil-dwelling bacteria, some members of this lineage are the only organisms capable of synthesizing lignin peroxidase and are therefore important in wood decomposition. Some basidiomycetes are ectomycorrhizal fungi (EMF) that improve the health and productivity of temperate forests via their mutualistic associations with tree roots. Temperate forests provide most of the hardwoods and softwoods used in construction, furniture-making, and papermaking.

Basidiomycetes can also be parasitic. One subgroup of the lineage consists entirely of parasitic forms called rusts, including species that cause serious infections in many crop plants. Plant parasites called smut fungi specialize in infecting grasses; a few parasitic basidiomycetes even infect other fungi.

Ascomycota

Ascomycetes are found in virtually every terrestrial habitat as well as some freshwater and marine environments. They are so named because most ascomycetes form mycelia and produce spores in structures called asci. Ascomycete mycelia also commonly reproduce asexually. In addition, single-celled yeasts are part of the Ascomycota lineage. Yeasts reproduce when daughter cells pinch off from the parent cell after mitosis. This asexual reproductive process is called **budding**.

About half of ascomycete species grow in mutualistic association with cyanobacteria and/or single-celled green algae, forming the structures called lichens (see Figure 29.13). Over 17,500 different lichens have been described to date; in most, the fungus involved is an ascomycete (although a few basidiomycetes participate as well).

Many ascomycetes form mutualistic EMF associations with tree roots. Ascomycetes are also the most common endophytic fungi on aboveground tissues. Many other ascomycetes are saprophytic and are abundant on forest floors and in grassland soils; parasitic forms are also common. In addition, several species are predatory—primarily on amoebae and other unicellular protists, but a few on larger prey. For example, the nematode-trapping fungus shown in Figure 29.6 is an ascomycete.

Some saprophytic ascomycetes can grow on jet fuel or paint and are used to help clean up contaminated sites. *Penicillium* is an important source of antibiotics, and *Aspergillus* produces citric acid used to flavor soft drinks and candy. Truffles are so highly prized that they can fetch $2000 per pound, and the multibillion-dollar brewing, baking, and wine-making industries would collapse without the yeast *Saccharomyces cerevisiae*. A few parasitic ascomycetes cause infections in humans and other animals. In land plants, parasites from this group cause several plant diseases, including Dutch elm disease and chestnut blight.

In summary, then, you can see that Fungi are a large and diverse group of organisms. They are economically important, and they also play important roles in the environment.

CHECK YOUR UNDERSTANDING

✔ If you understood this section, you should be able to . . .

1. Explain the role of the polar tube in microsporidial infections.
2. Explain how Glomeromycota can be beneficial to plants.
3. Explain whether or not budding produces genetic variation in yeasts.

Answers are available in Appendix A.

CHAPTER 29 Review

For media, go to Mastering Biology

29.1 Why Do Biologists Study Fungi?

- Fungi are used widely in the food processing, cheese-making, baking, and brewing industries. Fungi also provide many of our antibiotics.
- Parasitic fungi are responsible for devastating blights in crops and other plants.
- Many fungi species live in close association with land plants. The roots of most plants are infected by mutualistic fungi that provide the plant with nitrogen, phosphorus, and water. The fungi benefit by receiving sugars from photosynthesis along with other plant products.
- Once plants die, saprophytic fungi degrade the lignin and cellulose in wood and use nutrients from decaying plant material.
- Because they free up carbon atoms that would otherwise be locked up in wood, fungi speed up the carbon cycle in terrestrial ecosystems.

29.2 How Do Biologists Study Fungi?

- Morphological analyses of fungi have revealed adaptations that make fungi exceptionally effective at absorbing nutrients from the environment.

- Researchers have determined that fungi exhibit two growth forms—single-celled "yeasts," or mycelia composed of long, filamentous hyphae. Mycelia provide fungi with extremely high surface-area-to-volume ratios. Some fungal bodies are extremely large.
- Septa divide many hyphae into cells, but pores in the septa facilitate transport of nutrients and organelles through the hyphae.
- Scientists have determined from morphological studies that fungi possess four known, distinctive types of sexual reproductive structures: **(1)** chytrids produce motile gametes; **(2)** zygomycetes make tough zygosporangia; **(3)** basidiomycetes have club-shaped, spore-forming structures; and **(4)** ascomycetes have sac-like, spore-forming structures.
- Fungi can reproduce asexually, and many do so by forming conidia—asexual spores that grow into a new fungus.
- Recent analyses of DNA sequence data have revealed that fungi are more closely related to animals than they are to plants, and that the Basidiomycota, Ascomycota, and Glomeromycota each form monophyletic groups. Researchers are still trying to understand how the various groups of chytrids, zygomycetes, and microsporidians are related to each other and to the other fungal lineages.

29.3 What Themes Occur in the Diversification of Fungi?

- All fungi make their living by absorbing nutrients from living or dead organisms.
- Fungi often associate with other organisms, especially plants, in symbiotic relationships that can be mutualistic, parasitic, or commensal.
- Ectomycorrhizal fungi (EMF) and arbuscular mycorrhizal fungi (AMF) form mutualistic relationships with the roots of plants.
- Many fungi live in roots and shoots as endophytes. Some of these endophytic fungi secrete toxins that discourage herbivores; others live as commensals, providing no apparent benefit to the plant.
- Fungi form symbiotic relationships with many animal species and, in lichens, with various species of green algae and cyanobacteria.
- Extracellular digestion, in which enzymes are secreted into food sources, enables fungi to break down extremely large molecules before absorbing them.
- Lignin decomposes through a series of uncontrolled oxidation reactions triggered by the enzyme lignin peroxidase.
- Cellulose digestion occurs in a carefully regulated series of steps, each catalyzed by a specific cellulase.
- Many species of fungi have never been observed to reproduce sexually. Most can produce haploid spores either sexually or asexually.
- Sexual reproduction usually starts when hyphae from genetically different mating types fuse—an event called plasmogamy. If the fusion of nuclei, or karyogamy, does not occur immediately, a heterokaryotic mycelium forms, which in some species is dikaryotic. Heterokaryotic cells may eventually produce spore-forming structures where karyogamy and meiosis take place.

29.4 Key Lineages of Fungi

- Microsporidia are parasitic, single-celled fungi that use a unique structure called a polar tube to infect host cells.
- Chytrids are usually aquatic with motile, flagellated spores and gametes. Some are terrestrial decomposers and some are parasites.
- Zygomycetes are largely saprophytic on dead plant tissue. During sexual reproduction, they form zygosporangia after hyphae from different mating types fuse. Asexual reproduction is also common.
- The Glomeromycota is a monophyletic group of soil fungi that form arbuscular mycorrhizae with most land plants.
- The Basidiomycota is a monophyletic group of terrestrial fungi that includes mushrooms, bracket fungi, stinkhorns, and puffballs. During sexual reproduction, they form airborne spores in structures called basidia. Some can degrade wood. Others form ectomycorrhizae with certain trees.
- Ascomycota is a monophyletic group of terrestrial fungi. During sexual reproduction, they form airborne spores in structures called asci. Some form lichens in association with green algae or cyanobacteria.

Answers are available in Appendix A.

✔ TEST YOUR KNOWLEDGE

1. The mycelial growth habit leads to a body with a high surface-area-to-volume ratio. Why is this important?
   a. Mycelia have a large surface area for absorption.
   b. The hyphae that make up mycelia are long, thin tubes.
   c. Most hyphae are broken up into compartments by walls called septa, although some exist as single, gigantic cells.
   d. Hyphae can infiltrate living or dead tissues.
2. Evaluate the following statements about mushrooms. Select True or False for each statement.
   T/F A diploid zygote is formed when a male spore fertilizes a female spore.
   T/F Haploid spores are formed by meiosis in basidia.
   T/F A mushroom is a reproductive structure that consists largely of dikaryotic cells.
   T/F Mushrooms are part of the group that also includes bracket fungi, boletes, puffballs, and stinkhorns.
3. The hyphae of arbuscular mycorrhizal fungi (AMF) form bushy structures after making contact with the plasma membrane of a root cell. What is the function of these structures?
   a. They anchor the fungus inside the root, so the association is more permanent.
   b. They increase the surface area available for the transfer of nutrients.
   c. They produce toxins that protect the plant cells against herbivores.
   d. They break down cellulose and lignin in the plant cell wall.
4. What does it mean to say that a hypha is dikaryotic?
   a. Two nuclei fuse during sexual reproduction to form a zygote.
   b. Two independent nuclei, derived from different individuals, are present in each cell.
   c. The nucleus is diploid or polyploid—not haploid.
   d. It is extremely highly branched, which increases its surface area and thus absorptive capacity.

✔ TEST YOUR UNDERSTANDING

5. The Greek root *ecto* means "outer." Why are ectomycorrhizal fungi, or EMF, aptly named?
   a. Their hyphae form tree-like branching structures inside plant cell walls.
   b. They are mutualistic.
   c. Their hyphae form dense mats that envelop roots but do not penetrate the cell walls.
   d. They transfer nitrogen from outside their plant hosts to the interior.

6. Explain why fungi that degrade dead plant materials are important to the global carbon cycle. Do you accept the text's statement that, without these fungi, "Terrestrial environments would be radically different than they are today, and probably much less productive"? Why or why not?

7. Lignin and cellulose provide rigidity to the cell walls of plants. But in most fungi, chitin performs this role. Why is it logical that most fungi don't have lignin or cellulose in their cell walls?

8. **SOCIETY** Lawns are sometimes fertilized with nitrate that can be washed into neighboring woodlots by rain. If the trees in that woodlot are associated with EMF, what effect might the excess nitrate have on the fungi and/or trees?
   a. Fungal growth may be stimulated, causing the trees to transport more nitrogen to the fungi.
   b. The fungi may secrete more peptidases to break down the nitrate.
   c. The trees may take up the nitrate directly into their roots and rely less on the fungi.
   d. The excess nitrate may stimulate the fungi to transport more sugar to the trees.

✔ TEST YOUR PROBLEM-SOLVING SKILLS

9. **PROCESS OF SCIENCE** Table 29.1 mentions that chytrids are responsible for massive die-offs currently occurring in amphibians. Review Koch's postulates (see Ch. 26, Section 26.1); then design a study showing how you would use Koch's postulates to test the hypothesis that chytrid infections are responsible for frog deaths.

10. **PROCESS OF SCIENCE** Many mushrooms are extremely colorful. One hypothesis is that the colors serve as a warning to prevent animals from eating mushrooms, much like the bright yellow and black stripes on wasps help to deter potential predators. Design an experiment to test this hypothesis.

✔ PUT IT ALL TOGETHER: Case Study

Just how fast can fungal spores fly?

Some fungi have elaborate mechanisms for dispersing spores. For example, the "squirt gun" fungus *Pilobolus*, which grows in cow dung, forcibly shoots off tiny black sporangia filled with spores. Any sporangia that are flung onto fresh grass are likely to be eaten by a grazing cow, passed through its digestive system unharmed, and deposited in a new batch of dung. The spores carried within the sporangia are perfectly positioned to grow into a new mycelium.

11. *Pilobolus* is a zygomycete, and the sporangia just described produce asexual spores. In contrast, which of the following best describes what happens during the sexual phase of its life cycle?
   a. It produces eight ascospores in each ascus.
   b. It produces basidiospores that grow into mushrooms.
   c. It produces motile sperm and eggs.
   d. Hyphae from two compatible mating types fuse and form a zygosporangium.

12. **MODEL** Both asexual sporangia (shown in the photo) and zygosporangia can be found in cow dung. Make a simplified drawing that illustrates the events of plasmogamy, karyogamy, and meiosis that occur during sexual reproduction in *Pilobolus*. Be sure to indicate and label the zygosporangium.

13. Would you describe the association between *Pilobolus* and cows as a symbiotic relationship? Explain your answer.

14. Explain the process of extracellular digestion that occurs during the growth of *Pilobolus* mycelia through dung.

15. **QUANTITATIVE** Using high-speed video, researchers have measured several variables associated with sporangium discharge in various spore-shooting fungi. Based on the data in the table shown here, how do the launch speed and acceleration of *Pilobolus* compare with those of the other fungi? How does launch speed for each species compare with the top speed of 44 km/h reached by Jamaican athlete Usain Bolt during his 2009 world-record-setting race?

	Ascobolus immersus	*Podospora anserina*	*Pilobolus kleinii*
Launch speed (m/s)	14	21	9
Acceleration (m/s^2)	1,800,000	1,500,000	210,000
Range (m)	0.3	0.2	2.5
Turgor pressure (MPa)	0.40	0.40	0.55

Source: L. Yafetto et al. 2008. *PLoS ONE*, 3(9): e3237.

16. **PROCESS OF SCIENCE** Propose an experiment to test whether *Pilobolus* shoots its sporangia preferentially toward light. Hypothesize why such a preference might be advantageous.

Mastering Biology

Students Go to Mastering™ Biology for assignments, the eText, and the Study Area with animations, practice tests, and activities.

Professors Go to Mastering™ Biology for automatically graded tutorials and questions that you can assign to your students, plus Instructor Resources.

30 An Introduction to Animals

Jellyfish such as this hydromedusa belong to one of the most ancient lineages of animals.

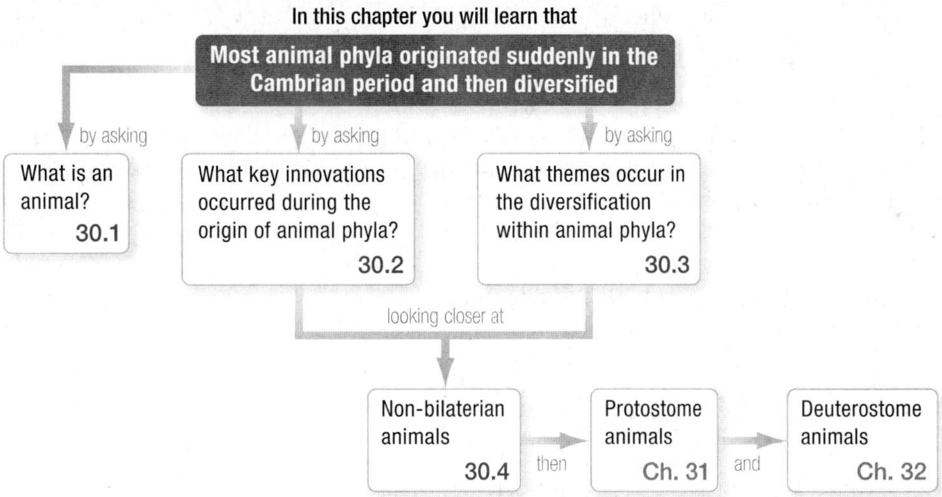

BIG PICTURE

This chapter is part of the Big Picture. See how on pages 722–723.

You're an animal, one of an extremely diverse and species-rich lineage on the tree of life. Estimates suggest that between 3 million and 10 million animal species exist on our planet today, some similar to you, others quite different. They are all your cousins, survivors of about 800 million years of evolution from a common eukaryotic ancestor.

Only about 1.4 million animal species have been described and given scientific names to date, but scientists are in a race to discover and describe more animal species before more of them go extinct due to human impacts (Ch. 54, Section 54.2). Scientists are eager to understand and preserve animal diversity not only due to our innate interest in our own lineage of life but also because humans depend on other animals in many ways. For example, humans rely on other animals for food, materials, labor, services such as pollination, and their major ecological roles in maintaining ecosystem function (Ch. 54, Section 54.3). Animals also play an important role in causing and transmitting human diseases and in helping scientists to cure them.

In this chapter you will find a broad overview of how the major groups of animals diversified, along with more detailed information on the characteristics of the lineages that appear first in the fossil record. (See Chapters 31 and 32 to focus on the most species-rich groups.) Let's start by considering the characteristics of animals.

30.1 What Is an Animal?

Like land plants and fungi, animals originated from single-celled eukaryotes (Ch. 27, Section 27.3). **Figure 30.1** reviews how animals fit in the tree of life. Animals occur in a lineage, called Opisthokonta, along with fungi, the single-celled or colonial eukaryote (protist) called choanoflagellates, and other protists (not shown in tree). Choanoflagellates are the closest living relatives, or sister group, to animals, sharing a common ancestor about 900 million years ago.

After you complete this section, you should be able to ...

- Analyze the characteristics that distinguish animals from other eukaryotes.

Despite their stunning diversity, **animals** form a monophyletic group, sharing key traits:

- *All animals are multicellular eukaryotes* whose cells lack cell walls but have an extensive **extracellular matrix (ECM)**; (Ch. 11, Section 11.1). The ECM includes proteins specialized for cell–cell adhesion and communication.
- *All animals are heterotrophs,* meaning that they obtain carbon compounds from other organisms (rather than via photosynthesis as autotrophs do). Most ingest ("eat") their food, rather than absorbing it across their exterior body surfaces.
- *All animals move* under their own power at some point in their life cycle.
- *All animals other than sponges have* **(1)** *neurons* (nerve cells) that transmit electrical signals to other cells; and **(2)** *muscle cells* that can change the shape of the body by contracting.

Multicellular fungi and animals are similar in that they are both multicellular heterotrophs that digest (break down) and absorb nutrients. However, animals are the only multicellular heterotrophs on the tree of life that usually ingest their food first, before they digest it. As a result, digestion in animals typically occurs within the space, or lumen, of the digestive tract rather than in the open exterior, as occurs with fungi.

In most animals, neurons connect to each other, forming a nervous system, and some neurons connect to muscle cells—which may contract in response to electrical signals from neurons. Muscles and neurons are adaptations that allow a large, multicellular body to move efficiently. What does the combination of multicellularity, heterotrophy, and efficient movement produce? Eating machines—animals are the largest detritivores, herbivores, and carnivores on Earth.

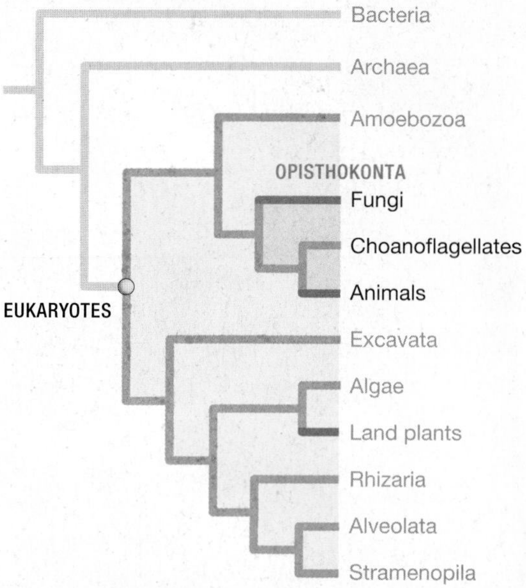

Figure 30.1 Choanoflagellates and Animals Are Sister Groups. The orange branches represent single-celled eukaryotes (protists).

The radiation of animals began around 550 million years ago during an event called the **Cambrian explosion** (Ch. 25, Section 25.3). After a slow start, diverse animals suddenly appear in the fossil record with not only muscles and nerves but also shells, exoskeletons, internal skeletons, legs, heads, tails, brains, eyes, antennae, jaw-like mandibles, and segmented bodies. The Cambrian explosion was arguably the most spectacular period of evolutionary change in the history of life. Today the descendants of these marine animals are key consumers in virtually every ecosystem, from the deep oceans to tropical forests.

Biologists currently recognize 30–35 **phyla**, or major lineages, of animals—the exact number is constantly being debated and revised as additional information comes to light. **Table 30.1** lists 28 phyla that are included in most published analyses.

How can we make sense of all this diversity? Let's first take a closer look at the key events that occurred during the *origin* of major animal lineages. Then let's examine the key themes of diversification that occurred *within* lineages after they originated.

CHECK YOUR UNDERSTANDING

✔ **If you understood this section, you should be able to ...**

List one trait that animals, plants, and fungi have in common, and one trait that is different between each pair.

Answers are available in Appendix A.

Table 30.1 An Overview of Major Animal Phyla

Group and Phylum	Common Name or Example Taxa	Number of Described Species
Non-bilaterian Groups		
Placozoa	Placozoans	1
Ctenophora	Comb jellies	190
Acoela	Acoelomate worms (other than flatworms)	350
Porifera	Sponges	8500
Cnidaria	Jellyfish, corals, anemones, hydroids, sea fans	11,500
Protostomes (possible sister group to Lophotrochozoa and Ecdysozoa)		
Chaetognatha	Arrow worms, pterobranchs	120
Protostomes: Lophotrochozoa		
Phoronida	Horseshoe worms	10
Gnathostomulida	Gnathostomulids	100
Entoprocta	Entoprocts, kamptozoans	170
Gastrotricha	Gastrotrichs	400
Brachiopoda	Brachiopods (lamp shells)	550
Acanthocephala	Acanthocephalans	1150
Nemertea	Ribbon worms	1200
Rotifera	Rotifers	2100
Bryozoa	Bryozoans, ectoprocts, moss animals	5700
Annelida	Segmented worms	16,800
Platyhelminthes	Flatworms	20,000
Mollusca	Mollusks (clams, snails, octopuses)	85,000
Protostomes: Ecdysozoa		
Priapulida	Priapulids	16
Kinorhyncha	Kinorhynchs	130
Onychophora	Velvet worms	165
Nematomorpha	Hair worms	330
Tardigrada	Water bears	1045
Nematoda	Roundworms	25,000
Arthropoda	Arthropods (spiders, insects, crustaceans)	1,200,000
Deuterostomes		
Hemichordata	Acorn worms	108
Echinodermata	Echinoderms (sea stars, sea urchins, sea cucumbers)	7000
Chordata	Chordates: tunicates, lancelets, sharks, bony fishes, amphibians, reptiles (including birds), mammals	65,000

✔ **QUANTITATIVE** Identify the five phyla with the most described species and list them in order from most to least diverse.

30.2 What Key Innovations Occurred during the Origin of Animal Phyla?

Biologists, of course, were not around to observe the origin of animals during the Cambrian explosion 550 million years ago (because biologists are themselves animals). What kind of data can a biologist use to study such ancient events? There are four main sources of evidence, each powerful in its own way:

After you complete this section, you should be able to . . .

- Analyze the significance of the origin of key innovations during the early radiation of animals.

1. *Fossils* are important because they provide the only direct evidence of what ancient animals looked like, when they existed, and where they lived. However, fossils are likely to occur only for animals that were abundant, had hard parts, lived in areas where sedimentation was occurring, and/or lived recently (Ch. 25, Section 25.2).

2. *Comparative morphology* provides information about which larval or adult morphological characteristics can be used to define the fundamental architecture, or **body plan**, of each animal lineage. In a phylogenetic context, these data can be used to infer which characteristics arose first during the evolution of animals, and which animal groups are more closely related.

3. *Comparative development* (or evolutionary developmental biology, often called *evo-devo*) provides information about patterns of gene expression and morphological change during development (Ch. 21, Section 21.5). This research can reveal when diverse structures are determined by similar genes and developmental processes (genetic and developmental homology), or when similar structures are determined by different genes and developmental processes (convergent evolution).

4. *Comparative genomics* provides information about the relative similarity of genes or whole genomes of diverse organisms (Ch. 20, Section 20.4). This relatively new source of data is providing dramatic insights into phylogenetic relationships and evolutionary history.

Sometimes these various sources of data on animal evolution suggest the same sequence of evolutionary events, providing *internal consistency* (see the case study on evolution of whales in Ch. 22, Section 22.2). However, on other occasions a new source of data can contradict insights gleaned from earlier studies, causing biologists to reevaluate their earlier conclusions.

Consider one of the most influential papers on animal phylogeny, published in 1997. Using sequences from a gene that codes for RNA in the small subunit of the ribosome, Anna Marie Aguinaldo and colleagues estimated the phylogeny of species from 14 animal phyla. The results were revolutionary, suggesting a different pattern of relationships among animal lineages than had been accepted for years based on morphological data.

The phylogenetic tree in **Figure 30.2** is an updated version of the 1997 result, based on further studies of many gene sequences. However, this tree still represents a hypothesis that lacks consensus in the scientific community.

Let's explore the tree by starting at the root and working toward the branch tips. (See **BioSkills 13** to review how to interpret phylogenetic trees.)

Origin of Multicellularity

Look at the root of the animal tree in Figure 30.2, and put your finger on the node that represents the common ancestor of all animals (indicated by a black dot). What does this part of the tree tell us?

The first important insight is that animals are a monophyletic group, meaning that all animals share a common ancestor that was multicellular. Data from fossils, comparative morphology, comparative development, and comparative genomics agree on a single origin of animals.

The second point to examine is which phylum branched off first during the diversification of animals, and thus is a sister group to all other animals. The prevailing hypothesis is that sponges (phylum Porifera) are the most ancient lineage of animals. Let's consider evidence for this hypothesis.

Fossil Evidence for the Sponges-First Hypothesis Sponges are the earliest animals to appear in the fossil record, more than 700 million years ago. The early presence of multicellular sponges and the absence of fossils of other multicellular organisms are both consistent with the basal position of sponges on the phylogeny.

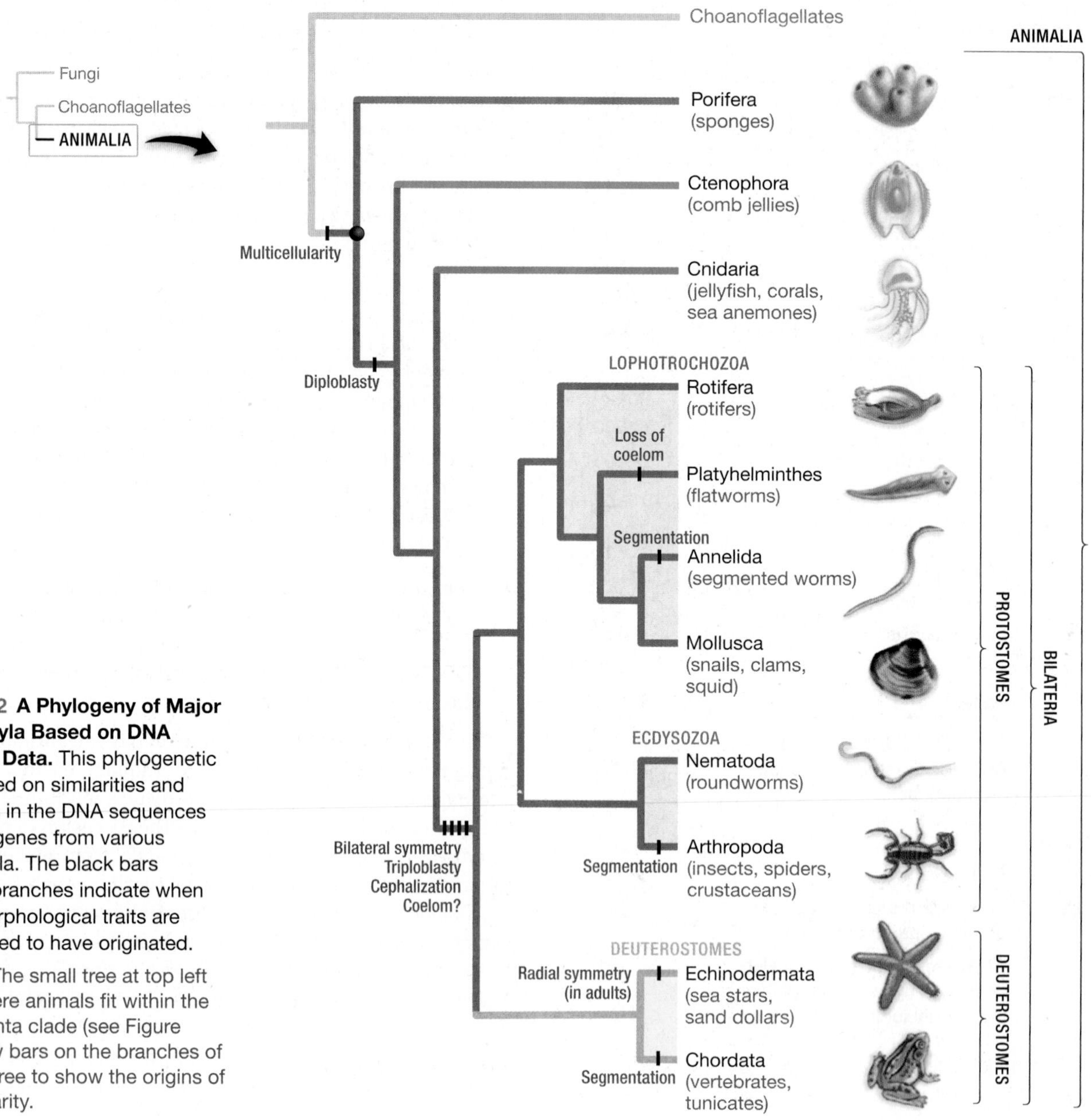

Figure 30.2 A Phylogeny of Major Animal Phyla Based on DNA Sequence Data. This phylogenetic tree is based on similarities and differences in the DNA sequences of several genes from various animal phyla. The black bars along the branches indicate when certain morphological traits are hypothesized to have originated.

✔ **MODEL** The small tree at top left shows where animals fit within the Opisthokonta clade (see Figure 30.1). Draw bars on the branches of this small tree to show the origins of multicellularity.

Further, geochemical evidence from ancient rocks suggests that the oxygen concentrations in the Precambrian atmosphere and oceans were a tiny fraction of the levels in the Cambrian and beyond. Recent measurements from some sponges living today shows that they can survive in an oxygen concentration as low as 0.5–4.0 percent of the present level, supporting the hypothesis that sponges could survive the low-oxygen conditions in ancient seas.

Morphological Evidence for the Sponges-First Hypothesis Sponges share several key characteristics with choanoflagellates:

- Both choanoflagellates and sponges are **benthic**, meaning that they live at the bottom of aquatic environments, and **sessile**, meaning that adults live permanently attached to a substrate rather than moving freely.
- Both feed in a similar way, using cells with similar morphology. As Figure 30.3 shows, the beating flagella of choanoflagellates and specialized cells in sponges called **choanocytes** trap bacteria and other organic debris. The trapped food particles are then digested inside the choanoflagellate cell body or within other cells of the sponge. That is, feeding occurs at the cellular level, unlike feeding in all other animals.

As you can see in Figure 30.3a, choanoflagellates sometimes form **colonies**—groups of individuals that are attached to each other. Some biologists once considered sponges to be colonies of single-celled protists due to the ability of sponge cells to re-aggregate after being dissociated. But sponges contain many specialized cell types that are dependent on each other, some of which occur in organized layers surrounded by extracellular matrix (ECM).

Recent research now shows that one group of sponges has true **epithelium**—a layer of tightly joined cells that covers the interior and/or exterior surface of the animal. Epithelium is essential to animal form and function (Ch. 39, Section 39.2).

Molecular Evidence for the Sponges-First Hypothesis Many molecular phylogenies show the sponges branching off first during the diversification of animals. This evidence supports the hypothesis that sponges are the sister group to all other animals.

What else can molecular evidence tell us? Comparative genomic studies suggest that despite the relative morphological simplicity of sponges, they possess a remarkably complex set of developmental **tool-kit genes** (Ch. 21, Section 21.3) necessary for all the basic molecular processes required by animals:

- Specialization of cell types
- Regulation of cell cycling and growth
- Adhesion among cells
- Adhesion between cells and extracellular matrix
- Recognition of self and nonself, thus innate immunity
- Developmental signaling and gene regulation
- Programmed cell death

Thus, a series of important genetic innovations appears to have occurred at the very root of the animal tree along with multicellularity. Subsequent duplication and diversification

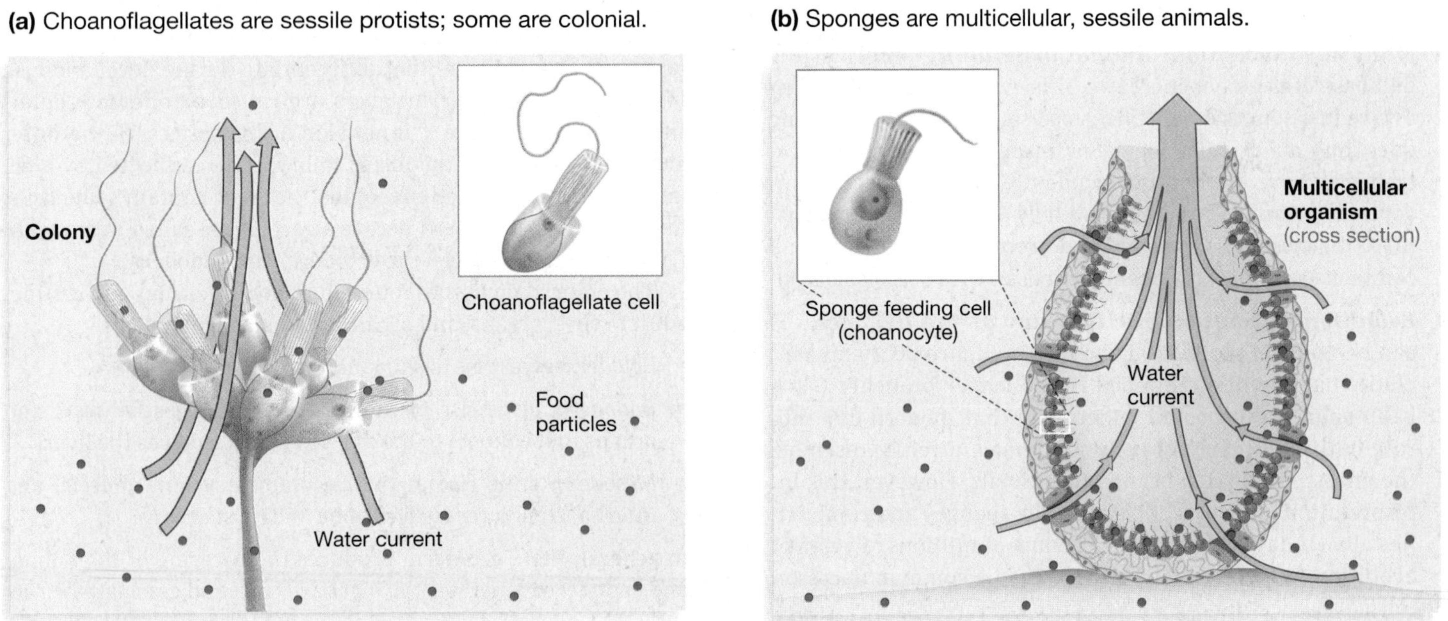

Figure 30.3 Choanoflagellates and Sponge Feeding Cells Are Similar in Structure and Function. Choanoflagellates **(a)** and sponges **(b)** are both suspension feeders. In both cases, specialized cells generate a current using flagella, then trap particles from the water flow.

of these genes contributed to the diversification of animal lineages.

Alternative Views: The Ctenophores-First Hypothesis Overall, there is compelling evidence that sponges are the sister group to all other animals. However, a flood of new genomic data has produced surprising alternative hypotheses. For example, some recent studies suggest that the Ctenophora (pronounced *ten-AH-for-ah*), or comb jellies, are the sister group to all other animals. This conclusion is based on the pattern of gains and losses of certain genes in ctenophores compared to other animals. For example, ctenophores are the only animal that lacks microRNAs, important regulators of gene expression (Ch. 19, Section 19.4).

Still, it's possible that ctenophores had an animal ancestor with microRNAs, and that this trait was subsequently lost. Remember that evolution involves not only the gain of new traits but also the loss of traits. Currently there is not sufficient evidence to reject the sponges-first hypothesis.

Insights from the Origin-of-Animals Debate No matter which hypothesis about the origin of animals ultimately attains the highest level of internal consistency (Ch. 22, Section 22.2), new genomic data have revealed a number of important insights that apply to the entire animal lineage:

- ***The evolution of animals is more complicated than a smooth transition from simple to complex.*** Two decades ago, many scientists predicted that more complex animals would have larger genomes than would simpler animals. But new genomic data have led to the rejection of this hypothesis. For example, the small starlet sea anemone (phylum Cnidaria) has almost as many genes as humans do. Many animal genes evolved early and were co-opted, rearranged, diversified, and sometimes lost in subsequent lineages.
- ***Many key innovations did not arise all at once.*** As the database of genomes increases, it is evident that traits often evolve in a sequence of "tinkering" steps, raising interesting questions about homology. For example, a morphological trait may have evolved independently in two lineages (convergent evolution) but may be built by the same (homologous) tool-kit genes and developmental processes inherited from a common ancestor. This phenomenon is called *deep homology*.
- ***Evolution did not stop within any of the lineages.*** It can be tempting to look at the tree in Figure 30.2 and conclude that evolution stopped in the longer branches (such as in sponges and comb jellies) and that modern-day representatives of early-branching animals directly resemble the ancestors of later-branching animals. However, this interpretation is incorrect because the sponges and comb jellies alive today represent the product of millions of years of evolution following their descent from a common ancestor. **Making Models 30.1** provides a tip for recognizing earlier-branching lineages as sister groups to other animals rather than as ancestors.

Let's continue our tour of the phylogeny in Figure 30.2 and see how these insights apply.

Making Models 30.1 Tips on Drawing Phylogenetic Trees (VI)

You can draw a dot on each node of a phylogenetic tree to represent the most recent common ancestor to the branches to the right of the node (see **BioSkills 13**). Drawing a dot can remind you that two branches emerging from a node are sister groups to each other; one is not the ancestor of the other.

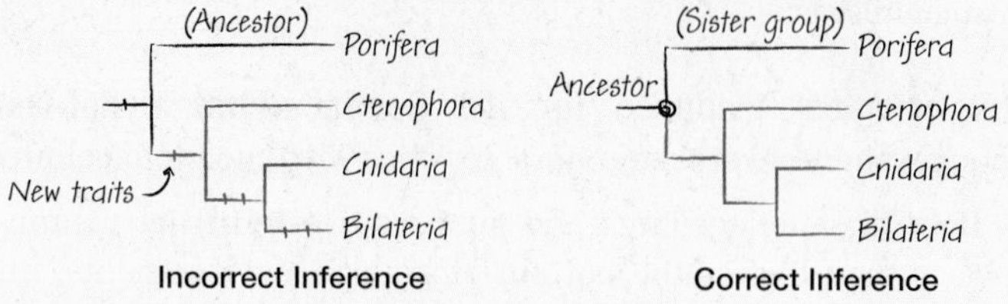

MODEL Draw several trait bars on the "Correct Inference" tree to show that evolution occurs in all branches, even Porifera.

To see this model in action, go to the Study Area of **Mastering Biology**

Origin of Embryonic Tissue Layers and Muscle

Sponges possess the basic genetic tool kit for cell–cell adhesion and cell–ECM adhesion, and a few sponges even have epithelium. However, sponges do not have complex **tissues**, groups of similar cells that are organized into tightly integrated structural and functional units.

Origin of Embryonic Tissue Layers Animals other than sponges are typically distinguished by the number of embryonic tissue layers they have. Animals whose embryos have two types of tissue are called **diploblasts** ("two-buds"); animals whose embryos have three types of tissue are called **triploblasts** ("three-buds"). (See **BioSkills 15** to review common Latin and Greek roots.)

The embryonic tissues are organized in layers, called **germ layers** ("sprout" layers that differentiate during development). In diploblasts these germ layers are called **ectoderm** ("outer-skin") and **endoderm** ("inner-skin"). In most cases the outer and inner "skins" of diploblast embryos are connected by a gelatinous material called **mesoglea** that may contain some cells. In triploblasts, however, there is a germ layer called **mesoderm** ("middle-skin") between the ectoderm and endoderm.

The embryonic tissues found in animals develop into distinct adult tissues, organs, and organ systems. In triploblasts:

- *Ectoderm* gives rise to skin and the nervous system.
- *Endoderm* gives rise to the lining of the digestive tract and organs that connect to the digestive tract, such as the liver.
- *Mesoderm* gives rise to the circulatory system, muscle, and internal structures such as bone and most organs.

In general, then, ectoderm produces the covering of the animal and endoderm generates the digestive tract and associated structures. Mesoderm gives rise to the tissues in between.

Look again at the animal phylogeny depicted in Figure 30.2. Traditionally, two groups of animals were recognized as diploblasts: the Ctenophora, or comb jellies, and the Cnidaria (pronounced *ni-DARE-ee-uh*), which include the jellyfish, corals, sea

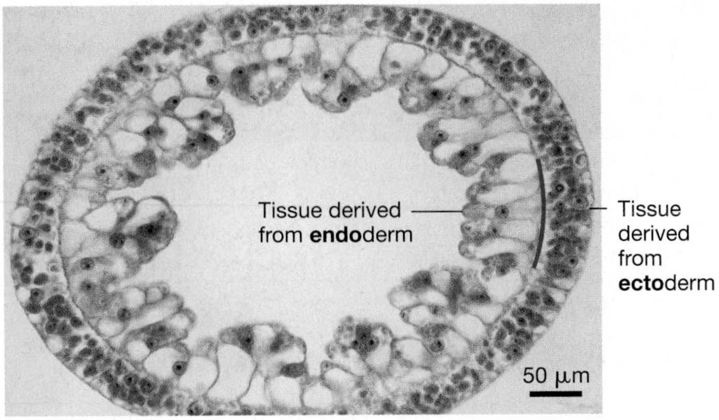

Figure 30.4 **Diploblastic Animals Have Bodies Derived from Ectoderm and Endoderm.** This is a cross section through the tube-shaped portion of a hydra's body (phylum Cnidaria; see Figure 30.5a). The cells are stained to make them more visible.

pens, anemones, and hydra. As an example of the diploblast body plan, Figure 30.4 shows a cross section of a hydra.

However, recent data suggest that some ctenophores and cnidarians have mesodermal cells in their mesoglea. Does this mean they are triploblasts? As it happens, some ctenophores and cnidarians contain genes coding for the structural components of mesodermal cells, but not the mesodermal specification genes that are present in true triploblasts. Here's the take-home message: Ctenophores and cnidarians share some components of the genetic tool kit for triploblasty, but not others—suggesting that the similarities among mesodermal cells in diploblasts and the mesoderm of triploblasts are the product of convergent morphological evolution based on a foundation of homologous genes, an example of deep homology.

Origin of Muscle Both larval and adult ctenophores swim primarily using cilia, but in some species, the adults also have contractile tissue that can change the body's shape and is used for swimming. Likewise, many cnidarians can change the shapes of their bodies, as seen in the rapid shortening of a sea anemone in response to a threat by a predator, and swimming by jet propulsion in jellyfish. If ctenophores and cnidarians lack true mesodermal muscle, how do these shape changes take place?

It turns out that all animals share homologous genes for contractile proteins. Even sponges, which lack muscle and are generally sessile as adults, possess the genes for key contractile proteins, such as actin and myosin. These proteins and others are used in ctenophores and cnidarians in contractile cells derived from endoderm and/or ectoderm called epitheliomuscular cells, which are functionally similar to true mesodermal muscle cells, but evolved independently.

The important point is that the entire genetic tool kit for muscles did not evolve all at once in the early Cambrian before the diversification of triploblasts. Some protein components of the muscle contractile apparatus evolved earlier and are homologous among animals, but these components were also co-opted and elaborated upon in different ways in different lineages. The result we observe today across animal groups is *functional* similarity (movement by contractile cells or tissues) achieved by independent evolutionary paths—another example of convergent evolution with deep homology.

Origin of Bilateral Symmetry, Cephalization, and the Nervous System

Body symmetry is a key morphological aspect of an animal's body plan. A body is said to be symmetrical if it can be divided by a plane so that the resulting pieces are nearly identical.

Ctenophores, many cnidarians, and some sponges have **radial symmetry** ("spoke symmetry"), which means they have at least two planes of symmetry. For example, almost any plane sectioned through the center of the hydra shown in Figure 30.5a produces two symmetrical halves. Many ctenophores are often described as having a special kind of radial symmetry called **biradial symmetry**—two planes of symmetry, not more. In this case, the two halves defined by the planes have rotational symmetry (as though spun 180 degrees) rather than resembling mirror images. Radial symmetry evolved independently in the phylum Echinodermata (pronounced *ee-KINE-oh-der-ma-ta*)—a group that includes sea stars, sea urchins, feather stars, and brittle stars.

In contrast, organisms with **bilateral symmetry** ("two-sided symmetry") have only one plane of symmetry and tend to have a long, narrow body. The annelid worm in Figure 30.5b, for example, has one plane of symmetry running lengthwise down its middle.

Based on the pattern of branches on the animal phylogeny in Figure 30.2, radial symmetry appears to have arisen earlier in the evolution of animals than bilateral symmetry. Bilateral symmetry occurs in all triploblastic lineages. Where on the tree did bilateral symmetry originate? The answer is more complex than it you might think. Let's see why.

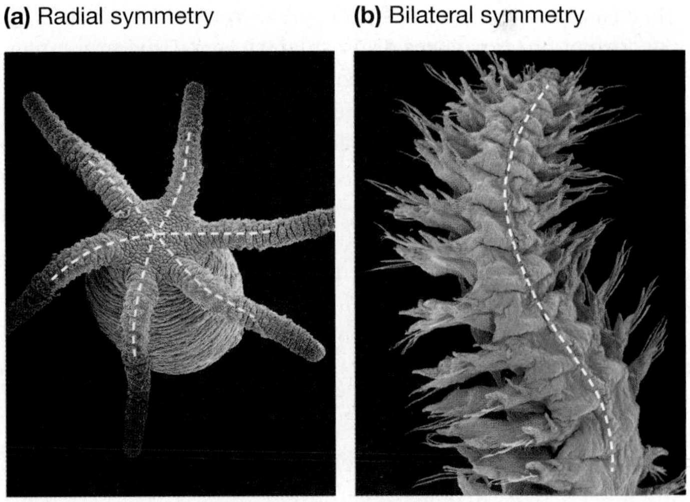

Figure 30.5 **Body Symmetry.** These colorized SEMs of **(a)** a hydra (looking down on oral surface; phylum Cnidaria) and **(b)** a polychaete worm (phylum Annelida) illustrate the difference between radial and bilateral symmetry.

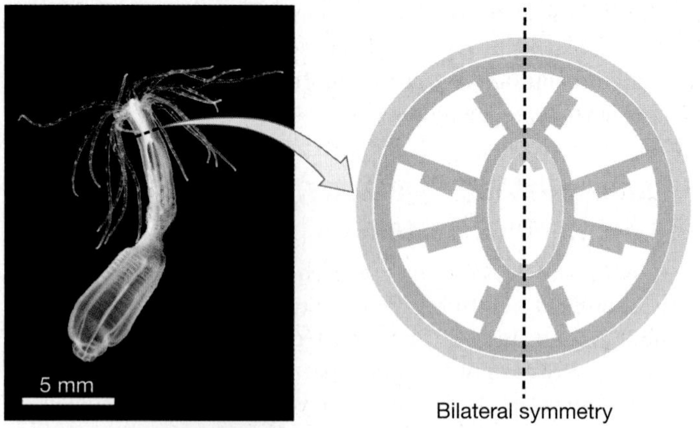

Figure 30.6 Morphological Evidence for Bilateral Symmetry in a Sea Anemone. Some corals and sea anemones have subtle bilateral symmetry despite the outward appearance of radial symmetry. *Nematostella*, shown here, is a small sea anemone that burrows in soft marine sediments.

Bilateral Symmetry: Homology or Convergent Evolution? Nearly all cnidarians appear radially symmetric at first glance, but a closer inspection of the internal anatomy reveals bilateral symmetry in some species, especially sea anemones. Examine the cross section of the sea anemone in Figure 30.6 and note the orientation of the gray rectangles representing retractor muscles. Only one plane of division, the one shown, results in mirror images on both sides of the plane. (To convince yourself, try to find others). Is the bilateral symmetry in sea anemones homologous to bilateral symmetry in triploblastic animals, or is it another example of convergent evolution?

Developmental biologists John Finnerty, Mark Martindale, and colleagues used developmental regulatory genes as a tool to address this evolutionary question. In triploblastic, bilaterally symmetric animals, called **bilaterians** for short, bilateral symmetry is achieved by the combination of anterior–posterior ("head–tail") axis formation and dorsal–ventral ("back–belly") axis formation. *Hox* genes are regulatory genes that are important in the development of the anterior–posterior axis—different *Hox* genes are expressed in different regions along this axis. In turn, they regulate the expression of other genes and thus establish regional identities (Ch. 21, Section 21.3). Similarly, *decapentaplegic* (*dpp*) genes are important in the development of the dorsal–ventral axis.

Finnerty and Martindale predicted that if bilateral symmetry in the starlet sea anemone *Nematostella vectensis* (an increasingly important model organism in biology; see **BioSkills 13**) is homologous to bilateral symmetry in bilaterians, the pattern of *Hox* and *dpp* gene expression would be similar in both. If convergent evolution had occurred, either *Hox* and *dpp* genes would not be expressed in *Nematostella* or their expression patterns would be unrelated to those in bilaterians.

Figure 30.7 shows Finnerty and Martindale's results. During *Nematostella* development, *Hox* genes are expressed along the anterior–posterior axis and *dpp* is expressed asymmetrically about the dorsal–ventral axis, a pattern of gene expression that

RESEARCH

QUESTION: Is bilateral symmetry in sea anemones homologous to bilateral symmetry in bilaterians?

HYPOTHESIS: Bilateral symmetry is homologous in sea anemones and bilaterians.

NULL HYPOTHESIS: Bilateral symmetry arose independently in sea anemones and bilaterians.

EXPERIMENTAL SETUP:

1. Stain gene products (proteins) of *Hox* gene in developing *Nematostella* embryos and larvae to reveal location of expression.

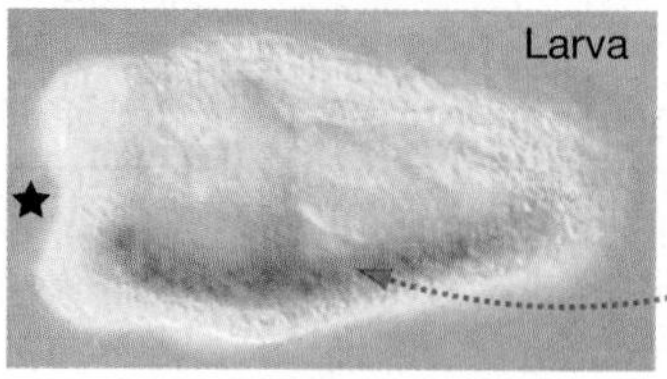

2. Repeat for other *Hox* and *dpp* gene products.

PREDICTION OF HYPOTHESIS: The pattern of expression of *Hox* and *dpp* genes in *Nematostella* will be similar to that found in bilaterians.

PREDICTION OF NULL HYPOTHESIS: *Hox* and *dpp* genes will not be expressed in *Nematostella,* or the expression pattern will be unrelated to that found in bilaterians.

RESULTS: Schematic summary of gene expression patterns:

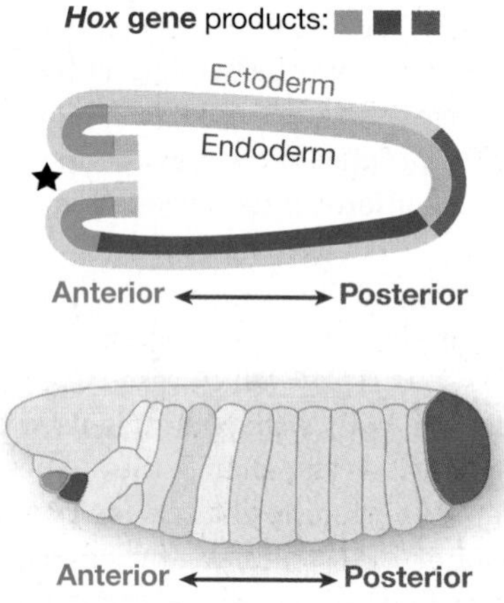

***Hox* genes** are expressed sequentially along the anterior–posterior axis in *Nematostella* larva (shown in longitudinal section)

Reference from prior research: ***Hox* genes** are expressed sequentially along the anterior–posterior axis in fruit fly larva (shown in surface view) and other bilaterians

Other results (not shown): ***dpp* genes** are expressed asymmetrically along the dorsal–ventral axis in *Nematostella*—as in bilaterians.

CONCLUSION: The genetic determination of bilateral symmetry has deep homology in *Nematostella* and bilaterians.

Figure 30.7 Genetic Evidence for Homology in Bilateral Symmetry in Cnidaria and Bilateria. The oral (mouth) side of the sea anemone is labeled as the anterior ("head end") and the aboral side is labeled as the posterior ("tail end") for comparison to bilaterians.

SOURCE: J. R. Finnerty, K. Pang, P. Burton, et al. 2004. Origins of bilateral symmetry: *Hox* and *dpp* expression in a sea anemone. *Science* 304: 1335–1337.

✔ **PROCESS OF SCIENCE** What results would have supported the null hypothesis?

is parallel to that observed in bilaterians. This evidence supports the hypothesis that bilateral symmetry in this sea anemone is homologous to bilateral symmetry in triploblastic animals—meaning that parts of the genetic tool kit that determine bilateral symmetry arose before the evolutionary split of cnidarians and bilaterians.

However, *Hox* and *dpp* gene expression in *Nematostella* is not identical to that in bilaterians. Further, subsequent studies on other cnidarians showed varying expression patterns of *Hox* and *dpp* genes. The take-home message is that the entire genetic tool kit for bilateral symmetry did not evolve all at once in the early Cambrian before the diversification of bilaterians. Some components of the tool kit evolved earlier—in the ancestor to cnidarians and bilaterians—and others evolved or were later co-opted differently, after the split of these two lineages.

The presence of *Hox, dpp,* and other developmental regulatory genes in cnidarians supports the hypothesis, shown in Figure 30.2, that cnidarians are a sister group to the bilaterians. Such genes are absent in sponges and ctenophores.

Origin of the Nervous System Over 99 percent of modern animals are overtly bilaterally symmetric. Is this body plan a "key innovation" that led to the diversification of the bilaterians during the Cambrian explosion? One hypothesis is that the evolution of the nervous system and the evolution of the head are tightly linked to bilateral symmetry and that together, these characteristics contributed to the radiation of bilaterians.

The function of neurons and nervous systems is to transmit and process information in the form of electrical signals (Ch. 43, Section 43.1). How are symmetry and nervous systems related?

- Sponges generally lack both nerve cells and symmetry. However, sponges do possess some of the essential tool-kit genes needed for the development and function of nerves.
- Cnidarians have nerve cells that are mostly organized into a diffuse arrangement called a **nerve net** (Figure 30.8a). These generally radially symmetric animals either float in water or live attached to a substrate. Radially symmetric organisms are more likely to encounter prey and other aspects of the environment in any direction. As a result, a diffuse nerve net can receive and send signals efficiently. In some species, clusters of nerves also occur adjacent to sensory cells, such as near the mouth.
- Ctenophores have a nerve net, nerves associated with the tentacles, and a specialized sensory organ used for sensing light and gravity. Some research shows that ctenophores also have clusters of nerves, called **ganglia**.
- The nervous systems of bilaterians are diverse, ranging from a nerve net to a more complex **central nervous system**, or **CNS**. In a CNS, some neurons are clustered into one or more large tracts or cords that project throughout the body; others are clustered into ganglia. Most of the bilaterally symmetric animals living today move through their environment. Bilaterally symmetric organisms tend to encounter prey and other aspects of the environment at the leading end. As a result, the animal benefits by having many neurons concentrated at that end, with nerve tracts that carry information down the length of the body (Figure 30.8b).

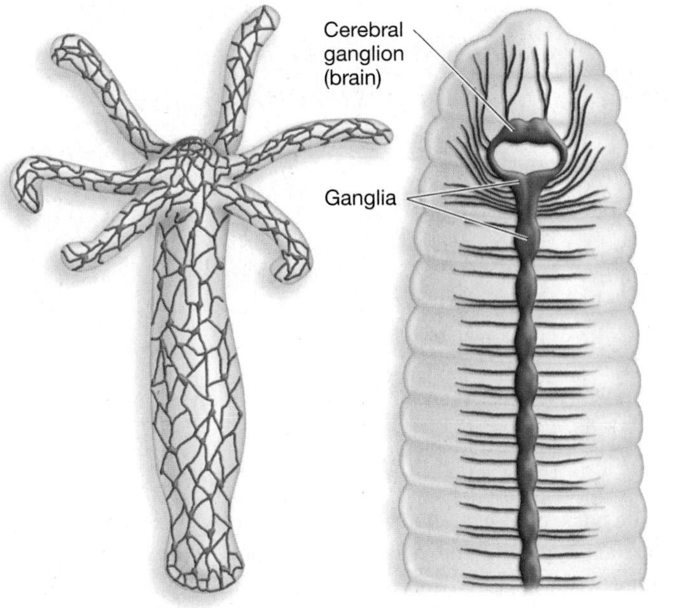

Figure 30.8 **Examples of How Body Symmetry Is Associated with the Nervous System. (a)** Radially symmetric animals, like this hydra, have a nerve net. **(b)** Bilaterally symmetric animals, like this earthworm, have a central nervous system.

It makes sense that the evolution of the CNS would coincide with **cephalization**: the evolution of a head, or anterior region, where structures for feeding, sensing the environment, and processing information are concentrated. The large mass of neurons that is located in the head, and that is responsible for processing information to and from the body, is called the cerebral ganglion or **brain**. The exact timing of the origin of the CNS and brain are hotly debated. Recent evidence supports the hypothesis that these structures did not arise just once in a common ancestor. Rather, the genetic tool kit for a nervous system evolved early and was co-opted multiple times in different lineages via convergent evolution.

The origin of a bilaterally symmetric body plan with cephalization and a CNS enabled rapid, directed movement and hunting. Lineages with this body plan had the potential to diversify into an array of formidable eating and moving machines.

Origin of the Gut and Coelom

Cnidarians have a body plan that looks like a sac. It has only one opening to the environment that is used for ingesting food, releasing gametes, and eliminating wastes. The space within the sac is called a gastrovascular cavity.

At a glance, it seems that ctenophores have the same sac-like body plan as cnidarians. However, recent videos taken by William Browne and colleagues have demonstrated clearly that ctenophores expel wastes through two tiny anal pores. This observation is significant because it suggests that at least some materials pass through the gastrovascular cavity in one direction.

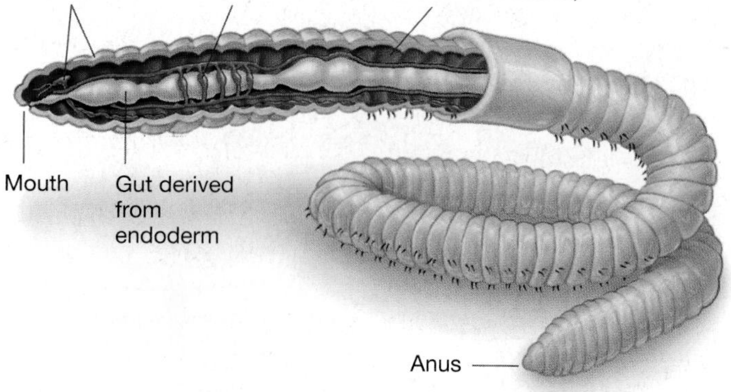

Figure 30.9 The Tube-within-a-Tube Body Plan Is Common in Bilaterians.

The basic bilaterian body shape is a tube within a tube where materials typically move in one direction. The inner tube is the individual's gut with a mouth at one end and an anus at the other, and the outer tube forms the skin and a nervous system (**Figure 30.9**). The mesoderm in between forms muscles and organs. The tube-like body plan with two openings enables specializations along the length of the gut and separates food and wastes.

Are the anal pores of ctenophores homologous to the anus of bilaterians? Researchers are seeking an answer to this question, but so far it seems that anal pores arose independently in ctenophores, yet co-opted some of the same genetic tool kit responsible for bilaterian posteriors.

Besides the gut cavity (which is lined by endodermally derived cells), some bilaterians have a fluid-filled cavity associated with the mesodermally derived tissues. That cavity is called the **coelom** (pronounced *SEE-lum*), as shown in the earthworm in Figure 30.9. The coelom provides a space for the circulation of oxygen and nutrients. It also enables the internal organs to move independently of each other and independently of the inner and outer tubes.

- Bilaterians whose coelom is completely lined with mesoderm-derived tissues are considered to have a true coelom and are known as **coelomates** (**Figure 30.10a**).
- Bilaterians that have no coelom, such as the flatworms (phylum Platyhelminthes), are called **acoelomates** ("no-cavity-form"; **Figure 30.10b**).
- Bilaterians whose coelom is only partially lined with mesoderm-derived tissues, such as roundworms (phylum Nematoda) and rotifers (phylum Rotifera), are known as **pseudocoelomates** ("false-cavity-form"; **Figure 30.10c**).

Earlier phylogenies based on morphological data predicted a gradual evolution from simple acoelomates to pseudocoelomates to coelomates. However, molecular data inspired an alternative view that the coelom arose in the ancestral bilaterian and was subsequently modified, reduced, or lost in many lineages. For example, the position of flatworms within the phylogeny in Figure 30.2 suggests that their ancestors had a coelom but that these traits were lost in the flatworm lineage. Further, the most diverse animals on Earth, the arthropods, have a highly reduced

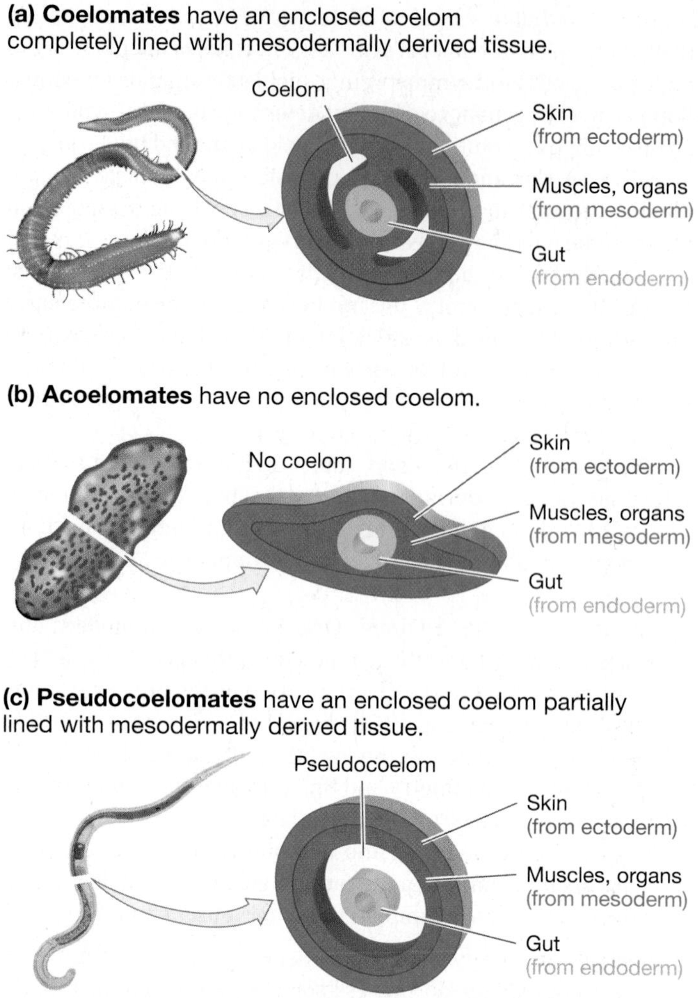

Figure 30.10 Animals May or May Not Have a Coelom.

coelom called a hemocoel (Ch. 31, Section 31.3). The evolutionary flexibility of the coelom has reduced its usefulness as a synapomorphy, or diagnostic character, for bilaterian animals. However, the presence or absence of a coelom is still an interesting component of the body plans of individual phyla.

Origin of Protostomes and Deuterostomes

Turn back to Figure 30.2 and put your finger on the node at the origin of the Bilateria, the monophyletic group of bilaterians. Now move your finger one node to the right. The common ancestor that existed at this node during the Cambrian was likely a bilaterally symmetric triploblast with a CNS, cephalization, and a coelom. This ancestor gave rise to a remarkable radiation of diverse animal lineages, only some of which are included in the tree.

What major groups occur within the bilaterian coelomates? Early studies of embryonic development inspired biologists to recognize two major subgroups:

1. **protostomes**, ("first-mouth"), named for the embryonic development of the mouth before the anus; and
2. **deuterostomes**, ("second-mouth"), named for the embryonic development of the anus before the mouth.

Embryonic germ layers form during a process called gastrulation, which begins when cells move from the surface into the interior of the embryo (Ch. 47, Section 47.4). These cell movements create a pore that opens to the outside, called a blastopore. In deuterostomes, the blastopore becomes the anus; the mouth forms later. Traditionally, biologists thought that in protostomes, the blastopore becomes the mouth and the anus forms later. However, recent studies of embryonic patterns of development in diverse bilaterians show that development is highly variable in protostomes—the blastopore may become the anus, the mouth, both the anus and the mouth, or neither.

Similarly, mesoderm was once thought to form by one developmental pathway in protostome embryos, and by a different developmental pathway in deuterostomes. But new data show that both types of mesoderm-forming processes occur in both groups.

Despite the morphological and developmental challenges to defining synapomorphies for the protostomes and deuterostomes, genomic data consistently support the formation of these two monophyletic groups within the Bilateria. Further, phylogenetic analyses consistently support two major groups within the protostomes: **(1)** the Lophotrochozoa (pronounced *loh-foh-troh-koh-ZOH-ah*), which includes the mollusks, annelids, flatworms, rotifers, and others; and **(2)** the Ecdysozoa (pronounced *eck-die-soh-ZOH-ah*), which includes the arthropods, nematodes, and others. **Lophotrochozoans** grow continuously when conditions are good. **Ecdysozoans** grow by shedding their external skeletons or outer coverings and expanding their bodies.

The vast majority of animal species are protostomes (Chapter 31). However, the largest animals and predators tend to be deuterostomes (Chapter 32).

Origin of Segmentation

Segmentation is defined as the division of the body or a part of the body into a series of similar structures. A segmented backbone is one of the defining characteristics of **vertebrates**, a monophyletic group within the Chordata that includes fishes, reptiles (including birds), amphibians, and mammals. Of the animals that are not vertebrates, called **invertebrates** (a non-monophyletic group; see Ch. 25, Table 25.2), segmentation is conspicuous in annelids (earthworms and other segmented worms) and arthropods (insects, spiders, and crustaceans)—both very prolific and diverse lineages.

Before molecular data were available, biologists grouped segmented animals such as annelids and arthropods together in the same lineage. Then Aguinaldo's landmark 1997 revision of animal relationships overturned this view, suggesting that segmentation arose independently in these groups (as indicated by separate trait bars in Figure 30.2).

Did segmentation in different phyla arise independently (by convergent evolution), or is it homologous? Applying the same logic that Finnerty and Martindale used in their study of bilateral symmetry in sea anemones, several researchers are investigating this question by comparing the genes that regulate segmentation in different phyla. For example, one recent study found that one of the same regulatory pathways that controls segmentation in arthropods, called the hedgehog pathway, is also important in regulating segmentation in annelid worms.

The conclusion so far is that some of the tool-kit genes for segmentation arose early in animal evolution, meaning that they are homologous in different phyla, rather than evolving independently. However, these genes were subsequently lost in some lineages and co-opted in different ways and elaborated upon in others, resulting in the convergent evolution of morphological segmentation in distantly related phyla. Research continues to test this hypothesis of deep homology.

Why have organisms with segmented bodies been so successful in terms of diversity? One leading hypothesis is that segmentation enables specialization. Small changes in the expression of certain tool-kit genes, such as *Hox* genes, along the length of a body can result in novel numbers, shapes, and sizes of body segments and appendages. For example, the evolution of snakes involved the loss of limbs (Ch. 21, Figure 21.14), and the evolution of arthropods involved the specialization of appendages on different segments (Ch. 31, Figure 31.4). Natural selection can then favor variations that are adaptive in certain aquatic or terrestrial environments—leading to diversification.

In summing up this tour of the animal phylogeny, a recurring theme has emerged that evolution is much more complex than a ladder-like, progressive acquisition of new genes and new structures. Many animal genes evolved early and were co-opted, rearranged, diversified, and sometimes lost in subsequent lineages. Thus, some of the most exciting questions in biology—such as what the earliest animals looked like—are still unanswered and the focus of ongoing research. These are questions that you could help to answer.

CHECK YOUR UNDERSTANDING

✔ **If you understood this section, you should be able to ...**

1. Cite two forms of evidence that support the hypothesis that sponges are the sister group to all other animals.
2. Describe which parts of your body relate to your tube-within-a-tube body plan.

Answers are available in Appendix A.

30.3 What Themes Occur in the Diversification within Animal Phyla?

After you complete this section, you should be able to ...

- Analyze the themes of diversification that occurred as animals evolved within phyla.

One of the great unsolved mysteries in the study of animal diversity is this: Why did almost all of the phylum-level body plans evolve so rapidly during the Cambrian, rather than more gradually over time? Animals diversified tremendously within phyla after the Cambrian, but no major new lineages arose during the last 500 million years.

SUMMARY Table 30.2 Diversification of Sensory Organs

Sense	Example
Sight *Stimulus*: light Flies use their compound eyes to find food and mates, and to escape predators	
Hearing *Stimulus*: sound Bats use their sense of hearing to find prey and to avoid obstacles in the dark	
Taste/smell *Stimulus*: molecules Some male moths have elaborate antennae to detect chemical signals in the air	
Touch *Stimulus*: contact, pressure Sea anemones detect and capture prey using their sense of touch	
Other senses	**Examples**
Stimulus: temperature (thermal energy)	Pit vipers
Stimulus: magnetic field	Sea turtles
Stimulus: electric field	Sharks
Stimulus: barometric pressure	Birds
Stimulus: gravity	Comb jellies

One mechanistic hypothesis is that gene regulatory networks are responsible for limiting the origin of new body plans. Gene regulatory networks determine patterns of animal development and thus lay down the body plan. The scientists studying these networks have observed that once the networks are established, they are very resistant to change—that is, changes tend to be fatal. However, a multitude of "switches" and "plug-ins" can be added to networks, providing a mechanism for diversification within lineages after the body plans are established.

As research continues on the origin of body plans, biologists' understanding of the subsequent diversification within lineages is also steadily improving. Several variables have been hypothesized to play a role in the adaptive radiation of animals (Ch. 25, Section 25.3):

- ***Higher oxygen levels*** The oxygen level increased around the start of the Cambrian. This change may have made the evolution of big, mobile animals possible due to the efficiency of aerobic respiration, which provides more energy for growth and movement than is available in the absence of oxygen (Ch. 9, Section 9.1).
- ***Rise of a higher-quality food source*** The rise of algae not only produced oxygen but also served as a higher-quality food source for early animals, meaning more energy and mass were available for growth and reproduction.
- ***The evolution of predation*** When predators evolved, they exerted selection pressure on other animals for shells, skeletons, rapid movement, and other adaptations for escaping capture, launching an evolutionary "arms race."
- ***New niches beget more new niches*** As animals diversified, they themselves created new niches that could support yet more ecological diversification.
- ***Modified genes, modified bodies*** As the animal genetic tool kit evolved—such as by duplication and diversification of *Hox* genes—the potential for morphological diversity increased.

Most likely a combination of all these variables was important to the diversification of animals. The result? Animals with the same body plan diversified in their strategies for finding food, ingesting food, and reproducing in different ecological niches. What are the major themes of diversification?

Sensory Organs

The evolution of a cephalized body was a major breakthrough in the evolution of animals. Along with a mouth and brain, a concentration of sensory organs in the head region—where the animal initially encounters the environment—is a key aspect of cephalization.

Certain senses are almost universal in animals (Ch. 44, Section 44.1). The common senses include sight, hearing, taste/smell, and touch (see examples in **Table 30.2**). At least some ability to sense temperature is also common. As animals diversified, a wide array of specialized sensory abilities also evolved. For example:

- ***Magnetic field*** Many birds, sea turtles, sea slugs, and other animals can detect Earth's magnetic field and use it as an aid in navigation (Ch. 50, Figure 50.9).

- ***Electric field*** Some aquatic predators, such as sharks, are so sensitive to electric fields that they can detect electrical activity in the muscles of passing prey (Ch. 44, Figure 44.21).
- ***Barometric pressure*** Some birds can sense changes in air pressure, which may aid them in avoiding storms.

Variation in sensory abilities is important: It allows different species of animals in aquatic and terrestrial environments to collect a wide array of information about the presence of food, predators, and mates.

Feeding

To organize the diversity of ways that animals find food, biologists distinguish *what* individuals eat from *how* they eat. Animals within a lineage often pursue different food sources and feeding strategies when they occupy different niches. For example, sea cucumbers and sea stars both have an echinoderm body plan, but they have different food sources and feeding strategies—the sea cucumber mops up detritus from the seafloor, and the sea star pries open clams and mussels and consumes them.

Conversely, animals from *different* lineages often pursue the *same* food sources and feeding strategies when they occupy similar niches. For example, sea cucumbers (phylum Echinodermata) and some burrowing polychaete worms (phylum Annelida) have different body plans but use a similar strategy of consuming detritus on the seafloor.

What Animals Eat: Diversification of Ecological Roles Animals can typically be classified as **(1) detritivores** that feed on dead organic matter, **(2) herbivores** that feed on plants or algae, **(3) carnivores** that feed on animals, or **(4) omnivores** that feed on a variety of organisms, including plants, animals, fungi, protists, archaea, and/or bacteria (**Table 30.3**). These ecological roles have important implications for whole ecosystems because consumption interactions move both energy and nutrients through food webs (Ch. 53, Section 53.1).

SUMMARY Table 30.3 Diversification of Ecological Roles

Ecological Role	Example
Detritivores Feed on dead organic matter Millipedes feed on decaying leaves	
Herbivores Feed on plants or algae Pandas eat vast amounts of bamboo	
Carnivores Feed on animals Owls hunt down and consume prey	
Omnivores Feed on a combination of plants, animals, fungi, protists, archaea and/or bacteria	*Example:* Humans

Another way to categorize the diverse ecological impacts of animal consumption is to consider the effect on the organisms that are consumed. Most carnivores are **predators** that kill their **prey**. Predators are often larger than their prey and kill them quickly using an array of mouthparts and hunting strategies. Most herbivores consume some plant tissue, such as leaves, without killing the whole organism.

However, some carnivores and some herbivores are **parasites**. They're much smaller than the hosts they feed on, so they can harvest nutrients from certain parts of their hosts. **Endoparasites** live inside their hosts and usually have simple, wormlike bodies. Tapeworms, of the phylum Platyhelminthes (flatworms), are endoparasites with no digestive system. Instead of a mouth, they have hooks or other structures on their head that attach to their animal host's intestinal wall. Instead of digesting food themselves, they absorb nutrients through their skin directly from their surroundings. Leaf miners, also endoparasites, are worm-shaped insect larvae that eat the nutrient-rich tissues inside leaves. **Ectoparasites** live on the outside of their **hosts**. Ectoparasites usually have limbs or mouthparts that allow them to grasp the host, and mouthparts that allow them to pierce their host's surface and suck the nutrient-rich fluids inside. Aphids and ticks are ectoparasites.

Whether animals consume other organisms completely or partially, they can drastically decrease the fitness of the consumed organisms. Thus, animals are an important agent of natural selection—influencing the evolution of the species they eat (Ch. 52, Section 52.1).

How Animals Feed: Four General Strategies Animal mouthparts vary, and the structure of an animal's mouthparts correlates closely with its method of feeding. Keep this in mind as you review the four general tactics that animals use to obtain food, summarized in **Table 30.4** on page 642.

SUMMARY Table 30.4 Diversification of Feeding Strategies

Strategy	Example
Suspension feeders (filter feeders) Capture food by filtering out particles floating in water or drifting through the air Barnacles use specialized legs to capture plankton	
Deposit feeders Ingest organic material that has been deposited within a substrate or on its surface Sea cucumbers use feeding tentacles to mop up detritus from the seafloor	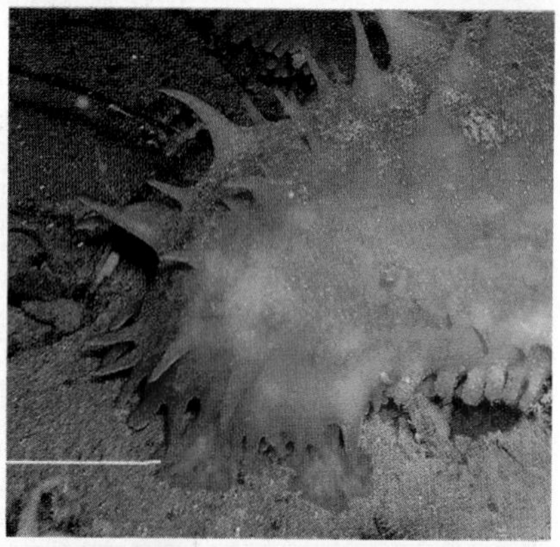
Fluid feeders Suck or mop up liquids like nectar, plant sap, blood, or fruit juice Butterflies and moths drink nectar through their extensible, hollow proboscis	
Mass feeders Take chunks of food into their mouths Lions bite off chunks of meat from prey carcasses	

Suspension feeders, also known as **filter feeders**, employ a wide array of structures to trap suspended particles—usually detritus or plankton, small organisms that drift in the currents. Sponges, like the one illustrated in Figure 30.3b, are suspension feeders. So are clams and mussels, which trap food by pumping water through their gills. Baleen whales suspension feed using sieve-like baleen plates that trap shrimp-like krill in their mouths. Barnacles capture particles with their feathery, jointed legs. Many sessile organisms are suspension feeders.

Suspension feeding occurs mostly in aquatic animals because particles float in water much more readily than in air. However, some terrestrial animals, such as flamingoes, suspension feed on aquatic organisms, and spiders suspension feed on flying insects.

Many **deposit feeders** digest organic matter in sediments; their food consists of sand- or mud-dwelling bacteria, archaea, protists, and fungi, along with detritus that settles on the surface of sediments. Earthworms, for example, swallow soil as well as leaves and other detritus on the surface of the soil. The seafloor is also rich in organic matter, which rains down from the ocean surface and collects in food-rich deposits. These deposits are exploited by a wide array of benthic animals, including many species of segmented worms (annelids) and sea cucumbers.

Unlike suspension feeders, which are diverse in size and shape and use various trapping or filtering systems, deposit feeders tend to be similar in appearance. They usually have simple mouthparts, and their body shape is wormlike.

Fluid feeders range from butterflies and moths that feed on nectar with a straw-like proboscis to vampire bats that feed on blood. Fluid feeders are found in a wide array of lineages and sometimes have mouthparts that allow them to pierce seeds, stems, skin, or other structures in order to withdraw the fluids inside. Other fluid feeders such as houseflies mop up liquids with sponge-like mouthparts.

Mass feeders ingest chunks of food. The structure of their mouthparts correlates with the type of food pieces that they harvest and ingest. Lions, for example, have razor-like teeth for tearing tough flesh into pieces small enough to be swallowed. Snails use a rasp-like structure called a radula to scrape bits of plant tissue or animal flesh into their mouths.

Movement

Animal locomotion provides an array of important functions: finding food, finding mates, escaping from predators, and dispersing to new habitats. Animals move in highly variable ways; they burrow, slither, swim, fly, crawl, walk, or run—mostly powered by muscle. Virtually all animals can be considered to have one (or more) of three types of skeletal systems that enable such complex movements (Ch. 45, Section 45.3):

1. **Hydrostatic skeletons** derive support from a flexible body wall in tension surrounding a fluid or soft tissue under compression. They occur mostly in soft-bodied organisms such as sea anemones, earthworms, and octopuses.
2. **Endoskeletons** derive support from rigid structures inside the body, such as bones in vertebrates and spicules in sponges.

3. **Exoskeletons** derive support from rigid structures on the outside of the body, such as the external armor of arthropods.

Coordinated contractions and relaxations of different muscles in the body enable various shape changes and forms of locomotion. Limbs contribute to this diversity of movement. They develop as outpockets of the body wall and can take a variety of forms, such as the lobe-like hydrostatic limbs of onychophorans (velvet worms), the jointed limbs of arthropods and vertebrates, the lobe-like and bristled parapodia of segmented worms, the flexible hydrostatic tube feet of echinoderms, and the arms and tentacles of mollusks (**Table 30.5**).

Traditionally, biologists have hypothesized that the major types of jointed and unjointed animal limbs were not homologous—that is, not descended from a common ancestor. Because the structures of animal appendages are so diverse, it seemed logical that they evolved independently of each other. As a result, biologists predicted that completely different genes are responsible for each major type of appendage.

Recent studies have challenged this view, however. For example, one experiment involves a gene called *Distal-less*, or *Dll*, that was originally discovered in fruit flies. The protein product of *Dll* seems to deliver a simple message as a fruit fly embryo develops: "Grow appendage out this way." (*Distal* means "away from the body.")

Biologists in Sean Carroll's lab set out to test the hypothesis that *Dll* was involved in the initial phase of limb or appendage formation in other animals. The team used a fluorescent marker that binds to the *Dll* gene product to locate tissues where the gene is expressed. They found that the fluorescent marker bound to *Dll* gene products in embryos of annelids, arthropods, echinoderms, and other phyla. More important, the highest concentrations of *Dll* gene products were found in cells that form appendages—even in phyla with wormlike bodies that have extremely simple appendages (**Figure 30.11**). Other experiments have shown that *Dll* is also involved in limb formation in vertebrates.

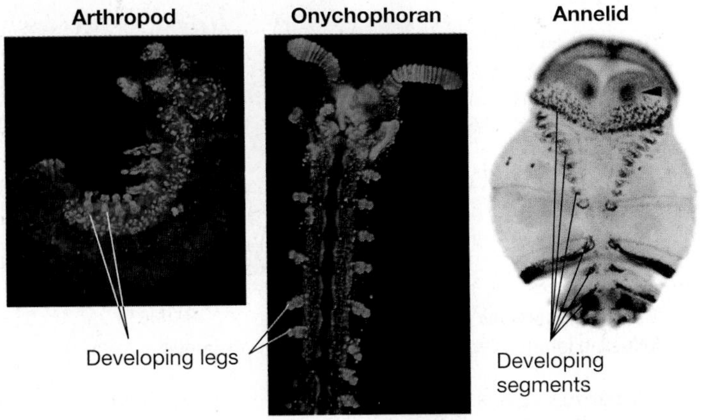

Figure 30.11 Evidence of Genetic Homology in Animal Appendages. In diverse organisms, *Dll* gene products (stained green and dark brown) are localized in areas of the embryos where appendages are forming. If the genetic tool kit for limbs was not homologous, *Dll* would be expressed in the limbs of some animals but not in others, or in a pattern unrelated to limbs.

DATA: G. Panganiban, S. M. Irvine, C. Lower, et al. 1997. *PNAS* 94: 5162–5166.

SUMMARY Table 30.5 Diversification of Limbs

Type of Limb	Example
Lobe-like limbs Onychophorans (velvet worms) use lobe-like limbs to crawl	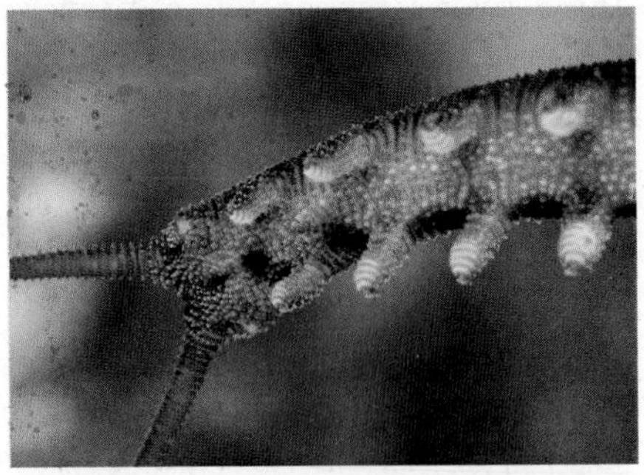
Jointed limbs The jointed limbs of arthropods (such as this crab) and vertebrates are used for locomotion and feeding	
Parapodia Polychaete worms use bristled parapodia to crawl and swim	
Tube feet Echinoderms like this sea star use tube feet to crawl and to grab prey	
Arms and tentacles Octopuses use muscular arms to crawl and to grab prey	

Based on these findings, biologists are concluding that all animal appendages have some degree of genetic homology, even if the limbs themselves evolved independently in different lineages. The idea is that the genetic tool kit of appendages evolved early in the history of the Bilateria and that, subsequently, evolution by natural selection produced the diversity of limbs observed today. In some lineages, such as snakes, the loss of limbs was an adaptation that increased diversification.

Reproduction

An animal may be efficient at moving and eating, but if it does not reproduce, the alleles responsible for its effective locomotion and feeding will not increase in frequency in the population. Evolution by natural selection occurs when individuals with certain alleles produce more surviving offspring than other individuals do (Ch. 22, Section 22.3).

Given the array of habitats and lifestyles of animals, it's not surprising that they vary tremendously in how they reproduce (**Table 30.6**). A few examples here will help illustrate the diversity of animal reproduction (see also Ch. 47, Section 47.1).

Asexual or Sexual Reproduction? In the phylum Rotifera, an entire lineage called the bdelloids (pronounced *DELL-oyds*) reproduces only asexually, through mitosis, by producing diploid eggs that can mature into adults without being fertilized—a process called **parthenogenesis**. Animals can engage in asexual reproduction by other methods, too, such as fission (splitting) as in coral polyps, and by budding, as in cnidarian polyps called hydra. Even certain fish, lizard, and snail species have been observed to undergo asexual reproduction only.

However, sexual reproduction via meiosis and fusion of gametes is by far the most common mode of reproduction in animals. Asexual reproduction tends to be more efficient than sexual reproduction, but sexual reproduction leads to greater genetic diversity, which is often favored by natural selection in variable or unfavorable environments (Ch. 13, Section 13.4). Some animals, like coral polyps, can reproduce either sexually or asexually depending on environmental circumstances.

Where Does Fertilization Occur? During sexual reproduction, fertilization may be external (in the environment) or internal (usually within the body of the female).

External fertilization is extremely common in aquatic species. Females lay eggs onto a substrate or release them into open water. Males shed swimming sperm on or near the eggs or into open water.

When internal fertilization takes place, a male typically inserts a sperm-transfer organ into the body of a female. In some animal species, males produce sperm in packets, which females then pick up and insert into their own bodies. But in seahorses, females insert eggs into the male's body, where they are fertilized. (The male is pregnant for a time and then gives birth to live young.)

Where Do Embryos Develop? Following internal fertilization, eggs and embryos may be retained in the female's body during development, or fertilized eggs may be laid outside of the body, where they develop independently or under the protection of the mother or father. "Live-bearing" and "egg-bearing" represent

SUMMARY Table 30.6 Diversification of Reproduction

Reproductive Strategy	Example
Asexual reproduction Polyps within a coral colony are genetically identical clones. They are produced by fission (splitting) or budding	
Sexual reproduction ***External fertilization*** Corals reproduce sexually by releasing eggs and sperm into the open water, where fertilization occurs	
Internal fertilization When damselflies copulate, the male holds the female behind her head with claspers and the female retrieves sperm from the male's sperm-transfer organ with the tip of her abdomen	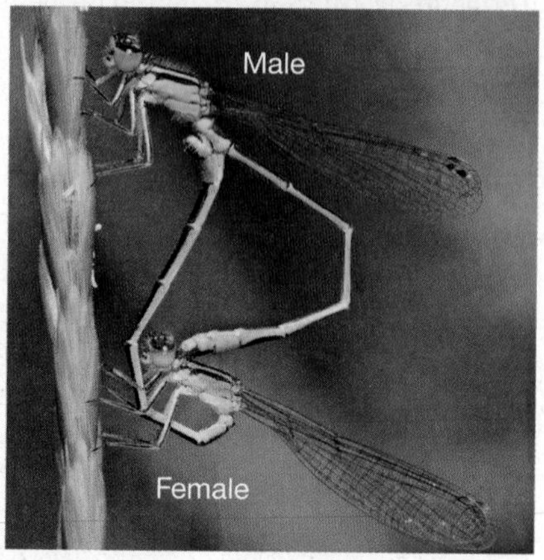

Modes of embryonic development following internal fertilization	Examples
Viviparous species nourish embryos internally and give birth to live young	Most mammals
Oviparous species deposit fertilized eggs; embryos are nourished by yolk	Most insects, birds
Ovoviviparous species retain eggs internally (embryos are nourished by yolk) and give birth to live young	Guppies, garter snakes

ends of a continuum with the gradations in between generally referred to as "egg-live-bearing":

- **Viviparous** ("live-bearing") species such as humans and most other mammals nourish embryos inside the body (nutrients are derived from the mother, generally via a highly vascularized organ called the placenta) and give birth to live young.
- **Oviparous** ("egg-bearing") species such as chickens and crickets deposit fertilized eggs. The embryos within are nourished by yolk.
- **Ovoviviparous** ("egg-live-bearing") species such as guppies and garter snakes retain eggs inside the body during early development, but the growing embryos are nourished by egg yolk instead of by nutrients transferred directly from the mother, as in viviparous species. Ovoviviparous females then give birth to well-developed young.

Most mammals and a few species of sea stars, velvet worms, fishes (including sharks), amphibians, and lizards are viviparous; some snails, insects, reptiles, and fishes are ovoviviparous. But the vast majority of animals are oviparous.

Life Cycles

Reproduction is just one component of the diverse life cycles of animals. The vast majority of sexually reproducing animals have diploid-dominant life cycles; in most cases the haploid stage is represented by relatively short-lived and single-celled gametes. But even this general pattern has interesting exceptions, such as haploid males in honeybee colonies. How else do animal life cycles vary?

Perhaps the most spectacular innovation in animal life cycles involves the phenomenon known as **metamorphosis** ("change-form"), a drastic change of form from one developmental stage to another. The young of some animals, such as humans, look recognizable as small versions of adults when they are born. Their development is said to be *direct*. However, other animals, such as sea urchins, undergo metamorphosis during their life cycle and are said to have indirect development (**Figure 30.12**).

During indirect development, embryogenesis produces **larvae** (singular: **larva**) that look radically different from adults, live in different habitats, and eat different foods. The sea urchin in Figure 30.12 has bilaterally symmetric and planktonic larvae, and it suspension feeds on single-celled algae. In contrast, the adult is radially symmetric and roams the seafloor or kelp forests, feeding on kelps and other species of macroalgae.

Through the process of metamorphosis, larvae transform into **juveniles**; they look like adults and live in the same habitats and eat the same foods as adults, but they are still sexually immature. After a period of growth and maturation, juveniles become **adults**, the reproductive stage in the life cycle.

Metamorphosis is extremely common in marine animals as well as in insects (Ch. 31, Section 31.3) and many amphibians. Why would natural selection favor such an indirect path to adulthood? One hypothesis emphasizes dispersal. In marine species that have limited or no movement as adults—such as sponges, corals, sea anemones, clams, and sea urchins—larvae function as a dispersal stage. They are a little like the seeds of many land plants—a life stage that allows individuals to move to new habitats. In insects such as butterflies, however, the larva is the main feeding stage and the adult is the dispersal stage.

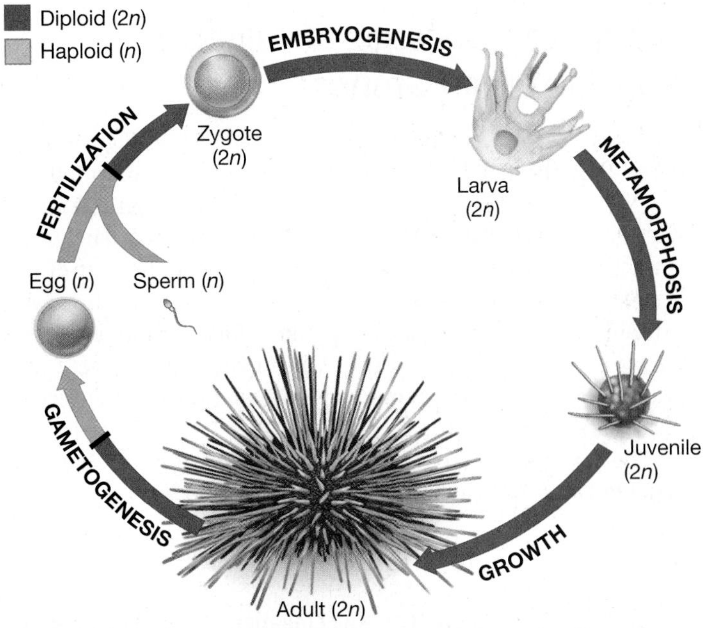

Figure 30.12 Many Animal Life Cycles Include Metamorphosis. Most sea urchins have a planktonic larva that is very different from the benthic adult. (Drawings are not to scale.)

✔ A few sea urchins produce especially large, yolk-rich eggs that develop directly into juveniles and then adults, bypassing the feeding larval stage. Circle the part of the life cycle that would be different in this case, and pose a hypothesis to explain why this difference might be adaptive under some conditions.

Another hypothesis for the occurrence of metamorphosis emphasizes feeding efficiency. Because larvae and adults feed on different foods in different ways, they do not compete with each other and can specialize to take advantage of available food sources. For example, the caterpillar larvae of butterflies can feast on nutritious leaves to acquire the energy needed for growth, while adult butterflies feast on flower nectar to acquire energy needed for migrating, finding mates, and reproducing.

The diversification of life cycles—as well as that of sensory structures, ecological roles, feeding strategies, limbs, and reproductive strategies—represent the consequences of evolutionary processes played out in ecological contexts over millions of years. Let's now examine how the general themes of diversification apply to individual phyla, starting with the earliest lineages to emerge during the origin of animals.

CHECK YOUR UNDERSTANDING

✔ If you understood this section, you should be able to . . .

1. Explain how the mouthparts of deposit feeders and mass feeders typically differ.
2. Suggest why external fertilization is more common in aquatic environments, and why internal fertilization is more common in terrestrial environments.

Answers are available in Appendix A.

30.4 Key Lineages of Animals: Non-Bilaterian Groups

The fossil record and phylogeny of animals (see Figure 30.2) agree that the Porifera (sponges), Ctenophora (comb jellies), and Cnidaria (jellyfish and others) were the earliest to diverge from the rest of the animals. Recall, however, that this finding does not mean that any of these animals (as we recognize them today) are ancestors of bilaterians (see Making Models 30.1).

After you complete this section, you should be able to ...

- Compare the three major non-bilaterian phyla of animals.

The key characteristics of non-bilaterian groups are summarized in **Table 30.7**.

Porifera (Sponges)

Sponges are a diverse group of asymmetric sessile, mostly marine, suspension-feeding animals. They occur from the deep ocean to coastal habitats of the world's oceans, providing food and habitats for other organisms. Some freshwater forms also occur. Fossil evidence shows that sponges have survived the dramatic changes on Earth for more than 700 million years.

Some traits are common among all sponges, such as a body plan composed of cell groups that create channels of water currents with small incurrent pores and larger excurrent pores (see Figure 30.3b). As water passes by feeding cells, organic debris and bacteria, archaea, and small protists are filtered out of the current and then digested at the cellular level. Some deep-sea sponges are predators, however—they capture small crustaceans on hooks that project from the body.

DIVERSITY **Table 30.7** **Key Lineages of Non-Bilaterian Animals**

Animalia >		Known Species	Feeding	Movement	Reproduction
	Porifera **Sponges** Multicellular but lacking true tissues, symmetry, muscles, and nerves; all are benthic	**8500** Marine Freshwater 0.1 mm–3 m	Most are suspension feeders; feeding cells capture bacteria and other food particles from water currents moving through cavities in the body; many species harbor photosynthetic symbionts	Most adults are sessile; limited shape-changes occur in some species due to coordinated changes in cells; dispersal is typically by ciliated larvae	Asexual reproduction occurs in a variety of ways; during sexual reproduction, fertilization is often internal
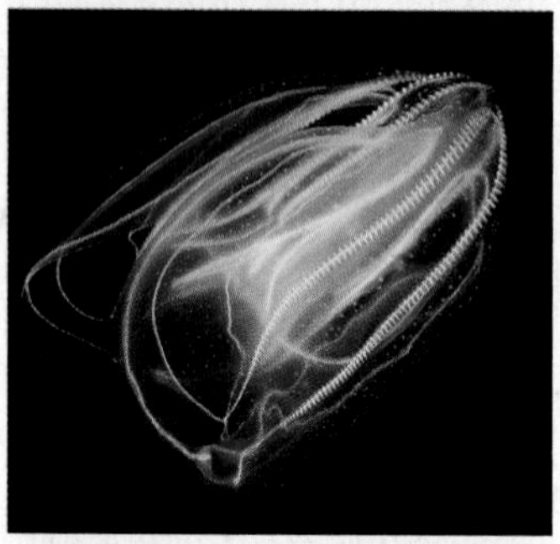	Ctenophora **Comb jellies** Biradially symmetric, diploblastic, gelatinous, and mostly planktonic with comb-like rows of cilia (bright lines in photo); no stinging cells; some are bioluminescent	**190** Marine 5 mm–1.5 m	Predators; some have two long tentacles with sticky cells that capture prey. Others capture prey using body mucus; have a mouth and two anal pores	Largest adult animals known to use cilia for locomotion; cilia occur in comblike plates that run the length of the animals; have a gravity-sensing organ	Most have both male and female organs and routinely self-fertilize, although fertilization is external; larvae are free-swimming
	Cnidaria **Jellyfish, corals, anemones, hydroids** Radially symmetric, diploblastic, gelatinous polyps (sessile) and medusae (planktonic) with stinging cells	**11,500** Marine Freshwater 5 mm–2 m (longest tentacles 75 m)	Predators; many have tentacles with stinging cells called cnidocytes that capture prey; have a single body opening; some harbor photosynthetic protists	Both polyps and medusae are capable of movement via contractions of muscle-like cells, enabling creeping and jet propulsion; larvae swim via cilia	Polyps produce new individuals asexually by budding, fission, or fragmentation; some polyps and medusae reproduce sexually; fertilization is usually external

For more go to https://goo.gl/TV2iww

Sponges are diverse in size, shape, and composition. They are distinguished partly by their **spicules**—stiff spikes of silica or calcium carbonate ($CaCO_3$) that, along with collagen fibers, give structural support to the ECM. Some sponges form thin encrusting layers on rocks, while others are spectacular "barrels" larger than a person. Sponges can also host a wide range of photosynthetic organisms, such as protists or cyanobacteria. Pigments produced by these microscopic symbionts—including green chlorophyll and orange carotenoids—are often why sponges are so colorful. What's more, sponges produce an array of toxins to defend themselves against predators and bacterial parasites. Researchers are interested in the possible use of these toxins in cancer chemotherapy.

Most adult sponges are sessile, though a few species are reported to creep over surfaces at rates of up to 4 millimeters per day. Recent research has confirmed that at least one species contracts its body to expel waste products, while others contract surface cells to change the rate of water flow through pores. These shape changes occur at the cellular level and do not involve muscles. Most species disperse via larvae that swim with the aid of cilia.

Asexual reproduction occurs in a variety of ways, such as fragmentation and budding, depending on the species. For example, a sponge fragment containing small groups of certain types of adult sponge cells can develop into a new adult. Although individuals of most species are hermaphrodites, producing both eggs and sperm, self-fertilization is rare because individuals release their male and female gametes at different times. Fertilization is often internal—choanocytes capture sperm from the water and deliver them to eggs held within with the sponge body.

Ctenophora (Comb Jellies)

Although a few species of comb jellies creep on the ocean floor, most are planktonic and predatory. Fewer than 200 species have been described to date, but some are abundant enough to represent a significant fraction of the total planktonic biomass. If you swim in the sea, you will likely bump up against comb jellies but not realize it because they are so transparent in the water.

Ctenophores lack the stinging cells that are characteristic of cnidarians. However, they have specialized sticky cells, called colloblasts, for trapping prey—a synapomorphy for this group. Despite their harmless, gelatinous appearance, ctenophores are efficient predators. For example, accidental introductions of the ctenophore *Mnemiopsis leidyi* (see Table 30.7), which preys on eggs and larvae of fishes and competes with other predators for food, have devastated fish production in the Black Sea, North Sea, and Caspian Sea.

Cnidaria (Jellyfish, Corals, Anemones, Hydroids)

Cnidarians are found in all of the world's oceans, occupying habitats from surface waters to benthic substrates. A key synapomorphy of the cnidarians is a specialized cell called a **cnidocyte** (pronounced *NIGH-doh-site*), which is used in prey capture. When a cnidocyte senses a fish or other type of prey, it forcibly ejects a barbed, spear-like structure that may contain toxins. The barbs hold the prey, and the toxins subdue it until it can be brought to the mouth and ingested. Cnidocytes are commonly located near the mouth or on elongated tentacles. A few cnidarian species produce toxins that are deadly to humans as well as to prey organisms; in Australia, twice as many people die each year from stings by box jellyfish as from shark attacks.

Many cnidarians have a life cycle that includes both a sessile **polyp** form that reproduces asexually and a free-floating **medusa** that reproduces sexually (see the hydrozoan medusa in the chapter-opening photo). Anemones and corals, however, exist only as polyps—never as medusae (see anemones and corals in Figure 30.6, Table 30.2, and Table 30.6).

Reef-building corals are ecosystem engineers—each polyp secretes a calcium carbonate endoskeleton that is left behind long after the polyp dies. Each generation of corals builds on the secretions of its predecessors, gradually creating the physical structure of a coral reef—one of the world's most productive habitats (Ch. 53, Figure 53.8).

Tropical waters are typically nutrient poor and hold little food for suspension feeders such as coral polyps. But corals have a mutualistic relationship with photosynthetic protists. The photosynthetic cells hosted by corals provide nutrients for the coral, and the corals provide habitat and protection to the cells. However, some coral reefs are now endangered due to global warming and ocean acidification (Ch. 54, Section 54.2). When the coral polyps become stressed, they expel the colorful symbionts and take on a "bleached" appearance that can be fatal.

While coral reefs are threatened by human activities in many parts of the world, jellyfish are thriving in some affected waterways. Researchers predict that jellyfish populations will soar in some areas over the next century, even as many other animal populations decline.

In this section you reviewed the non-bilaterian groups of animals. In the next two chapters, you'll get a closer look at bilaterians, starting with the most abundant and diverse animals on Earth, the protostomes (Chapter 31), and moving on to the largest predators on Earth, the deuterostomes (Chapter 32).

CHECK YOUR UNDERSTANDING

✔ If you understood this section, you should be able to . . .

MODEL Add bars to the phylogeny in Figure 30.2 to show the origin of unique traits (synapomorphies) for Porifera, Ctenophora, and Cnidaria.

Answers are available in Appendix A.

30.1 What Is an Animal?

- Animals are multicellular, heterotrophic eukaryotes with cells that lack cell walls and that ingest their prey.
- All animals move under their own power at some point in their life cycle. All animals other than sponges have specialized nerve and muscle cells that enable complex movements.
- Animals are eating machines, the largest consumers on Earth.
- Animals comprise 30–35 phyla and may number 3 million or more species.

30.2 What Key Innovations Occurred during the Origin of Animal Phyla?

- Animals are a monophyletic group that originated from a protist ancestor that was probably similar to choanoflagellates.
- The basic genetic tool kit for multicellular animals originated at the very root of the animal lineage. Tool-kit genes then diversified or were co-opted, networked, or lost in various ways in different lineages.
- Sponges can be complex and have many cell types, but lack well-defined tissues; their body plan is composed of groups of cells that form internal channels for water currents; most sponges lack symmetry.
- Most ctenophores have biradial symmetry, two embryonic germ layers, and neurons organized into a nerve net.
- Most cnidarians have radial symmetry, two embryonic germ layers, and neurons organized into a nerve net. However, subtle bilateral symmetry occurs in some cnidarians.
- Most animal species are bilaterians. They have bilateral symmetry, three embryonic germ layers, and a coelom (or evolved from an ancestor that had a coelom)—features that gave rise to a tube-within-a-tube body plan.
- Most bilaterians also have a centralized nervous system and cephalized bodies—meaning that a distinctive head region contains the mouth, brain, and sensory organs.
- The two main groups of bilaterians are the protostomes and deuterostomes.
- Segmentation is a key innovation in annelids, arthropods, and vertebrates.
- The evolution of animals is much more intricate than a smooth transition from simple to complex. Many key innovations did not arise all at once. Some evolved more than once. Some were lost. Evolution did not stop in any of the lineages.

30.3 What Themes Occur in the Diversification within Animal Phyla?

- Sensory abilities and sensory structures vary among species and correlate with their habitats and feeding methods.
- Animals fill four different ecological roles depending on their source of food: They are detritivores, herbivores, carnivores, or omnivores.
- Animals capture food in four ways: suspension feeding, deposit feeding, fluid feeding, and mass feeding.
- Although the types of appendages used in animal locomotion range from simple lobe-like limbs to complex jointed legs, some of the genes that indicate where appendages develop are homologous.
- Asexual reproduction occurs in many species, but sexual reproduction predominates. Fertilization may be external or internal, and embryos may complete development inside or outside the mother.
- Development may be direct or indirect. During indirect development, metamorphosis—a dramatic morphological transformation from one life stage to another—takes place.

30.4 Key Lineages of Animals: Non-Bilaterian Groups

- Poriferans (sponges) are diverse, mostly marine, benthic suspension feeders that pump water through internal chambers to feed.
- Ctenophores (comb jellies) are marine planktonic predators; they are the largest animals to use cilia for locomotion.
- Cnidarians (jellyfish, corals, anemones, hydroids) are diverse and abundant in ocean habitats around the world. The cnidocyte, or stinging cell, is a synapomorphy for this group.

Answers are available in Appendix A.

✔ TEST YOUR KNOWLEDGE

1. Which synapomorphy (shared, derived trait) distinguishes animals as a monophyletic group, distinct from choanoflagellates?
   a. multicellularity
   b. coloniality
   c. heterotrophy
   d. movement
2. In a tube-within-a-tube body plan, what is the interior tube derived from?
   a. ectoderm
   b. mesoderm
   c. the coelom
   d. endoderm
3. True or false? New major animal body plans have continued to evolve since the Cambrian about 550 million years ago.
4. Which of these characteristics does not apply to cnidarians?
   a. are diploblastic
   b. possess a gastrovascular cavity with one opening
   c. undergo metamorphosis
   d. have a central nervous system

✔ TEST YOUR UNDERSTANDING

5. Why is it ecologically significant that animals are heterotrophic *and* multicellular?
6. **QUANTITATIVE** To estimate the relative abundance of the major phyla, calculate how many named species of arthropods, mollusks, and nematode worms exist per named species of chordate (the phylum containing vertebrates, including humans; see Table 30.1). Do you think these calculations are likely to be underestimates or overestimates? Why?

7. **THINK CAREFULLY** Evaluate this statement: Animals evolved from simple to complex.

8. Which of these choices is an example of homology (similarity due to common ancestry)?
   a. suspension feeding in sponges and clams
   b. ectoparasite lifestyle in aphids and ticks
   c. cnidocytes (stinging cells) in jellyfish and sea anemones
   d. radial symmetry in cnidarians and echinoderms

✔ TEST YOUR PROBLEM-SOLVING SKILLS

9. **SOCIETY** Suppose that a gene originally identified in nematodes (roundworms) is found to be homologous with a gene that can cause developmental abnormalities in humans. Would it be possible to use fruit flies as a model organism to study this gene? Explain.

10. Humans possess which of the following traits? Select True or False for each trait.
    T/F triploblasty
    T/F parthenogenesis
    T/F viviparity
    T/F metamorphosis

✔ PUT IT ALL TOGETHER: Case Study

Why aren't comb jellies the most diverse animals on Earth?

The vast majority of animals that ever existed are now extinct (Ch. 25, Section 25.4), but Tereza Jezkova and John Wiens wondered which variables were most important in driving the diversification of species that exist today. Why are there so many species in some phyla, such as Cnidaria (see photo), but so few in others, such as Ctenophora?

11. **QUANTITATIVE** Draw a horizontal axis to represent the number of species within phyla using a logarithmic scale (1, 10, 100, 1000 species, etc.; see **BioSkills 5**). Then use Table 30.1 to map seven representative phyla from small to large at intervals of about an order of magnitude on this scale.

12. Based on your reading of this chapter, propose at least five traits that you think might have been most important in triggering diversification within phyla (examples: origin of hearing, origin of internal fertilization).

13. **MODEL** Jezkova and Wiens used a type of graph called a linear regression to find correlations between variables such as the proportion of species per phylum with legs (on the *y*-axis) and the diversification rate per phylum (on the *x*-axis). Sketch a graph to show what a strong positive correlation between these two variables would look like and what the absence of a correlation would look like.

14. **PROCESS OF SCIENCE** A sample of Jekova and Wiens' results is shown here. The R^2 value represents the strength of the correlation (where 0.00 is lowest and 1.00 is highest). The P value represents the statistical significance (see **BioSkills 3** to review how to interpret R^2 and P values). Which five traits look most important?

Presence of ...	R^2	P
Head	0.06	0.209
Coelom	0.01	0.672
Digestive system with two openings	0.09	0.111
Bilateral symmetry	0.06	0.214
Photoreceptors/eyes	0.38	< 0.001
Legs	0.08	0.153
Skeleton	0.14	0.047
Large size	0.02	0.516
Sexual reproduction	0.03	0.354
Separate sexes	0.17	0.027
Metamorphosis	0.04	0.300
Parasitic lifestyle	0.26	0.005
Terrestrial lifestyle	0.27	0.005
Mobile lifestyle	0.01	0.597

Source: T. Jezkova and J. J. Wiens. 2017. *American Naturalist* 189: 201–212.

15. Which traits do not correlate strongly with diversification rate *within* phyla, but are likely to have been important in the original diversification of animal phyla during the Cambrian? Select True or False for each trait.
    T/F presence of a head
    T/F mobile lifestyle
    T/F terrestrial lifestyle
    T/F bilateral symmetry

16. **PROCESS OF SCIENCE** The researchers know that correlation does not equal causation. However, can the *absence* of a correlation enable you to reject a hypothesis of causation? How would the R^2 values be different in a scenario where a single trait was important to diversification in many phyla versus a scenario where different traits were important to diversification in different phyla?

31 Protostome Animals

In species richness and numbers of individuals, protostomes, especially insects, are the most diverse and abundant animals. Shown here are just a few species from the most diverse group of insects on the planet: the beetles.

In this chapter you will learn that

Protostomes are the most diverse and abundant animals on Earth

starting with

What is a protostome? **31.1**

then asking

What is a lophotrochozoan? **31.2**

and examining

Key lineages: Flatworms, Annelids, Mollusks

then asking

What is an ecdysozoan? **31.3**

and examining

Key lineages: Roundworms, Arthropods

The five major animal lineages in the tree of life include the sponges (Phylum Porifera), the comb jellies (Phylum Ctenophora), the corals and jellyfish (Phylum Cnidaria), and the two bilaterian lineages, the protostomes and deuterostomes. Of these, the **protostomes** are the most diverse and abundant lineage of animals on Earth. To put this in perspective, consider that of the more than 30 animal phyla known to exist, at least 22 are protostomes (see Ch. 30, Table 30.1). You will likely recognize many of these phyla:

- *Annelida*, segmented worms including earthworms, polychaete worms, and leeches
- *Mollusca*, including snails, clams, octopuses, and squid
- *Arthropoda*, including insects, crustaceans (shrimp, lobsters, crabs), chelicerates (spiders and mites), and myriapods (millipedes, centipedes)

BIG PICTURE

This chapter is part of the Big Picture. See how on pages 722–723.

Some protostome phyla have little diversity, such as the phylum Priapula, which contains only 16 named species. On the other end of the spectrum, over 85,000 mollusks have been named so far, and about 1.2 million arthropod species (about a million of these are insects). Some scientists estimate that the actual number of arthropod species may be over 10 million.

Certain protostomes are not only diverse (in numbers of species) but also extremely abundant (in numbers of individuals). A typical acre of pasture in England is home to almost 18 *million* individual beetles and up to 9 *billion* roundworms. The world population of ants is estimated to be 1 million billion (1,000,000,000,000,000) individuals.

Given the diversity and abundance of protostomes, it's no surprise that they fill important ecological roles. As a group, protostomes live in virtually every aquatic and terrestrial habitat in the world and are detritivores, herbivores, or carnivores. If one of biology's most fundamental goals is to understand the diversity of life on Earth, then protostomes—particularly the mollusks and arthropods—demand our attention.

The ecological importance of protostomes extends to the health and welfare of humans in profound and diverse ways. For example:

- Some protostomes—such as clams, oysters, scallops, crabs, shrimp, lobsters—are a major direct source of food for humans, which makes them economically important.
- Humans rely on protostomes indirectly for their ecosystem services. For example, many farmers rely on earthworms to make soil fertile, on bees and other insects to pollinate their crops (Ch. 54, Section 54.3), and on insects and spiders that protect crops from herbivorous pests by eating or parasitizing them.
- Some protostomes (especially insects) can have a massive negative impact on food production by damaging crops.
- Some protostomes produce valuable materials such as silk (silkworm moth caterpillars) and pearls (oysters and mussels).
- Many protostomes play an important role in causing or transmitting human diseases, and they are common parasites.
- The protostomes include two of the most important model organisms in all of biological science: the fruit fly *Drosophila melanogaster* and the roundworm *Caenorhabditis elegans* (see **BioSkills 11**).

Biologists are urgently trying to identify protostome species and the roles they play in ecosystems. In this current age of humans, informally known as the Anthropocene (Ch. 25, Section 25.4), the rate of species extinction is escalating due to habitat loss and other threats (Ch. 54, Section 54.2). Many protostomes may become extinct before they are even discovered. Of the insect species monitored by the International Union for Conservation of Nature, 33 percent are in decline.

Let's take a closer look at what characterizes the protostomes and then explore the diversity within the major phyla.

31.1 What Is a Protostome?

The diversity of protostomes is hard to overstate. As the pie chart in **Figure 31.1** shows, the bright-colored wedges representing the major protostome phyla take up much more of the total number of named species than the non-protostome phyla, those labeled Chordata, Echinodermata, Cnidaria, and Porifera. It's easy to forget that most animals are protostomes, because the animals we tend to be most concerned about are ourselves and other vertebrates, in phylum Chordata.

After you complete this section, you should be able to ...

- Compare the protostomes to other animals.

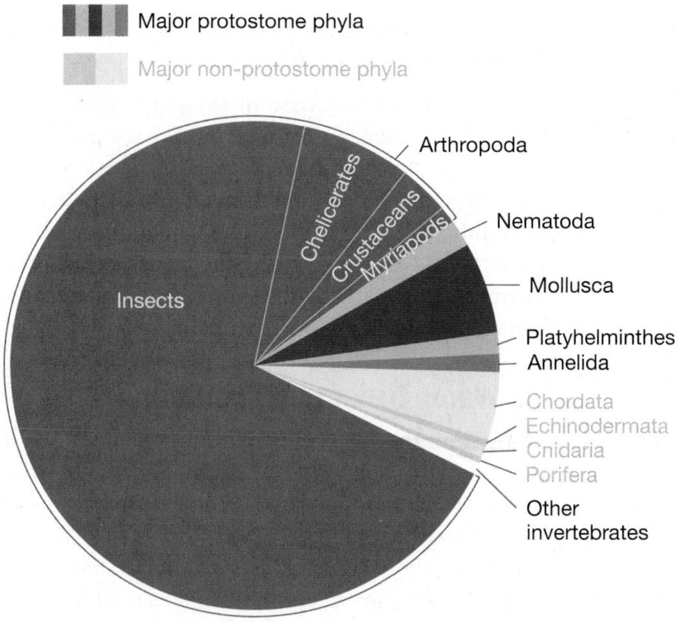

Figure 31.1 The Relative Diversity of Animal Lineages. Most animal species are protostomes. About 70 percent of all known species of animals on Earth are insects, most of them beetles. "Other invertebrates" include two dozen or so minor phyla, almost all of them protostomes.

DATA: The IUCN Red List of Threatened Species. Version 2017.2.

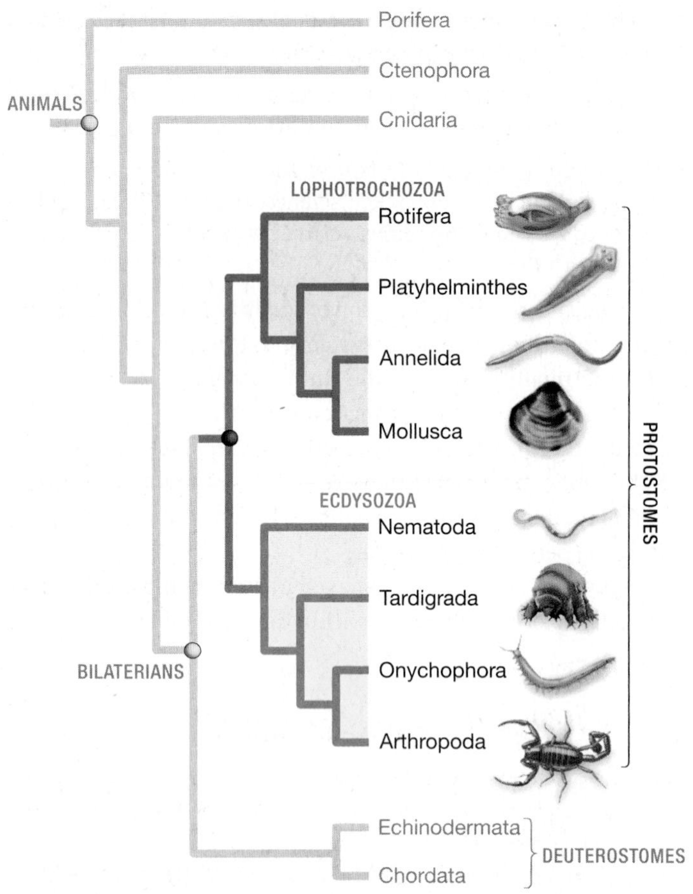

Figure 31.2 Protostomes Are a Monophyletic Group Comprising Two Major Lineages. There are 22 phyla of protostomes, but the 8 major phyla shown account for about 99.5 percent of the known species.

The phylogenetic tree in **Figure 31.2** provides some evolutionary context for examining the protostomes. To read the figure, start at the root of the tree (at top left) showing the origin of animals at the gray dot. The exact branching pattern is still under debate, but researchers agree that the Porifera, Ctenophora, and Cnidaria were the first lineages to split off during the early radiation of animals (Ch. 30, Section 30.2).

Now find the other gray dot (at bottom left) on the phylogeny in Figure 31.2 and put your finger on it. This node represents the most recent common ancestor of a monophyletic group of animals called the bilaterians (see **BioSkills 13** for help interpreting phylogenetic trees). The ancestor at this node was likely bilaterally symmetric and triploblastic (having three germ layers)—characteristics that gave rise to an array of body plans among bilaterians during the Cambrian explosion (Ch. 30, Section 30.2). These diverse bilaterians fall into two major monophyletic groups: the protostomes, indicated by colored branches, and the deuterostomes, represented in the tree by echinoderms and chordates.

Biologists traditionally distinguished the protostomes (literally, "first-mouth") from the deuterostomes ("second-mouth") by using developmental characteristics:

- Embryonic development of the mouth before the anus during gastrulation (rather than the anus before the mouth in deuterostomes)
- Inability of isolated early embryonic cells to develop into a complete embryo (rather than the ability of such cells to do so in deuterostomes)
- Formation of a coelom by the splitting of blocks of mesodermal cells (rather than by mesodermal pockets pinching off the gut in deuterostomes)

However, as researchers have examined the development of more diverse bilaterians, they have discovered many exceptions to the "protostome versus deuterostome" developmental patterns. During the 500 million years of evolution after the bilaterians diversified, some convergent evolution occurred, producing similar developmental characteristics in different lineages. For example, the anus develops before the mouth in some protostomes. Meanwhile, other characteristics thought to be synapomorphies—shared, derived traits within a lineage—were lost in some groups.

These insights seem to paint a messier picture at first, but they have opened up exciting areas of research that integrate genetics, development, evolution, and ecology. And despite the complexity of developmental data, molecular phylogenies consistently support the division of bilaterians into the protostomes and deuterostomes—there is widespread consensus on these major groupings within the Bilateria.

Now put your finger on the black dot in Figure 31.2. This node represents the most recent common ancestor of all the protostomes. The order of subsequent diversification within the protostomes has been greatly debated. Recent analyses of DNA sequence data have rejected many previous hypotheses based on morphological data. DNA sequence data consistently support two major subgroups within the protostomes:

1. the Lophotrochozoa (shown by purple branches), including annelid worms and mollusks, and
2. the Ecdysozoa (shown by red branches), including nematodes (roundworms) and arthropods.

Before considering each of these groups in more detail, let's examine two themes in the evolutionary diversification of the protostomes. Recall that all animal lineages evolved diverse ways of sensing the environment, feeding, moving, reproducing, and developing (Ch. 30, Section 30.3). In addition, diversification in protostomes was triggered by the transition from aquatic to terrestrial habitats, and by the compartmentalized and highly modifiable body plans of many protostomes.

The Water-to-Land Transition

Fossils from the Burgess Shale in Canada, Chengjiang in China, and other sites show that protostome lineages originated in the ocean (Ch. 25, Section 25.3). Today, protostomes are the most abundant animals in the world's oceans as well as in virtually every freshwater and terrestrial habitat. Like land plants and fungi, protostomes made the transition from aquatic to terrestrial environments.

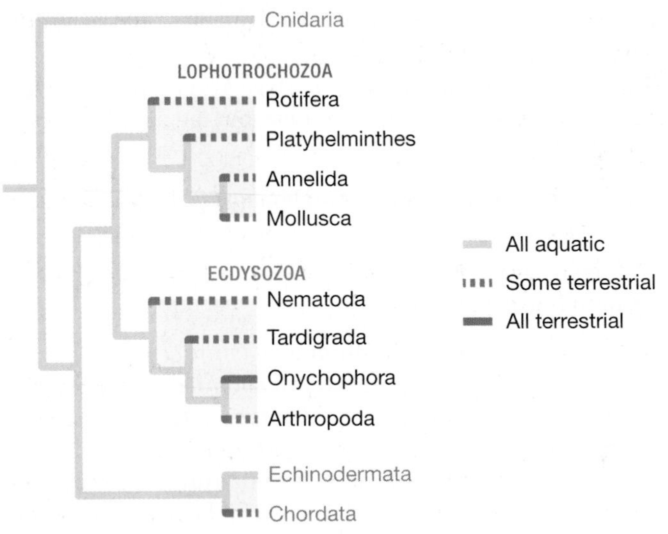

Figure 31.3 The Water-to-Land Transition Occurred Several Times in the Protostomes. The transition occurred only once in the deuterostomes, in phylum Chordata.

✔ According to this tree, terrestrial living arose independently eight different times in protostomes. Why wouldn't you conclude instead that terrestrial living was ancestral and the aquatic lifestyle arose multiple times independently?

To help put this achievement in perspective, note that green plants made the move from fresh water to land just once (Ch. 28, Section 28.3). It is not yet clear whether fungi moved from aquatic habitats to land once or several times (Ch. 29, Section 29.3). Only one lineage among deuterostomes moved onto land: tetrapod vertebrates (Ch. 32, Section 32.4). But protostomes made a water-to-land transition multiple times as they diversified (**Figure 31.3**).

Many protostome lineages, especially snails (Mollusca) and the insects and spiders (Arthropoda), survived on land. They also experienced an **adaptive radiation**—a period of rapid speciation resulting in a large number of descendant species that have adapted to a variety of habitats. Adaptive radiations tend to occur due to a combination of ecological opportunity and morphological, physiological, and/or behavioral innovations (Ch. 25, Section 25.3).

Ecological Opportunity Water-to-land transitions are important because they open up entirely new habitats and new types of resources to exploit. The transition of protostomes to land coincided in many cases with the adaptive radiation of plants on land. Land plants were able to thrive given the increased availability of sunlight and carbon dioxide on land compared to water. As a result, the high net primary productivity of land plants provided the energy and nutrients for entire terrestrial food webs (Ch. 53, Section 53.1) as well as physical niches where protostomes could survive (Ch. 49, Section 49.1).

Key Morphological, Physiological, and Behavioral Innovations Recall that an **adaptation** is a heritable trait that increases the fitness (reproductive success) of individuals in a particular environment. Several adaptations were required for protostomes to thrive on land, such as the ability to:

- exchange gases with the air,
- avoid drying out, and
- hold up their bodies under their own weight.

Land animals exchange gases with the atmosphere readily as long as they have a large, moist surface area that is exposed to the air (Ch. 42, Section 42.3). The bigger challenge is to prevent the gas-exchange surface and other parts of the body from drying out. How do terrestrial protostomes solve this problem?

- Roundworms, earthworms, and other terrestrial protostomes live in humid soils or other moist environments and exchange gases across their body surface. They have a high surface-area-to-volume ratio, which increases the efficiency of gas exchange.
- Some terrestrial arthropods and mollusks have tracheae, gills, or other respiratory structures located *inside* the body, minimizing water loss (Ch. 40, Section 40.4).
- Insects evolved a waxy layer that minimizes water loss from the body surface, with openings to respiratory passages that can be closed if the environment dries. Terrestrial snails (other than slugs) have a shell that minimizes water loss.
- Desiccation-resistant eggs covered with special membranes or shells evolved repeatedly in populations that made the transition to life on land. Further, some terrestrial protostomes engage in internal fertilization, which can help to prevent eggs (and sperm) from drying out.

Unlike land plants and fungi, many land animals can also move to moister habitats if the area they are in gets too dry.

What are some of the mechanical constraints of living on land? Without the upward buoyant force of water, animals on land require greater structural support to hold their bodies up against gravity and to move (Ch. 45, Section 45.4). This mechanical constraint has limited the size of many protostomes, including insects, on land.

Compartmentalized and Flexible Body Plans

Protostomes such as mollusks and arthropods have relatively compartmentalized body plans—their bodies are divided into different regions with different functions. This compartmentalization appears to have been a key innovation that enabled regions to evolve somewhat independently of each other in these lineages, a kind of evolutionary flexibility.

If you combine the concept of compartmentalization of body plans with the tendency for small changes in gene expression of regulatory genes to produce structural and functional innovation, you can begin to understand a mechanism for the dramatic diversification of protostomes in new ecological contexts.

For example, recall how *Hox* genes are known to play a role in establishing the anterior-to-posterior identities of body regions of most animals. Michalis Averof and Nipam Patel compared the expression of *Ubx-AbdA Hox* genes during embryonic development in four crustacean arthropods, which have segmented

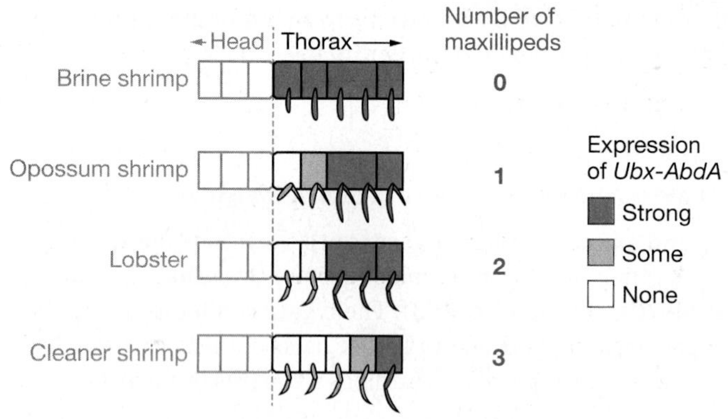

Figure 31.4 Small Changes in Gene Expression Can Lead to Morphological Innovation. In animals whose body plans are compartmentalized, such as segmented arthropods, evolution can occur in some body regions without affecting the function of other regions, providing evolutionary flexibility.

SOURCE: M. Averof and N. H. Patel. 1997. *Nature* 388: 682–686.

bodies. Figure 31.4 shows that although these crustaceans have the same number of thorax segments and express the same *Ubx-AbdA* gene (shown in red), a change in the expression pattern coincides with a change in the number of legs used for locomotion (red) compared to feeding appendages (blue; called maxillipeds because they occur on the thorax rather than the head). Notice how the change in functional identity of one segment does not necessarily change the functional identity of others.

Keep the concept of compartmentalization as a form of structural and functional flexibility in mind as you read more about the diversity of protostomes.

CHECK YOUR UNDERSTANDING

✔ **If you understood this section, you should be able to ...**

1. **QUANTITATIVE** Use Figure 31.1 and/or Table 30.1 to rank the five most diverse phyla of animals from largest to smallest number of species.
2. Propose one example of an adaptation for terrestrial living that may have arisen via convergent evolution in the protostomes that made the water-to-land transition independently.

Answers are available in Appendix A.

31.2 What Is a Lophotrochozoan?

The phylogeny in Figure 31.2 shows that the lophotrochozoans are a monophyletic group, meaning that they are all descendants of a common ancestor. The 13 phyla within the Lophotrochozoa (pronounced *loh-foh-troh-koh-ZOH-ah*) include the rotifers, flatworms (phylum **Platyhelminthes**; pronounced *plah-tee-hell-MIN-theez*), annelids, and mollusks (see the full list in Table 30.1).

After you complete this section, you should be able to ...

- Compare the lophotrochozoans to other protostomes.

The name Lophotrochozoa is derived from two distinctive morphological traits that occur in some, but not all, members of this lineage:

1. A feeding structure called a lophophore, which is found in three phyla
2. A type of larva called a trochophore, which is common to many of the phyla

As Figure 31.5a shows, a **lophophore** ("tuft-bearer") is a specialized structure that rings the mouth and functions in suspension feeding. Lophophores are found in bryozoans (moss animals), brachiopods (lamp shells), and phoronids (horseshoe worms).

Trochophores are a type of larvae common to marine mollusks, marine annelids, and several other phyla in the Lophotrochozoa. As Figure 31.5b shows, a **trochophore** ("wheel-bearer") larva has a ring of cilia around its middle. These cilia allow swimming and, in some species, sweep food particles into the mouth.

Trochophore larvae, like other larvae, occur in animals that undergo indirect development—where larvae often look radically different from adults, live in different habitats, and eat different foods (Ch. 30, Section 30.3). Recent analyses suggest that the trochophore larva originated early in the evolution of lophotrochozoans and later evolved into different larval types in some groups. However, larvae are not unique to lophotrochozoans—even the earliest known animals, the sponges, have larvae.

What other characteristics are unique to lophotrochozoans? Recent phylogenetic analyses have supported a long-standing hypothesis that the spiral pattern of cleavage in embryos is a synapomorphy for this group. When cells divide at oblique angles to each other during early embryogenesis, a spiraling pattern of cells in the blastula results (Figure 31.6a). This is in contrast to other patterns of cleavage, such as radial cleavage, where cells divide at right angles to each other (Figure 31.6b). Although spiral cleavage has been highly conserved in some lophotrochozoan phyla, it has been modified or lost in others. ✔ **You should be able to indicate the origin of spiral cleavage on Figure 31.2, as well as the loss of spiral cleavage in Rotifera.**

In sum, if you were to observe a lophophore, a trochophore larva, or spiral cleavage in an unfamiliar animal, you could be confident that the animal belongs to the Lophotrochozoa. However, not all lophotrochozoans have all three of these traits (current synapomorphies for the lophotrochozoa are all molecular).

Lophotrochozoans are morphologically diverse, ranging from tiny globular rotifers to giant clams that weigh as much as a person. Table 31.1 on page 656 summarizes the features of key lineages of lophotrochozoans.

Many lophotrochozoan phyla include species with long, thin, tubelike bodies that lack limbs. That is, they are **worms** with a basic tube-within-a-tube design (Ch. 30, Section 30.2). The outside tube is the skin, which is derived from ectoderm; the inside tube is the gut, which is derived from endoderm. Muscles and organs derived from mesoderm are located between the two tubes. Worms may or may not have a coelom (fluid-filled space lined

(a) Lophophores function in suspension feeding in adults.

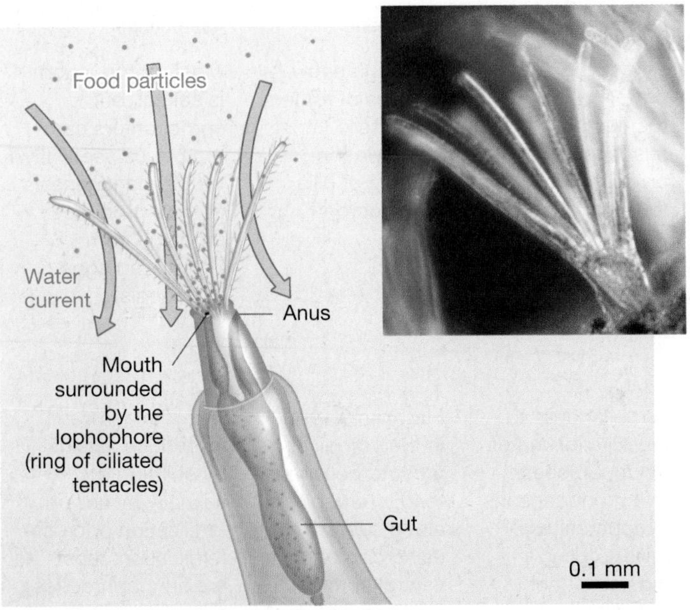

(b) Trochophore larvae swim and may feed.

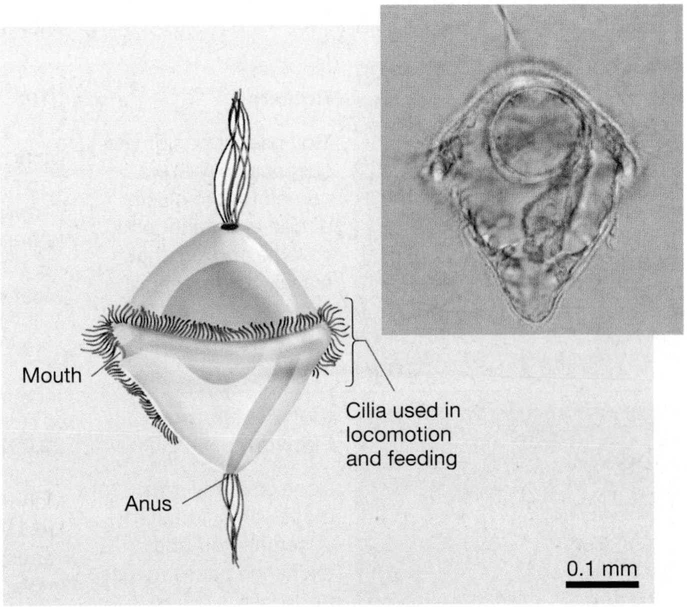

Figure 31.5 Lophotrochozoans Are Named after Two Traits. (a) Three phyla of lophotrochozoans, including the bryozoan shown here, have the feeding structure called a lophophore. **(b)** Many phyla of lophotrochozoans have the type of larva called a trochophore.

with mesodermally derived tissues; for example, flatworms do not). Either way, the entire body of a worm is usually involved in the hydrostatic skeleton, which enables shape changes and locomotion in soft-bodied organisms (Ch. 45, Section 45.3).

Despite their relatively simple body plan, wormlike lophotrochozoans have diverse feeding methods, including suspension, deposit, fluid, and mass feeding (Ch. 30, Section 30.3). Let's take a closer look at two of the most diverse groups of wormlike lophotrochozoans, the flatworms and the annelids. Then let's consider the most diverse phylum of lophotrochozoans—the mollusks.

What Is a Flatworm?

Flatworms are named for the broad, flattened shape of their bodies. (The Greek roots *platy* and *helminth* mean "flat-worm.") They lack a coelom and structures specialized for gas exchange and circulation of oxygen and nutrients. Before molecular data were available, many scientists concluded from this morphological simplicity that flatworms branched off earlier in the evolution of bilaterians. However, molecular data now confirms the placement of flatworms within the Lophotrochozoa, providing an important reminder that evolution is not always a tidy sequence of increasing complexity and that the *loss* of traits can be adaptive.

Biologists interpret the flattened bodies of flatworms as an adaptation that gives them an extremely high surface-area-to-volume ratio. This flat body plan allows nutrients and gases to diffuse efficiently to all of the cells inside the animal with minimal expenditure in complex internal structures. One physical constraint to this body plan is that the body surface must be moist for gas exchange to take place. Flatworms are thus restricted to aquatic and moist terrestrial environments. ✔ **You should be able to indicate the origin of the flattened, acoelomate body plan on Figure 31.2.**

(a) Spiral cleavage is unique to lophotrochozoans.

(b) Radial cleavage, for comparison

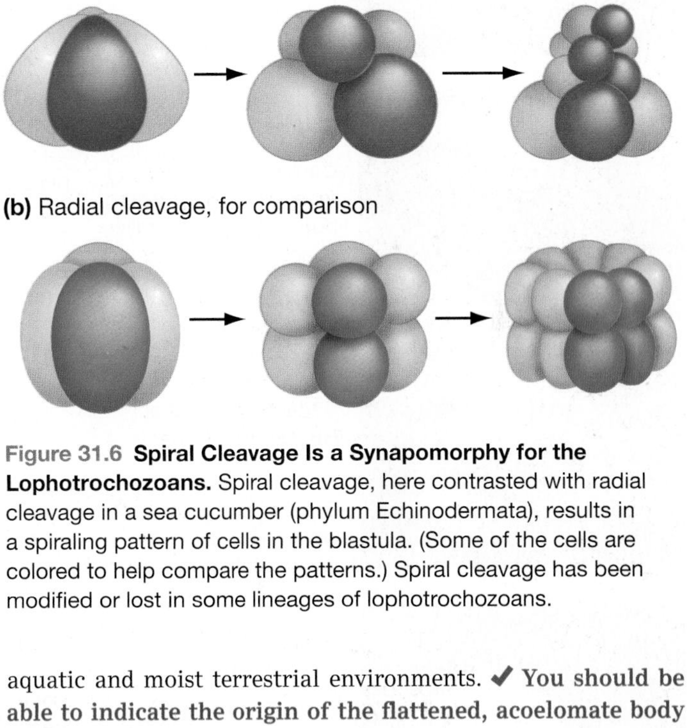

Figure 31.6 Spiral Cleavage Is a Synapomorphy for the Lophotrochozoans. Spiral cleavage, here contrasted with radial cleavage in a sea cucumber (phylum Echinodermata), results in a spiraling pattern of cells in the blastula. (Some of the cells are colored to help compare the patterns.) Spiral cleavage has been modified or lost in some lineages of lophotrochozoans.

The flatworms are a large and diverse phylum traditionally categorized into four major groups:

1. *Turbellaria* is a non-monophyletic collection of free-living flatworms. Although some turbellarian species live in moist terrestrial environments, most are freshwater or marine,

DIVERSITY Table 31.1 Key Lineages of Lophotrochozoans

Animalia > Protostomia > Lophotrochozoa >		Known Species	Feeding	Movement	Reproduction
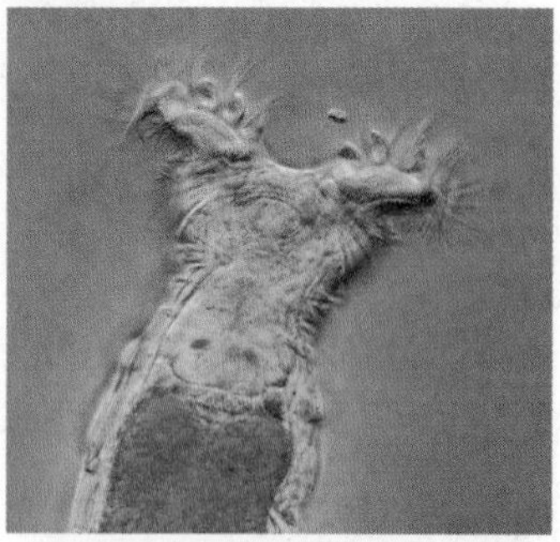	Rotifera **Rotifers** Tiny pseudocoelomate suspension feeders named for the cluster of cilia at anterior end, called a corona, that waves in a rotating pattern	**2100** Marine Freshwater Terrestrial (damp soils) Most < 1 mm	Cilia in the corona create a current that enables suspension feeding on microscopic food particles	Although a few species of rotifers are sessile, most swim via the beating of cilia in the corona	Most reproduction is sexual, but some rotifers can reproduce asexually via parthenogenesis (female produces embryos from unfertilized eggs via mitosis)
	Platyhelminthes **Flatworms** Acoelomate worms, most with a flattened, unsegmented body; exchange gases across body wall Traditional groups: **Turbellarians** **Cestodes** (tapeworms) **Trematodes** (flukes) **Monogeneans**	**20,000** Marine Freshwater Terrestrial (moist areas) 1 mm–20 m	Turbellarians (free-living), flukes (endoparasites), and monogeneans (ectoparasites) have a "blind" digestive tract with only one opening for ingestion of food and elimination of wastes; tapeworms (endoparasites) have no gut or mouth and absorb nutrients across their body wall	Flatworms vary in their mobility; most turbellarians tend to creep along substrates using cilia on their ventral surface; some can swim by undulating their bodies; tapeworms have hooks that attach them to their host	Turbellarians can reproduce asexually by splitting or sexually by self-fertilization or cross-fertilization; most are hermaphrodites; tapeworms and flukes reproduce sexually and have complex life cycles involving multiple hosts
	Annelida **Segmented worms** Coelomate worms, most with a segmented body and bristles called chaetae Traditional groups: **Polychaetes** (bristle worms) **Oligochaetes** (earthworms et al.) **Hirudinea** (leeches) Also: Echiuria and Sipuncula, which have lost their segmentation	**16,800** Marine Freshwater Terrestrial 1 mm–3 m	Polychaetes have diverse feeding strategies such as suspension feeding, mass feeding, and deposit feeding; almost all oligochaetes are deposit feeders in soils; about half of leeches are ectoparasites while others are predators or scavengers	Polychaetes and oligochaetes crawl or burrow, many using chaetae to improve traction; some polychaetes use appendages called parapodia that function as paddles; many polychaetes and leeches are good swimmers	Asexual reproduction occurs in some annelids via fragmentation or parthenogenesis; most polychaetes have separate sexes and external fertilization; some have a trochophore larva; oligochaetes and leeches are hermaphrodites with internal cross-fertilization
	Mollusca **Snails, clams, chitons, squid** Body plan includes muscular foot, visceral mass, and mantle Traditional groups: see **TABLE 31.2**	**85,000** Marine Freshwater Terrestrial 1 mm–14 m	Snails, chitons, and cephalopods have a rasping structure called a radula that is specialized for diverse feeding strategies; bivalves use gills for suspension feeding	The muscular foot is modified in many ways, enabling creeping in snails, slugs, and chitons; digging in clams; grasping in cephalopods	In many species, the sexes are separate, but many snails are hermaphrodites; fertilization is external in some, internal in others; some have a trochophore larva; sexual reproduction is the norm

For more go to https://goo.gl/GX4JUc

gliding over substrates using their ciliated ventral surface. Some marine flatworms are brightly colored and several centimeters long. However, the freshwater turbellaria common to biology labs, such as *Dugesia*, are more drab and only a few millimeters long (**Figure 31.7a**).

Most turbellarians prey on protists or small animals; others scavenge dead animals. You may be surprised to learn that they have a blind digestive tract—it has only one opening used for both ingesting food and eliminating wastes. The mouth is located near the middle of the body's ventral surface, at the end of a tubelike *pharynx*. In some flatworms, the pharynx can extend and pierce live prey or other food.

2. *Cestoda* are strictly endoparasitic tapeworms that parasitize diverse vertebrates, absorbing nutrients by diffusion across their body wall. Humans most often acquire tapeworms by eating undercooked pork, beef, or fish. This is why raw meats are often packaged with instructions for thorough cooking. Humans spend millions of dollars yearly on treating and preventing tapeworms in pets and farm animals.
3. *Trematoda* are endoparasitic or ectoparasitic flukes that parasitize vertebrates, arthropods, annelids, and mollusks. In humans, a fluke is responsible for schistosomiasis—a serious public health issue in many tropical and subtropical nations, infecting more than 200 million people worldwide (**Figure 31.7b**). Flukes are parasites with mouths; they feed by gulping host tissues and fluids, often causing blood in the host's urine or stool, kidney failure, and other problems.
4. *Monogenea* are tiny ectoparasites that parasitize specific tissues of particular species, usually the skin or gills of fishes.

Turbellarians can reproduce asexually by splitting in half, or sexually by self-fertilization or cross-fertilization during the copulation of hermaphroditic individuals (with both female and male organs). Monogeneans are also hermaphrodites. In contrast, the reproductive systems and life cycles of flukes and tapeworms are extremely complex. In many cases the life cycle involves two or even three distinct host species.

What Is a Segmented Worm?

Unlike flatworms, most **annelids** have a coelom, a fully developed digestive tract with a mouth and an anus, and a segmented body. See **Making Models 31.1** on page 658 for tips on using simple models to compare the body plans of different phyla.

Before molecular data were available, biologists grouped annelids and arthropods together in the same clade due to their segmentation. Molecular data have overturned this view, suggesting that segmentation arose independently in these groups. However, researchers are discovering that some of the same developmental tool-kit genes are involved in establishing segmentation in both phyla, suggesting convergent evolution at the morphological level but some degree of deep homology at the genetic level (Ch. 30, Section 30.2). ✔ **You should be able to indicate the origin of annelid segmentation on Figure 31.2.**

The common ancestor of the annelids had a key synapomorphy in addition to segmentation: bristle-like extensions called **chaetae** that extend from lobe-like appendages called **parapodia** (**Figure 31.7c**). Once again, although these appendages appear very different from the appendages of other bilaterian phyla, the homologous expression patterns of genes such as *Distal-less*, or *Dll*, in the appendages of diverse phyla suggest

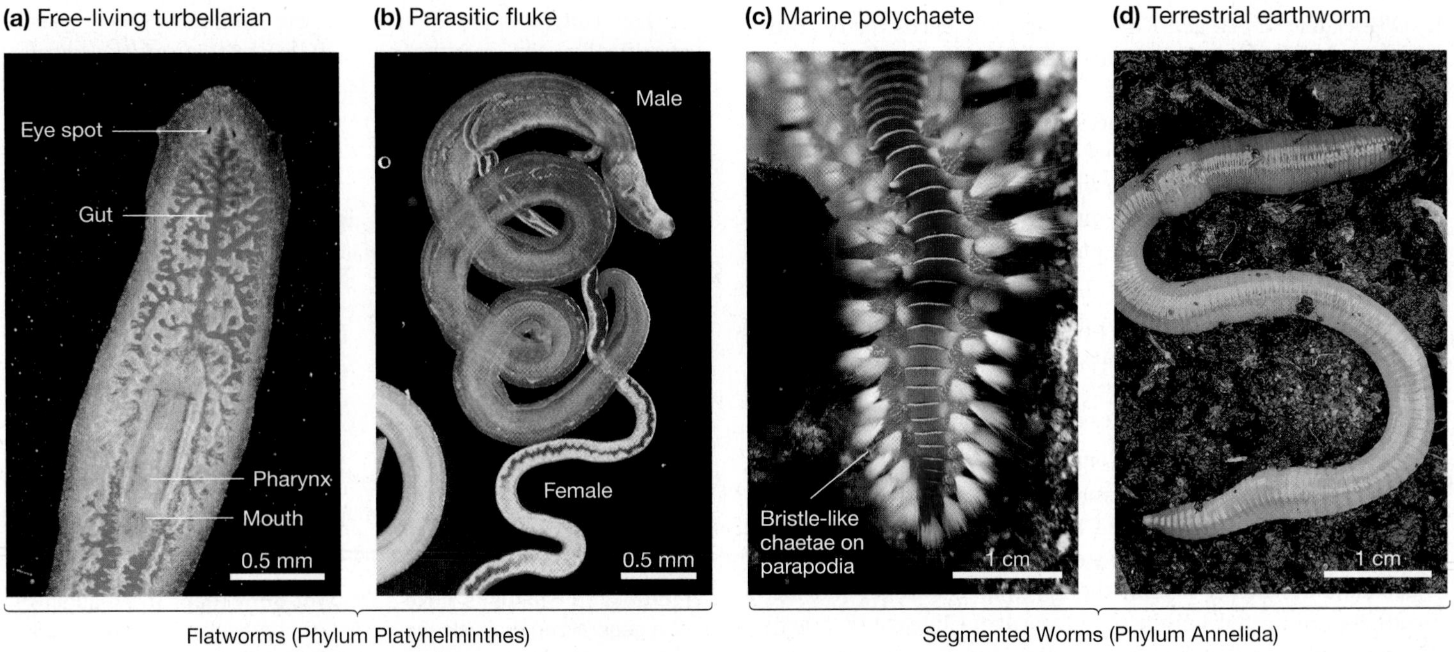

Figure 31.7 Wormlike Lophotrochozoans Are Diverse. (a) Flatworms in the genus *Dugesia* are common lab specimens. **(b)** The blood fluke *Schistosoma mansoni* sickens millions of humans in tropical regions. **(c)** The polychaete *Hermodice carunculata* has especially conspicuous bristles (chaetae). **(d)** The earthworm *Lumbricus terrestris* can be considered an ecosystem engineer due to its important role in the formation and aeration of soil.

> **Making Models 31.1 Tips on Drawing Animal Body Plans**
>
> Drawing simple diagrams of animal body plans can help you compare key features in different phyla. The amount of detail you include depends on how many features you would like to compare. Simple lines work best, and color can be very useful.
>
> 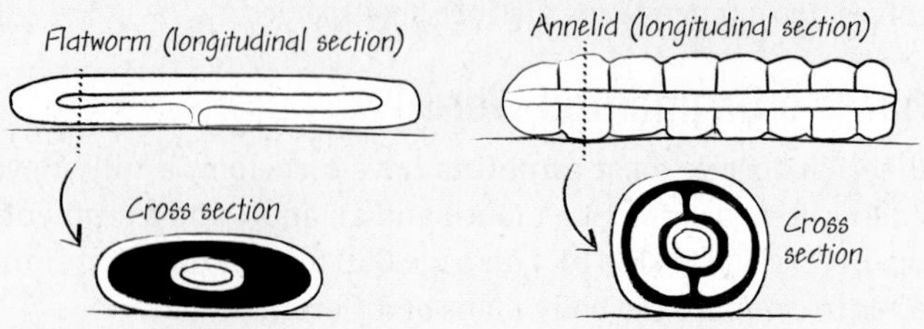
>
>
> **MODEL** Label the mouth in the flatworm. Label the mouth and anus in the annelid. In both cross sections, color the tissues derived from ectoderm blue and those from endoderm yellow (tissue derived from mesoderm is already shown in black). Draw small dots in the coelom of the annelid.
>
> To see this model in action, go to the Study Area of **Mastering Biology**

a common origin of the genetic tool kit for appendages (Ch. 30, Section 30.3).

Annelids were traditionally divided into three groups:

1. *Polychaeta* (*pol-ee-KEE-tah*), meaning "many-bristles," are diverse worms that live in a wide variety of marine habitats. Polychaetes that are highly mobile often have large parapodia and chaetae. More sedentary polychaetes have reduced parapodia and smaller, but still numerous, chaetae—adaptations for burrowing in sediments or living in tubes.
2. *Oligochaeta* (*ol-ee-goh-KEE-tah*) are the earthworms and other oligochaetes ("few-bristles"). Virtually all oligochaetes make their living by deposit feeding in soils (Figure 31.7d). Their tunnels are critically important in aerating soil, and their feces contribute large amounts of organic matter. Charles Darwin and many other biologists have remarked on the often unsung earthworm, which provides ecosystem services essential to natural ecosystems as well as agriculture.
3. *Hirudinea* (*hee-ru-DIN-ee-ah*) are the leeches. About half of leeches are ectoparasites that attach themselves to fish, humans, or other hosts and suck blood and other body fluids. The nonparasitic leech species are predators or scavengers.

Molecular data have added complexity to this three-group categorization. The Polychaeta and Oligochaeta are now known to be paraphyletic—that is, non-monophyletic groups including some but not all members of a clade (Ch. 25, Table 25.2); the Oligochaeta are nested within the Polychaeta lineage, and the Hirudinea are nested within the Oligochaeta lineage. Further, unsegmented worms called Sipunculida and Echiura are now grouped within Annelida—their ancestors were segmented, but this trait was lost over time. Research continues to clarify the phylogeny of the annelids.

What Is a Mollusk?

Mollusks (also spelled "molluscs") are a highly diverse monophyletic group of lophotrochozoans, comprising more than 85,000 species. One synapomorphy for mollusks is a body plan based on the following three major components (Figure 31.8):

1. the **foot**, a large muscle located at the base of the animal, used in movement;
2. the **visceral mass**, the region containing the main organ systems; and
3. the **mantle**, an outgrowth of the body wall that covers the visceral mass, forming an enclosure called the mantle cavity. In many species, the mantle secretes a hard calcium carbonate shell.

Once the molluscan body plan evolved, subsequent diversification was largely driven by adaptations that allowed the mollusks to move, feed, or reproduce in novel ways. The result was a dramatic radiation of mollusks into several major lineages, summarized in Table 31.2, including **chitons** (mollusks with dorsal shells made of plates), **bivalves** (clams and mussels), **gastropods** (slugs and snails), and **cephalopods** (squid and octopuses).

The Muscular Foot Is Separated from the Visceral Mass Snails and chitons have a large, muscular foot at the base of the body that works as a type of hydrostatic skeleton called a muscular hydrostat: While not containing a fluid-filled chamber or internal organs, it is composed of muscle tissue that is incompressible and can transmit shape changes (just like your tongue). How can an animal move with just one foot? Waves of muscle contractions sweep backward or forward along the length of the foot, allowing individuals to creep along a surface (see Figure 31.9 on page 660).

The compartmentalization of the molluscan body plan—specifically, the relative physical separation of the muscular foot and visceral mass (compared, for example, to the

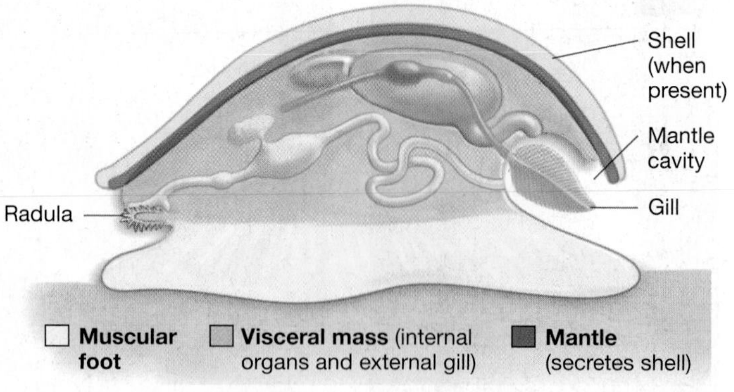

Figure 31.8 **Mollusks Have a Distinct Body Plan.** This diagram of a generalized mollusk shows that the body plan is based on a foot, a visceral mass, and a mantle. Gills are located inside a cavity created by the mantle.

✔ You should be able to indicate the origin of the molluscan body plan on Figure 31.2.

tube-within-a-tube body plan of annelids)—may have enabled greater diversification of the muscular foot across the phylum. For example:

- In bivalves such as clams, the foot is modified as a digging appendage.
- The foot of cephalopods such as squid and octopuses is modified as arms and **tentacles**: long, muscular extensions with suckers that aid in movement and prey capture. (Octopuses have eight arms. Squid and cuttlefish have eight arms and two longer tentacles specialized for grasping; look ahead to Figure 31.11)
- In sea butterflies (swimming sea snails), the foot forms winglike flaps that enable swimming.

With physical separation between the muscular foot and visceral mass, variations in the muscular foot could occur and be

DIVERSITY Table 31.2 Key Lineages of Mollusks

	Protostomia > Lophotrochozoa > Mollusca >	Known Species	Feeding	Movement	Reproduction
	Polyplacophora **Chitons** Named "many-plate-bearing" for their eight dorsal plates, or valves	**1000** Marine 2–33 cm	Most chitons are grazing herbivores; they use a radula to scrape algae and other organic matter off rocks	Waves of contractions along the length of the foot allow chitons to move by creeping, similar to how snails move	In most species the sexes are separate, and fertilization is external (or within mantle cavity); most species have a trochophore larva
	Bivalvia **Clams, mussels, scallops, oysters** Named "two-valves" for the two hinged parts of the calcium carbonate shell, which can be closed by a muscle in many species	**9200** Marine Freshwater 1 mm–1.2 m	Most are suspension feeders, creating a flow of water across gills within the mantle cavity; a single oyster can filter as much as 200 liters of water per day; no radula	Clams burrow into sediments using their muscular foot; scallops clap their valves together in a form of jet propulsion; mussels and oysters are sessile	Most reproduce sexually and have external fertilization; eggs develop into a trophophore larva, then another type of larva called a veliger
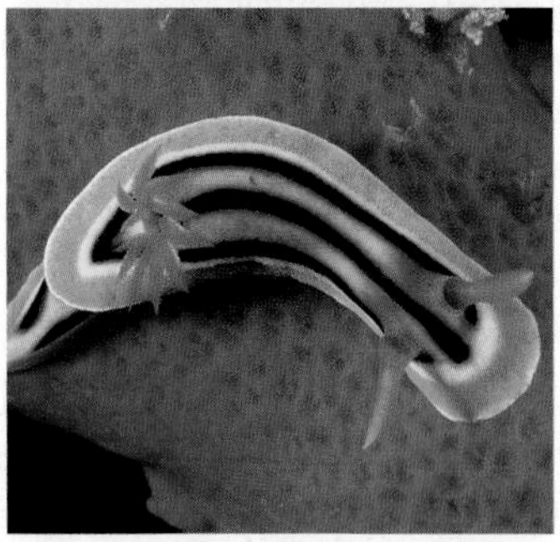	Gastropoda **Snails, slugs, nudibranchs** Named "belly-feet" for the large, muscular foot on ventral side; snails have a shell; slugs and nudibranchs have no shell	**70,000** Marine Freshwater Terrestrial 1 mm–1 m	Radula is diverse, enabling many feeding strategies; most gastropods are herbivores or detritivores, but some are predators (some can bore holes through shells of other mollusks)	Waves of contractions along the length of the foot allow gastropods to move by creeping; sea butterflies have winglike flaps that enable swimming	Most reproduce sexually; some reproduce asexually by parthenogenesis; most marine gastropods produce a veliger larva; terrestrial larvae are retained within egg case
	Cephalopoda **Squid, octopuses, nautiluses, cuttlefish** Named "head-feet" for large "head" that actually consists of the visceral mass; foot is modified as arms and tentacles; external shell only in nautiluses	**800** Marine 6 mm–14 m	Highly intelligent predators that hunt by sight and use arms or tentacles to capture prey; some cuttlefish and octopuses inject poison into prey; have a radula and a beak that can bite forcefully	Highly agile; squid swim by jet propulsion; octopuses crawl using arms; cuttlefish swim by undulating lateral fins; cephalopods can release a jet of "ink" to disorient predators	Sexes are separate, and some species have elaborate courtship displays; internal fertilization; no larvae

For more go to https://goo.gl/pcnv3Q

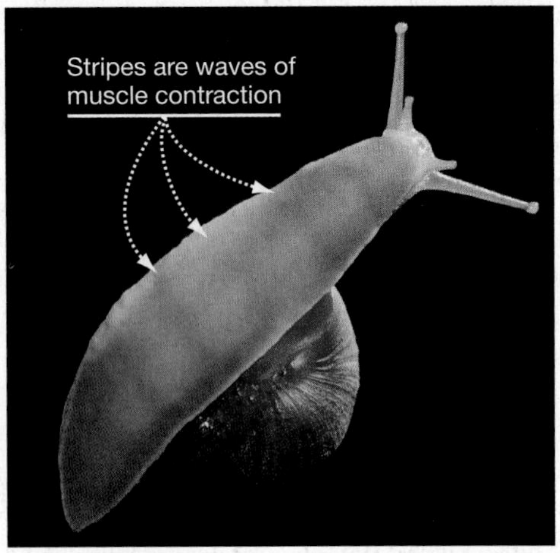

Figure 31.9 **How Snails Walk on One Foot.** Waves of muscle contractions, moving forward or backward along the length of the foot, allow mollusks like this snail to creep along a substrate.

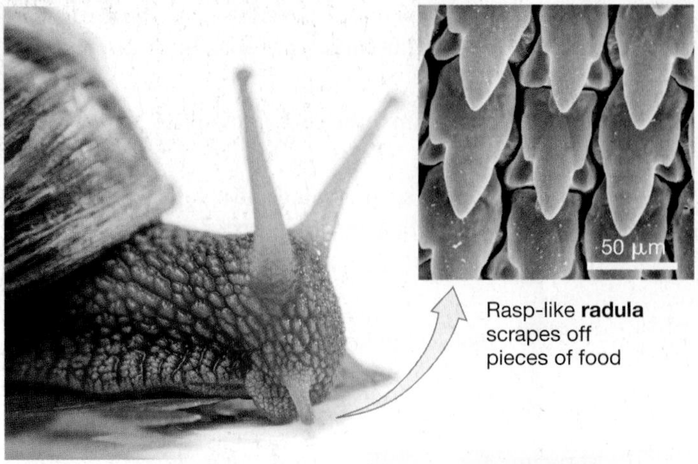

Figure 31.10 **The Radula Is Unique to Mollusks.** The size and shape of the radular teeth are highly correlated with the type of food eaten by the species.

honed by natural selection without compromising the function of the visceral mass. This compartmentalized molluscan body plan was thus a key innovation that enabled mollusks to exploit new ecological niches.

The Visceral Mass Includes the Radula and Gills The visceral mass is the region in all mollusks where organs and surrounding fluids are located. The coelom itself is highly reduced in most mollusks, functioning mostly in reproduction and excretion of wastes. The organs occupy a different type of body cavity called a **hemocoel** ("blood-hollow"; pronounced *HEE-mah-seal*), where body fluids bathe the organs directly in an open circulatory system (Ch. 42, Section 42.5). The hemocoel is different from a coelom because it is not lined in tissue derived from mesoderm and has a distinct developmental origin (usually originating from the blastocoel).

At the anterior end of the visceral mass, the mouth has a unique molluscan feeding structure called a **radula**, which functions like a rasp or file. The mollusk moves the radula back and forth over the food source, causing the many sharp plates to scrape material so that it can be ingested (Figure 31.10). Because such an unusual and complex structure is not likely to have evolved more than once, the radula probably evolved early in molluscan evolution. ✔ **You should be able to indicate the origin of the radula on Figure 31.2.** The radula was later lost in the bivalves, which acquire food by suspension feeding.

The gastropod radula is particularly diverse, varying in size and shape among species. Although most gastropods are herbivores or detritivores, specialized types of radular teeth allow some gastropods to act as predators. Species called drills, for example, use their radula to bore a hole in the shells of other mollusks.

The visceral mass often includes one or two external gas-exchange structures called **gills**. Gills are formed from a large number of tiny filaments that create a high surface area for gas exchange. In bivalves the gills are modified, forming a large suspension-feeding "net" that can capture small animals, protists, prokaryotes, and detritus from water pumped across the gills by cilia. A single oyster can filter the organic matter from 200 liters of seawater in a day, improving water clarity. Large restoration efforts are under way in places such as the Chesapeake Bay to restore oyster populations that help improve water quality.

Gills are highly vulnerable to drying out when exposed to air. Some terrestrial snails are limited to moist environments and keep their gills protected deep within their mantle cavities. Other terrestrial snails have lost their gills entirely and breathe using an internal lung instead, a novel organ derived from an infolding of the mantle.

The Mantle Has Different Functions in Different Groups In many species the mantle secretes a shell made of calcium carbonate. Some mollusk species have a shell with one, two, or eight parts, called valves; others have no shell at all. In bivalves, the protective shell is hinged and closes with the aid of muscles attached to the two valves. Many marine and terrestrial snails can retract into their shells when they are attacked or when their tissues begin to dry out. The shells of terrestrial snails are thinner and lighter than those of marine snails—an adaptation for life on land.

Numerous adaptations of the mantle allow diverse functions other than secreting shells. For example:

- In most terrestrial snails, the mantle forms an internal lung. This lung was a key innovation that led to a rapid diversification of snails on land during the Carboniferous period.
- In bivalves, the mantle is lined with muscle and forms tubes called **siphons**. For example, in clams, which burrow in soft sediments, two siphons extend upward into the water column, controlling incurrent and excurrent water flow over the gills.
- In cephalopods, the mantle forms a siphon that functions in jet propulsion. The mantle cavity fills with water and then the mantle muscles contract, forcing a stream of water out of a single moveable siphon and propelling the animal in the opposite direction (Figure 31.11).

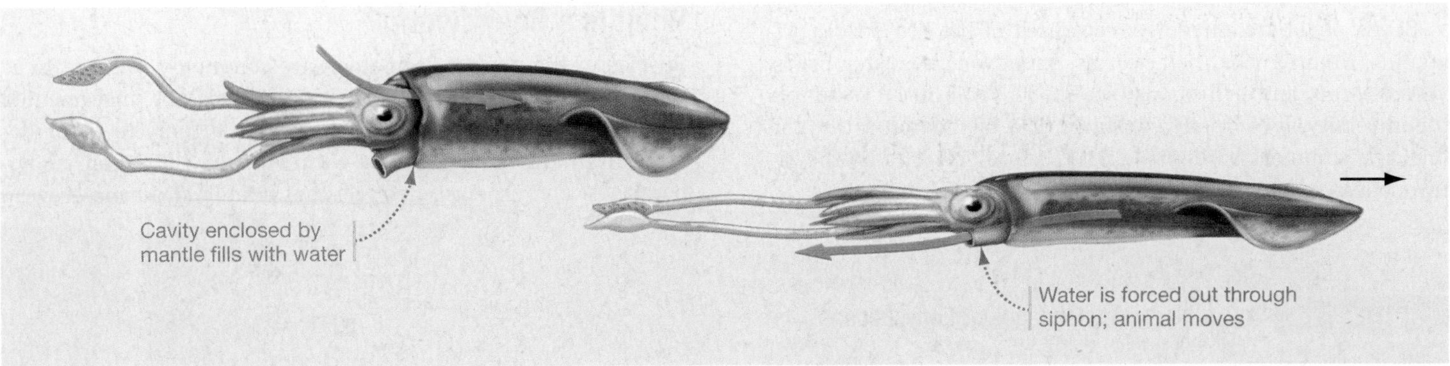

Figure 31.11 Jet Propulsion in Mollusks. In jet propulsion, muscular contractions of the mantle force the water out through a movable siphon formed by the mantle.

✔ Circle the part of the squid that is homologous to the foot of a snail.

The take-home message is that the compartmentalized molluscan body plan appears to have served as a morphological innovation that, in combination with ecological opportunity, enabled the spectacular diversification of mollusks. In other words, while all mollusks have a foot, visceral mass, and mantle, the sizes, shapes, and functions of each of these structures have diversified in ways that enable drastically different methods of feeding, moving, and reproducing. Other than the arthropods, mollusks are the most diverse phylum of animals on Earth.

CHECK YOUR UNDERSTANDING

✔ **If you understood this section, you should be able to ...**

1. Describe three ways you would expect a land-dwelling snail to differ from a snail that lives in water.
2. **MODEL** Draw and label a simple model of a squid, showing the three main features of the molluscan body plan.

Answers are available in Appendix A.

31.3 What Is an Ecdysozoan?

After you complete this section, you should be able to ...

- Compare the ecdysozoans to other protostomes.

Lophotrochozoans, like humans, increase their size continuously and incrementally during growth periods (Figure 31.12a)—the ancestral manner of growth for animals. In astonishing contrast, the Ecdysozoa (pronounced *eck-die-soh-ZOH-ah*) are characterized by a clear synapomorphy: All ecdysozoans grow intermittently by **molting**—that is, by shedding an exoskeleton or external covering.

The Greek root *ecdysis*, which means "to slip out, or escape," is appropriate because during a molt, an individual sheds its outer layer, or **cuticle**—called an **exoskeleton** if it is hard—and slips out of it (Figure 31.12b). Once the old covering is shed, the body is pumped up with fluid—causing it to expand—and a new cuticle or exoskeleton forms that is larger than the old one. As ecdysozoans grow and mature, they undergo a succession of molts, sometimes making dramatic morphological transformations along the way.

A stiff body covering is advantageous because it provides an effective structure for muscle attachment and affords protection. However, the trade-off is that during molting, the soft bodies of ecdysozoans are exposed and vulnerable to predators.

A hormone called ecdysone is important in regulating the molting cycle (Ch. 46, Section 46.3). Is ecdysone a key innovation for the ecdysozoans? No—studies show that this steroid hormone serves a variety of functions in many other animals. Thus, it appears that ecdysone was co-opted in the ecdysozoans, where it serves a novel function.

(a) Lophotrochozoans grow incrementally.

(b) Ecdysozoans grow by molting.

Figure 31.12 Lophotrochozoans and Ecdysozoans Differ in Their Mechanism of Growth. (a) Lophotrochozoans grow incrementally, as seen in the growth bands on this clam, which show periods of slower growth (narrow bands) and more rapid growth (wide bands). **(b)** In contrast, ecdysozoans such as this cicada undergo the process of molting, crawling out of their old exoskeletons and growing new, larger ones.

Seven phyla are currently recognized in the Ecdysozoa: priapulids, kinorhynchs, hair worms, roundworms, water bears, velvet worms, and arthropods (see Ch. 30, Table 30.1). To sample this diversity, let's briefly consider three of the non-arthropod lineages, summarized in **Table 31.3**, before delving into the arthropods in greater detail.

What Is a Roundworm?

Species in the phylum Nematoda are commonly referred to as **roundworms**, or **nematodes**. Roundworms are unsegmented worms with a pseudocoelom (lined only partially in tissue derived from mesoderm), a tube-within-a-tube body plan, no appendages, and an elastic cuticle that is molted during growth.

DIVERSITY **Table 31.3** Key Lineages of Ecdysozoans

	Animalia > Protostomia > Ecdysozoa >	Known Species	Feeding	Movement	Reproduction
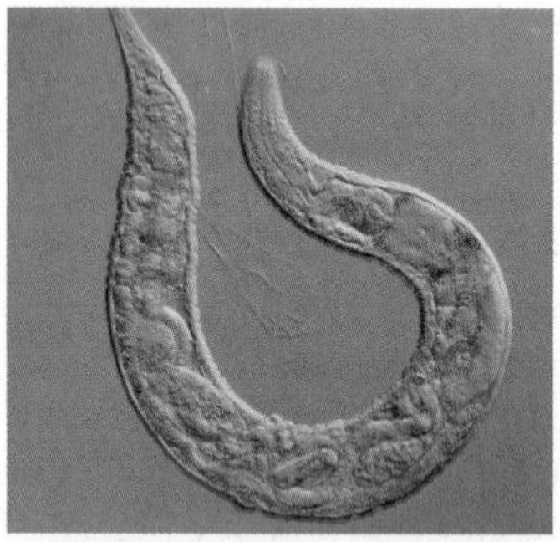	Nematoda **Roundworms** Unsegmented worms with no appendages, a pseudocoelom, and a thick, elastic cuticle that must be molted during growth; abundant worldwide	**25,000** Marine Freshwater Terrestrial <1 mm–9 m	Feed on wide variety of materials, including bacteria, archaea, fungi, plant roots, small protists, small animals, and detritus; most are free-living but some are parasitic; mouthparts vary	Move using an undulating motion of the hydrostatic skeleton; longitudinal muscles bend the body laterally on alternating sides (have no circumferential muscles)	Sexes are separate in most species (*C. elegans* is mostly hermaphroditic); internal fertilization leads to egg laying and direct development of offspring; individuals molt four times during lifetime
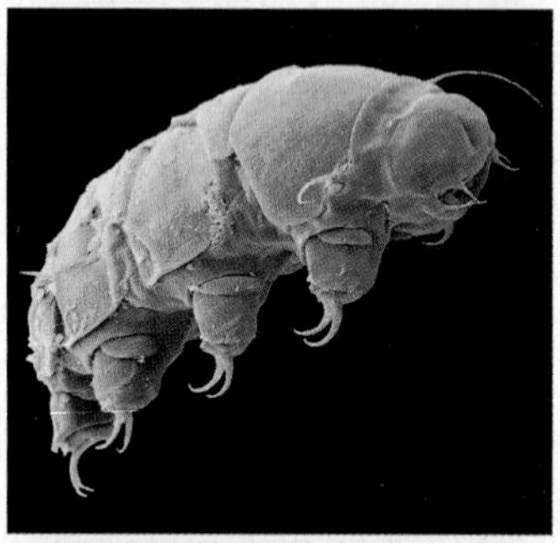	Tardigrada **Water bears** Microscopic animals with a segmented body, eight short legs, a hemocoel, and a flexible cuticle that must be molted during growth	**1045** Marine Freshwater Terrestrial (moist areas) 0.3–1.5 mm	Most feed by sucking fluids from plants or animals; others are detritivores; mouthparts vary; some can go without food or water for 10 years	Named "slow-walkers" for their slow walking movements underwater on three pairs of clawed, lobed legs; fourth pair of legs is used for grasping	Sexes are separate in most species; some are parthenogenic; some are hermaphroditic, able to self-fertilize; fertilization is usually external; may molt up to 12 times during growth
	Onychophora **Velvet worms** Caterpillar-like worms with dry, segmented bodies, lobe-shaped appendages, and a hemocoel; regularly molt a thin cuticle	**165** Terrestrial (humid areas) 0.5–20 cm	Nocturnal ambush predators; can squirt a sticky slime to ensnare prey, mostly small invertebrates; bite prey with jaws, then inject digestive saliva; suck up the digested, liquefied insides	Moves slowly using a hydrostatic skeleton; muscles enable parapodia to move independently; can squeeze its soft body into narrow crevices	Sexes are separate in almost all species, and females are usually larger than males; fertilization is usually internal; most species are ovoviviparous
	Arthropoda **Arthropods** Segmented animals with jointed appendages, a hemocoel, and an exoskeleton that must be molted during growth (see Table 31.4)	**1,200,000** Marine Freshwater Terrestrial 0.1 mm–4 m	Use specialized mouthparts to consume all types of foods from detritus and live prey to flower nectar and blood; serve diverse ecological roles	Most use jointed appendages for walking, running, jumping, and swimming; most also have wings; insect larvae such as maggots, caterpillars, and grubs move using hydrostatic skeletons	Sexes are usually separate, and sexual reproduction is the norm (parthenogenesis is rare); fertilization is usually internal; may or may not have larvae and metamorphosis

For more go to https://goo.gl/4BqcQa

Although some roundworms can grow to lengths of several meters, most are tiny—less than 1 mm long. They lack specialized systems for exchanging gases and circulating nutrients. Instead, gas exchange occurs across the body wall and nutrients move by diffusion from the gut to other parts of the body.

While around 25,000 species of roundworms have been named, the actual total may be over a million—they occur in virtually every habitat known, including deep below Earth's surface. The chemistry of the cuticle varies, enabling nematodes to withstand sometimes inhospitable environments such as alcoholic rotting fruit or acidic digestive tracts. Further, nematodes feed on a wide variety of material including bacteria, archaea, fungi, plant roots, small protists, and detritus. In most species, the mouthparts are structured in a way that increases the efficiency of feeding on a particular type of organism or material.

Roundworms are remarkably abundant. For example, biologists have found 90,000 roundworms in a single rotting apple and have estimated that rich farm soils contain up to 9 billion roundworms per acre. Based on these numbers, it is no surprise that roundworms play an important role in their ecosystems.

Most roundworms are free-living. Recall that one such roundworm, *Caenorhabditis elegans*, is among the most thoroughly studied model organisms in biology (**BioSkills 11**). Other roundworm species parasitize animals, including humans:

- Pinworms (*Enterobius vermicularis*) infect about 40 million people in the United States alone, causing itching of the anus.
- Advanced infections by the roundworm *Wuchereria bancrofti* result in the blockage of lymphatic vessels, causing massive swelling—the condition known as elephantiasis.
- The potentially fatal disease trichinosis is caused by species in the genus *Trichinella*. The parasites pass from other infected animals to humans when humans ingest raw or undercooked meat, primarily pork. An array of symptoms results, including discomfort caused by *Trichinella* larvae encasing themselves in cysts within muscle or other types of tissue.

Despite its relative simplicity, the nematode body plan has been extraordinarily successful in diverse environments.

What Are Water Bears and Velvet Worms?

Of the smaller phyla of ecdysozoans, Tardigrada and Onychophora (*on-ee-KOFF-er-uh*) are of special interest due to their close relationship to arthropods. They are similar to arthropods in having a segmented body and segmented limbs (see Table 31.3). Unlike arthropods, their cuticle is not hardened as an exoskeleton, and their limbs are not jointed. ✔ **You should be able to indicate the origin of the following traits on the ecdysozoan branches of Figure 31.2: molting, cuticle, segmented body, segmented limbs.**

The tardigrades, or water bears, are microscopic animals that live in diverse marine, freshwater, and terrestrial environments, such as in the film of water that covers plants in moist habitats. They are also famous for withstanding extreme conditions, such as outer space. The onychophorans, or velvet worms, are small, caterpillar-like organisms that live in moist leaf litter and prey on small invertebrates.

What Is an Arthropod?

Based on duration in the fossil record, species diversity, and abundance of individuals, arthropods are the most important phylum within the Ecdysozoa. Arthropods appear in the fossil record over 520 million years ago and have long been the most abundant animals observed in both aquatic and terrestrial environments. Well over a million living species have been described, and biologists estimate that millions or perhaps even tens of millions of arthropod species have yet to be discovered. In terms of species diversity, arthropods are easily the most successful phylum of eukaryotes. Further, arthropods are very well studied; recall that the fruit fly *Drosophila melanogaster* is an important model organism in biology.

The Arthropod Body Plan In addition to their bilateral symmetry, triploblastic tissue origins, and other protostome traits, **arthropods** ("joint-foot") are characterized by the following three key features:

1. ***A segmented body*** An athropod's body is organized into prominent regions, or **tagmata** (singular: **tagma**), where each tagma is formed from a few or many segments. For example, the body of a grasshopper and other insects is divided into a **head**, **thorax**, and **abdomen**, while the body of a spider or a crayfish is divided into the **cephalothorax** ("head-chest") and abdomen (see **Figure 31.13** on page 664).
2. ***An exoskeleton*** This hard external skeleton is made primarily of the polysaccharide chitin (Ch. 5, Section 5.2), strengthened by calcium carbonate ($CaCO_3$) in crustaceans; and
3. ***Jointed appendages*** Joints between segments in the legs and other appendages enable very rapid and precise movements (Ch. 45, Section 45.3).

The jointed appendages of arthropods are organized in pairs along the segmented body, one pair per segment—such as a pair of antennae on a head segment, or a pair of legs on a thorax segment. The appendages have an array of functions: exchanging gases, sensing the environment, feeding, reproduction (for example, grasping mates or holding egg masses) and locomotion (via swimming, walking, running, or jumping). ✔ **You should be able to indicate the origin of tagmata, exoskeleton, and jointed appendages on Figure 31.2.**

Arthropods have a highly reduced coelom. But, like mollusks, they possess an extensive body cavity called a hemocoel, which serves as part of the open circulatory system. Besides circulating gases and nutrients, the hemocoel serves as part of a hydrostatic skeleton that enables movements in caterpillars, grubs, maggots, and other types of soft-bodied arthropod larvae.

Why Has the Arthropod Body Plan Been So Successful? You have already learned that mollusk bodies are compartmentalized and flexible in evolutionary terms, meaning that the foot, visceral mass, and mantle diversified independently of each other. The compartmentalization and flexibility of the arthropod body plan, however, takes the prize for evolutionary potential.

Studies on the role of tool-kit genes—such as the *Hox* genes—during the development of different arthropods have contributed

(a) Insects have three tagmata.

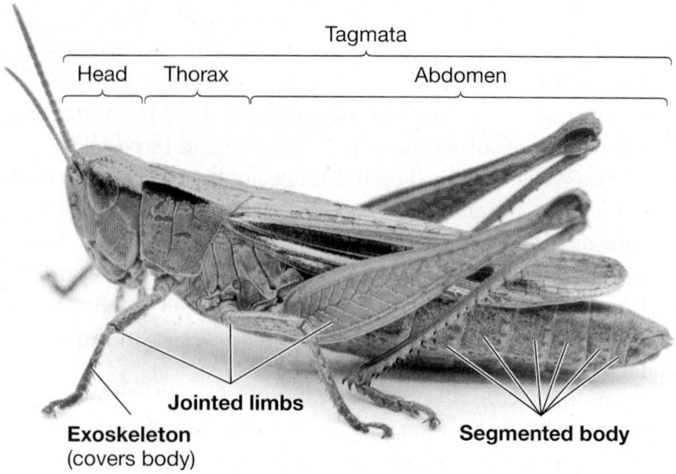

(b) Spiders have two tagmata.

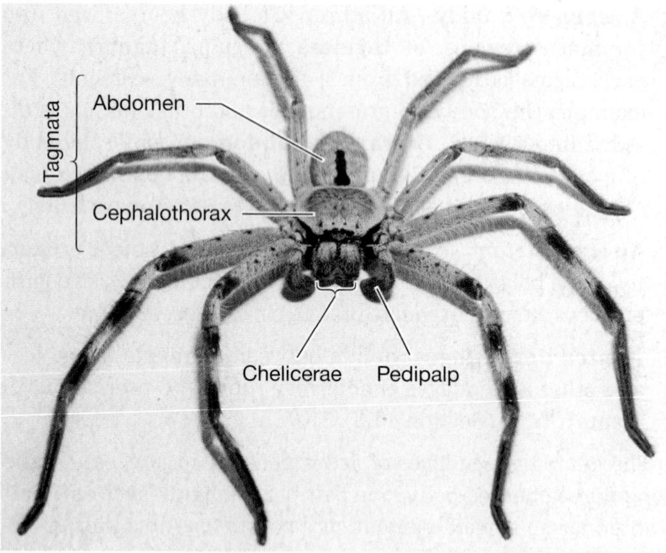

Figure 31.13 Arthropods Have a Specialized Body Plan. Arthropods have segmented bodies organized into two or three body regions, or tagmata. Arthropods also have jointed limbs, which enable these animals to move despite their hard outer covering, the exoskeleton.

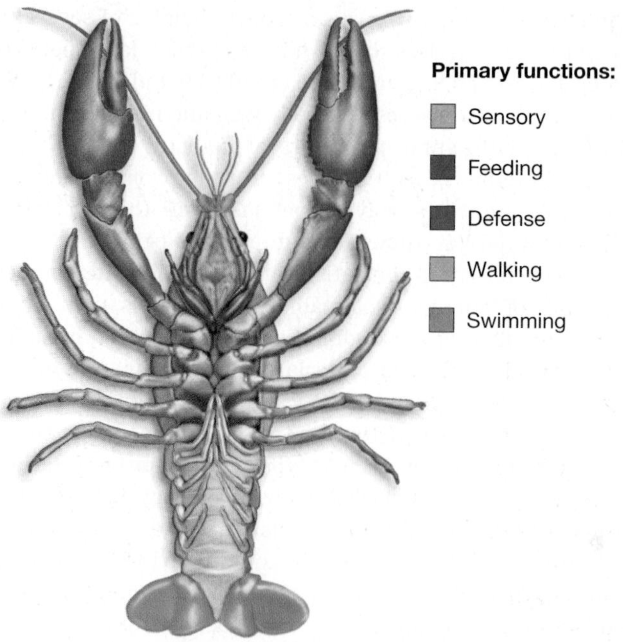

Figure 31.14 Arthropods Can Have Functional Specialization Among Segments in Their Body Plan. This ventral view of a crayfish illustrates how arthropod segments and appendages can be specialized for different functions.

enormously to our understanding of the process of arthropod diversification. These studies show that small changes in the timing and location of expression of certain tool-kit genes along the length of an arthropod can result in novel shapes, sizes, and functions of body segments and appendages.

To understand the evolutionary potential of the arthropod body plan, consider a hypothetical ancestral arthropod with many uniform body segments, each with a pair of identical appendages. The redundancy of this body plan can provide a fertile starting place for diversification because some of the segments can continue to function the old way, say for crawling, while other segments can become co-opted for new functions. For example, as shown in Figure 31.4, a pair of appendages could become specialized for feeding without compromising the animal's ability to crawl.

Genetic opportunity in the form of variation in the expression of tool-kit genes can combine with ecological opportunity by the process of natural selection, resulting in the diversification of arthropod body segments and appendages. **Figure 31.14** shows an example of how diverse, functionally specialized appendages can result in an arthropod body plan.

Origin of the Wing Insects were the first animals to achieve powered flight more than 350 million years ago, about 100 million years before their first vertebrate predators, the pterosaurs, launched into the air. This innovation enabled them not only to escape predators but also to search for new food resources and disperse to new habitats. Today there are more species of winged insects on the planet than all other animal species combined. How did wings evolve?

This question continues to inspire lively debate. The evidence is convincing that wings evolved only once before the adaptive radiation of insects on land, yet no fossils of transitional wing structures have yet been discovered. Developmental genetics has helped to test various transition hypotheses by looking for the expression of genes important to the development of wings in various body structures of other arthropods. However, the results have not yet been decisive or consistent across different species.

One thing is clear: wings did not evolve from the jointed limbs of arthropods. Rather, wings occur as unjointed extensions of the dorsal cuticle on the second and third segments of the insect thorax—in addition to the jointed legs on the ventral side. Further, wings have been secondarily lost or modified many times. For example, in flies the hindwings are greatly reduced to paddle-like halteres that function as gyroscopes during their highly acrobatic flight.

Arthropod Diversity

Arthropod taxonomy traditionally recognizes four major lineages of arthropods: myriapods, insects, crustaceans, and chelicerates (summarized in **Table 31.4**). Let's first examine these traditional groups and then consider a more recent view of their relationships.

Myriapods Have Long, Segmented Trunks **Myriapods** have relatively simple bodies—a head region and a long trunk featuring a series of many segments, each bearing either one pair of legs (in centipedes) or two pairs of legs (in millipedes, because pairs of segments became fused in the ancestor of millipedes). Some species have no eyes, but others have few to many **simple eyes**

DIVERSITY **Table 31.4 Key Lineages of Arthropods (Traditional Groupings, Separating Insects from Crustaceans)**

Protostomia > Ecdysozoa > Arthropoda >		Known Species	Feeding	Movement	Reproduction
	Myriapoda **Millipedes, centipedes** Relatively simple bodies with a head region and a long, segmented trunk with many legs	**16,000** Terrestrial 1–30 cm	Have mouthparts that can bite and chew; millipedes are detritivores; centipedes are predators, using poison-containing fangs to subdue prey (large centipedes can cause painful bites in humans)	Walk or run on their many legs; a few species burrow; millipedes have two pairs of legs per trunk section (due to the fusion of adjacent segments); centipedes have one pair of legs per segment	Sexes are separate, and fertilization is internal; males deposit sperm packets that are picked up by females or transferred to them by the male; juveniles undergo several molts
	Insecta **Insects** Remarkably diverse and abundant in terrestrial environments; body has three tagmata: head, thorax, and abdomen	**1,000,000** Terrestrial Freshwater 0.1 mm–20 cm	Most have four pairs of mouthparts that have diversified for feeding on almost every type of food source available on land; seek food using compound eyes and one pair of antennae	Most have three pairs of legs attached to the thorax, used for walking, running, jumping, or swimming; most have one or two pairs of wings that they use to fly	Sexes are separate, and sexual reproduction is the norm (parthenogenesis is rare); most mate via direct copulation; fertilization is internal; development may be direct or indirect
	Crustacea **Shrimp, lobsters, crabs, barnacles, isopods, copepods** Most diverse arthropods of the sea; body has three tagmata (head, thorax, and abdomen) or two (cephalothorax and abdomen)	**47,000** Marine Freshwater Terrestrial 0.1 mm–4 m	Most have 4–6 pairs of mouthparts (e.g., mandibles) that have diversified for feeding on almost every type of food source available; seek food using compound eyes and two pairs of antennae	Typically have pairs of legs with diverse forms enabling walking, running, and swimming; barnacles are sessile but use feathery legs to filter food from the water	Sexes are often separate, and sexual reproduction is the norm (parthenogenesis is rare); fertilization is usually internal; many have indirect development via a planktonic nauplius larva
	Chelicerata **Spiders, scorpions, horseshoe crabs, ticks, mites** Diverse on land; body has two tagmata: cephalothorax and abdomen	**100,000** Terrestrial Marine 0.1 mm–28 cm	Mites and ticks are ectoparasites; others are either active hunters or sit-and-wait predators, some make webs; spiders secrete digestive enzymes into prey and drink up the liquid	Most have four pairs of legs, one pair of chelicerae (used for diverse functions), and one pair of pedipalps; most walk, crawl, jump, or move through the air via silken threads	Sexes are usually separate, and sexual reproduction is the norm; fertilization is internal; sex organs fit in a "lock-and-key" fashion; development is direct, no metamorphosis

For more go to https://goo.gl/leMT41

DIVERSITY **Table 31.5 Key Orders of Insects**

Protostomia > Ecdysozoa > Arthropoda > Insecta >		Key Traits	Feeding	Life Cycle	Notes
	Coleoptera ("sheath-winged") **Beetles** **360,000** known species	Hardened forewings, called elytra, protect the membranous hindwings that power flight and act as stabilizers	Adults are important predators and scavengers; larvae often feed voraciously using chewing mouthparts	Complete metamorphosis (larvae are called grubs)	The most species-rich lineage on the tree of life—about 25 percent of all described insect species are beetles (see chapter-opening photo)
	Lepidoptera ("scale-winged") **Butterflies, moths** **180,000** known species	Wings are covered with tiny, often colorful scales; the forewings and hindwings often hook together and move as a unit	Adults often feed on nectar; larvae usually have chewing mouth-parts and either bore through food material or bite leaves	Complete metamorphosis (larvae are called caterpillars)	Most are important pollinators; some species, such as monarch butter-flies, migrate long distances
	Diptera ("two-winged") **Flies, including mosquitoes, gnats, midges** **150,000** known species	One pair of wings; reduced hindwings, called halteres, are club shaped and act as stabilizers during flight	Adults usually feed on nectar, blood, or other liquids; larvae may be parasitic or feed on rotting plant or animal tissues	Complete metamorphosis (larvae of some flies are called maggots and live in moist environments; some larvae are aquatic)	Are among the most agile insect fliers; wings beat at frequencies as high as 1000 beats per second
	Hymenoptera ("membrane-winged") **Ants, bees, wasps** **115,000** known species	Membranous forewings and hindwings lock together	Most ants feed on plant material; most bees feed on nectar; wasps are predatory and often have parasitic larvae	Complete metamorphosis; males are haploid (hatch from unfertilized eggs) and females are diploid; larvae are cared for by adults	Most live in colonies and many are eusocial, having a complex social structure with a division of labor and cooperative breeding

(Continued)

clustered on the sides of the head. (Human and cephalopod eyes are also simple, meaning they have just one lens.) Myriapods inhabit terrestrial environments all over the world.

Insects Have Three Tagmata, Unbranched Appendages, and One Pair of Antennae As illustrated in Figure 31.13a, **insects** have three tagmata—a head, thorax, and abdomen. Three pairs of walking legs extend from the ventral surface (belly) of the thorax. Most species have one or two pairs of wings on the dorsal surface (back) of the thorax.

Typically the head contains four pairs of mouthpart structures, one pair of slender **antennae** that are used to touch and smell, two or three simple eyes, and a pair of compound eyes. A **compound eye** contains many lenses, each associated with a light-sensing, columnar structure (Ch. 44, Figure 44.10). **Table 31.5** describes seven of the most prominent orders of insects.

About 1 million species of insects have been named so far, but biologists are sure that many more exist. Based on species diversity and numbers of individuals, insects dominate terrestrial environments. In addition, the larvae of some species are common in freshwater streams, ponds, and lakes.

Insects arose around the same time as land plants, about 480 million years ago. The close ecological associations between many insects and land plants suggest that the radiations of these two lineages were partly due to their coevolution—each provided ecological opportunities for the other.

Crustaceans Have Two or Three Tagmata, Branched Appendages, and Two Pairs of Antennae **Crustaceans** live primarily in marine and freshwater environments, where they play important ecological roles. Planktonic copepods, for example, are extremely common in surface waters around the world, where they

DIVERSITY Table 31.5 Key Orders of Insects (Continued)

Protostomia > Ecdysozoa > Arthropoda > Insecta >		Key Traits	Feeding	Life Cycle	Notes
	Hemiptera ("half-winged") **Bugs, including leaf hoppers, aphids, cicadas, scale insects** **85,000** known species	Called "half-winged" because the forewings of some bugs are hardened near the bases but membranous near the tips	Mouthparts are modified for piercing and sucking; most suck plant juices, but some are predatory	Incomplete metamorphosis (young are called nymphs and resemble adults)	"True bugs"; some are major pests of crops and gardens; others are predators of those pests, thus are important biological control agents
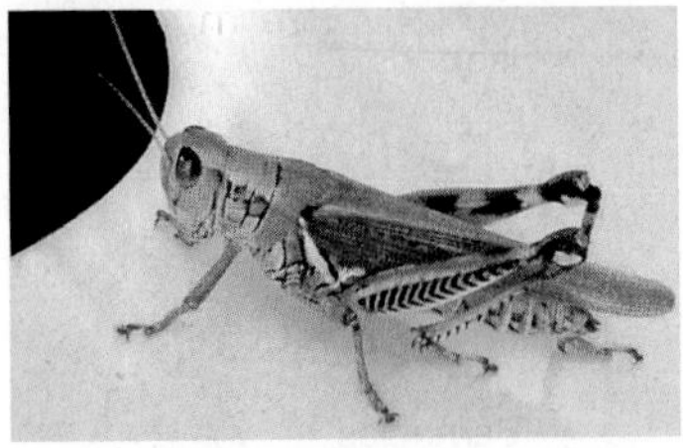	Orthoptera ("straight-winged") **Grasshoppers, crickets** **25,000** known species	Large, muscular hind legs enable jumping; at rest, the membranous hindwings are held folded like a fan beneath the forewings	Most have biting or chewing mouthparts and eat leaves	Incomplete metamorphosis (young are called nymphs and resemble adults)	Some males use sound to attract mates by rubbing their wings against their legs or against each other in a process called stridulation
	Odonata ("toothed") **Dragonflies, damselflies** **6,500** known species	Four membranous wings and long, slender abdomens	Nymphs and adults are predatory, often on flies; the "odon" refers to strong, toothlike structures on their mandibles	Incomplete metamorphosis: nymphs are aquatic, adults are terrestrial	They hunt by sight using enormous compound eyes with up to 28,000 lenses; have reduced antennae (used for touch and smell)

For more go to https://goo.gl/Y4ACVG

consume vast amounts of phytoplankton and, in turn, are an important food source for fish and marine mammals. Crustaceans are also important grazers and predators in shallow-water benthic environments. Just a few species of crabs and some isopods (known as sowbugs or pillbugs, commonly seen under wood piles in backyards) are terrestrial.

The segmented body of most crustaceans is divided into three tagmata: the head, thorax, and abdomen. However, some species have a **carapace**—a platelike section of their exoskeleton—that covers and protects the head and thorax, forming a fused cephalothorax. Crustacean appendages are usually branched, but some have lost their branch, becoming secondarily unbranched.

Crustaceans are the only type of living arthropod with two pairs of antennae. They also have sophisticated, compound eyes—usually mounted on stalks. Most crustaceans have 4–6 pairs of mouthparts that are derived from jointed appendages. Many species have mouthparts on the head called maxillae that manipulate food, and some have a pair of mouthparts called **mandibles** that can bite or chew. Many crustaceans also have feeding appendages on thoracic segments, called maxillipeds or claws, that can grab food and move it to the mouthparts. As a group, crustaceans are highly diverse in their feeding strategies, thanks in part to the evolutionary flexibility of their compartmentalized body plan.

Crustaceans are the oldest arthropod lineage still in existence. The tiny ostracods (seed shrimp) have a particularly rich fossil record dating back to the Cambrian. The only other arthropod lineage as ancient as the crustaceans are the trilobites, which expired in the end-Permian mass extinction about 252 million years ago (Ch. 25, Section 25.3).

Chelicerates Have Two Tagmata and Chelicerae The most prominent lineages of **chelicerates** are terrestrial and include spiders, scorpions, ticks, mites, and harvestmen (daddy long-legs). But some chelicerates are marine, such as the horseshoe crabs (which are not true crabs) and sea spiders.

The chelicerate body consists of two tagmata: the cephalothorax and abdomen. (Note that the cephalothorax and abdomen are also referred to as the prosoma and opisthosoma, respectively; these distinct terms are used to emphasize that the chelicerate tagmata are not homologous to those in other arthropods, even though they are functionally similar.) The cephalothorax usually has eyes.

Chelicerates are named for a pair of clawlike appendages called **chelicerae**, located near the mouth (see Figure 31.13b). Depending on the species, the chelicerae are used in feeding, defense, copulation, movement, or sensory reception. Besides four pairs of legs and one pair of chelicerae, chelicerates also have a pair of **pedipalps** located just behind the chelicerae, which may be used to manipulate food, transfer sperm, or carry out other functions. Chelicerates have no antennae.

All spiders are predators. Some species are active hunters that roam around in search of prey, but most spiders are sit-and-wait predators that either pounce on passing prey or create sticky webs to capture prey. In the latter case, the spider senses the vibrations of the struggling prey on the web, seizes it, and injects it with paralyzing venom via the chelicerae. No matter how they capture prey, spiders generally secrete digestive enzymes into the prey that enable digestion to begin externally, and then they drink up the nutritious liquid. Often all that's left of the prey is an empty exoskeleton.

Scorpions are also venomous predators; most feed on insects, but some occasionally take snakes and lizards. Although the venom of a scorpion is an effective defense against predators, it is primarily used to kill or paralyze prey. Unlike spiders and scorpions, mites and ticks are ectoparasitic; they use their piercing mouthparts to feed on host animals or, in the case of dust mites, on their hosts' dead skin (including yours).

Some of the most riveting discoveries about chelicerates involve their elaborate courtship displays, which may include both visual displays and the release of chemical odorants. In spiders, males may present a dead insect as a gift that the female eats as he uses his pedipalps to transfer sperm to her reproductive tract. In other species, the female eats the male as sperm is being transferred—the longer the meal lasts, the more successful the sperm transfer.

A New Phylogeny of Arthropods The relationships of lineages within the arthropods are currently undergoing a major revision. Insects and myriapods have long been considered sister groups within the arthropods due to shared morphological characteristics, such as their single pair of antennae and their unbranched appendages. However, recent molecular phylogenetic studies—as well as comparisons of new morphological data, developmental data, and evidence in the fossil record—strongly support placing the insect clade *within* the crustacean lineage, making the traditional concept of crustaceans paraphyletic (Figure 31.15).

While the exact relationships among arthropod lineages may take years of additional research to work out, many researchers currently support the hypothesis that crustaceans successfully invaded land as insects. In other words, insects *are* crustaceans.

Arthropod Metamorphosis

Many arthropods undergo a morphological transformation from a larva to a very different adult form. Others do not. Let's use insects as a case study to consider the adaptive implications of metamorphosis in arthropods.

Two Types of Insect Metamorphosis In insects, the presence or absence of a larval stage characterizes two distinct types of metamorphosis.

In **incomplete metamorphosis** (a form of direct development) juveniles called nymphs look like smaller versions of the adult. The aphid nymphs in Figure 31.16a, for example, shed their exoskeletons several times and grow—gradually changing from wingless, sexually immature nymphs to sexually mature adults, some of which can fly. But throughout their life, aphids

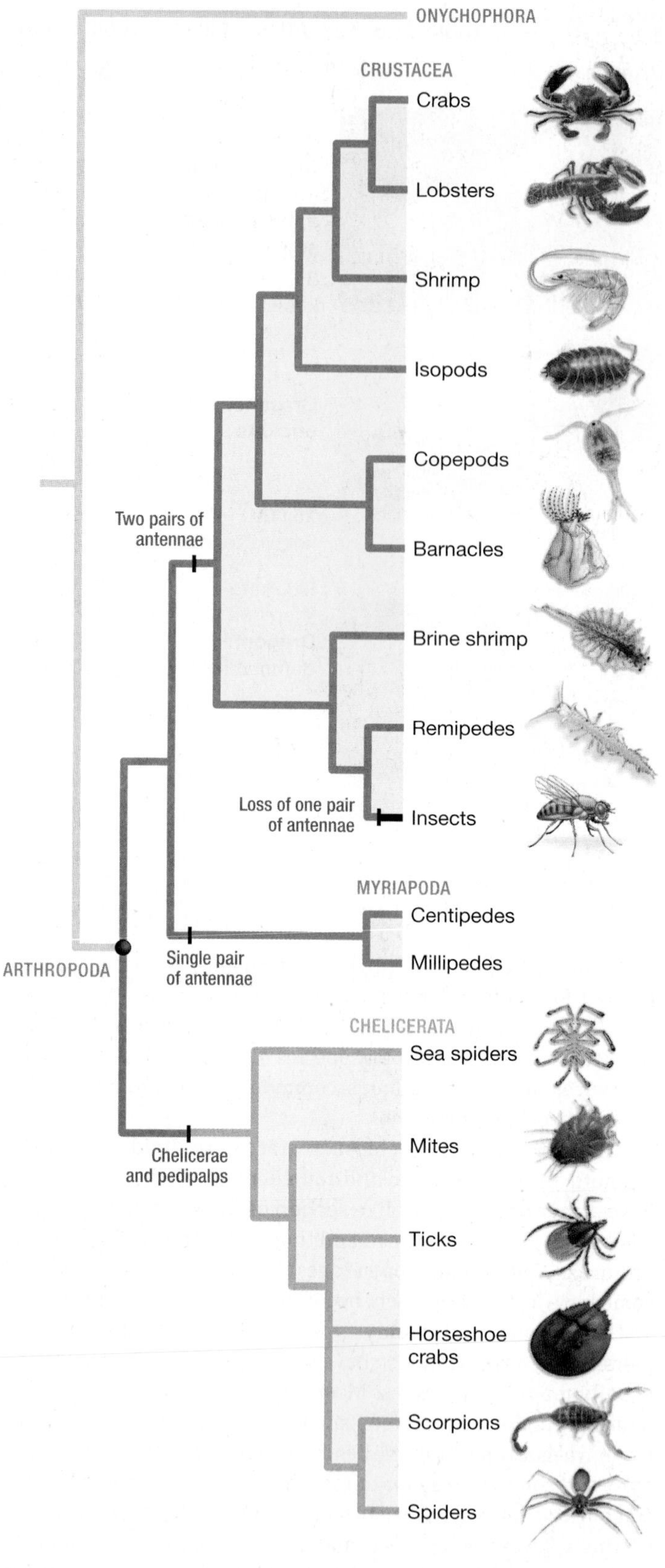

Figure 31.15 **Phylogeny of the Major Groups of Arthropods.** This phylogeny illustrates the current hypothesis that insects are a terrestrial clade within the crustaceans, overturning traditional classification, which is nonetheless still in widespread use due to its long history.

✔ **THINK CAREFULLY** Did insects evolve from remipedes (blind crustaceans found in coastal caves)?

(a) Aphid:
Incomplete metamorphosis

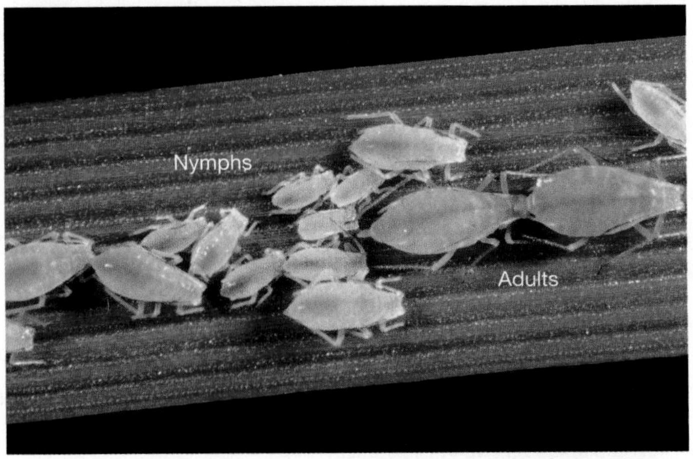

Nymphs look like smaller versions of adults and eat the same foods.

(b) Mosquito:
Complete metamorphosis

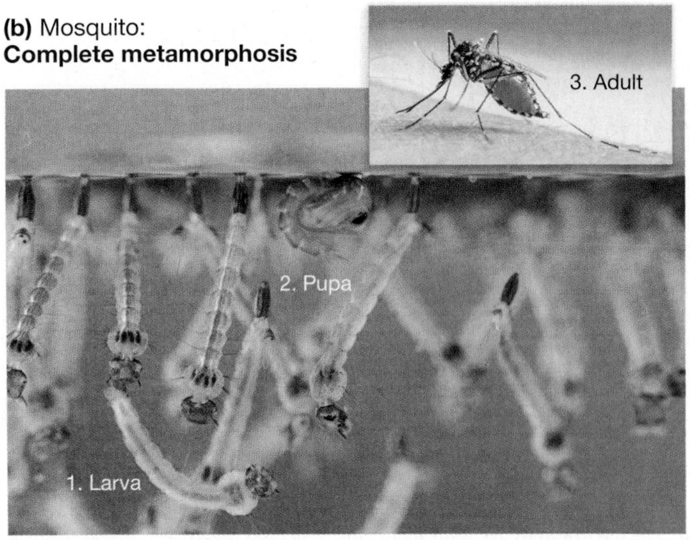

Larvae look substantially different from adults and eat different foods.

Figure 31.16 **Insect Metamorphosis Can Be Incomplete or Complete.**

live in the same habitats and feed on the same food source in the same way: They suck sap.

In **complete metamorphosis** (a form of indirect development) there is a distinct larval stage. An example is the life cycle of the mosquito, illustrated in Figure 31.16b:

1. Newly hatched mosquito larvae live in quiet bodies of fresh water, where they suspension feed on archaea, bacteria, algae, and detritus.
2. When a larva has grown sufficiently, it stops feeding and moving and secretes a protective case. The individual is now known as a **pupa** (plural: **pupae**). During pupation, the body is completely remodeled into a new, adult form.
3. The adult mosquito flies and feeds—females take blood meals from mammals, and males drink nectar from flowers.

What Is the Adaptive Significance of Metamorphosis? In insects, complete metamorphosis is 10 times more common than incomplete metamorphosis. How can this be explained?

Most adult insects are mobile, so using a larval stage for dispersal does not sufficiently explain this pattern. The leading hypothesis is based on efficiency in feeding. Because juveniles and adults in species with complete metamorphosis feed on different materials in different ways and sometimes even in different habitats, they do not compete with each other.

An alternative hypothesis is based on the advantages of functional specialization. In many moths and butterflies, larvae are specialized for feeding, whereas adults are specialized for mating and feed rarely, if ever. Larvae are largely sessile, whereas adults are highly mobile. If specialization leads to higher efficiency in feeding (as larvae) and reproduction (as adults) and thus higher overall fitness, then complete metamorphosis would be advantageous. These hypotheses are not mutually exclusive, however—both could be correct, depending on the species being considered.

Take-Home Messages

Protostome animals are a fabulously successful clade of animals, and millions of species of protostomes have yet to be discovered. Their multiple transitions to land and subsequent adaptive radiations—especially in snails, spiders, and insects—are particularly spectacular. A wealth of questions are left for biologists to explore, ranging from the evolutionary origin of one of the most significant morphological innovations in animal history—the insect wing—to the ecosystem services that protostomes supply to humans. Next, let's consider a lineage of animals that might be more familiar—the deuterostomes, including vertebrates like you.

CHECK YOUR UNDERSTANDING

✔ If you understood this section, you should be able to ...

1. List three lineages of arthropods that successfully made the transition from water to land.
2. Describe three ways that insects and spiders are similar, and three ways that they are different.

Answers are available in Appendix A.

CHAPTER 31 Review

For media, go to Mastering Biology

31.1 What Is a Protostome?

- Molecular evidence supports the hypothesis that protostomes are a monophyletic group divided into two major subgroups: the Lophotrochozoa and the Ecdysozoa.
- All protostomes are bilaterally symmetric and triploblastic.
- The ancestral protostome probably had a coelomate body plan, but the coelom has been reduced or lost in several phyla.
- The transition to living on land occurred several times independently during the evolution of protostomes, opening new habitats and driving diversification. The transition was facilitated by watertight shells and exoskeletons, the evolution of specialized structures to minimize water loss and manage gas exchange, and the ability to move to moist locations.
- Many protostome body plans are compartmentalized in such a way that some parts can evolve independently of others, facilitating diversification.

31.2 What Is a Lophotrochozoan?

- Lophotrochozoans grow incrementally (no molting). Some phyla have characteristic feeding structures called lophophores, and some have a distinctive type of larvae called trochophores. Spiral cleavage is a synapomorphy for the group but was modified or lost in some phyla.
- Flatworms (phylum Platyhelminthes), annelids (phylum Annelida), and mollusks (phylum Mollusca) are the most diverse phyla within the Lophotrochozoa. All three phyla made the water-to-land transition.
- Flatworms have a flattened, unsegmented, acoelomate body plan with no specialized structures for gas exchange or circulation. Some free-living turbellarian flatworms live in moist terrestrial environments, but most live in freshwater and marine environments. Tapeworms and flukes are parasitic.
- The annelids have a segmented, coelomate body plan. This diverse group includes both highly mobile and sedentary marine polychaetes, terrestrial earthworms, and leeches.
- The mollusks have a distinct body plan including a muscular foot, a visceral mass, and a mantle. These components diversified in many ways, enabling mollusks to occupy most aquatic—and some terrestrial—habitats. Mollusks include chitons, bivalves such as clams and scallops, gastropods such as snails and slugs, and cephalopods such as squid and octopuses.

31.3 What Is an Ecdysozoan?

- Species in the Ecdysozoa grow by molting—meaning that they shed their old external covering and grow a new, larger one.
- Roundworms (phylum Nematoda) and arthropods (phylum Arthropoda) are the most diverse phyla within the Ecdysozoa. The water bears (phylum Tardigrada) and velvet worms (phylum Onychophora) are also ecdysozoans.
- Roundworms have an unsegmented, pseudocoelomate, limbless body plan. Most are tiny, less than 1 mm long. But they live in virtually every habitat on the planet and can be exceptionally abundant.
- The arthropod body plan is segmented and protected by an exoskeleton made primarily of chitin. Appendages are jointed. The diversification of arthropod bodies is based on variation in the number and functional specialization of segments within body regions called tagmata.
- Arthropods are traditionally divided into four major groups: *myriapods* (millipedes, centipedes), *insects* (such as beetles, butterflies, flies, ants, bugs, grasshoppers, and dragonflies), *crustaceans* (such as shrimp, lobsters, crabs, barnacles, and copepods), and *chelicerates* (such as spiders, scorpions, horseshoe crabs, ticks, and mites). However, recent molecular data indicate that insects are a terrestrial subgroup of Crustacea.
- In incomplete metamorphosis, nymphs usually resemble adults and eat the same foods. Complete metamorphosis enables larvae and adults to specialize independently, creating opportunities for diversification.

Answers are available in Appendix A.

✔ TEST YOUR KNOWLEDGE

1. What is a lophophore?
   a. a specialized filter-feeding structure
   b. the single opening in species with a blind gut
   c. a distinctive type of larva with a band of cilia
   d. a synapomorphy that defines lophotrochozoans
2. What is the function of the arthropod exoskeleton?
   a. The presence of an exoskeleton has given arthropods a good fossil record, because hard parts fossilize more readily than do soft tissues.
   b. It has no well-established function. (Trilobites had an exoskeleton, and they went extinct.)
   c. It provides protection and functions in locomotion.
   d. It makes growth by molting possible.
3. One trait that is shared by the Lophotrochozoa and Ecdysozoa is __________.
4. Which protostome phyla successfully made the transition from water to land? Select True or False for each phylum.
   T/F Mollusca
   T/F Arthropoda
   T/F Annelida
   T/F Nematoda

✔ TEST YOUR UNDERSTANDING

5. Propose an explanation for why annelids and arthropods were thought to be closely related, before phylogenetic analyses in the late 1990s.
6. Use your understanding of insect evolution to predict two adaptations for terrestrial living that occurred in spiders.
7. Pose a hypothesis to explain why the evolution of the wing was such an important event in the evolution of insects.
8. **THINK CAREFULLY** Evaluate this statement: Evolution is said to occur when new traits accumulate over time, increasing complexity.

✔ TEST YOUR PROBLEM-SOLVING SKILLS

9. Brachiopoda is a phylum within the Lophotrochozoa. Even though they are not closely related to bivalve mollusks (such as clams or mussels), brachiopods look and act like bivalve mollusks. Specifically, brachiopods suspension feed, secrete calcium carbonate shells with two valves that hinge together in some species, and attach to rocks or other hard surfaces on the ocean floor. How is it possible for brachiopods and bivalves to be so similar if they did not share a recent common ancestor?

10. **SOCIETY** About a third of insect species measured are in decline, meaning that their populations are shrinking due to habitat destruction, pollution, and other factors. Predict how this change affects the bee pollination of crops such as apples and almonds.

✔ PUT IT ALL TOGETHER: Case Study

How could you count all the bees, beetles, and butterflies in one forest?

A team of 102 scientists spent a year surveying a small area of the San Lorenzo rain forest in Panama to count the number of species of arthropods living there. After collecting 129,494 specimens—using nets, traps, shovels, tree-climbing harnesses, helium balloons, and other creative gear—it took the team eight years to sort and identify the arthropods.

11. Which of the following nested groups best describes the taxonomic context for the San Lorenzo project?
   a. Animalia > Bilateria > Arthropoda > Ecdysozoa
   b. Protostomia > Lophotrochozoa > Ecdysozoa> Arthropoda
   c. Arthropoda > Protostomia > Ecdysozoa > Bilateria
   d. Bilateria > Protostomia > Ecdysozoa > Arthropoda

12. **QUANTITATIVE** Rather than measuring the entire 6000-hectare (ha) forest, the researchers sampled arthropod diversity by intensively collecting as many arthropods as they could in 12 plots that measured 20 m × 20 m square. If 1 ha $= 10{,}000\ m^2$, how many hectares of forest did they sample in all?
   a. $20\ m \times 20\ m \times 12 = 4800$ ha
   b. $4800\ m^2 \times 10{,}000\ m^2/1\ ha = 48{,}000{,}000$ ha
   c. $20\ m \times 20\ m = 400$ ha
   d. $4800\ m^2 \times 1\ ha/10{,}000\ m^2 = 0.48$ ha

13. **QUANTITATIVE** The graph below shows some of the data for the major arthropod groups collected. Notice that the scale on the *y* axis is logarithmic to make both small and large numbers legible on the same graph (see **BioSkills 5**). For example, there are about 400 species of spiders, but only 40 species of bees. About how many arthropods were found in total? About what percentage of these were beetles?

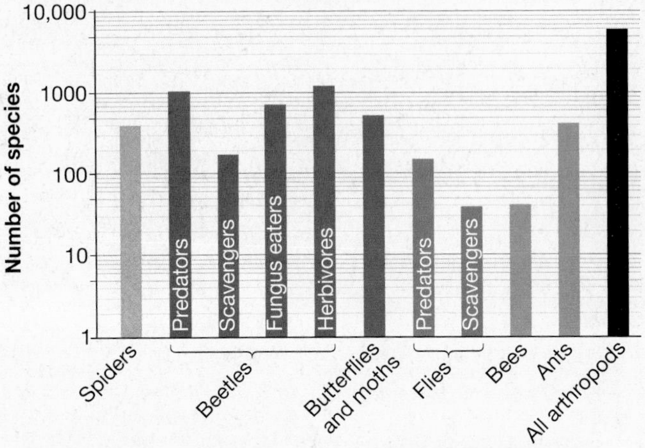

Source: Y. Basset et al. 2012. *Science* 388: 1481–1484.

14. Use the data in the graph to analyze the following statements. Select True or False for each statement, then correct any false statements to make them true.
   T/F All taxa shown in the graph are insects.
   T/F All taxa shown in the graph have segmented bodies and limbs.
   T/F All taxa shown in the graph are terrestrial.
   T/F If the researchers had sampled for mollusks, they probably would have found almost as many species as they did for arthropods.

15. **PROCESS OF SCIENCE** By comparing the number of arthropod species that were common among the 12 plots sampled, the researchers estimated that the total number of arthropods in the whole San Lorenzo forest is 25,000 species. To come up with this estimate, what's one assumption the researchers had to make about the forest?

16. The researchers observed a strong correlation between the number of arthropod species and the number of plant species in the San Lorenzo forest. Pose an evolutionary hypothesis to explain this correlation.

Mastering Biology

Students Go to Mastering™ Biology for assignments, the eText, and the Study Area with animations, practice tests, and activities.

Professors Go to Mastering™ Biology for automatically graded tutorials and questions that you can assign to your students, plus Instructor Resources.

32 Deuterostome Animals

In most ecosystems the "top predators"—animals that prey on other animals and aren't typically preyed upon themselves (such as this osprey)—tend to be deuterostomes.

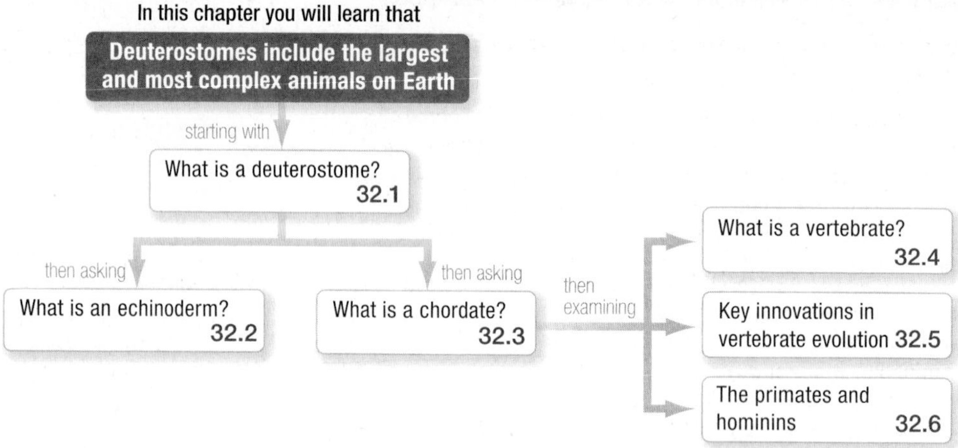

You may not think about yourself this way, but you are a deuterostome—a member of a lineage of animals that originated in the ocean during the Cambrian at least 550 million years ago and diversified to aquatic and terrestrial habitats all over the Earth, surviving five mass extinctions and constant global change (Chapter 25).

Of all the bilaterian animals (descendants of a common ancestor that had bilateral symmetry, triploblasty, and a coelom; Ch. 30, Section 30.2), deuterostomes are *not* the most abundant *nor* the most diverse. Protostome animals take the prize in this comparison—recall that about 70 percent of all known species of animals are insects (phylum Arthropoda), most of them beetles (see Ch. 31, Figure 31.1 and Table 31.5). But deuterostomes are notable because they include the largest-bodied and some of the most morphologically and behaviorally complex of all animals, including humans, and many of them are major consumers in food webs worldwide.

BIG PICTURE

This chapter is part of the Big Picture. See how on pages 722–723.

To explore deuterostome diversity, let's begin by taking a closer look at what characterizes the deuterostomes and then explore the echinoderms and chordates.

32.1 What Is a Deuterostome?

The **deuterostomes** (literally, "second-mouth") were initially grouped together because they appeared to undergo early embryonic development in a similar way: When a hummingbird, humpback whale, or human is just beginning to grow, the gut starts developing from posterior to anterior—the anus forms first and the mouth second. The coelom develops from outpockets of mesoderm (rather than the splitting of blocks of mesoderm).

After you complete this section, you should be able to ...

- Compare the deuterostomes to other animals.

More recently, as researchers have uncovered the developmental patterns of more and more animal species, they have concluded that some components of the "deuterostome pattern of development" are not unique to deuterostomes. For example, some protostomes also develop the anus first and the mouth second. Thus, the developmental distinction between deuterostomes and protostomes has become blurred.

However, phylogenies based on molecular evidence confirm that deuterostomes are a monophyletic group. Today, most biologists recognize three deuterostome phyla: the Echinodermata, Hemichordata, and Chordata (Figure 32.1):

- The echinoderms include the sea stars and sea urchins.
- The hemichordates include acorn worms and pterobranchs. Acorn worms burrow in marine sands or muds and make their living by deposit feeding or suspension feeding. Pterobranchs are small, tube-dwelling animals that typically live in colonies.
- The chordates include small, torpedo-shaped lancelets, sac-like tunicates, and the **vertebrates**, or animals with backbones. The vertebrates, in turn, comprise the hagfishes, lampreys, sharks and rays, bony fishes, amphibians, mammals, and reptiles (including birds).

Animals that are not vertebrates are collectively known as **invertebrates**. Over 95 percent of the known animal species are invertebrates (most of them protostomes; see Ch. 30, Table 30.1) but the vertebrates dominate the deuterostomes.

Although many deuterostomes share important features of embryonic development, their adult body plans and their feeding methods, modes of locomotion, and reproductive strategies are highly diverse. How did this diversity arise? Keep two themes in mind as you explore the chapter.

1. ***Evolution is not a linear progression from simple to complex.*** Recall from the exploration of sponges, comb jellies, cnidarians (Chapter 30), and protostomes (Chapter 31) that evolution tends to proceed more by tinkering with preexisting traits than by inventing new ones from scratch. The main gene regulatory networks that characterize animals originated in the Cambrian, then components were co-opted, rearranged, duplicated, and sometimes lost in subsequent lineages—producing the morphological traits that are characteristic of deuterostomes.
2. ***Ecological context is important for evolution.*** For example, plants and protostomes were already established on land when deuterostomes arrived, fueling an adaptive radiation in deuterostomes that ultimately produced the age of dinosaurs. When almost all dinosaurs and other large reptilian predators went extinct 66 million years ago, ecological opportunity fueled the radiation of mammals, birds, and ray-finned fishes.

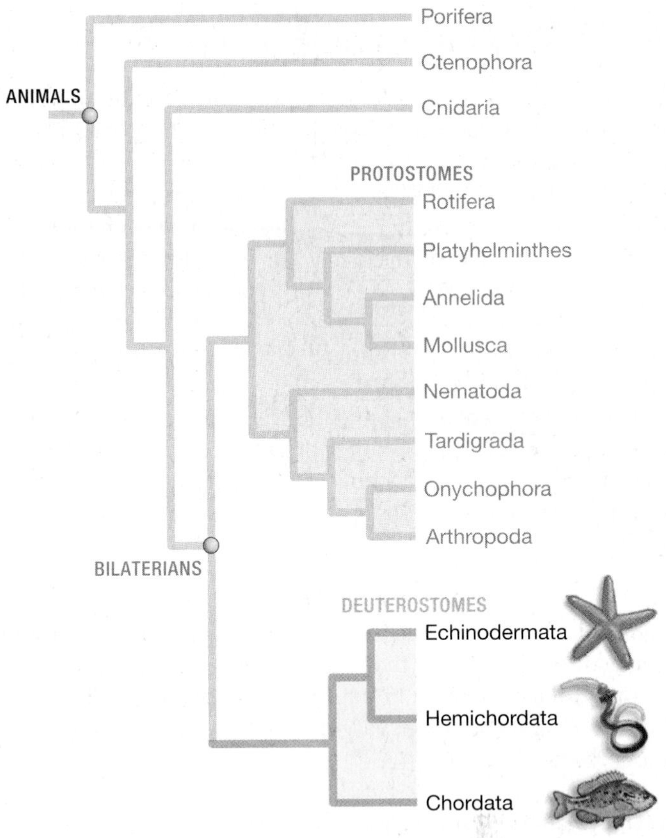

Figure 32.1 There Are Three Phyla of Deuterostomes. Vertebrates are in the phylum Chordata.

✔ Are invertebrates a monophyletic group? (Hint: See Ch. 25, Table 25.2, if you need help.)

Let's begin our tour of deuterostome diversity with the echinoderms and invertebrate chordates, then turn to vertebrates and our own closest relatives the hominins.

CHECK YOUR UNDERSTANDING

✔ **If you understood this section, you should be able to ...**

List four protostomes and four deuterostomes that you might see during your day today. Identify which of them are invertebrates and which are vertebrates.

Answers are available in Appendix A.

32.2 What Is an Echinoderm?

The echinoderms ("spiny-skins") were named for the spines observed in many species. Echinodermata is a large and diverse phylum that is exclusively marine. In numbers of species and range of habitats occupied, echinoderms are the second-most successful lineage of deuterostomes next to the vertebrates. To date, biologists have cataloged about 7000 species of echinoderms.

After you complete this section, you should be able to ...

- Compare the echinoderms to other deuterostomes.

Echinoderms are also abundant. In some deepwater environments, species in this phylum represent 95 percent of the total mass of organisms. The key lineages are described in **Table 32.1**.

How are the bodies of these animals organized?

The Echinoderm Body Plan

Echinoderms are deuterostomes that share several synapomorphies—traits that identify them as a monophyletic group: radial symmetry in adults, an endoskeleton of calcium carbonate, and the water vascular system. Let's consider each trait in turn.

Radial Symmetry in Adults All deuterostomes are considered bilaterians, because they evolved from an ancestor that was bilaterally symmetric (Ch. 30, Section 30.2). The ciliated planktonic larvae of echinoderms are also bilaterally symmetric (**Figure 32.2a**). But a remarkable event occurred early in the evolution of echinoderms: the origin of five-sided radial symmetry, also called pentaradial symmetry, in the benthic adults (**Figure 32.2b**).

Radially symmetric animals do not have heads (Ch. 30, Section 30.2). As a result, they tend to interact with the environment in all directions at once, or in varying directions, instead of facing the environment with a "head end." Adult echinoderms that are capable of movement tend to move equally well in all directions on the seafloor. Because echinoderms don't have anterior and posterior ends, the axes of the body are often described as oral (the side with the mouth) and aboral (the side without the mouth).

Endoskeleton of Calcium Carbonate A second noteworthy feature of the echinoderm body is its **endoskeleton**: a hard structure located just inside a thin layer of epidermal tissue, or skin, that protects and supports the body (shown without skin in Figure 32.2b). As an individual is developing, internal cells secrete plates of calcium carbonate. Depending on the species, the plates may remain separated and result in a flexible structure (as in sea stars), or they may fuse into a rigid case (as in sea urchins).

The echinoderm skeletal system also contains a unique type of "catch" collagen that can be reversibly stiff or flexible depending on signals from the nervous system. This innovation is adaptive because echinoderms can move their bodies into a specific shape, then "lock" into that position with no further energy investment.

(a) Echinoderm larvae are bilaterally symmetric.

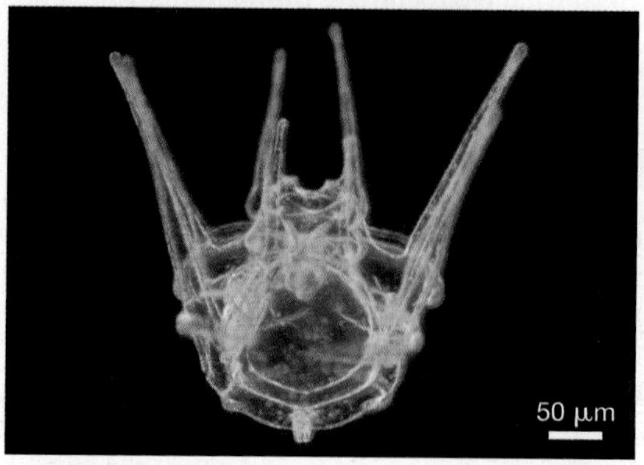

(b) Adult echinoderms are radially symmetric.

Figure 32.2 Body Symmetry Differs in Adult and Larval Echinoderms. (a) Bilaterally symmetric sea urchin larva. **(b)** Radially symmetric adult endoskeleton.

Water Vascular System and Tube Feet Echinoderms have a unique series of branching, fluid-filled tubes and chambers called the **water vascular system**. In most groups, the water vascular system is open to the exterior where it meets the body wall, and seawater flows into and out of the system. Inside, fluids move via the beating of cilia that line the tubes and chambers.

Figure 32.3a on page 676 highlights a particularly important part of the water vascular system called tube feet. **Tube feet** are elongated, fluid-filled appendages, each consisting of a balloon-like *ampulla* inside the body and a tube-like *podium* ("foot"; plural: *podia*) projecting on the outside (**Figure 32.3b**). The water vascular system forms a specialized hydrostatic skeleton that operates the tube feet. As tube feet extend and contract in a coordinated way, they alternately grab and release the substrate or prey using a combination of adhesion and suction, enabling movement and feeding. Tube feet also aid in gas exchange by increasing the surface area of soft tissue exposed to the aquatic environment.

Echinoderms Are Important Consumers

Depending on the lineage and species in question, echinoderms make their living by mass feeding on algae or other animals, suspension feeding, or deposit feeding. In most cases, an echinoderm's tube feet play a key role in obtaining food. Let's look at two examples of how echinoderms feed.

DIVERSITY Table 32.1 Key Lineages of Echinoderms

Deuterostomia > Echinodermata >		Known species	Feeding	Movement	Reproduction
	Crinoidea **Feather stars, sea lilies** Have 5–200 flexible arms held upward in the water	**600** Marine 3–150 cm	Most are suspension feeders; long arms with tube feet capture particles from water; cilia sweep particles to the upward-facing mouth	Sea lilies have long stalks and are sessile; feather stars hold on to substrate using flexible hooks and can swim short distances	Sexual reproduction with separate sexes; external fertilization
	Asteroidea **Sea stars** Have 5–40 arms radiating out from a central region; most spines occur as small knobs	**1900** Marine 1–100 cm	Predators or scavengers; some use tube feet to pull bivalves apart, then evert stomach into prey's visceral mass	Use tube feet to crawl over hard or soft substrates with oral side of the body facing down	Mostly sexual reproduction with separate sexes; external fertilization; most can regenerate arms; some asexual reproduction
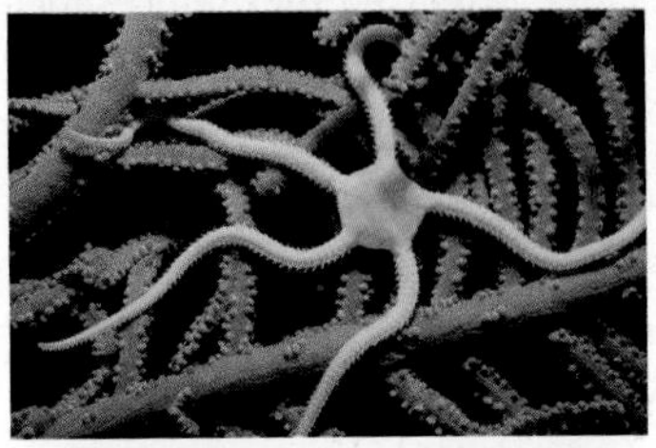	Ophiuroidea **Brittle stars, basket stars** Have 5–7 snake-like arms radiating out from a small, central disk	**2100** Marine 1–100 cm	Use flexible arms with delicate spines and tube feet for capturing prey, scavenging, deposit feeding, or suspension feeding	Rather than crawling on tube feet as in sea stars, ophiuroids crawl using shape changes in their flexible arms	Mostly sexual reproduction with separate sexes; external fertilization; most can regenerate arms; some asexual reproduction
	Echinoidea **Sea urchins, sand dollars** Sea urchins have globe-shaped bodies and long spines; sand dollars are flat	**1000** Marine 1–36 cm	Most urchins are herbivores; they graze using a jaw-like structure containing five teeth; most sand dollars are deposit feeders	Sea urchins crawl over hard surfaces using their tube feet and spines; sand dollars burrow in soft sediments	Mostly sexual reproduction with separate sexes; external fertilization; some asexual reproduction in larvae
	Holothuroidea **Sea cucumbers** Soft-bodied and elongated, with a mouth at one end and an anus at the other	**1250** Marine 2–200 cm	Suspension feed or deposit feed using modified tube feet called tentacles that are arranged in a whorl around their mouths	Burrow with wormlike movements using soft, muscular body wall and hydrostatic skeleton	Mostly sexual reproduction with separate sexes; external fertilization; some asexual reproduction by dividing body into two parts

For more go to https://goo.gl/KAfshM

Sea Stars Are Carnivores Consider the ochre sea star, *Pisaster ochraceus* (see Table 32.1), which is common along the rocky shores of the Pacific coast of North America. When this sea star wraps its arms around a bivalve, such as a mussel, the mussel responds by tightly closing the two halves of its shell. So how does the sea star feed?

The sea star is often able to pry the shell apart a few millimeters by clamping onto each half of the mussel's shell with its tube feet, making its endoskeleton rigid, and pulling. Once a gap exists, the sea star extrudes its stomach from its body and forces the stomach through the narrow opening in the bivalve's shell. Upon contact with the visceral mass of the bivalve, the stomach of the sea star secretes digestive enzymes, turning the mussel to a nutritious soup that the sea star stomach can then absorb. Eventually, only the shell of the prey remains.

Sea stars are important consumers in aquatic communities. For example, in the mid-1960s, Robert Paine famously discovered that when he removed sea stars from half of his study plots in rocky intertidal habitats, the number of species present in those communities plummeted compared to control plots (see Ch. 52, Figure 52.14). What had been diverse communities of algae and invertebrates became overgrown with solid stands of mussels.

(a) Echinoderms have a water vascular system.

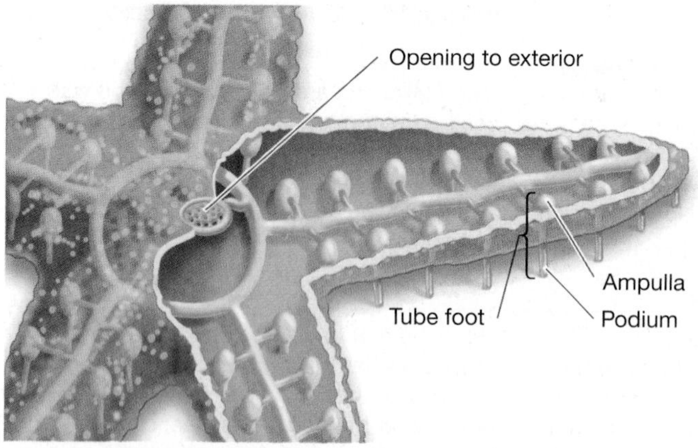

(b) Tube feet project from the underside of the body.

Figure 32.3 **Echinoderms Have a Water Vascular System with Tube Feet. (a)** The water vascular system is a series of tubes and chambers that radiates throughout the body. **(b)** Tube feet aid in movement because they extend from the body and can grab and release the substrate.

Sea Urchins Are Herbivores Sea urchins don't have flexible arms or eversible stomachs, but they do have a unique, jaw-like structure in their mouths called Aristotle's lantern. This feeding structure is made up of five calcium carbonate teeth that meet at the center of the mouth. Each tooth is operated by muscles, and the entire apparatus can be extended and retracted during feeding, increasing the animal's reach. The teeth scrape up material from the substrate or cut pieces from algae such as giant kelp.

In some areas of the world, urchins are extremely important grazers. In cases when humans have removed predators such as sea otters, and urchin populations have grown in response, the urchins' grazing can destroy kelp forests.

Sea stars (Asteroidea) and sea urchins and sand dollars (Echinoidea) are the most familiar echinoderms. Table 32.1 summarizes the key features of these groups along with the three other major lineages of echinoderms: the feather stars and sea lilies (Crinoidea), close relatives of sea stars called brittle stars and basket stars (Ophiuroidea), and close relatives of echinoids called sea cucumbers (Holothuroidea).

CHECK YOUR UNDERSTANDING

✔ **If you understood this section, you should be able to ...**

Explain how you would determine whether an animal you encounter along the seashore is an echinoderm.

Answers are available in Appendix A.

32.3 What Is a Chordate?

You are a chordate, yet chordates are so diverse that you would probably find it difficult to identify familiar features in some of your fellow chordates—such as the sea squirts, whose small, sessile, globular bodies hardly resemble animals at all. Which characteristics are shared within this phylum?

After you complete this section, you should be able to ...

- Compare the chordates to other deuterostomes.

Chordates are deuterostomes characterized by the presence of four key morphological features at some stage in their life cycles:

1. **pharyngeal slits or pouches**, which are openings into the throat or pouches in the throat;
2. a **dorsal hollow nerve cord**, which runs the length of the body;
3. a **notochord**, a stiff and supportive but flexible rod that runs the length of the body; and
4. a muscular, post-anal tail, meaning a tail that contains muscle and extends past the anus.

Electrical signals carried by the dorsal hollow nerve cord coordinate muscle movement, and the notochord stiffens the muscular tail. When this collection of traits first evolved, they created a long, streamlined animal that swam rapidly. Fossil evidence suggests that chordates diversified early during the Cambrian explosion of animals, about 550 million years ago.

The phylum Chordata is made up of three major lineages, or "subphyla": the **(1)** cephalochordates, **(2)** urochordates, and **(3)** vertebrates. The first two subphyla are sometimes referred to as the invertebrate chordates—meaning members of the Chordata that lack vertebrae (**Table 32.2**). They are of great interest to evolutionary biologists, who study the morphology, development, and genetics of these animals to better understand the evolutionary history of our own subphylum, the vertebrates.

Invertebrate chordates are also of increasing interest to ecologists due to their worldwide oceanic distribution, sometimes high density (thousands of individuals per cubic meter), and important role as consumers of small particles in food webs.

Let's take a closer look at the four chordate morphological features in each subphylum.

The Cephalochordates

Cephalochordates are also called lancelets, or amphioxus; they are small, mobile, torpedo-shaped animals with a "fishlike" appearance, and they make their living by suspension feeding

DIVERSITY Table 32.2 **Key Lineages of Invertebrate Chordates**

	Deuterostomia > Chordata >	Known species	Feeding	Movement	Reproduction
	**Cephalochordata** **Lancelets** Look like fish but are benthic invertebrates; have planktonic larvae	**25** Marine 4–7 cm	Suspension feed; adults extend head out of sediments, pump water through mouth and trap food particles in mucus on pharyngeal slits	Swim and burrow by contracting muscles on either side of the flexible, rod-like notochord, which functions as a stiff endoskeleton	Sexual reproduction with separate sexes; external fertilization; no asexual reproduction known
	**Urochordates** **Sea squirts, salps, and larvaceans** Sac-like adult body; most have non-feeding planktonic tadpole larvae	**2170** Marine 0.1–25 cm (colonies of salps to 2 m)	Suspension feed; capture particles by pumping water into incurrent siphon and across mucus sheet on pharyngeal slits, then out excurrent siphon	Adult sea squirts are sessile, but salps and larvaceans drift or propel themselves in open water; larvae swim with aid of notochord, which stiffens body	Most release both eggs and sperm, but some retain eggs for internal fertilization; some form large colonies via asexual budding

For more go to https://goo.gl/XhfrlB

(**Figure 32.4a**). Adult cephalochordates live on the ocean floor, where they burrow in sand and suspension feed with the aid of their pharyngeal slits.

The dorsal hollow nerve cord in cephalochordates runs parallel to a notochord, which extends all the way to the anterior end of the animal (thus the name "head-chords," although cephalochordates lack a well-defined head). Because the notochord stiffens a cephalochordate's body, muscle contractions on either side result in fishlike movement, used during dispersal or mating. (You can see an artist's depiction of an early cephalochordate called *Pikaia* in Ch. 25, Figure 25.10, at top left).

The Urochordates

Urochordates are also called tunicates. They have an external coat of polysaccharide, called a tunic, that covers and supports the body. The three major sub-lineages of urochordates include the sea squirts (or ascidians), the salps (and other thalaceans), and the larvaceans. All of the species described to date are marine—sea squirts live on the ocean floor; salps and larvaceans live in open water.

Sea squirts are by far the most diverse urochordates, with about 2000 named species. As **Figure 32.4b** shows, pharyngeal slits, a dorsal hollow nerve cord, notochord, and tail are present in sea squirt larvae. In these non-feeding larvae, the pharyngeal slits function in gas exchange, and the other three features enable swimming movements. As the larvae swim or float in the upper layers of the ocean, they drift to new habitats where food might be more abundant. As a result, larvae serve as a dispersal stage in the life cycle of sea squirts.

Adult sea squirts lack a dorsal hollow nerve cord, notochord, and tail, but they retain pharyngeal slits, which are used for gas exchange and feeding. These sessile animals suspension feed by using cilia to pump water into an incurrent siphon, across the mucus-covered pharyngeal slits, and out an excurrent siphon, trapping particles that are then digested in the U-shaped gut.

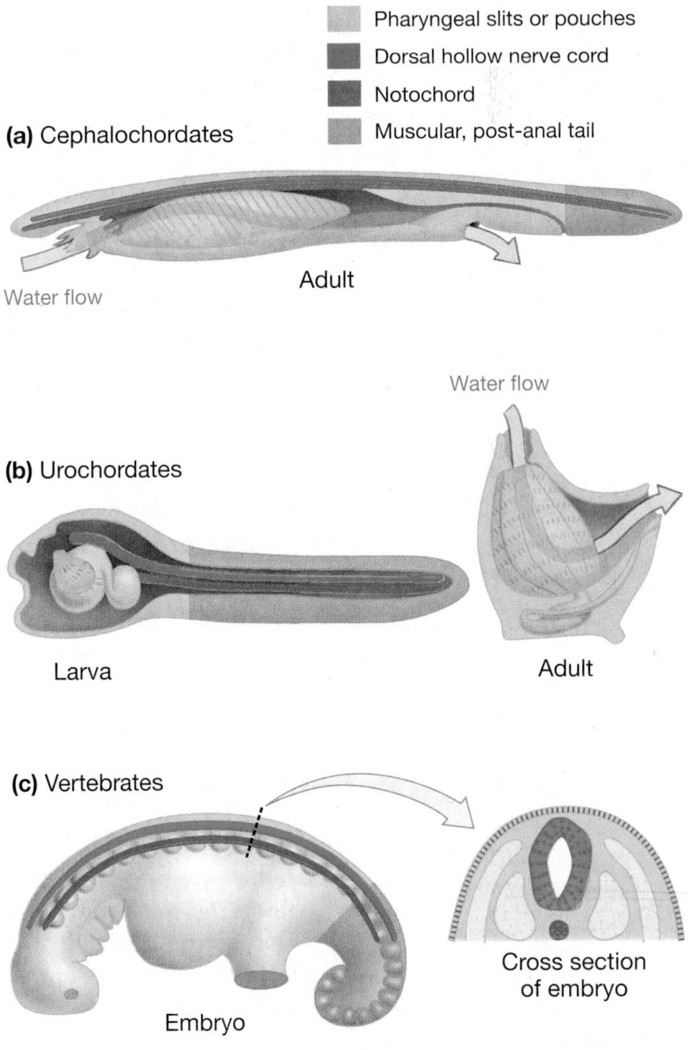

Figure 32.4 Four Key Characteristics Occur in All Chordates.

The Vertebrates

Vertebrates (Figure 32.4c) include the hagfishes, lampreys, sharks and rays, several lineages of bony fishes, amphibians, mammals (including humans), and reptiles (including birds). How do the chordate characteristics appear in this lineage?

In vertebrates, the dorsal hollow nerve cord is the spinal cord—a bundle of nerve cells that runs from the brain to the tail end of the body. Electrical signals from the spinal cord control movements and the function of organs (Ch. 43, Section 43.4).

Structures called pharyngeal pouches are homologous to pharyngeal slits and are present in all vertebrate embryos, including humans. In fishes and some amphibians, the creases between pouches open into slits and develop into part of the main gas-exchange organ—the **gills**. In terrestrial species, however, slits do not develop after the pharyngeal pouches form. In these species, the pharyngeal pouches are a vestigial trait, yet nonetheless an integral part of early vertebrate development.

The muscular post-anal tail is also a vestigial trait in some species. A tail occurs in all vertebrate embryos, but is lost during the course of development in many tailless species, including humans.

A notochord develops in all vertebrate embryos, where it functions in body support and helps organize the adult body plan. The notochord secretes proteins that help induce the formation of somites—segmented blocks of tissue that form along the length of the body (Ch. 47, Section 47.4). Cells in the somites later differentiate into the vertebrae, ribs, and skeletal muscles of the back, body wall, and limbs. In this way, the notochord is instrumental in the development of the trait that gave vertebrates their name. The notochord itself is retained into the adult stage only in hagfishes and lampreys, but persists as cells in the intervertebral disks in other vertebrates, including humans.

There are about 63,000 named species of vertebrates. Because they are so species rich and diverse, vertebrate chordates merit their own section in this chapter.

CHECK YOUR UNDERSTANDING

✔ **If you understood this section, you should be able to ...**

MODEL Sketch a simple model of a generic chordate, showing and labeling its four key features.

Answers are available in Appendix A.

32.4 What Is a Vertebrate?

After you complete this section, you should be able to ...

- Compare the vertebrates to other chordates.

What distinguishes vertebrates from other chordates? The vertebrates are a monophyletic group distinguished by the following synapomorphies:

- a column of cartilaginous or bony structural supports called **vertebrae**, which form along the dorsal side of the body; and
- a **cranium**, a bony, cartilaginous, or fibrous case that encloses the brain.

In species with bony skeletons, bones are cartilaginous first and become bony later in development. **Cartilage** is a strong but flexible tissue that consists of scattered cells in a gel-like matrix of polysaccharides and protein fibers (Ch. 39, Section 39.2). Cartilage is found throughout your body, including the outer ear and the tip of your nose. **Bone** is a dense tissue consisting of cells and blood vessels encased in a matrix made primarily of calcium phosphate, along with a small amount of protein fibers.

The vertebral column is important because it protects the spinal cord. The cranium is important because it protects the brain and sensory organs such as eyes. Together, the vertebrae and cranium protect the central nervous system (CNS) and key sensory structures.

The vertebrate brain develops as an outgrowth of the most anterior end of the dorsal hollow nerve cord, and it is important to the vertebrate lifestyle. Vertebrates are active predators and herbivores that must make rapid, directed movements in response to stimuli in their environment. More complex brains enable more complex interactions with the environment.

In early vertebrates, the brain was divided into three regions with important sensory functions:

1. The **forebrain** housed the sense of smell;
2. the **midbrain** was associated with vision; and
3. the **hindbrain** was responsible for balance and, in some species, hearing.

All vertebrates living today have retained these three regions, but the structures and functions of the regions have evolved over time. For example, in the jawed vertebrates, the hindbrain consists of enlarged regions called the **cerebellum** and **medulla oblongata**. Part of the forebrain also evolved into a large and elaborate structure called the **cerebrum**, especially in birds and mammals. (The structure and functions of the vertebrate brain are detailed in Ch. 43, Section 43.4.) The evolution of a large, three-part brain, protected by a hard cranium, was a key innovation in vertebrate evolution.

Among living vertebrates, the most species-rich and ecologically diverse lineages are the ray-finned fishes and the tetrapods. **Ray-finned fishes**, including familiar groups such as goldfish, tuna, and salmon, have fins supported by bony rods and occupy habitats ranging from deepwater environments, which are perpetually dark, to shallow ponds that dry up each year. **Tetrapods** ("four-footed"), including reptiles, amphibians, and mammals, tend to be the large herbivores and predators in terrestrial environments all over the world.

Today the number of species of ray-finned fishes is about the same as the number of tetrapods—both having over 30,000 species (Figure 32.5). How are the lineages of vertebrates related, and what are some of the key morphological innovations that arose during vertebrate evolution?

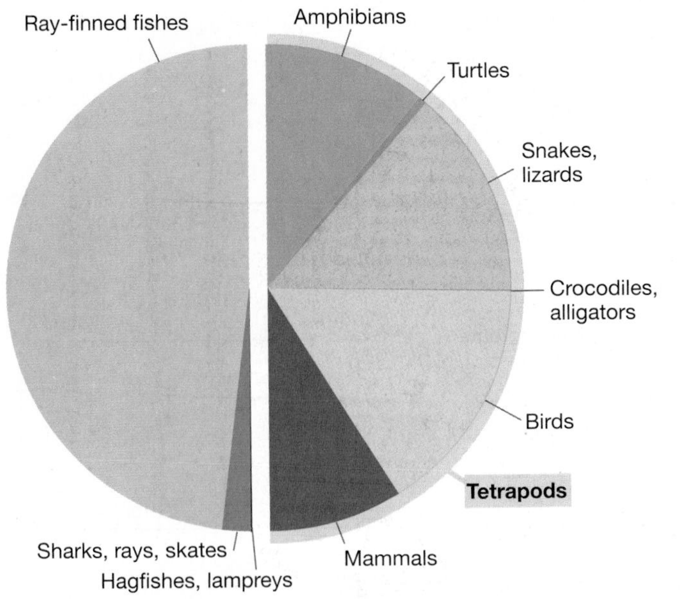

Figure 32.5 **Relative Species Abundance among Living Vertebrates.** Of the 63,000 species of described vertebrates, about half are ray-finned fishes and about half are tetrapods.

DATA: American Museum of Natural History; BirdLife.org; FishBase.org; IUCN Red List 2017; Reptile-Database.org.

CHECK YOUR UNDERSTANDING

✔ **If you understood this section, you should be able to . . .**

1. **QUANTITATIVE** Rank the top five groups of living vertebrates (see Figure 32.5) according to species abundance.
2. Name two chordate characteristics that cephalochordates and ray-finned fishes share, and two traits that occur in fishes but not cephalochordates.

Answers are available in Appendix A.

32.5 What Key Innovations Occurred during the Evolution of Vertebrates?

After you complete this section, you should be able to . . .

- Analyze the significance of key innovations during the evolution of vertebrates.

A major renaissance in the study of vertebrate evolution has come about in the last few decades due to synthesis of diverse sources of new data: **(1)** new fossil evidence continues to provide major insights; **(2)** phylogenetic analysis now incorporates fossil evidence and also a treasure trove of new molecular data; **(3)** evidence from developmental biology is enabling scientists to test for homology between structures in different vertebrate lineages; and (4) comparative genomics provides information about the relative similarity of genes or whole genomes of organisms (Chapter 20).

To explore some of the major events in vertebrate evolution revealed by the data, let's examine the phylogenetic tree of living chordates, especially vertebrates, in Figure 32.6 on page 680. This tree was constructed based on DNA and other molecular sequence data, but fossil evidence is essential in interpreting the pattern of traits plotted on the tree (indicated by labeled bars). Keep three general themes in mind while exploring vertebrate evolution:

1. ***Most vertebrates are extinct.*** The fossil record shows that the vertebrate tree of life was very "bushy" from the start, with many diverse branches that went extinct. Thus, the branches in Figure 32.6 represent only a small fraction of total diversity. Remember, none of the species alive today are ancestors of any other living species. (See Ch. 25, Section 25.1 and **BioSkills 13** to review how to read phylogenetic trees.)
2. ***Some traits evolved more than once.*** Convergent evolution has occurred in many lineages, creating a more complex picture of vertebrate evolution than was once appreciated. For example, flight evolved independently in pterosaurs, birds, and bats.
3. ***Traits are sometimes lost.*** Evolution is not just a progression from simple to complex, and it is not limited to the addition of new traits. The loss of traits is also common and adaptive, depending on the environment. For example, snakes are limbless reptiles that descended from reptiles with four limbs.

Although the phylogeny of chordates continues to be a topic of intense research, biologists are increasingly confident that the relationships described in Figure 32.6 are accurate. Let's start at the root of the tree at top left, then work our way to the bottom right.

Urochordates: Sister Group to Vertebrates

The tree in Figure 32.6 includes all the major lineages of living deuterostomes. Which lineage is most closely related to the vertebrates? Put your finger on the node identified by the first black dot. This node represents the most recent common ancestor of all chordates, possessing the key chordate traits described in the previous section: pharyngeal slits, a dorsal hollow nerve cord, a notochord, and a muscular, post-anal tail. The vertebrates are clearly a part of this monophyletic group.

Now move one node down to the second black dot, near the base of the purple Urochordata branch. According to the most recent molecular data, the urochordates are the closest living relatives of the vertebrates, not the cephalochordates as previously proposed. Thus, urochordates are considered a sister group to the vertebrates because these two lineages share a more recent common ancestor than either lineage does with the cephalochordates.

First Vertebrates: Origin of the Cranium and Vertebrae

Now move your finger down to the third black dot in Figure 32.6. The dot at this node represents the most recent common ancestor of all vertebrates. What might this ancestor have looked like?

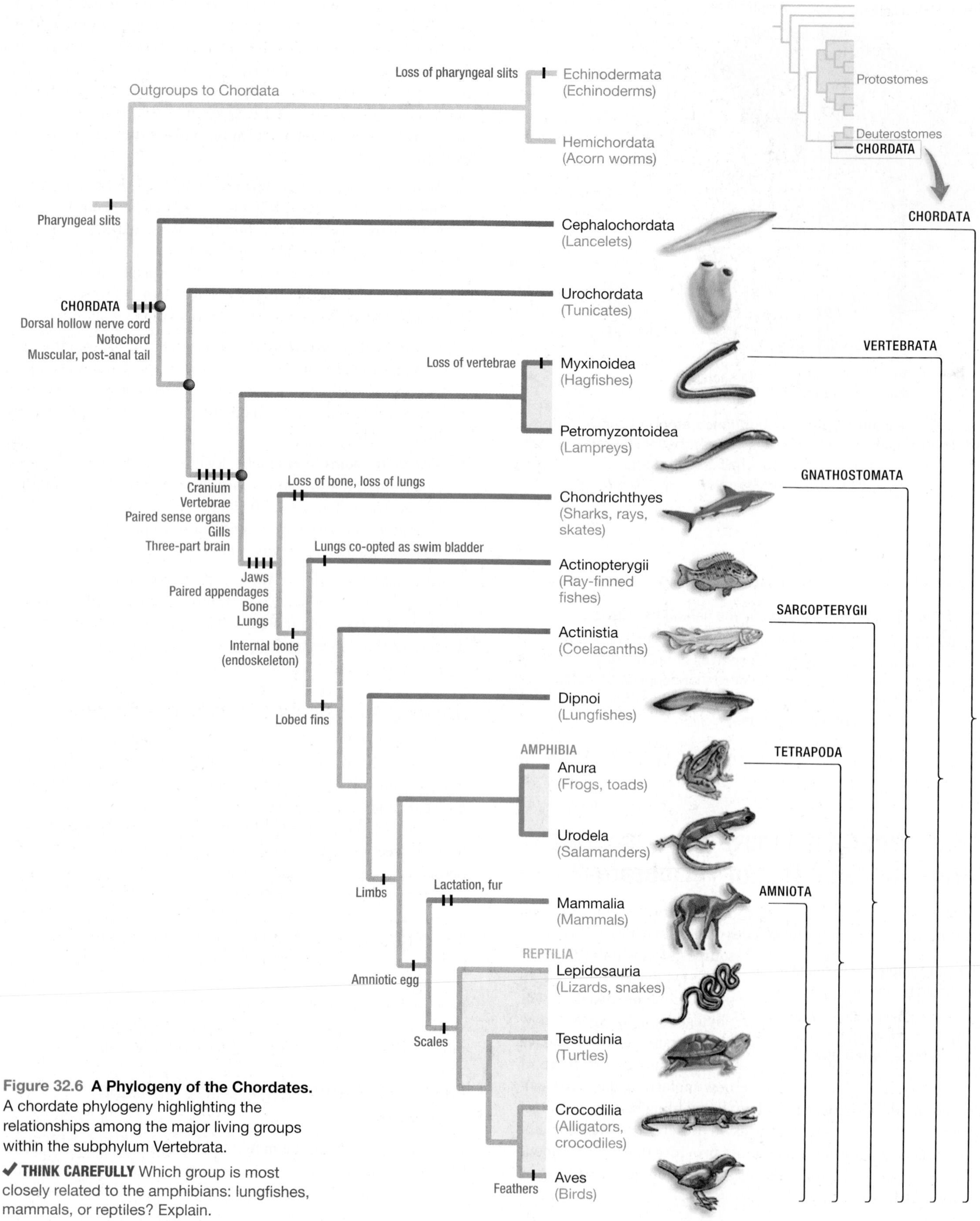

Figure 32.6 A Phylogeny of the Chordates. A chordate phylogeny highlighting the relationships among the major living groups within the subphylum Vertebrata.

✔ **THINK CAREFULLY** Which group is most closely related to the amphibians: lungfishes, mammals, or reptiles? Explain.

DIVERSITY Table 32.3 **Key Lineages of Jawless Fishes**

Chordata > Vertebrata >		Known Species	Feeding	Movement	Reproduction
	Myxinoidea **Hagfishes** Cranium, no vertebrae, no jaws, no paired appendages	**78** Marine 10–130 cm	Scavengers and predators; deposit feed on carcasses of dead fishes and whales, and may feed on buried prey in ocean sediments	Use notochord as structural support; swim by undulating; sometimes tie themselves in "knots" to remove copious defensive slime	External fertilization; do not have a larval stage; eggs hatch into juveniles that resemble miniature adults
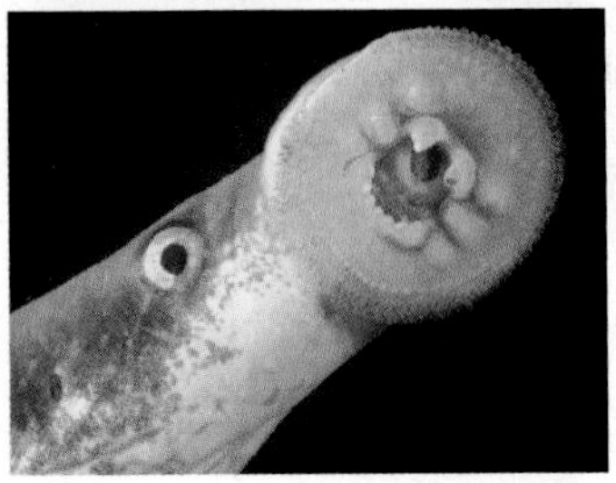	Petromyzontoidea **Lampreys** Cranium, some pieces of cartilage along nerve cord, no jaws, no paired appendages	**46** Marine Freshwater 10–100 cm	Most are ectoparasites; use spines in their mouth to rasp a hole in side of fishes and other hosts; suck blood and other body fluids	Use notochord as structural support; swim by undulating; can use suckers to loop upstream like an inchworm	Eggs hatch into larvae that look and act like lancelets (cephalochordates); burrow and suspension feed for years before metamorphosing into adults

For more go to https://goo.gl/M8TPOC

Evidence for Early Vertebrates The earliest known vertebrate fossils are present in the Chengjiang formation of China and the Burgess Shale and similar deposits in Canada, dated to the Cambrian explosion (Ch. 25, Section 25.3). The detailed vertebrate fossils show that the earliest members of this lineage lived in the ocean about 520 million years ago; they had streamlined, fishlike bodies, gills, a notochord, post-anal tail, and paired eyes. The fossils also suggest the presence of a cranium made of cartilage and possibly a series of cartilaginous reinforcements to the notochord—early vertebrae. Scientists will never know if any of the discovered fossil species are direct ancestors of today's vertebrates, but the fossils help scientists to pose and test hypotheses.

✔ **MODEL If you understand phylogenetic trees, you should be able to add the extinct Cambrian vertebrate *Haikouichthys* to the tree in Figure 32.6. (Note that there is no time scale in this tree.)**

Comparative developmental studies and comparative genomics are showing that complex traits such as the cranium and vertebrae don't just appear suddenly with no precedent. The specialized embryonic cells (such as neural crest cells) that are responsible for brain, cranium, and sensory cell formation are synapomorphies for vertebrates, yet the genes responsible for the differentiation of these cells were co-opted and modified from genes present in invertebrate chordate ancestors. Thus, tinkering of certain gene regulatory networks may have been central to key innovations in the evolution of the vertebrate head.

The Jawless Vertebrates Now turn again to the phylogenetic tree in Figure 32.6, and look at the first blue branch of living vertebrates to emerge near the root of the vertebrate tree: the hagfishes and lampreys (described in **Table 32.3**). Hagfishes and lampreys have pharyngeal slits, a dorsal hollow nerve cord, a notochord, and a muscular post-anal tail, plus a three-part brain, a cartilaginous cranium, and paired eyes. Lampreys have cartilaginous reinforcements (vertebrae) on their notochord, but hagfishes do not—the trait was lost.

The phylogenetic relationships among the hagfishes, lampreys, and gnathostomes (jawed vertebrates, described in the next section) have been hotly debated. Two hundred years ago, hagfishes and lampreys were grouped together as the class Cyclostomata ("round-mouthed"), based on their unique mouthparts (see Table 32.3) and absence of jaws. Then, starting in the 1970s, morphologists began to take a closer look and placed hagfishes and lampreys in two independent lineages—hagfishes representing the most basal lineage of vertebrates, and lampreys representing the sister group to gnathostomes.

So which is correct? Analysis of a new source of data—evolutionarily conserved microRNAs (Ch. 19, Section 19.4)—has decided in favor of the Cyclostomata hypothesis. This result is important because it emphasizes that some vertebrate characteristics appear to have been *lost* in the hagfish/lamprey lineages, including the loss of vertebrae in hagfishes.

Gnathostomes: Origin of the Vertebrate Jaw

The jawed vertebrates, or **gnathostomes** (pronounced *NATH-oh-stomes*), are a monophyletic group including five major living lineages: the cartilaginous fishes, ray-finned fishes, coelacanths, lungfishes, and tetrapods. The four lineages of fishes within this group are described in **Table 32.4** on page 682 (the tetrapods are described separately).

Note that unlike the amphibians, mammals, and reptiles, the "fishes" are not a monophyletic group. You can explore the

DIVERSITY Table 32.4 Key Lineages of Jawed Fishes

	Chordata > Vertebrata >	Known Species	Feeding	Movement	Reproduction
	Chondrichthyes **Sharks, rays, skates** Cartilaginous skeleton, jaws, and paired lateral fins	**1200** Marine (Some freshwater) 10 cm–13 m	Some suspension feed on plankton, but most are predators, often "top predator" because they are at the top of the food web	Sharks swim by undulating their bodies from side to side and beating their tails; rays and skates swim by flapping their wing-like pectoral fins	Internal fertilization; some are oviparous, some ovoviviparous, and some viviparous; a placenta evolved independently in certain shark lineages
	Actinopterygii **Ray-finned fishes** Bony skeleton, fins supported by long, bony rods arranged in a ray pattern; scales	**32,200** Marine Freshwater 8 mm–14 m	Can suck food toward their mouths, grasp it with their protrusible jaws, process it with teeth on jaws and with pharyngeal jaws in throat	Swim by undulating side to side and by flapping fins; many have a gas-filled swim bladder that prevents sinking	Most rely on external fertilization and are oviparous; others have internal fertilization and are viviparous; some have parental care; many undergo metamorphosis
	Actinistia **Coelacanths** Four fleshy, lobed fins supported by bones; thought to have gone extinct 65 mya, discovered alive in 1938	**2** Marine 1.5–2 m	Has a unique hinge in its skull that enables it to open its mouth extremely wide to consume large fish and cephalopods such as squid	Generally swim slowly and passively using four lobed fins in alternating pattern resembling use of limbs in tetrapods	Sexual reproduction; internal fertilization; ovoviviparous (retain embryos within eggs for a year or more before birth of fully developed young)
	Dipnoi **Lungfishes** Four lobed fins; use lungs to breathe air when oxygen level drops in their shallow aquatic habitats	**6** Freshwater 0.4–2 m	Omnivorous, eating algae and plant material as well as animals; teeth fuse to form tooth plates for crushing and chewing	Swim by undulating their bodies; some also use their lobed fins to walk along the bottom of ponds	Sexual reproduction; external fertilization; oviparous; eggs hatch into larvae that resemble juvenile salamanders

For more go to https://goo.gl/8WzoAO

distinction between monophyletic and paraphyletic groups in **Making Models 32.1**.

You might be tempted to look at the living lineages of jawed fishes in Figure 32.6 and conclude that the earliest jawed vertebrates must have been shark-like. Cartilaginous fishes, such as sharks and rays, have jaws; but unlike the other three lineages of jawed fishes, these jaws are made of reinforced cartilage rather than bone.

However, exciting new fossil evidence has disproved a shark-like origin of jawed fishes, pointing instead to jawed fishes called placoderms that show up in the Silurian, about 430 million years ago. Rather than being shark-like, these extinct armored fishes had heads covered with bony shields (**Figure 32.7**). It turns out that most of your skull is made up of this type of dermal ("skin") bone, making your skull homologous to the "head shields" of ancient fishes—evidence of your fish ancestry. Other remnants of placoderm armor occur today as dermal scales in sharks and rays, and as scales and fin rays in bony fishes.

The diverse lineages of extinct armored fishes are not included in the tree of living vertebrates in Figure 32.6. If pictured, the placoderms would occur after the jawless fishes diverged but before the origin of living jawed fishes.

Soon after the appearance of jawbones, teeth appear in the fossil record. With jaws and teeth, vertebrates became armed and dangerous. The evolution of jaws was significant because it improved the ability of fishes to capture prey, and also to bite, meaning that vertebrates were no longer limited to suspension or

Making Models 32.1 Tips on Drawing Phylogenetic Trees (VII)

Sometimes there is a mismatch between traditional taxonomic terms and evolutionary relationships, resulting in named groups that are not monophyletic, such as "fishes." Drawing models can help you understand the difference between monophyletic and paraphyletic groups. (You can review types of groups in Ch. 25, Table 25.2.)

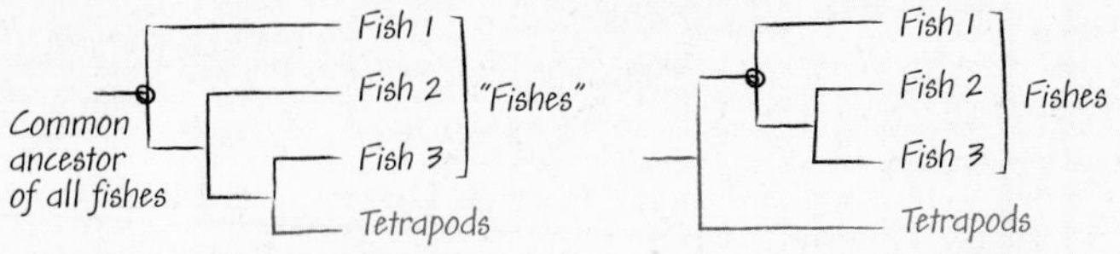

"Fishes" are not monophyletic — If fishes were monophyletic

MODEL Draw a third tree that has an accurate branching pattern, yet shows a monophyletic group called "fishes." (Hint: Some evolutionary biologists might consider you to be a fish.)

To see this model in action, go to the Study Area of **Mastering Biology**

deposit feeding. Instead, they could make a living as herbivores or predators. Other key traits also appeared in placoderms, such as paired fins and internal fertilization. The fossil record shows that soon after these key innovations evolved, a spectacular radiation of jawed vertebrates followed, filling marine and freshwater habitats.

Once jaws originated, changes in size and shape enabled a rapid diversification of feeding strategies, from suction feeding (caused by the negative pressure created in front of the mouth when the jaws are snapped open) to biting. Other modifications to the jaw also occurred:

- In most ray-finned fishes, the jaw is protrusible—meaning it can be extended to bite at food.
- Several species-rich lineages of ray-finned fishes have a second specialized jaw called a **pharyngeal ("throat") jaw**, consisting of modified gill arches. It is located in the back of the throat and makes food processing efficient (Ch. 41, Section 41.2).

To summarize, the origin of the jaw was a major evolutionary innovation, leading to a dramatic radiation of fishes during the Devonian period and the subsequent origin of tetrapods. More than 99 percent of living species of vertebrates are gnathostomes.

Origin of the Bony Endoskeleton

In some lineages of fishes from late in the Silurian period, the cartilaginous endoskeleton began to be stiffened by the deposition of bone. Unlike the dermal bone that had evolved earlier and was derived from neural crest cells (ectodermal in origin), the bony endoskeleton was derived from mesoderm and functioned to support swimming movements.

Figure 32.6 shows three living lineages of bony fishes: ray-finned fishes, coelacanths, and lungfishes. Their bodies are covered with interlocking scales that provide a stiff but flexible covering (remnants from armored ancestors). About 96 percent of all living fish species, including familiar groups such as the tuna, trout, cod, and goldfish, are teleosts, a sublineage of ray-finned fishes. Teleosts underwent an adaptive radiation about the same time that mammals did, after the end-Cretaceous mass extinction 66 million years ago when aquatic reptiles called ichthyosaurs and plesiosaurs went extinct.

Although coelacanths (*SEEL-uh-kanths*) and lungfishes represent independent lineages (thus do not form a monophyletic group), they are sometimes called **lobe-finned fishes**. Lobe-finned fishes were common and diverse in the fossil record in the Devonian period, about 400 million years ago. They are important because they represent a crucial evolutionary link between the fishes and the tetrapods. Only eight species are living today.

Origin of the Lungs

Look again at the phylogenetic tree in Figure 32.6, and pose a hypothesis for the timing of the origin of lungs. Did lungs first appear in lungfishes? You might be surprised to learn that lungs appeared earlier—in placoderms. Why would lungs evolve in an aquatic animal that already had gills?

Fossil evidence suggests that lungs originated as an auxiliary source of oxygen for the heart. Consider that the blood flow in the circulatory system of early fishes passed from the heart to the gills—where the blood picked up oxygen—then to the tissues of the body, then finally returned to the heart *low* in oxygen. The

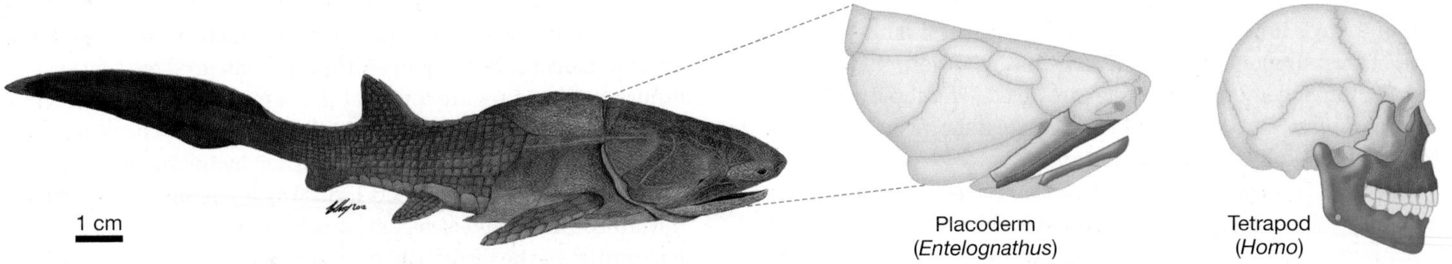

Figure 32.7 The Earliest Jaws Occurred in Extinct Armored Fishes Called Placoderms Placoderms were highly diverse. This artist's rendering illustrates a small placoderm dating to the boundary between the Silurian and Devonian periods, 419 million years ago. Four bones associated with the placoderm jaw have been color-coded to show their homology to human skull bones.

DATA: M. Zhu et al. 2013 *Nature* 502: 188–193; J. A. Long 2016 *Science* 354: 280–281.

low availability of oxygen to the heart would have limited metabolic activity.

Lungs arose as outpockets of the esophagus. The current origin-of-lung hypothesis predicts that early fishes filled this space with air either by gulping air from the surface or by secreting oxygen from tissues within. Then the close association between the lungs and heart was the precedent for the pulmonary circulation that occurs in living tetrapods (Ch. 42, Section 42.5).

The early appearance of the lungs has interesting implications. The lungs were lost in cartilaginous fishes and were coopted in ray-finned fishes as a gas-filled **swim bladder**, an important innovation that allowed ray-finned fishes to maintain neutral buoyancy and avoid sinking without a constant expenditure of energy.

Tetrapods: Origin of the Limb

The next great event in the evolution of vertebrates was the transition to living on land. The first vertebrates that had limbs and were capable of moving on land date to about 365 million years ago, late in the Devonian period. These were the first of the tetrapods—vertebrates with four limbs.

The tree in Figure 32.6 indicates that the transition from water to land probably occurred just once in the evolution of vertebrates (so far; several lineages of living fishes have a semi-terrestrial lifestyle), giving rise to three major lineages of living tetrapods: amphibians, mammals, and reptiles. Some lineages of tetrapods, such as snakes and whales, experienced a subsequent loss of limbs but are still considered members of the group tetrapods since they are descendants of a common ancestor that had limbs.

The Limbs-from-Fins Hypothesis To understand how tetrapods made the water-to-land transition, consider the morphology and behavior of their closest living relatives, the lungfishes. Most living species of lungfishes inhabit shallow, oxygen-poor water. True to their name, lungfishes breathe air with lungs, supplementing the oxygen taken in by their gills.

Some lungfishes also have fleshy fins supported by bones and are capable of walking short distances along watery mudflats or the bottoms of ponds. In addition, some species can survive extended droughts by burrowing in mud.

Lungfishes have a series of adaptations that allow them to survive on land for short periods. Could a lungfish-like organism have evolved into the first land-dwelling, limbed vertebrate?

Fossils provide strong links between the limbs of the ancestors of today's lungfishes and those of the earliest land-dwelling vertebrates, supporting a limbs-from-fins hypothesis. Figure 32.8 shows a phylogeny of the species involved. The fossil record is more complete than shown here—the figure shows just a sample of the species known from this interval.

The color coding in Figure 32.8 emphasizes that each fin or limb has a single bony element closest to the body (in blue) that articulates with two bones arranged side by side (in red), next to a series of elements farthest from the body (in orange). Because the structures are similar and no other animal groups have limb bones in this arrangement, the evidence for homology is strong.

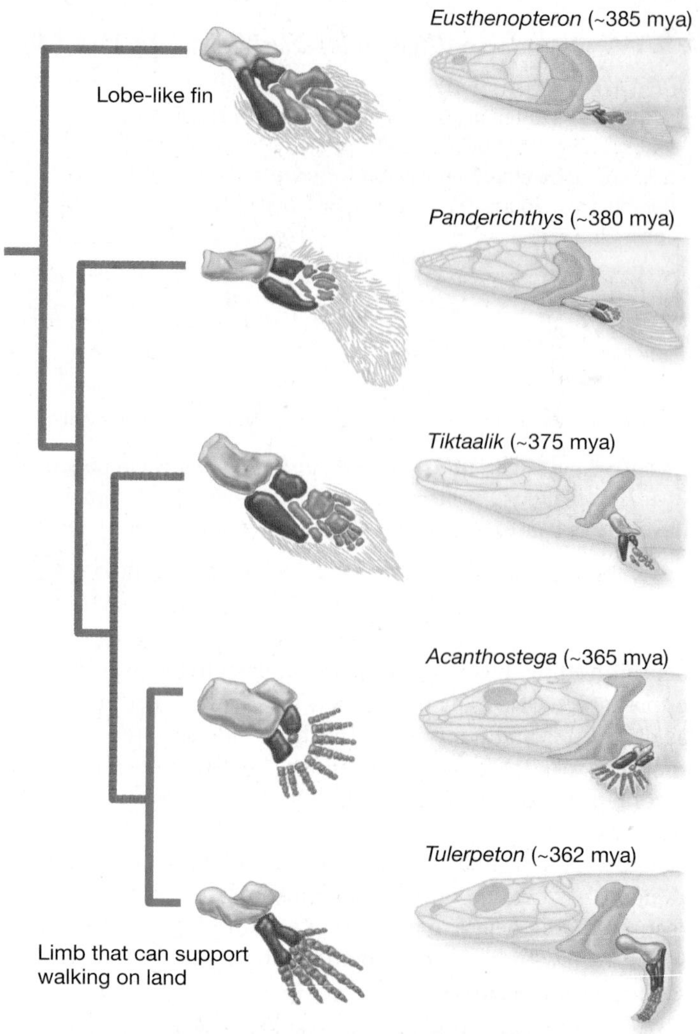

Figure 32.8 Fossil Evidence for the Limbs-from-Fins Hypothesis. The fossil record from the Devonian documents a series of species that show a gradual transition from a lobe-like fin to a limb that could support walking on land.

✔ **PROCESS OF SCIENCE** This phylogeny was estimated from traits other than the morphology of the limb. Why is this important in addressing how limbs evolved?

The hypothesis that tetrapod limbs evolved from fish fins has also been supported by molecular genetic evidence. Recent work has shown that several regulatory proteins involved in the development of fish fins and the upper parts of mammal limbs are homologous. For example, the proteins produced by several different *Hox* genes are found at the same times during development and in the same locations in fins and limbs. These data suggest that these appendages are patterned by the same genes.

The origin of the tetrapod limb was a key morphological innovation that enabled vertebrates to exploit the ecological opportunities of the terrestrial environment, including refuge from aquatic predators, access to new food sources, and an increased availability of oxygen. Just as the segmented body and limbs of the arthropod plan were key innovations that enabled insects to

DIVERSITY **Table 32.5** Key Lineages of Amphibians

	Chordata > Vertebrata > Amphibia >	Known Species	Feeding	Movement	Reproduction
	Anura **Frogs, toads** Stout-bodied; lacking tails as adults; large eyes	**7400** Freshwater Terrestrial 0.8–32 cm	Adults are carnivores; most are sit-and-wait predators that use their long, extensible tongues to capture prey	Four limbs; on land, frogs and toads kick their hind legs to jump or hop; in water, they kick their hind legs to swim	Sexual reproduction; usually external fertilization in water; aquatic tadpoles usually metamorphose into land-dwelling adults
	Urodela **Salamanders** Slender-bodied; tails in both larvae and adults; short limbs; large eyes	**675** Freshwater Terrestrial 1.5–180 cm	Most adults are carnivores; terrestrial species use their extensible tongues to capture prey; aquatic forms grab prey with their jaws	Four limbs; on land, salamanders walk; in water, they undulate their bodies to swim; some can regenerate limbs	Sexual reproduction; usually internal fertilization in water; aquatic larvae metamorphose into aquatic or terrestrial adults
	Gymnophiona **Caecilians** Slender-bodied; no limbs; small, skin-covered eyes detect light	**200** Freshwater Terrestrial 11–150 cm	Terrestrial caecilians prey on earthworms and other soil-dwelling animals; aquatic forms eat invertebrates and small fishes	Lack limbs, so resemble worms or snakes; burrow underground in wet tropical regions; in water, swim by undulating their bodies	Sexual reproduction; internal fertilization; most are viviparous; some have larvae, while others have direct development

For more go to https://goo.gl/BNXeQ7

diversify on land, the segmented body and limbs of vertebrates were key innovations that enabled them to diversify on land. The limbs of vertebrates diversified spectacularly—enabling running, gliding, crawling, burrowing, swimming, and flying.

Amphibians The first tetrapods to live on land were **amphibians** ("both-sides-living"). The name is appropriate because adults of most amphibians feed on land but lay their eggs in water. Most amphibians undergo metamorphosis from an aquatic larva to a terrestrial or semiterrestrial adult. Gas exchange occurs exclusively or in part across their moist, mucus-covered skin. Living amphibians represent a monophyletic group, Amphibia, which includes frogs and toads, salamanders, and the snakelike caecilians. The characteristics of these three groups are described in **Table 32.5**.

Amniotes: Origin of the Amniotic Egg

The fossil record shows clearly that internal fertilization and live birth (viviparity) evolved during the radiation of Devonian fishes, in placoderms—the males had structures called claspers that enabled copulation. However, internal fertilization was subsequently lost in several groups, including the ancestral tetrapod, which is hypothesized to have been egg-laying (oviparous).

About 20 million years after tetrapods appear in the fossil record, the first amniotes show up. The Amniota is a lineage of vertebrates that includes all tetrapods other than amphibians. The **amniotes** are named for a signature synapomorphy: the amniotic egg. Put your finger on the node of the tree in Figure 32.6 where the mammal and reptile branches meet. This node indicates that mammals and reptiles are sister groups that both inherited the amniotic egg from their common ancestor. Thus, mammals and reptiles are more closely related to each other than either of them is to the amphibians. Are you surprised?

Amphibians keep their jelly-coated eggs from drying out by laying them in water. In contrast, oviparous reptiles (including birds) and the few oviparous mammals produce an **amniotic egg**, which has a protective covering that reduces the rate of drying significantly. Species that produce amniotic eggs lay them outside of water.

As **Figure 32.9** on page 686 shows, the amniotic egg contains four internal membranes with distinct functions that enable survival in dry, terrestrial environments. The three inner membranes surround the embryo itself (the amnion), the yolk provided by the mother (the yolk sac), and the waste from the embryo (the allantois). These three membranes are in turn surrounded by a fourth membrane (the chorion), which is cushioned by a solution of water and protein called albumen. The multiple

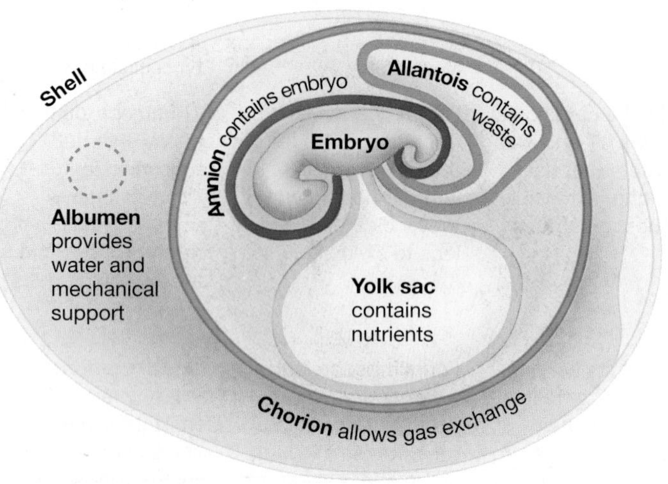

Figure 32.9 **An Amniotic Egg.** Amniotic eggs have four membrane-bound sacs within a leathery or hard shell.

membranes of the amniotic egg are thought to be advantageous because they:

- provide mechanical support, an important consideration outside of a buoyant aquatic environment; and
- increase the surface area available for exchange of gases and other materials. Efficient diffusion of molecules is important because it allows females to lay larger eggs that hatch into larger, more independent young.

In addition, amniotic eggs are surrounded by a shell. This layer is leathery in lizards and snakes. It is stiffened by some calcium carbonate deposits in turtles and crocodiles, and by extensive calcium carbonate deposits in birds. Lizards, snakes, turtles, and crocodiles often bury their eggs in moist soils, but bird eggs are more watertight and are laid in nests that are exposed to air.

In viviparous reptiles and mammals, the amniotic egg is modified and retained internally by the mother, an effective strategy for keeping embryos from drying out. The four membrane-bound sacs are still present in modified form, but there is no shell.

Mammals: Origin of Lactation and Fur

What distinguishes mammals from reptiles? **Mammals** are a monophyletic group of amniotes named for unique structures, called **mammary glands**, that produce milk. The milk nourishes developing young, by a process called **lactation**. Mammals are the only vertebrates with cheek muscles and lips—traits that make suckling milk possible.

Further, mammals are **endotherms** ("inside-heated") that maintain high body temperatures by oxidizing large amounts of food and generating large amounts of heat. Endothermy is hypothesized to be an adaptation that enables individuals to maintain high levels of activity—particularly at night or during cold weather. Mammals retain heat because the body surface is covered with layers of hair or fur made of the protein keratin. While fur is a synapomorphy for mammals, the keratin-secreting pathways responsible for this covering are homologous with the pathways producing the scales and feathers of reptiles.

The three major lineages of mammals alive today are described in Table 32.6: the egg-laying **monotremes**, the pouch-bearing **marsupials**, and the **placental mammals**, or **eutherians**, which are by far the most species-rich and morphologically diverse group of mammals. As in other lineages of fishes and reptiles that independently re-evolved viviparity, placental mammals produce a placenta within the uterus or oviduct during pregnancy.

The mammalian **placenta** is an organ combining maternal and embryonic tissues (Figure 32.10). It is rich in blood vessels that facilitate the flow of oxygen and nutrients from the mother to the developing embryo and that also remove nitrogenous wastes and carbon dioxide from the embryo. The embryo's contributions to the placenta are the allantois and chorion. The function of these structures is similar to the one they play in the amniotic egg of oviparous species—the diffusion of gases, nutrients, and wastes. After a development period called **gestation**, the offspring emerges from the mother's body.

What evolutionary advantages did viviparity and the placenta offer? Here are the leading hypotheses:

- Offspring develop at a more constant, favorable temperature.
- Offspring are protected.
- Offspring are portable—mothers are not tied to a nest.

However, all adaptive traits have trade-offs, and the benefits are environment specific (Ch. 22, Section 22.5). While the placenta may offer some advantages over laying eggs in some conditions, the placenta is energetically expensive to produce, and—as all human mothers are very aware—bearing live young is an energetically costly and sometimes risky endeavor.

The earliest mammals in the fossil record appear about 210 million years ago. While this date is highly debated, researchers

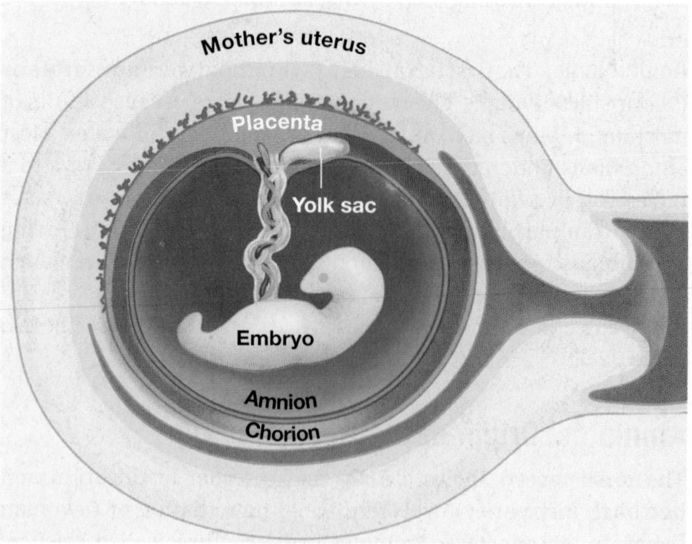

Figure 32.10 **The Mammalian Placenta Nourishes an Embryo Internally.**

✔ Compare the relative size of the yolk sac here with that found in the amniotic egg in Figure 32.9. Propose a hypothesis to explain your observation.

DIVERSITY Table 32.6 **Key Lineages of Mammals**

Chordata > Vertebrata > Mammalia >		Known Species	Feeding	Movement	Reproduction
	Monotremata **Platypus, echidnas** Egg-laying mammals; occur in nature only in Australia and New Guinea	**5** Freshwater Terrestrial 35–55 cm	Feed using leathery beak or bill; platypus feeds on small animals in streams; echidnas feed on ants, termites, and earthworms	Platypus swims and walks using webbed feet; echidnas walk; have lower metabolic rates than other mammals	Oviparous; platypus lays eggs in a burrow; echidnas keep eggs in a pouch; secrete milk for hatched young from skin glands
	Marsupiala **Marsupials** Pouch-bearing mammals; include kangaroos, koalas, opossums, and wallabies	**350** Terrestrial 0.1–2 m	Herbivores, carnivores, and omnivores; some appear very similar to certain placental mammals due to convergent evolution	Crawling, gliding, walking, running, or hopping (adaptations for swimming also occur but are rare)	Viviparous with a simple placenta; the young are born early and complete development in the mother's pouch, where they suck milk until mature
	Eutheria **Placental mammals** Well-developed placenta and extended pregnancy; distributed worldwide	**5100** Marine Freshwater Terrestrial 0.03–30 m	Herbivores, carnivores, and omnivores; the structure of teeth and digestive tracts correlates closely with diet	Walking, running, burrowing, gliding, flying, swinging, swimming	Viviparous with extensive placenta; after extended pregnancy, young are highly developed compared to marsupials

There are 29 orders of placental mammals. Here are the six most species-rich orders:
2255 species of **rodents** (rats, mice, squirrels, guinea pigs, gerbils, chipmunks, beavers, porcupines)
1150 species of **bats** (megabats, microbats)
500 species of **primates** (lemurs, monkeys, gibbons, and great apes, including humans)
450 species of **insectivores** (hedgehogs, moles, shrews)
329 species of **cetartiodactyls** (pigs, hippos, whales, deer, sheep, cattle, horses, giraffes, camels, llamas)
285 species of **carnivorans** (dogs, bears, cats, weasels, seals, raccoons, skunks)

For more go to https://goo.gl/NMPf1L

agree that mammals were widespread and ecologically diverse by about 165 million years ago, in the Jurassic period. This means that mammals began to diversify when dinosaurs and other reptiles were the dominant large herbivores and predators in terrestrial and aquatic environments. Many of these lineages subsequently died out, however, during the end-Cretaceous mass extinction 66 million years ago. The few lineages that survived underwent an adaptive radiation, giving rise to the many mammalian orders familiar today, filling the ecological roles that had once been dominated by dinosaurs and the ocean-dwelling, extinct ichthyosaurs and plesiosaurs.

One major diagnostic feature distinguishing mammals from their sister group, the reptiles, in the fossil record is the number and placement of openings in the side of the skull. These skull openings are important indications of function: Jaw muscles required for biting and chewing pass through them and attach to bones on the upper part of the skull. Whereas reptiles have multiple bones in their lower jaw and only one bone in their middle ear, mammals have only one bone in their lower jaw and three bones in their middle ear: the incus, malleus, and stapes.

Reptiles: Origin of Scales and Feathers Made of Keratin

The **reptiles** are a monophyletic group that represents the second major living lineage of amniotes besides mammals. The four major lineages of reptiles are summarized in **Table 32.7** on page 688: the lizards and snakes, turtles, crocodiles and alligators, and birds.

What Makes a Reptile a Reptile? What adaptations do reptiles have for life on land?

- The skin of reptiles is made watertight by a layer of scales or feathers made of keratin (homologous to mammalian fur).
- Reptiles breathe air through well-developed lungs.
- Reptiles lay shelled, amniotic eggs, or in some cases retain embryos internally and give birth to live young.

Most living non-avian reptiles are **ectotherms** ("outside-heated")—meaning that individuals do not use internally generated heat to regulate their body temperature (with some exceptions, such as tegu lizards that are endothermic during their

DIVERSITY **Table 32.7 Key Lineages of Reptiles**

Chordata > Vertebrata > Reptilia >		Known Species	Feeding	Movement	Reproduction
	Lepidosauria **Lizards, snakes** Elongated bodies and scaly skin, which they shed as they grow	**9800** Marine Freshwater Terrestrial 0.016–3 m	Small lizards are carnivores, while many large lizards are herbivores; snakes are carnivores that swallow prey whole, usually head first	Walking, running, swimming, crawling, burrowing, gliding	Mostly oviparous or ovoviviparous (some viviparity); mostly sexual reproduction but some parthenogenesis
	Testudinia **Turtles** Protective shell composed of bony plates that fuse to the vertebrae and ribs; horny beak	**340** Marine Freshwater Terrestrial 0.06–2 m	Most turtles are omnivores, eating a variety of foods (such as fish, frogs, worms, snails, insects, and/or plants) depending on their environment	Walking, burrowing, swimming; some migrate long distances to natal beaches with remarkable accuracy	All oviparous; other than digging nest, no parental care; some individuals are known to have lived over 150 years
	Crocodilia **Crocodiles, alligators** Long jaws filled with conical teeth; eyes on top of head and nostrils on top of long snouts	**25** Marine Freshwater 1.6–5.5 m	All carnivores; often sit in water for long periods of time, then ambush prey that have come to drink	Walking on land; swimming using side-to-side motion of large, muscular tail	Oviparous with extensive parental care; guard eggs laid in earth-covered nests or mounds on land
	Aves **Birds** Wings, feathers, large breast muscles, lightweight bodies, horny beaks, endothermy	**10,000** Marine Terrestrial 0.05–2.8 m	Carnivores, herbivores, detritivores; beak shape correlates with diet	Flapping flight, gliding, swimming, paddling, running; flightlessness has evolved repeatedly in certain groups	Oviparous, often with extensive parental care; one or both parents usually build a nest, incubate the eggs, and feed the young until they fledge

For more go to https://goo.gl/navR8J

reproductive phase). However, ectothermic reptiles bask in sunlight, seek shade, and perform other behaviors to keep their body temperature at an appropriate level.

The fossil record for reptiles includes the dinosaurs, ichthyosaurs, plesiosaurs, pterosaurs (pronounced *TARE-oh-sors*, flying reptiles with membranous wings), and other extinct lineages that flourished from about 250 million years ago until the end-Cretaceous mass extinction, 66 million years ago. The fossils of these diverse, extinct reptiles are particularly captivating in part because of their large sizes. The gargantuan, herbivorous sauropods take the prize for being the largest, reaching lengths of 40 meters and weights of 90 tons.

Birds and the Evolution of Flight Take note of the position of birds within the phylogenetic tree in Figure 32.6. Traditional classification separates birds from reptiles. However, phylogenetic evidence strongly supports the inclusion of birds within the monophyletic group called reptiles—meaning that birds are reptiles. The fossil record provides conclusive evidence that birds descended from one of several lineages of dinosaurs that had feathers. **Feathers** are specialized outgrowths of the skin composed of keratin. They provide insulation, are used for display, and furnish the lift and steering required for flight. The active lifestyles of birds are supported by endothermy, which evolved independently in birds and mammals. (The precise measurement of growth rates from fossilized dinosaur bones has recently revealed that many dinosaurs were mesotherms—having metabolic rates in between that of ectotherms and endotherms.)

Wings and flight evolved independently in three lineages of tetrapod amniotes: the pterosaurs (the largest having a wingspan over 10 m), the birds, and the bats (the second most diverse order of living mammals, following rodents; see Table 32.6). How did flight evolve?

The best data sets on this question involve feathered flight in birds. Since the early 2000s, paleontologists have discovered a spectacular series of feathered dinosaur fossils, including feathered dinosaurs that were capable of sophisticated flight 125 million years ago, long before the mass extinction of non-avian

Figure 32.11 **Feathers Evolved in Dinosaurs.** This artist's depiction of what *Anchiornis huxleyi,* a feathered dinosaur, might have looked like in life incorporates recent data on fossil feather colors. The colors may have been important in mating displays. Unlike modern birds, many feathered dinosaurs had large feathers on their legs.

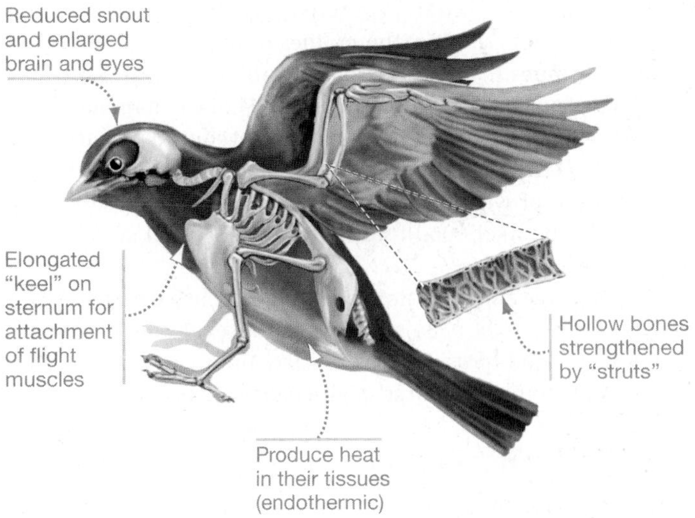

Figure 32.12 **In Addition to Feathers, Birds Have Several Features That Enable Flight.**

dinosaurs 66 million years ago. The newly discovered species and exceptionally well-preserved feathered specimens address key questions about the evolution of birds, feathers, and flight:

- ***Did birds evolve from dinosaurs?*** On the basis of skeletal characteristics, all of these recently discovered fossil species clearly show that birds are part of the monophyletic group called dinosaurs. That is, birds *are* dinosaurs, and thus they are the living representatives of this otherwise extinct clade of reptiles. (For a phylogenetic view of birds within dinosaurs, see Ch. 25, Case Study.)
- ***How did feathers evolve?*** The early fossils support a model of feathers evolving in a series of steps, beginning with simple projections from the skin (popularly called "dinofuzz") and culminating in the complex, branched structures familiar today. However, feathers occurred in multiple, non-flying dinosaur lineages, not just the ancestors of birds, suggesting that feathers originally may have been used for insulation and/or mating displays. Recent analysis of pigments in fossil feathers have shown striking color patterns in some dinosaurs (Figure 32.11).
- ***Did birds begin flying from the ground up or from the trees down?*** More specifically, did flight evolve with running species that began to jump and glide with the aid of feathers to provide lift? Or did flight evolve from tree-dwelling species that used feathers to glide from tree to tree or tree to ground? This question is still unresolved. Measurements of locomotion in present-day partridges have inspired an alternate hypothesis—that feathers may have helped early birds to achieve lift while running rapidly up steep slopes to high perches. The appearance of "flight" feathers on the legs of some dinosaurs adds an intriguing piece to the puzzle (see Figure 32.11).

Once feathers evolved in dinosaurs, the fossil record shows that a series of adaptations made powered, flapping flight increasingly efficient (Figure 32.12).

- Over time, birds have undergone a dramatic reduction in size compared to many of their theropod ancestors, enabling them to diversify into new niches. (Theropods are a diverse group of bipedal, carnivorous dinosaurs, including *Tyrannosaurus rex*.)
- Birds are extraordinarily light for their size, primarily because they have a drastically reduced number of bones, the largest of which are thin-walled and hollow except for bony "struts."
- The heads of birds have retained the proportions of the juveniles of their theropod ancestors—relatively shorter snouts but larger eyes and brains.
- Most dinosaurs have a flat sternum ("breastbone"), but the sternum in most birds has a projection called the keel (the flightless ratites do not), which provides a large surface area to which flight muscles attach.
- Birds are capable of sustained activity year-round because they are endotherms—they maintain a high body temperature by producing heat in their tissues.

From dinosaurs that jumped, glided from tree to tree, or ran up slopes, birds have evolved into extraordinary flying machines. Following the mass extinction of non-avian dinosaurs 66 million years ago, birds experienced a spectacular adaptive radiation resulting in the 10,000 species of birds distributed globally today.

Parental Care

As in other animal groups, numerous vertebrate lineages evolved a variety of methods of **parental care**—any physiological, morphological, or behavioral investment that improves the likelihood of offspring to survive. Here are some examples of parental care in vertebrates:

- Fanning aquatic eggs with oxygen-rich water
- Guarding eggs and/or newly hatched young from predators
- Keeping eggs and young moist (amphibians), or keeping them warm and dry (reptiles and mammals)
- Supplying young with food
- Teaching important survival skills to young

Mammals and birds give offspring particularly extensive care. In both groups, the mother (and often the father, in birds) continues to feed and care for individuals after birth or hatching—sometimes for many years. Female mammals also feed their offspring after birth with milk produced in mammary glands (Figure 32.13). These types of care are energetically expensive to provide, but can improve the animals' fitness by increasing the likelihood that their offspring will survive and reproduce.

Among large animals, the evolution of extensive parental care is hypothesized to be a major reason for the evolutionary success of mammals and birds. In both lineages, mothers produce relatively small numbers of large, high-quality offspring.

Take-Home Messages

Overall, as a result of the evolution of traits such as a three-part brain, cranium, jaw, bony endoskeleton, lungs, limbs, amniotic eggs, fur, and feathers, vertebrates became highly successful consumers in their aquatic and terrestrial ecosystems. However, vertebrate evolution was not a linear progression from simple to complex. Hagfishes have lost vertebrae. Snakes have lost legs. Some birds have lost the ability to fly. The loss of traits can be adaptive in certain environments and in certain lifestyles.

Remember that environments are constantly changing (Ch. 49, Section 49.2). Natural selection can favor traits that improve the fitness of animals adjusting to gradually changing environments. But major abrupt changes in the environment can be catastrophic. The dinosaurs were not poorly adapted to their natural environments—just unlucky to experience an asteroid impact of epic proportions (Ch. 25, Section 25.4).

Some evolutionary events, such as the vertebrate transition from water to land, appear to have occurred just once. But other events, such as the origin of flight and the origin of viviparity, have evolved independently in multiple lineages—they are examples of convergent evolution. Consider the remarkable diversity of snakes, lizards, turtles, mammals, and birds that have made the transition from land *back* to water.

Overall, fossil, developmental, morphological, and molecular evidence have demonstrated how novel structures, such as the vertebrate limb, were co-opted in multiple stages and in various ways from preexisting structures. Such syntheses of multiple lines of evidence have produced a wealth of insights in vertebrate evolution, answering many questions and raising many others.

Figure 32.13 **Parental Care Is Extensive in Mammals.** Female mammals, such as this timber wolf, produce highly nutritious milk to nourish their young.

CHECK YOUR UNDERSTANDING

✔ If you understood this section, you should be able to ...

1. Explain the adaptive significance of the cranium, vertebrae, jaws, tetrapod limb, and amniotic egg.
2. Map the "origin of endothermy" on the tree in Figure 32.6. (Hint: occurred independently in two lineages.)

Answers are available in Appendix A.

32.6 The Primates and Hominins

In this section you will examine the lineage of mammals called the primates, the fossil record of human ancestors, and data on the relationships among human populations living today. Who are your close vertebrate cousins?

After you complete this section, you should be able to ...

- Compare the primates to other vertebrates.
- Analyze the major insights about human evolution gleaned from fossil and molecular evidence.

The Primates

The mammalian lineage known as **primates** traditionally consists of two main groups: prosimians and anthropoids.

- The **prosimians** ("before-monkeys") typically consist of the lemurs, found in Madagascar (Figure 32.14a), and the lorises, pottos, and tarsiers of Africa and south Asia. Most prosimians living today are relatively small in size, reside in trees, and are active at night.
- The Anthropoidea, or **anthropoids** ("human-like"), include the New World monkeys found in Central and South America, the Old World monkeys that live in Africa and tropical regions of Asia, the gibbons of the Asian tropics, and the Hominidae, or **great apes**—orangutans, gorillas, humans, bonobos, and chimpanzees (Figure 32.14b).

The phylogenetic tree in Figure 32.15 shows the evolutionary relationships among these groups. Note that one of the prosimian lineages, tarsiers (the prosimian illustrated on the tree), appears to be more closely related to the anthropoids than to the other prosimians, indicating that the prosimians are not a monophyletic group.

What Makes a Primate a Primate? Primates tend to have hands and feet that are efficient at grasping, flattened nails instead of

(a) Prosimians

Lemur catta

(b) Anthropoids

Gorilla gorilla

Figure 32.14 **Traditional Taxonomy Divides Primates into Prosimians and Anthropoids.**

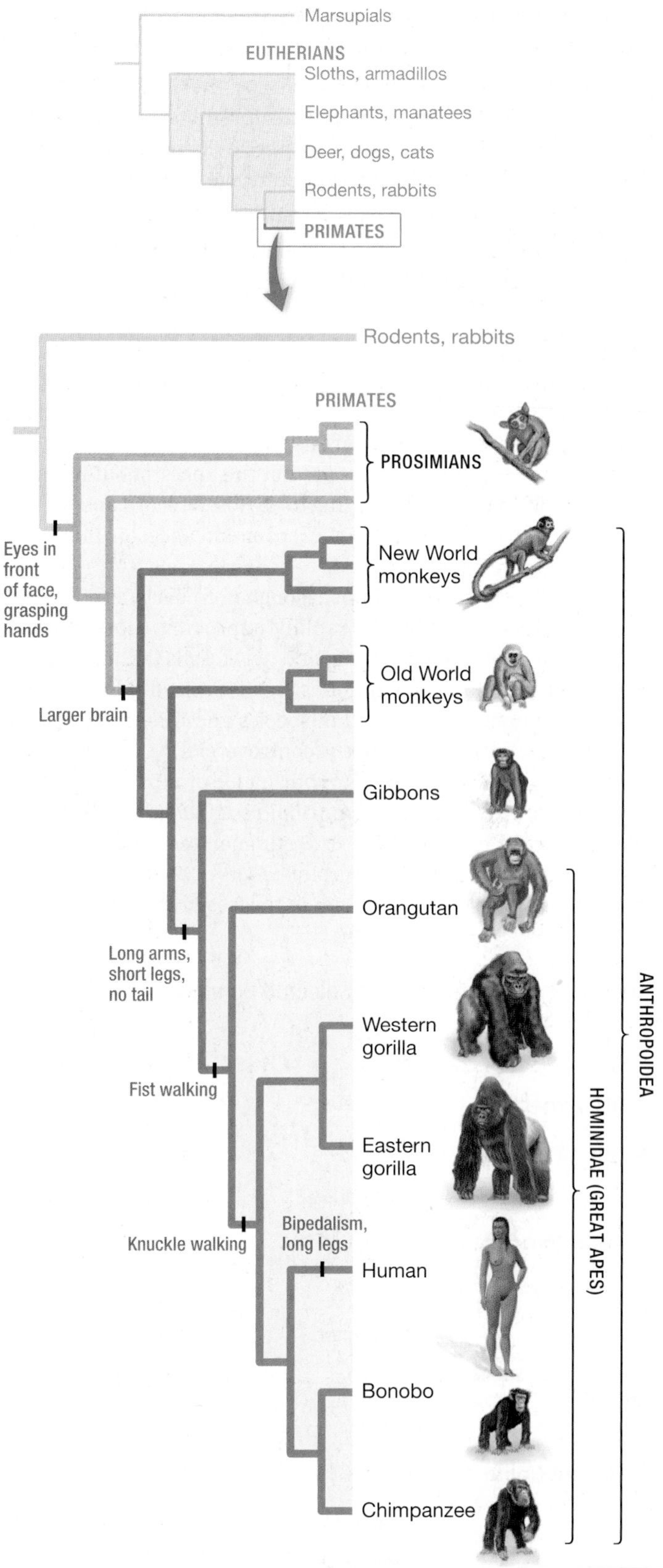

Figure 32.15 **A Phylogeny of the Primates.** Phylogenetic tree estimated from extensive DNA sequence data. Fossil and molecular data suggest that humans and chimps shared a common ancestor 6 to 7 million years ago.

claws on the fingers and toes, brains that are large relative to overall body size, color vision, complex social behavior, and extensive parental care of offspring.

Along with other mammal groups that live in trees and/or make their living by hunting, primates have eyes located on the front of the face. Eyes that look forward provide better depth perception than do eyes on the sides of the face. The hypothesis here is that good depth perception is important in species that run or swing through trees and/or attack prey.

What Makes a Great Ape a Great Ape? The great apes are also called **hominids**. Compared with most types of primate, the hominids are relatively large-bodied and have long arms, short legs (except for humans), and no tail.

All of the great ape species have distinct ways of walking.

- Orangutans rarely come down from trees to walk on the ground. But when they do, they typically walk with the backs of their hands pressed to the ground, or occasionally, with their knuckles pressed to the ground.

- Gorillas, bonobos, and chimps only knuckle walk. They rarely rise up on two legs—usually when displaying aggression.
- Humans are the only living great ape that is fully **bipedal** ("two-footed")—meaning they walk upright on two legs. Fossil footprints in Tanzania provide direct evidence that bipedalism occurred at least 3.6 million years ago.

Bipedalism is a synapomorphy that characterizes the hominins. The **hominins** are a monophyletic group comprising *Homo sapiens* and more than 20 extinct, bipedal relatives.

Fossil Humans

From extensive comparisons of DNA sequence data, it is clear that humans are most closely related to the chimpanzees and bonobos, and that our next nearest living relatives are the gorillas. Note, however, that chimps are not the ancestors of humans (see **Making Models 32.2**). According to the fossil record and molecular data, chimps and humans had a common ancestor that lived in Africa 6 to 7 million years ago.

The fossil record of hominins, though not nearly as complete as investigators would like, is rapidly improving. More than 20 species have been named to date, and new fossils that inform the debate over the ancestry of humans are discovered every year.

Although naming the hominin species and interpreting their characteristics remain intensely controversial, it is still traditional to organize hominins into four major groups that evolved after the recently characterized *Ardipithecus ramidus*—the oldest hominin known to date. **Table 32.8** summarizes key data for selected species within the four groups. **Figure 32.16** provides the time range of each of these species in the fossil record.

Making Models 32.2 Tips on Drawing Phylogenetic Trees (VIII)

It's a common misconception that humans evolved from chimps. Drawing phylogenetic trees can help to clarify why this interpretation is incorrect (see **BioSkills 13**). Chimps and humans are "sister groups" that share an extinct common ancestor that was neither chimp nor human.

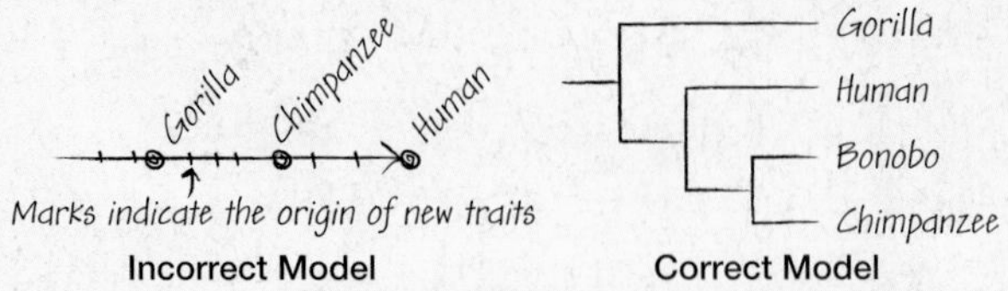

MODEL Draw a dot on the tree on the right that shows the common ancestor of humans and chimps. Then draw several trait bars representing synapomorphies on the branches for gorilla, human, bonobo, and chimp to show that evolution of new traits occurred in all branches, not just in humans.

To see this model in action, go to the Study Area of **Mastering Biology**

Gracile Australopithecines The small hominins called gracile australopithecines (pronounced *aw-STRALE-oh-PITH-eh-seens*), are referred to as *gracile*, or "slender," because they had a relatively slight build. Adult males were about 1.5 meters tall, and their estimated weight was about 36 kg.

The gracile australopithecines are placed in the genus *Australopithecus* ("southern ape"). The name was inspired by the earliest specimens, which came from South Africa. Two of the

Table 32.8 Characteristics of Selected Hominins

Genus	Species	Location of Fossils	Estimated Average Braincase Volume (cm^3)	Estimated Average Body Size (kg)	Associated with Stone Tools?
Ardipithecus	*A. ramidus*	Africa	325	40	no
Australopithecus	*A. afarensis*	Africa	460	38	no
	A. africanus	Africa	465	34	no
Paranthropus	*P. boisei*	Africa	480	41	no?
Early *Homo*	*H. habilis*	Africa	610	33	**yes**
	H. ergaster	Africa	760	64	**yes**
	H. erectus	Africa, Asia	1000	58	**yes**
	H. naledi	Africa	513	48	**yes**
Recent *Homo*	*H. heidelbergensis*	Africa, Europe	1200	71	**yes**
	H. neanderthalensis	Middle East, Europe, Asia	1430	72	**yes**
	H. floresiensis	Flores (Indonesia)	380	28	**yes**
	H. sapiens	Middle East, Europe, Asia	1480	64	**yes**

DATA: S. L. Robson and B. Wood. 2008. *Journal of Anatomy* 212: 394–425; L. R. Berger et al. 2015. *eLife* 4: e09560.

✔ **QUANTITATIVE** How much smaller is the average body size of *H. floresiensis* compared to the average *H. sapiens*? Express your answer as a percentage. How much smaller is the braincase volume of *H. floresiensis* compared to *H. sapiens*? What do these numbers suggest?

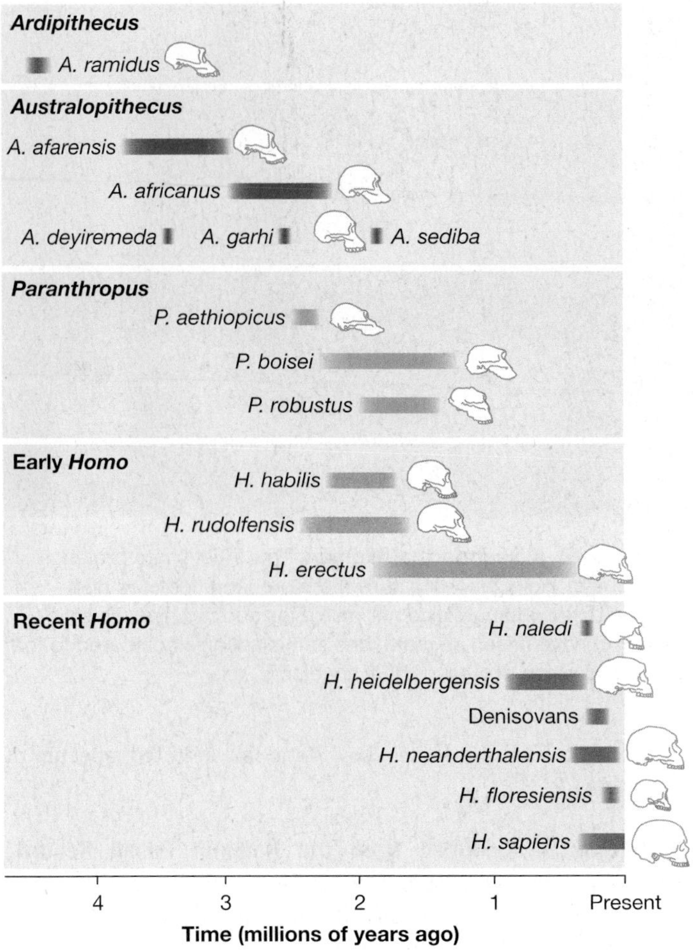

Figure 32.16 A Timeline of Hominin Evolution. The colored bars indicate approximate dates for the first appearance and last appearance in the fossil record of each species. (Not all named species are shown.)

✔ **QUANTITATIVE** Of hominin species shown, how many existed 1.8 million years ago, and how many 100,000 years ago?

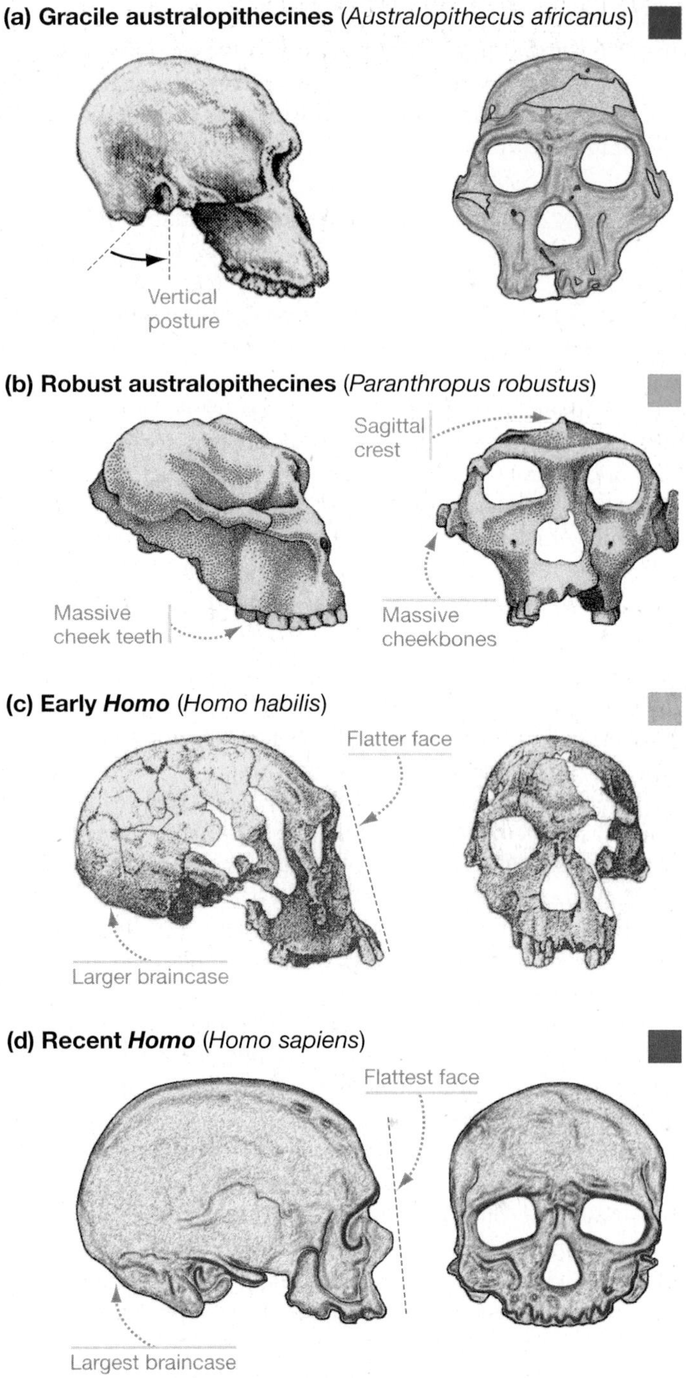

Figure 32.17 African Hominins Comprise Four Major Groups. The color key links these fossils with Figure 32.16 and Table 32.8.

✔ The skulls are arranged as they appear in the fossil record, from most ancient (a) to most recent (d). How did the forehead and brow ridge of hominins change through time?

most recent *Australopithecus* to be named are *A. sediba* (2010) from South Africa and *A. deyiremeda* (2015) from Ethiopia. The latter discovery suggests that multiple species of australopithecines coincided in East Africa more than 3 million years ago.

Several lines of evidence support the hypothesis that the gracile australopithecines were bipedal. The shape of the australopithecine knee and hip are consistent with bipedal locomotion, and the hole in the back of the skull where the spinal cord connects to the brain is oriented downward (Figure 32.17a), just as it is in *Homo sapiens*. In chimps, gorillas, and other vertebrates that walk on four feet, this hole is oriented backward (compare postures in Figure 32.15).

Robust Australopithecines The robust australopithecines are grouped in the genus *Paranthropus* ("beside-human"). The name "Paranthropus" was inspired by the hypothesis that they form a side branch during human evolution—an independent lineage that were not ancestors of *Homo*. Like the gracile australopithecines, these robust australopithecines were bipedal. They were much stockier than the gracile forms, however—about the same height but an estimated 8–10 kilograms (20 pounds) heavier on average. In addition, their skulls were much broader and more robust.

All three *Paranthropus* species had massive molars and jaws, very large cheekbones, and a sagittal crest—a flange of bone at the top of the skull (Figure 32.17b). Because muscles that work the jaw attach to the sagittal crest and cheekbones, researchers conclude that these hominins had tremendous biting power—inspiring the nickname "nutcracker man" for one species. Recent analysis of wear patterns in tooth enamel revealed that the biting power of *Paranthropus* was probably used to crush large quantities of tough grasses.

Early *Homo* Species in the genus *Homo* are called **humans**. As Figure 32.17c shows, species in this genus have flatter and narrower faces, smaller jaws and teeth, and larger braincases than the earlier hominins do. (The **braincase**, or cranium, is the portion of the skull that encloses the brain; it does not include the facial bones.)

The appearance of early members of the genus *Homo* in the fossil record coincides closely with the appearance of tools made of worked stone—most of which are interpreted as handheld choppers or knives. Although new fossil discoveries suggest the possibility that toolmaking originated at least 3.3 million years ago, many researchers favor the hypothesis that extensive toolmaking was a diagnostic trait of early *Homo*.

Recent *Homo* The recent species of *Homo* date from 1.2 million years ago to the present. As Figure 32.17d shows, these species have even flatter faces, smaller teeth, and larger braincases than the early *Homo* species do.

The skull in Figure 32.17d is that of a 30,000-year-old Cro-Magnon. The **Cro-Magnons** were a population of *Homo sapiens* who were accomplished painters and sculptors and who buried their dead in carefully prepared graves. There is also growing evidence that another recent *Homo* species, the **Neanderthal** people (*Homo neanderthalensis*), made art and buried their dead in a ceremonial fashion.

Figure 32.18 shows an artist's interpretation of what a Neanderthal woman might have looked like, based on recent fossil and genomic evidence (the latter of which enables identification of phenotypic alleles for hair and eye color). If she were showered, clothed, and walking down the street, you probably wouldn't pick her out as a different species from yourself, especially given the remarkable morphological diversity within our own species.

A striking recent *Homo* was discovered in 2003 on the island of Flores in Indonesia. *H. floresiensis* had braincases smaller than those of gracile australopithecines and were about a meter tall, inspiring the nickname "Hobbit." Fossil finds and stone tools suggest that *H. floresiensis* inhabited Flores from about 190,000 to 50,000 years ago. Did *H. floresiensis* and *H. sapiens* overlap? The answer is not yet known.

Perhaps the most notable new species of *Homo* is *H. naledi*, discovered in South Africa by cavers in 2013 and described to the world in 2015. The treasure trove of bones revealed what researchers first proposed was an early human with a mix of traits. For example, the general shape of the skull is modern, but the braincase is less than half the size of a modern human's. Then recent radiometric dating of the *H. naledi* fossils produced surprising results—the fossils are very young, between 236,000 and 335,000 years old, suggesting that it may be a recent, rather than early, *Homo*. Stay tuned for more details on this emerging story.

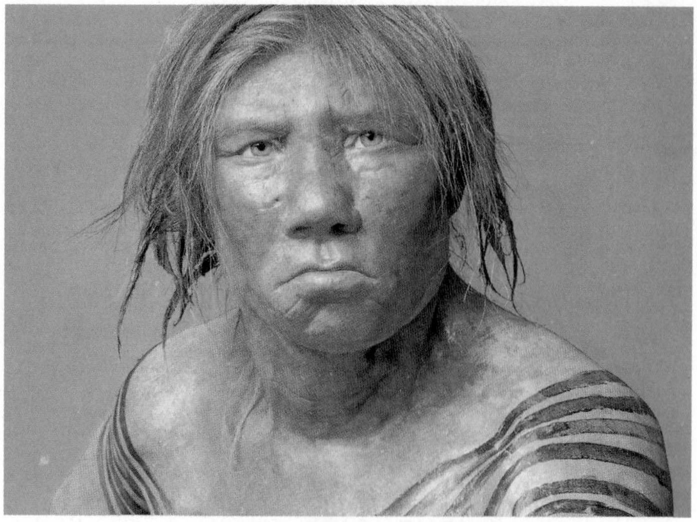

Figure 32.18 **A Neanderthal Woman.** This artist's interpretation of fossil and molecular data shows the probable features of a Neanderthal woman. Notice her pale eyes and red hair, which appear to have arisen as mutations in this species compared to the dark complexion of Neanderthal ancestors.

Table 32.8 summarizes key data for selected species of hominins.

What Can Be Deduced from the Hominin Fossil Record? Although researchers do not have a solid understanding of the phylogenetic relationships among the hominin species, several points are clear from the available fossil data:

- The shared, derived character of hominins is bipedalism.
- The hominin family tree has many branches. That is, there was not a linear progression from one species to another—multiple species of hominins coincided on Earth at the same time.
- Compared with the gracile and robust australopithecines and the other great apes, species in the genus *Homo* have extremely large brains relative to their overall body size. The cerebrum region, in particular, is greatly expanded.

Why did humans evolve such gigantic brains? The *H. sapiens* brain represents just 2 percent of an individual's body mass, but it uses about 20 percent of its energy—big brains are expensive to operate.

The leading hypothesis on this question is that early *Homo* began using symbolic spoken language and tools. The idea here is that increased language use and toolmaking triggered natural selection for the capacity to reason, plan, and communicate, enabling *Homo* to cooperate in complex social networks. This process favored individuals with larger brains.

To summarize, *Homo sapiens* is the sole survivor of an adaptive radiation of hominins that began in Africa 6–7 million years ago. Why all but one species went extinct is still a mystery, though climate change (causing oscillations in size and distribution of habitable areas) and competition for food and space are all likely to have played a part.

The Out-of-Africa Hypothesis

Before *H. sapiens* first appeared, hominins existed in Africa, Europe, and Asia. Where did our species originate?

The leading hypothesis for the evolution of *H. sapiens* is called the **out-of-Africa hypothesis**. It contends that *H. sapiens* evolved its distinctive traits in Africa and then dispersed throughout the world. Further, it claims that *H. sapiens* evolved independently of the earlier European and Asian species of *Homo*—meaning there was no interbreeding between *H. sapiens* and *H. neanderthalensis, H. erectus*, or *H. floresiensis*. Do the data support this hypothesis?

Fossil Evidence Until recently, the oldest *H. sapiens* fossils were located in East African rocks dating to about 195,000 years ago. However, new fossil discoveries in South Africa and Morocco suggest pushing back the fossil record for *H. sapiens* in Africa to about 300,000 years ago, representing populations that may or may not have led to descendants who are alive today. The relationships among these fossils remain unresolved.

Another startling new *H. sapiens* fossil discovery in Israel dates to about 180,000 years ago, providing evidence that modern humans may have migrated out of Africa earlier than once thought. But as predicted by the out-of-Africa hypothesis, younger *H. sapiens* fossils have been found in locations more distant from Africa—in Europe, Asia, Australia, and the Americas.

While the exact time and geographic origin of *H. sapiens* in Africa is actively debated, the fossil evidence provides strong support for the African origin of *H. sapiens* and subsequent migration. However, the fossil data cannot resolve the question of whether *H. sapiens* ever interbred with *H. neanderthalensis* where they overlapped in Europe and the Middle East, or possibly with *H. erectus* or *H. floresiensis* in Asia.

Molecular Evidence Many researchers have taken a molecular approach to understanding the history of our species, providing several dramatic insights. For example, Luigi Cavali-Sforza, Svante Pääbo, David Reich, and others have created phylogenetic trees based on genetic comparisons of diverse human populations living today. As Figure 32.19 shows, the first lineages to branch off led to descendant populations that live in Africa today. Later branches gave rise to lineages residing today in central Asia, Europe, East Asia, Polynesia, and the Americas. Based on this observation, it is logical to infer that the ancestral population of modern humans lived in Africa.

Did *H. sapiens* leave Africa in one wave? Multiple sources of data support the hypothesis that the human migration out of Africa occurred in *more* than one wave—an early wave moved east and south to Australia, and a later wave moved to Europe and the rest of mainland Asia. While the exact migration routes of *H. sapiens* are hotly debated as new data become available, researchers agree that the migration pattern was complex. Figure 32.20 on page 696 is a very generalized summary.

What about the question of interbreeding? There are now multiple lines of evidence that *H. sapiens* interbred with other species as they migrated out of Africa. First, Pääbo, Reich, and colleagues compared mitochondrial and nuclear gene sequences of extinct Neanderthals and diverse modern humans. They made the remarkable observation that 1 to 4 percent of the genome of indigenous Europeans and Asians—but not Africans—is derived from Neanderthals. These data suggest that *H. sapiens* and Neanderthals interbred in the Middle East after *H. sapiens* left Africa, but before migrating through Europe and Asia. Genetic analysis is revealing that some Neanderthal alleles are beneficial, while others are deleterious.

Second, sequence data from modern Asians and Aboriginal Australians suggest that modern humans interbred with a recently discovered cousin of Neanderthals, called the Denisovans (no species designation), in central Asia during their first migration out of Africa. About 5 percent of the genome of Aboriginal Australians and their island neighbors—descendants of the first migration—is derived from Denisovans.

In summary, the Neanderthal and Denisovan data reject the strict interpretation of the out-of-Africa hypothesis, because some gene flow definitely occurred on multiple occasions among *H. sapiens*, Neanderthals, and Denisovans between 40,000 to 180,000 years ago. However, other fossil and molecular evidence supports the general concept of the out-of-Africa hypothesis. That is, the overall data support an out-of-Africa hypothesis with "leakage."

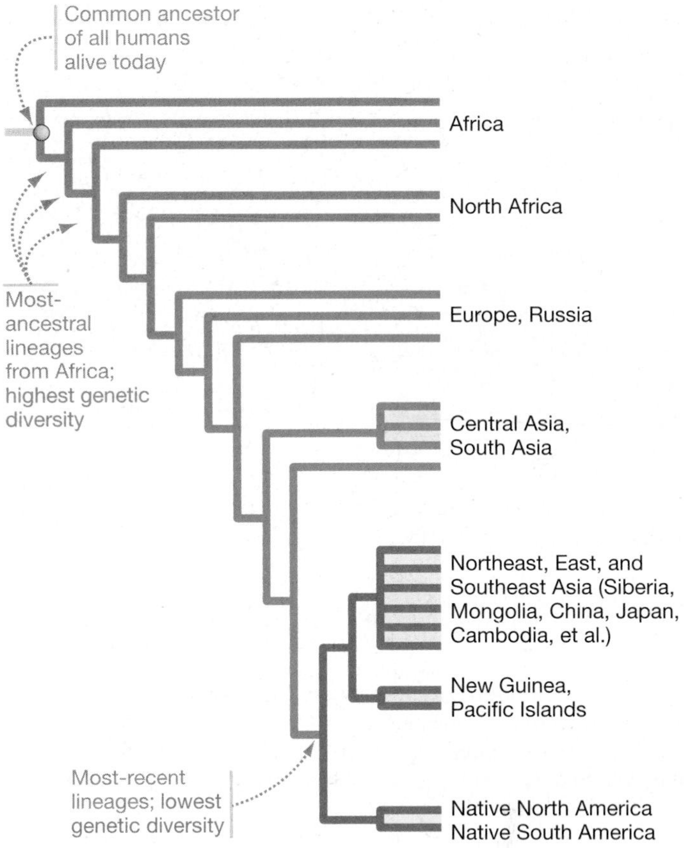

Figure 32.19 Phylogeny of Human Populations Living Today. The phylogeny was estimated from DNA sequence data. As predicted by the out-of-Africa hypothesis, the lineages with the highest genetic diversity and deepest history occur in Africa.

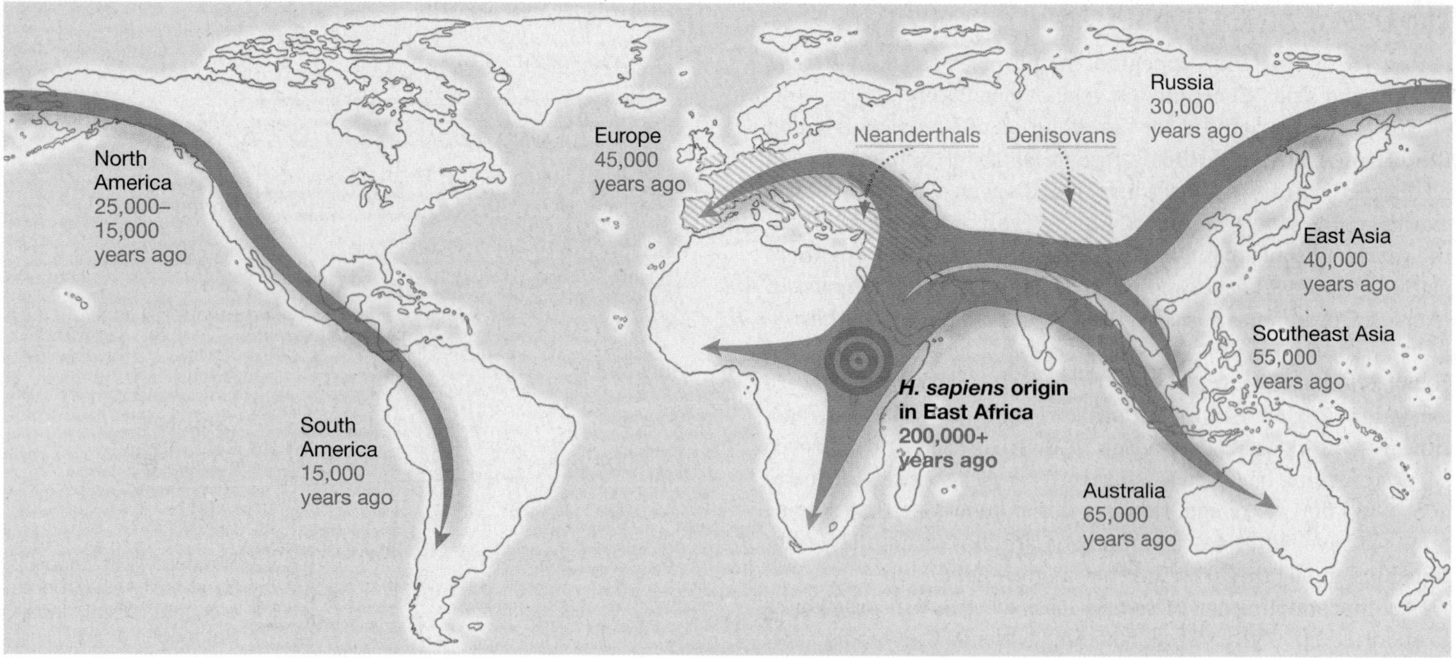

Figure 32.20 **Modern *Homo sapiens* Originated in Africa and Spread throughout the World.** The migration out of Africa occurred in multiple waves—first east and south to Australia, and later to Europe and the rest of Asia and the Americas (although in a manner that is much more complex than shown here). As modern humans migrated, they interbred with at least two other species of *Homo*, indicated on the map as Neanderthals and Denisovans.

Have Humans Stopped Evolving?

The study of human evolution is an exciting and dynamic field. The development of fast and inexpensive genome sequencing techniques has vastly increased biologists' ability to answer interesting questions about our species. All four processes of evolution—mutation, genetic drift, gene flow, and natural selection—are measurable in human populations today and are expected to continue. Consider just a few examples:

- Humans continue to evolve with their pathogens. For example, the frequencies of hemoglobin alleles vary in human populations according to the prevalence of malaria, a tropical mosquito-borne disease that kills about a million people each year.
- In the absence of Caesarian sections, the size of a newborn infant's head is limited by the size of the birth canal through the mother's pelvis. Infants whose heads are too large die during childbirth. How will the sharp rise in Caesarian sections affect average human head size at birth?
- A mutation that enabled adult humans to digest the milk sugar lactose evolved independently in multiple populations living with domesticated goats, sheep, and cows. The mutation is spreading worldwide.

Some researchers argue that humans may even be evolving more rapidly now than ever before in human history. Why? In part, the global human population has recently increased exponentially, increasing the rate at which new mutations are occurring compared to the pace of mutations during early human history. Plus, humans have changed the physical, chemical, and biological environment of Earth at an unprecedented rate, changing the abiotic and biotic contexts in which the processes of evolution take place.

In this chapter, you have learned that humans have a deep history. Our chordate ancestors emerged during the Cambrian explosion 540 million years ago, alongside almost all other major animal phyla. While the placoderms, pterodactyls, and many other animals went extinct, today's fruit flies, giant squid, kangaroos, and *H. sapiens* represent surviving branches of the legacy of life on a constantly changing Earth.

CHECK YOUR UNDERSTANDING

✔ **If you understood this section, you should be able to ...**

1. **THINK CAREFULLY** Correct this statement: "Humans evolved from gorillas."
2. Use the out-of-Africa hypothesis to explain why it is possible that some or none of your genome may be derived from Neanderthals.

Answers are available in Appendix A.

32.1 What Is a Deuterostome?

- Deuterostomes are a sister group to the protostomes. Both groups share a common ancestor that was bilaterally symmetric, triploblastic, and coelomate.
- Molecular evidence supports the hypothesis that deuterostomes are a monophyletic group divided into three major subgroups: Echinodermata, Hemichordata, and Chordata (which includes vertebrates).
- Many of the top consumers in ecosystems are deuterostomes.

32.2 What Is an Echinoderm?

- Echinoderm larvae are bilaterally symmetric, but most undergo a metamorphosis into five-sided, radially symmetric adults.
- Echinoderms have a calcium carbonate endoskeleton.
- The echinoderm water vascular system is composed of fluid-filled tubes and chambers; external projections of the system are called tube feet. Tube feet may be used for movement, feeding, or respiration.
- Echinoderms are important predators and herbivores in marine environments. They include the sea stars, brittle stars, sea urchins, sea cucumbers, and feather stars.

32.3 What Is a Chordate?

- Chordates are distinguished by the presence (at some point during development) of pharyngeal slits or pouches, a dorsal hollow nerve cord, a notochord, and a muscular, post-anal tail.
- Chordates include cephalochordates (lancelets), urochordates (ascidians, thalaceans, and larvaceans), and vertebrates.

32.4 What Is a Vertebrate?

- Vertebrates are distinguished by the presence of a cranium and vertebrae. In most groups of vertebrates, the body plan features an extensive endoskeleton composed of bone.
- Vertebrates have a brain with three regions, and paired sensory structures.
- Vertebrates are the most important large-bodied predators and herbivores in both marine and terrestrial environments.

32.5 What Key Innovations Occurred during the Evolution of Vertebrates?

- The vertebrates living today represent a small fraction of the diverse vertebrates that have thrived on Earth since their origin in the Cambrian about 540 million years ago.
- The earliest vertebrates had chordate characteristics, plus a cranium, paired eyes, and cartilaginous reinforcements of the notochord (vertebrae).
- The origin of jaws in armored fishes enabled them and their descendants to bite food and process it with teeth.
- The origin of limbs in tetrapods enabled walking and running on land, and flying. Amphibians, mammals, and reptiles (including birds) are all tetrapods. In some tetrapods, legs were lost (as in snakes, caecilians, whales), or co-opted as other structures (such as wings of birds, wings of bats).
- The evolution of the amniotic egg allowed tetrapods to lay large eggs on land.
- Lactation and fur were key innovations in mammals.
- Scales and feathers were key innovations in reptiles.
- Parental care expanded notably in some groups of ray-finned fishes and tetrapods—particularly mammals and birds.

32.6 The Primates and Hominins

- Humans are a tiny twig on the tree of life. Chimpanzees and humans diverged from a common ancestor that lived in Africa 6–7 million years ago.
- The fossil record of the past 4.5 million years contains many distinct species of hominins. *Homo sapiens* is the sole surviving representative of an adaptive radiation.
- During much of the last 2 million years, several species of humans were present in Africa or Europe at the same time. All lineages but *H. sapiens* went extinct without leaving descendant populations. However, some of the genomic DNA of *H. neanderthalensis* and an unnamed Denisovan species has been found in modern *H. sapiens*, suggesting interbreeding of *Homo* species in the distant past.
- The phylogeny of living humans, based on comparisons of DNA sequences, agrees with evidence in the fossil record that *H. sapiens* originated in Africa and later spread throughout Asia, Europe, and the Americas.
- Like all other animals on Earth, humans continue to evolve, and may even be evolving at an accelerated rate due to population growth and the unprecedented rate of environmental change caused by humans.

Answers are available in Appendix A.

✔ TEST YOUR KNOWLEDGE

1. True or false? Both echinoderms and vertebrates have endoskeletons. Explain.
2. What trait(s) characterize the vertebrates?
   a. vertebrae and a cranium
   b. jaws and a spinal cord
   c. bony endoskeleton
   d. notochord and post-anal tail
3. What two major lineages make up the living Amniota?
4. Most species of hominins are known only from Africa. Which species have been found in other parts of the world as well?
   a. early *Homo*—*H. habilis* and *H. ergaster*
   b. *H. erectus*, *H. neanderthalensis*, and *H. floresiensis*
   c. gracile australopithecines
   d. robust australopithecines

✔ TEST YOUR UNDERSTANDING

5. Which statement is true about endothermy in mammals and birds?
   a. They both inherited this trait from a common ancestor.
   b. Birds are more closely related to mammals than to reptiles.
   c. Endothermy arose independently in these two lineages.
   d. Birds are reptiles and all reptiles are ectothermic.

6. Compare amniotic eggs to placentas.
7. There is some evidence that pharyngeal slits occur in certain species of echinoderms that appear early in the fossil record. If confirmed, what do these data suggest?
   a. Echinoderms are chordates.
   b. Pharyngeal slits were present in the earliest echinoderms and lost later.
   c. Some lineages of echinoderms are more closely related to chordates than others.
   d. Pharyngeal slits should not be used as a trait in phylogenetic analysis.
8. **MODEL** Draw a phylogeny of major populations of modern humans that would reject the out-of-Africa hypothesis.

✔ TEST YOUR PROBLEM-SOLVING SKILLS

9. **PROCESS OF SCIENCE** Consider the evolution of protostomes (Chapter 31) and deuterostomes (this chapter): **(1)** Compare adaptations in protostomes and deuterostomes for living on land. **(2)** Recall that changes in the expression patterns of tool-kit genes in arthropods such as insects enabled a dramatic diversification of their segmented appendages and bodies. Design an experiment to test whether this concept applies to vertebrates.
10. **MODEL** Genetic diversity in living human populations is highest in Africa and decreases as a function of distance traveled by the human migration that left Africa many millennia ago (see Figure 32.20). Draw a graph to show this trend. (Hint: What is the independent variable? What is the dependent variable? What kind of graph is appropriate for this kind of data? See **BioSkills 2**.) Add a label where you would expect to find a data point for the Yanomamö tribe of the Amazon rain forest in South America. Why is it important to use indigenous people for this study?

✔ PUT IT ALL TOGETHER: Case Study

How did mammalian ear bones originate?

The size and shape of the vertebrate skull can reveal a great deal about an animal's lifestyle and evolutionary relationships. Consider your own skull. If you put your finger in your ear and move your jaw up and down, you can feel the space near the hinge of your jaw. Nestled in this space are the tiny bones that make your hearing possible: the malleus, incus, and stapes. All mammals have these three ear bones, but reptiles such as this *T. rex* don't. Where did ear bones come from?

11. Analyze the morphological data shown here and write a hypothesis to explain the origin of mammalian ear bones. (The cynodont shown is one of many extinct synapsid amniotes that lived early in the lineage that gave rise to mammals.)

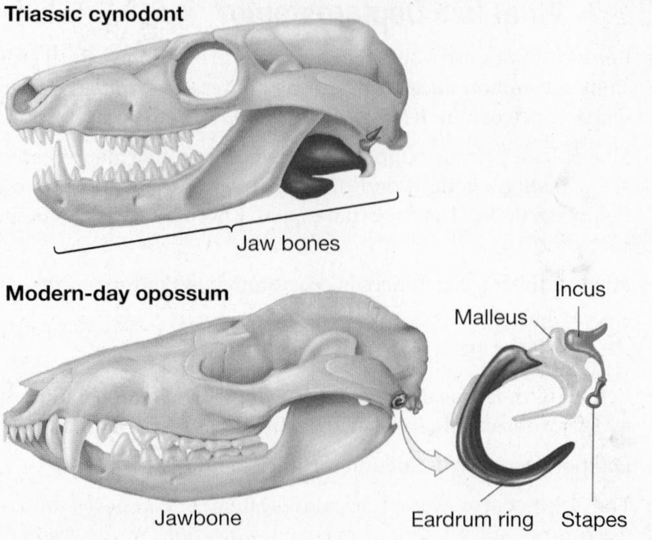

Source: X-X. Luo. 2011. *Annual Review of Evolution, Ecology, and Systematics* 42: 355–350.

12. **THINK CAREFULLY** How is the opossum related to you? Select True or False for each statement.
    T/F An opossum is an animal, but I am a human.
    T/F An opossum is a mammal, but I am a human.
    T/F An opossum is a marsupial, but I am a placental mammal.
    T/F The opossum and I are both tetrapods.
13. The illustration of the opossum skull shows that the ear bones are completely separated from the jawbone (as they are in all mammals). Pose a hypothesis to explain why this separation could be an adaptation that contributed to the radiation of mammals into diverse niches, including a nocturnal lifestyle.
14. Gene expression patterns can be used to test hypotheses based on morphology. For example, the regulatory gene *Bapx1* is expressed in the hinge of the developing lower jaw in fishes and reptiles. Where would you predict *Bapx1* expression to occur in mammals?
15. **PROCESS OF SCIENCE** Researchers studied mice embryos as a model organism to determine whether the cells of the ear bones originated from the same embryonic cells as the cells that form the jaw in other vertebrates (see **BioSkills 11**). Why would the researchers use mice instead of humans? Do the results from mice tell you something about your own ears? Why or why not?
16. Evolution often results in the co-option of a preexisting structure for a new use. Cite three examples from this chapter to support this statement. Evolution can also result in the loss of a trait. Cite three examples from this chapter.

Mastering Biology

Students Go to Mastering™ Biology for assignments, the eText, and the Study Area with animations, practice tests, and activities.

Professors Go to Mastering™ Biology for automatically graded tutorials and questions that you can assign to your students, plus Instructor Resources.

33 Viruses

This photomicrograph was created by treating seawater with a fluorescing compound that binds to nucleic acids. The smallest, most abundant dots are viruses. The larger spots and splotches are bacteria, archaea, and protists.

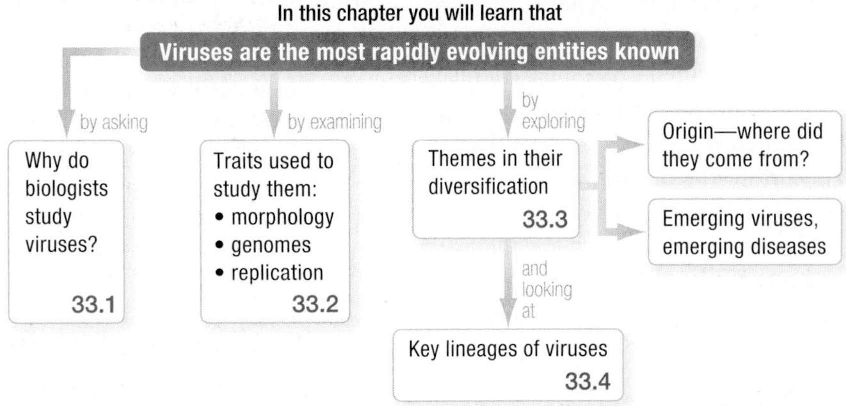

If you have ever been laid low by a high fever, cough, scratchy throat, body aches, and debilitating lack of energy, you may have wondered what hit you. The answer was probably a **virus**: an obligate, intracellular parasite.

Viruses enter a **host cell**, take over its biosynthetic machinery, and use that machinery to replicate—to manufacture a new generation of viruses. Viruses are termed "obligate" parasites because viral replication depends *completely* on host cells. Outside of a host cell, a virus exists as an infectious particle, called a **virion**. In this form it cannot replicate—a virion simply acts as a go-between to transmit the parasite from one host cell to another.

Viruses, unlike most other things studied in biology, are defined by what they are *not* and what they *cannot* do:

- They are not cells and are not made up of cells.
- They cannot manufacture their own ATP, amino acids, or nucleotides, and they cannot produce proteins on their own.

SUMMARY Table 33.1 Viruses versus Organisms

Characteristic	Viruses	Organisms
Hereditary material	DNA or RNA; can be single stranded or double stranded	DNA; always double stranded
Plasma membrane present?	No	Yes
Can carry out transcription independently?	No—even if a viral polymerase is present, transcription of viral genomes requires nucleotides provided by host cell	Yes
Can carry out translation independently?	No	Yes
Metabolic capabilities	Virtually none	Extensive—synthesis of ATP, reduced carbon compounds, vitamins, lipids, nucleic acids, etc.

When you have the flu, influenza viruses enter the cells that line your respiratory tract and replicate inside those cells. Every time you cough or sneeze, you eject millions or billions of infectious virions into the environment. If another person breathes in one of those virions, it may enter a respiratory tract cell in that person and start a new infection.

Because viruses are not cells, they are not considered organisms. Most biologists agree that viruses are not even alive, because they depend on their host cell to satisfy the five attributes of life (Ch. 1, Section 1.1). Yet viruses have a genome, they are superbly adapted to exploit the metabolic capabilities of their host cells, and they evolve. **Table 33.1** summarizes some characteristics of viruses and how they differ from organisms.

Nearly all organisms examined thus far are parasitized by at least one kind of virus. The surface waters of the world's oceans teem with bacteria and archaea, yet viruses outnumber them in this habitat by a factor of 10 to 1. If you leaned over a boat and filled a wine bottle with seawater, you would collect about 10 billion virions—nearly one and a half times the world's population of humans.

33.1 Why Do Biologists Study Viruses?

Any study of life's diversity would be incomplete unless it included a look at how viral parasites contribute to that diversity. Each type of virus infects a specific host-cell type, which may consist of one or more unicellular species or groups of cells in a multicellular species. The enormous diversity of potential host cells has resulted in the evolution of an equally diverse assortment of viruses. Do viruses, in turn, contribute to organismal evolution?

After you complete this section, you should be able to . . .

- Describe how viruses shape host evolution and cause disease.

Viruses Shape the Evolution of Organisms

Viruses directly contribute to organismal diversity in several ways. As agents of natural selection, viruses promote the evolution of organismal traits, such as physical barriers and immune defenses that reduce the occurrence or impact of infection (Chapter 48).

Viruses also directly influence the genetic makeup of organisms. They can introduce foreign genes into cellular genomes by **lateral gene transfer** (Ch. 20, Section 20.6)—picking up genes from one organism and shuttling them to another organism of the same species or even a different species.

Viruses also contribute their own genetic material to organisms. Researchers estimate that 5–8 percent of the human genome consists of remnants of viral genomes from past infections. Some of these viral genes are part of what makes us human. For example, a protein that is encoded by an abandoned viral gene is necessary for proper development and function of the human placenta. And a membrane protein that mediates gamete fusion during fertilization is strikingly similar to a viral fusion protein.

Viruses Cause Disease

Viruses are important from a practical standpoint. To health-care workers, agronomists, and foresters, these parasites are a persistent—and sometimes catastrophic—source of misery and economic loss. Much of the research on viruses is motivated by the desire to minimize the damage they can cause. In the human body, virtually every system, tissue, and cell can be infected by at least one kind of virus (**Figure 33.1**).

Viruses have caused the most devastating human epidemics in recent history. Physicians and researchers use the term **epidemic** (literally, "upon-people") to describe a disease that rapidly infects many individuals over a widening area. After Europeans conquered Mexico in the sixteenth century, for example, the native peoples of Mexico were decimated by epidemic diseases. It is estimated that 12 to 25 million people died from smallpox and a type of viral disease referred to as cocoliztli ("pest" in a local dialect; **Figure 33.2a**).

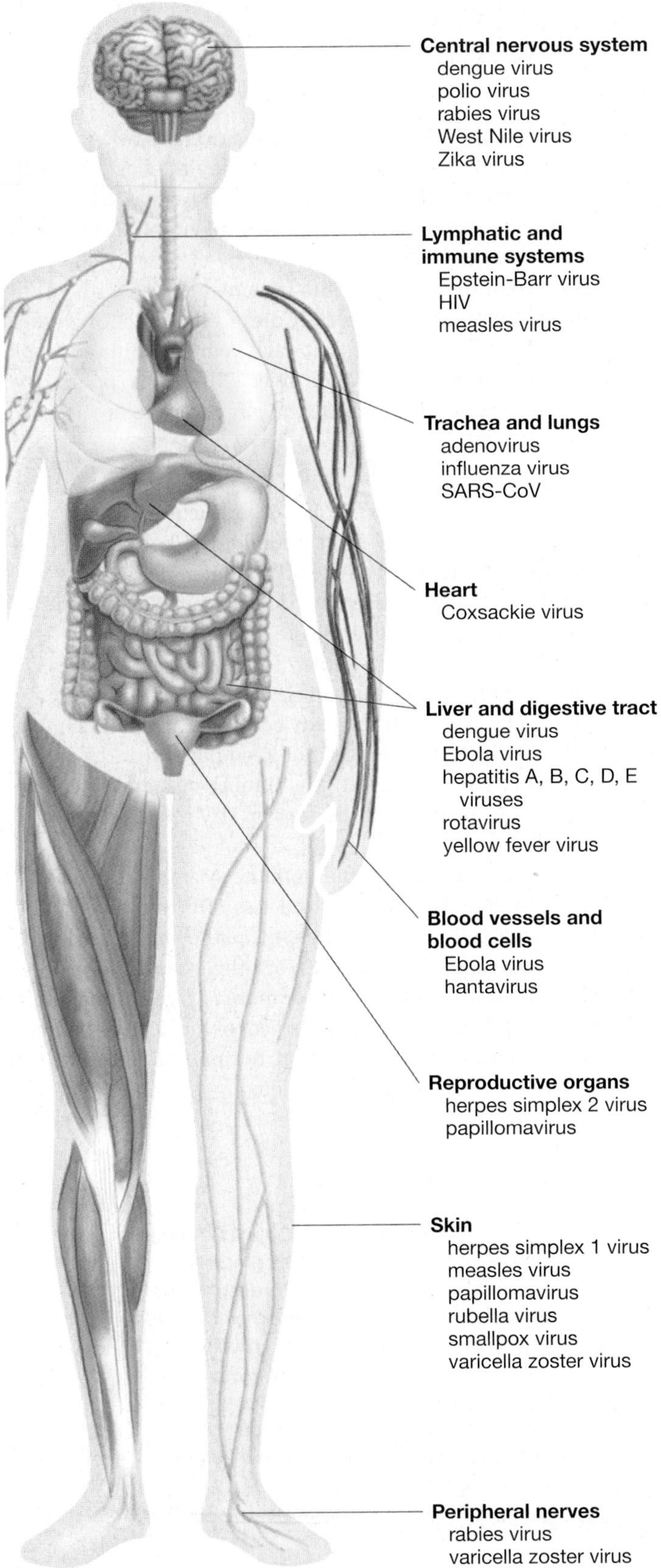

Figure 33.1 Human Organs and Systems That Are Parasitized by Viruses. Viruses listed in this figure may infect more than one tissue, but for simplicity, only one tissue is indicated for most of them.

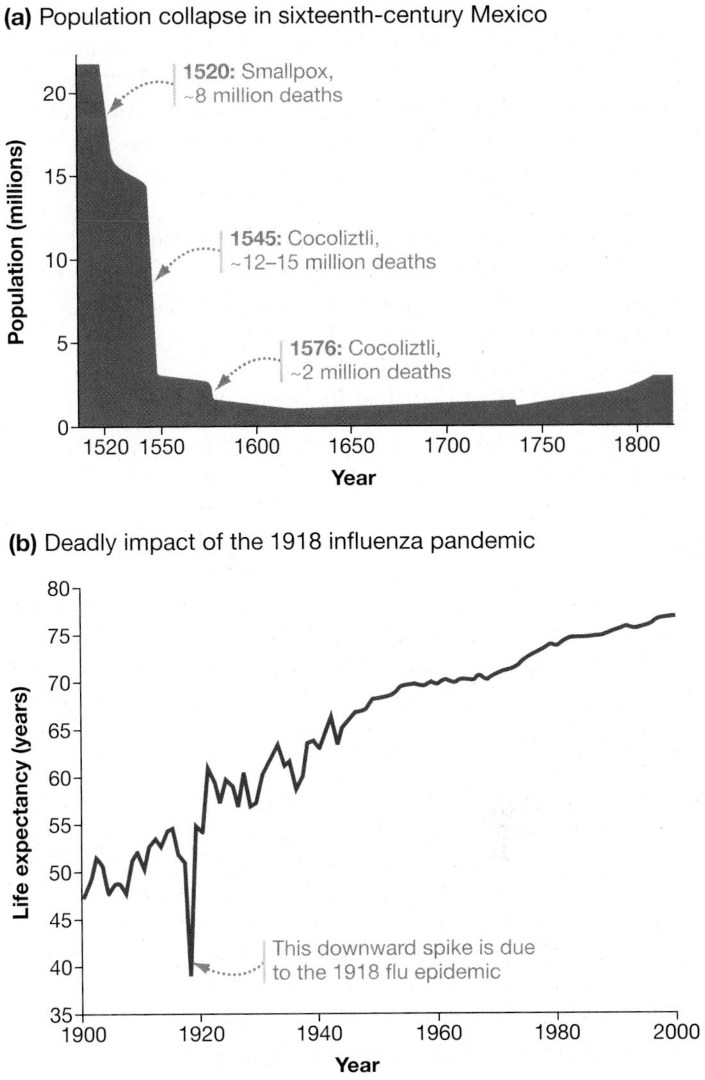

Figure 33.2 The Impact of Viral Epidemics on Human Populations. (a) This graph illustrates the catastrophic effects of the sixteenth-century smallpox and cocoliztli (viral hemorrhagic fever) epidemics on the native peoples of Mexico. **(b)** This graph illustrates the dramatic effect of the 1918 influenza pandemic on an otherwise upward trend.

DATA: (a) R. Acuna-Soto, D. W. Stahle, M. K. Cleaveland, et al. 2002. *Emerging Infectious Diseases* 8: 360–362. (b) E. Arias. 2010. United States life tables, 2006. *National Vital Statistics Reports* 58 (21): 1–40, Table 10. Hyattsville, MD: National Center for Health Statistics.

During the eighteenth and nineteenth centuries, it was not unusual for Native American tribes—whose population numbered fewer than 3 million to begin with—to lose 90 percent of their members over the course of a few years to measles, smallpox, and other viral diseases spread by contact with European settlers. To appreciate the impact of these epidemics, think of 10 close friends and relatives—then remove nine. In some cases, no one was left. A smallpox epidemic from 1837 to 1840 took the lives of up to 300,000 Native Americans from Texas to Alaska and drove the Mandan tribe of North Dakota to extinction.

An epidemic of worldwide scope is called a **pandemic**. The influenza outbreak of 1918–1919, called Spanish flu, qualifies as one of the most devastating pandemics ever recorded. The strain of influenza virus that emerged in 1918 was particularly **virulent**—meaning it tended to cause severe disease. The viral outbreak, which occurred just as World War I was ending, killed far more people than did the five-year conflict itself—an estimated 20 million in the first eight months. Worldwide, the Spanish flu is thought to have killed up to 50 million people. Figure 33.2b illustrates the impact this pandemic had on life expectancy in the United States during the twentieth century.

Current Viral Pandemics in Humans: AIDS

In terms of the total number of people affected, the measles and smallpox epidemics and the 1918 influenza pandemic mentioned earlier have been surpassed by the **acquired immune deficiency syndrome (AIDS)** pandemic. AIDS is an affliction caused by the **human immunodeficiency virus (HIV)**.

HIV is one of the most intensively studied of all viruses. Since the early 1980s, governments and private corporations from around the world have spent hundreds of millions of dollars on HIV research. Given this virus's current and projected impact on human populations around the globe, the investment is justified.

What Is the Scope of the AIDS Pandemic? Most pathogens afflict the very young and the very old, but because HIV is primarily sexually transmitted, young adults are most likely to contract the virus. The United Nations Programme on HIV and AIDS (UNAIDS) estimates that AIDS has already killed about 37 million people, and another 37 million people are infected with HIV worldwide. HIV infection rates remain highest in sub-Saharan Africa. In Swaziland, for example, blood-testing programs have confirmed that 28.8 percent of the adult population is HIV positive.

Often overlooked, however, is a "hidden" epidemic in the southern United States. If current rates continue as predicted by the Centers for Disease Control and Prevention (CDC), one in two southern African American gay and bisexual men will be infected with HIV—a rate higher than that of any other region in the world.

Treatment for AIDS Currently, there is no cure for AIDS, but drug treatments can now significantly delay the onset of AIDS-related diseases. For the millions of people who do not have access to these drugs, HIV infection will inevitably lead to full-blown AIDS after 8 to 12 years.

Without treatment, people who become infected with HIV in their late teens or twenties generally die of AIDS in their twenties or thirties. The UN estimates that each year 1.1 million people are being lost in the prime of their lives. Physicians, politicians, educators, and aid workers all use the same word to describe the epidemic's impact: staggering.

How Does HIV Cause AIDS? Like other viruses, HIV parasitizes specific types of cells—most notably, the helper T cells of the **immune system**, which is the body's defense system against disease. **Helper T cells** are crucial to the immune system's responses to invading pathogens (Ch. 48, Section 48.3).

If an HIV virion infects a helper T cell and replicates inside it, the cell may die as hundreds of new virions are released and infect more cells. Although the body continually replaces helper T cells, the production rate does not keep pace with the rate of destruction by HIV. As a result, the number of helper T cells in the bloodstream gradually declines as an HIV infection proceeds (Figure 33.3).

As the T-cell count drops, the immune system's responses to invading pathogens become less and less effective. Eventually, too few helper T cells are left to fight off pathogens, allowing infections by bacteria, fungi, protists, or other viruses to proceed unchecked. In almost all cases, one or more of these "opportunistic" infections proves fatal. HIV kills people indirectly—by making them susceptible to diseases that normally do not arise in those with a functioning immune system.

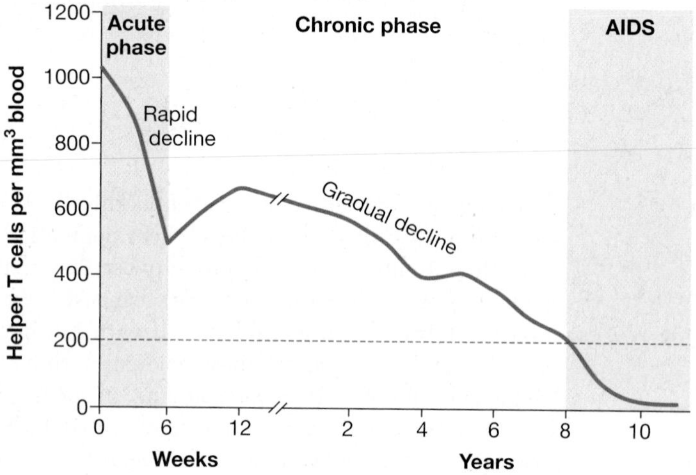

Figure 33.3 Helper T-Cell Count Declines during an HIV Infection. This graph of the number of helper T cells present in the bloodstream over time is based on data from a typical patient infected with HIV. The acute phase may be associated with symptoms such as fever. Few or no symptoms occur in the chronic phase. AIDS typically occurs when the T-cell count dips below 200 per mm^3 of blood.

DATA: G. Pantaleo and A. S. Fauci. 1996. *Annual Review of Microbiology* 50: 825–854.

CHECK YOUR UNDERSTANDING

✔ If you understood this section, you should be able to . . .

1. Discuss this statement: "Viruses are not alive, but they can shape the evolution of organisms."
2. Explain why the decimation of Native Americans by smallpox and other viral diseases in the eighteenth to nineteenth century is called an epidemic rather than a pandemic.
3. Explain why the clinical signs of AIDS take 8 to 12 years to appear.

Answers are available in Appendix A.

33.2 How Do Biologists Study Viruses?

After you complete this section, you should be able to . . .

- Explain how viruses are identified and describe how viruses infect host cells.

Biologists often use viruses as model systems to study processes that occur in cells. For example, during the last half of the twentieth century, key results in molecular biology originated in studies using viruses that infect bacterial cells, called **bacteriophages** ("bacteria-eaters"; BioSkills 11). In addition, many researchers who study animal and plant viruses focus on how to prevent or reduce the effects of the diseases that they cause. As it turns out, treating viral diseases is more complicated than treating diseases caused by other infectious agents. For example, the antibiotics used against bacterial infections often target metabolic activities that are specific to the pathogen. Recall that viruses depend on the metabolic capabilities of the host cell for their replication. This dependency means that traditional antibiotics are not useful for viral infections—if they disrupted the metabolism required for viral replication, they would end up killing the host cell.

To treat viral infections in animal cells, then, biologists must first identify virus-specific targets that may be used to **(1)** activate the immune response against them via **vaccines**, or **(2)** directly interfere with viral replication via antiviral drugs. The first step in identifying these targets is to isolate the virus in question to learn more about its structure and mode of replication.

Isolating viruses takes researchers into the realm of nanobiology, in which structures are measured in billionths of a meter. (One nanometer, abbreviated nm, is 10^{-9} meter; BioSkills 1.) Most viruses produce virions that range from about 20 to 300 nm in diameter. They are dwarfed by eukaryotic cells and—except for some "giant" viruses—even by most bacterial cells (Figure 33.4). Millions of virions can fit on the period at the end of this sentence.

If virus-infected cells can be grown in culture or harvested from a host individual, researchers can usually isolate the virus by passing solutions of infected cultures or patient samples (e.g., blood, saliva, or feces) through a filter. The filters used to study viruses have pores that are large enough for their virions to pass through but are too small to admit cells. If exposing susceptible host cells to the filtrate results in infection, researchers can conclude that a virus is likely to be the cause of infection. These steps are inspired by Koch's postulates, which established the criteria for linking a specific infectious agent with a specific disease (Ch. 26, Section 26.1).

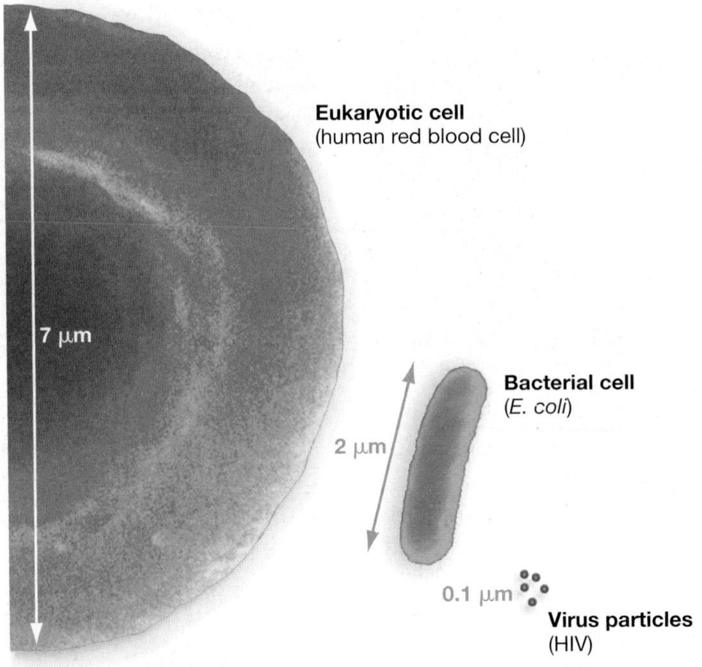

Figure 33.4 **Viruses Are Tiny.** HIV is nearly two orders of magnitude smaller than one of your red blood cells (one-seventieth the size) and one-twentieth the size of *E. coli*, a bacterial cell.

Once biologists have isolated a virus, they study and characterize it by analyzing **(1)** the structure (morphology) of the virion, **(2)** the nature of the genetic material that is transmitted from one host to another, and **(3)** the mechanism by which the virus replicates. Let's begin with morphological traits.

Analyzing Morphological Traits

Only the very largest viruses, such as the measles virus and the giant viruses, are visible with a light microscope. Therefore, to see the virion of a virus, researchers usually rely on transmission electron microscopy (BioSkills 9). Electron microscopy has revealed that viruses come in a wide variety of shapes, and many viruses can be identified by shape alone.

In their overall structure, most viruses fall into two broad categories: **(1)** those enclosed by just a shell of protein called a **capsid** and **(2)** those enclosed by both a capsid and one or more membranous **envelopes**. Figure 33.5 on page 704 provides an overview of some of the most common viral shapes and structures.

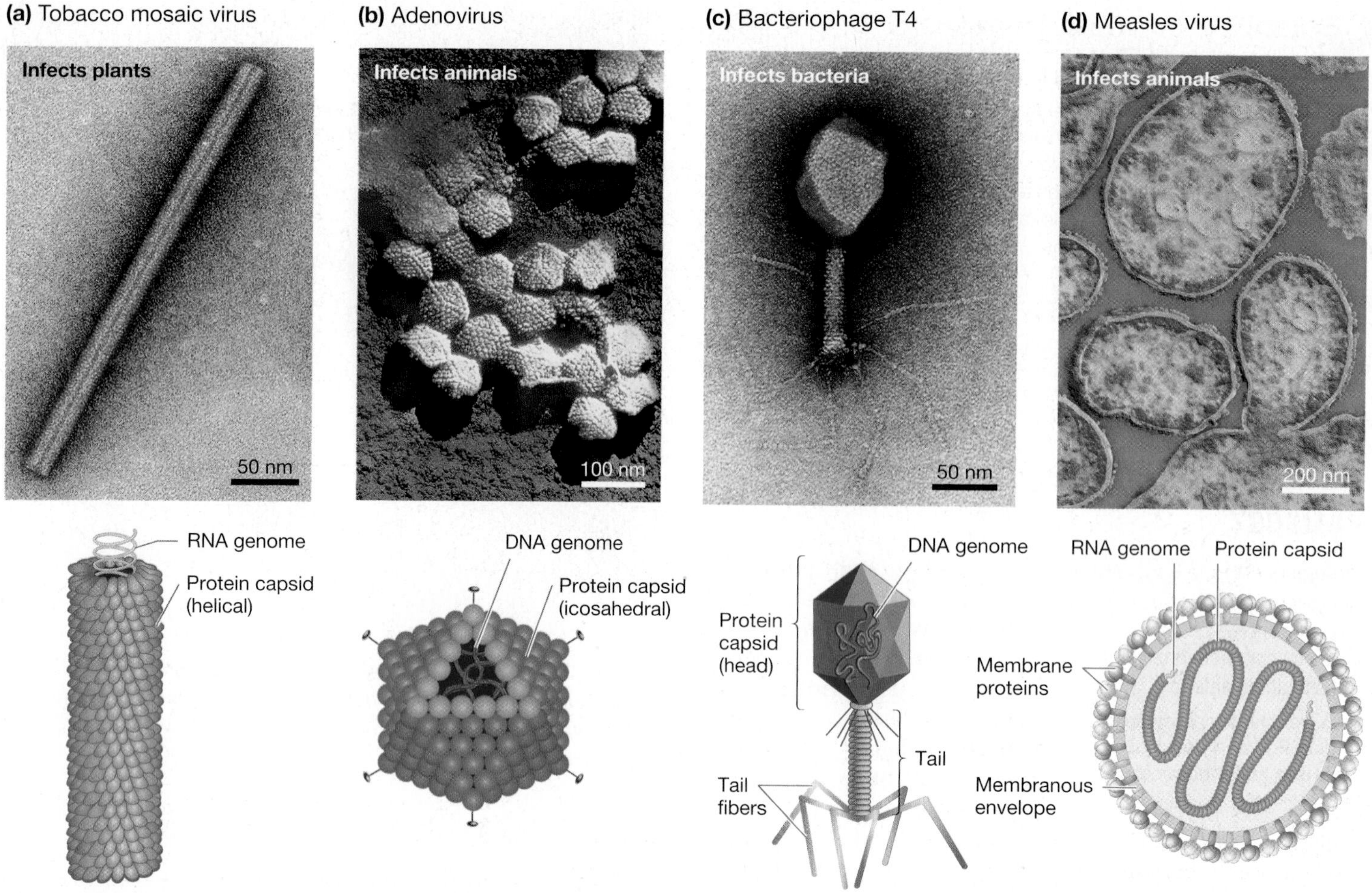

Figure 33.5 **Viruses Vary in Size and Shape and May Be "Naked" or Enveloped.**

Most viruses produce virions with capsid shapes that are either helical or icosahedral (Figures 33.5a and 33.5b; an icosahedron is a polyhedron with 20 triangular faces). But some viruses—such as the measles virus—have more complex capsid shapes, and the bacteriophage T4 has an elaborate "tail" (Figures 33.5c and 33.5d). A virion's capsid serves two functions: It protects the genome while outside the host and releases the genome when the virion infects a new cell.

Nonenveloped viruses, often called "naked" viruses, use only the capsid to protect their genetic material. The tobacco mosaic virus, adenovirus, and T4 virions illustrated in Figures 33.5a, 33.5b, and 33.5c are three different examples of naked viruses. The vast majority of bacteriophage virions fall in this category.

In enveloped viruses, such as the measles virus shown in Figure 33.5d, the envelope contains embedded viral proteins and is derived from a lipid bilayer in the host cell where the virion was manufactured. Many viruses that directly afflict animals—including humans—are enveloped. Later in this chapter you will learn how such viruses obtain their envelope from an infected host cell.

Analyzing the Genetic Material

Viruses can also be categorized based on the nature of their genome. This is not true for cells. DNA serves as the hereditary material in all cells, and information flows from DNA to mRNA to proteins (Chapters 1 and 16). Although all cells follow this pattern, which is called the central dogma of molecular biology, many viruses break it.

This conclusion traces back to work done in the 1950s, when Heinz Fraenkel-Conrat and colleagues separated the protein and nucleic acid components of a plant virus known as the tobacco mosaic virus, or TMV (see Figure 33.5a). They found that the nucleic acid portion of this virus consisted of RNA and that the RNA, by itself, could infect plant tissues and cause disease. This result was surprising because it showed that RNA—not DNA—functions as TMV's genetic material.

Subsequent research revealed an amazing diversity of viral genome types. In some groups of viruses, such as the agents that cause measles and flu, the genome consists of RNA. In others, such as the viruses that cause herpes and smallpox, the genome is DNA. Viral genomes may be linear or circular and may consist of a single molecule, or several molecules referred to as segments.

Further, the RNA and DNA genomes of viruses can be either single stranded or double stranded. The single-stranded RNA genomes can also be grouped into three different categories:

1. In **positive-sense** single-stranded RNA ([+]ssRNA), the genome contains the same base sequences as mRNAs and can be directly read by host-cell ribosomes to synthesize viral proteins.

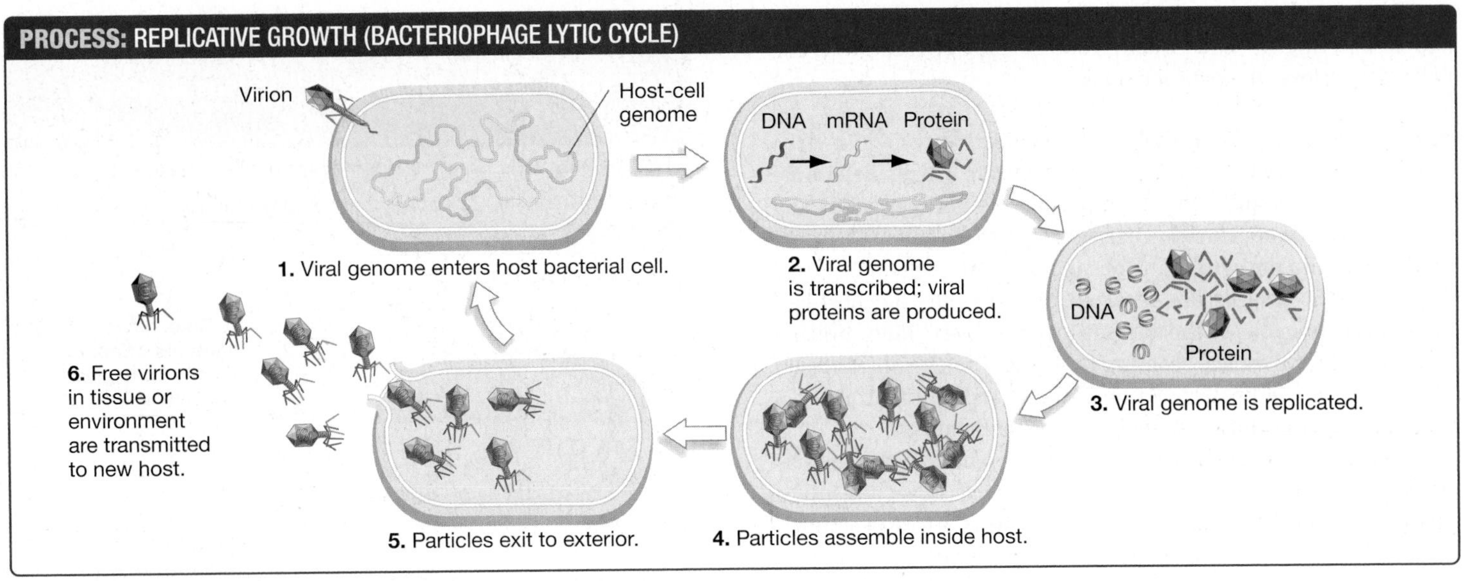

Figure 33.6 **Viral Replicative Growth In Bacteria.** Many bacterial viruses follow the same general cycle as this model, which shows the replication of bacteriophage T4. (An alternate strategy, termed lysogeny, postpones virion replication. It is shown in Figure 33.14.)

2. In **negative-sense** single-stranded RNA ([−]ssRNA), the base sequences in the genome are complementary to those in viral mRNAs.
3. In **ambisense** single-stranded RNA, part of the genome is positive sense and can be directly translated, and the other part of the same RNA is negative sense and must be copied before translation of viral proteins.

Although likely millions of types of virus exist, they all appear to infect their host cells in one of two general ways: via immediate **replicative growth**, which quickly produces the next generation of virions and often kills the host cell, or via a growth cycle that temporarily suspends production of virions and allows the viral genome to coexist with the host genome for a period of time. Let's start with the mode of infection that immediately manufactures virions.

Analyzing the Phases of Replicative Growth

Six phases are common to replicative growth in virtually all viruses: **(1)** attachment to a host cell and entry into the cytosol; **(2)** transcription of the viral genome and production of viral proteins; **(3)** replication of the viral genome; **(4)** assembly of a new generation of virions; **(5)** exit from the infected cell; and **(6)** transmission to a new host.

Figure 33.6 shows the replicative growth cycle of a DNA virus that infects bacterial cells—bacteriophage T4. This process is referred to as the **lytic cycle** because it ends with lysis (bursting) of the host bacterial cell.

Comparable replicative cycles also occur in plant and animal cells, but vary in their details: typically, nonenveloped viruses cause host cells to burst, while enveloped viruses bud from cells (look ahead to Figure 33.13). One thing should stand out as you examine the figure: In a single replicative cycle, one virion will produce many progeny.

Figure 33.7 shows how a bacteriophage's tremendous capacity for replication results in nonlinear growth that proceeds in a stepwise manner. This pattern of growth is very different from that observed in the bacterial host cells, where each individual produces only two progeny in each generation.

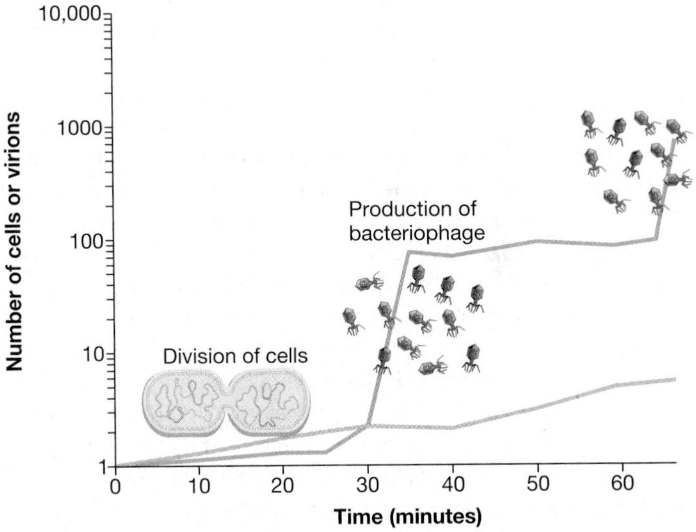

Figure 33.7 **Growth Patterns Differ between Viruses and Their Hosts.** Cells reproduce exponentially, as indicated by a linear pattern when plotted on a logarithmic scale. The production of virions is not exponential; instead, viruses such as bacteriophage T4 replicate in a stepwise pattern.

DATA: Courtesy of the Undergraduate Biotechnology Laboratory, California Polytechnic State University, San Luis Obispo.

✔ **QUANTITATIVE** Based on the information and trends in this graph, predict the number of extracellular virions that would be produced after 75 minutes if all of those released by 35 minutes infected new host cells. Compare this to the number of cells you would predict to be present after 75 minutes.

Each type of virus has a particular way of entering a host cell and completing the subsequent phases of the replicative cycle. Let's take a closer look at each phase.

How Do Viruses Enter a Cell? The replicative cycle of a virus begins when a free virion enters a target cell. This is no simple task. All cells are protected by a plasma membrane, and many also have a cell wall. How do viruses breach these defenses, insert themselves into the cytosol inside, and begin an infection?

Most plant viruses are inserted directly into the host-cell cytosol via abrasions or the damage incurred by sucking or biting insects that have viruses coating their mouthparts. In contrast, viruses that infect bacterial or animal cells must first attach to a specific molecule on the cell wall or plasma membrane.

The type of molecule used for attachment, often referred to as the virus receptor, varies among different viruses. As a result, a virus can infect only those host species, or tissues within a multicellular host, that have the correct receptor on their surface.

Bacteriophages will attach to proteins or carbohydrates on the cell wall, fimbriae, or flagella of their host cells (Ch. 7, Section 7.1). To gain access to the plasma membrane, they often use an enzyme called lysozyme to degrade a small region of the cell wall. The genome of the bacteriophage is then transferred to the cytosol in a process referred to as **uncoating**, which varies depending on the structure of the virus. The T4 bacteriophage shown in Figure 33.6, for example, uncoats its genome by injecting it through a protein structure similar to a hollow needle that is stabbed through the bacterial membrane.

Viruses that attack animal cells attach to one or more virus receptors in the host cell's plasma membrane. These molecules are typically either proteins or the carbohydrate portions of glycoproteins or glycolipids. The selectivity that viruses exhibit for their receptors is responsible for some of the organ-specific infections shown in Figure 33.1.

To appreciate how investigators identify virus receptors, consider research on HIV. In 1981—right at the start of the AIDS epidemic—biomedical researchers realized that people with AIDS had few or no helper T cells possessing a membrane protein called **CD4**. These cells, symbolized as $CD4^+$, have a key supportive role in the immune response (Ch. 48, Section 48.3).

Two research groups hypothesized that besides its normal function in the immune response, CD4 is co-opted to serve as the receptor for HIV attachment. To test this hypothesis, the researchers systematically blocked different membrane proteins on helper T cells and then determined whether that blockage prevented HIV infection. They used **antibodies** that bound specifically to CD4 and to other host membrane proteins.

The key assumption in their experiments was that antibodies directed against the receptor for viral attachment would prevent HIV from entering and infecting cells. After completing their work, both research teams reached the same result: Antibodies to CD4 prevented HIV from infecting host cells (Figure 33.8).

Although CD4 is necessary for HIV attachment, the viral genome will not be uncoated unless the virion also binds to a second membrane protein, a **co-receptor**. In most individuals, membrane proteins called CXCR4 and CCR5 are co-opted by the virus to act as co-receptors. Once the virion binds to both CD4

RESEARCH

QUESTION: Is CD4 the protein HIV uses to enter a host T cell?

HYPOTHESIS: CD4 is the membrane protein HIV uses to enter cells.

NULL HYPOTHESIS: CD4 is not the membrane protein HIV uses to enter cells.

EXPERIMENTAL SETUP:

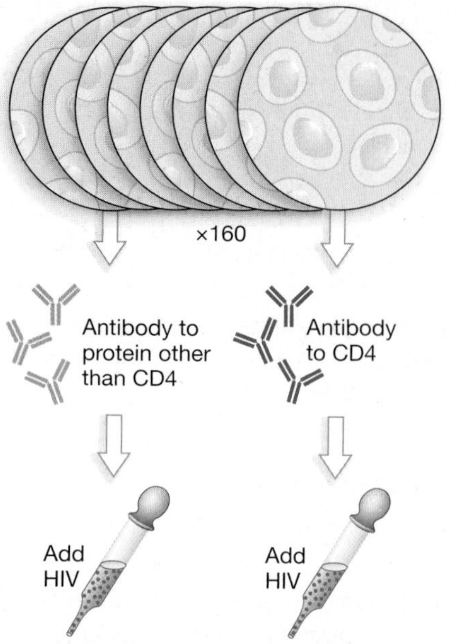

1. Take 160 identical samples of helper T cells from a large population of T cells growing in culture.

2. Add a different antibody to each sample of cells—each antibody will "block" a specific membrane protein.

3. Add a constant number of HIV virions to all samples. Incubate cultures under conditions optimal for virus entry.

4. Use microscopy to identify which samples contain cells infected by HIV.

PREDICTION OF HYPOTHESIS: HIV will not infect cells with antibody to CD4 but will infect other cells.

PREDICTION OF NULL HYPOTHESIS: HIV will infect cells with antibody to CD4.

RESULTS:

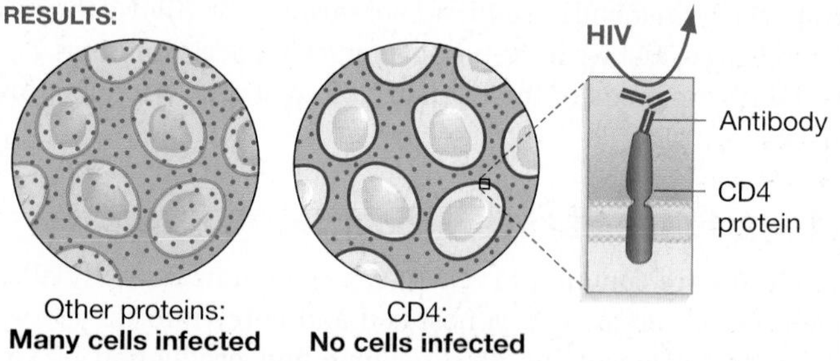

CONCLUSION: HIV uses CD4 as the receptor to enter helper T cells. Thus, only cells with free, unbound CD4 on their surface can be infected by HIV.

Figure 33.8 **CD4 Is the Receptor Co-opted by HIV to Attach to and Enter Host Cells.** In this experiment, only antibodies that bound to CD4 prevented HIV from gaining entry to the helper T cells.

SOURCE: A. G. Dalgleish, P. C. Beverley, P. R. Clapham, et al. 1984. The CD4 (T4) antigen is an essential component of the receptor for the AIDS retrovirus. *Nature* 312: 763–767. Also D. Klatzmann, E. Champagne, S. Chamaret, et al. 1984. T-lymphocyte T4 molecule behaves as the receptor for human retrovirus LAV. *Nature* 312: 767–768.

✔ Does this experiment prove that CD4 is the only membrane protein required for HIV entry? Explain why or why not.

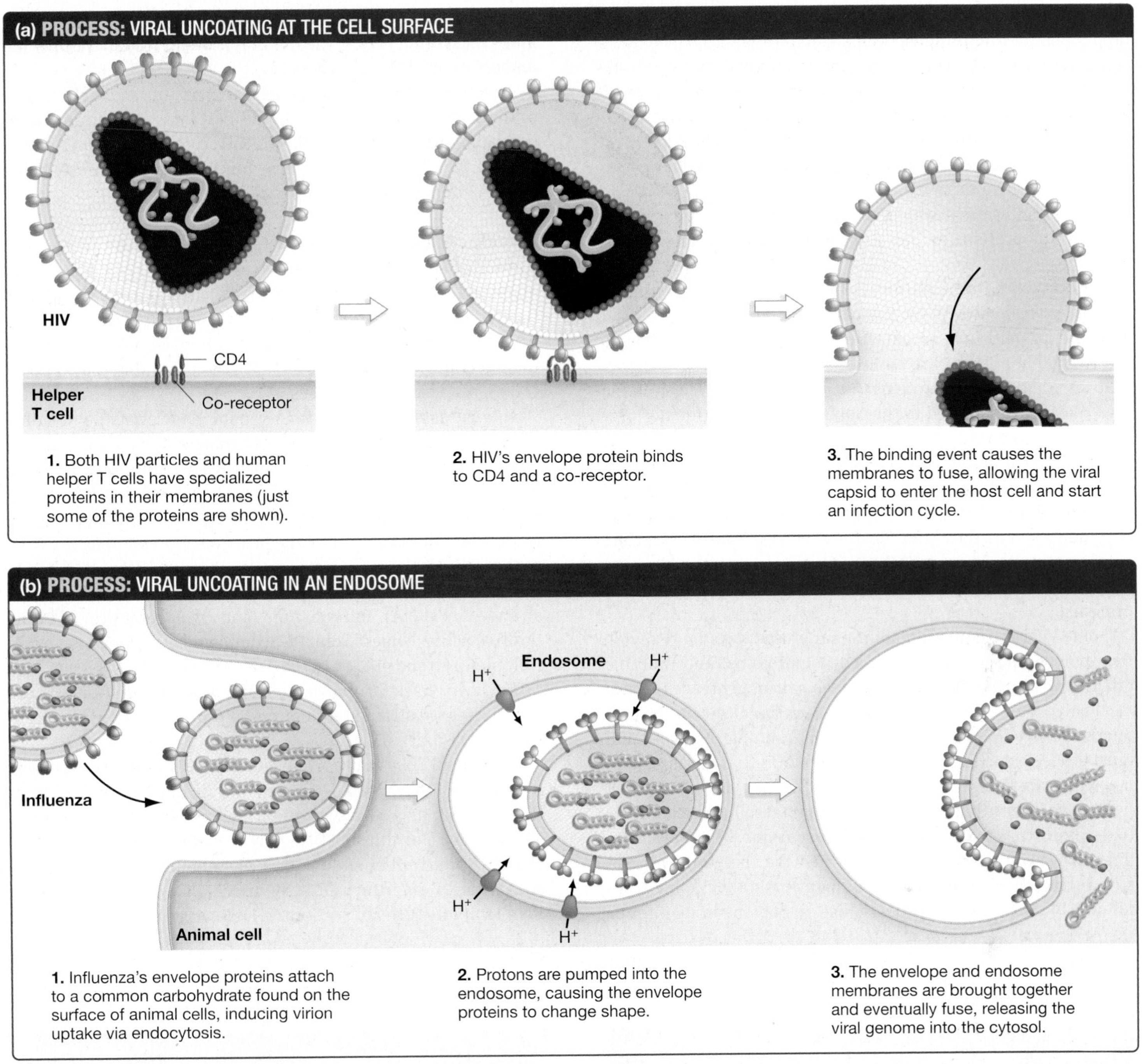

Figure 33.9 Enveloped Viruses Bind to Host Membranes and Uncoat via Fusion.

and a co-receptor, the lipid bilayers of the virion's envelope and the plasma membrane of the helper T cell fuse (Figure 33.9a). When fusion occurs, HIV has breached the cell boundary. The viral capsid then enters the cytosol and disassembles to release the genomic RNAs, and infection proceeds.

Another common entry process is shown in Figure 33.9b, where the influenza virion is first internalized via endocytosis (Ch. 7, Section 7.5). After the virion attaches to a carbohydrate component on the cell surface, it is pulled into the cell inside a vesicle called an endosome. When the endosome acidifies—a normal part of endocytosis—the viral attachment proteins change shape, promoting fusion of the viral envelope and the endosomal membrane.

Endocytosis is also commonly used by naked viruses to enter animal cells. As with enveloped viruses, acidification of the endosome serves as a signal for uncoating the genetic material. In this case, low pH alters the structure of the capsid proteins, which disrupts the endosomal membrane and allows the viral genetic material to enter the cytosol.

The discovery of virus receptors and requirements for uncoating has inspired a search for compounds that block these early events of the replicative cycle. This search has resulted in drugs that block attachment and uncoating of HIV by targeting the HIV envelope protein and host CCR5 co-receptor. Drugs that interfere with viral infection or replication are called **antivirals**.

How Do Viruses Produce Proteins? During the replicative cycle, production of viral proteins begins soon after a virus uncoats its genome. Depending on the virus, transcription of the viral genome may be accomplished by either host or viral RNA polymerases. In all viruses, however, translation of viral transcripts depends entirely on the host. Viruses lack the ribosomes, amino acids, ATP, and most of the other biosynthetic machinery required for translating their own mRNAs into proteins (Ch. 17, Section 17.3).

Viral proteins are produced in one of two ways, depending on whether the proteins end up embedded in a membranous viral envelope.

Viral mRNAs that code for envelope proteins are translated as if they were mRNAs for the cell's own membrane proteins. Envelope proteins are manufactured by ribosomes attached to the rough endoplasmic reticulum (rough ER), where carbohydrates are added to the proteins (Ch. 7, Section 7.5). The proteins then may be transported to the Golgi apparatus for further processing. In some viruses, such as HIV, the finished glycoproteins are transported to the plasma membrane, where they are assembled into new virions.

Viral mRNAs that code for proteins that don't end up in the envelope, such as those that form the virion's capsid, may take a different route. These mRNAs are translated by free ribosomes in the cytosol, just as if they were cellular mRNAs for cytosolic proteins.

Some viruses require an additional processing step to divide long polypeptide sequences, called polyproteins, into multiple distinct viral proteins. Viral enzymes called **proteases** cleave viral polyproteins at specific locations—a critical step in the production of functional proteins. In HIV, for example, a polyprotein must be cleaved to form the different proteins that assemble into new viral capsids.

The discovery that HIV produces a protease triggered efforts to develop drugs that would inhibit the enzyme. This work got a huge boost when researchers identified the three-dimensional structure of HIV's protease, using X-ray crystallographic techniques (BioSkills 6). The protease has an opening in its interior where the enzyme's **active site** is located (Figure 33.10a). Specific amino acid sequences in the HIV polyproteins fit into the opening and are cleaved at the active site.

Based on these findings, researchers immediately began searching for molecules that could fit into the opening and block the active site, thereby preventing the protease from functioning (Figure 33.10b). Several protease inhibitors are currently being used to interfere with HIV replication.

How Do Viruses Copy Their Genomes? Besides carrying out transcription and translation, viruses must copy their genetic material to make a new generation of virions. Viruses depend on the host cell for nucleotide monomers. Some DNA viruses also depend on the host cell's DNA polymerase machinery to replicate their genomes (Ch. 15, Section 15.3), requiring that the virus becomes localized to the nucleus.

In viruses that have an RNA genome, however, the invading virus brings along its own enzymes to make copies of the genome. Some of these viral enzymes, called **RNA replicases**, function as *RNA-dependent* RNA polymerases. In other words, they synthesize RNA from an *RNA* (rather than a DNA) template, using ribonucleotides provided by the host cell. For example, the RNA replicases of positive-sense single-stranded RNA ([+]ssRNA) viruses first convert the single-stranded genome into a double-stranded (dsRNA) intermediate and then produce multiple positive-sense copies from the complementary negative-sense (template) strand of the dsRNA (Figure 33.11).

To keep track of conversions between positive-sense and negative-sense strands, it is often helpful to model the process (see Making Models 33.1).

But not all RNA viruses replicate using RNA replicases. In certain animal RNA viruses, the genome is first transcribed from RNA to DNA by a viral enzyme called **reverse transcriptase**. This enzyme is an unusual DNA polymerase—one that can synthesize DNA from either an RNA or a DNA template. It first synthesizes a **complementary DNA**, or **cDNA**, using a single-stranded RNA template (Ch. 20, Section 20.1). Reverse transcriptase then

(a) HIV's protease enzyme

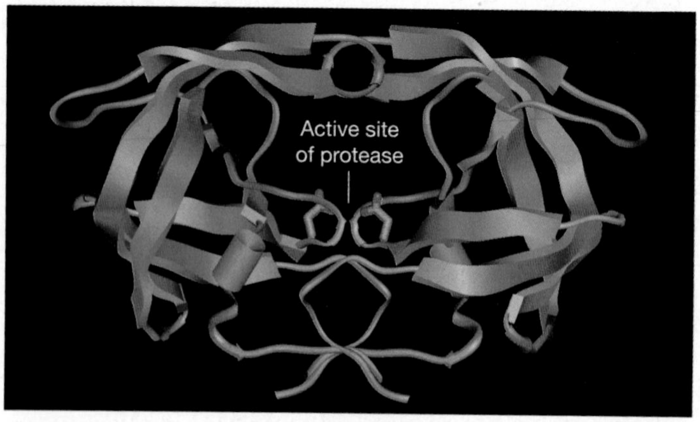

(b) A drug can block the active site

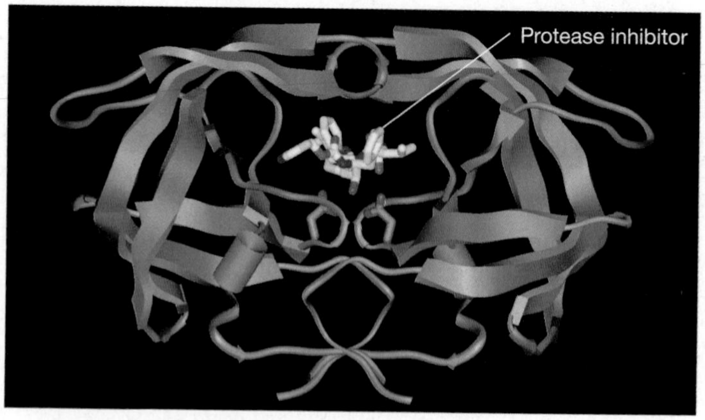

Figure 33.10 **The Active Site Is Located in the Interior of HIV's Protease. (a)** Ribbon diagram depicting the three-dimensional shape of HIV's protease enzyme. **(b)** After identifying protease's structure, researchers synthesized compounds that were predicted to fit into the active site and prevent the enzyme from working.

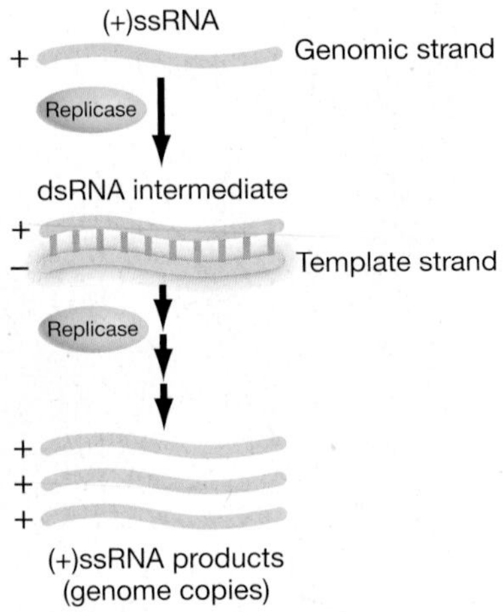

Figure 33.11 RNA Replicases Make Copies of Genomic RNA via a dsRNA Intermediate.

Making Models 33.1 Tips on Tracking Replication in RNA Virus Genomes

You can model genome replication in positive- and negative-sense RNA viruses by using simple symbols like those shown below. In models like these, it's not necessary to show nucleotide sequences; just draw simple arrows with plus and minus signs. The arrowheads mark the 3′ ends of the nucleic acid strands (Ch. 4, Section 4.1).

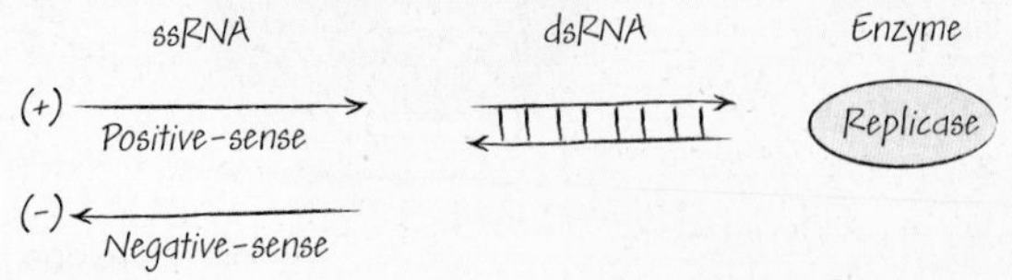

MODEL Draw all the steps in the replication of a negative-sense RNA viral genome. Indicate where RNA replicase is needed.

To see this model in action, go to the Study area of **Mastering Biology**

removes the RNA strand and catalyzes the synthesis of a second, complementary DNA strand, resulting in a double-stranded DNA (**Figure 33.12**). The dsDNA copy enters the cell nucleus and is then inserted into the host genome and used as a template for host cellular machinery—including an RNA polymerase—to produce viral mRNAs and viral genomic RNAs.

Viruses that reverse-transcribe their genome in this way are called **retroviruses** ("backward-viruses"). The name is apt because the initial flow of genetic information in this type of virus is RNA → DNA—the opposite of the central dogma (Ch. 16, Section 16.2). HIV is an example of a retrovirus.

The first antiviral drugs that were developed to combat HIV act by inhibiting reverse transcriptase. Logically enough, drugs of this type are known as reverse transcriptase inhibitors.

How Are New Virions Assembled? Once viral proteins have been produced and the viral genome has been replicated, assembly of a new generation of virions can take place.

During virion assembly, viral genomes are packaged into capsids. Some viruses also include copies of non-capsid proteins, often enzymes such as polymerases, inside the capsid. In many cases, the details of the assembly process are not well understood. Some viruses assemble the capsid first and then use motor proteins to pull the viral genome inside, while others assemble the capsid around the genomic material.

Enveloped viruses use the host endomembrane system (Ch. 7, Section 7.5) to transport their envelope proteins to the appropriate membrane for assembly. For example, virions of HIV and influenza assemble at the host cell's surface and acquire their envelope from its plasma membrane. Other enveloped viruses may assemble at the surface of internal membranes, such as the rough ER or Golgi apparatus.

In most cases, self-assembly occurs—though some viruses make proteins that provide scaffolding where new virions are put together. Researchers are in the process of testing new drugs designed to bind to HIV capsid proteins and inhibit their assembly during a viral infection.

How Do Progeny Virions Exit an Infected Cell? Most viruses leave a host cell in one of two ways: by budding from cellular membranes or by bursting out of the cell. In general, enveloped viruses bud, and nonenveloped viruses burst.

Viruses that bud from one of the host cell's membranes take some of that membrane with them. As a result, their envelope

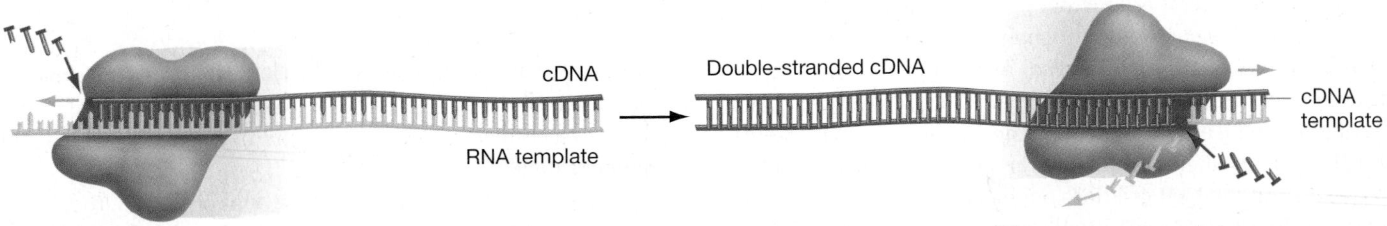

Figure 33.12 Reverse Transcriptase Catalyzes Synthesis of Double-Stranded DNA from an RNA Template. The single-stranded DNA produced initially by reverse transcriptase is called cDNA because its base sequence is complementary to that of the RNA template.

(a) A virion budding from a human host cell

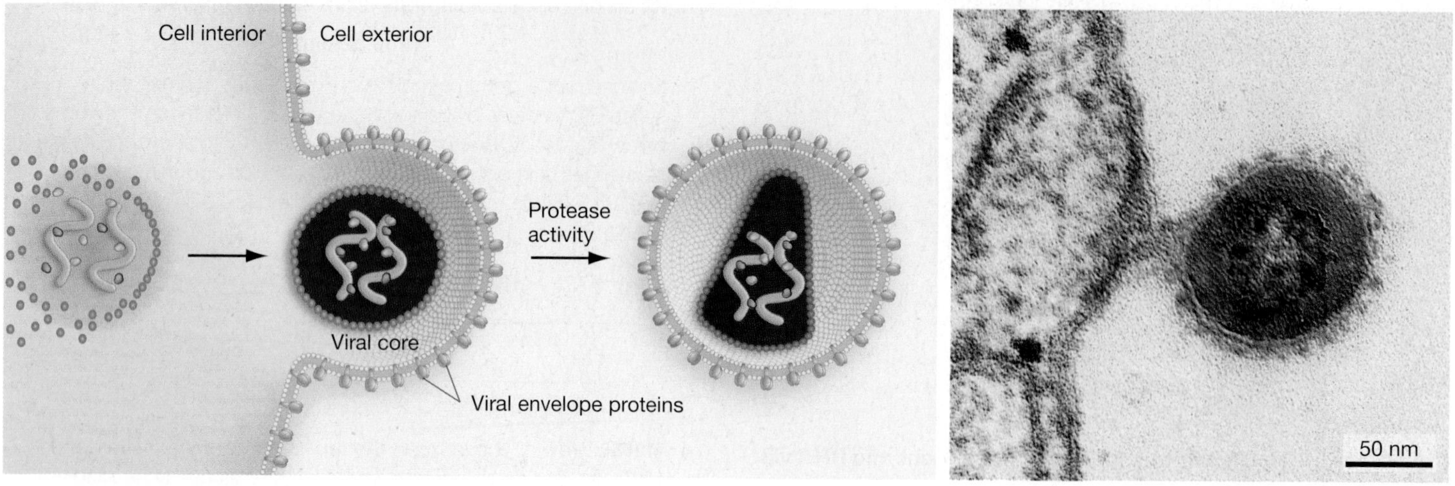

(b) Virions bursting from a human host cell

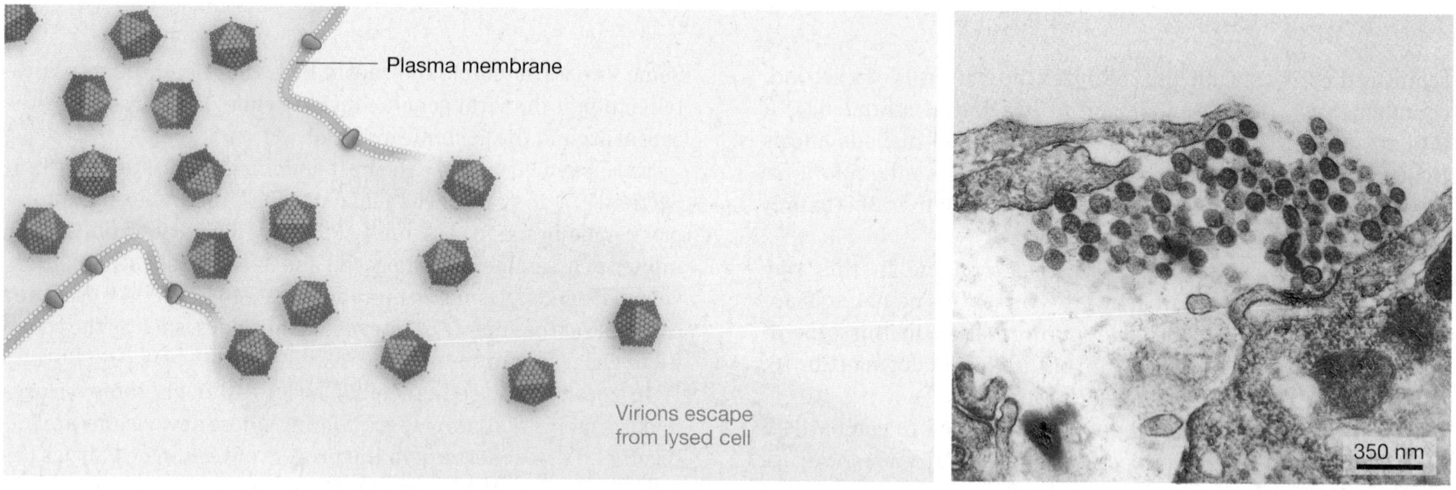

Figure 33.13 Viruses Leave Infected Cells by Budding or Bursting. (a) In enveloped viruses such as HIV, virions are formed as they bud from membranes of the host cell. **(b)** Virions of nonenveloped viruses normally exit the host cell by bursting the cell. Bursting kills the host cell; budding does not directly cause cell death.

✔ Suppose an HIV-infected cell and an adenovirus-infected cell were artificially lysed immediately before the exit of virions. How would that affect the number of infectious progeny released by each cell?

includes host-cell phospholipids along with envelope proteins encoded by the viral genome. In **Figure 33.13a**, HIV is shown budding from a helper T cell's plasma membrane. After the budding step, the capsid proteins are further processed by HIV's protease, giving the capsid its characteristic cone-like appearance.

Viruses that bud through internal membranes, such as those that enter the rough ER or Golgi apparatus, are secreted from the cell after being escorted through the endomembrane system. Regardless of the membrane used for producing enveloped viruses, exit from the host cell does not directly cause the death of the host. Cell death may eventually result from the virus depleting the cell of resources. In addition, HIV can trigger the death of both infected and uninfected T cells by inducing apoptosis (Ch. 48, Section 48.4).

In contrast, most nonenveloped viruses release their virions from the cell by lysing it—commonly referred to as the burst. For example, adenovirus produces proteins that help bring about cell lysis and so disperse a new generation of virions into the environment. **Figure 33.13b** shows virions bursting from an infected human cell.

How Are Virions Transmitted to New Hosts? Once the replicative cycle within a cell is complete, dozens to several hundred newly assembled virions are in the extracellular space. What happens next?

If the host cell is part of a multicellular organism, the new generation of virions may be transported through the body—often via the bloodstream or lymphatic system (Ch. 48, Section 48.2). In vertebrates, antibodies produced by the immune system may bind to the virions and mark them for destruction. But if a virion contacts an appropriate host cell before it encounters antibodies, then a new replicative cycle will start.

PROCESS: LYSOGENY (DORMANT STATE IN A BACTERIOPHAGE)

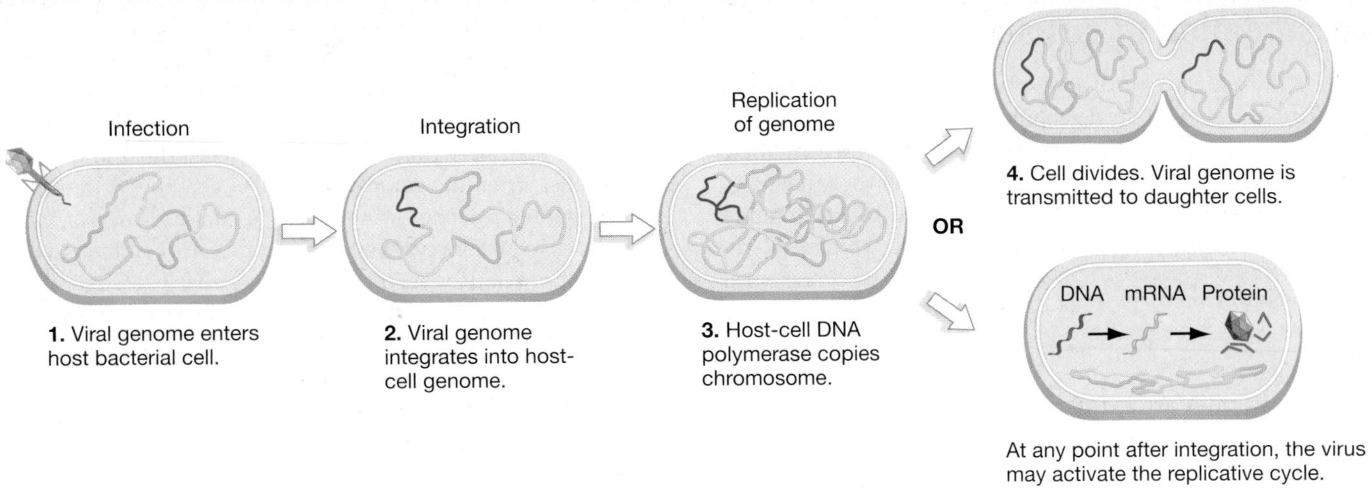

Figure 33.14 Some Bacteriophages Can Switch between the Lytic Cycle and Lysogeny. Molecular cues from the host can initiate this switch from the lytic cycle (see Figure 33.6) to lysogeny in viruses such as the bacteriophage lambda shown here.

The long-term success of a virus, however, depends on its ability to be transmitted through the environment from one organism to another. For example, when you cough, sneeze, wipe a runny nose, or defecate, you help rid your body of viruses. But you also expel them into the environment, where they may come into contact with a new host.

For a virus, a new host represents an unexploited habitat brimming with resources in the form of cells. The situation is analogous to that of an organism dispersing to a new habitat and colonizing it. The alleles carried by these successful colonists increase in frequency in the total population. In this way, natural selection (Ch. 22, Section 22.3) favors alleles that allow viruses to do two things: **(1)** replicate within a host and **(2)** be transmitted to new hosts. ✔ **If you understand how different types of viruses produce a new generation of virions, you should be able to compare and contrast the replicative cycles of bacteriophage T4 and HIV.**

Analyzing How Viruses Can Coexist with Host Cells

All viruses undergo replicative growth, but certain types of viruses can arrest the replicative cycle and enter a dormant state of coexistence with the host cell. In bacteriophages, this alternate mode of infection is called **lysogeny** (Figure 33.14).

The onset of lysogeny is triggered by molecular cues in the host that push the virus out of the replicative cycle. Instead of being actively transcribed and replicated, the viral DNA is incorporated into the host's chromosome, and expression of most of the viral genes is shut down.

Although no virions are produced during lysogeny, the host's DNA polymerase replicates the viral DNA each time the cell divides. Copies of the viral genes are passed on to daughter cells just as if they were the host's own genes. Some of these introduced genes are known to significantly alter bacterial hosts. For example, many of the pathogenic strains of *E. coli* that have recently emerged in humans are more virulent because they produce toxins encoded by bacteriophage genes that were introduced via lysogeny.

An integrated bacteriophage genome may be activated by different cues from the host cell, often associated with host stress. Activation leads to excision of the viral DNA from the host genome and reentry into the replicative growth cycle.

In viruses that infect animal cells, the dormant state is called **latency**. Certain animal viruses that enter a period of latency arrest their replicative growth and thus avoid a strong and effective immune response. Latency is a key adaptation that prolongs the infection and increases the likelihood that the virus will be transmitted to a new host. Depending on the virus involved, the viral genome may or may not be integrated into the host genome.

HIV is an example of an animal virus that requires latency. HIV is inefficiently transmitted—requiring direct contact with the T cells in body fluids such as blood, semen, or vaginal secretions—but the latency period prevents the host from clearing the virus before transmission to a new host occurs. The latency period begins at the start of many new infections. After the viral genes are integrated into the host genome, they are expressed only after the helper T cell host is activated by the immune response. Until that happens, HIV remains silent and hidden.

CHECK YOUR UNDERSTANDING

✔ **If you understood this section, you should be able to . . .**

1. Evaluate the claim that the viral replicative cycle more closely resembles the mass production of automobiles than the reproduction of cells.
2. **MODEL** Draw a model to compare and contrast the lytic and lysogenic pathways.

Answers are available in Appendix A.

33.3 What Themes Occur in the Diversification of Viruses?

The tree of life will never be free of viruses. Mutation and natural selection guarantee that viral genomes will continually adapt to the defenses offered by their hosts, whether those defenses are produced by an immune system or by biomedical researchers. Because most viral polymerases have high error rates and because viruses lack error repair enzymes, viral mutation rates are extremely high. Many viruses change constantly—giving them the potential to evolve rapidly.

After you complete this section, you should be able to . . .

- Discuss the relationship between rapid viral evolution and emerging human diseases.

Where Did Viruses Come From?

No one knows how viruses originated. Biologists working to answer this question are considering three hypotheses. Let's have a look.

Escaped-Genes Hypothesis: Origin of Viruses in Plasmids and Transposable Elements Like viruses, plasmids and transposable elements are acellular, mobile genetic elements that replicate with the aid of a host cell (Ch. 20, Section 20.6). Certain viruses are indistinguishable from plasmids except for one feature: They encode proteins that form a capsid and allow the genes to exist outside of a cell.

Some biologists hypothesize that simple viruses represent "escaped gene sets." This view proposes that mobile genetic elements are descended from clusters of genes that physically escaped from prokaryotic or eukaryotic chromosomes long ago.

According to this hypothesis, the escaped gene sets took on a mobile, parasitic existence because they happened to encode the information needed to replicate themselves at the expense of the genomes that once held them. In the case of viruses, the hypothesis is that the escaped genes included the instructions for making a protein capsid and possibly envelope proteins. Each type of virus may be the product of a distinct "escape event."

To support the escaped-genes hypothesis, researchers would need to discover a virus that had so recently derived from intact prokaryotic or eukaryotic genes that the viral genome still strongly resembled the DNA sequence of those genes.

Degeneration Hypothesis: Origin of Viruses in Symbiotic Bacteria Some researchers contend that DNA viruses with large genomes trace their ancestry back to free-living bacteria that once took up residence inside eukaryotic cells. The idea is that these bacteria degenerated into viruses by gradually losing the genes required to synthesize ribosomes, ATP, nucleotides, amino acids, and other compounds.

Although this idea sounds speculative, it cannot be dismissed lightly. There is evidence that the mitochondria and chloroplasts of eukaryotic cells originated as intracellular symbionts derived from ancestors that were independent, free-living cells (Ch. 27, Section 27.3). Investigators contend that, instead of evolving into intracellular symbionts that aid their host cell, the bacterial ancestors of DNA viruses became parasites capable of replication and transmission from one host to another.

To support the degeneration hypothesis, researchers have pointed to the giant virions of pandoraviruses and mimiviruses, which infect certain protists. These are among the largest known viruses, with sizes that approach or exceed those of many bacterial cells. When the equally large DNA genomes of these viruses were sequenced, they were found to contain some of the genes involved in protein synthesis—genes that are common to cells but had not previously been observed in viruses. It is still not clear, though, whether these genes are remnants from an ancestral cell or were picked up from a host cell during an infection. The discovery of a cell that possesses a genome similar to one of these large DNA viruses would support the hypothesis that they originated from the degeneration of a cell.

RNA-World Hypothesis: Origin of Viruses at the Origin of Life Some researchers propose a third alternative to explain the origin of viruses: that they trace their ancestry back to RNA-based life-forms that evolved early in Earth's history—life-forms postulated by the RNA-world hypothesis (Ch. 4, Section 4.4). If this hypothesis is correct, then the RNA genomes of some viruses would be descended from genes found in inhabitants of such an RNA world. It has even been suggested that retrovirus-like parasites may have been responsible for transforming the genetic material of cells from RNA to DNA.

To support this hypothesis, advocates point to the ubiquity of viruses—which suggests that they have been evolving along with organisms since life began. The RNA-world hypothesis also addresses a problem that weakens the other two hypotheses: Several proteins that are commonly expressed in many viruses are not expressed in any cell examined thus far. If viruses originated at the same time the cell did, then the genes coding for these proteins may have come from an RNA-world pool of genes instead of a cell.

Currently, there is no single widely accepted view of where viruses came from. Because scientists are almost certain that viruses originated more than once in the history of life, there is no such thing as the phylogeny of all viruses. All three hypotheses may be valid for the origin of different viruses.

Emerging Viruses, Emerging Diseases

The continuing diversification of viruses poses a never-ending challenge to their potential hosts. With alarming regularity, the news media carry accounts of deadly viruses that are infecting humans for the first time.

HIV is an example of a virus responsible for an **emerging disease**: a new illness that suddenly affects significant numbers of individuals in a host population. Another horrific example occurred in 1976, when the Ebola virus infected 318 individuals in the Democratic Republic of Congo and killed 280 of them. A more recent Ebola virus outbreak in West Africa claimed the lives of over 11,000 people in 2014–2016. There is hope that an experimental vaccine may thwart the virus in any future outbreaks.

Where does the Ebola virus hide between outbreaks? The virus is known to infect and kill antelopes, chimpanzees, and gorillas. Researchers suspect, however, that the virus resides harmlessly

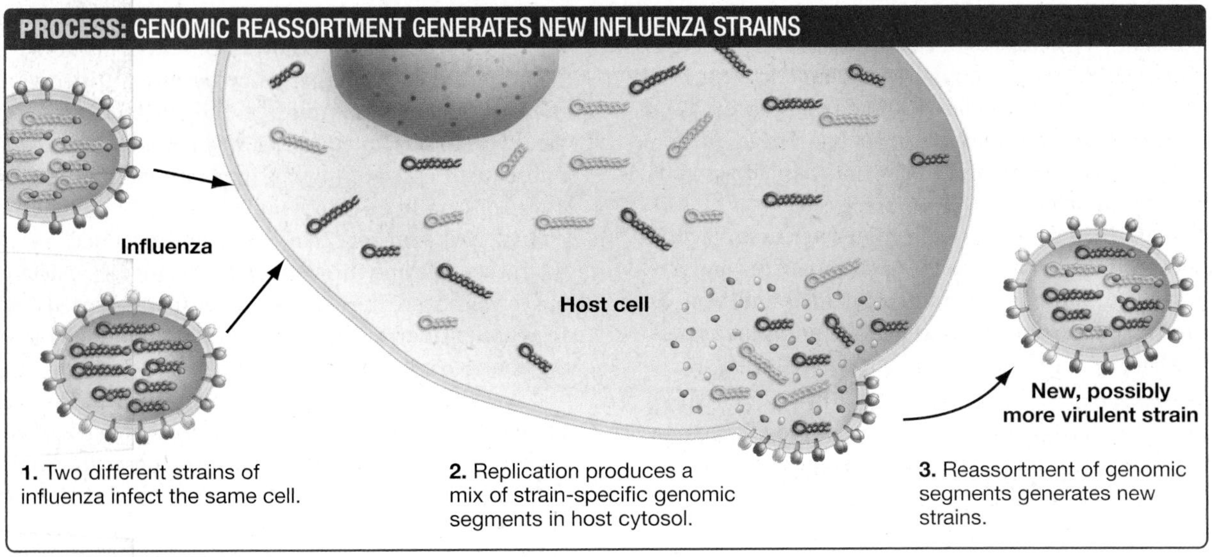

Figure 33.15 Influenza Can Generate New Strains via Genomic Reassortment.

in bats until conditions favor its reemergence. Both HIV and Ebola are considered emerging viruses because they "jumped" from their natural host species to a new host—humans.

Some Emerging Viruses Arise from Genome Reassortment

Each year brings the possibility of a new influenza pandemic, one that might resemble the devastating outbreak of 1918. The small changes in the influenza virus that arise from its error-prone RNA replicase are primarily responsible for the need for yearly vaccine updates. But influenza can also acquire entire alleles that are completely new to the strain that infects humans. (A virus **strain** consists of populations that have similar characteristics and is the lowest, or most specific, level of taxonomy for viruses.) How do new strains of influenza originate?

Influenza has a single-stranded RNA genome that consists of eight segments. If two virions infect the same cell, the replicated genomic segments are shuffled in the cytosol and randomly dealt out to the progeny during assembly.

The ability to randomly mix genomic segments becomes particularly significant when two different strains infect the same cell (Figure 33.15). For example, different strains of influenza can infect humans as well as other animals, such as birds and pigs. While the avian strains do not efficiently infect humans, they can infect pigs—and so can the human strains. This means that pigs can serve as mixing vessels to produce new recombinant strains. This type of reassortment is thought to have produced the 1998 pandemic strain of influenza, whose genomic segments came from human, pig, and avian strains.

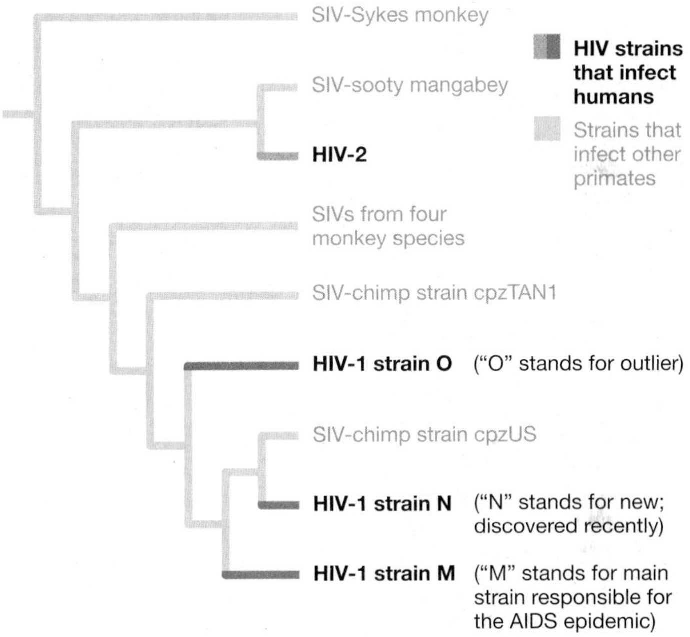

Figure 33.16 Phylogeny of HIV Strains and Related Viruses. Phylogenetic tree showing the evolutionary relationships among some of the immunodeficiency viruses that infect primates—including chimpanzees, humans, and several species of monkeys.

✔ Indicate where an SIV jumped to humans by drawing and labeling bars across the appropriate branches on the phylogenetic tree.

Using Phylogenetic Trees to Understand Emerging Viruses

How do researchers know that an emerging virus has jumped to a new host? The answer is to analyze the evolutionary history of the virus and then construct a phylogenetic tree for the virus (BioSkills 13) that includes its close relatives.

HIV, for example, belongs to a group of viruses that infect a wide range of mammals, including house cats, horses, goats, and primates. Several conclusions can be drawn from the phylogenetic tree of HIV, shown in Figure 33.16.

- ***There are simian immunodeficiency viruses (SIVs).*** Many of HIV's closest relatives parasitize cells that are part of the immune system in monkeys and chimpanzees. Several of these SIVs cause diseases with symptoms reminiscent of AIDS.

- ***There are two HIVs.*** There are two distinct types of human immunodeficiency viruses, called HIV-1 and HIV-2. Although both can cause AIDS, HIV-1 is far more virulent and is the better studied of the two. HIV-1's closest known relatives are SIVs isolated from chimpanzees that live in central Africa. In contrast, HIV-2's closest relatives are SIVs that parasitize monkeys called sooty mangabeys.

  In central Africa, where HIV-1 infection rates first reached epidemic proportions, contact between chimpanzees and humans is extensive. Chimps are hunted for food and kept as pets. Similarly, sooty mangabeys are hunted and kept as pets in western Africa, where HIV-2 infection rates are highest.
- ***Multiple "jumps" have occurred.*** Several strains of HIV-1 exist. The most important are called O for outlier (meaning the most distant group relative to other strains), N for new, and M for main. Each of these strains likely represents an independent origin of HIV-1 from a chimp SIV strain.

The last point is particularly important. The existence of distinct strains suggests that HIV-1 has jumped from chimps to humans several times. It may do so again in the future.

Responding to a New Virus Outbreak Physicians become alarmed when they see many patients with identical, and unusual, disease symptoms in the same geographic area over a brief period. Physicians report these cases to public health officials, who take on two urgent tasks: **(1)** identifying the agent that is causing the new illness and **(2)** identifying the origin of the outbreak.

As an example, in late 2002, an outbreak of an unknown respiratory syndrome—which became known as severe acute respiratory syndrome (SARS), rocked southern China and rapidly spread to several countries across the globe, causing over 8000 probable infections and killing 774 people. In 2003, the World Health Organization (WHO) led an international effort that identified the causative agent of SARs as a new coronavirus—a group of viruses known to be one of the causes of the common cold.

The next task, identifying the origin of a virus outbreak, is equally critical and takes old-fashioned detective work. By interviewing patients about their activities, researchers called epidemiologists decide how each patient could have acquired the virus.

In the case of the SARS outbreak, the horseshoe bat was found to be the natural host, or reservoir, of the virus. But the virus was also identified in the civet, a small mammal commonly sold in markets in southern China. This final observation provided the link between the bats and the human outbreak.

The rapid spread of the virus to different countries illustrates the impact of world travel on the transmission of disease. Fortunately, a major pandemic was avoided because SARS was mostly limited to people who were in close contact with infected individuals. To date, no additional SARS infections have been recorded, but a new and related virus—called the Middle East Respiratory Syndrome Coronavirus—caused a similar outbreak in Saudi Arabia in 2013 and produced a 30 percent death rate. There's little doubt that deadly viral diseases will continue to emerge as human expansion encroaches on wildlife habitat.

Reemergence of a More Virulent Virus The mosquito-borne Zika virus, once little known, is now considered an international public health emergency because of its explosive spread in South and Central America, the Caribbean, and Florida. Zika virus was first isolated in 1947 from a rhesus monkey in Uganda. Subsequent cases of human infection were infrequent and produced only mild flu-like symptoms.

In 2007 a major Zika virus outbreak led to epidemics in Southeast Asia and throughout the Pacific Ocean islands. Of growing concern were the rare reports of some infected adults developing Guillain-Barré syndrome, an autoimmune disorder (Ch. 48, Section 48.5) characterized by generalized muscle weakness.

Zika virus infection took a tragic turn in the outbreak that began in Brazil in 2013 and has spread to over 70 countries. Most human Zika virus infections are by mosquito bites, and in 80 percent of cases there are no symptoms. But the virus can also be spread directly through sexual contact and blood transfusion, and—with truly devastating effects—from infected mothers to their fetuses.

Public health officials have confirmed that the virus is associated with causing severe developmental defects in newborns, such as microcephaly (a small head circumference and decreased brain volume). Researchers now are working around the clock to develop an effective vaccine, antiviral treatments, and rapid detection methods.

CHECK YOUR UNDERSTANDING

✔ If you understood this section, you should be able to . . .

1. Compare and contrast the three hypotheses for the origin of viruses.
2. Predict what type of mutation in an avian flu virus would make it more dangerous to humans, and explain how this change might occur.

Answers are available in Appendix A.

33.4 Key Lineages of Viruses

Unlike the organisms discussed in previous chapters, viruses have no single phylogenetic tree that represents their evolutionary history. Instead, researchers most often focus on comparing base sequences in the genetic material of small, closely related groups of viruses to reconstruct the phylogenies of particular lineages, exemplified by the tree for HIV strains in Figure 33.16.

After you complete this section, you should be able to . . .

- Describe the lineages of viruses.

To organize the genomic diversity of viruses, researchers group them into seven general classes based on the nature of their genetic material and how they replicate. This approach, first proposed by David Baltimore in 1971, is called the Baltimore classification. Figure 33.17 summarizes the seven classes.

Notice that each of the Baltimore classes has its own unique strategy for transcribing and replicating the genetic material.

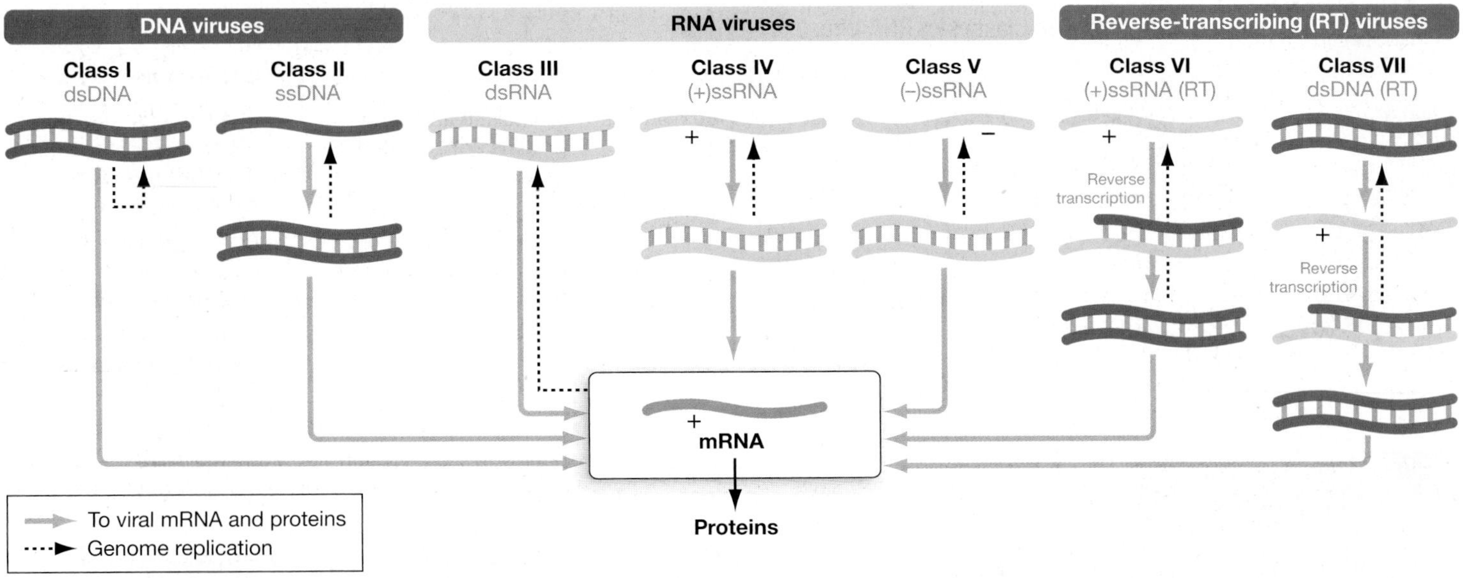

Figure 33.17 The Seven Strategies for Expression of Viral Genetic Material Converge on the Translation of mRNA. Notice that although Classes I and IV have the same types of genomes as Classes VII and VI, respectively, their genomes are replicated in different ways.

Recall that the genome of positive-sense single-stranded RNA viruses contains the same base sequences as mRNAs and can be directly translated by the host cell to make viral proteins. However, all the other strategies require additional steps to produce mRNA.

In the Baltimore classification system, researchers further distinguish viruses according to **(1)** virion and genome morphology, **(2)** nature of the host species, and **(3)** how the virus replicates within the host. Biologists have identified almost 100 virus families using these criteria, and the list keeps growing as new, emerging viruses are discovered. Viruses within families are further divided into distinct genera and species for convenience, although the phylogenetic relationships among viruses are not as clear as the relationships among organisms.

Tables 33.2, 33.3, and **33.4** (pages 715 through 717) summarize key characteristics of the seven Baltimore classes of viruses. As you review these tables, take time to consider how the

DIVERSITY Table 33.2 Baltimore Classes of DNA Viruses

		Morphology	Host Taxa	Replicative Cycle
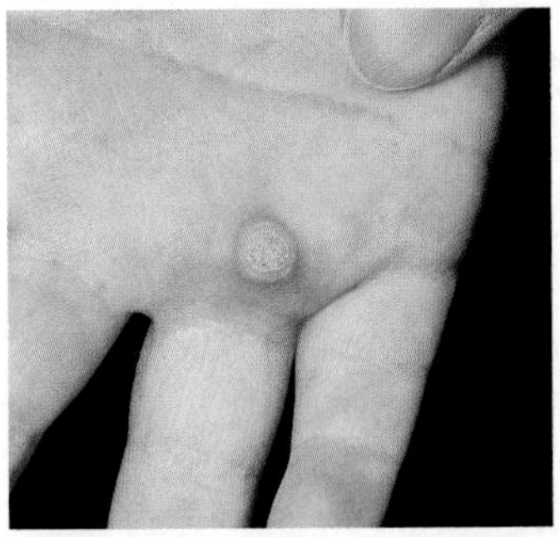	I. Double-Stranded DNA (dsDNA) Viruses **Largest group of viruses; 33 families, 142 genera** ◀ Warts are a symptom of infection by human papilloma virus	Virions are enveloped or naked. Capsids are icosahedral, helical, or complex. 4.5–2400 kb genomes, as a single circular or linear molecule or multiple linear molecules.	Most infect animals or bacteria; fewer infect protists or archaea. Not known to infect plants. Some genera may coexist with host via lysogeny or latency.	Genome replicated by a semiconservative mechanism. Genome used as template for transcription of mRNA. mRNA used to produce viral proteins.
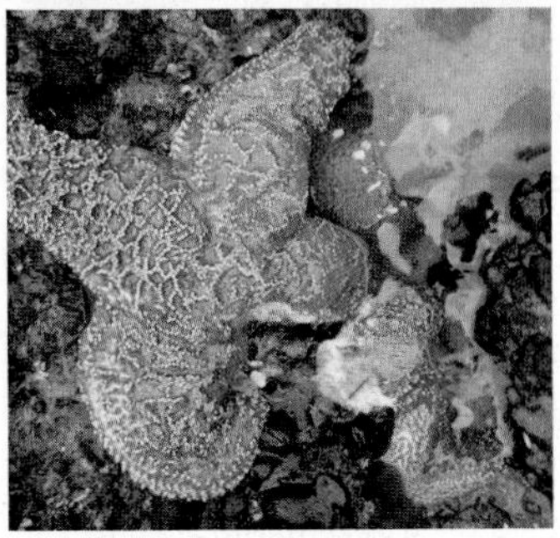	II. Single-Stranded DNA (ssDNA) Viruses **9 families, 44 genera** ◀ Sea star associated densovirus (SSaDV) causes wasting disease	Most virions are naked; a few are enveloped. Most capsids are icosahedral; a few are helical. 1.8–12.5 kb genomes, as single or multiple molecules that may be circular or linear.	Most infect animals; fewer infect bacteria or plants. Marine invertebrates recently identified as hosts.	Genome replicated via a dsDNA intermediate, where either strand is used in progeny virions. dsDNA intermediate used as the template for transcription of mRNA. mRNA used to produce viral proteins.

For more, go to https://goo.gl/nE1nPA

DIVERSITY Table 33.3 Baltimore Classes of RNA Viruses

	Morphology	Host Taxa	Replicative Cycle
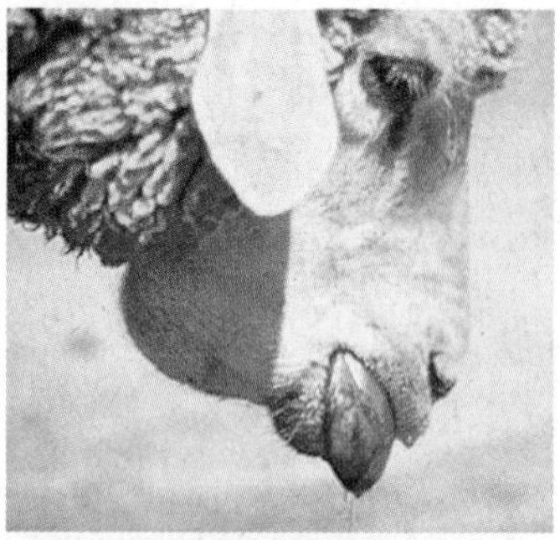 ◀ Sheep afflicted by the bluetongue virus **III. Double-Stranded RNA (dsRNA) Viruses** **12 families, 36 genera**	Most virions are naked; a few are enveloped. Capsids are icosahedral and often have multiple layers. 3.7–30 kb genomes, as one or more linear molecules.	Animals, plants, and protists are common hosts. Several known to infect hosts across different taxa (e.g., animals and plants).	Viral RNA replicase packaged in virion transcribes mRNA from the dsRNA genome. mRNA used as a template to make copies of the dsRNA genome. mRNA also used to produce viral proteins.
 ◀ Pea plant exhibiting mosaic virus leaf mottling **IV. Positive-Sense Single-Stranded RNA ([+]ssRNA) Viruses** **Largest group of RNA viruses; 32 families, 139 genera**	Virions are enveloped or naked. Most capsids are icosahedral; a few are helical. 3.4–31 kb genomes, as one or more linear molecules.	Plants and animals are most common hosts. Some can infect vertebrates and arthropods (e.g., mosquitoes), which are used as vectors for transmission.	Genomic RNA translated to produce a viral RNA replicase. Viral replicase makes copies of genomic RNA via dsRNA intermediate. Copies of viral genome used to produce polyproteins, from which viral proteins are generated via cleavage.
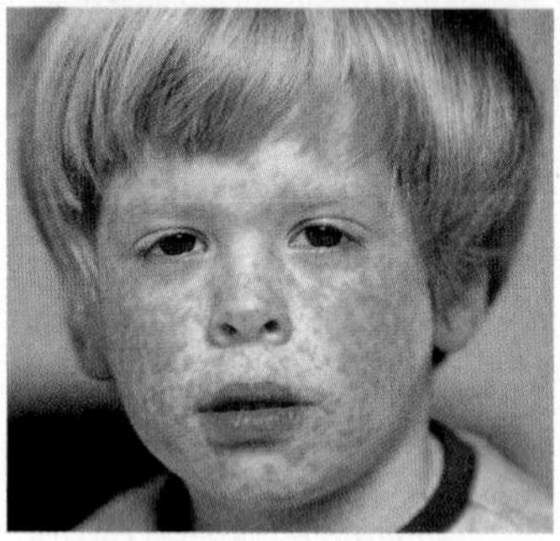 ◀ The rash on this young boy's face is one of the hallmark symptoms of measles virus infection **V. Negative-Sense Single-Stranded RNA ([−]ssRNA) Viruses** **9 families, 42 genera**	Most virions are enveloped; a few are naked. Capsids are helical. 10–25 kb genomes, as one or more linear molecules.	Most infect animals; fewer infect plants. Not known to infect bacteria or archaea	Viral RNA replicase packaged in virion transcribes mRNA from (−)ssRNA genome. Viral replicase makes copies of (−)ssRNA genomic RNA via dsRNA intermediate. mRNA used to produce viral proteins.

For more, go to https://goo.gl/HuZidO

different viruses have touched your life. Let's now survey each of the major classes to explore how viruses affect society.

Class I: Double-Stranded DNA (dsDNA) Viruses Smallpox is perhaps the most notorious of the dsDNA viruses. Although smallpox has caused millions of deaths throughout human history, it was eradicated by worldwide vaccination programs. In 1977, smallpox was declared to be extinct in the wild; the only remaining samples of the virus are stored securely in research labs. Other common animal viruses with dsDNA genomes include the human papilloma virus, herpesviruses, and adenoviruses. Viruses like human papilloma can induce their host cells to enter S phase (Ch. 12, Section 12.1) despite the absence of normal growth signals, leading to common warts. This ability is also responsible for the link between human papilloma virus infections and cervical cancer in women.

Class II: Single-Stranded DNA (ssDNA) Viruses The most familiar ssDNA virus is the canine parvovirus type 2, which causes a fatal disease in dogs if untreated. This disease is rare in domesticated dogs thanks to a widely used vaccine given during the first year of life. In 2014, a newly discovered densovirus triggered an epidemic of wasting disease in sea stars that is decimating populations from Baja California to southern Alaska.

Class III: Double-Stranded RNA (dsRNA) Viruses Many of the diseases in rice, corn, sugarcane, and other crops are caused by dsRNA viruses. The bluetongue virus, transmitted through midges and other biting insects, causes a deadly disease in ruminants (especially sheep) that has significantly affected the livestock industry. In humans, reovirus and rotavirus infections are the leading cause of infant diarrhea. Rotavirus alone is estimated by the WHO to be responsible for over 215,000 infant deaths each year.

DIVERSITY Table 33.4 Baltimore Classes of Reverse-Transcribing Viruses

		Morphology	Host Taxa	Replicative Cycle
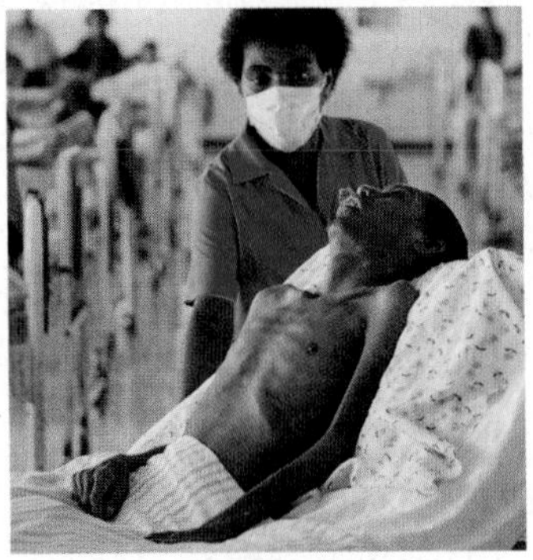	**VI. Reverse-Transcribing (+)ssRNA Viruses** Smallest group of viruses; only 1 family, 8 genera ◀ HIV compromises the immune system and physiology of this patient	Virions are enveloped. Capsids are icosahedral, but may be processed to change shape. 7–11 kb genomes, as a single linear molecule.	Known to infect only humans and other vertebrates.	Viral reverse transcriptase packaged in virion converts RNA genome to dsDNA, which is integrated into host genome. Integrated dsDNA transcribed to replicate (+)ssRNA genome and produce mRNA. mRNA used to produce viral proteins.
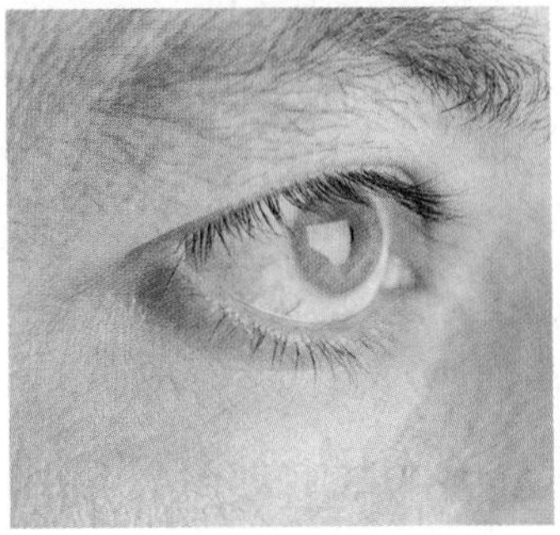	**VII. Reverse-Transcribing dsDNA Viruses** 2 families, 9 genera ◀ Jaundice is a symptom of infection by hepatitis B virus	Virions are enveloped or naked. Capsids are icosahedral. 3–8.3 kb genomes, as a single circular molecule, often incomplete dsDNA	Known to infect only vertebrates and plants.	Genome used as template for transcription of full-length (+)ssRNA and mRNA. Viral reverse transcriptase converts (+)ssRNA into dsDNA to replicate genome. mRNA used to produce viral proteins.

For more, go to https://goo.gl/torWbf

Class IV: Positive-Sense Single-Stranded RNA ([+]ssRNA) Viruses Most of the commercially important plant viruses belong to this group. Because they kill groups of cells in the host plant and turn patches of leaf or stem white, they are often named mottle, spotted, chlorotic (meaning lacking chlorophyll), or mosaic viruses. Viruses within this group that infect humans include the rhinovirus (cause of the common cold), polio, the SARS coronavirus, West Nile virus, and hepatitis A, C, and E. The West Nile virus is an example of a vector-borne virus; it can infect mosquitoes, which serve as vectors for transmitting the virus between birds and other animals, including humans. Other vector-borne viruses in this group include Zika, dengue, and yellow fever.

Class V: Negative-Sense Single-Stranded RNA ([−]ssRNA) Viruses A wide variety of plants and animals are parasitized by (−)ssRNA viruses. If you have ever suffered from the flu, mumps, or measles, then you are painfully familiar with these viruses. This group also includes some of the most deadly viruses on the planet, such as Ebola and rabies. Rabies is one of the few viral diseases for which getting vaccinated *after* a bite from a rabid dog can save your life. Left untreated, rabies is fatal and causes an estimated 59,000 deaths each year, primarily in Africa and Asia where few dogs are vaccinated.

Class VI: Reverse-Transcribing (+)ssRNA Viruses The Rous sarcoma virus (chickens), mouse mammary tumor virus, and murine (mouse) leukemia virus are well-studied retroviruses that have been shown to contribute to the development of cancer. Of the retroviruses, HIV is the most familiar and, in terms of the human population, by far the deadliest virus in this group.

Class VII: Reverse-Transcribing dsDNA Viruses The hepatitis B virus is among several viruses that cause hepatitis, but it is the only one that has a dsDNA genome. Symptoms of infection include yellowing of the eyes and skin. This condition, called jaundice, is caused by the accumulation of bilirubin in the blood as a result of liver damage. About 5 percent of adults infected with hepatitis B virus have a chronic infection that has been linked to the development of liver cancer.

Controlling the threat that viruses pose to society continues to be an intense area of research. For biologists, the evolution and emergence of novel viruses is both fascinating to study and at the same time terrifying, because of their potentially devastating consequences.

CHECK YOUR UNDERSTANDING

✔ **If you understood this section, you should be able to . . .**

Associate types of genomic characteristics used to describe viral diversity with each of the three hypotheses that have been posed to explain the origin of viruses (see Section 33.3).

Answers are available in Appendix A.

33.1 Why Do Biologists Study Viruses?

- Viruses have shaped the evolution of organisms by transferring genetic information between cells and serving as agents of natural selection via illness and death.
- Viruses cause disease—epidemics and pandemics.

33.2 How Do Biologists Study Viruses?

- Most viruses have a protein capsid that is either icosahedral or helical. Some viruses are covered by a membranous envelope.
- Viral genomes exhibit considerable diversity. The genetic material may consist of one or more molecules of DNA or RNA that is either double stranded or single stranded.
- The entry of viruses into host cells often depends on specific interactions between viral proteins and molecules on the host-cell surface that can function as virus receptors. The viral genome is then released into the cytosol by uncoating.
- Viruses use the host cell's chemical energy, substrates, and biosynthetic machinery to produce viral proteins and replicate their genomes.
- Enveloped viruses exit a cell by budding; naked viruses often exit by lysis, killing the host cell in the process.
- Some viruses may enter a dormant phase, in which they do not produce virions but instead coexist with the host cell. When in this form, viral genetic material will be transmitted to daughter cells when the host divides.

33.3 What Themes Occur in the Diversification of Viruses?

- Three hypotheses have been developed to explain the origin of viruses: **(1)** They are escaped gene sets from cells; **(2)** they are the products of degenerate cellular parasites; and **(3)** they coevolved with cells from the RNA world.
- Viruses continue to evolve. Factors influencing rates of viral evolution include errors during genome replication and genomic reassortment.
- Many emerging diseases—illnesses that suddenly affect significant numbers of individuals in a host population—are caused by viruses transmitted from other host species.

33.4 Key Lineages of Viruses

- In addition to their morphology and the nature of their genetic material, viruses vary in how their genomes are replicated and transcribed to produce mRNA.

Answers are available in Appendix A.

✓ TEST YOUR KNOWLEDGE

1. What do host cells provide for viruses?
   a. ATP
   b. nucleotides and amino acids
   c. ribosomes
   d. all of the above
2. What does reverse transcriptase do?
3. In the viral replicative cycle, what reaction do viral proteases catalyze? Select True or False for each statement.
   T/F They catalyze polymerization of amino acids into peptides.
   T/F They catalyze folding of long peptide chains into functional proteins.
   T/F They catalyze cutting of polyprotein chains into functional proteins.
   T/F They catalyze breaking of peptide bonds
4. What features distinguish the seven major categories of viruses?

✓ TEST YOUR UNDERSTANDING

5. The outer surface of a virion consists of either a membranous envelope or a protein capsid. How does the outer surface correlate with a virus's mode of exiting a host cell?
6. Compare and contrast the bacteriophage lytic cycle and lysogeny by addressing (1) the rate of replication of the viral genome, (2) production of virions, and (3) effect on the host cell.
7. **THINK CAREFULLY** If you come down with the flu, should your physician prescribe an antibiotic for you? Explain why or why not.
8. Of the viruses highlighted in Section 33.4, predict which of the following would be able to make viral proteins if nothing more than its genome were injected into a suitable host cell.
   a. pea mosaic ([+]ssRNA) virus
   b. bluetongue (dsRNA) virus
   c. measles ([−]ssRNA) virus
   d. human immunodeficiency (RNA reverse-transcribing) virus

✓ TEST YOUR PROBLEM-SOLVING SKILLS

9. **SOCIETY** If you were in charge of the government's budget devoted to stemming the AIDS epidemic, would you devote most of the resources to drug development or preventive medicine? Defend your answer.
10. Which of the following types of viruses would you expect to require periods of latency?
   a. viruses that have large genomes and require a long time for replication
   b. viruses that require a long time for transmission to new hosts
   c. viruses that require a long time for assembly into complex structures
   d. viruses that infect cells of the immune system

✔ PUT IT ALL TOGETHER: Case Study

How does Zika virus cause birth defects?

The mosquito-borne Zika virus (ZIKV) is one of the most feared viruses for pregnant women. Recent statistics justify this fear: Infants born to mothers infected with ZIKV during pregnancy face a risk of up to 42 percent of developing birth defects, including microcephaly (an abnormally small head and decreased brain volume). What do we know about how ZIKV causes damage to the developing brain?

11. ZIKV is a Baltimore class IV virus. Based on this classification, what do you know about the structure of its genome?

12. **PROCESS OF SCIENCE** Researchers have determined that host cells possess several different surface proteins that can serve as virus receptors for ZIKV. If you were to repeat the experiment described in Figure 33.8 using ZIKV instead of HIV, how would you expect the results to be different? Explain.

13. Several lines of evidence suggest that after ZIKV is transmitted to a pregnant woman via the bite of an infected mosquito, the virus then directly infects cells of the placenta (Ch. 47, Section 47.4), gaining access to the fetal brain. Following attachment, the virion would most likely enter the placental cell by:
    a. insertion via the mosquito's mouthparts
    b. endocytosis
    c. degrading the cell wall with lysozyme
    d. injection through a hollow, needle-like protein structure

14. **QUANTITATIVE** Researchers recently developed a mouse model for fetal brain defects associated with human ZIKV infection. To determine the effect of ZIKV on the number of neurons in the brain, ZIKV was injected into the brains of developing mouse embryos. Neurons were identified by staining sections of brain tissue with antibodies against NeuN, a neuron-specific protein (see **BioSkills 6** for more information on using antibody probes), and the number of NeuN-positive cells per mm^2 in specific regions was quantified. The results of three independent experiments are shown in the graph below. Use the P value provided to determine if the difference is significant or not (* means $P < 0.05$; see **BioSkills 3**). Based on these results, what can you conclude?

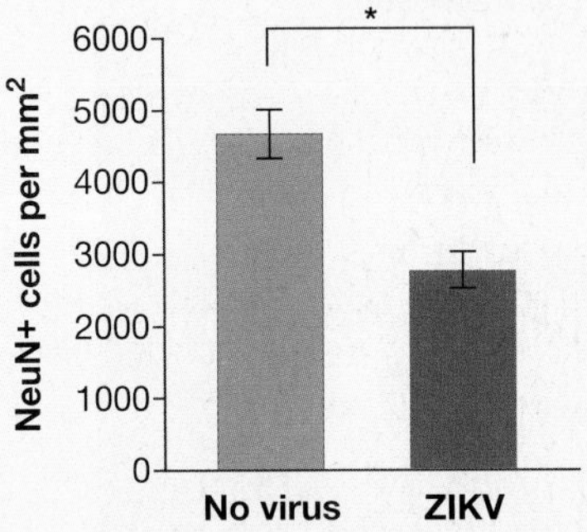

Source: Q. Shao et al. 2016. *Development* 143: 4127–4136.

15. **SOCIETY** Traveling and tourism have been badly affected in ZIKV-affected countries. What advice would you give a friend planning a trip to the Caribbean?

16. **MODEL** Draw a model to illustrate how the Zika virus genome is used to produce new virions. In your drawing, identify the steps required to replicate the genome and those used to produce mRNA. Also note where mRNAs that are used to produce capsid and envelope proteins are translated.

Mastering Biology

Students Go to Mastering™ Biology for assignments, the eText, and the Study Area with animations, practice tests, and activities.

Professors Go to Mastering™ Biology for automatically graded tutorials and questions that you can assign to your students, plus Instructor Resources.

■ For an introduction to the Mystery of the Newt case study, see page 17.

Now that you've learned about the diversification of life, let's return to the Mystery of the Newt case study introduced after Chapter 1. Recall the story about three hunters who were found dead at a campsite in Oregon next to a coffee pot containing a surprise ingredient: a rough-skinned newt (*Taricha granulosa*). Researchers have since discovered that these newts produce high levels of lethal tetrodotoxin (TTX).

✔ If you understand Unit 5, you should be able to apply your learning to this case study:

Are Newts Adapted to Kill Humans?

Humans do not typically eat newts or drink newt coffee. So, why do newts spend resources producing a toxin that is lethal to humans? Turns out they don't do it on purpose. Humans happen to be vulnerable to TTX because our voltage-gated sodium ion channels are homologous with other vertebrates'.

1. What does it mean that sodium channels in humans and other vertebrates are homologous?
   - **a.** The sodium channels are similar because they have the same function.
   - **b.** The sodium channels are adaptive.
   - **c.** The sodium channels arose independently in humans and other vertebrates.
   - **d.** The sodium channels are similar due to common ancestry.

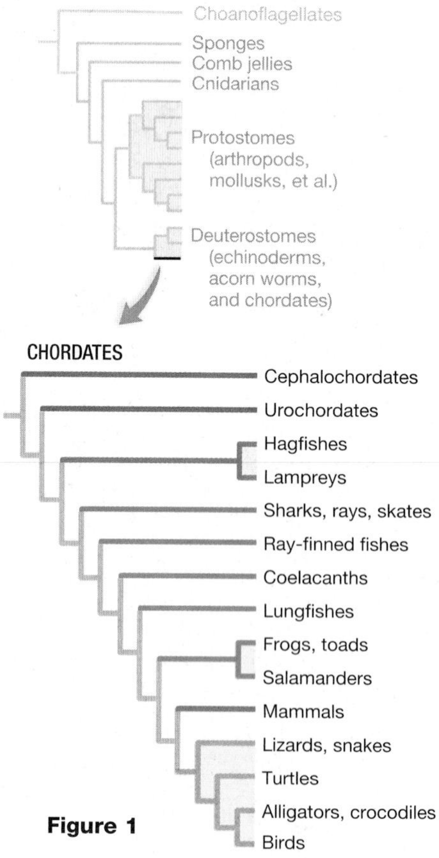

Figure 1

Researchers have discovered the remains of newts in the gut contents of a variety of predators, including trout, bullfrogs, otters, owls, hawks, ducks, and snakes. To understand your relationship to these other vertebrates, consider your place in the tree of life.

2. **MODEL** The tree of life is composed of smaller monophyletic groups nested within larger monophyletic groups. Circle the branches in the following monophyletic groups in the phylogentic trees in **Figure 1**. (See Sections 30.2 and 32.5)

   animals
   bilaterians
   deuterostomes
   chordates
   tetrapods
   amniotes
   mammals

3. **MODEL** Add a dot on the tree to indicate the most recent common ancestor of trout, bullfrogs, otters, owls, hawks, ducks, snakes, and humans. (See BioSkills 13)

When sodium channels are blocked by TTX in newt predators, nerve impulses cease and the animals becomes paralyzed, sometimes fatally (see Unit 1 case study on pages 142–143).

4. Voltage-gated ion channels specific to sodium arose independently in just three groups: bacteria, cnidarians, and bilaterians. These independent origins are an example of:

a. convergent evolution
b. homology
c. adaptive radiation

5. MODEL Draw bars on the branches of the phylogenetic tree (Figure 1) to show the following events in the evolution of voltage-gated sodium channels:

- **Voltage-gated ion channels** (not specific to sodium) appear in the comon ancestor of choanoflagellates and animals.
- **Na_V1** (the first voltage-gated ion channel specific to sodium) appears in the common ancestor of bilaterians.
- **Four types of Na_V1** appear in the common ancestor of ray-finned fishes and tetrapods due to gene duplication and diversification.
- **Nine types of Na_V1** appear in the common ancestor of amniotes due to further gene duplication and diversification.

Channel	Response to TTX	Main tissue location
$Na_V1.1$	Sensitive	Central nervous system
$Na_V1.2$	Sensitive	Central nervous system
$Na_V1.3$	Sensitive	Central nervous system
$Na_V1.4$	Sensitive	Skeletal muscle (including breathing muscles)
$Na_V1.5$	Resistant	Cardiac muscle (heart)
$Na_V1.6$	Sensitive	Central and peripheral nervous system
$Na_V1.7$	Sensitive	Neurons that perceive pain
$Na_V1.8$	Resistant	Neurons that perceive pain
$Na_V1.9$	Resistant	Neurons that perceive pain

The differences among the genes that code for Na_V1 channels arose by random mutations.

6. During evolution, some taxa, such as echinoderms, lost some of their Na_V1 channels. Can the loss of a trait be adaptive? Provide an example.

7. SOCIETY In humans and other amniotes, chance mutations occurred in some versions of Na_V1 channels making them resistant to TTX. Other channels remained sensitive (see the table). Predict one cause of death in humans exposed to TTX.

8. Some garter snakes are able to eat toxic newts and survive. Predict which Na_V1 channel is key to this accidental innovation.

Some humans, such as patients with cancer or addicts suffering heroin withdrawal, experience chronic pain due to the constant firing of their pain receptors. Researchers are testing the use of TTX as a drug to relieve pain in these patients.

9. SOCIETY Propose which Na_V1 channel would be a good target of TTX drug treatment in humans with chronic pain.

10. PROCESS OF SCIENCE Propose which model organism you would use to study the effectiveness of TTX for pain treatment (see BioSkills 11). Defend your choice by referencing the relationships in the tree of life.

a. *Arabidopsis thaliana*
b. *Drosophila melanogaster*
c. *Caenorhabditis elegans*
d. *Mus musculus*
e. *Saccharomyces cerevisae*

TTX is only one of many toxins and other active compounds that are important to the physiology, ecology, and evolution of organisms throughout the tree of life.

11. MODEL Mark the origin of the following toxins on the tree of life in the Big Picture of Diversity (pages 722–723).

Nicotine
Tobacco and other angiosperms
(See Unit 6 case study on pages 836–837)

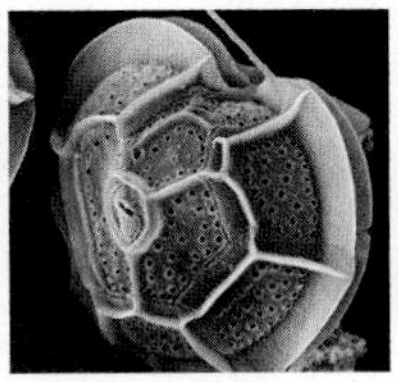

Saxitoxin
Cyanobacteria and dinoflagellates
Humans sometimes ingest by eating shellfish

Amatoxin
Death cap mushrooms and other fungi
Often mistaken for edible fungi

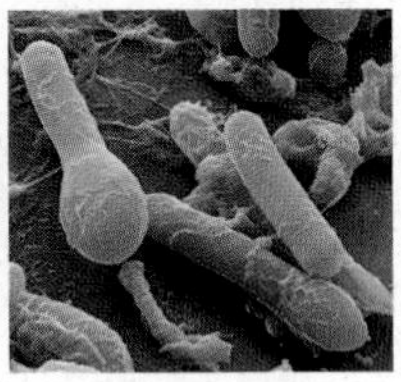

Botulinum toxin
Firmicute bacteria (Clostridium)
One "botox" variant is the most lethal toxin ever discovered

Your knowledge of the diversity of life can not only help you to solve the riddle of the toxic newt, but also help to solve problems that could have a big impact on human lives.

Each unit ends with a continuation of this story.

BIG PICTURE

This Big Picture shows the three-domain hypothesis, dividing life into the domains Bacteria, Archaea, and Eukarya. Most organisms on Earth are single-celled prokaryotes in the domains Bacteria and Archaea.

Only some of the many lineages of living organisms are included in this tree (see Chapters 26–32 for more details). You can use this Big Picture to practice your tree-thinking skills (see BioSkills 13). Also, be sure to do the blue exercises in the Check Your Understanding box below.

The Big Picture of Evolution (pp. 532–533) explains how the tree of life took shape. New branches are added when natural selection, genetic drift, and mutation occur in populations that are isolated by low levels of gene flow. Branches are "pruned" from the tree when extinction.

DIVERSITY OF LIFE

Viruses are enormously diverse and are important agents of organismal evolution, but are not themselves alive so are not included in the tree of life

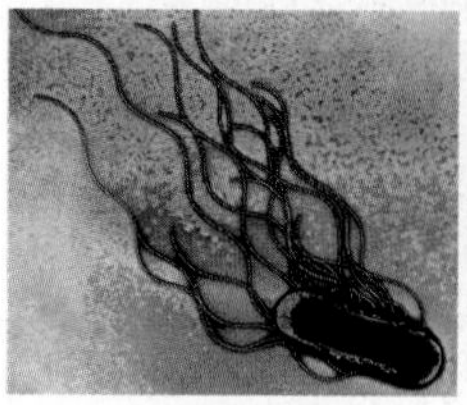

γ-Proteobacteria

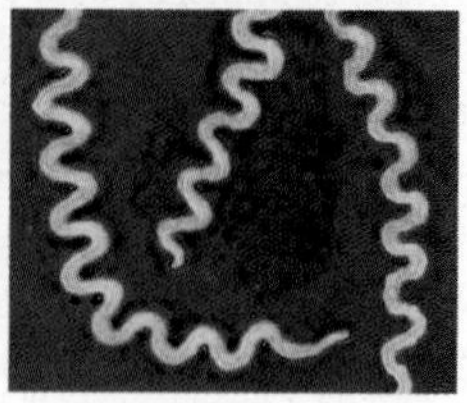

Spirochaetes

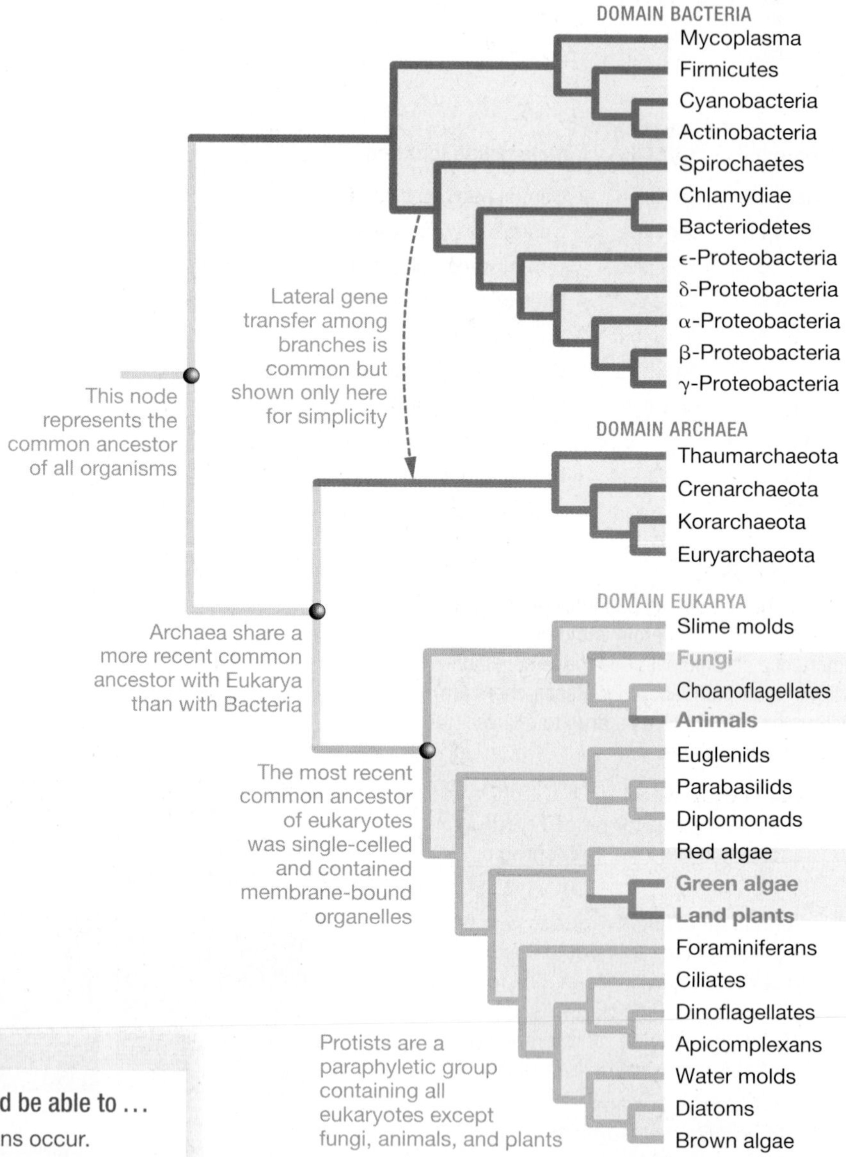

CHECK YOUR UNDERSTANDING

✔ **If you understand the big picture, you should be able to ...**

1. Circle the branches in the trees where humans occur.
2. In the tree on the left, draw an arrow from cyanobacteria to the root of plants to show the endosymbiosis event marking the origin of chloroplasts. Then draw an arrow from the α-proteobacteria to the root of Eukarya to show the origin of mitochondria.
3. Identify three examples of monophyletic groups in the trees and one example of a paraphyletic group.
4. Mark the origin of stinging cells in jellyfish (cnidarians).

Answers are available in Appendix A.

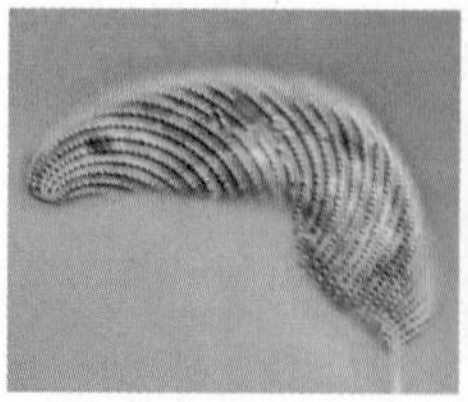

Euglenids

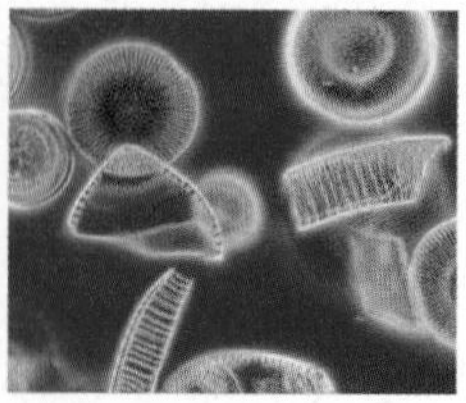

Diatoms

THE BIG PICTURE

Big Picture activities are available at **Mastering Biology**

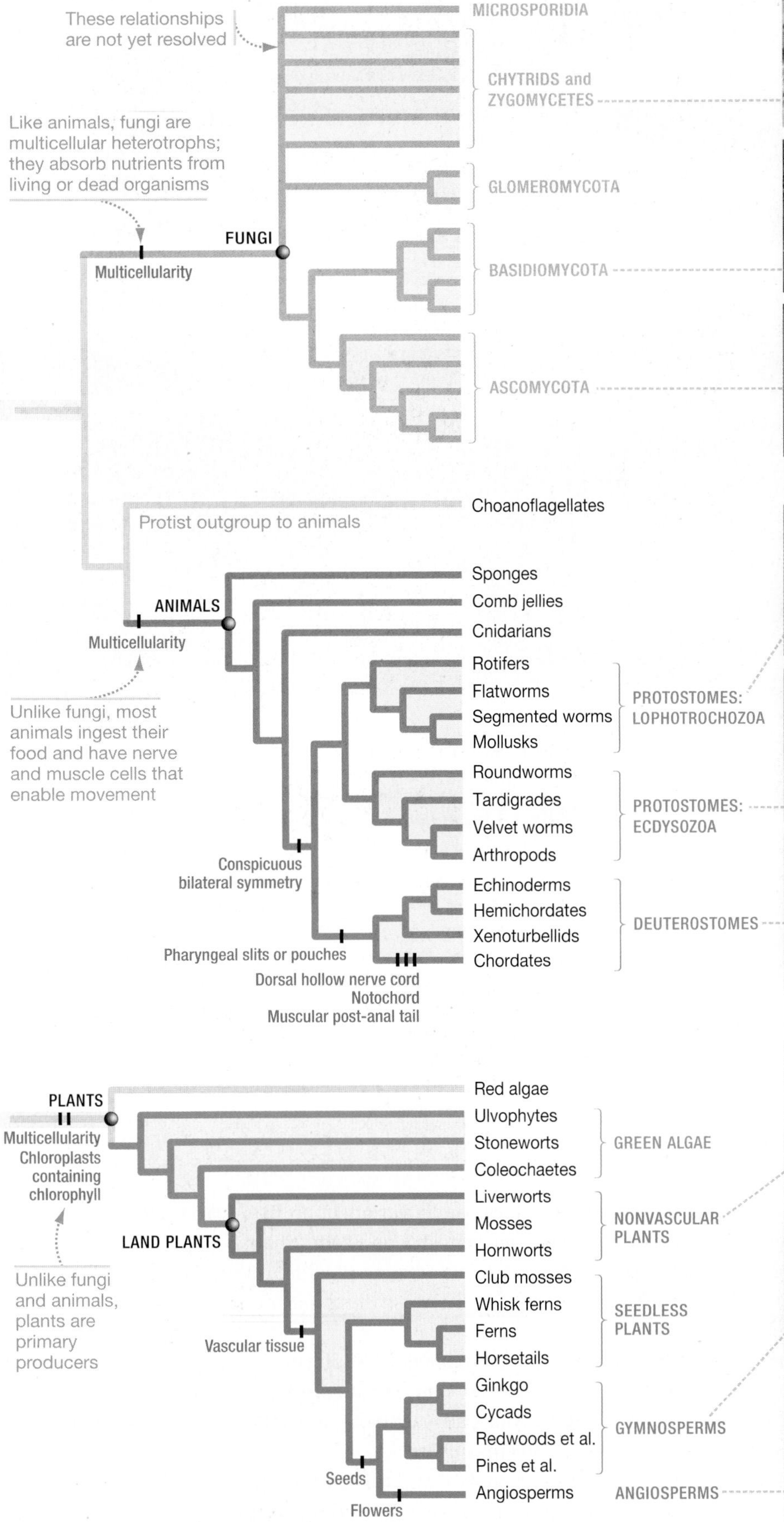

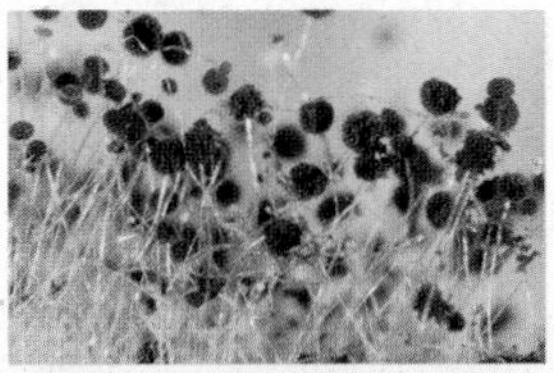

Zygomycetes
Have hyphae that yoke together and fuse; include many food molds

Basidiomycota
Terrestrial fungi that form spores on club-shaped basidia; include mushrooms, puffballs, and bracket fungi

Ascomycota
Form spores in a sac-like structure called an ascus; include morels, truffles, and yeast

Mollusks
The most diverse phylum of lophotrochozoans: about 85,000 described species including snails, clams, and octopuses

Arthropods
The most diverse phylum of ecdysozoans: over a million described species including millipedes, insects, lobsters, crabs, ticks, and spiders

Chordates
The most diverse phylum of deuterostomes: over 65,000 described species including vertebrates such as fishes, amphibians, reptiles, and mammals

Mosses
The most diverse lineage of nonvascular plants: over 12,000 described species, mostly in moist, terrestrial environments

Gymnosperms
An ancient group of seed plants: over 1000 described species including ginkgoes, cycads, redwoods, and pines

Angiosperms
The most diverse lineage of seed plants: about 300,000 described species including water lilies, roses, wheat, oak trees, and sunflowers

34 Plant Form and Function

Plants absorb water, nutrients, and sunlight from their surrounding environment in diverse ways. These giant water lilies (*Victoria amazonica*) are known for their exceptionally large leaves, which can be up to 9 meters in diameter.

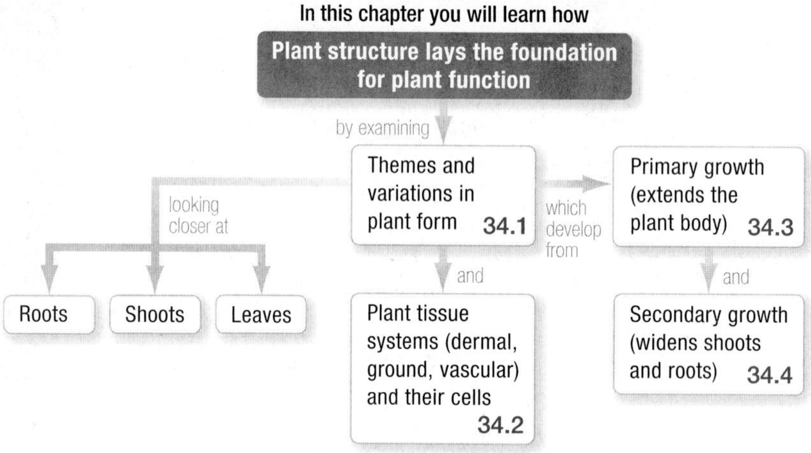

Land plants—along with algae, cyanobacteria, and a variety of protists—carry out photosynthesis, which is some of the most remarkable biochemistry on Earth (Chapter 10). Using the energy in sunlight and the simplest of starting materials—carbon dioxide, water, and specific key ions—plants synthesize thousands of different compounds. They use these compounds to build bodies that may live for thousands of years.

This feat is even more impressive when you consider that the starting materials that plants need to grow are tiny and diffuse, and are usually found at low concentrations over a large area. To gather the raw materials required for their sophisticated biosynthetic machinery, a plant's roots and shoots grow outward, extending the individual into the soil and atmosphere.

In essence, a plant's body harvests diffuse resources and concentrates them in cells and tissues. The structure of its body is dynamic, because most plants exhibit **indeterminate growth**; that is, unlike

BIG PICTURE

This chapter is part of the Big Picture. See how on pages 838–839.

animals, plants continue to grow throughout their lives. A bristlecone pine over 5000 years old has roots and shoots that are still growing. In response to favorable conditions, a plant sends shoots and roots in the most promising directions, seeking light and the simple compounds it requires.

The contrast between the plant and animal way of life is striking. Most animals move around, eat concentrated sources of food, and avoid stressful conditions. But plants stay in one place, extend their roots and shoots to harvest resources, make their own food, and cope with stress where they stand.

This chapter focuses on three fundamental questions:

1. How is the plant body organized?
2. Why are plants so diverse in size and shape?
3. How do plants grow throughout their lives?

Instead of surveying all of the major groups of plants, the focus here is on the flowering plants, or **angiosperms**. Recall that angiosperms are the most recent major group of plants to appear in the fossil record (Ch. 28, Section 28.4). With about 300,000 species, they are by far the most abundant, species rich, and geographically widespread plant group today. Also, their economic and medical importance to humans is difficult to overstate. Most of the food we eat and many of the medicines we use are derived from angiosperms.

If you look outside or down the produce aisle of a grocery store, you will see a wide diversity of plants and plant products. By the time you finish this chapter, you'll understand how plants are organized and how they grow. Exploring questions about the anatomy of flowering plants is vital to understanding the world at large as well as the other chapters in this unit. (You can review the importance of plant form and function in the Big Picture of plant and animal physiology on pages 838–839.)

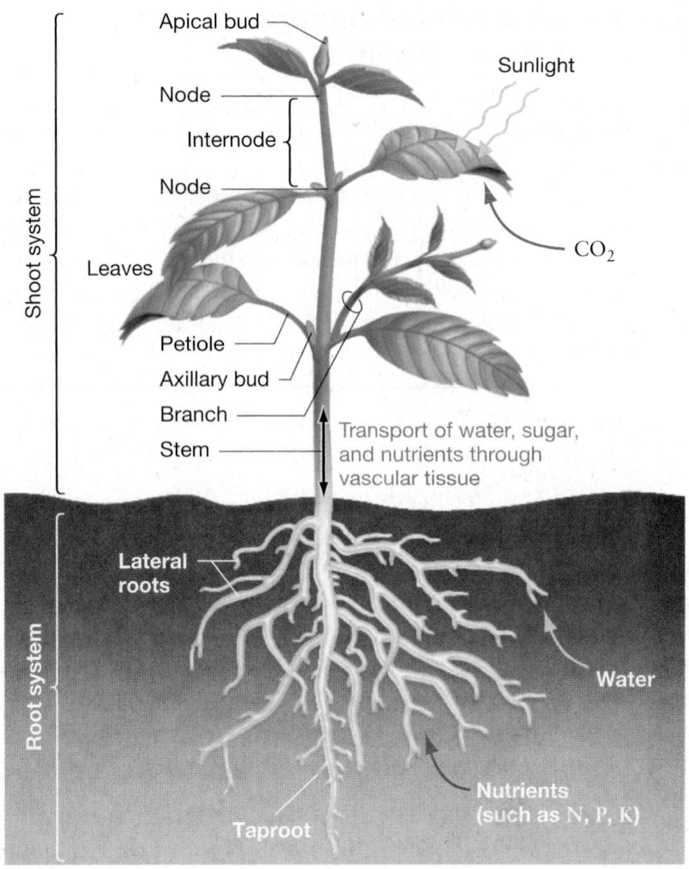

Figure 34.1 Typical Root and Shoot Systems Acquire and Transport Resources. Shoot systems are specialized for harvesting light and CO_2. Root systems absorb water and key nutrients such as nitrogen (N), phosphorus (P), and potassium (K).

✔ Suppose that this plant's growth was limited by access to light and nutrients, and that the plant had a new deposit of nutrient-rich soil to the left and much more sunlight suddenly available to the right. What would you expect this plant to look like in one month?

34.1 Plant Form: Themes with Many Variations

After you complete this section, you should be able to...

- Analyze the morphology of plant roots and shoots.

For photosynthesis to occur efficiently, plants need large amounts of light, carbon dioxide, and water. To synthesize the carbohydrates, proteins, nucleic acids, and lipids needed to build and maintain cells, plants must obtain nitrogen (N), phosphorus (P), potassium (K), magnesium (Mg), and a host of other nutrients. Most of these key elements exist in nature as ions dissolved in water found in soil.

Figure 34.1 identifies the major structures in the two basic systems that plants use to acquire the resources they need for photosynthesis. An aboveground portion called the **shoot system** harvests light and carbon dioxide from the atmosphere to produce sugars, while a belowground portion called the **root system** anchors the plant and takes in water and nutrients from the soil. As mentioned, both systems grow throughout the life of the individual, allowing the plant to increase in size, acquire resources, and reproduce. Collectively, the root and shoot systems make up the **plant body**.

Plant bodies can be quite diverse. Some plants, such as duckweed, an aquatic flowering plant, are small, typically just a few millimeters in length. Others are giants of the natural world. The Centurion tree, for example, is a eucalyptus tree with a height of nearly 100 m that is the tallest angiosperm in the world (redwoods are the tallest trees in the world, but they are gymnosperms, not angiosperms). Duckweed and the Centurion tree clearly have different challenges to growth and survival. The Centurion tree, for example, requires an extensive root system to support and sustain its shoot system, while duckweed does not.

In most plants, vascular tissue connects the root and shoot systems. Through vascular tissue, water and nutrients are transported from roots to shoots; sugars can be transported in both directions. As you learn about how plant form has evolved and serves necessary functions, compare and contrast what you learn with how animals have adapted to new environmental challenges. In many ways the two groups are remarkably similar.

The Importance of Surface Area to Volume Relationships

Before exploring the nature of root and shoot systems in more detail, it's important to recognize a key structural relationship that is critical to their function.

Root and shoot systems both function in absorption—roots absorb water and key nutrients, and shoots absorb light. Absorption takes place across a surface. But the cells that use the absorbed molecules and light occupy a volume. Thus, a plant body is more efficient as an absorption-and-synthesis machine when it has a large surface area relative to its volume.

Figure 34.2 illustrates this point. In this example, the cells in a plant are represented by cubes; the side of each cell is 50 μm long. Thus, each face of a cell has a surface area of

$$50 \times 50\ (\text{length} \times \text{width}) = 2500\ \mu\text{m}^2$$

and each cell has a volume of

$$50 \times 50 \times 50\ (\text{length} \times \text{width} \times \text{height}) = 125{,}000\ \mu\text{m}^3$$

Follow the calculations in the figure and note that the surface area to volume ratio of the entire structure changes depending on its shape. For example:

- If 64 cells are arranged in a cube, the surface area to volume ratio is 0.0300/μm.
- If 64 cells are arranged in a long tube, the surface area to volume ratio is 0.0425/μm.

This simple exercise has an important punch line: Tubes have a greater surface area to volume ratio than cubes with the same number and size of cells. If you calculate the ratio for the pictured flattened structure, you should see that this structure has even more surface area relative to volume than tubes.

It's no surprise, then, that the absorptive regions of a root system are tubelike, and the absorptive regions of a shoot system are flattened structures called leaves. Storage tissues such as tubers and seeds have a low surface area to volume ratio because they are not involved in absorption.

The Root System

Many root systems have a central **taproot**, as well as numerous **lateral roots** (see Figure 34.1). The root system anchors the plant in soil, absorbs water and ions from the soil, conducts water and selected ions to the shoot, obtains energy in the form of sugar from the shoot, and stores material produced in the shoot for later use.

Root systems can be impressive in extent. For example, in 1937 botanist Howard Dittmer grew a winter rye plant in a container full of soil for four months. He then unearthed the plant and meticulously measured its roots. The root system of this single individual contained more than 14 billion identifiable structures with a combined length of over 11,000 km—over one-fourth of Earth's circumference. The total surface area of the root system was almost 640 m^2, or about the size of 1.5 basketball courts. A root system like this is clearly adapted to absorbing diffuse resources located underground.

Other studies have shown that **(1)** the roots of trees routinely extend beyond the width of their aboveground canopy, and **(2)** it is not unusual for a plant's root system to represent over 80 percent of its total mass. Many plants devote a great deal of energy and resources to the growth of their root systems.

Although most root systems contain the same general structures, the root systems are diverse. This diversity can be analyzed on three levels:

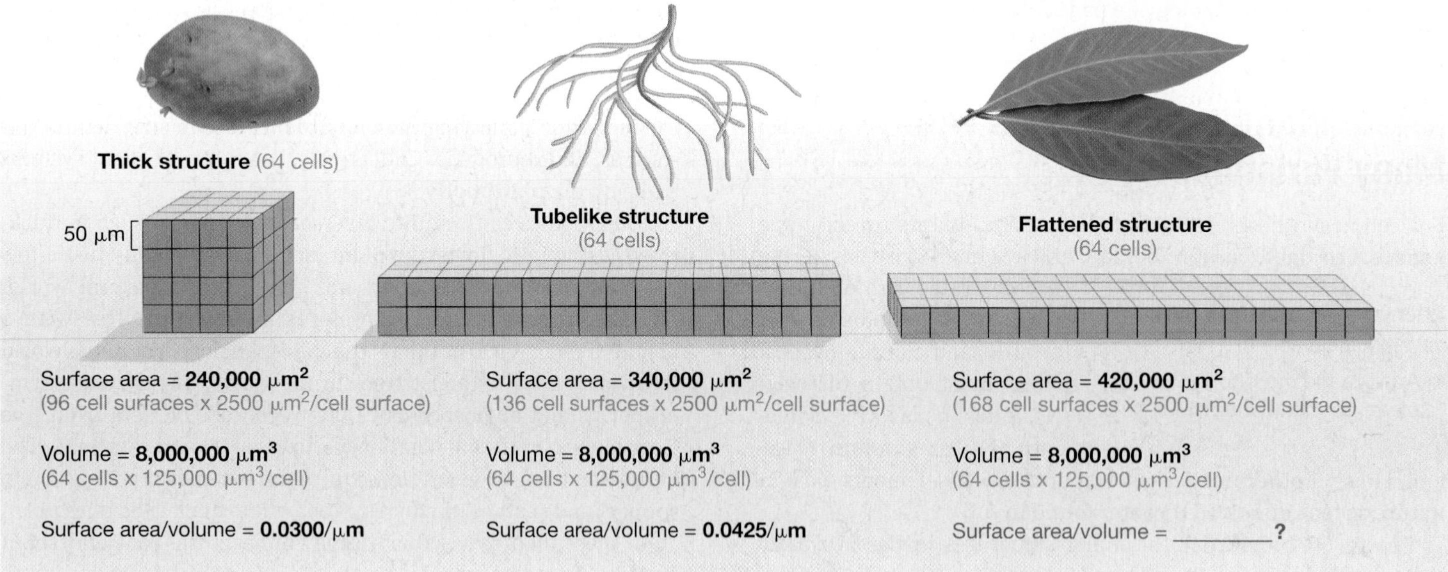

Figure 34.2 **The Morphology of Roots and Leaves Gives Them a High Surface Area to Volume Ratio.** In this example, the "thick structure" represents a potato-like storage organ; the "tubelike structure" represents a root; the "flattened structure" represents a leaf. Note that each schematic structure has the same number of cells (each represented as a small cube) and the same total volume—but a very different surface area.

✔ **QUANTITATIVE** What is the surface area to volume ratio for the flattened structure?

1. morphological diversity among species;
2. phenotypic plasticity, or changes in the structure of an individual's root system in response to the environment; and
3. modified roots that are specialized for unusual functions.

Let's consider each level in turn.

Morphological Diversity in Root Systems To appreciate the range of morphological diversity observed in the root systems of angiosperms, consider prairie plants.

Prairies are grassland ecosystems found in areas of the world such as central North America, the Serengeti Plain of East Africa, the Pampas region of Argentina, and the steppes of central Asia (Ch. 49, Section 49.4). Rain is abundant enough in these areas to support a lush growth of **herbaceous** plants—seed plants that lack woody tissue—yet scarce enough to exclude trees and most shrubs. The growth of woody species is also discouraged by fires that regularly sweep through these ecosystems.

Although the aboveground portions of prairie plants burn during fires and die back during the winter or dry season, their root systems are **perennial**, meaning that they live for many years. The root system sends up a new shoot system each spring and also after a fire. Figure 34.3 shows that the root systems of prairie plants can be very different, even if they live next to each other. For example, the dense, fibrous root systems of junegrass and switchgrass do not have a taproot, whereas the taproot of a compass plant can reach depths of over 4.5 m. As a result, competition for water and nutrients is minimized between these species.

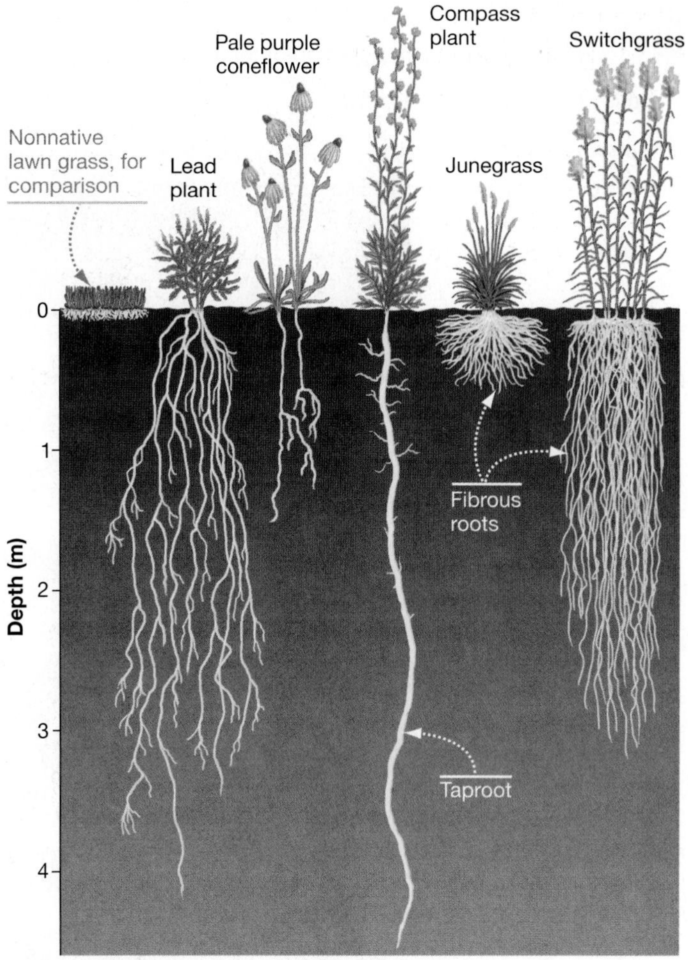

Figure 34.3 **Plants Have Diverse Root Systems.** The roots of prairie plants that live side by side can be very different.

✔ Which prairie plants would be more likely to absorb rain when it first enters soil? Which would be most likely to absorb water and continue growing during periods of drought?

Phenotypic Plasticity in Root Systems Morphological diversity in roots occurs within species as well as among species. Some of the within-species variation is due to genetic diversity among individuals, but some is due to how roots respond to the environment.

Roots show a great deal of **phenotypic plasticity**—meaning that their form is changeable, depending on environmental conditions. For example, spruce trees in waterlogged soils tend to have root systems less than a meter deep. Their roots are shallow because wet soil lacks oxygen, and plant root cells, like animal cells, suffocate in anoxic conditions. The same tree growing in drier soil would develop a root system several meters deep. The key point is that even genetically identical individuals may have very different root systems if they grow in different environments.

Phenotypic plasticity is particularly important in plants because they grow throughout their lives, and sometimes their environment changes. Roots actively grow into areas of soil where resources are abundant, such as nutrient-rich septic fields or sewer pipes that leak human waste. Roots stop growing or die back in areas where resources are used up or lacking.

Modified Roots The taproots and fibrous roots illustrated earlier do not begin to exhaust the types of roots found among plants. For example, some roots are **adventitious**—meaning they develop from an unusual source, the shoot system instead of the root system.

- In ivy, **anchor roots**—adventitious roots that grow from nodes in the shoot system (see Figure 34.1)—help anchor individuals to brick walls or other structures.
- The **prop roots** of corn are adventitious roots that help brace individuals in windy weather.

Even roots that are part of the root system can have specialized functions—meaning that they do things other than absorb water and nutrients and anchor the shoot system.

The **pneumatophores** of mangroves in the genus *Avicennia* are specialized lateral roots that function in gas exchange. These mangroves grow in submerged habitats where fine silt is deposited, cutting off the oxygen supply to their roots. Their root cells do not suffocate, however, because oxygen from the atmosphere can diffuse into the root system through the pneumatophores. These roots grow upward—not downward—in response to gravity.

The thick taproot of some biennial plants, such as carrots and beets, stores carbohydrates during the first of the plants' two growing seasons. These taproots are known as **storage roots**. Carrot roots—the "carrots" you might have eaten for lunch—are typically harvested at this point. Sugar beets are harvested to produce much of the sugar commonly found in grocery stores. If the roots were not harvested, the plants would resume growth after

SUMMARY Table 34.1 Form and Function of Modified Roots

Modified Root	Example
Anchor roots These adventitious roots anchor stems to walls and to other plants	
Prop roots These adventitious roots stabilize the stem	
Pneumatophores Pneumatophores allow gas exchange between roots and atmosphere	
Storage roots Sugar beets store carbohydrates and other nutrients in their taproot for future use	

✔ Why do root cells need oxygen?

the winter and use that stored energy during a second summer to "go to seed," producing a large flowering shoot. Carbohydrates in the taproot make the energy-intensive reproductive process possible.

See **Table 34.1** for some examples of modified roots.

The Shoot System

As Figure 34.1 indicates, the shoot system has an array of important anatomical features.

- The shoot system consists of one or more **stems**, which are vertical aboveground structures.
- A stem consists of **nodes**, where leaves are attached, and **internodes**, or segments between nodes.
- A **leaf** is an appendage that projects from a stem laterally on a stalk called the **petiole**. Leaves usually function as photosynthetic organs.
- The nodes where leaves attach to the stem are also the site of **axillary (or lateral) buds**, which form just above the site of leaf attachment.
- If conditions are appropriate, an axillary bud may grow into a **branch**—a lateral extension of the shoot system.
- The tip of each stem and branch contains an **apical bud**, where growth occurs that extends the length of the stem or branch.
- If conditions are appropriate, apical or axillary buds maydevelop into flowers or other reproductive structures.

As with root systems, diversity in shoots can be analyzed on three levels: morphological diversity among species, phenotypic plasticity within individuals, and modified shoots with specialized functions.

Morphological Diversity in Shoot Systems The shoot systems of angiosperms, essentially the visible part of the plant, range in size from species like the tiny ($<$ 5 mm diameter) duckweed that you may have seen growing on the surface of stagnant ponds to the Centurion tree with a height of nearly 100 m (300 ft). Even more impressive in stature are some gymnosperms: redwood trees reach heights of over 100 m (300 ft), and giant sequoias can have trunks that weigh 2.6 million kg (over 5.7 million lb)—about the same as 10 diesel locomotives.

Variation in the size and shape of the shoot system is important: It allows plants of different species to harvest light at different locations and thus minimize competition. It also allows plants to thrive in a wide array of habitats.

As an example of how the shape of a shoot system varies among species in different environments, consider the silversword plants native to Hawaii. You might recall that all silverswords are believed to be descended from the same ancestor—a species of tarweed that arrived in Hawaii from North America about 5 million years ago (Ch. 25, Section 25.3).

Silverswords represent an adaptive radiation: a lineage that rapidly split into many species occupying a wide array of habitats. Their shoot systems are particularly diverse in size, shape, and growth habit. Some silverswords grow low to the ground in dense mats; some form bunched rosettes of leaves; some are vines; others are woody shrubs or even small- to medium-sized trees.

Biologists interpret this diversity of shoot systems as a suite of adaptations for harvesting light and carbon dioxide in different environments.

- In lush habitats, where competition for light is intense, woody individuals that grow tall are favored by natural selection.
- In dry, windblown habitats, individuals with short stems or rosettes thrive because they require less water than taller individuals, and they don't blow over.

The adaptive radiation of silverswords has been based in part on diversification in their shoot systems.

Phenotypic Plasticity in Shoot Systems The size and shape of an individual's shoot system can vary dramatically based on variation in growing conditions: temperature, exposure to wind, and availability of water, nutrients, and light.

This conclusion was driven home in an experiment conducted by Jens Clausen and colleagues in the late 1930s. They transplanted several species of herbaceous plants between sites along an elevational gradient: from sea level to alpine habitats. In each case, the transplanted individuals were propagated from cuttings—meaning that they were genetically identical to individuals growing at the other locations. As the "Results" section in Figure 34.4 shows, the overall size and shape of the shoot system varied markedly among locations. These results indicate that genetics *and* environment affect overall plant form.

Because a plant's shoot system continues to grow over the course of its lifetime, it can respond to changes in environmental conditions just like the root system can. Experiments on phototropism, for example, established that shoot systems can bend toward light if an individual is shaded on one side (Ch. 37, Section 37.2). Plants also undergo differential growth, producing more branches and leaves in regions of the body that are exposed to the highest light levels. A plant's shoot system grows in directions that maximize its chances of capturing light.

Modified Stems As in root systems, the stems of many plants have modified structures to serve various functions.

- Many desert cacti have highly modified stems. Instead of functioning primarily to support leaves, cactus stems often enlarge into water-storage organs. Water accounts for up to 98 percent of the weight of a cactus stem. Cactus stems also contain the plant's photosynthetic tissue.

RESEARCH

QUESTION: How much does a plant's growth form depend on its environment?

HYPOTHESIS: (No explicit hypothesis—the goal of this experiment was to explore the interaction between genetic makeup versus environmental influence on size and shape.)

EXPERIMENTAL SETUP:

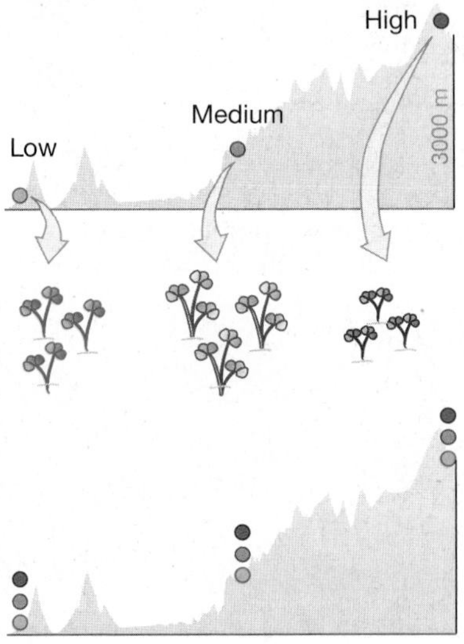

1. Take cuttings from individuals of *Potentilla glandulosa* growing in low-, medium-, and high-elevation habitats in the Sierra Nevada mountains.

2. Propagate genetically identical individuals.

3. Transplant individuals from each source population into each habitat (low, medium, and high elevation). Allow to grow and observe mature plants.

PREDICTION OF HYPOTHESIS: (No explicit predictions.)

RESULTS:

Examples of mature plants observed:

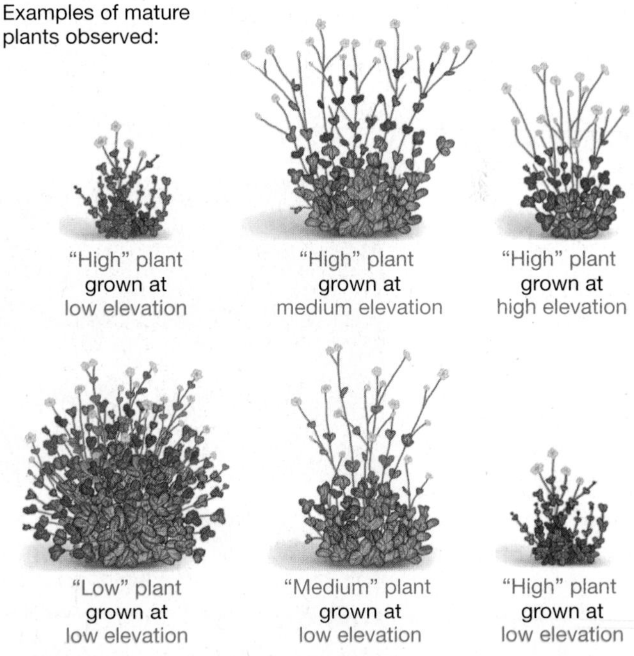

CONCLUSION: Environmental conditions have a profound influence on body size and shape (genetically identical plants look different at each site). But genetic makeup also has a large influence on plant morphology (plants from the different source populations looked different, even when all were grown in the same habitat).

Figure 34.4 Experimental Evidence Supports Phenotypic Plasticity in Shoot Systems.

SOURCE: J. Clausen, D. D. Keck, and W. M. Hiesey. 1940. *Experimental studies on the nature of species. I. Effect of varied environments on western North American plants*. Publication 520. Washington, DC: Carnegie Institution.

✔ **PROCESS OF SCIENCE** Why was it important for the researchers to propagate the individuals from cuttings before transplanting them?

SUMMARY Table 34.2
Form and Function of Modified Stems

Modified Stem	Example
Water-storage structures Cactus stems store water; the spines are modified leaves (the example shown is a longitudinal section)	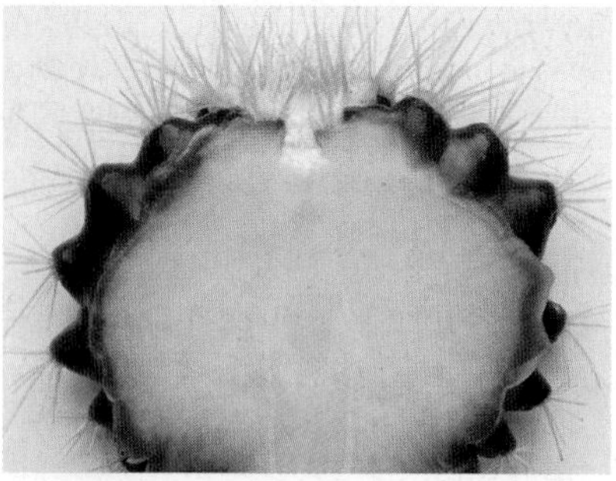
Stolons Strawberry stolons produce new individuals at nodes aboveground	
Rhizomes Rhizomes produce new individuals at nodes belowground, and store carbohydrates	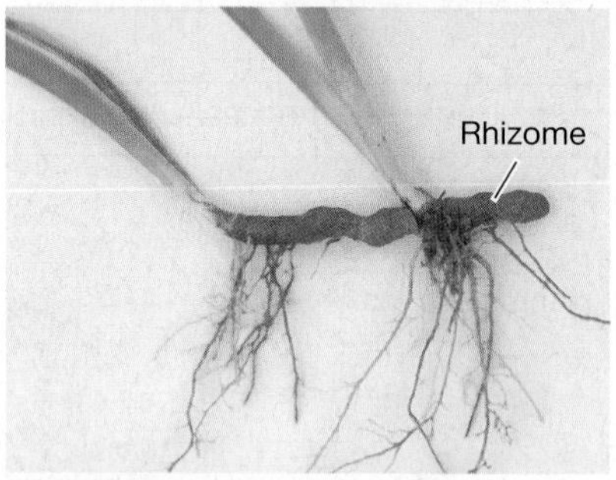
Tubers Tubers such as potatoes store carbohydrates such as starch	
Thorns Thorns provide protection from herbivores	

- **Stolons** are modified stems that grow horizontally along the soil surface, producing adventitious roots and leaves at each node. Because new plants form at these nodes, stolons function in asexual reproduction (Ch. 38, Section 38.1).
- Like stolons, **rhizomes** are stems that grow horizontally instead of vertically. They produce new plants at nodes and thus participate in asexual reproduction. But while stolons grow aboveground, rhizomes spread belowground. One quaking aspen plant in Utah was found to have 47,000 stems rising from a network of rhizomes, making it one of the largest organisms known. What we see as individual aspen trees are actually stems that are connected underground.
- **Tubers** are underground, swollen rhizomes that function as carbohydrate-storage organs. The eyes of a potato—a typical tuber—are nodes in the stem where new branches may arise.
- **Thorns** are modified stems that help protect the plant from attacks by large **herbivores**, or plant eaters, such as deer, giraffe, or cattle.

See **Table 34.2** for some examples of modified stems

The Leaf

In most plant species, the vast majority of photosynthesis occurs in the part of the shoot system called the leaf. The total area of leaf produced by a single plant can be enormous—a single tree can have hundreds of thousands of leaves with a total surface area equivalent to that of a football field. All of this area is available for absorbing photons and supporting photosynthesis.

A **simple leaf** (**Figure 34.5a**) is composed of just two major structures: an expanded portion called the **blade** and the stalk of the petiole. But leaves exhibit many variations on the central theme of a flattened structure specialized for performing photosynthesis. (Ch. 38, Section 38.5 explores leaf development more fully.)

Morphological Diversity in Leaves Glance outside or stroll through a garden, and you'll find many types of simple leaves with an easily recognizable blade and petiole. (Grass leaves will stump you, though, because they lack petioles entirely.) You will also find **compound leaves** that have blades divided into a series of leaflets (**Figure 34.5b**). You may even encounter doubly compound leaves, which have leaflets that are again divided (**Figure 34.5c**). Although a compound leaf easily can be mistaken for several individual leaves, the defining feature of an individual leaf (whether simple or compound) is the presence of an axillary bud where the petiole joins the stem (see Figure 34.1).

Not all leaf blades are thin with a large surface area, however. For example, plants that thrive in deserts or in cold, dry habitats tend to have needlelike leaves (**Figure 34.5d**). The leading hypothesis to explain this pattern is based on two observations: **(1)** Water is often in short supply in these environments because it is scarce in deserts, or frozen and thus unavailable in cold habitats; and **(2)** leaves with large surface areas lose large amounts of water through an evaporative process called **transpiration** (Ch. 35, Section 35.2). Thus, needlelike leaves

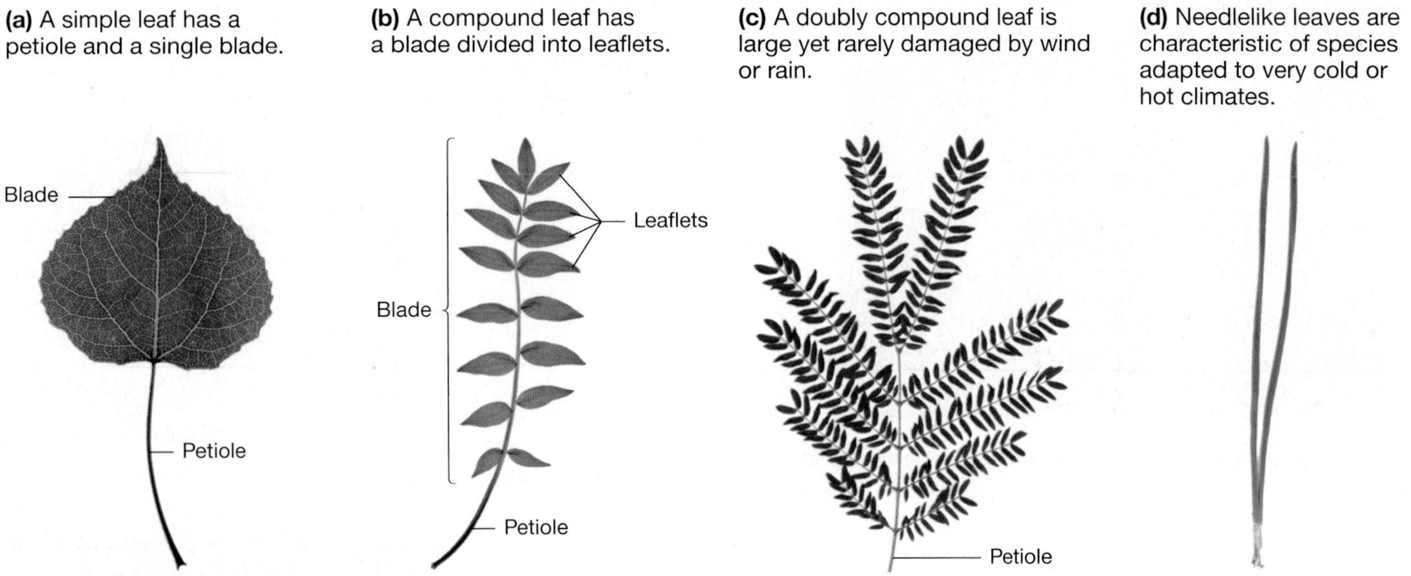

Figure 34.5 Leaves Vary in Size and Shape. The structure in **(a)** shows a simple leaf, while **(b)** and **(c)** show examples of compound leaves. Part **(d)** shows two needlelike leaves.

✔ **THINK CAREFULLY** How many leaves are shown in **(b)** and **(c)**?

are interpreted as adaptations that minimize transpiration in water-scarce habitats. Small, narrow leaves are also much less susceptible to wind damage than are large, broad leaves.

The arrangement of leaves on a stem can vary as much as leaf shape. For example, leaves can be

- arranged to alternate on either side of the stem (Figure 34.6a);
- paired opposite each other on the stem (Figure 34.6b);
- arranged in a whorl (Figure 34.6c);
- found in a compact arrangement where internodes are extremely short—leading to a rosette growth form (Figure 34.6d).

Clearly, leaves are arranged on stems in many ways. Despite this variation, the development and arrangement of leaves follow well-defined patterns that tend to be species specific. Leaf shape and the arrangement of leaves on a stem are usually determined genetically, so these characters are often used for plant identification.

Phenotypic Plasticity in Leaves Even though leaves do not grow continuously, they exhibit phenotypic plasticity just like root systems and stems do. Leaves from the same individual that grow in sun versus shade serve as a prominent example of phenotypic plasticity in leaf morphology. As the oak tree leaves in

Figure 34.6 The Arrangement of Leaves on Stems Varies.

Figure 34.7 **Leaves Exhibit Phenotypic Plasticity.** These leaves came from the same tree.

Figure 34.7 show, the leaf grown in the sun is much smaller than the leaf grown in the shade:

- *Sun leaves* typically are thicker and have a relatively small surface area, which reduces water loss in areas of the body where light is abundant.
- *Shade leaves* are relatively thin and broad, providing a high surface area that maximizes absorption of rare photons.

Water loss is less of a problem for shade leaves because temperatures are cooler in shade than in bright sun. ✔ **If you understand the concept of phenotypic plasticity, you should be able to explain why phenotypic plasticity in roots and shoots is expected to be more important (1) in environments where conditions are variable versus stable, and (2) in long-lived versus short-lived species.**

Modified Leaves Not all leaves function primarily in photosynthesis; some perform other roles (**Table 34.3**).

- Onion bulbs consist of thickened leaf bases that store nutrients, separated by highly condensed internodes.
- The thick leaves of plants called succulents, such as aloe vera, store water.
- The tendrils that enable plants, such as a grapevine, to climb are modified leaflets or leaves.
- The bright red leaves of poinsettias attract pollinators to the tiny yellow flowers that they surround.
- The hooded, tubelike leaves of the pitcher plant trap insects. When insects enter, they feed on the plant's nectar and appear to become dizzy. Eventually they fall into the bottom of the tube and drown in water that has accumulated. The insects are then digested by bacteria or enzymes secreted by the plant, and the plant absorbs the nutrients released.
- Cactus spines (the top image in Table 34.2) are modified leaves that protect the stem.

The variability of plant roots, stems, and leaves is impressive. Diversity, plasticity, and dynamism are recurring themes in the study of plant anatomy. In Section 34.2 you will explore the diversity of cells and tissues found within plants.

SUMMARY **Table 34.3**
Form and Function of Modified Leaves

Modified Leaves	Example
Bulbs Onion leaves store food	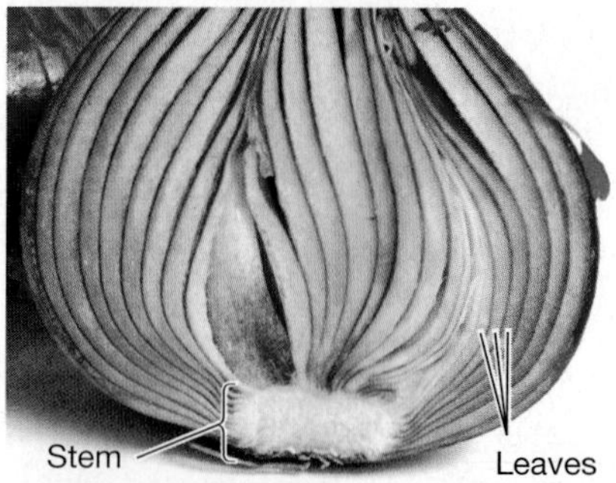
Succulent leaves Aloe vera leaves store water	
Tendrils Pea tendrils aid in climbing	
Floral mimics Red poinsettia leaves attract pollinators	
Traps Pitcher plant leaves trap entering insects, use their hood to discourage insects from flying out, and then digest them	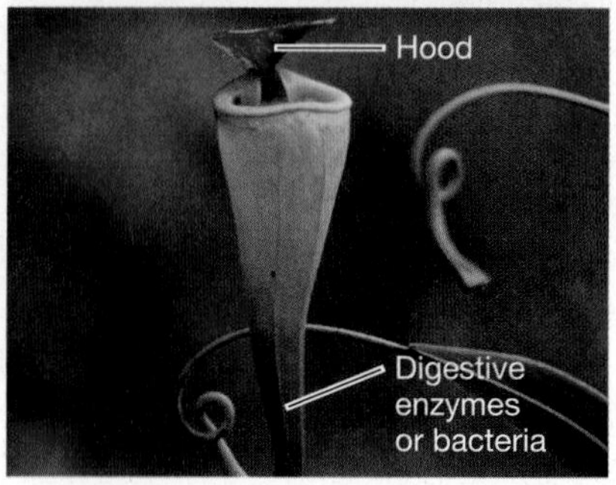

CHECK YOUR UNDERSTANDING

✔ **If you understood this section, you should be able to...**

1. **MODEL** Diagram a generalized version of the angiosperm (flowering plant) body, labeling each major part.
2. Provide two examples each of root systems, stems, and leaves that differ in structure and/or function from the generalized plant body shown in Figure 34.1.
3. *Bryophyllum daigremontianum* ("mother of thousands") produces hundreds of asexual baby plantlets that are genetically identical to the parent plant and can grow into mature plants. Predict whether plantlets obtained from a single parent plant, but grown under different environmental conditions, would likely show the same phenotypes.

Answers are available in Appendix A.

34.2 Plant Cells and Tissue Systems

After you complete this section, you should be able to...

- Identify the structural features that distinguish various plant tissues.

Recall that all eukaryotic cells, including those of plants and animals, share most of their key characteristics: chromosomes enclosed in a nuclear envelope, a plasma membrane studded with proteins that regulate the passage of materials in and out, mitochondria that produce ATP by oxidizing sugars, and an array of other organelles that synthesize or degrade key molecules (Ch. 7, Section 7.2).

In addition, plant cells have several features that are absent in animal cells (Figure 34.8):

- All plant cells are surrounded by a cellulose-rich **primary cell wall** that supports the cell and defines its shape. Furthermore, some plant cells go on to form a **secondary cell wall**, which is often hard and rigid, interior to the primary cell wall.
- The cytoplasm of adjacent plant cells is often connected via **plasmodesmata** (singular: **plasmodesma**) (Ch. 11, Figure 11.11b). Plasmodesmata consist of cytoplasm and segments of smooth endoplasmic reticulum (smooth ER) that run through tiny, membrane-lined gaps in the cell wall (see Figure 34.8b).
- Plant cells often contain several types of organelles that are not found in animals—specifically chloroplasts and a large, membrane-bound organelle called a vacuole, which fills most of the cell's volume.

Chloroplasts are, all importantly, the site of photosynthesis (Ch. 10, Section 10.1). Non-photosynthetic cells found in roots, seeds, flower petals, and other locations may have organelles that are similar to chloroplasts but are specialized for storing pigments, starch, oils, or proteins.

Vacuoles store various compounds, including wastes, and in some cases they digest wastes, as do animal lysosomes. In addition, plant vacuoles store water and nutrients. They may also hold pigments to yield color, or poisons to deter herbivores.

Another important distinction between plant cells and animal cells is that plant cells do not change position once they form—they remain "locked in place" during their entire existence. Some animal cells migrate within the body either early in the development of an individual or as mature (differentiated) cells.

Like animal cells, plant cells with similar structure are often associated and form distinct tissues. A **tissue** is a group of cells that functions as a unit. Plant tissues that consist of a single

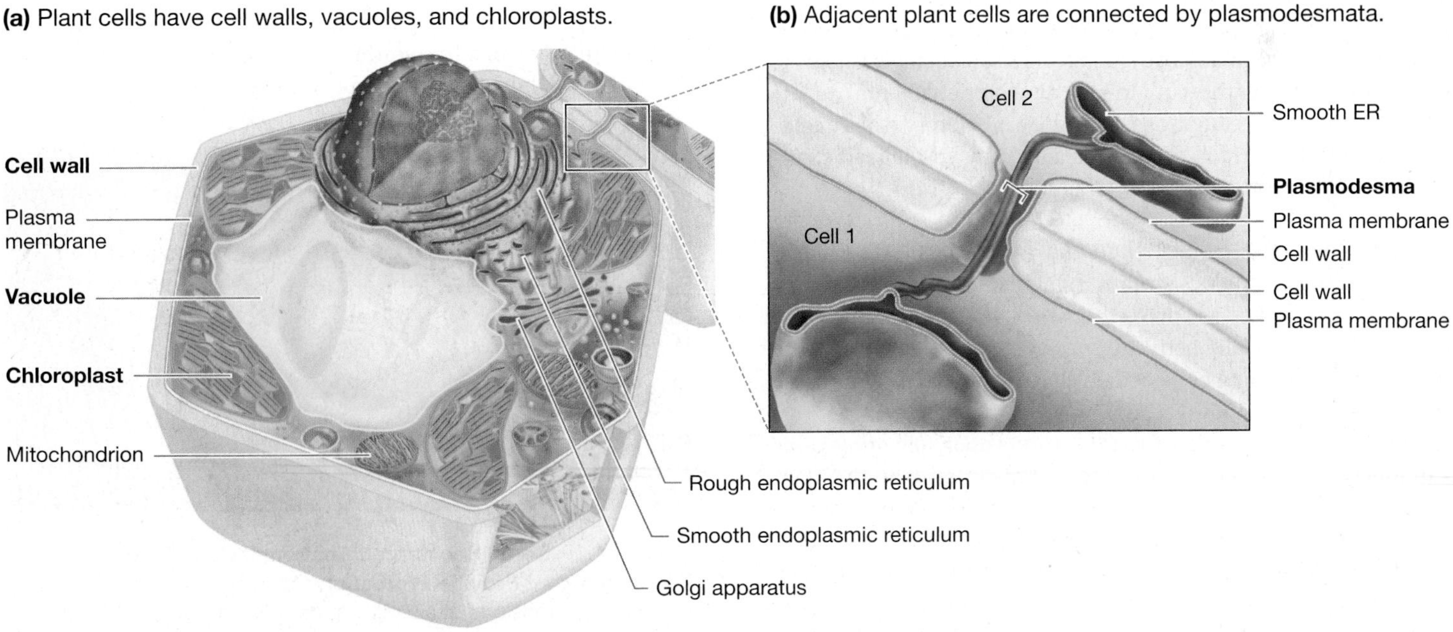

Figure 34.8 A Plant Cell Is Both Similar to and Different from an Animal Cell. Plant vacuoles are similar to animal lysosomes; however, the cell wall, chloroplasts, and plasmodesmata are unique to plants.

type of cell are called **simple tissues**; tissues that contain several cell types are termed **complex tissues**. Moreover, plant tissues are grouped together into broader categories, referred to as **tissue systems**, based on their structural features and location within the plant. The three tissue systems found in plants are **(1)** the dermal tissue system, **(2)** the ground tissue system, and **(3)** the vascular tissue system. Let's examine each of these in turn.

The Dermal Tissue System

The **dermal tissue system** consists of dermal tissue (literally, "skin" tissue). This tissue, also called the **epidermis**, is the outermost layer of cells that forms the interface between the organism and the external environment. In shoots, its primary function is to protect the plant—from water loss, disease-causing agents, and herbivores. In roots, the epidermis, which includes root hairs, functions primarily in absorbing water and nutrients.

The epidermis is made up of several different cell types, and therefore is a complex tissue. Let's consider the functional significance of the cell types found in dermal tissue.

Epidermal Cells Protect the Surface Most shoot epidermal cells are flattened and usually lack chloroplasts. Epidermal cells in the shoot system fulfill their protective role in part by secreting the **cuticle**—a waxy layer that forms a continuous sheet on the surface of leaves and stems (Ch. 28, Section 28.3). Waxes are lipids and are hydrophobic. As a result, the cuticle reduces the amount of water that stems and leaves lose by evaporation.

From a human perspective, the water-repellent properties of some cuticles make them a valuable ingredient in polishes and lipsticks. For example, carnauba wax, used in car and floor polishes, is secreted by epidermal cells in the leaves of carnauba palms native to Brazil.

Besides minimizing water loss, the cuticle forms a barrier to protect the plant from viruses, bacteria, and the spores or growing hyphae of parasitic fungi. In this way, the plant epidermis forms the first line of defense against disease-causing pathogens.

Waxes found in the cuticle can also be detrimental to the plant, however, by reducing gas exchange in photosynthetic cells. This potentially serious problem is solved by specialized structures, called stomata, in dermal tissue.

Stomata Regulate Gas Exchange and Water Loss Most plants have structures called **stomata** (singular: **stoma**). Typically found on leaves, they are pores that allow CO_2 to enter and O_2 to exit photosynthetically active tissues. A stoma is surrounded by two specialized **guard cells**, which change shape to open or close the pore (Figure 34.9).

When stomata are open, CO_2, O_2, water vapor, and other gases can move between the atmosphere and the interior of the plant by diffusion. Stomata open when adequate water is available and CO_2 is needed for photosynthesis. When stomata are open, water evaporates from the moist interior of the leaf to the surrounding atmosphere. Stomata close when conditions are dry. This prevents large amounts of water from being lost during transpiration.

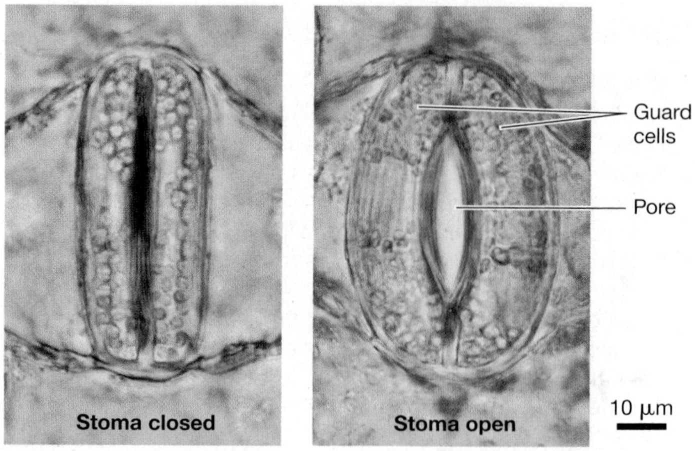

Figure 34.9 **A Stoma's Guard Cells Regulate the Opening of a Pore.**

Trichomes Perform an Array of Functions In addition to minimizing water loss and regulating gas exchange in shoots, cells in dermal tissue protect the individual from the damaging effects of intense sunlight and attacks by herbivores.

Trichomes (pronounced *TRY-komz*) are hairlike appendages made up of specialized epidermal cells. They are found in shoot systems and come in an array of shapes, sizes, and abundances. Depending on the species, trichomes may **(1)** keep the leaf surface cool by reflecting sunlight; **(2)** reduce water loss by forming a dense mat that limits transpiration; **(3)** provide barbs or store toxic compounds that thwart herbivores (Figure 34.10); or **(4)** trap and digest insects.

The Ground Tissue System

Most photosynthesis, as well as most carbohydrate storage, takes place in the most common of the tissue systems, the **ground tissue system**. Cells in ground tissue are also responsible for most of the synthesis and storage of specialized products such as colorful pigments, hormones, and toxins required for defense. Keep in mind, however, that some toxins are stored in trichomes,

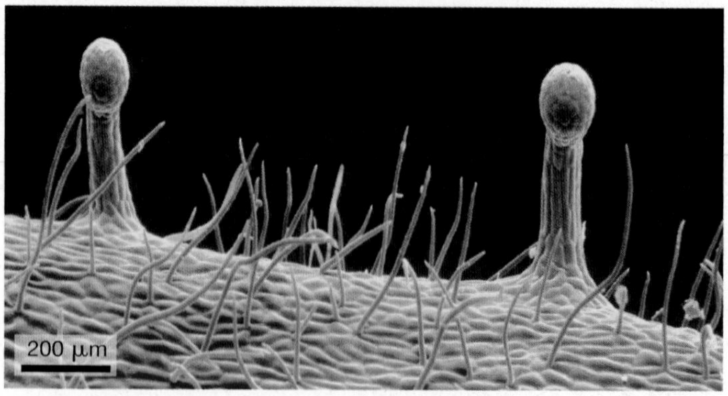

Figure 34.10 **Some Specialized Epidermal Cells Produce Trichomes That Provide Protection.** Some trichomes on this leaf are hairlike extensions of a single epidermal cell, while others are multicellular structures. The structures that are colored orange here hold toxins.

(a) In leaves: photosynthesis and gas exchange

(b) In roots: carbohydrate storage

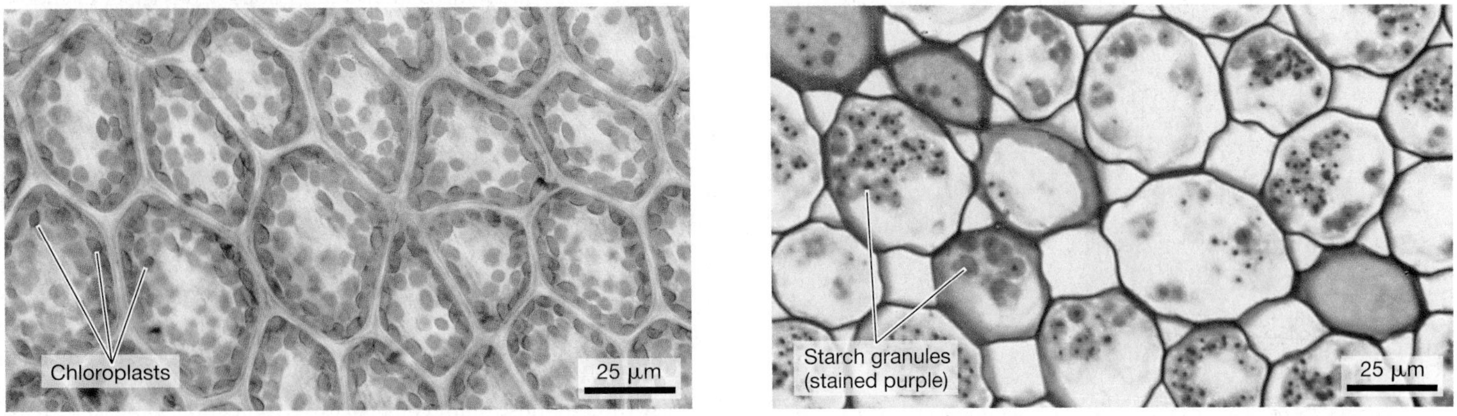

Figure 34.11 **Parenchyma Cells May Perform Photosynthetic or Storage Functions.**

✔ Give three possible examples of genes encoding proteins that are likely to be expressed in leaf cells, but not in root cells.

part of the dermal tissue system. Some ground tissue cells also play a large role in structural support of the shoot system.

The ground tissue system consists of three distinct tissues: parenchyma (pronounced *pa-REN-ki-ma*), collenchyma (*ko-LEN-ki-ma*), and sclerenchyma (*skle-REN-ki-ma*).

Parenchyma: Simple Tissue with Cells That Perform a Wide Array of Tasks **Parenchyma** tissue is composed of parenchyma cells, which have relatively thin primary cell walls and are the most common and versatile ground tissue cells. Parenchyma represents a simple tissue. The parenchyma tissue in leaves consists of parenchyma cells filled with chloroplasts, and it is the primary site of photosynthesis (Figure 34.11a). But in other organs, parenchyma cells store starch granules (Figure 34.11b). When you eat a potato or an apple, you are ingesting primarily parenchyma cells.

Many parenchyma cells are **totipotent**, meaning they retain the capacity to divide and develop into a complete, mature plant (much like embryonic stem cells in animals). The totipotency of parenchyma cells is important in healing wounds and in reproducing asexually via stolons or rhizomes. In each case, parenchyma cells may begin to divide, grow, and differentiate to form new roots and shoots.

The totipotency of parenchyma cells also allows gardeners to clone plants by making cuttings. For example, if you cut a piece of stem from a coleus plant and place it in water, parenchyma cells will divide to produce a mass of undifferentiated cells called a **callus**. Roots then develop from the callus, and the new individual can be planted in soil (see BioSkills 11).

Collenchyma: Simple Tissue with Cells That Function Primarily in Shoot Support The cells of **collenchyma** tissue are similar to parenchyma cells but are characterized by an unevenly thickened primary cell wall and are longer and thinner than parenchyma cells. Collenchyma is another example of a simple tissue.

Collenchyma tissue is often found just under the epidermis of stems, especially outside vascular bundles. Collenchyma cells are not rigid, and their ability to stretch allows stems to flex in the wind without tearing or breaking. These cells provide flexible structural support for shoots. They also support actively growing parts of the plant due to the expandable nature of their primary cell wall. The "strings" you may have peeled from a stalk of celery or rhubarb include many strands of collenchyma cells (Figure 34.12)

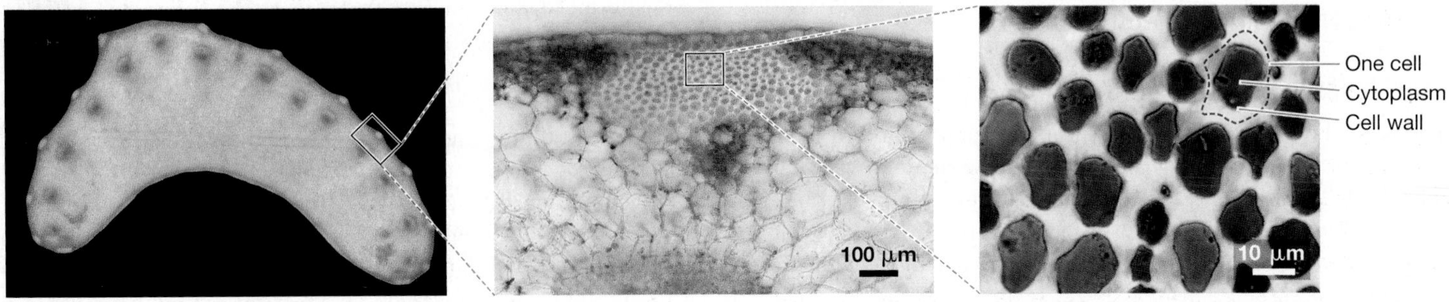

Figure 34.12 **Collenchyma Cells Support Growing Tissues.** A celery stalk is actually a petiole; the strands you can peel from it are columns of collenchyma cells.

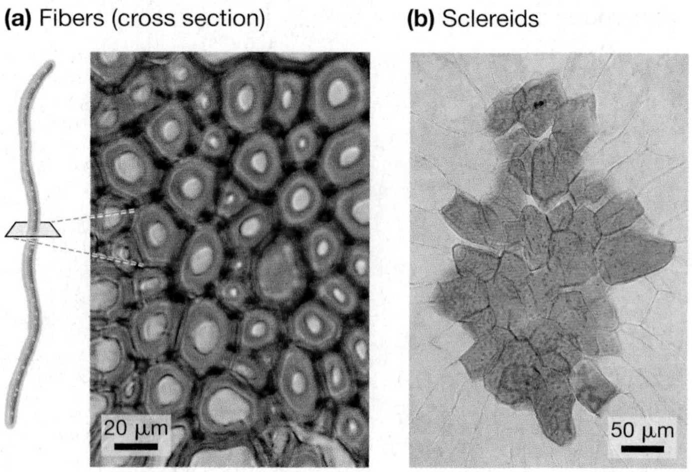

Figure 34.13 **Sclerenchyma Cells Support Mature Tissues. (a)** Fibers from *Tilia* (linden tree) and **(b)** sclereids from a pear (stained pink) have thickened secondary cell walls. These cells provide support and protection.

Sclerenchyma: Simple Tissue with Two Types of Specialized Support Cells The cells of **sclerenchyma** tissue are characterized by the presence of the thick, rigid secondary cell wall in addition to the relatively thin primary cell wall. Unlike the primary cell wall, which is composed primarily of **cellulose**, the secondary cell wall contains the tough, rigid compound **lignin** in addition to cellulose (Ch. 28, Section 28.3).

Unlike collenchyma cells, sclerenchyma cells have a nonexpandable secondary cell wall that specializes them for supporting stems and other structures after growth has ceased. Another key difference between collenchyma and sclerenchyma is that sclerenchyma cells are usually dead at maturity—meaning they contain no cytoplasm.

Two types of sclerenchyma cells are recognized: fibers and sclereids.

1. **Fibers** are extremely elongated. The fiber cells from ramie plants, a species of Asian nettle, can be over half a meter long. Fiber cells are important in the manufacture of paper, hemp or jute ropes, and linen and other fabrics (Figure 34.13a).
2. **Sclereids** are relatively short, have variable shapes, and often function in protection. The tough coats of seeds and the thick shells of nuts are composed of sclereids; these cells are also responsible for the gritty texture of pears (Figure 34.13b).

The Vascular Tissue System

The **vascular tissue system** functions in support and in long-distance transport of water and dissolved nutrients in vascular plants (Ch. 28, Section 28.3). It also moves the products of photosynthesis that are made and stored in ground tissue. The vascular tissue system consists of two complex tissues: xylem and phloem.

1. **Xylem** (pronounced *ZY-lem*) conducts water and dissolved nutrients in one direction: from the root system to the shoot system.
2. **Phloem** (*FLO-em*) conducts sugar, amino acids, hormones, and other substances in two directions: from roots to shoots and from shoots to roots.

Xylem Structure The xylem of many plants includes water-conducting cells as well as parenchyma cells and fibers. The two types of water-conducting cells in xylem are tracheids and vessel elements.

1. The xylem of all vascular plants contains **tracheids** (*TRAY-kee-idz*).
2. In angiosperms, as well as members of the Gnetophyta and a few other groups of vascular plants, xylem also contains **vessel elements** (Ch. 28, Section 28.3).

Although most plant cells perform their function while alive, tracheids and vessel elements—like sclerenchyma cells—are dead at maturity and contain no cytoplasm. The xylem cells have thick, lignin-containing secondary cell walls; in essence, the cell walls form pipelines for the movement of water and nutrients from roots to shoots. ✔ **THINK CAREFULLY A poison is injected directly into tracheids. This poison kills living cells by preventing the formation of ATP. What impact will this poison have on transport of water through the tracheids injected with poison?**

Tracheids are long, slender cells with tapered ends. The sides and ends of tracheids have **pits**, which are gaps in the secondary cell wall where only the primary cell wall is present (Figure 34.14a). When water is moving up a plant through tracheids, it moves from cell to cell both vertically and laterally through pits, because that is where resistance to flow is lowest.

Vessel elements, in contrast, are shorter and wider than tracheids. In addition to having pits, vessel elements have **perforations**—openings in the end walls that lack both primary and secondary cell walls (Figure 34.14b). In some species, the ends of vessel elements lack any cell wall at all, and stacked cells form open pipes called vessels. Vessel elements conduct water more efficiently than do tracheids, because their width and perforations offer less resistance to flow. Tracheids and vessel elements are often found adjacent to each other in those plants that contain both types of cells (Figure 34.14c).

You may have heard of **hardwood** and **softwood**. These designations don't really tell us how hard the wood is; instead, they indicate whether the wood contains vessel elements (hardwood—from flowering plants such as maple, oak, or balsa) or only tracheids (softwood—from conifers such as pine or spruce).

Phloem Structure In most vascular plants, phloem is made up primarily of two specialized types of cells: sieve-tube elements and companion cells (Figure 34.15). Both are alive at maturity and lack secondary cell walls. The phloem of many plants also includes fibers that aid in structural support.

- **Sieve-tube elements** are long, thin cells that have perforated ends called **sieve plates**. They are responsible for transporting sugars and other nutrients.

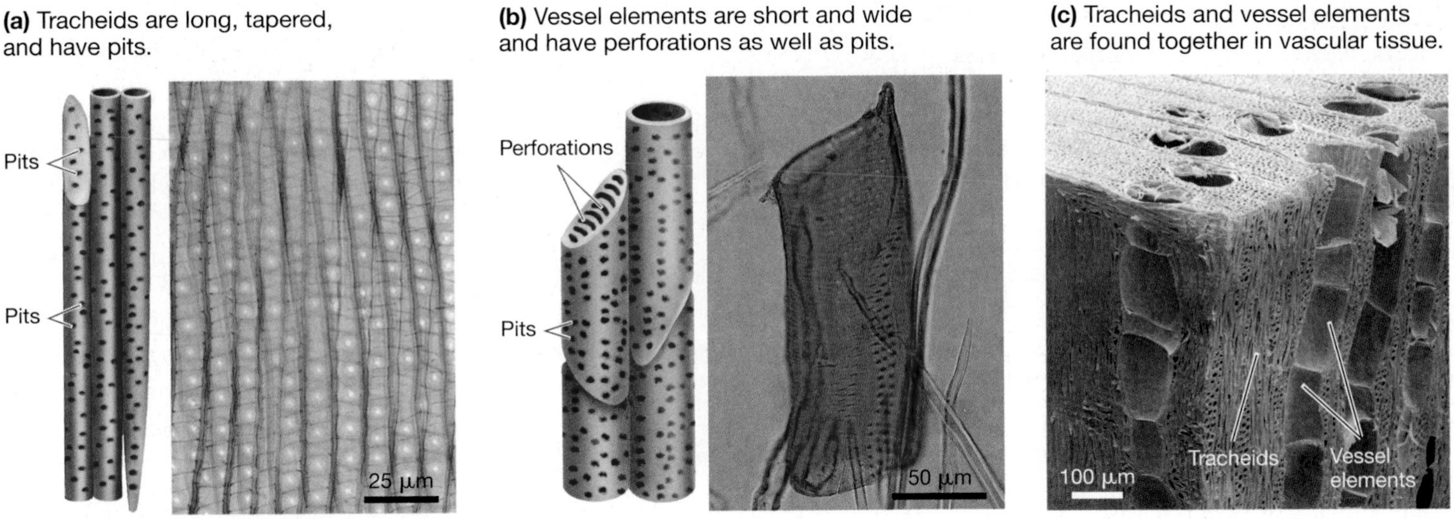

Figure 34.14 Xylem May Contain Two Types of Water-Conducting Cells. **(a)** Tracheids are long and thin compared to **(b)** vessel elements, which are much shorter and wider. **(c)** Both types of water-conducting cells are found in the vascular tissue of angiosperms.

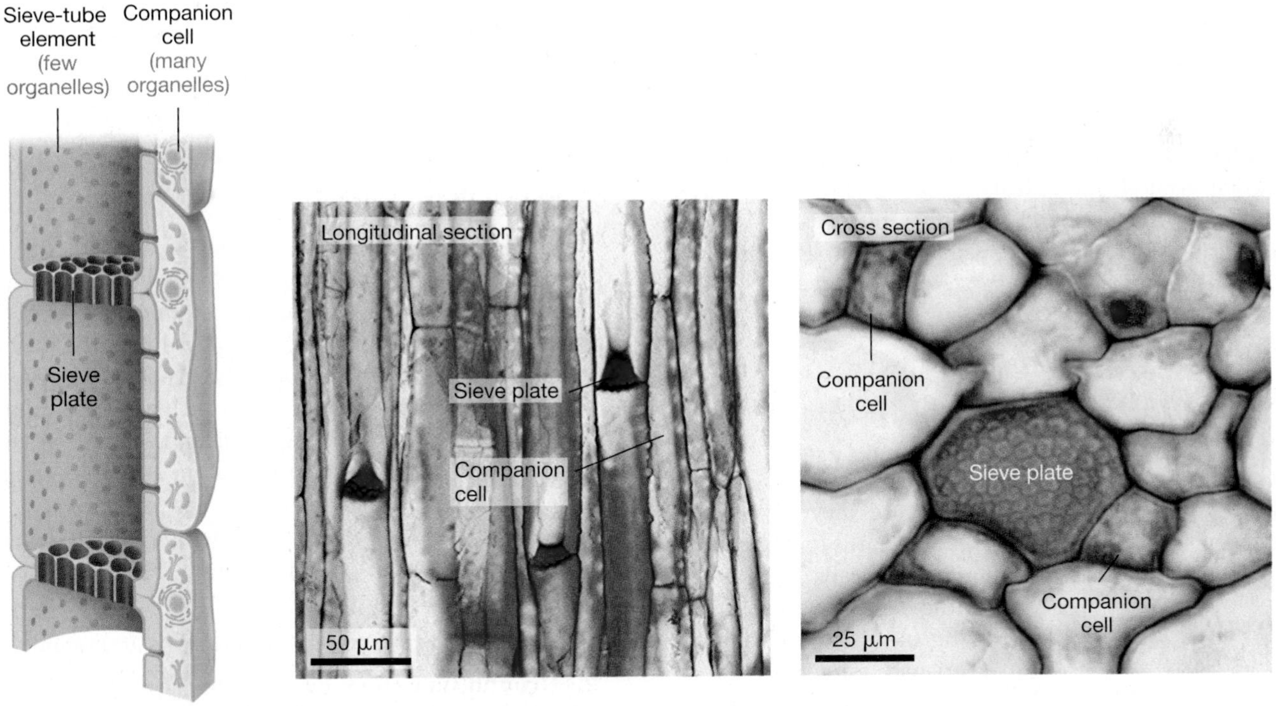

Figure 34.15 Phloem Consists of Sieve-Tube Elements and Companion Cells. Sieve-tube elements conduct sucrose throughout the body; companion cells support the metabolic activity of sieve-tube elements.

- **Companion cells** are not conducting cells, but instead provide materials to maintain the cytoplasm and plasma membrane of sieve-tube elements.

Sieve-tube elements lack nuclei and most other organelles, but they are directly connected to adjacent companion cells by numerous plasmodesmata (see Figure 34.8b). Companion cells contain most of the organelles normally found in plant cells and support the metabolic activity of sieve-tube elements.

CHECK YOUR UNDERSTANDING

✔ **If you understood this section, you should be able to...**

1. Explain the structural differences between plant and animal cells.
2. Describe the structural and functional differences between vessel elements and sieve-tube elements.

Answers are available in Appendix A.

SUMMARY Table 34.4 Plant Tissue Systems

Tissues Present	Description of Tissue	Function
Dermal Tissue System		
Epidermis	Complex tissue consisting of epidermal cells, guard cells, trichomes, and root hairs	Shoots: Protection, gas exchange Roots: Protection, water and nutrient absorption
Ground Tissue System		
Parenchyma	Simple tissue consisting of parenchyma cells (primary cell wall only)	Synthesis and storage of sugars and other compounds
Collenchyma	Simple tissue consisting of collenchyma cells (primary cell wall only with uneven thickenings; flexible cell walls)	Support (mostly in stems and leaves)
Sclerenchyma	Simple tissue consisting of sclerenchyma cells: sclereids or fibers (primary and secondary cell walls; dead at maturity)	Sclereids: protection (mostly in hard outer surfaces of seeds and fruits) Fibers: support (mostly in stems and leaves)
Vascular Tissue System		
Xylem	Complex tissue consisting of tracheids, vessel elements, and frequently parenchyma cells and fibers (tracheids, vessel elements, and fibers are dead at maturity; primary and secondary cell walls)	Transport of water and nutrients; structural support
Phloem	Complex tissue consisting of sieve-tube elements, companion cells, and frequently fibers (fibers are dead at maturity)	Transport of sugars, amino acids, hormones, etc.; support

Table 34.4 summarizes the major tissues and cell types found in the dermal, ground, and vascular tissue systems. Once you're familiar with the structure and function of plant tissues, you're ready to consider how these cells and tissues arise during primary and secondary growth.

34.3 Primary Growth Extends the Plant Body

Plants grow throughout their lives because they have many **meristems**—populations of undifferentiated cells that retain the ability to undergo mitosis. When meristematic cells divide, some of the daughter cells remain in the meristem, allowing the meristem to persist. Other cells, though, differentiate into distinct cell types with specific functions.

After you complete this section, you should be able to . . .

- Analyze the growth processes that allow shoots and roots to increase in length.

Apical meristems are located near the tip of each root and shoot. As cells in apical meristems divide, enlarge, and differentiate, root and shoot tips extend the plant body outward, allowing it to explore new space.

The division of apical meristem cells, and differentiation of those cells, is responsible for **primary growth**, which is common to all plants. The major consequence of primary growth is to increase the length of the root and shoot systems. All of the cells and tissues that are derived directly from apical meristems constitute the **primary plant body**. In general, most plants that do not produce woody tissues consist entirely of primary growth.

How Do Apical Meristems Produce the Primary Plant Body?

Whether located in the tips of roots or of shoots, apical meristems give rise to populations of cells that become three distinct **primary meristems**: protoderm, ground meristem, and procambium. These regions are partially differentiated but retain the character of meristematic cells because they keep dividing. The three types of primary meristems are important because they give rise to the dermal, ground, and vascular tissue systems (see Section 34.2). Meristem formation and tissue development are covered in more detail in Ch. 38, Section 38.5.

Figure 34.16 indicates where the apical meristems and the three primary meristems—protoderm, ground meristem, and procambium—are found in shoots and roots.

1. **Protoderm** gives rise to the dermal tissue system.
2. **Ground meristem** gives rise to the ground tissue system, which makes up the bulk of the primary plant body.
3. **Procambium** gives rise to the vascular tissue system.

Vascular tissue runs through ground tissue, so the cells that make up ground tissue are usually close to cells that conduct the water and nutrients they need. Figure 34.17 shows how the dermal, ground, and vascular tissue systems are organized in different parts of the primary plant body. The key point to remember is that the dermal, ground, and vascular tissue systems are derived from cells in primary meristems, which originated from apical meristems. Thus, they represent primary growth and the primary plant body. **Table 34.5** on page 740 summarizes the meristems and the components of primary growth that arise from them.

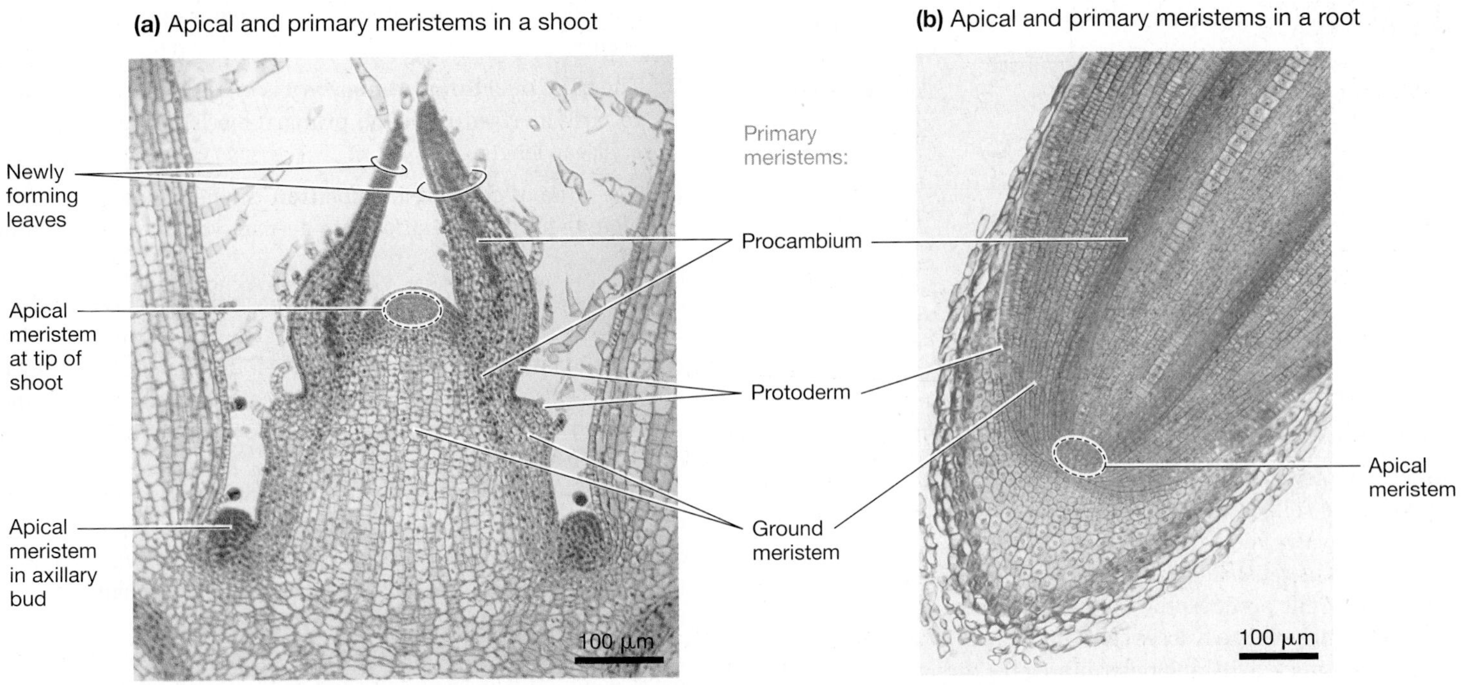

Figure 34.16 The Apical and Primary Meristems in the Shoot and Root Give Rise to Plant Tissues. Apical meristems consist of small, similar-looking cells that divide when conditions are favorable. Three types of primary meristems—protoderm, ground meristem, and procambium—are derived from the apical meristem and consist of partially differentiated cells that can still divide.

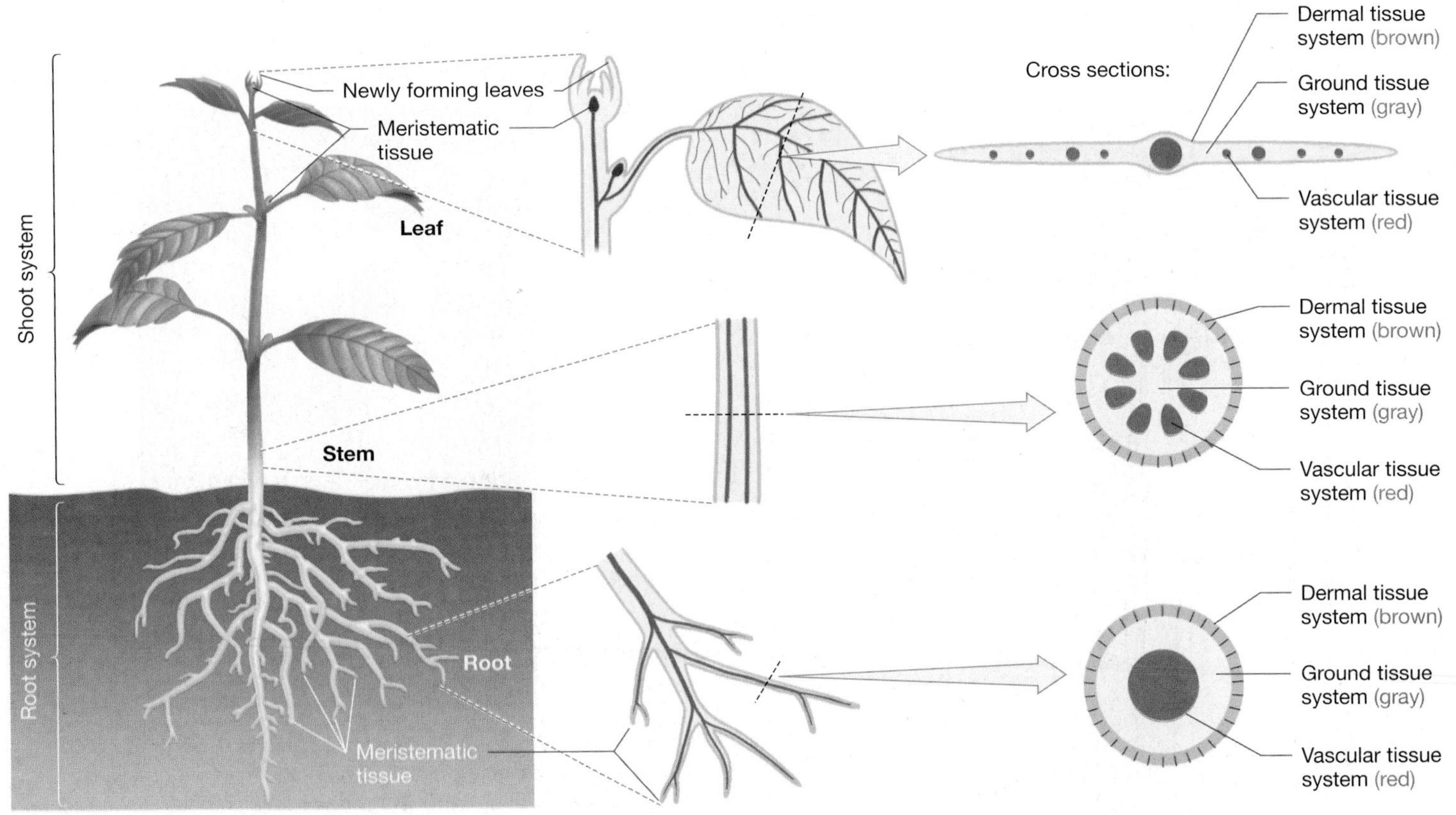

Figure 34.17 The Primary Plant Body Is Composed of the Dermal, Ground, and Vascular Tissue Systems.

✔ Are the youngest tissues found at the tip or the base of the shoot? Explain.

SUMMARY Table 34.5 Components of Primary Growth

Meristem	Primary Meristem	Primary Tissue System	Primary Tissues
Apical Meristem	Protoderm	Dermal tissue system	Epidermis
	Ground meristem	Ground tissue system	Parenchyma Collenchyma Sclerenchyma
	Procambium	Vascular tissue system	Xylem Phloem

How Is the Primary Root System Organized?

Roots have several features that allow them to grow into new regions of the soil, so they can furnish cells throughout the plant body with water and key nutrients.

A group of cells called the **root cap** protects the root apical meristem. Cells produced by the meristem constantly replenish the cap, which regularly loses cells. Root cap cells are also important in sensing gravity and determining the direction of growth. And they synthesize and secrete a slimy, polysaccharide-rich substance that helps lubricate the root tip, reducing friction and protecting the apical meristem as it is pushed through soil.

Three distinct populations of cells exist behind the root cap (Figure 34.18a):

1. The **zone of cellular division** contains the apical meristem, where cells actively divide, along with the protoderm, ground meristem, and procambium, where additional cell division occurs.
2. The **zone of cellular elongation** is made up of cells that are recently derived from the primary meristems and that increase in length.
3. The **zone of cellular maturation** is where older cells complete their differentiation into dermal, vascular, and ground tissues.

The zone of cellular elongation is the region most responsible for the growth of roots through the soil. The cells in this region increase in length by taking up water. Their expansion provides the force that pushes the root cap and apical meristem through the soil. When environmental conditions are good, roots can extend by as much as 4 centimeters per day.

The zone of cellular maturation is the most important root segment in terms of water and nutrient absorption. In this region, epidermal cells produce outgrowths called **root hairs**, which greatly increase the surface area of the dermal tissue. Root hairs furnish the actual sites of water and nutrient absorption (Figure 34.18b). The rest of the root system provides structural support for the root hairs, conducts water and nutrients to the shoot, stores the products of photosynthesis, and anchors the plant in the soil. Uptake of water and nutrients by root hairs is vital to plants.

The zone of cellular maturation is also where lateral roots begin to grow. In contrast to lateral branches in the shoot, which arise from meristems in axillary buds (see Figure 34.1), lateral roots arise within a ring of cells around the vascular tissue and then erupt through the surrounding ground tissue.

(a) Longitudinal section of a root

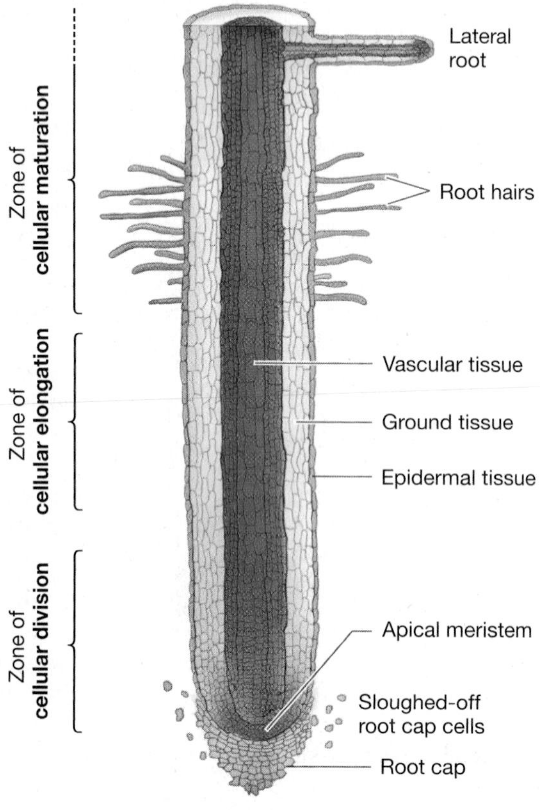

(b) Root of germinating radish seed

Figure 34.18 **Roots Extend into the Soil via Growth of Apical Meristems and Cell Elongation. (a)** A summary of the organization of the primary root system. The zone of cellular maturation is actually much larger than can be shown here. **(b)** The root of a germinating radish seed shows extensive growth of root hairs. Most absorption of water and nutrients occurs at root hairs.

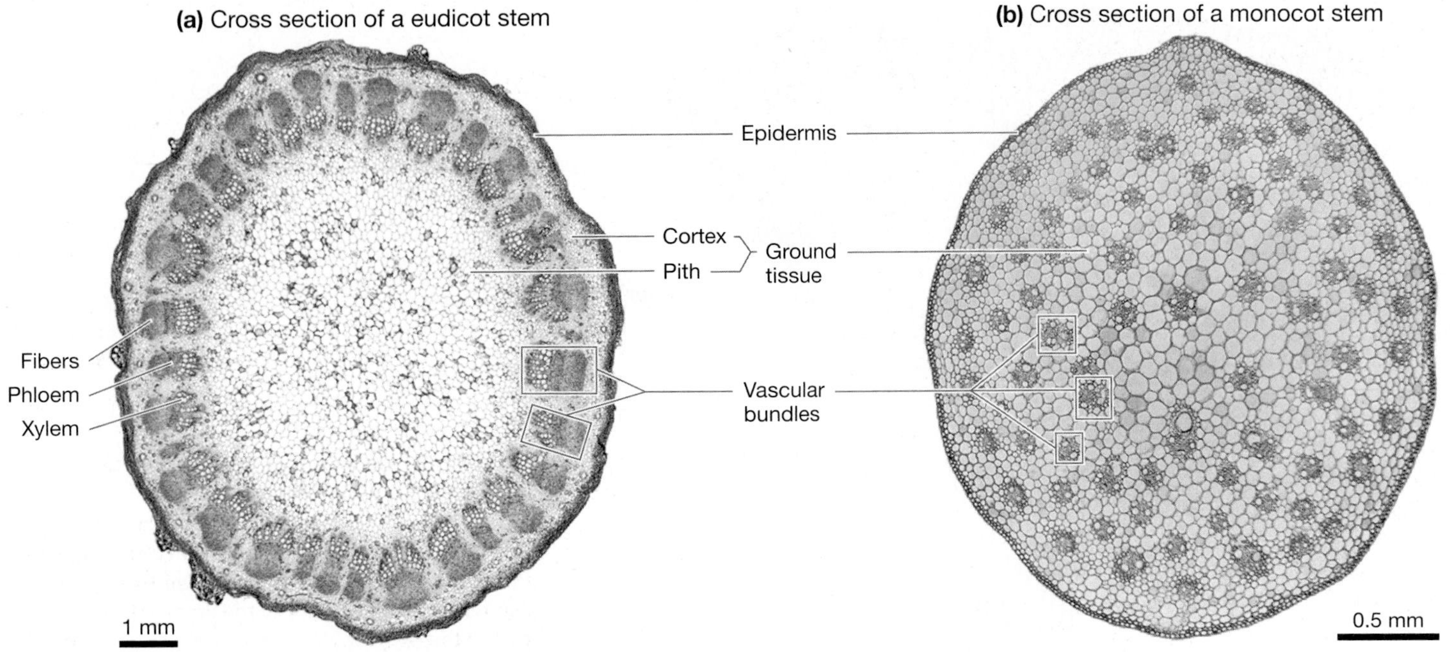

Figure 34.19 **Tissue Systems in Stems Have Distinct Arrangements.** Vascular bundles are **(a)** arranged in a ring near the perimeter of eudicot stems but **(b)** scattered throughout the ground tissue in monocot stems.

✔ **If you understand the organization of the primary root system, you should be able to explain which parts of roots are actively pushed through soil.**

How Is the Primary Shoot System Organized?

If you visit a garden regularly, you can imagine how the growing root tips push through the soil and expand to form complex networks underground. But even a casual observer can watch the growth of shoot systems over time, as the tips of stems extend and branch, and as new leaves form and expand.

Just behind each shoot apical meristem, the primary meristems give rise to dermal, ground, and vascular tissues. Figure 34.19a shows how these tissues are arranged in the stem of a mature sunflower. Note that vascular tissues are grouped into **vascular bundles**, which form strands running the length of the stem.

In sunflowers and other eudicots, the vascular bundles are arranged in a ring near the stem's perimeter (Ch. 28, Section 28.3). The ground tissue that the vascular tissue runs through is divided into two major regions: **pith**, the ground tissue that is toward the center of the stem, and **cortex**, the ground tissue that is between the vascular bundles and the epidermis. The roots of some plants also have distinct pith and cortex regions.

The arrangement of the vascular bundles and ground tissue is dramatically different in the stems of monocots such as grasses (including corn), lilies, and orchids. As Figure 34.19b shows, vascular bundles in corn stems are normally scattered throughout the ground tissue. (Review other comparisons of monocots and eudicots in Ch. 28, Section 28.3.)

The arrangement of vascular tissue in eudicots is significant and is associated with the ability to produce secondary growth. In Section 34.4 you will explore how secondary growth develops in trees and other woody plants.

CHECK YOUR UNDERSTANDING

✔ **If you understood this section, you should be able to...**

1. Explain the relationship between an apical meristem and the three primary meristems.
2. **MODEL** Draw a sketch that shows the arrangement of vascular tissue in a stem from an alfalfa plant (an example of a eudicot).

Answers are available in Appendix A.

34.4 Secondary Growth Widens Shoots and Roots

After you complete this section, you should be able to...

- Analyze the growth and development of woody plant tissues.

Primary growth increases the length of roots and shoots; its major function is to extend the reach of the root and shoot system and thus increase a plant's ability to absorb light and acquire carbon dioxide, water, and nutrients.

In trees and other woody plants, **secondary growth** increases the width of roots and shoots. Its major function is to increase the amount of conducting tissue available and provide the structural support required for extensive growth. Recall that the evolution of conducting tissues allowed the vascular plants to grow taller to compete for light (Ch. 28, Section 28.3). Without the support provided by secondary growth, however, roots would not be massive enough to anchor large shoot systems, and long stems would fall over or break. Only with the origin of secondary growth could vascular plants become the giants seen today.

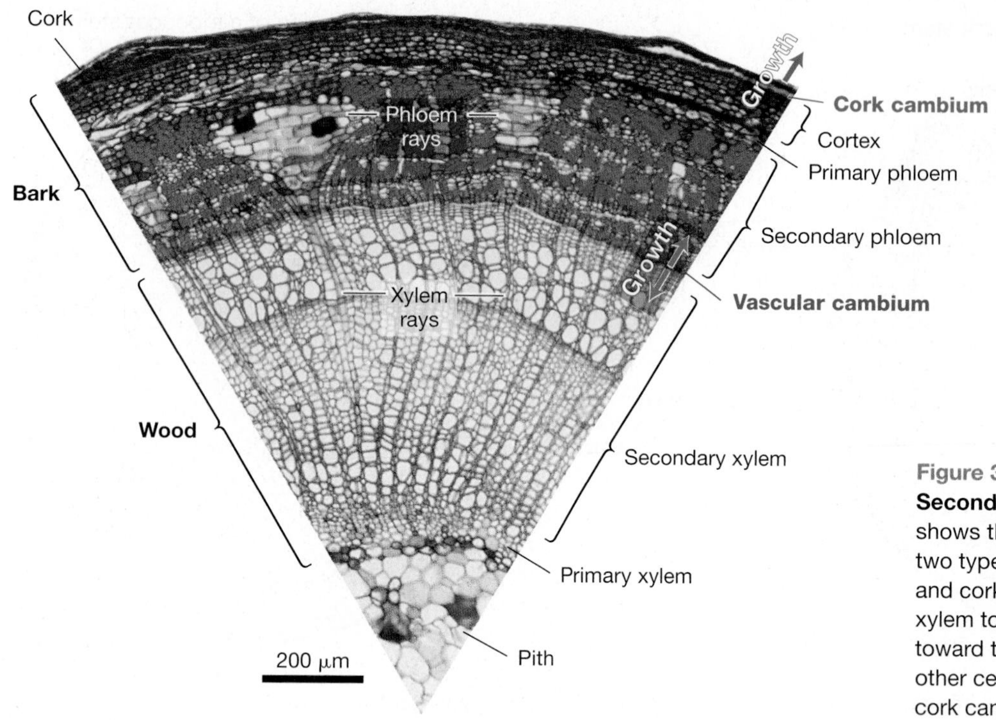

Figure 34.20 Cambia Are Responsible for Secondary Growth. This piece of a cross section shows that a linden tree (*Tilia* spp.) stem contains two types of cylindrical cambia: vascular cambium and cork cambium. The vascular cambium produces xylem toward the interior of the stem and phloem toward the exterior. Wood consists of xylem and other cells produced by the vascular cambium. The cork cambium produces cells associated with bark.

What Is a Cambium?

Secondary growth produces **wood** and occurs only in species that have a cambium in addition to apical meristems. A **cambium** (plural: *cambia*) is a special type of meristem (also called a lateral meristem) that differs from an apical meristem in two ways:

1. A cambium forms a cylinder that runs the length of a root, tree trunk, or branch and is made up of a single layer of meristematic cells. In contrast, apical meristems are clusters of cells localized at root tips and shoot tips.
2. In a cambium, and only there, cells divide in a way that increases the *width* of roots, trunks, and branches. Cells in an apical meristem, in contrast, divide in a way that extends the *length* of roots and shoots.

As Figure 34.20 shows, there are two distinct types of cambia in plants that undergo secondary growth.

1. A cylinder of meristematic cells called the **vascular cambium** in roots and stems produces xylem toward the interior of the cambium and phloem toward the exterior of the cambium. These vascular tissues are referred to as **secondary xylem** and **secondary phloem**, because they are formed as a result of secondary growth.
2. Another cylinder of meristematic cells called the **cork cambium** is located near the outer perimeter of roots, trunks, and branches. The cork cambium produces **cork**, the main tissue associated with tree bark.

One observation is critical to understanding how cambia work: The cork cambium produces new cells, called **cork cells**, primarily toward the exterior. (It also produces relatively few parenchyma cells toward the interior). The vascular cambium, in contrast, generates new layers of cells toward both the interior and exterior—more to the inside than the outside. The cells produced toward the outside become secondary phloem, while those produced toward the inside become secondary xylem (or wood). The new cells formed to the inside push all of the other cells (including the vascular cambium itself) toward the outside, causing an increase in girth. Tree trunks and branches grow wider each year as a result of cell divisions that occur exclusively in cambia located just underneath the bark. Table 34.6 summarizes the major types of cells that arise from each type of lateral meristem. Now let's explore how secondary growth is initiated in trees.

SUMMARY Table 34.6 Components of Secondary Growth

Meristematic Tissue	Mature Tissue	Direction of Growth	Mature Cell Composition	Mature Tissue Function
Cork Cambium	**Cork**	Produced to the outside	Cork cells	Protection (part of bark)
Vascular Cambium	**Secondary phloem**	Produced to the outside	Sieve-tube elements, companion cells, and sclerenchyma cells (fibers)	Transport of sugars, amino acids, hormones, etc.
	Secondary xylem (also called wood)	Produced to the inside	Tracheids, vessels, parenchyma cells (arranged in rays) and sclerenchyma cells (fibers)	Transport of water and ions; structural support

How Does a Cambium Initiate Secondary Growth?

Imagine a very young tree seedling. Initially, all of the seedling's tissues represent primary growth because they were derived from apical meristems. However, at some point the seedling will start to produce woody, secondary growth. As the tree matures, the woody shoot system is commonly referred to as the trunk and branches. Although a tree can get taller with each passing year, note that the position of any existing region of the trunk will stay at that same height, because new growth is initiated in the apical meristem. What occurs during the initiation of secondary growth?

Recall that the stems of eudicots contain vascular bundles arranged in a ring toward the periphery (see Figure 34.19a). Within each vascular bundle, xylem cells are located toward the interior while phloem cells are located toward the exterior. A single layer of cells located between the xylem and phloem becomes meristematic. This layer of cells represents the beginning of the vascular cambium.

Likewise, a ring of cells located just beneath the epidermis becomes meristematic. This represents the beginning of the cork cambium. The cork cambium and vascular cambium cells divide throughout the plant's life, producing secondary growth each year.

As a result, the woody tissues of tree trunks and branches are secondary growth and the wood we use for various purposes forms from vascular cambium activity (**Figure 34.21**).

Secondary growth in roots occurs similarly to that in shoots. A vascular cambium forms between the xylem and phloem, and a cork cambium forms interior to the epidermis. These lateral meristems increase the overall girth of roots.

What Do Vascular Cambia Produce?

Secondary phloem and xylem cells are not always produced simultaneously. In most cases, the vascular cambium produces many more secondary xylem cells than secondary phloem cells.

Structurally, primary and secondary phloem and primary and secondary xylem are complex tissues, made up of more than one cell type. Functionally, primary and secondary phloem are similar, as are primary and secondary xylem.

- Primary and secondary phloem function in sugar transport.
- Primary and secondary xylem function in water transport and structural support.

Secondary xylem makes up the structural material called wood. Secondary phloem makes up the inner part of a tree's bark.

Besides producing conducting cells such as sieve-tube elements, tracheids, and vessel elements, the vascular cambium produces sclerenchyma cells (fibers) for additional strength, along with some parenchyma cells. The parenchyma cells are formed in rows that radiate laterally across the xylem and phloem. These rows of cells are called **rays** (see Figure 34.20), and they form a living conduit that transports water and nutrients laterally across the trunk.

The results of cell division in a vascular cambium are highly asymmetrical, however. As the vascular cambium grows, all of the secondary xylem is retained and accumulates as wood, but the primary xylem eventually clogs and may rot away. In addition, the outermost secondary phloem becomes fragmented and compressed as the root or shoot increases in diameter. As a result, mature woody roots, trunks, and branches are dominated by secondary xylem.

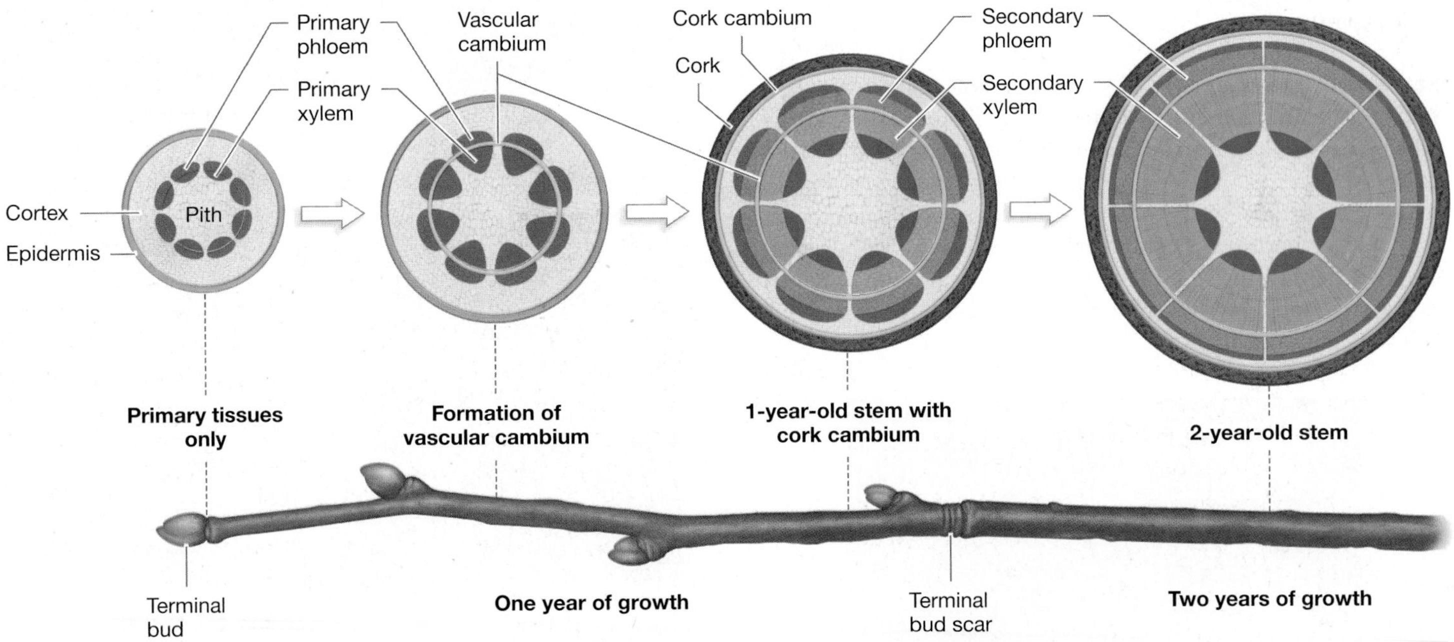

INTERACTIVE Figure 34.21 Initiation of Secondary Growth in Trees. Growth at the tips of tree trunks and branches begins as primary growth. Secondary growth is initiated when the vascular cambium and cork cambium become meristematic, thereby increasing the width, or girth, of the trunk or branch.

✔ **THINK CAREFULLY** The current diameter of a tree trunk is 10 inches. The trunk diameter increases at a rate of 2 inches per year. An insect burrows into the trunk and lays its eggs in the secondary xylem, 1 inch out from the center of the tree. How far from the center will the dormant eggs be located in two years?

What Do Cork Cambia Produce?

The main role of cork cambia is to produce cork cells toward the exterior that form the mature tissue called cork (see Figure 34.20). Together with the secondary phloem, the cork cambium and cork make up the **bark** of a tree trunk.

Bark provides a particularly tough barrier in species whose cork cells secrete a strong secondary cell wall containing lignin. Bark also helps prevent water loss because cork cells produce a layer of wax and other molecules inside their cell walls, making them impermeable to water and gases. Gas exchange can still occur between the atmosphere and living tissues inside the trunk, though—through small, spongy openings in the bark called **lenticels** (Figure 34.22). Cork cells die when they mature. As a trunk continues to widen, the cork layer often cracks and flakes, and the outer layers might even slough off the tree.

Bark is important because it protects the woody trunk from damage and pathogens as it increases in girth. As a trunk matures, the epidermal tissue produced by the apical meristem during primary growth is replaced by the bark, which takes over the role of preventing water loss and protecting from pathogens and herbivores. In some species, exceptionally thick bark can even protect the shoot system from fire damage. Redwood trees, for example, are adapted to fire-prone habitats and can have bark that is 61 cm (24 in.) thick.

The Structure of Tree Trunks

Trees are perennial plants that live for many years. As a tree matures and grows in width, the innermost xylem layers stop transporting water—only the xylem from the most recent years transports fluid.

Heartwood and Sapwood Xylem that no longer transports water begins accumulating protective compounds secreted by other tissues. These compounds form resins, gums, and other complex mixtures. The deposition of these molecules causes the oldest portions of secondary xylem to become darker than the younger portions. The darker-colored, inner xylem region is called **heartwood**, while the lighter-colored, outer xylem is called **sapwood** (Figure 34.23a). Such variations are often visible in wood floors and furniture.

Figure 34.22 **Lenticels Are Visible in the Bark of Paperbark Cherry (*Prunus serrula*).** Lenticels are natural breaks in cork tissue through which oxygen can enter a trunk.

(a) Heartwood and sapwood have different functions.

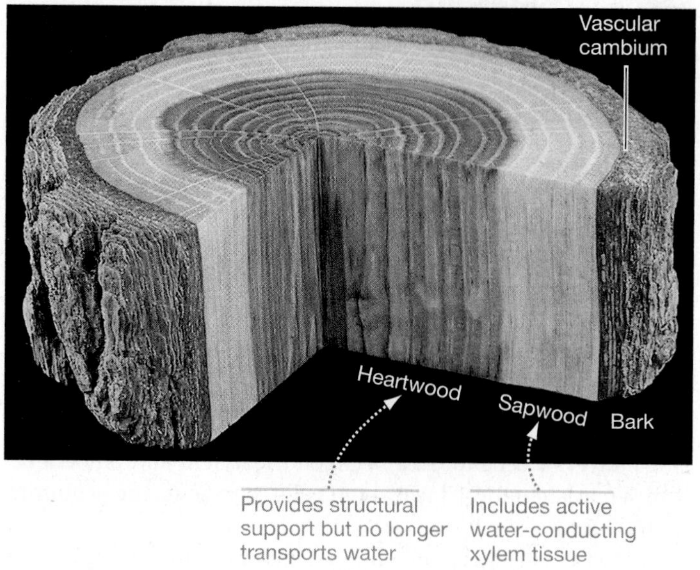

(b) Growth rings result from seasonal variation in cell size.

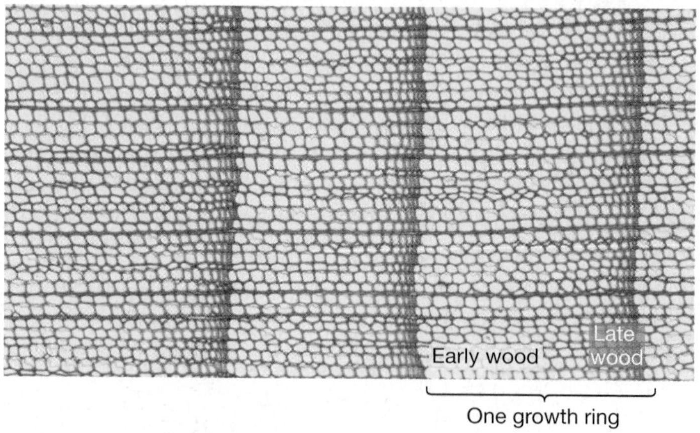

(c) Patterns in growth rings can tell a tree's history.

Figure 34.23 **Anatomy of a Tree Trunk. (a)** Transverse section of a tree trunk. **(b)** Section of wood, stained to show individual cells under a light microscope. **(c)** Unstained section through a fir tree from Germany's Black Forest.

✔ In part **(c)**, pick a thick growth ring and label the early wood and late wood.

Annual Growth Rings Another important phenomenon occurs in environments where the vascular cambium stops growing for a portion of each year. This period of **dormancy** takes place during the winter in cold climates and during the dry season in tropical habitats.

When the vascular cambium resumes growth in the spring or at the start of the rainy season, it produces large, relatively thin-walled cells, called early wood. As the growing season nears its end, conditions tend to dry out or become cooler; the secondary xylem cells that are produced at this time tend to be smaller, thicker walled, and darker and are called late wood. Thus, when growth is seasonal, regions of large, thin-walled cells alternate with layers of small, thick-walled cells, resulting in annual growth rings in the secondary xylem (Figure 34.23b).

Analyzing patterns in tree growth rings is an important field of study in biology. Because trees grow faster when moisture and nutrients are plentiful, wide tree rings are reliable indicators of wet years. In contrast, narrow rings signal drought years—or in the case of the tree shown in Figure 34.23c, years when abundant acid rain, due to air pollution, reduced growth. Years with generally less than optimal growth conditions could also be related to temperature or disease.

Dendrochronology is the science of dating and studying tree growth rings. By studying growth rings in fossil trees, dead trees, and extremely old living trees, biologists can often assemble a continuous record of growth dating back thousands of years. In doing so, they gain a better understanding of climate changes, fires, and other environmental events that occurred in the past. With continued research, scientists also hope to predict how forests might respond to the fluctuating temperatures associated with global climate change.

CHECK YOUR UNDERSTANDING

✔ **If you understood this section, you should be able to...**

1. **MODEL** Draw a three-year-old woody trunk in cross section, labeling secondary xylem and phloem, vascular and cork cambia, and cork. Add arrows showing the direction of growth due to the activity of vascular and cork cambia.
2. **MODEL** Draw the secondary xylem of a four-year-old woody trunk in cross section from a tree that experienced optimal growing conditions during its first two years of growth and drought conditions during the most recent two years.

Answers are available in Appendix A.

CHAPTER 34 REVIEW

For media, go to **Mastering Biology**

34.1 Plant Form: Themes with Many Variations

- The root and shoot systems of plants are specialized for harvesting the light, water, and nutrients required for performing photosynthesis. Structures involved in absorption have a high surface area to volume ratio.
- Roots extract water and nutrients such as nitrogen, phosphorus, and potassium from the soil.
- The shoot system consists of all aboveground portions of the plant, including stems and leaves.
- Leaves, which usually consist of a flattened blade that extends from a petiole, carry out photosynthesis by capturing light and carbon dioxide from the atmosphere.
- The morphology of root and shoot systems varies widely among species, allowing individuals to reduce competition for resources.
- Roots and shoots may be modified to perform a variety of other functions, including nutrient storage, water storage, protection, and asexual reproduction.
- Because roots and shoots grow throughout life, a plant is able to respond appropriately to changes in environmental conditions.

34.2 Plant Cells and Tissue Systems

- The dermal tissue system is usually one cell layer thick and plays a role in protection and water absorption by roots. In shoots the epidermis synthesizes the cuticle, which aids in water conservation.
- Stomata are pores formed by pairs of epidermal guard cells that open and close to control CO_2 uptake and water loss by leaves.
- The ground tissue system performs photosynthesis and stores carbohydrates and other compounds.
- The vascular tissue system transports materials throughout the plant. Within the vascular system, xylem tissue transports water and dissolved ions up the plant; phloem tissue transports sugars up and down.
- Ground tissue contains **(1)** parenchyma cells, which function in material synthesis, transport, and storage; **(2)** collenchyma cells, which provide structural support for growing regions; and **(3)** sclerenchyma cells—fibers and sclereid cells—that strengthen regions of the body that have stopped growing.

34.3 Primary Growth Extends the Plant Body

- Each apical meristem gives rise to three primary meristems: protoderm, ground meristem, and procambium.
- The primary meristems give rise to the dermal, ground, and vascular tissue systems, which extend throughout the plant body.
- During root growth, behind the zone of cellular division, cells become longer in the zone of cellular elongation and acquire specialized functions in the zone of cellular maturation.

34.4 Secondary Growth Widens Shoots and Roots

- In some plants, shoots and roots are widened by cylindrical meristems called vascular cambia that produce secondary xylem and secondary phloem, and cork cambia that produce cork cells.
- Cork cells produce waxes that protect stems from water loss.
- Wood consists of secondary xylem, while bark consists of all tissue outside of the vascular cambium.

Answers are available in Appendix A.

✔ TEST YOUR KNOWLEDGE

1. What is a sieve-tube element?
   a. the sugar-conducting cell found in phloem
   b. the widened, perforation-containing, water-conducting cell found only in angiosperms
   c. the nutrient- and water-absorbing cell found in root hairs
   d. the nucleated and organelle-rich support cell found in phloem
2. Evaluate the following statements regarding tracheids and vessel elements. Select True or False for each statement.
   T/F Both tracheids and vessel elements are specialized for water conduction.
   T/F Both tracheids and vessel elements have pits.
   T/F Vessel elements have perforation plates but tracheids do not.
   T/F Tracheids and vessel elements have to be alive in order to transport water.
3. Which statement best characterizes primary growth?
   a. It does not occur in roots, only in shoots.
   b. It leads to the development of cork tissue.
   c. It produces the dermal, ground, and vascular tissues.
   d. It produces rings of xylem, phloem, and cork tissue.
4. Which statement best characterizes secondary growth?
   a. It results from cell divisions in the vascular and cork cambia.
   b. It increases the length of the plant stem.
   c. It results from divisions in the apical meristem cells.
   d. It often produces phloem cells to the inside and xylem cells to the outside of the vascular cambium.

✔ TEST YOUR UNDERSTANDING

5. Describe the general function of the shoot and the general function of the root system. Which tissues are continuous throughout these two systems? Suggest a hypothesis to explain why the shoot and root systems of different species are so variable in size and shape.
6. Explain why continuous growth enhances the phenomenon known as phenotypic plasticity.
7. What does the cuticle do? What do stomata do? Predict how the thickness of the cuticle and the number of stomata differ in plants from wet habitats versus dry habitats.
8. **THINK CAREFULLY** You attached a birdfeeder to a tree 5 years ago at a position 6 feet above the ground. The tree has grown since at a rate of 1 foot per year. What is the current height of the birdfeeder?

✔ TEST YOUR PROBLEM-SOLVING SKILLS

9. Identify the structure you are consuming when you eat each of the following: asparagus, Brussels sprouts, celery, spinach, carrot, potato.
10. Trees can be killed by girdling—the removal of bark and vascular cambium in a ring all the way around the tree. Explain why.

Mastering Biology

Students Go to Mastering™ Biology for assignments, the eText, and the Study Area with animations, practice tests, and activities.

Professors Go to Mastering™ Biology for automatically graded tutorials and questions that you can assign to your students, plus Instructor Resources.

✔ PUT IT ALL TOGETHER: Case Study

How is plant growth affected by acid rain?

Plants experience a vast array of conditions that can alter their growth and development, including temperature and availability of water, nutrients, and light. They are also exposed to pollutants such as acid rain formed largely as a result of burning fossil fuels and volcanic eruptions. Acid rain can strip minerals and nutrients from the soil, and eat away the outer waxy layer of tissue that protects a plant. Acid rain is still a major environmental problem in certain regions. How does acid rain affect plant growth?

11. Develop a hypothesis on the likely effects of acid rain on primary growth in trees. Based on your hypothesis, make some predictions about the impact of acid rain on the height of tree trunks.
12. **QUANTITATIVE** Researchers tested the effects of acid rain (pH 2.0) on seedlings of two different species, camphor tree (*Cinnamomum camphora*) and chinaberry tree (*Melia azederach*), using distilled water as a control. Results of the experiments are shown here. Note that acid rain caused a significant reduction of growth in chinaberry trees (*** means $P < 0.001$, see **BioSkills 3**), but not in camphor trees. What was the approximate percentage of growth reduction observed in the chinaberry trees treated with acid rain?

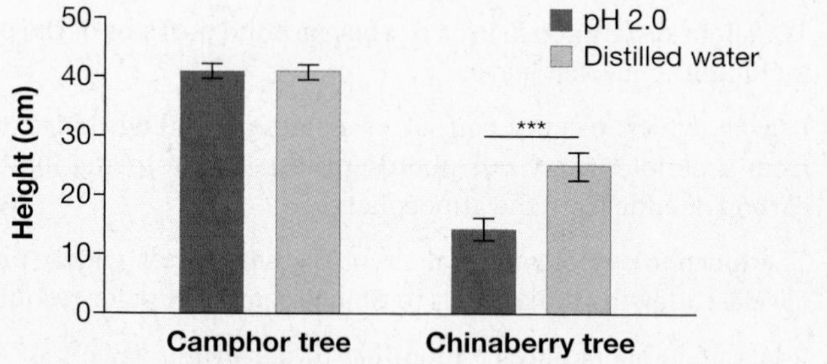

Source: H. B. Fan and Y. H. Wang. 2000. *Forest Ecology and Management* 126: 321–329.

13. Which meristem was likely affected in the chinaberry tree?
14. Give a plausible explanation for the differential effect of acid rain on height in camphor and chinaberry trees (i.e., what structural features might offer more protection from acid rain in one species versus another?).
15. Based on the results presented here, predict the impact of acid rain on the vascular cambium in the two species. How would that effect be apparent in the amount of wood produced in the tree trunks?
16. Beyond primary and secondary growth, what other aspects of plant growth might be negatively affected when plants are exposed to environmental stress?

35 Water and Sugar Transport in Plants

This chapter explores how plants move water from their roots to their leaves and how they transport sugars to all of their tissues—sometimes over great distances.

In this chapter you will learn how

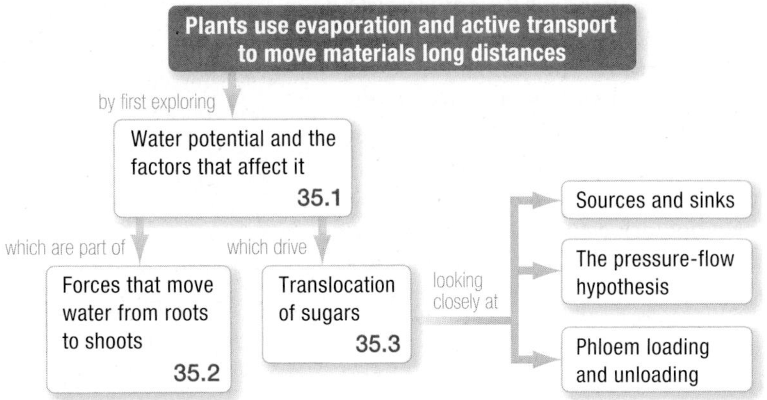

On a hot summer day, a large deciduous tree can lose enough water to fill three 55-gallon drums. To understand why, recall that the surfaces of leaves are dotted with many stomata (Ch. 34, Section 34.2). A recent study of trees in the Amazon discovered one species whose leaves have over 844 stomata per square millimeter. In most plants, stomata are open during the day so gas exchange can occur between the atmosphere and the cells inside the leaf. This exchange is crucial. For photosynthesis to take place, leaf cells must acquire carbon dioxide (CO_2).

There's a catch, however. While stomata are open, the moist interior of the leaf is exposed to the dry atmosphere. As a result, large quantities of water evaporate from the leaf. If the lost water is not replaced with water absorbed by roots, plant cells will dry out and die.

Evaporation from leaves is a challenge for plants, but it is also beneficial. Because water enters a plant through its roots and exits through its leaves, there is a regular flow of water from roots to shoots. Minerals that are absorbed by roots are carried to the leaves by this flow. Without this movement of water, shoots would be unable to receive the minerals they need to grow.

BIG PICTURE

This chapter is part of the Big Picture.See how on pages 838–839.

In addition, under hot conditions, evaporation cools the plant, just like sweating cools your body. Heavy rates of evaporation can lower leaf temperatures by as much as 10–15°C.

But the source of water can be a long distance from where it is needed. How do plants transport water against the force of gravity—from roots, through tree trunks, to distant branches and leaves? In certain redwood trees, for example, the leaves that lose water may be 100 meters (m) from the root hairs that absorb water. In addition, how do plants use phloem cells to move the sugar they produce in leaves to where sugar is stored or needed for growth? These questions are the heart and soul of this chapter. Answering them is a fundamental part of understanding how plants work.(You can review the critical role of water movement and transport in the Big Picture on pages 838–839.)

35.1 Water Potential and Water Movement

Before exploring the details of how water and sugars are transported through xylem and phloem, respectively, let's first consider the basic principles of diffusion and osmosis that play important roles in the movement of water through plants. Recall that solutes tend to diffuse from regions of higher concentration to regions of lower concentration. Also recall that water tends to move across membranes from regions of lower solute concentration to regions of higher solute concentration by **osmosis** (see Ch. 6, Section 6.3).

After you complete this section, you should be able to . . .

- Explain how water moves from regions of high water potential to regions of low water potential in plants.

Water can also move from a wet region to a dry region (think about water evaporating on a hot, dry day). This net movement of water—either across a membrane or into the surrounding air—occurs as a direct result of differences in water potential from one region to another.

What Is Water Potential?

Biologists use the term **water potential** to indicate the potential energy that water has in a particular environment compared with the potential energy of pure water at room temperature and atmospheric pressure. Recall that potential energy is stored energy (see Ch. 8, Section 8.1). Changes in potential energy are associated with changes in position, such as the position of a molecule or an electron.

Water potential is measured in units called **megapascals (MPa)**, where one MPa is equal to 10^6 pascals. A **pascal (Pa)** is a unit of measurement commonly applied to pressures—force per unit area. A car tire is inflated to about 0.2 MPa, and the water pressure in home plumbing is usually 0.2 to 0.3 MPa. Pure water at room temperature and atmospheric pressure has a water potential of 0 MPa. Solutions that consist of water and solutes have water potentials below 0 MPa (so they are negative).

Water potential is symbolized by the Greek letter Ψ (psi, pronounced *sigh*). Differences in water potential determine the direction that water moves. The net movement of water is always from areas of higher water potential to areas of lower water potential.

Water transport from roots to shoots, for example, occurs because of differences in water potential and is ultimately driven by evaporation of water from the stomata of leaves. This process, termed **transpiration**, surprisingly requires virtually no expenditure of energy by plants. Transpiration occurs whenever two conditions are met: **(1)** Stomata are open, and **(2)** the air surrounding leaves is drier than the air inside leaves. As a result of transpiration, water is pulled through plants under negative pressure (much like soda is pulled through a drinking straw).

Two major factors contribute to differences in water potential: solute potential and pressure potential. Let's examine each in turn.

What Factors Affect Water Potential?

To understand how water moves from cell to cell in a plant, first consider the cell in the beaker on the left in Figure 35.1a. Notice that it is suspended in a **solution**—a homogenous, liquid mixture containing several substances. In this case, the solution consists of water and dissolved substances, or **solutes**.

In the beaker on the left, the solute concentrations in the cell and in the surrounding solution are the same. Such a solution is said to be **isotonic** to the cell (Ch. 6, Section 6.3). When two regions separated by a membrane are isotonic with each other, water moves across the membrane in equal amounts in both directions. As a result, there is no *net* movement of water into or out of the cell.

The Role of Solute Potential What happens when the cell is transferred to the beaker on the right in Figure 35.1a? This beaker contains pure water, which has no solutes. As a result, the solution surrounding the cell is strongly **hypotonic** relative to the cell—that is, the surrounding solution has a lower solute concentration than the cell. Although water moves across the membrane in both directions, under these conditions, there will be a *net* movement of water into the cell by osmosis.

The tendency for water to move in response to differences in solute concentrations is determined by the **solute potential (Ψ_S)**. Solute potentials are always negative because they are measured relative to the solute potential of pure water. Solutions with *high* concentrations of solutes have *low* solute potentials.

The Role of Pressure Potential When an animal cell is placed in a hypotonic solution and water enters the cell via osmosis, the volume of the cell increases and the cell may even burst. This does not happen to plant cells, however.

If a plant cell swells in response to incoming water, its plasma membrane pushes against the relatively rigid cell wall. The cell wall resists expansion of the cell volume by pushing back, much as a bicycle tire pushes back against its inner tube as it is inflated.

The force exerted by the wall is called **wall pressure** (Figure 35.1b). As water moves into the cell, the pressure of the fluid contents inside the cell, known as **turgor pressure**, increases until wall pressure is induced. Cells that are firm and that experience wall pressure are said to be **turgid**.

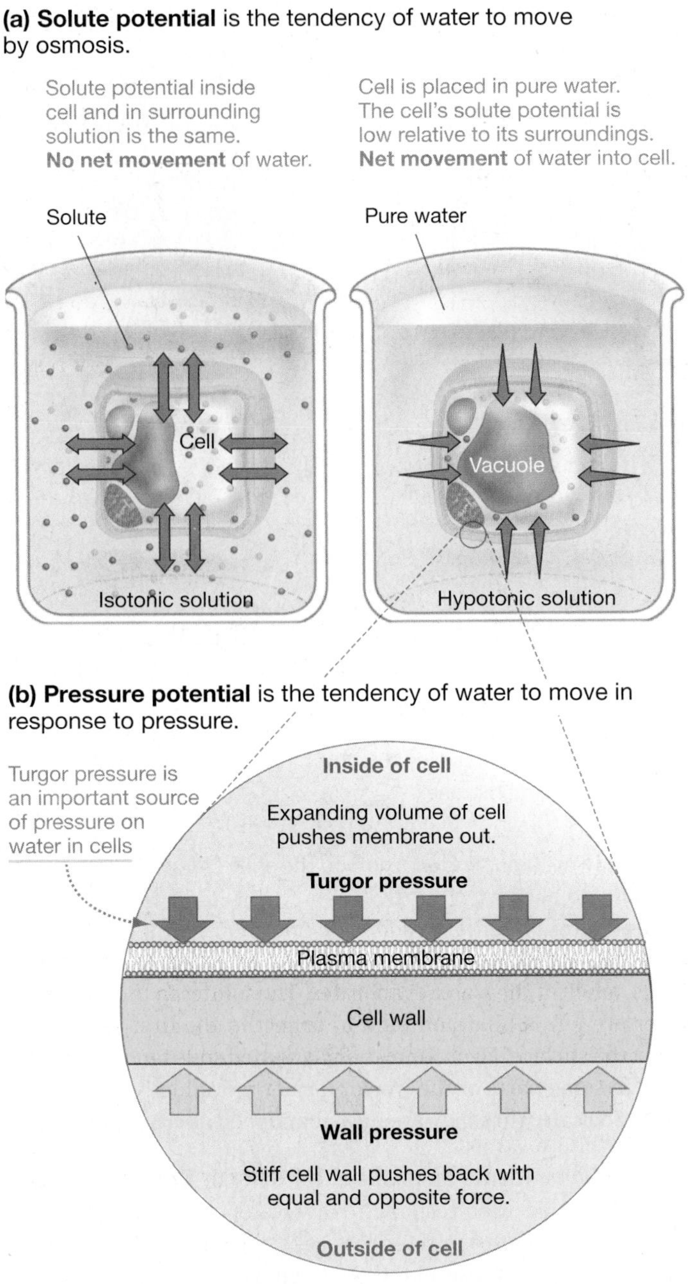

Figure 35.1 Two Major Factors Affect Water Potential: Solute Potential and Pressure Potential.

Turgor pressure is important because it counteracts the movement of water into cells by osmosis. In the example on the right in Figure 35.1a, the solute potential favors water moving into the cell. However, the rigid cell wall limits the amount of water that can enter the cell.

Pressure potential (Ψ_P) refers to any kind of physical pressure on water. Inside a cell, the pressure potential consists of turgor pressure and, in the opposite direction, wall pressure. While the solute potential of a solution can be only 0 MPa (pure water) or negative, pressure potential can be either positive or negative. When pressure is negative, it is called tension. Tension is what draws liquid up through a drinking straw.

When solute potential and pressure potential affect a cell at the same time, the combined effect of these two forces determines the net direction of water flow.

Working with Water Potentials

If you ignore the effects of gravity, which are negligible over short distances such as between cells, water potential is defined by the following equation:

$$\Psi = \Psi_S + \Psi_P$$

In other words, the potential energy of water in a particular location is the sum of the solute potential and the pressure potential existing at that site.

Although solute potentials (Ψ_S) are always negative, the pressure potential (Ψ_P) from turgor pressure is positive inside *living* cells. But as you will see later, the pressure potential in dead cells (e.g., xylem) can sometimes be negative. How do solute and pressure potentials combine to drive water into and out of living cells?

Water Movement in the Absence of Pressure In the U-shaped tube on the left side of Figure 35.2a, on page 750, two solutions are separated by a selectively permeable membrane. The system is open to the atmosphere and thus is not under any additional pressure, meaning $\Psi_P = 0$ MPa.

Note that the left side of the tube contains pure water, which has a Ψ_S of 0 MPa. The Ψ_S for the solution on the right side of the membrane is -1.0 MPa. Because water potential is higher on the left side of the tube than on the right side, the overall net movement of water is from left to right.

The right side of Figure 35.2a models the same situation with a cell that is initially **flaccid**—meaning it has no turgor pressure and therefore has a pressure potential of 0 MPa. Note that the cell has been placed in a solution of pure water. Because the cell has a low solute potential (-1.0 MPa) and the pure water has a higher water potential than the cell, water enters the cell via osmosis.

Water Movement in the Presence of Solute and Pressure Potentials On the left side of Figure 35.2b, the concentrations in the U-shaped tube are the same as in Figure 35.2a, but the solution on the right side of the tube experiences pressure exerted by a plunger. If the force on the plunger produces a pressure potential of 1.0 MPa on the right side, and if the Ψ_S for the solution on the right side of the membrane is still -1.0 MPa, then the water potential of the right side is $-1.0 \text{ MPa} + 1.0 \text{ MPa} = 0$.

In this case, the water potential on both sides of the membrane is equal and there will be no net movement of water. If the force on the plunger is greater than 1.0 MPa, the solution on the right side will have a *higher* water potential and water will flow from right to left.

The right side of Figure 35.2b models this situation in a living cell, where the incoming water creates turgor pressure. When the positive turgor pressure ($+1.0$ MPa) plus the cell's negative solute potential (-1.0 MPa) equals 0 MPa—the water potential of pure water—the system reaches equilibrium. At equilibrium, there is no additional net movement of water. In this way, turgor pressure acts like the plunger in Figure 35.2b.

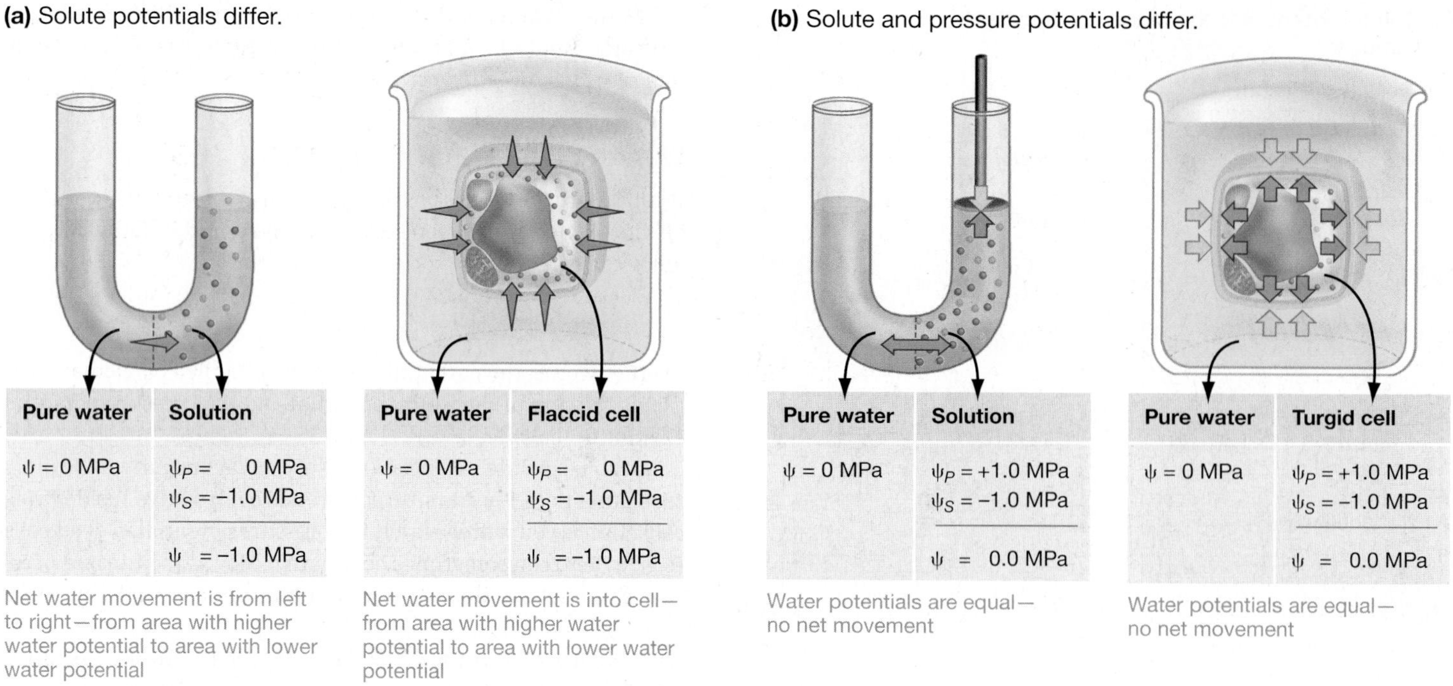

Figure 35.2 Solute Potential and Pressure Potential Interact.

✔ On the left side of part (a), does the solute potential on the right side of the tube increase, decrease, or stay the same when water flows from left to right by osmosis? Assume that solutes do not cross the membrane.

✔QUANTITATIVE If you understand the concept of water potential, you should be able to calculate the water potential of a cell with $\Psi_P = 0$ MPa and $\Psi_S = -2$ MPa. Indicate whether there will be an overall net movement of water into or out of the cell if that cell is placed into a solution in which $\Psi_S = -3$ MPa.

Water Potentials in Soils, Plants, and the Atmosphere

The water contained in leaf, stem, or root tissues has a pressure potential and a solute potential, just like the water inside a cell does. Likewise, both the soil surrounding the root system and the air around the shoot system have a water potential.

Water Potential in Soils In moist soil, the water that fills crevices between soil particles usually contains relatively few solutes and normally is under little pressure. As a result, its water potential tends to be high relative to the water potential found in a plant's roots, which is higher in solutes.

There are important exceptions to this rule, however:

- ***Salty soils*** Soils near ocean coastlines have water potentials as low as –4 MPa or less due to high solute concentrations. This value is much lower than the water potential typically found inside plant roots.
- ***Dry soils*** When soils are dry, water no longer flows freely in the spaces between soil particles. All of the remaining water adheres tightly to soil particles, creating a tension that lowers the water potential of soil water.

When the water potential in soil drops, water is less likely to move from soil into roots. If soil water potential is low enough, water may even move from roots to the soil. This situation would be deadly to plants.

Low water potential in soil is an enormously important issue for world agriculture. When soils are irrigated to boost crop yields, much of the water evaporates. The solutes in the irrigation water are left behind and tend to collect in the first few inches below the surface. Over time, then, irrigated soils tend to become salty. In some parts of the world, formerly productive soils have become so salty that they are now abandoned as cropland.

How Are Some Plants Specialized for Salty or Dry Soils? Salt-adapted species often respond to low water potentials in soil by accumulating solutes in their root cells, which then lowers their solute potential. These plants have enzymes that increase the concentrations of certain organic molecules in the cytoplasm. Thus they can keep the water potential of their tissues lower than that of salty soils and are able to absorb what little water is available.

Species that are adapted to dry soils cope by tolerating low solute potentials. For example, let's look at the changes in solute potential that were recorded in tissue from a woody shrub called ninebark (*Physocarpus malvaceus*). In Figure 35.3, each data point represents the solute potential recorded on a particular day; the lines between the data points are drawn simply to make the trend clear.

Ninebarks thrive on dry sites in the Rocky Mountains of western North America. Notice that the solute potential of their tissues is relatively high in June, at the start of the growing season. In the Rockies, June tends to be rainy and cool. July and August are progressively hotter and drier, however. As the data indicate, the solute potential in the tissue drops dramatically during this period.

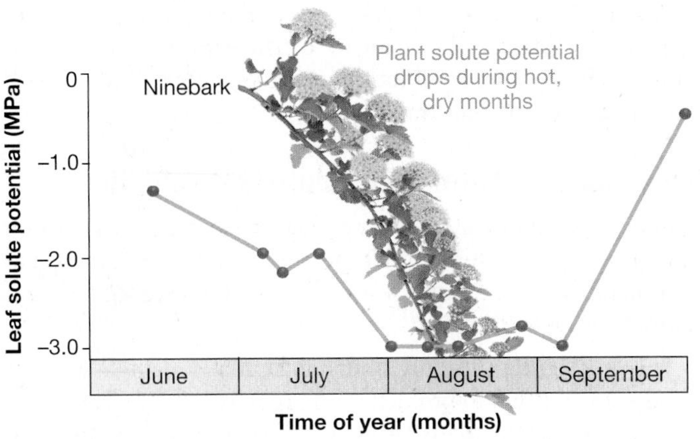

Figure 35.3 **Plants with Low Solute Potentials Can Grow in Dry Soils.** As soils dry, their water potential decreases. But if a plant's solute potential also drops, it can maintain a water-potential gradient that continues to bring water into the plant.

DATA: R. G. Cline and W. S. Campbell. 1976. *Ecology* 57: 367–373.

✔ **MODEL** Draw a line on the graph in Figure 35.3 that predicts the solute potential of soil, assuming that ninebark is able to absorb water from soil throughout the year.

Figure 35.4 **Wilting Occurs When Water Loss Leads to Loss of Turgor Pressure.** Wilting, though sometimes reversible, can be a life-threatening condition in plants, analogous to severe dehydration in humans.

This is a key observation. As the summer progresses and water potential in soil drops, ninebark shrubs are able to keep acquiring water and grow because the solute potentials of their tissues can drop to stay *below* the soil water potential.

Plants that are adapted to wetter sites cannot tolerate such low solute potentials in their tissues. When conditions get hot and dry, they have to close their stomata and stop photosynthesis—meaning that their growth will stop or slow down dramatically compared to dry-adapted plants. If plants keep their stomata open and leaf cells lose water faster than the water is replaced, the positive turgor pressure normally inside living cells drops to 0. If the cells do not regain turgor quickly, they are at risk of dehydration and death. When an entire tissue loses turgor pressure, it will **wilt** (Figure 35.4). Unless corrected, extensive wilting may lead to the death of the tissue and, eventually, the plant.

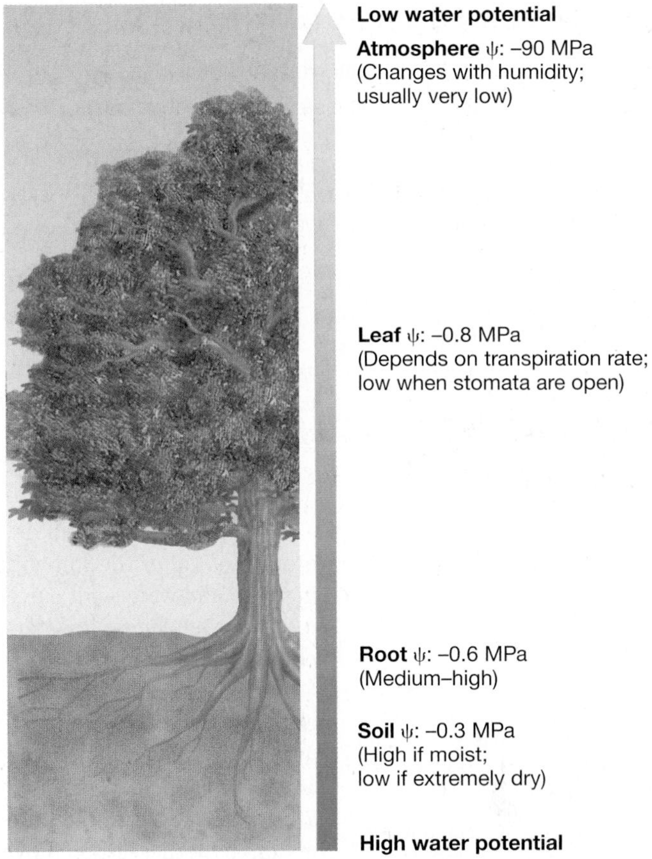

Figure 35.5 **A Water-Potential Gradient Exists between Soil, Plants, and Atmosphere.** Water moves from regions of higher water potential to regions of lower water potential.

Water Potential in Air In the atmosphere, water exists as a vapor with no solute potential. The pressure exerted by water vapor in the atmosphere depends on temperature and humidity. The lower the pressure potential, the faster liquid water evaporates into the atmosphere.

- When air is dry, few water molecules are present and the pressure they exert is low, increasing the rate of evaporation.
- When air is warm, water molecules move farther apart and also exert lower pressure.

Warm, dry air has an extremely low water potential, often approaching −100 MPa. When the weather is cool and rainy or foggy, however, the water potential of the atmosphere may be equal to the water potential inside a leaf. But normally, the water potential of the atmosphere is lower than the water potential inside a leaf, so water in the leaf evaporates quickly.

In most cases, water potential is highest in soil, medium to high in roots, low in leaves, and very low in the atmosphere. This situation sets up a **water-potential gradient** that causes water to move up through the plant. To move *up* a plant, water moves *down* the water-potential gradient that exists between the soil, its tissues, and the atmosphere. When it does so, it replaces the water lost through transpiration (Figure 35.5).

CHECK YOUR UNDERSTANDING

✔ **If you understood this section, you should be able to . . .**

1. Explain how grass roots can lose water instead of absorb water when a lawn is overfertilized.
2. Explain whether the roots of pickleweed (*Salicornia europaea*), a plant that grows well in salty soil, are likely to have a higher or lower water potential than the surrounding soil.

Answers are available in Appendix A.

35.2 How Does Water Move from Roots to Shoots?

After you complete this section, you should be able to . . .

- Explain how water moves into roots by osmosis, and how evaporation from leaves draws that water up through the shoot system.
- Describe the physical features that allow certain plants to conserve water.

Suppose that you are caring for the wilted plant in Figure 35.4. If you add water to the soil, the water potential of the soil will increase and water will move into the roots down a water-potential gradient.

Water moves into root cells by osmosis, but how does it move up to the shoots, against gravity? Biologists have tested three major hypotheses for how water could be transported to shoots:

1. Root pressure—a pressure potential that develops in roots—could drive water up against the force of gravity.
2. Capillary action could draw water up the cells of xylem.
3. Cohesion-tension, a force generated in leaves by transpiration, could pull water up from roots.

Which of these forces drives water to the top of trees? To answer this question, let's begin by considering how water and solutes move from the soil into the root **vascular tissue**, which contains xylem and phloem.

Movement of Water and Solutes into the Root

To understand how water enters a root, consider the cross section through a young buttercup root shown in Figure 35.6. Starting at the outside of the root and working inward, notice that several distinct tissues are present:

- The **epidermis** (literally, "outside skin") is a single layer of cells. In addition to protecting the root, some epidermal cells produce long, thin **root hairs**, which greatly increase the total surface area of the root.
- The **cortex** consists of ground tissue and stores carbohydrates.
- The **endodermis** ("inside skin") is a cylindrical layer of cells that forms a boundary between the cortex and the vascular tissue. The function of the endodermis is to control ion uptake and prevent ion leakage from the vascular tissue.
- The **pericycle** ("around circle") is a layer of cells located just interior to the endodermis that forms the outer boundary of the vascular tissue. The pericycle can produce lateral roots.
- Conducting cells of the vascular tissue, **xylem** and **phloem**, transport water and nutrients between roots and shoots and are located in the center of roots in buttercups and other eudicots. Notice that, in these plants, phloem is situated between each of four arms formed by xylem, which is arranged in a cross-shaped pattern.

Three Routes through Root Cortex to Xylem When water enters a root along a water-potential gradient, it does so through root hairs. As water is absorbed, it moves through the root cortex toward the xylem along three routes (Figure 35.7):

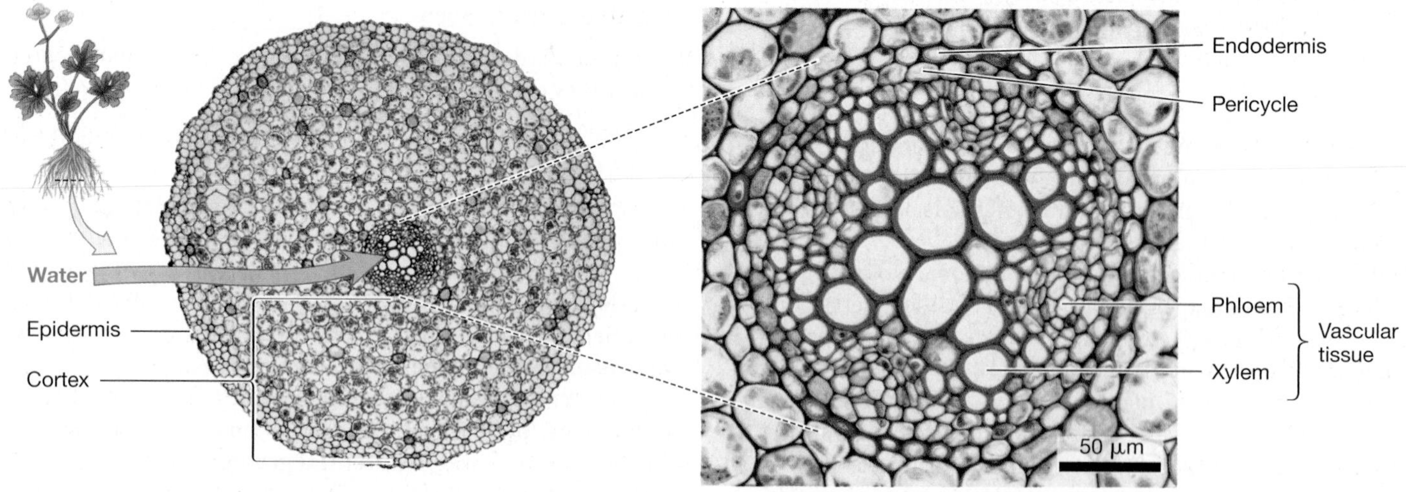

Figure 35.6 In Roots, Water Has to Travel through Several Tissue Layers to Reach Vascular Tissue. This cross section through a buttercup root shows the anatomy that is typical of roots in eudicots.

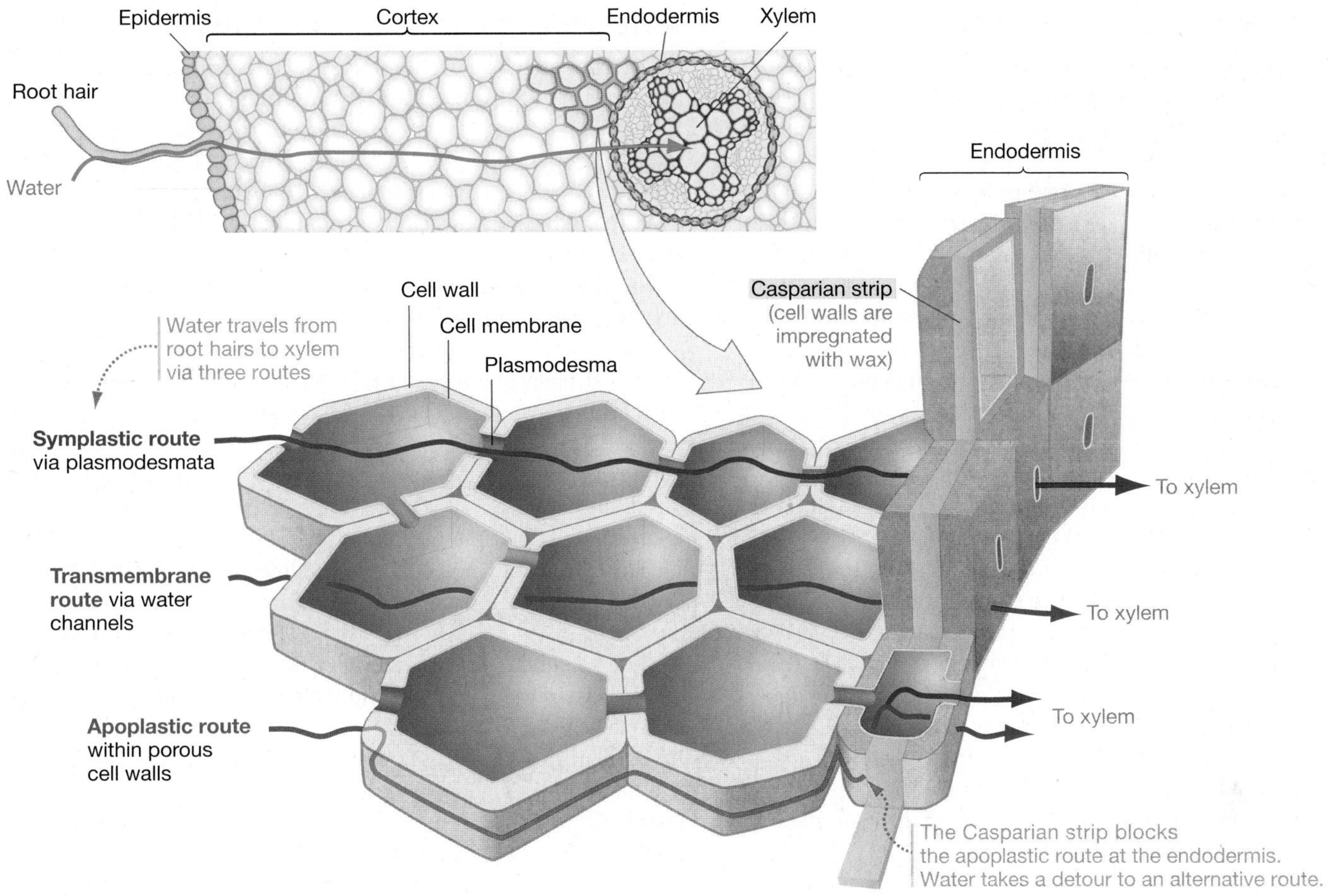

Figure 35.7 **Water Travels from Root Hair to Xylem along Three Routes.**

1. The *symplastic route* is inside the plasma membranes. The **symplast** consists of the cytosol and the continuous connections between cells via plasmodesmata (Ch. 11, Section 11.2).
2. The *transmembrane route* is based on water movement across plasma membranes, typically through aquaporin proteins—water channels located in the plasma membranes of many cells (Ch. 6, Section 6.4). Some water may also diffuse directly across plasma membranes.
3. The *apoplastic route* is outside the plasma membranes. The **apoplast** consists of cell walls, which are porous, and the spaces that exist between cells. Water moving along the apoplastic route *must* eventually pass through the cytoplasm of endodermal cells before entering xylem.

In essence, water can flow through tissues on its way toward xylem by passing through the symplast via plasmodesmata, crossing membranes via aquaporins, or passing around cells via the apoplast.

The Role of the Casparian Strip The situation changes when water reaches the endodermis. Endodermal cells are tightly packed and secrete a narrow band of wax called the **Casparian strip**. This layer is composed primarily of a compound called **suberin**, which forms a waterproof barrier where endodermal cells contact each other.

The Casparian strip blocks the apoplastic route by preventing water from moving through the walls of endodermal cells and into the vascular tissue (see Figure 35.7). The Casparian strip does not affect water and ions that move through the symplastic route.

Because of the Casparian strip, water and solutes have to move through the cytoplasm of an endodermal cell to reach vascular tissue. Endodermal cells, in turn, act as gatekeepers by regulating what enters the xylem.

By forcing water and ions to cross at least two membranes—one when entering the symplast outside the Casparian strip, another when leaving the symplast inside the Casparian strip—on their way from the soil to the xylem, plants can use specific channel and carrier proteins (Ch. 6, Section 6.4) to control what moves to the shoots. Endodermal cells allow ions that the plant needs, such as potassium (K^+), to pass through to the vascular tissue. In contrast, these cells can limit the passage of ions such as sodium (Na^+) or heavy metals that are not needed or may be harmful.

✔ **If you understand this concept, you should be able to predict what would happen if a plant had a mutation that prevented synthesis of suberin and formation of the Casparian strip.**

Water Movement via Root Pressure

Movement of ions and water into the root xylem is responsible for the process known as **root pressure**. Recall that root pressure is one of three hypothesized mechanisms for moving water up xylem, from root to shoot.

The Casparian strip in endodermal cells is essential for root pressure to develop. Without an apoplastic barrier between the xylem and the environment, ions and water would simply leak out of roots.

Stomata normally close during the night, when photosynthesis is not occurring and CO_2 is not needed. Their closure minimizes water loss and slows the movement of water through plants. But roots often continue to accumulate ions that their epidermal cells acquired from the soil as nutrients, and these nutrients are actively pumped into the xylem. The influx of ions lowers the water potential of xylem below the water potential in the surrounding cells. As a result, water from nearby cells enters xylem.

As water flows into xylem from other root cells in response to the solute gradient, a positive pressure is generated at night that forces fluid up the xylem. More water moves up xylem and into leaves than is being transpired from the leaves.

In certain low-growing plants, such as strawberries, enough water can move to force water droplets out of the leaves by a phenomenon known as **guttation**. If you have ever walked barefoot through the grass early in the morning, some of the water that gets your feet wet spent part of the night moving through plants (Figure 35.8).

At one time, positive root pressure formed the basis of a leading hypothesis to explain how water moves from roots to leaves in trees. However, follow-up research showed that over long distances, such as from the ground to the top of a large tree, the force of root pressure is not enough to overcome the force of gravity on the water inside xylem.

In addition, researchers demonstrated that cut stems, which have no contact with the root system, are still able to transport water to leaves. Biologists concluded that there must be some other mechanism involved in the long-distance transport of water.

Water Movement via Capillary Action

Researchers have also evaluated a hypothesis based on the phenomenon of **capillary action**, or movement of water up a narrow tube. When a thin glass tube (analogous to xylem cells) is placed upright in a pan of water, water creeps up the tube (Figure 35.9). The movement occurs in response to three forces: **(1)** surface tension, **(2)** adhesion, and **(3)** cohesion. Let's briefly consider each force in turn.

Surface tension is a force that exists among water molecules at an air–water interface. In the body of a water column, all the water molecules are surrounded by other water molecules and form hydrogen bonds in all directions. Water molecules at the surface, however, can form hydrogen bonds only with the water molecules beside and below them. Because they have fewer neighbors nearby, surface molecules share stronger attractive forces and bind together more tightly. This enhanced attraction results in tension that minimizes the total surface area.

Adhesion is a molecular attraction among dissimilar molecules. In this case, water interacts with a solid substrate—such as the glass walls of a capillary tube or the cell walls of tracheids or vessel elements—through hydrogen bonding. Water molecules are pulled upward as they bond to each other and adhere to the side of the tube.

Cohesion is a molecular attraction among like molecules, such as the hydrogen bonding that occurs among molecules in water (Ch. 2, Section 2.2). Because water molecules cohere, the upward pull by adhesion is transmitted to the rest of the water column. The water column rises against the pull of gravity.

The effects of adhesion, cohesion, and gravity are responsible for the formation of a concave surface boundary called a **meniscus** (plural: **menisci**). A meniscus forms at most air–water interfaces—including those found in narrow tubes, such as xylem. Menisci form because adhesion and cohesion pull water molecules up along the sides of the tube, while gravity pulls the water surface down in the middle.

When adhesion creates an upward pull at the water–container interface, surface tension creates an upward pull all across the

Figure 35.8 **Root Pressure Causes Guttation.** When ions accumulate in the xylem of roots at night, enough water enters the xylem via osmosis to force water up and out of low-growing leaves.

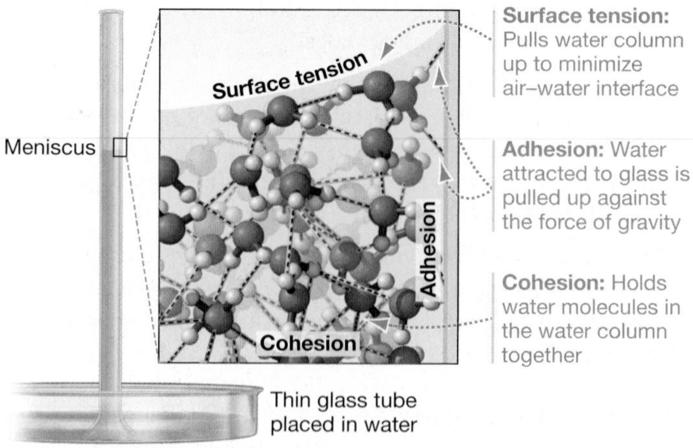

Figure 35.9 **Water Can Rise via Capillary Action.** Capillary action occurs through a combination of three forces, all of them generated by hydrogen bonding and acting together against gravity.

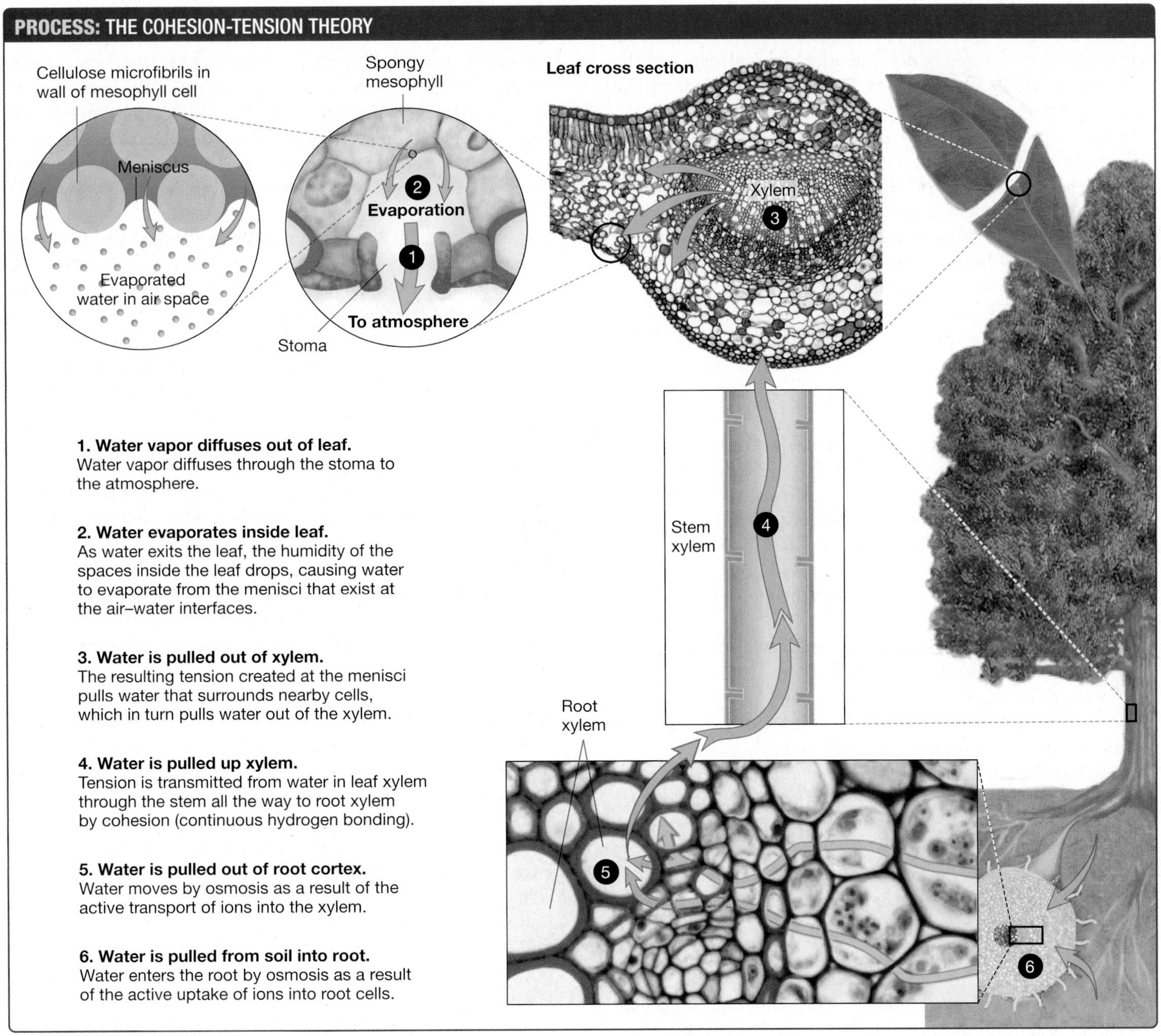

Figure 35.10 **Transpiration Creates Tension That Is Transmitted from Leaves to Roots.**

surface, and cohesion transmits both forces to the water below. All three forces counteract the effect of gravity, and the result is capillary action.

Like root pressure, capillary action can transport water only a limited distance. This process moves water along the surfaces of mosses and other low-growing, non-vascular plants, but it can raise the water in the xylem of a vertical stem only about 1 m.

Thus, root pressure and capillary action cannot explain how water moves from soil to the top of a redwood tree that can grow 5 to 6 stories higher than the Statue of Liberty. How then does it actually happen?

The Cohesion-Tension Theory

The leading hypothesis to explain long-distance water movement in vascular plants is the **cohesion-tension theory**, which states that water is pulled from roots to the tops of trees along a water-potential gradient, via forces generated by transpiration at leaf surfaces (see Figure 35.5). This process relies on two of the forces involved in capillary action, namely, cohesion and tension.

To understand how the cohesion-tension force works, start with step 1 in Figure 35.10. Notice that spaces inside the leaf are filled with moist air as a result of evaporation from the surfaces of surrounding cells. When a stoma opens, this humid air is exposed to the atmosphere, which in most cases is much drier. This contrast creates a steep water-potential gradient between the leaf interior and its surroundings. The steeper the gradient, the faster water vapor diffuses out through the stomata.

Step 2 shows that as water is lost from the leaf to the atmosphere, the humidity of the gas-filled space inside the leaf drops. In response, more water evaporates from the walls of the parenchyma cells. At the microscopic level, cell walls are not perfectly smooth, so menisci form at the air–water interface. As more water evaporates, the curvature of the menisci increases and the total area of the air–water interface increases.

In 1894 Henry Dixon and John Joly hypothesized that the formation of steep menisci produces a force capable of pulling water up from the roots, tens or hundreds of meters into the air. Is this really possible?

The Role of Surface Tension in Water Transport The key concept in the cohesion-tension theory is that the negative force or pull (tension) generated at the air–water interface is transmitted through the water outside of leaf cells (step 3 in Figure 35.10), to the water in xylem (step 4), to the water in the vascular tissue of roots (step 5), and finally to the water in the soil (step 6). Water can enter roots by osmosis even if transpiration is not occurring.

The transmission of pulling force from the leaf surface to the root is possible because **(1)** xylem forms a continuous network from roots to leaves, so there are continuous columns of water throughout the plant, and **(2)** all of the water molecules are held together through hydrogen bonds (cohesion).

Note that the plant does not expend energy to create the pulling force. The force is generated by energy from the Sun, which drives evaporation from the leaf surface. Water transport is solar powered.

In effect the cohesion-tension theory of water movement states that, because of the hydrogen bonding between water molecules, water is pulled up through xylem in continuous columns.

Creating a Water Potential Gradient You can also think about the cohesion-tension theory in terms of water potentials. Note that because tracheids and vessels are dead at maturity, the water in xylem does not cross plasma membranes. As a result, water does not move between cells by osmosis. In xylem, water movement is driven entirely by differences in pressure potential. The water in a column of xylem cells moves by **bulk flow**—a mass movement of molecules along a pressure gradient.

The pulling force generated at menisci in leaf cell walls lowers the pressure potential of water in leaves. Even though the tension created at each meniscus is relatively small, there are so many menisci in the leaves of an entire plant that the tension created by summing many small pulling forces is remarkable. It creates a water-potential gradient between leaves and roots that is steep enough to overcome the force of gravity and pull water up long distances.

To appreciate just how great the forces involved are, think of the vessel elements or tracheids in xylem as groups of drinking straws. When you use a straw, the vacuum that you create causes liquid to rise. Sucking on a drinking straw creates a pressure difference of about −0.1 MPa, which can draw water up a maximum of about 10 m. In contrast, the negative pressure exerted by the menisci in leaves can be as low as −2.0 MPa. The tension in xylem tissue may be 10 times the amount of pressure on a fully inflated car tire. (But note that the pressure on a car tire is positive, not negative.) The force is enough to draw water up 100 m.

The Importance of Secondary Cell Walls If you suck on a drinking straw hard enough, and the resistance at the end of the straw is high enough, then the pressure gradient between the inside of the straw and the atmosphere can overcome the stiffness of the straw and cause it to collapse. Similar forces are at work as water is pulled through xylem. How can vascular tissue withstand negative pressures as low as −2.0 MPa without collapsing?

The answer is found in a key adaptation: the secondary thickenings characteristic of the cell walls in tracheids and vessel elements. The cells in vascular tissue have walls that are reinforced with tough lignin molecules (Ch. 28, Section 28.3). As a result, xylem consists of rigid "pipes" that can withstand internal tension without collapsing, much like cartilage rings in your trachea prevent it from collapsing as you inhale.

The evolution of lignified secondary cell walls was an important event in the evolution of land plants because it allowed vascular tissue to withstand extreme negative pressures. The result? Tall trees.

What Evidence Do Biologists Have for the Cohesion-Tension Theory? If the cohesion-tension theory is correct, the water present in xylem should experience a strong pulling force. A simple experiment supports this prediction.

If you find a leaf that is actively transpiring and cut its petiole, the watery fluid in the xylem, the xylem sap, withdraws from the edge toward the inside of the leaf (Figure 35.11). According to the cohesion-tension theory, this observation is due to a transpirational pull at the air–water interface in leaf cells.

Although the observation that xylem sap is under tension was important, the cohesion-tension theory remained controversial. For example, several studies failed to document rapid changes in xylem pressure when temperature and humidity—and thus the magnitude of the water-potential gradient—were changed experimentally. Advocates of the theory blamed the negative results on instruments that could not document small, rapid changes in pressure potential. Who was right?

Chunfang Wei, Melvin Tyree, and Ernst Steudle answered the question in the late 1990s with an instrument called a xylem pressure probe (Figure 35.12). A xylem pressure probe includes a very thin glass tube that can be inserted directly into the xylem of a leaf. The probe allows researchers to record changes in xylem pressure instantly and directly.

To test the cohesion-tension theory, the researchers changed xylem pressure by raising the light intensity to alter transpiration rates in the leaves of corn plants. Their results?

- The graph in Figure 35.12 plots how xylem pressure changed as the researchers raised the light intensity over a 7-minute period. Note that as light intensity increased, the xylem

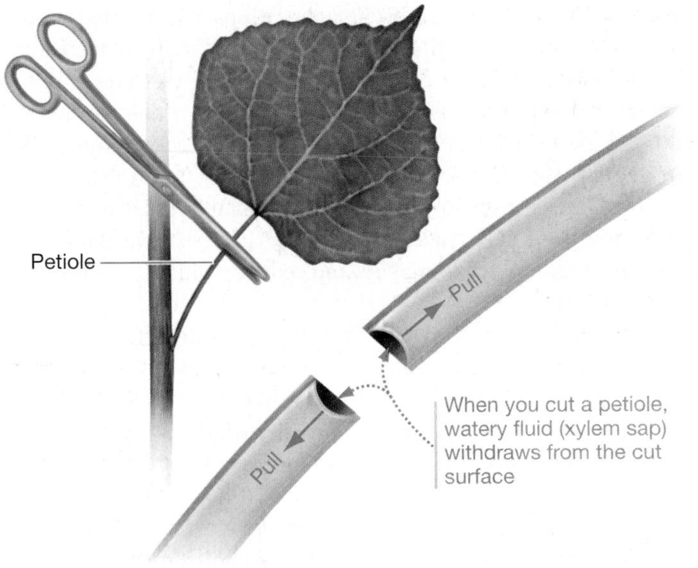

Figure 35.11 Xylem Sap in Cut Stems "Snaps Back."

pressure probe documented increased tension, or negative pressure.

- In addition, higher light intensity reduced the weight of the entire plant, suggesting that higher transpiration rates caused water loss.

Both observations are consistent with predictions that follow from the cohesion-tension theory.

These experiments convinced most biologists that increased transpiration leads to increased tension on xylem sap. Rising tension, in turn, lowers the water potential of leaves and exerts a pull on water in the roots, where the water potential is high. Based on these and other results, most biologists now accept the cohesion-tension theory.

Plant Features That Reduce Water Loss through Transpiration

One of the most important features of the cohesion-tension theory is that it does not require plants to expend energy to lift the water that the roots take in by osmosis. Xylem acts as a passive conduit—a set of narrow pipes that allows water to move from a region of higher water potential (the roots) to a region of lower water potential (the leaves). Water flows from roots to shoots as long as the water-potential gradient—from root to leaf to atmosphere—is intact.

When soils begin to dry, however, it becomes difficult for plants to replace water being lost via transpiration. If water is not replaced fast enough, the solute potentials of leaves drop, and leaves and branches begin to wilt. In response, stomata may close down partially or completely to reduce transpiration rates and conserve water. However, closing stomata affects the ability of plants to carry out photosynthesis, because CO_2 acquisition slows or stops.

The balance between conserving water and maximizing photosynthesis is termed the photosynthesis-transpiration

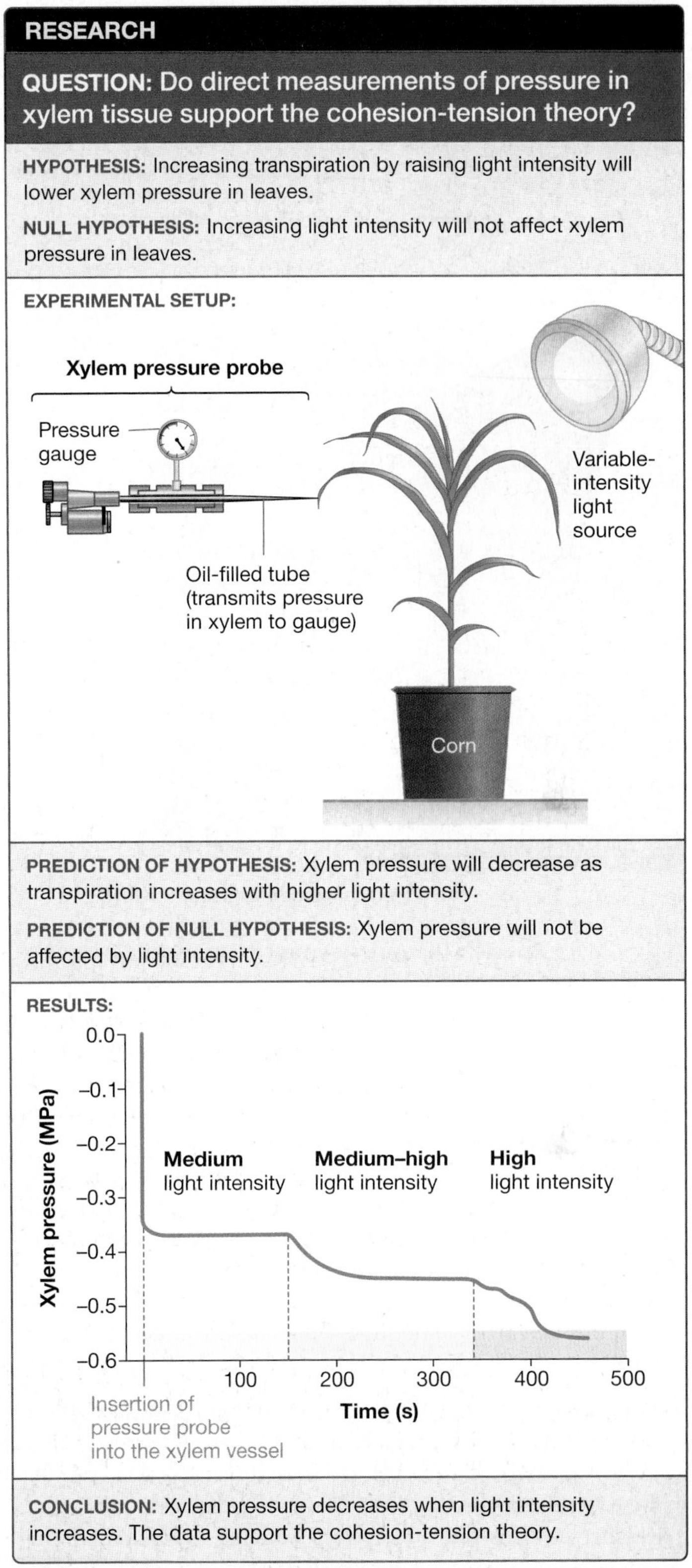

Figure 35.12 Changes in Pressure inside Xylem Can Be Measured.

SOURCE: C. Wei, M. T. Tyree, and E. Steudle. 1999. Direct measurement of xylem pressure in leaves of intact maize plants. A test of the cohesion-tension theory taking hydraulic architecture into consideration. *Plant Physiology* 121: 1191–1205.

✔**QUANTITATIVE** Suppose that during the same experiment, the researchers chose to plot changes in the water-potential gradient between the xylem in roots and leaves on the *y*-axis. What would the graph look like?

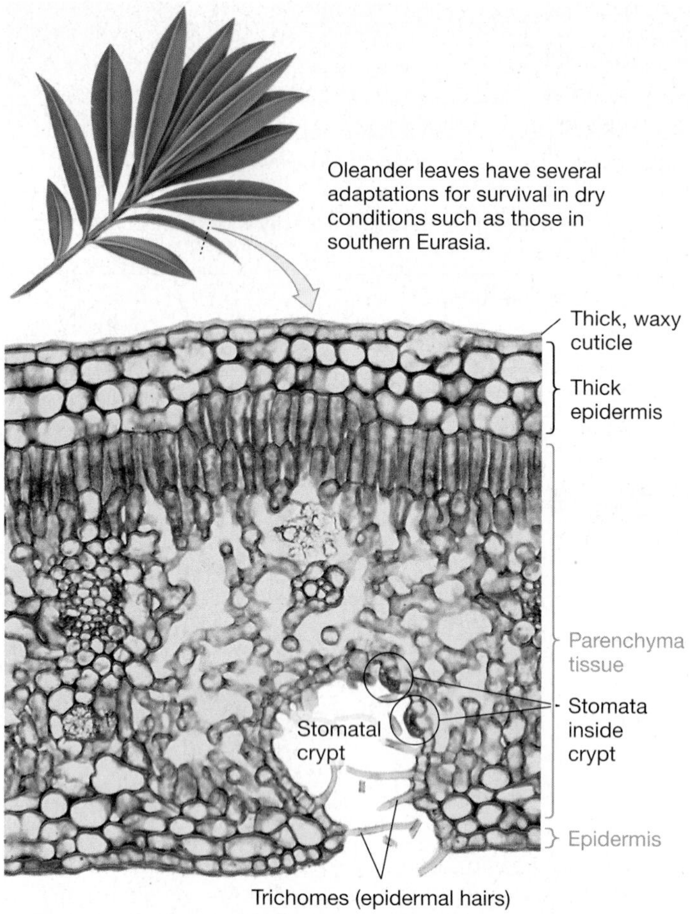

Figure 35.13 Species That Are Adapted to Dry Habitats Have Modified Leaf Structures.

compromise. This compromise is particularly delicate for species that grow on dry sites. How do they cope?

Plants that thrive in dry habitats have several adaptations that help them slow transpiration and limit water loss. Plants with adaptations for life in dry environments are called **xerophytes**. Consider the oleander plant, which is native to the dry shrub-grassland habitats of southern Eurasia. Figure 35.13 shows a cross section through an oleander leaf, which has the following special features:

- A particularly thick cuticle covers the upper surface of oleander leaves. This waxy layer minimizes water loss from cells, especially those that are directly exposed to sunlight. In general, species that are adapted to dry soils have much thicker cuticles than do species adapted to wet soils. They often have a thick, multilayered epidermis as well.
- The stomata of oleanders are located on the undersides of their leaves, inside deep **stomatal crypts** in the epidermis. Hairlike extensions of epidermal cells called **trichomes** (Ch. 34, Section 34.2) help shield the stomata from the atmosphere. The leading hypothesis to explain these features is that they slow the loss of water vapor from stomata to the dry air surrounding the leaf by minimizing air flow around the stomata.

Other species have different adaptations for limiting water loss. For example, recall that many species adapted to water-limited habitats—either cold environments where water is often frozen or deserts where rainfall is rare—have needlelike leaves (Ch. 34, Section 34.1). Needlelike leaves generally expose less surface area to the environment than do flat, blade-like leaves and thus have lower transpiration rates. Some plants, such as marram grass, produce curled leaves with stomata located toward the interior. This helps to trap moist air and reduce evaporation of water from stomata.

Many of the plant species that thrive in deserts and other hot, dry habitats can continue photosynthesizing even when soil moisture content is low. Recall that two novel biochemical pathways, **crassulacean acid metabolism (CAM)** and **C_4 photosynthesis**, allow plants to increase the CO_2 concentration in their leaves and conserve water (Ch. 10, Section 10.4). CAM plants open their stomata at night and store the CO_2 that diffuses into their tissues by adding the carbon dioxide molecules to organic compounds. C_4 plants minimize the extent to which their stomata open because they use CO_2 very efficiently. Mesophyll cells in C_4 plants take up CO_2 and add it to organic compounds. The CO_2 is then transferred to specialized cells called **bundle-sheath cells** (Ch. 10, Section 10.4), where the Calvin cycle operates. In effect, the C_4 cycle is a mechanism for concentrating carbon dioxide in cells deep inside the leaf, so stomata do not have to be fully open.

Like the cuticle and stomatal crypts of oleanders, CAM and C_4 plants have adaptations that help them conserve water by limiting transpiration.

CHECK YOUR UNDERSTANDING

✔ **If you understood this section, you should be able to . . .**

1. Predict changes in transpiration rates when each of the following occurs: (a) stomata close, (b) a rain shower starts, and (c) weather changes as dry air blows in.
2. Explain how water moves through xylem and often accumulates as droplets at the tips of grass leaves even when transpiration does not occur.
3. Explain whether stomata in pineapple (an example of a CAM plant) are likely to be closed or open during the daytime.

Answers are available in Appendix A.

35.3 Translocation of Sugars

After you complete this section, you should be able to . . .

- Explain how sugars are moved through phloem under positive pressure from leaves through the shoot and root systems.

Xylem and phloem are often found adjacent to each other, and both are specialized for transporting fluids. However, these two types of vascular tissue are radically different in their structure and in the way they transport fluids.

Translocation refers to the movement of sugars by bulk flow in multiple directions throughout a plant—but specifically, from sources to sinks. In vascular plants, a **source** is a tissue where sugar enters the phloem; a **sink** is a tissue where sugar exits the phloem.

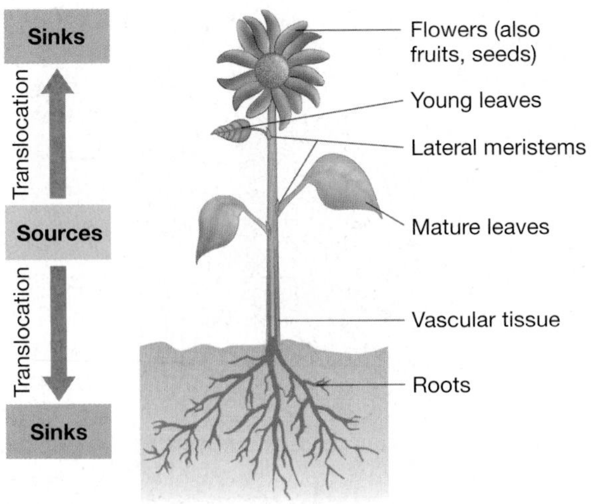

Figure 35.14 Sugars Move from Sources to Sinks.

Where do sources and sinks occur in a plant? The answer often depends on the time of year.

- ***During the growing season*** Mature leaves and stems that are actively photosynthesizing produce sugar in excess of their own needs. These tissues act as sources. Sugar moves from leaves and stems to a variety of sinks, where sugar use is high and production is low. Apical meristems, lateral meristems, developing leaves, flowers, developing seeds and fruits, and storage cells in roots all act as sinks (Figure 35.14).
- ***Early in the growing season*** When a plant resumes growth after the winter or the dry season, sugars move from storage areas to growing areas. Storage cells in roots and stems act as sources; developing leaves act as sinks.

Tracing Connections between Sources and Sinks

To explore the relationship between sources and sinks in more detail, consider research on sugar-beet plants that involved exposing their leaves to carbon dioxide molecules containing the radioactive isotope ^{14}C. The goal was to track where carbon atoms moved after they were incorporated into sugars via photosynthesis.

The location of ^{14}C atoms inside a plant can be documented in two ways: **(1)** by measuring the amount of radioactive CO_2 emanating from different tissues, or **(2)** by laying plant parts on X-ray film and allowing the radioactivity to expose and darken the film.

Many researchers have enclosed individual leaves of intact plants in a bag and introduced a known amount of radioactive CO_2 for a short period of time. One typical experiment documented that mature leaves retained just over 9 percent of the labeled carbon. In contrast, growing leaves retained 67 percent. These data are consistent with the prediction that fully expanded leaves act as sources of sugar, while actively growing leaves act as sinks.

In similar experiments, researchers exposed all the leaves on a growing plant to radioactive CO_2. One such experiment found that over 16 percent of the total carbon was translocated to root tissue within 3 hours. This result is consistent with the prediction that, during the growing season, roots also act as sinks.

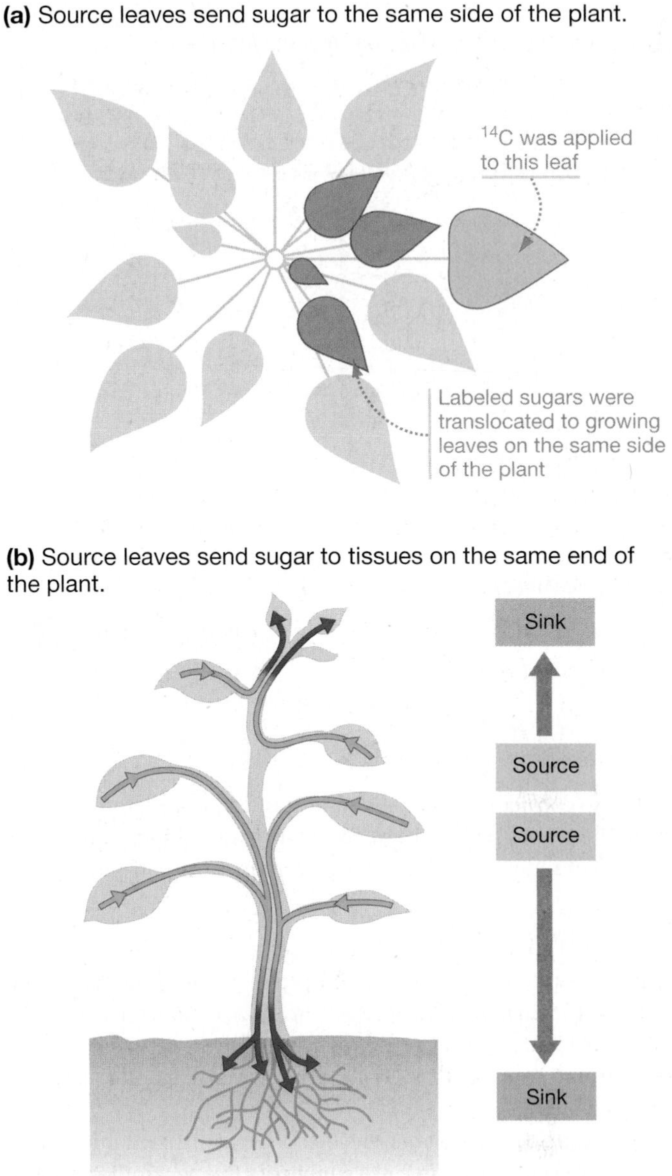

Figure 35.15 Sources Supply Sinks on the Same Side and Same End of the Plant.

Similar experiments support the following generalizations:

- Sugars can be translocated rapidly—typically 50–100 centimeters per hour (cm/hr).
- There is a strong correspondence between the physical locations of sources and sinks.

The second point is particularly interesting. Mature leaves that act as sources send sugar to tissues on the same side of the plant (Figure 35.15a). Experiments with tall plants also show that leaves on the upper part of the stem send sugar to apical meristems, but leaves on the lower part of the plant send sugar to the roots (Figure 35.15b). See **Making Models 35.1** on page 760 for help in mapping the movement of sugars between sources and sinks.

Making Models 35.1 Tips on Drawing Arrows (I)

You can use arrows in many ways in biological models—to show the passage of time, to point to something important, or to show movement of a substance from one place to another. Make sure you understand the meaning of arrows when you see them or draw them.

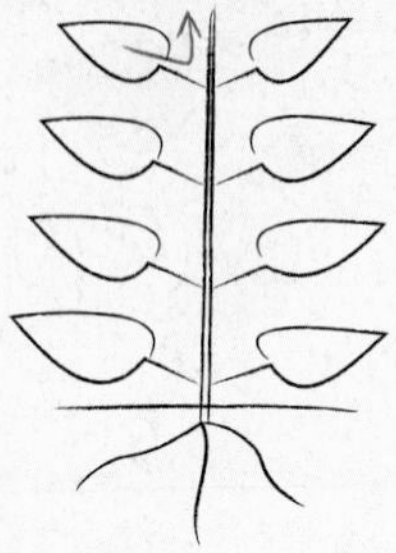

MODEL Use the model here to show the translocation of sugars from sources (leaves) to sinks (apical meristem and roots). One arrow is shown; add seven others and label one of them to show its biological meaning.

To see this model in action, go to the Study Area of **Mastering Biology**

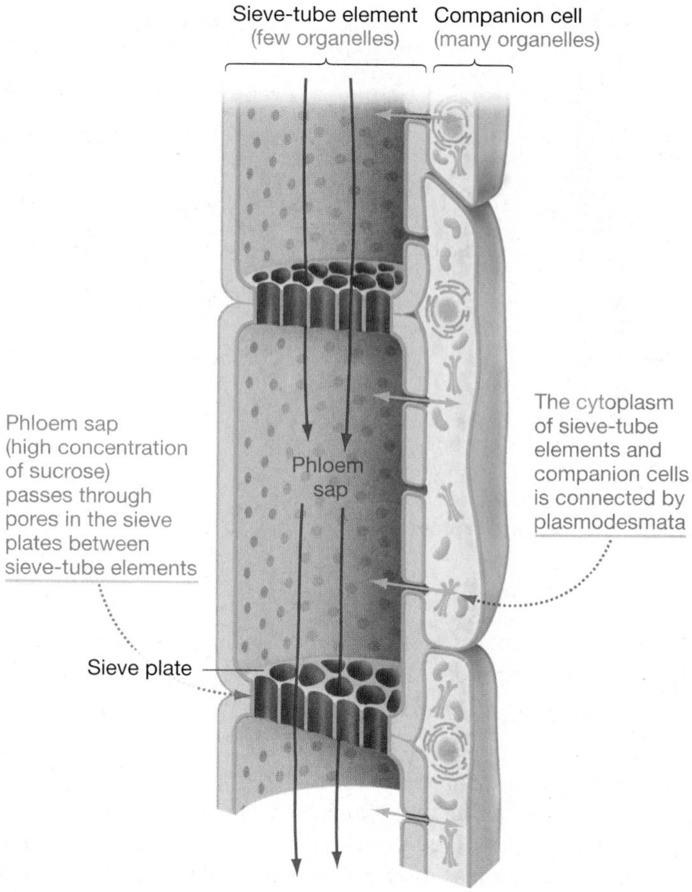

Figure 35.16 Sieve-Tube Elements Are Connected by Pores in Sieve Plates.

Why would leaves send sugar to tissues on a certain side or part of the plant's body? The answer hinges on understanding the structure of phloem and its arrangement in roots and shoots.

The Anatomy of Phloem

Phloem consists largely of two cell types, **sieve-tube elements** and **companion cells** (Ch. 34, Section 34.2). Unlike the tracheids and vessel elements that make up most of the xylem, sieve-tube elements and companion cells are alive at maturity and lack secondary cell walls.

Recall that, in most plants, sieve-tube elements lack nuclei and most other organelles. They are connected to one another, end to end, by perforated **sieve plates** (Figure 35.16). The pores, which are simply enlarged plasmodesmata, create a direct connection between the cytoplasms of adjacent cells. Companion cells, in contrast, have nuclei and a rich assortment of ribosomes, mitochondria, and other organelles. Companion cells are located adjacent to sieve-tube elements and function as their "support staff."

You might also recall that the phloem in primary and secondary vascular tissue is continuous throughout the plant—meaning that there is a direct anatomical connection between the phloem from the tips of shoots to the tips of roots (Ch. 34, Section 34.2). The sieve-tube elements in phloem represent a continuous system for transporting sugar throughout the plant body.

Vascular bundles, which contain xylem and phloem, run the length of plants, and certain bundles extend into specific branches, leaves, and lateral roots. Phloem sap does not move from one vascular bundle to another—instead, each bundle is separate and independent.

Based on these structural features, the physical relationships observed between sources and sinks in herbaceous plants are logical. For example, the phloem in the leaves on one side of a plant connects directly with the phloem of branches, stems, and roots on the same side, through a specific set of vascular bundles.

The phloem sap that flows through vascular tissue is often dominated by the disaccharide **sucrose**—table sugar. Phloem sap can contain small amounts of minerals, amino acids, mRNAs, hormones, and other compounds as well. How does this solution move? What mechanism is responsible for translocating sugars from sources to sinks?

The Pressure-Flow Hypothesis

In 1926 Ernst Münch proposed the **pressure-flow hypothesis**, which states that events at source tissues and sink tissues create a pressure potential gradient in phloem (Figure 35.17). The water in phloem sap moves down this pressure gradient, and sugar molecules are carried along by bulk flow.

Like the cohesion-tension theory for water transport in xylem, the pressure-flow hypothesis is based on movement down a water-potential gradient created by changes in pressure potential. Unlike the cohesion-tension model, however, transpiration does not provide the driving force to move phloem sap. Instead, differences between turgor pressure in the phloem near source tissues and turgor pressure in the phloem near sink tissues generate the necessary force. Creating these differences in turgor pressure usually requires an expenditure of energy.

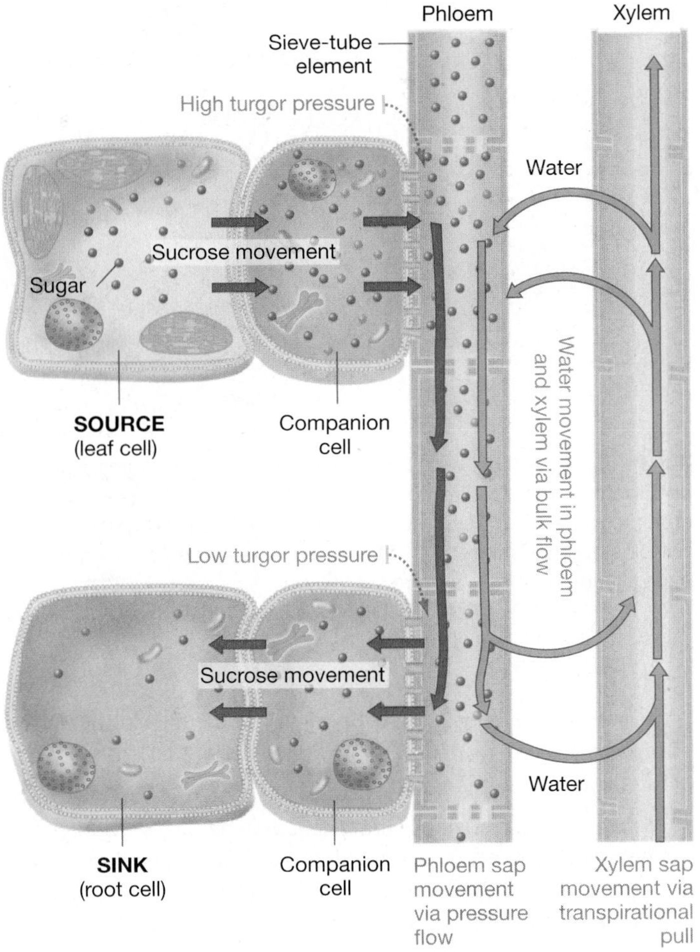

INTERACTIVE Figure 35.17 The Pressure-Flow Hypothesis: High Turgor Pressure Near Sources Causes Phloem Sap to Flow to Sinks. The pressure-flow hypothesis predicts that water cycles between xylem and phloem, and that water movement in phloem is a response to a gradient in pressure potential.

✔ **MODEL** This diagram shows how water moves between xylem and phloem in the middle of the growing season, when leaves are sources and roots are sinks. Add new arrows, in new colors, to indicate the direction of water and phloem sap flow in spring, when roots act as sources and leaves act as sinks.

Creating High Pressure Near Sources and Low Pressure Near Sinks To understand how Münch's model works, start with the source cell at the upper left in Figure 35.17. The red arrows reflect Münch's proposal that sucrose moves from source cells into companion cells and from there into sieve-tube elements.

Because of this phloem loading, the phloem sap in the source tissue has a high concentration of sucrose. Compared to the water in the adjacent xylem cells in a vascular bundle, the phloem sap has a very low solute potential.

As the blue arrows in the upper part of the diagram show, water moves down a water-potential gradient—flowing passively from xylem across the selectively permeable plasma membrane of sieve-tube elements. In response, turgor pressure begins to build in the sieve-tube elements in the source region.

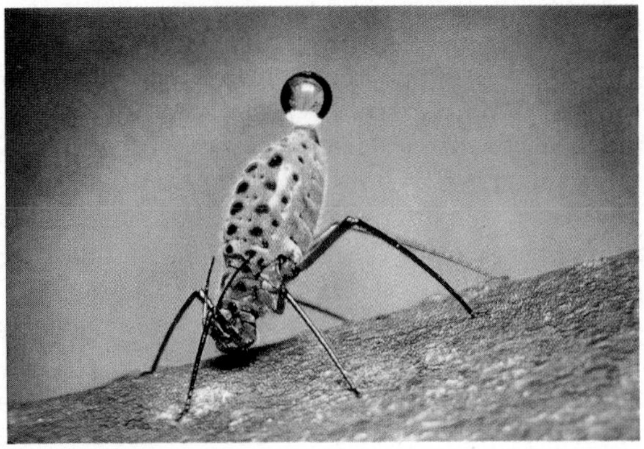

Figure 35.18 Aphids Feed on Phloem Sap. The tip of this aphid's mouthpart (the stylet) is in a sieve-tube element within the plant stem. The droplet emerging from the aphid's anus is honeydew, which consists of sugary phloem sap.

What is happening at the sink? Münch proposed that cells in the sink (bottom left in Figure 35.17) remove sucrose from the phloem sap by passive or active transport. As a result of this phloem unloading—a loss of solutes—the water potential in sieve-tube elements increases until it is higher than the water potential in adjacent xylem cells. As the blue arrows at the bottom of the figure show, water flows across the selectively permeable membranes of sieve-tube elements into xylem down a water-potential gradient. In response, turgor pressure in the sieve-tube elements in the sink drops.

The net result of these events is high turgor pressure in phloem at a source tissue and low turgor pressure at the sink, created by the loading and unloading of sugars, respectively. This difference in pressure drives phloem sap from source to sink via bulk flow. There is a one-way flow of sucrose and a continuous loop of water movement. Water returns to the source tissue via the xylem.

Testing the Pressure-Flow Hypothesis The pressure-flow hypothesis is logical, given the anatomy of vascular tissue and the principles that govern water movement. But has any experimental work supported the theory? Some of the best tests have relied on aphids—small insects that make their living ingesting phloem sap.

Aphids insert a syringe-like mouthpart, called a stylet, into sieve-tube elements. The pressure on the fluid in these cells forces it through the stylet and into the aphid's digestive tract. Excess water and sucrose that the aphid does not need is excreted out its anus as droplets of "honeydew" (Figure 35.18).

If the aphids are then severed from their stylets, sap continues to flow out through the stylets. This phenomenon allows researchers to collect phloem sap efficiently for analysis. It also confirms that the aphids do not actively suck the fluid. As predicted by the pressure-flow model, phloem sap is indeed under pressure, which forces it to enter the aphid.

This observation supports one of the fundamental predictions of the pressure-flow hypothesis. Now the question is, How does sucrose enter phloem at sources and leave phloem at sinks in a way that sets up the water-potential gradient?

Phloem Loading

In contrast to the cohesion-tension model of water movement in xylem, pressure flow often requires that plants expend energy to set up a water-potential gradient in phloem. To establish a high pressure potential in sieve-tube elements near source cells, large amounts of sugar have to be transported into the phloem sap—enough to raise the solute concentration of sieve-tube elements. This requirement is illustrated in Figure 35.17, top left.

In some cases, loading is active—it requires an expenditure of ATP and some sort of membrane transport system. But when the sucrose concentration in source cells is extremely high, movement of sucrose into sieve-tube elements can also occur via passive diffusion through plasmodesmata. How does phloem loading occur? What specific membrane proteins are involved?

To answer these questions, let's start by reviewing how transport proteins make it possible for sugars and other large or charged substances to cross a phospholipid bilayer.

How Are Sucrose and Other Solutes Transported across Membranes? **Passive transport** occurs when ions or molecules move across a plasma membrane by diffusion—that is, down their electrochemical gradient. The adjective "passive" is appropriate because no expenditure of energy is required for the movement to occur.

Recall that small, uncharged molecules diffuse across phospholipid bilayers rapidly (Ch. 6, Section 6.2). But ions and many large molecules diffuse across phospholipid bilayers slowly if at all, even when their movement is favored by a strong electrochemical gradient. To diffuse rapidly, they must avoid direct contact with the phospholipid bilayer by passing through a membrane protein.

Two types of membrane proteins—channels and carriers—facilitate the passive diffusion of specific ions or molecules (Ch. 6, Section 6.4).

1. **Channel proteins** form pores that selectively admit certain ions.
2. **Carrier proteins** undergo a conformational change that transports specific molecules across the lipid bilayer.

Channels and carriers are responsible for **facilitated diffusion.**

Active transport (Figure 35.19) occurs when ions or molecules move across a plasma membrane against their electrochemical gradient. The adjective "active" is appropriate because cells must expend energy in the form of ATP to move solutes in an energetically unfavorable direction.

Active transport always involves membrane proteins. **Pumps** are proteins that change shape when they bind ATP or a phosphate group from ATP. As they move, pumps transport ions or molecules against an electrochemical gradient.

Pumps establish an electrochemical gradient that favors the movement of an ion or molecule back across the plasma membrane. For example, the pump on the left side of Figure 35.19 has established a high concentration of ions or molecules (symbolized by the light gray balls) outside of the cell. As a result, these ions or molecules can then diffuse back into the cell. Why is this important?

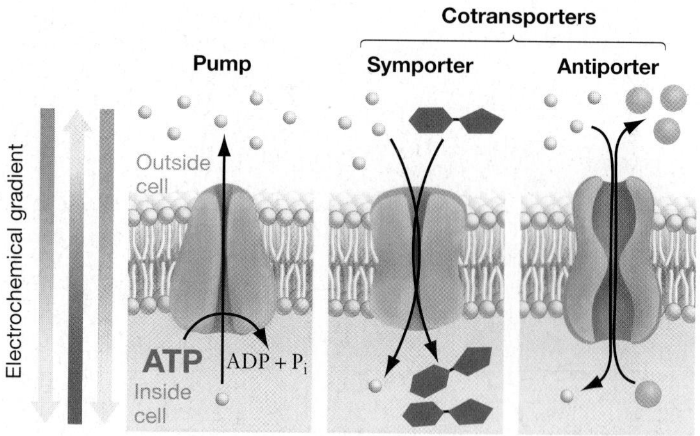

Figure 35.19 **Active Transport Moves Ions or Molecules against an Electrochemical Gradient.** All forms of active transport require an expenditure of ATP. Pumps use ATP directly; cotransport proteins use ATP indirectly. Cotransport depends on a previous expenditure of ATP by a pump.

In many cases, the electrochemical gradients established by pumps are used to transport other molecules or ions by two types of membrane proteins called **cotransporters.**

1. **Symporters** transport solutes *against* a concentration gradient, using the energy released when a different solute moves in the same direction *down* its electrochemical gradient. The red molecules in the middle of Figure 35.19 are moving through a symporter.
2. **Antiporters** work in a similar way, except that the solute being transported against its concentration gradient moves in the direction *opposite* that of the solute moving down its concentration gradient. The orange ions in Figure 35.19, right, are moving through an antiporter.

When solutes move through cotransport proteins, **secondary active transport** occurs.

Active transport, secondary active transport, and passive transport are all involved in translocating sugars in plants. Let's look first at events at source tissues, where the active transport of sucrose into sieve-tube elements results in a high pressure potential.

How Are Sugars Concentrated in Sieve-Tube Elements at Sources? Because sucrose may be more highly concentrated in companion cells than in photosynthetic cells where it is produced, researchers hypothesized that sucrose transport from source cells into companion cells may involve active transport. Another key observation—that strong pH differences exist between the interior and exterior of phloem cells—suggested that sucrose might enter companion cells with protons by secondary active transport.

Figure 35.20 illustrates the logic behind this hypothesis. Note this key claim: A membrane protein in companion cells hydrolyzes ATP and uses the energy that is released to transport protons (H^+) across the membrane to the exterior of the cell.

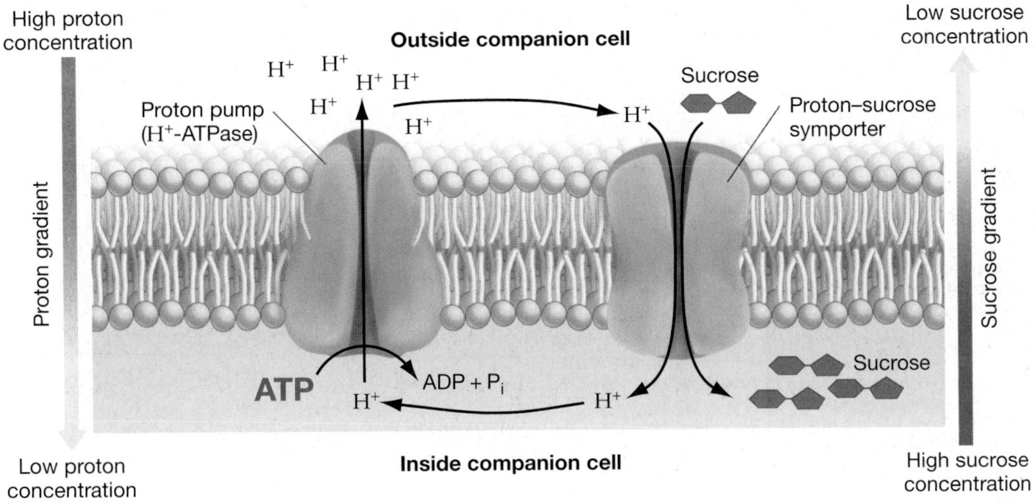

Figure 35.20 A Model for Cotransport of Protons and Sucrose. According to the model of cotransport, a proton pump hydrolyzes ATP to move hydrogen ions to the exterior of the cell. The resulting high concentration of H^+ outside the cell establishes an electrochemical gradient that allows the transport of sucrose into the cell against its concentration gradient.

Proteins like these are called **proton pumps**, or more formally, **H^+-ATPases**. Proton pumps establish a large difference in charge and in hydrogen ion concentration on the two sides of the membrane. The resulting electrochemical gradient favors the entry of protons into the cell.

The right side of Figure 35.20 shows the second key claim: A symporter acts as a conduit for protons and sucrose to enter the cell together. With this symporter, protons move down their electrochemical gradient, and sucrose moves against its concentration gradient.

If phloem loading depends on the activity of a proton–sucrose symporter, a mutation in the gene encoding this protein should result in a plant that cannot export sucrose from leaves to various sink tissues. What would this mutant plant look like?

Michael Sussman's lab used the model plant *Arabidopsis thaliana* (**BioSkills 11**) to isolate a mutant lacking a gene called *SUC2*. This gene was known to be expressed in vascular tissue and to encode a proton–sucrose symporter.

Mutant plants grew much more slowly than wild-type plants, indicating that something was indeed wrong with the mutant (**Figure 35.21a**). When the researchers looked at leaves under a microscope, they found that chloroplasts from mutant plants accumulated huge amounts of starch, an indication that they could not efficiently export sugar to the rest of the plant (**Figure 35.21b**). To test this hypothesis, the researchers then applied radioactive sucrose to a single source leaf on mutant and wild-type plants and tracked its location 5 hours later.

As **Figure 35.21c** shows, mutant leaves that were supplied with radioactive sucrose retained much more of the sucrose than did wild-type leaves. Also, sink tissues in the mutant plants

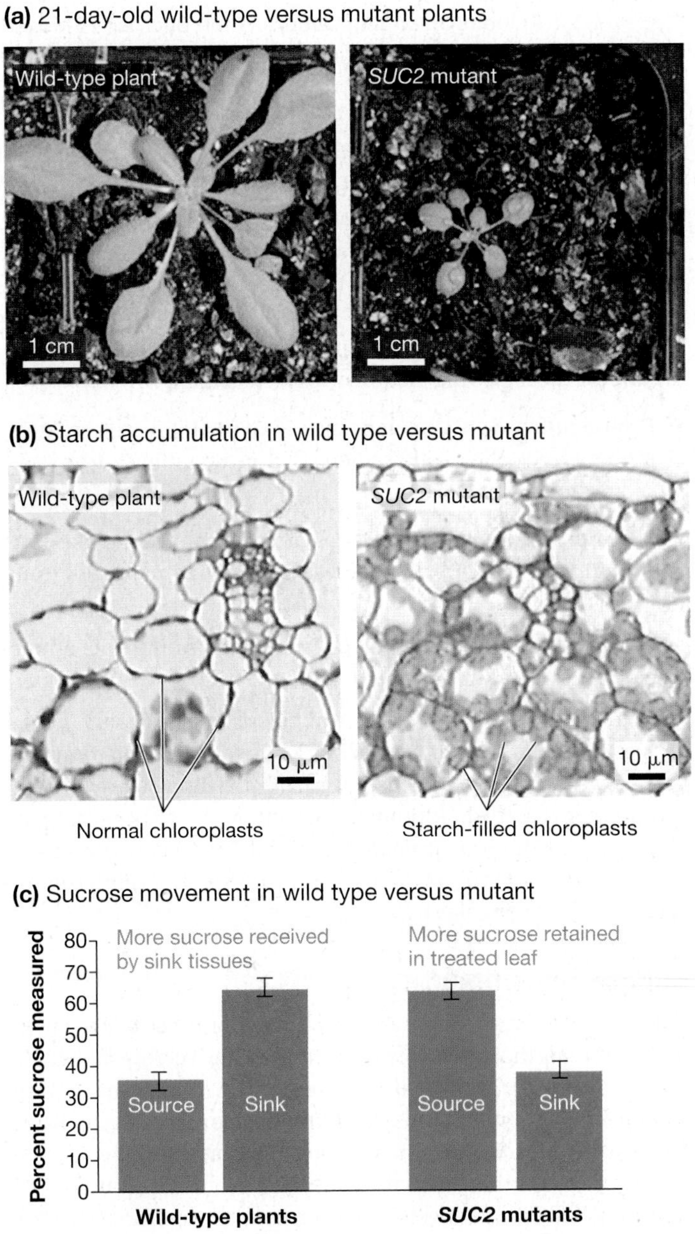

Figure 35.21 Proton–Sucrose Symporters Play a Role in Long-Distance Transport in Plants.

DATA: J. R. Gottwald, P. J. Krysan, J. C. Young, et al. 2000. *Proceedings of the National Academy of Sciences* 97: 13979–13984.

✔ **PROCESS OF SCIENCE** Why did the researchers measure radioactivity after only 5 hours, and not wait a longer period such as 24 hours?

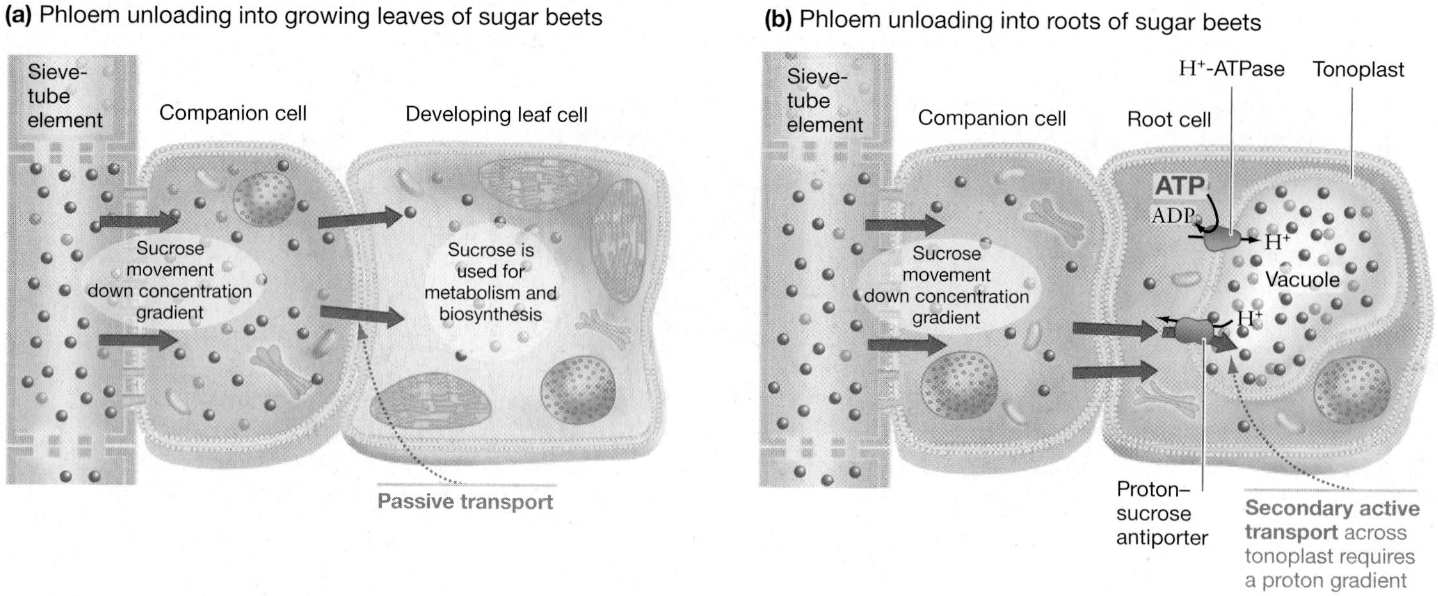

Figure 35.22 Phloem Unloading Occurs When Sucrose Is Moved Out of Phloem Tissue into Sinks. The mechanism of phloem unloading can vary from sink to sink, such as in **(a)** young leaves and **(b)** roots of the same plant.

received much less radioactive sucrose than wild-type sink tissues. Without the proton–sucrose symporter in companion cells, mutant plants were less able to load sucrose into phloem cells for long-distance transport, so that the excess sugar accumulated in leaves as starch.

There is now considerable genetic and biochemical evidence to support the following model for phloem loading:

- Proton pumps in the membranes of companion cells create a strong electrochemical gradient that favors a flow of protons into companion cells.
- A symporter in the membranes of companion cells uses the proton gradient to bring sucrose into companion cells from the source cells.
- Once inside companion cells, sucrose moves into sieve-tube elements via plasmodesmata.

More recent research with *Arabidopsis* suggests that members of a gene family called SWEET code for transport proteins that move sucrose out of cells—a crucial first step before phloem loading can proceed. Although such work on the mechanism of phloem loading in *Arabidopsis* and other species continues, most researchers are convinced that proton pumps and proton–sucrose symporters play a key role.

Phloem Unloading

Phloem unloading at sinks can also be active or passive. The membrane proteins that are involved in transporting sucrose molecules out of the phloem, as well as the mechanism of movement, vary among different types of sinks within the same plant. Mechanisms for phloem unloading also vary among different species.

To appreciate this diversity, consider how sucrose is unloaded in the phloem of sugar beets—a crop grown for the storage tissues in its root, which is a major source of the granulated and powdered sugar sold in grocery stores. Two major sinks are found in sugar beets: young, actively growing leaves, and the enlarged roots.

In young leaves of sugar beets, sucrose is unloaded down a concentration gradient by simple diffusion. This passive transport occurs because sucrose is rapidly used up inside the cells of young leaves to provide energy for ATP synthesis as well as carbon for the synthesis of cellulose, proteins, nucleic acids, and phospholipids needed by growing cells (Figure 35.22a).

In the roots of the same plant, however, an entirely different mechanism is responsible for unloading sucrose. Root cells in this species have a large vacuole that stores sucrose. The membrane surrounding this organelle is called the **tonoplast**. It contains two types of protein pumps that work together to accumulate sucrose in the vacuole, much like the phloem loading process described earlier (Figure 35.22b). As protons become concentrated in the tonoplast, they have a tendency to diffuse back out of the vacuole. A proton–sucrose cotransporter then uses the proton gradient to move sucrose from the cytosol into the vacuole against its concentration gradient. An important distinction in this case is that the cotransporter is an antiporter, while in source tissues it is a

CHECK YOUR UNDERSTANDING

✔ If you understood this section, you should be able to . . .

1. Describe how sieve plates facilitate the movement of sugars between adjacent sieve-tube elements.
2. Predict what would happen to the pressure gradient normally found in phloem if the companion cells were treated with a toxin that prevents proton pumps from working properly.

Answers are available in Appendix A.

symporter (see Figure 35.19). The active transport of sucrose into the vacuole allows sucrose to move passively from phloem into the storage cells, keeping the water potential of the phloem sap near the sink low—as required by the Münch pressure-flow hypothesis.

To summarize, more than eight decades of research provide convincing evidence that the pressure-flow hypothesis is fundamentally correct. Nevertheless, results of some recent studies indicate that the pressure-flow hypothesis may be incomplete and unable to explain the movement of sugars through phloem in all situations. This remains an active area of research.

The cohesion-tension theory for water movement and the pressure-flow hypothesis for phloem sap movement represent major advances in our understanding of how plants work.

CHAPTER 35 Review

For media, go to **Mastering Biology**

35.1 Water Potential and Water Movement

- Plants lose water as an inevitable consequence of exchanging gases with the atmosphere. The flow of water from soil to air via plant tissues transports minerals to the shoot and follows a water-potential gradient.
- Water potential (Ψ) is a measure of the tendency of water to move down its potential energy gradient and is expressed as megapascals (MPa).
- In plants, water potential has two components: **(1)** a solute potential, formed by the concentration of solutes in a cell or tissue; and **(2)** a pressure potential, provided by the cell wall and other factors. The water potential of a cell, tissue, or plant is the sum of its solute potential and pressure potential.
- When selectively permeable membranes are present, water moves by osmosis from areas of high solute potential to areas of lower solute potential.
- When no membranes are present, water moves by bulk flow from areas of high pressure to areas of lower pressure, independently of differences in solute potential.

35.2 How Does Water Move from Roots to Shoots?

- According to the cohesion-tension theory, water is pulled in one continuous column from roots to shoots, against the force of gravity, by the surface tension caused by transpiration from leaves.
- Surface tension occurs at menisci that form as water evaporates from the walls of leaf cells, and it is transmitted downward via hydrogen bonding between water molecules. In this way, evaporation of water from leaves is responsible for the movement of water from roots to shoots.
- Plants that occupy dry habitats have traits that limit the amount of water they lose to transpiration. These include cuticle, stomata that are located in stomatal crypts on the undersides of their leaves, trichomes, or narrow, needlelike leaves.
- CAM and C_4 photosynthetic cycles are adaptations in some plants that limit water loss in dry habitats.

35.3 Translocation of Sugars

- According to the Münch pressure-flow hypothesis, sugars are translocated from sources to sinks via bulk flow along a pressure gradient that develops in phloem.
- A pressure gradient is generated by the active transport of sugars into sieve-tube elements in source tissues, coupled with the transport of sucrose out of sieve-tube elements at sink tissues. Water moves by osmosis from xylem into sieve-tube elements near sources and cycles back to xylem near sinks.

Answers are available in Appendix A.

✔ TEST YOUR KNOWLEDGE

1. **THINK CAREFULLY** A cell is placed in a solution that is hypotonic to the cell. Which of the following best describes movement of water in this situation?
   a. Water will only flow into the cell.
   b. Water will only flow out of the cell.
   c. Water will flow into and out of the cell, but the overall net movement will be out of the cell.
   d. Water will flow into and out of the cell, but the overall net movement will be into the cell.
2. What important role does the Casparian strip play in the movement of water through plants?
   a. forces water to move through the cytoplasm of living endodermal cells as it makes its way from the soil to the xylem
   b. causes cells to shrink, thereby increasing pressure within cells
   c. loads sugars into xylem, thereby causing water to enter the xylem by osmosis
   d. acts as a filter that prevents salts, heavy metals, and other pollutants from entering root hairs
3. What is the role of companion cells in the movement of sugars through plants?
   a. They are the sites of sugar production by photosynthesis.
   b. They are the sites where starch is converted to sucrose.
   c. They secrete sucrose, which draws sugars through phloem under negative pressure.
   d. They accumulate sucrose, which is then transferred to adjacent sieve-tube elements.
4. Consider the following statements regarding the transport of phloem sap. Select True or False for each statement.
   T/F This is a passive process that is driven by the evaporation of water from leaves.
   T/F Sugars tend to move from sources to sinks.
   T/F Phloem sap moves through sieve-tube elements under positive pressure.
   T/F Sieve-tube elements and vessel elements are commonly involved in the transport of phloem sap.

✔ TEST YOUR UNDERSTANDING

5. The cells of a certain plant species can accumulate solutes to create very low solute potentials. Which of these statements is correct?
   a. The plant's transpiration rates will tend to be extremely low.
   b. The plant can compete for water effectively and live in relatively dry soils.
   c. The plant will grow most effectively in soils that are saturated with water year-round.
   d. The plant's leaves will wilt easily.
6. **MODEL** Draw a plant cell in pure water. Add dots to indicate solutes inside the cell. Now add more dots to indicate an increase in solute potential inside the cell. Add an arrow showing the *net* direction of water movement in response. Add arrows showing the direction of wall pressure and turgor pressure in response to water movement. Repeat the same exercise, but this time, add solutes to the solution outside the cell at a concentration that is greater than inside the cell.
7. A mutant plant lacking the ability to pump protons out of leaf companion cells will be unable to do which of the following?
   a. initiate transpiration
   b. load sucrose into sieve-tube elements
   c. carry out photosynthesis
   d. transport water through the xylem
8. **THINK CAREFULLY** Your friend claims that phloem always carries sugars down a plant. What, if anything, is wrong with that statement?

✔ TEST YOUR PROBLEM-SOLVING SKILLS

9. **THINK CAREFULLY** Consider a tree that is 50 m tall and is transpiring roughly 90 liters of water each day. Approximately how many calories will the tree use to transpire this quantity of water?
10. **SOCIETY** Salt is used to melt snow and keep roads clear during the winter in many cities. Land adjacent to de-iced roads often ends up with a high concentration of salt in the soil. Explain why plants growing near salted roads may appear wilted in the spring.

✔ PUT IT ALL TOGETHER: Case Study

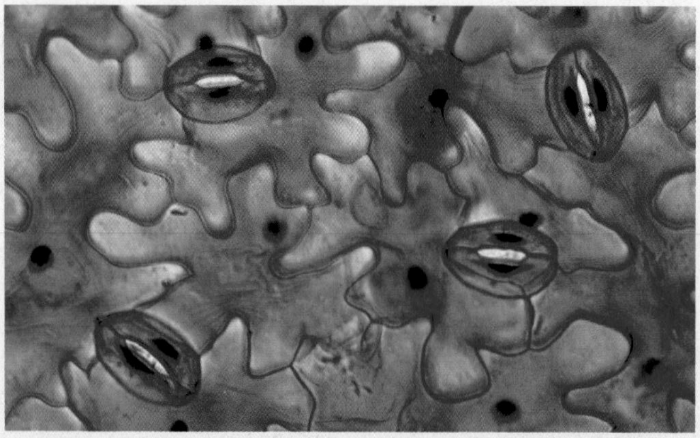

How do plant stomata respond to climate change?

Atmospheric CO_2 has been increasing rapidly since the late 1800s, largely due to human activities. Recall that CO_2 enters leaves through stomata and can then be used for photosynthesis. However, transpiration occurs as a result of water evaporating through stomata. How have plants responded to elevated CO_2 levels?

11. Which of these structural features can help to limit water loss in plants that occupy dry habitats?
    a. abundant companion cells and sieve-tube elements
    b. stomata that are located in pits on the undersides of their leaves, or narrow, needlelike leaves
    c. extensive networks of xylem and phloem
    d. stomata that are located on the top surface of leaves, or broad leaves
12. What impact, if any, do you predict elevated CO_2 levels will have on the number of stomata in leaves, and on the transpiration rate?
13. The amount of water that evaporates from stomata over a period of time is referred to as stomatal conductance, which is determined largely by the number of stomata in a given area of leaf surface. Researchers obtained specimens from preserved collections and measured stomatal conductance in leaves from oak trees and pine trees that grew at various times under different CO_2 levels. The data are shown in the following graph. In general, is the maximum stomatal conductance rate in plants more or less than it was a century ago?

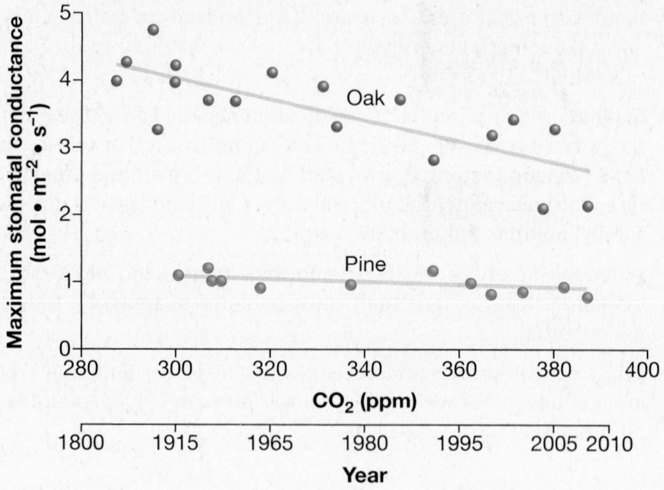

SOURCE: E. I. Lammertsma et al. 2011. *PNAS* 108: 4035–4040.

14. One prediction of global climate change is that there will be an increase in periods of drought in some regions. Given the data just presented, will plants be more or less likely to survive periods of drought as they are exposed to rising CO_2 levels?
15. **QUANTITATIVE** In the year 1915, the stomatal conductance of oak was approximately how many times higher than that of pine? How about in the year 2010?
16. Assuming that the CO_2 level continues to increase with time, how likely are plants to be able to continue to adapt by adjusting stomatal conductance?

Mastering Biology

Students Go to Mastering™ Biology for assignments, the eText, and the Study Area with animations, practice tests, and activities.

Professors Go to Mastering™ Biology for automatically graded tutorials and questions that you can assign to your students, plus Instructor Resources.

36 Plant Nutrition

In most plants, roots obtain the water and nutrients required for individuals to survive and thrive.

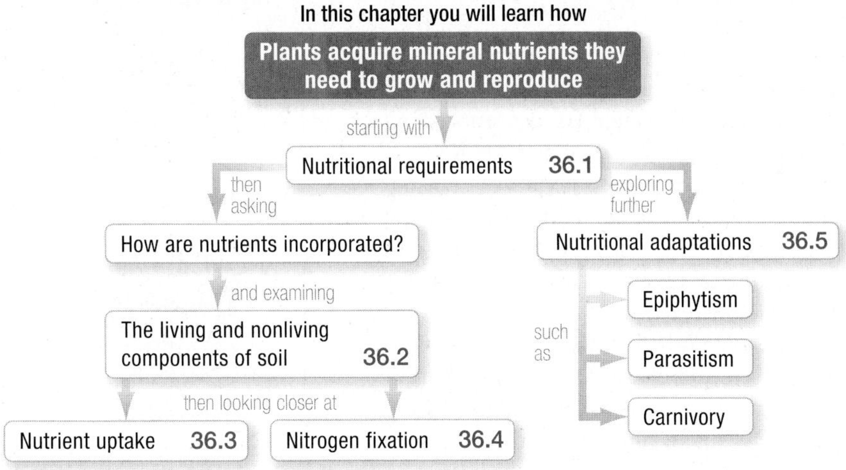

The most urgent tasks facing any organism are to acquire **(1)** carbon-containing molecules that will be used as cellular building blocks and **(2)** the chemical energy required to make ATP. Plants acquire both of these by producing sugar through photosynthesis. Yet plants cannot live on sugar alone. Plants synthesize all of their own nucleic acids, amino acids, enzymes and cofactors, chlorophylls, and other molecules necessary to grow and reproduce. Plants carry out some of the world's most impressive chemical syntheses.

A plant's ability to perform complex reactions depends on its capacity to harvest as raw materials a wide variety of elements in the form of ions and molecules. In addition to carbon dioxide and water, plants have to obtain nitrogen, phosphorus, potassium, and other elements. Most of these nutrients exist in soil, the majority as ions that are found at low—sometimes extremely low—concentrations. Once these ions are inside roots, the flow of water up the xylem carries the nutrients throughout the plant body.

The plant body is an efficient machine for harvesting these diffuse resources and concentrating them in cells and tissues. You have been introduced to how the organization and growth of the plant root and

BIG PICTURE

This chapter is part of the Big Picture. See how on pages 838–839.

shoot systems make resource acquisition possible (Chapter 34) and how water and nutrients are transported throughout the plant body (Chapter 35). In this chapter, you will concentrate on how plants take up elements from the soil so that these nutrients can be transported to the cells that need them.

Questions about nutrition are fundamental to understanding how plants work, increasing agricultural productivity, and maintaining the productivity of forests that supply lumber and fuel—and that also mitigate temperature increases associated with global climate change (Ch. 53, Section 53.3) by transforming atmospheric CO_2 into plant matter. Let's begin by analyzing the basic nutritional needs of plants.

36.1 Nutritional Requirements of Plants

After you complete this section, you should be able to ...

- Identify the nutrients required for plant growth.

What do plants need to live? In the early 1600s, Jean Baptiste van Helmont performed a classic experiment designed to answer this question. Van Helmont wanted to know where the mass of a growing plant comes from, and he used a willow tree as a study organism.

As **Figure 36.1** shows, he began by placing 90 kilograms (kg) of soil in a pot with a 2-kg willow sapling. He allowed the plant to grow for five years, adding only water during that time. At the end of the experiment, he reweighed the willow and the soil. The willow gained about 74 kg while the soil lost only 60 grams (g).

Where had the additional 74 kg of tree come from? Because he was not aware that gases have mass, van Helmont hypothesized that the new plant material came from water. He also ignored the loss of 60 g in the soil, chalking it up to measurement error.

As it turned out, van Helmont's measurements were not the problem—his conclusions were. Most of the added mass of the tree came from carbon dioxide in the atmosphere. The 60 g removed from the soil contained vital elements—the nutrients that are the focus of this chapter. What are they?

About half the elements in the periodic table—more than 60—can be found in the tissues of one or more plant species. The question that biologists, farmers, and foresters ask is, Which of these elements are essential for growth and reproduction in most species, and in what quantities?

Which Nutrients Are Essential?

Biologists define an **essential nutrient** as an element or compound that is required for normal growth and reproduction—meaning that the plant cannot complete its life cycle without it. Essential nutrients cannot be synthesized by the organism.

Researchers test whether a nutrient is essential by denying a specific element to plants and documenting what happens to the plants over time. For most vascular plants, 17 elements are essential. Just three of these—carbon, hydrogen, and oxygen—typically make up about 96 percent of the dry mass of a plant. The remaining 14 essential elements are sometimes called mineral nutrients, because they originate in soil.

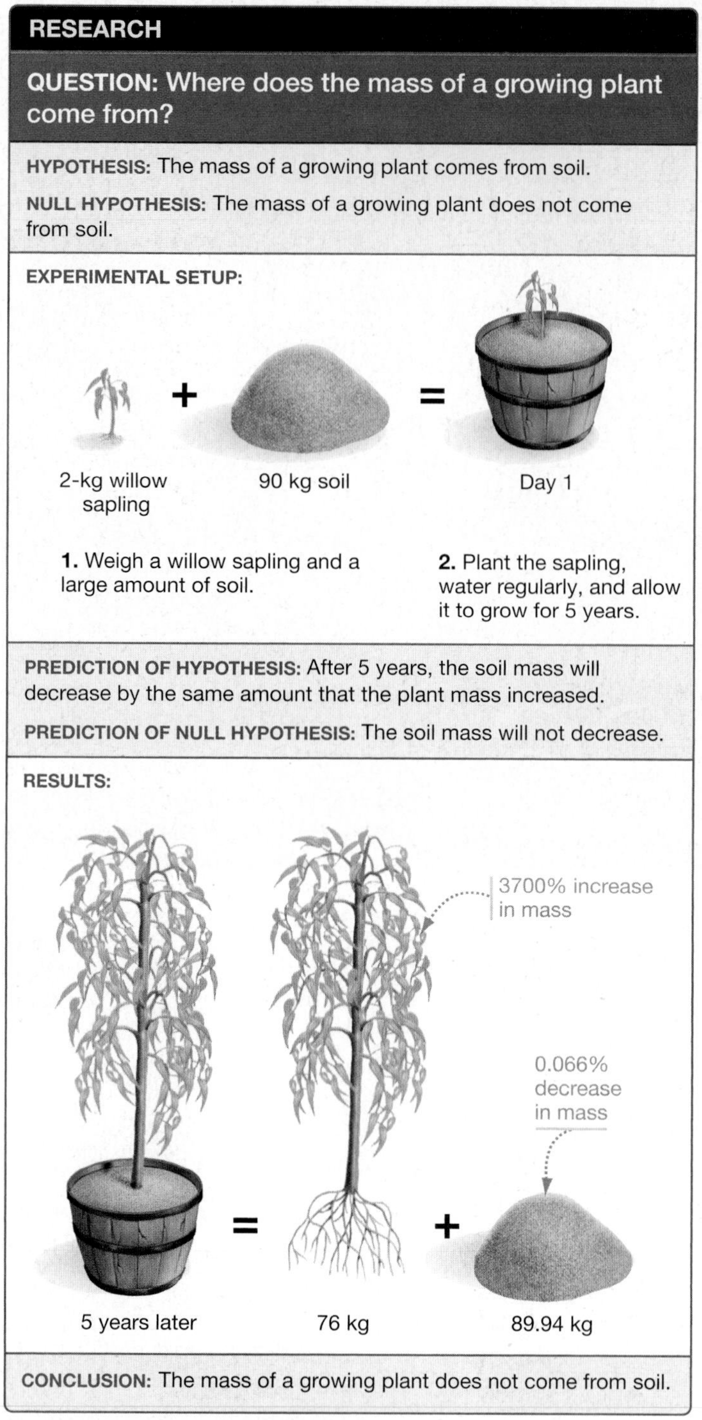

Figure 36.1 An Early Experiment Explored the Role of Soil in Plant Nutrition.

✔ **PROCESS OF SCIENCE** Many people think that most of a plant's mass comes from soil or water. Describe an experiment that would convince someone that most of a plant's mass comes from CO_2.

Although different classification schemes for the essential elements have been proposed, the most common is based on distinguishing nutrients that are obtained from water or air versus soil and then dividing soil nutrients into macronutrients and micronutrients based on their abundance in plants (**Table 36.1**).

Table 36.1 Essential Nutrients in Plants

Element	Form Available to Plants	Functions	Average % Dry Mass*	Deficiency Symptoms
Obtained from Water or Air: H_2O or CO_2				
Oxygen	O_2, H_2O	Electron acceptor in cellular respiration; major component of organic compounds	45	Usually affects roots: cells suffocate, leading to root rot and wilting
Carbon	CO_2	Substrate for photosynthesis; major component of organic compounds	45	Slow growth (starvation)
Hydrogen	H_2O	Major component of organic compounds; functions in electrical balance and establishment of electrochemical gradients	6	Slow growth due to cell death (desiccation)
Obtained from Soil: Macronutrients				
Nitrogen	NO_3^- (nitrate) NH_4^+ (ammonium ion)	Component of proteins, nucleic acids, ATP, chlorophyll, hormones, and coenzymes	1.5	Failure to thrive; chlorosis (yellowing of older leaves)
Potassium	K^+	Necessary for osmotic adjustment in cells; required for synthesis of organic molecules; cofactor for some enzymes	1.0	Chlorosis at margins of leaves or in mottled pattern; weak stems; short internodes
Calcium	Ca^{2+}	Important to regulatory functions, cell-wall structure, signal transduction; stabilizes membranes; enzyme cofactor	0.5	Necrosis (small spots of dead cells) in meristems; deformation of young leaves; stunted, highly branched root system
Magnesium	Mg^{2+}	Chlorophyll component; activates many enzymes	0.2	Chlorosis between leaf veins; premature leaf drop
Phosphorus	$H_2PO_4^-$ (dihydrogen phosphate ion) HPO_4^{2-} (hydrogen phosphate ion)	Component of ATP, nucleic acids, phospholipids, and several coenzymes	0.2	Stunted growth in young plants; dark green leaves with necrosis
Sulfur	SO_4^{2-} (sulfate ion)	Component of proteins containing methionine and cysteine, electron transport proteins, and coenzymes	0.1	Stunted growth; chlorosis
Obtained from Soil: Micronutrients				
Chlorine	Cl^- (chloride ion)	Necessary for water-splitting step of photosynthesis; functions in water balance and electrical balance	0.01	Wilting at leaf tips; general chlorosis and necrosis of leaves or development of bronze color
Iron	Fe^{3-} (ferric ion) Fe^{2-} (ferrous ion)	Necessary for chlorophyll synthesis; component of cytochromes and ferredoxin; enzyme cofactor	0.01	Chlorosis between veins of young leaves
Manganese	Mn^{2+}	Involved in photosynthetic O_2 evolution; enzyme activator; important in electron transfer	0.005	Chlorosis between leaf veins and small necrotic spots
Zinc	Zn^{2+}	Involved in synthesis of the plant hormone auxin, maintenance of ribosome structure, enzyme activation	0.002	Small internodes; stunted and distorted ("puckered") leaves
Boron	$H_2BO_3^-$ (borate ion)	Strengthens cell walls; required for pollen tube growth and normal membrane function	0.002	Black necrosis in young leaves and buds
Copper	Cu^+ (cuprous ion) Cu^{2+} (cupric ion)	Cofactor of some enzymes; present in lignin of xylem	0.0006	Light-green leaves with necrotic spots; twisted and malformed leaves
Nickel	Ni^{2+}	Cofactor for enzyme functioning in nitrogen metabolism	0.00001	Necrosis at leaf tips
Molybdenum	MoO_4^{2-} (molybdate ion)	Cofactor in nitrogen reduction; essential for nitrogen fixation	0.00001	Chlorosis between veins; necrosis of older leaves

*These percentages were obtained by drying vascular plants and documenting what proportion of the waterless mass consists of various elements.

Macronutrients Plants need relatively large quantities of certain elements in the soil; these elements are called **macronutrients**. Some of them are major components of nucleic acids, proteins, and phospholipids, all of which are plentiful in plants.

Among the macronutrients, nitrogen (N), phosphorus (P), and potassium (K) are particularly important because they often act as **limiting nutrients**, meaning their availability limits plant growth. If N, P, and/or K are added in appropriate quantities to soil as fertilizer, plant growth usually increases. This outcome explains why the leading ingredients in virtually every commercial fertilizer are N, P, and K.

Micronutrients In contrast to macronutrients, plants require **micronutrients** in small quantities. When plant tissues are dried and analyzed, micronutrients are typically present in only trace amounts. Instead of acting as components of macromolecules, micronutrients usually act as cofactors for specific enzymes—substances that are required for normal enzyme function (Ch. 8, Section 8.3).

It's important not to underestimate the role of micronutrients, even though plants need only tiny amounts of them. For example, a typical plant contains just one molybdenum atom for every 60 million hydrogen atoms, not including water. Yet plants die without molybdenum, because it functions as a cofactor for several enzymes involved in nitrogen processing.

What Happens When Key Nutrients Are in Short Supply?

In some cases, biologists can examine a plant that is growing poorly and diagnose a nutrient deficiency (Figure 36.2). For example, if older leaves are in poor condition, the plant probably needs more N, P, K, or magnesium. These elements represent **mobile nutrients**—meaning they are readily transported from older leaves to younger leaves when they are in short supply—so older leaves deteriorate first when these elements are scarce. **Immobile nutrients** such as iron or calcium, in contrast, stay tied up in older leaves. When they are in short supply, younger leaves are the first to show deficiency symptoms.

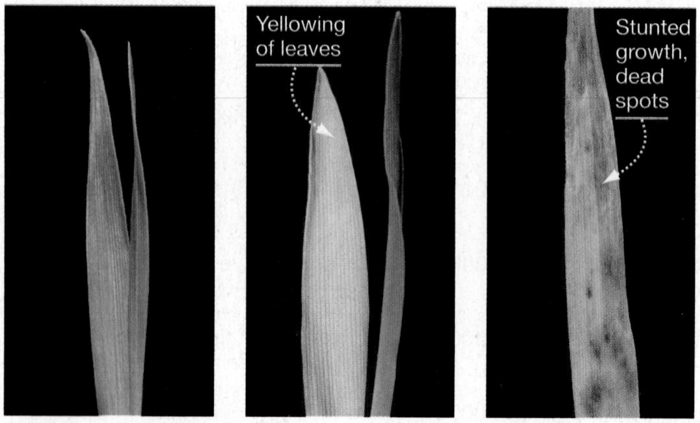

Figure 36.2 Nutrient Deficiencies Can Produce Distinctive Symptoms. Nutrient deficiencies produce symptoms specific to each nutrient. See Table 36.1 for more.

The ability to diagnose nutrient deficiencies is based largely on studies involving hydroponic growth systems. **Hydroponic growth** takes place in liquid cultures, without soil, so researchers can precisely control the availability of each nutrient. Individual essential elements can be omitted from hydroponic cultures to determine the effect of that deficiency. Incidentally, certain commercially available crops are often grown hydroponically as well (e.g., tomatoes, lettuce, and others).

Consider an experiment on copper deficiency in tomatoes (Figure 36.3). Researchers grew seedlings in two types of treatments. One treatment consisted of flasks containing water and

RESEARCH

QUESTION: How does copper deficiency affect plants?

HYPOTHESIS: Plants denied copper will grow poorly.

NULL HYPOTHESIS: Plants denied copper will grow normally.

EXPERIMENTAL SETUP:

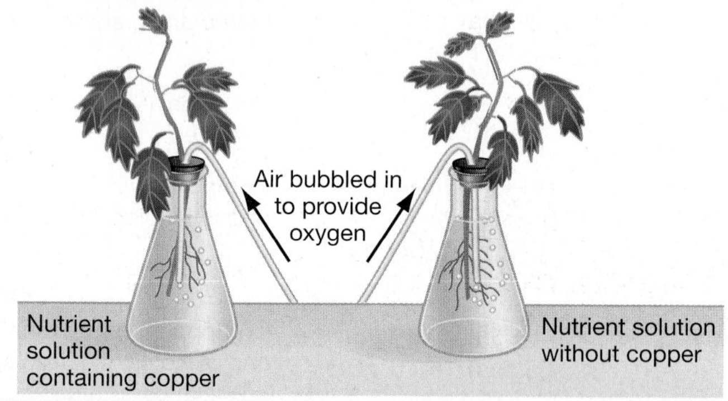

PREDICTION OF HYPOTHESIS: The copper-deficient plant will grow less than the normal plant.

PREDICTION OF NULL HYPOTHESIS: Both plants will grow the same.

RESULTS:

CONCLUSION: Copper deficiency leads to poor growth. All tissues appear to be affected adversely.

Figure 36.3 Hydroponics Can Be Used to Study Nutrient Deficiencies.

SOURCE: D. I. Arnon and P. R. Stout. 1939. The essentiality of certain elements in minute quantity for plants with special reference to copper. *Plant Physiology* 14: 371–375.

✔ **PROCESS OF SCIENCE** What problem would arise if this experiment had been done in soil with and without added copper?

all the essential nutrients in the relative concentrations that are optimal for tomato growth. The second treatment was identical, except that the nutrient solution lacked copper.

As Figure 36.3 shows, copper-deprived individuals have stunted shoots, unnaturally light-colored foliage, and curled leaves. Given copper's role as a cofactor or component of several enzymes involved in redox reactions required for ATP production, it is understandable that all tissues in the deficient plant were severely affected. Because copper is a micronutrient, it is reasonable to expect that a relatively small amount would cure the deficiency. In line with this prediction, researchers found that the symptoms were prevented if plants were cultured in a solution containing just 0.002 mg/L of copper. Analogous studies have been done on the other essential nutrients as well.

CHECK YOUR UNDERSTANDING

✔ If you understood this section, you should be able to ...

1. Explain why nitrogen is considered an essential macronutrient, while boron is considered an essential micronutrient.
2. Explain why a nitrogen deficiency might make the older leaves on a plant have a yellow color while the younger leaves have a healthy green color.

Answers are available in Appendix A.

36.2 Soil: A Dynamic Mixture of Living and Nonliving Components

After you complete this section, you should be able to ...

- Explain how the composition of soil affects nutrient availability.

The process of soil building begins with solid rock. As Figure 36.4 shows, weathering—the forces applied by rain, running water, temperature changes, and wind—continually breaks tiny pieces off large rocks. The weathering process is accelerated if small cracks develop in the rock. Plant roots that grow into a crack expand as they grow, widen the crack, and break off small flakes or pebbles. A similar effect occurs in high latitudes or at high elevations when water enters the cracks, freezes in winter, expands, and breaks off pieces.

Depending on their size and composition, the particles resulting from these processes are called gravel, sand, silt, or clay. These rock fragments are the first ingredient in soil. As organisms occupy the substrate, they add dead cells and tissues and feces. This decaying organic matter is called **humus** (pronounced *HEW-muss*).

With time, soil eventually becomes a complex and dynamic mixture of inorganic particles, organic particles, and living organisms. A single gram of soil normally contains dozens of small animals and plant roots; hundreds of thousands of protists, fungi, and microscopic animals; and hundreds of millions of bacteria and archaea (Figure 36.5 on page 772).

Both the parent rock that contributes inorganic soil components and the organisms and organic matter that occupy soils vary from one site to another. **Texture**—the proportions of gravel, sand, silt, and clay—and other soil qualities vary as well. Soil texture is important for several reasons:

- Texture of a soil affects the ability of roots to penetrate and obtain water and nutrients, as well as to anchor and support the plant body. For example, soil dominated by clay-sized particles tends to compact and resist root penetration.
- Texture affects a soil's ability to hold water and make it available to plants. Water tends to adhere to clay and silt particles but runs through sand and gravel.
- A soil's texture and water content dictate the oxygen available for cellular respiration. Oxygen used by root cells is found in air pockets among soil particles. This explains why overwatering a plant is just as detrimental as underwatering it: Overwatering drowns a plant's roots.

The best soils for many plants, called *loams*, contain roughly equal amounts of sand, silt, and clay, along with a high proportion of humus. Loams that have good texture and large amounts of organic matter can take thousands of years to develop through the weathering of rocks and the continual addition of humus.

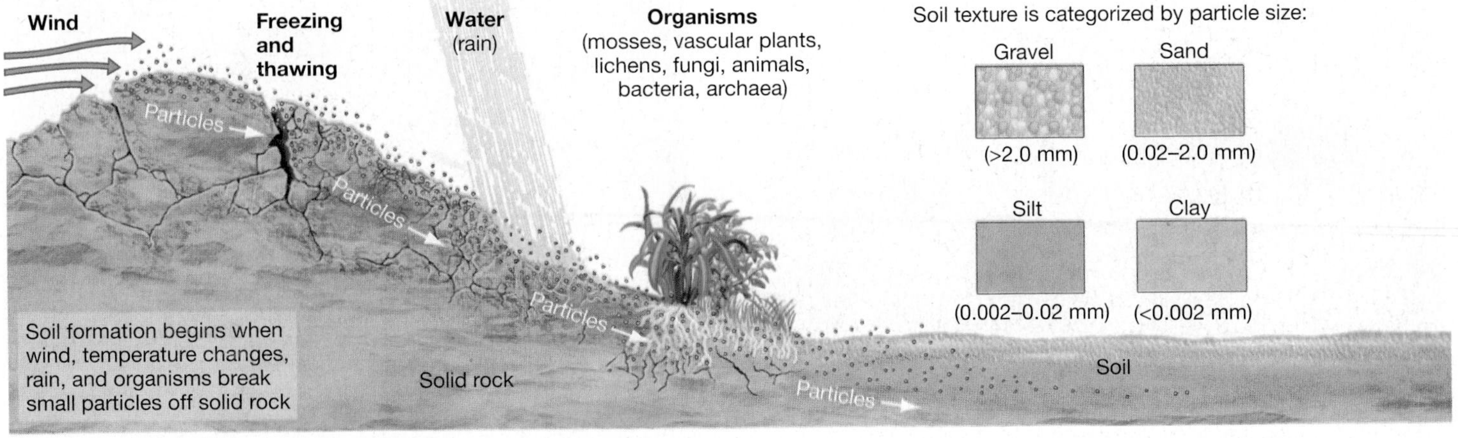

Figure 36.4 Soil Formation Begins with Weathering of Rock.

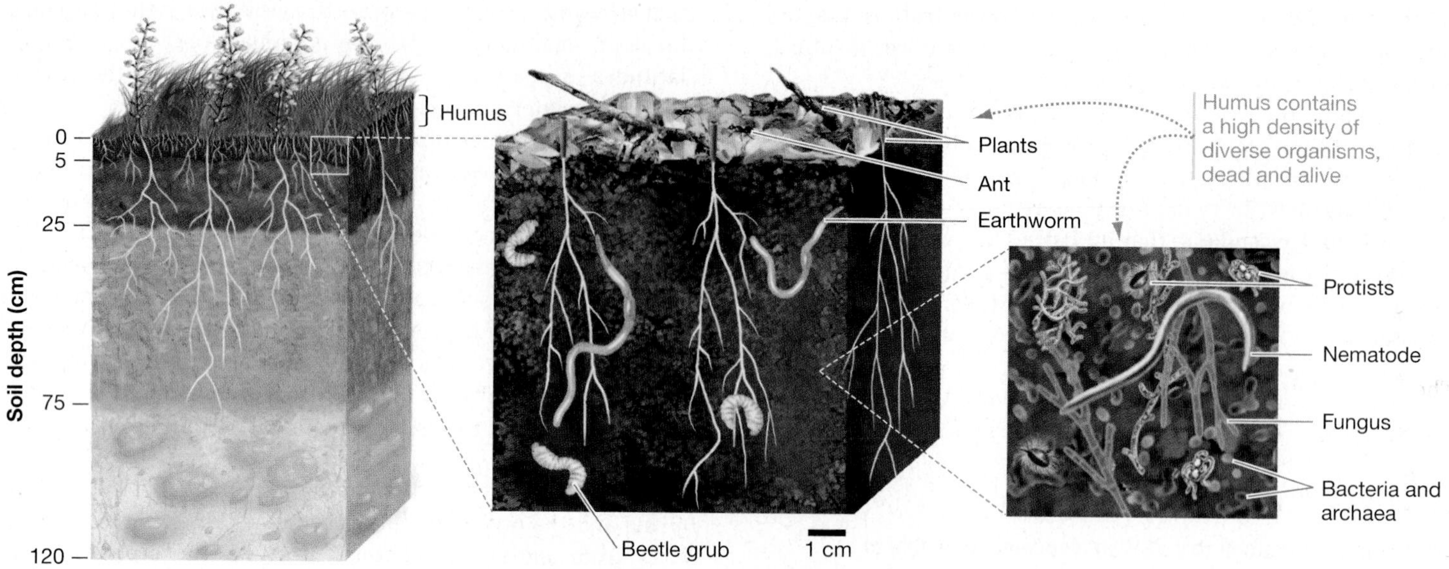

Figure 36.5 Mature Soils Are a Complex Mixture of Organic and Inorganic Components. In addition to mineral particles, soil contains humus—organic material derived from dead organisms—and a wide array of living organisms.

Unfortunately, it can take just a few years of abuse by humans for loams to blow or wash away.

The Importance of Soil Conservation

Soil erosion occurs when soil is carried away from a site by wind or water. It occurs naturally, such as when rivers cut away at their banks and carry material downstream. In most natural environments, though, the rate of soil formation exceeds the rate of soil erosion, so soils build up over time. Unfortunately, the situation can change dramatically when humans exploit an area.

When plant cover is removed through forestry, farming, or development, no stems and leaves remain to lessen the force of wind and rain, and plant roots no longer hold soil particles in place. The results can be devastating. At some locations in the Great Plains of the United States, 8–10 centimeters (cm) of topsoil were blown away during the Dust Bowl of the 1930s, when drought and poor farming practices left thousands of hectares of soil unprotected (**Figure 36.6a**).

U.S. soil erosion rates have declined dramatically since then. Still, researchers estimate that topsoil in almost 30 percent of all croplands in the United States is eroding ten times faster than it is replenished (**Figure 36.6b**). This loss of topsoil represents a major concern for food production in the future. Worldwide, an estimated 36 billion tons of soil are lost to erosion every year.

(a) Wind erosion in the United States, 1930s

(b) Runoff from a heavy rain carries topsoil from a crop field.

Figure 36.6 Soil Erosion Can Have Devastating Consequences.

If soil is managed carefully, however, it can be a renewable resource. Techniques that maintain long-term soil quality and productivity are the basis of **sustainable agriculture** and sustainable forestry. Farmers can reduce soil loss dramatically by

- planting rows of trees as windbreaks;
- using techniques that minimize the amount of plowing and tilling needed to control weeds; and
- planting crops in strips that follow the contour of hillsides.

What Factors Affect Nutrient Availability?

The elements required for plant growth are found in the soil not as atoms, but as **ions**. The second column of Table 36.1 lists the ion forms of some essential nutrients available to plants. Notice that some of these nutrients are available as elemental ions, such as K^+ or Cl^-, while others exist as molecular ions, such as HPO_4^{2-} or NO_3^-. The ionic charge of nutrients, as well as soil texture and soil pH, affect nutrient availability. Let's examine the relationships among these factors.

The Role of Ionic Charge and Soil Texture The ions present in soil typically behave in one of two ways, depending on their charge (Figure 36.7). Anions—ions with negative charges—usually dissolve in soil water, because they interact with water molecules via hydrogen bonding. However, phosphate ions are an exception to this rule, as they can form insoluble complexes with iron, aluminum, calcium, or other positively charged cations.

Because they exist as solutes, negatively charged anions are readily available to plants for absorption. They are also easily washed out of the soil by rain, however. The loss of nutrients via the movement of water through soil is called **leaching**.

Cations—ions with positive charges—dissolve in soil water but are not as immediately available as anions. In solution, cations interact with the negative charges found on two types of soil particles: **(1)** organic matter that is rich in negatively charged organic acids; and **(2)** the surfaces of clay particles that are rich in mineral anions (see Figure 36.7).

A soil's texture affects its interactions with ions. Organic soils that contain clay tend to retain nutrients, because cations often remain attached to the negatively charged clay, and because these soils hold water (and thus anions) better than sandy soils do. Sandy soils, however, are more prone to leaching since the soil particles are not negatively charged.

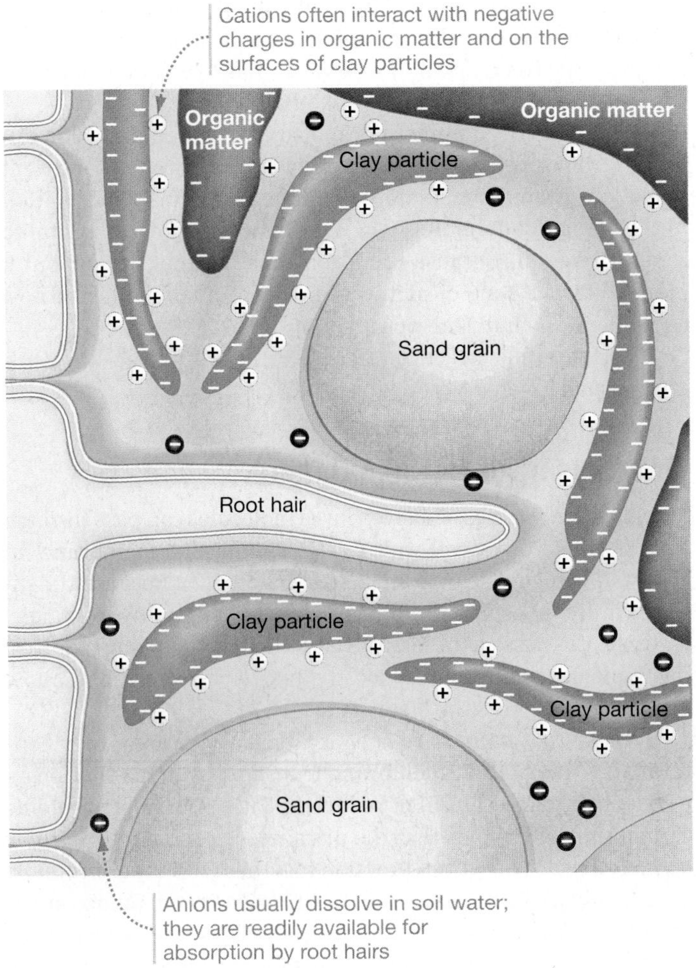

Figure 36.7 Cations Tend to Bind to Clay Particles and Organic Matter; Anions Stay in Solution.

The Role of Soil pH In addition to ionic charge of nutrients and soil texture, other factors influence the availability of essential elements. Perhaps the most important of these is soil pH.

Recall that the pH scale indicates the relative concentration of hydrogen ions in a solution, and that it ranges from 0 to 14 (Ch. 2, Section 2.2). Soils with low pH have a relatively high concentration of hydrogen ions and are considered acidic. Soils with a high pH contain relatively few hydrogen ions and are termed basic, or alkaline. Most plants thrive in soils with a relatively neutral pH—between 6 and 7. But some plants, such as blueberries, grow best in acidic soils (pH around 3 to 5); others such as lavender and fennel thrive on alkaline soils (pH above 8).

Acidic soils are found in regions such as conifer forests, where the decomposition of organic matter produces carbonic, phosphoric, or nitric acid. Alkaline soils, in contrast, are common in regions where limestone ($CaCO_3$) is abundant. Calcium ions released when limestone reacts with water replace protons clinging to soil particles. The protons then react with CO_3^{2-} to form bicarbonate ions (HCO_3^-), lowering the hydrogen ion concentration of the soil and raising its pH.

Soil pH affects the availability of plant nutrients in a number of ways. For example, the presence of protons in soil water can cause the release of cations that are bound to soil particles. The process responsible is called **cation exchange**. Cation exchange occurs when protons or other soluble cations bind to negative charges on soil particles and cause bound cations, such

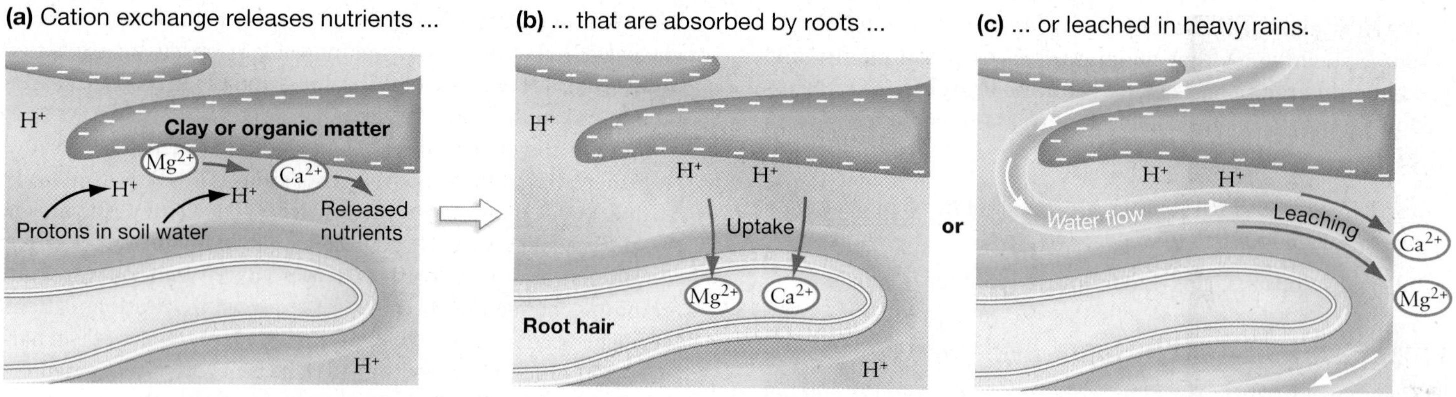

Figure 36.8 **Cation Exchange Releases Nutrients Bound to Soil Particles.** When cation exchange occurs, a proton binds to negative charges on clay or organic matter, releasing bound cations.

as magnesium or calcium, to be released from the soil (Figure 36.8a). Nutrients that are released by cation exchange become available for uptake by nearby plant roots (Figure 36.8b). Soil composition also affects the capacity for cation exchange.

Plants influence cation exchange because root cells release CO_2 as a by-product of cellular respiration (Chapter 9). The CO_2 reacts with H_2O to form carbonic acid, which releases protons. In addition, plants can actively transport protons out of root cells into the surrounding environment (that process is described in the next section). If soil is too acidic, however, rain may leach cations away before the roots can take up nutrients (Figure 36.8c). In many tropical rain forests around the world, deforestation has resulted in acidic, nutrient-poor soils that are not productive.

To summarize, anions stay in solution in soil water. They are readily available to plants but may wash away easily. Positive ions, in contrast, tend to bind to soil particles but can be released by cation exchange.

Nutrients are found at extremely low concentrations in soil but at high concentrations in plant cells. How are plants able to bring N, P, K, and other key elements into their bodies against a concentration gradient?

CHECK YOUR UNDERSTANDING

✔ **If you understood this section, you should be able to ...**

1. Explain whether negatively charged chloride ions or positively charged calcium ions are more readily absorbed by root hairs in soil with a neutral pH.
2. Explain how cation exchange enhances the availability of magnesium and calcium ions.

Answers are available in Appendix A.

36.3 Nutrient Uptake

In most species of plants, the root system is the site of nutrient uptake. You have been introduced to the general features of root anatomy (Ch. 34, Section 34.1) and the role of roots in water uptake (Ch. 35, Section 35.2). Now let's explore the detailed anatomy of roots and analyze events that occur in the plasma membranes of their epidermal and cortex cells.

After you complete this section, you should be able to ...

- Explain how plants absorb useful nutrients but prevent toxins from entering the shoot system.

Most nutrient uptake occurs just above the growing root tip, in the region called the **zone of maturation**. Recall that epidermal cells in this part of the root have extensions called **root hairs** (Figure 36.9). Root hairs dramatically increase the surface area available for nutrient and water absorption. For example, the root system of a single annual rye plant can have a total surface area the size of a basketball court.

Root hairs are so numerous and so efficient at absorbing nutrients from soil that, over time, they create a "zone of nutrient depletion" in soil immediately surrounding them. The formation of this mined-out region is why continued root growth is vital to a plant's health. Because roots continue to grow throughout a plant's life, the zone of maturation is continually entering new and potentially nutrient-rich areas of soil.

If an epidermal cell encounters soil where nutrients are available, what happens? How do ions enter?

Mechanisms of Nutrient Uptake

Ions, small molecules, and even large molecules can pass through the mesh-like plant cell wall freely. The plasma membrane, in contrast, is highly selective (Ch. 6, Section 6.2). Because the interior of the phospholipid bilayer is nonpolar and hydrophobic, it resists the passage of ions—even ions that the plant requires. Some membrane proteins span the bilayer, however, and allow specific ions to cross the membrane.

Because root hairs have such a large surface area, they contain large numbers of membrane transport proteins that bring nutrients into the cytosol of root cells. These transport proteins cannot import the ions that the plant needs completely on their own, however. They often function together with other proteins, specifically **proton pumps**, or H^+-ATPases. Proton pumps aid in the absorption of certain ions by establishing a relatively high concentration of protons on the outside of root epidermal cells

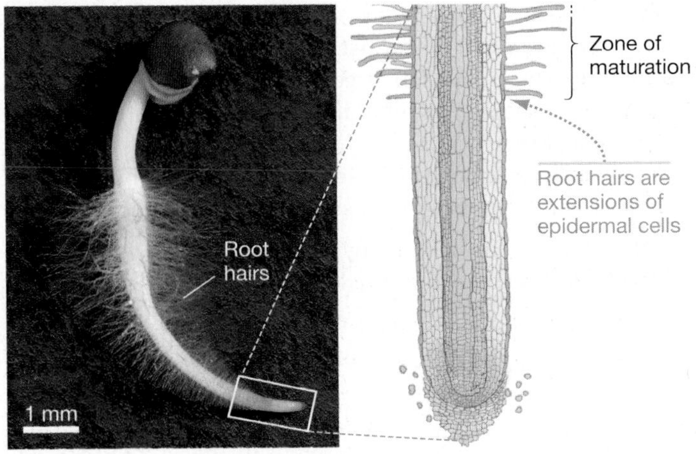

Figure 36.9 Root Hairs Provide Tremendous Surface Area for Nutrient Absorption.

(thus forming a relative negative charge inside the cells). As a result of the proton gradient, positively charged cations in soil are drawn into the negatively charged root cells through specific transport proteins. In addition, protons tend to diffuse back into the root cells, bringing anions along with them through cotransporter proteins. The next few sections describe the process of how proton pumps and transport proteins function in more detail.

Establishing a Proton Gradient Plants absorb nutrients and concentrate them in their tissues. In an epidermal cell, ions such as potassium (K^+), hydrogen phosphate (HPO_4^{2-}), and nitrate (NO_3^-) are many times more concentrated than they are in the soil water outside the cell. For these nutrients to be used by plants, they must enter root cells *against* a strong concentration gradient. How is this possible?

As Figure 36.10a shows, the answer hinges on proton pumps. Root cells use proton pumps to establish an **electrochemical gradient** that drives the import of essential ions into cells. These proteins are found in the plasma membranes of root epidermal and cortex cells, and they are similar to the pumps that make it possible for companion cells to load sucrose into phloem against a concentration gradient (Ch. 35, Section 35.4). Proton pumps in root cells transport protons to the exterior of the cells.

The activity of proton pumps leads to an excess of protons on the exterior of the plasma membrane relative to the interior. This results in a gradient favoring the movement of protons back into the cell. In addition, the outside of the cell becomes positively charged relative to the inside. Stated another way, there is a separation of charge—a **voltage**—across the membrane.

Proton pumps are found in all types of plant cells, and all plant cell membranes carry a voltage. Because voltage is a form of potential energy, the charges that are separated by the membrane create a **membrane potential**, or difference in electrical charge across a cell membrane. How is the membrane potential used to transport ions into cells *against* their concentration gradient?

Using a Proton Gradient to Import Cations Root cells establish a net negative charge as a result of pumping protons out of the cells. This negative charge draws cations into the cells as long as the electrical gradient is strong enough to overcome the concentration gradient. In essence, plant cells are batteries that are charged up to attract nutritionally necessary cations.

To drive this point home, consider potassium cations. Researchers who added radioactive potassium ions to the solution outside a root cell and followed their movement found that K^+ does indeed accumulate in cells due to an electrochemical gradient (Figure 36.10b). In follow-up experiments, biologists were able to isolate the membrane protein responsible for the uptake of K^+ and sequence the gene that encodes the protein.

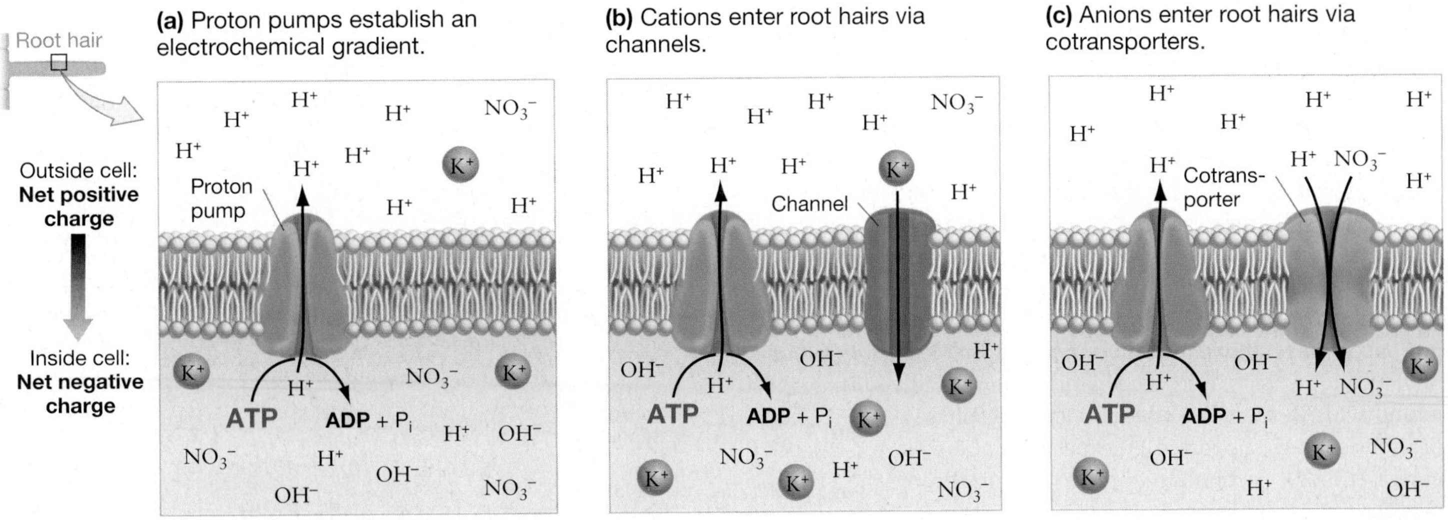

Figure 36.10 Ions Enter Roots along Electrochemical Gradients Created by Proton Pumps.

✔ Are cations and anions attracted to, or repelled from, the interior of the cell?

✔ **If you understand this concept, you should be able to explain what happens to cation uptake if (1) H^+-ATPases fail, and (2) a molecule blocks a cation channel.**

Cations such as K^+ enter root hairs through membrane proteins along an electrochemical gradient, but how is it possible for anions to enter? The negatively charged interior of the cell should repel these negatively charged ions.

Using a Proton Gradient to Import Anions Figure 36.10c shows how anions such as NO_3^- are able to enter root hairs against their electrochemical gradient. The key is that anions enter through membrane transport proteins called cotransporters (Ch. 35, Section 35.4), which transport two solutes at once. In this case, the cotransporter is a symporter: It brings NO_3^- into the cell along with a proton, and both solutes move in the same direction.

Here's the key observation: So much energy is released when a proton enters the cell along its electrochemical gradient that nitrate ions, phosphate ions, or other anions can be cotransported against their electrochemical gradients.

✔ **If you understand this concept, you should be able to explain what happens to anion uptake if (1) H^+-ATPases fail, and (2) a molecule blocks an anion symporter.**

Once ions have crossed the plasma membrane of an epidermal or cortex cell in the root, they continue to move via the symplast and through the endodermis toward the central vascular tissue. Recall that the endodermis is the site of the Casparian strip, which forces ions and water to pass through the plasma membrane and cytosol of endodermal cells (Ch. 35, Section 35.2). Once inside the endodermal layer, ions are actively transported out of cells, where they accumulate in the xylem sap. The ions are then transported passively by bulk flow to tissues throughout the plant.

Nutrient Transfer via Mycorrhizal Fungi To synthesize proteins and nucleic acids, plants need to extract large quantities of nitrogen and phosphorus from the soil. But concentrations of these nutrients are typically so low in soils that they limit plant growth. How do plants absorb enough N and P to satisfy their nutritional needs? In most cases, they rely on partnerships with fungi that live in close association with their roots.

You might recall that fungi and plant roots that live in association are called **mycorrhizae** (literally, "fungus-root"; Ch. 29, Section 29.1). Mycorrhizal fungi and plants are **symbiotic** ("living-together"), meaning that they live in physical contact with each other. Biologists estimate that more than 80 percent of all vascular plant species worldwide associate with mycorrhizal fungi.

In some habitats, plants receive large quantities of nitrogen from mycorrhizal fungi. The fungal symbionts in these associations are particularly efficient at digesting macromolecules (proteins and nucleic acids) in decaying organic material and absorbing the amino acids (a source of nitrogen) and phosphate ($H_2PO_4^-$) that are released. As supported by experimental evidence described earlier (Ch. 29, Section 29.3), the mycorrhizal fungi then transport the nutrients from soil to plant roots (Figure 36.11).

In exchange for these nutrients, the plant symbionts transfer sugars and other photosynthetic products to the fungi. Because there is a reciprocal exchange of nutrients, the symbiotic relationship is considered **mutualistic**, or mutually beneficial. Many horticulture companies sell mycorrhizal fungi products that can be applied to seeds. The fungi increase the overall growth and yield of major crop plants.

Fungi are particularly efficient at acquiring the nutrients required by plants, for two reasons:

1. Networks of filamentous hyphae increase the surface area available for absorbing nutrients by up to 700 percent.
2. Fungi can acquire nutrients from macromolecules in the soil that are unavailable to non-mycorrhizal plants.

Most plant species grow slowly and are overwhelmed by competitors if denied their mycorrhizal associates. Fungi clearly play a vital role in overall plant growth.

Mechanisms of Ion Exclusion

Plants have sophisticated systems for absorbing nutrients using membrane transporters and energy from the proton gradients they generate. However, not all ion uptake from soil is beneficial to plants. In some cases, plants are better off avoiding certain soil components.

Certain types of natural soils—as well as soils that have been contaminated by waste products from mining or smelting

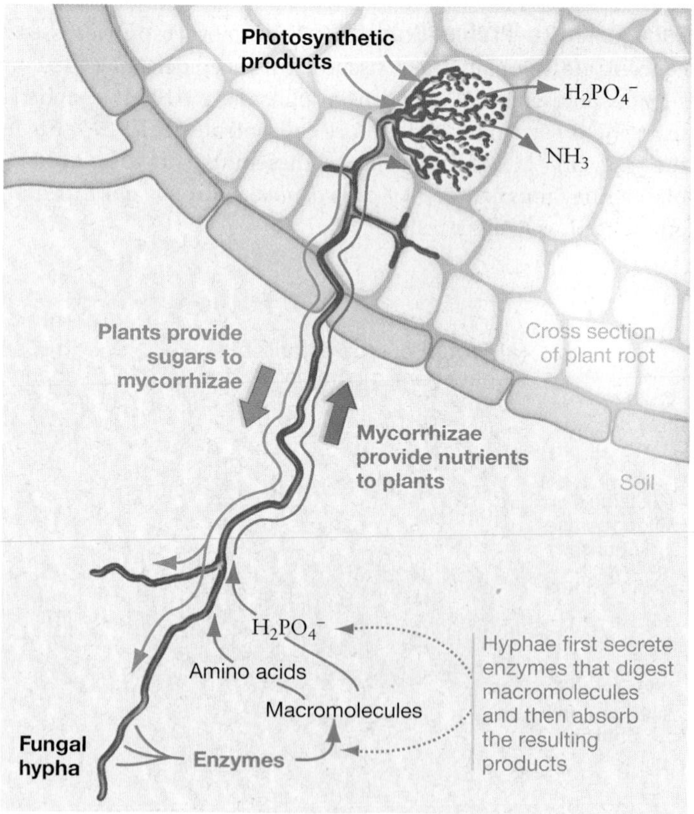

Figure 36.11 Most Mycorrhizal Fungi and Plants Are Mutualists. The fungus illustrated here provides nutrients to its host plant in exchange for photosynthetic products.

operations—contain enough cadmium, zinc, nickel, lead, or other metals to poison enzymes in most plants. Sodium is also detrimental at high concentrations: Too much Na^+ inside cells can disrupt enzyme function, and too much Na^+ in extracellular spaces can create a solute potential that pulls enough water out of cells to result in a loss of turgor.

Sodium poisoning is a key issue in many environments, including **(1)** ocean coastlines, **(2)** habitats near roads that are treated with salt to melt ice and snow, and **(3)** irrigated farmlands. When soils are irrigated, solutes are left behind when water evaporates from the surface—leading to salt buildup over time.

In light of these environmental factors, how do plants prevent detrimental ions from entering the xylem? There are several methods, which use either passive or active exclusion.

Passive Exclusion Ions move through roots following the same routes that water follows: the apoplastic, transmembrane, and symplastic pathways (Ch. 35, Section 35.2). Many ions do not enter the root symplast, simply because epidermal and cortex cells lack the requisite membrane transporters. This is a form of passive exclusion.

By having membrane transporters only for the ions that it requires, a plant can control which ions enter the symplast. Ions that reach the endodermis via the apoplastic pathway are blocked by the presence of the Casparian strip, so endodermal cells act as a selective filter—preventing unnecessary ions from reaching the xylem. The Casparian strip also qualifies as a mechanism for the passive exclusion of metals, sodium, and other unwanted ions (Figure 36.12).

The number of ion transporters present in root cells is also important in determining the toxicity of certain ions. For example, salt-tolerant strains of corn transport much less salt into their roots than do salt-intolerant strains—possibly because of genetic variation in the number of sodium channels found in root cells.

Active Exclusion by Metallothioneins and Phytochelatins Plants also have mechanisms for coping with toxins once they are inside their cells. This is important because even essential nutrients are toxic at high enough concentrations.

Copper channels, for example, admit the small amounts of copper ions required for normal cell function. But plants that grow on soils near copper-mining operations experience large concentration gradients that favor an influx of this nutrient, so a surplus is likely to build up inside the plant body. How do plants control the influx of copper and other types of excess nutrients to prevent them from poisoning key enzymes?

One mechanism for coping with toxic concentrations of metals involves small proteins called **metallothioneins**, and short peptides called **phytochelatins** that are synthesized by special enzymes. These macromolecules bind to metal ions and prevent them from acting as a poison. Producing them requires an expenditure of energy and represents a form of active exclusion.

Metallothioneins and phytochelatins have been found in a wide variety of organisms—bacteria, fungi, and animals as well as plants. Recent research on the mustard-family species *Arabidopsis thaliana* has shown that individuals from populations with a high tolerance for copper produce many more metallothionein proteins than do individuals from populations with a low

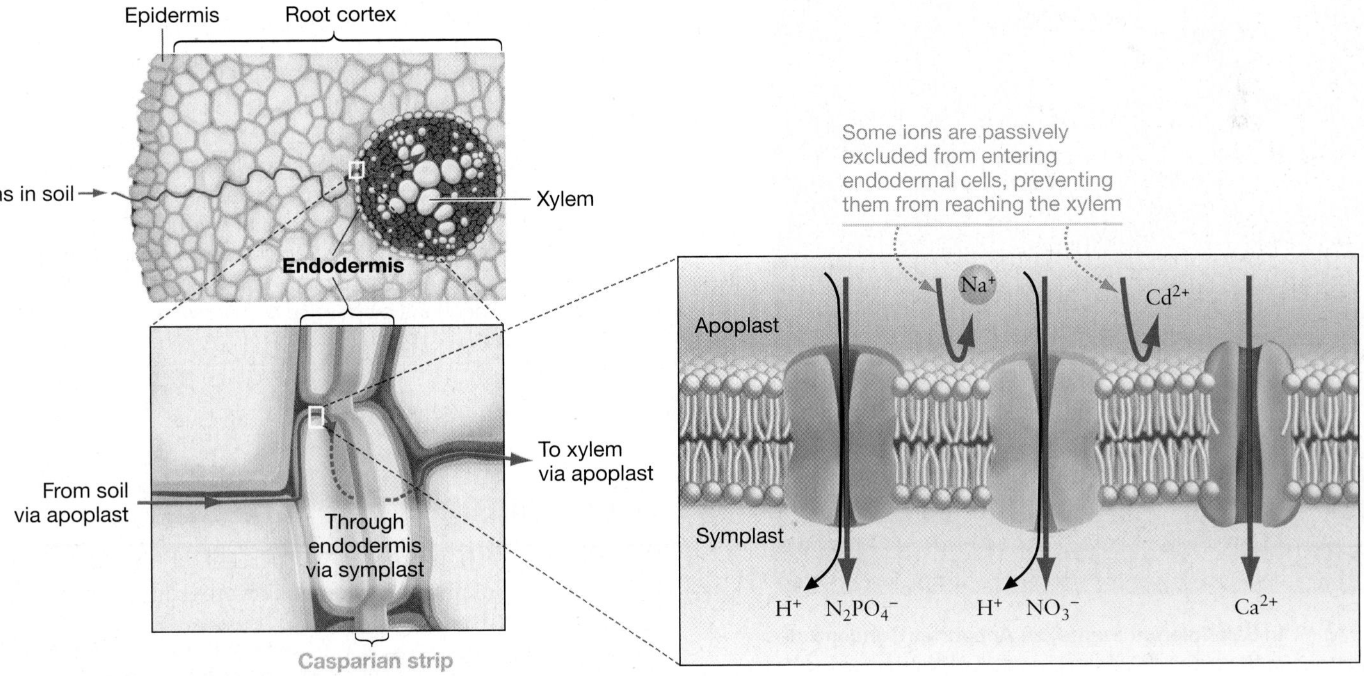

Figure 36.12 Passive Exclusion Occurs at the Plasma Membrane. Certain ions are excluded from cells because it is difficult for them to cross the plasma membrane.

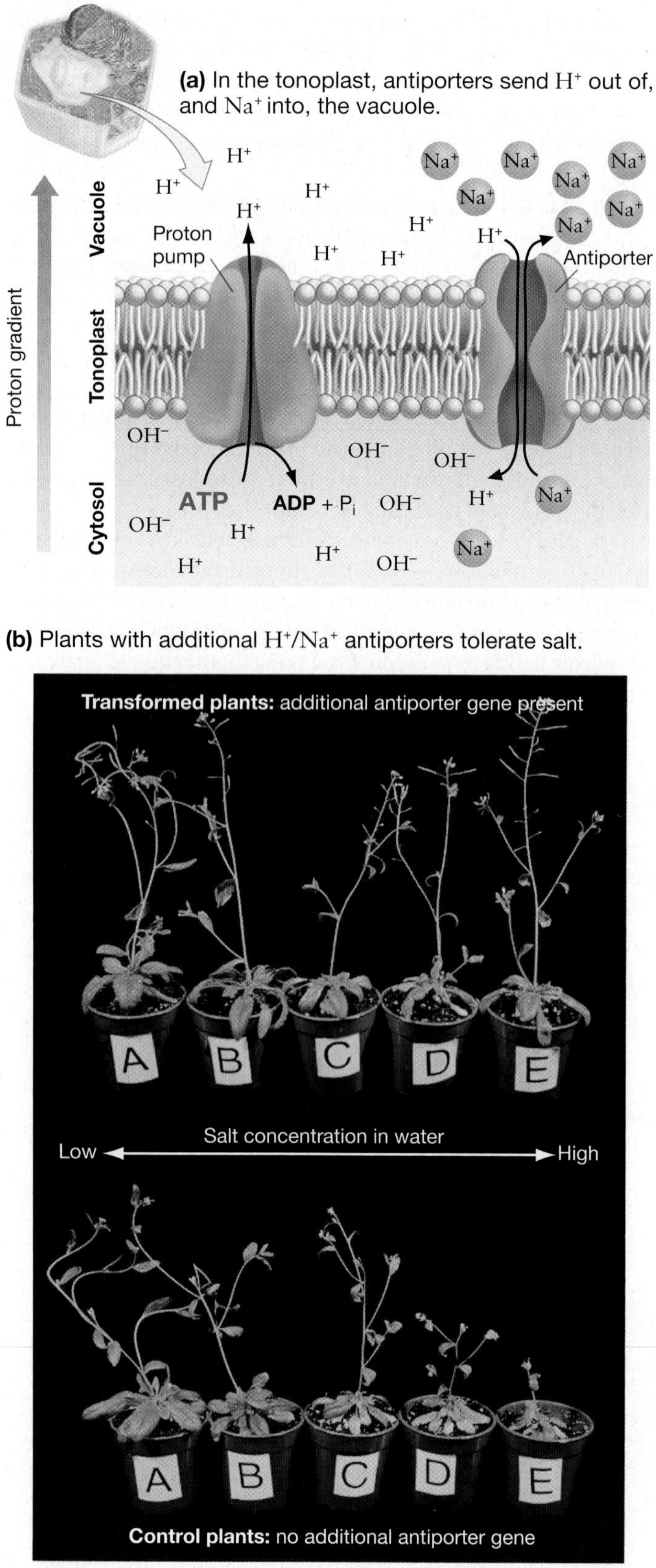

Figure 36.13 In Salt-Tolerant Plants, an Antiporter Concentrates Sodium Ions in Vacuoles. (a) In the tonoplast of salt-tolerant species, proton pumps establish a proton gradient that allows antiporters to concentrate sodium ions inside the vacuole where they cannot poison enzymes in the cytoplasm. **(b)** *Arabidopsis* plants transformed with the gene for the antiporter grow well even when given salty water.

tolerance for copper. (For more on the importance of *Arabidopsis thaliana* as a model organism, see **BioSkills 11**.)

Active Exclusion by Antiporters A second mechanism for actively neutralizing specific toxins involves transport proteins located in the **tonoplast**—the membrane surrounding the large central vacuole. Proteins in the tonoplast membrane allow plants to actively remove toxic substances from the cytosol and store them in the vacuole.

Perhaps the best-studied example involves salt-tolerant plants with proteins that move sodium ions from the cytosol into vacuoles, where they cannot poison enzymes (**Figure 36.13a**). Proton pumps in the tonoplast move protons into the vacuole, creating a gradient that favors the movement of protons back out of the vacuole. A transport protein that functions as an **antiporter** then uses this gradient to move sodium ions into the vacuole. The antiporter accomplishes this by carrying sodium ions *into* the vacuole as protons diffuse *out of* the vacuole. As a result, sodium ions are transported against their concentration gradient.

Knowledge of this H^+/Na^+ antiporter has created a great deal of excitement among biologists, because it offers a way to alter crop plants genetically so that they can grow well in salty soils created by poor irrigation practices. As an example, consider an experiment in which normal *Arabidopsis* plants and individuals transformed with extra copies of the *Arabidopsis* gene for the H^+/Na^+ antiporter were exposed to different levels of NaCl. The genetically transformed plants accumulated more sodium ions in their vacuoles, and as **Figure 36.13b** shows, they were able to grow efficiently even when exposed to high concentrations of salt. Similar results have been demonstrated in crops such as tobacco, rice, cotton, and others.

CHECK YOUR UNDERSTANDING

✔ **If you understood this section, you should be able to . . .**

1. Explain how plants generate and use charge differences across membranes to import cations and anions.
2. Explain whether pine seedlings grown in the presence of mycorrhizal fungi will likely grow faster or slower than pine seedlings grown in the absence of mycorrhizal fungi.
3. **MODEL** Make a diagram that traces a cadmium ion (Cd^{2+}) moving from the soil into a plant, showing how it is excluded at different locations by passive and active mechanisms of ion exclusion.

Answers are available in Appendix A.

36.4 Nitrogen Fixation

After you complete this section, you should be able to . . .

- Explain how plants obtain nitrogen.

Nitrogen gas (N_2) makes up 80 percent of the atmosphere. Unfortunately, plants and other eukaryotes cannot use nitrogen in this form. Nitrogen gas is unreactive because a great deal of energy is needed to break the triple bond between the two nitrogen atoms.

To synthesize amino acids, nitrogenous bases, and other nitrogen-containing compounds, plants normally absorb nitrogen in forms such as ammonium (NH_4^+) or nitrate ions (NO_3^-). But these ions are in short supply in many soils, meaning that plant growth is often limited by the availability of usable nitrogen. Most crop plants grow faster when they receive a nitrogen-containing fertilizer.

Nitrogen-based fertilizers have drawbacks, however. Fertilizer production is extremely energy intensive, and in many parts of the world, ammonia and other nitrogen-based fertilizers are too expensive for farmers. In more affluent regions, these fertilizers are used so extensively that they are causing serious pollution problems (Ch. 53, Section 53.2). For these and other reasons, there is intense interest in understanding the molecular basis of a phenomenon called biological nitrogen fixation.

The Role of Symbiotic Bacteria

Among all the organisms on the tree of life, only a few species of bacteria and archaea are able to absorb N_2 from the atmosphere and convert it to ammonia (NH_3), nitrites (NO_2), or nitrates. This process is called **nitrogen fixation**.

Nitrogen fixation requires a series of specialized enzymes and cofactors, including a large multi-enzyme complex called nitrogenase. The process is extremely energy demanding. An expenditure of 8 high-energy electrons and 16 ATP molecules is required for nitrogenase to reduce one molecule of N_2 to two molecules of NH_3. The process can be summarized as follows:

$$N_2 + 8\,e^- + 8\,H^+ + 16\,ATP \rightarrow 2\,NH_3 + H_2 + 16\,ADP + 16\,P_i$$

In some cases, bacterial cells that are capable of nitrogen fixation take up residence *inside* plant root cells. Although several different bacteria and plant hosts can be involved, the best-studied nitrogen-fixing bacteria are members of the genus *Rhizobium* that associate with plants in the **legume** family, which includes peas, soybeans, clover, and alfalfa. Members of the bacterial genus *Rhizobium* and closely related species are often called **rhizobia**.

Thanks to their ability to fix nitrogen, legumes are extremely important agricultural crops. Because it adds nitrogen to the soil, a legume crop is often planted after harvesting a grain crop such as corn or wheat. This type of crop rotation results in "free" nitrogen fertilizer that benefits the next grain crop. Because legumes are rich in nitrogen, they are an important part of the human diet in many parts of the world.

As Figure 36.14 shows, the infected root cells of legumes form distinctive structures called **nodules**, where nitrogen-fixing rhizobia are found. The nodules are pink because they include an iron-containing molecule called leghemoglobin (short for legume hemoglobin). **Leghemoglobin** is related to the hemoglobin that carries oxygen in your blood. Like hemoglobin, leghemoglobin binds oxygen.

Leghemoglobin is important because nitrogenase is poisoned by the presence of oxygen. Inside root nodules, oxygen molecules bind to leghemoglobin instead of binding to nitrogenase. In this way, leghemoglobin keeps the level of free oxygen low enough for nitrogenase to function while maintaining a source of oxygen for electron transport to occur. Intense cellular respiration and ATP synthesis are necessary to supply nitrogenase with the energy it requires.

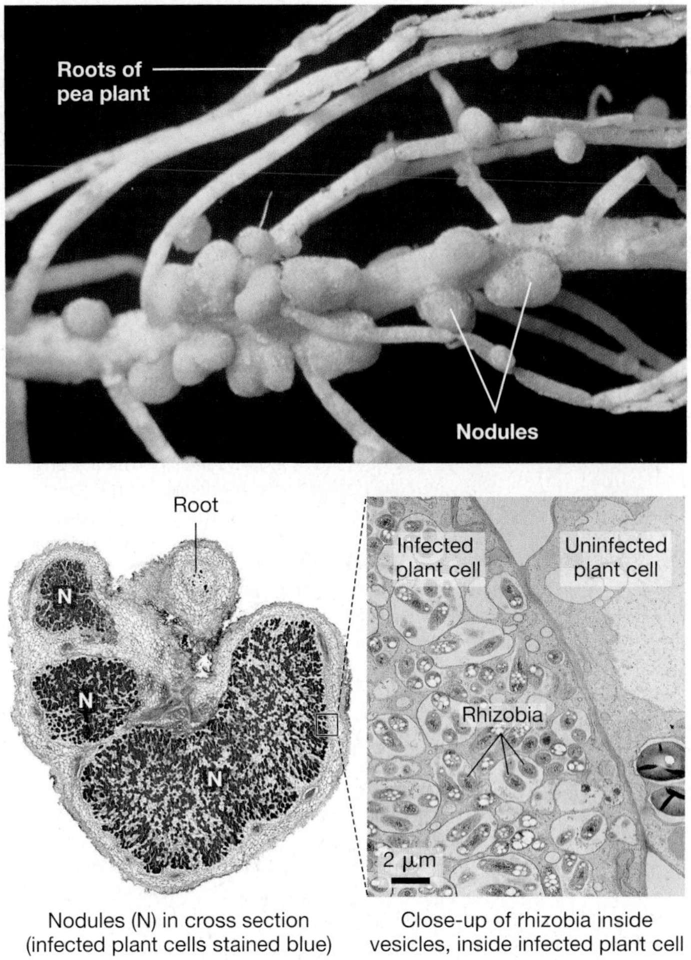

Figure 36.14 In Some Plants, Roots Form Nodules Where Nitrogen-Fixing Bacteria Live.

What Is the Relationship between Plants and Nitrogen-Fixing Bacteria?

Like mycorrhizal fungi and their host plants, legumes and rhizobia have a mutualistic relationship. The nitrogen-fixing bacteria provide the plant with ammonia, while the legume provides the bacteria with carbohydrates and protection. The association is costly for the host plant, however. For the bacteria to synthesize the enormous amounts of high-energy electrons and ATP required to manufacture ammonia, plants have to supply them with large quantities of sugar.

Given the high cost of maintaining this relationship with bacteria, does it make sense for legumes to facilitate the growth of nodules if nitrogen is readily available? Researchers addressed this question by growing soybean plants hydroponically in the presence and absence of nitrogen.

RESEARCH

QUESTION: Do legumes regulate root nodule growth based on demand for nitrogen?

HYPOTHESIS: Legumes invest resources to produce nodules only if there is a lack of nitrogen in their environment.

NULL HYPOTHESIS: Legumes invest resources to produce nodules regardless of whether they need additional nitrogen.

EXPERIMENTAL SETUP:

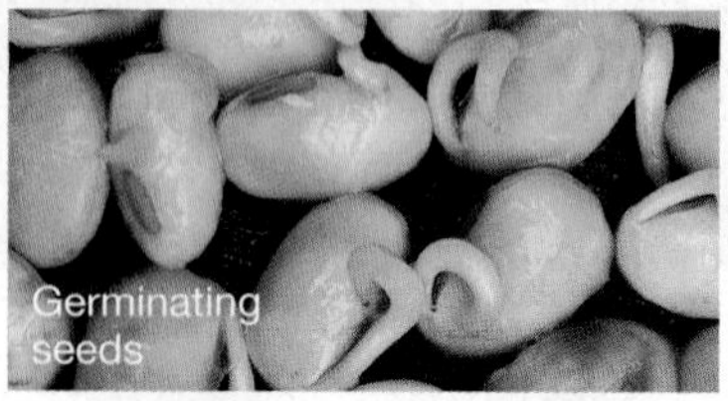

1. Inoculate soybean seeds with nitrogen-fixing bacteria (*Rhizobium* sp.).

2. Let seeds germinate and the seedlings initiate root nodule formation (10 days).

Treatment 1: nutrient solution with nitrogen

Treatment 2: nutrient solution without nitrogen

3. Transfer seedlings to hydroponic cultures containing either all essential nutrients (including nitrogen) or all essential nutrients except nitrogen.

4. Let plants grow for two weeks, measuring the size of the root nodules daily.

PREDICTION OF HYPOTHESIS: The nodules on plants grown with nitrogen will remain the same size, while nodules on plants grown without nitrogen will grow larger.

PREDICTION OF NULL HYPOTHESIS: Root nodules will enlarge in both treatments.

RESULTS:

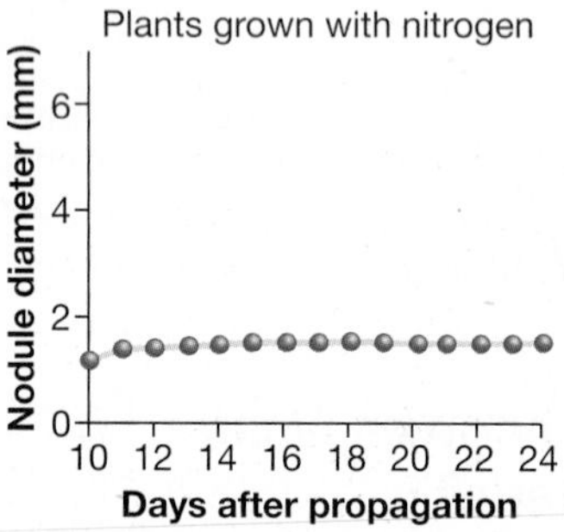

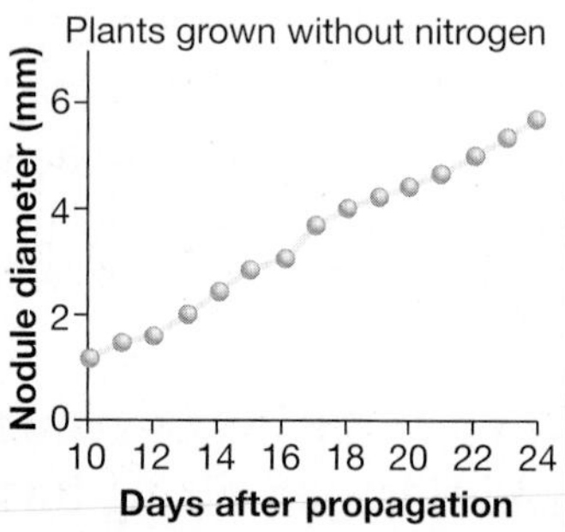

CONCLUSION: Legumes regulate nodule development by investing resources into nodules only when additional nitrogen is needed.

Figure 36.15 Legumes Regulate Growth of Root Nodules Based on Nitrogen Needs.

SOURCE: H. Fujikake et al. 2003. Quick and reversible inhibition of soybean root nodule growth by nitrate involves a decrease in sucrose supply to nodules. *Journal of Experimental Botany* 54 (386): 1379–1388.

✔ **PROCESS OF SCIENCE** Predict the results of a similar experiment in which nitrogen was absent during days 1–5 of hydroponic growth, added during days 6–10, then removed during days 11–14.

After inoculating seeds with nitrogen-fixing bacteria, researchers allowed the seeds to germinate and initiate nodule formation. The seedlings were then grown hydroponically, either with all essential nutrients except nitrogen or with all essential nutrients including nitrogen. Root nodules in the two treatments were then measured over a two-week period. As **Figure 36.15** shows, the nodules on soybean plants grown with nitrogen did not enlarge, but those on plants grown without nitrogen enlarged dramatically. This study indicates that legumes have the ability to fine-tune their relationship with nitrogen-fixing bacteria in order to meet their nitrogen needs while minimizing resources expended on the bacteria.

If roots allowed bacterial pathogens to enter, the plant would possibly die from the subsequent infection. How do roots distinguish beneficial bacteria from pathogens and allow only the rhizobia to infect the roots? Given the cost of the plant–bacterium interaction and its importance to both species, it's not surprising that the infection process is carefully regulated.

When a legume seed germinates under natural conditions, its roots do not contain a population of rhizobia. Instead, roots make contact with bacterial cells existing in the soil, the rhizobia infect root cells, and root cortex cells grow into a nodule.

The first event in infection is a recognition step that occurs between a legume-family plant and its symbiotic bacterium. First, young roots release compounds called flavonoids, which tend to attract rhizobia. When rhizobia contact the flavonoids, the bacteria respond by producing sugar-containing molecules called **Nod factors** (for *nod*ule formation). Nod factors, in turn, bind to signaling proteins on the membrane surface of root hairs.

The recognition step is specific to the species involved. Each legume species produces a different flavonoid that acts as a recognition signal, and each rhizobium species responds with one or more unique Nod factors. When researchers have switched recognition signals or Nod factors between species, the recognition step fails.

When Nod factors bind to the root-hair surface, they set off a chain of events that leads to dramatic morphological changes in the host legume. After the bacterial cells contact the root-hair surface, they move inside the root-hair cell wall and begin to multiply. This triggers an invagination of the root-hair plasma membrane called an **infection thread**. Rhizobia proliferate within the infection thread as it grows into the cortex of the root. Portions of the infection thread bud off, releasing rhizobia into membrane-bound clusters within cortex cells. A root nodule begins to form from infected cortex cells, which divide rapidly.

The interaction between rhizobia and legumes is one of the most complex and best studied of all mutualisms.

CHECK YOUR UNDERSTANDING

✔ **If you understood this section, you should be able to . . .**

1. Hypothesize why it is that not all plants have symbiotic bacteria that fix nitrogen.
2. Hypothesize why the interaction that establishes the infection by nitrogen-fixing bacteria is so complex.

Answers are available in Appendix A.

36.5 Nutritional Adaptations of Plants

Based on the data available so far, over 95 percent of vascular plant species take up nutrients from soil. And over 80 percent of all plants supplement their "diet" with nutrients acquired from mycorrhizal fungi, while a small but significant fraction associate with nitrogen-fixing bacteria. Perhaps 99 percent of all living plant species make their own sugar through photosynthesis—they are **autotrophs**, organisms that can synthesize their own food.

After you complete this section, you should be able to ...

- Identify examples of specialized methods some plants use to obtain nutrients.

What about the small number of plant species that don't follow these rules? Some parasitize other plants, some appear to live on air, and others catch insects and digest them.

Parasitic Plants

Parasites are organisms that live in close physical contact with individuals from another species and that lower the fitness of those individuals, usually by obtaining water or nutrients from the host. Biologists estimate that about 3000 species of angiosperms are parasitic—less than 1 percent of the plant species that have been studied and named to date.

Some parasitic plants are non-photosynthetic and obtain all of their nutrition by tapping into the vascular tissue of the host individual—they are **heterotrophs**, organisms that obtain food by consuming other organisms. Dodder ("strangle weed"), for example, is a plant that can wrap itself around other plants (Figure 36.16a) and produce structures called **haustoria** (singular: **haustorium**) that penetrate a host plant's vascular tissue and extract water and nutrients (Figure 36.16b).

Monotropa ("ghost plant") is also non-photosynthetic, and thus not green; these plants extract nutrients from fungi closely associated with their roots (Figure 36.16c). But most parasitic plants are photosynthetic and tap the xylem of other species for water and essential nutrients. For example, mistletoe (Figure 36.16d) is photosynthetic, but it forms haustoria and extracts water and ions from the host's xylem. Mistletoe is a serious cause of economic loss in the forests of western North America.

Epiphytic Plants

Species from a diverse array of plant lineages do not absorb nutrients from soil or from some other host plant. In fact, these species never even make contact with soil. As Figure 36.17a on page 782 shows, they often grow on the trunks or branches of trees. For this reason they are called **epiphytes** ("upon-plants"). Epiphytes are not parasitic.

In northern forests, epiphytism is common; mosses and ferns there often grow on tree trunks. The so-called Spanish moss that hangs from oak trees in the southeastern United States is another familiar epiphyte—although this species is actually not a moss but a bromeliad, a relative of the pineapple. In the tropics and subtropics, one-third of all ferns grow as epiphytes, and there are thousands of species of epiphytic orchids, bromeliads, and lycophytes.

Epiphytes absorb most of the water and nutrients they need from rainwater, dust, and particles that collect in their tissues or in the crevices of bark. As Figure 36.17b shows, some epiphytic bromeliads have leaves that grow in rosettes and form "tanks" that collect water and organic debris. In such cases, nutrients are actually absorbed through the leaves themselves.

Carnivorous Plants

Carnivorous plants trap insects and other animals, kill them, and then digest the prey to absorb their nutrients. Approximately 0.02 percent of plant species are carnivorous; they make their own carbohydrates via photosynthesis but use carnivory to supplement the nitrogen available in the environment. Most are found in bogs or other habitats where nitrogen is scarce or unavailable.

Figure 36.16 Parasitic Plants Extract Nutrients from Their Hosts.

(a) Epiphytes grow on trees.

(b) This epiphyte's leaves form water-holding "tanks."

Figure 36.17 Epiphytes Are Adapted to Grow in the Absence of Soil. The environments where many epiphytes grow—tree trunks and branches—are dry and nutrient poor.

✔ The mistletoe in Figure 36.16 and the bromeliads shown here both grow on other plants. How do they differ in their relationship to the trees that support them?

Modified Leaves Form an Array of Trapping Mechanisms Carnivorous plants use modified leaves or roots to trap insects. For example, the leaves of pitcher plants form tubes (see Ch. 34, Table 34.3). Prey are enticed to the tube by an attractive odor and then have difficulty climbing back out. Eventually they fall into the pool of water below, where they drown. Enzymes released by the plant digest the prey, and the plant absorbs the nutrients obtained from the insects' dead bodies.

Sundews have an alternative hunting strategy: They have modified leaves that function like flypaper. The leaf surfaces develop hairs that exude a sticky substance and trap insects (Figure 36.18). Glands near the sundew's trap release enzymes that digest the prey. The leaf then absorbs the nutrients that are released.

In the Venus flytrap, modified leaves trap insects mechanically. Sensory hairs protrude from the epidermis of each leaf. When an insect lands on the trap and bumps two or three of these hairs, the hair cells produce an electrical signal and the leaf responds by snapping shut (see Ch. 37, Figure 37.17). If the prey continues to stimulate hairs, the leaf secretes enzymes that digest the prey.

Figure 36.18 Sundews Have Modified Leaves with Sticky Surfaces That Catch Insects.

Costs and Benefits of Carnivory Initially, research on carnivorous plants focused on documenting that meat-eating was adaptive. Early experiments in natural environments supported this hypothesis by showing that, compared to individuals of the same species that are not fed fruit flies, carnivorous plants that were fed fruit flies grew faster and flowered more often.

Follow-up work explored the costs as well as the benefits of carnivory. For example, several teams of researchers showed that when carnivorous plants are provided with nitrogen-based fertilizers, they produce fewer of their specialized insect-trapping leaves and a higher proportion of leaves that function primarily in photosynthesis. These results suggest carnivory is a trait that shows phenotypic plasticity (Ch. 34, Section 34.1): Plants increase investment in prey-capture structures when nitrogen is rare, but they decrease investment in such structures when nitrogen is readily available.

Currently, some research groups are doing "evo-devo" research (Chapter 21) on carnivory. They are working to identify the novel alleles that made the evolution of carnivory possible in certain groups. Meat-eating may be relatively rare in plants, but it has inspired interesting new directions for research.

CHECK YOUR UNDERSTANDING

✔ If you understood this section, you should be able to . . .

1. Explain how "ghost plant" (*Monotropa*) can grow even though it doesn't carry out photosynthesis.
2. Explain why Venus flytrap plants are considered both autotrophs and heterotrophs.

Answers are available in Appendix A.

36.1 Nutritional Requirements of Plants

- About 96 percent of the dry mass of a typical plant consists of carbon, hydrogen, and oxygen. The vast majority of plants obtain these elements by absorbing carbon dioxide from the atmosphere and water from the soil.
- The other 4 percent of the plant body consists of a complex suite of elements, acquired from soil, that are required for normal growth and reproduction. These essential elements are usually absorbed in the form of ions.
- Nutrients are usually classified as macronutrients, needed in large supply, and micronutrients, needed in only trace amounts.
- When plants are lacking in an essential nutrient, they usually develop predictable deficiency symptoms.

36.2 Soil: A Dynamic Mixture of Living and Nonliving Components

- Soil is a complex and dynamic mixture of inorganic particles such as clay and sand, organic matter, and organisms. Soil provides plants with oxygen, water, and nutrients as well as a physical substrate for anchoring and supporting the plant body.
- Plants absorb nutrients in the form of ions. Anions in soil water are normally available to roots, but cations adhere to soil particles and are made available only after cation exchange.

36.3 Nutrient Uptake

- Nutrients are present at low concentrations in the soil surrounding roots.
- Plants import nutrients against a concentration gradient by pumping protons into the extracellular space. A large excess of protons outside the root hair creates a strong electrochemical gradient favoring the entry of positively charged ions through membrane transporters.
- The proton gradient in root-hair membranes also allows negatively charged ions to cross the plasma membrane via symporters.
- In addition to acquiring nutrients by establishing a proton gradient, many plants obtain nitrogen or phosphorus through a mutualistic relationship with mycorrhizal fungi in exchange for carbohydrates derived from photosynthesis.
- Passive and active systems also exist for excluding certain ions. This is important because all nutrients are toxic at high concentrations.

36.4 Nitrogen Fixation

- Certain plants, including those in the legume family, are capable of symbiosis with nitrogen-fixing bacteria. The bacteria colonize the interior of root cells after exchanging species-specific signals.
- Nitrogen-fixing bacteria receive protection and sugar from the host plant, in exchange for producing ammonia, a usable form of nitrogen for the plant, from nitrogen gas.

36.5 Nutritional Adaptations of Plants

- Most parasitic plants produce their own carbohydrates through photosynthesis, but all obtain water and nutrients by penetrating the xylem of host plants.
- Epiphytes grow on the trunks and branches of other plants and absorb mineral nutrients from rainwater, dust, or decaying matter trapped in the bark of their host plant.
- Carnivorous plants have evolved mechanisms for trapping and digesting insects and other animals. The primary benefit of carnivory is to obtain nitrogen in low-nitrogen environments.

Answers are available in Appendix A.

✓ TEST YOUR KNOWLEDGE

1. Consider the following statements regarding limiting nutrients. Select True or False for each statement.
   T/F Nitrogen (N), phosphorus (P), and potassium (K) are common examples.
   T/F Their presence limits the availability of micronutrients.
   T/F Their availability tends to limit plant growth.
   T/F Certain macronutrients and micronutrients can be considered limiting nutrients.
2. Where does most nutrient uptake occur in roots?
   a. at the root cap, where root tissue first encounters soil away from the zone of nutrient depletion
   b. at the Casparian strip, where ions must enter the symplast before entering xylem cells
   c. in the symplastic and apoplastic pathways
   d. in root hairs, in the zone of maturation
3. Why are proton pumps in root-hair plasma membranes important?
   a. They pump protons into cells, generating a membrane potential (voltage).
   b. They allow toxins to be concentrated in vacuoles, so the toxins do not poison enzymes in the cytoplasm.
   c. They set up an electrochemical gradient that makes it possible for roots to absorb cations and anions.
   d. They set up the membrane voltage required for action potentials to occur.
4. Why is the presence of clay particles important in soil?
   a. They provide macronutrients—particularly nitrogen, phosphorus, and potassium.
   b. They bind metal ions, which would be toxic if absorbed by plants.
   c. They allow water to percolate through the soil, making oxygen-rich air pockets available.
   d. The negative charges on clay bind to positively charged ions and prevent them from being leached out of the soil.

✓ TEST YOUR UNDERSTANDING

5. Suppose that certain root cells have an overall charge that is more negative than normal. What impact would this likely have on the uptake of anions such as NO_3^-?
   a. Anions would be less likely to enter roots.
   b. Anions would be more likely to enter roots.
   c. This would have no impact on the ability of anions to enter roots.
   d. This would make anions and cations equally likely to enter roots.

6. **THINK CAREFULLY** In a semester-long experiment tracking growth in plants, your lab partner—who often skips class—carefully records the mass of water added to a potted plant with the expectation that this addition will be fully accounted for in the mass gained by the plant. Is your lab partner right or wrong? Explain.

7. **THINK CAREFULLY** Your friend claims that all plants are autotrophs because they perform photosynthesis. Is that a correct statement? Explain.

8. Why is it important for plants to exclude certain ions? Summarize the difference between active and passive exclusion mechanisms.

✔ TEST YOUR PROBLEM-SOLVING SKILLS

9. **QUANTITATIVE** There is a conflict between van Helmont's data on willow tree growth and the data on essential nutrients listed in Table 36.1. According to the table, nutrients other than C, H, and O should make up about 4 percent of a willow tree's mass. Most or all of these nutrients should come from soil. But van Helmont claimed that the soil in his experiment lost just 60 g, while the tree gained 74,000 g. Calculate the percentage of the added mass accounted for by soil, and compare it to the predicted 4 percent. State at least one hypothesis to explain the conflict between expected and observed results. How would you test this hypothesis?

10. **PROCESS OF SCIENCE** Design an experiment, using radioactive carbon and the heavy isotope of nitrogen ($^{15}N_2$), that would test whether the rhizobia–pea plant interaction is mutualistic.

✔ PUT IT ALL TOGETHER: Case Study

What nutritional benefits can insects provide to carnivorous plants?

The carnivorous plant *Nepenthes bicalcarata* ("fanged pitcher plant") has a unique relationship with a species of ant—*Camponotus schmitzi* ("diving ant"). The diving ants are not digested by the pitcher plants, but instead live on the plants and consume nectar. Diving ants also dive into the digestive juices in the pitcher, swim to the bottom, and capture and consume trapped insects, leaving uneaten body parts and ant feces behind. What nutritional impact do the ants have on fanged pitcher plants? Do the pitcher plants derive any nutritional benefit from this relationship?

11. Based only on the information provided here, make a prediction on the effect of diving ants on overall pitcher plant growth.

12. **PROCESS OF SCIENCE** Researchers tested the hypothesis that the relationship between diving ants and pitcher plants is mutualistic (i.e., both species derive a nutritional benefit). To do so, they compared leaf surface area (as a measure of overall growth) in two sets of pitcher plants: plants with diving ants and plants without. The results are shown in the graph. The *P* values indicate whether there is a significant relationship between the size of the host plants and the surface area of the host plants' leaves. (For an explanation of statistical significance, see **BioSkills 3**.) Based on this graph, what conclusions can be drawn about the impact of diving ants on overall plant growth?

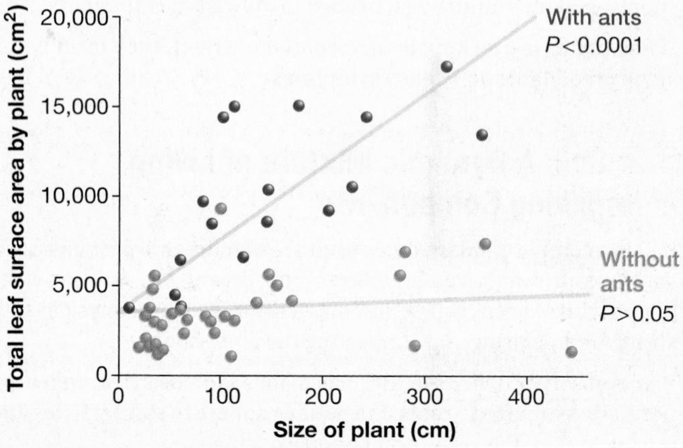

Source: V. Bazile et al. 2012. *PLoS ONE* 7: e36179.

13. Based on the background information presented here, does the relationship between diving ants and pitcher plants appear to be mutualistic? Explain.

14. Explain why carnivorous and parasitic plants are most common in nutrient-poor habitats.

15. Nitrogen is a key nutrient often obtained by carnivorous plants from the insects they digest. Are the results presented here what would be expected if nitrogen is a limiting nutrient? Explain.

16. Carnivorous plants and legumes (e.g., peas, soybeans) both absorb key nutrients directly from other organisms. How is nutrient acquisition in pitcher plants similar to that in legumes? How is it different?

37 Plant Sensory Systems, Signals, and Responses

Plants have sophisticated information processing systems and can respond to several environmental cues. The stem of this bean plant, for example, senses the pull of gravity and is growing away from it.

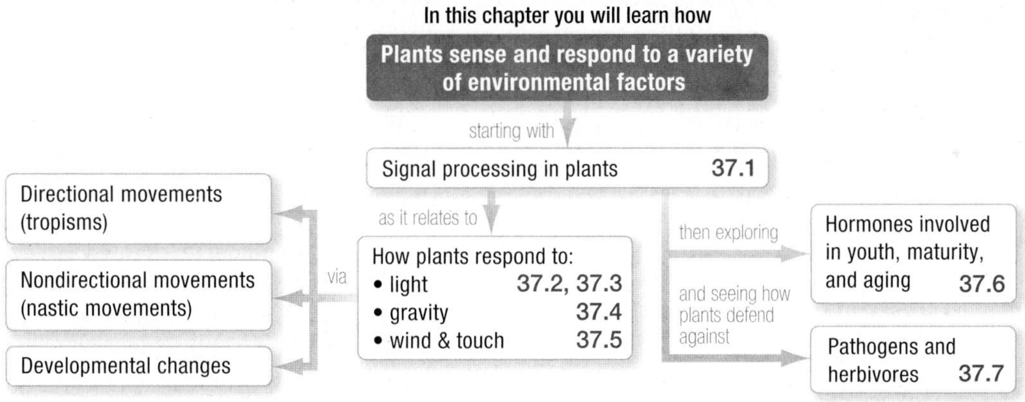

Imagine standing in place for several hundred years, like an oak tree. Each spring you produce flowers. All summer you absorb light from the Sun, carbon dioxide from the atmosphere, and water and nutrients from the soil. Water, ions, and sugars flow up and down your vascular tissue. For six months or more your body grows upward, outward, and downward. In fall, thousands of your offspring drop to the ground as acorns. Then as winter approaches, your metabolism slows and you stop growing. Like a hibernating animal, you spend the long, hard months of winter in a state of suspended animation, before resuming growth the following spring.

To stay alive, an oak needs to gather information about its environment. It has to sense the season of the year, the time of day, the pull of gravity, the force of wind, and attacks by enemies. It needs to sense which of its leaves are being shaded, and which leaves are receiving the wavelengths of light that are required to support photosynthesis.

Plants may not have eyes or ears, but they can sense and respond to light, gravity, pressure, and wounds. They have the equivalent of a sense of smell because they can perceive certain airborne molecules. It can

BIG PICTURE

This chapter is part of the Big Picture. See how on pages 838–839.

even be argued that they have a sense of taste, because their roots detect nutrients in the soil.

Besides gathering information about the conditions around them, plants have to respond appropriately. They do not jump or swim or run, but their shoot systems grow toward light or end up shorter and stockier in response to wind. In response to gravity, shoots grow up and roots grow down. In response to touch, the modified leaves of a Venus flytrap shut fast enough to catch live insects. If a plant is being attacked or if it senses that a neighboring individual is under attack, it may lace its tissues with toxic compounds or mobilize other defenses.

The message of this chapter is simple: Plants have sophisticated systems for collecting information about their environment and responding in ways that maximize their chances of surviving, thriving, and producing offspring. The ability to gather information and respond to it is one of the five fundamental attributes of life (Chapter 1). Let's begin by exploring how plant cells process information from the environment.

37.1 Information Processing in Plants

Every environment is full of information. But, like other organisms, plants primarily monitor aspects of the environment that matter most to them—those that affect their ability to stay alive and produce offspring.

After you complete this section, you should be able to ...

- Explain how plants detect and respond to environmental stimuli.

Figure 37.1 summarizes the three-step process by which plants gather, process, and respond to environmental stimuli:

Step 1 Sensory cells perceive an external stimulus and transduce, or change, it into an intracellular signal.

Step 2 A cell–cell signal is then sent by the sensory cells to target cells in other parts of the body that can respond to the information.

Step 3 Target cells receive this signal and change their activity in a way that produces an appropriate response.

Let's briefly consider how each of these steps works, then explore some details about how plants sense and respond to light, gravity, and other signals from the environment.

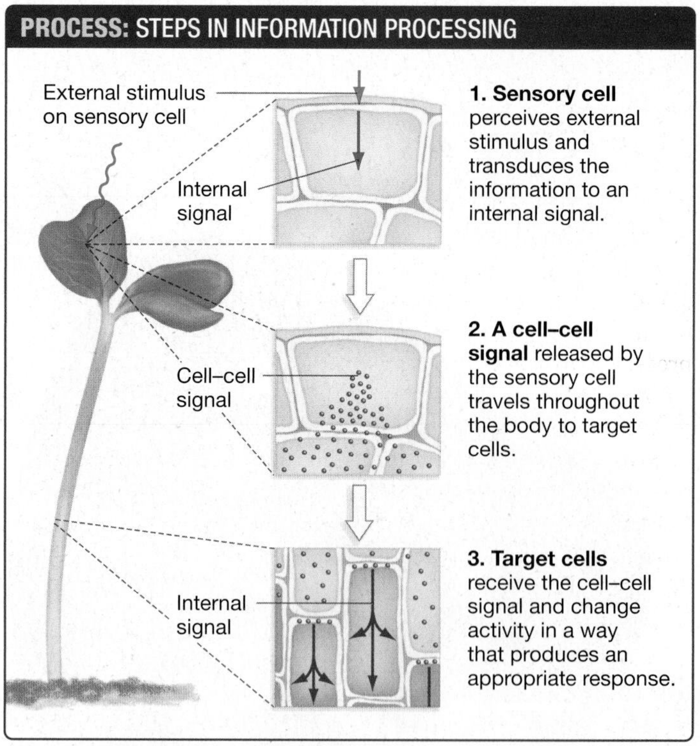

Figure 37.1 **In Many Cases of Plant Information Processing, Cell–Cell Signals Link a Stimulus and a Response.**

How Do Cells Receive and Process an External Signal?

When you text a friend, the signal travels from you to their receiver via airwaves. When the message arrives, the receiving cell phone changes the information in the airwaves into electrical signals and then into words that your friend can understand.

Plants work in much the same way. When light strikes a sensory cell, for example, the information it carries has to be changed into a form that is meaningful to that cell. Signals from the environment are usually detected by proteins specialized for that function. **Signal receptor** proteins change shape in response to an environmental stimulus, such as being struck by a particular wavelength of light, having pressure applied, or binding to a particular type of molecule.

When a signal receptor changes shape in response to a stimulus, the information changes form—from an external signal to an intracellular signal. The intracellular signal primes the sensory cell for action. In many cases, though, the cells that receive information from the environment are located in a part of the body that is distant from the cells that need to respond to the information. How does information from activated sensory cells get to target cells?

The answer usually involves a cell–cell signal called a **hormone**—an organic compound that's produced in small amounts in one part of a plant and transported to target cells, where it causes a physiological response. For example, a hormone produced in roots may cause stomata in leaves to close in response to dry soil conditions. Hormones may be transmitted from cell to cell by specialized transport proteins in cell membranes, in xylem sap or phloem sap, or by simple diffusion from the originating cell.

Plant cells routinely receive information from several different hormones at the same time, so it is common for different types of hormones to interact with each other and modulate the cell's response.

How Do Cells Respond to Cell–Cell Signals?

Cells are exposed to a constant stream of hormones and other types of cell–cell signaling molecules. Many of these signals have little to no effect on what is happening inside a given cell. The reason is simple: Signaling molecules elicit a response only if a cell has an appropriate receptor.

Recall that some signal receptors exist inside the cell, where they respond to signaling molecules that readily diffuse through the plasma membrane (Ch. 11, Section 11.3). However, most signal receptors are located in the plasma membrane, where they respond to external stimuli, or bind to signaling molecules that cannot or do not cross the membrane.

If the receptor resides inside the cell, the information carried by the signal is processed directly—with no intermediate steps. If the receptor is located in the plasma membrane, the subsequent process is called **signal transduction** (Figure 37.2); the verb "transduce" means "to convert from one form to another." In either case, the result may be a dramatic change in the cell's activity.

Signal transduction begins with stimulation of a receptor protein in the plasma membrane. You might recall that signal transduction pathways often involve two fundamental components: phosphorylation cascades and second messengers (Ch. 11, Section 11.3).

- **Phosphorylation cascades** are triggered when the change in the receptor protein's shape leads to the transfer of a phosphate group (PO_4^{3-}) from ATP to the receptor or an associated protein (see Figure 37.2). Phosphorylation activates proteins involved in signal transduction cascades, causing them to phosphorylate and activate a different set of proteins, which in turn catalyze the phosphorylation and activation of still other proteins.
- **Second messengers** are produced when receptor proteins trigger the production of intracellular signals or their release from storage areas (see Figure 37.2). Calcium ions (Ca^{2+}) stored in the vacuole, ER, or cell wall are one of several ions or molecules that function as the second messenger in plants.

In some cases, phosphorylation cascades and second messengers interact to modify the cell response. For example, second messengers can activate phosphorylation cascades in some signal transduction pathways.

To understand why a signaling molecule can have such significant effects on cell activity, look again at Figure 37.2. Activation of a signal transduction pathway results in the production of many phosphorylated proteins or the release of many second messengers. These events amplify the original signal many times. In this way, for example, low concentrations of plant hormones can have a large impact on target cells. Hormones are small molecules found in low concentrations, but because the information they carry is amplified during signal transduction, they produce big results.

As Figure 37.2 shows, the response to a signal binding with a receptor can include

- activation of membrane transport proteins (Ch. 6, Section 6.4), which produces a change in the membrane's electrical potential or the cell wall's pH. For example, plant cell walls develop the capacity to expand in response to the release of protons from the cell itself—which then can lead to a plant organ growing toward or away from a particular stimulus.
- changes in gene expression—via alterations in transcription activators or repressors (Ch. 19, Section 19.3) or the translation machinery—that result in new RNAs or proteins in the cell.

PROCESS: SIGNAL TRANSDUCTION

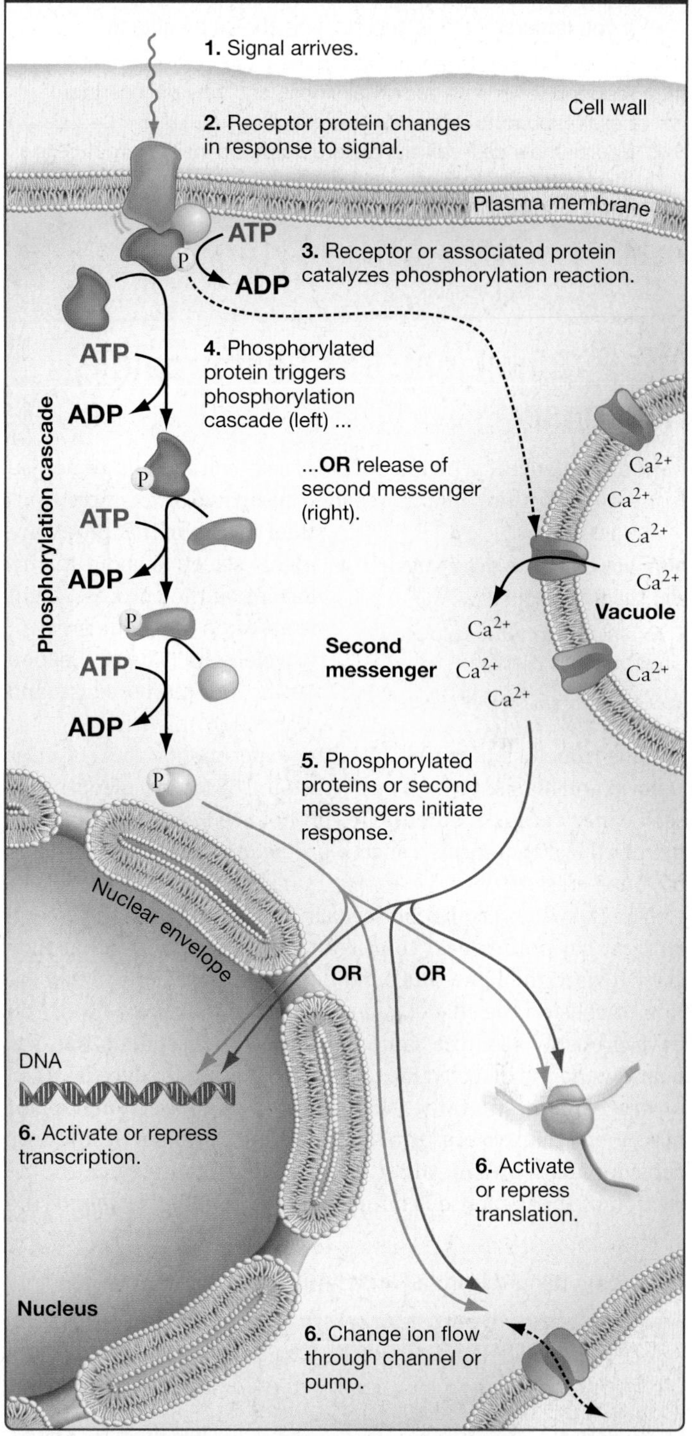

Figure 37.2 **Signal Transduction Changes an External Signal to an Internal Signal.** After an environmental or cell–cell signal is transduced to an intracellular signal, it triggers a change in the cell's activity.

For example, during seed germination, enzymes are produced that break down food reserves used by the developing embryo.

The response to a signal is important. When cells respond, the change in their activity helps the plant cope with the environmental change perceived by the sensory cell.

CHECK YOUR UNDERSTANDING

✔ If you understood this section, you should be able to . . .

1. Explain why only certain cells in the plant body—not all—respond to an environmental signal, and why only certain cells respond to a hormone.
2. Explain how cell–cell signals are amplified inside target cells by signal transduction.

Answers are available in Appendix A.

37.2 Blue Light: The Phototropic Response

After you complete this section, you should be able to . . .

- Explain the phototropic growth response in plants.

Most of the general principles of plant cell–cell signaling emerged from studies of how plants respond to light. For example, consider the claim that plants are highly selective about the information they process. Light is made up of a wide array of wavelengths (Ch. 10, Section 10.2), but plants sense and respond to only a few.

This conclusion traces back to experiments that Charles Darwin and his son Francis performed in 1881 with coleoptiles of reed canary grass. A **coleoptile** (pronounced *koh-lee-OP-til*) is a modified leaf that forms a sheath protecting the emerging shoots of young grasses.

The Darwins germinated seeds in the dark, placed the young, straight coleoptiles next to a light source, and noted that they grew toward the light. The bending response was due to elongation of cells on the shaded side of the plant (Figure 37.3a). You have probably seen the same response in houseplants that are near a window. Directed movement in response to light is called **phototropism** (literally, "light-turn"). Any directional plant movement that occurs in response to a directional stimulus is considered a **tropism**. Other plant movements in which the direction of the response is not dependent on the direction of the stimulus (e.g., closing of Venus flytrap leaves) are referred to as **nastic movements**. These are described later in this chapter.

When the Darwins exposed coleoptiles to light filtered through a solution of potassium dichromate, however, the coleoptiles did not bend toward the light. Potassium dichromate solutions filter out wavelengths in the blue part of the visible spectrum. The results of a follow-up experiment with coleoptiles show that when these plants are exposed to isolated wavelengths of blue, yellow, green, orange, and red light, bending occurs only toward blue light (Figure 37.3b).

The specificity of this response is important. Recall that **(1)** chlorophylls *a* and *b* are the primary photosynthetic pigments, and **(2)** these pigments absorb strongly in the blue and red parts of the spectrum (Ch. 10, Section 10.2). Plants exhibit a phototropic response if blue wavelengths are available, but show no response if blue wavelengths are not present. Plants bend toward blue light, which is important for photosynthesis.

Let's see what happens when blue light strikes a sensory cell, and follow the signal to examine the cells that undergo differential growth in response to blue light.

Phototropins as Blue-Light Receptors

Although biologists knew that the blue-light receptor must be a **pigment**—a molecule that absorbs certain wavelengths of light—it took decades to find it. A key breakthrough came in the early 1990s, when researchers found a membrane protein in the tips of emerging shoots that gains a phosphate group in response to blue light. Researchers hypothesized that the membrane protein becomes activated when it is phosphorylated in response to blue light, and that the activated protein then triggers the phototropic response.

Subsequent work succeeded in isolating the gene that codes for the membrane protein. The gene, named *PHOT1* (*PHOTOTROPIN 1*), was found by analyzing mutant *Arabidopsis thaliana* individuals that do not show a phototropic response to blue light (Figure 37.4). Further experiments showed that *PHOT1* codes for a blue-light receptor. An important question remained, however: Is the PHOT1 protein activated by phosphorylation, and does this trigger bending of the plant?

(a) Shoots bend toward full-spectrum light.

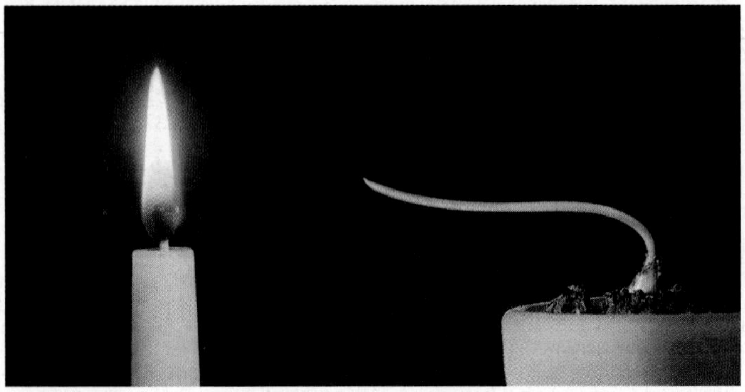

(b) Shoots bend specifically toward blue light.

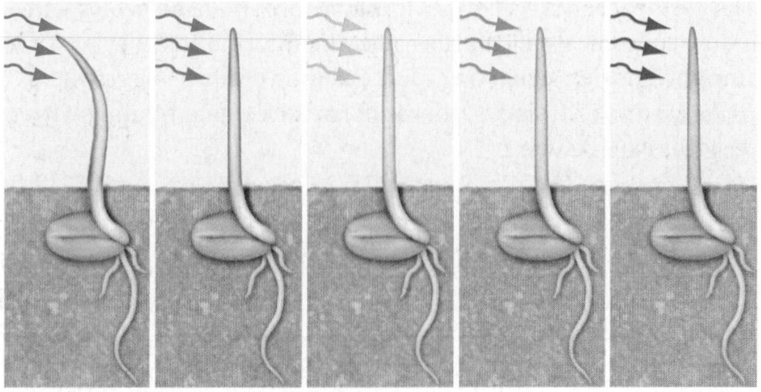

Figure 37.3 Experimental Evidence Suggests that Plants Sense Specific Wavelengths of Light. Though a candle is shown in part (a), the Darwins used paraffin lamps or other sources of artificial light in their original experiments. In part (b), each coleoptile is colored to indicate the color of light it was exposed to.

Figure 37.4 Phototropic Response Differs Greatly between a Wild-Type *Arabidopsis* and a Phototropin-Deficient Mutant.

As a first step in answering this question, researchers identified an amino acid in the PHOT1 protein that becomes phosphorylated after plants are exposed to blue light. During this analysis, they constructed a mutated *PHOT1* gene, called *phot1-m*, which coded for an altered protein with an amino acid that cannot be phosphorylated. This altered protein provided the researchers with a way to test whether phosphorylation of the PHOT1 protein was required for a phototropic response.

The key to the experiment was to compare bending in response to blue light in wild-type *Arabidopsis* with the response in mutant plants that lacked a PHOT1 protein. When the researchers introduced the *phot1-m* gene into a mutant plant, the altered protein failed to rescue phototropism. However, when the normal *PHOT1* gene was introduced into a mutant plant, phototropism was restored. The current consensus is that a phototropic response is initiated when the PHOT1 protein is phosphorylated.

Recent research indicates that *Arabidopsis* has a second blue-light receptor related to PHOT1, called PHOT2. Collectively, **photoreceptors** that detect blue light and initiate phototropic responses are known as **phototropins**.

Given its importance to photosynthesis, it's not surprising that blue light triggers an array of responses in addition to bending. The phototropins, for example, trigger signal transduction cascades that result in at least two other responses:

1. ***Chloroplast movements inside leaf cells*** These movements put chloroplasts in positions to optimize light absorption. In high light the chloroplasts move to the sides of cells to shade each other; in low light they spread out.
2. ***Opening of stomata*** As a result, carbon dioxide can diffuse into cells as blue light triggers photosynthesis.

Besides the phototropins, several other kinds of blue-light receptors have been identified in plants and also help control the opening of stomata, as well as stem elongation and flower production. Phototropism, however, ranks as the best studied of all blue-light responses. How is the signal from phototropins transmitted to the cells that cause bending?

Auxin as the Phototropic Hormone

Long before the phototropins were identified, biologists knew that sensory cells responded to blue light by releasing a hormone. The Darwins established this result when they followed up on their initial experiments.

RESEARCH

QUESTION: Where is light sensed to initiate phototropism in grass seedlings?

HYPOTHESIS: Light is sensed at the tip of a coleoptile.

NULL HYPOTHESIS: Light is not sensed at the tip of the coleoptile.

EXPERIMENTAL SETUP AND RESULTS:

Control: Bends toward light

Tip removed: No bending

Tip covered: No bending

Lower portion of coleoptile covered: Bends toward light

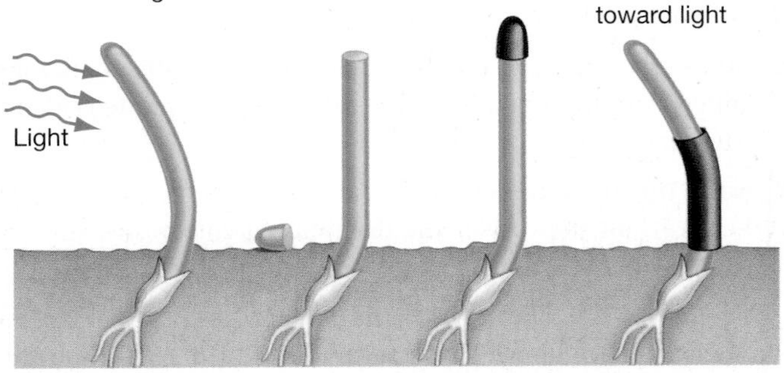

CONCLUSION: Light responsible for triggering phototropism is sensed at the coleoptile tip.

INTERPRETATION:

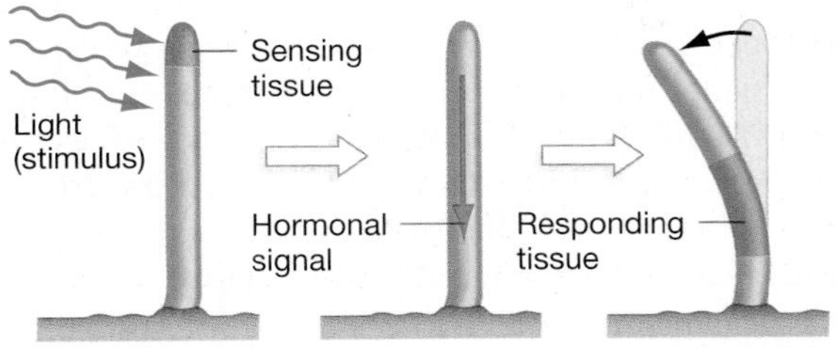

Figure 37.5 The Sensory and Response Cells Involved in Phototropism Are Not the Same.

SOURCE: C. Darwin and F. Darwin. 1897. *The Power of Movement in Plants*. New York: D. Appleton & Co.

✔ **PROCESS OF SCIENCE** A critic could argue that this experiment lacked appropriate controls for the treatments labeled "Tip removed" and "Tip covered." Suggest better controls for these treatments than the unmanipulated individual at the far left.

Figure 37.5 shows their experimental setup and results.

- If they removed the tips of coleoptiles, they found that the decapitated seedlings stopped bending toward the light.
- If they covered the tips of coleoptiles with opaque material, the seedlings did not bend toward light.
- If they put opaque collars below the tips, in the area where bending occurs, the seedlings bent toward light normally.

These data provided evidence that the blue-light sensors were located in the tips of the coleoptiles, and not in the part of the coleoptile that bends.

How did the sensory cells in the tip communicate with the cells that actually elongate? The Darwins proposed that phototropism depends on "some matter in the upper part which is acted on by light, and which transmits its effects to the lower part." Their hypothesis was that a substance produced at the tip of the coleoptile acts as a signal and is transported to the area of bending. This was the first explicit hypothesis stating that hormones—signaling molecules that can act at a distance—must exist.

The hormone hypothesis was not tested rigorously until 1913, when Peter Boysen-Jensen

1. cut the tips off young oat coleoptiles;
2. put a tip, and a porous block of the gelatinous compound called agar underneath it, on some of the decapitated coleoptiles; and
3. put a tip, and a piece of mica (a nonporous mineral) underneath it, on the rest of the decapitated coleoptiles and observed whether bending occurred.

As Figure 37.6a shows, only the coleoptiles treated with the porous agar block showed normal phototropism. Based on this observation, Boysen-Jensen concluded that the phototropic signal was indeed a chemical and that it could diffuse. He also inferred that the molecule was water soluble, because the agar that he used was a water-based gelatin.

Twelve years later, Frits Went extended these results. He placed the decapitated tips of oat coleoptiles on agar blocks, with the goal of collecting the hypothesized hormone controlling phototropism. Then he did something clever: He placed agar blocks that had been exposed to oat tips off-center on the decapitated coleoptiles of other individuals (Figure 37.6b). He also did the same with agar blocks that had not been exposed to oat tips, as a control (not shown).

Even though the coleoptiles were kept in the dark during the entire experiment, they responded by bending if their agar block had been exposed to oat tips. In this way, Went succeeded in producing the phototropic response without the stimulus of light.

Because it promotes cell elongation in the shoot, Went named the hormone **auxin** (from the Greek auxein, "to increase"). Auxin was the first plant hormone ever discovered.

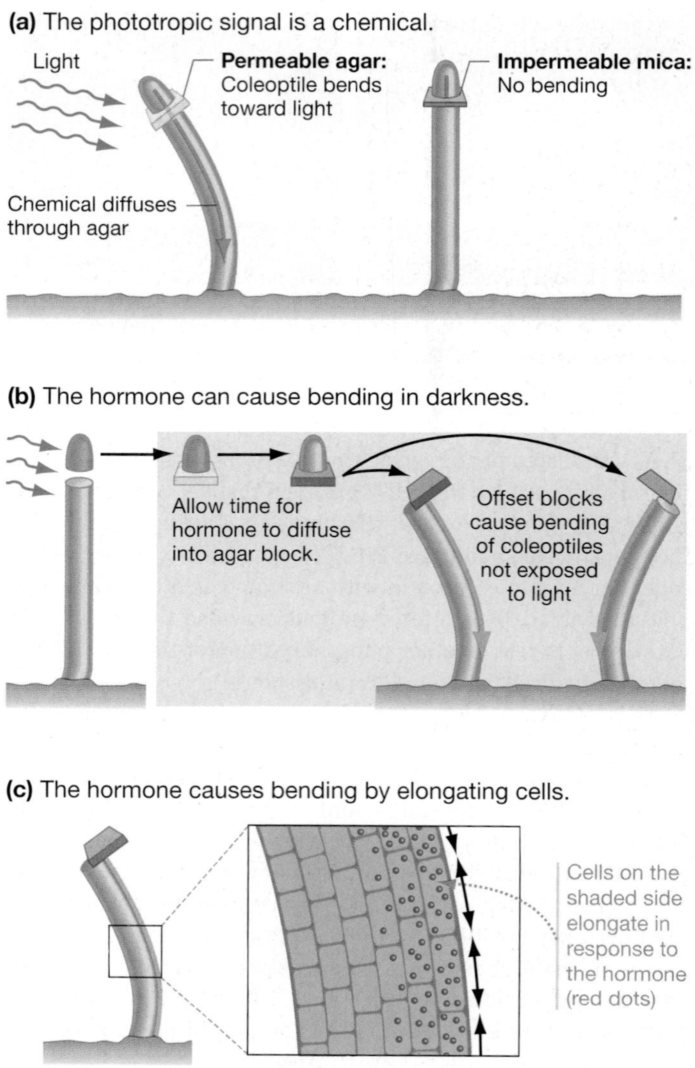

Figure 37.6 **Experimental Evidence Supports the Hormone Hypothesis for Phototropism. (a)** Coleoptiles bend in response to light if substances from the tip are allowed to move downward. **(b)** If bending can take place in darkness, then light is not directly required for the response. Only the hormone is required. **(c)** During the phototropic response, bending occurs because cells on the shaded side of the coleoptile elongate.

Isolating and Characterizing Auxin After years of effort, researchers in two laboratories independently succeeded in isolating and characterizing auxin. The hormone turned out to be indoleacetic acid, or IAA. It was difficult to identify because it is present in such low concentrations. Depending on the species and tissue involved, IAA concentrations range from 3 to 500 nanograms per gram of tissue. (The prefix *nano–* refers to billionths.)

Like the other plant hormones introduced in this chapter, auxin is a small molecule with a relatively simple structure. It is present in quantities so small that its concentration is difficult to measure. Yet its impact is huge. Auxin can bend stems—producing, in some cases, tree trunks that are permanently bowed.

The Cholodny–Went Hypothesis Went's experiments were a breakthrough in research on information processing: They confirmed the hormone hypothesis and led to the discovery and characterization of IAA. But Went's experiments also inspired an important hypothesis for *how* the hormone produces the bending response. Working independently, both N. O. Cholodny and Went proposed that phototropism results from an asymmetric distribution of auxin.

The Cholodny–Went hypothesis makes two predictions:

1. Auxin produced in the tips of coleoptiles is transported more down the shaded side than the illuminated side.
2. The asymmetric distribution of auxin causes cells on the shaded side of the coleoptile to elongate more than cells on the illuminated side (Figure 37.6c).

In essence, bending results from differential cell elongation.

To test the Cholodny–Went model, Winslow Briggs used thin sheets of mica inserted horizontally into coleoptile tips to block the movement of auxin. As predicted by the model, auxin

PROCESS: ACID-GROWTH HYPOTHESIS FOR CELL ELONGATION

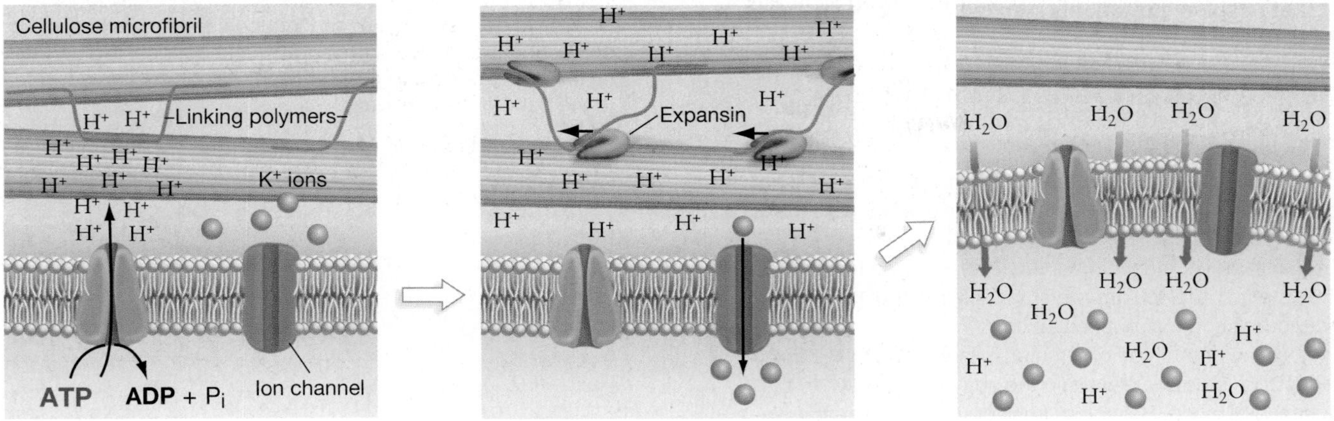

1. **Proton pumps acidify cell wall** outside the plasma membrane.

2. **Wall loosens** as activated expansins "unzip" hydrogen bonds connecting cellulose microfibrils to other cell-wall polymers. Electrochemical gradient brings K^+ or other ions into cell.

3. **Water follows by osmosis.** Increased turgor pressure pushes loosened wall out, elongating the cell.

Figure 37.7 **The Acid-Growth Hypothesis Requires Expansion of the Cell Wall and Intake of Water.** In response to auxin, activated proton pumps lower the pH outside the plasma membrane. This acidification causes a series of events that lead to elongation of the cell.

concentrations were the same on the lighted and shaded sides of divided tips. If a path for auxin movement was maintained, more of the auxin accumulated on the shaded side of the tip.

The Cholodny–Went asymmetric distribution model was correct. It explained how auxin leads to asymmetric cell elongation, and thus the bending response called phototropism. Biologists now know that the stems of eudicots and the coleoptiles of monocots respond to phototropic signals in essentially the same way.

The Cell Elongation Response How do cells in the stem respond to auxin? Experiments in corn plants and *Arabidopsis* suggested that auxin-binding proteins are found in stem and leaf cells. It took decades, however, for researchers to identify auxin receptors. In 2005 researchers provided strong evidence of an intracellular auxin receptor named TIR1. But the story was not complete—in 2014 other researchers showed that a different auxin-binding protein, named ABP1, is part of a receptor complex in the plasma membrane. Although initial investigations indicated that ABP1 may be required for auxin-based responses, more recent studies with mutants of the ABP1 gene show that plants that lack functional ABP1 exhibit normal auxin-based growth responses. The exact role of ABP1 remains unclear.

Researchers proposed that once auxin binds to receptors in stem cells, the signaling pathway that follows increases the activity of membrane H^+-ATPases, or proton pumps, in the plasma membrane. Recall that **proton pumps** use the energy in ATP to drive protons out of the cell against an electrochemical gradient (Ch. 35, Section 35.4). Because the pH of the cell wall decreases when H^+-ATPases are active, the idea that these pumps are responsible for cell elongation became known as the **acid-growth hypothesis**.

To understand the rationale behind the acid-growth hypothesis, keep in mind that two things have to happen for a plant cell to get larger:

1. The cell wall has to loosen up to allow for an increase in cell size.
2. Water has to enter the cell and generate turgor pressure on the cell wall to trigger an increase in cell volume.

As Figure 37.7 shows, both processes are triggered by pumping protons into the cell wall.

When proton pumping lowers the pH of the wall to 4.5, cell-wall proteins called **expansins** are activated. Expansins "unzip" the hydrogen bonds that form between cellulose microfibrils and other polymers in the cell wall, loosening the structure.

As protons are pumped out of the cell, an electrochemical gradient is established. The inside of the plasma membrane becomes much more negative than the outside, favoring the entry of potassium (K^+) or other positively charged ions. As the concentration of solutes increases inside the cell, water follows via osmosis.

This is a key point: Cells don't move water directly. Instead, they create an osmotic gradient that favors water movement. Pumping protons *out* of a cell is a way to bring water *into* the cell. The incoming water increases turgor pressure, which pushes out the loosened cell wall. The cell expands, with the direction of expansion depending in part on the orientation of the cellulose microfibrils. As a result, cells elongate in specific directions.

The upshot? When cell walls on one side of a stem are acidified in response to a signal from auxin, bending results through asymmetrical elongation. Auxin's role in phototropism is an exquisite example of information processing in plants.

Given the importance of light to plants, it shouldn't be surprising that they have photoreceptors that detect other wavelengths of light. Let's explore how plants respond to red and far-red light.

CHECK YOUR UNDERSTANDING

✔ **If you understood this section, you should be able to ...**

1. Explain why a plant with mutant phototropin genes will not bend toward light even when placed on a windowsill that receives bright sunlight.
2. Explain how you would manipulate a large bed of roses so their stems bend toward the east. Assume that you have an abundant supply of purified auxin.
3. Predict the phenotype of a mutant plant in which the auxin receptor in stems is always in the active state, even in the absence of auxin. Base your answer on the information in this section.

Answers are available in Appendix A.

37.3 Red and Far-Red Light: Germination, Stem Elongation, and Flowering

Plants are sensitive to wavelengths in the red and far-red portions of the visible spectrum, as well as to blue light. This sensitivity is interesting because red wavelengths (about 660 to 700 nm) and far-red wavelengths (over 710 nm) signal very different things to a plant.

After you complete this section, you should be able to ...

- Explain how red and far-red light interact with phytochrome to trigger seed germination, stem elongation, and flowering in some plants.

- Red light drives photosynthesis, just as blue light does.
- Far-red wavelengths are not absorbed strongly by photosynthetic pigments, so they tend to pass through leaves. As a result, far-red wavelengths are prominent in light that is filtered through tree leaves before it reaches the forest floor. Far-red light indicates shade caused by absorption of photosynthetic light wavelengths by leaves—in other words, it tells the plant that it's being shaded by other plants.

The Red/Far-Red "Switch"

The first hint that plants monitor red and far-red light emerged from studies by H. A. Borthwick and colleagues in the early 1950s on how lettuce seeds germinate. By exposing lettuce seeds to various wavelengths of light and plotting their frequency of germination, researchers discovered that germination rates peak when seeds receive red light (about 660 nm).

This observation made sense, because lettuce thrives best when it grows in bright sunlight. But the stimulatory effect of just 1 minute of red light disappeared if seeds were then exposed to 4 minutes of far-red light. This observation also made sense because far-red light indicates that the seeds are shaded. Wavelengths near 735 nm inhibit germination the most effectively.

Table 37.1 How Do Red Light and Far-Red Light Affect the Germination of Lettuce Seeds?

Biologists exposed moistened lettuce seeds to flashes of light containing one of two wavelengths in sequence: red and far-red (FR). After exposure to the last flash of light, the seeds were held in the dark for several days.

Light Exposure Sequence	Germination (%)
None (control)	9
Red	98
Red → FR	54
Red → FR → Red	100
Red → FR → Red → FR	43
Red → FR → Red → FR → Red	99
Red → FR → Red → FR → Red → FR	54
Red → FR → Red → FR → Red → FR → Red	98

SOURCE: H. A. Borthwick, S. B. Hendricks, M. W. Parker, et al. 1952. A reversible photoreaction controlling seed germination. *Proceedings of the National Academy of Sciences, USA* 38: 662–666, Table 1.

✔ **QUANTITATIVE** According to the data shown here, what is the average germination rate of lettuce seeds that were last exposed to red light? To far-red light? How would these values compare with the germination rate of seeds that are buried underground and receive no light at all?

Follow-up experiments showed that red and far-red light act like an on-off switch for lettuce seed germination (**Table 37.1**). Red light promotes lettuce germination; far-red inhibits it.

The key observation, though, is that the last wavelength sensed by the seed determines whether germination occurs at a high rate. How could this happen?

Phytochrome Is a Red/Far-Red Receptor

To interpret the red/far-red switch in seed germination, biologists hypothesized that the same pigment absorbs both wavelengths. Further, they suggested that the pigment exists in two shapes, or conformations: One shape absorbs red light, and one shape absorbs far-red light.

The idea was that switching behavior, or **photoreversibility**, occurs because light absorption makes the photoreceptor pigment change shape, like a light switch moving up or down in response to touch. Each conformation would be responsible for a different response.

Biologists called the hypothesized pigment **phytochrome** ("plant-color"). Later studies confirmed that phytochrome is a specialized light receptor, different from any of the pigments involved in absorbing light during photosynthesis or phototropism.

Figure 37.8 illustrates the photoreversibility hypothesis. One conformation of phytochrome, called P_r (phytochrome red), absorbs red light. Another conformation of the same molecule, called P_{fr} (phytochrome far-red), absorbs far-red light. Each conformation changes shape and switches to the other when it

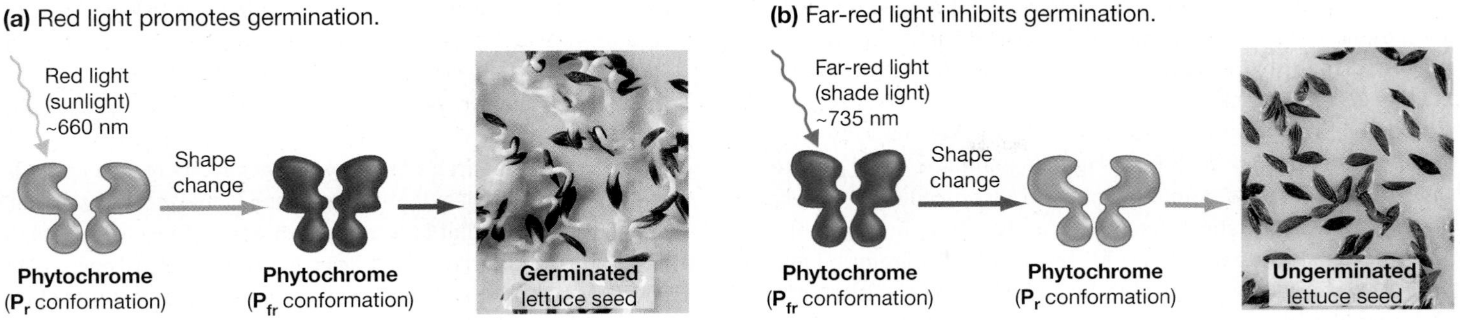

Figure 37.8 The Photoreversibility Hypothesis Addresses Phytochrome Behavior. Phytochrome switches between the P_r conformation and the P_{fr} conformation when it absorbs red or far-red light, respectively. Only the P_{fr} form—which is triggered by exposure to red light—promotes germination.

absorbs its specific wavelength. The result? Red light promotes seed germination; in far-red light, seeds remain ungerminated.

In addition to seed germination, phytochrome has been shown to play a role in other plant growth responses controlled by light, including etiolation and circadian rhythms. **Etiolation** describes the phenomenon in which plants have a pale yellow color and lengthen their stems and grow long, thin, and spindly when they are shaded or kept in dark conditions (Figure 37.9). You may have seen etiolated plant seedlings growing in the shade of a large tree. Also, the leaves of several plant species open and close in a slow but regular fashion determined by changes in light levels. These movements represent examples of *circadian rhythms*—events that occur roughly on a 24-hour cycle.

The photoreversibility hypothesis explains the red/far-red switch seen in seed germination and stem elongation. But can it also explain other plant responses? Let's consider another key event in plant growth—flowering—and the signals that trigger it.

Figure 37.9 Etiolation Affects a Plant's Appearance. Normal bean seedlings (left) and etiolated seedlings resulting from low light levels (right).

✔ What might be the adaptive advantage of etiolation?

Signals That Promote Flowering

Flowers develop from modified shoots. In essence, flower formation begins when an apical meristem stops making energy-harvesting stems and leaves and begins to produce the modified stems and leaves that make up flowers. Instead of making more food through photosynthesis, the flowering shoot commits to investing energy in sexual reproduction.

When does this happen? Early experiments on the environmental signals that promote flowering focused on the number of hours of light and dark during a day. As it turns out, the switch that triggers flowering in response to changes in day length also shows red/far-red photoreversibility. Phytochrome plays a key role in a phenomenon called photoperiodism.

Photoperiodism is any response by an organism that is based on photoperiod—the relative lengths of day and night. In plants, the ability to measure photoperiod is important because it allows individuals to respond to seasonal changes in climate—for example, to flower when pollinators are available and when resources for producing seeds are abundant. Let's take a closer look.

Responding to Changes in Photoperiod Experiments on photoperiodism in plants have shown that, with respect to flowering, plants fall into three main categories:

1. **Long-day plants** bloom in midsummer, when days are longest and nights shortest. Radishes, lettuce, spinach, corn, irises, and other long-day plants flower only when days are longer than a certain length—usually between 10 and 16 hours, depending on the species.
2. **Short-day plants** bloom in spring, late summer, or fall when days are shortest and nights longest. Asters, chrysanthemums, poinsettias, and other short-day plants flower only if days are shorter than a certain species-specific length.
3. **Day-neutral plants** flower without regard to photoperiod. Day and night length have no effect on flowering in roses, snapdragons, dandelions, tomatoes, cucumbers, and many **weeds**—plants that are adapted to grow in soils that have been disturbed enough to remove or damage existing vegetation.

How do plants sense changes in day or night length and initiate the chain of events that leads to flowering? Early experiments suggested that phytochrome is involved, and that night length, not day length, triggers flowering.

Figure 37.10 summarizes some of the key experiments. Researchers showed that for long-day plants, a night length *shorter* than a specific amount of time was needed to induce flowering; while for short-day plants, a night length *longer* than a specific time was needed (see Figure 37.10a). Therefore, long-day plants would more accurately be named "short-night" plants, and short-day plants are really "long-night" plants. The length of time in darkness that must be exceeded to initiate flowering in short-day plants, or not exceeded to initiate flowering in long-day plants, is called the **critical night length**.

Interestingly, interrupting the night period with a flash of specific wavelengths of light changed the flowering response (see Figure 37.10b). In short-day plants, for example, flowering was inhibited if the dark period was interrupted by a flash of red light—wavelengths around 660 nm. But a subsequent flash of far-red light—wavelengths around 735 nm—erased the effect. In some way, the phytochrome switch linked changes in night length to flowering.

More recent research has shown that phytochrome's effect is closely tied to the molecular mechanisms of timekeeping in plants. Plants have a clock that is reset each morning. Clock protein levels rise during the day and trigger expression of a gene called *CONSTANS* (*CO*). The CO protein is a transcription factor that affects the production of a flowering hormone. When phytochrome is activated by light, it stabilizes CO, so that CO accumulates in cells.

- In long-day (short-night) plants, a high level of CO stimulates production of the flowering hormone.
- In short-day (long-night) plants, a high level of CO *inhibits* production of the flowering hormone.

Many questions remain, however, and research is continuing. In the meantime, it's clear that a complex interaction takes place between clock proteins, phytochrome activation, and the production of the signal that initiates flowering. What is that signal?

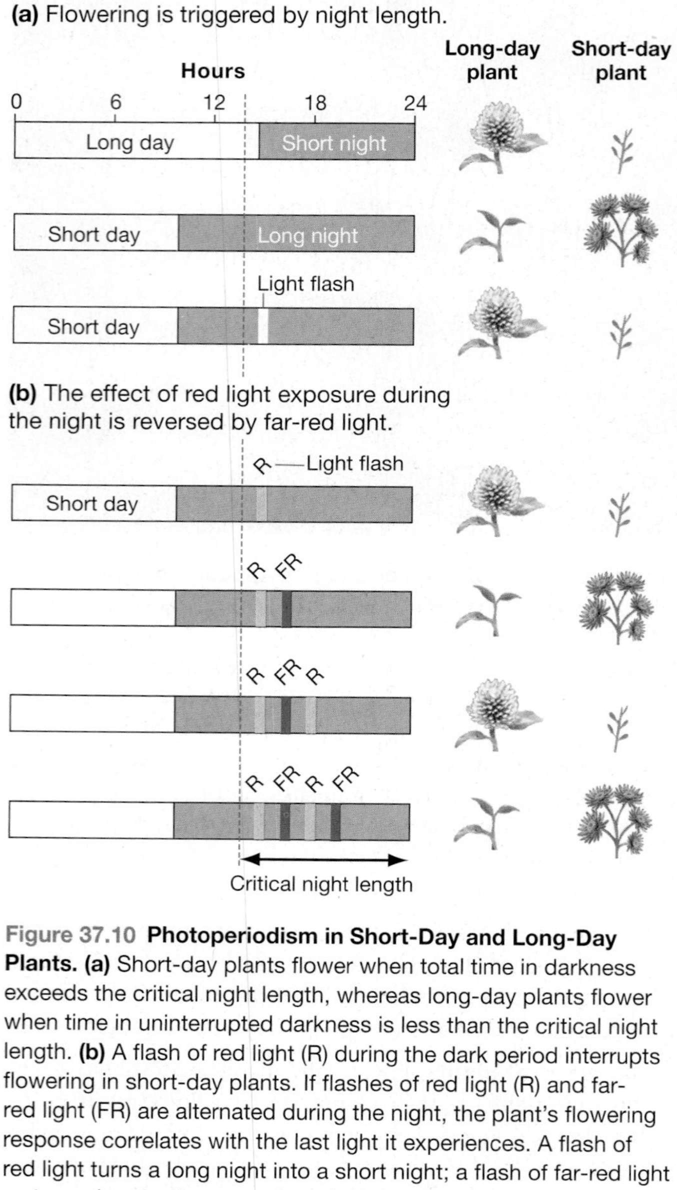

Figure 37.10 Photoperiodism in Short-Day and Long-Day Plants. (a) Short-day plants flower when total time in darkness exceeds the critical night length, whereas long-day plants flower when time in uninterrupted darkness is less than the critical night length. **(b)** A flash of red light (R) during the dark period interrupts flowering in short-day plants. If flashes of red light (R) and far-red light (FR) are alternated during the night, the plant's flowering response correlates with the last light it experiences. A flash of red light turns a long night into a short night; a flash of far-red light restores the plant's perception of a long night.

Discovery of the Flowering Hormone Since the 1930s, biologists have known that exposing even one leaf on a plant to the conditions necessary to induce flowering may result in the whole plant flowering. This result suggests that the signal to flower comes from leaves and travels to the apical meristem.

Grafting experiments, in which an organ from one individual is physically attached to a different individual, support this result. If you provide an experimental plant with the appropriate flower-triggering night length and then cut off a leaf or stem and graft it onto a second, experimental plant that has never been exposed to the correct photoperiod for flowering, the experimental plant will flower (Figure 37.11).

Like the experiments on phototropism with agar blocks (see Figure 37.6), this result supported the hypothesis that some substance produced in the transplanted leaf must travel up the recipient plant. This substance would then trip a developmental switch in the apical meristem, causing the change from vegetative growth to flowering. Biologists were so convinced that flowering must be induced by a hormone that they named it **florigen**, even though the actual hormone had not been discovered.

Almost 80 years later, researchers finally found the florigen molecule. This work began by focusing on the gene *FLOWERING LOCUS T* (*FT*) in *Arabidopsis*, which is known to promote flowering when activated. When researchers exposed leaves to short nights, they found that the gene was expressed in the leaf vascular tissue. Subsequently, some research indicates that the protein product of the *FT* gene is transported from leaves to the shoot apical meristem, and that the protein's presence triggers the activation of genes required for converting the stem to a flower.

Details are emerging on how the information present in the P_r–P_{fr} switch is translated into action. This affects germination, stem elongation, flowering, timekeeping, and other responses to red and far-red light.

Given their reliance on photosynthesis for both energy and reduced carbon compounds, it isn't surprising that plants are highly tuned to their light environment. They also need to tell which end is up. Next, let's explore how plants sense gravity.

CHECK YOUR UNDERSTANDING

✓ If you understood this section, you should be able to …

1. Explain why it is adaptive for red light to trigger germination in lettuce seeds, while far-red light inhibits it.
2. Explain why it is adaptive for changes in night length to affect the tendency to flower in some species.

Answers are available in Appendix A.

RESEARCH

QUESTION: Can a signal produced in one plant be used to induce flowering in another?

HYPOTHESIS: Signals from grafted leaves can induce flowering.

NULL HYPOTHESIS: Signals from grafted leaves cannot induce flowering.

EXPERIMENTAL SETUP:

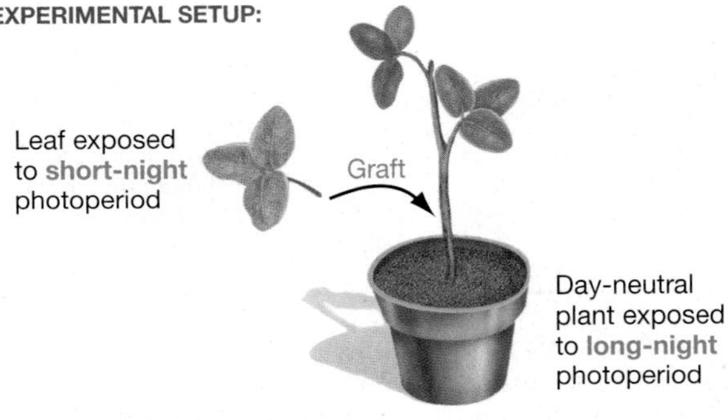

PREDICTION OF HYPOTHESIS: Plant will flower.

PREDICTION OF NULL HYPOTHESIS: Plant will not flower.

RESULTS:

CONCLUSION: A signal from a grafted leaf can induce flowering.

Figure 37.11 Experimental Evidence Supports the Hypothesis that a Hormone for Flowering Exists.

SOURCE: A. Lang, M. K. Chailakhyan, and I. A. Frolova. 1977. Promotion and inhibition of flower formation in a day-neutral plant in grafts with a short-day plant and a long-day plant. *Proceedings of the National Academy of Sciences, USA* 74: 2412–2416.

✓ PROCESS OF SCIENCE For this experiment to support the claim that a flowering hormone exists, a control treatment needs to be carried out. What is this control treatment? If the hypothesis is correct, what is the predicted outcome of the control treatment?

37.4 Gravity: The Gravitropic Response

After you complete this section, you should be able to …

- Explain the gravitropic response of plant roots.

The wavelengths, quantity, and direction of light that a plant receives change with the season, weather, time of day, and shading by other plants. However, gravity is constant and unidirectional. Light means food; gravity provides information about how the plant should orient itself in space.

Shoots usually respond to gravity by growing in an upward direction (see photo at the start of this chapter); roots usually respond by growing downward or laterally. How do plants sense gravity, and use it as a signal to orient their growth?

In 1881 Charles and Francis Darwin published one of the first experimental results on **gravitropism** ("gravity-turn")—the ability to move in response to gravity. Recall that the ends of root tips are covered by a protective tissue called the **root cap** (Ch. 34, Section 34.1). The Darwins found that roots stop responding to gravity if the root caps are removed. This observation suggested that gravity sensing occurs somewhere inside the root cap.

Recently biologists demonstrated precisely which cells are involved in gravity sensing in *Arabidopsis* roots. By killing tiny blocks of cells with laser beams, researchers showed that cells located at the center of the root cap are the most important for regulating the gravitropic response. Importantly, these cells always contain starch granules (Figure 37.12). Cells in the root cap respond to gravity and initiate gravitropism. Does starch play a role in sensing gravity?

The Statolith Hypothesis

The leading explanation for how plants sense gravity—the **statolith hypothesis**—is based on two interconnected ideas:

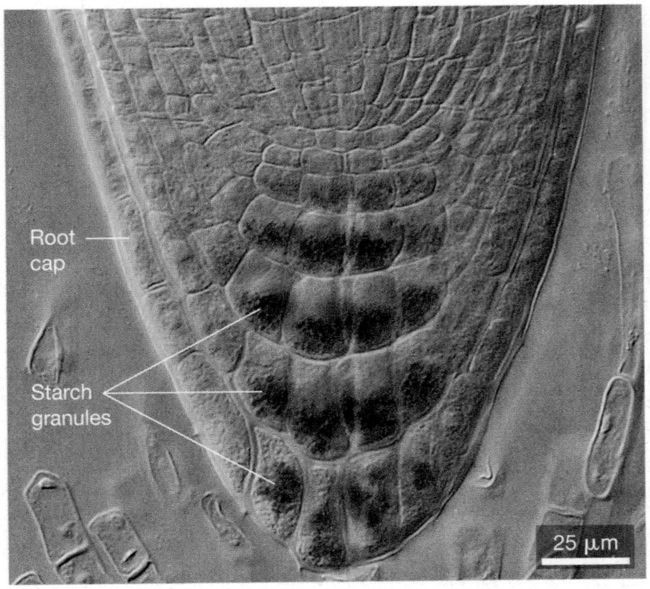

Figure 37.12 Gravity Sensing Occurs in the Root Cap. The root cap is a protective structure. Gravity-sensing cells in the center of the cap contain starch granules (stained dark purple).

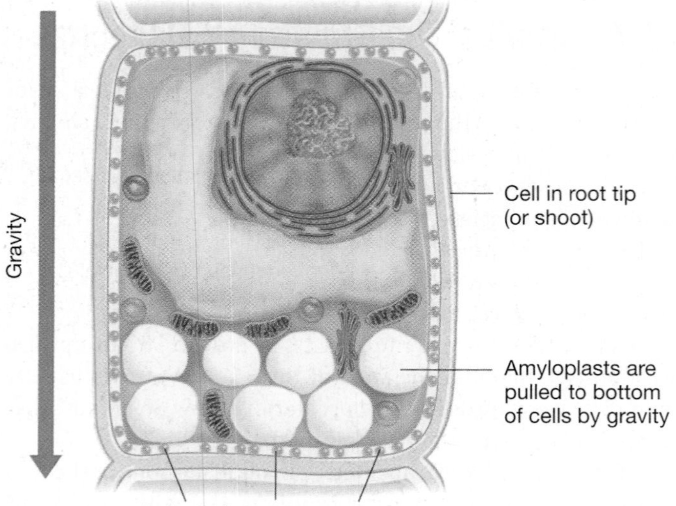

Figure 37.13 The Statolith Hypothesis States that Amyloplasts Stimulate Sensory Cells. Amyloplasts are filled with starch. They are dense, so they sink in response to gravity. The statolith hypothesis predicts that pressure receptors in the plasma membrane become activated as a result.

✔ **PROCESS OF SCIENCE** Assuming that the receptor is a protein, suggest a hypothesis to explain how signal transduction occurs in response to pressure from a statolith.

1. **Amyloplasts** are organelles that contain starch granules; starch is denser than water. These organelles are pulled to the bottom of root cap cells by the force of gravity (Figure 37.13).
2. The position of the amyloplasts activates pressure receptor proteins located in the plasma membrane. These receptors initiate the gravitropic response.

The statolith hypothesis was inspired by animals that use dense particles to sense gravity. Lobsters, for example, take up grains of sand that become positioned in specialized gravity-sensing organs in their antennae. The grains of sand are called **statoliths** ("place-stones"). When the animal tilts or flips over, the statoliths move in response to gravity. Inside the organ, the sand grain ends up pushing against a sensory cell. When this cell is activated, it indicates that the animal is no longer upright.

According to the statolith hypothesis, the same thing happens in root cap cells. If the wind tips a plant over, for example, the amyloplasts settle onto the new "lower" cell walls. The weight activates receptors, which signal that the root no longer faces in the correct direction.

Although recent experiments strongly support the statolith hypothesis, the search for the gravity receptor itself continues. In contrast, the second and third steps of information processing in response to gravity—the production of a cell–cell signal and the response of target cells—are much better understood.

Auxin as the Gravitropic Signal

Root cap cells that sense changes in the direction of gravitational pull respond by changing the distribution of auxin in the root tip. Figure 37.14 illustrates this chain of events, also described here.

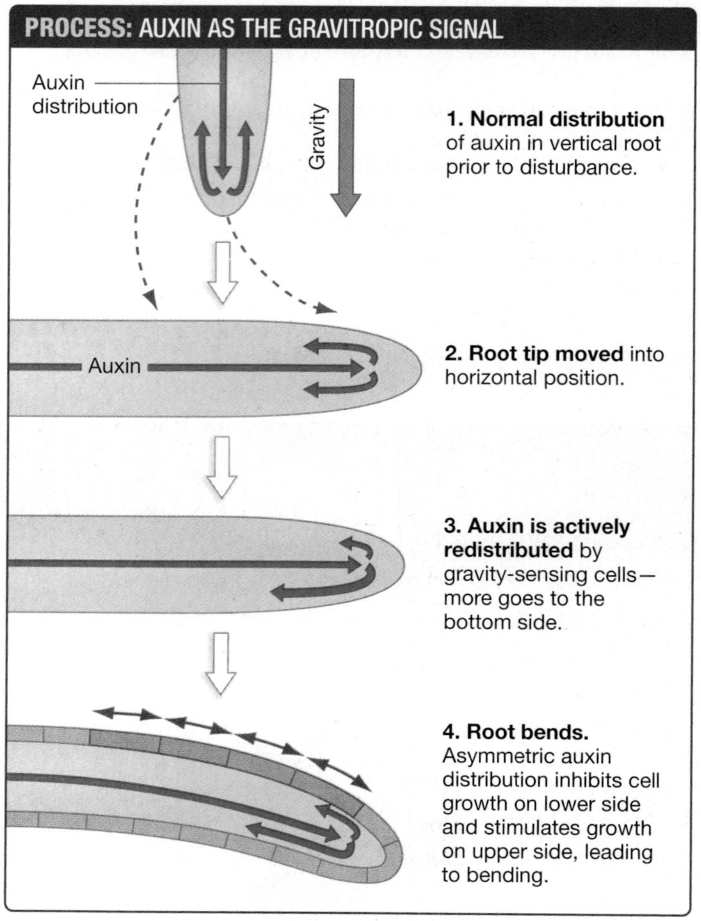

Figure 37.14 The Auxin Redistribution Hypothesis Explains Gravitropism. This sequence of events might begin when a growing root tip hits a rock and is displaced horizontally, or when a plant is tipped in a windstorm and partially uprooted.

Step 1 Under normal conditions, auxin flows down the middle of the root, then toward the perimeter, and finally away from the root cap.

Step 2 If the root is moved horizontally, sensory receptors in gravity-sensing cells trigger changes in the position of auxin transport proteins that redistribute auxin.

Step 3 Auxin is redistributed: The cells in the lower portion of the root receive increased concentrations of auxin; those in the upper portion receive lower concentrations.

Step 4 Because a high auxin concentration inhibits growth in roots, the differences in auxin concentration trigger differential growth. Cells in the lower portion of the root elongate more slowly compared with cells in the upper portion. The result is bending.

Note that the way root cells respond to auxin redistribution during the gravitropic response is opposite to the way that cells in the stem respond during phototropism. In stems, a high concentration of auxin leads to *increased* cell elongation and bending. In roots, a high concentration of auxin leads to *decreased* cell division and elongation. Roots bend as cells on the upper side of the zones of cellular division and elongation (Ch. 34, Section 34.3) continue to grow.

CHECK YOUR UNDERSTANDING

✔ If you understood this section, you should be able to ...

1. Explain how a root will grow in a downward direction even if placed in a horizontal position.
2. Explain the similarities and differences between the auxin redistribution in shoots and roots and the responses called phototropism and gravitropism.
3. Explain why auxin could be considered a gravitropic hormone.

Answers are available in Appendix A.

37.5 How Do Plants Respond to Wind and Touch?

Light and gravity are not the only physical forces that plants sense and respond to. Plants also react to mechanical stimuli such as wind and touch.

After you complete this section, you should be able to ...

- Explain how physical forces affect plant growth patterns and movement.

Let's analyze some of the best-studied responses to pushing or pulling forces acting on plants: growth responses that lead to exceptionally stout, stiff stems; directional movement toward or away from an object; and movement responses that make a Venus flytrap snap shut.

Changes in Growth Patterns

When plants are buffeted by wind, receptor cells transduce the mechanical force into an internal signal in the form of phosphorylated proteins or second messengers. Although the exact receptor signaling pathways are not known, studies in *Arabidopsis thaliana* have shown that transcription of a large suite of genes is turned up or down in response to touch or other mechanical stimuli that mimic the effect of wind. Some of the new protein products act to stiffen cell walls, resulting in plants that are shorter and stockier than plants that do not experience repeated vibrations or touching.

Figure 37.15 shows what this response—referred to as **thigmomorphogenesis**—looks like in *Arabidopsis*. In this experiment, plants were either touched lightly 2 times per day or left untouched. The flowering stalks of the touched plants were much shorter.

In response to wind or touch, then, plants change their growth patterns in ways that make them more likely to withstand the force, stay upright, and live long enough to produce flowers and fruit. Note that this altered growth is not directional—it merely results in sturdier plants.

Movement Responses

In some cases, plants respond to touch by moving in a specific direction. This response, **thigmotropism** ("touch-bending"), can be moderately fast. For example, species that grow by climbing up objects or other plants may have modified leaves or stems that form long, thin structures called *tendrils*. When a tendril makes contact with an object, it responds by wrapping itself around the item as fast as one or more times per hour (**Figure 37.16**). Plants that climb in this way are able to grow taller toward light without needing to invest considerable resources in making wood.

Figure 37.15 Plant Growth Changes in Response to Wind or Touch. The *Arabidopsis* plants on the right were touched several times daily, while plants on the left were untouched.

Movement is even faster in "touch-sensitive" plants. A Venus flytrap, for example, captures insects by closing quickly when an insect lands on its leaves. Rapid, nondirectional movements such as these are referred to as **thigmonastic movements**. These types of movements occur when a touch-receptor cell transduces the mechanical signal to an electrical signal.

To understand how plants use electrical signaling, recall that proton pumps in the plasma membrane of most plant cells give them a negative charge relative to the exterior environment (Ch. 35, Section 35.4). The separation of charges creates a membrane voltage, or **membrane potential**. If a receptor protein responds to touch by allowing ions to flow across the membrane—which changes the amount of charge on either side—then the membrane potential changes. In this way, the mechanical signal (touch) can be transduced to an electrical signal.

To travel from a sensory cell to a response cell, electrical signals are propagated in a characteristic form called an **action potential**. In animals, action potentials carry signals

Figure 37.16 Thigmotropism Is Directional Movement in Response to Touch. Portions of grape plants called tendrils wind around support structures after contacting them. The attachment provided by the tendrils allows plants to climb.

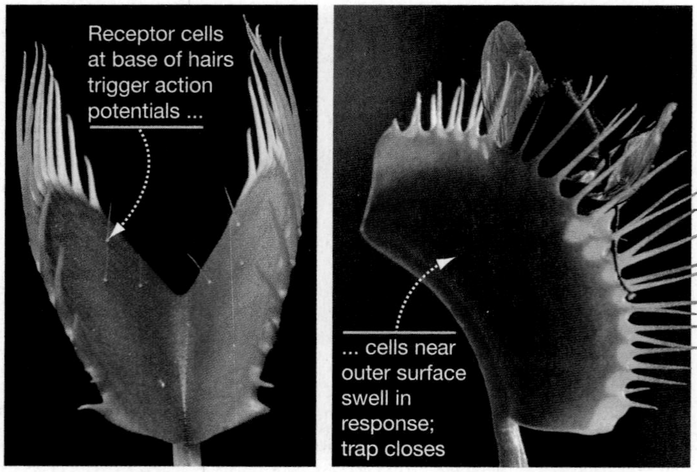

Figure 37.17 Venus Flytraps Close in Response to Action Potentials Generated by Sensory Hairs.

along the plasma membrane of nerve and muscle cells (Ch. 43, Section 43.2). How do plant action potentials move across tissues, when those tissues lack nerve cells? Recall that most plant cells are connected by membrane-lined **plasmodesmata** (Ch. 11, Section 11.2). Action potentials move more slowly in plants than in animals, but they can still carry information over long distances in a hurry. In a Venus flytrap, for example, action potentials generated by receptor cells at the base of sensory hairs inside the trap race across the leaf at a rate of about 10 cm/sec.

When the action potentials in a Venus flytrap reach cells on the outer surface of the trap, the cells change shape by swelling and push the trap shut (Figure 37.17). The response is rapid enough to resemble the way an animal's muscle contracts in response to an action potential.

CHECK YOUR UNDERSTANDING

✔ **If you understood this section, you should be able to ...**

1. Explain why it might be beneficial for plants grown in windy conditions to grow shorter and stockier than when grown in the absence of wind.
2. Explain how tendrils help some plants grow tall.

Answers are available in Appendix A.

37.6 Youth, Maturity, and Aging: The Growth Responses

After you complete this section, you should be able to ...

- Summarize how hormones regulate plant growth and responses to environmental stimuli.

Plants grow throughout their lives, from the time they germinate until the time they die. As the plant body grows, it matures into an efficient machine for absorbing sunlight, water, nutrients, and other diffuse resources. But growth is not constant. It speeds up when water, light, and nutrients are abundant and slows or stops when conditions are poor, or when it is time for leaves to drop or fruits to ripen.

Controlling growth in response to changes in age or environmental conditions is one of the most important aspects of information processing in plants. Hormones—appropriately referred to as plant growth regulators—play key roles in controlling growth. Let's consider six of the best-studied hormones involved in growth responses, starting with auxin—the molecule responsible for phototropism and gravitropism.

Auxin and Apical Dominance

When **apical dominance** occurs, growth is restricted to the main stems; the lateral buds in the axils of each leaf remain dormant. But if the apical bud dies or is removed (e.g., eaten by an herbivore or cut off by someone pruning their garden), the dormancy of the lateral buds is broken and lateral branches begin to grow. Figure 37.18 shows this phenomenon in action.

What caused the change? Because auxin is produced in shoot tips, researchers suspected that it might have a role in apical dominance as well as phototropism. This hypothesis was confirmed when it was shown that apical dominance could be sustained by applying auxin to a shoot's cut surface after its tip had been removed.

Auxin's role in apical dominance suggests that tip cells send a constant stream of information down to other organs and tissues. A break in the signal means that apical growth has been interrupted. In response, lateral branches sprout and begin to take over for the main shoot. Now the question is, how does this signal move?

Polar Transport of Auxin Auxin transport is **polar**, or unidirectional. If radioactively labeled auxin is added to the top end of a stem segment, the hormone emerges from the basal end. But if labeled auxin is added to the basal end, it is not transported toward the top. Auxin is the only plant hormone known to be transported through individual cells in one direction only.

Figure 37.18 When Apical Dominance Occurs, Growth of Lateral Buds Is Suppressed. **(a)** The stem of a coleus plant is shown still intact. **(b)** The same plant, several weeks after the stem was cut. The lateral shoots will orient themselves vertically.

Studies with auxin have also shown that the hormone is transported down the stem and through the root via parenchyma cells in the ground tissue and vascular tissue. Auxin enters the apical end of cells via a specialized membrane protein, diffuses to the other end of the cell, and then is transported out by proteins located only in the basal portion of the plasma membrane. Labeled auxin moves from cell to cell at about 10 cm/hr—roughly 10 times more slowly than substances traveling in phloem or xylem.

Because enzymes break down some auxin molecules as they travel down the long axis of the plant, polar transport sets up a strong gradient in auxin concentration. Auxin concentrations are higher near the tips of shoots than near the base of shoots.

What Is Auxin's Overall Role? Auxin clearly plays a key role in controlling growth via apical dominance, phototropism, and gravitropism. But this chemical messenger has other important effects as well:

- Auxin produced by seeds within the fruit promotes fruit development.
- Falling auxin concentrations are involved in the **abscission**, or shedding, of leaves and fruits, which is associated with the genetically programmed aging process called senescence.
- The presence of auxin in growing roots and shoots is essential not only for the proper differentiation of xylem and phloem cells in vascular tissue but also for the development of vascular cambium (Ch. 34, Section 34.1).
- Auxin stimulates the development of adventitious roots in tissue cultures and cuttings.

Auxin has so many different effects on plants that it has been difficult for biologists to understand its overall role. Recently, several investigators have proposed that one of the fundamental functions of auxin gradients is to signal where cells are located relative to the root–shoot axis. If conditions relating to the root–shoot axis change—for instance, a windstorm tips the plant or a deer eats the shoot apex—resulting changes in auxin gradients effectively signal how and where the individual's tissues should respond. Phototropism, gravitropism, apical dominance, and the production of adventitious roots are all ways of coping with changes along the root–shoot axis of the plant.

Cytokinins and Cell Division

Cytokinins are a group of plant hormones that promote cell division. (*Cyto* is the Greek root for cell; *kinin* refers to "kinesis," meaning movement or division.)

The Discovery of Cytokinins In their first attempts to grow plant cells and embryos in culture, biologists found that coconut milk, which stores nutrients used by growing coconut embryos, promoted cell division. This was the first hint that certain molecules can promote cell division in plants. Later experiments showed that molecules derived from the nitrogenous base adenine also stimulate the growth of cells in culture.

Eventually, naturally occurring adenine derivatives that stimulate growth were discovered in corn and apples, and were named cytokinins. Zeatin, the cytokinin found in the most species, is derived from adenine.

Cytokinins are synthesized in root tips, young fruits, seeds, growing buds, and other developing organs. But most of the zeatin and other cytokinins that are active in plants are synthesized in the apical meristems of roots and are then transported up into the shoot system via the xylem. Biologists still add cytokinins to plant cells growing in culture, to stimulate cell division (**BioSkills 11**).

How Do Cytokinins Promote Cell Division? After years of searching, biologists have now isolated and characterized a group of closely related proteins located primarily in the endoplasmic reticulum that act as cytokinin receptors. When cytokinins bind to these receptors in target cells, the receptors activate genes that regulate cell division.

Recent research on cytokinins has explored whether they affect molecules that regulate the cell cycle, including the cyclins and the cyclin-dependent kinases (Cdks; Ch. 12, Section 12.3). Recall that activated cyclins and Cdks allow cells to progress through checkpoints in the cell cycle and continue dividing.

To assess whether cytokinins affect cell-cycle genes, researchers grew *Arabidopsis* cells in culture so the nutrients and other molecules available could be carefully controlled. The biologists starved the cells of cytokinins for a day and then added the hormones again to half of the cells. When they assessed the level of mRNA from a cyclin gene called *CycD3*, they documented significant increases in the cells that were treated with cytokinins compared with the level in untreated cells.

This is strong evidence that cytokinins regulate growth by activating genes that keep the cell cycle going. In the absence of cytokinins, the cell cycle halts and cells stop dividing.

Gibberellins and ABA: Growth and Dormancy

In regions with temperate, seasonal climates, most seeds and mature plants start growing in spring. Conditions for growth are good at that time of year, because temperatures are warming and soil moisture levels are usually at their peak. Seedlings and mature plants continue to grow throughout the summer and early fall if moisture and nutrients are still available.

During drought conditions, however, growth stops, including in the embryos inside seeds. Embryos begin to develop as seeds mature, but cease this initial growth and remain dormant throughout the cold winter months. **Dormancy** is a temporary state of reduced metabolic activity.

Which signals initiate growth in response to changing environmental conditions, and which signals stop it? Two hormones provide the answer. **Gibberellins**, a large family of closely related compounds, stimulate growth in plants. **Abscisic acid**, commonly abbreviated **ABA**, inhibits growth. In at least some cases, the two hormones interact like start and stop signals.

The Discovery of Gibberellins Over 100 years ago, Japanese farmers noticed that some of their rice seedlings grew exceptionally quickly but fell over before they could be harvested. Biologists found that the plants with this "foolish seedling disease" were infected with the fungus *Gibberella fujikuroi*.

Researchers confirmed a causal connection between *Gibberella* infection and rapid stem elongation when they treated rice seedlings with an extract from the fungus. As predicted, the treated

Figure 37.19 **Gibberellin Enhances Stem Growth in Many Plants.** The two cabbage plants on the right were treated with gibberellin for five months. No gibberelin was applied to the control plant on the left.

seedlings produced abnormally long shoots. The same results have been replicated in other plant as well (Figure 37.19).

The active component in the fungus extract was eventually isolated and named gibberellic acid (GA), which is a gibberellin. Follow-up research showed that rice and other plants produce their own gibberellin but respond to applications of additional hormone by elongating their stems. In effect, infected seedlings suffer from a gibberellin overdose. Fungi that manipulate their host plants in this way release their spores higher off the ground, so they are likely to blow farther away from their parent.

Gibberellins are found in a wide array of fungi and plants. Most plant species produce several different gibberellins that function as active hormones.

Even though gibberellins have dramatic effects on growth, they are present in vanishingly small concentrations. In growing stems and leaves, active forms of gibberellin may be present in concentrations of about 10 nanograms per gram of tissue.

Defective Gibberellin Genes Cause Dwarfing To find the genes that are responsible for producing gibberellins, biologists analyzed mutant plants with abnormal stem length. Recall that Gregor Mendel analyzed the transmission of two alleles at a single gene that affected stem height in garden peas (Ch. 14, Table 14.2). One allele was associated with tall stems; the other was associated with dwarfed growth. The tall allele was dominant to the dwarf allele.

Dwarfed pea plants have much shorter stems than do normal individuals of the same age. However, when gibberellins were applied to dwarfed plants, they grew normally, suggesting that they lacked the means to synthesize the functional hormone. By analyzing dwarfed individuals, biologists were able to identify genes involved in gibberellin synthesis.

✔ **If dwarfed individuals receive the same amount of sun and thus perform as much photosynthesis as taller individuals, they often produce more flowers and seeds. Why?**

The gene responsible for the stem-length differences in garden peas is now known as *Le* (for *length*). Early work on dwarf mutants showed that they attain normal height if they are treated with the gibberellin called GA_1. This observation suggested that dwarf peas can respond to gibberellins normally—meaning that the problem is not with a hormone receptor.

In follow-up experiments, biologists treated dwarf peas with a radioactively labeled molecule used in the synthesis of GA_1. These plants did not produce radioactively labeled GA_1, even though plants with the normal allele did. Based on these results, researchers became convinced that the *Le* locus encodes an enzyme involved in GA synthesis.

Investigators recently identified the gene at the *Le* locus and confirmed that in dwarf plants, it contains a mutation that renders the enzyme inactive. More than 100 years after Mendel did his experiments, biologists finally understood the molecular basis of the dwarfing phenotype he studied.

In stems, gibberellins appear to promote both cell elongation and rates of cell division. But it is well established that auxin also promotes cell elongation, and that cytokinins also promote cell division. Research continues on how GAs, cytokinins, and auxin interact on the molecular level to control plant growth and development.

Gibberellins and ABA Interact during Seed Dormancy and Germination Many plants produce seeds that have to undergo a period of drying or a period of cold, wet conditions before they are able to germinate. A requirement for drying ensures that mature seeds will not sprout on the parent plant; an obligatory cold period prevents seeds from germinating just before the onset of winter, when young seedlings would be more likely to perish. Some seeds also have to receive a dose of red light, which indicates that they are in a sunny location.

In essence, then, seeds have an "off" setting that discourages germination and an "on" setting that initiates growth. The appropriate state for the on/off switch is determined by environmental cues such as temperature, moisture, and light.

By applying hormones to seeds, researchers learned that in many plants, ABA is the signal that inhibits seed germination and gibberellins are the signal that triggers germination. To understand how these messengers interact, researchers have studied a specific event: the production of an enzyme called α-amylase in germinating oat or barley seeds.

α-Amylase acts as a digestive enzyme that breaks the bonds between the sugar subunits of starch (Ch. 5, Section 5.2). (Your saliva contains an **amylase** that acts on the starch in food. In humans and other mammals, this enzyme initiates carbohydrate digestion in the mouth.)

PROCESS: GIBBERELLINS (GAs) ACTIVATE PRODUCTION OF α-AMYLASE

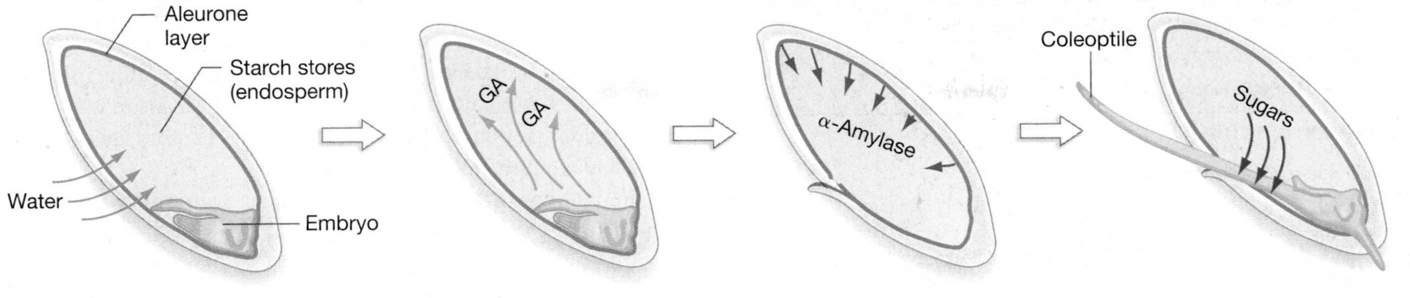

1. The seed absorbs water, and germination begins.

2. Gibberellins diffuse from the embryo to the aleurone layer.

3. Cells in the aleurone layer respond by releasing digestive enzymes, such as α-amylase.

4. The enzymes digest starch, releasing sugars and other molecules to the growing plant.

Figure 37.20 Gibberellin Plays a Critical Role in Seed Germination.

Figure 37.20 shows that, during the germination of a barley seed, α-amylase is released from a tissue called the aleurone layer. The enzyme diffuses into the seed's starch-filled storage tissue, called endosperm. Digestion of starch releases sugars that can be used by the growing embryo. Adding GA to the aleurone layer increases the production and release of α-amylase; adding ABA to that layer decreases the α-amylase level.

Research on the molecular interaction between GA and ABA carries several important messages:

- A cell's response to a hormone often occurs because specific genes are turned on or off.
- Hormones don't act on genes directly. Instead, a receptor on the surface of a cell's plasma membrane or in the cytosol receives the message and responds by initiating a signaling pathway, which activates specific gene regulatory proteins—the activators and repressors of transcription (Ch. 19, Section 19.3).
- Different hormones interact at the molecular level because they induce different gene regulatory proteins, which increase or decrease expression of key genes.

The logic runs as follows: Different hormones trigger the activation or production of different regulatory transcription factors. Hormone concentration affects the amount of each transcription factor activated or produced. Changes in transcription factors are responsible for changes in gene expression.

GA's role in seed germination has important commercial applications. For example, in the production of beer, brewers routinely use gibberellins in the malting process—the conversion of starches stored in barley seeds to sugars. The sugars that are released in response to GA treatment support fermentation by yeast and provide flavor and alcohol in the finished beer.

ABA Closes Guard Cells in Stomata One of the major challenges faced by land plants is replacing water that is lost to the atmosphere when **stomata** are open (Ch. 35, Section 35.3). Because stomata open in response to blue light, they allow gas exchange to occur while the plant is receiving the wavelengths of light used in photosynthesis.

However, if plant roots are unable to obtain enough water to replace what is lost from the leaves, stomata close. Closing stomata is an adaptive response when roots cannot find adequate water, because continued transpiration would lead to wilting and potential tissue damage.

Early work on the mechanism of stomatal closing suggested that ABA is involved. For example, applying ABA to the exterior of stomata causes them to close within seconds.

To explore the hypothesis that a hormone regulates stomatal closing, researchers performed the experiments summarized in Figure 37.21 on page 802. The fundamental idea was to grow plants whose roots had been divided. Only one side of the experimental plants was watered, while both sides of the control plants were watered. During this treatment, investigators documented that the water potential of the leaves remained the same in both control and experimental plants. Yet the stomata of experimental plants began to close while those of control plants remained open. This result suggested that roots from the dry side of the pot were signaling drought stress, even though the leaves were not actually experiencing a water shortage.

Follow-up experiments have supported two important predictions: ABA concentrations in roots on the dry side of the pot are extraordinarily high relative to the watered side, and ABA concentrations in the leaves of experimental plants are much higher than in the leaves of control plants.

These results suggest that ABA from roots is transported to leaves and that it serves as an early warning of drought stress. In doing so, ABA overrides the signal from the blue-light photoreceptors introduced earlier in this chapter and causes stomata to close to conserve water.

The Molecular Mechanism of Guard-Cell Closure To understand how stomata open and close, recall that a stoma consists of two **guard cells** (Ch. 34, Section 34.2). When the vacuoles of guard cells are filled with water, the cells are turgid. The shape of turgid guard cells forms an open pore between the two cells, which allows gas exchange between the atmosphere and the leaf interior. But when vacuoles lose water and guard cells become

RESEARCH

QUESTION: Can roots communicate with shoots?

HYPOTHESIS: Roots that are dry can signal shoots to close stomata.

NULL HYPOTHESIS: Roots cannot communicate with the shoot.

EXPERIMENTAL SETUP:

1. Divide roots of many plants into two sides.

2. In experimental group, water one side.

3. In control group, water both sides.

4. In both groups, measure water potential of leaves and observe stomata.

PREDICTION OF HYPOTHESIS: Stomata in experimental plants will close; stomata in control plants will stay open.

PREDICTION OF NULL HYPOTHESIS: Stomata in both experimental and control plants will stay open.

RESULTS: No difference between experimental and control plants in water potential of leaves.

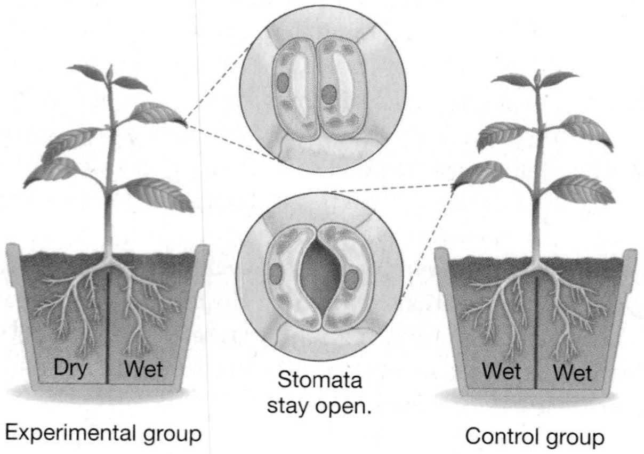

CONCLUSION: Roots can communicate with shoots. Dry roots signal the shoot and cause stomata to close, even though leaves are receiving sufficient water (from roots on the wet side of the plant).

Figure 37.21 Experimental Evidence Suggests that Roots Produce an "It's Too Dry" Signal.

SOURCE: P. G. Blackman and W. J. Davies. 1985. Root to shoot communication in maize plants of the effects of soil drying. *Journal of Experimental Botany* 36: 39–48.

✔ **PROCESS OF SCIENCE** Why was it important to show that the water potential of leaves was the same in the two treatments?

flaccid and lack turgor, cell shape changes in a way that closes the pore and stops both gas exchange and water loss via transpiration.

Based on these observations, pursuing the question of how stomata open and close is the same as asking how guard cells become turgid or flaccid—meaning, how does water flow into or out of vacuoles. Activation of PHOT by blue light leads to water entry and stomatal opening; activation of ABA receptors leads to water exit and stomatal closing.

Remember that cells do not transport water directly. Instead, they change ion concentrations, creating osmotic gradients that result in water movement. It shouldn't be surprising to learn that guard-cell opening and closing is based on changes in the activity of H^+-ATPases and ion transporters in the plasma membrane.

When photoreceptors are stimulated by blue light, large numbers of protons are pumped out of each guard cell. As a result, the interior of guard cells develops a negative charge. As **Figure 37.22a** shows, increased H^+-ATPase activity creates a strong electrochemical gradient that brings potassium and chloride ions into the interior of the guard cells. Water follows the incoming ions via osmosis. The upshot? The cells swell and pores open.

But as **Figure 37.22b** illustrates, guard cells respond to ABA in a very different way. When ABA reaches the guard cells, several things happen: **(1)** H^+-ATPases are inhibited, and **(2)** channels that allow chloride and other anions to leave along their electrochemical gradients are opened. When the anions leave guard cells, the change in membrane potential causes outward-directed potassium channels to open. Large amounts of K^+ leave the cells, and **(3)** water follows by osmosis. The result is a loss of turgor and closing of the pore.

Whether it acts on guard cells or seeds, ABA fulfills a general role in plants as a dormancy, or "no-growth," signal. In many cases, its action depends on input from other hormones and photoreceptors. To survive and reproduce successfully, plants have to integrate information from a variety of sources.

Brassinosteroids and Body Size

When you went through puberty, you experienced a growth spurt triggered by surges in steroid hormones called testosterone and estradiol. In plants, growth spurts are triggered by surges in similar steroid hormones called **brassinosteroids**.

The name "brassinosteroid" was inspired by two observations:

1. The hormones were initially discovered in 1979 in *Brassica napus*—a crop plant that is the source of the canola oil you may use in cooking.
2. They are **steroids**—part of a family of lipid-soluble compounds (Ch. 6, Section 6.1).

Brassinosteroids promote growth and are a key regulator of overall body size in plants. In the model organism *Arabidopsis thaliana* (**BioSkills 11**), for example, mutant individuals that cannot synthesize brassinosteroids are extremely dwarfed.

Recent research has highlighted an important difference between the brassinosteroids and the steroid hormones found in animals. Steroid hormones in animals act by entering cells, binding to receptors in the cytosol, and forming a hormone–receptor complex that enters the nucleus (Ch. 46, Section 46.2). This complex then binds to DNA and directly changes gene expression.

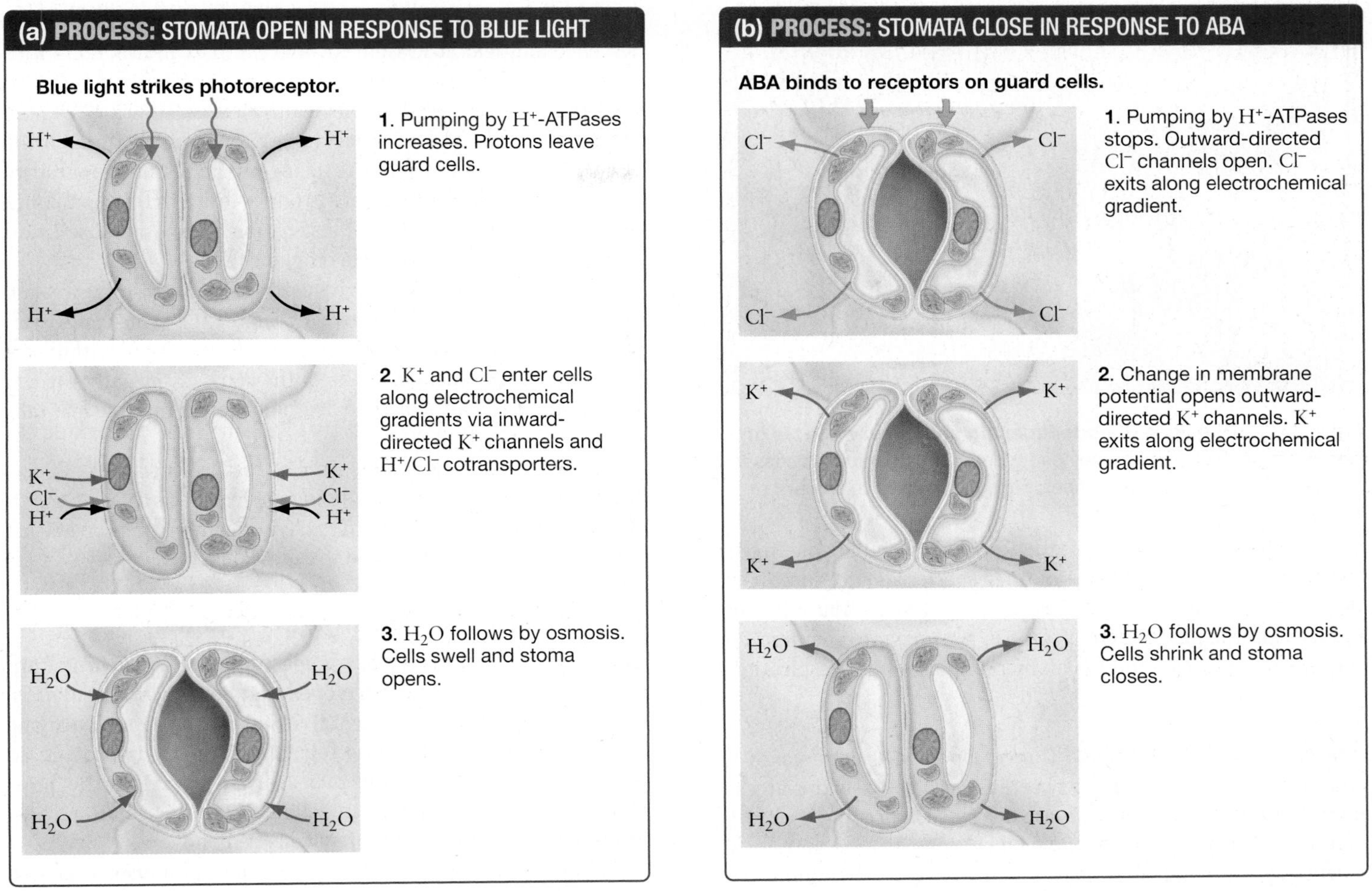

Figure 37.22 **Changes in Ion Flows Are Responsible for Opening and Closing Stomata. (a)** Activated blue-light receptors trigger ion flows into guard cells. The cells swell when water follows by osmosis. **(b)** Activated ABA receptors trigger ion flows out of guard cells. The cells shrink when water follows by osmosis.

Because steroids are lipid soluble, researchers were not surprised to find that testosterone and estradiol—animal hormones—cross the plasma membrane before binding to a receptor.

Brassinosteroids, in contrast, are more hydrophilic than animal hormones and never enter the cell. Instead, they bind to receptors on the plasma membrane and activate signal transduction events—phosphorylation cascades—that lead to changes in gene expression.

The genes for the synthesis of brassinosteroids appear to be homologous with the genes required for steroid hormone synthesis in animals, meaning that they are derived from a similar gene in the common ancestor of plants and animals. So how do brassinosteroids interact with auxin, cytokinins, and gibberellins to regulate growth and body size? These are questions for future research.

Ethylene and Senescence

Senescence is a regulated process of aging and eventual death of organs such as fruits and leaves or an entire organism. Like most aspects of plant growth and development, senescence is regulated by complex interactions between several hormones in response to changes in temperature, light, and other factors.

The hormone most strongly associated with senescence is **ethylene**. Like other plant hormones, ethylene is simple in structure and active at small concentrations. Unlike other plant hormones, however, ethylene is a gas at normal temperatures.

Ethylene is synthesized from the amino acid methionine and is involved in the following senescence events:

- fruit ripening, which hastens the aging and rotting of fruit;
- flowers fading; and
- leaf abscission—their detachment and fall.

In addition, ethylene influences plant growth, and it is a stress hormone induced by drought and other conditions. Ethylene regulates a surprisingly large range of physiological responses.

The Discovery of Ethylene Ethylene was initially discovered in ancient China, when fruit growers noticed that burning incense in closed rooms made pears ripen faster. Westerners made a similar observation in the late 1800s, when gas street lamps came into wide use in cities, and the plants growing near leaky gas lines dropped their leaves prematurely. Researchers showed that ethylene in lamp gas was the molecule responsible for the leaf loss; ethylene is likewise present in incense smoke.

In the 1930s ethylene was also found in the gases that are released by ripe apples. Subsequently, biologists documented

Figure 37.23 **Ethylene Speeds Ripening and Other Aspects of Senescence.** Exposure to ethylene triggers the ripening process in bananas. After ethylene treatment, they are shipped to stores.

sharp spikes in ethylene production during fruit ripening in tomatoes, bananas, and certain other species in addition to apples. Follow-up research on these species showed that ethylene induces **(1)** the production of some of the enzymes required for the ripening process, and **(2)** an increase in cellular respiration, which furnishes ATP.

Ethylene and Fruit Ripening During ripening, stored starch is converted to sugar, enhancing sweetness; protective toxins are removed or destroyed; cell walls are degraded, softening the fruit; chlorophyll is broken down; and pigments and aromas that signal ripeness are produced. Biologists interpret fruit ripening as an adaptation that enhances the fruit's attractiveness to birds, mammals, and other animals that disperse seeds to new locations.

Today, fruit sellers manipulate the level of ethylene to control fruit ripening. For example, they treat green bananas with ethylene after the bananas have been picked, to stimulate ripening (Figure 37.23). Conversely, apples are stored in warehouses with a high concentration of CO_2 and a low concentration of O_2, which inhibits ethylene production in the fruit. Apples stored under these conditions can be sold long after their original harvest date, when untreated fruits have rotted.

Ethylene and Leaf Abscission Ethylene's effects on leaf senescence and leaf abscission involve complex interactions with auxin and cytokinins. In addition to being synthesized in apical meristems, auxin is produced in healthy leaves (Figure 37.24, step 1). It is then transported from the leaf to the stem through the petiole. In response to age or to changes in ambient temperature or day length, leaves sometimes produce less auxin (step 2). As a result, cells in a region of the leaf petiole called the **abscission zone** become more sensitive to ethylene in the tissue.

Increased ethylene sensitivity activates enzymes that weaken the walls of cells in the abscission zone. At the same time, chlorophyll in the leaf is broken down, and nutrients are withdrawn and stored in parenchyma cells in the stem. Eventually the cell walls at the base of the petiole degrade enough that the leaf falls (step 3 in Figure 37.24).

Applications of cytokinins, in contrast, reverse these effects and dramatically extend the life span of leaves. As a result, ethylene and cytokinins are thought to have opposite effects on at least some of the processes involved in senescence.

An Overview of Plant Growth Regulators

Understanding how different signals interact is an exciting frontier in research on plant hormones that are growth regulators. Knowledge of the environmental signals that change hormone levels and the genes that they regulate is rapidly improving. In contrast, a great deal remains to be learned about how the information is carried by signal transduction pathways, how those pathways interact, and how plant cells respond.

Table 37.2 summarizes the functions of hormones discussed in this section. As you study this table, two key observations should emerge:

1. A single hormone often affects many different target tissues. As a result, there can be an array of responses to the same cell–cell signal. To interpret this pattern, biologists point out that hormones may carry a common message to a variety of tissues and organs. Auxin defines the root–shoot axis of the body; gibberellins trigger stem growth; cytokinins promote cell division; ABA slows or prevents growth; brassinosteroids increase overall mass; and ethylene signals senescence.

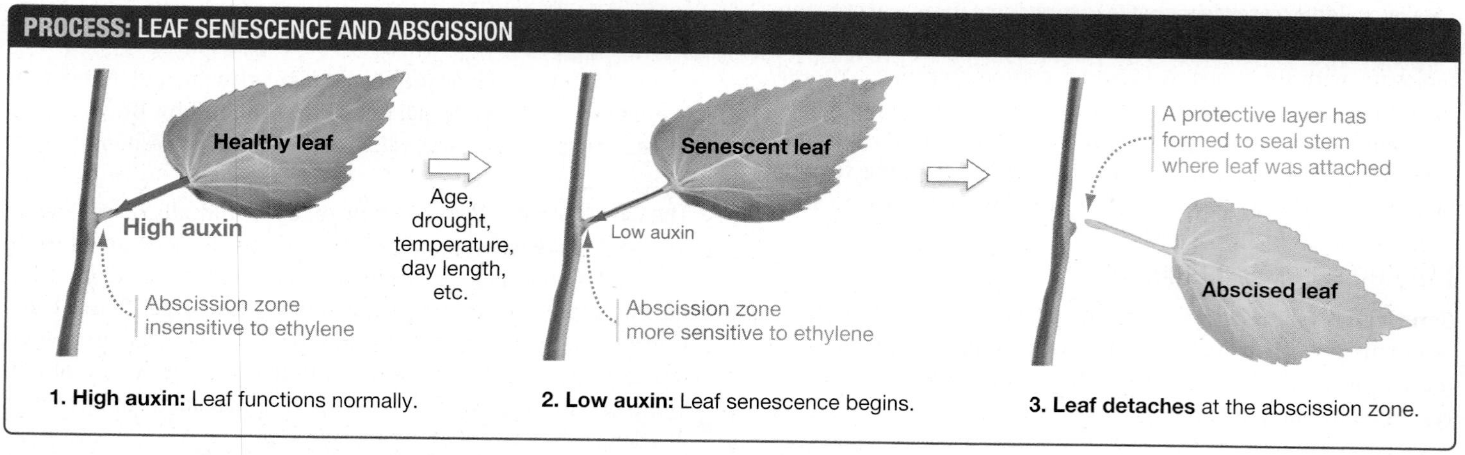

Figure 37.24 **Leaf Abscission Occurs in Response to Signals from Auxin and Ethylene.** Older (senescent) leaves produce much less auxin than young leaves do, leading to an increase in ethylene sensitivity and leaf abscission.

SUMMARY Table 37.2 **Plant Growth Regulators**

Hormone	Characteristics and Functions	Representative Effects
Auxin	• Involved in cell elongation and apical dominance • Promotes differentiation of xylem and phloem • Helps to define long axis of body (phototropism and gravitropism responses) • First plant hormone isolated and characterized • Produced in shoot apical meristems and young leaves	 Phototropism
Cytokinins	• Promote cell division in the presence of auxin • Promote chloroplast development and break lateral bud dormancy • Delay senescence (aging) • Produced in root apical meristems, many other tissues	 Budding
Gibberellins (GAs)	• Promote stem growth via both cell elongation and division • Promote seed germination • Produced in apical meristems, immature seeds, and anthers (pollen-producing organs)	 Germination
Abscisic acid (ABA)	• Inhibits bud growth and seed germination • Induces closure of stomata in response to water stress • Acts as a stress hormone analogous to cortisol in humans • Produced in almost all cells	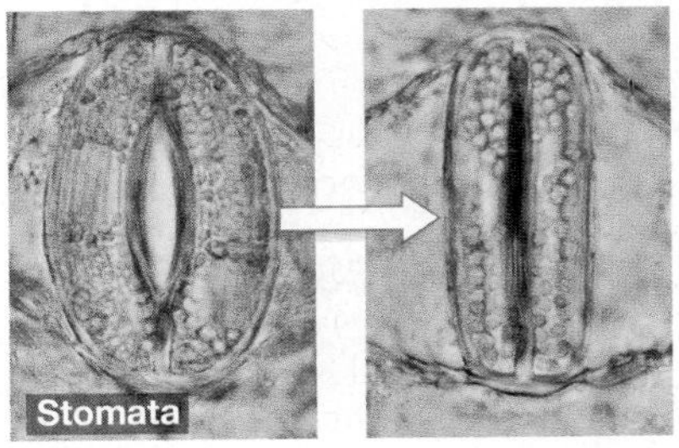 Stomata
Ethylene	• Exists in gas form • Involved in fruit ripening • Induces senescence of fruits, flowers, and leaves • Produced in all organs when plants are under stress	 Senescence
Brassinosteroids (BRs)	• Promote cell elongation in stems and leaves • Structurally related to steroid hormones in animals • Produced in almost all tissues	 Body size

2. In most cases, several different hormones can affect the same response. Stated another way, hormones do not work independently—they interact with each other. To make sense of this pattern, biologists point out that individual hormones tend to be produced in response to an environmental cue at a certain location, such as water availability at root tips. Many environmental cues may be changing at the same time, however. For plants to respond appropriately, they need to integrate information from various environmental cues perceived at various locations in the body.

Crosstalk is the interaction between the signaling pathways triggered by different hormones (Ch. 11, Section 11.3). The key insight to remember is that signaling systems form communication networks. Crosstalk is the molecular mechanism responsible for integrating information from many sensory cells and signals.

The complex interactions among hormones involved in the growth response have a purpose: allowing individuals to survive and thrive long enough to reproduce. The same can be said for the hormones involved in protecting plants from danger.

CHECK YOUR UNDERSTANDING

✔ If you understood this section, you should be able to . . .

1. Explain why pruning a shrub can cause it to become bushier.
2. Predict the effects of watering a large number of daisy plants, each in a pot in a greenhouse, with water that contains either ABA or GAs.
3. Write a hypothesis to explain the adage "one rotten apple spoils the whole bushel."

Answers are available in Appendix A.

37.7 Pathogens and Herbivores: The Defense Responses

After you complete this section, you should be able to . . .

- Explain how plants defend themselves against pathogens and herbivores.

Plants cannot run away from danger. Instead, they have to stand and fight. Like humans and other animals, plants are being constantly threatened by an array of disease-causing viruses, bacteria, and parasitic fungi. Furthermore, plant roots are susceptible to attacks by nematodes—soil-dwelling roundworms (Ch. 31, Section 31.3).

Disease-causing agents are termed **pathogens**. If plants were not able to sense attacks by pathogens and respond to them quickly and effectively, the landscape would be littered with dead and dying vegetation.

The waxy cuticle that covers epidermal cells is an effective barrier to most viruses, bacteria, fungi, and other disease-causing agents, and the structures called thorns, spines, and trichomes help protect leaves and stems from damage by herbivores (Ch. 34, Section 34.2). These represent examples of physical defense.

In addition, many plants lace their tissues with **secondary metabolites**—molecules that are closely related to compounds in key synthetic pathways but are not found in all plants. Some secondary metabolites function in plant defense by poisoning herbivores. These represent examples of chemical defense.

- The flavorful oils in peppermint, lemon, basil, and sage have insect-repellent properties.
- The pitch that oozes from pines and firs contains a molecule called pinene, which is toxic to bark beetles.
- The pyrethroids produced by *Chrysanthemum* plants are a common ingredient in commercial insecticides.
- Molecules called tannins are found in a wide array of plant species; when they are ingested by animals, tannins bind to digestive enzymes and make the herbivores sick.
- Compounds such as opium, caffeine, cocaine, nicotine, and tetrahydrocannabinol (THC) disrupt the nervous systems of plant-eating insects and vertebrates.

Although these defenses are effective, they are also expensive in terms of ATP and materials invested. So it's not surprising that plants often produce defenses or increase their existing defenses only in response to attacks by pathogens or herbivores.

Responses to attacks and infections in both plants and animals are called **inducible defenses** because they are triggered by the presence of a threat. Let's first consider how plants sense and respond to viruses and other pathogens and then explore what they do when attacked by insects and other herbivores.

How Do Plants Sense and Respond to Pathogens?

Like animals, plants have remarkably swift and diverse defense mechanisms to resist attacks by viruses, bacteria, and fungi. Plants use a wide variety of sensory proteins to detect invading pathogens and use signaling pathways to carry out their defense. Successful pathogens, in turn, have evolved elaborate mechanisms to evade these defenses.

An Evolutionary Arms Race Most plant pathogens are very specific, infecting only one or a few host species. This is due in part to the long history of coevolution of the pathogen and its host (Ch. 52, Section 52.1).

If a pathogen evolves a mechanism to exploit its host more effectively, the host may evolve new abilities to detect the pathogen and defend itself from attack. In response, the pathogen may then evolve an alternate strategy to evade detection by the host.

The complex interactions that have resulted from this back-and-forth "arms race" have proved challenging for biologists to understand. The human food supply is estimated to be reduced 50 percent worldwide due to plant disease. In the mid-1800s, Ireland experienced a potato famine in which millions of people faced starvation because of a potato pathogen. Pathogens can also be devastating to native species. The once-dominant American chestnut tree was nearly wiped out by a fungal pathogen in the early 1900s.

To cause disease, plant pathogens must first get past the surface of a host plant and begin growing in the apoplast, outside the plasma membranes of cells. Most pathogens enter through stomata or wounds. Once inside the plant, some pathogens kill host cells and feed on the cell debris. More commonly, a pathogen feeds on nutrients from a host cell while it is still alive. How do the plant cells respond to an attack in progress?

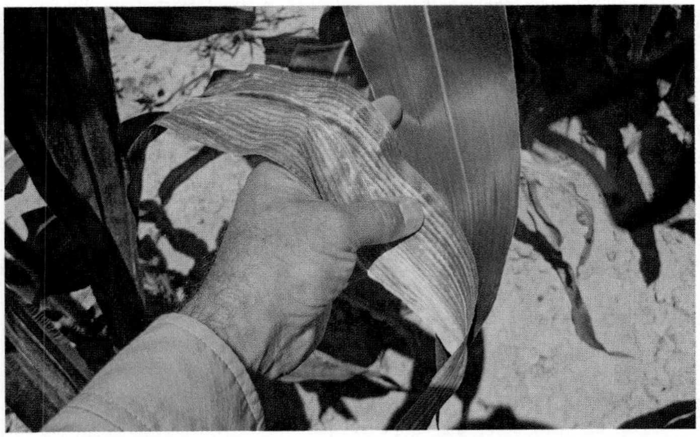

Figure 37.25 The Hypersensitive Response Protects Plants from Pathogens. The yellow, mottled appearance of the leaf on the left is due to the hypersensitive response.

The Hypersensitive Response Plant cells contain a variety of receptor proteins that can bind to molecules derived from pathogen cells. When contact is made, these "surveillance" proteins signal to the plant cell that a pathogen is present, which triggers a cascade of signals that lead to

1. stomatal closure, preventing more pathogen cells from entering the plant;
2. production of toxin molecules targeted to the pathogen;
3. reinforcement of the neighboring plant cell walls to limit movement of the pathogen; and
4. rapid suicide of cells in the infected region.

These events take place within hours of the start of an infection and are termed the **hypersensitive response (HR)**. The HR results in small brown spots in leaves—those cells dead by suicide—that accumulate over the growing season (Figure 37.25). By the end of the growing season, it is rare to find leaves lacking these brown spots. Plants that can detect an infection quickly and induce an HR remain healthy and show no further signs of disease. The HR is similar to the cell-mediated immune response in mammals, which leads to the death of infected cells (Ch. 48, Section 48.4). ✔ **If you understand this concept, you should be able to explain the adaptive value of the hypersensitive response despite the death of plant cells.**

In the 1990s biologists discovered that a mutant of *Pseudomonas*, a bacterial pathogen, could infect plant cells if it was artificially introduced inside leaves, but the pathogen was unable to cause disease if applied to the outside of leaves. This mutant could not produce a small molecule called coronatine, but the role coronatine played in infection was not clear.

Plants normally respond to *Pseudomonas* by closing their stomata within an hour of detecting the pathogen, effectively slamming the door on future invasions. The normal bacteria, however, are able to interfere with this defense by causing the stomata to re-open several hours later, allowing more bacteria to enter the leaf. In 2006, Maeli Melotto and coworkers discovered that coronatine is the molecule that triggers the stomata to re-open. Without the ability to make coronatine, the mutant bacteria couldn't re-open the stomata and were at a clear disadvantage.

Plant defenses are based on good communication not only from cell to cell, as in the HR, but also throughout the plant and even to neighboring plants.

An Alarm Hormone Extends the HR Once the HR is under way in a localized area of infection, a hormone produced at the infection site travels throughout the body and triggers a slower and more widespread set of events called **systemic acquired resistance (SAR)**. Over the course of several days, SAR primes cells throughout the root or shoot system for resistance to assault by a pathogen—even cells that have not been directly exposed to the disease-causing agent.

Figure 37.26 illustrates how the HR and SAR are thought to work together. In addition to triggering the HR, detection of a pathogen leads to the release of a hormone that initiates SAR. This signal acts globally as well as locally—that is, at the point of infection—and results in the expression of a large suite of genes called the pathogenesis-related (*PR*) genes. When expressed, these genes produce proteins that help prevent an infection from spreading.

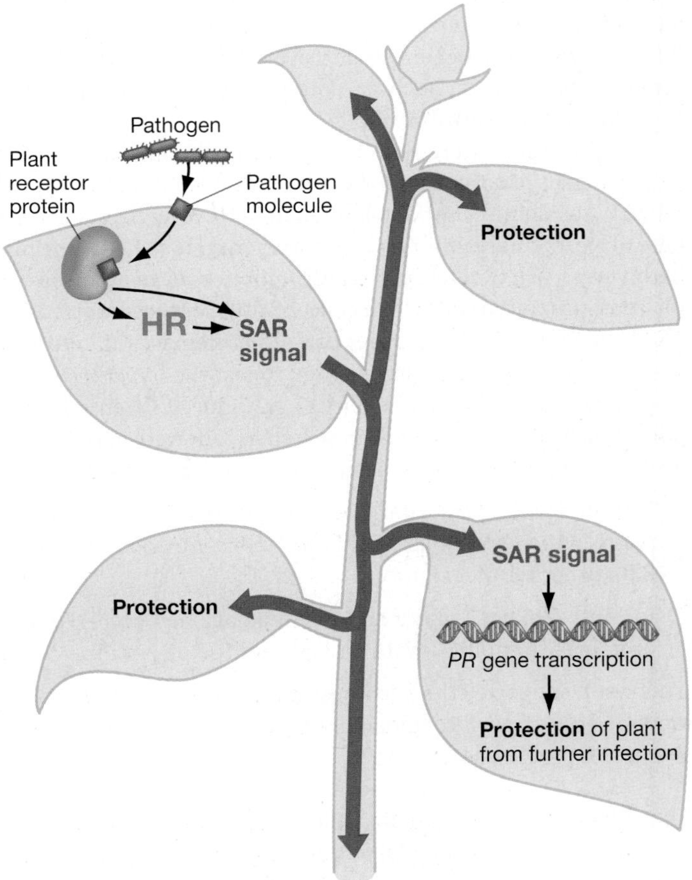

Figure 37.26 The Hypersensitive Response Produces a Signal That Induces Systemic Acquired Resistance (SAR). This diagram summarizes the current consensus on how the HR and SAR interact.

When biologists set out to locate the hormone responsible for SAR, they found that the level of **methyl salicylate (MeSA)**—a molecule derived from salicylic acid—increases dramatically after tissues are infected with a pathogen. Follow-up work revealed these findings:

- Phloem sap leaving infected sites has an elevated level of MeSA.
- Treatments that reduce MeSA reduce or abolish SAR.
- Adding MeSA to the lower leaves of tobacco plants leads to SAR in the upper, untreated leaves.

Biologists are working to understand these complex interactions in order to genetically engineer crop plants that can resist diseases. Despite these interventions, however, pathogens continue to evolve, in a never-ending arms race.

How Do Plants Sense and Respond to Herbivore Attack?

Over a million species of insects have been discovered and named so far. Most of them make their living by eating leaves, stems, phloem sap, seeds, roots, or pollen. Plants have effective induced defenses in response to pathogens such as viruses, bacteria, and fungi. But can they ramp up defenses in response to insect attack?

The Role of Proteinase Inhibitors When researchers started studying why some plant tissues are more palatable and digestible than others, biochemists discovered that many seeds and some storage organs, such as potato tubers, contain proteins called **proteinase inhibitors**.

Proteinase inhibitors block the enzymes—found in the mouths and stomachs of animals—that are responsible for digesting proteins. When an insect or a mammalian herbivore ingests a large dose of a proteinase inhibitor, the herbivore gets sick. As a result, herbivores learn to detect proteinase inhibitors by taste and avoid plant tissues containing a high concentration of these molecules.

Although many plant tissues contain proteinase inhibitors in low concentration, biologists wanted to test the hypothesis that these proteins might also be part of an induced defense by the plant. To evaluate this idea, researchers allowed herbivorous beetles to attack one leaf on each of several potato plants.

- In liquid extracted from the other leaves on the attacked plants, the proteinase inhibitor concentration averaged 336 μg per mL of leaf juice.
- In control plants, where no insect damage had occurred, the proteinase inhibitor level averaged just 103 μg per mL.

This result supported the hypothesis that a hormone produced by wounded cells travels to undamaged tissues and induces the production of proteinase inhibitors.

The Discovery of Systemin Biologists isolated the wound-response hormone by purifying the compounds found in tomato leaves and testing them for the ability to induce the synthesis of proteinase inhibitors. The hormone that is active in tomato plants and closely related species turned out to be **systemin**, a polypeptide just 18 amino acids long. Systemin was the first peptide hormone ever described in plants.

Researchers who labeled copies of systemin with a radioactive carbon atom, injected the hormone into plants, and then monitored its location confirmed that systemin moves from damaged tissues to undamaged tissues.

Currently, work on the production of systemin and proteinase inhibitors focuses on determining each step in the signal transduction pathway that alerts undamaged cells to danger (Figure 37.27). The data have indicated the following process:

Step 1 Systemin is released from damaged cells.

Step 2 Systemin travels through the plant body via phloem and binds to membrane receptors on target cells.

Step 3 The activated receptor triggers a series of chemical reactions that eventually synthesize another hormone called jasmonic acid.

Step 4 Jasmonic acid activates a cytosolic signaling pathway that induces the production of at least 15 new gene products, including proteinase inhibitors.

In this way, plants build potent concentrations of insecticides in tissues that are in imminent danger of attack.

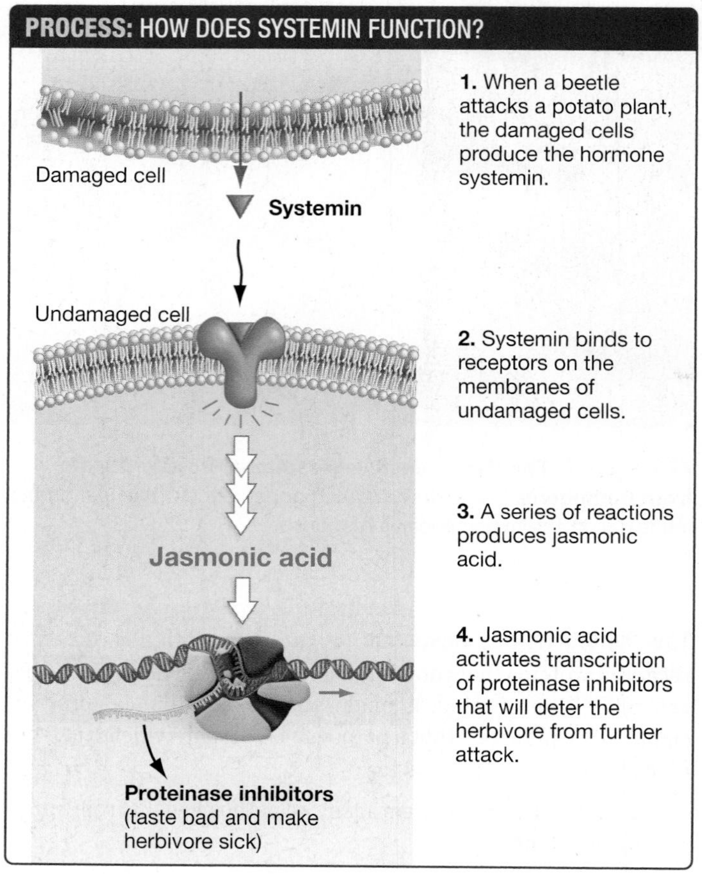

Figure 37.27 Signals from Insect-Damaged Cells Prepare Other Cells for Attack. Systemin, a hormone produced by herbivore-damaged cells, initiates a protective response in undamaged cells.

Figure 37.28 Parasitoids Kill Herbivores. When this plant was first attacked by this caterpillar, it produced pheromones that attracted a female wasp. The wasp laid her eggs in the caterpillar. As the wasp larvae grew and developed into pupae, they devoured the caterpillar.

"Talking Trees": Responses from Nearby Plants When an insect starts munching on a leaf, volatile compounds evaporate from the leaf's surface and travel through the air. In the 1980s, researchers began to suspect that plants growing near other plants under herbivore attack "eavesdrop" on these volatiles. In response, they increase their own defenses—even though they've yet to be attacked.

The "talking trees" hypothesis has received extensive support, primarily through experiments exposing plants to volatiles in the absence of any insects. Even if the volatiles come from an entirely different species, research has shown that plants can sense the presence of these chemicals and respond by increasing the production of proteinase inhibitors and other defense compounds.

In some cases, plants that are under attack even call for help from other organisms.

Pheromones Released from Plant Wounds Recruit Help from Wasps Caterpillars and other herbivorous insects have enemies of their own—often wasps that lay their eggs in the insects' bodies. When wasp eggs hatch inside a caterpillar, the wasp larvae begin eating the host from the inside out. An organism that is free living as an adult but parasitic as a larva, such as these wasps, is called a **parasitoid** (Figure 37.28). Unlike most parasites, parasitoids are similar in size to their hosts and must consume (and kill) their hosts to complete their development.

Biologists observed that parasitoids are particularly common when insect outbreaks occur in croplands. They wondered if wounded plants release compounds that actively recruit parasitoids. More specifically, they hypothesized that plants produce **pheromones**—chemical messengers that are synthesized by an individual and released into the environment, and that elicit a response from a different individual. Hormones act on cells inside the individual that produced them; pheromones act on another individual.

To explore this idea, researchers analyzed compounds that were released from corn seedlings during attacks by caterpillars. The insect-damaged leaves produced 11 molecules that were not produced by undamaged leaves. These compounds were not released by leaves that had been cut with scissors or crushed with a tool; only insect damage triggered their production.

To follow up on this result, the investigators put female wasps in an arena that contained leaves damaged by insects and leaves that had suffered only mechanical damage. In more than two-thirds of the tests performed, the wasps preferred to fly toward the insect-damaged leaves.

These results support the hypothesis that plants produce wasp attractants in response to attack by caterpillars. Biologists are increasingly convinced that plants can produce pheromones that recruit help in the form of egg-laden wasps.

From the research done on plant sensory systems, it is abundantly clear that plants don't just sit there. These organisms may be stationary, but they constantly monitor and respond to a wide array of information about their environment. Gaining a better understanding of phototropism, gravitropism, response to disease and herbivore attack, and other aspects of plant behavior is an exciting frontier in biology.

CHECK YOUR UNDERSTANDING

✔ **If you understood this section, you should be able to ...**

1. Explain the fitness advantage of having a wide array of receptor proteins in a single individual.
2. Explain why plants do not maintain a high proteinase inhibitor concentration in their tissues at all times.

Answers are available in Appendix A.

37.1 Information Processing in Plants

- In most cases, information processing in plants starts when a receptor protein changes shape in response to a stimulus.
- Receptor proteins are either embedded in the plasma membrane or intracellular (located inside the cytoplasm).
- When signal transduction occurs, an external signal is changed into an intracellular signal by a receptor embedded in the cell membrane.
- In sensory cells, signal transduction culminates in the production of hormones that are transported throughout the plant body.
- If a hormone binds to a receptor on a target cell, signal transduction occurs and culminates in changes in gene expression, altered translation rates, or changes in the activity of specific membrane pumps, channels, or ion carriers.

37.2 Blue Light: The Phototropic Response

- Phototropins are blue-light receptors that become phosphorylated and then initiate signal transduction pathways.
- In the phototropic response, cells near the tips of coleoptiles sense changes in blue light and respond by altering the distribution of the hormone auxin. Cells on one side of the coleoptile elongate in response to auxin much more than do cells on the other side of the coleoptile. In this way, plants bend toward sunlight.

37.3 Red and Far-Red Light: Germination, Stem Elongation, and Flowering

- Phytochrome is a red/far-red switch that allows plants to detect shade from other plants and regulates a wide variety of plant growth responses including germination, stem elongation, and flowering.
- The phytochrome protein changes shape when it absorbs red light. The same protein changes to an alternate shape when it absorbs far-red light.

37.4 Gravity: The Gravitropic Response

- Cells in the center of root caps contain amyloplasts with starch granules that function as statoliths, causing roots to grow down in response to gravity.
- Redistribution of auxin in response to gravity results in downward growth of roots and upward growth of shoots.

37.5 How Do Plants Respond to Wind and Touch?

- Thigmomorphogenesis is a change in growth in response to wind or touch whereby plants grow more slowly and become stocky.
- Thigmotropism is a directional response to touch, such as in tendril coiling.
- Thigmonastic movements are a rapid, nondirectional response to touch, such as in closing of Venus flytraps.

37.6 Youth, Maturity, and Aging: The Growth Responses

- Several key hormones regulate plant growth.
- Auxin establishes and maintains the long axis of the plant body, playing a key role in phototropism and gravitropism and maintaining apical dominance. All of these responses rely on the polar transport of auxin, which establishes a gradient in auxin from the plant's shoot tips to its root tips.
- Cytokinins promote cell division by regulating the cell cycle, and they delay senescence.
- Gibberellins (GAs) signal that conditions for growth are good and promote the initiation or continuation of growth and development.
- Abscisic acid (ABA) signals that environmental conditions are bad by suppressing growth, enforcing dormancy, and closing stomata.
- Regulation of dormancy and growth by ABA and GAs are examples of how hormones interact—allowing plants to integrate information from several different stimuli and respond appropriately.
- Ethylene is a gaseous hormone that triggers senescence and fruit ripening, and it plays a role in leaf abscission.
- Brassinosteroids are steroid hormones that regulate growth.

37.7 Pathogens and Herbivores: The Defense Responses

- Plants have receptor proteins that detect pathogen infection and trigger the hypersensitive response, which limits the spread of disease. The hypersensitive response precedes a slower, whole-plant response known as systemic acquired resistance (SAR).
- Herbivores trigger plant responses that lead to the synthesis of molecules that make plants less palatable.
- Localized infection or herbivory stimulates long-distance signaling and defense responses throughout the plant.
- Volatile signals triggered by herbivory attract parasitoids that attack the herbivores.

Answers are available in Appendix A.

✔ TEST YOUR KNOWLEDGE

1. Evaluate the following assertions about phototropism. Select True or False for each statement.
   T/F Cells on the illuminated side of a stem elongate more than cells on the shaded side.
   T/F Phototropism is triggered by blue light.
   T/F Phototropins play a significant role in phototropism.
   T/F The bending of the plant is due to cell elongation in response to auxin.
2. Which of the following statements about phytochrome is not correct?
   a. It is photoreversible.
   b. Its function was understood long before the protein itself was isolated.
   c. The P_{fr} form activates the responses to red light.
   d. It is involved in guard-cell opening.
3. Which of the following statements about hormones is correct?
   a. They tend to be large molecules.
   b. They exert their effects only on the same cells that produce them.
   c. They can exert strong effects only when they are present in high concentrations.
   d. They trigger a response by binding to target-cell receptors.
4. What evidence suggests that ABA from roots can signal guard cells to close?
   a. If roots are given sufficient water, guard cells close anyway.
   b. If roots are dry, guard cells begin to close—even though leaves may not be experiencing water stress.
   c. Applying ABA on guard cells directly causes them to close.
   d. If roots are dry, the ABA concentration in leaf cells drops dramatically.

✔ TEST YOUR UNDERSTANDING

5. Why was it logical to predict that amyloplasts function as statoliths?
   a. They are dense and settle to the bottom of gravity-sensing cells.
   b. They are present only in gravity-sensing cells.
   c. They make a direct physical connection with membrane proteins that have been shown to be gravity-receptor molecules.
   d. Their density changes in response to gravity.
6. Phytochromes can be considered "shade detectors," while phototropins such as PHOT1 can be considered "sunlight detectors." Explain why these characterizations are valid.
7. To explore how hormones function, researchers have begun to transform plants with particular genes. In one experiment, a gene involved in cytokinin synthesis was introduced into tobacco plants. Which one of the following results would be expected?
   a. Individuals produced more lateral branches.
   b. Stems grew extremely tall and slender.
   c. Roots were incapable of responding to gravity.
   d. Stomata were closed most of the time.
8. Suppose that a mutant plant is unable to make methyl salicylate. Explain why it is not likely to survive in the wild.

✔ TEST YOUR PROBLEM-SOLVING SKILLS

9. **PROCESS OF SCIENCE** In general, small seeds that have few food reserves must be exposed to red light before they will germinate. (Lettuce is an example.) In contrast, large seeds that have substantial food reserves typically do not depend on red light as a stimulus to trigger germination. State a hypothesis to explain these observations.
10. **PROCESS OF SCIENCE** In many species native to tropical wet forests, seeds do not undergo a period of dormancy. Instead, they germinate immediately. Predict the role of ABA in these seeds. How would you test your prediction?

✔ PUT IT ALL TOGETHER: Case Study

How fast can a plant move when tickled?

Leaflets of *Mimosa pudica* (common names: sensitive plant, touch-me-not) have a remarkable ability to close up in response to being touched or physically moved. How fast can the leaflets close? How does this occur? And more importantly, what benefit could this unusual response provide to the plant?

11. QUANTITATIVE Researchers studied the rate of closing and reopening of leaflets by measuring changes in the distance between opposing leaflets after being touched. Their results are shown in the accompanying graph (d/d_{max} is actual leaflet distance relative to maximal leaflet distance). How long does it take for leaflets to close? How long to reopen?

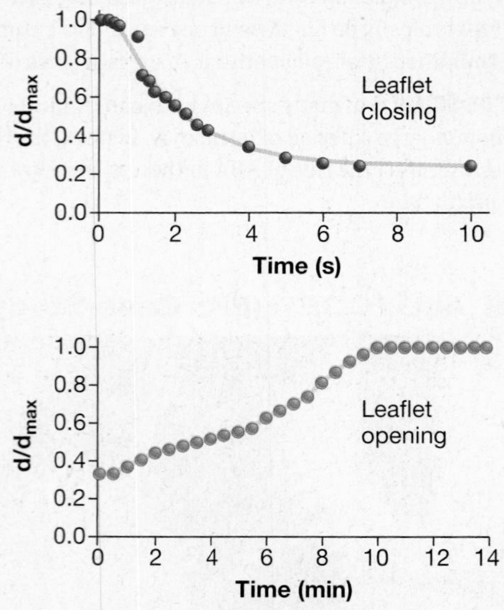

Source: A. G. Volkov et al. 2010. *Plant, Cell & Environment* 33: 163–173.

12. Which of the following terms best describes the leaflet movement?

a. thigmonastic movements
b. thigmotropism
c. thigmomorphogenesis
d. apical dominance

13. The mechanism of leaflet closure is similar to what happens during the shrinking of guard cells and closing of stomata (see Figure 37.22). When in the open position, special cells on the upper surface of *Mimosa* leaflets are filled with water and are under pressure. Explain how osmosis and flow of ions into and out of these special cells may be involved in leaflet closure and reopening.

14. If just the leaflets located toward the end of a leaf are touched, the adjacent leaflets close in fairly rapid succession until all leaflets on a leaf close up. Explain how electrical signaling may be involved in this response.

15. PROCESS OF SCIENCE Researchers have hypothesized that rapid leaf movements in *Mimosa* serve as a defense mechanism (e.g., closing leaflets may deter plant-eating insects). Propose an experiment to test this hypothesis.

16. *Mimosa* leaflets also close on their own at dusk and reopen at dawn (this cycle develops as a circadian rhythm). What receptor molecule mentioned in this chapter is likely involved in regulating this process?

38 Flowering Plant Reproduction and Development

In this chapter the focus is on the function of plant reproductive structures like this flower, and the developmental processes involved in transforming a seed into a mature flowering plant.

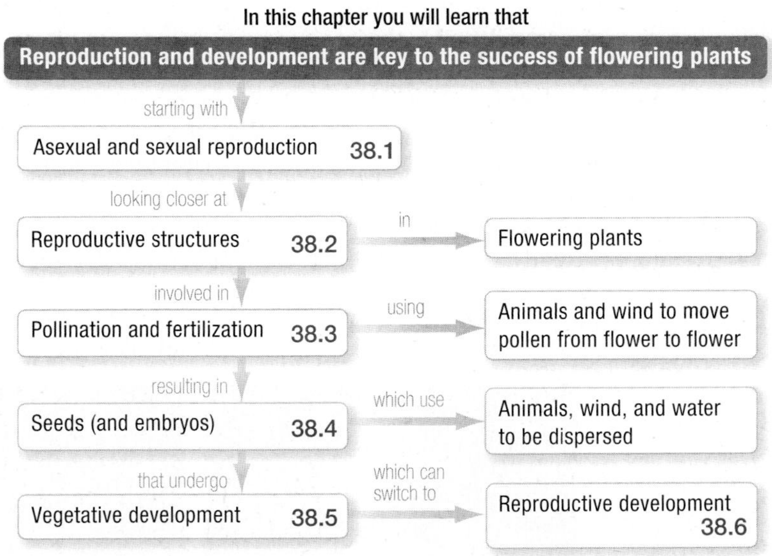

BIG PICTURE

This chapter is part of the Big Picture. See how on pages 838–839.

It is difficult to overemphasize the importance of the plant reproductive organs and developmental processes that are analyzed in this chapter—for plants, for biologists, and for you.

- For plants, every structure in the body and every physiological process—from water transport to photosynthesis—exist for one reason: to maximize the chances that the individual will produce offspring. Reproduction is the unconscious goal of everything that an organism does (Chapter 1).
- For biologists, plant reproduction and development are not only fundamental to understanding how plants work but also the basis for major industries. Agriculture, horticulture, forestry, biotechnology,

and ecological restoration draw extensively on what biologists know about plant reproduction and development.

- For you, plant reproduction and development for the most part means food. Human diets are based on consuming plant reproductive structures—primarily the seeds and fruits derived from flowers. A **flower** is a reproductive structure in angiosperms that produces gametes, attracts pollinators, receives gametes from other individuals, nourishes embryos, and develops seeds and fruits. **Seeds** consist of an embryo and nutrient stores surrounded by a protective coat. **Fruits** develop from the flower's seed-producing organ and contain seeds.

This chapter focuses on angiosperms, for three reasons: **(1)** Angiosperms represent over 85 percent of the land plants described to date; **(2)** virtually every important domesticated plant is an angiosperm; and **(3)** aspects of reproduction in other land plant lineages were described previously (Chapter 28). Before delving into how a seed can ultimately develop into a mature flowering plant, let's examine some of the general principles that underlie plant reproduction.

38.1 An Introduction to Flowering Plant Reproduction

Plant reproductive structures and processes vary among species. Consider just one aspect of reproductive organs—size. Flowers vary from microscopic to the size of a small child; seeds and fruits range from dust-like particles to coconuts.

Fortunately for students of plant biology, several basic principles unify this diversity of reproductive systems. Let's begin by exploring how flowering plants can reproduce either asexually or sexually.

After you complete this section, you should be able to ...

- Explain the role of sexual and asexual reproduction in flowering plants.

Asexual Reproduction

Asexual reproduction does not involve fertilization. It results in the production of **clones**—genetically identical copies of the parent plant.

Some plants extend their life indefinitely by asexual reproduction. The oldest of all known plants is a ring of creosote bushes in the Mojave Desert of California. The bushes comprise a clone that originated from a parent plant that started growing some 12,000 years ago.

Although all asexual reproduction is based on mitotic cell division, it includes a wide array of structures and mechanisms.

- Figure 38.1a shows shoots and roots emerging from a horizontal stem called a **rhizome**, which grows underground. If the emerging individuals become separated from the parent plant, they represent asexually produced offspring.
- The gladiolus plant in Figure 38.1b has propagated itself via modified stems called **corms**, which grow under the surface of the soil.
- The kalanchoe in Figure 38.1c produces "plantlets" (small plants) from meristematic tissue located along the margins of its leaves. When the plantlets mature, they drop off the parent plant and grow into independent individuals.
- In some species, including the dandelion in Figure 38.1d, mature seeds can form without fertilization. This phenomenon, known as **apomixis**, results in seeds that are genetically identical to the parent.

The key characteristic of asexual reproduction is efficiency. If an herbivore or a disease wipes out other plants that surround a grass plant, the grass can quickly send out rhizomes. Its asexually produced offspring are likely to fill the unoccupied space before seeds from competitors can establish themselves and grow. The parent plant can also nourish these progeny as they become established.

Although asexual reproduction is extremely common in plants, it does have an important downside: A fungus or other

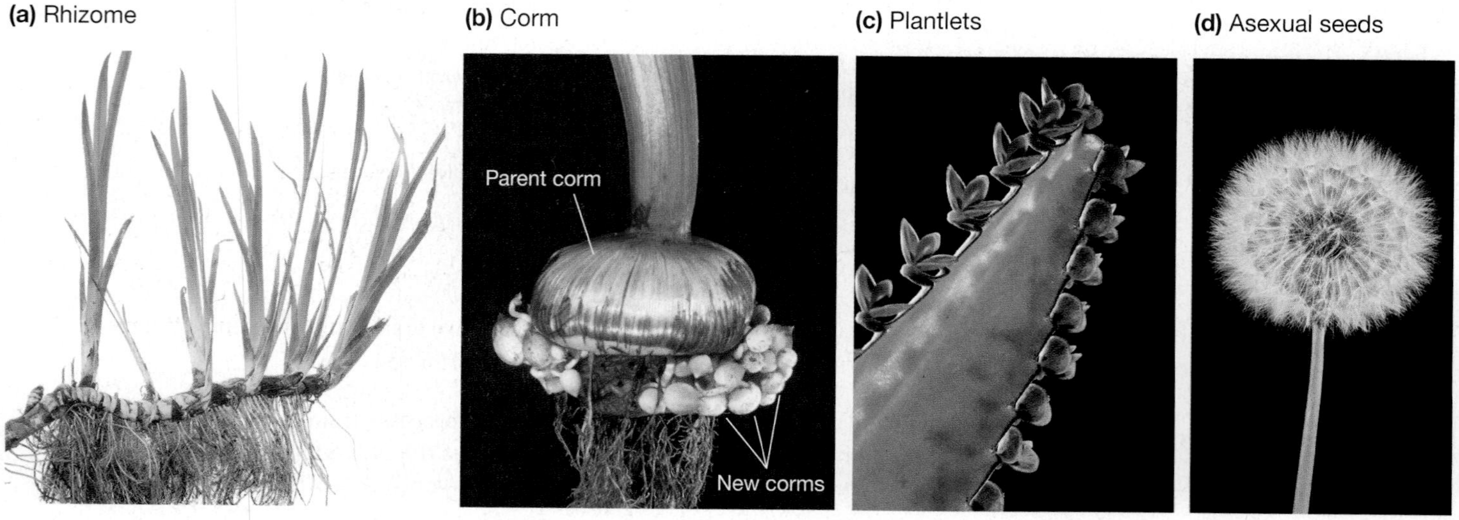

Figure 38.1 **The Mechanisms of Asexual Reproduction Are Diverse.** These are just a few mechanisms of asexual reproduction in flowering plants.

disease-causing agent that infects an individual plant will probably succeed in infecting the plant's cloned offspring as well, even if they are no longer physically connected.

This predicted outcome is based on the observation that plants fight disease with a wide variety of molecules (Ch. 37, Section 37.7). Populations of asexually produced individuals can lack the genetic diversity needed to generate an arsenal of effective disease-fighting compounds.

Evidence supporting this disadvantage of asexual reproduction comes from agriculture and horticulture. Asexually propagated apples, bananas, and other crops are known to be more susceptible to epidemics than are sexually propagated species. What are the fundamental events involved in sexual reproduction?

Sexual Reproduction and the Flowering Plant Life Cycle

Most flowering plants reproduce sexually. **Sexual reproduction** is based on meiosis and fertilization. Recall that **meiosis** is a type of nuclear division that results in four daughter cells, each made up of half the number of chromosomes present in the original parent cell, and **fertilization** is the fusion of haploid cells termed **gametes** (Ch.13, Section 13.1). Meiosis results in offspring that are genetically different from each other and from their parents.

The advantages of sexual reproduction are common to all eukaryotes that undergo meiosis. When and where meiosis occurs, however, is highly variable. In most animals, meiosis leads directly to the formation of gametes. In land plants, the situation is much different—meiosis and fertilization occur in alternate phases of a life cycle.

Land plants are characterized by a life cycle with two distinct multicellular forms—one diploid and one haploid. An individual in the diploid phase of the life cycle is called a **sporophyte**, while an individual in the haploid phase of the life cycle is called a **gametophyte**. A good way to keep these terms straight is to remember that sporophyte means "spore-plant," while gametophyte means "gamete-plant." Sporophytes produce spores by meiosis. Gametophytes produce gametes by mitosis.

You may recall that this type of life cycle, called **alternation of generations**, has evolved independently in various protists and land plants (Ch. 27, Section 27.3, and Ch. 28, Figure 28.12).

Figure 38.2 diagrams the life cycle of an angiosperm; five key stages are highlighted. Most of the flower structures shown in the figure, as well as the roots, stems, and leaves, are part of the sporophyte. Male and female gametophytes are produced in anthers and ovules respectively.

1. Meiosis occurs in sporophytes and results in the production of haploid **spores**. Meiosis and spore production occur inside structures called **sporangia** (singular, *sporangium*).
2. Spores undergo mitosis and develop into multicellular, haploid gametophytes. Male gametophytes are pollen grains, and female gametophytes are embryo sacs.
3. **Pollination** (transfer of pollen grains from an anther to a stigma) occurs, and gametophytes produce **sperm** (male gametes) and **eggs** (female gametes) by mitosis.
4. Fertilization occurs when two gametes fuse to form a diploid **zygote**.
5. The zygote undergoes mitosis and grows into a multicellular, diploid embryo (young sporophyte).

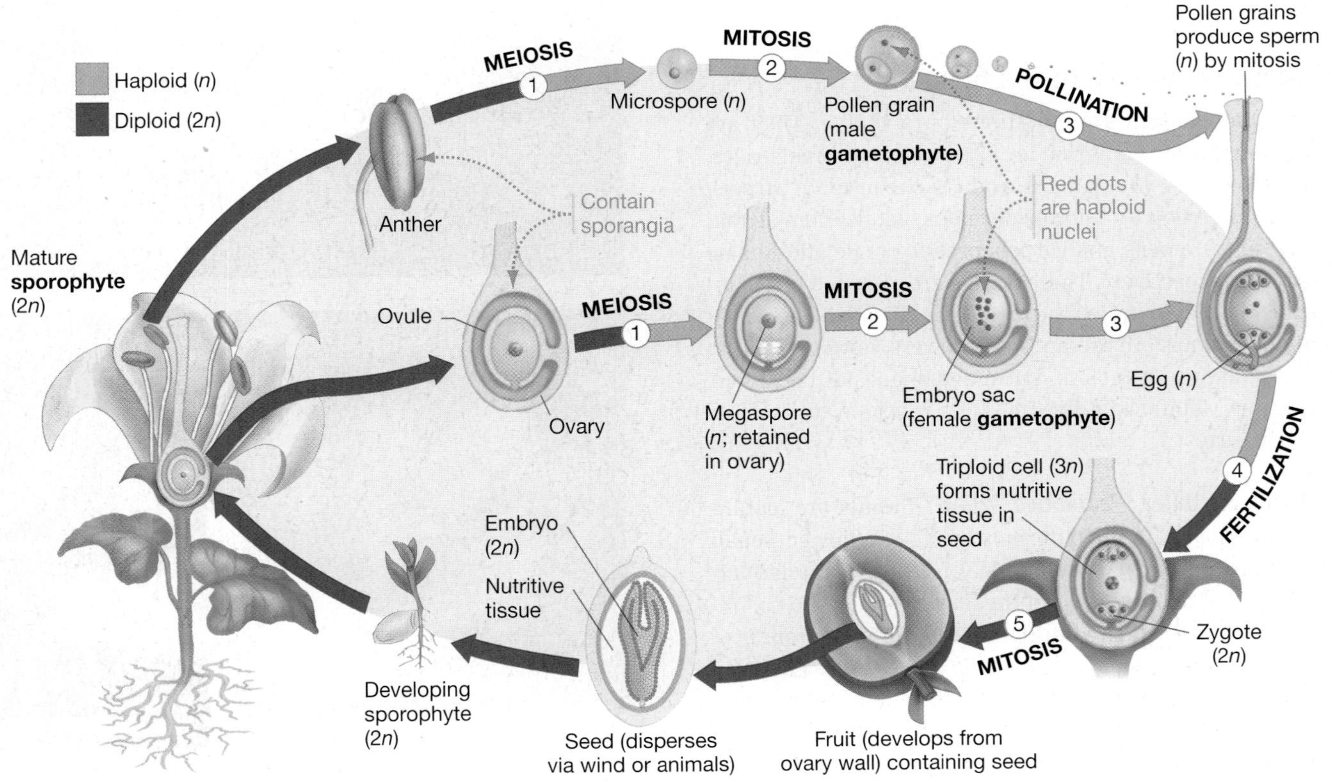

Figure 38.2 **All Angiosperms Exhibit Alternation of Generations.**

Details on the development of male and female gametophytes and the unique process of fertilization in angiosperms will be explored more fully in Sections 38.2 and 38.3.

✔ **If you understand the angiosperm life cycle, you should be able to (1) identify the male spore and the female spore in Figure 38.2, and (2) provide evidence to support the statement that in angiosperms, female gametophytes don't leave their parent plant.**

CHECK YOUR UNDERSTANDING

✔ **If you understood this section, you should be able to ...**

1. Explain an advantage and a disadvantage of asexual reproduction.
2. Evaluate whether the following statement is correct: "Sperm and eggs are formed by meiosis in flowering plants."

Answers are available in Appendix A.

38.2 Reproductive Structures

Each major group of plants, from mosses to angiosperms, has a unique variation on the theme of alternation of generations, as well as characteristic male and female reproductive structures (Ch. 28, Section 28.3). In this section the focus is on flowers, found only in angiosperms, and their role in the formation of male and female gametophytes.

After you complete this section, you should be able to ...

- Explain the functional role of various flower parts.

The General Structure of the Flower

Structurally, all flowers are variations on a theme. They are made up of four basic organs that are essentially modified leaves: **(1)** sepals, **(2)** petals, **(3)** stamens, and **(4)** one or more carpels. Sepals offer protection as a flower develops, petals often attract pollinators, and stamens and carpels produce male and female gametophytes, respectively. These organs are attached to a compressed portion of stem called the receptacle (**Figure 38.3a**).

Not all four organs are present in all flowers, however, and as **Figure 38.3b** shows, the colors, sizes, and shapes of these four components are fabulously diverse. Let's consider each of the four parts, in turn.

Sepals Form an Outer, Protective Whorl **Sepals** are leaflike structures that make up the outermost parts of a flower. Sepals are usually green and photosynthetic, and they are relatively thick compared with other parts of the flower.

Because they are attached to the receptacle in a circular or whorled arrangement, sepals enclose the flower bud as it develops and grows—protecting young buds from damage by desiccation, insects, or disease-causing agents. The entire group of sepals in the flower is collectively called the **calyx** (pronounced *KAY-liks*).

Figure 38.3 The Basic Structures in Flowers Are Highly Variable. (a) Flowers are made up of sepals, petals, stamens, and carpels. **(b)** The four parts vary greatly among species.

Figure 38.4 Insects See in the Ultraviolet Range. **(a)** The "flower" of a black-eyed Susan, seen by the unaided human eye. Note that this structure is a collection of dozens of individual flowers. **(b)** The same structure, photographed with a camera that records ultraviolet wavelengths that are visible to bees but invisible to humans.

Petals Furnish a Visual Advertisement Like sepals, **petals** are arranged around the receptacle in a whorl. Often brightly colored and scented, petals function to advertise the flower to bees, flies, hummingbirds, and other pollinators.

In some cases, the color of the petals correlates with the visual abilities of particular animals. Bees, for example, respond strongly to wavelengths in the blue and purple regions of the light spectrum, as well as yellow (they don't see red well). Flowers that attract bees, in turn, often have yellow, blue, or purple petals with ultraviolet patches.

The ultraviolet regions of petals in "bee flowers" frequently highlight the center of the flower (Figure 38.4). Why? In these flowers, the base of the petals contains a gland called a **nectary**. The nectary produces the sugar-rich fluid **nectar**, which many of the animals that visit flowers harvest along with pollen. In the process of collecting pollen or nectar, the visiting animal usually deposits on the female parts pollen from a different plant—accomplishing pollination.

The entire group of petals in a flower is collectively called the **corolla**. In some species, the petals within the corolla vary in size, shape, and function:

- Flattened petals may provide a landing pad for flying insects.
- Elongated, tubelike petals frequently have a nectary at their base that can be reached only by animals with a long beak or tongue-like proboscis.
- Some petals protect the reproductive organs located inside the corolla.
- Specialized cells in some petals synthesize and release molecules that provide a scent attractive to certain species of pollinating animals.

In contrast, wind-pollinated angiosperms such as oaks, birches, pecans, and grasses have flowers that have small petals or no petals at all, and they lack nectaries. These species do not invest in structures that aren't required for pollination.

Stamens Produce Pollen **Stamens** are reproductive structures that produce male gametophytes—also known as pollen grains. The male gametophytes, in turn, produce sperm that are eventually carried to eggs (see Figure 38.2).

Each stamen consists of two components:

1. a slender stalk termed the **filament**, and
2. the pollen-producing organs called **anthers** (see Figure 38.3a).

The anther is the business end of the stamen—where meiosis and pollen formation take place. The filament holds the stamen in a place where wind, insects, hummingbirds, bats, or other agents can make contact with the pollen grains produced in the anther.

Carpels Produce Ovules The fourth reproductive structure in flowers is the **carpel**, which produces female gametophytes. A carpel consists of three regions:

1. The **stigma** is a sticky tip that receives pollen.
2. The **style** is a slender stalk.
3. The **ovary** is an enlarged structure at the base of the carpel (see Figure 38.3a).

Inside the ovary, female gametophytes are produced in structures called **ovules**. An ovary may contain more than one ovule. When the female gametophytes that are produced inside ovules mature, they produce eggs (see Figure 38.2).

The "Sex" of Flowers Varies In most angiosperm species, stamens and carpels are produced on the same individual. Flowers that contain both stamens and carpels are referred to as **perfect flowers**.

There are also **imperfect flowers**, however, meaning they contain either stamens *or* carpels, but not both. Imperfect flowers that contain only stamens can be considered "male" flowers. Similarly, imperfect flowers that contain only carpels can be considered "female" flowers.[1]

In some cases, separate stamen- or carpel-producing flowers occur on the same individual. Species like these, including the

[1]Technically, flowers are not referred to as male and female. Instead, they are staminate or carpellate. Staminate flowers produce stamens, which produce pollen grains, which produce male gametes (sperm). Carpellate flowers produce carpels, which contain ovaries. Female gametophytes develop inside ovaries and produce female gametes (eggs). For convenience, though, the text sometimes refers to male and female flowers and reproductive structures.

Figure 38.5 **"Male" and "Female" Flowers Can Occur on the Same Individual or on Different Individuals. (a)** The tassels of corn contain male flowers; ears contain female flowers. **(b)** In *Cannabis sativa*, male and female flowers are found on different individuals.

corn plants illustrated in Figure 38.5a, are **monoecious** (pronounced *mon-oh-EE-shuss;* literally, "one-house"). In corn, the tassel is a collection of stamen-producing "male" flowers, and the ear contains a group of carpel-producing "female" flowers.

In contrast, some species with imperfect flowers are **dioecious** ("two-houses")—meaning that each individual plant produces either stamen-bearing flowers only or carpel-bearing flowers only. *Cannabis sativa* is a dioecious species (Figure 38.5b).

How Are Female Gametophytes Produced?

What purposes do the three parts of the carpel serve? The function of the stigma and style will become clear in Section 38.3; for now, let's concentrate on what happens inside the ovary.

Figure 38.6 illustrates a longitudinal section showing the inside of a typical angiosperm ovary. Notice that it contains one or more ovules. Each ovule contains a structure called the megasporangium, inside which there is a diploid cell called the megasporocyte. (The use of "mega" is appropriate, because these structures are much larger than their counterparts in the stamen.) The megasporangium is comparable to spore-producing structures found in other plants, such as those on the back of fern leaves (Chapter 28).

Figure 38.6 shows the steps that occur in the production of female gametophytes:

Step 1 The diploid megasporocyte divides by meiosis, producing four haploid **megaspores**.

Step 2 Three of the megaspores degenerate.

Step 3 The surviving megaspore divides by mitosis to produce a single large central cell with eight haploid nuclei.

Step 4 The haploid nuclei move to different positions, and cell walls form around most of them. One of the cells becomes the haploid egg. Typically, two **polar nuclei** stay together within the central cell—the largest cell in the ovule. The overall

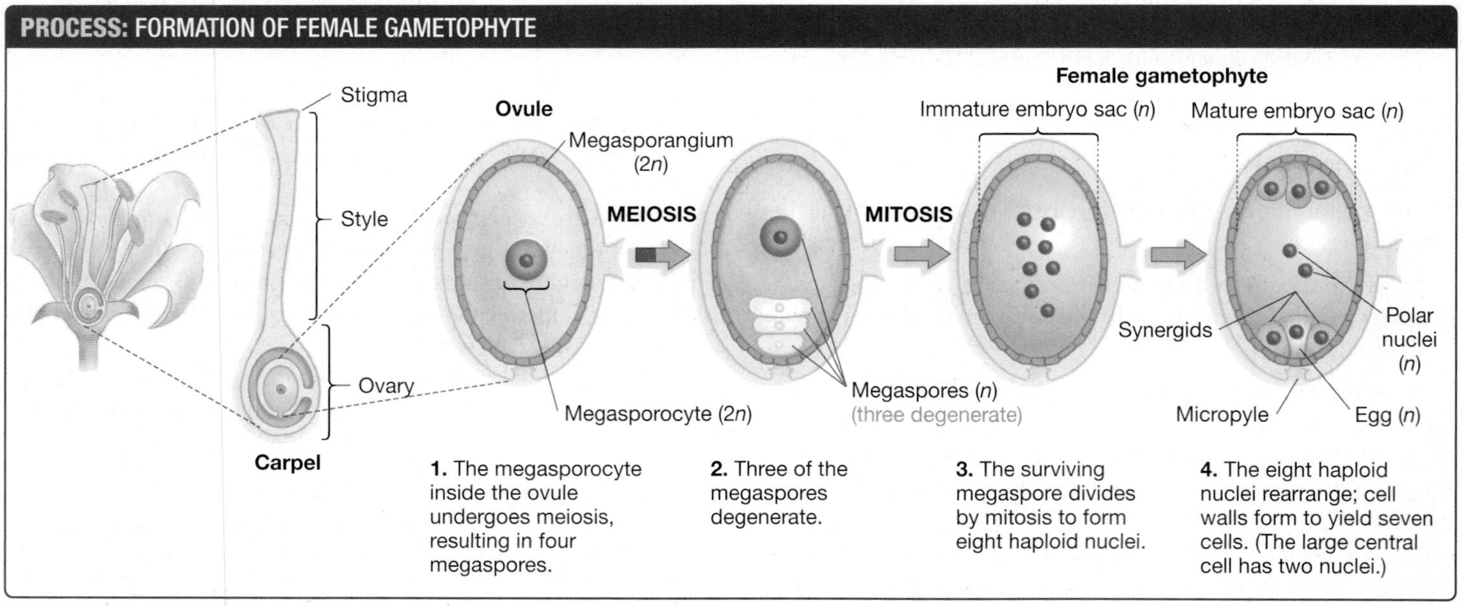

Figure 38.6 **In Angiosperms, Megaspores Produce Female Gametophytes.**

structure, with eight haploid nuclei and seven cells, is the female gametophyte—usually known as the **embryo sac**.

The number of polar nuclei varies among species. As you'll see later, the polar nuclei are involved in fertilization and play an important role in the development of embryo-nourishing tissue after pollination and fertilization take place.

In the carpel, then, a diploid megasporocyte divides by meiosis to form a megaspore, which then divides by mitosis to form the female gametophyte. Female gametophytes are encased in an ovary, retained in the flower, and produce an egg. The egg cell is located at one end of the female gametophyte, near an opening in the ovule called the **micropyle** ("little-gate"). The micropyle is where sperm will enter the ovule before fertilization. Two haploid cells called **synergids** lie close to the egg. You will soon learn more about the importance of synergids during the fertilization process.

How Are Male Gametophytes Produced?

Figure 38.7 provides a detailed look at the stamen and the steps that occur in the production of male gametophytes (that is, pollen). (Recall that a stamen consists of two major parts: an anther and a filament. Inside the anther, structures known as microsporangia contain diploid cells called microsporocytes.)

Step 1 Diploid microsporocytes undergo meiosis. Each haploid cell that results is a **microspore**.

Step 2 Normally, all of the microspores survive. Microspores then divide by mitosis and form a tube cell and generative cell.

Step 3 The tube cell and generative cells form a haploid, immature male gametophyte, also known as a **pollen grain**. Eventually the generative cell will divide by mitosis to form two sperm, and the tube cell will enlarge and carry sperm to a female gametophyte (described more fully in the next section).

In the anther, then, diploid microsporocytes divide by meiosis to form microspores, which then divide by mitosis to form male gametophytes (pollen grains). Each diploid microsporocyte ultimately yields four haploid pollen grains. Male gametophytes are dispersed from the flower and eventually produce sperm.

At the immature stage—before it has produced sperm—the male gametophyte consists of two cells: a small generative cell enclosed within a larger tube cell. The male gametophyte is considered mature when the haploid generative cell undergoes mitosis and produces two sperm cells.

In some species, this maturation step occurs while pollen is still in the anther. In other species, maturation and sperm production don't occur until after the pollen grain lands on a stigma and begins to grow. The reasons for this difference are not yet understood.

The wall of a pollen grain develops a tough outer coat that includes the watertight compound called **sporopollenin** (introduced in Ch. 28, Section 28.2). This coat protects the male gametophyte when the pollen is released from the parent plant into the environment. Depending on the species, pollen grains may be dispersed by animals, the wind, or water currents.

But then how does a pollen grain get to the mature carpel of the same species, where an egg cell is waiting?

CHECK YOUR UNDERSTANDING

✔ **If you understood this section, you should be able to ...**

1. Explain why stamens are considered the male part of flowers, while carpels are considered the female part of flowers.
2. Compare and contrast a female gametophyte and a male gametophyte.

Answers are available in Appendix A.

PROCESS: FORMATION OF MALE GAMETOPHYTE

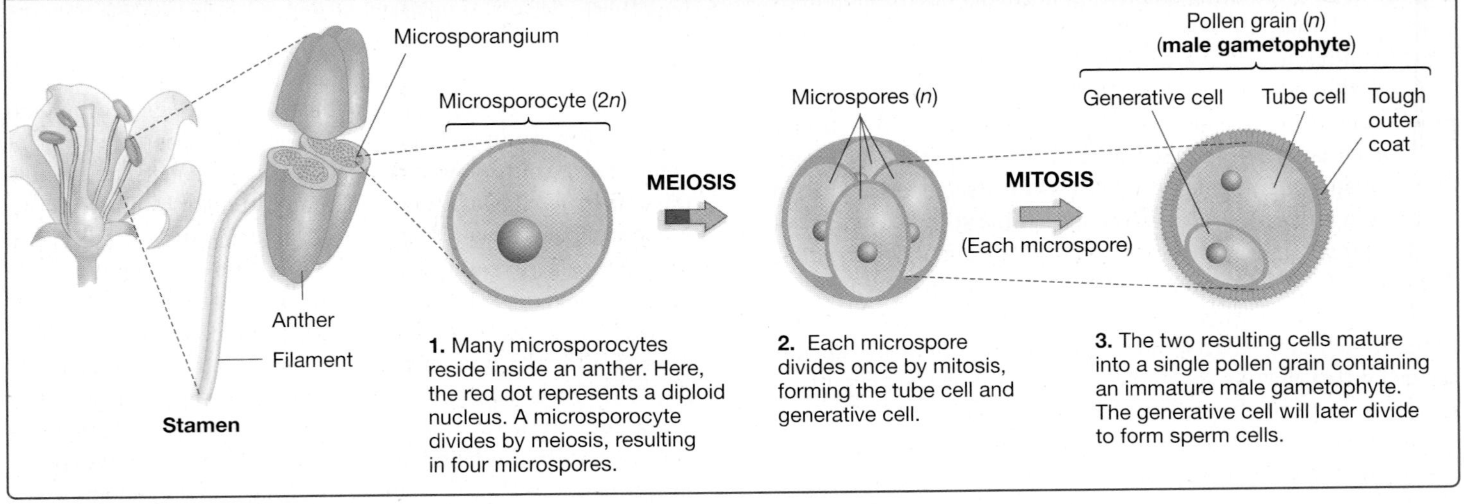

Figure 38.7 **In Angiosperms, Microspores Produce Male Gametophytes.**
✔ Define a gametophyte. Why are pollen grains considered male gametophytes?

38.3 Pollination and Fertilization

Pollination is the transfer of pollen grains from an anther to a stigma; fertilization occurs when a sperm and an egg actually unite to form a diploid zygote. The two events are separated in space and time.

After you complete this section, you should be able to . . .

- Explain the events that occur during pollination and fertilization in flowering plants.

Pollination is not restricted to angiosperms. Gymnosperms (Ch. 28, Section 28.3) also package their male gametes in pollen grains. Managing pollination and fertilization in angiosperms is a critical challenge for fruit growers and plant breeders. In addition, angiosperms' pollination and fertilization systems are thought to be key to their evolutionary success.

What aspects of pollination and fertilization allowed flowering plants to become so successful in terms of their numbers of species?

Pollination

Pollen can fall on the stigma of the same individual or the stigma of a different individual. **Self-fertilization**, or selfing, occurs when a sperm and an egg from the same individual combine to produce an offspring. In most cases, though, plants **outcross**—meaning that a sperm and an egg from different individuals combine to form an offspring. Outcrossing is the result of **cross-pollination**—when pollen is carried from the anther of one individual to the stigma of a different individual. As you'll see, many flower characteristics are correlated with the types of pollinators they attract.

Selfing versus Outcrossing: Costs and Benefits Selfing and outcrossing each have advantages and disadvantages. The primary advantage of selfing is that successful pollination is virtually assured—it doesn't depend on agents other than the plant itself.

Biologists have documented the benefit of pollination assurance by hand-pollinating plants that normally outcross. In most cases, the hand-pollinated individuals produce far more seed than do individuals that are pollinated naturally.

Other things being equal, self-fertilization should result in the production of many more seeds than outcrossing. Other things are not equal, however. Selfing has a distinct disadvantage: Selfed offspring are usually much less diverse genetically than outcrossed offspring are, even though selfing still involves meiosis (Chapter 13). In some cases, selfed offspring may also suffer from inbreeding depression (Ch. 23, Section 23.2).

Although outcrossing is riskier in terms of the reduced chance that pollination will occur, it results in more genetically diverse offspring that may be much more successful at warding off attacks from viruses, bacteria, and other pathogens.

Outcrossing is much more common than selfing. In many cases, plants have elaborate mechanisms to prevent selfing:

- ***Temporal avoidance*** In some species that have perfect flowers, male and female gametophytes mature at different times. Thus, self-fertilization does not occur.
- ***Spatial avoidance*** Selfing isn't possible in dioecious species and may be rare in monoecious species, unless pollinators transfer pollen between "different-sexed" flowers on the same individual. And in some species with perfect flowers, the anthers and stigma are so far apart that self-pollination is extremely unlikely—if pollen falls inside the flower, it has almost no chance of landing on the stigma.
- ***Molecular matching*** In many species that produce both pollen and ovules on the same plant, molecular interactions occur that prevent pollen grains from delivering sperm to the female gametophytes produced on the same plant. Such species are said to be **self-incompatible**. Many plants, just like animals, have the ability to recognize self and non-self cells and tissues.

Pollination Syndromes Cross-pollination can be accomplished in various ways: Pollen can be carried from flower to flower by physical agents such as wind or water, or by organisms such as insects, birds, or bats.

Recall that animals visit flowers to eat pollen grains, harvest nectar, or both. As an animal feeds from a flower, pollen grains adhere to its body. When the same individual visits another flower of the same species to feed, some of these grains are deposited on a stigma of the second flower.

In most cases, animal pollination is an example of **mutualism**: a mutually beneficial relationship between two species. Pollinators usually benefit by receiving food; flowering plants gain by having their pollen transferred to a different individual so that outcrossing takes place. How did these pollinator–plant relationships evolve?

Pollination syndromes are suites of flower characters that are associated with certain types of pollinators. For example, many bird-pollinated flowers tend to be red and unscented, and they open during the day when birds are active. In contrast, moths and bats are usually active at night. If they feed on nectar or pollen, the flowers they visit tend to be white (and thus more visible in low light), have a strong scent, and open at night. Observations such as these have led to the directed-pollination hypothesis, which proposes that natural selection has favored flower colors, shapes, and scents to attract specific pollinators (Ch. 28, Section 28.3).

The directed-pollination hypothesis has strong experimental support. For example, field observations led researchers to conclude that a white-flowered species of South American petunia (*Petunia axillaris*) was mainly pollinated at night by hawk moths, while a purple-flowered species (*P. integrifolia*) was pollinated during the day by bumblebees. (Note that white is highly visible at night, and purple is the color most easily seen by bees.)

Was flower color the key to flower choice by these insects? This is difficult to determine because of other differences in flower characteristics between two plant species, such as flower shape and scent. In this experiment, researchers identified a gene named *AN2* as being responsible for the pigmentation in the purple petunia. They introduced this gene into the genome of the white petunia, producing new plants that now had purple flowers. The researchers then documented that other than color, the wild-type and genetically altered flowers were identical.

RESEARCH

QUESTION: Does flower color influence pollinator preference?

HYPOTHESIS: Hawk moths prefer white petunias while bumblebees prefer purple, even when the flowers are otherwise identical.

NULL HYPOTHESIS: Neither insect has a clear preference for white- or purple-flowered petunias.

EXPERIMENTAL SETUP:

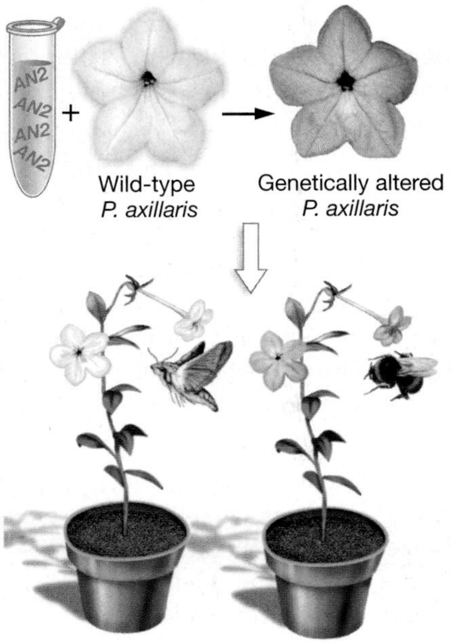

1. **Isolate *AN2* gene** for purple flower color from a related species (*P. integrifolia*) and insert it into *P. axillaris.*

2. **Grow** wild-type (white) and genetically altered (purple) *P. axillaris* in controlled greenhouse conditions.

3. **Count number of visits** by hawk moths and bumblebees to white and purple flowers.

PREDICTION OF HYPOTHESIS: Hawk moths will visit white-flowered *P. axillaris* more frequently, while bumblebees will visit purple-flowered *P. axillaris* more frequently.

PREDICTION OF NULL HYPOTHESIS: Both insects will visit white and purple flowers equally.

RESULTS:

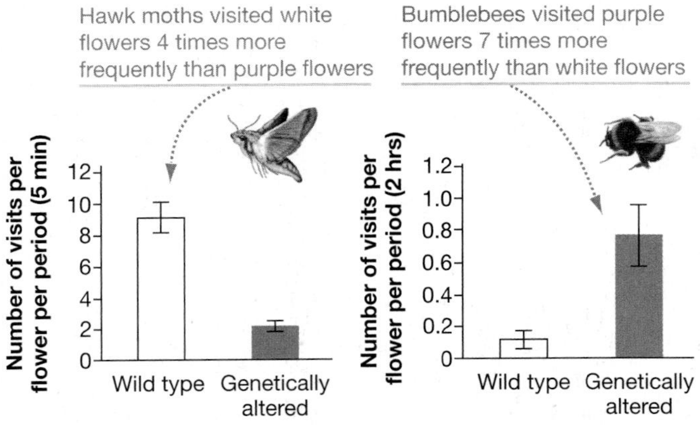

CONCLUSION: Petunia flower preference among bumblebees and hawk moths is significantly influenced by flower color.

Figure 38.8 **The Adaptive Significance of Flower Color.**

SOURCE: M. E. Hoballah T. Gübitz, J. Stuurman, et al. 2007. Single gene-mediated shift in pollinator attraction in *Petunia*. *Plant Cell* 19: 779–790.

✔ **PROCESS OF SCIENCE** Why did the researchers measure pollination in a greenhouse? Why did they plot flower visits on different scales?

Figure 38.9 **Pollinators Can Sometimes Be Inferred from Flower Structure and Color.** Charles Darwin predicted the existence of a moth that would be able to pollinate the orchid sometimes referred to as Darwin's orchid (left). Just such a pollinator (right) was discovered some 40 years after Darwin made his prediction.

In a controlled, greenhouse experiment, hawk moths and bumblebees were allowed to choose between the two *P. axillaris* plants. As predicted by the directed-pollination hypothesis, the results indicated that flower color was key to the preferences of these pollinators (Figure 38.8). These data strongly support the hypothesis that flower color is an adaptation that increases the frequency of pollination by particular pollinators.

Features associated with pollination syndromes are thought to be adaptations: traits that increase the fitness of individuals in a particular environment. In this case, flowers and pollinators have adaptations that increase pollination frequency and feeding efficiency, respectively. To capture this point, biologists say that **coevolution** has occurred (covered in Chapters 28 and 52).

One of the most famous examples of coevolution in pollination involves a species called Darwin's orchid (Figure 38.9), which is native to Madagascar. When first discovered by Western scientists, the orchid attracted a great deal of attention because it has a "spur" that can be as much as 28 cm (11 in) long, and there is a nectary at its base. Charles Darwin hypothesized that it must be pollinated by a moth with a tongue-like proboscis as long as the spur. The idea seemed preposterous at the time, but 40 years later, a hawk moth species with a proboscis that averages 38 cm in length was discovered pollinating the orchid in the wild.

Does Pollination by Animals Encourage Plant Speciation? In addition to affecting the fitness of individual plants, does pollination by animals make the formation of entirely new species more likely?

To answer this question, consider the following situation. A biologist has documented that two populations of a mountain-dwelling species called the alpine skypilot have flowers with different characteristics:

- Alpine skypilots that grow in forested habitats at or below the timberline have small flowers with short stalks and an aroma described as "skunky."
- Individuals that grow in the tundra habitats above the timberline have large flowers with long stalks and a sweet smell.

These differences are interesting, because different insects pollinate the two populations. Small flies are abundant at slightly lower elevations, are attracted to skunky odors, and pollinate the timberline individuals; large bumblebees are abundant at higher elevations, are attracted to sweet odors, and pollinate the tundra flowers.

Experiments have shown that bumblebees prefer to pollinate big flowers—probably because larger flowers can support their larger mass. Because flies and bumblebees prefer to visit different types of flowers, gene flow between the two skypilot populations is low and they are evolving distinct characteristics. The two populations may be on their way to becoming different species.

The message here is that evolutionary changes in the size or food-finding habits of a pollinator affect the angiosperm populations they pollinate. In return, changes in flower size and shape affect the insects pollinating that population. If a small population of Darwin's orchids evolved longer spurs, for example, the hawk moth pollinators living in that area would be under intense selection that favored the evolution of a longer proboscis.

Because mutation continuously introduces variations in traits, insect and angiosperm populations frequently change, diverge, and form new species. Changes in pollination can trigger the evolution of new species. It is no surprise that insects and angiosperms are exceptionally species-rich groups.

Pollination was obviously a crucial innovation during plant evolution. Now let's get down to specifics. What happens once a pollen grain is deposited on a stigma?

Fertilization

A pollen grain landing on the stigma of a mature flower of the same species initiates a series of events that lead to fertilization. Figure 38.10 walks you through the steps.

Step 1 After landing on the stigma of a mature flower from the same species, a pollen grain absorbs water and germinates. **Germination** is a resumption of growth and development. This step is blocked in many self-incompatible species if the pollen came from the same plant. When pollen germinates, the tube cell elongates and forms a structure called a **pollen tube** that grows through the stigma and down the length of the style. The direction of growth is affected by chemical attractants, called LUREs, which are small proteins released by the synergids.

Step 2 In most species, like the one illustrated in the figure, the tube-cell nucleus and the generative cell travel down the length of the pollen tube, and the generative cell divides to form two sperm. In other species, the generative cell forms sperm before the pollen is released from anthers.

Step 3 When the pollen tube reaches the micropyle of the ovule, it grows through it and enters a synergid within the female gametophyte. The synergid degenerates and two sperm are released.

PROCESS: POLLEN TUBE GROWTH AND FERTILIZATION

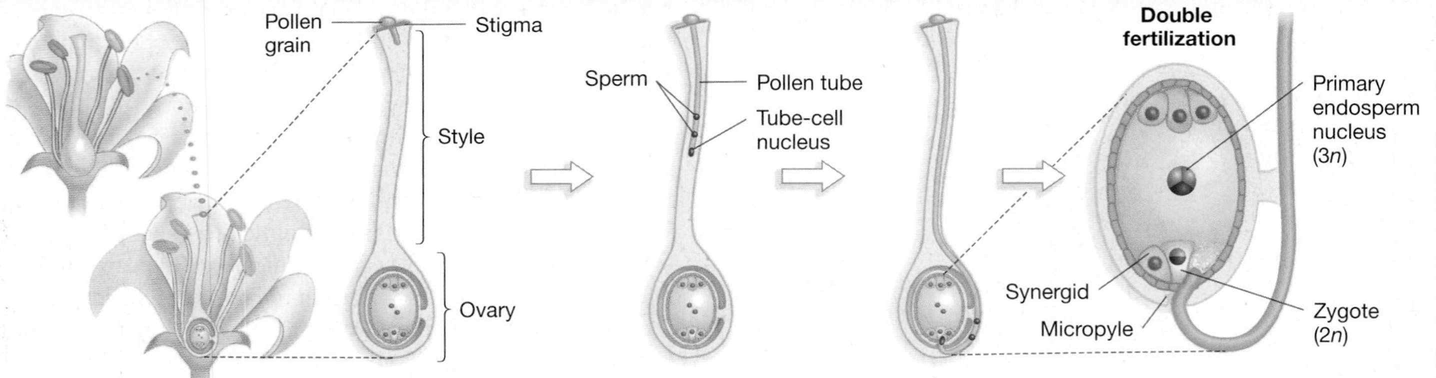

1. Pollen grain absorbs water and germinates on the stigma. Pollen tube begins growing down the style.

2. The tube-cell nucleus moves into pollen tube, and the generative cell nucleus divides by mitosis to form two sperm in pollen tube.

3. Pollen tube completes growth toward the egg by passing through micropyle and discharging the two sperm into a synergid.

4. One sperm unites with egg to form zygote. The other fuses with the two polar nuclei to form endosperm (nutrient tissue).

INTERACTIVE Figure 38.10 Double Fertilization Produces a Zygote and an Endosperm Nucleus. When the pollen tube reaches the female gametophyte, one sperm nucleus fertilizes the egg while the second fuses with the polar nuclei.

Step 4 In angiosperms an unusual event called **double fertilization** takes place. Each sperm released from a pollen tube has a different fate. One sperm unites with the egg to form the zygote. The other sperm moves through the female gametophyte and fuses with the polar nuclei in the central cell. In most cases, two polar nuclei are present and a triploid ($3n$) primary endosperm nucleus forms.

The triploid nucleus resulting from this second fertilization undergoes mitosis and cytokinesis to form the **endosperm** ("inside-seed") tissue. In most species, endosperm is triploid and stores nutrients for the embryo, including starch or oils (lipids) plus proteins and other nutrients. As one example, most of the cells in corn kernels are endosperm.

✔ **If you understand double fertilization, you should be able to identify in a female gametophyte, immediately after fertilization, which cells are haploid, diploid, and triploid.**

After fertilization, embryogenesis begins (see Section 38.5 for details) and seeds develop into mature, dormant structures that are adapted for dispersal by wind, water, or animals. What happens as seeds and fruits develop into mature structures ready for dispersal?

CHECK YOUR UNDERSTANDING

✔ **If you understood this section, you should be able to ...**

1. Explain why insects increase their fitness by visiting flowers, and why flowers increase their fitness by rewarding insects.
2. Describe the functional roles of the two products of double fertilization in angiosperms.

Answers are available in Appendix A.

38.4 Seeds and Fruits

A complex and coordinated series of events takes place as embryos develop within seeds and as seeds develop within fruits. Seeds and fruits are effective at dispersing embryos to new locations where they can grow into mature sporophytes. In this section you will explore some of the events in seed and fruit development in more detail.

After you complete this section, you should be able to ...

- Explain the roles of seeds and fruits in flowering plant reproduction.

Seed Maturation

As a seed matures, the embryo and endosperm develop inside the ovule and become surrounded by a covering called a **seed coat**. At the same time, the ovary around the ovule develops into a fruit, which encloses and protects the seed (or seeds, if a single ovary contains multiple ovules). Besides protecting the seeds, fruits often aid in dispersing seeds away from the parent plant. The mature seed consists of an embryo, a food supply—originating with endosperm—and a seed coat. In most cases, mature seeds leave the parent plant encased in a fruit (Ch. 28, Section 28.3).

Storage of Nutrients Along with pollen, the evolution of the seed was a crucial innovation as land plants diversified. Because seeds contain stored nutrients that the embryo will use for growth and development once the seed germinates, they allow offspring to be much more successful in colonizing habitats that are crowded with competitors than are offspring produced from spores, which are single cells. As a young plant emerges from the seed, it can subsist on stored nutrients until it is well enough established to absorb water from the soil and feed itself via photosynthesis. This also makes sense because many seeds germinate underground where light is not available and seedlings cannot produce organic compounds through photosynthesis.

Recall that one prominent lineage of angiosperms—the monocots—have just one **cotyledon** ("seed-leaf"; pronounced *kot-uh-LEE-don*), whereas eudicots have two (Ch. 28, Section 28.3). In most eudicots, the cotyledons take up the nutrients that were initially stored in the endosperm and store them again. In these species, there is no endosperm left by the time the seed matures—instead, the cotyledons function as the nutrient storage organ in mature seeds. **Figure 38.11** compares the seed structure in beans and corn—a representative eudicot and monocot, respectively.

The Role of Drying By the time a seed matures, the precursors of the root and shoot systems, along with the seed leaves, have formed. Once these events are accomplished, the seed tissues dry and the embryo becomes dormant—meaning it stops growing and waits.

The seeds of many species dry out, or desiccate, as they mature. Water makes up 90 percent of normal plant cells, but dried seeds contain just 5–20 percent water.

Loss of water is an adaptation that prevents seeds from germinating until after they are dispersed. The dry condition of seeds ensures that they will not germinate until water is available. This is adaptive in temperate species because water is crucial to the survival of germinated seedlings. Dry seeds are also less susceptible than wet seeds to damage from freezing.

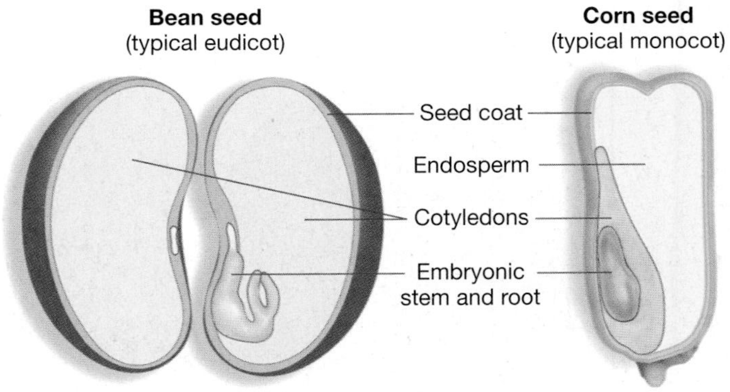

Figure 38.11 Seeds Contain an Embryo and a Food Supply Surrounded by a Tough Coat. In beans (left), a eudicot, the nutrients in the endosperm have been absorbed by the cotyledons and stored. In corn (right), a monocot, the endosperm is intact.

How do the cellular structures in the embryo and endosperm survive the drying process? When researchers reduce the amount of water surrounding isolated membranes or isolated proteins to the levels observed in extremely dry seeds, the membranes disintegrate and the proteins denature. Clearly, something is happening at the molecular level in seeds to keep these cell components intact.

Researchers established that one of these "somethings" involves sugars. As water leaves the seed during drying, sugars become concentrated and maintain the integrity of plasma membranes and proteins. If drying is extreme, the sugars form an extremely viscous liquid that contains little if any water. Substances such as this are considered vitrified, or glass-like. Biologists propose that this glassy state helps maintain the integrity of plasma membranes and proteins in seeds that experience extremely dry conditions. When seeds absorb water, the sugars dissolve and germination proceeds.

Drying is only one part of the seed maturation process, however. Equally important is the development of tissues surrounding the seed itself. In many cases, these tissues are required for the seed to be dispersed from the parent plant.

Fruit Development and Seed Dispersal

Fertilization in angiosperms initiates the development of the fruit as well as the seed and embryo (see Figure 38.2). After pollination and fertilization occur, the ovary begins to develop into the fruit while the ovules develop into seeds. As a fruit matures, the walls of the ovary thicken to form the **pericarp**, the part of the fruit that surrounds and protects the seed or seeds (Figure 38.12). Fruits can be dry when they are mature, as in nuts, or fleshy, as in cherries and tomatoes. Note that fruits are formed from tissues derived from the mother—not the embryo. Let's look at fruit structure and function in turn.

Fruit Structure Fruits come in three basic types (Figure 38.13):

1. **Simple fruits** such as the cherry develop from a single flower that contains a single carpel or several carpels that are fused together. This is the most common type of fruit.

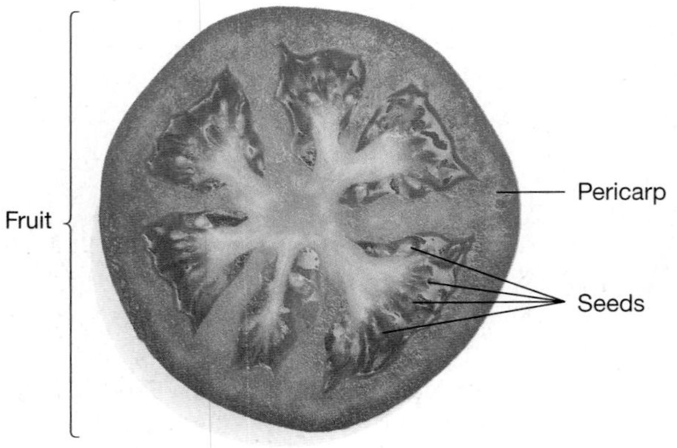

Figure 38.12 In Maturing Fruits, the Ovary Wall Develops into a Pericarp That Surrounds the Seeds. The seeds develop from ovules. A fruit consists of a pericarp and the enclosed seeds.

(a) Simple fruit (e.g., cherry):
Develops from a single flower with one carpel or fused carpels

(b) Aggregate fruit (e.g., blackberry):
Develops from a single flower with many separate carpels

(c) Multiple fruit (e.g., pineapple):
Develops from many flowers with many carpels

Figure 38.13 Three Major Types of Fruits. The structure of a fruit depends on the number of carpels found in each flower and whether the carpels are fused.

✔ **PROCESS OF SCIENCE** When fruits ripen, their color changes in a way that makes them more conspicuous to fruit eaters. State a hypothesis to explain why this color change might increase the fitness of an individual.

2. **Aggregate fruits** such as the blackberry also develop from a single flower, but one that contains many separate carpels.
3. **Multiple fruits** such as the pineapple develop from many flowers and thus many carpels.

In addition to the structures just mentioned, **accessory fruits** develop not from ovaries, but from other modified floral tissues. For example, the tiny seeds on the outside of strawberries are individual fruits, while the enlarged portion of the strawberry develops from a part of the stem called the receptacle (see Figure 38.3).

Fruit Function Fruits have two functions: They frequently aid in seed dispersal, and they protect seeds from physical damage and seed predators. Dispersal is important to the fitness of the young sporophyte. This is especially true in long-lived species, in which the offspring would compete with the parent plant for light, water, and nutrients if there were no dispersal. Fruits sometimes split open and release seeds to be dispersed directly. In many cases, however, seeds are dispersed to new locations while they are still enclosed in the fruit.

Dry fruits may simply fall to the ground or be dispersed by wind, propulsion, or animals. Some dry fruits have hooks or barbs that adhere to passing animals, while nuts are dispersed by seed predators. Fruits that are dispersed by wind often have external structures to catch the breeze and extend the distance they travel; the fruits of dandelions and maple trees are familiar examples. Fruits that float, such as coconuts, can disperse seeds in water.

Some plants actually disperse dry fruits via propulsion. The sandbox tree, for example, produces a seed pod that shrinks as it dries. Eventually the pod splits apart violently, spraying seeds in all directions with so much force that the plant is sometimes called the dynamite tree. The bursting seed pod sounds like a pistol shot, and seeds can be scattered as much as 40 m away from the parent plant. Similarly, the dwarf mistletoe fruit fills with sugars as it matures. Enough water follows via osmosis to make the fruit explode and shoot seeds as far as 5 m.

Animals are the most common dispersal agent for fleshy fruits. Just as plants have evolved to use animals for pollination, they have also evolved to use specific types of animals for seed dispersal. In cases like this, seed dispersal is an example of mutualism. The plant provides a fruit rich in sugars and other nutrients; in return, the animal carries the fruit to a new location and excretes the seeds along with a supply of fertilizer.

Mammals and birds are the most common seed dispersers. Mammals are often active at night and use their well-developed sense of smell to locate fruits. As you might imagine, fruits of mammal-dispersed seeds are usually dull colored and fragrant. In contrast, birds are usually active in the day and see well. Fruits of bird-dispersed seeds are usually brightly colored.

Some animals are seed predators that compete with dispersers. For example, birds usually swallow fruit whole and disperse seeds over long distances. Fruit-eating mice, however, chew their food, often killing the seeds inside. How can plants discourage seed predators such as mice from eating their fruits? The answer is one that might be familiar to you. Some plants such as chili peppers lace their fruits with a spicy-hot repellent called capsaicin. But does it work?

RESEARCH

QUESTION: Does the presence of capsaicin in chilies deter some predators but not others?

HYPOTHESIS: Capsaicin deters cactus mice (seed predators) but not birds (seed dispersers).

NULL HYPOTHESIS: Cactus mice and birds respond to capsaicin in the same way.

EXPERIMENTAL SETUP:

PREDICTION OF HYPOTHESIS: Both will eat hackberry, but only thrashers will eat pungent chilies.

PREDICTION OF NULL HYPOTHESIS: No difference between thrashers and mice in fruit consumed.

RESULTS:

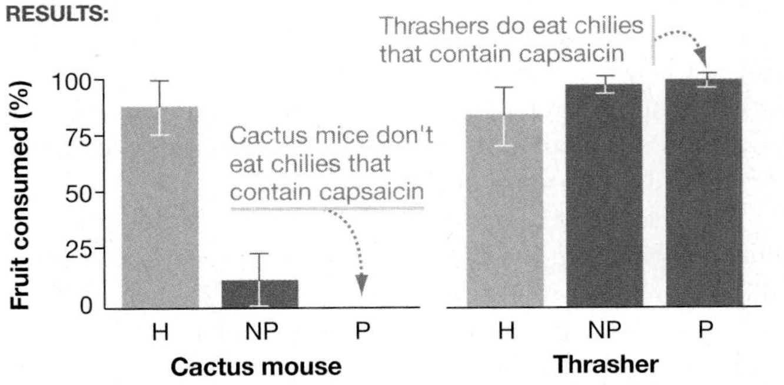

CONCLUSION: The presence of capsaicin deters cactus mice but not thrashers.

Figure 38.14 **Experimental Evidence Supports the Hypothesis that Capsaicin Deters Seed Predators but Not Seed Dispersers.**

SOURCE: J. J. Tewksbury and G. P. Nabhan. 2001. Directed deterrence by capsaicin in chilies. *Nature* 382: 403–404.

✔ **PROCESS OF SCIENCE** Why were hackberries used in this experiment?

An experiment by ecologists Joshua Tewksbury and Gary Nabhan tested the hypothesis that capsaicin promoted dispersal by the curve-billed thrasher and deterred predation by cactus mice (Figure 38.14). Each animal was offered three kinds of fruit: hackberries, fruits from a strain of chilies that can't synthesize capsaicin (non-pungent chilies), and pungent chilies that have lots of capsaicin. All three fruits looked similar and had equivalent nutritional value. For each animal tested, the researchers recorded the percentage of each fruit eaten during a specific time interval. They then calculated the average amount of fruit that was eaten by five test individuals from each species.

Curve-billed thrashers ate all three fruits equally, but the cactus mice ate fewer non-pungent chilies and avoided the capsaicin-laced fruits entirely. In a follow-up experiment, the researchers fed non-pungent chilies to each kind of animal. When the seeds had passed through the animal's digestive tract and were excreted, the researchers collected and planted the seeds. About 60 percent of the seeds that passed through the birds germinated, but none of those eaten by the mice germinated. Capsaicin appears to be an effective deterrent to seed predation.

Seed Dormancy

Once they have dispersed from the parent plant, seeds may not germinate for a period of time. This condition is referred to as **dormancy**. Dormancy is usually a feature of seeds from species that inhabit seasonal environments, where for extended periods conditions may be too cold or dry for seedlings to thrive. Based on this observation, dormancy is interpreted as an adaptation that allows seeds to remain viable until conditions improve. Consistent with this hypothesis, dormancy is rare or nonexistent in seeds produced by plants that inhabit tropical wet forests or other areas where conditions are suitable for germination year-round. What molecular mechanisms are responsible for dormancy?

The answer to this question is not well understood, but the hormone abscisic acid (ABA) plays important roles in seed development and dormancy (Ch. 37, Section 37.6). Mutants of some plants that cannot either make or respond to ABA exhibit a property called **viviparity**, in which the seeds germinate on the parent plant as soon as they are mature. In many species, ABA triggers the accumulation of storage compounds, desiccation tolerance, and the prevention of germination.

After a seed has been dispersed from the parent plant, it may remain dormant in the soil for years or even centuries before it will germinate. Over 130 years ago, botanist William Beal initiated an experiment that is still ongoing to address the question of just how long dormant seeds can remain viable. He collected seeds from soil in Michigan and filled 20 glass bottles with 50 seeds from each of 20 different plant species. The bottles were buried, and every 20 years one of the bottles is dug up and its seeds are planted. The most recent bottle excavation was in the year 2000, and most of the seeds were indeed still viable. The current record for longest seed dormancy is a date seed that germinated after being buried under ruins in Israel for 2000 years.

Seed Germination

What finally causes dormant seeds to germinate? Thanks to the broad range of habitats in which plants are adapted, germination depends on a wide variety of internal and external factors.

Requirements for Germination Most seeds need water, oxygen, and warm temperatures to germinate. The coats of some seeds are thick enough to prevent water and oxygen from physically reaching the embryo. For germination to occur, these seed coats must be disrupted, or **scarified**.

Crop seeds that require scarification are placed in large, revolving drums with pieces of sandpaper that abrade and scarify the seeds. When planted, these scarified seeds germinate quickly and uniformly. In nature, impermeable seed coats protect nutritious seeds from being killed by soil bacteria and fungi, and they often prevent seeds from germinating for many years.

To germinate, most seeds must experience particular environmental conditions. Species native to high latitudes or high elevations often produce seeds that must undergo cool, wet conditions before they will germinate. Without this level of control, germination before the ensuing winter would likely result in death of the seedling. Studies reveal that germination is regulated by two hormones, gibberellin and ABA, and that the levels of these hormones in seeds are affected by temperature (Ch. 37, Section 37.6). After the seed goes through a winter, its gibberellin level increases and germination begins.

Because small seeds have few nutrient reserves in their cotyledons or endosperm, many small-seeded species need to germinate near the soil surface, where individuals are exposed to light and can feed themselves via photosynthesis. Lettuce seeds and other small seeds must be exposed to red light before they will break dormancy and germinate (Ch. 37, Section 37.3). Red light is an important environmental cue, because wavelengths in the red portion of the light spectrum are used for photosynthesis. Red light and blue light signal that sunlight is abundant and that the seedling will not be shaded by other plants.

Finally, many of the seeds produced by species native to habitats where wildfires are frequent, as in the California chaparral and South African fynbos, have an unusual chemical requirement to break dormancy: They must be exposed to fire or smoke before they will germinate. In fact, the commercial food product "liquid smoke" induces germination in these seeds as well as actual smoke does. In fire-prone habitats, it is advantageous for seeds to germinate after fire has cleared away existing vegetation.

The message here is that dormancy can be broken in response to a wide variety of habitat-specific environmental cues. In general, the cue that triggers germination is a reliable signal that conditions for seedling growth are favorable for a particular species in a particular environment.

The Process of Germination The formation of a mature embryo is just one requirement to ensure the development of viable seeds. Even if specific environmental signals, such as light or heat, are required to break dormancy, seeds do not germinate without water. Water uptake is the first event in germination. Once the seed coat allows water penetration, water enters by moving along a steep water-potential gradient, because the seed is so dry.

Water uptake in a typical angiosperm seed has three distinct phases:

Phase 1 Germination begins with a rapid influx of water. Oxygen consumption and protein synthesis in the seed increase dramatically, but no new messenger RNAs are transcribed (Ch. 16, Section 16.2). Based on these observations, biologists have concluded that some of the key early events in germination are driven by mRNAs that are stored in the seed before maturation.

Phase 2 This phase is an extended period when water uptake slows or stops. New mRNAs are transcribed and translated into protein products. Mitochondria also begin to multiply. In effect, seeds take up enough water in phase 1 to hydrate

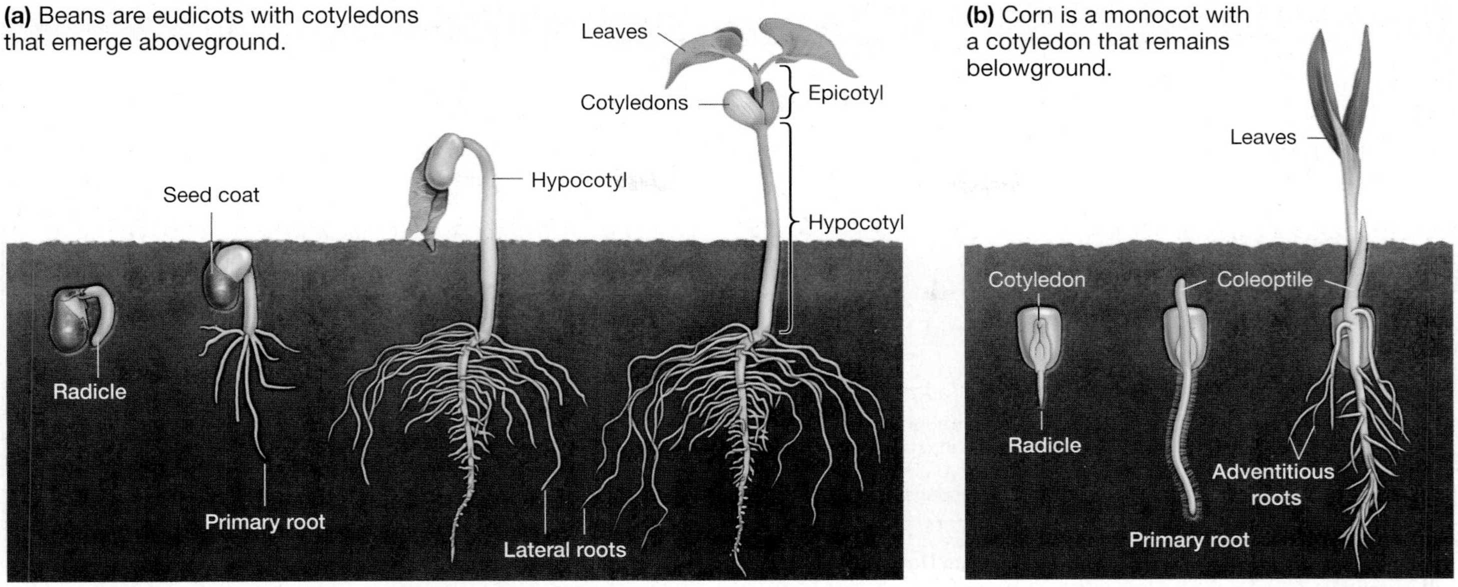

Figure 38.15 The Germination Sequence Varies among Species.

✔ In which of these species are cotyledons photosynthetic?

their existing proteins and membranes and then begin to manufacture the proteins and mitochondria needed to support growth.

Phase 3 Water uptake resumes as growth begins. This renewed phase of water uptake enables cells to develop enough turgor pressure to enlarge. Eventually, the seedling bursts from the seed coat.

Figure 38.15 shows what happens as eudicot and monocot embryos emerge from the seed. The **radicle**, or embryonic root, emerges first, and it then develops into the mature root system. This is important because the seedling must have a source of water in order to grow. Initially, leaves are less important because of the nutrients stored in the seed.

In eudicots, the shoot system with its cotyledons usually emerges shortly after the radicle appears. In monocots such as corn, the emerging shoot is covered by the protective coleoptile (Ch. 37, Section 37.2). Note that in eudicots, the emerging stem has a hook shape. As with the coleoptile of monocots, this trait is thought to protect the apical meristem from damage as the shoot works its way upward through rough soil particles. Recall that **meristems** are regions of active cell division that represent the ultimate source of all cells within plants (Ch. 34, Section 34.3). The role of meristems in plant development is described more fully in Section 38.5.

The next major event in the seedling's life occurs when either the cotyledons or the earliest leaves produced by the growing seedling begin photosynthesis. The seedling is said to be established when the young plant no longer relies on food reserves in its endosperm or cotyledons. Instead, it receives all of its nourishment on its own, from the compounds produced by photosynthesis.

Once established, the seedling embarks on a series of developmental events that ultimately leads to the formation of an adult plant with mature roots, stems, leaves, and eventually flowers. How does this transformation take place? What developmental events occur as a plant progresses from an immature embryo to a mature plant?

CHECK YOUR UNDERSTANDING

✔ **If you understood this section, you should be able to . . .**

1. Explain why beans, peas, and corn are classified as fruits.
2. Explain why it might be beneficial for seeds to remain dormant and viable for long periods of time.
3. Describe the events that occur at the cellular level when dormant seeds absorb water.

Answers are available in Appendix A.

38.5 Embryogenesis and Vegetative Development

After you complete this section, you should be able to . . .

- Explain how embryogenesis leads to the formation of meristems that serve as the ultimate source of cells in developing plants.

Plant development occurs as vegetative or reproductive development. **Vegetative development** produces the nonreproductive portions of the plant body—the roots, leaves, and stems. As a plant matures, some shoot meristems will produce reproductive structures, a process known as **reproductive development** (see Section 38.6).

Plants have to contend with varied environmental conditions throughout their developmental phases. Unlike most animals, they don't move around to find a place that suits their requirements. Instead, they adjust to their immediate surroundings, largely by continuously growing and developing roots, stems, and leaves. If an oak tree is heavily

PROCESS: EMBRYO DEVELOPMENT AND SEED MATURATION

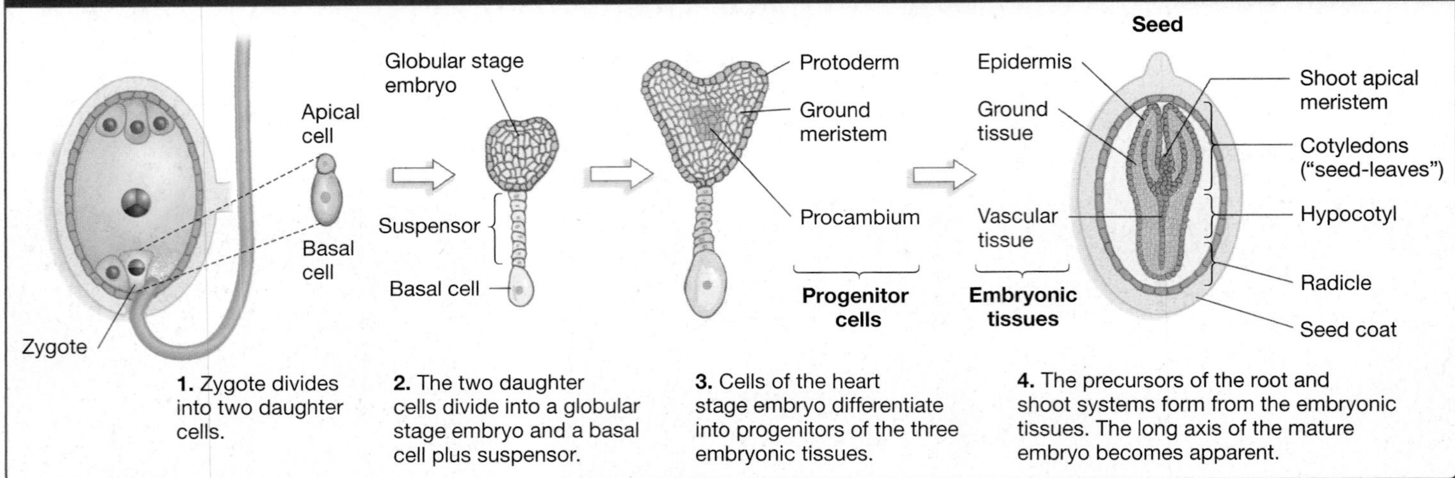

Figure 38.16 **Embryonic Tissues and Structures Develop Inside Seeds.** The embryo inside a seed has cotyledons and the beginnings of root and shoot systems. The embryonic epidermis and the ground and vascular tissues are organized in distinct layers.

✔ Label the apical–basal and radial axes on the globular stage embryo.

shaded on one side, it stops growing in that direction and extends branches on the other side. If it is heavily shaded on all sides, it directs its growth upward. This constant adjustment to changing environmental conditions is possible thanks to the meristems that are located at the tips of shoots and roots—meristems that are first initiated while the plant is developing as an embryo within a seed.

Let's begin our discussion of vegetative development by examining the earliest stages of development and differentiation that occur during embryogenesis.

Embryogenesis

Embryogenesis is the developmental process by which a single-celled zygote becomes a multicellular embryo. Although the details vary among species, the fundamental sequence of events during embryogenesis in all flowering plants is the same. The events illustrated in Figure 38.16 represent what occurs in the model plant *Arabidopsis* (**BioSkills 11**). The highlights are as follows:

- After fertilization, the zygote divides, and it produces daughter cells that are different in size, content, and fate. This type of cell division is called an asymmetric division and produces a smaller apical cell and larger basal cell; **apical** refers to the tip, while **basal** refers to the base.
- The asymmetry between the basal and apical cells helps establish one of the primary axes (directions) of the plant body: the **apical–basal axis** (also called the long axis). The basal cell divides parallel to the future apical–basal axis to produce a column of cells called the **suspensor**. The suspensor provides a route for nutrient transfer from the parent plant to the developing embryo.

The apical cell produces daughter cells that divide both perpendicularly to the apical–basal axis and parallel to it to produce a group of cells called the globular stage embryo. The regulation of cleavage plane orientation and the direction of cell expansion are critical for creating structures during plant development.

Another asymmetry arises as cells of the globular stage embryo continue to divide. Cells in the interior of the structure are completely surrounded by other cells of the embryo. In contrast, cells in the outermost layer contact surrounding tissues in the seed in addition to underlying embryo cells. This creates the second major body axis, the **radial axis**. The radial axis extends from the interior of the plant body out to the exterior.

Returning to the events in Figure 38.16:

- The heart stage embryo (named after its characteristic shape) forms shortly after the globular stage. During the heart stage, the three primary meristems form (protoderm, procambium, and ground meristem), giving rise to the dermal, vascular, and ground tissue systems, respectively (Ch. 34, Section 34.3).
- As the embryo continues to develop, the apical–basal axis of the plant continues to emerge and several important structures take shape: cotyledons, which in many species absorb nutrients from the endosperm and supply them to the rest of the embryo; the **hypocotyl** ("under-cotyledon"), which is the embryonic stem; and the radicle, the embryonic root. Some embryos also have an **epicotyl** ("above-cotyledon"), which is a portion of the embryonic stem that extends above the cotyledons (see Figure 38.15).

The initial events in embryogenesis illustrate a general point: The fate of a plant cell can be summed up in the old quip about the three keys to success in real estate—"location, location, location." Starting with the initial division that creates the apical and basal cells, plant cells differentiate based on where they are in the body.

✔ **If you understand the basic steps in embryonic development, you should be able to compare and contrast a cotyledon with a radicle in terms of their tissue composition and the progenitor cells they originated from.**

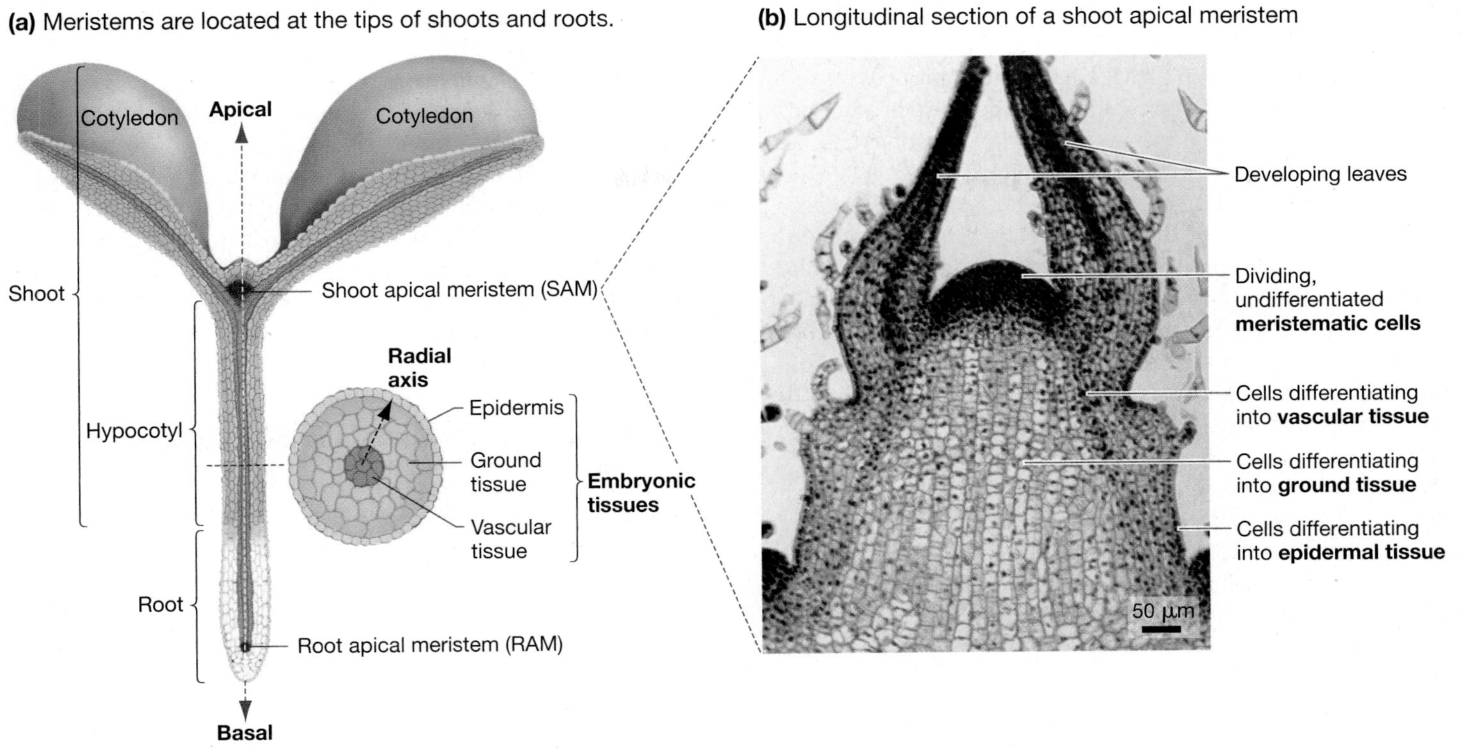

Figure 38.17 **Embryogenesis Establishes the Apical–Basal and Radial Axes as Well as Apical Meristems.** When cells in a meristem divide, the daughter cells either remain undifferentiated and continue to divide, or they differentiate into new epidermal, ground, or vascular cells.

Meristem Formation

The apical–basal and radial axes that are established during early embryogenesis are retained at later stages of development (Figure 38.17a). As the cotyledons, hypocotyl, and root begin to take shape, groups of cells called the **shoot apical meristem (SAM)** and **root apical meristem (RAM)** are specified. Figure 38.17b provides a close-up view of a shoot apical meristem.

The cells within the meristem are small and undifferentiated. Within each meristem, the rate of cell division is dictated by cell–cell signals produced in response to environmental cues, such as the arrival of spring, the presence of abundant water, or the amount of light striking the plant.

Just below and at the periphery of the meristem, daughter cells produced by mitosis and cytokinesis in the meristem grow in specific directions and initially differentiate into epidermal, ground, or vascular tissue. During embryogenesis, these tissues are produced and arranged along the radial axis (see Figures 38.16 and 38.17):

- The **epidermis** ("over-skin") is an outer covering of specialized cells that protects the individual.
- Inside the epidermal layer of cells is **ground tissue**, a mass of cells that may later differentiate into cells that are specialized for photosynthesis, food storage, or other functions.
- The **vascular tissue** in the center of the plant will eventually differentiate into specialized cells that transport food and water between root and shoot.

Careful microscopy allowed biologists to tease out the events that occur as meristems grow, and intense research continues to explore how interactions between cell–cell signals such as auxin influence the fate of cells produced by meristems. The root meristem can form all the underground portions of the plant, and the shoot meristem can form all the aerial portions, including reproductive structures. Throughout a plant's life, meristematic tissues continue to produce cells that can differentiate into adult tissues and structures (Chapter 34).

For the cotyledons and other embryonic structures to take shape, cell divisions need to occur in precise orientations. What's more, the resulting cells must exhibit differential growth. Some cells grow larger than others, and the direction of cell expansion along the apical–basal or radial axes is tightly controlled and often radically different.

Let's now examine the genetic control of axis determination in plants, which is initiated during embryogenesis.

Which Genes Determine Body Axes in the Plant Embryo?

The genetic approach to exploring development that began with research on *Drosophila melanogaster* has also proven to be a powerful tool for studying plant embryogenesis. Although the specific genes involved are different in plants than in animals, the basic mechanism by which genes direct the earliest events in development is similar.

Research on the genetics of early development in plants was pioneered by Gerd Jürgens and colleagues in the 1990s. This research group set out to identify genes that are transcribed in the zygote or embryo of *Arabidopsis* and that are responsible for establishing the apical–basal axis of the plant body. It's no coincidence that this effort was similar to the work on anterior–posterior pattern formation in *Drosophila* (Chapter 21), as Jürgens had participated in the work with flies.

The biologists' initial goal was to identify individuals with developmental defects at the seedling stage. More specifically, they were looking for patterning mutants that lacked particular regions along the apical–basal axis of the body. The team succeeded in finding several bizarre-looking mutants (**Figure 38.18**). Apical mutants lacked the first leaves, or cotyledons. Some mutants, called central mutants, lacked the embryonic stem, or hypocotyl. Other plants, dubbed basal mutants, lacked both hypocotyls and roots.

To interpret these results, the researchers suggested that each type of *Arabidopsis* mutant had a defect in a different gene, and that each gene played a role in specifying the position of cells along the apical–basal axis of the body. They hypothesized that these genes are analogous to the segmentation genes of fruit flies, which specify the fate of cells within well-defined regions along the anterior–posterior axis of insects.

What are these *Arabidopsis* genes, and how do they exert their effects? To answer these questions, consider the gene responsible for the basal mutants, which lack hypocotyls and roots. This gene has been cloned and sequenced and named *MONOPTEROS*. Because its DNA sequence indicated that its protein product has a DNA-binding domain, *MONOPTEROS* was hypothesized to encode a transcription factor that regulates the activity of target genes. This hypothesis was later shown to be correct.

The *MONOPTEROS* gene is activated in response to signals from auxin, a prominent plant hormone. Auxin is produced in the shoot apical meristem and transported toward the basal parts of the individual. This results in an auxin concentration gradient along the apical–basal axis of a plant. Much like the Bicoid concentration gradient in fruit fly embryos (Ch. 21, Section 21.3), the auxin concentration gradient provides positional information.

Together these observations indicate that auxin acts as a morphogen to trigger the production of the regulatory transcription factor MONOPTEROS. In turn, MONOPTEROS unleashes a regulatory cascade that determines which cells in the basal portion of the embryo will form hypocotyl and roots.

What is the take-home message here? Although the specific genes and proteins differ in animals and plants, independent evolution in the two groups has either converged or been constrained to adopt solutions for development that involve common principles. Cell-to-cell signals and regulatory cascades result in the step-by-step specification of a cell's position and fate.

Researchers have also taken up the question of which genes respond to these signals to direct the development of specific structures. Let's consider one example—the genetic control of leaf shape.

Which Genes Determine Leaf Structure and Shape?

Once a leaf begins to grow, its structure develops along three different axes: the proximal–distal, mediolateral ("middle-to-side"), and adaxial–abaxial ("upper–lower"; **Figure 38.19**). Proximal is toward the main body; distal is away from the main body. The mediolateral axis runs from the middle of a leaf toward its margin.

Researchers have begun to identify the genes responsible for specifying the three leaf axes. For example, analyses of mutant snapdragons and other species have shown that a gene called *PHANTASTICA* (abbreviated *PHAN*) may play a critical role in setting up the adaxial–abaxial axis of leaves.

The protein product of *PHAN* is a regulatory transcription factor. PHAN controls the expression of genes that cause cells to form the upper surface of leaves. It is part of a regulatory cascade that begins with auxin and other cell–cell signals and ends with the growth of a normal-shaped leaf.

All leaves are built along three axes, but the shape of leaves varies within an individual and between species (Chapter 34). What controls these differences between species? Experimentally altering regulatory genes involved in leaf development can lead to striking changes in leaf shape.

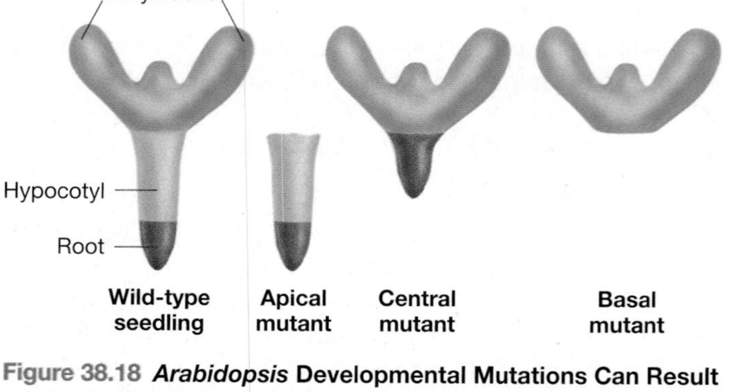

Figure 38.18 ***Arabidopsis*** **Developmental Mutations Can Result in Misshapen Embryos.** Researchers have identified *Arabidopsis* patterning mutants that are missing specific sections of the body along the apical–basal axis.

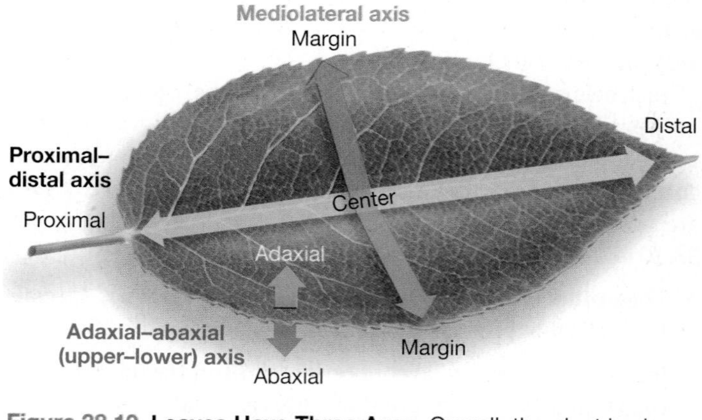

Figure 38.19 Leaves Have Three Axes. Overall, the plant body has just two axes: apical–basal and radial. Individually, however, every leaf has the three axes shown here. (To remember the difference between adaxial and abaxial, imagine that the b in abaxial stands for "below.")

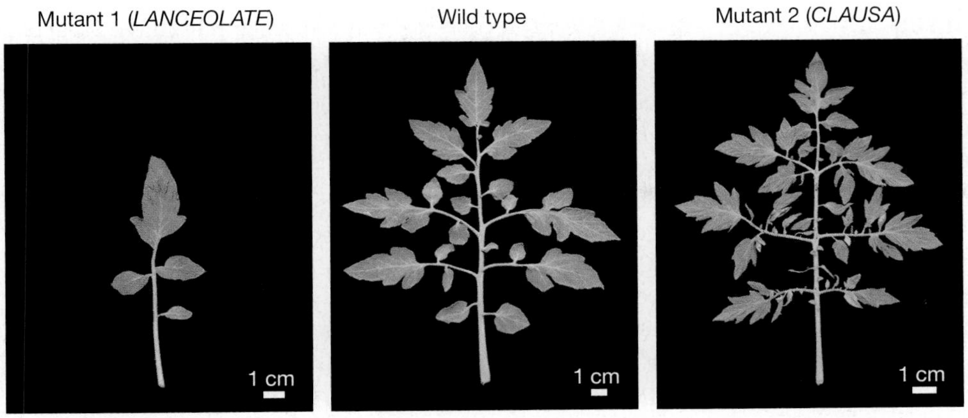

Figure 38.20 Regulatory Genes Control the Development of Leaf Shape. A wild-type tomato leaf is shown in the middle panel, surrounded by tomato leaves produced by two different regulatory gene mutations, *LANCEOLATE* and *CLAUSA*.

Figure 38.20 shows how mutations in two different regulatory genes can alter leaf shape in tomato plants. Studies like these are beginning to reveal the genes that control the development of particular leaf shapes and how changes in these genes may have led to the evolution of different types of leaves.

CHECK YOUR UNDERSTANDING

✔ **If you understood this section, you should be able to ...**

1. Identify differences between the globular stage, heart stage, and mature stage of embryogenesis.
2. Explain the statement "Meristems are perpetually embryonic."
3. Describe an experimental approach for potentially changing a plant that produces simple leaves into one that produces compound leaves.

Answers are available in Appendix A.

38.6 Reproductive Development

After you complete this section, you should be able to ...

- Explain how differential gene expression leads to the formation of floral organs.

Reproductive development is a distinguishing feature of plants. In animals, determination of cells that have the potential to form sperm or egg—the germ line—is one of the earliest events in development. In plants, there is no predetermined germ line. Instead, shoot meristems have the potential to switch from vegetative to reproductive development in response to environmental conditions such as length of day and night. Plants form reproductive structures—flowers—when environmental conditions trigger a shoot apical meristem to switch from vegetative to reproductive development (Ch. 37, Section 37.3). In some plants, the transition from vegetative to reproductive development occurs just once, while in others reproductive development can occur on a yearly basis.

The Floral Meristem and the Flower

Through a series of genetically controlled steps, a shoot apical meristem (SAM) transitions to a **floral meristem**; instead of vegetative structures, such as roots, stems, or leaves, the floral meristem produces flowers. The genes that take part in the regulatory cascade responsible for the maturation of a floral meristem are now well known.

Recall that a flower contains four kinds of organs: (1) sepals, (2) petals, (3) stamens, and (4) carpels. As Figure 38.21 shows, each of these organs is arranged in a characteristic pattern of whorls within whorls (a whorl is a circular arrangement). How does the floral meristem produce these four organs in their characteristic arrangement?

The first hint of an answer came over 100 years ago, when researchers discovered several types of **homeotic mutations** in flowers of popular garden plants. (Recall from Chapter 21 that homeotic genes regulate the development of various body parts in plants and animals.) In the mutants, one kind of floral organ was replaced by another. For example, one homeotic mutant had flowers with sepals, petals, another ring of petals, and carpels instead of having sepals, petals, stamens, and carpels. These mutants are similar to *Drosophila* homeotic mutants (Ch. 21, Section 21.3), where individuals have legs or antennae growing in the wrong location in place of the appropriate structure.

Just as an analysis of homeotic mutants in fruit flies led to understanding the genetic control of the body plan in animals, an analysis of homeotic flower mutants in *Arabidopsis* triggered a breakthrough in understanding the genetic control of flower structure.

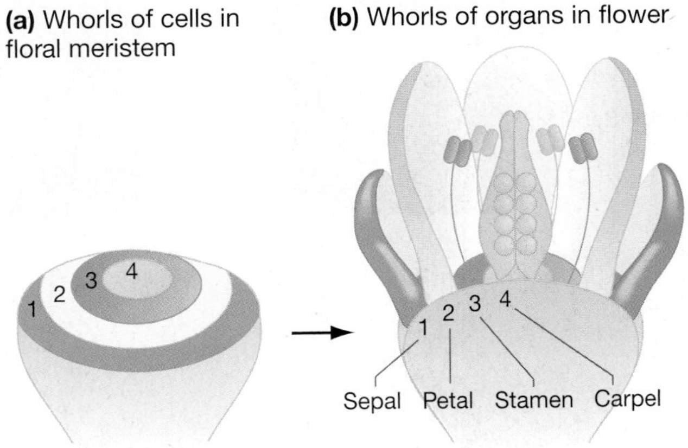

Figure 38.21 Whorls of Floral Meristem Cells Produce Circularly Arranged Flower Organs.

The Genetic Control of Flower Structures

Over 100 years after floral homeotic mutants were first described, Elliot Meyerowitz and colleagues assembled a large collection of *Arabidopsis* homeotic flower-structure mutants. The researchers' goal was to identify and characterize the genes responsible for specifying the four floral organs.

The group found three general classes of mutants, shown in Figure 38.22. Some mutants had only carpels and stamens; others had only sepals and carpels; still others had only petals and sepals. The key observation was that each type of mutant lacked the elements found in *two* of the four whorls.

What was going on? The biologists hypothesized that each class of homeotic mutant was caused by a defect in a single gene. They reasoned that if three genes set up the pattern of a flower, the mutant phenotypes suggested a hypothesis for how the three gene products interact. Because they referred to the three hypothetical genes as *A, B,* and *C*, the hypothesis is called the ABC model.

The ABC Model Three basic ideas underlie the ABC model (Figure 38.23a):

1. Each of the three genes is expressed in two adjacent whorls.
2. Because each gene is expressed in two adjacent whorls, a total of four different combinations of gene products can occur.
3. Each of these four combinations of gene products triggers the development of a different floral organ.

Specifically, the Meyerowitz group proposed that **(1)** the A protein alone causes cells to form sepals, **(2)** a combination of A and B proteins sets up the formation of petals, **(3)** B and C combined specify stamens, and **(4)** the C protein alone designates cells as the precursors of carpels.

Does this model explain how the three classes of homeotic mutants occur? The answer is yes, if two additional elements are added to the model:

4. The A protein inhibits production of the C protein.
5. The C protein inhibits production of the A protein.

With these ideas in mind, patterns of gene expression can be predicted that correspond to the mutant phenotypes. These are shown in Figure 38.23b.

For example, if the *A* gene is disabled by mutation, then it no longer inhibits the expression of the *C* gene, and all cells produce the C protein. As a result, cells in the outermost whorl express only C protein and develop into carpels, while cells in the whorl just to the inside produce B and C proteins and develop into stamens.

✔ If you understand the ABC model, you should be able to explain why mutants that lack expression of the *C* gene instead express *A* genes across all whorls.

Testing the Model Although the ABC model is plausible, elegant, and capable of explaining the data, it needed to be tested. To do so, Meyerowitz and co-workers mapped the genes responsible for the mutant phenotypes and cloned the genes using techniques you learned about earlier (Chapter 20). Once they had isolated the genes, they were able to obtain probes to perform in situ hybridizations. The goal was to learn the pattern of expression of the *A, B,* and *C* genes and see if that pattern corresponded to the model's predictions.

The mRNAs for the *A, B,* and *C* genes showed up in the sets of whorls predicted by the model. The *A* gene is expressed in the outer two whorls, the *B* gene is expressed in the middle two whorls, and the *C* gene is expressed in the inner two whorls. Later work showed that there weren't single *A, B,* and *C* genes, but groups of *A, B,* and *C* class genes. However, this finding didn't change the outline of the ABC model.

The result that *A, B,* and *C* class genes were expressed in the predicted regions supported the ABC model. Just as different combinations of *Hox* gene products specify the identity of fly segments (Chapter 21), different combinations of floral identity genes specify the parts of a flower.

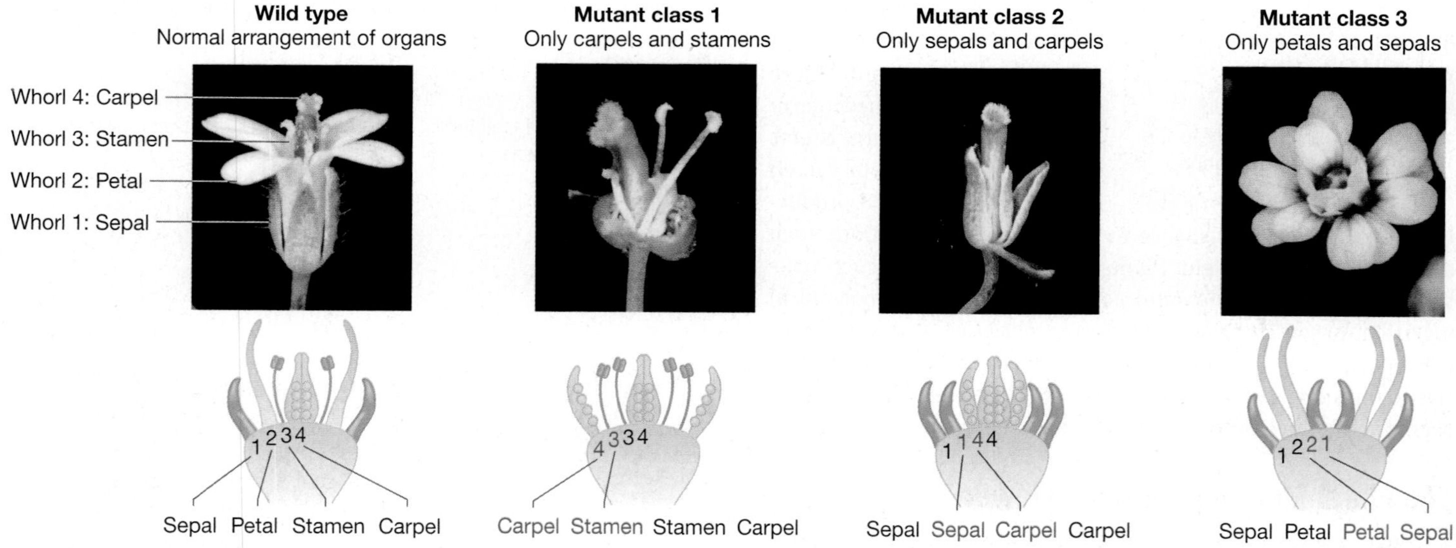

Figure 38.22 Homeotic Mutants of *Arabidopsis* Can Have Flower Organs in the Wrong Locations. Red labels in the mutants indicate a homeotic transformation (misplaced structure).

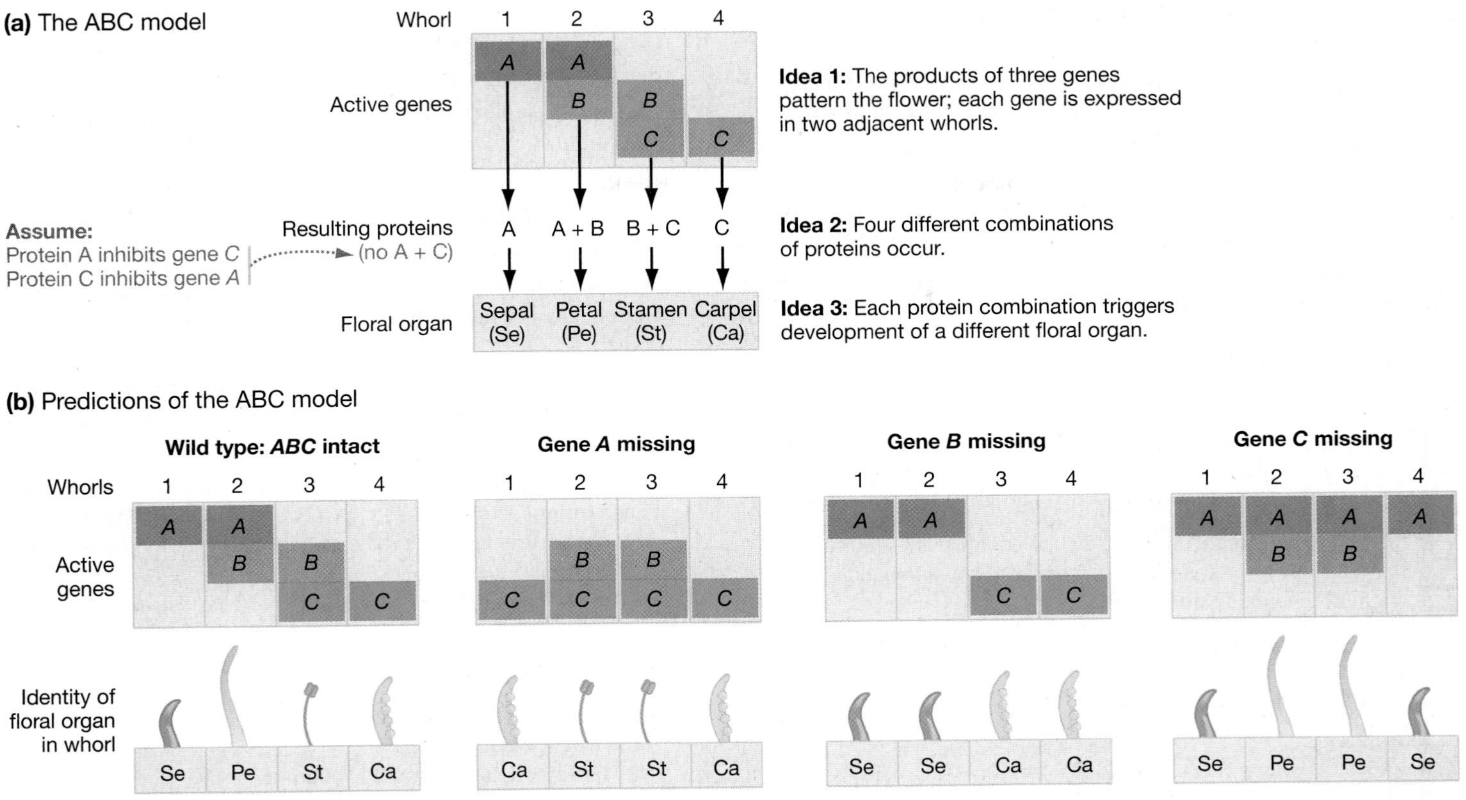

Figure 38.23 The ABC Model Illustrates Genetic Control of Flower Development. The ABC model is a hypothesis to explain the existence of three types of homeotic mutants in *Arabidopsis* flowers. The regions of gene expression were determined by in situ hybridization.

Plant Tool-kit Genes Like animals, plants have a genetic tool kit for development, although the tools have turned out to be different from those of animals. The differences, however, are not in the types of tools—hammer versus saw—but in their particular forms: nail hammer versus finishing hammer; circular saw versus band saw. For example, plants use a group of proteins called MADS-box transcription factors to specify structures, and animals use Hox transcription factors; plants use graded concentrations of auxin to specify position within the embryo, and animals use graded concentrations of other molecules, such as Bicoid. The details are different, but the broad mechanisms are the same.

Why should this be? There's likely a limit to the number of ways to build a multicellular organism. The different evolutionary paths taken in plants and animals to achieve this end have converged on these similar mechanisms, but each has used a different set of tools.

CHECK YOUR UNDERSTANDING

✔ **If you understood this section, you should be able to ...**

1. List the four floral organs in order of arrangement from the exterior to the interior of a flower.
2. Explain how homeotic mutations in plants are similar to those in animals.

Answers are available in Appendix A.

CHAPTER 38 Review

For media, go to Mastering Biology

38.1 An Introduction to Flowering Plant Reproduction

- Most plants are capable of asexual reproduction, which results in offspring that are genetically identical to the parent plant.
- Plants undergo alternation of generations, in which a diploid sporophyte phase alternates with a haploid gametophyte phase. Sporophytes produce spores by meiosis. Gametophytes produce gametes by mitosis.
- The relative size and life span of the gametophyte and sporophyte phases vary a great deal among plant groups.
- In angiosperms, or flowering plants, sporophytes are the large and long-lived phase where photosynthesis takes place; gametophytes consist of just a few cells.

38.2 Reproductive Structures

- In angiosperms, male and female gametophytes are microscopic and are produced inside flowers. Flowers are made up of sepals, petals, stamens, and one or more carpels.
- The lower part of the carpel, the ovary, contains one to many ovules. Within the ovule, a megasporocyte undergoes meiosis, producing a megaspore that develops into the female gametophyte.
- In the anthers of stamens, microsporocytes undergo meiosis. The resulting microspores develop into male gametophytes, which are enclosed in pollen grains.

38.3 Pollination and Fertilization

- Pollination occurs when pollen grains are transported to the stigma of the carpel. In most cases, the structure of a flower correlates with the morphology and behavior of its pollinator.
- If allowed to germinate on the stigma, a pollen grain sends a long pollen tube down the style. Two sperm travel down the pollen tube and enter the female gametophyte.
- In double fertilization, one sperm fuses with the egg to form a zygote, while the other fuses with polar nuclei within the female gametophyte. The fusion of sperm and polar nuclei produces endosperm—nutritive tissue that in most species is triploid.

38.4 Seeds and Fruits

- Seeds contain an embryo and a food supply surrounded by a coat.
- As an embryo develops, endosperm cells divide to form a nutrient-rich tissue. In addition, cells along the outside of the ovules form a protective seed coat, and the ovary develops into a fruit.
- In many cases, the mature fruit contains structures that help disperse the mature seed via wind, water, propulsion, or animals.
- Many seeds do not germinate immediately but instead experience a period of dormancy.
- A wide variety of conditions, ranging from scarification to exposure to red light, may break seed dormancy. In many cases, the event that triggers germination ensures that the seed germinates when environmental conditions are favorable.

38.5 Embryogenesis and Vegetative Development

- Vegetative development is the development and growth of all plant structures except flowers.
- Vegetative development begins with embryogenesis and occurs throughout the plant's life.
- Apical–basal and radial axes and apical meristems are established during early embryogenesis and retained in subsequent developmental stages.
- The development of an angiosperm embryo includes the formation of dermal tissue (epidermis), ground tissue, and vascular tissue layers and the development of the radicle, hypocotyl, and cotyledons.
- Vegetative development depends on meristems present at the tips of roots and shoots.
- The growth that occurs in vegetative development is finely tuned to the environment, allowing plants to adopt forms that maximize their survival and reproduction.
- Regulatory genes that control vegetative development are being discovered.

38.6 Reproductive Development

- Reproductive development is initiated when a shoot meristem switches from producing cells that will form shoots and leaves to producing cells that will form flowers, the plant's reproductive structure.
- The switch from vegetative to reproductive development in shoot meristems occurs in response to environmental conditions in many plants.
- Once a floral meristem is established, combinations of regulatory transcription factors encoded by *A*, *B*, and *C* classes of genes interact to specify the flower's sepals, petals, stamens, and carpels.

Answers are available in Appendix A.

✔ TEST YOUR KNOWLEDGE

1. Evaluate these statements about double fertilization. Select True or False for each sentence.
   T/F Two zygotes are formed, but only one survives.
   T/F One sperm fuses with an egg while the other sperm fuses with polar nuclei.
   T/F It results in the formation of a diploid embryo and triploid endosperm.
   T/F Both sperm involved in double fertilization form within the same pollen tube.
2. **THINK CAREFULLY** In angiosperms, are sperm and eggs produced by mitosis or meiosis? Which cells are spores? Which structures are gametophytes?
3. Which of the following does *not* occur during embryogenesis?
   a. formation of the radial axis
   b. production of the suspensor
   c. formation of the cotyledons and hypocotyl
   d. formation of the leaf lateral and proximal–distal axes
4. When does the apical–basal axis first become apparent?
   a. when the epidermal, ground, and vascular tissues form
   b. when the cotyledons, hypocotyl, and root form
   c. when the first cell division produces the apical cell and basal cell
   d. during the globular stage, when the suspensor is complete

✔ TEST YOUR UNDERSTANDING

5. Why is the emergence of the radicle an important first step in germination?
   a. Its hook helps protect the shoot that emerges later.
   b. It carries out photosynthesis to supply the embryo with food.
   c. It is important for establishing a supply of water to the growing embryo.
   d. It is necessary to break the seed coat.
6. What are the advantages and disadvantages of self-fertilization versus those of outcrossing?
7. In what sense are the tissues produced in the shoot and root apical meristems of a 300-year-old oak tree "embryonic"?

8. Human embryonic stem cells are capable of forming all types of cells in the human body. How are these cells similar to meristem cells in plants?

✔ TEST YOUR PROBLEM-SOLVING SKILLS

9. **PROCESS OF SCIENCE** Consider the following fruits: an acorn, a cherry, a burr, and a dandelion seed. Based on the structure of each of these fruits, predict how the seed is dispersed. Design a study that would estimate the average distance that each type of seed is dispersed from the parent plant.

10. **MODEL** Make a sketch of a simplified flower with all four organs, and indicate how the pattern of expression for just two genes (e.g., "*D*" and "*E*") could, hypothetically, regulate the development of each organ. Consider that what's important is whether a gene is on or off and that a structure can be specified when neither gene is expressed.

✔ PUT IT ALL TOGETHER: Case Study

How important are bees to pollination?

Most flowering plants can achieve pollination in several different ways. Those that produce pollen and carpels on the same plant may be self-pollinated, but they may also be cross-pollinated by insects or other pollinators. The cape gooseberry (*Physalis peruviana*) shown here is able to produce seed whether it is self- or cross-pollinated. Is one type of pollination better for reproductive success than the other?

11. Considering the gooseberry flower shown here, what types of cues might attract bees to the plant? What type of rewards do bees seek?

12. **QUANTITATIVE** Researchers conducted controlled experiments to test for effects of different types of pollination on cape gooseberry fruit size and seed production. The types of pollination included self-pollination, cross-pollination by bees, and manual cross-pollination (transferring pollen from one plant to another by hand). Results are shown in the graphs here. Compared to self-pollination, do bees have a significant impact on fruit mass? How about on seed formation? (* means $P < 0.05$; see **BioSkills 3**).

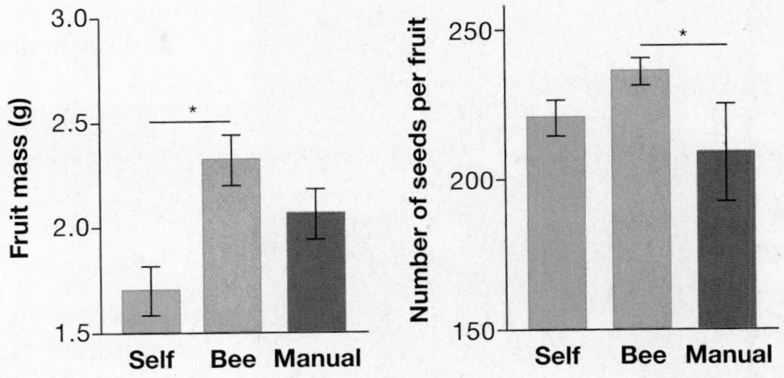

Source: A. Chautá-Mellizo, S. Campbell, M. Argenis Bonilla et al. 2012. *Basic and Applied Ecology* 13: 524–532.

13. Why might it be advantageous for plants to promote cross-pollination? What are the advantages of self-pollination?

14. **QUANTITATIVE** In general, larger fruits tend to earn higher profits for farmers who grow cape gooseberries as a crop. Compared to self-pollinated flowers, what is the approximate percentage of difference in fruit size of bee-pollinated flowers?

15. Based on the data in the graphs provided, approximately how many ovules are found in each gooseberry carpel?

16. **SOCIETY** Global bee populations have been declining over the past decade. What impact might this have on the world's human population?

END-OF-UNIT
CASE STUDY

UNIT 6

PUT IT ALL TOGETHER

▮ For an introduction to the Mystery of the Newt case study, see page 17.

Now that you've learned about plant form and function, let's consider how plants fit into the Mystery of the Newt case study. Recall the local legend about the rough-skinned newt and the three dead hunters. Tetrodotoxin (TTX), produced on the skin of the newt, was suspected of being responsible for the hunters' demise. Although TTX is toxic to humans and other animals, some garter snakes (*Thamnophis* sp.) are resistant to its effects.

✔ If you understand Unit 6, you should be able to apply your learning to this case study:

Can Plant Compounds Perform a Role Similar to Tetrodotoxin?

Unlike newts and other animals, plants do not have the ability to escape from potential predators. But that doesn't mean they are totally defenseless. Just like the rough-skinned newt, many plants produce various compounds—nicotine, caffeine, cocaine, and cyanide to name just a few—that are toxic to herbivores and appear to play a role in self-defense.

1. Nicotine, the addictive compound found in cigarettes and other tobacco products, interacts with the nervous system of animals and is considered a plant secondary metabolite. Which of the following best describes secondary metabolites? (See Section 37.7)

a. They are not directly required for plant growth, development, or reproduction, but may play a role in a plant's long-term survivability or fecundity.

b. They are produced in the secondary growth of a plant, such as the wood and bark formed by lateral meristems.

c. They represent metabolic waste products formed by the breakdown of primary metabolites.

d. They are synthesized in secondary cell walls, but not in primary cell walls.

Various herbivores, such as the beet armyworm (*Spodoptera exigua*) caterpillar, are able to use tobacco leaves as a source of food. However, studies have shown that nicotine causes reduced growth rates and increased mortality in *S. exigua*. Does nicotine help to minimize herbivory by insect pests?

Figure 1 illustrates results of a study that examined the role of nicotine in plant defense. Wild-type tobacco and tobacco that was genetically modified to be nicotine free were transplanted to a field plot in a native habitat with beet armyworms present. The amount of leaf area damage was recorded for a period of 16 days.

Figure 1

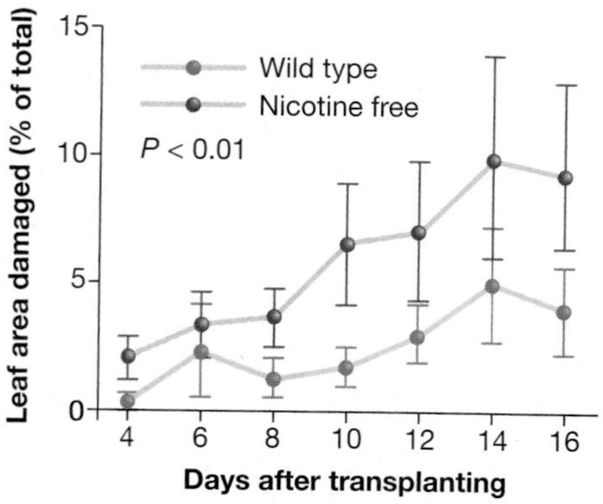

DATA: A. Steppuhn et al. 2004. *PLoS Biology* 2(8): e217, Figure 4.

2. QUANTITATIVE Do the experimental results support the hypothesis that nicotine plays a role in plant defense? Explain.

Although tobacco plants produce a baseline amount of nicotine on a normal basis, nicotine synthesis has also been shown to be an inducible defense.

3. Why might it be advantageous for tobacco to synthesize high levels of nicotine only when leaves are eaten by insects instead of producing high levels all of the time? (See Section 37.7)

How does leaf damage induce nicotine synthesis? Leaves damaged by herbivores produce jasmonic acid (see Figure 37.27), which is transported to roots and serves as a signal to increase nicotine production. Nicotine is synthesized in the roots and then transported to and stored in the leaves.

4. MODEL Sketch a generalized plant, including a simplified version of its vascular tissue. Based on your knowledge of plant vascular tissue, indicate: **(1)** the pathway by which jasmonic acid is likely transported from damaged leaves to roots, and **(2)** the pathway by which nicotine is transported from roots to leaves.

Interestingly, some insects are resistant to the toxic effects of nicotine (much like some garter snakes are resistant to TTX). A classic example is the tobacco hornworm, *Manduca sexta* (Figure 2), which can tolerate doses of nicotine that are fatal to other herbivores. Given their voracious appetites, hornworms can be very destructive to tobacco farms.

5. PROCESS OF SCIENCE Recall the experiment that shows leaf damage by armyworms on wild-type and nicotine-free tobacco (Figure 1). What do you predict a graph of the results might look like if tobacco hornworms were used as the herbivore in that experiment? Sketch your prediction using the blank graph below Figure 2 as a starting point.

Figure 2 Tobacco hornworm (*Manduca sexta*) caterpillar on damaged tobacco leaf.

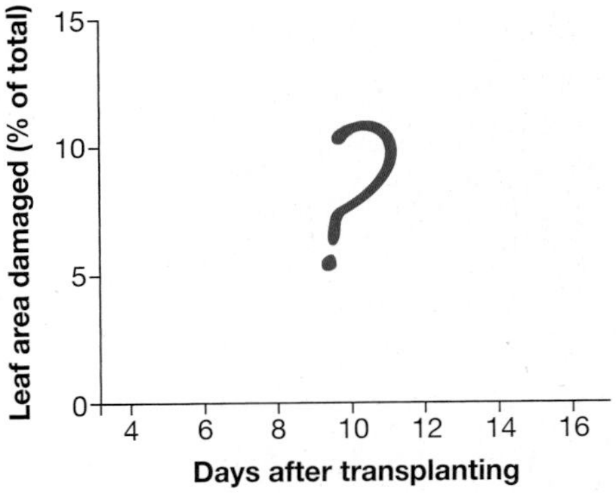

Figure 3 Big-eyed bugs (*Geocoris* sp., inset) feed on hornworm eggs and larvae.

The relationship between tobacco plants, hornworms, and their potential predators is quite complex. It turns out that hornworms use the nicotine from tobacco leaves as a defense against predators such as wolf spiders. However, tobacco plants release volatile compounds in response to being eaten that, when mixed with hornworm saliva, attract predators such as big-eyed bugs (*Geocoris*) that eat the eggs and larvae of tobacco hornworms (Figure 3).

6. Explain how compounds in tobacco plants are used as direct and indirect defenses.

7. Does nicotine in tobacco play a similar role to newt TTX? Explain.

The roles of plant-based secondary metabolites continue to be a major focus of research. Much like the newts and garter snakes, plants, herbivores, and their predators are in an "evolutionary arms race." The physiological response to toxins will be explored in Unit 7.

Each unit ends with a continuation of this story.

BIG PICTURE

THE BIG PICTURE

Plants and animals are diverse lineages of multicellular eukaryotes. They are different in important ways. Each lineage evolved independently from a different single-celled protist—plants with the ability to make their own food by photosynthesis, and animals reliant on obtaining energy from other organisms. Furthermore, plants are sessile, while most animals are capable of complex movements and locomotion.

Yet despite these differences, plants and animals face many of the same challenges to survive and reproduce in water and on land. Use this concept map to explore some of their similarities and differences in form and function.

Note that most boxes in the concept map indicate the chapters where you can go for more information. Also, be sure to do the blue exercises in the Check Your Understanding box below.

Plant and Animal FORM AND FUNCTION

are the product of

Evolutionary processes 22, 23

including

occurring in

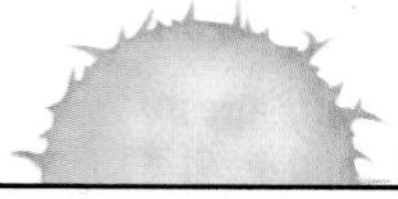

Abiotic environment
- Temperature
- Light
- Water availability
- pH, salinity, dissolved gases, nutrients
- Habitat structure

49

Biotic environment
- Parental care
- Competitors
- Predators, prey
- Parasites, hosts
- Mutualists

51, 52

create

Changing ecological contexts Unit 9

because plants and animals have

Key FUNCTIONS to survive and reproduce

correlate with

including

REGULATION OF WATER AND IONS

NUTRITION

PLANTS

- Water moves along a water potential gradient
- Turgor pressure provides structural support
- Terrestrial plants lose water evaporatively by transpiration, regulated by stomata and waxy surfaces

35

- Autotrophic; make their own food by photosynthesis

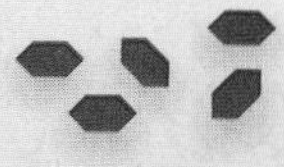

- Take in sunlight, CO_2, and minerals
- Obtain ions from soil, symbiotic fungi, or bacteria

36

ANIMALS

- Osmotic stress varies in marine, freshwater, and terrestrial habitats
- Urinary system maintains homeostasis of water and electrolytes while managing excretion of nitrogenous wastes
- Terrestrial animals limit evaporative loss

39

- Heterotrophic; must eat food to acquire energy
- Take in carbohydrates, fats, proteins, vitamins, and minerals
- Obtain nutrients by ingestion, digestion, and absorption

40

CHECK YOUR UNDERSTANDING

✔ If you understand the big picture, you should be able to . . .

1. Propose one mechanism that both plants and animals use on land to limit the evaporative loss of water.
2. Describe one difference between plants and animals regarding the haploid stage of their life cycles.
3. Give an example of an adaptation that enables large plants and animals to survive the physical constraints imposed by large body size.
4. Explain where cellular respiration fits into this map.

Answers are available in Appendix A.

Natural selection 22, 23 — produces →

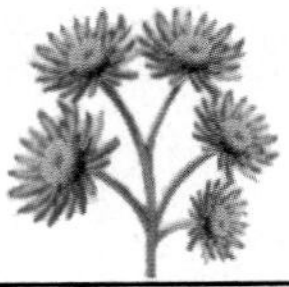

Adaptations
- Anatomical
- Physiological
- Behavioral

22, 34, 39

limited by →

Constraints
- Genetic
- Developmental
- Historical
- Mechanical 22

including →

Body size
- Length, mass, weight
- Surface area/volume ratio

34, 39

Fitness trade-offs 39

require ability to maintain → **Homeostasis** 39

influence → **Phenotypic plasticity** 34

changes → **Key STRUCTURES to survive and reproduce**

can be studied at different →

Levels of organization
- Populations
- Organisms
- Organ systems
- Organs
- Tissues
- Cells
- Molecules

1, 34, 39

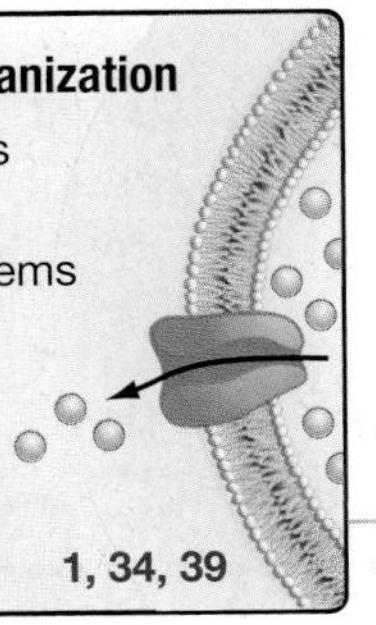

including →

Plant tissues
- Dermal
- Ground
- Vascular
- Meristem

34

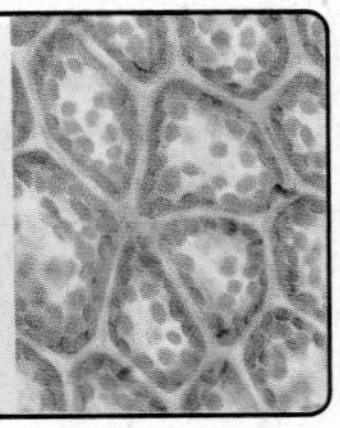

Animal tissues
- Nervous
- Muscle
- Epithelial
- Connective

39

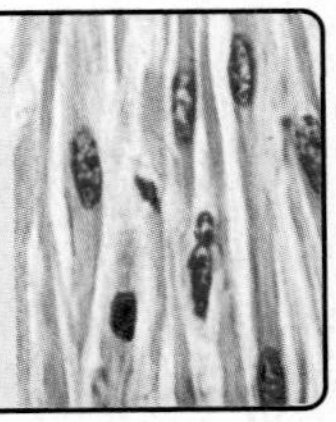

INTERNAL TRANSPORT

- Move water by bulk flow in xylem tissue, driven by transpiration

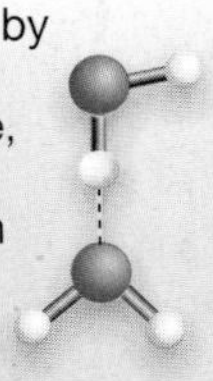

- Move sugars in phloem tissue along pressure gradient from sources to sinks

35

- Most move air or water through tracheae, lungs, or gills
- Circulatory systems may be open or closed
- In most, pumping heart moves blood through the body

42

SENSORY SYSTEMS, SIGNALS, & RESPONSES

- Detect light, gravity, heat, touch, wind, pathogens, herbivores, chemical signals
- Hormones (plant growth regulators) carry information and induce responses

37

- Detect light, gravity, heat, touch, sound, pathogens, predators, chemical signals
- Nervous system enables rapid response
- Endocrine system enables slower, longer-term responses via hormones

43, 44, 46

MOVEMENT

- Incapable of locomotion, but gametes and seeds can disperse
- Some can move slowly via differential growth or rapidly via cell-shape changes

37, 38

- Most are capable of movement using muscles

- Muscles work in combination with skeletal systems: hydrostatic skeletons, exoskeletons, and endoskeletons

45

REPRODUCTION

- Most can reproduce sexually, some asexually
- Undergo alternation of generations (haploid and diploid stages are both multicellular)
- Meristematic cells give rise to new structures throughout a plant's life

38

- Most can reproduce sexually, some asexually
- Multicellular stage is usually diploid; gametes are haploid
- Cells specialize early and contribute to embryonic germ layers that form all adult structures

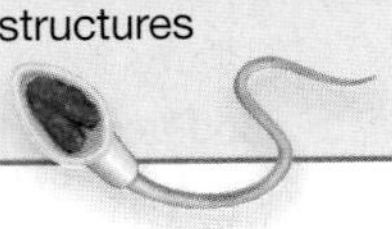

47

DEFENSE

- Have both physical and chemical defenses against pathogens and herbivores
- Infection launches a hypersensitive response

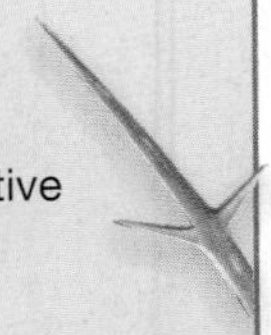

37

- Have both physical and chemical defenses against pathogens and predators
- Infection launches: (1) generic, innate immune response, and sometimes also (2) specific, adaptive immune response

48

39 Animal Form and Function

African elephants are adapted to living in hot environments. Their large ears facilitate heat loss to their surroundings.

In this chapter you will learn that

Animal structure is fundamentally related to function

starting with

- Adaptations and fitness trade-offs 39.1

then asking

- How does structure correlate with function? 39.2
  - exploring: Tissues, organs, and organ systems
- Why is body size important? 39.3
- How do animals regulate their internal environment (homeostasis)? 39.4
  - looking closer at: Heat exchange and thermoregulation 39.5

Elephants are symbolic of Africa. They inhabit a wide range of habitats there, from dense forests to deserts. But one thing that all these habitats have in common is high temperature. African elephants thrive in areas where the temperature can exceed 50°C (122°F).

Small animals avoid the midday heat by retreating to a cool burrow deep underground or the shade of a small shrub. But elephants and other large mammals have a harder time hiding from sunlight, so their survival depends on other traits that prevent them from overheating. For example, elephants regularly roll in muddy water to cool off. In addition, they lack a thick coat of insulating fur, which would retain body heat.

Elephants have another weapon in their arsenal against extreme heat—large ears. The ears are loaded with blood vessels that can carry warm blood to the surface, where body heat can be transferred to the

BIG PICTURE

This chapter is part of the Big Picture. See how on pages 838–839.

environment. Elephants regularly wiggle their ears to help promote heat transfer. As you might predict, African elephants have ears that are two to three times as large as those of closely related Asian elephants, which generally inhabit cooler environments.

In elephants, regulating heat exchange with the environment thus involves both structure (sparse fur and large ears) and function (shunting blood to the surface of the ears).

Anatomy refers to an organism's physical structure, or form. **Physiology** is how the physical structures in an organism function. You can review the importance of form and function in plants and animals in the Big Picture on pages 838–839. In this unit, starting with Chapter 39, you will focus on the relationship between anatomy and physiology in animals.

39.1 Form, Function, and Adaptation

Biologists who study animal anatomy and physiology are studying **adaptations**—heritable traits that make individuals more likely to survive and reproduce in a certain environment than individuals that lack those traits (Ch. 22, Section 22.3).

> **After you complete this section, you should be able to...**
>
> - Describe the relationship between the form and function of animal tissues, organs, and systems.

Recall that adaptation results from evolution by natural selection. Natural selection, in turn, occurs whenever individuals with certain alleles leave more offspring that survive to reproductive age than do individuals with different alleles. Because of this difference in reproductive success, the frequency of the selected alleles increases from one generation to the next.

For example, African elephants with alleles for larger ears have a better chance of surviving to produce more offspring than do elephants with alleles for smaller ears. The ability to use their ears to regulate body temperature is an adaptation that helps these elephants thrive in hot environments.

The Role of Fitness Trade-Offs

Adaptations increase fitness—the ability to produce viable offspring. But no adaptation is "perfect." Instead, adaptations are limited by which alleles are present in a population and by the nature of the traits that already exist—because all adaptations derive from preexisting traits.

The human spine, for example, is a highly modified form of the vertebral column in ancestors that walked on all four limbs (Ch. 32, Section 32.3). The modifications in the human spine can be considered adaptations that support our upright posture, but they are far from perfect—most adults experience back pain at some point during their lives. Evolution of the human spine has been constrained by the nature of the ancestral trait and by a lack of alleles that would improve its structure and function.

The most important constraint on adaptation, though, may be **trade-offs**—inescapable compromises between traits. For example, it takes a lot of energy to produce offspring through the process of reproduction and also to mount an immune response during an infection. Animals sometimes do not have enough energy to satisfy both needs. In these cases, a trade-off emerges: The animal may devote more energy to reproduction at the expense of strong immune function, or vice versa, or both traits might be negatively affected.

How do biologists study trade-offs in animal physiology? Let's consider experimental work on trade-offs in crickets. During mating, a male cricket produces a **spermatophore**, a packet of sperm surrounded by a large, gelatinous mass. The male deposits the spermatophore on the female's genital opening (Figure 39.1a). After mating, the female begins to eat the gelatinous mass (Figure 39.1b). The sperm packet remains behind, and the sperm slowly begin to enter her reproductive tract.

The longer it takes for the female to eat the mass, the more sperm are transferred to her, increasing the number of eggs fertilized by the male. Therefore, it might seem advantageous for the male to make as large a spermatophore as possible. However, males expend a great deal of energy in making spermatophores. In male crickets, is there a trade-off between producing

(a) Decorated crickets mating

(b) The female eats the gelatinous mass.

Figure 39.1 **Mating in Crickets Involves a Trade-off with Immune Function.** The spermatophore contains a gelatinous mass and a sperm packet. The female eats the gelatinous mass as the sperm from the packet enter her reproductive tract to fertilize her eggs.

RESEARCH

QUESTION: Is there a trade-off between reproductive and immune function in male crickets?

HYPOTHESIS: Male crickets need to make an energy trade-off between reproductive function and immune function.

NULL HYPOTHESIS: No energy trade-off between reproductive function and immune function is required.

EXPERIMENTAL SETUP:

1. Remove spermatophores from male crickets for 5 days:

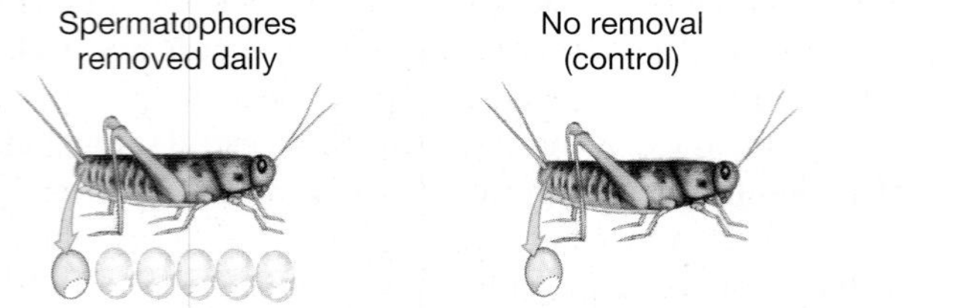

2. Draw hemolymph samples from both sets of crickets.

3. Add bacteria to samples; measure lysis of bacteria.

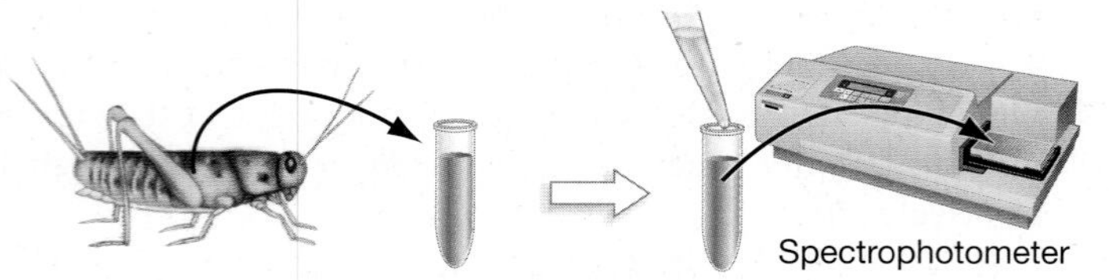

PREDICTION OF HYPOTHESIS: Hemolymph from males forced to produce more spermatophores will exhibit lower lytic activity than controls.

PREDICTION OF NULL HYPOTHESIS: There will be no difference in lytic activity between treated males and control males.

RESULTS:

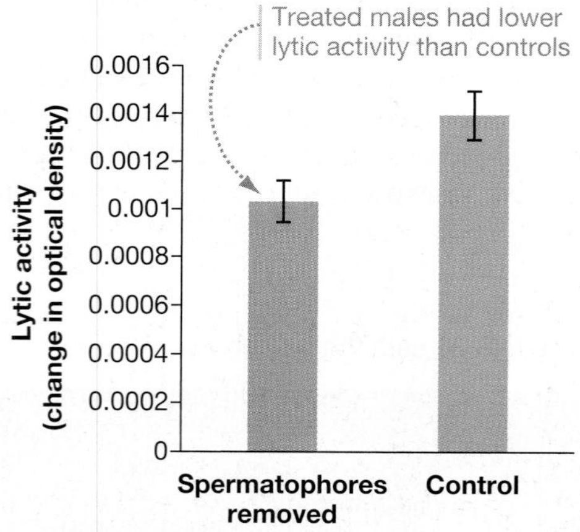

EXPERIMENTAL SETUP:

1. Inject male crickets:

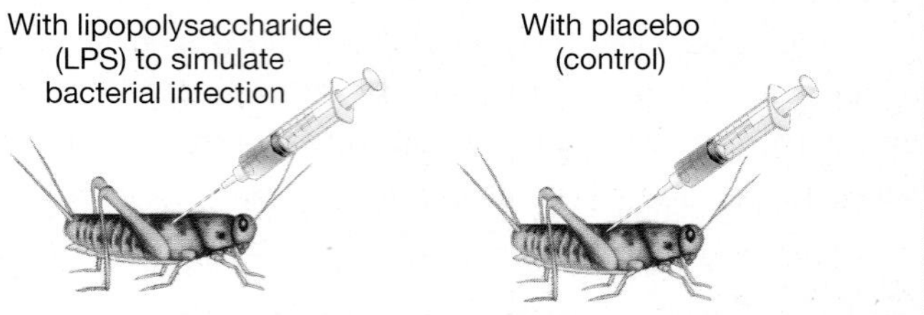

2. Remove spermatophores and measure size of gelatinous mass.

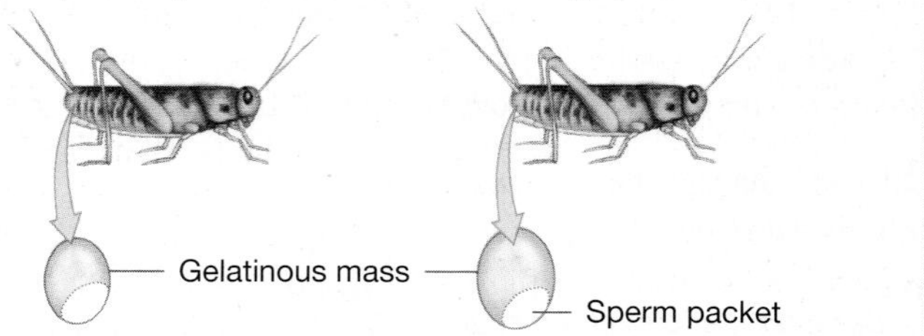

PREDICTION OF HYPOTHESIS: Spermatophores from LPS-injected males will have smaller gelatinous masses than those from control males.

PREDICTION OF NULL HYPOTHESIS: There will be no difference in gelatinous mass size between treated males and control males.

RESULTS:

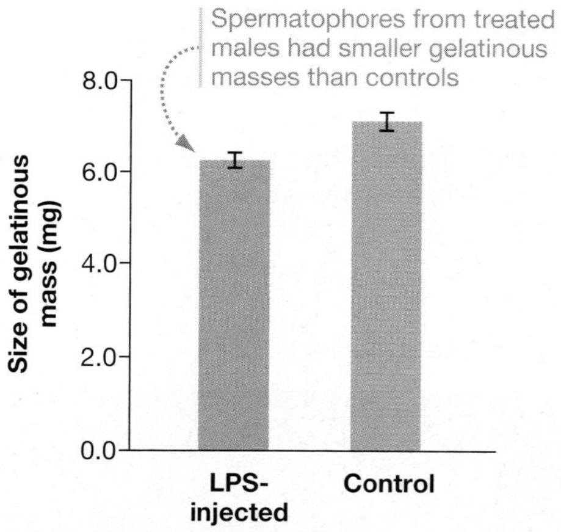

CONCLUSION: Male crickets must make an energy trade-off to support increased reproductive effort.

Figure 39.2 **Trade-offs between Reproduction and Immune Function in Male Crickets.** This study showed that males with experimentally increased investment in reproduction had less ability to kill bacteria than control males, and that males with experimentally stimulated immune function had lower reproductive function than control males.

SOURCE: A. M. Kerr, S. N. Gershman, and S. K. Sakaluk. 2010. Experimentally induced spermatophore production and immune responses reveal a trade-off in crickets. *Behavioral Ecology* 21: 647–654.

✔ **PROCESS OF SCIENCE** If the null hypothesis were supported by the results, what would each graph look like?

spermatophores and other costly processes, such as mounting an immune response against a pathogen?

To answer this question, biologists carried out an experiment using a powerful method—a reciprocal design. They altered the energy investment crickets made into each trait—reproduction and immune function—and observed the effect on the other trait (Figure 39.2).

First, the biologists removed spermatophores from male crickets daily, causing the crickets to spend more energy to make new ones. The biologists then took a sample of the crickets' hemolymph (a circulating fluid similar to blood) and added bacteria to find out whether the hemolymph could lyse, or rupture, the bacteria. As bacteria are lysed, the opaque hemolymph becomes clear. Lytic activity was measured using a spectrophotometer (BioSkills 8), which displayed the change in opacity, or optical density, of the hemolymph over time.

Compared to control crickets, which did not have their spermatophores removed, the experimental crickets had lower lytic activity (fewer bacteria lysed). This suggested that increased investment into making spermatophores was traded off against investment into immune function.

Next, the biologists injected male crickets with lipopolysaccharide (LPS), a component of the cell walls of bacteria, which caused the crickets to mount an immune response. The effect on spermatophore size was dramatic: Injected crickets produced smaller spermatophores than the control crickets.

When the researchers looked more closely, they realized that the size of the sperm packet remained largely unchanged, but the size of the gelatinous mass decreased in injected crickets. Smaller gelatinous masses are eaten more quickly by females, allowing less time for sperm transfer and decreasing male crickets' reproductive success. Taken together, these two sets of reciprocal experiments provide very strong evidence that there is a trade-off between reproductive and immune function in crickets.

Trade-offs are common in nature. Desert animals that sweat to cool off are threatened with dehydration. An eagle's beak is superbly adapted for tearing meat but not for weaving nesting materials together. In studying animal form and function, biologists study compromise and constraint as well as adaptation.

Adaptation and Acclimatization

In biology, adaptation refers to a genetic change in a population in response to natural selection exerted by the environment. Phenotypic change in an individual in response to environmental fluctuations is referred to as **acclimatization**. Acclimatization is reversible. It is similar to acclimation, which refers to changes that occur in an organism in a laboratory setting.

If you moved to Tibet, your body would acclimatize to the high elevation by making more of the oxygen-carrying pigment hemoglobin and more hemoglobin-carrying red blood cells. But human populations that have lived at high elevations in Tibet for many generations have adapted to this environment through genetic changes. At least five genes have been identified that help Tibetans live at high elevations, including some coding for proteins that assist hemoglobin in binding oxygen.

The ability to acclimatize is itself an adaptation. Light-skinned humans, for example, vary in the ability to tan in response to sunlight. Some individuals tan easily—they have alleles that allow them to acclimatize efficiently to environments with intense sunlight—while others do not. In this and many other cases, the ability to acclimatize is a genetically variable trait that can respond to natural selection.

CHECK YOUR UNDERSTANDING

✓ If you understood this section, you should be able to . . .

1. Use the terms "adaptation," "fitness," "natural selection," and "trade-off" to explain the results of the experiment on cricket reproduction and immune function.
2. Explain why an increase in hemoglobin concentration in runners training at high elevations would constitute acclimatization, not adaptation.

Answers are available in Appendix A.

39.2 Tissues, Organs, and Organ Systems: How Does Structure Correlate with Function?

After you complete this section, you should be able to . . .

- Analyze the relationship between the structure and function of tissues found in animals.

If a structure found in an animal is adaptive—meaning that it helps the individual survive and produce viable offspring—it is common to observe that the structure's size, shape, or composition correlates closely with its function. This applies to all structures, from cells to tissues to organs and organ systems.

For example, recall that biologists have documented extensive changes in beak size and shape in the medium ground finch (*Geospiza fortis*) on the Galápagos Islands (Ch. 22, Section 22.4). Such changes are due to natural selection. Individuals with deep beaks are better able to crack the large seeds that predominate during drought years, while individuals with small beaks are better able to harvest the small seeds that predominate during wet years.

As Figure 39.3 on page 844 shows, a strong correlation between diet and beak structure is also found among different species of Galápagos finches. Species with small, cone-shaped beaks eat small seeds; species with large, cone-shaped beaks eat large seeds; and species with long, tweezer-like beaks pick insects off tree trunks or other surfaces.

The mechanism responsible for these structure–function correlations is straightforward: If a mutant allele alters the size or shape of a structure in a way that makes it function more efficiently, individuals with that allele will produce more offspring than will other individuals. As a result, the allele will increase in frequency in the population over time.

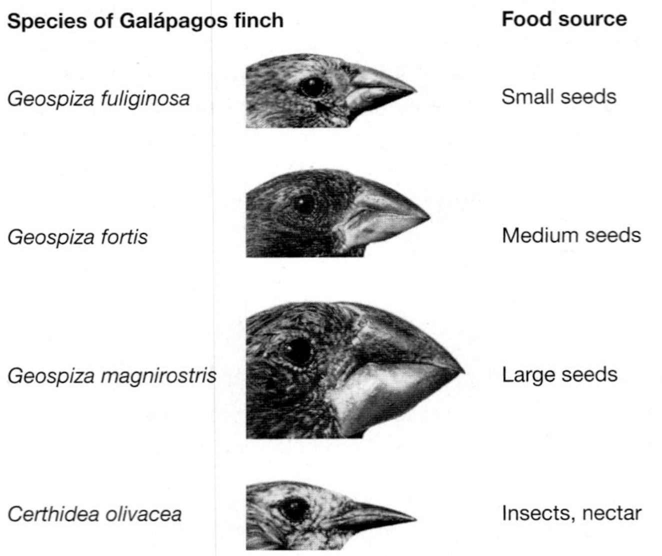

Figure 39.3 **In Animal Anatomy and Physiology, Form Often Correlates with Function.**

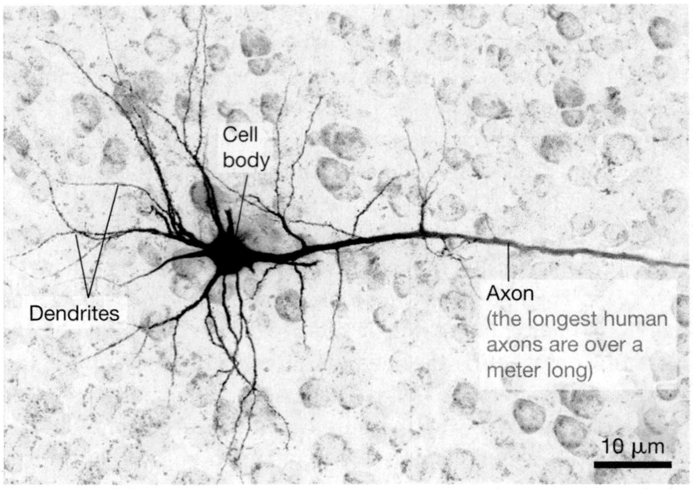

Figure 39.4 **In Nervous Tissue, Neurons Transmit Electrical Signals.** In a neuron, information is transmitted from dendrites to the cell body to the axon.

✔ Given a neuron's structure, is a neuron more likely to provide signals to specific cells and tissues or to broadcast signals widely throughout the body?

Structure–Function Relationships at the Molecular and Cellular Levels

Correlations between form and function start at the molecular level. For example, earlier chapters emphasized that the shape of proteins correlates with their role as enzymes, structural components of the cell, or transporters. The membrane proteins called channels form pores that allow specific ions or molecules to pass in or out of cells (Ch. 6, Section 6.4). The ends and interior of a channel are hydrophilic, which allows the protein to interact with the surrounding solution or the interior of the cell, while the perimeter is hydrophobic—allowing the protein to interact with the lipid bilayer. The protein's structure fits its function.

Similar correlations between structure and function occur at the level of the cell. In fact, a cell's specialized function can be predicted by examining its internal structure. Cells that manufacture and secrete large amounts of protein are packed with rough endoplasmic reticula (ER) and Golgi apparatuses; cells that store energy are dominated by large fat droplets; and cells that ingest and destroy invading bacteria have many lysosomes.

The overall shape of a cell can also correlate with its function. For example, cells that are responsible for transporting materials into or out of the body often have extremely large areas of plasma membrane. As a result, they have room to accommodate the thousands of membrane channels, carrier proteins, and pumps required for extensive transport.

Tissues Are Groups of Cells That Function as a Unit

Animals are multicellular—their bodies contain distinct types of cells that are specialized for different functions. Frequently, groups of animal cells work together to perform the same function. A **tissue** is a group of cells that function as a unit.

Most adult animals have four tissue types: **(1)** nervous tissue, **(2)** muscle tissue, **(3)** epithelial tissue, and **(4)** connective tissue. In each case, the structure of the tissue correlates closely with its function. Let's consider each type in turn.

Nervous Tissue **Nervous tissue** consists of nerve cells, which are also called **neurons**, and several types of supporting cells. Neurons transmit electrical signals, which are produced by changes in the permeability of the cell's plasma membrane to ions (Ch. 43, Section 43.1). Supporting cells have many functions, including regulating ion concentrations in the space surrounding neurons, supplying neurons with nutrients, or serving as scaffolding or support for neurons.

Although they vary widely in shape, all neurons have projections that approach other cells. As Figure 39.4 shows, most neurons have two distinct types of projections from the cell body, where the nucleus is located: **(1)** highly branched, relatively short processes called **dendrites**, and **(2)** a relatively long process called an **axon**. Dendrites facilitate transmission of signals from adjacent cells to the neuronal cell body; the axon carries electrical signals from the cell body to other cells.

Muscle Tissue **Muscle tissue** was a key innovation in the evolution of animals—like nervous tissue, it appears in no other lineage on the tree of life (Ch. 30, Section 30.2). Some of the functions of muscle include movement of the body, pumping of the heart, and mixing of food in the gastrointestinal tract. There are three types of muscle tissue (Figure 39.5); you, along with other vertebrates, have all three.

1. **Skeletal muscle** attaches to bones and exerts a force on them when it contracts. Skeletal muscle is responsible for most body movements. It has long cells with a *striated*, or striped, appearance produced by an overlapping arrangement of proteins.

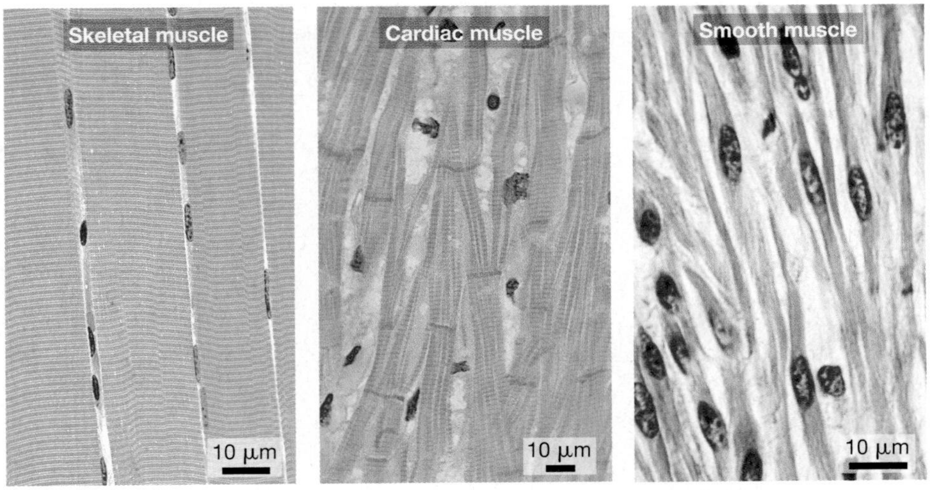

Long cells,
voluntary movement

Branched cells,
involuntary movement

Tapered cells,
involuntary movement

Figure 39.5 **Muscle Tissue Is Composed of Cells That Contract.** The three types of muscle tissue have distinctive structures and functions.

2. **Cardiac muscle** makes up the walls of the heart and is responsible for pumping blood throughout the body. The branching pattern of cardiac muscle allows electrical signals to spread throughout all cells of the heart, resulting in their coordinated contraction and relaxation.
3. **Smooth muscle** cells, which are tapered at each end, form a muscle tissue that lines the walls of the digestive tract and the blood vessels. Contraction and relaxation of smooth muscle help move food through the digestive tract and regulate blood pressure. (Muscle tissue and movement are explored in detail in Chapter 45.)

Epithelial Tissues **Epithelial tissues** are also called **epithelia** (singular: **epithelium**). Epithelia cover the outside of the body, line the inner surfaces of many organs, and form glands. An **organ** is a structure that serves a specialized function and consists of two or more tissues; a **gland** is an organ that secretes specific molecules or solutions such as hormones or digestive enzymes.

Perhaps no other tissues better exemplify the relationship between structure and function than epithelia. Epithelia can be divided into two major types: *simple epithelium* and *stratified epithelium* (Figure 39.6). Simple epithelia are only a single cell layer thick, allowing gases, water, nutrients, and other substances to move across them easily. Stratified epithelia, in contrast, consist of layers of closely packed cells, protecting body surfaces from the environment. Adjacent epithelial cells are joined by structures that hold them tightly together, such as tight junctions and desmosomes (introduced in Ch. 11, Section 11.2).

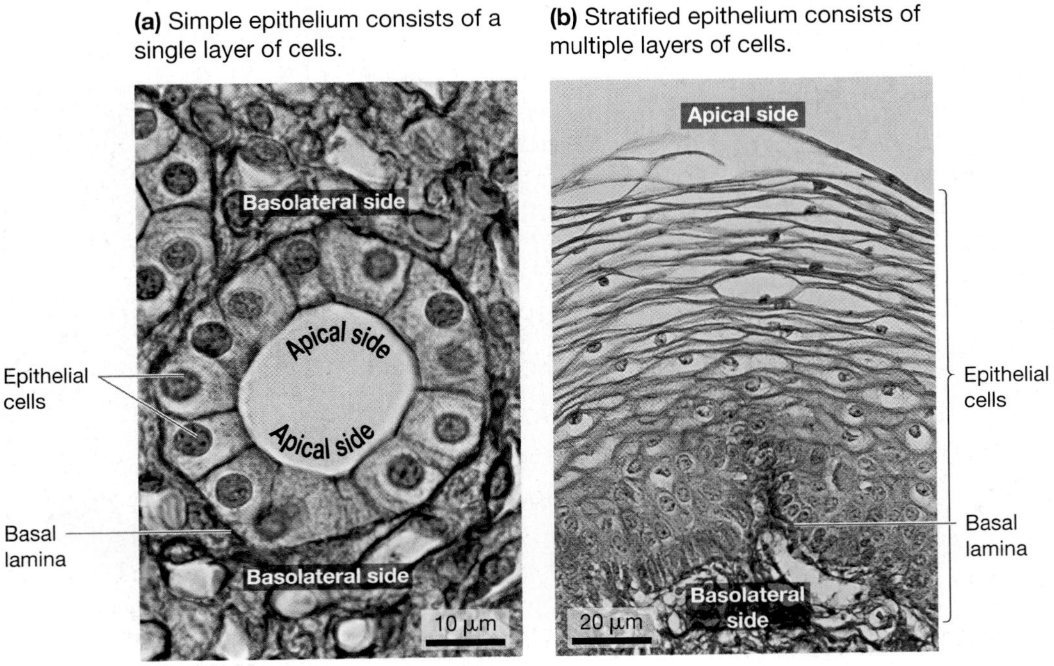

Figure 39.6 **In Epithelia, Epithelial Cells Regulate Which Materials Pass across Body Surfaces and Provide Protection.** **(a)** Simple epithelium is found in organs such as intestine and structures like kidney ducts (shown here), allowing easy transport of substances across organ boundaries. **(b)** Stratified epithelium is found in organs such as the vagina (shown here) and the skin, which are subjected to a lot of wear and tear.

Epithelial tissue has polarity, or sidedness. An epithelium has an **apical** side, which faces away from other tissues and toward the environment, and a **basolateral** side, which faces the interior of the animal and is attached to connective tissues. This connection is made by a layer of fibers called the **basal lamina**.

The apical and basolateral sides of an epithelium have distinct structures and functions. Epithelial cells, for example, line the surface of your trachea, or windpipe. The apical side of these cells secretes mucus and is covered with cilia that help sweep away dust, bacteria, and viruses. The basolateral side lacks these features and is cemented to the basal lamina.

Epithelial cells have short life spans. The cells that line your esophagus—the tube connecting your mouth and stomach—live for 2 to 3 days, while the cells that line your large intestine live for a maximum of 6 days. Muscle cells and neurons, in contrast, can live as long as you do. Epithelial cells are short lived because they are exposed to harsh environments, where they are likely to be killed or scraped away.

Epithelial tissue as a whole does not wear away, however, because it includes cells that actively undergo mitosis and cytokinesis—producing new epithelial cells to replace those lost on the apical side.

✔ If you understand the functions of epithelia, you should be able to predict which of the following is more likely to contain stratified epithelium: the inside surface of the mouth or the gas exchange surface of the lungs.

Connective Tissues **Connective tissues** consist of cells that are loosely arranged in a liquid, jellylike, or solid matrix. The matrix is composed of extracellular fibers and other materials, and it is secreted by the connective tissue cells themselves. The extracellular fibers are of three types: **(1)** *collagen*, which provides strength and structural integrity; **(2)** *reticular fibers*, which are arranged like a fine mesh net; and **(3)** *elastin*, a stretchy protein that allows the connective tissue to stretch and then easily return to its original shape.

There are four main types of connective tissue, and each secretes a distinct type of extracellular matrix (**Table 39.1**). The nature of the matrix determines the nature of the connective tissue.

1. **Loose connective tissue** contains an array of fibers in a soft matrix; it serves as a packing material holding organs and tissues together and as padding under the skin. Cells called **fibroblasts** make the extracellular matrix in loose connective tissue.
2. **Dense connective tissue** is found in the tendons and ligaments that connect muscles, bones, and organs. As Table 39.1 shows, the matrix in tendons and ligaments is dominated by collagen fibers.
3. **Supporting connective tissue** has a firm extracellular matrix. **Bone** and **cartilage** are connective tissues that provide structural support for the vertebrate body as well as protective enclosures for the brain and other components of the nervous system. The hard matrix of bone consists mainly of collagen and calcium phosphate, whereas cartilage is more flexible and is dominated by collagen and elastin. The

SUMMARY Table 39.1 Connective Tissues

Type	Example
Loose *Soft* extracellular matrix; holds tissue together loosely	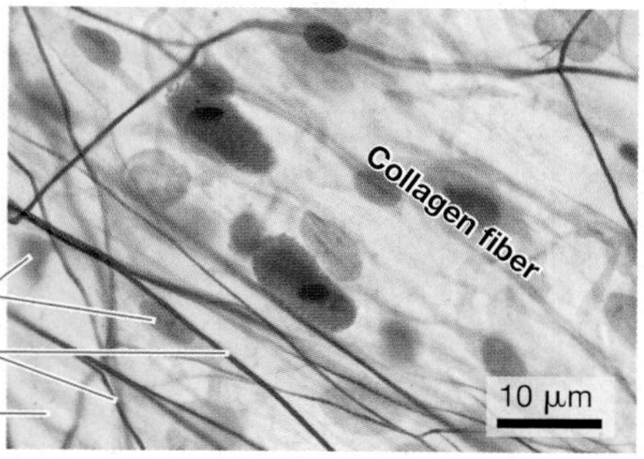
Dense *Fibrous* extracellular matrix; holds tissue together tightly	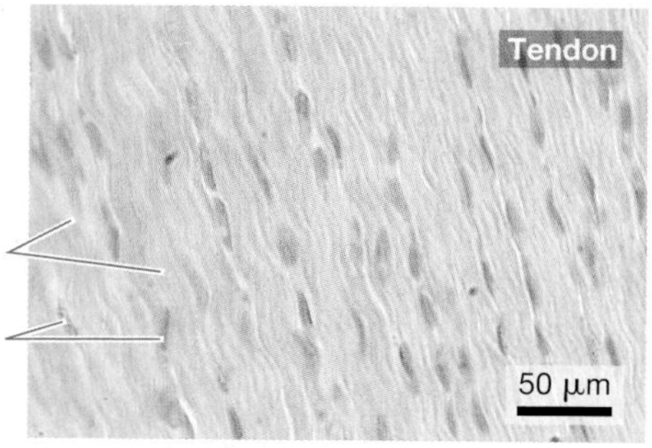
Supporting *Firm* extracellular matrix; functions in structural support and protection	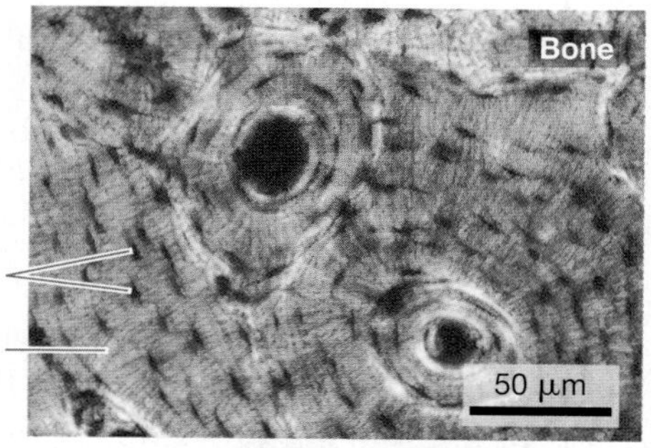 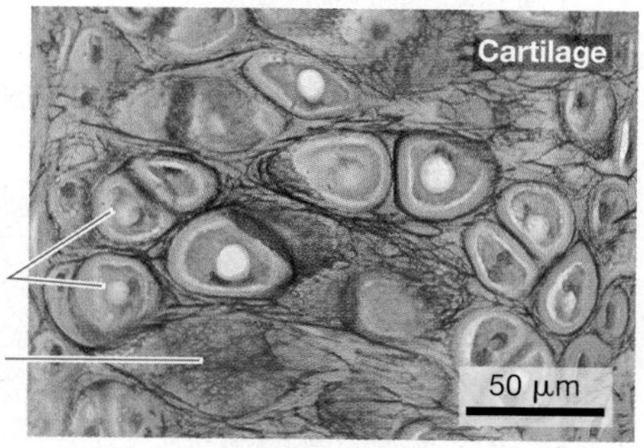
Fluid *Liquid* extracellular matrix; functions in transport	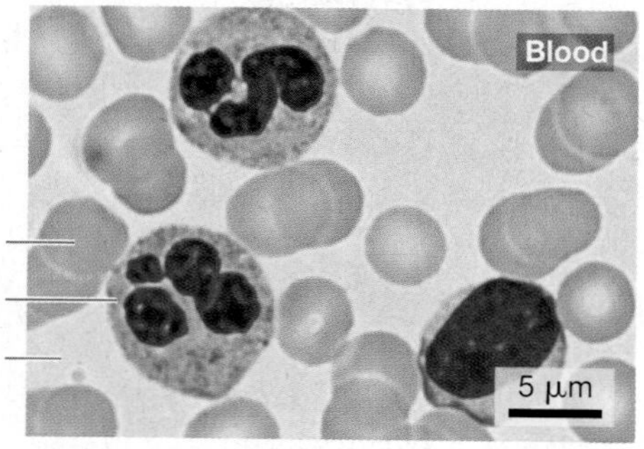

difference in flexibility between bone and cartilage reflects their functions: for example, bones are rigid to protect organs, and cartilage is flexible to provide resilient padding in joints.

4. **Fluid connective tissue** consists of cells surrounded by a liquid extracellular matrix. **Blood**, which transports materials throughout the vertebrate body, contains a variety of cell types and has a specialized extracellular matrix called plasma (Ch. 42, Section 42.4).

Organs and Organ Systems

Cells with similar functions are organized into tissues, and tissues are organized into specialized structures called organs. Recall that an organ is a structure that serves a specialized function and consists of at least two types of tissues. The small intestine, for example, consists of muscle, nervous, connective, and epithelial tissues (Figure 39.7a).

An **organ system** consists of groups of tissues and organs that work together to perform one or more functions. Using the digestive system as an example, Figure 39.7b illustrates how the structure of organs correlates with their function and how the components of an organ system work together in an integrated fashion.

Because an animal's body contains atoms, molecules, cells, tissues, organs, and organ systems, biologists who study animal anatomy and physiology must work at various levels of organization to understand how that body operates.

Figure 39.8 on page 848 illustrates these levels of organization, using the human nervous system as an example. Because the structure and function of each component in the body are integrated with those of other components, and because each level of organization is integrated with other levels of organization, the organism as a whole is greater than the sum of its parts. In other words, an organism is more than just a collection of individual organ systems, and each system is more than just a collection of individual cells, tissues, or even organs.

Each subsequent chapter in this unit focuses on a different organ system found in animals, beginning with the excretory system and ending with the immune system. Each of these systems can be interpreted as a suite of adaptations and trade-offs. Each system accomplishes a specific task required for survival and reproduction, and each works in conjunction with other systems.

Before delving into the various organ systems, however, it's essential to examine general phenomena that affect all systems in animals. Let's start by looking at how body size affects animal physiology.

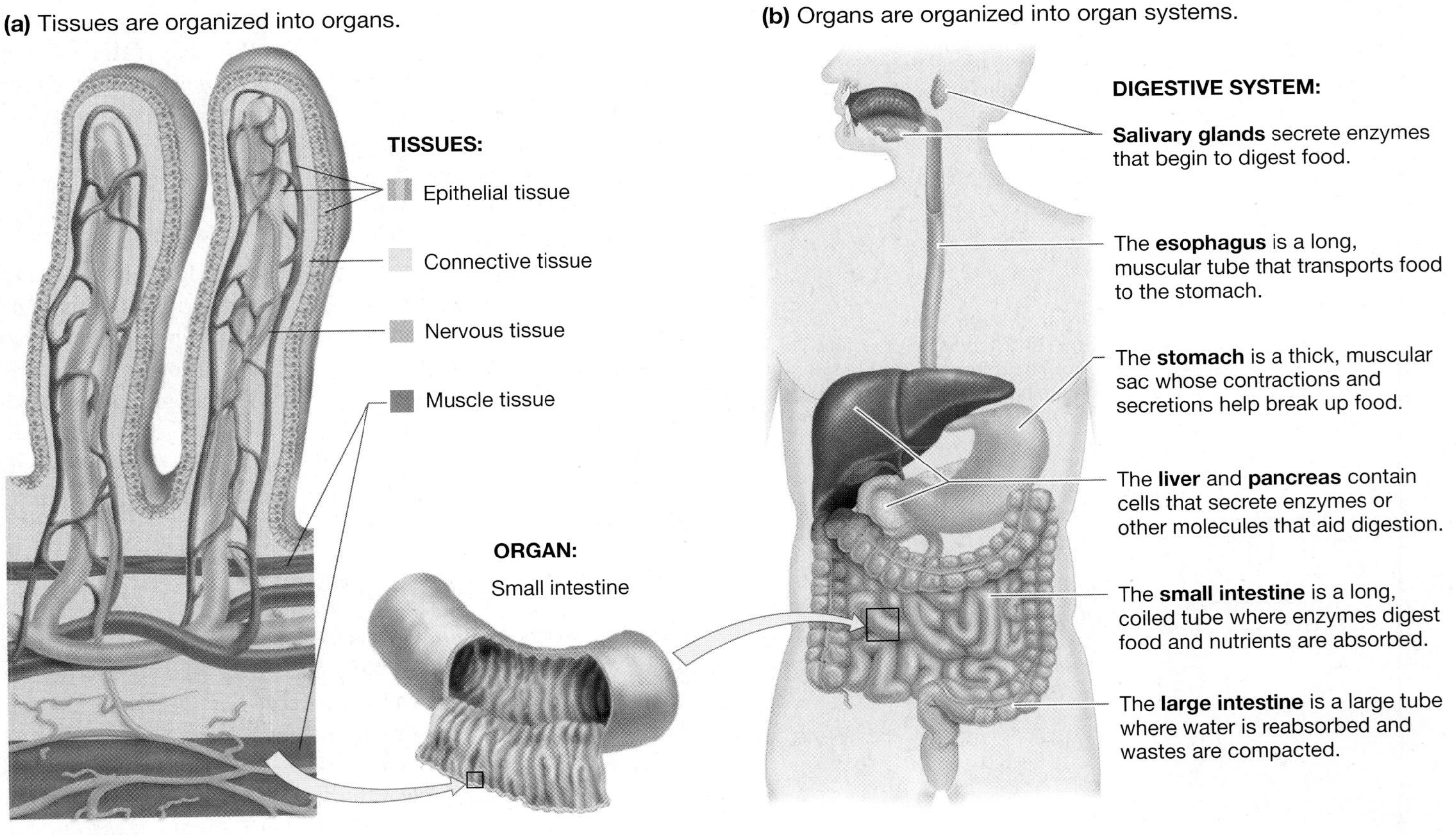

Figure 39.7 **Organs Are Composed of Tissues; Organ Systems Are Made Up of Organs. (a)** The human small intestine is an organ composed of all four major tissue types. **(b)** The human digestive system is essentially one long tube made up of different organs that form chambers where food is digested and nutrients are absorbed. The salivary glands, liver, and pancreas are organs that secrete specific enzymes or other compounds into the tube.

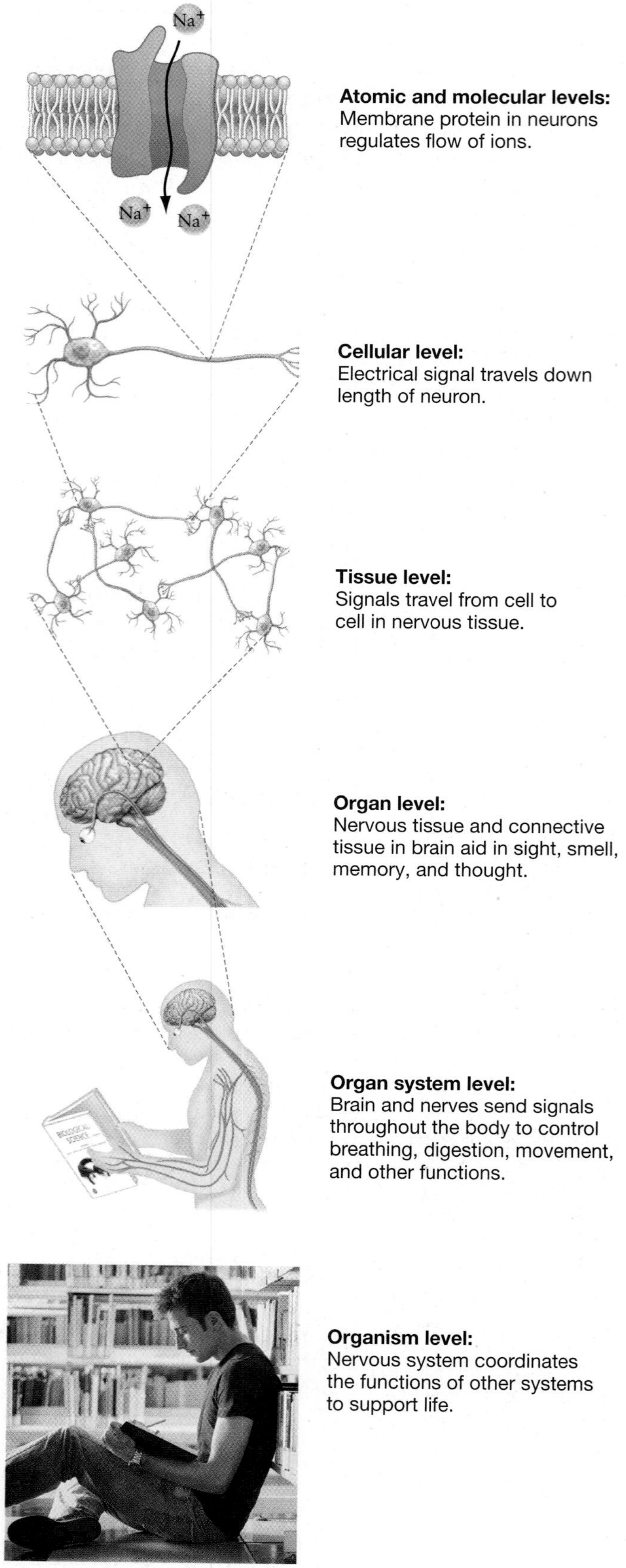

Figure 39.8 **Biologists Study Anatomy and Physiology at Many Levels.** The levels of organization within an organism are tightly integrated with each other in both structure and function.

CHECK YOUR UNDERSTANDING

✔ **If you understood this section, you should be able to ...**

1. Explain why the surface area of a simple epithelial cell tends to be larger than that of a stratified epithelial cell.
2. Identify the functions of each of the four tissue types in the small intestine.
3. For each of the following tissues, explain how its structure enables the tissue to perform a specific function: the meniscus, a flexible piece of cartilage in the knee joint; branched cardiac muscle cells; simple epithelium in the lungs.

Answers are available in Appendix A.

39.3 How Does Body Size Affect Animal Physiology?

Animals are living "machines," made up of molecules, cells, tissues, organs, and organ systems that have changed over time in response to natural selection.

After you complete this section, you should be able to...

- Explain how an animal's body size affects its physiology.

The laws of physics affect the anatomy and physiology of a living machine. The force of gravity, for example, limits how large an animal can be and still move efficiently. Or consider the forces exerted by the medium in which animals live. Because water is much denser than air, it is harder for animals to move through water. As a result, fish and aquatic mammals have much more streamlined bodies than terrestrial animals do.

Physical laws clearly affect body size. Just as clearly, body size has pervasive effects on how animals function. Large animals need more food than small animals do. Large animals also produce more waste, take longer to mature, reproduce more slowly, and tend to live longer. Conversely, small animals lose heat and water more rapidly than large animals do and are therefore more susceptible to damage from cold and dehydration. Juveniles and adults of the same species face different challenges simply because their body sizes are different.

Why is body size such an important factor in how animals work? How do biologists study the consequences of size? Let's consider each question in turn.

Surface Area to Volume Relationships: Theory

From microscopic roundworms to gigantic blue whales, animals span an incredible range of body masses—a total of 12 orders of magnitude. Many of the challenges posed by increasing size are based on the fundamental relationship between surface area and volume.

The relationship between the surface area and volume of the roots and shoots of plants affects water and light absorption by the plant (Ch. 34, Section 34.1). Similarly, surface area is important in animals because oxygen and nutrients must diffuse into an animal's cells, and waste products such as urea and carbon dioxide must diffuse out. The rate at which these and other

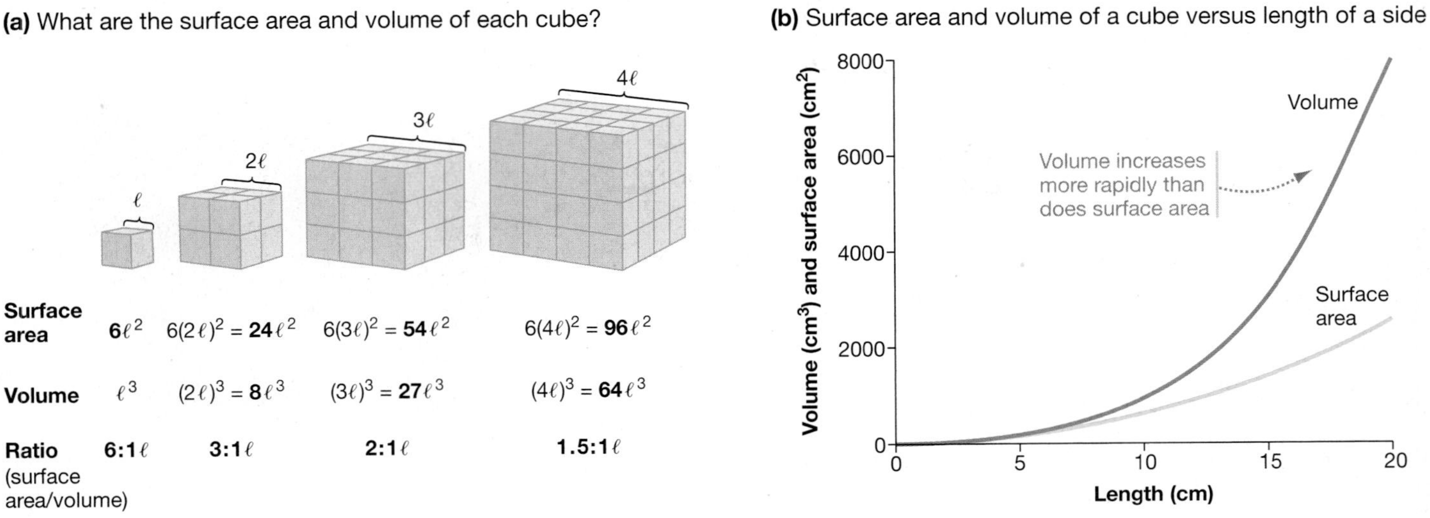

INTERACTIVE Figure 39.9 Surface Area and Volume Change as a Function of Overall Size. (a) The surface area of an object increases as the square of the length (ℓ). The volume increases as the cube of the length. **(b)** Volume increases much more rapidly than does surface area as linear dimensions increase.

molecules and ions diffuse depends in part on the surface area available for diffusion. In contrast, the rate at which nutrients are used and heat and waste products are produced depends on the volume of the animal.

The contrast between processes that depend on surface area and those that depend on volume is important for a simple reason. As an animal gets larger, its volume increases much more rapidly than its surface area does. This is true not just for animals, but for all organisms.

Reviewing a little basic geometry will convince you why this is so. As Figure 39.9a shows:

- The surface area of a cube increases as a function of its linear dimension *squared*. Because a cube has six sides, the surface area of a cube of length ℓ is $6\ell^2$ (six times the area of any one side).
- The volume of the same structure increases as a function of its linear dimension *cubed*. Hence, the volume of a cube of length ℓ is ℓ^3.

Figure 39.9b graphs the consequences of these relationships. The *x*-axis plots the length of a side in a cube; the *y*-axis plots the cube's volume (orange line) or surface area (yellow line). As a cube gets bigger, its surface area increases much more slowly than does its volume (or mass).

The same general relationship holds for cells, tissues, organs, organ systems, and whole organisms. Larger cells, for example, have lower surface area to volume ratios than smaller cells. How does the relationship between surface area and volume affect animal form and function?

Surface Area to Volume Relationships: Data

As an example of how surface area to volume relationships affect an animal's physiology, consider the metabolic rate of mammals. **Metabolism** is the sum of all chemical reactions that occur within an organism. Given that metabolism requires energy, the **metabolic rate** is the overall rate of energy consumption by an individual. Because consumption and production of energy in mammals depend largely on aerobic respiration, metabolic rate is often measured in terms of oxygen consumption.

Because an elephant is so much larger than a mouse, the elephant consumes a great deal more oxygen per hour than the mouse does. But what is going on at the levels of cells and tissues in these animals?

To compare metabolic rates in different species, biologists divide the overall metabolic rate by body mass and report a metabolic rate in units of milliliters of oxygen per gram per hour (mL O_2/g/hr). This "mass-specific," or relative, metabolic rate gives the rate of oxygen consumption per gram of tissue. Because an individual's metabolic rate varies dramatically with its activity, the accepted convention is to report the **basal metabolic rate (BMR)**—the rate at which an animal consumes oxygen while at rest, with an empty stomach, under normal temperature and moisture conditions.

Figure 39.10 on page 850 plots mass-specific BMR as a function of body mass for selected animals of different sizes. Notice that the *x*-axis on the graph has a logarithmic scale, making it easier to compare very small species with very large ones (for help with logarithms, see BioSkills 5).

What is the take-home message of this graph? On a per-gram basis, small animals have higher BMRs than do large animals. An elephant has more mass than a mouse, but a gram of elephant tissue consumes much less energy than a gram of mouse tissue does.

The leading hypothesis to explain this pattern is based on surface area to volume ratios. Many aspects of metabolism—including oxygen consumption, food digestion, delivery of nutrients to tissues, and removal of wastes and excess heat—depend on exchange across surfaces. As an organism's size increases, its mass-specific metabolic rate must decrease. Otherwise the surface area available for exchange of materials would fail to keep

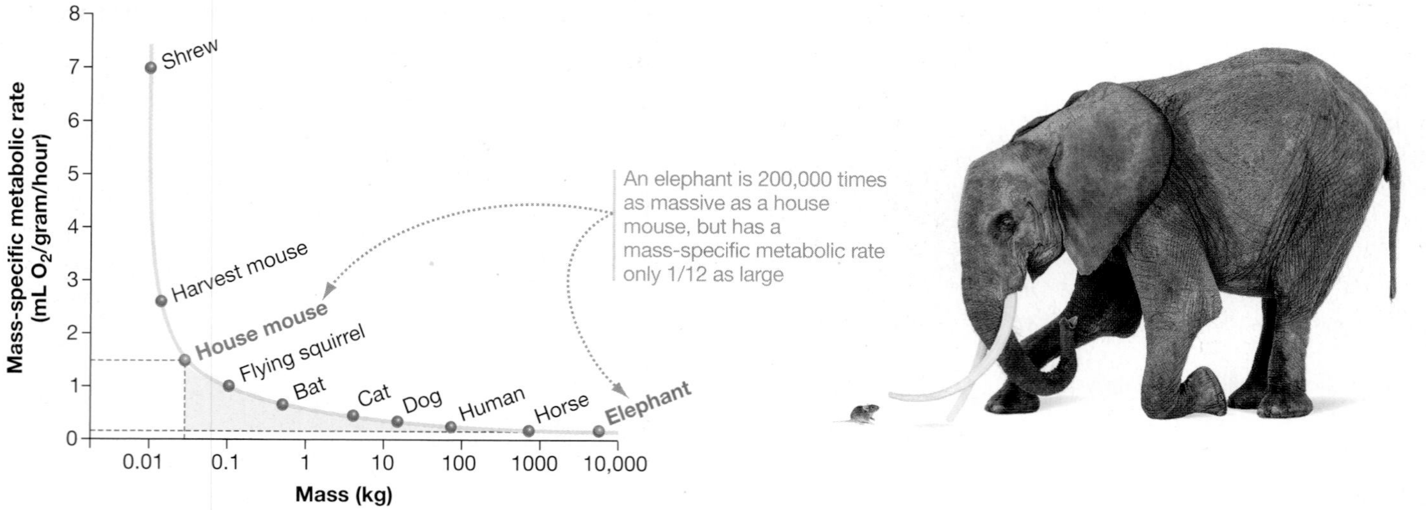

Figure 39.10 Small Animals Have Higher Relative Metabolic Rates than Large Animals Do.
✔ Which mammal must eat more to support each gram of its tissue: a Chihuahua or a Great Dane?

up with the metabolic demands generated by the organism's enzymes.

✔ **If you understand the relationship between body size and mass-specific metabolic rates of mammals, you should be able to predict what would happen to the body temperatures of mice and elephants if their mass-specific metabolic rates were swapped.**

Adaptations That Increase Surface Area

If the function of a biological structure depends on diffusion, that structure usually has a shape that increases its surface area relative to its volume. Flattening, folding, and branching are effective ways for structures to have a high surface area to volume ratio:

- ***Flattening*** In aquatic animals, **gills** are organs that allow the exchange of gases and dissolved substances between the animals' tissues and the surrounding water. Fish have **gill lamellae** (Figure 39.11a)—thin sheets of epithelial cells that provide the gill with an extremely high surface area relative to its volume. Because the surface area is so large, gases are able to diffuse across the gills rapidly enough to keep up with the growth in the volume of a developing fish.
- ***Folding*** In portions of the digestive tract where nutrients are transported into the body, the surface of the structure is folded. Extending from these folds are narrow projections called **villi** (Figure 39.11b). Together, the folds and villi make an extensive surface area available. Folded surfaces are common in diffusion-dependent organs.
- ***Branching*** The highly branched network shown in Figure 39.11c is a system of small, thin-walled blood vessels called **capillaries**. Capillaries are the sites where gases, nutrients, and waste products diffuse into and out of blood, and branching greatly increases their surface area. In general,

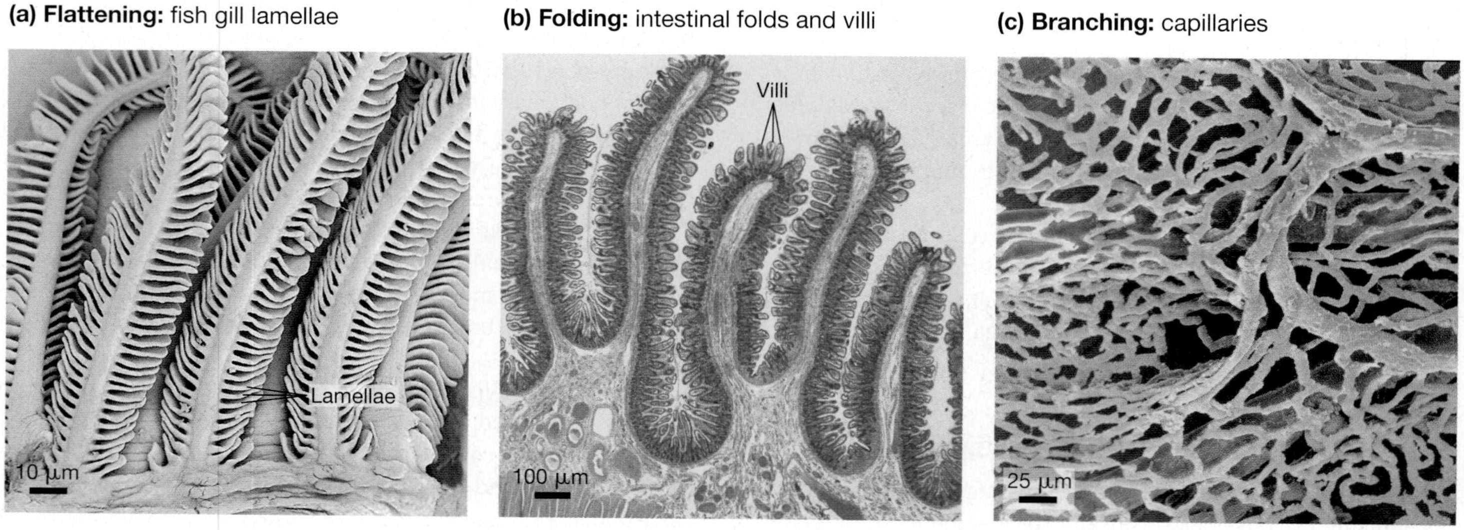

Figure 39.11 Certain Structures Increase the Surface Area of Tissues. The micrograph in part (c) has been colorized to highlight capillaries (pink).

highly branched structures increase the surface area available for diffusion.

The amount of surface area created by flattening, folding, and branching can be impressive. The highly branched capillaries in a human have a surface area of up to 1000 m^2; extensive folding gives a surface area of about 250 m^2 in your small intestine. For comparison, a doubles tennis court has a surface area of 261 m^2.

Surface area to volume relationships have a pervasive influence on the structure and function of animals. They will be an issue in almost every chapter in this unit.

CHECK YOUR UNDERSTANDING

✔ If you understood this section, you should be able to...

1. Predict whether salamanders that lack lungs and breathe entirely through their skin are small or large compared to salamanders that have lungs. Explain your reasoning.
2. Explain why fish gills have such a high surface area.

Answers are available in Appendix A.

39.4 Homeostasis

Adaptation and surface area to volume ratios are important themes in the analysis of animal form and function. So is homeostasis.

After you complete this section, you should be able to...

- Explain how homeostasis is maintained in animals.

Homeostasis (literally, "alike-standing") is defined as stability in the chemical and physical conditions within an organism's cells, tissues, and organs. Although conditions in an organism's environment can change, internal chemical and physical states vary slightly but are usually kept within a tolerable range.

Homeostasis: General Principles

Many of the structures and processes observed in animals can be interpreted as mechanisms for maintaining homeostasis with respect to some measure, such as pH, temperature, or calcium ion concentration. Let's review some important general ideas about homeostasis and then analyze how homeostasis can be maintained in the face of environmental fluctuations.

Homeostasis Is Achieved via Regulation Many organisms are able to **regulate** their internal conditions—to actively maintain relatively constant internal conditions even when the environment fluctuates. A dog maintains a body temperature of about 38°C whether it's cold or hot outside. If its body temperature rises, the dog might pant to cool off and maintain homeostasis. If its body temperature falls, it might shiver to bring its temperature back up to the target value. Because dogs have to spend energy on panting and shivering, it's reasonable to conclude that maintaining a relatively constant body temperature is very important.

However, maintaining a constant body temperature might not be feasible in some environments. For example, many aquatic animals do not maintain their body temperature above the ambient temperature because they would lose heat quickly to the surrounding water. Instead, these animals **conform** to environmental conditions. The body temperature of freshwater invertebrates, fishes, and turtles, for example, changes as the water warms or cools. These animals save energy by not using metabolic heat to regulate their body temperature, but they might experience trade-offs such as swimming too slowly to escape from predators in cold water.

Note that regulation and conformation lie at the two extreme ends of a spectrum, and most animals fall somewhere in between. For example, some animals conform to a range of mild ambient temperatures but expend energy to regulate their body temperature when their environment becomes dangerously cold or hot. These concepts also apply to homeostasis of other variables, such as water and ion concentration of body fluids (Chapter 40).

The Role of Epithelia Because epithelia are the interface between the internal and external environments, they play a key role in achieving homeostasis. Epithelia are responsible for forming an internal environment that can be dramatically different from the external environment and for allowing physical and chemical conditions inside an animal to be maintained at relatively constant levels.

As subsequent chapters will show, any epithelial cells are studded with membrane proteins that regulate the transport of ions, water, nutrients, and wastes. No molecule can enter or leave the body without crossing an epithelium. Homeostasis is possible because epithelia control this exchange.

Why Is Homeostasis Important? Much of the answer to this question is based on enzyme function. Recall that enzymes are proteins that catalyze chemical reactions within cells (Ch. 8, Section 8.3). Temperature, pH, and other physical and chemical conditions have a dramatic effect on the structure and function of enzymes. Most enzymes function best within a narrow range of conditions.

Other processes depend on homeostasis, too. Temperature changes affect membrane permeability and how quickly solutes diffuse. The expansion of water as it freezes can rip cells apart if tissues are allowed to cool much below 0°C. Conversely, extremely high temperatures can denature proteins—meaning that they lose their tertiary structure and cease to function.

When homeostasis occurs, conditions inside the body allow molecules, cells, tissues, organs, and organ systems to function at an optimal level. However, occasional departures from homeostasis can represent important adaptations. For example, a fever is a response to an infection by a pathogen. This increase in body temperature can help fight off the pathogen.

The Role of Regulation and Feedback

Most animals achieve homeostasis by using *regulatory systems* that monitor internal conditions such as temperature, blood pressure, blood pH, and blood glucose concentration. If one of these variables changes, a regulatory system acts quickly to modify it. Like the thermostat in a home heating system, each of these systems has a **set point**—a target range of values for the controlled variable.

Animals have a set point for blood pH, blood oxygen and nutrient concentrations, and other variables. In most mammals, the set point for body temperature is somewhere between 35°C and 39°C. How does a regulatory, or homeostatic, system allow an individual to maintain its tissues at the set point despite changes in activity and the environment?

A homeostatic system consists of three general components: a sensor, an integrator, and an effector. Figure 39.12 shows how these components interact to regulate temperature in mammals:

1. A **sensor** is a structure that senses some variable in the external or internal environment.
2. An **integrator** evaluates the incoming sensory information by comparing it to the set point and determines whether a response is necessary to achieve homeostasis.
3. An **effector** is any structure that helps restore the internal condition being monitored by the system.

In mammals such as the dog shown in Figure 39.12, the sensors are temperature receptors located throughout the body that constantly monitor information about body temperature. For example, receptors on neurons in a brain region called the **hypothalamus** sense cooling or heating, and they respond by altering the pattern of electrical signals that they send to adjacent neurons.

The electrical signals that originate with temperature receptors are transmitted to an integrator, also located in the hypothalamus. Current evidence indicates that separate centers in the hypothalamus sense and integrate changes in body temperature.

If a mammal is cold, cells in the hypothalamus send signals to effectors that return body temperature to the set point. Signals from the hypothalamus might induce shivering in skeletal muscles to generate warmth and fluffing of fur to improve insulation and retain heat. Signals from the same or nearby cells can also result in the release of blood-borne chemical signals that increase the rate of cellular respiration throughout the body, generating more body heat.

But if the same individual is too hot, the integrator in the hypothalamus sends signals that initiate sweating or panting—responses that cool the body. Other signals can induce behavioral changes that slow heat gain and production, such as seeking shade or a cool burrow and resting.

Homeostatic systems are based on negative feedback. When **negative feedback** occurs, effectors return internal conditions to set-point values. In response to either cooling or heating, behavioral and physiological responses move the body temperature back toward the set point via negative feedback. Note the following points about the effectors that maintain homeostasis:

- Feedback systems usually work in "antagonistic pairs": One set of responses increases a parameter while a corresponding set of responses decreases it.
- Input from sensors and integrators happens continuously, so feedback systems are constantly making fine adjustments relative to the set point.

Homeostatic systems are a key aspect of one of the five attributes of life (Ch. 1, Section 1.1): acquiring information from

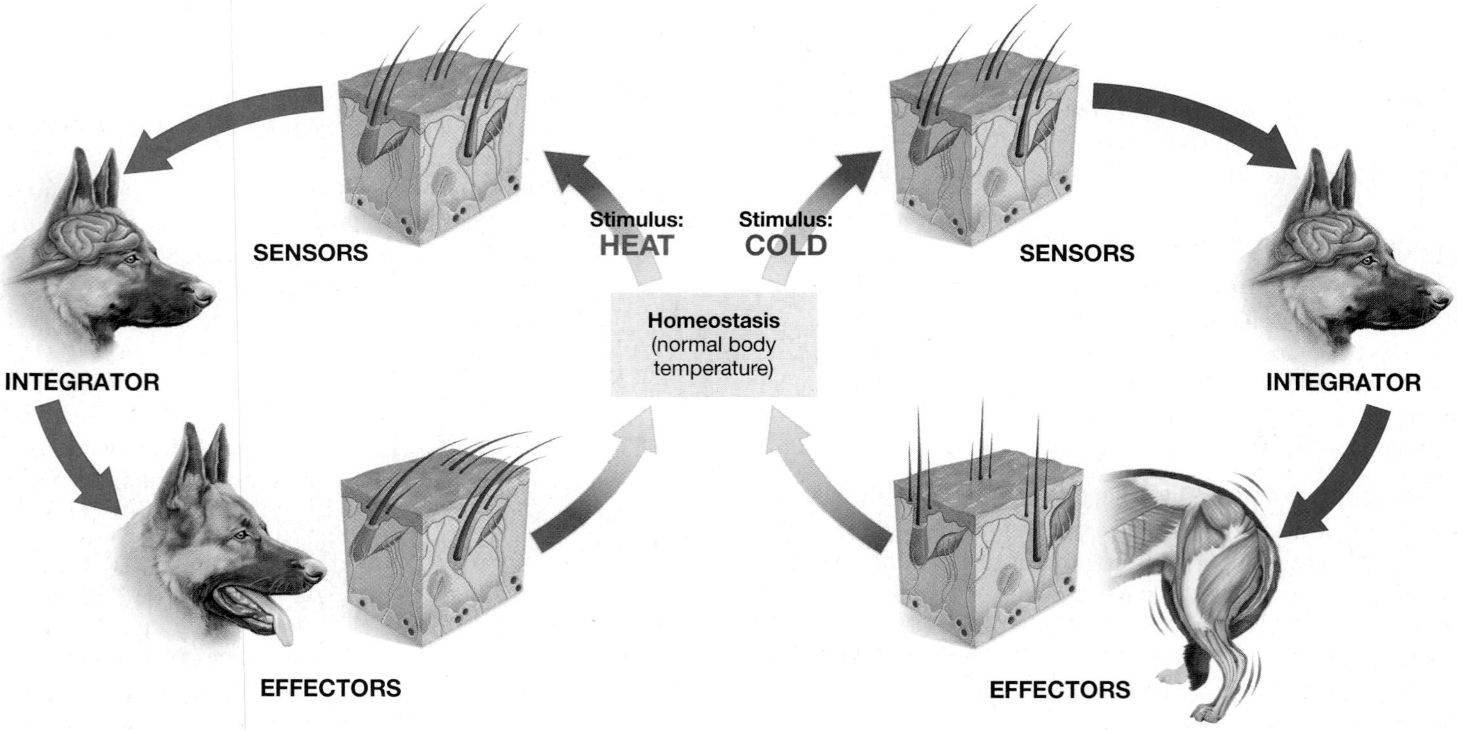

Figure 39.12 **In Mammals, a Homeostatic System Regulates Body Temperature through Negative Feedback.** In dogs, a set point for body temperature is maintained by a complex negative feedback system that includes sensors and integrators in the brain as well as sensors and effectors located throughout the body. The set point varies among species, from 30°C in monotremes to about 38°C in dogs and over 39°C in rabbits.

the environment and responding to it. Subsequent chapters in this unit explore how animals use such systems to achieve homeostasis with respect to the solute concentrations of their cells and tissues, their oxygen supply, and nutrient availability. In the rest of this chapter, you will focus in more detail on how different animals achieve homeostasis with respect to body temperature.

CHECK YOUR UNDERSTANDING

✔ **If you understood this section, you should be able to ...**

Use the example of thermoregulation in dogs shown in Figure 39.12 to explain why homeostatic negative feedback responses are often described as antagonistic.

Answers are available in Appendix A.

Figure 39.13 **There Are Four Methods of Heat Exchange.** The arrows indicate the direction of heat exchange.

39.5 Thermoregulation: A Closer Look

Let's take a closer look at how animals **thermoregulate**, or control their body temperature. All animals exchange heat with their environment. Heat flows "downhill," from regions of higher temperature to regions of lower temperature. If an individual is warmer than its environment, it will lose heat; if it is cooler than its surroundings, it will gain heat.

After you complete this section, you should be able to...

- Compare and contrast the methods by which animals thermoregulate.

How does heat exchange occur?

Mechanisms of Heat Exchange

As Figure 39.13 shows, animals exchange heat with the environment in four ways: conduction, convection, radiation, and evaporation.

1. **Conduction** is the direct transfer of heat between two physical bodies that are in contact with each other. For instance, when a turtle sits on a warm rock, heat is transferred from the rock to its body. The rate at which conduction occurs depends on the surface area of transfer, the steepness of the temperature difference between the two bodies, and how well each body conducts heat.
2. **Convection** is a special case of conduction. During conduction, heat is transferred between two solids; but during convection, heat is exchanged between a solid and a moving liquid or gas. For example, the heat loss that occurs when wind blows over your skin is due to convection. As the speed of the air or water flow increases, so does the rate of heat transfer.
3. **Radiation** is the transfer of heat between two bodies that are not in direct physical contact. All objects, including animals, radiate energy as a function of their temperature. The Sun radiates heat; so does your body, but to a much lesser degree.
4. **Evaporation** is the phase change that occurs when a liquid becomes a gas. Conduction, convection, and radiation can cause heat gain or loss, but evaporation leads only to heat loss. The turtle in Figure 39.13 is losing heat as water evaporates off its shell and skin. Because of the extensive hydrogen bonding in liquid water, a large amount of energy is needed to heat water and produce evaporation (Ch. 2, Section 2.2). If you get overheated on a summer day, splashing water on your skin and sweating will allow you to use evaporative heat loss to cool your body. Conversely, getting wet on a cold day can be deadly. The water on your skin absorbs so much heat from your body that your temperature may drop dangerously.

Heat exchange is critical in animal physiology because individuals that get too hot or too cold may die. You've already seen how mammals regulate body temperature through negative feedback. Let's now take a broader look at strategies for temperature regulation in animals.

Thermoregulatory Strategies

The ability of animals to thermoregulate varies widely. Two ways to organize this variation are by examining **(1)** how animals obtain heat, and **(2)** whether body temperature is held constant.

An animal that is an **endotherm** ("inner-heat") produces adequate heat to warm its own tissues, while an **ectotherm** ("outer-heat") relies principally on heat gained from the environment. Endotherms and ectotherms represent two extremes along a continuum of heat sources.

There are also two extremes on a continuum describing whether animals hold their body temperature constant: **Homeotherms** ("alike-heat") keep their body temperature constant, while in **poikilotherms** ("varied-heat"), body temperatures rise or fall depending on environmental conditions.

Humans, along with most birds and most other mammals, are strictly endothermic homeotherms. These species produce their own heat and maintain a constant body temperature. In contrast, most freshwater and terrestrial invertebrates, fishes, amphibians, and non-avian reptiles are ectothermic poikilotherms whose body temperatures change throughout the day and seasonally. But many animal species lie somewhere between these extremes:

- Some mammals, such as the African elephant featured in the chapter introduction, allow their body temperature to rise during the hotter part of the day—meaning they are somewhat poikilothermic.
- Small mammals that inhabit cold climates lose heat rapidly because their surface area is large relative to their volume. To survive when the ambient temperature is low, species such as dormice reduce their metabolic rate and allow their body temperature to drop, a form of poikilothermy. This condition is called **torpor**. Torpor during the winter that persists for weeks or months is called **hibernation**, while prolonged torpor to escape dry or hot conditions is called **estivation**.
- Naked mole rats are mammals that lack insulation because they have no fur. They live in underground tunnels and allow their body temperature to rise and fall with burrow temperatures. They are poikilothermic and intermediate between ectotherms and endotherms.
- Japanese honeybees exhibit poikilothermy when defending their hives from predatory hornets. The honeybees swarm an invading hornet and contract their flight muscles repeatedly to collectively produce heat endothermically (Figure 39.14). The temperature within the swarm rises to 47°C (117°F), killing the hornet but not the honeybees, which can tolerate temperatures up to 50°C (122°F).

Even in a homeothermic endotherm such as a mammal or bird, body temperature can vary widely in different body regions. When a Canada goose is standing on ice, its feet may be at a temperature of just 9°C, even though its body core is at 35°C. Similar variations exist in tuna and mackerel. These fishes are ectotherms but generate heat to warm certain sections of their bodies, such as their eyes or swimming muscles.

Comparing Endothermy and Ectothermy

Endotherms can warm themselves because their basal metabolic rates are extremely high—the heat given off by the high rate of chemical reactions is enough to warm the body. Mammals and birds retain this heat because they have elaborate insulating structures, such as fur or feathers.

Ectotherms can also generate heat as a by-product of metabolism. The amount of heat they generate is small compared with the amount generated by endotherms, however, because ectotherms have relatively low metabolic rates. The most important sources of heat gain in ectotherms are radiation and conduction: Ectotherms bask in sunlight or lie on warm rocks or soil.

You might have heard the terms "warm-blooded" and "cold-blooded" used to describe endotherms and ectotherms, respectively. But these terms are technically inaccurate, because the blood is not always warm in endotherms and cold in ectotherms. For example, an ectothermic lizard basking in the sun can attain a body temperature far above the ambient temperature, even higher than that of an endotherm.

Endothermy and ectothermy are best understood as contrasting adaptive strategies. Because endotherms maintain a high body temperature at all times, they can be active in winter and at night. Their high metabolic rates also allow them to sustain high levels of aerobic activities, such as running or flying.

These abilities come at a cost, however: To fuel their high metabolic rates, endotherms have to obtain large quantities of energy-rich food. The energy used to produce heat is then unavailable for other energy-demanding processes, such as reproduction and growth.

In contrast, ectotherms are able to thrive with much lower intakes of food. And because they are not oxidizing food to provide heat, they can use a greater proportion of their total energy intake to support reproduction.

What's the downside of ectothermy? Chemical reaction rates are temperature dependent, so muscle activity and digestion slow dramatically as the body temperature of an ectotherm drops. As a result, ectotherms are more vulnerable to predation in cold

(a) A hornet preys on a honeybee. **(b)** A swarm of bees surrounds a hornet ... **(c)** ... forming a hot defensive ball.

Figure 39.14 **Honeybees Use Heat to Kill Predators.** The infrared image in panel **(c)** shows the heat generated by the bees.

weather and in general are less successful than endotherms at inhabiting cold environments or remaining active on cool nights.

In short, each suite of adaptations has advantages and disadvantages. Like all adaptations, endothermy and ectothermy involve trade-offs.

Countercurrent Heat Exchangers

Homeothermic endotherms such as birds and mammals have sophisticated systems for thermoregulation that allow them to maintain high body temperatures even in cold environments, in part by minimizing heat loss from the body. Heat loss is a particularly important problem for aquatic mammals. If you've ever gone swimming in cold water, you can appreciate the problem faced by seals, otters, and whales. Water is such an effective conductor of heat that aquatic organisms lose metabolic heat rapidly. The dense, water-repellent fur of otters conserves heat by maintaining a layer of trapped air next to the skin. Seals and whales are insulated by thick layers of fatty blubber.

Some marine mammals have body parts containing specific arrangements of blood vessels that minimize heat loss. For example, the tongue of a gray whale, which is exposed to cold water during feeding, contains bundles of arteries and veins. Each bundle includes an artery that carries warm, oxygenated blood from the body core. The artery is encircled by smaller veins, which transport cool blood from the tongue surface back toward the body core (Figure 39.15a).

This type of arrangement, in which fluids flow through adjacent pipes in opposite directions, is called a **countercurrent exchanger**. The "exchanger" part of the name is apt because, in a case like the whale's tongue, heat is exchanged between the warm blood in the artery and the cool blood in the veins.

The countercurrent exchange system is key to minimizing heat loss in tissues exposed to the cold, such as the whale tongue. To see how it works, study the diagram on the left in Figure 39.15b. In this diagram, the fluid that enters the countercurrent heat exchanger is initially warm but steadily transfers heat to the adjacent, cooler fluid flowing in the opposite direction. There is a warmer-to-cooler gradient between the two fluids at every point along the length of the countercurrent exchanger.

If the two fluids ran in the same direction, as in the "concurrent" diagram on the right in Figure 39.15b, the gradient between them would disappear quickly as the arterial current cooled and the venous current heated. The countercurrent physical arrangement, in which an artery is tightly wrapped with several small veins, increases the rate of heat transfer beyond what would occur if there were a single large vein. Countercurrent exchangers, including heat exchangers, are effective because they maintain a gradient between the two fluids along their entire length, thereby maximizing heat transfer from the warm fluid to the cool fluid and minimizing heat loss to the environment.

Similar heat-conserving arrangements of arteries and veins are found in the flippers of whales and dolphins and in the legs of many mammals and birds, such as penguins that live in cold terrestrial environments.

Countercurrent exchangers are just one of many remarkable adaptations you'll encounter in this unit—structures and systems that allow animals to thrive in a wide array of environments.

CHECK YOUR UNDERSTANDING

✔ **If you understood this section, you should be able to ...**

1. Identify the main source of heat available to a lizard that lives underground.
2. Discuss the advantages and disadvantages of endothermy and ectothermy.
3. Explain what would happen to the body temperature of a Canada goose if the arteries and veins of its legs had concurrent rather than countercurrent blood flow.

Answers are available in Appendix A.

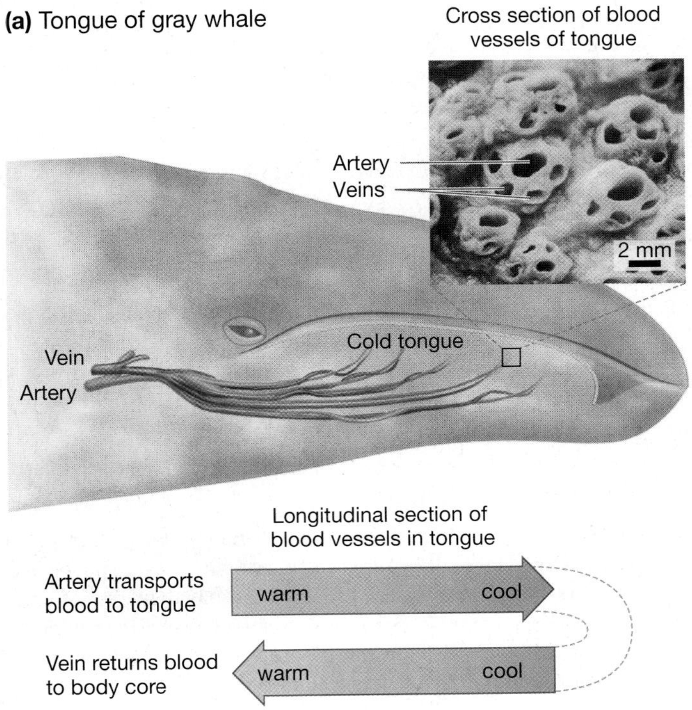

Figure 39.15 **Countercurrent Exchangers Conserve Heat.** **(a)** Bundles of arteries and veins in a whale tongue form heat exchangers that minimize heat loss from the tongue to the cold ocean water during feeding. **(b)** Countercurrent arrangements are much more efficient than "concurrent" arrangements. The data given here are hypothetical.

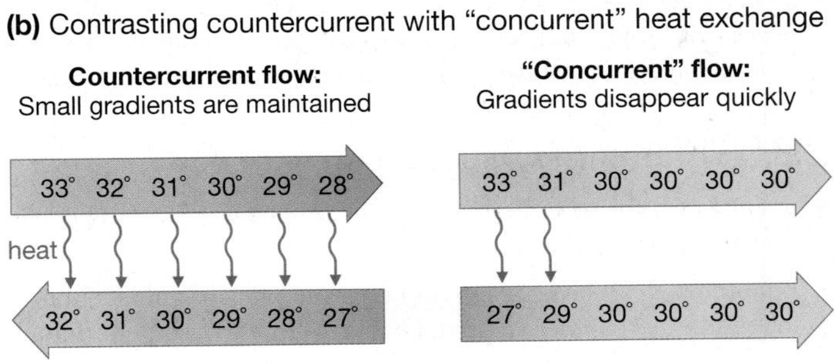

39.1 Form, Function, and Adaptation

- Animal structures and their functions represent adaptations, which are heritable traits that improve survival and reproduction in a certain environment.
- Adaptations involve trade-offs, or inescapable compromises between traits.
- Acclimatization is a reversible response to the environment that improves physiological function in that environment.

39.2 Tissues, Organs, and Organ Systems: How Does Structure Correlate with Function?

- Animal cells with a common function are grouped together into four general types of tissue: connective tissue, nervous tissue, muscle tissue, and epithelial tissue.
- Organs are structures that are made up of two or more tissues that together perform specific tasks.
- Organ systems consist of organs that work together in an integrated fashion to perform one or more functions.

39.3 How Does Body Size Affect Animal Physiology?

- Large animals have smaller surface area to volume ratios than small animals. As animals grow, their volume increases more rapidly than their surface area.
- Large animals have low mass-specific metabolic rates, in keeping with their relatively small surface area for exchanging the oxygen and nutrients required to support metabolism and the wastes and heat produced by metabolism.
- The relatively high surface area of small animals means that they lose heat extremely rapidly.

39.4 Homeostasis

- Homeostasis refers to relatively constant physical and chemical conditions inside the body.
- Homeostasis in a fluctuating environment is usually achieved by regulation carried out by homeostatic systems.
- Animals have set points, or target values, for various body parameters. When a regulated variable is not at its set point, negative feedback occurs. Responses to negative feedback return the parameter to the set point and result in homeostasis.
- Most animals have a set point for body temperature. If an individual overheats, it may pant, sweat, or seek a cool environment; if an individual is cold, it may shiver, bask in sunlight, or fluff its fur or feathers.

39.5 Thermoregulation: A Closer Look

- Animals vary from endothermic to ectothermic and from homeothermic to poikilothermic.
- Endotherms can be active in cold environments but must obtain a lot of energy to fuel their metabolism. Ectotherms do not require as much energy, but their activity depends on environmental temperature.
- Countercurrent heat exchangers have vessels in close contact that carry warm and cool fluids in opposite directions.

Answers are available in Appendix A.

✔ TEST YOUR KNOWLEDGE

1. True or False: The increase in red blood cell count in tourists visiting Tibet is an example of acclimatization.
2. Which of these examples best describes the concept that form facilitates function?
   a. Crickets must balance their resources between spermatophore production and immune defenses.
   b. A desert jackrabbit has large ears that help eliminate excess heat.
   c. An Antarctic fish maintains homeostasis by conforming to the stable, external temperature.
   d. Honeybees will swarm around a predatory wasp and contract their flight muscles to generate a lethal ball of heat.
3. Which of the following statements regarding surface area and volume in animals are correct? Select True or False for each statement.
   T/F As an animal grows, its volume increases more rapidly than its surface area.
   T/F A chihuahua has a higher surface area to volume ratio than a great dane.
   T/F Animals with high surface area to volume ratios heat and cool more slowly than animals with lower surface area to volume ratios.
   T/F As an animal's volume increases, its total surface area decreases.
4. Which of the following is an advantage that ectotherms have over endotherms of the same size?
   a. They require much less food.
   b. They are less vulnerable to predation during cold weather.
   c. They can remain active in cold weather or on cold nights.
   d. They have higher metabolic rates and grow more quickly.

✔ TEST YOUR UNDERSTANDING

5. For each of the following, explain how structure relates to function: absorptive sections of the digestive tract; capillaries; beaks of Galápagos finches; fish gills.
6. The metabolic rate of a frog in summer (at 35°C) is about eight times higher than in winter (at 5°C). Compare and contrast the frog's ability to move, exchange gases, and digest food at the two temperatures. During which season will the frog require more food energy, and why?
7. Explain why most endotherms are homeothermic and most ectotherms are poikilothermic.
8. **QUANTITATIVE** Consider three spheres with radii of 1 cm, 5 cm, and 10 cm. Calculate the surface area and the volume of each sphere, and plot the results on a graph with radius on the *x*-axis and surface area and volume on the *y*-axis. (Surface area of a sphere = $4\pi r^2$; volume of a sphere = $(4/3)\pi r^3$.) Explain how the graph shows the relationship between size and surface area to volume ratio.

✓ TEST YOUR PROBLEM-SOLVING SKILLS

9. Explain why it would be impossible for a gorilla the size of King Kong to have fur. (Your answer should explain how the surface area to volume ratio of a normal-sized gorilla would compare to Kong's; relate this to the role of surface area and volume in heat generation and heat transfer, and consider the function of fur.)

10. The dinosaur *Apatosaurus (Brontosaurus)* is one of the largest terrestrial animals that ever lived—over 20 m in length and weighing over 20 metric tons. Is it more likely that *Apatosaurus* was homeothermic or poikilothermic? Explain.

✓ PUT IT ALL TOGETHER: Case Study

How does gigantism affect the physiology of animals?

Many species of animals on islands are larger than related species on the mainland. Scientists hypothesize that this phenomenon, called island gigantism, evolved in response to the scarcity of competitors and predators on islands. Reduced competition and predation allows species to exploit more resources and frees them from the need to hide in small refuges.

11. QUANTITATIVE The graph shown here compares the average carapace (shell) length of mainland and island tortoises. Summarize the results (*** means $P < 0.001$, see **BioSkills 3**), then use the data to predict whether the surface area to volume ratio is higher in mainland or island tortoises.

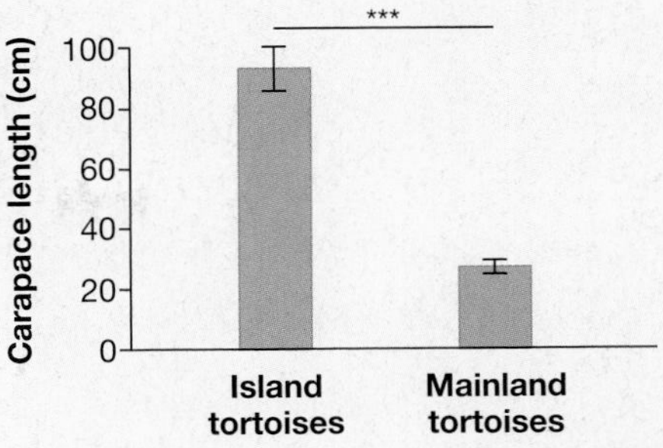

SOURCE: A. L. Jaffe, G. J. Slater, and M. E. Alfaro. 2011. *Biology Letters* 7: 558–561.

12. Which tortoises, mainland or island, need to eat more food per gram of their body mass?

13. Which of the following might be a trade-off of gigantism experienced by giant island tortoises?

a. They cool very rapidly during cold weather.

b. It would be difficult to sustain their high mass-specific metabolic rates on a diet of plants alone.

c. It could be more difficult to avoid thermally unfavorable conditions.

d. They could hide from nonnative predators more easily.

14. THINK CAREFULLY True or false: The body temperatures of island tortoises always closely match the temperatures in their environments.

15. Suppose that a small mainland tortoise and a large island tortoise are placed in the same pen at a zoo. Which tortoise will be more poikilothermic, the small or large tortoise? Why?

16. THINK CAREFULLY On a trip to the Galápagos Islands, you overhear a group of tourists refer to tortoises as "cold blooded." Explain why this word is not accurate to describe a giant tortoise.

40 Water and Electrolyte Balance in Animals

Terrestrial animals lose water every time they breathe, urinate, and defecate. For many animals, drinking is an important way to replace lost water and achieve homeostasis. This chapter explores how terrestrial and aquatic animals maintain water balance.

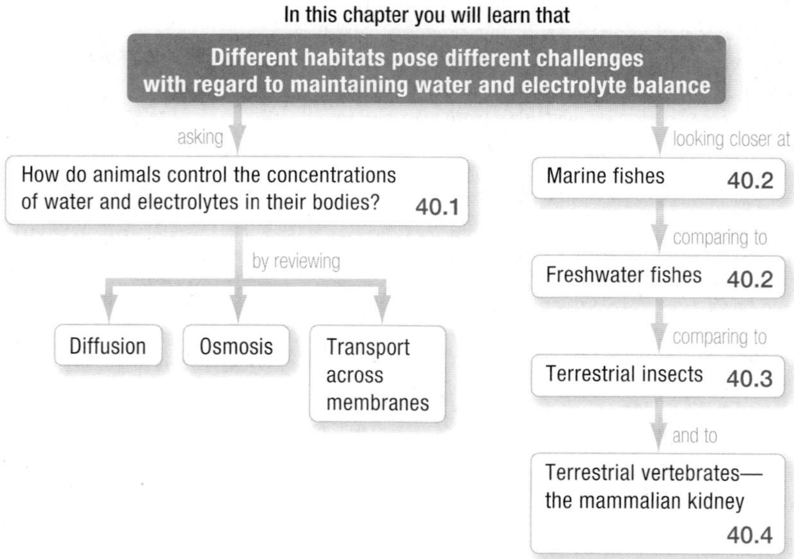

The chemical reactions that make life possible occur in an aqueous solution. If the balance of water and dissolved substances in the solution is disturbed, those chemical reactions—and life itself—may stop. Humans can survive just three days without drinking water. Many marine animals die if a hurricane introduces enough fresh water to the ocean shore to disrupt normal salt concentrations.

An animal achieves water balance when its water intake equals its water loss. Water balance is an important component of homeostasis—the maintenance of constant and favorable conditions in cells and tissues.

Water balance is intimately associated with sustaining balanced concentrations of electrolytes throughout the body. An **electrolyte** is a compound that dissociates into ions when dissolved in water. Electrolytes got their name because they conduct electrical current.

BIG PICTURE

This chapter is part of the Big Picture. See how on pages 838–839.

In many animals, the most abundant ions of electrolytes are sodium (Na^+), chloride (Cl^-), potassium (K^+), and calcium (Ca^{2+}). Cells require precise concentrations of these ions to function normally. In humans, imbalances can lead to muscle spasms, confusion, irregular heart rhythms, fatigue, paralysis, or even death.

Water and electrolyte balance is also associated with excretion. Animals produce urine to excrete wastes. Urine contains water, so excretion of urine inevitably leads to water loss. The amount of water an animal loses in its urine depends both on its hydration state and on the type of wastes it produces.

This chapter is focused on a single question: How do animals maintain water and electrolyte balance in marine, freshwater, and terrestrial environments? Answering it will introduce you to some of the most complex and important homeostatic systems known. You can review the importance of maintaining water and electrolyte balance in the Big Picture on pages 838–839.

40.1 Osmoregulation and Excretion

Recall that uncharged solutes move down their concentration gradients via **diffusion** (Ch. 6, Section 6.3). The movement of water is a special case of diffusion called **osmosis**. Osmosis occurs only when solutions are separated by a membrane that permits water to cross but selectively holds back some or all of the solutes (Chapter 6, Fig. 6.13).

After you complete this section, you should be able to …

- Explain the relationship between osmoregulation and excretion.

The concentration of solutes in a solution, measured in osmoles[1] per liter, is the solution's **osmolarity**. If the solutes are separated by a selectively permeable membrane and cannot cross that membrane, water moves from the side of lower osmolarity—that is, lower solute concentration—to the side of higher osmolarity—higher solute concentration.

Now let's examine how osmosis and diffusion affect water and ion balance in animals that live in different environments.

What Is Osmotic Stress?

Osmotic stress occurs when the concentration of dissolved substances in a cell or tissue is abnormal. It means that water and solute concentrations are different from their set points. Osmotic stress can be caused by loss or gain of water or electrolytes to or from the animal's environment.

Many organisms respond to osmotic stress by osmoregulating, just as they respond to heat or cold stress by thermoregulating (Ch. 39, Section 39.5). **Osmoregulation** is the process by which organisms control the concentrations of water and solutes in their bodies.

[1]The unit osmole is similar to the unit mole except that an osmole takes into account molecules that dissociate in solution. For example, because NaCl dissociates into Na^+ and Cl^- in solution, adding 1 mole of NaCl to water is equivalent to adding 2 osmoles of solute.

Not all animals encounter osmotic stress. For many marine invertebrates, such as sponges, jellyfish, and flatworms, achieving homeostasis with respect to water and solute balance is straightforward. Seawater is a fairly constant ionic and osmotic environment, and the concentrations of electrolytes and other solutes found in these animals nearly match those of the sea. Such animals are **osmoconformers**.

Seawater is **isosmotic** with the tissues of osmoconforming animals. Put another way, the solute concentrations inside and outside these animals are equal. Because the body fluids of osmoconforming marine invertebrates are isosmotic with seawater, osmosis doesn't alter water and solute balance and induce osmotic stress.

Osmotic Stress in Seawater, in Fresh Water, and on Land

In contrast to most marine invertebrates, marine and freshwater bony fishes and terrestrial animals are **osmoregulators**. These animals actively regulate osmolarity inside their bodies to achieve homeostasis. Osmoregulation in seawater, in fresh water, and on land involves very different challenges and solutions.

By osmoregulating, marine bony fishes keep the osmolarity of their tissues lower than that of seawater. The difference in osmolarity is most important in gills, which are organs involved in gas exchange. For gas exchange to occur with the environment, the epithelial cells on the surfaces of the gills must be in direct contact with seawater.

Seawater is **hyperosmotic** to the tissues of marine bony fishes—the solution outside the body has a higher solute concentration than the solution inside (Figure 40.1). Because there is a

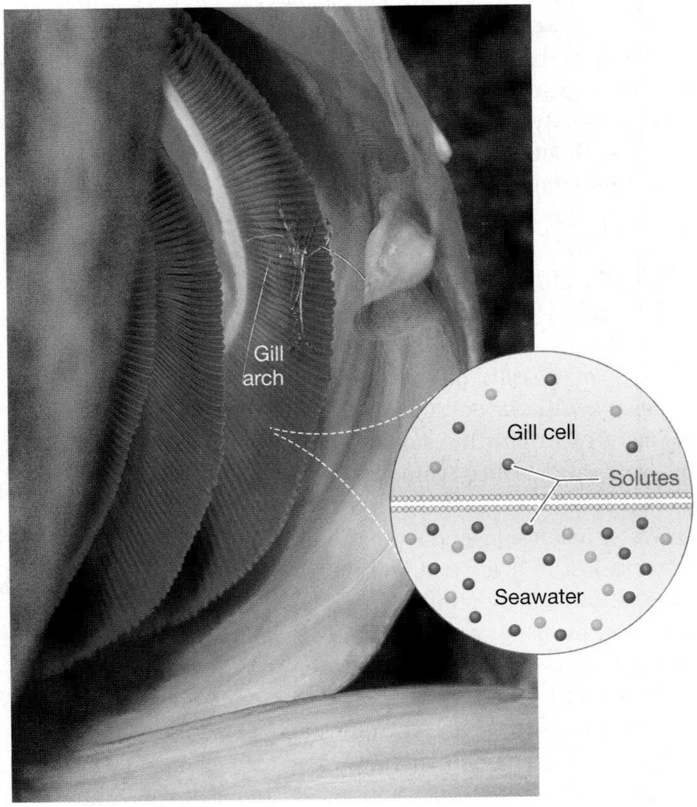

Figure 40.1 Seawater Is Hyperosmotic Compared to the Cells of Fish Gill Tissue.

large difference in osmolarity between the inside of each cell and the seawater outside, water tends to flow by osmosis out of the gill epithelium. If the water that these fishes lose across their gills is not replaced, their cells will shrivel and die. These animals face a trade-off between gas exchange and maintaining water and electrolyte balance.

Marine bony fishes replace the lost water by drinking large quantities of seawater. Drinking brings in excess electrolytes, however. Electrolyte balance is disrupted even further because ions diffuse into the gill epithelium, following a concentration gradient from seawater to tissue cells. To rid themselves of these excess electrolytes, marine bony fishes actively pump ions out of their bodies and back into the seawater, using membrane proteins located in the gill epithelium.

Freshwater fishes osmoregulate in an environment dramatically different from the ocean. Unlike their marine relatives, which lose water by osmosis and gain solutes by diffusion, freshwater fishes gain water and lose solutes.

Why? In the gills of freshwater fishes, epithelial cells have a higher solute concentration than the fresh water outside. The fresh water is **hyposmotic** to the fishes' tissues. As a result, these fishes gain water via osmosis across the gill epithelium, which puts them under osmotic stress. Just as in marine fishes, there is a trade-off between gas exchange and osmoregulation.

If a freshwater fish does not get rid of incoming water, its cells will burst and it will die. To achieve homeostasis and survive, freshwater fishes excrete large amounts of water in their urine and do not drink.

Freshwater fishes also undergo osmotic stress because electrolytes tend to diffuse out of the gill epithelium into the environment, down their concentration gradients. These animals must replace the lost electrolytes by obtaining them in food or by actively transporting them into the body from the surrounding water—usually across the gills.

What about land animals? In terms of water balance, terrestrial environments are similar to the ocean. Like marine bony fishes, land animals constantly lose water to the environment. On land, however, the process involved is not osmosis but evaporation (Ch. 39, Section 39.5).

The epithelial cells that line a turtle's lung and a fruit fly's gas-exchange structures have a moist surface, which protects the integrity of the cells' plasma membranes and promotes diffusion of gases across the epithelia. Because the atmosphere is almost always drier than this surface, terrestrial animals lose water by evaporation. Once again, there is a trade-off between gas exchange and osmoregulation.

Water balance in land animals is further complicated because water is lost in urine and, in some species, when they sweat or pant to lower body temperature. The lost water is replaced by drinking, ingesting water in food, or gaining metabolic water—water produced during cellular respiration (Ch. 9, Section 9.1). The relative importance of each of these methods of replacing water depends on the species. For example, many desert animals do not have access to drinking water throughout much of the year, so they rely more on water in food and metabolic water.

Making Models 40.1 illustrates how to use arrows to model the osmoregulatory challenges and solutions faced by animals.

Making Models 40.1 Tips on Drawing Arrows (II)

Using arrows to show gain and loss of water and ions is helpful in understanding the osmoregulatory challenges faced by animals as well as the ways they solve these problems. The model shown here uses filled arrows to illustrate the challenges faced by marine bony fishes (losing water to and gaining ions from the ocean) and open arrows to show the solutions (drink seawater, pump ions out across gills, low urine volume).

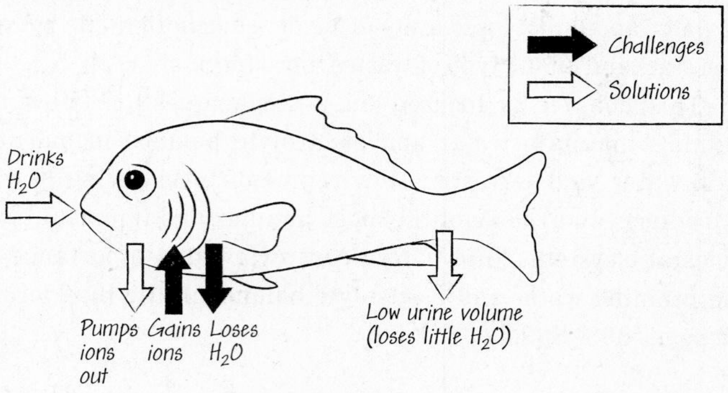

MODEL Draw models showing osmoregulatory challenges and solutions for (a) a freshwater fish and (b) a terrestrial turtle.

To see this model in action, go to the Study Area of **Mastering Biology**

How Do Electrolytes and Water Move across Cell Membranes?

What molecular mechanisms allow animals to cope with the diverse challenges they face in maintaining water and electrolyte balance? Recall that solutes move across cell membranes by passive or active transport (Ch. 6, Sections 6.3 and 6.4). *Passive transport* occurs by diffusion along an electrochemical gradient and does not require an expenditure of energy by the cell. *Active transport*, in contrast, occurs when a source of energy such as ATP powers the movement of a solute, establishing a concentration gradient or an electrochemical gradient.

Passive transport mechanisms include simple diffusion and facilitated diffusion. In *simple diffusion*, small, uncharged molecules such as gases move directly across the membrane. In *facilitated diffusion*, large and/or charged molecules move across the membrane through a channel.

In *primary active transport*, a source of energy such as ATP is used to move ions against their gradients. The sodium–potassium pump, or Na^+/K^+-ATPase, is important in active transport in animals. *Secondary active transport*, or *cotransport*, relies on membrane proteins that use an electrochemical gradient established by a pump during primary active transport. A cotransporter that moves different solutes in the same direction is called a *symporter*; a cotransporter that moves different solutes in opposite directions is called an *antiporter* (Ch. 35, Section 35.3).

How does water cross cell membranes? To date, there are no known mechanisms for actively transporting water across membranes. Instead, cells use pumps to transport ions and set up an osmotic gradient; water then follows by osmosis—often through the specialized membrane channels called **aquaporins** (Ch. 6, Section 6.4). In essence, cells move water by moving solutes.

Animals excrete excess solutes, along with waste products, using a urinary system. Because solutes and wastes often must be dissolved in water to be excreted, water balance is fundamentally related to excretion.

How Do Different Forms of Nitrogenous Waste Impact Water Balance?

Animal cells contain amino acids and nucleic acids that are used to synthesize proteins, RNA, and DNA. Excess amino acids and nucleic acids can be broken down in catabolic reactions that produce **ammonia (NH_3)**. Ammonia is toxic to cells because at high concentrations it raises the pH of intracellular and extracellular fluids enough to inactivate enzymes.

How do animals avoid these toxic effects? They get rid of the ammonia—by excreting it or by converting it to other nitrogen-containing compounds, which are then excreted. Ammonia and these other compounds are referred to as **nitrogenous wastes**.

Forms of Nitrogenous Waste Vary among Species Because nitrogenous wastes must be dissolved in water, their excretion inevitably leads to water loss. However, the amount of water an animal loses during excretion depends on the form of nitrogenous waste it excretes: ammonia, urea, or uric acid (**Table 40.1**).

- In freshwater fishes, ammonia is diluted to a low concentration and excreted in watery urine.
- In freshwater and marine bony fishes, ammonia diffuses across the gills into the surrounding water down its concentration gradient.
- In mammals (including humans) and adult amphibians, enzyme-catalyzed reactions convert ammonia to a much less toxic compound called **urea** (pronounced *yoo-REE-ah*), which is excreted in urine.
- In terrestrial arthropods, birds, and other reptiles, reactions convert ammonia to **uric acid**, the white, paste-like substance that you have probably seen in bird feces. Compared with urea and ammonia, uric acid is much less soluble in water (which explains why it is so difficult to wash bird droppings off a car). As a result, animals that excrete uric acid can get rid of excess nitrogen while losing little water.

Why Do Nitrogenous Wastes Vary among Species? The type of nitrogenous waste produced by an animal correlates with its lineage—its evolutionary history. For example, mammals excrete urea while reptiles (including birds) excrete uric acid (see Table 40.1).

Evolutionary history is not the entire story, however. Nitrogenous waste production is also related to the amount of osmotic stress that a species endures, which is influenced by the habitat it occupies.

- Terrestrial birds conserve water by excreting about 90 percent of their nitrogenous waste as uric acid and only 3–4 percent as ammonia. However, aquatic birds such as ducks excrete just 50 percent of their excess nitrogen as uric acid and 30 percent as ammonia.
- Tadpoles are aquatic and excrete ammonia, but many adult frogs and toads are terrestrial and excrete urea.
- Production of urea and uric acid is particularly common in animals—such as reptiles—that live in dry habitats.

To make sense of these observations, consider the fitness trade-off between the energetic cost of synthesizing each type of waste and the benefit of conserving water. Ammonia excretion requires a large water loss but little energy expenditure because the molecule isn't processed by enzymes. Uric acid excretion, in contrast, requires almost no loss of water but a series of enzyme-catalyzed, energy-demanding reactions. Different trade-offs are favored in different environments.

Now that you have a basic understanding of the osmoregulatory and excretory challenges facing animals, let's take a closer look at how marine, freshwater, and terrestrial animals maintain water and electrolyte homeostasis.

SUMMARY Table 40.1 Attributes of Three Forms of Nitrogenous Waste

Attribute	Ammonia	Urea	Uric Acid
Solubility in water	high	medium	very low
Amount of water required for excretion	high	medium	very low
Toxicity	high	medium	low
Groups in which it is the primary waste	most bony fishes, aquatic invertebrates	mammals, most adult amphibians, sharks, rays, skates	birds and other reptiles, most terrestrial arthropods (insects, spiders)
Method of synthesis	product of breakdown of amino acids and nucleic acids	synthesized in liver, starting with ammonia or amino groups from amino acids	synthesis starts with nucleic acids
Energy cost of synthesis	low	high	high
Method of excretion	in urine, and diffuses across gills	in urine (mammals); diffuses across gills (sharks)	with feces

CHECK YOUR UNDERSTANDING

✔ **If you understood this section, you should be able to…**

1. Explain why maintaining water homeostasis poses similar challenges to marine and terrestrial animals.
2. Explain the relationship between the osmotic stress experienced by a species and the main type of nitrogenous waste it excretes.

Answers are available in Appendix A.

40.2 Water and Electrolyte Balance in Marine and Freshwater Fishes

Osmoregulatory physiology in fishes largely depends on whether the fish lives in the ocean or a body of fresh water. Marine bony fishes and cartilaginous fishes (sharks, rays, and skates) experience severe osmotic stress because they live in water with very high osmolarity. Distinct strategies for dealing with osmotic stress have evolved in these vertebrate lineages since they diverged over 400 million years ago. In contrast, freshwater fishes have to cope with an osmotic stress that is the opposite of the challenge facing marine fishes. Let's start with marine fishes.

After you complete this section, you should be able to …

- Compare water and electrolyte balance in marine and freshwater fishes.

Osmoconformation versus Osmoregulation in Marine Fishes

Osmoregulation and osmoconformation are two strategies for living in the ocean, and each has its own costs and benefits. Marine bony fishes are osmoregulators (**Figure 40.2a**). Recall that they maintain a lower blood osmolarity than that of seawater by drinking seawater to replace water lost via osmosis and by actively transporting electrolytes out of the body. This process comes with a significant energetic cost.

Sharks, rays, and skates are osmoconformers (**Figure 40.2b**). However, the composition of their blood is quite different from that of seawater. Shark blood contains low concentrations of ions but a relatively high concentration of urea. This increases their blood osmolarity so that it is nearly isosmotic with seawater. The result? Sharks lose little water by osmosis. However, sharks must expend energy to make proteins that protect their cells from the toxic effects of the high urea concentration.

Even though they are osmoconformers, sharks still maintain a relatively low concentration of salt (NaCl) in their blood. To do so, sharks must excrete salt, because sodium and chloride ions diffuse into their gill cells from seawater along the ions' concentration gradients. Research on the molecular mechanism of salt excretion in sharks revealed the following key points:

- The mechanism is found in a wide array of species, including *Homo sapiens*. It is functioning in your kidneys right now.
- The mechanism represents a critically important concept in physiology. Plant and animal cells use active transport to set up a strong electrochemical gradient for one ion—typically Na^+ in animals and H^+ in plants. The sodium ion or proton gradient is then used to transport a variety of other substances without further expenditure of energy.

Salt excretion is fundamental to life. Let's dig in.

How Do Sharks Excrete Salt?

Research on salt excretion in sharks focused on an organ called the **rectal gland**, which secretes a concentrated salt solution into its rectum, where it is then excreted into the environment. To determine how this gland works, researchers studied it in vitro—meaning outside the shark's body, in a controlled laboratory environment. The basic approach was to dissect rectal glands, immerse them in a solution with a defined composition and osmolarity, and analyze the fluid that the glands produced.

Early experiments showed that normal salt excretion occurred only if the solution in the rectal gland contained ATP. This result supported the hypothesis that salt excretion involves

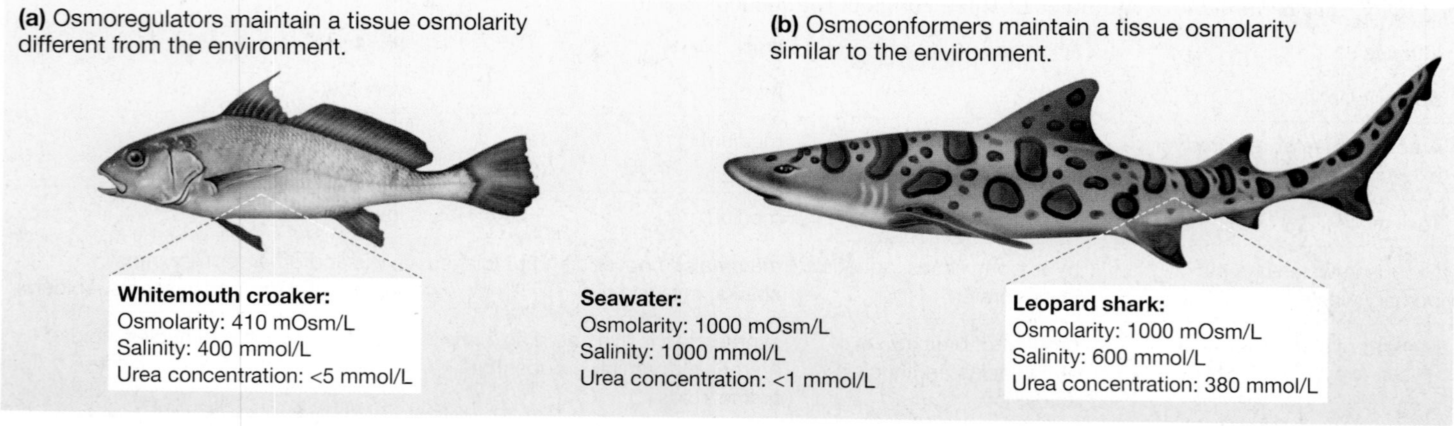

Figure 40.2 Osmoregulation and Osmoconformation Are Distinct Strategies for Marine Animals. (a) Marine bony fishes regulate their tissue osmolarity so it is lower than that of the ocean. **(b)** Marine cartilaginous fishes keep their tissue salinity lower than that of the ocean, but their high urea concentration keeps their osmolarity similar to that of the ocean.
DATA: A. G. Becker et al. 2011. *Neotropical Ichthyology* 9: 895–900; W. W. Dowd et al. 2010. *Journal of Experimental Biology* 213: 210–224.

active transport. A concentrated salt solution can be produced only if ions are actively transported against a concentration gradient. How do ions become concentrated?

The Role of Na^+/K^+-ATPase An energy-demanding mechanism for salt excretion implies that a protein in the plasma membrane of epithelial cells is actively pumping Na^+, Cl^-, or both. The best-characterized candidate was Na^+/K^+-ATPase.

To test the hypothesis that Na^+/K^+-ATPase is involved in salt excretion by shark rectal glands, biologists used a plant defense compound called **ouabain** (pronounced *WAH-bane*). This molecule is toxic to animals because it binds to Na^+/K^+-ATPase and prevents it from functioning.

Just as predicted, rectal glands that were treated with ouabain stopped producing a concentrated salt solution. This was strong evidence that Na^+/K^+-ATPase is essential for salt excretion.

A Molecular Model for Salt Excretion Subsequent work has shown that salt excretion in sharks is a multistep process, summarized in Figure 40.3.

1. Na^+/K^+-ATPase pumps Na^+ out of epithelial cells across the basolateral surface, into the **interstitial fluid**—the extracellular fluid surrounding the rectal gland. The pump also moves K^+ into the cell from the interstitial fluid, creating an electrochemical gradient that favors the diffusion of Na^+ into the cell and K^+ out of the cell.
2. The $Na^+/Cl^-/K^+$ cotransporter moves these three ions into the cell by secondary active transport, powered by the Na^+ gradient. Note that this cotransporter allows Na^+ to diffuse into the cell *down* its electrochemical gradient, causing Cl^- and K^+ to move into the cell *against* their electrochemical gradients. This entry of Cl^- and K^+ is possible only because these ions move through a cotransporter with Na^+.
3. As Cl^- builds up inside the cell, Cl^- diffuses down its electrochemical gradient out of the cell and into the lumen of the gland through a chloride channel located in the apical membrane. At the same time, K^+ diffuses out of the cell into the interstitial fluid through basolateral potassium channels.
4. Following its electrochemical gradient, Na^+ diffuses from the interstitial fluid into the lumen of the gland through spaces between the cells.

In many animals, epithelial cells that transport sodium and chloride ions contain the same combination of membrane proteins found in the shark rectal gland. These species include

- marine birds and other reptiles that drink seawater and excrete NaCl via salt glands in their nostrils;
- marine fishes that excrete salt from their gills; and
- mammals that transport salt in their kidneys.

Research on the shark rectal gland also produced an unforeseen benefit for understanding cystic fibrosis, the most common genetic disease in human populations of northern European extraction (Ch. 6, Section 6.4). Several years after the shark chloride channel was characterized, investigators identified a protein called cystic fibrosis transmembrane regulator (CFTR). Although the disease was known to be associated with defects in CFTR, no one knew what the protein did.

When investigators realized that the amino acid sequence of CFTR is 80 percent identical to that of the shark chloride channel, it was their first hint that CFTR is involved in Cl^- transport. Subsequent studies supported the hypothesis that cystic fibrosis results from a defect in a chloride channel. With this result, research on water and electrolyte balance in sharks shed light on an important human disease.

Research on the shark rectal gland and the gills of marine bony fishes has uncovered the molecular mechanisms of salt balance

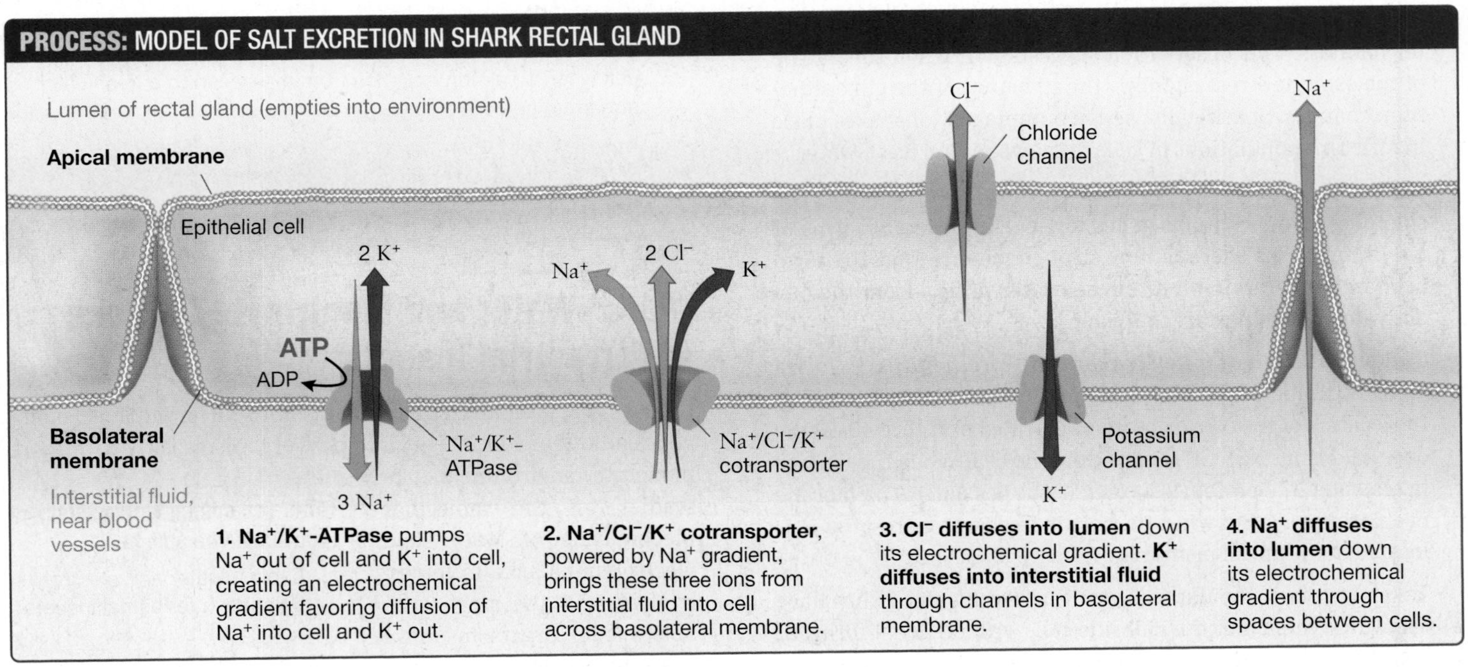

Figure 40.3 **The Rectal Gland Rids the Shark Body of Excess Salt.**

in these animals. How do freshwater fishes achieve homeostasis with respect to electrolytes?

How Do Freshwater Fishes Osmoregulate?

The osmotic stress facing freshwater fishes is the opposite of that facing marine fishes. Freshwater fishes lose electrolytes across their gill epithelium by diffusion. To maintain homeostasis, they must actively transport ions back into the body across the gill epithelium. How do they do this?

Sea Bass and Salmon as Model Systems To understand how freshwater fishes gain electrolytes, researchers have focused on sea bass and several species of salmon. In the course of a lifetime, individuals of these species move between seawater and fresh water. In doing so, they move between environments that present dramatically different osmotic stressors.

In marine bony fishes, specialized cells in the gill epithelium called chloride cells move salt using the same combination of membrane proteins used by epithelial cells in the shark rectal gland (Figure 40.3). When sea bass and salmon are in seawater, chloride cells are abundant and active. What happens to these cells when individuals move into fresh water? Do the changes that occur provide any insight into how these species acclimatize to their new environment and avoid dying of osmotic stress?

Is There a Freshwater Chloride Cell? The results of research on sea bass and salmon support the hypothesis that there is a freshwater version of the classical chloride cell—one that moves ions in the opposite direction of the seawater version. Instead of excreting salt, these cells import it.

The following lines of evidence have accumulated to date:

- ***Osmoregulatory cells may be in different locations.*** Salmon taken from fresh water and seawater have chloride cells in different locations on the gills. The same is true for other fish species that switch between freshwater and seawater habitats. This observation suggests that when the nature of the osmotic stress changes, the structure of the gill epithelium changes. Specifically, active pumping of ions takes place in different populations of cells in seawater and fresh water.
- ***Different forms of Na^+/K^+-ATPase may be activated.*** The salmon genome contains genes for several different forms of Na^+/K^+-ATPase. There is now strong evidence that the form activated when salmon are in seawater differs from the one activated when they are in fresh water.
- ***The orientation of a key transport protein "flips."*** In sea bass, researchers have been able to stain epithelial cells to reveal the location of the cotransporter illustrated in Figure 40.3—the one that brings Na^+, Cl^-, and K^+ into the cell. When the fish are in seawater, the protein is located in the basolateral membrane of chloride cells. But when they are in fresh water, the protein is located in the apical membrane (Figure 40.4).

Taken together, the data suggest that freshwater fishes have a freshwater version of the chloride cell, with different forms of Na^+/K^+-ATPase and transporters that result in the import rather than export of ions. Identifying the mechanisms of electrolyte uptake in freshwater fishes is an important challenge for researchers who want to know how aquatic organisms cope with osmotic stress.

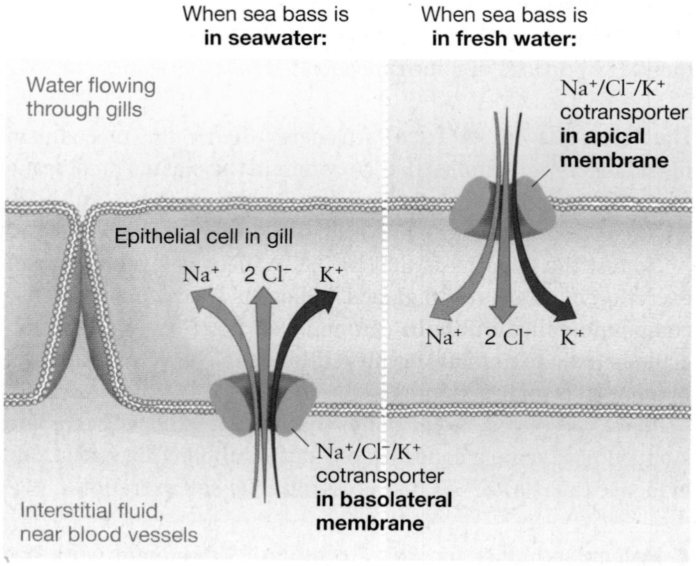

Figure 40.4 A Key Ion Cotransporter Can Change Locations in Gill Epithelial Cells of Sea Bass. Changing the location of the $Na^+/Cl^-/K^+$ cotransporter helps sea bass deal with the different osmotic stresses caused by seawater and freshwater environments. (Other membrane channels and pumps are present but not shown here.)

CHECK YOUR UNDERSTANDING

✔ **If you understood this section, you should be able to...**

1. List the advantages and disadvantages of being an osmoregulating fish versus an osmoconforming fish in a marine environment.
2. **MODEL** Diagram the changes that occur in a gill epithelial cell when a salmon migrates from the ocean to a freshwater stream to spawn.

Answers are available in Appendix A.

40.3 Water and Electrolyte Balance in Terrestrial Insects

After you complete this section, you should be able to ...

- Explain water and electrolyte balance in terrestrial insects.

By studying extreme situations or unusual organisms, biologists can often gain insight into how organisms cope with more moderate environments. In studies on the molecular mechanisms of water and electrolyte balance in terrestrial insects, the most valuable model organisms have been the desert locust and a common household pest called the flour beetle. (You may have seen the larvae of flour beetles, called mealworms, in bags of flour that were not shut tightly enough to keep adults from entering and breeding.)

Desert locusts and flour beetles live in environments where osmotic stress is severe. These insects rarely, if ever, drink—simply because little or no water is available in their habitats.

How do they maintain water and electrolyte balance? The answer has two parts: They minimize water loss from their body surface, and they carefully regulate the amount of water and electrolytes that they excrete in their urine and feces. Let's look at each issue in turn.

How Do Insects Minimize Water Loss from the Body Surface?

Terrestrial animals breathe by exposing an extremely thin respiratory epithelium to the atmosphere (Ch. 42, Section 42.3). Oxygen diffuses into this epithelium, and carbon dioxide diffuses out. Water also crosses the respiratory surface and is lost to the atmosphere via evaporation.

Evaporation from the rest of the body is another threat—this is a particular challenge to insects because they are small and, as a result, have a high surface area to volume ratio (Ch. 39, Section 39.3). In other words, insects have a relatively large surface area from which to lose water but a small volume in which to retain it.

Figure 40.5a shows how insects minimize evaporation from the surface of their bodies. This diagram is a cross section of an insect's exoskeleton, which consists of a layer of chitin—a tough, nitrogen-containing polysaccharide—and protein covered by a layer of wax. This combination of chitin, protein, and wax is known as the **cuticle**.

Recall that waxes, a type of lipid, are highly hydrophobic and thus highly impermeable to water (Ch. 6, Section 6.1). Researchers who removed the wax from insect exoskeletons found that the rate of water loss from the body surface increased sharply. This observation indicates that the wax layer is an adaptation that minimizes evaporative water loss.

In terrestrial insects, gas exchange occurs across the membranes of epithelial cells that line the **tracheae**, an extensive system of tubes. The insect tracheal system connects with the atmosphere at openings called **spiracles** (Figure 40.5b). Muscles just inside each spiracle open or close the opening, much as guard cells open or close the stomatal pores in plant leaves and stems (Ch. 37, Section 37.6).

When investigators manipulated assassin bugs (*Rhodnius*) so that their spiracles stayed open and then placed the animals in a dry environment, the insects died within three days. These data support the hypothesis that the ability to close spiracles is an important adaptation for minimizing water loss during respiration. If an insect is under osmotic stress, it may be able to close its spiracles, reduce its metabolic rate, and wait until conditions improve before resuming activity.

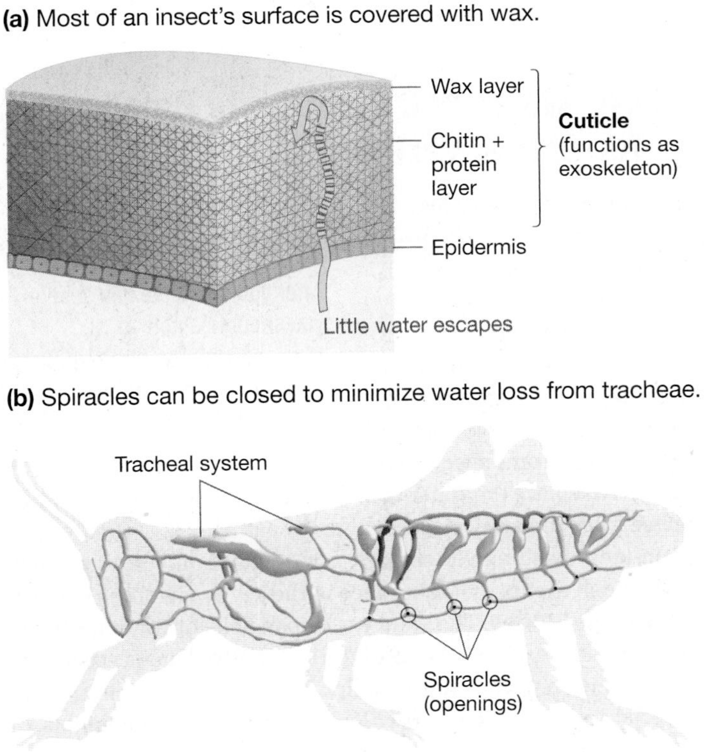

Figure 40.5 In Terrestrial Insects, Adaptations Limit Water Loss from the Body Surface and from the Respiratory System.

✔ In what season would a desert insect's cuticle likely have the most wax? Explain.

How Do Insects Regulate the Amount of Water and Electrolytes They Excrete?

For insects, minimizing water loss is only half the battle in avoiding osmotic stress. To maintain homeostasis, insects must also carefully regulate the composition of a blood-like fluid called **hemolymph**. Hemolymph is pumped by the heart and transports electrolytes, nutrients, and waste products, and is modified in a regulated process to produce urine.

The Malpighian Tubules Allow Insects to Make Concentrated Urine To maintain water and electrolyte balance, insects rely on excretory organs called **Malpighian tubules** and on their hindgut—the posterior portion of their digestive tract. As Figure 40.6 on page 866 shows, Malpighian tubules have a large surface area, are in direct contact with the hemolymph, and empty into the hindgut. The Malpighian tubules are responsible for forming a **filtrate**, a filtered liquid, from the hemolymph. This "pre-urine" then passes into the hindgut, where it is processed and modified before excretion.

How is the filtrate formed? Epithelial cells in the Malpighian tubules contain pumps that actively transport K^+ into the lumen of the organ. The resulting high concentration of K^+ brings water into the tubules by osmosis. Other electrolytes and nitrogenous wastes then diffuse into the filtrate down their concentration gradients.

The filtrate that accumulates inside the Malpighian tubules flows into the hindgut, where it joins digested food. If an insect is osmotically stressed, electrolytes and water from the filtrate are reabsorbed from the hindgut and returned to the hemolymph, while uric acid remains in the hindgut. Reabsorption results in formation of hyperosmotic final urine, conservation of water, and efficient elimination of nitrogenous wastes.

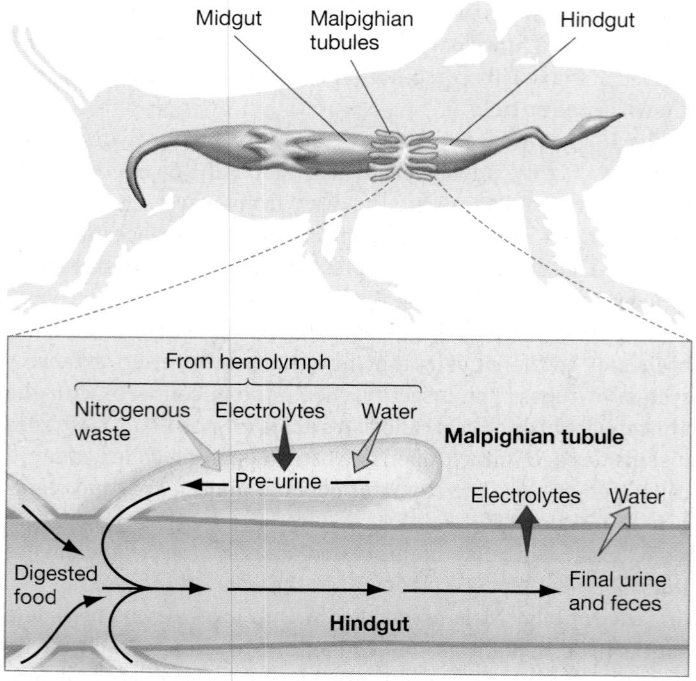

Figure 40.6 In Insects, Urine Forms in the Malpighian Tubules and Is Modified in the Hindgut. The isosmotic filtrate that forms in the Malpighian tubules empties into the hindgut, where it mixes with fecal material. Valuable substances such as electrolytes and water are selectively reabsorbed from the hindgut, leaving wastes to be excreted with feces.

In terrestrial insects, 80 to 95 percent of the water in the filtrate is reabsorbed and kept inside the body. The ability to recover this water allows these insects to live in extremely dry habitats such as deserts and flour bins.

How does reabsorption happen? The mechanism involves specific membrane pumps and channels, not unlike the system found in the chloride cells of fishes. To study it, researchers removed the hindgut epithelium from a desert locust and set it up as a sheet dividing two solutions. When they removed K^+ and Na^+ from the solution on the lumen side of the organ, water reabsorption stopped. This result established that the hindgut's ability to recover water from urine depends on ion movement: The epithelial cells in the hindgut transport specific ions out of the filtrate and into the hemolymph. Water follows by osmosis, forming concentrated urine.

General Principles of Water and Electrolyte Balance Regulation in Insects Several general principles have emerged from studies of insect excretion:

- Water is not pumped directly. Water moves between cells or body compartments via osmotic gradients that are set up by the active transport of ions.
- The formation of the filtrate is not particularly selective. Most of the molecules present in the hemolymph are also present in the Malpighian tubules.
- In contrast to filtrate formation, reabsorption is highly selective. The membrane pumps and channels involved in reabsorption are highly specific for certain ions and molecules. Waste products do not pass through the hindgut membrane. Instead, they remain in the hindgut and are eliminated along with the feces. Only valuable ions and molecules such as nutrients are reabsorbed.
- In contrast to filtrate formation, reabsorption is tightly regulated. The membrane pumps and channels involved in reabsorption are activated and deactivated in response to osmotic stress. If an insect is dehydrated, nearly all of the water in the filtrate is reabsorbed. But if it has plenty to drink, reabsorption does not occur and the urine is watery and hyposmotic to the hemolymph. The system is dynamic and allows precise control over water and electrolyte balance.

Given the numbers of species and individuals and the array of habitats that insects occupy, it is clear that their systems for maintaining water and electrolyte balance are remarkably effective. The general principles learned from studies of insect excretion turn out to be relevant to vertebrate systems as well.

CHECK YOUR UNDERSTANDING

✔ **If you understood this section, you should be able to...**

Explain how the following traits are involved in water retention in insects:

1. Excretion of nitrogenous waste in the form of uric acid.
2. Selective reabsorption of electrolytes in the hindgut.

Answers are available in Appendix A.

40.4 Water and Electrolyte Balance in Terrestrial Vertebrates

After you complete this section, you should be able to ...

- Explain water and electrolyte balance in terrestrial vertebrates.

In dealing with water loss, terrestrial vertebrates face the same hazards as terrestrial insects do. Crocodiles, turtles, lizards, frogs, birds, and mammals lose water from their body surfaces, from the surface of their lungs every time they breathe, and in their urine. Electrolytes are also lost in urine and, in some species, in sweat. To replace the water they lose, most terrestrial vertebrates drink. They replace lost electrolytes by eating.

In land-dwelling vertebrates, osmoregulation occurs primarily through events that take place in the key organ of the urinary system, the **kidney**. The kidney is responsible for water and electrolyte balance as well as the excretion of nitrogenous wastes. Its function is analogous to the Malpighian tubules and hindgut of insects. Let's first take a closer look at the kidney in mammals.

Structure of the Mammalian Kidney

Mammalian kidneys occur in pairs and tend to be bean shaped. A large blood vessel called the renal artery brings blood that contains nitrogenous wastes into the organ; another large blood

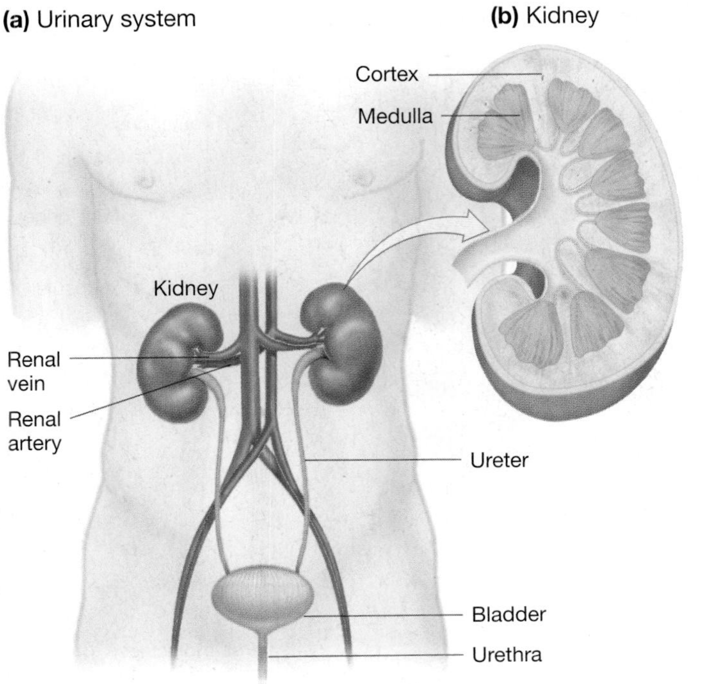

Figure 40.7 The Human Urinary System Consists of the Kidneys, Ureters, Bladder, and Urethra. **(a)** The kidneys are paired organs. The urine they produce is stored in the bladder and transported out of the body by the urethra. **(b)** Each kidney has an outer region called the cortex and an inner area called the medulla.

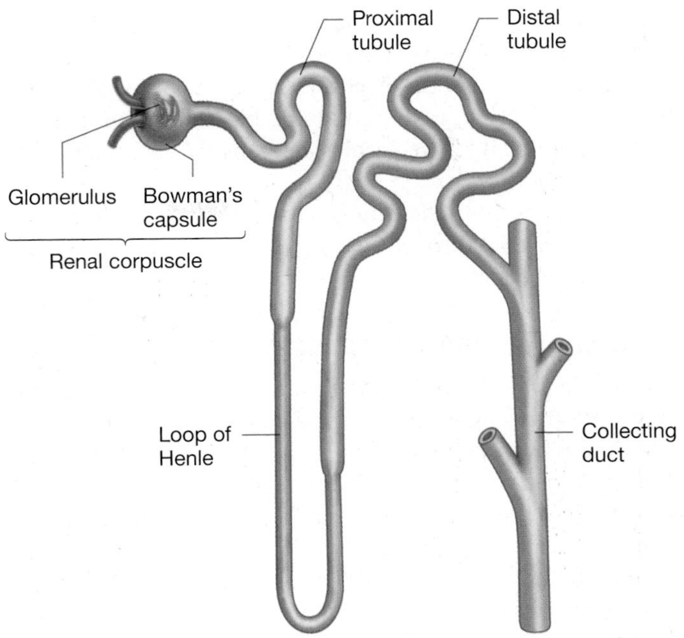

Figure 40.8 A Nephron Has Four Major Regions and Empties into a Collecting Duct. Urine formation begins in the renal corpuscle and ends in the collecting duct.

vessel, the renal vein, carries away blood that has been cleared of wastes.

The urine that forms in the kidney is transported via a long tube called the **ureter** to a storage organ, the **bladder**. From the bladder, urine is transported to the body surface through the **urethra** and then excreted. In most vertebrates, the kidneys are located near the dorsal (back) side of the body. Figure 40.7a summarizes the parts of the human urinary system.

Most of the kidney's mass is made up of small structures called nephrons. The **nephron** is the basic functional unit of the kidney. The work involved in maintaining water and electrolyte balance occurs in the nephron.

Most of the approximately 1 million nephrons in a human kidney are located almost entirely in the outer region of the organ, or **cortex** (Figure 40.7b). But some nephrons extend from the cortex into the kidney's inner region, or **medulla**.

Function of the Mammalian Kidney: An Overview

The nephron shares important functional characteristics with the insect excretory system:

- Water cannot be transported actively—it crosses membranes only by osmosis.
- To move water, cells in the kidney set up a strong osmotic gradient in the interstitial fluid surrounding the nephrons.
- By regulating these gradients and specific channel proteins, kidney cells exert precise control over loss or retention of water and electrolytes.

Figure 40.8 provides an overview of the nephron. Inside the kidney, nephrons are coiled and twisted, so the figure untwists a nephron for you to more easily see its components. A nephron has four major regions, the last of which is connected to a structure called the collecting duct. Each nephron is basically a tube that is closed at one end and open at the other. The closed end is the beginning of the nephron; the open end empties its contents into collecting ducts.

The major nephron regions and the collecting duct each have a distinct function:

- The *renal corpuscle* filters blood, forming a filtrate, or pre-urine, consisting of ions, nutrients, wastes, and water.
- The *proximal tubule* has epithelial cells that reabsorb nutrients, ions, and water from the filtrate into the blood.
- The *loop of Henle*, or *nephron loop*, establishes and maintains a strong osmotic gradient in the interstitial fluid surrounding the loop. Osmolarity of the interstitial fluid increases as the loop descends into the medulla.
- The *distal tubule* reabsorbs ions and water from the filtrate in a regulated manner—one that helps maintain water and electrolyte balance according to the body's needs.
- The *collecting duct* may reabsorb more water to maintain homeostasis. In addition, urea moves from the urine to the interstitial fluid at the base of the collecting duct and contributes to the medullary osmotic gradient set up by the loop of Henle.

Each of a nephron's four regions is surrounded by blood vessels. These are not shown in Figure 40.8 so that you can more easily see the nephron, but the blood vessels play a key role in kidney function: They bring waste-containing blood into the nephron and then take away the molecules and ions that are reabsorbed from the initial filtrate.

Now let's delve into the details. The sections that follow trace the flow of material through each region of the nephron and out of the collecting duct.

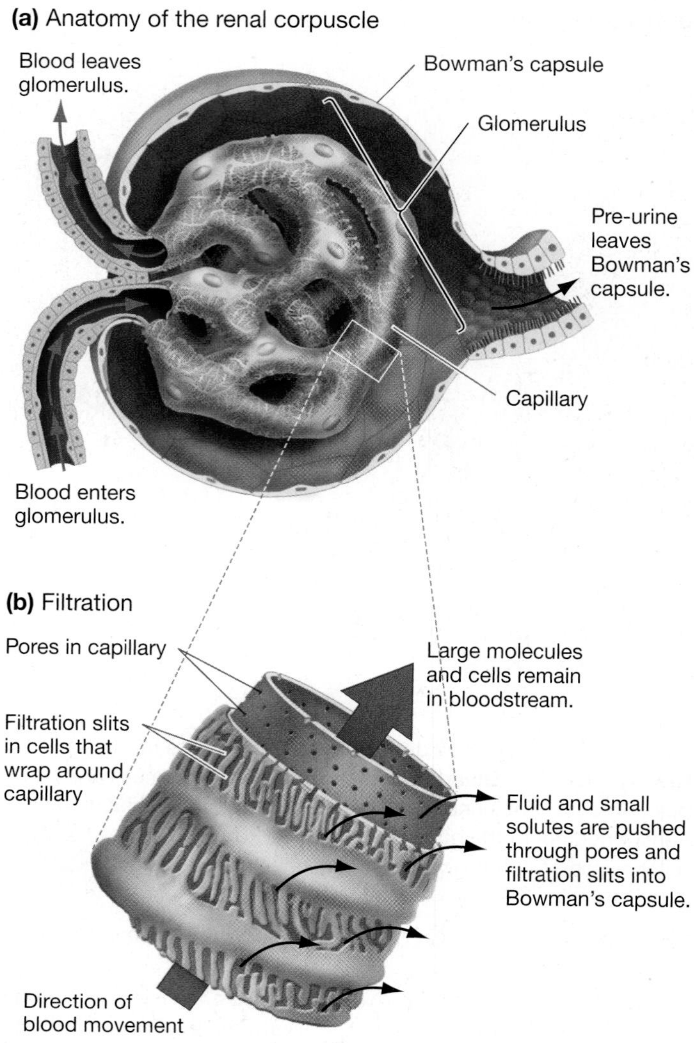

Figure 40.9 Urine Formation Begins when Blood Is Filtered in the Renal Corpuscle. (a) The renal corpuscle consists of Bowman's capsule and the glomerulus. **(b)** The capillaries in the glomerulus have pores and are surrounded by cells that have filtration slits. Blood pressure forces water and small molecules out of the capillaries, through the slits, and into Bowman's capsule.

Filtration: The Renal Corpuscle

In terrestrial vertebrates, urine formation begins in the **renal corpuscle** (literally, "kidney little-body"). As Figure 40.9a shows, the renal corpuscle forms a capsule that encloses a cluster of tiny blood vessels, or capillaries. These vessels bring blood to the nephron from the renal artery. Collectively, the cluster of capillaries is called the **glomerulus** ("ball of yarn"). The region of the nephron that surrounds the glomerulus is named **Bowman's capsule**, or the *glomerular capsule*. Together, the glomerulus and Bowman's capsule make up the renal corpuscle.

Figure 40.9b illustrates a key feature of the glomerular capillaries: They have large pores, or openings. In addition, they are surrounded by unusual cells whose membranes fold into a series of slits and ridges called filtration slits.

The structure of the renal corpuscle allows it to function as a **filtration** device. Water and small solutes from the blood pass through the pores and slits into Bowman's capsule. Filtration is based on size: Proteins, cells, and other large components of blood do not fit through the pores and do not enter the nephron. They remain in the blood instead.

Stated another way, urine formation starts with a size-selective filtration step—with blood pressure supplying the force required to perform filtration. In vertebrates, blood is under higher pressure than the surrounding tissues because it is pumped by the heart through a closed system of vessels. This pressure is enough to force water and small solutes through the pores in the glomerulus, allowing the renal corpuscle to strain large volumes of fluid without expending energy in the form of ATP.

Note the following critical facts about the filtration step:

- The renal corpuscles of a human kidney can produce about 180 liters of filtrate per day. This is an impressive volume—think of 90 two-liter bottles of soft drink arranged on a supermarket shelf.
- About 99 percent of the filtrate is reabsorbed—only a tiny fraction of the original volume is excreted as urine.

Filtering large volumes from the blood allows wastes to be removed effectively. Pairing this process with selective reabsorption allows waste excretion to occur with a minimum of water and nutrient loss.

Reabsorption: The Proximal Tubule

Where does the process of filtrate reabsorption begin? Filtrate leaves Bowman's capsule and enters a convoluted structure called the **proximal tubule**. The filtrate inside this tubule contains water and small solutes such as urea, glucose, amino acids, vitamins, and electrolytes. Some of these solutes are waste products; others are valuable nutrients.

Active Transport Occurs in Epithelial Cells The epithelial cells of the proximal tubule have a prominent series of small projections, called **microvilli** ("little shaggy hairs"), facing the lumen (Figure 40.10a). The microvilli greatly increase the surface area of this epithelium. A large surface area provides space for membrane proteins that act as pumps, channels, and cotransporters.

Epithelial cells in the proximal tubule are also packed with mitochondria, which suggests that these cells carry out extensive ATP-demanding active transport. Based on these observations,

(a) Microvilli increase the inner surface area of epithelium in proximal tubule.

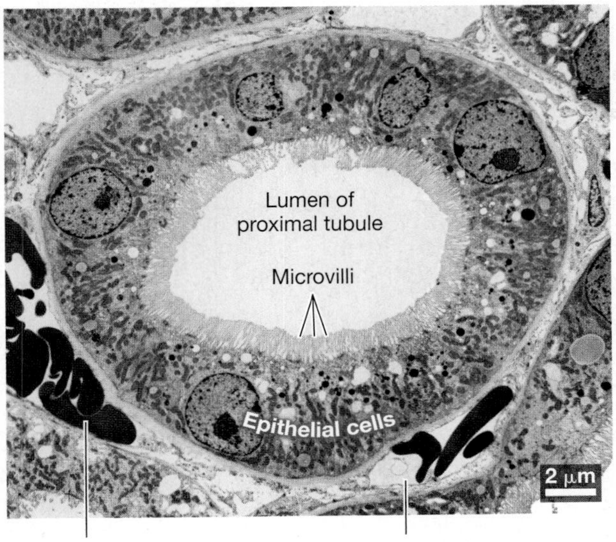

(b) Model of selective reabsorption in proximal tubule

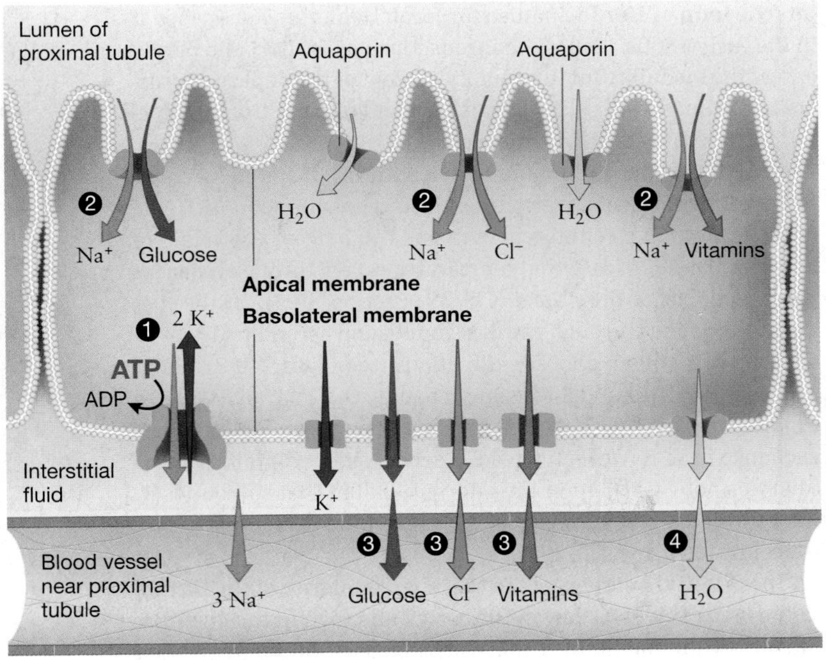

INTERACTIVE Figure 40.10 Water, Electrolytes, and Nutrients Are Reabsorbed in the Proximal Tubule. The step numbers in part (b) are explained in the following text section.

biologists hypothesized that the proximal tubule functions in the active transport of selected molecules out of the filtrate.

By injecting solutions of known composition into isolated rabbit and rat proximal tubules in the presence or absence of ATP, researchers confirmed that selected electrolytes and nutrients are actively reabsorbed from the filtrate that enters the tubules.

When solutes move from the proximal tubule into epithelial cells and then into the bloodstream, water follows along the osmotic gradient. In this way, valuable solutes and water are reabsorbed and returned to the bloodstream.

Ion and Water Movement Is Driven by a Concentration Gradient in the Interstitial Fluid Figure 40.10b summarizes the current model of the molecular mechanisms involved in selective reabsorption in the proximal tubule:

1. ***Active transport creates Na^+ gradient.*** Na^+/K^+-ATPase in the basolateral membrane moves Na^+ from the interior of epithelial cells surrounding the lumen of the proximal tubule to the interstitial fluid. The active transport of Na^+ out of the cells creates a concentration gradient favoring the entry of Na^+ from the lumen.
2. ***Na^+ gradient is used to remove solutes from filtrate.*** Na^+-dependent cotransporters in the apical membrane use the Na^+ gradient to remove ions and nutrients (e.g., Cl^-, glucose, vitamins) selectively from the filtrate in the lumen. The movement of Na^+ into the cell, *down* its electrochemical gradient, provides the means for moving other solutes *against* their gradients.
3. ***Removed substances diffuse into blood.*** The solutes that move from the lumen into the cell diffuse across the basolateral membrane into the interstitial fluid and then into nearby blood vessels.
4. ***Water moves into blood by osmosis.*** Water follows the movement of solutes from the proximal tubule into the cell and then out of the cell and into blood vessels. Recall that water moves by osmosis across the membranes of these epithelial cells through membrane proteins called aquaporins.

Almost all of the nutrients, along with about two-thirds of the NaCl and water that is originally filtered by the renal corpuscle, are reabsorbed in the proximal tubule. The osmolarity of the tubular fluid is unchanged despite this huge change in volume, however, because water reabsorption is proportional to solute reabsorption.

✔ **If you understand the way substances are reabsorbed, you should be able to assign one of the following transport mechanisms to each of the four steps just described: simple diffusion, facilitated diffusion, primary active transport, secondary active transport.**

In effect, then, the cells that line the proximal tubule act as a recycling center. The filtration step in the renal corpuscle is based on size; the reabsorption step in the proximal tubule selectively retrieves small substances that are valuable. The pumps and cotransporters in the proximal tubule reabsorb nutrients, electrolytes, and water but leave wastes. As the filtrate flows into the loop of Henle, it has a relatively high concentration of waste molecules and a relatively low concentration of nutrients.

Creating an Osmotic Gradient: The Loop of Henle

In mammals, the fluid that emerges from the proximal tubule enters the **loop of Henle**—named for Jacob Henle, who described it in the early 1860s. In most nephrons, the loop is short and barely enters the medulla. But in about 20 percent of the nephrons present in a human kidney, the loop is long and plunges from the cortex of the kidney deep into the medulla.

In 1942 Werner Kuhn offered a hypothesis, inspired by countercurrent heat exchangers, to explain what the loop of Henle does. Recall that a countercurrent heat exchanger is a system in which two fluids of different temperatures flow through adjacent pipes in opposite directions (Ch. 39, Section 39.5). In the medulla, fluid flows through the descending and ascending limbs of the loop of Henle in opposite directions (see Figure 40.8). Kuhn proposed that this opposite flow enables the loop of Henle to function as a countercurrent exchanger and multiplier. It doesn't exchange heat, however. Instead, it sets up and maintains a medullary osmotic gradient in the interstitial fluid that surrounds it. As you will learn next, this gradient is essential to creating highly concentrated urine.

Specifically, Kuhn proposed that the osmolarity of the filtrate in the loop of Henle is low in the cortex and high in the medulla. Further, Kuhn maintained that the osmolarity in the interstitial fluid surrounding the loop mirrors the gradient inside the loop. This is a key point.

Testing Kuhn's Hypothesis A series of papers published during the 1950s and early 1960s supplied important experimental support for the countercurrent exchange model. Figure 40.11 reproduces two particularly important data sets, obtained by comparing the osmolarity of kidney tissue slices. In both graphs, the *x*-axis shows the location in the kidney, from cortex to inner medulla. The *y*-axis indicates osmolarity, measured either as the percentage of the maximum observed or as solute concentration.

- Figure 40.11a shows data on the osmolarity of fluid inside the loop of Henle. The vertical lines represent the range of values observed at a particular location. As predicted by Kuhn's model, a strong gradient in osmolarity exists from the cortex to the inner medulla.
- Figure 40.11b shows that the concentrations of Na^+, Cl^- and urea in the interstitial fluid outside the loop of Henle also increase sharply from the cortex to the inner medulla.

These results suggested that the solutes responsible for the gradient outside the loop are Na^+, Cl^- and urea. The change in concentration of urea turned out to be particularly important.

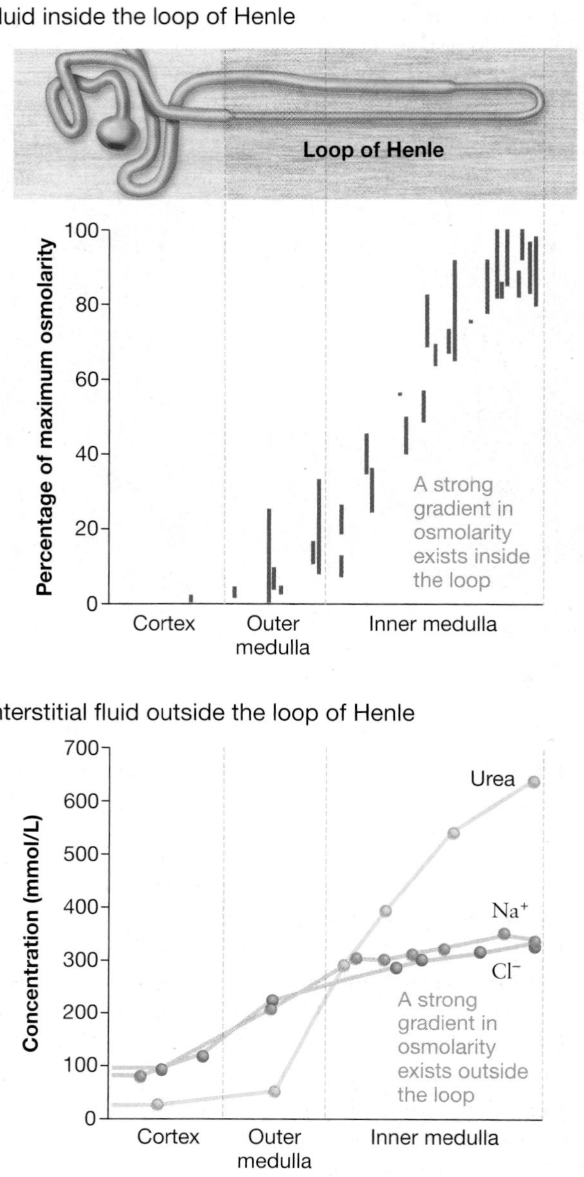

Figure 40.11 Data Confirm the Existence of a Strong Osmotic Gradient Both Inside and Outside the Loop of Henle. As the nephron plunges into the inner medulla, the concentration of dissolved solutes increases both inside **(a)** and outside **(b)** the loop of Henle.

DATA: K. J. Ullrich, K. Kramer, and J. W. Boyer. 1961. *Progress in Cardiovascular Diseases* 3: 395–431.

How Is the Medullary Osmotic Gradient Established and Maintained? The loop of Henle has three distinct regions: the descending limb, the thin ascending limb, and the thick ascending limb (Figure 40.12a). The thin and thick ascending limbs differ in the thickness of their walls. Do the three regions also differ in their permeability to water and solutes?

It took over 15 years of experiments performed in laboratories around the world to definitively answer that question. Researchers punctured the loop of Henle of rodents with a micropipette, analyzed the composition of the fluid inside, and compared it with the composition of the nephron's final product—urine.

In the ascending limb of the loop of Henle, Na^+ and Cl^- constituted at least 60 percent of the solutes; urea constituted about 10 percent. But in the distal tubule, urea was the major solute. These data suggested that Na^+ and Cl^-, but not urea, were being removed somewhere in the ascending limb. How?

Na^+ and Cl^- were also present at high concentrations in the tissue surrounding the thick ascending limb, so researchers hypothesized that Na^+ might be actively pumped out of this portion of the

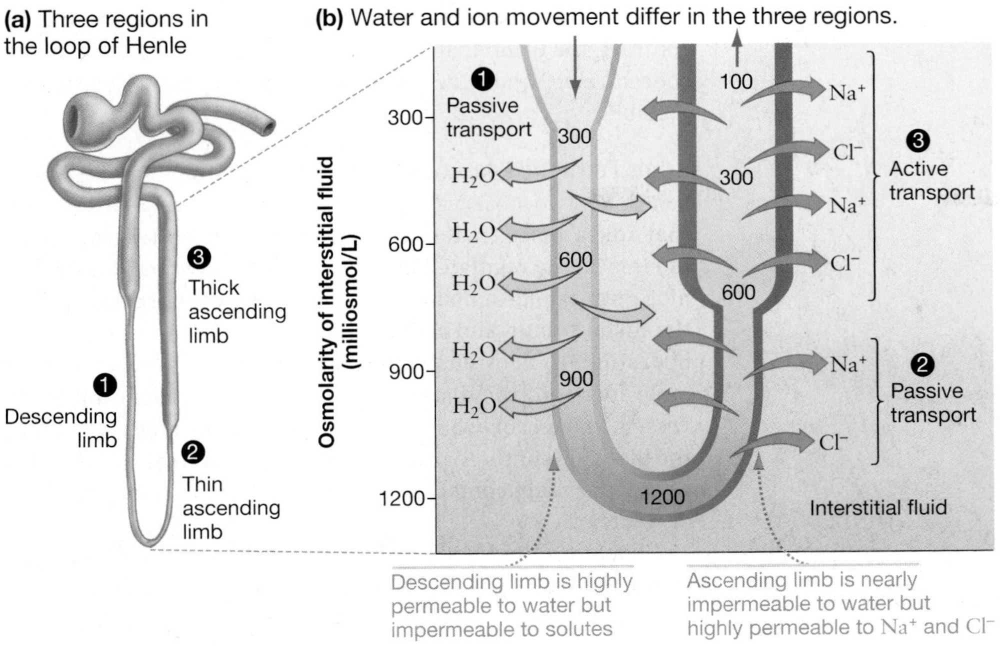

Figure 40.12 The Loop of Henle Maintains an Osmotic Gradient Because Water Leaves the Descending Limb and Salt Leaves the Ascending Limb. The values inside the loop in part (b) represent the osmolarity of the filtrate.

nephron. The hypothesis was that the active transport of Na^+ out of the filtrate in the thick ascending limb would create an electrical gradient that also favored the loss of Cl^-.

Follow-up experiments using ouabain and other poisons supported the hypothesis that Na^+ are actively transported out of the solution inside the thick ascending limb, and Cl^- follow along an electrochemical gradient. The epithelial cells responsible for salt excretion are configured almost exactly like the epithelium of the shark rectal gland (see Figure 40.3).

What is happening in the descending limb and the thin ascending limb of the loop of Henle? By injecting solutions of known concentration into the nephrons of rabbits, biologists documented that the descending limb is highly permeable to water but almost completely impermeable to solutes. The thin ascending limb of the loop, in contrast, is highly permeable to Na^+ and Cl^-, moderately permeable to urea, and almost completely impermeable to water.

The observations just summarized all came together in 1972 when two papers, published independently, proposed the same comprehensive model for how the loop of Henle works. The Na^+ and Cl^- pumped out of the ascending limb of the loop, along with urea reabsorption from the collecting duct, establish and maintain the medullary osmotic gradient. To understand this model, follow the events shown in Figure 40.12b:

1. ***Descending limb: Passive transport of water.*** As fluid flows down the descending limb, the fluid inside the loop loses water to the interstitial fluid surrounding the nephron. This movement of water is passive—it does not require an expenditure of ATP. The water follows an osmotic gradient created by the ascending limb. At the bottom of the loop—in the inner medulla—the fluids inside and outside the nephron have high osmolarity. The filtrate does not continue to lose water, though, because the membrane in the ascending limb is nearly impermeable to water.
2. ***Thin ascending limb: Passive transport of Na^+ and Cl^-.*** The fluid inside the nephron loses Na^+ and Cl^- in the thin ascending limb. The ions move passively, down their electrochemical gradients.
3. ***Thick ascending limb: Active transport of Na^+ and Cl^-.*** Near the cortex, the osmolarity of the surrounding interstitial fluid is low. Additional Na^+ and Cl^- are actively transported out of the nephron in the thick ascending limb.

The countercurrent flow of fluid, combined with changes in permeability to water and in the types of channels and pumps that are active in the epithelium of the nephron, creates a self-reinforcing system. The presence of an osmotic gradient stimulates water and ion flows that in turn maintain an osmotic gradient.

Here's how it works: Movement of NaCl from the ascending limb into surrounding tissue increases the osmolarity of the fluid outside the descending limb, which results in an outward flow of water across the water-permeable epithelium of the descending limb via osmosis. This loss of water in the descending limb increases the osmolarity of the fluid entering the ascending limb. The high concentration of salt in the fluid at the base of the ascending limb triggers a passive flow of ions out—reinforcing the osmotic gradient.

✔ **If you understand this concept, you should be able to predict what happens to the osmotic gradient when the drug furosemide inhibits membrane proteins that pump Na^+ and Cl^- ions out of the thick ascending limb. Specifically, how does this drug affect (1) water reabsorption in the descending limb, (2) the osmolarity of the filtrate at the bottom of the loop of Henle, and (3) reabsorption of salt in the thin ascending limb?**

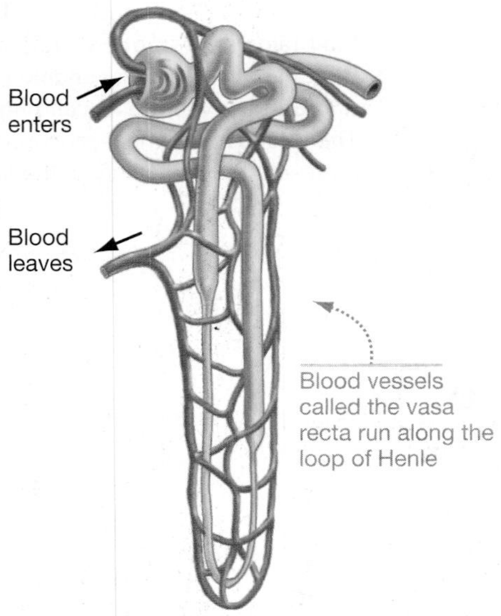

Figure 40.13 Blood Vessels Are Closely Associated with the Nephron. Water and solutes reabsorbed from the loop of Henle enter a system of blood vessels, the vasa recta.

The Vasa Recta Removes Water and Solutes That Leave the Loop of Henle What happens to the water and salt that move out of the loop from the filtrate into the interstitial fluid? They diffuse into the **vasa recta**, a network of blood vessels that runs along the loop. As a result, the reabsorbed water and electrolytes are returned to the bloodstream instead of being excreted in urine (Figure 40.13). The vasa recta joins up with small veins at the top of the medulla, which prevents the reabsorption of water and electrolytes from disrupting the concentration gradient in the medulla.

The removal of water that leaves the descending limb is particularly important. If it were not drawn off into the bloodstream, it would dilute the concentrated fluid outside the loop of Henle and quickly destroy the osmotic gradient.

Urea from the Collecting Duct Adds to the Medullary Osmotic Gradient Urea is also involved in creating the medullary osmotic gradient. The concentration of urea in the interstitial fluid is higher in the inner medulla and lower in the outer medulla. This gradient exists because the innermost section of the collecting duct is permeable to urea.

Although the system created by the nephron, vasa recta, and collecting duct may seem complex, its outcome is simple: it creates and maintains a strong osmotic gradient with the minimum possible expenditure of energy.

Regulating Water and Electrolyte Balance: The Distal Tubule and Collecting Duct

The first three steps in urine formation—filtration, reabsorption, and establishment of an osmotic gradient—result in a filtrate that is slightly hyposmotic to blood. Once the filtrate has passed through the loop of Henle, the major solutes that it contains are urea and other wastes along with a low concentration of ions.

The filtrate that enters the distal tubule is always dilute. In contrast, the urine that leaves the collecting duct can be highly concentrated when the individual is dehydrated. How is this possible?

Urine Formation Is under Hormonal Control The answer is based on two observations about any further reabsorption that takes place in the **distal tubule** and **collecting duct**: **(1)** It is highly regulated, and **(2)** it is altered in response to osmotic stress. The amount of Na^+, Cl^-, and water reabsorbed in the distal tubule and collecting duct varies with the animal's hydration. In addition, selective **secretion** of ions and other molecules into the distal tubule from the surrounding blood vessels may occur when levels of those molecules are high in the blood. All of these processes can lead to a final urine that is very different in composition from the filtrate leaving the loop of Henle.

Changes in the distal tubule and collecting duct are controlled by **hormones**—signaling molecules in the blood. Specifically:

- If the Na^+ level in the blood is low, the adrenal glands release the hormone **aldosterone**, which leads to activation of sodium–potassium pumps, causing reabsorption of Na^+ and secretion of K^+ in the distal tubule. Water is reabsorbed by osmosis. Aldosterone saves Na^+ and water and causes the excretion of K^+. It also activates sodium–proton antiporters that facilitate reabsorption of Na^+ from the distal tubule into the blood and secretion of H^+ from the blood into the distal tubule. The latter helps to regulate blood pH.
- If an individual is dehydrated, the brain releases **antidiuretic hormone (ADH)**. (The term "diuresis" refers to increased urine production, so antidiuresis means inhibited urine production.) ADH saves water.

How Does ADH Work? Epithelial cells of the collecting duct are joined by tight junctions (Ch. 11, Section 11.2), making the epithelium impermeable to water and solutes. ADH has the following important effects on epithelial cells in the collecting duct:

- ADH triggers the insertion of aquaporins into the apical membrane. As a result, cells become much more permeable to water and large amounts of water are reabsorbed.
- ADH increases the cells' permeability to urea, which is reabsorbed into the surrounding fluid. This helps create a concentration gradient favoring water reabsorption from the filtrate.

As Figure 40.14a shows, water leaves the collecting duct passively—following the medullary osmotic gradient maintained by the loop of Henle. When ADH is present, water is conserved by the body, and the urine is strongly hyperosmotic to the blood. The collecting duct is the final place where the composition of the filtrate can be altered.

When ADH is absent, however, few aquaporins are found in the epithelium of the collecting duct, and the structure is relatively impermeable to water (Figure 40.14b). In this case, a larger quantity of hyposmotic urine is produced. Urine exiting the collecting ducts moves from the kidneys into ureters and then is stored in the bladder until urination.

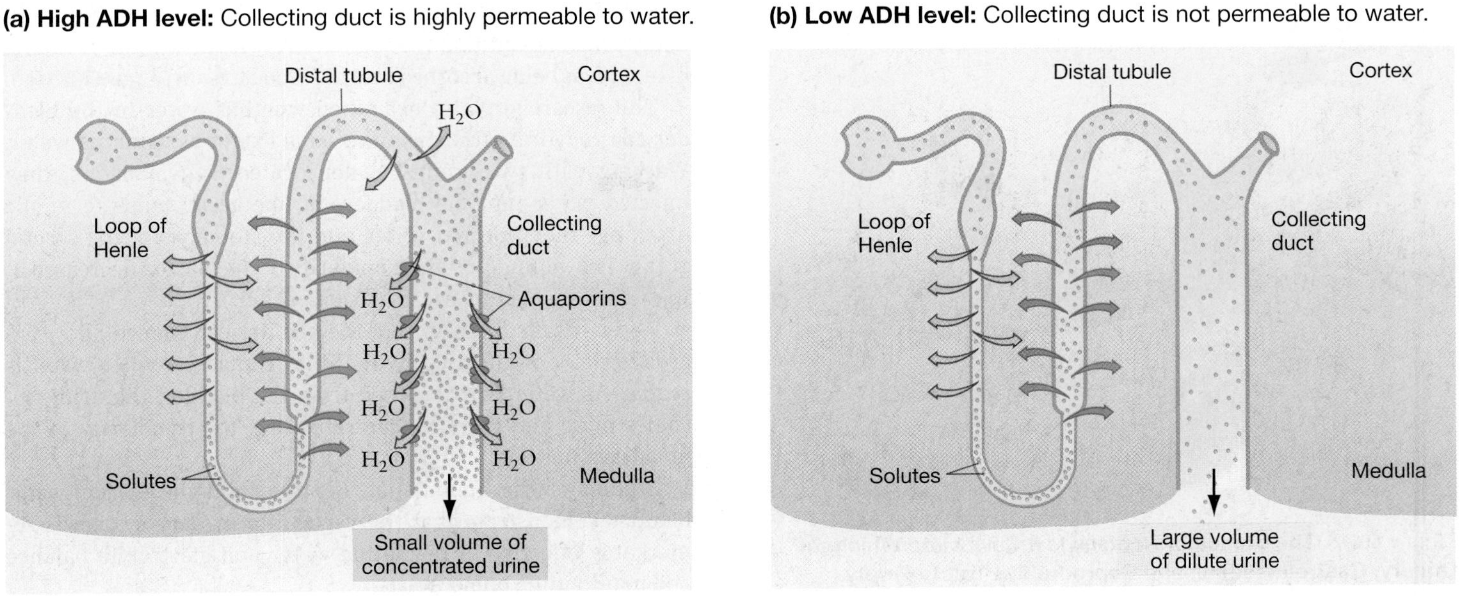

Figure 40.14 ADH Regulates Water Reabsorption by the Collecting Duct.

✔ If you understand ADH's effect on the collecting duct, you should be able to predict how urine formation is affected by ethanol, which inhibits ADH release, and by nicotine, which stimulates ADH release.

Table 40.2 reviews the functions of the four major regions of the nephron and the collecting duct.

Urine Formation in Nonmammalian Vertebrates

The loop of Henle is an important adaptation in mammals and some birds. Water loss is reduced because these animals can produce urine that is hyperosmotic to their blood. In contrast, the nephrons of fishes, amphibians, and non-avian reptiles lack loops of Henle, and their kidneys are therefore unable to produce concentrated urine.

Many fishes and amphibians do not need to produce concentrated urine. But conserving water is important in reptiles, especially those inhabiting deserts. Recall that reptiles produce nitrogenous wastes in the form of uric acid, which is excreted with very little water in urine that is hyperosmotic to their body tissues.

However, the kidneys of non-avian reptiles produce isosmotic urine. How, then, does it become hyperosmotic before it's excreted?

SUMMARY Table 40.2 Structure and Function of the Nephron and Collecting Duct

Structure	Function
Renal corpuscle (Bowman's capsule and glomerulus)	Size-selective filtration: forms filtrate from blood (water and other small substances enter nephron)
Proximal tubule	Reabsorbs electrolytes (active transport), nutrients, water
Loop of Henle	Establishes and maintains medullary osmotic gradient in interstitial fluid from outer to inner medulla
• Descending limb	• Permeable to water (passive transport out of filtrate)
• Thin ascending limb	• Permeable to Na^+, Cl^- (passive transport out of filtrate)
• Thick ascending limb	• Active transport of Na^+, Cl^- out of filtrate
Distal tubule	Aldosterone present: reabsorbs Na^+ No aldosterone present: does not reabsorb Na^+
Collecting duct	Regulates water retention
• Main portion	• High ADH level: water leaves filtrate; produces small volume of urine that is hyperosmotic to blood Low ADH level: water stays in filtrate; produces large volume of urine that is hyposmotic to blood
• Innermost portion	• Urea leaks out by passive transport, contributing to high osmolarity of inner medulla

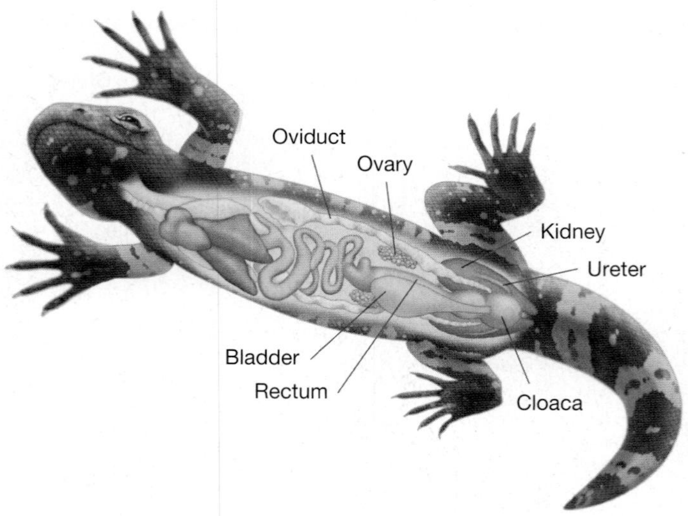

Figure 40.15 **The Cloaca of Reptiles Is a Cavity into Which the Urinary, Gastrointestinal, and Reproductive Tracts Empty.**

In most reptiles, the ureters empty isosmotic urine into the **cloaca**, a cavity into which the urinary, gastrointestinal, and reproductive tracts all empty (Figure 40.15). Reptiles are able to absorb water from urine across the wall of the cloaca into the bloodstream. Eventually, a semisolid uric acid paste is excreted along with the feces.

Some reptiles also have a bladder that collects the isosmotic urine from the ureters and stores it before emptying it into the cloaca. When water is available, these reptiles drink a lot, and their bladders fill up with dilute urine. Researchers hypothesized that the bladder acts as a "canteen" to store water for when it is unavailable.

To test this hypothesis, the investigators injected the bladders of dehydrated Gila monsters—large desert lizards—with radioactively labeled water and drew blood samples over time. The radioactivity of the lizards' blood plasma increased within a half hour, which meant that the radioactively labeled water from the bladder was being absorbed into the bloodstream (Figure 40.16a).

The researchers also examined whether water in the bladder can rehydrate lizards to the same extent as drinking water. Working with two groups of dehydrated Gila monsters, they injected water into the bladders of one group, and—to mimic drinking—they injected water into the stomachs of the second group. They found that the osmolarity of the plasma decreased at the same rate in both groups (Figure 40.16b).

These results indicate that the bladder can indeed allow lizards to carry a water supply that they can access when water is scarce. As water is reabsorbed from the bladder, the urine becomes more and more concentrated, but the osmolarity of the blood remains low.

Studies of the mammalian nephron and the Gila monster bladder demonstrate that the vertebrate urinary system is remarkably effective in regulating water and electrolyte balance and maintaining homeostasis.

CHECK YOUR UNDERSTANDING

✔ **If you understood this section, you should be able to...**

1. **MODEL** Diagram a mammalian nephron, labeling the major parts and showing where blood is filtered and where water, ions, nutrients, and urea are reabsorbed.
2. Predict how the following events would affect urine volume and concentration: (a) drinking massive amounts of water, (b) eating large amounts of salt, and (c) refraining from drinking water for 48 hours.
3. Suggest a hypothesis explaining how some desert rodents can produce urine that is even more concentrated than that of a dehydrated human. (Hint: This involves the loop of Henle.)

Answers are available in Appendix A.

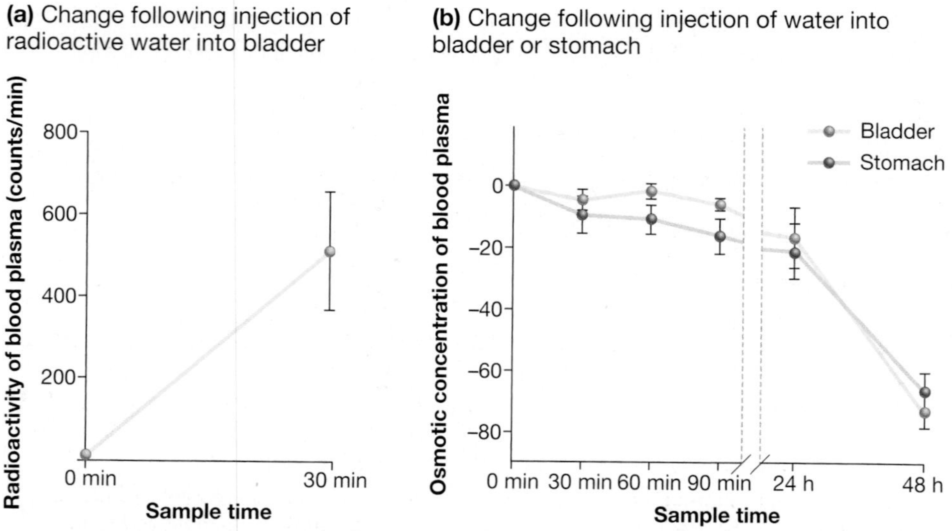

Figure 40.16 **Gila Monsters Can Reabsorb Water from Their Bladders. (a)** Radioactively labeled water passes from the bladder into the bloodstream. **(b)** Water reduces blood osmolarity at a similar rate in dehydrated lizards whether added to the stomach or the bladder.

DATA: J. R. Davis and D. F. DeNardo. 2007. *The Journal of Experimental Biology* 210: 1472–1480.

✔ What physiological trade-offs might be imposed by carrying a large amount of water in the bladder?

40.1 Osmoregulation and Excretion

- Solutes move across membranes via passive transport, facilitated diffusion, or active transport. Water moves across membranes by osmosis.
- In most animals, epithelial cells that selectively transport water and electrolytes are responsible for homeostasis.
- The mechanisms involved in regulating water and electrolyte balance vary widely among animal groups because different habitats present different types of osmotic stress.
- The type of nitrogenous waste excreted by an animal is affected by its phylogeny and its habitat type. Most fishes excrete ammonia; mammals and most adult amphibians excrete urea; and insects and most reptiles excrete uric acid.

40.2 Water and Electrolyte Balance in Marine and Freshwater Fishes

- Seawater is strongly hyperosmotic to the tissues of marine bony fishes, so they tend to lose water by osmosis and gain electrolytes by diffusion.
- Marine bony fishes are osmoregulators, whereas cartilaginous fishes including sharks are osmoconformers.
- Epithelial cells in the shark rectal gland and in the gills of marine bony fishes excrete excess salt using Na^+/K^+-ATPases and $Na^+/Cl^+/K^+$ cotransporters located in the basolateral membrane.
- Similar salt-excreting cells also exist in the salt glands of marine birds and other reptiles and in the kidneys of mammals.
- Fresh water is strongly hyposmotic to the blood of freshwater fishes, so they tend to gain water by osmosis and lose electrolytes by diffusion.
- Epithelial cells in the gills of freshwater fishes import ions using Na^+/K^+-ATPases located in the basolateral membrane and the $Na^+/Cl^-/K^+$ cotransporters located in the apical membrane.

40.3 Water and Electrolyte Balance in Terrestrial Insects

- A waxy coating on the insect exoskeleton limits evaporative water loss. Spiracles, the openings to the insect respiratory system, close when osmotic stress is severe.
- The Malpighian tubules of insects form a filtrate that is isosmotic with the hemolymph. If pumps in the epithelium of the hindgut are activated, then electrolytes and water are reabsorbed from the filtrate and returned to the hemolymph.
- Insects can form hyperosmotic urine that minimizes water loss during the excretion of nitrogenous wastes.

40.4 Water and Electrolyte Balance in Terrestrial Vertebrates

- Nephrons in the vertebrate kidney form a filtrate in the renal corpuscle and then reabsorb valuable nutrients, electrolytes, and water in the proximal tubule.
- A filtrate containing urea and electrolytes flows through the loop of Henle of mammalian kidneys, where changes in the permeability of epithelial cells to water and salt—along with active transport of salt—create a steep medullary osmotic gradient.
- Antidiuretic hormone increases the water permeability of the collecting duct, causing water to be reabsorbed along the osmotic gradient and hyperosmotic urine to be produced.
- The nephrons of fishes, amphibians, and non-avian reptiles do not have loops of Henle and therefore cannot produce urine that is hyperosmotic to the body fluids. However, some of these vertebrates can produce hyperosmotic urine by reabsorbing water from the cloaca or bladder.

Answers are available in Appendix A.

✔ TEST YOUR KNOWLEDGE

1. Which of the following statements regarding fishes that live in fresh water is/are correct? Select True or False for each statement.
   T/F Water moves across the gills via osmosis until equilibrium is established, at which time the water molecules stop moving.
   T/F They lose water to their environment primarily through the gills. They replace this water by drinking.
   T/F Water enters epithelial cells in their gills via osmosis. Electrolytes leave the same cells via diffusion.
   T/F They have specialized epithelia that actively pump electrolytes from the environment into the blood.
2. Which of the following organisms would lose the most water by osmosis across its gills?
   a. marine bony fish
   b. shark
   c. freshwater fish
   d. freshwater invertebrate
3. What effect does antidiuretic hormone (ADH) have on the nephron?
   a. It increases water permeability of the descending limb of the loop of Henle.
   b. It decreases water permeability of the descending limb of the loop of Henle.
   c. It increases water permeability of the collecting duct.
   d. It decreases water permeability of the collecting duct.
4. Fill in the blank: In Gila monsters, the organ in which water from urine is reabsorbed into the bloodstream is the ____________.

✔ TEST YOUR UNDERSTANDING

5. Compare and contrast the types of nitrogenous wastes excreted by animals. Identify which type can be excreted with the least water, which is most toxic, and which waste is excreted by bony fishes, by mammals, and by insects. Which type would you expect to be produced by embryos inside eggs laid on land?
6. The chloride cells of fish gills have a high density of mitochondria. How does this characteristic relate to the functional role of chloride cells? Would you expect other epithelial cells involved in ion transport to contain large numbers of mitochondria? Explain.

7. Explain why mammals would not be able to produce concentrated urine if their nephrons lacked loops of Henle.

8. Scientists have noted that marine invertebrates tend to be osmoconformers, while freshwater invertebrates tend to be osmoregulators. Suggest an explanation for this phenomenon.

✔ TEST YOUR PROBLEM-SOLVING SKILLS

9. Biologists have been able to produce mice that lack functioning genes for aquaporins. How would the urine of these mice compare to that of mice with normal aquaporins?
   a. lower volume and lower osmolarity
   b. lower volume and higher osmolarity
   c. higher volume and lower osmolarity
   d. higher volume and higher osmolarity

10. **QUANTITATIVE** To test the hypothesis that mussels are osmoconformers, researchers exposed mussels to water of varying osmolarities and then drew hemolymph samples from the mussels. Graph the data provided here. Put the independent variable on the *x*-axis and the dependent variable on the *y*-axis. Is the researchers' hypothesis supported by the data? Explain.

Water Osmolarity (milliosmol/L)	Hemolymph Osmolarity (milliosmol/L)
250	261
500	503
750	746
1000	992

✔ PUT IT ALL TOGETHER: Case Study

How does water pollution affect osmoregulation in fishes?

Fish and other aquatic organisms are exposed to many types of water pollutants, including metals such as aluminum. Although a low level of aluminum is found in unpolluted water, many lakes and streams have an increased level because of mining, sewage treatment, and accidental spills of toxic materials. Aluminum pollution can result in mass fish die-offs such as the one pictured here. How does this occur?

11. Which of the following is an osmoregulatory challenge that freshwater fishes need to overcome?
   a. diffusion of sodium ions out of the body
   b. diffusion of water out of the body
   c. active transport of sodium ions out of the body
   d. active transport of water out of the body

12. **QUANTITATIVE** In a laboratory, scientists exposed freshwater bony fish (*Prochilodus lineatus*) to water with a high level of aluminum and compared their blood osmolarity to that of fish exposed to water with a normal aluminum level (control). The results of the experiment are shown here (asterisks indicate $P < 0.05$ between control and treated groups at a given time; **BioSkills 3**). Do the data support the hypothesis that aluminum interferes with osmoregulation in freshwater fishes? Explain.

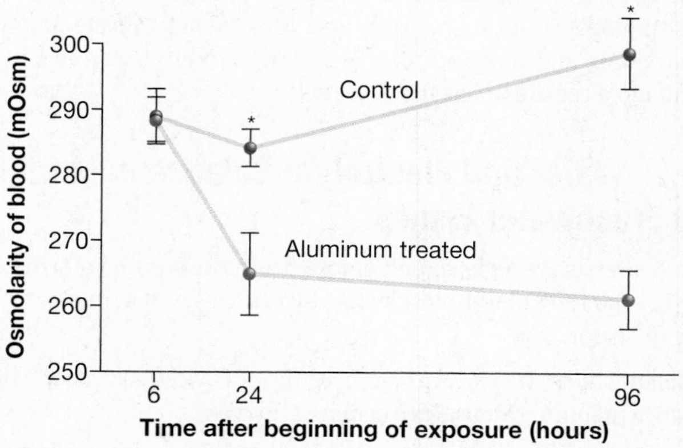

SOURCE: M. M. P. Camargo, M. N. Fernandes, and C. B. R. Martinez. 2009. *Aquatic Toxicology* 94: 40–46.

13. **PROCESS OF SCIENCE** Why did the scientists do this experiment in a laboratory instead of simply collecting fish from a river with a high aluminum level and documenting their osmoregulatory ability?

14. The scientists also measured the activity of Na^+/K^+-ATPase in the gills of the fish exposed to aluminum and compared it to that of the control fish. What do you suppose were their results? Explain.

15. **THINK CAREFULLY** True or false: Water moves by osmosis across a fish's gills to an area with a higher sodium ion concentration because water molecules are attracted to the sodium ions.

16. **MODEL** Draw a graph similar to the one here showing how the results would be different if the experiment had been performed on marine bony fish in seawater. (Assume that the osmolarity of seawater is 1100 mOsm and the set point osmolarity of marine bony fishes is 290 mOsm.)

41 Animal Nutrition

A snake has captured a meal. Animals obtain nutrients by ingesting food.

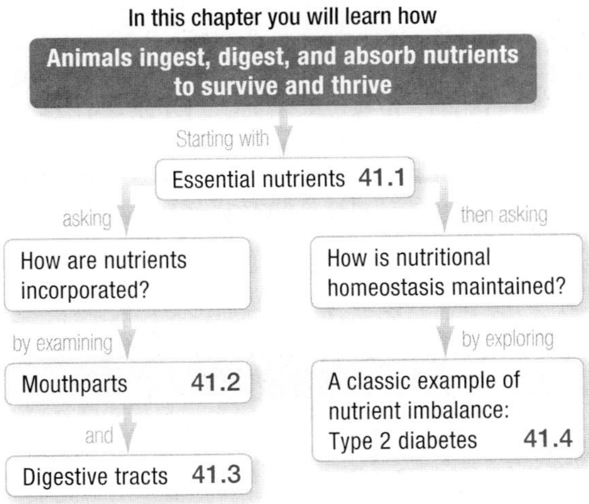

Animals get the two basic requirements for life—**(1)** chemical energy for synthesizing ATP and **(2)** carbon-containing compounds and minerals for building complex macromolecules—by ingesting other organisms. In short, animals are heterotrophs: They eat to live.

The types of food that are available to different animals vary widely, and food is often in short supply. From these observations, you might expect that many different ways of obtaining food have evolved in animals and that animals are under intense natural selection for making efficient use of the food they have. How do animals get their food, and how do they process it? Which substances in food are used as nutrients, and how do humans and other animals maintain appropriate levels of key nutrients in their bodies?

For you or any other animal to stay alive, food must be ingested, digested, and absorbed, and then the wastes must be eliminated. **Ingestion** is the process of bringing food into the **digestive tract**—also

BIG PICTURE

This chapter is part of the Big Picture. See how on pages 838–839.

known as the alimentary (literally, "nourishment") canal, or gastrointestinal (GI) tract. The digestive tract is a series of connected chambers and tubes where **digestion**, the mechanical and chemical breakdown of food, takes place. Various accessory glands secrete enzymes into the digestive tract that digest food into particles small enough for efficient **absorption**—the uptake of specific ions and molecules across the epithelium that lines the digestive tract. **Elimination** removes wastes from the animal's digestive tract. This process is summarized in **Figure 41.1** for a simple animal, a sea anenome.

Research on feeding and digestion is fundamental to understanding basic aspects of animal biology, but research on animal nutrition has important practical applications as well. For example, it addresses questions about why several nutrition-related diseases, including diabetes mellitus and obesity, are on the rise in many human populations.

Let's begin with a look at what animals must eat to live.

41.1 Nutritional Requirements

Humans and other animals get the chemical energy and carbon-containing building blocks they need from carbohydrates, proteins, and fats. All these substances are carbon compounds with high potential energy. (Chapters 5 and 6 analyzed the structures of carbohydrates and fats, and Chapter 9 detailed how these compounds are used to synthesize ATP and key macromolecules.)

After you complete this section, you should be able to ...

- Explain the nutrient requirements of animals.

A carbohydrate, protein, or fat is an example of a **nutrient**: a substance that an organism needs to remain alive. **Food** is any material that contains nutrients.

The amount of energy provided by foods is measured in kilocalories (on food labels, kilocalories are referred to as Calories). Because fats are rich in C—H bonds, they provide more energy than other nutrients: about 9 kcal/g versus about 4 kcal/g for carbohydrates and proteins.

Although all nutrients are necessary for growth and survival of animals, **essential nutrients** are those that cannot be synthesized and must be obtained from the diet.

Humans require four classes of essential nutrients:

1. **Essential amino acids** are amino acids that an animal cannot synthesize from simpler building blocks. The human diet requires nine essential amino acids, which must be obtained from food. (Recall from Ch. 3, Section 3.1, that all 20 amino acids are required to manufacture most proteins; humans can synthesize 11 of them.)
2. **Essential fatty acids** are fatty acids that an animal must obtain in its diet. (Fatty acids were introduced in Ch. 6, Section 6.1.) Humans can synthesize all fatty acids except two, which must be obtained from eating certain plants or fish.
3. **Vitamins** are organic, or carbon-containing, compounds that are vital for health but are required in only minute amounts. They have a variety of roles; several function as coenzymes in critical reactions (Ch. 8, Section 8.3). **Table 41.1** lists a few of the vitamins for which RDAs have been established, notes their functions, and indicates the problems that develop if they are missing in the diet.
4. **Minerals** are inorganic substances used as components of enzyme cofactors or structural materials (**Table 41.2**). Some, such as calcium and phosphorus, are needed in relatively large quantities. Others, such as iron and copper, are required in small or trace amounts. Minerals include ions of electrolytes, which influence osmotic balance and are required for normal membrane function (Chapter 40). Sodium (Na^+), potassium (K^+), and chloride (Cl^-) are the major ions of electrolytes in the human body.

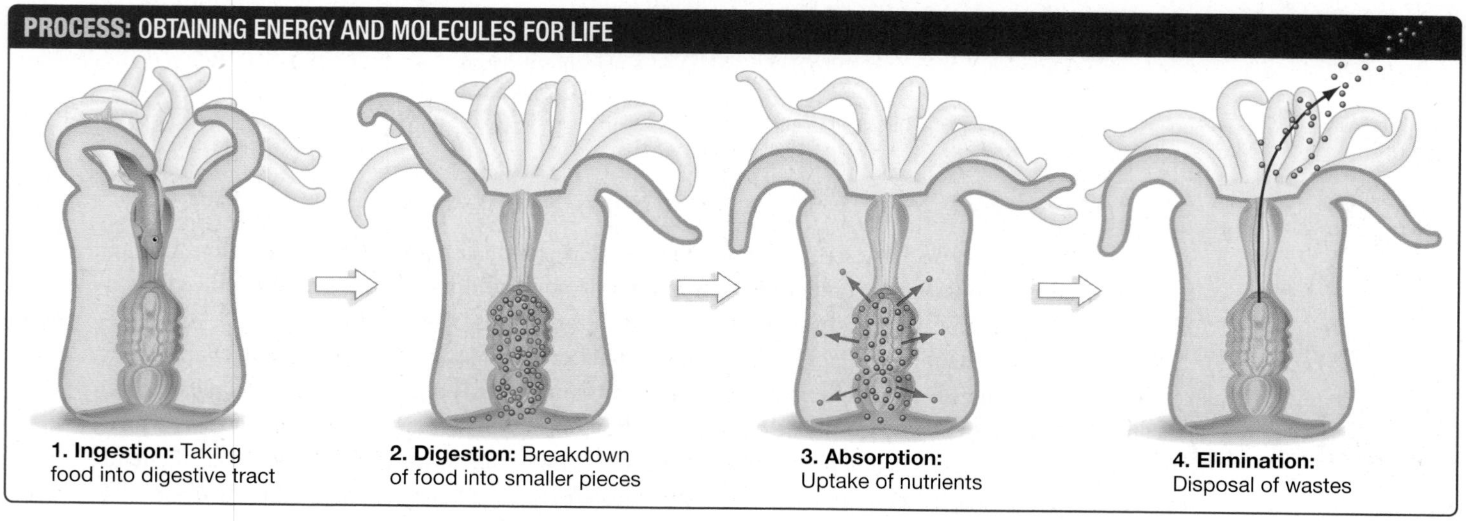

Figure 41.1 Animal Nutrition Is a Four-Step Process. Ingestion is followed by digestion, where food is broken down into absorbable components. Anything left after absorption is subject to elimination as waste.

To obtain nutrients, animals must ingest them, usually via a mouth. Because natural selection has resulted in dietary specialization in many animals (Ch. 39, Figure 39.3), the structure of animal mouthparts often strongly reflects the function of eating specific foods.

CHECK YOUR UNDERSTANDING

✔ **If you understood this section, you should be able to . . .**

Explain why somone who is a vegetarian and does not follow a careful diet might become anemic.

Table 41.1 Some Important Vitamins Required by Humans

	Source	Function	Effects if Deficient
Vitamin B_1 (thiamine)	legumes, whole grains, potatoes, peanuts	formation of coenzyme in citric acid cycle	beriberi (fatigue, nerve disorders, anemia)
Vitamin B_3 (niacin)	meat, whole grains	component of coenzymes NAD^+ and $NADP^+$	pellagra (digestive problems, skin lesions, nerve disorders)
Vitamin B_9 (folate)	green vegetables, oranges, nuts, legumes, whole grains; also synthesized by bacteria in intestine	coenzyme in nucleic acid and amino acid metabolism	anemia
Vitamin B_{12} (cobalamin)	red meat, eggs, dairy products; also synthesized by bacteria in intestine	coenzyme in synthesis of proteins and nucleic acids and in formation of red blood cells	anemia (fatigue and weakness due to low hemoglobin content in blood)
Vitamin C (ascorbic acid)	citrus fruits, tomatoes, broccoli, cabbage, green peppers	used in collagen synthesis, prevents oxidation of cell components, improves absorption of iron	scurvy (degeneration of teeth and gums)
Vitamin D_3 (cholecalciferol)	fortified milk, egg yolk; also synthesized in skin exposed to sunlight	aids absorption of calcium and phosphorus in small intestine	rickets (bone deformities) in children; bone softening in adults

Table 41.2 Major Minerals Required by Humans

	Source in Diet	Function	Effects if Deficient
Calcium (Ca)	dairy products, green vegetables, legumes	bone and tooth formation, nerve signaling, muscle response	loss of bone mass, slow growth
Chlorine (Cl)	table salt or sea salt, vegetables, seafood	fluid balance in cells, protein digestion in stomach (HCl), acid–base balance	weakness, loss of muscle function
Fluorine (F)	fluoridated water, seafood	maintenance of tooth structure	higher frequency of tooth decay
Iodine (I)	iodized salt, algae, seafood	component of the thyroid hormones thyroxine and T_3	goiter (enlarged thyroid gland)
Iron (Fe)	meat, eggs, whole grains, green leafy vegetables, legumes	enzyme cofactor; synthesis of hemoglobin and electron carriers	anemia, weakness
Magnesium (Mg)	whole grains, green leafy vegetables	enzyme cofactor	nerve disorders
Phosphorus (P)	dairy products, meat, grains	bone and tooth formation; synthesis of nucleotides and ATP	weakness, loss of bone mass
Potassium (K)	dairy products; meat; nuts; fruits; potatoes, legumes, and other vegetables	nerve signaling, muscle response, acid–base balance	weakness, muscle cramps, loss of muscle function
Sodium (Na)	table salt or sea salt, seafood	nerve signaling, muscle response, blood pressure regulation	weakness, muscle cramps, loss of muscle function, nausea, confusion
Sulfur (S)	any source of protein	amino acid synthesis	swollen tissues, degeneration of liver, intellectual disability

(a) The sharp teeth of mountain lions stab and slice prey.

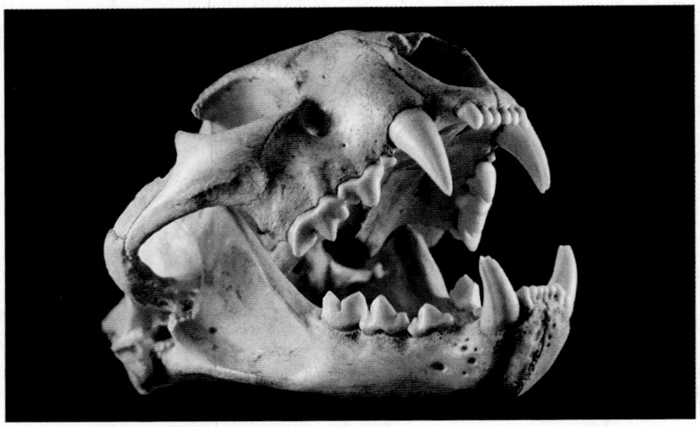

(b) A flexible skull allows snakes to swallow large prey whole.

Figure 41.2 **Mouthpart Structure Correlates with Function.**

41.2 Capturing Food: The Structure and Function of Mouthparts

Biologists assign animal feeding techniques to one of the following strategies (Ch. 30, Section 30.3):

After you complete this section, you should be able to ...

- Analyze the relationships between the structure and function of animal mouthparts.

- **Suspension feeders**, such as sponges and tubeworms, filter small organisms or bits of organic debris from water by means of cilia, mucus-lined "nets," or other structures.
- **Deposit feeders**, including earthworms and sea cucumbers, swallow sediments and other types of deposited material rich in organic matter.
- **Fluid feeders** suck or lap up blood, nectar, or other fluids.
- **Mass feeders** are the majority of animals. They seize and manipulate chunks of food.

Mouthparts as Adaptations

The types of food that animals harvest range from soupy solutions in decaying carcasses to nuts inside hard shells. Solutions have to be lapped up; nuts have to be cracked. Given the diversity of food sources that animals exploit, it is not surprising that they capture and process food using a wide variety of mouthpart structures, including jaws, teeth, beaks, and toxin-injecting organs.

Natural selection has closely matched the structure of animal mouthparts to their function in obtaining food. For example, most mammals chew their food and swallow distinct packets, or boluses. The meat-slicing teeth of carnivores such as the mountain lion shown in Figure 41.2a is one example of the many tooth shapes that evolved from the relatively simple and uniform teeth in the common ancestor of all mammals. Diversification of tooth shape has allowed mammals to exploit a wide range of foods. Another example is found in snakes, which have highly mobile skull bones and complex associated musculature that allow them to ingest large prey without chewing or biting off pieces (Figure 41.2b).

Let's pursue the correlation between mouthparts and food sources further, by analyzing the structure and function of jaws and teeth in what may be the most diverse lineage in any vertebrate family: the cichlid fishes of Africa.

A Case Study: The Cichlid Throat Jaw

The cichlids that inhabit the Rift Lakes of East Africa are a spectacular example of **adaptive radiation**—the diversification of a single ancestral lineage into many species, each of which lives in a different habitat or exhibits a distinct form (Ch. 25, Section 25.3). Lake Victoria, for example, is home to 300 cichlids that live nowhere else. Each Lake Victoria cichlid species feeds on a different specific item, but as a group they exploit almost every food source in the lake, from plankton to plants to other fishes.

How can a group of closely related species exploit so many different food sources? Many fish species have pharyngeal (throat) jaws located well behind the oral (mouth) jaws (Figure 41.3). Most non-cichlids use their pharyngeal jaws to move food down their throats, but cichlids and some other species can also use theirs to bite and process (e.g., crush, tear, or compact) their food. In

Figure 41.3 **Rift Lake Cichlids Have Two Sets of Biting Jaws.** This X-ray image shows the oral jaws, which capture food, and the pharyngeal jaws, which process it.

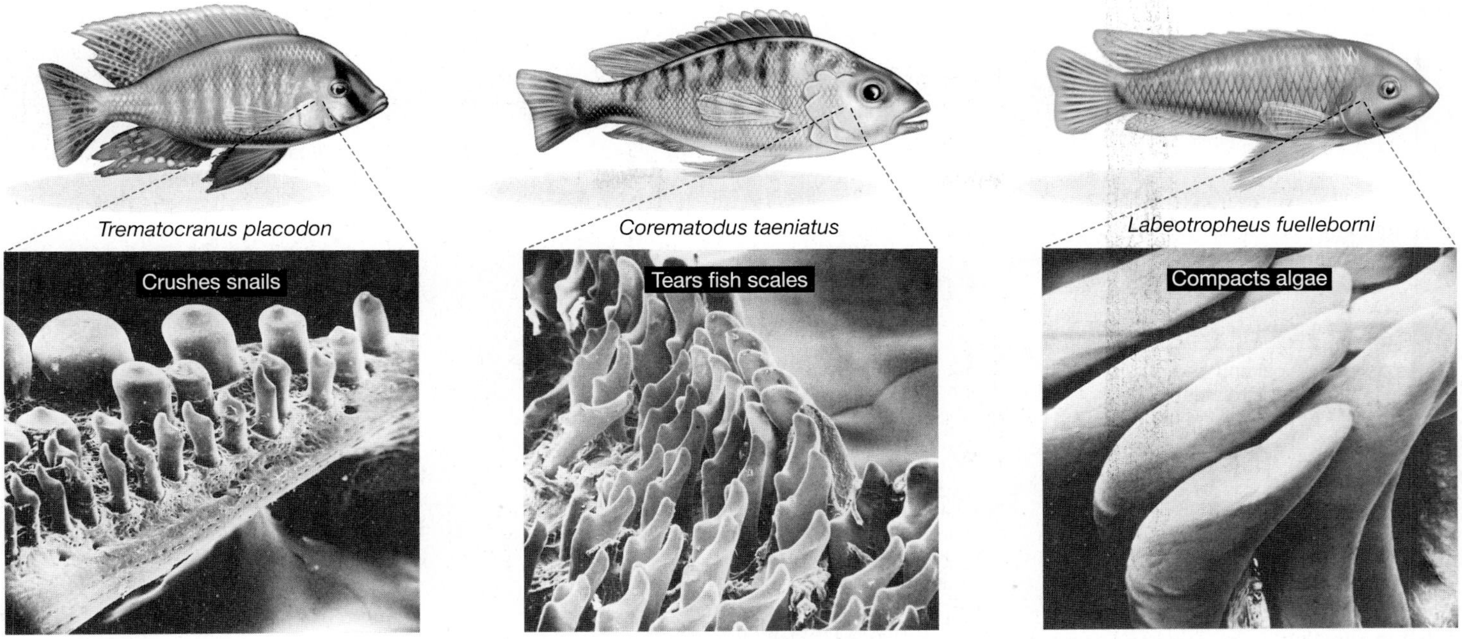

Figure 41.4 In Cichlids, the Structure of the Pharyngeal Jaw Correlates with the Type of Food Ingested.

cichlids this is possible because the upper pharyngeal jaw attaches to the skull, and because muscles connecting their lower pharyngeal jaw to the cranium allow it to move against the upper jaw.

Besides acting as a second set of biting jaws that make food processing more efficient, the pharyngeal jaws provide a more specialized set of toothlike structures. These protuberances vary in size and shape among cichlids, correlating with their function, such as crushing snail shells, tearing fish scales, or compacting algae (Figure 41.4). These observations are part of a large body of evidence supporting a general pattern in animal evolution: In response to natural selection, mouthparts have diversified, enabling animals to exploit a diversity of food sources. The structures of jaws, teeth, and other mouthparts correlate with their functions in harvesting and processing food.

CHECK YOUR UNDERSTANDING

✔ **If you understood this section, you should be able to ...**

Look in the mirror at your teeth, and explain how the shapes of the incisors and canines in front and molars in back reflect their functions in relation to the types of food humans are adapted to eat.

41.3 The Structure and Function of Animal Digestive Tracts

After you complete this section, you should be able to ...

- Analyze the roles of digestive organs in animals.

Digestion is a key process in animals because, unlike plants, unicellular organisms, and fungi, most animals do not acquire nutrients as individual molecules. (Some animals that live as internal parasites are exceptions.) Instead, most animals ingest packets of food that must be broken down into small pieces. Nutrients must be extracted from the small pieces, and waste materials must be eliminated. How and where does this processing occur?

An Introduction to the Digestive Tract

The following types of digestive tracts occur in animals:

1. **Incomplete digestive tracts** have a single opening, the mouth, through which the animal both ingests food and eliminates wastes. The mouth opens into a chamber, called a gastrovascular cavity, where digestion takes place (Figure 41.5).

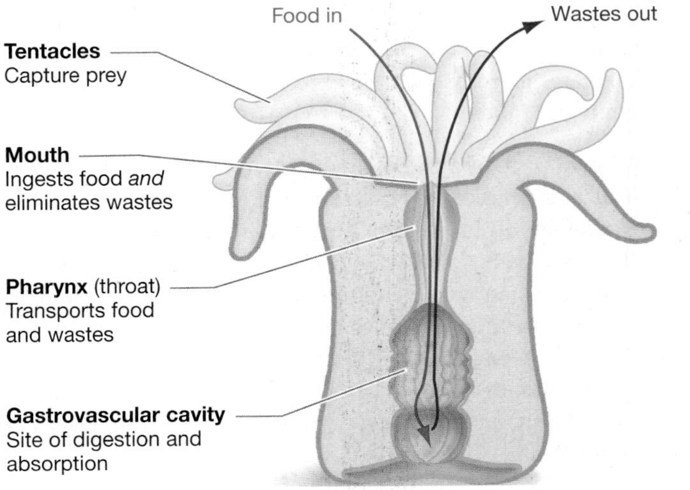

Figure 41.5 Sea Anemones Have an Incomplete Digestive Tract. Sea anemones use stinging cells located on their tentacles to capture small fishes, crustaceans, and other prey. Prey are ingested by the mouth and digested in the gastrovascular cavity, and the nutrients are absorbed; then wastes are eliminated through the mouth.

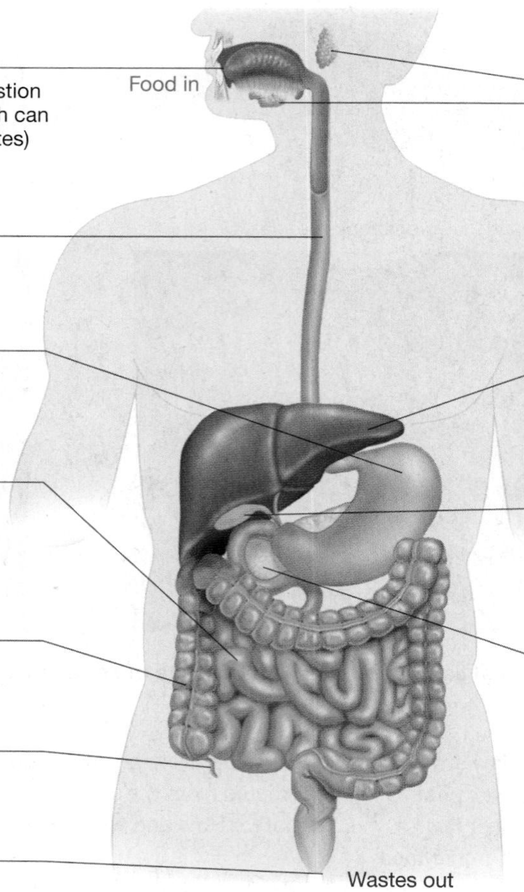

Figure 41.6 **In the Complete Human Digestive Tract, Food Materials Are Mechanically and Chemically Digested.** In humans, as in all vertebrates and most other animals, the digestive tract is a tube that runs from the mouth to the anus. Food materials are mechanically and chemically digested as they pass through the organs of this tract. The salivary glands, liver, gallbladder, and pancreas are not part of the tract itself. Instead, they are accessory organs that secrete material into the tract at specific points.

2. **Complete digestive tracts** have two openings—they start at the mouth and end at the anus. The interior of this tube communicates directly with the external environment via these openings (Figure 41.6).

One advantage of a complete digestive tract is that different chemical and physical processes can be confined to different compartments within the tract, so that they occur independently of each other and in a prescribed sequence. The stomach, for example, provides an acidic environment for digestion. However, in the small intestine, enzymes are specialized to function in a slightly alkaline environment. Also, thanks to the one-way flow of food and wastes, material can be ingested and digested without interruption, instead of alternating with waste removal as in an incomplete digestive tract.

The digestive tract is only one part of the digestive system, however. Several vital organs and glands are connected to the digestive tract. These accessory structures contribute digestive enzymes and other products to specific portions of the tract and therefore play a key role in digestion. In vertebrates, they include the salivary glands, liver, gallbladder, and pancreas (see Figure 41.6).

An Overview of Digestive Processes

Before analyzing the function of each component of the digestive system in detail, let's consider the general changes that happen to food as it is digested, both mechanically and chemically, on its way through the digestive tract. In this brief overview and in the detailed discussion that follows, humans will serve as a model species—simply because so much is known about human digestion.

In the process of mechanical digestion, humans break down large chunks of food into smaller pieces. We accomplish this by the chewing action of teeth, as well as by contraction of smooth muscle in the stomach and small intestine. Mechanical digestion increases the surface area of the food and mixes it into a watery slurry so that enzymes can more easily do their job—chemical digestion, or breaking of chemical bonds.

During chemical digestion, distinct chemical changes occur as food moves through each compartment in the digestive tract (Figure 41.7):

1. In the mouth, enzymes in the saliva begin the chemical breakdown of carbohydrates and lipids.

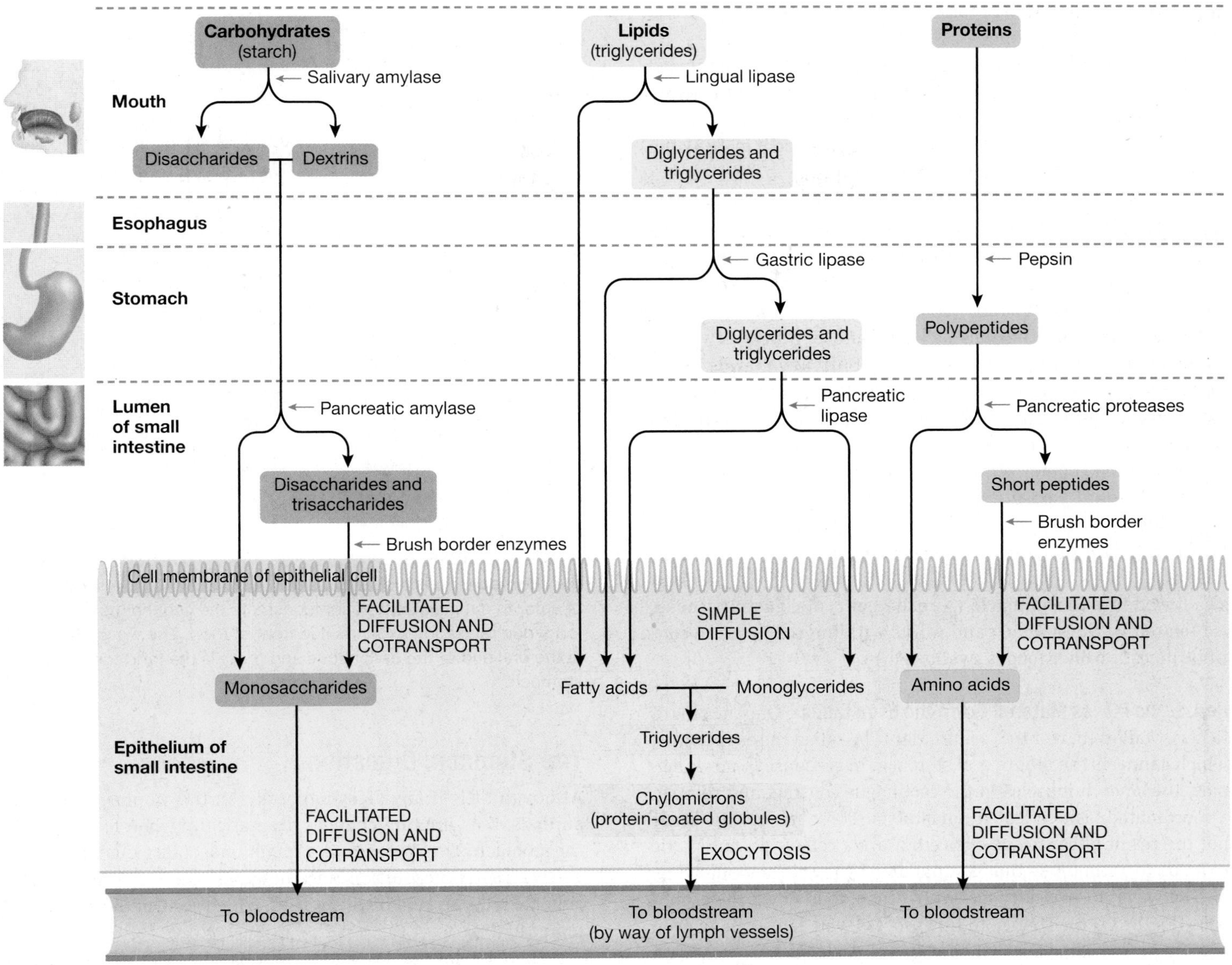

Figure 41.7 In Chemical Digestion, Carbohydrates, Lipids, and Proteins Are Processed in a Series of Steps. Three key types of macromolecules enter the digestive system (top of diagram). As they proceed through the digestive tract, they are broken apart by various enzymes. Simple sugars, fatty acids, monoglycerides, and amino acids then enter epithelial cells in the small intestine and are transported to the bloodstream.

2. Chemical digestion of protein begins in the acidic environment of the stomach. Meanwhile, lipids continue to be chemically digested in the stomach, but carbohydrates are not broken down further.
3. Chemical processing of the three major types of macromolecules—carbohydrates, proteins, and lipids—is completed in the small intestine (see Ch. 9, Section 9.1 for an overview of catabolic and anabolic pathways).
4. The small molecules that result from the digestion of these macromolecules are absorbed in the small intestine, along with water, vitamins, and ions.

After digestion and absorption are completed in the small intestine, the remaining food components enter the large intestine, where more water is absorbed and microbes digest plant material that was not digested in the small intestine. The waste material remaining in the large intestine is **feces**, which are eventually excreted. Because digestion is so important to understanding how animal bodies work, let's examine each step more closely. As the following sections track food materials from the mouth to the anus in humans, watch for notes highlighting the diversity of structures found in the digestive tracts of nonhuman animals.

The Mouth and Esophagus: Digestion and Ingestion

In mammals, digestion begins with the mechanical process of chewing—the tearing and crushing activity of teeth. Chewing reduces the size of food particles and softens them. Humans augment the mechanical digestion of food with their use of knives and cooking. Chewing and cooking both help prepare food for the chemical action of enzymes.

Starch breakdown was actually the first enzyme-catalyzed reaction ever discovered (see Chapter 8 for a review of enzyme function). In the early 1800s several researchers found that a component of certain plant extracts digested starch; in 1831 the same activity was discovered in human saliva.

Digestion Starts in the Mouth **Salivary amylase**, the enzyme responsible for carbohydrate digestion in the mouth, is one of the best-studied enzymes. Amylase cleaves bonds in starch to release dextrins, which are smaller carbohydrates of various lengths, as well as some disaccharides such as maltose, which is two glucose monomers joined by a single bond. **Salivary glands** in the mouth secrete amylase and also produce the slimy substance called **mucus**. The combination of water and mucus makes food soft and slippery enough to be swallowed.

Cells in the tongue synthesize and secrete another salivary enzyme, **lingual lipase** ("lingual" refers to the tongue), which begins the digestion of lipids by breaking triglycerides, a common form of fat in the diet, into diglycerides and fatty acids. Lingual lipase plays only a minor role in digestion, mostly once it has been swallowed into the stomach. (A related enzyme, **gastric lipase**, is released by the stomach and works with lingual lipase to continue digestion once food is swallowed.)

Peristalsis Moves Material Down the Esophagus Once a bolus of food is swallowed, it enters a muscular tube called the **esophagus**, which connects the mouth and stomach. In response to nerve signals, the smooth muscles in the esophagus contract and relax in a coordinated fashion called **peristalsis**. These nerve signals are not the result of conscious choice but are a reflex—an automatic reaction to a stimulus—that is stimulated by the act of swallowing.

The waves of peristalsis actively propel food down the esophagus (Figure 41.8). Thanks to this process, you can swallow even when your mouth is lower than your stomach, such as when you bend over to drink water from a drinking fountain. About 6 seconds after being swallowed, food reaches the bottom of the esophagus.

A Modified Esophagus: The Bird Crop In an array of bird species, the esophagus has a prominent, widened segment called the **crop** where food can be stored and, in some cases, processed.

The structure and function of the crop vary among bird species. In many groups, the crop is a simple sac that holds food and regulates its flow into the stomach. In these species, the crop is interpreted as an adaptation that allows individuals to eat a large amount in a short time; they then retreat to a safe location while digestion occurs. In addition, some birds store food in their crops and then regurgitate it into the mouths of their young.

The crop has independently evolved into a digestive organ in two leaf-eating species of bird, the hoatzin and the kakapo. Leaves are difficult to digest because they contain a large amount of cellulose (Ch. 5, Section 5.2). In the enlarged crop of these species, bacteria that are capable of breaking down cellulose perform digestion. The bacterial cells, along with the fatty acids that result from bacterial metabolism, leak out of the crop into the stomach and are used as food by the birds.

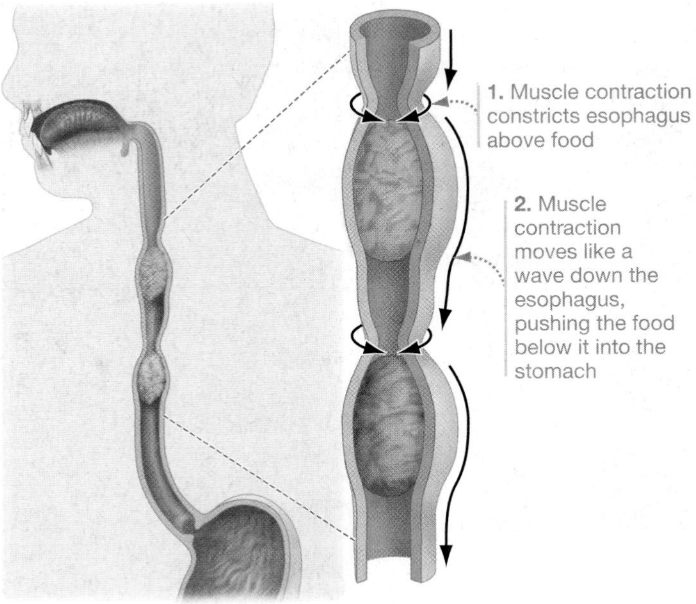

Figure 41.8 **In the Esophagus, Peristalsis Transports Food to the Stomach.** Peristalsis is a wave of contraction and relaxation of smooth muscle. Contraction constricts the esophagus behind food, and relaxation expands it in front of food. The wave begins at the oral end of the esophagus and propels the food toward the stomach.

The Stomach: Digestion

Although little if any digestion occurs in the esophagus of most animals, the situation changes dramatically when food reaches the stomach. The **stomach** is a tough, muscular pouch in the digestive tract, bracketed on both the superior and inferior ends by ringlike muscles called **sphincters**, which control the passage of material (Figure 41.9).

When a meal fills the stomach, muscular contractions churn and mix the stomach contents to a uniform consistency and solute concentration. A certain amount of mechanical digestion of food also results from this churning. The other main function of the stomach is the partial digestion of proteins and lipids.

Compared with the mouth or esophagus (or virtually any other tissue), the lumen of the stomach is highly acidic. Early researchers documented this fact by analyzing vomit or material collected by sponges that were tied to strings, swallowed, and pulled back up; chemists confirmed that the predominant acid in the stomach is hydrochloric acid (HCl).

Not long after this discovery, a physician named William Beaumont established that digestion of proteins takes place in the stomach. He reached this conclusion through an extraordinary series of experiments on a young man named Alexis St. Martin.

The Stomach as a Site of Protein Digestion In 1822, when St. Martin was 19 years old, a shotgun accidentally discharged into his abdomen, leaving a series of wounds. Despite repeated attempts, Beaumont was unable to close a hole in St. Martin's stomach. Eventually Beaumont inserted a small tube through the opening; the tube remained in St. Martin's body for the rest of his life.

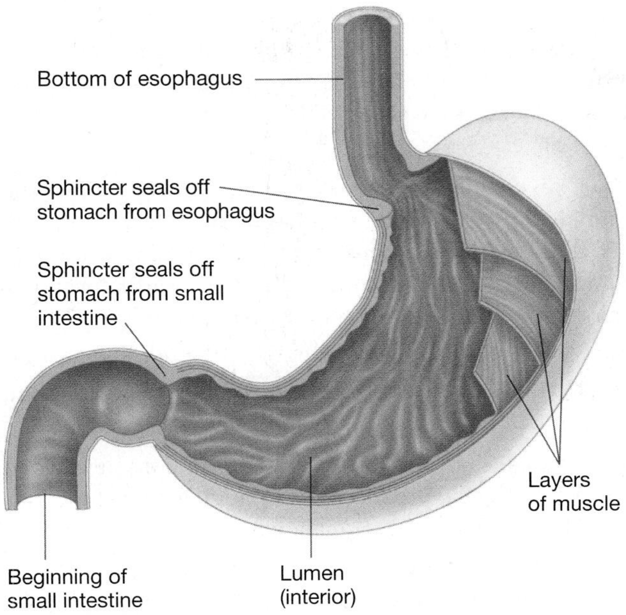

Figure 41.9 The Stomach Is a Muscular Pocket That Holds and Mixes Food as Well as Digesting Proteins. The stomach provides an acidic environment for protein digestion. Its muscular contractions mix food and break it into smaller pieces.

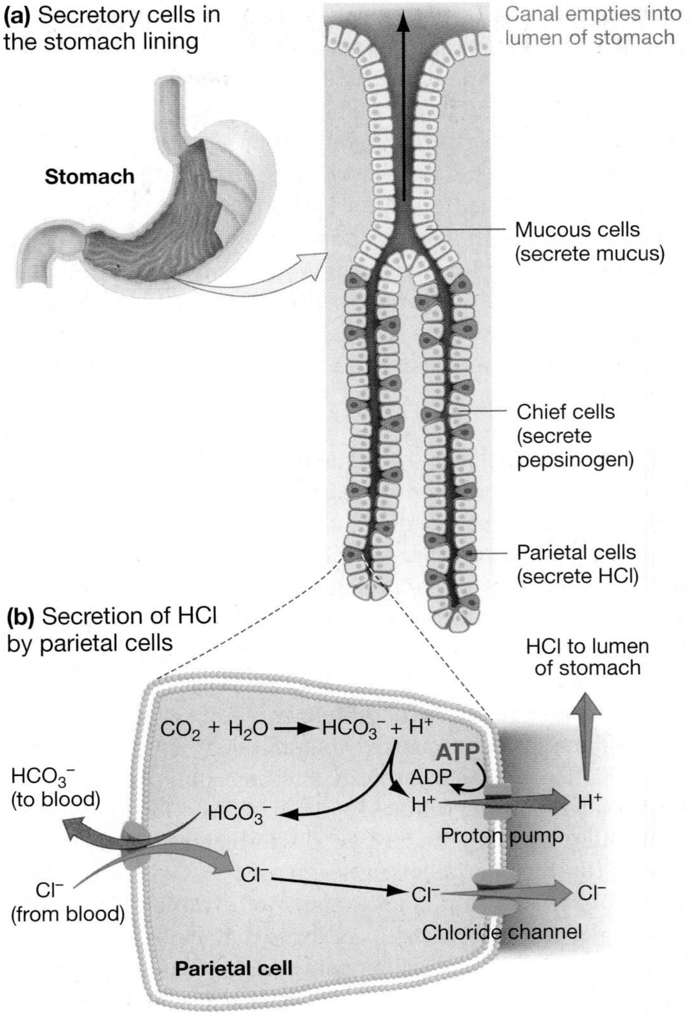

Figure 41.10 Cells in the Stomach Lining Secrete Mucus, Pepsinogen, and Hydrochloric Acid.

With the tube in place, Beaumont was able to tie a string onto small pieces of meat or vegetables, insert the food directly into St. Martin's stomach, and draw it out after various intervals. Beaumont also removed liquid from inside the stomach and observed how this gastric (stomach) juice acted on food in vitro. His experiments showed that gastric juice digests food—particularly meat.

The enzyme that is responsible for digesting proteins in the stomach was purified in 1836 and named **pepsin**. Because it breaks down proteins, biologists hypothesized that pepsin must be synthesized and stored in cells while it is in an inactive form—otherwise it would kill the cells that make it.

In 1870 a microscopist established the presence of granules in specialized stomach cells called chief cells. These granules were hypothesized to be a pepsin precursor. Follow-up work confirmed this hypothesis. The precursor compound, which came to be called pepsinogen, is converted to active pepsin by contact with the acidic environment of the stomach.

The acidic environment of the stomach disrupts the secondary and tertiary structures of proteins. In addition, pepsin cleaves the peptide bonds next to certain amino acids, reducing long polypeptides to relatively small chains of amino acids.

Which Cells Produce Stomach Acid? Researchers studying the anatomy of the stomach wall noticed clusters of distinctive **parietal cells** located in pits in the stomach lining (Figure 41.10a). An investigator also documented that the shape and activity of these cells appeared to vary as the digestion of a meal proceeded. Based on these observations, he inferred that parietal cells are the source of the HCl in gastric juice, which may have a pH as low as 1.5.

Earlier microscopists had shown that another type of cell, called a **mucous cell**, secretes additional mucus that is found in gastric juice. Mucus lines the gastric epithelium and protects the stomach from damage by HCl. To summarize, these anatomical studies showed that the epithelium of the stomach contains several types of secretory cells, each type specialized for a particular function.

How Do Parietal Cells Secrete HCl? The first clue to how parietal cells manufacture HCl emerged in the late 1930s, when a researcher found a high concentration of an enzyme called carbonic anhydrase in parietal cells.

This result was interesting because **carbonic anhydrase** catalyzes the formation of a proton (H^+) and a bicarbonate ion (HCO_3^-) from carbon dioxide and water:

$$CO_2 + H_2O \rightleftharpoons H^+ + HCO_3^-$$

A second clue to the formation of HCl came in the 1950s, when transmission electron microscopes allowed researchers to analyze parietal cells at high magnification (BioSkills 9). The resulting micrographs showed that parietal cells are packed with mitochondria. Because mitochondria produce ATP, the structure of parietal cells suggested that they might function in active transport.

Later work confirmed this hypothesis by showing that the protons formed by the reaction between carbon dioxide and water are actively pumped into the lumen of the stomach. Later studies showed that chloride ions from the blood enter parietal cells in exchange for bicarbonate ions, via a cotransport protein, and then move into the lumen through a chloride channel. Figure 41.10b diagrams the current model for HCl production. ✔ **If you understand this model, you should be able to explain why many heartburn (acid reflux) drugs contain proton pump inhibitors.**

What stimulates the parietal cells to secrete HCl? After being stimulated by nerves or the arrival of food, certain stomach cells produce the hormone **gastrin**. In response, parietal cells begin secreting HCl.

The Ruminant Stomach The structure and function of the stomach can vary, depending on the nature of the diet. In cattle, sheep, goats, deer, antelope, giraffe, and pronghorn—species that are collectively called **ruminants**—the stomach is specialized for digesting cellulose instead of proteins. Animals do not produce the enzymes required to digest cellulose. Yet cellulose is the main carbohydrate in the leaves, stems, and twigs that ruminants ingest.

Like the hoatzin and kakapo described earlier in this chapter, ruminants are able to harvest energy from cellulose thanks to a combination of specialized anatomical structures and symbiotic relationships with bacteria and unicellular protists. The microbes ferment the cellulose to produce food for themselves; meanwhile, other by-products of the fermentation, as well as some of the microbes themselves, are used as food by the ruminant. This relationship is an example of **symbiosis**, in which members of different species live in close contact with each other.

As Figure 41.11 shows, ruminants have four-chambered stomachs. The process of digestion in these stomachs is as follows:

Step 1 Food initially enters the largest chamber, the rumen, which serves as a fermentation vat. The rumen is packed with symbiotic bacteria and protists. These organisms produce **cellulase**, an enzyme capable of breaking the chemical bonds in cellulose, yielding glucose. The rumen is an oxygen-free environment, and the symbiotic organisms produce ATP from this glucose via fermentation, releasing fatty acids as a by-product (Ch. 9, Section 9.6). These fatty acids are absorbed by the ruminant and used as an energy source.

Step 2 The chamber adjacent to the rumen, called the reticulum, is similar in function. After plant material has been partially digested in the rumen and the reticulum, the animal regurgitates portions of that material into its mouth, forming a cud. The ruminant chews that regurgitated material further to enhance mechanical breakdown and then re-swallows it.

Step 3 Processed cud enters the third chamber, the omasum, where water and some minerals are absorbed.

Step 4 The final chamber, the abomasum, contains pepsin and other digestive enzymes produced by the ruminant and functions much like the stomachs of other mammals.

Most of a ruminant's food consists of **(1)** fatty acids and other compounds produced as waste products of fermentation reactions in symbiotic organisms, and **(2)** the symbiotic cells themselves.

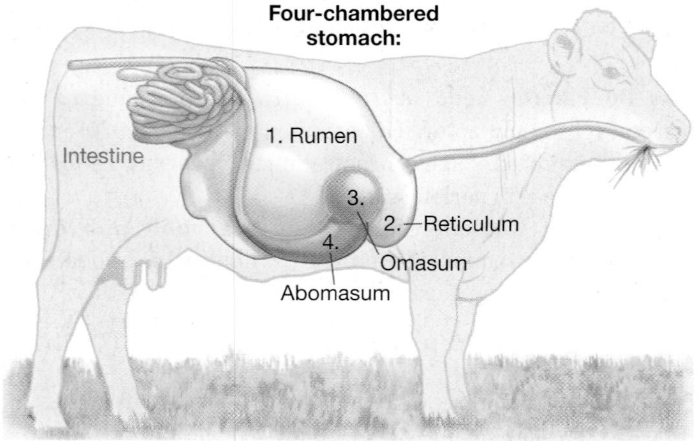

Figure 41.11 **Ruminant Stomachs Facilitate the Digestion of Cellulose by Symbiotic Organisms.** Ruminants obtain many of their nutrients from symbiotic bacteria and protists that live in two chambers of the stomach, the rumen and reticulum.

The Avian Gizzard The avian gizzard is another prominent type of modified stomach. Birds do not have teeth and cannot chew food into small pieces. Instead, most species swallow sand and small stones that lodge in the gizzard. As this muscular sac contracts, food is pulverized by the grit.

The gizzard is particularly large and strong in bird species that eat coarse foods such as seeds and nuts. The gizzard of a wild turkey, for example, can crack large walnuts.

Like the crop, the gizzard is interpreted as an adaptation that allows birds to ingest food quickly—without needing to chew—and digest it later. Biologists invoke the same hypothesis to explain why ruminants chew cud. The ability to regurgitate material and finish chewing while hiding in a place safe from predators is thought to increase fitness. ✔ **If you understand this concept, you should be able to explain why scavenging birds such as condors are often poisoned when eating carcasses of animals killed by hunters who use lead shot.**

The Small Intestine: Digestion and Absorption

In humans, as covered previously, the stomach breaks up food mechanically and starts the chemical digestion of proteins and lipids. Peristalsis in the stomach wall then moves small amounts of material through the sphincter at the base of the stomach and into the small intestine.

The **small intestine** is a long tube that is folded into a compact space within the abdomen. In the small intestine, partially digested food mixes with secretions from the pancreas, liver, and gallbladder and begins a journey of about 6 m (20 ft). When passage through this tubular structure is complete, digestion is mostly finished, and most nutrients—along with large quantities of water—have been absorbed.

Folding and Projections Increase Surface Area for Absorption

The surface area available for nutrient and water absorption in the small intestine is nothing short of remarkable (Ch. 39,

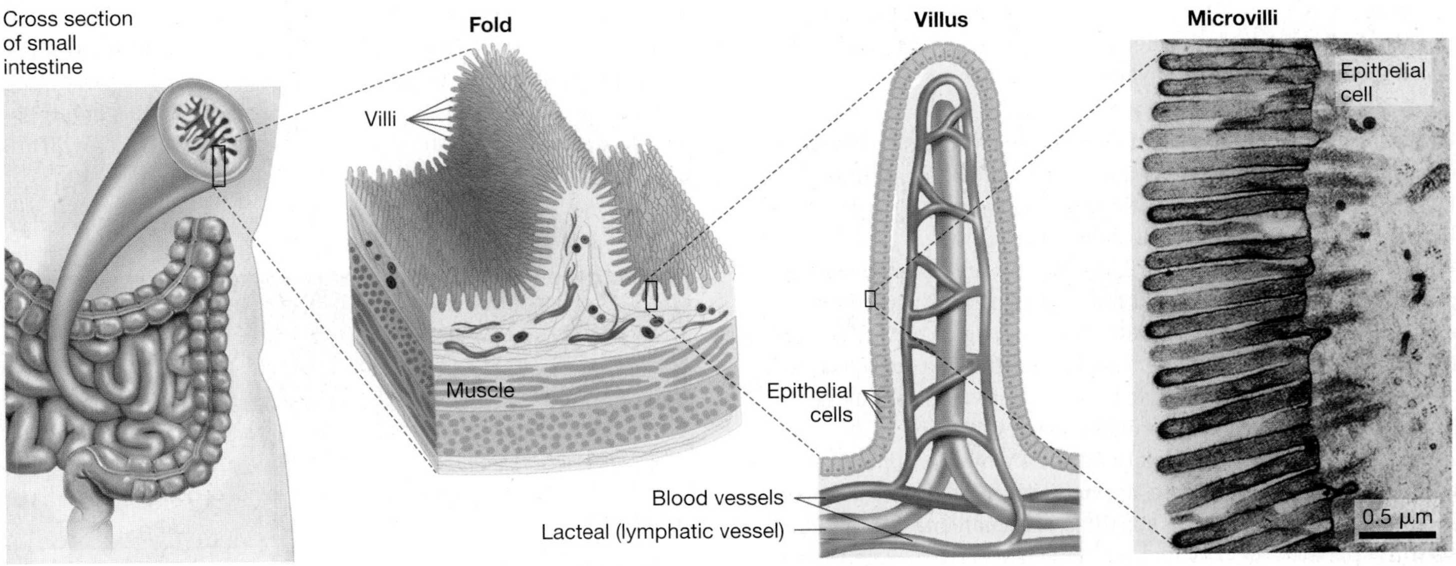

Figure 41.12 The Small Intestine Has an Extremely Large Surface Area. The villi that project from folds in the small intestine are covered with microvilli (colorized brown in the micrograph at the far right).

Section 39.3). As **Figure 41.12** shows, the organ's epithelial tissue is folded and covered with fingerlike projections called **villi** (singular: **villus**). In turn, the cells that line the surface of villi have tiny projections on their apical surfaces called **microvilli** (singular: **microvillus**). Microvilli project into the lumen of the digestive tract.

If the small intestine lacked folds, villi, and microvilli, it would have a surface area of about 3300 cm^2 (3.6 ft^2). Instead, the epithelium covers about 2 million cm^2 (over 2200 ft^2)—an area about the size of a tennis court.

The enormous surface area of the small intestine increases the rate of nutrient absorption. And because each villus contains blood vessels and a lymphatic vessel called a **lacteal**, nutrients pass quickly from epithelial cells into the body's transport systems (Ch. 42, Section 42.5).

To understand how digestion is completed and absorption occurs, let's explore what happens to proteins, lipids, and carbohydrates as they move through this section of the digestive tract. Again, humans are the model organism.

Protein Processing by Pancreatic Enzymes In the small intestine, the protein digestion that began in the stomach is completed, yielding individual amino acids that can enter the bloodstream and be transported to cells throughout the body.

How do these final stages of protein digestion occur? By the end of the nineteenth century, biologists had established that enzymes in the small intestine digest polypeptides to monomers. Later work showed that each of these protein-digesting enzymes, or **proteases**, is specific to certain types or configurations of amino acids in a polypeptide chain. Therefore, a suite of proteases is required to completely digest polypeptides to amino acid monomers.

In addition, by 1900 biologists had determined that proteases are synthesized in an inactive form in the **pancreas**, which is connected to the small intestine by the pancreatic duct. Like the production of inactive pepsinogen by chief cells in the stomach, the production of pancreatic enzymes in inactive conformations prevents pancreatic cells from digesting themselves.

How are pancreatic enzymes activated in the small intestine? A breakthrough in understanding came when Russian physiologist Ivan Pavlov showed that pancreatic enzymes become activated when they come into contact with juice from the upper part of the small intestine. Activation ceased to occur when he heated the intestinal juice, and because he knew that heat denatures enzymes (like all proteins), he hypothesized that the chemical that activated the pancreatic enzymes was also an enzyme.

Decades later, scientists figured out how this enzyme, called **enteropeptidase**, works. Other researchers demonstrated that enteropeptidase removes a short section from the N-terminus of trypsinogen, an inactive pancreatic protein, resulting in the active protease **trypsin**. Trypsin then triggers the activation of other pancreatic proteases, such as chymotrypsin, elastase, and carboxypeptidase (**Figure 41.13**). Once these enzymes are activated in the upper reaches of the small intestine, each of them

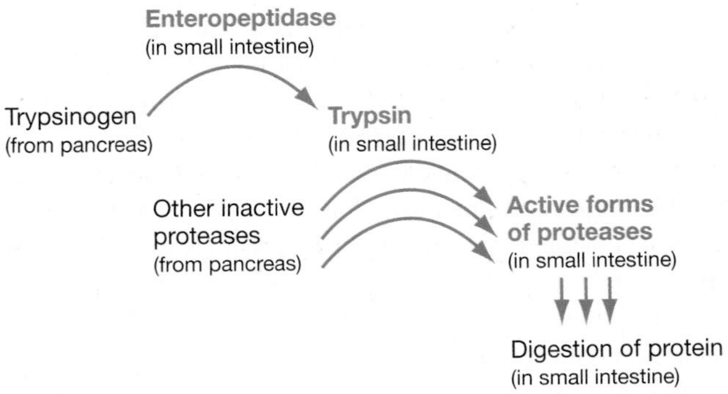

Figure 41.13 Enteropeptidase Triggers an Enzyme-Activation Cascade in the Small Intestine.

begins cleaving specific peptide bonds between amino acids. Eventually, enzymes embedded in the small intestine epithelium break these polypeptides into amino acid monomers.

What Regulates the Release of Pancreatic Enzymes? Digestive enzymes are needed only when food reaches the small intestine. Based on this simple observation, it was logical to predict that their release would be carefully controlled.

A classic experiment by William Bayliss and Ernest Starling, published in 1902, established how the release of pancreatic enzymes is controlled. Bayliss and Starling began by cutting the nerves between the pancreas and small intestine of a dog. Electrical signaling between the two organs via neurons was now impossible. But when the researchers simulated arrival of material from the stomach by introducing a weak HCl solution into the upper reaches of the animal's small intestine, its pancreas secreted enzymes in response. This observation was startling: The small intestine had successfully signaled the pancreas that food had arrived, even though the nerves connecting the two organs had been cut.

Starling hypothesized that a chemical messenger must be involved, and that the chemical messenger must originate in the small intestine and travel to the pancreas via the blood. He tested this hypothesis by cutting off a small piece of the small intestine, grinding it up, and injecting the resulting solution into a vein in the animal's neck. Minutes later, the pancreas sharply increased secretion.

Bayliss and Starling had discovered the first **hormone**—a chemical messenger that influences physiological processes at a very low concentration. The molecule they detected, which they called **secretin**, is produced by the small intestine in response to the arrival of food from the stomach.

Follow-up work showed that secretin's primary function is to induce a flow of bicarbonate ions (HCO_3^-) from the pancreas to the small intestine. Bicarbonate ions are important because they neutralize the acid arriving from the stomach (Ch. 2, Section 2.2).

Several years later, researchers discovered a second hormone produced in the small intestine. They called it **cholecystokinin** (pronounced *ko-la-sis-ta-KY-nin*). Cholecystokinin ("bile-bag-mover") stimulates the secretion of digestive enzymes from the pancreas and the secretion of molecules from the gallbladder that aid in processing lipids.

How Are Carbohydrates Digested and Transported? Besides manufacturing protein-digesting enzymes, the pancreas produces nucleases and an amylase that is similar to the salivary enzyme introduced earlier. **Nucleases** digest the RNA and DNA in food; **pancreatic amylase** continues the digestion of carbohydrates that began in the mouth by cleaving dextrins to release monosaccharides, disaccharides, and trisaccharides.

Digestion of carbohydrates does not end until they are completely broken down into monosaccharides. This is accomplished by a series of enzymes embedded in the epithelial cells of the small intestine—collectively known as brush border enzymes (because the microvilli resemble a brush; see Figures 41.7 and 41.12). For example, the brush border enzyme maltase breaks the bond in maltose to release two glucose molecules. ✔ **If you understand this concept, you should be able to predict the physiological reason that many people cannot digest the disaccharide lactose.**

What molecular mechanisms make it possible for epithelial cells to transport nutrients, such as glucose, from the lumen of the small intestine into the bloodstream? Two general principles apply to monosaccharide and amino acid absorption: **(1)** It is highly selective, in that proteins in the plasma membranes of microvilli are

RESEARCH

QUESTION: How is glucose transported into epithelial cells of the small intestine?

HYPOTHESIS: Glucose enters epithelial cells along with sodium ions via a Na^+-glucose cotransporter protein.

NULL HYPOTHESIS: Glucose transport does not depend on Na^+ transport.

EXPERIMENTAL SETUP:

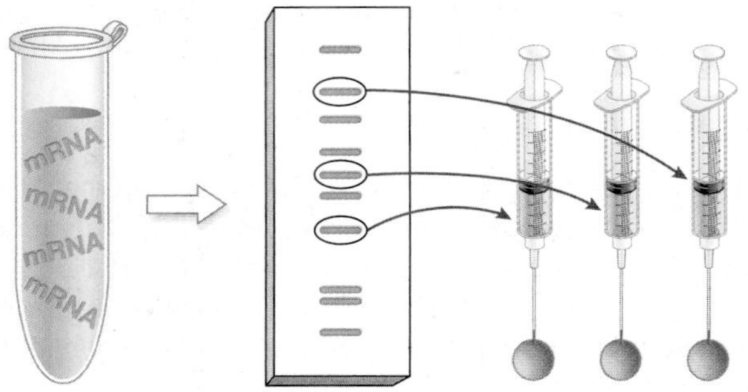

1. Purify mRNA from rabbit intestinal cells.

2. Separate mRNAs by size via gel electrophoresis.

3. Inject individual mRNAs into frog eggs. Test each egg—can it absorb Na^+ and glucose?

PREDICTION OF HYPOTHESIS: An egg will be able to absorb Na^+ and glucose because it received the mRNA that codes for the Na^+-glucose cotransporter.

PREDICTION OF NULL HYPOTHESIS: None of the eggs will be able to absorb Na^+ and glucose.

RESULTS:

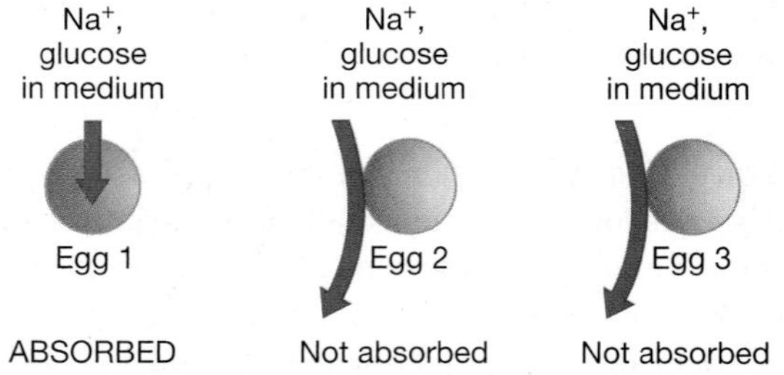

CONCLUSION: The egg that absorbs Na^+ and glucose received the mRNA from the Na^+-glucose cotransporter gene.

Figure 41.14 The Experimental Protocol for Locating the Na^+-Glucose Cotransporter Gene.

SOURCE: E. M. Wright. 1993. The intestinal Na^+/glucose cotransporter. *Annual Review of Physiology* 55: 575–589.

✔ **PROCESS OF SCIENCE** Why did the researchers inject the RNAs into frog eggs instead of into rabbit epithelial cells?

responsible for bringing specific nutrients into epithelial cells; and **(2)** it is usually active, meaning ATP is expended to transport nutrients into the cells against their concentration gradients.

Work over the past several decades has demonstrated both of these principles. One of the key results grew out of a series of experiments during the 1980s, which established that glucose absorption depends on the presence of an electrochemical gradient favoring an influx of sodium ions into the epithelium. Based on this finding, biologists hypothesized that the apical membranes of epithelial cells must contain a variety of cotransporters—membrane proteins that bring a nutrient molecule into the cell along with sodium ions. To confirm that a sodium–glucose cotransporter exists, investigators set out to find the gene that codes for it.

The researchers began by purifying mRNAs from rabbit intestinal cells (Figure 41.14), which presumably were transcribing the cotransporter genes. Then the team separated the mRNAs by size via gel electrophoresis (BioSkills 6) and injected one of each type of mRNA into a series of frog eggs—cells that do not normally transport glucose. The frog eggs translated the rabbit mRNAs into proteins.

One of the injected eggs was able to import Na^+ and glucose in tandem. The researchers concluded that this egg had received the mRNA for the rabbit Na^+-glucose cotransporter. They made a DNA copy of the mRNA (using techniques introduced in Ch. 20, Section 20.1), analyzed it to determine the sequence of the gene, and from that inferred the amino acid sequence of the membrane protein.

The discovery of the Na^+-glucose cotransporter inspired a three-step model for glucose absorption:

1. Na^+/K^+-ATPase (sodium–potassium pump) in the basolateral membrane of the epithelial cells creates an electrochemical gradient that favors the entry of Na^+.
2. Glucose from digested food enters the cell along with sodium ions via the Na^+-glucose cotransporter in the apical membrane.
3. Glucose diffuses into nearby blood vessels through a glucose carrier in the basolateral membrane.

If this configuration of pumps, cotransporters, and carriers sounds familiar, here's why: The same combination of membrane proteins occurs in the proximal tubule of the kidney, where the proteins are responsible for the reabsorption of sodium ions and glucose from urine (Ch. 40, Section 40.4).

Follow-up work showed that in the small intestine—just as in the proximal tubule—other cotransporters are responsible for absorbing other monosaccharides and amino acids, and that specific channels and carriers in the basolateral membrane are responsible for transporting each substance to the blood.

Digesting Lipids: Bile and Transport The pancreatic secretions include digestive enzymes that act on fats, in addition to enzymes that act on proteins and carbohydrates. When lipids arrive in the small intestine, they have been partially digested by lingual lipase from the mouth and gastric lipase from the stomach. In the small intestine, the enzyme **pancreatic lipase** completes digestion to release fatty acids and monoglycerides.

Recall that fats are insoluble in water (Ch. 6, Section 6.1). As a result, as shown in Figure 41.15 (step 1), they tend to form large globules as they are churned in the stomach. As these globules emerge from the stomach, they must be broken up in a process called **emulsification** before pancreatic lipase can act on them.

In the small intestine (step 2), emulsification results from the action of small molecules called bile salts. Bile salts mechanically digest fats by breaking up large fat golubles into smaller ones.

Bile salts are synthesized in the **liver**, an organ that performs an array of functions related to digestion, and secreted in a complex solution called **bile**, which is stored in the **gallbladder**. When bile enters the small intestine, it raises the pH and emulsifies fats. Once fats are broken into small globules (step 3), which increases their surface area, they can be attacked by pancreatic lipase and digested (step 4).

The monoglycerides and fatty acids released by lipase activity enter small intestine epithelial cells by simple diffusion. Once inside the cells, they are repackaged into triglycerides and

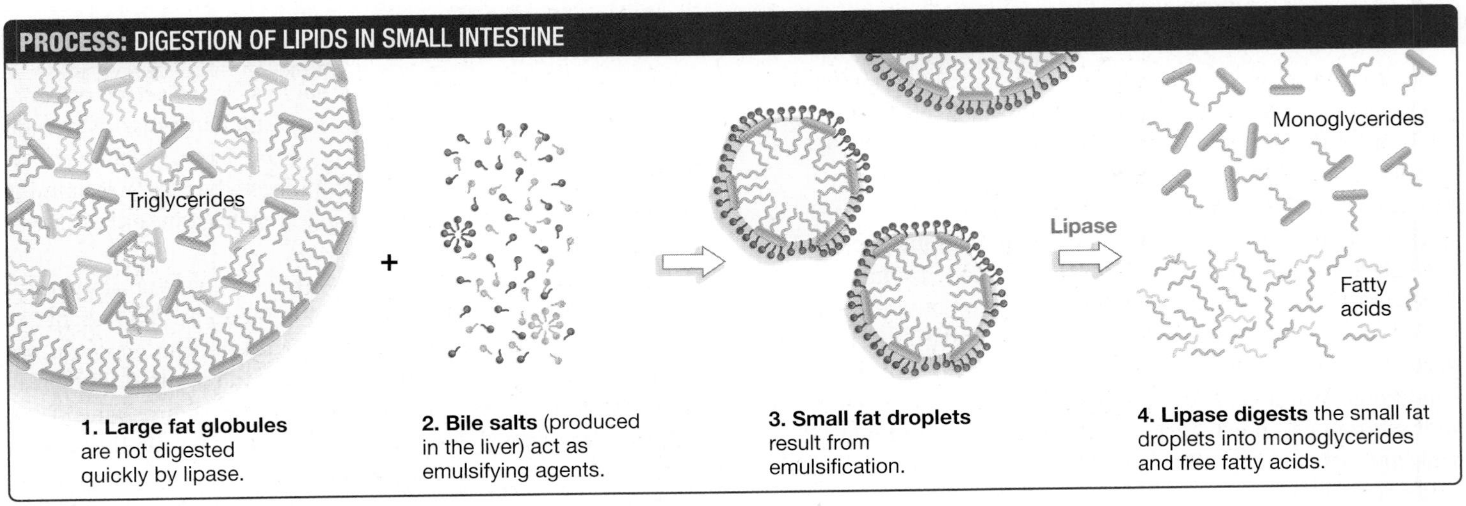

Figure 41.15 **Lipid Digestion in the Small Intestine Depends on Bile Salts and Pancreatic Lipase.** Bile salts break up large fat globules, enabling pancreatic lipase to digest fats quickly.

SUMMARY Table 41.3 **Digestive Enzymes in Mammals**

Digestion is accomplished by the enzymes listed here, by HCl produced in the stomach, and by bile salts made in the liver and stored in the gallbladder.

	Where Synthesized	Regulation	Function
Carboxypeptidase	Pancreas	Released in inactive form in response to cholecystokinin from small intestine; activated by trypsin	In small intestine, breaks peptide bonds in polypeptides, releasing amino acids
Chymotrypsin	Pancreas	Released in inactive form in response to cholecystokinin from small intestine; activated by trypsin	In small intestine, breaks peptide bonds in polypeptides, releasing amino acids
Elastase	Pancreas	Released in inactive form in response to cholecystokinin from small intestine; activated by trypsin	In small intestine, breaks peptide bonds in polypeptides, releasing amino acids
Gastric lipase	Stomach	Released in response to gastrin from stomach	In stomach, breaks bonds in fats, releasing fatty acids and diglycerides
Lingual lipase	Salivary glands	Released in response to taste and smell stimuli	In mouth and stomach, breaks bonds in fats, releasing fatty acids and diglycerides
Nucleases	Pancreas	Released in response to cholecystokinin from small intestine	In small intestine, break apart nucleic acids, releasing nucleotides
Pancreatic amylase	Pancreas	Released in response to cholecystokinin from small intestine	In small intestine, breaks apart carbohydrates, releasing smaller sugars
Pancreatic lipase	Pancreas	Released in response to cholecystokinin from small intestine	In small intestine, breaks bonds in fats, releasing fatty acids and monoglycerides
Pepsin	Stomach	Released in inactive form (pepsinogen); activated by low pH in stomach lumen	In stomach, breaks peptide bonds between certain amino acids in proteins, releasing polypeptides
Salivary amylase	Salivary glands	Released in response to taste and smell stimuli	In mouth, breaks apart carbohydrates, releasing dextrins and disaccharides
Trypsin	Pancreas	Released in inactive form (trypsinogen) in response to cholecystokinin from small intestine; activated by enteropeptidase from small intestine	In small intestine, breaks specific peptide bonds in polypeptides, releasing amino acids

processed into protein-coated globules called chylomicrons, which move by exocytosis into lacteals—look back at Figure 41.7—near the epithelial cells. The lacteals merge with larger lymph vessels, which then merge with large veins. In this way, fats enter the bloodstream without clogging small blood vessels. Because lipids are hydrophobic, packaging them into a chylomicron allows them to be carried in the aqueous environment of the blood until they are absorbed by cells for storage or use as energy.

Fat-soluble vitamins (vitamins A, D, E, and K) are also absorbed via simple diffusion and packaged for transport into chylomicrons. Water-soluble vitamins (including the B vitamins and vitamin C), in contrast, are absorbed via facilitated diffusion and cotransport, like the water-soluble monosaccharides and amino acids.

Table 41.3 summarizes the source and function of the major digestive enzymes found in mammals.

How Is Water Absorbed? When solutes from digested material are absorbed into the epithelium of the small intestine, water follows passively by osmosis. This is an important mechanism for **(1)** absorbing water that has been ingested and **(2)** reclaiming liquid that was secreted into the digestive tract in saliva, mucus, and pancreatic fluid.

This mechanism of water absorption inspired an important medical strategy called oral rehydration therapy. If a patient has diarrhea, clinicians frequently prescribe dilute solutions of glucose and electrolytes to be taken orally. When the solutes in the drink are absorbed in the small intestine through sodium-glucose cotransporters, enough water follows to prevent the life-threatening effects of dehydration. This simple treatment saves thousands of lives every year. ✔ **If you understand this strategy, you should be able to predict at least two digestive effects of a molecule that selectively blocks the sodium-glucose cotransporter.**

The Large Intestine: Absorption and Elimination

By the time digested material reaches the large intestine of a human, a large amount of water (approximately 5 liters per day) and virtually all of the available nutrients have been absorbed. The primary function of the **large intestine** is to form feces by absorbing additional water and compacting the wastes that remain.

These processes occur in the **colon**—the main section of the large intestine. Feces are held in the **rectum**, which is the final part of the large intestine, until they can be eliminated. Although the kidneys are responsible for maintaining water balance, water absorption in the large intestine is important for keeping the body well hydrated.

Variations in Structure and Function Sizes and functions of the large intestine vary dramatically among animals. In insects, the posteriormost portion of the digestive tract, called the hindgut, reabsorbs water and ions from "pre-urine," and excretes uric acid and feces (Ch. 40, Section 40.3). Among vertebrates, various lineages of fishes have no large intestine at all.

In some herbivorous vertebrates, the **cecum**, a blind sac at the anterior end of the large intestine, is greatly enlarged. These vertebrates include rabbits, many rodents, some marsupials, horses, elephants, tapirs, and leaf-eating primates. Like the crop of some birds and the modified stomach of ruminants, the cecum of these vertebrates contains symbiotic bacteria and protists that ferment cellulose. Rabbits and some other mammals are able to eliminate the cecum's contents as pellets, which the animal then re-ingests and passes through the digestive tract a second time. This particular example of **coprophagy**, or feces eating, allows the animal to absorb more nutrients from food.

In humans, a narrow pouch called the **appendix** emerges from the cecum. The function of the appendix has long been debated. It has often been described as vestigial—referring to a reduced or incompletely developed trait that is a vestige of evolutionary ancestry (Ch. 22, Section 22.2)—partly because it does not perform any obvious vital function. Indeed, if it becomes inflamed, it can be surgically removed from a patient with no ill effects. Although the appendix is not vital, it contains immune system cells and appears to act as a haven for symbiotic microorganisms that inhabit the colon. After an episode of diarrhea flushes symbiotic bacteria from the colon, the appendix may provide the additional bacteria needed for recolonizing the colon.

The Human Gut Microbiome Plays Important Roles in Digestion and Health Fermentation of cellulose is not as important to an omnivorous human as it is to an herbivore such as a ruminant or a rabbit, but our symbiotic gut bacteria play significant roles in our health. Fermentation by bacteria in the human colon breaks down the cellulose we eat to produce several important nutrients, such as vitamin K, that are absorbed into our bloodstream.

Exciting recent research has shown that our gut microbiota go far beyond simply producing vitamins. Human gut microbes play important roles in healthy metabolism and function of our immune systems. Imbalances in gut bacterial communities are associated with disorders ranging from inflammatory bowel disease to metabolic disorders, including obesity, type 2 diabetes mellitus, and even cancer.

A person's gut microbiota may become imbalanced due to use of antibiotics or poor diet. For example, the typical Western diet (high in animal protein and fat, low in fiber-rich plants) leads to gut microbial communities associated with increased risks of obesity and cardiovascular disease. In contrast, the typical Mediterranean diet (most fats and proteins from plants) favors gut microbes that may reduce the risk of these disorders.

Some physicians advocate fecal transplants—where feces of a healthy donor, containing large quantities of gut bacteria, are transplanted into the large intestine of a recipient—in treating certain chronic disorders such as inflammatory bowel disease. Studies have shown that this therapy often works very well, although the technique is proving slow to catch on—possibly due to the yuck factor.

CHECK YOUR UNDERSTANDING

✔ **If you understood this section, you should be able to ...**

1. Explain how each compartment in the human digestive tract aids the ingestion and digestion of food, absorption of nutrients, and elimination of wastes.
2. Predict the consequences of treating a person with a drug that inhibits the release of bile salts.
3. Predict how the relative volumes of the large intestines of herbivores and carnivores would compare, and explain your reasoning.

Answers are available in Appendix A.

41.4 Nutritional Homeostasis—Glucose as a Case Study

When digestion is complete, amino acids, fatty acids, ions, sugars, vitamins, and minerals enter the bloodstream and are delivered to the cells that need them. The body uses or stores these nutrients to maintain homeostatic levels in the blood and avoid imbalances. Too much or too little of a nutrient can be problematic or even fatal.

After you complete this section, you should be able to ...

- Analyze glucose homeostasis.

The illness **diabetes mellitus** is a classic example of nutrient imbalance. People with diabetes mellitus have an abnormally high level of glucose in their blood because cells cannot import the glucose. Over the course of a lifetime, a chronically elevated blood glucose level can lead to an array of complications, including blindness, impaired blood circulation, and heart failure. What causes the imbalance? More important, how is glucose homeostasis normally maintained?

Insulin's Role in Glucose Homeostasis

Insulin is a hormone that is secreted by cells in the pancreas when the blood glucose level is elevated. It travels through the bloodstream and binds to receptors on cells throughout the body. (See Chapter 46 for more detail on the structure and function of hormones and other chemical signals.) In response, cells that have insulin receptors increase their rate of glucose uptake and processing. Specifically, insulin stimulates cells in the liver and skeletal muscle to import glucose from the blood and synthesize glycogen from glucose monomers. As a result, the glucose level in the blood declines (Figure 41.16, top).

If the blood glucose level falls too low, as it does when an animal has not eaten for a while, other cells in the pancreas secrete a hormone called **glucagon**. In response to glucagon, liver cells catabolize stored glycogen and produce glucose via the process of **gluconeogenesis**, the synthesis of glucose from non-carbohydrate compounds. As a result, the glucose level in the blood rises (see Figure 41.16, bottom).

Insulin and glucagon interact to form a negative feedback system capable of achieving homeostasis with respect to the glucose concentration in the blood.

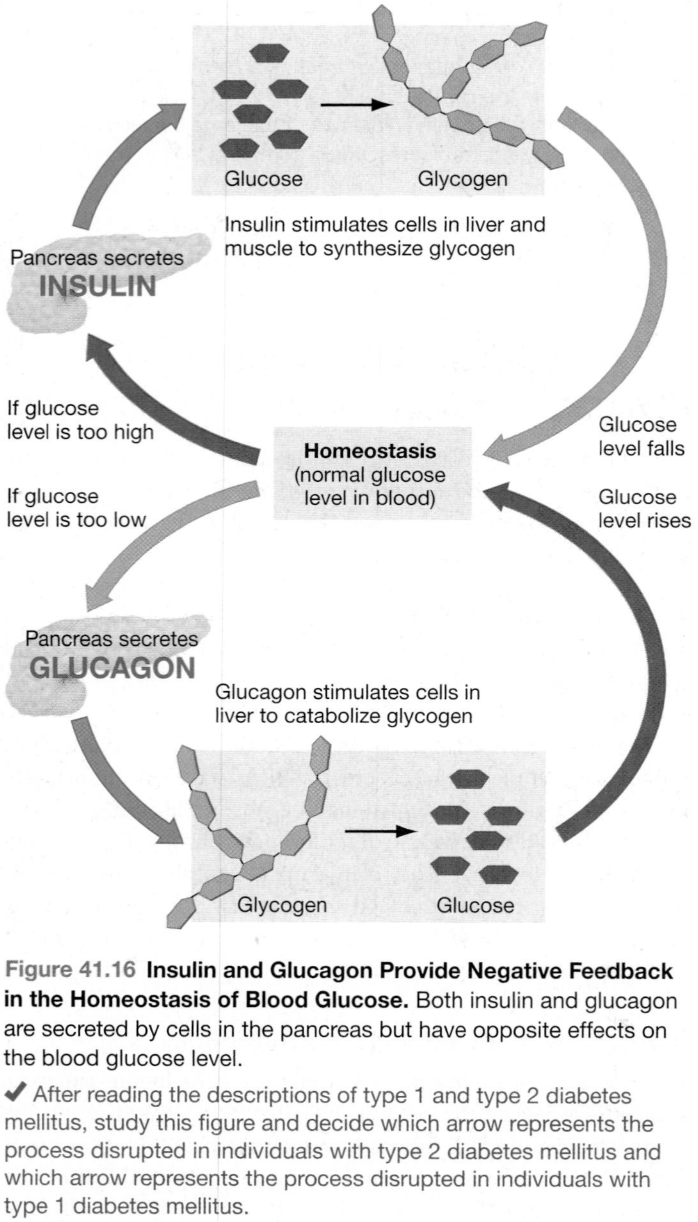

Figure 41.16 Insulin and Glucagon Provide Negative Feedback in the Homeostasis of Blood Glucose. Both insulin and glucagon are secreted by cells in the pancreas but have opposite effects on the blood glucose level.

✔ After reading the descriptions of type 1 and type 2 diabetes mellitus, study this figure and decide which arrow represents the process disrupted in individuals with type 2 diabetes mellitus and which arrow represents the process disrupted in individuals with type 1 diabetes mellitus.

Diabetes Mellitus Has Two Forms

Diabetes mellitus develops in people **(1)** who do not synthesize sufficient insulin or **(2)** whose cells are resistant to insulin, meaning that insulin does not effectively activate its receptor in target cells. The first condition is called type 1 diabetes mellitus; the second condition is type 2 diabetes mellitus. In both types, effector cells do not receive the signal that would result in uptake of glucose by cells and therefore a drop in blood glucose level.

Glucose imbalance has a direct effect on urine formation. Normally, signals from insulin keep the blood glucose level low enough that all of the glucose can be reabsorbed from the filtrate formed in the kidney. But when the blood glucose level is very high, so much glucose enters the nephron that it cannot all be reabsorbed. A high glucose level in the filtrate increases its osmolarity and decreases the amount of water reabsorbed from it. More water leaves the body, leading to high urine volume in both types of diabetes mellitus.

The word "diabetes" means "to run through"; water "runs through" people with diabetes. The word "mellitus" means "honeyed (sweet)." Before chemical methods of analyzing urine were available, physicians would taste the patient's urine. If the urine was sweet, then the patient very likely was suffering from diabetes mellitus.

How do these diseases develop? Type 1 diabetes mellitus is an autoimmune disease, meaning that the body's immune system mistakenly targets its own cells for destruction. In the case of type 1 diabetes mellitus, the insulin-producing cells of the pancreas are destroyed. Type 2 diabetes mellitus occurs when the receptors for insulin no longer function correctly or are reduced in number. The primary risk factors for developing type 2 diabetes mellitus are obesity, a high-sugar diet, lack of exercise, and genetic predisposition to the disease.

A high blood glucose level is not the only problem with diabetes. The level is high because glucose cannot be imported into cells, so cells that require glucose to function, such as brain neurons, can be starved unless the disease is treated.

Currently, type 1 diabetes mellitus is treated with insulin injections and careful attention to diet; type 2 diabetes mellitus is managed primarily through prescribed diets, exercise, and monitoring the blood glucose level, as well as taking drugs that increase cellular responsiveness to insulin. The challenge is to achieve homeostasis with respect to the blood glucose level in the absence of the body's normal regulatory mechanisms.

The Type 2 Diabetes Mellitus Epidemic

An epidemic of type 2 diabetes mellitus is currently under way in certain human populations. In the United States, 12 percent of persons aged 18 and over and 25 percent of persons aged 65 and over were diabetic in 2015. About 90–95 percent of these individuals had type 2 diabetes mellitus.

Because of a strong association between the prevalence of diabetes mellitus in parents and their children, researchers have long suspected that some individuals have a genetic predisposition for developing the disease. To date, alleles at dozens of genes that predispose individuals to type 2 diabetes mellitus

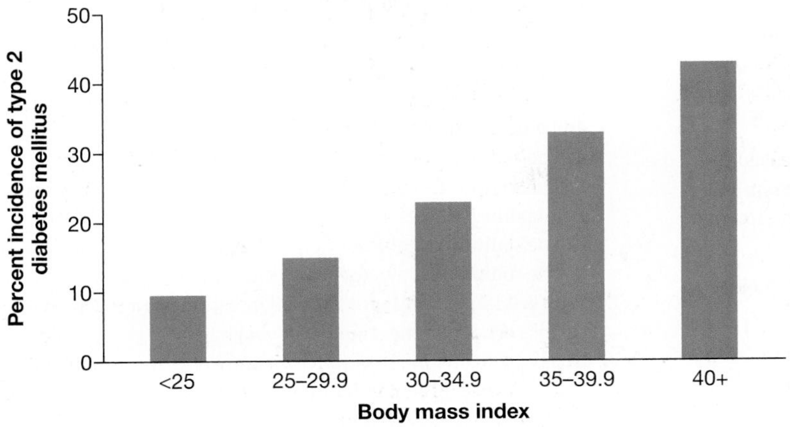

Figure 41.17 The Incidence of Type 2 Diabetes Mellitus Is Correlated with Obesity.
DATA: N. T. Nguyen et al. 2011. *Obesity Surgery* 21: 351–355.

✔ **QUANTITATIVE** A 1.8-m (6-ft) tall man has a BMI of 40. How much weight would he need to lose to reduce his risk for type 2 diabetes mellitus to less than 10 percent?

have been identified (using techniques introduced in Ch. 20, Section 20.5).

However, there is also evidence that obesity has an important impact on the incidence of type 2 diabetes mellitus. From 1999 to 2006, researchers collected data on more than 20,000 middle-aged and elderly Americans, 2900 of whom developed type 2 diabetes mellitus. As Figure 41.17 shows, the researchers found a strong relationship between the incidence of type 2 diabetes mellitus and **body mass index (BMI)**, a measure of obesity. BMI is calculated as body mass (in kilograms) divided by height (in meters) squared. A person with a BMI of 30 or higher is considered obese and has a far greater risk of developing type 2 diabetes mellitus than people with lower BMIs.

Type 2 diabetes mellitus used to be called adult-onset diabetes because the disease typically appeared in adults. The current epidemic of type 2 diabetes mellitus is not restricted to adults, however, and that name was abandoned as more and more young people were diagnosed with the disease. At present, one of every three children born in the United States will develop type 2 diabetes mellitus in their lifetime, and childhood diabetes rates are increasing rapidly every year worldwide. Two-thirds of American adults are overweight, and one-third are obese. Type 2 diabetes mellitus and obesity are clearly modern-day epidemics.

CHECK YOUR UNDERSTANDING

✔ **If you understood this section, you should be able to ...**

1. Identify the causes of each type of diabetes mellitus.
2. Explain what a person with type 1 diabetes mellitus should do when her or his blood glucose level is too high and when it is too low.

Answers are available in Appendix A.

CHAPTER 41 Review

For media, go to **Mastering Biology**

41.1 Nutritional Requirements

- The diets of animals include fats, carbohydrates, and proteins that provide energy; vitamins that serve as coenzymes and perform other functions; minerals that are used as components of enzyme cofactors or structural materials; and ions of electrolytes required for osmotic balance and normal membrane function.
- Fats contain more energy (about 9 kcal/g) than carbohydrates or proteins (about 4 kcal/g), making fats an efficient way to store energy in the body.

41.2 Capturing Food: The Structure and Function of Mouthparts

- Most animals are mass feeders that obtain food by seizing and manipulating it.
- Animal mouthparts include teeth, jaws, beaks, and toxin-injecting organs.
- Through natural selection, mouthparts in different animal species have become adapted for obtaining particular types of food.

41.3 The Structure and Function of Animal Digestive Tracts

- Most animals have a complete digestive tract that begins at the mouth and ends at the anus.
- Mechanical digestion usually begins in the mouth, where food is chewed, and continues in the stomach and small intestine, where food is mixed.
- In many animals, chemical digestion of food begins in the mouth. In mammals, salivary amylase hydrolyzes bonds in starch and glycogen, and lingual lipase hydrolyzes bonds in fats.
- Once food is swallowed, it is propelled down the esophagus by peristalsis.
- Digestion continues in the stomach. In the human stomach, a highly acidic environment denatures proteins, gastric lipase continues fat digestion, and pepsin begins the cleavage of peptide bonds that link amino acids.
- Food passes from the stomach into the small intestine, where it is mixed with secretions from the pancreas and liver.

- In the small intestine, carbohydrate digestion is continued by pancreatic amylase and intestinal brush border enzymes; fats are emulsified by bile salts and digested by pancreatic lipase; and protein digestion is completed by a suite of pancreatic proteases and intestinal brush border enzymes.
- Cells that line the small intestine absorb the nutrients released by digestion. In many cases, uptake is driven by an electrochemical gradient established by Na^+/K^+-ATPase that favors the diffusion of Na^+ into the cells.
- As solutes leave the lumen of the small intestine and enter cells, water follows by osmosis.
- Water reabsorption is completed in the large intestine, where feces form. Feces are expelled from the body in the process of elimination.
- The structure of organs in the digestive tract varies widely among species, in ways that support processing of the food each species ingests.

41.4 Nutritional Homeostasis—Glucose as a Case Study

- Diabetes mellitus is a condition in which the level of glucose in the blood is abnormally high.
- Type 1 diabetes mellitus is caused by a defect in the production of insulin—a hormone secreted by the pancreas that promotes the uptake of glucose from the blood by cells.
- Type 2 diabetes mellitus is characterized by a failure of cells to respond to insulin.
- The development of type 2 diabetes is correlated with obesity. The incidence of this disease has reached epidemic proportions in many populations.

Answers are available in Appendix A.

✔ TEST YOUR KNOWLEDGE

1. **QUANTITATIVE** Calculate and compare the caloric content of skim milk and whole milk. Per serving, skim milk contains 12 g carbohydrates, 8 g protein, and no fat; whole milk contains 12 g carbohydrates, 8 g protein, and 8 g fat.
2. Evaluate the following statements regarding digestion; select True or False for each statement.
   T/F Salivary amylase completes the digestion of starch into monosaccharides.
   T/F Pepsin digests proteins in the stomach.
   T/F Trypsin digests disaccharides into monosaccharides.
   T/F Pancreatic lipase performs the majority of chemical digestion of lipids.
3. Cellulose is fermented in which of the following structures in rabbits?
   a. small intestine
   b. cecum
   c. abomasum
   d. rumen
4. A hormone that reduces the blood glucose level is __________, and a hormone that increases the blood glucose level is __________.

✔ TEST YOUR UNDERSTANDING

5. Explain the role in nutrition of each of the following structures: bird crop, cow rumen, and elephant cecum.
6. Why is oral rehydration therapy with a solution of sodium chloride and glucose an effective treatment for dehydration?
   a. The sodium and glucose decrease urine output.
   b. The sodium and glucose facilitate water absorption by the small intestine.
   c. The sodium and glucose help kill intestinal bacteria.
   d. The sodium and glucose make the person thirsty.
7. Why is it important that the small intestine has a much greater surface area than the stomach or esophagus?
8. Explain why insulin injections are more effective in controlling the blood glucose level in individuals with type 1 diabetes mellitus than in those with type 2 diabetes.

✔ TEST YOUR PROBLEM-SOLVING SKILLS

9. When food is plentiful, animals tend to store most of what they eat as fat. Why is this?
10. Among vertebrates, the large intestine exists only in lineages that are primarily terrestrial (amphibians, reptiles, and mammals). Propose a hypothesis to explain this observation.

✔ PUT IT ALL TOGETHER: Case Study

What is the relationship between diet and the structure of the digestive tract?

Minnows are mainly carnivorous, eating insects and other small animals. However, herbivory has evolved independently in minnows several times. What changes in digestive structure and function are associated with the evolution of herbivory?

11. Like cichlids, minnows use their pharyngeal jaws to process food. Suggest some possible structural differences between the teeth on the pharyngeal jaws of carnivorous and herbivorous minnows.
12. Which of the following is true of the digestive tracts of minnows?
    a. They are incomplete but have both a mouth and an anus.
    b. They are complete, facilitating compartmentalization of digestion in different organs.
    c. They are incomplete, with no accessory organs.
    d. They are complete and include a large gastrovascular cavity.

13. **QUANTITATIVE** Researchers compared the relative gut length—the length of the digestive tract divided by body length—in four species of herbivorous minnows and four species of carnivorous minnows. The results are shown in the graph provided here (*** means $P < 0.001$; **BioSkills 3**). Based on these data, what conclusion can you draw about the relationship between diet and gut length?

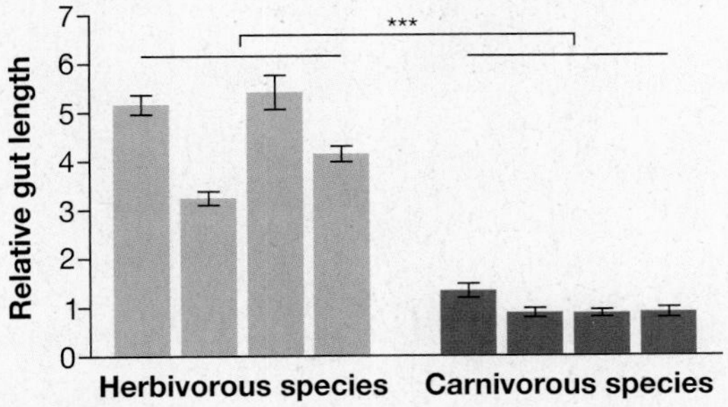

Source: D. P. German et al. 2010. *Physiological and Biochemical Zoology* 83: 1–18.

14. Suggest a function of the difference in relative gut lengths of herbivorous and carnivorous minnows.
15. Which minnows, herbivorous or carnivorous, should exhibit higher cellulase activity in the gut? Explain.
16. **PROCESS OF SCIENCE** Why did the researchers compare relative gut length instead of absolute gut length?

Mastering Biology

Students Go to Mastering™ Biology for assignments, the eText, and the Study Area with animations, practice tests, and activities.

Professors Go to Mastering™ Biology for automatically graded tutorials and questions that you can assign to your students, plus Instructor Resources.

42 Gas Exchange and Circulation

During intense exercise, animal circulatory systems deliver large amounts of oxygen to tissues and remove large amounts of carbon dioxide. This chapter explores how gas exchange occurs in animals that live in aquatic or terrestrial environments.

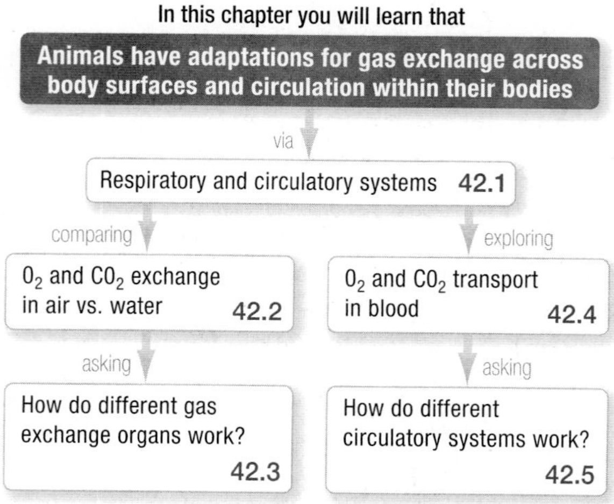

Animal cells are like factories that run 24 hours a day. Inside the plasma membrane, the chemical reactions that sustain life require a steady input of raw materials. Those reactions also produce a steady stream of wastes. Earlier in this unit, you learned how waste materials are excreted from the body (Chapter 40). You also examined how nutrients enter the body (Chapter 41). Let's now turn our attention to these major questions:

- How are two of the most important molecules in the economy of the cell—the oxygen (O_2) required for cellular respiration and the carbon dioxide (CO_2) produced by cellular respiration—exchanged with the environment?
- How are these gases—along with wastes, nutrients, and other types of molecules—transported throughout the body?

BIG PICTURE

This chapter is part of the Big Picture. See how on pages 838–839.

Understanding gas exchange and circulation is fundamental to understanding how animals work. If either process fails, the consequences are dire. Let's begin with an overview of animal respiratory and circulatory systems, and then plunge into the details of how gases are exchanged and transported.

42.1 The Respiratory and Circulatory Systems

When the mitochondria inside animal cells are producing ATP via **cellular respiration**, they use oxygen and produce carbon dioxide. (To review the details of this important metabolic pathway, see Chapter 9.) To support continued ATP production, cells have to obtain oxygen and expel excess carbon dioxide continuously (see the Big Picture, Energy for Life, on pages 236–237).

After you complete this section, you should be able to . . .

- Explain the relationship between the respiratory and circulatory systems.

How does this gas exchange occur between an animal's environment and its mitochondria? In most cases, gas exchange involves the five steps illustrated in Figure 42.1:

1. ***Ventilation***, the movement of air or water through a specialized gas exchange organ, such as a lung or gill.
2. ***Diffusion at the respiratory surface***, where O_2 moves from the air or water into the blood and CO_2 moves from the blood into the air or water, along their concentration gradients.
3. ***Circulation***, the transport of dissolved O_2 and CO_2 throughout the body—along with nutrients, wastes, and other types of molecules—via the circulatory system.
4. ***Diffusion at the tissues***, where O_2 moves from the blood into the tissues and CO_2 moves from the tissues into the blood, along their concentration gradients.
5. ***Cellular respiration***, the cell's use of O_2 and production of CO_2. In tissues, where cellular respiration has led to low O_2 levels and high CO_2 levels, gas exchange occurs between blood and cells.

Steps 1 and 2 are accomplished by the **respiratory system**, the collection of cells, tissues, and organs responsible for gas exchange between the animal and its environment. In essence, a respiratory system consists of structures for conducting air or water to a surface where gas exchange takes place.

In some animals the gas-exchange surface is the skin, but in most species it is located in a specialized organ such as the gills found in mollusks, arthropods, and fishes, the tracheae of insects, or the lungs of tetrapods. Section 42.3 analyzes the structure and function of gills, tracheae, and lungs in detail.

Step 3 in Figure 42.1 is usually accomplished by a **circulatory system**, which moves O_2, CO_2, and other materials around the body. In many cases, a muscular heart propels a specialized, liquid transport tissue throughout the body via a system of vessels.

Keeping in mind this broad overview of respiratory and circulatory systems, let's dive into the details by exploring how oxygen and carbon dioxide move between an animal's body and its environment.

CHECK YOUR UNDERSTANDING

✔ **If you understood this section, you should be able to . . .**

Explain why O_2 diffuses *into* the blood from alveoli, but diffuses *out of* the blood into tissues.

Answers are available in Appendix A.

42.2 Air and Water as Respiratory Media

Gas exchange between the environment and cells is based on diffusion. Under normal conditions, the oxygen concentration is relatively high in the environment (for example, in air that you inhale or in ocean water) and low in tissues, while the carbon dioxide level is relatively high in tissues and low in the environment. So oxygen tends to move from the environment

After you complete this section, you should be able to . . .

- Analyze the factors that affect oxygen and carbon dioxide levels on land and in the water.

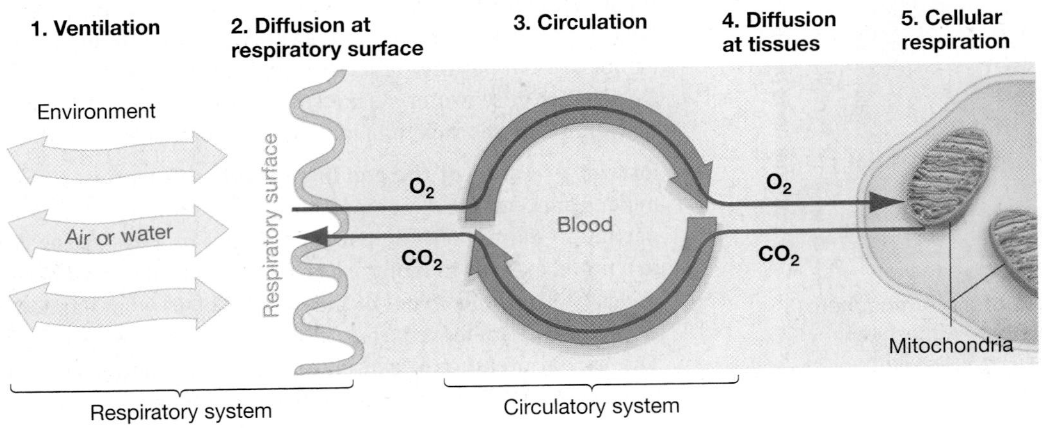

Figure 42.1 Gas Exchange Involves Ventilation, Diffusion, Circulation, and Respiration. In animals, oxygen and carbon dioxide are exchanged by diffusion across the surface of a lung, a gill, the skin, or some other gas exchange organ. In many species these gases are transported to and from tissue cells—where gas exchange also takes place—in a fluid connective tissue such as blood.

into tissues, and carbon dioxide tends to move from tissues into the environment.

How much oxygen and carbon dioxide are present in the atmosphere versus the ocean? What factors in air and water influence how quickly these gases move by diffusion?

How Do Oxygen and Carbon Dioxide Behave in Air?

The atmosphere is composed primarily of nitrogen (76 percent) and oxygen (21 percent) and has trace amounts of argon (0.93 percent) and CO_2 (0.04 percent). Nitrogen (N_2) and argon are not important to animals living at sea level and are usually ignored in analyses of gas exchange.

However, the data are a little misleading. To understand why, consider that the *percentage* of O_2 in the atmosphere does not vary with elevation. The atmosphere at the top of Mt. Everest is composed of 21 percent oxygen, just as it is at sea level. The key difference is that far fewer molecules of oxygen and other atmospheric gases are present per unit volume of air at high elevations than at sea level, because atmospheric pressure is lower at high elevations. To understand how gases move by diffusion, it is important to express their presence in terms of partial pressures instead of percentages. Pressure is force exerted per unit area. A **partial pressure** is the pressure of a particular gas in a mixture of gases.

To calculate the partial pressure of a particular gas, multiply the fractional composition (the fraction of air the gas makes up) by the total pressure exerted by the entire mixture (atmospheric pressure). The calculation is valid because the total pressure of a mixture of gases is the sum of the partial pressures of all the individual gases. This relationship is called Dalton's law.

For example, Figure 42.2 shows that the total atmospheric pressure at sea level is 760 mm Hg (millimeters of mercury). If you multiply this value by 0.21, which is the fraction of air that is O_2, you obtain a partial pressure of oxygen, abbreviated P_{O_2}, at sea level of 160 mm Hg. Because the atmospheric pressure is only about 250 mm Hg at the top of Mt. Everest, the P_{O_2}, there is only $0.21 \times 250 = 53$ mm Hg.

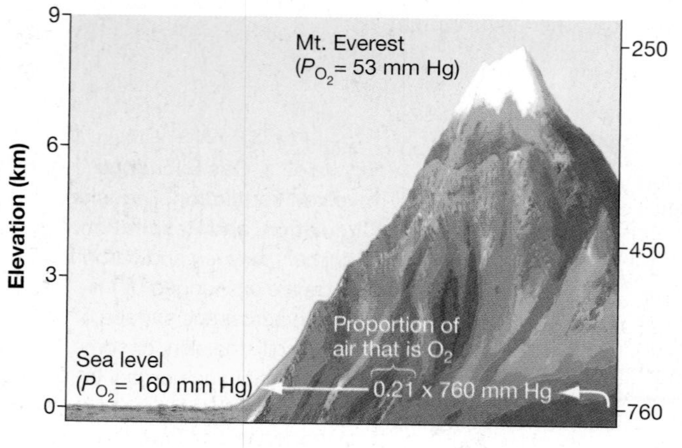

Figure 42.2 Oxygen Makes Up 21 Percent of the Atmosphere, but Its Partial Pressure Depends on Elevation. Atmospheric pressure, and thus oxygen partial pressure (P_{O_2}), falls with increasing elevation.

Oxygen and carbon dioxide diffuse between the environment and cells along their respective partial pressure gradients, just as solutes diffuse along their electrochemical gradients. In both air and water, O_2 and CO_2 move from regions of high partial pressure to regions of low partial pressure. It's hard to breathe at the top of Mt. Everest because the partial pressure of oxygen is low there—meaning that the diffusion gradient between the atmosphere and your lung tissues is small, so fewer molecules of O_2 diffuse into your tissues when you take a breath.

How Do Oxygen and Carbon Dioxide Behave in Water?

To obtain oxygen, water breathers face a much more challenging environment than air breathers do. Aquatic animals live in an environment that contains much less oxygen than the environments inhabited by terrestrial animals. At 15°C, a liter of air can contain up to 209 milliliters (mL) of O_2, while a liter of water may contain a maximum of only 7 mL of O_2. To extract a given amount of oxygen, an aquatic animal has to process 30 times more water than the amount of air a terrestrial animal breathes.

In addition, water is about a thousand times denser than air and much more viscous. As a result, water breathers have to expend much more energy to ventilate their respiratory surfaces than do air breathers.

What Affects the Amount of Gas in a Solution? Oxygen and carbon dioxide diffuse into water from the atmosphere, but the amount of gas that dissolves in water depends on several factors:

- ***Solubility of the gas in water*** Oxygen has very low solubility in water. Only 0.003 mL of oxygen dissolves in 100 mL of water for each increase of 1 mm Hg in oxygen partial pressure. Animals compensate for the low solubility of oxygen in water by having a molecule in blood that binds to oxygen and delivers it to tissues. Without this carrier molecule, the rate of blood flow to tissues would have to increase dramatically to meet oxygen demand.
- ***Temperature of the water*** As the temperature of water increases, the amount of gas that dissolves in it decreases. Other things being equal, warm-water habitats have much less oxygen available than cold-water habitats do. For a fish, breathing in warm water is comparable to a land-dwelling animal breathing at high elevation.
- ***Presence of other solutes*** Because seawater has a much higher concentration of solutes than fresh water does, seawater can hold less dissolved gas. At 10°C, up to 8.02 mL of O_2 can be present per liter of fresh water versus only 6.35 mL of O_2 per liter of seawater. As a result, freshwater habitats tend to be more oxygen rich than marine environments.
- ***Partial pressure of the gas in contact with water*** Gases move from regions of high partial pressure to regions of low partial pressure. So if the partial pressure of a gas dissolved in a liquid exceeds that of the gas in the adjacent atmosphere, the gas will bubble up out of the liquid. This is what happens when the cap is removed from a bottle of carbonated beverage. The partial pressure of carbon dioxide in the newly opened drink is much higher than it is in the atmosphere.

What Affects the Amount of Oxygen Available in an Aquatic Habitat? The partial pressure of oxygen varies in different types of aquatic habitats, just as it varies with altitude on land. Besides the factors just listed, other important considerations affect oxygen's availability in water. These include the presence of photosynthetic organisms and decomposers, the amount of mixing that occurs, and the surface area of the body of water. For example:

- Habitats with large numbers of photosynthetic organisms tend to be relatively oxygen rich. In contrast, oxygen content is extremely low in bogs and other stagnant-water habitats because oxygen is quickly depleted by decomposers that use it in cellular respiration.
- Unless currents mix water almost continuously, water near the surface has a much higher oxygen content than water near the bottom of the same habitat.
- Shallow ponds and streams tend to be much better oxygenated than deep bodies of water because shallower bodies have a higher ratio of surface area to volume.
- Rapids, waterfalls, and breaking waves are the most highly oxygenated of all aquatic environments because a large surface area is exposed to the atmosphere as water splashes and because air bubbles are incorporated into the water.

Now let's consider the structure and function of ventilatory organs. How do the gills of fishes, the tracheae of insects, and the lungs of mammals cope with the differences between air and water?

CHECK YOUR UNDERSTANDING

✔ If you understood this section, you should be able to . . .

Decide, for each of three aquaria (A, B, and C), whether a large or a small amount of air should be bubbled in to maintain oxygenation of the water. Aquarium A contains warm water and several fish; B contains cold water, several fish, and aquatic algae; and C contains warm water and sedentary animals.

Answers are available in Appendix A.

42.3 Organs of Gas Exchange

After you complete this section, you should be able to . . .

- Explain gas-exchange organs in animals.

Many small animals lack specialized gas-exchange organs, such as gills or lungs. Instead, they obtain O_2 and eliminate CO_2 by diffusion across the body surface. This is possible because their size and shape give them an extraordinarily high ratio of surface area to volume (Ch. 39, Section 39.3). In sponges, jellyfish, flatworms, and other species, diffusion across the body surface is rapid enough to fulfill their requirements for taking in O_2 and expelling CO_2.

Most of these animals are restricted to living in wet environments, however. Gas-exchange surfaces must be kept moist to permit diffusion of gases across them. The surfaces must also be thin, and thin tissues are prone to water loss. Living in wet or humid environments allows animals to exchange gases across their outer surface while avoiding dehydration.

In contrast, animals that are large or that live in dry habitats need some sort of specialized respiratory organ. Respiratory organs provide a greater surface area for gas exchange—enough to meet the demands of a large body filled with cells. In terrestrial animals, respiratory organs are located inside the body, which helps minimize water loss from the moist surfaces.

Biologists have long marveled at the efficiency of gills and lungs. To appreciate why, let's examine the physical factors that control diffusion rates and then look at the structure and function of these respiratory organs.

Physical Parameters: The Law of Diffusion

In 1855 Adolf Fick derived an equation regarding diffusion that was based on the results of experiments he had performed on the behavior of gases. **Fick's law of diffusion**, the equation shown in Figure 42.3, states that the rate of diffusion of a gas depends on five parameters:

1. Solubility of the gas in the aqueous film lining the gas-exchange surface
2. Temperature (k represents these first two parameters)
3. Surface area available for diffusion (A)
4. Difference in partial pressures of the gas across the gas-exchange surface ($P_2 - P_1$)
5. Thickness of the gas-exchange surface (D)

Fick's law identifies traits that allow animals to maximize the rate at which oxygen and carbon dioxide diffuse across respiratory surfaces. Specifically, Fick's law states that gases diffuse at the highest rates when three conditions are met:

1. A is large, meaning a large surface area is available for gas exchange. Given Fick's law, it is not surprising that the respiratory surface in the human lungs would cover about 140 m^2—about a third of a basketball court—if the epithelium were spread flat.

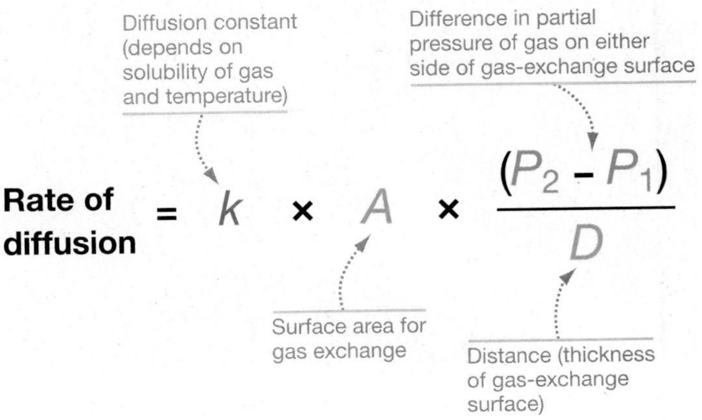

Figure 42.3 Fick's Law Describes the Rate of Diffusion.

2. D is small, meaning the gas-exchange surface is extremely thin. In the human lung, this barrier to diffusion is only 0.2 μm thick—about 1/200th of the thickness of this page.

3. $P_2 - P_1$ is large, meaning the partial pressure gradient of the gas across the surface is large. Large partial pressure gradients are maintained in part by having a circulatory system in close contact with the gas-exchange surface. When blood flows close to this surface, oxygen is rapidly taken away from the area where this gas is diffusing inward, and carbon dioxide is rapidly brought into the area where this gas is diffusing outward. As a result, $P_2 - P_1$ stays large.

What other aspects of gill and lung structure affect the diffusion rate? To answer this question, let's delve into the anatomy of these respiratory organs.

How Do Gills Work?

Gills are outgrowths of the body surface or throat in aquatic animals and are used for gas exchange as well as for osmoregulation (Ch. 40, Section 40.2). Gills are efficient solutions to the problems posed by water breathing, primarily because they present a large surface area for the diffusion of gases across a thin epithelium.

In some species of invertebrates, such as the red tubeworm (Figure 42.4a), gills project from the body surface and contact the surrounding water directly.

In other invertebrate species, such as the crayfish (Figure 42.4b), gills are covered by the exoskeleton. In animals with these "internal" gills, water must be driven over the gills by cilia, the limbs, or other specialized structures.

In contrast to the diversity of gills found in aquatic invertebrates, the gills of bony fishes are all similar in structure. Fish gills are located on both sides of the head, and in teleosts (Ch. 32, Section 32.5) the gills consist of four arches, as Figure 42.5a shows.

How Do Fishes Ventilate Their Gills? To move water through their gills so gas exchange can take place, most fishes open and close their mouth and **operculum**, the stiff flap of tissue that covers the gills. The pumping action of the mouth and operculum creates a pressure gradient that moves water over the gills.

In contrast, tuna and other fishes that are particularly fast swimmers force water through their gills by swimming with their mouths open. This process is called ram ventilation.

Regardless of how fish gills are ventilated, water flows in one direction through gills, passing over long, thin structures called **gill filaments** that extend from each gill arch. Each gill filament is composed of hundreds or thousands of **gill lamellae**. Gill lamellae are sheetlike structures, shown in detail in Figure 42.5b. Note that a bed of small blood vessels called capillaries runs through each lamella.

The Fish Gill Is a Countercurrent Exchange System The one-way flow of water through gill lamellae has a profound impact on gill function, for a simple reason: The flow of blood through the capillaries in each lamella is in the opposite direction to the flow of water. As a result, each lamella functions as a countercurrent exchanger (see Figure 42.5b).

Recall that countercurrent exchangers are based on two adjacent fluids flowing in opposite directions (Ch. 39, Section 39.5). Figure 42.6 illustrates why the countercurrent flow is so critical.

Note the following key points about Figure 42.6a, where water and blood flow in opposite directions:

- A slight gradient in percent oxygen in the water flowing over the gill lamellae and the blood (here, 10 percent) exists along the entire length of the lamellae.
- A large difference in oxygen percentage exists between the start and end of the system. In this example, the difference is 100% − 15% = 85% in the water and 90% − 5% = 85% in the blood.

The upshot? Most of the oxygen in the incoming water has diffused into the blood.

Now look at Figure 42.6b, where water and blood flow in the same direction.

Figure 42.4 Gills Can Be External or Internal. **(a)** Red tubeworms are marine polychaetes with gills that protrude from the body. **(b)** Crayfish gills are covered by the carapace (exoskeleton covering the head and thorax). In this photo, a portion of the carapace has been removed to expose the gills.

- A large gradient in partial pressure of oxygen (here, 100 percent) exists at the start of the system. The gradient in oxygen partial pressures declines rapidly and eventually disappears.
- A relatively small difference in oxygen partial pressure exists between the start and end of the system. In this example, the difference is 100% − 50% = 50% in the water and 50% − 0% = 50% in the blood.

With this arrangement, only half of the oxygen in the incoming water has diffused into the blood.

Countercurrent flow makes fish gills extremely efficient at extracting oxygen from water because it ensures that a difference in the partial pressure of oxygen in water versus blood is maintained over the entire gas-exchange surface.

The effect of countercurrent exchange is to maximize the $P_2 - P_1$ term in Fick's law of diffusion, averaged over the entire gill surface. Based on this observation, biologists cite countercurrent exchange as another example of how gills are optimized for efficient gas exchange.

How Do Insect Tracheae Work?

As noted earlier, air and water are dramatically different ventilatory media because they have different densities, viscosities, and abilities to hold oxygen and carbon dioxide. In addition, the consequences of exposing the gas-exchange surface to air versus water differ. On land, breathing leads to a loss of water by evaporation. How do terrestrial animals minimize water loss while maximizing the efficiency of gas exchange?

To answer this question, consider the tracheal system of insects. Recall that insects have an extensive system of air-filled tubes called **tracheae** located within the body (Ch. 40, Section 40.3). These tubes connect to the exterior through openings in the exoskeleton called **spiracles**, which can be closed to minimize the loss of water by evaporation (Figure 42.7). The interior ends of tracheae are tiny and highly branched. This structure allows the tracheal system to transport air close enough to cells that gas exchange can take place directly across their plasma membranes.

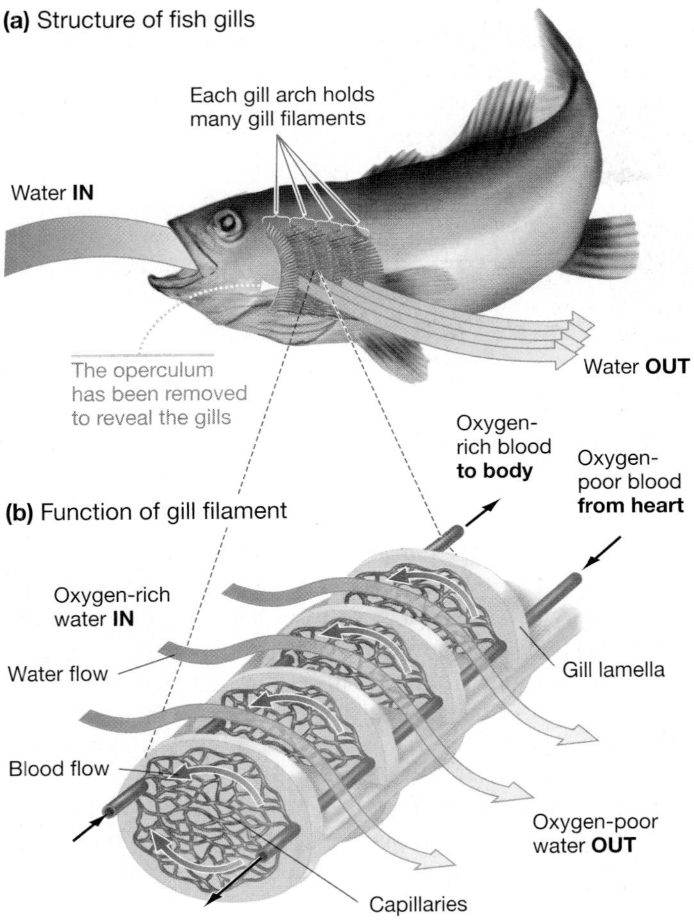

Figure 42.5 **Structure and Function of Fish Gills. (a)** Under the flap called the operculum (removed here), the four gill arches are found. **(b)** Fish gills are a countercurrent exchange system in which water and blood flow in opposite directions. In the blood vessels, red represents highly oxygenated blood, blue represents poorly oxygenated blood, and purple represents mixed blood.

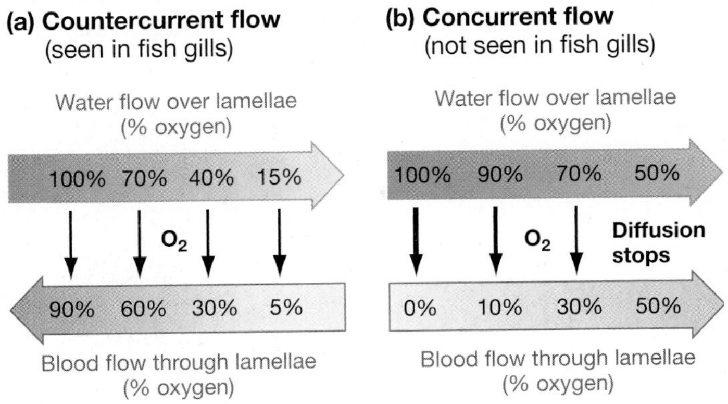

Figure 42.6 **Countercurrent Exchange Is Much More Efficient than Concurrent Exchange.** In the countercurrent exchange system of fish gills, oxygen is transferred along the entire length of the capillaries.

✔ If you understand this concept, you should be able to predict what would happen to oxygen transfer from water to blood if the flow were concurrent.

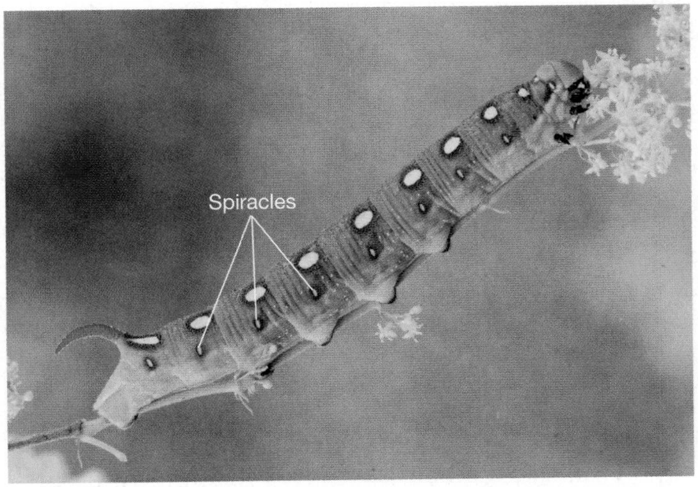

Figure 42.7 **Spiracles of the Hawkmoth Caterpillar.** Openings along the side of the caterpillar allow air to move in and out of the tracheae.

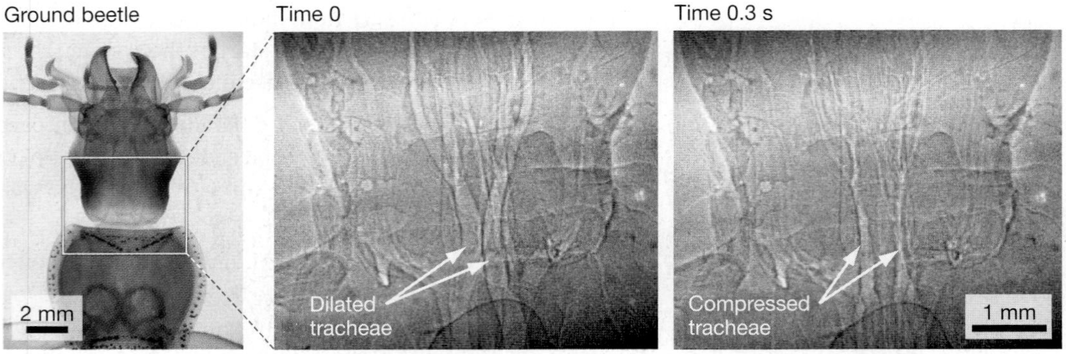

Figure 42.8 Tracheae Are Dilated and Compressed during Ventilation. Muscle relaxation and contraction during flight or other activity alternately dilates and compresses tracheae, causing pressure changes that promote air flow in and out of the tracheae. Beetles and other insects can inhale and exhale one-half the volume of their trachea, similar to humans who exchange about half their lung volume during mild exercise.

Air moves from the atmosphere into the spiracles and then through the tracheae to the tissues in the insect's body. Many insects are far too large to deliver gases to and from their tissues by simple diffusion alone. How do these large insects ventilate their tracheal system?

The answer lies in how the tracheae are alternately compressed and dilated as the muscles around them contract and relax. The muscle contractions and relaxations produce pressure changes that alter the volume of the tracheal system (Figure 42.8). This key point is an example of Boyle's law: If the volume occupied by a fixed amount of gas increases, the gas pressure decreases. If that volume declines, the gas pressure increases.

The volume of the tracheal system increases when muscles relax, causing pressure inside the system to go down and air from the atmosphere to rush in. What happens when muscles contract? The opposite—the volume of the tracheae decreases, pressure inside the system increases, and gas moves out of the tracheae into the atmosphere.

The action of the abdominal and flight muscles therefore stimulates air flow through the insect tracheal system, causing gases to move more quickly than they would by diffusion alone.

In large insects, the movement of gases is further promoted by larger tracheal diameters, which reduces the resistance to airflow. Researchers have shown that the diameter of the tracheae in large beetles is disproportionally larger than that in small beetles. Disproportionally large tracheae allow effective ventilation even in the largest beetles, which can weigh over 50 g and reach impressive lengths of over 15 cm (6 inches)—the size of a submarine sandwich!

Why don't beetles get even larger than this? Researchers hypothesize that if they did, the diameter of their tracheae would have to be so large that there would not be enough space in their bodies for much else. The beetles would lack sufficient muscles and other tissues to support their huge bodies.

In the Paleozoic era about 300 million years ago (mya), however, giant insects flourished. While the largest dragonflies today have a wingspan of about 15 cm, during the Paleozoic they had wingspans as large as 70 cm (2.3 feet)! How can this be explained?

One major hypothesis holds that giant insects evolved during the Paleozoic era because the atmospheric oxygen concentration was much higher than it is today (Figure 42.9). Recall that according to Fick's law, the rate of diffusion of a gas from one point to another is affected by the difference in partial pressures of

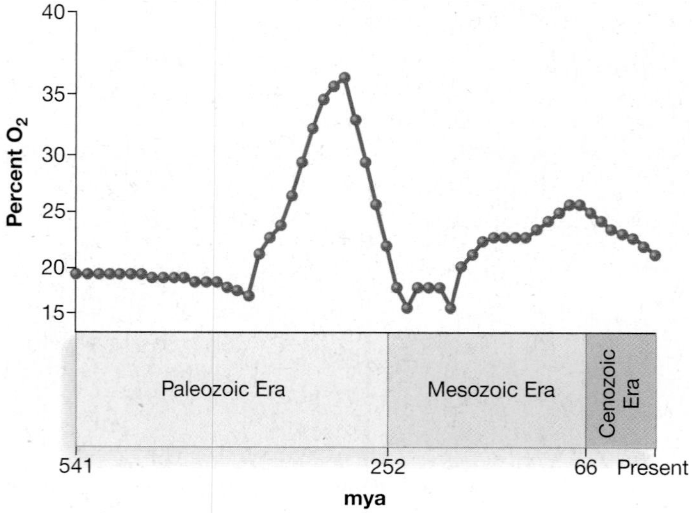

Figure 42.9 The Atmospheric Oxygen Level Peaked during the Paleozoic Era.

DATA: R. A. Berner. 1999. *Proceedings of the National Academy of Sciences* 96: 10955–10957.

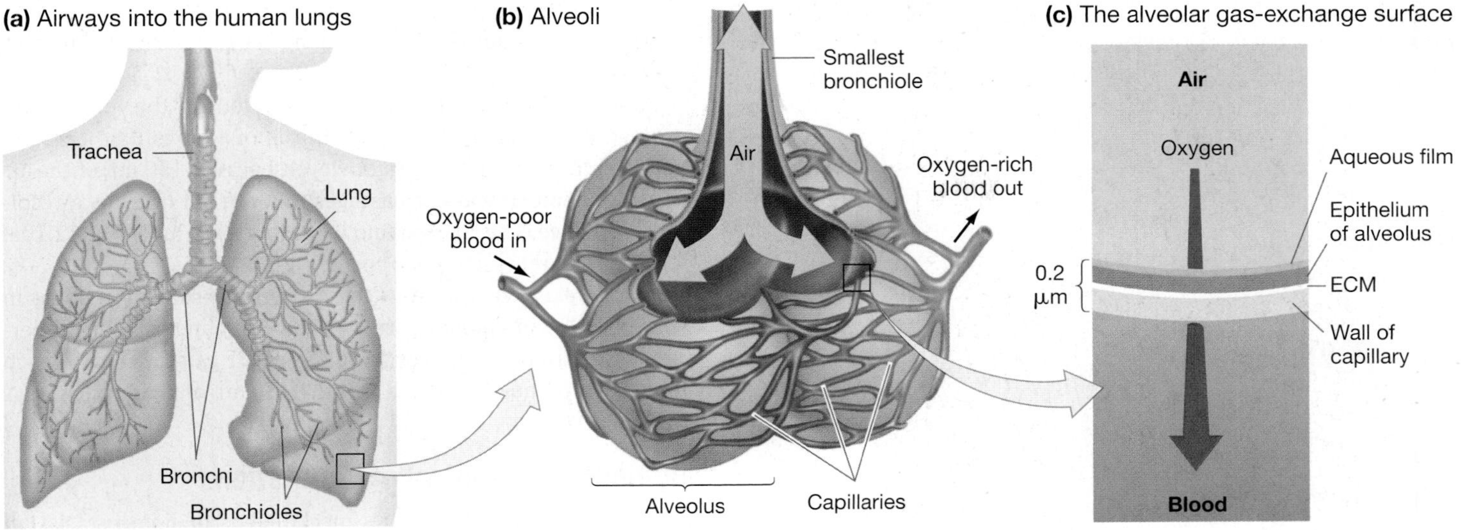

Figure 42.10 **Lungs Provide a Thin Membrane with a Large Surface Area for Gas Exchange between Air and Blood. (a)** The human respiratory tract branches repeatedly from the largest airway, the trachea, to the smallest, called bronchioles. The system of airways ends in clusters of tiny sacs called alveoli. **(b)** Alveoli are covered with capillary networks; together, these form **(c)** the gas-exchange surface.

the gas at the two points. During the latter part of the Paleozoic, the combination of high atmospheric oxygen level and relatively large tracheae allowed adequate ventilation even in extremely large insects.

By the start of the Mesozoic era about 50 to 100 million years later, the atmospheric oxygen concentration had declined precipitously. Giant insects went extinct, likely in part because they could no longer supply their tissues with enough oxygen.

How Do Vertebrate Lungs Work?

In most terrestrial vertebrates, air enters the body through both the nose and mouth. A tube known as the trachea (not to be confused with the tracheae of insects) carries the inhaled air to narrower tubes called **bronchi** (singular: **bronchus**). The bronchi branch off into yet narrower tubes, the **bronchioles**. The organs of ventilation, the lungs, enclose the bronchioles and part of the bronchi (Figure 42.10a).

Lungs are internal organs that are used for gas exchange. Terrestrial vertebrates—amphibians, reptiles (including birds), and mammals—have lungs, as do certain fishes and invertebrates. The amount of lung surface area available for gas exchange varies a great deal among species. In frogs and other amphibians, the lung is a simple sac lined with blood vessels. The lungs of mammals, in contrast, are finely divided into tiny sacs called **alveoli** (singular: **alveolus**; Figure 42.10b).

Each human lung contains approximately 150 million alveoli, which give mammalian lungs about 40 times more surface area for gas exchange than an equivalent volume of frog lung tissue. As Figure 42.10c shows, an alveolus and its associated capillary provide a very thin gas-exchange surface between air and blood. This surface consists of a thin aqueous film, a layer of alveolar epithelial cells, some extracellular matrix (ECM) material, and the wall of a capillary.

Apart from total surface area, the other major feature of lungs that varies among species is mode of ventilation. In the lungs of snails and spiders, air movement takes place primarily by diffusion. Vertebrates, in contrast, actively ventilate their lungs by pumping air via muscular contractions.

One mechanism for pumping air is **positive pressure ventilation**, used by frogs and some other amphibians. A frog lowers the floor of its throat, increasing the volume there and drawing in air from the atmosphere through the nasal passages and into the oral cavity. The animal then closes the nasal passages and contracts its throat muscles. These actions increase the air pressure in the oral cavity, forcing air into the lungs.

In effect, frogs push air into their lungs. In contrast, humans and other mammals pull air into their lungs. How does this **negative pressure ventilation** work?

The pressure inside the human chest cavity is about 5 mm Hg less than atmospheric pressure. This negative pressure surrounding the lung is just enough to keep the lung expanded. If a wound penetrates the chest wall and the pressure differential between the chest cavity and the atmosphere disappears, the lung on the side of the injury will collapse like a deflated balloon.

Humans ventilate their lungs by changing the pressure within their chest cavity between about −5 mm Hg and −8 mm Hg relative to the atmosphere. Inhalation is based on increasing the volume of the chest cavity and thus lowering the pressure. The change in volume is caused by a downward motion of the thin muscular sheet called the **diaphragm** and an expansion of the rib cage. As the pressure surrounding the lungs drops, the lungs expand, and the ensuing drop in pressure inside the lungs causes air to flow into the airways along a pressure gradient.

(a) Lungs expand and contract in response to changes in pressure inside the chest cavity.

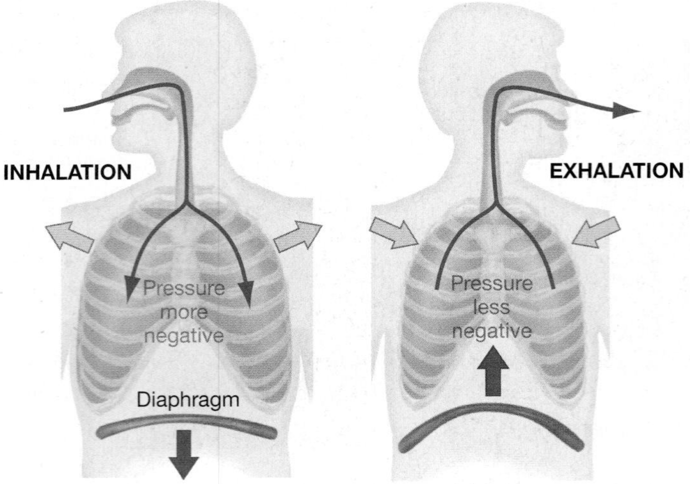

(b) Ventilatory forces can be modeled by a balloon in a jar.

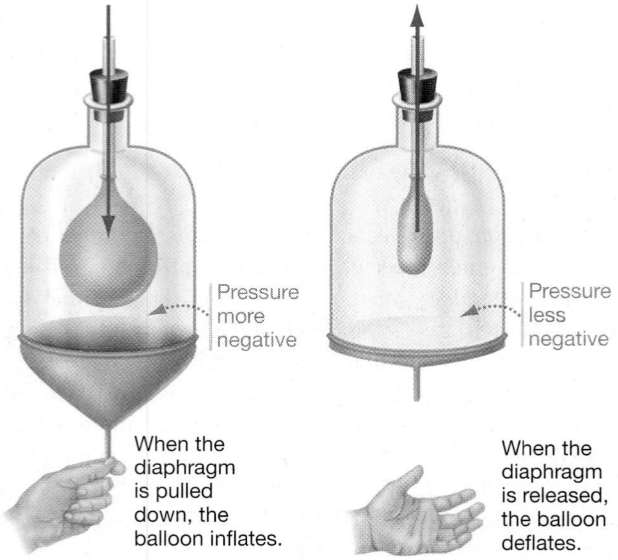

Figure 42.11 Changes in the Volume of the Chest Cavity Drive Negative Pressure Ventilation. (a) Inhalation: When the diaphragm and rib muscles contract, the volume of the chest cavity increases, lowering pressure within the lungs. In response, air flows into the lungs. Exhalation: When the diaphragm and rib muscles relax, the volume of the chest cavity decreases, causing internal pressure to increase. In response, lung volume decreases—due to elasticity of the lungs—and air flows out. **(b)** A model of negative pressure ventilation.

Figure 42.11a shows how contraction of the diaphragm causes the drop in pressure that results in inhalation.

Exhalation, in contrast, is a passive process—the volume of the chest cavity decreases as the diaphragm and rib muscles relax. Because the lung is elastic, it returns automatically to its original shape if it is not stretched or compressed. During exercise, though, exhalation is an energy-demanding, active process.

The changes in pressure that occur during negative pressure ventilation are analogous to changing the pressure within a jar, as shown in Figure 42.11b.

About 450 mL of air moves into and out of the lungs in an average breath. Only about two-thirds of this volume actually participates in gas exchange, however, because 150 mL of the air occupies **dead space**—air passages that are not lined by a respiratory surface. The trachea and bronchi shown in Figure 42.10a, for example, represent dead space.

During exercise, the chest cavity undergoes larger changes in volume, allowing much more gas to be exchanged. When a person is breathing hard, over 2500 mL of air can move with each inhalation–exhalation cycle, but the 150 mL of dead space stays the same.

Homeostatic Control of Ventilation

An animal is in trouble if its mechanisms of homeostasis fail to maintain blood oxygenation or to eliminate carbon dioxide. Adequate adenosine triphosphate (ATP) production depends on maintaining the partial pressures of oxygen and carbon dioxide within a narrow range, during both rest and vigorous exercise. How is ventilation controlled to achieve this critical homeostasis?

When mammals are resting, the rate of breathing is established by the medullary respiratory center, an area at the base of the brain, just above the spinal cord. This center stimulates the rib and diaphragm muscles to contract about 12 to 14 times per minute in humans.

But during exercise, things change. Active muscle tissue takes up more oxygen from the blood. As a result, the partial pressure of oxygen (P_{O_2}) in blood drops. Those same muscles release larger quantities of carbon dioxide to the blood, raising its partial pressure P_{CO_2} in blood.

The rise in P_{CO_2} in blood—not the drop in blood P_{O_2}—is the major factor controlling breathing rate in mammals. When carbon dioxide reaches the brain, it rapidly diffuses from the blood into the cerebrospinal fluid that bathes the brain. In both blood and cerebrospinal fluid, CO_2 reacts with water to form a hydrogen ion (H^+) and a bicarbonate ion (HCO_3^-):

$$CO_2 + H_2O \rightleftharpoons H^+ + HCO_3^-$$

The result is a slight drop in the pH of blood and cerebrospinal fluid. The change in blood pH is sensed by specialized neurons in the brain and in the large arteries that travel from the heart into the neck and to the base of the brain.

Signals from these neurons and from pH detectors in the medullary respiratory center are responsible for sustained increases in breathing rate and depth during exercise. The rise in ventilation rate and depth increases the rate of oxygen delivery to the tissues and the rate at which carbon dioxide is eliminated from the body, restoring P_{O_2} and P_{CO_2} to their resting levels. This control system is so effective that it can maintain stable blood levels of oxygen and carbon dioxide even during intense exercise.

Now let's look more closely at how blood transports oxygen and carbon dioxide between the gas-exchange surface and an animal's tissues.

CHECK YOUR UNDERSTANDING

✔ If you understood this section, you should be able to . . .

1. Identify two features that are common to gills, tracheae, and lungs, as well as one trait that is unique to each.
2. Describe what happens to the P_{O_2}, P_{CO_2}, and pH of your blood when you hold your breath.

Answers are available in Appendix A.

42.4 How Are Oxygen and Carbon Dioxide Transported in Blood?

Blood is a connective tissue that consists of cells in a watery extracellular matrix. Besides carrying oxygen and carbon dioxide between cells and the lungs or gills, blood transports nutrients from the digestive tract to other tissues in the body, moves waste products to the kidney and liver for processing, conveys hormones from glands to target tissues, delivers immune system cells to sites of infection, and distributes heat throughout the body.

After you complete this section, you should be able to . . .

- Explain how oxygen and carbon dioxide are transported in blood.

Given the wide variety of functions that blood serves, it is not surprising that it is a complex tissue. In an average human, 50–65 percent of the blood volume is composed of a liquid extracellular matrix called **plasma**. The remainder of the volume is made up of cells and cell fragments that are collectively called formed elements.

The formed elements in blood include platelets, several types of white blood cells, and red blood cells.

- **Platelets** are cell fragments that act to minimize blood loss from ruptured blood vessels. They do so by releasing material that helps form the blockages known as clots.
- **White blood cells** (also called **leukocytes**) are part of the immune system. They fight infections (described in detail in Chapter 48).
- **Red blood cells** transport oxygen from the lungs to tissues throughout the body. They also play a role in transporting carbon dioxide from tissues to the lungs. In humans, red blood cells make up 99.9 percent of the formed elements.

The human body synthesizes new red blood cells at the rate of 2.5 million per second to replace old red blood cells, which die at the same rate. New red blood cells last for about 120 days. Red blood cells, white blood cells, and platelets develop from stem cells located in the tissue inside bone (bone marrow).

Vertebrates other than mammals transport oxygen in red blood cells that retain their nuclei. But in mammals, red blood cells lose their nuclei as they mature, along with their mitochondria and most other organelles. Mammalian red blood cells are essentially bags filled with approximately 280 million copies of the oxygen-carrying molecule **hemoglobin**.

Structure and Function of Hemoglobin

Even though oxygen is not highly soluble in water, it is often found in a high concentration in blood. Blood has a high oxygen-carrying capacity because O_2 readily binds to the hemoglobin molecules in red blood cells.

The evolution of hemoglobin was a key event in the diversification of animals. By increasing the oxygen-carrying capacity of blood, hemoglobin made it possible for cellular respiration rates to increase. High rates of ATP production, in turn, support high rates of growth, movement, digestion, and other activities.

Hemoglobin is a tetramer, meaning that it consists of four polypeptide chains (Figure 42.12a). Each of the four polypeptide chains binds to a nonprotein group called **heme** (represented by black circles in the figure). Each heme group, in turn, contains an iron ion (Fe^{2+}) that can bind to an oxygen molecule. As a result, each hemoglobin molecule can bind up to four oxygen molecules. In blood, 98.5 percent of the oxygen is bound to hemoglobin; only 1.5 percent is dissolved in plasma.

What Is Cooperative Binding? Blood leaving the human lungs has a P_{O_2} of about 100 mm Hg, while at rest the blood in vessels in muscles and other tissues has a P_{O_2} of about 40 mm Hg. This partial pressure difference creates a diffusion gradient that unloads O_2 from hemoglobin to the tissues (Figure 42.12b).

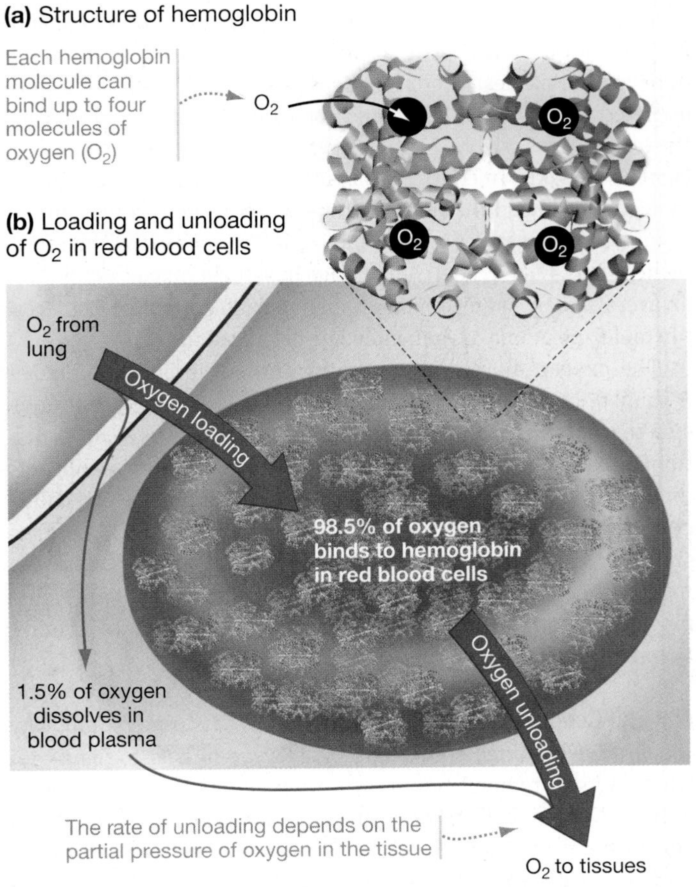

Figure 42.12 Hemoglobin Transports Oxygen to Tissues.

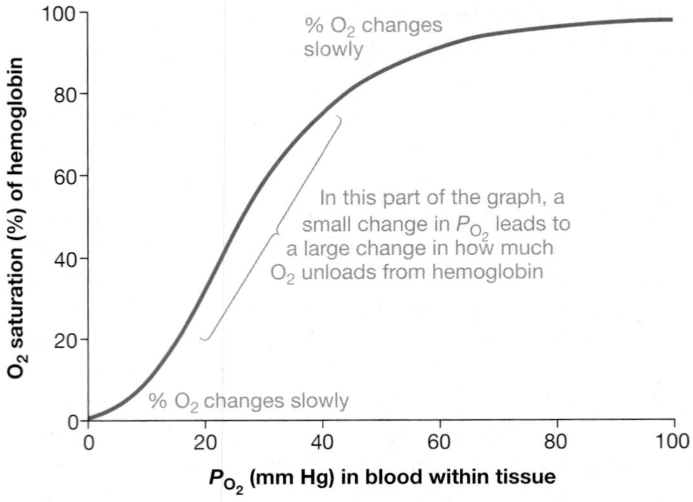

Figure 42.13 The Oxygen–Hemoglobin Equilibrium Curve Is Sigmoidal. A sigmoidal curve has three distinct regions.

Researchers who studied the dynamics of O_2 unloading in tissues found the pattern shown in **Figure 42.13**. This graph plots the percentage of O_2 saturation of hemoglobin in red blood cells versus the P_{O_2} in the blood within tissues. If the saturation of hemoglobin approaches 100 percent, it means that every possible binding site in hemoglobin contains an oxygen molecule.

The graph in Figure 42.13 is called an **oxygen–hemoglobin equilibrium curve**, an oxygen dissociation curve, or a hemoglobin saturation curve. Note that the *x*-axis plots the partial pressure of oxygen in tissues. In effect, this represents "demand." Oxygen-depleted tissues, where demand for oxygen is high, are toward the left on the horizontal axis; oxygen-rich tissues are toward the right. The *y*-axis, in contrast, plots the percentage of hemoglobin molecules in blood that are saturated with oxygen—a measure of "supply," or how many oxygen molecules on average are bound to hemoglobin. Each 25 percent change in saturation corresponds to an average of one additional oxygen molecule bound per hemoglobin molecule or one oxygen molecule delivered to tissues.

The oxygen–hemoglobin equilibrium curve is sigmoidal, or S-shaped, because the binding of each successive oxygen molecule to a subunit of the hemoglobin molecule causes a conformational change in the protein that makes the remaining subunits much more likely to bind oxygen. This phenomenon is called **cooperative binding**. (See **Making Models 42.1** for tips on understanding S-shaped curves.)

Conversely, the loss of a bound oxygen molecule changes hemoglobin's conformation in a way that makes the loss of additional oxygen molecules more likely.

Making Models 42.1 Tips on Drawing Graphs (I): Sigmoidal ("S-shaped") Curves

A sigmoidal curve results when the relationship between two variables is not constant. Comparing a sigmoidal curve (shown in the graph on the right here) to a straight line (graph on the left) can help you understand the S-curve. The dotted lines in the straight-line graph show how changes in values on the *x*-axis lead to equal changes in the *y*-axis values.

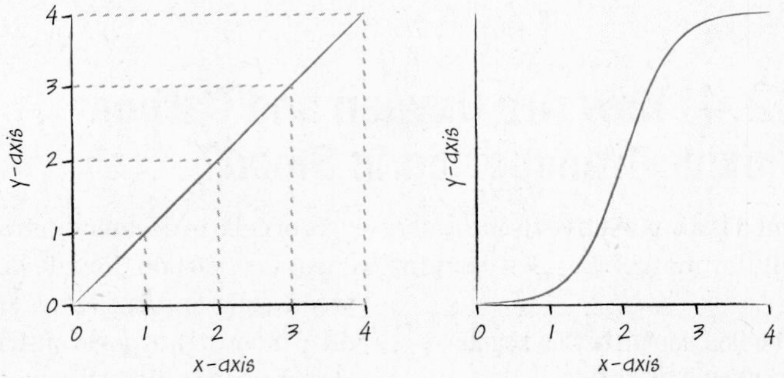

✔ **MODEL** Draw dotted lines on the sigmoidal curve that connect the equal changes in values on the *x*-axis to the associated changes in the *y*-axis values. Are the changes in the *y*-axis values equivalent or different? Describe the relationship between *x* and *y* in a sigmoidal curve in words.

To see this model in action, go to the Study Area of **Mastering Biology**

Why Is Cooperative Binding Important? To understand why cooperative binding is important, use **Figure 42.14a** to figure out what happens to hemoglobin saturation when oxygen demand in tissues changes. Let's begin with some basic observations. When blood arrives at tissues from the lungs, its hemoglobin saturation is close to 100 percent. At rest, tissue P_{O_2} is typically about 40 mm Hg. But during exercise, cells are using so much oxygen in cellular respiration that tissue P_{O_2} drops to about 30 mm Hg. Now do this:

- Put your finger on the *x*-axis at 40 mm Hg, trace the dashed line up until it hits the equilibrium curve, and check where this point is on the *y*-axis. The answer is about 75 percent. This means that when tissues are at rest, hemoglobin unloads 25 percent of its oxygen to the tissues.
- Next, put your finger on the *x*-axis at 30 mm Hg, trace the dashed line up, and check hemoglobin saturation at this point on the curve. The answer is about 58 percent. This means that when tissues are exercising, hemoglobin unloads about 42 percent of its oxygen to tissues.

So in response to a relatively small change in tissue P_{O_2} there is a relatively large change in the percentage of saturation of hemoglobin. The large change occurs because the equilibrium curve is extremely steep in the range of P_{O_2} values commonly observed in tissues. The curve is steep because of cooperative binding. Thus, cooperative binding is important because it makes hemoglobin exquisitely sensitive to changes in the P_{O_2} of tissues.

If cooperative binding did not occur, all four subunits of hemoglobin would load or unload oxygen independently of each other. They would lose or gain oxygen in direct proportion to the partial pressure of oxygen in the blood. The curve would not be as steep in the center—meaning that there would be only a small change in the percentage saturation of hemoglobin when tissue P_{O_2} changes.

Figure 42.14b shows how the oxygen–hemoglobin equilibrium curve might look if binding were noncooperative. As the dashed lines on this graph indicate, a relatively small change in oxygen delivery would occur when tissue P_{O_2} changes from its resting level of about 40 mm Hg to 30 mm Hg. Specifically, hemoglobin would

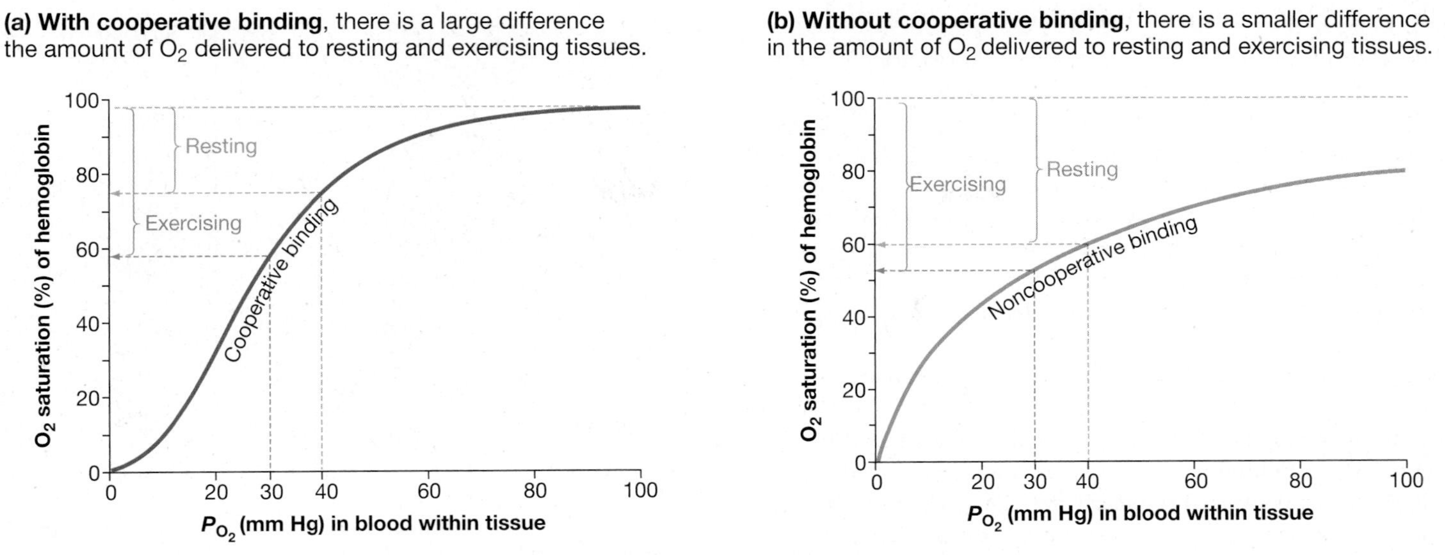

Figure 42.14 Cooperative Binding of O_2 by Hemoglobin Results in Greater O_2 Delivery than Noncooperative Binding. Hemoglobin is almost 100 percent saturated with oxygen until it arrives at tissues.

unload about 100% − 60% = 40% of its oxygen when tissues are at rest, and about 100% − 53% = 47% of its oxygen during exercise. This difference—about 7 percent—is much less than the 17 percent change observed with cooperative binding.

How Do pH and Temperature Affect Oxygen Unloading from Hemoglobin? Cooperative binding is only part of the story behind oxygen delivery. Hemoglobin—like other proteins—is sensitive to changes in pH and temperature.

As noted earlier, the partial pressure of CO_2 rises in active muscle tissue during exercise. The CO_2 produced by exercising muscle reacts with the water in blood to form bicarbonate and hydrogen ions. As a result, the pH of the blood in exercising muscle drops.

Decreases in pH alter hemoglobin's conformation. These shape changes make hemoglobin more likely to unload O_2 at any given value of tissue P_{O_2}. As Figure 42.15 shows, this phenomenon, known as the **Bohr shift**, causes the oxygen–hemoglobin equilibrium curve to shift to the right when pH declines.

The Bohr shift is important because it makes hemoglobin more likely to release oxygen during exercise or other conditions in which P_{CO_2} is high, pH is low, and tissues are under oxygen stress.

During exercise, active tissues also produce heat, causing their temperature to rise. Increasing temperature has the same result as decreasing pH: It shifts the oxygen–hemoglobin equilibrium curve to the right, representing a greater unloading of oxygen to tissues at any given P_{O_2}.

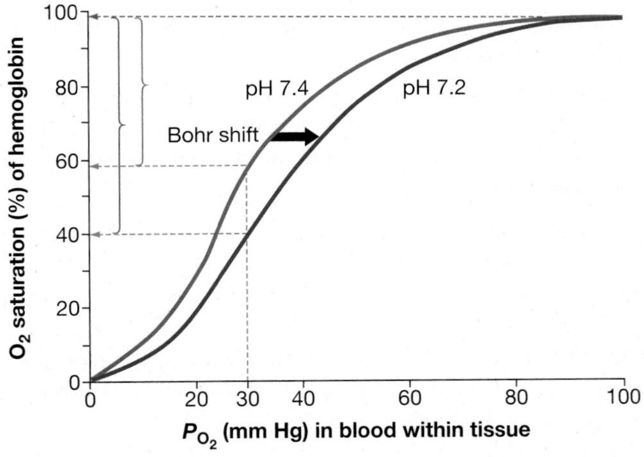

Figure 42.15 The Bohr Shift Makes Hemoglobin More Likely to Release Oxygen to Tissues with Low pH. As pH drops, oxygen becomes less likely to stay bound to hemoglobin at all values of tissue P_{O_2}. Exercising tissues have lower pH than resting tissues and so receive more oxygen from hemoglobin.

✔ **QUANTITATIVE** Estimate how much more oxygen is unloaded from hemoglobin at pH 7.2 than at pH 7.4 when the tissue has a P_{O_2} of 30 mm Hg.

Oxygen Delivery by Hemoglobin Is Extremely Efficient To appreciate cooperative binding and the Bohr shift in action, consider an experiment on how the oxygen transport system in rainbow trout responds to sustained exercise.

To begin, biologists had fish swim continuously against a current in a water tunnel. As the researchers increased the speed of the current and thus the swimming speed of the fish, they periodically sampled the O_2 content of arterial and venous blood. Arterial blood is freshly oxygenated and moving from the gills to tissues; venous blood is returning to the heart from the tissues. Not surprisingly, the biologists found that the arterial O_2 level remained fairly constant as swimming speed increased. This meant that the gills continued to saturate hemoglobin with oxygen.

In contrast, the O_2 content of venous blood, which had undergone gas exchange with the tissues, dropped steadily as swimming speed increased. When the fish had reached their maximum sustainable speed, virtually all the oxygen that had been available in the blood had been extracted. The data show that in hard-working tissues, the combination of increased temperature, lower pH, and lower P_{O_2} caused hemoglobin to become almost completely deoxygenated.

CO_2 Transport and the Buffering of Blood pH

The carbon dioxide that is produced by cellular respiration in the tissues enters the blood, where it reacts with water to form bicarbonate and hydrogen ions. Recall that the resulting drop in blood pH stimulates an increase in breathing rate. Rapid exhalation of CO_2 then counteracts the drop in blood pH.

Homeostasis with respect to blood pH is reinforced by a series of events that take place inside red blood cells. Biologists were able to work out what was happening when they discovered large amounts of the enzyme **carbonic anhydrase** in red blood cells.

The Role of Carbonic Anhydrase and Hemoglobin Recall that carbonic anhydrase catalyzes the formation of bicarbonate and hydrogen ions from carbon dioxide and water (Ch. 41, Section 41.3). Consequently, CO_2 that diffuses into red blood cells is quickly converted to bicarbonate ions and protons. The same reaction occurs in the plasma surrounding red blood cells, although much more slowly in the absence of the enzyme. Why is the carbonic anhydrase activity in red blood cells so important? The answer has two parts:

1. The protons produced by the enzyme-catalyzed reaction induce the Bohr shift, which makes hemoglobin more likely to release oxygen.
2. The partial pressure of CO_2 in blood drops when carbon dioxide is converted to soluble bicarbonate ions, maintaining a strong partial pressure gradient favoring the entry of CO_2 into red blood cells.

Thus, carbonic anhydrase activity promotes both O_2 delivery and CO_2 uptake in active tissues.

Once bicarbonate ions form in the red blood cell, they are transported into the blood plasma. The outcome is that most CO_2 is transported in blood (specifically in plasma) in the form of bicarbonate ions. In contrast, the protons produced by the reaction stay inside red blood cells (Figure 42.16).

What ultimately happens to these protons? When hemoglobin is carrying few oxygen molecules, it has a high affinity for protons. As a result, it takes up much of the H^+ that is produced when carbon dioxide reacts with water. The hemoglobin acts as a **buffer**—a compound that minimizes changes in pH.

What Happens When Blood Returns to the Lungs? When poorly oxygenated blood reaches the alveoli, its environment changes dramatically. In the lungs, a partial pressure gradient favors the diffusion of CO_2 from plasma and red blood cells to the atmosphere within the alveoli. As CO_2 diffuses from the blood into the alveoli, P_{CO_2} in the blood declines.

The drop in blood P_{CO_2} reverses the chemical reactions that occurred in tissues:

1. Hydrogen ions (protons) leave their binding sites on hemoglobin.
2. Protons react with bicarbonate ions to form CO_2.
3. CO_2 diffuses into the alveoli and is exhaled from the lungs.

In the meantime, hemoglobin has picked up O_2. Hemoglobin's affinity for oxygen is high in the alveoli because blood pH rises as P_{CO_2} declines. When blood leaves the lungs, it has unloaded its carbon dioxide, and its hemoglobin is saturated with oxygen. The cycle then begins anew.

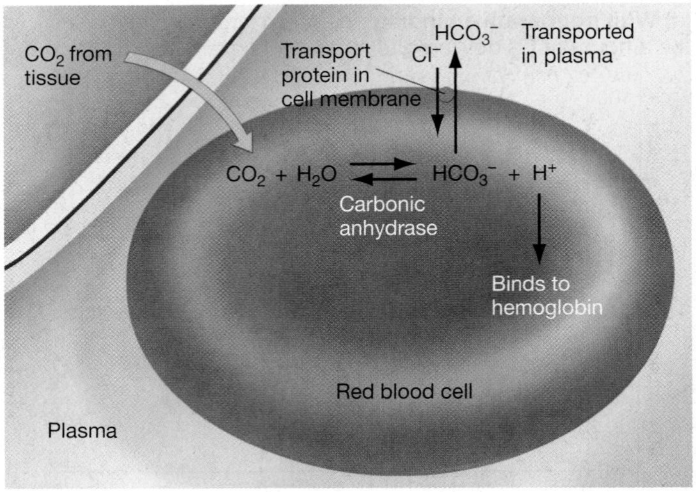

Figure 42.16 **Carbonic Anhydrase Is Vital to CO_2 Transport in Blood.** When CO_2 diffuses into red blood cells, carbonic anhydrase catalyzes a reaction to form a bicarbonate ion (HCO_3^-) and a proton (H^+). This reaction maintains the partial pressure gradient favoring the entry of CO_2 into red blood cells. The protons produced by the reaction bind to poorly oxygenated hemoglobin. Most CO_2 in blood is transported to the lungs in the form of HCO_3^-.

✔ This diagram shows the sequence of events in tissues. After reading the rest of Section 42.4, explain what happens when the red blood cell in the diagram reaches the lungs.

CHECK YOUR UNDERSTANDING

✔ **If you understood this section, you should be able to . . .**

1. Predict how the oxygen–hemoglobin equilibrium curves of Tibetan people, whose ancestors have lived at high elevations for thousands of years, compare to curves of people whose ancestors have lived at sea level for many generations.
2. Predict the effect of hyperventilation, in which gases are exhaled extremely rapidly, on blood P_{CO_2} and pH.

Answers are available in Appendix A.

42.5 Circulation

After you complete this section, you should be able to . . .

- Analyze how circulatory systems move materials through the body.

According to Fick's law, differences in the partial pressure of gases are only part of the story when it comes to understanding diffusion rates. Surface area and diffusion distance—*A* and *D* respectively in the Fick's law equation featured in Figure 42.3—also play a key role.

Animals without circulatory systems have various ways of minimizing the distance and maximizing the surface area available for diffusion of gases and other key solutes:

- Animals that are only a few millimeters in size, like rotifers and tardigrades, have a small enough volume that diffusion over their body surface is adequate to keep them alive.
- The flattened bodies of flatworms and tapeworms give these animals a high surface-area-to-volume ratio (Ch. 39, Section 39.3). In these species, too, molecules are exchanged with the environment directly across the outer body surface.
- Diffusion across the body wall also occurs in roundworms, where gas exchange is facilitated by muscular contractions in the body wall. Diffusion is enhanced as roundworms circulate fluids by sloshing them back and forth.
- Jellyfish and corals have a large, highly folded gastrovascular cavity that offers a large surface area for exchange of molecules with the environment.

In larger animals, however, the problem of providing a large enough surface area for diffusion is solved by a circulatory system. A circulatory system carries transport tissues called blood or hemolymph into close contact with every cell in the body. In this case, "close contact" is a distance of about 0.1 mm or less between the blood or hemolymph and cells within tissues. Diffusion is rapid at this scale.

To explore how circulatory systems work, let's start by distinguishing the two most basic types—open and closed. Open circulatory systems occur in most invertebrates, while all vertebrates have closed systems.

What Is an Open Circulatory System?

In an **open circulatory system**, a fluid connective tissue called hemolymph is actively pumped throughout the body in a limited system of vessels. The hemolymph is not confined exclusively to the vessels, however. Instead, hemolymph comes into direct contact with tissues. As a result, the molecules being exchanged between hemolymph and tissues do not have to diffuse across the wall of a vessel. Hemolymph transports wastes and nutrients and may also contain some cells, clotting agents, and oxygen-carrying pigments such as hemoglobin or a copper-containing pigment called hemocyanin.

Figure 42.17 illustrates the open circulatory system in a spider. Note that a muscular organ called the **heart** pumps hemolymph into vessels that empty into open, fluid-filled spaces. When the heart relaxes and its internal pressure drops below the pressure in these spaces, hemolymph enters the heart via little holes in its surface. General body movements also help hemolymph move to and from the heart.

Because it moves throughout the volume of the body, hemolymph is under relatively low pressure in open circulatory systems. As a result, hemolymph flow rates may also be low. These features make open circulatory systems most suitable for relatively sedentary organisms, which do not have high oxygen demand.

Insects, with their rapid movements and more active lifestyles, are an exception to this rule. In the open circulatory systems of insects, the limitations imposed by low hemolymph pressure are overcome by their tracheal respiratory system, which delivers oxygen directly to the tissues.

Another characteristic of open circulatory systems, because they lack discrete, continuous vessels, is that the flow of hemolymph cannot be directed toward tissues that have a high oxygen demand and CO_2 buildup. An open circulatory system moves hemolymph throughout an animal's body in much the same way that a ceiling fan moves air throughout a room in a house.

Crustaceans are an important exception to this rule, however. Even though their circulatory system is classified as open, these species have a network of small vessels that can preferentially send hemolymph to tissues with the highest oxygen demand.

What Is a Closed Circulatory System?

In a **closed circulatory system**, blood flows in a continuous circuit through a series of vessels in the body, under pressure generated by a heart. Because the blood is confined to vessels, a closed system can generate enough pressure to maintain a high flow rate.

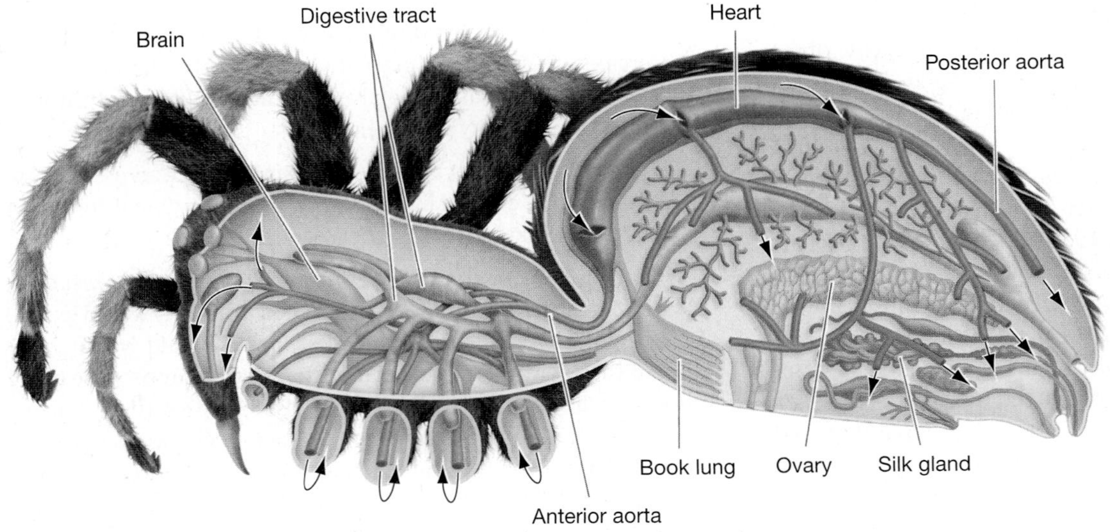

Figure 42.17 Spiders Have an Open Circulatory System. The arrows show the direction of hemolymph flow.

In a closed circulatory system, blood flow can also be directed in a precise way in response to the tissues' needs. For example, blood can be shunted to leg muscles during exercise, to the intestines after a meal, or to regions of the brain engaged in particular mental tasks.

Which Lineages Have Closed Circulatory Systems? Closed circulatory systems are found in vertebrates and a few other lineages where individuals tend to be active. Earthworms and other annelids, for example, have a closed circulatory system and exchange gases with the environment across their thin, moist skin, which has a dense supply of capillaries. As a result, annelids are able to obtain and circulate enough oxygen to support intense muscular activity. Most live as active burrowers and hunters.

A similar situation occurs in squid, octopuses, and other cephalopods that hunt down prey. The closed circulatory system of these mollusks generates high rates of blood flow, which oxygenates their muscles well enough to support rapid movements and a predatory lifestyle.

Closed circulatory systems contain various types of blood vessels, each having a distinct structure and function. Let's review the major types of blood vessels and then consider how the vessels of a closed circulatory system interact with the lymphatic system.

Types of Blood Vessels An enormous amount of tubing is required to distribute blood within "diffusion distance" of every cell in the body. If all the blood vessels in a human body were laid end to end, they would stretch about 100,000 km (over 60,000 miles).

Blood vessels are classified as follows:

- **Arteries** are tough, thick-walled vessels that take blood away from the heart. Small arteries are called **arterioles**.
- **Capillaries** are vessels whose walls are just one cell thick, allowing exchange of gases and other molecules between blood and tissues. Networks of capillaries are called **capillary beds**.
- **Veins** are thin-walled vessels that return blood to the heart. Small veins are called **venules**.

The structure of arteries, capillaries, and veins correlates closely with their functions in a closed circulatory system. For example, the heart ejects blood into a large artery, usually called the **aorta**. All arteries have both muscle fibers and elastic fibers in their walls, but elastic fibers dominate the walls of the aorta. As a result, the aorta can expand when blood enters it under high pressure from the heart.

When a contraction of the heart ends, the diameter of the aorta returns to its resting state. This elastic response propels blood away from the heart and augments the force generated by the heart contraction. Similar types of secondary pumping action occur to some extent in other arteries as well. This feature helps maintain forward blood flow in the period between heart contractions.

The walls of arteries and arterioles have a thick layer composed of smooth muscle fibers. When the muscle fibers relax, the vessel diameter increases, resistance to flow is reduced, and blood flow increases in the tissues served by the vessel. But when these muscle fibers contract, the vessel diameter decreases, increasing resistance to flow and slowing the flow of blood in the vessel. In this way, blood flow to specific tissues can be carefully regulated by signals from the nervous system to muscle fibers in the vessels supplying those tissues.

Capillaries are the smallest blood vessels. Their walls are only one cell layer thick, and they are just wide enough to let red blood cells through one at a time (Figure 42.18a). The extreme thinness of capillaries and the dense network they form throughout the body make them suitable sites for the exchange of gases, nutrients, and wastes between blood and the other tissues.

In some organs, such as the liver, the walls of capillaries contain many small openings that further diminish the barrier to diffusion between blood and the tissues. Despite their thinness, it is rare for capillaries to rupture because blood pressure drops dramatically as blood passes through arterioles on its way to capillary beds.

After blood from arteries and arterioles passes through capillaries, veins carry it back to the heart. Because blood is under relatively low pressure by the time it exits the tissues, veins have thinner walls and larger interior diameters than arteries do (Figure 42.18b).

Blood flow in veins is speeded by skeletal muscle activity in the extremities, which compresses large veins. Larger veins also contain one-way **valves**, which are thin flaps of tissue that prevent any backflow of blood.

All veins contain some muscle fibers, which contract in response to signals from the nervous system, decreasing the diameter and overall volume of the vessels. Blood pressure in a closed circulatory system is regulated, in part, by actively adjusting the volume of blood contained within the veins.

Exchange between Blood Plasma and Interstitial Fluid The relatively high operating pressure of closed circulatory systems, combined with the thinness of capillaries, produces a small but steady leakage of fluid from these blood vessels into the surrounding space. The area between cells is called interstitial space; the extracellular fluid that fills it is interstitial fluid. Blood cells are retained within capillaries, so interstitial fluid resembles plasma in its electrolyte composition.

Why does interstitial fluid build up? In 1896 Ernest Starling proposed that the following two forces were at work (Figure 42.19):

1. There is an outward-directed hydrostatic force in capillaries, created by the blood pressure generated by the heart. This force is analogous to the pressure that drives water through the wall of a leaky garden hose.
2. There is also an inward-directed osmotic force across the capillary walls, created by the higher concentration of solutes in the blood plasma than in the interstitial space.

Starling reasoned that at the end of the capillary nearest to an arteriole, the hydrostatic force (the blood pressure) would exceed the osmotic force. If so, then in that location fluid would move out of the capillary into the interstitial space. But because blood pressure drops as fluid passes through a long, thin tube, Starling proposed that at the venous end of the capillary, the inward-directed osmotic force would exceed the outward-directed hydrostatic force. Thus, the fluid that was lost at the arteriolar end of the capillary would be largely reclaimed at the venous end.

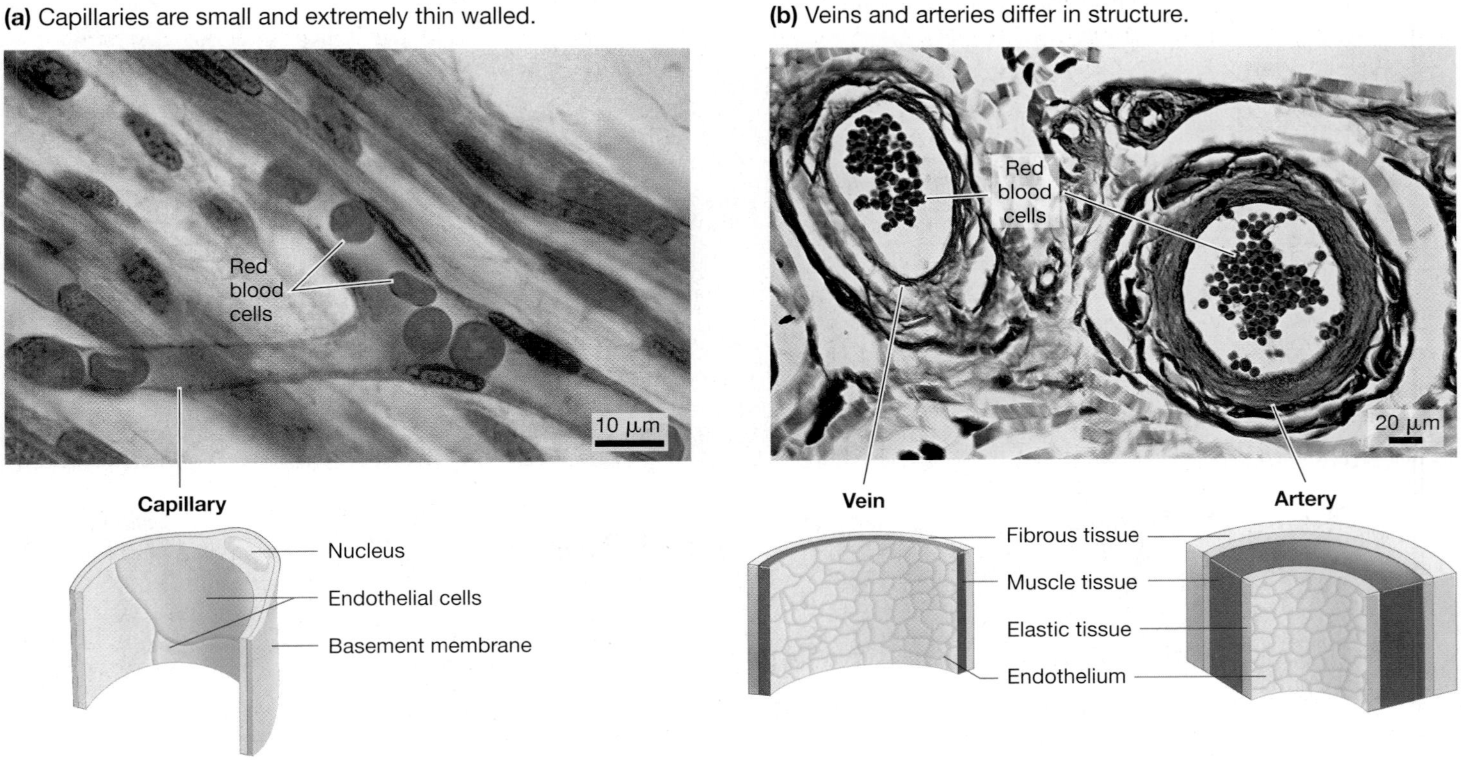

Figure 42.18 The Structures of Capillaries, Veins, and Arteries Reflect Their Different Functions. Notice the differences in relative wall thickness and overall size.

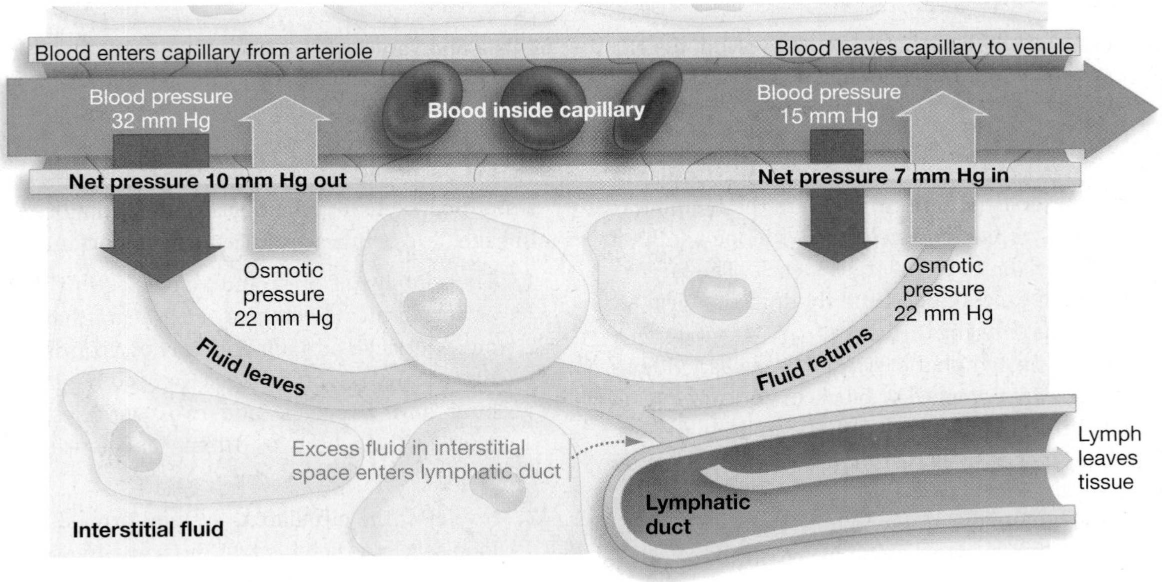

Figure 42.19 Pressure Differences in Capillaries Create Interstitial Fluid and Lymph. The balance of blood pressure and osmotic forces favors fluid loss from the beginning (inflow) end of capillaries and fluid recovery at the other (outflow) end. Fluid that is not recovered by the capillaries is transported out of the tissue as lymph, which eventually rejoins the blood circulation.

Note the word "largely," however—not all interstitial fluid is reabsorbed by capillaries. In Figure 42.19, not all of the fluid entering the interstitial space at the arteriolar end of the capillary bed has reentered the bloodstream at the venous end—there is a net buildup of interstitial fluid. What happens to this fluid?

The Role of the Lymphatic System Starling proposed that because interstitial fluid is continually added to the interstitial space, there must be a mechanism for draining the excess fluid. In fact, the fluid is collected in the **lymphatic system**: a collection of thin-walled, branching tubules called lymphatic vessels that

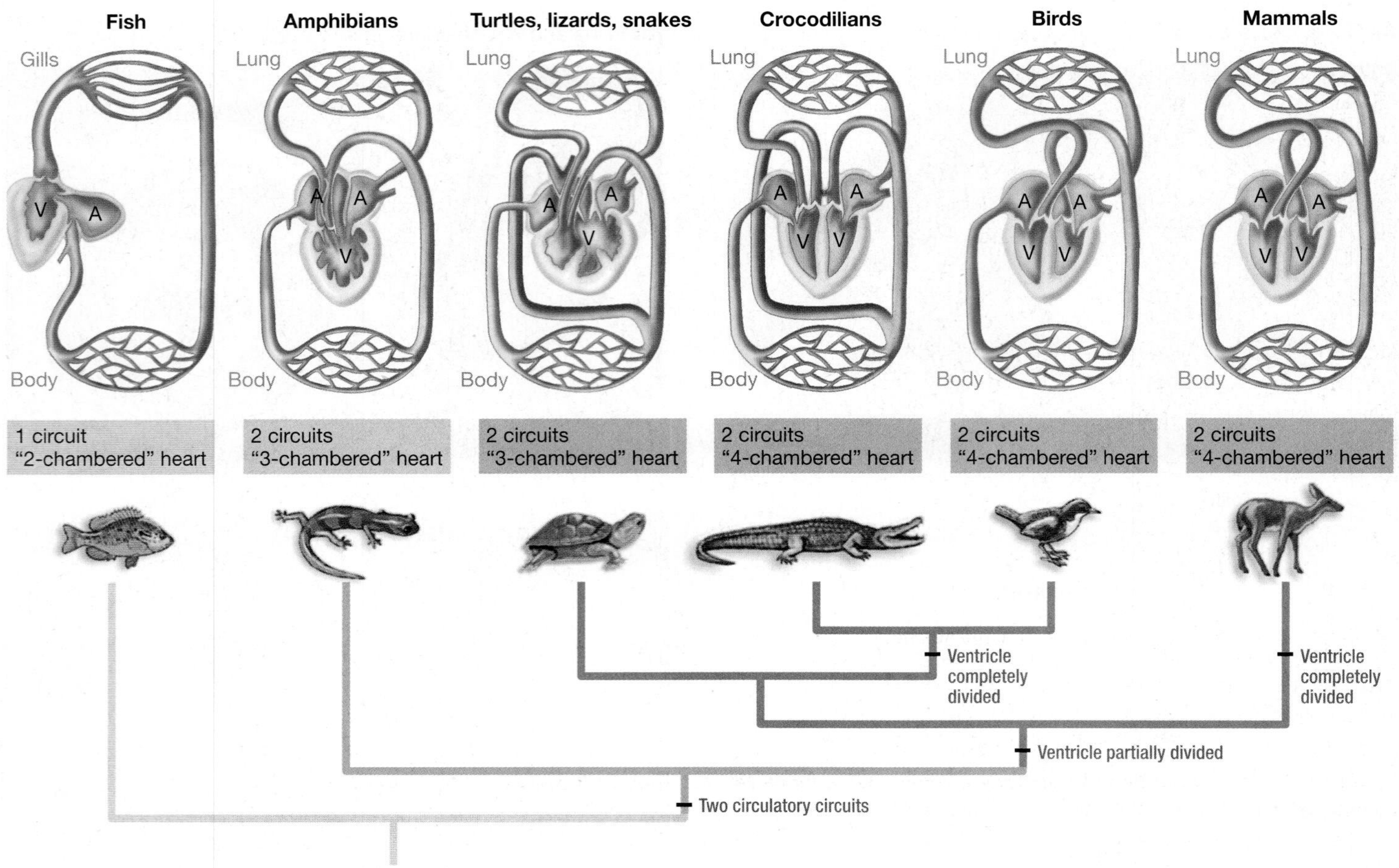

Figure 42.20 As Vertebrate Circulatory Systems Evolved, the Number of Atria and Ventricles Increased. "A" denotes the atria—chambers that receive blood coming into the heart from the body and the gills or lungs. "V" denotes the ventricles—chambers that pump blood out to the gills or lungs and the body.

permeate all tissues. Interstitial fluid that enters the lymphatic ducts is called **lymph**. Lymphatic vessels join with one another, like the tributaries of a river, to form larger vessels. The largest lymphatic vessels return excess interstitial fluid, in the form of lymph, to the major veins entering the heart.

The importance of the lymphatic system becomes evident when lymphatic vessels are damaged or blocked. For example, a disease called elephantiasis results when the lymphatic vessels in the extremities are blocked by parasitic worms that are transmitted from person to person via mosquito bites. The affected limbs swell dramatically because the lymph cannot be drained, and the skin thickens, cracks, and becomes very painful.

How Does the Heart Work?

In vertebrates, the heart contains at least two chambers: There is at least one thin-walled **atrium** (plural: **atria**), which receives blood, and at least one thick-walled **ventricle**, which generates the force required to propel blood out of the heart and through the circulatory system. Atria are separated from ventricles by atrioventricular (AV) valves.

The phylogenetic tree in **Figure 42.20** shows the evolutionary relationships among some major vertebrate lineages and a simplified sketch of the heart and circulatory system for each lineage. Two points are particularly important to note:

1. The number of atria and ventricles in the heart increased as vertebrates diversified. Fish hearts have one atrium and one ventricle; amphibians, turtles, lizards, and snakes have two atria and one ventricle; crocodilians, birds, and mammals have two atria and two ventricles. It is common to refer to these as two-, three-, and four-chambered hearts, respectively.
2. In fishes, the circulatory system forms a single circuit—one loop services the gills and the body. In other lineages, there are separate circuits to the lungs and to the body.

What Factors Favored the Evolution of Multichambered Hearts and Multiple Circulatory Circuits? Lungs evolved in some lineages of fishes about 400 million years ago, facilitating air-breathing and supplementing the water-breathing capacity of gills. In these animals, exemplified by the extant lungfish, a **pulmonary artery** carries blood to the lungs, and **pulmonary veins** return freshly oxygenated blood to the heart. Circulation is partially split into two circuits—a **pulmonary circuit** that takes blood to the lungs and gills, and a **systemic circuit** that takes blood to the body.

Because fishes live in the neutrally buoyant environment of water, gravity has little effect on blood flow. Even though blood pressure drops as blood passes through the gills—due to the mechanical resistance to flow that occurs in the gills' capillary beds—blood pressure stays high enough to move blood throughout the body.

The situation changed dramatically in those lineages that evolved into terrestrial vertebrates, whose lungs became their primary gas exchange organ. Gravity has a much larger effect on circulation in land-dwelling vertebrates than in aquatic vertebrates, especially in the case of blood flow to elevated portions of the body. To overcome gravity in terrestrial environments, blood must be pumped at high pressure. However, pressures must be lower in the capillaries of the lung because they are surrounded by alveolar air, not other tissues. A high pulmonary blood pressure would push liquid out of the capillaries into the alveoli and fill the alveoli with liquid. The evolution of pulmonary and systemic circuits allowed the best of both worlds—a high-pressure systemic circuit that pumps blood throughout the body, and a low-pressure pulmonary circuit that sends blood to the lungs.

Paired atria evolved in the ancestors of modern amphibians, but the pulmonary and systemic circulations are only partially separated in amphibians, turtles, lizards, and snakes. In these lineages, blood from the right and left atria may mix in the common ventricle before being expelled from the heart to the lungs or to the body. Turtles, lizards, and snakes, however, have partially divided ventricles that can limit the amount of mixing that occurs there (see Figure 42.20).

In addition, turtles, lizards, and snakes have a bypass vessel running from the right side of the ventricle directly into the systemic circulation. This bypass vessel is also observed in the unusual four-chambered hearts of crocodilians. The bypass vessels have an important function: They shunt blood from the pulmonary to the systemic circulation when the animal is underwater and not breathing. The result is a great reduction in blood flow to the lungs at those times.

The shunt might play other roles as well. Importantly, shunting blood from the right ventricle into the systemic circulation causes O_2-poor—but CO_2-rich—blood to be delivered to organs, including the stomach. As carnivores, alligators and other crocodilians eat protein-rich food, often including whole bones. Recall that digestion of protein relies on acid secretion by the stomach, and that acid production begins when CO_2 diffuses into gastric cells and reacts with water to produce H^+ (see Ch. 41, Figure 41.10). Could alligators use the bypass vessel to send extra CO_2, which could be used to make acid, to the stomach?

To test this hypothesis, researchers conducted a clever experiment (**Figure 42.21**). They surgically tied off the bypass vessel so that blood could no longer be shunted, effectively changing the alligator's heart anatomy to that of a mammal or bird. Control

RESEARCH

QUESTION: Why do alligators shunt blood when digesting?

HYPOTHESIS: Shunting increases flow of CO_2-rich blood to the stomach, facilitating acid production and bone digestion.

NULL HYPOTHESIS: Shunting plays no role in acid production or bone digestion.

EXPERIMENTAL SETUP:

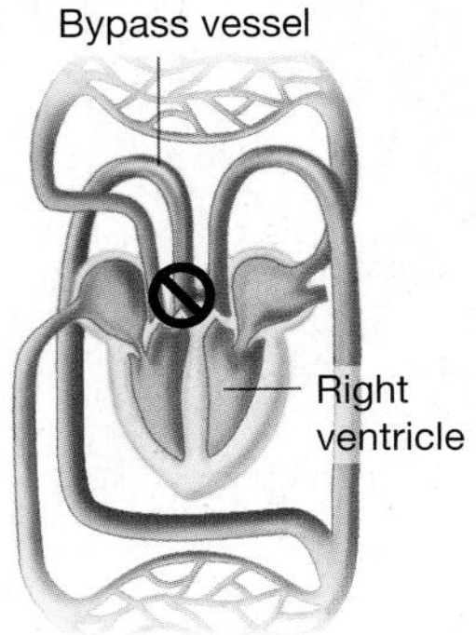

1. Surgically disrupt the shunt by tying off the opening from the right ventricle into the bypass vessel. Perform sham surgery on control alligators.

2. Feed the alligators.

3. Compare rates of gastric acid secretion (using pH electrode inserted into stomach) and bone digestion (using X-rays to measure bone size over time) in disrupted and control alligators.

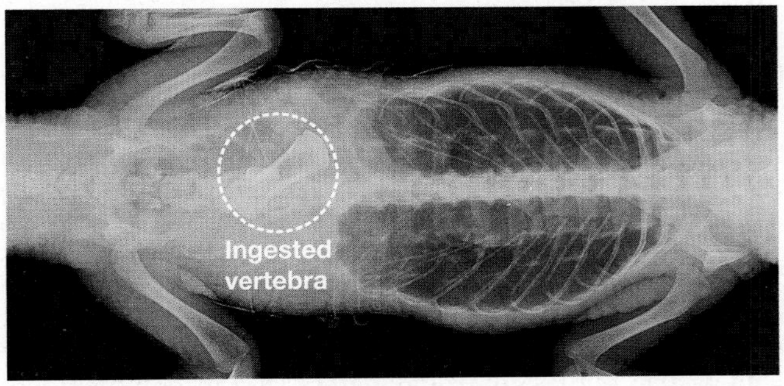

PREDICTION OF HYPOTHESIS: Alligators with disrupted shunts will produce less acid and digest bone more slowly than control alligators.

PREDICTION OF NULL HYPOTHESIS: There will be no differences between alligators with disrupted shunts and control alligators.

RESULTS:

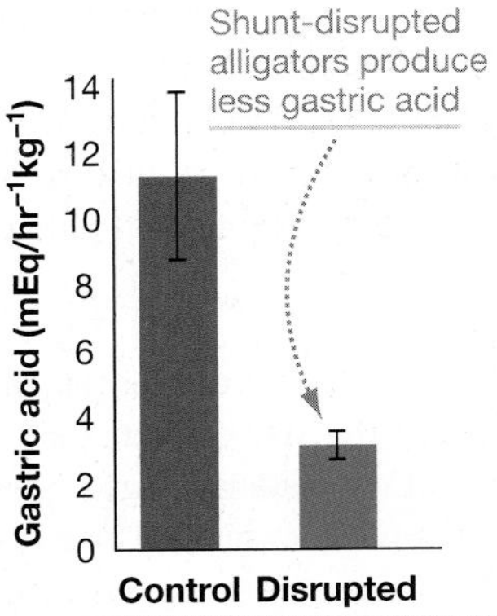

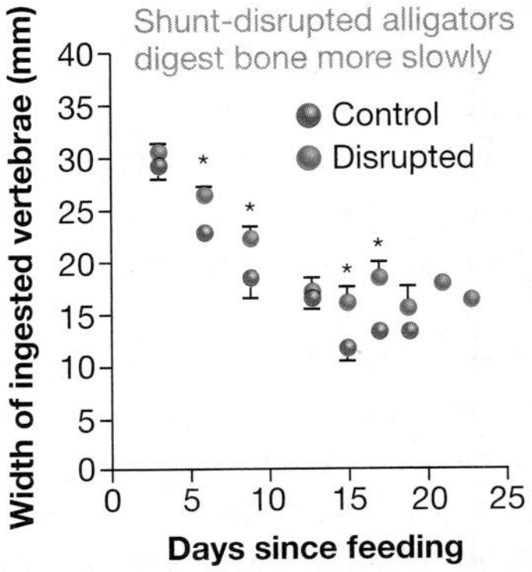

CONCLUSION: Alligators shunt O_2-poor, CO_2-rich blood from the right ventricle to systemic circulation when digesting, facilitating greater acid production and therefore bone digestion in the stomach.

Figure 42.21 The Circulatory Anatomy of Alligators Allows Them to Digest Bone.

SOURCE: C. G. Farmer, T. J. Uriona, D. B. Olsen, M. Steenblik, and K. Sanders. 2008. The Right-to-Left Shunt of Crocodilians Serves Digestion. *Physiological and Biochemical Zoology* 81: 125–137.

✔ **PROCESS OF SCIENCE** Why was a sham surgery performed instead of using intact, normal alligators as controls?

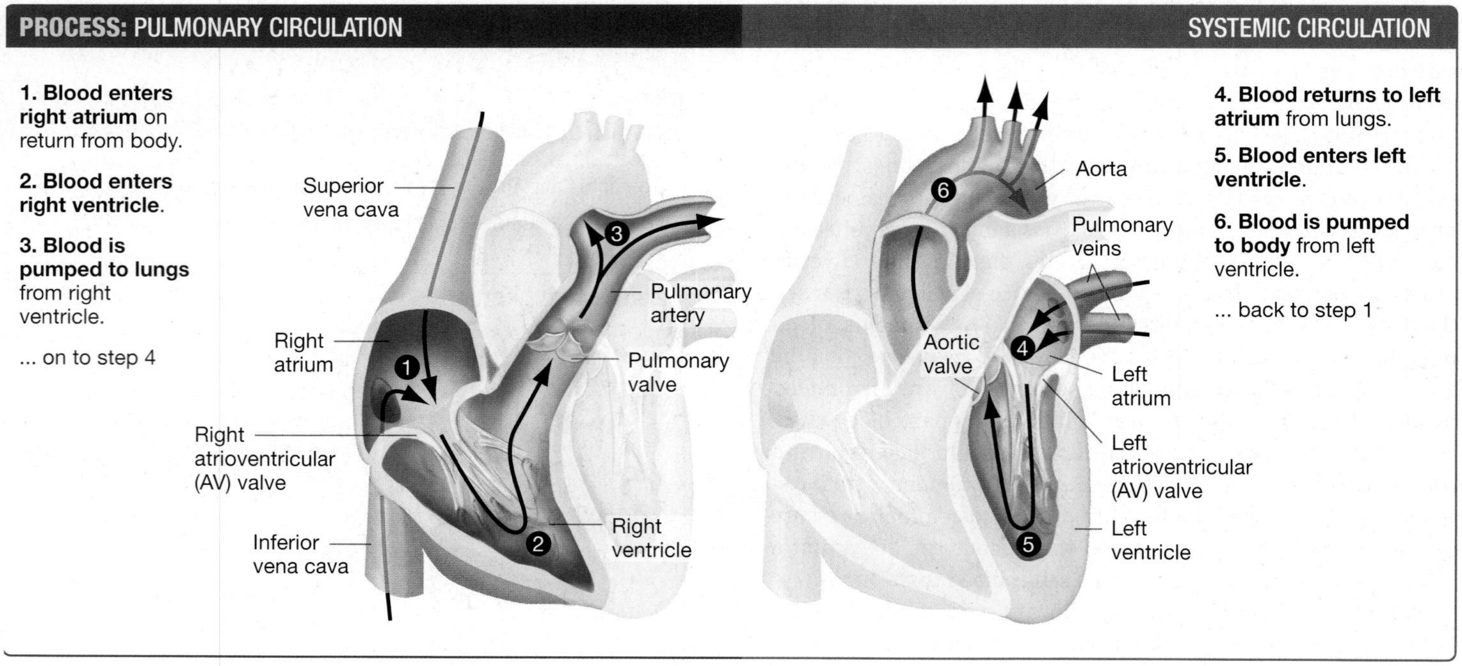

Figure 42.22 **The Human Heart Maintains Separation of Highly Oxygenated and Poorly Oxygenated Blood.** Blood flows through the chambers in the sequence shown.

alligators had the surgery, but the bypass vessel was not tied off (this is called a "sham" surgery). After the alligators had healed, the researchers fed them, measured their gastric acid production by inserting a pH electrode into the stomach, and took X-rays over time to measure the size of the bones in the stomach as the alligators digested them.

Just as the researchers predicted, the alligators that could no longer shunt CO_2-rich blood to the stomach did not produce as much gastric acid as control alligators, and it took them longer to digest the bones in their meals. Despite having completely divided ventricles, the shunt gives crocodilians circulatory flexibility depending on whether they are on land or in water, and whether they are fasting or digesting.

Unlike turtles, lizards, and snakes, birds and mammals have fully divided ventricles and lack a bypass vessel. This configuration completely separates the pulmonary and systemic circuits. Complete separation prevents mixing of blood from the two circuits, maximizing the P_{O_2} in the systemic circulation. As a result, tissues receive more oxygen and can produce more ATP to fuel energy-intensive processes such as aerobic exercise—flight, for example—and endothermy.

The Human Heart Your heart is located in your chest cavity, between your lungs, and is roughly the size of your fist. As Figure 42.22 shows, the human circulatory system returns blood from the body to the right atrium of the heart (step 1). This blood is low in oxygen, and it arrives via two large veins called the inferior (lower) and superior (upper) **venae cavae** (singular: **vena cava**).

When the muscles that line the right atrium contract, they send poorly oxygenated blood to the right ventricle (Figure 42.22, step 2). The right ventricle, in turn, contracts and sends blood out to the lungs, via the pulmonary artery. In this way, the right ventricle powers the movement of blood through the pulmonary circulation (step 3).

Blood flows from atrium to ventricle to artery in only one direction because one-way atrioventricular valves separate the heart's chambers from each other and from the adjacent arteries. As Figure 42.22 indicates, the valves are flaps, oriented to ensure a one-way flow of blood with little or no backflow. If heart valves are damaged or defective, the resulting backflow can be heard through a stethoscope and is called a **heart murmur**. The backflow reduces the organ's efficiency.

After blood circulates through the capillary beds in the lung's alveoli and becomes highly oxygenated, it returns to the heart through the pulmonary veins. The oxygenated blood enters the left atrium (Figure 42.22, step 4).

When the left atrium contracts, it pushes blood into the left ventricle (step 5). The walls of the left ventricle are so thick with muscle cells that their contraction sends oxygenated blood at high pressure through the aorta and into the arteries and capillaries that make up the systemic circulation (step 6).

Figure 42.23 summarizes the flow pattern through the human circulatory system and the partial pressures of O_2 and CO_2 at various points in the pulmonary and systemic circulations. Notice that blood vessels are called arteries or veins according to the direction of blood flow relative to the heart, not because of the oxygen content of the blood in them. So the pulmonary artery is

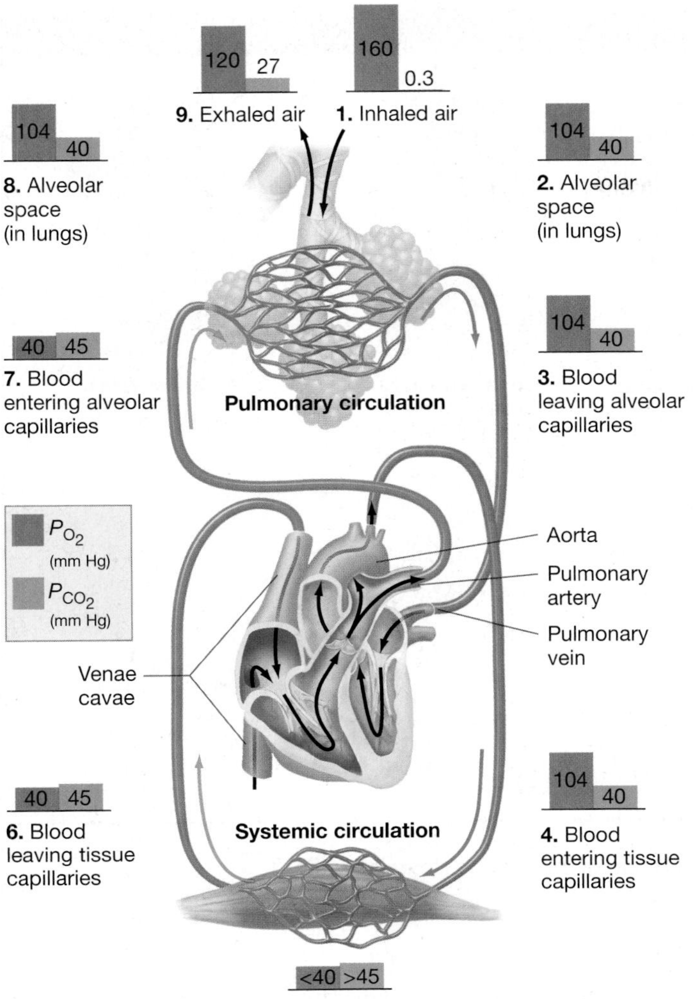

Figure 42.23 Partial Pressures of O_2 and CO_2 Vary throughout the Human Circulatory System.

✔ Why are the partial pressures of oxygen and carbon dioxide in exhaled air intermediate in magnitude between the partial pressures in inhaled and alveolar air?

called an artery because it takes blood away from the heart, even though this blood is low in oxygen.

Electrical Activation of the Heart Like other muscle cells, cardiac muscle cells contract in response to electrical signals. In invertebrates, the electrical signals that trigger heart contraction come directly from the nervous system. But a vertebrate heart will continue to beat even if all nerves supplying it are severed. Why? In vertebrates, a group of cells in the heart itself are responsible for generating the initial signal.

The cells that initiate contraction in the vertebrate heart are known as **pacemaker cells**. They are located in a region of the right atrium called the **sinoatrial (SA) node**.

The electrical signal generated in the SA node is rapidly conducted throughout the right and left atria. The signal spreads quickly from cell to cell thanks to a striking property of cardiac muscle cells: They form physical and electrical connections with each other.

All cardiac muscle cells branch to contact several other cardiac muscle cells, join end to end with these neighboring cells (see Ch. 39, Figure 39.5), and connect to them by specialized structures called **intercalated discs**. Because these discs contain many gap junctions (cell-to-cell connections; described in Chapter 11), electrical signals pass directly from one cardiac muscle cell to the next.

The electrical activation of the heart is reflected in the orange line on the graph at the bottom of Figure 42.24 on page 916. This line is called an **electrocardiogram**, or **EKG**—a recording of the electrical events that occur as the heart beats. An EKG recording is generated by amplifying the overall electrical signal conducted from the heart to the chest wall through the tissues of the body. By inspecting an EKG, physicians can diagnose disturbances of heart rhythm and detect damage to the heart muscle.

The drawings above the graph in Figure 42.24 show where the key electrical events are happening. The process is as follows:

1. The SA node generates an electrical signal.
2. The signal from the SA node quickly propagates to atrial muscle cells. As a result, the atria contract simultaneously and eject blood into the ventricles.
3. As the atria begin to contract, the signal is conducted to an area of the heart called the **atrioventricular (AV) node**. The AV node delays the signal slightly before passing it to the ventricles. The delay allows the atria to fill the ventricles with blood before the ventricles contract.
4. After the delay, the electrical impulse is rapidly transmitted through specialized conducting fibers in the muscular wall that separates the ventricles. The impulse spreads through both ventricles, causing them to contract as the atria relax. The ventricles empty because the signal and the resulting muscular contraction move from the bottom up to the top of each ventricle—toward the arteries that allow blood to exit.
5. The final electrical event occurs as the ventricles relax and their cells recover—restoring their electrical state before contraction.

✔ If you understand these concepts, you should be able to predict how the amount of blood ejected from the ventricles would change if there were no delay at the AV node.

The SA node and the muscle cells of the heart receive input from the nervous system and from hormones, chemical messengers carried in the blood. These inputs are important for regulating both the heart rate and the strength of ventricular contraction. In this way, the rate of blood flow through the circulatory system varies in response to electrical signals and hormones. During the "fight-or-flight" response (Ch. 46, Section 46.3), for example, a chemical signal called epinephrine causes both heart rate and contraction strength to increase—sending blood more quickly through the body in preparation for rapid movement.

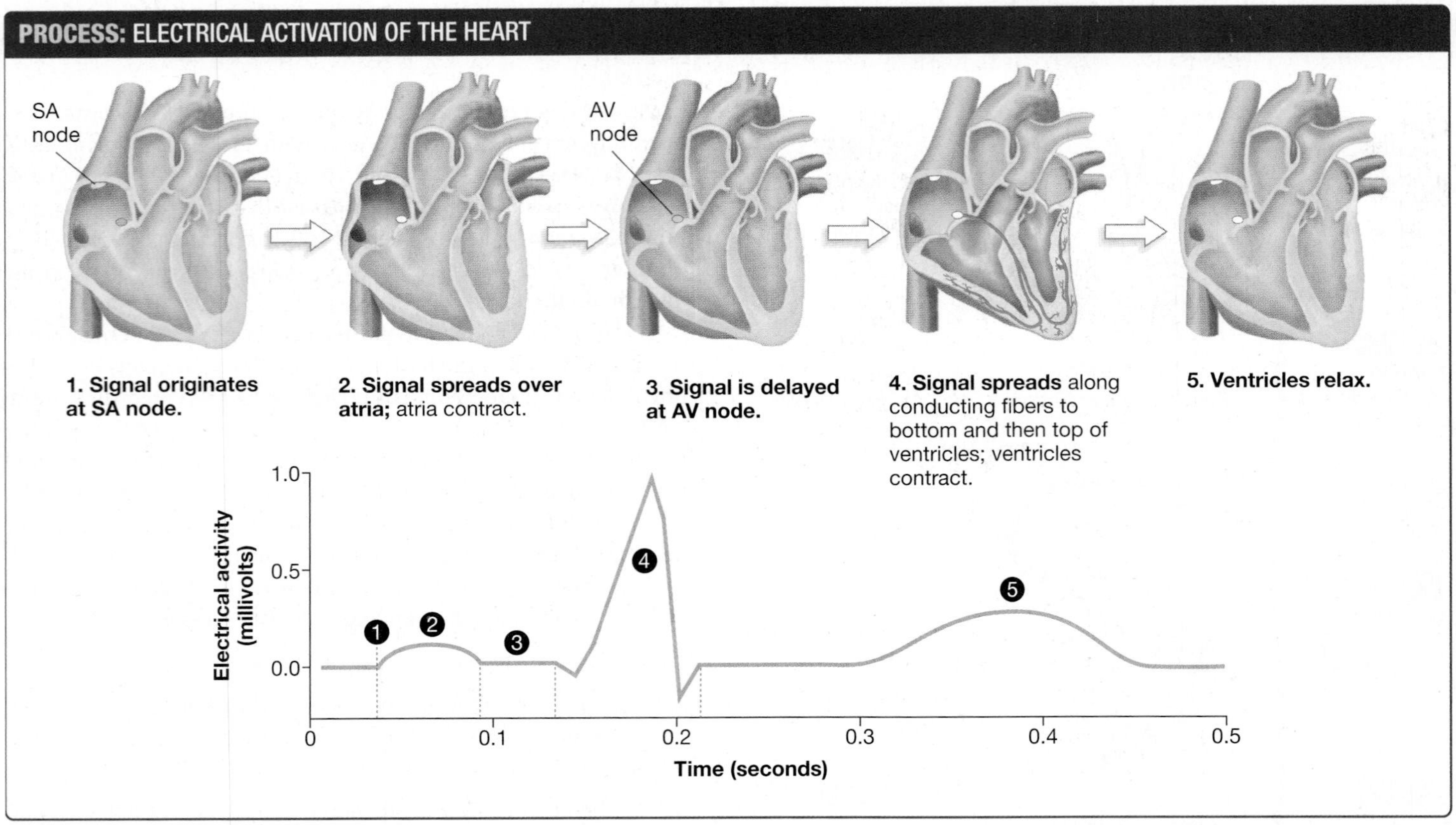

Figure 42.24 **Sequential Electrical Activation Leads to Coordinated Contraction of the Human Heart.** The rate and strength of contractions control the pressure in the heart chambers and arteries.

The Cardiac Cycle The electrical signals originating from the SA node ensure that the atria contract simultaneously, and the delay at the AV node ensures that the atria are relaxed by the time the ventricles contract. The contraction phases of the atria and the ventricles, called **systole** (pronounced *SIS-tuh-lee*), are therefore closely coordinated with their relaxation phases, or **diastole** (pronounced *dy-AS-tuh-lee*). This sequence of contraction and relaxation is called the **cardiac cycle**; the cycle consists of one diastole and one systole for both atria and ventricles.

Ventricular contraction (ventricular systole) leads to a rapid increase in pressure within both ventricles, as recorded in the dark purple line on the graph in Figure 42.25. Blood is ejected into the pulmonary artery and the aorta when ventricular pressure exceeds the pressure within each respective artery. Blood pressure measured in the systemic arterial circulation at the peak of ventricular ejection into the aorta is called the **systolic blood pressure**. Blood pressure measured just before ventricular ejection is called the **diastolic blood pressure**.

Clinicians report blood pressure measurements as two numbers separated by a slash; the first number is the systolic pressure and the second number is the diastolic pressure. People with blood pressures consistently higher than 130/90 mm Hg have high blood pressure, or **hypertension**.

Hypertension is a serious disease because it can lead to a variety of defects in the heart and circulatory system. Abnormally high blood pressure puts mechanical stress on arteries. If the walls of an artery fail, the individual may experience heart attack, stroke, kidney failure, or burst or damaged vessels.

Patterns in Blood Pressure and Blood Flow

As blood moves through capillaries, blood pressure drops dramatically—as the top graph in Figure 42.26 indicates. This happens because, as arteries branch, rebranch, and eventually form networks of capillaries, the total cross-sectional area of blood vessels in the circulatory system increases, as shown on the bottom graph. As the blood moves through greater and greater areas of space inside the vessels, the pressure that it exerts on the vessel walls drops.

As the line labeled "Velocity" in the bottom graph in Figure 42.26 indicates, the velocity of blood flow also decreases significantly in capillary beds relative to arteries and veins, because the same amount of fluid is passing through a much larger area. Recall that the slow flow of blood through capillaries is important: It provides sufficient time for gases, nutrients, and wastes to diffuse between tissues and blood.

Why Is Regulation of Blood Pressure and Blood Flow Important? The general patterns of blood pressure and blood flow diagrammed in Figure 42.26 don't tell the entire story, however. Blood movement is carefully regulated at an array of points throughout the circulatory system.

Recall that the walls of arterioles are made up partially of smooth muscle. Contraction or relaxation of this muscle can

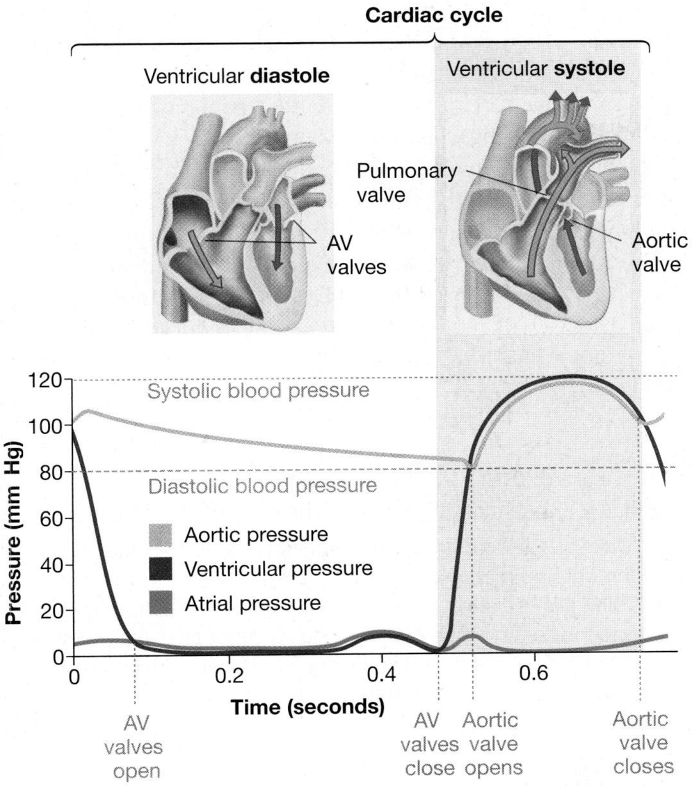

Figure 42.25 Blood Pressure Changes during the Cardiac Cycle. These data show the pressures in the left atrium, left ventricle, and aorta in the course of a cardiac cycle. In this example, the blood pressure measured in the upper arm would be 120/80 mm Hg. Right ventricle and pulmonary artery pressures would produce a similar pattern, but the blood pressure in the pulmonary artery would be much lower—closer to 25/8 mm Hg.

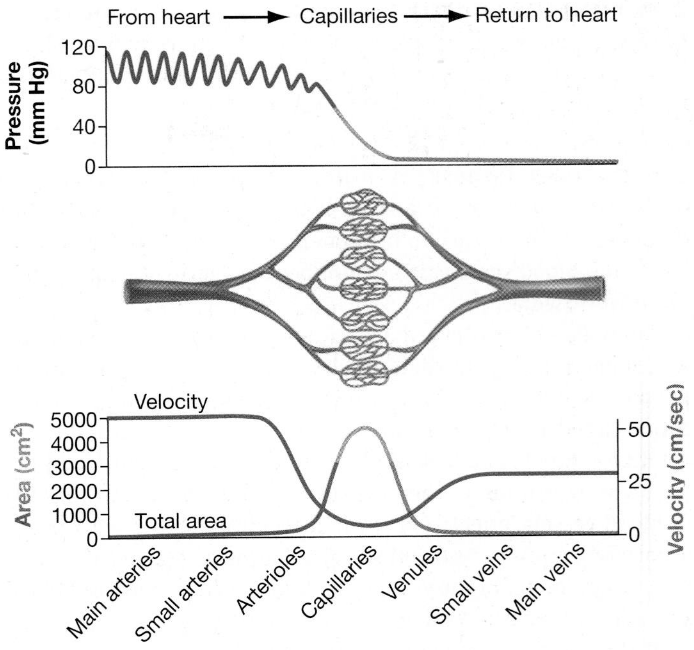

Figure 42.26 Blood Pressure Drops Dramatically in the Circulatory System. The top graph shows how blood pressure changes as blood leaves the heart and travels through arteries, capillaries, and veins, as in the branching pattern of the middle diagram. In arteries near the heart, each heartbeat causes fluctuations in blood pressure. These pressure pulses disappear in the capillaries, so blood flows there at a steady speed. The bottom graph plots the total area of blood vessels shown in the diagram, as well as the velocity of blood flow through the vessels.

restrict or allow blood flow to specific tissues. For example, arterioles in the skin dilate during exercise, diverting blood flow to the skin to eliminate excess heat. This accounts for the flushed facial appearance induced by vigorous exercise.

As another example of how blood pressure and blood flow are regulated, consider what happens if you sit long enough for the smooth muscle in the veins of your legs to relax. Then, if you stand up rapidly, blood pools in your legs under the influence of gravity and your blood pressure can drop enough to reduce blood flow to your brain and cause dizziness or even a blackout. More serious drops in blood pressure, due to severe dehydration or blood loss, can be fatal. Fortunately, decreases in blood pressure elicit a powerful homeostatic response.

Homeostatic Control of Blood Pressure Recall that all homeostatic responses involve **(1)** sensors that detect the change in condition, **(2)** an integrator that processes information about the change, and **(3)** effectors that diminish the impact of the change (Ch. 39, Section 39.4).

The body has specialized pressure-sensing receptors called **baroreceptors** that detect changes in blood pressure. Baroreceptors are found in the walls of the heart and some of the major arteries. When baroreceptors transmit nerve signals to the brain (the integrator) indicating a serious fall in blood pressure, a rapid, three-component effector response ensues:

1. Cardiac output—the volume of blood leaving each ventricle per minute—increases. This is due to an increase in heart rate and an increase in stroke volume, which is the amount of blood ejected from each ventricle during each cardiac cycle. (Cardiac output = heart rate × stroke volume.)
2. Arterioles serving the capillaries of certain tissues constrict to divert blood to more critical organs. (This occurs in tissues like the skin and intestines, which can endure short-term restrictions in their blood supply without damage.)
3. Veins constrict, decreasing their overall volume. Because more than half of the blood in the circulatory system is contained within the veins, constriction of these vessels shifts blood volume toward the heart and arteries to maintain blood pressure and flow to vital organs.

This coordinated response is mediated both by a portion of the nervous system called the sympathetic nervous system (Chapter 43) and by hormones produced by the adrenal glands (Chapter 46). Sympathetic nerves and the hormones involved in regulating blood pressure deliver their messages directly to **(1)** the SA node to increase heart rate, **(2)** the ventricles of the heart to increase stroke volume, and **(3)** the muscular walls of the arteries and veins to modify their total volume.

Cardiovascular Disease A healthy circulatory system is obviously critical to your overall well-being. Indeed, **cardiovascular disease**, which is a group of ailments collectively affecting the heart and blood vessels, is the number one cause of death in humans worldwide.

Many factors contribute to cardiovascular disease, including age, tobacco use, poor diet, obesity, inactivity, and genetics. As people age, their blood vessels harden and lose elasticity—a condition called **arteriosclerosis**.

High-fat diets and lack of physical activity can compound the problem by leading to the deposit of fatty plaques on the walls of blood vessels, effectively reducing their diameter. Nicotine in tobacco also constricts blood vessels. The loss of elasticity and the reduced diameter of the vessels combine to cause increased blood pressure, which can weaken the walls of arteries.

If the arteries that deliver blood to the heart muscle become completely blocked, a **myocardial infarction**, or heart attack, can occur. Contrary to popular thought, this condition is not defined as stopping of the heartbeat. In a myocardial infarction, a portion of heart tissue dies within minutes when it is deprived of oxygen. Depending on the location and extent of damage, myocardial infarction can affect heart function slightly, or it can cause rapid death. Timely medical intervention can improve the chances of surviving a heart attack.

Nearly 18 million people worldwide died from cardiovascular disease in 2017, and the number is projected to reach 23 million by the year 2030. It may get even higher because the incidences of obesity and diabetes mellitus have reached epidemic proportions (Ch. 41, Section 41.4). Effectively combating cardiovascular disease will require improved diet, reduced tobacco use, greater amounts of physical activity, and better access to health care across the globe.

CHECK YOUR UNDERSTANDING

✔ If you understood this section, you should be able to . . .

1. Make a labeled diagram showing how blood circulates through the mammalian heart.
2. **THINK CAREFULLY** In a medical drama on television, a patient's heart stops beating, and a doctor says the patient is in cardiac arrest. Is this the same thing as a myocardial infarction? Why or why not?

Answers are available in Appendix A.

CHAPTER 42 Review

For media, go to **Mastering Biology**

42.1 The Respiratory and Circulatory Systems

- Animal gas exchange involves ventilation, exchange of gases between the environment and the blood, and exchange of gases between blood and tissues.
- Animal circulation involves transportation of gases, nutrients, wastes, and other substances throughout the body.

42.2 Air and Water as Respiratory Media

- As media for exchanging oxygen and carbon dioxide, air and water are dramatically different.
- Compared with water, air contains much more oxygen and is much less dense and viscous. As a result, terrestrial animals have to process a much smaller volume of air to extract the same amount of O_2, and the amount of work required to do so is less than in aquatic animals.

42.3 Organs of Gas Exchange

- The structure of gills, tracheae, lungs, and other gas exchange organs minimizes the cost of ventilation while maximizing the diffusion rates of O_2 and CO_2.
- Consistent with predictions made by Fick's law of diffusion, respiratory epithelia tend to be extremely thin and to be folded to increase surface area.
- In fish gills, countercurrent exchange ensures that the differences in O_2 and CO_2 partial pressures between water and blood are favorable for gas exchange over the entire length of the ventilatory surface.
- Insect tracheae carry air directly to and from tissues.
- Breathing rate is regulated to keep the carbon dioxide content of the blood stable during rest and exercise.

42.4 How Are Oxygen and Carbon Dioxide Transported in Blood?

- The tendency of hemoglobin to give up oxygen varies as a function of the P_{O_2} in surrounding tissue in a sigmoidal fashion. As a result, a relatively small change in tissue P_{O_2} causes a large change in the amount of oxygen released from hemoglobin.
- Oxygen binds less tightly to hemoglobin when pH is low. Because CO_2 reacts with water to form protons, the existence of high CO_2 partial pressures in exercising muscle tissues lowers their pH and makes oxygen less likely to stay bound to hemoglobin and more likely to be unloaded into tissues.

- The CO_2 that diffuses into red blood cells from tissues is rapidly converted by the enzyme carbonic anhydrase into a bicarbonate ion and a proton. The protons that are released bind to poorly oxygenated hemoglobin. In this way, hemoglobin acts as a buffer that takes protons out of solution and prevents large fluctuations in blood pH.

42.5 Circulation

- In many animals, blood or hemolymph moves through the body via a circulatory system consisting of a pump (heart) and vessels.
- In open circulatory systems, overall pressure is low and tissues are bathed directly in hemolymph.
- In closed circulatory systems, blood is contained in vessels that form a continuous circuit. Containment of blood allows higher pressures and flow rates, as well as the ability to direct blood flow accurately to tissues that need it the most.
- In organisms with a closed circulatory system, a lymphatic system collects excess fluid that leaks from the capillaries and returns it to the circulation.
- In amphibians and some reptiles, blood from the pulmonary and systemic circuits may be mixed in the single ventricle.
- In mammals and birds, a four-chambered heart pumps blood into two circuits, which separately serve the lungs and the rest of the body. Crocodilians have a similar heart with a bypass vessel that can shunt blood from the pulmonary to the systemic circuit.
- In vertebrates, the cardiac cycle is controlled by electrical signals that originate in the heart itself.
- Heart rate, cardiac output, and constriction of arteries, arterioles, veins, and venules are regulated by chemical signals and by electrical signals from the brain.
- Cardiovascular disease is the leading cause of death in humans.

Answers are available in Appendix A.

✓ TEST YOUR KNOWLEDGE

1. Which of the following statements regarding the insect tracheal system is/are correct? Select True or False for each statement.
   T/F Tracheae dilate and constrict during flight or other types of movement, functioning as a "breathing" mechanism.
   T/F Spiracles open into the body cavity, allowing direct contact between the air and hemolymph.
   T/F Tracheae carry oxygen to alveoli; here the oxygen diffuses into the hemolymph to be delivered to tissues.
   T/F Spiracles can close to minimize water loss.
2. Identify at least two advantages of breathing air instead of water.
3. Which of the following promotes oxygen release from hemoglobin?
   a. a decrease in temperature
   b. an increase in O_2 level
   c. a decrease in pH
   d. a decrease in carbonic anhydrase activity
4. Describe the disadvantages of an open circulatory system relative to a closed circulatory system.

✓ TEST YOUR UNDERSTANDING

5. Explain how each parameter in Fick's law of diffusion is reflected in the structure of the mammalian lung.
6. Frog lungs have a smaller surface area for gas exchange than mammalian lungs. How do frogs compensate for this difference?
   a. Frog tissue absorbs more oxygen from the blood than mammalian tissue does.
   b. Frogs breathe more quickly than mammals.
   c. Frogs also obtain oxygen via diffusion across the skin.
   d. Frog lung tissue has a greater density of capillary beds than mammalian lung tissue.
7. **MODEL** Carp are fishes that thrive in stagnant-water habitats with low oxygen partial pressure. Compared with the hemoglobin of many other fish species, carp hemoglobin has an extremely high affinity for O_2. Draw an oxygen–hemoglobin equilibrium curve showing separate lines for carp and a fish that lives in water with a higher oxygen partial pressure. Explain why they differ.
8. Explain why a person who survives a myocardial infarction might need to have an artificial pacemaker implanted.

✓ TEST YOUR PROBLEM-SOLVING SKILLS

9. Predict how Antarctic icefish can transport enough oxygen in their blood to meet their needs even though they lack hemoglobin.
10. Why did separate systemic and pulmonary circulations evolve in species that have the high-pressure circulatory system required for rapid movement of blood?

✓ PUT IT ALL TOGETHER: Case Study

How do the cardiovascular systems of athletes adjust to strenuous exercise?

During exercise, the cardiovascular system must supply muscles with large amounts of oxygen and fuel and get rid of a lot of wastes. How do the cardiovascular systems of athletes respond to prolonged exercise?

11. During athletic training, the oxygen–hemoglobin dissociation curve
    a. shifts to the right, unloading more oxygen to tissues.
    b. shifts to the right, unloading less oxygen to tissues.
    c. shifts to the left, unloading more oxygen to tissues.
    d. shifts to the left, unloading less oxygen to tissues.

12. **THINK CAREFULLY** When athletes exercise, what is the primary physiological variable responsible for their sustained increase in ventilation rate?
   a. decreased blood P_{O_2}
   b. increased blood P_{CO_2}
   c. increased blood pH
   d. increased body temperature

13. **QUANTITATIVE** Researchers used echocardiography, a sonogram of the heart, to estimate the mass of the left ventricle in current athletes, non-athletes, and ex-athletes. The data are graphed below (*** means $P < 0.001$, and the P value comparing non-athletes and ex-athletes is > 0.05; **BioSkills 3**). What conclusion can be drawn from the graph?

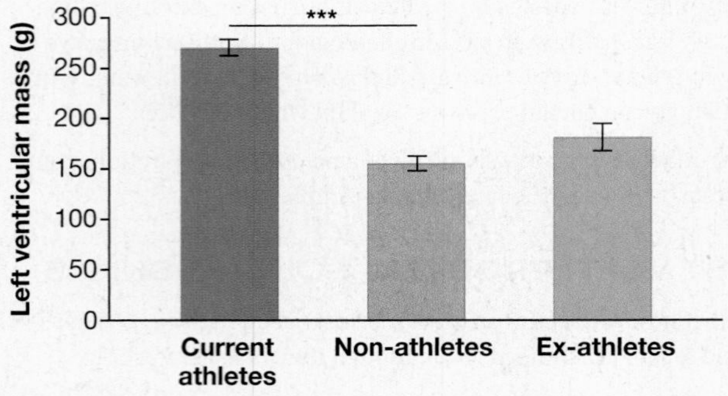

Source: L. M. Shapiro. 1984. *British Heart Journal* 52: 130–135.

14. Explain the advantage of the observed difference between current athletes and non-athletes in the graph shown here.

15. Researchers have also observed that athletes and non-athletes have the same mean resting cardiac output, even though athletes have a far lower resting heart rate. How is this possible?

16. **SOCIETY** Athletes are not the only people with enlarged hearts. Many patients with cardiovascular disease also have enlarged hearts. Suggest a cause of this enlargement.

Mastering Biology

Students Go to Mastering™ Biology for assignments, the eText, and the Study Area with animations, practice tests, and activities.

Professors Go to Mastering™ Biology for automatically graded tutorials and questions that you can assign to your students, plus Instructor Resources.

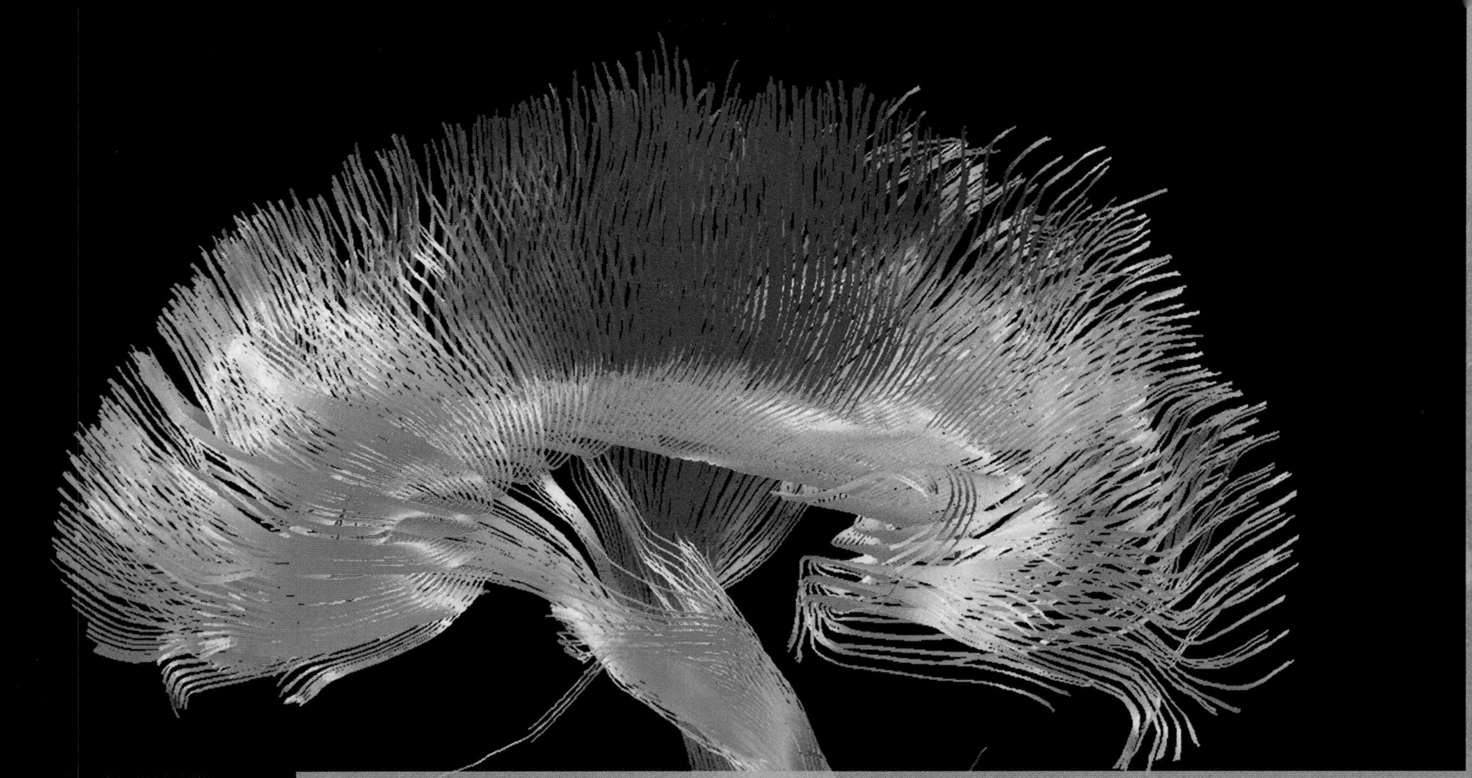

43 Animal Nervous Systems

Diffusion spectrum imaging reveals the trajectory of neural pathways in the brain. Such advances in brain imaging are allowing neurobiologists to study how neurons in the brain communicate with one another.

In this chapter you will learn how

Animals use electrical signaling along neurons for internal communication

beginning with

Principles of electrical signaling 43.1

then asking

How does the action potential work? 43.2

finally looking at

Synapses—the interface between neurons 43.3

then exploring

The vertebrate nervous system 43.4

including

The brain, learning, and memory

Most students and professional biologists are attracted to the study of neurobiology because they want to understand the human brain as well as higher-order processes such as consciousness, intelligence, emotion, learning, and memory. However, the human brain is very challenging to study as a whole structure. It contains billions of cells, interconnected through a myriad of neural pathways, only a tiny fraction of which are highlighted in the photo shown here.

Faced with this challenge, early researchers in neurobiology started simple: They focused on the function of individual nerve cells, or **neurons**, the cells mainly responsible for the working of the brain and the rest of the nervous system. Neurons are specialized to conduct information in the form of electrical signals from point to point in the body at speeds of up to 120 m/sec (270 mph). Thus, electrical signaling is a crucial aspect of information processing—one of the five attributes of life (introduced in Chapter 1).

BIG PICTURE

This chapter is part of the Big Picture. See how on pages 838–839.

Initial research on the electrical properties of single neurons laid a broad foundation for more recent studies of how the human brain works. This chapter proceeds in the same way. You'll begin by focusing on the neurons themselves and how they use electrical signaling to communicate with each other. In the last section of the chapter, you'll consider how the brain is organized and how phenomena such as memory work.

43.1 Principles of Electrical Signaling

The evolution of neurons was a key event in the diversification of animals, along with the evolution of muscles (Ch. 30, Section 30.1). All animals except sponges have neurons and muscle cells. Neurons transmit electrical signals; muscles can respond to signals from neurons by contracting.

After you complete this section, you should be able to . . .

- Explain the principles of electrical signaling in neurons.

Neurons are organized into two basic types of nervous systems:

1. The diffuse arrangement of cells called a **nerve net**, found in cnidarians (jellyfish, hydra, anemones) and ctenophores (comb jellies).
2. A **central nervous system (CNS)** that includes large numbers of neurons aggregated into clusters called ganglia.

Most animals with a CNS have a large cerebral ganglion, or brain, located in their anterior end. You also might recall that this phenomenon—the evolution of a bilaterally symmetric body with structures for information gathering and processing located at the head end—is known as cephalization (Ch. 30, Section 30.2).

Cephalization made most animals into efficient eating and moving machines: They face the environment in one direction, and sensory appendages take in information and send it to a nearby brain for processing. After integrating information from an array of sensory cells, the brain sends electrical signals to muscles and other organs that respond to the sensory stimuli.

Types of Neurons

The sensory cells are responsible for gathering information and can respond to light, sound, touch, or other stimuli. Sensory cells in an animal's skin, eyes, ears, mouth, and nose transmit information about the environment. Sensory cells inside the body monitor conditions that are important to keep in homeostatic balance, such as blood pH and temperature. In this way, sensory cells monitor conditions both outside and inside the body. Many sensory cells are **sensory neurons**, which carry information to the CNS. In vertebrates, the CNS consists of the brain and the spinal cord.

One function of the CNS is to integrate information from sensory neurons. Cells in the CNS called **interneurons** (literally, "between-neurons"), which pass signals from one neuron to another, perform this integration.

Some interneurons make connections to **motor neurons**, which are nerve cells that send signals to effector cells in glands or muscles. Recall that effectors are structures that bring about a physiological change in an organism (Ch. 39, Section 39.4). Motor neurons and sensory neurons are bundled together into long strands of nervous tissue called **nerves**.

All neurons and other components of the nervous system that are outside the CNS are considered part of the **peripheral nervous system**, or **PNS**. Section 43.4 describes the structure and function of the vertebrate PNS.

Typically, sensory information is transmitted via the PNS to the CNS, where it is processed. Then a response is transmitted back through the PNS to appropriate parts of the body. This reaction is an example of a **reflex**, an involuntary response to an environmental stimulus. For example, here's what happens if you prick your finger on a rose thorn, as shown in Figure 43.1:

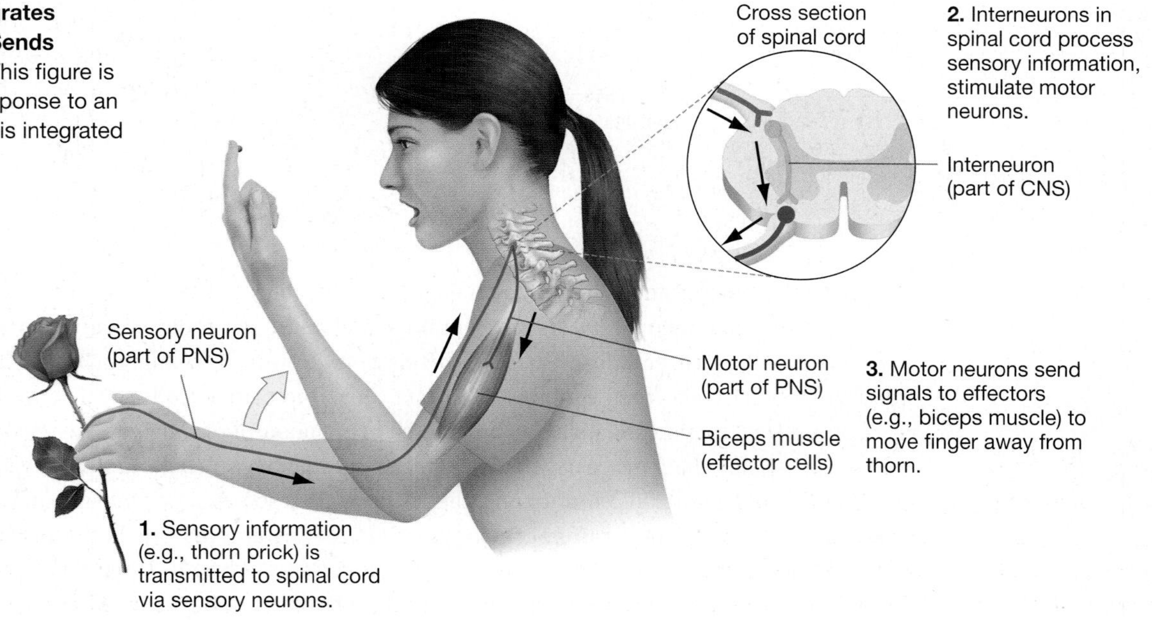

Figure 43.1 **The CNS Integrates Sensory Information and Sends Signals to Effector Cells.** This figure is an example of a reflex, a response to an environmental stimulus that is integrated in the spinal cord.

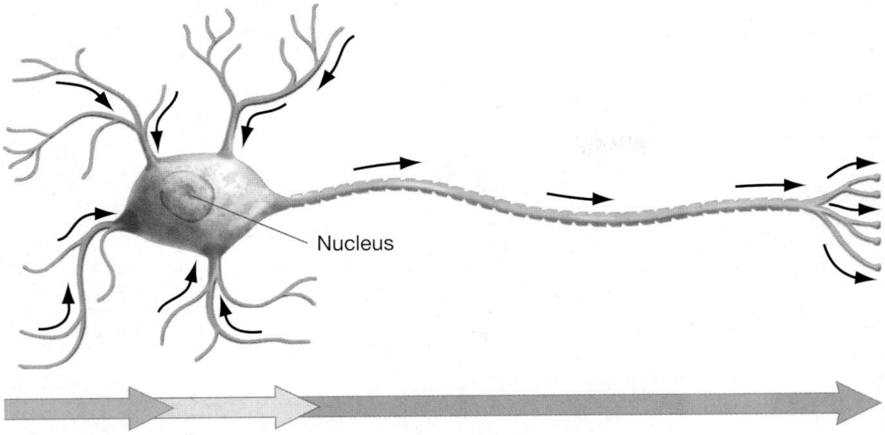

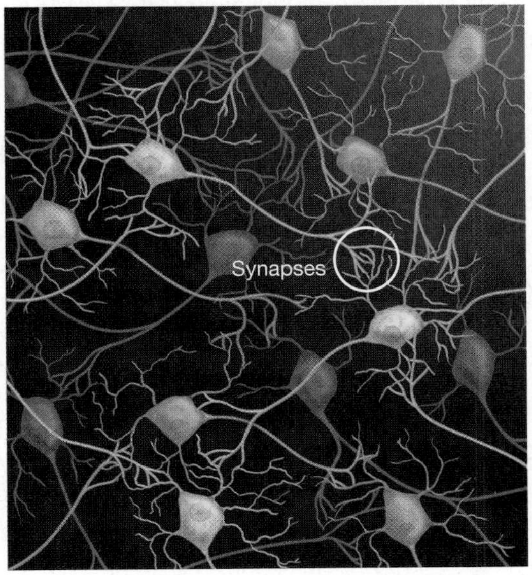

Figure 43.2 **How Does Information Flow in a Neuron? (a)** The three parts of a generalized neuron, and how information flows through them. **(b)** Most neurons receive inputs from and send projections to many other neurons.

Step 1 Pain sensors in your finger relay information to the spinal cord via sensory neurons.

Step 2 In the spinal cord, interneurons process the information and respond by stimulating motor neurons.

Step 3 The stimulated motor neurons make the biceps muscle contract, causing you to withdraw your hand even before your brain becomes aware of the pain.

The Anatomy of a Neuron

Neurons are difficult to study because they are small, transparent, and structurally complex. So when Camillo Golgi discovered that some neurons become visible when preserved nervous tissue is treated with a silver nitrate solution, his finding was a major advance. The year was 1873.

Through the early decades of the twentieth century, the work of Golgi and Santiago Ramón y Cajal revealed several important points about the anatomy of neurons. Most neurons have the same three parts, shown in Figure 43.2a:

1. A **cell body**, or **soma**, which contains the nucleus;
2. A highly branched group of relatively short projections called **dendrites**; and
3. One or more relatively long projections called **axons**.

Dendrites are rarely more than 2 mm long, but axons can be over a meter in length. The number of dendrites and their arrangement vary greatly from neuron to neuron.

Dendrites receive signals from the axons of other neurons; a neuron's axon sends signals to the dendrites and cell bodies of other neurons (Figure 43.2b). In short, dendrites and cell bodies receive signals; axons pass them on. Most neurons receive input from and send signals to many other neurons at connections called synapses, forming complex networks of neurons.

How do neurons produce the electrical signals that they transmit?

An Introduction to Membrane Potentials

Ions carry an electric charge. In virtually all cells, the cytoplasm and extracellular fluids adjacent to the plasma membrane contain unequal distributions of ions. As a result, there is a difference in charge across the membrane.

A difference in charge between any two points creates an **electrical potential**, or **voltage**. When an electrical potential exists across a plasma membrane, the separation of charges is called a **membrane potential**. If there is a large difference in charge across the membrane, the membrane potential is large.

Membrane potentials refer only to a separation of charge immediately adjacent to the plasma membrane, on either side of the membrane. Even if there is a large membrane potential, there may be no charge separation slightly farther from the membrane.

Units and Signs Membrane potentials are measured in units called millivolts. The **volt (V)** is the standard unit of electrical potential, and a **millivolt (mV)** is 1/1000 of a volt. As a comparison, an AA battery that you buy in a store has an electrical potential of 1500 mV between its positive and negative terminals. In resting neurons, the difference in electrical potential across a plasma membrane typically ranges from 65 to 80 mV.

By convention, membrane potentials are always expressed in terms of inside relative to outside, and the outside value is defined as 0 mV. Because there are usually more negatively charged ions and fewer positively charged ions on the inside surface of a membrane relative to its outside surface, membrane potentials are usually negative (for example, −65 mV).

Electrical Potential, Electric Currents, and Electrical Gradients Membrane potentials are a form of potential energy. Recall that potential energy is energy based on the position of matter.

To convince yourself that ions have potential energy when a membrane potential exists, consider what would happen if the membrane were removed. Ions would spontaneously move from the region of like charge to the region of unlike charge—causing a flow of charge. This flow of charge, which is called an **electric current**, would occur because like charges repel and unlike charges attract.

However, charge is not the only contributor to the potential energy of a membrane potential. Ions also have different concentrations across membranes. Therefore, a membrane potential also includes energy stored as the concentration gradients of charged ions on the two sides of the membrane. Recall that the combination of an electrical gradient and a concentration gradient is an **electrochemical gradient** (Ch. 6, Section 6.4).

What do all these facts have to do with neuron function? Neurons use the electrochemical gradient of ions across their membranes to power the signals that allow neurons to communicate with one another and with other cells.

How Is the Resting Potential Maintained?

When a neuron is not communicating with other cells, the difference in charge across its membrane is called the **resting potential**. To understand why the resting potential exists, consider the distribution of the various ions on the two sides of the neuron's plasma membrane, shown in Figure 43.3:

- The interior side of the membrane has relatively low concentrations of sodium (Na^+) and chloride (Cl^-) ions, a relatively high concentration of potassium ions (K^+), and some organic anions—proteins, amino acids, and other organic molecules that have dropped one or more protons and thus carry a negative charge.
- In the extracellular fluid, sodium and chloride ions predominate.

If each type of ion diffused across the membrane in accordance with its concentration gradient, organic anions and K^+ would leave the cell, while Na^+ and Cl^- would enter.

Ions cannot cross phospholipid bilayers readily, however. They cross plasma membranes mainly in one of three ways (Ch. 6, Section 6.4): by **(1)** primary or **(2)** secondary active transport against a gradient, or **(3)** by diffusion along a gradient through an **ion channel**—a protein that forms a pore in the membrane through which specific ions can diffuse. Let's examine how these types of transport are involved in establishing and maintaining a neuron's resting potential.

The sodium–potassium pump, Na^+/K^+-ATPase, actively pumps Na^+ out of the cell and K^+ into the cell. More specifically, the energy gained by the sodium–potassium pump when it receives a phosphate group from one ATP is used to move three

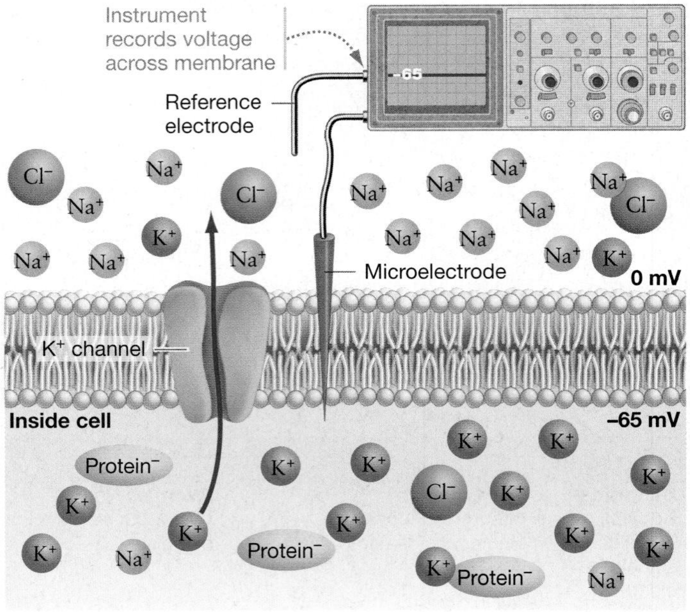

Figure 43.3 **Neurons Have a Resting Potential.** In resting neurons, the membrane is selectively permeable to K^+. As K^+ leaves the cell along its concentration gradient, the inside of the membrane becomes negatively charged relative to the outside. To measure a neuron's membrane potential, researchers use two electrodes: a reference electrode outside the cell and a microelectrode inserted into the cell.

✔ Will K^+ continue to leave the cell indefinitely? Explain why or why not.

Na^+ ions out of the cell and two K^+ ions into the cell (see Ch. 6, Figure 6.24). Active transport via Na^+/K^+-ATPase ensures that eventually the concentration of K^+ is much higher on the inside of the plasma membrane than the outside, while the concentration of Na^+ is lower on the inside than the outside.

At rest, the plasma membrane of a neuron is relatively impermeable to most cations. However, neurons have a relatively high number of potassium channels, called K^+ **leak channels**, which allow K^+ to leak across the membrane.

The K^+ concentration gradient established by Na^+/K^+-ATPase (high K^+ inside the cell, low K^+ outside) favors the net diffusion of K^+ out of the cell through leak channels. As K^+ moves from the interior of the cell to the exterior, the inside of the membrane becomes more and more negatively charged relative to the outside. This buildup of negative charge inside the membrane begins to attract K^+ and counteract the concentration gradient that had favored the movement of K^+ out.

These counteracting influences cause the membrane to reach a voltage at which equilibrium exists between the concentration gradient that favors movement of K^+ out and the electrical gradient that favors movement of K^+ in. At this voltage, called the **equilibrium potential** for K^+, there is no longer a net movement of K^+.

Although Cl^- and Na^+cross the plasma membrane much less readily than does K^+, some movement of these ions also occurs through a small number of leak channels that are selective for each of them. As a result, each type of ion has an equilibrium potential. The membrane as a whole has a membrane potential that combines the effects of the individual ions.

To summarize, the neuron has a negative resting membrane potential because Na^+/K^+-ATPase creates a concentration gradient across the cell that favors the exit of K^+ through leak channels. The resting potential represents energy stored as concentration and electrical gradients of specific ions. ✔ **If you understand this concept, you should be able to predict what would happen to the membrane potential if Na^+ or K^+ were allowed to diffuse freely across the membrane.**

Using Electrodes to Measure Membrane Potentials

During the 1940s, A. L. Hodgkin and Andrew Huxley focused on what has become a classic model system in the study of electrical signaling: the "giant" axons of squid.

Squid live in the ocean and are preyed on by fishes and whales. When a squid is threatened, electrical signals travel down the giant axons to muscle cells, causing the cells to contract. Their contraction expels water from a cavity in the squid's body. As a result, the squid lurches away from danger by jet propulsion (Ch. 31, Section 31.2). This extremely rapid response to threats is an adaptation that helps squid avoid predation.

Hodgkin and Huxley decided to study the squid giant axon simply because it is so large. Many of the axons found in humans are a mere 2 μm in diameter, but the squid giant axon is about 500 μm in diameter—large enough that the researchers could insert a wire down its length. By measuring the voltage difference between the wire inside the axon and another wire outside, Hodgkin and Huxley could record the membrane potential of the axon.

Later researchers developed glass microelectrodes with tips tiny enough to penetrate smaller axons and other parts of neurons. This development made it possible to record membrane potentials from a wide variety of neurons in many animal species.

Recordings made with electrodes inserted into neurons revealed that the resting potential is regularly disrupted by an event called the action potential when a neuron is stimulated.

What Is an Action Potential?

An **action potential** is a rapid, temporary change in a membrane potential. When stimulated, neurons produce action potentials that allow them to communicate with other neurons, muscles, or glands.

Although Hodgkin and Huxley initially studied the action potential in the squid giant axon, subsequent work has shown that action potentials have the same general characteristics in all species and in all types of neurons.

A Three-Phase Signal Figure 43.4 shows the form of the action potential that Hodgkin and Huxley recorded from the squid giant axon. The action potential has three distinct phases:

1. **Depolarization** of the membrane. In its resting state, a membrane is said to be polarized because the charges on the two sides are different. Depolarization means that the membrane becomes less polarized than before. During the depolarization phase, the membrane potential changes from highly negative, crosses zero, and then is briefly positive.
2. A rapid **repolarization**, which changes the membrane potential back to negative.
3. A **hyperpolarization** phase, when the membrane potential is slightly more negative than the resting potential.

Together, all three phases of an action potential occur within a few milliseconds.

For an action potential to begin in a squid giant axon, the membrane must depolarize from the resting potential (−65 mV in Figure 43.4) to about −55 mV. If the membrane depolarizes less than that, an action potential does not occur. But if this **threshold potential** is reached, certain channels in the axon membrane open, allowing ions to rush into the axon along their electrochemical gradients. The inside of the membrane becomes less negative and then positive with respect to the outside.

When the membrane potential reaches about +40 mV, an abrupt change occurs and the repolarization phase begins. The change is triggered by the closing of certain ion channels and the opening of other ion channels in the membrane.

To summarize, an action potential occurs because specific ion channels in the plasma membrane open or close in response

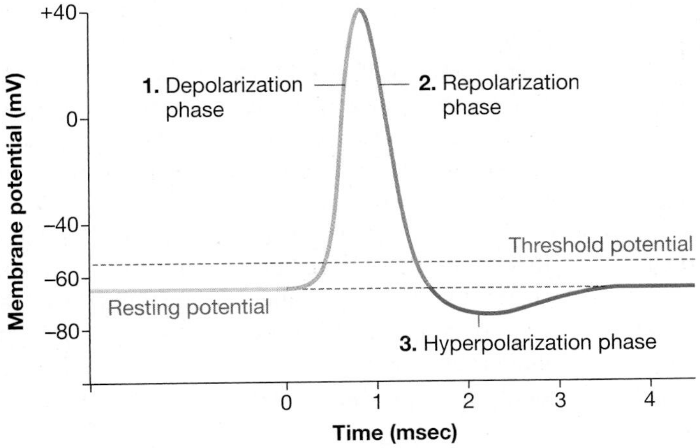

Figure 43.4 **Action Potentials Have the Same General Shape.** An action potential is a three-phase change in membrane potential that occurs the same way every time.

to changes in membrane voltage. An action potential always has the same three-phase form, even though the values of the resting potential, threshold potential, and peak depolarization may vary among species or even among types of neurons in one species.

An "All-or-None" Signal That Propagates Hodgkin and Huxley made other important observations about the action potential. Besides being fast and having three distinct phases, it is an all-or-none event.

- There is no such thing as a partial action potential.
- All action potentials for a given neuron are identical in magnitude and duration.
- Action potentials are always propagated down the entire length of the axon.

For example, when an action potential was recorded at a particular point on a squid axon, an action potential that was identical in shape and magnitude would be observed farther down the same axon soon afterward. Neurons are said to have **excitable membranes** because they can generate action potentials that propagate rapidly along the length of their axons.

Taken together, these observations suggested a mechanism for electrical signaling. In the nervous system, information is coded in the form of action potentials that propagate along axons. In the squid giant axon, action potentials signal muscles to contract. As a result, the animal escapes from danger.

CHECK YOUR UNDERSTANDING

✔ **If you understood this section, you should be able to ...**

1. **MODEL** Make a diagram similar to Figure 43.1 showing the neurons responsible for sensing, integrating, and responding to stepping on a nail. Label the key components in your drawing.
2. Predict what would happen to the resting potential of a squid axon if potassium leak channels were blocked.
3. Explain why the first phase of an action potential is called depolarization and why the second phase is called repolarization.

Answers are available in Appendix A.

43.2 Dissecting the Action Potential

After you complete this section, you should be able to ...

- Explain the steps in an action potential.

Of Na^+, Cl^-, and K^+, which ion or ions are most important in the electric currents that form the action potential? Are different ions responsible for the depolarization and repolarization phases of the event?

Hodgkin made a crucial start in answering this question when he realized that the peak of the action potential in the squid giant axon, about +40 mV, was close to the equilibrium potential for Na^+—recall that this is the voltage at which equilibrium exists between the concentration gradient that favors movement of Na^+ into the cell and the electrical gradient that favors movement of Na^+ out of the cell. If sodium channels opened early in the action potential, then Na^+ should flow into the neuron until the membrane potential reached this equilibrium potential. How could this hypothesis be tested?

Distinct Ion Currents Are Responsible for Depolarization and Repolarization

To understand the currents responsible for the action potential, Hodgkin and Huxley recorded electrical activity in squid giant axons that were bathed in solutions containing different concentrations of ions.

- Removing Na^+ from the solution surrounding the axon abolished the production of action potentials.
- When axons were bathed in solutions with various concentrations of Na^+, the peak of the action potential paralleled the concentration of Na^+. If the Na^+ concentration outside the axon was high, the peak was high. If the Na^+ concentration outside the axon was low, the peak was low. (Note that this change in the peak of the action potential occurred because the scientists altered the Na^+ concentration gradient. In a living animal, action potentials in the same neuron always have the same magnitude.)

These experiments furnished strong support for the hypothesis that the action potential begins when Na^+ flows into the neuron. In other words, sodium ions are responsible for the depolarization phase.

What happens during the repolarization phase? Using radioactive K^+, Hodgkin and Huxley showed that there was a strong flow of potassium ions out of the cell during the repolarization phase.

The action potential consists of a strong inward flow of sodium ions followed by a strong outward flow of potassium ions. ✔ **If you understand this concept, you should be able to add labels that read, "Sodium channels open, Na^+ enters" and "Potassium channels open, K^+ leaves" to Figure 43.4.**

How Do Voltage-Gated Channels Work?

The action potential depends on **voltage-gated channels**—membrane proteins that open and close in response to changes in membrane voltage. The shape of a voltage-gated channel, and thus its ability to admit ions, changes in response to the charges present at the inside of the membrane. Figure 43.5 shows a simple model of how voltage-gated sodium channels change shape as the membrane potential changes.

To confirm that voltage-gated channels exist, Hodgkin and Huxley used a technique called voltage clamping. **Voltage clamping** allows researchers to hold the voltage of a cell's plasma membrane at any desired value and record the electrical currents that occur at that voltage. When the researchers held the squid giant axon membrane at various voltages, different currents

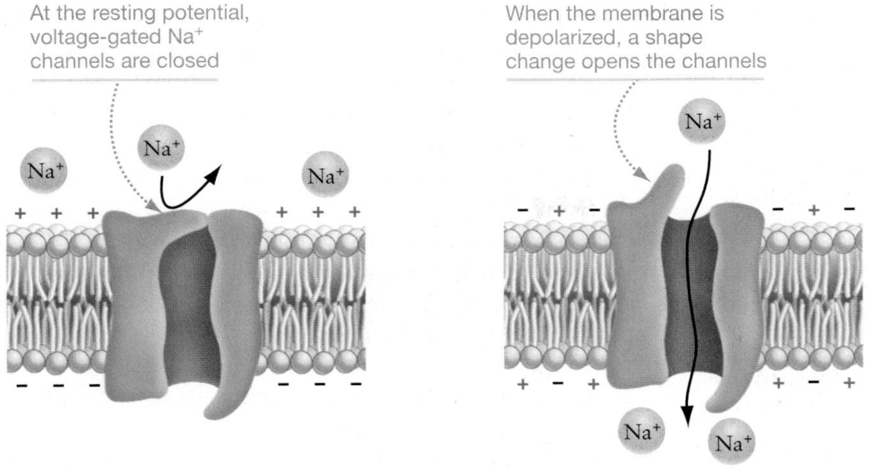

Figure 43.5 The Shape of a Voltage-Gated Channel Depends on the Membrane Potential. Changes in the shape of voltage-gated channels are responsible for changes in a neuron's permeability to ions.

resulted. These experiments supported the hypothesis that the membrane contained ion channels whose behavior depends on voltage.

Patch Clamping and Studies of Single Channels Studying individual ion channels became possible in the 1980s when Erwin Neher and Bert Sakmann perfected a variation of voltage clamping known as **patch clamping**. As Figure 43.6 shows, the researchers touched a tiny patch of axon membrane with an extremely fine-tipped microelectrode and applied suction to capture a single ion channel within the electrode's tip.

Using this technique, Neher and Sakmann were able to document for the first time the currents that flowed through individual channels. They showed that different ion channels behave differently:

- *Voltage-gated channels* are either open or closed. There is no gradation in channel behavior. This conclusion is based on the shape of the recorded current: Current flow starts and stops instantly, and the size of the current is always the same.
- *Sodium channels* open quickly when the membrane is depolarized. They stay open for about a millisecond, close, and remain inactive for 1 to 2 msec. That explains why the cell can repolarize: Once the sodium channels close, there is a lag before they can open again.
- *Potassium channels* open with a delay during depolarization. They continue to flip open and closed until the membrane repolarizes. Once the membrane returns to the resting potential, these channels remain closed.

Positive Feedback Occurs during Depolarization More detailed experiments on Na^+ channels also explained why the action potential is an all-or-none event. The key observation was that Na^+ channels become more likely to open as a membrane depolarizes. As a result, an initial depolarization leads to the opening of more Na^+ channels, which depolarizes the membrane further, which leads to the opening of additional Na^+ channels.

The opening of Na^+ channels exemplifies **positive feedback**—meaning that the occurrence of an event makes the same event more likely to recur. Positive feedback is rare in organisms: It cannot be employed as a regulatory mechanism under many circumstances, because it often leads to uncontrolled events. The opening of Na^+ channels during an action potential is one of the few examples known.

Using Neurotoxins to Study Channels and Currents In addition to using voltage clamping and patch clamping, researchers have used neurotoxins to explore the dynamics of voltage-gated channels. **Neurotoxins** are poisons that affect neuron function—often resulting in convulsions, paralysis, or unconsciousness. They come from sources as diverse as venomous snakes and foxglove plants.

For example, when biologists treated giant axons from lobsters with tetrodotoxin from puffer fish, they found that the resting potential in treated neurons was normal, but action potentials were abolished. More specifically, the flow of K^+ out of the cell was normal but the influx of Na^+ was wiped out. Researchers concluded that tetrodotoxin blocks the voltage-gated Na^+ channel, probably by binding to a specific site on the channel protein.

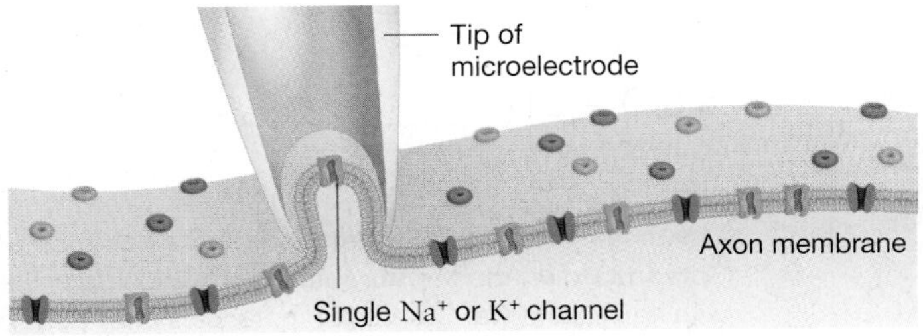

Figure 43.6 Patch Clamping Provides Insights into the Behavior of Ion Channels. In patch clamping, researchers use extremely fine-tipped microelectrodes to record electric currents through individual ion channels.

How Is the Action Potential Propagated?

To explain how action potentials propagate down an axon, Hodgkin and Huxley suggested the model that is illustrated in Figure 43.7a.

Step 1 At the start of an action potential, the influx of Na^+ attracts intracellular negative charges and repels positive charges, causing cations to spread away from the sodium channels.

Step 2 As cations are pushed farther from the initial sodium channels, they depolarize adjacent "downstream" portions of the membrane.

Step 3 Downstream voltage-gated Na^+ channels open when the adjacent membrane reaches threshold, resulting in a new action potential there.

In this way, new action potentials are continuously generated at adjacent areas of the plasma membrane (Figure 43.7b). Action potentials are all or none because new action potentials are always generated along the entire length of the axon.

Why don't action potentials propagate back up the axon in the direction of the cell body? To answer this question, recall that once Na^+ channels have opened and closed, they are less likely to open again for a short period. This condition is known as the **refractory** state. Action potentials propagate in one direction only because "upstream" sodium channels, in the direction of the cell body, are in the refractory state.

The hyperpolarization phase, in which the membrane is more negative than the resting potential, also keeps the positive charges that spread upstream from triggering an action potential in that direction. During the hyperpolarization phase, a much stronger stimulus would be necessary to raise the membrane potential to the threshold potential.

Axon Diameter Affects Propagation Speed Understanding how the action potential propagates helped researchers explain why the squid's axons are so large. When sodium ions enter the axon interior at the start of an action potential, they repel intracellular cations, causing them to spread along the inside of the membrane. Cations moving down axons with larger diameters meet less resistance than those moving down narrow axons. As a result, the charge spreads along the membrane more quickly.

The upshot is that the squid giant axon and other large-diameter axons transmit action potentials much more quickly than small axons can. The squid giant axon's large size is an adaptation that makes particularly rapid signaling possible.

Myelination Affects Propagation Speed Relatively few vertebrates have giant axons. Instead, vertebrates—and some invertebrates—have another adaptation that increases the speed

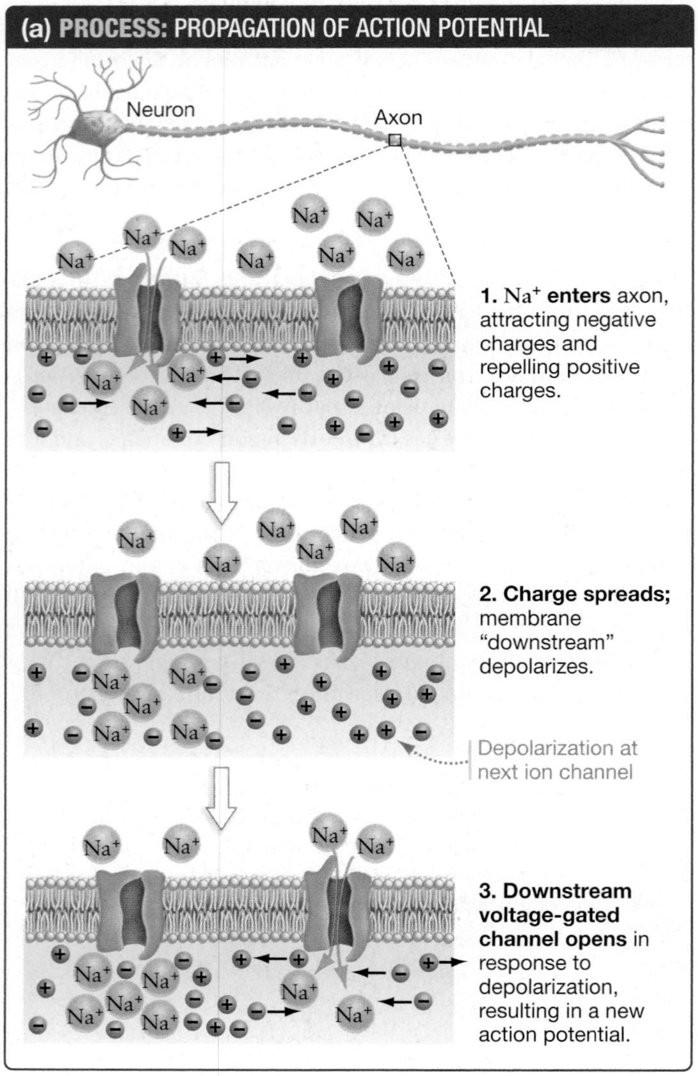

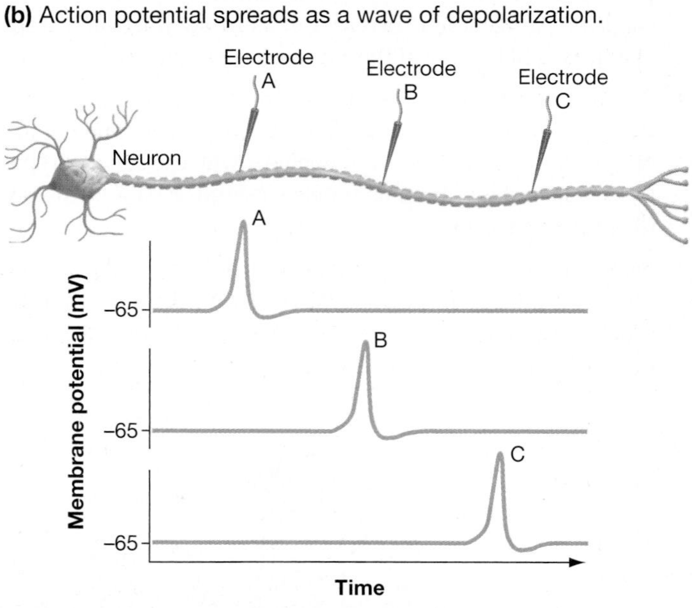

Figure 43.7 **Action Potentials Propagate because Charge Spreads Down the Membrane. (a)** An action potential starts with an influx of Na^+. **(b)** Action potentials propagate down the axon without getting smaller, because new action potentials are continuously generated along the axon membrane.

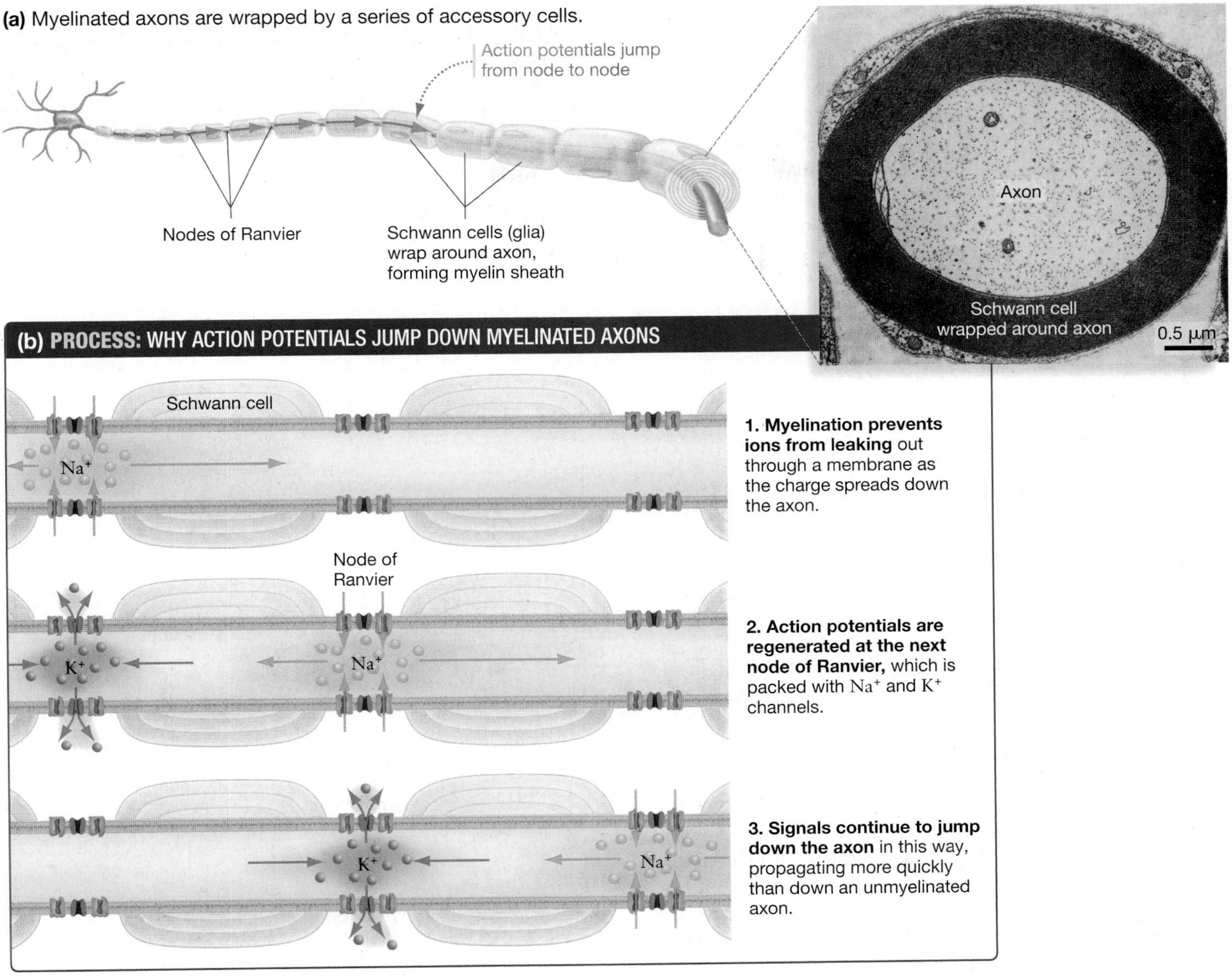

Figure 43.8 Action Potentials Propagate Quickly in Myelinated Axons.

of action potential propagation: specialized accessory cells whose membranes wrap around the axons of certain neurons.

In the central nervous system, these accessory cells are **oligodendrocytes**. In the peripheral nervous system, described in Section 43.4, they are **Schwann cells** (Figure 43.8a). Oligodendrocytes and Schwann cells are two examples of **glia**, which are nervous system cells that support neurons.

When oligodendrocytes or Schwann cells wrap around an axon, they form a **myelin sheath**, which acts as a type of electrical insulation. As intracellular cations spread down an axon during the propagation of action potentials, the myelin sheath prevents ions from leaking back out across the plasma membrane of the axon.

Consequently, the cations moving down the membrane are able to spread until they reach a gap in the myelin sheath, called a **node of Ranvier** (Figure 43.8b). The nodes have a dense concentration of voltage-gated Na^+ and K^+ channels, so new action potentials can be generated at the nodes.

Action potentials "jump" from node to node down a myelinated axon much more rapidly than they can move down an unmyelinated axon of the same diameter. In an unmyelinated axon, voltage-gated sodium and potassium channels are found all along its length, and action potentials propagate continuously down the axon. Myelination is an adaptation that makes rapid transmission of electrical signals possible in axons that have a small diameter. The effect is dramatic: Myelination can increase the speed of action potential propagation in neurons 15-fold.

To appreciate the importance of myelination, consider what happens when it is disrupted. If myelin degenerates, the transmission of action potentials slows considerably. The autoimmune disease **multiple sclerosis (MS)** develops when the immune system targets oligodendrocytes, destroying myelin in the CNS. As damage to myelin increases, electrical signaling becomes more impaired, affecting coordination among neurons and causing muscles to weaken. The symptoms of MS are highly variable; in severe cases, the disease can be crippling.

Once action potentials have been initiated in a neuron, they are propagated down the entire length of the axon. What happens when action potentials reach the end of the axon?

CHECK YOUR UNDERSTANDING

✔ **If you understood this section, you should be able to ...**

1. Explain why the action potential is an all-or-none phenomenon.
2. Explain why action potentials can propagate as quickly in the unmyelinated squid giant axon as in small, myelinated axons in vertebrates.

Answers are available in Appendix A.

43.3 The Synapse

After you complete this section, you should be able to ...

- Describe what happens in a synapse when neurons communicate.

Ramón y Cajal maintained that the plasma membrane of each neuron is separated from those of adjacent cells. This hypothesis was confirmed in the 1950s, when images from transmission electron microscopes revealed that most neurons are separated from one another at their junctions by tiny spaces called **synapses**. Despite this separation, neurons can communicate with one another. Therefore, some indirect mechanism must exist that transmits signals from cell to cell, across their plasma membranes.

In the 1920s, Otto Loewi showed that this indirect mechanism involves **neurotransmitters**. Neurotransmitters are chemical messengers that transmit information from one neuron to another neuron, or from a neuron to a target cell in a muscle or gland.

Loewi knew that signals from the vagus nerve slow the heart rate. To test the hypothesis that the signal from nerve to heart muscle is delivered by a chemical, he performed the experiment diagrammed in Figure 43.9.

First, Loewi isolated the vagus nerve and heart of a frog. As predicted, the heart rate slowed when he stimulated the vagus nerve electrically. Next, he took the solution that bathed the first heart and applied it to another frog heart—without stimulating the vagus nerve to that heart. He found that the second heart's rate slowed as well.

This result provided strong evidence for the chemical transmission of signals by the nervous system. The vagus nerve had released a neurotransmitter.

Synapse Structure and Neurotransmitter Release

As Figure 43.10 shows, **(1)** the membranes of axons and the cells they communicate with are separated by a tiny space, the **synaptic cleft**, and **(2)** the ends of axons contain numerous sac-like structures, called **synaptic vesicles**. Synaptic vesicles were hypothesized to be storage sites for neurotransmitters.

At a synapse, the neuron that contains the synaptic vesicles is called the **presynaptic neuron**, and the cell on the other side of the synaptic cleft is called the **postsynaptic cell**.

RESEARCH

QUESTION: How is information transferred from one neuron to another?

HYPOTHESIS: Molecules called neurotransmitters carry information from one neuron to the next.

NULL HYPOTHESIS: Information is not transferred between neurons in the form of molecules.

EXPERIMENTAL SETUP:

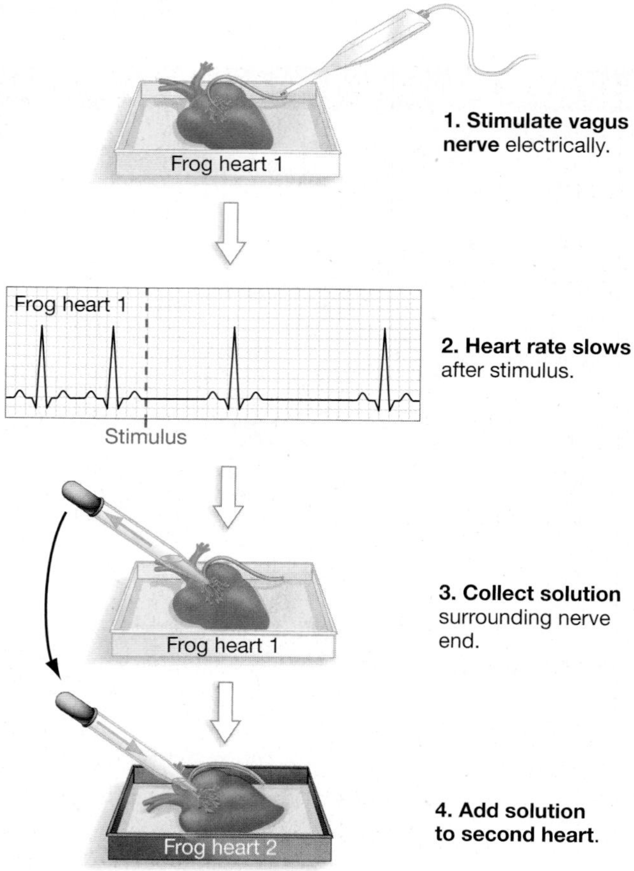

PREDICTION OF HYPOTHESIS: The rate of the second heart will slow.

PREDICTION OF NULL HYPOTHESIS: There will be no change in the second heart's rate.

RESULTS:

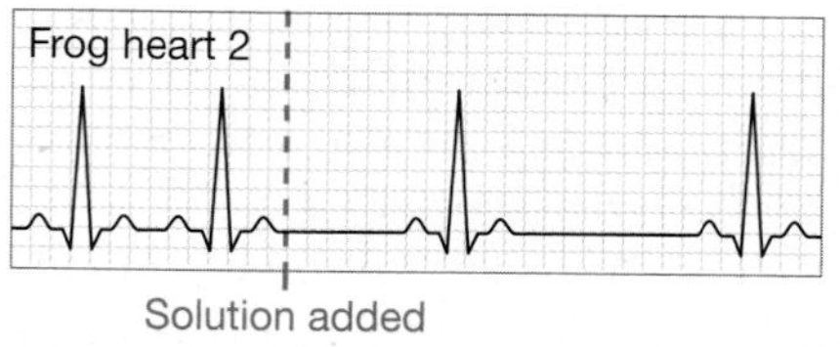

Heart rate slows after solution is added.

CONCLUSION: The vagus nerve releases molecules that slow heart rate. Neurotransmitters carry information.

Figure 43.9 Experimental Evidence for the Existence of Neurotransmitters.

SOURCE: O. Loewi. 1921. Über humorale Übertragbarkeit der Herznervenwirkung. *Pflügers Archiv European Journal of Physiology* 189: 239–242.

✔ **PROCESS OF SCIENCE** What would be an appropriate control for this experiment?

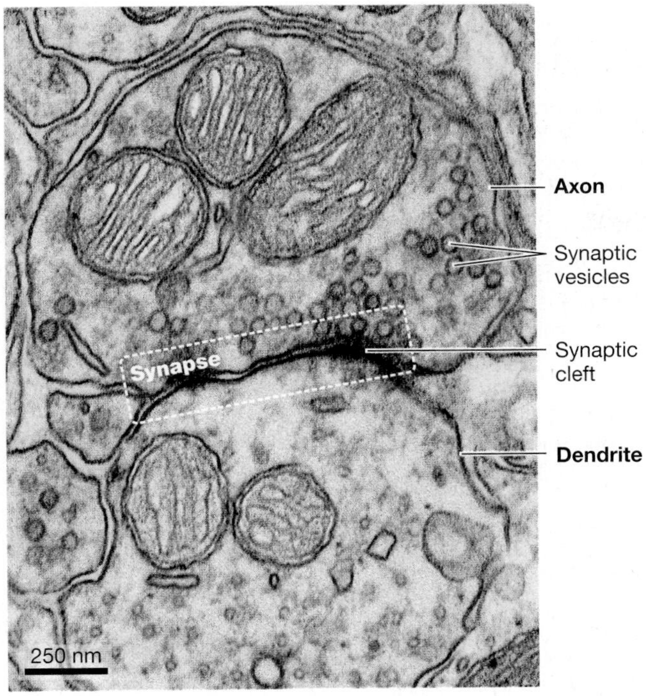

Figure 43.10 **Synaptic Vesicles Cluster Near Synapses.** A cross section of the site where an axon meets a dendrite.

Anatomical observations such as these, combined with chemical studies of the synapse, led to the model of synaptic transmission illustrated in Figure 43.11.

Step 1 An action potential arrives at the end of the axon, near the synaptic cleft.

Step 2 The depolarization created by the action potential opens voltage-gated calcium channels located near the synapse, in the plasma membrane of the presynaptic neuron. The electrochemical gradient for Ca^{2+} results in an inflow of calcium ions through the open channels.

Step 3 In response to the increased calcium ion concentration inside the axon, synaptic vesicles fuse with the presynaptic membrane and release neurotransmitters into the synaptic cleft. The delivery of neurotransmitters into the cleft is an example of exocytosis (Ch. 7, Section 7.5).

Step 4 Neurotransmitters bind to receptors on the postsynaptic cell. Thus, each neurotransmitter functions as a **ligand**, a molecule that binds to a specific site on a receptor molecule. Neurotransmitter-receptor binding causes ion channels in the postsynaptic membrane to open, leading to a change in the membrane potential of the cell. The combined effect on membrane potential of many neurotransmitters binding may trigger an action potential in the postsynaptic cell.

Step 5 The response ends when the neurotransmitters unbind from their receptors, causing the ion channels in the postsynaptic membrane to close.

After an action potential ceases, the neurotransmitters diffuse out of the synaptic cleft, and then are broken down or taken back up by the presynaptic cell. Let's take a closer look at how neurotransmitters allow neurons to communicate.

What Do Neurotransmitters Do?

Researchers can look for neurotransmitters by stimulating a neuron, collecting the molecules that are released, and analyzing them chemically. To find the receptor for a particular neurotransmitter, researchers can attach a radioactive atom or other type of label to the neurotransmitter and add the labeled

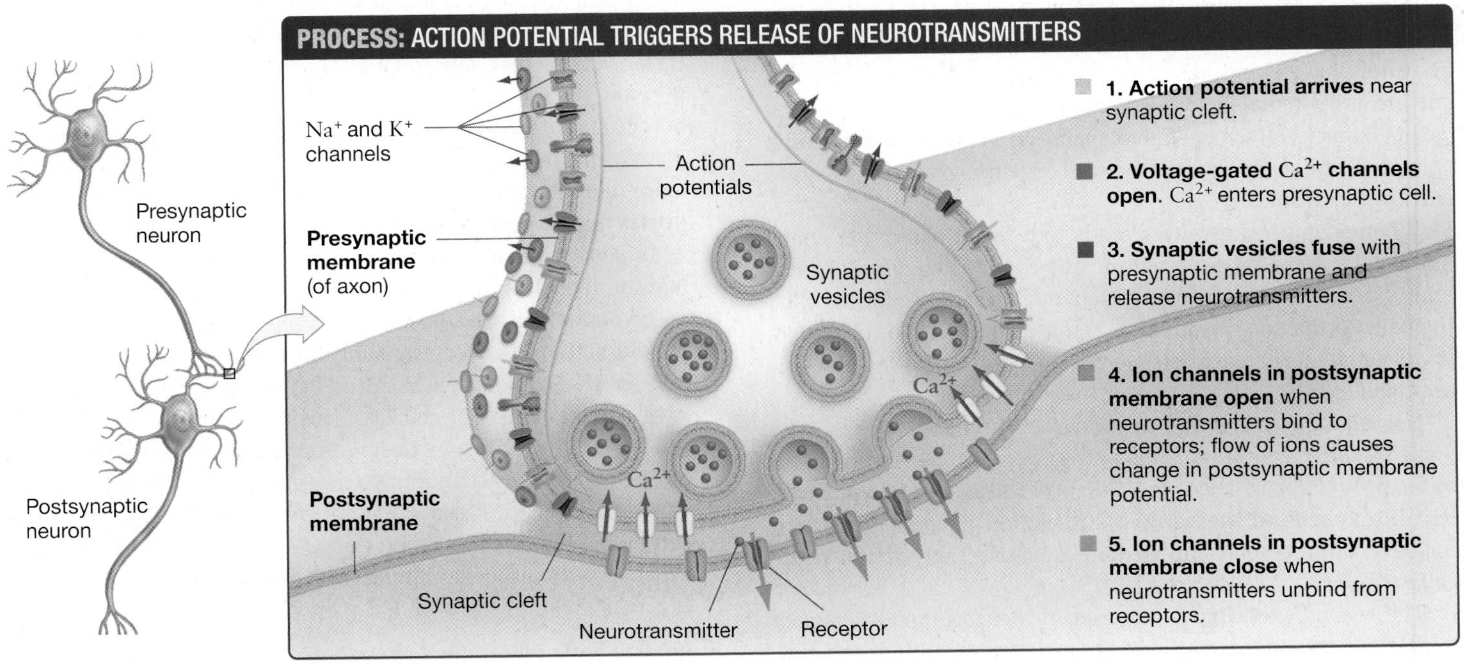

Figure 43.11 **Neurons Meet and Transfer Information at Synapses.** The sequence of events that occurs when an action potential arrives at a synapse.

Table 43.1 Types of Neurotransmitters and Their Characteristics

Neurotransmitter	Site of Action	Action*	Drugs That Interfere†
Acetylcholine	Neuromuscular junction, some CNS pathways	Excitatory (inhibitory in some parasympathetic neurons)	• Botulism toxin blocks release • Black widow spider venom increases, then eliminates, release • Alpha-bungarotoxin (in some snake venoms) binds to and blocks receptor
Monoamines			
Norepinephrine	Sympathetic neurons, some CNS pathways	Excitatory or inhibitory	• Ritalin (used for attention deficit hyperactivity disorder) increases release • Some antidepressants prevent reuptake
Dopamine	Many CNS pathways	Excitatory or modulatory	• Cocaine and amphetamine prevent reuptake
Serotonin	Many CNS pathways	Inhibitory or modulatory	• MDMA (ecstasy) increases level for several hours, then reduces level for days to weeks
Amino Acids			
Glutamate	Many CNS pathways	Excitatory	• PCP (angel dust) blocks receptor
Gamma-aminobutyric acid (GABA)	Some CNS pathways	Inhibitory	• Ethanol mimics response to GABA
Peptides			
Endorphins, enkephalins, substance P	Sensory pathways (pain)	Excitatory, modulatory, or inhibitory	• Natural and synthetic opiates (e.g., opium, heroin, morphine, and others) bind to and stimulate receptors • Naloxone blocks receptors; used to treat opiate drug overdose

*Excitatory neurotransmitters make action potentials more likely in postsynaptic cells; inhibitory neurotransmitters make action potentials less likely; modulatory neurotransmitters modify the response at other synapses.

†Drugs that prevent reuptake of neurotransmitters increase their activity.

neurotransmitter to nervous tissue. Receptors that are bound to the labeled neurotransmitter can then be isolated and analyzed.

Using techniques such as these, biologists have discovered and characterized a wide array of neurotransmitters and receptors. Some of them are listed in **Table 43.1**.

By patch-clamping receptors, biologists confirmed that many receptors are also ion channels. Such channels, called **ligand-gated channels**, have a binding site for a specific ligand on the part of the channel protein that faces the synaptic cleft. In contrast to voltage-gated channels, which open in response to a change in membrane voltage, ligand-gated channels open in response to binding by a specific neurotransmitter.

When a neurotransmitter binds to a ligand-gated ion channel in the postsynaptic membrane, the channel opens and allows ions to diffuse along their electrochemical gradient. In this way, the neurotransmitter's chemical signal is transduced to an electrical signal—a change in the membrane potential of the postsynaptic cell.

Not all neurotransmitters bind to ion channels, however. Some bind to receptors that activate enzymes whose action leads to the production of a second messenger in the postsynaptic cell. Recall that **second messengers** are chemical signals produced inside a cell in response to a chemical signal that arrives at the cell surface (Ch. 11, Section 11.3).

The second messengers induced by neurotransmitters may trigger changes in enzyme activity, gene transcription, or membrane potential. (In Chapter 46, you will explore the cellular role of second messengers in detail.)

Postsynaptic Potentials

What happens when a neurotransmitter binds to a ligand-gated ion channel in the postsynaptic cell?

Ligand-gated sodium channels on the membranes of dendrites are in particularly high concentration near synapses. When neurotransmitters bind, these channels open and allow cations such as sodium to enter the cell, causing depolarization (**Figure 43.12a**). In most cases, depolarization makes an action potential in the postsynaptic cell more likely. Changes in the membrane potential of a postsynaptic cell that make the cell more likely to produce an action potential are called **excitatory postsynaptic potentials (EPSPs)**.

If neurotransmitter-receptor binding leads to an outflow of potassium ions or an inflow of chloride ions or other anions in the postsynaptic cell, the postsynaptic membrane hyperpolarizes—making action potentials less likely to occur in the postsynaptic cell (**Figure 43.12b**). Changes in the membrane potential of a postsynaptic cell that make the cell less likely to produce an action potential are called **inhibitory postsynaptic potentials (IPSPs)**. If an EPSP and an IPSP occur at the same time in the same postsynaptic cell, they may cancel each other out (**Figure 43.12c**). Synapses also can be modulatory—meaning that their activity modifies a postsynaptic cell's response to input from other synapses.

Postsynaptic Potentials Are Graded It is critical to realize that, unlike action potentials, EPSPs and IPSPs are not all-or-none events. Instead, they are graded in magnitude.

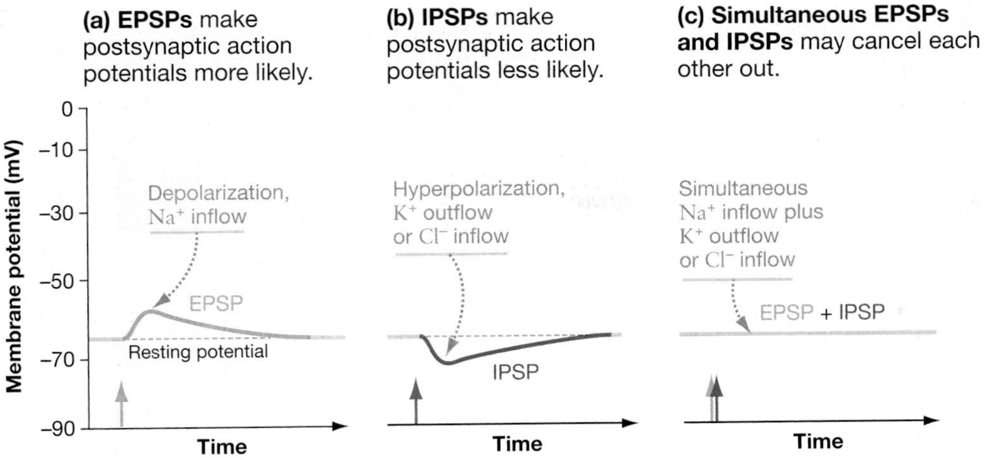

Figure 43.12 Events at the Synapse May Depolarize or Hyperpolarize the Postsynaptic Membrane. These recordings show what happens to the membrane potential of a postsynaptic neuron in response to the release of neurotransmitters that cause **(a)** depolarization or **(b)** hyperpolarization. In **(c)**, the simultaneous release of both neurotransmitters results in no change, because the depolarization and hyperpolarization cancel each other out.

The magnitude of an EPSP or IPSP depends on the amount of neurotransmitter that is released at the synapse at a given time. A higher concentration of neurotransmitter in the synaptic cleft leads to a larger EPSP or IPSP. Both types of signal are short lived because the effects cease when neurotransmitters unbind from their receptors and are recycled.

Anything that changes the amount or life span of neurotransmitters in the synaptic cleft may alter the normal functioning of neurons. The drugs cocaine and amphetamine, for example, exert their effects by inhibiting the uptake and recycling of the neurotransmitter dopamine (see Table 43.1).

Summation and Threshold How do EPSPs and IPSPs affect the postsynaptic cell? As Figure 43.13a shows, the dendrites and cell body of a neuron typically make hundreds or thousands of synapses with other neurons, as part of a neural network. Therefore, the postsynaptic neuron may receive a great many EPSPs and IPSPs at any instant.

If an IPSP and EPSP occur close together in space or time, the changes in membrane potential tend to cancel each other out. But if several EPSPs occur close together in space or time, they sum and make the neuron more likely to reach threshold and fire an action potential (Figure 43.13b). The additive nature of postsynaptic potentials is termed **summation**.

The sodium channels that trigger action potentials in a neuron are typically located near the place where the axon emerges from the cell body, a site called the **axon hillock** (see Figure 43.13a). As IPSPs and EPSPs are received throughout the dendrites and cell body, charge spreads to the axon hillock. If the membrane at the axon hillock depolarizes to the threshold potential, enough voltage-gated sodium channels open to trigger positive feedback and an action potential. Once an action potential starts at the axon hillock, it propagates down the axon to the next synapse.

Summation is critically important. Because neurons receive input from many synapses, and because IPSPs and EPSPs sum, information in the form of electrical signals is modified at the postsynaptic neuron before being passed along. An action potential in a presynaptic neuron does not always lead to an action potential in a postsynaptic neuron—the response by the postsynaptic cell depends on the information it receives from a wide array of neurons.

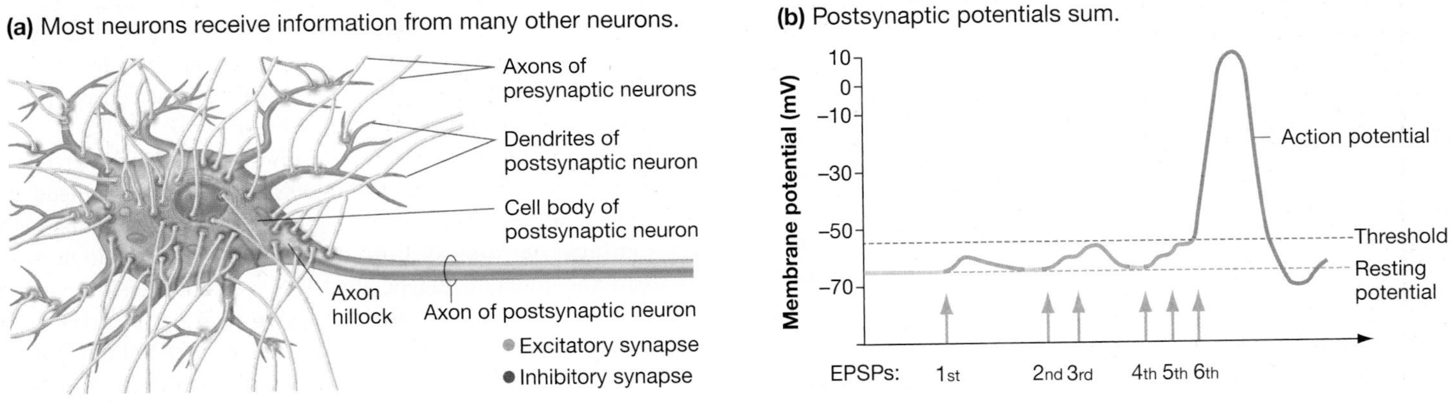

Figure 43.13 Neurons Integrate Information from Many Synapses. (a) The dendrites and cell body of a neuron typically receive signals from hundreds or thousands of other neurons. **(b)** When presynaptic action potentials arrive close together in time, the postsynaptic potentials summate. Here the first EPSP is insufficient to generate an action potential; two EPSPs arriving close together undergo summation but do not reach the threshold; the summation of three EPSPs arriving close together does reach the threshold. This example is simplified—in reality, hundreds or thousands of IPSPs and EPSPs sum to determine action potential frequency.

CHECK YOUR UNDERSTANDING

✔ **If you understood this section, you should be able to ...**

Predict the effect on an EPSP of an increase in the concentration of a synaptic enzyme that breaks down the neurotransmitter responsible for the EPSP.

Answers are available in Appendix A.

43.4 The Vertebrate Nervous System

In the first three sections of this chapter, you examined electrical signaling at the level of molecules, membranes, and individual cells. This section gives you a view of electrical signaling at the levels of tissues, organs, and systems.

After you complete this section, you should be able to ...

- Analyze the vertebrate nervous system.

To begin, let's consider the overall anatomy of the vertebrate nervous system. Recall from Section 43.1 that the vertebrate central nervous system (CNS) is made up of the brain and spinal cord and is concerned primarily with integrating information. The peripheral nervous system (PNS) is made up of neurons outside the CNS. You will first look at the PNS, and then at how researchers explore the function of the most complex part of the CNS: the human brain. The chapter concludes by returning to the molecular level (and to studies of invertebrates) to introduce recent work on learning and memory.

What Does the Peripheral Nervous System Do?

What functions do the cells of the peripheral nervous system control? Anatomical and functional studies indicate that the PNS consists of two divisions with distinct functions:

1. The **afferent division** transmits sensory information to the CNS.
2. The **efferent division** carries commands from the CNS to the rest of the body.

Neurons in the afferent division monitor conditions inside and outside the body. Once information from afferent neurons has been processed in the CNS, neurons in the efferent division carry signals that allow the body to respond appropriately to changed conditions.

As Figure 43.14 shows, the afferent and efferent divisions are part of a hierarchy of PNS functions. The efferent division is further subdivided into somatic and autonomic nervous systems:

- The **somatic nervous system** carries out voluntary responses, which are under conscious control. Skeletal muscles serve as the effectors.
- The **autonomic nervous system** carries out involuntary responses, which are not under conscious control. Autonomic neurons control internal processes such as digestion and heart rate. Smooth muscle, cardiac muscle, and glands serve as the effectors.

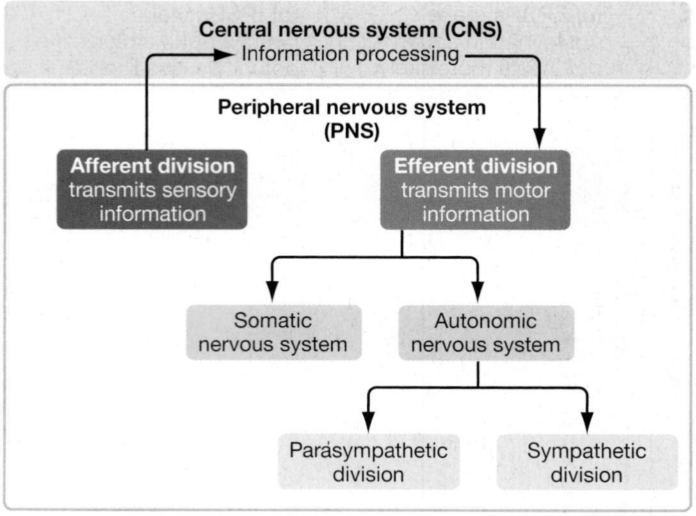

Figure 43.14 The Vertebrate Nervous System Has Several Functional Divisions.

Many organs are served by two functionally distinct types of autonomic nerves, summarized in Figure 43.15. The two types often have opposite effects on the same organ.

1. Nerves in the **parasympathetic nervous system** promote "rest-and-digest" functions that conserve or restore energy.
2. Nerves in the **sympathetic nervous system** typically prepare organs for stressful "fight-or-flight" situations.

To understand the autonomic nervous system, consider the changes that go on in your body when you begin exercising. During exercise, sympathetic neurons are activated and parasympathetic neurons are inhibited. Examine Figure 43.15 to see some of the effects this will have on the body: heart rate will increase; airways will dilate; digestive organs are inhibited; glucose is released from the liver into the bloodstream. All of these responses help the body maintain homeostasis during energetically demanding exercise.

Now suppose you've finished your workout and are making dinner. As you cook, the sight, smell, and thought of food begins to stimulate your parasympathetic neurons and inhibit sympathetic neurons. These effects are the opposite of exercising: your heart rate decreases; your airways constrict; your stomach, intestines, and accessory organs begin contracting or secreting digestive juices. You are now ready for that meal!

Besides the afferent and efferent divisions, some researchers have proposed a third division of the PNS: the neurons embedded in the wall of the gastrointestinal tract, from esophagus to anus. This **enteric nervous system** interacts with autonomic nerves but can also function independently (for example, if the vagus nerve is cut). The enteric nervous system plays a major role in regulating digestion, but it may also be important in immune function, mental health, cognition, and memory formation in the CNS.

Functional Anatomy of the CNS

Parasympathetic nerves originate at the base of the brain or the base of the spinal cord (see Figure 43.15). Most sympathetic nerves also originate in the spinal cord, but they emerge along the middle of its length. Similarly, most sensory neurons project

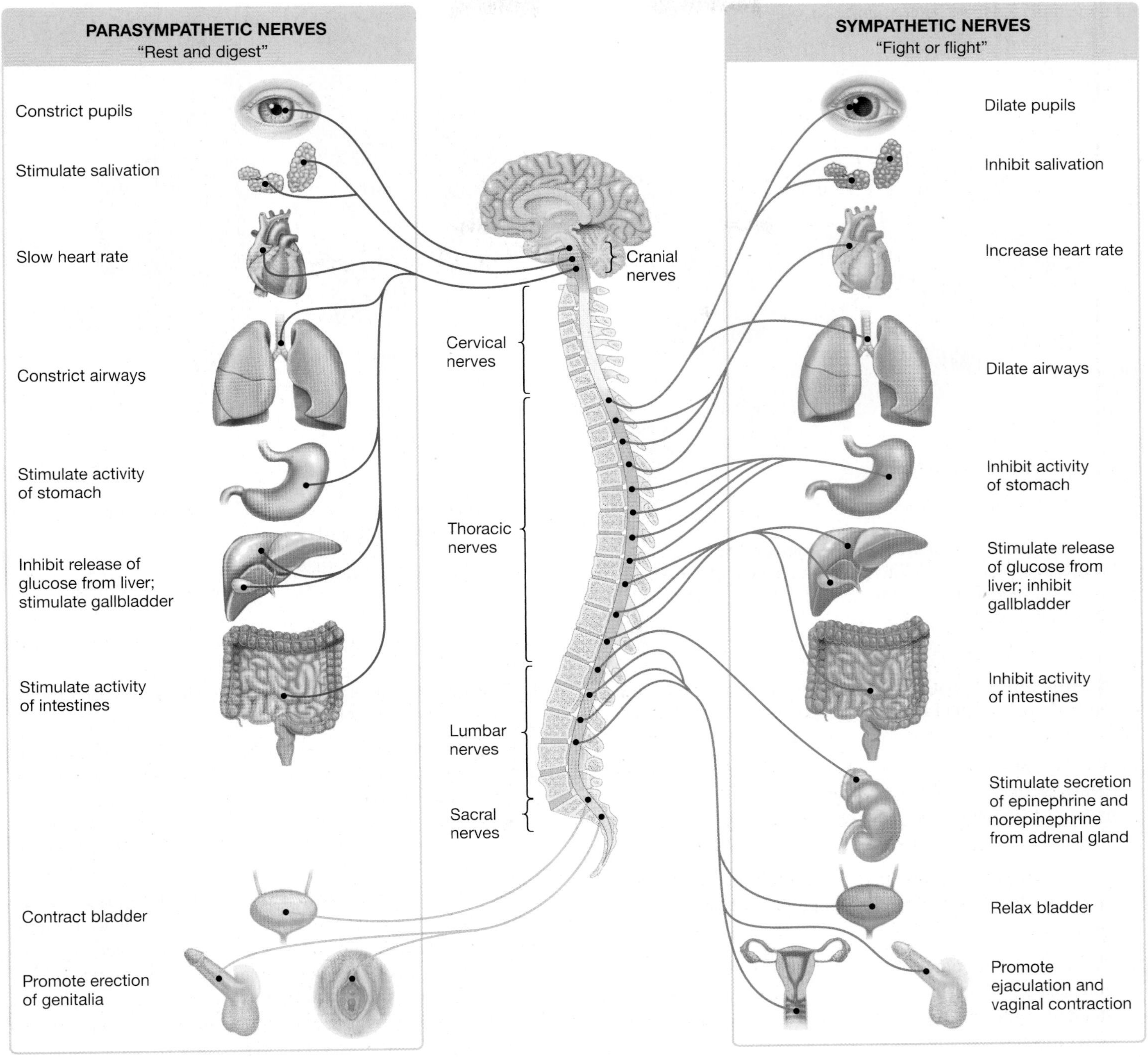

Figure 43.15 The Autonomic Nervous System Controls Internal Processes.

✔ Explain how the responses listed here for the pupils, heart rate, liver, and gallbladder are part of the rest-and-digest function or the fight-or-flight function.

axons to the spinal cord, and most somatic motor neurons project from the spinal cord.

In effect, then, the spinal cord serves as an information conduit. It collects and transmits information throughout the body. Virtually all the sensory information that enters the spinal cord must be sent to the brain for processing. The main exceptions involve spinal reflexes, such as the one illustrated in Figure 43.1, in which sensory neurons stimulate interneurons or motor neurons within the spinal cord itself.

What happens once sensory signals arrive at the brain? How are thousands of signals integrated to allow an animal to respond to stimuli? Let's begin our exploration by delving into the anatomy of the brain.

General Anatomy of the Human Brain The brains of all vertebrates can be functionally divided into three main parts based on sensory function: the forebrain (smell), midbrain (vision), and hindbrain (hearing and sometimes balance); Ch. 32, Section 32.4.

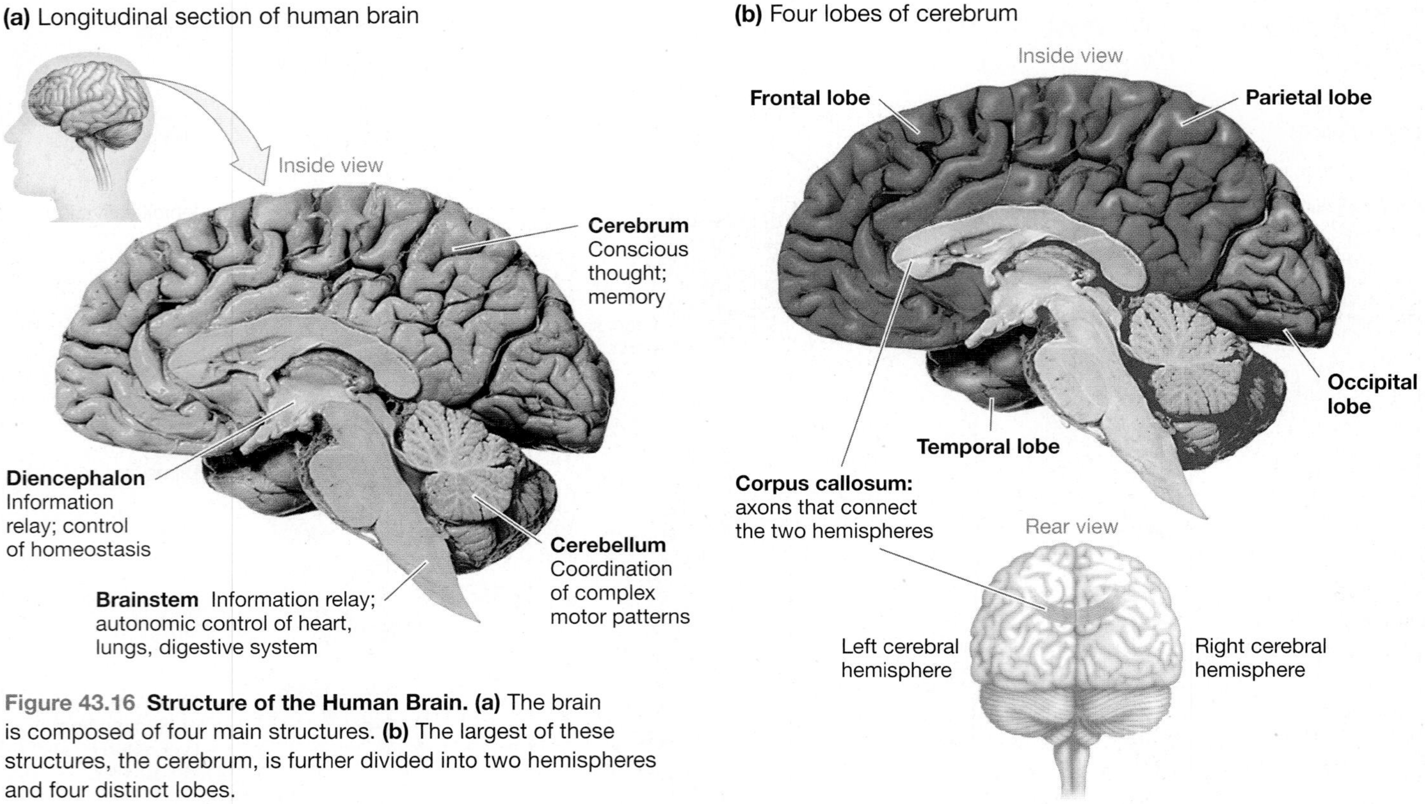

Figure 43.16 Structure of the Human Brain. (a) The brain is composed of four main structures. **(b)** The largest of these structures, the cerebrum, is further divided into two hemispheres and four distinct lobes.

Nineteenth-century anatomists further characterized the components of the human brain by emphasizing the four structures labeled in Figure 43.16a. Each structure has distinct functions:

1. The **cerebrum** accounts for the bulk of the human brain. It makes up most of the forebrain, is divided into left and right hemispheres, and is the seat of conscious thought and memory.
2. The **diencephalon** is also part of the forebrain. It relays sensory information to the cerebrum and functions in maintaining homeostasis.
3. The **brainstem** connects the brain to the spinal cord and contains the midbrain and parts of the hindbrain. It is the autonomic center for regulating cardiovascular, digestive, and other involuntary functions.
4. The **cerebellum** is a structure in the hindbrain that coordinates complex motor patterns.

Each cerebral hemisphere has four major areas, or lobes: **(1)** the **frontal lobe**, **(2)** the **parietal lobe**, **(3)** the **occipital lobe**, and **(4)** the **temporal lobe** (Figure 43.16b). The two hemispheres are connected by a thick band of axons called the **corpus callosum**.

The relative size of the entire brain, and of its component structures, varies greatly among vertebrates (Figure 43.17). For example, compare the size of the cerebrum in fishes and mammals. In fishes, the cerebrum is quite small and is involved mainly in the sense of smell. In mammals, the cerebrum is very large and contains regions specialized for memory and reasoning, in addition to the processing of multiple sensory and motor functions. Reflecting its role in these brain functions that are so prominent in humans, your own cerebrum is three times as large as those of

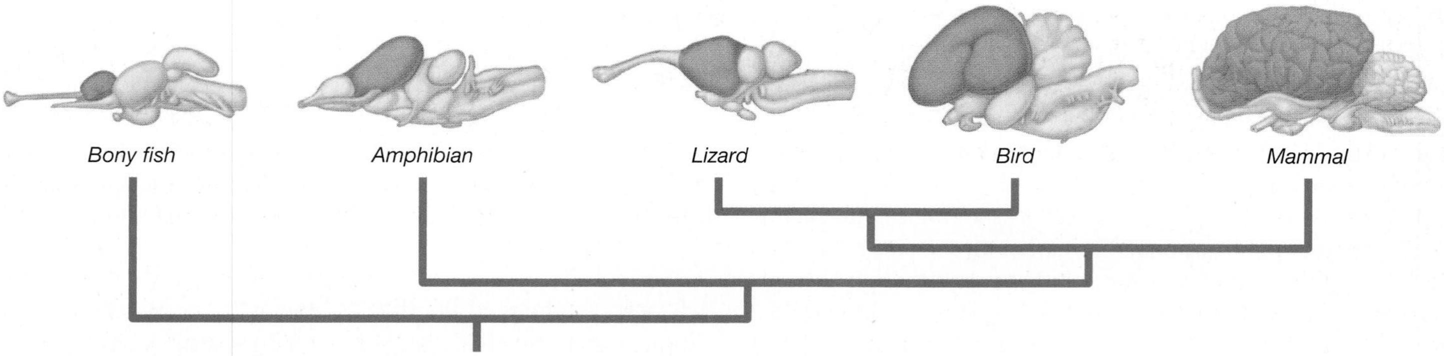

Figure 43.17 The Relative Sizes of Brain Regions Differ among Vertebrate Lineages. For example, the cerebrum (highlighted in brown) is larger in birds and mammals than in bony fishes, amphibians, and lizards.

comparably sized mammals. What methods do researchers use to explore the function of each area within the cerebrum?

Mapping the Brain I: Lesion Studies Early work on brain function focused on people with specific mental deficits caused by areas of brain damage or lesions. In 1848, for example, Phineas Gage was working on a railroad construction site when an iron rod over 3 cm in diameter was blasted through his skull. The rod entered beneath his left eye, exited through his forehead, and landed more than 20 m away. Miraculously, Gage survived this accident, but his personality did not fare so well. After the accident, Gage's physician reported him to be "fitful," "irreverent," and "obstinate," a dramatic change from his previous personality. The iron rod had damaged Gage's frontal lobe, providing some of the first clues that this part of the brain plays a role in personality and emotion.

The French physician Paul Broca formulated the hypothesis that specific regions of the brain are specialized for coordinating particular functions. Broca's hypothesis arose from his study of an individual who could understand language but could not speak. After the person's death in 1861, Broca examined the patient's brain and discovered a lesion in the left frontal lobe. Broca proposed that this region is responsible for speech.

Broca's hypothesis that functions are localized to specific brain areas has been verified through extensive efforts to map the cerebrum. Important advances were made by studying the behavior of people who had specific portions of their brains removed during surgery.

One such case is that of a 27-year-old man named Henry Gustav Molaison (referred to as "H.M." until his death). In 1953, surgeons treated him for life-threatening seizures by removing a small portion of his temporal lobe and about two-thirds of his **hippocampus**, a structure at the inner edge of the temporal lobe. Molaison recovered and lived for another 55 years. For the rest of his life, he had normal intelligence and vividly remembered his childhood, but he had no short-term memory. Brenda Milner, who studied Molaison for over 40 years, had to introduce herself to him every time they met; he could not even recognize a recent picture of himself.

Based on case histories such as Molaison's and studies of memory in laboratory animals, a consensus has emerged that several aspects of memory are governed by the hippocampus and interior sections of the temporal lobe. In particular, the hippocampus is responsible for the formation of new memories, which are then "stored" in the cerebrum to be retrieved later. How did scientists discover this role of the cerebrum?

Mapping the Brain II: Electrical Stimulation of Conscious Patients Wilder Penfield pioneered a different approach to studying brain function. Penfield studied people who were suffering from severe seizures and were scheduled to have seizure-prone areas of their brains surgically removed. While the patients were awake and under a local anesthetic, Penfield electrically stimulated portions of their cerebrums. His immediate goal was to map essential areas that should be spared from removal if possible.

When Penfield stimulated specific areas, patients reported sensations or experienced movement in particular regions of the body. From these responses, Penfield was able to map regions of the cerebrum involved in sensory and motor processing

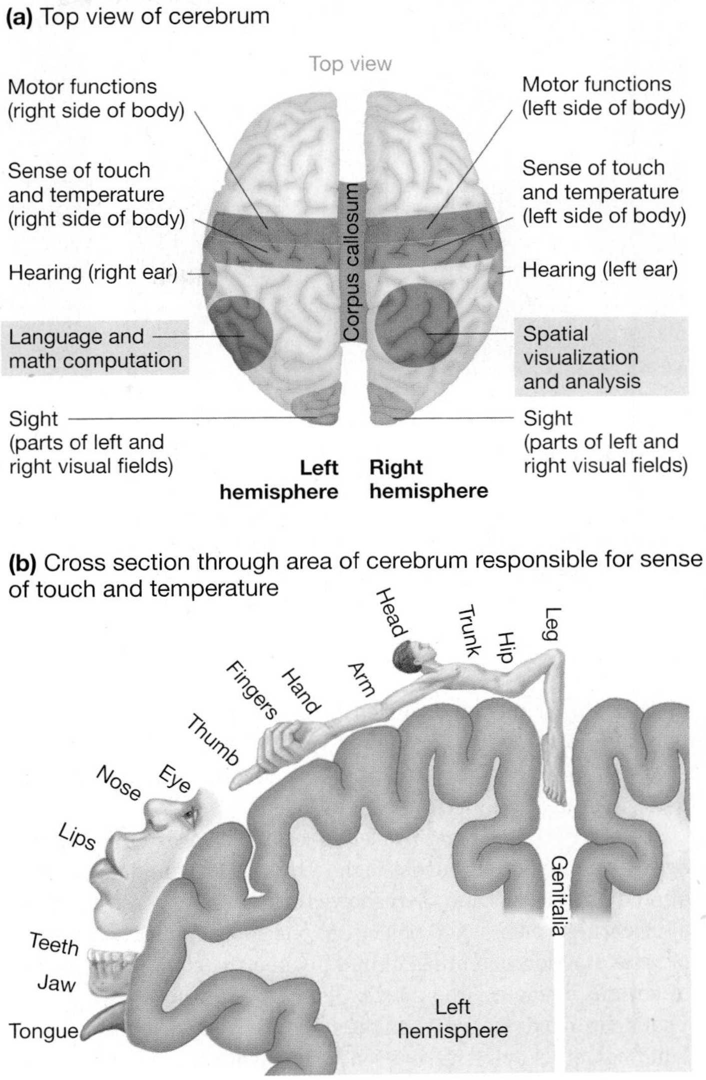

Figure 43.18 **Specific Brain Areas Have Specific Functions.** **(a)** A map of the cerebrum, in top view, showing the functions of some major regions. The map was compiled from studies of people with brain lesions or in whom brain regions had been removed surgically. (Note that the corpus callosum is not actually as wide as shown here.) **(b)** Researchers mapped the cerebral area responsible for the sense of touch and of temperature by stimulating neurons in the brains of patients who were awake. The size of the icons corresponds to the amount of cerebral area devoted to processing sensory information from those parts.

✔ Is there a correlation between the size of the cerebral area devoted to sensing a particular body part and the size of that body part? Explain.

(Figure 43.18). Brain surgeons still use this technique to map critical locations near tumors and seizure-prone areas.

Perhaps the most striking of Penfield's findings was that, on occasion, patients would respond to stimulation of their temporal lobe by having what appeared to be flashbacks. After one region was stimulated, a woman said, "I hear voices. It is late at night around the carnival somewhere—some sort of traveling circus. . . . I just saw lots of big wagons that they used to haul animals in."

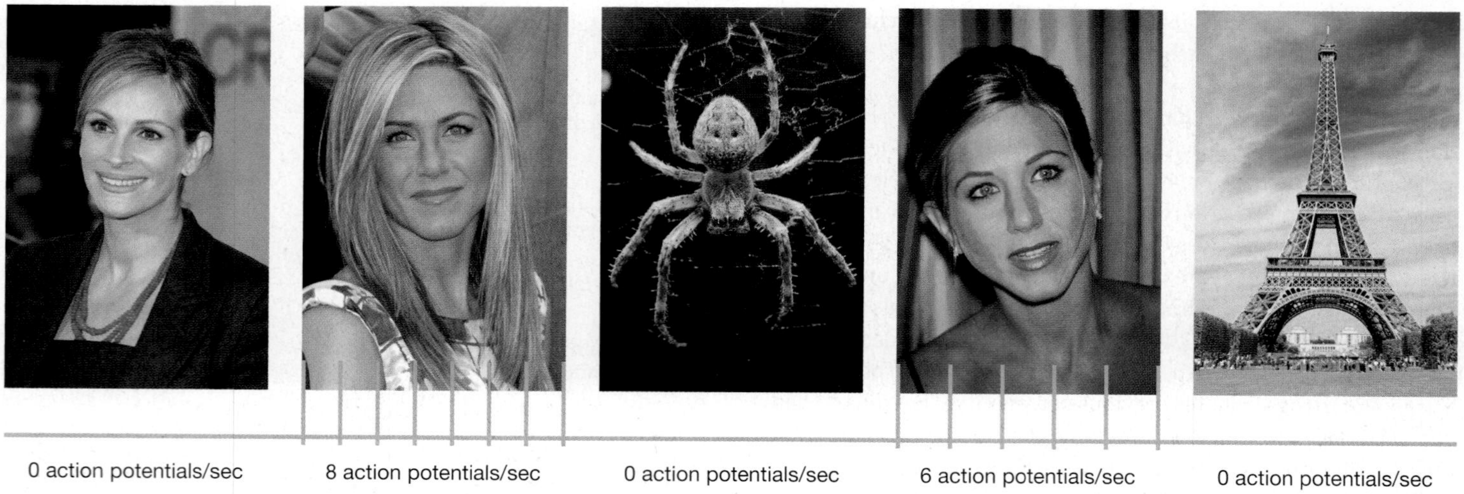

Figure 43.19 Single-Neuron Recording Reveals that Some Neurons in the Brain Recognize Specific Concepts. The recordings below each photograph show how a single neuron fires in response to images of actress Jennifer Aniston but not to other images.

DATA: R. Q. Quiroga, L. Reddy, G. Kreiman, et al. 2005. *Nature* 435: 1102–1107.

Was this a memory, stored in a small set of neurons that Penfield happened to stimulate? The hypothesis that memories are somehow stored in specialized cells is controversial. However, a recent study provided support for it.

Researchers attached tiny electrodes that recorded the electrical activity of individual neurons in the temporal lobes and hippocampi of study subjects. They then showed each subject a set of images of celebrities, places, and objects. In one subject, a specific neuron produced action potentials ("fired") when the subject viewed any of several images of the actress Jennifer Aniston (Figure 43.19). That neuron did not fire when the subject viewed images of spiders, buildings, other actresses, or even Aniston with another person. The neuron fired only when the subject viewed images of the actress alone, or even an image of her name spelled out. It appeared that, through experience, at least one of this person's neurons became singularly devoted to the concept of Jennifer Aniston.

Note that the researchers were testing only a tiny subset of the neurons present in the human brain, so there may be more than one "Jennifer Aniston neuron" in the brain. But the take-home message is that the formation or retrieval of memories associated with specific concepts involves specific neurons. Researchers are still examining exactly how these neurons interact with other parts of the brain during the processes of learning and remembering.

Methods such as lesioning and single-cell recording are powerful because they provide strong evidence for the functions of specific brain regions and even neurons. However, these methods merely establish a correlation between specific neurons or brain regions and a given function. In 1979, Francis Crick, known for his codiscovery of the double-helix structure of DNA but also an accomplished neurobiologist, suggested that major breakthroughs in neuroscience would come with the development of techniques that could stimulate or inhibit certain neurons without affecting others. In the early 2000s, shortly before Crick's death, this idea became a reality with the invention of optogenetics.

Mapping the Brain III: Optogenetics The technique known as **optogenetics** uses light to control the activity of targeted types of neurons in the brain. In one of the earliest applications of optogenetics, researchers inserted an algal gene that codes for a light-activated ion channel into the DNA of specific types of neurons in the brains of mice and then surgically implanted light-emitting probes in the brains of those mice (Figure 43.20a). By activating those probes, they could stimulate action potentials in the neurons that expressed the algal ion channel (Figure 43.20b). This technique made it possible to directly observe the role of specific neurons in specific behaviors.

In the past 15 years, optogenetics has exploded in popularity as a tool for studying the functions of neurons responsible for sleeping, running, mating, and many other behaviors. What's more, researchers are developing optogenetic tools to stimulate neurons that may be compromised in neurodegenerative diseases, including Parkinson's disease and Alzheimer's disease. Optogenetics is also being used to examine detailed neural mechanisms behind learning and memory, for example by elucidating the roles of specific subregions of the hippocampus in memory formation.

What other approaches to studying learning and memory have been productive?

How Do Learning and Memory Work?

Learning is an enduring change in behavior that results from a specific experience in an individual's life. **Memory** is the retention of learned information. Learning and memory are thus closely related and are often studied in tandem. Despite decades of intensive research on the mechanisms involved in learning and memory, scientists have barely scratched the surface of these complex phenomena, and major components of learning and memory remain mysteries that continue to be actively studied.

(a) A mouse with neurons genetically engineered to respond to light is implanted with a light-emitting brain probe.

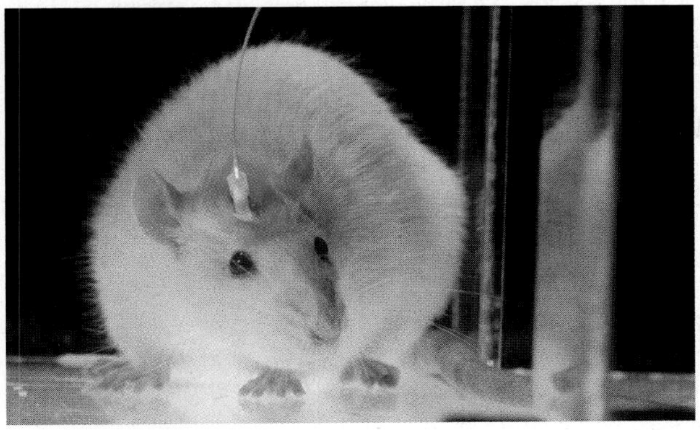

(b) Flashes of light from the probe cause light-sensitive neurons to respond with action potentials.

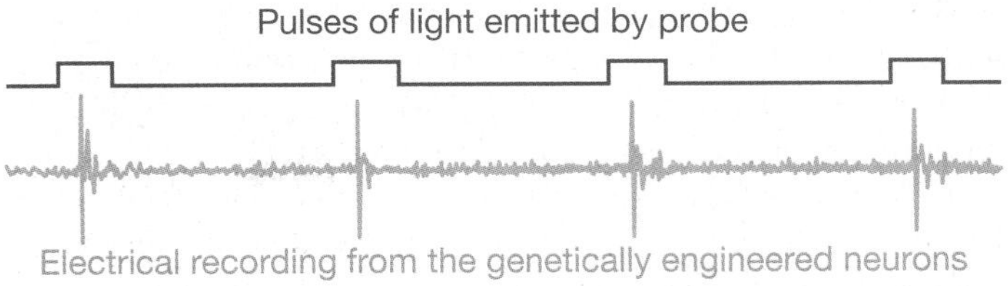

Figure 43.20 Optogenetics Allows Researchers to Stimulate Specific Neurons. With this technique, the roles played by these neurons can be studied in living animals.

As an introduction to how researchers explore these phenomena, let's first examine work that focuses on neurons and then review research at the molecular level.

Recording from Single Neurons during Memory Tasks How do the action potentials generated by a neuron change as learning and memory take place? Researchers have attempted to answer this question by recording from individual neurons in the temporal lobes of humans.

Physicians placed electrodes in specific brain regions of patients who were about to undergo surgery to remove seizure-prone areas of their brains. While the patients were still awake, the researchers projected words or names of objects on a screen and asked the patients to read them silently, read them aloud, or remember them and repeat them later. The data showed that individual neurons in the cerebrum's temporal lobe were relatively inactive while patients identified objects but extremely active when patients remembered the objects and repeated their names aloud.

What do such data mean? Neurons in the temporal lobe are most active during memory tasks. How can action potentials in particular neurons make memory possible?

Documenting Changes in Synapses Research on the molecular basis of memory is based on the idea that learning and memory must involve some type of short-term or long-term change in the neurons responsible for these processes. This change could be structural or chemical in nature. Structural changes might include modifications in the number of synapses that a particular neuron makes, the destruction of neurons, or the formation of new neurons. Chemical changes might involve alterations in the amount of neurotransmitter released at certain synapses or changes in the number of receptors present in postsynaptic cells.

To explore the molecular basis of learning and memory, Eric Kandel's group has focused on an animal much easier to study than any vertebrate—the sea slug *Aplysia californica* (**Figure 43.21a**). Much of their work has explored the reflex diagrammed in **Figure 43.21b**: When the siphon, a structure on the animal's back, is touched—for example, by a stream of water—the sea slug responds by withdrawing its gill. Withdrawing the gill protects it from predators. The reflex is produced by sensory neurons that are activated by touch and motor neurons that project to a gill muscle.

(a) Sea slug *Aplysia californica* (juvenile)

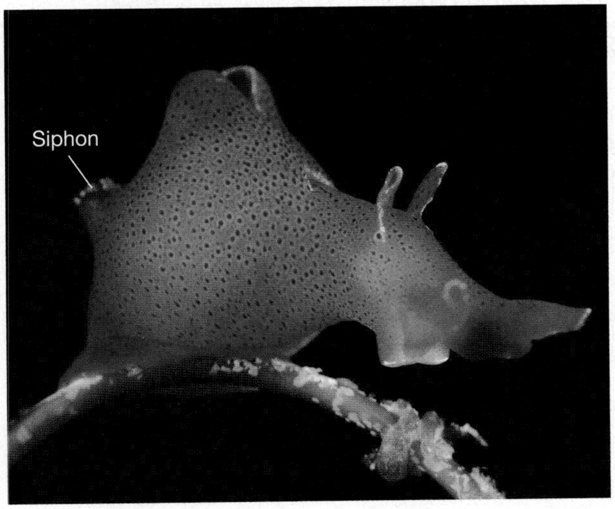

(b) Gill-withdrawal reflex protects the gills during an attack.

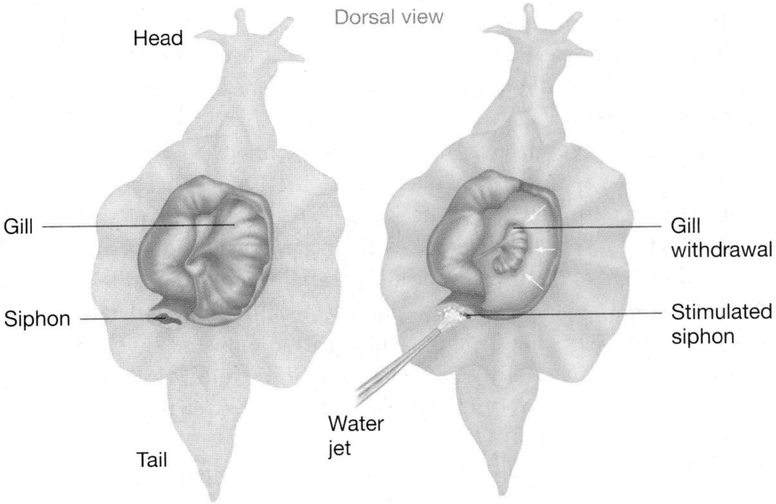

Figure 43.21 The Gill-Withdrawal Reflex in *Aplysia* Is a Model System in Learning and Memory.

Early work established that this simple reflex is modified by learning. For example, *Aplysia* also withdraw their gills when their tails are given an electrical shock. If a tail shock is repeatedly paired with a very light touch to the siphon—too light to normally produce a response on its own—an *Aplysia* will learn to withdraw its gill in response to a light siphon touch alone.

Follow-up studies showed that the neurons involved in learning in this reflex release the neurotransmitter **serotonin**. Repeated application of serotonin mimics what happens at the synapse during learning, when sensory neurons from the siphon and the tail fire repeatedly. These results suggest that in *Aplysia*, changes in the nature of the synapse form the molecular basis of learning and memory. A change in the responsiveness or structure of a synapse is termed **synaptic plasticity**.

To take a closer look at synaptic plasticity in *Aplysia*, Kandel's team replicated these results with sensory and motor neurons growing in culture. Figure 43.22a shows an *Aplysia* sensory neuron that synapses with two motor neurons on a culture plate. To mimic the learning process, the investigators applied serotonin to the synapse on one motor neuron five times over a short period.

When they stimulated the sensory neuron a day later, they found a huge increase in the size of EPSPs in the motor neuron postsynaptic to that synapse (Figures 43.22b and 43.22c). That motor neuron had also established additional synapses with the sensory neuron. The structure and behavior of the motor neuron exposed to serotonin had changed, based on its experience. The motor neuron that had not been exposed to serotonin had no increase in the size of its EPSPs or the number of its synapses.

Documenting Changes in Neurons Synapses change over time, but can new neurons form in the central nervous system? When early neurobiologists dissected brains from adult cadavers, they found that none of the brain neurons showed signs of mitosis. From these observations, they concluded that **neurogenesis**—the formation of new neurons—does not occur in adults.

In 1983, Steven Goldman and Fernando Nottebohm revisited this long-standing idea by examining the brains of songbirds. Songbirds were chosen because in their brains, specific regions that control singing behavior undergo dramatic seasonal changes in size. The researchers hypothesized that these changes result from an increased rate of neurogenesis during the reproductive season.

To test this hypothesis, the researchers injected canaries (Figure 43.23a) with radiolabeled thymidine, which becomes incorporated into newly synthesized DNA. The researchers later collected the birds' brains, cut them into very thin slices, mounted the slices on microscope slides, and brought the slides into contact with photographic film. In this technique, called autoradiography (BioSkills 6), any cells with radiolabeled thymidine in their DNA would expose the film, thereby indicating the presence and location of cells that were produced after the bird was injected.

The researchers found that adult songbirds are indeed able to produce large numbers of new neurons. Dozens of studies since

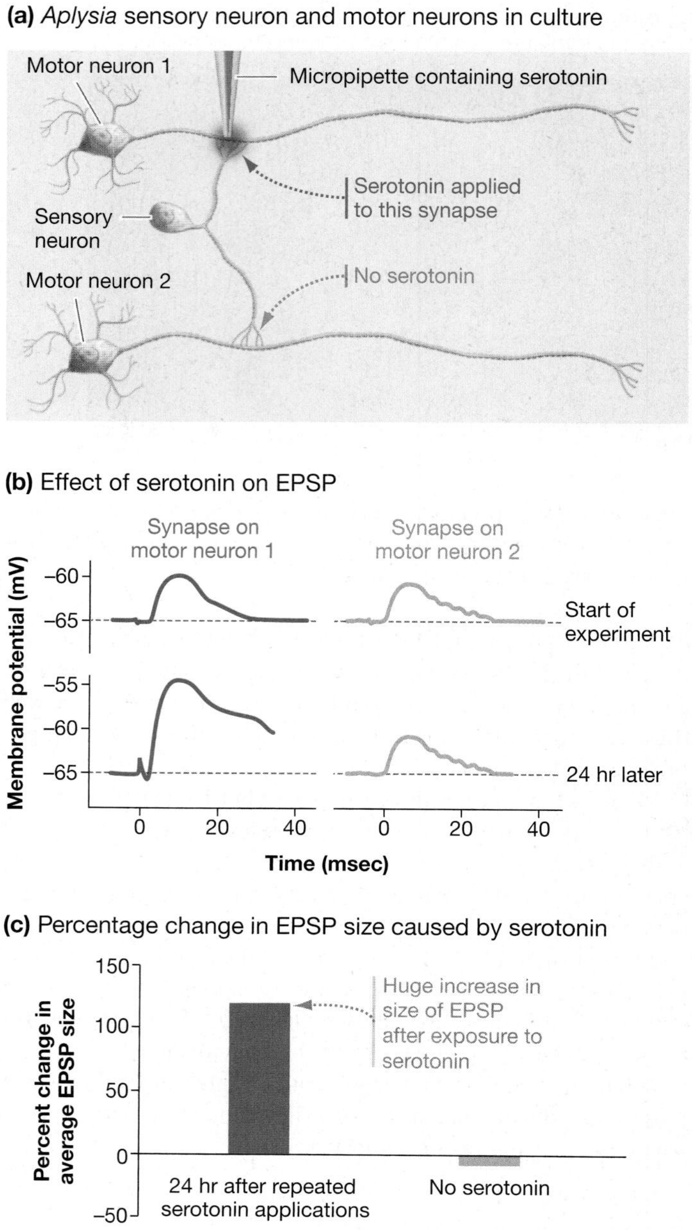

Figure 43.22 Learning and Memory Involve Changes in Synapses. (a) In this experiment, a single *Aplysia* sensory neuron was grown in culture with two motor neurons. A micropipet was used to apply serotonin to the synapse between the sensory neuron and one of the motor neurons. **(b)** Stimulating the sensory neuron produced EPSPs in both motor neurons. One day after repeated serotonin application, the EPSP was much larger at the serotonin-treated synapse but was unchanged at the untreated synapse. **(c)** Bar graph quantifying the results of 18 experiments.

DATA: K. C. Martin, A. Casadio, H. Zhu, et al. 1997. *Cell* 91: 927–938.

(a) Male canaries (left) sing to attract females (right).

(b) In songbirds, the size of the brain's song-control region increases in spring.

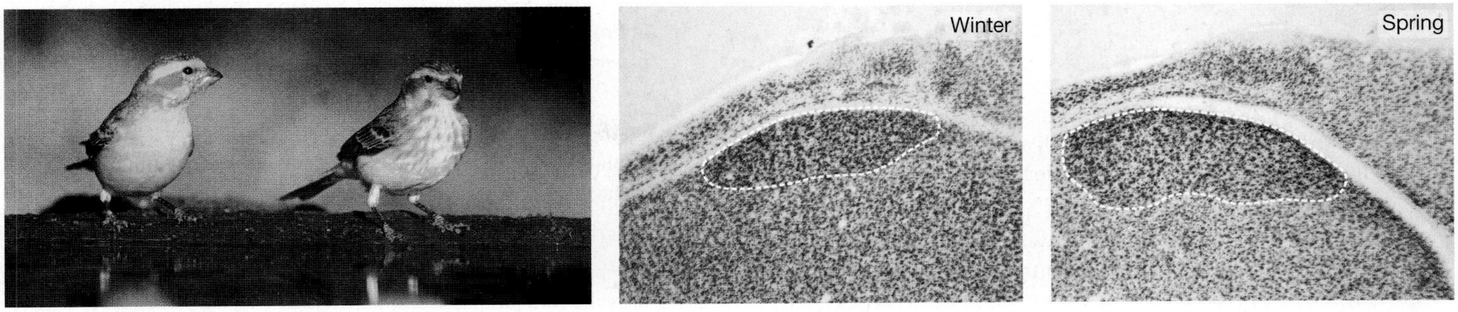

Figure 43.23 Neurogenesis Occurs in the Adult Songbird Brain.

Goldman and Nottebohm's experiment have confirmed that new neurons are incorporated into the song-control regions of the brain each spring, causing these regions to grow dramatically (**Figure 43.23b**). The new neurons promote learning and memory in the song-control system.

Other scientists around the same time and, indeed, even earlier obtained autoradiographic evidence of neurogenesis in adult mammals. However, the sheer number of new cells was much lower in mammals than in birds, so these reports did not convince all researchers. Skeptics initially suggested that the labeled cells in mammals might actually be glia rather than neurons.

A breakthrough came in the early 1990s with the development of new techniques that specifically label neurons. The markers used in these techniques bind to proteins that are found only in neurons. Since that time, neurogenesis has been definitively shown to occur in many adult vertebrates.

Scientists are only beginning to understand the functional roles of these new neurons. Do they have the same roles as neurons made during development? Can neurogenesis be used to help heal brain injuries? Some studies have even suggested that chronic stress during childhood (the stress of poverty or bullying, for example) can reduce the ability to make new neurons, thereby negatively affecting learning in adults.

Results such as these reinforce a growing consensus that learning, memory, and the control of complex behaviors involve not only molecular and structural changes in synapses but also changes in the number of neurons. Further, most researchers now agree that at least some aspects of long-term memory involve changes in gene expression. Chemical messengers called hormones also cause changes in gene expression in target cells (Chapter 46). But before investigating how hormones work, let's focus on the electrical signals involved in vision, hearing, taste, and movement (the subjects of Chapters 44 and 45).

CHECK YOUR UNDERSTANDING

✔ **If you understood this section, you should be able to ...**

1. **PROCESS OF SCIENCE** Describe the research strategies that allowed biologists to localize particular functions to specific regions in the brain.
2. **THINK CAREFULLY** Critique the concept that all brain damage is always permanent. What physiological mechanism(s) might be involved in brain repair after damage?

Answers are available in Appendix A.

CHAPTER 43 Review

For media, go to **Mastering Biology**

43.1 Principles of Electrical Signaling

- Most neurons have a cell body, multiple short dendrites that receive signals from other cells, and a single axon that transmits electrical signals to other neurons or to effector cells in glands or muscles.
- Studies of the squid giant axon established that neurons have a resting potential maintained by the sodium–potassium pump and potassium leak channels. When Na^+/K^+-ATPase hydrolyzes ATP, it transports 3 Na^+ out of the cell and 2 K^+ in, resulting in a concentration gradient that favors the exit of K^+ from the cell via leak channels. This makes the interior of the cell membrane negative compared to the outside.

43.2 Dissecting the Action Potential

- Studies of the squid giant axon established that the action potential is a rapid, all-or-none change in membrane potential.
- An action potential begins with an inflow of Na^+ that depolarizes the membrane. An outflow of K^+ follows and repolarizes the membrane.
- Both Na^+ and K^+ flow through voltage-gated channels.
- As Na^+ flows in, it repels cations, which spread along the inside of the cell from the site of the action potential, causing the adjacent portion of the membrane to depolarize enough to trigger an action potential there.

- Action potentials propagate most rapidly in axons that are large or myelinated.

43.3 The Synapse

- When an action potential arrives at a synapse, synaptic vesicles fuse with the axon's membrane and deliver neurotransmitters into the synaptic cleft. The neurotransmitters bind to receptors on the membrane of the postsynaptic cell.
- Some receptors are ligand-gated channels. These channels open in response to binding by a neurotransmitter, enabling ion flow that depolarizes or hyperpolarizes the postsynaptic cell's membrane.
- Postsynaptic potentials summate.
- If the membrane at the axon hillock of a postsynaptic neuron depolarizes to the threshold potential, an action potential is triggered.

43.4 The Vertebrate Nervous System

- The vertebrate CNS consists of the brain and spinal cord; the PNS consists of all nervous system components outside the CNS.
- In vertebrates, the PNS contains somatic and autonomic components. The somatic nervous system carries out voluntary responses by signaling skeletal muscles; the autonomic nervous system carries out involuntary responses by signaling effector cells that change internal conditions.
- Early efforts to map functional regions of the brain were based on analyzing deficits in individuals with brain lesions or on stimulating certain regions of the cerebrum.
- Research has established that learning and memory involve synaptic changes, including the release of more or less neurotransmitter and the formation of additional synapses.
- Neurogenesis (formation of new neurons) occurs in adult vertebrates and may be an important component of learning and memory.

Answers are available in Appendix A.

✔ TEST YOUR KNOWLEDGE

1. Which ion most readily leaks across a neuron's membrane, helping to establish the resting potential?
   a. Ca^{2+}
   b. K^+
   c. Na^+
   d. Cl^+
2. Which of these statements about myelination in neurons is/are correct? Select True or False for each statement.
   T/F It speeds propagation by increasing the density of voltage-gated channels all along the axon.
   T/F Multiple sclerosis is characterized by disrupted myelination of certain neurons in the central nervous system.
   T/F It speeds propagation by preventing cations from leaking out across the membrane as they spread down the axon.
   T/F It is more commonly observed in vertebrates than in invertebrates.
3. In a neuron, what creates the electrochemical gradient favoring the outflow of K^+ when the cell is at rest?
   a. Na^+/K^+-ATPase
   b. voltage-gated K^+ channels
   c. voltage-gated Na^+ channels
   d. ligand-gated Na^+/K^+ channels
4. Which of the following brain regions is responsible for formation of new memories?
   a. brainstem
   b. cerebellum
   c. frontal lobe
   d. hippocampus

✔ TEST YOUR UNDERSTANDING

5. Explain the difference between a ligand-gated K^+ channel and a voltage-gated K^+ channel.
6. Describe the role of summation in postsynaptic cells.
7. Compare and contrast the somatic nervous system and autonomic nervous system.
8. Why is memory thought to involve changes in particular synapses?
   a. At some synapses, more neurotransmitters are released after learning takes place.
   b. At some synapses, a different type of neurotransmitter is released after learning takes place.
   c. When researchers stimulated certain neurons electrically, individuals replayed memories.
   d. When researchers changed synapses in the brains of patients during surgery, the patients' memories changed.

✔ TEST YOUR PROBLEM-SOLVING SKILLS

9. Explain why drugs that prevent neurotransmitters from being taken back up by a presynaptic neuron have dramatic effects on the activity of postsynaptic neurons.
10. Alzheimer's disease is a common form of dementia affecting millions of people, especially the elderly. Two regions of the brain are particularly affected, often shrinking dramatically and accumulating large deposits of extracellular material. Based on your knowledge of memory, what two brain regions do you think these are? Explain.

✔ PUT IT ALL TOGETHER: Case Study

How can a frog kill with its skin?

Certain species of frogs in the genus *Phyllobates* have a powerful defensive adaptation—their skin can secrete a milky fluid that contains an extremely toxic compound called batrachotoxin (BTX). These frogs, which are found in Colombia, are known as poison dart frogs because some indigenous Colombian hunters coat the tips of their blowgun darts with the frogs' skin secretions. An animal hit by one of these darts dies quickly. What is the mechanism of action of BTX?

11. **QUANTITATIVE** The graph here shows the effect of BTX on the membrane potential of a squid giant axon.

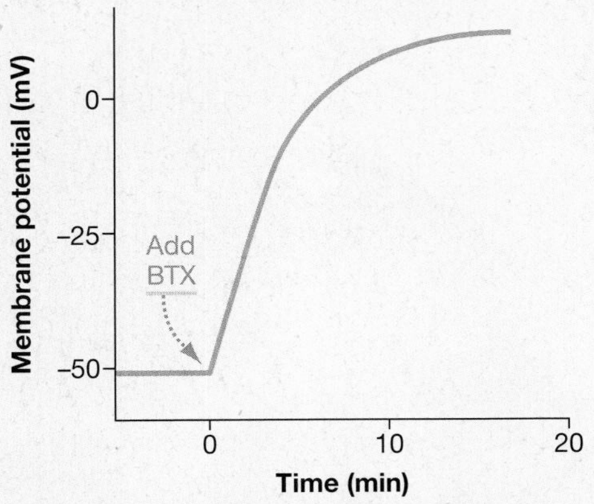

Source: T. Narahashi, E. X. Albuquerque, and D. Deguchi. 1971. *Journal of General Physiology* 58: 54–70.

Which of the following is the most likely explanation for the effect of BTX on the squid giant axon?

a. inactivation of Na^+/K^+-ATPase
b. closing of sodium channels
c. opening of sodium channels
d. opening of potassium channels

12. **PROCESS OF SCIENCE** Identify a research technique that could be used to discover how BTX affects specific membrane proteins. Based on the graph in Question 11, what would you expect this technique to show?

13. As the graph in Question 11 shows, BTX depolarizes the membrane and prevents repolarization. What effect would this have on electrical signaling by the nervous system?

14. **MODEL** Like neurons, cells in skeletal and cardiac muscle also produce action potentials. Create a concept map (**BioSkills 12**) showing how BTX could kill a mammal through its effects on nervous and muscle tissues.

15. Predict the effects of each of the following on the membrane potential of a neuron simultaneously poisoned with BTX: (a) removing extracellular sodium ions; (b) increasing the intracellular potassium ion concentration; and (c) adding tetrodotoxin from puffer fish.

16. Although BTX is a powerful antipredator poison, one snake species in Colombia eats poison dart frogs. Suggest a hypothesis that might explain how the snake is resistant to the toxin.

44 Animal Sensory Systems

In many species of moth, males have much larger antennae than females do. Receptor cells on the males' feathery antennae detect airborne chemical signals that are produced by sexually mature females. As a result, males can locate females in total darkness.

In this chapter you will learn how

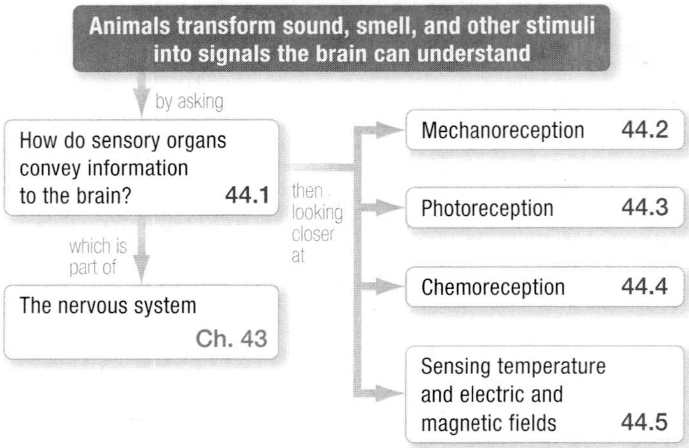

Many adult moths are active at night, when it is difficult or impossible to see. Instead of *looking* for a mate under these challenging circumstances, sexually mature female silk moths release a chemical attractant into the air. Male moths can detect even a single molecule of the attractant by using receptor cells located on their large, feathery antennae. Guided by an airborne gradient of attractant molecules, a male moth flies unerringly toward a female.

As they patrol in search of these airborne molecules, however, male moths are hunted by bats. Like moths, bats are active almost exclusively at night. Instead of hunting by sight, like a falcon or a cheetah, bats hunt with the aid of sonar: They emit a train of high-pitched sounds as they fly and then listen for echoes that indicate the direction and shape of objects in their path. If the object is a moth, the bat flies toward it, catches the moth in its mouth, and eats it.

BIG PICTURE

This chapter is part of the Big Picture. See how on pages 838–839.

But some moth species can hear the sounds bats emit. When moths detect sounds from an onrushing bat, they tumble out of the sky in chaotic escape flights.

If you were out at night as these dramas unfolded, you might be only dimly aware that bats and moths were flying about. Humans cannot smell moth attractants or hear the sounds that most bats emit when flying. It took decades of careful experimentation for biologists to understand how moths and bats sense the world around them and respond to the information they receive.

Sensing and interpreting changes in the environment are fundamental to how animals work. Indeed, everything from reproduction to feeding to detecting predators depends on sensory systems. To better understand sensory systems, let's begin with a basic question: How are sounds, smells, and other stimuli transformed into a signal that the brain can understand?

44.1 How Do Sensory Organs Convey Information to the Brain?

As a male moth flies through the night, its brain receives streams of signals from an array of sensory organs. Antennae collect information about the concentration of female attractant molecules; ears located on various parts of the body send data on the presence of high-pitched sounds; detectors for balance and gravity transmit signals about the body's orientation in space.

After you complete this section, you should be able to ...

- Explain how sensory organs communicate with the brain.

Each type of sensory information is detected by a sensory neuron or by a specialized receptor cell that makes a synapse with a sensory neuron. As Figure 44.1 shows, the moth's nervous system integrates the sensory input—the information from sensory neurons—and responds with motor output, via electrical signals, to specific muscle groups (effectors).

The ability to sense a change in the environment depends on two processes:

1. *Transduction*, the conversion of an external stimulus to an internal signal in the form of action potentials along sensory neurons, and
2. *Transmission* of the signal to the central nervous system (CNS).

The first process, **transduction**, requires a sensory receptor cell specialized for converting light, sound, touch, or some other signal into an electrical signal. Sensory receptors are membrane proteins that are categorized by type of stimulus:

- **Mechanoreceptors** respond to distortion caused by pressure.
- **Photoreceptors** respond to particular wavelengths of light.
- **Chemoreceptors** detect specific molecules.
- **Thermoreceptors** detect changes in temperature.
- **Nociceptors** sense harmful stimuli such as tissue injury.
- **Electroreceptors** detect electric fields.
- **Magnetoreceptors** detect magnetic fields.

With such a broad range of possible sensory receptors, it is no wonder that animals can monitor and respond to a wide array of changes in their environments.

Now, how do sensory cells receive information from the environment and report it to the brain, so an appropriate response can occur?

Sensory Transduction

During the resting state in most sensory cells, the inside of the plasma membrane is negative relative to the outside (Ch. 43, Section 43.1). When ion flows cause the inside to become less negative than the resting potential, the membrane is depolarized. When ion flows cause the inside to become more negative than the resting potential, the membrane is hyperpolarized.

Although sensory receptors can detect a remarkable variety of stimuli, they all transduce sensory input—such as light, sounds, touch, and odors—to a change in membrane potential. In this

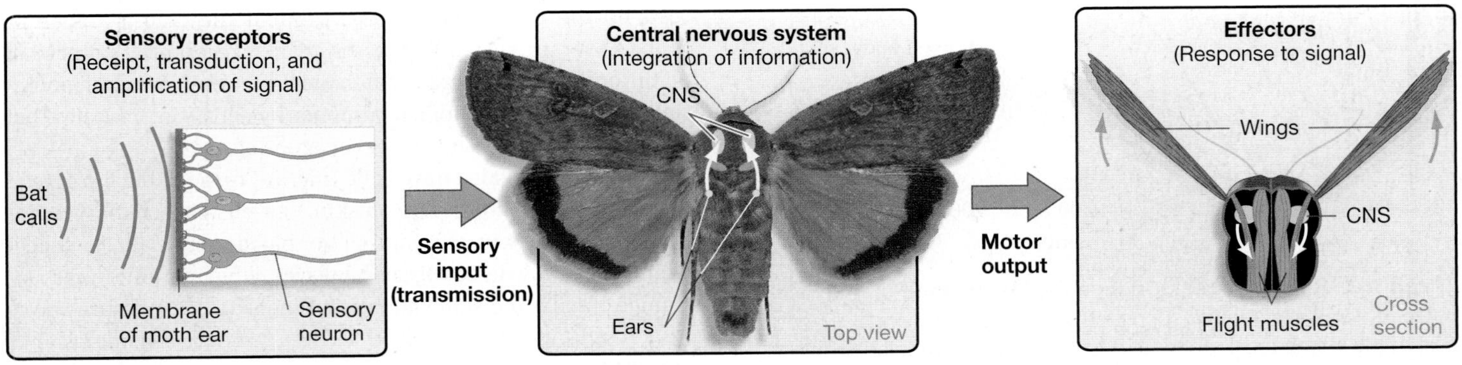

Figure 44.1 **Sensory Systems, the CNS, and Effectors Are Linked.** As shown here in a moth, sensory neurons relay information about conditions inside and outside an animal to the central nervous system. After integrating information from many sensory neurons, the CNS sends signals to muscles.

way, different types of information are transduced to a common type of signal—one that can be interpreted by the brain.

If a sensory stimulus induces a large change in a sensory receptor's membrane potential, there is a change in the firing rate of action potentials sent to the brain. The amount of depolarization that occurs in a sound-receptor cell, for example, is proportional to the loudness of the sound. If the depolarization passes threshold, enough voltage-gated sodium channels open to trigger action potentials that are relayed to the brain.

Figure 44.2a shows the membrane potential from a sound-receptor cell. When the experimenter played a sound, the sound-receptor cell depolarized for a short time in response. Other sensory cells work in a similar way.

Recall that all action potentials from a given neuron are identical in size and shape (Ch. 43, Sections 43.1 and 43.2). Figure 44.2b graphs the action potential "firing rate" recorded from a sensory neuron when sounds at various frequencies were played at two distinct intensities. Notice that loud sounds induce a higher rate of action potentials than do soft sounds. In this way sensory cells provide information about the intensity of a stimulus.

But if all types of external stimuli are converted to electrical signals in the form of action potentials, and if all action potentials are alike in size and duration, how does the brain interpret the incoming signals properly?

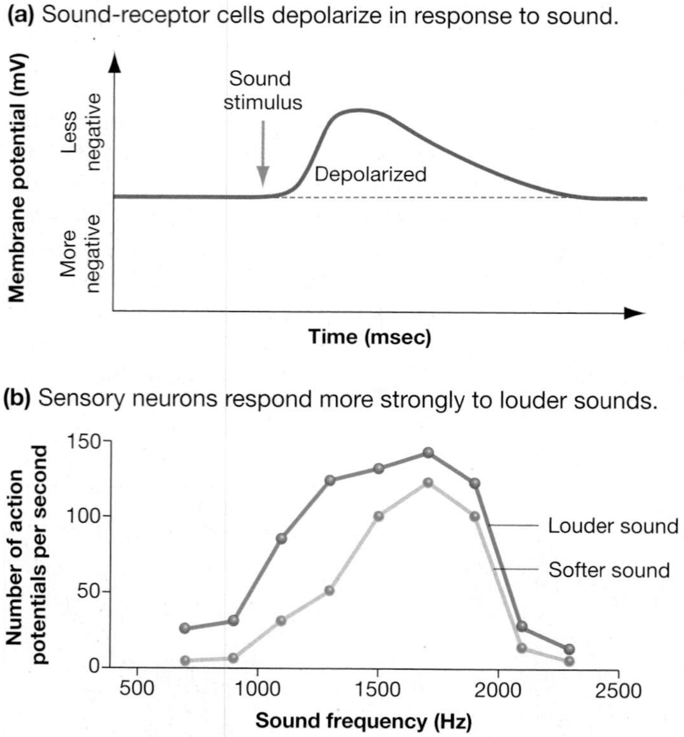

Figure 44.2 Sensory Inputs Change the Membrane Potential of Receptor Cells. (a) In response to sensory stimuli, ions flow across the membranes of receptor cells and either depolarize or hyperpolarize the membrane. **(b)** The rate at which action potentials occur in a sensory neuron provides information about the nature and intensity of the sensory stimulus. (The frequency of a sound, measured in hertz [Hz], determines its pitch.)

DATA: J. E. Rose, J. E. Hind, D. J. Anderson, et al. 1971. *Journal of Neurophysiology* 34: 685–699.

Transmitting Information to the Brain

There are two keys to understanding how the brain interprets sensory information. First, receptor cells tend to be highly specific. For example, each receptor cell in a human ear responds best to certain pitches of sound. Some receptors are more sensitive to low-pitched sounds, and others respond best to high-pitched sounds. The pattern of action potentials from a cell contains information about the pitch of the sound that is being detected, its intensity, and its duration.

The second key point is that each type of sensory neuron sends its signals to a specific portion of the brain. Axons carrying sensory information from the human ear project to the temporal lobes at the sides of the brain, but axons carrying sensory information from the eye deliver action potentials to the occipital lobe at the back of the brain. Different regions of the brain are specialized for interpreting different types of stimuli.

Now that the basic principles of sensory transduction and transmission have been introduced, let's delve into the details of the major sensory systems.

CHECK YOUR UNDERSTANDING

✔ **If you understood this section, you should be able to ...**

Contrast sensory transduction and transmission.

Answers are available in Appendix A.

44.2 Mechanoreception: Sensing Pressure Changes

After you complete this section, you should be able to ...

- Explain how animals sense pressure changes via mechanoreception.

Animals have a variety of mechanisms for **mechanoreception**—the detection of pressure changes. Crabs, for example, have a fluid-filled organ that helps them sense the pressure created by gravity. The organ, known as a **statocyst**, is lined with mechanoreceptor cells and contains a small calcium-rich particle that normally rests on the bottom of the organ. But if the crab is tipped or flipped over, this particle presses against receptors that are *not* on the bottom of the organ. When the brain receives action potentials from these receptors, it responds by activating muscles that restore the animal to its normal orientation.

Many animals also have cells that are responsible for detecting direct physical pressure on skin, as well as mechanoreceptor cells that monitor how far muscles or blood vessels are stretched. Animals hear by detecting sound waves, which produce pressure changes in air, and some aquatic animals detect pressure waves in water via a lateral line system. These pressure-sensing systems are all based on the same mechanism.

Let's briefly examine the general structure of a mechanoreceptor cell and its response to pressure, and then investigate the specific structures involved in mammalian hearing and, in fishes, the lateral line system.

How Do Sensory Cells Respond to Sound Waves and Other Forms of Pressure?

In mechanoreception, direct physical pressure on a plasma membrane or distortion of membrane structures by bending changes the conformation of ion channels in the membrane and causes the channels to open or close. The consequent change in ion flow through the channel proteins results either in a depolarization or a hyperpolarization. This changes the frequency of action potentials in a sensory neuron.

Some mechanosensing cells have receptors in their cell membranes. These mechanoreceptors are ion channels that open or close as the cell's shape alters when exposed to pressure changes. For example, when you feel someone tapping you on the shoulder, it's because the pressure from their finger distorts the cell membranes of your skin cells, changing the conformation of the ion channels. In other cases, mechanoreceptors in specialized cells called hair cells open or close with the assistance of specialized hair-like structures.

The Structure of Hair Cells **Hair cells** are mechanoreceptor cells, named for their stiff outgrowths called **stereocilia** (singular: **stereocilium**; Figure 44.3a). The "hairy-looking" stereocilia are microvilli that are reinforced by actin filaments.

Many hair cells also have a single **kinocilium**, a true cilium that contains a 9 + 2 arrangement of microtubules (Ch. 7, Section 7.6). Hair cells are found in the ears of land-dwelling vertebrates and the lateral line system in many species of fishes and some amphibians.

As Figure 44.3a shows, the stereocilia in a hair cell are arranged in order of increasing length; if a kinocilium is present, it is the longest of all the projections. These structures extend into a fluid-filled chamber.

Signal Transduction in Hair Cells As shown in Figure 44.3b, (1) if stereocilia are bent in the direction of the kinocilium in response to pressure from sound waves, (2) the distortion causes potassium (K^+) channels in the stereocilia to open. This is the common theme connecting pressure-sensing cells: Bending opens or closes ion channels.

Recall that the opening of K^+ channels usually causes an outflow of K^+ that hyperpolarizes neurons (Ch. 43, Section 43.1). Hair-cell plasma membranes respond differently, however, because they are bathed by extracellular fluid with an extraordinarily high K^+ concentration. As a result, (3) when the channels open, K^+ rushes in and causes the cells to depolarize.

In hair cells, (4) depolarization causes an inflow of calcium ions, which (5) triggers synaptic vesicles to fuse with the plasma membrane, leading to (6) an increase in the amount of neurotransmitter released at the synapse between the hair cell and a sensory neuron. The end result is excitation of the afferent sensory neuron, meaning that it becomes more likely to fire action potentials. You might recall that afferent neurons are part of the peripheral nervous system and conduct information to the CNS (Ch. 43, Section 43.4).

If sound-pressure waves bend stereocilia the other way, however, the K^+ channels close, and the cell hyperpolarizes. This decreases the amount of neurotransmitter released at the synapse and inhibits the postsynaptic sensory neuron, making it less likely to fire action potentials.

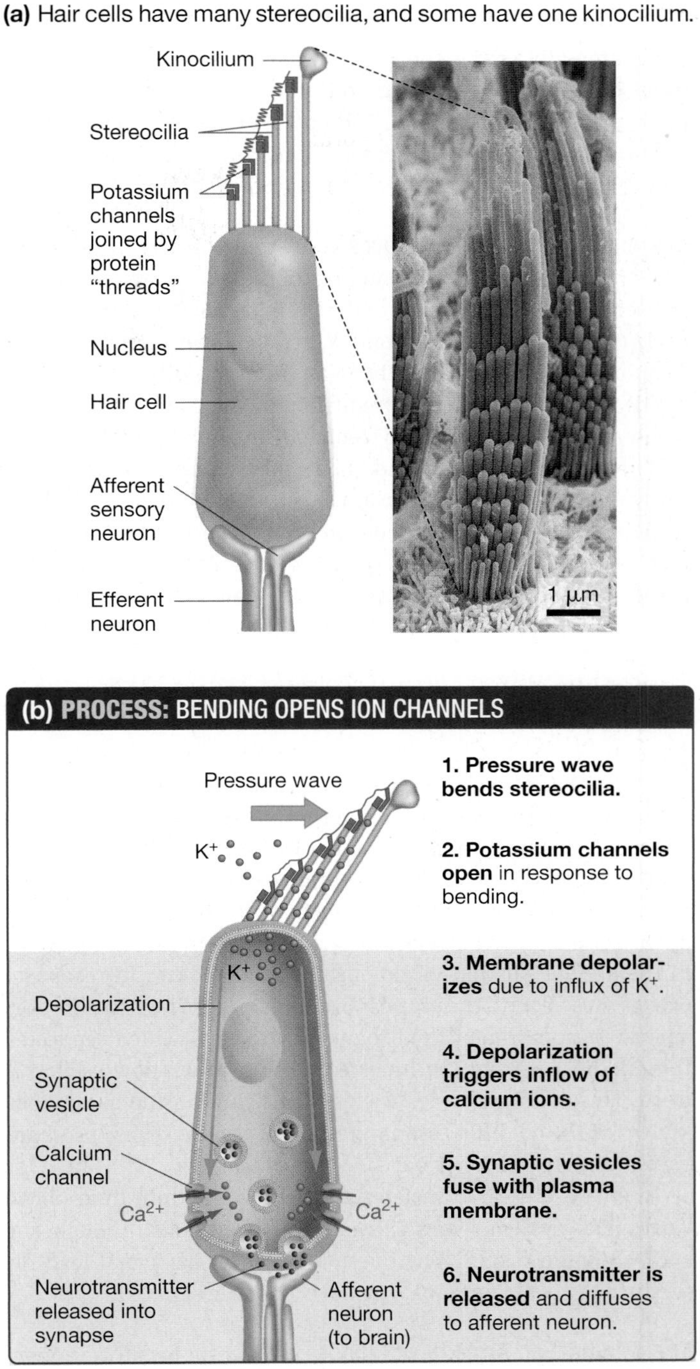

Figure 44.3 **Hair Cells Transduce Sound Waves to Electrical Signals.**

How can bending affect ion channels? Electron micrographs show that tiny threads connect the tips of stereocilia to each other. One hypothesis contends that when the stereocilia are bent, the threads connecting them somehow pull on the potassium channels in the membrane of the next longest stereocilium and open them like tiny trapdoors (see Figure 44.3b, step 2). Researchers still do not fully understand how the ion channels involved in pressure reception work.

The Mammalian Ear: Hearing

The mammalian ear is the site of two important mechanosensations: **hearing**, the sensation produced by the wavelike changes in air pressure called sound, and **equilibrium**, the sensation of orientation, rotation, and acceleration. Both of these senses use hair cells in similar ways to sense remarkably different stimuli.

Hearing Detects Sound Waves A sound consists of waves of pressure in air or water. The number of pressure waves that occur in 1 second is the **frequency** of the sound, reported in units called hertz (Hz), or cycles per second. When you hear different sound frequencies, you perceive them as different **pitches**. A high-pitched sound may have a frequency in the range of 8000 Hz, whereas a low-pitched sound frequency might be 1000 Hz.

The ear transduces sound waves into action potentials that carry information to the brain. To understand how changes in the membrane potential of hair cells result in hearing, let's focus on the human ear as a case study. The human ear has three sections: the **outer ear**, **middle ear**, and **inner ear**, as shown in Figure 44.4. A membrane separates each section from the next.

The path of sound through the ear is traced in Figure 44.4. The outer ear, which projects from the head, collects incoming pressure waves and funnels them into a tube known as the ear canal. At the inner end of the ear canal (see the lower part of Figure 44.4), the waves strike the **tympanic membrane**, or eardrum, which separates the outer ear from the middle ear.

The repeated cycles of air compression cause the tympanic membrane to vibrate back and forth with the same frequency as the sound wave. The vibrations are passed to three tiny bones in the middle ear that vibrate against one another in response. One of these bones, the **stapes** (pronounced *STAY-peez*), vibrates against a membrane called the **oval window**, which separates the middle ear from the inner ear. The oval window oscillates in response and generates waves in the fluid inside a chamber known as the **cochlea** (pronounced *KOK-lee-ah*). These pressure waves are sensed by hair cells in the cochlea.

In effect, the ear translates airborne waves into fluid-borne waves. The system seems extraordinarily complex, though, for such a simple result. Why doesn't the outer ear canal lead directly to the oval window? Why have a middle ear at all?

The Middle Ear Amplifies Sounds Biologists began to understand the function of the middle ear when they recognized two key aspects of its structure. First, the size difference between the tympanic membrane and the oval window is important. The tympanic membrane is about 15 times as large as the oval window, causing the amount of vibration induced by sound waves to increase by a factor of 15 when it reaches the oval window. This phenomenon is similar to taking the same amount of force used to bang on a very large door and applying it to a very small door.

In addition, the three middle ear bones act as levers that further amplify vibrations from the tympanic membrane. The overall effect in mammals is to amplify sound by a factor of 22—meaning that soft sounds are amplified enough to stimulate hair cells in the cochlea. Thus, biologists interpret the mammalian middle ear as an adaptation for increasing sensitivity to sound.

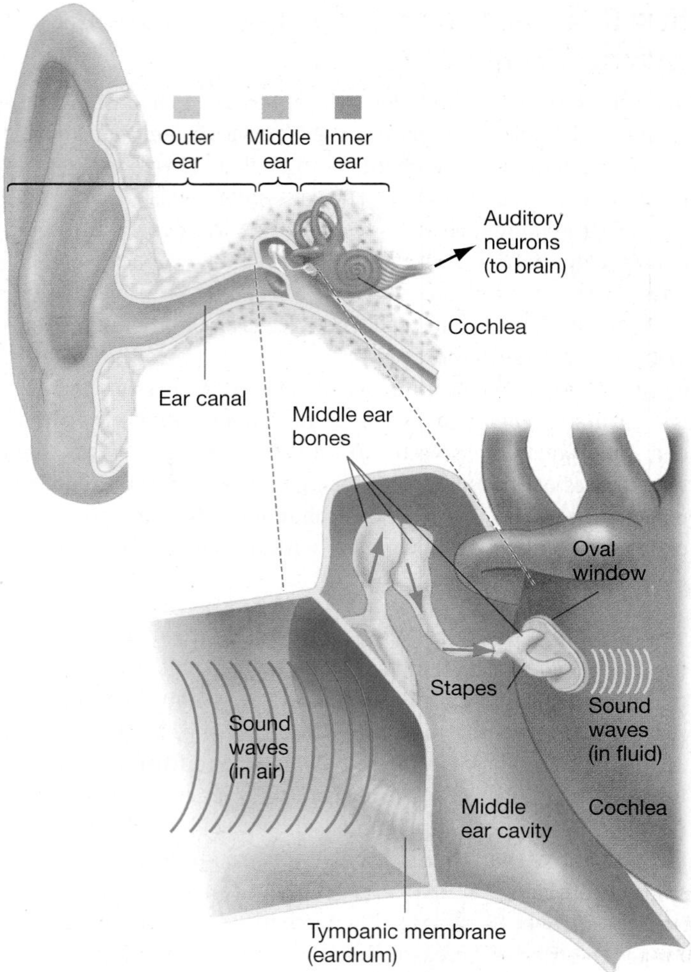

Figure 44.4 The Human Ear Detects Sound Waves. The middle ear starts with the tympanic membrane and ends at a cochlear membrane called the oval window. Sound waves cause the tympanic membrane to vibrate, which causes tiny bones in the inner ear to vibrate and bang against the oval window, resulting in pressure waves in the fluid inside the cochlea.

To summarize, the mammalian outer ear transmits sound waves from the environment to the middle ear; the middle ear amplifies these waves enough to stimulate the hair cells within the cochlea of the inner ear.

If all hair cells responded equally to all frequencies of sound, you would be able to perceive only one pitch. Everyone's voice—indeed, every noise—would sound the same. How can hair cells distinguish different frequencies?

The Cochlea Detects the Frequency of Sounds As Figure 44.5a shows, the cochlea is a coiled tube with a set of internal membranes that divide it into three chambers. Hair cells, forming rows in the middle chamber, are embedded in a tissue that sits atop the **basilar membrane** (Figure 44.5b). In addition, the hair cells' stereocilia touch yet another, smaller surface called the **tectorial membrane**. (The kinocilium is not present in a mature mammalian cochlear hair cell.) In effect, hair cells are sandwiched between membranes.

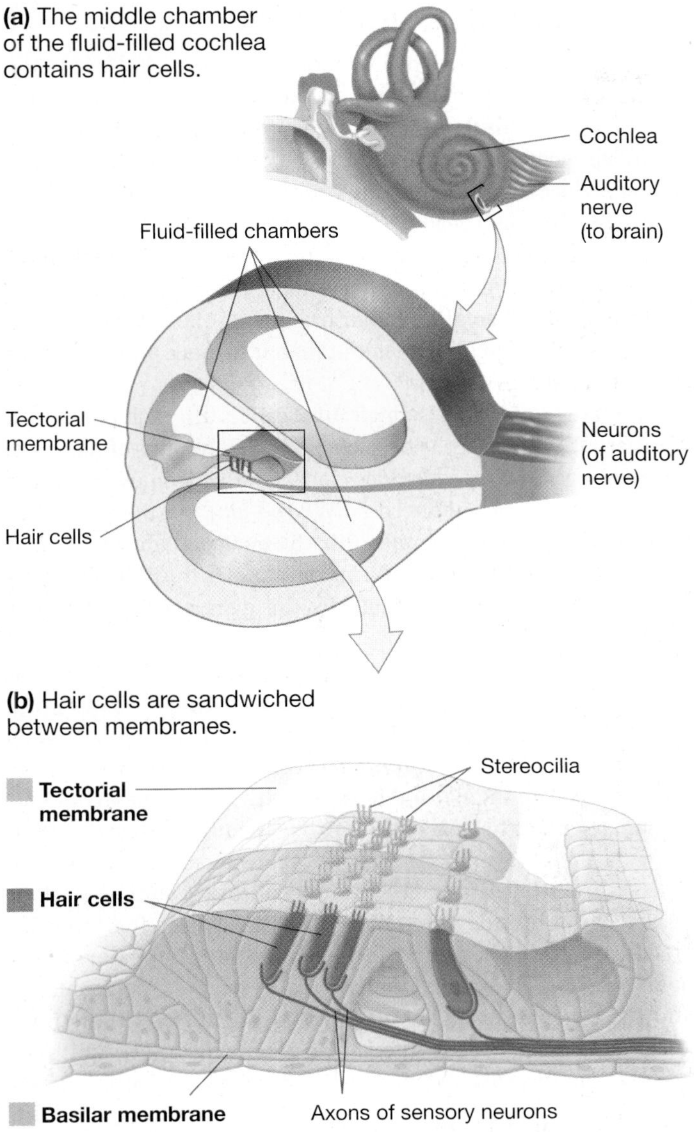

Figure 44.5 The Human Cochlea Contains Fluid-Filled Chambers Separated by Membranes.

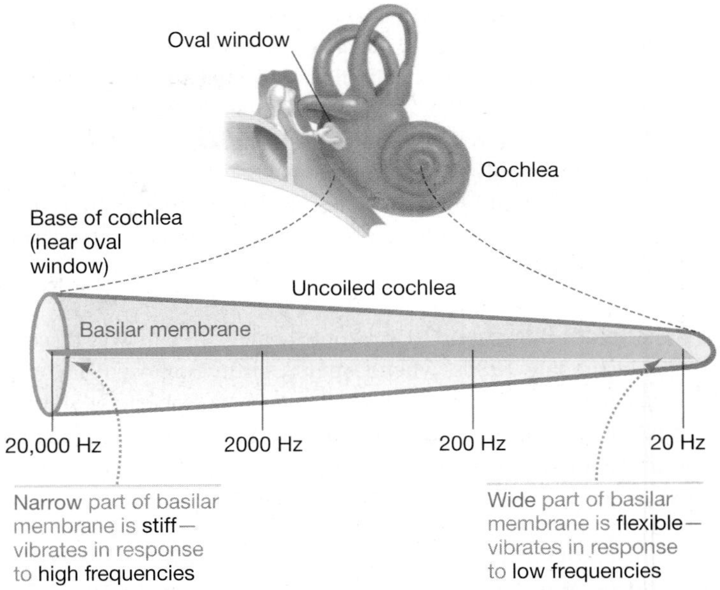

Figure 44.6 The Basilar Membrane Varies in Stiffness along Its Length. Different parts of the basilar membrane vibrate in response to different sound frequencies.

Researchers struggled for decades to understand how these membranes affect hair cell function. It is virtually impossible to study cochleas in living organisms, because the cochleas are tiny, complex, coiled, and buried deep inside the skull. During the 1920s and 1930s, however, Georg von Békésy pioneered work on the structure and function of these organs by performing experiments on cochleas that he had dissected from fresh human cadavers.

Von Békésy was able to vibrate the oval window and record how the cochlea's internal membranes moved in response. He found that when a pressure wave traveled down the fluid in the upper and lower chambers, the basilar membrane vibrated in response. His key finding, though, was that sounds of different frequencies caused the basilar membrane to vibrate maximally at specific points along its length (Figure 44.6). When the basilar membrane vibrated in a particular location, the stereocilia of the hair cells there were bent one way and then the other by the tectorial membrane.

Von Békésy also noted that the basilar membrane is stiff near the oval window and flexible at the other end. This is why each segment vibrates in response to a different frequency of sound. Just as a stiff drumhead produces a high-pitched sound and a loose drumhead yields a low-pitched sound, high-frequency sounds cause the stiff part of the basilar membrane to vibrate; low-frequency sounds cause the flexible part to vibrate.

To summarize, certain portions of the basilar membrane vibrate in response to specific frequencies and result in the bending of hair-cell stereocilia. In this way, hair cells in a particular place on the membrane respond to sounds of a certain frequency. When the occipital lobe of the cerebrum receives action potentials from neurons associated with specific hair cells, it interprets the action potentials as a particular pitch—meaning a specific frequency of sound. The result is the sense called hearing.

Complex sounds contain a wide variety of frequencies and trigger particular combinations of hair cells. Through experience, the brain learns which combinations of frequencies represent music, a fire alarm, or a best friend's voice.

Humans can hear sounds between 20 Hz and 20,000 Hz (20,000 Hz is equal to 20 kHz, or kilohertz). But some mammals can hear low-frequency infrasounds that are too low for humans to hear (*infra*– means "below or under"); others can hear high-frequency ultrasounds that are above the range of human hearing (*ultra*– means "beyond").

Elephants Detect Infrasound When Katherine Payne was observing elephants at a zoo in the mid-1980s, she noticed a subtle throbbing in the air. Payne knew that infrasound can produce such sensations.

To test the hypothesis that the elephants were producing infrasonic vocalizations, Payne returned to the zoo with a tape recorder and microphones that could pick up sounds at extremely low frequencies. Played at normal speed, the tape she made was silent. But when she raised the pitch of the sounds by speeding up the tape, she heard a chorus of cow-like noises. The elephants were calling to each other, using low-frequency sounds.

Follow-up research showed that elephants have the best infrasonic hearing of any land mammal. Because infrasound can travel exceptionally long distances, biologists hypothesize that infrasonic calls allow wild elephants to communicate when they are miles apart.

Bats Detect Ultrasound Ultrasonic hearing in bats was discovered in the late 1930s, when Donald Griffin borrowed the only ultrasonic apparatus then in existence from Robert Galambos, a fellow graduate student. Griffin used the machine to demonstrate that flying bats constantly emit ultrasounds. In subsequent experiments, he documented that a bat with cotton in its ears, or with its mouth taped shut, crashed into walls when released in a room. Blindfolded bats, in contrast, never crashed.

Griffin and Galambos concluded that bats use sound echoes (sonar) to navigate. This concept, termed **echolocation**, was an outlandish idea at the time. When Galambos described it at a meeting in 1940, another scientist shook him by the shoulders and said, "You can't really mean that!"

Bats generate high-frequency sound waves with their larynx, or voice box. These waves "bounce" off surfaces (including those of insects), producing echoes that the bat detects in its inner ear (Figure 44.7). Recall that hair cells on different sections of the basilar membrane sense sound waves of different frequencies. In bats, a huge area of the basilar membrane is specialized for sensing the high-frequency sounds of the returning echoes. Similarly, the part of the brain used to process sound is unusually large in bats, highlighting the extremely important role of echolocation in their navigation and hunting.

More recent research has shown that dolphins, shrews, and certain other animals besides bats use sonar. In fact, at least some of these species probably perceive shapes with their ears better than they do with their eyes.

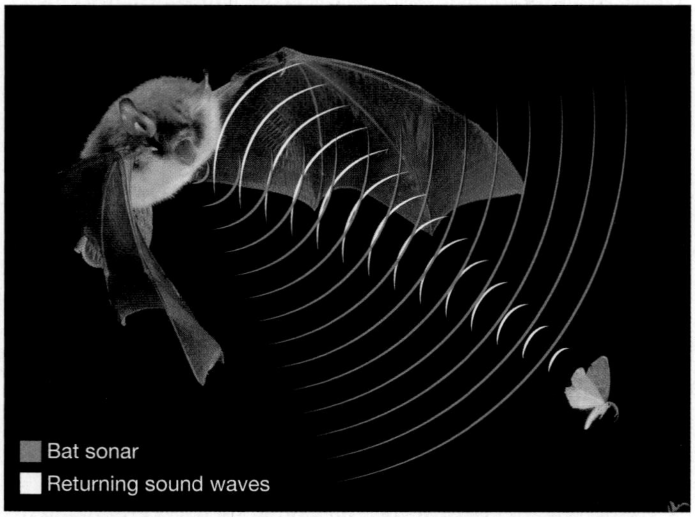

Figure 44.7 Bats Emit Ultrasonic Sound Waves That Bounce Off Surfaces. Returning sound waves are sensed by the bat's inner ear.

The Mammalian Ear: Equilibrium

The inner ear is responsible for more than hearing. Next to the cochlea are two other inner ear structures—the **semicircular canals** and **vestibule**—that allow us to maintain equilibrium. Equilibrium includes orientation (for example, up, down, and sideways), acceleration (moving your head in one plane), and rotation (moving your head in several planes).

Imagine a gymnast on the balance beam. As she jumps, flips, and dismounts, hair cells in her semicircular canals and vestibule transmit information to her brain about her changing position in space so that she can maintain her balance and posture as she performs (Figure 44.8).

The three fluid-filled semicircular canals are each positioned in different planes. At the base of each canal, groups of hair cells with stereocilia are embedded in a gel-like domed structure called a cupula (pronounced *KEWP-yoo-lah*). As the gymnast's head rotates during a complex flip, the movement of the fluid in the canals bends the stereocilia and stimulates the hair cells. This motion-specific stimulation allows the gymnast to detect her rotation.

The vestibule contains two dense clusters of hair cells called **maculae** (singular: **macula**) that detect linear acceleration and orientation. The kinocilia and stereocilia of the hair cells are embedded in a gelatinous matrix covered with a fibrous membrane. The membrane is filled with little crystals of calcium carbonate that make the fibrous membrane heavier than the gelatinous layer. Any rotation or acceleration causes the heavy fibrous layer to shift at a different rate than the gelationous layer containing the stereocilia. This causes the kinocilia and stereocilia to bend and the mechanoreceptors to open, depolarizing the maculae. One macula is situated horizontally, and the other is situated vertically, so the differential stimulation of their hair cells inform the gymnast's brain about her orientation. In addition, these cells are stimulated as she accelerates for dismount.

The importance of the inner ear's equilibrium system is evident when it is disrupted. **Vertigo** is a medical condition that makes affected people feel as if they are spinning—or their environment is. Vertigo has many causes, most associated with injury to the inner ear—for example, if a calcium carbonate crystal is dislodged from the vestibule and enters a semicircular canal. It is often debilitating unless treated.

Sensing Pressure Changes in Water

Hair cells in the ear allow mammals to sense changes in air pressure that are perceived as sound, or changes in fluid pressure inside the ear that are perceived as orientation. But in fishes and aquatic amphibians, hair cells in a different organ allow the perception of pressure changes in the water that surrounds them. In most fishes and larval amphibians, groups of hair cells are embedded in multiple cupulae inside water-filled canals that run the length of the body (Figure 44.9a), forming a sensory organ called the **lateral line system**.

Pressure changes in the surrounding water—whether resulting from waves, an animal swimming nearby, or some other force—cause changes in the pressure of water that enters and

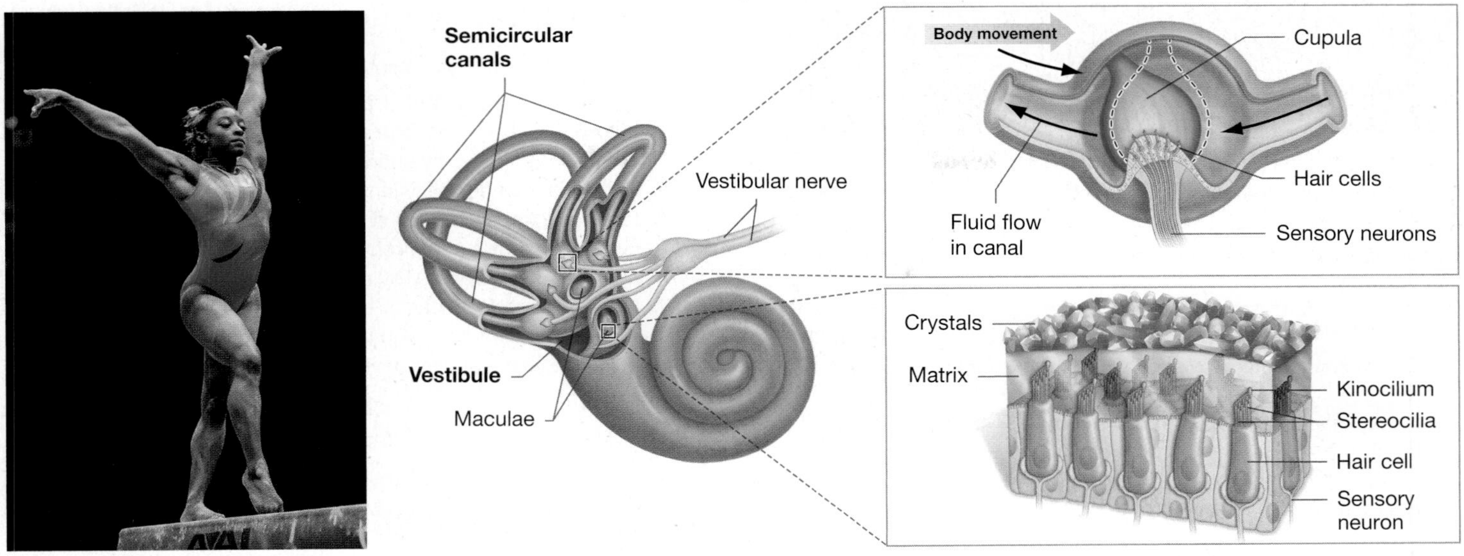

Figure 44.8 The Semicircular Canals and Vestibule of the Inner Ear Detect Orientation.

moves through the lateral line system (**Figure 44.9b**). Much like in the human inner ear, these changes cause kinocilia and stereocilia on the hair cells to bend, and the distortion leads to a change in the frequency of action potentials along sensory neurons that project to the brain. In this way, most aquatic vertebrates get information about pressure changes at specific points along the head and body.

What use do fishes make of the lateral line system? It seems reasonable that the lateral line system could be helpful for identifying mates, locating prey, or avoiding predators. However, fishes could also use vision, smell, or other senses for these functions. How important is the lateral line system?

To answer this question, researchers studied how nocturnal catfish locate prey. At night, catfish cannot use vision to hunt, so they must detect stimuli that persist in the wake of their prey after it has moved on. The researchers hypothesized that the nocturnal catfish hunt smaller fishes using the lateral line system.

To test this hypothesis, they conducted an ablation experiment on the catfish. Ablation is the removal or blocking of a structure or process. In this case, the researchers ablated the lateral line system in one group of catfish, using a chemical that blocks the hair cells from responding to pressure waves. In another group of catfish, they ablated the ability to smell or taste the water by surgically removing the lobes of the brain responsible for these senses. The set up is shown in **Figure 44.10** on page 952.

While the catfish whose smell and taste were ablated captured about the same percentage of guppies as the non-ablated control group, the catfish whose lateral line was ablated had very poor hunting success. The researchers concluded that the lateral line system is much more important than other senses in successful hunting by nocturnal catfish. Other studies have confirmed that fishes and aquatic amphibians use the lateral line system to detect anything moving in the surrounding water, including prey, predators, and mates.

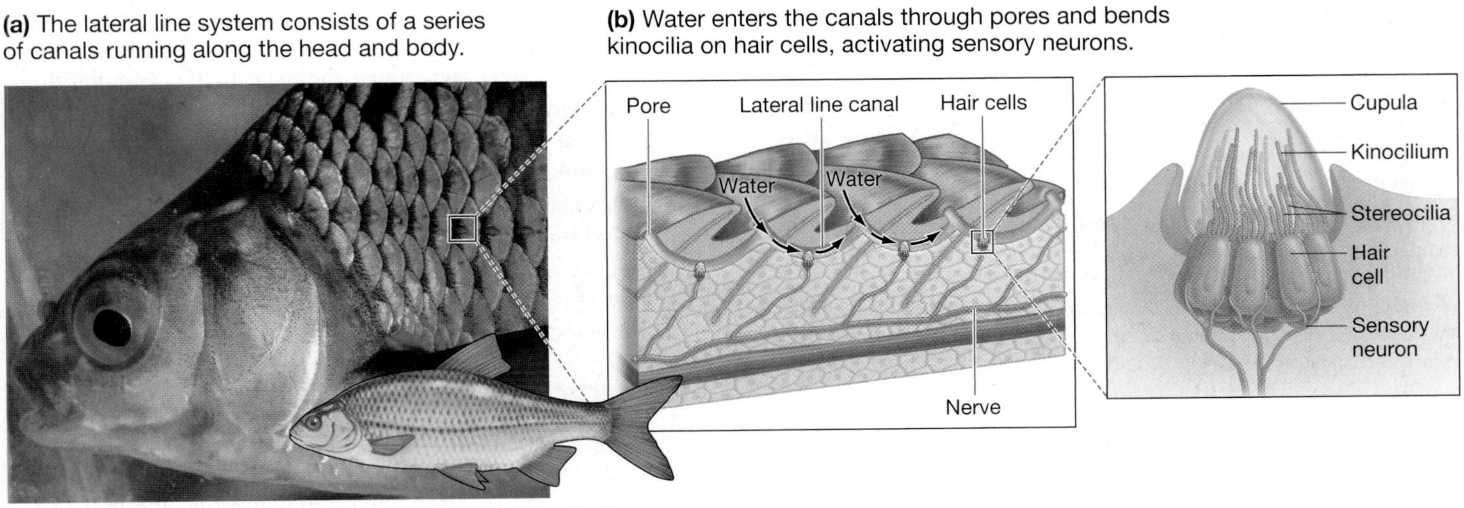

Figure 44.9 The Lateral Line System Detects Pressure Waves in Water. (a) Lateral line canals lie just under the fish's epidermis. **(b)** Hair cells within the canals have kinocilia and stereocilia embedded in a gel-like cupula.

RESEARCH

QUESTION: What cues do predatory catfish use to detect prey at night?

HYPOTHESIS: Predatory catfish detect prey using the lateral line system.

ALTERNATE HYPOTHESIS: Predatory catfish detect prey using smell and taste.

NULL HYPOTHESIS: Predatory catfish use none of these senses to detect prey.

EXPERIMENTAL SETUP:

1. Three experimental groups of catfish (n = 16 for each group):

Lateral line ablated	Smell and taste ablated	Control catfish

2. Acclimate catfish to dark tank at night.
3. Add guppies to tank and record capture success rate.

PREDICTION OF HYPOTHESIS: Catfish with their lateral line ablated will have lower capture success than control fish.

PREDICTION OF ALTERNATE HYPOTHESIS: Catfish with smell and taste ablated will have lower capture success than control fish.

PREDICTION OF NULL HYPOTHESIS: There will be no capture success difference among groups.

RESULTS:

Percentage of Guppies Captured

Lateral line ablated	Smell and taste ablated	Control catfish
17%	60%	65%

CONCLUSION: Nocturnal predatory catfish use the lateral line system to detect prey.

Figure 44.10 The Lateral Line System Is Used for Predation.

Source: K. Pohlmann, J. Atema, and T. Breithaupt. 2004. The importance of the lateral line in nocturnal predation of piscivorous catfish. *Journal of Experimental Biology* 207: 2971–2978.

✔ **PROCESS OF SCIENCE** If you wanted to conduct a study similar to this one on a diurnal species of catfish, what other sense would you need to control for, and how would you do it?

CHECK YOUR UNDERSTANDING

✔ **If you understood this section, you should be able to . . .**

1. Predict and explain the effect on hearing in each of the following cases: a punctured eardrum, a mutation that results in dramatically shortened stereocilia, and an age-related loss of flexibility in the basilar membrane.
2. Motion sickness can result when the brain receives conflicting information from senses about motion. Devise a hypothesis to explain why motion sickness is often worse when you read in a car.
3. Would you expect the aquatic clawed frog or the semiaquatic bullfrog to have a more developed lateral line system? Explain.

Answers are available in Appendix A.

44.3 Photoreception: Sensing Light

After you complete this section, you should be able to . . .

- Explain how animals sense light via photoreception.

Most animals have a way to sense light. The organs involved in **photoreception** range from simple light-sensitive eyespots in flatworms to the sophisticated, image-forming eyes of vertebrates, cephalopod mollusks, and arthropods.

Variation in the structure of light-sensing organs illustrates an important general principle about the sensory abilities of animals: In most cases, a species' sensory abilities correlate with the environment it lives in and its mode of life—how it finds food and mates. Eyes and other sensory structures are adaptations that allow individuals to thrive in a particular environment. Salamander species that live in meadows and forests have sophisticated eyes; those that live in lightless caves have no functional eyes at all. Keep this principle in mind as you delve into the details of how insects and vertebrates see.

The Insect Eye

Insects have **compound eyes** composed of hundreds or thousands of light-sensing columns called **ommatidia**. As Figure 44.11 shows, each ommatidium has a lens that focuses light onto a small number of receptor cells—usually four. The receptor cells, in turn, send axons to the brain. Each ommatidium acts like a single pixel on a computer monitor: It contributes information about one small piece of the visual field. Therefore, a compound eye with more ommatidia has higher resolution—meaning greater resolving power, or ability to distinguish objects.

In addition, the presence of many light-sensing columns makes species with compound eyes particularly good at detecting movement. Insects that hunt by sight, such as damselflies and dragonflies, have particularly large numbers of ommatidia.

The Vertebrate Eye

Compound eyes are found in insects, crustaceans, and certain other arthropods. Because compound eyes appear only in species that are part of the same monophyletic group (Ch. 25, Section 25.1), researchers conclude that this type of eye structure evolved just once—in an ancestor of today's arthropods.

(a) Ommatidia are the functional units of insect eyes.

(b) Each ommatidium contains receptor cells that send axons to the CNS.

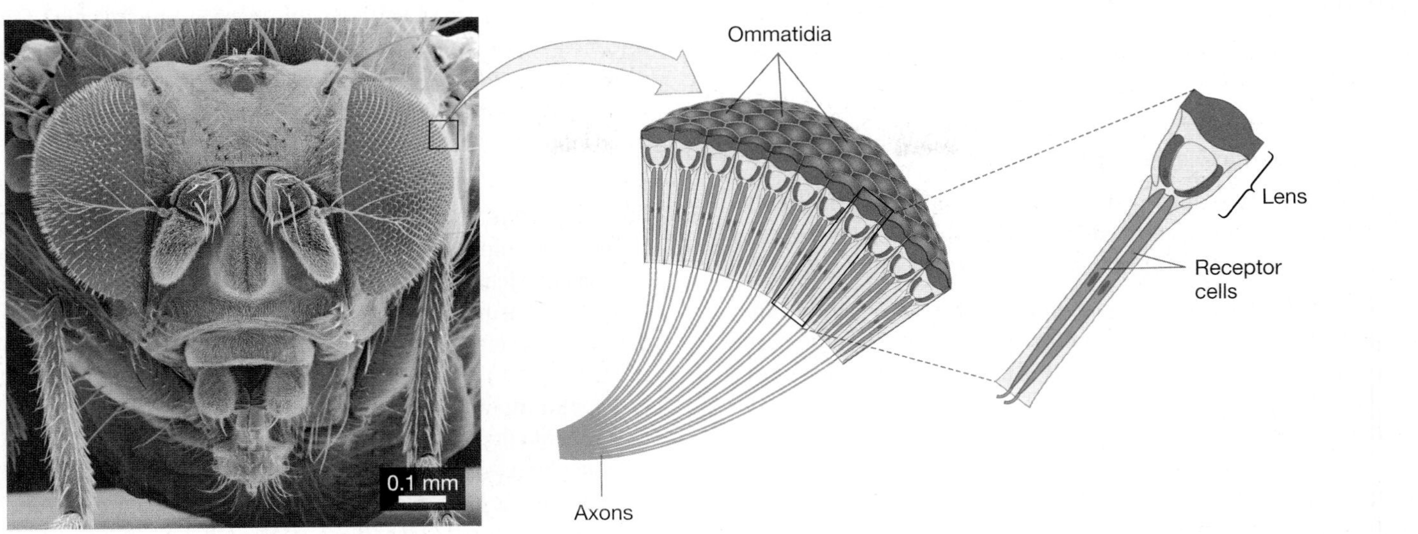

Figure 44.11 In the Compound Eyes of Insects, Each Ommatidium Sees Part of the Visual Field. The micrograph in part (a) is colorized to match the drawings in part (b).

In contrast, the **simple eye**—a structure with a single lens that focuses incoming light onto a layer of many receptor cells—evolved independently in several widely divergent groups, including annelids, cephalopod mollusks (squid and octopuses), and vertebrates. Let's examine the vertebrate version of the simple eye more closely.

The Structure of the Vertebrate Eye Figure 44.12a shows the major structures in the human eye:

- The outermost layer of the eye is a tough rind of white tissue called the sclera. This is the "white of the eye" in mammals.
- The front of the sclera forms the **cornea**, a transparent sheet of connective tissue.
- The **iris** is a pigmented, round muscle just inside the cornea. The iris can contract or expand to control the amount of light entering the eye.
- The **pupil** is the hole in the center of the iris.
- Light enters the eye through the cornea and passes through the pupil and a curved, clear **lens**.
- Together, the cornea and lens focus incoming light onto the retina in the back of the eye. The **retina** contains a layer of photoreceptors and several layers of neurons.

Figure 44.12b provides a closer look at the retina, which is attached to the rest of the eye by a single layer of pigmented

(a) The structure of the human eye

(b) In the retina, cells are arranged in layers.

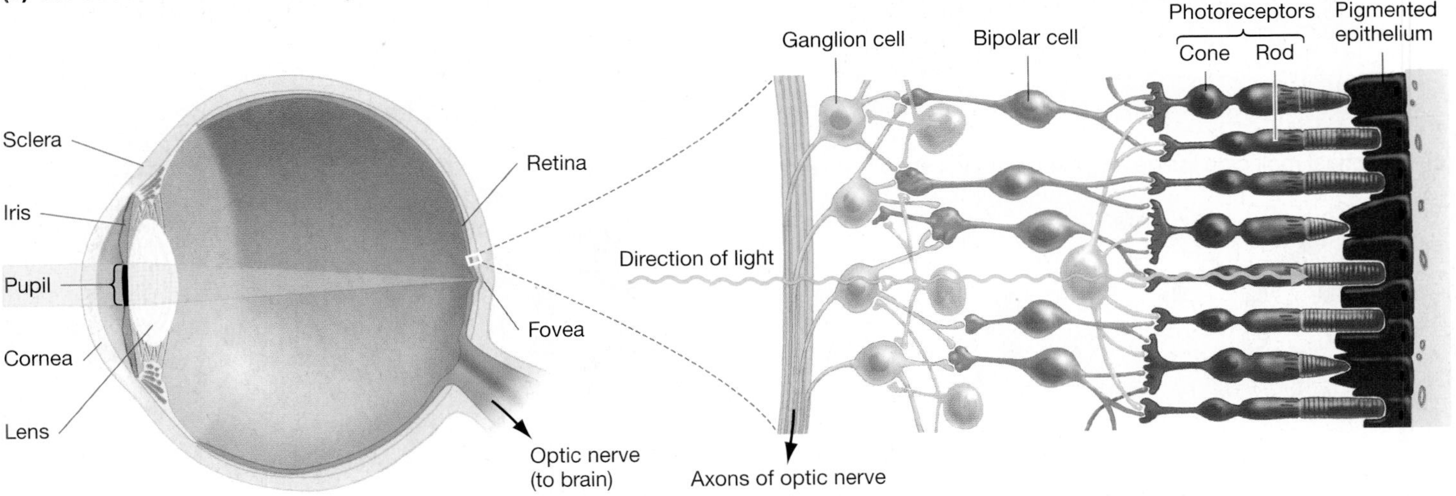

Figure 44.12 Simple Eyes Have a Single Lens That Focuses Incoming Light on Receptor Cells. (a) Light passes through the pupil of the eye and is focused onto the retina. **(b)** The photoreceptor cells (rods and cones), which respond to light, are in the "outermost" layer of the retina, farthest from the light source.

epithelial cells. From back to front, the retina is made up of three distinct cell layers:

1. The photoreceptors, sensory cells that respond to light, are held in place by the pigmented epithelium.
2. Photoreceptors synapse with an intermediate layer of connecting neurons called **bipolar cells**.
3. Bipolar cells synapse with neurons called **ganglion cells**, which form the innermost layer of the retina. The axons of the ganglion cells project to the brain via the **optic nerve**.

Note that vertebrate eyes—including yours—have a blind spot because there are no photoreceptor cells where the optic nerve leaves the retina. If light falls in this area, there are no sensory cells available to respond. No signal is sent to the brain, so the light isn't seen.

What Do Rods and Cones Do? Early anatomists established that the photoreceptors in vertebrate eyes come in two distinct types: small rod-shaped cells and cone-shaped cells, called **rods** and **cones**, respectively. When technical advances allowed changes in the membrane potentials of these cells to be recorded, it became clear that rods and cones differ in function as well as structure. Rods are sensitive to dim light but not to color. Cones, in contrast, are much less sensitive to faint light but respond to different wavelengths (i.e., colors). These discoveries explained why night vision is largely black and white—at night, the rods do most of the work.

How Do Rods and Cones Detect Light? As Figure 44.13a shows, rods and cones have segments that are packed with membrane-rich disks. The membranes contain large quantities of a transmembrane protein called **opsin**; each opsin molecule is associated with a molecule of the pigment **retinal** (Figure 44.13b). The two-molecule complex found in rod cells is called **rhodopsin**. Different complexes occur in cone cells.

Experiments with isolated retinal molecules confirmed that retinal changes shape when it absorbs light. Specifically, the number-11 carbon in the retinal molecule changes from the *cis* conformation to the *trans* conformation (Figure 44.13c). Retinal is a light switch. The shape change that occurs in retinal triggers a series of events that culminate in a stream of action potentials being sent to the brain.

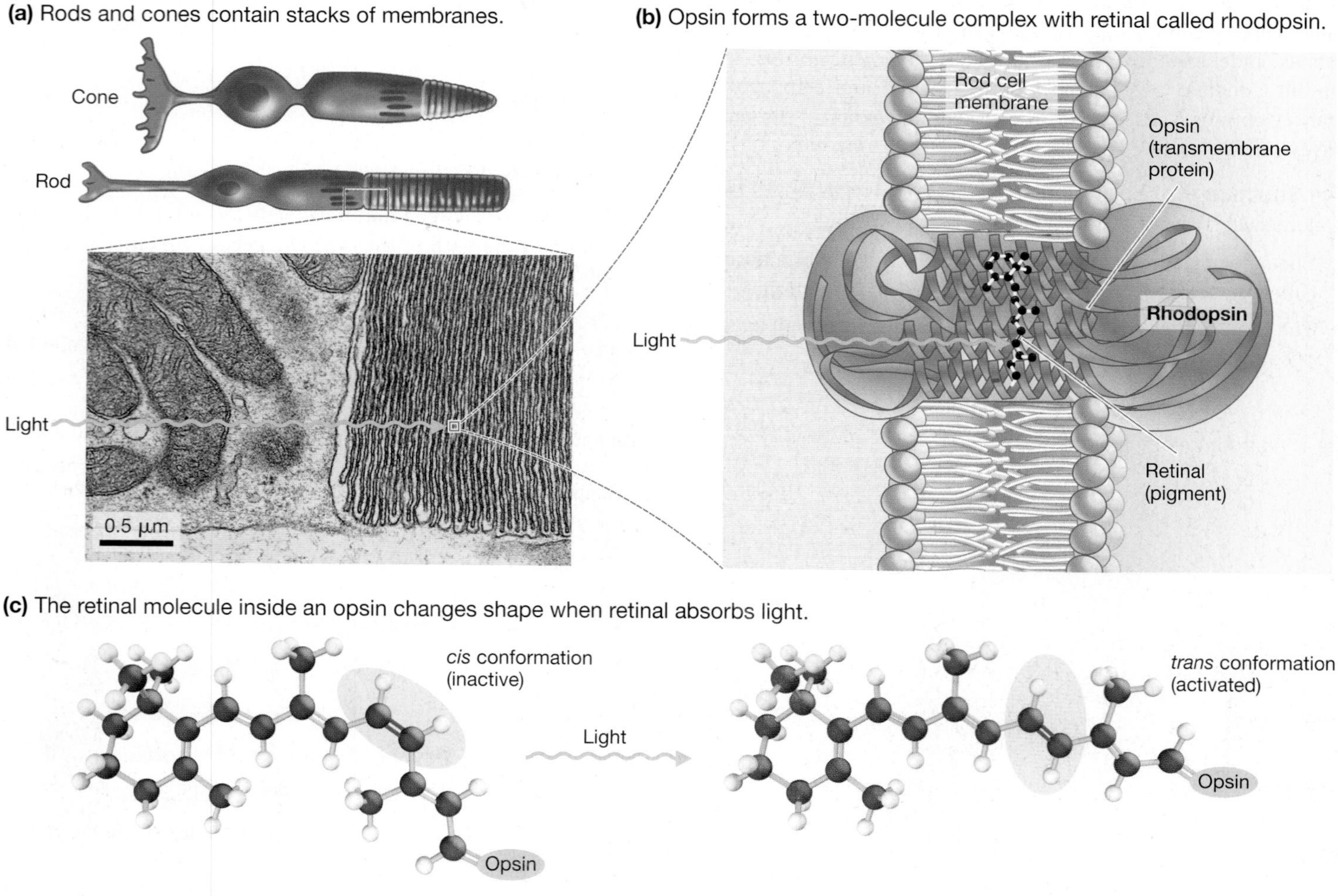

Figure 44.13 **Rods and Cones Are Packed with Transmembrane Opsin Proteins That Contain the Pigment Retinal. (a)** Rods and cones have membranous disks containing thousands of opsin and retinal molecules (together called rhodopsin in rods). **(b)** Each opsin holds one retinal molecule. **(c)** Retinal changes conformation when it absorbs light. In response, opsin also changes shape.

This system is exquisitely sensitive: Biologists have recorded a measurable change in the membrane potentials of rod cells in response to a single photon of light. But how do humans and other animals perceive color?

Color Vision: The Puzzle of Dalton's Eye To answer this question, consider the research program initiated by the physicist John Dalton[1] in the late eighteenth century. At the age of 26, Dalton realized that he and his brother saw colors differently than other people did. To them, red sealing wax and green laurel leaves appeared to be the same color, and a rainbow exhibited only two hues. Dalton and his brother could not differentiate the colors red and green. This condition is called red–green color blindness (Figure 44.14).

In a lecture delivered in 1794, Dalton explained his perceptions by hypothesizing that red wavelengths failed to reach his retinas. Further, he hypothesized that because a normal eyeball is filled with clear fluid, and because blue fluids absorb red light, his defective vision resulted from the presence of bluish fluid rather than clear fluid in his eyes.

To test this hypothesis, Dalton left instructions that his eyes should be removed after his death and examined to see if the fluid inside was blue. When he died 50 years later, an assistant dutifully removed the eyes from Dalton's corpse and examined them. The fluid inside the eye was not blue at all, however, but slightly yellow—the normal color for an older person. Further, when the back was cut off one eye and colored objects were viewed through the lens, the objects looked perfectly normal. Dalton's hypothesis was incorrect.

Color Vision: Multiple Opsins What, then, caused Dalton's color blindness? The key to answering this question was the discovery that the human retina contains three types of color-sensitive photoreceptors: blue, green, and red cones, named for the colors to which they are most sensitive.

To follow up on this result, biologists analyzed opsin molecules from the three cone types and found that each opsin had a distinct amino acid sequence. The three proteins are called the S, M, and L opsins, for short-, medium-, and long-wavelength sensitivity, respectively. Although retinal is the light-absorbing molecule in all photoreceptor cells, the different opsin molecules cause each type to respond to a different range of wavelengths of light. Figure 44.15a shows ranges for the three different opsins in cones.

Based on these results, biologists hypothesized that the brain distinguishes colors by combining signals initiated by the three classes of opsins. Figure 44.15b, for example, graphs how much light is absorbed by the S, M, and L opsins of humans across a

Figure 44.14 People with Red–Green Color Blindness Cannot Distinguish Red from Green. These images show what a person with red–green color blindness would see, compared to a person with normal color vision.

[1]Dalton was an accomplished physicist. He was the first proponent of the atomic theory of matter and formulated Dalton's law on the partial pressures of gases (introduced in Ch. 42, Section 42.2). Red–green color blindness is sometimes called daltonism in his honor.

(a) Different opsins respond to different wavelengths.

Opsin type	Absorption range	Peak absorption
S	400–500 nm	420–440 nm
M	450–630 nm	530–545 nm
L	500–700 nm	560–580 nm

(b) Opsin absorption ranges overlap.

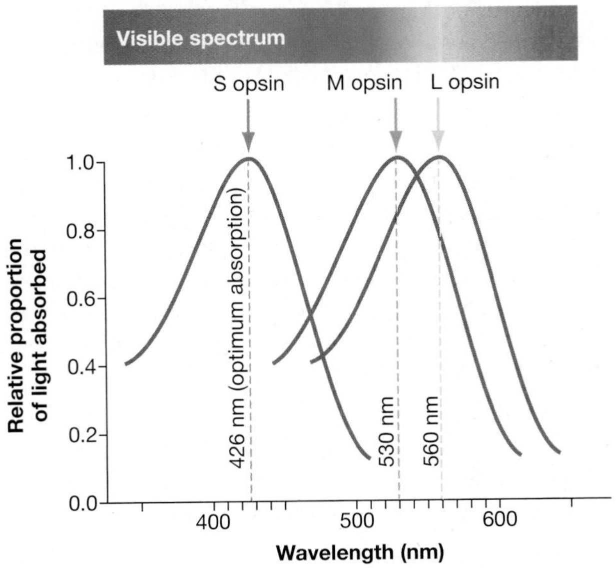

Figure 44.15 Color Vision Is Possible because Different Opsins Absorb Different Wavelengths of Light. Each human cone cell contains one of three different types of opsin. Each opsin absorbs a different range of wavelengths.

✔ The retinal molecules in S, M, and L opsins are identical. What is the likely reason that the opsins respond to different wavelengths of light?

range of wavelengths. Notice that light at a wavelength of 560 nanometers (nm) is absorbed strongly by L cones, to an intermediate degree by M cones, and not at all by S cones. In response to the corresponding signals from these cells, the brain perceives the color yellow.

Does this hypothesis explain Dalton's color blindness? According to the data in Figure 44.15, wavelengths from green to red are not absorbed by S opsin at all. It is thus unlikely that S opsin is involved in red–green color blindness. Did Dalton fail to distinguish red and green because his M or L cones were defective? Research has shown that red–green color-blind people lack either functional M or L cones, or both. Was the same true of Dalton?

This question was answered in the 1990s, when the genes for the M and L opsins were sequenced. Remarkably, Dalton's eyes had been preserved. Researchers managed to extract DNA from the 150-year-old tissue and analyze his M opsin genes and L opsin genes. They found that Dalton had a normal *L* allele but lacked a functional *M* allele. As a result, he did not have green-sensitive cones. The puzzle of Dalton's color vision was solved.

Red–green color blindness, or the inability to distinguish red and green due to an absence of either M or L cones, is estimated to affect about 5 to 10 percent of men and less than 0.5 percent of women. It is more prevalent in men than in women because it is an X-linked trait (Ch. 14, Section 14.4).

Do Other Animals See Color? What about other animals—do they see color the way humans do? The answer is, probably not. Animals that are active at night have relatively few cone cells and many rods, giving them high sensitivity to light but poor color vision. However, many vertebrate and invertebrate species have four or more types of opsins and probably perceive a world of colors that is much richer than ours.

In general, the types of opsins found in a species correlate with the environment it inhabits and its mode of life. For example,

- A marine fish called the coelacanth (pronounced *SEE-luh-kanth*), which lives in water 200 m deep, has two opsins that respond to the blue region of the spectrum (with absorption peaks at 478 nm and 485 nm). As a result, coelacanths perceive several distinct hues of blue that we would perceive as a single color. Presumably, these opsins offer an adaptive advantage to coelacanths because wavelengths in the yellow and red parts of the spectrum do not penetrate well into deep water—only blue light exists in the coelacanth's habitat.
- In humans and other primates that eat fruit, two of the three opsins are sensitive to wavelengths around 550 nm. The presence of these opsins allows individuals to distinguish between the greens, yellows, and reds of unripe and ripe fruits.
- Many animals (for example, some insects and birds) have opsins that are sensitive to ultraviolet (UV) light, which has shorter wavelengths than humans can see. Certain flowers have UV patterns that serve as signals for insect pollinators (Ch. 38, Figure 38.4). Also, many birds have strong UV patterns in their plumage that are important criteria used by females for selecting mates.

CHECK YOUR UNDERSTANDING

✔ **If you understood this section, you should be able to . . .**

1. Explain why retinal can be thought of as a "light switch."
2. Predict the type of vision loss that would result from each of the following: detachment of the retina following a car accident, a mutation that knocked out the gene for S opsin, and an age-related clouding of the lens.

Answers are available in Appendix A.

44.4 Chemoreception: Sensing Chemicals

After you complete this section, you should be able to . . .

- Explain how animals sense chemicals via chemoreception.

Chemoreception occurs when chemicals bind to chemoreceptors and thereby initiate action potentials in sensory neurons. The sense of taste, called **gustation**, and the sense of smell, called **olfaction**, originate in chemoreceptors. A chemoreceptor detects the presence of a specific molecule by undergoing a change in membrane potential when that molecule is present. In this way, information about the presence of a particular molecule is transduced to an electrical signal in the body.

Taste: Detecting Molecules in the Mouth

In humans, the chemoreceptor cells that sense taste are clustered in structures known as **taste buds**. Although taste buds are scattered around the mouth and throat, most taste buds are located on the tongue (Figure 44.16). A taste bud contains about 100 spindle-shaped taste cells, which make synapses with sensory neurons.

How do taste cells work on a molecular level, and how do they produce the sensation of taste? Early taste research focused on

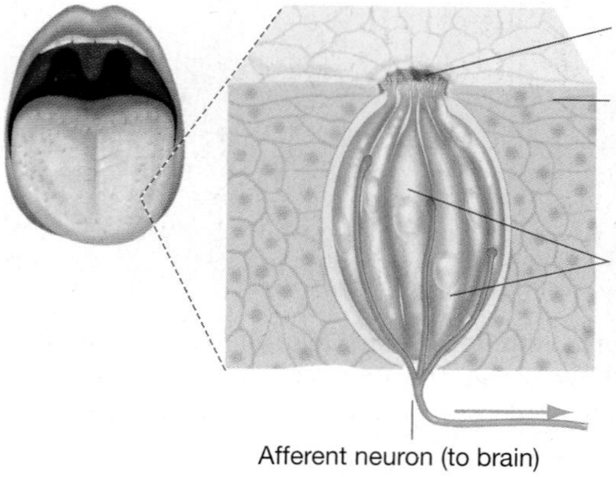

Figure 44.16 Taste Buds Contain Many Types of Chemoreceptors. A single taste bud, shown in this figure, is comprised of many taste cells.

the hypothesis that four "basic tastes" existed: salty, sour, bitter, and sweet.

Salty and Sour Researchers who analyzed the membrane proteins in taste cells found strong evidence that salty and sour sensations result from the activity of ion channels.

- The sensation of saltiness is due primarily to sodium ions (Na^+) dissolved in food. These ions flow into certain taste cells through open Na^+ channels and depolarize the cells' membranes.
- Sourness is due in part to the presence of protons (H^+), which flow directly into certain taste cells through H^+ channels and depolarize the membrane.

The sour taste of grapefruit and other citrus fruits, for example, results from the release of protons by citric acid. In general, the lower the pH of a food, the more it depolarizes a taste cell's plasma membrane, and the more sour the food tastes.

Compared to saltiness and sourness, the molecular mechanisms responsible for the sensations of bitterness and sweetness have been much more difficult to identify.

Why Do Many Different Foods Taste Bitter? Bitterness has been difficult to understand because molecules with very different structures, such as those in orange peel or unsweetened cocoa, are all perceived as bitter. How is this possible?

An answer began to emerge after researchers confirmed that some humans genetically lack the ability to taste certain bitter substances. In 1931, Arthur Fox was synthesizing phenylthiocarbamide (PTC) and accidentally released some of it into the air. A nearby colleague complained of a bitter taste in his mouth, but Fox could not taste anything. Follow-up research confirmed that the ability to taste PTC is inherited and polymorphic. About 25 percent of Americans cannot sense the molecule.

To find the gene responsible for this trait, biologists compared the distribution of genetic markers observed in "tasters" and "nontasters." The mapping effort recently narrowed down this gene's location to several candidate chromosomal regions. In one of the regions, researchers found a family of 40 to 80 genes that encode transmembrane receptor proteins.

Follow-up work has documented that each protein in the family binds to a different type of bitter molecule. A taste cell, however, can have many different receptor proteins from this family. As a result, many different molecules can depolarize the same cell and cause the sensation of bitterness.

Why are so many genes devoted to detecting bitterness? Many of the molecules that bind to these receptor proteins are found in toxic plants; most animals react to bitter foods by spitting them out and avoiding them in the future. In essence, bitterness indicates, "This food is dangerous; don't swallow it."

What Is the Molecular Basis of Sweetness and Other Tastes? Inspired by progress on bitter receptors, research teams used a similar approach—analyzing mutant mice that could not sense sweetness—to understand how sweet receptors work.

In humans and mice, three closely related membrane receptor proteins are responsible for detecting sweetness as well as glutamate and other amino acids. Glutamate triggers the sensation called **umami**, which is the meaty taste of the molecule monosodium glutamate (MSG). Glutamate is sensed by one particular pair of the three receptor proteins. Sweetness is sensed by a different pair.

Recent work has answered a long-standing question about the sweet sensation—why so many different types of sugars trigger the same sensation. As it turns out, a single receptor protein has binding sites for multiple types of sweet compounds, meaning that a variety of molecules can stimulate each sweet receptor cell.

The tongue appears to have receptors that are responsible for sensations other than salty, sour, bitter, sweet, and umami. For example, scientists are currently studying whether our tongues can detect calcium, other metals, and carbon dioxide. There is also evidence that mechanoreception in the tongue plays an important role in taste by giving foods a certain texture or "mouth feel." Although taste is beginning to reveal its secrets, the complete story will probably not be known for many years.

Olfaction: Detecting Molecules in the Air

Taste allows animals to assess the quality of their food before swallowing it. Olfaction, in contrast, allows animals to monitor airborne molecules that convey information. Wolves and domestic dogs, for example, can distinguish millions of different airborne molecules at vanishingly small concentrations. The molecules that constitute odor contain information about the movements and activities of prey and other members of an animal's own species.

Odorants Provide Information about the Environment Airborne molecules that convey information about food or the environment are called **odorants**. When they reach the nose, they diffuse into a mucus layer in the roof of the nose (Figure 44.17, page 958). There, they activate olfactory neurons by binding to membrane-bound receptor proteins. Axons from these neurons project to the **olfactory bulb**, the part of the brain where olfactory signals are processed and interpreted.

Understanding the anatomy of the odor-recognition system was a relatively simple task. Understanding how receptor neurons distinguish one molecule from another was much more difficult. Initially, investigators hypothesized that receptors respond to a small set of "basic odors," such as musky, floral, minty, and so on. The idea was that each basic odor would be detected by its own type of receptor, much like the way gustation works.

In 1991, Linda Buck and Richard Axel discovered a gene family in mice that is made up of hundreds of distinct coding regions and encodes receptor proteins on the surface of olfactory receptor neurons. Follow-up experiments confirmed that each receptor protein binds to a small set of molecules.

Further work established that most, if not all, vertebrates possess this family of genes. The number of olfactory receptor genes varies widely among mammals, from around 700 in some primates to over 4000 in elephants. But about half of the receptor genes in most mammal species have mutations that render them nonfunctional. For example, humans have about 800 receptor genes, but

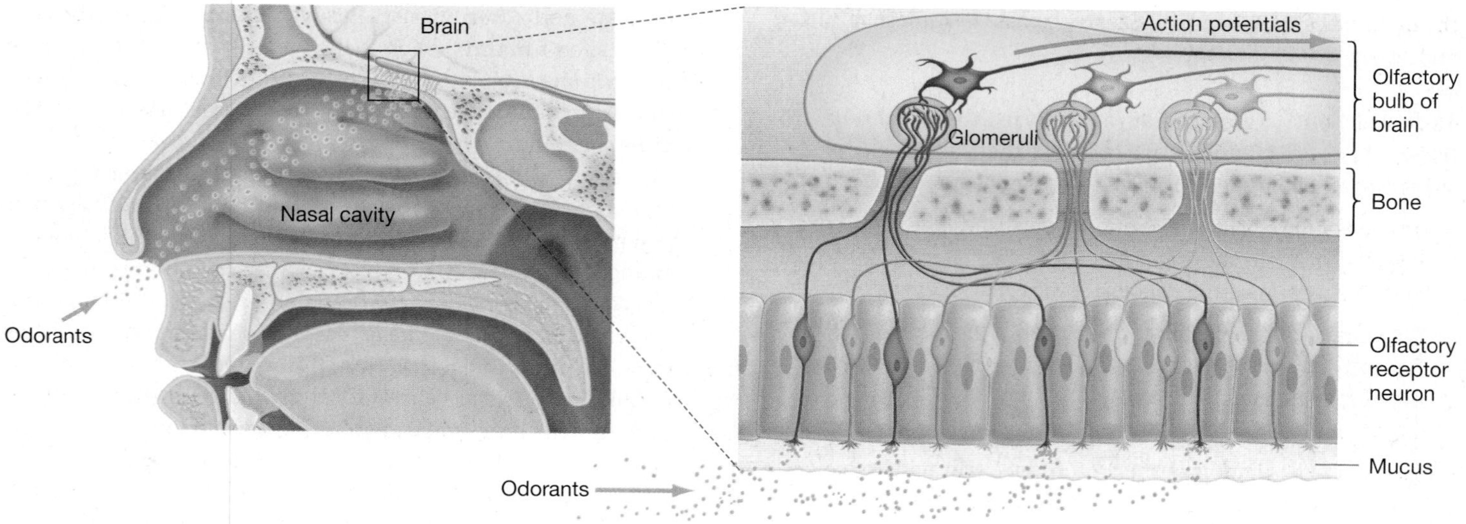

Figure 44.17 Chemoreceptor Cells in the Nose Respond to Specific Odorants. Each of the chemosensory neurons in the nose has one type of odor receptor protein on its dendrites. Sensory neurons with the same receptor protein project to the same glomerulus, or section within the olfactory bulb of the brain.

only half code for receptor proteins. This observation may explain why the sense of smell is so poor in humans compared with that of most other mammals. But the sense of smell is even worse in whales and dolphins—70% to 80% of their olfactory receptor genes are nonfunctional. These animals rely far more on hearing and vision than on smell to sense their environment.

In contrast, mammals such as rodents, horses, and elephants express at least 1000 types of functional receptor proteins, reflecting their reliance on olfaction to sense the environment. Humans take advantage of the strong olfactory sense of another mammal species, the dog, to sense odorants that we cannot detect, including explosives, drugs, and missing persons or criminals on the run. Because bloodhounds have the strongest sense of smell of any dog, they are often used in this capacity. The combination of many functional receptor protein types expressed in millions of olfactory cells spread over a vast surface area of olfactory epithelium makes the bloodhound an exquisite scent tracker. They can detect as few as one or two human skin cells left behind on a rock or branch in the woods.

Further research on olfaction centered around two questions. How many different receptor proteins occur in the membrane of each neuron involved in odor reception? How does the brain make sense of the input from so many different receptors?

Scientists determined that each olfactory neuron has only one type of receptor protein and that axons of neurons that respond to this receptor protein project to the same region in the olfactory bulb of the brain. These regions are called **glomeruli** (meaning "little balls"; singular: **glomerulus**). For example, each smell recognized by mice is associated with the activation of a different subset of the 2000 glomeruli in the olfactory bulb. Thus the activation of several specific glomeruli might be perceived as the smell "fresh bread." In essence, then, the sensing of odorants is similar to the eye's use of three cones to sense many colors; but odor reception works on a much larger scale.

Interestingly, odorant receptor proteins have recently been identified in locations other than the nose—for example, in tissues of the heart and pancreas. The proteins have also been identified in the cell membranes of sperm cells, where they appear to play an important role in guiding the sperm toward the egg. Research on this complex and impressive sense continues at a furious pace.

Pheromones Provide Information about Members of the Same Species Recall from the introduction to this chapter that male silk moths can locate female moths from miles away. Males have much larger antennae than females, suggesting that they use these antennae in a sex-specific manner.

These factors led scientists to hypothesize that female moths release a chemical into the environment that binds to chemoreceptors on the males' antennae and acts as an attractant. In 1959, scientists identified this chemical and named it bombykol, after the scientific name of the silk moth (*Bombyx mori*). Bombykol was the first chemically characterized **pheromone**, meaning a chemical that is secreted into the environment and that affects the behavior or physiology of animals of the same species.

Thousands of pheromones have since been identified in invertebrates and vertebrates alike, performing such roles as alerting other members of a beehive to an intruder or signaling a male rodent that a female is ovulating. In insects, pheromones typically bind to receptors on the antennae. In tetrapod vertebrates, pheromone receptors are often localized in the **vomeronasal organ (VNO)**, a sensory organ in the nasal cavity. The VNO is distinct from the nasal region containing sensory neurons that project to the olfactory bulb. The VNO and the olfactory bulb send signals to different parts of the brain, although some animals also sense odorants with their VNO. For example, a male snake may use its VNO to follow a pheromone trail of a female snake or a scent trail of a prey animal.

Do humans release pheromones? This is a hotly debated question. In 1971, Martha McClintock reported that the menstrual cycles of women living in close contact with one another become synchronized as a result of a secretion from the women's armpits. However, the responsible chemical was not identified. Furthermore, the methodology of this study has been criticized and its results have become controversial, as other researchers have failed to replicate them. Other studies have suggested that secretions from men's armpits can alter hormone levels in women, but both the amount of the chemical secreted and the reaction to it vary dramatically. Until scientists more fully understand the mechanisms by which these human "pheromones" achieve their effects, the jury is out.

CHECK YOUR UNDERSTANDING

✔ **If you understood this section, you should be able to ...**

1. Discuss why a loss in chemosensory ability occurs when you burn your tongue with extremely hot food.
2. **PROCESS OF SCIENCE** Develop a hypothesis to explain why the vomeronasal organ appears to be reduced or vestigial in most primates.

Answers are available in Appendix A.

44.5 Other Sensory Systems

The stimuli and senses discussed so far are the ones you are likely most familiar with. But animals can sense much more than pressure waves, light, and chemicals. All animals can sense temperature and painful stimuli, and some can even perceive electric or magnetic fields. Let's start with temperature and briefly examine each of these other stimuli and senses in turn.

After you complete this section, you should be able to ...

- Explain how temperature, electrical charge, and magnetic fields are detected.

Thermoreception: Sensing Temperature

Recall that many animals thermoregulate to maintain body temperature within an acceptable range (Ch. 39, Section 39.5). Virtually every physiological process, from digestion to metabolism, is temperature dependent, so the ability to sense temperature changes in the environment and respond accordingly is crucial.

Thermoreception Helps Animals Thermoregulate Animals detect heat energy by **thermoreception** and adjust their behaviors or physiological processes, such as shivering and sweating, in response.

Some thermoreceptors are located in the central nervous system. In mammals, the hypothalamus is the brain region that senses departures from homeostatic body temperature and sends signals to effectors to restore homeostasis (Ch. 39, Figure 39.12).

Thermoreceptors also are commonly found on skin and other outer surfaces of animals, so that changes in the temperature of the environment can be sensed. As an example, several types of thermoreceptors have been identified in mammals. Some receptors depolarize in response to cooling, and others depolarize in response to heating. Picking up a cold object stimulates "cold receptors" in your skin, resulting in an increase in the rate of action potentials in sensory neurons that inform your brain that the object is cold.

Interestingly, extreme temperatures are sensed by a different type of receptor, called a nociceptor, that also senses other painful stimuli such as those produced by certain chemicals, excessive pressure, and tissue damage. If you touch a hot stove burner, the pain you feel arises primarily via stimulation of nociceptors.

Pit Vipers Have Extremely Sensitive Thermoreceptors The pit vipers are a group of snakes that are named after the two temperature-sensitive pits located just beneath their nostrils (Figure 44.18a). Inside each pit is a membrane lined with exquisitely sensitive thermoreceptors—a rattlesnake's thermoreceptors can sense changes in temperature as small as 0.003°C.

Pit vipers use these thermoreceptors to sense the heat energy given off by prey or predators, and to detect potential burrows in which to hide. The brains of these snakes may combine visual and thermal stimuli into a "thermal image" that might look something like an image from an infrared camera (Figure 44.18b). Even in complete darkness, rattlesnakes can strike prey with deadly accuracy.

(a) Pit vipers have temperature-sensitive pits.

(b) Warm animals emit infrared radiation.

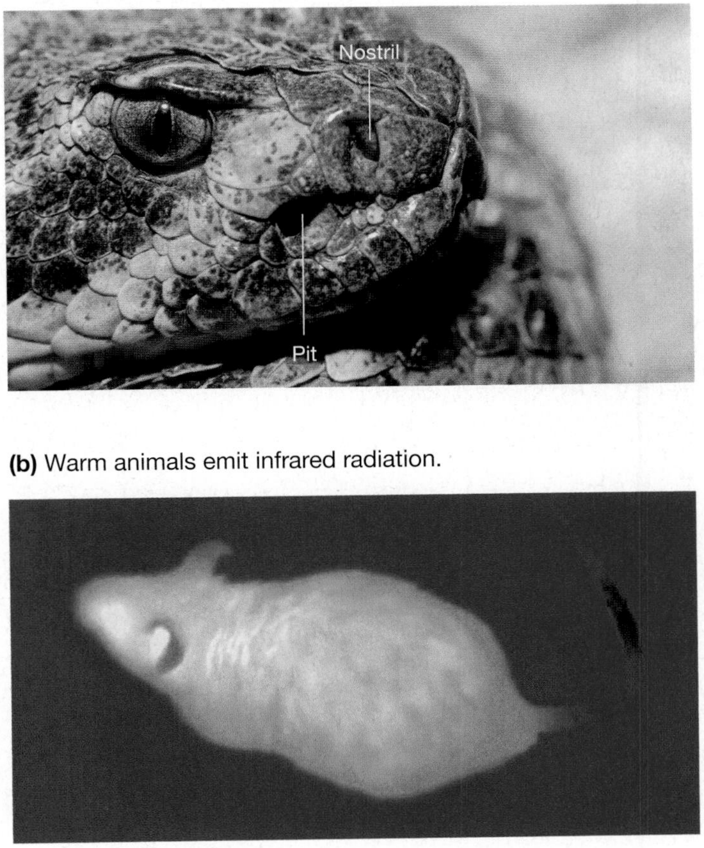

Figure 44.18 Pit Vipers Use Thermoreception to Detect Prey. (a) Pits are lined with extremely sensitive thermoreceptors. **(b)** Pit vipers can detect infrared radiation given off by rodents and other prey.

✔ **If you understand thermoreception, you should be able to predict whether a rattlesnake could strike effectively at its prey with (1) its eyes covered but pits exposed, (2) its eyes and pits covered with cotton cloth, and (3) its eyes and pits covered with an opaque, heat-blocking material.**

Electroreception: Sensing Electric Fields

All animals give off weak electrical impulses that arise from the activity of their nerves and muscles. Because water is a good conductor of these electrical impulses, many kinds of fishes use **electroreception**, or sensation of electric fields, to locate prey, detect predators, and navigate.

Sharks Use Electroreception to Hunt In sharks, some parts of the lateral line system are specialized to detect electric fields rather than pressure. Tiny pores scattered across a shark's head contain structures called **ampullae of Lorenzini** (Figure 44.19). These ampullae are lined with hair cells that detect electrical potentials in the water and send signals to the shark's brain via sensory neurons.

The sensitivity of these ampullae is remarkable—sharks can detect electrical potentials as small as a nanovolt (a billionth of a volt). This ability allows sharks to sense prey that are far away or even buried in the sand on the ocean floor. Combined with their exquisite sense of olfaction—sharks can detect a single drop of blood in a million drops of ocean water—electroreception makes them finely tuned predators.

Although data are scarce, scientists have some evidence to suggest that sharks also use electroreception to navigate. Ocean currents moving through Earth's magnetic field generate weak electrical currents. Sharks have been observed orienting themselves to these fields in the ocean as well as to artificially created fields in the laboratory.

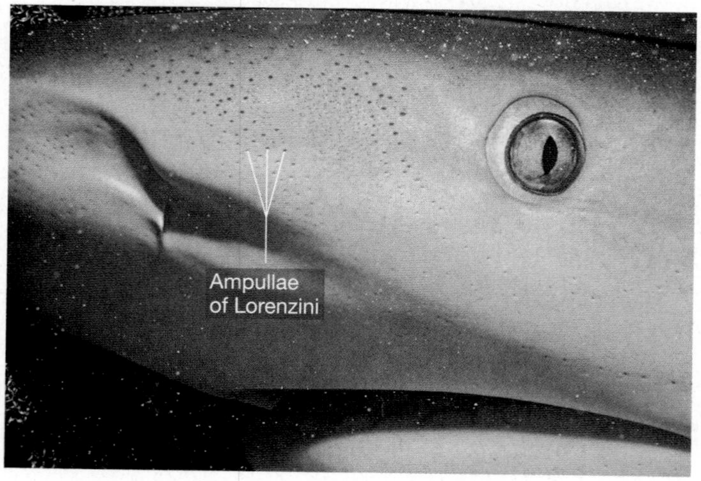

Figure 44.19 Ampullae of Lorenzini on a Shark's Head Detect Electric Fields to Help Locate Prey.

Electrogenic Fishes Generate Electric Fields **Electrogenic fishes** have specialized organs near their tails that generate electric fields stronger than those of regular nerves or muscles. The currents produced by these fields move in an arc through the water (Figure 44.20). Any item located within that arc will disrupt the currents, allowing the fish's electroreceptors to detect it. In this way, electrogenic fishes use their electric organs to locate prey, sense predators, navigate through murky water, and even communicate with other members of the same species.

Some electrogenic fishes have the ability to produce extremely strong currents that stun or kill their prey. The electric organs of electric eels take up over 80 percent of their body mass and can generate a 500-volt change in electrical potential and 1 ampere of current in the water around them. This amount of current is enough to kill a person swimming in water with an electric eel.

Magnetoreception: Sensing Magnetic Fields

The Earth produces a magnetic field as it rotates on its axis. Just as a compass responds to this magnetic field to indicate direction, animals may home in on magnetic fields while navigating. **Magnetoreception** has been described in many groups of organisms, including bacteria, fungi, invertebrates, and all vertebrate classes.

In general, studies of the mechanisms by which animals sense Earth's magnetic field are in their infancy. But scientists are confident that several distinct mechanisms have evolved. Recall that sharks can sense Earth's magnetic field indirectly via electric fields produced by ocean currents. Terrestrial animals, in contrast, are able to sense magnetic fields directly.

In 1968, German scientists noticed that European robins being kept in the laboratory with no visual cues to the outside began to sit at one end of their cages at the beginning of the migratory season. This behavior led the scientists to hypothesize that birds use magnetoreception to identify direction as they migrate.

Since then, support for this hypothesis has accumulated. For example, when European robins were placed into circular chambers with artificial magnetic fields, the birds always oriented themselves in relation to the artificial field. What's more, disruption of the magnetic field prevents birds from navigating properly. When scientists fitted homing pigeons with little caps that reversed the polarity of the magnetic field, the pigeons flew in the direction opposite to the one they were trained to fly in.

What enables birds to sense magnetic fields? A combination of factors are likely at work. One hypothesis is that deposits of iron inside sensory neurons in the beak play a role in the response to changes in the magnetic field. In support of this hypothesis, cutting the axons of these neurons prevents birds from responding to artificial changes in the magnetic field in the lab.

To complicate matters, however, magnetoreception in birds apparently also depends on vision. Covering the right eye—but not the left eye—of migrating birds interferes with their ability to navigate using magnetic cues. Scientists have recently found evidence for magnetoreceptors in the retinas of birds.

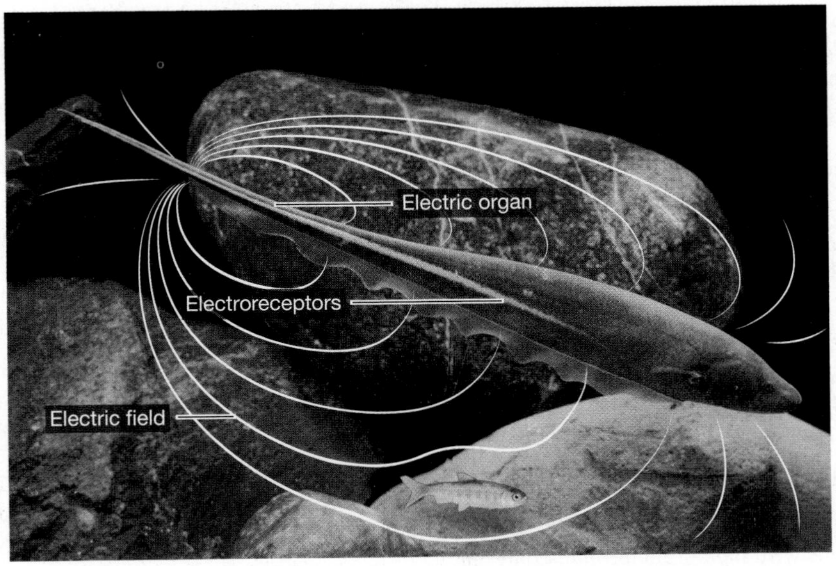

Figure 44.20 Electrogenic Fishes Can Create a Strong Electric Field. The field, generated by the fish's electric organ, produces an electrical current in the surrounding water. When a prey animal disrupts the current, the fish detects its presence using electroreceptors.

This complication highlights a key feature of sensory perception: Animals do not use individual senses in isolation, but rather combine sensations of many types when locating prey, evading predation, or communicating with other individuals. Whether it is a homing pigeon using magnetoreception and photoreception to find its way home, or a rattlesnake using a visual and thermal image to strike at a mouse, all animals depend on sensory systems that work together to collect the information needed to survive.

CHECK YOUR UNDERSTANDING

✔ **If you understood this section, you should be able to ...**

1. Suggest a sensory adaptation that could help vampire bats, which feed exclusively on the blood of large mammals such as cattle and humans, to locate prey.
2. Propose a mechanism by which a female sea turtle migrates to lay eggs on the beach where she was born decades earlier.

Answers are available in Appendix A.

CHAPTER 44 Review

For media, go to **Mastering Biology**

44.1 How Do Sensory Organs Convey Information to the Brain?

- Sensory stimuli as different as sound and light are transduced to changes in membrane potential in receptor cells. These changes alter the pattern of action potentials that sensory neurons send to the brain.
- The brain is able to distinguish different types of stimuli because axons from different types of sensory neurons project to different regions of the brain.

44.2 Mechanoreception: Sensing Pressure Changes

- Pressure receptors detect direct physical stimulation, including stimulation from sound.
- Hair cells, the major sensory receptors in the vertebrate ear, undergo a change in membrane potential in response to bending of their stereocilia.
- Sound waves of a certain frequency cause a certain part of the cochlea's basilar membrane to vibrate. Hair cells at this location stimulate action potentials in sensory neurons in response to the vibration.
- Hair cells in the semicircular canals and vestibule of the mammalian inner ear signal the brain about equilibrium.
- Hair cells in the lateral line systems of fishes and aquatic amphibians are stimulated by pressure changes in the water.

44.3 Photoreception: Sensing Light

- In the vertebrate eye, photoreceptors (rods and cones) contain light-sensitive molecules that consist of retinal paired with an opsin protein.
- The rhodopsin found in rods is stimulated by even the faintest light.
- Color vision is possible because cones contain opsins that respond to specific wavelengths of light absorbed by retinal.
- Humans distinguish colors based on the pattern of stimulation of three types of opsins found in cones. People who lack one of the functional cone opsins are color blind, meaning they cannot distinguish as many colors as people with all three opsins can.

44.4 Chemoreception: Sensing Chemicals

- Chemoreceptors detect the presence of specific chemicals.
- Taste buds contain taste cells with membrane proteins that play key roles in the response to chemicals. Sodium ions and protons enter taste cells via channels and depolarize the membrane directly, producing the sensations of saltiness and sourness, respectively. Sugars and some toxic compounds bind to membrane receptors, resulting in action potentials in sensory neurons that are interpreted by the brain as sweet and bitter flavors, respectively.
- Smell, or olfaction, is used to detect molecules from the outside environment. Airborne chemicals are detected by hundreds of different odor-receptor proteins located in the membranes of receptor cells in the nose.

44.5 Other Sensory Systems

- Thermoreceptors respond to changes in temperature.
- Nociceptors respond to painful stimuli, including extreme temperatures, certain chemicals, excessive pressure, and tissue damage.
- Electroreceptors contain modified hair cells that respond to electric fields.
- Magnetoreceptors respond to magnetic fields and are often used in navigation and orientation.

Answers are available in Appendix A.

✔ TEST YOUR KNOWLEDGE

1. In the human ear, how do different hair cells respond to different frequencies of sound?
   a. Waves of pressure move through the fluid in the cochlea.
   b. Hair cells are "sandwiched" between membranes.
   c. Receptor proteins in the stereocilia of each hair cell are different; each protein responds to a certain range of frequencies.
   d. Because the basilar membrane varies in stiffness, it vibrates in certain places in response to certain frequencies.
2. Evaluate the following statements according to information presented in this chapter. Select True or False for each statement.
   T/F Most human eyes have one type of rod and three types of cones.
   T/F Rods are more sensitive to dim light than cones are.
   T/F Nocturnal animals have fewer rods than diurnal animals.
   T/F Both rods and cones use retinal and opsins to detect light.
3. Which of these statements about taste is *true*?
   a. Sweetness is a measure of the concentration of hydrogen ions in food.
   b. Sodium ions from foods can directly depolarize certain taste cells.
   c. All bitter-tasting compounds have a similar chemical structure.
   d. Sourness is detected when hydrogen ions bind to membrane receptors.
4. What type of sensory system do migrating birds use to detect direction?

✔ TEST YOUR UNDERSTANDING

5. Considering that sounds and odors both trigger changes in the patterns of action potentials in sensory neurons, how does the brain perceive which sense is which when the action potentials reach the brain?
   a. The action potentials stimulated by sounds are different in size and shape from those stimulated by odors.
   b. The axons from different sensory neurons go to different areas of the brain.
   c. Mechanoreception is not consciously perceived by the brain, whereas chemoreception is.
   d. Chemoreception is not consciously perceived by the brain, whereas mechanoreception is.
6. Give three examples of how the sensory abilities of an animal correlate with its habitat or method of finding food and mates.
7. Compare and contrast the lateral line system of fishes with electroreception in sharks.
8. Scientists generally think that a "good hypothesis" is one that is reasonable and testable and inspires further research into a phenomenon. Using these criteria, explain whether Dalton's hypothesis about color vision was a good hypothesis. Was it correct?

✔ TEST YOUR PROBLEM-SOLVING SKILLS

9. **QUANTITATIVE** Scientists collected data on the date of onset of the menstrual cycles in a group of women who moved into a college dormitory together in the fall. The *y*-axis of the graph shows the mean difference (in days) between the onset of a woman's cycle and the average onset date of the rest of the women. Evaluate whether these data provide evidence for the existence of a human pheromone.

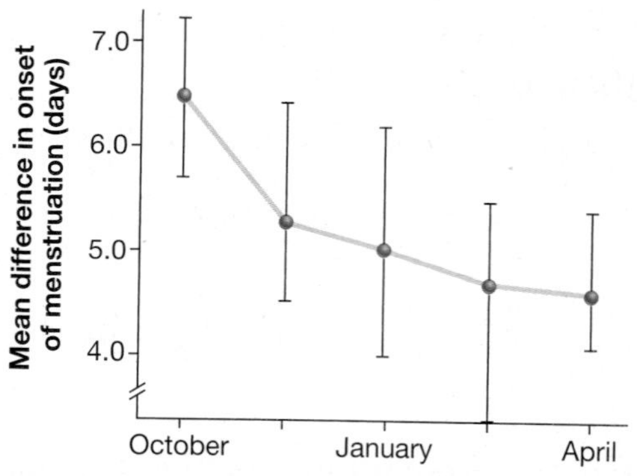

DATA: M. K. McClintock. 1971. *Nature* 229: 244–245, Figure 1.

10. **PROCESS OF SCIENCE** Design experiments to test the hypothesis that electric eels are both electrogenic and electroreceptive.

✔ PUT IT ALL TOGETHER: Case Study

How do pheromones keep a hive of honeybees buzzing?

Honeybees live in social groups consisting of a queen, up to several hundred male drones, and thousands of infertile female workers. The drones mate with the queen only, and the workers protect the hive, forage, and feed and groom the queen. The health of the hive depends on the female workers performing these duties instead of reproducing. What roles do pheromones play in maintaining a functional beehive?

11. Scientists have identified dozens of pheromones used by honeybees for communication. Which type of sensory system uses pheromones?
a. mechanoreception
b. photoreception
c. chemoreception
d. thermoreception
e. electroreception

12. Honeybees produce an alarm pheromone when their hive is molested. This pheromone stimulates the bees to protect the hive. If you were to count the number of alarm pheromone receptors in honeybee tissues, which type of bee would likely have the most—a queen, a drone, or a worker? Why?

13. Why might an alarm pheromone be more effective for triggering a protective response in a hive than signals that involve other senses, such as vision or hearing?

14. **QUANTITATIVE** Researchers observed that the queen produces a pheromone that attracts both drones and workers. They hypothesized that this pheromone inhibits ovarian development in the workers, making the workers infertile. To test this hypothesis, they exposed workers to a synthetic version of the queen pheromone and then recorded their "ovary development score." (Higher scores indicate more fully developed ovaries.) The results are shown in the graph here. Do these results support the researchers' hypothesis? Why or why not? (*** signifies $P < 0.001$; see **BioSkills 3** for more on statistical significance.)

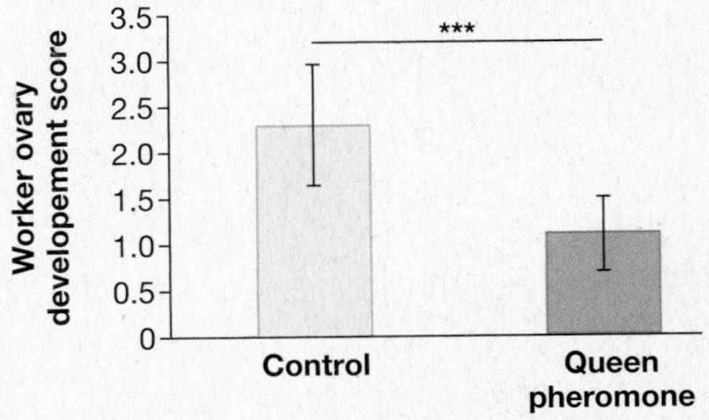

Source: S. E. R. Hoover, C. I. Keeling, M. L. Winston, et al. 2003. *Naturwissenschaften* 90: 477–480.

15. **PROCESS OF SCIENCE** In the experiment described in Question 14, the researchers dissolved the queen pheromone in diethyl ether, a chemical that helps volatilize the pheromone, making it easier for the workers to detect it in the air. The control treatment consisted of plain diethyl ether. Why did they use this as the control, instead of simply not exposing the workers to any chemical?

16. **SOCIETY** Beekeepers carefully manage their beehives by adding mated queens, drones, and workers at appropriate times. However, sometimes the bees abandon a new hive before the queen can mature, and the result is inefficient hive management and honey production. Suggest a potential use of queen pheromone in controlling this problem.

Mastering Biology

Students Go to Mastering™ Biology for assignments, the eText, and the Study Area with animations, practice tests, and activities.

Professors Go to Mastering™ Biology for automatically graded tutorials and questions that you can assign to your students, plus Instructor Resources.

45 Animal Movement

Basilisk lizards are able to run on water—literally. This impressive escape strategy demonstrates the extent to which muscle-generated movements have diversified among animals. For most animals, complex muscle movements make the difference between life and death.

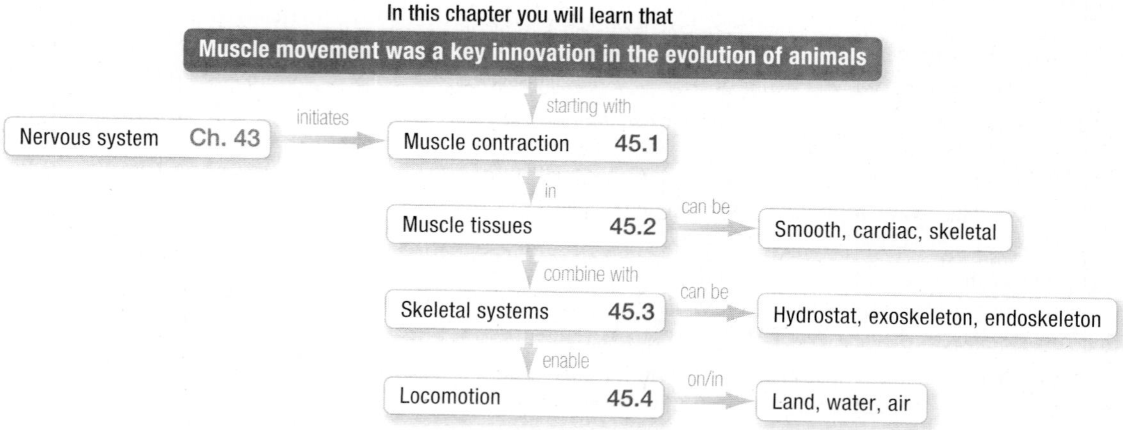

You may have discovered while studying biology that plants and animals are more similar to each other than they initially appear. For example, plants and animals both require water and nutrients in specific quantities, have highly specialized tissues and complex reproductive structures, and launch defenses against parasites and predators. You can see a comparison of these and other traits in the Big Picture of Plant and Animal Form and Function on pages 838–839.

However, animals possess a quality that clearly distinguishes them from plants and other organisms: movement by virtue of muscle contractions. Muscle-generated movements were a key innovation in animal evolution. Rapid movements, along with sophisticated sensory structures (Chapter 44) and complex information processing systems (Chapter 43), were vital to animal diversification (Chapter 30), and these attributes made animals efficient eating machines in diverse ecosystems.

Muscles generate movement by exerting force and causing shape changes. Movement falls into two general categories:

BIG PICTURE

This chapter is part of the Big Picture. See how on pages 838–839.

1. ***Movement of the entire animal relative to its environment.*** **Locomotion**, the movement of an organism or cell from place to place under its own power, enables animals to avoid predators and to seek food, water, mates, and shelter. Modes of animal locomotion include undulating, jetting, swimming, walking, running, jumping, gliding, and flying.
2. ***Movement of one part of the animal relative to other parts (not involved in locomotion).*** This type of movement also has important functions—for example, to ventilate gills for gas exchange and to grasp prey. Even sessile animals such as corals and barnacles use complex, muscle-generated movements to survive and reproduce.

How do animals accomplish their spectacular movements? In this chapter, you'll start by probing into the mechanism of muscle contraction, which serves as the "engine" for most animal movements. Then you'll consider how muscle and skeletal systems work together to produce locomotion. And finally, you'll explore a research field called **biomechanics**, in which the principles of physics and engineering are applied to questions about the mechanical structure and function of organisms. Let's jump in.

45.1 How Do Muscles Contract?

The mechanism responsible for the contraction of muscle has fascinated and perplexed scientists for many centuries. Before the advent of microscopes and modern research techniques, scientists could only speculate about what makes muscles contract and relax.

After you complete this section, you should be able to . . .

- Analyze how muscles cause movement in animals.

Early Muscle Experiments

Put your hand on your biceps muscle and flex it. You'll notice that the muscle seems to swell. What is actually happening in the muscle? In the second century CE, Roman physician and philosopher Galen proposed that spirits flowed from nerves into muscles, inflating them and increasing their diameter. This "inflation" hypothesis persisted into the seventeenth century, when French philosopher René Descartes suggested that nerves carry fluid from the pineal gland—a part of the brain then considered to be the seat of the soul—to muscles, making them shorten and swell.

Later that century, Dutch anatomist Jan Swammerdam tested Descartes' inflation hypothesis with a simple yet elegant experiment. He placed a piece of frog thigh muscle into a glass syringe with the nerve protruding through an airtight hole in the side and with a drop of water in the tip of the syringe. He then stimulated the nerve, causing the muscle to contract. If the muscle's volume changed during contraction, the drop of water should move. But it did not. The volume of the muscle remained constant.

Swammerdam's experiment demonstrated an important point: The contraction mechanism is inherent to the muscle itself—muscle is not like a balloon, and the nerve is not like a water hose filling a balloon. This insight was confirmed by Italian scientist Luigi Galvani in the 1790s, when he cut the nerve to a frog's leg muscle and then connected the two cut sides with a metal conductor. The muscle contracted. He concluded that the nerve and muscle possess "animal electricity" that can induce contraction.

If the shape of a muscle does not change by inflation, what is the mechanism of muscle contraction? What does electricity have to do with it?

The Sliding-Filament Model

Early microscopists established that the muscle tissue in vertebrate limbs and hearts is composed of slender fibers. A **muscle fiber** is a long, thin muscle cell. Within each muscle cell are many threadlike, contractile structures called **myofibrils**. The myofibrils inside muscle fibers often look striped, or striated, due to the alternating light–dark units called **sarcomeres**, which repeat along the length of a myofibril (Figure 45.1).

The microscopists observed that sarcomeres shorten as myofibrils contract. Sarcomeres then lengthen when the cell relaxes and an external force stretches the muscle. Based on these

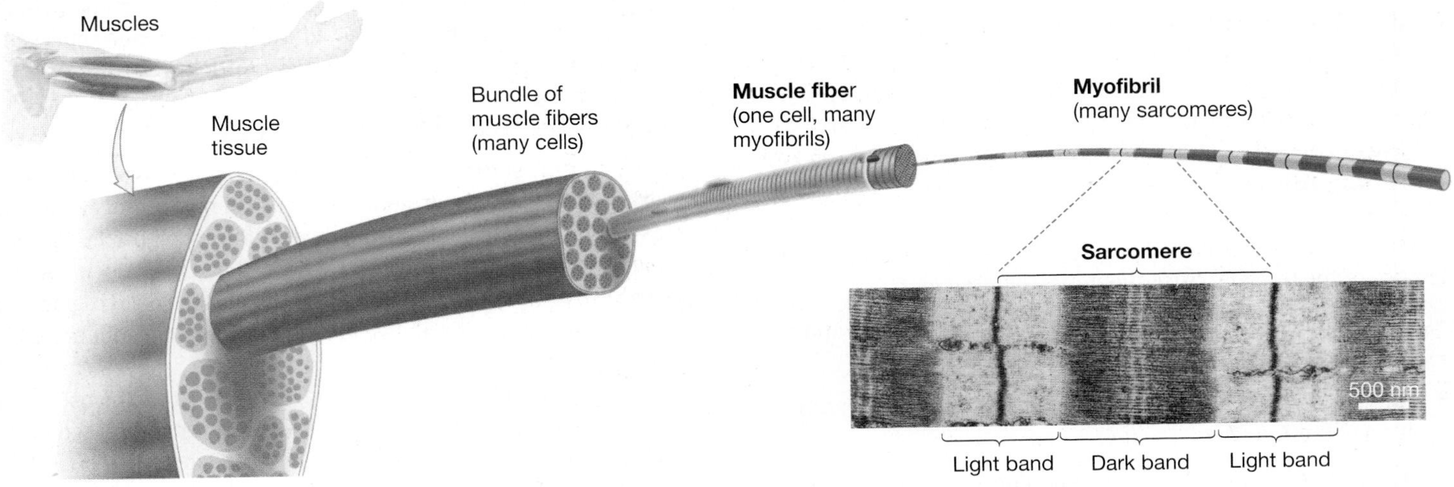

Figure 45.1 Muscle Cells Contain Many Myofibrils, Which Contain Many Sarcomeres. Skeletal muscle cells (fibers) have a striped appearance due to repeating sarcomeres, which are units of alternating light–dark bands.

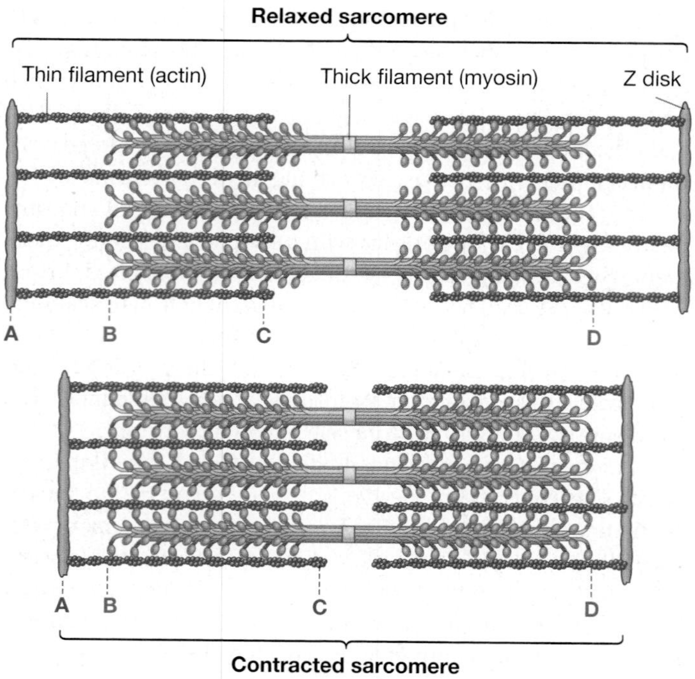

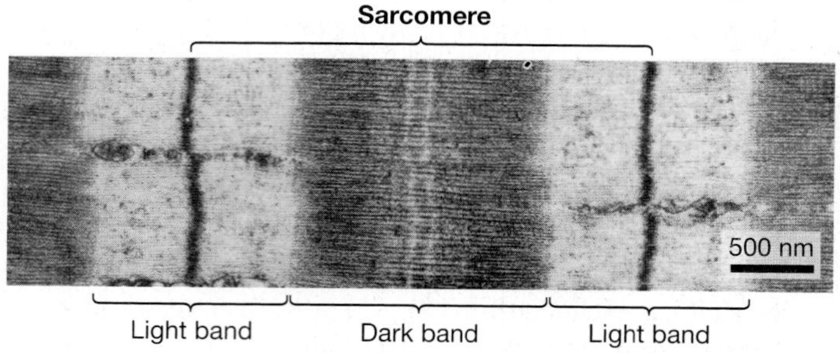

Figure 45.2 The Sliding-Filament Model Explains Sarcomere Contraction. When a sarcomere contracts, the lengths of the thin filaments (distance from A to C) and thick filaments (distance from B to D) do not change. Rather, the filaments slide past one another.

✔ According to the model shown here, why is the dark band in a sarcomere dark and the light band light?

observations, it became clear that the question of how muscles contract simplifies to the question of how sarcomeres shorten.

Biologists knew that the answer must involve the two types of protein that had been found in sarcomeres: **actin** and **myosin**. But they did not know the shapes of these molecules or how they were arranged within the sarcomere. Did both types of molecules span the entire length of the sarcomere? Or were they restricted to certain bands within the sarcomere?

In 1952, biologist Hugh Huxley examined cross sections of sarcomeres using electron microscopy. He observed that there were two types of filaments, **thin filaments** and **thick filaments**, and that these filaments overlapped in the dark bands but not in the light bands. Huxley and his collaborator Jean Hanson also observed that sarcomeres stripped of their myosin had no dark bands. They concluded that the thick filaments must be composed of myosin, and the thin filaments must be composed of actin.

How did myosin and actin interact to shorten the sarcomere? In 1954, Huxley and Hanson hit on the key insight when they observed how the light and dark bands in sarcomeres changed when a muscle contracted. Overall, the width of the dark bands did not change during a contraction, but the light bands became narrower.

To explain these observations, Huxley and Hanson proposed the **sliding-filament model** illustrated in Figure 45.2. They hypothesized that the filaments slide past one another during a contraction. That is, the sarcomere shortens with no change in the lengths of the thin and thick filaments themselves:

- The distance from point A to point C does not change, and the distance from point B to point D does not change.
- Points A and B move closer to each other during contraction, as do points C and D.

Interestingly, another pair of researchers, Andrew F. Huxley (no relation to Hugh) and Rolf Niedergerke, published the same result at the same time and in the same issue of *Nature*. In the years that followed, research has confirmed that the Huxley–Hanson model is correct in almost every detail.

Each thin filament is composed of two coiled chains of actin, a common component of the cytoskeleton of eukaryotic cells (Ch. 7, Section 7.6). One end of a thin filament is anchored to a structure called the **Z disc**, which forms the wall between neighboring sarcomeres. The other end of a thin filament is free to interact with thick filaments. Thick filaments are composed of multiple strands of myosin. They span the center of the sarcomere and are free at both ends to interact with thin filaments.

To appreciate how the sliding-filament model works, consider the following analogy: Two large trucks are parked 50 m apart, facing each other. Each has a long rope attached to the front bumper. Six burly weightlifters stand in a line on a platform in front of each truck, grab onto the rope, and pull, hand over hand, so that the two trucks roll toward one another. ✔ **If you understand the sliding-filament model, you should be able to explain which elements in this analogy represent the Z discs, which represent the thin filaments, and which represent the thick filaments.**

How Do Actin and Myosin Interact?

How does this sliding action occur at the molecular level? Early work on the three-dimensional structure of myosin revealed that a myosin molecule contains a pair of subunits whose "tails" are coiled around one another and whose "heads" are bent to the side. Each myosin head can bind to actin, and the head region can catalyze the hydrolysis of ATP into adenosine diphosphate (ADP) and a phosphate ion.

In addition, electron microscopy revealed that myosin and actin are locked together shortly after an animal dies, when its muscles enter the stiff state known as rigor mortis. Because ATP is unavailable in dead tissue, the data suggested that ATP is required for myosin to release from actin once the two molecules have bound to each other.

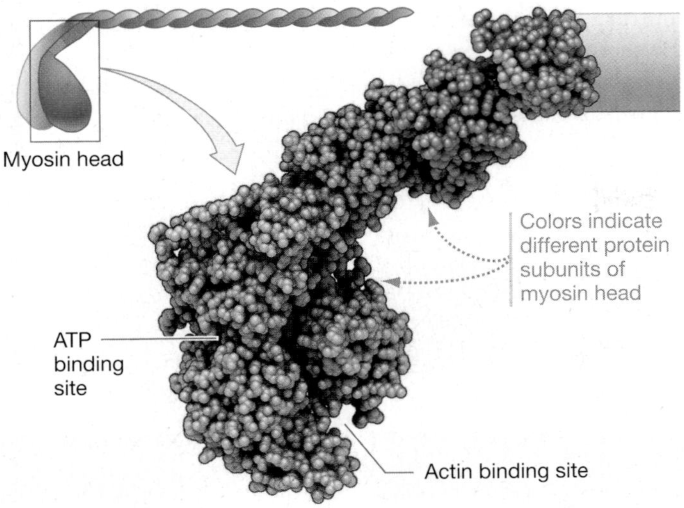

Figure 45.3 Myosin's "Head" Binds ATP and Actin. A myosin molecule includes two myosin proteins, each of which has a head that contains binding sites for ATP and actin.

Later, Ivan Rayment and colleagues solved the detailed three-dimensional structure of the myosin head (**Figure 45.3**). Using X-ray crystallography (**BioSkills 6**), Rayment's group determined the location of the actin-binding site. They also showed that myosin's conformation was significantly different when ATP was bound to the head than when ADP was bound.

Based on these data, Rayment and co-workers proposed a four-step model for actin–myosin interaction (**Figure 45.4**):

Step 1 ATP binds to the myosin head, causing a conformational change that releases the head from the actin in the thin filament.

Step 2 ATP is hydrolyzed to ADP and inorganic phosphate. The myosin head then pivots and binds to a new actin subunit farther down the thin filament (toward a Z disc). A myosin head bound to actin forms what is known as a cross-bridge. In this position, the myosin head is "cocked" in its high-energy state, ready for the power stroke.

Step 3 When inorganic phosphate is released, the head pivots back to its original conformation. This bending, called the power stroke, moves the entire thin filament relative to the thick filament.

Step 4 After ADP is released, the myosin head is ready to bind to another molecule of ATP.

As ATP binding and hydrolysis continue, myosin continues to bind actin, move the thin filament, release actin, and bind again, much like when you swing along on monkey bars. This movement of the thin filament along the thick filament causes the two Z discs to be pulled closer together, as long as ATP is available and the muscle is being stimulated. (Note that the transition from step 4 to step 1 cannot occur when ATP supplies run out, which explains why rigor mortis sets in after death.)

The same basic ratcheting mechanism between actin and myosin is responsible for the amoeboid movement observed in amoebae and slime molds (Ch. 27, Section 27.3) and the streaming of cytoplasm observed in algae and land plants. Actin and myosin have played a critical role in the diversification of eukaryotes

PROCESS: MODEL FOR ACTIN–MYOSIN INTERACTION

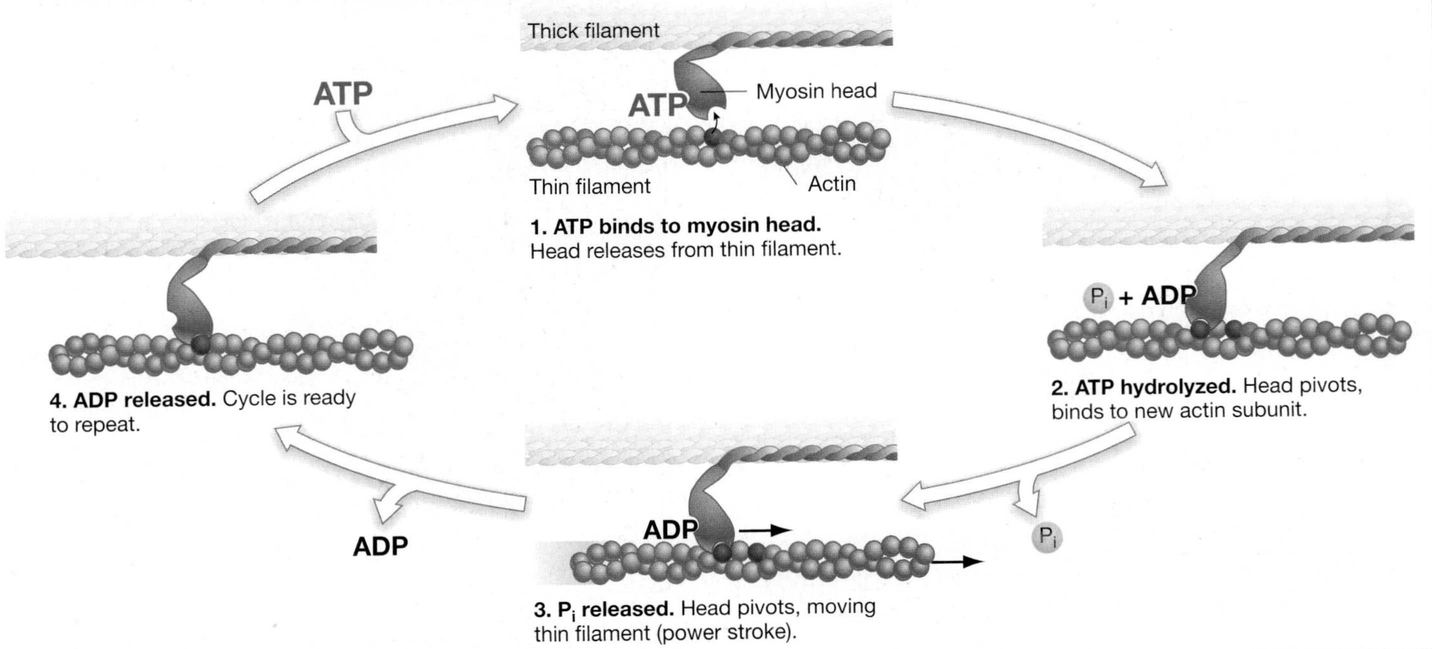

INTERACTIVE Figure 45.4 Myosin and Actin Interact during Muscle Contraction. Summary of the current model of how myosin and actin interact as a sarcomere contracts. The four steps repeat rapidly. (For simplicity, only one head of the myosin molecule is shown.)

because they make movement possible in the absence of cilia and flagella.

Considering that ATP is almost always available in living muscles, how do muscles ever stop contracting and relax? Besides actin, thin filaments contain two key proteins called **tropomyosin** and **troponin**, which work together to block the myosin binding sites on actin. When these sites are blocked, the myosin–actin interaction cannot occur, and thick and thin filaments cannot slide past each other. As a result, the muscle relaxes (Figure 45.5a).

But when calcium ions bind to troponin, the resulting troponin–tropomyosin complex moves in a way that exposes the myosin binding sites on actin. As Figure 45.5b shows, myosin then binds, and contraction can begin.

How are calcium ions released so that contraction can begin? The process begins when the muscle cell is stimulated by a neuron.

How Do Neurons Initiate Contraction?

You are probably sitting as you read. If so, contract your calf muscles to point your toes. Your nervous system just played a critical role in controlling the timing of your muscle contractions. First, your central nervous system—your brain—received input from an array of sensory cells in your peripheral nervous system, such as the ones in the retinas of your eyes as you read this paragraph. Then your brain integrated this information and triggered action potentials in the motor neurons that caused your calf muscles to contract. (See Chapter 43 to review the structure and function of the nervous system.)

Figure 45.6 summarizes what happens when an action potential from a motor neuron arrives at a muscle cell:

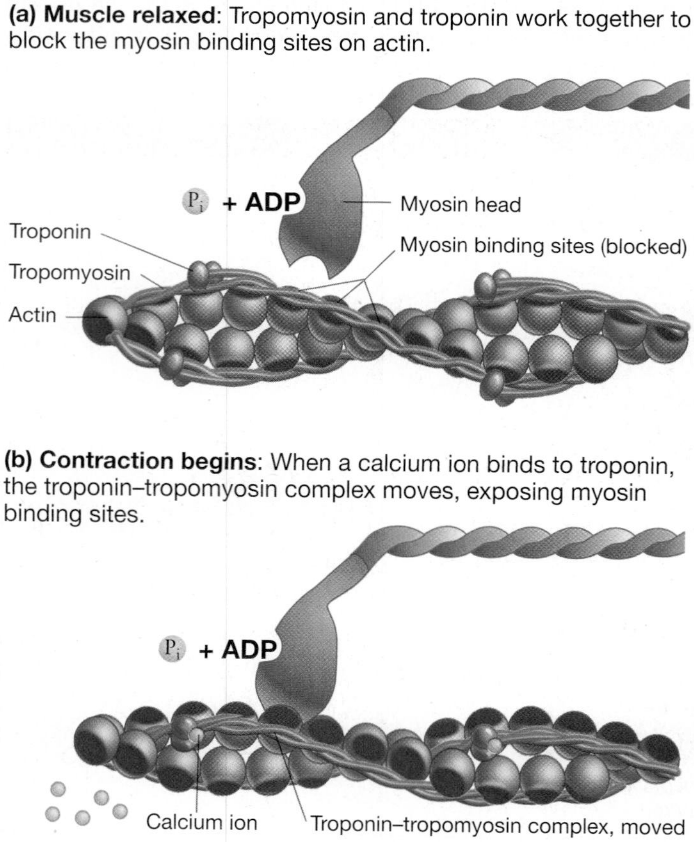

Figure 45.5 Troponin and Tropomyosin Regulate Muscle Activity. Note that the myosin head is in its energized state when a muscle is relaxed.

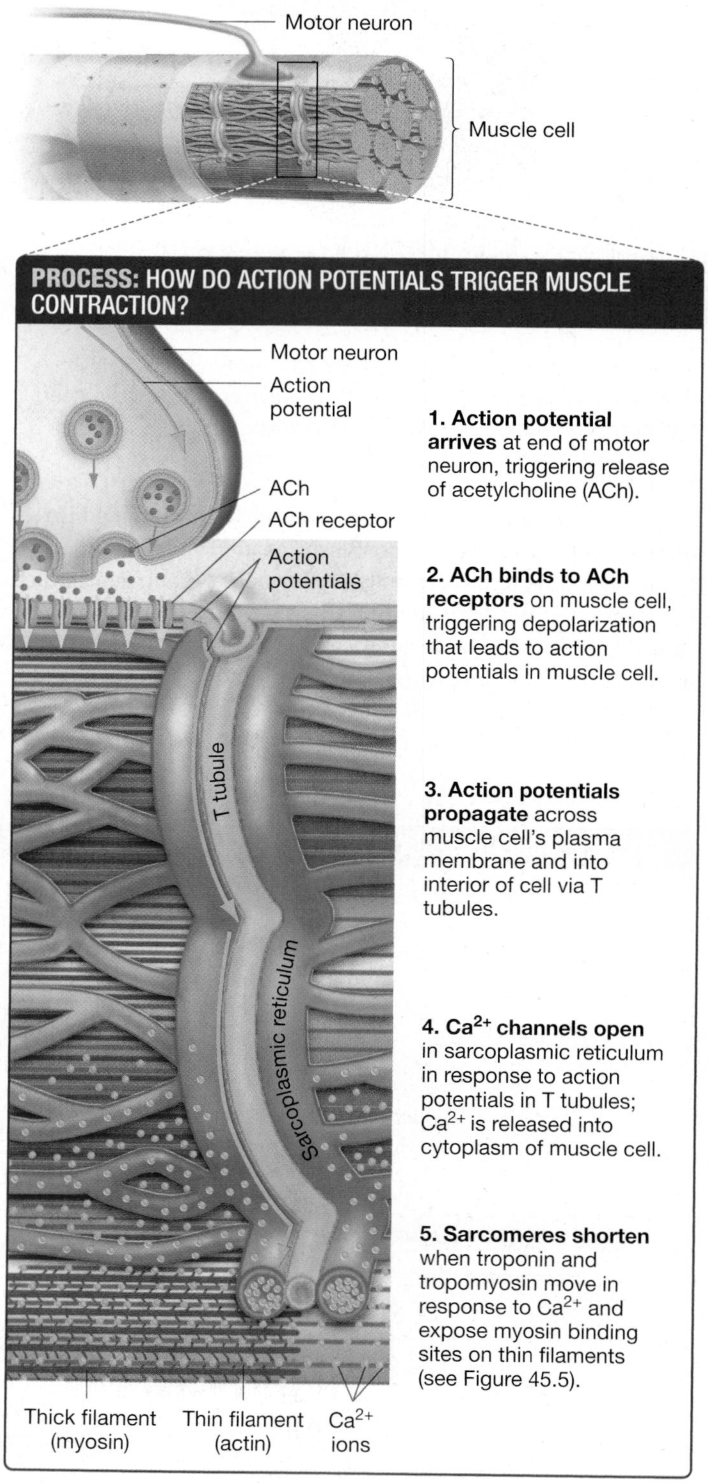

Figure 45.6 Action Potentials Trigger Ca^{2+} Release. Action potentials at the neuromuscular junction trigger the release of Ca^{2+} inside the muscle cell. Ca^{2+} binds to troponin, moving the troponin–tropomyosin complex and allowing myosin to form a cross-bridge with actin.

Step 1 When an action potential arrives at the end of the motor neuron, it triggers the release of the neurotransmitter **acetylcholine (ACh)** into the synaptic cleft between the neuron and the muscle cell (see Ch. 43, Figure 43.11).

Step 2 ACh diffuses across the synaptic cleft and binds to receptors on the plasma membrane of the muscle cell. Binding of ACh opens a ligand-gated ion channel in the receptor protein, resulting in depolarization of the muscle cell. If enough ACh is released by the motor neuron, the depolarization triggers action potentials in the muscle cell.

Step 3 The action potentials propagate along the length of the muscle cell and spread into the interior of the cell via invaginations of the plasma membrane called **T tubules**. (The T stands for *transverse*, meaning "extending across.")

Step 4 T tubules intersect with extensive sheets of smooth endoplasmic reticulum called the **sarcoplasmic reticulum**. When an action potential passes down a T tubule and reaches one of these intersections, it causes nearby calcium channels in the sarcoplasmic reticulum to open. Calcium ions diffuse from the sarcoplasmic reticulum through these channels into the cytoplasm, where the sarcomeres are located.

Step 5 Calcium ions bind to troponin, causing tropomyosin to move and expose the myosin binding sites on the actin filaments. The muscle fiber can now contract.

By this series of events, the interaction of the nervous system and muscle tissue at the neuromuscular junction precisely regulates muscle contractions—and thus complex movement.

CHECK YOUR UNDERSTANDING

✔ **If you understood this section, you should be able to ...**

1. Describe the sliding-filament model.
2. Predict the effect on muscle function of drugs that have the following actions: increase acetylcholine release at the neuromuscular junction; prevent conformational changes in troponin; and block uptake of calcium ions into the sarcoplasmic reticulum.

Answers are available in Appendix A.

45.2 Classes of Muscle Tissue

After you complete this section, you should be able to ...

- Analyze muscle tissue in animals.

How do muscle cells and muscle tissues vary? After years of careful anatomical study, biologists concluded that animals have three classes of muscle tissue: **(1)** smooth muscle, **(2)** cardiac muscle, and **(3)** skeletal muscle. You, and all other vertebrates, have all three.

Smooth, cardiac, and skeletal muscle share several properties. They all contract as described by the sliding-filament model, and they all contract in response to electrical stimulation. However, the three classes of muscle also differ in important ways, summarized in **Table 45.1** on page 970:

- ***Voluntary versus involuntary*** **Voluntary muscles** can contract in response to conscious thought (and also by unconscious reflexes) and are stimulated by neurons in the somatic division of the peripheral nervous system. **Involuntary muscles** contract only in response to unconscious electrical activity and are stimulated and inhibited by neurons in the autonomic division of the peripheral nervous system. (See Chapter 43 for a review of the somatic and autonomic divisions of the nervous system.)
- ***Multinucleate versus uninucleate*** Muscle cells may have one or many nuclei depending on the size of the cells.
- ***Striated versus unstriated*** As shown in Figure 45.1, the actin and myosin filaments in some muscle cells are aligned in rows forming sarcomeres, giving the cells and tissues a banded appearance when viewed through a microscope; for this reason, such muscle tissue is often called **striated muscle**. Other muscle cells are unstriated.

Let's apply these characteristics to each class of muscle tissue in more detail.

Smooth Muscle

Smooth muscle cells are unbranched, tapered at each end, and often organized into thin sheets. Contraction of smooth muscle cells occurs when actin and myosin bind and move past one another, but these proteins are not arranged into sarcomeres as they are in skeletal and cardiac muscle. Hence, smooth muscle cells are unstriated and appears smooth. Smooth muscle cells are relatively small and have a single nucleus.

Smooth muscle is essential to the function of the lungs, blood vessels, digestive system, urinary bladder, and reproductive system. Bronchioles in the lungs have a layer of smooth muscle that controls the size of airways; similarly, smooth muscles in blood vessels can contract or relax to alter blood-flow patterns and blood pressure. Layers of smooth muscle in the gastrointestinal tract help mix and move food, and uterine smooth muscle is responsible for expelling the fetus during birth.

What causes smooth muscle to contract? Smooth muscle is autorhythmic, meaning that it can spontaneously contract, with no stimulation by the nervous system. However, the autonomic nervous system can also stimulate and inhibit contraction in smooth muscle. Because smooth muscle is innervated only by autonomic neurons, it is involuntary.

In the digestive system, ACh released by parasympathetic ("rest-and-digest") neurons stimulates contraction of smooth muscle in the stomach and intestine, aiding digestion (Chapter 43). In contrast, sympathetic ("fight-or-flight") neurons release the neurotransmitter norepinephrine, and the adrenal glands adjacent to the kidneys release the hormone epinephrine (also called adrenaline). Norepinephrine and epinephrine have the opposite effect to that of acetylcholine: They inhibit contraction of smooth muscle in the gut.

Cardiac Muscle

Cardiac muscle makes up the walls of the heart and is responsible for pumping blood throughout the body. Unlike smooth muscle cells, cardiac muscle cells contain sarcomeres and are

SUMMARY Table 45.1 Classes of Muscle Tissue in Vertebrates

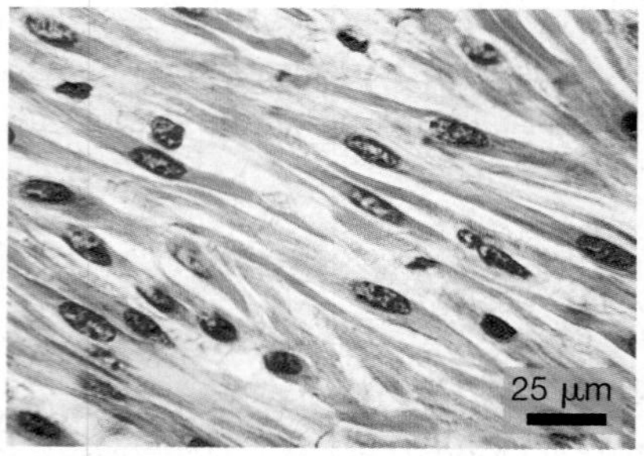

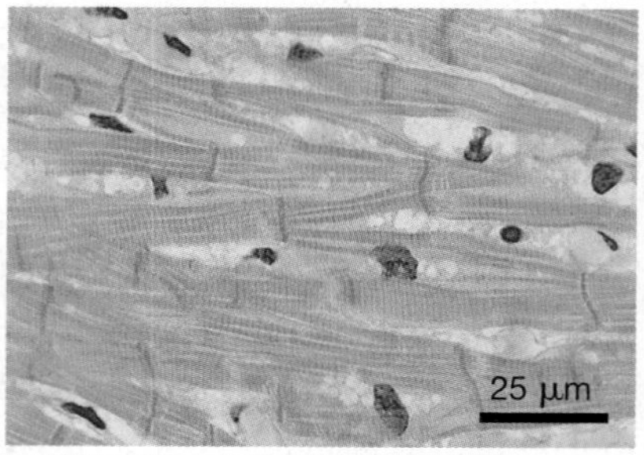

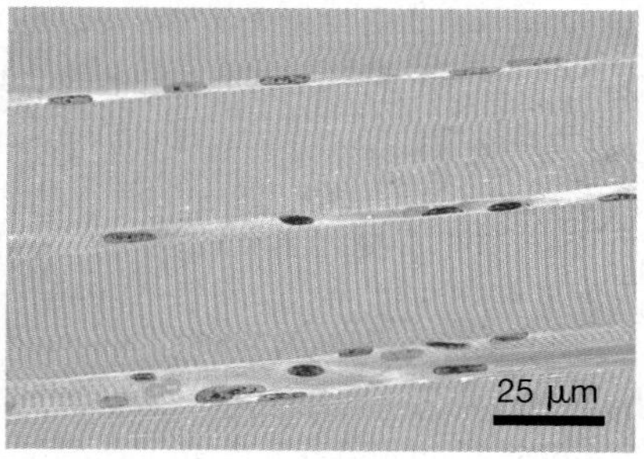

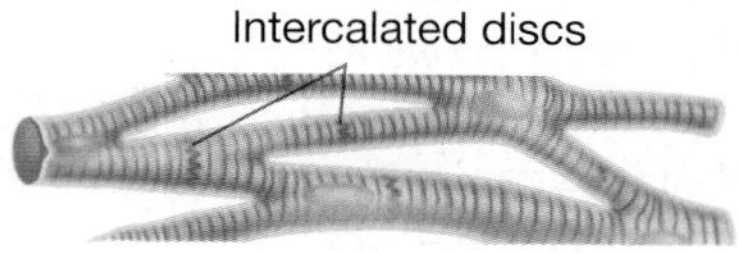

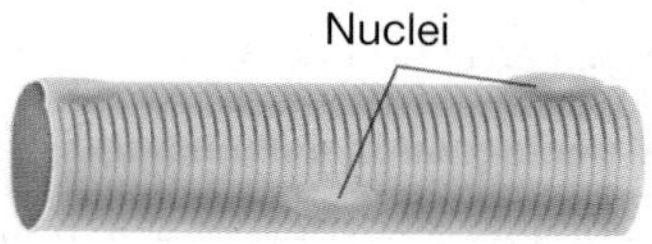

	Smooth Muscle	Cardiac Muscle	Skeletal Muscle
Location	Intestines, arteries, other	Heart	Attached to the skeleton
Function	Move food, help regulate blood pressure, etc.	Pump blood	Move skeleton
Cell characteristics	Single nucleus	1 or 2 nuclei	Multinucleate
	Unstriated	Striated	Striated
	Unbranched	Branched; intercalated discs form direct cytoplasmic connections between cells	Unbranched
	No sarcomeres	Contains sarcomeres	Contains sarcomeres
	Activity is “involuntary,” meaning that signal from motor neuron is not required	Activity is “involuntary,” meaning that signal from motor neuron is not required	Activity is “voluntary,” meaning that signal from somatic motor neuron is required

striated. Further, cardiac muscle cells have a unique branched structure, and they are directly connected end to end via specialized regions called intercalated discs. These discs are critical to the flow of electrical signals from cell to cell and thus to the coordination of the heartbeat.

Like smooth muscle, cardiac muscle is autorhythmic and involuntary—it contracts following spontaneous depolarizations, and the rate and strength of contractions are influenced by autonomic neurons. During rest, parasympathetic neurons release acetylcholine onto the heart. This neurotransmitter slows the rate of depolarization of cardiac cells. The result is a lower heart rate. During exercise, or when an animal is frightened, stressed, or otherwise stimulated, sympathetic neurons release norepinephrine onto the heart, and the adrenal glands release epinephrine. Norepinephrine and epinephrine increase heart rate and strengthen the force of cardiac muscle contraction. The result is that more blood is pumped from the heart—an essential component of the fight-or-flight response.

Skeletal Muscle

Skeletal muscle consists of exceptionally long, unbranched muscle fibers. For example, a skeletal muscle fiber of a cat may be 0.4 mm wide and 40 mm long—enormous compared to most cells. These large cells result from the fusion of many smaller embryonic cells during development, accounting for the multiple nuclei spread out along the cell. Each muscle fiber is packed with myofibrils, each of which may contain thousands of sarcomeres, giving skeletal muscle its striated appearance.

Skeletal muscle is so named because it usually attaches to the skeleton. When it contracts, skeletal muscle exerts a pulling force on the skeleton, causing it to move—powering the sprint of cheetahs, the flight of hummingbirds, and the pinch of crab claws. In addition, skeletal muscle encircles the openings of the digestive and urinary tracts and controls swallowing, defecation, and urination.

A significant fraction of the body of many animals is composed of skeletal muscle. For example, 63 percent of the body weight of trout is skeletal muscle, and mammals—including humans—of all sizes are 40 to 45 percent muscle. When animals engage in load-bearing exercise, such as weight-lifting by humans, existing muscle fibers synthesize additional contractile proteins and become larger; no new cells are formed. The increased size of the muscle fibers allows a muscle to do more work—for example, to pull on bones with greater force to lift heavier weights. Clearly, skeletal muscle plays an important role in animal biology. Some of the major skeletal muscles in the human body are shown in Figure 45.7.

Skeletal muscle is distinguished from cardiac and smooth muscle by being voluntary. Skeletal muscle must be stimulated by somatic motor neurons to contract. If these motor neurons are damaged, as can occur with a spinal cord injury, skeletal muscle cannot contract and becomes paralyzed.

Although all muscles contract as described by the sliding-filament model, not all skeletal muscle fibers have the same contractile properties. The force output of skeletal muscles depends on **(1)** the relative proportion of different fiber types, **(2)** the organization of fibers within the muscle, and **(3)** how the muscle is used. Let's take a closer look at these sources of variation in muscle performance.

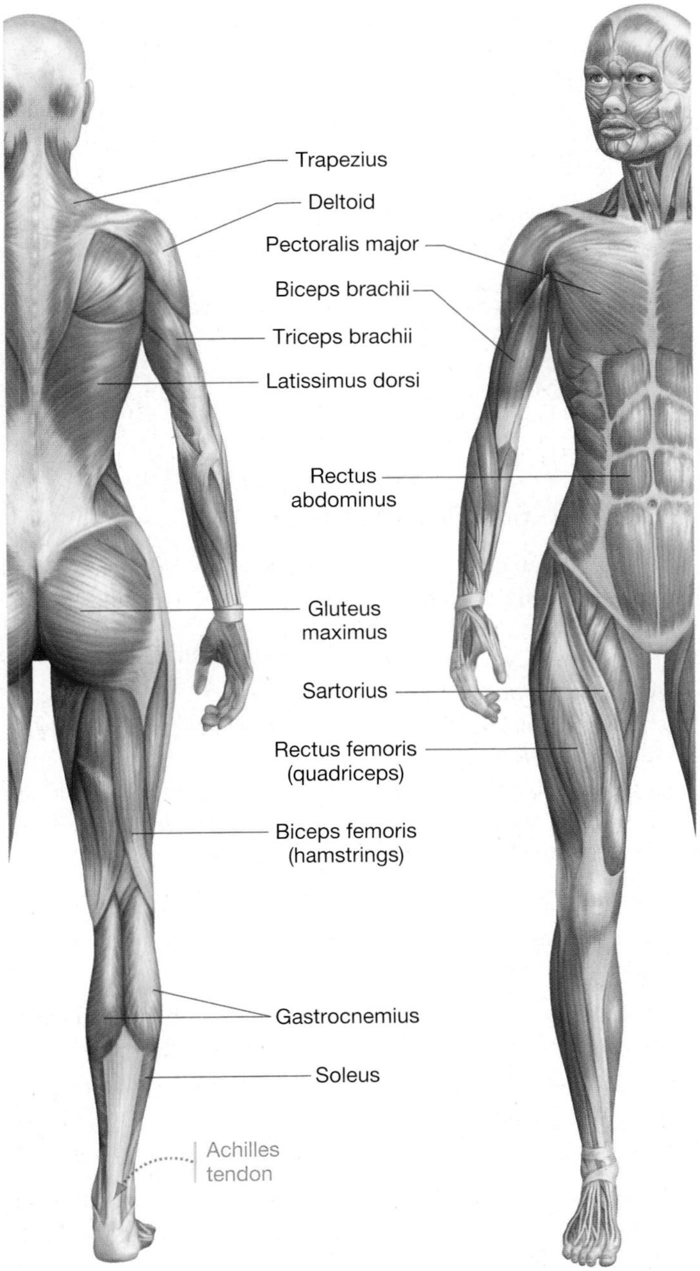

Figure 45.7 The Major Skeletal Muscles of the Human Body. Skeletal muscles, including those illustrated here, make up a large portion of the human body.

Skeletal Muscle Fiber Types Skeletal muscle fibers can be divided into general types based on their structural and functional characteristics, summarized in **Table 45.2**:

- **Slow muscle fibers** (slow oxidative fibers) appear red because they contain a high concentration of myoglobin, an iron-bearing pigment that carries oxygen (similar to but distinct from hemoglobin in the blood). Slow fibers contract slowly because their myosin hydrolyzes ATP at a slow rate. They also fatigue slowly because they have many mitochondria and can generate steady quantities of ATP using oxidative phosphorylation—that is, aerobic respiration (Chapter 9)—thanks to the plentiful supply of oxygen delivered by myoglobin.
- **Fast muscle fibers** (fast glycolytic fibers) appear white because they have a low myoglobin concentration. They contract rapidly because their myosin hydrolyzes ATP quickly, but they also fatigue rapidly because their primary source of ATP is glycolysis rather than aerobic respiration.
- **Intermediate muscle fibers** (fast oxidative/glycolytic fibers) appear pink or red. Their contractile properties vary, but they are intermediate between those of slow and fast fibers, because intermediate fibers derive ATP from both glycolysis and aerobic respiration.

SUMMARY **Table 45.2 Skeletal Muscle Fiber Types**

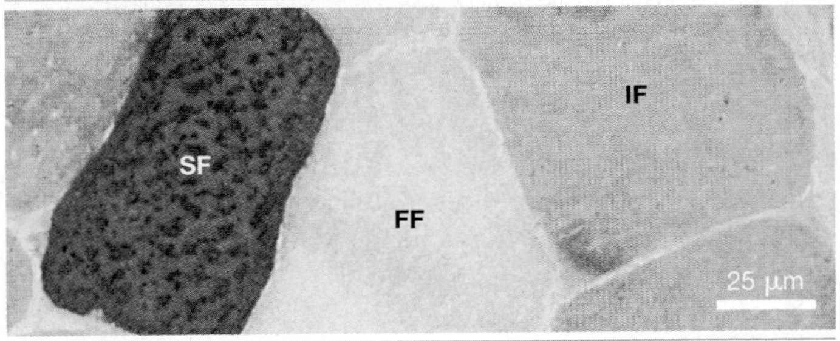

Slow fiber	Fast fiber	Intermediate fiber
Red	White	Pink or red
High myoglobin concentration	Low myoglobin concentration	High myoglobin concentration
Derives most ATP via aerobic respiration (slow oxidative)	Derives most ATP via glycolysis (fast glycolytic)	Derives ATP from glycolysis and aerobic respiration (fast oxidative/ glycolytic)
Many mitochondria	Few mitochondria	Many mitochondria
Slow twitch	Fast twitch	Intermediate twitch
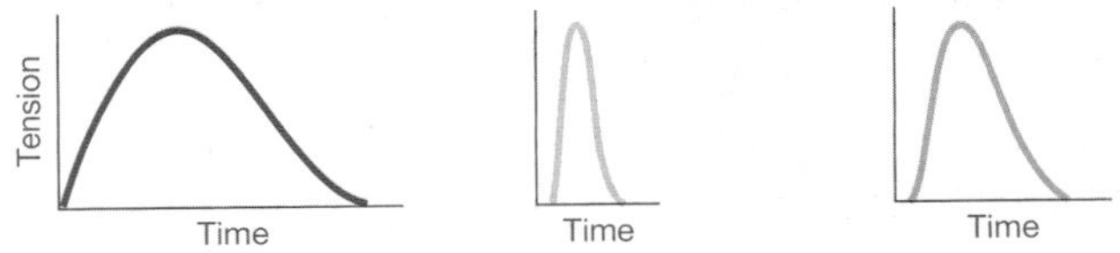		
Fatigues slowly	Fatigues quickly	Intermediate fatigue

The different fiber types are present in all skeletal muscles, but their relative abundances differ from muscle to muscle. Slow fibers are abundant in muscles specialized for endurance—such as the leg muscles of birds that swim or walk a lot (the "dark meat" of chicken legs). In humans, the soleus muscle in the back of the calf is an example of a muscle with a high proportion of slow fibers—it helps to keep you upright when you stand.

Fast fibers contract and relax up to three times as fast as slow fibers, making fast fibers well suited for bursts of activity. The "white meat" of chicken breasts, specialized only for quick bursts of flight to escape predators, is made primarily of fast fibers. The muscles that control your eye movements are another example.

✔ **If you understand muscle fiber types, you should be able to predict the most abundant fiber type in the postural muscles of the human neck.**

Can humans change their muscle fiber types through training? Experiments have shown that endurance training can increase the density of mitochondria and myoglobin in muscle fibers as well as the number of blood vessels in muscles, enabling athletes to improve their muscle performance. However, training does not change slow fibers to fast fibers, nor the reverse.

Skeletal Muscle Fiber Organization How much the length of a muscle changes is determined by the length of its muscle fibers—how many sarcomeres are lined up in a row in each fiber. By contrast, the force exerted by a muscle is proportional to the cross-sectional area of the muscle—the number of sarcomeres lined up side by side, working together to exert a pull. Thus, the arrangement of fibers within a muscle influences the contractile properties of the muscle.

For a given muscle volume, some muscles are organized to maximize length change, because the fibers are parallel to each other in long bands (**Figure 45.8**, left)—a longer chain of sarcomeres in a myofibril produces a greater change in length. For example, the sartorius muscle in the human thigh, the longest muscle in our bodies, has parallel fibers.

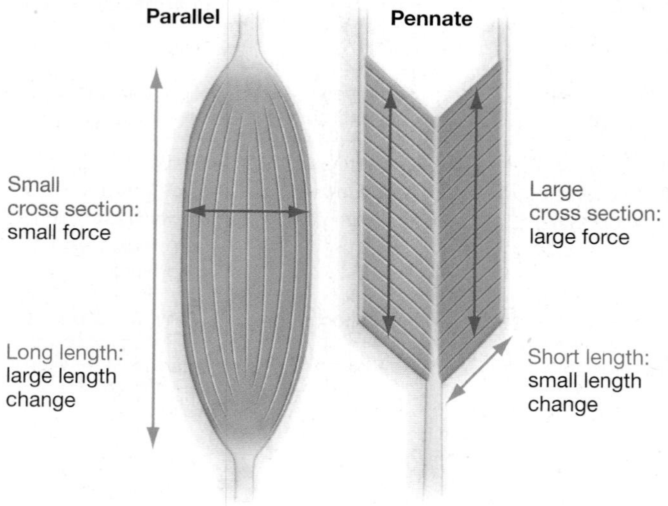

Figure 45.8 Patterns of Muscle Fiber Arrangement Affect Muscle Contractile Properties. Most muscles have fibers arranged in either a parallel or pennate pattern. The orientation of muscle fibers affects the contractile properties of the muscle.

Other muscles are organized to maximize force, because the fibers are in a diagonal, or pennate, pattern ("penna" means feather; see Figure 45.8, right)—more sarcomeres pulling in parallel produce a greater force. The gastrocnemius muscle in the human calf is a pennate muscle.

Context of Muscle Contraction The relative abundances of fiber types and organization of muscle fibers are not sufficient to account for the diverse contraction properties of skeletal muscles. Muscle force also varies according to how extended the muscle is when it contracts and how rapidly it is allowed to shorten—if at all. Muscles like the quadriceps in your thigh can even exert a force while they are lengthening, such as when you ease your weight down a step.

These circumstances depend on the interaction between the muscle and the skeleton. Let's take a closer look at how the muscle and skeletal systems interact to produce movement.

CHECK YOUR UNDERSTANDING

✔ **If you understood this section, you should be able to . . .**

1. Contrast the structures of smooth, cardiac, and skeletal muscle.
2. Explain how a person's skeletal muscles might change when they train for a marathon.

Answers are available in Appendix A.

45.3 Skeletal Systems

All a muscle can do is pull. How can complex movements be accomplished using an engine that can only pull? Also, muscles are limited in how much they can shorten. How, then, can they cause dramatic shape changes in animal bodies?

After you complete this section, you should be able to . . .

- Analyze skeletal systems in animals.

Muscle forces and shape changes are transmitted to other parts of the body and to the environment via the skeleton. Skeletal systems perform the following functions:

- *Protection* from physical and biological assaults.
- *Maintenance of body posture* despite the downward pull of gravity and the vagaries of wind and waves.
- *Re-extension of shortened muscles* because if no mechanism of re-extension existed, muscles would shorten only once.
- *Transfer of muscle forces* to other parts of the body and to the environment, enabling a much greater range of force production and shape change than can be accomplished by muscle alone.

The relative importance of these roles varies among animals according to their lifestyles and environments. For example, natural selection has favored turtles with robust shells in some environments—a skeletal adaptation for protection. In other environments, natural selection has favored highly reduced shells—an adaptation for rapid locomotion. There are trade-offs between protection and mobility.

(a) Annelids (e.g., earthworms) have **hydrostatic skeletons**.

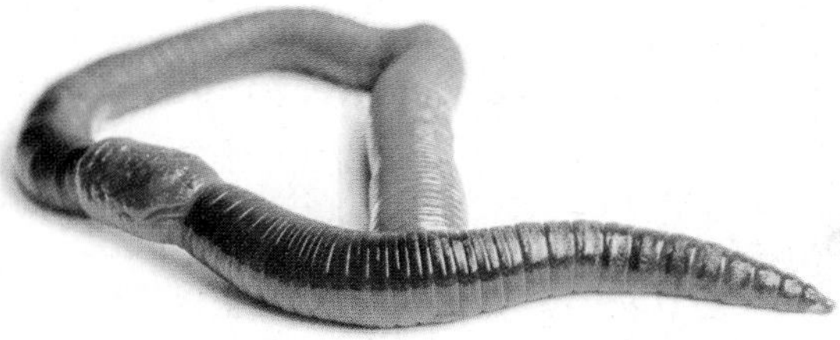

(b) Vertebrates (e.g., frogs) have internal **endoskeletons**.

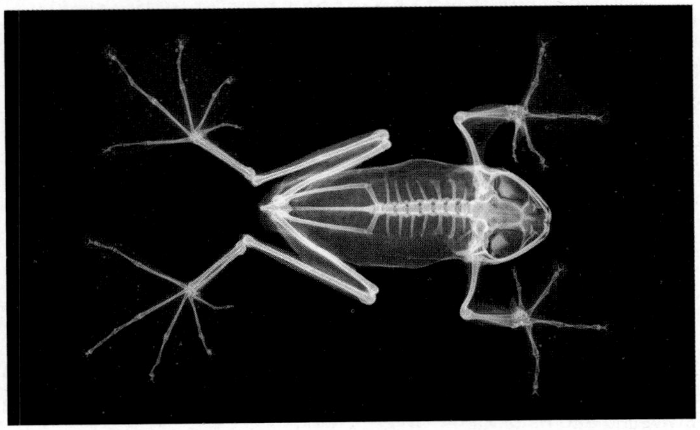

(c) Arthropods (e.g., crabs) have external **exoskeletons**.

Figure 45.9 **Three Types of Skeletal Systems Allow Animals to Move.**

Despite the stunning diversity in animal bodies, virtually all animals can be considered to have one (or more) of three types of skeletal systems (Figure 45.9):

1. *Hydrostatic skeletons* use the hydrostatic pressure of enclosed body fluids or soft tissues to support the body (see Figure 45.9a).
2. *Endoskeletons* have rigid structures inside the body (see Figure 45.9b).
3. *Exoskeletons* have rigid structures on the outside of the body (see Figure 45.9c).

Let's consider how each skeletal system transmits muscle forces and shape changes.

Hydrostatic Skeletons (Hydrostats)

Despite their squishy appearance, soft-bodied animals do have skeletons—hydrostatic skeletons. First let's look at how they are built, and then consider how they function.

Hydrostat Structure **Hydrostatic skeletons** ("still-water skeletons"), or hydrostats, are constructed of an extensible body wall in tension surrounding a fluid or deformable tissue under compression. When fluid is under compression, its pressure increases. The pressurized internal fluid, rather than a rigid structure, enables soft-bodied animals to maintain posture, re-extend muscles, and transfer muscle forces to the environment.

Hydrostatic skeletons occur in diverse animals, from sea anemones and jellyfish to mollusks and many types of worms (Chapter 31). Hydrostats also support *parts* of animals, such as the tongues and penises of humans and the tube feet of echinoderms (Ch. 32, Section 32.2).

The structures of hydrostatic skeletons are diverse as well. The body wall of hydrostats may include different numbers and orientations of muscle layers and fiber-reinforced cuticles or connective tissues. The interior may include seawater, coelomic fluid, blood, or soft organs such as intestines. Hydrostatic skeletons composed mostly of muscle, such as tongues and tentacles, are called muscular hydrostats.

Hydrostat Function How do animals with hydrostatic skeletons move? Consider an earthworm. Its body wall consists of a cuticle reinforced with collagen fibers as well as two layers of muscle—longitudinal muscles, oriented along the length of the animal, and circumferential muscles, oriented in bands around each segment. Longitudinal and circumferential muscles in earthworms make up an **antagonistic muscle group**, a group of two or more muscles that re-extend one another via the skeleton. When the circumferential muscles contract, they make the segments narrower and squeeze the coelomic fluid and internal tissues, thus increasing internal pressure. The pressure pushes outward in all directions, extending the relaxed longitudinal muscles and lengthening the segment (Figure 45.10).

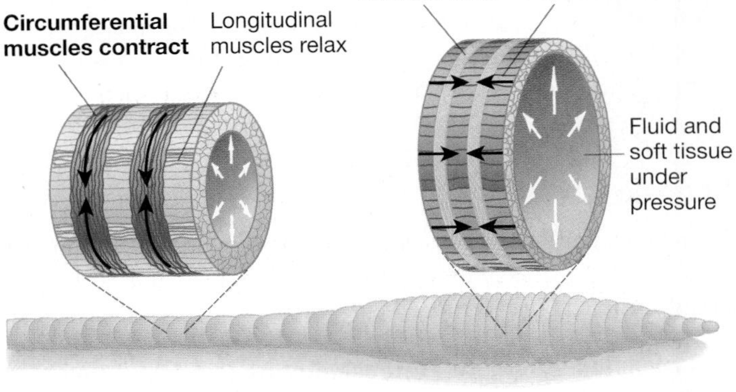

Figure 45.10 **Antagonistic Muscle Groups Cause Shape Changes in Hydrostatic Skeletons.** The pressure of the internal fluid or tissue transmits forces between muscle groups and between muscles and the environment.

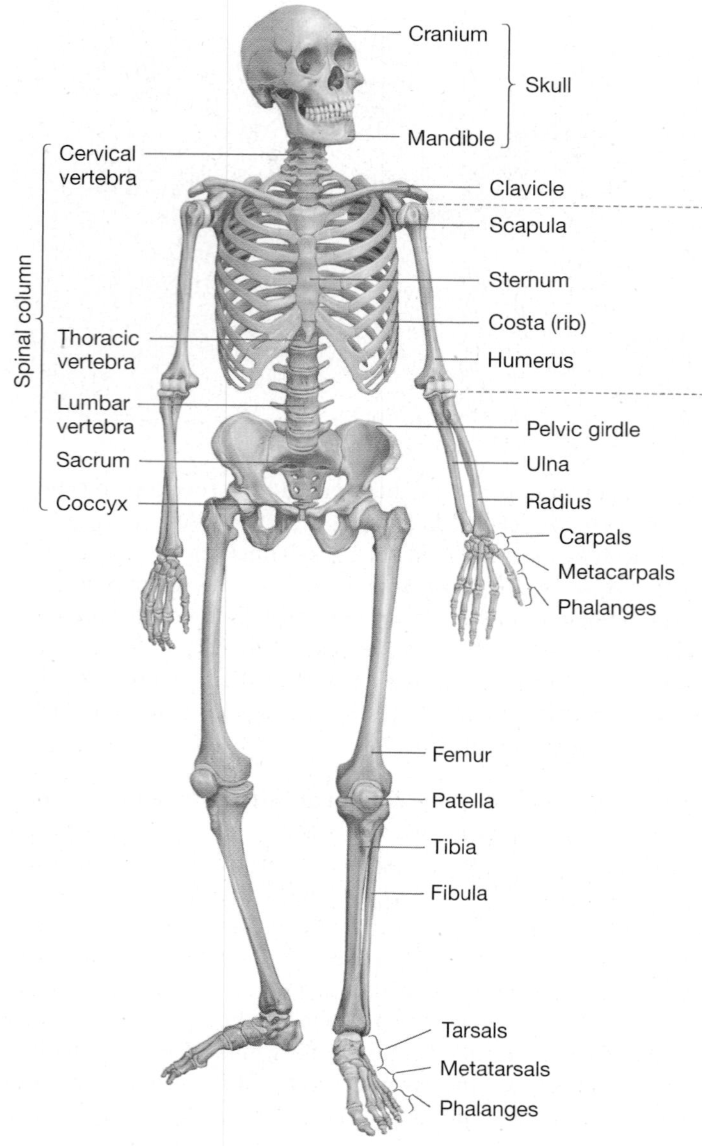

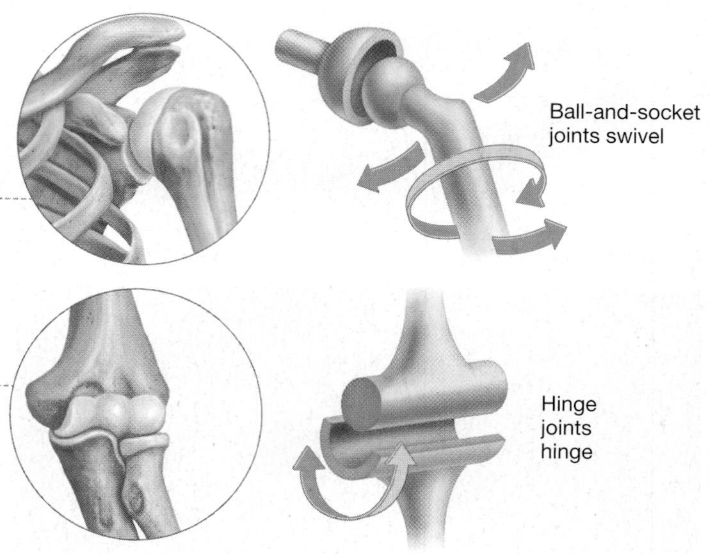

Figure 45.11 **Bones of the Human Endoskeleton.** Bones are rigid and cannot change shape themselves; instead they articulate at joints that make specific types of movement possible, such as swiveling and hinging.

When the longitudinal muscles contract and the circumferential muscles relax, the reverse occurs—the segments become wider and shorter, pushing sideways against the soil. Alternating contractions of longitudinal and circumferential muscles pass down the earthworm in waves, called **peristalsis**. In this way, earthworms move forward or backward within their underground burrows.

Endoskeletons

Even though parts of you, such as your tongue, are supported by a hydrostatic skeleton, your endoskeleton is what keeps you standing up and on the move.

Endoskeleton Structure **Endoskeletons** ("inside skeletons") are rigid structures that occur within the body. Even the most ancient of animal lineages, the sponges, secrete spicules—stiff spikes of silica or calcium carbonate—that provide structural support for the body. In echinoderms, the endoskeleton consists of calcium carbonate plates just beneath the skin—fused into a rigid case in sea urchins, but suspended in a flexible matrix that enables bending of the arms in sea stars.

The vertebrate endoskeleton differs from those of most sponges and echinoderms—and from hydrostatic skeletons—in that it is composed of rigid levers (the bones) separated by joints. Vertebrates change the shapes of their bodies largely by changing the *joint angles* between bones in the limbs and between the limbs and the rest of the body, rather than changing the shapes of body segments themselves.

Vertebrate skeletons are composed of three main elements:

1. **Bones** are made up of cells in a hard extracellular matrix of calcium phosphate with small amounts of calcium carbonate and protein fibers. The adult human body contains 206 bones (Figure 45.11a). The meeting places where adjacent bones interact are called **articulations**, or **joints**. Bones articulate in ways that limit the range of motion, for example enabling a swivel in the shoulder joint but a hinge in the elbow joint (Figure 45.11b).
2. **Cartilage** is made up of cells scattered in a gelatinous matrix of polysaccharides and protein fibers. Cartilage can be quite rigid, as in the clam-crushing jaws of some stingrays, or more rubbery, as in the pads that cushion the joints in your knees and back.
3. **Ligaments** are bands of fibrous connective tissue, primarily collagen, that bind bones to other bones. Ligaments stabilize the joints.

Bones attach to skeletal muscle via bands of fibrous connective tissue called **tendons**. The ropelike structure of tendons transmits muscle forces to precise locations on the bones, sometimes quite a distance away from the muscle itself.

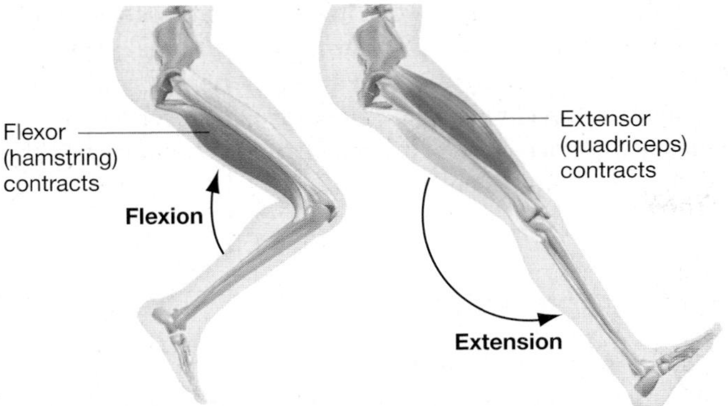

Figure 45.12 **Endoskeletons Move by Contraction and Relaxation of Flexor and Extensor Muscles.**
The muscles are attached to the outside of the skeleton.

✔ Use this figure, along with Figure 45.7, to assign the roles of flexor or extensor to the biceps brachii and triceps brachii muscles.

Endoskeleton Function: Movement Bones and cartilage do a good job of resisting compression (pushing) and bending, whereas tendons and ligaments do a good job of resisting tension (pulling). These structures interact with muscle in ways that enable the efficient transmission of muscle forces and shape changes.

Vertebrate skeletons move by means of changes in joint angles controlled by antagonistic muscle groups. For example, consider how your thigh muscles flex and extend your knee joint. The hamstring muscles in the back of your thigh are **flexors**, muscles that pull bones closer together, decreasing the joint angle between them. They swing your lower leg back toward your thigh, reducing the angle of your knee joint. The quadriceps muscles in the front of your thigh are **extensors**, muscles that increase the angle of a joint. They straighten your leg at the knee joint (Figure 45.12).

The hamstrings and quadriceps muscles accomplish the large swing of the lower leg by inserting into different locations on the tibia (shin bone). The articulation of the tibia with the femur (thigh bone) serves as the pivot point for this lever. Along with enabling a shape change, the bone transmits the forces exerted by the thigh muscles to the foot—such as when the extension of your leg enables you to kick a ball.

Endoskeleton Function: Calcium Homeostasis Besides their mechanical functions, bones serve several physiological functions. Chief among these is storage of calcium and other minerals. Calcium is necessary for the proper functioning of many physiological processes, including cell division, release of neurotransmitters, and muscle contraction (see Section 45.1). The calcium in bone is the main stored source of calcium for these functions. When the blood calcium ion level falls, the bones release calcium ions, maintaining blood–calcium homeostasis so that essential physiological functions can occur.

Several cell types work together to keep the blood calcium ion level relatively constant. **Osteoblasts** are bone-building cells—they secrete the protein- and calcium-rich extracellular matrix that hardens to form bone. **Osteoclasts**, however, are bone-resorbing cells. When the blood calcium ion level is low, osteoclasts secrete acid onto bone tissue, causing small amounts of minerals to be resorbed into the blood.

The relative rates of bone building and bone resorption are affected by many factors, including hormones, aging, diet, and exercise. In healthy adults, bone building and breakdown are balanced, so bone mass remains constant even though the turnover rate is high. Weight-bearing exercise helps to increase bone mass, while aging and malnutrition can reduce it. **Osteoporosis** ("bone pores"), a disease in which reduced bone mass can make bones brittle and susceptible to fracture, is a common affliction of elderly people.

Bones perform other physiological functions as well. The interior of long bones, called bone marrow, is the source of the red blood cells needed to carry oxygen in the blood (Ch. 42, Section 42.4) and the white blood cells needed for the immune system (Chapter 48).

Exoskeletons

The mechanical function of rigid cuticle in exoskeletons is similar in many ways to the function of endoskeletons. Exoskeletons occur primarily in arthropods, including insects, crustaceans, and arachnids (spiders, ticks, scorpions).

Exoskeleton Structure An **exoskeleton** ("outside skeleton") is an exterior skeleton that encloses and protects an animal's body. The origin of the exoskeleton was a key innovation that preceded the spectacular diversification of arthropods, the most diverse and abundant animals on Earth (Ch. 31, Section 31.3).

The material composition of exoskeletons varies. Insect exoskeletons consist of a cuticle formed from a composite of proteins and the polysaccharide chitin (see Ch. 5, Table 5.1). Chitinous ingrowths of the skeleton form **apodemes**, where muscles attach. Crustaceans such as crabs and lobsters have a cuticle that is mineralized with calcium carbonate, making their exoskeletons relatively thick and hard—and heavy. Most crustaceans are aquatic, and their buoyancy in water helps support their weight.

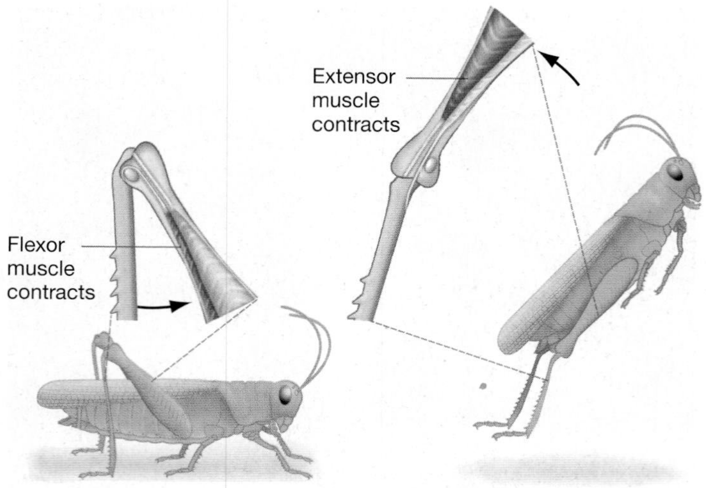

Figure 45.13 Exoskeletons Move by Contraction and Relaxation of Flexor and Extensor Muscles. The muscles are contained within the skeleton.

Exoskeleton Function Like vertebrates, arthropods have paired flexor–extensor muscles that operate their jointed skeletons, causing movements that are based on changes in joint angles rather than changes in the dimensions of the segments themselves (**Figure 45.13**). Unlike vertebrate muscles, however, the muscles of arthropods must be packed *within* the skeleton.

In many arthropods, the problem of the interior placement of muscles is solved by having the pennate, or feather-like, arrangement of muscle fibers illustrated in Figure 45.8. This arrangement boosts force output by effectively increasing the muscle cross-sectional area but not the muscle width during a contraction.

The disadvantage of pennate muscles is that their length change is small, so they have a limited range of motion. Arthropods compensate for this constraint in part by the placement of their apodemes, which can convert a small shortening of muscle to a large change in joint angle.

The rigid levers of vertebrate skeletons can grow continuously as the rest of the body grows. But because the rigid exoskeletons of arthropods encase the growing soft tissue like a suit of armor, they must be shed—molted—periodically and replaced with a bigger one. Arthropods are vulnerable to predation during molts because their skeleton is soft and dysfunctional at that time.

Hydrostatic skeletons, endoskeletons, and exoskeletons all transmit muscle forces and shape changes to other parts of the body and to the environment. While hydrostatic skeletons are the most widespread in terms of the number of animal phyla that contain species with skeletons of this type (virtually all—even arthropods, echinoderms, and vertebrates), the lever-based, segmented, jointed skeletons of arthropods and vertebrates win the prize for overall functional diversity.

CHECK YOUR UNDERSTANDING

✓ If you understood this section, you should be able to …

1. Compare and contrast the structure and function of hydrostatic skeletons, endoskeletons, and exoskeletons.
2. Predict what would happen if neurons simultaneously stimulated the biceps brachii and triceps brachii muscles to contract.

Answers are available in Appendix A.

45.4 Locomotion

After you complete this section, you should be able to …

- Explain how animals locomote.

The most spectacular capability conferred on animals by the combination of muscle contractions and skeletal systems is efficient locomotion. Some animals migrate thousands of miles every year, and others perform astonishing feats of acrobatics. Locomotion has been shaped by natural selection and has been central to complex ecological relationships since the radiation of animals during the Cambrian explosion more than 500 million years ago (Ch. 30, Section 30.1).

Many of the diverse modes of locomotion are already familiar to you. Here are some examples:

- ***On land*** Crawling, walking, running, climbing, hopping, jumping, burrowing
- ***In water*** Undulating, jetting, swimming, rowing
- ***In air*** Flying, gliding

What variables do biologists analyze to unlock the secrets of animal locomotion?

How Do Biologists Study Locomotion?

The number of experimental studies on locomotion has increased exponentially in recent years, partly due to the conceptual breakthrough offered by the field of biomechanics—applying the principles of engineering to quantify the mechanics of organisms. Biomechanics studies the physical act of locomotion at different levels, including material properties, structures, motions, forces, and energetics. Let's examine each of these levels, starting with materials.

How Are the Material Properties of Tissues Important to Locomotion? A great deal of research has been devoted to analyzing the active contractile properties of muscle because they are central to understanding how forces are generated in locomotion. However, the passive material properties of the skeletal elements are essential to understanding the transmission of forces.

Consider how you walk and run. A pioneer of biomechanics, R. McNeill Alexander, observed that, in terms of the movement and energy exchange of the center of mass, walking is mechanically similar to the swinging of an upside-down pendulum, whereas running is mechanically similar to a bouncing ball. Based on this insight, Alexander hypothesized that the large tendons of the lower legs of terrestrial animals—for example, the Achilles

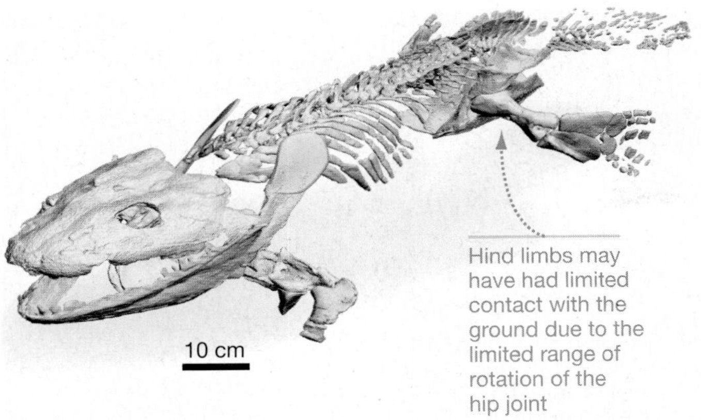

Figure 45.14 The Relationship between Structure and Function Can Be Studied Using Computer Models. This computer image of the early tetrapod *Ichthyostega* was built using high-resolution scans of fossil bones. The 3-D model measures the range of rotation of the major limb joints based on skeleton geometry, determining the overall range of motion of the limbs. The model found that the range of rotation of the shoulder and hip joints was more limited than previously thought, rejecting the hypothesis that these animals could walk on all four limbs.

tendon at the back of your ankle (see Figure 45.7)—work as springs when animals run.

To test this hypothesis, scientists measured the elastic properties of tendon by clamping a piece of tendon in a device that measures the pulling force of the tissue as it is stretched to different lengths and released. The tendon returned 93 percent of the energy invested, losing only 7 percent to heat. This high rate of elastic-energy storage explains how tendons add a spring to the steps of runners, reducing the amount of muscle-generated power that must be generated from step to step.

The importance of the material properties of muscles and skeletal tissues is most obvious when they fail, whether in subtle or catastrophic ways. You may have firsthand experience with the debilitating consequences of broken bones or strained tendons.

How is Musculoskeletal Structure Adapted for Locomotion? Many biologists begin their study of locomotion by examining the size and shape of the skeletal elements. Careful measurements of skeleton geometries can reveal a great deal about the posture of the animal, the range of motion of joints, and skeletal function in general. For example, a team of biologists recently used structural analysis to test the hypothesis that an early tetrapod, called *Ichthyostega*, could walk on all four limbs like today's salamanders (**Figure 45.14**).

Similar mechanical principles can be applied to relate the shapes of wings to flying ability and the shapes of aquatic animals to swimming ability. Sometimes, however, you just have to watch the action itself to understand how an animal uses its body to locomote.

How is Locomotion Studied in Living Animals? All the motions of different parts of the skeleton, such as the angular rotation of limbs and the pattern of footfall on the ground, together produce locomotion.

Photographer Eadweard Muybridge is famous for his pioneering photo sequences of locomotion. His work with animals reportedly began in 1872 when he was commissioned by a racehorse owner to settle a wager on whether horses are ever completely airborne during a gallop. His results are shown in the top sequence of **Figure 45.15**. High-speed video and digital images of many other animals have since been recorded, providing insights into many forms of locomotion, such as:

- The gait of human sprinters (see Figure 45.15, bottom)
- The complex wing-beat patterns of hovering bees and hummingbirds
- The upright, bipedal (two-footed) gait of basilisk lizards running across the surface of water (see chapter opening image)
- The footfall pattern that prevents centipedes from tripping on their own legs
- The aerial undulating of snakes that glide down from treetops
- The limb-like use of fins in lungfish "walking"
- Peristalsis in the muscular feet of crawling snails

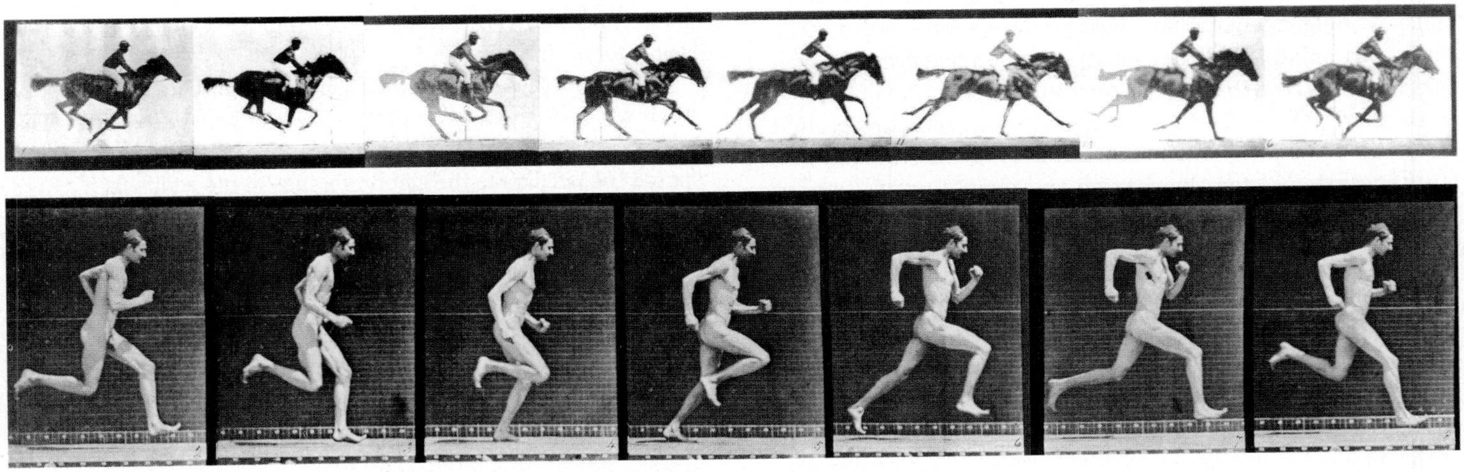

Figure 45.15 The Motions of Locomotion Can Be Captured on Film. Eadweard Muybridge shot many photo series of animals during locomotion to enable precise analysis of limb and body motions over time.

For most animals, the pattern of movement during locomotion varies with speed. For example, horses walk at slow speeds, trot at intermediate speeds, and gallop at fast speeds—each gait has a distinct pattern of leg motions.

Computers facilitate the analysis of the many images captured in motion studies. The results are themselves insightful, but they also serve as an important stepping-stone to understanding the forces involved in locomotion.

What Forces Are Involved in Locomotion? If an animal wants to move forward, it must push something backward, as predicted by Newton's third law of motion. Otherwise, the animal could move but not get anywhere, like a person with smooth shoes on slick ice. The types of forces that are important in locomotion vary according to whether the animal is locomoting on land, in water, in air, or some combination of the three.

On land, gravitational and inertial forces dominate. The gravitational force experienced by an animal is its weight, which is the product of its mass and the acceleration due to gravity (9.8 m/sec^2 on Earth). Weight is important on land because most terrestrial animals must hold themselves up to move forward. Inertial forces are proportional to mass and velocity, and they represent the resistance of bodies and limbs to acceleration and deceleration. Note in the horse and human photo sequences in Figure 45.15 that the arms and legs must swing back and forth dramatically—an energy-intensive process of acceleration and deceleration.

In water, gravitational forces are less important than on land, due to the counteracting buoyant forces supporting the animal's weight in water. However, aquatic animals must overcome drag, the force that resists forward motion through fluids. Convergent evolution of torpedo-shaped bodies has occurred in diverse aquatic animals, from tuna to dolphins and ichthyosaurs, due to the strong selection for bodies that minimize drag during rapid locomotion (Ch. 25, Section 25.1). Aquatic animals that move more slowly face less drag and thus are morphologically more diverse.

Water and air are both fluids, but air is a thousand times less dense than water. Buoyant forces are therefore negligible in air, making gravitational forces very important to fliers—most animals that locomote by flying or gliding have adaptations that make them lightweight. They must produce a force called lift to counteract gravity, and they must also minimize drag. As a result, fast fliers tend to have very streamlined shapes.

How are forces measured? Scientists can measure ground force—the force with which a terrestrial animal strikes the substratum during a step—by coaxing an animal to walk, run, or hop on an instrument called a force plate. Quantifying forces in fluids is more nuanced. It often requires indirect measurement by visualizing the flow of air or water around the animal (**Figure 45.16**).

Figure 45.16 Visualization of Airflow Is Used to Analyze the Forces Involved in the Hovering of a Bat. The lift and drag forces acting on flying and swimming animals can be measured by observing fluid flow around the animal. The arrows represent the velocity of tiny water droplets illuminated by a laser in front of a high-speed camera.

What is the Cost of Locomotion? Animals must spend energy to find food and mates and to escape from predators. However, the more energy an animal spends on locomotion, the less it can spend on producing offspring. There is strong selection pressure to minimize the cost of locomotion. How do biologists measure this cost?

To get a sense of variables that determine the cost of locomotion, consider the classic studies by Alexander on the gait transition from walking to running in humans. Alexander discovered that walking is an efficient mode of transport because you exchange potential energy at the top of your stride with kinetic energy midstride. However, the resulting pendulum-like motion of your center of mass is not efficient at higher speeds, when it becomes more cost effective to run using your spring-like tendons and other skeletal tissues to store energy between strides.

Alexander hypothesized that animals locomote using the most energy-efficient gait at each speed. To test this hypothesis in horses, physiologists Dan Hoyt and Richard Taylor trained horses to walk, trot, and gallop at a range of unnatural speeds—for example, trotting at a speed where the horse would normally have preferred to gallop. Hoyt and Taylor fitted each horse with an oxygen mask and ran it on a treadmill, so that they could measure oxygen consumption and speed simultaneously. The rate of oxygen consumption is a measure of energy use because oxygen consumption is proportional to ATP production during aerobic respiration. The researchers then plotted energy use versus speed for each gait.

Hoyt and Taylor also filmed the horses moving freely around their paddock and measured the speeds at which the horses used different gaits when given free choice. The researchers then compared the lab data to the gait preference in the paddock.

The graph in **Figure 45.17** shows Hoyt and Taylor's results. The three curves at the top represent the energy used per distance traveled at the three gaits—the dips in the curves indicate the speeds at which the gaits were most efficient. The bars at the bottom of the graph show the speeds and gaits chosen by the horses when they were able to locomote freely in the paddock. The data support the hypothesis that the horses use the most energy-efficient gaits at different speeds and avoid intermediate speeds where the cost of locomotion is higher.

Similar studies have been conducted for diverse animals running, swimming, and flying, with similar results. Natural

RESEARCH

QUESTION: Do horses minimize the cost of locomotion?

HYPOTHESIS: Horses choose gaits that minimize energy use at different speeds.

NULL HYPOTHESIS: Horses do not choose gaits based on the cost of locomotion.

EXPERIMENTAL SETUP:

1. Measure oxygen consumption of horses trained to walk, trot, and gallop at a range of speeds on a treadmill. Calculate energy used per distance traveled at different speeds.

2. Videotape the same horses locomoting freely in the paddock, and measure the gaits and speeds they choose to use naturally.

PREDICTION OF HYPOTHESIS: For each gait, there is a range of speeds at which energy use is minimized. Horses will favor these gaits and speeds.

PREDICTION OF NULL HYPOTHESIS: There will be no correlation between chosen gaits and energy consumption.

RESULTS:

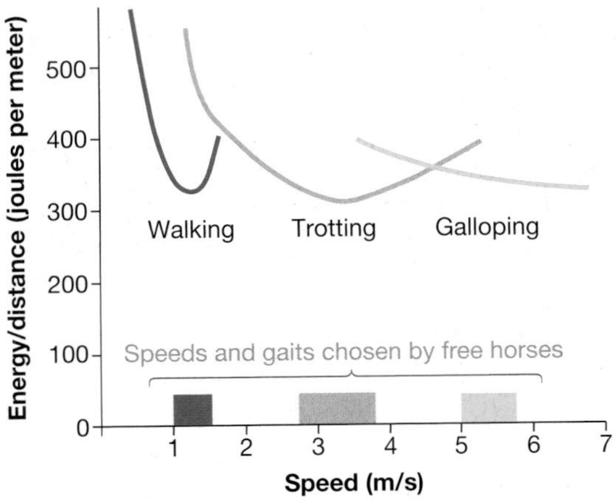

CONCLUSION: Horses choose gaits that minimize energy use at different speeds and avoid speeds with high energy consumption.

Figure 45.17 **Horses Minimize the Cost of Locomotion by Choosing Appropriate Gaits.**

SOURCE: D. F. Hoyt and C. R. Taylor. 1981. Gait and the energetics of locomotion in horses. *Nature* 292: 239–240.

✔ **QUANTITATIVE** Use the graph to estimate the relative cost of galloping rather than trotting at 3.5 meters/second (m/s).

selection favors animals that locomote efficiently because they have more energy available for other vital activities.

Size Matters

Animals that use muscles to power locomotion span a vast range of sizes—an astonishing 10 orders of magnitude (10,000,000,000)—from tiny ants to giant whales. Many of these animals also grow over a large size range during their development. As is true for many other aspects of animal structure and function, size matters.

Two organisms may be geometrically similar—that is, they may have exactly the same proportions—but if they are different sizes, they are more different than they seem. To start, the ratio of surface area to volume decreases as the organism gets larger, because surface area is proportional to length squared, while volume is proportional to length cubed (Ch. 39, Section 39.3). This concept has far-reaching implications for physiology. It also has important mechanical implications.

The weight of an animal is proportional to its volume, and the ability of leg bones to support the weight is proportional to their cross-sectional area. Thus, large terrestrial animals must have disproportionately hefty skeletal elements to avoid breaking their legs—something that Galileo observed 400 years ago when he compared skeletons of small and large animals (**Figure 45.18**).

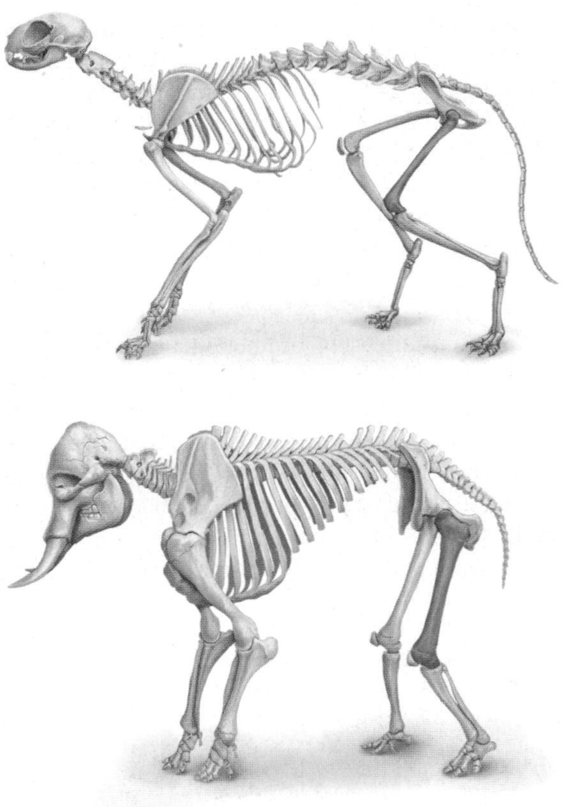

Figure 45.18 **Size Influences Skeleton Geometry.** When scaled to the same size, an elephant has thicker bones and a more upright stance than a cat does, due to the disproportionate burden of gravity on larger animals. The femurs are highlighted for comparison.

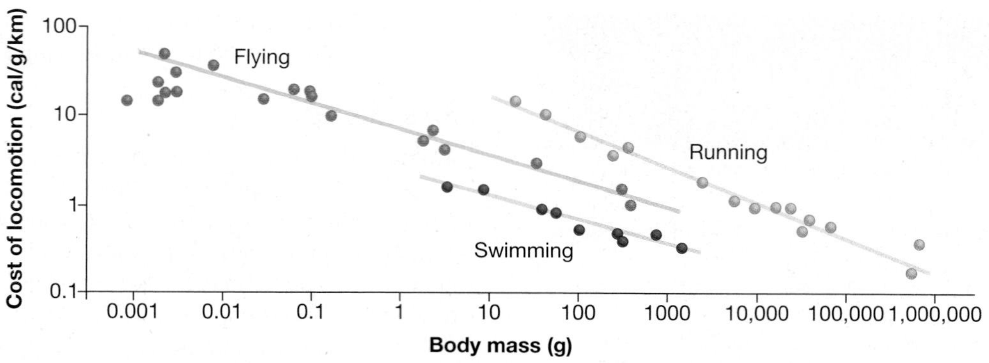

Figure 45.19 The Cost of Locomotion for Running, Flying, and Swimming Animals Decreases with Body Mass.

DATA: K. Schmidt-Nielsen. 1972. Locomotion: Energy cost of swimming, flying, and running. *Science* 177: 222–228.

✔ **QUANTITATIVE** About how much more costly is it to run than to swim for animals with a body mass of 100 g?

So, in the late 1980s, it was surprising when Andrew Biewener discovered that animals of different sizes maintain similar stresses in their skeletal tissues. How do they do that?

Biewener observed that posture and behavior are important variables. Small animals locomote in a more crouched posture and make great leaps, while large animals such as elephants are more straight-legged, so that their skeletons, rather than their muscles, can support their body weight. Larger animals also locomote more gently. If a house cat were enlarged to the size of an elephant, it would break its own bones when trying to pounce.

Size is also of paramount importance in determining how animals locomote through fluids. Fluids have an inherent viscosity, or stickiness, that is of minor concern to animals that are very large and/or fast (and thus have high inertia), such as dolphins, which can glide through water with little effort. But viscosity is of enormous concern to animals that are very small and/or slow (and thus have low inertia), such as plankton. If you were the size of a grain of rice and went for a swim in a lake, the water would feel like corn syrup.

Size also affects the cost of locomotion. In the 1970s, physiologist Knut Schmidt-Nielsen compared the energy costs of different modes of locomotion. To account for widely different body sizes and distances traveled, he normalized all data to the energy cost in calories per gram of body tissue per kilometer traveled and then plotted the data on a log–log plot (**BioSkills 5**).

Schmidt-Nielson's data showed that larger animals use less energy per gram of body tissue per distance traveled than smaller animals—all the lines slope down to the right (**Figure 45.19**). The data also showed that for a given body size, running is a more expensive mode of transport than flying, and swimming is the least expensive.

Many other aspects of locomotion can also be plotted as a function of body size. The results reveal general principles that can be used to make predictions for diverse animals, even those that are extinct.

The take-home message? The laws of physics establish definite constraints on the possible sizes, shapes, and modes of locomotion in animals. Understanding these principles helps illuminate themes and variations among the diversity of animals.

CHECK YOUR UNDERSTANDING

✔ **If you understood this section, you should be able to . . .**

1. Consider locomotion on land, in water, and in the air. For which type of locomotion is gravity most difficult to overcome? For which type of locomotion is drag most difficult to overcome?
2. Movies such as *Jurassic Park* often show *Tyrannosaurus rex* running swiftly and jumping onto its prey, including humans. Evaluate the accuracy of this portrayal in terms of biomechanics.

Answers are available in Appendix A.

45.1 How Do Muscles Contract?

- The muscles of vertebrate limbs are composed of cells, called muscle fibers, that contain contractile elements called myofibrils, each divided into contractile units called sarcomeres.
- Sarcomeres appear striated, or banded, due to the aligned arrangement of thick filaments (myosin) and thin filaments (actin).
- Sarcomeres shorten when thick filaments of myosin slide past thin filaments of actin in a series of binding events mediated by the hydrolysis of ATP.
- Calcium ions play an essential role in muscle contraction by making the actin in thin filaments available for binding by myosin.
- Acetylcholine released from somatic motor neurons is the neurotransmitter that stimulates contraction of skeletal muscle.

45.2 Classes of Muscle Tissue

- Smooth muscle lines bronchioles, blood vessels, the gastrointestinal tract, and certain reproductive organs. Smooth muscle cells are small and unstriated and have a single nucleus. Contractions are involuntary.
- Cardiac muscle is found in the heart and forces blood through the circulatory system. Cardiac muscle cells are striated, branched, and connected to one another by intercalated discs. Contractions are involuntary.
- Most skeletal muscles are attached to the skeleton and are responsible for voluntary movement of the body. Skeletal muscle cells, called fibers, are long, striated, and multinucleate.
- Skeletal muscle fibers are specialized to contract slowly or quickly and to have a high or low endurance. These properties depend on the concentration of myoglobin present and the use of aerobic respiration and/or glycolysis for the production of ATP.
- Skeletal muscle fibers are mostly organized in a parallel arrangement, which maximizes shortening, or a pennate pattern, which maximizes force production.

45.3 Skeletal Systems

- Hydrostatic skeletons, or hydrostats, are composed of a body wall in tension surrounding fluid or soft tissue under compression.
- Endoskeletons are internal skeletons, surrounded by soft tissue. In vertebrates, the jointed endoskeleton is composed of bones, cartilage, and ligaments.
- Bones are involved in movement, protection of organs, storage of calcium, and production of blood cells.
- Exoskeletons are external skeletons, enclosing soft tissue. In arthropods, the muscles occur within the rigid, jointed cuticle composed of chitin, proteins, and sometimes minerals. Exoskeletons must be shed to enable growth.
- Movement is based on antagonistic muscle groups that act on a skeleton. Examples include flexors and extensors, which change the joint angle between rigid skeletal segments—especially in limbs.

45.4 Locomotion

- Locomotion is movement relative to the environment and requires the transmission of muscle forces to the land, water, or air surrounding the animal.
- Locomotion can be studied at different levels: material properties, structures, motions, forces, and energetics.
- Locomotion on land is usually dominated by gravitational and inertial forces. Swimmers must overcome drag. Fliers must overcome drag and must generate enough lift to counteract gravity.
- Body size is important to the mechanics of locomotion.

Answers are available in Appendix A.

✔ TEST YOUR KNOWLEDGE

1. Which of the following statements regarding control of muscle tissue is/are correct? Select True or False for each statement.
   T/F All skeletal muscles are voluntary.
   T/F Cardiac muscle is involuntary.
   T/F Some smooth muscle tissues are voluntary and others are involuntary.
   T/F Parasympathetic and sympathetic neurons innervate skeletal, cardiac, and smooth muscle tissues.
2. In muscle cells, myosin molecules continue moving along actin molecules as long as
   a. ATP is present and troponin is not bound to Ca^{2+}.
   b. ADP is present and tropomyosin is released from intracellular stores.
   c. ADP is present and the intracellular acetylcholine level is high.
   d. ATP is present and the intracellular Ca^{2+} concentration is high.
3. Which of the following is critical to the function of most exoskeletons, endoskeletons, and hydrostatic skeletons?
   a. Muscles interact with the skeleton in antagonistic groups.
   b. Muscles attach to each of these types of skeleton via tendons.
   c. Muscles extend joints by pushing skeletal elements.
   d. Segments of the body or limbs are extended when paired muscles relax in unison.
4. True or false: A large animal will experience twice the gravitational force of a small animal half its length if their geometries are the same.

✔ TEST YOUR UNDERSTANDING

5. How did data on sarcomere structure inspire the sliding-filament model of muscle contraction? Explain why the observation that muscle cells contain many mitochondria and extensive smooth endoplasmic reticulum turned out to be logical once the molecular mechanism of muscular contraction was understood.
6. Rigor mortis is the stiffening of a body after death that occurs when myosin binds to actin but cannot unbind. What prevents myosin from unbinding?

7. R. McNeill Alexander discovered that the arch of the human foot operates like a spring during running. Predict how a runner's oxygen consumption would change if a runner wore shoes that prevented the arches from changing shape. Explain your reasoning.

8. Explain why the energetic cost of swimming decreases as a fish grows.

✔ TEST YOUR PROBLEM-SOLVING SKILLS

9. Atropine is a compound found in many poisonous nightshade plants. It blocks acetylcholine receptors in the heart. Predict the effect of ingestion of atropine on heart rate. Explain your logic.

10. **QUANTITATIVE** The speed at which you switch from a walk to a run can be predicted using what is called the Froude number, which is based on the relative importance of gravitational and inertial forces of your pendulum-like walking gait.

$$\text{Froude number} = \frac{(\text{speed of locomotion})^2}{\text{gravitational acceleration} \times \text{leg length}}$$

Most mammals change from a walk to a run at a Froude number of 0.5. If gravitational acceleration is $9.8\ m/s^2$, and your leg length is 0.9 m, at what speed are you likely to switch from a walk to a run?

a. 2.0 m
b. 1.9 m/s
c. 2.1 m/s
d. $2.0\ m/s^2$

11. Compare and contrast the structure and function of the three types of skeletal muscle fibers.

12. Predict who would likely have a greater proportion of fast glycolytic fibers in their gastrocnemius (calf) muscle—an elite distance runner or an elite sprinter. Explain.

13. **THINK CAREFULLY** Predict the effect of training for a marathon on the number of muscle cells in the gastrocnemius. Explain.

14. **QUANTITATIVE** To discover the relationship between muscle-fiber types and performance, researchers obtained tiny biopsies of the gastrocnemius of 14 elite distance runners, 18 trained but non-elite distance runners, and 19 untrained subjects. They categorized the fiber types as slow or fast. (At the time of the study, intermediate fibers had not been identified as a third type.) Some of their data are shown here (* means $P < 0.05$; **BioSkills 3**). What conclusions can you draw from these data?

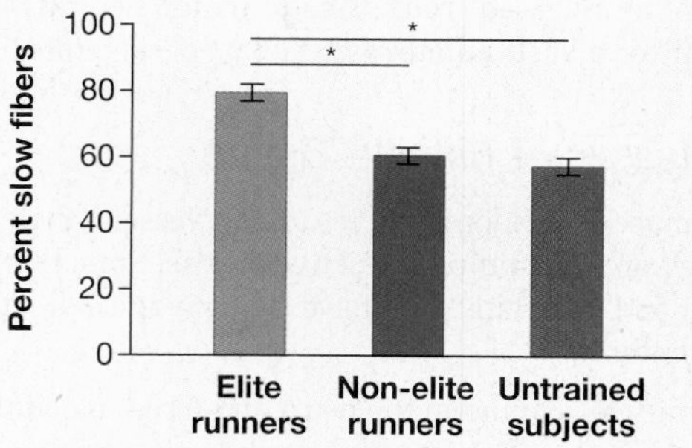

Source: W. J. Fink, D. L. Costill, and M. L. Pollock. 1977. *Annals of the New York Academy of Sciences* 301: 323–327.

15. The researchers looked more closely at the data within the group of elite runners. Although the mean proportion of slow fibers was 79 percent in this group, individual values ranged from 27 percent in one runner to 98 percent in another. How does this finding affect your interpretation of the relationship between athletic performance and muscle-fiber types?

16. Imagine that Paula Radcliffe is racing against a bird and a fish, each with the same mass as Paula. Which organism would have the highest cost of locomotion during the race?

✔ PUT IT ALL TOGETHER: Case Study

Is athletic performance related to the relative proportions of muscle fiber types?

Distance runner Paula Radcliffe has won dozens of long-distance races and held the women's world record for the marathon since 2003. Scientists, trainers, and athletes alike have wondered about the extent to which muscle structure and function contribute to success in athletes such as Radcliffe. What makes elite distance runners so good? Are their muscles somehow different from those of less successful athletes and non-athletes?

Mastering Biology

Students Go to Mastering™ Biology for assignments, the eText, and the Study Area with animations, practice tests, and activities.

Professors Go to Mastering™ Biology for automatically graded tutorials and questions that you can assign to your students, plus Instructor Resources.

46 Chemical Signals in Animals

The spectacular transformation that occurs during insect metamorphosis is triggered by chemical signals called hormones.

In this chapter you will learn that

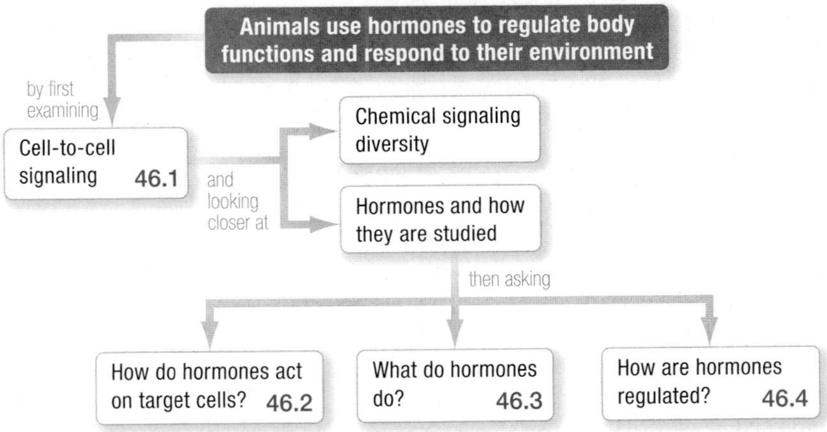

In response to sights, sounds, and other sensory stimuli, an animal's nervous system sends rapid messages, in the form of action potentials, to precise locations in the body. In many cases, these messages result in immediate, temporary responses such as muscle contractions and movement.

In response to changes in external or internal conditions, cells in the nervous system or the endocrine system also release molecules that produce longer-term responses in a broad range of tissues and organs. The **endocrine system** is a collection of organs and cells that secrete chemical signals into the bloodstream. A chemical signal that circulates through body fluids and affects distant target cells is called a **hormone**.

The goal of this chapter is to explore how hormones and other types of internal chemical signals work in animals. Together, animal nervous and endocrine systems process information about the internal and external environments—a function that is one of the five key attributes of life (Chapter 1).

Let's begin with an overview of chemical signaling systems and then plunge into analyzing how hormones regulate the activity of target cells.

BIG PICTURE

This chapter is part of the Big Picture. See how on pages 838–839.

46.1 Cell-to-Cell Signaling: An Overview

Animal chemical signals are present in extremely low concentrations but can have enormous effects on their target cells. Unlike action potentials, which are electrical impulses that have a short-term effect on a single cell or on a small population of adjacent cells, the messages that chemical signals carry can have a relatively long-lasting effect.

> After you complete this section, you should be able to . . .
>
> - Analyze cell-to-cell signaling in animals.

In combination, electrical and chemical signals allow animals to coordinate the activities of cells throughout the body. They are the mechanism responsible for maintaining trillions of cells as an integrated unit called an individual.

Major Categories of Chemical Signals

The chemical signals found in animals have diverse structures and functions. **Table 46.1** summarizes how biologists organize the diversity of chemical signals, based on where the molecules originate and where they act. Notice that the names for most of the categories use the Greek word root *crin*, meaning "separated." Its use captures something essential about how chemical signals act: They are released from cells and thus are separated from them.

SUMMARY **Table 46.1**
Five Categories of Chemical Signals in Animals

Type of Chemical Signal	Source and Target
Autocrine signals act on the same cell that secretes them.	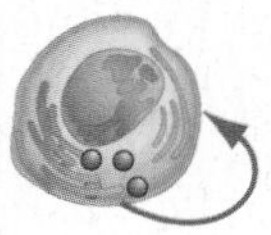
Paracrine signals diffuse locally and act on nearby cells.	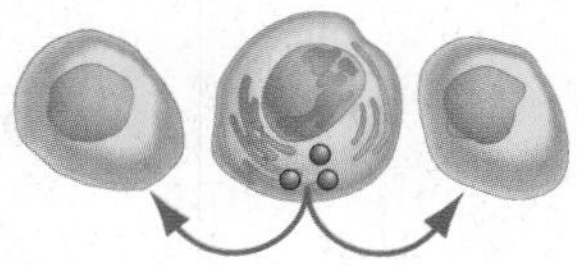
Endocrine signals are hormones carried between distant cells by blood or other body fluids.	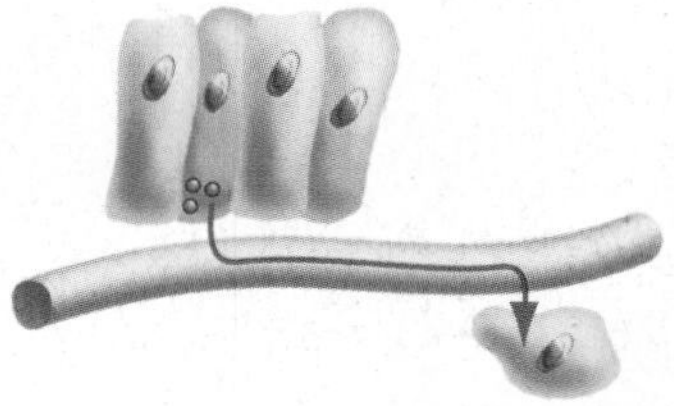
Neural signals (neurotransmitters) diffuse a short distance between neurons.	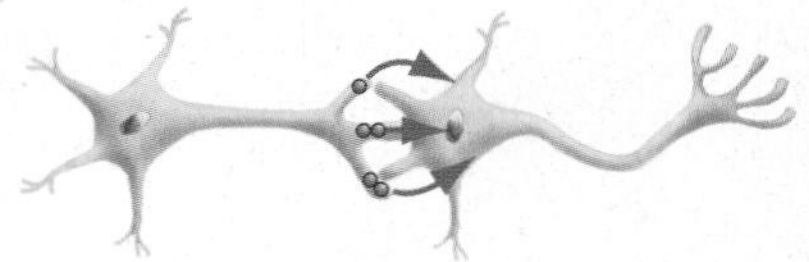
Neuroendocrine signals (neurohormones) are hormones released from neurons.	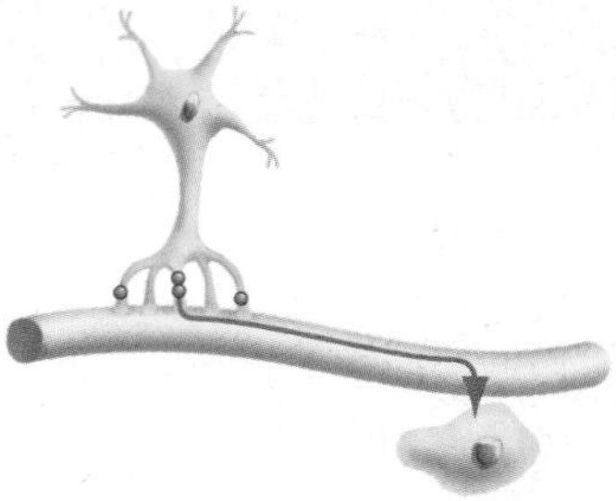

Note that these five classes of chemical messenger are not necessarily structurally distinct classes of molecules. For example, the endocrine signals found in a particular organism routinely belong to several families of chemical compounds, ranging from amino acid derivatives to lipids. And a particular family of molecules—say, peptides or the lipids called steroids—may function as endocrine, autocrine, and paracrine signals in the same individual.

Autocrine Signals Act on the Same Cell That Secretes Them

Autocrine signals affect the same cell that releases them. Perhaps the best-studied autocrine signals are **cytokines** (literally, "cell-movers"). Most cytokines amplify the response of a cell to a stimulus. An example is interleukin 2, which in the course of fighting an infection is synthesized and released by a type of white blood cell called a T cell. Interleukin 2 activates T cells to help eliminate the infection. It also causes the cells to divide repeatedly, producing more activated T cells for host defense (Chapter 48).

Paracrine Signals Act on Neighboring Cells

Paracrine signals diffuse locally and act on target cells near the source cell. Cytokines, for example, may act as paracrine signals as well as autocrine signals, because they can trigger responses by other nearby cells of the immune system.

In fact, a single chemical messenger can often be assigned to more than one category of signal, based on its mode of action. Like some cytokines, the cell–cell signals named **insulin** and **glucagon** cross categories. These molecules are produced by two distinct populations of cells within the regions of the **pancreas** called the islets of Langerhans. The molecules act on nearby pancreatic cells as paracrine signals and ensure a smooth, steady response to changing blood glucose levels. But they also act as hormones, in that they are released into the blood or other body fluids and affect distant cells—in this case, controlling the concentration of glucose in the blood.

Endocrine Signals Are Hormones

Endocrine signals are carried to distant cells by blood or other body fluids. The cells that produce endocrine signals may be organized into discrete organs called **glands** or may be interspersed among the cells of other organs—as are the islets of Langerhans in the pancreas.

Because hormones are well studied and particularly important to understanding how animals work, they are the focus of this chapter. Many or most of the principles discussed are also relevant to the other categories of animal cell–cell signals, however.

Neural Signals Are Neurotransmitters

You might recall that when an action potential arrives at a synapse, it triggers the release of neurotransmitters, which bind to receptors on the postsynaptic cell and induce a change in membrane potential—altering the tendency of the postsynaptic cell to fire action potentials (Ch. 43, Section 43.3).

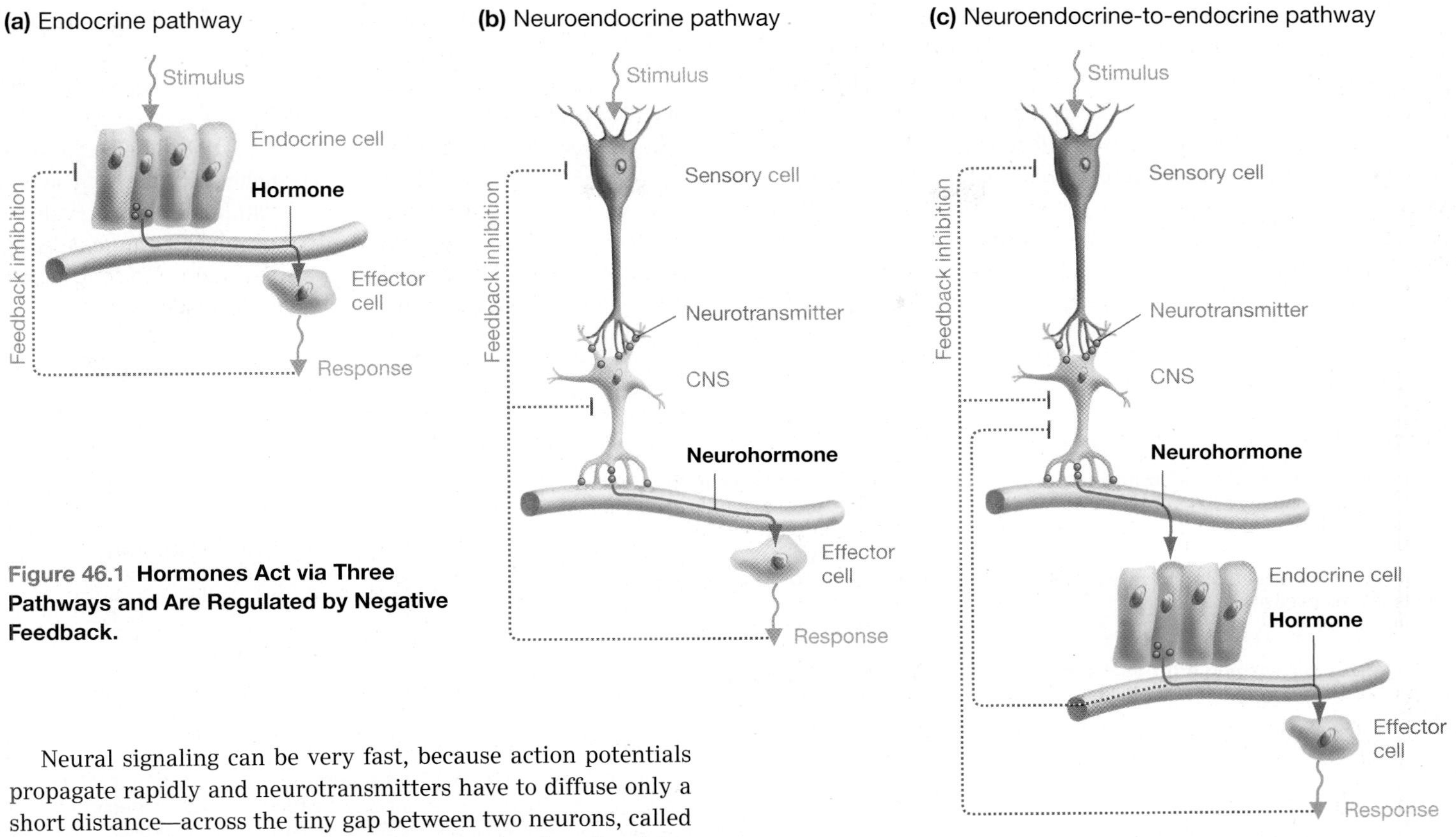

Figure 46.1 Hormones Act via Three Pathways and Are Regulated by Negative Feedback.

Neural signaling can be very fast, because action potentials propagate rapidly and neurotransmitters have to diffuse only a short distance—across the tiny gap between two neurons, called the synaptic cleft. Neural signals also tend to be short lived, because the signaling molecules are quickly broken down or taken back up by the presynaptic cell.

Neuroendocrine Signals Act on Distant Target Cells Contrary to popular belief, the endocrine and nervous systems do not function independently of each other. Even though they are released from neurons, **neuroendocrine** ("nerve-inside-separated") signals share a key attribute with endocrine signals: They act on distant cells. For this reason, they are called **neurohormones**.

Antidiuretic hormone (**ADH**; also called vasopressin) is a particularly well-studied neuroendocrine signal. ADH is produced by neurons that have their cell bodies in a brain region called the **hypothalamus**. But instead of acting as a neural signal, ADH acts on cells in the collecting duct of the kidney to help regulate water excretion (Ch. 40, Section 40.4).

Hormone Signaling Pathways

In plants, sensory cells perceive changes in the environment and broadcast a hormonal signal that triggers an appropriate response from effector cells, which are cells that will produce a response. Cells that can respond to a hormonal stimulus can also be called target cells for that hormone. In animals, some endocrine cells respond directly to an environmental stimulus by secreting hormones (Figure 46.1a). But frequently, hormonal signaling in animals involves additional steps.

In many cases, information about external or internal conditions is gathered by sensory receptors and then integrated by neurons in the central nervous system (CNS) before the production of a hormonal signal. Neurons in the CNS respond by releasing neurohormones, which act on effector cells directly (Figure 46.1b) or stimulate cells in the endocrine system, which respond by producing hormones (Figure 46.1c).

All three types of signaling pathway—endocrine, neuroendocrine, and neuroendocrine-to-endocrine—are regulated by **negative feedback**. In negative feedback, the product of a process inhibits its production. **Positive feedback**, in contrast, occurs when the product of a process *stimulates* its production, resulting in greater and greater production of that product. (For help in diagramming feedback loops, see **Making Models 46.1**.)

Making Models 46.1 Tips on Drawing Feedback Loops

Using arrows to show stimulation and bar-headed lines to show inhibition is a useful way to conceptualize feedback loops. The model here shows a negative feedback system in which endocrine gland A releases hormone A, which stimulates endocrine gland B to release hormone B. Hormone B inhibits the release of hormone A.

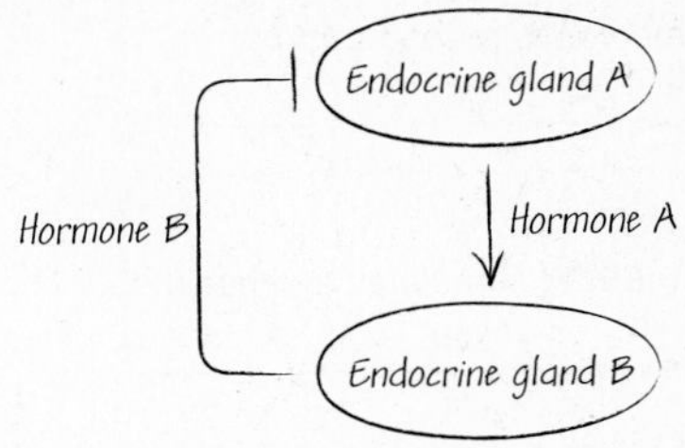

MODEL Draw a model showing a positive feedback loop. What will happen to the secretion of hormone B over time?

To see this model in action, go to the Study Area of **Mastering Biology**

Negative feedback is key to homeostasis (Ch. 39, Section 39.4). In the endocrine pathway, the hormone produced by the effector cells feeds back on the endocrine cells that secrete it, lowering production of the hormone and down-regulating the response (see Figure 46.1a). The response of effector cells also feeds back to cells that initiate the neuroendocrine and neuroendocrine-to-endocrine pathways. A change in input from these initiating cells then lowers production of the neurohormone and reduces the response (see Figures 46.1b and 46.1c). Neuroendocrine-to-endocrine signaling pathways have an additional layer of regulation, because the hormonal signal usually inhibits production of the neurohormone (see Figure 46.1c).

Positive feedback, in contrast, leads a physiological variable *away* from homeostasis. Release of a hormone causes a response in a tissue that causes it to release more of that hormone, stimulating a greater response, still higher hormone secretion, and so on. Why would it be beneficial to depart from homeostasis? Positive feedback is typically observed when a rapid and extreme change in physiological function is needed, as with the increased contraction of the uterine muscles during childbirth (Ch. 47, Section 47.6).

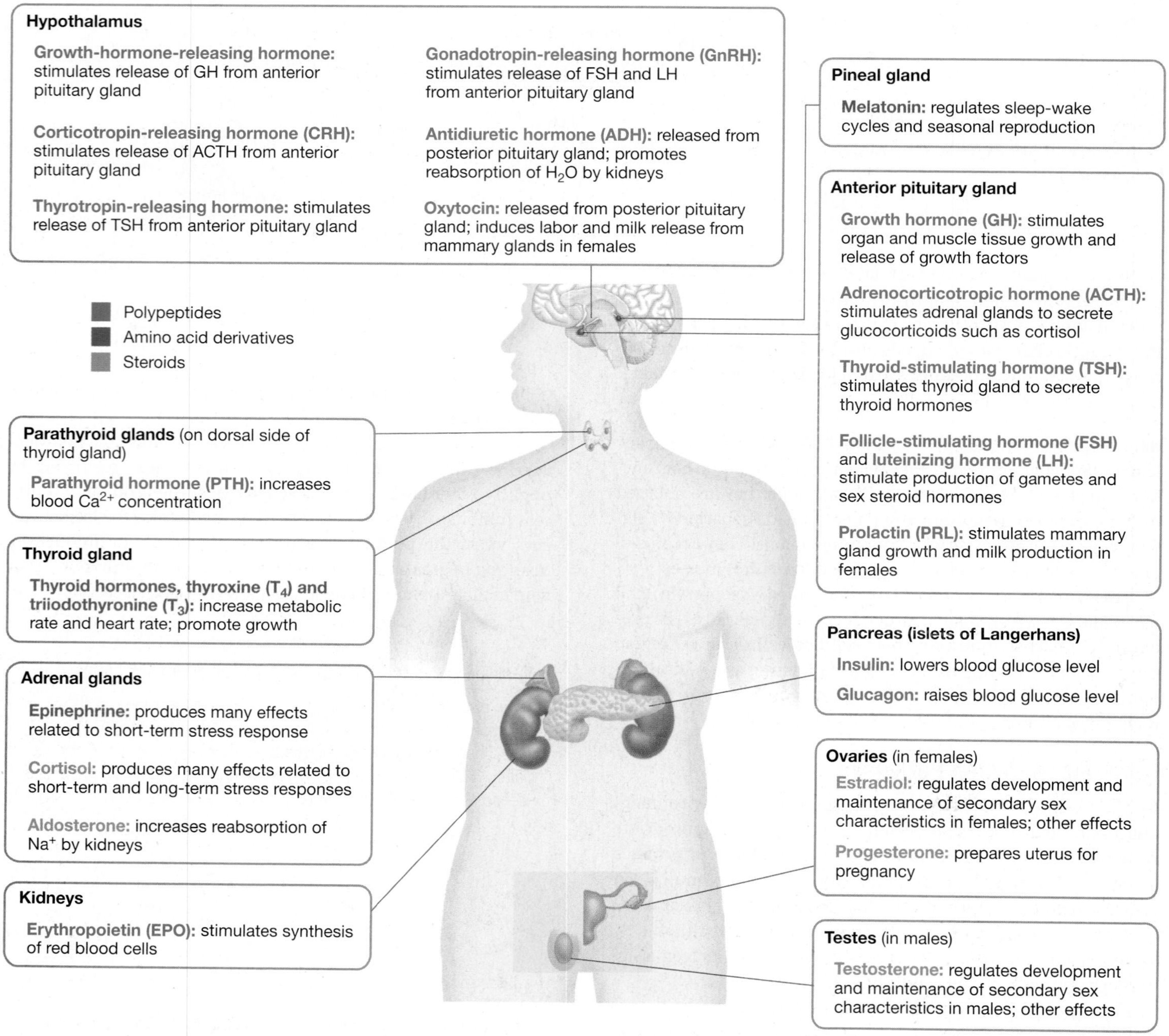

Figure 46.2 Humans Possess a Diverse Array of Endocrine Glands and Hormones. This list is only partial. The heart, gastrointestinal tract, adipose (fat) tissue, and many other organs and cells also produce hormones.

The take-home message? The nervous system and endocrine system are tightly integrated into a neuroendocrine system. Endocrine signals are released in response to electrical signals; in turn, endocrine signals modulate the electrical signals transmitted by the nervous system.

Negative feedback in the endocrine system is analogous to temperature control by a heat-sensitive thermostat. If the temperature is too high, the thermostat sends a signal that turns the furnace off; if the temperature is too low, the thermostat sends a signal that turns the furnace on. The result is a relatively constant air temperature. In animal cell–cell signaling, negative feedback reduces the production or secretion of a hormone, or both.

What Structures Make Up the Endocrine System?

The endocrine system is the collection of cells, tissues, and organs responsible for hormone production and secretion. Organs that secrete a hormone into the bloodstream are called **endocrine glands**.

The tissues and organs that make up the endocrine system vary widely among animals. For example, neurons that manufacture and secrete hormones are important in insects, where they regulate molting, metamorphosis, and other processes. Salmon have an unusual gland that secretes a hormone responsible for regulating calcium ion concentration.

Even within one species, the diversity of endocrine system components can be impressive. For example, Figure 46.2 shows major human glands with endocrine functions. As mentioned earlier, hormone-secreting cells are not always organized into discrete glands. In many cases, they are located in other kinds of organs. The islets of Langerhans in the pancreas are one example.

It's important to note that not all glands in the body are part of the endocrine system. **Exocrine glands**, in contrast to endocrine glands, deliver their secretions through outlets called ducts into a space other than the circulatory system. Most of the digestive glands are either exocrine glands—an example is the salivary glands—or mixed endocrine and exocrine glands, such as the pancreas (Ch. 41, Section 41.3). The exocrine cells of the pancreas secrete digestive enzymes through ducts into the intestine. The endocrine portion of the pancreas consists of cells that secrete insulin and glucagon directly into the bloodstream.

Researchers have identified hundreds of hormones involved in every physiological system, and new hormones are still being discovered. However, scientists initially thought that all internal communication was accomplished via the nervous system, because nerves are visible during dissection and hormones are not. How did scientists discover hormones?

How Do Researchers Identify a Hormone?

Research on animal hormones began in the mid-1800s, when the Swiss-German biologist Arnold Berthold performed an experiment on roosters. Long before the term "hormone" was coined, Berthold hypothesized that the testes release chemical signals that causes male roosters to act aggressively toward one another. His hypothesis stemmed from the observation that when young male roosters are castrated (have their testes removed so that more tender meat develops), they do not act aggressively toward one another and do not grow the large combs and wattles typical of roosters (Figure 46.3).

RESEARCH

QUESTION: What causes typical male anatomy and behavior in roosters?

HYPOTHESIS: A chemical signal released from the testes is responsible for aggression and comb and wattle growth in roosters.

NULL HYPOTHESIS: A chemical signal released from the testes is not responsible for aggression and comb and wattle growth in roosters.

EXPERIMENTAL SETUP:

Group 1: Male chicks castrated (testes removed)

Group 2: Male chicks castrated, testis re-implanted in abdominal cavity

Group 3: Male chicks castrated, testes transplanted between chicks

PREDICTION OF HYPOTHESIS: Castrated males will fight less and have smaller wattles and combs than castrated males with testes implanted in the body.

PREDICTION OF NULL HYPOTHESIS: There will be no difference in fighting behavior and comb and wattle size between groups.

RESULTS:

Group 1: Birds have small wattles and combs; don't fight.

Groups 2 and 3: Birds develop large, red wattles and combs; fight aggressively.

Necropsy shows that implanted testes reestablished blood vessels but not nerves.

CONCLUSION: A chemical signal released from the testes into the bloodstream causes males to fight and stimulates comb and wattle growth.

Figure 46.3 **The Testes Produce a Chemical Signal Required for Male Anatomy and Behavior in Roosters.** This hormone was later named testosterone.

SOURCE: A. A. Berthold. 1849. Transplantation der Hoden. *Arch. Anat. Physiol. Wiss. Med.* 16: 42–49.

✔ **PROCESS OF SCIENCE** Now that testosterone is commercially available for use in research, how could Berthold's experiment be modified to yield more specific results?

To test his hypothesis, Berthold castrated a group of young male chickens. In some, he re-implanted one of the testes he had just removed into the birds' own abdominal cavities or those of other castrated chickens. As these birds matured, Berthold found that they acted aggressively and grew large wattles and combs, like normal roosters, whereas the control castrated birds did not.

The genius of Berthold's experiment was that it separated hormonal from neural influences. The testes that he implanted in the birds' abdominal cavities did not form new neural connections with the body, but they did establish new blood vessels. When the castrated birds with re-implanted testes developed typical male anatomy and behavior, Berthold concluded that these characteristics were stimulated not by nerves, but by one or more blood-borne chemical signals. It would be over 50 years until the official term for these chemical signals—hormones—was coined. Although this chemical was not isolated and named until years after his death, Berthold had discovered **testosterone**, the hormone that controls development of male reproductive anatomy and stimulates reproductive behavior.

A Breakthrough in Measuring Hormone Levels

Documenting an association between a particular gland or hormone and an effect in the body is just a first step. To understand hormone action, researchers have to figure out how these signals help animals stay alive and produce offspring.

The key to this goal is the ability to quantify the levels of hormones circulating in the bloodstream. This ability eluded researchers for decades, however, because hormones are present in the blood in extremely small concentrations.

The breakthrough came in the 1950s when Rosalind Yalow developed the **radioimmunoassay**, a feat for which she later received the Nobel Prize in Medicine. In a radioimmunoassay, the quantity of hormone in a blood sample can be estimated by adding a radioactively labeled version of that hormone to the sample. The labeled and unlabeled hormones will compete with each other for binding to an antibody (BioSkills 6).

The technique of radioimmunoassay has revolutionized the study of hormones and the treatment of endocrine diseases, allowing the concentrations of hormones to be measured precisely in blood samples drawn from animals or patients. Currently, radioimmunoassays are used every day in endocrine research and medical laboratories. Much of what researchers know about hormone function was learned in studies employing these assays.

CHECK YOUR UNDERSTANDING

If you understood this section, you should be able to ...

1. Discuss why it is often difficult to differentiate between the functions of the nervous and endocrine systems.
2. Classify each of the following as autocrine, paracrine, endocrine, or neuroendocrine: (a) testosterone stimulates aggression in roosters; (b) antidiuretic hormone binds to collecting duct cells in the kidney; and (c) a cell releases a chemical signal into the extracellular fluid, where it diffuses to bind to a neighboring cell.

Answers are available in Appendix A.

46.2 How Do Hormones Act on Target Cells?

Hormones have an astonishing array of functions, but they all share certain characteristics. Chief among these are that they can have strong effects even at very low concentrations, and they exert these effects by binding to proteins called **hormone receptors** in target tissues. Let's take a look at how these characteristics are related to the functions of various chemical classes of hormones.

After you complete this section, you should be able to ...

- Explain how hormones trigger responses in target cells.

Hormone Concentrations Are Small, but Their Effects Are Large

Hormones have profound effects on individuals, even though they are present at vanishingly small concentrations. As an example, consider research that led to the discovery of **growth hormone (GH)**.

Several researchers noted that rats and other laboratory animals stopped growing when their pituitary glands were removed. Based on this observation, it was hypothesized that the **pituitary gland**, located at the base of the brain, produces a chemical signal that promotes cell division and other aspects of growth.

To test this hypothesis, a research group purified a polypeptide from cow pituitary glands, injected the polypeptide into lab rats, and documented rapidly accelerated growth. When the researchers injected 0.01 mg of the polypeptide each day for nine days into rats that lacked pituitary glands, the growth plates in the rats' leg bones widened by 50 percent. Those rats also gained an average of 10 g more than rats that lacked pituitary glands and did not receive the polypeptide injections. Stated another way, an additional 0.09 mg of polypeptide—now known to be GH—led to a weight gain of 10,000 mg.

Further, 1 kg of cow pituitary tissue contains a mere 0.04 g of GH. By mass, GH makes up just 4 one-thousandths of 1 percent of the cow pituitary gland, but it has a dramatic effect on animal growth.

The Three Chemical Classes of Hormones

Figure 46.4 illustrates the three major classes of chemicals that can act as hormones in animals, based on chemical structure:

- Peptides and polypeptides, which are chains of amino acids linked by peptide bonds (Chapter 3)
- Amino acid derivatives
- Steroids, which are a family of lipids distinguished by a four-ring structure (Chapter 6)

Figure 46.4 shows one example of each class of signal. Secretin, a hormone produced in the small intestine that stimulates the exocrine portion of the pancreas, is a polypeptide. The hormone **epinephrine**, also known as **adrenaline**, is synthesized in the medulla of the **adrenal glands** from the amino acid tyrosine. The hormone **cortisol** is synthesized in the cortex of the adrenal glands from the steroid cholesterol. (The Greek word roots

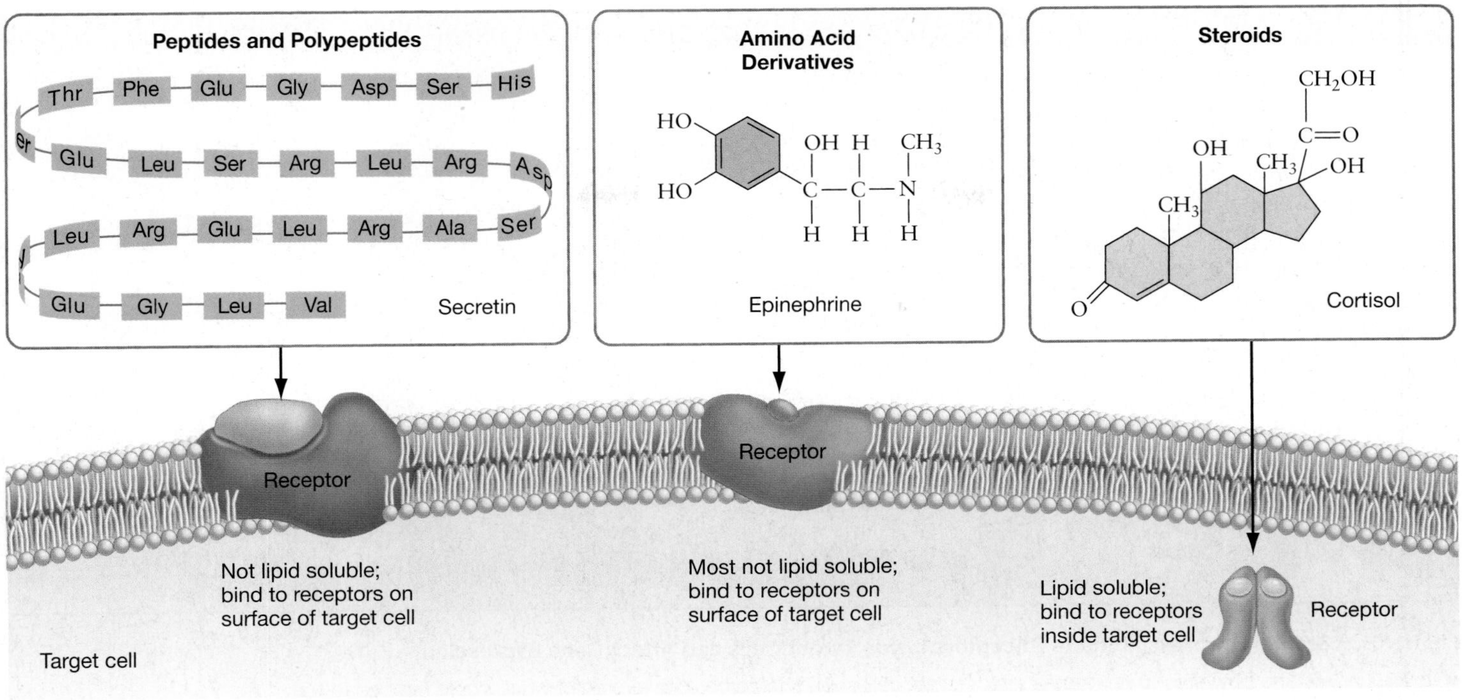

Figure 46.4 **Most Animal Hormones Belong to One of Three Chemical Classes.**

epi and *nephron* mean "top-kidney"; the Latin word roots *ad* and *renal* also mean "top-kidney.")

Given that small amounts of polypeptide, amino-acid-derived, and steroid hormones have large effects on the activity of cells, organs, and systems, how do the three classes of hormones differ? The major difference is that steroids are lipid soluble, but polypeptides and most amino acid derivatives are not (see Figure 46.4).

Important exceptions to this rule are the **thyroid hormones**, **triiodothyronine** (also known as T_3) and **thyroxine** (also known as T_4), which are produced by the **thyroid gland**. The thyroid hormones are derived from the amino acid tyrosine but are lipid soluble.

Differences in solubility are important because steroids and thyroid hormones cross plasma membranes much more readily than do other types of hormones. To affect a target cell, all polypeptides and most amino acid derivatives bind to a receptor on the cell surface. Lipid-soluble hormones, in contrast, can diffuse through the plasma membrane and bind to receptors inside the cell.

To compare these two distinct paths of hormone action, let's consider how estradiol and epinephrine affect target cells. As a steroid and a nonsteroid, respectively, they serve as model systems for target cell responses to hormonal signals.

Steroid Hormones Bind to Intracellular Receptors

Estrogens are steroid hormones that direct the development of female secondary sex characteristics in many animal species. In humans and other mammals, the most important estrogen is the molecule **estradiol** (formally, 17 β-estradiol).

Because of estradiol's importance in reproduction by humans and domesticated animals, its mode of action has been the topic of intense investigation for over 50 years. How do target cells receive the signal carried by estradiol?

Estradiol receptors are primarily located in the nuclei of cells in the uterus, hypothalamus, and mammary glands. The presence of estradiol receptors in these cells indicates that they are target cells for estradiol, whereas cells that lack these receptors are not. This is a crucial point: Hormones are broadcast throughout the body via the bloodstream, but they act only on target cells that express the appropriate receptor.

During the 1970s and 1980s, work in several laboratories suggested that estradiol and other steroid hormones affect gene transcription after they bind to their receptors. For example, researchers injected laboratory animals with estradiol or other steroid hormones and documented changes in the mRNAs and proteins produced in target cells. These data showed that steroid hormones can cause dramatic changes in the amount or timing of mRNA production by a large number of genes.

How do steroid hormones accomplish this? The estradiol receptor, like other members of the steroid-hormone receptor family, has two copies of a distinctive DNA-binding domain. DNA-binding domains are sections of a protein that make physical contact with DNA. The presence of these domains in the estradiol receptor suggested that once estradiol binds to it, the hormone–receptor complex might affect gene expression by binding directly to DNA.

Follow-up work confirmed that steroid hormone–receptor complexes bind to specific sites in DNA called **hormone-response elements**. Hormone-response elements are located just "upstream" (in the 5′ direction) from the start of target genes. Gene expression changes when a regulatory molecule such as a steroid hormone–receptor complex binds to the hormone-response element for that gene.

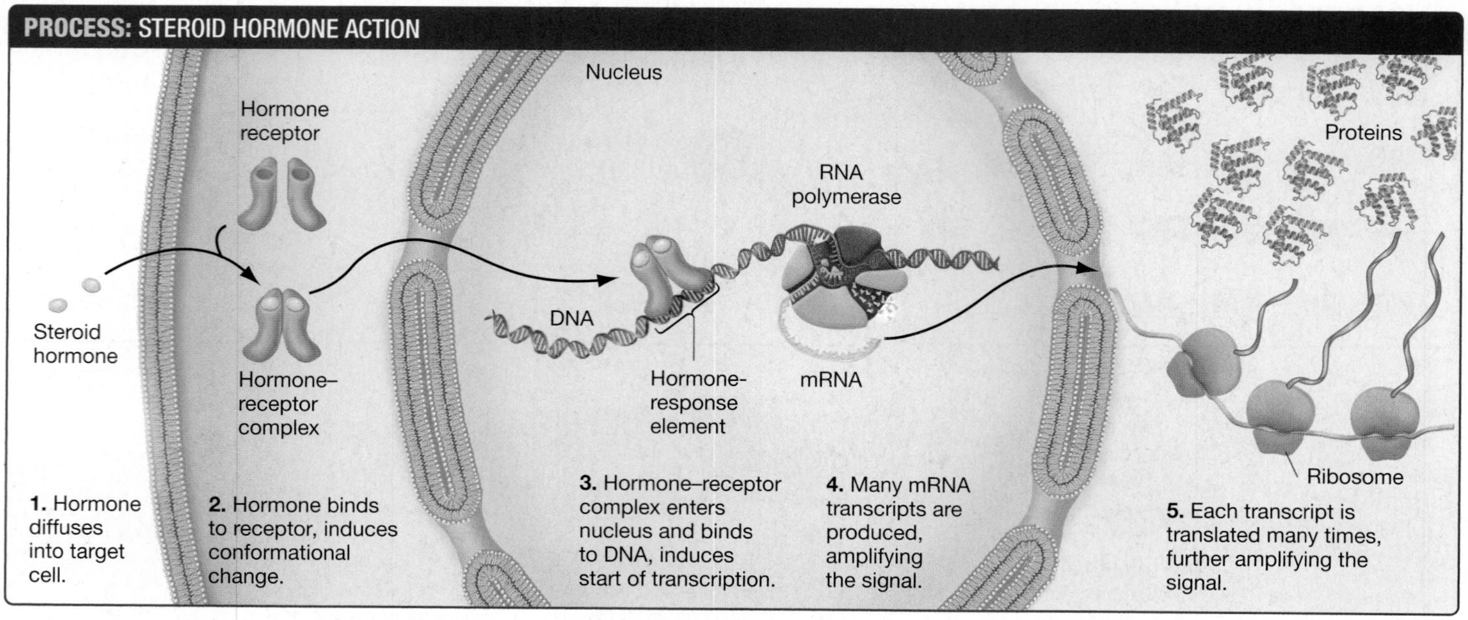

Figure 46.5 Steroid Hormones Bind to Receptors Inside Target Cells and Affect Gene Expression.

✔ If you understand how steroid hormones affect target cells, you should be able to predict what would happen to the estradiol response if an individual had mutations that changed the DNA sequence of its hormone-response element.

Figure 46.5 summarizes the current model of how steroid hormones affect target cells. Because each hormone–receptor complex leads to the production of many copies of the gene product, the signal from the hormone is amplified. In this way, a small number of hormone molecules produce a large change in the activity of target cells and tissues.

In contrast to steroid hormones, polypeptide hormones and most amino-acid-derived hormones are not lipid soluble and therefore cannot enter cells. How do they stimulate their target cells?

Polypeptide Hormones Bind to Receptors on the Plasma Membrane

For water-soluble messengers such as polypeptide and most amino-acid-derived hormones to affect a cell, they must bind to receptors on the cell surface. Because the messenger never enters the target cell, its message must be transduced—changed into a form that is active inside the cell. This phenomenon is known as **signal transduction** (Ch. 11, Section 11.3).

In 1948 a biologist named Raymond Ahlquist published an exhaustive set of studies on how epinephrine—a water-soluble, amino-acid-derived hormone—affects dogs, cats, rats, and rabbits. The responses fell into two distinct categories, depending on the tissue being considered. To explain this observation, Ahlquist hypothesized that epinephrine binds two types of receptor. He called these hypothetical receptors the alpha receptor and the beta receptor.

Follow-up work with molecules that block epinephrine receptors documented that there are actually four types of epinephrine receptors: two types of alpha receptors and two types of beta receptors. Each is found in a distinct tissue type, and each induces a different response from target cells.

The discovery of four epinephrine receptors reinforces the concept of tissue specificity observed in experiments on the estradiol receptor. Hormones are transmitted throughout the body, not unlike cell phone signals that are broadcast through the atmosphere. But a hormone's message is received only by cells with the appropriate receptor—just as a cell phone signal is received only by a phone with the correct number. Because there are four epinephrine receptors, the same hormone can trigger different effects in different cells. What happens once epinephrine binds to one of these receptors?

What Acts as the Second Messenger in Signal Transduction?

Signal transduction occurs when a chemical messenger (the first messenger) binds to a cell-surface receptor, activating an intracellular second messenger and triggering a response inside the cell. Recall that a **second messenger** is a nonprotein signaling molecule that increases in concentration inside a cell in response to a signaling molecule that binds at the surface (Ch. 11, Section 11.3).

One effect of epinephrine is an increase in the glucose level in the blood. How does a signal from epinephrine trigger this effect? To answer that question, biologists focused on the enzyme **phosphorylase** (Figure 46.6a), which catalyzes a reaction that cleaves glucose molecules off glycogen (Figure 46.6b). Phosphorylase exists in active and inactive states; the enzyme switches between these states when it is phosphorylated or dephosphorylated by another enzyme.

Phosphorylase is present in liver cells—the primary source of blood glucose during the short-term stress response. As predicted, when researchers added epinephrine to extracts from homogenized (ground up) liver cells, much larger amounts of phosphorylase were activated than in cell extracts that did not receive epinephrine (Figure 46.6c).

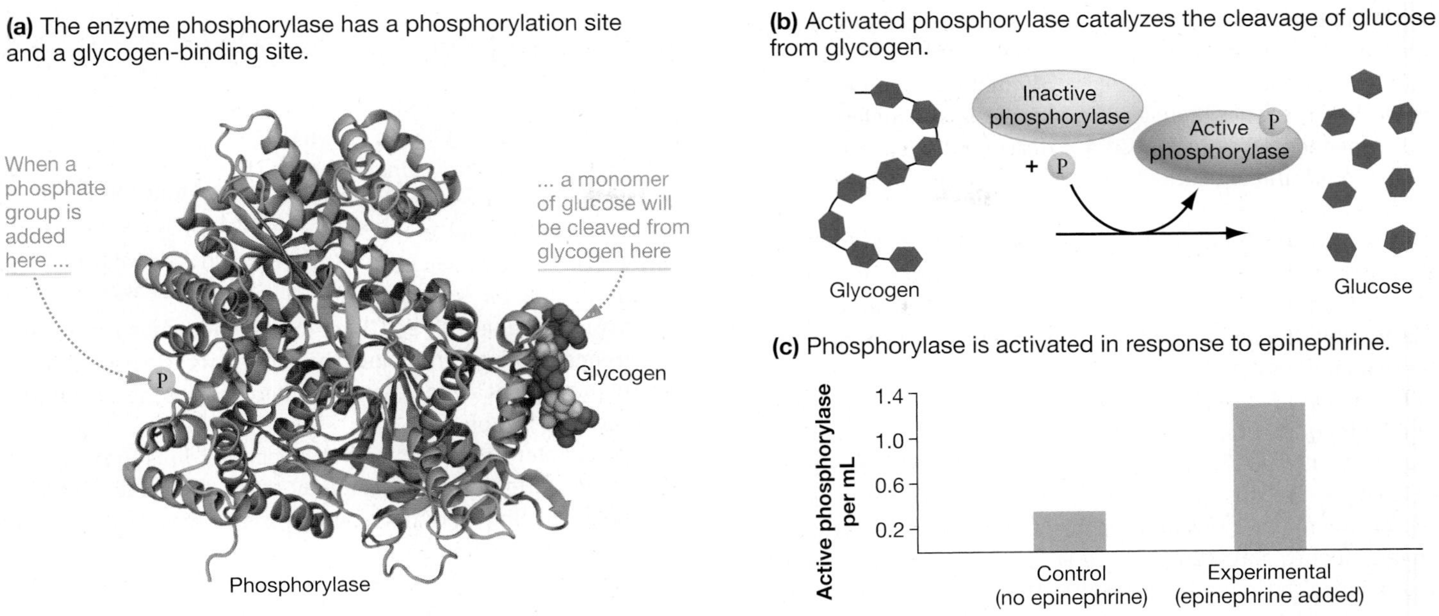

Figure 46.6 The Enzyme Phosphorylase Catalyzes the Release of Glucose Monomers from Glycogen.
(a) Activation of phosphorylase requires phosphorylation. **(b)** Glucose cleaved from glycogen is released into the blood.
(c) When epinephrine is added to extracts from liver cells, the amount of activated phosphorylase increases dramatically.

DATA: T. W. Rall, E. W. Sutherland, and J. Berthet. 1957. *Journal of Biological Chemistry* 224: 463–475.

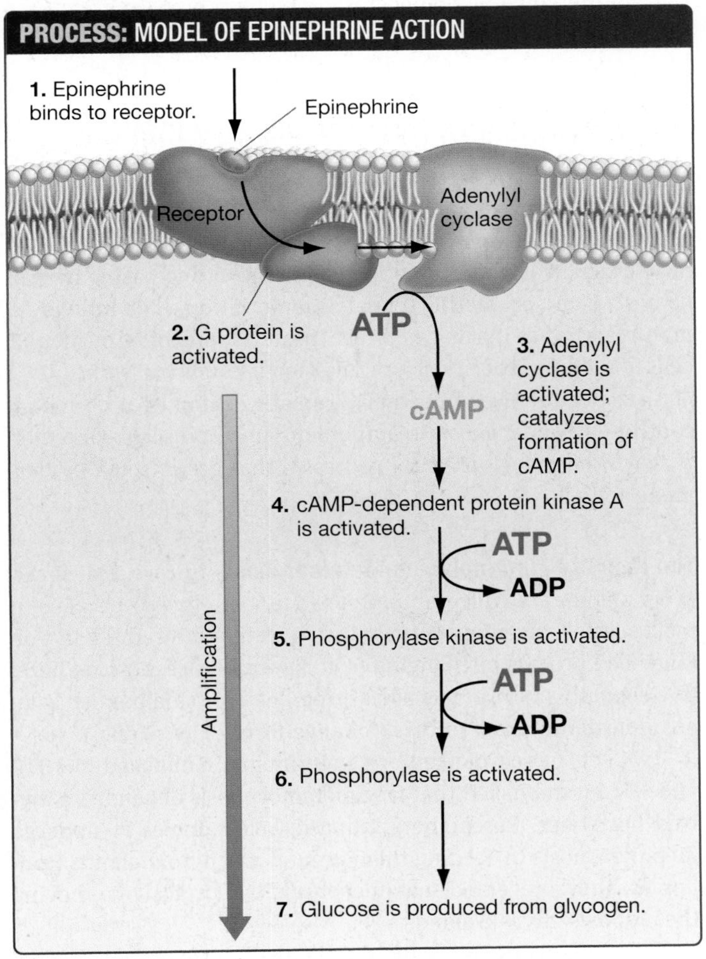

Figure 46.7 Epinephrine Triggers a Signal Transduction Cascade That Results in the Activation of Phosphorylase. Epinephrine's signal is amplified at each of steps 3–6.

This observation suggested that something in the homogenized cells activated the phosphorylase when epinephrine was present. By purifying components of the liver cell extracts and testing them one by one, researchers eventually found the key ingredient in the activation of phosphorylase: a molecule called cyclic adenosine monophosphate, or **cyclic AMP (cAMP)**.

The role of cAMP in epinephrine signaling was confirmed when researchers studied epinephrine's effects on the rat heart. During the short-term stress response, the heart beats faster and with more force, sending more blood to the tissues. When researchers injected rats with epinephrine, they found that the cAMP level increased inside heart cells *before* the phosphorylase level increased. This supported the hypothesis that cAMP is a second messenger in this system.

A Signal Transduction Cascade Amplifies the Hormonal Signal

How does cAMP transfer the hormonal signal on the cell surface to phosphorylase inside the cell? Follow-up work revealed the mechanism of epinephrine action on liver cells (Figure 46.7).

When epinephrine binds to its receptor, it activates a G protein, which then activates the enzyme adenylyl cyclase. Adenylyl cyclase catalyzes a reaction that converts ATP to cAMP. Next, cAMP initiates a chain of events called a **signal transduction cascade** (also called a **phosphorylation cascade**; Ch. 11, Section 11.3) by binding to an enzyme called cAMP-dependent protein kinase A. This enzyme responds by phosphorylating the enzyme phosphorylase kinase, which then phosphorylates phosphorylase.

In studying this model, focus on the following overall points:

- cAMP transmits the signal from the cell surface to the signal transduction cascade.
- Together, cAMP production and the subsequent phosphorylation events amplify the original signal from epinephrine.

To appreciate the second point, consider that, in response to stimulation by the activated G protein, adenylyl cyclase is thought to catalyze the formation of at least 100 molecules of cAMP. In turn, each of these cAMP molecules activates many molecules of cAMP-dependent protein kinase A. Subsequently, each protein kinase A molecule activates many molecules of phosphorylase kinase, and so on.

In this way, the binding of just a single molecule of epinephrine may trigger the release of millions or even billions of glucose molecules. Amplification through a signal transduction cascade explains why tiny amounts of hormones can have such huge effects on an individual.

The model in Figure 46.7 was inspired by experiments on the type of epinephrine receptor called the beta-1 receptor. But other researchers showed that a completely different signal transduction event occurs when epinephrine binds to a different type of receptor, the alpha-1 receptor. In this and many other receptor systems, calcium ions (Ca^{2+}) serve as the second messenger in conjunction with a molecule called IP_3. Diacylglycerol (DAG) and 3′, 5′-cyclic GMP (cGMP) are also common second messengers in hormone response systems.

Why Do Different Target Cells Respond in Different Ways?

Researchers are increasingly impressed with the diversity and complexity of signal transduction cascades. For example, target cells that have the same receptor protein may have different second messengers or different enzyme systems that are available for activation. As a result, the same hormone and receptor can give rise to different responses in different target cells.

This finding helps explain one of the most fundamental observations about hormones: The same chemical messenger can trigger different responses in cells from different organs or in cells at different developmental stages. The reason is that the cells contain different receptors, second messengers, amplification steps, protein kinases, enzymes, or transcriptionally active genes.

To summarize this section, steroid and thyroid hormones tend to exert their effects through changes in gene expression; they activate transcription factors that lead to the production of new proteins. In contrast, polypeptide and most amino-acid-derived hormones tend to trigger signal transduction cascades that activate existing proteins, usually by phosphorylation.

CHECK YOUR UNDERSTANDING

If you understood this section, you should be able to ...

1. **MODEL** Draw a cell, showing (a) a polypeptide hormone binding to its receptor and resulting in activation of an intracellular protein, and (b) a steroid hormone binding to its receptor and resulting in production of a new protein.
2. Explain why an animal's response to a polypeptide hormone is typically rapid and short lived compared to its response to a steroid hormone.

Answers are available in Appendix A.

46.3 What Do Hormones Do?

At the beginning of this chapter, you read that hormones are chemical messengers. If that's so, what do hormones "say"?

> **After you complete this section, you should be able to ...**
> - Analyze the roles of hormones.

A first step in answering this question is to recognize that a single hormone can exert various effects in an animal. In humans, for example, the thyroid hormones T_3 and T_4 stimulate metabolism and thus oxygen consumption throughout the body. But they also promote growth, increase heart rate, and stimulate the synthesis of many important macromolecules.

A second step in grasping what hormones do is to recognize that several different hormones may affect the same aspect of physiology. Insulin, glucagon, epinephrine, and cortisol all influence the glucose level in the blood.

Some hormones have extremely diverse effects; other hormones have functions that appear to overlap. These observations begin to make sense when hormone action is viewed in the context of the whole organism. Hormones coordinate the activities of cells in three arenas: **(1)** development, growth, and reproduction; **(2)** response to environmental challenges; and **(3)** maintenance of homeostasis. Let's analyze each of these actions in turn.

How Do Hormones Direct Developmental Processes?

In animals, as in plants, hormones play a key role in regulating growth and development. Growth hormones and sex hormones have crucial roles in promoting cell division, increasing overall body size, and promoting sexual differentiation as an individual matures; certain hormones direct the development of particular cells and tissues at critical junctures in an individual's life.

Let's explore two of the most dramatic examples of hormonal control—metamorphosis in amphibians and in insects—and then survey other developmental processes that are affected by hormone action.

The Role of T_3 in Amphibian Metamorphosis Frogs, toads, and salamanders are called amphibians ("double-lives") because in most species, juveniles (larvae) live in water while adults live on land. The process of changing from an immature, aquatic larva to a sexually mature, terrestrial frog, toad, or salamander is an example of **metamorphosis** ("change-form"; Figure 46.8).

Two sets of complementary experiments, published in 1912 and 1916, established that frog metamorphosis depends on thyroid hormones. Researchers induced frog tadpoles to undergo metamorphosis by feeding them ground-up thyroid glands from horses; they prevented metamorphosis by surgically removing the tadpoles' thyroid glands.

Follow-up work showed that T_3 is the hormone that triggers many of the changes observed in metamorphosis. In response to a signal from the brain, the pituitary gland secretes **thyroid-stimulating hormone (TSH)**. TSH stimulates the thyroid gland

Larval frogs (tadpoles)	
Locomotion	Swim via movements of muscular tail
Respiration	Gas exchange occurs in gills, across skin, and in lungs
Nutrition	Most are herbivorous; have a long gut specialized for digesting algae and plants
Excretion	Most nitrogenous wastes excreted as ammonia

Adult frogs	
Locomotion	Tailless; walk, jump, or swim using limbs
Respiration	Gas exchange occurs across skin and in lungs
Nutrition	Most are carnivorous; have a large mouth or large tongue for catching prey; have a short gut specialized for digesting proteins
Excretion	Most nitrogenous wastes excreted as urea

Figure 46.8 **Amphibian Metamorphosis Is a Continuous Process.** When metamorphosis begins in a frog, toad, or salamander, the individual stays active and continues to feed. The continuous and gradual transition from juvenile to adult is mediated by T_3.

to produce T_4, and T_4 is then converted to the more active T_3 at target tissues.

In juvenile amphibians, an increase in the level of T_3 stimulates the growth of new structures, such as legs. Other structures—such as a tadpole's tail—disintegrate or are absorbed. Some tissues change structure and function. For example, changes in existing cells are responsible for the switch from a tadpole's long intestine, specialized for digesting plant material, to an adult frog's short intestine, specialized for digesting insects and other prey. In the liver, cells respond to T_3 by manufacturing the enzymes required to excrete urea—the nitrogenous waste product released by frogs—instead of the ammonia produced by tadpoles.

✔ **If you understand the basic principles of hormone action, you should be able to suggest a hypothesis explaining why different frog cells can respond to T_3 in such different ways.**

Hormone Interactions Regulate Insect Metamorphosis Some insects show a remarkable type of juvenile-to-adult transition called complete metamorphosis (Ch. 31, Section 31.3). In species that undergo this process, juveniles are called larvae. As in amphibians, insect larvae look completely different from adults, live in different habitats, and eat different food.

As insect larvae grow, they undergo a series of molts in which they shed their old exoskeleton, expand their bodies, and produce a new exoskeleton. After a specific number of these juvenile molts, however, they secrete a tough covering called a pupal case and become a pupa. Inside the pupal case, specific populations of larval cells give rise to a completely new adult body. The rest of the larval body disintegrates.

In insects, metamorphosis depends on interactions between two hormones. If **juvenile hormone (JH)** is present at a high concentration in the larva, surges of the hormone **ecdysone** induce the growth of the larva via molting. But if the JH level is low, ecdysone triggers a complete remodeling of the body—metamorphosis—and the transition to adulthood and sexual maturity (Figure 46.9).

Figure 46.9 **Insect Metamorphosis Is Triggered by Hormones.** When complete metamorphosis begins in an insect, the individual enters a resting stage called the pupa. Pupation is triggered by low levels of specific hormones.

Sexual Development and Activity in Vertebrates In mammals and other vertebrates, long-distance cell-to-cell signals play key roles as embryos develop. Hormones also direct anatomical and physiological changes that occur later in life. Some of the most important of these changes involve the reproductive organs.

Events early in development dictate whether the sex organs, or **gonads**, of a vertebrate embryo become male (**testes**) or female (**ovaries**). This process is called primary sex determination. In mammals, primary sex determination depends on genes located on sex chromosomes. Expression of these genes dictates whether testes or ovaries develop, and these early gonads then begin producing different hormones. In human males, the early testes produce two hormones:

1. The steroid hormone testosterone induces early development of the male reproductive tract.
2. A polypeptide hormone called **Müllerian inhibitory substance** inhibits development of the female reproductive tract.

In females, the ovaries produce the steroid hormone estradiol, which is required for further development of the female reproductive tract.

Sex hormones also play a key role in the juvenile-to-adult transition in mammals. When humans reach early adolescence, for example, surges of sex hormones lead to the physical and emotional changes associated with **puberty**. These developmental changes create the adult phenotype and the ability to produce offspring.

In boys, surges of testosterone are responsible for changes that include enlargement of the penis and testes and growth of facial and body hair. In girls, an increased concentration of estradiol leads to the enlargement of breasts, the onset of menstruation, and other changes. In both sexes, a growth surge begins at puberty. This growth is stimulated by GH produced in the pituitary gland (see Figure 46.2). GH regulates growth factors, which are signaling molecules that control the cell cycle (Ch. 12, Section 12.4). Growth originates in the epiphyseal plates, which are small pieces of cartilage separating the shaft from the end at both extremities of long bones.

Puberty is associated with a growth spurt because the effect of GH on the human skeleton is enhanced by the action of sex hormones, which surge during adolescence. Although growth and sex hormones continue to be produced long after puberty, growth in humans stops when the epiphyseal plates are replaced with bone, making further growth impossible.

The sex hormones continue to play a key role in adults. In humans, sex hormones are instrumental in regulating sperm production, the menstrual cycle, and reproductive behavior (Chapter 47). The result of the regular release of sex hormones is that humans can mate year-round. In many animals, reproductive behavior is instead confined to specific times of the year. In these species, environmental cues such as increasing day length, warmth, or onset of seasonal rains trigger the release of sex hormones.

How Photoperiod Affects Sex Hormone Release The increase in day length—or increasing **photoperiod**—during spring is particularly important in stimulating the release of sex hormones in seasonally reproducing mammals, lizards, and birds. The lengthening photoperiod is sensed by photoreceptors, which are sensory receptors that respond to light (Ch. 44, Section 44.3).

The location of these photoreceptors depends on the animal. In mammals, photoreceptors in the retinas of the eyes send signals to the **pineal gland** via a pathway leading through the brain and spinal cord. The pineal gland secretes the hormone **melatonin**, which relays photoperiodic information to the hypothalamus, a brain region that initiates a series of signals directing production of sex hormones. The pineal gland also regulates sleep–wake cycles.

Maximal melatonin secretion occurs in the dark, so stimulation of the pineal gland by photoreceptors *reduces* melatonin secretion. Animals therefore experience a daily rhythm in melatonin level, which is highest at night. As photoperiod increases in spring, there are fewer hours of darkness, resulting in a lower *overall* level of melatonin (Figure 46.10). This decline in melatonin level "informs" the hypothalamus to stimulate the testes and ovaries to make sex hormones, readying the body for reproduction.

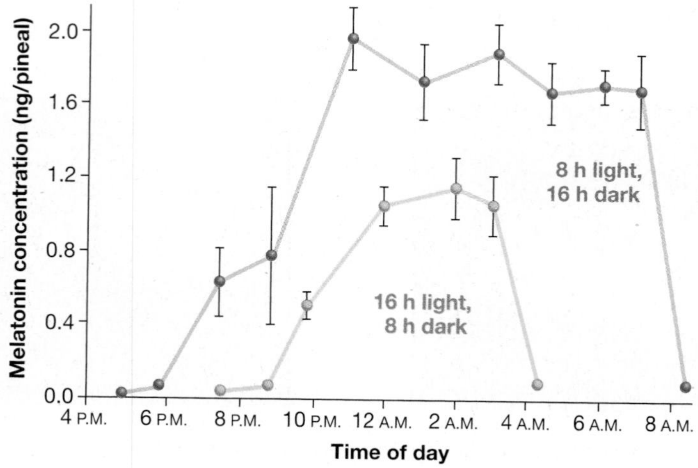

Figure 46.10 Melatonin Level Is Affected by Time of Day and Photoperiod in Seasonal Breeders. Melatonin level is highest at night, and is higher overall when photoperiod is short (8 h light, 16 h dark) than when it is long (16 h light, 8 h dark).

DATA: H. Illnerova, K. Hoffmann, and J. Vanecek. 1984. *Neuroendocrinology* 38: 226–231.

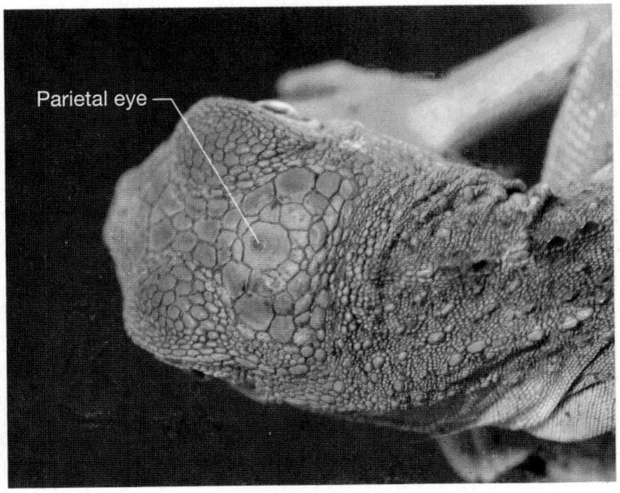

Figure 46.11 **Some Diurnal Reptiles Have a Parietal Eye.** A tiny hole in the skull allows sunlight to directly stimulate photoreceptors in the parietal eye, a brain region associated with the pineal gland.

Many diurnal (day-active) lizard species do not rely solely on retinal photoreceptors. These lizards also have a small hole in the top of their skull, covered only by a thin layer of skin (Figure 46.11). Light passes through this hole and stimulates photoreceptors in the parietal eye, a part of the brain associated with the pineal gland. Interestingly, in birds, neither retinal nor pineal photoreceptors are responsible for sending photoperiodic information to the hypothalamus. It appears that this information originates in photoreceptors located diffusely throughout the brain.

The common thread among mammals, lizards, birds, and other vertebrates is that information about photoperiod is transduced from a signal detected by photoreceptors into a hormonal signal originating at the hypothalamus. The role of the hypothalamus is discussed in more detail in Section 46.4.

Some Chemicals Can Disrupt Hormone Signaling In 1962, Rachel Carson stunned the world with her book *Silent Spring*, in which she described how commercially produced synthetic chemicals—even at low concentrations—can adversely affect humans and wildlife. Since then, evidence has continued to accumulate that many pesticides, industrial chemicals, and other pollutants interfere with normal endocrine function in many animal species.

In the early 1990s, scientists coined the term **endocrine disruptor** to describe chemicals that interfere with normal hormonal signaling. This interference can happen in a number of ways, including binding of the chemical to hormone receptors and altering the metabolism of hormones, thereby affecting physiological processes and behavior. Many endocrine disruptors are **xenoestrogens**—foreign chemicals that bind to estrogen receptors and induce estrogen-like effects. For example, exposure to the widely used herbicide atrazine causes feminization and even sex change in male frogs, induces reproductive-tissue abnormalities and reduced spawning in fishes, and is associated with low sperm counts and increased risk of birth defects in humans.

Another well-known xenoestrogen is bisphenol A (BPA), an industrial chemical used in many plastic and metal products, including containers for food and water. Laboratory animals exposed to BPA show abnormal development of reproductive and brain tissues, along with increased risk of some types of cancers.

As research continues into the possible health effects of xenoestrogens and other endocrine disruptors, many states and countries are limiting or banning their use. For example, atrazine is currently banned in the European Union, and BPA is banned in some countries for use in products such as baby bottles.

How Do Hormones Coordinate Responses to Stressors?

When an animal is thrust into a dangerous or unpredictable situation, hormones play a part in both the short-term and long-term responses. Let's explore each of these response categories in turn.

Short-Term Responses to Stress The short-term reaction, called the **fight-or-flight response**, is triggered by the sympathetic nervous system (Ch. 43, Section 43.4). If you were being chased by a grizzly bear, action potentials from your sympathetic nerves would stimulate your adrenal medulla and lead to the release of epinephrine.

To determine how epinephrine affects the body, researchers injected human volunteers with epinephrine or a saline solution—as a control—and then measured five physiological parameters in the volunteers. As shown in Figure 46.12 on page 996, epinephrine caused dramatic increases in the concentrations of free fatty acids and glucose in the blood, pulse rate, blood pressure, and oxygen consumption by the brain. In addition, the volunteers who were injected with epinephrine reported strong subjective feelings of anxiety and excitement.

Other experiments showed that epinephrine redirects blood away from the skin and digestive system and toward the heart, brain, and muscles. Epinephrine does this by relaxing smooth muscle surrounding blood vessels in the heart, brain, and muscles—increasing blood delivery.

Taken together, the responses to epinephrine lead to a state of heightened alertness and increased energy use that prepares the body for rapid, intense action such as fighting or fleeing. Epinephrine coordinates the activities of cells in many organs and systems throughout the body to prepare an individual to cope with a life-threatening situation.

RESEARCH

QUESTION: How does epinephrine affect the body?

HYPOTHESIS: Epinephrine causes changes involved in the fight-or-flight response.

NULL HYPOTHESIS: Epinephrine is not involved in the fight-or-flight response.

EXPERIMENTAL SETUP:

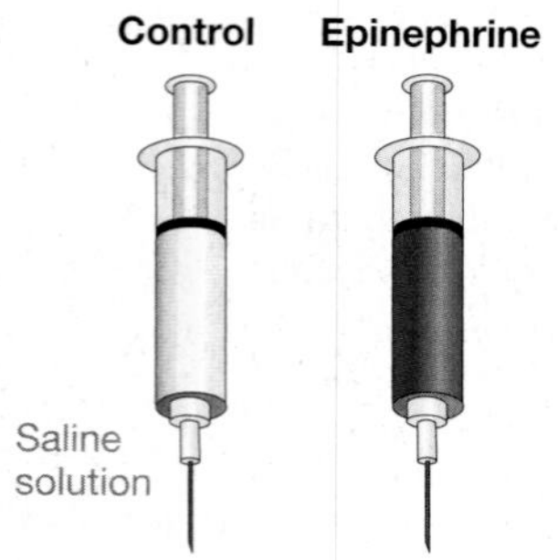

1. Inject human volunteers with saline solution or epinephrine.

2. Document changes in fatty-acid and glucose concentrations in blood, pulse rate, blood pressure, and oxygen consumption in brain.

PREDICTION OF HYPOTHESIS: The epinephrine-treated group will have higher fatty-acid and glucose concentrations in blood, pulse rate, blood pressure, and brain oxygen consumption than the control group.

PREDICTION OF NULL HYPOTHESIS: There will be no differences in the physiological state of individuals based on the substance injected.

RESULTS:

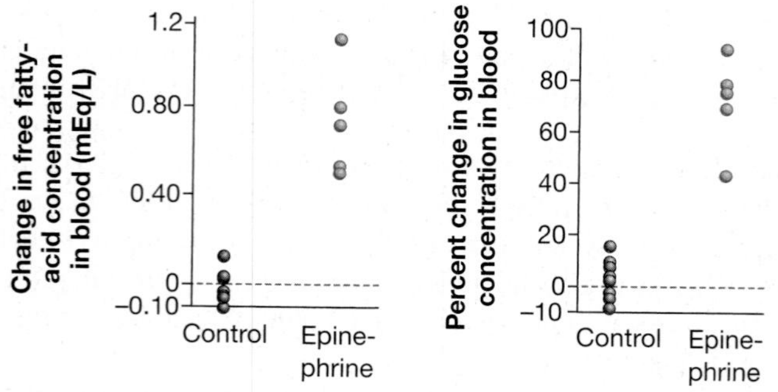

	Control	Epinephrine
Pulse rate (beats/min)	78.3	89.6
Blood pressure (mm Hg)	90.9	108.7
O_2 consumption in brain (cc O_2/100 g/min)	3.41	4.16

CONCLUSION: Epinephrine causes an array of changes associated with the fight-or-flight response.

Figure 46.12 **Epinephrine Prepares the Body for Action.** Each point in the graphs represents data from one volunteer; the data in the table are average values of the study participants.

SOURCES: B. D. King, L. Sokoloff, and R. L. Wechsler. 1952. The effects of *l*-epinephrine and *l*-norepinephrine upon cerebral circulation and metabolism in man. *Journal of Clinical Investigation* 31: 273–279. P. S. Mueller and D. Horwitz. 1962. Plasma free fatty acid and blood glucose responses to analogues of norepinephrine in man. *Journal of Lipid Research* 3: 251–255.

✔ **PROCESS OF SCIENCE** Why did researchers bother to inject volunteers in the control group with saline? Why did they inject them with anything?

Long-Term Responses to Stress If you have ever experienced the fight-or-flight response, you may recall that the state is short lived. Once an epinephrine "rush" wears off, most people feel exhausted and want to rest and eat.

What happens if the stress continues and turns into a long-term condition? In the course of their lifetime, many people experience periods of starvation or fasting, prolonged emotional distress, or chronic illness. How do hormones help humans and other animals cope with extended stress?

Early studies of long-term stress in human subjects suggested a role for cortisol, which is produced in the adrenal cortex. An increased level of cortisol was found in airplane pilots and crew members during long flights, athletes who were training for intense contests, parents of children undergoing treatment for cancer, and college students who were preparing for final exams. Why?

What Does Cortisol Do? In humans, cortisol's primary role is to ensure the continuing availability of glucose for use by the brain during long-term stress. Cortisol manages the following main processes that maintain glucose production:

- Cortisol induces the synthesis of liver enzymes that make glucose from amino acids and other chemical precursors.
- Cortisol makes adipose tissue—fat tissue—and resting muscles resistant to insulin. Insulin normally stimulates **adipocytes** and resting muscle cells to remove glucose from the bloodstream. But when cortisol makes these cells resistant to insulin, glucose is reserved for use by the brain and exercising muscles.
- Cortisol promotes the release of fatty acids—the body's major fuel molecules—from adipose tissue, for use by the heart and muscles.

Because of its importance in regulating blood glucose concentration, cortisol is classified as a **glucocorticoid**. Corticosterone, produced by some other vertebrates, is a glucocorticoid similar to cortisol.

The long-term stress response comes at a high price, however, as any victim of a serious injury or illness knows. Glucocorticoids make amino acids available for glucose synthesis by promoting the degradation of contractile proteins in muscle. The resulting loss of muscle mass may cause severe weakness. Also, glucocorticoids impair wound healing and suppress immune and inflammatory responses. These processes are costly in terms of energy use, but suppressing them makes the body more susceptible to infection.

The overall concept here is that the long-term stress response is a compromise—a fitness trade-off (Ch. 39, Section 39.1). The fuel requirements of the brain are met at the expense of other tissues and organs.

How Are Hormones Involved in Homeostasis?

Recall that homeostasis is the maintenance of relatively constant physical and chemical conditions inside the body. Homeostatic systems depend on three components (see Ch. 39, Figure 39.12):

1. A sensory receptor that monitors a condition
2. An integrator that processes information from the sensor and compares it to a normal value, or set point
3. Effector cells that return the condition to the set point

In homeostatic systems, messages often travel from integrators to effectors in the form of hormones.

Insulin, Glucagon, and Blood Glucose Homeostasis The regulation of blood glucose concentration is a good example of homeostasis regulated by hormones. The blood glucose level must be maintained within a narrow range of values. If the level gets too low, the brain does not have enough fuel to function. High blood glucose levels, however, are toxic to neurons and blood vessels and, when chronic, can even result in organ failure and amputation of extremities (Ch. 41, Section 41.4).

When an animal eats, a rising blood glucose level stimulates the release of insulin from the pancreas (see Ch. 41, Figure 41.16). Insulin stimulates effector cells throughout the body to import glucose from the blood for storage or use in metabolism, causing the blood glucose level to drop to normal values.

Hours after a meal, the blood glucose level declines as glucose is used for cellular respiration, which causes glucagon to be released from the pancreas. Glucagon has basically the opposite effect of insulin: It causes glucose-storing cells in the liver to export glucose into the blood, thereby increasing the blood glucose level.

ADH, Aldosterone, and Water and Electrolyte Balance Recall that when an individual is dehydrated, antidiuretic hormone (ADH) is released from the pituitary gland (Ch. 40, Section 40.4). ADH increases the permeability of the kidney's collecting ducts to water, causing water to be reabsorbed from urine and saved.

ADH is instrumental in achieving homeostasis with respect to water balance. For example, the ethanol in alcoholic beverages inhibits the release of ADH from the pituitary. Consequently, people who imbibe large quantities of these beverages produce large quantities of dilute urine. The resulting water loss can lead to dehydration and nausea—symptoms associated with an alcoholic hangover.

When the sodium ion concentration in body fluids is low, **aldosterone** is released from the adrenal cortex (Ch. 40, Section 40.4). Because aldosterone increases reabsorption of sodium ions in the distal tubules of the kidney, it plays a key role in homeostasis with respect to electrolyte concentrations and overall volume of body fluids. Adrenal hormones with this effect are called **mineralocorticoids**.

ADH saves water; aldosterone saves sodium ions. Together, they are key players in maintaining water and electrolyte balance.

EPO and Oxygen Availability **Erythropoietin (EPO)** is a crucial component in the homeostatic system that maintains the blood oxygen level. When the blood oxygen level falls, the kidneys and other tissues release EPO, which stimulates the production of red blood cells. The more red blood cells there are, the higher the oxygen-carrying capacity of blood is. If you moved to a high elevation and experienced chronic oxygen deficit, your body would respond by releasing EPO and increasing your red blood cell count. This explains why many endurance athletes live or train at high altitudes—the high red blood cell count may give them an athletic advantage (Figure 46.13).

Some athletes have turned to EPO injections as a way to increase the oxygen-carrying capacity of their blood and give themselves a competitive edge. This practice is dangerous as well as illegal. The increased viscosity of blood in EPO abusers is accentuated during exercise, when the blood plasma volume drops due to dehydration. The combination of high blood viscosity and low blood volume can impair blood flow through capillaries, increasing the risk of tissue damage and blood clotting. If clots form in blood vessels that lead to the heart or brain, a heart attack or stroke may occur. EPO abuse is thought to be responsible for the collapse and death of several cyclists during races in the mid-1990s.

Figure 46.13 **High Altitude Training Can Aid Athletic Performance.** Low oxygen levels in the air cause release of EPO, which stimulates red blood cell production.

CHECK YOUR UNDERSTANDING

If you understood this section, you should be able to...

1. Compare and contrast the roles of the hormones involved in metamorphosis of amphibians and insects.
2. Explain how the various changes induced by an elevated cortisol level result in a response to long-term stress.

Answers are available in Appendix A.

46.4 How Is the Production of Hormones Regulated?

Most hormones are released in response to an environmental cue or a message from an integrator in a homeostatic system. Often, the nervous system is closely involved. For example, environmental cues that signal the onset of the breeding season or the presence of a predator are received by sensory receptors and interpreted by the brain. Similarly, integration in most

After you complete this section, you should be able to ...

- Analyze the regulation of hormone production.

homeostatic systems is done by neurons in the central nervous system (CNS)—the brain and spinal cord (Ch. 43, Section 43.4).

Based on these observations, the short answer to the question posed in the title of this section is simple: In many cases, hormone production is directly or indirectly controlled by the nervous system.

The Hypothalamus and Pituitary Gland

The pituitary gland is directly connected to the hypothalamus. This physical link is the basis of the connection between the CNS and the endocrine system. As you read in Section 46.2, removing the entire pituitary gland in laboratory animals caused them to stop growing. The animals also lost the ability to maintain a normal body temperature and suffered atrophy (shrinkage) of their genitals, thyroid glands, and adrenal cortexes. Not surprisingly, their life span shortened dramatically.

These experiments suggested that, in addition to secreting GH, the pituitary secretes hormones that regulate the production of a wide variety of other hormones. As a case study, let's look at the pituitary hormone that acts on the adrenal glands.

Controlling the Release of Glucocorticoids Early work on rats suggested that a molecule from the pituitary gland affects the adrenal gland. This molecule soon came to be called **adrenocorticotropic hormone**, or **ACTH** (*adreno* refers to the adrenal glands; *cortico* refers to the outer portion, or cortex, of each gland; and *tropic* means "affecting the activity of"). ACTH is also known as corticotropin.

ACTH was purified and characterized in 1943. When human volunteers were injected with ACTH, the level of the glucocorticoid hormone cortisol rose in their blood. This result provided evidence that ACTH is a regulatory hormone. The adrenal cortex secretes cortisol in response to ACTH released from the pituitary. Similar experiments have established that ACTH stimulates production of the entire class of glucocorticoids.

What regulates ACTH release? Biologists from two laboratories independently showed that ACTH is released in response to a molecule produced by the hypothalamus. After years of effort, a different team of researchers succeeded in purifying a peptide—just 41 amino acids long—called **corticotropin-releasing hormone (CRH)**. When the hypothalamus releases CRH, it stimulates cells in the pituitary to secrete ACTH into the bloodstream.

Negative Feedback by Glucocorticoids What *stops* glucocorticoid secretion? The key is to recognize that glucocorticoids themselves suppress ACTH production by the pituitary gland. Glucocorticoids accomplish negative feedback—they suppress their own production.

When human volunteers were injected with cortisol, the ACTH level in their bloodstream dropped dramatically. Cortisol also inhibits release of CRH from the hypothalamus. Thus, if glucocorticoid levels become too high, the ACTH level falls. But if glucocorticoid levels become too low, the ACTH level rises and drives a compensatory increase in glucocorticoid production. Figure 46.14 summarizes the relationships among CRH, ACTH, and the glucocorticoid cortisol.

What happens when negative feedback fails? Certain pituitary tumors diminish the ability of cortisol to suppress ACTH production, leading to persistently high blood levels of ACTH and cortisol. The result is **Cushing's disease**, an unrelenting stress response that depletes the body's protein reserves. It is fatal if not treated.

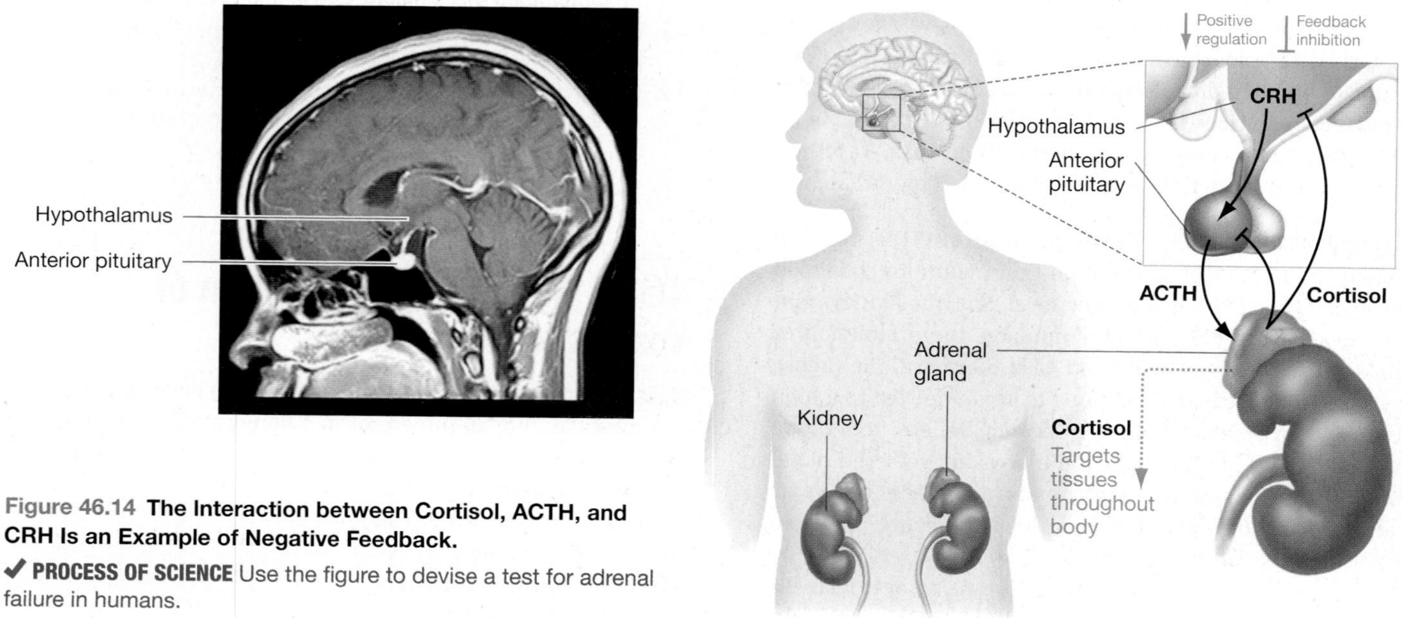

Figure 46.14 The Interaction between Cortisol, ACTH, and CRH Is an Example of Negative Feedback.

✔ **PROCESS OF SCIENCE** Use the figure to devise a test for adrenal failure in humans.

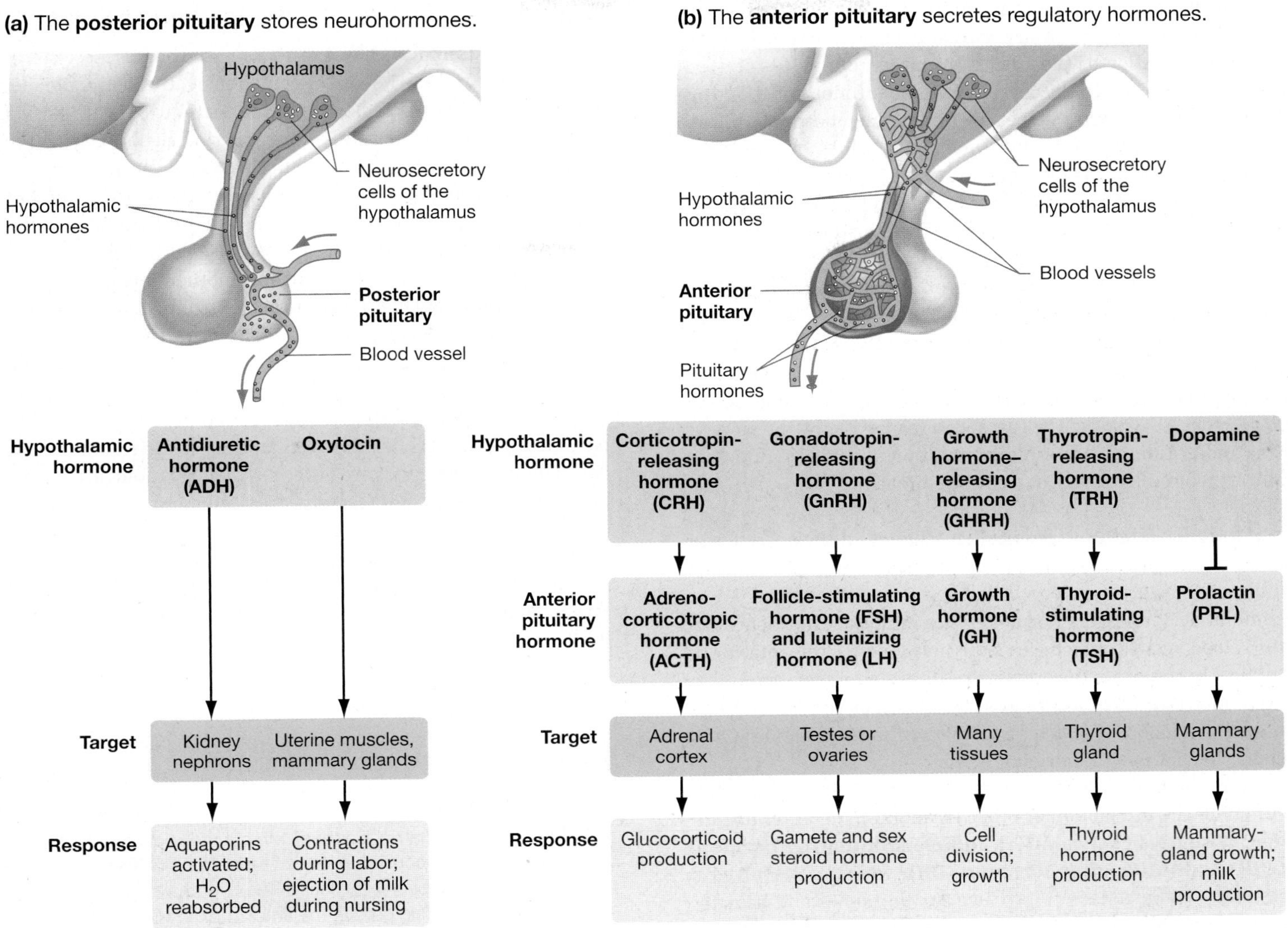

Figure 46.15 **The Hypothalamus and Pituitary Interact Closely. (a)** Developmentally and anatomically, the posterior pituitary is an extension of the hypothalamus. Neurosecretory cells in the hypothalamus extend directly into the posterior pituitary and secrete ADH (vasopressin) and oxytocin. **(b)** The hypothalamus and the anterior pituitary communicate indirectly, via blood vessels. Hormones produced by other populations of neurosecretory cells in the hypothalamus travel in the blood to the anterior pituitary, where they control the release of pituitary hormones. The RH in the names of some of the hypothalamic hormones stands for "releasing hormone" because it stimulates release of another hormone from the anterior pituitary gland.

Patterns in Glucocorticoid Release Recall that cortisol is responsible for regulating the level of glucose in the blood (Section 46.3). Under ordinary circumstances, the production of CRH by the hypothalamus displays a daily rhythm and reaches its highest level in the early morning hours. This pattern drives a corresponding daily rhythm in ACTH production and blood cortisol level.

The morning peak in blood cortisol typically coincides with arousal and initiation of the day's activities—and the effect is saving glucose for use by the brain. The unpleasant symptoms of jet lag are due in part to the daily cortisol rhythm being out of synchrony with local time for several days after you arrive in a new time zone.

When the brain processes stimuli that produce pain or anxiety, it initiates a long-term stress response and a sustained increase in CRH production. Increased CRH production causes blood ACTH and cortisol levels to remain much higher throughout the day than they are in the unstressed state.

The Hypothalamic–Pituitary Axis: An Overview The CRH-ACTH-glucocorticoid relationship is just one of many hormone systems based on interactions among the hypothalamus, pituitary, and target glands or cells. The **hypothalamic–pituitary axis** forms two anatomically distinct systems because the pituitary gland has two segments: the **anterior pituitary** and the **posterior pituitary** (Figure 46.15). The anterior pituitary develops from cells in an embryo's mouth and throat lining; the posterior pituitary is an extension of the brain.

The hypothalamic neurons responsible for hormone secretion by the posterior pituitary are distinct from those that control hormone secretion by the anterior pituitary. Both types of hypothalamic neurons synthesize and release neurohormones and are therefore called **neurosecretory cells**. Their hormone release is under the control of brain regions responsible for integrating information about the external or internal environment. For example, information about an upcoming exam or athletic contest might trigger action potentials that lead to the release of CRH.

The Posterior Pituitary Neurosecretory cells that project from the hypothalamus produce the hormones ADH and oxytocin, which are then stored in the posterior pituitary. From there, ADH and oxytocin are released into the bloodstream. This is an example of the neuroendocrine pathway of hormone action. Recall that ADH aids in the reabsorption of water by the kidneys. **Oxytocin** stimulates uterine contractions during birthing and subsequent milk release in female mammals.

The Anterior Pituitary Unlike the situation in the posterior pituitary, the hypothalamus and anterior pituitary are connected indirectly. Neurosecretory cells in the hypothalamus secrete stimulatory ("releasing hormones") or inhibitory hormones into tiny blood vessels, which carry the hormones to the anterior pituitary. In response, the anterior pituitary alters its secretion of hormones that enter the general bloodstream and act on target tissues or glands. This is an example of the neuroendocrine-to-endocrine pathway of hormone action.

Many of the hormones produced by the anterior pituitary stimulate the production of other hormones. The anterior pituitary hormones include ACTH; **follicle-stimulating hormone (FSH)** and **luteinizing hormone (LH)**, which are involved in stimulating the gonads to produce sex hormones and gametes; GH; **prolactin**, which stimulates mammary gland growth and milk production in mammals; and TSH.

Control of Epinephrine by Sympathetic Nerves

When biologists analyze how the nervous and endocrine systems interact to control the release of epinephrine, the distinction between the two systems begins to blur. In Section 46.2 you learned about how epinephrine acts as an endocrine signal. During the fight-or-flight response, sympathetic nerves trigger the release of epinephrine from the adrenal medulla into the bloodstream. But in addition, some sympathetic nerves release the related molecule **norepinephrine** directly onto target cells. Epinephrine and norepinephrine, which differ from one another only by the presence of an additional methyl group on epinephrine, are members of the family of molecules called **catecholamines**.

During the fight-or-flight response, the endocrine system broadcasts epinephrine by secreting it into the bloodstream, whereas the nervous system delivers norepinephrine directly to particular cells. Although the mechanisms for distributing these two chemical messengers are different, their physiological effects during the fight-or-flight response are similar.

Catecholamines function as neurotransmitters as well as hormones. This dual role exemplifies how the nervous and endocrine systems are not separate systems, but rather are coordinated components of organisms' physiological strategies for maintaining homeostasis.

CHECK YOUR UNDERSTANDING

If you understood this section, you should be able to . . .

1. Explain how negative feedback affects the release of ACTH.
2. Discuss the relationship between processing centers in the brain, neurosecretory cells in the hypothalamus, and hormone-secreting cells in the anterior pituitary.

Answers are available in Appendix A.

CHAPTER 46 Review

For media, go to **Mastering Biology**

46.1 Cell-to-Cell Signaling: An Overview

- Chemical signals in animals fall into five categories: autocrine, paracrine, endocrine, neural, and neuroendocrine.
- Hormones are chemical messengers that are released from neurons or cells of the endocrine system, circulate in the blood or other body fluids, and trigger a response in target cells containing an appropriate receptor.

46.2 How Do Hormones Act on Target Cells?

- Hormones have a variety of chemical structures. Most animal hormones are peptides or polypeptides, amino acid derivatives, or steroids.
- Animal hormones have two basic modes of action. Steroid and thyroid hormones are lipid soluble, cross plasma membranes readily, and usually bind to receptors inside cells. Most polypeptide and amino-acid-derived hormones are not lipid soluble; they bind to receptors located in the plasma membranes of target cells.
- Most steroid and thyroid hormones act by inducing a change in gene expression.

- Polypeptide and most amino-acid-derived hormones trigger signal transduction cascades that activate one or more target proteins by phosphorylation.
- Although they are produced in tiny concentrations, hormones have large effects because they trigger gene expression or because their message is amplified through a signal transduction cascade.

46.3 What Do Hormones Do?

- Together with the nervous system, hormones coordinate the activities of diverse cells and tissues. A single hormone may affect a wide array of cells and tissues and induce a variety of responses.
- Estradiol is an example of a hormone that regulates development and sexual maturation. Estradiol is required for the maturation of female secondary sex characteristics in adolescence.
- Melatonin is a hormone that regulates reproductive physiology in response to seasonal changes in day length.
- Epinephrine and cortisol are examples of hormones that help individuals cope with environmental changes. Epinephrine activates the short-term response to stressors by triggering the fight-or-flight response. Cortisol triggers the long-term response to stressors by inducing changes that conserve glucose for use by the brain.
- Hormones are involved in a wide array of homeostatic interactions. For example, hormones are involved in directing cells that modify the concentrations of glucose, water, sodium ions, and other substances in the blood and interstitial fluid.

46.4 How Is the Production of Hormones Regulated?

- In many cases, the release of a hormone is regulated by hormones from the anterior pituitary.
- Hormone-secreting cells in the anterior pituitary are regulated by hormones released by the hypothalamus.
- The long-term stress response is a well-studied example of hormone regulation. The brain responds to long-term stress by triggering the release of the hypothalamic hormone CRH. CRH activates the anterior pituitary to release ACTH, which stimulates the production of cortisol by cells in the adrenal cortex. Because cortisol inhibits the production of ACTH and CRH, the chain of events is regulated by negative feedback.
- Some hormones, such as epinephrine, are released in response to sympathetic nervous system activation.

Answers are available in Appendix A.

✔ TEST YOUR KNOWLEDGE

1. How do steroid hormones differ from polypeptide hormones and most amino-acid-derived hormones?
   a. Steroid hormones are lipid soluble and cross plasma membranes readily.
   b. Polypeptide and amino-acid-derived hormones are longer lived in the bloodstream and thus exert greater signal amplification.
   c. Polypeptide hormones are the most structurally complex and induce permanent changes in target cells.
   d. Only steroid hormones bind to receptors in the plasma membrane.
2. What is a hormone-response element?
   a. a receptor for a steroid hormone
   b. a receptor for a polypeptide hormone
   c. a segment of DNA where a hormone–receptor complex binds
   d. an enzyme that is activated in response to hormone binding and produces a second messenger
3. Which of the following assertions regarding hormones is correct? Select True or False for each statement.
   T/F Growth and metamorphosis are controlled by hormones.
   T/F Cortisol stimulates the production of ACTH.
   T/F Hormones produced by the hypothalamus are considered neurohormones.
   T/F Hormonal changes during puberty and pregnancy are forms of endocrine disruption.
4. True or False: In hormone systems, negative feedback occurs when the presence of a hormone inhibits release of the hormone.

✔ TEST YOUR UNDERSTANDING

5. Compare and contrast the modes of action of lipid-soluble and water-soluble hormones.
6. Why is the observation that one hormone may bind to more than one type of receptor important?
7. Compare and contrast the structure and function of the anterior and posterior pituitary glands.
8. **PROCESS OF SCIENCE** Design a study to test the hypothesis that the symptoms of jet lag are caused by disruption of normal daily cortisol rhythms.

✔ TEST YOUR PROBLEM-SOLVING SKILLS

9. **PROCESS OF SCIENCE** Suppose that during a detailed anatomical study of a marine invertebrate, you found a small, previously undescribed structure. How would you test the hypothesis that the structure is a gland that releases one or more hormones?
10. **QUANTITATIVE** Scientists set out to test the hypothesis that the herbicide atrazine is an endocrine disruptor that feminizes male amphibians. They treated male amphibians with atrazine and then compared their circulating testosterone concentration with those of males and females that were not treated with atrazine (controls). The results are shown here (* signifies $P < 0.05$; see **BioSkills 3** for more on statistical significance). Was the hypothesis supported? Why or why not?

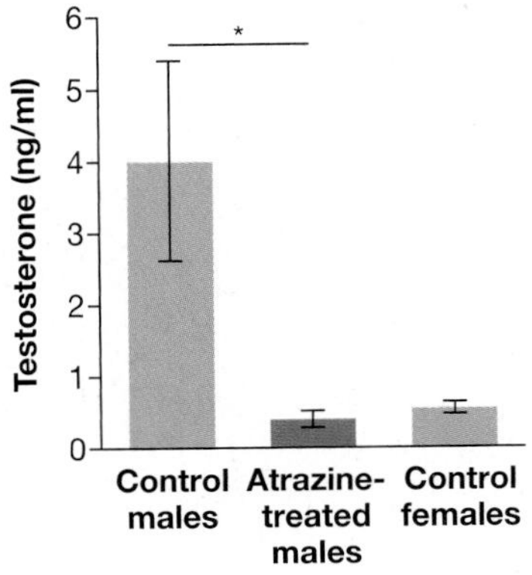

DATA: T. B. Hayes, A. Collins, M. Lee, et al. 2002. *Proceedings of the National Academy of Sciences, USA* 99: 5476–5480.

✔ PUT IT ALL TOGETHER: Case Study

Is ecotourism stressful to animals?

Ecotourism helps conserve wildlife by increasing the value of wildlife conservation for local people. In Argentina, the world's largest breeding colony of Magellanic penguins is visited by thousands of tourists every year, causing the penguins to become habituated to the presence of people. Has ecotourism affected the penguins' ability to respond to stressors? Has it affected the penguins' secretion of corticosterone?

11. Corticosterone is a(n):
- **a.** neurohormone
- **b.** steroid hormone
- **c.** amino-acid-derived hormone
- **d.** polypeptide hormone

12. In what part(s) of a target cell would you expect to find corticosterone receptors?

13. If penguins in a colony are habituated to the presence of tourists, how would their corticosterone response to being visited by tourists compare to that of penguins in a colony not previously exposed to tourists?

14. QUANTITATIVE Scientists tested the hypothesis that penguins habituated to tourists have a blunted stress response overall. They examined the corticosterone response of habituated and non-habituated penguins to another stressor—that of being captured and restrained for 30 minutes. They found that both habituated and non-habituated penguins secreted corticosterone in response to being captured and restrained, but that the level of corticosterone after 30 minutes differed between the groups. The results are shown here (* means $P < 0.05$; **BioSkills 3**). Was the hypothesis supported? Explain.

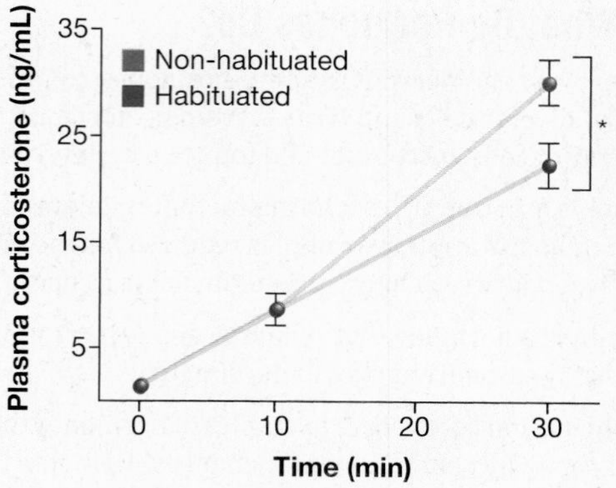

Source: B. G. Walker, P. D. Boersma, and J. C. Wingfield. 2006. *Conservation Biology* 20: 146–154.

15. In response to a visit by tourists, which penguins—habituated or non-habituated—would show a greater change in expression of genes for liver enzymes involved in converting amino acids to glucose? Explain.

16. Suggest at least two distinct physiological mechanisms that could be responsible for the difference in corticosterone response to a stressor in habituated and non-habituated penguins. (Hint: Refer to Figure 46.14.)

Mastering Biology

Students Go to Mastering™ Biology for assignments, the eText, and the Study Area with animations, practice tests, and activities.

Professors Go to Mastering™ Biology for automatically graded tutorials and questions that you can assign to your students, plus Instructor Resources.

47 Animal Reproduction and Development

The swollen red rump of this female Hamadryas baboon indicates that she is about to produce an egg. She will probably mate with several males before the egg is fertilized.

In this chapter you will learn that

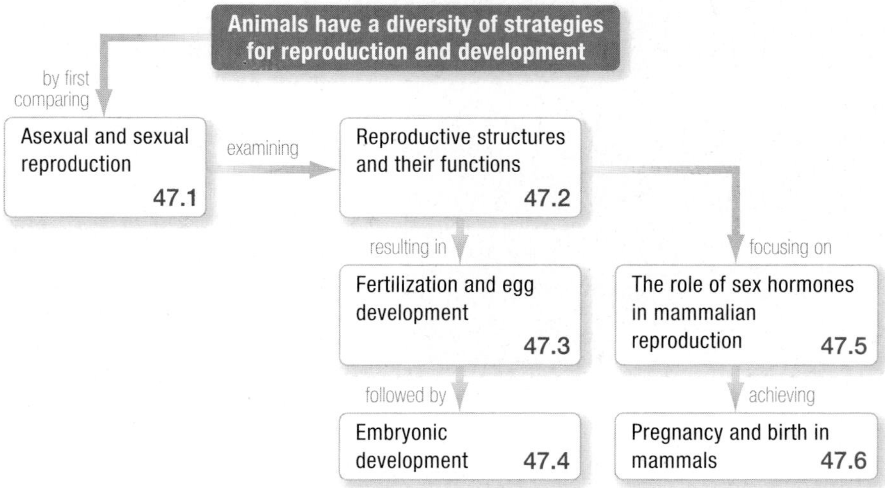

All the cells, tissues, organs, and systems introduced in Unit 7 exist for one reason: They allow animals to survive long enough and gather enough resources to reproduce. Put another way, adaptations allow organisms to produce offspring. Reproduction is the underlying purpose of virtually everything that an animal does. Replication is a fundamental attribute of life (Chapter 1).

Although evolution by natural selection explains *why* animals reproduce, the goal of this chapter is to explore *how* reproduction occurs and how an animal develops before it is born. Understanding and manipulating reproductive systems is an important issue for physicians, veterinarians, farmers, zookeepers, conservation biologists, and many others. In this chapter you'll discover some of the diversity of reproductive modes that occur in animals, but the focus is on mammalian reproduction, using humans as the primary model organism.

BIG PICTURE

This chapter is part of the Big Picture. See how on pages 838–839.

47.1 Asexual and Sexual Reproduction

After you complete this section, you should be able to...

- Explain asexual reproduction, sexual reproduction, and gametogenesis in animals.

In earlier chapters you explored how asexual and sexual reproduction differ. **Asexual reproduction** occurs without fusion of gametes, results in offspring with genetic material from only one parent, and often yields offspring that are genetically identical to the parent. **Sexual reproduction**, in contrast, typically involves the fusion of haploid gametes and results in offspring that are genetically different from each other and from their parents.

Figure 47.1 **Mechanisms of Asexual Reproduction in Animals Are Diverse.**

Let's first consider how animals engage in asexual reproduction and then review the mechanisms responsible for sexual reproduction.

How Does Asexual Reproduction Occur?

For thousands of animal species, females can reproduce with no genetic contribution from males. There are three main mechanisms of asexual reproduction:

1. In **budding**, an offspring begins to form within or on a parent, such as on the hydra shown in Figure 47.1a. The process is complete when the offspring—a miniature version of the parent—breaks free and begins to grow on its own.
2. In **fission**, an individual simply splits into two or more descendants, as shown in anemones in Figure 47.1b.
3. In **parthenogenesis** (literally, "virgin-origin"), females develop eggs and produce offspring without any genetic contribution from a male. Parthenogenesis occurs in a diversity of lineages, including certain invertebrates, fishes, lizards, snakes, and birds; Figure 47.1c shows parthenogenesis in a whiptail lizard.

In budding and fission, females **clone** themselves—that is, produce large numbers of identical copies of themselves asexually. Parthenogenetic offspring may be produced by mitosis, in which case they are also clones of their mother. They can also be produced by meiosis after a doubling of chromosome number, or by the fusion of two of the products of meiosis. In these cases, recombination during meiosis can create offspring with variable genetic composition.

Some animals can reproduce only asexually or sexually, but many animal species regularly switch between modes. Why?

Switching Reproductive Modes in *Daphnia*: A Case History

Daphnia are crustaceans that live in freshwater habitats throughout the world. In a typical year, *Daphnia* produce only diploid, female offspring throughout the spring and summer, via parthenogenesis. The offspring produced by parthenogenesis develop in a structure called a brood pouch (Figure 47.2), and the young *Daphnia* are released when the female molts her exoskeleton.

In late summer or early fall, however, many of the parthenogenetically produced offspring develop into males. Once the males have matured, sexual reproduction ensues: by meiosis, males produce haploid **sperm** and females produce haploid **eggs**, and these gametes fuse to create diploid offspring. **Fertilization** is the fusion of sperm and egg to form a **zygote**.

Biologists try to explain observations such as switching reproductive modes at two levels (Ch. 50, Section 50.3):

1. **Proximate causation** addresses *how* a trait is produced. When researchers identify the genetic, developmental, hormonal, or neural mechanisms responsible for a phenotype, they are working at the proximate level.
2. **Ultimate causation** addresses *why* a trait occurs, in terms of its effect on fitness. Researchers who work at the ultimate level try to understand the evolutionary history of traits.

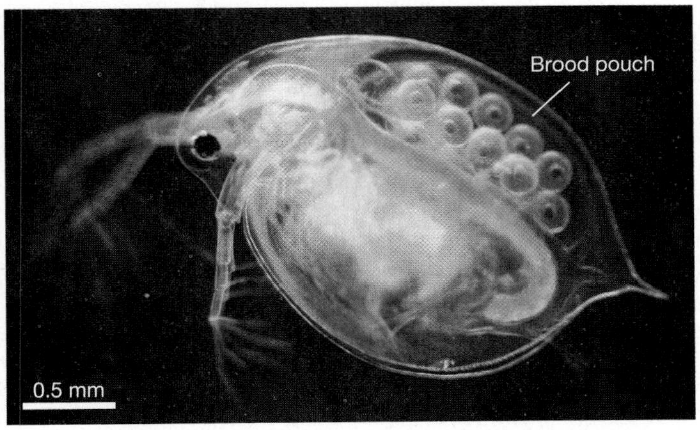

Figure 47.2 **Female *Daphnia* Can Produce Offspring by Parthenogenesis.**

Let's consider work on proximate aspects of the asexual–sexual switch in *Daphnia* and then consider ultimate causation.

What Environmental Cues Trigger the Switch? For decades, most researchers contended that day length triggered the asexual–sexual switch. The idea was that the shortening days of late summer or fall affected sensors in the nervous system (Chapter 43), and these receptors produced electrical or hormonal signals that induced the production of males and haploid eggs.

In 1965, however, biologists showed that high population densities are also a factor. Researchers who brought *Daphnia* populations into the lab and kept day length constant found that the highest proportion of sexually reproducing females occurred at the highest population density (Figure 47.3a).

Another group of investigators built on this result by pinpointing the specific aspects of crowding that affected the animals. These biologists brought *Daphnia* into the laboratory and altered day length, the amount of food available to individuals, and the quality of the water. To vary water quality, the investigators used either clean water or "crowded" water taken from tanks where *Daphnia* were being maintained at high density.

As Figure 47.3b shows, individuals in the study population switched to sexual reproduction only if they were exposed to water from crowded populations, low food availability, *and* short day lengths. In short, *Daphnia* needs three different cues from the environment to switch to sexual reproduction. Two of these cues are associated with high population density; the third is associated with the onset of winter.

Why Do *Daphnia* Switch between Asexual and Sexual Reproduction? *Daphnia* appear to start sexual reproduction when conditions worsen. Why?

The leading hypothesis to answer this question points out that sexually produced offspring are genetically diverse. Recall that genetic diversity increases fitness in environments with rapidly evolving parasites, deteriorating physical conditions, or other types of rapid environmental change (Ch. 13, Section 13.4). When the environment changes, genetically diverse offspring are likely to include individuals that will survive better and reproduce more than offspring that are identical to their parents.

Mechanisms of Sexual Reproduction: Gametogenesis

The mitotic cell divisions, meiotic cell divisions, and developmental events that produce male and female gametes are collectively called **gametogenesis**. **Spermatogenesis** is the formation

(a) Sexual reproduction is more common in crowded populations of *Daphnia pulex* than in sparse populations.

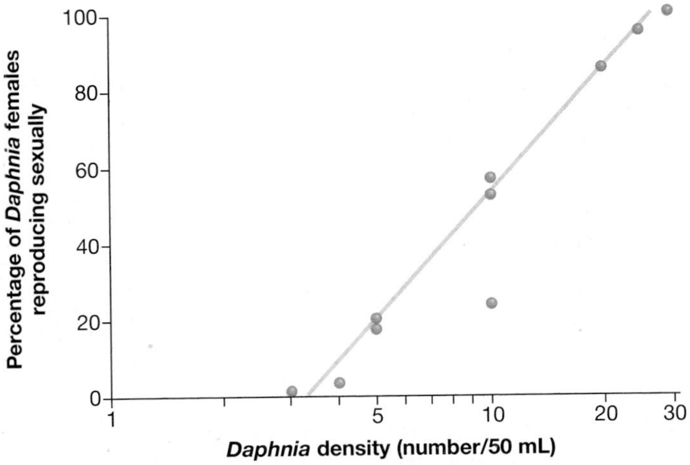

(b) In *Daphnia magna*, a combination of environmental cues triggers the switch to sexual reproduction.

Water quality	Food concentration	Day length	Sexual broods (%)
Clean	Low	Short	0
Crowded	Low	Short	44
Clean	Low	Long	0
Crowded	Low	Long	0
Clean	High	Short	0
Crowded	High	Short	0
Clean	High	Long	0
Crowded	High	Long	0

Figure 47.3 **In *Daphnia*, Environmental Cues Signal the Switch from Asexual to Sexual Reproduction.** **(a)** There is a strong positive correlation between the percentage of females that reproduce sexually and the density of the population (plotted on a log scale here; BioSkills 5). **(b)** Environmental conditions were varied experimentally for *Daphnia*. "Crowded" water was taken from tanks containing dense populations.

DATA: (a) R. G. Stross and J. C. Hill. 1965. *Science* 150: 1462–1464. (b) O. T. Kleiven, P. Larsson, and A. Hoboek. 1992. *Oikos* 65: 197–206.

✔ **PROCESS OF SCIENCE** Design an experiment to determine which molecule or molecules in "crowded" water serve as a signal that triggers sexual reproduction.

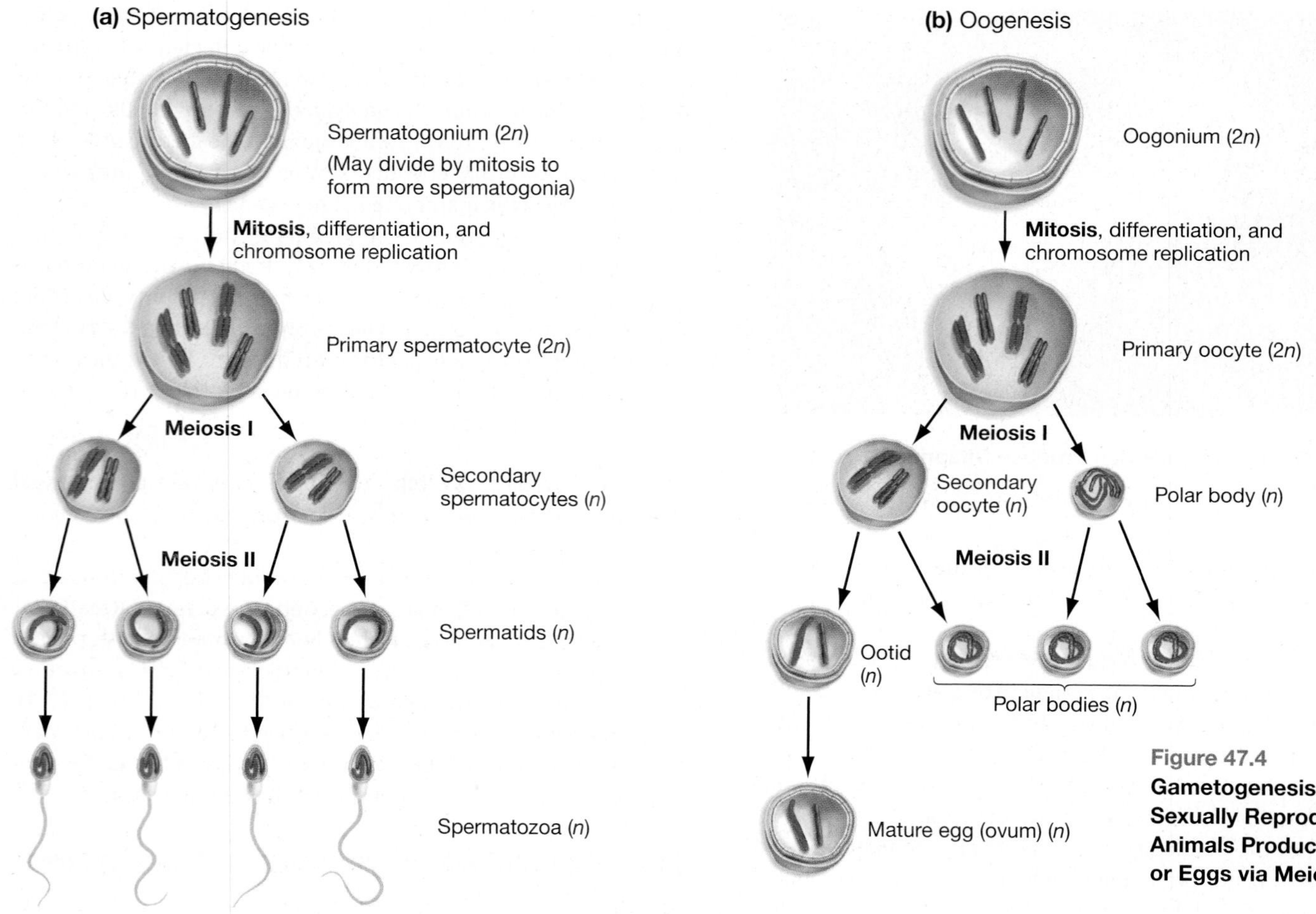

Figure 47.4 Gametogenesis in Sexually Reproducing Animals Produces Sperm or Eggs via Meiosis.

of sperm; **oogenesis** is the formation of **ova** (singular: **ovum**) (Figure 47.4). An ovum is also known as a mature egg.

In the vast majority of animals, gametogenesis occurs in a sex organ, or **gonad**. Male gonads are called **testes**; female gonads are called **ovaries**. Early in development, reproductive cells known as germ cells enter the testes and ovaries and give rise to diploid cells that will undergo gametogenesis.

Spermatogenesis in Mammals Figure 47.4a summarizes the events that take place during spermatogenesis inside tiny tubules in mammalian testes. Note that in the male gonad, diploid cells called **spermatogonia** (singular: **spermatogonium**) divide by mitosis. Some of the resulting cells continue to function as spermatogonia; others differentiate to form specialized cells that are committed to developing into sperm. As spermatogenesis continues, the cells move closer to the lumen of the tubules.

The specialized diploid cells produced by spermatogonia are called **primary spermatocytes**. They undergo meiosis I and produce two haploid **secondary spermatocytes**, which then undergo meiosis II. The result is four haploid cells called spermatids.

Each haploid spermatid develops into a **spermatozoan** (plural: **spermatozoa**), also commonly known as sperm—a cell that is specialized for carrying a haploid genome from the male through the female reproductive tract and fertilizing an egg. The spermatozoa are shed into the lumen of the tubules. The production of spermatogonia, primary spermatocytes, and sperm occurs continuously throughout a male's adult life.

Structure and Function of Sperm As a mammalian sperm develops, it acquires the four main compartments shown in Figure 47.5: the head, neck, midpiece, and tail.

- The head contains the nucleus and an enzyme-filled structure called the **acrosome**. The enzymes stored in the acrosome allow the sperm to penetrate the barriers surrounding the egg.
- The neck encloses a centriole that will combine with a centriole contributed by the egg to form a centrosome. The centrosome is required for formation of the spindle apparatus during mitosis (Ch. 12, Section 12.2).
- The midpiece is packed with mitochondria, which produce the ATP required to power motility (movement).
- The tail region consists of a **flagellum**—a long structure, composed of microtubules and surrounded by plasma membrane, that whips back and forth to make motility possible (Ch. 7, Section 7.6).

Sperm are stripped-down, streamlined cells that are specialized for racing other sperm to the egg. Eggs, in comparison, are larger, far less mobile storage containers that are packed with valuable materials to fuel embryonic development.

Oogenesis in Mammals Figure 47.4b highlights an important similarity between spermatogenesis and oogenesis: In the female gonad, diploid cells called **oogonia** (singular: **oogonium**) divide by mitosis. Some of the resulting cells continue to function

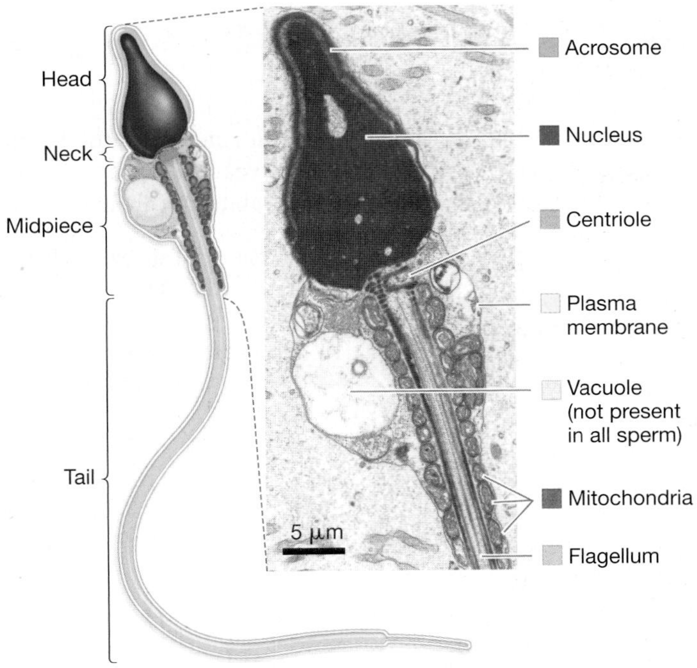

Figure 47.5 Mammalian Sperm Are Specialized for Motility and Fusion with an Egg. The morphology of human sperm is typical of many mammal species.

as oogonia; others differentiate to form specialized cells that are committed to developing into an egg.

However, subsequent steps in gametogenesis are markedly different in females. When the specialized cells produced by an oogonium, the **primary oocytes**, undergo meiosis, only one of the haploid cells produced can mature into an egg. This **secondary oocyte** is arrested in meiosis II until it is fertilized by a sperm. Only then does it complete meiosis to become an ootid, which matures into an ovum, or a mature egg.

To concentrate the cytoplasm and organelles into the secondary oocyte, the other cells produced by meiosis in females have a tiny amount of cytoplasm and do not mature into eggs. Because this distribution of cytoplasm is so unequal, the smaller cells are called **polar bodies**. Polar bodies usually degenerate shortly after their formation.

Another difference between gametogenesis in male and female mammals is that the production of primary oocytes stops early in development; in fact, it stops during fetal development, before females are born, and remains arrested for months, years, or decades, depending on the species. At sexual maturity, some of these primary oocytes will successively reenter the meiotic cycle to produce secondary oocytes and ova.

Structure and Function of Eggs Eggs are larger than sperm because eggs contain the organelles and nutrients required for the embryo's early development. In the vast majority of species, the mature egg cell is a membrane-bound structure consisting of a haploid nucleus, other organelles, and a large supply of nutrients provided by **yolk**—a fat- and protein-rich cytoplasm.

Just outside the plasma membrane of an egg, a fibrous, matlike sheet of glycoproteins called the **vitelline envelope** forms

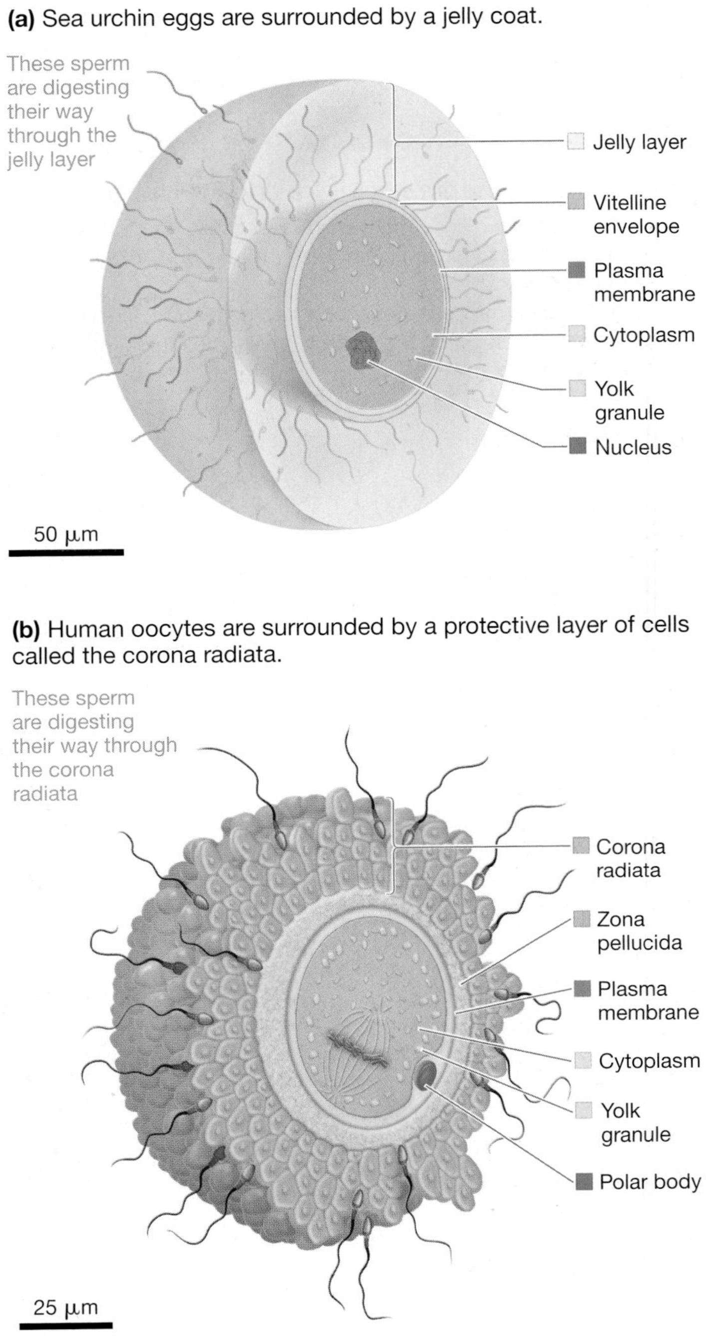

Figure 47.6 Eggs Are Surrounded by Protective Structures.

and surrounds the egg. In many aquatic animals, such as the sea urchin, a large, gelatinous matrix known as a jelly layer surrounds the vitelline envelope, further enclosing and protecting the egg (Figure 47.6a). The plasma membrane and surrounding layers play a key role in binding sperm from the same species and initiating fertilization.

In the eggs of humans and other mammals, the vitelline envelope is unusually thick and is called the **zona pellucida**. This structure is surrounded by a layer of cells known as the corona radiata, which a sperm must penetrate before it can fertilize the oocyte (Figure 47.6b).

The evolution of the amniotic egg, with its four membrane-bound sacs (Ch. 32, Section 32.5), was a key innovation. It allowed tetrapods to lay large eggs that do not dry out on land.

CHECK YOUR UNDERSTANDING

✔ **If you understood this section, you should be able to . . .**

1. Predict the conditions that would favor asexual reproduction in an animal that is capable of both asexual and sexual reproduction.
2. **MODEL** Diagram spermatogenesis and oogenesis, labeling the cells and showing their ploidy (designate diploid as $2n$ and haploid as n).

Answers are available in Appendix A.

47.2 Reproductive Structures and Their Functions

In the first section of this chapter, you considered the broad contrast between asexual and sexual reproduction, and you explored how animals produce sperm and eggs during sexual reproduction. Now let's explore the mechanics of sexual reproduction in more detail, starting with the anatomy of the male and female reproductive systems.

After you complete this section, you should be able to . . .

- Explain how reproductive structures in males and females assist fertilization.

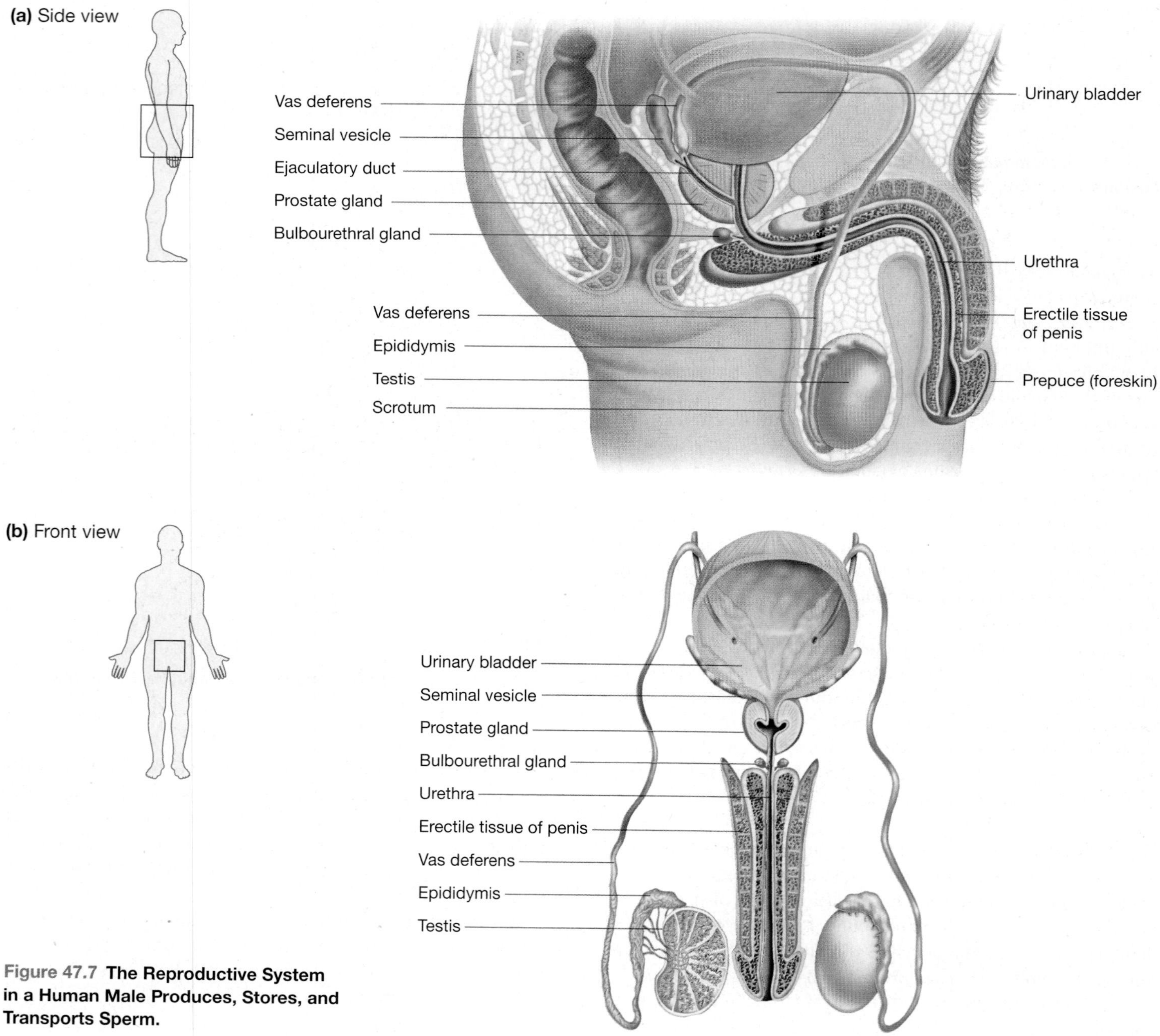

Figure 47.7 The Reproductive System in a Human Male Produces, Stores, and Transports Sperm.

The Male Reproductive System

Male reproductive anatomy consists of structures that produce sperm or accessory fluids, along with structures to deliver the sperm to the female reproductive tract. Let's examine these structures, focusing on human males.

Functional Anatomy of the Human Male Reproductive System

In humans, the external anatomy of the male reproductive system, or **genitalia**, consists of the scrotum and the penis. The saclike **scrotum** holds the testes; the **penis** functions as the organ of **copulation**, or sexual intercourse, necessary for internal fertilization.

Although it consists of many structures, the internal reproductive system in human males has just three basic functional components. Figure 47.7 shows the relevant structures in side view and front view.

1. ***Spermatogenesis and sperm storage*** Sperm are produced in the testes, and they are stored and become mature nearby in the **epididymis**.
2. ***Production of accessory fluids*** Complex solutions form in the **seminal vesicles**, **prostate gland**, and **bulbourethral glands**. These solutions, called accessory fluids, are added to sperm to form **semen**. Table 47.1 lists some components of the accessory fluids.
3. ***Transport and delivery*** A **vas deferens** is a tube with smooth muscle in its walls that transports sperm from the epididymis to the short **ejaculatory duct** within the prostate gland, where the sperm are mixed with accessory fluids. The resulting semen then enters the **urethra**, a longer tube that passes through the penis and services both the reproductive and urinary systems. The semen is expelled from the body during **ejaculation**.

Other Male Animals Exhibit Diverse Reproductive Anatomy

The composition of the accessory fluids varies widely among animals. In many insects, spiders, and vertebrates, molecules in the accessory fluids cause the semen to congeal after it arrives in the female reproductive tract, forming a plug. Experiments have shown that these copulatory plugs can serve as an effective deterrent to future matings. In some species, though, females or other males actively remove the plugs.

SUMMARY Table 47.1 **Accessory Fluids in Human Semen**

Source	Content	Function
Seminal vesicles	Fructose (a sugar)	Source of chemical energy for ATP synthesis fueling sperm motility
	Prostaglandins	Stimulate smooth muscle contractions in uterus
Prostate gland	Antibiotic compound	Prevents urinary tract infections in males
	Citric acid	Nutrient used by sperm
Bulbourethral glands	Alkaline mucus	Lubricate tip of penis; neutralize acids in urethra

Another diverse aspect of male reproductive anatomy has to do with a bone inside the penis called the **baculum**, which helps stiffen the penis during copulation. Some mammal species, including humans, lack a baculum. But in rodents, the variable shape of the baculum can be used to distinguish among species. And in seals, baculum size correlates with mating system—the mating practices observed in a species. For example, in seal species in which females routinely mate with several males before becoming pregnant, males have not only large testes for their size but also a large baculum. The testes and baculum are much smaller in species in which females mate with a single male.

The Female Reproductive System

In animals, oogenesis occurs in the ovaries. Let's consider two highly specialized examples of the functional anatomy of female reproductive systems, those of birds and mammals. Birds are **oviparous** ("egg-bearing") animals that lay an amniotic egg protected by a hard shell; most mammals are **viviparous** ("live-bearing"), and embryonic development takes place entirely within the mother's body.

Anatomy and Function of the Reproductive Tract of Female Birds

Figure 47.8 describes the events that take place as an egg

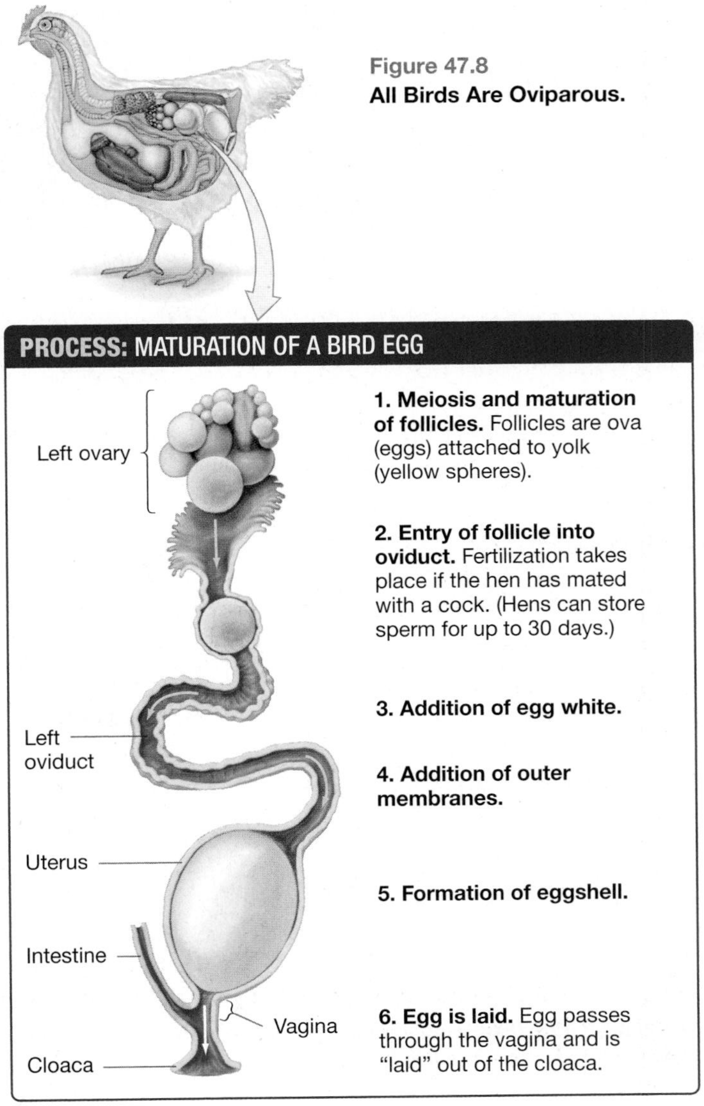

Figure 47.8
All Birds Are Oviparous.

moves through the reproductive tract of a female bird. The result is a hard-shelled egg, such as the chicken eggs you cook for breakfast. The egg is "laid" from the **cloaca** (pronounced *kloh-AY-kuh*)—a chamber that the reproductive, digestive, and excretory systems flow into and that opens to the environment (Ch. 40, Figure 40.15).

From the time the egg is released from the ovary until the zygote undergoes mitosis and the embryo begins to develop, a bird egg is a single cell. The ostrich egg, which can be over 15 cm (6 inches) in diameter, contains one of the largest single cells known in animals. A bird egg stores enough nutrients and water to sustain development until hatching.

Note that although male birds have two testes, females of most bird species have just one functional ovary. The presence of a single working ovary is thought to be an adaptation that reduces weight to make flight more efficient.

Functional Anatomy of the Human Female Reproductive System Figure 47.9 shows side and front views of the human female reproductive system. Female genitalia include the **labia minora** (singular: **labium minus**) and the **labia majora** (singular: **labium majus**), the clitoris, the opening of the urethra, and the opening of the vagina.

- The labia are folds of skin that cover the urethral and vaginal openings.
- The **clitoris** is an organ that functions in sexual arousal; it develops from the same population of embryonic cells that gives rise to the penis in males. Note that the medial section perspective of Figure 47.9 makes the clitoris appear small, but it actually consists of extensive internal tissue. The clitoris becomes erect during sexual stimulation and is covered with a protective sheath called the *prepuce*.

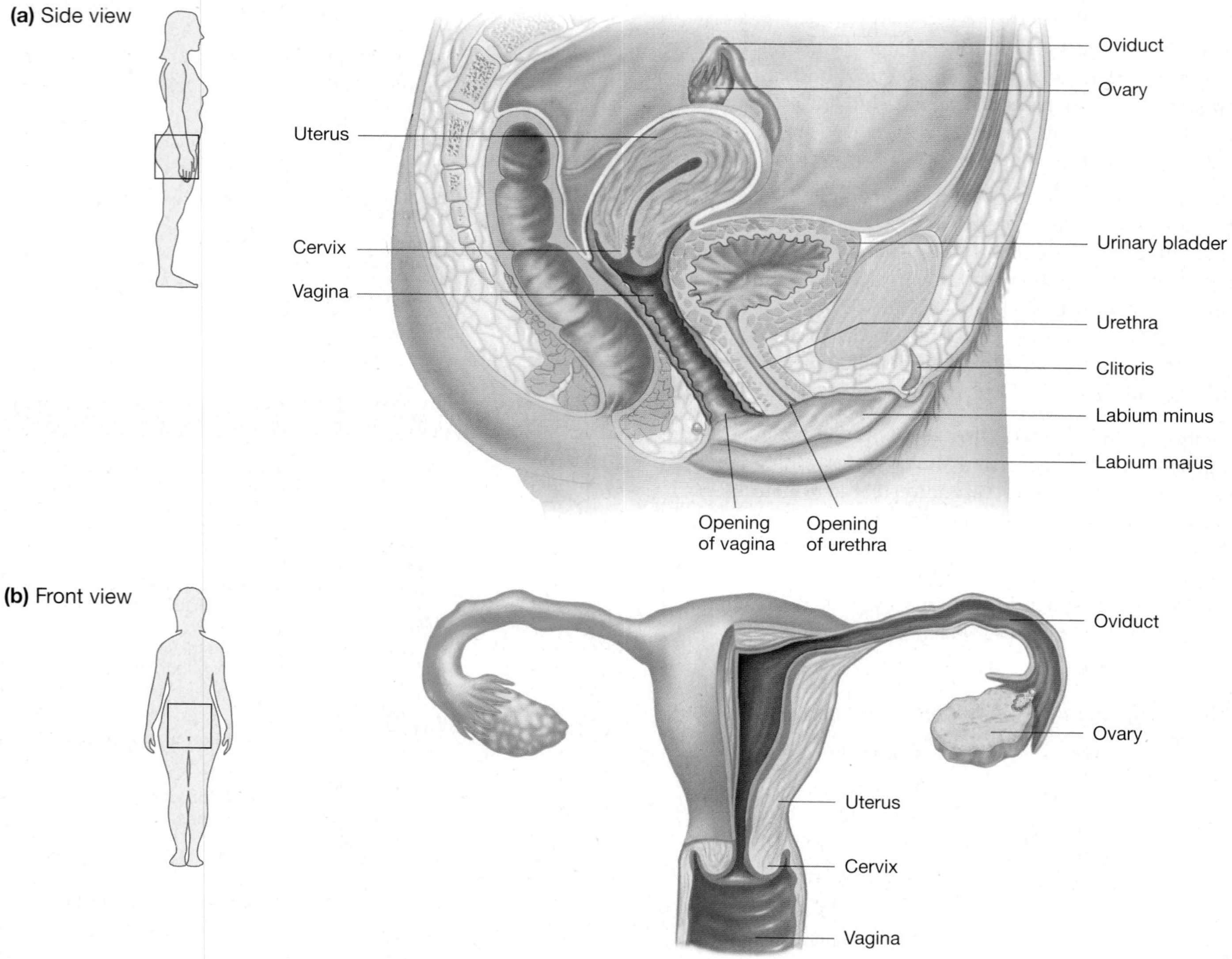

Figure 47.9 **The Reproductive System in a Human Female Produces Eggs and Nurtures the Embryo and Fetus.**

✔ **PROCESS OF SCIENCE** Generate a hypothesis to explain why male and female gonads are paired in mammals. How would you test your hypothesis?

- The urethral opening, where urine is expelled, is separate from the reproductive structures.
- The **vagina** is the muscular tube where semen is deposited during sexual intercourse. The vagina is also referred to as the birth canal because the baby moves out of the uterus and through the vagina during childbirth.

The internal structures of the female reproductive system in humans and other mammals serve two basic functions:

1. ***Production and transport of eggs*** Eggs are produced in the paired ovaries. During **ovulation**, a secondary oocyte is expelled from the ovary and enters the **oviduct**, also known as the **fallopian tube** or uterine tube, where fertilization may take place. The fertilized egg is transported through the oviduct to the muscular sac called the **uterus**, where it may implant.
2. ***Development of offspring*** The uterus is where embryonic development takes place. During childbirth, the baby passes through an opening in the **cervix**—the bottom part of the uterus—and into the vagina.

How do sperm reach the egg, and how is the egg fertilized? These questions are addressed in the next section.

CHECK YOUR UNDERSTANDING

✔ If you understood this section, you should be able to ...

1. List in order the structures through which sperm pass from spermatogenesis to ejaculation.
2. List in order the structures through which an ovum and subsequently a fetus pass from oogenesis to childbirth.

Answers are available in Appendix A.

47.3 Fertilization and Egg Development

After you complete this section, you should be able to...

- Explain fertilization and egg development.

Fertilization marks the beginning of the remarkable events of animal development, the set of processes that allow the zygote to develop into a multicellular organism. Development is part of a cycle that connects one generation to another: Gametes formed by adults of one generation unite in fertilization, creating a new individual that develops into an adult of the next generation. In this way, development and reproduction are inextricably linked.

Let's begin by examining different strategies for fertilization, the cellular mechanisms of fertilization, and different approaches to supplying the developing embryo with nutrients. Many species use external fertilization, the release of gametes into the environment followed by fertilization outside the mother. In other species, sperm are deposited into the reproductive tract of the female, and internal fertilization occurs.

External Fertilization

Most animals that rely on external fertilization live in aquatic environments. This is logical, because gametes and embryos must be protected from drying. If external fertilization occurred in a terrestrial environment, either the gametes or the resulting zygote would likely die of desiccation.

Animals that use external fertilization also tend to produce huge numbers of gametes. For example, female sea stars release millions of eggs into the surrounding seawater during spawning (Figure 47.10), and males release many times that number of sperm. This strategy increases the probability of a sperm and an egg meeting in an ocean or lake, and it ensures that some offspring will survive even after most of them fall prey to predators.

Sperm and eggs from different individuals must be released into the environment synchronously by males and females for external fertilization to work. How is gamete release coordinated? The answer has two parts.

1. Gametogenesis in both the male and female usually occurs in response to environmental cues, such as lengthening days and warming water, that indicate a favorable season for breeding.
2. Gametes are released in response to specific cues from individuals of the same species.

In fishes and other aquatic animals with well-developed eyes, external fertilization by spawning is often the culmination of an elaborate courtship ritual between a male and a female. In other species, chemical messengers called pheromones (Ch. 44, Section 44.4) are involved in synchronizing gamete release.

Figure 47.10 External Fertilization Occurs in Aquatic Environments. Female sea stars in the genus *Asterias* can release as many as 100 million eggs at once.

Internal Fertilization

The vast majority of terrestrial animals as well as a significant number of aquatic animals use internal fertilization. It occurs in one of two ways:

1. After copulation, in which males deposit sperm directly into the female reproductive tract with the aid of a copulatory organ, usually called a penis.
2. After males package their sperm into a structure called a **spermatophore**, which is then picked up and placed into the female's reproductive tract by the male or the female. In some salamander species, for example, the male places the spermatophore on the ground within its territory. Later the female picks it up with her cloaca. In this case, the result is internal fertilization without copulation.

Sperm Competition and Second-Male Advantage An important insight about internal fertilization came from experiments on dung flies conducted by Geoff Parker. In 1970 Parker confirmed the existence of **sperm competition**—competition between sperm from different males to fertilize the eggs of the same female.

Parker's experiments consisted of a series of matings of one female with two males. In each experiment, Parker selected the two males in such a way that he could distinguish between their offspring. He found that the proportion of offspring fathered by each male was not 50:50. Instead, whichever male was last to copulate fathered an average of 85 percent of the offspring produced.

Besides suggesting that sperm were competing to fertilize eggs, these data indicated that, in this experiment, the second male won. Follow-up research has confirmed that **second-male advantage** is widespread, although not universal, in insects and some other animal groups. How does it occur?

To explore why second-male advantage occurs, biologists turned to the fruit fly *Drosophila melanogaster*, a model organism for studies of reproduction and development (BioSkills 11). One research group introduced a gene into male fruit flies that caused them to produce sperm with green fluorescent tails (Figure 47.11). When a female mated with a green-spermed male and then mated with a male with normal-colored sperm, the researchers found that the number of green sperm observed in the female's sperm-storage area was far lower than when no second mating took place.

To interpret this finding, the biologists suggested that the sperm of the second male physically dislodged the first male's sperm from the female's sperm-storage area. The researchers also demonstrated that the fluid that accompanies sperm during fertilization can displace stored sperm deposited by other males. Together, these mechanisms result in the second male's sperm fertilizing most of the eggs laid.

Why Is Testis Size Variable among Species? Research on sperm competition has recently contributed another major finding. In species where females routinely mate with multiple males before laying eggs or giving birth, males have extraordinarily large testes for their size and produce huge numbers of sperm.

Fertilization is similar to a lottery in which sperm represent tickets. The more tickets a male enters in the competition, the

RESEARCH

QUESTION: Does the "second-male advantage" occur in sperm competition?

HYPOTHESIS: In sperm-storage areas, sperm from the second male displace sperm from the first male.

NULL HYPOTHESIS: The mechanism does not involve sperm displacement from sperm-storage areas.

EXPERIMENTAL SETUP:

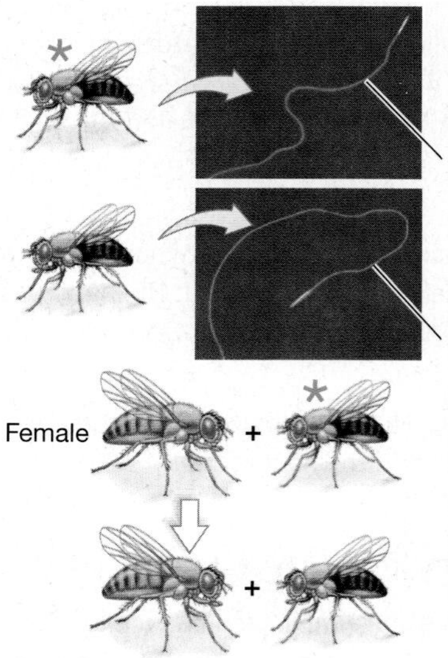

1. Introduce a gene into some male fruit flies to mark their sperm with green fluorescent protein.

2. Allow female fruit flies to mate with green-spermed males.

3. Allow some of those females to mate with males having normal-colored sperm. Observe sperm in all females' storage areas.

PREDICTION OF HYPOTHESIS: When females mate twice, little sperm from the first male remains in storage.

PREDICTION OF NULL HYPOTHESIS: When females mate twice, most or all of the sperm deposited by the first male is still present.

RESULTS:

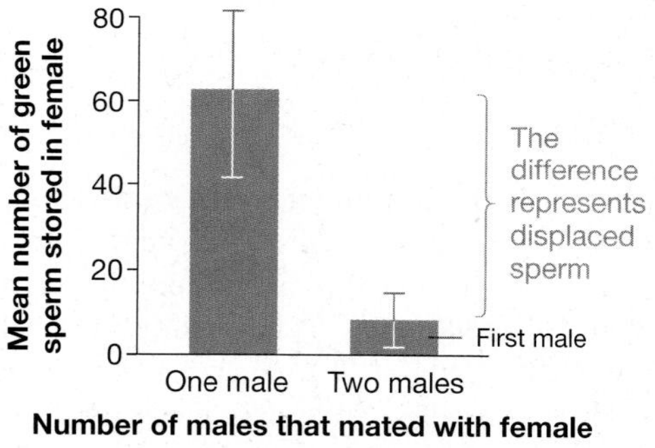

CONCLUSION: If a female mates a second time, most sperm from the first male she mated with disappears.

Figure 47.11 Experimental Evidence Supports Second-Male Advantage in *Drosophila*. The graph shows the average number of green sperm stored in females when males with green sperm were the only male to mate or the first of two males to mate.

SOURCE: C. S. C. Price, K. A. Dyer, and J. A. Coyne. 1999. Sperm competition between *Drosophila* males involves both displacement and incapacitation. *Nature* 400: 449–452.

✔ Do these data show that sperm from the second male physically displace green sperm from the first male?

(a) Long genital spines in males of the seed beetle *Callosobruchus maculatus*

(b) During sperm competition, males with longer genital spines father more offspring.

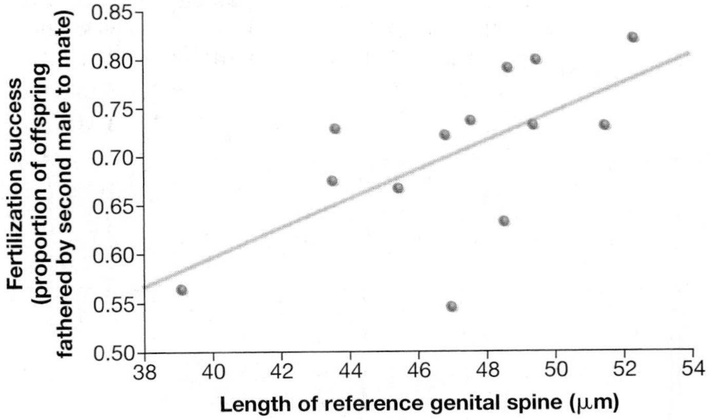

Figure 47.12 **In Some Insects and Spiders, Individual Variation of Male Genitalia May Affect Reproductive Success. (a)** An elaborate male reproductive structure in a seed beetle, used to transfer sperm to the female. The spikes at the top of the inset photo are called genital spines. **(b)** Data on the size of reference genital spines from experiments on reproductive success during sperm competition.

DATA: C. Hotzy and G. Arnqvist. 2009. *Current Biology* 19: 404–407.

higher is his chance of "winning" fertilizations and passing his alleles on to the next generation. The larger the testes, the more sperm produced, and the more likely they are to win the lottery.

The lottery model has been challenged, however, by evidence that females often store sperm and exert control over which sperm are successful in fertilization. In other words, females do not always accept the results of sperm competition passively. They do this by physically ejecting sperm from less desirable males. This phenomenon has been dubbed cryptic female choice, an appropriate name because the selection of sperm by females is hidden from males.

How Does External Anatomy Affect Sperm Competition? The structure of male genitalia varies greatly among animal species. In many groups of insects and spiders, for example, closely related species are morphologically identical except for their distinctive genitalia. Why are these organs so diverse?

The leading hypothesis is that certain shapes of genitalia give males an advantage in sperm competition. To test this hypothesis, biologists measured the size of the spines found on the tips of male genitalia in a species of seed beetle (Figure 47.12a) and then documented which males were most successful at fertilizing eggs when females mated with two males.

The *x*-axis in Figure 47.12b plots the length of genital spines; the *y*-axis plots the proportion of eggs fertilized by the second male to copulate with a female during experimental matings. The line through the data points shows that second-to-mate males with longer genital spines tend to father a higher percentage of offspring. The long spines stick in the female's reproductive tract and prolong copulation time, even if the female is ready for copulation to end. The upshot? Size matters, in seed beetles. Natural selection that occurs during sperm competition may explain why genitalia are so diverse among insect and spider species.

Now that you have explored how the size and shape of male external genitalia are related to sperm competition and reproductive success, let's examine what happens when sperm are released into the female reproductive tract and begin their hunt for the egg.

The Cell Biology of Fertilization

The first challenge of fertilization is for a sperm to find the egg. In many species, sperm accomplish this by following a concentration gradient of chemical signals secreted by the egg. In species that use external fertilization, there's an additional challenge: ensuring that only sperm and egg of the same species come together. Once in contact, the sperm and egg have to fuse, something that few other cells in the body ever do.

What's more, in most species, thousands of sperm can compete to fertilize an egg, but fusion must be limited to a single sperm so the zygote does not receive extra chromosomes and exceed the diploid number ($2n$). Finally, the fusion of the two gametes has to trigger the onset of development.

Research on fertilization began in earnest early in the twentieth century, when biologists started to study the sperm–egg interaction that occurs during external fertilization in sea urchins. Figure 47.13 on page 1014 outlines major events of this process.

1. Sea urchin sperm find the egg by following a gradient of a chemical released from the egg's jelly layer.
2. The head of the sperm initially binds to the jelly layer.
3. Binding triggers the acrosome reaction, a release of the contents of the sperm's acrosome. The enzymes that were contained in the acrosome digest nearby portions of the jelly layer and the vitelline envelope.

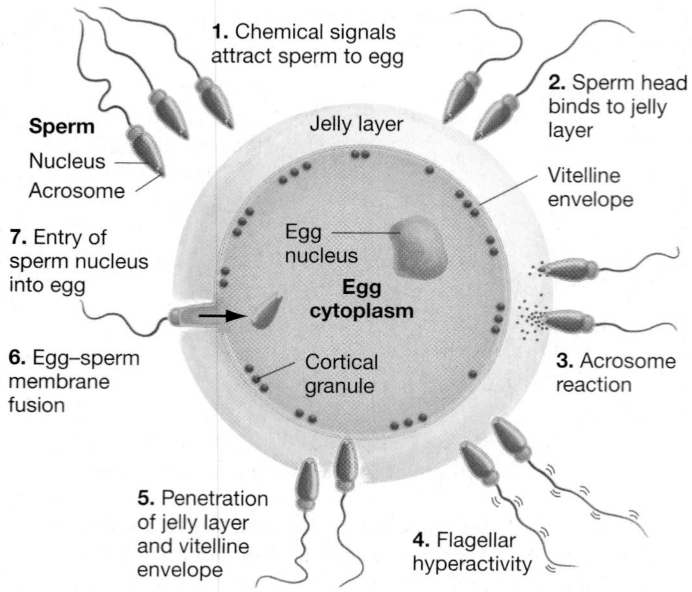

Figure 47.13 Fertilization Is a Complex Process. Steps in the fertilization of a sea urchin egg are shown in this example.

4. Initial contact with the jelly layer also ramps up the beating of the sperm's flagellum, powering the sperm toward the egg's plasma membrane.
5. The sperm head contacts the egg's plasma membrane.
6. Proteins in the plasma membranes of the egg and sperm induce membrane fusion.
7. The sperm nucleus then enters the egg, and the sperm and egg nuclei fuse, thereby reestablishing the diploid number. The egg is now activated and ready for development—fertilization is complete.

In marine habitats, sperm and eggs from different species can mix together in seawater. What prevents cross-species fertilization and the consequent production of dysfunctional hybrid offspring?

How Do Gametes from the Same Species Recognize Each Other? In the 1970s, Victor Vacquier and co-workers succeeded in identifying a protein on the head of sea urchin sperm that binds to the plasma membrane of sea urchin eggs in a species-specific manner. They called this protein bindin.

Follow-up work showed that bindin proteins from even very closely related species of sea urchin are distinct, and that egg plasma membranes contain receptors that bind only to bindin from sperm of the same species. Similar types of specific protein–protein interactions occur between the sperm and eggs of mammals.

✔ **If you understand the importance of protein–protein interactions in fertilization, you should be able to explain why adding a molecule that binds to a bindin-like protein on the human sperm head would be an effective contraceptive.**

What Prevents More than One Sperm from Entering the Egg? Researchers found that even when hundreds of sperm are clustered around an egg, only one sperm participates in fertilization. This observation makes sense: Multiple fertilization, or polyspermy, would result in a zygote that had extra copies of each chromosome, and embryos with the wrong number of chromosomes usually die. How do animals avoid polyspermy?

Sea urchins have two mechanisms that block polyspermy. The first begins immediately after sperm–egg contact and involves depolarization of the egg's plasma membrane—changing the membrane potential from negative to positive (Ch. 43, Section 43.1). Additional sperm cannot fuse with the egg while it is depolarized, so polyspermy is prevented. However, the depolarization lasts for only about one minute, so this mechanism is short lived.

The second block to polyspermy, one seen in most animals, is erection of a physical barrier to sperm entry. This process begins when entry of the sperm nucleus causes calcium ions (Ca^{2+}) to be released from the egg's endoplasmic reticulum into the cytoplasm. As Figure 47.14a shows, a wave of Ca^{2+} release starts at the point of sperm nucleus entry and spreads throughout the egg.

The egg contains thousands of vesicles called cortical granules that lie just beneath the plasma membrane (see Figure 47.13). The release of Ca^{2+} causes the cortical granules to fuse with the plasma membrane and release their contents by exocytosis to the exterior of the cell. This is the same mechanism of Ca^{2+}-mediated exocytosis as that used when neurons release neurotransmitters at synapses (Ch. 43, Section 43.3).

The cortical granules contain proteases that digest the exterior-facing portion of the receptors for sperm. This prevents any new sperm from binding to the egg surface. In addition, ions and other compounds released by the cortical granules accumulate between the egg's plasma membrane and the vitelline envelope. These concentrated solutes cause water to diffuse into the space between the plasma membrane and the vitelline envelope. The influx of water lifts the vitelline envelope away from the cell. Compounds from the cortical granules cross-link molecules in the vitelline envelope to form a tough **fertilization envelope** (Figure 47.14b). This impenetrable barrier keeps additional sperm from reaching the sea urchin egg.

Although mammalian eggs do not produce a fertilization envelope, they do experience a wave of Ca^{2+} ions that triggers the release of enzymes from cortical granules. Similar to events in the sea urchin egg, these enzymes destroy the egg's receptors for sperm and modify proteins on the egg surface and on a structure similar to the vitelline envelope, blocking polyspermy by preventing the binding of additional sperm once the egg is fertilized.

Now that you've read about how fertilization takes place, let's look at where development occurs.

Why Do Some Females Lay Eggs, while Others Give Birth to Live Offspring?

In many oviparous species that release or deposit their eggs into the environment, such as sea stars, sea urchins, and most insects, the parents provide no further care. Birds, however, incubate their eggs and feed the young after hatching; fishes may guard their eggs from predators and fan the eggs to oxygenate them.

In viviparous species, the embryo attaches to the reproductive tract of the mother and receives nutrition and exchanges gases directly with her circulatory system. In **ovoviviparous** species, offspring also develop inside the mother's body but are nourished by nutrient-rich yolk stored in the egg.

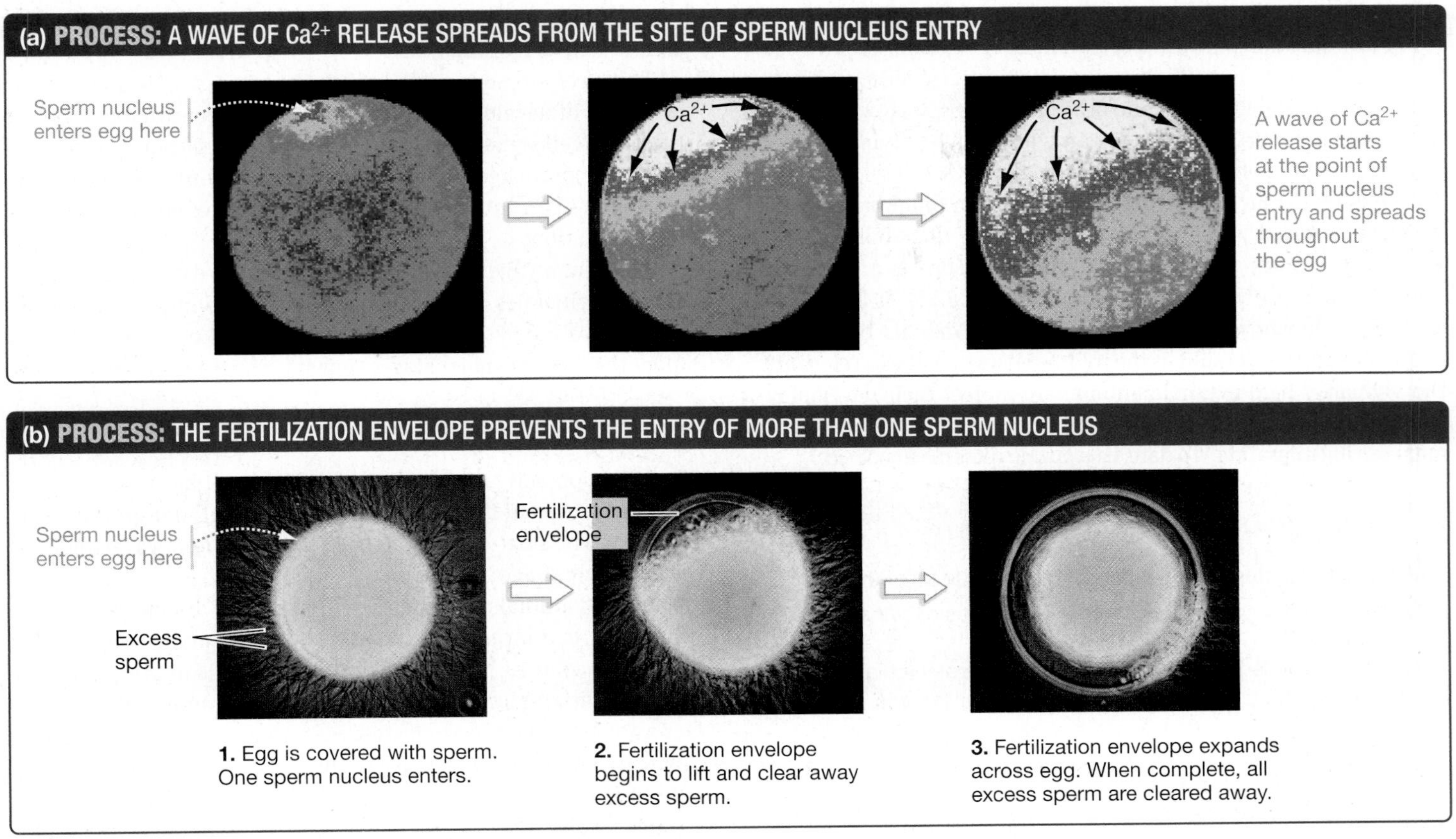

Figure 47.14 A Physical Barrier Erected After Fertilization Prevents Polyspermy. The events of sea urchin fertilization are shown here. **(a)** During fertilization, a wave of Ca^{2+} release begins at the point of sperm nucleus entry and spreads under the plasma membrane across the egg in about 30 seconds. Colors represent Ca^{2+} concentration, where white is highest and blue lowest. The Ca^{2+} concentration was measured with an injected reagent that fluoresces in the presence of Ca^{2+}. **(b)** In response to the increased Ca^{2+} concentration, a fertilization envelope arises in about 40 seconds and clears away excess sperm.

Why does oviparity exist in some groups and viviparity or ovoviviparity in others? Biologists tackled this question by studying the lizard genus *Sceloporus*. Some *Sceloporus* populations are oviparous; others are ovoviviparous. The biologists analyzed a phylogenetic tree—based on molecular and morphological data—of many *Sceloporus* species (**Figure 47.15**). Two conclusions should make sense to you (see **BioSkills 13** for help with interpreting phylogenetic trees):

1. Because the basal branches of the tree represent oviparous species, egg laying probably represents the original, or ancestral, condition.
2. As the red branches on the tree show, ovoviviparity evolved independently in two groups.

Biologists have also found that ovoviviparous populations of sea stars have evolved from oviparous populations in some instances.

Why did natural selection favor these changes between egg laying and live birth? Researchers have hypothesized that natural selection should favor live birth in cold habitats. Low temperatures slow the development of embryos. Therefore, in cold habitats, it might be advantageous for females to retain eggs inside their bodies so that they can behaviorally thermoregulate to maintain the offspring at a more favorable temperature for development (Ch. 39, Section 39.5). This hypothesis is supported by the fact that ovoviviparous *Sceloporus* species live in the highlands of the southwestern United States and central Mexico, where temperatures can be very low.

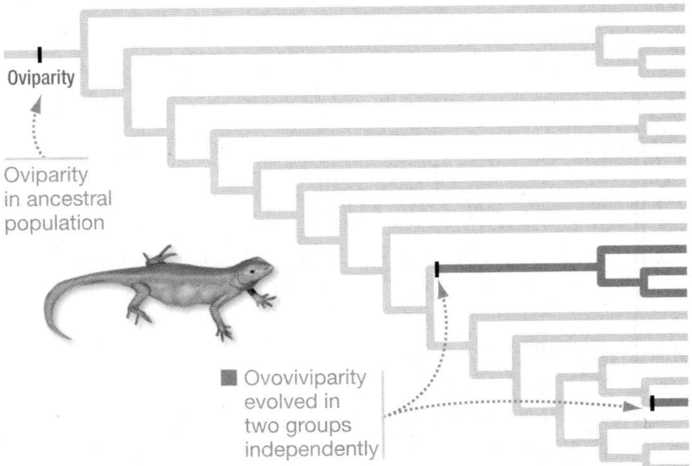

Figure 47.15 Ovoviviparity Has Evolved More than Once in *Sceloporus* Lizards. Each branch on this phylogenetic tree represents a *Sceloporus* species or population from central Mexico.

Sceloporus lizards have been a productive group in which to study this question because the trait has evolved more than once in extremely closely related species—making it easier to find correlations between live birth and other characteristics, such as living in cold habitats. Indeed, in lizards as a whole, viviparity has evolved independently at least 100 times.

The issue is much more difficult to study in mammals, where viviparity evolved just once. Monotremes (the duck-billed platypus and echidnas) are oviparous, while marsupials and eutherian mammals are viviparous (Ch. 32, Section 32.5). It is still not clear why, and how, viviparity evolved in mammals before the marsupial–eutherian split. Is the cold-habitats hypothesis relevant? Did viviparity help mammalian mothers protect their developing young from parasites and predators, or control the fetal environment during migration? To date, this puzzle remains unsolved.

CHECK YOUR UNDERSTANDING

✔ If you understood this section, you should be able to . . .

1. Predict some advantages and disadvantages of oviparity and viviparity.
2. Explain why fertilization is blocked when a sea urchin egg is treated with a drug that allows Ca^{2+} to enter the egg.

Answers are available in Appendix A.

47.4 Embryonic Development

After you complete this section, you should be able to . . .

- Explain the stages of embryonic development.

"Making Babies: 25,000 Genes, Some Assembly Required." This wry chapter title written by evolutionary developmental biologist Sean Carroll highlights one of the most remarkable events of life. How *is* a baby—or any other animal—assembled from a single cell?

You have already learned about the basic genetic and cellular processes of development (Chapter 21). The central message was that cell–cell signaling causes different sets of transcription factors to be produced in various cells throughout the embryo, resulting in differential gene expression and thereby enabling cell differentiation.

The goal of this section is to look in more detail at how these general principles apply to embryonic development—in essence, what's involved in "some assembly required." Let's follow development through the major stages shown for humans in Figure 47.16.

Cleavage

Fertilization activates development. One cell becomes two, two become four, and the beginnings of a multicellular organism are in place. The stage of rapid cell divisions that follows fertilization is called **cleavage**.

In most animals, cleavage distributes the cytoplasm present in the egg into a larger and larger number of smaller and smaller daughter cells. As a result, the number of cells in the embryo increases without any overall growth of the embryo. The cell divisions of cleavage are the fastest that ever occur. For example, in fruit fly embryos, mitotic divisions occur every 10 minutes, resulting in the production of about 50,000 cells in half a day.

The cells that are created during cleavage are called **blastomeres** (literally, "bud-parts"). When cleavage is complete, the embryo in many animals consists of a mass of blastomeres called a **blastula** ("little-sprout" or "bud").

In humans and almost all other mammals, fertilization takes place in the oviduct, and cleavage occurs as the embryo moves down the oviduct toward the uterus (Figure 47.17). After implantation, the embryo completes its development in the uterus.

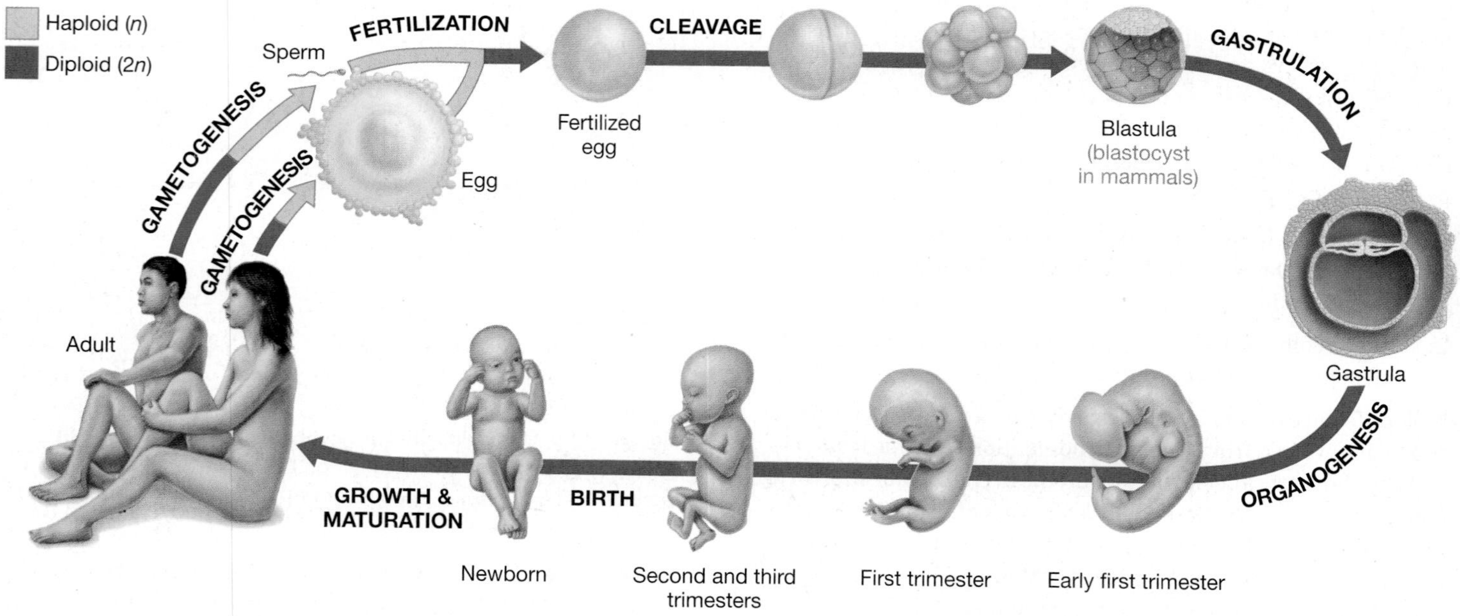

Figure 47.16 Human Development Proceeds in Ordered Phases.

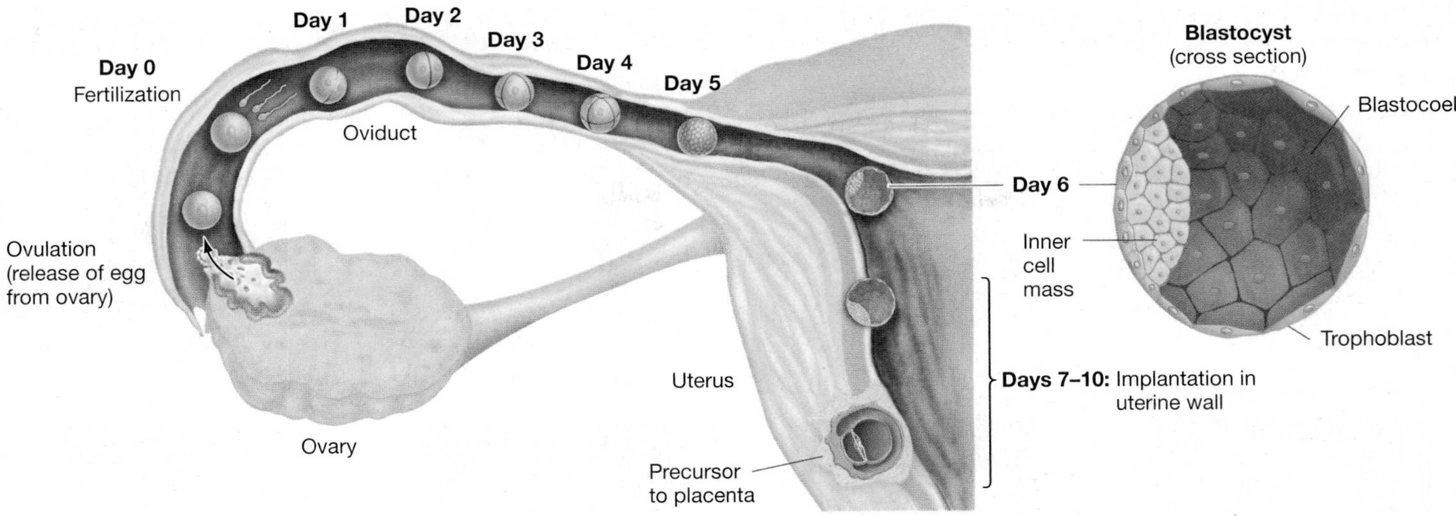

Figure 47.17 **In Most Mammals, Cleavage Occurs before the Embryo Reaches the Uterus.** Human cleavage and a human blastocyst are shown here.

Cleavage in mammals results in a type of blastula called a **blastocyst** ("sprout-bag"), which has two populations of cells. The exterior of the blastocyst is a thin sheet of cells called the **trophoblast** ("feeding-sprout"). Inside the trophoblast is a fluid-filled cavity (**blastocoel**) and a cluster of cells called the **inner cell mass (ICM)**. There is a distinction between the trophoblast and the ICM—the embryo develops from the ICM, and the trophoblast forms part of an organ called the **placenta**. Cells from the mother's uterus also form part of the placenta. A key innovation in the evolution of mammals, the placenta exchanges nutrients, wastes, and gases between the mother and the embryo.

Gastrulation

As cleavage nears completion, cell division slows and cell movement accelerates. In this stage of development, called **gastrulation**, extensive and highly organized cell migrations and changes in cell shape rearrange the embryonic cells into a structure called the **gastrula**.

Intensive research on this phase of development started in the 1920s with efforts to follow the migration of individual cells during gastrulation in newts and frogs. In these early experiments, tiny blocks of agar (a gelatinous compound) were soaked with a nontoxic dye. The dyed blocks were then pressed against the surface of blastula-stage embryos so that a small number of blastomeres became marked with dye. By allowing marked embryos to develop and then examining them at intervals during gastrulation, researchers were able to follow the migration of cells. Today, more sophisticated ways are used to mark individual cells and track them throughout the embryo, but the principle of marking and following cells remains the same.

During gastrulation, the basic patterns and structures of the animal body are formed. Gastrulation produces the three germ layers that form the basis of later structures and creates the body axes that determine directional patterns. Figure 47.18 on page 1018 shows the dramatic reshaping of a frog embryo during gastrulation.

Formation of Germ Layers The pattern of gastrulation varies widely among animal species, but the general outcome is the same: Gastrulation forms three embryonic tissue layers. A tissue is a group of cells that function as a unit (Ch. 39, Section 39.2).

Most animal embryos have three primary (first) tissue layers: **(1) ectoderm** ("outside-skin"), **(2) mesoderm** ("middle-skin"), and **(3) endoderm** ("inner-skin"). These embryonic tissue layers are called **germ layers** because they give rise to all the organs and tissues of the adult. Figure 47.18 shows two views of how cell migrations during gastrulation form the three germ layers in a frog embryo. Cells that will become ectoderm are shown in blue; cells destined to form mesoderm are shown in pink, and cells that will form endoderm are shown in yellow.

Step 1 The fates of different regions of the frog blastula are specified by cytoplasmic determinants and signals. This blastula contains a fluid-filled interior space called the blastocoel, which is present in most animal embryos.

Step 2 As gastrulation begins, an invagination (indentation) forms on the outer surface as cells change their shape. In frogs, this invagination starts out as a slit that eventually forms a circular opening known as the **blastopore**.

Step 3 Cells from the surface fold into the interior of the embryo through the blastopore, forming a tube that extends across the embryo. The tube will become the gut, or digestive tract. The blastocoel is displaced and eventually disappears.

Step 4 The migration of cells into and across the embryo ultimately results in the formation of the three germ layers: endoderm on the inside, mesoderm in the middle, and ectoderm on the outside.

✔ **If you understand gastrulation in a frog embryo, you should be able to describe how the ectoderm comes to completely cover the embryo.**

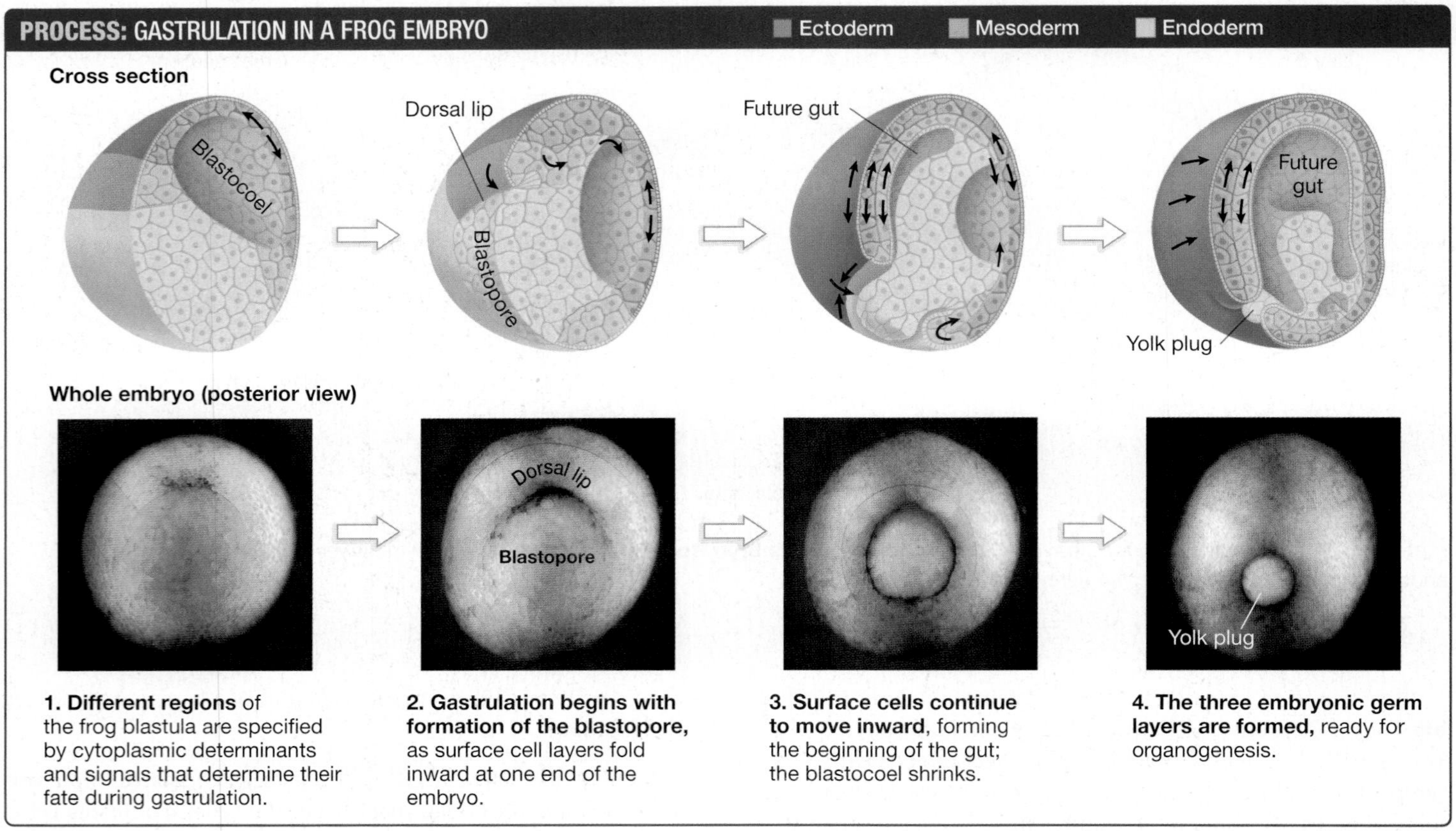

Figure 47.18 **Gastrulation Creates Head-to-Tail and Back-to-Belly Axes and Three Germ Layers in the Embryo.** Gastrulation in a frog is shown here. Important elements vary between species, but in every animal embryo, gastrulation requires the coordination of a remarkable set of cell migrations and cell-shape changes.

Each germ layer forms certain tissues and organs. Remarkably, the correspondence between germ layer and organ type is the same in most animals, even those as distantly related as people and flies. Figure 47.19 shows some of the organs that the germ layers produce in a human. Ectoderm forms the outer covering of the body and the nervous system; mesoderm produces muscle, most internal organs, and connective tissues such as bone and cartilage; endoderm produces the inner lining of the digestive tract and of the many organs that develop from the gut, such as the liver and lungs.

Creating Body Axes Besides establishing the germ layers, gastrulation has another critical role: creating the body axes. In frogs, for example, the blastopore becomes the anus (posterior), and the opposite end of the gut tube becomes the mouth (anterior); the region where cells first move into the blastopore (see Figure 47.18) defines the dorsal, or back, side of the embryo, and the opposite region becomes the ventral side. In this way, the anterior–posterior and dorsal–ventral axes of the body become apparent as gastrulation proceeds.

However, well before they are revealed, the major body axes in frogs and many other animals are already at least partially determined. Determination occurs either through the action of regulatory molecules called cytoplasmic determinants and/or through interactions between cells, a process known as induction. Determination ultimately results from differential gene expression (Ch. 21, Section 21.2).

Organogenesis

When gastrulation ends, the outside, inside, and middle layers are in place, and the positions of the head, tail, back, and belly are apparent—but this is only a start. For one thing, there are no organs within the embryo. The heart, brain, liver, and lungs all need to be formed from the germ layers, properly positioned, and connected with other organs. **Organogenesis** ("organ-origin") gets these jobs done. During organogenesis, cells divide, move, differentiate, and assemble into tissues and organs using instructions encoded in the genome, with guidance from signals sent by other cells (Chapter 21).

Initial Formation of the Notochord, Neural Tube, and Somites

To understand the genesis of any organ, it is necessary to trace back through many earlier structures, some of which exist only in the embryo. Let's begin with the assembly of key embryonic structures in chordates (the group of animals that includes humans and all other vertebrates): the notochord, neural tube, and somites.

Figure 47.20 shows the developmental path to these structures. Although this figure illustrates what happens in a frog embryo, similar events occur in the embryos of chickens, humans, and all other vertebrates.

First, a rod-like element called the **notochord** forms from mesodermal cells soon after gastrulation is complete (Figure 47.20, step 1). Molecular signals produced in the

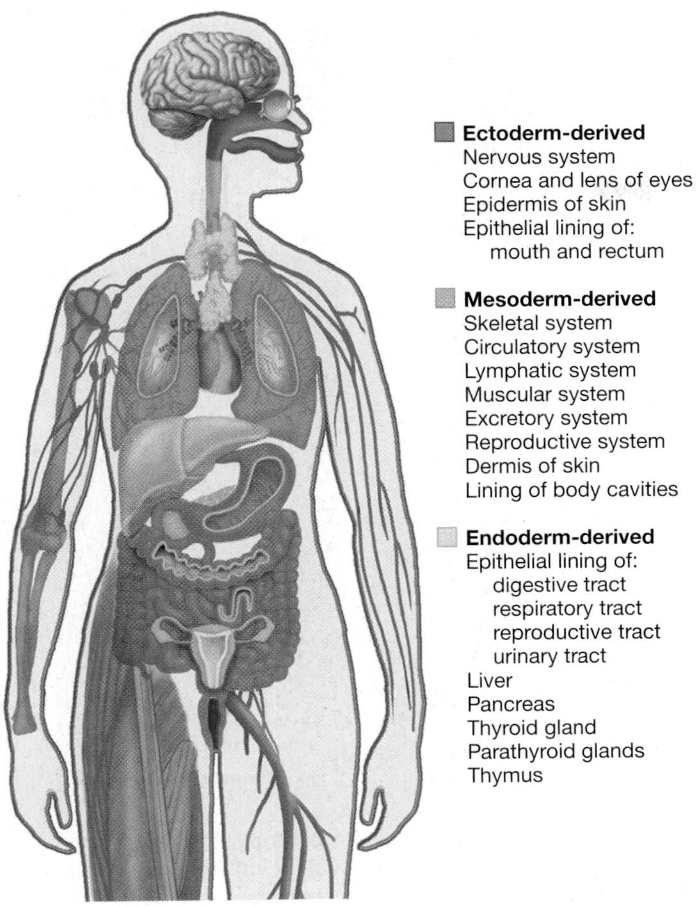

Figure 47.19 **The Three Embryonic Germ Layers Give Rise to Different Adult Tissues and Organs.** Each germ layer forms the same types of organs in all animals.

notochord induce the dorsal ectoderm to invaginate, or fold (Figure 47.20, step 2). This folding forms the **neural tube**, a tube of ectoderm that runs along the dorsal midline (middle back) of the embryo and gives rise to the brain and spinal cord (Figure 47.20, step 3). As organogenesis continues, mesodermal cells near the notochord become organized into **somites** (Figure 47.20, step 4).

Somites are paired blocks of mesodermal tissue that extend along either side of the dorsal midline of the embryo (Figure 47.21, page 1020). Somite formation is a response to changes in the cell adhesion molecules that keep mesodermal cells attached to each other (Ch. 11, Section 11.2).

Note that the notochord shown in Figures 47.20 and 47.21 is unique to the chordates. In some species of chordates, the notochord is a long-lasting structure that functions as a simple internal skeleton—it stiffens the body and makes efficient swimming movements possible. But in vertebrates, such as chickens, frogs, and humans, the notochord is transient: It appears only in embryos. As organogenesis proceeds in vertebrates, many of the cells in the notochord undergo programmed cell death by apoptosis (Ch. 21, Section 21.4).

Somites Differentiate into Skin, Bone, and Muscle Somites are transient structures, just like the notochord is. But unlike the notochord, somites produce many important structures of the adult. By marking somite cells and following them over time, researchers discovered that somites give rise not only to muscle but also to the lower (dermal) layer of the skin and much of the skeleton (Figure 47.22, page 1020).

As organogenesis proceeds, somite cells break away in distinct groups that migrate to their final locations in the developing embryo. Cell migrations such as these are critical to organogenesis. For example, once in their new locations, skeletal muscle precursor cells divide, begin synthesizing muscle-specific proteins, and eventually differentiate. The muscle cells later fuse to form long fibers within each muscle.

Differentiation of the Neural Tube and Central Nervous System Organogenesis of the central nervous system begins when the ectoderm along the dorsal surface of a vertebrate embryo begins folding to form the neural tube (see Figure 47.20).

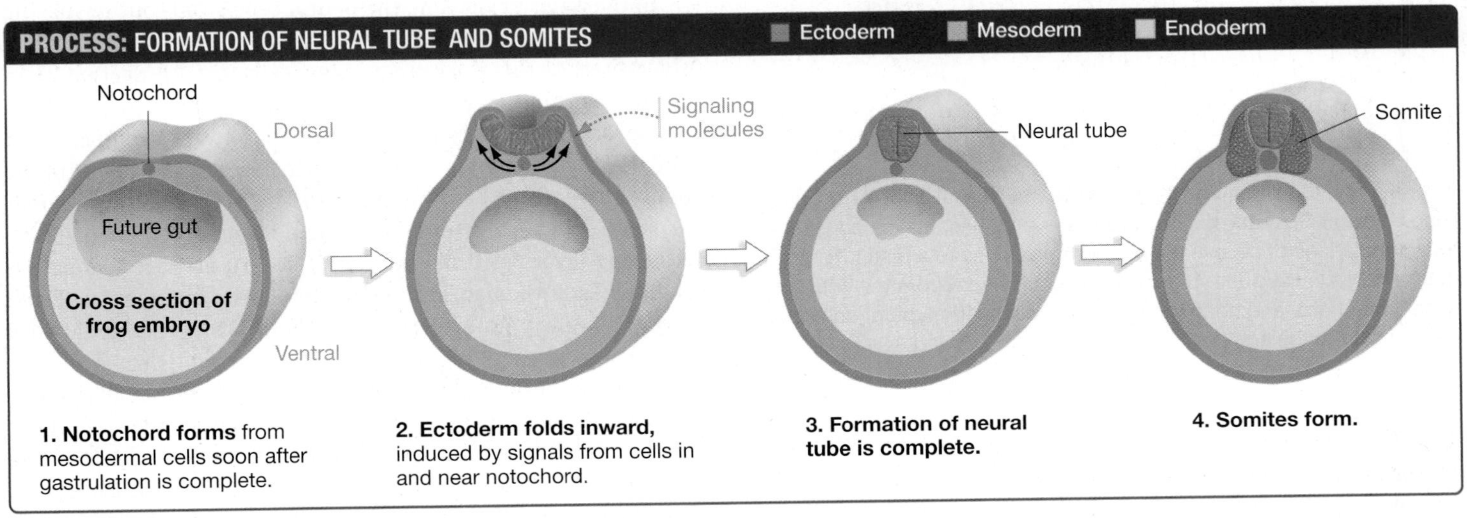

Figure 47.20 **The Notochord, Neural Tube, and Somites Form Early in Organogenesis.** The figure shows a cross section of an early frog embryo at different stages of development.

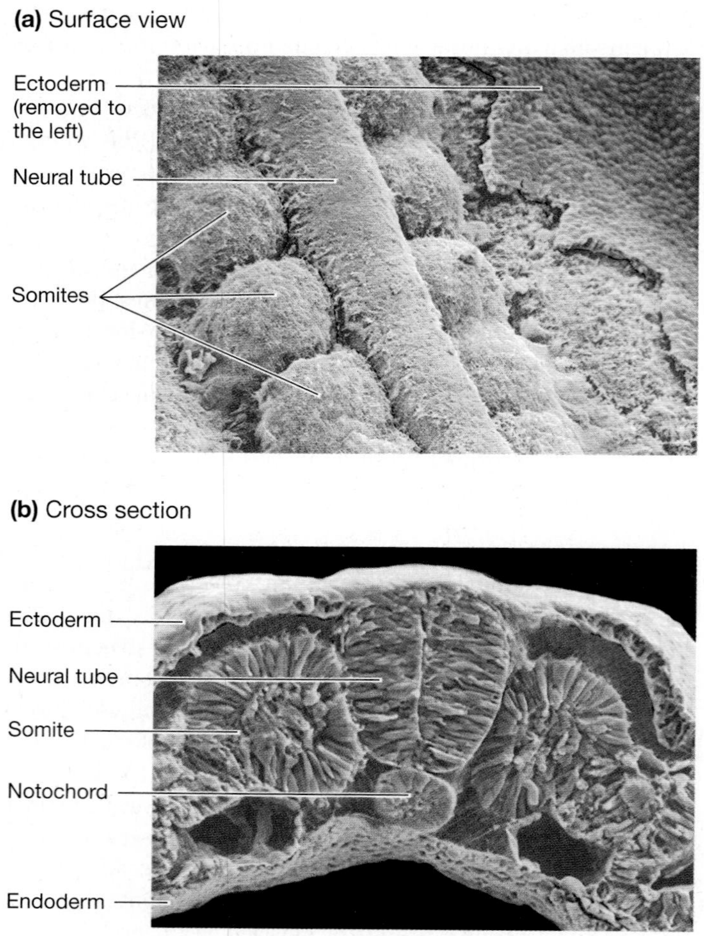

Figure 47.21 Somites Develop from Mesodermal Cells. **(a)** Surface view of chick embryo. **(b)** Cross section of chick embryo.

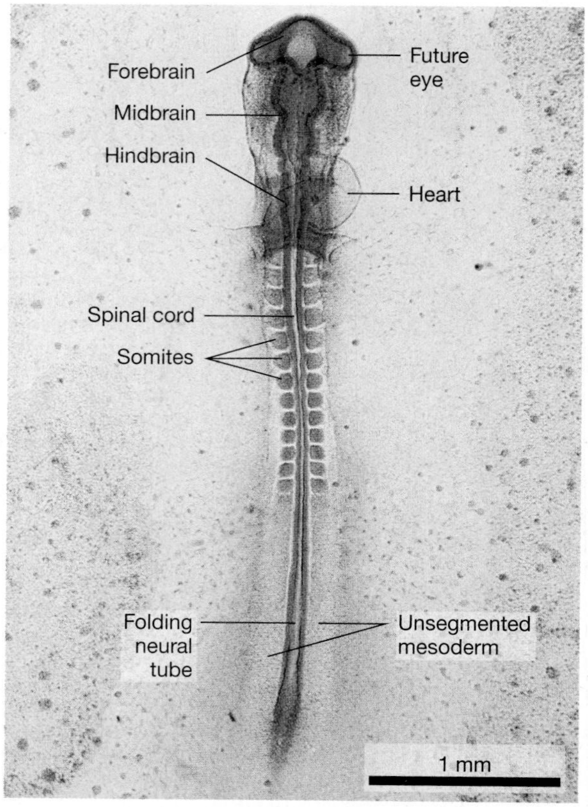

Figure 47.23 Development of the Central Nervous System and Somites. This micrograph of a chick embryo was taken 33 hours into development. Development advances more rapidly in the anterior than the posterior. Brain regions have formed in the anterior even before the neural tube is fully formed in the posterior. Similarly, somites form from unsegmented mesoderm in an anterior-to-posterior wave.

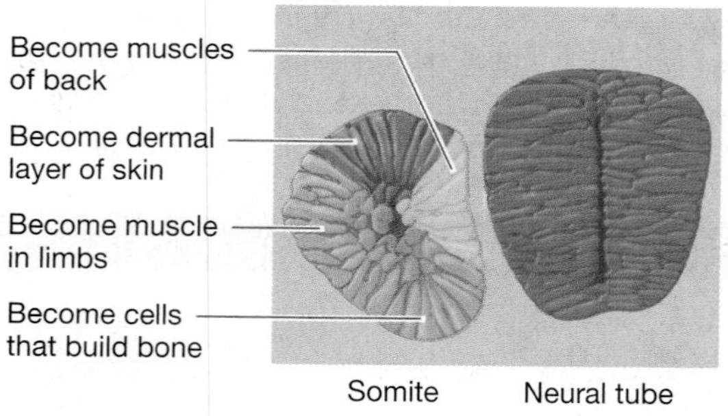

Figure 47.22 Somites Form Alongside the Neural Tube and Give Rise to Adult Structures. Each somite eventually breaks up into four populations of cells, each of which gives rise to a distinct set of structures in the adult. Signaling molecules sent from the neural tube, notochord, and nearby mesoderm specify the type of cell in each region of the somite.

As with somite development, molecular signaling from the notochord is crucial in determining which region of ectoderm will become part of the neural tube.

Once the neural tube is in place, organogenesis proceeds along two axes: the anterior–posterior axis and a radial axis that runs from the center of the neural tube to the periphery. The anterior portion of the neural tube contributes to the brain, and the posterior portion contributes to the spinal cord (Figure 47.23).

In the anterior portion, the neural tube swells out in specific regions and then is folded back onto itself to form different brain structures. As this is occurring, stem cells (Ch. 21, Section 21.3) in the innermost layer of the neural tube divide. In every cell division, one daughter cell remains as a stem cell in the inner layer, and the other daughter cell migrates to the outermost layer of the neural tube, where it differentiates. The type of cell it becomes depends on when during development the stem cell divided. The resulting layers of different cell types are arranged along the radial axis of the neural tube like the layers of an onion. These layers of cell types are critical for both brain and spinal cord function, with each layer made up of a different type of cell.

Organogenesis is an elaborate, multistep process. Cells need to divide, specialize, change their shapes, migrate, and sometimes die for an organ to form. All these behaviors are ultimately the outcome of reading a shared set of genetic instructions in cell-specific ways by distinct sets of regulatory transcription factors. The whole process is controlled by signaling molecules and genetic regulatory cascades (Ch. 21, Section 21.2).

Once the rudiments of organs are in place, development proceeds largely by their growth and refinement. But development doesn't occur in a vacuum. Particularly in mammals, the environment provided by the mother is critical. One class of signaling molecule—hormones—plays a major role in creating an environment that supports development and reproduction. Let's take a look at how hormones control the maturation and function of the tissues that produce sperm and eggs.

CHECK YOUR UNDERSTANDING

✔ **If you understood this section, you should be able to ...**

1. State when you would first be able to point out the future posterior and dorsal regions of a frog embryo, and explain what clues you would use to identify these regions.
2. **PROCESS OF SCIENCE** Design an experiment using chick or frog embryos to test the hypothesis that the notochord is needed for formation of the neural tube.

Answers are available in Appendix A.

47.5 The Role of Sex Hormones in Mammalian Reproduction

After you complete this section, you should be able to...

- Explain the role of sex hormones in mammalian reproduction.

Recall that the sex hormones testosterone and estradiol are steroids; the latter belongs to a class of hormones known as estrogens (Ch. 46, Section 46.2). Testosterone and estradiol bind to receptors within the cytoplasm or nucleus of target cells. The resulting hormone–receptor complexes bind to DNA and trigger changes in gene expression.

Testosterone and estradiol are classified as gonadal hormones because they are produced in the gonads. Most testosterone is synthesized in specialized cells inside the testes; most estradiol and other estrogens are synthesized by cells that surround each developing egg in the ovaries. These surrounding cells form a structure called a **follicle**.

The sex hormones play a key role in three events:

1. Development of the reproductive tract and brain in embryos
2. Maturation of the reproductive tract during the transition from childhood to adulthood
3. Regulation of spermatogenesis and oogenesis in adults

To begin exploring the action of sex hormones, let's take a closer look at the transition from childhood to adulthood in humans.

Which Hormones Control Puberty?

In amphibians, the juvenile-to-adult transition is triggered by the hormone T_3 (triiodothyronine); in insects, it occurs in response to ecdysone (Ch. 46, Section 46.3). But in humans, this transition—called **puberty**—is directed by increased levels of testosterone in boys and estradiol in girls.

Gonadal hormone production is well regulated by the hypothalamic–pituitary axis (Ch. 46, Section 46.4). Recall that chemical signals from the hypothalamus control the release of regulatory hormones from the anterior pituitary gland, which then cause the release of hormones from other glands.

Puberty begins when the hypothalamus releases a hormone called **gonadotropin-releasing hormone (GnRH)**. This hormone stimulates the anterior pituitary gland to release two hormones: luteinizing hormone (LH) and follicle-stimulating hormone (FSH), which enter the bloodstream and stimulate the testes and ovaries to secrete testosterone and estradiol, respectively (Figure 47.24). LH and FSH also stimulate gametogenesis.

The model in Figure 47.24 raises a question: What triggers increased GnRH secretion at the appropriate age? Although this question remains unanswered, some evidence indicates that nutritional state is involved. For example:

- The current average age for the onset of puberty in females in the United States is slightly over 12 years. During the eighteenth and nineteenth centuries, when the general nutritional state of the population was poorer, the average age was 17 years.
- Girls who have large fat stores tend to enter puberty earlier than do girls who are thin.

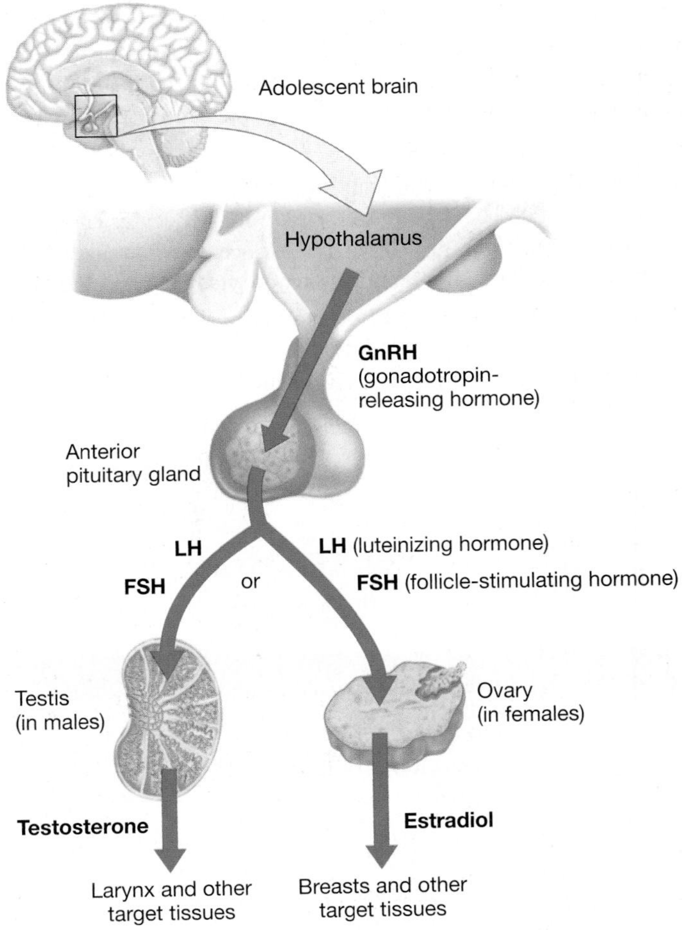

Figure 47.24 In Humans, Puberty Is Triggered by Hormones from the Hypothalamus and Anterior Pituitary.

✔ How does control of testosterone and estradiol secretion compare with control of cortisol release by the adrenal gland? (See Ch. 46, Figure 46.13.)

- Girls who exercise intensively and have little body fat, such as elite gymnasts and ballerinas, often have delayed onset of puberty.

If you recall how secretion of the adrenal hormone cortisol is controlled, however, you might suspect that the model of sex-hormone regulation in Figure 47.24 is simplified. Many hormones participate in negative feedback—also called feedback inhibition—meaning that the presence of the hormone inhibits the factor that triggers its release.

Do sex hormones participate in negative feedback? The short answer to this question is yes. To appreciate the details, let's investigate hormonal control of the human menstrual cycle.

Which Hormones Control the Menstrual Cycle in Humans?

Figure 47.25 illustrates the sequence of events in the human ovary during the **menstrual cycle**, a monthly reproductive cycle. Although the cycle's length varies among women, it averages about 28 days. In conjunction with changes in the ovary (ovarian cycle), as illustrated in the figure, the lining of the uterus undergoes a dramatic thickening followed by regression (uterine cycle). Ultimately, if fertilization does not occur, part of the uterine lining sloughs off and is expelled through the vagina.

Day 1 in the menstrual cycle is marked by the beginning of **menstruation**—the expulsion of the uterine lining. The remainder of the ovarian cycle has two distinct phases:

1. ***Follicular phase*** A follicle matures during the **follicular phase**, which lasts about 14 days. Primary oocytes complete meiosis I during this phase. Ovulation occurs when the follicle is mature and releases its secondary oocyte into the oviduct.
2. ***Luteal phase*** The **luteal phase** begins with ovulation and also averages 14 days in length. Its name refers to the **corpus luteum** ("yellowish body"), a hormone-secreting structure that forms from the ruptured follicle and degenerates if fertilization does not occur.

The regular occurrence of ovulation throughout the year makes humans extremely unusual among mammals. Although some mammals ovulate multiple times during the year, most ovulate only during a single prescribed breeding season—often in response to environmental cues such as changing photoperiod, and less often in response to cues from males.

Moreover, only humans and other great apes menstruate. In the vast majority of mammals, the lining of the uterus is reabsorbed if pregnancy does not occur. The females of these mammals have an **estrous cycle** and are sexually receptive only during estrus—when they are said to be "in heat."

Whether an estrous cycle or a menstrual cycle occurs, the basic sequence of ovarian cycle events, involving a follicular phase preceding ovulation and a luteal phase following ovulation, is shared among mammals. Hormonal control of estrous and menstrual cycles is also similar.

How Do Pituitary and Ovarian Hormones Interact during a Menstrual Cycle? By monitoring hormone concentrations in the blood or urine of a large number of women over the course of the menstrual cycle, researchers were able to document dramatic changes in the secretion of estradiol and several other hormones:

- LH and FSH are secreted by the anterior pituitary gland in response to GnRH.
- The corpus luteum in the ovary secretes the steroid hormone **progesterone** along with estrogens, including estradiol.
- In general, estradiol secretion surges during the follicular phase, and progesterone secretion surges during the luteal phase.

How are these changes regulated? Experiments helped establish that variations in the concentrations of estradiol and progesterone affect the release of the pituitary hormones LH and FSH. Researchers worked with three volunteers whose ovaries had

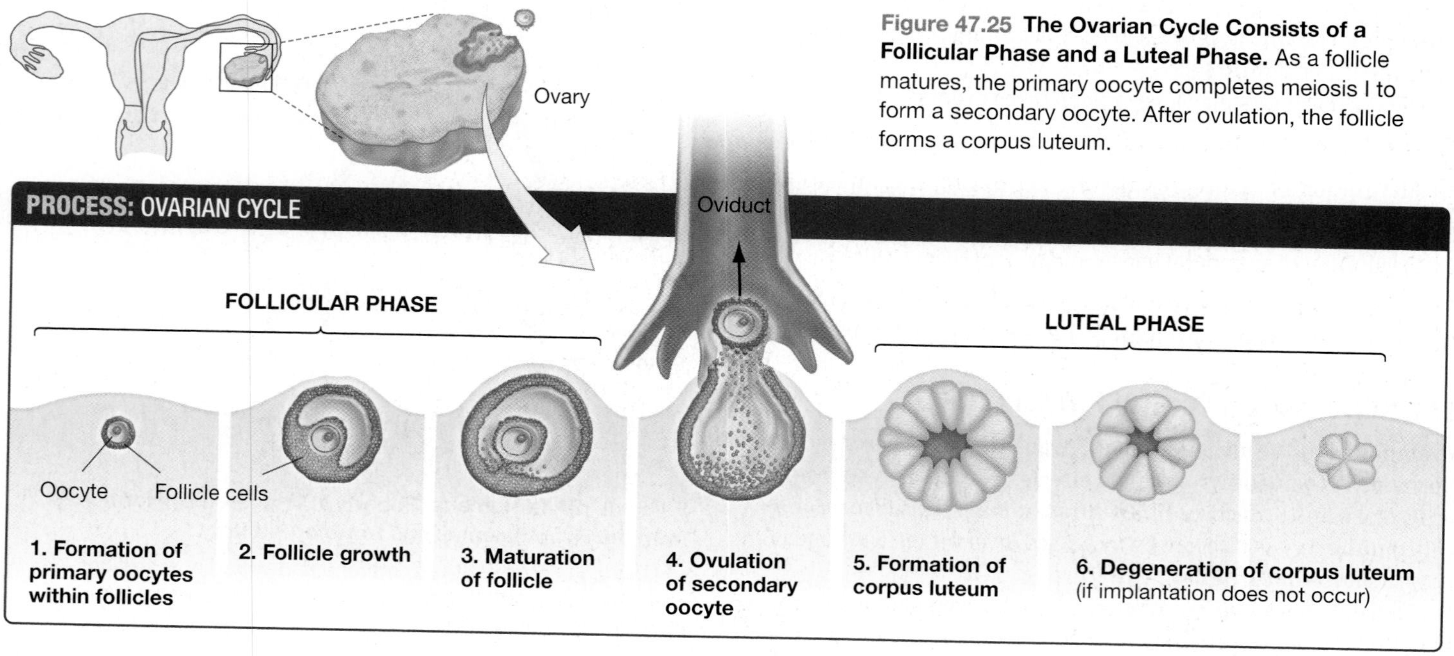

Figure 47.25 The Ovarian Cycle Consists of a Follicular Phase and a Luteal Phase. As a follicle matures, the primary oocyte completes meiosis I to form a secondary oocyte. After ovulation, the follicle forms a corpus luteum.

been removed due to cancerous growths or other problems. The women were receiving low doses of estradiol, which appeared to exert negative feedback on LH and FSH release.

But when the investigators injected the women with larger doses of estradiol or with progesterone, dramatic changes took place. For example, a large dose of estradiol stimulated a dramatic spike in LH level. This result suggested that positive feedback was occurring and that estradiol's effect on anterior pituitary hormone secretion depends on the dose: A high level of estradiol increases the release of LH, whereas a low level of estradiol suppresses it. In contrast, progesterone injections appeared to inhibit FSH and LH secretion, indicating that progesterone exerts only negative feedback on the pituitary hormones.

To summarize the interplay between LH, FSH, estradiol, and progesterone, let's start at day 1 in **Figure 47.26** and follow key events as the cycle progresses.

Days 1–7

- As the uterus sheds much of its lining, a follicle begins to develop in one ovary under the influence of FSH.
- The follicle secretes estradiol and a small amount of progesterone.
- While its level is still relatively low, estradiol suppresses LH secretion through negative feedback.

Days 8–14

- As the follicle grows, its secretion of estradiol gradually increases. The increase in estradiol concentration stimulates mitosis and an increase in cell number in the uterine lining.
- The enlarged follicle produces large quantities of estradiol, which begin to exert positive feedback on LH secretion.
- Positive feedback results in a spike in the LH level, just after the estradiol concentration peaks.
- The LH spike triggers ovulation and ends the follicular phase.

Days 15–21

- As the corpus luteum develops from the remains of the ruptured follicle, it secretes large amounts of progesterone and small quantities of estradiol, in response to LH.
- The rise in progesterone concentration inhibits secretion of LH and FSH and activates the thickened uterine lining, creating a spongy tissue with a well-developed blood supply. In this way, progesterone fosters an environment that supports embryonic development if fertilization occurs.

Days 22–28

- If fertilization does not occur, the corpus luteum degenerates.
- The progesterone level falls as the corpus luteum shrinks.

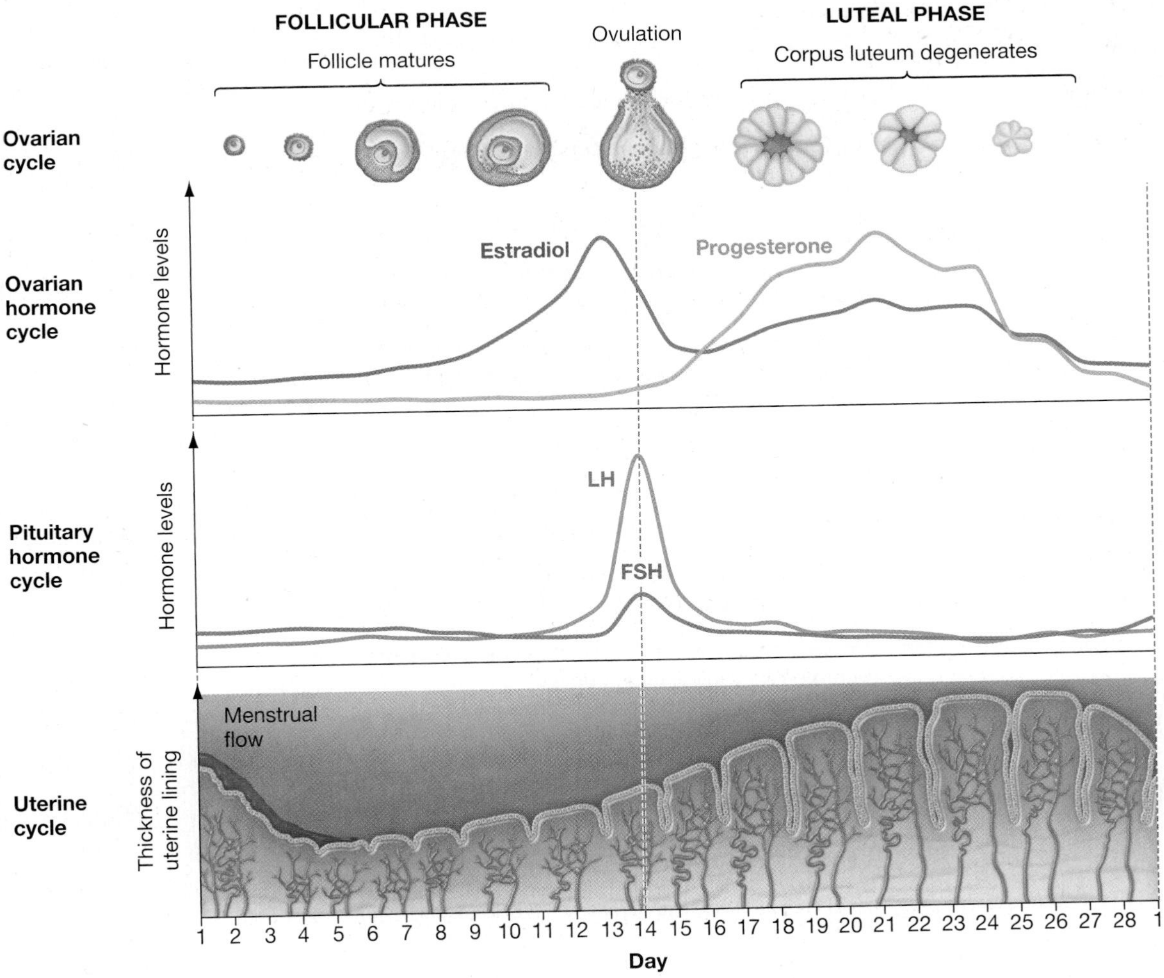

Figure 47.26 **Hormones Regulate Events in the Human Menstrual Cycle.**

- The decline in progesterone level causes the thickened lining of the uterus to degenerate. This in turn causes the menstrual bleeding that marks the first day of the next cycle.
- GnRH, LH, and FSH are released from the inhibitory control that progesterone exerts.
- The FSH level rises, and a new menstrual cycle begins.

The interplay between ovarian and pituitary hormones is similar in other mammal species that have been studied to date.

Manipulating Hormone Levels to Prevent Pregnancy Data on hormonal control of the menstrual cycle opened new avenues in birth control research. Specifically, researchers found that manipulating levels of progesterone and estradiol can prevent ovulation and serve as a relatively safe and effective method of **contraception**, or preventing unwanted pregnancies.

Hormone-based contraceptive methods deliver synthetic versions of progesterone or of progesterone and estradiol. These hormones suppress the release of GnRH, FSH, and LH through negative feedback.

In the United States, birth control pills are the most widely used contraceptive method. Hormone-containing pills are taken for three weeks and then stopped for one week to allow menstruation to occur. Other popular hormonal contraceptives include injections and hormone-secreting patches and implants. All hormonal contraceptives *prevent* pregnancy by preventing ovulation, but they do not terminate pregnancies that have already begun; contrary to popular belief, this statement is also true for emergency contraception (the "morning-after pill").

As **Table 47.2** indicates, hormone-based methods are just one approach to preventing pregnancy. Most other methods work mainly by preventing sperm from contacting the oocyte.

The "Percent Effectiveness" column on the far right of the table indicates the average percentage of women who do not become pregnant during one year of typical use of that method. Effectiveness usually increases dramatically if couples use a method exactly as specified for optimal efficacy during every episode of sexual intercourse.

CHECK YOUR UNDERSTANDING

✔ If you understood this section, you should be able to ...

1. Describe the function of FSH, and explain why the FSH level increases at the end of the menstrual cycle.
2. Predict the consequences of a drug that inhibits the release of FSH.

Answers are available in Appendix A.

SUMMARY Table 47.2 Comparing Methods of Birth Control

Type	Name	Mode of Action	Percent Effectiveness*
Barrier methods	Condom	Covers penis and prevents sperm from entering uterus.	85
	Female condom	Covers labia, vagina, and cervix and prevents sperm from entering uterus.	79
	Diaphragm	Covers cervix and prevents sperm from entering uterus.	84
	Sponge	Covers cervix and prevents sperm from entering uterus; also contains a molecule that immobilizes sperm.	84
	Spermicide	Foam or jelly covers cervix and prevents sperm from entering uterus; contains a molecule that immobilizes sperm.	71
Behavioral methods	Rhythm method	Couple refrains from vaginal intercourse around time of ovulation.	80
	Withdrawal	Man withdraws penis before ejaculation.	73
Hormone-based methods	The pill, the patch, the ring, the shot, the implant	Provides continuous or cyclical delivery of progesterone or progesterone plus estradiol.	92 to 99.9†
	Emergency contraception	Delivers progesterone or progesterone plus estradiol after unprotected vaginal intercourse.	92
Pregnancy termination	Mifepristone	Blocks progesterone receptors so menstruation occurs even after fertilization and implantation.	92
Other	Intrauterine device (IUD)	Small, T-shaped structure inserted into uterus; induces uterus to produce substances hostile to sperm and eggs. Some IUDs also secrete hormones.	99

*"Percent Effectiveness" indicates the average percentage of women who do not become pregnant during one year of typical use.
†Depends on delivery system used.

When sperm and egg do unite successfully, the menstrual cycle is interrupted. The corpus luteum does not degenerate, progesterone and estradiol levels stay high, and menstruation does not occur. Instead, the woman is now pregnant.

47.6 Pregnancy and Birth in Mammals

Viviparity allows the mother to provide a warm, protected environment for offspring during early development. Oviparous species that guard or incubate their eggs also provide warm, safe surroundings for their young. Any investment that parents make in an offspring comes at a cost, however: The more a mother invests in each offspring, the fewer offspring she can produce.

After you complete this section, you should be able to...

- Identify the major events in pregnancy and birth in mammals.

Pregnancy and **lactation**—providing milk that nourishes offspring after birth—represent some of the most extreme forms of parental care known in animals. And in some mammal species, parental care continues long after lactation ends. Humans, for example, are largely or completely dependent on their parents for protection and nutrition until puberty or young adulthood.

Let's examine how mammals make this investment, starting with marsupials and then turning to eutherian mammals.

Gestation and Development in Marsupials

Marsupials are one of the three major monophyletic groups of mammals (Ch. 32, Section 32.5). In marsupials, the corpus luteum is not maintained, and the young are ejected from the mother's body at the end of the estrous cycle. As a result, they are far less developed at birth than are the young of eutherian mammals, which undergo a lengthier **gestation**—the developmental period that takes place inside the mother's uterus.

However, the jaws, gut, lungs, and forelimbs of a newly born marsupial are relatively well developed at birth. That allows the offspring to climb from its mother's vagina to a nipple, which is usually enclosed in a pouch created by a flap of skin (Figure 47.27). The offspring clamps onto the nipple and continues to develop, fed by the mother's milk.

Even after growing large enough to leave the pouch and begin moving and feeding on its own, the offspring will return to the pouch for protection. Marsupial mothers invest a great deal in their offspring, even though a relatively short period of development takes place in the uterus.

Major Events during Human Pregnancy

Marsupials and eutherians differ sharply in terms of how long the offspring is retained inside the mother's uterus. Let's consider humans as a model organism in eutherian reproduction.

When a secondary oocyte is released from the human ovary, the cell is viable for less than 24 hours. Human sperm, in contrast, remain capable of fertilizing an egg for up to five days. Therefore, sexual intercourse in humans must occur less than five days before ovulation or immediately after ovulation for pregnancy to result.

Although an ejaculation may deposit hundreds of millions of sperm in the female reproductive tract, most die as they travel through the uterus. Only 100 to 300 sperm actually succeed in reaching the oviduct, where fertilization takes place.

Implantation of the Embryo Smooth muscle contractions in the oviduct gradually move the developing zygote toward the uterus. When the blastocyst arrives at the uterus, it undergoes **implantation**—meaning that it becomes embedded in the thickened, vascularized wall of the uterus. It will stay in the uterus for approximately 270 days (9 months).

Once the embryo is implanted in the uterine lining, its cells begin synthesizing and secreting the hormone **human chorionic gonadotropin (hCG)**, which prevents the corpus luteum from degenerating; this hormone is later produced in larger quantities by the placenta. As long as hCG is present, the ovary continues secreting progesterone, and the menstrual cycle is arrested. Some hCG is excreted in the mother's urine and is used as an indicator of pregnancy in pregnancy tests.

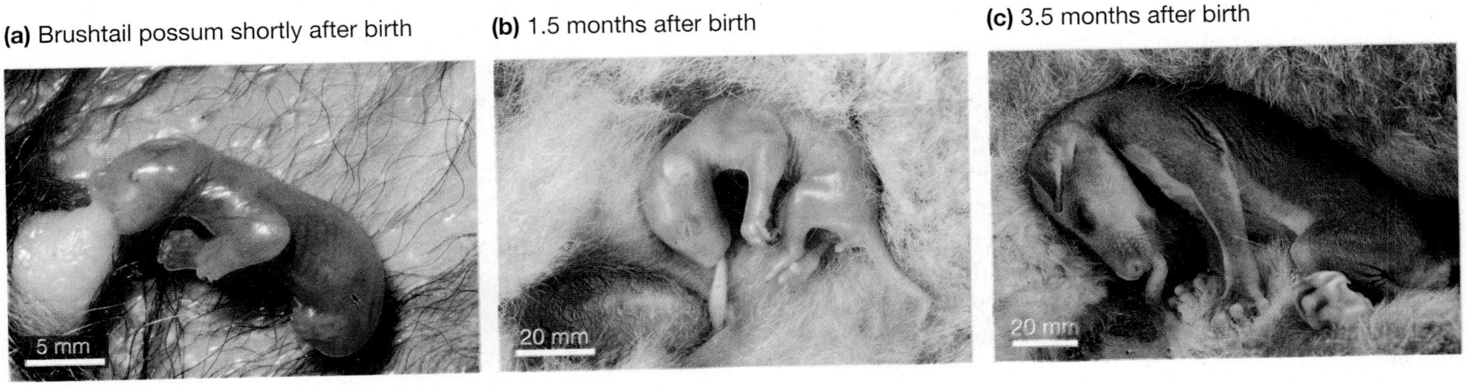

Figure 47.27 **Marsupials Trade a Long Gestation Period for a Long Lactation Period.** Compared to eutherians, marsupials—such as this brushtail possum—spend a short time developing in the uterus and a relatively long time being fed milk after birth.

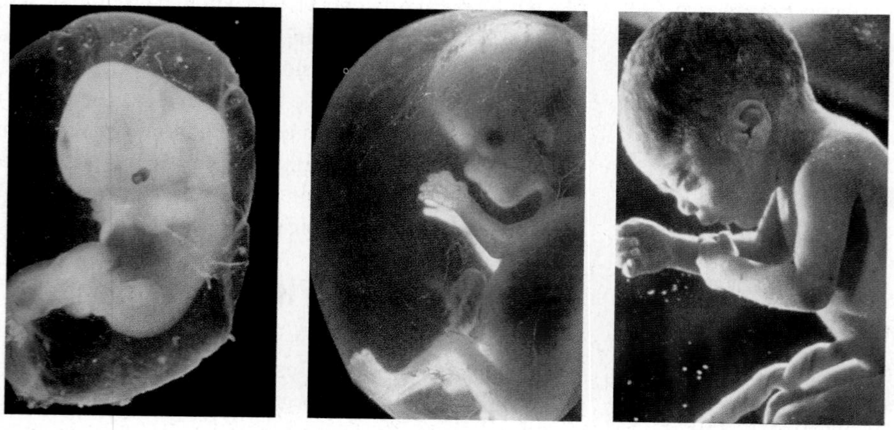

Figure 47.28
Human Gestation Is Divided into Three Trimesters.

The First Trimester Human gestation is divided into three 3-month stages called trimesters (**Figure 47.28**). Not long after implantation is complete, mass migrations of cells during gastrulation result in the formation of the three germ layers. By 8 weeks into gestation, these tissue layers have differentiated into the various organs and systems of the body. Also by this time, the heart has begun pumping blood through the circulatory system. The embryo at this stage is called a **fetus**.

Early in the first trimester, the trophoblast cells (see Figure 47.17) contribute to several important membranes. One of these membranes, the **amnion**, completely surrounds the embryo. The amnion eventually fills with amniotic fluid, which provides the embryo with a protective cushion.

Another key event in the first trimester is the formation of the placenta, which starts to develop on the uterine wall at the time of implantation. The placenta contains a dense supply of blood vessels from the mother, which provide nutrition for the growing fetus. Arteries transport blood from the circulatory system of the fetus, through the **umbilical cord**, to an extensive capillary bed in the placenta. This capillary bed provides a large surface area for the exchange of gases, nutrients, and wastes between maternal and fetal blood, even though the maternal and fetal blood do not commingle.

The placenta secretes a variety of hormones, including large amounts of progesterone and estrogens, into the maternal bloodstream. Because these hormones suppress the release of GnRH, LH, and FSH through negative feedback, they prevent the maturation and ovulation of additional follicles. By the end of the first trimester, the placenta is producing more than enough progesterone to replace the amount that had been produced by the corpus luteum, which has begun degenerating by this time. In essence, the placenta takes over from the corpus luteum in secreting the hormones required to maintain the pregnancy.

✔ **If you understand how hormones influence pregnancy, you should be able to explain why women who produce a low level of progesterone from the corpus luteum are prone to miscarriage.**

The Second and Third Trimesters After the fetal organs and placenta form during the first trimester, the rest of development consists mainly of growth (see Figures 47.28b and 47.28c). During the last weeks of pregnancy, the brain and lungs undergo particularly dramatic growth and development. If a baby is born prematurely, intervention may be required to keep the baby alive until the lungs can complete their development.

The machinery and level of hospital care required by premature infants emphasize just how superbly adapted mothers are for nourishing a growing fetus in the uterus. It costs hundreds of thousands of dollars for health care providers to do what mothers do naturally in the last trimester. Let's take a closer look at this critical aspect of pregnancy.

How Does the Mother Exchange Materials with the Fetus?

In oviparous and ovoviviparous species, mothers produce relatively large eggs that contain all of the nutrients and fluids the embryo needs for development until hatching. But in some viviparous species, eggs are relatively small and contain almost no nutrients.

In species such as humans, the developing embryo depends on the mother's body for oxygen; chemical energy in the form of sugars, amino acids, and other raw materials for growth; and waste removal. What physiological changes occur in human mothers to meet these demands?

Oxygen and Nutrient Exchange between Mother and Fetus During pregnancy, a mother's respiratory and circulatory systems change in ways that increase the efficiency of gas exchange and nutrient transfer with the fetus. For example:

- A woman's total blood volume expands by as much as 50 percent during pregnancy.
- To accommodate the increase in blood volume, maternal blood vessels dilate (widen) and blood pressure drops.
- The mother's heart enlarges and beats faster, increasing her cardiac output by almost 50 percent.
- The mother's breathing rate and volume increase to meet the fetus's demand for oxygen and removal of carbon dioxide.

In addition, important adaptations in the placenta increase the efficiency of gas exchange between the mother and the fetus. In many species, such as sheep, maternal and fetal blood flow through the placenta in a countercurrent fashion. Countercurrent flows maintain gradients that increase the efficiency of diffusion or other types of exchange (Ch. 39, Section 39.5).

Countercurrent flow does not occur in the human placenta. Instead, another mechanism increases the efficiency of exchange between mother and fetus. Maternal arteries in humans empty into a space at the junction of the maternal and fetal portions of the placenta. This space is packed with small projections called villi, which contain the fetal blood vessels. Thus, a large surface area of fetal tissue is bathed with highly oxygenated maternal blood. The fetal villi are analogous to the villi of the small intestine (Ch. 39, Section 39.3, and Ch. 41, Section 41.3), which provide a large surface area for nutrient absorption.

Toxic Chemicals Can Be Transferred from Mother to Fetus Mothers and embryos exchange more than nutrients and wastes—they can also exchange dangerous substances. For example, children of mothers who drink alcohol during pregnancy have a higher risk for hyperactivity, severe learning disabilities, and depression. Collectively, these disorders are termed **fetal alcohol syndrome (FAS)**. Brain scans of babies born with FAS reveal numerous structural abnormalities and reduced brain volume. These irregularities are thought to be responsible for dramatic reductions in IQ and other measures of intelligence and learning ability. On the basis of results such as these, public health officials strongly advise pregnant women not to drink *any* alcohol.

Birth

Although the mechanisms responsible for initially triggering the birthing process are not completely understood, the posterior pituitary hormone oxytocin is important in stimulating smooth-muscle cells in the uterine wall to begin contractions (Ch. 46, Section 46.4). The contractions that expel the fetus from the uterus constitute **labor**. Figure 47.29 shows the three stages of the birthing process in humans.

Step 1 The uterus initially contracts at relatively low frequency. The opening in the cervix begins to dilate. Once it is fully dilated, uterine contractions become more forceful, longer lasting, and more frequent.

Step 2 The baby is expelled through the cervix and the vagina.

Step 3 After the baby is delivered, the placenta remains attached to the uterine wall. At this point, caregivers clamp and cut the umbilical cord, which connects the child and the placenta. When the mother delivers the placenta and accompanying membranes, birth is complete.

Although this description sounds straightforward, in reality a large number of complications are possible. For example, one study on Swedish mothers showed that in the 1700s, approximately 1.4 mothers died for every 100 infants successfully delivered. In most cases, the cause of death was blood loss or infection following delivery. In the late 1800s, the introduction of hand-washing practices by midwives caused this number to drop dramatically. The advent of antibiotics and blood transfusions in the twentieth century further reduced maternal mortality, which is now less than

PROCESS: BIRTHING SEQUENCE

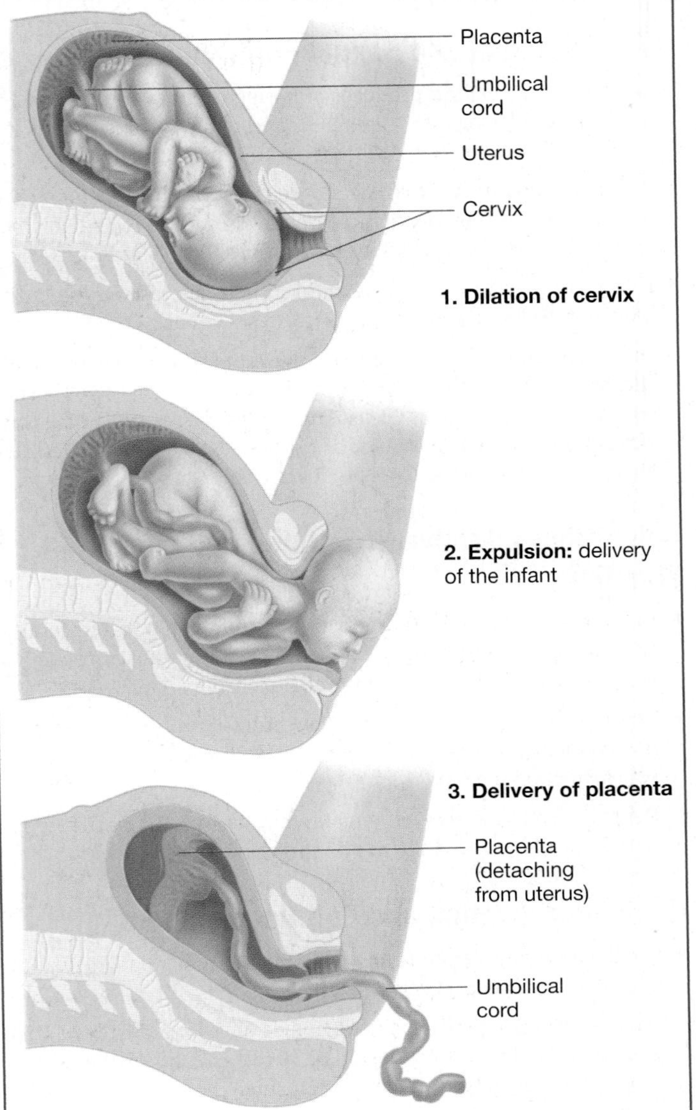

Figure 47.29 **Human Birth Occurs in Three Stages.**

0.007 percent. Improved nutrition, sanitation, and medical care have also reduced infant mortality rates in many countries.

The huge decline in the rate of death associated with childbirth qualifies as one of the great triumphs of modern medicine. However, mortality rates for mothers and infants remain high in many developing nations that lack sterile facilities and antibiotics.

CHECK YOUR UNDERSTANDING

✔ **If you understood this section, you should be able to ...**

1. Identify an advantage associated with the short gestation time of marsupials. Identify an advantage associated with the long gestation time of eutherians.
2. Describe the major events that occur over the course of a woman's pregnancy.

Answers are available in Appendix A.

47.1 Asexual and Sexual Reproduction

- Asexual reproduction produces offspring that have genetic material from only one parent. Sexual reproduction involves the fusion of haploid gametes from different parents during fertilization.
- Asexual reproduction may be favored in constant environments, whereas sexual reproduction may be favored in changing environments, or when food availability is low and availability of mates is high.
- The main mechanisms of asexual reproduction are budding, fission, and parthenogenesis.
- In human males, spermatogenesis is continuous throughout adult life, but in human females, all primary oocytes are formed by oogenesis early in development. Meiosis of oocytes is arrested for long periods of time, and cell division during meiosis is so unequal in females that just one egg—not four—is produced from each primary oocyte.

47.2 Reproductive Structures and Their Functions

- In humans, the male reproductive system includes structures specialized for producing and storing sperm, synthesizing components of semen, or transporting and delivering semen.
- The female reproductive system includes structures specialized for producing eggs, receiving sperm, and nourishing offspring during early development.
- Depending on their species, females may lay eggs (oviparity) or retain them and give birth to live offspring (viviparity).

47.3 Fertilization and Egg Development

- Fertilization is external in many aquatic animals but internal in almost all terrestrial species.
- When sperm competition occurs, males have large testes relative to their body size, and the last male to mate usually fathers a disproportionately large number of offspring.
- Fertilization is a multistep process that allows haploid sperm and eggs to fuse, forming a diploid zygote.
- Multiple mechanisms prevent polyspermy, fertilization by more than one sperm.

47.4 Embryonic Development

- The major stages of embryonic development are cleavage, gastrulation, and organogenesis.
- Cleavage is the stage of rapid cell divisions that follows fertilization and changes the fertilized egg (zygote) into a mass of cells.
- During gastrulation, coordinated cell migrations and shape changes transform the cleavage-stage embryo into a gastrula.
- Gastrulation establishes the three germ layers (ectoderm, mesoderm, and endoderm) as well as the anterior–posterior and dorsal–ventral body axes.
- The germ layers give rise to distinct organs in the adult, and each germ layer produces organs of similar types in all animals.
- Organogenesis is the formation of tissues and organs from the germ layers.
- Early in vertebrate organogenesis, cells in the notochord release signals that induce the formation of two structures: (1) the neural tube—precursor to the brain and spinal cord—from overlying ectoderm, and (2) somites—precursors to muscle, bone, and the lower layer of skin—from nearby mesoderm.

47.5 The Role of Sex Hormones in Mammalian Reproduction

- In mammals, GnRH from the hypothalamus triggers the release of FSH and LH from the anterior pituitary gland. FSH and LH regulate the production of the gonadal hormones testosterone and estradiol in the testes and ovaries, respectively.
- During the human menstrual cycle, progesterone exerts negative feedback and estradiol exerts both positive and negative feedback on the production of FSH and LH. Interactions between the pituitary and ovarian hormones are responsible for regulating cyclical changes in the ovaries (ovarian cycle) and uterus (uterine cycle).

47.6 Pregnancy and Birth in Mammals

- If fertilization occurs, the developing embryo and placenta secrete the hormone hCG, which arrests the menstrual cycle and allows pregnancy to continue.
- During the first trimester, the embryo becomes implanted in the uterine wall, the placenta forms, and organs develop.
- During pregnancy, the mother's blood volume, heart rate, and breathing rate increase. Nutrients and gases are exchanged efficiently between mother and fetus in the placenta.

Answers are available in Appendix A.

✓ TEST YOUR KNOWLEDGE

1. What term describes the mode of asexual reproduction in which offspring develop from unfertilized eggs?
   a. parthenogenesis
   b. budding
   c. regeneration
   d. fission
2. In sperm competition, what is "second-male advantage"?
   a. the observation that when females mate with two males, each male fertilizes the same number of eggs
   b. the observation that when females mate with two males, the second male fertilizes most of the eggs
   c. the observation that females routinely mate with at least two males before laying eggs or becoming pregnant
   d. the observation that accessory fluids prevent matings by second males—for example, by forming copulatory plugs
3. Which of the following statements regarding animal development is/are correct? Select True or False for each statement.
   T/F The neural tube forms after organogenesis is complete
   T/F The blastocyst is formed during cleavage
   T/F During cleavage, the zygote divides rapidly without growth, forming a mass of cells
   T/F Animals have two germ layers
4. True or false: The corpus luteum is retained upon implantation due to the presence of the hormone human chorionic gonadotropin (hCG).

✔ TEST YOUR UNDERSTANDING

5. Summarize the experimental evidence that *Daphnia* require three cues to trigger sexual reproduction. Discuss what these cues indicate about the environment.
6. Many frogs and mice are similar in size, yet a frog egg is vastly larger than a mouse egg. Propose a plausible explanation for this difference in the egg size.
7. How do spermatogenesis and oogenesis in humans differ with respect to numbers of cells produced, gamete size, and timing of the second meiotic division?
8. Give examples of negative and positive feedback in hormonal control of the human menstrual cycle. Why can a high estradiol level be considered a "readiness" signal from a follicle?

✔ TEST YOUR PROBLEM-SOLVING SKILLS

9. **PROCESS OF SCIENCE** Propose an experiment to test the hypothesis that cells from only one region of a frog blastula form the ectoderm. What results from this experiment would support this hypothesis?
10. **QUANTITATIVE** The BMI z-score is a relative measure of body mass index (BMI; Ch. 41, Section 41.4) that takes into account age. Higher values represent heavier individuals for a given height. The table here shows the BMI z-score of pre- and post-pubertal girls at three ages.

	BMI z-score	
Age	**Pre-pubertal**	**Post-pubertal**
11	−0.22	0.75
12	−0.28	0.52
13	−0.56	0.34

DATA: S. E. Anderson, G. E. Dallal, and A. Must. 2003. *Pediatrics* 111: 844–850.

Which of the following conclusions can you draw from the data?
- **a.** At a given age, there are more girls with low BMI z-scores than with high BMI z-scores.
- **b.** At a given age, girls with high BMI z-scores are more likely to have begun puberty than girls with low BMI z-scores.
- **c.** Girls 11, 12, and 13 years of age are equally likely to have begun puberty.
- **d.** There is no relationship between BMI z-score and age of beginning puberty.

✔ PUT IT ALL TOGETHER: Case Study

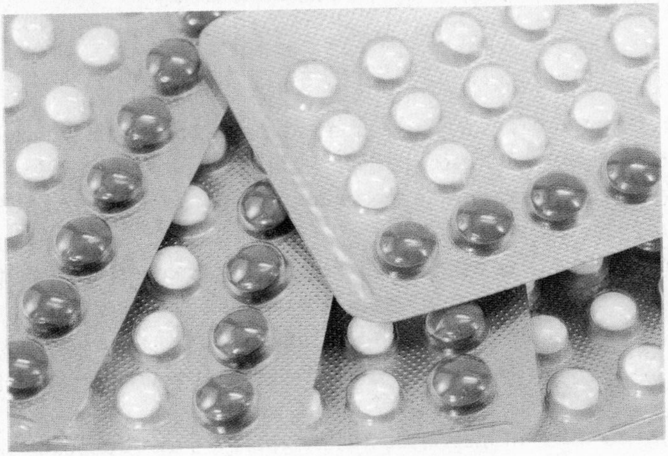

How does the birth control pill prevent pregnancy?

In the 1960s, the U.S. Food and Drug Administration approved a contraceptive that allowed women to plan desired pregnancies and prevent unwanted pregnancies. Oral hormonal contraception ("the pill") uses synthetic hormones similar in structure to progesterone and/or estradiol. What is the pill's mechanism of action?

11. Which of the following is the most effective form of contraception?
    - **a.** condom
    - **b.** diaphragm
    - **c.** withdrawal
    - **d.** the pill
12. Use your knowledge of the hormonal regulation of reproduction to predict the effect of a daily synthetic progesterone pill on (a) pituitary secretion of LH and FSH, and (b) ovarian secretion of estradiol and progesterone.
13. **QUANTITATIVE** Scientists confirmed the pill's mechanism of action by measuring plasma hormone levels in women before and after they went on the pill (* means $P < 0.05$, ** means $P < 0.01$, and *** means $P < 0.001$; **BioSkills 3**). Do the data shown here support the hypothesis that the pill affects hormonal signaling?

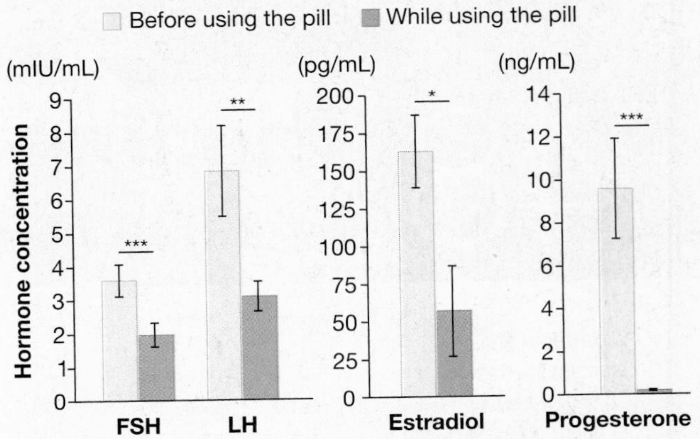

Source: U. J. Gaspard, M. A. Romus, D. Gillain, et al. 1983. *Contraception* 27: 577–590.

14. Use the information in the graph to explain how the pill affects each of the following: (a) maturation of a follicle, (b) thickening of the uterine lining during the follicular phase, (c) probability of ovulation, and (d) volume of menstrual fluid.
15. **THINK CAREFULLY** Use Table 47.2 to compare and contrast the mechanisms of action of emergency contraception and mifepristone to that of the pill. Which methods act as contraception and which act to terminate a pregnancy? Explain.
16. **SOCIETY** Imagine that four different contraceptives are under development. One blocks ovulation, one blocks fertilization, one blocks cleavage, and one blocks implantation. In the United States, which contraceptive is likely to be the least controversial to bring to market? Why?

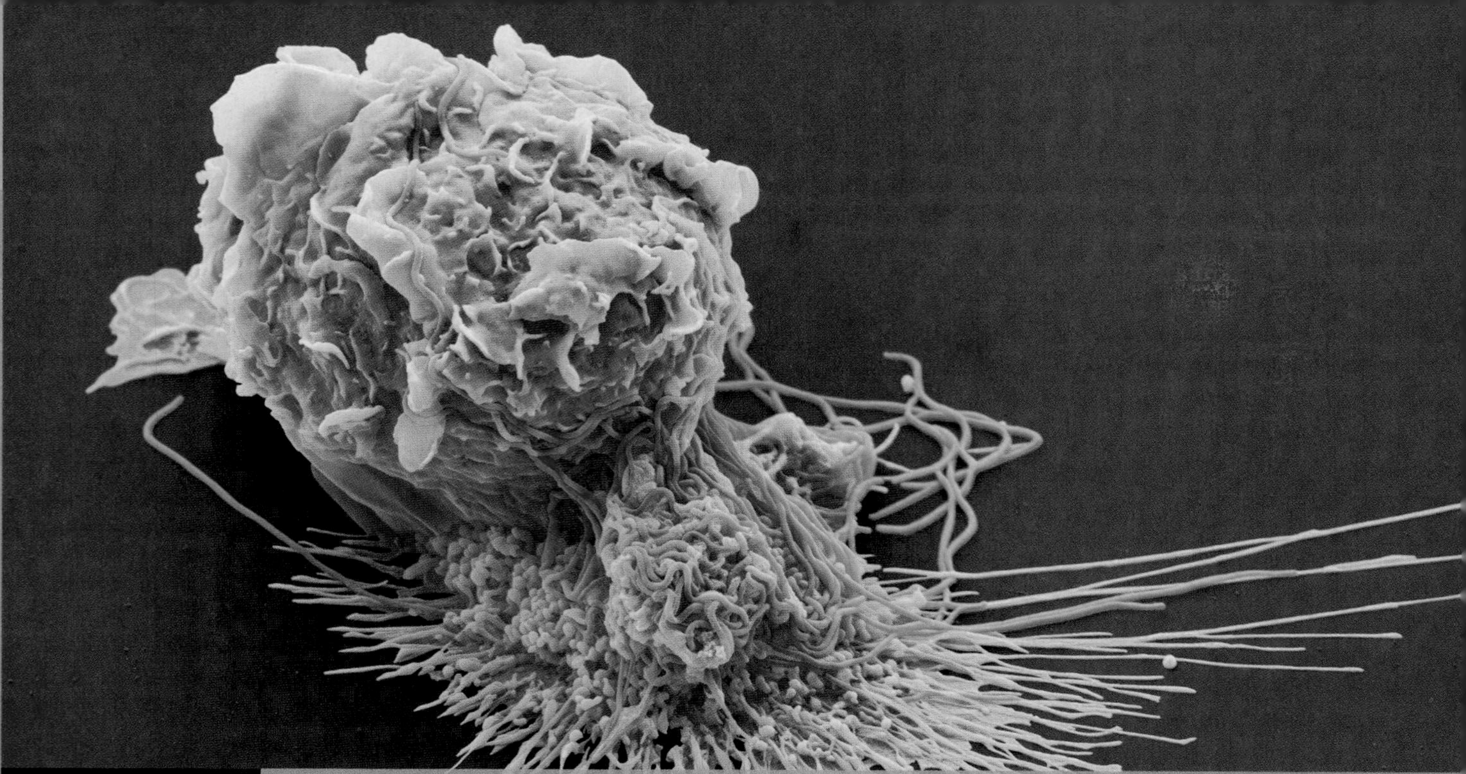

48 The Immune System in Animals

Colorized scanning electron micrograph of an immune system cell (a neutrophil) attacking *Borrelia* bacteria (blue), which can cause Lyme disease. In this chapter you'll explore how the immune system recognizes and eliminates foreign cells and viruses.

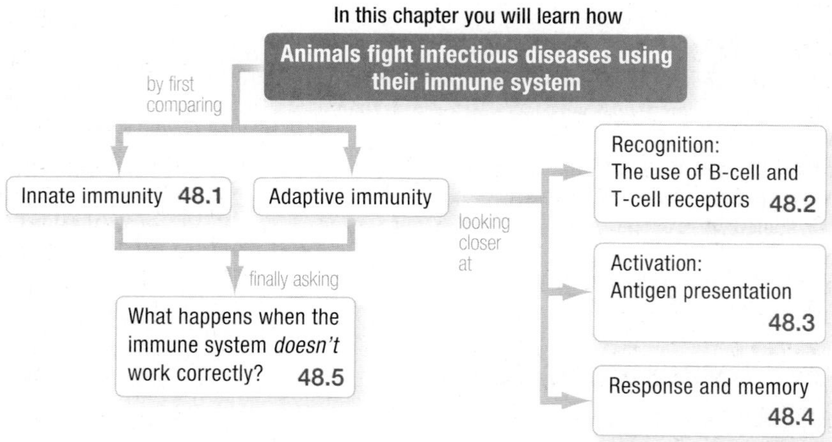

Infectious disease threatens the survival of every organism. For example, animals alone may be infected by thousands of different disease-causing bacteria, viruses, parasitic worms, fungi, and protists—collectively referred to as **pathogens**. Given the ability of pathogens to cause illness and death, it is remarkable that so many animals stay healthy for most of their lives.

If an animal contracts a bacterial or viral illness, however, it may eventually recover, even without medical intervention. Frequently, after recovering from an infection, the animal will be immune to (literally, "exempt" from) the disease—that is, it will not become ill when infected by the same pathogen in the future. **Immunity** is a resistance to or protection from the symptoms caused by a disease-causing pathogen that invades the body.

The **immune system** is responsible for defending animals against pathogens. Its success depends on three key processes: **(1)** preventing the entry of potential pathogens; **(2)** detecting the presence of a pathogen by distinguishing it from the animal's own body; and **(3)** destroying the pathogen.

BIG PICTURE

This chapter is part of the Big Picture. See how on pages 838–839.

SUMMARY **Table 48.1 Innate Versus Adaptive Immunity**

Innate Immunity	Adaptive Immunity
Occurs in all animals	Occurs only in vertebrates
Has both cell-mediated and secreted components	Has both cell-mediated and secreted components
Rapid response	Slow response
Generic response against any type of pathogens	Specific response against pathogen strains
No adaptation: Type of response does not vary when infections reoccur	Adaptation and memory: Response is more rapid and efficient when infections reoccur

When biologists began analyzing the immune system, they found that certain immune system cells are ready to respond to foreign invaders at all times, while others must be activated first. Cells that are *always ready* to respond confer **innate immunity**. Cells that must first be *selectively activated* to carry out their tailored responses to a specific pathogen confer **adaptive immunity**.

Table 48.1 compares the innate and adaptive immune responses. In combination, they form a powerful suite of defenses against a formidable and ever-changing array of pathogens. Let's begin by exploring how immune system cells and their secreted cellular products collaborate to battle infections.

48.1 Innate Immunity: First Response

Innate immunity is so named because it is inherent in all animals and is ready to go from the moment of birth. In contrast, adaptive immunity occurs in only 1 percent of animals—the vertebrates—and in humans, it is not fully developed until 6 months after birth. Clearly, innate immunity on its own has succeeded in protecting a spectacular abundance and diversity of animals, both invertebrates and vertebrates. It is the first line of defense and includes exterior anatomical structures that protect animals from invading pathogens as well as systems for interior detection and responses.

After you complete this section, you should be able to ...

- Explain how the innate immune response defends animals against broad groups of pathogens.

To launch an investigation into innate immunity, let's first focus on how the body prevents entry by foreign invaders. Then let's consider what happens if some do get in.

Barriers to Entry

The most effective way for animals to avoid getting an infection is to prevent pathogens from entering their bodies in the first place. In humans and many other animals, the most important deterrent to infection is the exterior surface. For example,

- The armored bodies of insects are covered with a tough layer called cuticle, which includes a layer of wax (Ch. 40, Section 40.4).
- Soft-bodied invertebrates like slugs, snails, and earthworms are covered with a protective layer of **mucus** (the adjective is "mucous"), a slimy mix of glycoproteins and water that traps pathogens and sloughs off.
- Human skin, an epithelial tissue (Ch. 39, Section 39.2), has an outer layer of dead cells that are reinforced with tough fibers of the protein keratin (Ch.7, Section 7.6).

Besides providing a physical barrier, outer surfaces also ensure a restrictive chemical environment. For example, the oil secreted by your skin cells is converted to fatty acids by bacteria that live harmlessly on your body. The fatty acids lower the pH of the skin's surface to about 5, creating an acidic environment that prevents the growth of most pathogens.

Unless the protective exterior surface is broken by an injury, the places in animal bodies that are most vulnerable to pathogen entry are the openings in the surface where the digestive tract, reproductive tract, gas-exchange surfaces, and sensory organs contact the environment.

How Are Openings in the Body Protected? As **Figure 48.1** shows, openings in the outer surface of animal bodies are protected by an array of specialized secretions and structures that discourage pathogen entry.

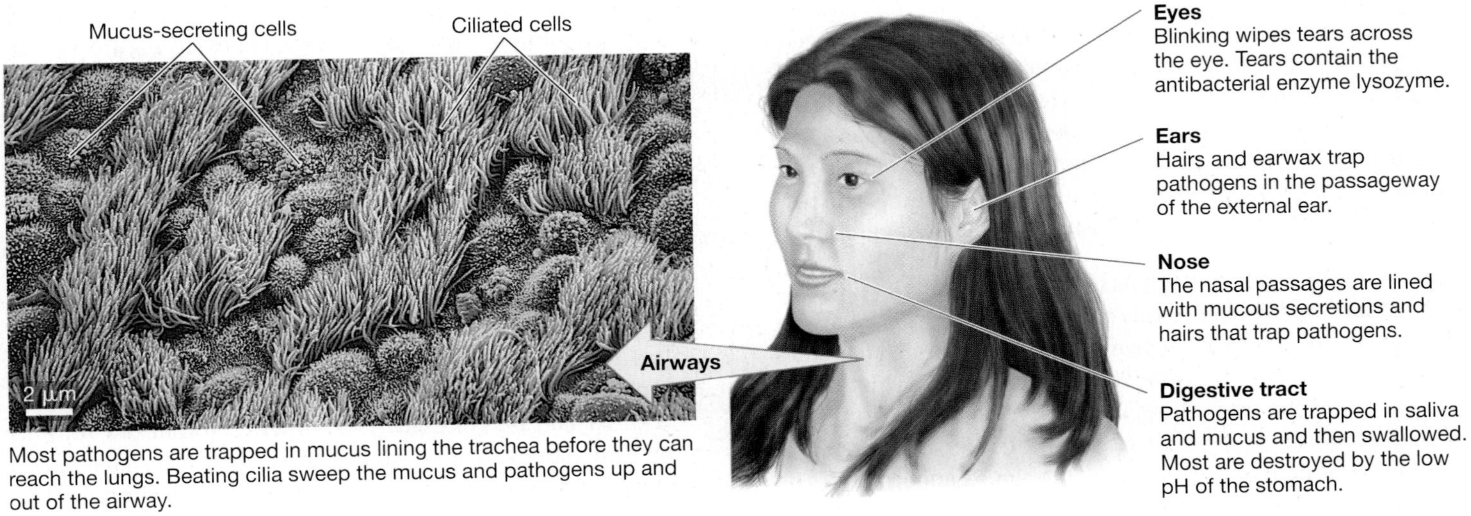

Most pathogens are trapped in mucus lining the trachea before they can reach the lungs. Beating cilia sweep the mucus and pathogens up and out of the airway.

Figure 48.1 The Body Protects Openings in a Variety of Ways.

The mucus that protects the surface of soft-bodied invertebrates is equally important in protecting the surface openings in vertebrates. For example, many of the pathogens that you breathe in or ingest while eating or drinking stick to the mucus that lines your airways and digestive tract. Pathogens that are stuck in mucus cannot come in contact with the plasma membranes of epithelial cells. Many of these pathogens are swept out of the respiratory tract by the beating of moveable cellular structures called cilia (Ch. 7, Section 7.6). They are then either coughed out or swallowed and killed in the acidic environment of the stomach.

Gaps in the body that are not covered with mucous layers are often protected by other types of secretions. For example, your ears are protected by waxy secretions, and your eyes are protected by tears that contain the enzyme **lysozyme**, which catalyzes the hydrolysis of the molecules that make up bacterial cell walls (Ch. 5, Section 5.2).

How Do Pathogens Gain Entry? When preventive measures fail, as they sometimes do, pathogens gain entry to animal tissues. Flu viruses, for example, have an enzyme on their surface that disrupts the mucous lining of the respiratory tract. When the outer surface of the virus contacts a host cell beneath the mucous layer, the virus can enter the cell and begin an infection. Wounds provide another important mode of entry. When the skin is broken, bacteria and other pathogens gain direct access to the tissues inside.

To viruses, bacteria, and fungi, your body is an ecological paradise, brimming with resources. Given that a single bacterium could give rise to a population of 100 trillion in a day, something must be done, and fast, when the outer defenses of the body are penetrated. What happens then?

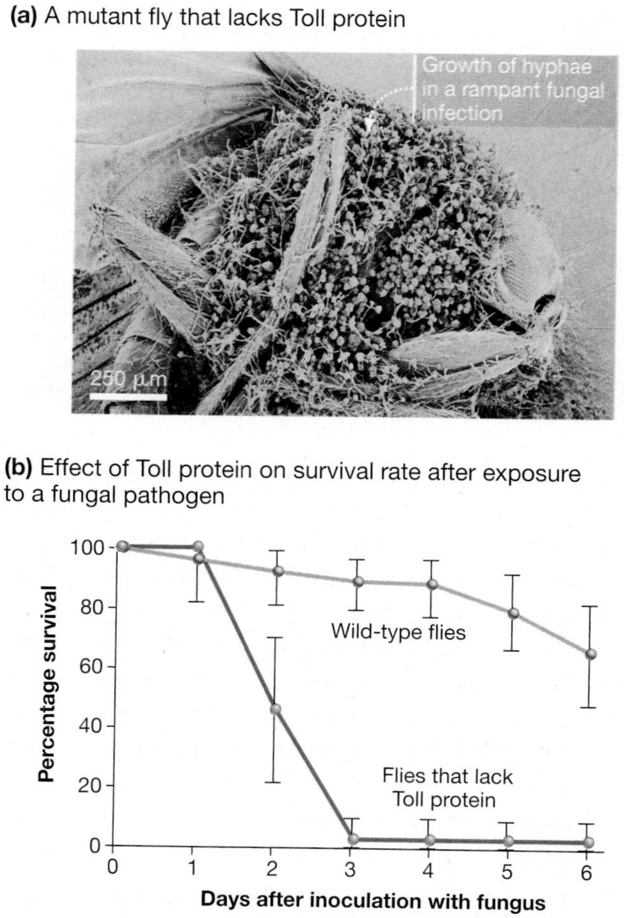

Figure 48.2 **Toll Protein Is Critical to the Innate Immune Response.** **(a)** A mutant fruit fly with a fungal infection. **(b)** Researchers compared the survival rate of wild-type flies and mutant flies that lack Toll protein after both groups were exposed to a fungal pathogen. The error bars represent the 95 percent confidence interval for each point (see BioSkills 3).

DATA: B. Lemaitre et al. 1996. *Cell* 86: 973–983.

The Innate Immune Response

If foreign invaders penetrate the body's protective barrier, the innate immune response is triggered—the body's first response to pathogens. The cells responsible for this response are a class of blood cells known as **white blood cells**, or leukocytes ("white-cells"), to distinguish them from the red blood cells that transport oxygen in vertebrates (Ch. 42, Section 42.4).

The white blood cells involved in innate immunity provide an immediate, *generic* response that is directed against the general type of pathogen encountered. Don't let the term "generic" mislead you into thinking that the response is indiscriminant. Instead, innate immunity is considered generic because it is directed against broad groups of pathogens. For example, the innate immune response can distinguish between fungi and bacteria but cannot identify a specific strain within either group.

How Are Pathogens Recognized by the Innate Immune Response? The answer to this question emerged from a breakthrough in the late 1990s, when results from unrelated studies came together. Earlier, German researchers had identified a protein that is important in the early development of fruit flies. When certain fly genes were mutated, the flies no longer produced this protein and their larvae developed abnormally. The strange appearance of the larvae inspired the researchers to name the protein Toll (in German, "toll" means amazing).

A link between Toll protein and immune function in fruit flies was revealed in 1996 when Bruno Lemaitre and Jules Hoffmann found that fungal infections were more severe in flies lacking Toll protein (Figure 48.2). Interestingly, the immune system defect in these mutant flies appeared to be specific to fungi, since the flies were fully capable of defending themselves against other types of pathogens.

Just a year later, two labs independently identified mammalian analogs of Toll in humans. Following this discovery, Bruce Beutler and his colleagues showed that a bacterial antigen did not trigger an immune response in mice with a defective Toll-like protein. An **antigen** is any foreign molecule that can elicit an immune system response. The antigen used by Beutler's team, called lipopolysaccharide (LPS), is present on the surface of Gram-negative bacteria (Ch. 26, Section 26.3).

Beutler and his colleagues had discovered that the Toll-like protein functions as a receptor that receives the signal that a pathogen is present, in this case a Gram-negative bacterium. This breakthrough was key to understanding how the innate immune response could defend against a particular type of pathogen.

Toll-like receptors (TLRs) are a subset of a larger group of proteins called **pattern-recognition receptors**, which serve as sentinels that detect the presence of molecules associated with pathogens and relay an alert signal to the cell. TLRs have also been observed in fungi and plants, suggesting that they arose in a common ancestor of all eukaryotes. Eleven TLRs (TLR1–TLR11) have been identified in humans, each one responding to one or more antigens.

What all the antigens have in common is that they are ubiquitous within the broad group in which they have been identified (e.g., *all* of the vast number of Gram-negative bacteria produce LPS), yet they do not occur in the host animal. Thus, these antigens serve as reliable signals of attack for the innate immune system. A mere 11 TLRs in humans is enough to detect virtually any type of invasion.

Pattern-Recognition Receptors Transduce Signals When a pattern-recognition receptor, such as a TLR, on the surface of a white blood cell binds to its antigen, it triggers a signal cascade within the cell that will orchestrate the most appropriate innate immune response (see Ch. 11, Section 11.3, for an introduction to signal transduction). For example,

- When the Toll protein in wild-type fruit flies or TLR2 in humans binds to zymosan, a common molecule in fungi, a signal cascade activates the production and secretion of antimicrobial peptides, which destroy the fungal pathogens.
- When human TLR4 is activated by LPS, a signal cascade leads to the production and secretion of **cytokines** ("cell-movers"). Cytokines are a class of diverse molecules that signal other immune system cells in various ways, such as increasing white blood cell production, attracting other immune cells to the site of infection, or stimulating other immune cells into action.
- When TLR7 on a human white blood cell binds single-stranded viral RNA, the cell may produce and secrete a specific type of cytokine called an interferon. Interferons stimulate neighboring cells to produce proteins that interfere with viral replication, enabling those cells to resist viral infection.

✔ If you understand the role of innate immune responses, you should be able to return to Figure 48.2 and predict how the same two groups of flies would respond to a Gram-negative bacterial pathogen.

Once the general signal of an invasion is received by TLRs, the first response is sent out, followed by a cascade of further actions that result in a fully engaged immune response. For example, consider what would happen if you were to trip on the sidewalk while running across campus and scrape your elbow. You would rapidly observe redness, swelling, and pain in the affected area. What is responsible for these changes?

The Inflammatory Response in Humans The appearance of your scraped elbow is a direct result of the innate immune response acting to promote tissue healing and repair, and to defend against infection. Figure 48.3 focuses on what happens if a wound becomes infected, and summarizes the major steps in an **inflammatory** ("in-flames") **response**—a multistep, innate immune response to trauma or infection observed in an array of animals. Note that this overview simplifies the situation—many other cell types and cell–cell signals are involved in responding to pathogens at a site of infection.

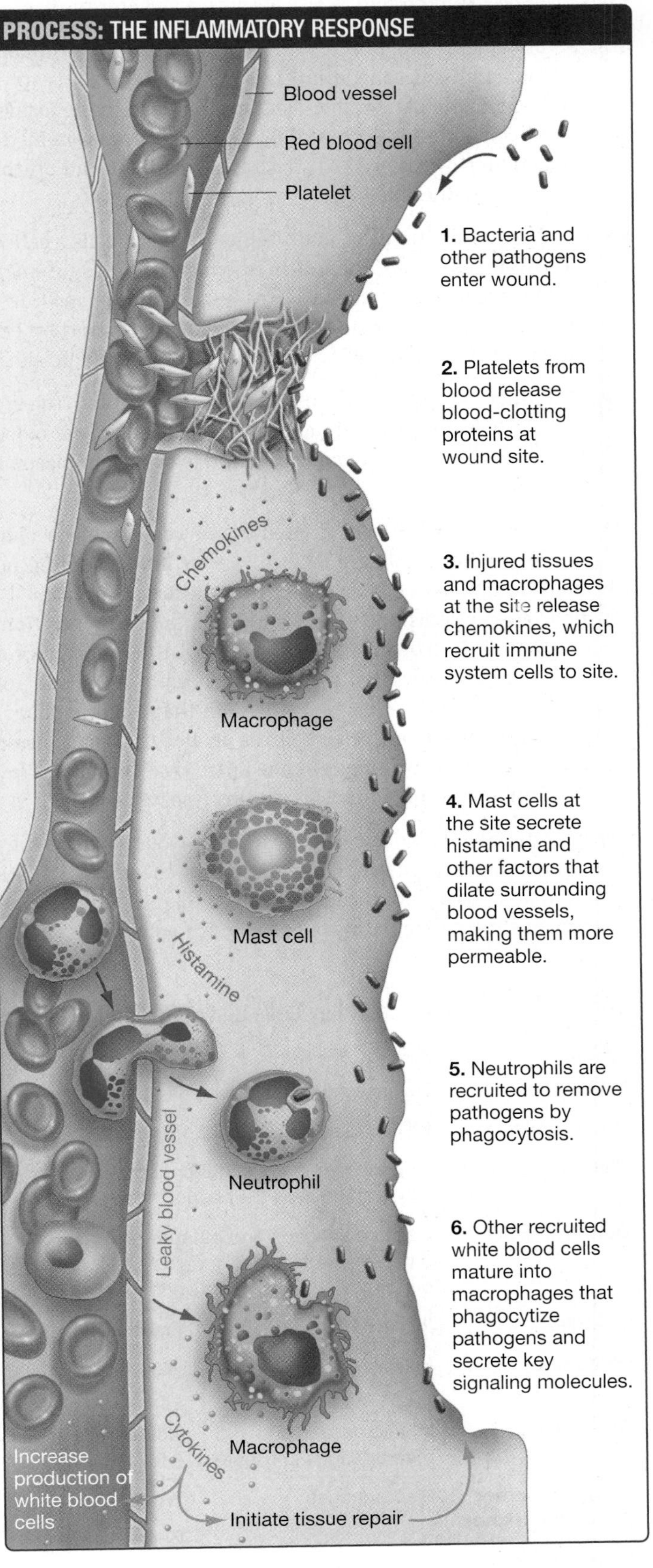

Figure 48.3 The Inflammatory Response of Innate Immunity Has Many Components.

✔ Which of the components depicted in the figure would cause the three classical signs of an infected wound: redness, swelling, and heat?

Step 1 A break in the skin allows pathogens to enter the body.

Step 2 If capillaries and other small blood vessels are broken, blood components called **platelets** immediately release proteins that participate in the reactions that lead to the formation of clots and lessen bleeding. Other clotting proteins in the blood form cross-linked structures that help wall off the wound and reduce blood loss.

Step 3 Wounded tissues and white blood cells called **macrophages** secrete **chemokines**, which are a class of small cytokines that recruit other cells to the site of injury and infection. The localized production of chemokines is important because it forms a gradient that marks a path to the wound site.

Step 4 Other white blood cells, called **mast cells**, release chemical messengers such as **histamine** that induce blood vessels slightly farther from the wound to dilate and become more permeable.

Step 5 The combination of dilated blood vessels and a chemokine gradient is like a 911 call reporting a building on fire. White blood cells called **neutrophils** move out of dilated blood vessels and migrate to the site of the infection. Neutrophils destroy invading cells by engulfing them, a process called phagocytosis (see the micrograph at the start of the chapter, and Ch. 7, Section 7.5). Cells that perform phagocytosis are collectively referred to as phagocytes. Once inside neutrophils, the invading cells are killed by a complex array of antimicrobial compounds, including lysozyme.

Step 6 Other white blood cells arrive at the wound, where they mature into macrophages. Besides secreting chemokines, these new macrophages produce additional cytokines that have an array of effects: stimulating bone marrow to make and release additional white blood cells, inducing fever—an elevated body temperature that aids in healing—and activating cells involved in tissue repair and wound healing. Macrophages also act as phagocytes, helping to clear invading cells from the area.

The inflammatory response continues until all foreign material is eliminated and the wound is repaired. But what happens when the innate immune response fails to contain and eliminate invading pathogens? Most invertebrates would be overcome by the infection and die, but vertebrates are protected by additional defenses.

White blood cells known as **dendritic** ("tree-like") **cells** capture antigens and debris from the site of infection by endocytosis (Ch. 7, Section 7.5), in particular by **macropinocytosis** ("large-cell-drinking")—uptake of liquids and small particles into a vesicle—phagocytosis, and receptor-mediated endocytosis. They present this information to cells that confer adaptive immunity, acting as messengers between the innate immune response and the adaptive immune response (see Section 48.3). Without this transfer of information, the adaptive immune response would not be activated to respond to the infection.

Table 48.2 summarizes key cells and molecules involved in innate immunity.

CHECK YOUR UNDERSTANDING

✔ If you understood this section, you should be able to ...

1. Explain the role of Toll-like receptors in innate immunity.
2. Describe how cleaning a wound to remove dirt and applying bandages to halt blood flow mimic events in the innate immune response.

Answers are available in Appendix A.

SUMMARY Table 48.2 Key Cells and Signaling Molecules of Innate Immunity

Cells	Primary Function		
Mast cells	Release signals that increase blood flow to wound area		
Neutrophils	Kill invading cells via phagocytosis		
Macrophages	Release cytokines that recruit other cells to wound site and stimulate a variety of activities; kill invading cells via phagocytosis		
Dendritic cells	Capture antigens and other foreign debris by macropinocytosis, phagocytosis, and receptor-mediated endocytosis. Connect innate and adaptive immune responses by activating certain cells of the adaptive immune response (antigen presentation; see Section 48.3).		
Molecules	**Produced By**	**Received By**	**Message/Function**
Histamine	Mast cells	Blood vessels	High concentration dilates blood vessels near wound site, among other activities
Chemokines*	Injured tissues and macrophages in tissues	Neutrophils and macrophages	Mark path to wound; promote dilation and increased permeability of blood vessels
Cytokines other than chemokines	Macrophages	White blood cells	Mark path to wound
		Bone marrow	Increase production of macrophages and neutrophils
		Central nervous system	Induce fever by raising set point for body temperature
		Local tissues	Stimulate cells involved in wound repair

*Note that chemokines are a subset of cytokines.

48.2 Adaptive Immunity: Recognition

In addition to the innate immune response, vertebrates have evolved the ability to recognize specific antigens and to differentiate between different species and even different strains of pathogens. The response of white blood cells involved in this ability is customized to particular invaders, so this arm of the immune system is often referred to as the adaptive immune response.

After you complete this section, you should be able to ...

- Describe how the adaptive immune response recognizes pathogen-specific antigens.

Given the array of pathogens that exist, an animal is almost certain to be exposed to an enormous variety of antigens during its lifetime. Is there a limit to how many different antigens its adaptive immune response can recognize?

Experiments conducted in the early 1920s answered this question. Researchers synthesized organic compounds that do not exist in nature, injected the novel molecules into rabbits, and observed whether they activated the animals' adaptive immune response. To the scientists' amazement, the rabbits could recognize and respond to every novel antigen tested. Each rabbit produced proteins in its blood, called **antibodies**, that specifically bound to the particular antigen that was injected.

The take-home message is that the adaptive immune response can recognize a seemingly limitless array of antigens. This observation, and subsequent research on antibody production, led to the identification of four key characteristics of adaptive immunity:

1. ***Specificity*** Antibodies and other components of the adaptive immune response bind only to specific sites on specific antigens.
2. ***Diversity*** The adaptive immune response can recognize and be activated by virtually any type of antigen.
3. ***Memory*** Adaptive immune responses are stronger and quicker when an individual is exposed to antigens encountered in previous infections.
4. ***Self–nonself recognition*** Molecules that are produced by the individual do not normally trigger a response, meaning that adaptive immunity can distinguish between self and nonself. Nonself molecules serve as antigens; self molecules do not.

The next question is, What cells and organs are responsible for the adaptive immune response?

An Introduction to Lymphocytes

The white blood cells that carry out the major features of the adaptive immune response are called **lymphocytes**. In contrast to the diversity of white blood cells involved in the innate immune response, lymphocytes are primarily divided into two distinct cell types that differ in their role in the adaptive immune response and their site of maturation.

The Discovery of B Cells and T Cells In 1956 a group of researchers gained an important insight into the adaptive immune response, by accident. The biologists wanted to produce and isolate antibodies against a particular toxic antigen, so they injected the antigen into a large number of chickens to induce an immune response and then collected the antibodies.

Besides injecting many normal chickens, the biologists also happened to include nine chickens that had undergone experimental removal of an organ—present only in birds—called the bursa of Fabricius. Six of the chickens that lacked a bursa died, and the three surviving chickens failed to produce antibodies to the antigen. In contrast, all of the chickens with an intact bursa survived and produced large numbers of antibodies. To make sense of these results, the researchers proposed that the bursa of Fabricius is critical for antibody production and that antibodies are important in neutralizing toxic antigens.

Not long after this observation was published, three groups of scientists independently conducted a related experiment in mice. To explore the function of the **thymus**—an organ located in the upper part of the chest of vertebrates—these scientists removed the organ from newborn mice. Mice lacking a thymus developed pronounced defects in their immune systems. For example, when they received skin grafts from unrelated mice, their immune systems did not recognize the skin as foreign. In contrast, mice with an intact thymus quickly mounted an immune response that killed the foreign skin cells.

The results of these experiments and follow-up experiments showed that lymphocytes from the bursa of Fabricius and the thymus have different functions. The bursa-dependent lymphocytes, or **B cells**, produce antibodies. The thymus-dependent lymphocytes, or **T cells**, are involved in graft rejection along with other immune functions, including recognizing and killing host cells that are infected with a virus.

Later work showed that in humans and other species that have no bursa of Fabricius, B cells mature in bone marrow. T-cells mature in the thymus.

The requirement for maturation in a particular organ distinguishes lymphocytes from the other white blood cells of the innate immune response introduced in Section 48.1. But what is the origin of lymphocytes? Where do they go after they mature? What is their life cycle?

Where Are Lymphocytes Found? The colored structures in Figure 48.4 on page 1036 mark the major sites in the human body that play key roles in the life of lymphocytes.

- ***Lymphocyte origin*** All lymphocytes originate in **bone marrow**. The bone marrow is the major blood-forming organ in the human body, responsible for the production of red blood cells and white blood cells. It consists of soft lymphoid tissue that fills the internal cavities in bones.
- ***Lymphocyte maturation*** B cells mature in the bone marrow in humans and many other animals. T cells mature in the thymus, which is located just behind the sternum (breastbone) in humans.
- ***Lymphocyte activation*** Lymphocytes have receptors that allow them to recognize antigens and become activated in the spleen and lymph nodes. The **spleen**, an organ located near the stomach in the abdominal cavity, is also involved in destroying old red blood cells. **Lymph nodes** are small, oval

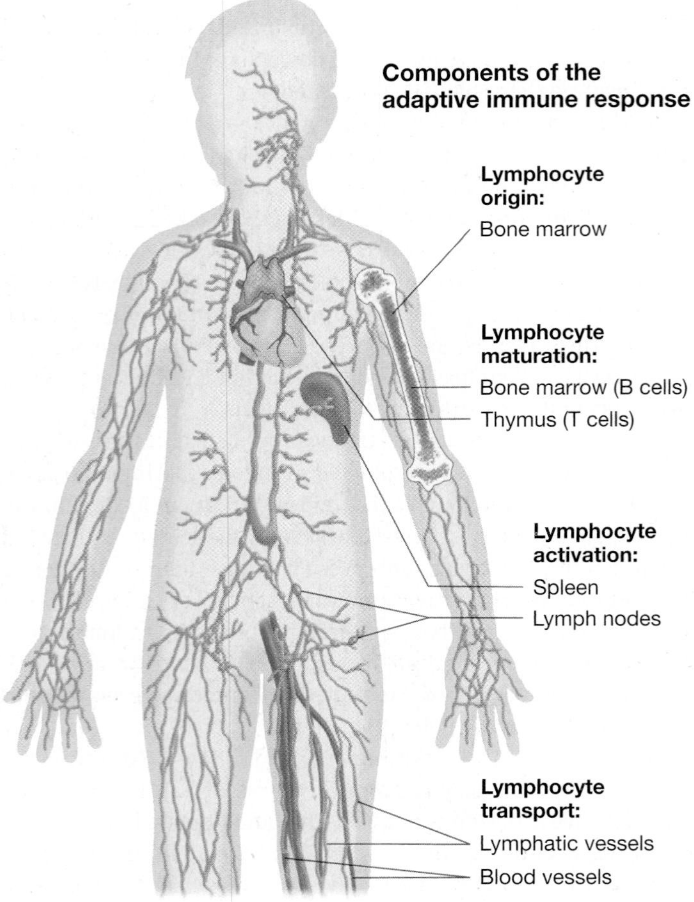

Figure 48.4 Lymphocytes Are Formed, Activated, and Transported in the Human Immune System.

organs that are located all around the body. Lymph nodes filter the lymph passing through them. Recall that **lymph** is a mixture of fluid and lymphocytes (Ch. 42, Section 42.5). The liquid portion of lymph originates in fluid that is forced out of capillaries by blood pressure.

- ***Lymphocyte transport*** Lymphocytes circulate through the **lymphatic system**, which consists of the blood, bone marrow, lymph nodes, and other organs, such as the spleen, that are involved in the production, maturation, and activation of lymphocytes. Lymph is transported throughout the body via lymphatic vessels, which are thin-walled branching tubules.

Large numbers of lymphocytes, as well as other white blood cells, are associated with the skin and with epithelial tissues that secrete mucus—primarily in the digestive and respiratory tracts. Collectively, the immune system cells found in these mucus-secreting tissues are called **mucosa-associated lymphoid tissue (MALT)**. White blood cells in the skin and MALT are important because they surveil points of pathogen entry.

Now let's turn to one of the most fundamental questions in immunology: How do B cells and T cells recognize so many different antigens? Let's begin with a look at B-cell and T-cell receptors—the molecules that initiate the adaptive immune response.

Lymphocytes Recognize a Diverse Array of Antigens

By the 1960s, biologists understood that B cells can produce antibodies to many different antigens, that each kind of B cell can synthesize only one kind of antibody, and that each antibody is specific to a particular antigen. The next question was: How do B cells receive the message to start making antibodies? Researchers hypothesized that each B cell formed in the bone marrow has thousands of copies of a receptor on its surface that, like its antibody, recognizes only one message—the antigen.

The Discovery of B-Cell Receptors To test the hypothesis that B cells have antigen-specific receptors on their surfaces, researchers injected experimental animals with radioactively labeled antigens. This strategy was similar to that of experiments with labeled estradiol that allowed biologists to isolate the estradiol receptor (Ch. 46, Section 46.2).

As it turned out, the labeled antigens bound to a protein on the surface of only those B cells that produced antibodies to the antigen. Chemical analysis revealed that these surface proteins, now called **B-cell receptors (BCRs)**, had the same overall structure as the antibodies that the B cells produced and secreted into the blood.

A BCR consists of two distinct polypeptides (Figure 48.5a). The smaller polypeptide is called the **light chain**. The larger polypeptide is roughly twice the size of the light chain and is called the **heavy chain**. Each BCR has two copies of the light chain and two copies of the heavy chain that are all held together by disulfide bonds (Ch. 3, Section 3.2). Each heavy chain includes a transmembrane domain that anchors the BCR in the plasma membrane of the B cell.

Further research showed that the light and heavy chains each consist of two regions: one whose amino acid sequence is virtually identical among different BCRs, and another whose sequence varies, forming a unique antigen-binding site. These regions are known as the **constant (*C*) region** and **variable (*V*) region**, respectively.

B Cells Also Produce Antibodies The antibodies produced by a B cell are identical in structure to its BCR, except that they lack the transmembrane domains. Instead of being inserted into the plasma membrane, antibodies are secreted from the cell and circulate via blood and lymph throughout the body (Figure 48.6).

Both the BCRs and the antibodies produced by B cells belong to the **immunoglobulin (Ig)** family of proteins. Immunoglobulins are crucial to the adaptive immune response.

Table 48.3 on page 1038 shows the five classes of immunoglobulins that act as antibodies: IgG, IgE, IgD, IgA, and IgM. Each class is distinguished by unique amino acid sequences in the heavy chains, and each has a distinct function in the immune response. Besides acting as antibodies, IgD and monomeric forms of IgM also serve as BCRs.

The Discovery of T-Cell Receptors It took much longer for researchers to isolate and characterize the **T-cell receptor (TCR)**. The technique used to identify BCRs was not useful for the receptors on T cells, suggesting that TCRs cannot bind antigens on their own. It turns out that T cells require other cells to **(1)** process the

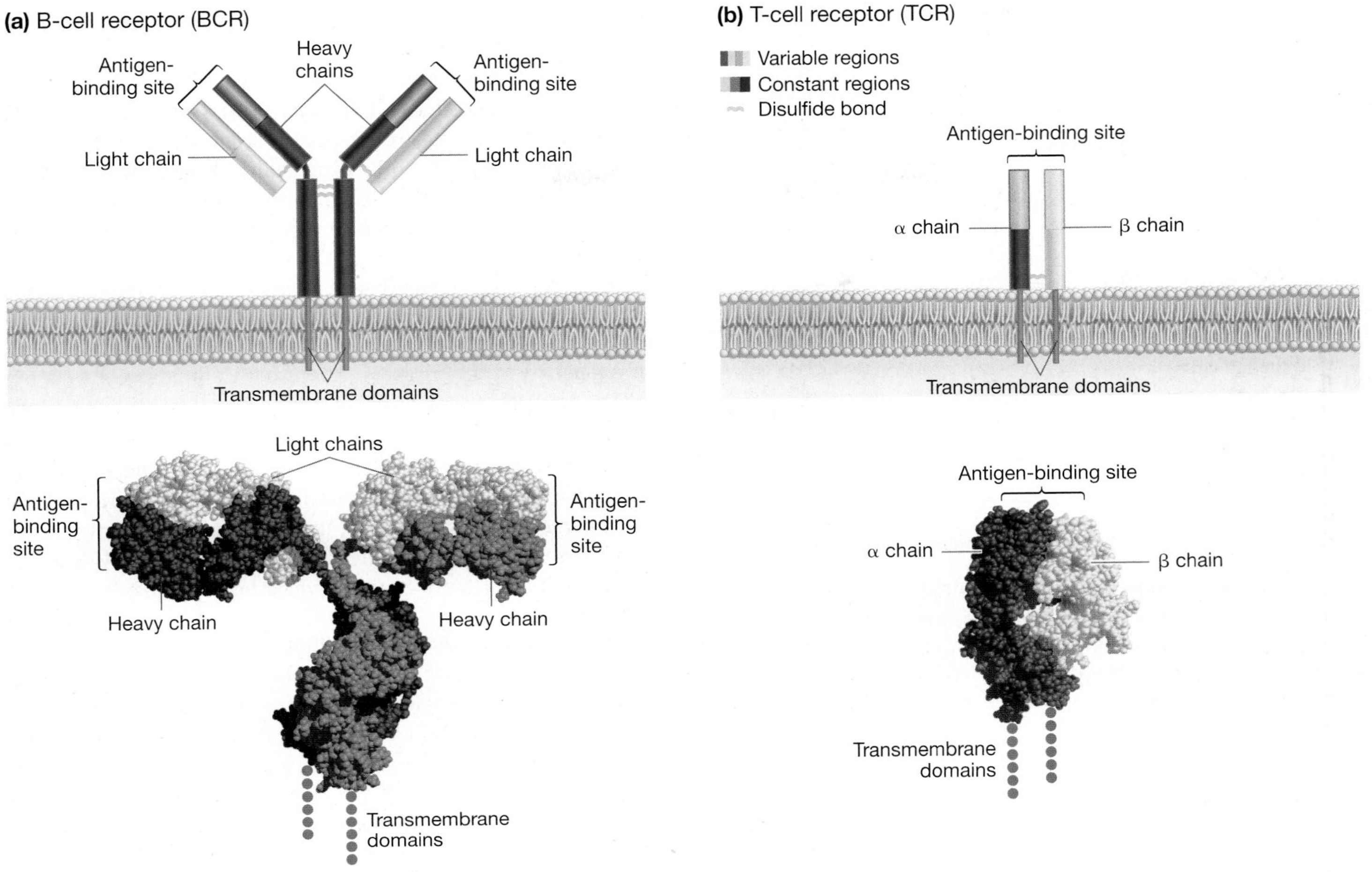

Figure 48.5 B-Cell Receptors and T-Cell Receptors Have Transmembrane Domains, Constant (*C*) Regions, and Variable (*V*) Regions That Form the Antigen-Binding Sites. Schematic and space-filling models of **(a)** the B-cell receptor, which is shaped like a Y, and **(b)** the T-cell receptor, which resembles one "arm" of the B-cell receptor.

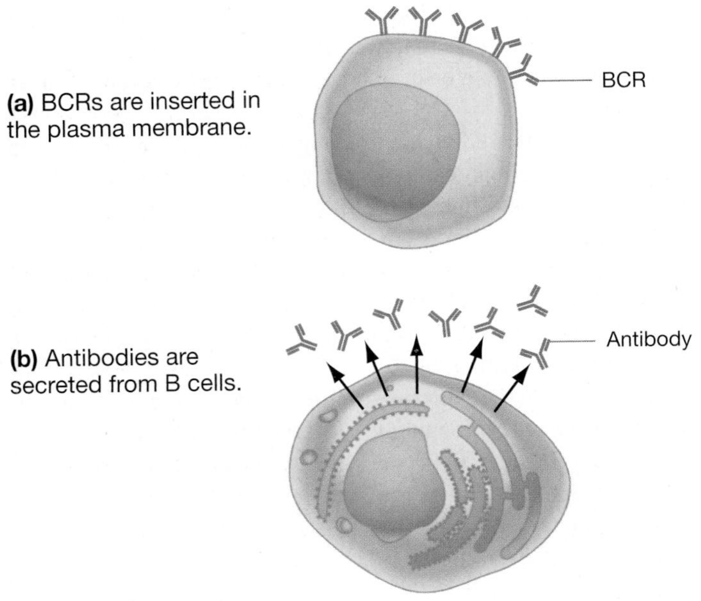

Figure 48.6 Antibodies Are Identical in Structure to B-Cell Receptors, Except They Lack the Transmembrane Domains.

antigens and **(2)** present them to the TCRs. This means that for a TCR to recognize an antigen, the foreign molecule must first undergo a complex process called **antigen presentation**.

This is a fundamentally important distinction. B cells bind to antigens directly; T cells bind only to antigens that are displayed by other cells of the immune system or cells infected by a pathogen.

Other data showed that a TCR is composed of two protein chains: an alpha (α) chain and a beta (β) chain (**Figure 48.5b**). TCRs belong to the immunoglobulin family of proteins: their overall shape is similar to the shape of one of the two "arms" of an antibody or BCR, and the *V* and *C* regions of TCRs are arranged like those in BCRs.

Receptors and Antibodies Bind to Epitopes Immunoglobulins (antibodies, BCRs, and TCRs) do not bind to entire antigens. Instead, they bind to a selected region of the antigen called an **epitope**.

To understand the relationship between an antigen and an epitope, consider that every bacterium, virus, fungus, and protist is made up of a large number of different molecules. Many of these molecules serve as antigens because they would be recognized as being foreign to your cells. In other words, your cells do not synthesize such molecules. In turn, each antigen may have

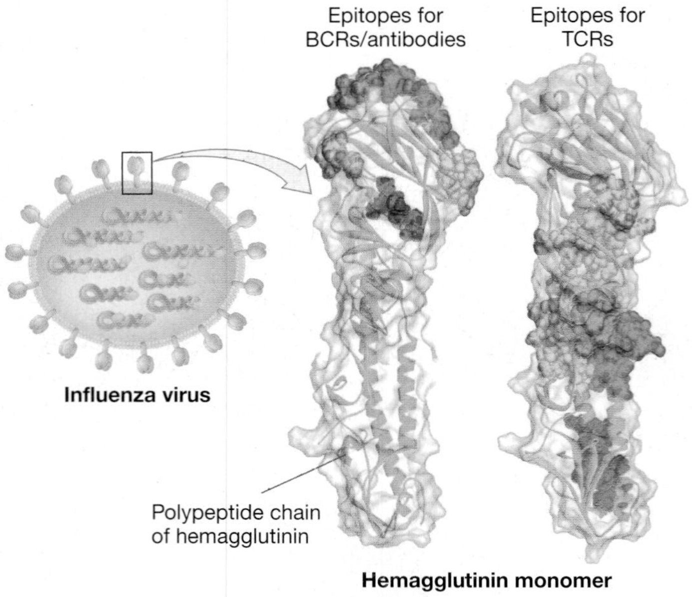

Figure 48.7 Most Antigens Have Multiple Epitopes. The envelope of the influenza virus (left) includes the protein hemagglutinin. As shown in the two space-filling models (right), each hemagglutinin monomer has several epitopes (indicated by different colors) for BCRs and antibodies, which are distinct from those for TCRs.

Table 48.3 Five Classes of Immunoglobulins

Name	Structure (Secreted Form)	Function
IgG	Monomer	The most abundant type of secreted antibody. Circulates in blood and interstitial fluid. Protects against bacteria, viruses, and toxins.
IgE	Monomer	Secreted in minute amounts. Involved in response to parasitic worms. Also responsible for hypersensitive reaction that produces allergies.
IgD	Monomer	Present on membranes of immature B cells; rarely secreted. Serves as a B-cell receptor (BCR).
IgA	Dimer	Most common antibody in breast milk, tears, saliva, and the mucus lining the respiratory and digestive tracts. Prevents bacteria and viruses from attaching to mucous membranes; helps immunize breastfed newborns.
IgM	Pentamer	First type of secreted antibody to appear during an infection. Binds many antigens at once; effective at clumping viruses and bacteria so that they can be phagocytosed. Monomeric form also serves as BCR.

many different epitopes, where binding by lymphocyte receptors and antibodies takes place.

Figure 48.7 illustrates the structure of a protein called hemagglutinin, which is found on the surface of the influenza virus (Ch. 33, Section 33.3). This protein is an antigen with several distinct epitopes. In hemagglutinin, as in many antigens, the epitopes recognized by BCRs and antibodies are different from those for TCRs because of how epitopes are presented: BCRs and antibodies bind to epitopes that are part of an intact antigen, while TCRs bind to epitopes that have been processed and presented by other cells. Each epitope is recognized by a particular BCR, antibody, or TCR. It is not unusual for an antigen to have between 10 and 100 epitopes.

How do BCR, antibody, and TCR proteins recognize specific epitopes? The presence of unique amino acid sequences in the *V* regions of every BCR, antibody, and TCR (see Figure 48.5) explains why each of these proteins binds to a unique epitope. Your body can respond to an almost limitless number of antigens because the number of different BCRs, antibodies, and TCRs is virtually limitless. How does all this variation come to be?

The Discovery of Gene Recombination In 1965, W. J. Dryer and J. Claude Bennett proposed a fantastic-sounding explanation for how immunoglobulin genes code for so many different variable regions and thus so many different proteins. They hypothesized that as a lymphocyte is maturing, a segment from a variable-region gene is cut and combined with a segment from the constant-region gene. Further, they proposed that what is cut and pasted differs in each lymphocyte. The result is that each lymphocyte has novel "*V* + *C* genes".

At the time, this type of genetic recombination in cells had not been observed, and most researchers considered the gene-recombination hypothesis wildly implausible.

Eleven long years passed before the first experimental evidence was obtained to support Dryer and Bennett's hypothesis. In 1976, Susumu Tonegawa showed that the length of DNA in the part of the genome that codes for the *V* and *C* regions is shorter in mature lymphocytes than it is in immature lymphocytes. This is exactly what would be predicted from the gene-recombination hypothesis.

The flurry of studies inspired by Tonegawa's result showed that the gene for light chains has dozens of different variable (*V*) segments, several different joining (*J*) segments, and a single constant (*C*) segment. The heavy-chain gene includes diversity (*D*) segments along with a set of *V* and *J* segments similar to those in the light-chain gene. The genes that encode the α and β chains of the TCR have a similar arrangement of distinct segments, each with multiple versions.

As a lymphocyte matures, the various gene regions are randomly mixed and matched to produce unique receptors. Figure 48.8 illustrates the steps involved in the production of a BCR light chain. A similar process occurs in the DNA encoding a BCR heavy chain. Consider how these recombination events lead to BCR diversity:

1. In the light-chain gene, one of 40 *V* segments recombines with one of 5 *J* segments. This step can produce $40 \times 5 = 200$ different light chains.

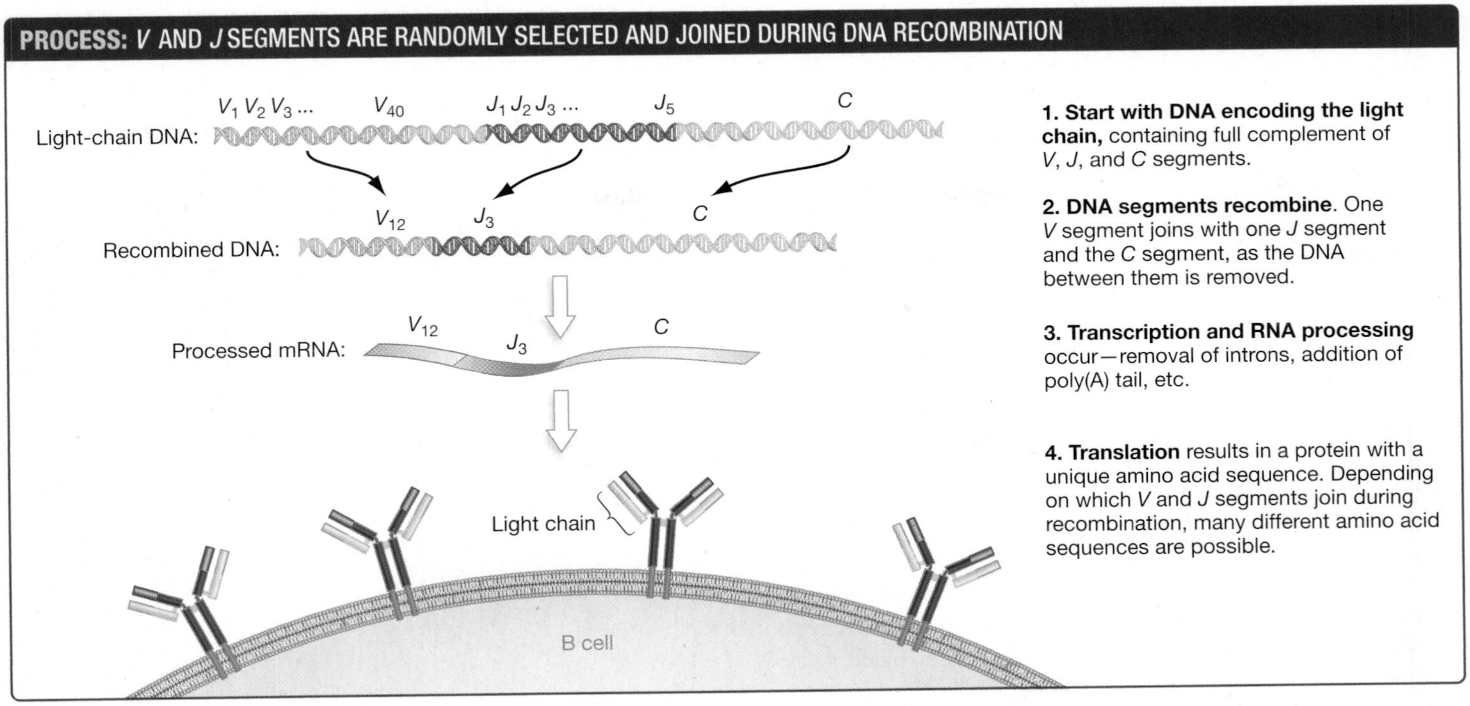

Figure 48.8 As B Cells Mature, Immunoglobulin-Gene Segments Recombine to Form a Single Gene. In the mature B cell depicted here, the final light-chain gene consists of the V_{12}, J_3, and C segments joined together; the final heavy-chain gene might consist of the V_{48}, D_{22}, J_1, and C segments joined together (not shown).

2. In the heavy-chain gene, any one of 51 *V* segments, 27 *D* segments, and 6 *J* segments can recombine.
3. The light-chain and heavy-chain gene rearrangements occur independently. When the polypeptides from each gene are assembled in the BCR, they form a specific antigen-binding site.

What's more, gene segments do not always join precisely during recombination. Some variation occurs as to where the *V* and *D* and the *D* and *J* segments join. As a result, an estimated 10^{10} to 10^{14} different BCRs can form in a single individual. TCR production is just as diverse.

✔ **QUANTITATIVE If you understand how unique genes for the light and heavy chains of the BCR are made, you should be able to calculate how many different antigen-specific BCRs could exist in (1) a human (you don't need to take into account the variation that occurs when segments don't join precisely) and (2) a single human B cell.**

Gene recombination is the molecular mechanism responsible for the specificity and diversity of the adaptive immune response. This process allows each lymphocyte to produce a unique BCR, antibody, or TCR—enabling it to recognize a unique epitope. The surface geometry of each variable region makes the interaction between an immunoglobin and its epitope extremely specific.

How Does the Immune System Distinguish Self from Nonself?

If the immune system has the remarkable ability to generate specific, targeted responses against virtually any substance, what keeps it from reacting against an individual's own molecules? If a receptor responded to a **self molecule**—that is, a molecule belonging to the host—the receptor would trigger an immune response. Because this type of response is extremely rare compared to responses against foreign antigens, biologists hypothesized that there must be some mechanism for eliminating self-reactive B cells and T cells.

To test this hypothesis, researchers injected B cells and T cells possessing anti-self receptors into mice and found that the injected lymphocytes were eliminated. Follow-up work showed that if B cells and T cells maturing in the bone marrow and thymus have anti-self receptors, the cells are likely to be destroyed or permanently inactivated before they leave these organs. The conclusion? Some form of "self-education" occurs during the maturation of lymphocytes.

Researchers have also identified another type of T cell that can regulate immune responses. Regulatory T cells limit the intensity of normal responses by suppressing certain parts of the immune system, and they may help inhibit any self-reactive cells that slip through the self-education system. Defective or insufficient numbers of regulatory T cells may be partly responsible for the development of immune disorder diseases (see Section 48.5).

Even after self-reactive lymphocytes are inactivated or removed, the structural diversity that remains in the receptors of mature B and T cells is sufficient for recognizing virtually any foreign antigen. But this diversity alone is not enough to mount an effective immune response. The enormous repertoire of possible BCRs and TCRs means that only a few cells will possess any particular epitope-specific receptor. To engage the invading pathogen successfully, these few lymphocytes must proliferate. How does this selective proliferation take place?

CHECK YOUR UNDERSTANDING

✔ If you understood this section, you should be able to . . .

1. Cite two key events that must occur during lymphocyte maturation to ensure an effective and safe adaptive immune response.
2. **QUANTITATIVE** Explain the benefit of having discrete gene segments encoding the variable region rather than having multiple genes that each encode whole variable regions.

Answers are available in Appendix A.

48.3 Adaptive Immunity: Activation

After you complete this section, you should be able to . . .

- Compare and contrast the steps required for activating B and T cells.

Lymphocytes are normally in a resting, or inactive, state. Over the course of a day, an inactive lymphocyte may hang out in the skin or mucosa-associated lymphoid tissue (MALT), enter the lymphatic vessels, migrate through a lymph node, cross over into the blood, pass through the spleen, return to the blood, and so on.

If an inactive lymphocyte does not encounter the epitope that it is programmed to respond to, the cell eventually dies. As it turns out, this is the fate of most of the lymphocytes that originate from the bone marrow. During an infection, however, some of these inactive lymphocytes will go on to mount a massive attack that is targeted against the specific invader.

The Clonal Selection Theory

The diversity of the adaptive response is useful only if it can be directed against an infection. In the 1950s, Frank Bernet and colleagues developed the **clonal selection theory** to explain how only the most useful lymphocytes are activated during an infection. Their theory proposed four key points about the adaptive immune response (illustrated in Figure 48.9 using a B cell model):

1. ***Antigens are recognized by receptors on B cells and T cells.*** Each lymphocyte that matures in the bone marrow or thymus expresses a unique receptor on its surface that binds to a unique epitope in an antigen.
2. ***Lymphocytes require receptor–epitope binding and cross-linking to become activated.*** When receptors on a lymphocyte bind to epitopes, the receptors are cross-linked and "throw a switch" to change the metabolic activities in the lymphocyte that ultimately pushes the cell from a resting to an activated state.
3. ***Activated lymphocytes are cloned.*** An activated lymphocyte divides, reenters the cell cycle, and thus makes many identical copies of itself. In this way, specific cells are selected and cloned in response to an infection.
4. ***Activated lymphocytes endure.*** Some of the cloned cells descended from an activated lymphocyte persist long after the pathogen is eliminated. As a result, the cloned cells can respond quickly and effectively if the infection reoccurs.

PROCESS: THE CLONAL SELECTION THEORY

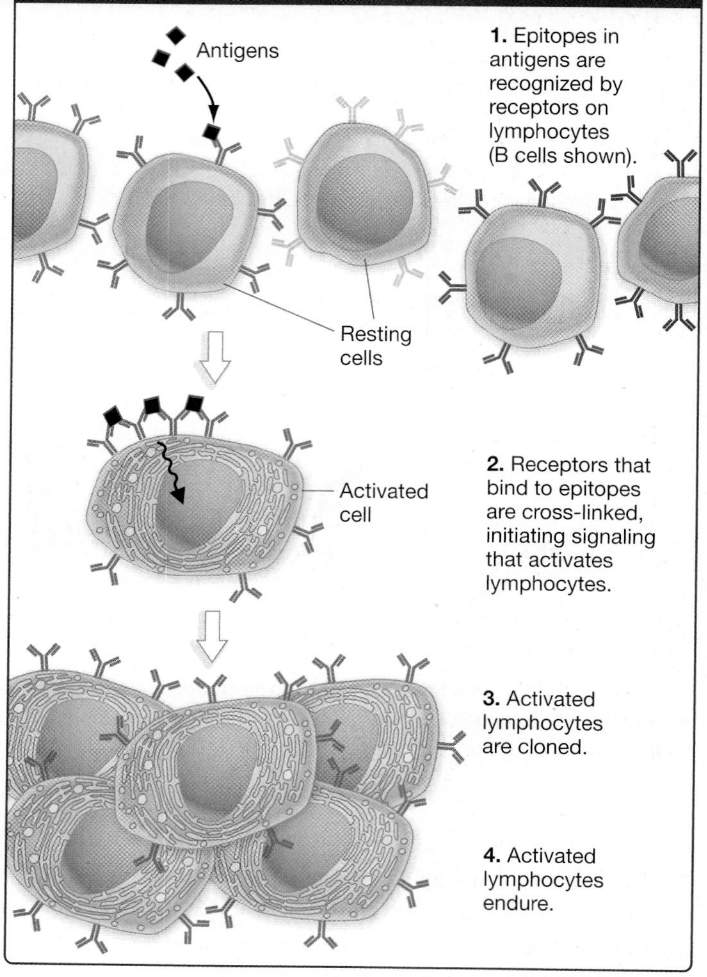

Figure 48.9 The Clonal Selection Theory. Certain lymphocytes (in this case, B cells) are "selected" when they bind to epitopes in antigens. This binding signals those cells to proliferate.

✔ If you understand the clonal selection theory, you should be able to explain what the words "clonal" and "selection" refer to.

It's important to note that activation of relevant B cells and T cells is a tightly controlled process. The mechanism is reminiscent of the precautions that nations with powerful missiles take to avoid accidental launches. For the most dangerous weapons, the signal to launch is checked and cross-checked, using a series of codes and signals. In the immune system, the checking and cross-checking occur through protein–protein interactions on the surfaces of cells, and the release and receipt of cytokines and other signaling molecules.

Let's first take a look at how T cells are activated and then examine how they play a role in the activation of B cells.

T-Cell Activation

Recall that as the innate immune response battles invaders at the site of an infection, dendritic cells gobble up antigens and debris via endocytosis (see Section 48.1). Dendritic cells collect information from the battle scene and then report to the lymph nodes, where they present antigens to T cells. Antigen presentation is a key event *that links the innate and adaptive arms of the immune system.*

To understand how the activation system works, let's explore how antigens are taken up, processed, and presented to naive T cells—inactive T cells that have not yet encountered antigen—by dendritic cells. (Section 48.4 examines other types of immune cells that are involved in presenting antigens to *activated* T cells.)

Antigen Presentation to Naive T Cells via MHC Proteins Recall that the receptors on T cells can bind only to epitopes that have been processed and presented on the surface of other cells. The surface proteins responsible for presenting these epitopes are called **major histocompatibility (MHC) proteins**. MHC proteins have a groove that binds to small epitope-containing antigen fragments that are typically 8 to 20 amino acids in length (see Figure 48.10).

MHC proteins come in two types, called **class I** and **class II MHC proteins**. Dendritic cells present peptides to naive T cells via both classes of MHC proteins, but the origins of the peptides differ for the two classes: The antigens that are processed and loaded onto class I MHC proteins are derived from the cell's interior, while those loaded onto class II MHC proteins are obtained from outside the cell. Class I MHC loading takes place inside the endoplasmic reticulum (ER); class II MHC proteins are loaded inside endosomes.

Figure 48.11 shows how dendritic cells process extracellular protein antigens and load the peptide fragments onto class II MHC proteins. (A similar loading process takes place for class I MHC proteins in the ER, but with antigens derived directly from the cytosol.)

Step 1 Dendritic cells ingest antigens at a site of infection by endocytosis.

Step 2 In one common pathway, the antigen is moved from the endocytic vesicle to an endosome, where an enzyme complex catalyzes the hydrolysis of antigen proteins into small peptide fragments (Ch. 7, Section 7.5).

Step 3 Some of the peptides are bound to the grooves of class II MHC proteins, which were made in the rough ER and transported to the endosome.

Step 4 The MHC–peptide complexes are exported from the endosomes into vesicles for transport to the cell surface.

Step 5 The vesicles fuse with the plasma membrane, resulting in MHC–peptide complexes being displayed on the cell surface.

It's important to note that humans have multiple genes encoding class I and class II MHC proteins. As a result, humans can produce several distinct proteins of each class that vary in the type of peptide that is presented. In addition, the MHC genes are among the most polymorphic of any genes known—meaning that many different alleles (Ch. 14, Section 14.2) exist in the population. Because so many distinct alleles exist, a wide array of peptides can be bound and presented—allowing dendritic cells to activate a response to many different pathogens. Also keep in mind that MHC class I proteins occur on all types of nucleated cells, not just cells of the immune system.

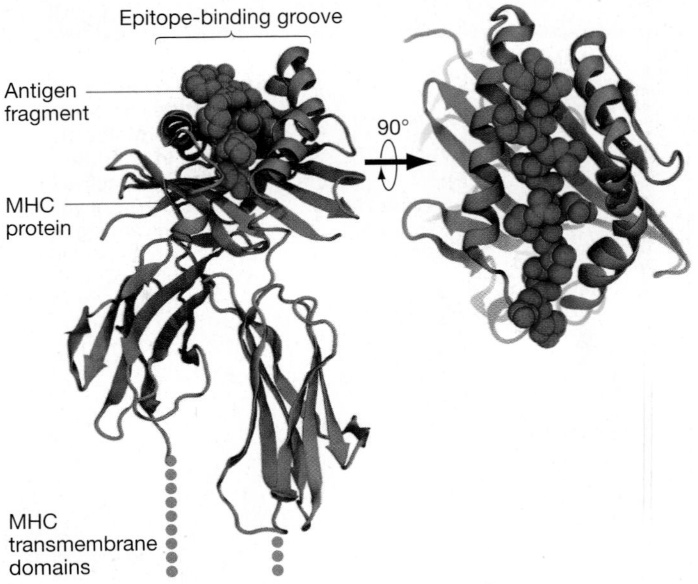

Figure 48.10 MHC Structure Promotes Peptide Binding. In this class II MHC protein, an epitope-containing fragment processed from the hemagglutinin antigen (see Figure 48.7) fits into the epitope-binding groove like a hot dog in a bun. Two alpha helices (Ch. 3, Section 3.2) flank the epitope-binding groove. Class I MHC proteins have a similar binding groove.

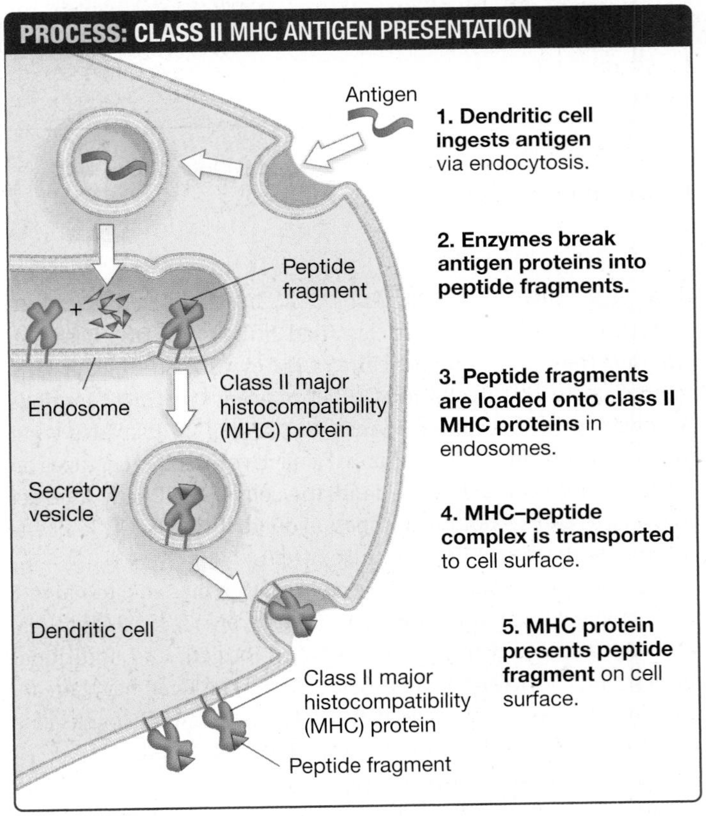

Figure 48.11 Dendritic Cells Present Antigens. Dendritic cells take in antigens, break them into fragments, and present the fragments in the grooves of MHC proteins. This figure shows the process for class II MHC proteins.

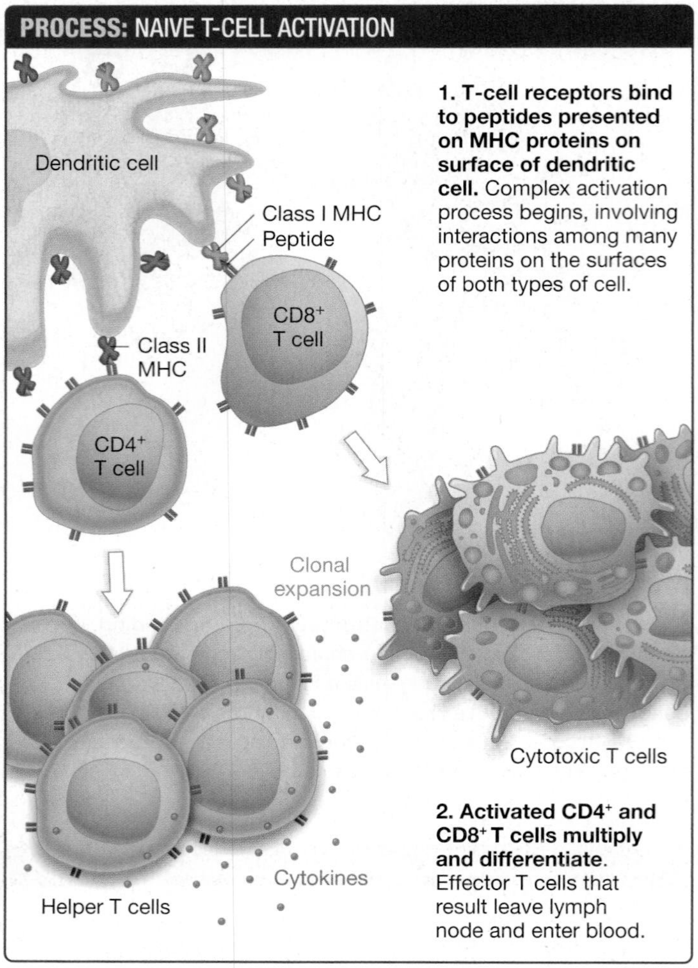

Figure 48.12 Naive T Cells Are Activated by Interacting with MHC–Peptide Complexes.

How Are Naive T Cells Activated by Antigen-Presenting Dendritic Cells? T cells are classified as $CD4^+$ or $CD8^+$, based on whether they have proteins called **CD4** or **CD8** on their plasma membranes. $CD4^+$ T cells and $CD8^+$ T cells have distinct functions in the adaptive immune response. Figure 48.12 illustrates what happens when either type of naive (inactive) T cell recognizes an MHC–peptide complex on a dendritic cell. $CD4^+$ T cells interact with class II MHC–bound epitopes on dendritic cells; $CD8^+$ T cells interact with class I MHC–bound epitopes.

Antigen-presenting dendritic cells carry a message to naive T cells saying, "I've found something. Is it foreign?" If a TCR binds, then the antigen is recognized as being foreign, and additional interactions between the dendritic cell and the T cell begin an activation process. In most cases, full activation of $CD8^+$ T cells also requires cytokines produced by activated $CD4^+$ T cells.

An activated T cell divides repeatedly to produce genetically identical daughter cells. This event, called **clonal expansion**, is a crucial step in the adaptive immune response. It leads to a large population of lymphocytes (in this case, T cells) capable of responding specifically to the antigen that has entered the body. During clonal expansion, in addition to replication, the daughter T cells differentiate to become **effector cells**, undergoing many morphological changes that prepare them for their specific functional roles in the immune response.

Cytotoxic T Cells and Helper T Cells When activated $CD8^+$ T cells undergo clonal expansion, the daughter cells develop into effector cells called **cytotoxic** ("cell-poison") **T cells**, also known as cytotoxic T lymphocytes (CTLs) or killer T cells (bottom right of Figure 48.12). The adjectives "cytotoxic" and "killer" are appropriate: Cytotoxic T cells kill cells that are infected with an intracellular pathogen.

In contrast, the daughter cells of activated $CD4^+$ T cells differentiate into effector cells called **helper T cells** (bottom left of Figure 48.12). The adjective "helper" is also appropriate: Helper T cells assist with the activation of other cells involved in the immune response. There are two types of helper T cells, designated T_H1 and T_H2, and they have distinct functions: T_H1 cells help activate cytotoxic T cells; T_H2 cells help activate B cells. During the activation phase, the dendritic cell will often direct which type of helper T cell the $CD4^+$ lymphocyte becomes. The outcome usually depends on which types of Toll-like receptors (TLRs) were activated in the dendritic cell at the site of infection.

✔ **If you understand the role of antigen presentation by MHC proteins, you should be able to create a table that summarizes the roles of the class I and class II MHC proteins with regard to the origin of the antigens that are loaded onto them, the type of T cells that bind to them, and the activity that is stimulated in these T cells via the interaction.**

Following activation and clonal expansion, effector T cells leave the lymphatic system, enter the blood, and migrate to the site of infection.

B-Cell Activation and Antibody Secretion

Like T cells, B cells undergo clonal expansion: They are replicated and undergo significant morphological changes when they are activated. As Figure 48.13a shows, inactive B cells have a large nucleus, little cytoplasm with few mitochondria, and a ruffled plasma membrane. Upon full activation, B cells produce a massive amount of rough ER and a large number of mitochondria (Figure 48.13b).

Many of the proteins synthesized in a cell's rough ER are inserted into the plasma membrane or secreted from the cell (Ch. 7, Section 7.5). The increased amount of rough ER in activated B cells is required for manufacturing and secreting antibodies.

The activation process that leads to these changes involves several steps and stimulation by T_H2 helper T cells (Figure 48.14):

Step 1 The BCRs on B cells interact directly with antigens that are floating free in lymph or blood. This interaction results in the first part of B-cell activation. The bound antigen is internalized and digested into fragments, which are then loaded onto class II MHC proteins. As a result, a B cell that encounters its antigen displays peptide epitopes of the antigen cradled in class II MHC proteins on its own surface.

Step 2 When a T_H2 helper T cell with a complementary receptor arrives, it binds to the MHC–peptide complex on the antigen-presenting B cell. This binding triggers activation signals that stimulate the helper T cell.

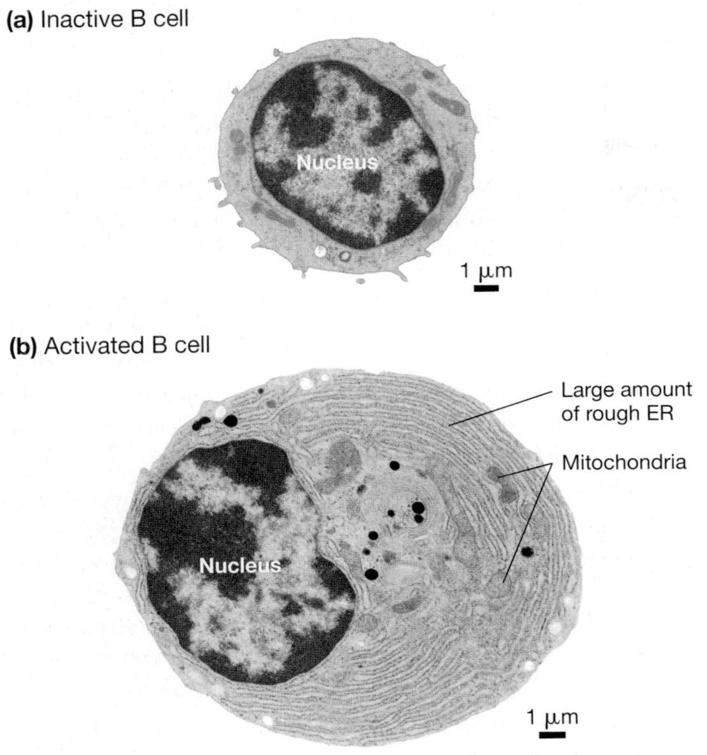

Figure 48.13 **B Cells Change Their Morphology When Activated.** **(a)** An inactive B cell has a small amount of cytoplasm with few organelles. **(b)** An activated B cell, called a plasma cell, has extensive rough endoplasmic reticulum (ER) and many mitochondria—suggesting that a great deal of protein synthesis is taking place.

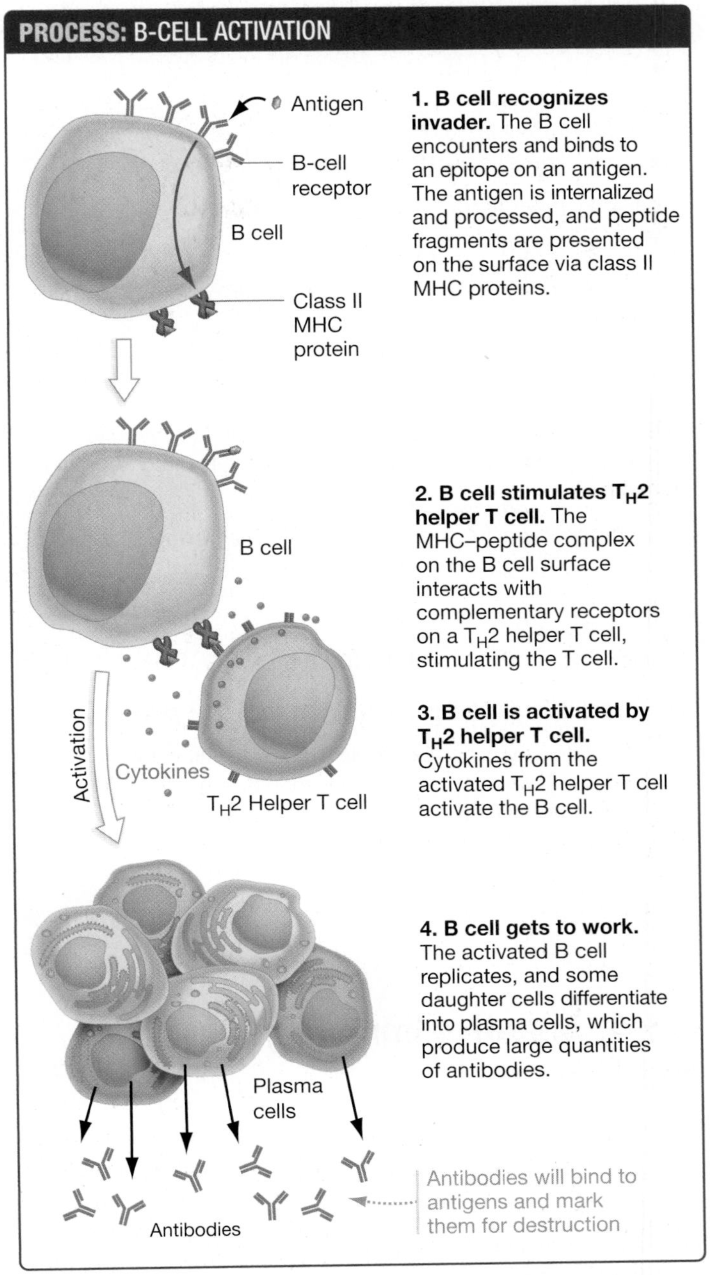

Figure 48.14 **B Cells Are Activated by Binding to Antigens and Interacting with T_H2 Helper T Cells.**

Step 3 The T_H2 helper T cell responds by releasing cytokines that trigger the second part of B-cell activation.

Step 4 The now fully activated B cell replicates, and some of the daughter cells differentiate into effector B cells called **plasma cells**. (Other activated B cells become memory cells, see Section 48.4). Plasma cells produce and secrete large numbers of antibodies. Recall that antibodies are identical to B-cell receptors, except that they lack a transmembrane domain and are secreted instead of being inserted into the plasma membrane (see Figure 48.6).

When B cells are fully activated, they migrate to specialized areas in the lymph nodes and spleen called germinal centers. There, the DNA sequences that code for immunoglobulins (BCRs and antibodies) undergo rapid mutations that modify the variable regions. This process, called **somatic hypermutation**, is responsible for fine-tuning the adaptive immune response. Somatic hypermutation can generate BCRs that bind to the antigen more tightly than BCRs formed during B-cell maturation in the bone marrow. B cells with receptors that bind best to the free antigen live and produce daughter cells; those that bind to the antigen less effectively die. It's important to note that unlike germline mutation, somatic hypermutation affects only individual B cells, and the mutations are not transmitted to offspring.

Once effector B cells and T cells have been activated, the adaptive immune response is in full swing. Cytotoxic and helper T cells move into the site of infection, and plasma B cells begin releasing antibodies specific to the invading pathogen to circulate in the blood and lymph. The immune system has recognized the invaders and initiated its response.

To learn how scientists use models to think about the cells of the immune system and how they interact, see **Making Models 48.1** on page 1044. Then let's look at what happens once the secreted antibodies and activated cells are recruited to the site of infection. For pathogens, the results are usually devastating.

Making Models 48.1 Tips on Drawing Immune System Processes

When modeling how cells of the immune system arise and interact, scientists often draw simple shapes to represent different cell types, antibodies, receptors, and cytokines. You can use arrows to indicate the clonal expansion of cells or the effect of one cell on another.

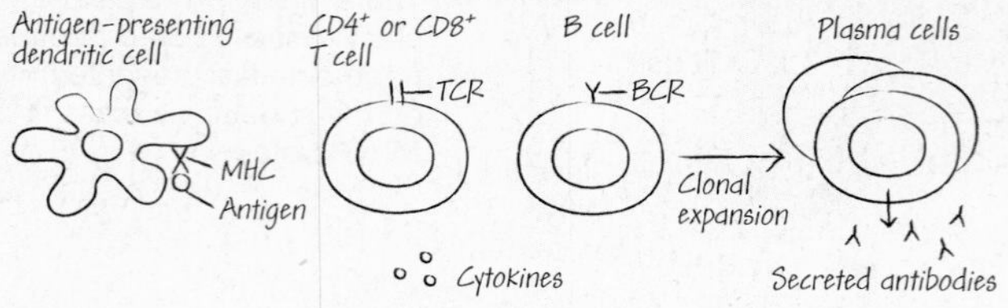

MODEL Make a drawing that shows how the innate immune response is required for activating the adaptive immune response.

To see this model in action, go to the Study Area of **Mastering Biology**

CHECK YOUR UNDERSTANDING

✔ **If you understood this section, you should be able to ...**

Generate a hypothesis to explain the observation that individuals who are heterozygous for the genes encoding MHC proteins tend to be healthier than individuals who are homozygous for these genes.

Answers are available in Appendix A.

48.4 Adaptive Immunity: Response and Memory

In combination with the white blood cells involved in innate immunity, the cells of the adaptive immune response are almost always successful in eliminating threats from bacteria, fungi, viruses, and other parasites.

After you complete this section, you should be able to ...

- Compare and contrast how the adaptive immune system responds to extracellular and intracellular pathogens.

The many different mechanisms used by adaptive immunity to dispose of foreign invaders are broadly grouped into two responses:

1. The **humoral (immune) response** is promoted by T_H2 cells and involves the production of antibodies and other proteins secreted into the blood and lymph by activated B cells. (The Latin root *humor* means "fluid.")
2. The **cell-mediated (immune) response** is promoted by T_H1 cells and involves the activation of phagocytic cells and cytotoxic T cells (activated $CD8^+$ cells), among others. This response primarily takes place via cell–cell contact.

Let's examine in turn how these responses deal with extracellular invaders and those invaders that take up residence within cells.

How Are *Extracellular* Pathogens Eliminated? The Humoral Response

Recall that once a B cell recognizes an invader and is activated by T_H2 helper T cells, the activated B cell replicates and differentiates into plasma cells (see Figure 48.14). Antibodies from the plasma cells then begin attaching to extracellular bacteria, fungi, viruses, and other foreign material. These bound antibodies interfere with the infection in four ways (Figure 48.15):

1. ***Opsonization ("preparation for eating")*** Antibodies from plasma cells coat pathogens at the infection site. Pathogens that are coated with antibodies are readily destroyed by phagocytes.
2. ***Neutralization*** Coated pathogens are blocked from interacting with—and thus infecting—host cells. Their participation in the infection is neutralized.
3. ***Agglutination ("gluing together")*** In many cases, antibodies cause the clumping of antigens, including those on cells and viruses, via a process called **agglutination**. Each antibody has at least two binding sites (see Table 48.3), so a single antibody can bind epitopes on cells or viruses and cross-link them. Clumped cells and viruses cannot infect the cells of the body and are easy targets for phagocytes.
4. ***Co-stimulation of complement proteins*** Antibodies that are bound to pathogens also activate a lethal group of proteins called the **complement system**. Complement proteins circulate in the bloodstream and assemble at antigen–antibody complexes. When complement proteins are activated, they participate in activities that result in punching deadly holes in the plasma membranes of pathogens.

Within a few days, this combination of killing mechanisms—armies of phagocytic cells and complement proteins that home in on antibody-tagged material—usually eliminates all of the extracellular pathogens. But what about those pathogens that reside within the cells of the body?

How Are *Intracellular* Pathogens Eliminated? The Cell-Mediated Response

Recall that the innate immune response includes dendritic cells and macrophages, both of which can ingest some of the invaders at the site of infection (Section 48.1). Antigen-presenting dendritic cells then activate naive T cells in the lymph nodes, which differentiate into helper T cells, including cytotoxic T cells (see Figure 48.12).

Macrophages kill foreign cells and, at the same time, process and present antigens to activate the adaptive immune response via class II MHC proteins. Macrophages display the processed peptides on their surfaces at the site of infection. If the class II MHC–peptide complex is recognized by a T_H1 helper T cell, two things happen. First, the phagocytic activity of the macrophages is enhanced. Second, the T_H1 cells secrete cytokines that recruit additional phagocytic cells to the site and activate cytotoxic T cells—increasing the inflammatory response.

Along with the class II MHC proteins, class I MHC proteins also are involved in the cell-mediated response. Recall that class I MHC proteins display peptides processed from cytosolic proteins

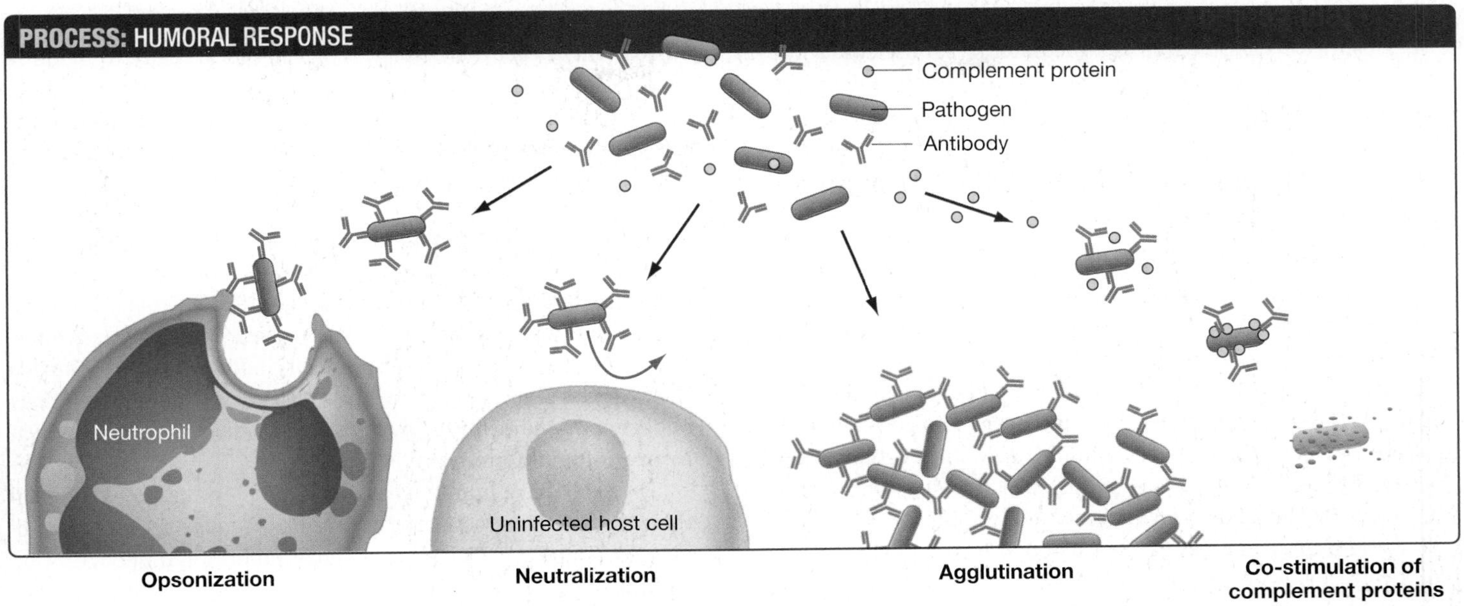

Figure 48.15 **The Humoral Response Eliminates Extracellular Pathogens.**

(Section 48.3). This means that if a cell were infected, peptides from the foreign proteins present inside the cell would be loaded onto some of the class I MHC proteins and presented to circulating cytotoxic T cells.

Cells that display the antigens of intracellular pathogens are effectively waving a flag that says, "I'm infected. If you destroy me, you'll destroy the infection." Elimination of the intracellular pathogens involves several steps (Figure 48.16).

Step 1 As cytotoxic T cells migrate into the area, those that recognize the class I MHC–peptide complex are stimulated to respond to the signal.

Step 2 After cytotoxic T cells recognize an infected target, they form a tight attachment that directs the secretion of molecules from the T cell to the target cell's surface. Adjacent cells are not exposed. Some of the molecules are proteins that assemble into pores in the target cell's plasma membrane. These pores allow other proteins from the T cell to pass directly into the cytoplasm of the target cell.

Step 3 Once in the cytoplasm, the T cell proteins activate a signaling cascade that causes the target cell to self-destruct via **apoptosis** (Ch. 21, Section 21.4). The result of apoptosis is the death and fragmentation of a cell into smaller vesicles, called apoptotic bodies, which are ingested by phagocytes like macrophages. Once the cytotoxic T cell has delivered this signal, often referred to as the "kiss of death," the T cell releases the dying cell and binds to another infected cell. Over time, all the infected cells are eliminated.

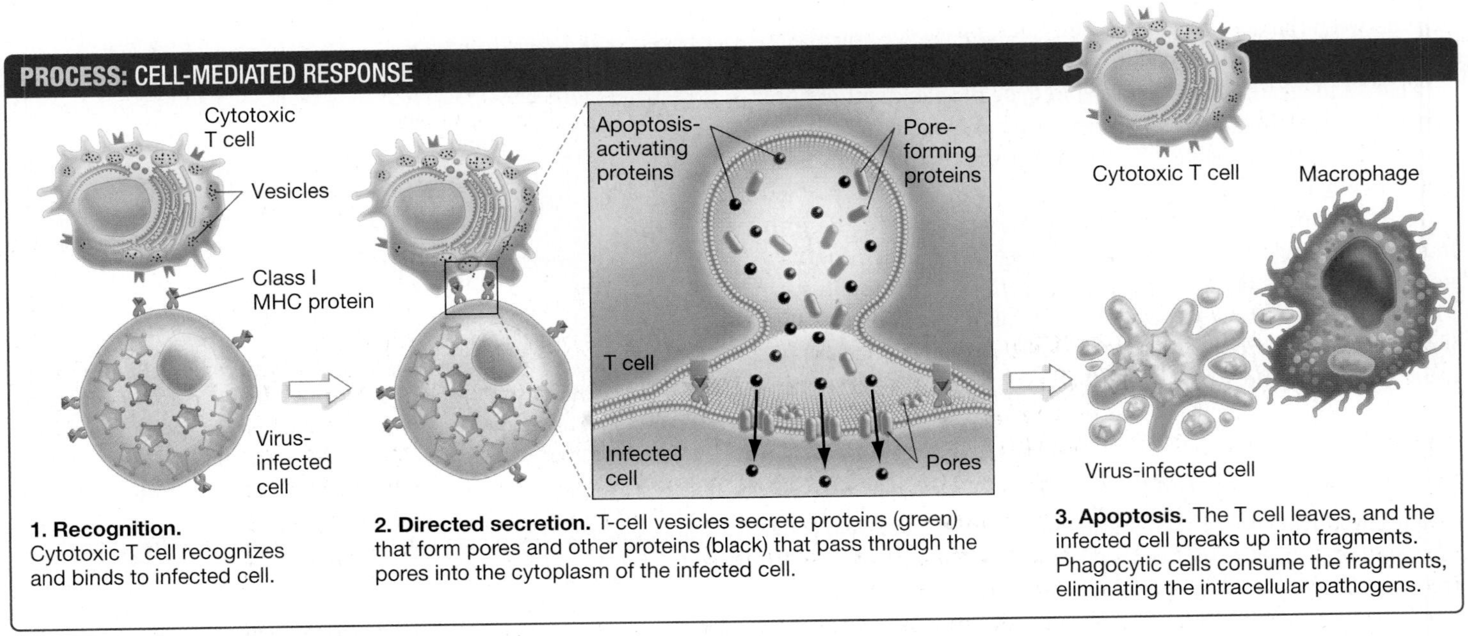

Figure 48.16 **The Cell-Mediated Response Eliminates Intracellular Pathogens.**

If the pathogen is able to replicate only inside host cells, as is the case with viruses, this cell-mediated response limits the spread of the infection by preventing the production of new generations of pathogens.

Why Does the Immune System Reject Foreign Tissues and Organs?

The innate and adaptive immune responses have devastating effects on invading pathogens. Unfortunately, they are equally deadly in their responses to tissues or organs that are introduced into a patient to heal a wound or cure a disease.

Consider the problems that can arise with blood transfusions. You might recall that certain individuals have red blood cells with membrane glycoproteins called A and B (Ch. 14, Section 14.5). These molecules act as antigens if they are introduced into a person whose own blood cells lack those glycoproteins.

For example, if you have type A blood, it means that your red blood cells have the A glycoprotein. If your red blood cells are transfused into a person who lacks the A glycoprotein—meaning someone who has type B or type O blood—the recipient's immune system will recognize the A glycoprotein as an antigen and mount a response against it. For a transfusion to be successful, the recipient must be given blood that contains the same glycoproteins found on his or her own red blood cells or that lacks the A and B glycoproteins entirely (type O).

Similar problems arise in tissue and organ transplants, except that the molecules directing rejection are the class I MHC proteins. There is an adaptive immune response against the foreign MHC proteins, and the innate immune response is activated based on the *absence* of "self MHC" signals. If transplanted cells do not display the same class I MHC proteins as host cells, components of innate immunity kill them.

To prevent strong immune reactions against transplanted organs or tissues, physicians seek donors who have MHC proteins that are extremely similar to those of the recipient, as is often the case for siblings. Even with close relatives, however, molecular differences will exist between the donor and recipient. Thus, physicians must also treat the recipient with drugs that suppress the immune response.

Thanks to steady improvements in drug development and in systems for matching MHC types between donors and recipients, the success rate for tissue and organ transplants has improved dramatically in recent years.

As the transfusion and transplant examples show, the immune system rejects foreign cells because they either contain nonself molecules or lack self molecules. To your immune system, a mismatched blood transfusion or an organ transplant is indistinguishable from a massive influx of bacteria, viruses, or other foreign invaders.

Responding to Future Infections: Immunological Memory

Besides producing the cells that implement the humoral and cell-mediated responses, activated B cells and T cells produce specialized daughter cells called **memory cells**. Memory cells do not participate in the initial adaptive immune response—the **primary immune response**. Instead, they provide surveillance after the original infection has been cleared. Memory cells remain in the spleen and lymph nodes for years or decades, ready to mount a rapid response should an infection with the same antigen reoccur.

The production of memory lymphocytes is a hallmark of the vertebrate immune response. How do memory cells protect an individual from future infections?

The Secondary Response Is Strong and Fast If the same antigen enters the body a second time, memory cells are able to recognize certain epitopes of the antigen and will trigger a second adaptive immune response, or **secondary immune response**. Figure 48.17 compares the rate of antibody production during the first and second exposures to a virus. The launching of a secondary immune response by means of memory cells is known as **immunological memory**.

The secondary immune response is faster and more efficient than the primary response. It is faster because the presence of memory T and B cells increases the number of lymphocytes that already have the correct antigen-specific receptors, thus decreasing the lag time for activating the adaptive immune response. It is more efficient because some of the memory B cells pass through another round of somatic hypermutation, the same process that occurred at the start of the primary immune response (see Section 48.3). Memory B cells with receptors that bind best to the antigen's epitopes live and produce daughter cells.

To review the adaptive immune response, including how B cells and T cells are activated, expanded, and differentiated into the various effector cells, study Table 48.4.

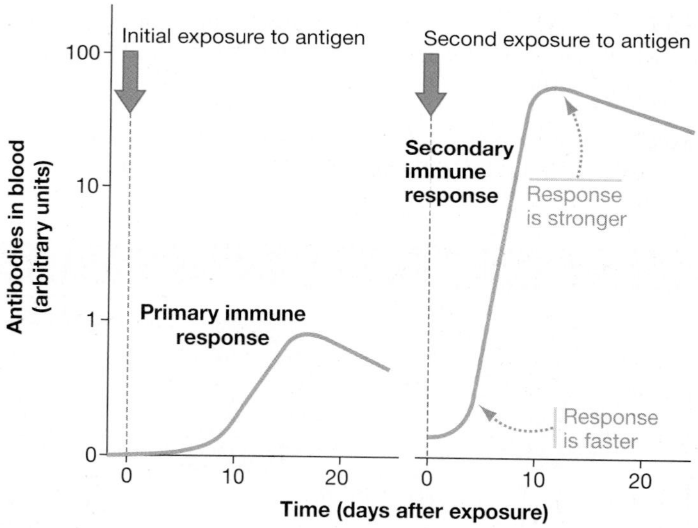

Figure 48.17 The Secondary Immune Response Is Faster and Stronger than the Primary Response. The two curves summarize results that are commonly observed when biologists inject the same antigen into a mouse at two different times and measure the concentration of antibodies to that antigen in the blood. The change in antibody concentration (plotted on a logarithmic scale) over time indicates the strength and speed of the response to the antigen.

✔ Imagine you injected a new antigen at the same time you began the second exposure to the original antigen. Draw a line on the graph to show what you predict the response to this new antigen would be in terms of the concentration of antibodies to it.

SUMMARY Table 48.4 Cells of the Adaptive Immune Response

Type of Lymphocyte	Method of Activation	Cells That Result from Activation and Clonal Expansion	Function of Resulting Cells
B cells	Receptor binds to free antigen; class II MHC–peptide complex then interacts with TCR of T_H2 cell	Plasma cells	Secrete antibodies
		Memory B cells	Participate in secondary immune response
$CD4^+$ T cells	Receptor binds to class II MHC–peptide complex on dendritic cell or other antigen-presenting cell	T_H1 helper T cells	Activate cytotoxic T cells; regulate inflammatory response; activate the cell-mediated response
		T_H2 helper T cells	Activate B cells; activate the humoral response
		Memory T cells	Participate in secondary immune response
$CD8^+$ T cells	Receptor binds to class I MHC–peptide complex on dendritic cell or other antigen-presenting cell; T cell is then stimulated by cytokines secreted by T_H1 cells	Cytotoxic T cells	Kill infected host cells that present complementary class I MHC–peptide complexes
		Memory T cells	Participate in secondary immune response

Vaccination Leads to Immunological Memory The production and effectiveness of memory cells explain phenomena that have been observed throughout recorded history. For example, in the middle ages, Chinese and Turkish practitioners protected people from the smallpox virus (Ch. 33, Section 33.4) by intentionally exposing them to the dried crusts of smallpox pustules taken from infected individuals. This procedure is an example of **immunization**—the conferring of immunity to a particular disease.

In the late 1700s Edward Jenner refined this immunization technique. In Jenner's day, milkmaids' faces were not pockmarked with scars from smallpox infections. Jenner knew that cows often had a smallpox-like disease called cowpox. He hypothesized that milkmaids became immune to smallpox because they had been exposed to cowpox while milking cows.

To test this hypothesis, Jenner exposed a boy to fluid from a cowpox pustule. Later he exposed the same child to fluid from a smallpox pustule. As predicted, the boy did not contract smallpox. Jenner's technique was named **vaccination**, from the Latin root *vacca*, which means "cow" (Figure 48.18). Vaccination is the introduction of a **vaccine**, a preparation containing antigens from a weakened or altered pathogen. The vaccine mimics an infection, priming the body's immune system to effectively fight off later encounters with the unaltered pathogen.

Figure 48.18 **Vaccination with Fluid from a Cowpox Pustule.** (1901 drawing from the French weekly magazine *Le Petit Journal*).

The data in Figure 48.17 explain why vaccination is an effective defense against certain pathogens—it speeds up the body's response to an infection. The antigens used in a vaccine are usually components of a pathogen's exterior because antibodies can readily attach to them.

Consider the three general types of vaccines against viruses:

1. *Subunit vaccines* consist of isolated viral proteins. Familiar examples include vaccines against hepatitis B and influenza.
2. *Inactivated viruses* have been damaged by chemical treatments—often exposure to formaldehyde—or exposure to ultraviolet light. They do not cause infections but are antigenic. If you have been vaccinated for hepatitis A or polio, you may have received an inactivated virus. Human papilloma virus (HPV) vaccines are based on virus-like particles that are formed by HPV surface proteins. They are not infectious, because they lack the virus's DNA.
3. *Attenuated viruses* are also called "live" virus vaccines because they consist of complete virus particles that infect cells. Researchers make these viruses harmless by culturing them on cells from species other than the normal host. In adapting to the atypical cells, the viruses lose the ability to replicate rapidly in their normal host cells and are eliminated by the immune response before causing disease. The smallpox and measles vaccines consist of attenuated viruses.

After vaccination, the body mounts a primary immune response that produces memory cells. If a second infection occurs later, these populations of memory cells respond quickly and eliminate the threat before illness develops. Vaccinations function like fire drills or earthquake preparedness exercises—they prepare the immune system for a specific threat.

Vaccines have been powerful tools for preventing disease. Smallpox was eradicated from the human population by a global vaccination effort, and we are close to doing the same with polio. But other battles are under way as researchers work around the clock to develop vaccines against emerging diseases such as Ebola and Zika virus (Ch. 33, Section 33.3).

Unfortunately, the genomes of some viruses such as the human immunodeficiency virus (HIV) mutate rapidly, and thus they can code for antigens with modified amino acid sequences. Consequently, traditional vaccines have been ineffective for immunization against such diseases. Memory cells generated from HIV vaccination are unlikely to recognize the changed epitopes in the rapidly evolving strains. Currently, the only protection against HIV is preventing infection.

CHECK YOUR UNDERSTANDING

✔ **If you understood this section, you should be able to ...**

1. Explain how viruses are eliminated by both humoral and cell-mediated immune responses.
2. Compare and contrast inactivated and attenuated viruses, and predict which would result in a more effective cell-mediated immune response.

Answers are available in Appendix A.

48.5 What Happens When the Immune System *Doesn't* Work Correctly?

After you complete this section, you should be able to ...

- Use examples to discuss the problems that can arise when the human immune system doesn't function properly.

The vertebrate immune system is a marvel of adaptation. A healthy immune system can defeat the vast majority of infections without medical intervention and with little impact on the body. The immune system is a formidable threat to pathogens, but if its response is dysfunctional, it can also be a liability to an animal's health and even survival.

To appreciate this point, let's first examine what happens when the immune system activates inappropriate responses and then look at the consequences when it fails to respond.

Allergies

For some people, walking through a field of ragweed or petting a cat is a prescription for developing a runny nose, itchy eyes, and labored breathing. For other people, being stung by a bee or eating a peanut is a life-threatening experience. These are examples of **allergies**, or allergic reactions—abnormal immune responses to antigens.

For reasons that are still not clear, certain people produce the IgE class of antibodies (see Table 48.3) in response to specific molecules found in cat dander, nuts, plant pollen, or other substances. Molecules that trigger allergic reactions are called **allergens**.

Allergic reactions are considered inappropriate responses because the normal role of IgE antibodies is to defend against infections by parasitic worms. Recent research has shown that in areas where infections with intestinal worms are common, allergies are almost nonexistent. This finding suggests that allergies don't occur if IgEs are expressed normally—that is, in response to worm infections.

The production of IgE antibodies in an allergic reaction triggers a series of events known as the **hypersensitive reaction**. When a susceptible person is first exposed to an allergen, the IgE antibodies that are produced bind to receptors on mast cells and certain other white blood cells. This binding occurs at the constant region of the IgE heavy chains (see Figure 48.5). Once IgE is bound, the cells (and the person) are said to be sensitized.

If the person is later exposed to the same allergen, the variable regions of these previously bound IgE antibodies bind to the allergen molecules and cross-link them, signaling the cell to rapidly secrete histamine, cytokines, and other compounds. In response to these substances, blood vessels dilate and become more permeable, smooth muscle cells in the airways contract, and other cells secrete mucus (to review the inflammatory response, see Figure 48.3). These responses are common manifestations of the hypersensitive reactions known as hay fever, hives, and asthma.

In severe hypersensitive reactions, blood vessels can dilate to the point where blood pressure plummets, oxygen delivery to the brain is reduced dramatically, and the person loses consciousness. In addition, severe smooth muscle contractions in the digestive and respiratory tracts can induce vomiting, diarrhea, and complete constriction of the airways. This combination of events, known as anaphylactic shock, is lethal unless the person receives immediate medical attention.

Although allergic reactions are considered abnormal, they are still directed against substances that are foreign to the body. What happens when the immune system turns against the body's own tissues?

Autoimmune Diseases

An immune response directed against molecules or cells that normally exist in the host is known as **autoimmunity**. Autoimmune reactions often result in disease due to the destruction of the body's own cells and structures by lymphocytes of the adaptive immune response.

- Multiple sclerosis (MS) results from the production of cytotoxic T cells that attack the myelin sheath of nerve fibers (Ch. 43, Section 43.2). Because damage to myelin reduces the efficiency of nerve signaling, coordination problems result.
- Rheumatoid arthritis develops when self-reactive antibodies alter the lining of joints, causing painful inflammation.
- Type 1 diabetes mellitus occurs when cytotoxic T cells attack and kill insulin-secreting cells in the pancreas, resulting in a lack of insulin and inability to regulate blood glucose levels (Ch. 41, Section 41.4).
- Celiac disease is an autoimmune disorder of the small intestine triggered by ingested gluten proteins (Ch. 3, Case Study), resulting in damage to the villi, small fingerlike projections that line the small intestine (Ch. 41, Section 41.3). When the villi are damaged, nutrients cannot be absorbed properly into the body.

The mechanisms behind these diseases are still poorly understood, but an interesting connection is emerging for some. Studies conducted since the late 1960s suggest that individuals who live in homes with a high level of sanitation have an increased risk of developing both allergies and autoimmune diseases.

This observation has led to the development of the **hygiene hypothesis**, which states that autoimmune and allergic responses arise in individuals who, because of hygienic practices, have experienced less exposure to parasites. Recall that allergies—which involve inappropriate IgE-based hypersensitivity reactions—are virtually absent in people who live where intestinal worms, the normal target of IgE antibodies, are common. Such people also rarely have autoimmune diseases, even though autoimmune diseases do not involve the IgE-based hypersensitivity reaction.

This correlation between hygiene and immune disorders points to the close ties that have been forged during the coevolution of the immune system and the invaders it defends us against. Low infant mortality and other benefits of sanitary lifestyles appear to come at the cost of inappropriate immune responses.

But what happens when the immune system fails to respond at all?

Immunodeficiency Diseases

Children who are born with a genetic disorder called severe combined immunodeficiency (SCID) lack a normal immune system and are unable to fight off infections. They are even susceptible to "opportunistic" pathogens that would normally be incapable of causing disease. If not given a bone-marrow transplant or kept in a completely sterile environment, children with SCID typically will die before they are 2 years old.

In a similar way, people who are infected with the **human immunodeficiency virus (HIV)** suffer from a progressive failure of the immune system (Ch. 33, Section 33.1). HIV infects and kills $CD4^+$ T cells. As the infection continues, populations of $CD4^+$ T cells gradually decline to a point where the immune system can no longer mount an effective adaptive immune response to infection. Eventually, HIV-infected people develop **acquired immune deficiency syndrome (AIDS)**. Like those with SCID, people with AIDS will succumb to illnesses that physicians almost never see in people with healthy immune systems.

Over the past three decades, HIV infections have destroyed the immune systems of millions of people. This fact, coupled with the marked increase in the incidence of allergies and autoimmune diseases in the developed world, illustrates the importance of continued research into the many nuances of the immune system.

CHECK YOUR UNDERSTANDING

✔ **If you understood this section, you should be able to . . .**

Predict the consequences of producing self-reactive IgE antibodies.

Answers are available in Appendix A.

CHAPTER 48 Review

For media, go to Mastering Biology

48.1 Innate Immunity: First Response

- Animals protect themselves from infection by establishing barriers that prevent the entry of pathogens.
- Innate immunity provides a rapid, generic response to broad classes of pathogens based on the recognition of pathogen-associated molecules by pattern-recognition receptors.
- During an inflammatory response, white blood cells of the innate immune response react to an infection by taking up foreign material and cells by endocytosis and by releasing cytokines that stimulate neighboring cells to respond to the infection and recruit more phagocytic cells into the area.

48.2 Adaptive Immunity: Recognition

- The vertebrate adaptive immune response possesses T cells and B cells that recognize specific epitopes on antigens via unique receptors found on the cells' surfaces.
- The genes that encode T-cell and B-cell receptor proteins are generated through gene recombination—a rearrangement of gene segments.
- Because every receptor that results from gene recombination is slightly different, the adaptive immune response can recognize and respond to an enormous variety of antigens.

48.3 Adaptive Immunity: Activation

- The clonal selection theory explains how the most appropriate cells for controlling an infection are selected and replicated to mount an effective adaptive immune response.
- T cells are activated when their receptors recognize epitopes displayed in MHC proteins on antigen-presenting cells. Full activation and differentiation of T cells often requires secondary signals from molecules called cytokines.
- B cells are activated when their receptors bind to free antigens. Most B cells differentiate into plasma cells and memory cells only when fully activated by helper T cells.

48.4 Adaptive Immunity: Response and Memory

- Extracellular pathogens are eliminated via phagocytosis and substances secreted in the humoral response, including antibodies. Antibodies contribute to the response by opsonizing, neutralizing, or agglutinating pathogens or by co-stimulating the lytic activity of complement proteins.
- Intracellular pathogens are eliminated via the cell-mediated response, in which infected host cells are induced to self-destruct (undergo apoptosis) by cytotoxic T cells.

- The immune system rejects blood transfusions and tissue transplants based on the presence of nonself molecules—often cell-surface glycoproteins or MHC proteins—or on the absence of self molecules.
- Memory lymphocytes produced during a primary infection or vaccination allow the immune system to respond rapidly and effectively to future infections by the same pathogen.

48.5 What Happens When the Immune System *Doesn't* Work Correctly?

- Allergies are abnormal immune responses to antigens called allergens.
- Autoimmune diseases occur when lymphocytes initiate an immune response against the body's own tissues.
- Immunodeficiency diseases (e.g., SCID and AIDS) arise from genetic mutations or viral infections that disrupt the function of key components of the immune system.

Answers are available in Appendix A.

✔ TEST YOUR KNOWLEDGE

1. What is the primary difference between the innate and adaptive immune responses?
a. The innate immune response does not distinguish between pathogens, while the adaptive immune response does.
b. Only the innate immune response is activated by antigens.
c. The adaptive immune response generates immunological memory and is more specific than the innate immune response.
d. The innate immune response does not kill cells; the adaptive immune response does.

2. The overall role of the inflammatory response is to
a. contain and eliminate foreign cells and material at the site of infection.
b. increase heat at the site of infection to activate enzymes used in the immune response.
c. produce antibodies that bind to and eliminate invading cells.
d. increase blood flow at the site of a wound to flush out invading pathogens.

3. What is the difference between an epitope and an antigen?

4. What is one of the differences between $CD4^+$ and $CD8^+$ T cells?
a. $CD4^+$ cells are immature, and $CD8^+$ cells are mature.
b. $CD4^+$ cells are activated, and $CD8^+$ cells are not.
c. $CD4^+$ cells interact with class II MHC proteins, and $CD8^+$ cells interact with class I MHC proteins.
d. $CD4^+$ cells activate cell-mediated responses, and $CD8^+$ cells activate humoral responses.

✔ TEST YOUR UNDERSTANDING

5. Explain how gene recombination leads to the production of vast numbers of different B-cell receptors.

6. What steps are required for most B cells to become fully activated and differentiate into plasma cells?

7. Why is clonal selection necessary for the adaptive immune response but not the innate immune response? Select True or False for each statement.
T/F The adaptive immune response uses receptors to recognize pathogens, and the innate immune response does not.
T/F There is more receptor diversity in the adaptive immune response than in the innate immune response.
T/F Cells in the innate immune response do not require activation, and those in the adaptive immune response do.
T/F Clonal selection is used for targeting pathogens, and the innate immune response is used only to stop blood flow from the wound.

8. What would a vaccine have to contain to protect a patient from chicken pox? Explain why we don't have vaccines for HIV.

✔ TEST YOUR PROBLEM-SOLVING SKILLS

9. Which of the following outcomes would be expected if somatic hypermutation did not occur?
a. The diversity of pattern-recognition receptors would be significantly lowered.
b. B and T lymphocytes would not be able to produce receptors that recognize antigens.
c. The adaptive immune response would not be activated by pathogens.
d. The secondary immune response to a repeat infection would produce the same antibodies as those made in the primary immune response.

10. Propose a hypothesis to explain how self-reactive B cells are identified and eliminated during maturation.

✔ PUT IT ALL TOGETHER: Case Study

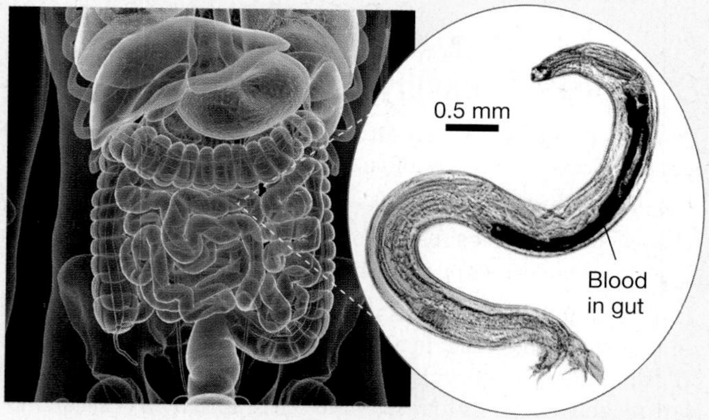

Are parasitic worms good for your immune system?

In developed countries, an enormous change has occurred within the human body over the past century—the loss of parasitic worms. Due to improvements in sanitation, roundworms that have inhabited human intestines (such as the hookworm above) and challenged our immune system for millions of years are no longer a threat. Does the end of this long-term relationship come at a cost?

11. What is the hygiene hypothesis? What correlation does it attempt to explain?

12. Explain how the adaptive and innate immune responses work together to defend the human body against infection by parasitic worms.

13. Pattern-recognition receptors have been identified that recognize worm products. Explain how these receptors are used to activate the adaptive immune response most appropriate for worm infection.

14. Parasitic worms modulate their host's immune response by inducing the proliferation of regulatory T cells. What role do these cells play in the adaptive immune response? How could this effect on regulatory T cells be linked to the hygiene hypothesis?

15. **QUANTITATIVE** The roundworm *Heligmosomoides polygyrus* is a natural intestinal parasite of mice, and it offers an excellent model of the immunology of worm infections in humans. Scientists evaluated the impact of parasitic roundworms on immune disorders using mice prone to developing type 1 diabetes mellitus. Five-week-old mice were infected with *H. polygyrus* (Hp). Two weeks later, half of the mice were cured of the infection (Rx). When the mice were 40 weeks old, scientists calculated the percentage of mice that developed diabetes in both groups: those exposed to roundworms and those in uninfected control groups (** means $P < 0.01$; see **BioSkills 3**). What two conclusions are supported by the results shown below?

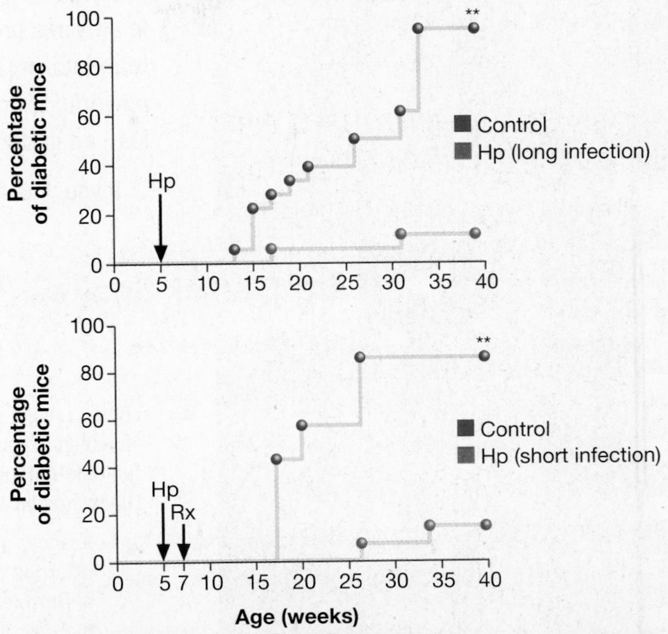

Source: P. K. Mishra et al. 2013. *Mucosal Immunology* 6: 297–308.

16. **SOCIETY** There is a strong association between the prevalence of diabetes mellitus in parents and their children. If you were a physician, would you ask parents who have type 1 diabetes to intentionally infect their children with intestinal worms? What additional information would you need to make this decision?

Mastering Biology

Students Go to Mastering™ Biology for assignments, the eText, and the Study Area with animations, practice tests, and activities.

Professors Go to Mastering™ Biology for automatically graded tutorials and questions that you can assign to your students, plus Instructor Resources.

END-OF-UNIT
CASE STUDY

UNIT 7

PUT IT ALL TOGETHER

■ For an introduction to the Revenge of the Newt case study, see page 17.

Now that you've learned the details of animal structure and function, it's time to check back in on the newts and garter snakes case study. As you've learned in previous units, many rough-skinned newt populations produce high levels of tetrodotoxin (TTX), but some garter snake populations have evolved resistance to this powerful toxin, allowing them to eat newts without dying. Let's use the material you learned in Unit 7 to better understand this fascinating predator-prey relationship.

✔ If you understand Unit 7, you should be able to apply your learning to this case study:

Do Garter Snakes Resistant to TTX Experience Trade-Offs?

Unit 7 revolves around the idea that the structure of an anatomical part is closely related to its function. Likewise at the molecular level, structural change imparted by mutations in voltage-gated sodium channels functionally allows resistant snakes to eat newts because TTX does not bind as strongly to the mutant channels.

1. Identify the method by which sodium moves through voltage-gated channels in neurons and muscles. (See Section 40.1)
   a. Simple diffusion
   b. Facilitated diffusion
   c. Primary active transport
   d. Secondary active transport
2. **MODEL** Use Figure 43.4 as a model to diagram how the action potentials would compare in the tissues of a resistant garter snake and a non-resistant garter snake when consuming a toxic newt. (See Section 43.1)
3. Describe the process by which some garter snake populations evolved resistance to newt TTX. (Chapter 22)
4. Predict a possible trade-off associated with the evolution of TTX resistance. (See Section 39.1)

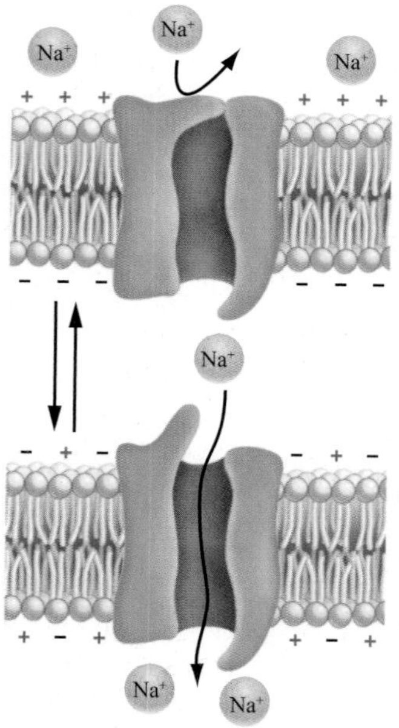

Changes in the shape of voltage-gated channels are responsible for changes in a membrane's permeability to sodium ions.

You learned earlier that TTX-resistant garter snakes have an advantage over TTX-sensitive garter snakes when they consume TTX (Unit 4 case study on pages 530–531). But what about when they consume other, non-toxic prey? Can we observe a trade-off in function associated with the structural difference in the sodium channel?

5. **PROCESS OF SCIENCE** Design a study to test the hypothesis that TTX-resistant garter snakes experience a trade-off in physiological function. (See Section 39.1)

Researchers Chris Feldman and Normand Leblanc, and their graduate students Robert del Carlo and Jessica Reimche, recently sought to answer this question by comparing neuronal and muscular function in TTX-sensitive and TTX-resistant garter snakes in the absence of TTX.

First, the team measured the flow of sodium ions through voltage-gated sodium channels in TTX-sensitive and TTX-resistant garter snakes when exposed to a standardized stimulus. Results are shown in **Figure 1**. Sodium ion flow is expressed in units of picoamperes (pA) per picofarad (pF); the more negative the value, the more ions are flowing across the membrane per unit of time.

6. **QUANTITATIVE** Contrast ion flow through voltage-gated sodium channels in TTX-sensitive (blue line) and TTX-resistant (red line) garter snakes.

Figure 1

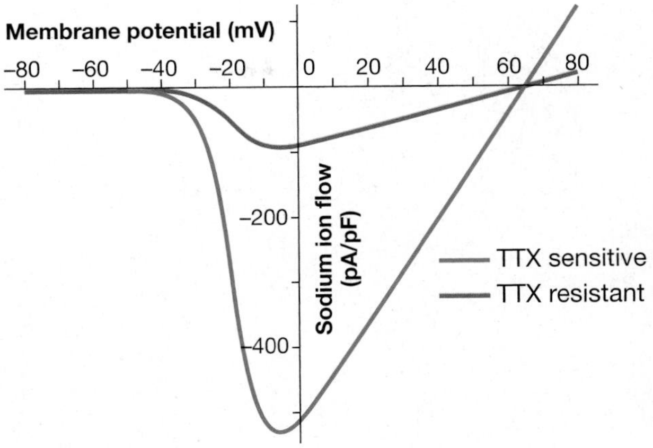

DATA: R. del Carlo, J. Reimche, N. Leblanc, and C. Feldman. 2018. Personal communication.

The research team also examined whether muscle function differs in TTX-sensitive and TTX-resistant garter snakes. They took muscle samples from both types of snake and measured the force the muscle produced in response to standardized stimuli. Resulting data are shown in **Figure 2**.

7. **QUANTITATIVE** Contrast the mean force generated by the muscle tissue of TTX-sensitive (blue line) and TTX-resistant (red line) garter snakes.
8. Propose a hypothesis to explain what is responsible for the difference in ion flow through the channels (Figure 1) and in force generated by muscle in the TTX-sensitive and TTX-resistant garter snakes (Figure 2). (See Sections 43.2 and 45.1)
9. Given both sets of data, suggest at least two additional physiological systems whose functions could be impacted by TTX-resistance. Describe the specific mechanisms by which the mutant sodium channel could affect function.

Figure 2

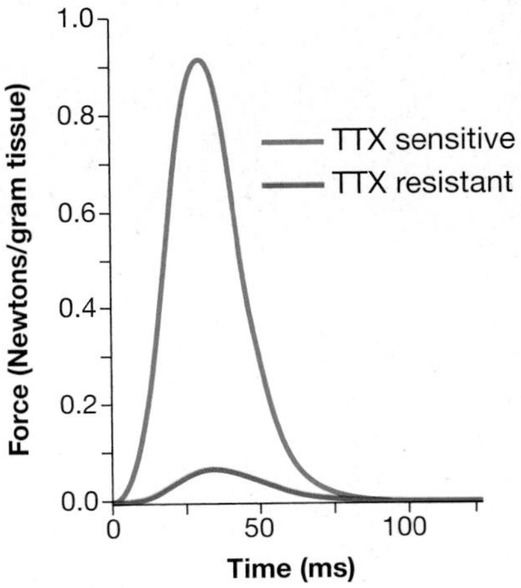

DATA: R. del Carlo, J. Reimche, N. Leblanc, and C. Feldman. 2018. Personal communication.

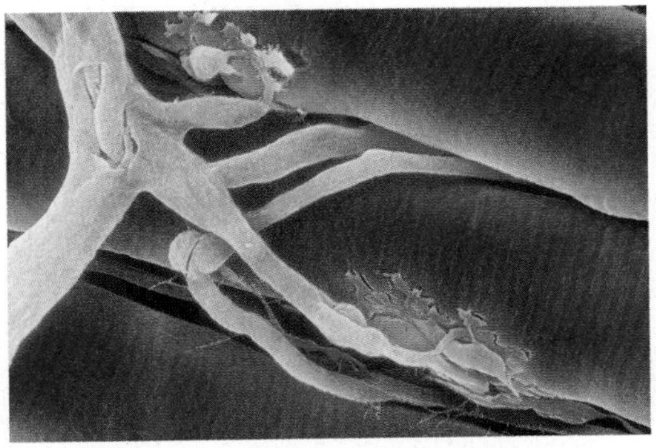

Colored scanning electron micrograph of nerve synapses (orange) and muscle fibers (blue).

10. Based on the evidence, do you think that garter snakes that are resistant to TTX experience trade-offs? Explain your reasoning.

How do these insights play out in garter snakes in the wild? Researchers do not yet know. It is possible that the weaker neural responses and muscle strength could mean that TTX-resistant garter snakes experience more difficulty capturing prey, are less able to escape from predators, tire more easily when competing for mates, or many other possibilities. So, when eating newts, resistant snakes have the upper hand, but otherwise they could be outcompeted by TTX-sensitive snakes. As you read the next unit (Unit 8), think about ways in which this this population variability could play out in an ecological setting.

Each unit ends with a continuation of this story. ■

49 An Introduction to Ecology

Toucans eat palm fruits, dispersing the seeds. Humans also consume these fruits (açaí berries) yet threaten many forests with clear-cutting. Increasingly, ecology requires the study of how humans affect—yet depend on—healthy ecosystems for life.

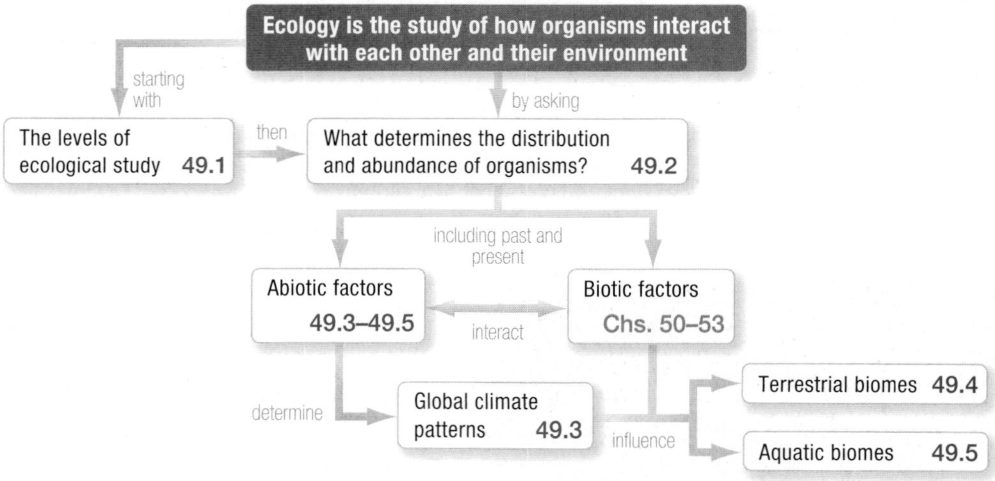

The study of how organisms interact with each other and their environment is called **ecology**. Just as molecules interact with each other in complex systems within cells, and cells and tissues interact with each other in complex systems in multicellular organisms, so too do organisms interact with each other and the nonliving environment in complex ecological systems.

The German biologist Ernst Haeckel first defined "oekologie" less than a decade after Charles Darwin published *On the Origin of Species by Means of Natural Selection* in 1859 (Chapter 22). The timing was no coincidence—evolution and ecology are intertwined and codependent. The diversity of life on Earth (Unit 5) is the result of evolutionary processes (Unit 4) that played out in a changing ecological context over 3.5 billion years.

BIG PICTURE

This chapter is part of the Big Picture. See how on pages 1190–1191.

Other fields of biology are increasingly integrating with ecology as well. In this unit, you'll be applying what you've learned about biochemistry, genetics, development, physiology, evolution, and the diversity of life to ecological questions. Ecologists work in subdisciplines ranging from ecological developmental biology to physiological ecology and conservation genetics.

A major trend in ecology is the increasing focus on human impacts. Except for the episode 66 million years ago, when a mountain-sized asteroid struck Earth, environments are changing more rapidly right now than they have at any time in the past 3.5 billion years. Humans are causing widespread habitat loss and degradation, pollution, introduction of exotic organisms, and climate change—transforming Earth in unprecedented ways.

Yet at the same time, human health and welfare depend on healthy ecosystems. Thus it is no exaggeration to say that the work of ecologists is essential to human endeavors that range from food security and economics to the avoidance of war. As the evolutionary developmental biologist and educator Sean Carroll wrote in his 2016 book, *Serengeti Rules*:

> If the motto for the twentieth century was "Better Living through Medicine," the motto for the twenty-first century must be "Better Living through Ecology."

Let's start by reviewing the levels of ecological study, then consider the types of variables that determine the distribution and abundance of organisms in terrestrial and aquatic environments.

49.1 Levels of Ecological Study

To understand how organisms interact with each other and their environment, ecologists work at five main levels: **(1)** organisms, **(2)** populations, **(3)** communities, **(4)** ecosystems, and **(5)** the biosphere. The levels aren't rigid or exclusive; researchers routinely draw on multiple levels—including lower levels—to explore a particular question. (You can review all the levels of organization from atoms to the biosphere in The Big Picture of Doing Biology on pages 18–19). To get started, let's consider how the levels of ecology apply to the example of sockeye salmon (**Table 49.1**).

After you complete this section, you should be able to . . .

- Compare the levels of ecological study.

Organismal Ecology

At the organismal level, researchers explore morphological, physiological, and behavioral **adaptations**—heritable traits that increase the fitness of an individual in a particular environment relative to individuals lacking the trait (Chapter 50).

For example, after spending four or five years feeding and growing in the ocean, sockeye salmon travel hundreds or thousands of kilometers to return to the freshwater stream where they hatched. Females create nests in the gravel stream

SUMMARY Table 49.1
Five Main Levels of Ecological Study

Level	Sockeye Salmon Example
Organismal ecology How do individuals interact with each other and their physical environment? Salmon migrate from saltwater to freshwater environments to breed	
Population ecology How and why does population size change over space and time? Each female salmon produces thousands of eggs. On average, only a few offspring will survive to return to the same stream to breed	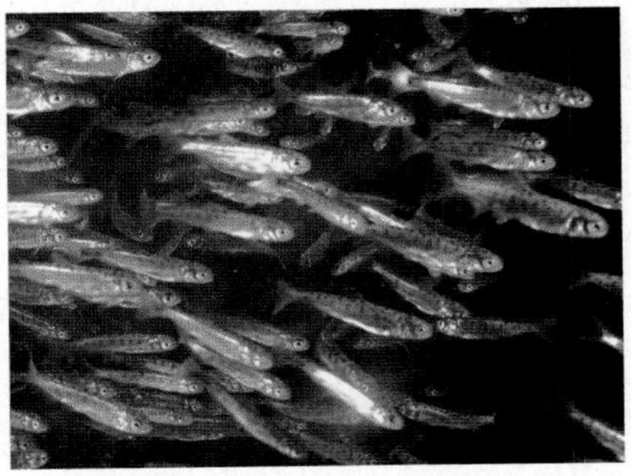
Community ecology How do species interact, and what are the consequences? Salmon are prey as well as predators	
Ecosystem ecology How does energy flow and how do nutrients cycle through the local environment? Salmon die and then decompose, releasing nutrients that are used by other organisms	
Global ecology How is the biosphere affected by global changes in nutrient cycling and climate? Worldwide populations of salmon are affected by changes in water flow and temperature due to global climate change	

bottom and lay eggs. Nearby males compete for the chance to fertilize eggs as they are laid. When breeding is finished, all the adults die.

What variables might a biologist measure at this level?

- Physiological function of sodium pumps in gills as salmon move between saltwater and freshwater environments
- Migration pathways of individuals
- Behavior of males competing for access to eggs

Population Ecology

A **population** is a group of individuals of the same species that lives in the same area at the same time. When biologists study population ecology, they focus on how the number and distribution of individuals in a population change over time (Chapter 51).

For example, researchers use mathematical models to predict the future of salmon populations. Many of these populations have declined due to overharvesting and degradation of habitat caused by pollution and dam construction. If the factors that affect population size can be described accurately enough, mathematical models can assess the impact of proposed dams, changes in weather patterns, altered harvest levels, or specific types of protection efforts.

What variables might a biologist measure at this level?

- Number of salmon spawning in streams each year
- Number of eggs produced by females
- Geographic range of salmon catches or sightings

Community Ecology

A biological **community** consists of populations of different species that interact with each other within a particular area. Community ecologists ask questions about the nature of the interactions between species and the consequences of those interactions. Research might concentrate on predation, parasitism, and competition, or explore how communities respond to fires, floods, and other disturbances (Chapter 52).

As an example, consider the interactions among salmon and other species in the marine and stream communities where they live. When they are at sea, salmon eat smaller fish and are themselves hunted and eaten by orcas, sea lions, humans, and other mammals; when they return to fresh water to breed, they are preyed on by bears and bald eagles. In both marine and freshwater habitats, salmon are subject to parasitism and disease. They are also heavily affected by disturbances—particularly changes in their food supply and the quality of their breeding streams.

What variables might a biologist measure at this level?

- Prey species in gut contents of salmon
- Density of gill parasites in salmon in different streams
- Fish biodiversity in different stream communities

Ecosystem Ecology

Ecosystem ecology is an extension of organismal, population, and community ecology. An **ecosystem** consists of all the organisms in a particular region along with nonliving components. These physical and chemical, or **abiotic** (literally, "not-living"), components include air, water, and the nonliving parts of soil. At the ecosystem level, biologists study how nutrients and energy move among organisms and through the surrounding atmosphere and soil or water (Chapter 53).

Salmon are interesting to study at the ecosystem level because they link marine and freshwater ecosystems. They harvest nutrients in the ocean and then, when they migrate, die, and decompose, they transport those molecules to streams and streamside forests. In this way, salmon transport chemical energy and nutrients from one habitat to another.

What variables might a biologist measure at this level?

- Decomposition rate of salmon in different streams
- Levels of soil nitrogen in forests with and without salmon streams
- Rate of sedimentation of streams in deforested watersheds

Global Ecology

Biologists define the **biosphere** as a thin zone surrounding the Earth where all life exists—the sum of all terrestrial and aquatic ecosystems. It extends from about 5 km below the land surface and 10 km down in the deepest trenches of the sea to over 10 km up into the atmosphere. This may sound like a thick layer, but it represents only 0.002 percent of Earth's diameter—thinner by comparison than the peel on an apple—and most of life occupies a much smaller zone within the biosphere.

The young field of global ecology is growing rapidly as ecologists scramble to quantify the effects of human impacts on the biosphere. Humans are modifying landscapes and releasing massive amounts of energy and nutrients to the biosphere, causing changes that transcend individual ecosystems.

Since the biosphere is practically a closed system, the actions of people in one part of the world may alter distant ecosystems, creating difficult challenges for education and public policy. For example, salmon are sensitive to changes in water temperature and water flow in streams. Who is accountable for the impacts of climate change on salmon?

Which variables might a biologist measure at this level?

- Frequency of floods and droughts in global salmon habitats
- Northward shift of salmon due to climate change

Conservation Biology Applies All Levels of Ecological Study

The five levels of ecological study are synthesized and applied in conservation biology. **Conservation biology** is the effort to study, preserve, and restore threatened genetic diversity in populations, species diversity in communities, and ecosystem function (Chapter 54). Conservation biologists are like physicians. But instead of prescribing treatments for sick people, they prescribe remedies for threatened environments to preserve biodiversity, clean air, pure water, and productive soils.

Each chapter in this unit focuses on a different level of ecological study, ending with the study of biodiversity and conservation. Turn to the Big Picture of Ecology (pages 1190–1191) to visualize how the different levels of ecology interact. In this introductory chapter, let's take a global look at factors that determine the distribution and abundance of organisms. In particular, what abiotic factors set the stage for life on Earth?

CHECK YOUR UNDERSTANDING

✔ **If you understood this section, you should be able to . . .**

Indicate the main level of ecological study addressed by this question: How does the decline of bee pollinators affect fruit production in apple trees?

Answers are available in Appendix A.

49.2 What Determines the Distribution and Abundance of Organisms?

When you look around at the plants and animals in your neighborhood, you can tell right away that you are not located deep in the Amazon rain forest—no scarlet macaws, jaguars, anacondas, blue morpho butterflies, or açaí palms out there. Different species live in different places. Why?

After you complete this section, you should be able to . . .

- Compare the abiotic and biotic factors that affect the geographic distribution and abundance of organisms.

The range of conditions that a species can tolerate and the range of resources it can use is referred to in ecology as its ecological **niche** (pronounced *nitch*). A niche, in turn, is a product of many abiotic and biotic factors, both present (that you can go outside and measure) and past (whether a century ago or 3 billion years ago).

Let's consider a tropical plant, the açaí palm (see the chapter-opening photo) as a case study. You may be familiar with açaí berries (pronounced *ah-SIGH-ee*), a popular ingredient of fruit bowls and smoothies. Why do açaí palms, *Euterpe oleracea*, live where they do? Use the summary in **Table 49.2** as a framework to compare abiotic and biotic variables that determine the geographic range of this plant.

Present Abiotic Factors

No one species can survive the full array of environmental conditions present on Earth due to **fitness trade-offs**, the evolutionary compromise that must be made between traits that cannot be optimized simultaneously (see Ch. 22, Section 22.5). Organisms tend to be adapted to a limited set of abiotic conditions, such as a particular temperature and moisture regime on land or a particular salinity, water depth, and flow regime in water.

Açaí palms live in the tropics where it's hot and wet. Ecologists have studied açaí palms in their natural habitat, making several observations:

- Açaí palms cannot tolerate freezing temperatures.
- Açaí palms can tolerate heat as long as they are not dry.
- Açaí palms require wet soils that are occasionally flooded.
- Açaí palms cannot tolerate salt water.

SUMMARY Table 49.2
Abiotic and Biotic Factors Interact

ABIOTIC FACTORS | **BIOTIC FACTORS**

Present Abiotic
Which nonliving factors affect the survival and reproduction of açaí palms?

- Sunlight
- Temperature
- Precipitation
- Flood frequency
- Salinity of floodwater

Present Biotic
Which organisms interact with açaí palms?

- Toucans consume and disperse seeds
- Palms compete with many other plants for access to sunlight
- Fungal parasites can kill palm trees

Past Abiotic
Which past abiotic factors influenced the distribution of palms?

- Continental drift
- Rising of Andes mountain range
- Sea-level changes
- Closing of Isthmus of Panama

Past Biotic
Which past biotic factors influenced the distribution of açaí palms?

- Interactions with organisms such as the giant sloth, now extinct
- Settlement of Amazon rain forest by humans about 15,000 years ago

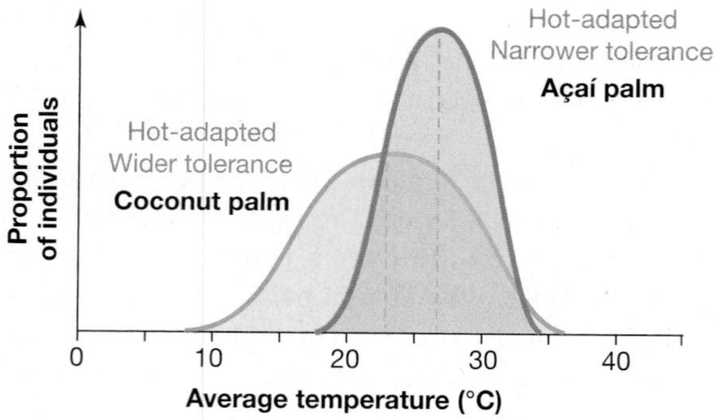

Figure 49.1 Niche Models Show the Range of Conditions Tolerated by a Species.

Temperature is particularly important in determining where organisms can live because temperature has a major impact on physiology, especially for organisms like palms that have limited ability to regulate their own temperatures. Enzymes are particularly sensitive to temperature and tend to function optimally in a narrow temperature range (Ch. 8, Section 8.4). That's why temperature can control the metabolic rate of the organisms and thus all aspects of their biology, including growth and reproduction—and survival.

Water availability can also be a matter of life and death. About 75 percent of the volume of a typical cell is water, which serves as a solvent, a source of physical support, a temperature buffer, and many other roles (Ch. 2, Section 2.2). Organisms vary tremendously in their adaptations for acquiring and retaining water.

How can you visualize the tolerance of a species to temperature, water, and other abiotic variables? One of the great contributors to modern ecology, G. Evelyn Hutchinson, proposed that a species' ecological niche could be visualized using a model. Consider the simplest case of the model by focusing on a single variable, temperature.

Figure 49.1 shows the proportion of coconut palms and açai palms that can tolerate different average temperatures. The resulting model is a set of curves that represents the temperature niche (thermal niche) for each species. In this example, you can see that the coconut palm and açai palm are both tropical species preferring relatively high average temperatures (27°C translates to about 81°F; see **BioSkills 1**). The açai palm, however, has a narrower temperature niche with a greater heat tolerance. Note that even though a species can survive at a certain temperature, it will not necessarily thrive there—the fitness of organisms tends to be lower near the extremes of the niche.

Niche models are a useful tool for quantifying and simplifying complex ecological relationships—they can be used to explore theoretical principles, to make predictions at the start of a study, or to visualize data at the end of a study, and they can include one or many variables. For more practice with niche models, see **Making Models 49.1**.

It is important to remember that temperature, precipitation, and other abiotic variables can vary from place to place within a geographic range, and from hour to hour, day to day,

Making Models 49.1 Tips on Drawing Graphs (II): Niche Models and Abiotic Tolerance

There are three elements to draw in the simplest niche model: (1) the *x*-axis is the environmental variable of interest; (2) the dependent variable on the *y*-axis is the number or frequency of individuals in the species; and (3) the curve shows the range of the condition tolerated and the relative abundance of individuals at each part of the range.

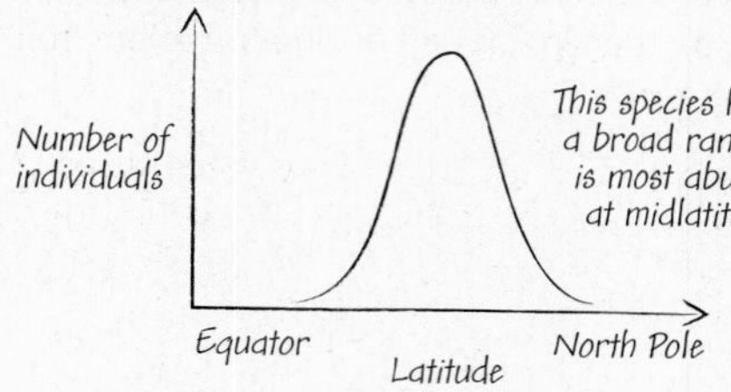

MODEL Draw your own niche model for a species of palm tree that is most abundant near the equator, rare at midlatitudes, and absent farther north.

To see this model in action, go to the Study Area of **Mastering Biology**

and year to year. This variation is biologically important—not an "error" in measurement—and can make the ecologist's job of quantifying ecological processes more interesting and challenging, especially because climate change is causing the variability of temperature and precipitation to increase.

Present Biotic Factors

Recall that the definition of niche refers to the range of conditions that a species can tolerate and the range of resources it can use. What kind of resources? These resources can include not only abiotic factors such as sunlight but also **biotic** ("living") factors related to other organisms, such as access to pollinators or protection from herbivores.

What kinds of biotic interactions matter to açaí palms? Consider that one important characteristic determining the geographic distribution of species is its **dispersal**, the movement of individuals from their place of origin to the location where they live and breed as adults. For organisms like açaí palms that can't move, how do they disperse?

You may not think that when a toucan *eats* açaí berries (see photo in Table 49.2), it is helping the açaí palm. When birds like toucans consume the fruits and then fly away and poop out the seeds elsewhere, they are unintentionally dispersing the seeds. In fact, these seeds have a *higher* rate of germination after passing through the bird's gut than they do if the fruit falls directly to the ground. Palms and toucans have coevolved in a mutualistic relationship, each influencing the survival and geographic distribution of the other.

But not all biotic interactions are beneficial. For example, some species of fungus can parasitize palms, stealing their nutrients and lowering their fitness—even causing death. Also, palms compete with many other plants for access to resources. All interactions among species in a community, whether beneficial or harmful, can influence the distribution of the species involved (explored in detail in Chapter 52).

Past Abiotic Factors

The Earth is constantly changing. According to the theory of plate tectonics, the landforms and oceans that appear static now have been in constant flux for the entire history of life. Mountains rise and fall and whole continents crash together and break apart due to *continental drift*, the slow movement of Earth's surface plates over millions of years (Figure 49.2).

What's more, fluctuations in climate cause average temperatures and patterns of precipitation to change, glaciers and ice caps to form and melt, sea levels to rise and fall, and islands and lakes to form and disappear. The scope of change is difficult to fathom just by looking out the window wherever you are today.

Alfred Russel Wallace was among the first to recognize the connection between historic geographic changes and the present distribution of organisms. In addition to codiscovering evolution by natural selection (Chapter 22), Wallace founded the study of **biogeography**—the study of how organisms are distributed geographically and through geologic time.

How could past abiotic changes affect the distribution of the 2600 described species of palms today? Here are some examples of research questions that you could study:

- Did palms occur on the ancient continent of Pangaea before it broke apart?
- Are there palms in western Africa that are closely related to the palms in eastern South America?
- Did the formation of the Isthmus of Panama—the land bridge connecting North America and South America—enable palms to disperse north or south?
- When the Andes Mountains arose along the western rim of South America causing the massive Amazon river system to flow east, did the rivers fragment the palm populations, leading to speciation (see Ch. 24, Figure 24.6)?

Past Biotic Factors

As major changes occurred in abiotic factors over geologic time, they did so in concert with biotic factors. Which of these biotic factors might have influenced the distribution of açaí palms?

You can consider this question at different time scales—such as from the evolution of plants from water onto land about 475 million years ago (Ch. 28, Section 28.3) to the more recent extinction of species that were once part of tropical forest communities. For example, giant sloths—herbivores the size of elephants (see image in Table 49.2)—may have shaped species interactions in South America for thousands of years, but most had disappeared from South America by about 10,000 years ago.

Without a doubt, interactions with humans have dominated the species interactions of the recent past. Modern humans began dispersing around the globe from Africa about 80,000 years ago, probably reaching South America about 15,000 years ago (Ch. 32, Section 32.6). Over thousands of years, humans

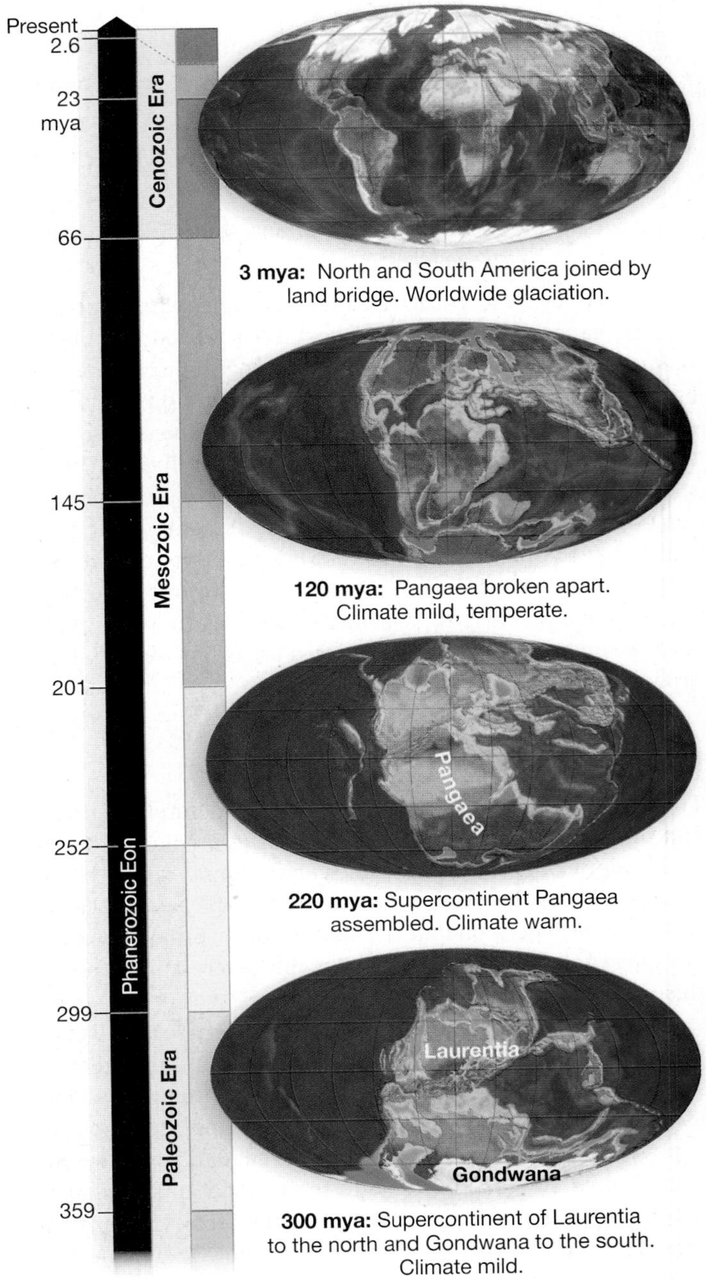

Figure 49.2 Continental Drift Causes a Radical Change in the Shapes and Positions of Continents and Oceans over Time. The movement of continents throughout the history of life has affected the total amount of land area, the relative amounts of land in the tropics versus northern latitudes, and the nature of ocean currents—and thus the distribution of organisms over time.

DATA: R. Blakey. Colorado Plateau Geosystems, Arizona USA.

have influenced the distribution of species like açaí palms in many ways:

- Land-use changes, such as clearing forests and damming streams
- Hunting, fishing, and other forms of harvesting, especially of large mammals such as giant sloths
- Dispersing organisms to new locations, sometimes purposely and sometimes by accident

It should be no surprise that ancient peoples who valued açaí berries, bananas, and other tropical foods traded and planted them in new areas, increasing their geographic range. Such dispersal sounds like it would always be a good thing. But if an **exotic species**—one that is not native—is introduced into a new area, spreads rapidly, and competes successfully with native species, it is said to be an **invasive species**. Since açaí palms are somewhat finicky in their growing requirements, their transplantation to other areas, such as Hawaii, is not considered a threat. However, other invasive species introduced to South America are a threat to açaí palms, such as invasive grasses that become established in deforested areas, preventing the regrowth of native palms.

Looking to the Future

The study of ecology involves not only consideration of present and past abiotic and biotic factors but also predictions for the future. For example, how will the current rapid rate of climate change affect the geographic distribution of açaí palms? Açaí berries have become economically important in Brazil, so beyond biological interest, there is an economic interest in preserving this species for the future.

Figure 49.3 is an example of how ecologists Úrsula Vaz and João Nabout approached the climate question. First, they observed the current distribution of açaí palms to make a model of the species' abiotic niche in terms of temperature and precipitation. Then the researchers entered data on published climate predictions for 2080 into their model to see what the geographic distribution of açaí palms might look like in the future. The maps in the "Results" section show the present and predicted future geographic ranges of suitable habitat for açaí palms. What do the data say? And what is not included in this data set?

Here are three take-home messages from this research that you can bear in mind for the rest of the chapter.

1. ***Start with the basics, then add more complexity.*** Vaz and Nabout's study results suggest that the range of suitable areas for palms increases by 2080, contracting in some areas but expanding in others. This looks like good news! Even so, the researchers point out that their model did not yet include some important variables. For example, some of the predicted growth areas occur in parts of Brazil where deforestation rates are highest. Also, biotic interactions are not included. Toucans and other large birds are being hunted out of some fragmented forests, reducing the survivorship and dispersal ability of palms.

RESEARCH

QUESTION: Will climate change affect the geographic distribution of açaí palms?

HYPOTHESIS: Climate change will cause a change in the distribution of açaí palms.

NULL HYPOTHESIS: Climate change will not affect the distribution of açaí palms.

STUDY DESIGN:

1. **Identify 95 locations** in South America where açaí palms have been described.
2. **Niche modeling:** Use a computer program to measure these five climatic variables in the 95 locations based on a database of recorded climate records:
   - Annual **mean temperature**
   - Annual **range** of **temperatures**
   - **Precipitation** in **wettest** month
   - **Precipitation** in **driest** month
   - **Precipitation** in **warmest** month
3. Obtain **climate change predictions** for South America in 2080 based on published mathematical models.
4. **Construct maps** of suitable growth areas for açaí palms today and in 2080 and compare their distributions.

PREDICTION OF HYPOTHESIS: The areas of suitable growth for açaí palms will shrink, grow, or move to new locations by 2080.

PREDICTION OF NULL HYPOTHESIS: The areas of suitable growth for açaí palms will be the same in 2080 as they are now.

RESULTS:

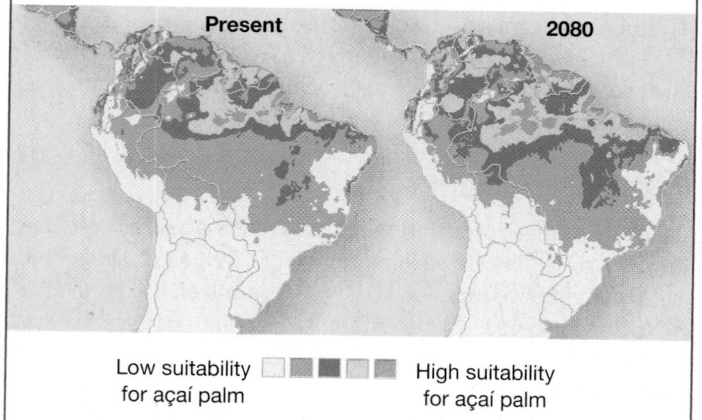

CONCLUSION: The suitable area for the growth of açaí palms increases slightly, contracting in some areas and expanding in others, but factors not included in the model may reduce the range of palms.

Figure 49.3 **Where Will Açaí Palms Grow in 60 Years?**

SOURCE: Ú. L. Vaz and J. C. Nabout. 2016. Using ecological niche models to predict the impact of global climate change on the geographical distribution and productivity of *Euterpe oleracea* Mart. (Arecaceae) in the Amazon. *Acta Botanica Brasilica* 30(2): 290–295.

✔ **PROCESS OF SCIENCE** The map was divided into small squares and then the data were analyzed for each square. Evaluate the benefits of using small squares versus large squares for the analysis.

2. ***Ecology is interdisciplinary.*** Like many studies in ecology, the work by Vaz and Nabout is collaborative. Increasingly, ecology involves international teams of scientists from diverse fields including geography, mathematics, climatology, geology, and chemistry. It also requires collaboration with private and public organizations such as local governments and farming corporations.
3. ***Humans degrade ecological systems yet depend on ecological systems.*** Climate change, deforestation, hunting, pollution, and other processes are abiotic and biotic factors that threaten the açaí palms, yet humans benefit from the açaí palms. Preserving ecosystems is ultimately a win for humans.

So the next time you enjoy an açaí smoothie, consider the magnificence of this treat—it's the product of 3.5 billion years of complex interactions between organisms and their environment. Since climate has such a major effect on the distribution of life on the planet, let's take a closer look at what causes climate patterns. For example, why are the tropics hot and wet?

CHECK YOUR UNDERSTANDING

✔ If you understood this section, you should be able to . . .

Identify a past and present abiotic factor and a past and present biotic factor that could be important to the distribution and abundance of sockeye salmon.

Answers are available in Appendix A.

49.3 Climate Patterns

After you complete this section, you should be able to . . .

- Analyze the factors that establish climate patterns.

Weather consists of the specific *short-term* atmospheric conditions of temperature, precipitation, sunlight, and wind. Although your local meteorologist might not always make a good prediction on the night's snowfall, **climate**—the prevailing, *long-term* weather conditions found in an area—occurs in relatively predictable global patterns.

Why Are the Tropics Warm and the Poles Cold?

In general, areas of the world are warm if they receive a large amount of sunlight (solar radiation) per unit area; they are cold if they receive a small amount of sunlight per unit area. Over the course of a year, regions at or near the equator receive much more sunlight per unit area—and thus much more energy in the form of light and heat—than regions that are closer to the poles.

- At the equator, the Sun is often directly overhead at midday. As a result, sunlight strikes the surface at or close to an angle of 90 degrees. At these angles, Earth receives a maximum amount of solar radiation per unit area (**Figure 49.4**).
- Because Earth's surface slopes away from the equator, the Sun strikes the surface at lower and lower angles moving toward the poles. When sunlight arrives at a low angle, even at midday, much less energy is received per unit area.

The pattern of decreasing average temperature with increasing latitude is a result of Earth's spherical shape.

Why Are the Tropics Wet?

One of the most striking climate patterns on Earth involves precipitation. When average annual rainfall is mapped for regions around the globe, it is clear that areas along the equator receive the most moisture. In contrast, locations about 30° latitude north and south of the equator are among the driest on the planet. Why do these patterns occur?

Simple questions often have surprising answers. It turns out that a major cycle in global air circulation, called a **Hadley cell**, is responsible for making the Amazon River basin wet and the Sahara dry. Hadley cells are named after George Hadley, who in 1735 conceived of the idea of enormous air circulation patterns.

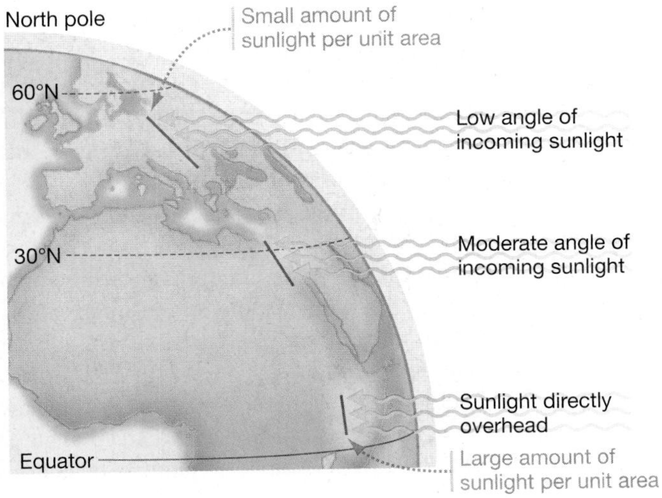

Figure 49.4 Solar Radiation per Unit Area Declines with Increasing Latitude. Over the course of a year, the Sun is frequently almost directly overhead at the equator at midday. As a result, equatorial regions receive a large amount of solar radiation per unit area compared to high-latitude regions.

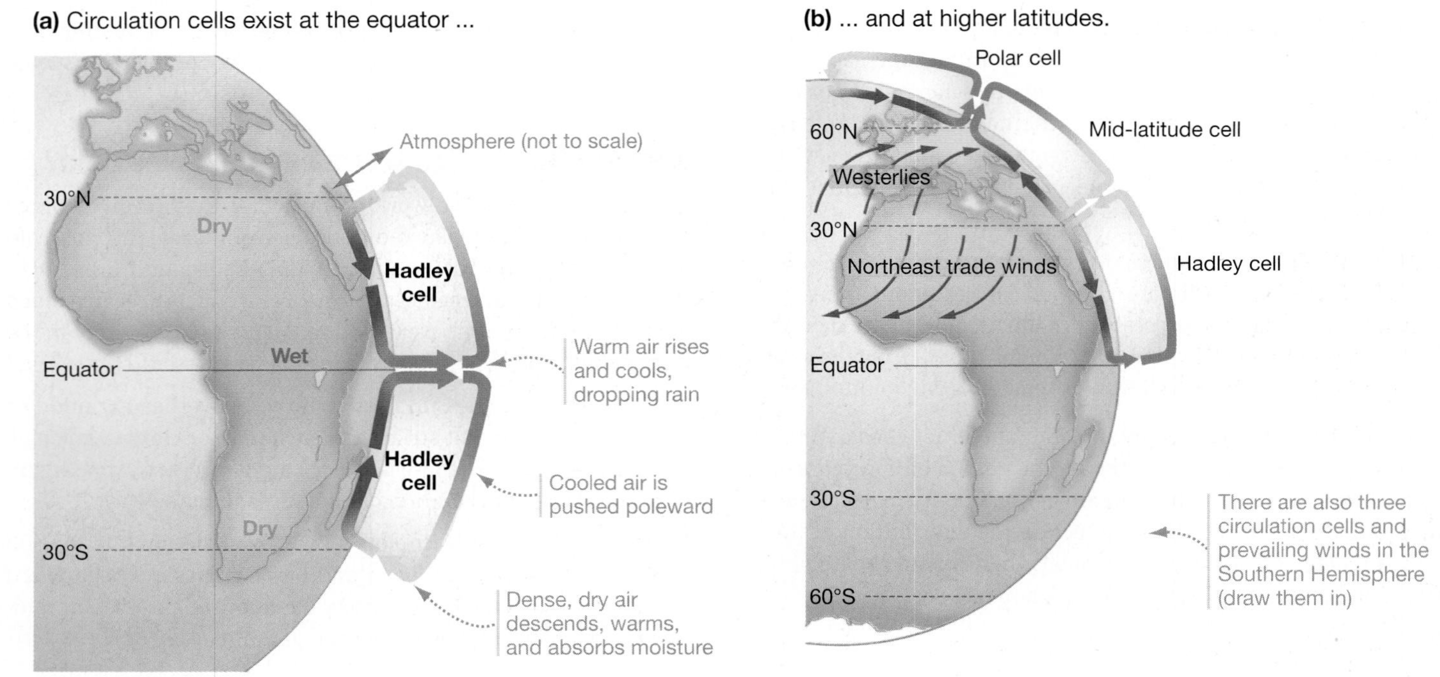

Figure 49.5 **Global Air Circulation Patterns Affect Rainfall.** Hadley cells explain why the tropics are much wetter than regions near 30° latitude north and south. Regions with rising air, such as those at the equator, tend to be wetter than regions with descending air.

✔ **MODEL** In part (b), draw arrows to show the three circulation cells and prevailing wind directions in the Southern Hemisphere.

To understand how Hadley cells work, put your finger on the equator in Figure 49.5a and follow the arrows.

- Air that is heated by the intense sunlight along the equator expands, lowering its pressure and causing it to rise. Note that this warm air holds a great deal of moisture, because warm water molecules tend to stay in vapor form instead of condensing into droplets.
- As air rises above the equator, it cools. Its ability to hold water declines, producing rain.
- As more air is heated along the equator, the cooler air above Earth's surface is pushed poleward. When the air mass has cooled enough, its density increases and it begins to sink.
- As the air sinks, it absorbs more and more solar radiation reflected from Earth's surface and begins to warm, gaining water-holding capacity. Thus the air approaching the Earth "holds on" to its water, and little rain occurs where it returns to the surface. These areas are bathed in warm, dry air.

The result? High levels of precipitation occur along the equator and a band of deserts occurs in the vicinity of 30° latitude north and south.

Similar air circulation cells occur between 30° latitude and 60° latitude (midlatitude cells) and between 60° latitude and the poles (polar cells), north and south (Figure 49.5b). The result is six major cells of moving air encircling the globe.

To make matters more complex, the air within each cell does not move in a straight north–south or south–north direction as you might expect from a static diagram. The spinning of the Earth on its axis causes a **Coriolis effect**, the tendency for moving air or water to be deflected from a straight path, instead swerving in a clockwise pattern in the Northern Hemisphere and in a counterclockwise pattern in the Southern Hemisphere. This effect accounts for the different directions of the prevailing winds and ocean currents at different latitudes (see Figure 49.5b).

What Causes Seasonality in Weather?

The striking global patterns in temperature and moisture that you've just reviewed are complicated by the phenomenon of seasons. Seasons are defined as regular, annual fluctuations in temperature, precipitation, or both.

Why does Earth experience seasons? One common misconception is that seasons change because Earth's distance from the Sun varies as it moves around the Sun. Actually, Earth's distance from the Sun changes very little. Also, if distance from the Sun alone did cause seasons, all of Earth would experience the change in temperatures at the same time.

As Figure 49.6 shows, starting on the left side, seasonality instead occurs because Earth is tilted on its north-to-south axis by 23.5 degrees. Due to this incline, the orbiting Earth's Northern Hemisphere is tilted toward the Sun in June. In this position, it faces the Sun most directly and receives the largest amount of solar radiation per unit area.

The Southern Hemisphere, in contrast, is tilted away from the Sun in June. As a result, it presents a steep angle for incoming sunlight and receives its smallest quantity of solar radiation per unit area.

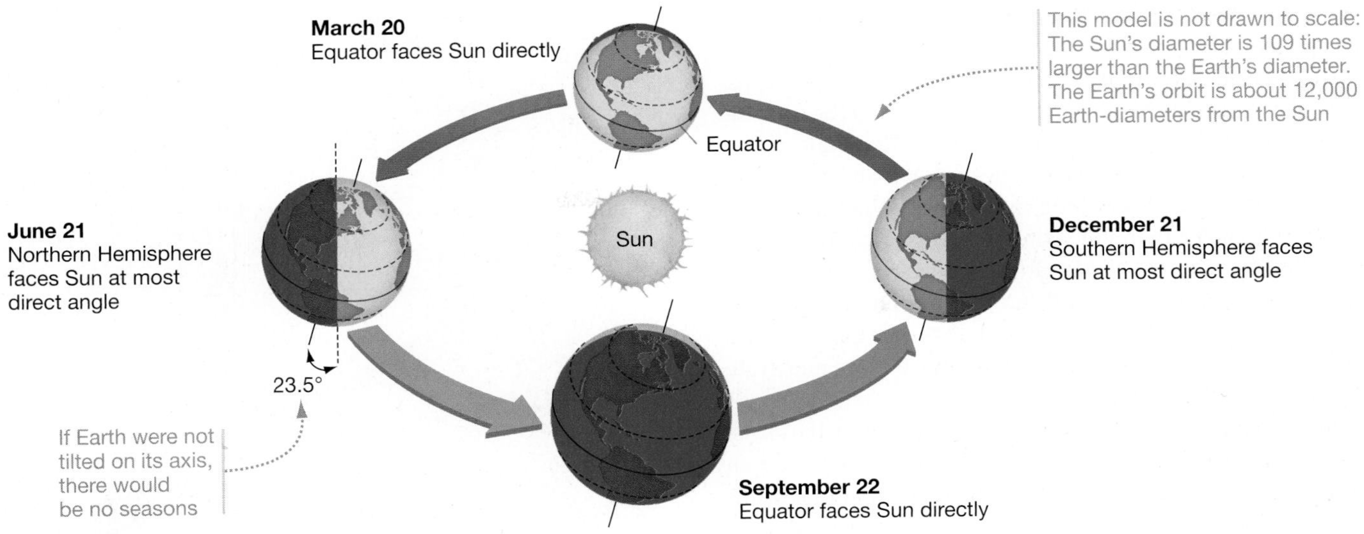

Figure 49.6 Earth's Orbit and Tilt Create Seasons at High Latitudes. Seasons occur due to annual variation in the amount of solar radiation that different parts of the Earth receive.

✔ **THINK CAREFULLY** Models are abstractions of reality, not direct representations. If the Sun's diameter is 13 mm in this diagram, how large would the Earth's diameter be if it were drawn to scale? How far would Earth have to be from the Sun to show an accurate orbital diameter? Why do you think the model is not shown to scale?

The upshot? In June it is summer in the Northern Hemisphere and winter in the Southern Hemisphere. Conversely, in December it is summer in the Southern Hemisphere and winter in the Northern Hemisphere. In March and September, the equator faces the Sun most directly and the tropics receive the most solar radiation. If Earth were not tilted on its axis, there would be no seasons.

What Regional Effects Do Mountains and Oceans Have on Climate?

The broad patterns of climate that are dictated by global heating patterns, air circulation cells, and seasonality are modified by regional effects. The most important of these are due to the presence of mountain ranges and proximity to an ocean.

Mountain Ranges Affect Regional Climate Mountains affect regional temperature as well as precipitation. **Elevation**, the height of land above sea level, is an important factor on its own because temperature decreases as a function of altitude—it's colder in the mountains than in the lowlands, even in the tropics.

The cooler temperatures at high elevation also affect precipitation patterns. Figure 49.7 shows the Cascade Mountain range along the Pacific Coast of North America as an example.

- In these northern latitudes, the prevailing winds are from the west (comparable to the westerlies in Figure 49.5b). These winds bring moisture-laden air from the Pacific Ocean onto the continent.
- As the air masses begin to rise over the mountains, the air cools and releases large volumes of water as rain.
- Once cooled air has passed the crest of a mountain range, the air is relatively dry because much of its moisture content has already been released. Areas that receive this dry air are said to be in a **rain shadow**.

The upshot? The areas along the western slopes of the Cascade Mountains are covered in lush rain forests while the areas to the east are deserts. Similar patterns occur at other mountain ranges.

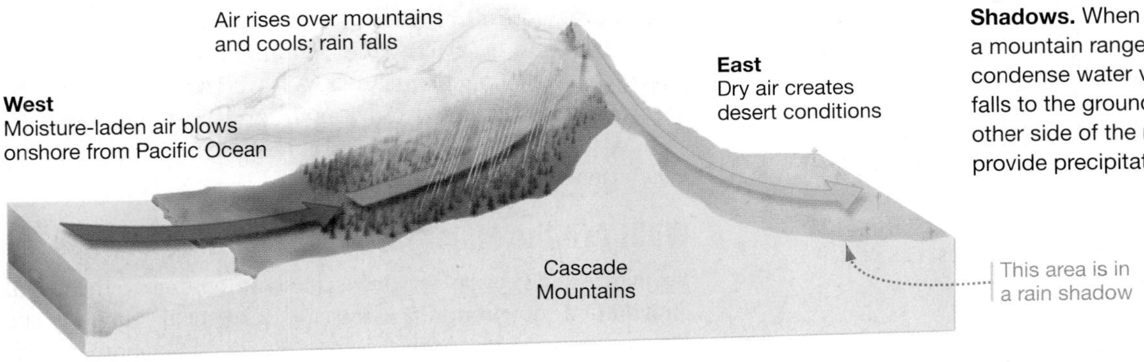

Figure 49.7 Mountain Ranges Create Rain Shadows. When moisture-laden air rises over a mountain range, the air cools enough to condense water vapor. As a result, moisture falls to the ground as precipitation. On the other side of the mountain, little water is left to provide precipitation.

Oceans Affect Regional Climate While the presence of mountain ranges tends to produce extremes in precipitation, the presence of an ocean has a moderating influence on temperature. Water has an extremely high **specific heat**, meaning that it has a large capacity for storing heat energy (Ch. 2, Section 2.2).

Because water has a high specific heat, it can absorb a great deal of heat from the atmosphere in summer, when the water temperature is cooler than the air temperature. As a result, the ocean moderates summer temperatures on nearby landmasses. Similarly, the ocean releases heat to the atmosphere in winter, when the water temperature is warmer than the air temperature. As a result, islands and coastal areas have much more moderate climates than do inland areas.

Further, ocean currents can influence local climate patterns. Massive ocean currents flow in cycles, or **gyres**, that move in a clockwise direction in the Northern Hemisphere and in a counterclockwise direction in the Southern Hemisphere, due to the Coriolis effect. The result is that warm water is delivered to colder latitudes and colder water is delivered to warmer latitudes.

For example, Great Britain lies at about the same northern latitude as Newfoundland, Canada, but the climate in Great Britain is much milder than Newfoundland—it rarely snows in London. Why? The currents within the North Atlantic Gyre move warm water up the western Atlantic toward Great Britain to the east (Figure 49.8).

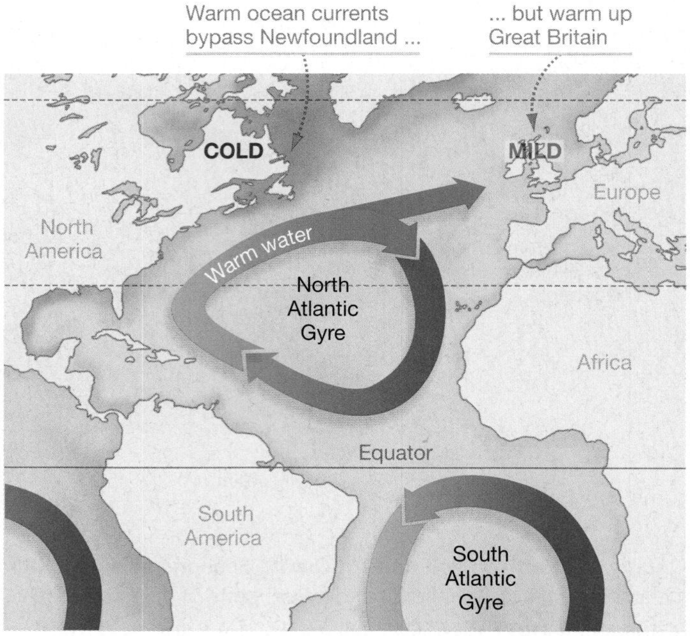

Figure 49.8 Ocean Currents Have Regional Effects on Climate. The ocean currents in the North Atlantic move in a clockwise pattern due to the Coriolis effect. This current moves warm water northeast toward Europe, causing the climate there to be much more moderate than predicted by latitude alone.

Do Biotic Factors Affect Climate?

While it is tempting to think that climate is determined solely by abiotic factors, organisms can also influence weather and climate (Ch. 53, Section 53.3). For example:

- The transpiration and evaporation of water from tropical forests increases cloud formation and regional rainfall. Thus, deforestation by humans decreases rainfall.
- Organisms produce and absorb greenhouse gases such as carbon dioxide and methane, which affect global temperature by influencing how much heat is retained in the atmosphere (see Ch. 53, Figure 53.19).
- The burning of fossil fuels (ancient organisms) by humans is vastly accelerating the release of greenhouse gases into the atmosphere, causing rapid change to global climate.

How do all these climate patterns combine to determine the distribution of organisms on Earth? Let's look at types of terrestrial biomes first, then turn to aquatic biomes.

CHECK YOUR UNDERSTANDING

✔ If you understood this section, you should be able to . . .

1. Use the niche concept to explain this pattern: Douglas fir trees are abundant on the western slopes of the Cascade Mountains but not the eastern slopes.
2. **SOCIETY** Predict where on the globe the highest risk of skin cancer would occur, and explain why.

Answers are available in Appendix A.

49.4 Types of Terrestrial Biomes

After you complete this section, you should be able to . . .

- Compare the characteristics of the major terrestrial biomes.

If you could walk from the equator in South America to the North Pole, you would notice startling changes in the organisms around you. Lush tropical forests with broad-leaved evergreen trees would give way to seasonally dry forests and then to dry shrubland and deserts. The deserts would yield to the vast grasslands of central North America, which terminate at the boreal forests of the subarctic. If you pressed on, you would reach the end of the trees and the beginning of the most northerly community—the arctic tundra.

If you walked from sea level up to the top of the Cascade Mountains of British Columbia, you would experience similar types of changes. Wet forest along the coast would give way to drier forest at higher elevations. Near the peaks, trees would thin and eventually yield to alpine tundra filled with hardy, low-growing plants. Broad-leaved evergreen forests, deserts, grasslands, and tundra are **biomes**: regions characterized by distinct abiotic characteristics and dominant types of vegetation.

What Are the Major Natural Terrestrial Biomes?

Ecologists and geographers have collaborated to map the global distribution of naturally occurring terrestrial biomes. The most common of these are shown in Figure 49.9. Ignore, for the

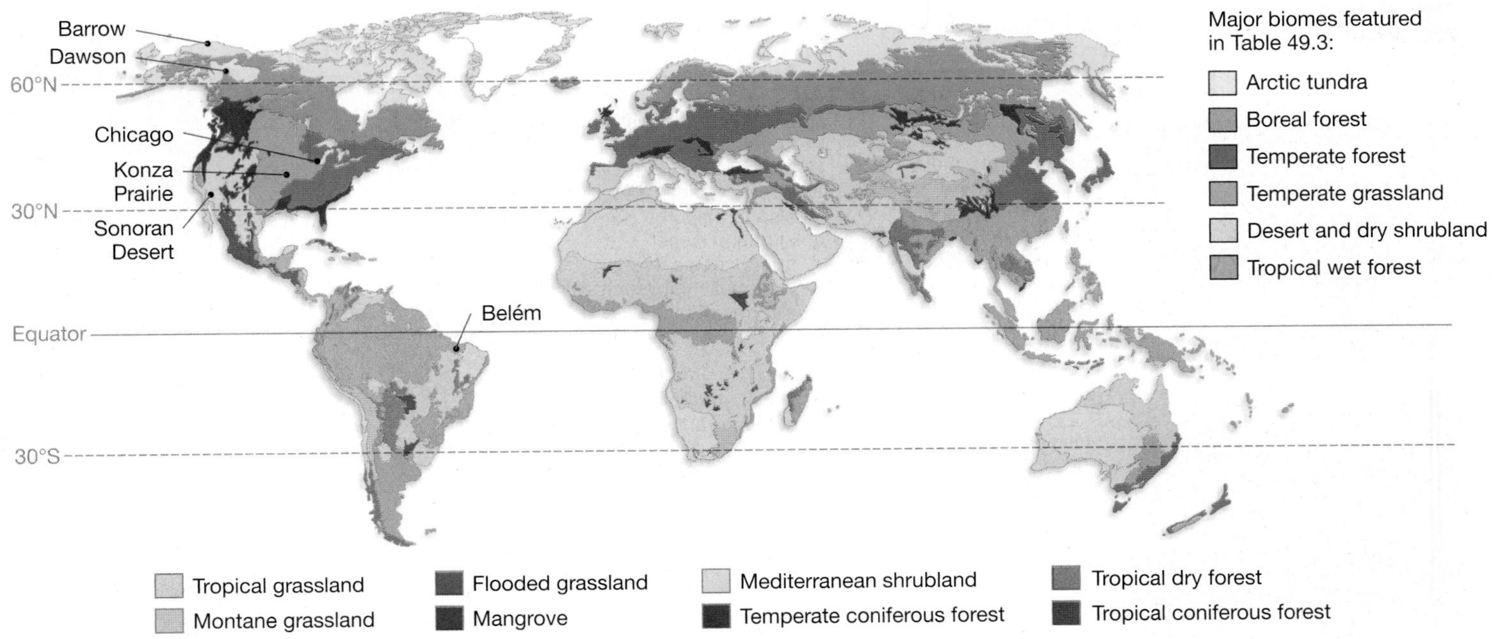

Figure 49.9 Distinct Terrestrial Biomes Are Found throughout the World. The six major biomes listed to the right of the map are featured in this chapter. (Table 49.3 cites data from the labeled locations on the left.)

DATA: D. M. Olson, E. Dinerstein, E. D. Wikramanayake, et al. 2001. *BioScience* 51: 933–938.

moment, the effects of humans on the landscape. Notice that different ecosystems may belong to the same biome. For example, the Sonoran Desert in North America and the Sahara in Africa are both part of the same desert biome, but each is a unique ecosystem with its own community of indigenous organisms due to historical biotic and abiotic factors.

Each of the terrestrial biomes found around the world is associated with a distinctive set of abiotic conditions determined largely by climate:

- *Temperature* is critical because the enzymes that make life possible work at optimal efficiency only in a narrow range of temperatures (Ch. 8, Section 8.3). Temperature also affects the availability of moisture: Water freezes at low temperatures and evaporates rapidly at high temperatures.
- *Moisture* is significant because it is required for life, and because terrestrial organisms constantly lose water to the environment through evaporation or transpiration. To stay alive, they must reduce water loss and replace lost water.
- *Sunlight* is essential because it is required for photosynthesis.
- *Wind* is important because it exacerbates the effects of temperature and moisture. Wind increases heat loss due to evaporation and convection, and it increases water loss due to evaporation and transpiration. Wind also has a direct physical impact on organisms such as birds, flying insects, and plants—it pushes them around.

Of the four interdependent components of climate, variation in temperature and moisture are the most important in determining plant distribution and abundance. More specifically, the type of terrestrial biome that develops in a particular region is governed by **(1)** average annual temperature and precipitation, and **(2)** annual variation in temperature and precipitation. Each biome contains species that are adapted to a particular temperature and moisture regime.

Biologists are particularly concerned with how temperature and moisture influence **net primary productivity (NPP)**. NPP is defined as the total amount of biomass generated by the fixation of carbon per year minus the amount that is oxidized during cellular respiration, where **biomass** refers to the total mass of organisms, of primary producers in this case. Carbon compounds that are oxidized during cellular respiration release energy to the organism and carbon to the atmosphere, so they are not available for the production of biomass (Ch. 53, Section 53.1).

NPP is crucial because it represents the organic matter that is available as food for other organisms. In terrestrial environments, NPP is often estimated by measuring **aboveground biomass**—the total mass of living plants, excluding roots.

Photosynthesis, plant growth, and NPP are maximized on land when sunlight is available, temperatures are warm, and conditions are wet. Conversely, photosynthesis cannot occur efficiently in low sunlight, at low temperatures, or under drought stress. Rates of photosynthesis are maximized in warm temperatures, when enzymes work efficiently, and in humid weather, when stomata can remain open and CO_2 is readily available.

To help you understand how temperature and precipitation influence biomes, let's take a closer look at the six terrestrial biomes that represent Earth's most extensive natural vegetation types: arctic tundra, boreal forest, temperate forest, temperate

BIOMES Table 49.3 Featured Terrestrial Biomes

	Biome Description	Dominant Vegetation	Avg. Precipitation (cm)	Avg. Temperature (°C)
	**Arctic tundra** Occurs throughout the arctic regions of the Northern Hemisphere. Has low productivity, low species diversity, and low aboveground biomass. Soil can be permanently frozen (called permafrost).	Ground is completely covered by low plants or lichens, but trees are absent. Animal diversity also tends to be low, although insect abundance can be extremely high (such as swarms of biting flies).	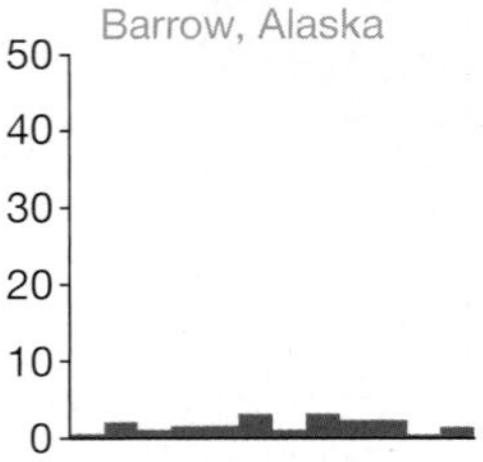	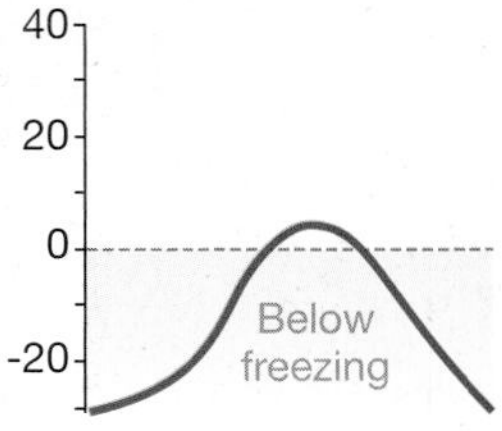
	**Boreal forest** Also called taiga. Occurs across most of Canada, Alaska, Russia, and northern Europe. Has low productivity and exceptionally low species diversity due to low temperatures and low precipitation.	Dominated by highly cold-tolerant conifers including pines, spruce, fir, and larch trees. Aboveground biomass is high because many of the slow-growing species are long lived. These forests have low structural diversity.	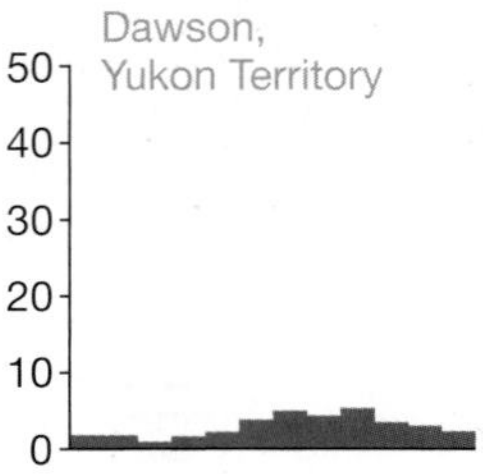	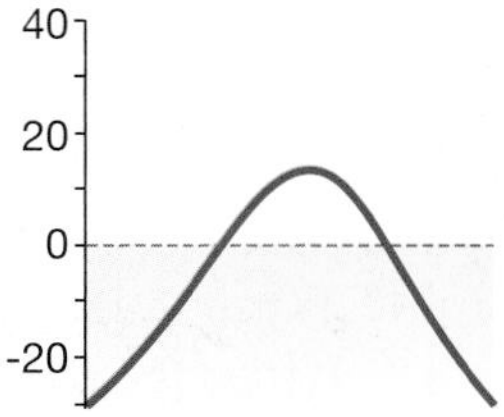
	**Temperate forest** Occurs in eastern North America, Europe, and eastern Asia. Productivity and diversity are moderate because temperatures do not support photosynthesis year-round.	Dominated by deciduous trees, which are leafless in winter and grow new leaves each spring and summer. Needle-leaved evergreens also common. Moderate structural diversity.	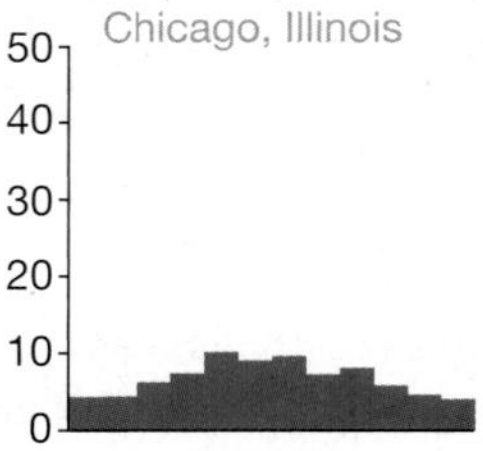	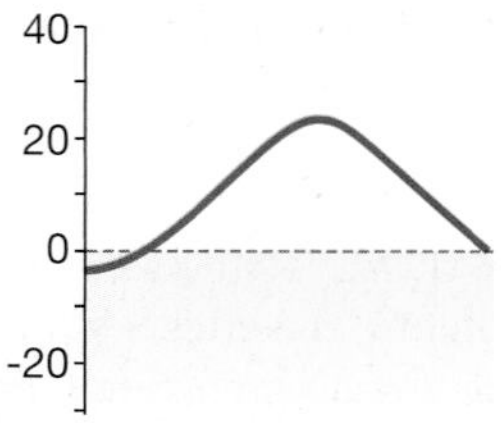
	**Temperate grassland** Also called prairie in North America and steppe in central Eurasia. Productivity is generally lower than in forests, but grassland soils are often highly fertile.	Grasses dominate because conditions are too dry to enable tree growth and/or because trees are burned out by fires. Plant life is extremely dense both aboveground and belowground (roots).	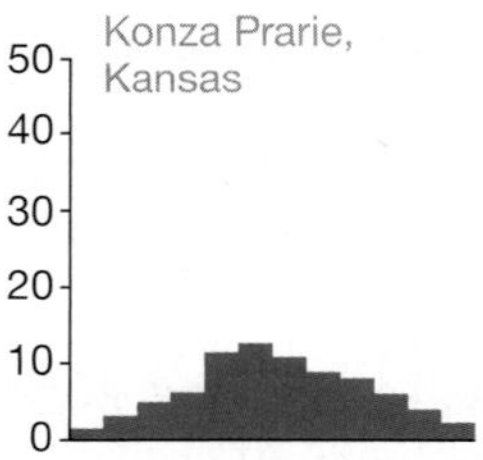	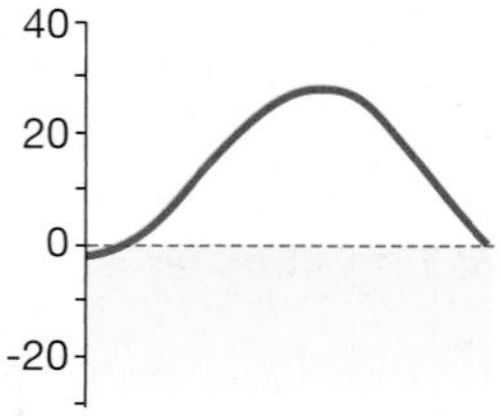
	**Desert and dry shrubland** Occurs at 30 degrees latitude north and south (see Figure 49.5). Distinguished by extreme temperatures, low moisture, and extremely low productivity.	Plants such as cacti have no leaves, leaves modified as spines, or leaves with a thick, waxy coating (adaptations for reducing water loss). Growth is slow year-round or occurs during episodes of rainfall.	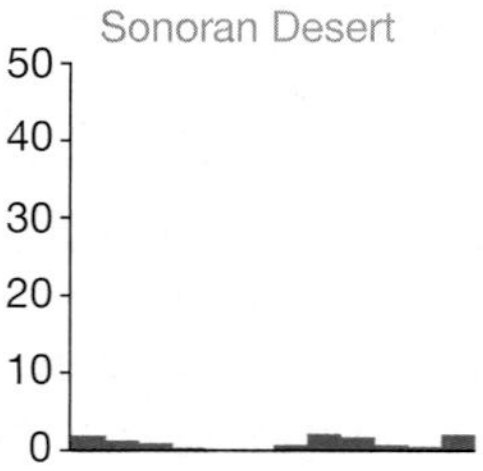	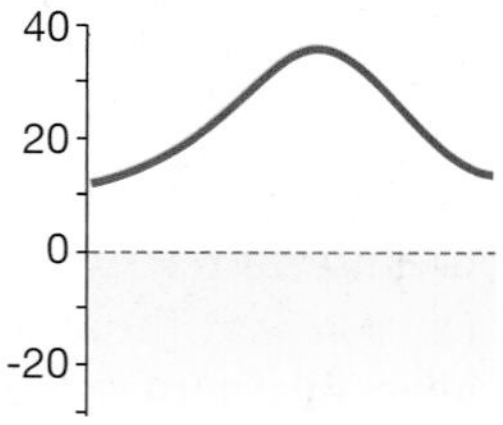
	**Tropical wet forest** Also called tropical rain forest. Occurs in equatorial regions. Favorable year-round growing conditions produce high aboveground biomass, high species diversity, and high productivity.	Plants have broad leaves, are evergreen, and are diverse in size and growth form. High structural diversity (tall trees, short trees, vines, shrubs, etc.) presents wide array of habitat types for animals.	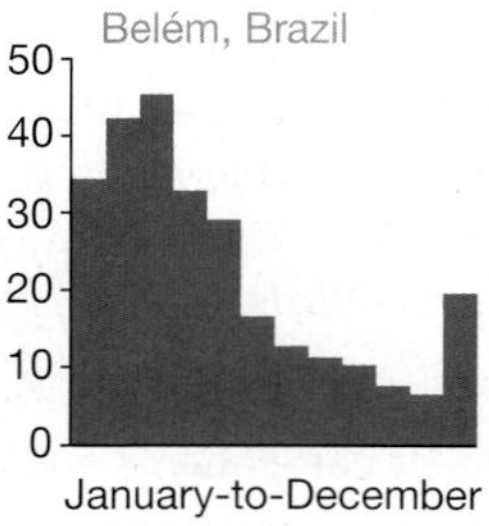	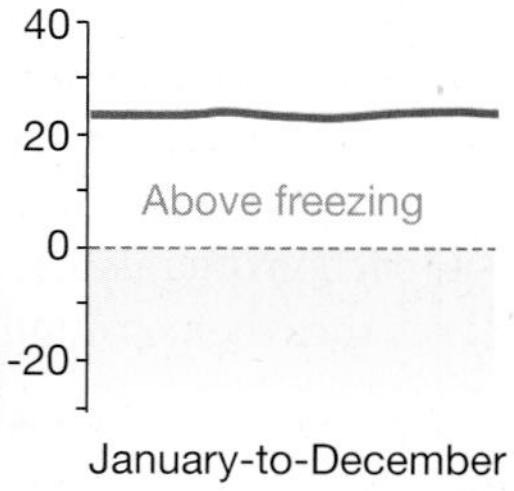

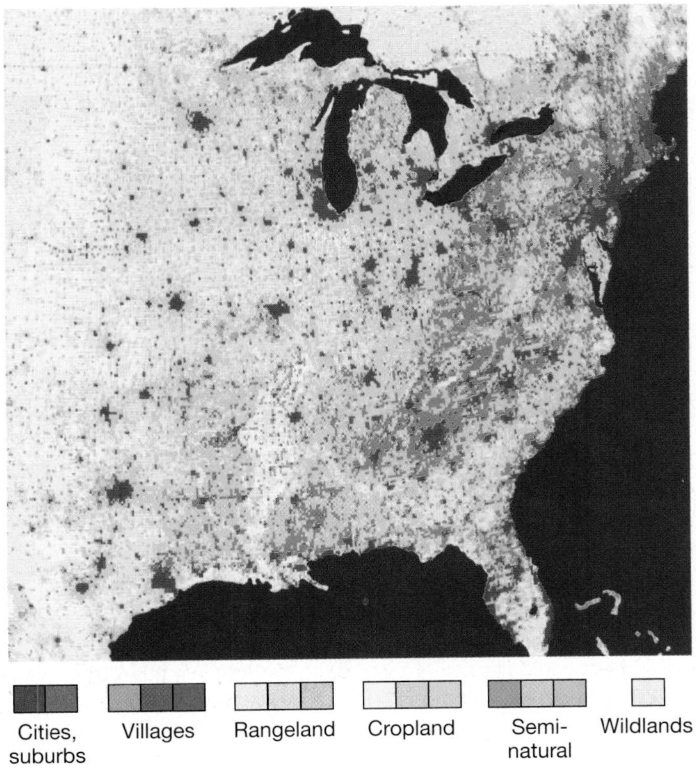

Figure 49.10 Anthropogenic Biomes Dominate Eastern North America. Several color subcategories have been consolidated in the key for simplicity. For example, cropland can be irrigated or rain-fed, residential, sparsely populated, or remote.

DATA: E. C. Ellis, K. Klein Goldewijk, S. Siebert, et al. 2010. *Global Ecology and Biogeography* 19: 589–606.

✔ Where do the remaining wildlands occur?

grassland, desert and dry shrubland, and tropical wet forest (**Table 49.3**). The table includes an example of temperature and precipitation data from a specific location within each biome (the locations are labeled in Figure 49.9). Data plotted in purple indicate the average monthly precipitation at those locations, graphed throughout the year; data plotted in red indicate the daily average temperature at those locations, graphed throughout the year.

✔ **If you understand the importance of temperature and precipitation in biomes, you should be able to explain why NPP is higher in tropical wet forests than in temperate forests.**

Human Land Use Is Displacing Natural Biomes

The map in Figure 49.9 shows the distribution of terrestrial biomes as they would occur naturally. This map is important because it represents the product of 3.5 billion years of interplay between geology, climate, evolution, and ecology.

Now for the rest of the story. The exponential increase in the world's human population (Ch. 51, Section 51.5) is driving land-use changes and other impacts that are so significant that some researchers have called for a new epoch of history—the Anthropocene (Ch. 54, Section 54.2). Consider these data:

- More than 75 percent of Earth's ice-free land shows evidence of direct alteration by humans—for example, farming, logging, and urban development (*not* including the effects of pollution, invasive species, or climate change).
- The remaining regions, or wildlands, account for just 11 percent of terrestrial net primary productivity.

What does the distribution of biomes look like with human impact factored in? Biogeographer Erle Ellis and colleagues have pioneered efforts to map anthropogenic biomes, or "anthromes" worldwide, with eye-opening results. **Figure 49.10** zooms in on one part of the world, eastern North America, for a close-up of the distribution of anthropogenic biomes. The remaining wildlands (pale gray on the map) are so sparse that they are difficult to locate. The resulting patchwork of fragments of natural biomes and anthromes has inspired a branch of study called **landscape ecology**, which measures the structure of these patchwork landscapes and their ecological processes, such as the flow of nutrients through watersheds.

As you can see from Figure 49.10, humans are directly affecting terrestrial biomes. Such changes to the landscape lead to widespread impacts on the distribution and abundance of organisms.

How Is Global Climate Change Affecting Terrestrial Biomes?

It is now well established that human-caused emissions of greenhouse gases like CO_2 are causing climate change (see data and discussion in Ch. 53, Section 53.3). Climate has changed frequently and radically throughout the history of life, but recent increases in average temperatures around the globe, combined with a projected increase of up to 6°C over the next 100 years, represent one of the most rapid periods of climate change since life began.

How Do Scientists Study the Effects of Climate Change on Biomes? Biologists collaborate with scientists in many other disciplines worldwide to understand how global climate change is affecting the distribution of life on Earth. In general, these teams use four kinds of strategies:

1. *Simulation studies* are based on computer models of weather patterns in local regions. By increasing average temperatures in these models, researchers can predict how wind and rainfall patterns, storm frequency and intensity, and other aspects of weather and climate will change. In effect, simulation studies predict how temperature and precipitation profiles, like those in Table 49.3, will change with increasing temperatures—and how the distribution of biomes might change as a result.
2. *Observational studies* are based on long-term monitoring at fixed sites around the globe. Researchers are documenting changes in key physical variables—temperature and precipitation—as well as changes in the distribution and abundance of organisms, such as the timing of insect emergence and bird migrations. By extrapolation, they can predict how change may continue at these sites.

3. *Historical studies* examine the relationship between CO_2 levels, climate change, and the distribution and abundance of organisms during events that occurred millions of years ago. For example, researchers have been able to measure the pattern of change in the biodiversity of the Amazonian rain forest as a function of past climate fluctuations. These patterns provide valuable data for models designed to predict future changes.
4. *Experiments* are designed to simulate changed climate conditions and to record responses by the organisms present.

Even though the scientific community did not become alert to the reality of global climate change until the 1980s, a large number of multiyear experiments have already been done. Because arctic regions are projected to experience particularly dramatic changes in temperature and precipitation, many of these experiments have been conducted in the tundra biome. Let's consider how the boundaries of biomes are shifting.

Figure 49.11 **Experimental Increases in Temperature Change Species Composition in Arctic Tundra.** The transparent, open-topped chambers shown here are an effective way to increase average annual temperature. Inside the chambers, species composition changes relative to otherwise similar sites nearby.

✔ **PROCESS OF SCIENCE** What other variable besides temperature might be affected by the transparent chambers? What kind of experiment could the researchers do to determine if this variable is affecting the species composition in the chamber?

Climate Change Is Causing Terrestrial Biomes to Shift Figure 49.11 shows one example of how these multiyear experiments are set up. Transparent, open-topped chambers are placed at random locations within a study site, and the characteristics of the vegetation inside the chambers are recorded annually. The same measurements are made in randomly assigned study plots that lack chambers.

Across experiments, the average annual temperature inside the open-topped chambers increases by 1–3°C. This increase is consistent with conservative projections for how temperatures will change over the next century.

In an analysis of data from 11 such study sites scattered throughout the arctic, several strong patterns emerged:

- Overall, species diversity decreases.
- Compared with control plots, grasses and shrubs increase inside the chambers, and mosses and lichens decrease.

These experimental results support simulation and observational studies predicting that arctic tundra environments are giving way to boreal forest. Biomes are shifting right before our eyes.

Tipping Points Cause Abrupt Changes in Biomes Note that not all areas are expected to transition smoothly and incrementally from one biome to another based on changes in temperature and/or rainfall. There is growing evidence that small perturbations in climate in some areas will cause abrupt changes to biomes.

For example, scientists predict that decreased rainfall over the Amazon will cause a rapid transition from forest to grassland due to positive feedback loops in the system. Decreases in forest cover will decrease the rate of rain-cloud formation, which will further decrease the rainfall. Decreased rainfall will increase the prevalence of fires, which will further decrease forest cover. The predicted result is a rapid change—a tipping point—from a forested stable state to a grassland stable state, with dramatic effects on biodiversity.

Climate Variability Is Increasing within Biomes Other simulation and observational studies have converged on a remarkable conclusion: Increases in average global temperature are increasing variability in temperature and precipitation. Stated another way, global climate change is making climates more extreme. Average temperature and average annual precipitation in many regions may not be changing substantially, even though variation is.

For example, so far the frequency of monsoon rains, cyclones, hurricanes, and other types of storm events is not changing. However, the amount of precipitation and the wind speeds associated with these events are increasing. The severity of droughts is also on the rise.

Since organisms have adaptations to certain abiotic conditions, increased variability poses physiological challenges. For example, during a recent cold snap in Texas, some anole lizards in a southern population had heritable traits (enzyme properties) that enabled them to survive the cold temperatures. Others died—that is, natural selection caused evolution in the lizard population, increasing the frequency of cold-tolerant individuals. But now think ahead to summer: What will happen during a severe summer heat wave?

Both human land-use patterns and climate change dramatically alter the composition and distribution of terrestrial biomes. These factors also affect aquatic biomes. Let's take a closer look.

CHECK YOUR UNDERSTANDING

✔ **If you understood this section, you should be able to . . .**

1. **SOCIETY** Identify which anthropogenic biome, or anthrome, you live in. What natural biome occurred there before?
2. Predict how the projected 2080 range of açaí palms would differ from that shown in Figure 49.3 if tipping points in biomes were factored in.

Answers are available in Appendix A.

49.5 Types of Aquatic Biomes

About 70 percent of Earth's surface is covered in water. What types of biomes occur there? And what are the characteristics that distinguish aquatic biomes?

After you complete this section, you should be able to ...

- Compare the characteristics of the major aquatic biomes.

Let's think back to the sockeye salmon from the beginning of the chapter (see Table 49.1). Most salmon begin their lives in streams, travel to the ocean to feed and grow, then return to their natal streams to mate and die. Thus salmon migrate between a freshwater aquatic biome (a stream) and a marine aquatic biome (the ocean). The major abiotic factors that distinguish streams from oceans include salinity, water depth, water flow, and nutrient availability. Let's consider each factor in turn.

Salinity

Water is an excellent solvent—one reason that it is so essential to life (Ch. 2, Section 2.2). The water in clouds is nearly pure, but as rainwater percolates through soils and flows across rocks, it picks up solutes. The fresh water of ponds, lakes, and rivers has very low solute concentrations. However, as this fresh water steadily flows into the ocean, it continually adds solutes. Water evaporates from the ocean, forming rain that returns fresh water to land, but the solutes stay in the ocean. Over millions of years, this process has resulted in a relatively high concentration of solutes in seawater.

The proportion of solutes dissolved in water determines its **salinity**, generally measured as the number of grams of solute per kilogram of water—a unitless number described as parts per thousand. The salinity of fresh water varies from about 0.06 to 0.3 parts per thousand, whereas the salinity of the open ocean is fairly constant at 35 parts per thousand.

What makes the ocean "salty"? Solutes with a positive charge—like sodium, potassium, magnesium, and calcium—combine with solutes with a negative charge—like chlorine, bromine, and fluorine—to form salts. The concentration of different salts in the ocean is determined by their solubility in water. For example, calcium carbonate has a low solubility in water, so it precipitates as limestone at the bottom of the ocean. But sodium chloride (familiar to you as table salt) has such a high solubility in water that it makes up about 86 percent of sea salt.

Because salinity dramatically affects osmosis and water balance in organisms (Chapters 35 and 40), species have physiological adaptations that allow them to cope with a specific range in salinity. Therefore, salinity is a major determinant of the distribution of organisms in aquatic biomes.

Note that 97.5 percent of Earth's water is salt water in the ocean. Most of the remaining 2.5 percent of fresh water is locked up in the ice caps and glaciers, leaving only about 0.5 percent in groundwater and in ponds, lakes, and rivers. Further, while there are many ponds and rivers on Earth, most of the volume of Earth's fresh water is contained within a few very large lakes such as Lake Baikal in Russia, Lake Tanganyika in East Africa, and Lake Superior in North America.

Water Depth and Sunlight Availability

Water absorbs and scatters light, so the amount and types of wavelengths available to organisms change dramatically as water depth increases. Light availability has a major effect on productivity.

At the water surface, all wavelengths of light are equally available. But as Figure 49.12a shows, ocean water absorbs light in the red region of the visible spectrum, resulting in the familiar blue hue of underwater scenes (Figure 49.12b). This is important because key photosynthetic pigments like chlorophyll have high absorption peaks for red wavelengths (Ch. 10, Section 10.2). Since red wavelengths are unavailable underwater, the productivity of photosynthetic organisms there is dramatically constrained.

The quantity of light available to organisms also diminishes rapidly with increasing depth. In pure seawater, the total amount of light available at a depth of 10 m is less than 40 percent of what it is at the surface; virtually no light reaches depths greater than 40 m (131 ft).

Since water depth and sunlight availability are so important to the distribution and abundance of organisms, biologists have used these abiotic variables to define zones within oceans and

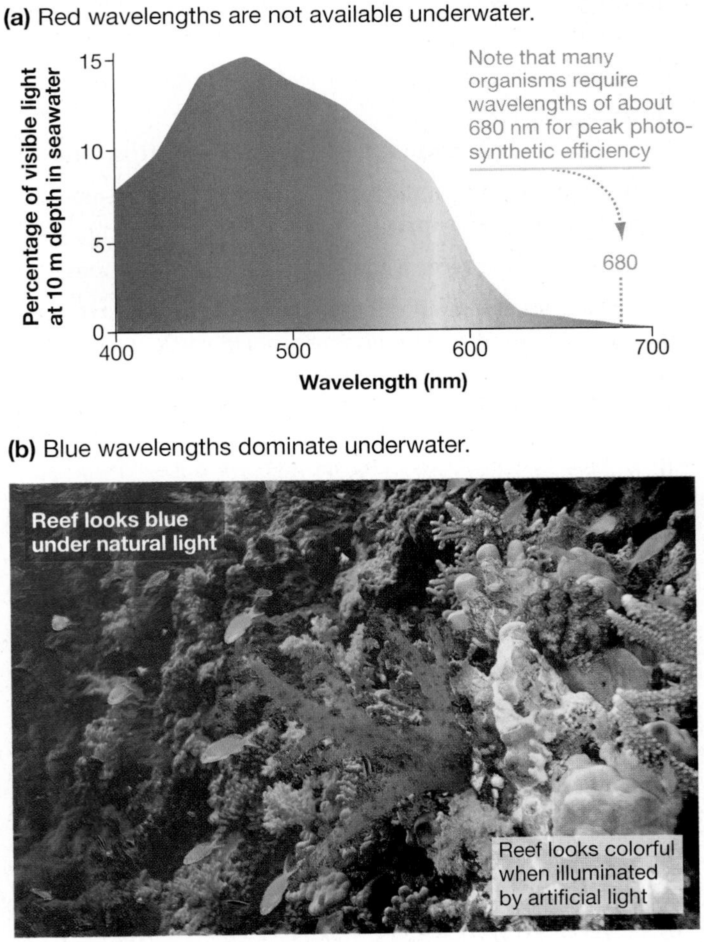

Figure 49.12 Light Is Limited Underwater.
DATA: M. B. Saffo. 1987. *BioScience* 37: 654–664.

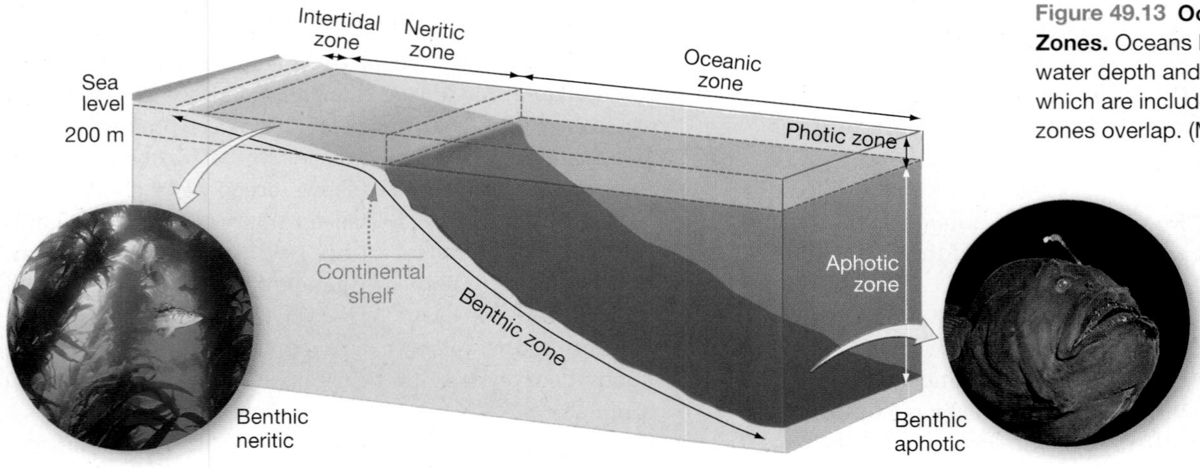

Figure 49.13 Oceans Have Distinctive Zones. Oceans have zones defined by water depth and light availability, some of which are included here. Note that some zones overlap. (Model not to scale.)

lakes. For example, Figure 49.13 shows six of the most commonly referenced zones in the ocean, some of which overlap.

- The **intertidal zone** consists of a rocky shoreline, sandy beach, or mud flat that is exposed to the air at low tide but submerged at high tide.
- The **neritic zone** extends from the intertidal zone to depths of about 200 m. Its outermost edge is defined by the end of the **continental shelf**—the gently sloping, submerged portion of a continental plate.
- The **oceanic zone** is the "open ocean"—the deepwater region beyond the continental shelf.
- The bottom of the ocean at all depths is the **benthic zone**.
- The intertidal and sunlit regions of the neritic, oceanic, and benthic zones make up a **photic zone**.
- Areas that do not receive sunlight are in an **aphotic zone**.

In the tropics, shallow portions of the neritic zone host the most productive environments in the world, **coral reefs**—large assemblages of colonial marine corals (related to sea anemones). Coral reefs provide habitat for many other organisms. The autotrophs in coral-reef ecosystems include not only algae that live on the reef but also mutualistic protists called dinoflagellates that live within the corals—the colors of most corals are derived from the photosynthetic pigments of these protists. Because the water is warm and sunlight penetrates to the ocean floor in these habitats all year, primary productivity and biodiversity are extremely high.

Although lakes are much smaller than oceans, they also have zonation based on water depth and sunlight availability. Figure 49.14 shows five of the most commonly referenced lake zones, some of which overlap. The photic, aphotic, and benthic zones are comparable in definition to those in the ocean.

- The **littoral zone** consists of the waters along the shore that are shallow enough for plants to take root.
- The **limnetic zone** is offshore and comprises water that receives enough light to support photosynthesis but that is too deep for plants to take root.

Organisms tend to occupy specific zones depending on their adaptations. For example, drifting organisms called **plankton** are abundant in the photic zone, comprising photosynthetic algae and cyanobacteria as well as the small crustaceans (organisms related to shrimp and crabs), fish, and other animals that eat them. Invertebrates and fish that consume dead organic matter, or **detritus**, are particularly abundant in the benthic zone.

Note that in the photic zones of both oceans and lakes, water is rarely pure and may contain high concentrations of sediments and other light-reducing components. The **turbidity**, or cloudiness, of water is an important determinant of light penetration. Turbidity is sometimes caused by natural processes such as erosion of river sediments by floodwaters and the erosion of coastal sediments by wave action. Turbidity is also caused by many human activities, such as runoff from agricultural fields and algae blooms caused by nutrient pollution.

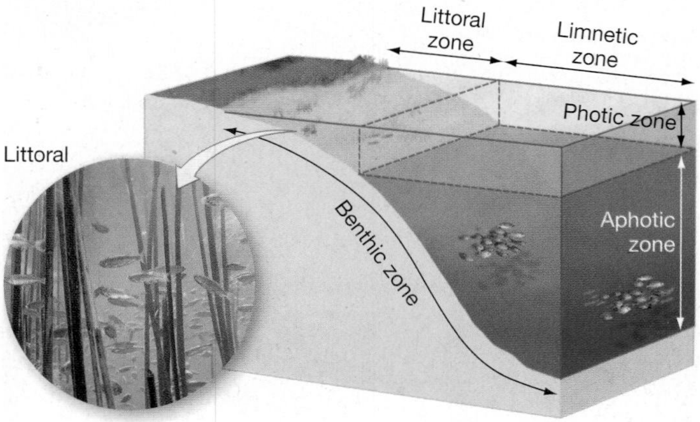

Figure 49.14 Lakes Have Distinctive Zones. Like oceans, lakes have zones defined by water depth and light availability, as depicted in this simplified model. (Model not to scale.)

Water Flow

Water flow is important because it influences the availability of oxygen, light, and nutrients. For example, the fast-moving water near the source of streams tends to be high in O_2, clear, and low in nutrients. The slower-flowing water downstream tends to be lower in O_2, more turbid, and more nutrient rich.

Water movement also presents a physical challenge. It can literally sweep organisms away. Organisms that live in fast-flowing streams, like mayfly larvae, have streamlined shapes and behavioral adaptations that help them maintain their position in the stream. Marine organisms that live in intertidal regions are also exposed to water flow—sometimes to violent wave action during storms—and to tidal currents. A growing field of biomechanics studies the mechanical implications of water flow to the physiology, evolution, and ecology of aquatic organisms.

Nutrient Availability

In many aquatic ecosystems, nutrients such as nitrogen and phosphorus are in short supply. Nutrient levels are important: The scarcity of nutrients limits growth rates in the photosynthetic organisms that provide food for other species.

Consider the vast oceanic zone, which is as sparsely populated as a desert. Sunlight is abundant in the photic zone of the open ocean, but nutrients are extremely scarce. When the photosynthetic organisms and the animals that feed on them die, their bodies drift downward out of the photic zone and thus are no longer available. In the open ocean, the benthic zone can be several kilometers below the surface. The benthic organisms in the aphotic zone survive by consuming this detritus that sinks down from the photic zone.

Water flow can carry nutrients from one area to another. Let's consider three types of water flow that are critical sources of nutrients in aquatic biomes: coastal runoff, ocean upwelling, and lake turnover.

Coastal Runoff Few nutrients are available in rushing mountain streams. But as the water flows downhill and the streams grow wider and slower, nutrients gather and tend to sink to the bottom, where they collect in the form of debris. For this reason, nutrients tend to be plentiful in the shallow **estuaries** where freshwater creeks and rivers meet the sea. Since both nutrients and sunlight are plentiful, estuaries are among the most productive environments on Earth.

Ocean Upwelling In certain coastal regions, nutrients are brought up to the surface by currents that cause upwelling. Figure 49.15 shows how this happens off the coast of Peru. Here the prevailing winds blow water to the north. Because the Earth rotates constantly, this wind-driven water current is moved offshore and replaced by water moving up from the ocean bottom.

The upwelling water is nutrient rich. In effect, it recycles nutrients that earlier had fallen to the ocean floor. The nutrients support luxuriant growths of photosynthetic cells, which feed an array of small grazing animals, which in turn feed small fish called anchoveta. When currents are favorable and upwelling is steady, the anchoveta fishery is the most productive in the world. However, due to periodic lulls in the trade winds—called El Niño events—warm water stays near the shore and upwelling stalls, devastating the fishery.

Lake Turnover Unless you live at northern latitudes, most of the ponds and lakes you see are made by humans. Most natural ponds and lakes occur at the world's higher latitudes, where they were scoured by glaciers as recently as 10,000 years ago. Each year in temperate regions, these bodies of water undergo remarkable changes known as the spring and fall **turnovers**. The spring and fall lake turnovers occur in response to changes in air temperature, as shown in Figure 49.16 on page 1072.

PROCESS: OCEAN UPWELLING

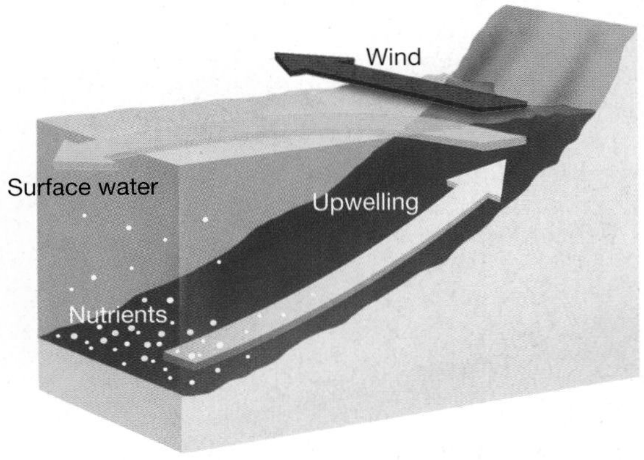

1. Winds blow: Along the coast of Peru, the prevailing winds blow north, moving water at the surface.

2. Surface water moves: As the Earth rotates, the moving surface water is forced offshore.

3. Upwelling: As surface water leaves, it is replaced by nutrient-laden water welling up from the bottom.

Figure 49.15 **Ocean Upwelling Brings Nutrients to the Surface.**

Note that in winter, water at the surface of a northern lake is locked up in ice at a temperature of 0°C. The water just under the ice is slightly warmer and is relatively oxygen rich because it was exposed to the atmosphere as winter set in. Because water is densest at 4°C, the water at the bottom of the lake is at 4–5°C. The bottom water is also oxygen poor because organisms that feed on detritus use up available oxygen in the process. A gradient in temperature such as this is called a **thermocline** ("heat-slope"), or thermal stratification.

The ice begins to melt, however, when spring arrives. The temperature of the water on the surface rises until it reaches 4°C. The water at the surface of the lake is now heavier than the water below it. As a result, the water on the surface sinks, aided by wind. The water at the bottom of the lake is displaced and comes to the surface, completing the spring turnover.

During the spring turnover, water at the bottom of the lake carries sediments and nutrients from the bottom up to the surface. This flush of nutrients triggers a rapid increase in the growth of photosynthetic algae and bacteria; biologists call it the spring bloom. The lake then stratifies again in the summer.

When temperatures cool in the fall, the water at the surface reaches a temperature of 4°C and sinks, a process often aided by wind, displacing water at the bottom and creating the fall turnover. The fall turnover is important because it brings oxygen-rich water from the surface down to the bottom, and because it again brings nutrients from the bottom up to the sunlit surface waters.

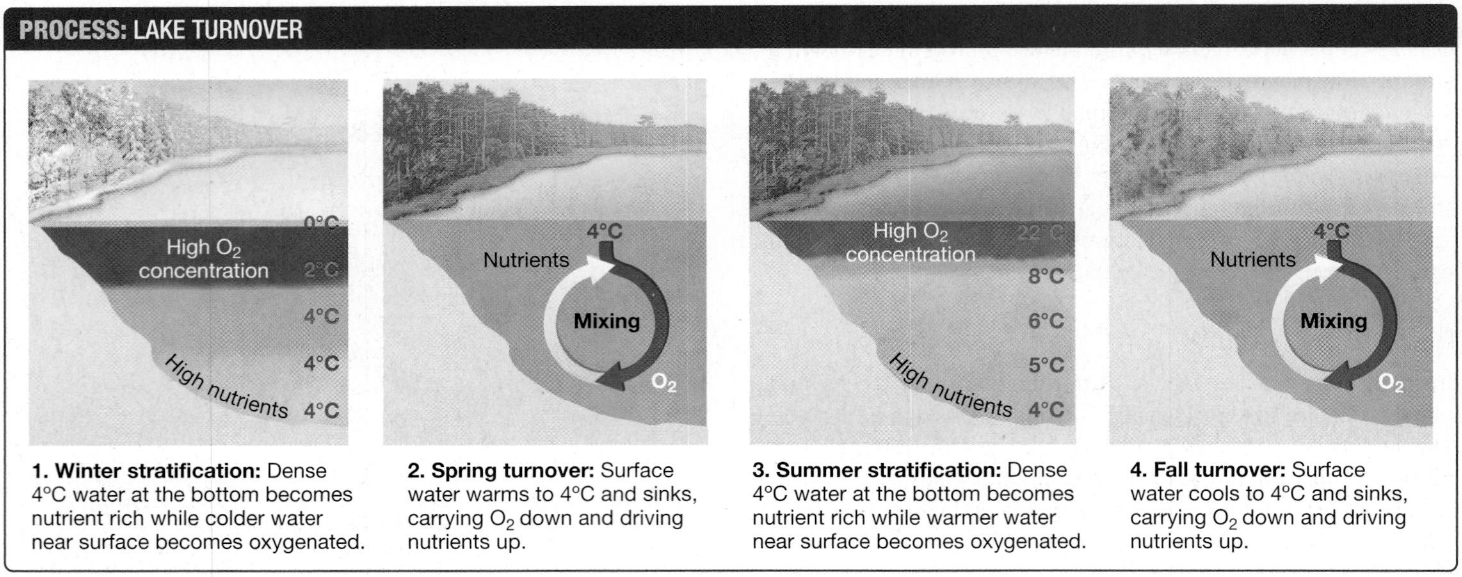

Figure 49.16 **In Temperate Regions, Lakes Turn Over Each Spring and Fall.**

Without the spring and fall turnovers, most freshwater nutrients would remain on the bottom of lakes, isolated from the photic zone. These aquatic ecosystems would be much less productive as a result.

Table 49.4 summarizes the major aquatic biomes in terms of their water depth, sunlight availability, water flow, nutrient availability, and characteristic organisms. As you review the table, notice how these abiotic factors interact. ✔ **If you understand the variables that are important in aquatic biomes, you should be able to predict which zones of oceans and lakes have the highest NPP and to explain your logic.**

How Are Aquatic Biomes Affected by Humans?

Perhaps one of the most common misconceptions about aquatic biomes is that water is unlimited and the aquatic biomes are vast—so waterways can "absorb" human impacts ranging from kitchen garbage to industrial pollution. While it's true that the ocean is very deep from a human perspective (average depth = 3.7 km), it is very shallow given its breadth—like a piece of paper. If all the water on Earth were balled up into one large droplet, it would be about the diameter of the Gulf of Mexico (Figure 49.17).

Humans have a long history of treating aquatic biomes as unlimited resources, causing both direct and indirect changes to their physical, chemical, and biotic properties. Consider a few examples:

- Humans eliminate aquatic biomes by filling or draining wetlands, damming streams, and removing water from ponds and rivers for agricultural irrigation.
- Humans also create new aquatic anthromes such as municipal reservoirs, decorative or recreational ponds and lakes, stormwater drainage systems, and agricultural ditches.
- Humans are disrupting aquatic communities. They transport organisms around the world and can introduce new species, sometimes with profound effects—such as the invasive zebra mussel becoming a scourge in the Great Lakes of North America. Humans also overexploit marine resources, especially top predators such as tuna and sharks, causing cascading effects on food webs (Chapters 53 and 54).
- Humans intentionally and unintentionally dispose of debris into aquatic biomes, clogging up creeks and ultimately creating enormous masses of garbage at the center of ocean gyres. Humans also release chemical pollutants, ranging from nutrient runoff to mercury, into aquatic biomes.
- Global climate change influences aquatic ecosystems by the same mechanisms discussed for terrestrial ecosystems. For example, increased rainfall on land causes increased flow in streams. In addition, global warming is causing sea level to rise, the pattern of ocean currents to change, and increased stratification (preventing upwelling and turnovers). Further, increased atomospheric CO_2 is causing ocean acidification (the lowering of

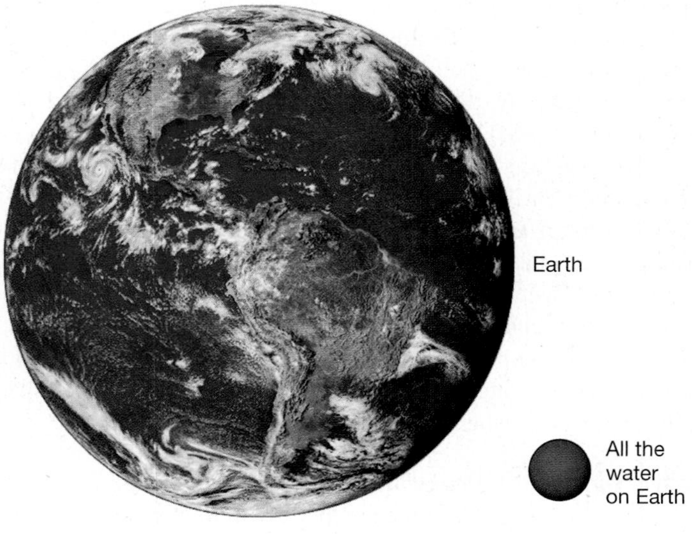

Figure 49.17 **Aquatic Biomes Are Finite Systems.**

BIOMES Table 49.4 Featured Aquatic Biomes

	Biome Description	Water Depth/Sunlight	Water Flow/Nutrients	Organisms
	**Lakes and Ponds** **Freshwater** Most natural lakes and ponds occur at high latitudes, formed in depressions created by the scouring action of glaciers thousands of years ago.	Biologists describe the structure of lakes and ponds by naming zones according to water depth and sunlight availability: e.g., littoral zone, limnetic zone, benthic zone, photic zone, and aphotic zone (see Figure 49.13).	Water movement is driven by wind and temperature. Seasonal turnovers mix the warmer, more oxygenated surface water with the deeper, more nutrient-rich benthic water (see Figure 49.15).	Organisms are adapted to certain zones. For example, cyanobacteria and algae—and animals that eat them—are abundant in the photic zone. Invertebrates and fish that eat dead organic matter are abundant in the benthic zone.
	**Freshwater Wetlands** **Freshwater** Freshwater wetlands are shallow-water habitats where the soil is saturated for at least part of the year. Include marshes, swamps, and bogs.	Unlike lakes and ponds, wetlands have only shallow water. All or most of the water in wetlands receives sunlight. Emergent vegetation (plants that grow above the surface of the water) capture sunlight before it strikes the water.	Bogs have low or no water flow and are oxygen poor and nutrient poor. Marshes and swamps have higher water flow, higher oxygen levels, and more nutrients, and so are more productive than bogs.	Bogs are dominated by peat moss and other plants adapted to acidic soils and low nutrients. Marshes are dominated by grasses and reeds. Swamps are dominated by trees and shrubs.
	**Streams** **Freshwater** Streams are bodies of water that move constantly in one direction from an area of higher elevation to an area of lower elevation. Creeks are small streams and rivers are large streams.	Small streams are shallow enough that sunlight reaches the bottom. Availability of sunlight is usually not a limiting factor for organisms, except when streams are deep and/or turbidity is high.	Streams tend to be colder, narrower, faster, oxygen rich, and nutrient poor near their origin at higher elevation and wamer, wider, slower, oxygen poor, and nutrient rich near their endpoint at lower elevation.	The same stream can contain different types of organisms upstream and downstream. For example, fish and insect larvae with adaptations for fast-moving water are common upstream, and algae and plants are more common downstream.
	**Estuaries** **Freshwater/Marine** Estuaries form where rivers meet the ocean, meaning that fresh water mixes with salt water. They include saline wetlands as well as adjacent open, brackish waters that vary in salinity.	Most estuaries are shallow enough that sunlight reaches the bottom. However, water depth and turbidity may fluctuate in response to tides, storms, and floods.	Water flow fluctuates daily due to tides, storms, and floods, causing fluctuations in salinity, turbidity, and nutrients. Estuaries are nutrient rich because nutrient-laden sediments are deposited when river water slows as it enters the ocean.	Estuaries are among the most productive environments on Earth, with high species diversity. They serve as a nursery for young fish, which feed on abundant vegetation, benthic invertebrates, and plankton while hiding from predators.
	**Oceans** **Marine** Oceans form a vast, continuous body of salt water that is remarkably uniform in chemical composition. Oceans are deep (average depth = 3.7 km) but depth is small compared to breadth.	Biologists describe the structure of oceans by naming zones according to depth and sunlight penetration: e.g., intertidal zone, neritic zone, oceanic zone, benthic zone, photic zone, and aphotic zone (see Figure 49.13).	Tides and wave action are major influences in the intertidal zone. Upwelling currents are important in the neritic zone. Large-scale currents circulate water in the oceanic zone in response to winds and the Earth's rotation.	Each zone of the ocean is populated by distinct species that are adapted to the physical conditions present. Coral reefs in the neritic zone have high species diversity, but the oceanic zone is sparsely populated.

pH) to occur, limiting the ability of organisms like corals to build skeletons and triggers many other effects (Ch. 53, Section 53.3).

The importance of climate and other abiotic factors to the distribution and abundance of organisms was the main focus of this chapter. Next, let's take a closer look at the biotic factors. How do interactions among organisms of the same species and interactions among different species affect the overall distribution and abundance of organisms?

CHECK YOUR UNDERSTANDING

✔ **If you understood this section, you should be able to . . .**

SOCIETY Predict how a lower rate of turnover and more stratification in Lake Tanganyika, caused by climate change, could affect its sardine fishery. Explain.

Answers are available in Appendix A.

49.1 Levels of Ecological Study

- Biologists study ecology at five main levels: **(1)** organisms, **(2)** populations, **(3)** communities, **(4)** ecosystems, and **(5)** the biosphere.
- The goal of organismal ecology is to understand the morphological, physiological, and behavioral adaptations of individuals to their abiotic and biotic environment.
- Population ecology focuses on how and why populations grow or decline over time and space.
- Community ecology is the study of how different species interact and how these interactions change over time.
- Ecosystem ecology analyzes interactions between communities and their abiotic environment—particularly the flow of energy and the cycling of nutrients.
- Global ecology studies processes that transcend ecosystems, such as the effects of climate change.
- To preserve endangered species and communities, conservation biologists apply methods and results from all levels.

49.2 What Determines the Distribution and Abundance of Organisms?

- For each species, a unique combination of abiotic and biotic factors determines where individuals live and the size of populations.
- Understanding historical events, such as the movement of entire continents, is important to interpreting current patterns of species distributions.
- Abiotic and biotic factors often interact to produce a different effect on species distributions than either type of factor would have on its own.

49.3 Climate Patterns

- Climate varies around the globe because sunlight is distributed asymmetrically—equatorial regions receive on average more solar radiation than do regions toward the poles.
- Air circulation cells such as Hadley cells create bands of wet and dry habitats, and the tilt of Earth's axis causes seasonality in the amount of sunlight that non-equatorial regions receive.
- The presence of mountains can create local areas of wet or dry habitats, and oceans moderate temperatures in certain terrestrial habitats.

49.4 Types of Terrestrial Biomes

- Terrestrial biomes are defined by climatic regimes and distinct types of terrestrial vegetation.
- The major terrestrial biomes include tropical wet forest, desert and dry shrubland, temperate grassland, temperate forest, boreal forest, and arctic tundra.
- The productivity of biomes varies. In general, photosynthesis is most efficient where average temperatures are high and moisture is ample.
- Humans are changing the terrestrial landscape, creating new anthropogenic biomes—such as cities, suburbs, farms, and ranchlands—in place of natural biomes.
- Global climate change is causing temperature profiles and precipitation patterns to change within terrestrial biomes and is also changing the geographic boundaries of biomes.

49.5 Types of Aquatic Biomes

- Salinity, water depth, water flow, and nutrient availability are key factors determining types of aquatic biomes.
- Aquatic biomes include lakes and ponds, wetlands, streams, estuaries, and oceans.
- Humans are affecting aquatic biomes by physically altering them (such as by damming a river), by changing their chemistry (such as by polluting and causing ocean acidification), and by changing species compositions (such as by introducing invasive species).
- Global climate change is affecting aquatic biomes by changing factors such as water flow, sea level, ocean circulation patterns, and patterns of upwelling and turnovers.

Answers are available in Appendix A.

✔ TEST YOUR KNOWLEDGE

1. Name the five main levels of study in ecology from smallest to largest.

2. Where do rain shadows exist?
a. the part of a mountain that receives prevailing winds and heavy rain
b. the region beyond a mountain range that receives dry air
c. the region along the equator where precipitation is abundant
d. the region near 30°N and 30°S latitude that receives dry air

3. What is the main type of vegetation in a tropical wet forest?
a. shrubs and bunchgrasses
b. herbs, grasses, and vines
c. broad-leaved deciduous trees
d. broad-leaved evergreen trees

4. What is one expected consequence of global climate change?
a. Average rainfall will increase.
b. Average rainfall will decrease.
c. Variability in rainfall will increase.
d. We cannot make predictions about future rainfall.

✔ TEST YOUR UNDERSTANDING

5. Which of these statements about aquatic biomes is true? Select True or False for each statement.
T/F The ocean is so vast that it is not influenced by human impacts.
T/F Water depth, water flow, salinity, and nutrient availability are important abiotic factors in aquatic biomes.
T/F Once nutrients sink to the bottom of lakes, they are no longer available to organisms.

6. Explain how the open ocean is similar to the desert, and how it is not.

7. Temperate forest in China and temperate forest in North America are part of the same biome but are distinct ecosystems. Explain.

8. Compare the distribution of the natural terrestrial biomes of eastern North America with the distribution of anthropogenic biomes.

✔ TEST YOUR PROBLEM-SOLVING SKILLS

9. **THINK CAREFULLY** Like Earth, Mars has seasons—its Northern Hemisphere experiences winter when its Southern Hemisphere experiences summer. Which statement is the best explanation for seasonality on Mars?
   - **a.** Mars is tilted on its axis.
   - **b.** Mars has an elliptical orbit that brings it closer to the Sun in summer than in winter.
   - **c.** Mars has a circular orbit and maintains a constant distance from the Sun.
   - **d.** The Southern Hemisphere of Mars consistently receives direct solar radiation.

10. Scientists predict that global climate change will cause a greater increase in average temperature at higher latitudes than in the tropics. This seems like good news since most of the world's biodiversity occurs in the tropics. However, recent data suggest that a smaller temperature change in the tropics will cause a *larger* physiological response in organisms there than the larger temperature change will in organisms in temperate areas. Based on what you have learned about thermal niches in this chapter, propose a hypothesis to explain this result.

✔ PUT IT ALL TOGETHER: Case Study

How could such a pretty fish be such an ecological problem?

The red lionfish, *Pterois volitans*, has beautiful red stripes, streaming fins, and a fearless disposition, and it is deadly. Native to the Pacific Ocean, the red lionfish was first discovered on coral reefs in the Bahamas in 1985. The species has spread to over 3 million square kilometers of the western Atlantic and Gulf of Mexico. Lionfish have a voracious appetite for native fish, are armed with venomous spines, and have no natural predators in their new range.

11. Why are species like the lionfish considered "invasive"?
    - **a.** They are found in areas where they are not native.
    - **b.** They were accidentally introduced by humans.
    - **c.** They spread aggressively and displace native species.
    - **d.** They are predators of other fish.

12. **MODEL** Lionfish are usually found in full-salinity seawater (35 parts per thousand). However, lab studies show that lionfish can tolerate salinities as low as 4 parts per thousand for several days, suggesting that the low-salinity water of estuaries (such as at the mouth of the Amazon River) may not be a barrier to lionfish dispersal along the coasts. Sketch a niche model predicting lionfish abundance as a function of salinity.

13. DNA studies suggest that the lionfish invasion began with just a few females released off the coast of Florida, probably from a fish tank. Based on your knowledge of ocean currents, predict the direction in which the lionfish dispersed most quickly.

14. **PROCESS OF SCIENCE** List the five main levels of ecological study, and write a question about the ecology of lionfish that you could research at each level.

15. **QUANTITATIVE** Biologists are scrambling to study the reproductive biology of lionfish to inform management decisions. For example, Alexander Fogg and colleagues studied lionfish females in the northern Gulf of Mexico and estimated that an average female could produce 2,332,490 eggs per year. What time of year is reproductive rate the highest? To answer this question, the researchers collected an average of 200 lionfish each month and measured ovary size relative to body size, an indicator of reproductive capacity. Describe the take-home message of the graph. How do the error bars affect your confidence in this message (see **BioSkills 3**)?

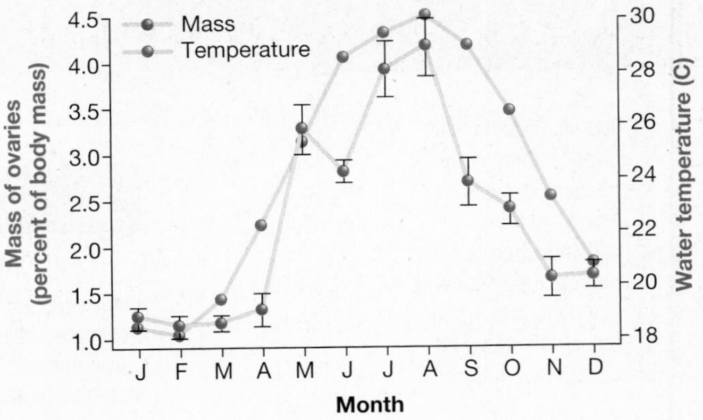

Source: A. Q. Fogg, N. J. Brown-Peterson, and M. S. Peterson. 2017. *Bulletin of Marine Science* 93: 1–23.

16. **SOCIETY** Volunteer conservation organizations have hosted "lionfish derbies" and lionfish cookouts to promote the capture of thousands of lionfish from reefs in Florida and the Bahamas. Conservationists are also trying to promote a fishery for these small but tasty fish. Give one reason that it would be easy to promote the consumption of lionfish. Describe one possible challenge to this effort.

Mastering Biology

Students Go to Mastering™ Biology for assignments, the eText, and the Study Area with animations, practice tests, and activities.

Professors Go to Mastering™ Biology for automatically graded tutorials and questions that you can assign to your students, plus Instructor Resources.

50 Behavioral Ecology

Two male cuttlefish perform threatening behavioral displays. The encounter will end when one cuttlefish swims away in defeat, leaving the other with improved access to a mate.

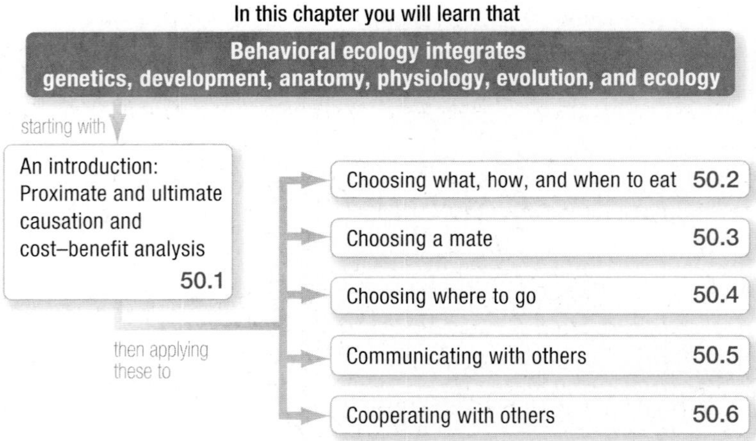

To biologists, **behavior** is action—specifically, the response to a stimulus. Pond-dwelling bacteria swim toward drops of blood that fall into the water. Time-lapse movies of growing sunflowers document the steady movement of their flowering heads throughout the day as they turn to face the shifting sunlight. Filaments of the fungus *Arthrobotrys* form loops that ensnare worms, which the fungus then digests.

Among animals, action may be frequent and even spectacular. Peregrine falcons reach speeds of up to 320 km/hr (200 mph) when they dive in pursuit of ducks or other flying prey. Hognose snakes go belly-up and feign death—mouths agape—when harassed by predators. Deep inside a hive, honeybees communicate about food sources by "dancing."

Organismal ecology (introduced in Ch. 49, Section 49.1) is the study of how morphological, physiological, and behavioral adaptations enable individual organisms to live in their physical and biological

BIG PICTURE

This chapter is part of the Big Picture. See how on pages 1190–1191.

environments. **Behavioral ecology** focuses on a subset of organismal ecology—the behavioral adaptations that have evolved in response to ecological selection pressures. You can see how behavior fits into the Big Picture of Ecology on pages 1190–1191.

Although all organisms respond in some way to signals from their environment, most behavioral research is performed on large, familiar, and behaviorally complex animals—especially arthropods, mollusks, and vertebrates—including humans. In this chapter, let's draw from these commonly studied organisms to explore how and why organisms behave the way they do.

Figure 50.1 **One Elephant Consoles Another.** Elephants have close social bonds and are known to comfort group members who have lost a family member or experienced other distress.

50.1 An Introduction to Behavioral Ecology

Behavior is a particularly integrative field in biological science. While behavioral ecology focuses primarily at the level of the organism, this field integrates levels from molecules up to ecosystems, depending on the question being asked (see the levels of biological organization, illustrated on page 18). To understand why animals and other organisms do what they do, researchers have to ask questions about genetics, development, anatomy, physiology, evolution, and ecology.

> **After you complete this section, you should be able to ...**
> - Compare proximate and ultimate causation.
> - Compare innate and learned behaviors.

In making sense of this broad approach, it's helpful to recognize that biologists ask questions and test hypotheses at two fundamental levels—proximate and ultimate. Let's first explore these levels and then consider different types of behaviors.

Proximate and Ultimate Causation

The eminent evolutionary biologist Ernst Mayr wrote an influential paper on "cause and effect in biology" in 1961. He clearly identified two types of biological causation:

1. **Proximate** (or mechanistic) **causation** explains *how* actions occur in terms of the genetic, neurological, hormonal, and skeletal–muscular mechanisms involved.
2. **Ultimate** (or evolutionary) **causation** explains *why* actions occur—based on their evolutionary consequences and history. Behavior is just like any other phenotype in that it can evolve by natural selection.

It's important to recognize that efforts to explain behavior at the proximate and ultimate levels are complementary. To understand what an organism is doing, biologists want to know how the behavior happens and why.

To grasp the idea of proximate and ultimate levels of causation, think about how you would react if your friend walked into the room seeming visibly distressed. What would you do? Personalities and relationships vary, but in general you would probably try to comfort your friend. This response occurs not only in humans but also in other highly social animals such as dolphins, dogs, other primates, and elephants (Figure 50.1). What are the proximate and ultimate causes of this consolation behavior?

Proximate Causes: How Does Consolation Behavior Occur? When you see and/or hear a friend in distress, neurons in your brain process this sensory information and cause some interesting changes. First, the brain signals your adrenal glands (above your kidneys) to produce more of the stress hormone cortisol, causing you to experience some of your friend's stress—an empathy response. At the same time, neural signals in the posterior pituitary in your brain signal the hormone oxytocin, which activates another part of your brain, the anterior cingulate cortex, which stimulates a caregiving response. These responses include consolation behaviors such as speaking in a soft voice and physical caresses.

What happens if the distressed person you encounter is not a friend, but a stranger? The neuronal and physiological responses that regulate empathy and consolation are reduced.

Other highly social animals show similar patterns of consolation behavior. For example, James Burkett and colleagues measured physiological aspects of consolation behavior in prairie voles. Voles will lick and groom group members that experience a stressful event, but not strangers. When the researchers blocked the oxytocin receptors in their study population, the voles no longer exhibited consolation behaviors, even though they continued normal self-grooming. This type of experiment is a powerful strategy used by biologists to demonstrate proximate causation: to remove or supply a hypothesized molecular or cellular mechanism then see if the behavior occurs in response.

Ultimate Causes: Why Does Consolation Behavior Occur? Individuals in human populations vary in their heritable traits, including the production and reception of oxytocin signals that mediate consolation behaviors. Since humans are highly social animals, traits that strengthen social bonds with friends are generally favored by natural selection because individuals engaged in these behaviors have, on average, increased health, increased access to resources, and decreased vulnerability to threats, leading to higher evolutionary fitness.

Note that a behavior that is adaptive in one species in one environment may not be adaptive in another species in another environment. For example, Burkett and colleagues observed oxytocin-mediated consolation behavior in the highly social prairie voles, but not in the less social meadow voles.

Consolation behavior illustrates how behavioral ecologists work at the interface of many fields and at both proximate and ultimate levels to address *how* and *why* questions about organisms in their environments. Before considering some of the most prominent questions in behavioral ecology, let's take a closer look at types of behavior.

Types of Behavior: An Overview

Scientists often use a framework to organize complex phenomena like behaviors. For example, behaviors differ in the extent to which they are learned and their flexibility. You can visualize this variation as a continuum, as shown in Figure 50.2. Let's take a closer look.

Some Behaviors Are More Innate and Inflexible Did anyone ever teach you how to yawn or sneeze? Because these behaviors require no learning, they are said to be innate. **Innate behavior** is inherited—meaning that it is passed genetically from parents to offspring (see left side of Figure 50.2).

Consider what happens to a kangaroo rat that is feeding at night and hears the sound of a rattlesnake's rattle. It jumps back—away from the direction of the stimulus. The jump-back behavior takes place the same way every time—it's hardwired. Highly inflexible, stereotyped behavior patterns like a kangaroo rat's jump-back behavior are called fixed action patterns, or FAPs. In humans, yawns and sneezes are also FAPs.

You can imagine why the jump-back behavior in kangaroo rats is adaptive: The faster they escape from predators, the more likely they are to survive. But innate behaviors have a disadvantage: They tend to be inflexible and thus may not align well to the real world if the environment changes—for example, if the jump-back behavior is repeatedly triggered by the buzz of airplanes passing over.

When the pioneering behavioral researchers Konrad Lorenz, Niko Tinbergen, and Karl von Frisch first began to explore animal behavior experimentally, they focused on these types of inflexible, innate actions—such as the way goose chicks imprint on their mother and follow her around. This research strategy is common in biological science. If you are addressing a new question—whether it's how DNA is transcribed or how ions cross plasma membranes—it's helpful to start by studying the simplest possible system. Once simple systems are understood, investigators can delve into more complex situations and questions. Lorenz, Tinbergen, and von Frisch won a joint Nobel Prize in 1973 for their behavioral work.

Some Behaviors Are More Learned and Flexible Were you born knowing how to drive a car? Clearly not; people have to learn how to drive. **Learning** is an enduring change in behavior that results from a specific experience in an individual's life. Behaviorist B. F. Skinner is well known for his pioneering experiments with rats and mice showing that their behaviors are subject to conditioning. Today, studies of learning in the lab are often complemented by observational studies in the field.

For example, chimpanzees learn to use different types of raw materials like bark, twigs, and leaves to fashion tools to forage for food and water. The chimpanzee in Figure 50.2 has used a specially prepared plant stem to sponge up water from a narrow hole in a rock. Chimps make different tools to fish for stinging bees, stingless bees, ants, and honey in nests of different sizes and shapes. They learn what works by trial and error.

Chimpanzees also learn by watching others. Unlike innate behaviors, learned behaviors can spread through populations. Behavioral ecologists Sylvain Alem, Clint Perry, and colleagues measured this phenomenon by tracking the spread of learned foraging behaviors through bee colonies. The researchers trained individual bumblebees (*Bombus terrestris*) to pull a string to retrieve a nectar reward. When the researchers introduced a single trained bee into a colony of naïve bees, the naïve bees learned the string-pulling behavior by interacting with the trained bee, then served as trainers to other bees (Figure 50.3). In control colonies that contained the reward apparatus but no trained bees, none of the bees learned how to retrieve the reward.

- Fixed, stereotyped
- Low variation in population
- Little influence by environment
- Can be adaptive because occurs in individuals without training

A lion yawns

- Flexible
- High variation in population
- Highly influenced by environment
- Can be adaptive because flexible in response to changing environment

A chimpanzee uses a stem to sponge water from a hole

Figure 50.2 Behavioral Traits Vary in Flexibility and Dependence on Learning.

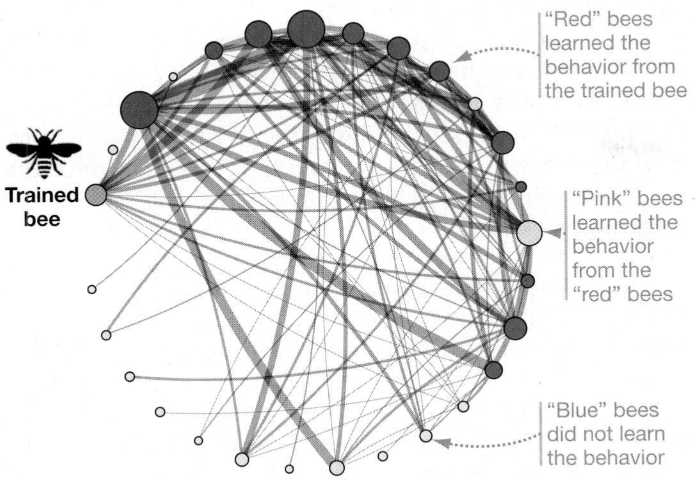

Figure 50.3 Learning Spreads Through Social Networks. In this network model, each dot represents a bumblebee, the lines represent social interactions between pairs of bees, and the size of the lines and circles represent the relative number of social interactions, which vary among bees. In this colony, only one bee was trained to pull a string to receive a nectar reward, but this behavior spread through the social network.
DATA: S. Alem, C. J. Perry, et al. 2016. *PLOS Bioloy* 14: e1002564.

Learning is an adaptive behavioral strategy in changing or unpredictable environments. While the learned behaviors themselves are not genetically determined and cannot be passed on to offspring via gametes, the capacity to learn has a genetic basis and varies among individuals and species. Behavioral flexibility is a hot topic of research in the context of climate change: Individuals and species whose behaviors have higher flexibility are predicted to be more likely to survive rapid changes in temperature, precipitation, and shifting interactions with other species.

Choices Involve Trade-Offs

Flexible, learned behaviors often involve choices and trade-offs. Most animals have a range of actions that they can perform in response to a situation. Animals take in information from the environment and, based on that information, respond in a certain way. But each behavior has different costs and benefits, resulting in **fitness trade-offs**, inescapable compromises between two traits that cannot be optimized simultaneously (Ch. 22, Section 22.5). For example, it's impossible for an animal to simultaneously maximize the energy invested in finding food and the energy invested in finding mates.

Behavioral ecologists use a framework borrowed from economics called **cost–benefit analysis** to understand and quantify the behavioral choices that animals make. Costs and benefits are measured in terms of their impact on fitness—their success in producing viable, fertile offspring.

The next five sections of this chapter introduce some of the most prominent challenges to organismal behavior, each of which requires cost–benefit analysis. These sections present research on how each challenge is met at the proximate level and at the ultimate level. As you read each section, keep two important concepts in mind:

1. ***Choices are not usually conscious.*** Nonhuman animals are not known to be self-aware of the choices and decisions they make. An elephant does not discover a distressed fellow elephant and then ask herself, "Should I caress my friend to make her feel better?" Instead, generations of natural selection on these systems result in neural and endocrine processes that are capable of taking in information and directing behavior that is likely to pass that genome on to the next generation.
2. ***The behavior of individuals varies within populations.*** Like other heritable traits, behavioral traits vary among individuals—some individuals are more fearful, social, or active than others. While it is convenient to study the average behaviors in a population, biologists are increasingly aware that variation among individuals is important to ecological interactions and evolution.

Let's begin with an introduction to research on a resource that every animal has to find in its environment to stay alive: food.

CHECK YOUR UNDERSTANDING

✔ **If you understood this section, you should be able to . . .**

1. Predict the proximate and ultimate causes of flower choice by nectar-feeding bees.
2. Predict whether innate preference for one species of flower or learned preference for multiple flowers would be favored by natural selection in bees during climate change.

Answers are available in Appendix A.

50.2 Choosing What, How, and When to Eat

After you complete this section, you should be able to . . .

- Compare the proximate and ultimate causes of foraging decisions.

When animals seek food, they are **foraging**. In some cases, the food source that a species exploits is extremely limited. Giant pandas, for example, subsist almost entirely on the shoots and leaves of bamboo. But individuals must still make decisions about which bamboo plants to harvest, as well as which parts of a particular plant to eat and which to ignore.

In most cases, animals have a relatively wide range of foods that they exploit over the course of their lifetime. To understand why individuals eat the way they do, let's consider **(1)** research on genetic variation in foraging behavior and **(2)** the concept of optimal foraging.

Proximate Causes: Foraging Alleles in *Drosophila melanogaster*

When Marla Sokolowski was working as an undergraduate research assistant, she noticed that some of the fruit fly larvae she was studying tended to move after feeding at a particular

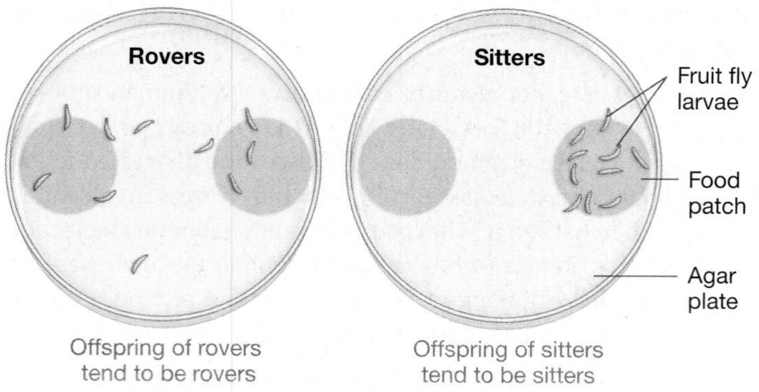

Figure 50.4 Foraging Behavior in Fruit Fly Larvae Is Heritable.

location, while others tended to stay put. She reared these "rovers" and "sitters" to adulthood, bred them, and found that the offspring of rovers also tended to be rovers, while offspring of sitters tended to be sitters (**Figure 50.4**). Thus, this aspect of behavior is at least partly due to genetic variation among individuals.

Using information from the *Drosophila melanogaster* genome sequence and genetic mapping techniques (Chapter 20), Sokolowski found and cloned a gene associated with the rover–sitter difference. She named it *foraging* (*for*).

Follow-up work showed that the rover allele is dominant over the sitter allele, and that the protein product of the *for* gene is involved in a signal transduction cascade in the nervous system—meaning it is involved in the response to cell–cell signals in neurons. Rovers and sitters tend to behave differently when they're foraging because they have different alleles of the *for* gene.

Researchers have recently discovered that adult flies express the same foraging alleles as larval flies—rover flies disperse more often and farther than sitter flies. Further, genes that are homologous to *for* have now been found in other animals such as honeybees and the roundworm *Caenorhabditis elegans*.

Ultimate Causes: Optimal Foraging

By altering population density experimentally and documenting the reproductive success of rovers and sitters, Sokolowski has shown that the rover allele in fruit fly larvae is favored by natural selection when population density is high and food is in short supply. In contrast, the sitter allele reaches high frequency in low-density populations, where food is abundant. Rovers do better at high density because they are more likely to find unused patches of food; sitters do better at low density because they don't waste energy moving around.

These results are exciting because they link a proximate mechanism with an ultimate outcome. In this case, the presence of certain alleles is responsible for a difference in fitness in specific types of habitats.

In *Drosophila*, foraging is a relatively inflexible and innate behavior—individuals don't "decide" to be rovers or sitters. It is the variability of food supplies that maintains both alleles in the population. But in most animal species, foraging behavior is much more flexible and dependent upon learning.

Introduction to Optimal Foraging When biologists set out to study why animals forage in a particular way, they usually start by assuming that individuals make decisions that have the effect of maximizing the amount of usable energy they take in, given the costs of finding and ingesting their food and the risk of being eaten while they're at it. This hypothesis—that animals maximize their feeding efficiency—is called **optimal foraging**. The concept can be expressed as

$$\text{Fitness} \propto \text{Feeding efficiency} \propto \frac{\text{Benefits}}{\text{Costs}}$$

where $\propto$ means "is proportional to." The logic behind optimal foraging is that when animals maximize benefits and minimize costs, they will have more time and energy available for reproduction—thus, higher fitness. Animals that forage optimally are expected to be favored by natural selection, so they will pass on their heritable behavior patterns to the next generation.

Note that truly "optimal" behavior is rarely achieved in biology. Organisms within populations vary and are never "perfect" due to fitness trade-offs, genetic and historical constraints, and constantly changing environments (Ch. 22, Section 22.5).

The specific costs and benefits of foraging depend on the feeding strategy and the environmental conditions of the organism. Different animals face different types of choices and therefore different variables subject to optimization. For example:

- Antarctic fur seals optimize their foraging by adjusting the duration and location of each dive depending on the depth and density of patches of food.
- Bumblebees optimize their flight paths among flowers to minimize the overall distance traveled while foraging.
- Plants called Venus fly traps optimize their foraging by waiting for several touches by trapped insects before producing digestive enzymes.

You can draw models of feeding scenarios to help you visualize the problem and identify the variables of interest (see **Making Models 50.1**). Optimal foraging is a powerful hypothesis because it can be quantified. How do biologists test for optimal foraging?

Making Models 50.1 Tips on Modeling Optimal Foraging

Drawing a model of optimal foraging can help to reveal the costs and benefits of a particular feeding scenario. The model below shows two possible flight patterns of a bee foraging among five flowers. The benefits received from the five flowers (energy in the form of pollen and nectar) are the same in both cases.

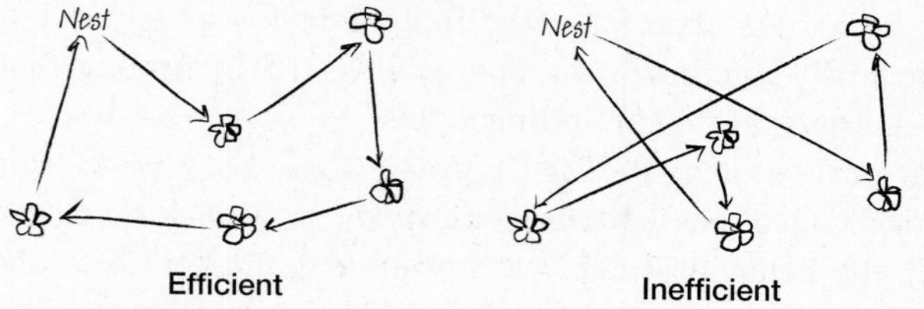

MODEL Use a ruler to measure the distances traveled in the two scenarios. Then use the ratio of distances to calculate the relative costs of foraging. (Express your cost ratio as a percentage.)

To see this model in action, go to the Study Area of **Mastering Biology**

A Test of Optimal Foraging in Cuttlefish Consider one study of optimal foraging in the pharaoh cuttlefish (*Sepia pharaonis*; pictured in the chapter-opening photo), a cephalopod cousin of octopuses and squid. You might not think of mollusks as intelligent, but cephalopods have the most complex brains of all invertebrates and complex cognitive behaviors to match.

Behavioral ecologists Tsang-I Yang and Chuan-Chin Chiao posed an interesting question: Can cuttlefish "count"? More specifically, can cuttlefish distinguish different numbers, sizes, and quality of prey—as well as the costs of capturing them—and then make an optimal choice about what to eat?

To answer this question, Yang and Chiao set up an experiment with 54 juvenile pharaoh cuttlefish raised from eggs in aquaria in the lab. The researchers designed a two-chamber tank (**Figure 50.5**) that would present the cuttlefish with a visual choice between two shrimp meals that varied in number (1 to 5), size (small or large), and/or quality (alive or freshly dead). They presented a particular choice of meals to each cuttlefish 18 times and replicated the experiment with 10 to 15 cuttlefish. The cuttlefish were not fed for four hours before data collection, so that they would be hungry during the trials.

The "Results" section in Figure 50.5 shows a sample of Yang and Chiao's data. The researchers made several observations:

- ***Cuttlefish can "count" and chose the number of their prey.*** Hungry cuttlefish chose more prey over a lesser number.
- ***Cuttlefish chose the size of their prey.*** Cuttlefish chose larger prey over smaller ones of the same quality.
- ***Cuttlefish chose the quality of their prey.*** Cuttlefish chose fewer live prey over a greater number of dead prey.

The cuttlefish showed that they could weigh the number, size, and quality of two meal options at the same time. On average, the cuttlefish chose the meals that offered them the highest net benefit: the highest quantity of quality food. How confident are the researchers with the results? Statistical tests comparing meal choices consistently produced $P < 0.01$, indicating that the differences were statistically significant (see **BioSkills 3**).

What about the costs? It takes time and energy for cuttlefish to capture and subdue prey—especially large prey—and there is a risk of being injured by prey. To see if cuttlefish weigh this cost when they are hunting, Yang and Chiao repeated their two-chamber experiments with a twist: This time they compared cuttlefish that were hungry versus satiated. What did the researchers find? Hungry cuttlefish are more willing to risk capturing larger shrimp than satiated cuttlefish, suggesting that they weigh the cost of capture in their meal choices as well as the benefits.

RESEARCH

QUESTION: Do cuttlefish make optimal choices when they hunt for prey?

HYPOTHESIS: Cuttlefish can distinguish the number, size, and/or quality of their prey to choose an optimal meal.

NULL HYPOTHESIS: Cuttlefish do not choose meals based on the number, size, and/or quality of prey.

EXPERIMENTAL SETUP:

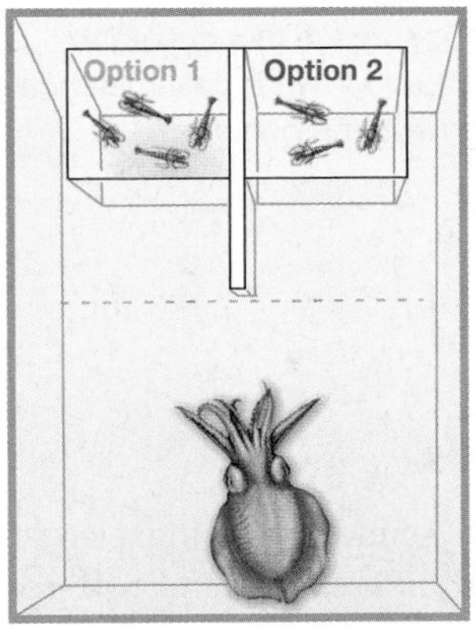

View of tank from above

1. **Prepare tanks** in the lab containing single young cuttlefish (*Sepia pharaoanis*, 2–4 cm).
2. **Present meal options** consisting of 1, 2, 3, 4, or 5 shrimp that are either small or large and either alive or dead.
3. **Record** which option was chosen.
4. **Repeat** 18 times per cuttlefish and with 10–15 cuttlefish.
5. **Repeat** using different combinations of meal options.

PREDICTION OF HYPOTHESIS: Cuttlefish will choose more shrimp over less, large shrimp over small, and live shrimp over dead.

PREDICTION OF NULL HYPOTHESIS: Cuttlefish will show no pattern of preference for meals based on number, size, or quality.

RESULTS: Three examples from many combinations of meal options:

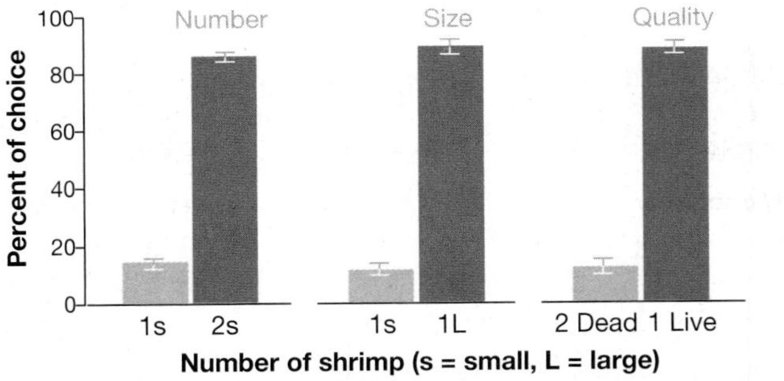

CONCLUSION: Cuttlefish can "count" and typically select optimal choices when hunting for prey.

Figure 50.5 Cuttlefish Exhibit Optimal Foraging.
SOURCE: T-I. Yang and C-C. Chiao. 2016. Number sense and state-dependent valuation in cuttlefish. *Proceedings of the Royal Society B* 283: 20161379.

✔ **PROCESS OF SCIENCE** During replicate measurements, the researchers varied the side of the meal chamber (right or left) that would contain the most or largest shrimp. In what way was this a control?

Introduction to Evolutionary Game Theory Notice how the choice between a small and a large meal is relatively simple. But what happens if the cuttlefish has to consider whether to compete with another cuttlefish for access to large prey, or whether seeking the meal will expose the cuttlefish to a hungry shark?

Evolutionary game theory is a mathematical modeling approach to predicting the outcome of natural selection on behaviors when multiple "players" are interacting. The core idea of game theory is that the behavioral strategy of one player depends on how it perceives the behavioral strategy of others. For example, if two cuttlefish are competing for access to a large shrimp, the (unconscious) choice of each cuttlefish to either attack the competitor or swim away depends on the likelihood that the competitor is likely to attack or swim away.

Evolutionary game theory, which was first published in biology by John Maynard Smith and George Price in 1973, and economic game theory, which was advanced by mathematician John Nash in 1950, arose somewhat independently. Today, however, these theories are very much cross-pollinating—an example of the power of interdisciplinary thinking. In biology, game theory applies not only to the evolution of foraging behaviors but also to other important behaviors such as choosing a mate.

CHECK YOUR UNDERSTANDING

✔ **If you understood this section, you should be able to ...**

Predict what kind of scenario might result in a cuttlefish declining to take a meal—that is, under what circumstances might the costs of foraging outweigh the benefits?

Answers are available in Appendix A.

50.3 Choosing a Mate

Choosing a mate is not as simple as it sounds. For context, consider that different species have evolved different general patterns of mating between males and females. These patterns, called mating systems, are summarized in **Table 50.1**. Notice that the systems are organized according to the number of males and the number of females that mate during a mating season.

After you complete this section, you should be able to ...

- Compare the proximate and ultimate causes of mate choice.

Bald eagles, like many birds, engage in **monogamy** by forming lasting pair bonds between one female and one male. In contrast, many cuttlefish engage in **promiscuity**, where each individual may have multiple mates. **Polygamy** lies in between, called either polygyny (literally, "many-females") or polyandry ("many-males") depending on whether one male is mating with many females (as in red deer) or one female is mating with many males (as the queen bee does in some honeybee species).

This mating system framework is a good starting point for understanding mating behavior, but the real world is a lot messier than this tidy table suggests. Keep in mind three important insights about mating systems:

1. ***Mating systems vary a lot among species, and even in individuals.*** For example, a marmoset (monkey) female may engage in polyandry when she is young and then in monogamy when she is older.
2. ***Mating systems are not always as they appear.*** For example, you might observe elephant seals to be socially polygynous (see Ch. 23, Figure 23.11), only to discover (via DNA sequencing of offspring) that some females have had furtive matings with nondominant males.
3. ***Mating behaviors can serve functions other than producing offspring.*** For example, mating behavior in humans and other highly social animals can function to create and maintain social bonds—both between sexes and within sexes.

Why would one mating system be adaptive over another? The answer to this question is complex and depends on a number of factors. Biologists seek answers to questions like these:

SUMMARY Table 50.1 Mating Systems

Mating System	Participants	Description	Examples of Groups Where this Mating System Occurs	
Monogamy	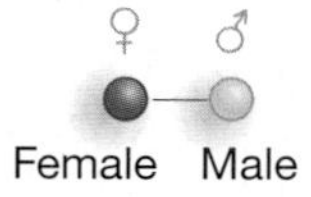	One female mates with one male, forming a pair bond; common in birds (which are oviparous) but rare in mammals (which are mostly viviparous and have lactation)	• Bald eagles • Mute swans • Osprey • Sandhill cranes • Humans	
Polygamy				
Polygyny	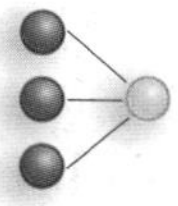	One male mates with two or more females (a group that is sometimes called a harem); often occurs in situations where a male defends a territory of resources	• Red deer • Elephant seals • Horses • Gorillas • Humans	
Polyandry	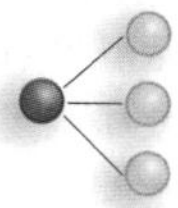	One female mates with two or more males; the rarest of the mating systems; some females choose multiple mates while others are coerced	• Honeybees • Marmosets and tamarins • Sandpipers • Field crickets • Humans	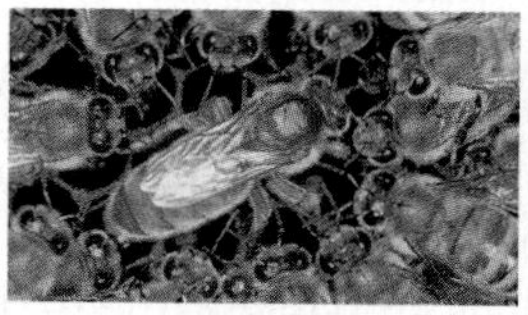
Promiscuity	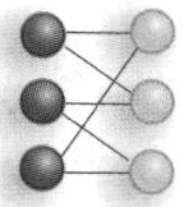	Males and females each have two or more sexual partners; some females choose multiple mates while others are coerced	• Cuttlefish • Ladybugs (ladybird beetles) • Snowshoe hares • Bonobos • Humans	

- How many *potential mates* are available, and how are they distributed?
- How much *competition* is there for access to mates?
- How much *parental care* can the mates contribute?
- What quantity and quality of *resources* are available?
- What are the *costs and benefits* of mating for each individual?

Let's take a specific example, *Anolis* lizards, and consider the proximate mechanisms responsible for mate choice and the ultimate question of why animals choose the mates they do.

Proximate Causes: How is Sexual Activity Triggered in *Anolis* Lizards?

Mate choice typically involves a *chooser* and a *courter* and some kind of **signal**, an information-containing behavior or characteristic. Remember: The "choice" in "mate choice" is rarely conscious; rather, it is a complex neural, hormonal, and motor response to sensory inputs. How do lizards choose mates?

The lizard *Anolis carolinensis* lives in the woodlands of the southeastern United States. It is socially polygynous; each male (the courter) typically defends a feeding area, or **territory**, sufficient to support two to six females (the choosers). Males signal sexual readiness to females by extending a brightly colored patch of skin called a dewlap and bobbing up and down as if doing push-ups (Figure 50.6a). However, sexual behavior is seasonal, lasting only from April to August, as indicated by the size of sexual organs in males and females relative to overall body size (Figure 50.6b).

What causes the onset of sexual behavior? The proximate answer is sex hormones—testosterone in males and estradiol in females. Testosterone is produced in the testes of males, and estradiol in the ovaries of females (Ch. 47, Section 47.5). The evidence for the effects of sex hormones is direct. Testosterone injections induce courtship activity in castrated males with no prior courtship activity; estradiol injections induce sexual activity in females whose ovaries have been removed.

But a full proximate explanation must also consider environmental and social cues. How do male and female *Anolis* lizards synchronize their sexual readiness? In a pioneering study, David Crews set out to answer this question by exposing a large number of lizards to controlled light conditions and social combinations in the laboratory. He observed three main results:

1. Females that were exposed to springlike conditions began producing eggs; females in the field that were not exposed to springlike conditions did not.
2. Females that were exposed to breeding males began producing eggs much earlier than did females that were alone, with other females, or with castrated males.
3. Females that were exposed to breeding males whose dewlaps had been surgically removed were slow to produce eggs—as slow as females grouped with castrated males.

These results support the hypothesis that two types of signals are necessary to produce the hormonal changes that lead to sexual behavior in *Anolis* lizards. Females must be exposed to springlike conditions *and* to breeding males with dewlaps.

(a) Display behavior of a male *Anolis* lizard

(b) Changes in sexual organs through the year

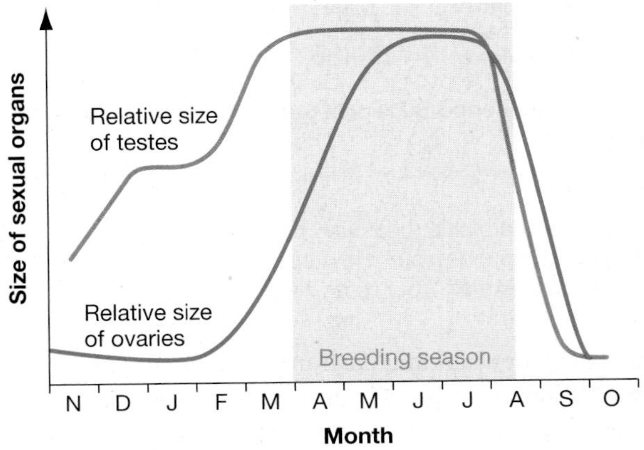

Figure 50.6 Sexual Behavior in *Anolis* Lizards Is Seasonal.
DATA: D. Crews. 1975. *Science* 189: 1059–1065.

Ultimate Causes: Sexual Selection

A type of selection called **sexual selection**(introduced in Ch. 23, Section 23.3) favors individuals possessing traits that increase their ability to obtain mates.

There are two types of sexual selection. When an individual of one sex choses an individual of another sex as a mate, ***inter*sexual** ("between-sexes") **selection** is said to occur (see Ch. 23, Figure 23.12a). Why would it be adaptive to choose a particular mate? In some cases, individuals choose mates that will provide *direct* benefits to them or their offspring, such as protection or food. Individuals can also choose mates that will provide *indirect* benefits by contributing "good alleles" to their offspring. In both cases, traits such as long tails, bright colors, and energetic courtship displays can serve as honest signals of mate quality. Sometimes, however, attractive traits are chosen just because they are attractive; if heritable, both the signal and the response are passed on to offspring.

When two individuals of the same sex compete with one another for mates, ***intra*sexual** ("within-sexes") **selection** is taking place (see Ch. 23, Figure 23.12b). For example, red deer males that win physical battles with other males tend to have higher mating success, as do pharaoh cuttlefish that successfully intimidate other males using threatening displays (see chapter-opening photo). Competing for mates can be costly in terms of energy expense, risk of injury, and other factors, but it can also have benefits—increased fitness due to reproductive success.

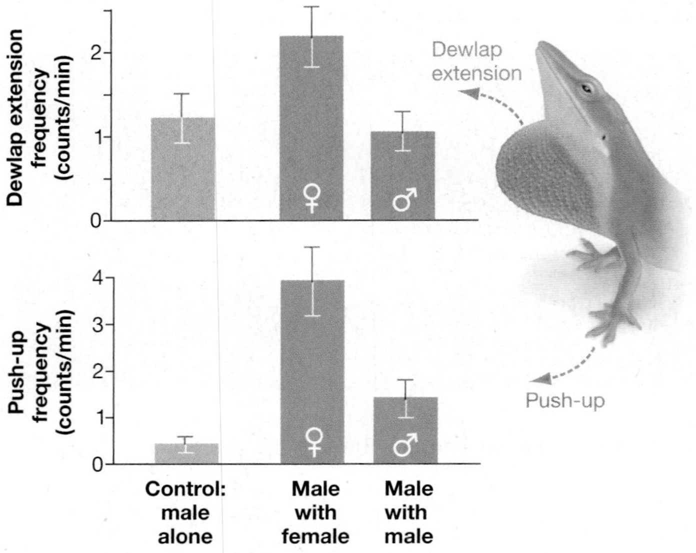

Figure 50.7 Evidence Supports Sexual Selection in *Anolis* Lizards.

DATA: T. Driessens, B. Vanhooydonck, and R. Van Damme. 2014. *Behavioral Ecology and Sociobiology* 68: 173–184.

✔ The researchers also measured frequencies of dewlap extensions and push-ups in the presence of predators. Predict the results and explain your reasoning.

Are bright dewlaps and push-ups in *Anolis* lizards signals that attract females, or ward off other males? Raoul Van Damme and colleagues set out to answer this question by measuring the rate of male signaling in various laboratory contexts. The researchers placed individual male lizards in laboratory habitats by themselves, in the presence of females, and in the presence of other males. Then they measured the frequency of dewlap extensions and push-ups in each case.

The graph in **Figure 50.7** shows some of Van Damme's results. Both dewlap extensions and push-ups are signals that woo females, supporting the intersexual selection hypothesis. However, push-ups are also used as territorial displays in the presence of males. In this species of *Anolis* under lab conditions, both intersexual and intrasexual selection appear to be in play.

And there's a twist: while *Anolis* lizards have long been recognized as socially polygamous, recent genetic evidence has revealed that many females are also mating with males from other territories—a genetically promiscuous mating pattern. Why is this mating behavior adaptive? One hypothesis is that it increases the fitness of females by increasing the genetic diversity of their offspring, who will have a higher average likelihood of survival in unpredictable environments.

CHECK YOUR UNDERSTANDING

✔ **If you understood this section, you should be able to ...**

1. **SOCIETY** Propose one proximate cause of human sexual readiness that is similar to that in *Anolis* lizards.
2. **SOCIETY** Propose one ultimate cause of human mate choice that is similar to that in *Anolis* lizards.

Answers are available in Appendix A.

Data on mate choice in *Anolis* lizards reinforce a central theme of this chapter: Animals usually make unconscious decisions in a way that maximizes their fitness.

50.4 Choosing Where to Go

Biologists have explored a wide array of questions related to the movement of individuals in search of food, mates, shelter, and other resources. Let's take a closer look at one dramatic example of these "lifestyle" decisions: the proximate and ultimate mechanisms responsible for migration. In ecology, **migration** is defined as the long-distance movement of a population associated with a change of seasons.

After you complete this section, you should be able to ...

- Compare the proximate and ultimate causes of migration.

At the proximate level, how do animals know where to go in moving from one seasonal "home" and another? At the ultimate level, what are the fitness benefits of having more than one home range that could outweigh the costs of long-distance migrations?

Proximate Causes: How Do Animals Navigate?

To organize research into how animals navigate during migratory movements, biologists distinguish three categories of navigation:

1. **Piloting** is the use of familiar landmarks.
2. **Compass orientation** is movement that is oriented in a specific direction.
3. **True navigation (map orientation)** is the ability to locate a specific place on Earth's surface.

Imagine that you asked someone to blindfold you, drive you to a remote area 100 miles from home, and drop you off by the side of the road without your cell phone. After removing your blindfold, how would you find your way home?

Piloting If you were lucky, you might recognize some landmarks and remember how to return home. The offspring of some migratory birds and mammals navigate in a similar way. They learn the route when following their parents one or more times—south in the fall and north in the spring—and later are able to pilot the route on their own.

Compass Orientation If you were dropped off and did not recognize any specific landmarks, which way would you start walking? If you had a compass and knew which direction to head, that would be a big help. Once you got closer to home and found a familiar landmark, you could pilot the rest of the way.

How do other animals perform compass orientation? To date, most research on compass orientation has been done on migratory birds such as robins, fish such as salmon, and insects such as butterflies. To determine where north is, these animals appear to use the Sun, the stars, and Earth's magnetic field.

The Sun is difficult to use as a compass reference because its position changes during the day. It rises in the east, is due south at noon (in the Northern Hemisphere), and sets in the west. To use the Sun as a compass reference, then, an animal must have an internal clock that defines morning, noon, and evening. The **circadian clock** that exists in organisms maintains a 24-hour rhythm of chemical activity. The clock is set by the light–dark transitions of day and night. It tells individuals enough about the time of day that they can use the Sun's position to find magnetic north.

The situation is actually simpler on clear nights because migratory birds in the Northern Hemisphere can use the North Star to find magnetic north and select a direction for migration. Dung beetles can use the band of light caused by the Milky Way galaxy to navigate. But what if the weather is cloudy?

Under these conditions, migratory birds and fish appear to orient using Earth's magnetic field. There are two main hypotheses for how animals detect magnetism. One hypothesis contends that animals can detect magnetism by their visual system, through a chemical reaction that involves electron transfer among molecules. An alternative hypothesis maintains that individuals have small particles of magnetic iron—the mineral called magnetite—in their bodies, such as in the beaks of birds and "noses" of fish. Changes in the positions of magnetic particles, in response to Earth's magnetic field, could then be detected and provide reliable information for compass orientation. So far, the data suggest that some animals may be capable of both mechanisms.

True Navigation (Map Orientation) If you were dropped off on the side of the road with a compass *and a map* of your location, finding your way home would be easy.

Do any animals have the equivalent of map orientation? Neurobiologist and ecologist Kenneth Lohmann and his colleagues tested this question in green sea turtles. Sea turtles have the remarkable ability to travel around vast areas of the ocean and then return to a specific coastal nesting or feeding site with pinpoint accuracy. How do they do it?

The Earth's magnetic field varies around its surface in predictable gradients of both intensity and angle, or inclination. Lohmann hypothesized that the turtles use this magnetic field as a source of precise positional information, much like we use latitude and longitude coordinates to locate ourselves on a map.

To test this hypothesis, Lohmann captured green sea turtles at their coastal feeding grounds in Melbourne Beach, Florida, fitted them with soft harnesses attached to a tracking device, and then let them swim one at a time in a circular pool (Figure 50.8). A coil system around the pool was used to change the magnetic field to replicate the magnetic signature of locations to the north and south of Melbourne Beach.

The data at the bottom of Figure 50.8 show how the turtles responded. Each dot represents the average swimming direction for one turtle. The bold arrow represents the average for the whole group. When the magnetic field in the pool simulated a location 337 km north of Melbourne Beach, the turtles swam south, the direction of "home." When the magnetic field in the pool simulated a location 337 km south of Melbourne Beach, the turtles swam north. Note that the direction of magnetic north did not change in either case, so compass orientation alone could not account for the ability of the turtles to navigate. Rather, the turtles were able to locate themselves on the map—they possess map orientation. Similar results have been discovered in salmon, which navigate through complex river systems between the ocean and natal streams.

RESEARCH

QUESTION: How do sea turtles find their way?

HYPOTHESIS: Green sea turtles use magnetic map orientation to navigate.

NULL HYPOTHESIS: Green sea turtles use compass orientation and piloting to navigate.

EXPERIMENTAL SETUP:

1. Capture green sea turtles in Melbourne Beach, Florida.

2. Place turtles in soft cloth harnesses attached to electronic tracking device in a swimming arena.

3. Monitor direction of swimming in different magnetic fields simulating locations to the north and south of Melbourne Beach.

PREDICTION OF HYPOTHESIS: Sea turtles will swim south in a simulated north location, and swim north in a simulated south location.

PREDICTION OF NULL HYPOTHESIS: Sea turtles will swim in random directions with respect to the two simulated locations.

RESULTS:

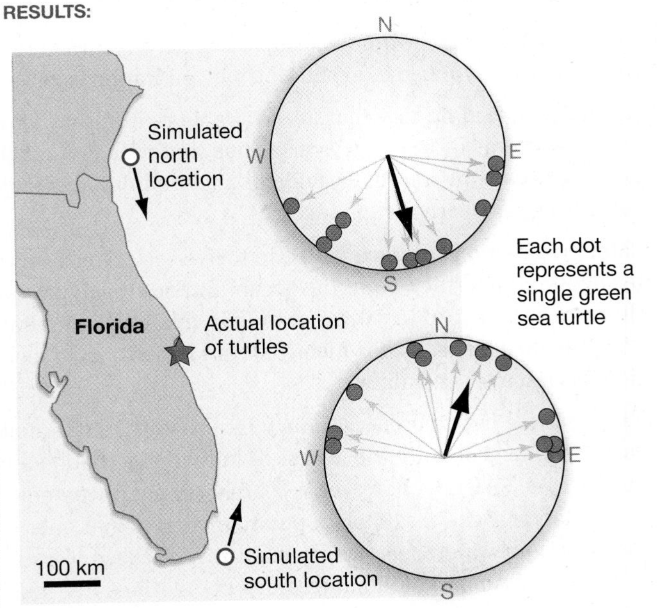

CONCLUSION: Green sea turtles use magnetic map orientation to navigate.

Figure 50.8 Sea Turtles Navigate Using Map Orientation.

SOURCE: K. J. Lohmann, C. M. F. Lohmann, L. M. Ehrhart, et al. 2004. Geomagnetic map used in sea-turtle navigation. *Nature* 428: 909–910.

✔ Describe what the data for the two sites would have to look like to reject Lohmann's hypothesis.

Neural Maps So far, the context of this discussion is long-distance migration. However, you may be surprised to learn that some insects (such as the ants in Ch. 1, Section 1.5) and other arthropods have elaborate pathway integration abilities that enable them to monitor their speed and direction during foraging trips and return home without referring to landmarks, stars, or magnetic fields.

Mammals like rats and humans take this neural processing a step further by forming a neural map of the physical space they experience. This neural navigation system occurs in a part of the brain called the entorhinal cortex. People with Alzheimer's disease lose cells in this area early in their illness, which is why they tend to wander and get lost.

Ultimate Causes: Why Do Animals Migrate?

Why do some animals have more than one home range? Migratory movements can be spectacular in their extent and in the navigation challenges they pose:

- Arctic terns nest during summer in tundra in the far north. One population flies south along the coast of Africa to wintering grounds off Antarctica and then flies back north along the eastern coast of South America (**Figure 50.9**). Chicks follow their parents south but make the return trip on their own. The journey totals over 32,000 km (20,000 miles).
- Salmon that hatch in rivers along the Pacific Coast of North America and northern Asia migrate to the ocean when they are a few months to several years old, depending on the species. After spending several years feeding and growing in the North Pacific Ocean, they return to the stream where they hatched. There they mate and die (Ch. 49, Section 49.1).

In most cases, at the ultimate level, it is relatively straightforward to generate hypotheses explaining why migration exists.

- Arctic terns feed on fish that are available in different parts of the world at different seasons. Thus, individuals that do not migrate might starve; individuals that do migrate should achieve higher reproductive success.
- Salmon eggs and young are safer and thrive better in freshwater habitats than they do in the ocean. But adult salmon can find much more food in saltwater than in freshwater habitats and grow to a bigger size, which confers a benefit in competition for nest sites and mates.

Testing these hypotheses rigorously is more difficult because researchers must compare the fitness of individuals that do and do not migrate. Recently, however, researchers have documented the evolution of sockeye salmon populations that do not migrate to the ocean—in watersheds where migratory sockeye were introduced by humans. The nonmigratory populations are much smaller bodied than the migratory populations—presumably because less food is available in the freshwater habitats.

The fitness of migrant and nonmigrant sockeye salmon appears to be similar in this case, due to fitness trade-offs. Although the larger-bodied migratory salmon are more successful in competing for nesting sites and mates, they also experience higher costs in terms of time, energy, and predation risk.

At the ultimate level, researchers are also struggling to understand how a nonmigratory population can evolve into a migratory one that travels to a completely different part of the world. It's likely that habitat changes occurring in response to global climate change will furnish an example (see Ch. 53, Section 53.3). Changes in the range and arrival times of migratory species have already been documented; it may be only a matter of time before a nonmigratory population begins to migrate, to cope with habitat shifts caused by climate change.

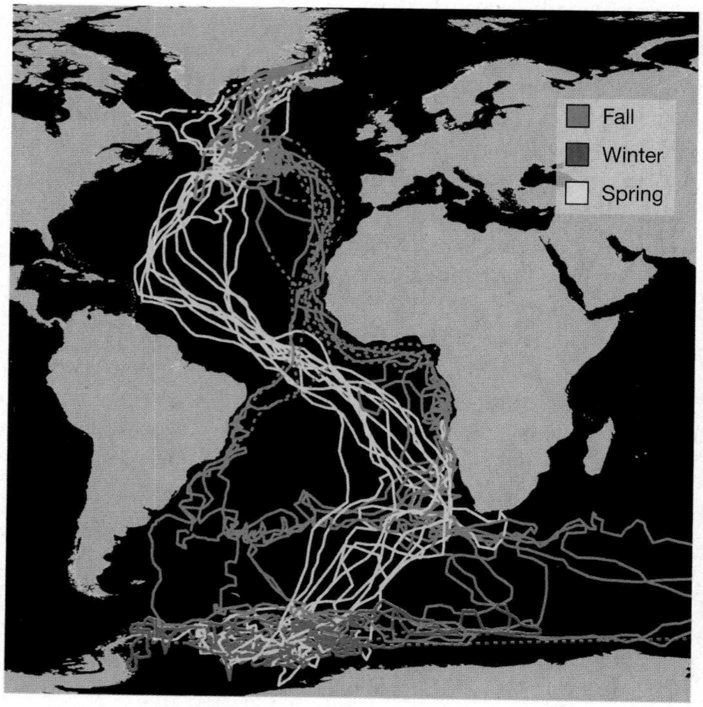

Figure 50.9 Arctic Terns Have Spectacular Migrations. The map shows the migratory pathways of 11 arctic terns, based on one year's data collected using tiny geolocators.

CHECK YOUR UNDERSTANDING

✔ **If you understood this section, you should be able to . . .**

Predict whether light pollution in urban areas disrupts bird migration by affecting piloting, compass orientation, and/or map orientation.

Answers are available in Appendix A.

50.5 Communicating with Others

After you complete this section, you should be able to . . .

- Compare the proximate and ultimate causes of communication.

The song of a bird, the bobbing dewlap of a male lizard, and the sentences in this biology text all have the same overall goal: communication. In biology, **communication** is defined as any process in which a signal from one individual modifies the behavior of a recipient individual. Communication is a crucial component of animal behavior. It creates a stimulus that elicits a response.

By definition, communication is a social process. For communication to occur, it is not enough that a signal is sent; the signal must be received and acted on. Let's see how this process plays out in honeybees.

Proximate Causes: How Do Honeybees Communicate?

Honeybees are highly social animals that live in hives. Inside the hive, a queen lays eggs that are cared for by workers. Besides caring for young and building and maintaining the hive, workers obtain food for themselves and other members of the colony by gathering nectar and pollen from flowering plants.

Biologists noticed that bees appear to recruit to food sources, meaning that if a new source is discovered by one or a few individuals, many more bees begin showing up over time. This observation inspired the hypothesis that food-finders have some way of communicating the location of food to others. How do bees "talk" to each other?

In the 1930s Karl von Frisch began studying bee communication by observing bees that built hives inside the glass-walled chambers he had constructed. He found that if he placed a feeder more than 100 m away from the hive, the bees moved in a circular pattern with straight runs down the middle accompanied by vigorous waggles of the abdomen. Von Frisch named this a "waggle dance" (Figure 50.10). Other workers appeared to follow the progress of the dance, first by touching the displaying bee as it danced and then by flying away in search of the food source.

Figure 50.10 **Honeybees Perform a Waggle Dance.** Successful foragers move in a figure-eight pattern, waggling their abdomens during the straight run in the middle.

To investigate the function of these movements further, von Frisch placed feeders containing sugar water at progressively greater distances from the hive. By catching bees at the feeders and dabbing them with paint, he could individually mark successful food-finders. Using these methods, von Frisch made three key observations:

1. The length of the waggle run was proportional to the distance the foragers had to fly to reach the feeder.
2. The direction of the waggle run correlated with the direction of the food source from the hive.
3. The direction of the food source was communicated as relative to the current position of the Sun.

For example, if food is directly away from the Sun's current position, marked bees do the waggle portion of their dance directly downward (Figure 50.11a). However, if the food is 90 degrees to the right of the Sun, marked bees waggle 90 degrees to the right of vertical (Figure 50.11b).

These results are nothing short of astonishing. Honeybees do not have large brains, yet they are capable of symbolic language.

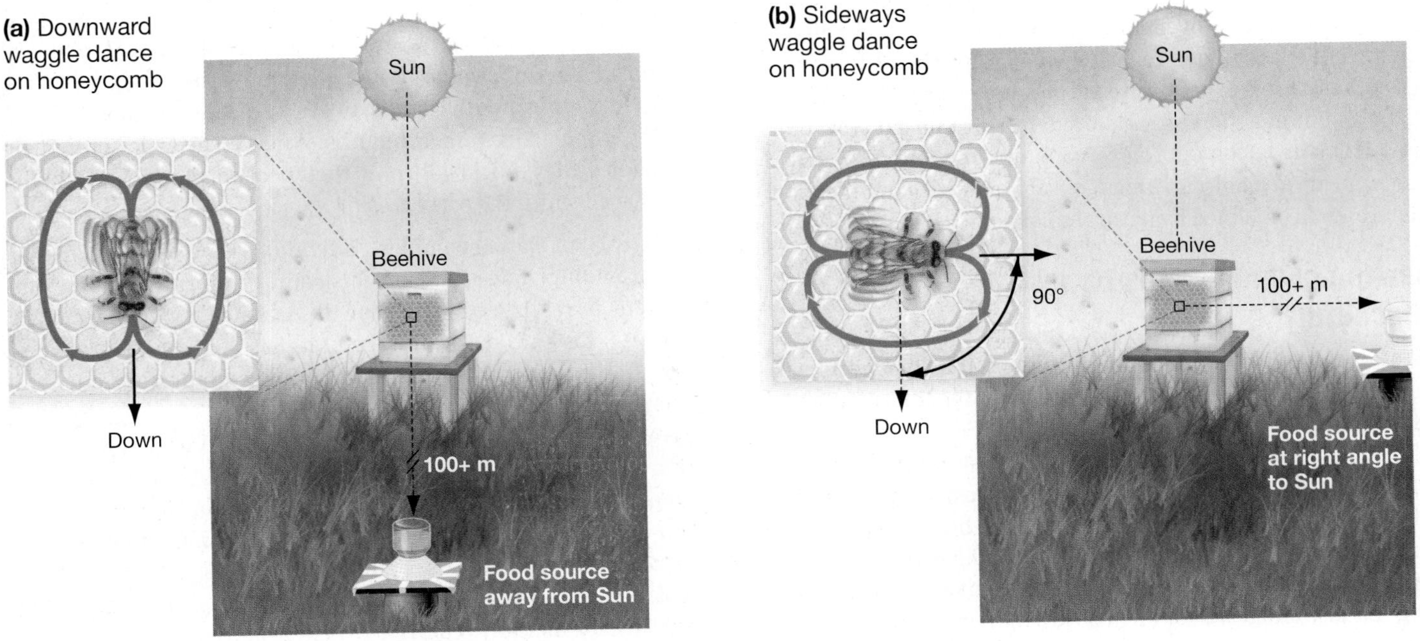

Figure 50.11 **The Direction of the Waggle Dance Indicates the Location of Food Relative to the Sun.** Waggle dances are done only when the food is at least 100 m from the hive.

What's more, they are able to interpret the angle of the waggle dance performed on a vertical surface and to respond by flying horizontally along the corresponding angle.

Further work has confirmed that the dance language of bees includes several modes of communication that are effective in the dark hives: tactile information in the movements themselves, sounds made during the dance, and scents that indicate the nature of the food source.

Ultimate Causes: Why Do Honeybees Communicate the Way They Do?

At the ultimate level, it's straightforward to come up with a hypothesis for why honeybees use a complex method of communication to share information about food sources. If more workers find a rich food source quickly, then the total amount of resources harvested as well as the number of offspring produced by the hive should increase. Thus the frequency of bee colonies with successful communication should increase over time.

In general, why do animals use the modes of communication that they do? One of the most general observations about communication is that the type of signal used by an organism correlates with its habitat. Bees, ants, and termites that live in dark nests rely on olfactory and tactile communication. Similarly, bats, wolves, and other animals that are active at night communicate via sound or scent. In contrast, animals that are active during the day and that live in open or treeless habitats tend to rely on visual communication.

Sound travels much farther than light in aquatic habitats. Thus, it is logical to observe that humpback whales rely on songs for long-distance communication. In some cases, humpback whale songs can travel hundreds of kilometers. Groups of whales use acoustic communication to keep together during migrations, and individual males sing to attract mates and warn rivals.

Each mode of communication has trade-offs. Songs and calls can carry information over long distances but are short lived. To be effective, they have to be repeated, and frequent repetition requires a large expenditure of time and energy. In addition, acoustic communication attracts predators. It's no surprise that when a hawk or falcon approaches a marsh inhabited by courting red-winged blackbirds, things get very quiet. Communication systems have been honed by natural selection.

When Is Communication Honest or Deceitful?

At the ultimate level, one of the questions that biologists ask about communication concerns the quality of the information. Is the signal reliable?

Current evidence suggests that honeybees are always honest. Stated another way, honeybee dances consistently "tell the truth" about the location of food sources. This observation is logical, because the honeybees that occupy a hive are closely related and cooperate extensively in the rearing of offspring. As a result, it is advantageous for an individual to convey information accurately. If a food-finder provided inaccurate or misleading information to its hivemates, fewer offspring would be reared and the food-finder's fitness would be reduced.

Figure 50.12 Deceitful Communication Is Most Common between Species. By playing dead, hognose snakes can avoid being eaten.

In some cases, however, natural selection has favored the evolution of deceitful communication. Consider two of the hundreds of fascinating examples:

- ***Deceiving individuals of the same species*** Male cuttlefish typically produce threating color displays toward rivals when competing for mates (see chapter-opening photo). However, in certain cases, some males flash a courtship display on the side of their bodies facing the females, but then flash a display that *mimics a female* on the side of their bodies facing rival males. This deceitful communication helps males to avoid detection by their rivals during courtship efforts.
- ***Deceiving individuals of another species*** Hognose snakes are famous for their theatrical ability to play dead (**Figure 50.12**). When perturbed by a predator, hognose snakes turn belly-up, produce a foul odor, hang their tongues out of their gaping mouths, and may even produce a couple of droplets of blood from their mouths. When turned back onto their bellies, they quickly flip onto their backs again and sustain the charade until the coast is clear.

In each of these examples, individuals increase their fitness by providing inaccurate or misleading information. It's important to realize, though, that the signaler is not consciously "lying." Instead, natural selection has simply favored certain behavioral traits that effectively communicate deceitful information.

Also, as you might expect, cheating works best in a population when it is relatively rare. If deceit becomes extremely common, then natural selection will strongly favor individuals that can detect and avoid or punish liars. But if liars are rare, then natural selection will favor individuals that are occasionally fooled but are more commonly rewarded by responding to signals in a normal way. This is an example of frequency-dependent selection (Ch. 23, Section 23.3).

CHECK YOUR UNDERSTANDING

✔ If you understood this section, you should be able to ...

1. Make a table comparing the costs and benefits of auditory, olfactory, and visual communication.
2. Predict why natural selection favors individuals that can detect deceitful communication and either avoid "liars" or punish them.

Answers are available in Appendix A.

50.6 Cooperating with Others

After you complete this section, you should be able to ...

- Compare explanations for apparent altruistic behavior.

The behaviors reviewed so far all have a key common element: They help individuals respond to environmental stimuli in a way that increases their fitness. But one type of behavior appears to contradict this pattern—altruism.

Altruism is behavior that has a fitness cost to the individual exhibiting the behavior and a fitness benefit to the recipient of the behavior. It is the formal term for self-sacrificing behavior. Altruism decreases an individual's ability to produce offspring but helps others produce more offspring.

In terms of ultimate causation, altruistic behavior appears to be paradoxical. If certain alleles make an individual more likely to be altruistic, those alleles should be selected against. Meanwhile, individuals with selfish alleles should be more likely to survive and produce more offspring. Over time, selfish alleles should increase in frequency while self-sacrificing alleles should decrease in frequency. Does altruism actually occur? Let's consider four explanations for apparent altruistic behavior.

Kin Selection

There is no question that self-sacrificing behavior occurs in nature. For example, black-tailed prairie dogs perform a behavior known as alarm calling. These burrowing mammals live in large communities, called towns, throughout the Great Plains region of North America. When a badger, coyote, hawk, or other predator approaches a town, some prairie dogs give alarm calls that alert other prairie dogs to scan for the threat.

Giving these calls is risky. In several species of ground squirrels and prairie dogs, researchers have shown that alarm-callers draw attention to themselves by calling and are in much greater danger of being attacked than non-callers are.

Hamilton's Rule How can natural selection favor the evolution of self-sacrificing behavior? William D. Hamilton answered this question by creating a mathematical model to assess how an allele that contributes to altruistic behavior could increase in frequency in a population.

To model the fate of altruistic alleles, Hamilton represented the fitness *cost* of the altruistic act to the actor as C and the fitness *benefit* to the recipient as B. Both C and B are measured in units of offspring produced. His model showed that the allele could spread if

$$Br > C$$

where r is the **coefficient of relatedness**. The coefficient of relatedness is a measure of how closely the actor and recipient are related. Specifically, r measures the fraction of alleles in the actor and recipient that are identical by descent—that is, inherited from the same ancestor (see **Quantitative Methods 50.1** on page 1090 to see how r is derived).

The formulation $Br > C$ is called **Hamilton's rule**. It states that altruistic behavior is most likely when three conditions are met:

1. The fitness benefits are high for the recipient (high B).
2. The altruist and recipient are close relatives (high r).
3. The fitness costs to the altruist are low (low C).

When Hamilton's rule holds, alleles associated with altruistic behavior will be favored by natural selection—because close relatives are very likely to have copies of the altruistic alleles. These alleles will increase in frequency in the population.

Inclusive Fitness Hamilton's rule is important because it shows that individuals can pass on their alleles to the next generation not only by having their own offspring but also by helping close relatives produce more offspring. To capture this point, biologists refer to direct fitness and indirect fitness.

- *Direct fitness* is derived from an individual's own offspring. Parents can increase their direct fitness by spending resources to ensure the welfare of their offspring. This behavior is called parental care (Ch. 32, Section 32.4).
- *Indirect fitness* is derived from helping relatives produce more offspring than they could produce on their own. For example, a grandmother can increase her indirect fitness by caring for her daughter's child so that her daughter will have more resources available to have a second child.

The combination of direct and indirect fitness components is called **inclusive fitness**.

$$\text{inclusive fitness} = \text{direct fitness} + \text{indirect fitness}$$

Biologists use the term **kin selection** to refer to natural selection that acts through benefits to relatives at the expense of the individual. While kin selection decreases the direct fitness of an individual, it increases the indirect fitness, resulting in an overall increase in an individual's inclusive fitness.

✔ **SOCIETY If you understand kin selection, you should be able to explain why it is likely to be common in humans.**

Testing Hamilton's Rule Does Hamilton's rule work? Do animals really favor relatives when they act altruistically? To test the kin-selection hypothesis, behavioral ecologist John Hoogland studied which of the inhabitants of a black-tailed prairie dog town were most likely to give alarm calls.

Within a large prairie dog town, individuals live in small groups called coteries that share the same underground burrow. Members of each coterie defend a territory inside the town.

Quantitative Methods 50.1 Calculating the Coefficient of Relatedness

The coefficient of relatedness, *r*, varies between 0.0 and 1.0. If two individuals have no identical alleles that were inherited from the same ancestor, then their *r* value is 0.0. Because every allele in pairs of identical twins is identical, their coefficient of relatedness is 1.0.

What about other relationships? **Figure 50.13a** shows how *r* is calculated between half-siblings. (To review what the boxes, circles, and lines in a pedigree mean, see Ch. 14, Figure 14.22.) Half-siblings share one parent. Therefore, *r* represents the probability that half-siblings share alleles due to inheriting alleles from their common parent. It is critical to realize that in each parent-to-offspring link of descent, the probability of any particular allele being transmitted is ½. This is so because meiosis distributes alleles from the parent's diploid genome to their haploid gametes randomly. Thus half the gametes produced by a parent get one of the alleles present at each gene, and half the gametes produced get the other allele. Half-siblings are connected by two such parent-to-offspring links. The overall probability of two half-siblings sharing the same allele by descent is $\frac{1}{2} \times \frac{1}{2} = \frac{1}{4}$ (To review rules for combining probabilities, see **BioSkills 4**.)

To think about this calculation in another way, focus on the red arrows in Figure 50.13a. The left arrow represents the probability that the mother transmits a particular allele to her son. The right arrow represents the probability that the mother transmits the same allele to her daughter. Both probabilities are ½. So the probability that the mother transmitted the same allele to both her son and daughter is $\frac{1}{2} \times \frac{1}{2} = \frac{1}{4}$.

Figure 50.13b shows how *r* is calculated between full siblings. The challenge here is to calculate the probability of two individuals sharing the same allele by inheriting it through their mother or through their father. The probability that full siblings share alleles as a result of inheriting them from one parent is ¼. So the probability that full siblings share alleles inherited from either their mother or their father is $\frac{1}{4} + \frac{1}{4} = \frac{1}{2}$.

✔ **QUANTITATIVE** If you understand the coefficient of relatedness, you should be able to calculate the *r* between these relatives: parent and child, grandparent and grandchild, and first cousins.

(a) What is the probability that half-siblings inherit the same allele from their common parent?

***r* between half-siblings:**
$\frac{1}{2} \times \frac{1}{2} = \frac{1}{4}$

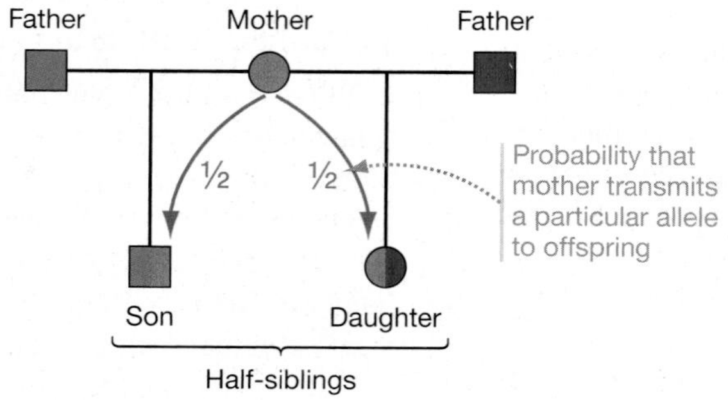

(b) What is the probability that full siblings inherit the same allele from their father or their mother?

Probability that they inherit same allele from **father**:
$\frac{1}{2} \times \frac{1}{2} = \frac{1}{4}$

Probability that they inherit same allele from **mother**:
$\frac{1}{2} \times \frac{1}{2} = \frac{1}{4}$

***r* between full siblings:**
$\frac{1}{4} + \frac{1}{4} = \frac{1}{2}$

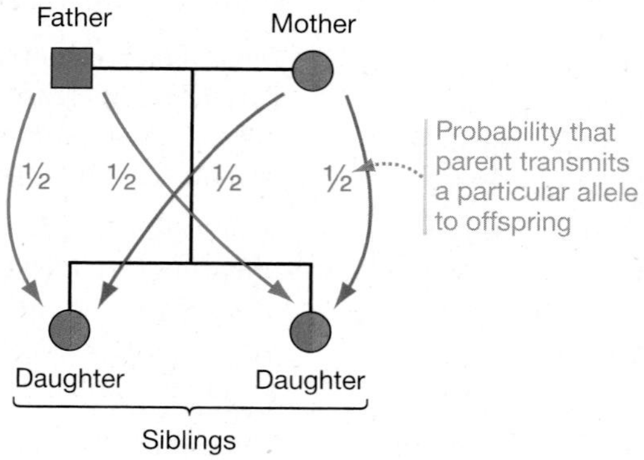

Figure 50.13 The Coefficient of Relatedness (*r*) Is Calculated from Information in Pedigrees.

By tagging offspring that were born over several generations, Hoogland identified the genetic relationships among individuals in the town. Hoogland determined whether each individual had

- no close genetic relatives in its coterie;
- no offspring in the coterie but at least one sibling, cousin, uncle, aunt, niece, or nephew; or
- at least one offspring or grandoffspring in the coterie.

The kin-selection hypothesis predicts that individuals who do not have close genetic relatives nearby will rarely give an alarm call. To evaluate this prediction, Hoogland recorded the identity of callers during 698 experiments. In these studies, a stuffed badger was dragged through the colony on a sled.

Were prairie dogs with close relatives nearby more likely to call, or did kinship have nothing to do with the probability of alarm calling? The bar graph in **Figure 50.14** shows the average proportion of times that individuals in each of the three categories called during an experiment. There is no statistical difference between the two red bars (kin but no offspring in coterie, and offspring in coterie). However, both red bars are significantly higher than the orange bar (no kin in coterie). The data indicate that black-tailed prairie dogs are much more likely to call if they live in a coterie that includes close relatives.

This same pattern—of preferentially dispensing help to kin—has been observed in many other species of social mammals and birds. Most cases of self-sacrificing behavior that have been analyzed to date are consistent with Hamilton's rule and are hypothesized to be the result of kin selection.

Manipulation

Note that altruism refers to a type of behavior with specific fitness consequences to an organism, not to the generosity of an organism as you might expect from the everyday use of the word in human culture. Sometimes altruistic behavior is involuntary.

For example, one of the most famous applications of Hamilton's rule is in explaining the complex social structure of organisms such as the bees discussed earlier, as well as ants and wasps. In these societies, workers sacrifice most or all of their direct reproduction to help rear the queen's offspring; this phenomenon is called **eusociality**.

Are eusocial insects altruistic? In insect societies that have a single, diploid queen who mates once with a *haploid* male, all the workers (who are all female) are highly related, with a coefficient of relatedness of 0.75—much higher than the 0.5 of their daughters. These numbers suggest that it would be more adaptive for workers to raise sisters than to raise their own offspring, supporting Hamilton's rule.

However, the queens in some insect societies mate with diploid males, or with multiple males, reducing the coefficient of relatedness of workers. Inclusive fitness is still important, but it is not sufficient to explain why less than 1 in 1000 workers would lay eggs. What else is going on?

One factor is that workers are sometimes manipulated into altruistic behavior. The queen and/or other workers police the nest and kill any eggs laid by workers. The more effective the policing, the less likely the workers are to continue expending energy to lay eggs. In other cases, the queen suppresses reproduction in the workers via pheromones or other chemical signals. Manipulation, or "enforced altruism," makes sense at the ultimate level because efficient colonies will be favored by natural selection, while rampant "cheating" can cause colonies to collapse.

Reciprocal Altruism

During long-term studies of highly social animals such as horses, elephants, dolphins, lions, chimpanzees, rats, and vampire bats, biologists have observed nonrelatives helping each other. Chimps and other primates, for example, may spend considerable time grooming unrelated members of their social group—cleaning their fur and removing ticks and other parasites from their skin. In vampire bats, individuals that have been successful in finding food are known to regurgitate blood meals to non-kin that have not been successful and are in danger of starving.

How can self-sacrificing behavior like this evolve if kin selection is not acting? The leading hypothesis to explain altruism among nonrelatives is called **reciprocal altruism**—an exchange of fitness benefits that are separated in time. Reciprocal altruists help individuals who have either helped them in the past or are likely to help them in the future.

RESEARCH

QUESTION: Do black-tailed prairie dogs prefer to help relatives when they give an alarm call?

HYPOTHESIS: Individuals give an alarm call only when close relatives are near.

NULL HYPOTHESIS: The presence of relatives has no influence on the probability of alarm calling.

EXPERIMENTAL SETUP:

1. Determine relationships among individuals in prairie dog coterie.

2. Drag stuffed badger across territory of coterie.

3. From observation tower, record which members of coterie give an alarm call.

4. Repeat experiment 698 times. Each prairie dog coterie is tested 6–9 times over a 3-year period.

PREDICTION OF KIN-SELECTION HYPOTHESIS: Individuals in coteries that contain a close genetic relative are more likely to give an alarm call than are individuals in coteries that do not contain a close genetic relative.

PREDICTION OF NULL HYPOTHESIS: The presence of relatives in coteries will not influence the probability of alarm calling.

RESULTS:

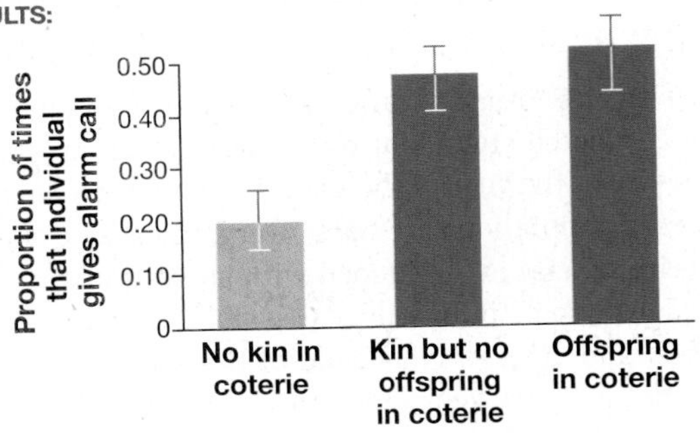

CONCLUSION: Alarm calling usually benefits relatives.

Figure 50.14 **Black-Tailed Prairie Dogs Are More Likely to Give Alarm Calls If Relatives Are Nearby.**

SOURCE: J. L. Hoogland. 1983. Nepotism and alarm calling in the black-tailed prairie dog (*Cynomys ludovicianus*). *Animal Behaviour* 31: 472–479.

✔ **PROCESS OF SCIENCE** How can you be confident that the differences in the data were not due to chance? (See BioSkills 3.)

Data that have been collected so far support the reciprocal altruism hypothesis in some instances.

- Among vervet monkeys, individuals are most likely to groom unrelated individuals that have groomed or helped them in the past (**Figure 50.15**).
- Vampire bats are most likely to donate blood meals to non-kin that have previously shared food with them.

Reciprocal altruism is also widely invoked as an explanation for the helpful and cooperative behavior commonly observed among unrelated humans.

Cooperation and Mutualism

Some behaviors may appear altruistic, but they actually may be selfish because they increase, rather than decrease, the fitness of the animal exhibiting the behavior. That is, the benefits of assisting others can exceed the costs.

- African wild dogs cannot hunt large prey alone, but they can succeed by hunting as a pack. The fitness of all the dogs increases when they work together. This is an example of cooperation among members of the same species.
- Female horses that spend more time in grooming relationships with other female horses tend to suffer less harassment from males—and to have more foals.

When cooperation occurs among individuals of different species, mutualism is said to occur. The relationship between a cleaner shrimp and a fish is mutualistic because both the shrimp and the fish benefit when the shrimp clears off—and eats—parasites from the fish's gills and jaws (Ch. 52, Section 52.1).

Individuals Do Not Act for the Good of the Species

Some apparent acts of cooperation are misunderstood, such as the widely circulated story that rodents called lemmings sacrifice themselves for the good of their species. The story claims that when lemming populations are high, overgrazing is so extensive that the entire species is threatened with starvation and extinction. In response, some individuals are said to throw themselves into the sea and drown. The good-of-the-species hypothesis maintains that even though some individuals die, the behavior evolved because the group benefits.

To understand why this type of self-sacrificing behavior does not occur, suppose that there are "self-sacrificing" alleles and "selfish" alleles. Individuals with self-sacrificing alleles die and do not produce offspring. But individuals with selfish alleles survive and produce offspring. As a result, selfish alleles increase in frequency while self-sacrificing alleles decrease in frequency. Thus, it is not possible for self-sacrificing alleles to evolve by natural selection.

Figure 50.15 Vervet Monkeys Engage in Reciprocal Altruism by Exchanging Help with Grooming.

Take-Home Messages

To summarize, organisms behave altruistically only when doing so either increases their direct fitness or their inclusive fitness, or when they are coerced. As with feeding, mating, finding a place to live, and communicating, helping others has costs and benefits that must be weighed by the organism. This "decision making" occurs at the proximate level through genetic and physiological mechanisms, and at the ultimate level through evolutionary processes.

Biologists can use evolutionary game theory (introduced in Section 50.2) to quantify and predict social interactions among individuals given that the behavioral strategy of one player depends on how it perceives the behavioral strategy of others. Since behaviors vary among individuals in populations and in different environments, the study of behavioral ecology requires systems thinking about complex networks of interactions.

This chapter focused on the behavioral adaptations of organisms within populations. Next, let's consider the distribution and abundance of entire populations (Chapter 51).

CHECK YOUR UNDERSTANDING

✔ **If you understood this section, you should be able to ...**

Assess whether an elephant is being altruistic when it comforts a distressed fellow elephant. Explain your reasoning.

Answers are available in Appendix A.

50.1 An Introduction to Behavioral Ecology

- Proximate causation: Experiments and observations focus on understanding how specific gene products, neuron activity, and hormonal signals cause behavior.
- Ultimate causation: Researchers seek to understand the adaptive significance of behavior, or why it enables individuals to survive and reproduce, thus increasing their fitness.
- Behaviors vary in the extent to which they are innate and inflexible versus learned and flexible.
- Animals appear to weigh the costs and benefits of responding to a particular situation in various ways. Costs and benefits are measured in terms of their impact on fitness.

50.2 Choosing What, How, and When to Eat

- Proximate causation: Fruit flies will be sitters or rovers depending on which alleles they carry for a certain gene.
- Ultimate causation: The sitter and rover alleles are maintained in fruit fly populations because the fitness consequences of the alleles depend on fluctuating environmental circumstances.
- Optimal foraging theory predicts that animals will make foraging decisions that maximize energy gain and minimize energy loss, time spent, and predation risk, among other factors. For example, cuttlefish are choosy about the number, size, and quality of their prey.
- Evolutionary game theory predicts the outcome of natural selection when the behavioral strategy of one player depends on how it perceives the behavioral strategy of others.

50.3 Choosing a Mate

- Proximate causation: The onset of sexual behavior in *Anolis* lizards depends on surges of the sex hormones testosterone and estradiol that are triggered by springlike temperatures *and* courtship displays from males.
- Ultimate causation: *Intra*sexual selection favors individuals that select mates that improve their fitness and/or the fitness of their offspring. *Inter*sexual selection favors individuals that are able to compete with others of the same sex for access to mates. Sexual selection in *Anolis* lizards is determined by both female choice and by male–male competition.

50.4 Choosing Where to Go

- Many animals migrate between different habitats rather than staying in one place due to the seasonal availability of food and the location of good nesting and mating places.
- Proximate causation: Animals can find their way by following familiar landmarks (piloting); by getting compass information from the Sun, the stars, and Earth's magnetic field; and/or by using the magnetic field for true navigation (map orientation).
- Ultimate causation: The benefits of migrating to food sources and mating and nesting areas must outweigh the energetic costs and danger of the migration.

50.5 Communicating with Others

- Animals communicate with touch, sounds, scents, and/or sight depending on the environment.
- Proximate causation: Honeybees have a symbolic language—their dancing movements communicate accurate information about the direction and distance of food sources.
- Ultimate causation: The ability to communicate effectively increases the fitness of animals by enabling them to find mates and avoid predators.
- Communication can be honest or deceitful. Deceitful communication can be adaptive when it is rare.

50.6 Cooperating with Others

- Proximate causation: Altruistic behavior can be voluntary or coerced.
- Ultimate causation: Altruistic behavior appears paradoxical because it lowers the direct fitness of an individual.
- When altruistic behavior is directed toward close relatives, alleles that lead to self-sacrificing may increase in frequency due to kin selection (increasing the inclusive fitness of the altruist).
- Some animals that live in close-knit social groups engage in reciprocal altruism—meaning they exchange help over time.
- Other animals cooperate because everyone benefits, such as in cooperative foraging. Cooperation among species is called mutualism.
- Individuals do not act "for the good of the species."

Answers are available in Appendix A.

✓ TEST YOUR KNOWLEDGE

1. What do proximate explanations of behavior focus on?
   a. how displays and other types of behavior have changed through time, or evolved
   b. the "adaptive significance" of a behavior
   c. genetic, neurological, and hormonal mechanisms of behavior
   d. appropriate experimental methods when studying behavior
2. What do ultimate explanations of behavior focus on?
3. **THINK CAREFULLY** A behavior is considered adaptive if it increases an individual's fitness. How is fitness measured? Select True or False for each statement.
   T/F strength
   T/F body size
   T/F speed
   T/F number of viable offspring
4. **THINK CAREFULLY** Why does altruism seem paradoxical?
   a. Sometimes altruistic behavior is actually selfish.
   b. Altruism does not actually help others.
   c. Alleles that cause an organism to behave altruistically should be selected against since these alleles should lower the organism's fitness.
   d. Animals behave altruistically to help the species, but sometimes their behavior harms the species.

✔ TEST YOUR UNDERSTANDING

5. Is it true that all organisms forage optimally? Why or why not?
6. The male cuttlefish in the chapter-opening photo can rapidly change their skin colors (under nerve control) to flash warning patterns to rivals. Predict the proximate and ultimate causes of this behavior.
7. Propose an evolutionary hypothesis to explain the observation that some bird populations do not migrate if people supply food for them in feeders.
8. **QUANTITATIVE** Hamilton's rule states that an altruistic allele could spread in a population if $Br > C$, where B represents the fitness benefit to the recipient, r is the coefficient of relatedness between altruist and recipient, and C represents the fitness cost to the altruist. If $r = 0.5$ between the altruist and the recipient, what would the ratio of costs to benefits have to be for the altruistic allele to spread?
   a. $C/B > 0.5$
   b. $C/B > 0$
   c. $C/B < 0.5$
   d. $C/B < 0$

✔ TEST YOUR PROBLEM-SOLVING SKILLS

9. **PROCESS OF SCIENCE** Evolutionary biologist Hopi Hoekstra and colleagues have hypothesized that the burrow-digging behavior of mice (and the resulting shape of their underground burrows) is heritable—innate and not learned. Design an experiment to test this hypothesis.
10. J. B. S. Haldane once remarked that he'd be willing to lay down his life to save two brothers or eight cousins. Explain what he meant.

✔ PUT IT ALL TOGETHER: Case Study

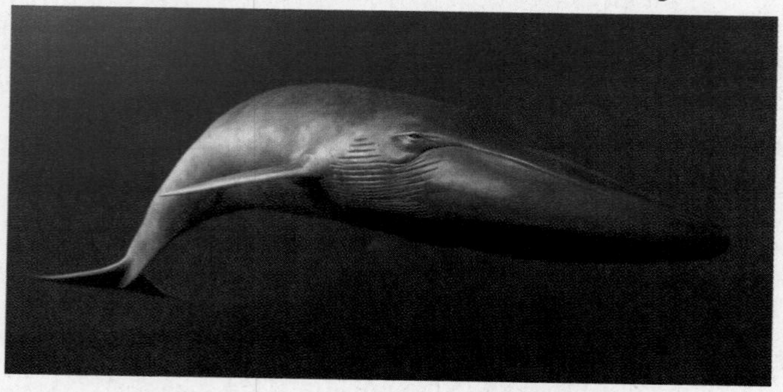

How does military sonar affect whale behavior?

Mass strandings of whales occur on beaches near military exercises where sonar is used, raising concerns about the effects of human-generated underwater sounds on animal behavior. Scientists are collecting behavioral data on several species of whales to find out how sonar affects them.

11. Whales communicate with one another using sound. What is one benefit and one cost to whales of using sound to communicate underwater?
12. Researchers followed tagged blue whales to observe how they respond to simulated military sonar—using sound levels much lower than those typically used during military exercises. Analyze the sample of data below for one individual blue whale and summarize the behavioral effect of the sound exposure.

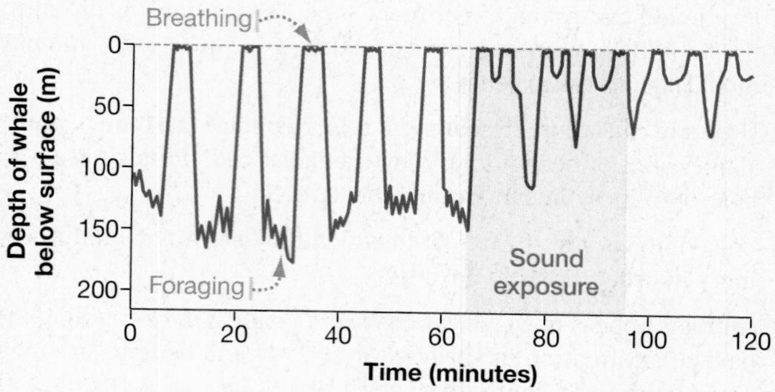

Source: J. A. Goldbogen, B. L. Southall, S. L. DeRuiter, et al. 2013. *Proceedings of the Royal Society B* 280: 2013.0657.

13. **QUANTITATIVE** Using the graph, estimate the number of minutes of foraging per hour before and after the sound exposure. Then predict the effect of sonar on the fitness of blue whales. Explain your reasoning.
14. Predict why the whale foraged at a depth of 100–170 m.
   a. The whale learned to forage at this depth from its mother.
   b. The whale had an innate instinct to feed at this depth.
   c. The whale's food was most plentiful at this depth.
   d. The whale could not dive any deeper than this depth.
15. **PROCESS OF SCIENCE** The researchers also measured the speed and direction of whale swimming in response to the sound exposure. Whales increase their speed and swim away from the direction of sound. Design a study to test the hypothesis that this behavior leads to beach strandings. Note that you will not receive permission to conduct the study if your actions are likely to cause strandings to occur.
16. **SOCIETY** There are debates about what action the military should take to avoid harming the whales. How could you apply cost–benefit analysis to address this problem?

Mastering Biology

Students Go to Mastering™ Biology for assignments, the eText, and the Study Area with animations, practice tests, and activities.

Professors Go to Mastering™ Biology for automatically graded tutorials and questions that you can assign to your students, plus Instructor Resources.

51 Population Ecology

This vivid school of crescent-tail bigeyes off the coast of Papua New Guinea demonstrates how individuals associate together in populations. The distribution and abundance of individuals in populations changes over space and time.

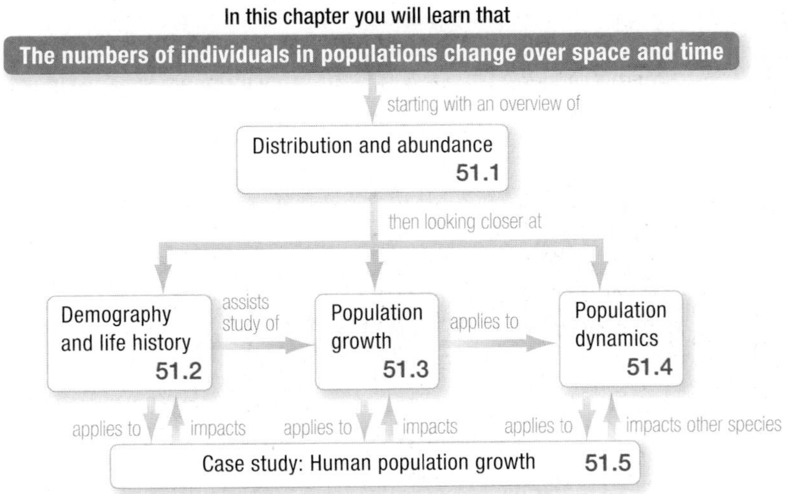

If you asked a biologist to name two of today's most pressing global issues, she might say global climate change and extinction of species. If you were asked to name two of the most pressing issues facing your region, you might say traffic and the price of housing. All four problems have a common cause: recent and dramatic increases in the size of the human population.

In 1940, for example, Mexico City had a population of about 1.6 million. The city grew to 5.4 million in 1960, 13.9 million in 1980, and over 21 million in 2017. How much bigger will it get over the next 50 years? And how large will the entire human population be by the time you retire?

A **population** is a group of individuals of the same species that live in the same area at the same time; **population ecology** is the study of how and why the number of individuals in a population changes over time and space. Population ecologists often integrate multiple levels of organization—for example, studying how the morphological, behavioral, or physiological adaptations of organisms affect population size

BIG PICTURE

This chapter is part of the Big Picture. See how on pages 1190–1191.

during climate change. (You can see how population ecology fits into the Big Picture of Ecology on pages 1190–1191.)

Further, population ecology serves as an important link between evolution and ecology. Recall that one of Charles Darwin's greatest insights was **population thinking**, which holds that variation among individuals in a population is not irrelevant; rather, understanding this variation is key to understanding how populations function and change through time in their abiotic and biotic environments (Ch. 22, Section 22.1).

With the explosion of human populations across the globe, the massive destruction of natural habitats, changes in climate, and the resulting threats to species throughout the tree of life, population ecology has become a vital field in biological science. The mathematical and analytical tools introduced in this chapter help biologists predict changes in population size and design management strategies to save threatened species.

Let's start by considering the geographic distribution of individuals within populations, then examine the variables that influence population size over time.

51.1 Distribution and Abundance

If you were interested in studying the population biology of sockeye salmon, açaí palms, pharaoh cuttlefish, humans, or any other species alive today, you might want to start by asking, "Where do they live, and how many are there?"

After you complete this section, you should be able to ...

- Explain how biologists measure the geographic distribution of individuals within populations.

Geographic Distribution

Distribution and abundance are two fundamental concepts in ecology. Most species identification guides show the **range**, or geographic distribution, of different species. Two types of factors determine range:

1. *Abiotic factors*, such as temperature, rainfall, the presence of geographical structures like mountains and oceans, and large-scale ongoing and historical processes such as continental drift (discussed in detail in Chapter 49);
2. *Biotic factors*, such as the past and current interactions with other species that provide habitat, food, or competition (discussed in detail in Chapter 52).

Ranges are dynamic—in constant flux as abiotic and biotic factors change over time.

The range of a species can also be considered at different scales. For example, the map in Figure 51.1 shows the vast global range of the common lizard, *Zootoca vivipara*, from Ireland in western Europe to Japan in eastern Asia. This map indicates that *Z. vivipara* can live in a variety of habitats and is cold adapted—it lives farther north than other reptiles, and only at high altitudes at the southern extent of its range.

Lizards are not distributed uniformly within this range, however. The **population density**—the number of individuals per unit area—of *Z. vivipara* varies throughout its range. If you zoom in on one region, such as the British Isles, you can observe that *Z. vivipara* is remarkably widespread, yet absent from some areas. Zooming in further still, the distribution of *Z. vivipara* is clumped according to the distribution of suitable habitats, such as meadows and open woodland. And if you zoom in all the way to where *Z. vivipara* individuals are visible within a certain

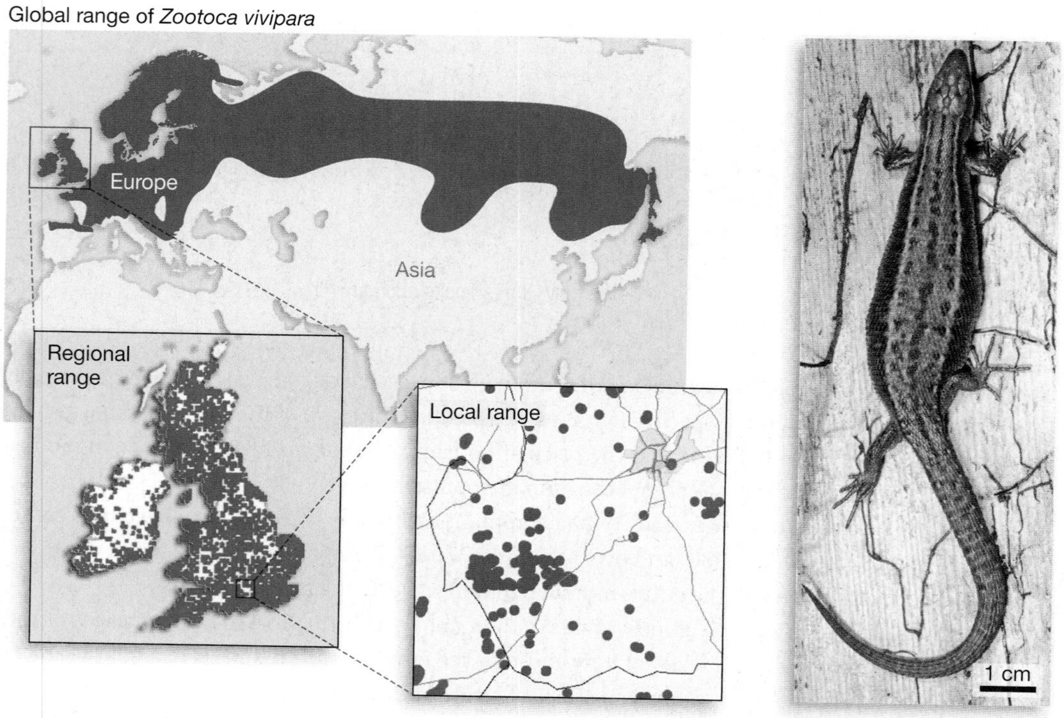

Figure 51.1 The Geographic Range of *Zootoca vivipara* Can Be Evaluated at Different Scales.

SUMMARY Table 51.1 Spatial Distribution of Individuals in Populations

Spatial Distribution		Proximate Explanations	Ultimate Explanations	Examples
Random	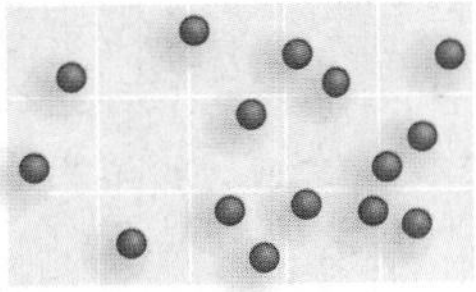	The dispersal of seeds, gametes, or larvae is random due to variation in factors such as wind and currents.	The average fitness of populations increases in variable environments when dispersal distributes individuals.	 Dandelions are randomly distributed in a meadow due to dispersal by wind.
Clumped	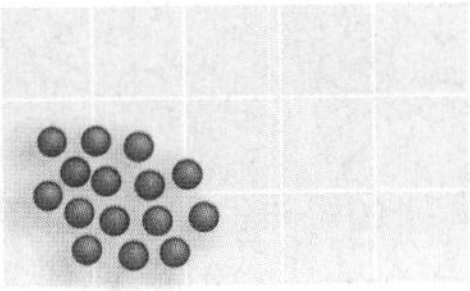	Individuals associate in social groups for feeding, mating, and/or avoiding predators; or resources are patchy.	Individuals experience higher fitness when they associate in groups; selection favors individuals that find patchy resources.	 Bigeyes associate in schools where they are more likely to avoid predation.
Uniform	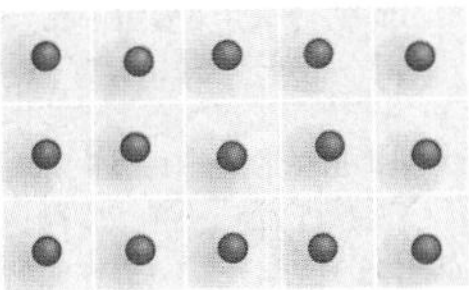	Individuals distance themselves from each other as they compete for nutrients, nesting space, or other resources.	Competition reduces the fitness of individuals; selection favors traits that minimize competition.	 Gannets compete for suitable nesting areas along rocky coasts.

patch of habitat, the distribution of individuals may appear relatively uniform due to their maintenance of feeding and mating territories.

In general, individual organisms can be arranged in different patterns within populations at the local scale, as summarized in **Table 51.1**. The arrangement may be

- *random*, if the position of each individual is independent of the others, as may occur when seeds are dispersed by the wind;
- *clumped*, if the organisms associate in social groups (such as schools of fish) or the quality of the habitat is patchy; or
- *uniform*, if negative interactions occur among individuals, such as competition for space, water, or other resources.

The particular dispersion pattern of a population has both proximate causes, in terms of physiological and behavioral mechanisms, and ultimate causes, in terms of evolutionary adaptations (see Ch. 50, Section 50.1 for an overview of proximate and ultimate causation).

If a species range is small, it may consist of a single population of interbreeding individuals. If a range is large, it may consist of many populations—sometimes contiguous, and sometimes isolated in space. If individuals from a species occupy many small patches of habitat, so that they form many independent populations, they are said to represent a **metapopulation** ("beyond population")—a population of populations connected by dispersal. Because humans are reducing large, contiguous areas of forest and grasslands to isolated patches or reserves, more and more species are experiencing metapopulation structure.

Sampling Methods

Distribution maps are always a work in progress—the accuracy of a species' range depends on the extent and currency of the field sampling efforts for that species. How can the population size, density, or distribution pattern of individuals be determined?

It depends on the species. For sedentary or sessile (mostly immobile) organisms such as trees, kelp, dandelions, mushrooms, and barnacles, researchers sample habitats by counting the individuals that occur along lines of known position and length—called *transects*—or inside rectangular plots—called *quadrats*—set up at random locations in the habitat or at regular intervals along transects. These counts can be extrapolated to the entire habitat area to estimate the total population size. They also can be compared to later sampling efforts to document trends over time.

In contrast, counting organisms that are mobile, such as lizards, rabbits, butterflies, fishes, and snakes, is much more challenging. For example, the number of lizards inside quadrats or along transects changes constantly as individuals move. Further, it can be difficult to track whether a particular individual has already been counted. But if individuals can be captured and then tagged or marked in some way, the total population size of a mobile species can be estimated by using a sampling method called mark–recapture (**Quantitative Methods 51.1**, page 1098).

Populations are composed of individuals of different ages. To get a sense of how the age structure and size of a population change through time, let's examine a field of study in ecology called demography.

CHECK YOUR UNDERSTANDING

✔ **If you understood this section, you should be able to ...**

PROCESS OF SCIENCE Propose a sampling method to measure the distribution of açaí palms in a tropical forest. Explain your reasoning.

Answers are available in Appendix A.

Quantitative Methods 51.1 Mark–Recapture Studies

To begin a mark–recapture study, researchers catch individuals and mark them with leg bands, ear tags, or some other method of permanent or temporary identification (**Figure 51.2**). After the marked individuals are released, they are allowed to mix with the unmarked animals in the population for a period of time. Then a second trapping effort is conducted, and the numbers of unmarked and marked individuals that were recaptured are recorded.

To estimate the total population from these data, researchers make a number of assumptions:

- The percentage of marked individuals (*M*) in the total population (*N*) is equal to the average percentage of marked individuals (*m*) found in each recapture sample (*n*). This relationship can be expressed algebraically:

Number of marked animals → M; Total population size → N; Number of marked animals in the recapture → m; Total number of animals (marked and unmarked) in the recapture → n

$$\frac{M}{N} = \frac{m}{n}$$

(Eq. 51.1)

- Individuals are not moving in and out of the study area.
- Individuals mix between captures.
- No bias exists regarding which individuals are caught in each recapture.
- Individuals do not learn to avoid or seek out traps after being caught once.
- Individuals do not change their behavior, attract and/or deter predators or mates, or die as a result of being marked or trapped.

Having measured *M*, *n*, and *m*, the researcher can estimate *N*. Since the number of marked individuals in each recapture will vary by chance, researchers often conduct multiple recaptures and calculate the average *N*.

Total population size: $$N = \frac{Mn}{m}$$

(Eq. 51.2)

Figure 51.2 Ecologists Can Use Marked Study Animals to Estimate Population Size. This seal has been captured, marked with a tag on its hind flipper, and released.

✔ If you understand the mark–recapture method, you should be able to answer the following questions:

1. **QUANTITATIVE** Suppose you trap 255 stickleback fish in a lake and mark them by clipping the first spine of their dorsal fins. One month later you recapture a total of 162 individuals, 78 of them marked. Estimate the total population size.
2. **PROCESS OF SCIENCE** Suppose that you re-trapped the fish population every month and noticed that the percentage of marked individuals decreased each time. What are two possible explanations for the decline?

51.2 Demography and Life History

The number of individuals present in a population depends on four processes: birth, death, immigration, and emigration.

After you complete this section, you should be able to . . .

- Compare the variables that determine the size and structure of a population over time.
- Explain the trade-offs that occur in life-history patterns.

- Populations grow as a result of births, here meaning any form of reproduction, and through **immigration**, which occurs when individuals enter a population by moving from another population.
- Populations decline due to deaths and **emigration**, which occurs when individuals leave a population to join another population.

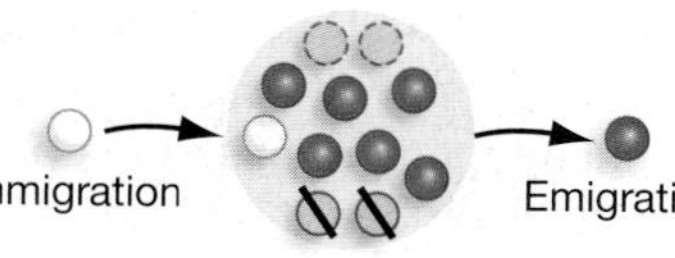

Analyzing birth rates, death rates, immigration rates, and emigration rates is fundamental to **demography**: the study of factors that determine the size and structure of populations through time.

To predict the future of a population, biologists have to know something about its makeup. They need to know the population's **age structure**—how many individuals of each age are alive. They also need to know how likely individuals of different ages are to survive to the following year, how many offspring are produced by females of different ages, and how many individuals of different ages immigrate and emigrate during each **generation time**—the average time between a mother's first offspring and her daughter's first offspring.

If a population consists primarily of young individuals with high reproductive rates and a short generation time, the population size is likely to increase. But if a population comprises chiefly old individuals with low reproductive rates and a long generation time, then the population size is almost certain to decline. Life tables are useful tools for analyzing this type of information.

Life Tables

A **life table** summarizes the probability that an individual will survive and reproduce in any given time interval over the course of its lifetime. Life tables were invented almost 2000 years ago; ancient Romans used them to predict food needs. In modern times, life insurance companies use life tables to predict the likelihood of a person dying at a given age. Biologists use life tables to study diverse questions, sometimes focusing on survival and sometimes accounting for both survival and reproduction.

There are two general approaches to collecting data for life tables. For relatively short-lived species such as lizards and sunflowers, researchers can track particular individuals over their entire lifetimes. For longer-lived species such as panthers and redwood trees, researchers must take a "snapshot" of the age structure of a population at one particular time—which is easier to do but less insightful. Let's consider an example of the first approach.

***Zootoca vivipara*: A Case Study** To understand how researchers use life tables, consider again the lizard *Z. vivipara*. As their name suggests ("vivipara" means live birth), most populations are ovoviviparous and give birth to live young, unlike most lizards, which lay eggs (Ch. 30, Section 30.3).

European biologists Henk Strijbosch and Raymond Creemers set out to construct a life table of a low-elevation population of *Z. vivipara* in the Netherlands, with the goal of comparing the results to data that other researchers had collected from *Z. vivipara* populations in the mountains of Austria and France and in lowland habitats in Britain and Belgium. They wanted to know whether populations that live in different environments vary in basic demographic features.

To begin the study, Strijbosch, Creemers, or their students visited their study site daily throughout the seven months that these lizards are active during the year. Each day, the team captured, measured, and marked as many individuals as possible.

Because this program of daily monitoring continued for seven years, the biologists were able to document the number of young produced by each female in each year of her life. If a marked individual was not recaptured in a subsequent year, they assumed that it had died sometime during the previous year. The data allowed researchers to calculate the number of individuals that survived each year in each particular age group as well as how many offspring each female produced. What did the numbers reveal?

Table 51.2 shows the data Strijbosch and Creemers collected for females. In most cases, biologists focus on females when calculating life-table data, because in sexually reproducing species like lizards and humans, only the females bear young (and the availability of males for fertilization is not a limiting factor). This means that growth rates depend entirely on females. Let's walk through Table 51.2 step by step.

Age Class and Number Alive The first column in Table 51.2 lists the age classes of *Z. vivipara*. An **age class** is a group of individuals of a specific age—for example, all female lizards between 0 (birth) and 1 year old. These data show that *Z. vivipara* females have a maximum life span of about seven years.

But not all females live to be this old. The second column in the table shows how many female lizards survived to reach each age class. Suppose 1000 female *Z. vivipara* are born in a particular year. These individuals represent a **cohort**—a group of the same age that can be followed through time. How many individuals would survive to age 1, age 2, and so on? On average, 424 would still be alive one year later, and 308 would still be alive two years later.

Table 51.2 Life Table for *Zootoca vivipara* Females in the Netherlands*

Age Class (x)	Number of Survivors (N_x)	Survivorship (l_x)	×	Age-Specific Fecundity (m_x)	=	Average Births/Year/Original Female ($l_x m_x$)
0	1000	1.000		0.00		0.00
1	424	0.424		0.04		0.02
2	308	0.308		1.47		0.45
3	158	0.158		2.07		0.33
4	57	0.057		2.44		0.14
5	10	0.010		3.25		0.03
6	7	0.007		3.25		0.02
7	2	0.002		3.25		+ 0.01
						R_0 = 1.00 = Net reproductive rate

*See **QUANTITATIVE METHODS 51.2** for a walk-through of the calculations.

DATA: H. Strijbosch and R. C. M. Creemers. 1988. Comparative demography of sympatric populations of *Lacerta vivipara* and *Lacerta agilis*. *Oecologia* 76: 20–26. (Note: *Lacerta vivipara* has since been renamed as *Zootica vivipara*.)

Survivorship The third column in Table 51.2 converts the data in the second column to proportions. **Survivorship**—a key component of a life table—is defined as the proportion of offspring produced that survive, on average, to a particular age (the opposite of mortality). For example, the *Z. vivipara* survivorship from birth to age 1 was 0.424 in the Netherlands population (that is, about 42 percent survived to age 1 and about 58 percent died). Survivorship from birth to age 2 was 0.308 (about 31 percent survived to age 2).

Life tables do not usually include data bars—the bars in this table, such as the orange bars for survivorship, are intended to help you visualize the patterns in the numbers in the different columns. Biologists typically visualize survivorship by plotting a graph with the number of survivors on a logarithmic scale on the *y*-axis (that is, increasing by a factor of 10 to enable a large range of data to be legible at both ends of the scale). Age is plotted on the *x*-axis. If survivorship is constant over time, the slope of the line on this semi-logarithmic plot will be constant and the line will be straight; if survivorship changes with age, the line will be curved (see BioSkills 5). The resulting graph is called a **survivorship curve**. Studies on a wide variety of species indicate that three general types of survivorship curves exist (Figure 51.3a).

1. ***Type I*** Humans and many other mammals have what biologists call a Type I survivorship curve. In this pattern, survivorship throughout life is high and then drops dramatically in old age, indicating that most individuals approach the species' maximum life span.
2. ***Type II*** Type II survivorship curves occur in species where individuals have about the same probability of dying in each year of life. Blackbirds and other songbirds have this type of curve—they generally experience parental care early in life, preventing high rates of early mortality, but are vulnerable to predation or adverse weather events at any age.
3. ***Type III*** Many insects and plants have Type III curves—a pattern defined by extremely high death rates for larvae or seedlings (which are produced in relatively large numbers) but high survival rates for individuals that make it past this risky stage.

Figure 51.3b provides a graph for you to plot survivorship of *Z. vivipara*, using the data reported in Table 51.2. Note the logarithmic scale on the *y*-axis of the graph. What type of survivorship curve does *Z. vivipara* have?

Fecundity The fourth column in Table 51.2 shows the data for **fecundity**, the number of female offspring produced by each female in the population. Because the researchers documented the reproductive output of the same *Z. vivipara* lizard females year after year, they were able to calculate a quantity called **age-specific fecundity**: the average number of female offspring produced by a female in each age class. The blue data bars in Table 51.2 illustrate the pattern of fecundity in lizards of different ages: Older female lizards have a higher age-specific fecundity than younger females. Note that when researchers are not able to determine the sex of offspring, they assume a 1-to-1 ratio of males and females unless the data from earlier studies suggest otherwise.

Reproductive Rate If survivorship is higher in young lizards and fecundity is higher in older lizards, what is the outcome? The purple bars in the fifth column of Table 51.2 show the product of survivorship multiplied by fecundity—most of the offspring in the population are produced by females of intermediate age.

The sum of the purple bars is a measure of **net reproductive rate**, which indicates whether the population is increasing or decreasing (as long as immigration and emigration are insignificant). Since this sum is equal to 1, this population of *Z. vivipara* is stable; when the sum > 1, the population is increasing, and when

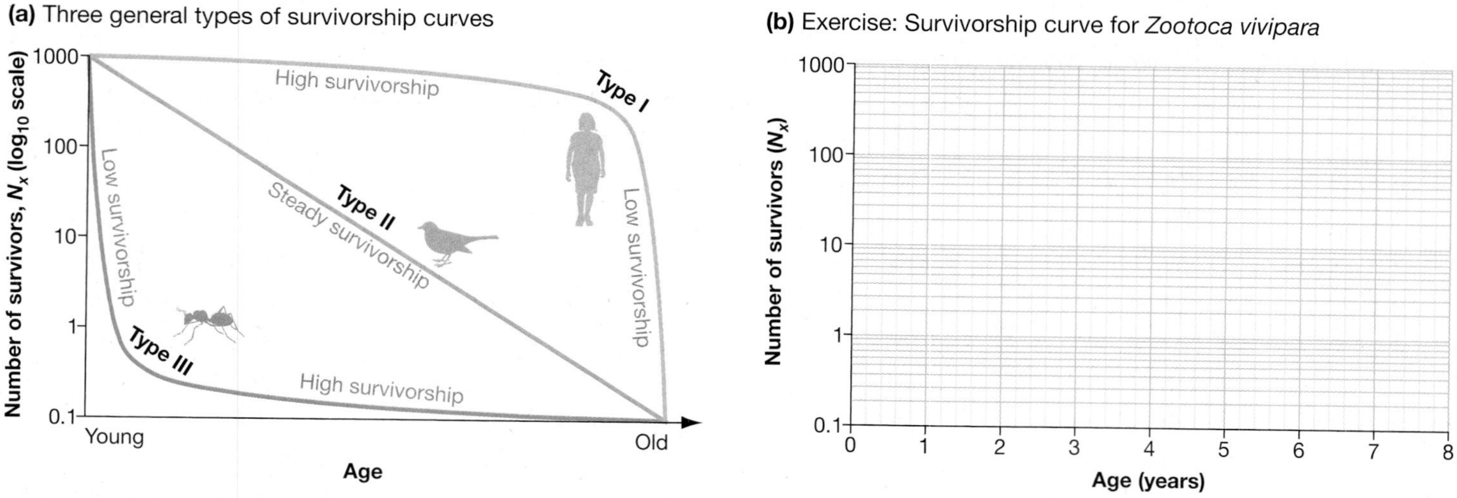

Figure 51.3 Survivorship Curves Identify When Survival Rates Are Low, Steady, or High.

✔ **QUANTITATIVE** Fill in the graph in part (b) with data on survivorship of *Zootoca vivipara* given in Table 51.2 (note the logarithmic scale on the *y*-axis). Compare the shape of the curve to the generalized graphs in part (a).

the sum < 1, the population is decreasing. **Quantitative Methods 51.2** walks through the calculations in the life table in more detail. How do the data on the Netherlands populations of *Z. vivipara* compare with populations in different habitats?

The Role of Life History

The life-table data in Table 51.2 are interesting because they contrast with results from other populations of *Z. vivipara*. Consider data on fecundity:

- *In the Netherlands*, almost no 1-year-old female *Z. vivipara* reproduce.
- *In Brittany, France*, 50 percent of 1-year-old female lizards reproduce.
- *In the mountains of Austria*, females don't begin breeding until they are 4 years old.

The contrasts between lizard populations are equally stark when looking at survivorship. In the Austrian population, most females live much longer than do the lowland individuals in the Netherlands and France.

In Brittany, fecundity is high but survivorship is low; in Austria, fecundity is low but survivorship is high. The population in the Netherlands is intermediate. In this species, key aspects of the life table vary dramatically among populations located in different abiotic and biotic environments.

What Are Fitness Trade-Offs? Why isn't it possible for *Z. vivipara* females to have both high fecundity and high survival? The answer is **fitness trade-offs**, the inescapable compromise between two traits that cannot be optimized simultaneously (Ch. 23, Section 23.5).

Fitness trade-offs occur because every individual has a restricted amount of time and energy at its disposal—meaning that its resources are limited. If a female lizard devotes a great deal of energy to producing a large number of offspring, she is not able to devote that same energy to her immune system, growth, nutrient stores, or other traits that increase survival. That is, a female can maximize fecundity, maximize survival, or strike a balance between the two.

Life History Patterns Result from Fitness Trade-Offs An organism's **life history** describes how an individual allocates resources to growth, reproduction, and activities or structures that are related to survival. Traits such as survivorship, age-specific fecundity, age at first reproduction, and growth rate are all aspects of an organism's life history. Understanding variation in life history is all about understanding fitness trade-offs.

In almost all cases, biologists find that life history is shaped by natural selection in a way that maximizes an individual's fitness in its environment. Further, certain life history traits tend to occur together, forming life history patterns that are adaptive in certain

Quantitative Methods 51.2 Using Life Tables to Calculate Population Growth Rates

If immigration and emigration are insignificant, a life table containing both survivorship and fecundity data can be used to calculate a population's growth rate. Let's take a closer look at the calculations in Table 51.2. Recall that life tables usually focus exclusively on females.

Survivorship is symbolized as l_x, where x represents the age class being considered. Survivorship for age class x is calculated by dividing the number of females in that age class (N_x) by the original number of females in that cohort (N_0):

$$\text{Survivorship: } l_x = \frac{N_x}{N_0} \quad \text{(Eq. 51.3)}$$

(N_x: Number of females in age class x; N_0: Number of original females)

Age-specific fecundity is symbolized m_x, where x again represents the age class being considered. Age-specific fecundity is calculated as the total number of female offspring produced by females of a particular age, divided by the total number of females of that age class present. It represents the average number of female offspring produced by a female of age x.

Documenting age-specific survivorship and fecundity allows researchers to calculate the average number of female offspring produced by females during each age of life, as $l_x m_x$. For example, if the frequency of females in a population living to be 2 years old is 0.308, and if 2-year-old females have an average of 1.47 female offspring, then the average number of female births at age 2, per original female born into the population, is $l_x m_x = 0.308 \times 1.47 = 0.453$.

Summing $l_x m_x$ values over the entire life span of a female gives the net reproductive rate, R_0, of a population:

$$\text{Net reproductive rate: } R_0 = \sum_{i=0}^{x} l_x m_x \quad \text{(Eq. 51.4)}$$

(Sum of all ages; Survivorship at age x; Fecundity at age x)

Here the Greek letter sigma (Σ) stands for "sum." The expression on the right-hand side of the equals sign is read, "sum the $l_x m_x$ values from each year of life, starting from birth or hatching ($i = 0$), through age x."

The net reproductive rate represents the growth rate of a population per generation. The logic behind the equation for R_0 is that the growth rate of a population per generation equals the average number of female offspring that each female produces over the course of her lifetime. A female's average lifetime reproduction, in turn, is a function of survivorship and fecundity at each age class. In Table 51.2, R_0 is the sum of the survivorship multiplied by the fecundity values in the right column.

If R_0 is greater than 1, then the population is increasing in size. If R_0 is less than 1, then the population is declining.

✔ **QUANTITATIVE** If you understand these calculations, you should be able to determine whether the population of fish represented by the following data set is growing, stable, or declining.

x	l_x	m_x
1	0.650	0.00
2	0.520	0.15
3	0.225	1.65
4	0.005	3.50

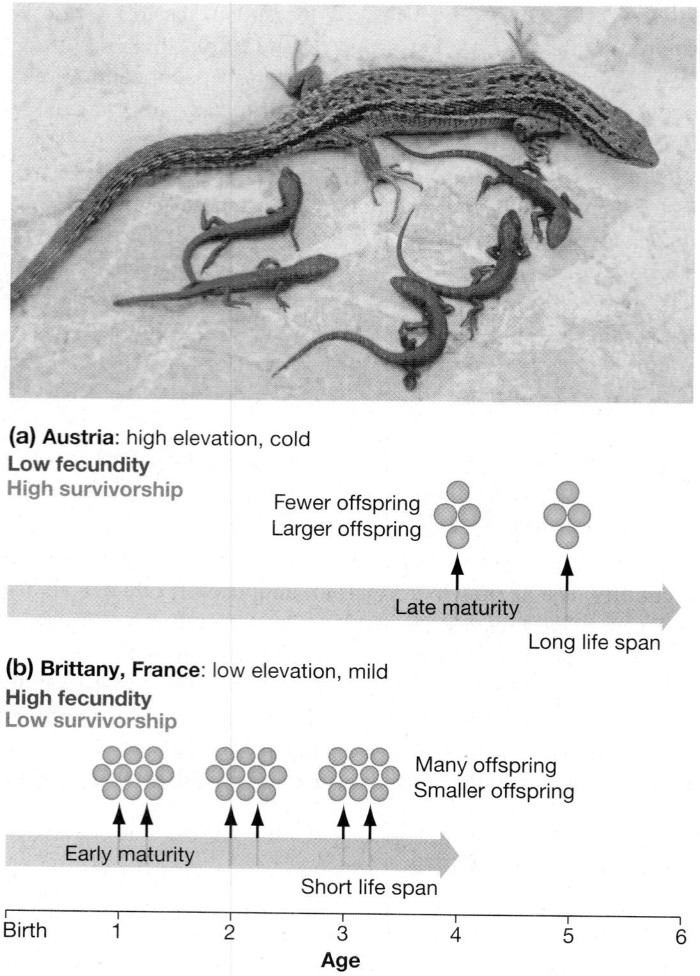

Figure 51.4 **Life-history patterns in *Zootoca vivipara* differ in populations that experience harsh and mild conditions.**

✔ **PROCESS OF SCIENCE** Pose a hypothesis to address this research question: Will the warming of Europe due to climate change cause evolution in *Z. vivipara* life-history traits? (Be sure to reference at least one of the traits shown in the models.)

conditions. For example, in *Z. vivipara* populations, biologists contend that females who live a long time but mature late and have few offspring each year have high fitness in cold, high-elevation habitats such as Austria (see the model in Figure 51.4a). In these habitats, females have to reduce their reproductive output and put more energy into traits that increase survival in a harsh environment. In contrast, females who have short lives but mature early and have large numbers of offspring twice each year do better in warm, low-elevation habitats such as Brittany (Figure 51.4b).

Life-history traits evolve by natural selection just like other traits do: heritable variation in individuals within populations leads to differential survival and reproductive success. As with evolution in other traits, there are some common misconceptions about the evolution of life-history traits (introduced in Ch. 22, Section 22.5). Here are a few key points to keep in mind:

- ***Evolutionary change occurs in populations, not organisms.*** The life-history traits of an individual lizard cannot evolve, but the average traits in a population can.
- ***Evolution is not goal directed.*** Lizards cannot evolve traits on purpose because they need or want them. Selection filters existing variation that originates from chance mutations.
- ***Evolution does not perfect organisms.*** Some traits cannot be optimized simultaneously due to fitness trade-offs or genetic or historical constraints. Also, the evolutionary processes of genetic drift and gene flow cause changes that are random with respect to fitness (Chapter 23).

The trade-offs in life-history patterns are important not only in comparing individuals or populations within a species but also in comparing species.

Life-History Patterns Vary across Species In general, individuals from species with high fecundity tend to grow quickly, reach sexual maturity at a young age, and produce many small eggs or seeds. The mustard plant *Arabidopsis thaliana*, for example, germinates and grows to sexual maturity in just 4 to 6 weeks. In this species, individuals usually live only a few months but may produce as many as 10,000 tiny seeds. They live fast and die young. Fruit flies, rabbits, guppies, and dandelions have similar life-history patterns.

In contrast, individuals from species with high survivorship tend to grow slowly and invest resources in traits that reduce damage from enemies and increase their own ability to compete for water, sunlight, or food. A coconut palm, for example, may take a decade to mature but live 60–70 years and produce offspring each year. Coconut palms invest resources in making stout stems that allow them to grow tall, a relatively extensive root system, large fruits (coconuts), molecules that make herbivores sick, and enzymes that reduce infections from disease-causing fungi. These traits increase survivorship but decrease fecundity. Giant pandas, humans, and whales have similar life-history patterns.

An *Arabidopsis thaliana* plant and a coconut palm represent two ends of a broad continuum of life-history characteristics (Figure 51.5). Most species fall somewhere in between.

Applying Life-History Data to Solve Problems The life-history continuum serves as an important framework that can help biologists solve problems such as the conservation of endangered species or the containment of invasive species. Consider a few examples:

- ***Which life stages are conservation priorities?*** Whooping cranes, sea turtles, spotted owls, and many other endangered species have high juvenile mortality, low adult mortality, and low fecundity—that is, they are on the high-survivorship, low-fecundity end of the continuum. In these species, the fate of a population is extremely sensitive to increases in adult mortality. Based on this insight, conservation biologists have begun an intensive campaign to reduce the loss of adult female sea turtles in fishing nets (Figure 51.6). Previously, most conservation action had focused on protecting eggs and nesting sites.
- ***Where do invasive species lie on the life-history continuum?*** It seems intuitive that invasive species would have high reproductive capacity and would appear on the high-fecundity and low-survivorship end of the life-history continuum, but extensive research on invasive species has

Figure 51.5 **Life-History Traits Form a Continuum.** Every organism, population, or species can be placed somewhere on this life-history continuum.

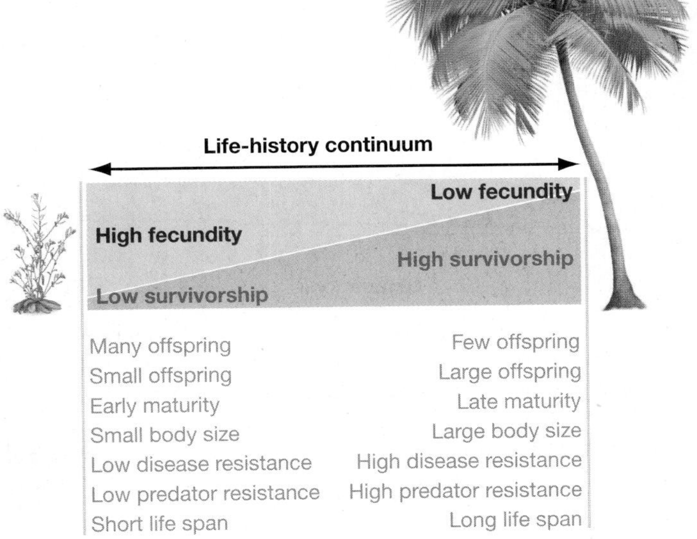

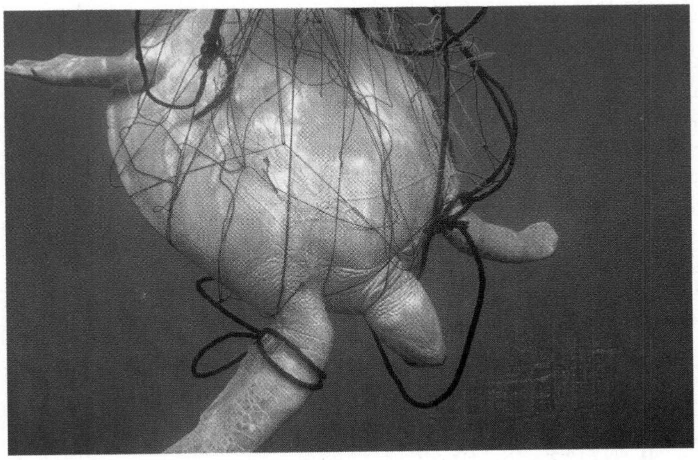

Figure 51.6 **The Mortality of Adult Sea Turtles in Fishing Nets Dramatically Affects Sea Turtle Populations.** Life-history data helps to prioritize conservation action.

shown that a variety of life-history patterns are successful. Further, many invasive species appear to have adaptive capacity, that is, sufficient genetic variation in life-history traits to evolve quickly in new habitats. This research is critical for thwarting invasive species, such as lionfish in the western Atlantic (see Case Study at the end of Chapter 49, page 1075) and Burmese pythons in the Florida everglades (see Case Study at the end of this chapter).

- ***How are species affected by climate change?*** Direct measurements of life-history traits of populations in the field are essential for predicting future population sizes. For example, climate change has reduced the thickness of the insulating snow layer at high elevations in Colorado. The fitness of *alpine marmots* has declined because they use more energy staying warm at the expense of fecundity. In contrast, the fitness of *yellow-bellied marmots* has increased due to the earlier availability of high-quality food.

CHECK YOUR UNDERSTANDING

✔ **If you understood this section, you should be able to . . .**

1. Predict the possible consequences of global warming to the net reproductive rate of sea turtles given that the sex ratio of many sea turtles and other reptiles is determined by the temperature of the eggs in the nest; typically, warmer eggs become female and cooler eggs become male.
2. Describe one fitness trade-off that is taking place in the life-history pattern of the North Pacific giant octopus (*Enteroctopus dofleini*): The female lives three to five years, then lays thousands of eggs in a single spawning event, cares for them until they hatch, and then dies.

Answers are available in Appendix A.

Life tables reveal the pattern of survivorship and fecundity of a population, linking life-history traits to population growth. Let's take a closer look at changes in population size over time.

51.3 Population Growth

After you complete this section, you should be able to . . .

- Compare exponential and logistic population growth models.

Recall that four processes affect a population's size: Births and immigration add individuals to the population; deaths and emigration remove them. It follows that a population's overall growth, or change in population size (ΔN) per unit time (Δt), is a function of birth rates, death rates, immigration rates, and emigration rates. Here, the Greek symbol Δ (delta) means "change."

$$\frac{\Delta N}{\Delta t} = Births - Deaths + Immigrants - Emigrants$$

Let's begin with the simplest scenario and assume that a population is isolated, with no immigration or emigration.

$$\frac{\Delta N}{\Delta t} = Births - Deaths$$

Exponential Growth

What would happen if *Arabidopsis* or coconut palms established a new population with plenty of resources, and it began to grow?

An Exponential Growth Equation If no immigration or emigration is occurring, then a population's growth is equal to the number of individuals (N) in the population times the difference between the birth rate per individual (b) and death rate per individual (d).

$$\frac{\Delta N}{\Delta t} = (b - d)N$$

The difference between the birth rate and death rate per individual is called the **per capita rate of increase** and is symbolized **r**. (Per capita means "for each head.")

$$\frac{\Delta N}{\Delta t} = rN$$

If $b > d$, then $r > 0$ and the population is growing. But if $b < d$, then $r < 0$, and the population is declining. If $r = 0$, then the population size does not change.

The expression $\Delta N/\Delta t$ represents a change in population size over a discrete time interval such as a year. Sometimes ecologists prefer to calculate the so-called instantaneous change in population size over a time interval that is so small that it approaches zero. Differential calculus can be used to create an expression dN/dt for instantaneous growth:

$$\frac{dN}{dt} = rN$$

This equation is a model of **exponential population growth**, which occurs when r does not change over time but the instantaneous growth dN/dt increases as population size N increases due to the product of r times N. Since this is a differential equation, r becomes an *instantaneous rate of increase* rather than *per capita rate of increase* (although the two terms are often used interchangeably).

When conditions are optimal for a particular species—meaning birth rates per individual are as high as possible and death rates per individual are as low as possible—then r reaches a maximal value called the **intrinsic rate of increase, r_{max}**. In species such as *Arabidopsis* and fruit flies, which breed at a young age and produce many offspring each year, r_{max} is high. In contrast, r_{max} is low in species such as giant pandas and coconut palms, which take years to mature and produce few offspring each year. Stated another way, r_{max} is a function of a species' life-history pattern.

Each species has a characteristic r_{max} that does not change. But at any specific time, a population has an instantaneous growth rate, or per capita rate of increase, r, that tells you what the growth rate is at a particular time. So r is always less than or equal to r_{max}. The instantaneous growth rate is often much lower than r_{max} because conditions for growth are rarely optimal. However, no matter the value of r, as long as r is constant and greater than 0, then growth increases exponentially when resources are unlimited. Quantitative Methods 51.3 explores other useful derivations of the exponential growth model.

Graphing Exponential Growth The graph in Figure 51.7 plots exponential population growth for various positive values of r. You can see why these curves are typically described as being J-shaped.

It's important to emphasize that even though r is constant, exponential growth adds an increasing number of *individuals* as the total number of individuals, N, gets larger. As an example, an r of 0.02 per year in a population of 1 billion adds over 20,000,000 individuals in a year. The same r in a population of 100 adds just over 2 individuals in a year. The r is the same, but the number of individuals added is not. See Making Models 51.1 to explore this important distinction.

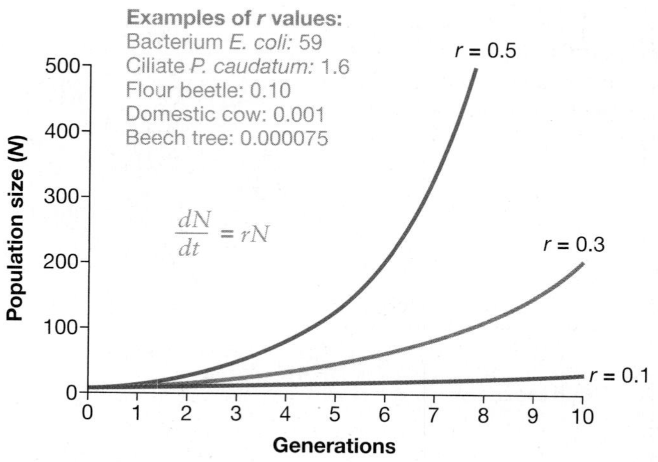

Figure 51.7 Exponential Growth Occurs When Resources Are Unlimited. Population size may increase slowly or rapidly, depending on the size of r [shown in units of individuals/(individual × day)].

DATA: T. Fenchel. 1974. *Oecologia* 14: 317–326.

Making Models 51.1 Tips on Drawing Graphs (III): Exponential ("J-shaped") Growth Curves

It can be counterintuitive to accept that the instantaneous growth rate r is constant in exponential curves, because the number of individuals added to the population grows as N gets larger. You can draw slopes on a model to understand this distinction.

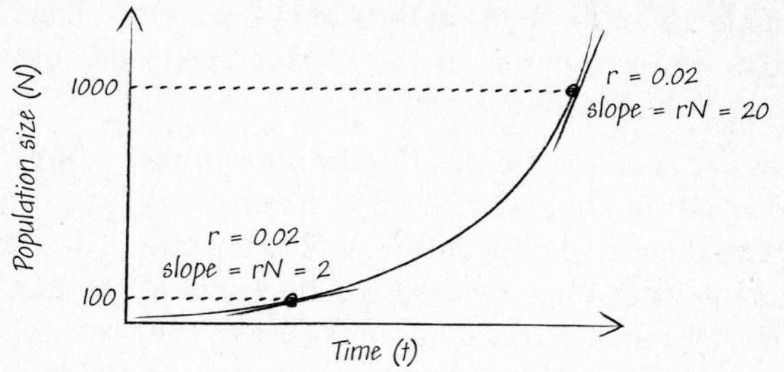

MODEL Draw a dot on the curve at $N = 500$ and calculate the slope at this point (the number of individuals being added to the population at that time) as rN, noting that r is constant at 0.02.

To see this model in action, go to the Study Area of **Mastering Biology**

Exponential Growth Is Density Independent The key point about exponential growth is that the population size does not limit growth. Biologists say that this type of population growth is **density independent**. In nature, exponential growth is common in two types of circumstances:

Quantitative Methods 51.3 Using Growth Models to Predict Population Growth

Population ecologists use different versions of growth equations to answer different kinds of questions about populations. These equations differ in the variables employed and in the assumptions made. Here are a few examples from the population ecologist's mathematical toolbox.

How can I calculate population growth rate from life-history data?

The per capita (or instantaneous) rate of increase, r, is related to the net reproductive rate, R_0 (see Table 51.2 and Quantitative Methods 51.2), which means that r can be estimated from life-table data:

Natural logarithm · Net reproductive rate · Generation time

Per capita growth rate: $$r = \frac{\ln R_0}{g}$$

(Eq. 51.5)

(**BioSkills 5** explains how to calculate a natural logarithm using your calculator.) r is generally a more useful measure of growth rate than R_0, because r is independent of generation time.

How can I predict the size of a population in the future if growth is continuous?

For species that breed continuously (not at defined intervals), like humans and bacteria, the expected size of a population at some point in the future can be estimated by using an integrated form of the exponential growth equation (modified using the rules of calculus):

Population size at time zero · Per capita growth rate · Time · Base of natural logarithm

Population size at end of time t: $$N_t = N_0 e^{rt}$$

(Eq. 51.6)

where e is about 2.72. This equation can be applied in numerous ways, such as to predict the recovery of an endangered species following conservation efforts or to predict the future population size of an invasive species that has just begun to colonize a new habitat. For example, if 10 rats were introduced from a ship onto the Galápagos Islands $N_0 = 10$ and the r for rats is about 0.015 individuals/(individuals × day), in one year ($t = 365$ days), the predicted rat population (N_{365}) would be about 2400 rats (rounded from 2386). (Notice how the time units, days in this case, cancel out when r is multiplied by t.) Keep in mind that these predictions assume that resources are unlimited and that birth rates and death rates are constant over time.

How can I predict the size of a population in the future if growth is *not* continuous?

Many species breed only once per year, especially insects and annual plants that live for only one season or year, and then die (with non-overlapping generations). In these cases, population growth occurs in discrete events rather than continuously.

The parameter λ (lambda) is called the **finite rate of increase**. (In mathematics, a *finite rate* refers to an observed rate over a given period of time.) In equation form, λ is the ratio of the population size during the next breeding season divided by the starting population size:

Population size one breeding interval later · Population size at time zero

Finite rate of increase: $$\lambda = \frac{N_1}{N_0}$$

(Eq. 51.7)

The finite rate of increase can thus be used to estimate the size of the population at the end of time t when breeding takes place seasonally:

Population size at time zero · Finite rate of increase to the power of t

Population size at end of time t: $$N_t = N_0 \lambda^t$$

(Eq. 51.8)

The size of the population at time t is equal to the starting size, times the finite rate of increase multiplied by itself t times.

The parameters λ (the finite rate of increase) and r (the per capita growth rate or instantaneous rate of increase) have a simple mathematical relationship:

Per capita growth rate · Base of natural logarithm

Finite rate of increase: $$\lambda = e^r$$

(Eq. 51.9)

The relationship between any finite rate and any instantaneous rate is given by *finite rate* $= e^{\text{instantaneous rate}}$.

✔ If you understand how to apply population growth equations, you should be able to solve these problems:

1. **QUANTITATIVE** Calculate r for a population of *Z. vivipara* lizards with $R_0 = 1.00$ and $g = 3$ years.
2. **QUANTITATIVE** Conservation biologists are trying to save a population of orchids in Hawaii from extinction. If the population currently contains about 50 individuals and is expected to sustain an r of 0.00048, predict the size of the population in 10 years.
3. **QUANTITATIVE** For two years, the students in an introductory biology class have used the mark–recapture technique to estimate the growing population size of gray squirrels on campus. They have determined that $\lambda = 1.2$. If the population size of squirrels is about 225 now, predict the size of the population in 5 years.

1. ***Colonization of a new habitat*** A few individuals start a new population in a new habitat with plentiful resources, such as when microorganisms colonize a new nutrient source, or when rats escape from a ship and colonize an island.
2. ***Recovery after a disaster*** A population has been devastated by a storm, flood, deforestation, or some other type of catastrophe and then begins to recover, starting with a few surviving individuals.

But it is not possible for exponential growth to continue indefinitely. Resources are never infinite. What happens when resources like food, water, sunlight, oxygen, or nesting areas start to run out?

Logistic Growth

If *Arabidopsis* populations grew exponentially for a long period of time, they would eventually fill all available habitat. When population density gets very high, the population's per capita birth rate b will decrease and the per capita death rate d will increase, causing the per capita growth rate r to decline. Stated another way, growth is often **density dependent**.

To analyze what happens when growth is density dependent, biologists use a parameter called carrying capacity. **Carrying capacity, *K*,** is defined as the maximum number of individuals in a population that can be supported in a particular habitat *over a sustained period of time*.

A Logistic Growth Equation If a population of size N is below the carrying capacity K, then the population should continue to grow. However, r is expected to decrease in proportion to the ratio $(K - N)/K$. If you substitute $r[(K - N)/K]$ for r in the exponential growth equation, the result for density-dependent growth is:

$$\frac{dN}{dt} = rN\left[\frac{K - N}{K}\right]$$

This expression is called a logistic growth equation.

In the expression $(K - N)/K$, the numerator defines the number of additional individuals that can be accommodated indefinitely in a habitat with carrying capacity K. Dividing $K - N$ by K turns this number of individuals into a proportion. Thus $(K - N)/K$ describes the proportion of "unused resources and space" in the habitat. It can be thought of as the environment's resistance to growth.

- When N is small, then $(K - N)/K$ is close to 1 and growth should be close to exponential. For example: $(1000 - 10)/1000 = 0.99$.
- As N gets larger, $(K - N)/K$ gets smaller. For example: $(1000 - 900)/1000 = 0.10$.
- When N is at carrying capacity (meaning that $K = N$), then $(K - N)/K$ is equal to 0 and growth stops.

Thus, as a population approaches a habitat's carrying capacity, its growth is predicted to slow. Note that this is a mathematical model based on theory. Real populations are not prevented from surpassing their carrying capacity, as you will soon see.

The logistic growth equation describes **logistic population growth**, the decrease in growth as a population size approaches the carrying capacity. Whereas exponential growth is density independent, logistic growth is density dependent.

Graphing Logistic Growth Figure 51.8a illustrates the S-shaped density-dependent growth in a hypothetical population. The graph shows changes in population size over time, and has three sections representing changes in growth.

1. Initially, growth is rapid because resources are abundant.
2. With time, N increases to the point where competition for resources or other density-dependent factors becomes important. As a result, growth begins to slow.
3. When the population is at the habitat's carrying capacity, the growth is 0—the graph of population size versus time is flat.

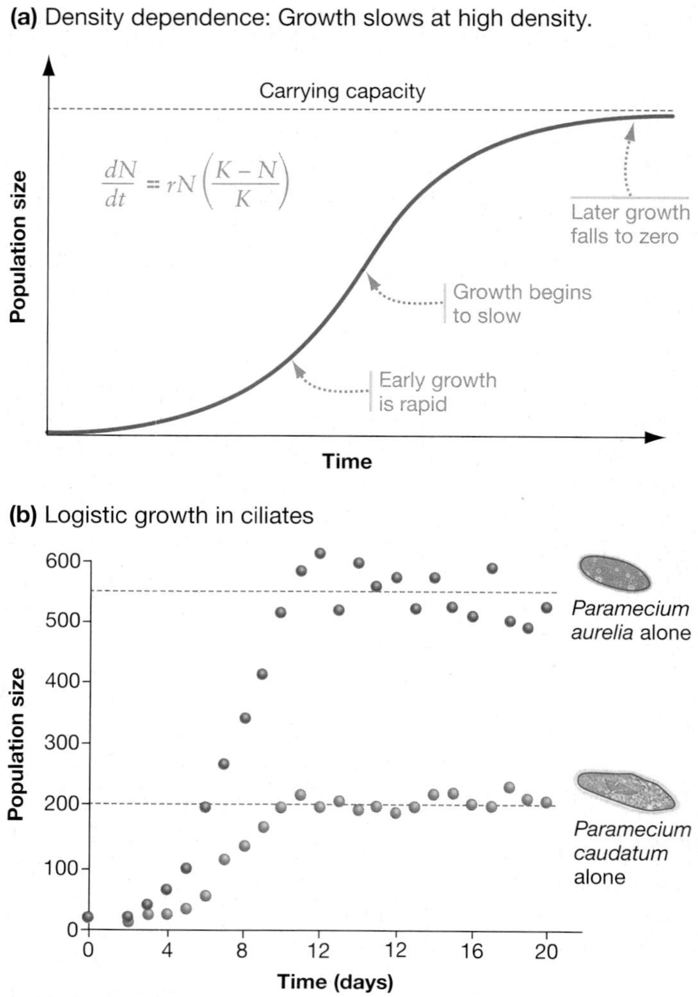

INTERACTIVE Figure 51.8 Logistic Growth Is Dependent on Population Size. (a) An S-shaped curve illustrates the pattern predicted by the logistic growth equation. **(b)** Data from laboratory experiments with two species of *Paramecium* (grown separately).

DATA: G. F. Gause. 1934. *The Struggle for Existence*. New York: Hafner Press.

This is what happened in an experiment on laboratory populations of single-celled ciliates, *Paramecium aurelia* and *P. caudatum*. Russian biologist Georgii Gause placed 20 individuals from one of the *Paramecium* species into 5 mL of a solution. He created many replicates of these 5-mL environments for each species separately. He kept conditions as constant as possible by adding the same amount of food to each, washing the solution to remove wastes, and maintaining the pH at 8.0.

In the graphs in Figure 51.8b, each data point represents population size for one of the species—the average of the replicates in the experiment. Note that both species exhibited logistic growth in this environment. The carrying capacity differed in the two species, however. The population size of *P. aurelia* averaged about 550 individuals, but that for the larger *P. caudatum* averaged about 200 individuals. (When the two species are grown together, one species gets eliminated; see Ch. 52, Section 52.1).

In both of the *Paramecium* species in this experiment, exponential growth could be sustained for only about ten days. What factors regulate population size and limit carrying capacity?

What Factors Limit Population Size?

Population sizes are regulated by two general types of factors:

1. *Density-independent factors* alter birth rates and death rates irrespective of the number of individuals in the population, and they usually involve changes in the abiotic environment—variation in weather patterns, or catastrophic events such as cold snaps, hurricanes, volcanic eruptions, or drought.
2. *Density-dependent factors* change in intensity as a function of population size, and they are usually biotic.

Let's take a closer look at density-dependent factors.

Density-Dependent Factors Limit Population Size Density-dependent factors are responsible for the declining growth in logistic population growth. **Table 51.3** lists some of the major types of density-dependent factors that reduce survivorship, fecundity, or both. For example, when a new population of *Arabidopsis* is established, a high density of seedlings will thrive at first. But as the population becomes more crowded, each seedling has reduced access to water, nutrients, and sunlight due to competition, and survivorship declines. Competition also reduces fecundity in surviving plants by limiting the resources available for investment in seeds.

Crowded populations can also be more vulnerable to infectious disease and parasites, exposure to a buildup of toxic waste, increased predation rates, and crowd-induced behavioral changes that decrease the probability that individuals will survive and produce viable offspring. Together, these factors increase death rates and decrease birth rates.

Note that density-dependent factors can be based on *intraspecific* ("within-species") interactions, such as competition among members of a cohort for food, or *interspecific* ("between-species") interactions, such as predation, parasitism, or competition among species for food (discussed in detail in Chapter 52). Both types of density-dependent interactions define a particular habitat's carrying capacity.

Carrying Capacity Is Not Fixed and Can Be Overshot It's important to recognize that the logistic growth model assumes a constant K. In reality, K changes over time due to fluctuations in food availability, space, and other density-dependent factors.

In general, variability in growth data is common and expected due to the complexity of biological systems. If you revisit the *Paramecium* data in Figure 51.8, you will notice that—even in highly controlled laboratory conditions—the data points do not perfectly follow the predictions of the mathematical model. In natural populations, variability is even more pronounced. Further, the logistic growth model might lead you to expect that growing populations will naturally slow and peak at carrying capacity. But in the real world, populations may dramatically overshoot the carrying capacity and then crash, or they may cycle in regular intervals above and below the carrying capacity. In the next section, you'll explore these possible population dynamics.

CHECK YOUR UNDERSTANDING

✔ **If you understood this section, you should be able to ...**

1. Predict the shape of the population growth curve of Norway rats after they were introduced to the resource-abundant Hawaiian Islands from ships in the 1700s. Explain your reasoning.
2. **PROCESS OF SCIENCE** Propose a density-dependent factor that limits growth in the lizard *Zootoca vivipara* and design an experiment to test this hypothesis.

Answers are available in Appendix A.

Table 51.3 Density-Dependent Factors That Limit Population Size

Factor	Example
Competition for resources Food, Territory, Water, Light, Nesting sites, Nutrients, Oxygen Dense tree saplings compete for limited light, water, and nutrients	
Disease and parasitism Stress-related degradation of health Infectious disease Parasitism Pigs in dense confinement are prone to illness	
Predation Increased predation as prey density increases Hare density influences lynx predation rates	
Toxic wastes Ammonia Uric acid Alcohol Carbon dioxide Densities of fruit fly larvae are limited by ammonia buildup from their feces	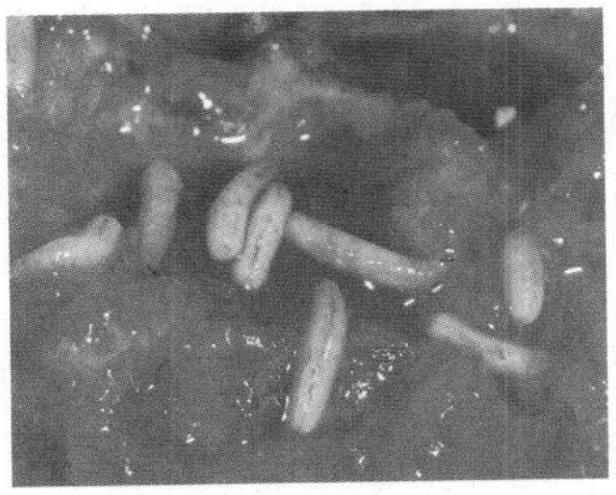
Social behavior Stress-mediated behavior Dominance behavior Mating behavior Parental-care behavior Predator-avoidance behavior Blue crab cannibalism increases with density	

51.4 Population Dynamics

The tools introduced in the previous two sections provide a foundation for exploring how biologists study **population dynamics**—changes in populations through time and space. Research on population dynamics has revealed a wide array of patterns in natural populations other than exponential and logistic growth. Let's start by considering how population size can crash or cycle over time.

After you complete this section, you should be able to ...

- Compare patterns in natural populations other than exponential and logistic growth.

Why Do Some Populations Crash?

In a classic study, population ecologist Victor Scheffer recorded the overshoot and collapse of a reindeer population on a remote Alaskan island called St. Paul. After 40 reindeer were placed on the island in 1911 as a food source for people, the reindeer population grew exponentially to more than 2000 animals in just 30 years. Twelve years later, only eight were left (Figure 51.9).

What happened on St. Paul Island? Scheffer examined several variables, including density-independent variables like harsh winters (which cover the lichens and low plants with snow and ice) and density-dependent variables like disease and hunting. Multiple factors can interact at the same time, but Scheffer concluded that the key cause of the crash was over-foraging of the food supply. The reindeer consumed the lichen and plants at an unsustainable rate—much faster than the resources could regrow.

How many reindeer could the island have *sustained* over time? Scheffer estimated from food availability that the carrying capacity of St. Paul Island was *one-third* of the maximum population size reached, about 700 reindeer—if they had not overexploited their food reserves. Other studies have shown that once carrying capacity is overshot, the carrying capacity itself declines, as indicated by the dashed line in Figure 51.9.

Was this overshoot and crash an exceptional event? No. A comparable overshoot and collapse of a reindeer population was documented by David Klein on St. Matthew Island between 1944 and 1966.

If the overshoot of carrying capacity is not as dramatic, a more cyclical pattern can occur. Let's take a closer look.

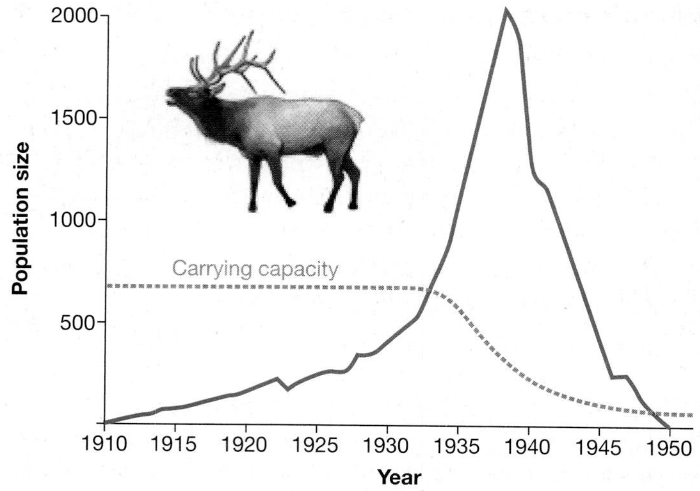

Figure 51.9 The Reindeer on St. Paul Island Overshot the Island's Carrying Capacity. The carrying capacity of the island was estimated to be about 700. But once the food sources were depleted by the reindeer, the carrying capacity declined.

DATA: V. B. Scheffer. 1951. *Scientific Monthly* 73: 356–362.

✔ **SOCIETY** Pose a counterargument for this claim: The human carrying capacity of Earth must be higher than the current human population size—because after all, everyone alive today must have enough resources to survive.

Why Do Some Populations Cycle?

Analyzing population cycles has been an especially productive way to understand how intraspecific and interspecific density-dependent factors interact. Figure 51.10 shows a classic case of population cycling that involves two species: snowshoe hare and lynx in northern Canada. The *x*-axis on this graph plots a 50-year interval; the *y*-axes show changes in hare population density (on the left, with a logarithmic scale) and changes in lynx density (on the right). Data for hares are plotted in tan; data for lynx are plotted in blue. Note three key points:

1. The scales on the *y*-axes are different—there are many more hares than lynx. (The *y*-axis for hares is logarithmic to compress the values so they can be compared more easily to changes in the lynx population.)
2. The hare and lynx populations cycle every 10 years on average.

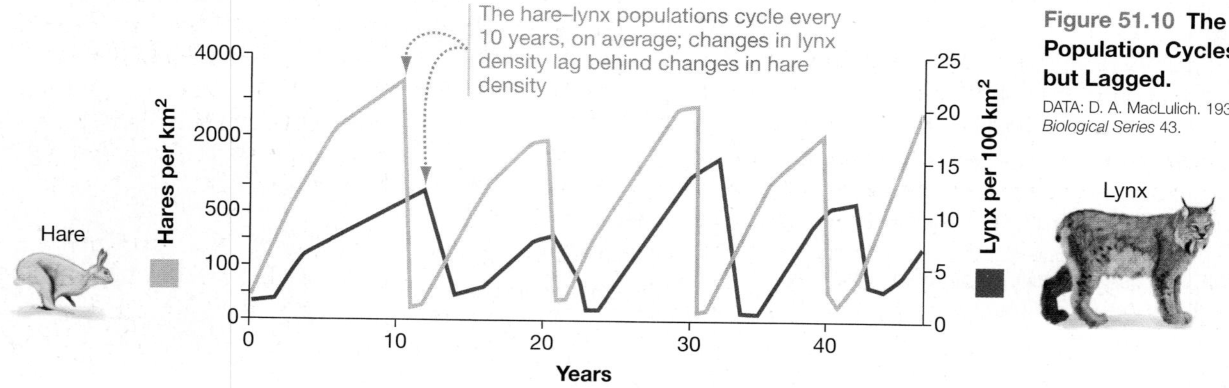

Figure 51.10 The Hare and Lynx Population Cycles Are Synchronous, but Lagged.

DATA: D. A. MacLulich. 1937. *University of Toronto Studies Biological Series* 43.

3. Changes in lynx density lag behind changes in hare density by about two years.

Snowshoe hares are herbivores—they subsist on leaves and stems that grow close to the ground. Lynx are predators, and they subsist mainly on snowshoe hares.

Is It Food or Predation? To explain the hare–lynx cycle, biologists posed two hypotheses based on density-dependent factors:

1. ***Bottom-up hypothesis*** When their populations reach high density, hares use up all their food and starve; in response, lynx also starve.
2. ***Top-down hypothesis*** Lynx populations reach high density in response to increases in hare density. At high density, lynx eat so many hares that the prey population crashes.

Stated another way, either hares control lynx population size, or lynx control hare population size. The cycle could be controlled by intraspecific competition for food or interspecific interactions—or an interaction of both factors.

A Field Experiment To test these hypotheses, Charles Krebs and colleagues set up a series of 1-km^2 study plots in boreal forest habitats that were as identical as possible (**Figure 51.11**).

RESEARCH

QUESTION: What factors control the hare–lynx population cycle?

BOTTOM-UP HYPOTHESIS: Food availability for the hares controls the hare–lynx cycle.

TOP-DOWN HYPOTHESIS: Predation controls the hare–lynx cycle.

INTERACTION HYPOTHESIS: The interaction of food availability and predation controls the hare–lynx cycle.

NULL HYPOTHESIS: The hare–lynx cycle isn't driven by predation, food availability, or a combination of those two factors.

EXPERIMENTAL SETUP:

Measure the hare survival rate in seven study plots (similar boreal forest habitats, each 1 km^2) from 1987 to 1994 (most of a population cycle).

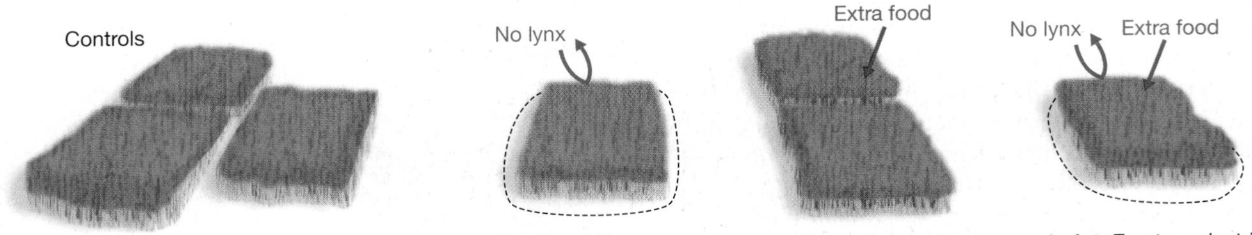

3 plots: Unmanipulated controls

1 plot: Erect an electrified fence that excludes lynx but allows free access by hares.

2 plots: Supply extra food for hares.

1 plot: Erect an electrified fence that excludes lynx; supply extra food for hares.

PREDICTION OF EXPERIMENTAL HYPOTHESES: Hare populations in at least one type of manipulated plot will be higher than in the control plots.

PREDICTION OF NULL HYPOTHESIS: Hare populations in all of the plots will be the same.

RESULTS:

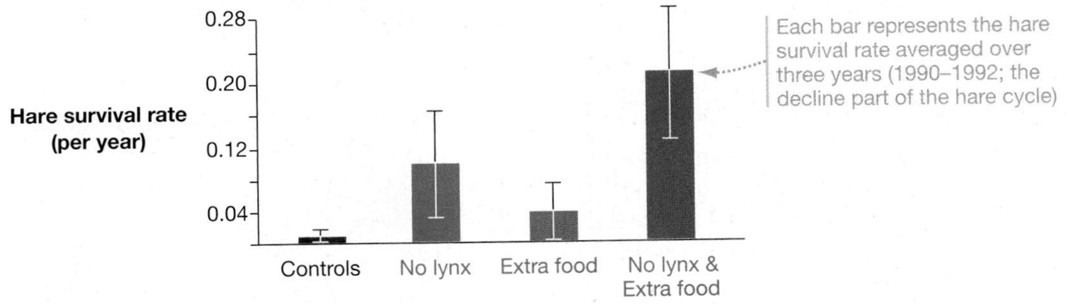

CONCLUSION: Hare populations are limited by both predation and food availability. When predation and food limitation occur together, they have a greater effect than either factor does independently.

Figure 51.11 **Experimental Evidence Indicates that Both Predation and Food Availability Drive the Hare–Lynx Cycle.**

SOURCE: C. J. Krebs, S. Boutin, R. Boonstra, et al. 1995. Impact of food and predation on the snowshoe hare cycle. *Science* 269: 1112–1115.

✔ **PROCESS OF SCIENCE** Some of the experimental plots were not replicated. What is one benefit and one cost of having replicates in this particular experiment?

- Three plots were left as unmanipulated controls.
- One plot was ringed with an electrified fence made up of a mesh that excluded lynx but allowed hares to pass freely.
- Two plots received additional food for hares year-round.
- One plot had a predator-exclusion fence and was also supplemented with food for hares year-round.

The biologists then used the mark–recapture method (see Quantitative Methods 51.1) and radio collars to monitor the size of the hare and lynx populations over an 8-year period.

The graph in Figure 51.11 plots average hare survival rate in the lynx-exclusion, food-addition, and combination plots over three years, in the decline part of the hare cycle from 1990 to 1992. The graph should convince you that plots with predators excluded showed higher hare survival rates than did control plots. This result supports the hypothesis that predation by lynx reduces hare populations. Plots with supplemental food also had higher hare survival rates than controls. But the plot with supplemental food *and* predators excluded had the highest hare survival rate of all.

These data support the hypothesis that hare populations are limited by availability of food as well as by predation and that food availability and predation intensity interact—meaning the combined effect of food and predation is larger than the effect of both in isolation. One hypothesis to explain this combined effect is that when hares are at high density, individuals are weakened by nutritional stress and are more susceptible to predation. An alternate hypothesis is that hares experience higher stress in the presence of lynx and compromise their foraging behavior to avoid predation.

Note that although the interaction of food availability and predation was supported by the data, other factors may also affect population growth in hares. For example, recent climate models suggest that density-independent factors such as temperature, rainfall, and snow may interact with density-dependent factors such as the concentrations of defense compounds in plants to influence the hare–lynx cycles.

Let's now consider multiple populations, and bring immigration and emigration into the discussion. One of the most important dynamics observed in population ecology is the "blinking on and off" of fragmented metapopulations.

How Do Metapopulations Change through Time?

Glanville fritillaries are an endangered species of butterfly native to the Åland islands off the coast of Finland (Figure 51.12a). Population ecologist Ilkka Hanski chose to study this species as a research model for metapopulations because it occupies isolated patches of habitat within its geographic range. Also, the population patches were small enough to be quantified.

Model Predictions: Metapopulations Should Be Dynamic Research on Glanville fritillaries began with a survey of meadow habitats on the islands. Because fritillary caterpillars feed on just two types of host plant, *Plantago lanceolata* and *Veronica spicata*, Hanski was able to pinpoint potential butterfly habitats.

Hanski's large research group estimated the fritillary population size within each patch of habitat by counting the number of

(a) Glanville fritillaries live in patches of meadow in Finland.

(b) PROCESS: MODEL OF METAPOLULATION DYNAMICS

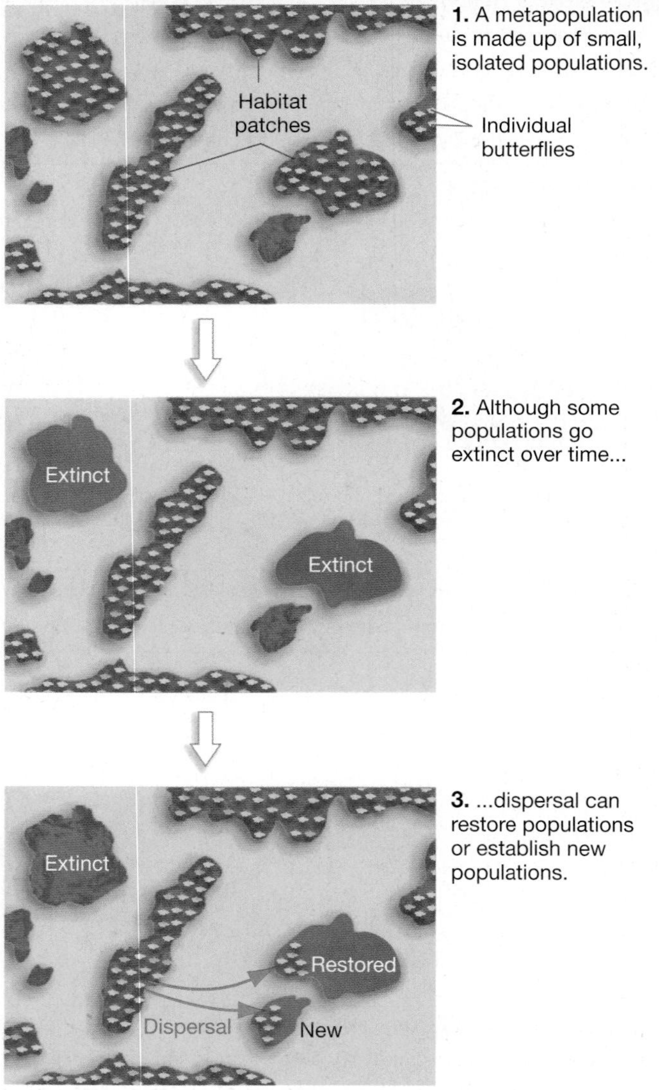

Figure 51.12 Metapopulation Dynamics Depend on Extinction and Recolonization. The overall size of a metapopulation stays relatively stable even if individual populations go extinct. These populations may be restored by dispersal, or unoccupied habitats might be colonized.

larval webs in the patch. Of the 1502 meadows that contained the host plants, 536 had Glanville fritillaries. The patches ranged from 6 m^2 to 3 ha in area. Most had only a single breeding pair of adults and one larval group, but the largest population contained hundreds of pairs of breeding adults and 3450 larvae.

These data represent the starting conditions. How would the metapopulation change over time? Hanski proposed a model of metapopulation dynamics illustrated in Figure 51.12b.

- Given enough time, each population within the larger metapopulation might go extinct, especially if the population is small. The cause could be a catastrophe, such as a storm; it could also be a disease outbreak or a sudden influx of predators.
- Dispersal from nearby populations would reestablish populations in empty habitat fragments.

This metapopulation model predicts that there will be a balance between extinction and recolonization. That is, even though populations blink on and off over time, the overall metapopulation is maintained at a stable number of individuals.

An Experimental Test Does this metapopulation model correctly describe the population dynamics of fritillaries? To answer this question, Hanski's group conducted a mark–recapture study. They caught and marked many individuals, released them, and revisited fritillary habitats two years later to recapture as many individuals as possible (see Quantitative Methods 51.1).

Just as the metapopulation model predicted, some populations had gone extinct and others had been created. On average, butterflies were lost from 200 patches each year; 114 unoccupied patches were colonized and newly occupied each year. The overall population size was relatively stable even though constituent populations came and went.

To summarize, the history and future of a metapopulation are driven by the birth and death of populations, just as the dynamics of a single population are driven by the birth and death of individuals. Further, dispersal is an important source of individuals to recolonize patches of vacated habitats. These dynamics have important implications for conserving biodiversity.

Metapopulation Dynamics Are Important to Conservation Habitat destruction caused by suburbanization, farming, highway construction, and other human activities leaves small populations isolated in pockets of intact habitat (Figure 51.13; see also anthromes in Ch. 49, Section 49.4 and habitat destruction in Ch. 54, Section 54.2). Work on Glanville fritillaries and other species has shown that a small, isolated population—even one within a nature preserve—is unlikely to survive over the long term. What can be done?

Data collected by Hanski's group have identified four attributes of Glanville fritillary subpopulations that increase their likelihood to persist:

1. They have larger subpopulation sizes.
2. They occupy larger geographical ranges.
3. They are closer to neighboring subpopulations (and hence more likely to be colonized).
4. They have higher genetic diversity.

Figure 51.13 **Human Activities Are Turning Once Intact Populations into Metapopulations by Dividing Habitats into Fragments.**

Results like these have important messages for conservation biologists:

- Areas that are being protected for threatened species should be big enough to maintain large subpopulations that are unlikely to go extinct in the near future. Large subpopulations are more likely than small subpopulations to endure disasters such as storms, and they are much less susceptible to the damaging effects of inbreeding.
- When it is not possible to preserve large tracts of land, an alternative is to establish systems of smaller tracts that are connected by corridors of habitat, so that dispersal between patches is possible.
- It is crucial to preserve at least some patches of unoccupied habitat to provide future homes for immigrants.

These results are made more complex by climate change. Data on existing populations show that changes in temperature and precipitation are already altering the ranges of species. For some species with large ranges, such as the lizard *Zootoca vivipara* (see Figure 51.1), local extinctions may occur in some populations, such as in the mountains of Spain, while the range expands northward elsewhere. Other species with smaller ranges or the inability to relocate rapidly are likely to go extinct.

The ultimate driver of habitat fragmentation, as well as overexploitation, pollution, climate change and other human impacts, is the staggering growth of the human population and correlated resource use. Let's apply what you have learned to this important case study.

CHECK YOUR UNDERSTANDING

✔ **If you understood this section, you should be able to ...**

Predict whether African savanna elephants, now pushed into many isolated patches of habitat by habitat destruction and fragmentation, have metapopulation structure. Explain your reasoning.

Answers are available in Appendix A.

51.5 Case Study: Human Population Growth

Fossil and molecular evidence indicates that *Homo sapiens* originated in Africa about 200,000 years ago, then migrated into Asia, Europe, Australia, and eventually North and South America (Ch. 32, Section 32.6). The geographic range of our species now consists of most of the land surface area of the planet—and the current population size approaches 8 billion. How did our population size change over that time?

After you complete this section, you should be able to ...

- Analyze the changes in human population growth over time.

It is tempting to think that humans follow different rules than other animals, but the same principles of population ecology that you learned earlier in the chapter apply to humans as well. For example, as the life-table data presented in Section 51.2 showed, age dramatically affects the probability that an individual will reproduce and survive to the following year. A population's age structure, in turn, dramatically influences the population's growth over time. Let's take a closer look at age structure in human populations, then examine human population growth curves.

Age Structure in Human Populations

To study age structure in our species, researchers typically use stacks of horizontal bars to plot the number of males and the number of females in each age cohort—often in five-year intervals between birth (0) and 100. The resulting graph is called a population pyramid or age pyramid.

Age Pyramids In countries where industrial and technological development took place generations ago, like Sweden, survivorship has been high and fecundity low for several decades. An age pyramid like that in Figure 51.14a results. The most striking pattern in these data is that most of the age classes contain similar numbers of people. The evenness occurs because about the same number of infants are being born each year, and because most of them survive to old age.

In contrast, the age distribution is bottom heavy in developing nations like Honduras in Central America (Figure 51.14b). Because they are currently undergoing rapid growth, these populations are dominated by the very young. The number of children being born recently is higher than the number of people existing in older age classes.

Analyzing an age pyramid can give biologists important information about a population's history. But studying age distributions can also help researchers predict a population's future. Look again at Figure 51.14a, and note that the lighter bars show what the age structure of Sweden is projected to be in 2050. Some changes in the age distribution will occur because survivorship has increased while fecundity remains the same. But the population is not expected to grow quickly, because there is not a large number of people reaching reproductive age. The projections highlight a major public policy concern in the industrialized countries: how to care for an increasingly aged population.

Now consider Figure 51.14b again. Thanks to dramatic improvements in health care, most young Hondurans survive to reproductive age. The projected age distribution in 2050 reflects the resulting population surge but does not continue to flare further out, because fecundity is expected to decline. The projections illustrate a major public policy concern in developing countries: providing education and jobs for an enormous group of young people who will be reaching adulthood during the lifetimes of preceding generations.

Population "Momentum" The data in Figure 51.14b make another important point: Because of recent and rapid population

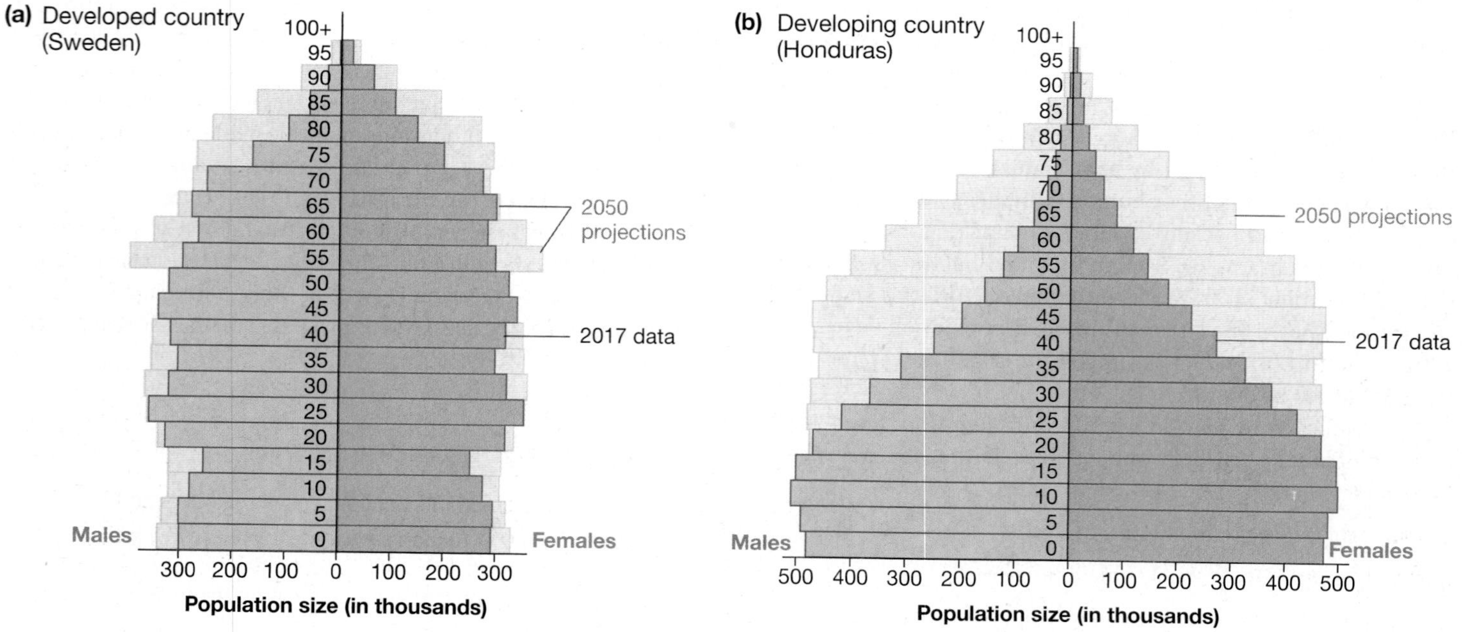

Figure 51.14 The Age Structure of Human Populations Varies Dramatically.
DATA: United Nations, *World Population Prospects: The 2017 Revision*.

growth in developing countries, overall population size will increase dramatically in these nations over the course of your lifetime. Honduras will have many more people in 2050 than it does now. A large part of this increase will be due to increased survivorship. But the number of offspring being born each year is also expected to stay high, even though fecundity is predicted to *decline*. This result seems paradoxical, but it is based on a key observation: These populations now contain so many young women that the overall number of births will stay high, even though the average number of children *per female* is much lower than it was a generation ago.

To capture this point, biologists say that these populations have momentum or inertia. Combined with high survivorship, age structures like these make continued increases in the total human population almost inevitable.

Analyzing Change in the Growth Rate of Human Populations

It is almost impossible to overemphasize just how dramatically the human population has grown recently. Figure 51.15 plots changes in the human population over the past 2000 years. Although the curve's shape looks superficially similar to exponential growth, the growth rate for humans has—until recently—actually *increased* over time, leading to a curve that rises *more steeply* than exponential growth. It took all of human history to reach a population size of 1 billion in 1804, and only 123 years to reach 2 billion. Most recently, it took only 12 years to add 1 billion people to the population to reach 7 billion.

Consider that until your grandparents' generation, no individual had lived long enough to see the human population double. But many members of the generation born in the early 1920s lived to see the population *triple*. And to drive home the impact of continued population doubling, consider this: If you were given a penny on January 1, then \$0.02 on January 2, \$0.04 on January 3, and so on, you would be handed \$10,737,418.00 on January 31.

The astonishing growth rate of the human population has raised a sense of alarm that we will overshoot—or have already overshot—the carrying capacity of our planet. This concern is not new. For example, Thomas Robert Malthus's book, *An Essay on the Principle of Population*, was widely influential in the early 1800s. Malthus warned that the population cannot continue to increase unchecked—competition for resources will eventually slow population growth, whether by famine, war, or voluntary reduction of family size. Malthus was an important source of inspiration for Charles Darwin in his formulation of the theory of evolution by natural selection (Ch. 22, Section 22.3).

How Large Is the Current Human Population? The world population approaches 8 billion. Further, about 80 million additional people—more than double the population of California—are being added each year.

The consequences of recent and current increases in human population are profound (as you will see in Chapter 54). Aside from being the primary cause of habitat loss and species extinctions, overpopulation is linked by researchers to declines in living standards, mass movements of people, political instability, and acute shortages of water, fuel, and other basic resources in many parts of the world. Overpopulation is the driver of many ecological and social ills.

The one encouraging trend in the data is that the growth rate of the human population has already peaked and begun to slow. The highest growth rates occurred between 1965 and 1970, when populations increased at an average of 2.04 percent per year. Between 1990 and 1995, the overall growth rate in human populations averaged 1.46 percent per year; the current growth rate is 1.09 percent annually.

In humans, r may be undergoing the first long-term decline in history. The question is: Will the human population stabilize or decline in time to prevent global—and potentially irreversible—damage?

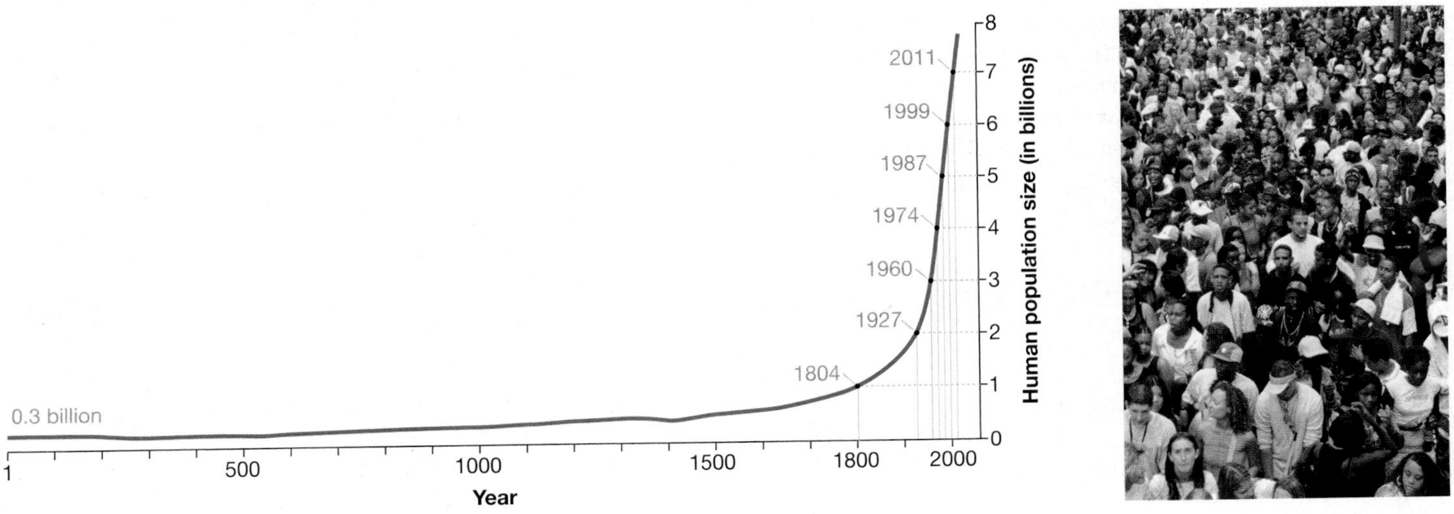

Figure 51.15 The Human Population Has Been Growing Rapidly. Graph based on estimates from historical and census data.

DATA: G. Vince. 2011. *Science* 334: 32–37; updated with current data from the United Nations, *World Population Prospects: The 2017 Revision*.

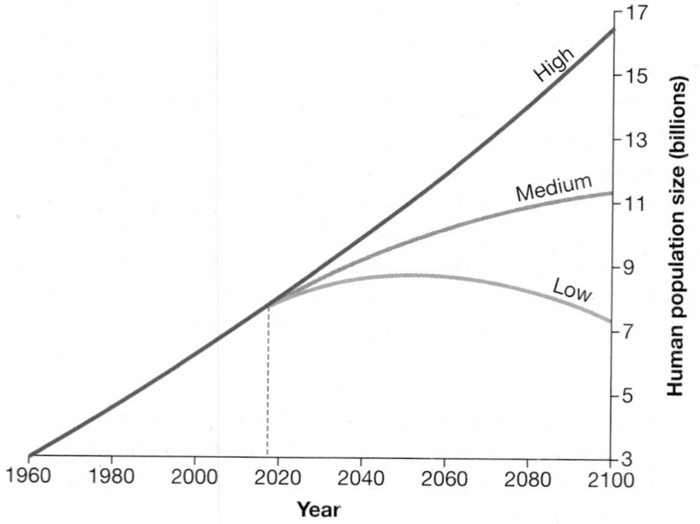

Figure 51.16 Projections Forecast Human Population Growth. The "High," "Medium," and "Low" labels refer to fertility rates.
DATA: United Nations, *World Population Prospects: The 2017 Revision.*

Will Human Population Size Peak in Your Lifetime? The United Nations Population Division makes regular projections for how human population size will change between now and 2100. The UN projections are based on three scenarios, which hinge on different values for **fertility** rates—the average number of surviving children that each woman has during her lifetime. Currently, the worldwide average fertility rate is 2.5 children per woman. This represents an enormous reduction in fertility from the 1950s, which averaged 5.0 children per woman. Figure 51.16 shows how total population size is expected to change by 2100 if average fertility continues at 2.5 (high), or declines to 2.0 (medium) or 1.5 (low) children per woman.

The middle number in the projections is statistically the most likely scenario. Its value is close to the replacement rate of the population, 2.1. The **replacement rate** is the average fertility required for each woman to produce exactly enough offspring to replace herself and her offspring's father. (It is slightly above 2.0 to account for the mortality of some women before they bear children.) When this fertility rate is sustained for a generation, $r = 0$, and there is **zero population growth**.

A glance at Figure 51.16 should convince you that the three scenarios are starkly different. If average fertility around the world stays at its current level (assuming that average survivorship does not change), world population will be closing in on 17 billion by the end of the century—more than doubling today's population with no signs of peaking. The low-fertility projection, in contrast, predicts that the total human population in 2100 will be similar to today's size after having peaked near 9 billion in 2050. If survivorship rates decline, whether from density-independent factors such as droughts or density-dependent factors such as famines, then the population will decline more rapidly than predicted from fertility rates alone.

Take-Home Messages

Humans experience the same population ecology principles as other organisms, such as fitness trade-offs between survivorship and fecundity and density-dependent limits to population growth. Will the human population level off at carrying capacity, as the *Paramecium* populations did in laboratory studies? Or will the human population crash like the reindeer population on St. Paul Island? Or perhaps something in between?

The human population boom is important not only because it affects virtually all aspects of human society from traffic jams to the price of housing to the spread of infectious diseases. The human population boom is also central to the population ecology of other organisms because the growing human population dramatically affects them via habitat destruction, pollution, overharvesting of resources, and climate change. As the human population rises, the impact on other species increases.

You have learned in this chapter that populations do not exist in isolation—the population growth of one species is often influenced by interactions with other species within complex biological communities. Biological communities are explored in greater detail in Chapter 52.

CHECK YOUR UNDERSTANDING

✔ **If you understood this section, you should be able to ...**

1. **SOCIETY** Evaluate whether the following report is evidence of density-independent or density-dependent growth in humans: A 2017 study estimated that 16 percent of early deaths were caused by pollution—three times more deaths than AIDS, malaria, and tuberculosis combined.
2. **SOCIETY** Explain why the human population is projected to continue its rapid growth, even though fertility rates are declining.

Answers are available in Appendix A.

CHAPTER 51 Review

For media, go to Mastering Biology

51.1 Distribution and Abundance

- The geographic ranges of species vary in size and are determined by a combination of abiotic and biotic factors.
- Individuals in a population have random, clumped, or uniform density distributions depending on the availability of resources, dispersal ability, social behavior, and other factors.
- There are many methods for measuring population size, population density, and distribution patterns—the best method to use depends on the behavior of the species.

51.2 Demography and Life History

- Life tables summarize how likely it is that individuals of each age class in a population will survive and reproduce.

- Three general types of survivorship curves have been observed: **(I)** high survivorship throughout life, **(II)** a constant probability of dying at each year of life, and **(III)** high mortality in juveniles followed by high survivorship in adults.
- There is a trade-off between survival and reproduction due to the limited availability of resources.
- In comparing the life-history traits of populations or species, there is a continuum between those with high survivorship and low fecundity and those with low survivorship and high fecundity.
- Life-history data helps biologists to design effective conservation strategies for diverse species.

51.3 Population Growth

- The growth rate of a population can be calculated from life-table data or from the direct observation of changes in population size over time.
- Exponential growth (J-shaped growth curve) occurs when resources are abundant; growth is density independent.
- Logistic growth (S-shaped growth curve) occurs when growing populations approach the carrying capacity, K, of their environment; growth is density dependent.
- Density dependence in population growth is due to competition for resources, disease, predation, or other factors that increase in intensity when population size is high. At high density, survivorship and fecundity decrease.
- Carrying capacity can change over time and is sometimes surpassed.

51.4 Population Dynamics

- If resources are used unsustainably, a population can overshoot the carrying capacity of its environment and crash.
- Food availability and predator–prey interactions are density-dependent factors that drive the population cycle of the snowshoe hare and other species.
- In some metapopulations, populations may blink on and off over time even though the overall metapopulation size is stable. Dispersal can reestablish populations in empty habitat fragments.
- Humans are fragmenting natural populations via habitat destruction, forcing some populations into metapopulation structure.

51.5 Case Study: Human Population Growth

- Age structure profoundly affects population dynamics. A population with few juveniles and many adults past reproductive age, like the human populations of developed nations, may be declining or stable in size. In contrast, a population with a large proportion of juveniles, like the human populations of developing nations, is likely to increase rapidly in size.
- The total human population has been increasing rapidly and approaches 8 billion.
- Based on various scenarios for average female fertility, the total human population in 2100 is expected to be between 9 and 17 billion.
- The future of the human population hinges on fertility rates and survivorship rates.

Answers are available in Appendix A.

✔ TEST YOUR KNOWLEDGE

1. **THINK CAREFULLY** What is usually true of exponential growth? Select True or False for each statement.
   T/F The population is growing.
   T/F The per capita growth rate is constant.
   T/F The per capita growth rate increases rapidly over time.
   T/F The per capita growth rate is very high.
2. In what populations does exponential growth tend to occur?
   a. populations that colonize new habitats
   b. populations that experience intense competition
   c. populations that experience high rates of predation
   d. populations that have surpassed their carrying capacity
3. If most individuals in a population are young, why is the population likely to grow rapidly in the future?
   a. Death rates will be low.
   b. The population has a skewed age distribution.
   c. Immigration and emigration can be ignored.
   d. Many individuals will begin to reproduce soon.
4. Why is there a trade-off between survivorship and fecundity?

✔ TEST YOUR UNDERSTANDING

5. Which of the following species is most likely to have a Type III survivorship curve?
   a. humans, *Homo sapiens*
   b. common lizards, *Zootoca vivipara*
   c. thale cress plants, *Arabidopsis thaliana*
   d. lynx cats, *Lynx canadensis*
6. **SOCIETY** Pose a hypothesis to explain why the human population has undergone especially rapid growth over the past 200 years. Describe two examples of density-dependent factors that limit human population growth. Is it possible that humans have surpassed Earth's carrying capacity?
7. Explain why biologists want to maintain (a) "habitat corridors" that connect populations in a metapopulation, and (b) unoccupied habitat that is appropriate for the species in question.
8. **THINK CAREFULLY** Indicate what is correct and incorrect about this statement: If lizards want to survive during climate change, they can evolve new life-history traits.

✔ TEST YOUR PROBLEM-SOLVING SKILLS

9. **SOCIETY** When wild plant and animal populations are logged, fished, or hunted, only the oldest or largest individuals tend to be taken. Many of the commercially important species are long lived and are slow to begin reproducing. If harvesting is not regulated carefully and exploitation is intense, what impact does harvesting have on a population's age structure? How might harvesting affect the population's life table and growth rate?
10. **MODEL** Make a rough sketch of the age distribution in a developing versus a developed country, and explain why the shapes of the diagrams are different. How is AIDS, which is a sexually transmitted disease, likely affecting the age distribution in countries hard hit by the epidemic?

✔ PUT IT ALL TOGETHER: Case Study

Are Burmese pythons taking over Florida?

Burmese pythons (*Python molurus bivittatus*) are constricting snakes that can reach enormous sizes (up to 7 meters in length). They are native to Southeast Asia but were released into southern Florida from the pet trade. Many other snakes occur naturally in this area. Are the introduced pythons a problem?

11. QUANTITATIVE Burmese pythons were first found in the wetlands of Everglades National Park in the 1980s, but only rarely. The accompanying graph shows what happened next. Most of the data are derived from chance encounters of pythons on roads (pythons are notoriously difficult to find). How would you describe the type of population growth in pythons from 2000 to 2008?

a. logistic
b. exponential
c. linear
d. logarithmic

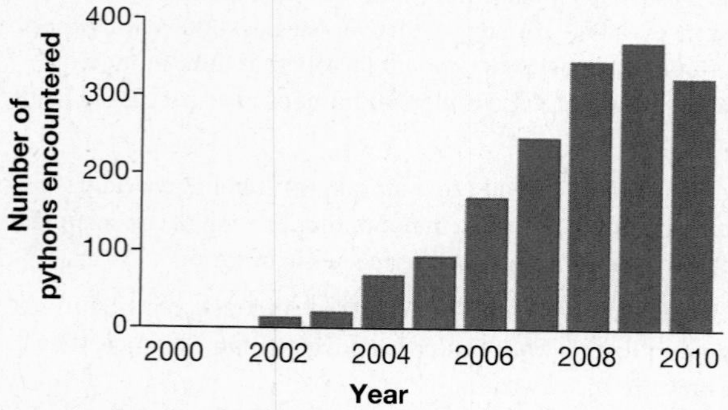

DATA: M. E. Dorcas, J. D. Willson, R. N. Reed, et al. 2012. *Proceedings of the National Academy of Sciences* 109: 2418–2422.

12. Researchers hypothesize that the severe winter of 2010 caused the decline in the number of pythons encountered that year. Is cold weather typically a density-dependent factor or a density-independent factor? Explain the difference.

13. PROCESS OF SCIENCE The data in the graph show the number of pythons that were encountered, not the total number of pythons in the population. Design a mark–recapture experiment to estimate the total population size of pythons, and list at least two assumptions you would be making in your experiment.

14. A life table would help researchers make predictions about python population growth. If you were to track a cohort of pythons over time to construct a detailed life table, what kind of data would you want to collect?

15. Pythons are generalist predators, consuming a wide variety of prey including birds, mammals ranging from rodents to deer, and alligators. A road survey showed a 99 percent decrease in the frequency of raccoon and opossum observations within python areas from 2003 to 2011. Rabbits and foxes have also disappeared from the area. Do you think food availability limits, or could limit, the carrying capacity of Burmese pythons in the Everglades?

16. The range of Burmese pythons in Florida is expanding rapidly. List at least one abiotic factor and one biotic factor that might limit the range of pythons.

Mastering Biology

Students Go to Mastering™ Biology for assignments, the eText, and the Study Area with animations, practice tests, and activities.

Professors Go to Mastering™ Biology for automatically graded tutorials and questions that you can assign to your students, plus Instructor Resources.

52 Community Ecology

In this chapter you'll explore how populations of different species interact with one another, and how these communities of organisms—including the coral reef community shown here—change over time.

In this chapter you will learn that

Communities consist of complex and dynamic interactions among species

starting with

Types of species interactions:
- Commensalism + / 0
- Competition − / −
- Consumption + / −
- Mutualism + / +

52.1

important to → Community structure 52.2 (*examining*)

Community structure *changes over time, causing* → Community dynamics: disturbance and succession 52.3

Community structure *reveals* → Geographic patterns in species richness 52.4

Community dynamics *partly explains* → Geographic patterns in species richness

Like other biological systems, communities are exciting to study because they are more than the sum of their parts. For example, you could describe a coral reef community as a coexistence of corals, flatworms, sea anemones, sharks, and so on—because a biological **community** consists of all the populations of different species that interact in a certain area. Yet the thrill of a coral reef comes not from the list of members, but from the mind-boggling complexity and function of the interacting network as a whole. These interactions are dynamic, self-regulating, and the product of ongoing evolution.

Ecologists want to know how communities work, and how to manage them in a way that will preserve biodiversity, ecosystem function, and an environment that people want to live in. Community ecologists study a wide range of questions, for example: What is the relationship between corals (a type of colonial animal related to sea anemones) and the photosynthetic protists that they host? How many species depend directly or indirectly on corals? Can coral reefs recover from the increasingly massive bleaching events caused by extreme warming in the oceans?

You can see how community ecology fits into the Big Picture of Ecology on pages 1190–1191. Let's start by taking a closer look at types of species interactions that can exist within a biological community.

BIG PICTURE

This chapter is part of the Big Picture. See how on pages 1190–1191.

52.1 Species Interactions

To study species interactions, biologists begin by analyzing the effects of one species on the fitness of another. Recall that the **fitness** of an individual is defined as its ability to produce viable, fertile offspring (Ch. 22, Section 22.3). Does the relationship between two species provide a fitness benefit to members of one or both species (a "+" interaction), does it incur a fitness cost (a "–" interaction), or does the association have no effect on the fitness of a participant (a "0" interaction)? Let's explore the four most common types of interactions:

After you complete this section, you should be able to ...

- Compare the major types of species interactions.

1. **Commensalism** occurs when one species benefits but the other species is unaffected (+ / 0).
2. **Competition** occurs when individuals use the same resources—resulting in lower fitness for both (– / –).
3. **Consumption** (including herbivory, predation, and parasitism) occurs when one organism eats or absorbs nutrients from another. The interaction increases the consumer's fitness but decreases the victim's fitness (+ / –).
4. **Mutualism** occurs when two species interact in a way that confers fitness benefits to both (+ / +).

As you analyze each type of interaction in the following discussion, watch for these key themes:

- ***Species interactions can affect the distribution and abundance of a particular species.*** One species can increase or decrease the distribution and/or abundance of another species.
- ***Species act as agents of natural selection when they interact.*** When two species influence each other's adaptations over time, biologists say that **coevolution** is occurring.
- ***The outcome of interactions among species is dynamic and conditional.*** Just as two people may alternately help or compete with each other depending on circumstances, the interactions between species may change over time and space.

Species Interaction: Commensalism

Of the four general types of interactions, commensalism (+ / 0) is the least studied. Commensalism can be challenging to demonstrate, because an absence of fitness benefits or costs (0) can be difficult to quantify. Commensalism is also very conditional.

An example is the relationship between epiphytic orchids and their host trees (Figure 52.1). Epiphytes (literally, "upon-plants") grow on the branches of trees. Their roots are adapted for grabbing onto the surface of tree bark. Epiphytes absorb most of the water and nutrients they need from rainwater, insect feces, dust, and other particles that collect in and around them. The epiphytes benefit by being held aloft in the sunlight, but the fitness of the tree is neither increased nor reduced.

However, if a large number of orchids or other epiphytes populate the branches of a tree, then the relationship becomes competitive—the epiphytes and the tree begin to compete for access to sunlight. The outcome of the interaction depends on the number and types of epiphytes present and may change over time.

Figure 52.1 Commensals Gain a Fitness Advantage but Don't Affect the Species They Depend On.

Species Interaction: Competition

Competition (– / –) lowers the fitness of both individuals involved for a simple reason: Time and energy are limited, so any amount of these resources expended in competition reduces the amount available for foraging, finding mates, reproducing, and other activities that increase fitness. Since some individuals in a competition may "win" and some may "lose," it can seem counterintuitive that the fitness of *both* would decline. The key insight here is that competing is energetically costly, so even the "winner" would have been better off not having to compete at all.

Competition that occurs between members of the same species is called **intraspecific** ("within-species") **competition**. Intraspecific competition for space, sunlight, food, and other resources intensifies as a population's density increases. As a result, intraspecific competition is a major factor in density-dependent growth (Ch. 51, Table 51.3).

Interspecific ("between-species") **competition** occurs when individuals from different species use the same resources. Note that species may compete *directly* for the same resources, such as a lion fighting off a hyena to control access to a zebra carcass. Species may also compete *indirectly* for the same resources, such as when a finch consumes all of the available seeds on a bush, leaving no seeds for species visiting the bush later.

Using the Niche Concept to Analyze Competition

Early work on interspecific competition focused on the concept of the **niche**, the range of resources that the species is able to use, or the range of conditions it can tolerate. Making Models 52.1 reviews G. Evelyn Hutchinson's proposal (Ch. 49, Section 49.2) that a niche could be envisioned by plotting a species' habitat requirements along a series of axes, such as the species' food or temperature requirements. Remember that individuals in populations vary: The "population thinking" perspective (Ch. 22, Section 22.1) is important to interpreting the horizontal range of the curve in niche models.

Making Models 52.1 Tips on Drawing Graphs (IV)

In the simplest form of a niche model, there are three elements to draw: (1) the *x*-axis is the habitat requirement of interest; (2) the dependent variable on the *y*-axis is the number or proportion of resources used; and (3) the curve shows the number or proportion of resources used at each part of the range.

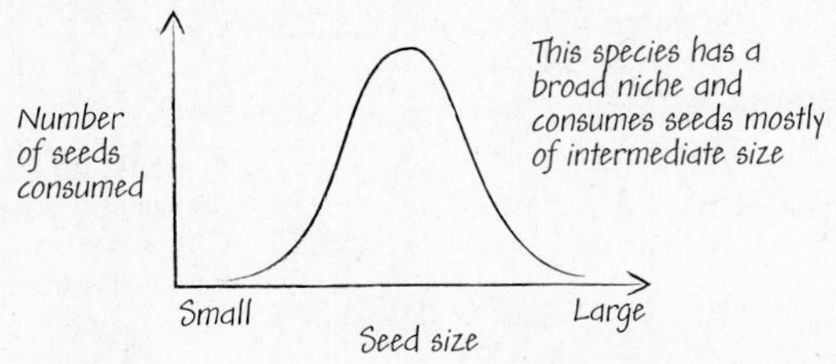

MODEL Draw your own niche model for a species that consumes mostly the smallest seeds, only rarely medium-sized seeds, and never large seeds.

To see this model in action, go to the Study Area of **Mastering Biology**

What happens when the niches of two species overlap? Interspecific competition occurs, where the extent of competition depends on the extent of niche overlap.

What Happens When Complete Niche Overlap Occurs? In the 1930s, G. F. Gause conducted a series of laboratory experiments with two species of the unicellular pond-dweller *Paramecium*.

- When *P. caudatum* and *P. aurelia* grew in separate cultures, both species exhibited logistic growth—the number of individuals increased rapidly at first, then leveled off as the population reached its carrying capacity (Ch. 51, Section 51.3).
- When the two species grew in the same culture together, sharing the same food supply, only the *P. aurelia* population exhibited a logistic growth pattern, although with a lower carrying capacity than when alone. *P. caudatum*, in contrast, was driven to extinction (**Figure 52.2a**).

The stronger competitor drove out the weaker competitor because the two laboratory populations had completely overlapping niches (**Figure 52.2b**). Based on these data, Gause proposed that two species that occupy the exact same niche (in all of its biotic and abiotic dimensions) cannot coexist, a hypothesis called the **competitive exclusion principle**. Some biologists invoke this principle to explain the disappearance of Neanderthals (*Homo neanderthalensis*) after thousands of years of coexistence with our species, *Homo sapiens* (Ch. 32, Section 32.6).

What Happens When Partial Niche Overlap Occurs? In the more typical case where niches partially overlap, individuals that use joint resources are at a disadvantage relative to individuals that use other resources. This is because competing individuals are likely to acquire fewer resources or to expend extra time and energy to acquire the resources. If both species experience a similar decrease in fitness due to the overlap of their niches, then both species may persist in the area of overlap, even if in reduced numbers. However, if one species is the stronger competitor, then the weaker

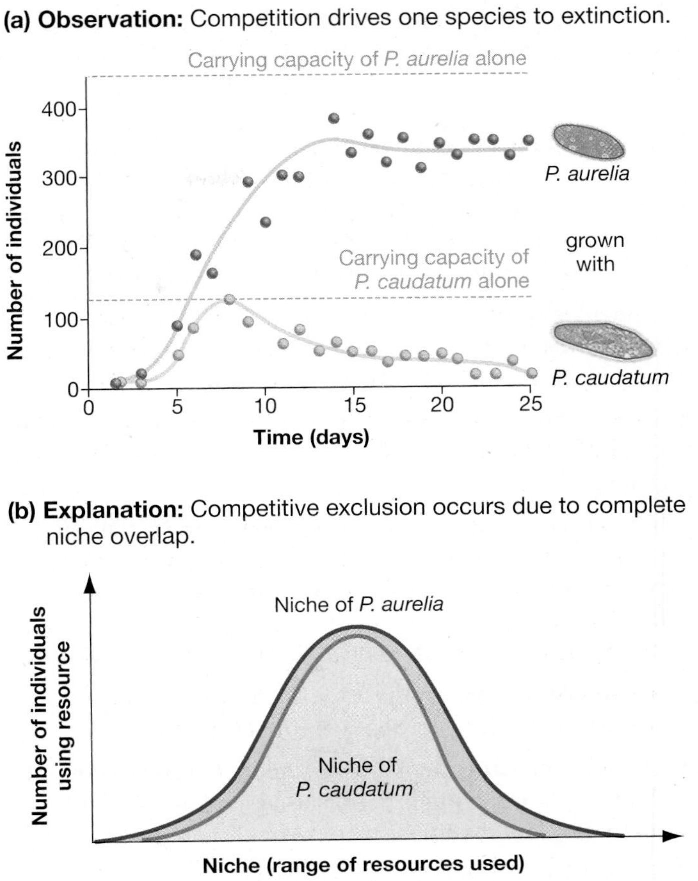

Figure 52.2 Competitive Exclusion Is Evident in *Paramecium*.
DATA: G. F. Gause. 1934. *The Struggle for Existence*. New York: Hafner Press.

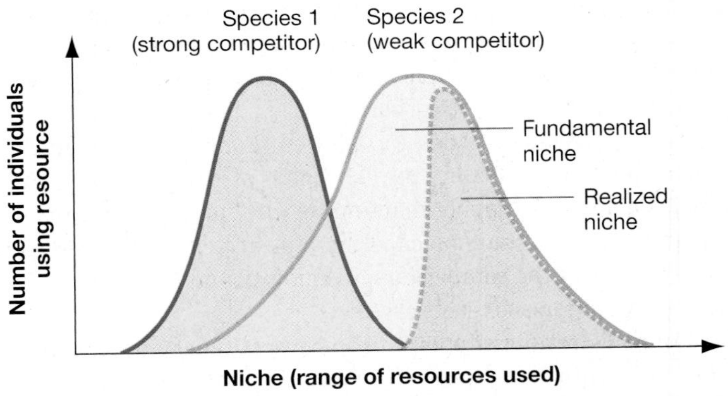

Figure 52.3 Fundamental Niches Are Broader than Realized Niches. In this example, Species 2 occupies a smaller range of resources (its realized niche) than it would if it had no competition.

competitor should be able to retreat to, or persist in, the area of non-overlap. Cases like these bring up an important distinction:

- A **fundamental niche** is the total theoretical range of environmental conditions that a species can tolerate.
- A **realized niche** is the portion of the fundamental niche that a species actually occupies, given limiting factors such as competition with other species (**Figure 52.3**).

Let's take a look at an important field experiment that demonstrated the difference between a fundamental and realized niche.

Experimental Studies of Competition in Nature Joseph Connell was a graduate student in the late 1950s when he started a side project on competition. Connell had observed an interesting pattern along an intertidal rocky shore in Scotland. He noticed that two species of barnacles had distinctive distributions:

- ***Upper intertidal zone*** The adults of one barnacle species, *Chthamalus stellatus* (pronounced *THAM-al-us*) occurred in an upper intertidal zone.
- ***Lower intertidal zone*** The adults of the other species, *Semibalanus balanoides*, were restricted to a lower intertidal zone.

Although adult barnacles live attached to rocks, their planktonic larvae are mobile. The young of both species were found together in the lower intertidal zone. Why aren't *Chthamalus* adults found there as well?

To answer this question, Connell considered two hypotheses:

1. ***Competition hypothesis*** Adult *Chthamalus* are outcompeted in the lower intertidal zone by *Semibalanus*.
2. ***Abiotic tolerance hypothesis*** Adult *Chthamalus* are absent from the lower intertidal zone because they do not thrive in the physical conditions there.

Connell tested these hypotheses by performing the experiment shown in Figure 52.4. He began by removing a number of rocks that had been colonized by *Chthamalus* from the upper intertidal zone at multiple study sites and transplanting them into the lower intertidal zone of each site. After moving the rocks, he allowed *Semibalanus* larvae to colonize them. Once the spring colonization period was over a year later, Connell divided each rock into two sides. In one half, he removed all *Semibalanus* that were in contact with or next to a *Chthamalus*.

This experimental design allowed Connell to document *Chthamalus* survival in the absence of competition with *Semibalanus*, and compare that finding with survival during competition. This is a common experimental strategy in competition studies: One of the competitors is removed, and the response by the remaining species is observed.

Connell's results supported the competition hypothesis. On the rocks with both species present, *Semibalanus* killed many of the *Chthamalus* by growing against them and lifting them off the substrate. As the orange bar in the "Results" section of Figure 52.4 shows, *Chthamalus* survival was much higher when all of the *Semibalanus* were removed. Connell observed similar results at the different study sites.

The fundamental niches of these two species partially overlap. Under the conditions tested in this experiment, *Semibalanus* is a superior competitor for space, but *Chthamalus* is able to thrive in a smaller realized niche where *Semibalanus* cannot survive. ✔ **MODEL If you understand this result, you should be able to sketch and label a graph of the fundamental and realized niches of the two barnacle species as a function of depth in the intertidal zone.**

RESEARCH

QUESTION: Why is the distribution of adult *Chthamalus* restricted to the upper intertidal zone?

COMPETITION HYPOTHESIS: Adult *Chthamalus* are outcompeted in the lower intertidal zone.

ABIOTIC TOLERANCE HYPOTHESIS: Adult *Chthamalus* do not thrive in the physical conditions of the lower intertidal zone.

EXPERIMENTAL SETUP:

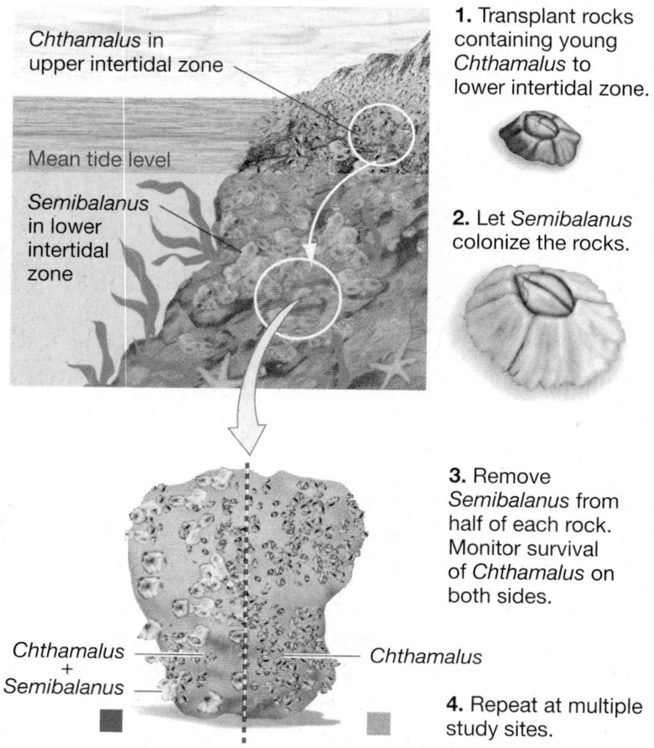

PREDICTION OF COMPETITION HYPOTHESIS: *Chthamalus* will survive better in the absence of *Semibalanus*.

PREDICTION OF ABIOTIC TOLERANCE HYPOTHESIS: *Chthamalus* survival will be low in both the presence and absence of *Semibalanus*.

RESULTS:

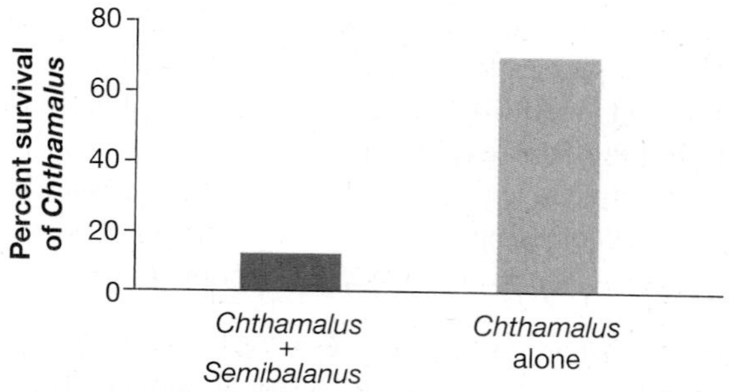

CONCLUSION: *Chthamalus* do not occur in the lower intertidal zone because they are outcompeted by *Semibalanus*.

Figure 52.4 Experimental Evidence Shows Competition.

SOURCE: J. H. Connell. 1961. The influence of interspecific competition and other factors on the distribution of the barnacle *Chthamalus stellatus*. *Ecology* 42: 710–723.

✔ **PROCESS OF SCIENCE** Why was it important to carry out the experiment with multiple rocks at multiple study sites?

Fitness Trade-Offs in Competition Why haven't *Semibalanus* and other successful competitors taken over the world? Biologists answer this question by invoking the concept of **fitness trade-offs**—inevitable compromises in adaptation (Ch. 22, Section 22.5). The key insight is that the ability to compete for a particular resource—like space on rocks or edible seeds of a certain size—can require time and energy that can no longer be invested in other traits that contribute to fitness, such as enduring drought conditions, warding off disease, or preventing predation.

In the case of *Chthamalus* and *Semibalanus* growing in the intertidal zone, the fitness trade-off is rapid growth and success in competing for space versus the ability to endure the harsh physical conditions of the upper intertidal zone. *Chthamalus* grow slowly and are small, but can survive long exposures to the air and to intense sun and heat; *Semibalanus* grow fast and are large, but cannot tolerate harsh physical conditions.

Also, competitive ability is not an inherent quality of species—it depends on environmental conditions. For example, Burmese pythons have unremarkable competitive abilities in their native Southeast Asia, due to the checks and balances provided by thousands of years of species interactions. But when these large snakes were released to the everglades of Florida in the 1990s, they began to consume large quantities of prey (see Ch. 51 case study, p. 1116) and outcompete native species such as the endangered indigo snake. The remarkable competitive success of invasive species poses a threat to communities around the world—and conservation biologists are very concerned.

Mechanisms of Coexistence: Niche Differentiation Because competition is a −/− interaction, there is strong natural selection on both species to avoid it. Figure 52.5 shows the predicted outcome of competition over many generations: An evolutionary change in traits reduces the amount of niche overlap, and thus the amount of competition. Note that *individuals* are not deciding to change; rather, the frequency of certain traits is changing in the *population* over time as individuals with lower fitness die or have fewer offspring than those with higher fitness.

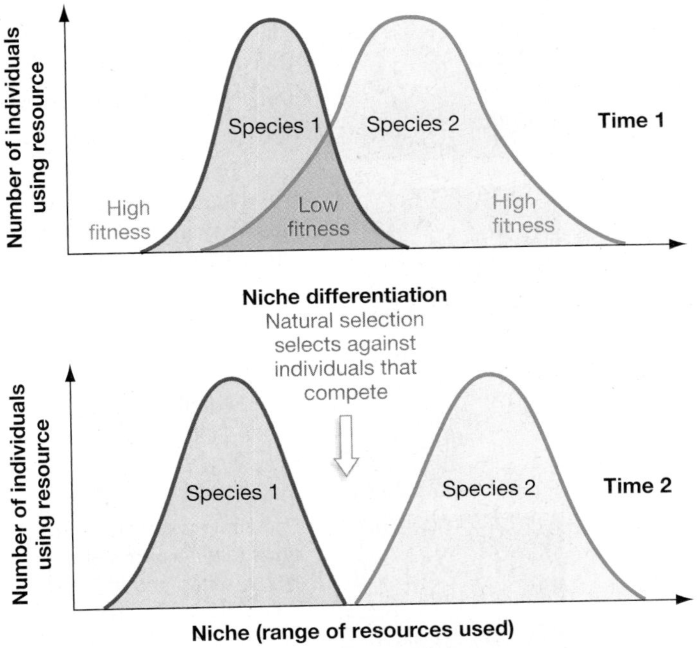

Figure 52.5 Competition Can Lead to Niche Differentiation.

An *evolutionary* change in resource use, caused by competition over generations, is called **niche differentiation** or resource partitioning. The evolutionary change that occurs in species traits, and that enables species to exploit different resources, is called **character displacement**. Character displacement makes niche differentiation possible.

One spectacular example of character displacement involves the Galápagos finches, famous for their diversity in beak shapes (Ch. 22, Section 22.4). Specializations in beak shape are adaptations for harvesting particular foods, such as small or large seeds.

Figure 52.6 shows the frequency of beak depths (top half of the beak only) in a population of medium ground finch, *Geospiza fortis*, on the island of Daphne Major in the 1940s. You can see on the *x*-axis that beak depth, a functionally important measure of beak shape, ranged from 3.5 to 5.6 mm. However, on the island of Santa Cruz at around the same time, *G. fortis* lived alongside the small ground finch, *G. fuliginosa*, and had deeper beaks on average, from 4.0 to 7.5 mm. From these data you could hypothesize either that competition causes character displacement in *G. fortis* or that the difference between populations is due to other factors such as predation or food availability.

More recently, a natural experiment enabled Peter and Rosemary Grant to confirm that character displacement was taking place in *G. fortis* on Daphne Major. The Grants noticed that the large ground finch, *G. magnirostris*, had established a breeding population on Daphne Major in 1982, living alongside

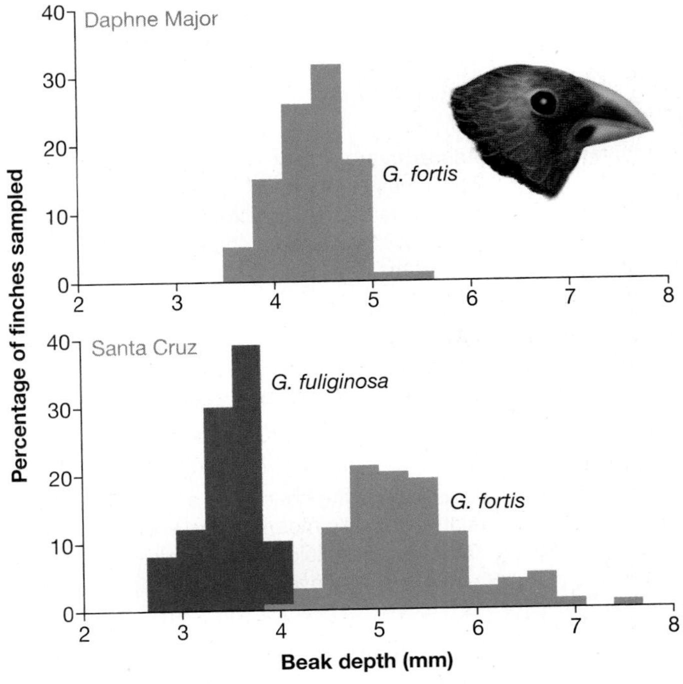

Figure 52.6 Niche Differentiation Is Evident in Galápagos Finches. Average beak depths in *Geospiza fortis* were higher on the island with a competitor present.

DATA: P. Grant. 1999. *Ecology and Evolution of Darwin's Finches*. Princeton, NJ: Princeton University Press.

or "sympatric" with *G. fortis. G. magnirostris* has a large beak capable of opening large, hard, drought-tolerant seeds. What effect did competition with *G. magnirostris* have on *G. fortis*?

At first the Grants observed a slow but steady decline in average beak size in *G. fortis* after the arrival of *G. magnirostris*. Then in 2004, during a severe drought when competition with *G. magnirostris* was at its most intense, the average beak size of *G. fortis* declined sharply.

The Grants had seen evolution in action—directional selection on *G. fortis* beak size causing character displacement (reduced average beak size) and niche differentiation (beak sizes adapted for eating different seed sizes). That is, natural selection had favored *G. fortis* individuals that did not compete with *G. magnirostris*.

✔ **MODEL If you understand niche differentiation, you should be able to sketch the frequency of beak depths in *G. fortis* before (see top graph in Figure 52.6) and many years after the arrival of *G. magnirostris* on Daphne Major.**

Species Interaction: Consumption

Consumption is a +/− interaction that occurs when one organism eats part or all of another. **Table 52.1** compares three types of consumption interactions—**predation, herbivory,** and **parasitism**—according to the type and size of species involved, the extent of consumption, and the duration of the interaction.

To explore how biologists analyze the impact of consumption, let's consider a series of insights about predators, herbivores, parasites, and their victims.

Consumption Interactions Can Affect the Distribution and Abundance of Species Biologists can employ a number of approaches to measure the consequences of consumption interactions to the abundance of species involved. For example, they can conduct an experiment by excluding a predator from an area and measuring changes to the abundance of prey compared to control areas. This was the strategy used by Charles Krebs and colleagues when they sought to understand the cycling of snowshoe hare and lynx populations in boreal forests (see Ch. 51, Figure 51.10). The researchers gained a number of insights using this approach:

- The lynx reduced the population size of snowshoe hares.
- The size of the effect of lynx on the hare population was conditional: It was more dramatic when food supplies for the hares were limited.
- The population size of the hares affected the population size of the lynx.

That is, each species in the interaction could affect the distribution and abundance of the other species, and the size of the effect was influenced by other biotic and abiotic variables.

SUMMARY **Table 52.1** Consumption Interactions

Interaction	Description	Consequence to Victim	Examples	
Predation	A **predator** kills and consumes all or most of the **prey**. Predators are typically **carnivores** ("meat-eaters") but seed predators consume and kill plant embryos.	Usually fatal; all or most of prey tissues are consumed		• Lynx eat hares • Orcas eat seals • Robins eat earthworms • Spiders eat flies
Herbivory	An **herbivore** ("plant-eater") consumes plant and algal tissues such as leaves, stems, fruits or roots, but typically not entire organisms.	May or may not be fatal depending on the extent of damage		• Pandas eat bamboo leaves • Sea urchins eat kelp • Butterflies eat flower nectar • Toucans eat palm fruits
Parasitism	An **endoparasite** consumes a relatively small amount of tissue or nutrients from *inside* a host, typically over a long period.	Often not fatal; parasites are usually small relative to their host	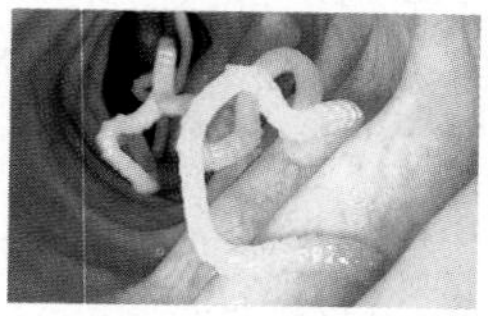	• Tapeworms and hookworms absorb nutrients from within gut of humans
A **parasite** is an organism that lives in or on a host, taking resources from it.	An **ectoparasite** consumes a relatively small amount of tissue or nutrients (often fluids) from *outside* a host, typically over a long period.	Often not fatal; parasites are usually small relative to their host	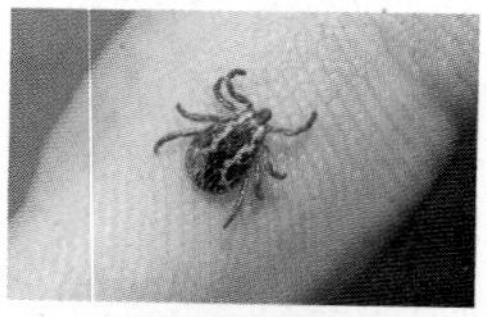	• Ticks drink blood • Aphids drink plant fluids
	A **parasitoid** is free living as an adult but endo- or ectoparasitic as a larva, hatching from an egg laid inside or on a host and consuming the host, eventually killing it.	Usually fatal to host	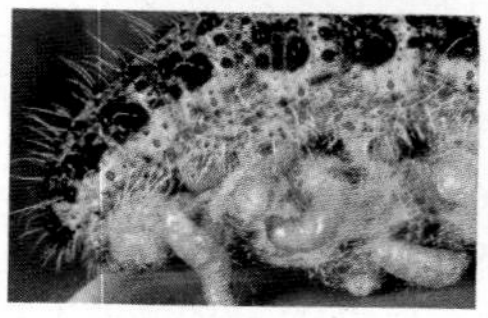	• Parasitoid wasps lay their eggs in caterpillars; the larvae hatch and eat the caterpillar from the inside out (shown emerging)

Biologists can also measure the effects of consumption interactions by looking at correlations in natural environments. For example, in a study of a thousand women living without contraceptives in the Amazonian lowlands:

- Women infected with *hookworms* experienced lower fecundity than women without hookworms due to delayed first pregnancy and longer interbirth intervals.
- Women infected with *roundworms* experienced higher fecundity than women without roundworms due to earlier first pregnancy and shorter interbirth intervals.

These results show that parasites can affect the abundance of host populations by affecting not only survivorship but also fecundity (Ch. 51, Section 51.1).

Research on the effects of consumers on the distribution and abundance of organisms has paid off in practical benefits: In some cases, biologists have been able to control pests by introducing consumers. In agriculture and forestry, the use of predators, herbivores, and parasites as biocontrol agents is a key part of **integrated pest management**: strategies to maximize crop and forest productivity while using a minimum of insecticides or other types of potentially harmful compounds.

For example, the invasive plant purple loosestrife had succeeded in taking over large expanses of wetlands in North America. Through careful experimentation, however, biologists were able to find four insect species from the native purple loosestrife range in Europe and Asia that feed exclusively on purple loosestrife. By introducing the herbivores into wetlands, researchers have helped to dramatically reduce purple loosestrife infestations.

This approach has enormous benefits but also serious risks—introducing species to communities to limit one species can have unexpected consequences to others. Ecologists encourage proceeding with extreme caution.

Consumption Is an Agent of Natural Selection Over the long term, how do species that interact via consumption affect each other's evolution? When predators and prey, herbivores and plants, or parasites and hosts interact over time, a **coevolutionary arms race** often results—a repeating cycle of reciprocal adaptation. Traits that increase feeding efficiency evolve in predators, herbivores, and parasites. In response, traits evolve that make prey and hosts unpalatable or elusive. In counter-response, selection favors consumers with traits that overcome the prey adaptation. And so on. Coevolution is a major driver of evolutionary change.

The consequences of coevolution are diverse, profound, and sometimes astonishing. Let's sample the evolutionary consequences of consumption by comparing two general types of adaptations for defense.

Constitutive Defenses Are Present in the Absence of Consumers Natural selection strongly favors traits that allow individuals to avoid being eaten. Prey species may hide, flee, poison or threaten to poison their consumers, school together to confuse their consumers, or deploy armor or weapons in defense (Table 52.2).

SUMMARY Table 52.2 Constitutive Defenses

Type of Defense	Example
Cryptic coloration and object resemblance Resemblance to a leaf enables this grasshopper to escape detection	
Escape behavior Prey may have adaptations to detect predators and run, fly, jump, or swim away	
Toxins and other defense chemicals Prey may have tissues laced with toxic compounds. Many advertise their toxicity with warning coloration	
Schooling/flocking Predators can become confused by groups of prey	
Defense armor and weapons Prey may have sharp thorns, protective shells, or other defense structures	

Traits like these are called **constitutive defenses**, or standing defenses, because they are present even in the absence of consumers.

The diversity and specificity of constitutive defenses is nothing short of mind-boggling. Consider the discovery made by naturalist Henry Walter Bates as he collected butterflies in the Amazonian rain forest in the mid-1800s. Bates found more species of butterflies in his study area than in all of Europe—and there was a surprising pattern among the diversity. He noticed pairs of species from different taxonomic groups that were not closely related, yet had very similar wing patterns and occupied the same parts of the forest, a pattern of resemblance called **mimicry**.

Based on what you have learned about niche differentiation, you might have expected that the traits of species occupying similar niches would become more *different* over time, not more alike. What could explain this seeming paradox in butterflies?

Bates had an insight when he observed that some of the butterflies exuded foul substances when he handled them and that these butterflies were passed over by would-be predators. Concluding that the foul substance thwarted predators, he hypothesized that natural selection favored mimic species that, although nontoxic themselves, resembled the unpalatable species—a form of mimicry now known as **Batesian mimicry**. Follow-up experimental studies supported this hypothesis—and recent genetic studies have discovered some of the molecular mechanisms responsible for the shared color patterns.

Bates also noticed that some pairs of similar but distantly related butterflies are *both* unpalatable. What was the fitness advantage of this pattern? Several hypotheses were proposed, but German naturalist Fritz Müller hit on the key idea—that the existence of similar-looking unpalatable prey in the same habitat increases the likelihood that predators will learn to avoid them. Thus, co-mimicry boosts the fitness of both species. The adaptive advantage of this type of mimicry, now known as **Müllerian mimicry**, was subsequently supported by field experiments.

Butterflies are not the only organisms to engage in mimicry. Bates and Müller both observed diverse examples, some of them quite shocking, such as a caterpillar with a close resemblance to a venomous snake. Figure 52.7 provides a comparison of Batesian and Müllerian mimics in the case of stinging and nonstinging insects with the familiar yellow and black warning coloration.

One of the take-home messages from studies of mimicry is that the + / – consumption interaction between two species can have unexpected consequences for other species in the community. That is, natural selection in the form of predator–prey interactions can result in other types of interactions among prey species, such as competition and mutualism.

Inducible Defenses Are Triggered by Consumers Although constitutive defenses can be extremely effective, they can also be expensive due to the energy and resources that must be devoted to producing and maintaining them. Based on this observation, it should not be surprising that many prey species have **inducible defenses**: physical, chemical, or behavioral defensive traits that are induced in the prey in response to the presence of a consumer.

Figure 52.7 Prey Species Display Two Basic Forms of Mimicry.

Plants are famous for their inducible chemical toxins such as cyanide, nicotine, caffeine, cocaine, and morphine.

Inducible defenses decline in prey species if consumers leave the habitat. They are efficient energetically, but they are slow—it takes time to produce them. To see how inducible defenses work, consider research on the blue mussels that live in an estuary along the coast of Maine. Biologists had documented that blue mussels had thicker shells in the part of the estuary with many crabs. Did the thicker shells occur in response to the presence of crabs, or for some other reason?

To test the inducible-defense hypothesis, Mark Bertness and colleagues carried out the experiment diagrammed in Figure 52.8. The tank on the right allowed the researchers to measure shell growth in mussels that were "downstream" from crabs, thus exposing them to the scent of crabs. As predicted by the inducible-defense hypothesis, the mussels exposed to a crab in this way developed significantly thicker shells than did mussels that were not exposed to the scent of a crab. These results suggest that, even without direct contact, mussels can sense the presence of crabs and increase their investment in defenses.

Inducible defenses are favored by natural selection because prey individuals with effective inducible defenses survive predation and pass these heritable responses to offspring, increasing the frequency of those defenses in the population over time.

Parasite–Host Coevolution Can Be Particularly Complex Predators tend to be larger or similar in size compared to their prey. In contrast, parasites tend to be much smaller than their hosts. Further, many parasites have complex life cycles that involve moving between more than one host. Thus, to thrive, parasites do not just have to invade tissues, consume them, and reproduce while evading defensive responses by their host. They also have to be transmitted to new hosts. What kind of adaptations do parasites have to enable transmission to new hosts?

RESEARCH

QUESTION: Are mussel defenses induced by the presence of crabs?

HYPOTHESIS: Mussels increase investment in defense in the presence of crabs.

NULL HYPOTHESIS: Mussels do not increase investment in defense in the presence of crabs.

EXPERIMENTAL SETUP:

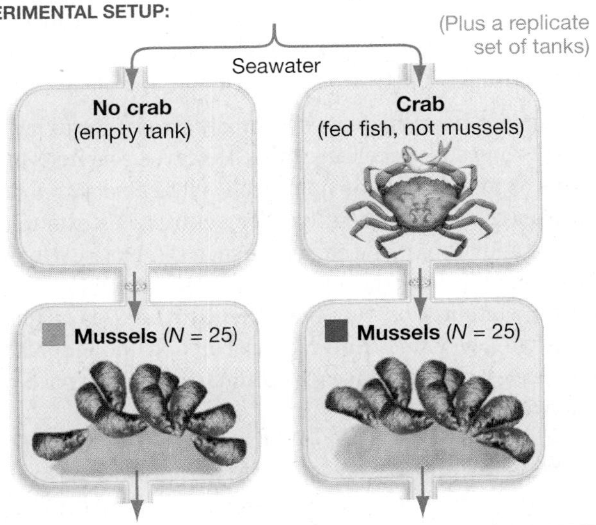

PREDICTION OF HYPOTHESIS: Mussels downstream of the crab tank will have thicker shells than mussels downstream of the empty tank.

PREDICTION OF NULL HYPOTHESIS: Mussels in the two tanks will continue to have shells of equal thickness.

RESULTS:

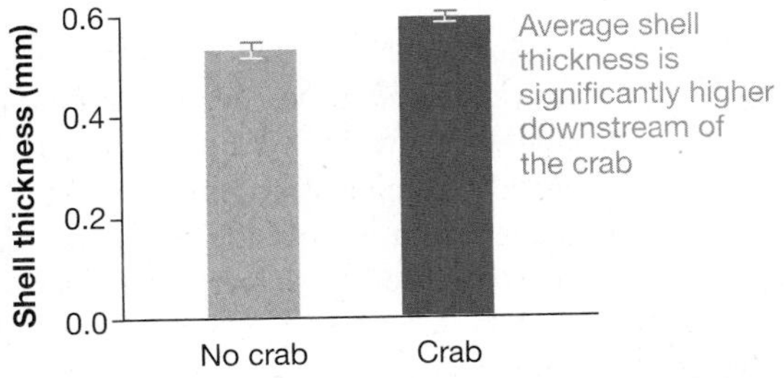

CONCLUSION: Mussels increase investment in defense when they detect crabs. Shell thickness is an inducible defense.

Figure 52.8 Experimental Evidence Shows Inducible Defenses.

SOURCE: G. H. Leonard, M. D. Bertness, and P. O. Yund. 1999. Crab predation, waterborne clues, and inducible defenses in the blue mussel, *Mytilus edulis*. *Ecology* 80: 1–14.

✔ **PROCESS OF SCIENCE** Why did the researchers feed fish rather than mussels to the crabs?

Figure 52.9 Some Parasites Manipulate Host Behavior. When the tropical ant *Cephalotes atratus* becomes infected with parasitic nematodes, its appearance and behavior attract bird predators.

To answer this question, consider species of tree-dwelling ants (**Figure 52.9a**) that are parasitized by nematodes (roundworms). When nematodes have matured inside a host, they lay eggs in the ant's abdomen. The exoskeleton in the area becomes translucent, and the yellow eggs inside make the structure look reddish instead of the normal black (**Figure 52.9b**). Infected ants also hold the abdomen up in a "flagging" posture, making them look like the berries that grow on the trees occupied by the ants (**Figure 51.9c**). Experiments have shown that fruit-eating birds are much more likely to pluck the abdomen from infected ants than uninfected ants, clearly reducing the fitness of infected ants.

Why is this adaptive for nematodes? To complete their life cycle, the nematodes have to grow inside birds, before being shed in bird feces in other trees where they are subsequently eaten by ants. The morphological and behavioral changes in the ants make the parasite much more likely to be transmitted to a new host, increasing the fitness of the nematodes. Biologists have now cataloged a large number of case studies like this, from an array of parasites and hosts.

Notice how quickly pairs of species interactions can become complex in communities. This one example of parasitism involves a network of interactions between a nematode, an ant, a plant, and a bird. Changes in the abundance and/or adaptations of one species can affect all the others.

Species Interactions: Mutualism

Mutualisms are +/+ interactions that involve a wide variety of organisms and rewards. Fitness benefits vary from factors related to survival, such as access to nutrients or protection from predators, to factors related to reproduction, such as dispersal of pollen or seeds. Let's consider a sample of this diversity.

(a) Mutualism between flowers and pollinators

(b) Mutualism between two animals providing protection

Figure 52.10 Mutualisms Take Many Forms. **(a)** As this bat reaches for nectar with its tongue, it will pick up pollen on its head and carry it to the next flower. **(b)** This clown anemonefish and sea anemone protect each other from predators.

- Orange nectar bats visit flowers to slurp up nectar with their long tongues (**Figure 52.10a**). The bats benefit by using nectar as a food source; flowering plants benefit because in the process of visiting flowers, bats carry pollen from one plant to another and accomplish pollination. Coevolution between flowering plants and pollinators is common and spectacular (Ch. 38, Section 38.3).
- One of the most important of all mutualisms occurs between mycorrhizal fungi and plant roots. Experimental evidence indicates that mycorrhizal fungi receive sugars and other carbon-containing compounds in exchange for nitrogen or phosphorus acquired by the plant partner (Ch. 29, Section 29.3). The plant feeds the fungus; the fungus fertilizes the plant.
- Mutualism occurs between nitrogen-fixing bacteria and the plant species that house them in their tissues (Ch. 36, Section 36.4). The host plants provide sugars and protection to the bacteria; the bacteria supply ammonia or nitrate in return.
- Mutualism is also common between bacteria and animals, particularly in the microbiome community in digestive systems of the hosts. The host provides a suitable habitat for the bacteria; the bacteria provide digestive and other metabolic favors to the host. For example, bacteria in the guts of termites and cows enable the digestion of cellulose. Interest in research on the health benefits provided by bacteria in humans is skyrocketing (Ch. 26, Section 26.1).
- **Figure 52.10b** illustrates an association that provides mutual aid and protection. The sea anemone's stinging tentacles defend the clown anemonefish, which has coevolved a mucus that protects it from the sting. Meanwhile, the fish chases away predators of the anemone. The fish feces also provide nitrogen to the single-celled, photosynthetic protists hosted within the anemone. Thus, this example represents a three-way mutualism between fish, anemone, and algae.

You might hear the term **symbiosis** ("together-living") used to refer to mutualistic relationships. However, scientists typically use this term to refer to any physically close association between two species, whether good, bad, or neutral. For example, both parasitic and mutualistic bacteria live in symbiosis with you.

You may also hear the term "cooperation" used to refer to mutualistic relationships. However, scientists use this term to refer to mutually beneficial interactions among individuals of the same species, such as wolves that hunt together in a pack (Ch. 50, Section 50.6).

Mutualists Are Not Trying to Be "Nice" Even though mutualisms benefit both species, the interaction does not involve individuals from different species being altruistic, or "nice," to each other. Researchers since Darwin have recognized that each individual in the interaction tends to behave in a way that maximizes the benefits while minimizing the costs. Thus, a mutualism interaction is the by-product of two individuals pursuing their own self-interest to maximize fitness. Some researchers think of mutualisms as "mutual parasitism."

In this light, it's not surprising that some species "cheat" on mutualistic systems. For example, deceit pollination occurs when a plant produces a showy flower but no nectar. Pollinators receive no reward for making a visit and carrying out pollination. Conversely, some would-be pollinators are "nectar robbers" that pierce a hole in the side of the flower and retrieve the nectar without pollinating the flower. In both cases, +/+ interactions have become +/− interactions.

Mutualisms Are Context Dependent Researchers have measured the fitness consequences of mutualisms to participating species in a wide variety of contexts—varying in both abiotic and biotic factors. While some mutualisms are quite robust, others are rather context dependent. Consider the effects of nutrient availability:

- ***In soils with low nitrogen*** Plants such as soybeans invest resources in root nodules to support large colonies of nitrogen-fixing bacteria. The benefits of the imported nitrogen compounds to the plant outweigh the energetic costs of supporting the bacteria. The mutualism benefits both the plant and the bacteria.
- ***In soils with high nitrogen*** The costs of supporting the nitrogen-fixing bacteria colonies may outweigh the benefits, turning a +/+ interaction into a +/− interaction.

Biologists are increasingly interested in the nuances of context-dependent mutualisms because the environment is rapidly changing due to human impacts. For example, since coral health is so important to coral-reef communities, many researchers are studying the effects of nutrient runoff, ocean warming, and ocean acidification on the stability of the mutualistic relationship between corals and the photosynthetic protists that they harbor.

Mutualisms Can Be Specific or General Just as some predators specialize in eating one kind of prey and others are generalists—eating whatever is available—mutualisms too can be specific or general. Figure 52.11 shows a model of this concept for the case of flower–pollinator mutualisms.

As you might predict, there are fitness trade-offs associated with different lifestyles. For example, a specialist pollinator might have adaptations that enable it to collect nectar very efficiently from a certain species of flower, increasing its fitness. But if the environment changes and the flower is no longer available, the specialist pollinator could go extinct. Generalists, on the other hand, may not be as efficient in their feeding as the specialist, but are more likely to persist in a variable environment.

Once again, biologists are increasingly interested in studying the extent to which mutualisms are specific or general in light of rapidly changing environments. For example, climate change is causing the mismatch of geographic ranges of interacting species and/or the mismatch of the timing of events like flowering or insect emergence. Specialists are less likely to survive these mismatches than generalists.

Take-Home Messages

Commensalism, competition, consumption, and mutualism have different fitness consequences on the species that are engaged in these interactions, and these consequences change over time and space—which is why biologists are so eager to study the effects of human impacts on species interactions. The quality of your life—such as the food you eat and the parasites you harbor—depend on this research. Table 52.3 summarizes the fitness effects, short-term impacts on population size, and long-term evolutionary consequences of species interactions.

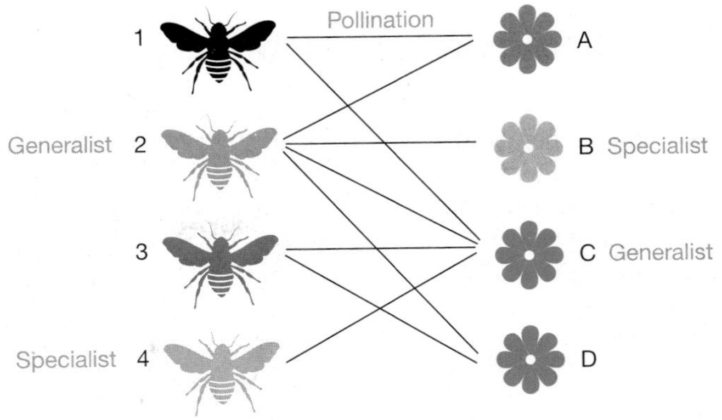

Figure 52.11 Pollinators and Flowers Can Be Specialists or Generalists in Their Mutualisms.

✔ **SOCIETY** Pesticides are causing declines in many bee species. Use the model to predict which type of crop plant is mostly likely to survive this environmental change.

CHECK YOUR UNDERSTANDING

✔ **If you understood this section, you should be able to . . .**

1. **THINK CAREFULLY** Explain whether niche differentiation (resource partitioning) involves a conscious choice by the species that are interacting.
2. **PROCESS OF SCIENCE** Hypothesize what type of species interaction is occurring when toucans consume palm fruits and pass the palm seeds in their feces, and design an experiment to test your hypothesis.

Answers are available in Appendix A.

SUMMARY Table 52.3 Species Interactions

Type of Interaction	Fitness Effects	Short-Term Impact: Distribution and Abundance	Long-Term Impact: Coevolution
Commensalism	+/0	Population size and range of commensal may depend on population size and distribution of host.	Strong selection on commensal to increase fitness benefits in relationship; no selection on host
Competition	−/−	Reduces population size of both species; if competition is asymmetric, the weaker competitor may become locally extinct, or it may retreat to a part of its niche that does not overlap with the stronger competitor.	Niche differentiation via selection to reduce competition
Consumption Herbivory Predation Parasitism	+/−	Impact on prey population depends on predator density, prey density, and effectiveness of defenses.	Strong selection on prey for effective defense; strong selection on consumer for traits that overcome defenses; coevolutionary arms races result
Mutualism	+/+	Population size and range of each species depend on those of the other species.	Strong selection on both species to maximize fitness benefits and minimize fitness costs of relationship

52.2 Community Structure

So far, the focus has been primarily on interactions that occur between two species. But in a community, each species interacts with dozens, hundreds, or even thousands of other species. The resulting complexity of interactions is mind-boggling. How can biologists begin to characterize communities? In general, a community structure has four key attributes:

> **After you complete this section, you should be able to ...**
>
> - Describe the factors that determine community structure.

1. the total number of species;
2. the relative abundance and distribution of those species;
3. the sum of interactions among all species; and
4. the physical attributes of the community, which may include abiotic factors (such as water depth or altitude gradient), and biotic factors (such as the physical structure provided by corals in a coral reef or trees in a forest).

Let's take a closer look at how biologists measure these attributes of community structure.

How Many Species Occur in Communities?

The number and relative abundance of species are key characteristics of community structure. Ecologists use two main methods of quantifying the number of species in a community:

1. **Species richness** is a count of how many species are present in a given community.
2. **Species diversity** is a weighted measure that incorporates both the number of species and their relative abundance, or *evenness*.

For a given species richness, species diversity is high if evenness is high, meaning that all the species have comparable abundance, but species diversity is low if just one or a few species dominate the community. This distinction can be visualized using simple models (see **Making Models 52.2**) and/or quantified using mathematical methods such as the Shannon index (explained in **Quantitative Methods 52.1**). Be aware that since relative abundance data are often not available, biologists and teachers sometimes use the terms "species richness" and "species diversity" interchangeably.

Making Models 52.2 Tips on Modeling Species Diversity

Simple models can help to distinguish the meaning of species richness and species diversity. You can use any type of symbol to represent different species in a community, such as letters, numbers, geometric shapes, or cartoons of organisms. Don't spend a lot of time on lifelike drawings—it's the concept that matters.

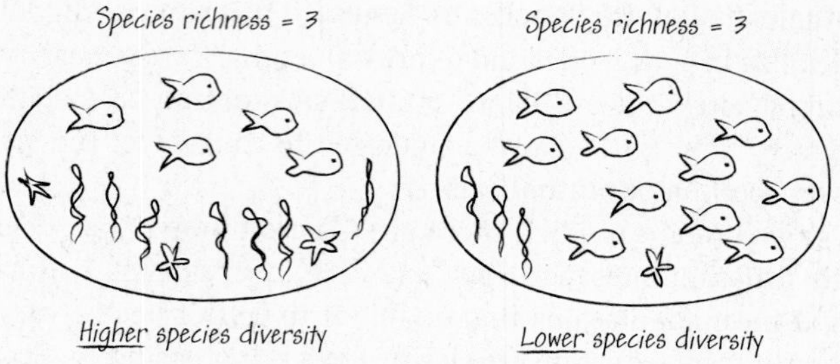

MODEL Using your own symbols, draw a model showing two communities with the same species richness (= 4) but different species diversity.

To see this model in action, go to the Study Area of **Mastering Biology**

Making Models 52.3 Tips on Drawing Food Webs

In food web models, pairwise species interactions are assembled into networks where the autotrophs such as plants and algae are typically placed at the bottom, the consumers of those organisms (primary consumers) are one level up, then the consumers of those animals (secondary consumers) are still higher, and so on. The arrows indicate the flow of energy and nutrients from one level to another.

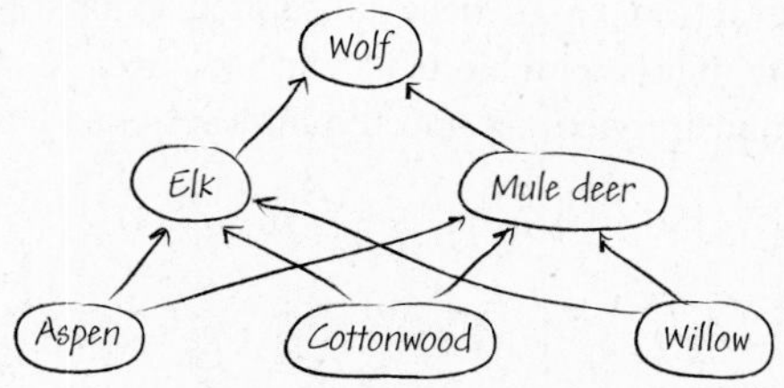

MODEL Add the following species interactions to the food web: Grasses are autotrophs that are eaten by elk and mice; mice are eaten by coyotes; coyotes are eaten by wolves.

To see this model in action, go to the Study Area of **Mastering Biology**

How Do Species Interactions Form Networks?

Biologists can study interactions between pairs of species, then piece interactions together into more complex multispecies interaction networks. An example of this approach was shown in Figure 52.11, which revealed an overlap of mutualistic relationships of generalist and specialist flowers and their generalist and specialist pollinators. That overlap led indirectly to competition among flowers and competition among pollinators.

You might also be familiar with linking consumption interactions sequentially into a **food chain** such as "aspen is eaten by elk, and elk are eaten by wolves." Overlapping food chains can then be combined into a **food web**, a summary of some or all of the consumption interactions in a community. The structure of food webs is based on trophic ("feeding") levels where, by convention, the producers are on the "bottom" and the consumers are on the "top" as energy and nutrients move up through the web's levels (detailed in Ch. 53, Section 53.1). **Making Models 52.3** illustrates one simple example of how a visual model can help to assemble pairwise consumption relationships in a community.

Quantitative Methods 52.1 Measuring Species Diversity

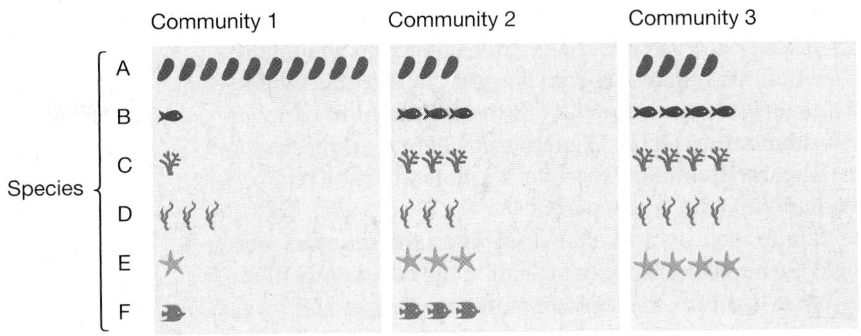

	Community 1	Community 2	Community 3
Species richness:	6	6	5
Evenness:	Low	**High**	**High**
Species diversity:	1.350	**1.794**	1.610

Figure 52.12 Species Diversity Can Be Quantified.

Consider the composition of the three hypothetical communities shown in **Figure 52.12**. This simple model shows the difference between two measures of diversity, *species richness* and *species diversity*. Species richness is the number of species found in a community. In this case, communities 1 and 2 have equal species richness, and community 3 is lower in richness by one species. It's important to note, however, that communities 2 and 3 have similar relative abundances of each species, or what biologists call high *evenness*. Community 1, in contrast, is highly uneven. Fifty-five percent of the individuals in community 1 belong to species A, and other species are relatively rare. An uneven community has lower effective diversity than its species richness would indicate.

To take evenness into account, other diversity indices have been developed. A simple example is the Shannon index of species diversity, given by the following equation:

$$H' = -\sum_{i=1}^{n} p_i \ln p_i$$

In this equation, p_i is the proportion of individuals in the community that belong to species i. The index is summed over all the species in the study (n).

The Shannon index of species diversity for the hypothetical communities shown in Figure 52.12 was thus calculated by **(1)** computing the proportion of individuals in each community that belong to each species, **(2)** taking the natural logarithm of each of these proportions (see **BioSkills 5** for a review of logarithms), **(3)** multiplying each natural logarithm times the proportion for each species, and **(4)** summing the total across the number of species in the community (5 or 6 in these examples) and multiplying by −1.

✔ **QUANTITATIVE** If you understand the equation, you should be able to do these calculations and get the same result given in Figure 52.12.

Notice that while communities 1 and 2 have the same species richness, community 2 has higher diversity because of its greater evenness. Community 3 has lower species richness than community 1 but higher diversity.

As you might imagine, adding more species and species interactions to models of community structure dramatically increases the complexity of the network. Fortunately, scientists from multiple disciplines have collaborated to develop powerful computer models to process and mine the complex community data, an example of systems biology and bioinformatics.

For example, consider the computer-generated network diagram of species interactions in a temperate forest in Japan, shown in **Figure 52.13**. The only pairwise species interactions included (represented by lines) are those that occur underground between species of plants and fungi (represented by dots). You can probably imagine how complex the network becomes when you add other underground and aboveground organisms to the model. While it is easier to study smaller networks, biologists are discovering that to fully understand the structure and function of communities, it is important to quantify their full complexity.

Notice in Figure 52.13 that some species (dots) are engaged in many more species interactions than others. Let's take a closer look at how certain species can play an especially important role in structuring communities.

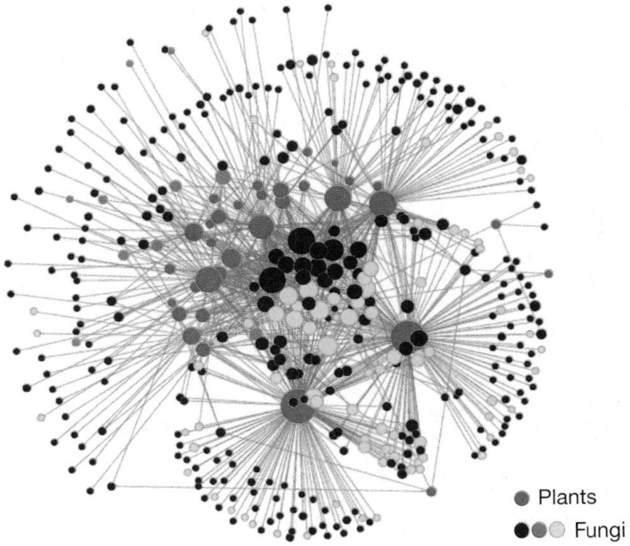

Figure 52.13 Network of Underground Plant–Fungus Species Interactions in a Temperate Forest. Each dot represents a species. The size of the dots represents relative abundance. The lines represent commensal, consumption, and mutualistic interactions.

DATA: H. Toju et al. 2014. *Nature Communications* 5: 5273.

Why Are Some Species More Important than Others in Structuring Communities?

In the mid-1960s, ecologist Robert Paine hypothesized that the sea star *Pisaster ochraceus* had a greater impact on community structure than its numbers would suggest. To test this hypothesis, Paine measured the number of species present in rocky intertidal communities on the Pacific coast where *P. ochraceus* was present and also in similar experimental plots where he removed *P. ochraceus*. What did he discover?

The data in **Figure 52.14** show that when the sea stars were removed, the number of species present in the community plummeted. What had been diverse communities of algae and invertebrates became overgrown with solid stands of mussels. The mussel *M. californianus* is a dominant competitor for space on the rocky shore, but its populations had been held in check by sea-star predation.

To describe the effect that a predator such as *P. ochraceus* can have on a community, Paine coined the term "keystone species." A **keystone species** has a much greater impact on the distribution and abundance of the surrounding species than its abundance and total biomass would suggest. In subsequent studies of keystone species, a couple of insights have emerged:

- Keystone species are not always predators. Their influence can be derived from many types of species roles and interactions in communities.
- Keystone species are not a discrete group. Rather, they represent one end of a continuum of influence on other species.

Biologists generally recognize two different types of effects of influential species on community structure: Some have a "bottom-up" effect and others have a "top-down" effect. Let's consider each in turn.

Bottom-Up Influences on Community Structure When the amount of nutrients, sunlight, water, and other abiotic factors determine the abundance of primary producers, biologists say that *bottom-up control* is occurring. The presence of a particular species at the "bottom" of the food web can have a major effect on community structure.

For example, the photosynthetic protists that live inside coral polyps are the main primary producers of coral reefs. When extreme warming events cause the protists to be expelled (and the corals to become "bleached"), the loss of the protists and their coral hosts is devastating to the entire coral-reef community.

In this example, corals are considered "ecosystem engineers" because they create the physical structure that provides habitat for other organisms in the community. When structural species such as these are removed by harvesting, disease, competition with an invasive species, climate change, or other factors, the community structure can change dramatically.

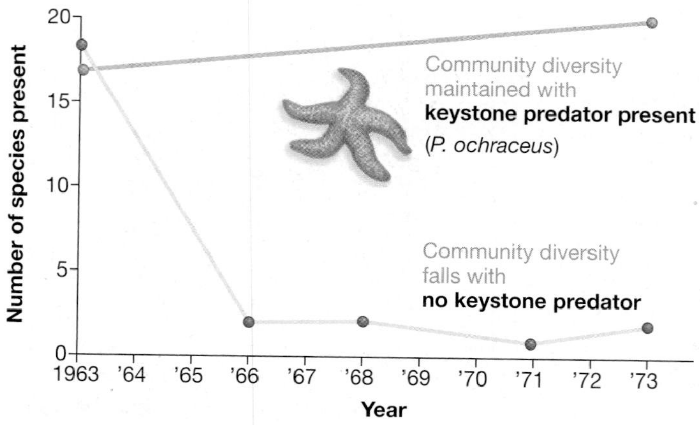

Figure 52.14 Keystone Predation Alters Community Structure in a Rocky Intertidal Habitat. Each data point represents the average number of species present in several study plots.

DATA: R. T. Paine. 1966. *American Naturalist* 100: 65–75.

Top-Down Influences on Community Structure As you have learned, the presence of certain consumers like *P. ochraceous* can also significantly change the number of species present in a community and the abundance of each. When a consumer limits a prey population in this way, biologists say that *top-down control* is occurring.

It is rarely practical to remove a top predator from a community as Paine was able to do with sea stars, but sometimes events present an opportunity for a natural experiment. That's exactly what happened in the Greater Yellowstone ecosystem of western North America, where wolves were hunted out in the early 1900s but reintroduced in 1995.

What effect did the absence of wolves have on the Greater Yellowstone communities? During the 1990s, ecologists William Ripple and Robert Beschta had been trying to determine why aspen trees had been declining for most of the last century. Then they made an important connection—the number of aspen trees started to decline at the same time that the wolves had disappeared. Ripple and Beschta hypothesized that there was an *indirect* link between aspen trees and wolves.

How could they test their hypothesis? One way was to see if the number of aspen trees would begin to increase as the reintroduced wolf population increased. Although the Greater Yellowstone wolf population has grown to only about 400 individuals since reintroduction, their presence has already led to an increase in aspen—and to many other changes in the food web (**Figure 52.15**).

- In Yellowstone, wolves primarily feed on elk. As elk numbers declined and individuals changed their feeding habits due to fear of being in the open, populations of favorite elk foods such as aspen, cottonwood, and willow have increased.
- The changes in willow and other plant species triggered an increase in beavers, which compete with elk for these plants.
- Beavers dam streams, forming ponds that provide habitat for frogs, turtles, and an array of other species, whose numbers have increased.

When changes in top-down control cause conspicuous effects two or three links away in a food web, biologists say that

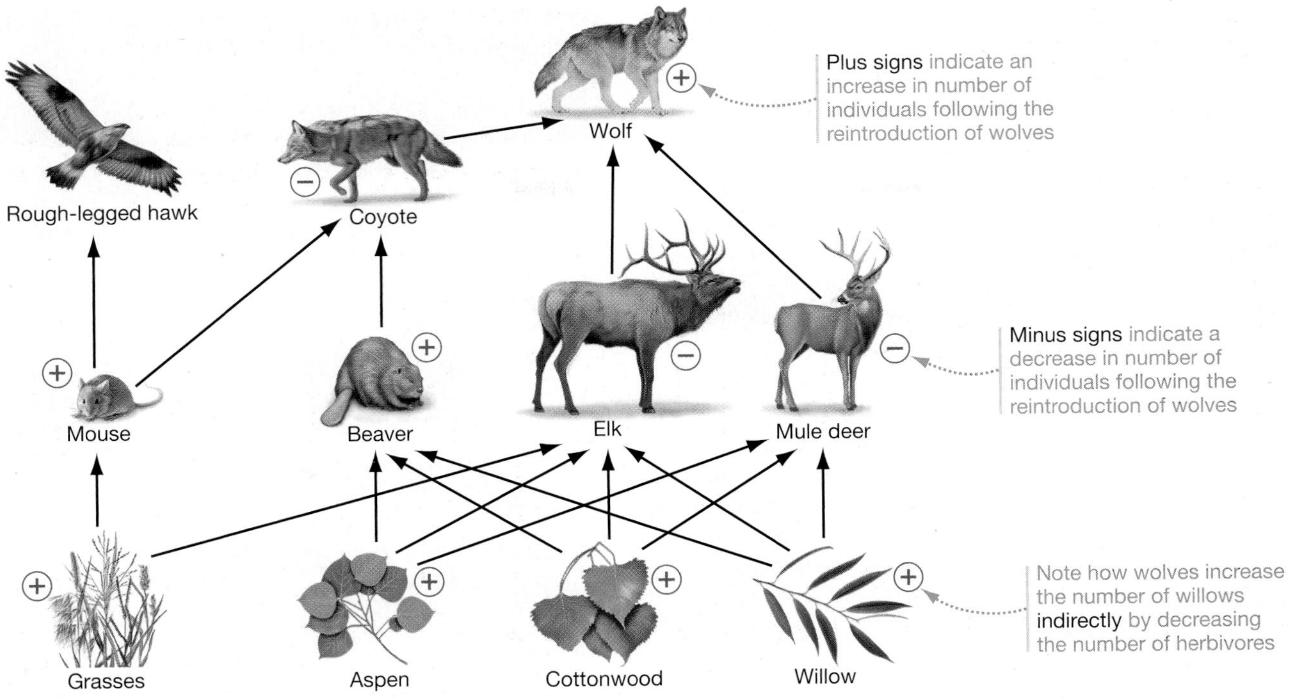

Figure 52.15 The Reintroduction of Wolves to Yellowstone Caused a Trophic Cascade. Wolves directly affect the species they eat, but they also indirectly affect other species in the food web. Only a few of the hundreds of species affected by the reintroduction of wolves are shown.

✔ **MODEL** Use the model to predict how the reintroduction of wolves affected the number of hawks. Draw a (+) or (–) to show your prediction.

a **trophic cascade** has occurred. Trophic ("feeding") cascades have been observed in many communities. The overfishing of sharks, for example, has caused a trophic cascade in some coral reefs; similarly, the overhunting of sea otters has triggered a trophic cascade in some kelp forests.

The message from these studies is clear: The removal of top predators from communities can have widespread, and sometimes unexpected, effects on the abundance of many species. This effect is increasingly important in conservation biology as many top predators are disappearing due to fishing, hunting, and habitat destruction.

When top predators are in their place, however, and autotrophs are in abundance, is the structure of communities generally predictable or unpredictable?

How Predictable Are Communities?

Do biological communities have a tightly prescribed organization and composition, or are they merely loose assemblages of species?

The Clements–Gleason Dichotomy Beginning with a paper published in 1936, Frederick Clements hypothesized that biological communities are stable and orderly entities with a highly predictable composition dependent on climate. His reasoning was that species interactions are so extensive—and coevolution so important—that communities have become highly integrated and interdependent units in nature. Stated another way, the species within a community cannot live without each other.

In developing this hypothesis, Clements likened the development of a plant community to the development of an organism. He argued that communities develop over time by passing through a series of predictable stages dictated by extensive interactions among species. Eventually, this developmental progression culminates in a final stage, known as a **climax community**. The climax community is stable—it does not change over time.

Henry Gleason, in contrast, contended that the community found in a particular area is neither stable nor predictable. He claimed that plant and animal communities are ephemeral associations of species that just happen to occur together due to chance historical events—such as the random dispersal of seeds to an area following a disturbance.

Which viewpoint is more accurate? Let's consider experimental and observational data.

An Experimental Test To explore the predictability of community structure experimentally, ecologists David Jenkins and Arthur Buikema constructed 12 identical ponds. They filled the ponds at the same time with water that contained enough chlorine to kill any preexisting organisms. (The setup is shown in Figure 52.16 on page 1132.) The Clements and Gleason hypotheses make clear and contrasting predictions:

- If community structure is predictable, then, once the chlorine has vaporized and made the water habitable, each pond should develop the same community of species through colonization (species dispersed via wind, waterfowl, and other means).
- If community structure is unpredictable, then each pond should develop a different community.

To test these predictions, the researchers sampled water from the ponds repeatedly for one year. They measured temperature, chemical makeup, and other physical characteristics of the water, and identified the species present in each pond by examining the samples of plankton under the microscope.

Their results? The rows of colored bars on the graph in Figure 52.16 indicate the species identified in each of the 12 ponds. Note that the researchers measured a total species richness of 61 planktonic species in all of the ponds. However, individual ponds had only 31 to 39 species apiece. Several species occurred in most or all 12 of the ponds, but each pond had a unique species assemblage. Thus, both predictions were partially supported. This type of result is common in biology—while it is desirable to receive a clear yes or no answer to a question, the real world is often more complex than conceptual models, with data falling along a gradient between two hypothetical extremes.

Mapping Current and Past Species' Distributions If communities were predictable assemblages, then the ranges of species that make up a particular community would be congruent—meaning that the same group of species would almost always be found living together. Is this what researchers have observed?

Consider the forests of North America. When biologists began documenting the ranges of tree species, they found that the boundaries of individual species ranges did not always coincide. For example, oaks and hickories often grow together in oak–hickory forests. Yet with increasing elevation up a mountain, hickories tend to drop out of the species mix as white pines start to appear, creating a gradient from oak–hickory to oak–pine.

Data on the historical composition of plant communities support these observations. Species do not come and go in the fossil record in tightly integrated units. Instead, the ranges of individual species tend to change independently of one another.

Overall, Clements's position on predictability was too extreme and Gleason's view of the importance of randomness may be closer to being correct. Although both biotic interactions and climate are important in determining which species exist at a certain site, chance and history also play a large role.

CHECK YOUR UNDERSTANDING

✔ If you understood this section, you should be able to ...

1. **SOCIETY** Identify two variables that you could measure to compare the community structure of your own intestinal microbiome (microbial community) after following different diets.
2. Explain why the overfishing of sharks could be responsible for the growth of algae that now threaten some coral reefs, even though sharks don't eat algae.

Answers are available in Appendix A.

RESEARCH

QUESTION: Are communities predictable or unpredictable?

COMMUNITIES-ARE-PREDICTABLE HYPOTHESIS: The group of species present at a particular site is highly predictable.

COMMUNITIES-ARE-UNPREDICTABLE HYPOTHESIS: The group of species present at a particular site is highly unpredictable.

EXPERIMENTAL SETUP:

1. Construct 12 identical ponds. Fill at the same time and sterilize water so that there are no preexisting organisms.

2. Examine water samples from each pond. Identify each plankton species present in each sample.

COMMUNITIES-ARE-PREDICTABLE PREDICTION: Identical plankton communities will develop in all 12 ponds.

COMMUNITIES-ARE-UNPREDICTABLE PREDICTION: Different plankton communities will develop in different ponds.

RESULTS (AFTER 1 YEAR):

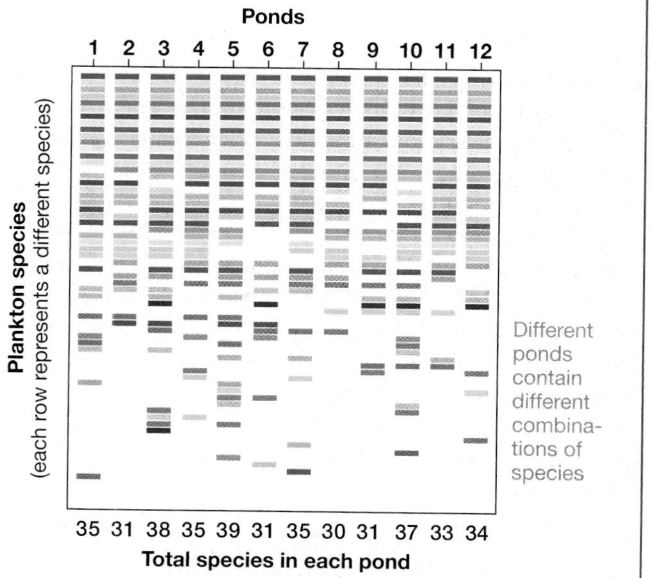

CONCLUSION: Although about half of the species present appear in all or most ponds, each pond has a unique composition. Both hypotheses are partially correct.

Figure 52.16 Experimental Evidence Indicates that Identical Communities Do Not Develop in Identical Habitats.

SOURCE: D. G. Jenkins and A. L. Buikema. 1998. Do similar communities develop in similar sites? A test with zooplankton structure and function. *Ecological Monographs* 68: 421–443.

✔ QUANTITATIVE What percentage of the 61 species were present in all of the ponds? Does this measure suggest that the community composition is more predictable or unpredictable?

52.3 Community Dynamics

Like cells, individuals, and species, communities can be described as dynamic. Communities can change slowly during gradual geologic events, such as deglaciation 11,000 years ago, or more rapidly due to sudden events, such as fires.

After you complete this section, you should be able to ...

- Describe the factors that determine community dynamics.

Disturbance and Change in Ecological Communities

Community composition and structure may change radically in response to changes in abiotic and biotic conditions. Biologists have become particularly interested in how communities respond to disturbance. A **disturbance** is any strong, short-lived disruption to a community that changes the distribution of living and/or nonliving resources.

Forest fires, windstorms, floods, the fall of a large-canopy tree, disease epidemics, and short-term eruptions in herbivore numbers all qualify as disturbances. These events are important because they alter light levels, nutrients, unoccupied space, or some other aspect of resource availability.

Biologists have come to realize that the impact of disturbance is a function of three factors: **(1)** the type of disturbance, **(2)** its frequency, and **(3)** its severity—for example, the speed and duration of a flood, or the intensity of heat during a fire. Biologists also now realize that most communities experience a characteristic type of disturbance. In most cases, disturbances occur with a predictable frequency and severity.

To capture this point, biologists refer to a community's **disturbance regime**. For example, fires kill all or most of the existing trees in a boreal forest every 100 to 300 years, on average. In contrast, many small-scale tree falls, usually caused by windstorms, occur in temperate and tropical forests every few years.

To appreciate why biologists are so interested in understanding disturbance regimes, consider a recent study on the fire history of giant sequoias in California (Figure 52.17a). Giant sequoias grow in small, isolated groves on the western side of the Sierra Nevada mountain range. Individuals live more than a thousand years, and many have been scarred repeatedly by fires.

Thomas Swetnam, a specialist on tree-ring research, obtained samples of cross sections through the bases of 90 giant sequoias in five different groves. As Figure 52.17b shows, the cross sections contained many growth rings that had been scarred by fire. To determine the date of each disturbance, he counted tree rings back from the present.

The graph in Figure 52.17c plots the average number of fires that occurred each century in one of the groves, over the past 2000 years. Swetnam found that in most of the groves, 20–40 fires had occurred each century. The data indicated that each tree had been burned an average of 64 times.

This study established that fires had been extremely frequent in these forests for most of recorded history. Because not enough time would pass for large amounts of fuel to accumulate between fires, the fires were probably of low severity. However, recent human intervention has reduced the frequency of forest fires—with unintended and unfortunate consequences. When fires do occur, the buildup of fuel causes massive conflagrations that are often devastating to the forest community.

Partly because of this work, the biologists responsible for managing sequoia groves now set controlled fires or let low-intensity natural fires burn instead of suppressing them immediately. To maintain communities in good condition, biologists have to ensure that the normal disturbance regime occurs. Otherwise, community composition changes dramatically.

Similarly, studies of disturbance regimes along the Colorado River in southwestern North America inspired land managers to begin releasing at intervals a huge pulse of water from the

Figure 52.17 **A Fire-Prone Community Shows a History of Disturbance.** Because trees form one ring (light band/dark band) every year, researchers can count the rings to determine how often fires have occurred during the last 2000 years in giant sequoia groves.

DATA: T. W. Swetnam. 1993. *Science* 262: 885–889.

reservoirs behind dams. The floods that have resulted were designed to mimic the natural disturbance regime. According to follow-up studies, the artificial floods appear to have benefited the plant and animal communities downstream by depositing nutrients and creating additional sandbar habitat.

Succession: The Development of Communities after Disturbance

During a disturbance, a community may have low or high *resistance* to the disturbance, meaning that it experiences a large or small amount of change. Following a disturbance, a community may have low or high *resilience*, meaning that it recovers slowly or quickly (see Ch. 54, Figure 54.12). In general, experimental evidence suggests that communities with higher species richness tend to have both higher resistance and higher resilience (Ch. 54, Section 54.3).

But what if a severe disturbance removes all or most of the organisms from a terrestrial or aquatic area? The recovery that follows is called **succession**. The name was inspired by the observation that species with certain types of life-history patterns tend to succeed others over time.

- **Primary succession** occurs when a disturbance removes the soil and its organisms as well as organisms that live above the soil surface. Glaciers, floods, volcanic eruptions, and landslides often initiate primary succession.
- **Secondary succession** occurs when a disturbance removes some or all of the organisms from an area but leaves the soil intact, including the seeds and microorganisms within. Fire and logging are examples of disturbances that initiate secondary succession on land. Hurricane storm surge and dead zones caused by nutrient pollution are examples of disturbances that initiate secondary succession in aquatic environments.

How does recovery proceed after a severe disturbance? Imagine a landscape before primary succession begins, such as the area surrounding the Kilauea volcano in Hawaii. There are no animals, no plants, and no soil. The process of soil building begins with solid rock, which weathers over time to produce gravel, sand, silt, and clay (Ch. 36, Section 36.2). As organisms occupy the substrate, they add dead cells and tissues and feces. With time, soil eventually becomes a complex and dynamic mixture of inorganic particles, organic particles, and living organisms that enable plants to take root and add structure to the landscape.

During secondary succession, soil is already present, so the process is much more rapid than primary succession. Figure 52.18 shows a typical sequence of terrestrial plant communities that develop as succession proceeds. In both terrestrial and aquatic environments, early successional communities are dominated by short-lived species that are small in stature and disperse their seeds, larvae, or other dispersal agents over long distances. Late successional or climax communities are dominated by species that tend to be long lived, large, and good competitors for resources such as light and nutrients (see life-history trade-offs in Ch. 51, Section 51.2).

The particular sequence of species that appears over time is called a successional pathway. What determines the pattern and rate of species replacement during succession at a certain time and place? To answer this question, biologists focus on three factors:

Figure 52.18 Secondary Succession in Temperate Forests. Succession leads to the development of a temperate forest from a disturbed state (in this case, an abandoned agricultural field).

1. the particular traits of the species involved;
2. how the species interact; and
3. historical and environmental circumstances, such as the size of the area involved and weather conditions.

Let's consider each factor in turn, focusing on terrestrial succession.

The Role of Species Traits Species traits, such as dispersal capability and the ability to withstand extreme dryness, are particularly important early in succession. As common sense would predict, recently disturbed sites tend to be colonized by plants and animals with good dispersal ability. When these organisms arrive, however, they often have to endure harsh environmental conditions.

Pioneering species tend to have "weedy" life histories. In ecology, a **weed** is a plant that is adapted for growth in disturbed soils. Because soil is often disturbed when biomass is removed, weeds tend to thrive at the start of secondary succession.

Early successional species devote most of their energy to reproduction and little to competitive ability. They are at the "high fecundity, low survivorship" end of the life-history continuum (see Ch. 51, Figure 51.5). More specifically, pioneering species have small seeds, rapid growth, and a short life span, and they begin reproducing at an early age. As a result, they have a high reproductive rate (high r_{max}—see Ch. 51, Section 51.3). But in addition, they can tolerate the severe abiotic conditions common in the absence of other organisms, such as high light levels, extreme temperatures, poor nutrient availability, and desiccation.

The Role of Species Interactions Once colonization is under way, the course of succession tends to depend less on how species cope with aspects of the abiotic environment and more on how they interact with other species. This change occurs because plants that grow early in succession change abiotic conditions in a way that makes the conditions less severe.

For example, because plants provide shade, they reduce temperatures and increase humidity. Their dead bodies also add organic material and nutrients to the soil. During succession, existing species can have one of three effects on subsequent species:

- **Facilitation** takes place when the presence of an early-arriving species makes conditions more favorable for the arrival of certain later species, such as by providing shade or nutrients. But the facilitation is not purposeful or altruistic; it is just a by-product of the way of life of the early-arriving species.
- **Tolerance** means that existing species do not affect the probability that subsequent species will become established. Once again, tolerance is not purposeful; it is just an absence of influence of early-arriving species on late-arriving species.
- **Inhibition** occurs when the presence of one species inhibits the establishment or regrowth of another. For example, a plant species that requires high light levels to germinate may be inhibited late in succession by the presence of mature trees that prevent sunlight from reaching the forest floor. Alternately, an established species may produce a chemical that inhibits the growth of other species.

The Role of Chance and History Besides species traits and species interactions, the pattern and rate of succession depend on historical and environmental context. For example, the communities that developed after forest fires disturbed Yellowstone National Park in 1988 depended on the size of the burned patch and how hot the fire had been at that location.

Succession is also affected by the particular weather or climate conditions that occur during the process. Variation in weather and climate causes different successional pathways to occur in the same place at different times.

Analyzing species traits, species interactions, and the historical–environmental context provides a useful structure for understanding why particular successional pathways occur. To see this theoretical framework in action, let's examine data on the course of succession that has occurred in Glacier Bay, Alaska.

A Case Study: Glacier Bay, Alaska An extraordinarily rapid and extensive glacial recession is taking place at Glacier Bay, mapped out in Figure 52.19 on page 1136. In just 200 years, glaciers that once filled the bay have retreated approximately 100 km, exposing tracts of barren glacial sediments to colonization. Because of this event, Glacier Bay has become an important site for studying succession.

Figure 52.19 shows the plant communities found in the area.

- Locations that have been ice free for 20 years or less do not have a continuous plant cover. Instead, they host scattered individuals of willow and small shrubs called mountain avens.
- Areas that have been deglaciated for about 100 years are inhabited by dense thickets of a shrub called Sitka alder as well as by scattered Sitka spruce trees.
- The oldest sites, ranging from 150 to 200 years of soil exposure, are dense forests of Sitka spruce and western hemlock.

These observations inspired a hypothesis for the pattern of succession in Glacier Bay: With time, the youngest communities of willow and mountain avens succeed to alder thickets, which subsequently become dense spruce–hemlock forests. The key claim is that the entire bay is following a single successional pathway.

Research has challenged this hypothesis, however. Biologists who reconstructed the history of each community by studying tree rings found that distinct successional pathways have occurred in different parts of the bay. These data raise an important question: How do species traits, species interactions, and dispersal patterns interact to generate the observed successional pathways?

- ***Species traits*** Western hemlock is abundant at older sites but largely absent from young ones. This pattern is logical for this species, because its seeds germinate and grow only in soils containing a substantial amount of organic matter. In addition, young hemlock trees can tolerate deep shade but not bright sunlight. Thus Western hemlock does not tolerate early successional conditions, but thrives later in succession.
- ***Species interactions*** Sitka alder facilitates the growth of Sitka spruce. The facilitating effect occurs because symbiotic bacteria that live inside nodules on the roots of alder convert atmospheric nitrogen (N_2) to nitrogen-containing molecules that alder use to build proteins and nucleic acids. When alder

Figure 52.19 Successional Stages Can Be Seen in Glacier Bay.

leaves fall and decay or roots die, the nitrogen becomes available to spruce. Although spruce trees are capable of invading and growing without the presence of alder, they grow faster when alder stands have added nitrogen to the soil. Note that spruce gains from the association. Alder is either unaffected or negatively affected if it becomes shaded by spruce.

Competition is another important species interaction. For example, shading by alder trees reduces the growth of spruce trees until spruce trees are tall enough to protrude above the alder thicket. Once the spruce trees breach the alder canopy, however, alder trees die out, unable to compete with spruce trees for light.

- ***Historical and environmental context*** According to geological evidence, glacial ice was more than 1100 m thick in the upper part of Glacier Bay during the mid-1700s. Because forests grow to an elevation of only 700 m or 800 m in this part of Alaska, the glacier had eliminated all of the existing forests. But in the lower part of the bay, the ice was substantially thinner. As a result, some forests remained on the mountain slopes beyond the ice, even at the height of glaciation.

  The difference is crucial. In the lower part of the bay, nearby forests provided a ready source of Sitka spruce and western hemlock seed. But in the upper part of the bay, no forests remained to contribute seed. As a result, dramatically different successional pathways were set in motion.

To summarize, successional pathways are determined by the adaptations that certain species have to their abiotic environment, interactions among species, and the history of the site. Species traits and species interactions tend to make succession predictable; history and chance events contribute a degree of unpredictability to succession.

CHECK YOUR UNDERSTANDING

✔ If you understood this section, you should be able to . . .

1. Explain how early successional species alter the environment in ways that make growing conditions more difficult for themselves.
2. Predict one life-history trait that is likely to be common among bacteria in the early succession following this disturbance: Antibiotics wipe out the bacterial community in your gut (including not only parasitic species, but mutualistic and commensal species).

Answers are available in Appendix A.

52.4 Geographic Patterns in Species Richness

After you complete this section, you should be able to . . .

- Explain geographic patterns in species richness.

Why are some communities more species rich than others? Ecologists have observed a general correlation between species richness and two abiotic variables: the geographic area occupied by the community, and the latitude of the community. Let's take a closer look at each pattern in turn.

Predicting Species Richness: The Theory of Island Biogeography

When researchers first began counting how many species are present in various communities, a strong pattern emerged: Larger patches of habitat contain more species than do smaller patches of habitat. The observation is logical because large areas should contain more niches and thus support higher numbers of species.

Early work on species richness highlighted another pattern, however—one that was harder to explain. Islands in the ocean have smaller numbers of species than do areas of the same size on continents. Why?

As it happens, answering this question has had far-reaching implications in ecology and conservation biology. This situation is quite common in biological science: Research on a seemingly esoteric question—often done for the sheer joy of discovery—turns out to have important implications for unrelated issues, or applications to practical problems that once were intractable.

The Role of Immigration and Extinction Robert MacArthur and Edward O. Wilson tackled the question of why islands in the ocean have smaller numbers of species than do areas of the same size on continents (see the species-area plot in Ch. 54, Figure 54.9). They assumed that speciation occurs so slowly that the number of species on an island is a product of just two processes, immigration and extinction, which should vary with the number of species already present.

- Immigration rates should decline as the number of species on the island increases, because individuals that arrive are more likely to be members of a species that is already there. In addition, competition should prevent new species from becoming established when many species are already present on an island—meaning that immigration should seldom result in establishment of a new population.
- Extinction rates should increase as species richness increases, because niche overlap and competition for resources will be more intense.

The curves in Figure 52.20a illustrate these two predictions. Note that when the two processes interact, the result is an equilibrium—a balance between the arrival of new species and the extinction of existing ones. If species richness changes due to a hurricane or fire or some other disturbance, then continued immigration and extinction should restore the equilibrium value.

The Role of Island Size and Isolation MacArthur and Wilson also realized that immigration and extinction rates should vary as a function of island size and how far the island is from a continent or other source of immigrants.

- Immigration rates should be higher on large islands that are close to mainlands, because immigrants are more likely to find large islands that are close to shore than small ones that are far from shore.
- Extinction rates should be highest on small islands that are far from shore, because fewer resources are available to support large populations, and because fewer individuals arrive to keep the population going.

Their theory makes two predictions: Species richness should be higher on **(1)** larger islands than smaller islands, and **(2)** nearshore islands versus remote islands (Figure 52.20b).

Applying the Theory The MacArthur–Wilson model, called the theory of island biogeography, is important for several reasons:

- It is relevant to a wide variety of island-like habitats such as mountaintops, lakes and ponds, and caves.
- It made specific predictions that could be tested in experimental and observational studies.
- It is relevant to species that have a metapopulation structure, such as Glanville fritillary butterflies (Ch. 51, Section 51.4) and an increasing number of species that have experienced habitat fragmentation.

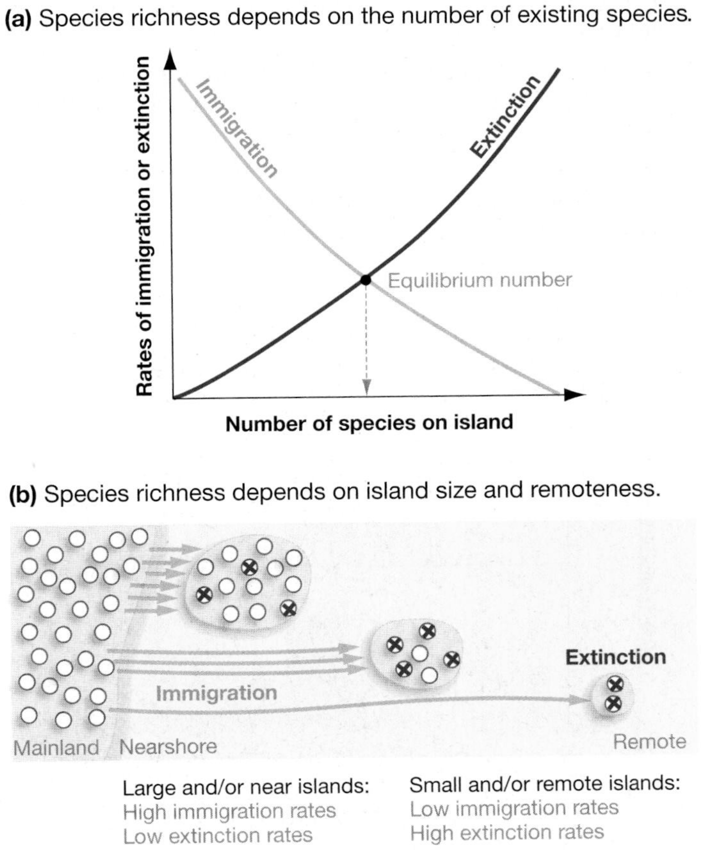

Figure 52.20 Theoretical Models Predict Species Richness as a Function of Island Characteristics.

✔ **MODEL** Suppose the mainland habitats closest to an island were wiped out by suburbanization. Sketch a new blue line in part (a) that shows lower immigration rates, then predict how the equilibrium number of species on the island will be affected.

- It can help inform decisions about the design of natural preserves that are surrounded by human development. In general, the most species-rich reserves should be ones that are **(1)** relatively large, and **(2)** located close to other relatively large habitat areas.

Global Patterns in Species Richness

Biologists have long understood that large habitat areas tend to be species rich, and the theory of island biogeography has been successful in predicting how species richness should vary among island-like habitats. But researchers have had a much more difficult time explaining what may be the most famous geographic pattern in species richness in the history of biology: There are more species in the tropics than at higher latitudes.

In birds, mammals, fish, reptiles, many aquatic and terrestrial invertebrates, and trees, species diversity declines as latitude increases. A sample of the data for birds and mammals is shown in **Figure 52.21**. This pattern is widespread, although not universal—a number of marine groups, as well as shorebirds, show a positive relationship between latitude and diversity. Why does a latitudinal diversity gradient in communities occur?

Over 30 hypotheses have been proposed to explain the latitudinal diversity gradient. These hypotheses generally fall into four broad categories according to the type of variables that are referenced:

1. ***Abiotic: spatial hypotheses*** Earth's surface area is higher near the equator than near the poles; the larger the area, the more space and potential geographic diversity. Also, the tropical areas are the oldest, most stable, and least harsh (for example, have experienced less scouring by ice sheets) and thus have had the longest time to diversify.

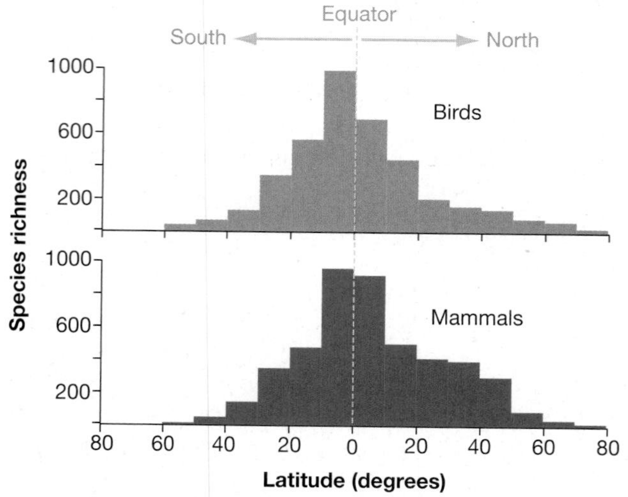

Figure 52.21 Species Richness of Birds and Mammals Have a Strong Latitudinal Gradient. The bird data include terrestrial birds from the Western hemisphere; the trend is similar in terrestrial birds from the Eastern hemisphere, but not in shorebirds.

DATA: J. Rolland et al. 2014. *PLOS Biology* 12: e1001775; D. A. Duchêne and M. Cardillo. 2015. *Global Ecology and Biogeography* 24: 1203–1362.

2. ***Abiotic: energy hypotheses*** The higher availability of sunlight and rain in the tropics (Ch. 49, Section 49.3) provides a higher net primary production—that is, a larger amount of energy and nutrients in the form of biomass that can support many organisms in a community.
3. ***Biotic: ecological interactions hypotheses*** A higher complexity of species interactions like competition, predation, parasitism, and mutualism occurs in the tropics, which leads to higher rates of niche differentiation. New niches beget more new niches.
4. ***Biotic: evolutionary history hypotheses*** The higher species diversity in the tropics is a product of relatively high speciation rates, low extinction rates, high immigration rates (colonization), and/or low emigration rates (dispersal).

How do biologists test these hypotheses? The most common approach is to look for correlations between each variable and species richness using different taxonomic groups (living today and/or extinct) in different latitudinal ranges (as they occur today and/or as they lived in the past). For example, phylogenic analysis of butterfly species alive today enabled researchers to discover that the speciation rate of butterflies has been higher in the tropics than in temperate areas.

What can you conclude from all this? While there is currently no consensus on which variables are the most important in determining the latitudinal diversity gradient in communities, one insight is very clear: Multiple abiotic and biotic factors interact in important ways that vary among taxonomic groups and in aquatic versus terrestrial environments. It's human nature to seek a simple answer to questions, but the reality is that the causes of species diversity gradients are fundamentally complex and require interdisciplinary systems thinking to understand.

In sum, evolution and ecology are closely intertwined. Species interactions in communities are an important driver of evolutionary change, which in turn affects species interactions in communities. What's more, individual species interactions combine to create a complex web of interactions that change over time and space. These ideas have many practical implications as climate change and other human impacts are causing changes to communities worldwide. In the next chapter, let's take a closer look at the role of communities in the flow of energy and the cycling of nutrients through ecosystems.

CHECK YOUR UNDERSTANDING

✔ **If you understood this section, you should be able to . . .**

Define species richness and predict which nature preserve is likely to have the highest species richness: One that is large or small? One that is near other preserves or remote? One that is near the tropics or at higher latitudes?

Answers are available in Appendix A.

52.1 Species Interactions

- Two interacting species may each experience a net fitness cost (–) or benefit (+) from the interaction. These costs and benefits depend on the conditions that prevail at a particular time and place and may change through time.
- Commensalism (+ / 0) occurs when one species benefits but the other is unaffected.
- Competition (– / –) occurs when the niches of two species overlap—meaning they use the same resources. Competition may result in the complete exclusion of one species (the weaker competitor). It may also result in niche differentiation, in which competing species evolve traits that allow them to exploit different resources.
- Consumption (+ / –) occurs when consumers eat part or all of other individuals, which resist through constitutive defenses or inducible defenses.
- Parasites are consumers that generally spend all or part of their life cycle in or on their host (or hosts). They usually have traits that allow them to escape host defenses and even manipulate host behavior to increase their likelihood of transmission to a new host. In turn, hosts have evolved counter-adaptations that help fight off parasites.
- Mutualism (+ / +) provides individuals with food, shelter, transport of pollen, defense against predators, or other benefits. For each species involved, the costs and benefits of mutualism may vary over time and from place to place.
- Interactions among species have two main outcomes: (1) They affect the distribution and abundance of the interacting species, and (2) they are agents of natural selection and thus affect the evolution of the interacting species, by a process called coevolution.

52.2 Community Structure

- The attributes of community structure include the number of species present (species richness), their relative abundance, the sum of their interactions, and the physical attributes of the community as determined by a combination of abiotic and biotic factors.
- Food webs comprise all the consumption relationships in a community. Food webs are structured based on trophic levels, with producers at the "bottom" and consumers at the "top."
- Keystone species are those that have a larger effect on their communities than their numbers would suggest.
- Bottom-up factors such as the primary productivity of producers can affect the structure of communities.
- Top-down factors such as consumption by keystone species and trophic cascades can affect the structure of communities.
- In some cases, the structure of an entire community can change dramatically if a single species is removed from it (as from overhunting or fishing) or added to it (such as an invasive species).
- Historical and experimental evidence supports the view that communities are dynamic rather than static, and that their composition is neither entirely predictable nor stable over time.
- The assemblage of species found in a community is primarily a function of climate and chance historical events.

52.3 Community Dynamics

- Each community has a characteristic disturbance regime—meaning a type, severity, and frequency of disturbance that it experiences.
- Three types of factors influence the pattern of succession that occurs after a disturbance: **(1)** A species' physiological traits influence when it can successfully join a community; **(2)** interactions among species influence if and when a species can become established; and **(3)** the historical and environmental context of the site affects which species are present.

52.4 Geographic Patterns in Species Richness

- Species richness is higher in large islands near continents than in small, isolated islands, due to differences in immigration and extinction.
- The theory of island biogeography is relevant to a wide variety of isolated habitats.
- Species richness is higher in the tropics and lower toward the poles due to the interaction of several abiotic variables (such as land area and sunlight availability) and biotic variables (such as species interactions and speciation rates).

Answers are available in Appendix A.

✔ TEST YOUR KNOWLEDGE

1. What is niche differentiation?
   a. the evolution of traits that reduce niche overlap and competition
   b. interactions that allow species to occupy their fundamental niche
   c. the degree to which the niches of two species overlap
   d. the claim that species with the same niche cannot coexist
2. The relationship between corals and the photosynthetic protists they harbor is an example of
   a. commensalism.
   b. competition.
   c. consumption.
   d. mutualism.
3. What is true about inducible defenses? Select True or False for each statement.
   T/F They are always present; thus, an individual is always able to defend itself.
   T/F They make it impossible for a consumer to launch surprise attacks.
   T/F They result from a coevolutionary arms race.
   T/F They make efficient use of resources, because they are produced only when needed.
4. Pioneer species tend to have high ________ and lower survivorship.

✔ TEST YOUR UNDERSTANDING

5. Cite an example to explain why species interactions are conditional and dynamic.
6. **THINK CAREFULLY** Evaluate this statement: Species want to increase their fitness by helping friendly species and hurting enemy species.
7. What is a disturbance? Consider the role of fire in a forest. Compare the consequences of high-frequency versus low-frequency fire, and high severity versus low severity of fire.
8. **MODEL** Draw a vertical food chain (part of a food web) showing that sea urchins eat kelp and that sea otters eat sea urchins. Use this model to (1) show or explain how a trophic cascade works; and (2) offer an explanation for why otters are considered a keystone species.

✔ TEST YOUR PROBLEM-SOLVING SKILLS

9. **SOCIETY** In some circumstances, the use of probiotics, which stimulate the rapid growth of bacteria that are mutualistic or commensal with humans, can eliminate the need to use antibiotics, which can wipe out helpful bacteria along with harmful bacteria. The use of probiotics is an example of which process?
   a. succession
   b. competitive exclusion
   c. parasitism
   d. niche differentiation
10. Suppose that a two-acre lawn on your college's campus is allowed to undergo succession. Describe how species traits, species interactions, and the site's history might affect the community that develops.

✔ PUT IT ALL TOGETHER: Case Study

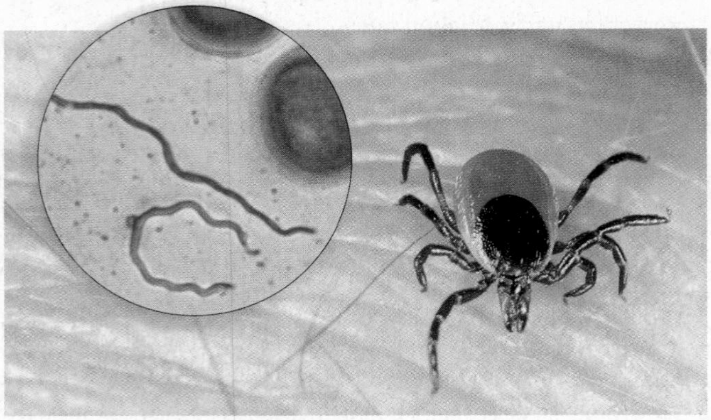

How is forest fragmentation linked to the rise of Lyme disease?

Lyme disease is caused by infections of the bacterium *Borrelia burgdorferi* (stained blue in inset, among red blood cells) which is transferred to humans via blood-sucking bites from the tick, *Ixodes scapularis*. Lyme disease can cause flu-like symptoms in the short term and more serious illnesses in the long term, if not treated. Why is Lyme disease on the rise in eastern North America?

11. In relation to humans, *B. burgdorferi* is a/an _____ and the tick is a/an _____.
    a. endoparasite/ectoparasite
    b. endoparasite/host
    c. parasitoid/host
    d. parasite/predator
12. **QUANTITATIVE** Newly hatched ticks do not harbor *B. burgdorferi*. Rather, they pick it up from certain hosts that are reservoirs of the bacterium. When a larval tick crawls onto a host, it may get groomed off and killed by the host, or it may feed successfully, in which case it may or may not then become infected with *B. burgdorferi*. Felicia Keesing and others measured the rate of tick feeding and infection on different hosts. What percentage of ticks on mice were infected with *B. burgdorferi*? What percentage of ticks on opossums were groomed off and killed?

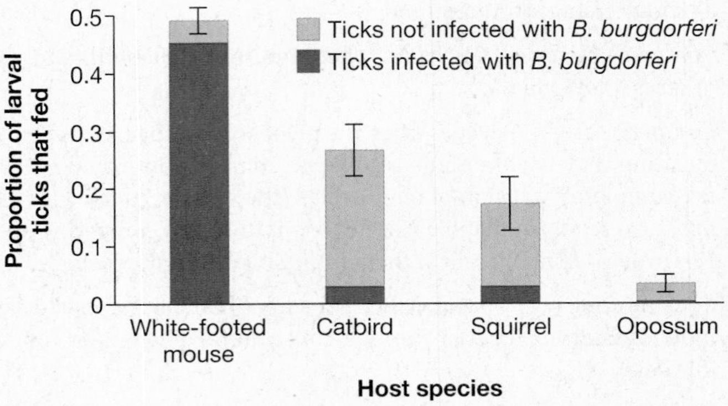

Source: F. Keesing, J. Brunner, S. Duerr, et al. 2009. *Proceedings of the Royal Society B* 276: 3911–3919.

13. Deer are hosts of ticks but are not reservoirs of *B. burgdorferi*—their immune systems detect and kill the bacterium. But a tick with a prior bacterial infection can consume a blood meal from a deer and then bite and infect a human. The fitness effects of a deer on *B. burgdorferi* can be summarized as:
    a. + because the deer supplies the bacterium with food.
    b. − because the deer kills the bacteria in its blood.
    c. + because the deer feeds and disperses the tick that can then infect a human with the bacterium.
    d. − or + depending on whether the deer is the tick's last host.
14. Researchers have measured a positive correlation between forest fragmentation and incidence of Lyme disease. According to the theory of island biogeography, how do you predict fragmentation will affect species richness?
15. White-footed mice occupy a broad ecological niche—they occur in most communities regardless of habitat quality. Many other species, like the opossum, are absent from low-quality forest fragments. Based on this information and the data in Question 12, propose a hypothesis to explain the observed relationship between increased forest fragmentation and increased incidence of Lyme disease.
16. **SOCIETY** A letter to the editor in a local newspaper asserts that "we cannot afford to preserve natural forests, because we need to use these resources to help humans." How could you respond to this comment in terms of Lyme disease?

Mastering Biology

Students Go to Mastering™ Biology for assignments, the eText, and the Study Area with animations, practice tests, and activities.

Professors Go to Mastering™ Biology for automatically graded tutorials and questions that you can assign to your students, plus Instructor Resources.

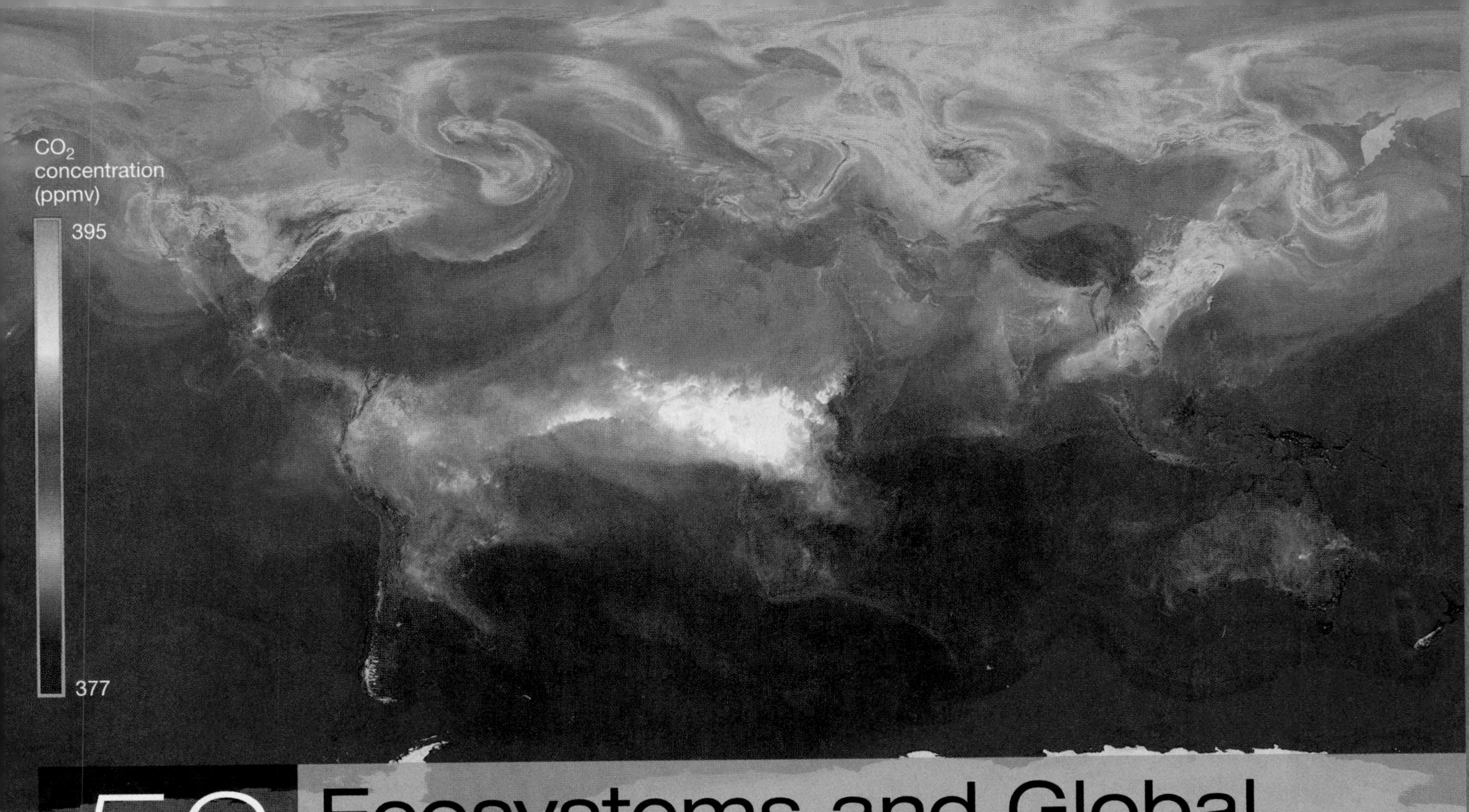

53 Ecosystems and Global Ecology

This NASA computer model shows the global circulation of carbon dioxide, a gas that is essential to photosynthesis, yet is also one of the greenhouse gases causing climate change.

In this chapter you will learn how

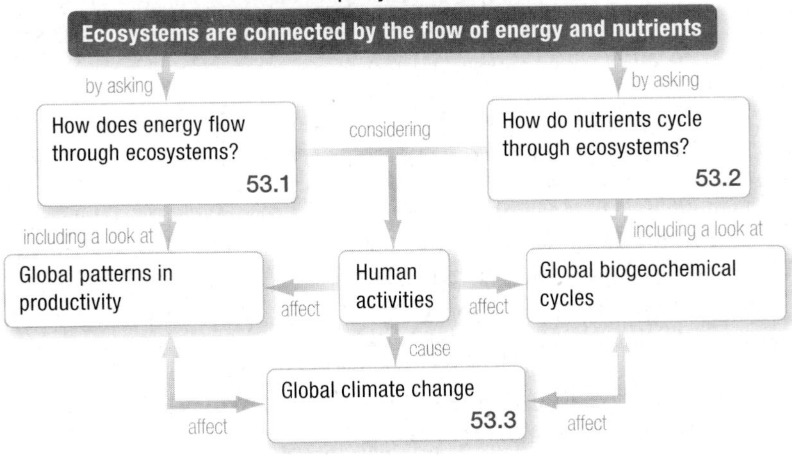

In this chapter you will find a simple, but important, message: In nature, everything is connected. This is because the components of an **ecosystem**—the community of interacting species present in a region, along with the abiotic components of the soil, water, and atmosphere—are linked by the flow of energy and the cycling of nutrients needed to sustain life. Further, ecosystems are themselves linked by the global exchange of energy and nutrients throughout the **biosphere**, the thin zone of soil, water, and atmosphere surrounding Earth where all life exists.

There once was more of a divide between studies of natural ecosystems and studies of human impacts. But because humans are removing species, introducing new species, changing landscapes, polluting ecosystems, and changing the climate, ecologists must increasingly incorporate human impacts into their work. Truly "natural" ecosystems (free of human influence) no longer exist, not even in remote places like the Antarctic.

BIG PICTURE

This chapter is part of the Big Picture. See how on pages 1190–1191.

The biological, chemical, and geological consequences of human impacts have been so dramatic that scientists have proposed naming a new epoch in the geological timescale—the Anthropocene ("new human epoch"; Ch. 25, Section 25.2). This epoch is marked by a decline in biodiversity as well as a decline in the ecosystem services that humans depend on for survival (explored in detail in Chapter 54).

The good news is that an unprecedented level of collaboration is now going on among scientists around the world, and they have increasingly powerful tools to measure and model the mind-boggling complexity of ecosystems. The work of ecologists is essential to the decision-making process in human societies that will determine the quality of life on Earth for centuries to come.

Let's explore large-scale ecology by considering how energy flows through ecosystems, how nutrients cycle, and how both processes are central to understanding global climate change. You can see how ecosystems fit into the Big Picture of Ecology on pages 1190–1191.

53.1 How Does Energy Flow through Ecosystems?

One of the defining characteristics of life is the use of energy (Ch. 1, Section 1.1). To stay alive and reproduce, organisms must acquire and use energy.

After you complete this section, you should be able to ...

- Describe the flow of energy through ecosystems.

Energy enters ecosystems via **primary producers**, also called **autotrophs** (literally, "self-feeders")—organisms that can synthesize their own food from inorganic sources. In most ecosystems, primary producers use solar energy to manufacture food via photosynthesis (Chapter 10). But in deep-sea hydrothermal vents and iron-rich rocks deep below Earth's surface, primary producers use the chemical energy contained in inorganic compounds such as methane (CH_4) or hydrogen sulfide (H_2S) to make food (Ch. 26, Section 26.3).

Do primary producers *create* energy? No. But they *transform* the energy in sunlight or inorganic compounds into the chemical energy stored in sugars and other organic (carbon-containing) molecules. Recall that the first law of thermodynamics states that energy cannot be created or destroyed, but only transferred and transformed (Ch. 8, Section 8.1).

The total amount of chemical energy produced in a given area and time period is called **gross primary productivity (GPP)**. The primary producers use this chemical energy in two ways:

1. ***Cellular respiration*** It takes energy just to stay alive. Primary producers have to use chemical energy to fuel cellular respiration (or fermentation) to produce the energy currency, ATP (Chapter 9). ATP, in turn, is used to fuel a diverse array of metabolic processes.
2. ***Growth and reproduction*** Chemical energy that isn't used for cellular respiration can be applied to growth and reproduction. Energy that is invested by primary producers in building new tissue or offspring is called **net primary productivity (NPP)**.

Because cellular respiration, growth, and reproduction are not perfectly efficient (second law of thermodynamics), some energy is "lost" (released) from primary producers during these processes as molecular waste and heat without being used. Linking these concepts together:

$$\mathbf{NPP = GPP - R}$$

(where R = energy either used in cellular respiration or "lost")

NPP represents the total amount of chemical energy that is stored in organic material, or **biomass**. NPP is critically important because it is the amount of energy available via primary producers to the other organisms of an ecosystem, like you.

How Efficient Are Autotrophs at Capturing Solar Energy?

How much of the energy that enters an ecosystem is captured as NPP? To answer this question, let's look at data from the world's most intensively studied ecosystem. Since the early 1960s, ecologists Gene Likens, Herbert Bormann, and colleagues have been studying how energy and nutrients flow through a temperate forest ecosystem in the northeastern United States: the Hubbard Brook Experimental Forest in New Hampshire.

During one year, from June 1, 1969, to May 31, 1970, 1,254,000 kcal of solar radiation per square meter (m^2) entered this ecosystem in the form of sunlight. If this amount of energy were available in the form of electricity, it would easily power two 150-watt lightbulbs per square meter of land, burning continuously for one year.

By documenting rates of photosynthesis in a large sample of forest plants, Likens and Bormann calculated that the plants used 10,400 kcal/m^2 of this energy in photosynthesis—a mere 0.8 percent of the incoming sunlight (**Figure 53.1**).

Why so inefficient? For comparison, the best solar panels produced today have an efficiency of about 22 percent in normal outdoor conditions. The answer includes several aspects of photosynthesis (Chapter 10).

- Even when conditions are ideal, the pigments that drive photosynthesis can absorb only a fraction of the light wavelengths available and thus a fraction of the total energy received.
- Plants in temperate biomes have drastically reduced photosynthetic rates in winter due to the reduction in sunlight.
- The efficiency of enzymes used in photosynthesis is very temperature dependent.
- If conditions get dry during the summer, stomata close to conserve water. Photosynthesis stalls due to a lack of CO_2.

How much of the gross primary production goes to the production of new biomass? By measuring plant growth, Likens and Bormann estimated that only about 45 percent of GPP went to

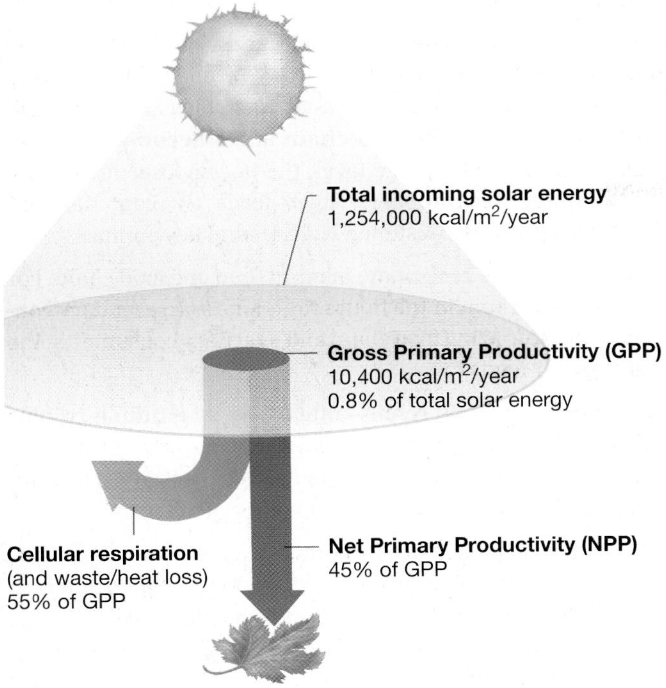

Figure 53.1 Ecologists Measure the Energy That Enters Ecosystems. These data are from the Hubbard Brook Experimental Forest.

DATA: J. R. Gosz et al. 1978. *Scientific American* 238: 93–102.

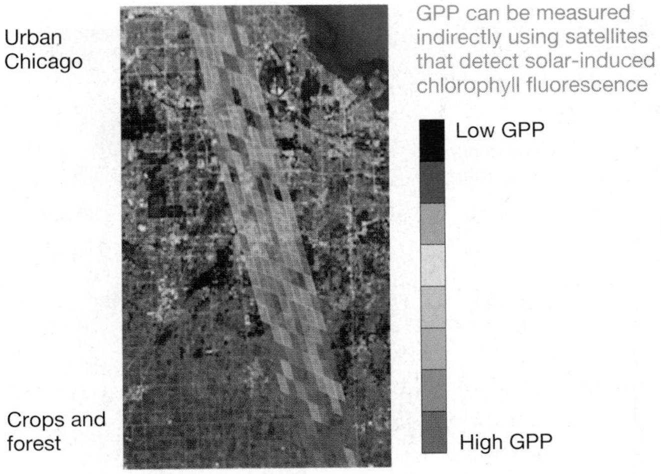

Figure 53.2 Satellites Can Measure GPP from Space. When chlorophyll absorbs sunlight during photosynthesis, a small amount of the energy is released as fluorescence, color coded green in this satellite scan through the Chicago area.

DATA: Y. Sun/NASA/JPL-Caltech.

NPP. The other 55 percent either was used for cellular respiration or lost from the organisms as heat.

Other biologists have done similar studies in other ecosystems using a variety of new tools to measure photosynthesis, such as satellite analysis of chlorophyll fluorescence (Figure 53.2). The general pattern is that only a tiny fraction of incoming sunlight is converted to chemical energy, and only a fraction of this gross primary productivity is used to build biomass.

What Happens to the Biomass of Autotrophs?

Graduate student Raymond Lindeman made a key insight about the fate of autotroph biomass in the 1940s, based on his research at Cedar Creek Bog in Minnesota. Lindeman proposed the *energy flow model* for ecosystems, whereby energy could flow from autotrophs through other organisms in the form of biomass.

- **Consumers** eat living organisms. **Primary consumers** eat primary producers; **secondary consumers** eat primary consumers; **tertiary consumers** eat secondary consumers; and so on.
- **Decomposers**, or **detritivores**, obtain energy by feeding on the remains of other organisms or waste products. Dead animals and dead plant tissues ("plant litter") are collectively known as **detritus**. Many fungi and bacteria are decomposers, but humans and other animals rarely act as decomposers.

Energy flows when one organism eats another. However, a key insight is that energy constantly dissipates as it flows through ecosystems and is ultimately "lost" as heat (meaning that it is no longer usable), whereas nutrients continue to cycle (Figure 53.3). Consider that all the *energy* stored in your tissues right now ultimately will be released into the atmosphere, but all the carbon, nitrogen, and other *nutrients* present in your tissues will persist in other organisms and the abiotic environment for millennia.

Fifteen years after Lindeman's model was published, Howard Odum carried the concept of energy flow to the next level by quantifying energy flow in an aquatic ecosystem. Odum categorized organisms according to their roles. Organisms that obtained energy from the same type of source were said to occupy the same **trophic** ("feeding") **level**.

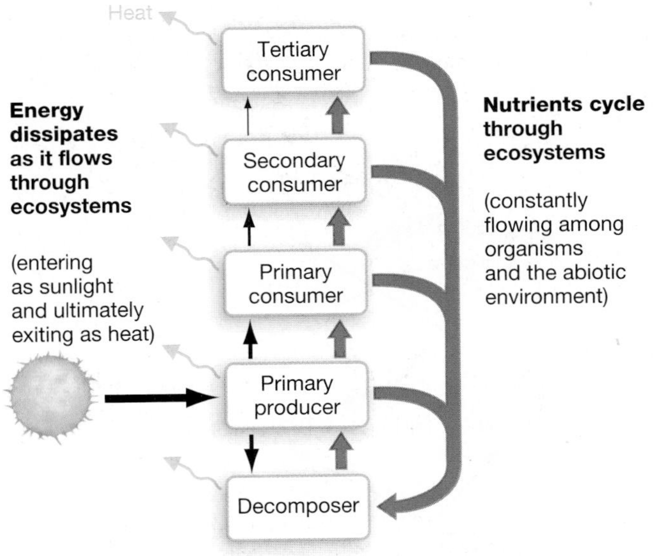

Figure 53.3 Energy Eventually Dissipates, but Nutrients Cycle.

Figure 53.4 Trophic Levels Identify Steps in Energy Transfer.

Trophic Structure: Food Chains and Food Webs Biologists often begin with a simple model and add complexity. A **food chain** focuses on one possible pathway of energy flow among trophic levels in an ecosystem. **Figure 53.4** shows two examples of food chains: a **grazing food chain** and a **decomposer food chain**. At the second trophic level, the decomposer food chain has a **primary decomposer**, which feeds on plant detritus, rather than a primary consumer. Note several key points:

- Some consumers participate in more than one food chain. For example, the robin in the figure functions as a secondary consumer in the grazing food chain and a tertiary consumer in the decomposer food chain.
- Consumers—such as robins—often feed at multiple trophic levels. You are a primary consumer when you eat a green salad, a secondary consumer when you eat beef, and a tertiary consumer when you eat tuna (a predator of other fish).

Because most consumers feed on a wide array of organisms and at multiple trophic levels, food chains are embedded in more complex **food webs**. The food web shown in **Figure 53.5** is a more detailed description of the trophic relationships among the organisms in Figure 53.4, as well as those among other organisms in the ecosystem. A complete food web would include *all* the organisms interacting in this ecosystem, along with estimates of the amount of energy transferred at each link. Food webs are a way to summarize energy flows and to document the complex trophic interactions that occur in ecosystems. For tips on drawing food webs, see **Making Models 52.3** on page 1128.

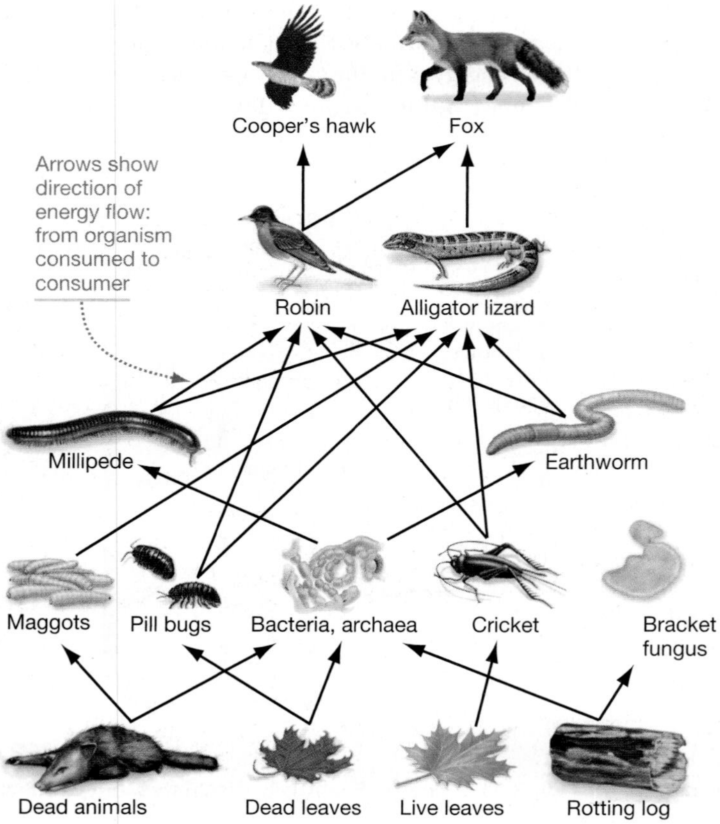

Figure 53.5 Food Webs Describe Trophic Relationships in an Ecosystem. Food webs offer a more in-depth analysis of feeding relationships than food chains do (see Figure 53.4). This food web shows only a few of the feeding relationships and species of a temperate deciduous forest. Organisms are not drawn to scale.

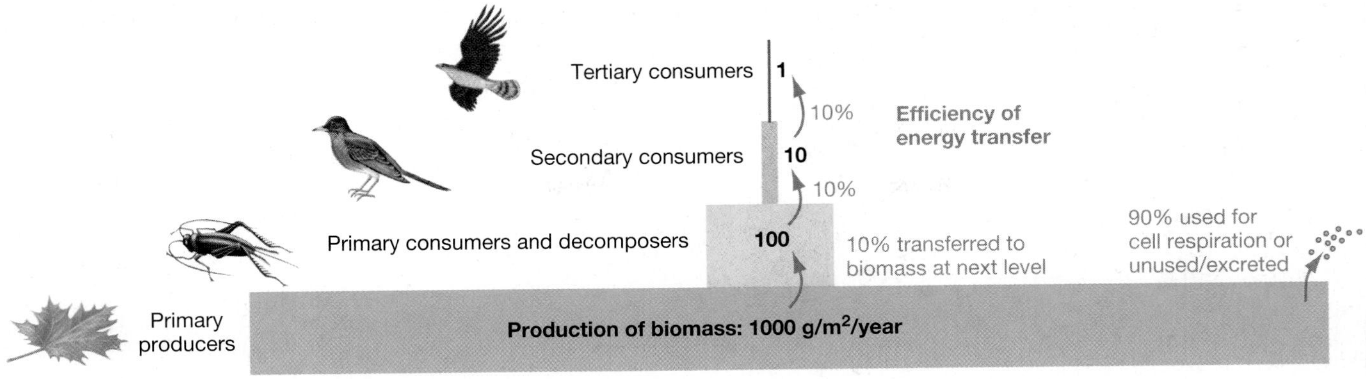

INTERACTIVE Figure 53.6 Productivity Declines at Higher Trophic Levels. This generalized example of a pyramid of productivity shows a 10 percent efficiency of transfer at each level.

Energy Flow to Grazers versus Decomposers Figures 53.4 and 53.5 show that the primary producers—maple leaves in this case—are a source of energy whether alive or dead. How much biomass gets eaten alive versus dead?

The answer to this question varies among habitats. In forest ecosystems, the percentage of NPP consumed alive can be as low as 5–7 percent, because indigestible wood isn't transferred to other organisms until it decays. But in marine habitats, where algae account for most of NPP, the percentage of NPP that gets consumed alive is much higher—often 35–40 percent. The "dead versus alive" percentage can also vary from season to season or even week to week according to ambient temperatures, light levels, nutrient levels, and species interactions.

Regardless of whether they are eaten dead or alive, note that primary producers are not growing *on purpose* to feed the other trophic levels. On the contrary, many primary producers have defenses to deter consumers since such adaptations increase their evolutionary fitness (Ch. 52, Section 52.1).

Energy Transfer between Trophic Levels

All ecosystems share a characteristic pattern: The total biomass produced each year declines from lower trophic levels up to the higher levels. To understand why, let's go back to the Hubbard Brook Experimental Forest.

Chipmunks are small rodents that are seed predators at Hubbard Brook. On average, these primary consumers harvest 31 kcal/m^2 of energy each year. Of that total, how much is used to create biomass? Careful measurements revealed that:

- 80.7 percent is used for cellular respiration.
- 17.7 percent is unused—meaning that it is not digested and absorbed. Instead, it is excreted.
- 1.6 percent goes into the production of new chipmunk tissue by growth and reproduction.

The same pattern would be seen in a secondary consumer—a Cooper's hawk, for example, that eats chipmunks and other primary consumers. Consider the energy that a hawk expends to catch prey. This energy cannot be used to make hawk biomass.

The general point is simple: Because most of the total energy consumed is used for cellular respiration rather than growth and reproduction, the amount of biomass produced at the second trophic level must be less than the amount at the first trophic level; productivity at the third trophic level must be less than that at the second.

The Pyramid of Productivity The data in Figure 53.6 depict the biomass produced at trophic levels 1–4 in a temperate deciduous forest. A separate horizontal bar represents each trophic level. Note that the graph reports two numbers:

1. *Productivity* is a rate, measured here in units of grams of biomass produced by growth and reproduction per square meter of area each year (g/m^2/year). (Note that these numbers do *not* contain information about the nutritional value of the biomass, such as relative amounts of carbohydrates or proteins.)
2. *Efficiency* is a dimensionless ratio—the units cancel out. (It is reported here as a percentage.) Efficiency is the fraction of biomass transferred from one trophic level to the next.

Biomass production at each trophic level varies widely among ecosystems. But as a rough rule of thumb, the efficiency of biomass transfer from one trophic level to the next is only about 10 percent. Where does the *biomass* go that is not transferred? It is released as carbon dioxide (from cellular respiration) and wastes such as feces and urine. Where does the *chemical energy* go that is not transferred? Ultimately, it is dissipated as heat.

To help you unpack the meaning of the pyramid of productivity, consider three hypothetical food choice scenarios for humans compiled in Figure 53.7 on page 1146. How much rice, chicken, or sardines can humans harvest given the same starting biomass of primary producers? This simple model makes a number of assumptions, for example:

- Humans eat only one type of food in each scenario.
- The foods all have the same nutritional value per gram.
- Humans spend the same amount of energy on activities such as collecting and digesting food in each scenario.

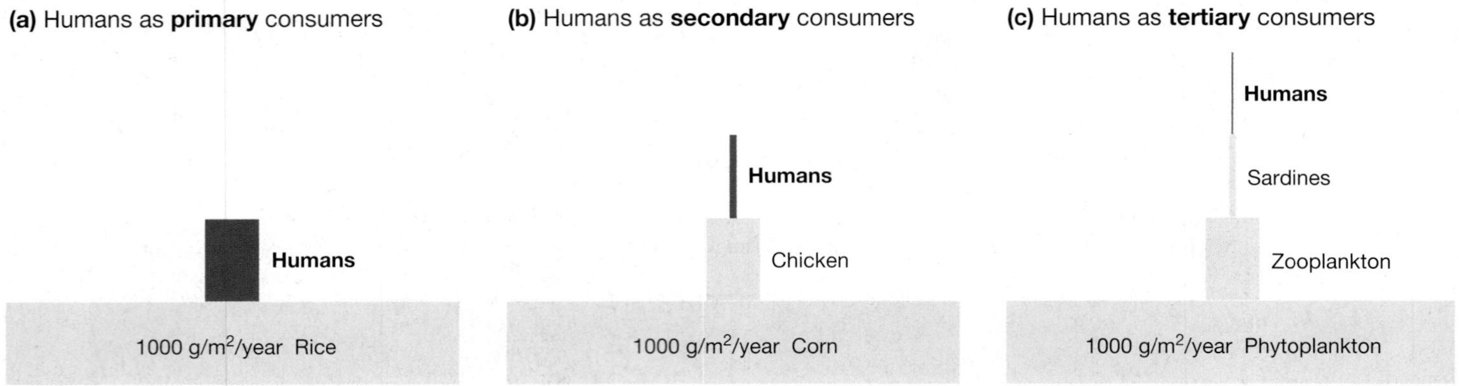

Figure 53.7 It Is More Efficient for Humans to Feed at Lower Trophic Levels.

✔ **QUANTITATIVE** Calculate how many grams/m^2/year of human biomass can be supported in each scenario. Then do the same for a fourth scenario where humans eat only tuna (which eat fish such as mackerel, which eat small fish like sardines). How much more human biomass can be supported by rice compared to tuna, given the assumptions of the model?

- The energy transfer between levels is always 10 percent.

Based on the models of productivity and efficiency shown in Figure 53.7, which food choice can support the most humans? The lower the trophic level of humans, the more humans that can be supported by a given biomass of primary producers.

Efficiency Varies The "10 percent rule" in ecological efficiency masks a great deal of interesting variation. For example:

- Large mammals are more efficient at producing biomass than small mammals, because large animals have a smaller surface-area-to-volume ratio and lose less heat (Ch. 39, Section 39.3).
- Biomass production is much more efficient in ectotherms than in endotherms. Ectotherms rely principally on heat gained from the environment and do not oxidize sugars to keep warm (Ch. 39, Section 39.5). As a result, they devote much less energy to cellular respiration than endotherms do. Chipmunks devote just 1.5 percent of their energy intake to growth; lizards and other ectotherms may devote 20–40 percent. Many insects devote over 40 percent of their energy intake to growth.

✔ **SOCIETY If you understand this concept, you should be able to explain why it may be more efficient to eat lobster than chicken, even though they feed at the same trophic level.**

Biomagnification The pyramid of productivity helps explain how some pollutants become concentrated in top consumers. Some types of molecules do not break down quickly in the environment and are neither digested by organisms nor excreted efficiently. These include heavy metals like mercury, emitted mostly from coal-fired power plants, and synthetic (human-made) organic compounds called POPs—persistent organic pollutants—that are used for herbicides, pesticides, and other industrial applications.

Mercury and POPs undergo **biomagnification**—meaning that they increase in concentration in consumers at higher levels in a food chain. Why?

Biomagnification occurs in the following steps:

1. Even when the concentration of the pollutant in the environment is low, a large quantity can be taken up by the abundant primary producers at the bottom of the productivity pyramid.
2. The primary consumers cannot metabolize or excrete the pollutants they consume. Thus, even though only about 10 percent of the biomass consumed is incorporated into consumer biomass, *all* of the pollutants are retained—a tenfold increase in concentration of pollutants.
3. The same process occurs in secondary consumers, causing another tenfold increase in the concentration of pollutants.
4. And so on. The more trophic levels there are in a food web, the higher the biomagnification of pollutants.

The upshot? Even when a pollutant occurs in low concentrations in an environment, it can become highly concentrated in top consumers.

One POP, called DDT (dichlorodiphenyltrichloroethane), was made famous by ecologist Rachel Carson in her 1962 book, *Silent Spring*. Carson was alarmed that many birds were dying in areas that were sprayed with DDT to control mosquitoes—birds with high concentrations of DDT produced thin-shelled eggs that got crushed in the nest. Carson's resulting campaign for responsible chemical use spurred the beginning of the environmental movement. DDT was banned in the United States in 1972. Populations of top predators like bald eagles are back on the rise.

Humans are top predators, too, and are vulnerable to biomagnification. ✔ **SOCIETY If you understand biomagnification, you should be able to explain how you could get mercury poisoning from eating tuna every day, but not from eating the same quantity of sardines every day.**

Global Patterns in Productivity

The pyramid of productivity in Figure 53.6 illustrates that the productivity of organisms at higher trophic levels ultimately depends on the NPP at the bottom level. Where is most NPP produced, and what limits its production to those areas?

Is Productivity Higher on the Land or in the Sea? Figure 53.8 summarizes satellite data on NPP from around the globe. Note that the highest NPP is indicated in red, yellow, and green, and that these colors occur only on land. In general, NPP on land is much higher than it is in the oceans. Why?

The leading hypothesis to explain this pattern is simply that much more light is available to drive photosynthesis on land than in marine environments. Water quickly absorbs many of the light wavelengths that are used in photosynthesis (see Ch. 49, Figure 49.12).

Which Terrestrial Ecosystems Are Most Productive? The terrestrial ecosystems with highest productivity are located in the wet tropics. Except in the world's major deserts (such as the large black area representing the Sahara, a desert in northern Africa), and high mountains, NPP on land tends to decline from the equator toward the poles. Why? The availability of sunlight and average temperature decreases with latitude. The low productivity in deserts is an exception attributed to the low annual rainfall caused by air circulation patterns (Ch. 49, Section 49.3).

You might recall that the equator-to-poles gradient in NPP is thought to be important in explaining the latitudinal gradient in species richness that occurs in most types of plants and animals (Ch. 52, Section 52.4). Areas of high productivity tend to be "hotspots" for high species richness.

Which Marine Ecosystems Are Most Productive? Productivity patterns in marine ecosystems contrast strongly with those of terrestrial environments. Marine productivity is highest along coastlines, and it can be as high near the poles as in the tropics. Why? Nutrients are the main limiting factor of marine productivity, and the shallow water along coasts receives more nutrients:

- Rivers carry nutrients from terrestrial ecosystems and deposit them on the coasts.
- Nearshore ocean currents bring nutrients from the cold, deep water of the oceanic zone up to the surface, a process called upwelling (Ch. 49, Section 49.5).

The coastal areas colored deep blue in Figure 53.8 include most of the world's most important fisheries.

In contrast, the oceanic zones, colored black in Figure 53.8, have extremely low NPP. They lack both river inputs and upwelling. Further, nutrients in organisms near the surface of the open ocean—where sunlight is abundant—constantly fall to dark, deeper waters in the form of dead cells, where they are no longer available to organisms near the surface. In productivity per square meter, the open ocean is a desert.

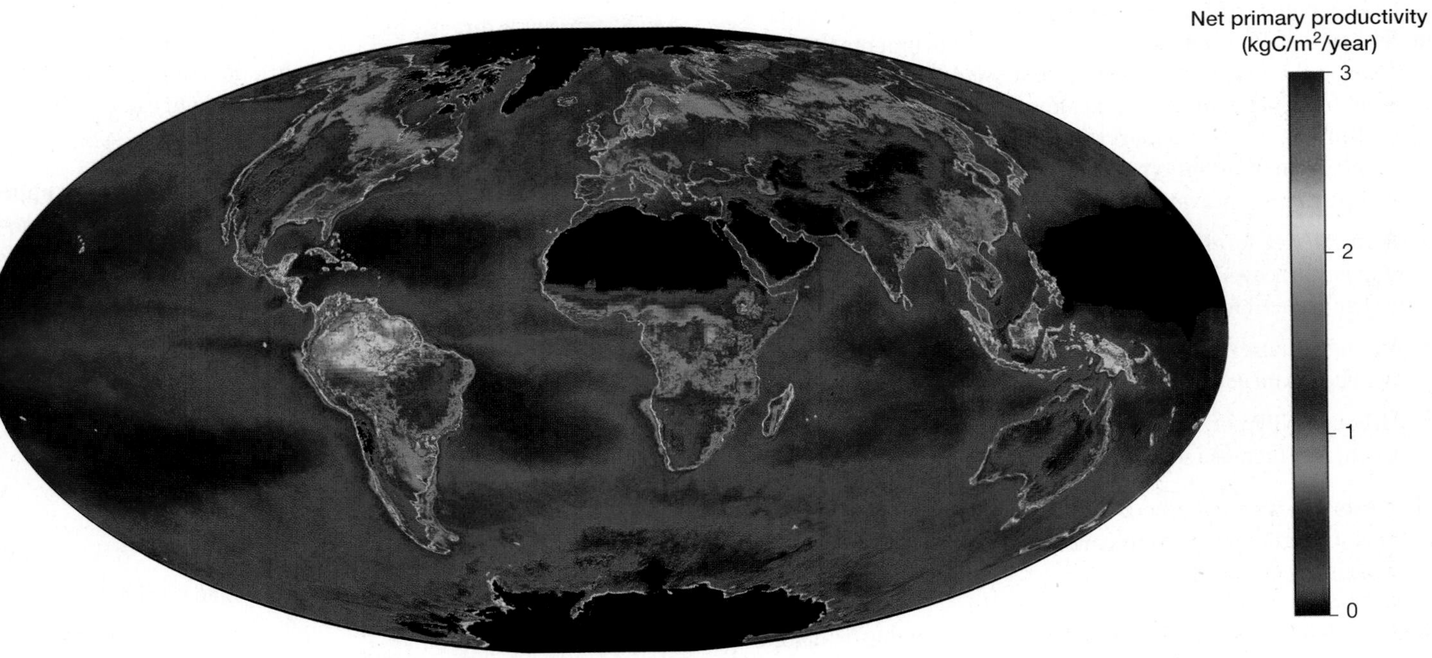

Figure 53.8 **Net Primary Productivity Varies among Geographic Regions.** The terrestrial ecosystems with the highest NPP are in the tropics; tundra and deserts have the lowest. The highest productivity in the oceans occurs in nutrient-rich coastal areas.

DATA: http://earthobservatory.nasa.gov.

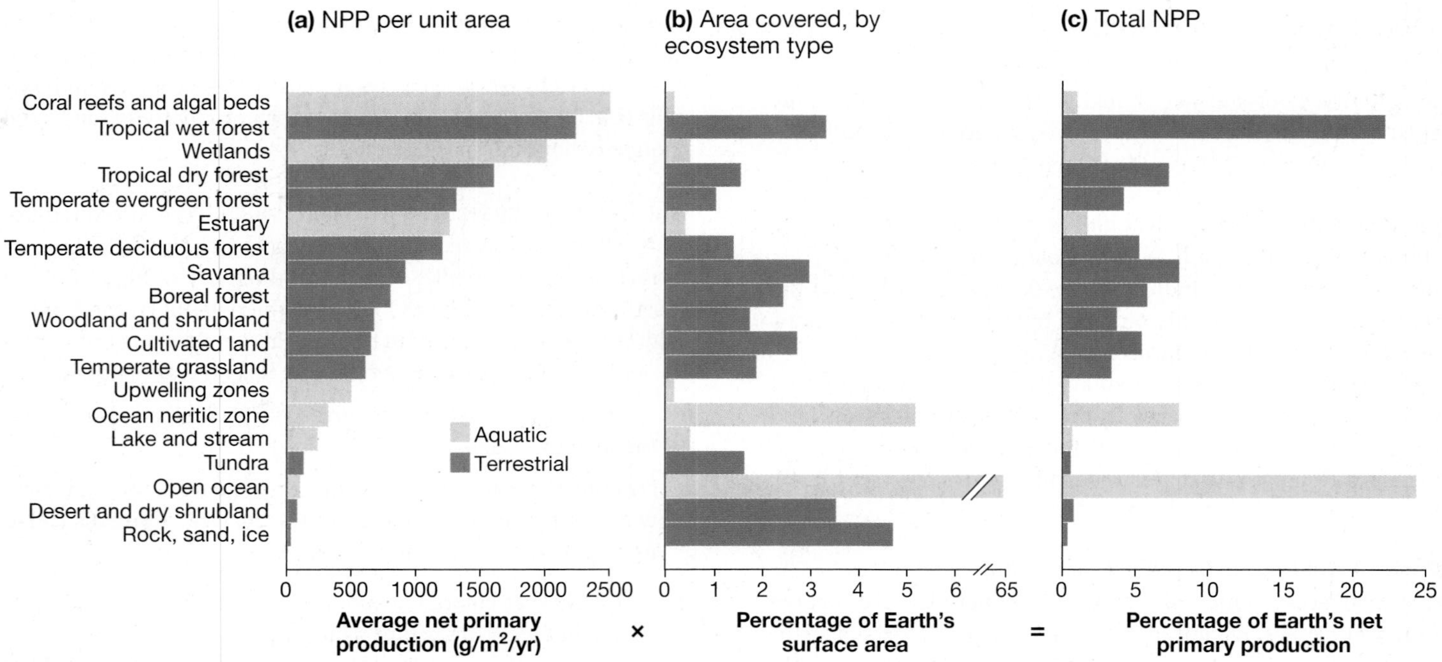

Figure 53.9 Net Primary Productivity Varies among Biomes.

DATA: H. Leith and R. H. Whittaker (eds.). 1975. *Primary Productivity in the Biosphere*. New York: Springer-Verlag.

✔ **QUANTITATIVE** Estimate the average NPP per unit area for the open ocean and the total area covered by the open ocean. How do these numbers explain the results in column (c)?

Which Biomes Are Most Productive? Figure 53.9 presents NPP data in a different way—organized by biome (Ch. 49, Sections 49.4 and 49.5). Figure 53.9a provides data on average NPP per square meter per year for each biome; Figure 53.9b documents the total area covered by each type of ecosystem; and Figure 53.9c presents the percentage of the world's total productivity—the result of multiplying the data in part (a) by the data in part (b). In each case, data from different types of biomes are plotted as horizontal bars. Note these observations:

- Tropical wet forests and tropical dry forests (which have a dry season) cover less than 5 percent of Earth's surface but together account for over 30 percent of total NPP.
- Among aquatic ecosystems, the most productive habitats are algal beds and coral reefs, wetlands, and estuaries.
- Although NPP per square meter is extremely low in the open ocean, this biome is so extensive that its total NPP is high.

Biologists are very concerned about the destruction of tropical wet forests and coral reefs, in part because these are the world's most productive habitats.

How Much of Earth's Total NPP Are Humans Using? If total NPP is so important to ecosystems, then ecologists must quantify how much of it humans use—a challenging task. A research group led by Helmut Haberl approached this challenge in terms of *potential* NPP—meaning, NPP that would exist without human activity. Using data from a wide array of sources, Haberl's team estimated that humans are currently preventing or appropriating 24 percent of potential NPP. Of this percentage,

- 53 percent is directly harvested and used (timber, food, etc.);
- 40 percent is prevented through land-use changes (such as parking lots and other forms of development); and
- 7 percent is destroyed due to human-induced fires.

This analysis is astonishing. One species—our own—is appropriating almost a quarter of the planet's biomass. Given increases in human population projected to occur over your lifetime (Ch. 51, Section 51.5), the demand for NPP is likely to reach the physical limit of what can be appropriated, with potentially dire consequences (Chapter 54).

Humans are affecting the flow of energy through ecosystems as well as the cycling of nutrients within and among ecosystems. Let's take a closer look at nutrient cycling.

CHECK YOUR UNDERSTANDING

✔ **If you understood this section, you should be able to . . .**

1. **SOCIETY** Explain why it takes about 10 times more energy to grow a kilogram of beef than a kilogram of wheat. Which food source is more sustainable?
2. **THINK CAREFULLY** Evaluate this statement: Plants create energy so that animals will have food.

Answers are available in Appendix A.

53.2 How Do Nutrients Cycle through Ecosystems?

Energy is transferred when one organism eats another. But tissues also contain carbon (C), nitrogen (N), phosphorus (P), sulfur (S), calcium (Ca), and other elements considered **nutrients** because they are essential for normal metabolism, growth, and reproduction. The transfer of both energy and nutrients links the biotic and abiotic components of ecosystems.

After you complete this section, you should be able to ...

- Describe how nutrients cycle within ecosystems.
- Compare the major biogeochemical cycles.

Atoms are constantly reused as they move through trophic levels, but they also spend time suspended in air, dissolved in water, or held in soil. The path an element takes as it moves from abiotic systems through producers, consumers, and decomposers and back again is referred to as its **biogeochemical** ("life-Earth-chemical") **cycle**. Because humans are disturbing biogeochemical cycles on a massive scale, research on the topic is exploding.

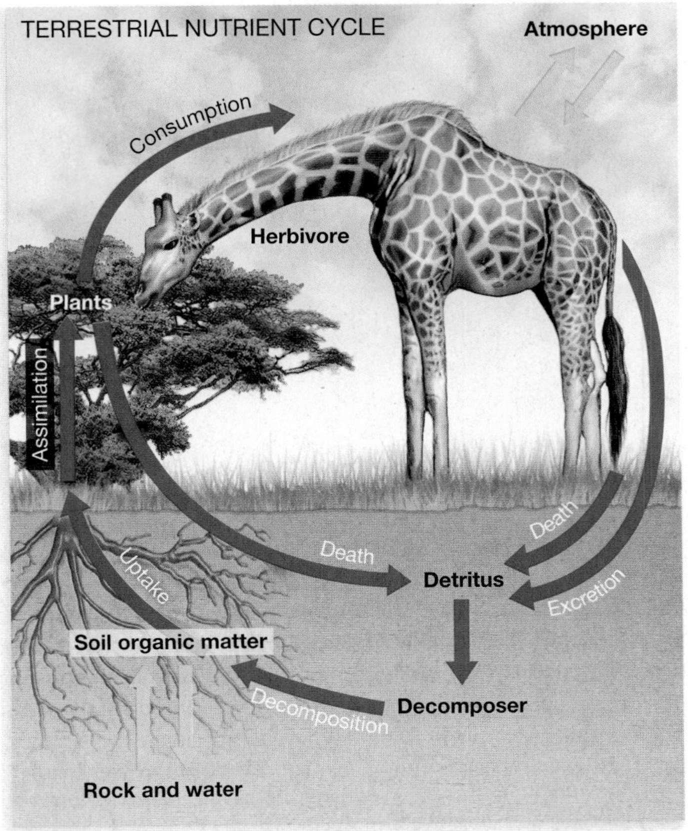

Figure 53.10 **Terrestrial Nutrients Share a Generalized Cycle.** Nutrients cycle from organism to organism in an ecosystem via assimilation, consumption, and decomposition (red arrows). Nutrients are also imported and exported from rock, water, and the atmosphere (yellow arrows).

Nutrient Cycling within Ecosystems

To understand the basic features of a biogeochemical cycle, study the generalized terrestrial nutrient cycle in Figure 53.10. The red arrows indicate the general pathways of nutrients as they cycle among organisms in a terrestrial ecosystem. The yellow arrows indicate nutrient imports and exports to and from rock, water, and the atmosphere.

Start with the "Uptake" arrow on the bottom left, which indicates that nutrients such as nitrogen and phosphorus are taken up from the soil by plants and assimilated into plant tissue. Note that if the plant tissue is eaten, the nutrients pass to the consumer food web; if the plant tissue dies, the nutrients enter the region labeled "Detritus," which contains wastes and dead organisms.

What happens as detritus decomposes? Nutrients that reside in plant litter, animal excretions, and dead animal bodies are used by bacteria, archaea, roundworms, fungi, and other primary decomposers. These microscopic decomposers and the carbon-containing compounds that they release combine to form what biologists call **soil organic matter**, a complex mixture of partially and completely decomposed detritus. When the detritus becomes completely decayed, it is called **humus** because it is rich in a family of carbon-containing molecules called humic acids. (Ch. 36, Section 36.2, describes other components of the soil.)

Eventually, decomposition converts the nutrients in soil organic matter to an inorganic form. For example, cellular respiration by soil-dwelling bacteria and archaea converts the nitrogen present in amino acids to ammonium (NH_4^+) or nitrate (NO_3^-) ions, which are then available for uptake by plants. Cellular respiration also converts carbon to CO_2, which is released to the atmosphere.

What Factors Control the Rate of Nutrient Cycling? The decomposition of detritus most often limits the overall rate at which nutrients move through an ecosystem. Until decomposition occurs, nutrients stay tied up in intact tissues. The decomposition rate, in turn, is influenced by three types of factors:

1. Abiotic conditions such as oxygen availability, temperature, and precipitation
2. The quality of the detritus as a nutrient source for the fungi, bacteria, and archaea that accomplish decomposition (e.g., lignin, a highly fibrous constituent of wood, can be digested only by basidiomycete fungi and a few bacteria)
3. The abundance and diversity of detritivores present

To appreciate the importance of abiotic conditions for the decomposition rate, consider the difference in detritus accumulation between boreal forests and tropical wet forests—two moist biomes that differ in temperature. Boreal forests are cool, so the rate of decomposition is slow. As a result, the top layer of soil accumulates a thick layer of partially decomposed detritus and organic matter.

Figure 53.11 Temperature and Moisture Affect Decomposition Rates.

In contrast, tropical wet forests are warm all year long. Conditions are so favorable for fungi, bacteria, and archaea that decomposition keeps pace with inputs. As a result, the top layer of soil remains very thin, storing little decomposed detritus and organic matter (Figure 53.11). Whereas nutrients cycle slowly through boreal forests, they cycle rapidly through wet tropical forests.

✔ SOCIETY If you understand this concept, you should be able to explain why farms created by clearing tropical forests tend to perform poorly.

Sources of Local Nutrient Export and Import Although nutrients are constantly cycled within individual ecosystems, they can also be lost from one ecosystem and exported to another.

- Carbon dioxide is released to the atmosphere by cellular respiration in primary producers, consumers, and decomposers.
- If an herbivore eats a plant and moves out of the ecosystem, the nutrients are exported.
- Nutrients leave ecosystems when flowing water or wind removes particles or inorganic ions and deposits them somewhere else.

Several of the major impacts that humans have on ecosystems—such as farming, logging, burning, and soil erosion—accelerate nutrient export. Farming and logging remove biomass in the form of plants; burning releases nutrients to the atmosphere; soil erosion moves nutrients that are bound to soil particles or dissolved in water. Further, these processes are often interrelated, as when logging leads to soil erosion and runoff.

For an ecosystem to function normally, nutrients that are exported must be replaced by imports. There are four major mechanisms to replace nutrients:

1. Ions that act as nutrients are released when rocks weather.
2. Nutrients can blow in on soil particles or arrive in streams.
3. Carbon is added when primary producers fix carbon via photosynthesis.
4. Nitrogen is added when nitrogen-fixing bacteria convert molecular nitrogen (N_2) in the atmosphere to usable nitrogen in ammonium or nitrate ions.

In ecosystems that humans manage, nutrients usually have to be actively replaced to maintain productivity. People accomplish this by adding fertilizers or planting nitrogen-fixing crops.

An Experimental Study of Nutrient Cycling The ideas presented thus far are logical and provide a strong conceptual foundation for thinking about nutrient cycles. But to explore nutrient movement in more depth, researchers have turned to experiments.

One of the first large-scale experiments on nutrient cycling took place at the Hubbard Brook Experimental Forest, introduced earlier in the chapter. A group of researchers led by Likens and Bormann began by choosing two similar **watersheds**—areas drained by a single stream—for study. Before starting the experiment, they documented that 90 percent of the nutrients in the watersheds was in soil organic matter; a mere 9.5 percent was in plant biomass. The nutrients in the soil were either dissolved in water or attached to small particles of sand or clay.

To begin the experimental treatment, the researchers cut all vegetation, including the trees, in one of the two watersheds (Figure 53.12). In each of the following three years, the clear-cut watershed was treated with an herbicide to prevent vegetation from regrowing. As a result, the experimental watershed was devegetated. Researchers left the other watershed untreated to serve as a control.

The graph in Figure 53.12 documents the amount of total dissolved substances that washed out of the streams in each watershed over the course of the four years of devegetation. Losses from the devegetated site were typically 15 times as high as they were from the control site.

Why? Plant roots were no longer able to hold soil particles in place, so some of the soil itself washed away. Further, plant roots no longer took up and recycled nutrients that were dissolved in the remaining soil water. Instead, large quantities of nitrate and

RESEARCH

QUESTION: How does the presence of vegetation affect the rate of nutrient export in a temperate-forest ecosystem?

HYPOTHESIS: Presence of vegetation lowers the rate of nutrient export because it increases soil stability and recycling of nutrients.

NULL HYPOTHESIS: Presence of vegetation has no effect on the rate of nutrient export.

EXPERIMENTAL SETUP:

1. Choose two similar watersheds. Document nutrient levels in soil organic matter, plants, and streams.

2. Devegetate one watershed, and leave the other intact.

3. Monitor the amount of dissolved substances in streams.

PREDICTION OF HYPOTHESIS: Amount of dissolved substances in stream in devegetated watershed will be much higher than amount of dissolved substances in stream in control watershed.

PREDICTION OF NULL HYPOTHESIS: No difference will be observed in amount of dissolved substances in the two streams.

RESULTS:

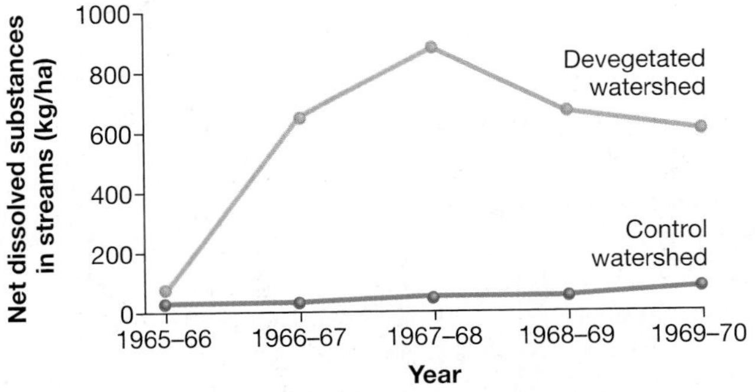

CONCLUSION: Presence of vegetation limits nutrient export. Removing vegetation leads to large increases in nutrient export.

Figure 53.12 **Experimental Evidence Shows that Deforestation Increases the Rate of Nutrient Export from Ecosystems.**

SOURCE: G. E. Likens, F. H. Bormann, N. M. Johnson, et al. 1970. Effects of forest cutting and herbicide treatment on nutrient budgets in the Hubbard Brook watershed-ecosystem. *Ecological Monographs* 40: 23–47.

✔ **PROCESS OF SCIENCE** Propose a hypothesis to explain why the level of nutrient export from the experimental site appears to peak and then decline.

other nutrients washed out of the soil and were exported from the ecosystem in runoff water.

Long-term devegetation of this type has occurred in formerly forested areas of the Middle East, North Africa, and elsewhere, due to intensive farming and grazing. Nutrients that used to occur within those soils—and the soils themselves—were exported. Today, the productivity of these regions is a tiny fraction of what it once was. Once soil is lost, it takes a very long time to reform (Ch. 36, Section 36.2).

Global Biogeochemical Cycles

When nutrients leave one ecosystem, they enter another. In this way, the movement of ions and molecules among ecosystems links local biogeochemical cycles into one global system. To explore how these global biogeochemical cycles work, let's consider the global water, nitrogen, phosphorus, and carbon cycles.

To understand how biogeochemical cycling works on a global scale, researchers focus on four fundamental questions:

1. ***Major reservoirs:*** What are the major reservoirs where elements are stored for a period of time?
2. ***Natural processes:*** What natural processes move elements from one reservoir to another, and how fast do elements move?
3. ***Human impacts:*** How are human activities affecting the movement of elements among reservoirs?
4. ***Interactions:*** How does one biogeochemical cycle interact with another?

Biogeochemical cycles can be very complex. Biologists often make visual models of the cycles to summarize the major reservoirs and processes involved. See **Making Models 53.1** for an introduction to interpreting and making these models.

Making Models 53.1 Tips on Drawing Nutrient Cycles

Global biogeochemical ("nutrient") cycles are constructed from two components: reservoirs, and processes that connect the reservoirs. The complexity of the models depends on how many elements are included, and whether the relative sizes of reservoirs and processes are indicated by color, size of arrows, numbers, or other strategies. Practice drawing your own cycles by using boxes to represent reservoirs and arrows to represent processes.

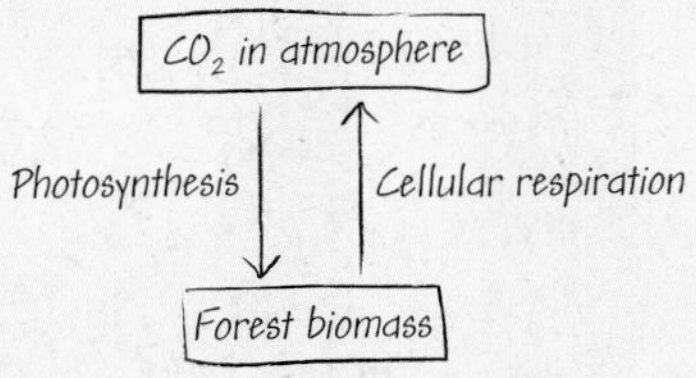

MODEL Increase the complexity of the above model of the carbon cycle by adding a box and arrow to show how the burning of fossil fuels moves carbon to the atmosphere as CO_2.

To see this model in action, go to the Study Area of **Mastering Biology**

The Global Water Cycle A simplified version of the **global water cycle** appears in Figure 53.13. The diagram shows the estimated amount of water that moves between major reservoirs over the course of a year, powered by energy from the Sun.

To analyze this cycle, begin at the left with evaporation of water out of the ocean, and the subsequent precipitation of water back into the ocean. For the marine component of the cycle, evaporation exceeds precipitation—meaning that, over the oceans, there is a net gain of water to the atmosphere.

When this water vapor moves over the continents, it is joined by a small amount of water that evaporates from lakes and streams and a large volume of water that is transpired (lost) by plants (Ch. 35, Section 35.2). However, precipitation exceeds evaporation and transpiration on land—meaning that there is a net gain of water on land. The cycle is completed by the water that moves from the land back to the oceans via streams and **groundwater**—water that is underground. Overall, there is no net gain or loss of water in the global water cycle.

Note that 97.5 percent of Earth's water is salt water in the ocean. Most of the remaining 2.5 percent of fresh water is locked up in the ice caps and glaciers, leaving about 0.5 percent available in lakes, rivers, and in groundwater for drinking. Most groundwater is stored in **aquifers**—layers of porous rock, sand, or gravel that are saturated with water. Aquifers are said to be closed (contained) when nonporous rock layers overlie them, and open (uncontained) when they can be recharged by water percolating down from the surface (see Figure 53.13). If depleted, closed aquifers may take thousands of years to recharge.

What effects are humans having on the water cycle? Perhaps the most direct impacts concern rates of groundwater depletion.

- Asphalt and concrete surfaces reduce the amount of water that percolates from the surface to deep soil and rock layers.
- When grasslands and forests are converted to agricultural fields, root systems that hold water are lost. More water runs off into streams and less percolates into groundwater.
- Irrigated agriculture, household use, and industrial applications such as hydraulic fracturing (fracking) are removing unprecedented amounts of water from groundwater storage. The Ogallala–High Plains Aquifer is the largest underground reservoir of fresh water in the United States. Billions of gallons are withdrawn each day, supplying 30 percent of the irrigation water for the country. The water is estimated to be 10,000 years old, and in much of its range it is being depleted much faster than it is being recharged (Figure 53.14), raising concerns about the future supply.

The **water table**—the upper limit of the underground layer of soil that is saturated with stored water—is dropping on every continent. About 40 percent of the world's population has insufficient water resources, meaning that many populations have already grown *beyond* carrying capacity for water. Water scarcity tends to lead to strife within and among populations.

Why not just remove the salt from salt water? Desalination plants are on the rise worldwide, but industrial desalination requires a tremendous amount of energy, making this source of fresh water expensive and an additional source of pollution.

The Global Nitrogen Cycle Figure 53.15 illustrates the **global nitrogen cycle**, which tracks the movement of nitrogen among terrestrial and aquatic ecosystems. To follow the cycle, start at the top in the atmosphere. This represents the main reservoir for nitrogen in the form of nitrogen gas (N_2)—a biologically inert molecule that comprises about 78 percent of the air you breathe. Nitrogen is usable by organisms only when it is reduced, or "fixed," meaning that it is converted from unusable atmospheric N_2 to usable ammonium (NH_4^+) or nitrate ions (NO_3^-; Ch. 36, Section 36.4).

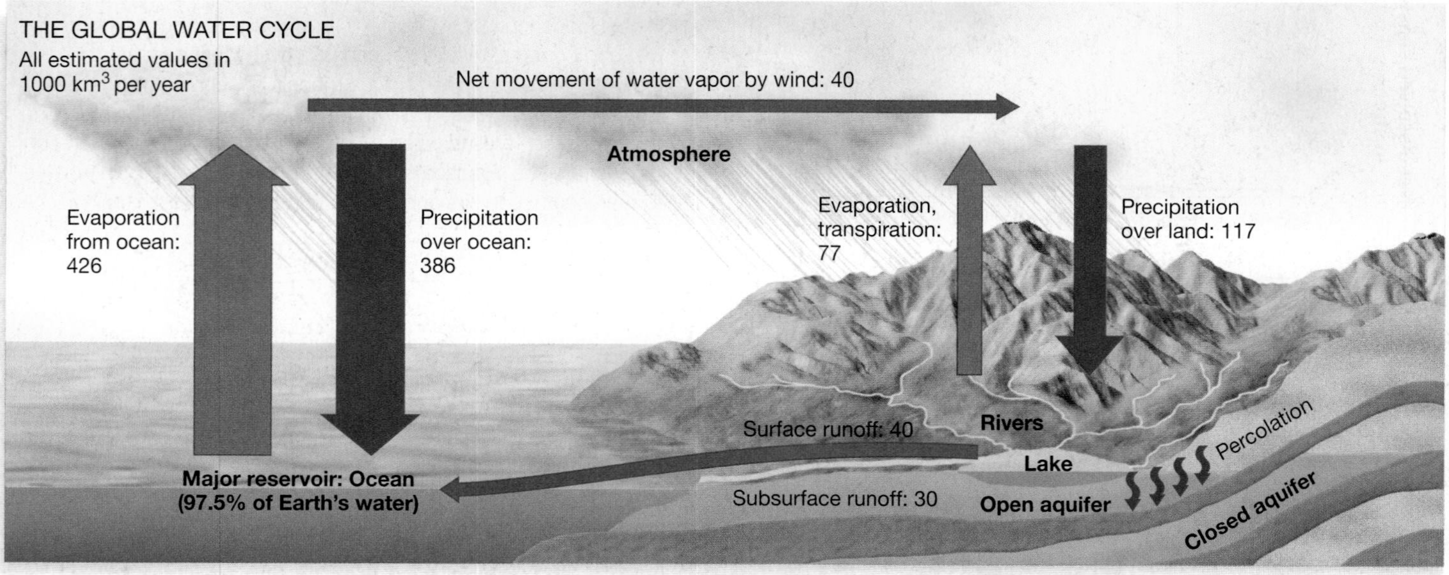

Figure 53.13 The Global Water Cycle.

DATA: U. Schneider et al. 2016. *Proceedings of the International Association of Hydrological Sciences* 374: 29–34.

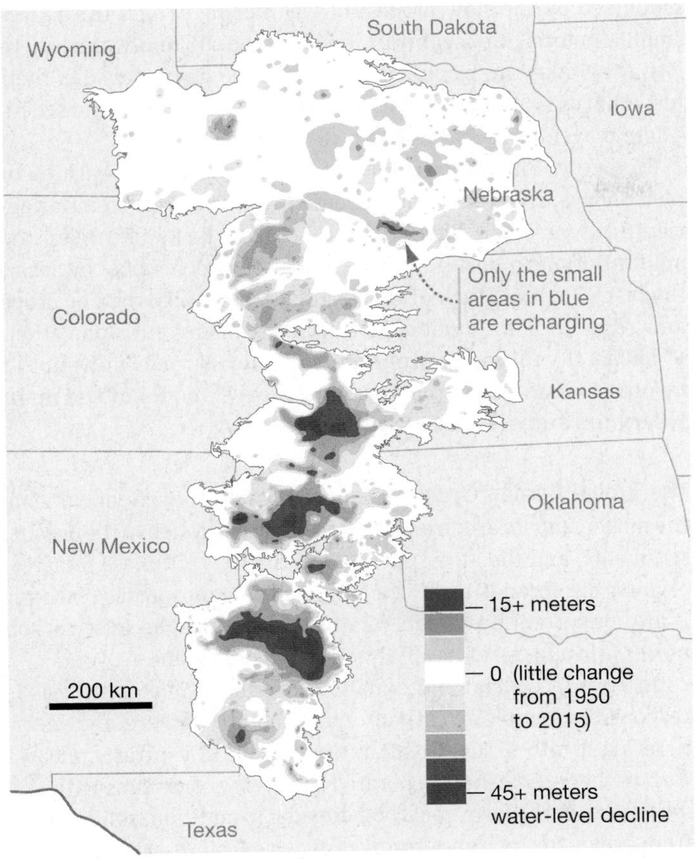

Figure 53.14 Water Levels in the Ogallala–High Plains Aquifer Declined from 1950 to 2015.

DATA: U.S. Geological Survey.

The fixation of nitrogen is usually a slow process, so nitrogen often limits primary productivity in terrestrial and aquatic ecosystems. Reduced nitrogen builds up slowly in soils and water over time and then cycles relatively rapidly among soil, microorganisms, fungi, plants, and animals within both terrestrial and aquatic ecosystems. Ultimately, bacteria use nitrogen-containing molecules as electron acceptors in cell respiration and return N_2 to the atmosphere, completing the cycle.

Nitrogen fixation occurs naturally due to lightning-driven reactions in the atmosphere and from enzyme-catalyzed reactions in bacteria that live in the soil and oceans. New evidence shows that some nitrogen also enters soils via weathering of rocks. However, the nitrogen cycle has been profoundly altered by human activities:

- Nitrogen pollution from the burning of fossil fuels causes acid rain. It is also a source of greenhouse gases that contribute to climate change (see Section 53.3) and is a source of gas that depletes the ozone layer in the upper atmosphere (a layer that protects the Earth from ultraviolet radiation from the Sun).
- When nitrogen is added to experimental plots in grasslands, a few competitively dominant species grow rapidly and displace other species that do not respond to nitrogen inputs as strongly. Productivity increases in the short term, but species diversity decreases, sometimes causing a long-term decrease in productivity (Ch. 54, Section 54.3).
- Nitrogen-laced runoff from agricultural fields causes algal blooms in aquatic ecosystems—an overfertilization effect called **eutrophication**—with serious consequences

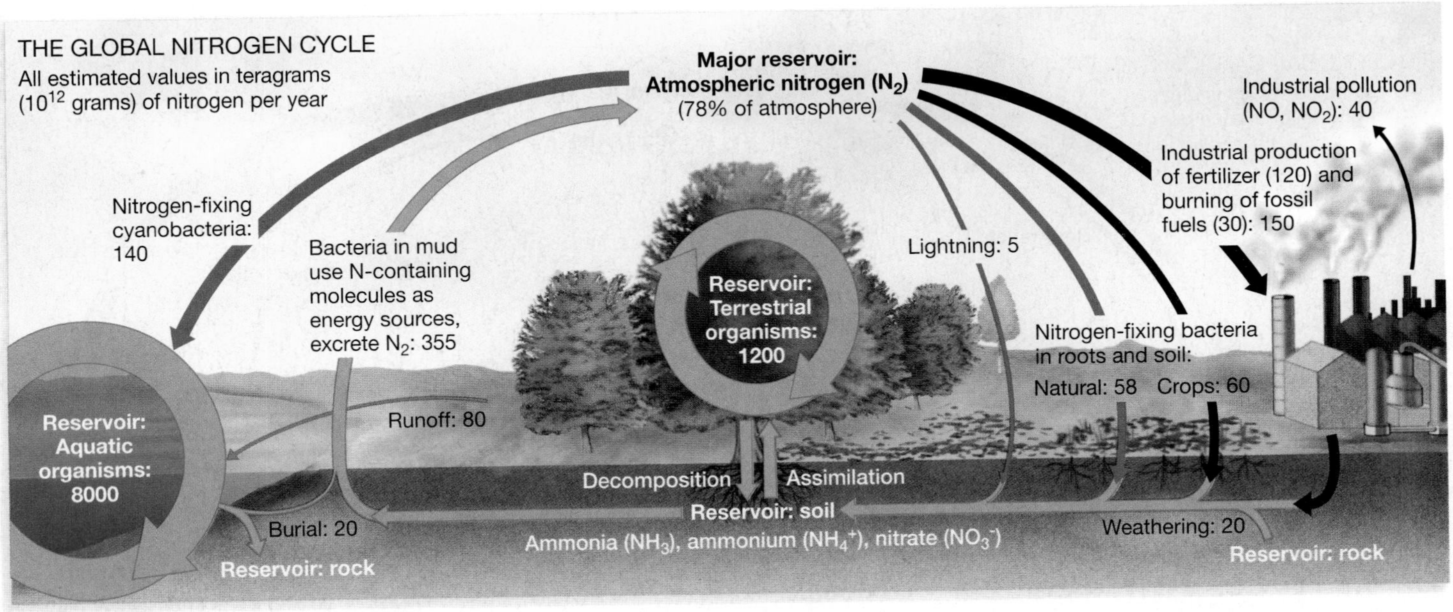

Figure 53.15 The Global Nitrogen Cycle. Nitrogen enters ecosystems as ammonia or nitrate via fixation from atmospheric nitrogen. It is exported in runoff and as nitrogen gas given off by bacteria.

DATA: D. Fowler et al. 2013. *Philosophical Transactions of the Royal Society B* 368 (1621): 20130165.

✔ **QUANTITATIVE** Calculate the percentage of total nitrogen fixation (all downward-pointing arrows) that is caused by human activities (black arrows).

Figure 53.16 **Nutrient Runoff Causes Harmful Algal Blooms.** Blooms, in turn, can lead to oxygen-starved dead zones.

(Figure 53.16). When the algae die and are decomposed by microbes, oxygen is depleted from the water, causing the formation of oxygen-free "dead zones" (see Ch. 26, Figure 26.15).

Human activities have doubled the rate of nitrogen fixation in the nitrogen cycle, causing widespread effects on ecosystems.

The Global Phosphorus Cycle The **global phosphorus cycle** tracks the movement of phosphorus among terrestrial and aquatic ecosystems. Unlike nitrogen, the main reservoir of phosphorus is in rock in the Earth's crust; this element does not occur in gas form in the atmosphere. Phosphorus is normally mobilized by the slow process of weathering, which introduces small amounts of phosphorus ions into the soil and water where it may be taken up by organisms. When the organisms die, their phosphorus is recycled or is carried by water runoff to the ocean. There it eventually settles and returns to rock.

However, humans are mining unprecedented quantities of phosphorus from rocks and introducing it into ecosystems as agricultural fertilizer. The fertilizer industry has quadrupled the amount of phosphorus entering global biogeochemical cycles in the last 75 years. Much of this fertilizer is not absorbed by crops and runs off into aquatic ecosystems, causing eutrophication. **✔ MODEL If you understand the phosphorus cycle, you should be able to make a simple "box and arrow" model of the main reservoirs and processes involved.**

The Global Carbon Cycle The **global carbon cycle** documents the movement of carbon among terrestrial ecosystems, the oceans, sediments, and the atmosphere (Figure 53.17). Of these four reservoirs, the ocean is by far the largest. The atmospheric reservoir is also important despite its relatively small size, because carbon moves into and out of it rapidly—through organisms.

In both terrestrial and aquatic ecosystems, photosynthesis is responsible for taking carbon out of the atmosphere and incorporating it into tissue. Cellular respiration, in contrast, releases carbon that has been incorporated into living organisms to the atmosphere, in the form of carbon dioxide. In addition, some carbon from organisms becomes trapped in geological reservoirs such as oil and coal (remains of ancient organisms) or sedimentary rock such as limestone (remains of skeletons of ancient organisms).

How have humans changed the carbon cycle? For thousands of years, humans have altered the global carbon cycle by deforesting

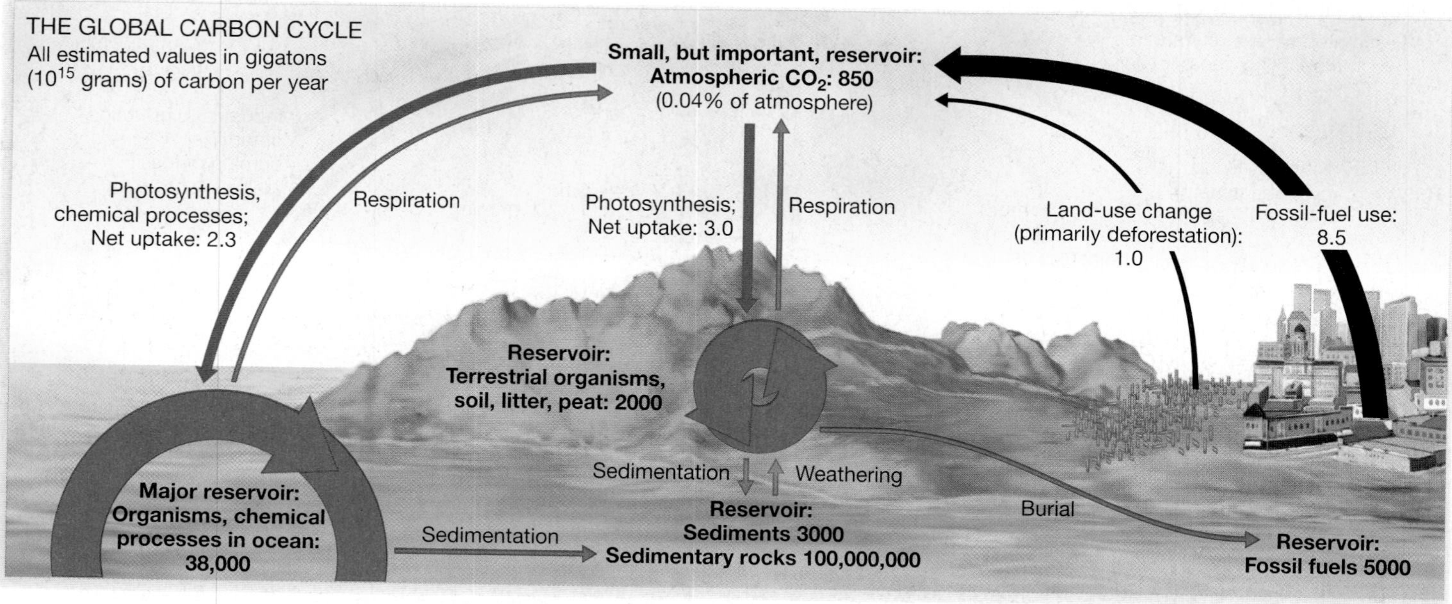

Figure 53.17 **The Global Carbon Cycle.** Photosynthesis and cellular respiration move carbon in and out of the atmosphere in the form of CO_2 (red arrows). Human activities such as deforestation and the burning of fossil fuels move carbon from inactive reservoirs, such as oil and coal, to an active reservoir, the atmosphere (black arrows).
DATA: B. Hönisch, A. Ridgwell, D. N. Schmidt, et al. 2012. *Science* 335: 1058–1063.

land for agriculture and settlements—releasing the carbon stored in trees to the atmosphere.

But the human influence on the carbon cycle increased sharply during the industrial revolution. Burning fossil fuels moves gigatons of carbon from an inactive geological reservoir, in the form of oil or coal, to an active reservoir—the atmosphere. When you burn gasoline, you are releasing carbon atoms that have been locked up in petroleum reservoirs for millions of years.

Figure 53.18a highlights the dramatic increase in carbon released from fossil-fuel burning over the past century; Figure 53.18b shows corresponding changes in atmospheric CO_2 concentrations. Note that CO_2 levels in the atmosphere have now risen above 400 parts per million (ppm). Analysis of CO_2 levels in air bubbles trapped in Antarctic ice reveals that CO_2 fluctuated within a range of about 170 to 300 ppm for the extent of the record—800,000 years. This means that current CO_2 levels are 30 percent higher than the highest level measured over the last 800,000 years. One recent study estimates that the current rise of CO_2 is steeper than any other time since the last mass extinction, 66 million years ago.

Why is the increase in atmospheric CO_2 so important? Because it changes the climate.

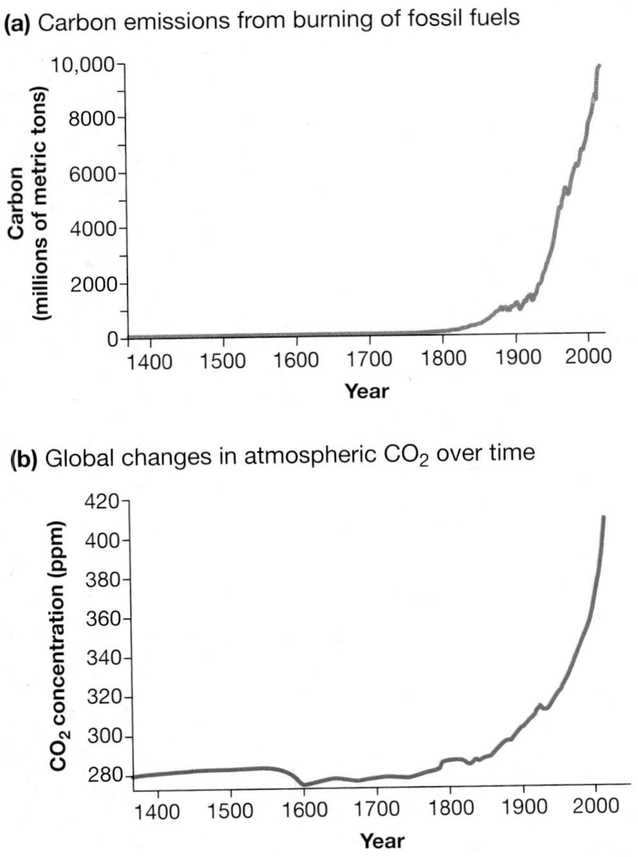

Figure 53.18 Humans Are Causing Increases in Atmospheric Carbon Dioxide. **(a)** Rates of carbon flow from fossil-fuel burning have increased as human populations have increased. **(b)** For centuries, average CO_2 concentrations in the atmosphere were about 280 parts per million (ppm).

DATA: Carbon Dioxide Information Analysis Center.

CHECK YOUR UNDERSTANDING

✔ **If you understood this section, you should be able to . . .**

Make a table comparing the following aspects of the water, nitrogen, phosphorus, and carbon cycles: (1) important reservoirs, (2) primary natural processes that move the elements among reservoirs, (3) main impacts by humans.

Answers are available in Appendix A.

53.3 Global Climate Change

The current human population is experiencing the most dramatic episode of environmental change in human history. The epic event has three main interacting drivers: changes in land use (Chapters 49 and 54), the rapid loss of species (Chapter 54), and global climate change, also called global warming.

After you complete this section, you should be able to . . .

- Describe how human activities cause global climate change.
- Analyze the effects of climate change on organisms in aquatic and terrestrial ecosystems.

Global warming and global climate change refer to the same general phenomenon, but with different emphases:

- **Global warming** refers to the increase in the *average* temperature of the planet. The average temperature is very important in determining physical, chemical, and biological processes.
- **Global climate change** refers to the sum of all the changes in local temperature and precipitation patterns that result from global warming—including the frequency and intensity of storms, droughts, and other events. The effects of global warming vary depending on geography, local weather patterns, ocean currents, and other factors. But averaged over the entire planet and over time, the Earth is already much warmer than it was just a few decades ago and is projected to get even warmer.

Note that "weather" and "climate" do not have the same meaning: **weather** refers to the *short-term*, highly variable atmospheric conditions such as temperature, moisture, sunlight, and wind at a specific place and time, whereas **climate** refers to the *long-term* average pattern of regional or global weather. For example, on a July morning in San Francisco, the weather can be foggy and cool on the coast and sunny and warm inland, but both areas experience the same climate.

Why is global warming such a problem? Let's explore the underlying cause of global climate change and then consider the consequences.

What Is the Cause of Global Climate Change?

Climate change is normal. Throughout Earth's history, the average temperature of the atmosphere has fluctuated. Scientists today are not alarmed because there is change; they are alarmed because the *rate* of change is unprecedented, and because it is caused by human activities. What is going on?

The Greenhouse Effect In the 1820s, French mathematician Joseph Fourier wondered why Earth's atmosphere was warmer than predicted by calculations of incoming solar radiation. Using a vase and some panes of transparent glass, he made a physical model that convinced him that the atmosphere could be trapping heat, thus insulating the Earth.

Three decades later, Eunice Foote and John Tyndall demonstrated experimentally that CO_2 plays a role in this warming effect. Carbon dioxide is a **greenhouse gas**: It traps heat that has been radiated from Earth and keeps it from being lost to space, similar to the way the glass of a greenhouse traps heat, causing a **greenhouse effect** (Figure 53.19). Be careful not to confuse the greenhouse effect with the sunscreen effect of the ozone (O_3) layer, which prevents some of the Sun's ultraviolet radiation from reaching Earth in the first place.

The greenhouse effect is a natural process that predates humans, and there are other greenhouse gases besides CO_2, such as methane (CH_4), water vapor, and nitrous oxide (N_2O). So what's the problem? Around 1900, eminent physical chemist Svante Arrhenius posed a stunning hypothesis: The atmospheric CO_2 would rise above natural levels due to *human-caused* CO_2 emissions, which would lead to global warming.

By the 1950s, the common reassuring argument was that the vast ocean would be able to absorb excess CO_2 produced by humans. However, Roger Revelle, a pioneer in oceanography, and his colleague Hans Suess reported in 1957 that the complex chemistry of the ocean limits how much CO_2 it can absorb. Their data supported the hypothesis made by Arrhenius.

How could the global warming hypothesis be tested? At the time of Revelle and Suess's research, there were no reliable records of atmospheric CO_2 levels—local CO_2 measurements varied widely due to fluctuating emissions from cities and other variables. So Revelle hired a young geochemist, Charles David Keeling, to gather precise measurements of CO_2 concentrations in remote, pristine locations, one of which was the Mauna Loa Observatory in Hawaii. By 1961, Keeling was able to confirm that global CO_2 was rising. Despite many obstacles, he persevered in collecting invaluable records for more than 40 years.

Figure 53.20 shows CO_2 data from Mauna Loa over the last four decades, along with temperature and solar energy data.

- There is no indication that the rate of increase of CO_2 is slowing, despite an array of recent regulations and incentives to reduce CO_2 emissions.
- The atmospheric temperature fluctuates, but there is a clear increase in average temperature over time. (The relatively flat trend between 2000 and 2010 was caused in part by the increased average temperature of the *ocean* during this interval.)

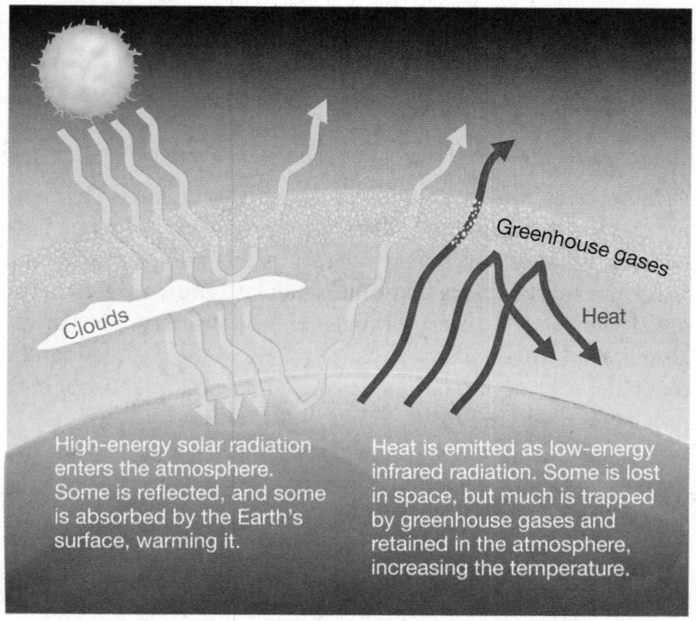

Figure 53.19 The Greenhouse Effect Explains Atmospheric Warming. Greenhouse gases in the atmosphere have an effect that is similar to glass in a greenhouse—high-energy radiation in the form of short-wavelength light can easily pass through the gases, but low-energy infrared (long-wavelength) radiation gets trapped.

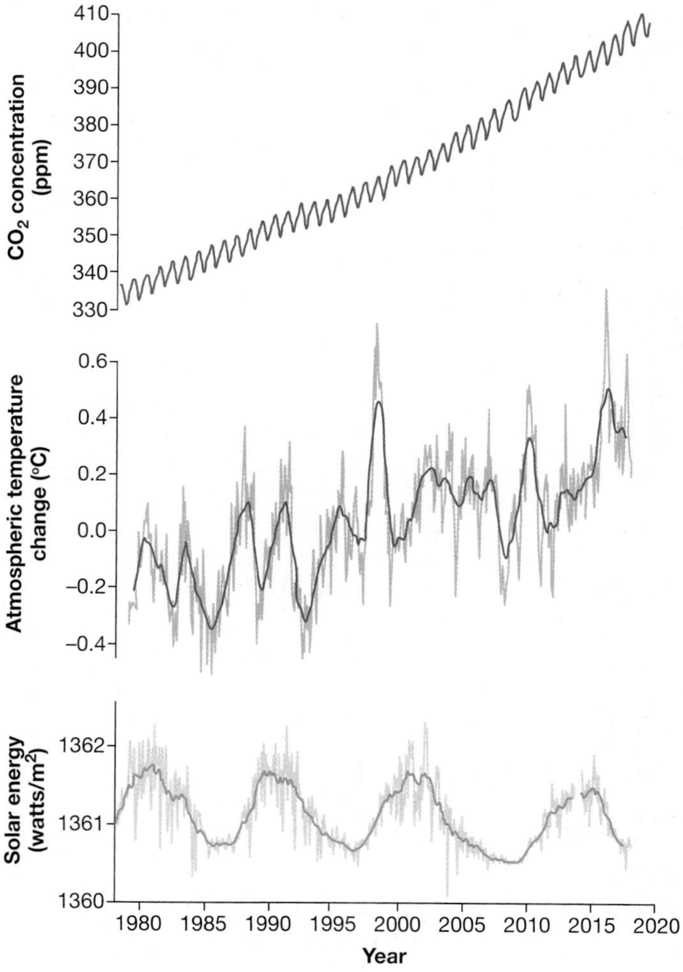

Figure 53.20 Recent Data Confirm an Overall Rise in Atmospheric CO_2 and Temperature, but Not in Solar Radiation.

DATA: NOAA/ESRL; T. R. Karl et al. (eds.). 2009. *Global Climate Change Impacts in the United States*. Cambridge UP; GISS/NASA.

✔ Notice the annual cycles in CO_2 concentrations. Predict why atmospheric CO_2 concentrations are low in the Northern Hemisphere in summer and high in winter.

- Even though the amount of solar energy cycles regularly, there is no net increase over time. Thus, fluctuations in solar energy are not sufficient to account for the marked increase in temperature, as one alternative hypothesis had predicted.

Based on all the long-term data available today, scientists agree that in the absence of humans, the climate would currently be in a cooling trend rather than warming trend. So the present rapid increases in atmospheric CO_2 and temperature are astonishing. Why are they increasing so fast?

Why Is the Climate Changing So Rapidly? There are two main reasons for the rapid rise in CO_2. First, the human population has exploded in size. The planet now has over 7 billion people, a 700 percent increase over a mere two centuries—a blink of an eye in Earth's history (Ch. 51, Section 51.5). But an equally important part of the answer involves resource use. The average per capita (per person) use of fossil fuels is skyrocketing—especially in industrialized countries.

The data in Figure 53.21 illustrate the consequences of population size and resource use. By multiplying the total population size of a country in part (a) by the quantity of CO_2 emissions per person in that country in part (b), the total emissions per country can be calculated in part (c). For example, even though the average person in China causes half the CO_2 emissions that an average person from the United States does, the total emissions from China surpass all others due to its enormous population size.

Conversely, although the United States represents less than one-twentieth of the world population, Americans produce one-seventh of the global CO_2 emissions. Looking at the data in another way, the population of the United States was almost twice that of Nigeria in 2016, but the average American produced 35 times as much CO_2 as the average Nigerian that year.

The bottom line is that the recent exponential increase of the human population, plus the steep increase in per capita fossil fuel use—as well as land-use changes such as deforestation—have caused a steep increase in the emission of CO_2 and other greenhouse gases. These changes, in turn, have caused the recent and dramatic increase in the average temperatures and climate variability around the globe. The rapid rate of climate change is ultimately rooted in human behavior.

How Much Is the Climate Changing?

In 1988, an international group composed of hundreds of scientists, called the Intergovernmental Panel on Climate Change (IPCC), was formed to evaluate the consequences of increases in CO_2 and other heat-trapping gases to climate. The group has since produced a series of reports summarizing the state of scientific knowledge on the issue. How has climate already changed, and what changes can we expect in the future?

Average Temperature The latest IPCC report released in 2018 concludes that the average global surface temperature reached approximately 1°C (1.8°F) above pre-industrial levels in 2017—due to human-induced warming. Four independent records of temperature summarized by NASA concur that the last decade has been the warmest in recorded history.

Predicting the future state of a system as complex and variable as Earth's climate is extremely difficult. Global climate models are the primary tool that scientists use to make these projections. A global climate model is based on a large series of

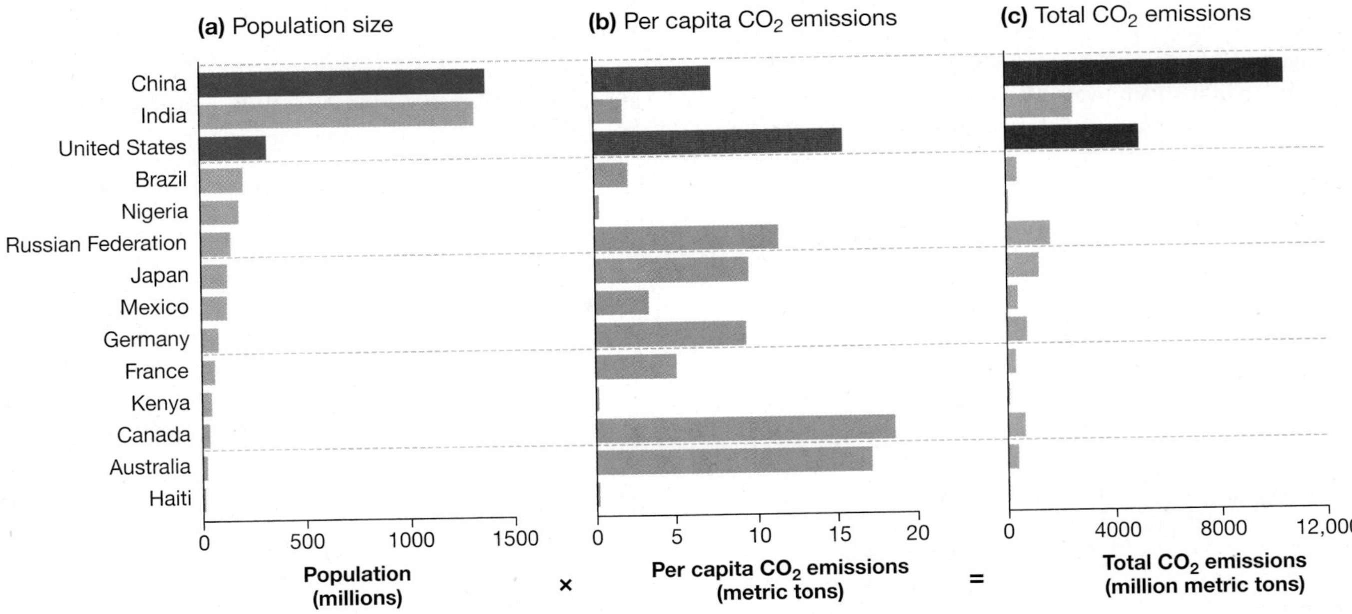

Figure 53.21 **Total Emissions Depend on Population Size and Resource Use.** These data are from 2016. The bars for China and the United States are emphasized for comparison.

DATA: Emission Database for Global Atmospheric Research, Version 4.3, 2016.

✔ **QUANTITATIVE** Estimate what the total emissions in China will be if the average release of CO_2 per person doubles due to increased use of fossil fuels.

equations that describe how the concentrations of various gases in the atmosphere, solar radiation, transpiration rates, and other parameters interact to affect climate.

The IPCC 2014 report predicts that average global temperature will increase another 1.5–2.0°C (2.7–3.6°F) by the year 2100, and 0.5–7.0°C (0.9–13°F) by the year 2300. The low numbers are based on models that assume that greenhouse gas emissions will peak and then decline below current levels by 2100; the high numbers are based on models that assume continued intensive use of fossil fuels and increases in greenhouse gas emissions.

For a sense of perspective on what these numbers mean, consider the predicted temperature scenarios in the context of temperature data from the last 5 million years (**Figure 53.22**). Note that the *x*-axis compresses the data on the far left and stretches the data on the far right to enable a closer look at recent and future trends. The *y*-axis shows the change in atmospheric temperature over time compared to today's average atmospheric temperature. The data offer a couple of take-home messages:

- Even under the most conservative scenario (purple curve), the average temperature over the next few hundred years will be higher than it has been over the past 5 million years.
- With no changes in human activities, the rise in average temperature could be extreme (red curve).
- Even moderate changes in human behavior result in less extreme increases (orange and blue curves).

Continued global warming is inevitable. But the extent of warming is highly dependent on current decisions about fossil fuel use and land use. In 2015, an unprecedented global collaboration of 196 countries created the Paris climate agreement to try to cap global temperature rise well below 2°C this century. So far, the major industrialized countries are failing to meet their emissions pledges.

Temperature Variation The average temperature is expected to keep rising, but that doesn't mean it will rise the same amount everywhere, or in a steady way. The polar regions will experience a larger average change in temperature than the tropics, and some ecosystems at the same latitudes will experience higher temperature increases than others due to wind patterns, ocean currents, and other factors (Ch. 49, Section 49.3). Further, the average temperature in some seasons will change more than in others.

Climate models also predict that temperature extremes will increase. Some areas will experience more extreme freezes in the winter and/or more extreme heat waves in the summer. Extreme temperatures can have a critical effect on ecosystems. For example, severe heat waves can foster intense fires that abruptly change biomes from forests to grasslands. How does temperature affect precipitation?

Effects of Climate Change on the Water Cycle Rising temperatures affect the water cycle, changing where, when, and how much water falls as precipitation, as well as how quickly water evaporates. The interaction of temperature and water determine how much soil moisture is available in ecosystems, a critical variable driving primary productivity.

The data in **Figure 53.23** show predictions for the percentage change in soil moisture expected during this century. The areas in green and blue, such as central Asia and the Sahara of Africa, are predicted to have increased soil moisture. However, the vast red areas in North America, South America, Europe, southern Africa, and Australia are predicted to experience drier soils. Reduced soil moisture reduces primary productivity and increases the risk of fires.

Climate models predict that precipitation will continue to become more variable. For example, the severity of extreme weather events such as droughts, monsoon rains, tornados, and hurricanes is increasing.

Warmer temperatures are also causing changes to the vast oceans. For example, the sea level is rising due to the thermal expansion of the warmer oceans and due to the melting of glaciers and polar ice caps, which reduces the Earth's fresh-water reservoirs and floods coastal areas. The melting of the ice caps, in turn, affects ocean currents and cloud cover. In essence, global warming has a domino effect on other factors.

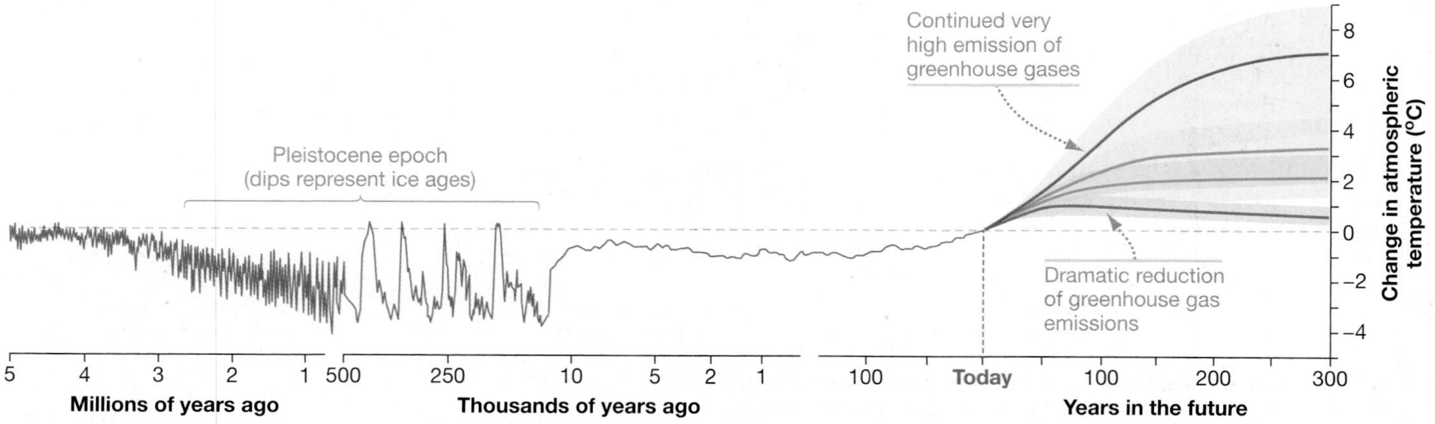

Figure 53.22 Scientists Expect Continued Warming in the Future. The scale on the x-axis varies.

DATA: C. Moritz and R. Agudo. 2013. *Science* 341: 504–508; IPCC, *Climate Change 2014: Synthesis Report*. 2014. Core Writing Team, R. K. Pachauri and L. A. Meyer (eds.). Intergovernmental Panel on Climate Change, Geneva, Switzerland, 151 pp.

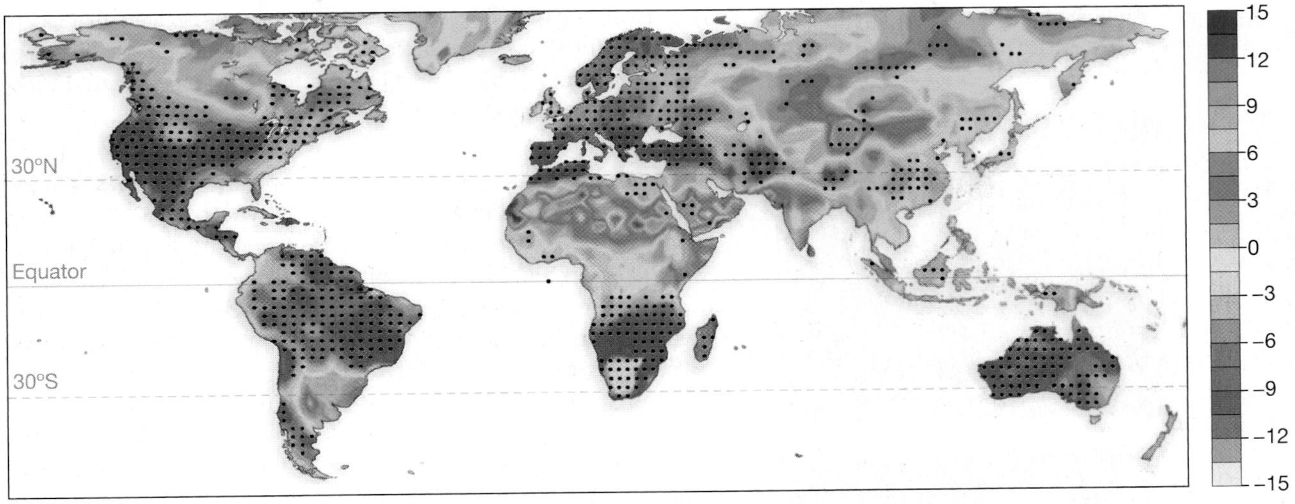

Figure 53.23 Reductions in Soil Moisture Are Predicted for Many Areas. The map shows a summary of 11 models predicting the percentage change in soil moisture by the end of the century, assuming the temperature scenario depicted by the blue curve in Figure 53.22. Black stippling indicates areas for which at least 9 of the 11 models agree on the direction of change.

DATA: A. Dai. 2013. *Nature Climate Change* 3: 52–58.

Documenting Positive and Negative Feedback Positive and negative feedback loops can make predicting the future effects of temperature changes very difficult. Ecosystems can respond to warming in ways that increase *or* decrease the concentrations of CO_2 and other greenhouse gases and that affect other drivers of global climate.

Positive feedback in climate change occurs when changes due to global warming result in the further acceleration of warming. ("Positive" in this context means "increasing" instead of "good.") Many types of positive feedback occur. For example:

- Warmer and drier climate conditions have increased the frequency of forest fires. Forest fires release CO_2, which leads to even more warming.
- Tundra sequesters almost twice as much carbon in the form of soil organic matter as occurs in the atmosphere. Traditionally, decomposition rates are extremely low. However, decomposition rates are increasing sufficiently to produce a net flow of carbon to the atmosphere—including CH_4 and N_2O, greenhouse gases that are less abundant than CO_2 but have a more potent effect on warming.
- Melting of the polar ice caps exposes more open water to sunlight. The darker open water absorbs more solar radiation than ice, increasing the rate of warming there.

Negative feedback in climate change occurs when changes due to global warming result in increased uptake and sequestration of CO_2 and other greenhouse gases—meaning that global warming should then be reduced. ("Negative" in this context means "decreasing" rather than "bad.") For example, recent experiments have shown that the growth rates of some plant species increase in direct response to increasing atmospheric CO_2, thus reducing atmospheric CO_2. However, other plant species, including crops such as wheat, experience decreased growth in warmer, more variable, or more extreme conditions.

Will ecosystem responses increase or decrease global warming, on average? While the predictions of different models vary, the overall trend is the same—it's going to keep getting hotter.

Climate is an abiotic component of ecosystems. How does climate change affect organisms?

Biological Effects of Climate Change

Even though global temperatures have risen only slightly in comparison with projections for the next 50–100 years, biologists have already documented dramatic impacts on the biosphere

- the development, physiology, and behavior of organisms;
- the growth, decline, and evolution of populations;
- the interaction of species within communities;
- the dynamics of communities over time; and
- the flow of energy and cycling of nutrients in ecosystems.

Consider some of the drivers of these biotic changes, summarized in **Table 53.1** on page 1160.

Geographic Range Shifts The geographic ranges of many species are shifting as their "climate envelopes" of suitable habitat move toward the poles, toward higher altitudes, or toward areas of adequate precipitation.

- Until recently, the range of mountain pine beetles was limited in North America by freezing temperatures. Milder temperatures have triggered deadly infestations of beetles in areas where they had not been abundant before, threatening the boreal forest ecosystem (see Table 53.1).
- Copepods are small crustaceans that serve as key primary consumers in marine plankton. A 40-year study in the North Atlantic showed that a southern species of copepod has moved north, while the range of cold-water species has declined.

SUMMARY Table 53.1 Biological Effects of Climate Change

Type of Effect	Example
Geographic Range Shifts The geographic redistribution of species to more favorable climate envelopes can change ecological interactions. Mountain pine beetles are spreading north and east, devastating pine forests (which turn red before they die).	
Phenology Shifts The change in the timing of seasonal events such as migration, flowering, hatching, and feeding, can result in mismatches, such as the persistence of white winter coats in hares even after the snow has melted in spring.	
Evolutionary Adaptation Change in allele frequencies can increase fitness in new conditions. In warmer environments, the red morph of the ladybird beetle is more adaptive than the heat-absorbing black morph.	
Extinction Species that are not able to survive and/or reproduce under new physical or biological conditions go extinct. Some desert lizards are not able to cope with the increasing intensity of heat.	
Ocean Acidification and Deoxygenation The ocean becomes more acidic as it absorbs CO_2, causing a wide range of consequences, such as the erosion of mollusk shells (white spots on the shell in the image) and coral skeletons.	

Shifts in the geographic distribution of species can lead to changes in species interactions and mismatches between species and resources, both of which can affect the structure and function of ecosystems.

Phenology Shifts The timing of seasonal events, or **phenology**, is changing in many biomes. Spring arrives in the United States 10 days to two weeks earlier than it did just 20 years ago. As a result, many biological activities associated with spring occur earlier. Recent studies have confirmed that migratory bird species are arriving on their breeding grounds in the northeastern United States an average of 13 days earlier than in previous decades, and fruit trees are ripening as much as 18 days earlier.

Note, however, that different species in an ecosystem do not necessarily synchronize changes in their timing. Mismatches often result, disrupting trophic interactions and other types of species interactions (Chapter 52). For example:

- Snowshoe hares use daylight length as a cue to change their coat color from white to brown in the spring. Earlier snowmelts in spring are causing a mismatch between coat color and the ground color, increasing the vulnerability of hares to predation by lynx (see Table 53.1).
- Insectivorous birds like great tits (*Parus major*) rely on the abundance of caterpillars to feed their hatchlings. However, a 20-year study has shown that the peak caterpillar biomass now occurs earlier due to the warmer spring, but the reproductive timing of great tits has not changed. Thus, there is a mismatch that has resulted in fewer and lower-weight great tit fledglings.

Because each species is a component of a community, mismatches between two species can have a domino effect on other species and the whole ecosystem.

Evolutionary Adaptation In some populations, changing climate and species interactions are already causing allele frequencies to change—meaning that some species are evolving in response to global climate change. There are clearly both evolutionary winners and losers, depending on the ecological context and adaptive capacity of the organisms. For example:

- Increasing temperatures in the Netherlands have caused evolution in the two-spot ladybird beetle, *Adalia bipunctata*. There are two heritable color patterns of these beetles: black with red spots and red with black spots (see Table 53.1). The frequency of the black morph has declined from 60 percent to 20 percent in one area in just 30 years. It turns out that the black morphs have lower fitness under warmer conditions because they absorb more solar energy and thus overheat.
- Similarly, tawny owls, *Strix aluco*, have heritable variations in color, ranging from pale gray to brown. Increasingly mild winters over the last 30 years have reduced the average snow cover. This change has reduced selection against the brown owls, which are more conspicuous in snow than the gray owls, resulting in a dramatic increase in the frequency of the brown owls across the country over time.

Once again, because each species is a component of the community within each ecosystem, evolution within a species can have a domino effect on other species.

Extinctions Biologists are concerned that rapid changes in climate, coupled with other stressors such as land-use changes, will cause high extinction rates (discussed in detail in Chapter 54).

- A team of ecologists determined that 12 percent of local lizard populations have gone extinct in Mexico since 1975 because the temperature is changing too rapidly for the lizard populations to adapt (see Table 53.1).
- As the climate warms, alpine tundra habitats are disappearing. Some plants and animals that have restricted geographic ranges and grow on mountaintops are running out of habitat.
- A marsupial mammal called the Bramble Cay melomys has gone extinct in Australia because its island habitat has been eliminated by sea-level rise.

Coral reefs are the subject of intensive research because they are among the most productive and species-rich ecosystems in the world. Research has shown that the increasing frequency of extreme warming events is causing many corals to bleach—the "whitening" phenomenon that occurs when corals lose their mutualistic photosynthetic protists, the primary producers in the coral-reef ecosystem. Corals have slow generation times (5–100 years) and tend to be very sensitive to temperature changes, so biologists fear that many corals are at risk of extinction—they may be unable to acclimatize, adapt, and/or disperse to more favorable environments rapidly enough. To complicate matters, temperature combines with other stressors, such as ocean acidification, to threaten coral ecosystems.

Ocean Acidification and Deoxygenation Although the ocean's capacity to absorb CO_2 is not limitless, it is still vast—it absorbs about one-third of global CO_2 emissions per year. This absorption has complex consequences for marine ecosystems.

When atmospheric CO_2 dissolves, it reacts with water to form bicarbonate (H_2CO_3), which then drops a proton to form carbonic acid (HCO_3^-). Increased CO_2 levels in the atmosphere are increasing the rate of this reaction, and the protons that are released are lowering the pH of the oceans. Since the beginning of the industrial revolution, the pH has decreased by 0.1, which translates to a 26 percent increase in acidity. ✔ **If you understand this concept, you should be able to predict what happens to the pH of water when you blow bubbles into it with a straw.**

Seawater is now more acidic than it has been for 65 million years. How does acidity affect organisms?

- Increased acidity reduces the availability of calcium carbonate ions to organisms that build calcium carbonate skeletons—such as corals, mollusks, and echinoderms (see Table 53.1). The more acidic the water, the slower the skeletons form. In some conditions, skeletons can dissolve altogether.
- Increased acidity reduces the ability of some larval fish to smell predators.
- Increased acidity reduces the ability of some sperm to swim.

While aquatic environments are becoming more acidic, they are also becoming deoxygenated. The oxygen in the oceans has declined about 2 percent since 1960. Warmer water has a reduced capacity to store dissolved O_2 and also a reduced tendency to mix with deeper, denser water. Deoxygenated water reduces the metabolic capacity of organisms in aquatic environments and has potentially far-reaching effects on ecosystems.

Biological Effects Interact in Ecosystems Understanding the complex interaction of biotic and abiotic factors in ecosystems—and predicting future changes—are among the most challenging and important tasks that ecologists face today.

Ecologists are discovering that some ecosystems are resilient. That is, climate change and other factors cause changes to individual species or species interactions, but ecosystem structure and function remain intact overall. However, some ecosystems are not as resilient. Factors sometimes combine in ways that cause a "tipping point" to occur, a sudden shift in the ecosystem that is not reversible, such as the sudden shift of a forest to a grassland due to the combined effects of climate change, forest fires, and invasive grasses, or the sudden shift of a coral reef to an algal bed due to the combine effects of rising temperatures, ocean acidification, and pollution.

Ultimately, the flow of energy and the cycling of nutrients through ecosystems depends heavily on net primary productivity. Let's see how climate change affects NPP.

Consequences to Net Primary Productivity

As emphasized earlier in this chapter, NPP is the basis of both energy flow and nutrient cycling in ecosystems. Thus, ecologists are closely monitoring effects of climate change on NPP.

How Is NPP Changing on Land? A study from 1982 to 1999 showed an overall increase in global terrestrial productivity. However, a study from 2000 to 2014 revealed a different trend. The green areas of the map shown in **Figure 53.24** on page 1162 indicate increased productivity, while the blue areas indicate decreased productivity. Overall, the global terrestrial NPP in the Southern Hemisphere declined.

This result does not seem intuitive at first, because an increase in atmospheric CO_2 would be expected to increase rates of photosynthesis. The researchers of this study point out that large-scale droughts in the Southern Hemisphere have decreased NPP in some of the world's most productive forests, counteracting the overall increase in NPP in the Northern Hemisphere.

More recent studies have supported this conclusion. For example, a drying trend in the Amazon rain forest, combined with clear-cutting and burning, has turned one of the world's largest **carbon sinks**, or long-term carbon reservoirs, into a carbon source. These data illustrate the complex interaction of effects of CO_2, temperature, and precipitation on productivity. Further, many terrestrial ecosystems are nutrient limited. Therefore, even if CO_2 increases, there may be insufficient N and P to support increased productivity, especially in areas where intense rain events increase nutrient export from the soil to aquatic ecosystems.

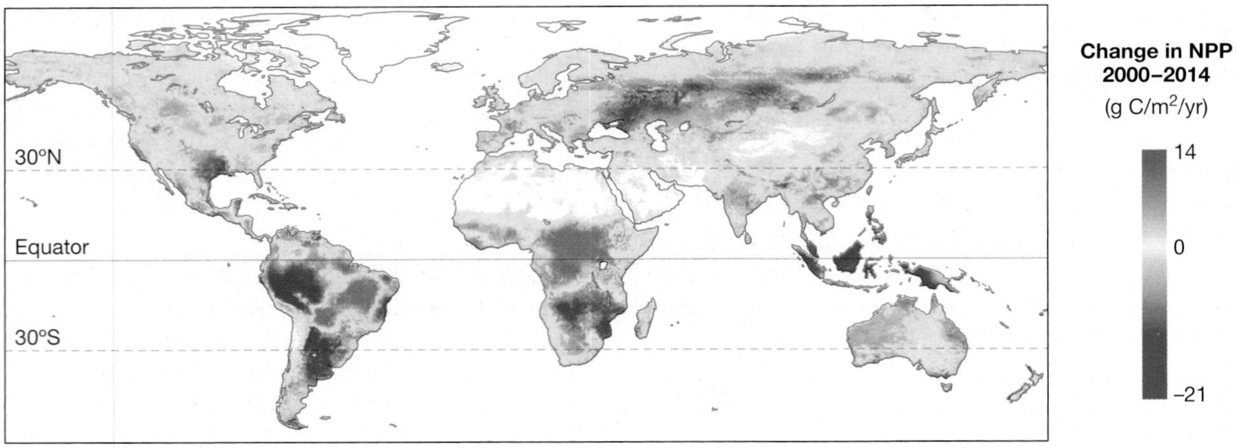

Figure 53.24 **Average Terrestrial NPP Has Declined in the Southern Hemisphere.** The data represent the average annual change in terrestrial NPP from 2000 to 2014. NPP has increased in some areas and declined in others.

DATA: Z. Li et al. 2016. *Hydrology and Earth System Sciences* 20: 2169–2178.

How Is NPP Changing in the Oceans? As on land, there is considerable variation in local productivity in the oceans. For example, productivity has dropped dramatically in large areas of the pelagic zone, but it has increased in a number of other regions. Overall, scientists agree that there is a general correlation between surface water temperatures and decreased productivity—NPP is decreasing in the oceans. What's going on?

One of the leading explanations for the drop in marine productivity hinges on changes in water density. To understand the logic behind this hypothesis, recall that lakes can become stratified—that is, have layers of different densities (Ch. 49, Section 49.5). Stratification occurs because water reaches its highest density at 4°C and becomes much less dense at higher temperatures.

Seawater also becomes stratified. This layering is important because the 4°C water in the benthic zone is nutrient rich due to the rain of decomposing organic material from the surface. When warmer surface water becomes lighter, water currents are much less likely to be strong enough to overcome the density difference and bring nutrient-rich, 4°C water all the way up to the surface, where nutrients can spur the growth of photosynthetic bacteria and algae (Figure 53.25). In this way, the warming global climate is causing surface waters to become more nutrient poor.

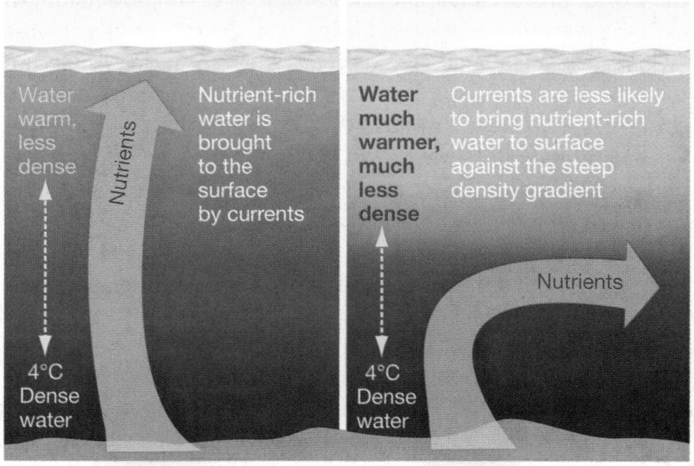

Figure 53.25 **Temperature Stratification in the Ocean Affects Productivity.**

Local and Global Consequences What are the consequences of these unprecedented global changes in NPP? There are two main take-home messages:

1. Changes in NPP cause feedback on the rate and pattern of climate change. If terrestrial and marine NPP continue to decline, the rate of absorption of CO_2 will decline, exacerbating climate change with far-reaching effects to humans and other organisms.
2. Changes in NPP combine with other factors, such as changes in precipitation patterns and ocean acidification, to alter food web dynamics—and thus ecosystem function. Maintaining ecosystem function is critical to the preservation of biodiversity and the ecosystem services that humans rely upon for survival (described in detail in Chapter 54).

In this chapter, you considered the theme that everything in nature is connected. As you've seen, humans are very much a part of these connections. Understanding how humans fit into the big picture is key to taking actions to improve ecosystem health and slow global climate change.

CHECK YOUR UNDERSTANDING

✔ **If you understood this section, you should be able to ...**

1. Explain to a fourth-grade student, an elderly neighbor, a parent, or your best friend why the climate is changing.
2. **SOCIETY** Provide at least two hypotheses to explain why increased CO_2 levels in the atmosphere are decreasing, rather than increasing, the productivity of some crops.

Answers are available in Appendix A.

53.1 How Does Energy Flow through Ecosystems?

- Net primary productivity (NPP) is the foundation of ecosystems because it represents the energy and nutrients, in the form of biomass, available for consumption by other organisms. Humans are appropriating a quarter of NPP.
- Energy is dissipated as it flows from producers to consumers and decomposers in a food web.
- Because energy transfer from one trophic level to the next is inefficient, ecosystems have a pyramid of productivity: Biomass production is highest at the lowest trophic level and lower at each higher trophic level.
- Unlike nutrients, persistent toxins can be magnified in concentration as they are passed up trophic levels in a food web, a process called biomagnification.
- Among terrestrial ecosystems, NPP is highest in tropical wet forests and tropical dry forests. The productivity of terrestrial ecosystems is limited by temperature and moisture.
- Among aquatic ecosystems, productivity is highest in coral reefs, wetlands, and estuaries. The productivity of aquatic ecosystems is limited by the availability of nutrients and sunlight.
- Although productivity is extremely low in the open ocean, the area covered by this ecosystem is so extensive that it accounts for the highest percentage of Earth's overall productivity.

53.2 How Do Nutrients Cycle through Ecosystems?

- Unlike energy that enters an ecosystem and is ultimately dissipated as heat, nutrients such as carbon and nitrogen move indefinitely through ecosystems in biogeochemical cycles.
- To analyze nutrient cycles, biologists focus on the reservoirs where elements reside and the processes that move elements between reservoirs.
- The rate of nutrient cycling is strongly affected by the rate of decomposition of detritus.
- Nutrients "leak" from one ecosystem to another when biomass is exported. Experiments have shown that vegetation prevents nutrient export by reducing runoff.
- Humans are harvesting fresh water from reservoirs faster than it can be replenished by natural processes.
- Human activities have doubled the amount of usable nitrogen that enters ecosystems each year, leading to increased productivity and to pollution, as well as to loss of biodiversity.
- The burning of fossil fuels by humans releases carbon from an inactive geological reservoir to the atmosphere. Current CO_2 levels are 30 percent higher than the highest level measured over the last 800,000 years.

53.3 Global Climate Change

- Average global temperatures are increasing rapidly because land-use changes and burning of fossil fuels have increased the flow of carbon in the form of CO_2 into the atmosphere. CO_2 is one of the greenhouse gases that trap heat in the atmosphere. (Note that the blocking of UV light by the ozone layer and the trapping of heat by greenhouse gases are different atmospheric phenomena.)
- Both positive (increasing) and negative (decreasing) feedback loops are occurring in response to global warming.
- Average temperature is increasing, but local temperatures and precipitation vary. The variability of temperature and rainfall is also increasing, including extreme weather events.
- Biologists have documented an array of biological consequences of climate change, including changes to geographic ranges, phenology, adaptations, extinctions, and ocean acidification and deoxygenation.
- The warming associated with climate change is causing a decrease in average NPP both on land and in the oceans, influencing both climate and ecosystem function.
- The climate is certain to continue changing over the next centuries, but the degree of change depends on human behavior.

Answers are available in Appendix A.

✔ TEST YOUR KNOWLEDGE

1. What is the main global reservoir of nitrogen?
2. True or False: Most of the net primary productivity that is consumed is used for growth by primary consumers. Explain.
3. Which of the following is the longest-lived reservoir for carbon?
   a. atmospheric CO_2
   b. marine plankton (primary producers *and* consumers)
   c. fossil fuels
   d. wood
4. Devegetation has what effect on ecosystem dynamics?
   a. It increases belowground biomass.
   b. It increases nutrient export.
   c. It increases NPP.
   d. It increases soil organic matter.

✔ TEST YOUR UNDERSTANDING

5. **QUANTITATIVE** If the GPP of a grassland is 5000 kcal/m^2/year and 55 percent is used up by cellular respiration, what is the NPP?
   a. 2250 kcal/m^2/year
   b. 2750 kcal/m^2/year
   c. 5000 kcal/m^2/year
   d. need more data
6. Explain why decomposition rates in a field in Nebraska would differ from the decomposition rates in a field in the Amazon. How do decomposers regulate nutrient availability in ecosystems?

7. Why are the open oceans nutrient poor? Why are coastal areas and intertidal habitats relatively nutrient rich?

8. Explain why it is more energy efficient (in terms of the amount of total NPP required) to eat a pound of tofu (bean curd) than a pound of hamburger.

✔ TEST YOUR PROBLEM-SOLVING SKILLS

9. **THINK CAREFULLY** Suppose that record snows blanket your campus this winter. Your friend says this is proof that global warming isn't really occurring. What is the flaw in your friend's logic?
   a. The average temperature of the Earth is not actually increasing.
   b. Global warming refers to temperatures, but snow is a type of precipitation.
   c. While the average global temperature is increasing, local temperatures and precipitation (weather) will vary.
   d. Your friend is confusing global warming and global climate change.

10. During the Carboniferous period, plant growth was extensive but rates of decomposition slowed (probably due to the formation of vast, oxygen-poor swamp habitats). As a result, large amounts of biomass accumulated in terrestrial environments (much of this biomass is now coal). The fossil record indicates that atmospheric oxygen increased, atmospheric carbon dioxide decreased, and global temperatures dropped. Explain why.

✔ PUT IT ALL TOGETHER: Case Study

Is the Amazon rain forest doomed?

Scientists around the world are collaborating to understand how deforestation, climate change, and natural processes will interact to affect one of the most productive and biodiverse ecosystems on Earth.

11. Which of the following types of ecosystem typically has the highest net primary productivity?
   a. temperate forest
   b. tropical dry forest
   c. tropical grassland
   d. tropical wet forest

12. Select True or False for each statement about the effects of clear-cutting on the Amazon rain forest, then explain your reasoning.
   T/F Nutrient export is likely to decline.
   T/F Atmospheric CO_2 is likely to decline.
   T/F Soil moisture is likely to decline.
   T/F Species diversity is likely to decline.

13. **PROCESS OF SCIENCE** Researchers have measured the effects of periodic forest fires on primary productivity in Amazon rain forest plots, comparing years with average precipitation and years with severe drought. Propose which controls would be necessary for such studies.

14. **MODEL** This box-and-arrow model summarizes some of the feedback links observed in the Amazon rain forest. Select True or False for the statements that follow, based on whether they are represented by the model. (Note that the boxes and arrows in this model are used differently than those in the nutrient cycle models; see **BioSkills 12**.)
   T/F The burning of fossil fuels increases atmospheric CO_2.
   T/F Tree growth reduces atmospheric CO_2.
   T/F The death of trees promotes the invasion of grasses.
   T/F An increase in CO_2 increases the frequency of droughts.
   T/F Drought increases the frequency of forest fires.

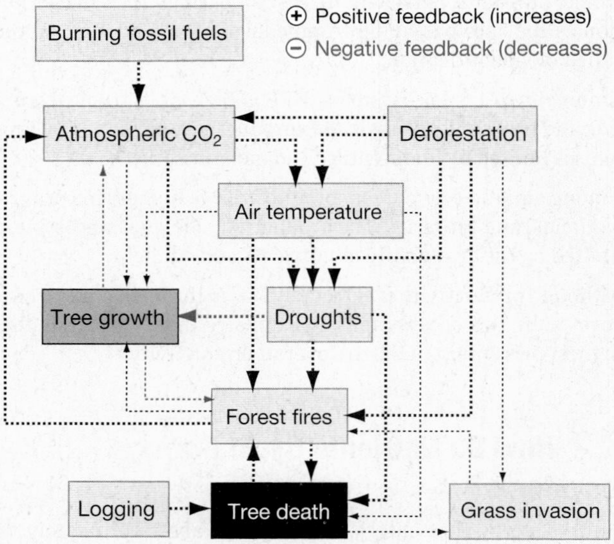

Source: P. M. Brando et al. 2014. *Proceedings of the National Academy of Sciences* 111: 6347–6352.

15. Many studies have raised the concern that positive feedback loops among numerous variables in the Amazon will cause an ecosystem tipping point—a rapid and irreversible transition from forest to grassland. Use the model in Question 14 as a tool to summarize a possible sequence of effects that could cause a transition from forest to grassland.

16. **SOCIETY** Journalists must be concise when reporting science news. Why might journalists and the public struggle to understand and discuss ecosystems ecology and global warming?

54 Biodiversity and Conservation Biology

The island of Borneo contains some of the most spectacular rain forests in the world, yet it is being cleared at an alarming rate to make way for oil palm plantations. This action threatens both biodiversity and ecosystem function. Fortunately, positive change is under way thanks to the actions of many educated people like you.

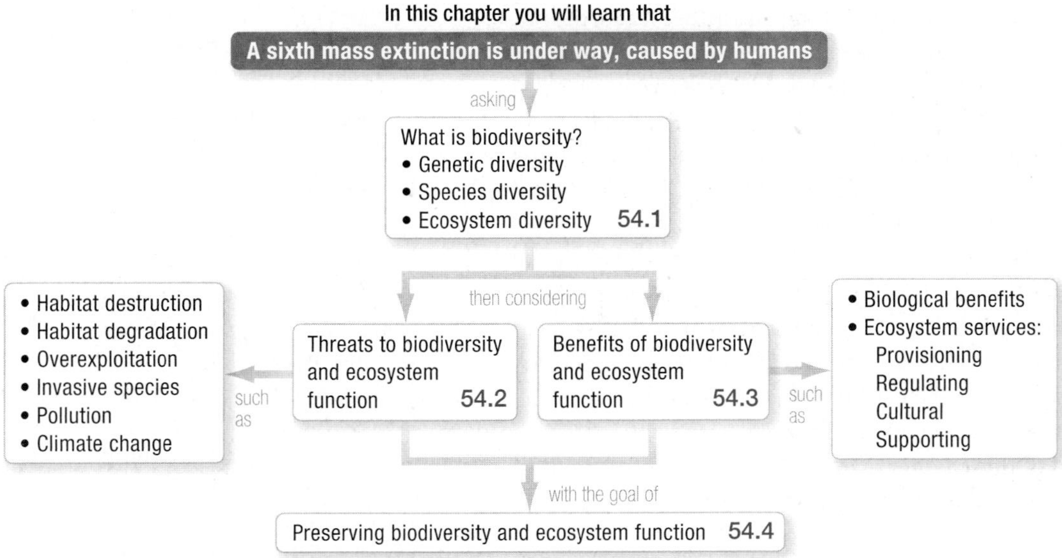

Borneo is famous for having some of the most luxuriant rain forests on the planet—15,000 species of plants, 222 species of mammals, and hundreds of birds, amphibians, and fishes. Six thousand of these species occur nowhere else on Earth. But when conservation biologist Willie Smits gazed out upon an area near Samboja on the Southeast Asian island of Borneo in 2002, he saw a devastated wasteland:

- almost total elimination of plant and animal life;
- depleted soil fertility and failing crops;
- unusually severe climate: droughts, floods, fires; and
- a destitute human population: 50 percent unemployed, high crime rate, poor nutrition, low life expectancy.

BIG PICTURE

This chapter is part of the Big Picture. See how on pages 1190–1191.

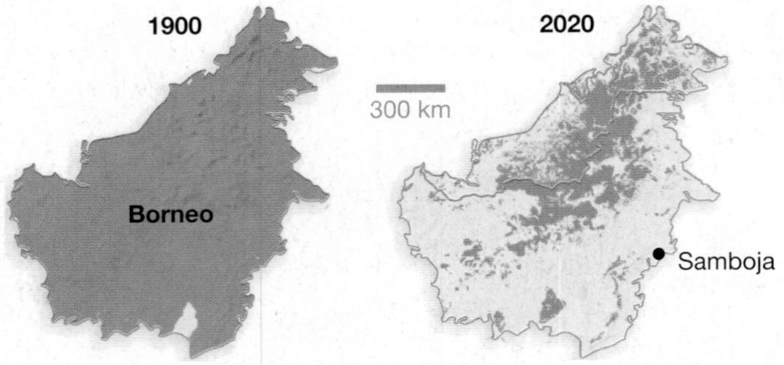

Figure 54.1 **Forest Is Disappearing in Borneo.** Borneo, a large island in Southeast Asia (see map in Figure 54.4), was once completely forested. By 2020, only a third of the forest will remain.
DATA: World Wildlife Fund; D. L. A. Gaveau et al. 2016. *Scientific Reports* 6: 32017.

What happened in Samboja? Starting in the 1950s, loggers started cutting down the valuable hardwood trees. Following the eventual clear-cut, the remnants were burned to make room for oil palm plantations to supply the world's demand for cooking oil, products such as lipstick, and biodiesel. In just four decades, Borneo lost more forest than would fit in Florida—a quarter of its land area. Figure 54.1 shows the loss of forest over the last century.

The orangutans have been one major casualty of this land transformation, which is how Smits became interested in Samboja (see chapter-opening photo). For these endangered apes, already threatened by poaching for the pet trade, the massive loss of habitat was the last straw. Smits realized that if he wanted to save the orangutans, he would have to save the forest.

The scenario at Samboja is not unique. As the human population soars (Ch. 51, Section 51.5) and the available resources plummet, more wild areas on Earth are being tamed, often with unintended consequences for ecosystems and humans.

So is the situation hopeless? No. The growing field of **conservation biology** is making progress in studying, preserving, and restoring biodiversity. Further, an increasing number of people around the world are realizing that preserving the environment does not go against human interests. On the contrary, preserving the environment helps people.

The goal of this chapter is to size up the threats to, and benefits of, biodiversity, and then explore some conservation success stories. Let's start by defining biodiversity.

54.1 What Is Biodiversity?

In essence, **biodiversity** means biological diversity. But this concept is more complex than it might seem. Let's take a closer look at the different levels of biodiversity, then see how biodiversity maps out globally.

After you complete this section, you should be able to ...

- Explain how biodiversity is measured.

Biodiversity Can Be Measured and Analyzed at Several Levels

To get a complete understanding of the diversity of life, biologists recognize and analyze biodiversity at the genetic, species, and ecosystem levels. Let's consider each level in turn.

Genetic Diversity **Genetic diversity** is the total genetic information contained within all individuals of a population, species, or group of species. It is measured as the number and relative frequencies of all genes (and their alleles) present in a sample.

At the genetic level, biodiversity is everywhere around you—and even inside you in a flourishing community of microbes that are essential to your health. Recently, efforts to catalog genetic diversity have been advanced by technical breakthroughs (Ch. 20, Section 20.3):

- *Genome sequencing* is a research protocol that sequences the entire genome of multiple members of the same species.
- *Environmental sequencing* is a research protocol that sequences all or most of the genes and/or alleles present in a sample taken from the soil or water present in a habitat (or your gut).

Environmental sequencing is a way to document genetic diversity even if the species present are unknown.

Genetic diversity is important because it represents the adaptive capacity of a population or higher taxonomic group—the ability of that group to persist over time despite changes in the environment.

Species Diversity The diversity of species present is a key feature of biological communities. It is usually quantified in two ways (introduced in Ch. 52, Section 52.4):

1. **Species richness** is a count of how many species are present in a defined area (which may vary in size from local to regional).
2. **Species diversity** is a weighted measure that incorporates both species richness and *evenness*, the relative abundance of each species present.

For a given species richness, species diversity is high if evenness is high, meaning that all the species have comparable abundance, but species diversity is low if just one or a few species dominate the community. This distinction can be quantified using mathematical methods such as the Shannon index (explained in Quantitative Methods 52.1 on page 1129) and/or visualized using models (see Making Models 52.2 on page 1128). Be aware that people sometimes use the terms "species richness" and "species diversity" interchangeably (which can be confusing).

Ecologists traditionally measure species diversity by spending hours at their field sites employing transects, traps, quadrats, mark-and-recapture techniques, and other methods to quantify the number of species present and the abundance of individuals of different species (Ch. 51, Section 51.1). This important work can be painstaking, especially in biodiverse areas such as the forests of Borneo—where, for example, a lamp can be used to attract hundreds of moth species in a single night.

One new technique for identifying species is called **DNA barcoding**—the use of well-characterized gene sequences to distinguish species. The idea is that variation in one gene is higher between species than within species. For example, variation in the gene for mitochondrial cytochrome *c* oxidase subunit 1 can be used to distinguish among animal species. Thus, when biologists take a tissue sample from a specimen in the field, they can find the match for the sequence of a target gene by comparing it to those in a reference library of known sequences. To date, tens of thousands of species have been barcoded.

Supporters of barcoding say this technique is revolutionary because it helps nonspecialists to identify species. Also, it enables species to be identified even when the entire organism is not available, such as when a piece of bone is found.

Other scientists caution against the use of DNA barcoding to replace traditional taxonomy. One argument is that sequences of one gene do not accurately represent the genetic variation at the species level—which can lead to ambiguous or incorrect results. Moving forward, DNA barcoding methods will continue to improve, and different combinations of barcoding and traditional methods will be used to identify species depending on the specimen available and the questions being asked.

When quantifying species diversity, it's also important to remember that counting species is not like counting beans in a bowl—the evolutionary and ecological contexts of all species are not equal. Thus, ecologists increasingly turn to newer measures of diversity at the species level:

- **Phylogenetic diversity** In terms of evolution, some species give a community evolutionary distinctiveness because they are phylogenetically distant from other species. For example, a conservationist might prioritize the survival of the single living species of red panda over one of the 15 species of raccoons because if a species like the red panda is wiped out, its entire lineage could be lost forever. (See Ch. 25, Section 25.1, for an introduction to phylogenetic trees.)
- **Functional diversity** In terms of ecology, some species have functional traits that play a key role in the function of an ecosystem: They may be key primary producers, important ecosystem engineers creating habitat for other species, keystone predators, or nitrogen fixers (Chapters 52 and 53). When functional diversity is a priority, preservation of species with unique functional traits is important.

Table 54.1 uses simple models to illustrate the differences between species diversity, phylogenetic diversity, and functional diversity. Ecologists can employ one or all of these measures depending on the data available and the questions being asked.

SUMMARY Table 54.1 Diversity Measures of Species

	High Diversity	Low Diversity
Species Diversity Number of species (**species richness**) and the **evenness** of the species (see Quantitative Methods 52.1)	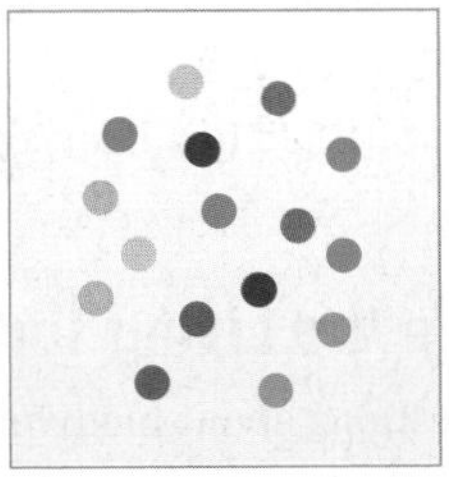 Species richness = 8 Evenness = high	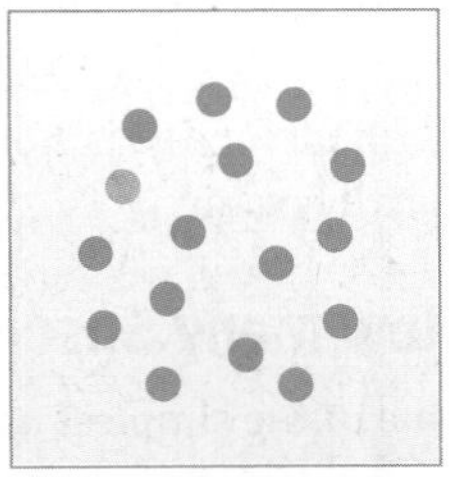 Species richness = 2 Evenness = low
Phylogenetic Diversity Evolutionary distinctiveness, measured using branch lengths among species in a phylogeny	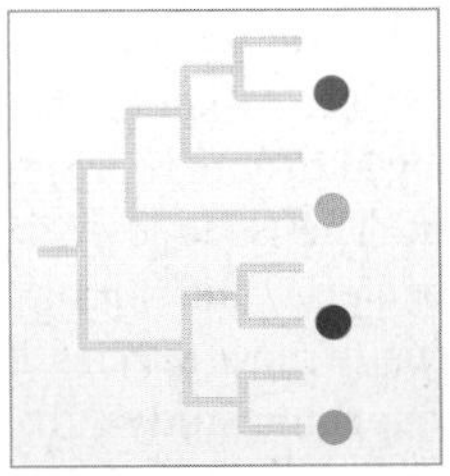 Species richness = 4 Relationships = distant	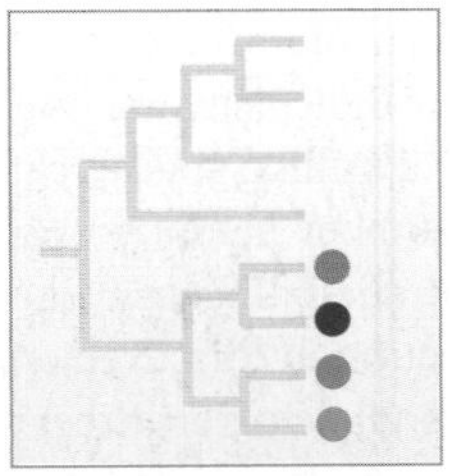 Species richness = 4 Relationships = close
Functional Diversity Ecological distinctiveness, measured by categorizing and counting the functional traits of species	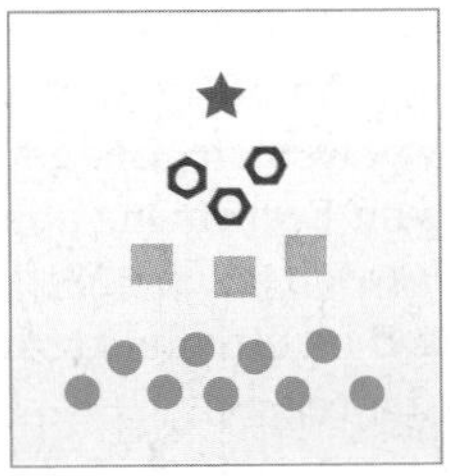 Species richness = 4 Functional groups = 4	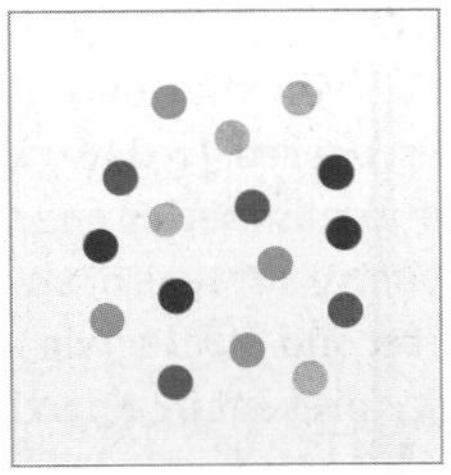 Species richness = 4 Functional groups = 1

✔ **MODEL** Draw a simple model comparing two communities that have the same species richness (4 species) and the same functional diversity (2 functional groups) but different species diversity.

Ecosystem Diversity Communities vary in the number of species that occur within each trophic level (horizontal diversity) and the number of trophic levels (vertical diversity). **Ecosystem diversity** is a measure of these complex factors plus interactions with the abiotic environment. It is the most difficult level of biodiversity to measure because the complexity of ecosystems is so vast.

Intertwined with the concept of ecosystem diversity is **ecosystem function**, which refers to the sum of biological and chemical processes that are characteristic of a given ecosystem—such as primary production, nitrogen cycling, decomposition, and carbon storage. These functions emerge from the sum of the feeding, growing, moving, respiring, excreting, and decomposing processes of the interacting member organisms within their abiotic context. Preserving ecosystem function is an increasingly urgent priority in conservation biology.

Note that the three levels of biodiversity are not independent of each other. For example, consider what might happen if the sole species of nitrogen-fixing bacteria disappeared from a grassland community: The genes that code for the nitrogen-fixing enzymes would be lost, followed by loss of the species that depend on that nitrogen as well as loss of important nutrient cycling functions of the ecosystem.

Change through Time Biodiversity can be recognized and quantified on several distinct levels, but it is also dynamic.

- Mutations that create new alleles increase genetic diversity; natural selection, genetic drift, and gene flow may eliminate certain alleles or change their frequencies, leading to an increase or decrease in overall genetic diversity (Chapter 23).
- Speciation increases species diversity; extinction decreases it (Ch. 25, Sections 25.3 and 25.4).

- Changes in climate or other physical conditions can result in the formation of new ecosystems or ecosystem functions—such as the formation of coral reef atolls from eroding volcanic islands. Disturbances such as volcanic eruptions and human activities can destroy ecosystems or ecosystem functions.

Biodiversity is not static. It has been changing since life on Earth began.

How Many Species Are Living Today?

One of the simplest questions about biodiversity is also one of the most difficult to answer: How many species are there on Earth? The answer is not known, but educated guesses range from 5 to 100 million. Given the massive effort that it would take to document every form of life on Earth, the answer will probably never be known.

Biologists are well aware, though, that the approximately 1.5 million species cataloged to date represent a tiny fraction of the number actually present. The issue is even more acute in poorly studied lineages like bacteria and archaea, where researchers routinely discover dozens of new species in each environmental sequencing or direct sequencing study (Ch. 20, Section 20.3). For example, a direct sequencing study estimated that over 500 species of bacteria live in the human mouth, but only about half had been named already. New species are discovered every year, even among well-studied groups like birds and mammals.

Given that only a fraction of the organisms alive have been discovered to date, how can biologists go about estimating the total number of species on Earth? One approach is based on intensive surveys of species-rich taxonomic groups in defined sites; a second strategy is based on attempts to identify all of the species present in a particular region.

Taxon-Specific Surveys In a taxon-specific survey, biologists conduct an intensive search for a specific taxon in a well-defined area. For example, Philippe Bouchet and colleagues conducted a massive survey of marine mollusks in coral-reef habitats along the west coast of New Caledonia, a tropical island in the southwest Pacific Ocean. The team spent more than a year collecting mollusks at 42 sites over a total area of almost 300 km^2. The effort produced more than 127,000 individuals representing 2738 species. These numbers far exceed the mollusk diversity recorded for any comparable-sized area.

In reporting their findings, Bouchet and his team emphasized that 20 percent of the species found were represented by a single specimen. This observation suggests that many species are exceedingly rare and thus likely to be missed by less intensive sampling efforts. And when the investigators compared their data with a survey in progress at a second site—different in reef structure but only 200 km away from the original site—they found that only 36 percent of species were shared between the two sites. Taken together, the results support the hypothesis that the 85,000 known mollusk species may represent just a third to a half of the actual total.

All-Taxa Surveys As the name suggests, in an all-taxa survey, biologists attempt to find and catalog all of the species present in an area. The most spectacular example of an all-taxa survey so far is the Census of Marine Life, a massive international effort to record the global diversity of life in the oceans. About 2700 scientists from more than 80 countries launched 540 expeditions over a 10-year period, ending in 2010, that surveyed the number and distribution of marine species around the globe—ranging from deep-sea vents to coral reefs (Figure 54.2) to intertidal mudflats.

The participants collected over 30 million records, including 6000 potentially new species. The total number of marine species is currently estimated by the census to be 250,000.

Where Is Biodiversity Highest?

Both taxon-specific surveys and all-taxa surveys provide data on the total number of species as well as on their geographic distributions. What patterns of biodiversity emerge?

Mapping Species Richness Researchers use various strategies for mapping global biodiversity data. The most common strategy is to divide the world's land area into a grid of parcels and plot the number of species (species richness) that occurs in each parcel. The accuracy of this method relies on the degree to which search effort is distributed across parcels worldwide.

Sampling efforts have been most extensive for terrestrial vertebrates, so these animals are often used as an indicator of

Figure 54.2 **Species Richness Can Be Estimated via All-Taxa Surveys.** Scuba divers survey coral reef diversity on the Great Barrier Reef in Australia.

Figure 54.3 Species Richness Is Not Evenly Distributed Globally. The map shows areas where the number of species of birds, mammals, and amphibians are highest.
DATA: P. D. Mannion et al. 2014. *Trends in Ecology and Evolution* 29: 42–50.

general biodiversity patterns. Figure 54.3, for example, shows a map of global biodiversity of birds, mammals, and amphibians, combining data from many sources. The dark red area on the map shows that biodiversity is exceptionally high along the western rim of the Amazon basin.

The areas of highest biodiversity are different for different taxonomic groups. However, some trends do appear.

- The most prominent geographic pattern in the distribution of biodiversity corresponds to latitude: In most, but not all, taxonomic groups, species richness is highest in the tropics and declines toward the poles (Ch. 52, Section 52.4).
- Biodiversity is also higher on land than in the sea. Even though land covers only about 30 percent of Earth's surface, about 9 out of 10 species (microbes aside) are terrestrial.
- Areas with greater geographical variation tend to have higher biodiversity than areas that are flat or featureless.

On land, tropical rain forests are particularly species rich. Even though they represent just 7 percent of Earth's land area, they are thought to contain at least 50 percent of all species present. For example, half of a square kilometer of tropical forest in Borneo contains 1175 tree species, which is more than all of the temperate forest in Europe, Asia, and North America combined. In the sea, coral reefs are especially species rich.

An active research area today involves understanding *why* tropical forests and coral reefs are so species rich and *why* the other biodiversity patterns occur. The mechanisms that determine these patterns are complex because they combine local abiotic factors, biotic factors, and historical context (Ch. 49, Section 49.2; Ch. 52, Section 52.4).

For example, one hypothesis to explain the high biodiversity of tropical forests involves both productivity and evolutionary time: Net primary productivity (NPP; see Ch. 53, Figure 53.6) is high in the tropics due to the abundance of sunlight and precipitation, and the large biomass of producers supports many consumers. Further, given that tropical ecosystems have persisted for a relatively long time (without being scoured by ice sheets), there has been plenty of time for speciation and the creation of new niches, increasing the diversity of communities. Researchers continue to test this hypothesis and others.

Mapping Endemism, Phylogenetic Diversity, and Functional Diversity Increasingly, biologists are looking beyond patterns of species richness alone to see how other metrics of diversity map out. Here are some variables being studied:

- Numbers of **endemic species**—meaning, species that are found in a particular area and nowhere else
- Phylogenetic diversity (see Table 54.1)
- Functional diversity (see Table 54.1)

As you might expect, sometimes these variables are highly correlated. An area that has high species richness is also likely to have high phylogenetic diversity and high functional diversity. However, the maps of each variable do not necessarily overlap exactly. Ecologists can use powerful computer models to identify areas where interesting exceptions occur, and areas where biodiversity is highest in *all* of its measures.

Mapping Biodiversity Hotspots So far, the focus has been on identifying where biodiversity is highest. But if conservation is the goal, then it helps to consider where diversity is highest *and* where it is most threatened.

In 1988, conservation biologist Norman Myers pioneered the term **biodiversity hotspot** to characterize regions that are in

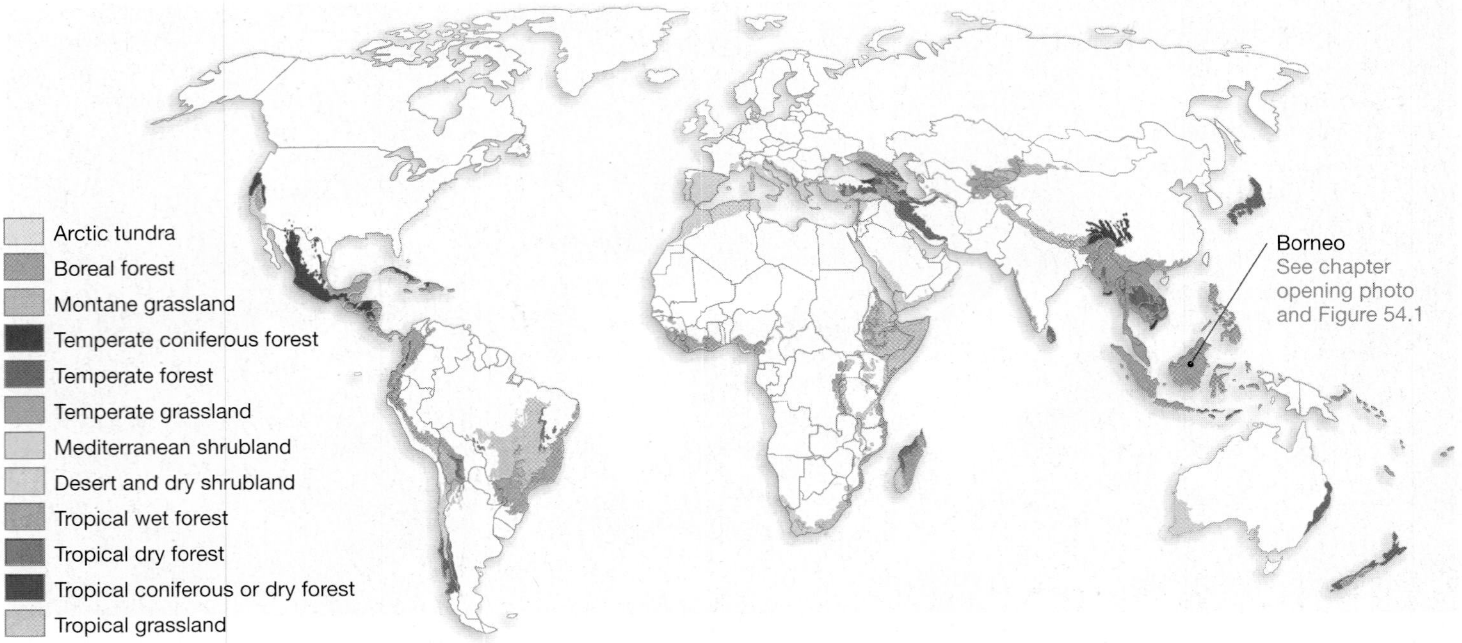

Figure 54.4 Biodiversity Hotspots Are a High Priority for Conservation. Areas that have the same color belong to the same biome (shown in Ch. 49, Figure 49.9).

DATA: S. Sloan et al. 2014. *Biological Conservation* 177: 12–24.

most urgent need of conservation action—areas where efforts to preserve habitat would have the highest return on investment. Since 1988, Myers and many collaborators, including organizations such as Conservation International, have identified 35 hotspots, shown in Figure 54.4, that meet two main criteria:

1. They contain at least 1500 endemic vascular plant species (more than 0.5 percent of the world's total).
2. They have lost at least 70 percent of their primary vegetation.

Vascular plants were used as a metric in identifying hotspots due to their importance in determining NPP. Although the hotspots defined by these criteria represent just 17 percent of Earth's land area (excluding Antarctica), they support

- 50 percent of all plant species
- 77 percent of all terrestrial vertebrate species
- 42 percent of all endemic terrestrial vertebrate species

One criticism of the second criterion of biodiversity hotspots is that it considers cumulative past loss, not current rate of loss. Most notably, hotspots do not include the rain forests found in the Amazon, the Congo River basin, and New Guinea—because those environments have not *yet* lost 70 percent of their primary vegetation. Effectively preserving these areas along with the hotspots would afford protection for over 65 percent of all land plants, found in just 5 percent of Earth's land surface.

Another criticism of hotspots is that they focus on species diversity and endemism but do not consider phylogenetic diversity and functional diversity. These variables will be easier to incorporate on a global scale as more data become available.

Maps of conservation priorities are needed only because the threats to biodiversity are profound. Let's take a closer look.

CHECK YOUR UNDERSTANDING

✔ If you understood this section, you should be able to . . .

Explain how you could measure biodiversity in the rain forests of Borneo at the genetic, species, and ecosystem levels.

Answers are available in Appendix A.

54.2 Threats to Biodiversity and Ecosystem Function

No species lasts forever. Extinction, like death, is a fact of life. If extinction is natural, why are biologists so concerned about habitat and species conservation? The answer is rate.

After you complete this section, you should be able to . . .

- Explain how human activities threaten biodiversity and ecosystem function.

Today, species are vanishing faster than ever before in Earth's history—other than when a gigantic asteroid smashed into Earth and wiped out most of the dinosaurs 66 million years ago (Ch. 25, Section 25.4). Modern rates of extinction are 100 to 1000 times greater than the average, or "background," rate recorded in the fossil record over the past 550 million years. For example:

- During the last 3000 years, Earth has lost *half* of its large terrestrial mammals.
- According to data on the current rates of extinction in birds and other well-studied groups, 60 percent of all remaining species are likely to be wiped out within 500 years.

The sixth mass extinction in the history of life is occurring now.

Either directly or indirectly, current extinctions are being caused by the demands of a rapidly growing human population—currently at over 7 billion and increasing by about 80 million people per year (Ch. 51, Section 51.5). The island we call planet Earth is far more crowded than it used to be.

How do biologists track extinctions that are under way? Beginning in 1994, the International Union for the Conservation of Nature (IUCN) began compiling the Red List to quantify the risk of extinction of different species. For example, the data in Figure 54.5 show that an astonishing 13 percent of birds, 25 percent of mammals, and 41 percent of amphibians are threatened—that is, they are either Critically Endangered, Endangered, or Vulnerable. An **endangered species** is a species whose numbers have decreased so drastically that it is almost certain to go extinct unless effective conservation programs are put in place. Orangutans, for example, are critically endangered.

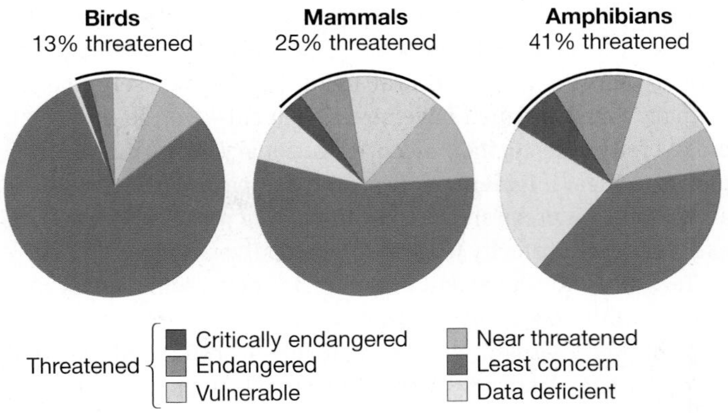

Figure 54.5 IUCN Designations Identify Threatened Species. The total number of threatened species in each group was estimated assuming that the species in the gray (data-deficient) category have the same rate of being threatened as the others. Extinct species are not included.

DATA: IUCN 2017. The IUCN Red List of Threatened Species. Version 2017.3.

Multiple Interacting Threats

Why are so many species endangered? Threats can be challenging to categorize and quantify because multiple threats often overlap and interact. Further, there is woefully insufficient data for most species, especially plants and invertebrates. Thus, most reviews of threats focus on species that are most intensively studied, mostly vertebrates.

In one review of nearly 2000 threatened vertebrate species, researchers tallied the types of threats experienced by each species, then calculated the proportions of the total that fell into five major categories. The results in Figure 54.6 reveal several patterns:

- So far, habitat loss is the single most important factor in the decline of the species sampled, especially in terrestrial and freshwater environments.
- Overexploitation is the dominant problem for marine species.
- So far, invasive species and pollution have had a larger impact on freshwater species than terrestrial and marine species.
- So far, climate change has had a larger impact on marine species than freshwater and terrestrial species.

These data give you a sense of general trends. As you might imagine, the importance of different types of threats varies dramatically among taxonomic groups. For example, overexploitation is by far the major threat to marine fishes, but it is only a minor threat to marine birds.

Are you curious why eating a burger is shown as a threat in Figure 54.6? This symbol emphasizes that human food systems alone are major contributors to all of the types of threats listed. Keep this in mind as you explore each threat in more detail.

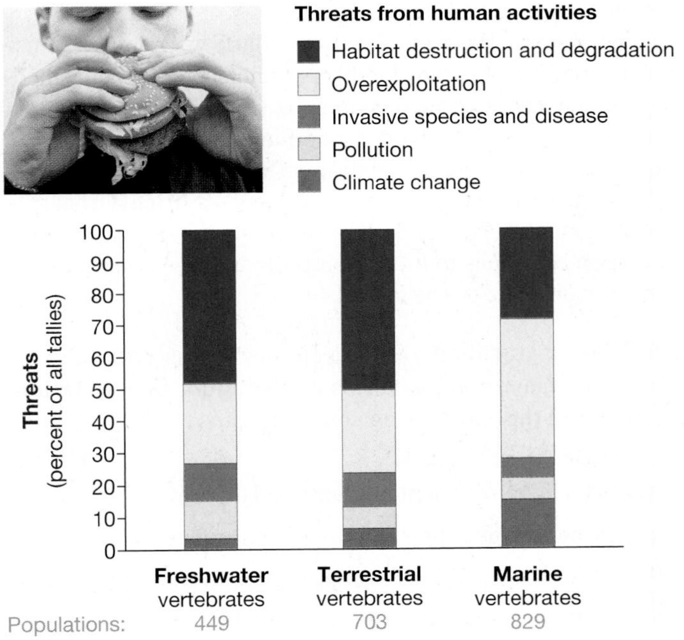

Figure 54.6 Endangerment Is Caused by Multiple Factors. Since each type of organism can experience more than one threat, the results show proportions of all threats documented for all the populations sampled.

DATA: WWF 2016. *Living Planet Report.*

✔ **QUANTITATIVE** What proportion of threatened terrestrial species experience habitat destruction or degradation?

Habitat Destruction Humans cause **habitat destruction**, the physical and functional loss of habitats, by many methods:

- Logging and burning forests
- Damming rivers
- Dredging or filling estuaries and wetlands
- Plowing prairies and grazing livestock
- Excavating minerals and fuels
- Building housing developments, golf courses, shopping centers, office complexes, airports, roads, and other structures

Today, less than a quarter of Earth's land is in its wild state (see map in Ch. 49, Figure 49.9), and most of the wild land is near the poles, where biodiversity is lowest.

On a global scale, one of the most important types of habitat destruction is deforestation—especially the conversion of primary forests to agricultural fields and human settlements. Primary forests—which have never been cut—tend to have high biodiversity due to their deep evolutionary history and complex ecological interactions. Research comparing primary forest and reforested areas shows that preserving primary forest is an extremely high priority for sustaining biodiversity.

To appreciate the extent of deforestation in some areas, consider satellite images of wet tropical rain forest in the Brazilian state of Mato Grosso that were made in 1992 and in 2006 (Figure 54.7). Analyses of satellite photos such as these have shown that as many as 3 million hectares (ha) were deforested in the Amazon each year during the 1990s. (A hectare is 100 meters by 100 meters—approximately the size of two football fields.)

More recent analyses suggest that the global rate of deforestation has now slowed compared to an average net loss of 16 million ha/year in the 1990s. But conservationists greet this news with caution. Net losses dropped because of extensive forest regrowth in China, North America, and Europe, even though over 40 million ha of primary forest have been lost in South America, Southeast Asia, and Africa since 2000.

Forest losses in the tropics are particularly worrisome because these forests contain many biodiversity hotspots, and they moderate the effects of climate change by storing carbon. Further, when forests are cut, cloud formation—thus precipitation—is reduced and soil is eroded, setting off a positive feedback loop that can trigger a tipping point in the biome—shifting it from rain forest to savanna (Ch. 53, Section 53.3). Meanwhile, the human population continues to increase rapidly in these areas, putting even more pressure on the forests.

Figure 54.7 Rates of Deforestation in Tropical Wet Forests Are High. In the top satellite image from Mato Grosso, Brazil, the green areas are intact primary forest. In the lower image of the same location 14 years later, the light areas have been burned or logged and converted to agricultural fields and pastures.

Habitat Degradation Aside from destroying natural areas outright, human activities can also cause **habitat degradation**, a reduction in the quality of a habitat. Humans can degrade habitats in many ways, for example:

- Artificial lights disorient migrants and disrupt night foraging.
- Traffic noise interrupts communication and causes stress.
- Pet cats have contributed to at least 63 vertebrate extinctions.

One of the most pervasive patterns of habitat degradation is **habitat fragmentation**—dividing contiguous areas of natural habitats into small, isolated fragments. For some species, a narrow pipe intruding into their habitat is a sufficient barrier to limit dispersal. But by far the most serious primary cause of fragmentation is the millions of kilometers of roads and the subsequent degradation of habitat that their access enables.

Habitat fragmentation concerns biologists for several reasons. First, like habitat destruction, fragmentation can reduce habitats to a size that is too small to support some species. This situation is especially true for top predators such as wolves, mountain lions, and grizzly bears, which need vast natural spaces for feeding, finding mates, and reproducing successfully.

When top predators in food webs are lost, trophic cascades may occur, affecting species throughout the food web in complex ways (Ch. 52, Section 52.2). Trophic cascades often have unexpected impacts on many aspects of ecosystem function, ranging from susceptibility to fire, disease, and invasive species to changes in primary productivity and nutrient cycling.

Another concern about habitat fragmentation is that it can force species into a metapopulation structure (Ch. 51, Section 51.4). In species that lack adaptations for this type of population structure, the small, isolated populations are much more likely than large populations to be wiped out, for two reasons:

1. Catastrophic events such as storms, disease outbreaks, or fires exterminate small populations more readily than large populations. Biologists refer to episodes like this as stochastic—meaning, chance—events.
2. In the absence of gene flow, small populations suffer from inbreeding depression—because so few potential mates are available—and random loss of alleles due to genetic drift.

Genetic drift leads to the random loss or fixation of alleles, so it reduces genetic variation in populations (Chapter 23). This result is a concern, because if the environment changes due to the emergence of a new disease or climate change, the population may lack alleles that confer high fitness in the new environment.

Habitat fragmentation is also a concern because it creates an "edge effect" that is being documented in a long-term experiment in a tropical wet forest. In an area near Manaus, Brazil, that was slated for clear-cutting, ecologists William Laurance, Thomas Lovejoy, and their colleagues set up 66 square, 1-ha experimental plots that remained uncut. Thirty-nine of these study plots were located in fragments designed to contain 1, 10, or 100 ha of intact forest. Twenty-seven of the plots were set up nearby, in continuous wet forest. As **Figure 54.8** shows, the distribution of the study plots allowed the research team to monitor changes inside forest fragments of different sizes and to compare these changes with conditions in unfragmented forest. Note that many of the study plots were located on forest edges.

When the research group surveyed the plots at least 10 years after the initial cut, they recorded two predominant effects:

1. A startling drop in biomass, or the total amount of fixed carbon, took place in the forest edge plots (colored blue in the map). As the graph in the "Results" section of Figure 54.8 shows, most of the loss occurred within 2–4 years of the fragmentation event.
2. On average, biomass did not decrease in the forest interior plots that were used as a control (colored green in the map). But the closer the forest interior plots were to an edge, the more likely they were to experience biomass loss.

In tropical wet forests, most of the biomass is concentrated in large trees. Due to exposure to high winds and dry conditions, the edges of the experimental plots contained many downed trees. Follow-up work showed that slow-growing trees typical of undisturbed forests are being replaced by early successional "weedy" species in edge habitats (Ch. 52, Section 52.3). A large number of bird and understory plant species have disappeared as well. A study of 1700 vertebrate species found that 85 percent were affected by forest edges.

The major insight of this experiment and other research is clear: When habitats are fragmented, the quality *and* quantity of habitats decline drastically. Besides losing over 8 million ha of primary forest in South America, Southeast Asia, and Africa each year to clear-cutting, we are losing large amounts of high-quality habitat throughout the world to fragmentation.

Overexploitation **Overexploitation** refers to any unsustainable removal of organisms from the natural environment for use by humans. Overexploitation is especially important in marine environments (see Figure 54.6). Two-thirds of harvestable marine species are now considered fully exploited or depleted, and a recent study of the global fishing industry concluded that 90 percent of individuals in large-bodied fish species from bottom or open-water habitats have already been removed from the world's oceans (**Table 54.2**, page 1174). The affected species include tuna, swordfish, marlin, cod, halibut, and flounder—and many species of sharks, which are harvested for their fins. Unless stricter fishing limits are established quickly, these popular species will go extinct. Further, removal of these top predators has an immense impact on species at lower trophic levels, causing a cascade of effects on food webs (Ch. 52, Section 52.2).

RESEARCH

QUESTION: How does fragmentation affect the quality of tropical wet forest habitats?

HYPOTHESIS: Fragmentation reduces the quality of wet forest habitats.

NULL HYPOTHESIS: Fragmentation has no effect on the quality of wet forest habitats.

EXPERIMENTAL SETUP: Measure biomass in 66 study plots.

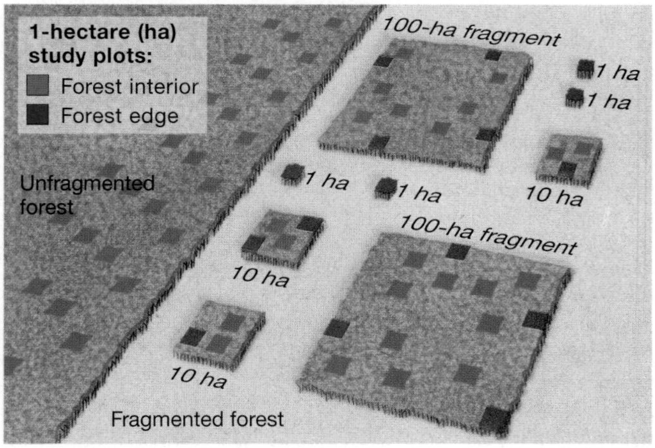

PREDICTION OF HYPOTHESIS: Biomass will decline in forest edge plots compared with those of the forest interior.

PREDICTION OF NULL HYPOTHESIS: Biomass will be the same inside forest fragments and along edges as in the forest interior.

RESULTS:

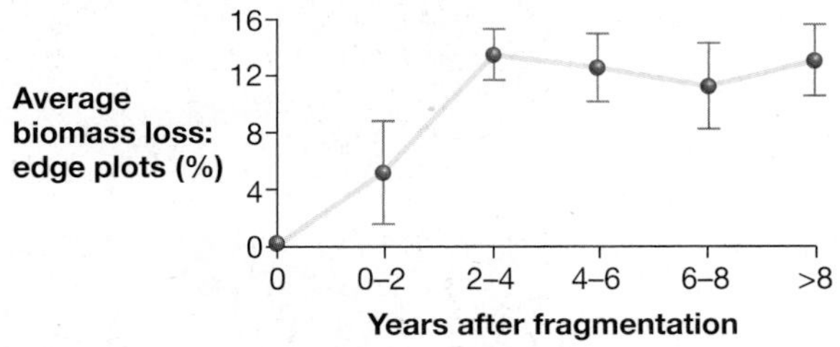

CONCLUSION: Biomass declines sharply along edges of forest fragments.

Figure 54.8 Experimental Evidence Shows Edge Effects in Fragmented Forests. Researchers tracked 66 study plots among forest fragments of different sizes.

SOURCE: W. F. Laurance, S. G. Laurance, L. V. Ferreira, et al. 1997. Biomass collapse in Amazonian forest fragments. *Science* 278: 1117–1118.

✔ **PROCESS OF SCIENCE** The locations of the fragments and study plots were assigned randomly. Why was this step important?

SUMMARY Table 54.2 Major Threats to Biodiversity

Threat	Example
Habitat destruction and degradation The conversion of primary forest to agricultural fields is a particularly devastating cause of habitat destruction.	
Overexploitation Overexploitation is the dominant threat for marine species, especially for large predators in top trophic levels, like these endangered bluefin tuna.	
Invasive species and disease Invasive species, like this Burmese python in the Florida Everglades, are introduced to a new area, multiply rapidly, and threaten native species.	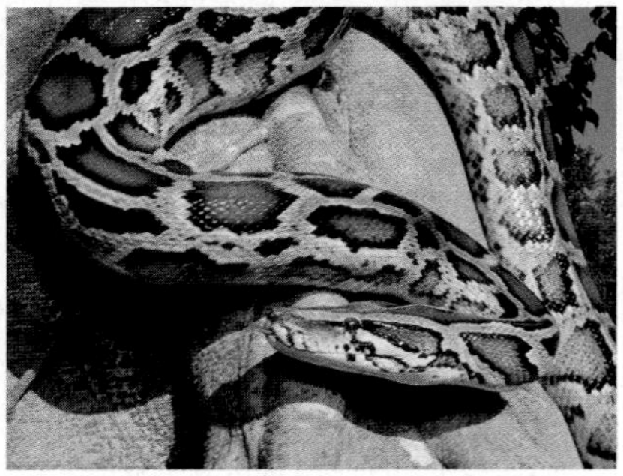
Pollution Chemical pollutants have reached every corner of the globe, but are particularly threatening to aquatic species.	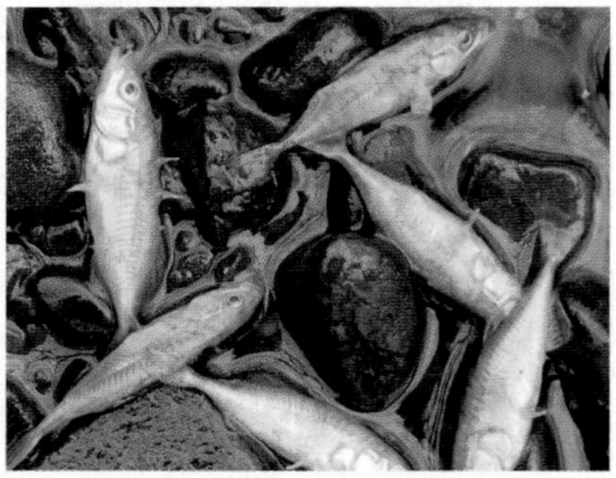
Climate change Climate change poses different types of threats to different species. For example, some corals become bleached (lose their symbiotic photosynthetic protists) when water temperature warms.	

Overexploitation also affects terrestrial environments.

- Overhunting is a major threat to many mammal populations. Aside from the traditional hunt for game species such as bison, elk, and moose, the growth of the "bushmeat" trade has devastated populations of primates and other mammals in Africa and elsewhere. Adding to the problem, many animals are hunted not for their meat, but for their ivory, skins, horns, or other body parts valued in folk medicines or decorations.
- The capture of exotic animals for the pet trade continues at a rapid rate despite many laws to prevent it. Exotic pets range widely from primates like tamarins and orangutans to parrots, iguanas, and frogs.
- Overexploitation can also threaten populations of plants (such as ginseng) and fungi (such as truffles).

Overexploitation is a grave concern for marine and terrestrial species worldwide due to the increasing demand and decreasing supply. Ten thousand years ago, only 0.1 percent of Earth's mammalian biomass consisted of humans and their domesticated animals. Today this number has risen to 90 percent due to the overexploitation of wild animals combined with the industrial-scale production of domesticated animals. To help you visualize numbers like these, see **Making Models 54.1**.

Invasive Species and Disease Global trade and global travel have vastly increased the rate of exchange of plants, animals, fungi, and microorganisms around the world. A nonnative species that is introduced into a new area is called an **exotic species**. Exotic species do not necessarily pose problems, and they can even be beneficial in some circumstances. But some exotic species directly threaten native species by eating them, competing with them, causing disease, or other types of deleterious interactions. For example, a lethal fungus is spreading from Asia in the exotic salamander pet trade, endangering native

Making Models 54.1 Tips on Drawing Graphs (V): Visualizing Numbers

When you're reading, it's easy to gloss over numbers and miss their meaning. Sometimes it helps to stop and draw a simple graph so that you can visualize the data and process what the numbers represent (see **BioSkills 2** on graphing). The graph below shows the numbers you just read in the text.

MODEL Select an appropriate style of graph to show this trend: Only 1411 Bengal tigers remained in India in 2004, but the number of tigers increased to 1706 in 2010 and 2226 in 2014, thanks to efforts to limit poaching.

To see this model in action, go to the Study Area of **Mastering Biology**

salamanders worldwide. Exotic species can also pose indirect threats by changing the local biotic or abiotic resources.

If an exotic species is introduced to a new area, grows to large population size, and disrupts species native to the area, it is called an **invasive species**. For example, lionfish originated in the Pacific Ocean but are now a scourge in coral reef communities in the western Atlantic (see end-of-chapter Case Study in Chapter 49). Similarly, Burmese pythons were set free by pet owners into the Florida Everglades and are now dramatically affecting native wildlife there (see Table 54.2 and the end-of-chapter Case Study in Chapter 51).

Pollution Humans change both the biotic environment and the abiotic environment by releasing chemicals into ecosystems. Many pollutants are persistent. They last in organisms and the environment for a long time and may continue to cause harm long after their initial release—and at great distances from their initial release. Pollution tends to cause the most harm in freshwater ecosystems where the pollution concentration from terrestrial runoff can be particularly high.

- Industrial pollutants include sulfur (causing acid rain), arsenic, mercury, and lead. Even in the distant Arctic, polar bear milk contains industrial pollutants called PCBs.
- Pharmaceutical pollutants (drugs) like oral contraceptives and ibuprofen are excreted in human urine and pass through wastewater-processing facilities into streams and rivers, where they are still biologically active.
- Pesticides and herbicides are a type of persistent organic pollutant (POP) that have a direct effect where they are used, but also affect species far away by dispersing in waterways and undergoing biomagnification in food webs (Ch. 53, Section 53.2). Honey samples worldwide contain traces of neonicotinoid pesticides.
- Nutrients like nitrogen and phosphorus run off of farms and fields, causing eutrophication (Ch.53, Section 53.3).
- Garbage such as plastics fills landfills, waterways, and ocean gyres, posing both physical and chemical threats.

Note that humans are not immune from the effects of pollutants. A recent study in *The Lancet* estimated that 9 million people were killed by causes directly linked to pollution in 2015—three times more deaths than were due to malaria, AIDS, and tuberculosis combined.

Climate Change The ongoing change to global climate caused by the burning of fossil fuels, deforestation, and other human activities has already affected species and led to extinctions (see a discussion of the causes and consequences of climate change in Ch. 53, Section 53.4). Climate change is affecting biodiversity in a variety of ways:

- As the global temperature increases, polar ice and alpine habitats are disappearing, endangering species native to arctic and high-elevation tundra.
- Coastal habitats are threatened by sea-level rise caused by the melting of polar ice caps and glaciers plus the thermal expansion of the oceans. Several recent models project that sea levels will rise an additional 0.8 to 2.0 m by the year 2100, inundating many coastal habitats (and cities), including areas with biodiversity hotspots.
- Biomes are becoming redistributed due to changes in both temperature and rainfall, causing mismatches between species distributions and abiotic conditions.
- Average temperature and precipitation are changing, and so are the extremes, causing more frequent and severe droughts, floods, deep freezes, and heat waves. These extreme events can be a matter of life or death for some organisms, leading to evolution or extinction in populations.
- The species compositions of communities are changing because trees and other slowly dispersing species are unable to keep up with rapid changes in climate and because migrations and other timed events are out of sync.
- Coral reefs are suffering from bleaching induced by high temperatures (see Table 54.2); this change is causing a cascade of effects in those very diverse ecosystems.
- Ocean acidification—due to the rise of atmospheric carbon dioxide concentrations and the consequent increase in carbonic acid formation in the oceans (and subsequent dissociation to carbonate and H^+)—is interfering with many aspects of organismal physiology, including the ability of corals, crustaceans, and other animals to make calcium carbonate skeletons.

Although climate change is a natural phenomenon, the climate is currently changing at an extreme rate due to human behavior.

How Will These Threats Affect Future Extinction Rates?

Developing projections of future biodiversity is a rapidly growing area of research, due to major advances in computer modeling, increased availability of paleontological and current data to input into the models, and increased demand for projections as the world is faced with important decisions about conservation.

Formulation of biodiversity scenarios is a highly integrative field of ecology because it must use information from multiple levels, including

- the effects of evolutionary processes on genetic variation, adaptation, and extinction in populations (Chapter 23);
- the impact of abiotic factors such as temperature and precipitation on the distribution and abundance of organisms (Chapter 49);
- the behavior of organisms in response to changing conditions (Chapter 50);
- the basic demographic characteristics of populations, such as fecundity and survivorship (Chapter 51);
- the consequences of species interactions at the level of communities (Chapter 52) and ecosystems (Chapter 53); and
- the effect of biodiversity threats at all these levels.

You can see how all these levels fit together in the Big Picture of Ecology on pages 1190–1191. Different models integrate different variables. Let's consider three examples of models ranging from single populations up to global species distributions.

Estimating the Probability that a Population Will Go Extinct A population viability analysis, or PVA, is a model that estimates the likelihood that a population will avoid extinction for a given time period. In most cases, a PVA attempts to combine data on

- age-specific survivorship and fecundity (Ch. 51, Section 51.2),
- geographic structure (Chapter 49), and
- rate and severity of habitat disturbance (Ch. 52, Section 52.3).

Typically, a population is considered viable if the analysis predicts that it has a 95 percent probability of surviving for at least 100 years. This type of analysis is particularly useful for creating management plans for species like the northern spotted owl and island foxes, which are listed as endangered in the IUCN Red List and the U.S. Endangered Species Act.

Estimating the Effect of Habitat Area on Species Richness One strategy for predicting the future of Earth's biodiversity focuses on the consequences of the most urgent problem: habitat loss. Given reasonable projections of how much habitat will be lost over a given time period, biologists can estimate rates of extinction based on **species–area relationships**. The concept here is to track backward on a species–area curve (see **Quantitative Methods 54.1**) to determine the species richness of a given habitat if its area were to decline due to habitat destruction or climate change.

To understand how biologists use the species–area curve in Quantitative Methods 54.1 to project extinction rates, ask yourself this question: If sea-level rise wiped out 90 percent of the habitats in part of the Bismarck Archipelago—for example, if *A* were reduced from 10,000 km^2 to 1000 km^2—what percentage of species would disappear? According to the graph, the number of bird species present would drop from about 75 to about 53. Thus, roughly 30 percent of species would vanish. The slope of the line in species–area curves is species dependent.

A drawback of this method is that using the curve to extrapolate backward can overestimate extinction rates due to assumptions inherent to the model. Thus, the technique must be used with caution and cross-referenced with other projection methods.

Quantitative Methods 54.1 Species–Area Plots

The goal of species–area plots is to quantify the relationship between species richness and habitat area. **Figure 54.9** gives an example of a species–area curve—a graph that plots habitat area on the *x*-axis and the number of species present on the *y*-axis. In this case, the habitat areas are islands in the Bismarck Archipelago near New Guinea, in the South Pacific, and the species documented are birds. Each data point represents the number of species found on a different island. Note that the relationship between habitat area and the number of species present is linear when both axes are logarithmic. (For help interpreting logarithms, see **BioSkills 5**.)

The "log–linear" relationship turns out to be typical for other habitats and taxonomic groups as well. When biologists have plotted species–area relationships for plants, butterflies, mammals, or birds from islands or continental habitats around the globe, the relationship is consistently described by the function $S = cA^z$, where:

- S is the number of species.
- A represents the habitat area.
- The c term is a constant that scales the data. Its value is high in species-rich areas, such as coral reefs or tropical wet forests, and low in species-poor areas, such as arctic tundra.
- The exponent z represents the slope of the line on a log–log plot. Thus, z describes how rapidly species numbers change with area.

For example, the solid line drawn through the points in Figure 54.9 is described by the function $S = (18.9)\,A^{0.15}$. This equation can be used to predict the species richness of other areas of a given size, such as an island that has not yet been sampled. If an unsampled island is known to have an area of 8 km^2, the species richness would be predicted as $S = 18.9 \times 8^{0.15} = 26$.

✔ **QUANTITATIVE** If you understand species–area curves, you should be able to use the equation describing the line in Figure 54.9 to predict the species richness of an island in the Bismarck Archipelago with an area of 5000 km^2.

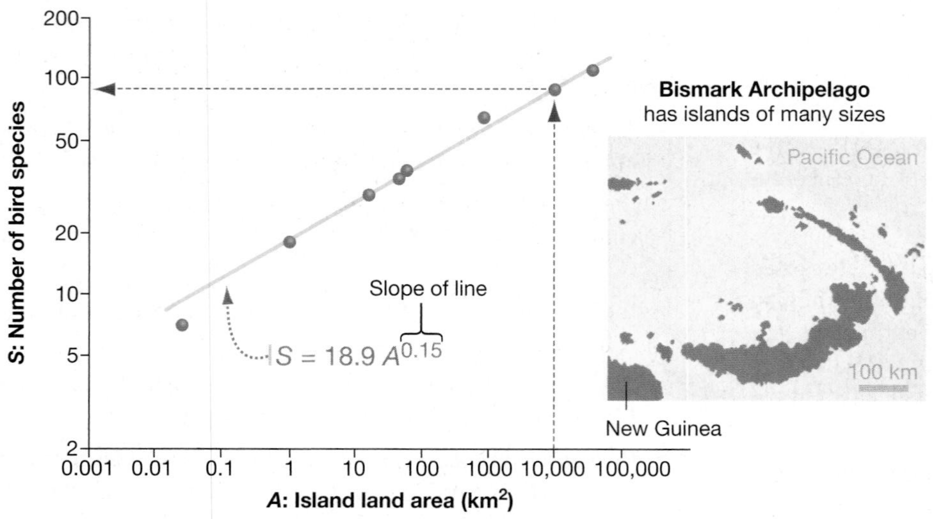

Figure 54.9 Species–Area Plots Quantify the Relationship between Species Richness and Habitat Area. The dashed arrows on this graph show the number of bird species that are expected to live on an island with an area of 10,000 km^2.

DATA: J. M. Diamond. 1975. *Biological Conservation* 7: 129–146.

Estimating the Effect of Climate Change on Species Distributions

How will climate change affect the global distribution of species? Can reliable predictions be made? Bioclimate envelope models, also called niche-based models, use the relationship between current species distributions and abiotic variables such as temperature and precipitation to project the future distributions of species under climate change.

For example, behavioral ecologist Barry Sinervo and colleagues took a two-step approach. First, they collected extensive data from one part of the globe, measuring extinction rates of 48 species of lizards at 200 sites in Mexico since 1975. The results were shocking: Twelve percent of local populations had already gone extinct because both the average and extreme temperatures are changing too rapidly for the lizard populations to adapt (Ch. 53, Section 53.4).

Sinervo and his international team then used their data to predict worldwide responses to global warming given projected climate changes and the relationship between lizard extinctions and climate (Figure 54.10). They estimated that 4 percent of local lizard populations have already gone extinct worldwide and that lizard species extinctions could reach 20 percent by 2080. These alarming projections were validated with local data, suggesting that climate change is not a problem of the future—it is already a matter of life and death for lizard populations today.

Take-Home Messages about Making Biodiversity Predictions It is important to note that at all levels of study, a great deal of variation exists in projected future biodiversity. And this variation has different sources:

- Projections must make assumptions about future levels of threats to biodiversity, such as the extent of climate change, but these levels are unknown and depend on how rapidly humans are able to reduce emissions of greenhouse gases and improve land-use practices.
- Variation in some projections is due to knowledge gaps in basic biology. More accurate projections will be possible as more data are collected.
- Some of the variation is due to differences in how the models are constructed. Scientists address this issue by using multiple models to cross-check projections, and by validating their models with empirical data from the real world.

One of the greatest challenges of making projections is the ability to measure how different factors interact. For example, the precipitous worldwide decline in amphibians is due to a combination of factors ranging from habitat loss, pollution, and climate change to infection by a chytrid fungus. These factors affect each other in complex feedback loops, requiring a systems biology approach that quantifies the network of interactions.

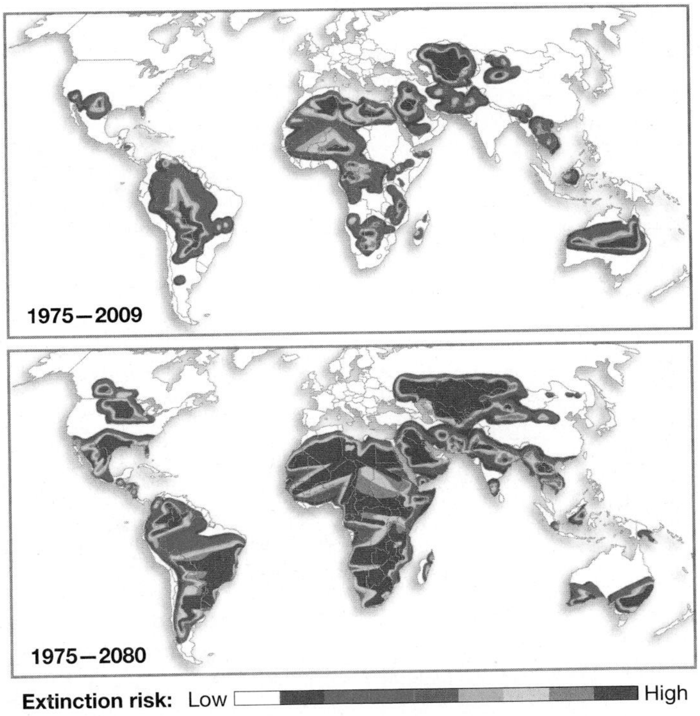

Figure 54.10 Global Extinction Rates of Lizards Can Be Predicted. The projections are based on current extinction rates in Mexico and projections of global climate change, and were validated using observed extinction rates in regions of four other continents.

DATA: B. F. Sinervo et al. 2010. *Science* 328: 894–899.

CHECK YOUR UNDERSTANDING

✔ **If you understood this section, you should be able to . . .**

1. Explain why fragmentation reduces habitat quality and can lead to the reduction of genetic diversity.
2. **SOCIETY** Propose how the industrial food system contributes to each type of threat in Table 54.2.
3. Describe how you would determine whether each of the threats in Table 54.2 was affecting the orangutans of Borneo.

Answers are available in Appendix A.

54.3 Why Are Biodiversity and Ecosystem Function Important?

After you complete this section, you should be able to . . .

- Explain the importance of biodiversity and ecosystem function.

One of the questions biologists have to address about biodiversity and ecosystem function is one that you may have heard from friends or relatives: Who cares? People are concerned about their grades or job or health or car or love life. Why should they worry about something as abstract as biodiversity and ecosystems?

The answer has two parts, one biological and the other economic and cultural. Let's consider each part in turn.

Biological Benefits of Biodiversity and Ecosystem Function

Do biological communities and ecosystems function better when species diversity is high? The effort to answer this question began in the mid-1980s and has blossomed since.

Biodiversity Increases Productivity Figure 54.11 summarizes a classic experiment on how species richness affects one of the most basic aspects of an ecosystem function: the production of biomass. Net primary productivity (NPP) is important because it is the source of chemical energy used by species throughout a food web (Ch. 53, Section 53.1).

As the "Experimental Setup" portion of the figure shows, David Tilman and colleagues divided 32 grassland plant species into five functional categories. The functional groups differ by traits such as the timing of their growing season, allocation of resources to woody stems versus seeds, C_3 versus C_4 photosynthetic pathways, and whether they fix nitrogen. The researchers planted 13-m by 13-m plots with a mixture of between 0 and 32 randomly chosen species, representing from 0 to 5 functional categories (where 0 species were used as controls), and measured the amount of aboveground biomass produced in each plot.

RESEARCH

QUESTION: Do high species richness and high functional diversity of species increase aspects of ecosystem function such as net primary productivity (NPP)?

HYPOTHESIS: NPP increases with increasing species richness and with increasing functional diversity of species.

NULL HYPOTHESIS:

EXPERIMENTAL SETUP:

1. Plant a total of 289 experimental plots, each with up to 32 species and up to **5 functional groups:**

2. Measure the amount of aboveground biomass produced in each plot.

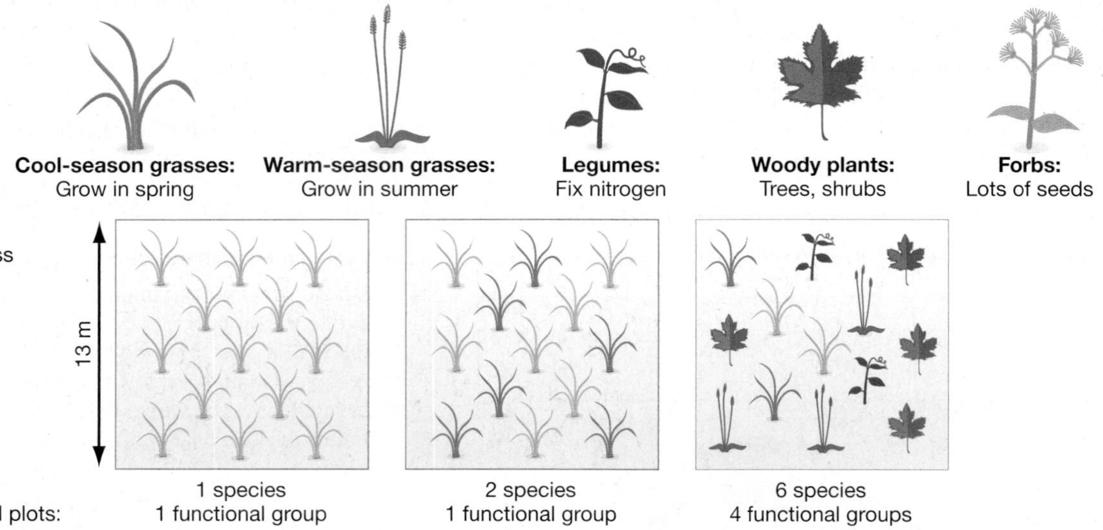

PREDICTION OF HYPOTHESIS:

PREDICTION OF NULL HYPOTHESIS:

RESULTS:

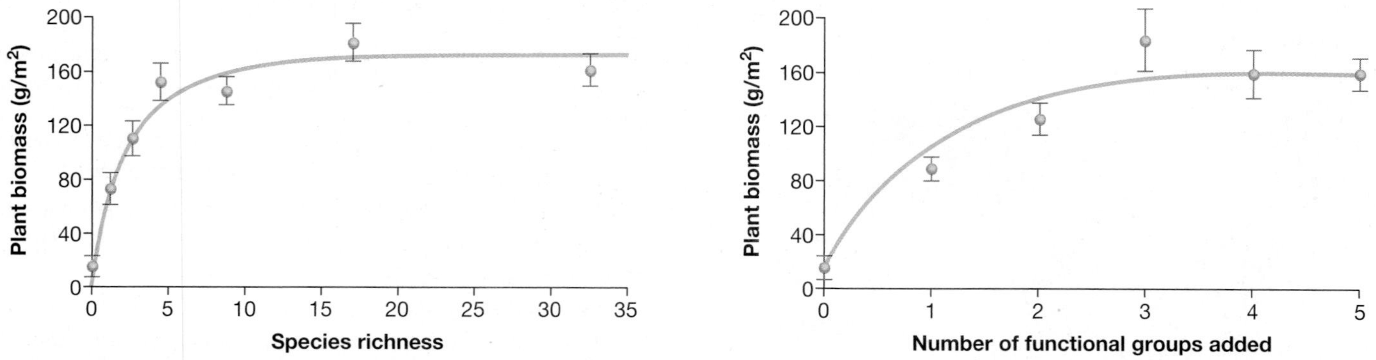

CONCLUSION: In this plant community, NPP increases with increasing species richness and increasing functional diversity of plants, at least up to a point.

Figure 54.11 **Evidence Shows that the Productivity of Ecosystems Depends on the Number and Types of Species Present.**

SOURCE: D Tilman, J. Knops, D. Wedin, et al. 1997. The influence of functional diversity and composition on ecosystem processes. *Science* 277: 1300–1302.

✔ **PROCESS OF SCIENCE** Write in the null hypothesis and both sets of predictions.

The results graphed in Figure 54.11 indicate that both the number and type of species present had important effects on biomass production. Experimental plots with more species and with a wider diversity of functional groups were more productive. More than 500 controlled experiments and 67 field studies performed in various plant communities have since supported this conclusion.

One aspect of the result that has varied among studies is the shape of the curves. Note in Figure 54.11 that total biomass leveled off as species richness and functional diversity increased. This observation suggests that increasing species diversity improves ecosystem function only up to a point. However, other studies discovered a more linear relationship that continues to increase as species richness increases. The difference has to do with the amount of total standing biomass in different types of plant communities (such as forest versus grassland).

Why would species richness have a positive impact on NPP? Several causal mechanisms may be at work.

- ***Resource use efficiency*** Diverse assemblages of plant species make more efficient use of the sunlight, water, and nutrients available and thus lead to greater overall productivity. For example, some prairie plants extract water near the soil surface, while others use water available a meter or more below the surface, a result of niche differentiation (Ch. 52, Section 52.1). When species diversity is high, more overall water is used and more photosynthesis can occur.
- ***Facilitation*** Certain species provide nutrients, partial shade, or other benefits that happen to make the environment more favorable for other species (Ch. 52, Section 52.3). In prairies, the presence of plants in the onion family may discourage herbivores, or the decaying roots, stems, and leaves from nitrogen-fixing species may fertilize other species.
- ***Chance inclusion of high-productivity species*** In many habitats, it is common for one or two species to be extremely productive. If the number of species in a study plot is low, it is likely that the "big producer" will be missing and NPP will be low. But if the number of species in a study plot is high, then likely the big producer is present and NPP will be high. Simply due to chance—a phenomenon called the sampling effect—high-species plots will tend to outproduce low-species plots.

The three mechanisms listed are not mutually exclusive—several can operate at the same time. Further, long-term and large-scale experiments have shown that the reason for the pattern can change over time, that the slope of the productivity curve can continue to rise over time, and that the strength of the diversity–productivity relationship depends on the evolutionary history of the species involved.

Does Biodiversity Lead to Stability? When biologists refer to the stability of a community, they mean its ability to

1. maintain productivity and other aspects of ecosystem function as conditions change over time and
2. recover to former levels of productivity or species richness after a disturbance.

Resistance is a measure of the extent to which a community remains unchanged during a disturbance. **Resilience** is a measure of how quickly a community recovers following a disturbance. The model in Figure 54.12 shows one way to visualize the relationship between resistance and resilience in communities.

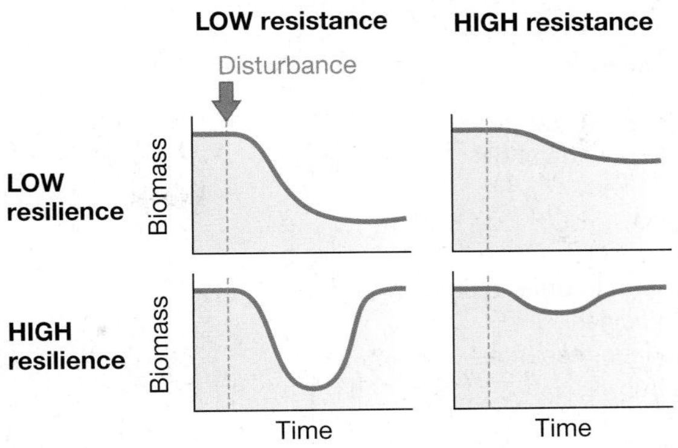

Figure 54.12 **Diversity Increases Resistance and Resilience.**

To test the hypothesis that diversity increases resistance and resilience, Tilman's team analyzed data from a natural experiment. A **natural experiment** occurs when comparison groups are created by an unplanned, unmanipulated change in conditions.

In 1987–1988, a severe drought hit their study site. When the drought ended, the team asked whether species richness affected resistance. As predicted, drought resistance was higher in more diverse communities than in less diverse ones.

In addition, the group analyzed the change in biomass in each plot four years after the drought versus the biomass before the drought. The data indicated that most plots containing five or fewer species showed a significant lowering of biomass after the disturbance, whereas all of the plots that contained more than five species regained the same biomass that they had before the disturbance. This is strong evidence that species-rich areas are more resilient following a disturbance.

Numerous follow-up experiments in ecosystems around the world have shown that biodiversity provides resistance and resilience in the face of climate extremes and other stressors, such as invasion by exotic species. And biologists are increasingly confident that species richness has a strongly positive effect on how ecosystems function.

Ecosystem Services: Economic and Social Benefits of Biodiversity and Ecosystems

The biological benefits of biodiversity—increased productivity, resistance, and resilience—benefit humans too. The prosperity, health, and happiness of humans completely depend on the preservation of biodiversity and functioning ecosystems. It is a common misconception that humans are outside of or apart from nature, as exemplified by the common phrase "humans and animals" in popular culture (see **BioSkills 17**). But humans are a small, recent twig on the animal branch of the tree of life (Ch. 32, Section 32.5) and are very much a part of ecosystems and dependent on their function.

Considered collectively, all the direct and indirect benefits that humans derive from organisms and the ecosystems they compose

SUMMARY Table 54.3 Ecosystem Services

Type of Service	Example
Provisioning services provide raw materials • Food • Fuel • Fiber and other materials • Medicines • Genetic resources • Water	
Regulating services are part of Earth's life-support system • Climate moderation • Soil formation • Erosion control • O_2 and CO_2 regulation • Water capture • Water purification • Air cleaning • Flood control • Storm mitigation • Waste decomposition	
Cultural services enrich quality of life • Aesthetics • Recreation • Education • Spiritual value • Human mental and physical health	
Supporting services enable all the other ecosystem services • Primary productivity • Nutrient cycling • Pollination • Pest control	

are called **ecosystem services**. A team of ecologists and economists led by Robert Costanza has estimated the value of ecosystem services at \$125 trillion per year—more than twice the gross national product of all nations combined. Let's take a closer look at some of the services that nature provides, summarized in **Table 54.3**.

Provisioning Services Wild species have provided the raw material to fuel the development of human societies ever since the first large-scale, highly organized cultures began to develop about 10,000 years ago. People felled vast tracts of forest for building materials and fuel; selectively bred wild plants to yield food and fibers; domesticated animals to provide food, labor, and material goods; harvested the oceans for protein; and processed plants, animals, and fungi as sources of medicines. Today as much as ever, humans rely on other organisms for most of their food and material needs.

The value of provisioning services is relatively easy to quantify because food, fuels, wood, and other materials are traded in the marketplace. For example, mushrooms have an annual food value of more than \$1 billion, the fishing industry has a value of \$80–100 billion, and raw materials provided by tropical forests are valued at \$300 billion.

Bioprospecting—the exploration of bacteria, archaea, protists, plants, fungi, and animals as novel sources of drugs or ingredients in consumer products—has benefited from the recent explosion of genetic information. Biologists can now search genomes from a wide array of species to find alleles with desired functions. Recent advances are impressive: Researchers have developed a potent painkiller from the paralyzing sting of tropical cone snails and a blood anticoagulant from the saliva of vampire bats.

Regulating Services Services provided by functioning ecosystems—such as water purification, flood control, and waste decomposition—are part of Earth's life-support system. Yet, these regulating services are often taken for granted until human actions disrupt their function.

This was the case in New York City in the 1990s. New York City is known for its excellent drinking water, piped in from the watersheds of the Catskill Mountains. But the water quality declined as those watersheds became developed—and something had to be done. A cost–benefit analysis revealed that it was far cheaper to restore the ecosystem function of the watersheds (\$1 billion) than to filter the water artificially (\$6–8 billion up front plus \$300 million annually). The "free" water purification services that had been provided by the forested Catskill ecosystem suddenly became quite valuable.

Cultural Services People integrate nature in their lives in diverse cultural ways, ranging from artistic celebrations to recreation, education, and spiritual contemplation. Eminent biologist and author Edward O. Wilson coined the term "biophilia" to refer to the innate, unconscious bond between humans and other organisms as a result of our deep coevolutionary history.

Thousands of studies are showing that humans who are removed from natural environments—such as people who spend most of their time in apartments, office buildings, factories, and jails—often suffer both physical and psychological setbacks, a

Figure 54.13 Hand Pollination Is Labor Intensive and Expensive. Following a decline of bees due to pesticide use, workers in Maoxian County, China, hand pollinate entire orchards.

syndrome called "nature-deficit disorder" by nature advocate Richard Louv. Conversely, people who spend more time outdoors in nature tend to have reduced stress; reduced rates of obesity, diabetes, and depression; improved problem-solving skills; enhanced creativity and curiosity; improved sleep patterns; improved social qualities such as empathy; and improved vision, among other traits.

On average, people who spend more time in nature tend to be healthier and happier. (How much time do you spend outdoors?) Government agencies and businesses are taking note because human health and happiness also have economic value.

Supporting Services Some ecosystem services support all the other categories of services. Primary productivity is a key example because the generation of biomass determines the provisioning, regulating, and cultural services that ecosystems can provide.

In a similar way, pollination and pest control are critical to multiple ecosystem services. The value of pollination became clear in a region in China where the excessive use of pesticides led to a particularly sharp decline of bees. Human workers must now pollinate apple, pear, and peach orchards by hand, at great expense (Figure 54.13). In the United States alone, insect-pollinated crops like almonds, apples, and watermelons produce $40 billion worth of products annually.

The air you breathe, the water you drink, and the soils that grow the food you eat are services that ecosystems provide. The quality of your life is tied to the quality of those ecosystems.

An Ethical Dimension

The data reviewed in this section suggest that there are compelling biological and economic reasons to preserve species and ecosystem function. Ecosystems work better, and people live better, when biodiversity is high. But in addition, an array of biologists, philosophers, and religious leaders argue that humans have an ethical obligation to preserve species and ecosystems. There are at least three arguments for this position:

1. Organisms have intrinsic worth. Humans diminish the world by extinguishing species and destroying ecosystems.
2. Industrialized nations are responsible for most of the environmental harm, yet poor countries are disproportionately affected by the consequences of these actions. It is unethical for rich countries to inflict hardship on poor countries.
3. It is unethical for the current generation to deprive future generations of ecosystem services.

From this perspective, extinction and disruption of ecosystem function are moral issues as well as biological and economic problems. One of the most important reasons to preserve biodiversity and ecosystem function may simply be that it is the right thing to do.

CHECK YOUR UNDERSTANDING

✔ **If you understood this section, you should be able to ...**

1. Explain why the high biodiversity of tropical forests leads to high productivity, in terms of efficient resource use, when compared to oil palm plantations.
2. **PROCESS OF SCIENCE** Design an experiment to test the hypothesis that the tropical forest of Borneo is more effective at building soil, retaining soil nutrients, and minimizing soil erosion than the oil palm plantations are.
3. **SOCIETY** Explain how the planting of trees in poor city neighborhoods could decrease the inequality in physical and mental health among people in poor and wealthy neighborhoods.

Answers are available in Appendix A.

54.4 Preserving Biodiversity and Ecosystem Function

Extinction is irreversible—and biodiversity loss is not slowing down. But there are reasons for hope.

After you complete this section, you should be able to ...

- Analyze strategies for preserving biodiversity and ecosystem function.

Solving any global problem requires a common goal to be defined. In the case of preserving biodiversity and ecosystem function, the objective is to sustain diverse communities in natural landscapes while supporting the extraction of resources required to maintain the health and well-being of the human population. A growing, evidence-based, international movement called Nature Needs Half (or Half-Earth) is calling for at least half of the Earth to be allocated for conservation to achieve a sustainable future.

How close are we to this objective? The map in Figure 54.14 shows the current status of the world's 846 ecoregions, which are subdivisions of terrestrial biomes (Ch. 49, Figure 49.9). So far, 12 percent of the ecoregions, shown in green, currently meet the half-protected status. This is discouragingly small compared to the target of 50 percent.

The good news is that 37 percent of the ecoregions, shown in light blue, have the *potential* to reach half-protected status. These areas, including much of the forests of the Amazon and Borneo, have enough quality habitat remaining to still be protected. The other half of the world's ecoregions, shown in orange and red, are more imperiled. The take-home message: There is a lot of work ahead, but there remains hope for a sustainable future. Even in imperiled areas, small changes can make big differences. What's working? And what can you do to help?

Addressing the Ultimate Causes of Loss

To begin, most biologists recognize that it will be impossible to preserve high species diversity and high-functioning ecosystems if two things happen:

1. The human population grows to 10 billion or more—from its current 7+ billion—over the next half-century, as projected by the UN's "high-fertility" scenario (Ch. 51, Section 51.5).
2. The rate of consumption of fossil fuels and other resources does not decline, especially in industrialized nations and growing developing nations (Ch. 53, Section 53.3), because consumption threatens ecosystems not only directly but also indirectly via climate change.

While it is complex and controversial to pinpoint a human carrying capacity for the planet (Ch. 51, Section 51.5), clearly humans are nearing, or have already surpassed, some of the recoverable limits of available resources. The way forward involves both reducing population growth and improving **sustainability**—the managed use of resources at a rate only as high as the rate at which they are replaced.

You may be familiar with efforts to develop renewable sources of energy, promote agricultural and forestry practices that maintain productivity without the addition of synthetic fertilizers or groundwater, limit fishing catches so that populations do not crash irrecoverably, and recycle metals and other raw materials. In essence, the challenge is to create ways for humans to live off the resources that are being produced continuously on Earth rather than using species to extinction and exhausting resources like petroleum, groundwater, and topsoils that have been stored for centuries or millennia. Sometimes small changes, like eating less beef, can make a big difference (Figure 54.15).

It may take decades for human population size and resource use to stabilize and for sustainable development practices to become widespread. In the meantime, thousands of species and untold ecosystem services may be lost. What specific approaches do conservationists use to protect natural resources?

Conservation Strategies to Preserve Genetic Diversity, Species, and Ecosystem Function

Despite the discouraging persistence of biodiversity loss, conservation efforts have managed a number of successes. Let's consider several examples.

Education Campaigns In some cases, public education campaigns are the key to major breakthroughs. For years, conservationists have used so-called flagship, or umbrella, species to rally support for conservation measures. Charismatic species such as pandas, tigers, and elephants tend to mobilize more effort than

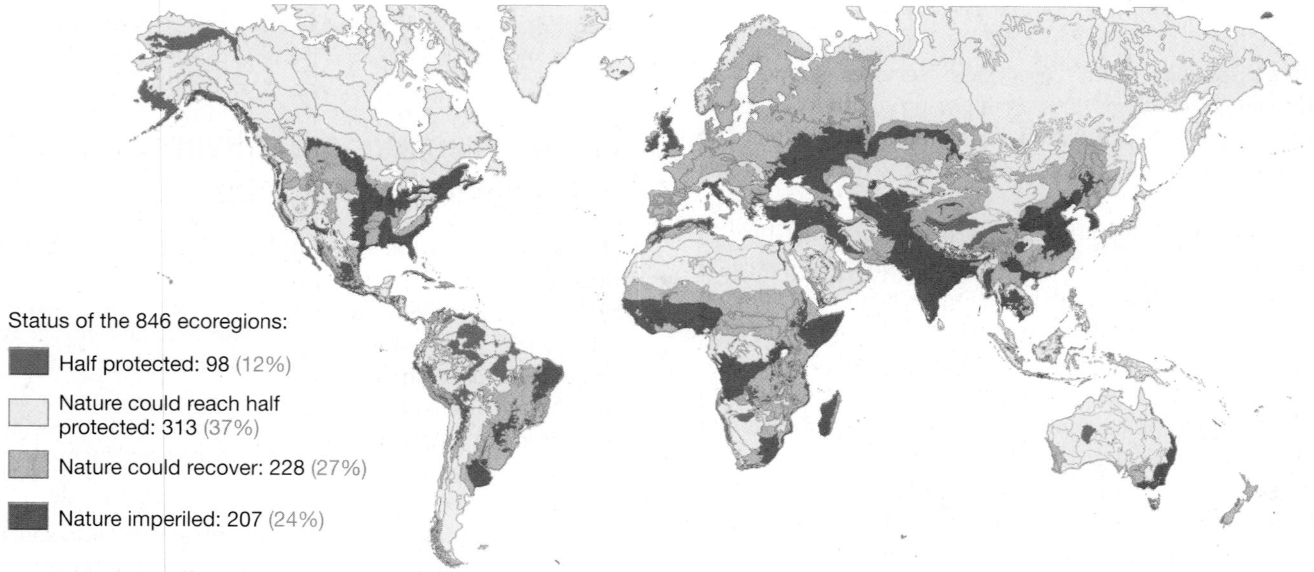

Figure 54.14 **Nature Needs Half: Current Protection Status of 846 Ecoregions on Earth.** The goal of the Half-Earth movement is to work toward a sustainable future where at least 50 percent of the land area is in the half-protected status (shaded dark green).
DATA: E. Dinerstein et al. 2017. *BioScience* 67: 534–545.

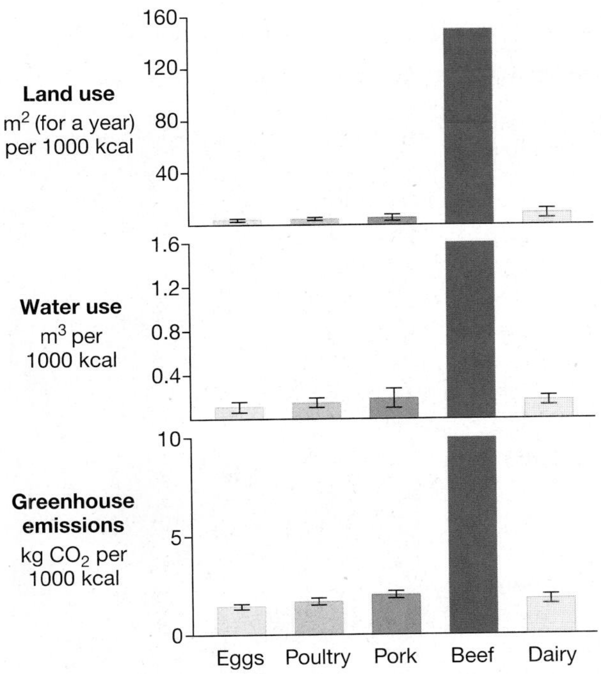

Figure 54.15 Beef Is a Resource Hog. The data show how many resources are used to produce 1000 kilocalories (kcal) of food (about half the daily intake for one person) in five different categories. Greenhouse emissions include methane, but are quantified in CO_2-equivalent units. (Note: In everyday English, 1000 kcal are equivalent to 1000 Calories.)

DATA: G. Eshel, A. Shepon, T. Makov et al. 2014. *Proceedings of the National Academy of Sciences* 111: 11996–12001.

✔ **QUANTITATIVE** If you decided to eat 1000 kcal of chicken (poultry) instead of beef today, about how much less land would be required to support your feast?

unfamiliar species such as copepods and forbs. By preserving these familiar species, however, conservationists can succeed in preserving the ecosystems where they live.

Consumers are usually unaware of the environmental impacts of their purchases, but can cause positive change in the marketplace when they become educated. For example, public education campaigns have succeeded in raising awareness that when fins are cut off of sharks and the sharks are thrown back to sea to die, not only does the shark population plummet, but organisms throughout the ecosystem, including those in coral reefs, are affected due to trophic cascades (Ch. 52, Section 52.2). Public demand for shark-fin soup has declined (**Table 54.4**, page 1184).

Genetic Restoration When habitat destruction or fragmentation result in small, isolated populations that are vulnerable to the loss of genetic diversity from chance events, genetic restoration can sometimes be accomplished by artificial gene flow.

Consider the case of a remnant population of the endangered Florida panther in the swampland of southern Florida (see Table 54.4). By the early 1990s, only about 25 adults remained, and they exhibited several recessive disorders and low genetic diversity—both signs of inbreeding. In an effort to prevent imminent extinction, the U.S. Department of the Interior moved eight panthers (locally called pumas) from Texas to Florida to introduce new alleles into the Florida population. The number of panthers quadrupled in 10 years. The offspring of hybridized animals were larger and had reduced incidents of defects.

Seed Banks About 95 percent of the world's calories come from just 30 crops. Agricultural scientists are racing to preserve diverse strains of crop and wild plants in **seed banks**—long-term storage facilities—to avoid the loss of genetic diversity in local and global populations (see Table 54.4). Unfortunately, the seeds of some plants, such as coffee, are not viable in seed banks.

Management Plans for Invasive Species Many new measures are arising to limit the global exchange of species. For example, ships are now required to exchange ballast water out at sea rather than in port, cargoes are inspected at border crossings, and nurseries are selling fewer invasive plants. The invasive wetland plant, purple loosestrife, whose seeds are transported in ship ballast water, is now a target for all three strategies (see Table 54.4). Prevention is more effective than treatment.

Management Plans for Threatened Species The Endangered Species Act was signed into law in the United States in 1973 to prevent species from becoming extinct. It is designed to protect threatened organisms as well as the ecosystems they depend on. One well-known success story is the bald eagle (see Table 54.4), which was placed on the list as an endangered species in 1967 due to multiple threats, such as habitat loss, hunting, and the effects of DDT, a pesticide that caused the catastrophic thinning of eggshells. A drastic reduction of these threats enabled the bald eagle population to recover dramatically.

Ex Situ Conservation and Reintroduction of Animals Some species—including Pere David's deer and the northern white rhinoceros—are extinct in the wild and exist only in captivity. These animals are examples of **ex situ conservation**, the preservation of species in zoos, aquaria, or other artificial settings. (Seed banks are an example of ex situ conservation of plants.) "Ex situ" means out of place. The name was inspired by a contrast with in situ ("in place") conservation, the establishment of protected areas where populations can be maintained in their natural habitat. Ex situ conservation is not ideal, but it is preferred over extinction.

In some cases, captive breeding of ex situ individuals and reintroduction into the wild have succeeded in creating wild populations. Examples include the whooping crane, peregrine falcon, California condor, and Arabian oryx (see Table 54.4).

Wildlife Corridors One long-term solution to enabling gene flow among isolated populations is the establishment of **wildlife corridors**, strips of undeveloped habitat that connect preserved areas. In some cases, wildlife corridors are as simple as a walkway under or over a major highway, so that animals such as salamanders and grizzly bears can cross without being killed (see Table 54.4). In other cases, a corridor can consist of 1000 km of makeshift paths across private and public lands, such as the corridors used by tigers in southern India.

SUMMARY **Table 54.4** Examples of Conservation Strategies to Preserve Genetic Diversity, Species, and Ecosystem Function

Conservation Strategy	Example
Education campaigns Global outcry about shark finning has helped to reduce the demand for shark-fin soup by more than 80% in China since 2012. Anti-finning regulations are enabling shark populations to recover.	
Genetic restoration After eight Texas pumas were introduced to Florida in 1995, the endangered panther population there quadrupled in size. Gene flow can counteract the effects of genetic drift and inbreeding in small populations.	
Seed banks Several organizations are storing seeds worldwide to avoid the permanent loss of genetic diversity in both domesticated and wild plants.	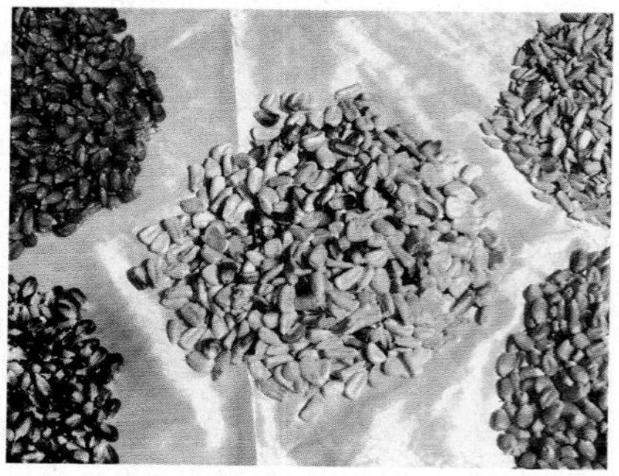
Management plans for invasive species New barriers thwart the movement of invasive species like purple loosestrife around the globe, such as the requirement that ships exchange their ballast water mid-ocean.	
Management plans for threatened species The bald eagle was placed on the endangered species list in 1967, then recovered so successfully that it was delisted in 2007.	
Ex situ conservation and reintroduction Once classified by the IUCN as extinct in the wild, the Arabian oryx has now reestablished wild populations after reintroduction from captive populations.	
Wildlife corridors Wildlife corridors, such as the vegetated overpass in Banff National Park in Canada, reconnect fragmented habitats, enabling gene flow and access to resources.	
Designating effective protected areas No-catch zones on the Great Barrier Reef of Australia have enabled the recovery of reef communities, boosting the numbers of commercially valuable fish in adjacent areas.	
Ecosystem restoration The restoration of clear-cut forests in Brazil (shown here), Borneo, and elsewhere has successfully produced dramatic increases in both species diversity and ecosystem services.	
Quantifying ecosystem services Costa Rica pays some landowners not to deforest their land because the ecosystem services provided by the intact forest, including ecotourism, are so valuable.	

By facilitating the movement of individuals, corridors allow areas to be recolonized if a species or a population was lost. Also, by improving gene flow, corridors can introduce new alleles that counteract the deleterious effects of genetic drift and inbreeding in small populations.

Designing Effective Protected Areas By far the most effective strategy for preserving biodiversity is to protect entire areas of land and sea from exploitation and degradation. Fortunately, the total amount of protected area is increasing at a steady rate globally. By strategically protecting prime habitat areas, countries worldwide can realize many benefits. For example, the creation of marine protected areas on the Great Barrier Reef of Australia—where fishing is forbidden—increases species diversity, resilience during storm events, and the abundance of commercially valuable fish in nearby fishing areas (see Table 54.4).

Ecosystem Restoration In many areas of the world, ecosystems are already heavily degraded or lost. The state of Illinois, for example, was estimated to contain 22 million acres of prairie when European settlers arrived; today only about 2000 acres (0.009 percent) of this biome remain. When natural areas are badly degraded or gone, biologists turn to restoration.

Thousands of large-scale and small-scale ecosystem restoration projects are now occurring around the globe, ranging from river dam removal to reforestation. Recall the dire scenario in Borneo described at the beginning of this chapter. Willie Smits had been working in Borneo for two decades in a desperate attempt to slow the rate of deforestation and save orangutans from the brink of extinction. Frustrated by his lack of progress, he decided to try a new approach—the restoration of 2000 ha of clear-cut rain forest in Samboja (see Table 54.4).

While Smits would be the first to agree that primary forest is of greater ecological value than restored forest, his progress just since 2002 is remarkable. His efforts have resulted in

- over 1600 tree species in a multilayered canopy;
- 137 species of birds (up from 5); 9 species of primates;
- rainfall up 25 percent; temperature down 3–5°C; and
- 3000 people earning income.

The cornerstone of Smits's Samboja Lestari project is integrating the ecological restoration of the forest along with the economic restoration of the local Dayak tribe—a win-win scenario. Although the size of Samboja Lestari is relatively small and the clear-cutting of rain forest continues in Borneo at a rapid pace, Smits's success offers reason for hope. How successful are ecosystem restoration projects in general?

Researchers have analyzed hundreds of such restoration projects to determine the resilience of ecosystems after human disturbances, with surprising results. Recovery of biomass following disturbances was much slower in forests than in other ecosystems, but unexpectedly fast overall—decades rather than centuries. Biomass recovery is not equivalent to biodiversity or ecosystem function, but many of the restoration projects show positive outcomes in terms of species richness and other variables. It is not too late to restore many degraded habitats.

In general, restoration success is positively correlated with the size of the area restored, its connectedness, remoteness, and the intactness of surrounding areas (Ch. 52, Section 52.4). However, another important insight is that an accumulation of many small changes can also have a big effect. If you and all your neighbors planted several native trees, shrubs, or other plants in your yards or neighborhoods, you could have a very positive effect on biodiversity and ecosystem function right where you live.

Quantifying Ecosystem Services Many ecologists are convinced that the traditional approaches to conservation are doomed to fail on their own—nature reserves are too small and isolated to succeed in preserving sufficient biodiversity and ecosystem function. Some of these scientists advocate for an ecosystem services approach to conservation—a quantification of ecosystem services to make explicit in the marketplace that long-term human prosperity depends on the preservation of these services.

The idea is that the cost of replacing ecosystem services is at least 10 to 100 times higher than the cost of maintaining them—if they are replaceable at all (if not, they are priceless). Therefore, it is an economic imperative to protect these services before they are lost. The problem is that quantifying ecosystem services is not an easy task, and incorporating these services into the marketplace requires a complicated collaboration among scientists, governments, nongovernmental organizations (NGOs), businesses, educators, communities, and local leadership. Further, policy mechanisms can be challenging to implement.

Successful experiments are under way, however. For example, the Costa Rican government has made policy changes that allow landowners to be compensated for the ecosystem services provided by their properties, such as carbon storage, biodiversity conservation, and scenic beauty. The payments are funded largely from transportation taxes. The logic here is that the Costa Rican economy will profit more in the long term from **ecotourism**—recreational visits to their wild places—if their wild places are preserved from logging by private owners in the short term (see Table 54.4). But the private owners must be compensated. Numerous other similar programs are in development.

In general, although social and political changes are difficult to implement, outcomes that boost biodiversity and ecosystem function are a boon to humans.

Take-Home Message

Your generation is facing the most serious global environmental crisis in the history of our species. However, biologists are showing that there is enormous potential for positive change. As someone with a background in biology, you have the intellectual tools to make an important, positive impact. In addition to studying life, biologists can help save it.

CHECK YOUR UNDERSTANDING

✔ **If you understood this section, you should be able to . . .**

SOCIETY Describe some of the things *you* could do to help preserve biodiversity and ecosystem function.

Answers are available in Appendix A.

54.1 What Is Biodiversity?

- Biodiversity is the diversity of life, quantified at the level of genetic diversity, species diversity, and ecosystem diversity.
- Only about 1.5 million species have been named out of an estimated 5 to 100 million species on Earth.
- Biodiversity surveys focus either on a specific taxon within an area, or on all species within an area.
- On average, biodiversity tends to be higher in the tropics than near the poles, higher on land than in the sea, and higher in areas with geographic diversity.
- Tropical wet forests (on land) and coral reefs (at sea) are particularly species rich.
- Biodiversity hotspots can be mapped based on species richness, areas of endemism, degree of threat, and/or unique evolutionary or ecological characteristics.

54.2 Threats to Biodiversity and Ecosystem Function

- The sixth mass extinction in the history of life is occurring due to human impacts. Species are vanishing faster than at any other time in Earth's history except the mass extinction event 66 million years ago.
- Habitat destruction and degradation are currently the leading causes of extinction. Conversion of primary forest to agricultural fields and housing developments is particularly devastating.
- Although overexploitation also occurs on land, it is the dominant threat for marine species.
- Many types of pollutants are found virtually everywhere on Earth, but they are particularly threatening to aquatic species.
- Climate change is a significant threat to biodiversity and ecosystem function due to multiple impacts including changes to average and extreme temperature and precipitation, melting of polar ice caps, sea-level rise, and ocean acidification.
- To estimate how many species will go extinct in the future, biologists make calculations based on many variables including current extinction rates, habitat sizes, predicted shifts in bioclimate envelopes, demographics, and species interactions.

54.3 Why Are Biodiversity and Ecosystem Function Important?

- At the ecosystem level, experiments have shown that high species richness increases aspects of ecosystem function such as productivity, resistance to disturbance, and resilience following a disturbance.
- There are many types of ecosystem services. Humans gain direct economic benefits from fishing, forestry, agriculture, and other activities that harvest materials from ecosystems.
- Humans also benefit from the ability of functioning ecosystems to build and hold soil, moderate local climates, retain and cycle nutrients, retain surface water and recharge groundwater, prevent flooding, and produce oxygen.
- Human happiness and health depend on healthy ecosystems.

54.4 Preserving Biodiversity and Ecosystem Function

- The major threats to biodiversity—habitat destruction and climate change—will not lessen until the human population stabilizes and resource use becomes sustainable.
- Solutions to the biodiversity crisis include preserving genetic diversity, protecting key habitats, restoring ecosystems, and integrating conservation with public policy so that protecting ecosystem services is a greater part of both global and local decision-making processes.
- Ecosystem restoration efforts are having a positive impact on biodiversity around the world. Conservation efforts make a difference. *You* can make a difference!

Answers are available in Appendix A.

✔ TEST YOUR KNOWLEDGE

1. What does species richness refer to? Select True or False for each statement.
   T/F the number of species in an area
   T/F the evenness of species in an area
   T/F the functional diversity of a species in an area
   T/F the phylogenetic diversity of species in an area
2. **MODEL** What community property does the red arrow in the model represent?
   a. low resilience
   b. high resilience
   c. low resistance
   d. high resistance

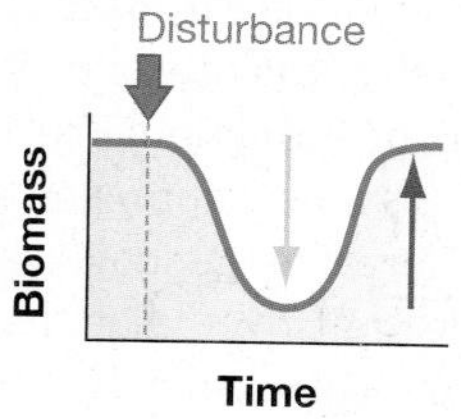

3. What is a biodiversity "hotspot"?
   a. an area where an all-taxon survey is under way
   b. an area where an environmental sequencing study has been completed
   c. a habitat with high NPP
   d. an area with high species richness and high threat to those species
4. What is the primary threat in marine environments?

✔ TEST YOUR UNDERSTANDING

5. **THINK CAREFULLY** Why is the common phrase "humans and animals" not accurate in biology? Suggest one way that a more scientific outlook could boost conservation efforts.
6. **PROCESS OF SCIENCE** Biologists claim that the all-taxa survey now under way at the Great Smoky Mountains National Park in the United States will improve their ability to estimate the total number of species living today. Discuss the benefits and limitations that this data set will provide in understanding the extent of global biodiversity.
7. Some biologists prefer to focus efforts on preserving endangered species while others prefer to focus on preserving ecosystems. What is your advice to biologists, based on the evidence from this chapter?
8. Explain why the construction of wildlife corridors can help maintain genetic diversity in a fragmented landscape.

✔ TEST YOUR PROBLEM-SOLVING SKILLS

9. The population size of mountain pine beetles in the American Northwest has long been held in check by freezing temperatures during the winters. As winters warm, populations of pine beetles are increasing, killing whitebark pine forest. Grizzly bears rely on the nuts of whitebark pine trees. Do you think climate change is affecting the fitness of the grizzly bear population?
   a. No, because climate change is not a problem yet.
   b. No, because climate change is affecting the beetles, not the bears.
   c. Yes, because climate change is decreasing the food supply of the bears, which means they will be less strong.
   d. Yes, because climate change is decreasing the food supply of the bears, which is likely to reduce their ability to survive or reproduce.

10. **SOCIETY** The maps shown here chronicle the loss of old-growth forest (more than 200 years old) that occurred in the United States. In your opinion, under what conditions is it ethical for conservationists who live in the United States to lobby government officials in Brazil, Indonesia, and other tropical countries to slow the rate of loss of old-growth forest?

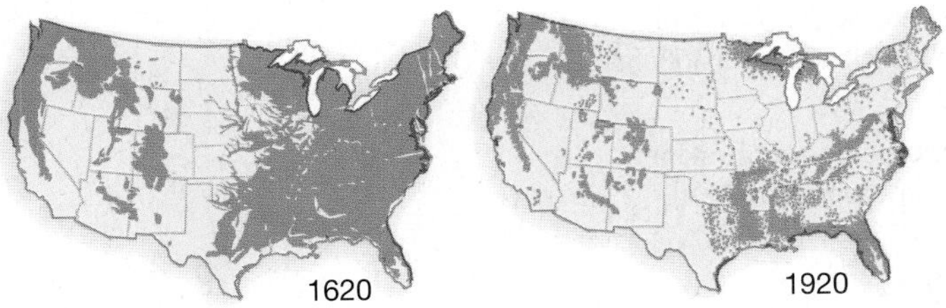

✔ PUT IT ALL TOGETHER: Case Study

How can biodiversity help coffee crops?

Do you drink coffee? A lot of people do—coffee is a major tropical crop, valued at over $100 billion per year. The most popular species of coffee, *Coffea arabica*, originated in Africa and is now planted extensively in Central and South America. The pest called the coffee berry borer beetle (*Hypothenemus hampeii*) moved along with the coffee and is a major problem for coffee farmers, sometimes destroying half of the coffee crop in mature plantations.

11. Based on what you have learned in this chapter about global trends in terrestrial ecosystems, what is currently the largest threat to the wet tropical forests in Central America?
   a. habitat destruction, such as the clearing of forests for coffee plantations
   b. exotic species, such as the introduction of coffee berry borers
   c. pollution, such as the use of pesticides in coffee plantations
   d. climate change, such as changes in precipitation patterns

12. Pesticides are generally ineffective in killing coffee berry borers. Ecologist Daniel Karp and colleagues conducted a study in Costa Rica to determine if natural predators like birds reduce the beetle population. The researchers selected 12 control plots and used nets to exclude birds from 12 similar treatment plots. What is the take-home message of the graph? (Remember, * means $P < 0.05$; see **BioSkills 3**.)

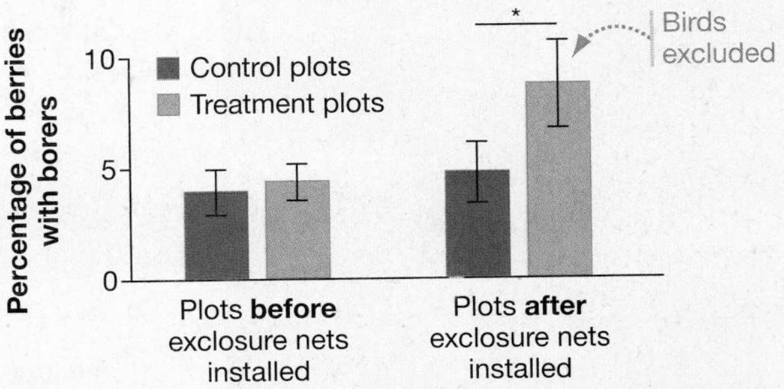

Source: D. S. Karp et al. 2013. *Ecology Letters* 16: 1339–1347.

13. **PROCESS OF SCIENCE** Look again at the graph in Question 12. Why did the researchers bother to collect data from the coffee shrubs before adding the nets to treatment plots to exclude birds?

14. **QUANTITATIVE** The abundance of birds and other predators of borer beetles depends on how much natural forest is left within and around the coffee plantations. What is the approximate percentage increase in borer predators per hectare if forest cover is doubled from 15 to 30 percent?

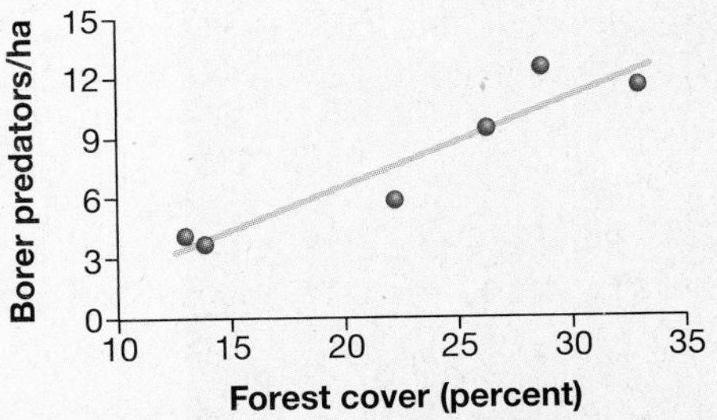

Source: D. S. Karp et al. 2013. *Ecology Letters* 16: 1339–1347.

15. **MODEL** Draw a simple model to compare the relative species richness and species diversity of these two coffee plantations: one is clear-cut and planted with only coffee shrubs, the other retains some natural forest (see **Making Models 52.2** on page 1128). For symbols, use a few letters to represent different species of plants and a few numbers to represent different species of animals in a square 1-ha plot.

16. **SOCIETY** Evaluate this statement: Leaving some natural forest in and around coffee plantations is a "win-win" situation—a win for ecosystems and a win for farmers.

Mastering Biology

Students Go to Mastering™ Biology for assignments, the eText, and the Study Area with animations, practice tests, and activities.

Professors Go to Mastering™ Biology for automatically graded tutorials and questions that you can assign to your students, plus Instructor Resources.

END-OF-UNIT
CASE STUDY

UNIT 8

PUT IT ALL TOGETHER

▮ For an introduction to the Mystery of the Newt case study, see page 17.

Now that you've learned about ecological processes and patterns, it's time to return to the Mystery of the Newt case study introduced after Chapter 1 (page 17). Recall the story about three hunters who were found dead at a campsite in Oregon next to a coffee pot containing a surprise ingredient: a rough-skinned newt (*Taricha granulosa*). This was an unusual human–newt species interaction!

✔ If you understand Unit 8, you should be able to apply your learning to this case study:

What Is the Larger Ecological Context of Toxic Newts?

Are toxic newts flourishing in your neighborhood? It depends where you live! One way to begin thinking about the larger ecological context of rough-skinned newts is to look at their distribution and abundance.

1. Which terrestrial biome do rough-skinned newts occupy? (Compare the map in **Figure 1** to Figure 49.9)
   a. temperate coniferous forest
   b. tropical wet forest
   c. montane grassland
2. Predict which abiotic factor limits the eastern expansion of the newt range. (See Section 49.3)
   a. Coriolis effect
   b. Hadley cell
   c. rain shadow
3. **PROCESS OF SCIENCE** Suggest which sampling method you could use to estimate the population size of newts and explain how it works. (See Section 51.1)

Figure 1

DATA: IUCN, Conservation International, and NatureServe. 2004. *Global Amphibian Assessment.*

Rough-skinned newts have a small amount of the toxin TTX in their skin at all times, but secrete large amounts of toxin when threatened by a predator. Further, they advertise their toxicity by arching in an "Unken reflex" posture (**Figure 2**), flashing their yellow or orange underbelly.

4. Is the Unken reflex an honest signal? Why would it be adaptive? (See Section 50.5)
5. Is TTX a standing or induced defence? What are the fitness trade-offs of this type of defense? (See Section 52.1)

Figure 2

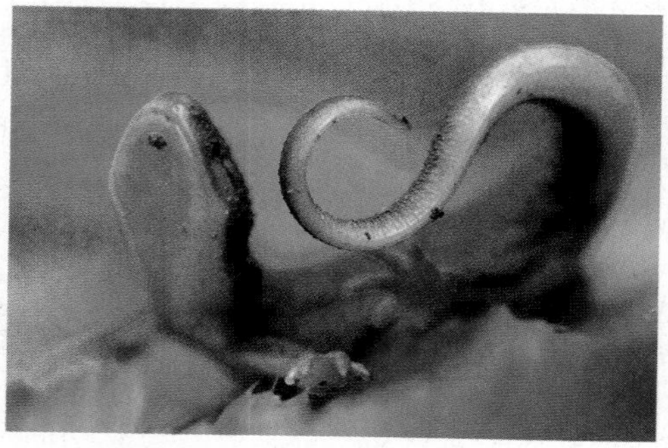

The Unken reflex flashes the orange underbelly of rough-skinned newts, warning predators of the toxin.

Newts are amphibians. They can live in moist terrestrial habitats as adults but require freshwater ecosystems (such as streams where hunters gather water for coffee!) to find a mate, deposit eggs, and grow as larvae. Thus, the food web of newts includes both terrestrial and aquatic organisms, only a few of which are shown in **Figure 3.**

6. **MODEL** Draw a model that shows how the energy that a newt expends in its Unken reflex ultimately came from the Sun. (See Section 53.1)
7. **THINK CAREFULLY** Explain how it is possible that carbon that you exhale in your breath could be incorporated into TTX in newts. (See Section 52.2)
8. What type of species interaction are newts and garter snakes engaged in? What are the fitness consequences to each species? (See Section 52.1)

Figure 3

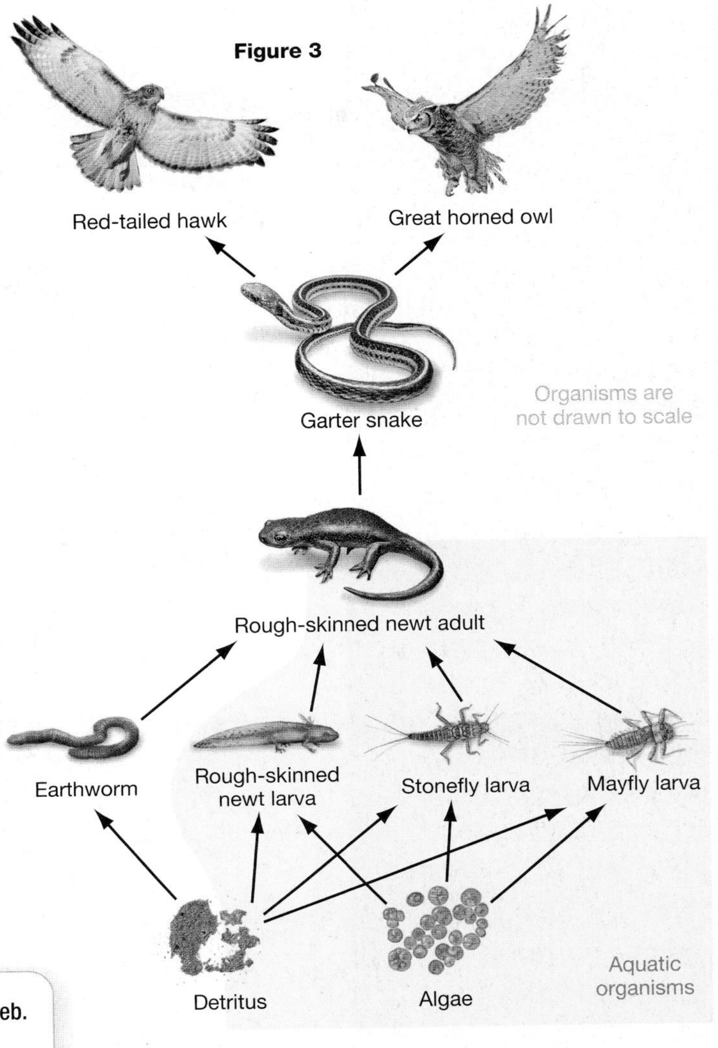

Recall that newts vary in their toxicity and that snakes tend to have higher resistance to TTX in areas of high newt toxicity (see Unit 5 case study on pages 720–721). Further, TTX-resistant garter snakes survive when they eat newts but they tend to be slower than non-resistant snakes and may become toxic to their predators.

9. Predict how the size of the hawk population would vary in areas of high and low newt toxicity. (See Section 52.2)

The nutritious eggs of newts could be added to the aquatic food web. But guess what? Newts secrete TTX in their eggs.

10. **PROCESS OF SCIENCE** List the five main levels of ecological study, and write a question about the ecology of newt eggs that you could research at each level. (See Section 49.1)

Female newts attach clumps of eggs to underwater vegetation. The eggs contain the toxin TTX.

In what ways to newts and humans normally interact?

11. **SOCIETY** Provide three examples of ecosystem services that healthy newt habitat may provide to humans. (See Section 54.3)
12. **SOCIETY** Predict three types of threats that the newt populations are experiencing due to human activities. (See Section 54.2)

Your knowledge of ecology can not only help you to solve the riddle of the toxic newt, but also to understand the connections among all living things. The end-of unit case studies in this text happen to follow the story of the newt, but every organism has its own story to tell, with plenty of unsolved mysteries ranging in scale from molecules to whole ecosystems.

BIG PICTURE

Ecology's primary goal is to understand the distribution and abundance of organisms. To reach this goal, ecologists study how organisms interact with each other and their abiotic environments.

Human activities are causing radical changes in organisms, populations, communities, and whole ecosystems. As a result, virtually every box and arrow in this concept map is being affected.

In addition to documenting changes, biologists are educating the public and policy makers about the value of biodiversity and healthy ecosystems, and researching the most effective ways to preserve species and keep ecosystems functioning normally.

Note that most boxes in the concept map indicate chapters and sections where you can go for review. Also, be sure to do the blue exercises in the Check Your Understanding box below.

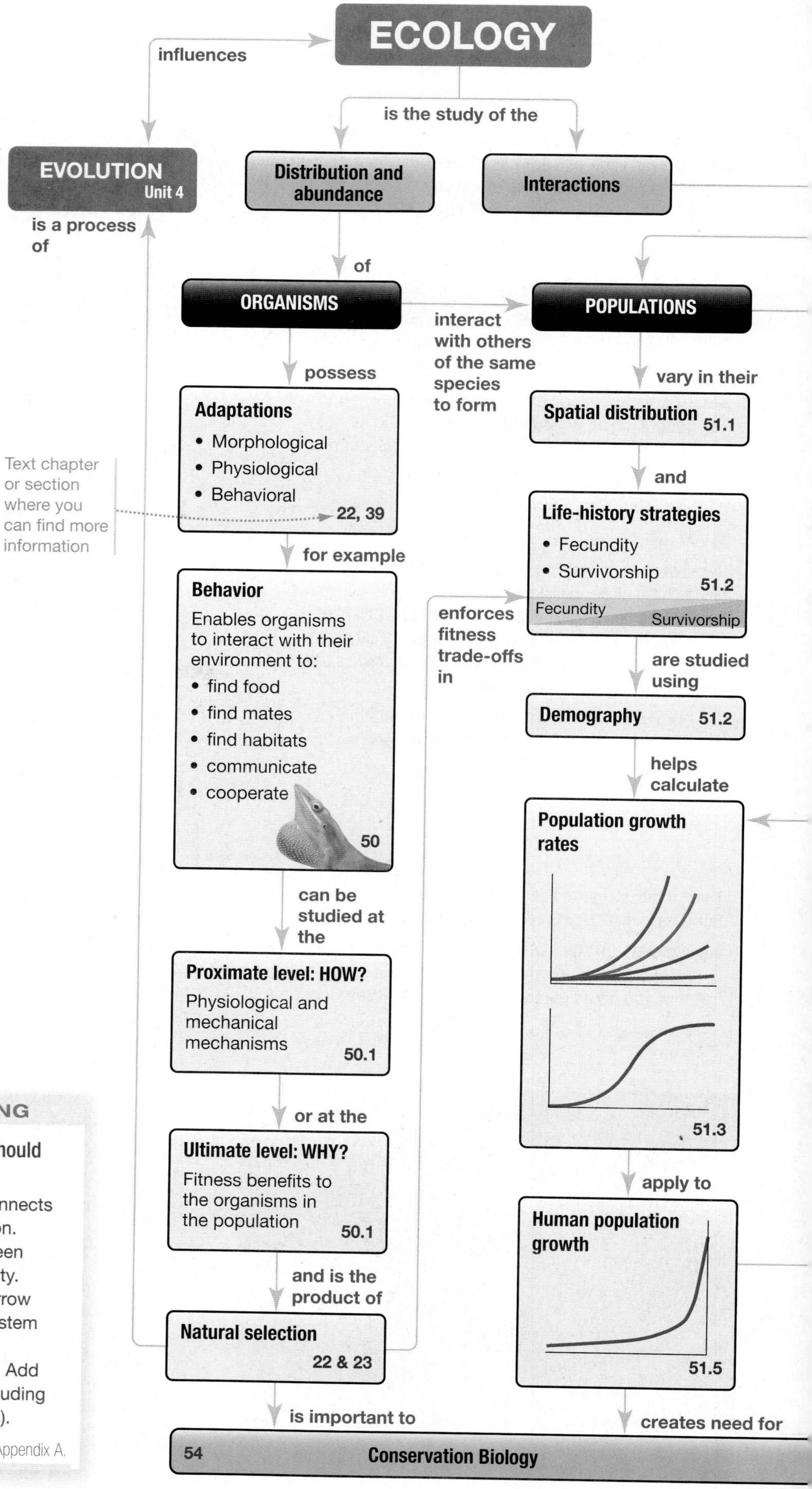

CHECK YOUR UNDERSTANDING

✔ **If you understand the big picture, you should be able to . . .**

1. Add an arrow and linking phrase that connects species interactions with natural selection.
2. Explain why there is a relationship between human population growth and biodiversity.
3. Give a specific example to explain the arrow between primary productivity and ecosystem services.
4. Label the greenhouse effect on the map. Add a box for CO_2 emissions to the map, including appropriate arrow(s) and linking phrase(s).

Answers are available in Appendix A.

THE BIG PICTURE

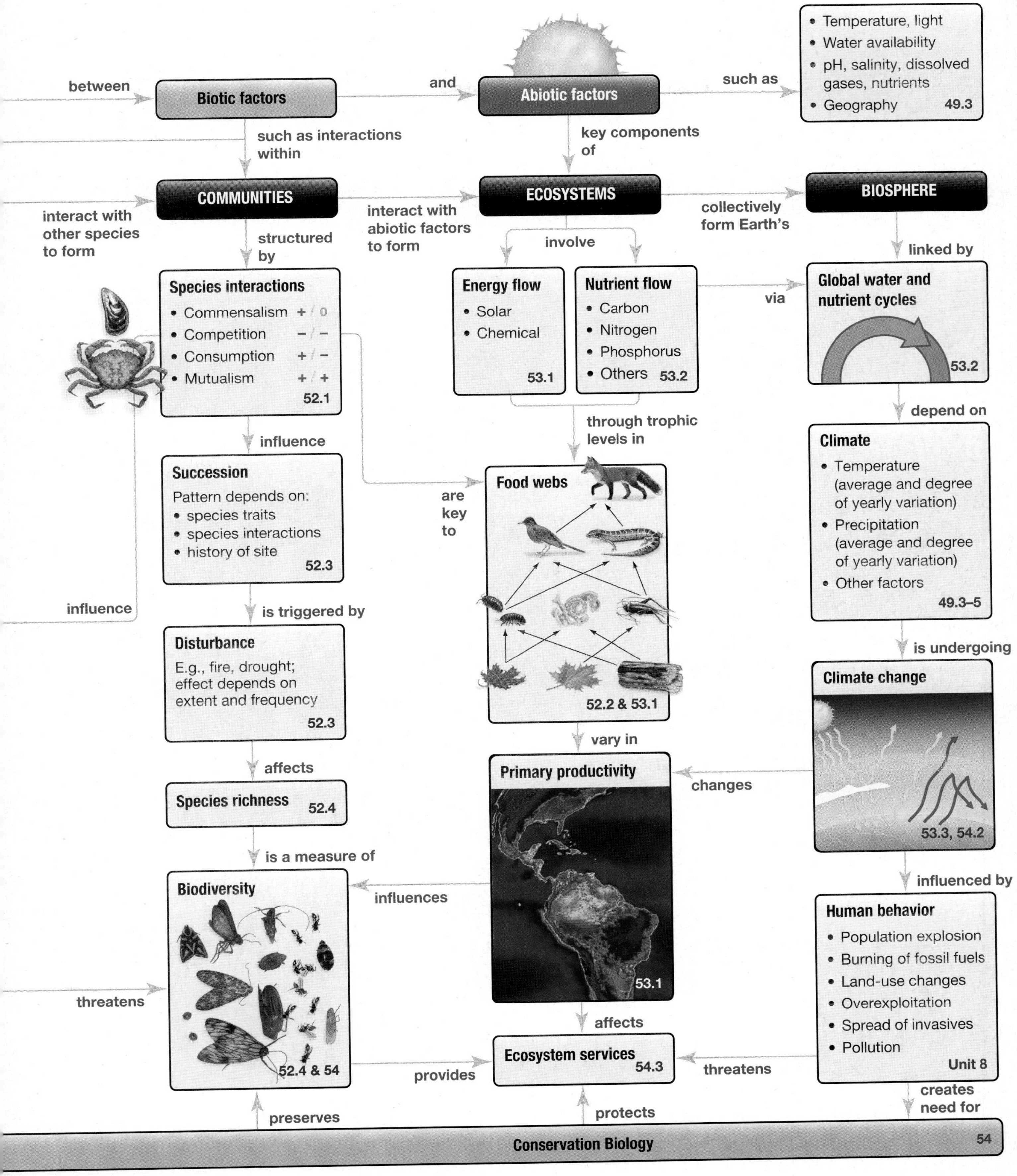

between
Biotic factors
and
Abiotic factors
such as
• Temperature, light
• Water availability
• pH, salinity, dissolved gases, nutrients
• Geography 49.3
such as interactions within
key components of
COMMUNITIES
ECOSYSTEMS
BIOSPHERE
interact with other species to form
interact with abiotic factors to form
collectively form Earth's
structured by
involve
linked by
Species interactions
• Commensalism +/0
• Competition −/−
• Consumption +/−
• Mutualism +/+
52.1
Energy flow
• Solar
• Chemical
53.1
Nutrient flow
• Carbon
• Nitrogen
• Phosphorus
• Others 53.2
via
Global water and nutrient cycles
53.2
influence
through trophic levels in
depend on
Succession
Pattern depends on:
• species traits
• species interactions
• history of site
52.3
are key to
Food webs
52.2 & 53.1
Climate
• Temperature (average and degree of yearly variation)
• Precipitation (average and degree of yearly variation)
• Other factors
49.3–5
influence
is triggered by
is undergoing
Disturbance
E.g., fire, drought; effect depends on extent and frequency
52.3
Climate change
53.3, 54.2
vary in
changes
affects
Primary productivity
53.1
Species richness 52.4
is a measure of
influences
influenced by
Biodiversity
52.4 & 54
Human behavior
• Population explosion
• Burning of fossil fuels
• Land-use changes
• Overexploitation
• Spread of invasives
• Pollution
Unit 8
threatens
affects
Ecosystem services 54.3
provides
threatens
creates need for
preserves
protects
Conservation Biology 54

APPENDIX A Answers

Chapter 1

IN-TEXT QUESTIONS AND EXERCISES

p. 2 CYU (1) apply Because viruses are not cells, they are not considered organisms. Most biologists agree that viruses are not even alive, because they depend on their host cell for replication and information processing, which are two additional characteristics of life. **(2)** analyze Your diagram can be in the form of a table or a flowchart, and it should show that the three unifying theories depend on various combinations of the five characteristics of life, e.g.: (1) The cell theory—cells, replication, energy; (2) The chromosome theory of replication—replication, information, energy; (3) The theory of evolution—evolution.

p. 4 Fig. 1.2 analyze If Pasteur had done either of the things listed, he would have had more than one variable in his experiment. Thus critics could claim that he got different results because of the differences in broth types or flask types and not the difference in exposure to preexisting cells. The results would not be definitive.

p. 4 CYU (1) understand The cell theory states only that all organisms are made of cells, and all cells come from preexisting cells. In multicellular organisms, the cells are not all identical because they can become specialized for particular functions. **(2)** analyze Your diagram should start with one circle (one cell) and use two arrows to show it dividing to result in two circles (two cells), and so on.

p. 6 CYU (1) analyze DNA (gene) ⟶ RNA ⟶ protein. **(2)** apply To stay alive, the flamingo requires energy. By feeding on aquatic organisms, it obtains molecules that can be used as building blocks for DNA, RNA, proteins, etc., and for making ATP.

p. 7 CYU understand Over time, traits such as a beak shaped to retrieve a certain food type that is especially abundant in a particular habitat will become increasingly frequent in the population.

p. 8 Making Models 1.1 apply See **Figure A1.1**. Molds and other fungi appear not to be closely related to green algae or land plants, because they differ from both green algae and land plants at two positions (5 and 8, counting from left to right).

p. 9 Fig. 1.8 apply The eukaryotic cell is roughly 10 times the size of the prokaryotic cell.

p. 10 CYU analyze From the sequence data provided, species A and B differ only in one letter of the DNA sequence (position 10 from left). Species C differs from species A and B in four letters (positions 1, 2, 9, and 10). Species A and B would be closest on a phylogenetic tree since they appear to be more closely related, while species C is more distantly related.

p. 11 Fig. 1.9 apply Male giraffes spend most of their time feeding on vegetation about 3.25 meters high; females spend most of their time feeding at about 2.5 meters.

p. 13 Fig. 1.11 apply If you used just one ant, the interpretation of the experiment would not likely change, but you would have less confidence in the conclusions drawn.

p. 14 CYU (1) create The key here is to test predation rates during the hottest part of the day (when desert ants actually feed) versus other parts of the day. The experiment would best be done in the field, where natural predators are present. One approach would be to capture a large number of ants, divide the group in two, and measure predation rates (number of ants killed per hour) when they are placed in normal habitat during the hottest part of the day versus an hour before (or after). You would need to include a control group—ants outside during the hottest part of the day. **(2)** evaluate If you didn't include a control, critics could argue that predation did or did not occur because of your experimental setup or manipulation, not because of differences in temperature. You also would need to ensure that there is no difference in ant body size or walking speed, how ants were captured and maintained, or other traits that might make the ants in the two groups more or less susceptible to predators. The ants should also be put out in the same habitat, so the presence of predators is the same in the two treatments.

✔ TEST YOUR KNOWLEDGE

1. understand d **2.** remember F, F, F, T **3.** remember b **4.** analyze Yes. Long necks could be advantageous for more than one reason.

✔ TEST YOUR UNDERSTANDING

5. apply They would need to show that the entity they discovered is cellular, replicates, processes information, acquires and uses energy, and has evolving populations. **6.** understand a **7.** understand Individuals with certain advantageous traits are selected, in the sense that they produce the most offspring. **8.** understand A null hypothesis specifies what a researcher should observe when the hypothesis being tested isn't correct.

✔ TEST YOUR PROBLEM-SOLVING SKILLS

9. evaluate In everyday English, the word "theory" is often used to mean a hunch or speculation about how something works. But in science, the meaning is different. A scientific theory is not just an educated guess, hunch, or speculation—it is an idea whose validity can be tested through data collection. The theory of evolution has been validated by large bodies of observational and experimental data. This theory is a scientific explanation that is so well established that no new evidence is likely to alter it. **10.** analyze Yes, they are likely to evolve differently. In areas of the world where HIV infection rates are high, the genes (heritable traits) that confer resistance to HIV should increase in the population over time.

✔ PUT IT ALL TOGETHER: Case Study

11. analyze The flow of information is from DNA to RNA to protein. Physical traits like leaf shape and size are a product of the proteins produced. **12.** analyze The Latin root *tri* means "three" and *foli* means "leaf." The name is appropriate because, as the photo shows, the vine consists of three-leaf clusters along its length. **13.** analyze Statistically, there is significantly less leaf damage in the vines on leafy host trees compared to that in vines creeping on the ground or vines on bare tree trunks. This result suggests that growing among other leaves protects the vine from predation by plant eaters. But the data do not directly show that leaf mimicry reduces herbivory. To test whether mimicry reduces herbivory, researchers would need to move vines to another host and measure herbivory on vines with similar versus dissimilar host-leaf shapes. **14.** analyze Exposing different vines to different light levels would add another variable to the study. Under these varying conditions, changes in the herbivory index might be due to light availability. **15.** apply The study included bare tree trunks as a control to show that the vine was not protected merely by climbing up a support to avoid predators on the ground. **16.** understand "Fitness" refers to an individual's ability to survive and reproduce.

Figure A1.1

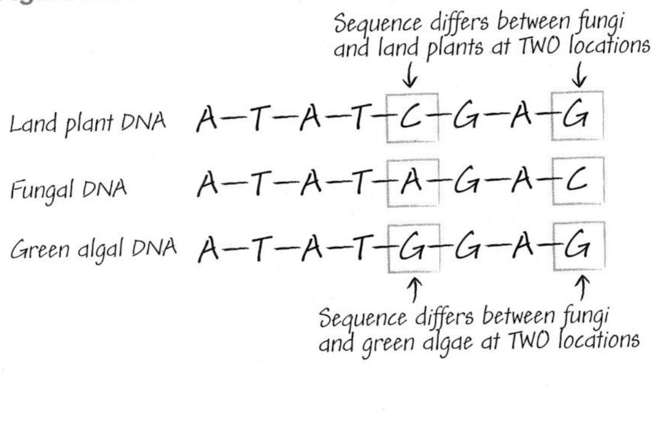

END-OF-UNIT CASE STUDY

Mystery of the Newt

p. 17 (1) create One possible hypothesis is: Rough-skinned newts contain a toxin in their skin. **(2)** create There are many possible approaches. You could design an experiment similar to what Butch Brodie, Jr. did as an undergraduate: Collect newts and trap mice (potential predators) in the woods. Grind up the newt skin into a fine powder and mix with water in different concentrations. Inject the newt skin solution or water as a control into mice ($n = 10$). Observe whether the mice live or die.

BIG PICTURE Doing Biology

pp. 18–19 CYU (1) understand Biologists design and carry out a study, either observational or experimental, to test their ideas. As part of this process, they state their ideas as a hypothesis and null hypothesis and make predictions. They analyze and interpret the data they have gathered, and determine whether the data support their ideas. If not, they revisit their ideas and come up with an alternative hypothesis and design another study to test these new predictions. **(2)** understand There are many possible examples. Consider, for example, the experiment on navigation in foraging desert ants (Chapter 1). In addition to testing how the ants use information on stride length and number to calculate how far they are from the nest (multicellular organism and population levels), researchers also could test how the "pedometer" works at the level of cells and molecules. **(3)** analyze A hypothesis is a testable statement to explain a specific phenomenon or a set of observations. The word "theory" refers to proposed explanations for very broad patterns in nature that are supported by a wide body of evidence. A theory serves as a framework for the development of new hypotheses. **(4)** analyze The next step is to relate your findings to existing theories and the current scientific literature, and then to communicate your findings to colleagues through informal conversations, presentations at scientific meetings, and eventually publication in peer-reviewed journals.

BioSkills

BIOSKILLS 1: p. 23 CYU (1) apply $5.0 \text{ km} \times 0.62 \text{ mile/km} = 3.1 \text{ miles}$. **(2)** apply $5/9(98.6°F - 32) = 37°C$. **(3)** apply Multiply your weight in pounds by 1 kg/2.2 pounds (0.45). **(4)** apply 9.2×10^{-7}; 2.3×10^{7}. **(5)** apply The answer (4.6) has 2 significant figures. When you multiply, the answer can have no more significant figures than the least accurate measurement—in this case, 1.6.

BIOSKILLS 2: p. 26 CYU (1) apply About 18 percent. **(2)** apply A dramatic drop (almost 10 percent). **(3)** analyze No, they would not be different. Because the data values in a bar chart are discrete or categorical rather than continuous, their order of presentation does not matter (though it's convenient to arrange the bars in a way that reinforces the overall message). **(4)** apply The most common height is 68 inches. $68 \text{ inches} \times 2.54 \text{ cm/inch} = 170 \text{ cm}$. **(5)** apply The dependent variable is the percentage of children. The independent variable is the type of response. See **Figure AB.1**.

BIOSKILLS 3: p. 29 CYU (1) analyze Test 2, the estimate based on the larger sample, is likely to have a smaller standard error. The more replicates or observations you have, the more precise your estimate of the average should be. **(2)** analyze The hummingbirds displayed a statistically significant preference for sucrose over both water and aspartame (*** means $P < 0.001$), but they consumed solutions of sucrose and erythritol with equal preference (no asterisk means no significant difference).

BIOSKILLS 4: p. 29 CYU
(1) apply $1/2 \times 1/2 \times 1/2 \times 1/2 = 1/16$
(2) apply $1/6 + 1/6 + 1/6 = 1/2$

BIOSKILLS 5: p. 30 CYU (1) understand Exponential **(2)** apply $\ln N_t = \ln N_0 + rt$. **(3)** apply $\text{pH} = -\log_{10}(2.75 \times 10^{-4}) = 3.56$. **(4)** apply $[\text{H}^+] = 10^{-5.43} = 0.00000372 = 3.72 \times 10^{-6}$.

BIOSKILLS 6: p. 31 Fig. B6.1 understand DNA and RNA are acids that tend to lose a proton in solution, giving them a negative charge.

p. 33 Fig. B6.4 apply The probe must be single stranded so that it will bind by complementary base pairing to the target DNA, and it must be labeled so that it can be detected. The probe will base-pair only with fragments that include a sequence complementary to the probe's sequence. A probe with the sequence 5′-AATCG-3′ will bind to the region of the target DNA that has the sequence 5′-CGATT-3′ as shown here:

5′-AATCG-3′
3′-TCCGGTTAGCATTACCATTTT-5′

p. 34 CYU (1) analyze The lane with no band comes from a sample where no PCR product was amplified. The faint band has very few copies of PCR product, while the bright band has many. **(2)** evaluate Understanding a molecule's structure is often critical to understanding how the molecule functions in cells.

BIOSKILLS 7: p. 35 CYU (1) understand Size, shape, and/or density. **(2)** apply Mitochondria, because they are larger than ribosomes.

BIOSKILLS 8: p 36 CYU understand A decrease in transmittance usually results in an increase in absorbance.

BIOSKILLS 9: p. 38 CYU analyze It doesn't necessarily mean that the cell lacks mitochondria. It's possible that no mitochondria happened to be present in this extremely thin section sliced through the cell.

BIOSKILLS 10: p. 39 Fig. B10.1 apply No—each type of cDNA would be represented many times, because many copies of each type of mRNA were present in the cells, and many bacterial cells were used to prepare the library.

p. 40 Fig. B10.3 analyze The polymerase will begin at the 3′ end of each primer. On the top strand in part (b), it will move to the left; on the bottom strand, it will move to the right. As always, synthesis is in the 5′ → 3′ direction.

p. 44 CYU (1) apply ddNTPs lack the —OH (hydroxyl) group on the 3′ carbon of deoxyribose that is required to extend the DNA chain during synthesis. **(2)** analyze (a) Primer 1b binds to the top right strand and would allow DNA polymerase to synthesize the top strand across the target gene. Primer 1a, however, binds to the top left strand and would allow DNA polymerase to synthesize the top strand *away* from the target gene. Primer 2a binds to the bottom left strand and would allow DNA polymerase to synthesize the bottom strand across the target gene. Primer 2b, however, binds to the bottom right strand and would allow DNA polymerase to synthesize the bottom strand away from the target gene. (b) Tell her to use primer 1b with primer 2a. **(3)** apply Start with a microarray containing exons from a large number of human genes. Isolate mRNAs from brain tissue and liver tissue, and make labeled cDNAs from each. Probe the microarray with both cDNAs, and record where binding occurs. Binding events identify genes that are transcribed in each type of tissue. Compare the results to identify genes that are expressed in brain but not liver, or in liver but not brain.

BIOSKILLS 11: p. 48 CYU (1) analyze Because the artificial conditions of cell culture differ from natural conditions, it may not be clear how the results apply to noncancerous cells that are not growing in cell culture. **(2)** analyze (a) *Caenorhabditis elegans* would be a good possibility, because the cell fates are known for each cell in a 33-cell embryo. You could find mutant individuals that lacked normal development, and compare the resulting embryos with normal embryos to identify the cells that change and examine how they change. (b) Any of the multicellular organisms described in BioSkills 11 would be a candidate, but *Dictyostelium discoideum* might be particularly interesting because cells stick to each other only during certain points in the life cycle.

BIOSKILLS 12: p. 50 CYU (1) analyze This model focuses on the alleles for one gene on one pair of homologous chromosomes. Other genes and other chromosomes are not shown, nor are any of the other contents of the cell. **(2)** understand The red balls represent oxygen atoms. These atoms have no color in real life. **(3)** understand DNA molecules. **(4)** apply The arrow represents the movement of the sodium ion through a protein channel to the other side of the membrane. **(5)** analyze The lizard is probably about the size of a human hand (and would be too small to see clearly if drawn to the same scale as the human and the dog). **(6)** understand The stomata cells are too small to be seen on the leaf. **(7)** analyze One way is to add an arrow that points from "genes" to "alleles" and is labeled "have different versions called."

BIOSKILLS 13: p. 51 CYU (1) apply See **Figure AB.2**. **(2)** understand Mammals, lizards/snakes, turtles, alligators/crocodiles, and birds have amniotic eggs. **(3)** analyze See **Figure AB.2**.

Figure AB.1

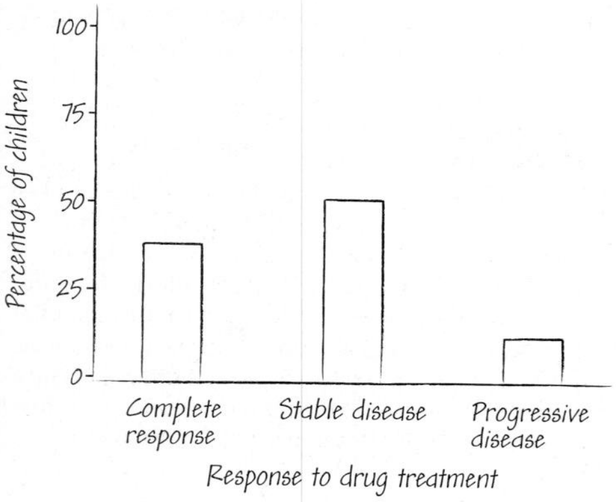

Figure AB.2

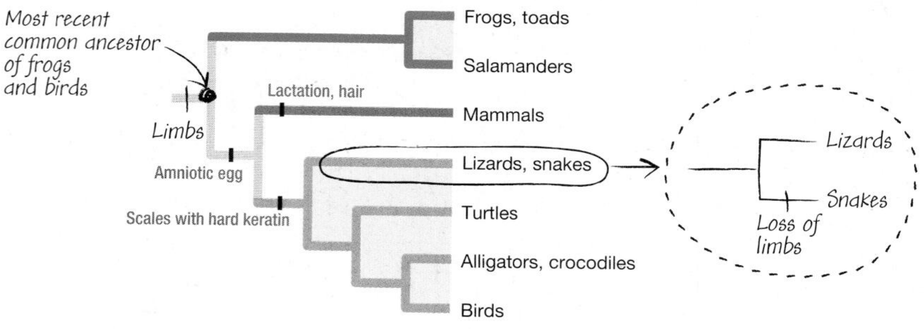

(4) analyze See Figure AB.3. (5) apply Mammals are equally related to lizards and turtles, because the most recent common ancestor of mammals and lizards is the same as the most recent common ancestor of mammals and turtles.

BIOSKILLS 14: p. 53 CYU analyze See Figure AB.4.

BIOSKILLS 15: p. 53 CYU (1) apply "Different yoked-together." **(2)** apply "Sugary loosened." **(3)** apply "Study of form." **(4)** apply "Three bodies."

BIOSKILLS 16: p. 55 CYU create Many examples are possible. See Ch. 11, Figure 1.11 as an example of the format to use for your Research box.

BIOSKILLS 17: p. 55 CYU (1) evaluate The type of misconception is goal-oriented thinking. Evolution is not goal directed. Legs did not evolve in fish because fish wanted or needed them. (See Chapter 22 for more information on evolution by natural selection.) **(2)** evaluate In science, a theory is an explanation for a broad class of phenomena that is supported by a wide body of evidence. In everyday use, a theory is a hunch or speculation about how something works. This difference in meaning could cause confusion because someone might dismiss a scientific theory (such as the theory of evolution or the theory of gravity) as speculation rather than treat it as a scientific explanation that is so well established that no new evidence is likely to alter it.

BIOSKILLS 18: p. 57 CYU (1) understand Your answer should include the six steps to study success presented in BioSkills 18, but written in your own words. **(2)** analyze The Bloom's level for question 1 would be *Understand*.

Chapter 2

IN-TEXT QUESTIONS AND ANSWERS

p. 60 Fig. 2.3 apply In phosphorus, there are 15 electrons, 3 electron shells, and the outer shell contains 4 orbitals. Since isotopes of an element only vary in neutrons, the number of protons and electrons will not differ between them, so all of these values will be the same.

p. 62 Fig. 2.7 understand Oxygen and nitrogen have high electronegativities. They hold shared electrons more tightly than C, H, and many other atoms, resulting in polar bonds.

p. 63 Fig. 2.8 apply *Water*: arrows pointing from hydrogens to oxygen atom; *ammonia*: arrows pointing from hydrogens to nitrogen atom; *methane*: double arrows between carbons and hydrogens; *carbon dioxide*: arrows pointing from carbon to oxygens; *molecular nitrogen*: double arrows between nitrogens.

p. 64 CYU (1) apply It has changed from being a hydrogen isotope to being a helium isotope. **(2)** apply See Figure A2.1. **(3)** analyze No, NaCl is formed by an ionic bond and CCl_4 would be formed by covalent bonds. Carbon is more electronegative than sodium and far less likely to lose its valence electrons to generate a stable +4 cation capable of ionic bonding with four chloride anions.

p. 65 analyze (1) The difference in electronegativity between the C and O atoms in CO_2 is similar to the difference between H and O atoms in H_2O. Both consist of three atoms bonded together with polar covalent bonds, but CO_2 is a linear molecule with double bonds and no lone electron pairs, whereas H_2O is a bent molecule with single bonds and two lone electron pairs. (2) If water had a molecular shape similar to CO_2, the partial negative charge on oxygen would have partial positive charges on either side. Compared to the actual, bent molecule, the partial negative charge would be much less exposed and less able to participate in hydrogen bonding.

p. 66 Fig. 2.13 apply Oils, like the octane in this figure, must be nonpolar. When an oil is poured into water, the polar water molecules would interact with each other via hydrogen bonding rather than with the oil molecules. The oil is thus separated from the water to form a distinct layer via hydrophobic interactions.

p. 68 Table 2.2 understand *Row 1:* Most reactions important for life take place in aqueous solution. *Row 2:* Ice floats, insulating bodies of water and preventing them from freezing solid. *Row 3:* The temperature of aqueous solutions changes slowly; oceans moderate coastal climates. *Row 4:* Evaporation of water from an organism cools the body.

p. 69 Making Models 2.1 apply See Figure A2.2.

p. 70 Fig. 2.16 apply The concentration of protons would decrease because milk is more basic (pH 6.5) than black coffee (pH 5).

p. 70 apply The bicarbonate concentration would increase. The protons (H^+) released from carbonic acid would react with the hydroxide ions (OH^-) dissociated from NaOH to form H_2O, leaving fewer protons free to react with bicarbonate to produce carbonic acid.

p. 70 CYU (1) understand When NaCl is dissolved, water disrupts the ionic bond between sodium and chloride and surrounds each of the ions. When glucose is dissolved, the covalent bonds are not disrupted, but water surrounds it by hydrogen-bonding with different parts of the molecule. **(2)** apply The heat of vaporization is used to cool off the pig as the water in mud takes up a large amount of heat from the body to change from liquid to gas. **(3)** apply The proton concentration in this solution would be 0.5 M, which is pH 0.3. To determine the number of protons, you would need to know the volume of the solution.

p. 73 Fig. 2.19 remember See Figure A2.3. The amount of potential energy in bonds is inversely related to the strength of the bonds. For example, the bonds in H_2 would be weaker and easier to break than those in H_2O.

p. 73 CYU (1) understand The electrons are shifted farther from the nuclei of the carbon and hydrogen atoms and closer to the nuclei of the more electronegative oxygen atoms. **(2)** analyze The reactants have higher chemical energy than the products. The entropy, however, is not increased or decreased based only on the number of molecules involved. The heat given off from this reaction would lead to increased entropy in the environment. Overall, the reaction would be spontaneous based on the change in potential energy. **(3)** understand The potential energy from the bonds is converted into kinetic energy, which is often in the form of thermal energy. The first law of thermodynamics is preserved because the energy is not destroyed, but only transformed.

p. 74 Fig. 2.21 remember The water-filled flask is the ocean; the gas-filled flask is the atmosphere; the condensed water droplets are rain; the electrical sparks are lightning.

p. 74 CYU analyze The conditions used in Miller's experiment would need to accurately mimic the environment of early Earth to support this theory.

p. 76 Table 2.3 apply All the functional groups in Table 2.3, except for the sulfhydryl group (—SH), are considered polar. The sulfhydryl group is very slightly polar, since sulfur's electronegativity is only slightly higher than that of hydrogen. When present on the amino acid cysteine, the sulfhydryl group is classified as nonpolar.

p. 77 CYU (1) apply The number of valence electrons in silicon is the same as carbon, which means that the number of covalent bonds formed would be the same. **(2)** analyze The carbonyl functional group is present on formaldehyde. This group is known to react with other componds to produce larger molecules.

✔ TEST YOUR KNOWLEDGE

1. remember b **2.** remember a **3.** remember c **4.** understand potential energy and entropy

Figure AB.3

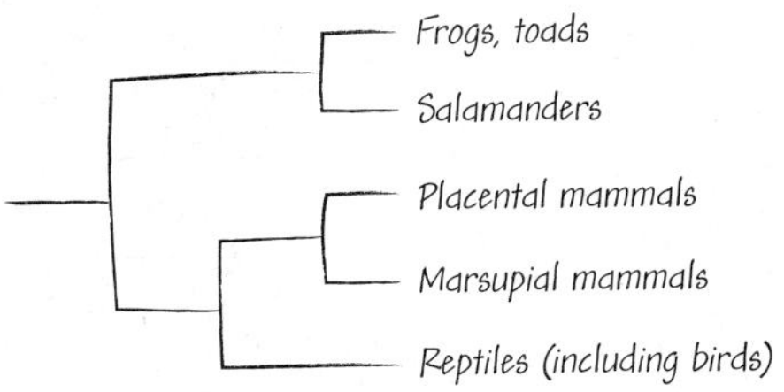

Figure AB.4

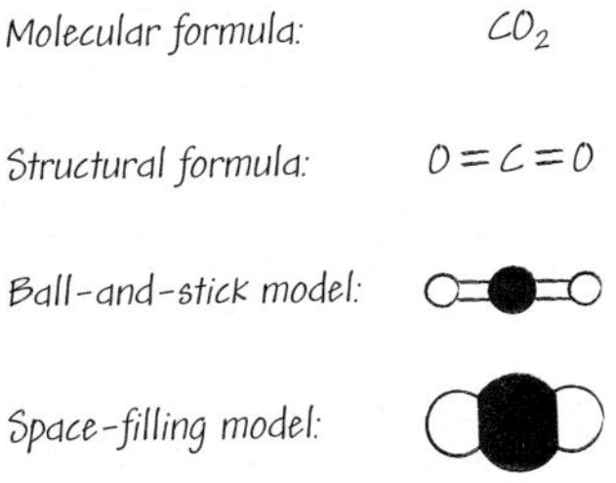

Figure A2.1

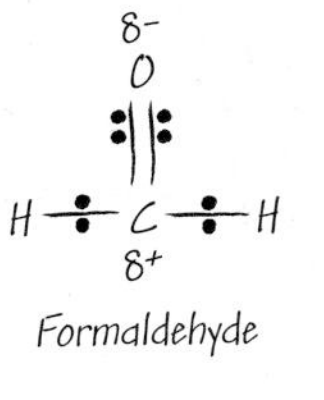

Figure A2.2

$CH_4 + 2\,O_2 \longrightarrow CO_2 + 2\,H_2O$

Figure A2.3

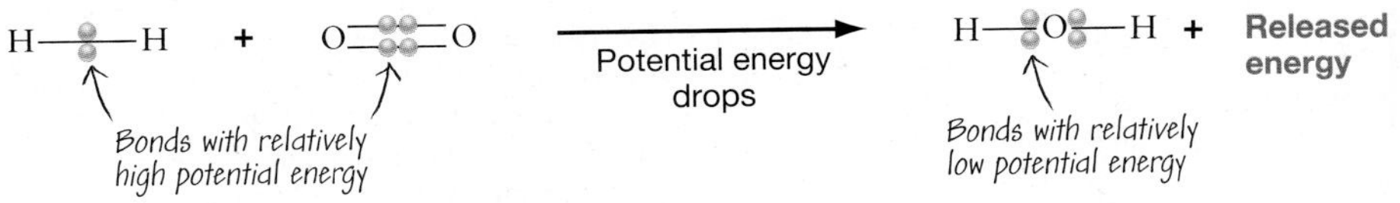

✔ TEST YOUR UNDERSTANDING

5. *apply* c. Acetic acid has more highly electronegative oxygen atoms than the other molecules. When oxygen is bonded to carbon or hydrogen, a polar covalent bond forms. **6.** *apply* Relative electronegativities would be F > O > H > Na. One bond would form with sodium, and it would be ionic. **7.** *apply* The proton concentration of a pH 8.5 solution is $1.0 \times 10^{-8.5}$ M, or 3.2×10^{-9} M. This is about 32 times less concentrated than a solution with pH 7.0 (1.0×10^{-7}) **8.** *apply* If additional CO_2 were added, then the sequence of reactions would be driven to the right and make the ocean more acidic. The dissociation reaction of carbonic acid lowers the pH of the solution by releasing extra H^+ into the solution.

✔ TEST YOUR PROBLEM-SOLVING SKILLS

9. *analyze* b **10.** *analyze* In hot weather, water absorbs large amounts of heat due to its high specific heat and high heat of vaporization. In cold weather, water releases the large amount of heat that it has absorbed.

✔ PUT IT ALL TOGETHER: Case Study

11. *apply* Water is denser in its liquid form than its solid form. This results in ice floating on the surface and serving as a blanket to insulate the liquid water from the colder temperature in the air. **12.** *analyze* The water will taste fresh. The liquid water in the ocean forms the crystalline structure of ice when water molecules interact with each other, excluding the salt ions. **13.** *apply* F, T, T, F. The freezing process is exothermic, like the condensation of water vapor into liquid. Even though the molecules have less entropy in ice, the second law of thermodynamics is preserved because the heat energy that is released will increase disorder in the environment (review Section 2.3). **14.** *create* One hypothesis might be that AFP binds to ice crystals and forms a nonpolar shell that prevents hydrogen bonding between water molecules to form a larger crystal. By limiting crystal growth, the cell would not freeze solid. **15.** *evaluate* To address the effectiveness of AFP, you might ask, "How long did the investigators keep the plants at each temperature?" **16.** *evaluate* There is no "correct" response, but some may prefer carrot AFP expressed in tomatoes because they do not like the idea of expressing an animal protein in plants. A pro may be the prevention of lost crops due to abnormally cold weather that may occur due to climate change. A con may be the unexpected consequences in growth, taste, or nutritional value of foods that contain AFP.

Chapter 3

IN-TEXT QUESTIONS AND EXERCISES

p. 82 Fig. 3.2 *understand* The R-groups shaded in green contain mostly C and H, which have roughly equal electronegativities. Electrons are evenly shared in C—H bonds and C—S bonds, so the groups are nonpolar. All of the R-groups shaded in pink have a highly electronegative oxygen atom with a partial negative charge, making them polar.

Figure A3.1

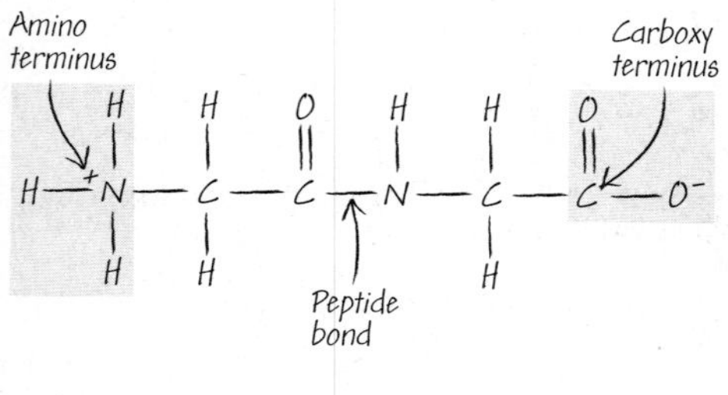

p. 84 CYU (1) *apply* See Figure A3.1. **(2)** *apply* From most hydrophilic to most hydrophobic: (1) aspartate, (2) asparagine, (3) tyrosine, (4) valine. The most hydrophilic amino acids will have side chains with full charges (ionized), like aspartate, followed by those with the largest number of highly electronegative atoms, like oxygen or nitrogen. Highly electronegative atoms form polar covalent bonds with carbon or hydrogen. The most hydrophobic will not have oxygen or nitrogen in their side chains, but instead will have the largest number of C—H bonds, which are nonpolar covalent. **(3)** *understand* Unlike the other single covalent bonds in the backbone, peptide bonds behave like double bonds in terms of stability and lack of movement of atoms around the bond. This difference is due to electron sharing between the pair of valence electrons on the nitrogen and the carbon of the peptide (C—N) bond.

p. 88 Fig. 3.10 *analyze* The tertiary structure that is rich in disulfide bonds would be most stable. The covalent bonds formed between cysteines are stronger than noncovalent bonds holding together the structures rich in α-helices and β-pleated sheets.

p. 88 CYU (1) *understand* Protein structure is hierarchical: Secondary, tertiary, and quaternary structure all depend on bonds and other interactions between amino acids that are covalently linked in a chain in a specific order (primary structure). **(2)** *apply* Nonpolar amino acid residues would be found in the interior of a globular protein like trypsin, grouped with other nonpolar residues due to hydrophobic interactions.

p. 91 CYU *analyze* The released calcium ions bind to certain proteins to change their folded shape and adopt a functional form, which alters cellular activity. In prion proteins, shape changes occur from protein–protein interactions and not via interactions with ions. In addition, an altered prion is infectious, meaning that it will promote additional normally folded proteins to adopt the infectious form.

p. 93 CYU (1) *remember* The six tasks performed by proteins: catalysis, structure, movement, signaling, transport, and defense. **(2)** *apply* Enzymes bind to specific substrates based on the structure of their active site. An active site is formed when the polypeptide of an enzyme is fully folded into its tertiary structure. The information required for directing the polypeptide into this folded state is in its primary structure. The primary structure is also responsible for the specific amino acid residues located in the active site that interact with the substrate.

✔ TEST YOUR KNOWLEDGE

1. *remember* d **2.** *remember* b **3.** *understand* The information present in the order and type of amino acids that make up the polypeptide (i.e., the primary structure). **4.** *remember* T, F, F, T. (The active site is the place where substrates bind and react. The second and third responses are false because they are only involved in folding the enzyme and not the catalytic process.)

✔ TEST YOUR UNDERSTANDING

5. *apply* The protein diversity would significantly decrease. Using 20 different amino acids, a total of 20^5 (3.2×10^6 or 3,200,000) different peptides can be generated. If only 10 different amino acids were available, then the number of peptides would drop to 10^5 (1×10^5 or 100,000). This would be a 32-fold decrease in diversity. **6.** *understand* Molecular chaperones facilitate folding by keeping unfolded proteins from clumping together so that they can fold into the shapes that are determined by the information in their primary structures. **7.** *remember* c **8.** *apply* The ability to regulate protein activity would be impaired. The proteins would be either permanently active or inactive, depending on how they are folded.

✔ TEST YOUR PROBLEM-SOLVING SKILLS

9. *analyze* Proline's side chain is covalently bonded to the nitrogen in the core amino group as well as to the central carbon. This would restrict the movement of the side chain relative to the core nitrogen and would further restrict the backbone when the nitrogen participated in a peptide bond with a neighboring amino acid. **10.** *create* See Figure A3.2.

✔ PUT IT ALL TOGETHER: Case Study

11. *apply* There are many possible answers. Some common foods prepared with wheat flour, for example, are pizza, bagels, pasta, batter-fried chicken, and burritos. Many other condiments, such as soy sauce in Asian foods, also contain wheat. **12.** *understand* a **13.** *apply* You would expect proline to appear once or twice in the chain (at a frequency of 1/20 at each of the 33 positions, or 1.65 times). **14.** *analyze* Amino acid differences would be expected in the active site or in regions that affect the folded structure of this site. Either of these changes could result in a different active site that is better at either binding to the peptides or catalyzing the reaction to cleave the peptide bonds. **15.** *analyze* The AN-PEP system has fewer proline-rich peptides in comparison to the negative control. The increase in peptide concentration in the negative control may result from the digestion of gluten by the normal gut enzymes, which would release more of the proline-rich peptide being evaluated. **16.** *create* Administering AN-PEP in pill form with a gluten-containing meal might digest the peptides before they cause an immune response, so the patient with celiac disease could at least occasionally not adhere to such a strict diet.

Figure A3.2

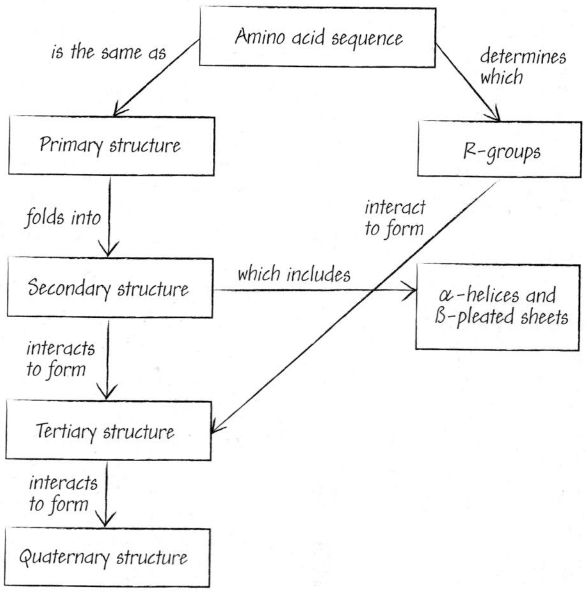

Chapter 4

IN-TEXT QUESTIONS AND EXERCISES

p. 97 Making Models 4.1 understand See Figure A4.1.

p. 98 Fig. 4.3 apply 5′-UAGC-3′

p. 98 CYU (1) apply See Figure A4.2. **(2)** understand All of the sugars would be deoxyribose instead of ribose, so the 2′ carbon would be bonded to a hydrogen instead of a hydroxyl. In addition, the uracil base at the 5′ end would be replaced with a thymine base. **(3)** understand Nucleotides are activated by linking additional phosphates to an existing 5′ phosphate. Activation increases the chemical energy in the nucleotides enough to offset the decrease in entropy that will result from the polymerization reaction so that the polymerization proceeds spontaneously.

p. 101 Making Models 4.2 apply See Figure A4.3.

p. 102 Fig. 4.7 apply It is not spontaneous—energy must be added (as heat) for the reaction to occur.

p. 103 CYU (1) apply DNA primary structure is based on covalent bonds and has greater variability compared to the secondary and tertiary levels. Secondary structure is based on complementary base pairing and involves two strands of DNA. The tertiary structure of DNA often includes proteins. **(2)** analyze In the G-T pair, only one hydrogen bond could form—between the bottom N-H in guanine and one of the carbonyl groups (C=O) of thymine. The other two potential H-bonding sites in guanine would repel the similar partial charges in the aligned groups extending from the thymine base. No hydrogen bonding would be possible in an A-C pair. (3) understand The complementary base pairing in the secondary structure of DNA allows each strand to serve as a template. After separating the strands of a DNA double helix, each template strand can be copied to make two identical double-stranded DNA molecules.

p. 105 CYU (1) analyze Most RNA is single-stranded, so its nucleotide sequence (primary structure) affects the ability of complementary base-pairing interactions to form stems and loops (secondary structures) and fold the molecule into a specific higher-order shape (tertiary structure). Because most DNA exists in double-stranded form, the secondary and tertiary levels of structure do not demonstrate the sequence-based variability seen in RNA. **(2)** understand Like proteins, single-stranded RNA molecules will fold into different shapes that have different properties. The additional hydroxyl on the ribose sugar also makes RNA more reactive, such that it can support catalytic functions.

p. 107 CYU (1) understand RNA provides a template that can be copied and can also catalyze the polymerization reaction required for its own replication. **(2)** understand Examples would include (1) the production of nucleotides, and (2) polymerization of RNA. It is thought that nucleotides were scarce during chemical evolution, so their catalyzed synthesis by a ribozyme would have been advantageous. Catalysis by an RNA replicase would have dramatically increased the reproductive rate of RNA molecules.

✔ TEST YOUR KNOWLEDGE

1. remember c **2.** remember c **3.** understand F, T, F, T. The first response is false because ribonucleotides are used for RNA, not DNA synthesis. The third response is false because pairing occurs between nitrogenous bases, not sugars. **4.** remember One end has a free phosphate group on the 5′ carbon; the other end has a free hydroxyl group bonded to the 3′ carbon.

✔ TEST YOUR UNDERSTANDING

5. understand DNA is a more stable molecule than RNA because it lacks a hydroxyl group on the 2′ carbon and is therefore more resistant to cleavage. **6.** apply a; if 30 percent is adenine, then 30 percent would be thymine, since they are base-paired together. This means that 40 percent consists of G-C base pairs, which would be equally divided between the two bases. **7.** apply The DNA sequence of the new strand would be 5′-ATCGATATC-3′. The RNA sequence would be the same, except each T would be replaced by a U. **8.** evaluate No. Catalytic activity in ribozymes depends on the tertiary structure generated from folding single-stranded molecules. Fully double-stranded forms of the RNA would not form the same tertiary structure.

✔ TEST YOUR PROBLEM-SOLVING SKILLS

9. create See Figure A4.4 for a sample concept map.
10. apply See Figure A4.5.

Figure A4.1

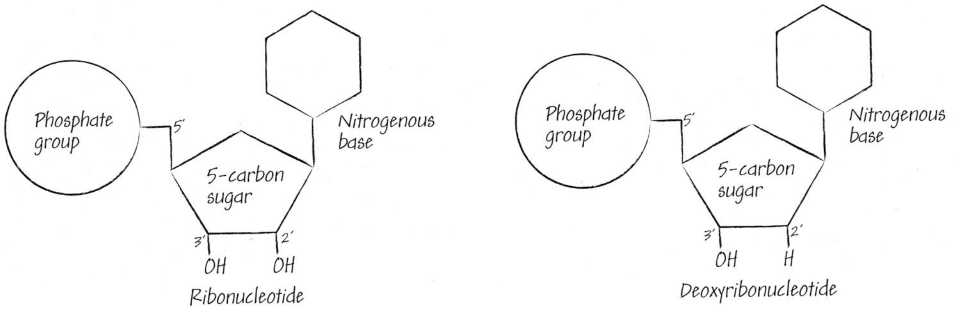

Figure A4.2

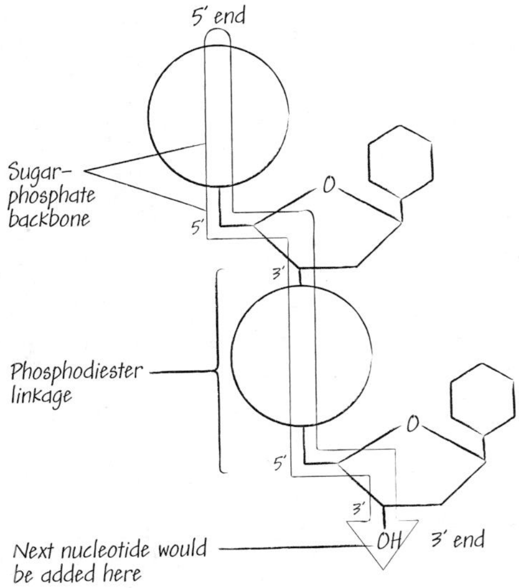

Figure A4.3

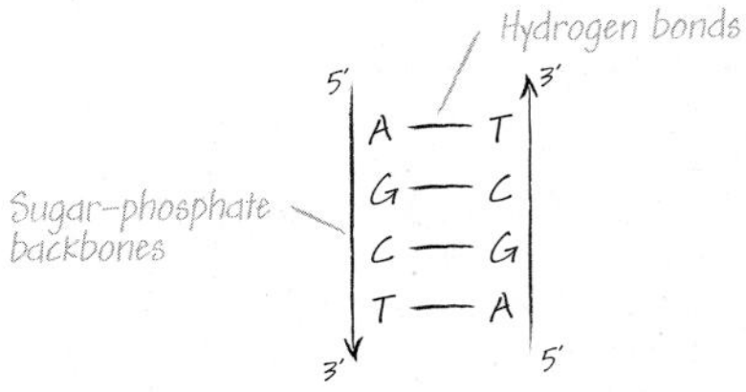

Figure A4.4

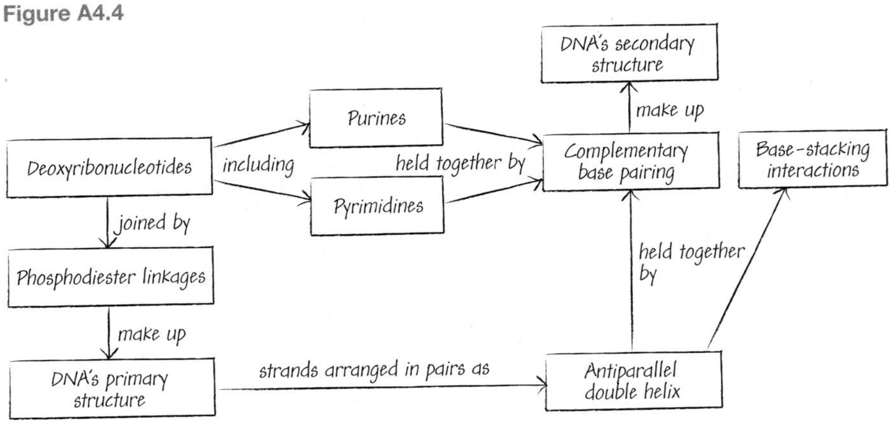

Figure A4.5

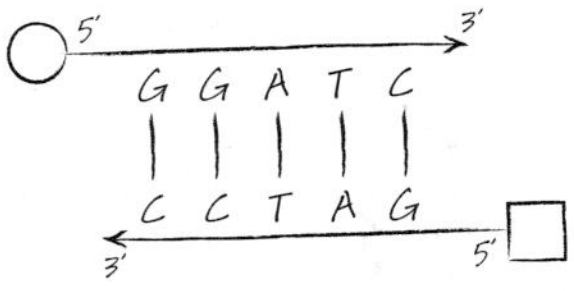

✔ PUT IT ALL TOGETHER: Case Study

11. analyze The capacity for storing information would be severely limited. It would be like trying to express ideas with a language that consisted of only one 4-letter word that is repeated over and over. According to this hypothesis, DNA would not serve as an effective information storage molecule. **12.** analyze Watson and Crick's model had the sugar–phosphate backbones oriented toward the exterior and based on the interior. Pauling's structure would not likely exist in cells because the nonpolar bases would be more exposed to water and the negatively charged sugar–phosphate backbones would repel one another. **13.** analyze The amount of water would affect the influence of hydrophobic interactions, which push the mostly nonpolar nitrogenous bases away from water and cause the DNA to twist into a double helix. **14.** analyze Within each DNA sample tested, the molar ratios of A:T and that of G:C were close to one, meaning that for every adenine there is a thymine, and for every guanine there is a cytosine. A key observation was that the ratios between A:G and T:C were not close to one, meaning that all four nucleotides were not present in equal ratios. These data are contrary to what Levene's tetranucleotide model would have predicted, which is that each nucleotide would be present in the same molar amounts. In addition, since the molar ratios of A:G and T:C were quite different in the organisms tested, the primary structure of DNA appears to vary among them. **15.** analyze Chargaff's data show that an approximately 1:1 molar ratio exists between adenine and thymine bases, and between guanine and cytosine bases. Watson and Crick used these data to come up with complementary base pairing, which requires that every adenine pairs with a thymine and every guanine pairs with a cytosine. In RNA, Chargaff's rules do not apply since RNA is single-stranded and the pairing is not consistent throughout the molecule. **16.** evaluate Science is seldom advanced in isolation. Watson and Crick could not have arrived at their model of the double helix without Levene's discovery of the structure of nucleotides and how they are linked together, Franklin's X-ray crystallography data, or Chargaff's biochemical analysis of the molar proportions of nucleotides in double-stranded DNA.

Chapter 5

IN-TEXT QUESTIONS AND EXERCISES

p. 111 Fig. 5.2 apply See the structure of mannose in Figure A5.1.

p. 111 CYU apply See Figure A5.2.

p. 113 Making Models 5.1 understand See Figure A5.3.

p. 115 CYU (1) apply They could differ in (1) location of linkages (e.g., 1,4 or 1,6); (2) types of linkages (e.g., α or β); (3) the sequence of the monomers (e.g., two galactose and then two glucose, versus alternating galactose and glucose); and/or (4) whether the four monomers are linked in a line or whether they branch. **(2)** analyze The α-glucose residues would all be oriented the same (no longer flipped across the glycosidic linkages) and the molecule would coil into a helix. The hydrogen bonds that were present between adjacent β-glucose polymers would no longer occur with the α-glucose polymer, so multi-strand fibers would not be formed.

p. 117 Fig. 5.7 apply The percentage of inhibition would not change for the intact glycoprotein bar. The purified carbohydrate bar would be at zero inhibition, and the purified protein bar would be similar to the intact glycoprotein bar.

p. 118 Fig. 5.8 understand All of the C—C and C—H bonds should be circled.

p. 118 CYU (1) understand *Aspect 1:* The β-1,4-glycosidic linkages in these molecules result in insoluble fibers that most organisms cannot break down with enzymes. *Aspect 2:* When individual molecules of these carbohydrates align, bonds form between them and produce fibers or sheets that resist pulling and pushing forces. **(2)** evaluate This claim is likely correct because carbohydrates have the potential for greater structural variation than proteins. Glycosidic linkages between sugar monomers can vary more in location and geometry than can peptide-bond linkages between amino acid residues, which are completely standardized. The different monosaccharide monomers vary extensively in their size, the position of the carbonyl group, the orientations of hydroxyl groups, and the presence of modifications in the polymerized forms (e.g., modified sugar residues in chitin and peptidoglycan). **(3)** apply The energy-storage molecules, like starch, are being hydrolyzed to release glucose. Disaccharides like the lactose in milk would be hydrolyzed to release glucose and galactose. These sugars may be further broken down to produce ATP and raw materials for building other molecules, such as glycolipids and glycoproteins. The insoluble cellulose that makes up dietary fiber does not get broken down, but it will help retain water and support the digestion and passage of fecal material.

✔ TEST YOUR KNOWLEDGE

1. understand Monosaccharides can differ from one another in (1) the location of their carbonyl group; (2) the number of carbon atoms they contain; and (3) the orientations of their hydroxyl groups. **2.** remember a **3.** remember c **4.** remember a

✔ TEST YOUR UNDERSTANDING

5. understand T, F, F, T. The second and third responses are false because the changes would result in different monosaccharides. **6.** understand The α-1,4-glycosidic linkages in starch form independent helical structures that are easily hydrolyzed by enzymes found in most organisms. In contrast, the β-1,4-glycosidic linkages result in large fibers and sheets that exclude water, so they resist hydrolysis. **7.** analyze Glycogen consists of glucose residues that are connected by α-1,4-glycosidic linkages with branches arising from α-1,6-glycosidic linkages. Chitin is made up of modified glucose residues (NAG) that are connected by β-1,4-glycosidic linkages with no branching. While chitin polymers interact with one another via hydrogen bonds, glycogen molecules do not. **8.** apply When bacteria contact lysozyme, the peptidoglycan in their cell walls begins to degrade, leading to the death of the bacteria. Lysozyme therefore helps protect humans against bacterial infections.

✔ TEST YOUR PROBLEM-SOLVING SKILLS

9. apply d; Lactose is a disaccharide of glucose and galactose, which can be cleaved by enzymes expressed in the human gut to release galactose. **10.** analyze Amylase breaks down the starch in the cracker into glucose monomers that taste sweet.

✔ PUT IT ALL TOGETHER: Case Study

11. apply Because the reacting hydroxyl at C-1 in glucose was in the α position (see Figure 5.4a), the monosaccharides in sucrose are said to be joined by an α-1,2-glycosidic linkage. (Note: Since this linkage is formed with β form of fructose, it is often referred to as being an α-β-1,2-glycosidic linkage.) **12.** apply Fructose is a hexose based on having six carbons, and it is a ketose

Figure A5.1

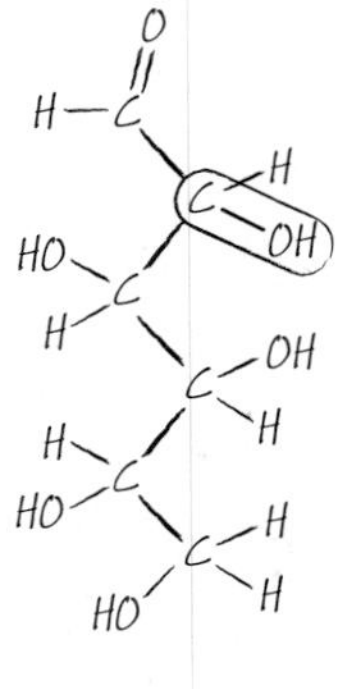

Figure A5.2

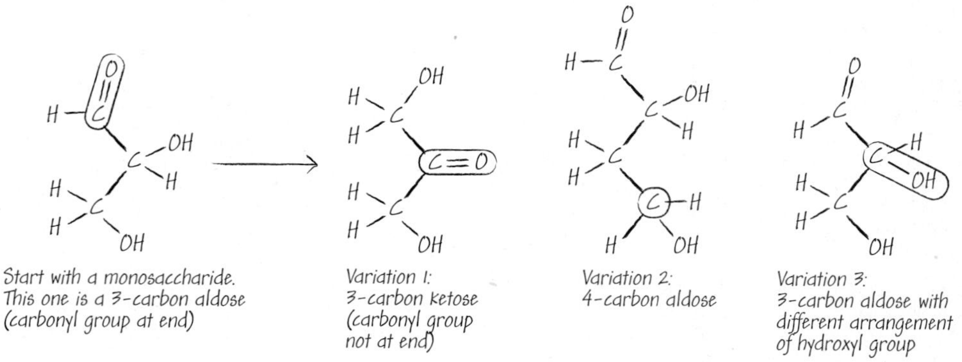

Figure A5.3

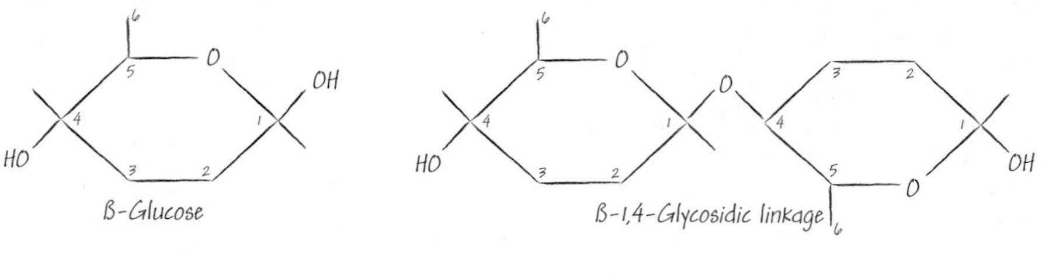

because the carbonyl is within the carbon chain (C-2, based on how ring structures are formed using the carbonyl group and the convention for numbering carbon atoms). **13.** analyze The position of the carbonyl is the most striking structural difference between these sugars (glucose is an aldose while fructose is a ketose) and so it is most likely responsible for the disparity in taste perception. **14.** apply To convert the starch polymer into monosaccharides, it must be hydrolyzed by an enzyme such as amylase. Since fructose is not in starch, the second event must be to convert some of the glucose sugars into fructose. **15.** analyze The escape times of fructose-fed rats were always slower than those of the rats not fed fructose. This result suggests that fructose impaired the rats' ability to recall what they had learned about finding the escape chamber. The rats that were fed a diet enriched in omega-3 fatty acids demonstrated an enhanced memory compared to rats on diets deficient in omega-3, even when they also consumed fructose, suggesting that omega-3s help offset the deleterious effects of dietary fructose. **16.** evaluate These results suggest that the best diet for performance in your classes will be rich in omega-3 fatty acids and low in fructose. Maybe it's time to replace your soda with fish oil!

Chapter 6

IN-TEXT QUESTIONS AND EXERCISES

p. 125 Fig. 6.5 remember A circle should be drawn around the phosphate and polar or charged group extending from the top of the molecule.

p. 125 CYU (1) understand In general, unsaturated lipids are more fluid than saturated lipids at a given temperature. **(2)** analyze Steroids have a distinctive four-ring structure with variable side groups attached; fats consist of three fatty acids linked to glycerol; many phospholipids also have a glycerol linked to fatty acids, but instead of three fatty acids, they have two plus a hydrophilic, phosphate-containing "head" region. **(3)** understand Free fatty acids are amphipathic because their hydrocarbon tails are hydrophobic but their carboxyl functional groups are hydrophilic. In fats, the charged carboxyl groups of fatty acids are converted to ester linkages. This change reduces the difference in polarity across the molecule, making it more uniformly nonpolar. As a result, fats are not considered amphipathic.

p. 127 analyze Amino acids have amino and carboxyl groups that are ionized in water, and nucleotides have negatively charged phosphates. Due to their charge and larger size, both of these compounds would be placed below the small ions at the bottom of the permeability scale ($< 10^{-12}$ cm/sec).

p. 128 Fig. 6.11 apply Increasing the number of phospholipids with polyunsaturated tails would increase permeability of the liposomes. Starting from the left, the first line (no cholesterol) would represent liposomes with 50 percent polyunsaturated phospholipids, the second line would be 20 percent polyunsaturated phospholipids, and the third line would contain only saturated phospholipids.

p. 129 CYU (1) understand Phospholipids are amphipathic, so their hydrophilic end will interact with water and their hydrophobic end will not. Water molecules are more ordered when surrounding the hydrophobic regions. When phospholipids are brought together to form a membrane, the hydrophobic regions are tucked away from water and thus the entropy in water increases. Events that increase entropy tend to be spontaneous. **(2)** apply See **Table A6.1**.

Table A6.1

Factor	*Effect on permeability*	*Reason*
Temperature	*Decreases as temperature decreases.*	*Lower temperature shows movement of hydrocarbon tails, allowing more interactions (membrane is more dense).*
Cholesterol	*Decreases as cholesterol content increases.*	*Cholesterol molecules interact with the hydrocarbon tails, making them more tightly packed.*
Length of hydrocarbon tails	*Decreases as length of hydrocarbon tails increases.*	*Longer hydrocarbon tails have more interactions (membrane is more dense).*
Saturation of hydrocarbon tails	*Decreases as degree of saturation increases.*	*Saturated fatty acids have straight hydrocarbon tails that pack together tightly, leaving few gaps.*

p. 130 apply Since temperature is a measure of thermal motion, increasing temperature would increase the rate of diffusion and thus the rate of passive transport to achieve equilibrium across a membrane.

p. 130 apply No, the right side of the membrane will have a higher concentration of solute at equilibrium. If you said yes, recall that pressure from the downward pull of gravity will push water molecules back to the left, against continued transport of water toward the higher solute concentration. This opposing force would prevent the solutions separated by the membrane from achieving the same concentration.

p. 132 CYU create See **Figure A6.1**.

p. 133 Fig. 6.19 create Repeat the procedure using a lipid bilayer that is free of membrane proteins, such as synthetic liposomes constructed from only phospholipids. If proteins were responsible for the pits and mounds, then this control would not show these structures.

p. 134 apply Your arrow should point from below to above the membrane. There is no concentration gradient for chloride, but the upper side has a net positive charge, which favors the import of negative ions like chloride.

p. 136 Fig. 6.24 analyze No, the results would be the same. The channel protein allows movement of ions in either direction, so the orientation of the channel would not affect the rate or directionality of the chloride ion transport.

p. 137 Making Models 6.1 apply See **Figure A6.2**. In this drawing, the lipid bilayer is represented by two lines because it does not require the structural details of how the lipids are organized.

p. 139 CYU (1) Peripheral membrane proteins are associated with one of the two hydrophilic surfaces of the membrane, but do not cross through the hydrophobic interior. Integral membrane proteins completely pass through the membrane via a series of adjacent nonpolar amino acids. **(2)** understand Passive transport does not require an input of energy—it happens because of energy already present in an existing gradient. Active transport is active in the sense of requiring an input of

Figure A6.1

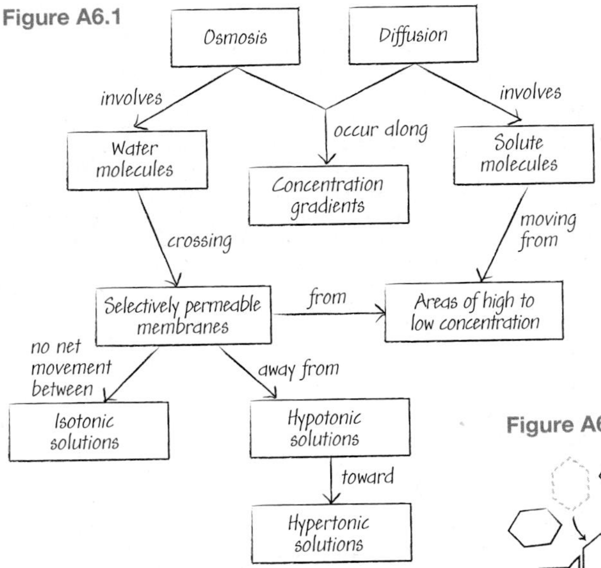

Figure A6.2

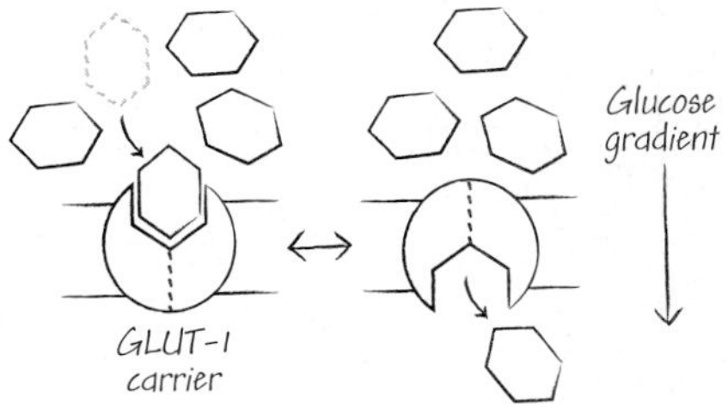

energy from, for example, ATP. **(3)** *create* See **Figure A6.3.**

p. 139 Fig. 6.27 *understand* *Diffusion:* Description as given; no proteins involved. *Facilitated diffusion:* Passive movement of ions or molecules along an electrochemical gradient, through transmembrane proteins; made possible by channel or carrier proteins. *Active transport:* Active movement of ions or molecules in a single direction, often against their gradient; made possible by pump proteins powered by an energy source such as ATP.

✔ TEST YOUR KNOWLEDGE

1. *understand* c **2.** *understand* b **3.** *understand* For osmosis to occur, a concentration gradient and a membrane that allows water to pass, but not the solute, must be present. **4.** *understand* d

✔ TEST YOUR UNDERSTANDING

5. *apply* No, because they have no polar end to interact with water. Instead, these lipids would float on the surface of water, or collect in droplets suspended in water, reducing their interaction with water to a minimum. **6.** *apply* See **Figure A6.4. 7.** *analyze* Channel proteins form pores in the membrane, some of which have different closed and open conformations, and carrier proteins undergo conformational changes to shuttle molecules or ions across the membrane. **8.** *apply* b

✔ TEST YOUR PROBLEM-SOLVING SKILLS

9. *analyze* F, F, T, T. The first two responses are false because the interior solution in both sets of liposomes is the same. The third is true because aquaporins allow for an increased rate of water transport. The fourth is true because if the Na or Cl ions can cross the membrane in frog eggs, then osmosis may be reduced or even be prevented from occurring. **10.** *apply* NaCl will dissociate into 1 M of Na^+ and 1 M of Cl^- on the left with only 1.5 M of K^+ on the right. Since there are more ions on the left side, water would move by osmosis from the right to the left. If the CFTR protein were added to the membrane, Cl^- ions would move along an electrochemical gradient from left to right, resulting in a higher concentration of ions on the right side of the membrane. This new ion gradient would reverse the direction of osmosis, resulting in the movement of water from the left side to the right.

✔ PUT IT ALL TOGETHER: Case Study

11. *apply* The saturated fat in meat undergoes a change in its physical state as it is heated—from a semisolid consistency to a liquid. When meat is grilled, the liquefied fat drips off, resulting in less saturated fat in the final product. When the meat is raw, all the saturated fat remains. **12.** *understand* Saturated fats have fatty acid tails with only C—C single bonds while unsaturated fats have one or more C═C double bonds in the tails. C═C bonds normally result in kinks that increase the spacing between fats and thus make unsaturated fats more fluid than saturated fats. **13.** *apply* To convert an oil into a semisolid compound, the hydrocarbons would need to become more saturated. The term "hydrogenation" is used because this process would involve adding hydrogen atoms to the vegetable oil to convert C═C double bonds into C—C single bonds (i.e., CH—CH). **14.** *analyze* Trans fats are unsaturated because they have one or more double bonds, but they have physical characteristics similar to saturated fats due to the *trans* double bond that straightens the hydrocarbon chain. **15.** *evaluate* No. The data show a correlation between blood levels of trans fats and atherosclerosis (i.e., these two events appear to occur together), but these data do not show the cause. "Causation" means that one event is responsible for the occurrence of the other (e.g., studying and doing well on an exam). **16.** *create* Other factors that may affect heart disease include gender, age, weight, heart disease in the family, other dietary habits, and so on.

END-OF-UNIT CASE STUDY What's So Toxic About Tetrodotoxin?

pp. 142–143 (1) *understand* TTX would need to be hydrophilic so that it is soluble in water. Since it is still active in the boiled water, it must also be heat resistant. **(2)** *understand* c **(3)** *apply* The hydroxyl is a polar group that forms hydrogen bonds with water molecules. **(4)** *analyze* Sodium ion channels transport sodium ions through a pore across the membrane in either direction, but the net movement would be along the electrochemical gradient. Carriers also transport substances based on the gradient, but they do so by binding to the substance and undergoing a conformational change to move it across the membrane. Pumps differ from channels and carriers by using an energy source outside of the gradient, such as ATP, to bind and transport substances across the membrane in one direction. **(5)** *understand* The Na^+/K^+-ATPase establishes the gradient by using ATP to undergo conformational changes that result in moving three sodium ions out of the cell and two potassium ions into the cell. **(6)** *remember* The extracellular side of the membrane will be more positively charged. **(7)** *analyze* As the electrical gradient is reduced, the channels open and sodium ions pass through the membrane along their electrochemical gradient. **(8)** *apply* For the deactivated conformation, more plus symbols should be drawn outside the membrane and more minus symbols inside. The reverse should be drawn for the activated confornation. **(9)** *apply* The helix would contain basic (positively charged) amino acids, such as lysine, arginine, and histidine. **(10)** *analyze* TTX prevents the transport of sodium ions under conditions that would normally cause the channels to be open. **(11)** *create* One drawing would show TTX binding to the channel and causing it to be stuck in the inactivated state. The other drawing would show TTX plugging the pore of an activated channel to prevent ion transport.

BIG PICTURE The Chemistry of Life

pp. 144–145 CYU (1) *understand* Oxygen is much more electronegative than hydrogen, so within water, the electrons are unequally shared in the O—H covalent bonds. The resulting partial negative charge around the oxygen and partial positive charges around the hydrogen atoms allow for hydrogen bonds to form among water molecules. **(2)** *analyze* Unlike other macromolecules, nucleic acids can serve as templates for their own replication. RNA is generally single-stranded and can adopt many different three-dimensional structures. The flexibility in structure, combined with the presence of reactive hydroxyl groups, contribute to the formation of active sites that catalyze chemical reactions. One or more of these catalytic RNA molecules may have evolved the ability to self-replicate. DNA is not likely to have catalyzed its own replication, as it is most often double-stranded, with no clear tertiary structure, and it lacks the reactive hydroxyl groups. **(3)** *remember* In the amino acid, the nitrogen in the amino (NH_3^+) group and the carbon in the carboxyl (α) group should be circled. In the nucleotide, the oxygen in the hydroxyl (OH) group and the phosphorus in the phosphate (PO_4^{2-}) group on the nucleotide should be circled. **(4)** *understand* A line representing a protein should be drawn such that it completely crosses the lipid bilayer at least once. The protein could be involved in a variety of different roles, including transport of substances across the membrane in the form of a channel, carrier, or pump.

Chapter 7

IN-TEXT QUESTIONS AND EXERCISES

p. 149 CYU *understand* **(1)** The ribosomes are macromolecular machines that synthesize all of the proteins in the cell. **(2)** Photosynthetic membranes increase food production by providing a large surface area to hold the pigments and enzymes required for photosynthesis. **(3)** Flagella propel cells through liquid using a motor that rotates a long, rigid filament. **(4)** The external layer of thick, strong material supports the cell membrane and provides protection from mechanical damage.

p. 154 Fig. 7.11 *create* Toxins are stored in membrane-enclosed vacuoles to prevent them from contacting and damaging components of the cytosol or other organelles in the cell.

Figure A6.3

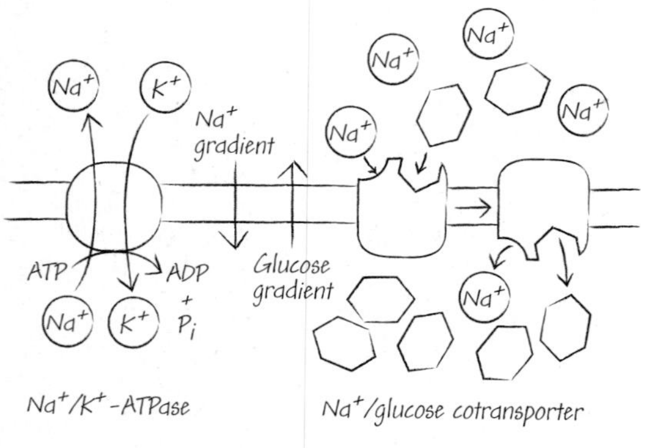

Figure A6.4

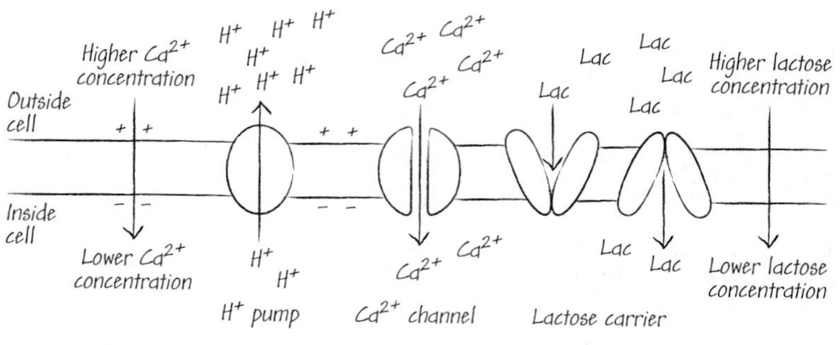

p. 158 CYU (1) understand The two different organelles contain specific sets of enzymes. Lysosomal enzymes digest macromolecules in the acidic lumen of this organelle, releasing monomers that can be recycled into new macromolecules. Peroxisomes contain catalase and other enzymes that process fatty acids and toxins via oxidation reactions. **(2)** understand From top to bottom of Table 7.1: administrative/information hub, protein factory, large molecule manufacturing and shipping (protein synthesis and folding center, lipid factory, protein finishing and shipping line, waste processing and recycling center), warehouse, fatty-acid processing and detox center, power station, food-manufacturing facility, support beams, perimeter fencing (used for cell wall and ECM).

p. 158 Fig. 7.15 remember See Figure A7.1.

p. 159 CYU (1) apply Such cells would likely have many lysosomes to digest and recycle components that are consumed, such as bacteria. **(2)** apply Secretory cells, like the pancreatic cell in Figure 7.15a, would have an increased amount of the rough ER and Golgi because most of the proteins would be made in the rough ER and then pass through the Golgi before secretion.

p. 161 Fig. 7.17 apply The researchers would have used immunostaining and fluorescence microscopy. The pyruvate kinase appears green because the researchers used an antibody against this protein that is tagged with a green fluorescent molecule.

p. 161 CYU analyze **(1)** Nucleotides are small enough that they would diffuse through the nuclear pore complex along their gradients—a passive process that would not require energy. **(2)** Large proteins must be escorted through the nuclear pore complex in a directional manner—an active process that requires energy and results in the protein being concentrated inside the nucleus.

p. 163 Fig. 7.19 analyze During the chase period, proteins appear to have first entered the Golgi apparatus after 7 minutes and then started to move into secretory granules after 37 minutes. This means that in this experiment, it took approximately 30 minutes for the fastest-moving proteins to pass through the Golgi apparatus.

p. 166 Making Models 7.1 apply See Figure A7.2. Step 1: Ribosomes at the rough ER finish the polypeptide as it is moved into the ER lumen. Step 2: Polypeptide folds and is packaged into a vesicle. Step 3: Vesicle is transported to the *cis* Golgi. Step 4: Protein is processed as it moves through the Golgi apparatus and cargo receptors package it into a transport vesicle. Step 5: Secretory vesicle is transported to the plasma membrane. Step 6: Vesicle fuses with the plasma membrane and releases protein to the outside of the cell.

p. 167 CYU (1) analyze Proteins that enter the nucleus are fully synthesized and have an NLS that interacts with another protein to get it into the organelle. The NLS is not removed. Proteins that enter the ER have a signal sequence that interacts with an SRP during protein translation by a free ribosome. The ribosome + signal sequence + SRP is moved to the ER and attaches to an SRP receptor. Protein synthesis continues through a translocon channel, moving the protein into the ER. The signal is removed once it enters the organelle. **(2)** analyze The protein would end up in a lysosome. The ER signal would direct the protein into the ER before it is completely synthesized. The mannose-6-phosphate tag will direct the protein from the Golgi to a late endosome, which matures into the lysosome. Thus the complete protein is never free in the cytosol, where the NLS could direct it into the nucleus.

p. 172 Fig. 7.29 apply You should have drawn a pair of doublets that have moved completely past one another. The doublets would not bend without the links holding them together.

p. 172 CYU (1) analyze Actin filaments are made up of two strands of actin monomers, intermediate filaments are made up of a number of different protein subunits, and microtubules are made up of tubulin protein dimers that form a tube. Actin filaments and microtubules exhibit polarity (or directionality), and new subunits are constantly being added or subtracted at either end (but added faster to the plus end). **(2)** understand The three motor proteins are myosin, kinesin, and dynein. Myosin is used to separate animal cells during cell division, for muscle contraction, and for cytoplasmic streaming. Kinesin is used for moving vesicles to the plus ends of microtubules. Dynein is used for moving vesicles to the minus ends of microtubules and for bending microtubules in flagella and cilia.

✔ TEST YOUR KNOWLEDGE

1. understand They have their own small, often circular chromosomes; they produce their own ribosomes; and they replicate independently of cellular division. **2.** understand b **3.** remember a **4.** understand The binding and subsequent hydrolysis of ATP molecules cause conformational changes in motor proteins that result in their moving along the filament.

Figure A7.1

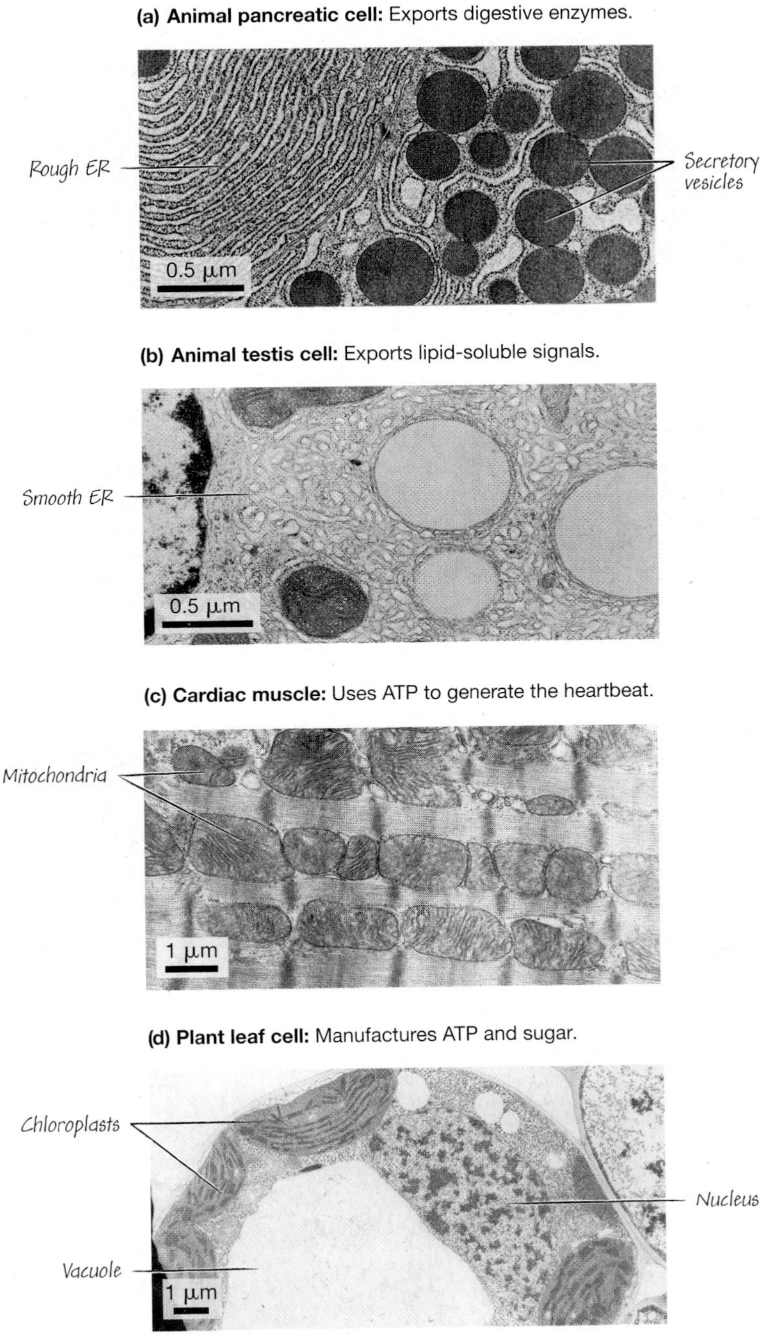

✔ TEST YOUR UNDERSTANDING

5. apply b; fimbriae are involved in bacterial attachment to surfaces and other cells, which would be important in the ability to grow on teeth. **6.** apply a; the endoplasmic reticulum is responsible for synthesizing the membrane proteins required for the transport of solutes across the plasma membrane. **7.** apply The NLS actively imports the protein into the nucleus, leaving very little of the protein in the cytoplasm. Diffusion alone would not drive all the protein into the nucleus. **8.** evaluate F, T, F, T. The first response is false because myosin walks along actin filaments. The third statement is false because kinesin moves vesicles to the plus ends, which are directed away from the microtubule-organizing center.

✔ TEST YOUR PROBLEM-SOLVING SKILLS

9. analyze The patient may have a defect in (1) a hydrolytic enzyme that breaks down the undigested material, (2) the process of adding a mannose-6-phosphate signal to a lysosomal hydrolase, or (3) the mannose-6-phosphate receptor in the Golgi apparatus. Any of these three defects would result in lysosomes that are missing an enzyme necessary to break down the undigested material. **10.** analyze The radiolabeled proteins would likely be found in the cytosol and in mitochondria. Cytosolic proteins include proteins like the actin and myosin involved in muscle contraction. Because of the high energy demand, you would also expect many proteins needed for ATP production to be imported into mitochondria.

✔ PUT IT ALL TOGETHER: Case Study

11. apply Lysosomes originate from late endosomes, which receive enzymes exported from the Golgi apparatus (all part of the endomembrane system). If melanosomes have a similar origin, then they would arise from an endosome-like compartment that also receives cargo from the Golgi apparatus. **12.** apply See **Figure A7.3**. The melanosomes (dark dots in the drawing) are moved along microtubule tracks by motor proteins. Microtubules originate near the nucleus of animal cells, and the positive ends project outward. Kinesin would be responsible for moving melanosomes outward to generate dark-colored cells. Dynein carries melanosomes back toward the nucleus in light-colored cells. **13.** analyze The melanosomes secreted by melanocytes would be available for uptake from the extracellular space by keratinocytes. The uptake process would likely be phagocytosis. But phagocytosis normally sends imported cargo to the lysosome for destruction, so if the organelles are to avoid destruction and remain in the keratinocyte, this final step must be inhibited. **14.** create *Hypothesis:* Exposure to UV radiation causes melanocytes to produce more melanosomes, which are then taken up by neighboring keratinocytes to darken the skin. *Experiment:* Expose part of the skin of an individual to UV light and keep another part of the skin protected. Take skin samples from both locations and use microscopy to count the number of melanosomes produced in melanocytes under the two conditions. **15.** analyze Autophagy is the process of recycling damaged organelles and other cytoplasmic components by wrapping portions of the cytoplasm in a membrane and fusing it with lysosomes. If melanosomes were removed by autophagy, then you would expect to find a higher rate of autophagy in keratinocytes in light-skinned individuals than in dark-skinned individuals. **16.** analyze When autophagy was induced, the skin appeared lighter, and melanin decreased significantly compared to the control. When autophagy was inhibited, the skin appeared darker, and melanin increased significantly compared to the control and the trial with autophagy induced. Conclusion: Autophagy can affect the appearance of skin and the amount of melanin in skin cells.

Chapter 8

IN-TEXT QUESTIONS AND ANSWERS

p. 177 apply (1) If ΔS is positive (products have more disorder than reactants), then according to the free-energy equation, ΔG is more likely to be negative as temperature (T) increases even if ΔH is positive. The increased temperature represents added thermal energy that may be used to drive an endothermic reaction to completion, making the reaction spontaneous. (2) Exothermic reactions may be nonspontaneous if they result in a large decrease in entropy—meaning that the products are more ordered than the reactants (ΔS is negative).

p. 178 Fig. 8.4 understand The slope of the line is approximately 0.05, which is calculated by determining the change in rate ($Y_2 - Y_1$) and dividing it by the change in temperature ($X_2 - X_1$) along the line. This means that the reaction rate increases 0.05 units (1/time to completion) for each 1°C increase in temperature.

p. 179 CYU (1) apply The potential energy that is lost as the enthalpy is decreased in the products is transformed into kinetic energy, which is most likely in the form of heat or light. **(2)** understand If the reactant concentration increases, so too will the number of collisions between reactant molecules. As a result, the rate of the chemical reaction would increase. If the concentration decreases, the rate would also decrease.

p. 181 apply In part (a), the electron donor is AH_2 and the electron acceptor is FAD. In part (b), the electron donor is BH_2 and the electron acceptor is NAD^+.

p. 182 Fig. 8.10 apply The ΔG in the uncoupled reaction is positive (> 0), and each of the steps in the coupled reaction have a negative (< 0) ΔG.

p. 183 analyze Redox reactions transfer energy between molecules or atoms via electrons. When oxidized molecules are reduced, their potential energy increases. ATP hydrolysis is often coupled with the phosphorylation of another molecule. This phosphorylation increases the potential energy of the target molecule.

p. 183 CYU (1) understand Electrons in C—H bonds are not held as tightly as electrons in C—O bonds, so they have higher potential energy. **(2)** understand The three phosphate groups in ATP have four negative charges in close proximity. The repulsive forces from the clustered negative charges result in weak, unstable bonds that have high potential energy.

p. 184 Making Models 8.1 apply See **Figure A8.1**. The specificity results from the shape of the binding pocket

Figure A7.2

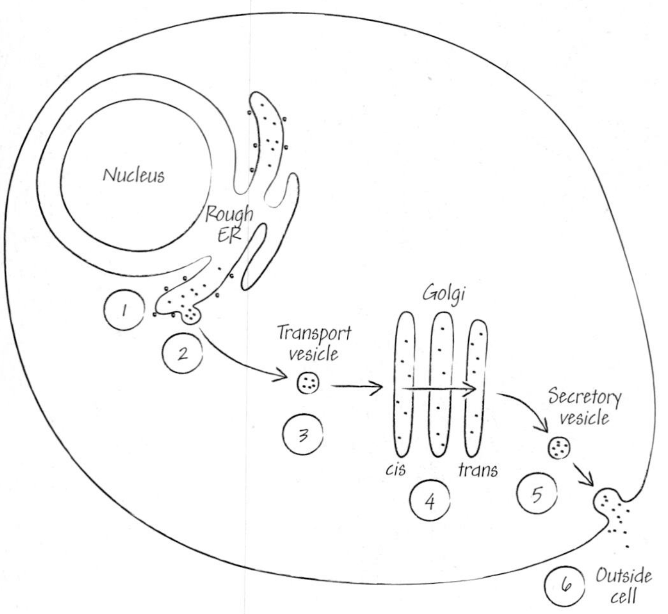

Figure A7.3

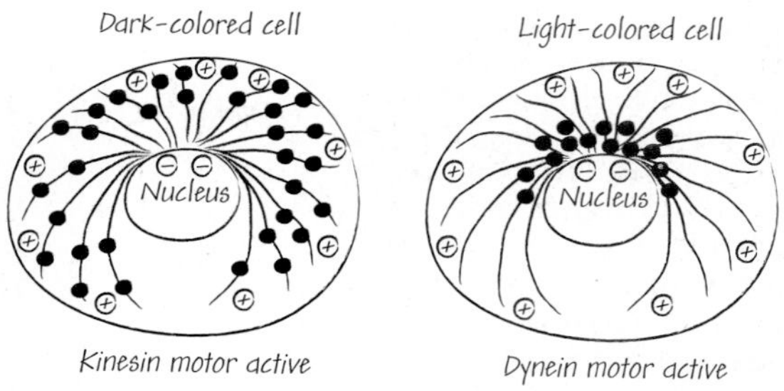

Figure A8.1

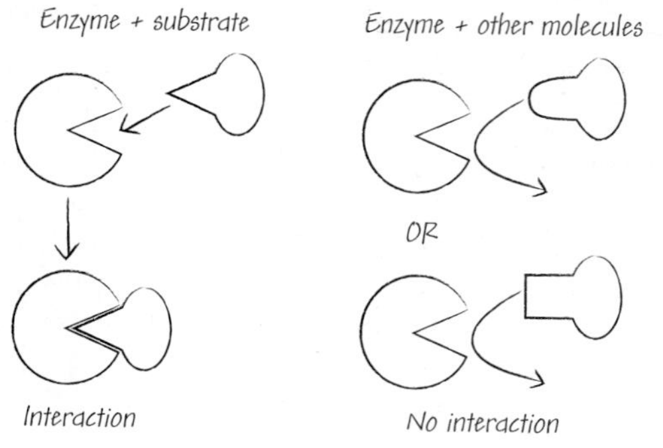

relative to the substrate. Another factor that enhances specificity is the unique set of amino acid residues in the active site, which form the correct interactions to tightly bind the substrate.

p. 185 Fig. 8.13 understand No—a catalyst affects only the activation energy, not the overall change in Gibbs free energy (ΔG).

p. 185 remember (1) binding substrates, (2) R-groups, (3) structure

p. 186 Fig. 8.15 apply Each enzyme molecule is saturated with substrate at the maximum rate of product formation, so to increase the rate, you would need to add more enzyme.

p. 186 CYU (1) understand Enzymes lower activation energy by destabilizing bonds in the substrates and promoting the formation of the transition state. **(2)** create The Mg^{2+} ion carries positive charges that will interact with the negative charges on ATP. This interaction helps the enzyme transfer a phosphate from ATP to its substrate.

p. 188 CYU (1) understand Reaction rate is based primarily on the activity of the enzyme. Once the temperature reaches a level that causes unfolding and inactivation of the enzyme, the rate decreases to the uncatalyzed rate. **(2)** analyze Both are mechanisms that regulate the interaction between enzymes and their substrates. In competitive inhibition, the regulatory molecule binds directly to the active site and interferes with substrate binding. In allosteric regulation, the regulatory molecule binds to a different region and causes the enzyme to change conformation and either open or close the active site.

p. 190 apply If enzyme 2 is absent, then there would be a buildup of product B because it is not being used to make C. At equilibrium, the concentration of A and B would be higher than expected in the fully functional pathway. If C is not being produced, it cannot serve as a reactant for the enzyme 3 catalyzed reaction. As a result, both C and D would be lower than expected from the functional pathway.

p. 190 CYU (1) understand If the regulated enzyme is at the start of a metabolic pathway, then when it is turned off, there will be no product to serve as a substrate for the following reaction. As a result, none of the subsequent reactions would occur, including the formation of the final product. **(2)** analyze The retro-evolution hypothesis starts with a single key reaction and then evolves additional enzyme-catalyzed reactions to generate more of the substrate for the original reaction. In the patchwork hypothesis, the evolution of a new pathway occurs by repurposing enzymes from other already existing pathways. New activities may arise in the process.

✔ TEST YOUR KNOWLEDGE

1. understand T, T, F, F. (The third response is false because exergonic reactions can occur if the products have lower entropy so long as there is a large drop in enthalpy in the reaction. The fourth response is false because the reaction rate is based on the activation energy, which may be high and result in an exergonic reaction occurring slowly, if at all.) **2.** remember c **3.** remember d **4.** understand The final product of a pathway inhibits the activity of an enzyme early in the same pathway, thus reducing the activity of all subsequent steps in the pathway.

✔ TEST YOUR UNDERSTANDING

5. understand Reactions proceed when the reactant molecule (the key) fits snugly into the active site of an enzyme (the lock). Fischer's original model assumed that enzymes were rigid; in fact, enzymes are flexible and often change their shape after binding to substrates. **6.** apply d. (Energy, such as the thermal energy in fire, must be provided to overcome the activation energy barrier before the reaction can proceed.) **7.** apply The uncoupled reaction ($A + B \rightarrow AB$) has a ΔG of about +1.3 kcal/mol. For the coupled reaction, step 1 ($A + B + ATP \rightarrow AB{-}P + ADP$) has a ΔG of about −3 kcal/mol and step 2 ($AB{-}P + ADP \rightarrow AB + P_i + ADP$) also has a negative ΔG (about −3 kcal/mol). **8.** apply Catabolic reactions will often have a negative ΔG based on a decrease in enthalpy and increase in entropy. Anabolic reactions are the opposite—a positive ΔG that is based on an increase in enthalpy and decrease in entropy.

✔ TEST YOUR PROBLEM-SOLVING SKILLS

9. apply See Figure A8.2. **10.** apply b. (Binding of the sugar likely induces a conformational change in the enzyme that forms a functional active site.)

✔ PUT IT ALL TOGETHER: Case Study

11. analyze If this amino acid cannot be synthesized by cells, then it must be obtained by breaking down proteins into amino acids, which would include phenylalanine. The amount and type of protein consumed in your diet would affect the amount of phenylalanine available in your body. **12.** create See Figure A8.3.

Figure A8.2

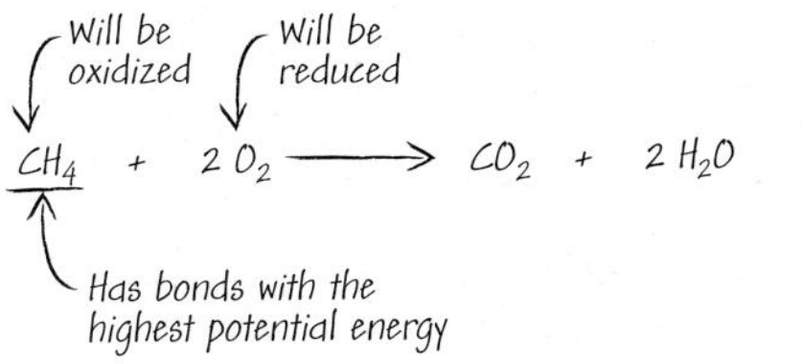

Figure A8.3

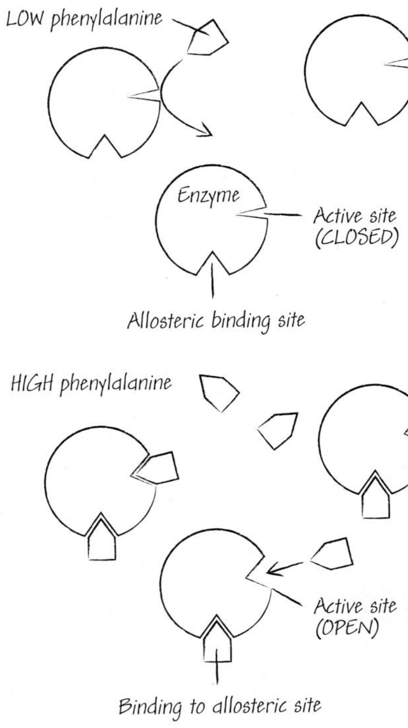

13. analyze These results support the hypothesis that phenylalanine is an allosteric regulator of PAH. This is shown by the different protein products generated after exposure of PAH to trypsin under the two conditions—intact PAH in the absence of phenylalanine versus PAH fragments in the presence of phenylalanine. The data indicate that PAH assumes a different conformation in the presence of phenylalanine than in its absence. A conformational change is what allowed PAH to be cut by trypsin in the presence of the amino acid. **14.** analyze (1) You could look for a defect in the production of the coenzyme. If the the coenzyme were lacking, PAH would not be able to break down phenylalanine. (2) Another defect could be in the reactions with NADH that are responsible for regenerating the reduced form of the coenzyme. **15.** analyze The functional form of PAH consists of four identical protein subunits, each providing the same active site. **16.** create Two possible areas of research are (1) development of drugs that would help a defective PAH assume the catalytically active conformation and (2) administration of the coenzyme to compensate for possible defects in its production.

Chapter 9

IN-TEXT QUESTIONS AND EXERCISES

p. 195 Fig. 9.2 understand *Glycolysis:* "What goes in" = glucose, NAD^+, ADP, P_i; "What comes out" = pyruvate, NADH, ATP. *Pyruvate processing:* "What goes in" = pyruvate, NAD^+; "What comes out" = acetyl CoA, NADH, CO_2. *Citric acid cycle:* "What goes in" = acetyl CoA, NAD^+, FAD, ADP or GDP, P_i; "What comes out" = CO_2, NADH, $FADH_2$, ATP or GTP. *Electron transport and oxidative phosphorylation:* "What goes in" = NADH, $FADH_2$, O_2, ADP, P_i; "What comes out" = NAD^+, FAD, H_2O, ATP.

p. 195 Making Models 10.1 understand When glucose is completely oxidized, all the carbons end up in CO_2. The processes represented by the arrows are shown in Figure A9.1.

p. 197 CYU apply The radioactive carbons in glucose can be fully oxidized by cellular respiration to generate radiolabeled CO_2. Other molecules, like lipids and amino acids, would end up radiolabeled since they are made in other anabolic pathways using intermediates from the central pathways.

p. 200 CYU (1) understand Reactions in the energy-investment phase require ATP to produce phosphorylated products. Without ATP, the reactions are endergonic and would not occur. The energy-payoff phase consists of exergonic reactions that harvest the chemical energy to produce ATP, so no input of energy is required. **(2)** apply See Figure A9.2. If the regulatory site had a higher affinity for ATP than the active site, then when ATP is present, it would first bind to the regulatory site and inhibit the enzyme. As a result, the rate of ATP production in glycolysis would be low at all concentrations of ATP.

p. 201 CYU remember The three molecules in the figure that speed up the reaction are pyruvate, NAD^+, and CoA (reaction substrates). The two molecules that slow down the reaction are acetyl CoA and NADH (reaction products). These molecules regulate the activity of pyruvate dehydrogenase indirectly by affecting other enzymes that cause it to be phosphorylated (inactive) or dephosphorylated (active).

p. 204 Fig. 9.13 apply NADH would be expected to have the highest amount of chemical energy since its production is correlated with the largest change in free energy in the graph.

p. 204 CYU (1) understand See Figure A9.3. **(2)** understand The enzymes in the first and fourth reactions are regulated by ATP, which inactivates the enzymes when ATP levels are high by binding to an allosteric regulatory site. The enzyme in the third reaction is regulated by high levels of NADH, which serves as a competitive inhibitor by binding to the active site.

p. 206 apply The O_2 at the bottom of the graph would have the highest redox potential, and the FMN in complex I would have the lowest redox potential. The higher the redox potential, the greater the ability to accept electrons.

p. 207 Fig. 9.15 analyze The proton gradient arrow should start in the intermembrane space and point down across the membrane into the mitochondrial matrix. *Complex I:* "What goes in" = NADH, H^+; "What comes out" = NAD^+, e^-, transported H^+. *Complex II:* "What goes in" = $FADH_2$; "What comes out" = FAD, e^-, H^+. *Complex III:* "What goes in" = e^-, H^+; "What comes out" = e^-, transported H^+. *Complex IV:* "What goes in" = e^-, H^+, O_2; "What comes out" = H_2O, transported H^+.

p. 208 Fig. 9.17 create They could have placed the vesicles in a basic solution that has a pH above that of the solution inside the vesicle. Doing this would set up a proton gradient across the membrane to test for ATP synthesis.

p. 208 analyze When glucose is oxidized in glycolysis, pyruvate processing, and the citric acid cycle, reduced electron carriers (NADH or $FADH_2$) are produced. The electron carriers pass electrons to the ETC in mitochondria, which then produces a proton gradient across the inner membrane. This proton gradient powers the synthesis of ATP via the ATP synthase.

p. 210 CYU (1) understand ATP synthase consists of a membrane-associated F_o unit and an F_1 unit joined by a rotor shaft. When protons flow through the F_o unit, it spins the rotor shaft within the fixed F_1 unit. This spinning shaft causes structural changes in the F_1 that drives the synthesis of ATP from ADP and P_i. **(2)** analyze The ETC in aerobic respiration uses oxygen (O_2) as the terminal electron acceptor; in anaerobic respiration it uses a different acceptor, like nitrate (NO_3^-) or sulfate (SO_4^{2-}). The redox potential is higher in O_2 than in NO_3^- or SO_4^{2-}, so there is a larger change in free energy in the aerobic ETC. For each glucose (or equivalent), there would be more protons transported and more ATP generated from aerobic respiration compared to anaerobic respiration.

p. 212 CYU analyze Both fermentation and the ETC regenerate NAD^+ from NADH. In fermentation, the only ATP produced comes from glycolysis, while the proton gradient formed by the ETC is used to make many more ATP per glucose. Your cells would not survive using fermentation alone due to the low yield of ATP from this process.

✔ TEST YOUR KNOWLEDGE

1. remember d **2.** understand c **3.** remember Most of the energy is stored in the form of NADH. **4.** understand T, T, T, F (Both the first and third responses are true because glycolysis is part of fermentation and it produces NADH while the follow-up reactions regenerate NAD^+. The fourth response is false because fermentation does not produce electron acceptors for the ETC.)

✔ TEST YOUR UNDERSTANDING

5. analyze Both processes produce ATP from ADP, but substrate-level phosphorylation occurs when enzymes transfer a phosphate group from a substrate to ADP, while oxidative phosphorylation forms ATP from ADP and P_i. In substrate-level phosphorylation, the energy driving the reaction is in the phosphorylated substrate, while oxidative phosphorylation uses energy in the proton-motive force generated by an ETC.
6. apply b **7.** analyze Both phosphofructokinase and isocitrate dehydrogenase are regulated by feedback inhibition, where the product of the reaction or series of reactions inhibits the enzyme activity. They differ in that phosphofructokinase is regulated by allosteric inhibition while isocitrate dehydrogenase is controlled by competitive inhibition. **8.** understand Oxidative phosphorylation is possible via a proton gradient that is established by redox reactions in the ETC. Uncoupling proteins disrupt this relationship by allowing protons to freely pass through the inner membrane, thus reducing the proton gradient and the production of ATP by oxidative phosphorylation.

✔ TEST YOUR PROBLEM-SOLVING SKILLS

9. create When complex IV is blocked, electrons can no longer be transferred to oxygen, the final acceptor, and cellular respiration stops. Fermentation could keep glycolysis going, but it is unable to fuel a cell's energy needs over the long term. The low production of ATP would result in systemic cell death, with those cells that lack the capacity for fermentation dying first. **10.** apply For each glucose molecule, two ATP are produced in glycolysis and two ATP are produced in the citric acid cycle via substrate-level phosphorylation. A total of 10 NADH and 2 $FADH_2$ molecules are produced from glycolysis, pyruvate oxidation, and the citric acid cycle. If each NADH were to yield 3 ATP, and each $FADH_2$ were to yield 2 ATP, then a total of 34 ATP would be produced via oxidative phosphorylation. Adding these totals would result in 38 ATP. A cell will not produce this much ATP, because the proton-motive force is used in other transport steps and because of other issues that may reduce the overall efficiency.

✔ PUT IT ALL TOGETHER: CASE STUDY

11. create The active transport of protons against the proton gradient requires energy released from redox reactions in the ETC. If the strength of the proton motive force increases above the energy released by the redox reactions, then transport of protons will be halted. When this occurs, electron transport will slow down or stop. **12.** analyze The greatest difference in survival occurs around week 120. This is determined by drawing a vertical line at different time points and measuring the difference in survival between the two conditions. **13.** analyze An additional 5.5 years would be added to the life span of the U.S. population. One way to calculate this is to first divide the life span in

Figure A9.1

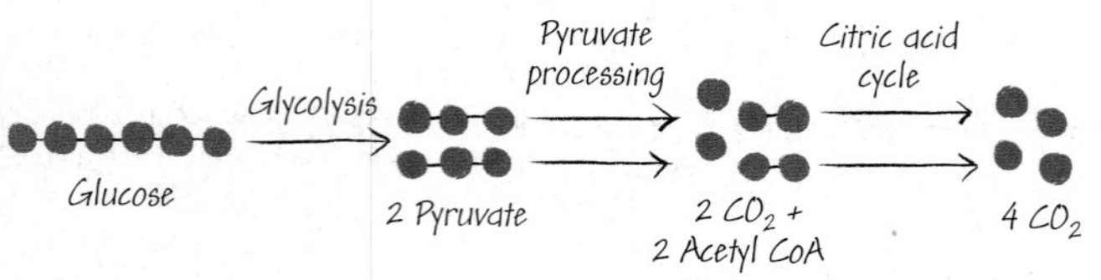

Figure A9.2

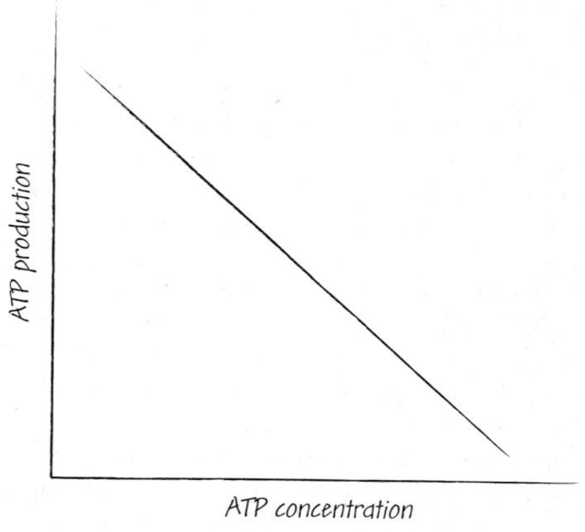

Figure A9.3

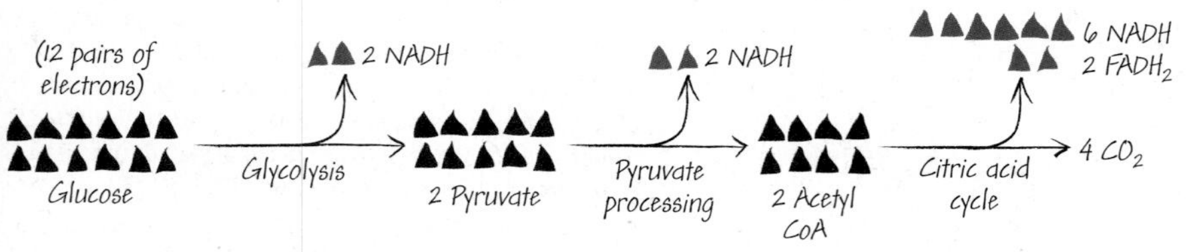

DNP-treated mice (770 days) by the life span of the control (719 days), then multiply this factor (1.07) by the average life span of the U.S. population (79 years). The difference between this value (84.5 years) and the current average resulted in an increase of 5.5 years. **14.** create Isolated mitochondria from either DNP-treated mice or control mice could be treated with the same amount of pyruvate. If DNP is reducing the proton gradient, and thus efficiency of oxidative phosphorylation, you would expect less ATP to be produced from these mitochondria per pyruvate compared to those obtained from the control mice. **15.** analyze To meet the ATP demands of the cell, more high-energy molecules, like glucose, would be broken down in DNP-treated mice instead of using them in anabolic reactions that contribute to increased body mass. **16.** analyze The increased respiration likely arises from an elevated demand for O_2 to drive electron transport to address the reduced efficiency of ATP synthesis. Fever would result from the increased production of heat as protons passively move across the membrane, similar to what is observed in brown adipose tissue. Death is likely due to insufficient ATP being produced to keep cells alive.

Chapter 10

IN-TEXT QUESTIONS AND EXERCISES

p. 215 apply See Figure A10.1. This reaction requires an input of energy because there are more high-energy chemical bonds in the products compared with the reactants, and there is a decrease in entropy.

p. 215 Fig. 10.1 apply See Figure A10.1.

p. 216 Making Models 10.1 Figure A10.2 shows one example of a completed model.

p. 217 CYU apply The Calvin cycle requires the ATP and NADPH produced in the light-capturing reactions. To continue producing ATP and NADPH, the light reactions require the ADP, P_i, and $NADP^+$ that is regenerated by the Calvin cycle.

p. 219 Fig. 10.6 analyze See Figure A10.3.

p. 220 Fig. 10.9 apply The energy state corresponding to a photon of blue light would be the same as shown, but the energy state corresponding to green light would be located between the energy states corresponding to red and blue photons.

p. 222 CYU apply The outer pigments would be more likely to absorb blue photons (short wavelength, high energy), and interior pigments would absorb red photons (long wavelength, low energy). This organization establishes a pathway to direct photon energy toward the reaction center since resonance energy is transferred from higher to lower energy levels.

p. 224 understand Light → Antenna pigments → Reaction center (electrons from water splitting are accepted here) → Pheophytin → ETC → Proton gradient → ATP synthase.

p. 226 remember See Figure A10.4.

p. 227 CYU analyze In mitochondria, electrons at a high-energy state are donated by NADH or $FADH_2$ (primary donors) and passed through an ETC to generate a proton-motive force. The electrons at a lower energy state at the end of the chain are accepted by O_2 (terminal acceptor) to form water. In chloroplasts, electrons at a lower energy state are donated by H_2O (primary donor), excited by photons or resonance energy, and passed through an ETC to generate a proton-motive force. These electrons are then excited a second time by photons or resonance energy, and the electrons at a higher energy state are accepted by $NADP^+$ (terminal acceptor) to form NADPH.

p. 228 Fig. 10.20 analyze The researchers had no basis for predicting these intermediates. They needed to perform the experiment to identify them.

p. 231 Fig. 10.25 apply The highest concentration of organic acids in the vacuoles of CAM plants would be found in the morning, since these acids are made during the night and used up during the day.

p. 231 CYU (1) understand In photorespiration, O_2 is fixed to RuBP by rubisco instead of CO_2. The reactions that follow use ATP and result in the release of CO_2, which decreases the overall rate of CO_2 fixation and the production of sugar. **(2)** understand (a) In C_3 plants, CO_2 is delivered to rubisco by diffusion through stomata. (b) In C_4 plants, the CO_2 that diffuses through stomata is first fixed by PEP carboxylase into four-carbon organic acids in mesophyll cells. These organic acids are then transported into bundle-sheath cells, where they release CO_2 to rubisco. (c) In CAM plants, CO_2 is delivered to rubisco in a two-step process similar to C_4 plants, but in the same cell. CAM plants take up CO_2 at night and the four-carbon organic acids are stored in

Figure A10.1

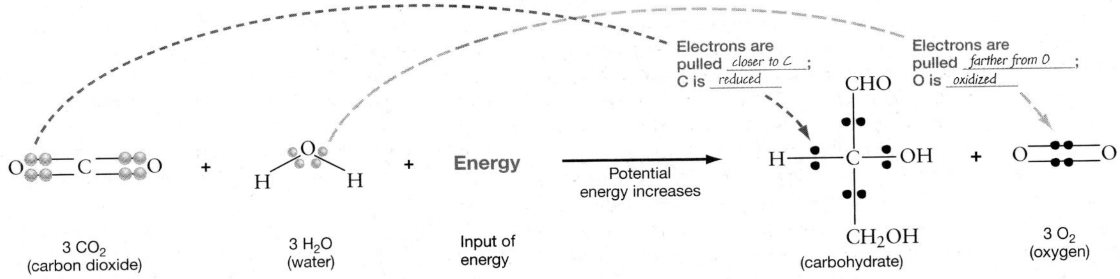

Figure A10.2

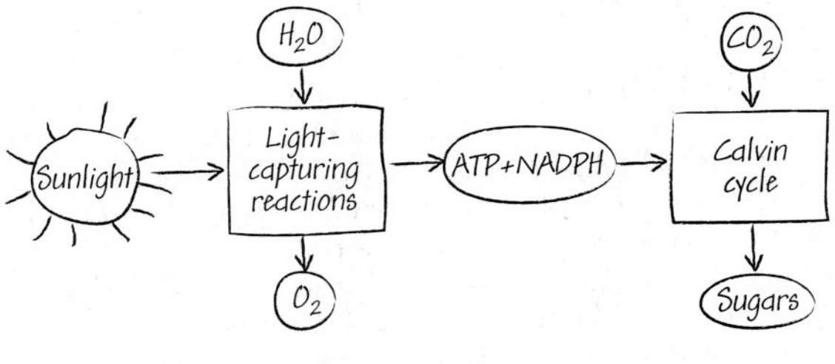

Figure A10.3

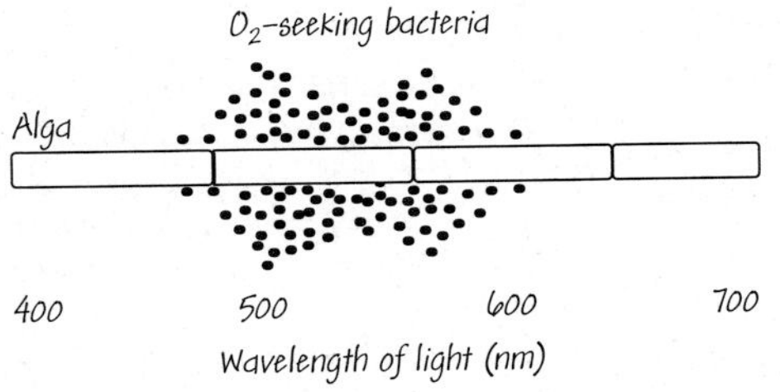

Figure A10.4

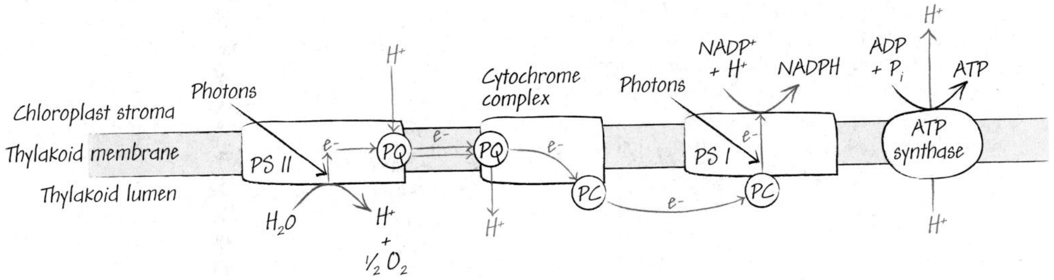

the central vacuoles. During the day, the organic acids are processed to release CO_2 to rubisco.

p. 232 apply Each complete cycle requires 3 ATP and 2 NADPH molecules. To complete six runs through the cycle, a total of 18 ATP and 12 NADPH molecules are needed. By following the number of carbons, it is apparent that only three RuBP molecules are required, since they are fully regenerated every three cycles: 3 RuBP (15 carbons) fix and reduce 3 CO_2 to generate 6 G3P (18 carbons), yielding 1 G3P (3 carbons); the other five G3P are used to regenerate 3 RuBP (15 carbons).

p. 233 CYU (1) understand When light is present, there is an increase in photosynthetic proteins and regulatory molecules activate rubisco. The production of photosynthetic proteins is decreased when sugar levels are high. Rubisco activity is inhibited when CO_2 levels are low. **(2)** apply The concentration of starch would be highest at the end of the day and lowest at the start of the day. Starch is made and stored in the chloroplasts of leaves during periods of high photosynthetic activity during the day. At night, it is broken down to make sucrose, which is transported throughout the plant to drive cellular respiration. (Cellular respiration also occurs during the day, but the impact is minimized due to the photosynthetic production of sugar.)

✔ TEST YOUR KNOWLEDGE

1. remember d **2.** remember d **3.** remember a **4.** understand Light energy is first converted to chemical energy when electrons are transferred from excited pigments to an electron carrier in a photosystem reaction center.

✔ TEST YOUR UNDERSTANDING

5. understand Most of the energy captured by chlorophyll in chloroplasts is converted into chemical energy by reducing electron acceptors in ETCs. When pigments are extracted, the antenna complexes, reaction centers, and ETCs have been disassembled, so the energy is given off as fluorescence and/or heat. **6.** understand The fixation phase occurs when CO_2 is fixed to RuBP by rubisco to form 3-phosphoglycerate (3PGA). The reduction phase uses ATP to phosphorylate the carbons and NADPH to reduce them with high-energy electrons to form G3P. The regeneration phase uses more ATP to convert some of the G3P to RuBP to continue the cycle. **7.** apply It would require 36 photons. To produce G3P from 3 CO_2 molecules and regenerate RuBP, a total of 6 NADPH and 9 ATP are required. Each NADPH is made by exciting two electrons from water in photosystem II and then again in photosystem I. This means that 24 photons are required for 6 NADPH, which would also produce 6 ATP. To make 9 ATP, three more pairs of electrons are required for a total 18 electrons that must be excited by 36 photons. **8.** understand F, F, T, T (The first and second responses are false because ATP is not directly used for CO_2 fixation by rubisco and the ATP produced by photosynthesis is used only in the stroma of chloroplasts for activities in the Calvin cycle. The third and fourth responses are correct and reflect the reduction and regeneration phases of the Calvin cycle.)

✔ TEST YOUR PROBLEM-SOLVING SKILLS

9. apply (1) O_2, ATP, and NADPH would be formed by noncyclic electron flow. (2) No O_2 or NADPH would be formed, but ATP may be made by cyclic electron flow. (3) Initially, O_2 and NADPH would be formed by noncyclic electron flow, but no ATP would be made. Without ATP, the Calvin cycle would halt and, once all the $NADP^+$ is reduced to NADPH, noncyclic electron flow would switch to cyclic electron flow. **10.** analyze b (The wavelength of light would fully excite PS I, but is less likely to excite PS II, resulting in cyclic electron flow since few electrons could be harvested from water by PS II.)

✔ PUT IT ALL TOGETHER: CASE STUDY

11. analyze It is possible that some aquatic plants and algae require the C_4 pathway. If you said that it was not possible, you may be thinking that the role of the C_4 pathway is to avoid desiccation, which is not the case. The C_4 pathway is used to increase the concentration of CO_2 to effectively drive carbon fixation by the Calvin cycle. Low CO_2 concentrations may (and do) exist in certain aquatic environments. **12.** apply The vast majority of the energy stored in the organic molecules present in fossil fuels, whether it is in plant or animal matter, originated from sunlight. **13.** analyze Biofuels are produced using CO_2 present in the atmosphere, so there would be no net release of CO_2 from the combustion of these molecules. **14.** apply At maximal photosynthetic activity, the cells will convert most of the available G3P into starch, which is stored in granules. **15.** create The strain with small antenna complexes has a higher rate of photosynthesis over time and reaches its maximum growth rate earlier as compared to the control. The reduced antenna complexes would capture less light, allowing more light to pass through the water to reach microalgae at the bottom of the pool. **16.** create Increased photosynthetic activity could be achieved by bubbling CO_2 through the culture to increase its concentration and promote the Calvin cycle.

BIG PICTURE Energy for Life

p. 236 CYU (1) understand Photosynthesis uses H_2O as an electron donor and releases O_2 as a by-product; cellular respiration uses O_2 as an electron acceptor and releases H_2O as a by-product. **(2)** understand Photosynthesis uses CO_2 as a substrate to fix and reduce it to produce sugar; cellular respiration oxidizes sugar and releases CO_2 as a by-product. **(3)** analyze CO_2 fixation would essentially stop, but CO_2 would continue to be released by cellular respiration. CO_2 levels in the atmosphere would increase rapidly, and production of new plant tissue would cease—meaning that most animals would quickly starve to death. **(4)** analyze ATP "is used by" the Calvin cycle; photosystem I "yields" NADPH; $FADH_2$ "donates high-energy electrons" to the ETC.

Chapter 11

IN-TEXT QUESTIONS AND EXERCISES

p. 240 apply If the cellulose microfibrils were digested, the cell wall's ability to resist tension would be diminished. As a result, turgor pressure would cause the plant cell to swell until its plasma membrane burst, and the cell would die.

p. 242 CYU analyze Plant cell walls and animal ECMs are both fiber composites, but they differ in the types of molecules used. Plant cell walls consist mostly of carbohydrates: cellulose microfibrils are the fibrous component, and the polysaccharides like pectins are the ground substance. Animal ECMs consist mostly of proteins: collagen fibrils are the fibrous component, and proteoglycans are the ground substance.

p. 245 apply All of the cells in the embryo would be capable of attaching to one another, regardless of the cell type. This would severely affect the development of different tissues and organs. The embryo would likely die.

p. 245 Fig. 11.10 apply *Prediction of hypothesis:* Cells treated with an antibody that blocks membrane proteins involved in adhesion will not adhere. *Prediction of null hypothesis:* Cells in all samples will adhere.

p. 247 CYU (1) analyze The three structures differ in composition, but their function is similar. The middle lamella in plants is composed of pectins that glue adjacent cells together. Tight junctions are made up of membrane proteins that line up and "stitch" adjacent cells together. Desmosomes are rivet-like structures composed of proteins that link the cytoskeletons of adjacent cells. **(2)** understand At plasmodesmata, the plasma membranes of adjacent plant cells are continuous, and the cells share portions of the smooth endoplasmic reticulum. Gap junctions connect adjacent animal cells by forming protein-lined pores. Both structures are openings between cells that allow cytosol, including ions and small molecules, to be shared.

p. 249 Fig. 11.13 apply The conformational change would likely expose a nuclear localization signal on the receptor, which is required for the protein to be transported into the nucleus.

p. 251 analyze The spy is the signaling molecule that arrives at the cell surface (the castle gate). The guard is the G-protein-coupled receptor in the plasma membrane, and the queen is the G protein. The commander of the guard is the enzyme that is activated by the G protein to produce second messengers (the soldiers).

p. 252 apply After adrenaline binds to the receptor in both heart cells and liver cells, the resulting signals are processed via distinct transduction pathways that activate different proteins and thus trigger different responses in the different cells.

p. 253 Fig. 11.17 analyze (1) cell responses A, C, and possibly B at a low level; (2) cell responses A, B, and C; (3) cell responses B and C.

p. 253 CYU (1) understand Each signaling molecule binds to a specific receptor protein. A cell can respond to a signaling molecule only if it has the appropriate receptor. Only certain cell types will have the appropriate receptor for a given signaling molecule. **(2)** understand Signals are amplified if one or more steps in a signal transduction pathway results in the production of multiple copies of a new signal. This signal may be in the form of a second messenger (small molecule or ion) or an altered protein (e.g., phosphorylation cascade). Signals are diversified when different downstream molecules are affected by a single step in the signaling pathway. **(3)** apply The intracellular signal transduced by abscisic acid includes the release of calcium ions as second messengers in guard cells. To deactivate the signal, these ions could be pumped back into the smooth ER.

p. 254 CYU (1) analyze The signal transduction pathways—consisting of signaling molecules, receptors, and second messengers—are similarly organized in both unicellular and multicellular organisms. There is more variety in the means of transmitting the signal between cells in multicellular organisms than in unicellular organisms. For example, there are no gap junctions or plasmodesmata for direct intercellular signal transmission between unicellular organisms. **(2)** understand Quorum sensing occurs when signaling molecules secreted by cells reach a high concentration, which is used as a measure of population density. Examples of activities coordinatd by quorum sensing include the aggregation of *Dictyostelium* amoebae, the production of light by *Vibrio fischeri* in bobtail squid, and the production of biofilms by bacterial communities.

✔ TEST YOUR KNOWLEDGE

1. understand A fiber composite consists of cross-linked filaments that withstand tension and a ground substance that withstands compression. The cellulose microfibrils in plants and collagen fibrils in animals functionally resemble the steel rods in reinforced concrete. The pectins and other gelatinous polysaccharides in plants and proteoglycans in animals functionally resemble the concrete ground substance. **2.** remember a (Recall that all proteins in the ECM are synthesized in the rough ER by ribosomes and then secreted into the extracellular space.) **3.** understand T, T, T, T (All of these responses reflect roles of various types of intercellular connections; see Section 11.2.) **4.** understand c

✔ TEST YOUR UNDERSTANDING

5. analyze b **6.** understand Dissociated sponge cells gradually began to aggregate, adhering to other cells of the same tissue type. Selective adhesion results from the presence of specific types of cell adhesion proteins in the plasma membrane, including cadherins. These proteins can bind only to the same or complementary adhesion proteins present on other cells. **7.** apply For each Ras-GTP, you would need 10 pennies for kinase 1 ($0.10), 100 nickels for kinase 2 ($5.00), and 1000 dimes for kinase 3 ($100.00). Constructing the model would therefore require $105.10. **8.** evaluate Information from different signals may conflict or be reinforcing. Crosstalk between signaling pathways allows cells to integrate information from many signals at the same time instead of responding to each signal in isolation.

✔ TEST YOUR PROBLEM-SOLVING SKILLS

9. analyze (a) No amplification can occur, because the number of signaling molecules dictates the size of the response. (b) Only one step is available for regulation, which includes either blocking the intracellular receptor or making it more responsive to the signaling molecule. (c) As with a signal transduction pathway, gene expression may be altered. Other types of responses, however, such as muscle contraction or the rapid mobilization of glucose, are not possible. **10.** analyze Antibody binding to the receptor may cause the two parts of the receptor to dimerize, which normally occurs when binding to a signaling molecule. The result would be activation of the phosphorylation cascade even in the absence of the signaling molecule.

✔ PUT IT ALL TOGETHER: CASE STUDY

11. understand The pheromone–receptor complex activates the G protein by triggering the replacement of bound GDP with GTP. Activation breaks the G protein into two subunits, which then go on to affect the activity of other proteins. **12.** apply The protein related to Ras initiates a phosphorylation cascade that phosphorylates and activates a series of mitogen-activated protein kinases (MAPKs). The terminal active MAPK from this cascade phosphorylates other proteins, triggering the cell response. **13.** analyze The scaffold protein, Ste5, increases the speed of the response by grouping a series of MAPKs together. Since the number of proteins bound to Ste5 is limited, amplification at each step in the phosphorylation cascade is decreased compared to a pathway without a scaffold protein. **14.** analyze In *ptc1*Δ cells, the amount of binding between Fus3 and Ste5 remains high at concentrations of α-factor that cause Fus3 to be released from Ste5 in wild-type cells. This suggests that Ste5 must be dephosphorylated by Ptc1 to release Fus3. Since the mating response requires release of Fus3, it would not be expected to occur in cells with a *ptc1*Δ mutation. **15.** create One possible hypothesis is that proteins involved in signaling the assembly of microfilaments are grouped together with the receptors and have a limited ability to move beyond the area of signal reception. **16.** create Many cell signaling pathways interact with one another by crosstalk. This interaction between distinct pathways could suppress the effect of the drug, or it could result in the drug interfering with other unrelated cell responses (i.e., side effects).

Chapter 12

IN-TEXT QUESTIONS AND EXERCISES

p. 259 Making Models 12.1 apply See **Figure A12.1**. Recall that after the chromosomes have replicated, each chromosome contains two attached chromatids.

p. 260 apply If there were no gap between the S and M phases, the graph would show an immediate increase in the percentage of radiolabeled cells in mitosis at the end of the pulse and start of the chase period.

p. 260 CYU (1) analyze The gap phases are periods of the cell cycle when there is no DNA synthesis (S phase) and no separation of the chromosomes or cytoplasm (M phase). In the G_1 phase, cells perform their normal functions and "decide" when or if the cell undergoes division. In the G_2 phase, cells contain replicated chromosomes as they grow and prepare for M phase. **(2)** apply The G_2 phase is the time between the end of the pulse and the starting point when radiolabeled cells are undergoing mitosis. Based on the data in the graph, this phase is 4 hours long.

p. 261 understand (1) DNA is the genetic material in chromosomes. (2) Chromosomes are made of chromatin, which is a complex of DNA and histone proteins. (3) Sister chromatids are identical copies of the same chromosome that are joined together.

p. 262 Fig. 12.5 apply (1) The prophase cell has replicated its DNA, so it will have twice as much DNA (120 picograms [pg]). Each chromosome will have two chromatids, but the number of chromosomes will remain at 22. (2) The anaphase cell still has 120 pg of DNA, but when the chromatids separate, each is defined as a daughter chromosome, so the cell has 44 chromosomes. (3) After cytokinesis, each daughter cell will have as much DNA (60 pg) and as many chromosomes (22) as the parent cell in G_1 phase. If you stated that the daughter cells have more or less DNA or chromosomes than the parent cell, recall that mitotic cell division does not change the number of chromosomes or the amount of DNA.

p. 264 apply See **Table A12.1**.

p. 265 Fig. 12.6 apply Daughter chromosomes will move toward the pole at the same rate as the darkened sections.

p. 267 CYU (1) apply See **Figure A12.2**. **(2)** understand *Type 1:* The shrinkage of kinetochore microtubules transports daughter chromosomes to the opposite poles of the spindle. *Type 2:* The push against polar microtubules by motor proteins and the pull on astral microtubules by other motors on the plasma membrane move the spindle poles to opposite sides of the cell. **(3)** analyze Both cell types use cytokinesis to divide the cytoplasm into two daughter cells. Plant cells produce a cell plate that grows in the middle of the spindle and fuses with the plasma membrane. Animal cells use a cleavage furrow in the middle of the spindle to draw in the plasma membrane, which eventually fuses with itself.

p. 268 Fig. 12.10 create Remove a sample of the cell extract at different times after the cyclin mRNA was added. Inject each of the samples into different frog oocytes that are arrested in the G_2 phase. Use light microscopy to determine if M phase occurs based on morphological changes.

p. 269 Fig. 12.11 analyze If the cyclin concentration did not decline, MPF would remain active and the cell would be stuck in M phase.

p. 269 understand MPF activity requires a cyclin and a Cdk, and it is turned on by phosphorylation at the Cdk's activating site and dephosphorylation by a phosphatase at the Cdk's inhibitory site. Enzymes that degrade cyclin reduce MPF activity.

p. 271 CYU understand The four checkpoints are (1) the G_1 checkpoint between G_1 and S; (2) the G_2 checkpoint

Figure A12.1

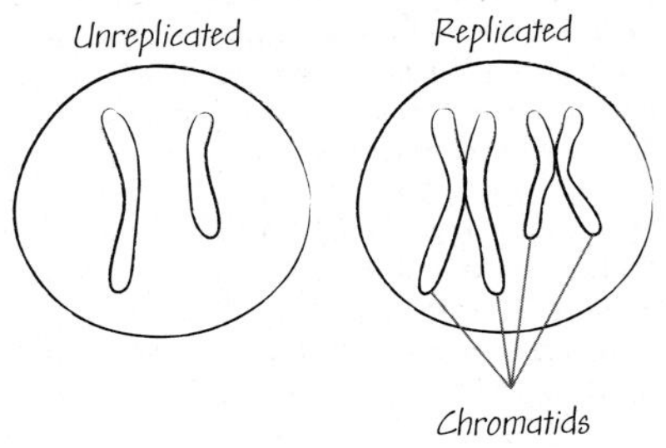

Figure A12.2

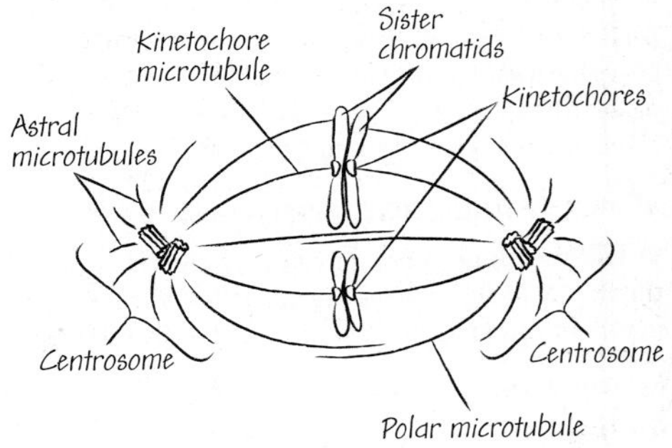

Table A12.1

	Prophase	Prometaphase	Metaphase	Anaphase	Telophase
Spindle apparatus	Starts to form	Contacts and moves chromosomes	Anchors poles to membrane and produces tension at kinetochores	Pulls chromatids apart	Defines site of cytokinesis
Nuclear envelope	Present	Disintegrates	Nonexistent	Nonexistent	Re-forms
Chromosomes	Condense	Attach to microtubules	Held at metaphase plate	Sister chromatids separate into daughter chromosomes	Collect at opposite poles

between G_2 and M; (3) the first M checkpoint between metaphase and anaphase; and (4) the second M checkpoint between anaphase and telophase. Without these four checkpoints, cells could be overproduced in multicellular organisms, and cell division could result in daughter cells that are inviable or have defects in their genetic material.

p. 273 CYU (1) apply Both types of tumors have unregulated cell division. However, benign tumors remain localized to a single mass and do not invade other tissues, while malignant tumors spread from the original tumor and may metastasize to distant locations. **(2)** understand Growth factors activate the expression of E2F and cyclins. The cyclins bind to a Cdk to phosphorylate Rb, which releases E2F to promote expression of S-phase specific genes.

✔ TEST YOUR KNOWLEDGE

1. remember b **2.** remember d **3.** remember c **4.** remember The sister chromatids on replicated chromosomes separate, and the spindle poles are pushed farther apart.

✔ TEST YOUR UNDERSTANDING

5. apply All of the chromosomes must be replicated during S phase, the spindle apparatus must connect with the kinetochores of each sister chromatid in prometaphase, the sister chromatids of each replicated chromosome must be partitioned in anaphase, and the daughter chromosomes must be divided between two daughter cells by cytokinesis. **6.** understand Daughter chromosomes were observed to move toward the poles faster than the marked regions of fluorescently labeled kinetochore microtubules. **7.** analyze F, T, T, F. (If the Rb protein is nonfunctional or G_1 cyclin is overexpressed, E2F expression would immediately cause cells to enter S phase.) **8.** analyze The absence of growth factors in normal cells would cause the cells to arrest in G_1 phase, and eventually all the cells in the culture would be in G_1. The cancerous cells would be unlikely to depend on these growth factors, so those cells would not arrest and would continue through the cell cycle.

✔ TEST YOUR PROBLEM-SOLVING SKILLS

9. analyze a (Adding the length of time spent in each phase allows you to determine that the cell cycle is 8.5 hours long. After 9 hours, all radiolabeled cells would have passed through a full cycle and be in either S phase or G_2—none would be in M phase.) **10.** apply The embryo passes through multiple rounds of the cell cycle, but cytokinesis does not occur during the M phases.

✔ PUT IT ALL TOGETHER: CASE STUDY

11. apply Microtubules are actively polymerized during the start of M phase, when the mitotic spindle is produced. During anaphase, the kinetochore microtubules are depolymerized as daughter chromosomes move to opposite poles of the cell. At the end of M phase, the spindle microtubules are depolymerized to remove the spindle before entering G_1. **12.** apply The second M-phase checkpoint. If microtubule depolymerization were inhibited, the chromosomes would not completely separate in anaphase, and the cells would arrest in M phase. **13.** analyze See **Figure A12.3**. The asynchronous culture would arrest in M phase, so after the labeled cells pass through G_2, they would remain in M phase along with other cells that were in G_2 during the pulse (resulting in only 50 percent of the M-phase cells being labeled). The percentage of labeled mitotic cells would decrease over time as additional unlabeled cells accumulate in M phase. **14.** create The stathmin protein decreases the stability of microtubules. When stathmin is absent, the microtubule-stabilizing effect of Taxol is enhanced, and the cells are more likely to arrest. As a result, the tumors do not grow. **15.** apply The cyclin-dependent kinase subunit of MPF is the enzyme responsible. **16.** create Inactivating genes for G_1 cyclins or E2F could serve as an alternative therapy to arrest cancerous cells in G_1.

END-OF-UNIT CASE STUDY How Did the Newt Become So Toxic?

pp. 276–277 (1) understand Bacteria present in a host animal may secrete signaling molecules that accumulate as the population increases. When the signaling molecules reach a certain level based on the density of the population, or quorum, a signaling pathway is activated resulting in a cellular response. In this case, the response would be the production of TTX. **(2)** analyze The TTX appears to be obtained from the diet of the puffer fish, which would support the exogenous hypothesis. **(3)** create Similar to what was done in the pulse–chase experiment, the TTX in the diet could be radiolabeled. If the TTX present in the liver is the same as what was provided in the diet, then the liver TTX would be radiolabeled. **(4)** apply TTX may circulate throughout the body, but bind specifically to liver and skin cells, causing it to accumulate around these tissues. There would likely be a receptor protein on the cells of these tissues that either binds TTX or a TTX carrier protein. **(5)** understand Secretory vesicles are formed by the secretory pathway, which includes the rough ER and Golgi apparatus. The rough ER produces cargo that is transported by vesicles to the Golgi apparatus. Receptors at the trans cisterna of the Golgi then concentrate cargo and form secretory vesicles, which eventually fuse with the plasma membranes of granular gland cells. **(6)** create TTX might be loaded into secretory vesicles by a pump protein that actively imports TTX from the cytosol. Alternatively, TTX might be manufactured by enzymes residing in the ER or Golgi, then recruited by receptors on the surface of secretory vesicles as the vesicles are formed in the *trans*-Golgi cisterna. **(7)** analyze After electric shock, there is significantly less TTX in the skin tissue. This would suggest that the shock is causing TTX to be secreted out of the granular glands onto the skin surface. After nine months, the skin tissue appears to regenerate some of the TTX. **(8)** understand When the G-protein-coupled receptor binds to its signal, it causes a G protein to be activated by replacing GDP with GTP. The active G protein then binds to an enzyme and induces it to produce a second messenger. The production of the second messenger leads to a cellular response, in this case the release of TTX from secretory vesicles. **(9)** apply b **(10)** analyze The newt's toxicity does not appear to be dependent on the presence of the toxin in its diet. This is not the case in puffer fish, where the absence of TTX in the diet resulted in loss of toxicity. This difference suggests that newts either make their own TTX (endogenous hypothesis), or harbor bacteria that produce the toxin (symbiotic hypothesis).

Chapter 13

IN-TEXT QUESTIONS AND ANSWERS

p. 280 apply Haploid number is $n = 4$; the organism is diploid, and $2n = 8$.

p. 280 Making Models 13.1 apply See **Figure A13.1**.

p. 280 apply See **Figure A13.2**. Because sister chromatids come from replication of the single double-helical molecule of DNA in the unreplicated chromosome, there is no difference in the genetic information present in the unreplicated and replicated chromosomes, so it makes sense to call both these structures one chromosome even though they look quite different from one another.

p. 282 apply 39. This is because there are 78 replicated chromosomes at the start of meiosis I, each with two molecules of DNA. The number of replicated chromosomes is reduced to 39 at the end of meiosis I. During meiosis II, the sister chromatids of the replicated chromosomes are separated to become individual daughter chromosomes, each with one molecule of DNA. This means that each of the daughter cells at the end of meiosis contains 39 unreplicated chromosomes and 39 molecules of double-stranded DNA.

Figure A12.3

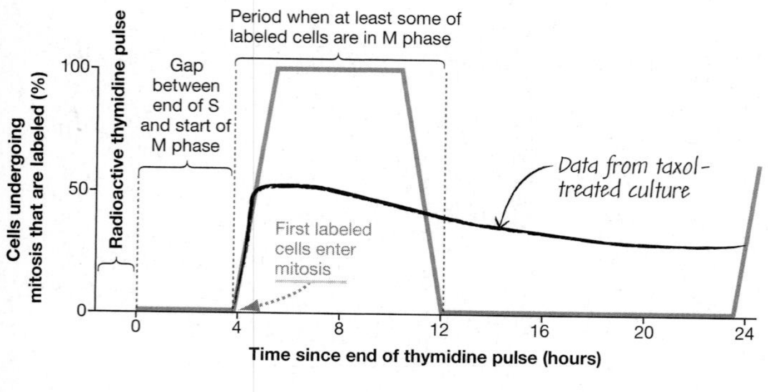

Figure A13.1

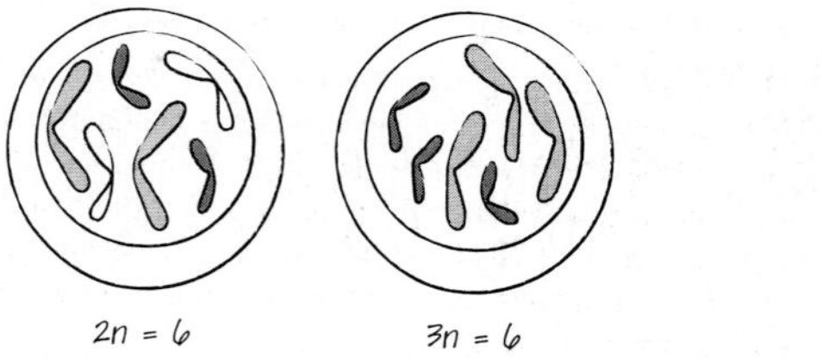

Figure A13.2

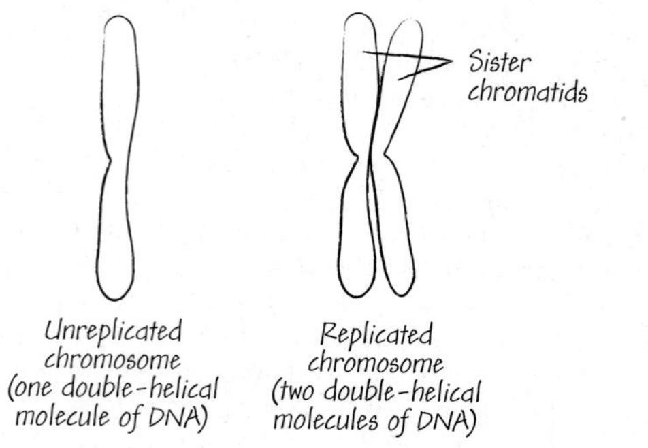

p. 284 apply Because mitosis separates sister chromatids of single chromosomes and there is no need for homologous chromosomes to pair, mitosis in a triploid cell, or a cell of any level of ploidy, can occur easily. In contrast, chromosome pairing is essential for proper separation of homologous chromosomes in meiosis. Because it's impossible to pair three of anything, the three copies of each chromosome in a triploid cannot be separated into daughter cells in a reproducible way.

p. 286 CYU (1) analyze The separation of homologous chromosomes (not sister chromatids, as in mitosis) during anaphase I of meiosis I reduces both chromosome number and DNA amount in each of the haploid daughter cells. In mitosis, sister chromatids of each replicated chromosome are separated and placed into separate cells. This halves the DNA content of each chromosome, but does not change chromosome number.
(2) analyze Each daughter cell would contain 5 units of DNA. This is because the number of duplicated chromosomes is reduced by half in meiosis I, bringing the DNA content to 10 units per cell. In meiosis II, sister chromatids are separated, reducing the DNA content per cell to 5 units.

p. 287 apply Each gamete would inherit either all maternal or all paternal chromosomes. This would limit genetic variation in the offspring by preventing many possible combinations of maternal and paternal chromosomes in gametes.

p. 288 Making Models 13.2 apply See Figure A13.3.

p. 288 CYU (1) analyze False. The minimum number of chromosomes required to observe independent assortment is 4. This is because independent assortment is based on the independent alignment of one homologous pair of chromosomes relative to at least one other homologous pair. **(2)** understand True. Crossing over combines different segments of non-sister chromatids, whereas independent assortment is based on the independent alignment across the metaphase I plate of maternal and paternal versions of different homologous pairs of chromosomes. **(3)** analyze True. Each $2n = 4$ parent can produce gametes with 4 different combinations of chromosomes. Assuming the parents are not genetically identical, the 4 different gametes in each parent can come together at fertilization to produce $4 \times 4 = 16$ different combinations of chromosomes in offspring.

p. 290 CYU (1) create See Figure A13.4.
(2) analyze Aneuploidy for most autosomes arrests development. For this reason, aneuploidy for autosomes other than 21 can be common in gametes and early embryos, but not seen at birth.

p. 291 Fig. 13.12 apply There would be 240 more. (Notice that the number of child-producing individuals in the sexual population doubles at each generation, but in the asexual population, these individuals quadruple at each generation. Therefore, after 4 generations of reproduction [this is what's needed to create generation 5], there will be $2^4 = 16$ child-producing offspring in the sexual population and $4^4 = 256$ in the asexual population. The base in each equation is the fold increase at each generation, and the exponent is the number of generations; look back at Figure 13.12 to see why these numbers make sense.) This diagram shows that slight differences in population growth rate can quickly lead to big differences in the number of individuals over many generations.

p. 292 Fig. 13.13 apply The rate of outcrossing is predicted to rise initially, as the pathogen selects for resistant worms, and then to fall as the worms in the population gain resistance and take advantage of the increased numbers of offspring possible with self-fertilization of hermaphrodites.

p. 292 CYU (1) evaluate One aspect of the purifying selection hypothesis is that sexual reproduction is advantageous for allowing a way around having a harmful mutation inherited by all offspring of an individual with the mutation. If mutation rate (the frequency of new mutations) were high, then sexual reproduction is predicted to be favored. If mutation rate were low, then the relative advantage of sexual reproduction would be lower, and at sufficiently low mutation rates, be lost. **(2)** evaluate According to the changing-environment hypothesis, the advantage of sexual reproduction stems from the production of genetically diverse offspring, some of whom may better be able to resist evolving pathogens or parasites. Since sexual reproduction among genetically identical individuals will produce offspring genetically identical to the parents and to one another, there should be no advantage of sexual reproduction. In fact, in this case, a disadvantage is predicted because of the twofold cost of males.

✔ TEST YOUR KNOWLEDGE

1. remember a (If you answered c, be careful not to mistake sister chromatids for a homologous chromosome pair.)
2. understand b **3.** remember 1/2 (This is because half of a person's chromosomes come from the father and half from the mother, regardless of the sex of the individual.)
4. remember c

✔ TEST YOUR UNDERSTANDING

5. understand Homologous chromosomes are similar in size and shape, they have the same genes (but often different alleles), and they originate from different parents. Sister chromatids are found in replicated chromosomes. They contain exact copies of the single double-helical molecule of DNA that was present in the unreplicated chromosome and replicated during S phase. **6.** analyze At the end of meiosis I. (This is because in a kidney cell before DNA replication, there are pairs of each chromosome, with one double-helical molecule of DNA per chromosome; at the end of meiosis I, each cell has only one of each type of chromosome, but each chromosome has two double-helical molecules of DNA.) **7.** apply a **8.** analyze Tetraploids produce diploid gametes, which combine with a haploid gamete from a diploid individual to form triploid offspring.

✔ TEST YOUR PROBLEM-SOLVING SKILLS

9. apply c (Because $n = 3$, independent assortment creates $2^n = 2^3 = 8$ different combinations of chromosomes. One of these combinations is made up of paternal chromosomes only. Therefore, a haploid cell has a 1/8 chance of receiving only paternal chromosomes.) **10.** create (a) Such a study might be done in the laboratory, controlling conditions in identical populations of rotifers infected with fungus. One population would be kept moist; the other population of rotifers would be allowed to dry out. After various periods of time, water would be added to each population and then the rotifers would be observed to see if fungal infections reappeared. (b) Wind disperses the rotifer to new and often pathogen-free areas. In this case, the ticket to a sex-free existence

Figure A13.4

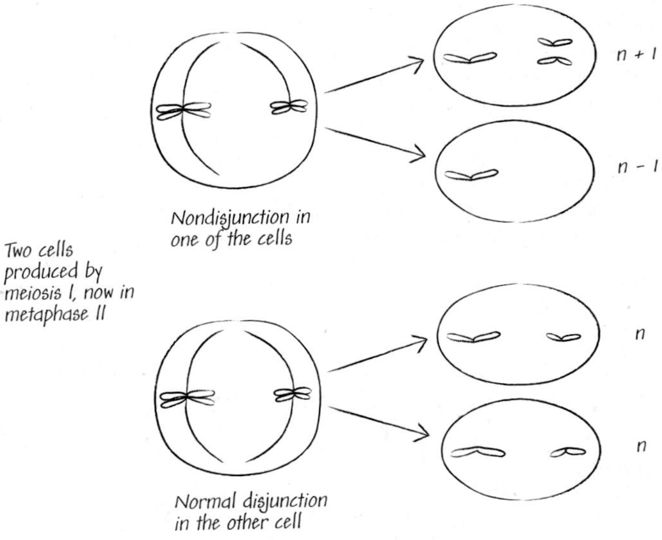

Figure A13.3

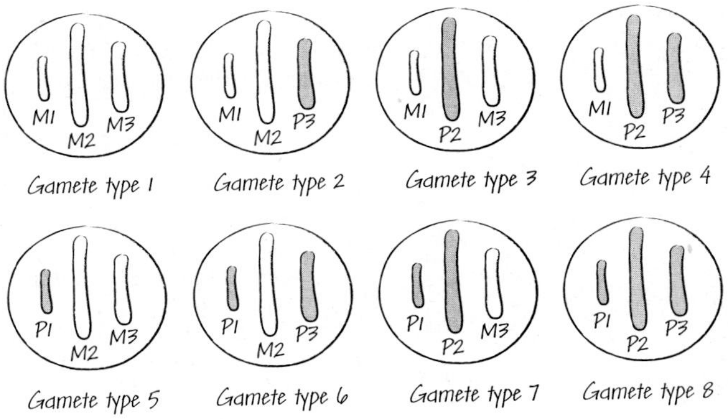

is not genetic diversity but the evolution of an alternative means of evading pathogens. Fungus-infected rotifers that can dry out and survive effectively rid themselves of the pathogen.

✔ PUT IT ALL TOGETHER: Case Study

11. evaluate The results are consistent with the hypothesis because they show that as females age, the amount of cohesin falls, the separation between kinetochores increases, and the percentage of aneuploid eggs goes up. **12.** understand This relationship is logical because of cohesin's role in holding sister chromatids together. As the cohesin level falls, there is less "glue" to hold the sister chromatids together and therefore greater separation between the kinetochores on each replicated chromosome. **13.** analyze This attachment is necessary to pull each homologous pair of chromosomes apart, moving one homolog to one pole and the other homolog to the other pole. This event accomplishes segregation of the homologs to different daughter cells. **14.** analyze One possible explanation is that the greater separation between kinetochores seen in older females allows spindle fibers from each pole to attach to a chromosome. The smaller separation between kinetochores in younger females may help to prevent this error. **15.** analyze There is a sharp increase in the percentage of aneuploid eggs in mice between 12 and 15 months of age. Women in their forties see a sharp increase in aneuploid children. Therefore, a 15-month-old mouse would correspond roughly to a woman in her mid-forties. **16.** analyze After examining the information presented in Figure 13.11, you can tell the woman that while she's right about having a greater chance of having a child with Down syndrome than if she had conceived at a younger age, her chances are still roughly 59 out of 60 of having a child without Down syndrome. You can say this because a 42-year-old woman has a 1-in-60 chance of having a child with Down syndrome.

Chapter 14

IN-TEXT QUESTIONS AND EXERCISES

p. 298 CYU (1) understand Blending inheritance hypothesis: Offspring would have intermediate-sized necks. Acquired characters hypothesis: Offspring would have longer necks than either parent. **(2)** apply Pea plants have a relatively short generation time, can be grown inexpensively in a small area, and produce many offspring, which is essential in analyzing genetic hypotheses. Disadvantages of elephants as a model organism include a long generation time and small numbers of offspring, not to mention their immense size, which would make studies with them exhorbitantly expensive and impractical.

p. 299 apply A cross of two hybrids of the same type involves a cross of a male and a female; self-fertilization involves male and female gametes that should contain exactly the same genetic determinants as those from the cross between hybrids. Thus, it makes no difference whether the male and female gametes come from the same individual or different individuals if the individuals are the same type of hybrid.

p. 299 Fig. 14.3 evaluate An experiment is a failure only if you don't learn anything from it. That is not the case here.

p. 302 Making Models 14.1 apply See Figure A14.1. All three Punnett squares show a 1 *RR*: 1 *Rr* ratio of genotypes.

p. 302 evaluate Filling in the top and side of a Punnett square requires writing out the types of gametes. To do this, you have to understand how different forms of a gene (alleles) separate from one another—as described by the principle of segregation. The offspring phenotype ratios will be 1 round : 1 wrinkled; the offspring genotype ratios will be 1 *Rr* : 1 *rr*.

p. 302 CYU (1) evaluate Mendel came to this conclusion for two reasons: (a) The results could not be explained if each individual had only one copy of a gene. With only a single gene for each trait, the offspring would inherit one of the parental genotypes, and it would lead to a mix of F_1 offspring phenotypes, depending on which parental allele was inherited. The reappearance of the recessive trait in 1/4 of the F_2 offspring could also not be explained if offspring had inherited only one allele of each gene from one or the other of their parents. (b) While more complicated models with many copies of each gene might possibly be fit to the data, he applied the principle that the simplest model is the most likely explanation for the results. **(2)** evaluate If there was a biased transmission of one allele over another, say the dominant allele was present in a greater proportion of gametes than the recessive allele, then Mendel would have obtained ratios different from a 3 : 1 ratio. Only when there is a 1 : 1 mix of gametes with the dominant and the recessive alleles will there be a 3 dominant: 1 recessive ratio.

p. 303 Making Models 14.2 apply The square on the right is wrong because it shows alleles for only one kind of trait in each gamete. See Figure A14.2 for the *PPRr* × *PPRr* cross.

p. 303 Making Models 14.3 analyze See Figure A14.3. (Notice that there are two identical copies of the *P* allele, so *all* gametes will receive a copy of this allele.)

p. 305 CYU (1) apply Because independent assortment considers the independence of the segregation of alleles of two or more distinct genes. At least two genes need to be considered, not one. **(2)** evaluate In a testcross, one of the parents is always a homozygous recessive strain. The only type of gamete produced by this strain contains recessive alleles of every gene considered in the cross. Given this, coupled with the finding that an *RrYy* strain produces 4 gamete genotypes in equal proportions, indicates that 4 different offspring phenotypes, each in equal amounts, are expected. In contrast, self-fertilization of an *RrYy* strain involves 4 different gamete genotypes, each in equal amounts, present in both sperm and egg cells. With dominant and recessive alleles for each gene, the offspring phenotypes will occur in a 9 : 3 : 3 : 1 ratio.

p. 308 CYU (1) apply Independent assortment involves independent segregation of the alleles of two separate genes on separate chromosomes. Therefore, the simplest model of independent assortment requires drawing at least 4 chromosomes shown as two homologous pairs. Conversely, because segregation considers the separation of the two copies of a single gene carried on a single type of chromosome, only the two chromosomes of one homologous pair need to be drawn. **(2)** apply See Figure A14.4.

p. 309 Fig 14.11 apply The four genotypes would be X^Y/X^w, X^y/X^W, X^Y/X^W, and X^y/X^w.

p. 310 Fig.14.12 apply Neither hypothesis is fully supported. However, linked genes do not assort independently but instead tend to be inherited together.

p. 310 Fig. 14.13 apply This observation is most likely due to sampling (or random) error. Even though a 1 : 1 ratio is expected, some variation occurs when sample size is less than infinite. Another explanation is that these genotypes differ in fitness: White-eyed, yellow-bodied flies have a survival advantage over red-eyed, gray-bodied flies. These two explanations are not mutually exclusive.

p. 311 Fig. 14.15 analyze The gene colored orange is *ruby eyes*; the gene colored blue is *miniature wings*.

p. 314 Fig. 14.18 analyze In this case, the 9 : 3 : 3 : 1 ratio comes from four different forms of one trait (comb shape), whereas in a standard dihybrid cross, the four different phenotypes come from two different traits that each come in two different forms.

p. 315 Fig. 14.20 analyze Very light or very dark wheat kernels are produced by a small number of genotypes, but intermediate-colored kernels are produced by a much larger number of genotypes. For example, white kernels are produced by only 1 genotype (*aabbcc*), but intermediate-colored kernels are produed by 7 different genotypes, many of which are predicted to occur relatively frequently in self-fertilization of an F_1 individual heterozygous for all three genes.

p. 316 CYU (1) apply The form of dominance of an allele is relative to that of another particular allele. This means that if one allele is paired with another, say I^A with *i*, I^A may be dominant to *i*, but if I^A is paired with a different allele, say I^B, I^A may be codominant with I^B. **(2)** create There are many ways to construct this study, but the basic idea is to set up two different situations: produce through crosses a set of individuals that share the same genotype at the gene in question (for example, *aa*) but are genetically variable at other genes. Raise and maintain these individuals in the same environment. If there are differences in phenotype, then these differences are likely due to gene interaction. Contrast this experimental strategy with one in which individuals are genetically identical across all genotypes and are raised and maintained in a set of different environments. If there is strong variation in phenotype in this case, then differences in phenotype are due to environmental effects. Gene interaction and environmental effects are not mutually exclusive, and both may play roles

Figure A14.1

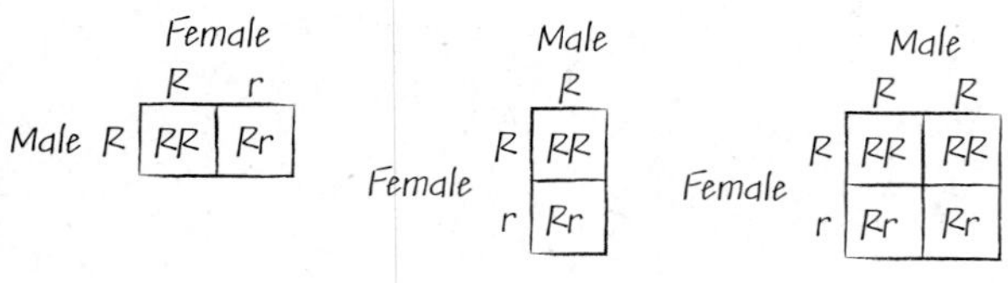

Figure A14.2

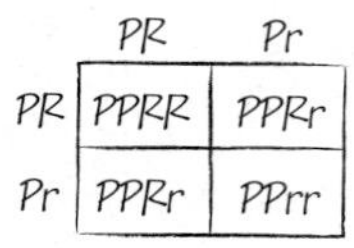

Figure A14.3

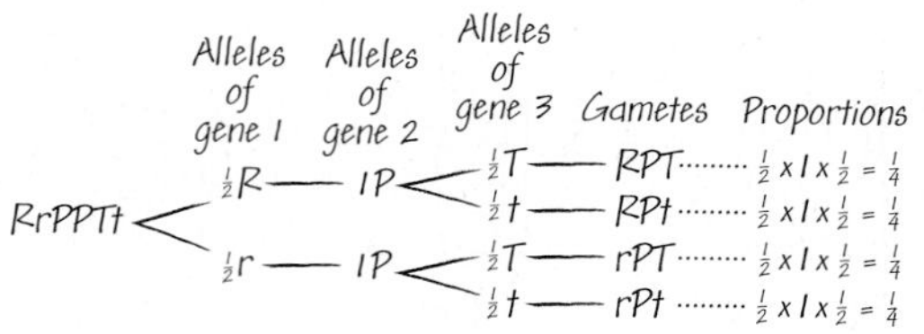

in the phenotypic variation of individuals who share a genotype at one gene. **(3)** analyze The lightest (white) and darkest color phenotypes are each associated with single genotypes (*aabbcc* or *AABBCC*). Given that the F_2 was produced by self-fertilization of a fully heterozygous F_1 (*AaBbCc*), 8 gamete types are expected to be produced with independent assortment. One of these gamete types is *abc*, and another is *ABC*. The chance that a sperm and an egg carrying identical gamete genotypes come together at fertilization to produce either of the *aabbcc* or *AABBCC* genotypes is given by the product of the gamete frequencies, or $1/8 \times 1/8 = 1/64$.

p. 318 Fig. 14.22 analyze A heterozygous female and a color-blind male. 1 color-blind male : 1 color-blind female.

p. 318 CYU (1) evaluate X-linked recessive. (Unaffected parents can have affected children [this indicates the trait is recessive], and only males express the trait [this makes the trait likely to be X-linked recessive]). **(2)** analyze There is a 2/3 chance that the first unaffected daughter in generation III is a carrier. (The possibility of this daughter being homozygous recessive is eliminated because she does not express the trait. This leaves a 1/3 chance that she is homozygous dominant and a 2/3 chance that she is heterozygous.) There is a 1/2 chance that the new child will be a carrier. (The probability is different in this case because the phenotype is unknown. Because both parents are heterozygous, any future child has a 1/2 chance of being heterozygous for the recessive allele.)

✔ TEST YOUR KNOWLEDGE

1. understand a **2.** understand b (Don't make the mistake of thinking that alleles must be recessive if they are either harmful or rare.) **3.** understand a (Remember that dominance and recessiveness are relative terms. Only by comparing the phenotype of a particular heterozygote to phenotypes of the two different homozygotes can dominance and recessiveness be learned. This is impossible to do with a haploid organism.)

✔ APPLY PROBLEM-SOLVING STRATEGIES

4. analyze **Example Solution** Here you are given parental and offspring phenotypes and then asked to infer the parental genotypes. As a starting point, assume that the coat colors are due to the simplest genetic system possible: one autosomal gene with two alleles, where one allele is dominant and the other recessive. Because female II produces only black offspring, it's logical to suppose that black is dominant to brown. Let's use B for black and *b* for brown. Then the male parent is *bb*. To produce offspring with a 1 : 1 ratio of black : brown coats, female I must be *Bb*. But to produce all black offspring, female II must be BB. This model explains the data, so you can accept it as correct.

5. analyze **Example Solution** The near 9 : 3 : 3 : 1 ratio of offspring indicates that this is a dihybrid cross. Starting with this hypothesis, let *O* stand for the allele for orange petals and *o* the allele for yellow petals; let *S* stand for the allele for spotted petals and *s* the allele for unspotted petals. The spotted to unspotted plants are roughly in a 3 : 1 ratio, and the orange to yellow plants are also roughly in the same 3 : 1 ratio. Therefore, start with the hypotheses that *O* is dominant to *o*, that *S* is dominant to *s*, that the two genes are found on different chromosomes so they assort independently, and that the parent's genotype is *OoSs*. If you do a Punnett square for the *OoSs* × *OoSs* mating, you'll find that progeny phenotypes are predicted to be in a 9 : 3 : 3 : 1 proportion. This is very close to what's observed.

✔ TEST YOUR PROBLEM-SOLVING SKILLS

6. apply The ratio of genotypes in the F_1 is expected to be 1 *Pp* : 1 *pp* and the ratio of phenotypes is expected to be 1 purple flower : 1 white flower. The F_2 offspring phenotypes are expected to be in a 3 purple : 1 white ratio. **7.** apply *Ss* for both parents. **8.** apply Your answer to the first three parts should conform to the F_1 and F_2 crosses diagrammed in Figure 14.5a, except that different alleles and traits are being analyzed. The recombinant gametes would be *Yi* and *yI*. **9.** analyze Albinism indicates the absence of pigment, so let *b* stand for an allele that produces the absence of blue pigment and *y* for an allele that produces the absence of yellow. If blue and yellow pigment blend to give green, then both green parents are *BbYy*. The green phenotype is found in *BBYY, BBYy, BbYY*, and *BbYy* offspring. The blue phenotype is found in *BByy* or *Bbyy* offspring. The yellow phenotype is observed in *bbYY* or *bbYy* offspring. Albino offspring are *bbyy*. The phenotypes of the offspring should be in the ratio 9 : 3 : 3 : 1 as green : blue : yellow : albino. **10.** analyze *Cross 1 genotypes:* non-crested (*Cc*) × non-crested (*Cc*) = 22 non-crested (*CC* or *Cc*), 7 crested (*cc*). *Cross 2 genotypes:* crested (*cc*) × crested (*cc*) = 20 crested (*cc*). *Cross 3 genotypes:* non-crested (*Cc*) × crested (*cc*) = 7 non-crested (*Cc*), 6 crested (*cc*). Non-crested (*C*) is the dominant allele. **11.** evaluate Let *D* stand for the normal allele and *d* for the allele responsible for Duchenne-type muscular dystrophy. The woman's family has no history of the disease, so her genotype is almost certainly *DD*. The man is not afflicted, so he must be *DY*. (The trait is X-linked, so he has only one allele; the "Y" stands for the Y chromosome.) Their children are not at risk. The man's sister could be a carrier, however—meaning she has the genotype *Dd*. If so, then half of the second couple's male children are likely to be affected. **12.** apply Your stages of meiosis should look like a simplified version of Figure 13.6, with $2n = 4$ instead of $2n = 6$. The *A* and *a* alleles could be on the red and blue versions of a long chromosome, and the *B* and *b* alleles could be on the red and blue versions of a shorter chromosome, similar to the way the genes associated with sickle cell disease and cystic fibrosis are shown in Figure 13.8. The *A* and *B* genes can be located anywhere on their respective chromosomes, but each chromosome should carry only one allele. Each pair of red and blue chromosomes is a homologous pair. Sister chromatids bear the same allele (e.g., both sister chromatids of the long blue chromosome might bear the *a* allele). Chromatids from the longer and shorter chromosomes are not homologous. There would be 4 types of gametes: 1 *AB* : 1 *Ab* : 1 *aB* : 1 *ab*. To identify the events that result in the principles of segregation and independent assortment, see Figures 14.7 and 14.8 and substitute *A, a* and *B, b* for *R, r* and *Y, y*. **13.** understand F (linked genes are separated by crossing over in a fraction of meioses); F (genetic map distances measure the frequency of crossing over between a pair of genes, not nucleotides between genes); T (this is the basis for genetic mapping); T. **14.** analyze According to Mendel's model, palomino individuals should be heterozygous at the locus for coat color. If this is so, if you mated palomino individuals, you would expect to see a 1 : 2 : 1 combination of chestnut, palomino, and cremello offspring. If blending inheritance occurred, however, all the offspring should be palomino.

15. analyze Because this is an X-linked trait, the father who has hemophilia could not have passed the trait on to his son. Therefore, the mother in couple 1 must be a carrier and must have passed the recessive allele on to her son, who is XY and affected. To educate a jury about the situation, you should draw what happens to the X and Y during meiosis and then make a drawing showing the chromosomes in couple 1 and couple 2, with a Punnett square showing how these chromosomes are passed to the affected and unaffected children. **16.** analyze The curved-wing allele is autosomal recessive; the lozenge-eyed allele is sex-linked (specifically, X-linked) recessive. Let *L* be the allele for long wings and *l* be the allele for curved wings; let X^R be the allele for red eyes and X^r the allele for lozenge eyes. The female parent is LlX^RX^r; the male parent is LlX^RY. **17.** analyze The chance that their first child will have hemophilia is 1/2. This is because all sons will have the disease and the chance of having a firstborn son is 1/2. The chance of having a carrier as their first child is also 1/2. This is because all daughters and none of the sons will be carriers, and there is a 1/2 chance of having a firstborn daughter. (Recall that males cannot carry an X-linked recessive trait—with only one X chromosome, males either have the trait or do not.)

Figure A14.4

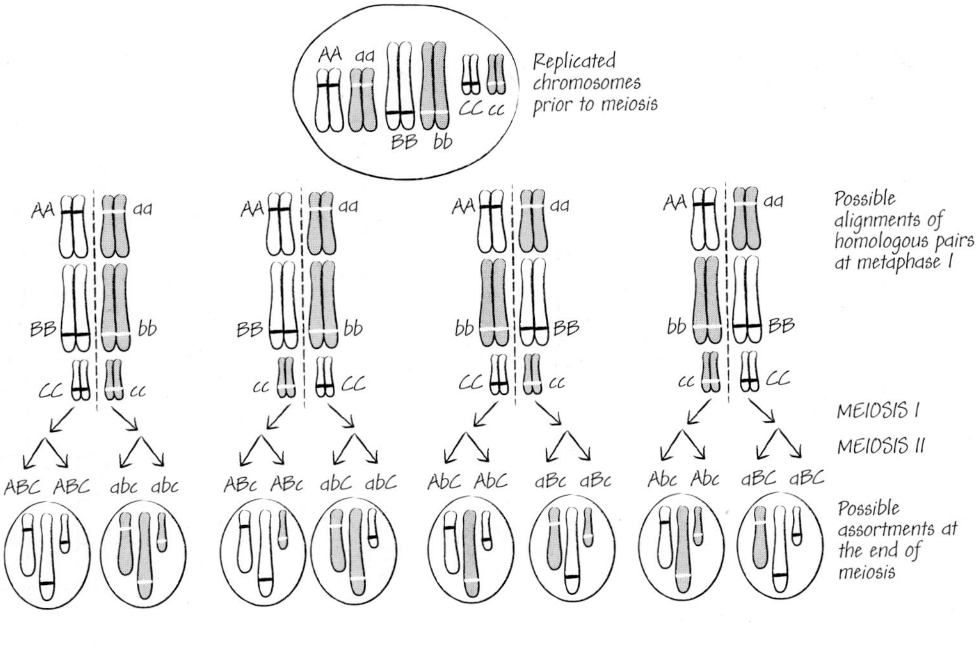

✔ PUT IT ALL TOGETHER: Case Study

18. apply a (Recall that pleiotropy is the ability of a single gene to affect more than one trait, something that's seen in ALD.) **19.** analyze c (Note that the trait skips a generation, only males show the trait, and about half the sons of a woman with affected offspring show the trait.) **20.** apply c (Unless there is new mutation, the female in the first generation must have been a carrier who transmitted the ALD allele to her daughter.) **21.** apply 1 (It's a certainty because ALD is an X-linked recessive disorder. Therefore, the son of a man affected by ALD will get a Y chromosome from his father and an X chromosome from his mother. Since the mother has no family history of ALD, she can be assumed to be homozygous for the wild-type, non-ALD allele of the gene.) **22.** apply The color-blind son without ALD must have inherited an X chromosome from his mother with the color blindness allele and a wild-type allele of the ALD-associated gene. His mother has the color blindness allele and an acquired ALD-causing allele on one of her X chromosomes and wild-type alleles of these genes on the other X chromosome. If a crossover occurred in this woman between the loci (sites on the chromosome) of the color blindness and ALD associated genes, this would produce two recombinant X chromosomes. One of these would have the color blindness allele and a wild-type allele at the ALD locus. The son inherited this recombinant X chromosome, and that is why he is color-blind but without ALD. **23.** apply "Any child you have has a 1/8 chance of ALD. However, there are two parts to the story: If you have a daughter, she will not have ALD. If you have a son, there is a 1/4 chance that he will have ALD. Therefore, there's a 1/8 chance that your firstborn child will have ALD." (This advice comes from knowing that the mother of this daughter must be a carrier of ALD because she has affected sons. Thus, there is a 1/2 chance that this mother transmitted the ALD allele to her daughter. If this daughter does have the allele, she must be heterozygous. In that case, none of her daughters will have ALD, but 1/2 of her sons are likely to have it. To calculate the overall chance that the first child of this couple will have ALD, you need to consider that the woman planning to have children has a 1/2 chance of being a carrier, that she has a 1/2 chance of having a son as a first child, and that 1/2 of sons born to this carrier female will have ALD. Multiplying these probabilities reveals a 1/8 chance that the first child of this couple will have ALD. See **BioSkills 4** on how to work with probabilities.)

Chapter 15

IN-TEXT QUESTIONS AND EXERCISES

p. 324 evaluate This finding would have indicated that the virus injected both DNA and protein into the cell. This result would force a reconsideration of the model that only DNA entered the cell during T2 infection and would have made it impossible to distinguish between protein and DNA as the genetic material.

p. 325 CYU (1) evaluate One explanation is that the procedure used to remove viral capsids was not perfectly efficient, allowing some of the viral capsids that contain radioactive sulfur in the proteins to remain attached to the cells. **(2)** analyze Because a molecule is a group of atoms held together with chemical bonds, each separate strand of DNA meets this definition, as do the two hydrogen-bonded strands of DNA. **(3)** understand The 5′ end of a DNA strand has a phosphate group attached to one of the carbons of the sugar that is at one end of the chain, and the 3′ end of a DNA strand has a hydroxyl (—OH) group attached to a different carbon of the sugar that is at the other end of the chain.

p. 326 CYU (1) understand In the conservative replication hypothesis, the parental double-stranded DNA molecule would remain intact after replication and a daughter DNA molecule would be created with both strands new. By contrast, in the semiconservative replication hypothesis (the correct hypothesis), each of the DNA molecules found after replication would contain one parental (old) strand and one daughter (newly replicated) strand of DNA. **(2)** understand The mass difference allowed the researchers to distinguish DNA strands before any rounds of replication began from DNA strands that were replicated after the study started. Making this distinction is at the heart of the Meselson–Stahl experiment.

p. 327 Fig. 15.5 analyze The same two bands should appear; but the upper band (DNA containing only ^{14}N) should get wider and more visible, and the lower band (hybrid DNA) should get narrower and less visible, because each succeeding generation has relatively less heavy DNA.

p. 330 apply Helicase, topoisomerase, single-stranded DNA-binding proteins, primase, the sliding clamp, and DNA polymerase are all required for leading-strand synthesis. If any one of these proteins is nonfunctional, DNA replication will not occur.

p. 331 apply See **Figure A15.1**.The leading strand would be continuous if either DNA polymerase I or DNA ligase were defective. However, on the lagging strand, if (a) DNA polymerase I was defective, there would be many unjoined Okazaki fragments that begin with RNA primers (because DNA polymerase I works to remove these primers); and if (b) ligase was defective, the lagging strand would have Okazaki fragments without RNA primers but with nicks separating each fragment (because ligase works to join Okazaki fragments).

p. 332 Fig. 15.11 apply About 2300 seconds. This comes from knowing that replication proceeds bidirectionally, so replication from each fork will replicate half the chromosome. This is 4.6 million base pairs ÷ 2 = 2.3 million base pairs. At 1000 base pairs per second, this requires 2.3 million base pairs ÷ 1000 base pairs per second = 2300 seconds.

p. 333 CYU (1) apply A loss of DNA helicase activity would be expected to cause an immediate arrest of the replication fork, because helicase is responsible for opening the two strands of DNA. However, the arrest of the replication fork from loss of topoisomerase wouldn't be far behind—the replication fork would soon stop moving as the DNA strands ahead of the replication fork became twisted. **(2)** understand A combination of two things causes this: DNA has antiparallel strands, and DNA can only be synthesized in the 5′ → 3′ direction. When these factors are put together, then one of the replicating strands follows the replication fork and the other moves away from it. **(3)** understand The leading (continuous) strand is a long strand that requires only one primer, but the lagging (discontinuous) strand is made of a set of short DNA fragments (Okazaki fragments), each of which requires an RNA primer and therefore primase.

p. 334 Fig. 15.12 analyze The left-hand end of this DNA would be the mirror image of the right-hand end of the molecule at the bottom of Figure 15.12, with a 3′ single-strand overhang extending to the left. (This is because neither end of double-stranded DNA can be replicated to the end; the strand that becomes shorter is

Figure A15.1

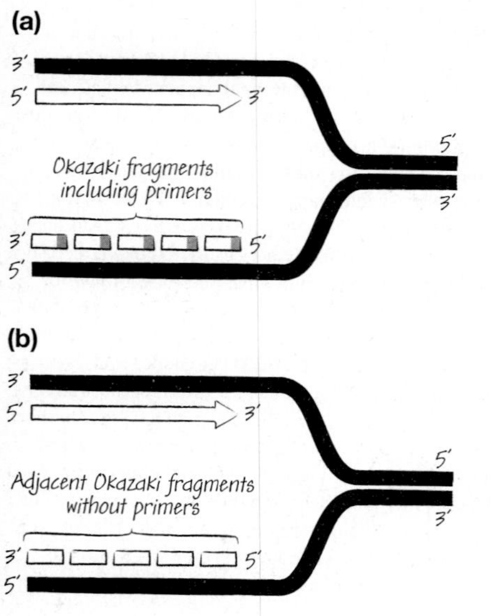

Figure A15.2

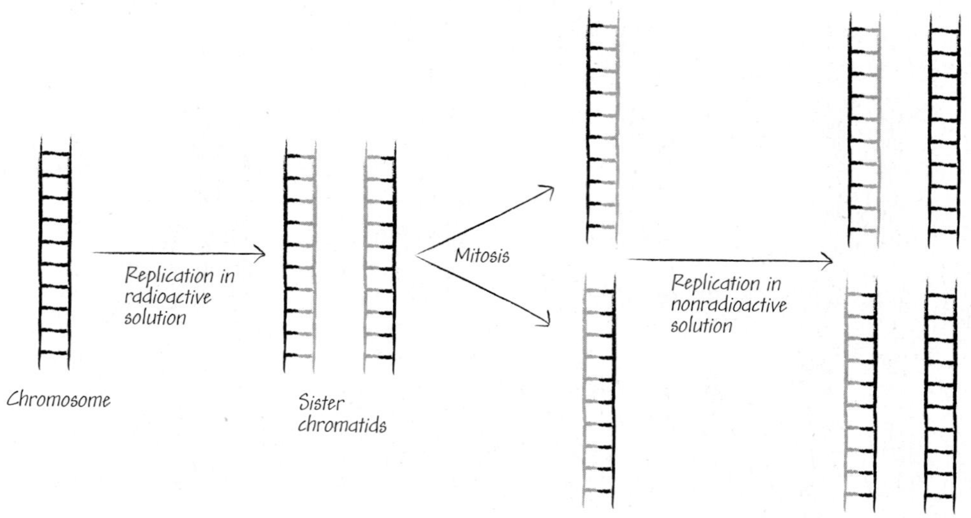

reversed on each end due to the antiparallel orientation of the two DNA strands.)

p. 335 CYU (1) understand The strand with a 3′ end cannot be replicated to its end because when it is used as a template, the complementary newly synthesized strand must be made in a direction that moves from the region near the 3′ end toward the other end of the chromosome. Given that an RNA primer is needed to start DNA synthesis, this means that this template strand isn't replicated to its end. In contrast, when the strand terminating with a 5′ end is used as a template, the newly replicated strand grows toward the end of the chromosome and can be extended all the way to the 5′ end. **(2)** understand The short repetitive DNAs come from repetitively copying the RNA template in telomerase that codes for this repeat. Doing this over and over creates the many copies.

p. 338 CYU (1) analyze The mutation rate would increase because there would be no way to distinguish the strand with the error and the strand to use as a template in correcting this error. **(2)** apply The mutation rate would increase because DNA repair systems use distortions in the DNA helix to recognize where there's a problem in the DNA sequence. **(3)** apply This evidence comes from human DNA repair deficiencies, notably xeroderma pigmentosum (XP). People with XP cannot effectively repair UV light damage and have a vastly elevated cancer risk. The implication is that normally functioning repair systems protect against cancer.

✔ TEST YOUR KNOWLEDGE

1. understand c **2.** remember b **3.** remember c **4.** remember 5′ → 3′

✔ TEST YOUR UNDERSTANDING

5. analyze Labeling either DNA or proteins. **6.** analyze T, T, T, F **7.** understand Telomerase binds to the overhang at the end of a chromosome. Once bound, and using its RNA template as a guide, it begins catalyzing the addition of deoxyribonucleotides to the overhang in the 5′ → 3′ direction, lengthening the overhang. This creates a single strand of DNA that is then used as a template on which primase, DNA polymerase, and ligase add deoxyribonucleotides to the lagging strand in the 5′ → 3′ direction, restoring the lagging strand to its original length. **8.** analyze d (The regularity of DNA's structure allows irregularities such as damaged bases to stand out and be recognized by DNA repair proteins.)

✔ TEST YOUR PROBLEM-SOLVING SKILLS

9. analyze (a) In Figure A15.2, the gray lines represent DNA strands containing radioactivity. (b) After one round of replication in radioactive solution, one double-stranded DNA would be radioactive in both strands and the other would not be radioactive in either strand. After another round of DNA synthesis, this time in nonradioactive solution, one of the four DNA molecules would be radioactive in both strands and the other three DNA molecules would contain no radioactivity in any strand.

10. analyze (a)The double mutant of *uvrA* and *recA* is most sensitive to UV light, the single mutants are in between, and the wild type is least sensitive. (b) The *recA* gene contributes more to UV light repair through most of the UV dose levels. But at very high doses, the *uvrA* gene is somewhat more important than the *recA* gene.

✔ PUT IT ALL TOGETHER: Case Study

11. analyze b (For an antibiotic to be effective, it has to spare some cells and harm others. If ciprofloxacin killed either eukaryotic cells alone or all types of cells, it would act as a powerful poison in people being treated. An inhibitor of telomerase would not block DNA replication, and it would have no effect on bacteria because they do not have this enzyme.) **12.** apply See Figure A15.3. **13.** apply Ligase activity. The presence of DNA fragments suggests that DNA gyrase cut DNA but was unable to ligate (join) it again. **14.** analyze Naladixic acid. (This can be concluded because the plot for naladixic acid's effect on DNA synthesis rate falls most rapidly—it's the steepest drop.) **15.** analyze Roughly twice as effective. (You can discover this by estimating the relative rates of DNA synthesis 10 minutes after addition of naladixic acid [~25 percent] and norofloxacin [~50 percent].) That the rate of DNA synthesis with naladixic acid is half the rate of norofloxacin suggests that naladixic acid is twice as effective as norofloxacin at inhibiting DNA synthesis. **16.** create There's no one right answer. What you need to communicate is that knowing how this "wonder drug" works requires understanding something about DNA's structure and how it is copied in bacteria. This includes the double-helical structure of DNA; the need to separate its strands during replication; the consequent introduction of twists that need to be relaxed; the enzyme called DNA gyrase that keeps DNA from being hopelessly knotted; and the fact that ciprofloxacin stops DNA gyrase from working and in that way blocks DNA synthesis and stops the deadly bacteria from growing. Because this is an article for a general audience, you need to keep it interesting and accessible.

Chapter 16

IN-TEXT QUESTIONS AND EXERCISES

p. 342 CYU (1) evaluate Srb and Horowitz's job would been complicated. Their results would show that two genes were needed for each enzyme. Unless they knew they were working with multi-polypeptide enzymes, and which gene coded for which polypeptide, they might not have come to a simple interpretation of what genes do. **(2)** create One definition is that a gene codes for one or more related RNAs or polypeptides.

p. 343 Fig. 16.1 apply (1) No, because it could not make citrulline from ornithine without enzyme 2. (2) Yes, because it would no longer need enzyme 2 to make citrulline.

p. 344 Fig. 16.3 analyze Accounting for separation in eukaryotes of where DNA is housed (the nucleus) and where proteins are synthesized (the cytoplasm) demands some intermediate or messenger molecule to carry information in DNA to where the information is used in protein synthesis. In bacteria, which lack a nucleus, there isn't a clear physical separation between where DNA is located and where proteins are synthesized. This means there isn't an obvious demand to hypothesize some carrier of information—mRNA—between DNA and the site of protein synthesis.

p. 346 CYU (1) understand Messenger RNA (mRNA) is an apt name because this RNA carries the information in DNA to where it's used in protein synthesis.**(2)** understand The arrow between DNA and RNA represents information being transferred from one molecule (DNA) to another (mRNA) during transcription. The arrow between RNA and protein represents information in one molecule (mRNA) being used to direct protein synthesis in the process of translation. **(3)** understand By explaining how information in DNA (genotype) can be used to direct the synthesis of proteins, which in turn are responsible for phenotypes of the cell and organism.

p. 348 Fig. 16.6 apply See Figure A16.1.

p. 348 analyze **(1)** The codons in Figure 16.4 are CGC, AAC, CUG (part a) and UGC, AAC, CUG (part b); according to the codon table, they are translated correctly. **(2)** There are many possibilities (just pick alternative codons for one or more of the amino acids); here is one mRNA sequence (running 5′ → 3′): 5′-GCG-AAC-GAU-UUC-CAG-3′. To get the corresponding DNA sequence, write this sequence but substitute T's for U's: 5′-GCG-AAC-GAT-TTC-CAG-3′. Now write the complementary bases, which will be in the 3′ → 5′ direction: 3′-CGC-TTG-CTA-AAG-GTC-5′. When this second strand is transcribed by RNA polymerase, it will produce the given mRNA, with the proper 5′→ 3′ orientation.

p. 348 CYU (1) analyze They would have found that one-base-pair insertions or deletions would disrupt protein function because of changing the reading frame, but that adjacent two-base-pair insertions or deletions would not destroy protein function because the reading frame would be maintained. **(2)** apply

Figure A15.3

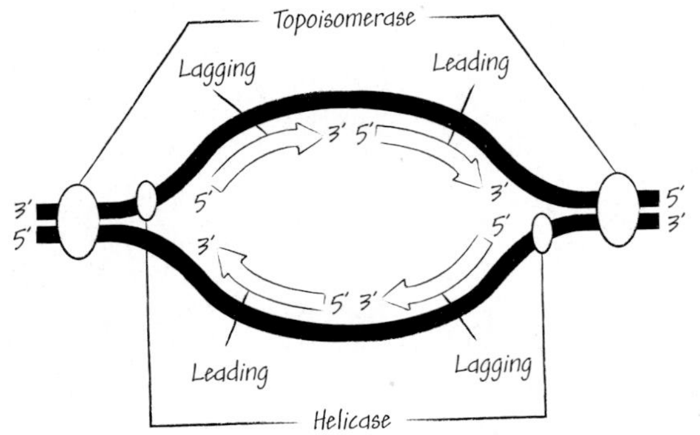

Figure A16.1

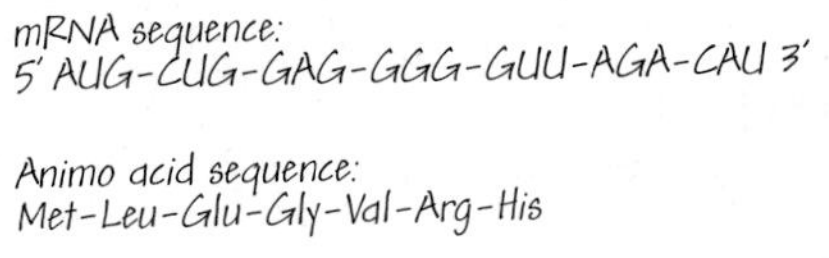

5′-UAUCCAUGGCACUUUAAAC-3′ (AUG is start; UAA is stop.) **(3)** apply Six. (This is because Met and Trp are coded by one codon, Cys is coded by two different codons, and Stop is encoded by three different codons. Therefore, $2 \times 3 = 6$ different mRNA sequences are possible.)

p. 350 apply The ATA → ATC mutation is a silent mutation because both sequences specify isoleucine. The ATA → TTA mutation is a nonsense mutation because it creates a stop codon. The ATA → ACA mutation is a missense mutation because the original sequence specifies isoleucine and the mutated sequence specifies threonine.

p. 351 Fig. 16.9 analyze Chromosomes 2, 3, 6, 10, 13, 14, 15, 18, 19, 21, 22, and the X chromosome show aneuploidy. Virtually every chromosome has structural rearrangements, and translocations are the most obvious. These are seen when two or more different colors occur on the same chromosome.

p. 351 CYU (1) analyze Position is predicted to play a larger role in the case of nonsense mutations because a nonsense mutation near the start of a reading frame is certain to destroy protein function; a nonsense mutation near the end of the reading frame may not affect protein function, because most of the polypeptide is properly specified. In missense mutations, position is unlikely to have any effect. **(2)** apply One way is for the mutation to be a silent mutation that changes the codon sequence but not the amino acid that's specified. Another way is for the mutation to be a missense mutation that changes the specified amino acid to one that has no effect on protein function. **(3)** understand A mutation is defined as a heritable change in DNA sequence, so a chromosome structural change will obviously change DNA sequence, given that chromosomes contain a characteristic sequence of DNA. Changes in chromosome number affect the number of copies of the DNA sequence. Because this change is heritable, it is a mutation.

✔ TEST YOUR KNOWLEDGE

1. understand a **2.** remember a **3.** understand The sequence of bases codes the information needed to create proteins. **4.** understand d.

✔ TEST YOUR UNDERSTANDING

5. evaluate Your friend has a point because it's primarily the set of proteins in a given cell at a given time that determine phenotype. But she's missing the idea that maintaining these proteins or adjusting their levels in response to changing conditions is impossible unless the information needed for their synthesis is stored in DNA. **6.** analyze An advantage of having more than the minimum number of codons is error tolerance. Because of the redundancy in an expanded code, not every base-pair change will change the meaning of a codon. **7.** apply T; base pairs can be changed anywhere. F; because a reading frame is the proper register of codons, frameshift mutations must occur in a reading frame, or protein-coding sequence of DNA. F; "neutral" refers to the effects on fitness of a mutation, and this does not depend on the degeneracy of the genetic code. F; deleterious mutations can occur anywhere a functional DNA sequence exists, such as noncoding sequences involved in regulation. **8.** understand The problem lies in the second part of the statement. All mutations do change the genotype, if genotype is defined as the genetic constitution of an individual. However, many mutations do not change phenotype (e.g., think of neutral and silent mutations).

✔ TEST YOUR PROBLEM-SOLVING SKILLS

9. analyze Substrate 3 would accumulate. Hypothesis: The individuals have a mutation in the gene for enzyme D. **10.** analyze c (This is because in a non-overlapping code, there are four codons [beginning with a first codon of AUG]. In a maximally overlapping code, the second codon would begin with the second base in the sequence, and the third codon would begin with the third base in the sequence. In this case, the codons would be AUG UGU GUU and so on to the end of the mRNA sequence. Following this pattern, there would be 10 complete codons in this mRNA sequence.)

✔ PUT IT ALL TOGETHER: Case Study

11. analyze b (A frameshift mutation will change the meaning of all codons that follow the mutation. Therefore, protein function is likely to be lost, and this creates a null allele. A missense mutation may or may not cause loss of the protein function, and neutral and silent mutations will not eliminate protein function.) **12.** analyze Transcription. This is because the end product of transcription is RNA, and no RNA is produced in individuals homozygous for the light-color allele. (The actual answer is a bit more complicated because nonsense mutations can destabilize mRNA; the light-colored mutant allele actually does have a nonsense mutation in the coding region that leads to destruction of the mRNA.) **13.** analyze Because there is a change from Ala to Thr in the human protein, the sequence of the non-template strand must have been altered from a G in the first position (a GC followed by any base codes for Ala) to an A in the first position (AC followed by any base codes for Thr). The base-pair change must have been from a GC base pair to an AT base pair in the DNA coding for first position of the codon. **14.** apply No, because the redundancy of the genetic code makes it possible for different codons to specify the same amino acids. (In fact, comparisons of amino acid and base sequences for related genes show that base sequences vary more than amino acid sequences.) **15.** create One of several plausible explanations is that just as for the mainland and beach mice, a mutation must have occurred in the population ancestral to modern Europeans either before or after their migration to Europe. This mutation gave a reproductive advantage to people who carried it in the environments of Europe. For this reason, over generations, the allele became more and more frequent until it predominated in this population. **16.** create There are many possible approaches, but all should center on the fact that the underlying genetics of a significant human trait, skin color, was revealed by studies of a model organism, the zebrafish.

Chapter 17

IN-TEXT QUESTIONS AND EXERCISES

p. 355 Fig. 17.1 understand RNA and DNA strands are antiparallel. The DNA template is "read" 3′ → 5′ (while RNA is being synthesized 5′ → 3′).

p. 358 CYU (1) analyze Terminator sequences require the formation of a hairpin loop in the RNA. Any DNA sequence that creates two complementary sequences in RNA separated by a short non-complementary sequence will create a hairpin loop, and a huge number of DNA sequences will do this. Promoters, in contrast, must have a specific sequence that allows a sigma protein to bind. For this reason, promoters can have only a limited number of sequences. **(2)** analyze Like sigma, the general transcription factors recognize the promoter sequence but are not directly involved in the synthesis of RNA, which is the role of the bacterial core enzyme.

p. 359 Fig. 17.5 apply There would be no loops. This is because the molecules would match up exactly if the DNA lacked introns.

p. 360 CYU (1) apply The 5′ and 3′ UTRs, which do not code for amino acids, are either exons or parts of exons because they are present in the mRNA. **(2)** understand Base pairing between sequences in the primary transcripts and small nuclear RNAs (snRNAs) of snRNPs that make up the spliceosome allows recognition of the regions to be spliced. **(3)** The mRNA could not be used to start a new round of translation (because the ribosome couldn't begin translation without the cap), and the mRNA would be degraded from the 5′ end (because the protective cap is no longer present).

p. 361 Making Models 17.1 apply See **Figure A17.1**.

p. 362 CYU (1) understand Because polyribosomes occur when two or more ribosomes are simultaneously translating a single mRNA, this can occur in both the cytoplasm of a eukaryotic cell or a bacterial cell. In contrast, linked transcription and translation requires both of these processes to occur in the same place and at the same, something that's impossilble in a eukaryotic cell that transcribes and processes RNA in the nucleus, but translates RNA in the cytoplasm. **(2)** understand There is no clear way to match particular amino acids to particular codons, but there is an easy way to specifically read codons by having a base sequence in a hypothesized adapter molecule that is complementary to the codon being read.

p. 363 Fig. 17.10 analyze If the amino acids stayed attached to the tRNAs, the gray line in the graph would stay high and the green line low. If the amino acids were transferred to some other cell component, the gray line would decline but the green line would remain low.

p. 364 understand (1) The amino acid attaches on the 3′ end of the arm of the L-shaped structure that is farthest from the anticodon loop. (2) The anticodon is antiparallel in orientation to the mRNA codon, and it contains the complementary bases.

p. 364 CYU (1) understand Two parts of the tRNA let it act as an adapter: The anticodon can read codons, and the amino acid attachement site allows the tRNA to bring the amino acid that's specified by the codon to the

Figure A17.1

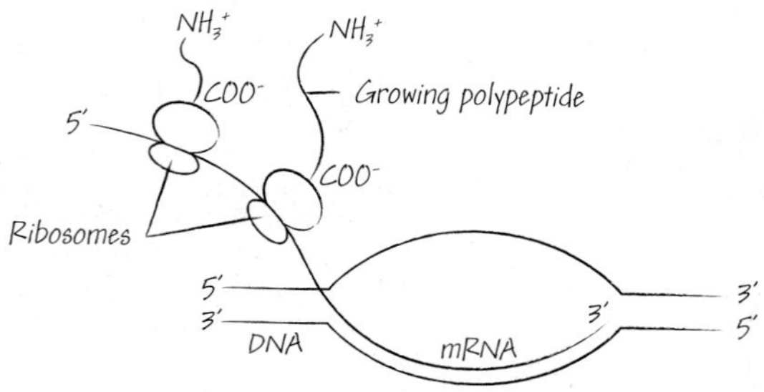

ribosome. tRNA essentially acts as a bridge that links codons to amino acids. **(2)** analyze Two. (Because there are about 40 different types of tRNA and 20 different aminoacyl transferases.) **(3)** analyze 61. (This answer comes from understanding that wobble pairing allows one tRNA to read more than one codon. Without wobble pairing, there would have to be one type of tRNA for each of the 61 amino acid-specifying codons.)

p. 369 CYU (1) analyze Having the initiator tRNA in the P site is important because if the polypeptide chain is to grow, the A site must be empty to allow the next aminoacyl tRNA to come into the ribosome. **(2)** understand After transfer of the polypeptide from the tRNA in the P site to the tRNA in the A site, the ribosome moves (translocates) one codon down the mRNA. The tRNAs remain bound to their codons as the ribosome moves. This event shifts the position of the tRNAs within the tRNA binding sites of the ribosome so that the tRNA that was in the A site is now in the P site. **(3)** analyze Because the release factor binds to the same A site in the ribosome normally occupied by an aminoacyl tRNA, the shape of the release factor and aminoacyl tRNA must be nearly identical to allow for a good fit.

✔ TEST YOUR KNOWLEDGE

1. remember c **2.** remember c (It's easy to confuse the start codon used in translation with the promoter used in transcription, and it's also easy to incorrectly think that the start codon is at the start—the 5′ end—of the mRNA. In fact, the start codon is within the mRNA at the very 3′ end of the 5′ untranslated region, 5′ UTR.) **3.** understand a **4.** apply b

✔ TEST YOUR UNDERSTANDING

5. apply b **6.** analyze F, F, T, T. **7.** analyze c (The reasoning is that sigma is needed only to initiate transcription. This mutant can continue elongation and termination but cannot start new transcripts. This effect is consistent with a loss of sigma activity at the elevated temperature.) **8.** analyze A promoter and a start codon are similar in that both are sequences of nucleic acids that are required to start important processes, and both determine the sites where the process will begin. The key differences are that the promoter is needed to start *transcription* and the start codon is needed to start *translation*. Also, the promoter is a DNA sequence only, and the start codon is a sequence found within an mRNA.

✔ TEST YOUR PROBLEM-SOLVING SKILLS

9. analyze Cordycepin has a structure very similar to a ribonucleoside triphosphate, yet it terminates transcription. Examination of cordycepin's structure reveals that it lacks the 3′ OH. This 3′ OH is critical to support the addition of another nucleotide to the growing RNA chain, so a reasonable prediction is that after cordycepin is incorporated as a ribonucleoside monophosphate into the RNA chain, its lack of a 3′ OH prevents the further addition of ribonucleotides. **10.** apply Twenty amino acids per second. The reasoning is that codons are 3 nucleotides long, so to match the rate of a polymerase moving at 60 nucleotides per second along the template DNA, the ribosome could translate up to 20 codons per second. At this rate, it would require 90 seconds to synthesize a polypeptide containing the 1800 amino acids that would be produced from an mRNA with 1800 codons (1800 amino acids ÷ 20 amino acids/sec = 90 seconds).

✔ PUT IT ALL TOGETHER: Case Study

11. understand b (Because α-amanitin is an inhibitor of transcription, it is expected to prevent RNA synthesis.) **12.** analyze The toxin could bind to any of the sites other than the catalytic site that are important for RNA polymerase to work in transcription. Examples of these are DNA entry and exit channels, the RNA exit channel, the nucleotide entry channel, and the protrusions shown within the polymerase that separate DNA strands and guide the exit of the RNA. **13.** apply d (Pre-mRNAs are transcribed by pol II, and this is the enzyme inhibited by α-amanitin. You can see this in the graph by the sharp fall in RNA pol II activity and the virtually flat line for RNA pol I and III.) **14.** analyze A bit less than 1 μM. Shutting down 95 percent of transcription means that transcription occurs at 5 percent of the normal level. If you trace a line from the *y*-axis at 5 percent until you intersect the curve showing pol II activity, then trace a line down from this point to intersect the *x*-axis, you'll see that the α-amanitin concentration is less than 1.0 μM. A reasonable estimate on this logarithmic scale would be ~0.6 μM. **15.** analyze It would be hard to follow how fast a pre-mRNA is spliced if new pre-mRNAs were constantly made during the study. Adding a toxin to block pre-mRNA production allows biologists to focus on the rate of splicing from start to finish. **16.** create Many possible experiments could be used to test this idea. One is to measure levels of protein synthesis in liver cells exposed to α-amanitin to learn if protein production is decreased. This trial would need to be followed by an experiment in which protein synthesis is caused to decrease in liver cells not exposed to α-amanitin, to see if those cells die just like the ones exposed to α-amanitin did. Treating the cells with a compound that blocks protein synthesis without affecting transcription or other processes could do this.

Chapter 18

IN-TEXT QUESTIONS AND EXERCISES

p. 376 Fig. 18.3 analyze β-galactosidase would be expressed in treatments 2 and 3. This is because glucose is the preferred carbon source, and for this reason, only glucose is predicted to inhibit expression of β-galactosidase when lactose is present.

p. 376 CYU (1) understand Given that proteins, the most common gene product, are produced in bacteria following transcription, translation, and often post-translational modifications, controlling whether a gene is or isn't transcribed introduces a long delay into the decision of whether an active protein is or isn't present. Regulating gene expression at the level of activating or inactivating an already synthesized protein (post-translational control) causes a near immediate response. Even so, transcriptional control provides a huge advantage by allowing cells to avoid investing the energetic and material resources needed to synthesize proteins poised for action but often never called to act. **(2)** understand This question relates to the idea of the previous one—a cell that always produces proteins, whether needed or not, is investing energy and material resources inefficiently—and in a competitive environment, inefficient cells lose.

p. 377 Making Models 18.1 apply See Figure A18.1. Your answer doesn't need to include all the labels and can show the DNA, the gene, DNA regulatory sequence, and repressor many different ways, so long as you can explain to others what your symbols mean.

p. 379 Fig. 18.7 understand Put the "repressor protein" on the operator. No transcription will take place. Then put the "RNA polymerase" on the promoter. No transcription will take place. Finally, put "lactose" on the repressor protein and then remove the resulting lactose–repressor complex from the operon. Transcription will begin.

p. 380 understand The same outcome can be achieved by either (a) preventing lactose from ever entering the cell (inducer exclusion) or (b) preventing high levels of transcription when the CAP–cAMP complex fails to form (even when the repressor is off the operator).

p. 381 CYU (1) understand In this way, the genes that encode proteins needed to metabolize lactose will be transcribed only when lactose is available to be metabolized. Otherwise, energy and resources, such as nucleotides and amino acids, would be wasted. **(2)** understand See Figure A18.2. **(3)** understand This is possible if operons for using sugars other than glucose have a binding site for CAP near their promoters. In this way, these operons would be transcribed efficiently only when the glucose level is low—exactly like the *lac* operon. **(4)** understand It makes sense to transcribe the *trp* operon, which functions to synthesize tryptophan, only when this amino acid is needed. Tryptophan works as a co-repressor to allow the *trp* repressor to bind to the operator, so transcription of *trp* operon genes is turned off when tryptophan is present, and transcription is turned on when the tryptophan level is low.

p. 382 CYU (1) analyze Because the same repressor protein is needed to regulate all genes in the regulon, a mutant

Figure A18.1

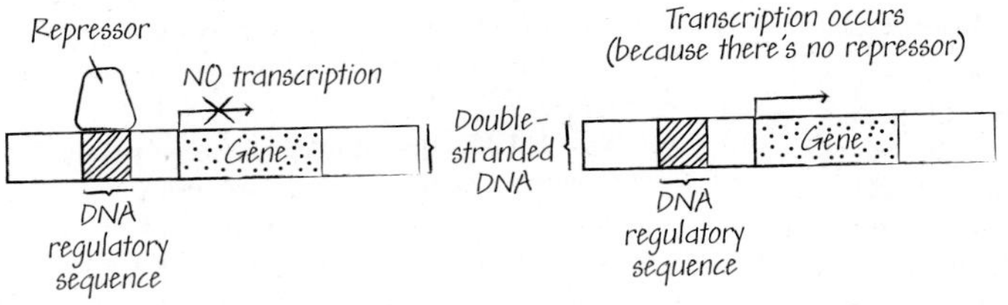

Figure A18.2

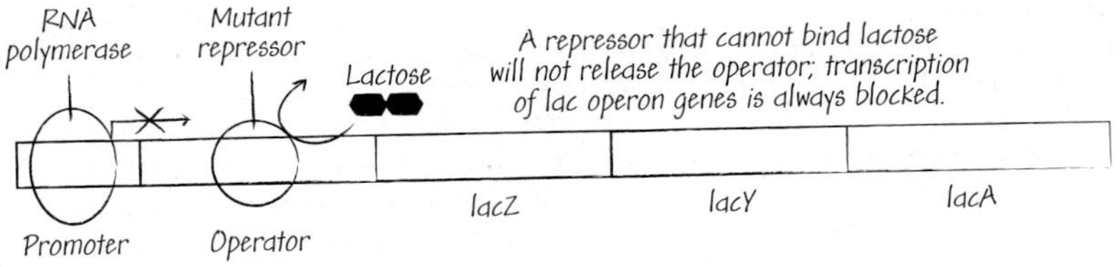

repressor would be unable to bind to any operator in the regulon. This result is predicted to affect the expression of every regulon gene. In contrast, mutation of a single operator should affect only the gene associated with that operator and no others. **(2)** analyze This allows negative feedback control of the SOS regulon. Having the *lexA* gene as part of the regulon allows for a high level of LexA repressor protein synthesis when there is extensive DNA damage. As the DNA damage is repaired, the LexA cleavage signal grows weaker, allowing the LexA protein to increase in concentration. This protein then turns off the SOS regulon genes. If the *lexA* gene were not part of the regulon, an indirect mechanism to restore the LexA repressor would be required.

✔ TEST YOUR KNOWLEDGE

1. remember d **2.** remember b **3.** remember DNA; RNA; **4.** remember a

✔ TEST YOUR UNDERSTANDING

5. apply F, F, F, T. **6.** Regulation of the *lac* operon should be normal. The location of the *lacI* gene isn't important, because the gene produces a protein that diffuses within the cell to the operator. **7.** analyze b (Because constitutive expression of the *lexA* gene allows the LexA repressor protein to build up after DNA damage has been repaired by SOS regulon gene products. The increased LexA level turns off the regulon after the DNA damage–repairing genes have been expressed.) **8.** apply The *lac* operon would be strongly induced. Once inside the cell, the IPTG will bind to the repressor, causing it to release from DNA. IPTG cannot be broken down, so its concentration will remain high. Finally, because glucose is absent, there will be no inducer exclusion to inhibit IPTG transport through the galactoside permease transporter and no cAMP to activate CAP.

✔ TEST YOUR PROBLEM-SOLVING SKILLS

9. analyze c (Under this condition, the mutant is predicted to have low levels of transcription because cAMP would not be available to activate CAP, but in the wild type, transcription would be at maximum level [because the CAP–cAMP activator would be in place and the repressor would be off the operator]. Under all other conditions, expression of the *lac* operon in the mutant and wild-type cells is predicted to be indistinguishable.) **10.** analyze Cells with functioning β-galactosidase will produce blue colonies; cells with *lacZ* mutations or *lacY* mutations will not produce β-galactosidase and will produce white colonies. The *lac* promoter could be mutated so that RNA polymerase cannot bind.

✔ PUT IT ALL TOGETHER: Case Study

11. understand a (Transcription produces an mRNA copy from a template strand of DNA. Because the genes are regulated by controlling production of their mRNA, this must be transcriptional control.) **12.** understand Bound. (In positive control, activator proteins bind to DNA sequences near the promoter and then interact with RNA polymerase to promote transcription.) **13.** analyze When there are relatively few bacterial cells (low cell density), the amount of inducer (the secreted signaling molecule) will be low. Thus, there will be little or no *lux* operon transcription. As cell density increases to high levels, the concentration of inducer builds up to a threshold level needed to activate transcription of the light-producing genes. This level of inducer will be found only when cells are at high density. **14.** understand This means that LuxR activity is regulated by a change in shape. (LuxR is found in two shapes, one without the inducer and the other with the inducer. Only the shape with the inducer allows LuxR to bind to DNA and serve as an activator.) **15.** analyze One regulatory protein controls the transcription of more than one gene or operon. Regulons, however, typically contain many genes or operons, but in this case only one gene (*luxR*) and one operon (*luxICDABE*) are controlled by the LuxR activator. **16.** create Pathogenic bacteria may benefit from mounting a full frontal assault, releasing toxin only when there are enough cells to produce sufficient toxin to alter the metabolism of the infected host.

Chapter 19

IN-TEXT QUESTIONS AND EXERCISES

p. 386 CYU (1) understand The existence of chromatin means that genes are often inaccessible for transcription, but the fact that mechanisms must be in place to open chromatin to access genes offers the possibility of regulating whether genes are ready for, or resistant to, transcription. **(2)** understand RNA splicing opens the possibility of splicing one primary transcript in multiple, regulated ways. This can generate different proteins to meet the unique demands of particular cell types. Because splicing is rare in bacteria, these organisms lack the possibilities for regulation offered by alternative splicing in eukaryotes.

p. 389 Fig. 19.5 analyze Acetylation of histones decondenses chromatin and allows transcription to begin, so HATs are involved in positive control. Deacetylation condenses chromatin and inactivates transcription, so HDACs are involved in negative control.

p. 390 Fig. 19.6 create They could do something to change histone modifications to see how this affects gene transcription instead of just making the observation that certain histone modifications and low rates of transcription go together.

p. 391 CYU (1) apply Many more genes than normal are predicted to be expressed because the inability to methylate DNA would lead to more decondensed chromatin. **(2)** understand Addition of acetyl groups to histones or methyl groups to DNA can cause chromatin to decondense or condense, respectively. Different patterns of acetylation or methylation will determine which genes in muscle cells versus liver cells can be transcribed and which genes are not available for transcription. **(3)** understand Genetic inheritance is based on differences in DNA sequences, essentially different alleles that lead to differences in phenotype. Epigenetic inheritance is inheritance that is due to anything else. In reality, however, the mechanism of epigenetic inheritance appears to be the inheritance of different patterns of chromatin condensation.

p. 392 Fig. 19.7 apply A typical eukaryotic gene usually contains introns and many regulatory sequences, including enhancers, and each particular mRNA produced from the gene usually codes for one protein. Bacterial operons usually lack introns and enhancers, and produce an mRNA that codes for two or more distinct proteins.

p. 393 apply The general transcription factors found in muscle and nerve cells are similar or identical; the sets of regulatory transcription factors found in the two cell types are different.

p. 394 analyze DNA forms loops when distant regulatory regions, such as silencers and enhancers, are brought close to the core promoter through binding of regulatory transcription factors to Mediator.

p. 395 CYU (1) apply Bacterial regulatory sequences are found close to the promoter; eukaryotic regulatory sequences can be close to the promoter or far from it. Bacterial regulatory proteins interact directly with RNA polymerase to initiate or prevent transcription; eukaryotic regulatory proteins influence transcription by altering chromatin structure or binding to the general transcription complex through Mediator proteins. **(2)** understand Regulatory sequences are predicted to be the same because different cells of an individual contain the same DNA sequence. In contrast, transcription factors are predicted to differ because it is largely differences in transcription factors that allow gene expression patterns to differ between cell types. **(3)** analyze Promoter-proximal elements are binding sites for regulatory transcription factors. If the DNA sequence in a promoter-proximal element changed, this would be likely to alter how tightly, or if at all, the transcription factor associated with this element could bind to DNA. In turn, this change in binding would increase or decrease the amount of time the transcription factor was associated with DNA, therefore changing how efficiently transcription initiation was promoted by this transcription factor.

p. 396 apply See Figure A19.1. (There are many possibilities because alternative splicing allows any exon or intron to be skipped or retained in the final mRNA.)

p. 398 apply A microRNA can recognize a specific target whenever it can form complementary base pairs with the target. One way that a miRNA could recognize more than one target mRNA would be if different mRNAs shared a common sequence recognized by the miRNA. Another way would be if the miRNA did not pair perfectly with its target sequence. In this case, two different mRNAs could have related but distinct sequences that the same miRNA could bind to.

p. 398 CYU (1) understand If one species made more extensive use of alternative splicing than another, then more gene products would be produced in the species with more extensive alternative splicing, even with the same number of genes. **(2)** understand The name "RNA interference" is apt because small RNAs, such as miRNAs, interfere with mRNAs by targeting them for destruction or preventing them from being translated. **(3)** apply Regulation at the post-transcriptional level is costly in terms of energy and materials because either an mRNA or a protein already has to have been synthesized for post-transcriptional control to function. The advantage of post-transcriptional control is that it allows a more rapid response than transcriptional control because there are fewer steps between an mRNA or a newly synthesized protein and the final, active gene product (often an active protein). Another advantage

Figure A19.1

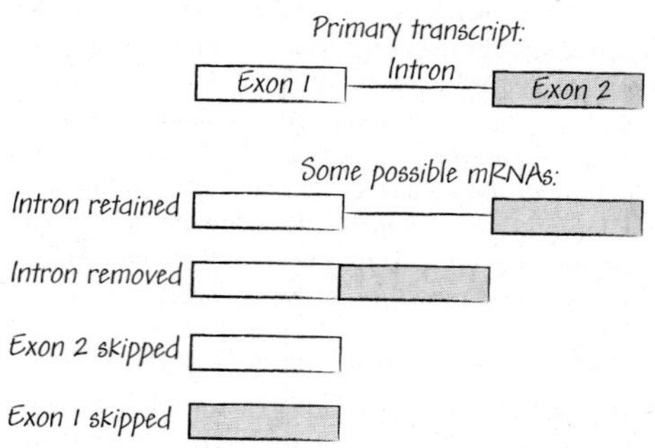

of post-transcriptional control is that it can fine-tune transcriptional regulation. For example, if an mRNA is produced in crude cuts of a lot or a little, then miRNA or translational control can work to more closely regulate the precise amount of the molecule.

p. 400 CYU (1) apply Many different types of mutations can disrupt control of the cell cycle and initiate cancer. These different mutations can, however, result in the same pattern of uncontrolled cell growth. **(2)** understand Tumor suppressor genes can be compared crudely to brakes and proto-oncogenes to accelerators. To run out of control, the accelerator must be active when it shouldn't be, and at the same time, the ability to put on the brakes must be lost. This is why tumor suppressor genes—the brakes—must not produce an active product, and proto-oncogenes—the accelerator—must produce an overly active product for cell division to run out of control. **(3)** understand An important role of p53 is to shut down DNA replication if there is DNA damage. With a loss-of-function mutation, p53 would fail to perform this role and cells would continue replicating damaged DNA. Replication of damaged DNA would lead to errors, which in turn would lead to mutations in many genes.

p. 401 CYU analyze **(1)** The number of regulatory elements would be predicted to decrease. This is because operons group genes that function in the same pathway into co-transcribed groups, and transcription of the entire set is controlled by the same set of DNA regulatory elements. In this way, a given set of regulatory elements could control many genes. Without this clustering, a separate set of regulatory elements is needed for each gene. analyze **(2)** miRNA-mediated control is based on controlling the presence of mRNAs (through cleavage) or the ability to translate mRNAs. This system presupposes that mRNAs are stable enough to bother controlling, whether it's useful to remove them or prevent their use. In the case of mRNAs with a fleeting existence, controlling mRNA stability or translatability takes away most of its advantages.

✔ TEST YOUR KNOWLEDGE

1. remember c **2.** understand d **3.** understand The set of regulatory transcription factors present in a particular cell, not differences in DNA sequence, is largely responsible for which genes are expressed and therefore for making one cell different from another. In the same individual, except for differences due to mutations, cells contain the same DNA sequence. **4.** remember b

✔ TEST YOUR UNDERSTANDING

5. analyze (a) Enhancers and the CAP binding site are similar because both are sites in DNA where regulatory proteins bind. They are different because enhancers generally are located at great distances from the promoter, whereas the CAP binding site is located adjacent to the promoter. (b) Promoter-proximal elements and the *lac* operon operator are both regulatory sites in DNA located close to the promoter. There usually is more than one promoter-proximal element but only one *lac* operator. (c) General transcription factors and sigma are proteins that must bind to the promoter before RNA polymerase can initiate transcription. They differ because sigma is part of the RNA polymerase holoenzyme, while the general transcription complex recruits RNA polymerase to the promoter. **6.** analyze The mutation is likely in a regulatory sequence, either in the promoter, a proximal control element, or an enhancer. Because this is a loss-of-function mutation, the mutation is probably acting to reduce or prevent transcription initiation. The mutation illustrates the important idea that alterations of regulatory sequences outside the coding region often have major effects on gene function. **7.** apply T, F, F, F. **8.** apply The cell is predicted to arrest in the cell cycle and most likely will undergo cell death through activation of apoptosis genes because of the continually active p53.

✔ TEST YOUR PROBLEM-SOLVING SKILLS

9. analyze c (This is because promoters and enhancers are brought into close physical proximity when transcription begins [see Figure 19.9]. Because rats of malnourished mothers initiate *Hnf4a* gene transcription infrequently, the promoter and enhancer will be together less often in these animals compared with rats born to well-nourished mothers.) **10.** create See Figure A19.2. The value for the normal diet should be shown as 1.0, and the value for the low-protein diet should be shown as 0.64 (0.64 comes from the ratio of the cpm of the low-protein diet divided by the cpm of the normal diet, or 7368/11,478).

✔ PUT IT ALL TOGETHER: Case Study

11. understand c (See Figure 19.9.) **12.** understand DNA looping, as shown for an idealized gene in Figure 19.9, step 3, could bring the enhancer bound by an activator (transcription factor) physically close to the promoter even if the enhancer is a huge distance away along the DNA molecule. **13.** analyze The transcription factor could bind with different affinities to the two closely related sequences, with the sequence promoting stronger binding leading to higher rates of transcription initiation. **14.** analyze Both enhancers drive much higher expression than the *Kitl* gene without the enhancer. On average, the enhancer associated with dark hair drives higher levels of transcription than the one associated with blond hair. Although the values of *Kitl* mRNA levels overlap, the *P* value indicates that there is a significant difference in expression. **15.** evaluate The results show only *Kitl* mRNA expression levels, and the relationship between these levels and dark versus blond hair would need to be established. The idea that the two forms of the enhancer help to determine dark versus blond hair can be tested in many ways. One approach would be to express the two different forms of the engineered gene in white mice to see if the blond allele form changed coat color to blond and the dark allele form changed the coat to a dark color. Whatever approach is taken, the goal should be to obtain results that are more informative about the phenotype of blond versus dark hair rather than levels of an mRNA. **16.** analyze Adding the microRNA precursor to mouse embryos should lead to its being processed into a mature, single-stranded miRNA. The miRNA–RISC complex would be capable of base pairing to the target *Kitl* mRNA, leading to its cleavage. This would reduce levels of functional *Kitl* mRNA in the embryo. Postnatal mice could be examined to see if their coat color were lighter than normal. If so, then *Kitl* is important for normal coat color development.

Figure A19.2

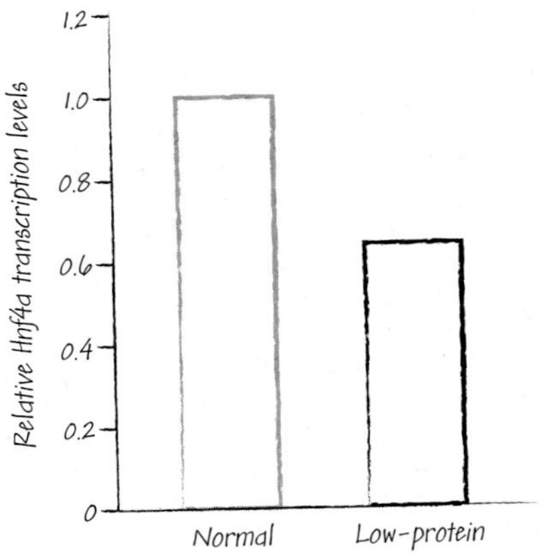

BIG PICTURE Genetic Information

pp. 404–405 CYU (1) remember Star = DNA, mRNA, proteins. **(2)** understand RNA "is reverse-transcribed by" reverse transcriptase "to form" DNA. **(3)** analyze E = splicing, etc.; E = meiosis and sexual reproduction (along with their links). **(4)** analyze Chromatin "makes up" chromosomes; independent assortment and recombination "contribute to" high genetic diversity.

Chapter 20

IN-TEXT QUESTIONS AND EXERCISES

p. 410 CYU (1) apply Without an antibiotic gene, it wouldn't be possible to distinguish cells that had taken up plasmid DNA from the vast majority of cells that did not take up the plasmid. Researchers would need to examine many more colonies to find one with the recombinant plasmid. **(2)** analyze The primary differences are that in laboratory-based genetic engineering, (a) scientists introduce particular genes into a Ti plasmid that becomes incorporated into the plant cell, and (b) the Ti plasmid is modified so it does not lead to uncontrolled growth in the plant cell.

p. 411 CYU (1) analyze If a band or bands on a gel used for paternity analysis does not match a potential father, then he can be excluded as the father. If there is a match across all bands, there is a possibility, even if it's very small, that the match is coincidental rather than due to paternity. **(2)** analyze Primers that bind to sequences outside the region of interest allow for amplification of sequences with any number of STR copies. Primers that bind inside an STR locus would not reveal the total number of STR copies within the locus.

p. 413 Fig. 20.4 apply For genes made of exons separated by introns, the ORF would extend across an exon, but once into flanking introns, there is no true reading frame. Extending the reading frame of the exon into introns is like searching for codons in a random sequence of RNA; eventually sequences for stop codons may be encountered, but these are not stop codons that are used. A functional start codon would be on only the 5′-most exon in a gene.

p. 414 CYU (1) analyze If Humpty Dumpty could have been repaired after his fall, it would have required piecing together all the broken bits of shell into their original places. This is similar to the challenge faced in de novo genome assembly when the randomly generated fragments of the genome must be put back together again in the right order. **(2)** evaluate There are no objectively right or wrong answers, but a point on the pro side is that having this knowledge may allow a lifestyle change that reduces the chance of developing the disease or the severity of its symptoms and promotes living life to its fullest. A point on the con side is that this knowledge could cause overwhelming worry that ruins the quality of life.

p. 417 Fig. 20.7 apply *Chromosome 1:*

cβ2-ε-Gγ-Aγ-cβ1-δ-cβ2-ε-Gγ-Aγ-cβ1-δ-β

Chromosome 2: β

p. 418 CYU (1) apply Parasites don't need genes that code for enzymes required to synthesize molecules they acquire from their hosts. **(2)** apply Humans are predicted to make more use of alternative splicing (because they are more complex than roundworms, yet have about the same number of genes).

p. 421 Fig. 20.14 analyze The predicted effect on the mosquito population is that it would no longer be able to carry the malaria parasite.

p. 422 CYU (1) apply The base-pairing between an extensive region of the sgRNA and the target DNA sequence. **(2)** analyze Gene drive technologies could be used in weaponized forms to deliberately cause harm, for example by destroying a critical food source, or they could be started with good intentions but spiral out of control to cause harm.

p. 423 CYU (1) analyze An ex vivo approach introduces the therapeutic gene into only a limited number of cells. This approach could work well to restore a missing secreted protein such as a hormone or enzyme in the blood that travels throughout the body. But ex vivo gene therapy wouldn't work well in the case of a defective intracellular protein, because any cells with the defect would remain defective and continue to cause the disease. **(2)** evaluate For a nonfunctional allele, all that's required is to add a functioning allele anyplace in the genome or even on a vector that remains separate from the genome. In contrast, a misfunctioning allele requires either replacement with a functional allele or inactivation of the mis-functioning allele *and* addition of a functioning allele.

p. 424 apply Homeostasis is a property that emerges from the interaction of molecules within the cell. Homeostasis is a new property, not exhibited by any single molecule, and it cannot be easily predicted from knowing which molecules are present.

p. 425 CYU (1) apply The purpose of RNA-seq is to learn which genes are transcribed and how often they appear as RNAs. This is done by comparing the sequences of the cDNAs to sequences in the genome. Without a genome sequence, this analysis would be impossible. **(2)** create (Many answers are possible.) One argument is that the scientists did not create life—but came close—because the synthetic genome had to be introduced into a preexisting cell that lacked a chromosome in order to support replication. The researchers did not fully synthesize a free-living organism. A counter-argument is that they did synthesize a new form of life, and in this sense, life was created.

✔ TEST YOUR KNOWLEDGE

1. understand They cut DNA at specific sites, known as recognition sites, and produce DNA fragments useful for cloning. **2.** remember b **3.** understand a **4.** apply d (This is because there is no relationship between the complexity of an organism and the size of its genome.)

✔ TEST YOUR UNDERSTANDING

5. understand cDNAs can be made (using reverse transcriptase) from RNA isolated from particular cells or tissues. The population of cDNAs can then be amplified by PCR and then sequenced and analyzed using a deep sequencing approach to reveal the relative levels of expression of particular genes. **6.** remember T, F, F, T **7.** apply Water flea gene density is about 31,000 genes/200 Mbp = 155 genes/Mbp. In humans, gene density is about 21,000 genes/3000 Mbp = 7 genes/Mbp. The relative gene density of water flea/human = 155/7 = 22. **8.** evaluate SNPs are used as genetic markers to find particular genes, but they are rarely disease-causing mutations. Often SNPs fall outside coding and regulatory regions, so there's only a slight chance that the SNP she found associated with the disease is the disease-causing mutation.

✔ TEST YOUR PROBLEM-SOLVING SKILLS

9. apply If the grave were authentic, it might include two very different parental patterns along with five children whose patterns each represented a 50/50 mix between the two parents. The unrelated individuals would have patterns not shared by anyone else in the grave. **10.** analyze Given the genetic, physiological, and morphological similarities between humans and chimps, you would expect that the livers and blood of chimpanzees and humans would function similarly. However, given the cognitive differences between these species, you might expect differences in gene expression in the brain. The RNA-seq data support this prediction.

✔ PUT IT ALL TOGETHER: Case Study

11. understand Isolate DNA from a potato plant resistant to potato blight. Cut the DNA into fragments with a restriction endonuclease that cuts on either side of the gene and that leaves sticky ends. Cut copies of a plasmid with the same restriction endonuclease. Mix the fragments and plasmids to allow complementary base pairing by sticky ends of fragments and plasmids. Use DNA ligase to catalyze formation of phosphodiester bonds and seal the sequences. The large set of recombinant plasmids that contain potato DNA would include some recombinant DNAs with the gene that allows resistance to potato blight. **12.** analyze Isolates 1, 5, and 7 contain only the potato gene and no plasmid. The absence of a 396-bp PCR product on the gel indicates that they do not contain any plasmid DNA. **13.** analyze If the DNA at each cycle steadily doubles, it is predicted to yield a 33.6-million-fold increase (a 2^{25}-fold increase). **14.** apply The positive control shows that the PCR conditions are right for amplifying the plasmid DNA. This means that the absence of a product from a particular sample is unlikely to be caused by technical difficulties. The negative control shows that plasmid DNA must be present for the PCR to generate a product. **15.** create If a search of human gene sequence databases revealed a gene that was similar in base sequence and coded for a similar product, and if follow-up work confirmed that the potato and human genes were similar in their pattern of exons and introns, then the researchers could claim that they are homologous. **16.** evaluate Some possible reasons for concern are whether it is safe for the environment or human health to have a foreign gene such as a bacterial gene in a GM crop plant. The researchers in this study avoid the problem of foreign DNA by using PCR to select plants having the allele that confers resistance to potato blight, but not having plasmid DNA.

Chapter 21

IN-TEXT QUESTIONS AND EXERCISES

p. 431 CYU (1) understand Because a cloned animal is derived from a single cell, it means that all the genetic information needed to create the organism during development is contained in the genome of that cell. Given that different cells can be used for cloning, this means that these cells are genetically equivalent and equivalent genetically to the zygote. **(2)** understand Understanding development can largely be reduced to understanding how genes are regulated in different ways in different cells and at different times during development. Therefore, knowing how genes are expressed is essential to understanding development.

p. 432 Fig. 21.3 analyze It made gene expression possible in any type of cell, including including fibroblasts.

p. 433 CYU (1) evaluate These results indicate that the cells were not determined to be eye cells in the early embryo but were determined in the later embryo. **(2)** understand "Self-renewal" means that the population of stem cells is maintained, even when stem cells divide. This occurs because one of the daughter cells becomes committed to a pathway of differentiation, and the other daughter cell remains a stem cell. **(3)** understand Transcription factors act as master regulators that change the gene expression patterns of adult cells to an embryonic-like state in which the cells are able to differentiate into any cell of the body—they are pluripotent.

p. 435 Fig. 21.5 apply The "green" cytoplasmic determinant would now be found in all daughter cells, and all of these cells would be specified as the same cell type.

p. 436 CYU (1) analyze Induction, because muscle forms only when one group of cells is exposed to another group of cells; this suggests the use of a muscle-specifying signal. **(2)** apply The mutation is more likely to affect specification by induction because induction depends on cell–cell signaling. **(3)** understand Because certain mutations resulted in overproduction of cells, this suggests that the normal gene product is involved in killing off cells that are produced during development.

p. 439 analyze Since higher-than-normal levels of the anterior-specifying Bicoid protein would be present in posterior regions, anterior structures would be predicted to be shifted farther back in the embryo.

p. 441 Fig. 21.13 evaluate If the genes they have similar DNA sequences, they would be considered related.

p. 442 CYU (1) analyze Because *bicoid* is a maternal effect, this means that the mother's genotype controls the phenotype of her offspring. Offspring of a *bicoid* mutant homozygote mother are expected to have patterning defects, whereas offspring of a *bicoid* mutant homozygote father are expected to have normal patterning if the mother has at least one wild-type allele of the *bicoid* gene. **(2)** evaluate For Bicoid to be a morphogen, different anterior structures would need to be specified at different Bicoid concentrations, with more anterior structures specified at higher concentrations. If Bicoid were an inducer but not a morphogen, then there's no requirement for a dependence on concentration. It's likely that above a threshold to induce anterior structures, different concentrations of Bicoid would exert the same effect—induction of anterior structures. **(3)** evaluate If large regions were specified before specifying individual segments, then loss of a gene that acts earlier and is higher in the cascade would cause the loss of large regions—no segments would form. In turn, loss of a gene that acts later and lower in the cascade would be predicted to alter a particular segment, but not affect the formation of adjacent segments.

p. 443 analyze Researchers created a limbless knockout mouse because interpreting how well a particular *sonic hedgehog* limb enhancer works to direct limb development would have been nearly impossible starting from a mouse that already had limbs.

p. 443 CYU (1) evaluate A mutation in a regulatory region can change where, when, and to what level an important regulator of development is expressed. This can lead to changes in the form of an organism with no need for changes to the coding sequence of the gene. (Shifts in where *Hoxc8* is expressed in snakes is an example.) **(2)** evaluate Because evolution involves the change in form of organisms over time, knowing that form is achieved through development and how form can be modified are essential to understanding evolution and evolutionary change.

✓ TEST YOUR KNOWLEDGE

1. remember b **2.** remember stem cells **3.** remember a **4.** understand a conserved gene that can be expressed at different times and places during development to produce different types of structures in different organisms.

✓ TEST YOUR UNDERSTANDING

5. analyze You can tell her that it's fine to start her work because all cells of an individual should be genetically equivalent, meaning they have the same sets of genes, regardless of the cell type or whether the gene is expressed (used to produce a final gene product, typically a protein) in that cell. (A common misunderstanding is that only cells that express, or turn on, particular genes contain those genes.) **6.** analyze T, F, T, F, **7.** apply Differentiation is triggered by the production (or activation) of regulatory transcription factors, which induce other transcription factors, and so on—a sequence that constitutes a regulatory cascade—as development progresses. At each step in the cascade, a new subset of genes is activated—resulting in a step-by-step progression from undifferentiated to fully differentiated cells. **8.** evaluate c. This result shows that a mouse tool-kit gene is similar enough (conserved) to a fly gene to be able to take over its function.

✓ TEST YOUR PROBLEM-SOLVING SKILLS

9. apply c. Since the morphogen concentration is halved every 100 μm away from the posterior pole, a drop to $\frac{1}{16}$ the highest concentration would require four "halvings" of the initial concentration ($\frac{1}{2} \times \frac{1}{2} \times \frac{1}{2} \times \frac{1}{2} = \frac{1}{16}$), and this would take place in four steps of 100 μm each—or 400 μm from the morphogen source. This is where the leg is predicted to be formed. **10.** create *Hypothesis 1:* The mutation falls in a regulatory region of the *Pitx1* gene that controls *Pitx1* expression in the spine-forming region during development. *Hypothesis 2:* If the mutation prevents *Pitx1* expression where spines form, then spine formation will not occur during development, resulting in spineless fish, even though the *Pitx1* coding region is normal.

✓ PUT IT ALL TOGETHER: Case Study

11. remember c. **12.** analyze a. (This is because molecules outside the cell can bind to cell-surface receptors and trigger differentiation; an extracellular signal protein is a common type of signaling molecule.) **13.** apply As cells differentiate, they express different sets of proteins. Therefore, looking for specific new proteins is a logical way to assess differentiation (and often easier and more sensitive than looking for changes visible under the microscope). **14.** create The gene expression patterns in the potential iPS cell lines and in the conventionally produced iPS cells are virtually identical. This finding supports the idea that the cells produced by this method are indeed iPS cells. You could test this idea further by examining the biological properties of the cells created by the new procedure, especially properties related to the ability to cause them to differentiate into various cell types in ways that parallel conventionally made iPS cells. **15.** analyze It's not a surprise because many important regulators in development are conserved across species. **16.** evaluate Using iPS cells would be preferable because these cells do not require the destruction of a human embryo. People disagree on whether obtaining cells from human embryos is ethical, but some people believe it is unacceptable, so it is better to avoid the issue entirely.

END-OF-UNIT CASE STUDY How Can Mutations Save a Snake?

pp. 446–447 (1) analyze
(TTX-sensitive) 5′-GGU UCC GAC GGC UUG-3′
(TTX-resistant) 5′-GUU UCC AAC GGC UUG-3′
(2) analyze
(TTX-sensitive) Gly-Trp-Asp-Gly-Leu
(TTX-resistant) Gly-Trp-Asn-Gly-Leu
(3) apply c **(4)** understand a **(5)** analyze c (In a heterozygote, there will be an approximately 50/50 mix of channel proteins, some that bind TTX well and others that bind TTX poorly. With this mixture of channel proteins, the expectation is that that influx of sodium will be approximately 50% of normal levels—an amount between what would be observed in either homozygote. Therefore, the "resistant" form of the gene would act as an incompletely dominant allele relative to the wild type allele.) **(6)** analyze If the mutations are missense mutations, these could lead to amino acid substitutions that allow the pore to function as a sodium channel but reduce its ability to bind TTX (and therefore influence TTX sensitivity). **(7)** evaluate False (Note that the last entry to the phylogeny, *Heterodon platirhinos*, is highly resistant, but has no amino acid substitutions in pore-forming regions 3 or 4.) True (The fact that *H. platirhinos* is highly resistant, but has no amino acid substitutions in pore-forming regions 3 or 4 suggests that either a mutation in another portion of the sodium channel gene or a completely different gene must have lead to TTX resistance.) True (Indeed, this trend is seen in numerous cases, for instance between the moderately and highly resistant entries 3 [*T. couchi*] and 4 [*T. atratus*] in the phylogeny.) True **(8)** understand In this case, mutations in the coding sequence are more likely to lead to resistance. This is because regulatory mutations are expected to lead to changes in the amount of the protein. If a hypothetical regulatory mutation lowered the amount of the normal channel protein, the snake might be TTX resistant but it would also likely be dead because it couldn't move (or could only move very slowly). Conversely, if a regulatory mutation increased the amount of the protein, no TTX resistance would be provided. In contrast, mutations in the coding sequence that alter the amino acid sequence of the protein might allow sodium movement through the channel but prevent TTX binding—and therefore provide resistance.

Chapter 22

IN-TEXT QUESTIONS AND ANSWERS

p. 450 CYU analyze From the viewpoint of typological thinking, a meadow of flowers would represent numerous copies of one type. From the viewpoint of population thinking, a meadow of flowers would represent a population of individuals with characteristics that vary. For example, some flowers might be slightly larger or smaller than others, vary in color, or have shorter or longer petals.

p. 452 Fig. 22.4 analyze According to Lamarck's early model, a fish species evolved into a more complex and "better" tetrapod species as it acquired legs so that it could walk on land. In contrast, Darwin and Wallace's model is not goal oriented. They proposed that traits varied among individuals in aquatic, semiaquatic, and terrestrial environments. Individuals with favorable heritable traits produced more offspring than did others without those traits, resulting in changes in the population over time.

p. 453 Fig. 22.5 apply If vestigial traits resulted from inheritance of acquired characteristics, some individuals must have lost the traits during their own lifetimes and passed the reduced traits on to their offspring. For example, a certain monkey's long tail might have been bitten off by a predator. The new traits would then somehow have passed to the individual's eggs or sperm, resulting in shorter-tailed offspring, until humans with a coccyx resulted.

p. 454 Making Models 22.1 apply See Figure A22.1.

p. 456 Table 22.2 explain Many possible correct answers; answers for items 1, 2, and 6 are already provided in the table. 3. Discovery of fishes that have limbs intermediate between fins and legs. 4. Tiny tailbone of humans. 5. Origin of antibiotic resistance in bacteria. 7. Same bone pattern in human arm and bat wing. 8. Killer whales are in the process of speciation today.

p. 457 CYU (1) apply If birds evolved from dinosaurs, you would expect to find transitional fossil dinosaurs with feathers (such fossils have been found). You would also expect birds and dinosaurs to share many homologous traits (and they do). **(2)** apply The DNA sequences of chimpanzees and humans are so similar because we share a recent common ancestor.

p. 458 Making Models 22.2 apply See Figure A22.2.

p. 459 CYU (1) apply The ability of flowers with long petals to produce more surviving offspring than do flowers with short petals. **(2)** apply See Figure A22.3.

p. 460 Fig. 22.9 apply **(1)** Relapse occurred because the few bacteria remaining after drug therapy were not eliminated by the patient's weakened immune system and began to reproduce quickly. **(2)** No—almost all of the cells present at the start of step 3 would have been resistant to the drug, so family members or heath-care workers would not respond to drug therapy.

p. 462 Fig. 22.12 apply *Prediction of Hypothesis:* Beak measurements were different before and after the drought. *Prediction of Null Hypothesis:* No difference was found in beak measurements before and after the drought.

Figure A22.1

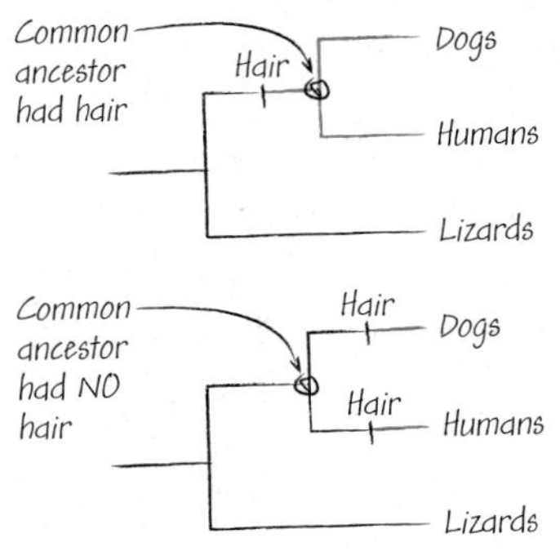

Figure A22.2

Figure A22.3

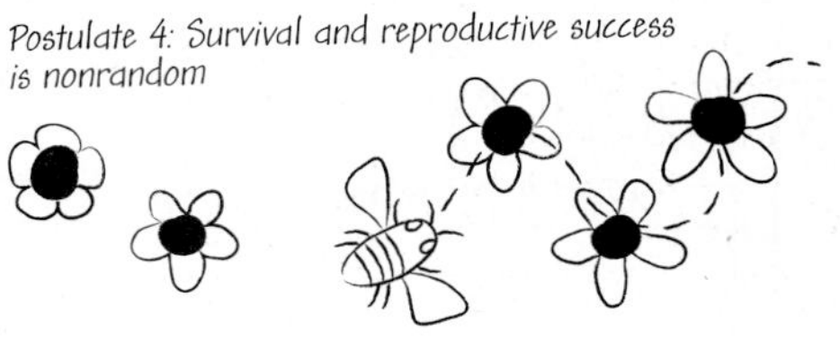

p. 463 Fig. 22.13 apply One way to label the graph would be to draw a vertical line intersecting with the *x*-axis at 1977, and add the label "Drought."

p. 463 evaluate The data suggest that the statement is true only some of the time, depending on environmental conditions. For example, finches had deeper beaks in 1980 compared with those measured in 2000, presumably because beaks that were less deep had higher fitness in 2000. Because most of the error bars are small, the fluctuation in the traits of the finches over time appears to be real and not a sampling error. This means that none of the traits always increases or decreases fitness—it depends on the situation.

p. 463 CYU (1) apply *Postulate 1:* Traits vary within the population? Yes, finches have beaks of different depths. *Postulate 2:* Some of the trait variation is heritable? Yes, beak depth is genetically determined. *Postulate 3:* There is variation in survival and reproductive success (some individuals produce more offspring than others)? Yes, some finches died and produced no offspring, while others lived and produced offspring. *Postulate 4:* Individuals with certain heritable traits produce the most offspring? Yes, fiches with deep beaks survived the drought and produced the most offspring. Natural selection occurred. **(2)** create Many solutions are possible, as long as all four of the ideas in answer **(1)** are represented. See Figure A22.4.

p. 464 Table 22.3 create Many possible correct answers; answers for items 1, 5, and 9 are already provided in the table. 2. In the caves, fish with no eyes arose by chance and had higher fitness, on average, than fish with eyes, increasing the frequency of eyeless fish over time. 3. The frequency of finches with deeper beaks increased over time because finches with deeper beaks could open tough seeds, increasing their fitness. 4. I acclimatized to the cold temperature in the classroom. 6. A mutation in the *rpoB* gene arose by chance, enabling some *M. tuberculosis* cells to survive the antibiotics. 7. A mutation in the *rpoB* gene arose by chance, enabling some *M. tuberculosis* cells to survive the antibiotics. 8. The loss of a digestive system is an adaptation in tapeworms that improves their efficiency in absorbing nutrients from their animal host. 10. No one beak shape is perfectly adapted for opening seeds because different shapes are more effective at opening seeds of different sizes. 11. The ear bones of humans are an adaptation for transmitting sound, but are not ideal due to the constraints of evolutionary history (having evolved from jawbones.)

p. 466 Making Models 22.3 apply See Figure A22.5.

p. 466 CYU (1) analyze In biology, an adaptation is any heritable trait that increases an individual's ability to produce offspring in a particular environment. In everyday English, "adaptation" is often used to refer to an individual's nonheritable adjustment to meet an environmental challenge, a phenomenon that biologists call acclimatization. The phenotypic changes resulting from acclimatization are not passed on to offspring. **(2)** apply By chance, a mutation occurred in some roses that resulted in sharp bumps on the stems. Individuals with these early prickles survived herbivore attacks and produced more offspring than roses without the mutation, increasing the frequency of the prickled roses in the population over time. Note that roses cannot cause a mutation to occur because they want or need it (see Table 22.3).

✔ TEST YOUR KNOWLEDGE

1. understand False **2.** understand F, F, F, T (If you answered true to one or more of the first three statements, remember that species, not individuals, change over time, and that evolution does not perfect organisms or necessarily make them more complex and better adapted over time.) **3.** remember Homologous **4.** remember F, T, F, F (If you answered true to the first, third, or fourth statement, remember that evolutionary fitness refers to the ability of an organism to pass its genes on to the next generation, not to its speed, strength, or longevity.)

✔ TEST YOUR UNDERSTANDING

5. understand F, F, T, F (If you answered true to the first, second, or fourth statement, review Table 22.3 and associated text for help with common misconceptions. The heritable characteristics of individuals do not change in response to the environment. Instead, change occurs in populations. But populations do not become more perfect over time due to the many constraints on natural selection.) **6.** understand Mutation produces new genetic variations, at random, with no forethought about which variations might prove adaptive in the future. Individuals with mutations that are disadvantageous won't produce many offspring, but individuals with beneficial mutations will produce many offspring. The beneficial mutations will thus increase in frequency through selection. **7.** understand b (In some environments, being big and strong lowers fitness; remember, fitness refers to number of healthy offspring produced, not strength or speed.) **8.** apply When antibiotics are overused, susceptible bacteria are constantly being killed off, even when an infection is not present. However, any individual bacteria that are resistant to these antibiotics will survive and multiply, reducing the likelihood that antibiotic treatment will be effective in the future.

✔ TEST YOUR PROBLEM-SOLVING SKILLS

9. apply If an *individual* coral colony can acclimatize to the warming water, this means it might survive as the ocean continues to warm. Individuals that cannot acclimatize will die. Over time, natural selection will favor corals that have heritable characteristics that enable them to tolerate warm water. That is, coral *populations* will adapt to warm water (as long as the rate of warming does not surpass the temperature tolerance of all corals, which are not able to disperse rapidly to new habitats). **10.** evaluate The theory of evolution fits the six criteria as follows. (1) and (2) It provides a common underlying mechanism responsible for puzzling observations such as homology, geographic proximity of similar species, the law of succession in the fossil record, vestigial traits, and extinctions. (3) and (4) It suggests new lines of research to test predictions about the outcome of changing environmental conditions in populations, about the presence of transitional forms in the fossil record, and so on. (5) It is a simple idea that explains the tremendous diversity of living and fossil organisms and why species continue to change today. (6) The realization that all organisms are related by common descent and that none are higher or lower than others was a surprise.

✔ PUT IT ALL TOGETHER: Case Study

11. analyze Evolution by inheritance of acquired characteristics predicts that the mice in the different environments needed to change color, so they did so and then passed their traits on to their offspring. The theory of evolution by natural selection is not goal oriented. It predicts that white and brown mice are descendants of an ancestral population that varied in color. The white mice had higher fitness in the sand dunes environment, where they were more likely to escape the notice of predators. Likewise, the brown mice had higher fitness in the mainland environment. **12.** understand a (If you answered b, c, or d, note that evolutionary adaptations are very different from adaptation in the everyday sense. Because adaptations are heritable traits, they cannot change during an organism's lifetime, and they cannot change in response to "want" or "need." Review Table 22.3 and associated text, as well as **BioSkills 17**, for help with common misconceptions). **13.** apply (1) Fur color varies within mouse populations. (2) Fur color is heritable. (3) More mice are produced than can survive, and some produce many surviving offspring while others produce few or none. (4) Mice with certain heritable traits, such as the ability to avoid predators by blending in their environment (camouflage), survive and produce more offspring.
14. analyze [Many possible correct answers.] One advantage of models is that all the mice are exactly alike except for color, enabling the control of all variables except the one being studied. Models are also easier to work with than live mice. One disadvantage of using models is that they do not have the same smell and behaviors as real mice. **15.** analyze The data show that mice that do not match the color of the soil are attacked more often than mice that do match. This supports the hypothesis that fur color is an adaptation in mice, because mice that avoid detection by predators are more likely to survive and produce offspring. **16.** apply Biologists consider a *P* value less than 0.05 to be statistically significant, that is, showing a real difference and not just sampling error. Since $0.01 < 0.05$, you can be confident that the mice that were camouflaged were attacked less than those that were not.

Figure A22.4

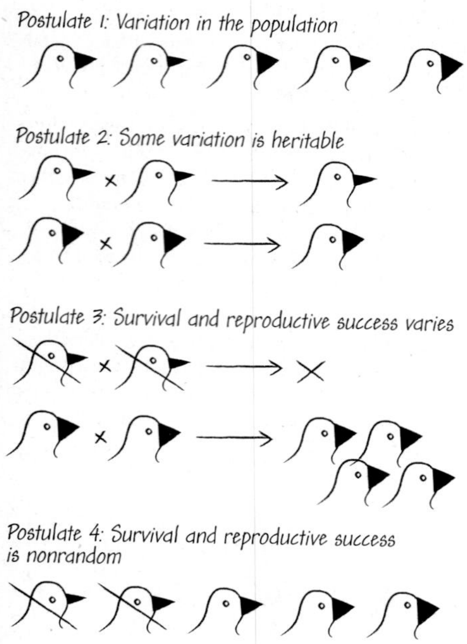

Figure A22.5

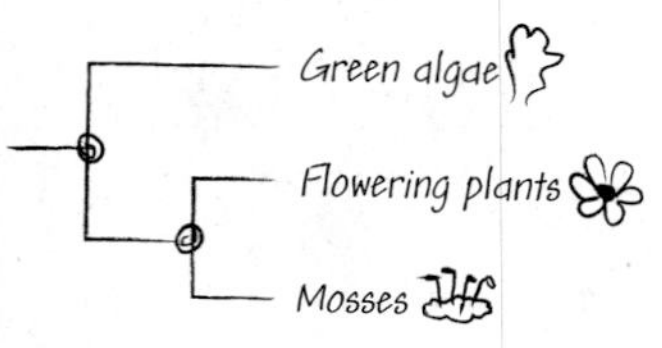

Chapter 23

IN-TEXT QUESTIONS AND EXERCISES

p. 472 apply **(1)** The frequencies of the three genotypes are shown by $p^2 + 2pq + q^2 = 1$:

$$\text{Freq}(AA) = p^2 = (0.6)^2 = 0.36$$

$$\text{Freq}(Aa) = 2pq = 2(0.6)(0.4) = 0.48$$

$$\text{Freq}(aa) = q^2 = (0.4)^2 = 0.16$$

(2) Allele frequencies in the offspring gene pool:

$$\text{Freq}(A) = p = 0.36 + \frac{1}{2}(0.48) = 0.60$$

$$\text{Freq}(a) = q = \frac{1}{2}(0.48) + 0.16 = 0.40$$

p. 473 Table 23.1 apply *(Step 2)* The observed allele frequencies, calculated from the observed genotype frequencies, are 0.43 for *M* and 0.57 for *N*. *(Step 3)* The expected genotypes, calculated from the observed allele frequencies under Hardy–Weinberg equilibrium, are 0.185 for *MM*, 0.490 for *MN*, 0.325 for *NN*. *(Step 4)* The observed homozygote and heterozygote frequencies are nearly identical to the expected values (each one is off by less than 3 percent).

p. 474 CYU apply Given the observed genotype frequencies, the observed allele frequencies are:

$$\text{Freq}(H) = p = 0.574 + \frac{1}{2}(0.339) = 0.744$$

$$\text{Freq}(h) = q = \frac{1}{2}(0.339) + 0.087 = 0.256$$

Given these allele frequencies, the genotype frequencies expected under Hardy–Weinberg equilibrium are *HH*: $0.744^2 = 0.554$; *Hh*: $2(0.744 \times 0.256) = 0.381$; *hh*: $0.256^2 = 0.066$. There are 4 percent too few heterozygotes observed, relative to the expected frequency ($0.381 - 0.339 = 0.042$, or 4.2 percent). One of the assumptions of the Hardy–Weinberg principle is not met at this gene in this population, at this time.

p. 475 CYU apply If inbreeding alone occurs, the genotype frequencies would change (*AA* and *aa* would become more common, *Aa* less common) but the allele frequencies would remain $A = 0.99$ and $a = 0.01$. However, if inbreeding plus natural selection occurs, then the frequency of the *a* allele likely will decrease in the population (and the frequency of *A* will increase), since the white sea turtles are easier for predators to find.

p. 476 apply If the population were under Hardy–Weinberg equilibrium then the allele frequencies would be $q = 0.7$ (the square root of $q^2 = 0.5$) and $p = 0.3$ (because $p + q = 1$), and these allele frequencies would not change over time. However, directional selection is likely to occur because albino sea turtles are more visible and thus more vulnerable to predators than sea turtles with normal pigmentation. The frequency of the *A* allele is likely to increase and the frequency of the *a* allele is likely to decrease over time.

p. 480 Fig. 23.10 create The researchers are trying to control for as many variables as possible other than beak color. Brothers are more likely to share more heritable traits than unrelated males.

p. 481 Fig. 23.11 apply The most successful males produce 100 offspring and the most successful females produce 10 offspring. Thus, the reproductive success of these males is ten times higher than the females.

p. 482 CYU analyze Sexual selection for large horns will favor the $Ho^+ Ho^+$ genotype (directional selection toward large horns), while ecological selection for small horns will favor the $Ho^P Ho^P$ genotype (directional selection toward small horns). This fitness trade-off will likely result in a high frequency of heterozygotes with intermediate horns. (This is, in fact, what researchers have observed.)

p. 483 Fig. 23.13 apply See Figure A23.1. Your graph should show that for a larger population size of 4000, the fluctuation of allele frequencies is less dramatic over 100 generations than it is for a population of 400.

p. 483 analyze Differences in allele frequencies in Table 23.1 (e.g., frequencies of $MM = 0.292$, 0.025, and 0.179 in the three populations) could be due to chance fluctuation like that in the simulated populations in Figure 23.13.

p. 484 Fig. 23.14 apply Original population: freq(A) = p = $(9 + 9 + 11)/54 = 0.54$; New population: freq(A) = p = $(2 + 2 + 1)/6 = 0.83$. The frequency of *A* has increased dramatically due to genetic drift.

p. 485 Making Models 23.1 apply See Figure A23.2. Your drawing should use lizard icons, circles, or other symbols with variation in a trait to show a larger source population producing a smaller new population. The frequency of the trait (shown by the icons and/or marked alleles) should be different in the new population than in the source population, due to genetic drift.

p. 485 Fig. 23.15 apply Original population at Time 1: freq(a) = q = $(5 + 2 + 2)/54 = 0.17$; new population at Time 2: freq(a) = q = $(1 + 1 + 1)/6 = 0.50$. The frequency of *a* has increased dramatically due to genetic drift.

p. 485 CYU apply Genetic drift is likely to occur in palms in the forest fragments due to the bottleneck effect. The allele frequencies within each of the surviving forest fragments are likely to differ from the allele frequencies in the original, intact population by chance.

p. 487 CYU apply Gene flow between the two corn populations (for example, via wind blowing pollen from one field to another) could result in the introduction of the herbicide-resistance allele from the GM crop to the organic crop and/or for the introduction of non-herbicide-resistance alleles from the organic crop to the GM crop. Assuming that some of the seeds from each crop are used to plant the next generation in each population, the allele frequencies of the two crops will become more similar over time.

p. 490 CYU evaluate **1.** The statement suggests that aphids mutated their genes on purpose, but mutations just happen by accident. **2.** Some mutations are deleterious, but others can be beneficial or neutral.

✔ TEST YOUR KNOWLEDGE

1. understand It defines what genotype and allele frequencies are expected if evolutionary processes and nonrandom mating are *not* occurring. **2.** understand b **3.** understand T, F, F. Note that genetic drift and gene flow sound similar but are distinct processes. If you answered *true* for the second statement, you were probably thinking of gene flow. Recall the mnemonic "Genetic drift drops and lifts (referring to allele frequencies); gene flow goes to and fro (referring to geographic movement)." **4.** understand True. If gene flow into a population introduces a beneficial allele, fitness will go up; but if gene flow introduces a deleterious allele, fitness will go down.

✔ TEST YOUR UNDERSTANDING

5. apply d; freq(A) = p = 0.9, so freq(a) = q = 0.1. Freq (aa) = $q^2 = (0.1)^2 = 0.01$. The number of babies with cystic fibrosis = 2500(0.01) = 25. **6.** apply Marrying close relatives is a form of inbreeding that would have increased the homozygosity of recessive alleles (inbreeding tends to shift alleles from heterozygotes, where they don't cause disease, to homozygotes, where they do). As a result, the royal families were plagued by genetic diseases. **7.** evaluate Mutations do not occur because an organism wants or needs them. They just happen by accident and can be beneficial, neutral, or deleterious. **8.** create Your concept map should have linking verbs that relate the following information: Selection may decrease, maintain, or increase genetic variation. Genetic drift tends to reduce it by causing random loss or fixation of alleles. Gene flow may increase or decrease variation (depending on whether immigrants bring new alleles or emigrants remove alleles). Mutation increases it. See the Big Picture of Evolution on p. 532–533.

✔ TEST YOUR PROBLEM-SOLVING SKILLS

9. apply If we let *q* stand for the frequency of the loss-of-function allele, we know that $q^2 = 0.0001$; therefore $q = \sqrt{0.0001} = 0.01$. By subtraction, the frequency

Figure A23.1

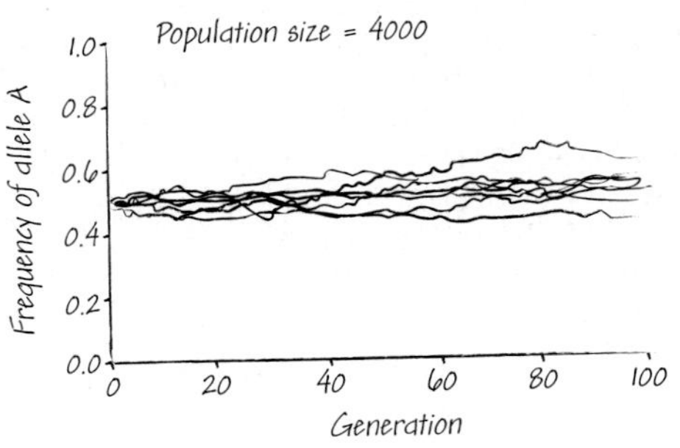

Figure A23.2

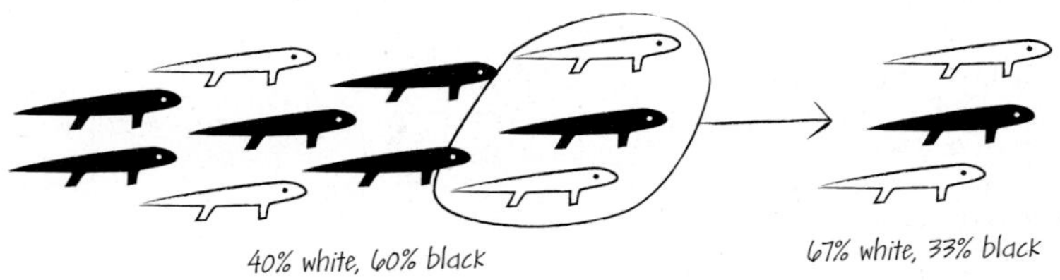

of normal alleles is 0.99. Under the Hardy–Weinberg principle, the frequency of heterozygotes is $2pq$, or $2 \times 0.99 \times 0.01$, which is 0.02 (about 2 people per hundred). **10.** analyze For the species whose males never help raise offspring, the fundamental asymmetry of sex is pronounced, and sexual dimorphism is likely to be high. If males invest a great deal in raising offspring, then the fundamental asymmetry of sex is small and sexual dimorphism is likely to be low.

✔ PUT IT ALL TOGETHER: Case Study

11. apply b. Genetic diversity in forest fragments is expected to decrease, because toucans are no longer carrying seeds (and their alleles) among palm populations. **12.** analyze Average seed size is lower in the forest fragment with toucans absent (about 9.5 mm) than in the forest with toucans present (about 11 mm). **13.** apply Directional selection. The average seed size of palms has declined in the absence of toucans, presumably because smaller seeds have higher fitness in the absence of toucans. **14.** apply If larger palm seeds have higher fitness than smaller palm seeds (in the presence of toucans), then the fitness of palms will decrease over time in the absence of toucans because the large seeds will no longer be dispersed and will have a lower chance of survival. **15.** apply Forest fragments may vary in slope, moisture, species composition, and other factors. If the researchers compare just two fragments, they may happen to get a result that is not typical of the other fragments, by chance. By comparing many fragments, the researchers can be more confident that the trend in their samples represents a real trend in the rest of the forest. **16.** evaluate This statement is false. You could point out that the data show clearly that evolution in palm seed size has occurred over the few decades since humans have caused forest fragmentation and local extinction of toucans. Evolution, a change in allele frequencies in a population, can occur as quickly as in a single generation. In this case, human fragmentation of the forest has already caused measurable evolution in the forest—and this effect is likely to increase over time as the fitness of the palms declines, causing a cascade of effects on other species.

Chapter 24

IN-TEXT QUESTIONS AND EXERCISES

p. 494 Making Models 24.1 apply See **Figure A24.1** for one possible answer. Note that in trees like this, the shapes of the branches at the node are abitrary—there are many ways to make one line split into two. Data about the rate of divergence of species are rarely available and are rarely incorporated into the shape of the tree branches.

p.495 (1) apply The monkeyflowers are experiencing habitat isolation because they live at different elevations. They also have behavioral isolation because they are attracting different pollinators, which means the sperm in the pollen from one population will not be carried to the eggs in the ovules in the other population. **(2)** apply The fruit flies are experiencing hybrid inviability because the male larvae are dying. Hybrid female flies cannot reproduce without males, so the populations of flies will remain reproductively isolated.

p. 497 Fig. 24.4 create See **Figure A24.2.**

p. 498 CYU (1) apply Habitat isolation is likely to be most relevant in the case of trumpeters; if trumpeters cannot cross rivers, then they would be able to mate with trumpeters on the same side of the river but not on the other side—a form of prezygotic isolation. **(2)** apply All three species concepts could be employed in the case of trumpeters in the Amazon. Reproductive isolation could be assessed as a measure of the biological species concept; the birds are alive today so their ability to produce viable offspring could be tested, although the birds would have to be brought together into the same habitat because they are geographically isolated. Morphological differences could be assessed as a measure of the morphospecies concept, although researchers might debate how different the differences would have to be. A phylogenetic tree based on DNA sequences or other traits could be assessed as a measure of the phylogenetic species concept. All three concepts have advantages and disadvantages; the results of the three tests could be compared to see if they are congruent.

p. 499 apply The small population on the right represents the new population of large ground finches on Daphne Major. The red color represents the increasingly large beak size on the island. In step 2, add a small arrow connecting the original and new populations to show that some gene flow still occurs.

p. 499 CYU (1) analyze This is an example of vicariance—chance events (land-use changes such as habitat destruction) physically separated the elephants into geographically isolated populations. **(2)** apply Allopatric speciation would likely occur again, because gene flow among the elephant populations is limited. The small elephant populations have experienced bottleneck events where large populations were reduced in size; this is a type of genetic drift in which the allele frequencies in some remnant populations are likely to be different than in others by chance (see Ch. 23, Section 23.4). These differences are often exaggerated by continued genetic drift, mutation, and natural selection, causing the populations to diverge over time to the point where they are genetically isolated and may eventually form new species (if they do not go extinct).

p. 500 Fig. 24.6 analyze The two clusters of species in the tree represent the species to the north and south of the Amazon River, as this was the earliest vicariance event to subdivide the trumpeters.

p. 502 apply The gray and green balls represent individuals with different feeding preferences. For example, the gray balls could represent resident killer whales that eat fish, while the green balls represent transient killer whales that eat marine mammals. In the beginning, there is gene flow between the different ecotypes; but at the end, the two ecotypes are reproductively isolated.

p. 503 apply A diploid grape plant sustains a defect in meiosis that results in the formation of diploid gametes. The individual self-fertilizes, producing tetraploid offspring. The tetraploid grape plants self-fertilize or mate with other tetraploid individuals, producing a tetraploid population.

p. 504 apply A tetraploid ($4n$) species (in this case, wheat) gives rise to diploid ($2n$) gametes, and a diploid wheat species gives rise to haploid (n) gametes. When a haploid gamete (n) fertilizes a diploid gamete ($2n$), a sterile triploid offspring results. If an error in mitosis occurs that doubles the chromosome number before meiosis, then hexaploid wheat is formed.

p. 504 CYU analyze The isolation of trumpeters is allopatric (occurs in different geographic areas), whereas the isolation of killer whales is sympatric (occurs in the same area). However, in both cases genetic isolation was caused by a lack or reduction of gene flow among populations. Also in both cases, genetic divergence occurred within populations due to mutation, selection, and drift.

p. 505 Figure 24.10 create After many generations, the average gill raker number of the whitefish is likely to be low. The graph would look similar to the bell-shaped curve of gill raker number in the original benthic population (with no whitefish having a high gill raker number).

p. 506 Fig. 24.12 analyze The hypothesis would have been rejected if orange and red gene regions in the experimental hybrid (corresponding to the regions inherited from the two parent species) did not align with those regions in the genome of the natural hybrid, *H. anomalus*.

p. 507 CYU analyze When reinforcement occurs, hybrids have low fitness; this is why avoiding hybridization improves the fitness of reproductive individuals (their reproductive efforts are not wasted). In contrast, the hybrids in the case of speciation by hybridization have higher fitness, usually because they happen to have adapatations for certain ecological conditions.

✔ TEST YOUR KNOWLEDGE

1. remember a **2.** remember b **3.** understand T, T, F, F; the biological species concept can only be applied to living species with sexual reproduction. **4.** analyze The statement is false; gene flow reduces the divergence of populations.

✔ TEST YOUR UNDERSTANDING

5. understand a **6.** apply Sample answer: Variation in male calls could occur, such as a lower-pitched call at the cool forest floor and a higher-pitched call in the warm forest canopy. Females in the forest floor could start to prefer males with lower-pitched calls, and females in the canopy could start to prefer males with higher-pitched calls, creating genetic isolation. Over time, genetic divergence could occur between the two populations. **7.** apply c **8.** apply False. Speciation is sometimes a slow process, as in the formation of 8 species of trumpeters over 3 million years. But speciation can occur rapidly, as in the formation of a new species of *Tragopogon* by allopolyploidy in a single generation.

✔ TEST YOUR PROBLEM-SOLVING SKILLS

9. apply The sticklebacks are geographically isolated in the different lakes. This is a classic setting for allopatric speciation to occur. Whether the fish in the different lakes would be considered different species

Figure A24.1

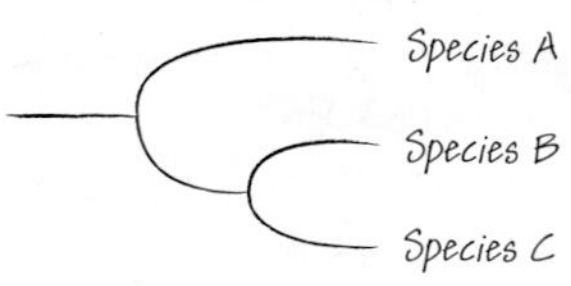

Figure A24.2

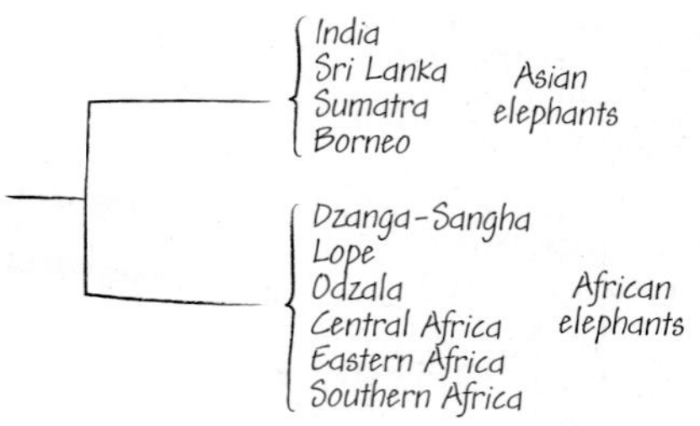

today depends on how much time has passed since the populations diverged and which species concept is used. **10.** analyze If the isolated populations and habitat fragments are small enough, species are likely to dwindle to extinction due to inbreeding and loss of genetic variation, or due to catastrophes such as a severe storm or a disease outbreak. If the isolated populations survive, they are likely to diverge into new species via allopatric speciation due to these vicariance events, because they are genetically isolated and because the habitats may differ.

✔ PUT IT ALL TOGETHER: Case Study

11. apply Biological species concept: Human populations are not considered separate species because all human populations can successfully interbreed. Morphological species concept: Although people of different races and from different populations look superficially different, they are not separate species because they have virtually identical anatomy and physiology. Phylogenetic species concept: Human populations are not separate species because they all arose from a very recent common ancestor and cluster together as one branch in a phylogenetic tree of species. (DNA comparisons have revealed that human races are remarkably similar genetically and do not differ enough to qualify for species status.) **12.** apply The morphospecies concept was used to distinguish fossil species. It was very difficult to determine whether morphological differences among fossils represented normal variation within populations (e.g., difference between a male and female) or meaningful differences among species. **13.** apply a **14.** analyze Yes. Modern humans and Neanderthals interbred in Europe after modern humans migrated out of Africa. This explains why modern Africans have no Neanderthal genes, but many humans in other parts of the world do carry Neanderthal genes—they are the result of hybridization events between the two species. **15.** evaluate To determine if there is a significant difference between groups, you could separate the data by eye color then run a statistical test such as a T-test (**BioSkills 3**) to compare the means and variance of the two groups. If $P < 0.05$ the difference is said to be significant. **16.** apply Extinction of one population. Modern humans may have been stronger competitors (better able to obtain food or other resources), which likely contributed to the Neanderthals' extinction.

Chapter 25

IN-TEXT QUESTIONS AND EXERCISES

p. 511 apply See **Figure A25.1**. The node is the point where the chimpanzee and human branches meet. This is the inferred ancestor of chimpanzees and humans (which is neither chimpanzee nor human).

p. 512 Making Models 25.1 apply See **Figure A25.2** for one possible solution.

p. 512 Making Models 25.2 analyze See **Figure A25.3**. The arctic grayling is equally related to all the salmon because all the salmon share a common ancestor.

p. 513 Fig 25.1 apply See **Figure A25.4**.

p. 516 Fig. 25.4 analyze See **Figure A25.5**. SINEs group (19, 20) identifies peccaries and pigs as a monophyletic group.

p. 517 CYU (1) apply See **Figure A25.6**. In a fish fraud analysis like this, each fish sample is represented by its own branch in the tree so that the question of origin can be answered. However, in typical species trees (e.g., see Figure 24.4), all samples with identical sequences are lumped into one branch. **(2)** understand Whales did not evolve directly from hippos. Rather, whales and hippos share a common ancestor (which was neither a whale nor a hippo) at the node where their two branches meet. **(3)** analyze Hair and limb structures in humans and whales are examples of homology because they are traits that can be traced to a common ancestor. All mammals have hair and similar limb-bone structure. However, extensive hair loss and advanced social behavior in whales and humans are examples of homoplasy. These traits are not common to all mammalian species and likely arose independently during the evolution of specific mammalian lineages.

p. 519 apply Intact fossils are the most likely to yield DNA evidence because the organic remains are preserved intact. Also, the younger the fossil, the more likely it is to contain intact DNA.

p. 519 apply While the fossil record does have limitations (habitat, taxonomic, tissue, temporal, and abundance biases), the presence of feathers in a non-avian dinosaur such as *Tyrannosaurus* is sufficient evidence to reject the hypothesis that feathers are unique to birds.

p. 521 CYU (1) apply Trace fossil. **(2)** apply Mollusks have lived on Earth since the Cambrian, about 541 million years. **(3)** analyze Predation is a behavior that is difficult to observe in the fossil record, but drill holes are a good choice because they occur in clamshells that fossilize well. Also, clams live in areas where sediments are likely to be deposited, have hard parts that are likely to fossilize well, and they occur over a long time span up to the present.

p. 522 create In habitats on the mainland, complex communities have been established for a long time, so tarweeds in California experience greater competition for resources and are limited to specific niches.

p. 525 CYU apply **(1)** Ecological opportunity: Many resources were available because no other animals (or other types of organisms) existed to exploit them. Also, the evolution of new species in new niches made new niches available for predators. **(2)** Morphological innovation: changes in form such as limbs and complex mouthparts were important because they enabled animals to live in habitats other than the benthic area and to consume new types of food.

p. 525 Fig. 25.11 analyze The era of the dinosaurs ended during the end-Cretaceous. About 15 percent of plant and animal families went extinct at that time. [Because species can be difficult to identify (Ch. 24, Section 24.1), families are a more practical and reliable metric of ancient taxonomic diversity; plants and animals were used because they have a more complete fossil record than other taxonomic groups.]

p. 527 Fig. 25.13 analyze The data lines would be flatter, similar to the background extinction line.

p. 527 CYU (1) apply No, pterosaurs were probably well adapted to their natural environment before an asteroid stuck 66 million years ago. During the mass extinction, most organisms, including pterosaurs, were affected by the toxic environment and/or the drop in primary productivity which caused whole ecosystems to collapse. Pterosaurs were unlucky. **(2)** analyze Overfishing is more similar to a mass extinction event than to background extinction, because sharks are harvested no matter how well adapted they are to their environment.

✔ TEST YOUR KNOWLEDGE

1. understand c; although teeth, shells, and other hard parts are most likely to be preserved, fossils of soft tissues

Figure A25.1

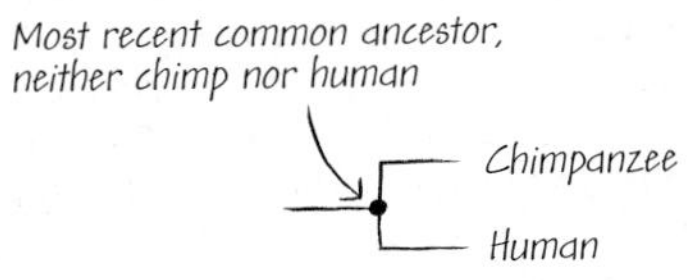

Figure A25.2

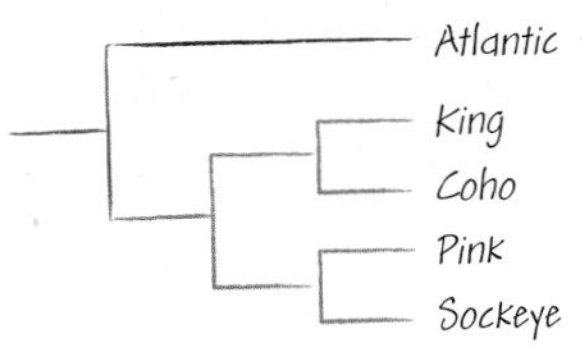

Figure A25.3

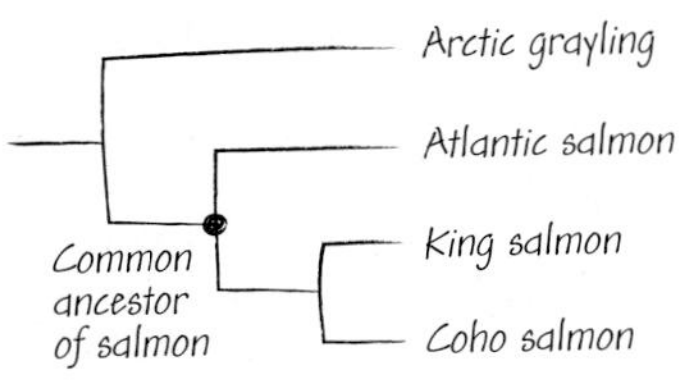

Figure A25.4

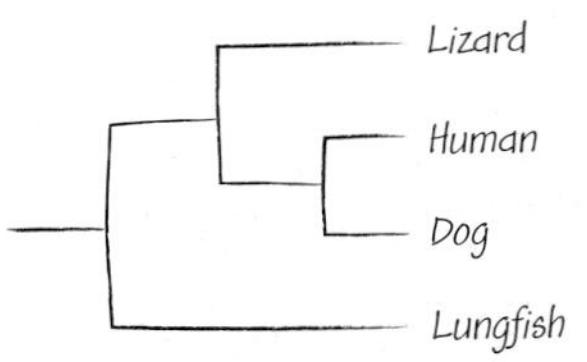

Figure A25.5

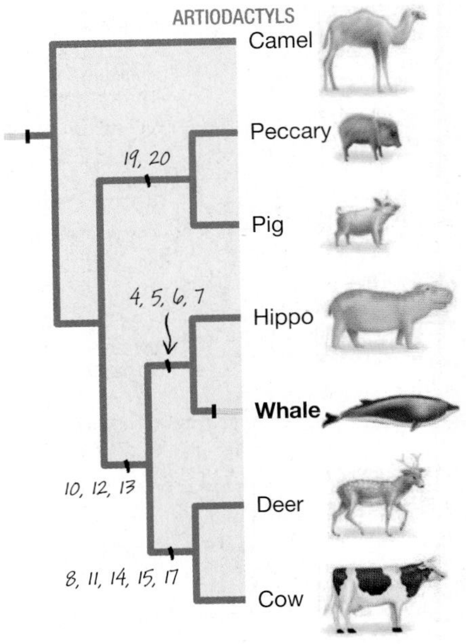

and other traces of organisms also occur and are not always converted into rock. **2.** understand The statement is false; the loss of a trait, such as the loss of legs in snakes or whales, can be a valuable synapomorphy for a clade. **3.** understand b **4.** understand d

✔ TEST YOUR UNDERSTANDING

5. understand Parsimony assumes that the tree with the fewest number of character changes is the most likely to represent evolutionary history. In the case of the artiodactyl astragalus, parsimony was misleading because the astragalus was lost when whales evolved limblessness—creating two changes (a loss following a gain) instead of one (a gain). **6.** evaluate One snip gets an ancestor, all its descendants, and nothing else. **7.** apply b **8.** apply Winged insects appear earlier, in the Devonian, whereas the first birdlike reptiles don't appear until much later in the Jurassic (and flight at some point after). Flight evolved independently in these groups, so it's an example of convergent evolution.

✔ TEST YOUR PROBLEM-SOLVING SKILLS

9. create See one solution in **Figure A25.7** (your drawing may use different branch rotations). The most parsimonious tree has the fewest number of changes (five in this case). Adding more characters to the matrix can improve your confidence in the results.

10. create *P. falciparum* originated from *Plasmodium* in gorillas, because *P. falciparum* shares a most recent common ancestor with *Plasmodium* in gorillas. This information could possibly be used to understand how *P. falciparum* is so lethal (e.g., researchers could look for a sequence change present in a human parasite that might be linked to lethality, see what the sequence codes for, and make a drug that targets this product).

Figure A25.6

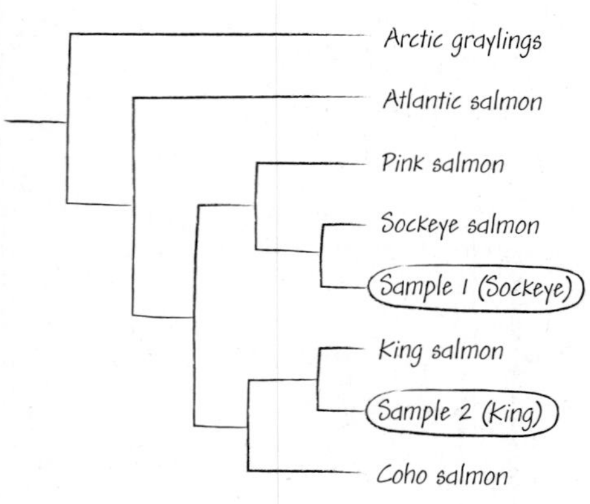

Figure A25.7

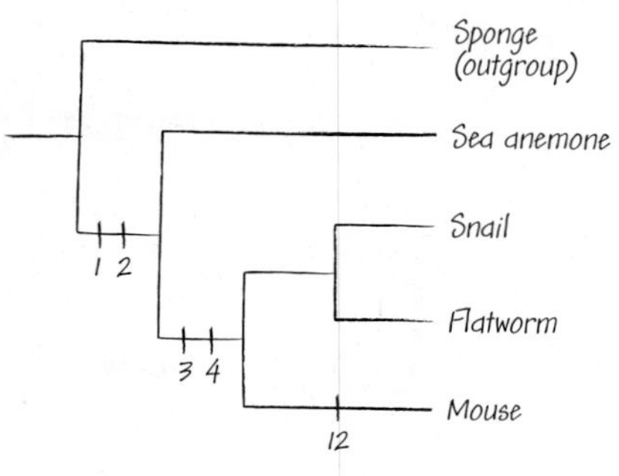

✔ PUT IT ALL TOGETHER: Case Study

11. analyze The end of the Jurassic (vertical line at 145 mya), entire Cretaceous (vertical line at 66 mya) within the Mesozoic era, and the beginning of the Paleogene in the Cenozoic era. **12.** analyze b; because birds are part of the monophyletic group called theropod dinosaurs and birds still live today, dinosaurs are not extinct. (The original definition of dinosaurs did not include birds and therefore was not a monophyletic group.) **13.** analyze The tree shows that most of the lineages died out at the K–Pg boundary, when an asteroid struck Earth and caused a mass extinction. If the Phanerozoic started at "noon," then the mass extinction of dinosaurs would have occurred at $66/541 = 0.12 =$ the moment when 12 percent of the time was left $= 0.12 \times 12$ hours $= 1.5$ hours before midnight, or 10:30 p.m. **14.** analyze F; while *Archaeopteryx* and *Aurornis* are next to each other in the phylogeny, *Archaeopteryx* and living birds share a more recent common ancestor (at the node at the base of the *Archaeopteryx* branch) than either does with *Aurornis* (see **BioSkills 13** for more practice reading trees). T; birds are more closely related to Enantiornithes. F; *Archaeopteryx* is assumed to be a sister group to the clade that includes living birds, not a direct descendant; the two taxa are assumed to have shared a common ancestor at the node where the branches meet. F; *T. rex* and living birds share a common theropod ancestor, but *T. rex* is not a direct ancestor of birds. **15.** apply See **Figure A25.8**. If feathers and wings existed before the origin of living birds, then these traits cannot be considered a synapomorphy for living birds, because they are not unique to this group; the mere presence of feathers and wings is not sufficient to explain the adaptive radiation of birds. **16.** evaluate Many possible answers. Researchers are likely relatively confident about the age and geographic location of the fossils found, the morphological features of the fossils, and the sudden extinction of the long-standing lineages. The researchers are probably not confident about the overall diversity of feathered dinosaurs, as new species are still being discovered (and only some of the discovered species are included in the illustration). Researchers are probably not confident about some of the divergence times of the phylogeny. The phylogeny is likely to change depending on the number of taxa and traits included and the type of analysis used.

END-OF-UNIT CASE STUDY Are Garter Snakes and Newts Engaged in an Arms Race?

pp. 530–531 (1) apply False, True, True, True **(2)** apply Two alleles (versions of a gene) are represented. New alleles arise via random mutation. **(3)** apply *Newts—Benefit of TTX production:* Deters predators, increasing newt survival and reproduction. *Newts—Cost of TTX production:* Requires expense of energy which cannot be used for other aspects of survival and reproduction. *Snakes—Benefit of TTX resistance:* Enables snakes to eat newts without dying, increasing snake survival and reproduction. *Snakes—Cost of TTX production:* Snakes move more slowly and are more vulnerable to predators, decreasing their chance of survival and reproduction. **(4)** apply No, evolution does not make species perfect because there are trade-offs to many traits. For example, resistance to TTX and speed of locomotion cannot both be optimized simultaneously. Also, the environment is variable over space and changing over time; no traits are optimal in all conditions. **(5)** analyze There is an overall correlation between areas of high newt toxicity and areas of high snake resistance, although the overlap is not perfect. Hunters would be most likely to die from newt-infused coffee in the two dark red spots in western Oregon. **(6)** apply *T. atratus, T. couchii, T. sirtalis 1*. **(7)** apply See **Figure EOU A4.1**. **(8)** apply No, even if the branches are rotated, there is no way to get all the colored branches together without changing the relationships in the tree. **(9)** apply It is more parsimonious to conclude that the most recent common ancestor was TTX-sensitive and that TTX-resistance arose independently in multiple lineages. **(10)** apply Homoplasy **(11)** apply It looks like an evolutionary arms race is occurring because there is evidence that snakes and newts are evolving adapatations and counter-adaptations in response to interactions with each other. Where newt toxicity is higher, there is natural selection for TTX-resistance in snakes. Where TTX-resistance occurs in snakes, there is natural selection for higher toxicity in newts. In this way, the toxicity of newts can increase over time and the TTX-resistance in snakes can also increase over time.

Figure A25.8

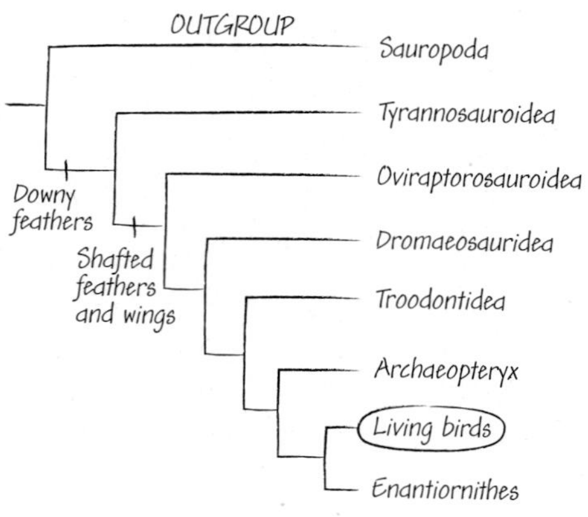

BIG PICTURE Evolution

pp. 532–533 CYU (1) understand Circle = inbreeding, sexual selection, nonrandom mating, natural selection, genetic drift, mutation, and gene flow. **(2)** analyze Adaptation "increases" fitness; synapomorphies "identify branches on" the tree of life. **(3)** analyze Several answers possible, for example, adaptive radiations and mass extinctions → can be studied using → the fossil record. **(4)** analyze Genetic drift, mutation, and gene flow "are random with respect to" fitness.

Figure EOU.A4.1

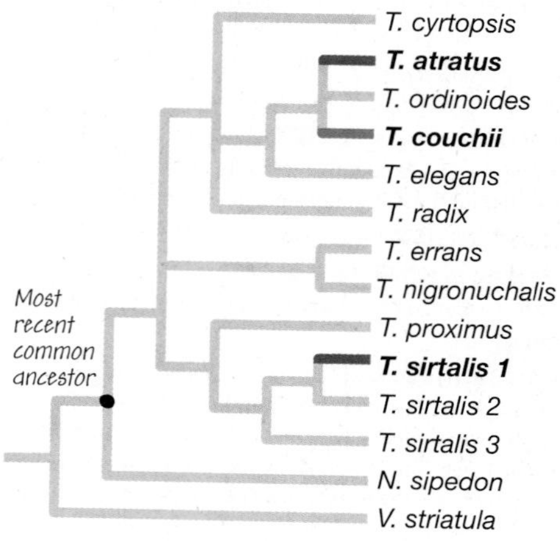

Chapter 26

IN-TEXT QUESTIONS AND EXERCISES

p. 539 CYU (1) remember Extremophiles includes those microbes that thrive in harsh conditions such as very high temperatures and very low or high pH. **(2)** understand Endospores provide a way for prokaryotes to survive conditions that are unsuitable for growth. The bacterial DNA is walled off from the rest of the cell, forming a resistant endospore that can resume growth when conditions are favorable. **(3)** apply Certain microbes are able to metabolize pollutants and convert them into harmless by-products. Bioremediation can be an effective and cost-efficient way to reduce certain pollutants.

p. 539 Fig. 26.3 evaluate Weak—different culture conditions may have revealed different species.

p. 541 CYU (1) create Conditions should mimic a spill—sand or stones with a layer of crude oil, or seawater with oil floating on top. Add samples, from sites contaminated with oil, that might contain cells capable of using molecules in oil as electron donors or electron acceptors. Other conditions (temperature, pH, etc.) should be realistic. **(2)** create Use metagenomic analysis. After isolating DNA from a soil sample, fragment and sequence the DNA. Compare the sequences to those from known organisms and use the data to place the species on the tree of life.

p. 543 Fig. 26.8 analyze Approximately 500 times larger. The diameter of *Thiomargarita* is approximately 250 microns. The diameter of *Mycoplasma* is approximately 0.5 microns. $250 \div 0.5 = 500$.

p. 548 apply If bacteria and archaea did not exist, then (1) the atmosphere would have little or no oxygen, and (2) almost all nitrogen would exist in molecular form (the gas N_2).

p. 548 Fig. 26.13 analyze Approximately 3.5 times as much. Species A will obtain roughly 200 kcal/mol glucose, while species B will obtain roughly 700 kcal/mol glucose: $700 \div 200 = 3.5$.

p. 549 Fig. 26.14 create Add a label for "Animals" in the upper right quarter of the circle, and draw arrows leading from "Organic compounds with amino groups" to "Animals" and from "Animals" to "NH_3."

p. 550 CYU (1) understand Prokaryotes can take up DNA (including plasmids) from their immediate environment through the process of transformation. **(2)** understand *E. coli* is a rod-shaped bacterium whose cell wall includes an outer membrane. *Staphylococcus aureus* is a spherically shaped bacterium arranged in clusters whose cell wall consists of a relatively thick layer of peptidoglycan. **(3)** understand Photoautotrophic prokaryotes perform photosynthesis like plants do (they use the energy in light and the carbon from carbon dioxide to produce their own organic compounds). Chemoorganoheterotrophic prokaryotes rely on absorbing organic molecules as a source of energy and cellular building blocks.

p. 552 CYU (1) understand Cyanobacteria produce oxygen as a by-product of photosynthesis and nitrogen in a form useable by other organisms. **(2)** remember *Sulfolobus* would likely be found in hot, acidic, high-sulfur enironments.

✔ TEST YOUR KNOWLEDGE

1. remember b **2.** remember d **3.** remember peptidoglycan **4.** understand T, T, T, F (the microbe *should* be able to be isolated from the diseased experimental animal).

✔ TEST YOUR UNDERSTANDING

5. c **6.** An electron donor provides the potential energy required to produce ATP. **7.** Large amounts of potential energy are released and ATP produced when oxygen is the electron acceptor, because oxygen is so electronegative. Large body size and high growth rates are not possible without large amounts of ATP. **8.** This result supports their hypothesis because the drug poisons the enzymes of the electron transport chain and prevents electron transfer to Fe^{3-}, which is required to drive magnetite synthesis. If magnetite had still formed, another explanation would have been needed.

✔ TEST YOUR PROBLEM-SOLVING SKILLS

9. create Hypothesis: A high rate of tooth cavities in Western children is due to an excess of sucrose in the diet, which is absent from the diets of East African children. To test this hypothesis using an animal system, you could feed a group of mice a diet that contains sucrose and another group a diet that lacks sucrose, then monitor the abundance of *S. mutans* in both groups. Although a similar study in Western and East African children might be theoretically feasible, it would be difficult to avoid unintended negative consequences. **10.** create Sample water or soil polluted with benzene-containing compounds. Put samples from these environments in culture tubes where benzene is the only source of carbon. Monitor the cultures and study the cells that grow efficiently.

✔ PUT IT ALL TOGETHER: Case Study

11. understand b **12.** understand Plasma membranes and cell walls of Bacteria are composed of different molecules than those in Archaea. These two groups also have different types of DNA polymerases, RNA polymerases, and ribosomes. **13.** evaluate The concept of prokaryotes and eukaryotes representing the two primary lineages of life is not well supported by either model. Each hypothesis presents prokaryotes as two separate groups. **14.** apply See **Figure A26.1**. **15.** evaluate According to the two-domain hypothesis, Archaea includes prokaryotes and eukaryotes. Although the prokaryotes lack nuclei, eukaryotic cells possess nuclei. **16.** create You would want to know as much as possible about the cell biology, DNA sequence data, and even the fossil record (if available) for each of the major groups (especially members at the base of each group). The discovery of early eukaryotic cells with features similar to intermediate lineages of Archaea would be consistent with the eocyte hypothesis. Early eukaryotic cells with features similar to early Archaea would be consistent with the three-domain hypothesis.

Chapter 27

IN-TEXT QUESTIONS AND EXERCISES

p. 556 apply Protists are not considered monophyletic because they do not include the plants, animals, and fungi. (See Figure 27.1.)

p. 559 CYU (1) understand *Plasmodium* species are transmitted to humans by mosquitoes. If mosquitoes can be prevented from biting people, they cannot spread the disease. The insecticide-treated nets provide a physical barrier as well as a means to kill the mosquitoes. **(2)** create Iron added → primary producers (photosynthetic protists and bacteria) bloom → more carbon dioxide taken up from atmosphere during photosynthesis → consumers bloom, eat primary producers → bodies of primary producers and consumers fall to bottom of ocean → large deposits of carbon-containing compounds form on ocean floor.

p. 562 CYU (1) understand Opisthokonts (meaning "base-tail") have a flagellum at the base or back of the cell; stramenopiles (meaning "straw-hairs") have straw-like hairs on their flagella. **(2)** understand In direct sequencing, DNA is isolated directly from the environment and analyzed to place species on the tree of life. It is not necessary to see the species being studied.

p. 562 understand See **Figure A27.1**.

p. 563 Fig. 27.8 apply Two—one derived from the original bacterium and one derived from the host cell that engulfed the bacterium.

p. 565 understand Membrane infoldings observed in bacterial species today support this hypothesis—they confirm that the initial steps could have occurred. The continuity of the nuclear envelope and ER are consistent with the hypothesis, which predicts that the two structures are derived from the same source (infolded membranes).

p. 566 Fig. 27.13 understand If euglenids could take in food via phagocytosis (ingestive feeding), then that would have provided a mechanism by which a smaller photosynthetic protist could have been engulfed and incorporated into the cell via secondary endosymbiosis.

Figure A26.1

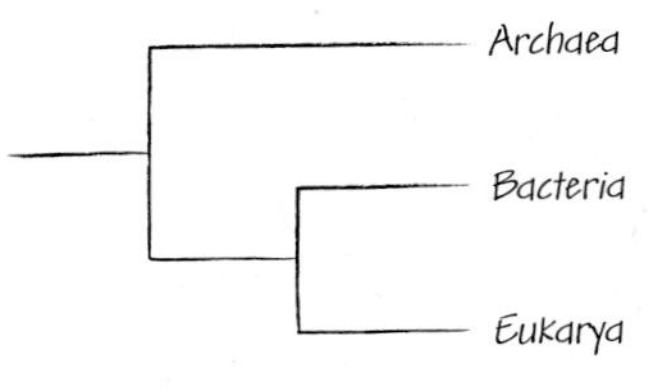

Figure A27.1

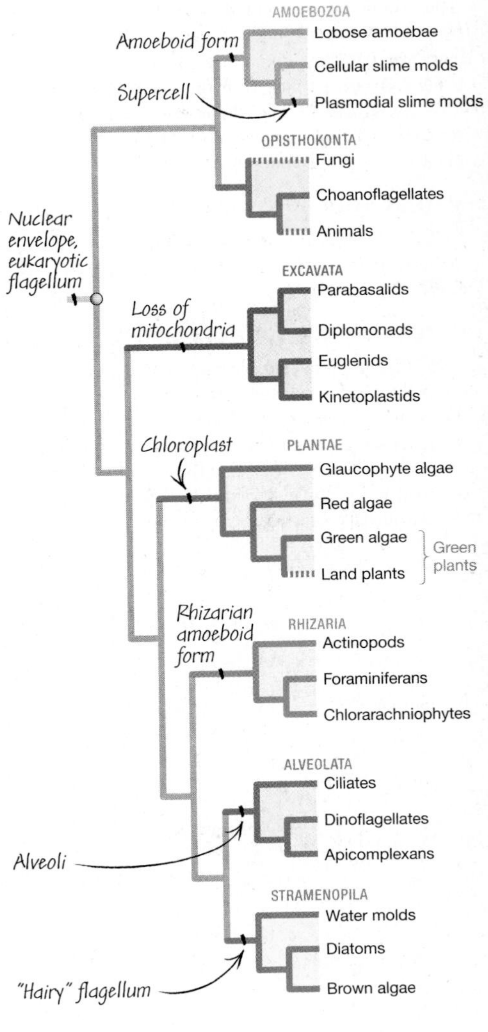

p. 568 evaluate When food is scarce or population density is high, the environment is changing rapidly (deteriorating). Offspring that are genetically unlike their parents may be better able to cope with the new and challenging environment.

p. 568 Making Models 27.1 apply See **Figure A27.2**.

p. 570 Fig. 27.17 understand In the brown algae species shown, which display alternation of generations, gametes are formed by mitosis. If you answered "meiosis," you were probably thinking about how humans and other animals produce gametes (egg and sperm) by meiosis. Take another look at Figure 27.17 and notice that gametophytes and gametes are haploid (there is no change in chromosome number). Meiosis results in the formation of spores that contain one-half the number of chromosomes that sporophytes have.

p. 571 CYU (1) apply The acquisition of the mitochondrion and the chloroplast represent the transfer of entire genomes, and not just single genes, to a new organism. **(2)** evaluate The gametes are the only haploid stage in humans. Therefore, the life cycle of humans is more comparable to *Thallassiosira* (which is dominated by diploid cells) than *Gyrodinium* (which is dominated by haploid cells). The diploid and haploid stages in humans look radically different from each other. Therefore, the human life cycle is more comparable to *Laminaria* than to *Ectocarpus*.

p. 571 a & b, **p. 572** c, **p. 573** d & e, **p. 574** f apply See Figure A27.1.

p. 574 Fig. 27.18 analyze If the startle hypothesis were not supported, then the graph representing the number of bioluminescent dinoflagellate cells consumed per copepod would look essentially identical to that of the non-bioluminescent dinoflagellates.

p. 574 CYU (1) analyze The amoeboid cells of Amoebozoa likely evolved separately from the amoeboid cells in Rhizaria. **(2)** analyze Dinoflagellates are very distantly related to opisthokonts. Therefore, bioluminescence likely evolved independently in these separate lineages.

✔ TEST YOUR KNOWLEDGE

1. understand b **2.** remember diatoms **3.** remember F (amoeboid motion is found in protists that lack cell walls), T, T, T **4.** remember b

✔ TEST YOUR UNDERSTANDING

5. understand Because all eukaryotes living today have cells with a nuclear envelope, it is valid to infer that their common ancestor also had a nuclear envelope. Because bacteria and archaea do not have a nuclear envelope, it is valid to infer that the trait arose in the common ancestor of eukaryotes. **6.** understand The host cell provided a protected environment and carbon compounds for the endosymbiont; the endosymbiont provided increased ATP from the carbon compounds. **7.** evaluate It confirmed a fundamental prediction made by the hypothesis and could not be explained by any alternative hypothesis. **8.** understand All alveolates have alveoli, which are unique structures that function in supporting the cell. Among alveolates there are species that (1) are ingestive feeders, photosynthetic, or parasitic, and (2) move using cilia, flagella, or a type of amoeboid movement.

✔ TEST YOUR PROBLEM-SOLVING SKILLS

9. analyze If the apicoplast that is found in *Plasmodium* (the organism that causes malaria) is genetically similar to chloroplasts, and if glyphosate poisons chloroplasts, it is reasonable to hypothesize that glyphosate will poison the apicoplast and potentially kill the *Plasmodium*. Because humans have no chloroplasts, this would be a good treatment strategy for malaria, provided that the glyphosate produces no other effects that would be detrimental to humans. **10.** evaluate Primary producers usually grow faster when the CO_2 concentration increases, but to date they have not grown fast enough to make the CO_2 level drop—the level of this gas has been increasing steadily over decades.

✔ PUT IT ALL TOGETHER: Case Study

11. remember c **12.** analyze The plasmodial stage of *Physarum* is one large supercell consisting of thousands of nuclei. The migrating slug of *Dictyostelium* is an aggregation of hundreds of individual cells, each with its own plasma membrane. Both of these structures, however, are able to move about their environment via amoeboid motion. **13.** analyze Because $P < 0.05$, there was a statistically significant increase in the amount of time that it took the slime mold to navigate the trap and reach the food when the agar was pretreated with extracellular slime. **14.** create To determine if there are speed differences, you could design an experiment with two conditions—one with slime mold placed on plain agar and one with slime mold placed on agar coated with extracellular slime. For both conditions, a food source could be placed a specified distance away from the slime mold and the time required to reach the food source recorded. **15.** create (1) To test light versus dark environments, place some slime mold in the middle of a chamber that has one half exposed to light and the other half covered up and, therefore, dark. (2) To test dry versus moist environments, place some slime mold in the middle of a chamber with one side kept dry and the other side moist. (3) To test oats versus sugar, place some slime mold in the middle of a chamber with oats on one side and sugar on the opposite side. For each setup, observe mold growth to see which treatment is preferred. To ensure robust results, replicate these experiments several times. **16.** evaluate Slime molds find the shortest route possible from one food item to another, therefore conserving energy. This decision-making process could be used to design roads or rail lines that connect major cities.

Chapter 28

IN-TEXT QUESTIONS AND EXERCISES

p. 580 CYU (1) apply Deforestation would have negative consequences including lower oxygen and higher carbon dioxide levels, erosion and loss of soil, and habitats that are drier and more subject to temperature swings. **(2)** analyze Most of the food that we eat with caloric value comes from plants dirctly (e.g., fruits, vegetables, wheat, rice, corn, sugar) or indirectly (e.g., an animal that fed on grass or other plants or fish that ate other fish that ultimately relied on photosynthetic algae as a food source).

p. 583 Fig. 28.6 apply If you find the common ancestor of all the green plants (at the base of the Ulvophyceae), the lineages that are collectively called green algae don't include that common ancestor and all of its descendants—only some of its descendants. The same is true for nonvascular plants (the common ancestor here is at the base of the Hepaticophyta) and seedless vascular plants (the common ancestor here is at the base of the Lycophyta).

p. 584 CYU understand Green algae and land plants share an array of morphological traits that are synapomorphies, including the chlorophylls they contain. Green algae also appear long before land plants in the fossil record. Phylogenetic trees estimated from DNA sequence data show that green algae and land plants share a common ancestor—green algae are the initial groups to diverge, and land plants diverge subsequently.

p. 586 Fig. 28.8 evaluate Water flows more easily through a short, wide pipe than through a long, skinny one because there is less resistance from the walls of the pipe. In addition, the end walls of vessels have perforations that allow water to move freely between vessels.

p. 589 Fig. 28.12 understand Mitosis. Haploid cells in the gametophyte produce haploid gametes. There is no change in number of chromosomes, and the ploidy level remains constant. The formation of sperm and eggs in plants is different from that in animals, whose gametes are formed directly by meiosis.

p. 589 understand Cells in the sporophyte undergo meiosis, producing haploid spores, which undergo mitosis, becoming haploid gametophytes. Cells in the gametophyte then undergo mitosis, producing gametes, which fuse to become diploid zygotes. Zygotes grow by mitosis into a mature diploid sporophyte.

p. 589 Making Models 28.1 create See **Figure A28.1**. It is different from the generic alga life cycle in that both the

Figure A27.2

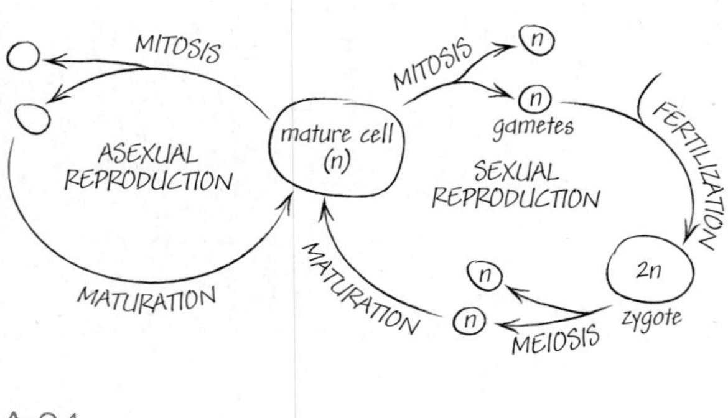

Figure A28.1

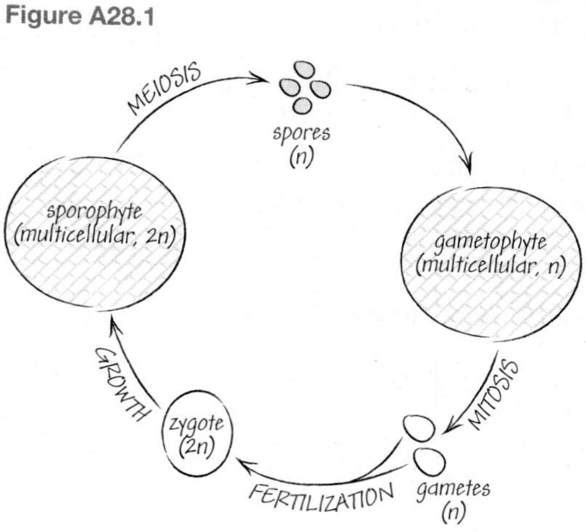

diploid and haploid stages are multicellular.

p. 590 Fig. 28.14 understand There are multicellular haploid stages (gametophytes) and multicellular diploid stages (sporophytes) in these plants.

p. 591 apply In the hornwort photo, the sporophyte is the spike-like green and yellow structure; the gametophyte is the leafy-looking structure underneath. The horsetail gametophyte is the microscopic individual on the left; the sporophyte is the much larger individual on the right.

p. 592 analyze Gymnosperm gametophytes are microscopic, so they are even smaller than fern gametophytes. The gymnosperm gametophyte is completely dependent on the sporophyte for nutrition, while fern gametophytes are not. Fern gametophytes are photosynthetic and even supply nutrition to the young sporophytes. Fern gametophytes are bisexual, while separate male and female gametophytes are found in gymnosperms.

p. 593 Fig. 28.19 analyze Consistent—the fossil data suggest that gymnosperms evolved earlier, and gymnosperms have larger gametophytes than angiosperms.

p. 595 Fig. 28.21 analyze To eliminate all other variables except for the one being tested, such as the possibility of pollination by other insects in the field. Scales differ because the two insect species visited flowers at very different frequencies.

p. 596 Fig. 28.23 analyze Flagellated sperm were likely lost once in the lineage leading to the three groups mentioned (in the single line leading up to the split of those three groups in Figure 28.6).

p. 597 CYU (1) apply Cuticle prevents water loss from the plant; UV-absorbing compounds allow plants to be exposed to high light intensities without damage to their DNA; vascular tissue moves water up from the soil and moves photosynthetic products down to the roots. **(2)** understand Pollen originated in the lineage leading to the splitting of gymnosperms and angiosperms. Flowers and fruits originated in the angiosperms.

p. 599 CYU (1) understand Spores are produced on the underside of fern leaves. Therefore, the fern is the sporophyte. **(2)** understand Gymnosperms and angiosperms are the two major groups of seed plants. Gymnosperms produce seeds that are not enclosed in a fruit (they do not produce flowers), whereas angisosperms do produce flowers; as a result, their seeds are enclosed within a fruit.

✔ TEST YOUR KNOWLEDGE

1. remember c **2.** remember c **3.** remember a **4.** remember T, T, F (seeds of angiosperms contain triploid endosperm cells along with a diploid embryo; gymnosperm seeds contain haploid female gametophyte cells that are used to nourish the embryo), F [spores are single-celled haploid structures that grow into gametophytes; seeds are multicellular structures that contain an embryo (sporophyte)].

✔ TEST YOUR UNDERSTANDING

5. understand In a gametophyte-dominant life cycle, the gametophyte is larger and longer lived than the sporophyte and produces most of the nutrition. In a sporophyte-dominant life cycle, the sporophyte generation is the larger, longer-lived, and photosynthetic phase of the life cycle. **6.** understand Plants build and hold soils required for human agriculture and forestry, and they increase water supplies that humans can use for drinking, irrigation, or industrial use. Plants release oxygen that we breathe. **7.** analyze Cuticle prevents water loss from leaves but also prevents entry of CO_2 required for photosynthesis. Stomata allow CO_2 to diffuse but can close to minimize water loss. Liverwort pores allow gas exchange but cannot be closed if conditions become dry. **8.** understand Homosporous plants produce a single type of spore that develops into a gametophyte that produces both egg and sperm. Heterosporous plants produce two different types of spores that develop into two different gametophytes that produce either egg or sperm. In a tulip, the microsporangium is found within the stamen, and the megasporangium is found within the ovule. Microspores divide by mitosis to form male gametophytes (pollen grains); megaspores divide by mitosis to form the female gametophyte.

✔ TEST YOUR PROBLEM-SOLVING SKILLS

9. apply A "reversion" to wind pollination might be favored by natural selection because it is costly to produce a flower that can attract animal pollinators. Because wind-pollinated species grow in dense clusters, they can maximize the chance that the wind will carry pollen from one individual to another (less likely if the individuals are far apart). Wind-pollinated deciduous trees flower in early spring before their developing leaves begin to block the wind. **10.** create Alter one characteristic of a flower, and present the flower to the normal pollinator. As a control, present the normal (unaltered) flower to the normal pollinator. Record the amount of time the pollinator spends in the flower, the amount of pollen removed, or some other measure of pollination success. Repeat for other altered characteristics. Analyze the data to determine which altered characteristic affects pollination success the most.

✔ PUT IT ALL TOGETHER: Case Study

11. analyze Springtails facilitate the transfer of sperm, as their presence led to significantly higher mean numbers of sporophytes. Their impact was especially important when male and female gametophytes were separated from each other. The mean number of sporophytes formed in the presence and absence of springtails when male and female gametophytes were united with each other (i.e., separated by 0 cm) was not significantly different (even though the trend was toward more sporophytes in the springtail treatment). **12.** understand Moss sperm have flagella, as do the sperm of green algae. **13.** understand It would be helpful to know whether springtails are actually found on mosses in the natural environment and, if so, whether they are present when gametophytes are producing sperm and eggs. It would also be helpful to know whether male and female gametophytes grow right next to each other or some distance apart. **14.** evaluate Spores do not fertilize eggs; sperm do. Spores and sperm are distinct structures with different life cycle roles. Spores grow into gametophytes. **15.** understand Other types of animals may help transfer sperm. In addition, rain may help splash sperm from male to female gametophytes. **16.** analyze Given the results presented here, animal-mediated fertilization likely originated well before the origin of flowering plants. Mosses were among the earliest land plants; they evolved well before flowering plants.

Chapter 29

IN-TEXT QUESTIONS AND EXERCISES

p. 609 CYU (1) remember A few examples are athlete's foot, vaginitis, diaper rash, ringworm, pneumonia, and thrush. **(2)** understand Plants that grow in association with mycorrhizal fungi are likely to grow faster and more robustly than plants without the fungi. **(3)** understand By digesting large carbon-containing compounds (such as lignin and cellulose) that would otherwise be tied up in dead trees and other organisms, fungi produce small organic molecules and release CO_2 through cellular respiration. This CO_2 is later absorbed by plants and fixed during photosynthesis into sugars (and later processed back into lignin, cellulose, and other large organic compounds).

p. 610 Fig. 29.4 understand Like other eukaryotes, yeasts contain membrane-bound nuclei and organelles such as mitochondria. Bacteria, however, lack membrane-bound nuclei and organelles.

p. 613 Fig. 29.8 understand Because conidia are formed by mitosis, all conidia formed from a single fungus are genetically identical.

p. 614 CYU (1) understand Swimming spores and gametes, zygosporangia, basidia, and asci. **(2)** understand Many types of mold consist of large numbers of conidia (asexual spores). These spores are released and can easily start to grow on nearby fruit. **(3)** evaluate Ascomycetes and basidiomycetes share a recent common ancestor. Therefore, ascomycetes are more closely related to basidiomycetes than they are to zygomycetes.

p. 616 Fig. 29.11 analyze Labeled nutrient experiments identify which nutrients are exchanged and in which direction. They explain why plants do better in the presence of mycorrhizae, and how the fungus also benefits.

p. 620 analyze Human sperm and eggs undergo plasmogamy followed by karyogamy during fertilization, but cells like the heterokaryotic cells of fungi do not occur in humans or other eukaryotes.

p. 620 Fig. 29.15 analyze The haploid mycelium.

p. 621 CYU (1) understand When birch tree seedlings are grown in the presence and in the absence of EMF, individuals denied their normal EMF cannot acquire sufficient nitrogen and phosphorus. The EMF acquire sugars from the birch. Therefore, both organisms benefit. **(2)** understand Spores are haploid cells that can be distributed long distances. When environmental conditions are favorable, they can germinate and grow into fungal hyphae. **(3)** analyze Meiosis and the production of haploid spores occur in each of these structures. In a zygosporangium, meiosis occurs in a multinucleate cell; in basidia and asci, there is a single diploid nucleus. In asci, meiosis is followed by one round of mitosis.

p. 626 CYU (1) remember The polar tube can penetrate the cell membrane of a host cell and provide a passageway for the microsporidian to enter the host cell. **(2)** analyze Members of the Glomeromycota form mycorrhizal relationships with plants. As a result, the plants can acquire more nitrogen and phosphorus than they are able to without the fungi. **(3)** analyze Budding produces new yeast cells by mitosis. Therefore, the newly formed yeast cells are genetically identical and budding does not lead to genetic variation.

✔ TEST YOUR KNOWLEDGE

1. understand a **2.** remember F (Spores grow into fungal hyphae; they are not involved in fertilization, and they are not designated as male and female.), T, T, T. **3.** understand b **4.** remember b

✔ TEST YOUR UNDERSTANDING

5. understand c **6.** evaluate *Sample answer*: Accept. Along with a few bacteria, fungi are the only organisms that can digest wood completely. If the wood is not digested, carbon remains trapped in wood. Without fungi, CO_2

would be tied up and unavailable for photosynthesis, and the presence of undecayed organic matter would reduce the space available for plants to grow. **7.** apply Fungi produce enzymes that degrade cellulose and lignin. **8.** apply c

✔ TEST YOUR PROBLEM-SOLVING SKILLS

9. create Study design: (1) Confirm that the chytrid fungus is found only in sick frogs and not healthy frogs. (2) Isolate the chytrid fungus and grow it in a pure culture. (3) Expose healthy frogs to the cultured fungus and see if they become sick. (4) Isolate the fungus from the experimental frogs, grow it in culture, and test whether it is the same as the original fungus. **10.** create Study design: (1) Collect a large array of colorful mushrooms that are poisonous; also capture mushroom-eating animals, such as squirrels. (2) Present a hungry squirrel with a choice of mushrooms that have been dyed or painted a drab color versus others treated with a solution that is identical to the dye or paint used but uncolored. (3) Record which mushrooms the squirrel eats. (4) Repeat the test with many squirrels and many mushrooms.

✔ PUT IT ALL TOGETHER: Case Study

11. apply d **12.** create See **Figure A29.1**. 13. evaluate The relationship between *Pilobolus* and cows might better be described as commensal; while *Pilobolus* benefits from this relationship by being dispersed by the cow and provided with a nutrient-rich environment in which to grow, cows may not derive much benefit from this relationship. **14.** understand Fungal mycelia release enzymes as they grow throughout dung. These enzymes break down undigested material into usable molecules, which are absorbed directly into the fungal mycelium. **15.** analyze Launch speed and acceleration of *Pilobolus* sporangia are less than that of the other fungi. Usain Bolt's top speed of 44 km/h translates to about 12.2 m/s (since 1 km = 1000 m and 1 hour = 3,600 seconds). Therefore, Usain Bolt's top speed is slower than the launch speeds of *Ascobolus* and *Podospora*, and just slightly faster than the launch speed of *Pilobolus*. **16.** create One possible approach would be to grow *Pilobolus* in a controlled room with a single light source at a distinct location. After sporangia are discharged, their location could easily be determined relative to the location of the light source. *Sample hypothesis*: Sporangia shot toward light may be more likely to land on turf that is fed on (and defecated on) by cows, because cows feed on photosynthesizing grass.

Chapter 30

IN-TEXT QUESTIONS AND EXERCISES

p. 630 CYU analyze The main trait that animals have in common with plants and fungi is that they are all multicellular eukaryotes. Unlike plants, animals and fungi are heterotrophs. And unlike both plants and fungi, animals move under their own power and have neurons and muscles.

p. 631 Table 30.1 analyze

1. Arthropoda (1,200,000 described species)
2. Mollusca (85,000)
3. Chordata (65,000)
4. Nematoda (25,000)
5. Platyhelminthes (20,000)

p. 632 Fig. 30.2 apply You should have added one bar and "multicellularity" label to the fungi branch and another bar and "multicellularity" to the animal branch; multicellularity arose independently in these two groups.

p. 634 Making Models 30.1 apply See **Figure A30.1**.

p. 636 Fig. 30.7 analyze Stained *Hox* and *dpp* gene products either would not be found in *Nematostella* at all or would not occur in the same anterior–posterior and dorsal–ventral pattern as observed in bilaterians.

p. 639 CYU (1) understand *Possible evidence:* Sponges are the earliest animals to appear in the fossil record; the feeding cells of sponges (choanocytes) have a strong resemblance to choanoflagellates, the protists that are most closely related to animals; sponges lack complex tissues; many molecular phylogenies support sponges as the sister group to other animals. **(2)** apply Your tube-within-a-tube body plan consists of a bilaterally symmetrical, elongated body with an inner tube (gut and associated organs) derived from endoderm; an outer tube (mostly skin) derived from ectoderm; muscles and organs derived from mesoderm in between; and one end (head) cephalized with sensory organs and brain (derived from ectoderm).

p. 645 Fig. 30.12 create The larva and metamorphosis arrow should be circled in the life cycle. It might be adaptive to skip the feeding larval stage if more food resources are available for the mother (who needs energy to produce the yolk) than for the larva; if predation pressure on larvae is high; or if other environmental factors for larvae are unfavorable, resulting in high mortality of the larvae.

p. 645 CYU (1) analyze The mouthparts of deposit feeders are relatively simple because these animals usually gulp soft material. The mouthparts of mass feeders are more complex because they have to tear off and process chunks of relatively hard material. **(2)** create Gametes that are shed into aquatic environments can float or swim. This cannot happen on land, so internal fertilization is more common there. Also, gametes are more at risk of drying out on land, and internal fertilization eliminates that risk.

p. 647 CYU analyze You should have drawn a mark and label on each horizontal branch of Figure 30.2 as follows: (Porifera) body plan composed of cell groups forming pores that channel water currents; (Ctenophora) sticky feeding cells called coloblasts, or anal pores; (Cnidaria) stinging cells called cnidocytes.

✔ TEST YOUR KNOWLEDGE

1. remember a **2.** remember d **3.** remember False—almost all the major body plans arose rapidly during the Cambrian. **4.** understand d

✔ TEST YOUR UNDERSTANDING

5. understand Many unicellular organisms are heterotrophic, but they can consume only small bits of food. Animals are multicellular, so they are larger and can consume larger amounts of food—making them important consumers in food webs. **6.** evaluate There are about 18.5 named arthropods for every named species of chordate (1,200,000/65,000), about 1.3 named mollusks for every named species of chordate (85,000/65,000), and about 0.4 named species of nematode worm for every named species of chordate (25,000/65,000). These numbers are likely to be underestimates because there are probably many more arthropods, mollusks, and nematode worms than have been described, yet relatively few of the larger, more familiar chordates (mostly vertebrates) left to be described. **7.** evaluate While it is generally true that early animals were simpler than later animals, evolution is more nuanced than a simple-to-complex story suggests. For example, many essential tool-kit genes evolved right at the base of the animal tree and were later co-opted in various ways in different lineages. Also, sometimes traits are lost over time rather than gained, such as a coelom in flatworms and legs in snakes. Also, some traits—such as segmentation—evolved independently in different groups. Moreover, the lineages that were earliest to diverge from other animals did not stop evolving. **8.** apply c

Figure A29.1

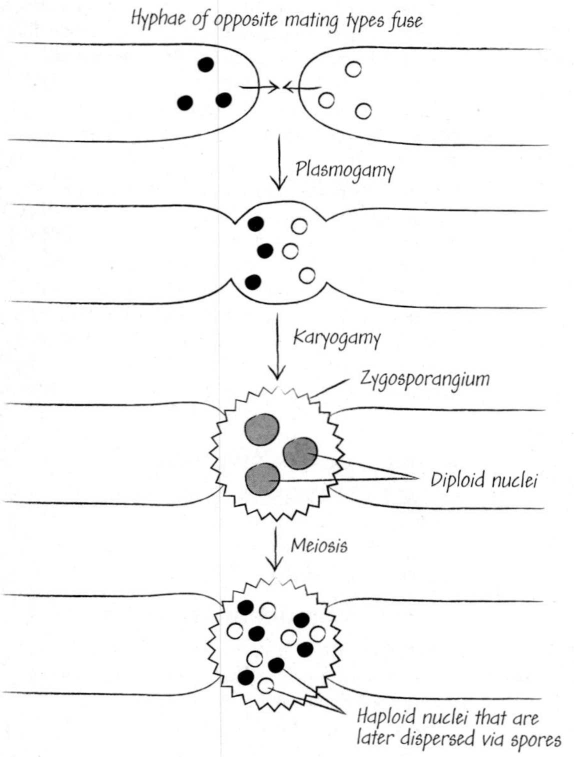

Figure A30.1

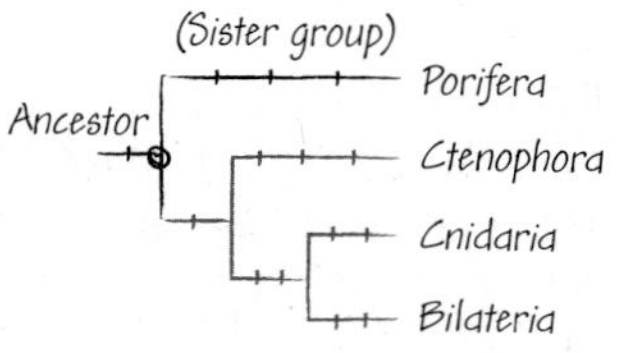

✔ TEST YOUR PROBLEM-SOLVING SKILLS

9. evaluate Yes—if the same gene is found in nematodes and humans, it was likely found in the common ancestor of protostomes and deuterostomes. Therefore, fruit flies are also likely to have this gene. However, sometimes genes are lost during evolution, so it's possible that the ancestors to flies had the gene, but flies do not. **10.** apply T, F, T, F.

✔ PUT IT ALL TOGETHER: Case Study

11. analyze See Figure A30.2.

12. evaluate Many answers are possible, but yours should include traits either mentioned as key innovations in Section 30.2 (such as central nervous system, bilateral symmetry, segmentation) or in the themes of diversification in Section 30.3 (certain types of sensory organs, ecological roles, or feeding strategies). **13.** create See Figure A30.3. It's fine if you showed the linear trend without including data points, but this answer has data points for clarity. In general, the strength of a correlation (indicated with an R^2 value between 0.0 and 1.0) is highest when there is a steep slope with the data points close to the line, rather than a flat slope with data points that are scattered. **14.** analyze Photoreceptors/eyes, skeleton, separate sexes, parasitic lifestyle, terrestrial lifestyle. You are identifying the traits with a high R^2 but a low P value. **15.** apply T, T, F (because all animals in the Cambrian lived in the ocean), T; the traits in the "true" answers are discussed in Section 30.2 focusing on the origin of animal phyla. **16.** analyze Yes—whereas a positive correlation between y and x does not necessarily mean that an increase in x *causes* an increase in y (there might be a third causative variable), the absence of a correlation between y and x suggests the absence of causation. The R^2 value would be very high (close to 1.000) if a single trait was important to diversification in many phyla, but very low (closer to 0.000) if different traits were important to diversification in different phyla—because each phylum represents one data point in each regression.

Chapter 31

IN-TEXT QUESTIONS AND EXERCISES

p. 653 Fig. 31.3 understand Aquatic living was the ancestral trait because the outgroups (e.g., Cnidaria, Porifera) were aquatic. There is also fossil evidence that the aquatic lifestyle was ancestral.

p. 654 CYU (1) analyze 1. Arthropoda (1,200,000 described species), 2. Mollusca (85,000), 3. Chordata (65,000), 4. Nematoda (25,000), 5. Platyhelminthes (20,000). **(2)** analyze *Several possible answers:* Internal respiratory structures; body surface that minimizes water loss; eggs with a covering that minimizes water loss; internal fertilization.

p. 654 apply See Figure A31.1.

p. 655 apply See Figure A31.1.

p. 657 apply See Figure A31.1.

p. 658 Making Models 31.1 apply See Figure A31.2.

p. 658 Fig. 31.8 apply See Figure A31.1.

p. 660 apply See Figure A31.1.

p. 661 Fig. 31.11 understand You should have drawn a circle around all the arms (the shorter appendages) and tentacles (the two longer ones).

p. 661 CYU (1) analyze Snails on land would likely have gills or lungs inside the body or otherwise be protected from drying out, thinner shells that are less heavy, and desiccation-resistant eggs. **(2)** apply See Figure A31.3.

p. 663 apply See Figure A31.1.

p. 663 apply See Figure A31.1.

p. 668 Fig. 31.15 understand No. Put your finger on the node where the remipede branch and the insect branch meet. This node represents the most recent common ancestor of both lineages; it was neither a remipede nor an insect as they exist today. Note: Researchers do not yet agree on whether the remipedes are the true sister group to insects; this is a leading hypothesis. (See BioSkills 13 for more practice reading phylogenetic trees.)

p. 669 CYU (1) analyze Myriapods (millipedes and centipedes), isopod crustaceans, and spiders. **(2)** apply [Many possible answers] *Similarities:* Both have an arthropod body plan with segmented bodies, exoskeletons, and jointed appendages; both live on land. *Differences:* An insect has three body regions, and a spider only has two; some insects eat plants, and all spiders are predators; most insects have wings, but no spiders have wings; most insects have six legs, and spiders have eight.

Figure A30.2

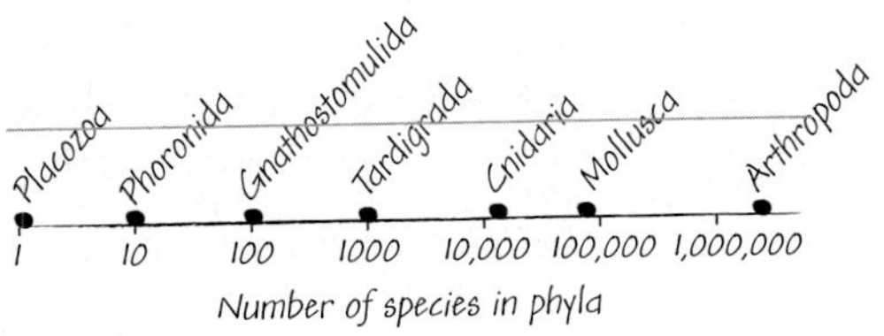

Figure A30.3

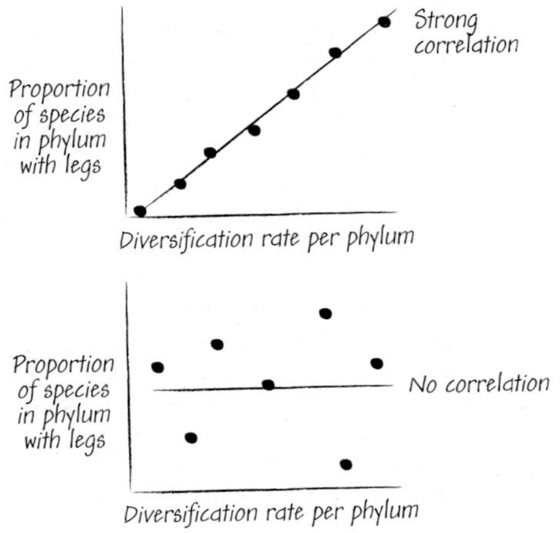

Figure A31.1

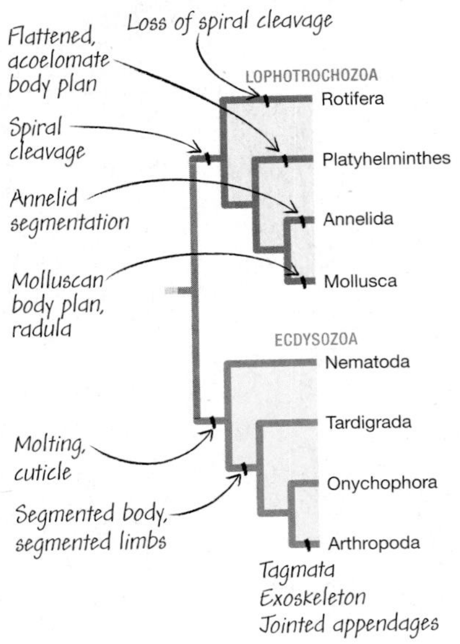

Figure A31.2

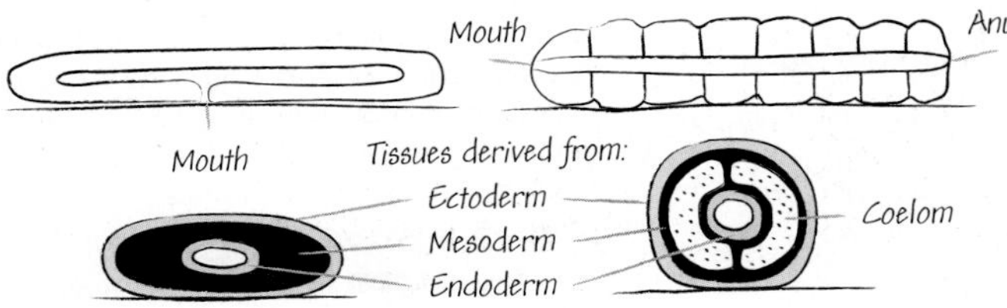

Figure A31.3

✔ TEST YOUR KNOWLEDGE

1. understand a **2.** understand c **3.** understand Both groups are bilaterally symmetric triploblasts with the protostome pattern of development. **4.** remember T, T, T, T

✔ TEST YOUR UNDERSTANDING

5. analyze Both annelids and arthropods have segmented bodies, unlike other protostomes such as flatworms and nematodes. **6.** apply Spiders have internal respiratory structures that minimize water loss, a waxy cuticle that minimizes water loss, and internal fertilization. **7.** create The ability to fly allowed insects to disperse to new habitats and find new food sources efficiently. **8.** evaluate It is true that evolution occurs when new traits accumulate, but it is *not* true that evolution *only* occurs when new traits accumulate. Evolution can also occur when traits are lost, such as the loss of a coelom in flatworms, loss of a digestive system in tapeworms, and loss of a second pair of antennae in insects. Some lineages adapt to their environments by becoming less complex over time.

✔ TEST YOUR PROBLEM-SOLVING SKILLS

9. apply If the distinct ancestors of brachiopods and mollusks lived in similar habitats and experienced natural selection that favored similar traits, then they would have evolved to have similar forms and habitats. This is called convergent evolution (Ch. 25, Section 25.1). **10.** analyze If bee populations are declining at the same or similar rate as for insects overall, losing a third of bee species could seriously affect the success of crops like apples and almonds. If these crops are not pollinated, they will not bear fruit for humans to eat.

✔ PUT IT ALL TOGETHER: Case Study

11. apply d **12.** analyze d **13.** analyze The total number of arthropods is about 6000 species. The total number of beetles is about 1000 + 200 + 700 + 1000 = 2900, which is about half, or 50 percent, of the species counted (rough estimate). **14.** analyze F (Spiders are not insects.), T, T, F (There are far fewer mollusk species than arthropod species, as shown in Figure 31.1.) **15.** create *One of several possible answers:* The researchers would have to assume that the diversity of species and number of unique species per area in parts of the forest that were *not* surveyed were comparable to the plots that were surveyed. **16.** analyze Most of the arthropod species are insects that experienced an adaptive radiation on land, thanks in part to their coevolution with plants, which were also experiencing an adaptive radiation on land. The more plants, the more types of habitats and food sources the insects could use, and the greater their ecological opportunity for speciation.

Chapter 32

IN-TEXT QUESTIONS AND EXERCISES

p. 673 Fig. 32.1 apply Invertebrates are not a monophyletic group. The group does not include all the descendants of the common ancestor, because vertebrates are excluded (see Ch. 25, Table 25.2).

p. 673 CYU analyze [Many possible answers] *Protostomes:* fly, beetle, moth, spider, snail, earthworm; all are invertebrates. *Deuterostomes:* dog, cat, chicken, robin, sparrow, squirrel, human; all are vertebrates.

p. 676 CYU understand If it's an echinoderm, it should have five-part radial symmetry, a calcium carbonate endoskeleton just underneath the skin, and a water vascular system (e.g., visible tube feet).

p. 678 CYU understand Your model (body indicated by an oval or similar shape) should have a dorsal hollow nerve cord just beneath the top, dorsal edge; another line representing the notochord below the nerve cord; a post-anal tail labeled at one end; and a series of lines or bumps representing pharyngeal slits near the other end (see **Figure A32.1**). The realism of your model does not matter as long as it is tidy and includes these four labeled elements.

p. 679 CYU (1) analyze According to Figure 32.5: 1. ray-finned fishes; 2. birds; 3. snakes, lizards; 4. amphibians; 5. mammals. **(2)** analyze Cephalochordates and fishes share pharyngeal slits, dorsal hollow nerve cord, notochord, and post-anal tail at some point in their life cycles; but fishes have vertebrae, cranium, and brain with three regions.

p. 680 Fig. 32.6 apply Mammals and reptiles are equally related to amphibians because mammals and reptiles share a common ancestor, and this ancestor shares a common ancestor with amphibians. Amphibians are more closely related to mammals and reptiles than to lungfishes because the tetrapods share a more recent common ancestor with each other than any of them do with lungfishes. Note that the proximity of the labels at the tips of the tree is not a good indicator of relationships, because you can rotate the branches around each node without changing the meaning of the tree (see **BioSkills 13**).

p. 681 apply Because this tree has no time scale, the label for this extinct vertebrate would be added to a branch tip that's vertically aligned with the others, for example, between Urochordata and Myxinoidea (hagfishes). This species would be a descendant of the node representing the most recent common ancestor of vertebrates but would not be labeled as the ancestor itself, because it cannot be known whether this fossil species is a direct ancestor.

p. 683 Making Models 32.1 apply See **Figure A32.2**.

p. 684 Fig. 32.8 apply If the goal of a tree is to test a hypothesis about a trait such as a limb, that trait should *not* be included when drafting the tree; doing so would result in circular reasoning (for example, if you sort your socks by their color, you should not be surprised to find that socks of the same color are in the drawer together).

p. 686 Fig. 32.10 analyze The yolk sac is smaller in the mammal—its function in an amniotic egg (source of nutrients) has been taken over by the placenta.

p. 690 CYU (1) understand The cranium protects the brain and sensory organs. The vertebrae protect the spinal cord. Jaws allow animals to capture food efficiently and process it by crushing or tearing. The ability to process food more efficiently increased the importance of jawed vertebrates as herbivores and carnivores. The tetrapod limb enables vertebrates to move on land. The amniotic egg enables vertebrates to lay eggs on land; amniotic eggs are resistant to drying out. **(2)** apply You should have added a label indicating the origin of endothermy on the red Mammalia branch, along with "Lactation, fur." You also should have added a trait bar indicating the origin of endothermy on the orange Aves branch. (Endothermy has also been discovered in a few ray-finned fishes and a few lizards.)

p. 692 Making Models 32.2 apply See **Figure A32.3**.

p. 692 Table 32.8 analyze 44 percent, 26 percent; relative to body size, *H. floresiensis* has a smaller brain than *H. sapiens,* suggesting that it might have had reduced cognitive capacity compared to a *H. sapiens* of the same body size.

p. 693 Fig. 32.16 apply About six hominin species existed 1.8 mya (of the ones included in the figure), and three existed 100,000 years ago.

p. 693 Fig. 32.17 analyze The forehead became much larger, and the face became "flatter"; the brow ridges are less prominent in later skulls than in earlier skulls.

p. 696 CYU (1) apply You might be tempted to correct this statement by saying "Humans evolved from chimps," which is a common misconception. Humans and chimps are sister groups that share a common ancestor that was neither chimp nor human. Similarly, humans, chimps, and gorillas share a common ancestor that was neither a gorilla, chimp, nor human (see **Making Models 32.2**). **(2)** apply *Homo sapiens* originated in Africa, then migrated to Europe and interbred with Neanderthals, and then migrated to the rest of the world. If you have purely African heritage, you probably do *not* have any Neanderthal in your genome, but if your heritage is from anywhere else or mixed, you are probably 1–4 percent Neanderthal.

✔ TEST YOUR KNOWLEDGE

1. remember True; echinoderms have a calcium carbonate endoskeleton, and vertebrates have a bony endoskeleton made of calcium phosphate. **2.** understand a **3.** remember reptiles and mammals **4.** understand b

✔ TEST YOUR UNDERSTANDING

5. understand c **6.** analyze Amniotic eggs and placentas are similar in that they both nourish and protect growing embryos; they are both made of a chorion, amnion, and yolk sac; and they are both adaptive strategies for life on land. The main difference is that amniotic eggs can be deposited outside the mother's body, protected in a shell filled with albumin, while the placenta is retained within the mother's uterus and relies on the exchange of nutrients, wastes, and gases with the mother. **7.** apply b **8.** evaluate A phylogeny of modern humans would reject the out-of-Africa hypothesis if the earliest branches to diverge from others in the phylogenetic tree were from any other region but Africa, such as Asia or Europe.

✔ TEST YOUR PROBLEM-SOLVING SKILLS

9. (1) analyze Protostomes and deuterostomes living on land both have mechanisms for preventing their eggs from drying out (e.g., thick membranes in snail and

Figure A32.1

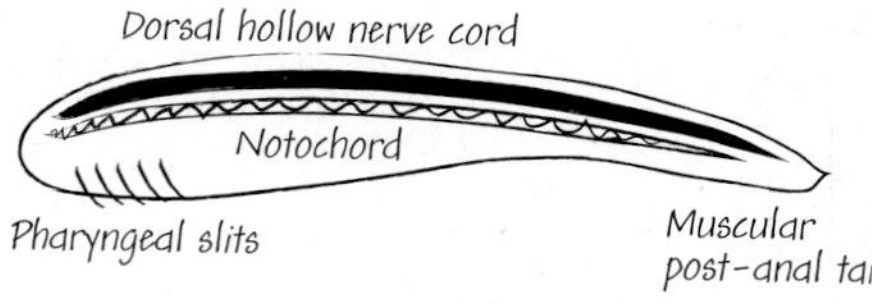

Figure A32.2

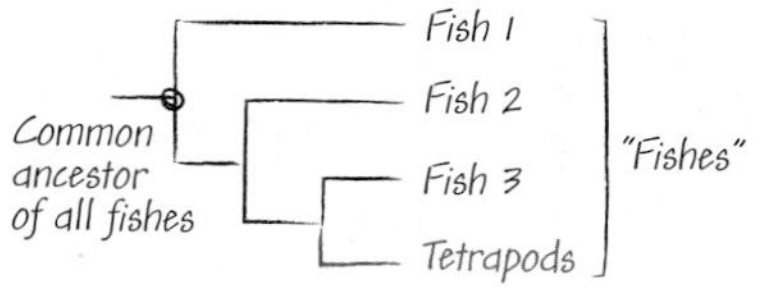

Figure A32.3

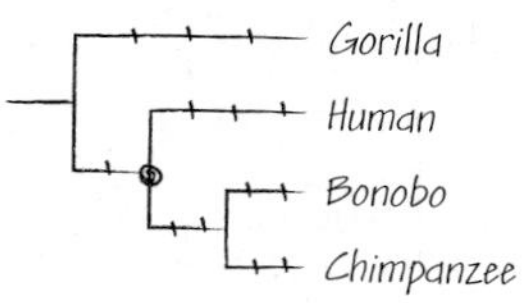

insect eggs; amniotic egg in tetrapods), preventing their skin from drying out (e.g., waxy cuticle in insects, scaly skin in reptiles), and keeping their respiratory surfaces from drying out (e.g., lungs in land snails and tracheal system in insects; lungs in tetrapods) (2) create Vertebrate limbs are diverse in form and function (they can serve as arms, flippers, wings, etc.) but have homologous bones within. Molecular probes could be used to visualize the expression patterns of tool-kit genes such as *Hox* in the embryonic limbs of different vertebrate species to show how changes in the expression of homologous genes can cause changes to the structure and function of limbs. **10.** create See **Figure A32.4**. It is important to use indigenous people for this study to minimize the amount of gene flow (transfer of alleles from one population to another) that occurred during the history of the study populations due to immigration and interbreeding.

✔ PUT IT ALL TOGETHER: Case Study

11. analyze *Hypothesis:* The bones of the mammalian ear evolved from the jawbones of an early amniote. **12.** evaluate F, F, T, T; in everyday English it's common to refer to "animals and humans" or "mammals and humans," but these phrases do not make sense in a biological context because humans are mammals, which are animals (see **BioSkills 17**). **13.** create It makes sense that a separation of the jawbone (feeding function) from the ear bones (hearing function) would be adaptive, because it releases each structure from the constraint of the other and allows both to evolve independently. Ear bones could then have evolved to be different sizes or shapes, allowing different sensitivity to pitch or volume depending on the animal's niche. For example, acute hearing would be very adaptive in a nocturnal environment where vision is a poor source of information. **14.** apply In the developing ear bones of mammals. (This is where *Bapx1* expression was indeed found.) **15.** evaluate Model organisms such as mice are used because they are short lived, easy to maintain, and easy to study (see **BioSkills 11**). It would be impractical to do similar experiments on human embryos, plus it is not culturally acceptable to conduct experiments on human embryos. Although early development in mice and humans is not identical, most processes are similar because they are both placental mammals that share a common ancestor with a similar developmental pattern. Therefore, the results from mice would likely be a good proxy for the patterns expected in humans. **16.** apply [Many possible answers] *Some examples:* Lungs were co-opted for use as swim bladders. Fins were co-opted for use as limbs. Arms were co-opted for use as wings. Scales made of keratin were co-opted for use as feathers. Jawbones were co-opted for use as ear bones. Vertebrae were lost in hagfishes. Fins were lost in tetrapods. Tails were lost in adult frogs. Limbs were lost in caecilians and snakes. Amniotic eggs were lost in marsupials and placental mammals. Tails (after the embryonic stage) were lost in great apes, including humans.

Chapter 33

IN-TEXT QUESTIONS AND EXERCISES

p. 703 CYU (1) analyze Viruses are not considered to be "alive," because they depend on their host cell to satisfy the five attributes of life; but they contribute to organismal evolution by acting as agents of natural selection, introducing foreign genes by lateral gene transfer, and contributing their own genetic material to host cells. **(2)** analyze An epidemic is a disease that rapidly infects many individuals over a widening area, and a pandemic is an epidemic that is worldwide in scope. In this example, the disease spread only in North America in native populations, not worldwide. **(3)** understand HIV infects and kills helper T cells, which play a central role in the human immune response. Over several years, the steady decline in helper T cell numbers leads to a compromised immune response that eventually allows other infections to occur, resulting in the clinical signs of AIDS.

p. 705 Fig. 33.7 analyze Starting with 1 virion, approximately 80 virions were produced in 35 minutes. If all of these virions infected new cells, you could expect each of the 80 infected cells to produce 80 virions by 75 minutes, or a total of 6400 virions. The cells appear to replicate once every 25 minutes, which means they will have replicated three times during this period. The total number of cells at 75 minutes would be approximately 2^3, or 8 cells. There would be 800 times as many virions as cells.

p. 706 Fig. 33.8 analyze No—it shows only that CD4 is required. HIV may need to use other proteins not identified in this experiment. (Subsequent work showed that other proteins are required.)

p. 709 Making Models 33.1 apply See **Figure A33.1**.

p. 710 Fig. 33.13 apply The number of infectious HIV virions would drop to zero if the cell were artificially lysed before budding could occur. There would be no difference in the number of infectious adenovirus virions released from an adenovirus-infected cell.

p. 711 analyze In T4, the viral genome is injected into the host, whereas HIV inserts its genome via membrane fusion. HIV must convert its RNA genome into double-stranded DNA before replication, which is not necessary for T4, which has a DNA genome. Both viruses use DNA to make mRNA that is translated into viral proteins. T4 virions are made in the cytosol of the cell and are released when the cell bursts. HIV virions bud from the cell surface, a process that does not require cell death.

p. 711 CYU (1) evaluate Like cars in the assembly line of an automobile plant, new virions are assembled from premanufactured parts, and large numbers of progeny are produced per generation. Cells, on the other hand, reproduce by division of a single integrated unit, resulting in only two progeny per generation. **(2)** apply See **Figure A33.2**.

p. 713 Fig. 33.16 apply Your drawing should have the following bars and labels: on branch to HIV-2 (sooty mangabey to human); on branch to HIV-1 strain O (chimp to human); on branch to HIV-1 strain N (chimp to human); on branch to HIV-1 strain M (chimp to human).

p. 714 CYU (1) analyze The escaped-genes and degenerate-cell hypotheses state that viruses originated from cells, while the RNA-world hypothesis suggests that viruses originated in parallel with, and maybe even influenced, the origin of cells. The escaped-genes and RNA-world hypotheses state that the viruses originated from parasitic molecules. The degenerate-cell hypothesis states that viruses originated from parasitic organisms. **(2)** apply A mutation that made transmission between humans more efficient would make the virus more dangerous. Such a mutation could occur via genomic reassortment in pigs, where an avian virus and a human

Figure A32.4

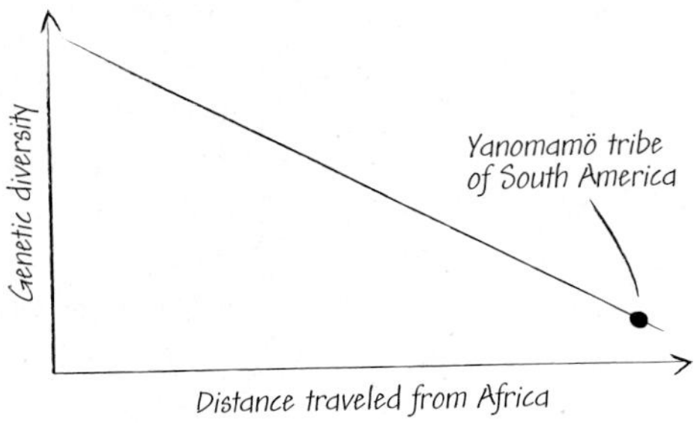

Figure A33.1

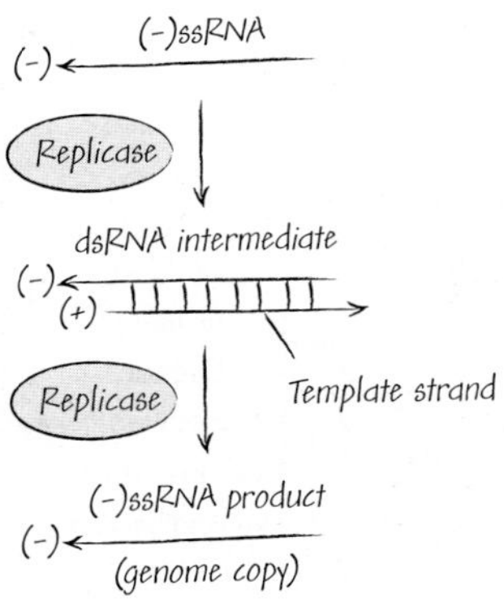

Figure A33.2

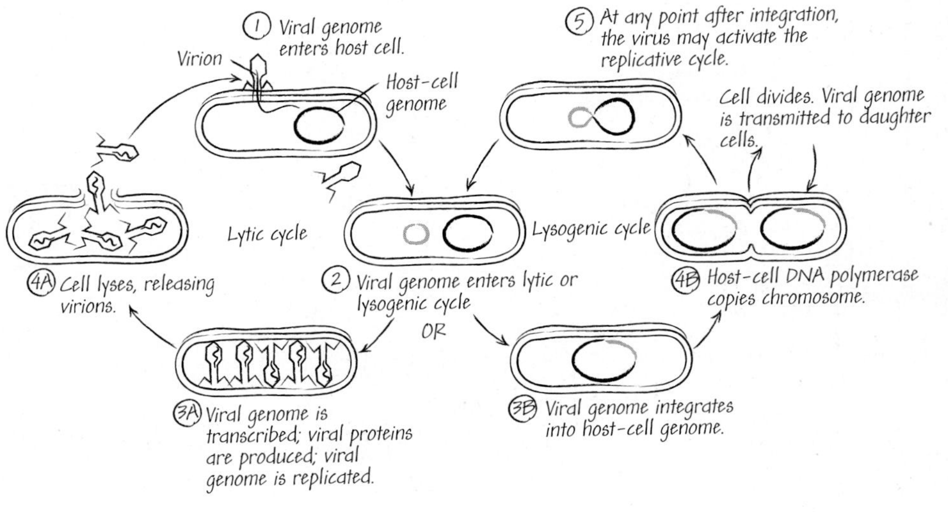

virus could co-infect cells and produce recombinants.

p. 717 CYU evaluate Each hypothesis can be associated with types of genomes as follows: *Escaped-genes hypothesis:* single and double-stranded DNA or possibly single-stranded RNA. *Degenerate-cell hypothesis:* double-stranded DNA. *RNA-world hypothesis:* single- or double-stranded RNA.

✔ TEST YOUR KNOWLEDGE

1. remember d **2.** understand Reverse transcriptase converts positive-sense single-stranded RNA to double-stranded DNA in class VI and class VII viruses. **3.** remember F, F, T, T **4.** understand The type of genetic material included in the virion (classes I–V) and how the genome is replicated (compare classes I and IV with classes VII and VI, respectively).

✔ TEST YOUR UNDERSTANDING

5. apply A virus with an envelope generally exits a host cell by budding. A virus that lacks an envelope generally usually exits by bursting the host cell (lysis). **6.** analyze (1) The rate of viral genome replication is much higher in the lytic cycle. In lysogeny, the viral genome can replicate only when the host cell replicates. (2) Only the lytic cycle produces virions. (3) The lytic cycle results in the host cell's death, while lysogeny allows the host cell to survive. **7.** apply No. Most prescribed antibiotics are designed to disrupt the metabolic activity of unicellular infectious agents, like bacteria. Viruses rely on host-cell metabolism to replicate, so antibiotics would not be useful against them without also killing your cells. **8.** apply a; the (+)ssRNA genome of the pea mosaic virus can directly serve as mRNA to produce viral proteins, while all of the other viruses require viral enzymes to transcribe their genome into mRNA.

✔ TEST YOUR PROBLEM-SOLVING SKILLS

9. evaluate Prevention is currently the most cost-effective program, but it does not help people who are already infected. Treatment with effective drugs not only prolongs lives but also reduces virus loads in infected people, so that they have less chance of infecting others. **10.** apply b; by prolonging the infection in this manner, these viruses are more likely to be transmitted to a new host.

✔ PUT IT ALL TOGETHER: Case Study

11. remember As a class IV virus, ZIKV has a linear, positive-sense, single-stranded RNA genome. **12.** apply All of the samples treated with different antibodies would contain infected cells. If ZIKV can use more than one type of receptor, then samples that are treated with antibodies that specifically target one virus receptor will leave the other receptors available for attachment. **13.** apply b **14.** analyze The total number of neurons (NeuN+ cells) after ZIKV infection decreases significantly in comparison with the control, in which no virus was injected. These results suggest that embryonic infection causes significant neuronal loss in the developing brain. **15.** evaluate If your friend is pregnant, she should probably defer her travel plans due to the risk of mother-to-fetus transmission. Otherwise, advise your friend to take preventative measures such as using mosquito netting, protective clothing, and insect repellent. **16.** apply See **Figure A33.3**.

END-OF-UNIT CASE STUDY Are Newts Adapted to Kill Humans?

pp. 720–721 (1) apply d **(2)** apply See **Figure EOU A5.1**. **(3)** apply See **Figure EOU A5.1**. **(4)** apply a. **(5)** analyze See **Figure EOU A5.1**. **(6)** understand Yes, the loss of a trait can be adaptive. For example, the loss of legs is adaptive in snakes and whales. **(7)** analyze Asphyxiation when breathing muscles become paralyzed. **(8)** analyze Nav1.4. **(9)** analyze Nav1.7. **(10)** apply d, Mus musculus (mouse) because of the options (mustard plant, fruit fly, nematode, mouse, yeast; see **BioSkills 11**) the mouse is the most closely related to humans, sharing a more recent common ancestor. **(11)** apply See **Figure EOU A5.2**, showing excerpts from the Big Picture of Diversity.

BIG PICTURE Diversity of Life

p. 722 CYU (1) understand In the tree on the left, you should have circled the Animals branch. In the tree on the right, you should have circled the Chordates branch. **(2)** understand Your first arrow should have a tail that originates on the purple cyanobacteria branch and a head that points to the gray branch just to the left of the clade containing red algae, green algae, and land plants. Your second arrow should have a tail that originates on the purple α-proteobacteria branch and a head that points to the gray branch to the left of the third black node in the tree, which is the root of Eukarya. **(3)** analyze Monophyletic groups (passing the "one-snip test" described in **BioSkills 13**) include Bacteria, Archaea, Eukarya, Fungi, Animals, Plants, Ascomycota, Basidiomycota, Protostomes, Deuterostomes, Gymnosperms, Angiosperms, and many others (including each individual branch tip, such as Chordates). Paraphyletic groups include Protists, Green Algae, Nonvascular Plants, and Seedless Plants. **(4)** apply You should have made a vertical tick mark on the horizontal red branch labeled Cnidarians.

Chapter 34

IN-TEXT QUESTIONS AND ANSWERS

p. 725 Fig. 34.1 apply New branches and leaves should develop to the right (the plant will also lean that way); new lateral roots will develop to the left.

p. 726 Fig. 34.2 analyze Surface area to volume = 420,000 $\mu m^2/8{,}000{,}000\ \mu m^3 = 0.0525/\mu m$.

p. 727 Fig. 34.3 understand Grasses that are shallow rooted would absorb water when it first enters soil. Plants with long taproots can reach water deep underground, which would be beneficial during periods of drought.

Figure EOU A5.1

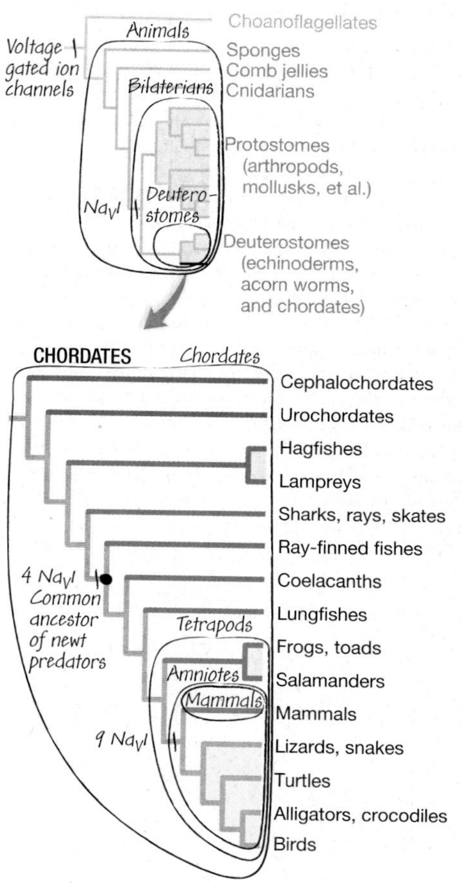

Figure A33.3

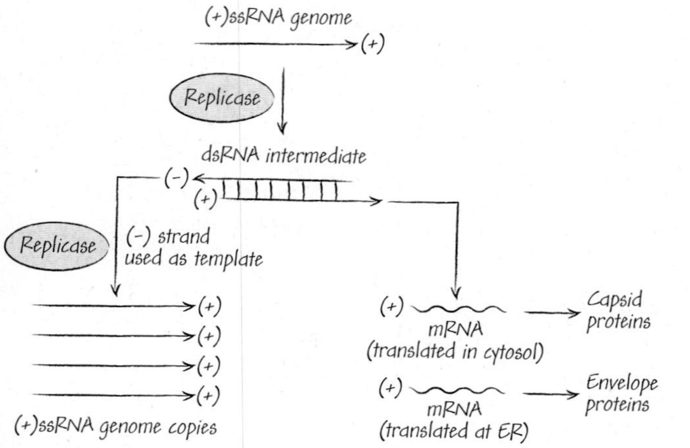

Figure EOU A5.2

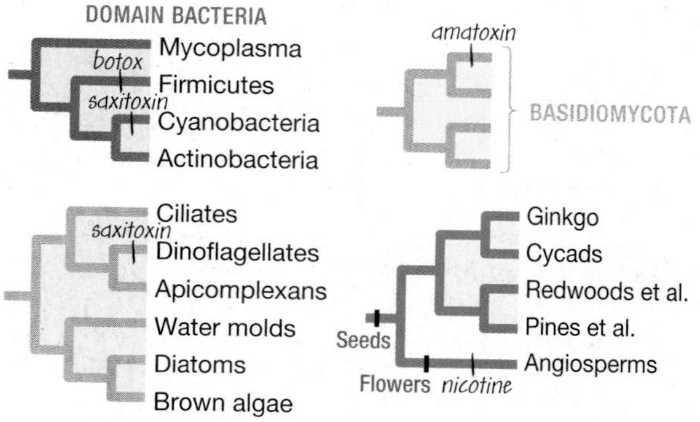

Figure A34.1

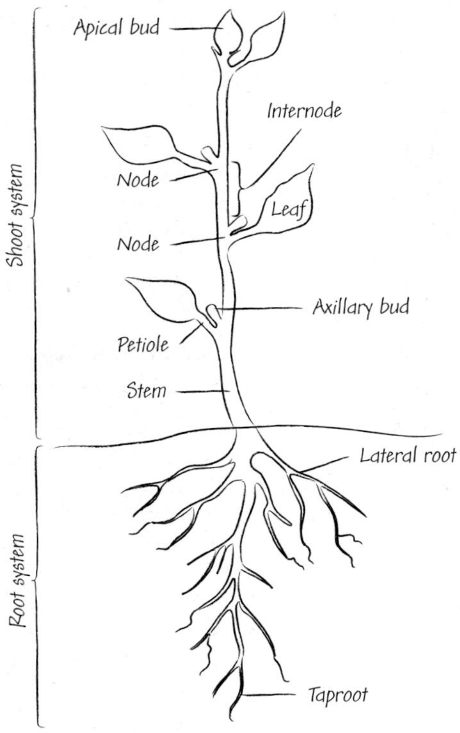

p. 728 Table 34.1 understand Root cells need oxygen to carry out aerobic cellular respiration, which generates ATP needed to keep cells alive.

p. 729 Fig. 34.4 understand The individuals are genetically identical—thus, any differences in size or shape in the different habitats are due to phenotypic plasticity and not genetic differences.

p. 731 Fig. 34.5 apply Although parts (b) and (c) may look like multiple leaves, they each show just one leaf composed of multiple leaflets. Each compound leaf represents just one leaf defined by a single axillary bud.

p. 732 apply Phenotypic plasticity is more important (1) in environments where conditions vary because it gives individuals the ability to change the growth pattern of their roots, shoots, and stems to access sunlight, water, and other nutrients as the environment changes; and (2) in long-lived species because it gives individuals a mechanism to change their growth pattern as the

Figure A34.2

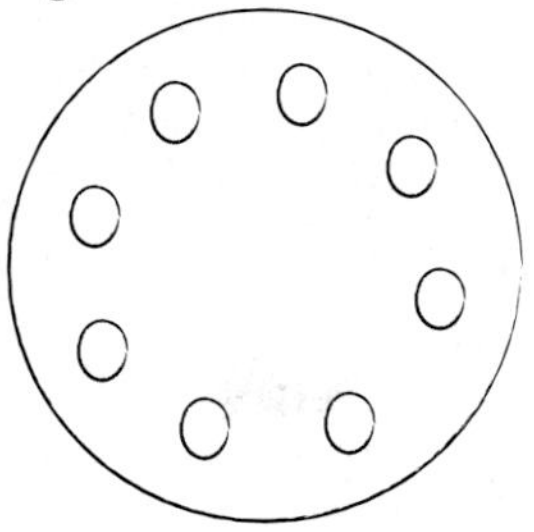

environment changes throughout their lifetime.

p. 733 CYU (1) remember See Figure A34.1. **(2)** analyze The generalized body of a flowering plant has a taproot with many lateral roots and broad leaves. Examples of deviations from this generalized body structure follow. Modified roots: Unlike taproots, *fibrous roots* do not have one central root, and *adventitious roots* arise from stems. Modified stems: *Stolons* grow along the soil and grow roots and leaves at each node, and *rhizomes* grow horizontally underground. Both of these modified stems function in asexual reproduction. Modified leaves: The *needlelike leaves* of cacti don't lose water to transpiration compared with typical broad leaves, and they protect the plant from herbivory. *Tendrils* on climbing plants are modified leaves specialized for wrapping around trees or other substrates to facilitate climbing. **(3)** apply Even though the plantlets are genetically identical, they would likely show different phenotypes because of phenotypic plasticity.

p. 735 Fig. 34.11 understand Many possible answers; for example, in leaf cells, genes encoding proteins involved in photosynthesis and formation of stomata, the cuticle, and trichomes would be expressed. These genes wouldn't be expressed in the root cells.

p. 736 apply Unlike many other plant cells, tracheids (and vessel elements) perform their function after they die. The remaining secondary cell walls serve as conduits through which water flows, by a process that does not require ATP. Therefore, the poison will have no effect on water transport through the injected tracheids.

p. 737 CYU (1) analyze Plant cells contain a cell wall found exterior to the cell membrane, while animal cells lack a cell wall. Plant cells are connected via plasmodesmata and may contain chloroplasts, while animals lack these structures. **(2)** analyze Vessel elements are dead at maturity and consist of cell walls whose function is to carry water typically from roots throughout the rest of the plant. Water is pulled through xylem under negative pressure. Sieve-tube elements are alive at maturity but contain relatively few organelles. These cells transport sugars throughout the plant under positive pressure.

p. 739 Fig. 34.17 understand The youngest tissues are found at the tip of the shoot, while the oldest tissues are found at the base. The apical meristem is the source of new growth and is located at the tip of shoots.

p. 741 understand The zones of cellular division and cellular elongation are actively pushed through soil. The zone of maturation remains in place (this is why root hairs can emerge here without being torn off as the root grows through soil).

p. 741 CYU (1) understand The apical meristem gives rise to the three primary meristems. The apical meristem is a single mass of cells localized at the tip of a root or shoot; the primary meristems are localized in distinctive sites behind the apical meristem. **(2)** apply See Figure A34.2.

p. 743 Fig. 34.21 apply Tree trunks grow wider because of cell divisions that occur only in the vascular and cork cambia (both located just underneath the bark). Secondary xylem (wood) remains in place once it is formed. Therefore, the eggs will remain in their current position (1 inch out from the center).

p. 744 Fig. 34.23 understand You should label a light band as early wood, and a dark band as late wood; both bands together comprise one growth ring.

p. 745 CYU (1) understand See Figure A34. 3. (2) apply See Figure A34.4.

✔ TEST YOUR KNOWLEDGE

1. remember a **2.** understand T, T, T, F. If you selected True for the fourth statement, remember that tracheids and vessel elements transport water only after they are dead. They lack a living cytoplasm and water is transported through the hollow cell walls that remain. **3.** understand c **4.** understand a

✔ TEST YOUR UNDERSTANDING

5. create The general function of both systems is to acquire resources: The shoot system captures light and carbon dioxide; the root system absorbs water and nutrients. Vascular tissue is continuous throughout both the shoot and root systems. Diversity in roots and shoots enables plants of different species to live together in the same environment without directly competing for resources. **6.** understand Continuous growth enhances phenotypic plasticity because it allows plants to grow and respond to changes or challenges in their environment (such as changes in light and water availability). **7.** apply Cuticle reduces water loss; stomata facilitate gas exchange. Plants from wet habitats should have a relatively large number of stomata and thin cuticle. Plants living in dry habitats should have relatively few stomata and thick cuticle. **8.** understand Stems, branches, and roots grow from their tips. Although a tree can get taller with each passing year, the position of any existing region of the trunk will stay at that same height because new growth is initiated in the apical meristem. Thus, the birdfeeder will remain at 6 feet.

Figure A34.3

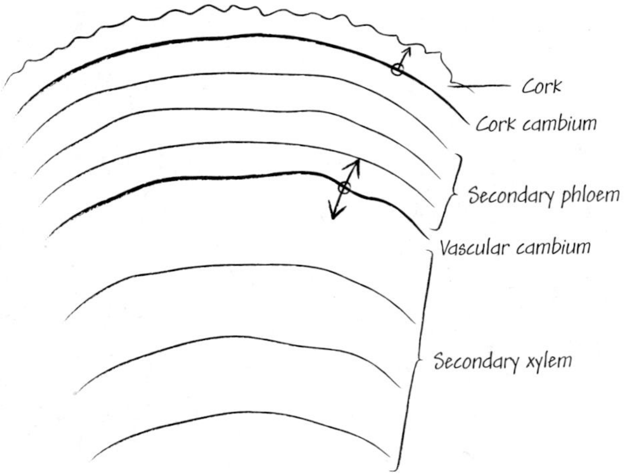

Figure A34.4

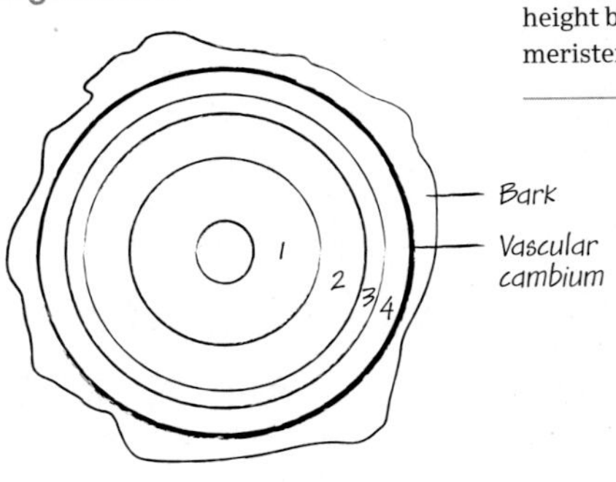

✔ TEST YOUR PROBLEM-SOLVING SKILLS

9. apply Asparagus—stem; Brussels sprouts—lateral buds; celery—petiole; spinach—leaf (petiole and blade); carrots—taproot; potato—modified stem. **10.** apply Girdling disrupts transport of solutes in secondary phloem. The tree's root system starves.

✔ PUT IT ALL TOGETHER: CASE STUDY

11. create Acid rain may damage apical meristems and likely have a negative effect on primary growth. Trees exposed to acid rain will probably be shorter than trees exposed to neutral rain. **12.** analyze Chinaberry trees treated with acid rain had about a 40 percent reduction in height. Trees treated with acid rain were approximately 15 cm tall while those treated with distilled water were approximately 25 cm tall—a difference of about 10 cm. Thus, 10 cm/25 cm = 40%. **13.** analyze The apical meristem was negatively affected. **14.** evaluate Strongly acidic rain had little effect on height (apical meristem activity) in *C. camphora*, but it had a major impact in *M. azedarach*. The *C. camphora* may have been protected from the effects of acid rain by having thick, resistant leaves with a thick cuticle; tough, resistant bark; or roots that could still function in acidic soil. **15.** create Acid rain may cause damage to the vascular cambium and will likely have a negative impact on secondary growth in chinaberry trees (but probably not in camphor trees based on the primary growth results presented). Chinaberry trees exposed to acid rain will probably produce less wood than trees exposed to rain with a neutral pH. **16.** apply Possibilities include negative effects on leaf production, number and quality of flowers and seeds, root formation, and any other component of plant growth.

Chapter 35

IN-TEXT QUESTIONS AND EXERCISES

p. 750 Fig. 35.2 apply The solute potential increases (becomes less negative), which reduces the solute potential gradient between the two sides.

p. 750 evaluate Ψ cell $= 0 + (-2) = -2$ MPa; Ψ solution $= -3$ MPa. There will be an overall net movement of water out of the cell.

p. 751 Fig. 35.3 create Any line is acceptable as long as it always runs *above* the existing line for ninebark leaves.

p. 752 CYU (1) apply If the concentration of solutes in the soil is greater than in the root cells (as a result of extra fertilizer), then water will tend to leave the roots and enter the soil by osmosis. **(2)** apply The roots of pickleweed will likely have a lower water potential than the surrounding soil. As a result, water would tend to enter the roots.

p. 753 apply The plant would not be able to exclude toxic solutes from the xylem, because water would not be forced through living cytoplasm of endodermal cells on its way to the xylem.

p. 757 Fig. 35.12 create The line on the graph would go up, and large jumps would occur each time the light level was turned up, because increased transpiration rates would increase negative pressure (tension) at the leaf surface and thus increase the water-potential gradient.

p. 758 CYU (1) apply Transpiration rates would tend to (a) decrease when stomata close, (b) decrease when it rains (because very little evaporation would occur), and (c) increase as dry air blows in. **(2)** understand Ions and other molecules are transported into roots, and water then follows by osmosis. The influx of water builds pressure in the xylem, which then forces water through the xylem and out of the leaves by the process of guttation. **(3)** apply CAM plants tend to close their stomata during the daytime to minimize water loss by transpiration. Therefore, pineapple would likely have closed stomata during the daytime.

p. 760 Making Models 35.1 apply See **Figure A35.1**.

p. 761 Fig. 35.17 create See **Figure A35.2**.

p. 763 Fig. 35.21 understand With time, the plant will metabolize the radioactive sucrose, and the level of radioactivity will decline.

p. 764 CYU (1) understand Sieve plates are enlarged pores between adjacent sieve-tube elements. These openings represent regions with no cell wall. As a result, phloem sap can easily move between cells. **(2)** understand Proton pumps are used to move sugars into companion cells. If they are not working properly, the pressure gradient would be diminished and, as a result, sugars would not be transported from sources to sinks.

✔ TEST YOUR KNOWLEDGE

1. understand d. It might be tempting to choose "a" because water will flow into the cell, but don't forget water can flow both into and out of the cell. **2.** remember a **3.** remember d **4.** understand F (This process requires energy and relies on actively transporting sugars into and out of cells.), T, T, F (Sieve-tube elements are involved in transport of phloem sap, but vessel elements are not. They are involved in movement of water through xylem by transpiration.)

✔ TEST YOUR UNDERSTANDING

5. understand b **6.** apply See **Figure A35.3**. **7.** understand b **8.** evaluate Phloem can carry sugars down a plant (e.g., if they are moving from leaves down to roots). But phloem can carry sugars up as well (e.g., if sugars are moving to flowers or fruits that are developing at the tip of a stem).

✔ TEST YOUR PROBLEM-SOLVING SKILLS

9. apply The tree would use virtually *no* calories to transpire the water. Water transport is driven by evaporation of water from leaves, which then pulls water from roots to shoots. The tree does not expend energy during this process. **10.** apply The high concentration of solutes in the soil, formed from the salt and melting snow, would result in a low water potential in the soil.

Figure A35.1

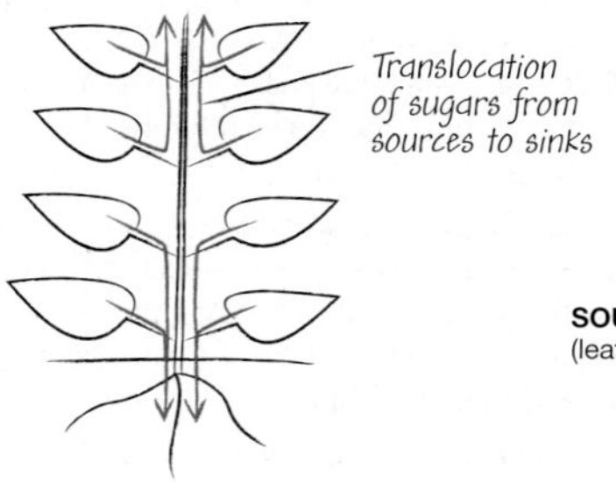

Figure A35.2

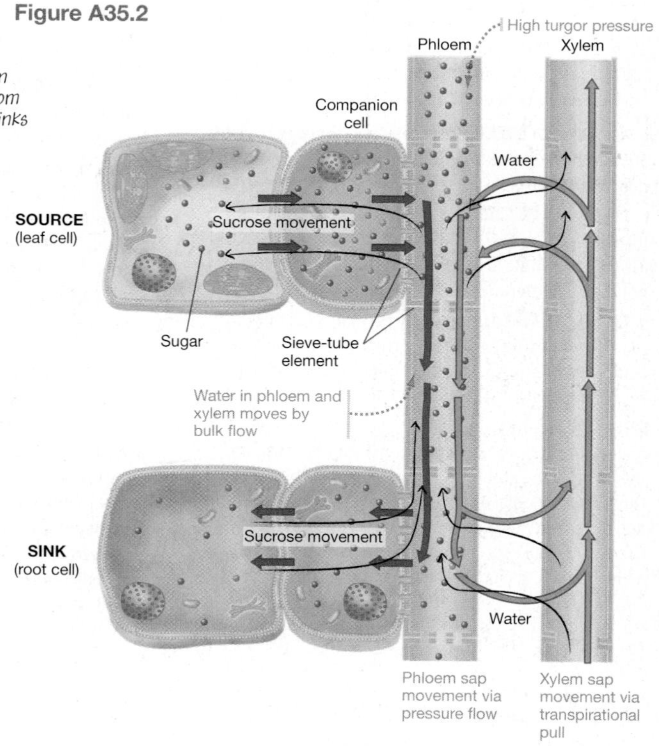

Figure A35.3

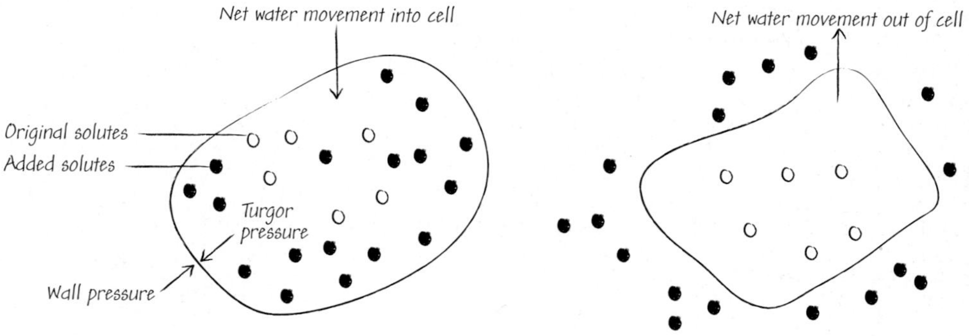

This would make it difficult for plants growing in that region to absorb water.

✔ PUT IT ALL TOGETHER: CASE STUDY

11. understand b **12.** create As the CO_2 level increases, one might predict a decrease in the number of stomata (as plants could absorb sufficient CO_2 even with fewer stomata) and therefore a decrease in transpiration rate. **13.** analyze The plants illustrated in this study have lower stomatal conductance now than they did a century ago. **14.** evaluate Because plants are transpiring less (and thus conserving water), they would probably be more likely to survive a drought under elevated CO_2 levels. **15.** analyze In 1915, stomatal conductance in oak was roughly four times that in pine. In 2010, oak was roughly three times that of pine. **16.** apply At some point, plants would likely reach a limit in their phenotypic flexibility; they would not be able to decrease the number of stomata any further without compromising CO_2 uptake and decreasing photosynthesis.

Chapter 36

IN-TEXT QUESTIONS AND EXERCISES

p. 768 Fig. 36.1 create To test the hypothesis that most of a plant's mass comes from CO_2 in the atmosphere, plants could be grown in the presence or absence of CO_2 and growth could be compared between the two groups. Or, plants could be grown in the presence of labeled CO_2, which could be analyzed in plant tissues.

p. 770 Fig. 36.3 analyze Because soil is so complex, it would be difficult to assume that the soils with and without added copper are identical except for the difference in copper concentration. Also, it would be better to compare soils that had no copper versus normal amounts of copper, instead of comparing those with and without added copper, because even soil without added copper might already contain trace amounts of copper that could have a big impact on plant growth.

p. 771 CYU (1) understand While both nitrogen and boron are needed for normal plant growth, nitrogen is considered a macronutrient because it is found in a relatively high level within plants, but boron is a micronutrient because it is found in only trace amounts. **(2)** apply Since nitrogen is a mobile element, it can be moved from older regions to younger regions when it is in short supply.

p. 774 CYU (1) understand Because anions such as Cl^- are less likely to bind to soil particles in neutral soil, they are more available and more readily absorbed by root hairs than cations such as Ca^{2+}. **(2)** understand Magnesium and calcium are both positively charged ions. When protons are released to the soil via cation exchange, they displace magnesium and calcium ions from soil particles and make them more readily available for absorption.

p. 775 Fig. 36.10 apply Since the interior of the cell is negative, cations are attracted to the interior, while anions are repelled from the interior.

p. 776 apply (1) If there is no voltage across the root-hair membrane, there is no electrical gradient favoring entry of cations, and absorption stops. (2) There is no route for cations to cross the root-hair membrane, so absorption stops.

p. 776 apply (1) If there is no proton gradient across the root-hair membrane, there is no gradient favoring entry of protons, so symport of anions stops. (2) There is no route for anions to cross the root-hair membrane, so absorption stops.

p. 778 CYU (1) understand Proton pumps establish an electrical gradient across root-hair membranes, and the inside of the membrane is much more negative than the outside. Cations follow this electrical gradient into the cell. Anions are transported with protons via symporters. **(2)** apply Pine seedlings grown with mycorrhizal fungi will likely grow faster than seedlings without fungi, since mycorrhyzae help plants facilitate the absorption of nutrients from the soil. **(3)** apply See Figure A36.1.

p. 780 Fig. 36.15 create Nodule diameter would increase steadily during days 0–5, then level off during days 6–10, then increase steadily again during days 11–14.

p. 780 CYU (1) create Many possible answers, including these: If N-containing ions are abundant in the soil, the energetic cost of maintaining N-fixing bacteria may outweigh the benefits of increased nitrogen availability. If mycorrhizal fungi provide a plant with N-containing ions, the energetic cost of maintaining N-fixing bacteria may outweigh the benefits of increased nitrogen availability. Plants that grow near species with N-fixing bacteria might gain nitrogen by absorbing N-containing ions after root nodules die, or by "stealing it" (roots of different species will often grow together). There could be a genetic constraint (Chapter 22): Plant species may simply lack the alleles required to manage the relationship. Note: To be considered correct, your hypothesis should be plausible (based on the underlying biology) and testable. **(2)** create Many answers are possible, including this: Because the bacterial cell enters root cells, the plant needs to have reliable signals that it is not a parasitic bacterium. It is advantageous for the plant to be able to reject N-fixing bacteria if usable nitrogen is already abundant in the soil.

p. 782 Fig. 36.17 understand Mistletoe is parasitic (it extracts water and certain nutrients from host trees); bromeliads are not (they are epiphytes and don't affect the fitness of their host plants).

p. 782 CYU (1) understand *Monotropa* absorbs carbon and other nutrients from fungi that grow in close association with its roots. Since it lacks chloroophyll, *Monotropa* does not produce its own organic compounds through photosynthesis. **(2)** apply Venus flytraps are autotrophic because they can produce organic compounds through photosynthesis. They are also heterotrophs because they absorb organic compunds from the insects they digest.

✔ TEST YOUR KNOWLEDGE

1. understand T, F (their presence has nothing to do with limiting the availability of micronutrients), T, T **2.** remember d **3.** understand c **4.** remember d

✔ TEST YOUR UNDERSTANDING

5. understand a **6.** understand Your lab partner is incorrect. Only a small portion of water absorbed by a plant is actually retained in the plant. Most of a plant's organic matter comes from conversion of CO_2 into organic compounds through photosynthesis. **7.** understand The statement is incorrect. If you answered "correct," remember that some parasitic plants cannot perform photosynthesis and therefore must absorb all their nutrients from other organisms. These represent examples of heterotrophic plants. **8.** understand Some metal ions are poisonous to plants, and high levels of other ions are toxic.

Figure A36.1

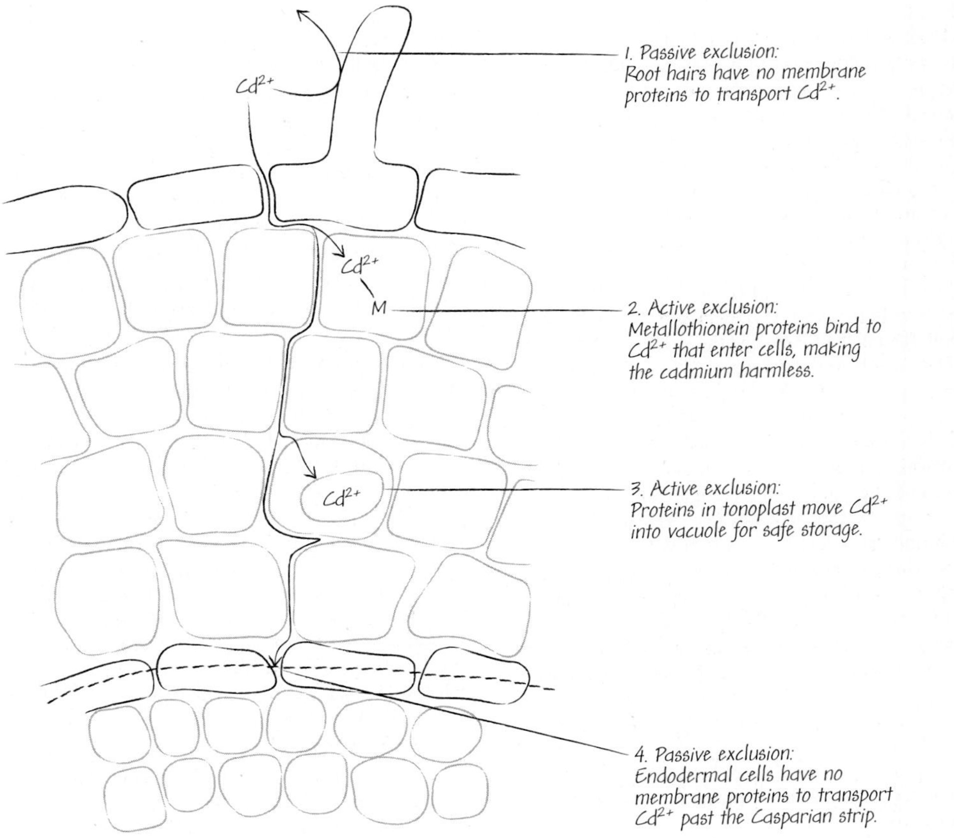

Passive mechanisms of ion exclusion—such as a lack of ion channels that allow passage into root cells—do not require an expenditure of ATP. Active mechanisms of exclusion—such as production of metallothioneins—require an expenditure of ATP.

✔ TEST YOUR PROBLEM-SOLVING SKILLS

9. create The soil lost 60 g while the tree gained 74,000 g. Dividing 60 by 74,000 and converting to a percentage equals 0.08 percent, far less than 4 percent. The water that van Helmont used may have contained many of the macro- and micronutrients that were incorporated into plant mass. To test this hypothesis, conduct an experiment of the same design but have one treatment with pure water and one treatment with water containing solutes ("hard water"). Compare the percentage of soil mass incorporated into the willow tree in both treatments. *Another explanation:* Willow may be unusual. To test this hypothesis, repeat the experiment with willow and several other species under identical conditions, and compare the percentage of soil mass incorporated into the plants. *A third hypothesis:* Van Helmont's measurements were inaccurate. Repeat the experiment. **10.** create Grow pea plants in the presence of rhizobia, exposed to air with (1) N_2 containing the heavy isotope of nitrogen and (2) radioactive carbon dioxide. Allow the plant to grow; then analyze the rhizobia and plant tissues. If the rhizobia–pea plant interaction is mutualistic, then ^{15}N-containing compounds and radioactive-carbon-containing compounds should be observed in both rhizobia and plant tissues. As a control, grow pea plants in the presence of the labeled compounds but without rhizobia. If the mutualism hypothesis is correct, the plant should contain labeled carbon but no heavy nitrogen.

✔ PUT IT ALL TOGETHER: Case Study

11. create It is reasonable to predict that the ants might have an overall *negative* impact on plant growth (because they consume nectar and other insects that would otherwise be digested and absorbed by the pitcher plant). However, the preferred answer would be an overall *positive* impact on plant growth since the ants do leave behind feces and body parts of insects. **12.** analyze Plants with diving ants show significantly larger total leaf surface area than plants without ants. Plants without ants grow taller with no significant increase in leaf surface area. In plants with ants, however, leaf surface increases significantly as the plants grow taller. Therefore, diving ants have a positive impact on overall plant growth. **13.** evaluate Yes, both the pitcher plants and the ants benefit from the relationship. Ants eat nectar and insects that visit the pitcher plants. Pitcher plants absorb nutrients from feces and insect body parts. **14.** understand If nutrients in soil are scarce, there is intense natural selection favoring alternative ways of obtaining nutrients—for example, by digesting insects or stealing nutrients from other plants. **15.** evaluate Yes. The availability of limiting nutrients limits plant growth. When plants are provided with additional limiting nutrients (often as fertilizer, but in this case as insects), growth usually increases. **16.** evaluate Both pitcher plants and legumes acquire nutrients from soil and directly from other organisms. Nitrogen tends to be the key nutrient absorbed from other organisms in both types of plants. Bacteria grow in root nodules of legumes and provide nitrogen to the plant (the plants do not trap the bacteria). Pitcher plants, however, must lure their prey, capture them, and digest them to obtain nutrients.

Chapter 37

IN-TEXT QUESTIONS AND EXERCISES

p. 788 CYU (1) understand The only cells that can respond to a specific environmental signal or hormone are cells that have an appropriate receptor for that signal or hormone. **(2)** understand Once a receptor in the plasma membrane is activated, it triggers the activation of many proteins in a phosphorylation cascade or the production of many second messengers.

p. 789 Fig. 37.5 create For "Tip removed," cut the tip and replace it, to control for the hypothesis that the wound itself influences bending (not the loss of the tip). For "Tip covered," cover the tip with a transparent cover instead of an opaque one, to control for the hypothesis that the cover itself (not excluding light) affects bending.

p. 792 CYU (1) apply Phototropins are pigments that detect blue light. If plants are unable to produce functional phototropins, they will not exhibit an appropriate phototropic growth response. **(2)** apply Using a syringe or other device, apply auxin to the west side of each stem, just below the tip. **(3)** apply Stems would become very long and would not respond to directional light.

p. 792 Table 37.1 analyze The average germination rate of lettuce seeds last exposed to red light is about 99 percent. The average for seeds last exposed to far-red light is about 50 percent. Both of these values are much higher than the germination rate of buried seeds that receive no light at all; their germination rate is only 9 percent.

p. 793 Fig. 37.9 evaluate By causing plants to grow longer or taller, etiolation increases their chances of finding light, when light is otherwise limited.

p. 795 CYU (1) understand Lettuce grows best in full sunlight. If a seed receives red light, it indicates that the seed is in full sunlight—meaning that conditions for growth are good. But if a seed receives far-red light, it indicates that the seed is in shade—meaning that conditions for growth are poor. **(2)** understand Night length changes very predictably throughout the year as seasons change, so it is a reliable signal for plants to perceive.

p. 795 Fig. 37.11 create An appropriate control treatment would be to graft a leaf from a plant that had never been exposed to a short-night photoperiod onto a new plant that also had never been exposed to the correct conditions for flowering. If the hypothesis is correct, the grafted leaf should not induce flowering.

p. 796 Fig. 37.13 create *One possibility:* Direct pressure from a statolith changes the shape of the receptor protein. The shape change triggers a signal transduction cascade, leading to a gravitropic cell response.

p. 797 CYU (1) apply Statoliths will resettle toward the bottom of root cells when the orientation of those cells is altered. Roots will detect the new position of the statoliths and adjust growth in a downward position so that statoliths become positioned on the bottom of the cells. **(2)** analyze In both roots and shoots, auxin is redistributed asymmetrically in response to a signal. In phototropism, auxin is redistributed to the shaded side of the shoot tip in response to blue light. In gravitropism, auxin is redistributed to the lower part of the root or shoot in response to gravity. **(3)** understand In root cells, auxin concentrations indicate the direction of gravity. The gravitropic response is triggered by changes in the distribution of auxin in root-tip cells.

p. 798 CYU (1) understand Short, stocky plants are less likely to be damaged by wind than long, tall plants of the same species. **(2)** understand Tendrils help anchor certain plants, so they support the plant as it increases in height.

p. 800 apply The dwarfed individuals can put more of their available energy into reproduction because they are using less energy for growth than taller individuals.

p. 802 Fig. 37.21 evaluate Without these data, a critic could argue that the stomata closed in response to water potentials in the leaf, not a signal from the roots.

p. 806 CYU (1) apply When shrubs are pruned, apical buds are likely to be cut off. Removing apical buds removes the hormones that prevent axillary buds from growing. As a result, more axillary buds grow into branches and the shrub becomes bushier. **(2)** apply *ABA:* Stomata should close and photosynthesis should stop compared to individuals that do not receive extra ABA. *GA:* The stems should elongate rapidly, compared to plants that do not receive extra GA. **(3)** create The rotten apple produces ethylene, which, being a gas, stimulates all of the other apples in the bushel basket to ripen too fast.

p. 807 apply Sacrificing a small number of cells due to the hypersensitive response is far less costly for the plant than death.

p. 809 CYU (1) understand Because specific receptor proteins recognize and match specific proteins produced by pathogens, having a wide array of these proteins allows individuals to recognize and respond to many types of pathogens. **(2)** understand If herbivores are not present, synthesizing large quantities of proteinase inhibitors wastes resources (ATP and substrates) that could be used for growth and reproduction.

✔ TEST YOUR KNOWLEDGE

1. remember F (cells on the shaded side elongate more than cells on the illuminated side), T, T, T **2.** remember d **3.** remember d **4.** understand b

✔ TEST YOUR UNDERSTANDING

5. understand a **6.** evaluate The P_{fr}–P_r switch in phytochromes lets plants sense whether they are in shade or sunlight, and it triggers responses that allow sun-adapted plants to avoid shade. Phototropins are activated by blue light, which indicates full sunlight, and trigger responses that allow plants to photosynthesize at full capacity. **7.** apply b **8.** apply When one leaf on the mutant plant is chewed on by a herbivore, the damaged leaf would not be able to signal to the rest of the plant through systemic acquired resistance (SAR), so it would not be able to fend off future herbivore attacks elsewhere on the plant.

✔ TEST YOUR PROBLEM-SOLVING SKILLS

9. create Small-seeded plants need to perform photosynthesis early in seedling development or they will starve. Therefore, they need to germinate in direct sunlight. The food reserves in large-seeded plants can support seedling growth for a relatively long time without photosynthesis. Therefore, they do not need to germinate in direct sunlight. **10.** create In these seeds, (a) little or no ABA is present, or (b) ABA is easily leached from the seeds, so its inhibitory effects are eliminated. Test hypothesis (a) by determining whether ABA is present in the seeds. Test hypothesis (b) by comparing the amount of ABA present before and after running water—enough to mimic the amount of rain that falls in a tropical rain forest over a few days or weeks—over the seeds.

✔ PUT IT ALL TOGETHER: CASE STUDY

11. analyze Approximately 7 seconds to close. This is deceiving, however, because leaflets are almost entirely closed within the first 3 seconds. It takes about 10 minutes for leaflets to reopen. **12.** remember a **13.** apply Leaflets are kept in the open position due to turgor pressure in cells on the upper surface. Leaflet closure is due to a rapid loss of pressure in those cells. The cells lose pressure due to ions flowing out of the cells, followed by loss of water by osmosis. **14.** apply Stimulating the first set of leaflets generates an action potential that causes those leaflets to close. The action potential is then propagated along the rest of the leaf, causing all leaflets to close in sequence. **15.** create One possibility would be to present *Mimosa pudica* and other closely related species that don't display rapid leaf movements to herbivores and measure the leaf consumption of each species. **16.** apply Phytochrome.

Chapter 38

IN-TEXT QUESTIONS AND EXERCISES

p. 816 (1) evaluate The microsporocyte divides by meiosis to generate male spores (i.e., microspores); the megasporocyte divides by meiosis to generate female spores (i.e., megaspores). **(2)** In angiosperms, the female gametophyte stays within the ovary at the base of the flower even after it matures and produces an egg cell. Fertilization and seed development take place in the same location.

p. 816 CYU (1) understand An advantage of asexual reproduction is that the organism does not need to rely on finding gametes from another individual (and therefore reproduction is assured). A disadvantage is that all offspring are genetically identical (and therefore less likely to be able to deal with diverse stresses from disease, parasites, or other environmental factors). **(2)** evaluate That statement is incorrect. Spores are produced by meiosis. Gametes are formed by mitosis.

p. 819 Fig. 38.7 evaluate A gametophyte is the multicellular individual that produces gametes by mitosis. The pollen grain is a male gametophyte because it is multicellular and produces sperm by mitosis.

p. 819 CYU (1) understand Stamens are considered male because they produce pollen (then ultimately produce sperm). Carpels are considered female because they produce ovules (which ultimately produce eggs). **(2)** analyze Both are multicellular individuals that produce gametes by mitosis. The female gametophyte is larger than the male gametophyte and produces an egg; the male gametophyte produces sperm.

p. 821 Fig. 38.8 analyze To eliminate all other variables except for the one being tested, such as the possibility of pollination by other insects in the field. Scales differ because the two insect species visited flowers at very different frequencies.

p. 823 apply The endosperm nucleus in the central cell is triploid; the zygote is diploid; the synergid and other cells remaining from the female gametophyte are haploid.

p. 823 CYU (1) evaluate Insects feed on nectar and/or pollen in flowers. Flowers provide food rewards that encourage insects to visit them. The individuals that attract the most pollinators produce the most offspring. **(2)** understand One product of double fertilization is the diploid zygote, which eventually grows by mitosis into a mature sporophyte. The other product is the triploid endosperm nucleus, which will grow by mitosis to form a source of nutrients for the embryo.

p. 824 Fig. 38.13 create *Hypothesis:* Fruit changes color when it ripens as a signal to fruit eaters. The color change is advantageous because seeds are not mature in unripe fruit and are unlikely to survive. Fruit eaters disperse mature seeds in ripe fruit, and these are more likely to survive and eventually germinate.

p. 825 Fig. 38.14 understand Hackberries served as a control to demonstrate that animals that avoided chilies were hungry.

p. 827 Fig. 38.15 analyze Beans—their cotyledons are aboveground and green.

p. 827 CYU (1) evaluate Beans, peas, and corn are classified as fruits because they develop from the female parts of flowers. Vegetables, however, come from roots, stems, or leaves. **(2)** evaluate Long periods of dormancy allow seeds to remain viable through conditions that are less than ideal for growth. When adequate water, oxygen, and temperatures are available, seeds can then germinate and grow into a mature plant. **(3)** remember Oxygen consumption and protein synthesis increase. Water uptake eventually slows or stops. At this point, new mRNAs are produced and mitochondria multiply. Water uptake then resumes and enables cells to enlarge as the seedling emerges from the seed coat.

p. 828 Fig. 38.16 apply See Figure A38.1.

p. 828 apply A cotyledon and a radicle develop into organs with distinct functions, but both are composed of the same tissues: Epidermis is derived from protoderm, ground tissue is derived from ground meristem, and vascular tissue is derived from the procambium.

p. 831 CYU (1) analyze The globular stage embryo is a mass of undifferentiated cells that lacks embryonic tissues. The heart stage embryo shows the early stages of cotyledons and the progenitors of the three embryonic tissues. The mature embryo shows distinct embryonic tissues including epidermis, ground tissue, and vascular tissue. **(2)** understand As long as a meristem exists, it has the potential to divide and produce new cells, tissues, and organs. **(3)** create If the plant could be genetically engineered with a mutant form of the *CLAUSA* gene, then that would likely result in a shift in leaf type from simple to compound.

p. 832 understand Mutual inhibition between the *A* and *C* genes helps create separate regions of gene expression. If *C* gene expression is lost, then the inhibition of *A* gene expression is also lost, and *A* genes are expressed in their normal location and where they are normally inhibited by *C* gene expression.

p. 833 CYU (1) remember Floral organs, listed from the exterior toward the interior, are sepals, petals, stamens, carpels. **(2)** analyze In both plants and animals, mutations in homeotic genes result in structures growing in the wrong location in place of the appropriate structure (e.g., by forming petals where stamens should be or developing legs in place of antennae).

✔ TEST YOUR KNOWLEDGE

1. remember F (double fertilization results in the formation of a zygote and endosperm), T, T, T **2.** remember Sperm and eggs are formed by mitosis, whereas megasporocytes (female) and microsporocytes (male) undergo meiosis to generate megaspores and microspores, respectively. These divide mitotically to give rise to female and male gametophytes—the embryo sac and pollen grain, respectively. **3.** remember d **4.** remember c

✔ TEST YOUR UNDERSTANDING

5. understand c **6.** analyze Outcrossing increases genetic diversity among offspring, making them more likely to survive if environmental conditions change from the parental generation. However, outcrossing requires cross-pollination to be successful. Self-fertilization results in relatively low genetic diversity among offspring but ensures that pollination succeeds. **7.** understand Just like the three tissue layers in plant embryos, the tissues produced in the shoot and root apical meristems of a 300-year-old oak tree can differentiate into all of the specialized cell types found in a mature plant. **8.** understand Human embryonic stem cells and meristem cells are both capable of extensive cell divisions that ultimately give rise to all of the cells within the adult person or plant, respectively.

✔ TEST YOUR PROBLEM-SOLVING SKILLS

9. evaluate Acorns have a large, edible mass and are usually animal dispersed (e.g., by squirrels that store them and forget some). Cherries have an edible fruit and are animal dispersed. Burrs stick to animals and are dispersed as the animals move around. Dandelion seeds float in wind. To estimate the distance that each type of seed is dispersed from the parent, (1) set up "seed traps" to capture seeds at various distances from the parent plant; (2) sample locations at various distances from the parent and analyze young individuals—using genetic techniques introduced in Chapter 20 to determine if they are offspring from the parent being studied; (3) mark seeds, if possible, and find them again after dispersal. **10.** create See Figure A38.2 for one possible solution. Notice that by using just two genes, each with two possible states (either on or off), four different outcomes, or floral structures, are possible.

✔ PUT IT ALL TOGETHER: Case Study

11. analyze Bees likely use visual cues such as color and patterns (possibly including ultraviolet patterns), and they also use scent when locating gooseberry flowers. Bees are likely looking for nutrient rewards such as pollen or nectar. **12.** analyze Flowers that were pollinated by bees did form fruits with significantly higher mass. However, there was no significant difference between number of seeds per fruit of self- and bee-pollinated flowers. **13.** evaluate Cross-pollination increases genetic

Figure A38.1

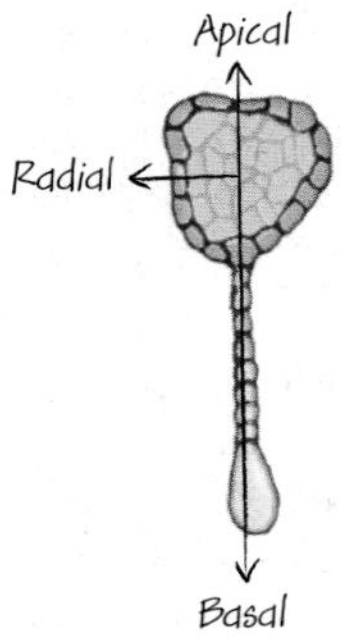

Figure A38.2

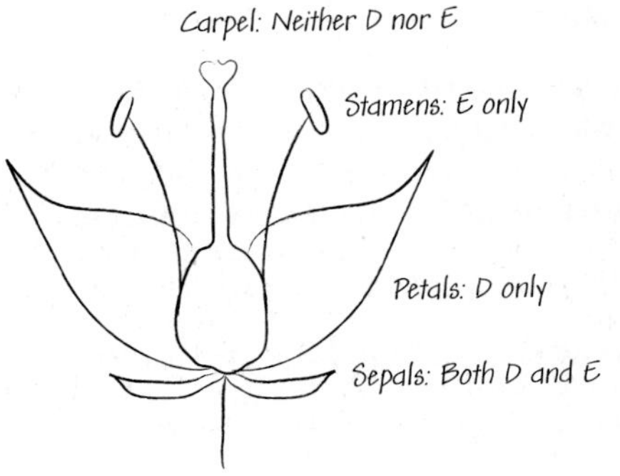

diversity in the offspring (seeds). Self-pollination requires only one plant and therefore is advantageous because it doesn't depend on insects or other pollinators. **14.** analyze Fruits produced by bee-pollinated flowers are approximately 35 percent larger. **15.** analyze At least 240 ovules per carpel (because fruits from bee-pollinated flowers contain roughly that many seeds). **16.** evaluate The recent decline in bee populations is indeed a major concern for food production on a global scale. Fewer bees means lower crop production. This problem will make it more difficult to feed and sustain the increasing number of people worldwide.

END-OF-UNIT CASE STUDY Can Plant Compounds Perform a Role Similar to Tetrodotoxin?

pp. 836–837 (1) understand a **(2)** analyze Yes. Although the armyworms ate leaves on both types of plants, there was more damage to the nicotine-free plants than the wild-type plants. **(3)** apply Nicotine synthesis may require a lot of resources. Therefore, the plant is better off using its energy and resources for nicotine synthesis only when needed. **(4)** understand Your sketch should indicate that jasmonic acid is transported through phloem while nicotine is transported through xylem. **(5)** apply One might predict equal damage in the wild type and nicotine-free plants since the hornworms can tolerate nicotine. **(6)** understand Nicotine acts as a direct defense by deterring certain pests (e.g., armyworms) but other compounds act as an indirect defense by attracting predators that feed on hornworms. **(7)** evaluate Yes. Nicotene and TTX are both toxic and are used to deter potential predators and, in both cases, there are predators that have evolved resistance to the toxins.

BIG PICTURE Plant and Animal Form and Function

pp. 838–839 CYU (1) analyze Plants and animals may accumulate compounds in the epidermis and skin (such as waxes and other lipids) that reduce water loss. Both plants and animals evolved closable openings in this barrier to allow regulated gas exchange. Plants and animals may also change the amount of the body exposed to the sun (in plants by moving leaves, in animals by changing posture or taking shelter) to reduce evaporative water loss. **(2)** analyze Plants have alternation of generations including a multicellular haploid stage, whereas the haploid stage in animals is single celled. **(3)** analyze Very large animals and plants have strong supporting structures such as thick bones or trunks that withstand the force of gravity. They also need elaborate internal transport systems to bring nutrients to each cell and carry wastes away. **(4)** analyze Cellular respiration fits into one of the key functions for survival—nutrition. Both plants and animals depend on sugar (produced by plants via photosynthesis) as their energy source to make ATP via cellular respiration.

Chapter 39

IN-TEXT QUESTIONS AND ANSWERS

p. 842 Fig. 39.2 create In each graph, the bars for treated and control males would have the same height.

p. 843 CYU (1) understand The gelatinous mass of the spermatophore of male crickets is an adaptation because males with larger gelatinous masses father more offspring, giving them higher fitness. Therefore, the large gelatinous mass has evolved by natural selection. Production of a large mass is a trade-off because males who invest a lot of energy into masses have reduced immune function, and because males who have increased immune function cannot make very big masses. **(2)** apply If runners train at high elevation where there is a low level of oxygen, their bodies might produce more hemoglobin to help deliver more oxygen to tissues. This is a reversible change to an individual's phenotype, which is acclimatization. Adaptations cannot occur in individuals but must evolve by natural selection in populations.

p. 844 Fig. 39.4 apply Most neurons have an axon, which transmits signals to specific cells and tissues rather than widely throughout the body.

p. 846 apply The inside surface of the mouth contains stratified epithelium; it is subjected to wear and tear from food. The gas exchange surface of the lungs contains simple epithelium; its thinness (a single cell layer) facilitates the diffusion of gases across it.

p. 848 CYU (1) understand The surface area of a simple epithelial cell is large because it functions in the movement of molecules in or out of the body, so it must contain many membrane proteins to assist with this transport. In contrast, a stratified epithelial cell is part of a thick layer of cells responsible for protection and therefore does not need a large surface area. **(2)** understand In the small intestine, neurons help the organ respond to stimuli from food and brain, muscles contract to mix food, epithelia absorb nutrients, and connective tissue holds the other tissues together. **(3)** apply The meniscus is made of cartilage, which is flexible and provides padding between the bones of the upper and lower leg; cardiac muscle cells are branched, allowing rapid communication between cells, which ensures coordinated contraction of the heart; the simple epithelial cells of tbe lungs are very small, allowing rapid diffusion of gases across them.

p. 850 Fig. 39.10 apply The Chihuahua must eat relatively more because it has a higher mass-specific metabolic rate than the Great Dane.

p. 850 apply The mouse's body temperature would drop because it would lose heat rapidly across its high surface area. The elephant's body temperature would rise because it cannot lose heat fast enough across its small surface area.

p. 851 CYU (1) apply Lungless salamanders that breathe through their skin are all small compared to salamanders with lungs. Their surface area to volume ratios are high enough that gas exchange across the skin can support the salamanders' metabolism. **(2)** understand Fish gills contain highly flattened structures that increase the surface area, helping gases and nutrients move across them.

p. 853 CYU create A dog's maintenance of homeostatic body temperature is antagonistic because when one response (e.g., warming the body by shivering) is activated, the other response (e.g., cooling the body by panting) will always be inhibited.

p. 855 CYU (1) apply The main source of heat for a lizard living underground is conduction of heat from the warm soil. **(2)** analyze Endotherms can remain active during the winter and at night and sustain high levels of aerobic activities such as running or flying, but they require large amounts of food energy. Ectotherms need much less food and can devote a larger proportion of their food intake to reproduction, but they have a harder time maintaining high activity levels at night or in cold weather. **(3)** apply If the legs of a Canada goose had concurrent flow, not much heat would be transferred from the arteries to the veins and the goose's body temperature would drop as it loses heat to the cold snow.

✔ TEST YOUR KNOWLEDGE

1. understand True **2.** understand b **3.** understand T, T, F, F **4.** understand a

✔ TEST YOUR UNDERSTANDING

5. understand *Absorptive sections* have many folds and projections that increase their surface area for absorption. *Capillaries* have a high surface area because they are thin and highly branched, making exchange of substances more rapid. *Beaks of Galápagos finches* have sizes and shapes that correlate with the type of food each species eats. Large beaks are used to crack large seeds; long, thin beaks are used to pick insects off surfaces; etc. *Fish gills* contain thin, flattened structures with a large surface area, which facilitates the exchange of gases and wastes. **6.** analyze A frog can move, breathe, and digest much faster at 35°C than at 5°C because enzymes work faster (rates of chemical reactions increase) at higher temperatures. The frog will need much more food energy during the summer to support its higher metabolic rate. **7.** understand Because endotherms generate much of their body heat metabolically, they can adjust the amount of heat produced to maintain their body temperature near the set point. Ectotherms cannot produce as much heat metabolically, so their body temperatures will decline when ambient temperatures go down and will rise when ambient temperatures go up or when they can bask in the sun. **8.** understand See **Figure A39.A1**. The graph shows that as size (radius) increases, the volume increases more rapidly than the surface area.

✔ TEST YOUR PROBLEM-SOLVING SKILLS

9. apply King Kong is endothermic, and his huge mass would generate a great deal of heat. His relatively small surface area compared to that of a normal-sized gorilla would not be able to dissipate the heat—especially if the surface were covered with insulating fur. **10.** apply *Apatosaurus* (*Brontosaurus*) was likely homeothermic, given its huge size. Due to its low surface area to volume ratio, its body temperature would not change much between day and night because it takes a long time to lose or gain heat across such a relatively small surface area.

✔ PUT IT ALL TOGETHER: Case Study

11. apply The surface area to volume ratio is higher for mainland tortoises because their body size is smaller. **12.** understand The mainland tortoises need to eat more food per gram of body mass because their mass-specific metabolic rate is higher. **13.** understand c **14.** understand False. The body temperature of ectotherms does not necessarily match the ambient temperature. If a tortoise is basking in a sunny field, the solar radiation might

Figure 39.A1

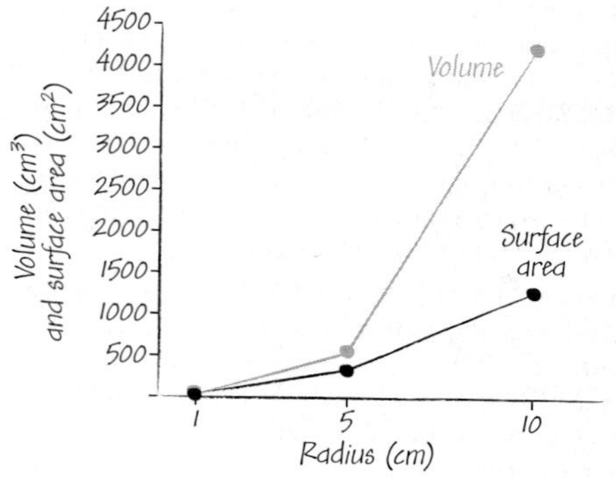

raise its body temperature well above ambient, and the tortoise's low surface area to volume ratio would cause it to retain that heat for a long time. **15.** apply The small tortoise from the mainland would be more poikilothermic (the temperature would rise higher during the day and drop lower at night) because heat would be transferred more readily between the environment and body of the small tortoise due to its high surface area to volume ratio. **16.** understand Tortoises are ectotherms, but it is inaccurate to describe them as cold blooded because their blood could actually be quite warm. This is especially true of giant tortoises, whose low surface area to volume ratios would cause them to lose heat so slowly that their body temperature could be elevated well above ambient temperature.

Chapter 40

IN-TEXT QUESTIONS AND EXERCISES

p. 860 **Making Models 40.1** create See **Figure A40.1**.

p. 862 CYU (1) understand Marine animals face dehydration because the high osmolarity of seawater causes them to lose water by osmosis. Terrestrial animals face dehydration because they constantly lose water to the air via evaporation from their body surfaces. **(2)** understand Marine and terrestrial species are under severe osmotic stress in the form of water loss to the environment. For this reason, they often excrete urea or uric acid because these require less water excretion. Freshwater species, in contrast, gain too much water from the environment, so they excrete ammonia along with large volumes of water.

p. 864 CYU (1) understand Marine cartilaginous fishes osmoconform, so they do not lose water to their environment. Because they do not need to drink seawater, they do not absorb as much NaCl from drinking. However, they must expend energy to make proteins that protect their cells from the toxic effects of maintaining a high urea concentration in their tissues. Marine bony fishes osmoregulate, so they do not need to make those proteins. However, they must drink seawater, which increases the amount of NaCl they absorb, so they must expend energy to excrete the NaCl.

(2) understand See **Figure A40.2**.

p. 865 Fig. 40.5 analyze The cuticle would likely have the most wax in the summer, when hot, dry weather would accelerate evaporation from the body surface.

p. 866 CYU (1) understand Uric acid has very low solubility in water and can be excreted without much water loss. **(2)** understand When electrolytes are reabsorbed in the hindgut epithelium, water follows along an osmotic gradient.

p. 869 understand In step 1, the Na^+/K^+-ATPase performs primary active transport. In step 2, the Na^+-dependent cotransporters perform secondary active transport. In steps 3 and 4, solutes and water move via facilitated diffusion. Simple diffusion does not occur here.

p. 871 apply (1) Less water will be reabsorbed because the osmotic gradient in the interstitial fluid will not be as steep. (2) Filtrate osmolarity will be lower because less water will have been reabsorbed. (3) Salt reabsorption will be reduced because the concentration of NaCl in the filtrate will be lower.

p. 873 apply Ethanol inhibits water reabsorption, leading to a larger volume of less-concentrated urine. Nicotine increases water reabsorption, leading to a smaller volume of more-concentrated urine.

p. 874 Fig. 40.16 create Carrying a large amount of water in the bladder may be energetically costly to the animal. The animal also might not be able to run as quickly and therefore might have an increased predation risk.

p. 874 CYU (1) understand See **Figure A40.3**. This is an example of a nephron model showing reabsorption and secretion of the substances you learned about in this chapter. **(2)** understand (a) Water intake leads to lowered electrolyte concentrations in the blood and filtrate and the production of large volumes of dilute urine. (b) Eating large amounts of salt results in concentrated, hyperosmotic urine. (c) Water deprivation triggers ADH release and the production of small volumes of concentrated, hyperosmotic urine. **(3)** create Several hypotheses are possible, including that desert rodents would have a greater number of nephrons with loops of Henle penetrating deep into the medulla, or that these loops would have a greater concentration of active transporters pumping NaCl into the interstitial fluid.

✔ TEST YOUR KNOWLEDGE

1. understand F, F, T, T **2.** understand a **3.** remember c **4.** remember bladder

✔ TEST YOUR UNDERSTANDING

5. analyze Ammonia is the most toxic and must be diluted with large amounts of water to be excreted safely. Urea and uric acid are less toxic and do not have to be excreted with large amounts of water but require more energy expenditure to produce. Uric acid can be excreted with the least water. Bony fishes excrete ammonia; mammals excrete urea; and insects excrete uric acid. You would expect the embryos inside terrestrial eggs to produce uric acid, because it is the least toxic and has very low solubility in water. **6.** analyze Mitochondria produce ATP in cellular respiration. A key component of salt transport by chloride cells

Figure A40.1

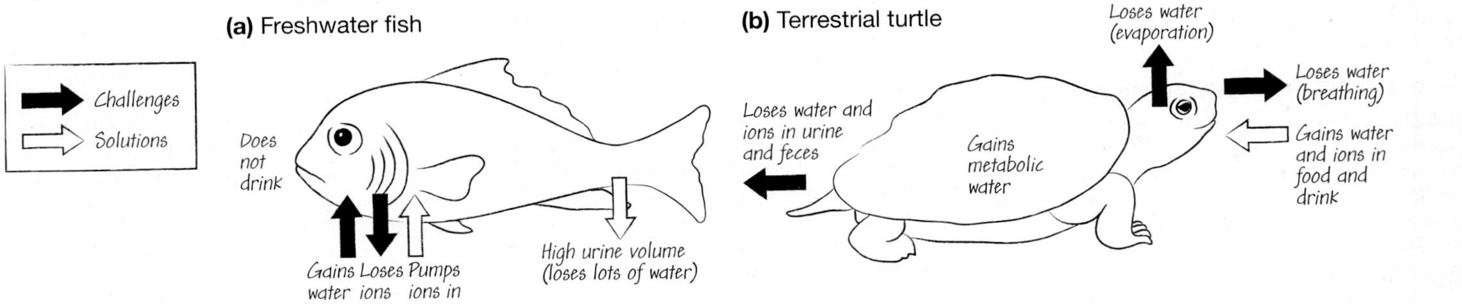

Figure A40.2

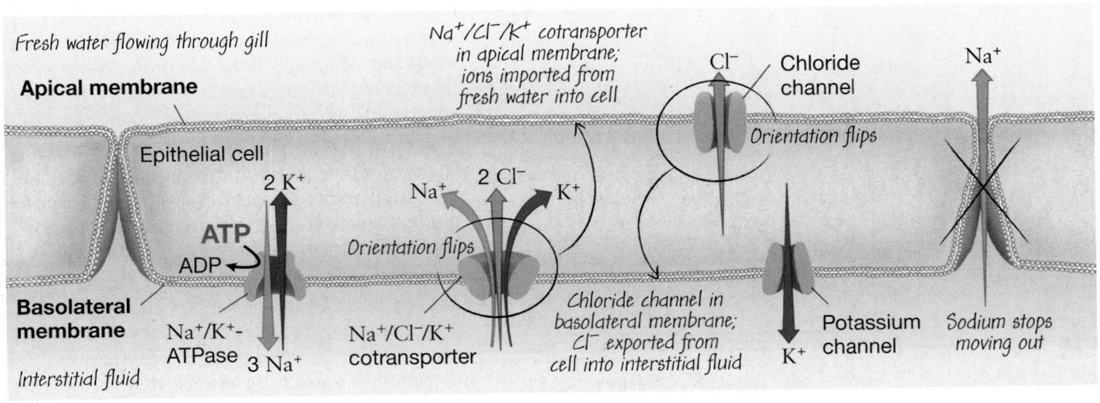

Figure A40.3

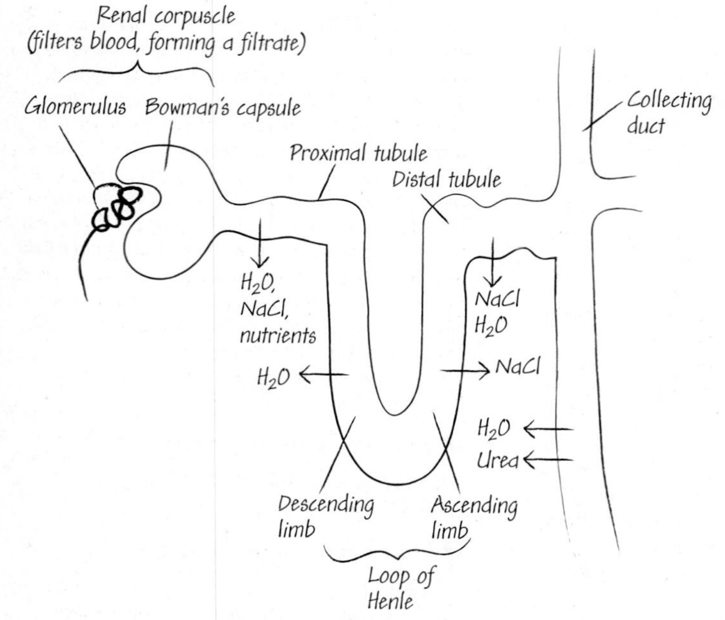

is Na^+/K^+-ATPase, which requires ATP to function. Because ATP fuels establishment of the ion gradients necessary for the cotransport mechanisms used by chloride cells as well as other epithelial cells involved in ion transport, an abundance of mitochondria would be expected in these cells. **7.** understand Without loops of Henle, there would be no concentration gradient in the interstitial fluid of the kidney's medulla, so water could not be absorbed from the pre-urine in the collecting duct; as a result, concentrated urine could not be formed. **8.** understand Fresh water has such low osmolarity that if invertebrates osmoconformed to it, the ion concentrations in their interstitial fluid would be too low to conduct electrical currents in their excitable tissues.

✔ TEST YOUR PROBLEM-SOLVING SKILLS

9. analyze c. Without aquaporins in the collecting duct, water cannot be reabsorbed, which would result in increased urine volume and decreased urine osmolarity. **10.** analyze The hypothesis that the mussels are osmoconformers is supported because the osmolarity of their hemolymph is very close to the wide range of osmolarities of the water they were exposed to. See **Figure A40.4**.

✔ PUT IT ALL TOGETHER: CASE STUDY

11. remember a **12.** evaluate Yes, the data support the hypothesis because the blood osmolarity of the aluminum-treated fish dropped between 6 hours and 24 hours after aluminum exposure began, whereas the blood osmolarity of the control fish did not change during that period. **13.** understand The researchers did a laboratory study to establish cause and effect by comparing two groups (the aluminum-treated fish and the control fish) that differed in only one variable (the aluminum content of the water). If they had simply sampled fish from water polluted with aluminum, they could not have attributed any observed effects on osmoregulation to aluminum with certainty because of the uncontrolled variables. **14.** create The activity of Na^+/K^+-ATPase was lower in the aluminum-treated fish than in the control fish. As a result, fewer sodium ions were imported from the water by active transport, leading to the observed reduction in blood osmolarity. **15.** understand False. Water moves by osmosis because there is a net diffusion of water molecules from solutions of low osmolarity to solutions of high osmolarity, not because water molecules are attracted to ions such as sodium. **16.** create See **Figure A40.5**.

Figure A40.4

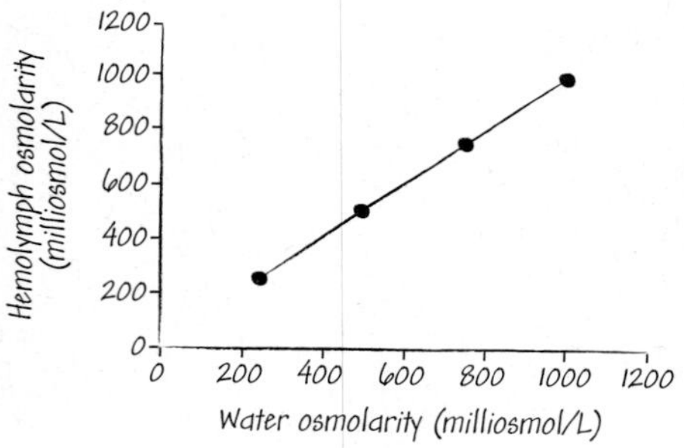

Figure A40.5

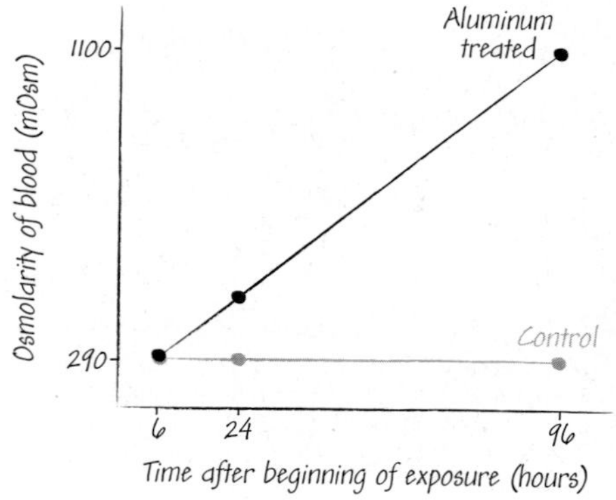

Chapter 41

IN-TEXT QUESTIONS AND EXERCISES

p. 879 CYU analyze Vitamin B_{12} and iron are found in animal products such as meats and are needed for production of red blood cells, so a vegetarian whose diet is low in vitamin B_{12} or iron may become anemic.

p. 881 CYU apply Humans have sharp canines and incisors, which may help to bite and tear meat. Humans also have flat molars, which help in grinding grains and other plant material.

p. 886 apply Proton pump inhibitors reduce the amount of HCl that enters the lumen of the stomach, therefore easing the symptoms of heartburn.

p. 886 analyze If lead shot is used to kill game, and condors feed on the game, they are apt to consume the lead shot and retain it in their gizzards as though it were rocks, to assist in grinding of food. The lead can then poison the birds.

p. 888 analyze People who cannot digest lactose do not produce sufficient quantities of the enzyme lactase, which is required to cleave the bond between the two monosaccharides in lactose.

p. 888 Fig. 41.14 understand Frog eggs don't normally make the sodium-glucose cotransporter protein, so the researchers could be confident that if the protein appeared in the eggs, it was from the injected RNA. They would not be confident of this if they had injected RNAs into rabbit epithelial cells, where the protein was probably already present.

p. 890 apply **(1)** Decreased energy yield from foods due to reduced glucose absorption, **(2)** increased glucose content in feces due to the passing of unabsorbed glucose into the colon, and **(3)** watery feces due to decreased water reabsorption in the large intestine (lower osmotic gradient). (An additional effect would be increased flatulence due to the metabolism of unabsorbed glucose by bacteria that produce gases as waste products.)

p. 891 CYU (1) understand *Mouth:* Ingestion; teeth mechanically digest food into smaller particles; salivary amylase begins to chemically digest carbohydrates; lingual lipase initiates the chemical digestion of fats. *Esophagus:* Food is moved to the stomach via peristaltic contractions. *Stomach:* Churning of food provides mechanical digestion. HCl denatures proteins and activates pepsin; pepsin begins to chemically digest proteins. *Small intestine:* Pancreatic enzymes complete the chemical digestion of carbohydrates, proteins, lipids, and nucleic acids. Most of the water and all of the nutrients are absorbed here. *Large intestine:* More water is absorbed; feces are formed and accumulate in the rectum. *Anus:* Feces are expelled out the anus in the process of elimination. **(2)** apply If the release of bile salts is inhibited, fats would not be digested and absorbed quickly, and they would pass into the large intestine. The person would likely produce fatty feces and lose weight over time. **(3)** apply Herbivores would have higher-volume intestines (longer and wider) than carnivores to house the extensive microbial communities necessary for fermenting cellulose.

p. 892 Fig. 41.16 apply In type 2 diabetes mellitus, the top black arrow from glucose to glycogen (inside liver and muscle cells) is the disrupted process. In type 1 diabetes mellitus, the green arrow on the top left, indicating insulin produced by the pancreas, is the disrupted process.

p. 893 Fig. 41.17 apply A 1.8-m tall man with a BMI of 40 weighs 130 kg. To reduce his risk for type 2 diabetes mellitus to less than 10 percent, he would need to lower his BMI below 25, so he would have to reduce his weight to less than 84 kg. Therefore, he would need to lose at least 46 kg.

p. 893 CYU (1) understand Type 1 diabetes mellitus is an autoimmune disease, meaning that the body's immune system mistakenly kills the insulin-producing cells of the pancreas. Type 2 diabetes mellitus is caused by a combination of factors, including genetic predisposition and obesity. **(2)** apply When the blood glucose level of a person with type 1 diabetes mellitus is too high, he or she should get an injection of insulin, which will trigger the absorption and storage of glucose by their cells. When their blood glucose level is too low, which might happen if they inject too much insulin, they should eat something with a high concentration of sugar (such as orange juice or a candy bar) to increase their blood glucose level quickly. (If their blood glucose level is dangerously low, a glucagon shot may be administered.)

✔ TEST YOUR KNOWLEDGE

1. understand One serving of skim milk has approximately $(12 \times 4) + (8 \times 4) = 80$ kcal, while one serving of whole milk has about $(12 \times 4) + (8 \times 4) + (8 \times 9) = 152$ kcal. **2.** understand F, T, F, T **3.** remember b **4.** remember insulin, glucagon

✔ TEST YOUR UNDERSTANDING

5. understand The bird crop is an enlarged sac that can hold quickly ingested food; in certain leaf-eating species, it is filled with symbiotic bacteria and functions as a fermentation vessel. The cow rumen is an enlarged portion of the stomach, and the elephant cecum is a blind sac at the anterior end of the large intestine. Both structures are filled with symbiotic organisms and function as fermentation vessels. **6.** understand b **7.** understand Nutrient absorption occurs in the small intestine but not in the stomach or esophagus, and the rate of absorption increases with surface area. **8.** understand Individuals with type 1 diabetes mellitus do not produce enough insulin, so insulin injections can compensate for the missing insulin. In contrast, individuals with type 2 diabetes produce insulin but have insulin receptors that do not function correctly or are reduced in number, so insulin injections are not as effective.

✔ TEST YOUR PROBLEM-SOLVING SKILLS

9. understand Fat contains more than twice the energy per gram as protein or carbohydrate, so fat storage minimizes the energetic cost of carrying stored food energy. **10.** create Terrestrial vertebrates are exposed to increased risk of water loss, and water reabsorption is the primary function of the large intestine. Most aquatic vertebrates do not need to reabsorb large amounts of water from their feces.

✔ PUT IT ALL TOGETHER: Case Study

11. understand The pharyngeal jaws of carnivorous minnows have sharp teeth specialized for puncturing animal prey, such as insects. The pharyngeal jaws of herbivorous minnows have flattened teeth specialized for grinding and compacting vegetation and algae. **12.** understand b **13.** evaluate The data show that the relative gut length of herbivorous minnows is greater than that of carnivorous minnows. **14.** understand The longer gastrointestinal tracts of herbivores allow more time to digest plants and algae, and provide more area to house symbiotic microbes that aid in digesting vegetation. **15.** apply Cellulase activity is higher in herbivorous minnows because their diets contain more cellulose than the carnivores' diets. Cellulase is produced by gut microbes. **16.** analyze The minnows likely varied in overall body size, so larger fish would have had longer guts regardless of diet. Dividing gut length by body length allowed the researchers to rule out differences in gut length due solely to differences in body size.

Chapter 42

IN-TEXT QUESTIONS AND EXERCISES

p. 897 CYU analyze Oxygen diffuses into the blood from the alveoli because the concentration of oxygen is higher in the inhaled air in the alveoli than it is in the blood in lung capillaries. However, when oxygen arrives at tissues, the tissues have a lower concentration of oxygen than the blood because they are using oxygen in cellular respiration, so oxygen diffuses from blood into tissues. Oxygen diffuses along its concentration gradient.

p. 899 CYU create A: Large amount of air, because the oxygen-carrying capacity of warm water is low. B: Small amount of air, because the oxygen-carrying capacity of cold water is higher and because algae contribute oxygen to the water through photosynthesis. C: Small amount of air, because sedentary animals require relatively little oxygen.

p. 901 Fig. 42.6 apply If concurrent flow occurred, less oxygen would be transferred from water to blood because the partial pressure gradient driving diffusion would fall to zero partway along the length of the capillary.

p. 905 CYU (1) understand Common features include large surface area, short diffusion distance (a thin gas-exchange membrane), and a mechanism that keeps fresh air or water moving over the gas-exchange surface. Only fish gills use a countercurrent exchange mechanism; only tracheae deliver oxygen directly to cells without using a circulatory system; only mammalian lungs contain alveoli—small air sacs surrounded by capillaries where gas exchange occurs. **(2)** apply The P_{O_2} decreases as oxygen is used up, the P_{CO_2} increases as CO_2 diffuses into the blood from tissues but cannot be exhaled, and the pH drops as the CO_2 dissolves in blood to form bicarbonate ions and H^+.

p. 906 Making Models 42.1 create See Figure A42.1. The equal changes in x are not associated with equal changes in y in a sigmoidal curve. At low values of x (zero to x_1), y increases slowly (zero to y_1). At intermediate levels of x (x_1 to x_2), y increases much more rapidly (y_1 to y_2). At high levels of x (x_3 to x_4), y increases slowly again (y_3 to y_4).

p. 907 Fig. 42.15 analyze According to the data in the figure, when the tissue is 30 mm Hg, the oxygen saturation of hemoglobin is about 40 percent for blood at pH 7.2 and about 58 percent at pH 7.4. Therefore, about 60 percent of the oxygen is released from hemoglobin at pH 7.2, but only about 42 percent of the oxygen is released at pH 7.4.

p. 908 Fig. 42.16 analyze In the lungs, a strong partial pressure gradient favors diffusion of dissolved CO_2 from blood into the alveoli. As the partial pressure of CO_2 in the blood declines, hydrogen ions leave hemoglobin and react with bicarbonate ions to form more CO_2, which then diffuses into the alveoli and is exhaled from the lungs.

p. 908 CYU (1) apply The curves of Tibetans should be shifted to the left relative to the curves of people adapted to sea level—meaning that Tibetans' hemoglobin should have a higher affinity for oxygen at all partial pressures. **(2)** apply Hyperventilation is extremely rapid ventilation. Carbon dioxide will be exhaled faster than tissues produce it, and so the P_{CO_2} level in the blood will decline, causing an increase in pH.

p. 913 Fig. 42.21 analyze Sham surgeries were performed on control alligators to make sure that something about the act of surgery, rather than the treatment itself, did not affect the variables measured in the study.

p. 915 Fig. 42.23 analyze Air from the alveoli mixes with air in the dead space in the bronchi and trachea on its way out of the body. This dead-space air is from the previous inhalation ($P_{O_2} = 160$ mm Hg; $P_{O_2} = 0.3$ mm Hg), so when the alveolar air mixes with it, the partial pressures in the exhaled air reach levels intermediate between those of inhaled and alveolar air.

p. 915 understand Without the delay at the AV node, the ventricles would not have the chance to fully fill with blood from the atria. Consequently, the volume of blood ejected from the ventricles would be smaller.

p. 918 CYU (1) understand See Figure A42.2.

(2) remember A myocardial infarction occurs when some of the heart tissue dies due to a lack of oxygen. It is not the same as cardiac arrest, which is when the heart stops beating. Only a massive infarction could interfere with proper pumping of the heart and stop it from beating.

✔ TEST YOUR KNOWLEDGE

1. understand T, F, F, T **2.** understand Air is less dense and viscous than water, so it takes less energy to ventilate with it. In addition, oxygen tends to be more abundant in air than in water, making it easier to obtain. **3.** remember c **4.** understand An open circulatory system cannot direct the hemolymph toward specific organs. Also, because the pressure is lower in an open circulatory system, there is a slower flow of hemolymph.

Figure A42.1

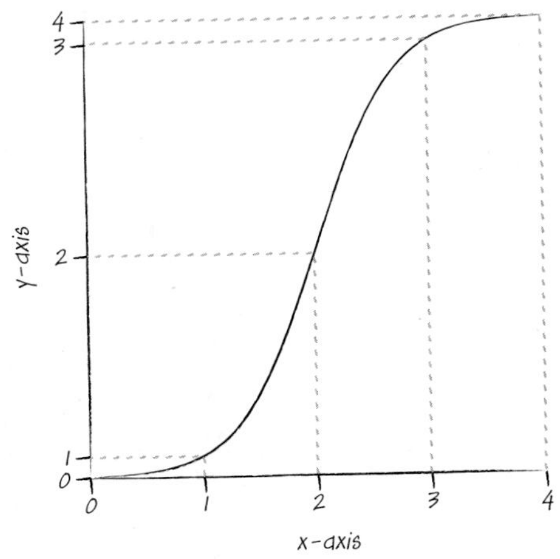

Figure A42.2

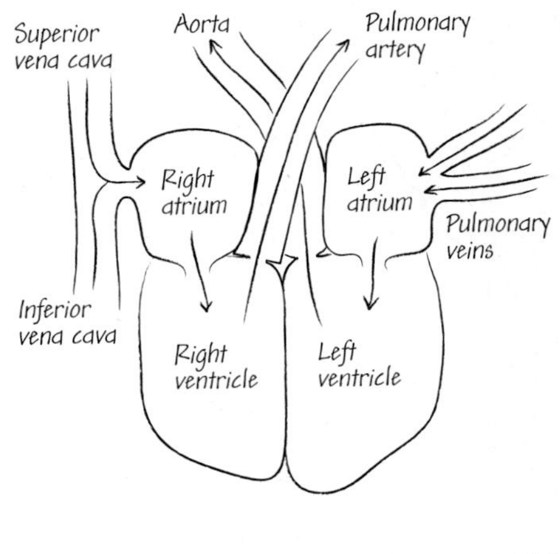

Figure A42.3

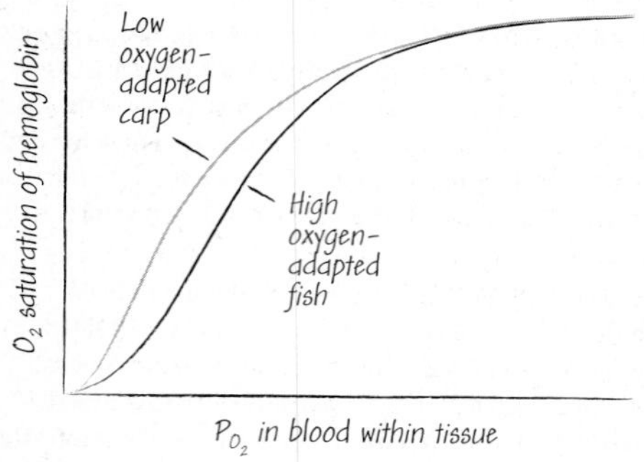

✔ TEST YOUR UNDERSTANDING

5. analyze Lungs increase the temperature of the air and are moist, allowing greater solubility of gases in the respiratory surface (increasing *k*); alveoli present a large surface area (large *A*); the epithelium of alveoli is thin (small *D*); and constant delivery of poorly oxygenated blood to alveoli maintains a steep partial pressure gradient, favoring diffusion of oxygen into the body ($P_2 - P_1$ is high). **6.** understand c **7.** create The carp's oxygen–hemoglobin equilibrium curve is shifted left compared to that of a typical fish that lives in high-oxygen water. This means that at a given water P_{O_2}, more oxygen is bound to hemoglobin, showing the increased affinity of hemoglobin for oxygen in species such as carp. See Figure A42.3. **8.** analyze Because myocardial infarction is death of heart muscle cells, an infarction could lead to scar tissue and thereby disrupt the flow of electricity from the sinoatrial (SA) node across the heart. An artificial pacemaker could solve this problem by stimulating heart muscle cells to contract at the correct time.

✔ TEST YOUR PROBLEM-SOLVING SKILLS

9. apply Cold water carries more oxygen than warm water does, so icefish blood can carry enough oxygen to supply the tissues with oxygen even in the absence of hemoglobin. The oxygen is simply dissolved in the blood. **10.** analyze If the pulmonary circulation were under pressure as high as that found in the systemic circulation, large amounts of fluid would be forced out of capillaries in the lungs. There is a trade-off between the thin surface required for rapid gas exchange and the thickness needed in blood vessels to withstand high pressure.

✔ PUT IT ALL TOGETHER: Case Study

11. understand a **12.** understand b (Contrary to popular belief, high P_{CO_2} plays a much more important role than low P_{O_2} in the increased ventilation rate during exercise.) **13.** analyze Current athletes have a higher left ventricular mass than non-athletes; this difference is statistically significant, because $P < 0.001$. However, this increase in mass is not permanent, because the left ventricular masses of ex-athletes and non-athletes are not significantly different ($P > 0.05$). **14.** understand The increased left ventricular mass in athletes allows their hearts to pump more blood per minute to exercising muscles to meet the muscles' demand for oxygen delivery and carbon dioxide removal. **15.** analyze If athletes and non-athletes have the same resting cardiac output (volume of blood pumped per ventricle per minute) but athletes have a lower resting heart rate (number of beats per minute), then athletes must have a higher stroke volume (volume of blood pumped per ventricle per beat). The increased ventricular muscle mass facilitates an increased force of contraction, resulting in more blood being ejected per beat. **16.** create In people with cardiovascular disease, the heart muscle often enlarges as it works harder to pump blood through hardened arteries.

Chapter 43

IN-TEXT QUESTIONS AND EXERCISES

p. 924 Fig. 43.3 apply No—as K^+ leaves the cell along its concentration gradient, the interior of the cell becomes more negative. As a result, an electrical gradient favoring movement of K^+ into the cell begins to counteract the concentration gradient favoring movement of K^+ out of the cell. Eventually, the two opposing forces balance out, and there is no net movement of K^+.

p. 925 apply If Na^+ were allowed to diffuse freely (and other ions could not diffuse), the membrane potential would equal the equilibrium potential for Na^+. If K^+ were allowed to diffuse freely (and other ions could not diffuse), the membrane potential would equal the equilibrium potential for K^+.

p. 926 CYU (1) analyze See Figure A43.1. **(2)** apply The resting potential would be much less negative because K^+ could no longer leak out of the cell. **(3)** understand The first phase of an action potential is called depolarization because the difference in charge across the membrane is reduced (the membrane becomes "less polarized"). The second phase is called repolarization because the difference in charge increases again.

p. 926 understand See Figure A43.2.

Figure A43.1

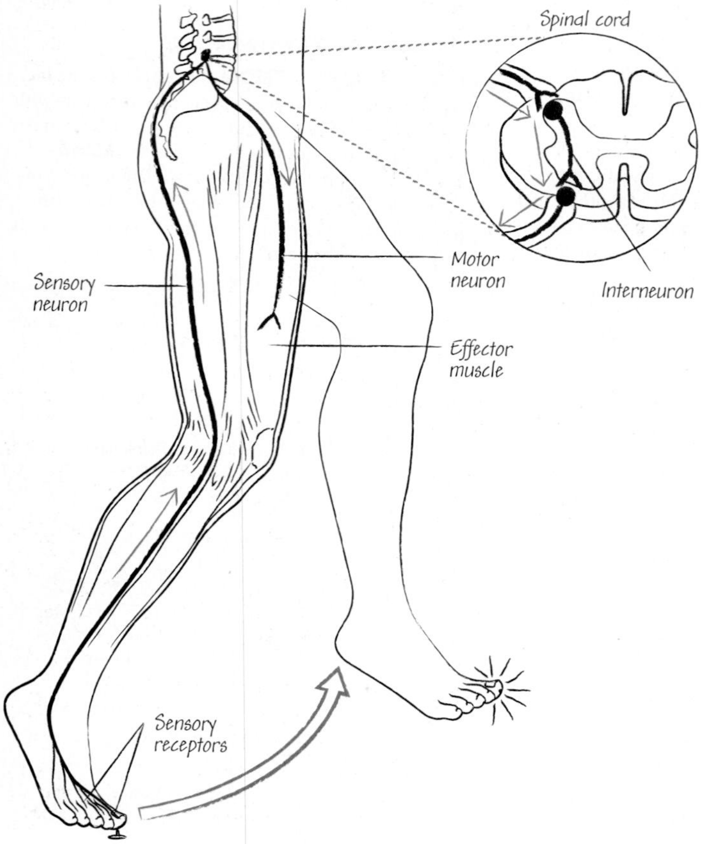

Figure A43.2

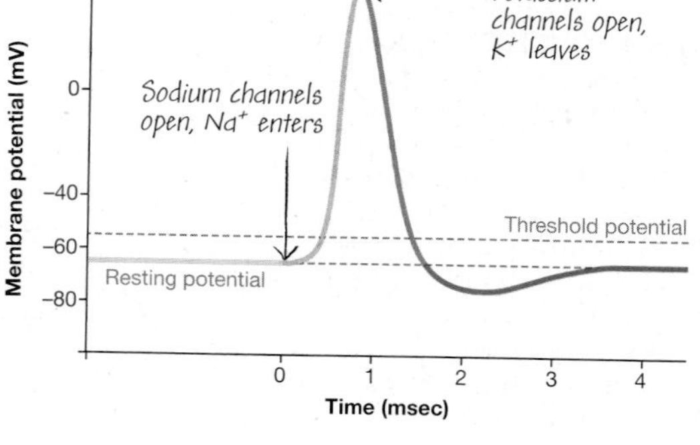

p. 930 CYU (1) understand If the threshold potential is attained, the probability that the voltage-gated sodium channels will open approaches 100 percent, and an action potential is produced. If the depolarization does not reach the threshold potential, the massive opening of Na^+ channels does not occur, and neither does an action potential. This is why the action potential is all or none. **(2)** understand Action potential propagation in the unmyelinated squid giant axon is fast because the axon's diameter is so large that cations experience less resistance as they flow down the axon. In small, myelinated axons of vertebrates, the small diameter increases resistance, but the myelination speeds propagation.

p. 930 Fig. 43.9 apply Collect a solution from the end of the vagus nerve at the heart *without* the nerve being stimulated. Expose a second heart to this solution. There should be no change in heart rate.

p. 934 CYU apply The EPSP would likely be smaller and briefer because there would be less neurotransmitter in the synaptic cleft.

p. 935 Fig. 43.15 analyze In the rest-and-digest mode, the pupils constrict, allowing less light to stimulate the eyes; the heart rate decreases, which conserves energy; the liver stores glucose, which enters the blood when food is digested; and the gallbladder releases products that promote digestion. In the fight-or-flight mode, the pupils open to admit more light; the heart rate increases to support muscle activity; the liver releases glucose into the blood, fueling increased muscle activity; and the gallbladder retains its contents, which are not used when food is not being digested.

p. 937 Fig. 43.18 analyze No—for example, part (b) indicates that the size of the cerebral area devoted to the trunk is no bigger than that devoted to the thumb.

p. 941 CYU (1) understand One strategy is to study individuals with known brain lesions and correlate the location of the lesion with a deficit in mental or physical function. Another is to directly stimulate brain areas or specific neurons in conscious animals and record the responses. **(2)** evaluate The discovery that synapses can change over time and that new neurons can be produced in adult vertebrates suggests that brain damage may not always be permanent. Following brain injury, repair might occur through synaptic plasticity or neurogenesis. The extent of repair would likely depend on many factors, including the region that was injured, the extent of the injury, and rehabilitative processes.

✔ TEST YOUR KNOWLEDGE

1. remember b **2.** understand F, T, T, T **3.** remember a **4.** remember d

✔ TEST YOUR UNDERSTANDING

5. analyze The ligand-gated channel opens in response to binding by a neurotransmitter; the voltage-gated channel opens in response to membrane depolarization. **6.** understand Postsynaptic potentials are additive. If a postsynaptic cell receives multiple EPSPs in a short period, it may be depolarized to the threshold potential and fire an action potential. IPSPs counteract the effect of EPSPs, making the postsynaptic cell less likely to fire an action potential. **7.** analyze The somatic nervous system controls voluntary skeletal muscle activity, such as movement of arms and legs. The autonomic nervous system controls internal involuntary activities, such as digestion, heart contraction, and gland secretion. **8.** understand a

✔ TEST YOUR PROBLEM-SOLVING SKILLS

9. apply The neurotransmitters will stay in the synaptic cleft longer, prolonging their binding to ligand-gated channels in the postsynaptic membrane. Ion flows across that membrane will be augmented, affecting the postsynaptic cell's membrane potential and likelihood of firing action potentials. **10.** create Alzheimer's disease is characterized by shrinkage of the hippocampus and the outer part of the cerebrum (called the cerebral cortex). The hippocampus is responsible for the formation of new memories; the outer part of the cerebrum is involved in the storage of memories, as indicated by electrical stimulation of outer cerebral areas during brain surgery and by recordings of activity in cerebral neurons of subjects viewing photographs.

✔ PUT IT ALL TOGETHER: Case Study

11. analyze c (BTX opens sodium channels, causing the membrane to depolarize. Unlike what happens during an action potential, however, the sodium channels remain open, and the membrane does not repolarize.) **12.** apply Patch clamping could be used to study the effect of BTX on individual ion channels. If the channel inside the tip of the microelectrode was a sodium channel, you would expect BTX to cause current to flow through the channel and to continue flowing as long as BTX was present, accounting for the prolonged depolarization shown in the graph. **13.** apply An action potential consists of a depolarization followed by a repolarization. By preventing repolarization, BTX stops action potentials from being produced and thus abolishes electrical signaling by the nervous system. **14.** create See **Figure A43.3**. **15.** apply (a) Depolarization would not occur, because there would be no gradient for inward flow of sodium ions. (b) The peak depolarization would be less positive because both Na^+ channels and K^+ leak channels are open in the presence of BTX, and increasing the intracellular K^+ concentration shifts the equilibrium potential for K^+ to a more negative value. (c) Depolarization would not occur, because tetrodotoxin blocks voltage-gated sodium channels. **16.** create One hypothesis is that in this snake species, a voltage-gated sodium channel evolved that is not affected by BTX. For example, a mutation in the gene for the sodium channel could reduce the effect of the toxin on the snake's nervous system, giving individuals with this mutation higher reproductive success and thereby causing the allele for this sodium channel to become fixed in the population by natural selection.

Figure A43.3

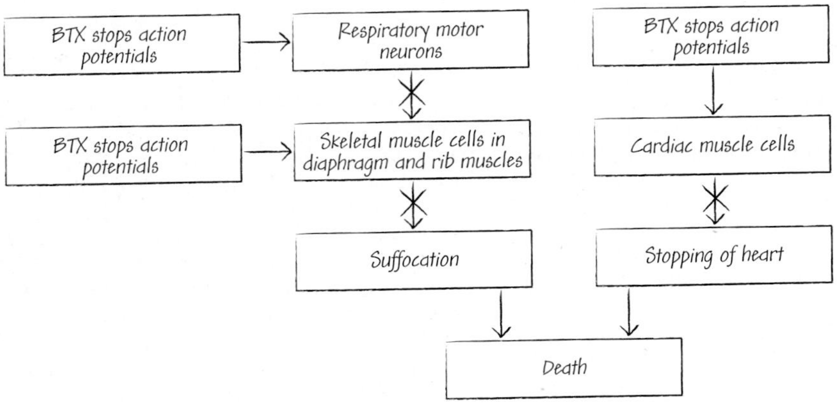

Chapter 44

IN-TEXT QUESTIONS AND EXERCISES

p. 946 CYU understand Sensory transduction is when sense receptors convert a stimulus such as a sound wave, photon, or chemical into action potentials, whereas transmission refers to how sensory neurons send the action potentials to the central nervous system.

p. 952 Fig. 44.10 create Vision could be used to detect prey in diurnal catfish. To control for this, you could "blindfold" catfish with an opaque material over their eyes.

p. 952 CYU (1) apply A punctured eardrum wouldn't vibrate correctly and would result in hearing loss at all frequencies in the affected ear. If the stereocilia were too short to reach the tectorial membrane, vibration of the basilar membrane would not cause them to bend, and sound would not be detected. A loss in basilar membrane flexibility would result in the inability to hear lower-pitched sounds, such as those in human speech. **(2)** create When you're reading in a car, the nerves from your eye signal your brain that you are stationary, whereas the nerves from the inner ear signal the brain that you are moving. This mismatch can make you feel queasy and dizzy. **(3)** apply The clawed frog would have a more developed lateral line system than the bullfrog, because the clawed frog is entirely aquatic whereas the bullfrog spends much of its time on land. The lateral line system functions only in water.

p. 955 Fig. 44.15 apply The S, M, and L opsin proteins are each different in structure. These structural differences affect the ability of retinal to absorb specific frequencies of light and change shape.

p. 956 CYU (1) understand Retinal acts like an on–off switch that indicates whether light has fallen on a rod cell. When retinal absorbs light, it changes shape. The shape change triggers events that result in a change in action potentials, signaling that light has been absorbed. **(2)** apply Retina detachment would separate the retina from the optic nerve, resulting in blindness. The mutation would produce blue–purple color blindness because the S opsin absorbs wavelengths in that region of the spectrum. A clouded lens would reduce the amount of light that reaches the retina, reducing visual sensitivity.

p. 959 CYU (1) apply Extremely hot food damages taste receptor proteins, making them temporarily unable to respond to their chemical triggers. **(2)** create One hypothesis is that other complex methods of communication, including gestures, facial expressions, and language, have evolved in primates and made sensation with the vomeronasal organ less important.

p. 960 apply (1) A rattlesnake with its eyes covered could strike effectively because its pits could sense the prey. (2) A rattlesnake with its eyes and pits covered with cotton cloth could strike effectively. The cloth would block vision, but heat energy from the prey would penetrate the cloth and stimulate thermoreceptors in the pits. (3) A rattlesnake with its eyes and pits covered with an opaque, heat-blocking material would not be able to strike effectively, because the material would block vision and heat energy.

p. 961 CYU (1) create Vampire bats have thermoreceptors in their noses that allow them to detect heat from their endothermic prey. **(2)** create Female sea turtles likely use magnetic cues to locate the beaches where they were born.

✔ TEST YOUR KNOWLEDGE

1. remember d **2.** understand T, T, F, T **3.** understand b **4.** remember magnetoreception

✔ TEST YOUR UNDERSTANDING

5. understand b **6.** understand The many possible examples include infrasound hearing allowing elephants to hear over long distances, ultrasonic hearing allowing bats to hunt by echolocation and moths to avoid bat predators, several blue opsins allowing coelacanths to distinguish the hues of blue light present in their deep-sea habitat, and red and yellow opsins allowing fruit-eating primates to distinguish ripe from unripe fruit. **7.** analyze Both the lateral line system and electroreception use hair cells to sense changes in the water. Hair cells in the lateral line system are mechanoreceptors that depolarize in response to changes in water pressure. In contrast, electroreceptive hair cells depolarize or hyperpolarize in response to changes in the electric field. **8.** evaluate Dalton's hypothesis was good. It was reasonable because blue fluid absorbs red light and would prevent red light from reaching the retina, and this prediction could be tested by dissecting his eyes after his death. Although the hypothesis was not correct, it inspired a rigorous test and led researchers to think of alternative explanations.

✔ TEST YOUR PROBLEM-SOLVING SKILLS

9. evaluate The data show that the mean time between onset of a woman's menstrual cycle and that of the rest of the women declined over time after they moved into the dormitory. It declined most rapidly after they first moved in together. The leveling of the graph line over time shows that the onset dates became more synchronized. The data are consistent with the hypothesis that a pheromone acted to synchronize the women's menstrual cycles, but they provide no direct evidence for the existence of such a pheromone. **10.** create To determine whether electric eels are electrogenic, place an eel in a tank of water and measure the electrical currents in the water. To determine whether they are electroreceptive, place an eel in a tank in the dark and add an object the size and shape of a prey animal that emits no scent or taste. Block mechanoreceptors in the lateral line system with a chemical, as described for Figure 44.10. If the eel locates the object, it would support the hypothesis that electric eels are electroreceptive.

✔ PUT IT ALL TOGETHER: Case Study

11. remember c **12.** analyze A worker should have the most alarm pheromone receptors because workers protect the hive. **13.** analyze Visual signals could not be detected easily at night or inside a dark hive. Audible signals could be hard to interpret if they were produced by thousands of workers, especially if the bees made similar sounds under other conditions. In contrast, an alarm pheromone would quickly spread throughout a hive even in darkness if each worker responded to the pheromone by secreting more of it, and the specificity of receptor proteins would enable bees to associate that pheromone with a particular type of stimulus. **14.** evaluate The results support the hypothesis because ovaries were less fully developed in workers exposed to the queen pheromone than in workers in the control group. **15.** analyze The researchers used diethyl ether as the control treatment because it was present as a solvent in the queen pheromone treatment. It's possible that diethyl ether itself could affect the bees' ovarian development. By including diethyl ether in both treatment groups, the researchers could attribute any difference in ovarian development to the presence of the pheromone. **16.** create Beekeepers starting a new hive can add a strip of material treated with queen pheromone to that hive to prevent bees from abandoning it before the queen is mature.

Chapter 45

IN-TEXT QUESTIONS AND EXERCISES

p. 966 Fig. 45.2 understand The dark band includes thin filaments as well as thick filaments; the light band consists of thin filaments only.

p. 966 analyze The trucks represent the Z discs, the ropes represent the thin filaments, and the weightlifters represent the thick filaments.

p. 969 CYU (1) understand In a sarcomere, thick myosin filaments are sandwiched between thin actin filaments. When the heads on myosin contact actin and change conformation, they pull the actin filaments toward the center of the sarcomere, shortening the whole sarcomere. **(2)** apply Increased acetylcholine release would result in an increased rate of action potentials in muscle cells and more forceful muscle contraction. Preventing conformational changes in troponin would prevent muscle contraction. Blocking uptake of calcium ions into the sarcoplasmic reticulum would lead to sustained muscle contraction.

p. 972 apply The postural muscles of the human neck are composed mainly of slow oxidative fibers, which are specialized for endurance.

p. 972 CYU (1) analyze Smooth muscle fibers have one nucleus, cardiac muscles fibers have 1–2 nuclei, while skeletal muscle fibers have multiple nuclei. Smooth muscle fibers are short and tapered at the ends, and cardiac muscle fibers are branched, whereas skeletal muscle fibers are long and cylindrical. **(2)** apply Training for a marathon would likely cause many of the person's muscle fibers to become redder in color due to increased growth of blood vessels and storage of oxygen-storing pigments. The cells would also have more mitochondria.

p. 975 Fig. 45.12 apply The biceps brachii is a flexor because it reduces the angle of the arm joint when it contracts. The triceps brachii is an extensor because it increases the angle of the arm joint.

p. 976 CYU (1) analyze Hydrostatic skeletons are made of soft tissues that vary from animal to animal; endoskeletons are made of calcium phosphate, calcium carbonate, and proteins; and exoskeletons are made of chitin, proteins, calcium carbonate, and other substances. Hydrostatic skeletons and endoskeletons are on the inside of animals, whereas exoskeletons are on the outside. Shape changes in hydrostatic skeletons occur from changes in body segments, whereas shape changes in endo- and exoskeletons occur from changes in the angles of joints between rigid levers. All types of skeletons attach to skeletal muscle and serve to transmit muscle forces. **(2)** apply No movement of the arm would occur because, for these antagonistic muscles to produce movement, one of them must be contracted while the other is relaxed.

p. 979 Fig. 45.17 apply Galloping at 3.5 m/s would use about 33 percent more energy than trotting (about 400 J/m versus 300 J/m).

p. 980 Fig. 45.19 apply Running is about 10 times as costly as swimming (about 8 cal/g/km versus 0.8 cal/g/km).

p. 980 CYU (1) analyze Gravity is most difficult to overcome for flying animals. Drag is hardest for aquatic animals. **(2)** evaluate It is unlikely that a *T. rex* could rapidly chase prey and jump on them. Because of their large size and therefore high inertia, it would be difficult to run fast, and jumping would cause bones to break because the skeleton would not be able to support the body's weight. *T. rex* likely moved relatively slowly and fed on slow-moving prey.

✔ TEST YOUR KNOWLEDGE

1. understand T, T, F, F **2.** understand d **3.** understand a **4.** understand False.

✔ TEST YOUR UNDERSTANDING

5. understand The key observation was that the banding pattern of sarcomeres changed during contraction. As the entire sarcomere became shorter, the width of the dark bands did not change, but the light bands became narrower. This observation suggested that some portions of the structure slid past other portions. The many mitochondria in muscle cells supply the large amounts of ATP needed to power the movement of myosin heads along actin filaments; the extensive smooth ER stores the calcium ions that initiate contraction by binding to troponin. **6.** understand ATP is required for myosin and actin to unbind. Because no ATP is made after death, rigor mortis occurs when myosin and actin remain bound. **7.** apply The oxygen consumption of the runner would increase because the arches of his or her feet would no longer store as much elastic energy, which normally reduces the energetic cost of running. **8.** understand The viscosity of water has a greater effect on small fishes than large fishes, making it easier for a large fish to move through the water.

✔ TEST YOUR PROBLEM-SOLVING SKILLS

9. apply In cardiac muscle, the binding of acetylcholine to its receptors causes the heart rate to slow. Ingestion of atropine would increase heart rate because it blocks acetylcholine receptors. **10.** apply c:

$$0.5 = v^2/(9.8\ \text{m/s}^2 \times 0.9\ \text{m})$$

$$0.5 = v^2/8.8\ \text{m}^2/\text{s}^2$$

$$v^2 = 0.5 \times 8.8\ \text{m}^2/\text{s}^2$$

$$v = \text{square root of } 4.4\ \text{m}^2/\text{s}^2$$

$$v = 2.1\ \text{m/s}$$

✔ PUT IT ALL TOGETHER: Case Study

11. analyze Slow oxidative fibers are red because they have a lot of myoglobin, which holds the oxygen necessary for the cells' high rates of aerobic respiration. These fibers contract slowly and do not fatigue readily. Fast glycolytic fibers are white due to their low myoglobin concentration; they contract and fatigue rapidly. Fast oxidative/glycolytic (intermediate) fibers are intermediate in color (pink) and physiological properties.

12. apply The sprinter would likely have a greater proportion of fast glycolytic fibers, because these cells are specialized for bursts of activity. **13.** apply There would likely be no difference in the number of muscle cells, even though training could cause the entire muscle to enlarge. Muscle cells can become larger with training; new cells are not added. **14.** evaluate The elite runners on average had a significantly higher proportion of slow fibers than the non-elite runners and untrained subjects. **15.** evaluate Although the elite runners had a higher proportion of slow fibers on average, that proportion does not necessarily predict the athletic performance of any given athlete, as evidenced by the low percentage in some individuals who performed at the elite level. **16.** analyze Radcliffe would have the highest cost of locomotion, because running is more energetically costly than flying or swimming.

Chapter 46

IN-TEXT QUESTIONS AND EXERCISES

p. 985 Making Models 46.1 create See Figure A46.1. Hormone B will increase over time because hormone B stimulates hormone A, which then stimulates an increase in hormone B, and so on.

p. 987 Fig. 46.3 create Rather than re-implanting testes, researchers could now castrate animals and inject them with testosterone. Doing this would show that any subsequent changes were due specifically to testosterone rather than another chemical released into the blood by the testes.

p. 988 CYU (1) understand It is difficult to differentiate between the nervous and endocrine systems because some neurons secrete hormones and some endocrine glands respond to neural signals. **(2)** apply (a) endocrine, (b) neuroendocrine, (c) paracrine.

p. 990 Fig. 46.5 apply The steroid hormone–receptor complex would probably fail to bind to the hormone-response element. Then gene expression would not change in response to the hormone—the arrival of the hormone would have little or no effect on the target cell.

p. 992 CYU (1) understand See Figure A46.2. **(2)** analyze Steroid hormones stimulate production of new proteins, which takes time to occur and results in long-lasting effects. In contrast, polypeptides rapidly stimulate activation of already existing proteins, which is transitory because the proteins can become inactivated just as quickly.

p. 993 create The cells could differ in receptors that, when bound to T_3, would induce changes in the expression of different genes.

p. 996 Fig. 46.12 understand The saline injection controlled for any stress induced by the injection procedure and for introducing additional fluid into the body.

p. 997 CYU (1) analyze In amphibians, an increase in the concentration of thyroid hormones stimulates metamorphosis from a larva to an adult without a pupal stage. In holometabolous insects, a decrease in the concentration of juvenile hormone and an increase in the concentration of ecdysone stimulate metamorphosis during a pupal stage. **(2)** understand By suppressing immune system function, promoting the release of fatty acids from storage cells and the release of amino acids from muscle cells for energy production, and preventing the uptake of glucose by adipocytes and resting muscle cells in response to signals from insulin, cortisol conserves glucose supplies for use by the brain.

p. 998 Fig. 46.14 create To test for adrenal failure, researchers could inject ACTH and monitor the cortisol level. If the cortisol level doesn't increase, adrenal failure is likely.

p. 1000 CYU (1) understand ACTH triggers the release of cortisol, but cortisol inhibits ACTH release by blocking the release of CRH from the hypothalamus and by suppressing ACTH production by the anterior pituitary. **(2)** understand Processing centers in the brain are responsible for integrating a wide array of sensory input. To start a response to this sensory input, they stimulate neurosecretory cells in the hypothalamus. Neurohormones from these cells travel to the anterior pituitary, where they trigger the production and release of other hormones.

✓ TEST YOUR KNOWLEDGE

1. understand a **2.** remember c **3.** understand T, F, T, F. **4.** remember True.

✓ TEST YOUR UNDERSTANDING

5. analyze Lipid-soluble hormones such as steroids usually bind to receptors inside the cell, forming a hormone–receptor complex that binds to DNA and alters transcription. Water-soluble hormones such as polypeptides bind to receptors on the cell surface and trigger production of a second messenger and, in some cases, a signal transduction cascade, ending in activation of proteins already present in the cell. **6.** analyze This is one way that the same hormone can trigger different effects in different tissues. For example, epinephrine binds to four different types of receptors in different tissues and elicits a different response from each. **7.** analyze The posterior pituitary is an extension of the hypothalamus and a storage area for hypothalamic hormones. The anterior pituitary develops from nonneural tissue; it synthesizes and releases an array of hormones in response to releasing hormones from the hypothalamus. **8.** create Subject volunteers to travel and jet lag, measure their ACTH and cortisol levels, and correlate these results with the subjects' perceived level of jet lag symptoms.

✓ TEST YOUR PROBLEM-SOLVING SKILLS

9. create There are two basic strategies: (1) Remove the structure from some individuals and compare their behavior and condition to sham-operated individuals in the same environment, or (2) make a liquid extract from the structure, inject it into some individuals, and compare their behavior and condition to individuals in the same environment that were injected with a saline solution. **10.** evaluate Yes—the finding that testosterone concentrations in atrazine-treated male frogs were significantly lower than those of control males and similar to those of control females supports the hypothesis that atrazine is an endocrine disruptor that feminizes male amphibians.

✓ PUT IT ALL TOGETHER: CASE STUDY

11. remember b **12.** remember Corticosterone receptors would be found in the cytoplasm or nucleus (intracellular). **13.** apply In response to a visit by tourists, corticosterone levels should show a smaller increase in penguins habituated to tourists than in penguins not previously exposed to tourists. **14.** evaluate The hypothesis was supported because the corticosterone response to capture and restraint for 30 minutes was significantly higher in non-habituated penguins than in habituated penguins. **15.** apply Corticosterone is a glucocorticoid similar to cortisol, which induces the synthesis of liver enzymes that make glucose from amino acids. If corticosterone has the same effect as cortisol, and if non-habituated penguins have

Figure A46.1

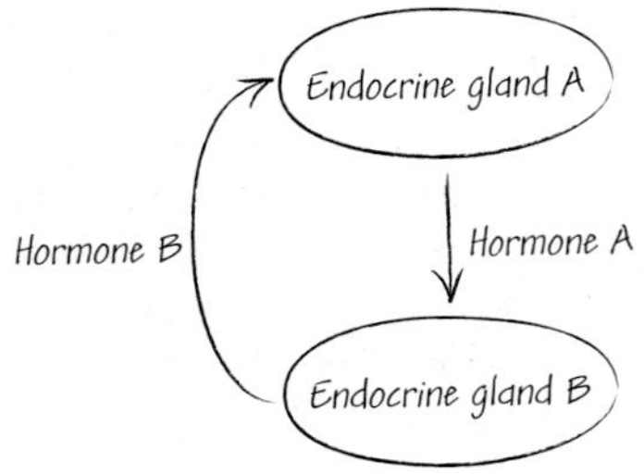

Figure A46.2

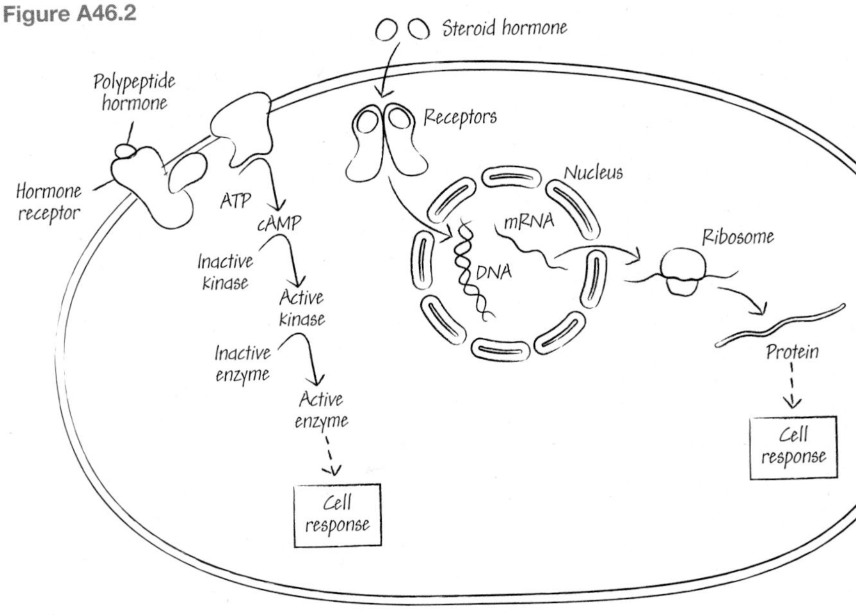

higher corticosterone levels than habituated penguins, then non-habituated penguins should show a greater increase in expression of genes for those enzymes. **16.** create There are several possibilities. In habituated penguins, the brain could become less sensitive to the stimulus of a tourist's presence, resulting in release of less CRH by the hypothalamus. Alternatively, the anterior pituitary gland could become less sensitive to CRH, secreting less ACTH. Finally, the adrenal gland could become less sensitive to ACTH, secreting less corticosterone.

Chapter 47

IN-TEXT QUESTIONS AND EXERCISES

p. 1005 Fig. 47.3 create Isolate and identify the molecules found in "crowded" water. Test each molecule by adding it, at the same concentration found in crowded water, to clean water occupied by a single *Daphnia* and recording whether the female produces a male-containing brood. Repeat with many test females and for each moleecule identified. As a control, record the number of male-containing broods produced in clean water.

p. 1008 CYU (1) apply Asexual reproduction would be expected in environments where conditions change little over time. **(2)** understand See Figure A47.1.

p. 1010 Fig. 47.9 create (More than one possible answer—this is an example.) Having two gonads is an "insurance policy" against loss or damage. To test this hypothesis, surgically remove one gonad from a large number of male and female rats. As a control, do a similar operation on a large number of similar male and female rats, but do not remove either gonad. Once the animals have recovered, place them in a barn or other "natural" setting and let them breed. Compare reproductive success of control rats with that of rats with unpaired gonads.

p. 1011 CYU (1) comprehend Testis to epididymis to vas deferens to ejaculatory duct to urethra. **(2)** comprehend Ovary to oviduct to uterus to cervix to vagina.

p. 1012 Fig. 47.11 evaluate No—the data are consistent with the displacement hypothesis, but do not provide direct evidence for it. There are other plausible explanations for the data, such as females ejecting sperm.

p. 1014 analyze If the added molecule blocked the bindin-like protein on the human sperm head, that protein would not be able to bind to its receptor on the egg. Fertilization would not take place.

p. 1016 CYU (1) analyze Oviparity usually requires less energy input from the mother after egg laying, and mothers do not have to carry eggs around as long—meaning that they can lay more eggs and be more mobile. But, before egg laying, mothers provide the egg with all the nutrition the embryo will require, and eggs may not be well protected after they are laid. Viviparity usually increases the likelihood that the developing offspring will survive until birth, but the space available in the mother's reproductive tract limits the number of young that can be produced. If viviparous young can be nourished longer than oviparous young, they may be larger and more able to fend for themselves. apply **(2)** The increase in Ca^{2+} concentration within the egg triggers formation of the fertilization envelope, which in turn blocks fertilization.

p. 1017 apply Future ectodermal cells are initially at one pole of the spherical embryo. During gastrulation, they spread out as an expanding sheet that eventually covers the entire embryo as the future endodermal and mesodermal cells move into the interior of the embryo.

p. 1021 CYU (1) understand The future posterior and dorsal regions could be identified as the blastopore forms and cells begin folding into the embryo. The blastopore marks the future posterior region of the embryo, and cells that are the first to be folded into the embryo mark the future dorsal side. **(2)** create In either type of embryo, remove a portion of the notochord. The prediction is that the neural tube would not form in the region without a notochord. Alternatively, part of a notochord could be transplanted from one embryo to a region under the ectoderm in another embryo where the neural tube does not form. In this case, the prediction is that a second neural tube would form above the transplanted notochord.

p. 1021 Fig. 47.24 analyze In both cases, the hypothalamus produces a releasing hormone (GnRH or CRH) that stimulates the release of regulatory hormones by the anterior pituitary (LH and FSH or ACTH). The pituitary hormones travel via the bloodstream and act on the gonads or adrenal glands, inducing the release of hormones from these glands.

p. 1024 CYU (1) understand FSH triggers maturation of an ovarian follicle. Its level rises at the end of the menstrual cycle because its secretion is no longer inhibited by progesterone, which stops being produced at a high level when the corpus luteum degenerates. **(2)** apply The drug would keep the FSH level low, meaning that follicles would not mature and would not begin producing estradiol and progesterone. The uterine lining would not thicken.

p. 1026 analyze If the level of progesterone isn't high enough, the uterine lining will not be maintained adequately, and a miscarriage is likely.

p. 1027 CYU (1) understand The short gestation time of marsupials allows them to immediately conceive again, so they can have offspring more frequently. The long gestation time of eutherians allows mothers to provide offspring with nutrients and optimal developmental conditions for an extended period. **(2)** understand During the first trimester, gastrulation creates three cell layers, organogenesis occurs, the heart begins pumping blood, and the placenta and umbilical cord form. During the second and third trimesters, a huge amount of fetal growth occurs, and the brain and lungs are completely developed.

✔ TEST YOUR KNOWLEDGE

1. remember a **2.** understand b **3.** understand F, T, T, F **4.** remember True

✔ TEST YOUR UNDERSTANDING

5. create *Daphnia* females switched to sexual reproduction only when they were exposed to short day lengths, water from crowded populations, *and* low food concentration. These conditions are likely to occur in the fall. **6.** apply Frogs are oviparous, and the egg must contain everything the embryo needs to develop. This makes the egg large. Mice are viviparous, and the embryo obtains almost everything it needs from the mother. Thus there is no need for large eggs provisioned with everything required for development. **7.** analyze Spermatogenesis generates four haploid sperm from each primary spermatocyte; oogenesis produces only one haploid egg from each primary oocyte. Eggs are much larger than sperm because eggs contain more cytoplasm. In spermatogenesis, the second meiotic division occurs right after the first meiotic division; in oogenesis, the second meiotic division is delayed until fertilization. **8.** understand *Negative feedback:* LH triggers secretion of estradiol, but at low levels estradiol inhibits further release of LH; LH and FSH trigger secretion of progesterone, but progesterone inhibits further release of LH and FSH. *Positive feedback:* High levels of estradiol trigger release of more LH. The follicle can produce high levels of estradiol only if it has grown and matured; in that sense, high levels of estradiol can be considered a signal that the oocyte is ready for ovulation to occur.

✔ TEST YOUR PROBLEM-SOLVING SKILLS

9. analyze Cells from a variety of locations in many frog blastulas could be labeled with a dye. The position of these cells could be assessed relative to the blastocoel, which is displaced to one "side" of the embryo. The descendants of these labeled cells could then be followed through later

Figure A47.1

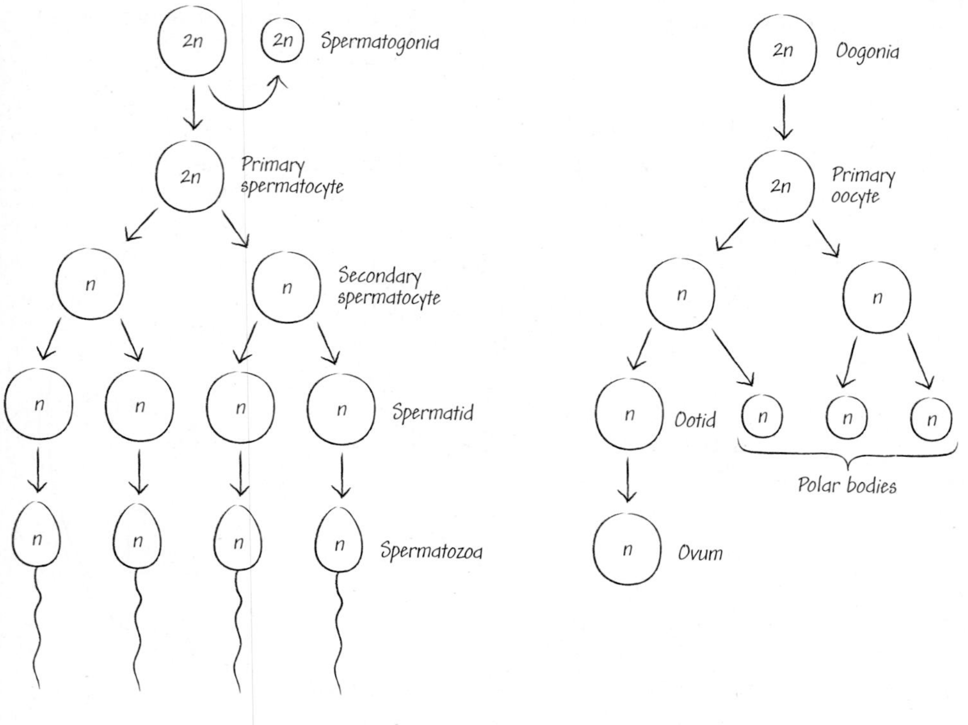

stages of development. If cells from only one location came to cover the outside of the embryo and contributed to structures normally formed by the ectoderm, then the hypothesis would be supported. **10.** analyze b

✔ PUT IT ALL TOGETHER: Case Study

11. remember d **12.** apply (a) Pituitary secretion of FSH and LH would decline due to negative feedback by progesterone. (b) Because of the decline in FSH and LH secretion, follicles would not develop, and therefore the ovaries would secrete little estradiol and progesterone. **13.** evaluate The data from the study do support the hypothesis that the pill interferes with normal hormonal signaling because the women showed statistically significant declines in all four major reproductive hormones (FSH, LH, estradiol, and progesterone) after they started using the pill. **14.** analyze (a) The follicle would not develop, because the FSH level is low. (b) The uterine lining would not thicken much, because the estradiol level during the follicular phase would be low. (c) Ovulation would be unlikely to occur because the estradiol level would never be high enough to stimulate the LH surge needed for ovulation. (d) The volume of menstrual fluid would be lower than that of women not on the pill, because the uterine lining would not thicken as much. **15.** analyze Both the pill and emergency contraception contain progesterone or progesterone plus estradiol, and both work by preventing ovulation; they would not terminate a pregnancy. In contrast, mifepristone is a progesterone receptor blocker that does terminate a pregnancy. **16.** analyze A contraceptive that blocked either ovulation or fertilization is likely to be the least controversial because some people contend that a new life begins at the moment of fertilization. Contraceptives that block cleavage or implantation are likely to stir more controversy because these events occur after fertilization.

Chapter 48

IN-TEXT QUESTIONS AND EXERCISES

p. 1033 apply Each Toll-like protein is a pattern-recognition receptor that responds to a different group of pathogens. Since bacteria and fungi are very different pathogens, you would not expect a receptor that recognizes fungi to be required for a response against bacteria, so flies lacking Toll protein would survive the bacterial infection as well as wild-type flies.

p. 1033 Fig. 48.3 analyze Mast cells would cause the three classical signs of infection. Mast cells release histamine, which triggers dilation of surrounding blood vessels, resulting in increased blood flow into the area, which causes redness, swelling, and heat.

p. 1034 CYU (1) understand Toll-like receptors recognize molecules associated with broad groups of pathogens, such as viruses, fungi, and bacteria. Upon binding to these molecules, the receptors cause signals to be transduced that activate cellular responses directed at eliminating these groups of pathogens. **(2)** apply Cleaning the wound to remove dirt mimics the phagocytic activity of macrophages and neutrophils by eliminating foreign cells and material. Applying bandages to halt blood flow mimics the effect of chemokines released from injured tissues and macrophages, which attract immune system cells to facilitate wound repair.

p. 1039 apply (1) In a human, 200 different light chains are possible (see text). The recombination of the *V*, *D*, and *J* segments results in $51 \times 27 \times 6 = 8262$ possible heavy chains. Since these chains combine to produce BCRs, a total of $200 \times 8262 = 1{,}652{,}400$ BCRs are possible in a human. (2) Only one type of BCR can be produced by a single human B cell.

p. 1040 CYU (1) understand The production of BCRs and TCRs via DNA recombination is required for an effective response. Elimination or inactivation of self-reactive lymphocytes is required for a safe response. **(2)** analyze The cell can make a larger number of differing variable-region genes per unit of DNA if the variable-region DNA is segmented. For example, 3 *V* segments and 3 *J* segments can randomly assemble into 9 different light-chain genes, whereas only 6 light-chain genes would be possible with the same amount of DNA if the genes were preassembled.

p. 1040 understand "Clonal" refers to the cloning—producing many exact copies—of cells that are "selected" by the binding of their receptor to a specific antigen (or MHC–peptide complex, which is introduced later).

p. 1042 understand See **Table A48.1**.

p. 1044 Making Models 48.1 apply See **Figure A48.1**.

p. 1044 CYU create Individuals who are heterozygous for MHC genes have greater variability in their MHC proteins than do individuals who are homozygous for these genes. The increased variability in MHC proteins allows a greater variety of peptides to be presented to T cells, which can then recognize, attack, and eliminate a wider range of pathogens. As a result, heterozygous individuals likely are better able to fend off infections than homozygous individuals would be.

p. 1046 Fig. 48.17 create See **Figure A48.2** on page A:56. The response to the new antigen would resemble the primary response that occurred after the first injection of the original antigen.

p. 1048 CYU (1) understand Extracellular viruses are opsonized, neutralized, and agglutinated by antibodies produced in the humoral response. Intracellular viruses are eliminated when their host cells are destroyed by cytotoxic T cells in the cell-mediated response. **(2)** analyze Attenuated viruses consist of complete virus particles; they can infect the cells of a host, but not well enough to cause disease. Inactivated viruses have been damaged by chemical treatments and cannot infect cells at all; they remain as extracellular particles. Since attenuated viruses can infect cells, their antigens will be processed and presented by class I MHC–peptide complexes, which promotes a stronger cell-mediated response than can inactivated viruses.

p. 1049 CYU apply IgE antibodies trigger hypersensitive reactions. If self-reactive IgE antibodies were produced, you would expect a strong response wherever the self molecules are present, leading to chronic inflammation or, in severe cases, anaphylactic shock.

✔ TEST YOUR KNOWLEDGE

1. remember c **2.** understand a **3.** remember An epitope is the part of an antigen that is recognized by a B-cell or T-cell receptor or an antibody produced by a B cell. **4.** remember c

✔ TEST YOUR UNDERSTANDING

5. understand In each B cell, different *V* and *J* segments are recombined to make the light-chain gene, and different *V*, *D*, and *J* segments are recombined to make the heavy-chain gene, giving the B cell a unique sequence in both chains of the BCR. **6.** understand B cells are first activated to divide via interactions between the BCRs and the antigens they recognize. Most B cells are then fully activated to differentiate and continue to divide when one of their class II MHC–peptide complexes interacts with an activated $CD4^+$ T cell that has a complementary TCR. **7.** analyze F, T, F, F **8.** analyze The vaccine would have to contain a chicken pox virus antigen that can stimulate an appropriate primary adaptive immune response. Vaccines have not worked for HIV, because the antigens on this virus are constantly changed through mutation, rendering it unrecognizable to memory cells generated following vaccination.

Table A48.1

MHC Type	Origin of Antigen	Type of T Cell That Binds	Activity Stimulated
Class I MHC	Host-cell cytosol	$CD8^+$	T cell is activated to kill infected cells
Class II MHC	Extracellular environment	$CD4^+$	T cell is activated to secrete cytokines that support the immune response of other cells

Figure A48.1

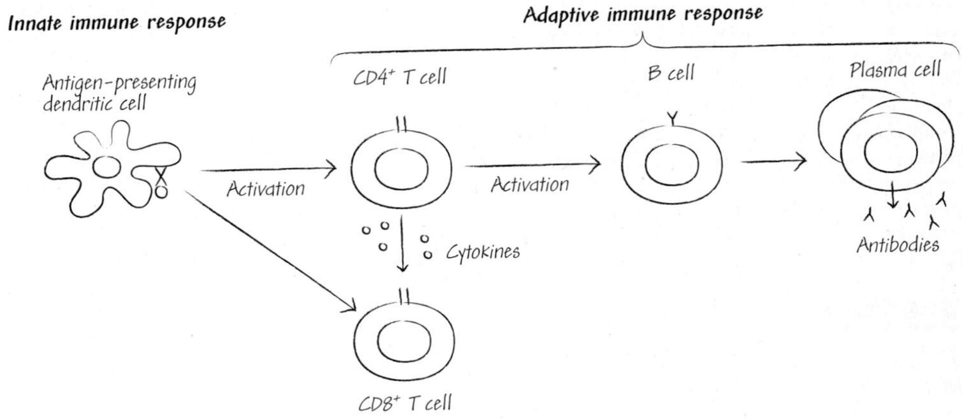

Figure A48.2

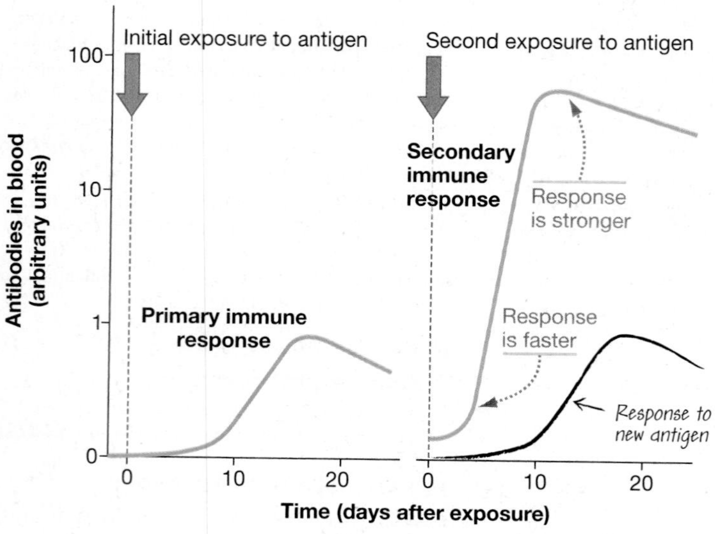

✔ TEST YOUR PROBLEM-SOLVING SKILLS

9. apply d **10.** create When the BCRs in self-reactive B cells bind to self molecules during maturation, the cells respond by either becoming permanently inactivated or undergoing apoptosis.

✔ PUT IT ALL TOGETHER: Case Study

11. understand The hygiene hypothesis states that allergic and autoimmune responses arise in individuals who, because of hygienic practices, have experienced less exposure to pathogens and parasites. It attempts to explain the positive correlation between the risk of developing these disorders and the level of sanitation. **12.** apply Defense against infection with parasitic worms involves IgE antibodies, which are part of the humoral response of adaptive immunity. Mast cells that are sensitized by IgE are part of the innate immune response; they respond to worm antigens by releasing chemicals like histamine that promote inflammation. **13.** apply Pattern-recognition receptors inform antigen-presenting cells to direct $CD4^+$ T cells to become T_H2 helper cells, which promote the humoral immune response. **14.** create Regulatory T cells suppress parts of the immune system, limiting the magnitude of the adaptive immune response. They also may help inhibit self-reactive lymphocytes, which are involved in autoimmunity. In hygienic areas, parasitic worm infections are rare, so regulatory T-cell numbers would be lower, self-reactive lymphocytes would not be as inhibited, and the incidence of autoimmunity would be higher. **15.** analyze First, infection with *H. polygyrus* inhibits the onset of type 1 diabetes in mice that are prone to developing the disease. Second, this inhibition does not require lifelong worm infection, because a two-week infection early in life still results in significant protection. **16.** evaluate To make this decision, you would need to know if adverse effects are associated with worm infections in children, the level of risk for the children to develop type 1 diabetes, and if the worms' modulation of the immune response would affect responses to other pathogens or vaccines.

END-OF-UNIT CASE STUDY Do Garter Snakes Resistant to TTX Experience Trade-Offs?

pp. 1052–1053 (1) understand b **(2)** create See Figure EOU A7.1. **(3)** understand A garter snake with a mutation in the gene for the voltage-gated sodium channel that made it resistant to TTX was able to eat toxic newts when other garter snakes were not. This provided a major food resource that allowed the snake to gain weight and grow rapidly, have more offspring, and pass on the gene for TTX resistance to more offspring than snakes without the mutation. The mutation spread rapidly through a population over time. **(4)** apply Many answers are possible. One might predict that the mutant sodium channel might not allow as much sodium ion transport, or might allow too much sodium ion transport, altering the dynamics of the action potential and therefore impacting neuron and muscle function. **(5)** create Scientists could compare the physical performance of TTX-sensitive and TTX-resistance snakes in terms of muscle strength, crawling speed, endurance, or other variables. They could also measure ion flow through channels in the muscle of each type of snake. **(6)** analyze When exposed to a standardized stimulus, the sodium channels of TTX-sensitive newts allow passage of over five times as many sodium ions compared to the channels in TTX-resistant newts. **(7)** analyze When exposed to a standardized stimulus, the muscle tissue of TTX-sensitive newts exerts a force that is nine times stronger than the force exerted by the tissue of TTX-resistant newts. **(8)** create The structure of the sodium channel in TTX-resistant garter snakes is different than that in the sensitive snakes. Perhaps the channel's pore is smaller, or contains a positively charged moiety, either of which situations would allow less sodium flow per unit time. Reduced sodium flow in both the motor neurons and muscle tissue likely leads to a reduced frequency of action potentials in the plasma membranes of muscle cells, reducing the amount of calcium secreted by the sarcoplasmic reticulum into the cytoplasm, reducing the number of sarcomeres stimulated to contract. This is one way that the muscles in TTX-resistant newts would produce less force. **(9)** create Many answers are possible. Any system that involves muscular function or voltage-gated sodium channels could potentially be affected. For example, there could be an impact on digestive function if contraction of the muscle in the gastrointestinal tracts of TTX-resistant snakes is affected. Similarly, there could be an impact on the cardiovascular systems of TTX-resistant snakes, since cardiac muscle and the smooth muscle in the blood vessels would have mutant sodium channels. **(10)** evaluate The evidence presented shows that TTX-resistant garter snakes do experience significant trade-offs in their neural and muscular systems. While the reduced functionality has been demonstrated in the laboratory, however, it remains to be seen whether this has a significant impact on wild snakes.

Figure EOU A7.1

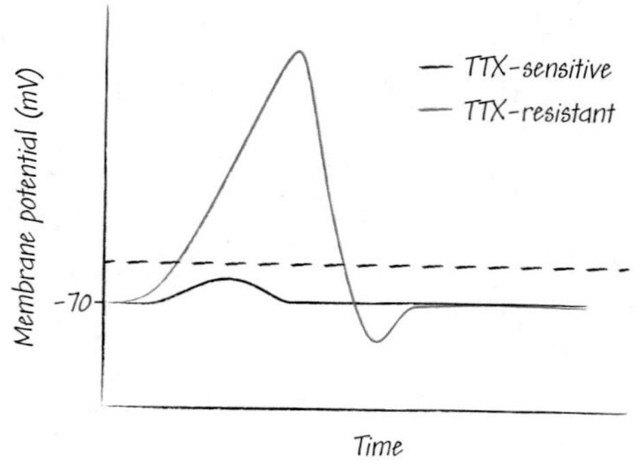

Chapter 49

IN-TEXT QUESTIONS AND EXERCISES

p. 1057 CYU apply This question focuses on the interaction between species, thus addressing the community level of ecology.

p. 1058 Making Models 49.1 apply See Figure A49.1.

p. 1060 Fig. 49.3 evaluate Using small squares would give more resolution to the map and show more information about variation in the landscape. Larger squares would be easier to work with but would not contain as much information.

p. 1061 CYU apply Many possible answers. *Present abiotic*: temperature, salinity of water. *Past abiotic*: sea-level change. *Present biotic*: predation by bears. *Past biotic*: fishing by humans as they migrated to North America more than 15,000 years ago.

p. 1062 Fig. 49.5 apply See Figure A49.2.

p. 1063 Fig. 49.6 analyze If drawn to scale, the Earth would be 13 mm/109 = 0.12 mm in diameter (tiny, about the size of a point made by a very sharp pencil), which would make it hard to see the pattern of daylight and

Figure A49.1

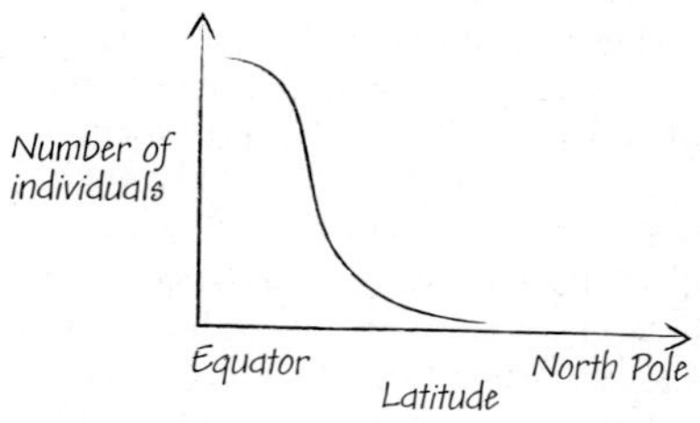

Figure A49.3

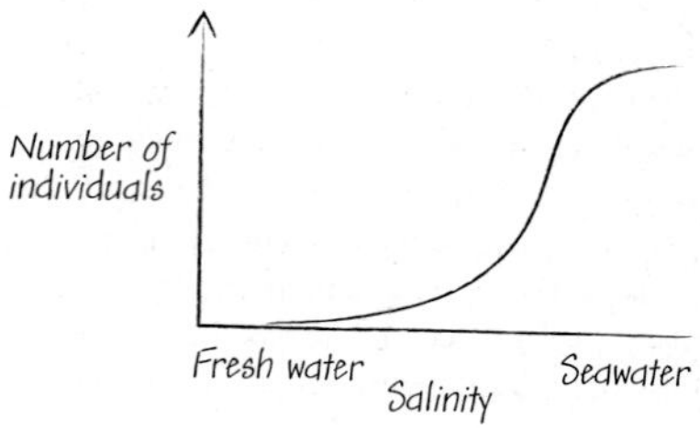

night on Earth. Earth would also have to be drawn $12{,}000 \times 0.12$ mm = 1400 mm, or 1.4 m away from the Sun, which would be impossible to do in a textbook or on a computer screen.

p. 1064 CYU (1) apply The western slopes of the Cascade Mountains are wet and the eastern slopes are dry (a rain shadow). Even if fir trees disperse to the dry areas, they die because they cannot tolerate the dry conditions beyond their wetter niche. **(2)** apply The highest risk of skin cancer is in the tropics, where the Sun is often directly overhead during midday and there is a large amount of solar radiation per unit area.

p. 1067 Fig. 49.10 analyze Most of the remaining wildlands occur in Canada, at the top of the image.

p. 1067 understand NPP is higher in tropical wet forests than in temperate forests because the tropics experience more sunlight, warmer temperatures, and more rain all year, creating a more favorable environment for photosynthesis than at higher latitudes, where winter limits the growth season.

p. 1068 Fig. 49.11 create The transparent chambers might reduce the amount of rain that enters the chambers (due to the sloped sides) or might reduce the wind, the consumption of plants by herbivores, or the dispersal of wind-pollinated seeds. Researchers could measure these variables inside and outside the chambers to see if there was a difference.

p. 1068 CYU (1) analyze To answer this question, you would first examine the map and key in Figure 49.10 to decide which anthrome you live in, such as cities, suburbs, or croplands. You would then examine Figure 49.9 and find your natural biome location there, such as temperate forest. Humans have removed the natural vegetation in many areas to use land for other purposes. **(2)** apply If the Amazon rain forest experienced a tipping point and abruptly transitioned to grassland (due to a positive feedback loop between reduced rain, increased fire, and decreased forest area), then the expected range of açaí palms in 2080 would definitely be reduced.

p. 1072 apply The neritic zone of oceans has the highest NPP because nutrients are available from runoff and upwelling, and light is also available. The littoral zone of lakes has the highest NPP because both nutrients and light are abundant.

p. 1073 CYU apply If the lake is more stratified all year, preventing turnovers, this means that the nutrients from deep water will not be returned to the photic zone at the surface and the oxygen at the surface will not be mixed to deep water. This trend is likely to reduce the sardine population because they will have a reduced food supply (due to reduced productivity) and reduced oxygen availability (due to lack of mixing with the surface). (These consequences have indeed already occurred, affecting the humans who rely on the sardines for food and income.)

✔ TEST YOUR KNOWLEDGE

1. understand Organismal, population, community, ecosystem, and biosphere/global. **2.** remember b **3.** remember d **4.** remember c

✔ TEST YOUR UNDERSTANDING

5. apply F, T, F (Ocean upwelling and lake turnover are examples of how nutrients can be returned to surface waters.) **6.** analyze Like deserts, open oceans have plenty of sunlight but very low productivity. Unlike deserts, which are water limited, open oceans are nutrient limited. **7.** apply Temperate forests in China and North America are part of the same biome because they are both dominated by deciduous forests and have similar temperature and precipitation profiles. However, the particular species found in each are different due to historical factors, making each ecosystem unique. **8.** analyze The natural biome of eastern North America is primarily temperate forest. However, most of this area is now occupied by human-influenced biomes such as cities, suburbs, and farmland.

✔ TEST YOUR PROBLEM-SOLVING SKILLS

9. apply a; As with Earth, the tilt of Mars's axis causes annual variation in the amount of solar radiation received by the northern and southern hemispheres of the planet. If you chose answers b or c, keep in mind that if a planet's distance from the Sun alone determined seasons, the entire planet would experience the same season at the same time. **10.** create Temperature is relatively constant in the tropics all year, whereas temperature fluctuates dramatically at higher latitudes. Organisms that are physiologically adapted to constant temperatures may not have as much genetic variation for traits that enable adaptation to higher temperatures, compared to organisms at higher latitudes, which are adapted to tolerate different temperatures at different times of the year.

✔ PUT IT ALL TOGETHER: Case Study

11. understand c (All answers are true statements about lionfish, but answer c defines the term "invasive.") **12.** apply See Figure A49.3. The curve should be highest in full-salinity seawater (35 ppt) and lower as salinity declines. **13.** apply The North Atlantic Gyre flows in a clockwise direction, so the prevailing current on the Florida coast would be south to north, and it would disperse lionfish up the coast. **14.** create Examples of possible questions (you can use Table 49.1 as a model): *Organismal*: What range of salinities can lionfish tolerate? *Population*: What is the growth rate of the lionfish population? *Community*: How are the lionfish affecting other species in the coral-reef community? *Ecosystem*: How do lionfish affect the net primary productivity of the coral reef? *Global*: How will climate change and ocean acidification affect the invasive behavior of lionfish? **15.** analyze The mass of lionfish ovaries is lowest in January and February and highest in July and August. The error bars increase confidence in the month-to-month trend because the bars are small relative to the fluctuation in ovary size over time. The graph emphasizes that ovary size is correlated with water temperature—the trends in both curves are very similar. **16.** evaluate There are many possible answers. For example, it would be easy to convince people about the devastating effects of invasive species like lionfish on coral-reef health; most people love coral reefs and would be highly motivated to eat lionfish if it helped protect coral reefs. On the other hand, lionfish are small, pretty, and have venomous spines that may not sound appetizing and could discourage handling. Also, it is not intuitive for humans to kill one pretty thing (lionfish) to protect another pretty thing (coral reefs).

Figure A49.2

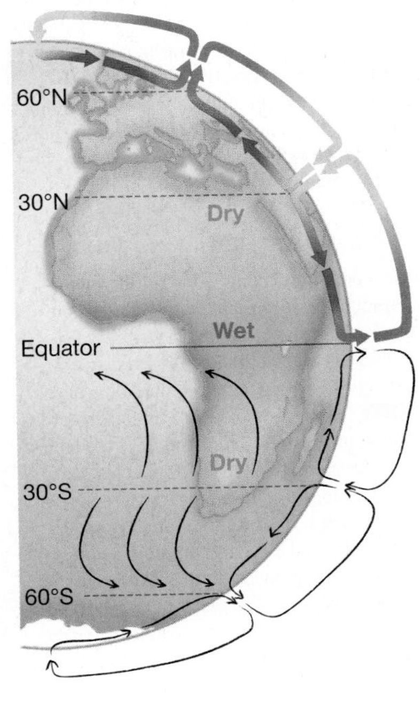

Chapter 50

IN-TEXT QUESTIONS AND EXERCISES

p. 1079 CYU (1) apply Proximate cause: Bees are attracted to flowers with a certain color, shape, and/or scent. Ultimate: Bees with adaptations that enable them to choose flowers with high nectar rewards will have more energy and thus higher fitness than other bees. **(2)** apply In a changing environment where the number and types of flowers might be unpredictable, bees with a capacity for learning to find new nectar sources will have higher fitness than bees with an innate attraction to only one type of flower.

p. 1080 Making Models 50.1 analyze Energy costs are proportional to distance traveled in this simple case, so the distance traveled by the bee can be used as a proxy for energy costs. Further, since relative costs are being used to compare efficiency, the distances in the model can be used as a proxy for the actual distances between flowers. This analysis shows that inefficient foraging is 40 percent more costly than efficient foraging. See Figure A50.1 on page A:58.

p. 1081 Fig. 50.5 analyze By swapping the side of the meal chamber where the most or larger shrimp are placed, the researchers are making sure that the cuttlefish is really evaluating the two meals rather than being trained to always select the meal on a certain side.

p. 1082 CYU analyze Several possible scenarios might result in the costs of a meal outweighing the benefits: If a cuttlefish is already satiated, it might not expend the energy to capture more prey; if the prey is very large, the time and energy needed to subdue and process it might outweigh the benefits; or if the prey is dangerous (such as a crab with claws) it might pose a risk to the health of the cuttlefish. Also, a cuttlefish might decline to take a meal if it has to compete with a competitor (such as another cuttlefish) or expose itself to a predator (such as a shark) to access the meal.

p. 1084 Fig. 50.7 analyze The frequency of dewlap extensions and push-ups would probably decrease in the presence of predators because these behaviors might attract predators (which is what the researchers observed). The increased risk of predation would offset the benefits of attracting mates. The lizards can't mate if they are dead.

p. 1084 CYU (1) analyze As in lizards, a rise in sex hormones in humans is a proximate cause of sexual readiness. **(2)** analyze Humans, like lizards, tend to choose a mate who will increase their fitness. For example, women often prefer men who will be able to help provide for offspring.

p. 1085 Fig. 50.8 analyze The data would either be distributed randomly around the circles, showing no

Figure A50.1

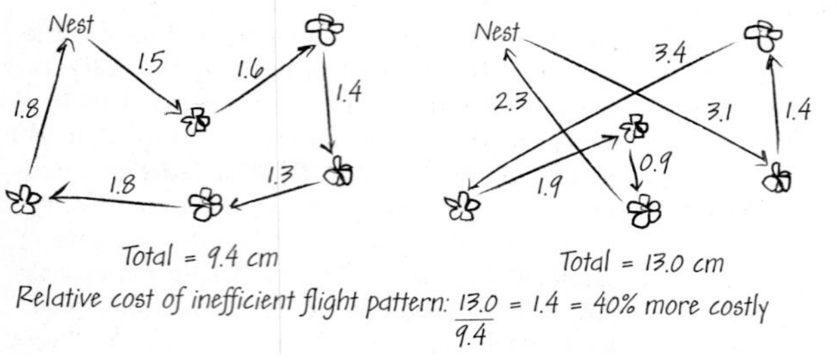

preference for direction, or the turtles would swim the same direction in both cases, suggesting that they are using compass orientation rather than map orientation.

p. 1086 CYU apply The piloting of birds could be disrupted by light if the light changes the appearance of landmarks. Light could also disrupt compass orientation by making the stars difficult to see at night. The ability of the birds to sense the magnetic field in true (map) navigation would be less likely to be affected.

p. 1089 CYU (1) analyze Your table should contain at least one cost and one benefit for each category. For example: Auditory communication allows individuals to communicate over long distances (benefit) but can be heard by predators (cost). Olfactory communication is effective in the dark (benefit), and scents can continue to carry information long after the signaler has left (benefit), but scents do not carry long distances (cost). Visual communication is effective during the day (benefit) but can be seen by predators (cost). **(2)** apply The ability to detect deceit protects the individual from fitness costs (e.g., being eaten, having a mate's eggs fertilized by someone else); avoiding or punishing "liars" should also lower the frequency of deceit (because it becomes less successful). Alleles associated with detecting and avoiding or punishing liars should be favored by natural selection and increase in frequency.

p. 1089 apply Humans often live near kin, are good at recognizing kin, and in many cases can give kin resources for protection that increase fitness.

p. 1090 apply Between parent and child, $r = ½$; between grandparent and grandchild, $r = ½ \times ½ = ¼$; between first cousins, $r = ½ \times ½ \times ½ = ⅛$. (Explanation: The r between a child's mother and uncle $= ½$ because they are full siblings, the r between the child and his mother $= ½$, and the r between that child's cousin and uncle $= ½$ because the cousin is the child of the uncle; multiply these three probabilities to get the answer, ⅛ .)To review rules of combining probabilities, see **BioSkills 4**.

p. 1091 Fig. 50.14 create You could check the figure legend or results section of the study to see the results of a statistical test comparing averages. For example, if the differences in averages is statistically significant, then P will be less than 0.05 (see **BioSkills 3**).

p. 1092 CYU apply The consolation behavior in elephants is probably not altruistic, because the caregiver probably receives fitness benefits from the interaction over time, such as closer social ties and the possibility of reciprocal care in the future (an example of reciprocal altruism).

✔ TEST YOUR KNOWLEDGE

1. understand c **2.** understand Adaptive significance of behavior; how behavior affects fitness, how behavior has evolved. **3.** understand F, F, F, T (Recall that fitness has a different meaning in the context of evolution—the ability to survive and reproduce viable offspring—than in everyday use, where it refers to strength, body size, and/or speed.) **4.** remember c (If you answered d, remember that individuals do not act for the good of the species; see explanation near the end of Section 50.6.)

✔ TEST YOUR UNDERSTANDING

5. evaluate Evolution favors organisms that forage optimally, thus maximizing benefits and minimizing costs. But individuals are not making conscious choices, individuals vary in populations, and environmental circumstances vary. As a result, some individuals will forage suboptimally and suffer reduced fitness due to evolutionary trade-offs, environmental constraints, and genetic and historic contraints. **6.** apply Proximate: When a male cuttlefish sees another male, especially if it is near a potential mate, its brain sends a signal to its skin to change to a warning display. Ultimate: Males that have higher fitness if they happen to have adaptations that enable them to intimidate rival males and thus produce more offspring with females. **7.** apply Migration is condition dependent in some birds. If enough food is available locally, the costs of migration outweigh the benefits, so the birds' fitness will be higher if they stay put. **8.** analyze c

✔ TEST YOUR PROBLEM-SOLVING SKILLS

9. create There are many possible answers, but a good experiment would assess the shape of underground burrows in an environment that excludes the possibility of learning. For example, you could measure burrow shapes of mice that were raised in the absence of other mice and of mice raised around other mice. (This is what Hoekstra did, making a cast of excavated burrows and then measuring the depth, length, and angles of each.) **10.** analyze Individuals have an r of ½ with full siblings and an r of ⅛ with first cousins. If the biologist will lose his life, he needs to save two siblings, or eight first cousins, to keep the "lost copy" of his altruism alleles in the population.

✔ PUT IT ALL TOGETHER: Case Study

11. understand One benefit of using sound underwater is that it travels much farther than light, enabling communication over long distances where visibility is poor. One disadvantage is that sound is short lived and must be repeated to be effective, which requires time and energy. **12.** analyze The whale stopped foraging when the sound started. It immediately surfaced, breathed more frequently, and reduced the depths of its dives. **13.** analyze Before the sound exposure, the whale was foraging for about 30 minutes per hour. After the sound exposure, the whale either stopped foraging completely or reduced the foraging rate to less than about 5 minutes per hour. If the sound was repeated or ongoing, it could affect the whale's fitness because the whale would stop feeding, and thus stop obtaining energy, and so it would have fewer resources to produce offspring. The sonar might also disrupt whale mating behavior, further reducing fitness. **14.** evaluate c (According to optimal foraging theory, whales probably spend the most time foraging where the density or quality of the food source is highest.) **15.** create The most convincing study would be an experiment in an area that had two groups of whales: One group would be subjected to sonar levels used by the military while the other group would be somehow shielded from this sonar. Then the behavior of the two groups could be compared. However, because the first group might strand itself on the beach, this experiment could not be conducted. Another approach would be to use an observational study rather than an experiment (Chapter 1), and collect data on sonar use and locations of beach strandings to see if the correlation between them is higher than you would expect to see by chance. This approach would not prove the cause, but it would provide convincing evidence of a cause. **16.** apply There are a number of benefits to conducting military exercises and a number of costs to whales. By studying the pattern of sonar use by the military, the behavioral patterns of whales, and the effects of sonar on whales, it could be possible to reduce the costs to whales with or without reducing the benefits to the military. For example, the military might be able to limit sonar use to locations that are far away from known whale migration routes and/or to times of the year when whales are not nearby.

Chapter 51

IN-TEXT QUESTIONS AND EXERCISES

p. 1097 CYU apply A transect is a good choice for trees because they do not move; you could count the number of açaí palms that occur along several transects throughout the forest and then use these numbers to calculate an estimate for the whole forest.

p. 1098 (1) apply $N = Mn/m = (255 \times 162)/78 = 530$. **(2)** analyze *Possible explanations:* Some marked individuals died as a result of the marks (fin clips); marked individuals learned to avoid the traps; some of the marked individuals moved out of the population; unmarked individuals moved into the population; unmarked individuals were "born" into the population (both marked and unmarked individuals produce unmarked offspring).

p. 1100 Fig. 51.3 analyze See **Figure A51.1**. The shape of the data is most similar to the Type II line in part (a).

p. 1101 apply $R_0 = 0.47$; the population is declining.

p. 1102 Fig. 51.4 create [Many possible answers.] *Example:* An increase in average temperature will cause evolution in *Zootoca vivipara* populations such that average life span will decrease and average number of offspring per female will increase. [In general, the life-history patterns of lizards in Brittany, France, are likely to become more common than those in the mountains of Austria as the climate in Europe warms. However, the weather conditions in particular regions may vary widely due to increased variability in climate, as described in Chapter 49].

p. 1103 CYU (1) create If the average temperature of sea turtle nests increases in a population, a higher proportion of offspring are likely to be female. If the number of females in the population increases and there are still enough males to fertilize the eggs, then the net reproductive rate could increase. But if all hatchlings become female and there are no males at all, then the population could become extinct. **(2)** apply The female

Figure A51.1

(b) Exercise: Survivorship curve for *Zootoca vivipara*

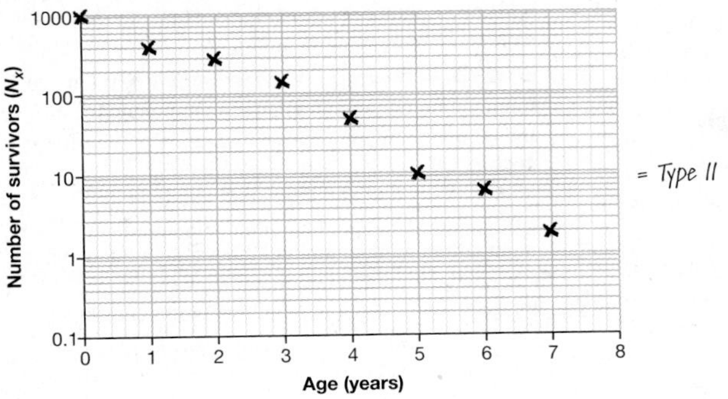

Figure A51.2

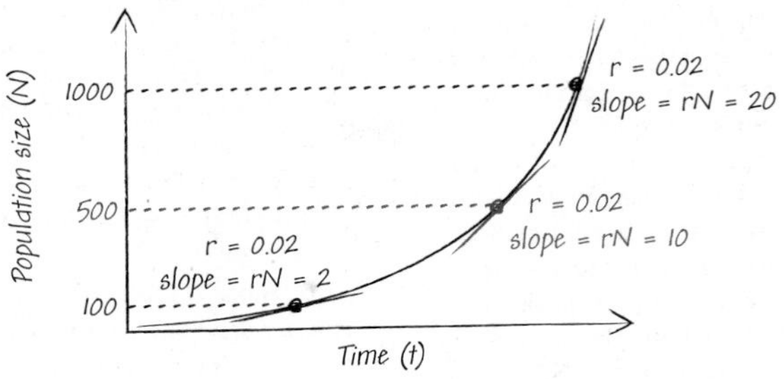

octopus is allocating so much of her resources to her fecundity (thousands of eggs) that she has insufficient resources for her own survival after spawning them. Also, since the octopus produces thousands of eggs, the eggs must be relatively small and thus have fewer resources than if she laid fewer, larger eggs.

p. 1104 Making Models 51.1 apply See **Figure A51.2**.

p. 1105 (1) apply Using Equation 51.5, $r = (\ln 1)/3 = 0$, which means that the population is neither growing nor declining. **(2)** apply Using Equation 51.6, $N_t = 50e^{(0.00048)(3650)} = 290$; the orchid population will grow to only about 290 individuals in 10 years (3650 days). **(3)** apply Using Equation 51.8, $N_t = (225)(1.2)^5 = 560$; the squirrel population will reach about 560 individuals in 5 years, more than doubling (if the assumptions of the model are met, such as the availability of unlimited resources).

p. 1107 CYU (1) apply If conditions were suitable for rats and there were no natural predators on the island, the rat population would be likely to increase exponentially at first. But resources would eventually become limiting, and population size would be limited by the carrying capacity. (Note, however, that sometimes populations exceed carrying capacity and then crash). **(2)** create One possible density-dependent factor is competition for food. To test this idea, compare carrying capacity in identical fenced-in areas that differ only in the amount of food added. (You could also test a hypothesis of space limitation by doubling the size of the enclosure but keeping the amount of food the same.)

p. 1108 Fig. 51.9 apply If the human population uses resources unsustainably (faster than they can be replaced), then the population can grow higher than the carrying capacity. Thus, our current population may be higher than the carrying capacity, meaning that we could not sustain this number of people on Earth over time. Density-dependent factors will cause the population to decline.

p. 1109 Fig. 51.11 evaluate One major benefit of replicated plots is that researchers can control for variables that are peripheral to the study, such as differences in natural food sources for the hares in different forest plots or differences in other predators besides lynx. One cost of creating replicate plots is the expense and labor associated with installing 4 km of fencing per plot and/or providing extra food. Another cost is the increased labor of measuring hare density in additional plots. (Large-scale studies such as these can be expensive and labor intensive.)

p. 1111 CYU apply No, the elephants do not have metapopulation structure if they are isolated. In metapopulations, dispersal can occur among habitat patches.

p. 1114 CYU (1) analyze Pollution is a form of toxic waste, and AIDS, malaria, and tuberculosis are major infectious diseases. Both are examples of density-dependent factors that limit human population growth. **(2)** understand Fewer children are being born per female, but there are many more females of reproductive age due to high fecundity rates in previous years combined with increased survival of infants and children.

✔ TEST YOUR KNOWLEDGE

1. understand T, T, F, F (Remember that during exponential growth, the per capita growth rate, r, is constant no matter whether the population is small or large.) **2.** understand a **3.** understand d **4.** understand Resources such as time and energy are limited, so survivorship and fecundity cannot both be maximized simultaneously.

✔ TEST YOUR UNDERSTANDING

5. analyze c (Many seeds do not survive, resulting in an early, steep drop in survivorship.) **6.** apply The population has undergone rapid growth because advances in nutrition, sanitation, and medicine have allowed humans to live at high density without suffering from decreased survivorship and fecundity. Also, human innovations such as machines have enabled the rapid extraction of resources. However, growth will slow more rapidly as density-dependent effects such as disease, famine, and water scarcity increase death rates and lower birth rates. It's possible (some researchers say likely) that human population has already surpassed carrying capacity. **7.** understand (a) Corridors allow individuals to move between populations, increasing gene flow and making it possible to recolonize habitats where populations have been lost. (b) Maintaining unoccupied habitat makes it possible to recolonize the habitat. **8.** apply It is true that the life-history traits of lizards may evolve in response to climate change, but it is incorrect to say that they will evolve because lizards want or need them to. Natural selection merely sorts existing variation; lizards with heritable traits that increase their fitness will increase in frequency in populations over time.

✔ TEST YOUR PROBLEM-SOLVING SKILLS

9. apply Fewer older individuals will be left in the population; there will be relatively more young individuals. If too many older individuals are taken, population growth rate may decline sharply as reproduction stops or slows. (But if relatively few older individuals are taken, more resources are available to younger individuals and their survivorship and fecundity, and the population's overall growth rate, may increase.) **10.** apply As a sexually transmitted disease, AIDS will reduce the number of sexually active adults. If the epidemic continues unabated, the numbers of both reproductive-age adults and children will decline, causing a top-heavy age distribution dominated by older adults and the elderly. See **Figure A51.3**.

✔ PUT IT ALL TOGETHER: Case Study

11. apply b **12.** understand Cold weather is typically a density-independent factor because it affects the pythons the same way no matter how many pythons there are.

Figure A51.3

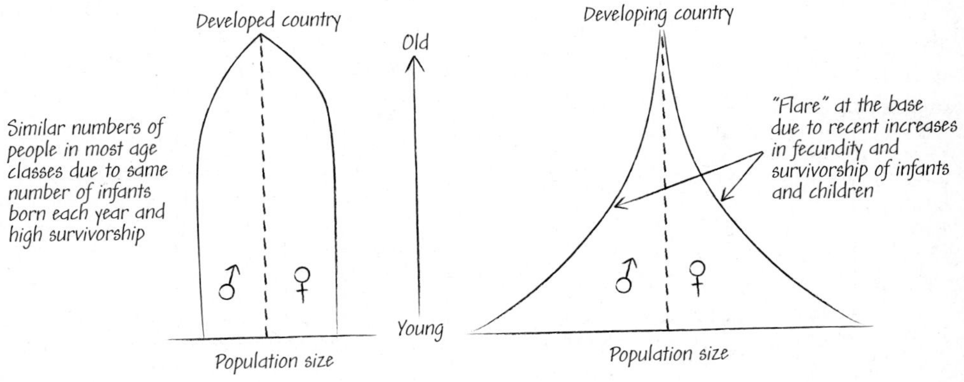

A density-dependent factor, such as competition for food, increases with increasing python density. **13.** create First, trap a number of pythons and mark them with a tattoo or other kind of mark that will not be lost when the snakes shed. Release the marked snakes, wait for a month (or longer), and re-trap the pythons. Use the equation $N = Mn/m$ to estimate the total population size (N), where M is the number of snakes that were marked, n is the number of snakes caught during the recapture, and m is the number of marked snakes in that recapture. Resample as many times as practical and calculate the average N. Repeat in other areas. *Assumptions:* The snakes are not moving into or out of the study area; the marked and unmarked snakes mix between sampling attempts; marked and unmarked snakes are equally likely to be trapped; the fitness of the marked individuals is not affected by the marks. **14.** apply It would be ideal to measure both the survivorship (the proportion of the cohort that survive to be 1 year old, 2 years old, etc.) and the age-specific fecundity (the average number of offspring produced by each female in each age class). **15.** evaluate Since pythons are generalists, it would seem that food availability would not be a primary factor in determining the carrying capacity. However, the dramatic disappearance of raccoons and opossums indicates that predation rates are very high and at least some food sources are disappearing. In the long run, the size of the python population will be food-limited (the population cannot increase exponentially forever). **16.** apply *Abiotic factors:* cold temperatures to the north, the occurrence of deep seawater (Atlantic Ocean and Gulf of Mexico) surrounding southern Florida. *Biotic factors:* competition for food, the occurrence of parasites, or the spread of disease; increased predation on eggs or juveniles.

Chapter 52

IN-TEXT QUESTIONS AND EXERCISES

p. 1119 Making Models 52.1 apply See Figure A52.1. The exact shape of the curve is unimportant, just its range and trend.

p. 1120 apply Your graph should look like Figure 52.3, with *Semibalanus* on the left, *Chthamalus* on the right, and the x-axis indicating depth in the intertidal zone. The realized niche of *Chthamalus* is in the upper intertidal zone.

p.1120 Fig. 52.4 evaluate If Connell had used only one or a few rocks at only one study site, critics might conclude that his results occurred because of something specific to the rock or the site rather than due to competition between species.

p. 1122 apply See Figure A52.2.

p. 1125 Fig. 52.8 apply This experimental design tested the hypothesis that mussels can sense the presence of crabs. If the crabs had been fed mussels, a critic could argue that the mussels detect the presence of damaged mussels—not crabs. (As it turns out, the mussels can do both. The experimenters did the same experiment with broken mussel shells in the chamber instead of a fish-fed crab.)

p. 1127 Fig. 52.11 apply Crop plants that are generalists in terms of pollination (like flower C in the model) are more likely to survive in an environment where pesticides are killing many bees.

p. 1127 CYU (1) evaluate Resource partitioning is an *evolutionary* change in resource use. The individuals of each species do not choose or try to have traits that reduce competition—they simply have those traits (or not). Individuals with traits that allow them to exploit different resources than the other species produce more offspring, which inherit those traits. (Recall from Chapter 22 that natural selection occurs on individuals, but adaptive responses are properties of populations.) **(2)** create *Hypothesis:* Toucans and palms are in a mutualistic interaction. *Alternate hypothesis:* Toucans and palms are in a consumption interaction (herbivory). To test your hypothesis, you must collect data to determine how the fitness of the toucans and palms changes as a result of the interaction. For example, you could measure the number of surviving offspring produced by toucans that are fed palm fruits versus other typical foods, or use an indirect measure of fitness such as body mass. (Toucans benefit from eating palm fruits.) Then you could determine the rate of survival of seeds in the feces of toucans relative to seeds that were not consumed by toucans. If the seeds are killed by the toucans, then the interaction is a consumption interaction. If the seeds in the toucan feces have a higher rate of survival and reproduction than the controls, then the interaction is a mutualistic interaction (which is what researchers discovered).

p. 1128 Making Models 52.2 apply Your drawing can use any type of simple symbols as long as you put four different symbols in each community to represent four species (species richness = 4). To indicate a difference in species diversity, one community must show more evenness (even abundance) of each species while the other community must be dominated by one or two species. Figure A52.3 is one example of an appropriate answer model (using geometric symbols to represent species).

p. 1128 Making Models 52.3 apply You should have added grass, mice, and coyote to the food web such that arrows point up from grass to mice, grass to elk, mice to coyote, and coyote to wolf. Grass should be on the bottom trophic level; mice should be on the second level, and coyote should be in a new level between mice and wolf (see Figure 52.15).

p. 1129 analyze The calculations for community 1 are as follows. (Note that your final answer is sensitive to the number of digits you use in your calculations; see **BioSkills 1**.)

Species A	Species B
$p_A = 10/18 = 0.555$	$p_B = 1/18 = 0.0555$
$\ln 0.555 = -0.589$	$\ln 0.0555 = -2.891$
$-0.589 \times 0.555 = \mathbf{-0.327}$	$-2.891 \times 0.0555 = \mathbf{-0.160}$
Species C	**Species D**
$p_C = 1/18 = 0.0555$	$p_D = 3/18 = 0.167$
$\ln 0.0555 = -2.891$	$\ln 0.167 = -1.790$
$-2.891 \times 0.0555 = \mathbf{-0.160}$	$-1.790 \times 0.167 = \mathbf{-0.299}$
Species E	**Species F**
$p_E = 2/18 = 0.111$	$p_F = 1/18 = 0.0555$
$\ln 0.111 = -2.198$	$\ln 0.0555 = -2.891$
$-2.198 \times 0.111 = \mathbf{-0.244}$	$-2.891 \times 0.0555 = \mathbf{-0.160}$

Summing the values of $p \times \ln p$ for each species and multiplying by -1 gives the Shannon index of species diversity for community 1:

$$(-1)[(\mathbf{-0.327}) + (\mathbf{-0.160}) + (\mathbf{-0.160}) + (\mathbf{-0.299}) + (\mathbf{-0.244}) + (\mathbf{-0.160})] = 1.350$$

Similar calculations would give species diversity values of 1.794 for community 2 and 1.610 for community 3.

p. 1131 Fig. 52.15 apply You should have drawn a (+) symbol by the hawk. A reduction in coyotes increases the population of mice, which increases the population of hawks.

p. 1132 CYU (1) apply You could measure the total number of species in your gut microbiome, the relative abundance of each species, the type of species interactions present, and the change in any of these variables with diet, age, medicine use, and other factors. **(2)** apply A trophic cascade might be occurring, where a decrease in the number of sharks has caused an increase in the number of fish that eat the herbivores of algae, causing an increase in the growth of algae.

p. 1132 Fig. 52.16 analyze The percentage of 61 species present in all of the ponds (top of the chart) is $14/61 = 0.23 = 23$ percent. Since this measure is less than half, it suggests that the community composition was more unpredictable than predictable.

Figure A52.1

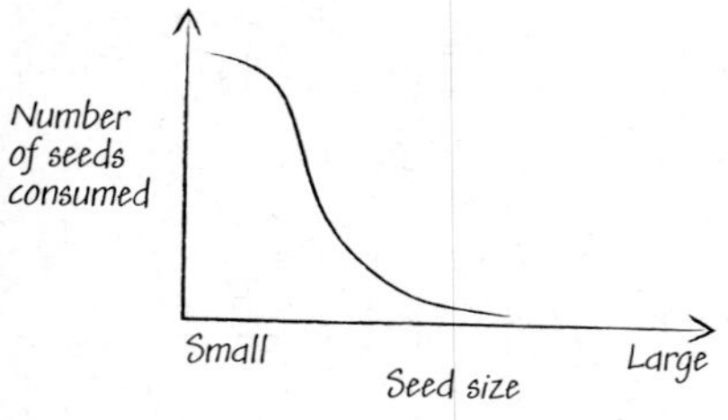

Figure A52.2

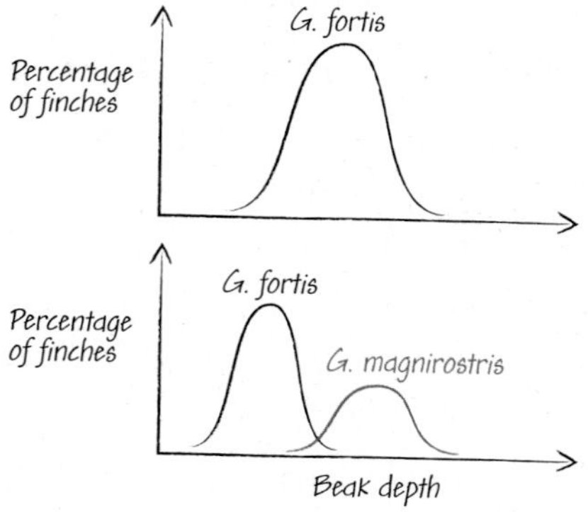

Figure A52.3

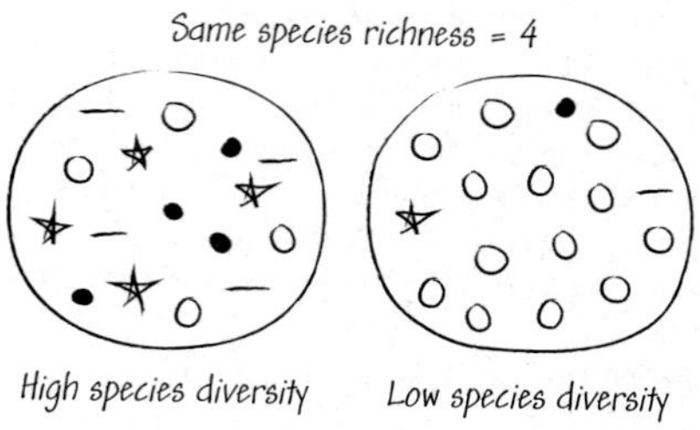

p. 1136 CYU (1) understand The shade provided by early successional species increases humidity, and decomposition of their tissues adds nutrients and organic material to the soil. These conditions favor growth by later successional species, which can outcompete the early successional species. **(2)** apply An early succession species is likely to have rapid growth, a short life span, and low competitive abilities—that is, it will likely occur at the high-fecundity, low-survivorship end of the life-history continuum (see Ch. 51, Figure 51.5). It is also likely to have good dispersal abilities.

p. 1137 Fig. 52.20 apply Lowering the rate of immigration would move the whole immigration curve downward. The equilibrium number of species (determined by the intersection of the immigration and extinction curves) would decrease. See Figure A52.4.

p. 1138 CYU apply Species richness is the number of species in a community. The theory of island biogeography predicts that species richness is likely to be higher in a nature preserve that is large and close to other preserves. The latitudinal species gradient predicts that species richness is likely to be higher in a preserve in the tropics.

✔ TEST YOUR KNOWLEDGE

1. remember a **2.** understand d **3.** understand F, F, T, T **4.** remember fecundity

✔ TEST YOUR UNDERSTANDING

5. understand (Many examples.) A pollination relationship between pollinator and flower is mutualistic if the pollinator receives a nectar reward and the flower receives pollination services. However, if the pollinator takes nectar without delivering pollen, or if the flower does not produce nectar even though the pollinator delivers pollen, then the (+/+) relationship becomes a parasitic (+/−) relationship. **6.** evaluate This statement is incorrect. Unlike individual humans, species do not have conscious thought and cannot make decisions. Species are made of diverse individuals with varying traits. Individuals either succeed or fail at surviving and reproducing under various conditions, including interacting with other species. Over time, the frequencies of traits change in the population. **7.** analyze Disturbance is any short-lived event that changes the distribution of resources. Compared to low-frequency fires, high-frequency fires would tend to be less severe (less fuel builds up) and would tend to exert more intense natural selection for adaptations to resist the effects of fire. Compared to low-severity fires, high-severity fires would open up more space for pioneering species and would tend to exert more intense natural selection for adaptations to resist the effects of fire. **8.** analyze See Figure A52.5. Note that per convention, kelp should be on the bottom, sea urchins in the middle, and sea otters on the top. Arrows should point up to show the flow of nutrients and energy from one trophic level up to the next. (1) Trophic cascades occur when a consumer has indirect effects on other species lower in the food web. Otters do not directly affect the kelp, but they eat the sea urchins that eat the kelp, therefore controlling the size of the kelp population. (You may have shown this on your model and/or described it in a sentence or two.) (2) Otters are considered keystone species because their presence has a large effect on the structure of the kelp community; in the absence of otters, the sea urchin population would soar and overgraze the kelp, devastating the kelp forest and many of the other organisms living in that community.

✔ TEST YOUR PROBLEM-SOLVING SKILLS

9. apply b **10.** analyze The exact answer will depend on the location of the campus. The first species to appear must have good dispersal ability, rapid growth, rapid reproduction, and tolerance for severe conditions. The two-acre plot is likely to be colonized first by pioneer species that have "weedy" characteristics. But once colonization is under way, the course of succession will depend more on how the various species interact with each other. The presence of one species can inhibit or facilitate the arrival and establishment of another. For example, an early-arriving species might provide the shade and nutrients required by a late-arriving species. The site's history and nearby ecosystems may influence which species appear at each stage; for instance, an undisturbed community nearby could be a source for native species. The pattern and rate of this succession is also influenced by the overall environmental conditions affecting it. Only species with traits appropriate to the local climate are likely to colonize the site.

✔ PUT IT ALL TOGETHER: Case Study

11. apply a **12.** analyze Mice: about 45 percent (red part of the bar); opossums: about 95 percent (since only about 5 percent fed successfully, about 95 percent must have been groomed off and killed). **13.** analyze d **14.** apply Forest fragmentation will subdivide large areas of habitat into many small islands of habitat, causing species richness to decline. **15.** create As forest fragmentation increases, species richness will decline. As species richness declines, white-footed mice will still be present in the population (due to the species' broad niche), but species like opossums will be absent (due to their narrow niches). Since mice are huge reservoirs of the Lyme bacterium and species like opossums are not, a greater percentage of animals present in forest fragments will host the Lyme bacterium. Thus, the incidence of Lyme disease will increase. **16.** evaluate The best prevention for Lyme disease is to preserve natural forests with high species richness—when the community is more diverse, there will be fewer white-footed mice and humans will have a lower exposure to Lyme disease. The forest provides important resources—including health benefits—to humans when it is intact.

Figure A52.4

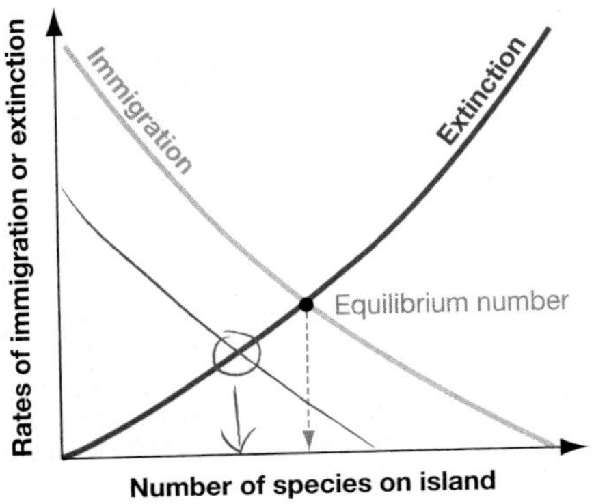

Figure A52.5

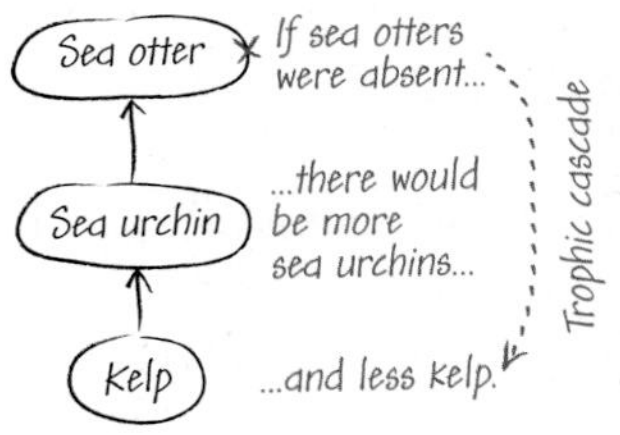

Chapter 53

IN-TEXT QUESTIONS AND EXERCISES

p. 1146 Fig. 53.7 analyze Given that each trophic level has only 10 percent of the biomass of the level below it: (1) Rice scenario supports 100 g/m^2/year of humans; (2) chicken scenario supports 10 g/m^2/year of humans; (3) sardine scenario supports 1 g/m^2/year of humans; (4) tuna scenario supports 0.01 g/m^2/year of humans. So, given the assumptions of this simple model, a rice diet can support 10,000 times more human biomass (= 100/0.01) than a tuna diet.

p. 1146 apply Crustaceans and most fish are ectothermic, so they are much more efficient at converting primary production into the biomass in their bodies than are endothermic birds and mammals.

p. 1146 apply Tuna are top predators in marine food webs. Mercury occurs in low concentrations in the water, but accumulates in higher concentrations at each trophic level. By eating a lot of tuna, you will accumulate mercury over time and potentially get mercury poisoning. Sardines are primary consumers, a low trophic level, and so they have a hundredfold or thousandfold lower concentrations of mercury. (Note that mercury and other pollutants that organisms cannot digest or excrete become more concentrated at higher trophic levels, but regular organic molecules such as carbohydrates and proteins *do not*.)

p. 1148 Fig. 53.9 analyze The average NPP is about 125 g/m^2/year. Area = 65 percent. Although NPP in the ocean is very small, the ocean is so vast that it ends up being the largest contributor to total NPP.

p. 1148 CYU (1) apply To grow a kilogram of beef, first you have to grow about 10 kg of grain or grass and feed it to the cow. Only 10 percent of this 10 kg will be used for growth and reproduction—the other 9 kg is mostly used for cellular respiration to keep the cattle alive, which results in the release of carbon in the form of CO_2. For efficient use of NPP, eating wheat as a primary food source is more sustainable than eating beef, because 10 times more wheat biomass can be harvested than beef biomass. **(2)** evaluate While it is true that plants are an important source of food for animals, this statement is incorrect for two reasons: (1) According to the first law of thermodynamics, plants cannot *create* energy—rather, they *transform* light energy into chemical energy; (2) it is common for humans to apply our goal-oriented perspective to other organisms, but plants do not act purposefully to help other organisms (see **BioSkills 17**).

p. 1150 apply In the tropics, nutrients cycle rapidly through forests, meaning that most of the nutrients are tied up in biomass rather than slowly decomposing in the soil. When the trees of tropical rain forests are hauled away, the nutrients are exported from the site, leaving only poor soil to nourish the crops.

p. 1151 Fig. 53.12 create One logical hypothesis is that the total amount exported increases as tree roots and other belowground organic material decay and begin to wash into the stream; the amount exported should begin to decline as nutrient reserves become exhausted, until eventually there are no more nutrients to wash away.

p. 1151 apply See **Figure A53.1**.

p. 1153 Fig. 53.15 analyze Total nitrogen fixation = 140 + 5 + 58 + 60 + 150 = 413 teragrams of nitrogen per year. Of this, 60 + 150 = 210 teragrams is caused by human activities. Thus, human activities fix about 210/413 teragrams = 51 percent of the nitrogen.

p. 1154 analyze See **Figure A53.2**.

p. 1155 CYU analyze See **Table A53.1**.

p. 1156 Fig. 53.20 analyze Photosynthesis increases in summer in the Northern Hemisphere, resulting in removal of CO_2 from the atmosphere.

p. 1157 Fig. 53.21 analyze $2 \times 10{,}000 = 20{,}000$ million metric tons of CO_2, double the current total.

p. 1161 apply The pH of the water will decrease (acidity will increase) because CO_2 from your breath reacts with water to form carbonic acid.

p. 1162 CYU (1) explain explain [One example of an explanation in non-technical language.] The climate is changing mainly because humans have been burning more and more fossil fuels like coal and oil. These fuels release large amounts of gases to the atmosphere that trap heat. The trapped heat causes the average climate to be warmer and also more variable. **(2)** apply [Many possible answers.] *Hypothesis 1:* The increased CO_2 is causing climate to change, including an increase in average temperature that is too warm for some crops. *Hypothesis 2:* The increased CO_2 is causing climate to change, including a change to precipitation that reduces growth in some crops. *Hypothesis 3:* The increased CO_2 is causing climate to change, including an increase in variability such as severe droughts, freezes, floods, or heat waves that reduce growth in some crops. *Hypothesis 4:* The increased CO_2 is causing climate to change; thus, the ranges of some important species in the community, such as pollinators, are shifting, causing reduction of growth in some crops.

✔ TEST YOUR KNOWLEDGE

1. remember the atmosphere (N_2 gas) **2.** remember False; most NPP is used for cellular respiration in primary consumers. **3.** remember c **4.** understand b

✔ TEST YOUR UNDERSTANDING

5. apply a **6.** analyze Warmer, wetter climates (as in the Amazon) speed decomposition; cool temperatures (as in Nebraska) slow it. Decomposition regulates nutrient availability because it releases nutrients from detritus and allows them to reenter the food web. **7.** understand The open ocean has almost no nutrient input from the land and has little upwelling to supply nutrients from the deep ocean. In contrast, intertidal and coastal areas receive large inputs of nutrients from rivers as well as from upwellings from ocean depths. **8.** apply Tofu derives from a plant, a primary producer at the base of the productivity pyramid. Cows are primary consumers, one level up. It would be about 10 times more efficient to eat tofu, because 10 times the quantity of plants would be necessary to provide one pound of beef.

✔ TEST YOUR PROBLEM-SOLVING SKILLS

9. evaluate c (The friend is confusing local weather variation with global climate. Even though Earth's average temperature is increasing, this does not mean that every season everywhere will be warmer; in some places, weather will be more extreme, including hotter heat waves and sometimes colder and/or wetter winters.) **10.** apply Atmospheric oxygen would increase due to extensive photosynthesis, but carbon dioxide levels would decrease because the rate of photosynthesis was high but the rate of decomposition was low. The temperature would drop because fewer greenhouse gases would be trapping heat reflected from Earth's surface.

✔ PUT IT ALL TOGETHER: CASE STUDY

11. understand d **12.** analyze F (nutrient export is likely to increase because most of the nutrients were stored in the biomass that was removed, and further erosion of soil will occur); F (trees absorb CO_2 from the atmosphere, so removal of trees will remove this carbon sink); T (soil moisture will decline because evaporation from the clear-cut land will increase); T (with the loss of the main primary producers and structural components of the ecosystem, biodiversity will plummet). **13.** create For measuring the effects of forest fires on primary productivity in forest plots, a good control would be forest plots that were comparable to the treatment plots in every way possible (species diversity, temperature, precipitation, topography, etc.) except that they did not burn or were prevented from burning. **14.** analyze T, T, T, T, T (These true statements are all represented by one of the feedback arrows in the model.) **15.** analyze Positive feedback loops can cause acceleration of destructive processes. For example, this model shows that increased atmospheric CO_2 leads to warmer temperatures and increased droughts, which increases the risk of fires, which kills trees, which increases atmospheric CO_2, which exacerbates the problem. Climate change, deforestation, and forest fires are combining to greatly increase the likelihood of fires that would destroy the trees and replace them with grasses. Once this tipping point is reached, the same processes would maintain the grassland, preventing forest regrowth. **16.** evaluate One reason people have a hard time understanding and discussing ecosystems and global warming is that so many important variables are interacting in complex ways over such large scales. It is difficult to convey this complexity to nonscientists in a simple yet accurate way.

Figure A53.1

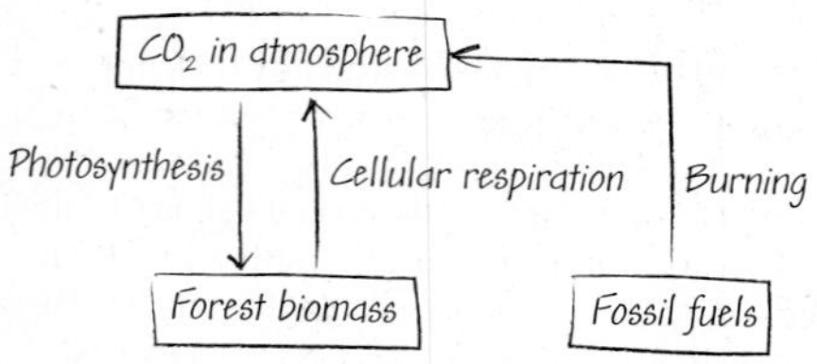

Figure A53.2

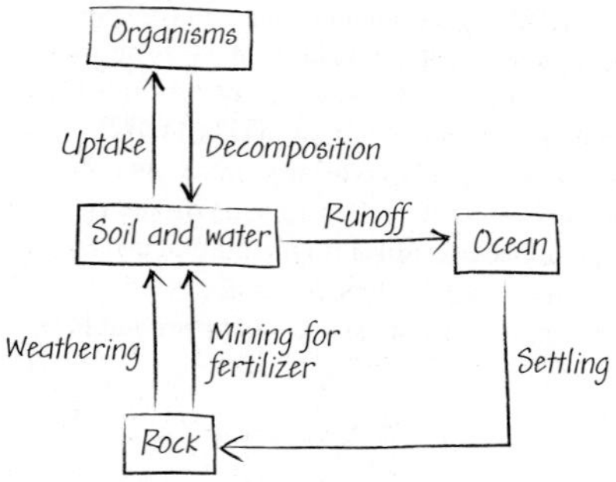

Table A53.1

	Important reservoirs	Natural processes	Human impacts
H_2O	Ocean; lakes and rivers; aquifers	Evaporation; precipitation	Pumping of groundwater; runoff (depletes fresh water)
N	Atmosphere; organisms	N-fixation; decomposition; assimilation; denitrification	Industrial N-fixation; burning of fossil fuels (adds N to cycle)
P	Rock	Weathering; uptake; decomposition	Industrial mining (adds P to cycle)
C	Organisms; atmosphere; ocean; fossil fuels	Photosynthesis; respiration	Burning of fossil fuels; deforestation (adds C to cycle)

Chapter 54

IN-TEXT QUESTIONS AND EXERCISES

p. 1167 Table 54.1 apply **Figure A54.1** is one example of an appropriate answer model (using geometric symbols to represent species). Note that in this sample answer, the symbols are arranged in rows to make them easier to compare, but this organization does not represent the geographic location of species.

p. 1170 CYU apply At the genetic level, you could measure the genetic diversity of the orangutan population by comparing DNA sequences obtained from blood or feces samples. Alternately, you could use environmental sequencing to look at the genetic diversity of soil organisms. At the species level, you could monitor the species richness of trees, birds, mammals, and other groups, and identify any changes in species richness. At the ecosystem level, you could measure the ability of the growing forest to retain nutrients in the soil, prevent erosion, trap carbon, increase the prevalence of cloud cover and rain, and other functions.

p. 1171 Fig. 54.6 analyze About 50 percent.

p. 1173 Fig. 54.8 evaluate Because the treatments and study areas were assigned randomly, no bias is involved in picking certain areas that have unusual biomass or species diversity. They should represent a random sample of biomass and species diversity in fragments of various sizes versus intact forest.

Figure A54.1

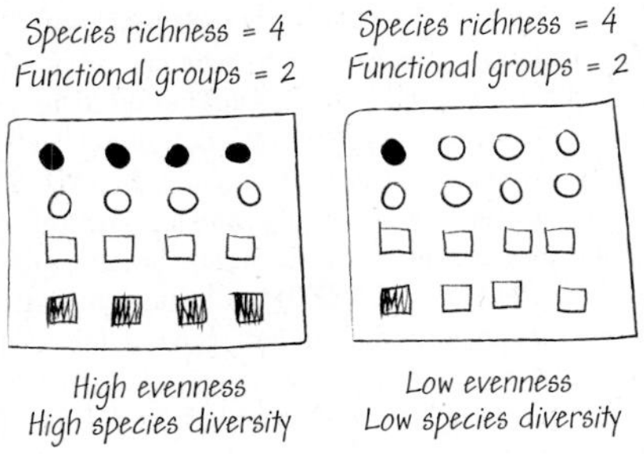

Figure A54.2

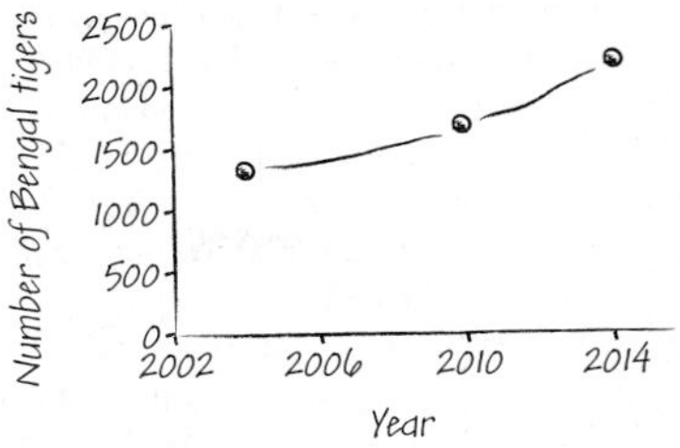

p. 1174 Making Models 54.1 create One example of an appropriate graph is shown in **Figure A54.2**. A scatter plot is a good choice for these data because both variables are continuous (see **BioSkills 2**). The graph helps to show that the tiger population has increased about a third within a decade.

p. 1176 Quantitative Methods 54.1 apply Using the equation $S = (18.9)A^{0.15}$, plug in 5000 km^2 for A and solve for S such that $S = 18.9 \times 5000^{0.15} = 68$ species.

p. 1177 CYU (1) understand Fragmentation reduces habitat quality by creating edges that are susceptible to invasion and loss of species, due to changed abiotic conditions. Genetic diversity can be reduced inside fragments as inbreeding occurs, exposing rare deleterious alleles to natural selection, and/or as populations lose genetic diversity via genetic drift. **(2)** analyze [Many possible answers.] The industrial food system that produces burgers (as shown in Figure 54.6) contributes to threats in these ways: (1) Forests, grasslands, and other types of habitats are destroyed to make way for growing crops such as wheat and for grazing cattle; (2) the overexploitation of other animals, like bison, has made way for cattle; (3) the transport of foods around the world enables the accidental introduction of invasive plants, insects, and other organisms; (4) the pesticides, herbicides, and fertilizers used in food systems are major pollutants; (5) the production and transport of food around the world produces a major amount of greenhouse gases, contributing to climate change. **(3)** create Habitat destruction and fragmentation in Borneo are obvious from the clear-cut areas. To assess overexploitation, you could interview local citizens and law enforcement officers on the history of poaching orangutans for the pet trade or for other uses. To assess effects from invasive species, you could use published or local accounts of community structure over time. To assess pollution, you could take direct measurements of soil, water, air, and possibly tissue samples (in dead orangutans, feces, or blood samples). To assess the effects of climate change, you could study local long-term temperature and precipitation data, which could affect the species present in the orangutan forest habitat.

p. 1178 Fig. 54.11 analyze *Null hypothesis:* NPP is not affected by species richness or functional diversity of species. *Prediction of hypothesis:* NPP will be greater in plots with more species and more functional groups. *Prediction of null hypothesis:* There will be no difference in NPP based on species richness or number of functional groups.

p. 1181 CYU (1) understand If many species are present, and each uses the available resources in a unique way, then a larger proportion of all available resources should be used—leading to higher biomass production. **(2)** create Establish several study plots in an area of Borneo where some of the forest has been converted to oil palm plantation but the adjacent forest (of similar elevation and topography) is still intact—half of the plots in the plantation and half in the native tropical forest. Monitor soil depth, soil nutrient levels, precipitation, and stream sediment levels regularly, and compare after several years. **(3)** apply Exposure to nature has a number of physical and mental health benefits, such as reduced stress and depression and reduced rates of obesity and diabetes. Since poor neighborhoods tend to have less access to natural areas than do wealthy city residents, planting trees in poor neighborhoods could help to decrease health inequality among poor and wealthy people.

p. 1183 Fig. 54.15 analyze It takes about 150 m^2 for a year to produce 1000 kcal of beef, and about 5 m^2 for a year to produce 1000 kcal of chicken, so you would use about 145 m^2 less land by eating chicken instead of beef today.

p. 1185 CYU apply [Many possible answers.] You could consume fewer resources and produce less waste; drive less and buy local; eat less beef; stop eating foods that threaten species, such as shark fin soup; buy only native plants for your yard and plant more to help create corridors among habitats; resist the temptation to release unwanted bait like earthworms or pets like snakes into nature; keep your cat inside; support local and global efforts to preserve natural areas; encourage others to do the same.

✔ TEST YOUR KNOWLEDGE

1. remember T, F, F, F **2.** remember b **3.** remember d **4.** remember overexploitation

✔ TEST YOUR UNDERSTANDING

5. understand In biology, humans are a type of animal, so saying "humans and animals" is incorrect, like saying "apples and fruit." Switching from a "humans and animals" to a "humans and other animals" worldview could encourage conservation in a number of ways. For example, if people feel more closely related to other animals, we may be more motivated to prevent extinctions. Likewise, if people feel more closely related to other animals, it may seem more obvious that humans are part of ecosystems rather than separate from nature, increasing motivation to reduce our negative impacts on ecosystems. **6.** apply By comparing the number of species estimated to reside in the park before and after the survey, biologists will have an idea of how much actual species diversity is being underestimated in other parts of the world—where research was at the level of Great Smoky Mountains National Park before the survey. The limitation is that it may be difficult to extrapolate from the data—no one knows if the situation at this park is typical of other habitats. **7.** evaluate No single approach is correct; there are good arguments for both efforts. In a case like the orangutans of Borneo, it is not possible to save the orangutans without saving their ecosystem. On the other hand, it's much easier to attract money and volunteers to help save a charismatic primate like an orangutan than it is to save a patch of forest. Both types of efforts are needed; their relative importance depends on the circumstance. **8.** understand Wildlife corridors facilitate the movement of individuals. Corridors allow areas to be recolonized if a species is lost and enable the introduction of new alleles that can counteract genetic drift and inbreeding in small, isolated populations, which tend to reduce genetic diversity.

✔ TEST YOUR PROBLEM-SOLVING SKILLS

9. apply d **10.** evaluate There is no "correct" answer. One argument is that conservationists could approach officials in tropical countries, including Brazil and Indonesia, and offer to help them learn from the mistakes made in developed nations. It is better to preserve enough forests to maintain biodiversity and keep ecosystem services intact, thus avoiding the expenditures that developed countries face in cleaning up pollution and restoring ecosystems and endangered species.

✔ PUT IT ALL TOGETHER: Case Study

11. understand a **12.** apply When birds are excluded, the percentage of berries with borers approximately doubled. The difference is considered statistically significant (not just sampling error), because the P value is less than 0.05. **13.** apply It was important for the researchers to demonstrate that there was no difference between control and treatment plots at the start of the experiment, so they could be sure that any change seen was due to the exclusion of birds and not some other factor such as a difference in shade, water availability, abundance of beetles, and so on. **14.** analyze There are about 4 borer predators per hectare at 15 percent cover and about 11 borer predators per hectare at 30 percent cover, a difference of $11/4 \times 100 = 2.75$ or 275 percent. That is, the abundance of borer predators is about three times higher. **15.** apply One example of an appropriate answer model is shown in **Figure A54.3**. In this example, species richness = 3 in the clear-cut plantation and = 12 in the plantation including native forest. The species diversity in the clear-cut plantation is very low because it is dominated by one species, coffee. The species diversity is higher in the second plot, due not only to the higher number of species (including birds) but also to the more even distribution of abundance among species.

16. evaluate It does look like a win-win situation. Leaving more natural forest increases the biodiversity at the genetic, species, and ecosystems levels, increasing the chance of resilience and ecosystem function. Leaving some of the natural forest is a win for farmers because the forest provides ecosystem services, such as pest management, that farmers rely on for a successful business.

Figure A54.3

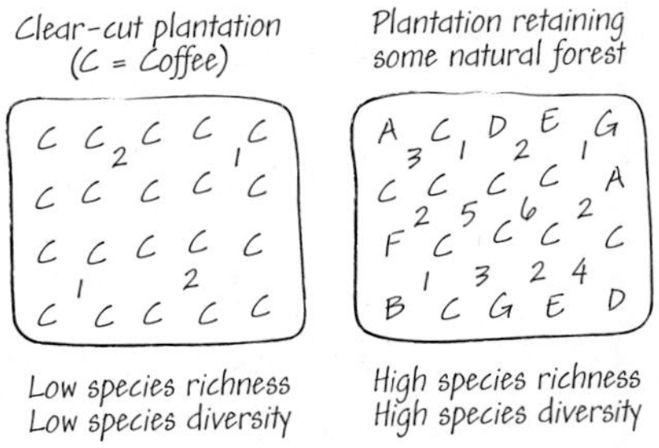

END-OF-UNIT CASE STUDY What Is the Larger Ecological Context of Toxic Newts?

pp. 1188–1189 (1) apply a **(2)** apply c. **(3)** apply Since newts are mobile, the mark-recapture method would be appropriate. First, you could capture a sample (*M*) of newts (for example by setting up traps on land or by using a net in a pond), then mark them in a harmless way, such as by applying a small tattoo. After a certain period time during which newts are allowed to mix, such as a week, you could recapture a sample of newts (*n*) and count how many of them have a tattoo (*m*). You would then use the equation $N = Mn/m$ to estimate the population size. You would want to conduct multiple resamples and calculate the average *N*, keeping the assumptions of the method in mind as you form your conclusions. This process would have to be repeated for different geographic areas if you wanted to estimate the size of the newt population throughout its range. **(4)** apply Yes, the Unken reflex is an honest signal because it communicates danger and the newt is actually dangerous. This signal is adaptive to the newt because the fitness of the newt will be higher if a predator does not eat it or attempt to eat it. **(5)** apply The small amount of TTX present in the skin at all times represent a small standing defense and the larger amounts of toxin that are secreted under threat represent a larger induced defense. The newt must expend energy to maintain a standing defense, however the benefit is that it is ready at all times. The induced defense is more efficient in terms of expending energy only when there is a threat, but the trade-off is that the defense may not be effective rapidly enough during a sneak attack by a predator. **(6)** apply Your model should show that the energy from the sun cannot be absorbed directly by the newt. Instead, plants or algae must first absorb the light energy and transform it into chemical energy. See Figure EOU A8.1. **(7)** apply The CO_2 from your breath could not be absorbed by the newt and incorporated into TTX directly. Instead, the CO_2 would have to be absorbed by plants or algae and incorporated into organic molecules via photosynthesis. Then, newts could either consume these autotrophs directly, or consume other animals that ate the plants or algae. Metabolic reactions within the newt could then transform food molecules they consumed into TTX. **(8)** apply Newts and garter snakes are engaged in a consumption (predation) interaction. The fitness of the snakes increases as a result of the interaction when the snakes are resistant to the TTX toxin and survive. If the snakes are not resistant to TTX and consume newts with high toxicity, the fitness of the snakes would decline. Whether the snakes are resistant to TTX or not, the newts never benefit from the interaction because they expend energy to produce the toxin. **(9)** apply In areas of high newt toxicity, snakes are likely to have high resistance to TTX and survive eating toxic newts. The abundance of snakes could support a large hawk population unless the snakes retain the TTX, in which case the size of the hawk population would decline. In areas of low newt toxicity, the newt population could support a large snake population, which could in turn support a large hawk population. **(10)** apply *Organism:* How much TTX is present in newt eggs in different parts of their range? *Population:* Is there a relationship between the fecundy of female newts and the amount of TTX in their eggs? *Community:* Do predators avoid newt eggs that contain TTX? *Ecosystem:* How does the presence of TTX in newt eggs affect the flow of nutrients in aquatic food webs? *Biosphere:* Is human-induced climate change affecting the survival and fecundity of newts, for example by affecting average temperature or size of freshwater ponds? **(11)** apply Healthy newt habitat, such as freshwater ponds and streams, can provide humans with food (provisioning service), can purify water (regulating service), and can provide recreation areas (cultural service). **(12)** apply Human activities affecting newts may include damming of their freshwater streams; polluting their ponds and streams; deforesting their terrestrial habitat; overfishing trout and other species in their food web, causing indirect effects; and causing climate change, which changes the temperature and precipitation patterns in their habitat.

Figure EOU A8.1

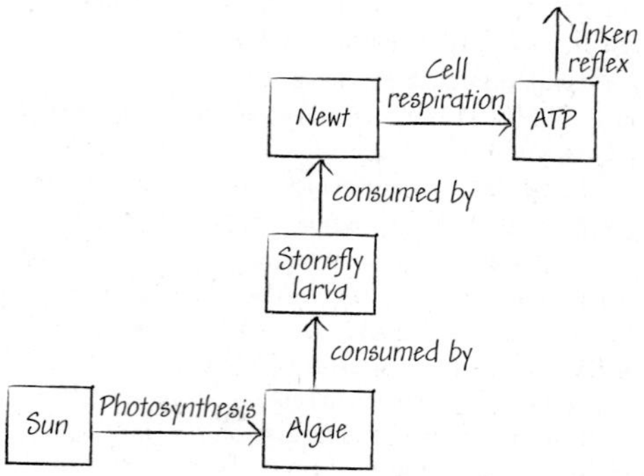

BIG PICTURE Ecology

p. 1190–1191 CYU (1) analyze Species interactions "are agents of" natural selection. **(2)** analyze As the human population climbs, more resources are needed to support the population. The harvesting of resources affects other species both directly (e.g., overfishing) and indirectly (e.g., climate change), resulting in decreased biodiversity. **(3)** understand Many possible answers. For example, primary productivity provides th e food we eat, both directly (plants) and indirectly (the consumers in the food web that rely on the primary productivity of producers). Primary productivity in forests stores carbon. **(4)** apply The greenhouse effect is shown by the diagram in the Climate Change box. One way to add a "CO_2 emissions" box to the concept map: [Human behavior] "increases" [CO_2 emissions]; and [CO_2 emissions] "cause" [greenhouse effect/climate change].

Periodic Table of Elements

Key	
1	Atomic number
H	Atomic symbol
Hydrogen	Name
1.008	Atomic weight*†

- Most common elements in living things
- Other major elements (minerals) found in living things
- Important trace elements (minerals) found in living things
- Elements mostly found in non-living things
- Elements not found in nature (synthesized by scientists)

1	2	3	4	5	6	7	8	9	10	11	12	13	14	15	16	17	18
1 **H** Hydrogen 1.008																	2 **He** Helium 4.003
3 **Li** Lithium 6.941	4 **Be** Beryllium 9.012											5 **B** Boron 10.81	6 **C** Carbon 12.01	7 **N** Nitrogen 14.01	8 **O** Oxygen 16.00	9 **F** Fluorine 19.00	10 **Ne** Neon 20.18
11 **Na** Sodium 22.99	12 **Mg** Magnesium 24.31											13 **Al** Aluminum 26.98	14 **Si** Silicon 28.09	15 **P** Phosphorus 30.97	16 **S** Sulfur 32.07	17 **Cl** Chlorine 35.45	18 **Ar** Argon 39.95
19 **K** Potassium 39.10	20 **Ca** Calcium 40.08	21 **Sc** Scandium 44.96	22 **Ti** Titanium 47.87	23 **V** Vanadium 50.94	24 **Cr** Chromium 52.00	25 **Mn** Manganese 54.94	26 **Fe** Iron 55.85	27 **Co** Cobalt 58.93	28 **Ni** Nickel 58.69	29 **Cu** Copper 63.55	30 **Zn** Zinc 65.38	31 **Ga** Gallium 69.72	32 **Ge** Germanium 72.64	33 **As** Arsenic 74.92	34 **Se** Selenium 78.96	35 **Br** Bromine 79.90	36 **Kr** Krypton 83.80
37 **Rb** Rubidium 85.47	38 **Sr** Strontium 87.61	39 **Y** Yttrium 88.91	40 **Zr** Zirconium 91.22	41 **Nb** Niobium 92.91	42 **Mo** Molybdenum 95.96	43 **Tc** Technetium [98]†	44 **Ru** Ruthenium 101.1	45 **Rh** Rhodium 102.9	46 **Pd** Palladium 106.4	47 **Ag** Silver 107.9	48 **Cd** Cadmium 112.4	49 **In** Indium 114.8	50 **Sn** Tin 118.7	51 **Sb** Antimony 121.8	52 **Te** Tellurium 127.6	53 **I** Iodine 126.9	54 **Xe** Xenon 131.3
55 **Cs** Cesium 132.9	56 **Ba** Barium 137.3	*	72 **Hf** Hafnium 178.5	73 **Ta** Tantalum 180.9	74 **W** Tungsten 183.8	75 **Re** Rhenium 186.2	76 **Os** Osmium 190.2	77 **Ir** Iridium 192.2	78 **Pt** Platinum 195.1	79 **Au** Gold 197.0	80 **Hg** Mercury 200.6	81 **Tl** Thallium 204.4	82 **Pb** Lead 207.2	83 **Bi** Bismuth 209.0	84 **Po** Polonium [209]	85 **At** Astatine [210]	86 **Rn** Radon [222]
87 **Fr** Francium [223]	88 **Ra** Radium [226]	**	104 **Rf** Rutherfordium [265]	105 **Db** Dubnium [268]	106 **Sg** Seaborgium [271]	107 **Bh** Bohrium [270]	108 **Hs** Hassium [277]	109 **Mt** Meitnerium [276]	110 **Ds** Darmstadtium [281]	111 **Rg** Roentgenium [280]	112 **Cn** Copernicium [285]	113 **Nh** Nihonium [286]	114 **Fl** Flerovium [289]	115 **Mc** Moscovium [289]	116 **Lv** Livermorium [293]	117 **Ts** Tennessine [293]	118 **Og** Oganesson [294]

*Lanthanides	57 **La** Lanthanum 138.9	58 **Ce** Cerium 140.1	59 **Pr** Praseodymium 140.9	60 **Nd** Neodymium 144.2	61 **Pm** Promethium [145]	62 **Sm** Samarium 150.4	63 **Eu** Europium 152.0	64 **Gd** Gadolinium 157.3	65 **Tb** Terbium 158.9	66 **Dy** Dysprosium 162.5	67 **Ho** Holmium 164.9	68 **Er** Erbium 167.3	69 **Tm** Thulium 168.9	70 **Yb** Ytterbium 173.1	71 **Lu** Lutetium 175.0
**Actinides	89 **Ac** Actinium [227]	90 **Th** Thorium 232.0	91 **Pa** Protactinium 231.0	92 **U** Uranium 238.0	93 **Np** Neptunium [237]	94 **Pu** Plutonium [244]	95 **Am** Americium [243]	96 **Cm** Curium [247]	97 **Bk** Berkelium [247]	98 **Cf** Californium [251]	99 **Es** Einsteinium [252]	100 **Fm** Fermium [257]	101 **Md** Mendelevium [258]	102 **No** Nobelium [259]	103 **Lr** Lawrencium [262]

DATA: IUPAC Commission on Isotopic Abundances and Atomic Weights. Atomic Weights of the Elements, 2017 (http://www.sbcs.qmul.ac.uk/iupac/AtWt/)

*Atomic weights are reported to four significant figures.

†For elements with a variable number of protons and/or neutrons, the mass number of the longest-lived isotope of the element is reported in brackets.

Glossary

5′ cap A modified guanine (G) nucleotide added to the 5′ end of eukaryotic mRNAs. Helps protect the mRNA from being degraded and promotes initiation of translation.

abdomen A region of the body; in arthropods, one of the three prominent body regions (tagmata), located posterior to the thorax.

abiotic Not alive (e.g., air, water, and some components of soil). Compare with **biotic**.

aboveground biomass The total mass of living plants in an area, excluding roots.

abscisic acid (ABA) A plant hormone that inhibits growth; it stimulates stomatal closure and triggers dormancy.

abscission In plants, the normal (often seasonal) shedding of leaves, fruits, or flowers.

abscission zone The region at the base of a petiole where cell-wall degradation occurs; results in the dropping of leaves.

absorption In animals, the uptake of ions and small molecules, derived from food, across the lining of the digestive tract.

absorption spectrum The amount of light of different wavelengths absorbed by a pigment. Usually depicted as a graph of light absorbed versus wavelength. Compare with **action spectrum**.

accessory fruit A fruit-like structure (e.g., strawberry) that develops not from an ovary, but from some tissue exterior to the carpel.

acclimation A change in a study organism's phenotype that occurs in response to laboratory conditions.

acclimatization A change in an individual's phenotype that occurs in response to a change in natural environmental conditions.

acetylation Addition of an acetyl group (—$COCH_3$) to a molecule. Acetylation of histone proteins is important in controlling chromatin condensation.

acetylcholine (ACh) A neurotransmitter that triggers contraction of vertebrate skeletal muscle cells but slows the rate of contraction of cardiac muscle cells.

acid Any compound that gives up protons or accepts electrons during a chemical reaction or that releases hydrogen ions when dissolved in water.

acid-growth hypothesis The hypothesis that auxin triggers elongation of plant cells by increasing the activity of proton pumps, making the cell wall more acidic and leading to expansion of the cell wall and an influx of water.

acoelomate A bilaterian animal that lacks an internal body cavity (coelom). Compare with **coelomate** and **pseudocoelomate**.

acquired immune deficiency syndrome (AIDS) A human disease characterized by death of immune system cells (in particular helper T cells) and subsequent vulnerability to other infections. Caused by the human immunodeficiency virus (HIV).

acrosome A caplike structure, located on the head of a sperm, that contains enzymes capable of digesting the outer coverings of an egg.

actin A globular protein that can be polymerized to form filaments. Actin filaments are part of the cytoskeleton and constitute the thin filaments in muscle cells.

actin filament A long fiber, about 7 nm in diameter, composed of two intertwined strands of polymerized actin protein; one of the three types of cytoskeletal fibers. Involved in cell structure and movement. Also called a *microfilament*. Compare with **intermediate filament** and **microtubule**.

action potential A rapid, temporary change in electrical potential across a membrane, from negative to positive and back to negative. Occurs in cells, such as neurons and muscle cells, that have an excitable membrane.

action spectrum The relative effectiveness of different wavelengths of light in driving a light-dependent process such as photosynthesis. Usually depicted as a graph of some measure of the process, such as O_2 production, versus wavelength. Compare with **absorption spectrum**.

activation energy The amount of kinetic energy required to initiate a chemical reaction; specifically, the energy required to reach the transition state.

activator A protein that binds to a DNA regulatory sequence to increase the frequency of transcription initiation by RNA polymerase. See also **transcriptional activator**.

active site The location in an enzyme molecule where substrates (reactant molecules) bind and react.

active transport The movement of ions or molecules across a membrane in a single direction, often against a gradient. Requires energy (e.g., from hydrolysis of ATP) and assistance of a transport protein (e.g., pump).

adaptation Any heritable trait that increases the fitness of an individual with that trait, compared with individuals without that trait, in a particular environment.

adaptive immunity Immunity to a particular pathogen or antigen conferred by activated B and T cells in vertebrates. Characterized by specificity, diversity, memory, and self–nonself recognition. Also called *adaptive immune response*. Compare with **innate immunity**.

adaptive radiation Rapid evolutionary diversification within one lineage, producing many descendant species that have adapted to a wide variety of habitats.

adenosine triphosphate (ATP) A molecule consisting of an adenine base, a sugar, and three phosphate groups that are linked together with covalent bonds that have high potential energy. Universally used by cells as a monomer for RNA synthesis and to store and transfer chemical energy.

adhesion The tendency of certain dissimilar molecules to cling together due to attractive forces. Compare with **cohesion**.

adipocyte A fat cell.

adrenal gland Either of two small endocrine glands, one above each kidney. The outer portion (cortex) secretes several steroid hormones; the inner portion (medulla) secretes epinephrine and norepinephrine.

adrenaline See **epinephrine**.

adrenocorticotropic hormone (ACTH) A peptide hormone, produced and secreted by the anterior pituitary, that stimulates release of steroid hormones (e.g., cortisol and aldosterone) from the adrenal cortex. Also known as *corticotropin*.

adult A sexually mature individual.

adventitious root A root that develops from a plant's shoot system instead of from the plant's root system.

aerobic Referring to any metabolic process, cell, or organism that uses oxygen as an electron acceptor. Compare with **anaerobic**.

afferent division The part of the nervous system that transmits information about the internal and external environment to the central nervous system. Consists mainly of sensory neurons. Compare with **efferent division**.

age class All the individuals of a specific age in a population.

age-specific fecundity The average number of female offspring produced by a female in a certain age class.

age structure The proportion of individuals in a population that are of each possible age.

agglutination The clumping of cells, viruses, or other particles by cross-linking molecules (e.g., antibodies).

aggregate fruit A fruit (e.g., raspberry) that develops from a single flower that has many separate carpels. Compare with **multiple fruit** and **simple fruit**.

alcohol fermentation Catabolic pathway in which pyruvate produced by glycolysis is converted to ethanol in order to oxidize NADH to NAD^+.

aldosterone A hormone that stimulates the kidney to conserve salt and water and promotes retention of sodium ions; produced in the adrenal cortex.

allele A particular version of a gene.

allergens Any molecule (antigen) that triggers an allergic response (an allergy).

allergy An IgE-mediated abnormal response to an antigen (allergen), usually characterized by dilation of blood vessels, contraction of smooth muscle cells in the airways, and increased activity of mucus-secreting cells.

allopatric speciation Speciation that occurs when populations of the same species become geographically isolated, often due to dispersal or vicariance. Compare with **sympatric speciation**.

allopatry Condition in which two or more populations live in different geographic areas. Compare with **sympatry**.

allopolyploidy (adjective: *allopolyploid*) The state of having more than two full sets of chromosomes (polyploidy) due to the hybridization between species. Compare with **autopolyploidy**.

allosteric regulation Regulation of an enzyme's (or other protein's) activity by binding of a regulatory molecule at a site distinct from the active site. Binding often results in a change in the protein's shape that affects the function of the active site.

α-amylase See **amylase**.

α-helix (alpha-helix) A secondary structure in proteins formed when the polypeptide backbone coils into a spiral shape stabilized by hydrogen bonding.

alternation of generations A life cycle involving alternation of a multicellular haploid stage (gametophyte) with a multicellular diploid stage (sporophyte). Occurs in most plants and some protists.

alternative splicing In eukaryotes, the splicing of primary RNA transcripts from a single gene in different ways to produce different mature mRNAs and thus different polypeptides.

altruism Any behavior that has a fitness cost to the individual (lowered survival and/or reproduction) and a fitness benefit to the recipient. See **reciprocal altruism**.

alveolus (plural: *alveoli*) Any of the tiny, air-filled sacs of a mammalian lung.

ambisense Describing a single strand of RNA that contains a region with the same sequence as an mRNA required to produce viral proteins and another region that is complementary to the sequence of a different mRNA. Compare with **positive-sense** and **negative-sense**.

amino acid A small organic molecule with a central carbon atom bonded to an amino group ($-NH_3$), a carboxyl group ($-COOH$), a hydrogen atom, and a side chain. When amino acids are linked together to form proteins, they are referred to as *residues*.

aminoacyl tRNA A transfer RNA molecule that is covalently bound to an amino acid.

aminoacyl-tRNA synthetase An enzyme that catalyzes the addition of a particular amino acid to its corresponding tRNA molecule.

ammonia (NH_3) A small molecule, produced by the breakdown of proteins and nucleic acids, that is very toxic to cells. It is a strong base that gains a proton to form the ammonium ion (NH_4^+). The major nitrogenous waste of bony fishes and aquatic invertebrates.

amnion The innermost of the membranes surrounding the embryo in an amniotic egg.

amniotes A major lineage of vertebrates (Amniota) that reproduce with amniotic eggs. Includes all reptiles (including birds) and mammals—that is, all tetrapods except amphibians.

amniotic egg An egg that has a watertight shell or case enclosing a membrane-bound water supply (the amnion and chorion), food supply (yolk sac), and waste sac (allantois).

amoeboid motion A sliding movement observed in some protists, accomplished by the formation of cytoplasmic extensions of the cell called pseudopodia. One form of *cell crawling*.

amphibians A lineage of vertebrates, many of which breathe through their skin and feed on land but lay their eggs in water; includes frogs, salamanders, and caecilians.

amphipathic Containing hydrophilic and hydrophobic regions.

ampullae of Lorenzini Structures on the heads of sharks that contain electroreceptors.

amylase Any enzyme that can break down starch by catalyzing hydrolysis of the α-glycosidic linkages between glucose residues.

amyloplasts Starch-storing organelles (plastids) in plants. In root cap cells, they settle to the bottom of the cell and may be used as gravity detectors.

anabolic pathway Any set of chemical reactions that synthesizes large molecules from smaller ones. Generally requires an input of energy. Compare with **catabolic pathway**.

anaerobic Referring to any metabolic process, cell, or organism that uses an electron acceptor other than oxygen, including fermentation or anaerobic respiration. Compare with **aerobic**.

anaphase The stage in mitosis or meiosis during which chromosomes are moved to opposite poles of the spindle apparatus.

anatomy The physical structure of organisms.

ancestral trait A trait found in the ancestors of a particular group.

anchor root Adventitious root that helps anchor individual plants to surfaces, including other plants, walls, and other structures.

aneuploidy (adjective: *aneuploid*) The state of having an abnormal number of copies of a certain chromosome.

angiosperm A flowering vascular plant that produces seeds within mature ovaries (fruits). The angiosperms form a single lineage. Compare with **gymnosperm**.

animal A member of a major lineage of eukaryotes (Animalia) whose members typically have a complex, multicellular body with cells that lack cell walls; eat other organisms; and move under their own power at some point in their lives.

animal model A non-human animal used by medical researchers that develops a disease with parallels to a disease of humans. Studied in hopes that findings may apply to human disease.

anion A negatively charged ion.

annelids Members of the phylum Annelida (segmented worms). Distinguished by a segmented body and a coelom that functions as a hydrostatic skeleton. Annelids belong to the lophotrochozoan branch of the protostomes.

annual Referring to a plant whose life cycle normally lasts only one growing season—less than one year. Compare with **perennial**.

anoxygenic Referring to any process or reaction that does not produce oxygen. Photosynthesis in purple sulfur bacteria is anoxygenic because it does not use water as an electron donor, so no O_2 is produced. Compare with **oxygenic**.

antagonistic muscle group A set of two or more muscles that re-extend one another by transmitting their forces via the skeleton.

antenna (plural: *antennae*) A long appendage of the head that is used to touch or smell.

antenna pigments Part of a photosystem; chlorophyll molecules and accessory pigments organized and modified by proteins to capture light and direct energy to a central reaction center during photosynthesis.

anterior Toward an animal's head and away from its tail. The opposite of **posterior**.

anterior pituitary The part of the pituitary gland containing endocrine cells that produce and release a variety of peptide hormones in response to other hormones from the hypothalamus. Compare with **posterior pituitary**.

anther The pollen-producing structure at the end of a stamen in flowering plants (angiosperms).

antheridium (plural: *antheridia*) The sperm-producing structure in most land plants except angiosperms.

Anthropocene Proposed name of a new, current epoch in the geologic time scale to reflect the dramatic physical, chemical, and biological changes that humans are causing on Earth; not yet officially adopted by the International Commission on Stratigraphy.

anthropoids A major lineage of primates, including humans and the other great apes, gibbons, and all monkeys. Compare with **prosimians**.

antibiotic Any substance, such as penicillin, that can kill or inhibit the growth of bacteria.

antibody A protein produced by B cells that can bind to a specific part of an antigen, tagging it for removal by the immune system. All monomeric forms of antibodies consist of two light chains and two heavy chains, which vary between different antibodies. Also called *immunoglobulin*.

anticodon The sequence of three bases (a triplet) in a transfer RNA molecule that can bind to an mRNA codon with a complementary sequence.

antidiuretic hormone (ADH) A peptide hormone, secreted from the posterior pituitary gland, that stimulates water retention by the kidney. Also called *vasopressin*.

antigen Any foreign molecule, often a protein, that can stimulate an innate or adaptive response by the immune system.

antigen presentation A process by which small peptides, derived from ingested or intracellular antigens, are complexed with MHC proteins and transported to the cell surface, where they are displayed and can be recognized by T-cell receptors.

antiparallel Describes the opposite orientation of nucleic acid strands that are hydrogen-bonded to one another, with one strand running in the $5' \rightarrow 3'$ direction and the other in the $3' \rightarrow 5'$ direction.

antiporter A carrier protein that allows an ion to diffuse down an electrochemical gradient, using the energy of that process to transport a different substance in the opposite direction *against* its concentration gradient. Compare with **symporter**.

antiviral Any drug or other agent that can interfere with the transmission or replication of viruses.

aorta In terrestrial vertebrates, the major artery carrying oxygenated blood away from the heart.

aphotic zone In an aquatic habitat, deep water receiving no sunlight. Compare with **photic zone**.

apical Toward the top. In plants, at the tip of a branch. In animals, on the side of an epithelial layer that faces the environment and not other body tissues. Compare with **basal** and **basolateral.**

apical–basal axis The shoot-to-root axis of a plant.

apical bud A bud at the tip of a stem or branch, where growth occurs to lengthen the stem or branch.

apical dominance Inhibition of lateral bud growth by the apical meristem at the tip of a plant branch.

apical meristem A group of undifferentiated plant cells, at the tip of a shoot or root, that is responsible for primary growth. Compare with **cambium**.

apodeme Any of the chitinous ingrowths of an exoskeleton to which muscles attach.

apomixis The formation of mature seeds without fertilization; a type of asexual reproduction.

apoplast In plants, the region outside plasma membranes consisting of the porous cell walls and the intervening extracellular air space. Compare with **symplast**.

apoptosis A series of tightly controlled changes in cellular activity that lead to the self-destruction of a cell. Occurs frequently during embryological development and as part of the immune response to remove infected or cancerous cells. Apoptosis is the most common form of **programmed cell death**.

appendix A narrow pouch that extends from the cecum in some mammals.

aquaporin A type of channel protein that facilitates the movement of water (osmosis) across a plasma membrane.

aqueous In a watery or in a water-based environment. Often used in describing a solution where water serves as the solvent.

aquifer An underground layer of porous rock, sand, or gravel that is saturated with water.

arbuscular mycorrhizal fungi (AMF) Fungi from the Glomeromycota lineage whose hyphae enter the root cells of their host plants. Also called *endomycorrhizal fungi*.

Archaea One of the three taxonomic domains of life, consisting of unicellular prokaryotes distinguished by cell walls made of certain polysaccharides not found in bacterial or eukaryotic cell walls, plasma membranes composed of unique isoprene-containing phospholipids, and ribosomes and RNA polymerase similar to those of eukaryotes. Compare with **Bacteria** and **Eukarya**.

archegonium (plural: *archegonia*) The egg-producing structure in most land plants except angiosperms.

arteriole A small vessel that carries blood to a capillary bed.

arteriosclerosis Hardening and loss of elasticity of arteries.

artery Any thick-walled vessel that carries blood (oxygenated or not) under relatively high pressure away from the heart to organs of the body. Compare with vein.

arthropods Members of the phylum Arthropoda. Distinguished by a segmented body; a hard, jointed exoskeleton; paired, jointed appendages; and an extensive body cavity called a hemocoel. Arthropods belong to the ecdysozoan branch of the protostome animals.

articulation A point of contact between two rigid components of a skeleton, such as between bones of a vertebrate endoskeleton or between segments of cuticle in an arthropod exoskeleton. Also called **joint**.

artificial selection Deliberate manipulation by humans, as in animal and plant breeding, of the genetic composition of a population by allowing only individuals with desirable traits to reproduce.

ascus (plural: *asci*) Specialized spore-producing cell found at the ends of hyphae in "sac fungi" (Ascomycota).

asexual reproduction Any form of reproduction in which offspring inherit DNA from only one parent. Includes binary fission, budding, and parthenogenesis. Compare with **sexual reproduction**.

astral microtubule A microtubule that arises from the one of the spindle poles in mitosis or meiosis and interacts with proteins on the plasma membrane.

atom The smallest identifiable amount of a particular element.

atomic number The number of protons in the nucleus of an atom, giving the atom its identity as a particular chemical element.

atomic weight The average mass of an element that is based on the relative proportions of all its naturally occurring isotopes.

ATP synthase A large membrane-bound protein complex that uses the potential energy stored in a proton gradient to couple proton transport and ATP synthesis.

atrioventricular (AV) node A region of the heart between the right atrium and right ventricle where electrical signals from the atrium are delayed briefly before spreading to the ventricle. This delay allows the ventricle to fill with blood before contracting. Compare with **sinoatrial (SA) node**.

atrium (plural: *atria*) A thin-walled chamber of the heart that receives blood from veins and pumps it to a neighboring chamber (the ventricle).

autocrine Relating to a chemical signal that affects the same cell that produced and released it.

autoimmunity A pathological condition in which the immune system recognizes self molecules and attacks the cells or tissues of an individual's own body.

autonomic nervous system The part of the vertebrate peripheral nervous system that controls internal organs and involuntary processes, such as stomach contraction, hormone release, and heart rate. Includes parasympathetic and sympathetic nerves. Compare with **somatic nervous system**.

autophagy The process by which damaged organelles and other cytoplasmic components are surrounded by a membrane and delivered to a lysosome to be recycled.

autopolyploidy (adjective: *autopolyploid*) The state of having more than two full sets of chromosomes (polyploidy) due to a mutation that doubled the chromosome number. All the chromosomes come from the same species. Compare with **allopolyploidy**.

autosomal inheritance The inheritance patterns that occur when genes are located on autosomes rather than on sex chromosomes.

autosome Any chromosome other than a sex chromosome (i.e., any chromosome other than the X or Y in mammals).

autotroph Any organism that can synthesize reduced organic compounds from simple inorganic sources such as CO_2 or CH_4. Most plants and some bacteria and archaea are autotrophs. Also called *primary producer*. Compare with **heterotroph**.

auxin Indoleacetic acid (IAA), a plant hormone that stimulates phototropism and cell elongation.

axillary bud A bud that forms at a node and may develop into a lateral (side) branch. Also called *lateral bud*.

axon A long projection of a neuron that can propagate an action potential.

axon hillock The site in a neuron where an axon joins the cell body and where action potentials typically are first triggered.

axoneme A structure found in eukaryotic cilia and flagella and responsible for their motion; composed of two central microtubules surrounded by nine doublet microtubules (9 + 2 arrangement).

B cell A type of lymphocyte that originates and matures in the bone marrow and, with T cells, is responsible for adaptive immunity. Produces antibodies and also functions in antigen presentation. Also called *B lymphocyte*.

B-cell receptor (BCR) A transmembrane protein in the plasma membrane of a mature B cell that can bind to a specific antigen. Its antigen-binding sites are identical to those in the antibodies produced by the same cell.

background extinction The average rate of low-level extinction that has occurred continuously throughout much of evolutionary history. Compare with **mass extinction**.

Bacteria One of the three taxonomic domains of life, consisting of unicellular prokaryotes distinguished by cell walls composed largely of peptidoglycan, plasma membranes similar to those of eukaryotic cells, and ribosomes and RNA polymerase that differ from those in archaea or eukaryotes. Compare with **Archaea** and **Eukarya**.

bacteriophage Any virus that infects bacteria.

baculum A bone inside the penis; usually present in mammals that have a penis lacking erectile tissue.

balancing selection A mode of natural selection in which no single allele is favored over time and across locations, on average. An overall balance of fitness and frequency is maintained among alleles.

ball-and-stick model A representation of a molecule where atoms are shown as balls—colored, and sometimes scaled, to indicate the atom's identity—and covalent bonds are shown as rods or sticks connecting the balls in the correct geometry.

bark The protective outer layer of woody plants, composed of cork cells, cork cambium, and secondary phloem.

baroreceptor One of the specialized nerve cells in the walls of the heart and certain major arteries that detect changes in blood pressure and trigger appropriate responses by the brain.

basal Toward the base. In plants, toward the root or at the base of a branch where it joins the stem. In animals, on the side of an epithelial layer that abuts underlying body tissues. Compare with **apical**.

basal body The microtubule-organizing center for cilia and flagella in eukaryotic cells. Consists of nine triplets of microtubules arranged in a circle and establishes the structure of axonemes. Structurally identical to a centriole.

basal lamina A thick, collagen-rich extracellular matrix that underlies most epithelial tissues in animals and attaches it to connective tissue.

basal metabolic rate (BMR) The total energy consumption by an organism at rest in a comfortable environment. For aerobes, often measured as the amount of oxygen consumed per hour.

base Any compound that acquires protons or gives up electrons during a chemical reaction or accepts hydrogen ions when dissolved in water.

base stacking Describes the interactions that form between adjacent base pairs in nucleic acid strands of a double helix. Consist of hydrophobic interactions and van der Waals interactions that tightly pack and stabilize the base pairing in nucleic acid secondary structure.

basidium (plural: *basidia*) Specialized spore-producing cell at the ends of hyphae in "club fungi" (Basidiomycota).

basilar membrane One of the membranes in the vertebrate cochlea between which the hair cells are located. Vibrates when sound waves travel through the fluid in the cochlea.

basolateral Toward the bottom and sides. In animals, the side of an epithelial layer that faces other body tissues and not the environment.

Batesian mimicry A type of mimicry in which a harmless or palatable species resembles a dangerous or poisonous species. Compare with **Müllerian mimicry**.

behavior In biology, an action in response to a stimulus.

behavioral ecology A discipline that focuses on behavioral adaptations that have evolved in response to ecological selection pressures; a subset of organismal ecology, which also includes morphological and physiological adaptations.

beneficial In genetics, referring to any mutation, allele, or trait that increases an individual's fitness. Compare with **deleterious** and **neutral**.

benign tumor A mass of abnormal tissue that develops due to unregulated growth but does not spread to other organs. Benign tumors are not cancers. Compare with **malignant tumor**.

benthic Living at the bottom of an aquatic environment.

benthic zone The area along the bottom of an aquatic environment.

β-pleated sheet (beta-pleated sheet) A secondary structure in proteins, formed when the polypeptide backbone folds into a sheetlike shape stabilized by hydrogen bonding.

bilateral symmetry An animal body pattern in which one plane of symmetry divides the body into a left side and a right side. Typically, the body is long and narrow, with a distinct head end and tail end. Compare with **radial symmetry**.

bilaterian A member of a major lineage of animals (Bilateria) that are bilaterally symmetrical at some point in their life cycle, have three embryonic germ layers, and have a coelom (or evolved from ancestors that had a coelom). All protostomes and deuterosomes are bilaterians.

bile A complex solution produced by the liver, stored in the gallbladder, and secreted into the small intestine. Contains steroid derivatives called bile salts that are responsible for emulsifying fats during digestion.

binary fission The process of cell division used for asexual reproduction of many prokaryotic cells. The genetic material is replicated and partitioned to opposite sides of a growing cell, which then divides in half, creating two genetically identical cells.

biodiversity The diversity of life considered at three levels: genetic diversity; species diversity; and ecosystem diversity.

biodiversity hotspot A region that is especially rich in species and also highly threatened by human activities, thus a high priority for conservation.

biofilm A complex community of bacteria enmeshed in a polysaccharide-rich, extracellular matrix that allows the bacteria to attach to a surface.

biogeochemical cycle The pattern of circulation of an element or molecule among living organisms and the environment.

biogeography The study of how species and populations are distributed geographically and through geologic time.

bioinformatics A discipline that blends biology and computer science to manage, analyze, and interpret biological sequence information, particularly DNA and amino acid sequences.

biological species concept The definition of a species as a population or group of populations that are reproductively isolated from other groups. Members of a species have the potential to interbreed in nature to produce viable, fertile offspring but cannot interbreed successfully with members of other species. Compare with **morphospecies concept** and **phylogenetic species concept**.

bioluminescence The emission of light by a living organism via an enzyme-catalyzed reaction.

biomagnification In animal tissues, an increase in the concentration of heavy molecules (such as mercury) or persistent organic pollutants (such as DDT) as they are passed up a food chain.

biomass The total mass of all organisms in a given population or geographical area; usually expressed as total dry weight.

biome A large terrestrial or aquatic region characterized by distinct abiotic characteristics and dominant types of vegetation.

biomechanics A field of biology that applies the principles of physics and engineering to analyze the mechanical structure and function of organisms.

bioprospecting The effort to find commercially useful compounds by studying organisms—especially species that are poorly studied to date.

bioremediation The use of living organisms, usually bacteria or archaea, to degrade environmental pollutants.

biosphere The thin zone surrounding Earth where all life exists; the sum of all terrestrial and aquatic ecosystems.

biotechnology Applications of biological knowledge, particularly in the areas of molecular genetics and genomics, to technological applications, particularly in research, agriculture, and medicine.

biotic Living, or produced by a living organism. Compare with **abiotic**.

bipedal Walking primarily on two legs; characteristic of hominins.

bipolar cell A type of cell in the vertebrate retina that receives information from one or more photoreceptors and passes it to ganglion cells.

biradial symmetry An animal body pattern that has two planes of symmetry (such as in ctenophores). Typically, the body is in the form of a cylinder or disk. Compare with **radial symmetry** and **bilateral symmetry**.

bivalent A pair of synapsed homologous chromosomes during prophase of meiosis I.

bivalves A lineage of mollusks, including clams and mussels, that have shells made of two parts, or valves.

bladder An organ that holds urine until it can be excreted.

blade The wide, flat part of a plant leaf.

blastocoel A fluid-filled cavity in the blastula of many animal species.

blastocyst The mammalian blastula. A roughly spherical structure composed of trophoblast cells on the exterior, a fluid-filled cavity (the blastocoel), and a cluster of cells (the inner cell mass) that fills part of the interior space.

blastomere A cell created by cleavage in early animal embryos.

blastopore An opening (pore) in the surface of some early embryos that cells move through during gastrulation.

blastula In animals, a ball of cells (blastomeres) formed by cleavage of a zygote and typically containing a fluid-filled cavity (the blastocoel). See **blastocyst**.

blood A type of connective tissue consisting of red blood cells and leukocytes suspended in a liquid extracellular matrix called plasma. Transports materials throughout the vertebrate body.

body mass index (BMI) A mathematical relationship used to assess obesity in humans. Calculated as body mass (in kg) divided by the square of height (in m^2).

body plan The basic architecture of an animal's body, including the number and arrangement of limbs, body segments, and major tissue layers.

Bohr shift The rightward shift of the oxygen–hemoglobin equilibrium curve that occurs with decreasing pH. It results in hemoglobin being more likely to release oxygen to tissues with low pH, such as exercising muscle.

bone A type of vertebrate connective tissue consisting of living cells and blood vessels within a hard, extracellular matrix composed of calcium phosphate ($CaPO_4$) and small amounts of calcium carbonate ($CaCO_3$) and protein fibers.

bone marrow The major blood-forming organ consisting of soft tissue filling the inside of large bones; contains stem cells that develop into red blood cells and white blood cells throughout life.

Bowman's capsule The hollow, double-walled, cup-shaped portion of a nephron that surrounds a glomerulus in the vertebrate kidney. Also called *glomerular capsule*.

brain A large mass of neurons, located in the head region of an animal, that is involved in information processing; may also be called the cerebral ganglion.

braincase Portion of the skull that encloses the brain. See **cranium**.

brainstem The most posterior portion of the vertebrate brain, connecting to the spinal cord; responsible for autonomic body functions such as regulating heart rate, respiration, and digestion.

branch (1) A part of a phylogenetic tree that represents populations through time. (2) A lateral extension of a plant's shoot system.

brassinosteroids A family of steroid hormones found in plants; they stimulate growth.

bronchiole Any of the small tubes in mammalian lungs that carry air from the bronchi to the alveoli.

bronchus (plural: *bronchi*) In mammals, one of a pair of large tubes that lead from the trachea to each lung.

budding A form of asexual reproduction in which an outgrowth from the parent breaks free as an independent individual; occurs in yeasts and some invertebrates and plants.

buffer A substance that, in solution, acts to minimize changes in the pH of that solution when acid or base is added.

bulbourethral gland In male mammals, one of a pair of small glands at the base of the urethra that secrete alkaline mucus (part of semen), which lubricates the tip of the penis and neutralizes acids in the urethra during copulation. In humans, also called *Cowper's gland*.

bulk flow The directional mass movement of a fluid due to pressure differences, such as movement of water through plant xylem and phloem, and movement of blood in animals.

bundle-sheath cell A type of cell found around the vascular tissue (veins) of plant leaves.

C3 pathway The most common form of photosynthesis in which atmospheric CO_2 is fixed by rubisco to form 3-phosphoglycerate, a three-carbon molecule. Used in first phase of the Calvin cycle.

C4 pathway A variant type of photosynthesis in which atmospheric CO_2 is first fixed by PEP carboxylase into four-carbon molecules, rather than the three-carbon molecules of the classic C_3 pathway. Used to concentrate CO_2 to reduce photorespiration in the Calvin cycle while stomata are closed to prevent water loss.

C_4 photosynthesis A variant type of photosynthesis in which atmospheric CO_2 is first fixed into four-carbon sugars, rather than the three-carbon sugars of classic C_3 photosynthesis. Enhances photosynthetic efficiency in hot, dry environments by reducing loss of oxygen due to photorespiration.

cadherin Any of a class of cell-surface proteins involved in selective cell—cell adhesion. Important for coordinating movements of cells and the establishment of tissues during embryological development.

callus In plants, a mass of undifferentiated cells that can generate roots and other tissues necessary to create a mature plant.

Calvin cycle In photosynthesis, the set of reactions that uses NADPH and ATP formed in the light-capturing reactions to drive the fixation of CO_2, reduction of the fixed carbon to produce sugar, and regeneration of the substrate used to fix CO_2.

calyx All of the sepals of a flower.

CAM See **crassulacean acid metabolism**.

cambium (plural: *cambia*) In woody plants, tissue that consists of two types of cylindrical meristems (also called *lateral meristems*) that increase the width of roots and shoots through the process of secondary growth. See **vascular cambium** and **cork cambium**.

Cambrian explosion The rapid diversification of animal body types and lineages that occurred during a 50-million-year period about 541 mya at the start of the Proterozoic eon.

cancer General term for a disease caused by cells that grow in an uncontrolled fashion, invade nearby tissues, and spread to other sites in the body.

capillary In a closed circulatory system, any of the many small, thin-walled blood vessels that permeate all tissues and organs and allow exchange of gases and other molecules between blood and body cells.

capillary action The tendency of water to move up a narrow tube due to adhesion, cohesion, and surface tension.

capillary bed A network of capillaries.

capsid A shell of protein enclosing the genome of a virus particle.

carapace In crustaceans, a large platelike section of the exoskeleton that covers and protects the cephalothorax (e.g., a crab's "shell").

carbohydrate Any of a class of molecules that contain a carbonyl group, several hydroxyl groups, and multiple carbon-hydrogen bonds. Synonymous with *sugar*. See **monosaccharide, disaccharide, oligosaccharide,** and **polysaccharide**.

carbon cycle, global The movement of carbon among abiotic and biotic reservoirs in terrestrial and aquatic ecosystems.

carbon fixation The process of converting gaseous carbon dioxide into an organic molecule; often associated with photosynthesis. See also **PEP carboxylase** and **rubisco**.

carbon sink A reservoir that stores carbon-containing compounds for an indefinite period of time.

carbonic anhydrase An enzyme that catalyzes the formation of carbonic acid (H_2CO_3) from carbon dioxide and water.

carboxylic acids Organic acids with one or more carboxyl groups (R-COOH).

cardiac cycle One complete heartbeat cycle, including systole and diastole.

cardiac muscle The muscle tissue of the vertebrate heart; responsible for pumping blood. Consists of long, branched fibers that are electrically connected and that initiate their own contractions; not under voluntary control. Compare with **skeletal muscle** and **smooth muscle**.

cardiovascular disease A group of diseases of the heart and blood vessels that may be caused by poor diet, obesity, inactivity, genetics, tobacco use, age, and other factors.

carnivore (adjective: *carnivorous*) An animal whose diet consists predominantly of meat, or other animals. Some plants are carnivorous, trapping and killing small animals and then absorbing nutrients from the prey's body. Compare with **herbivore** and **omnivore**.

carotenoid Any of a class of accessory pigments, found in chloroplasts, that absorb wavelengths of light not absorbed by chlorophyll; typically appear yellow, orange, or red. Includes carotenes and xanthophylls.

carpel The female reproductive organ in a flower. Consists of the stigma, to which pollen grains adhere; the style, through which the pollen tube grows; and the ovary, which houses the ovule. Compare with **stamen**.

carrier A heterozygous individual carrying a normal allele and a recessive allele for an inherited trait. The individual does not display the phenotype of the recessive trait but can pass the recessive allele to offspring.

carrier protein A transmembrane protein that facilitates diffusion of a small molecule (e.g., glucose) across a membrane by a process involving a reversible change in the shape of the protein. Also called *carrier* or *transporter*. Compare with **channel protein**.

carrying capacity (*K*) The maximum population size of a certain species that a given habitat can support over a sustained period of time.

cartilage A type of connective tissue that consists of relatively few cells scattered in a stiff matrix of polysaccharides and protein fibers. Provides structural support. Found mostly in vertebrates.

Cas protein A *C*RISPR-*as*sociated protein; an endonuclease that cuts DNA at sites complementary to an RNA (either crRNA in nature or sgRNA engineered for biotechnological applications) bound by the Cas protein. Cas proteins are a core element of the CRISPR-Cas system of genome editing.

Casparian strip In plant roots, a waxy layer containing suberin, a water-repellent substance that prevents movement of water through the walls of endodermal cells, thus blocking the apoplastic pathway of water and ion movement into the vascular tissue.

catabolic pathway Any set of chemical reactions that breaks down large, complex molecules into smaller ones, releasing energy in the process. Compare with **anabolic pathway**.

catabolite activator protein (CAP) A bacterial protein that activates the transcription of operons involved in the use of sugars other than glucose.

catalysis (verb: *catalyze*) Acceleration of the rate of a chemical reaction due to a decrease in the free energy of the transition state, called the activation energy.

catalyst Any substance that increases the rate of a chemical reaction without itself undergoing any permanent chemical change.

catecholamine Any of a class of small compounds, derived from the amino acid tyrosine, that are used as hormones or neurotransmitters. Catecholamines include epinephrine, norepinephrine, and dopamine.

cation A positively charged ion.

cation exchange The release (displacement) of cations, such as magnesium and calcium from soil particles, by protons in acidic soil water. The released cations are available for uptake by plants.

CD4 A protein on the plasma membrane of some human T cells. $CD4^+$ T cells can give rise to helper T cells.

CD8 A protein on the plasma membrane of some leukocytes in humans. $CD8^+$ T cells can give rise to cytotoxic T cells.

cecum A blind sac (having only one opening) between the small intestine and the colon. Enlarged in some species (e.g., rabbits) that use it as a fermentation chamber for digestion of cellulose.

cell A highly organized compartment bounded by a thin, flexible structure (plasma membrane) and containing concentrated chemicals in an aqueous (watery) solution. The basic structural and functional unit of all organisms.

cell body The part of a neuron that contains the nucleus; where incoming signals are integrated. Also called the *soma*.

cell crawling A form of cellular movement involving actin filaments in which the cell produces bulges in the membrane that stick to the substrate and are used to pull the cell forward.

cell cycle An ordered sequence of events in which a eukaryotic cell increases in size, replicates its chromosomes, evenly partitions the chromosomes to two daughter nuclei, and then undergoes division of the cytoplasm.

cell-cycle checkpoint Any of several points in the cell cycle at which progression of a cell through the cycle can be regulated.

cell division Formation of new cells by division of preexisting cells.

cell-mediated (immune) response The type of immune response that involves generation of cytotoxic T cells from $CD8^+$ T cells. Defends against pathogen-infected cells, cancer cells, and transplanted cells. Compare with **humoral (immune) response**.

cell membrane See **plasma membrane**.

cell plate A flattened, sac-like structure formed in the middle of a dividing plant cell from Golgi-derived vesicles containing cell-wall material; ultimately divides the cytoplasm into two separate cells.

cell theory The theory that all organisms are made of cells and that all cells come from preexisting cells.

cell wall A fibrous layer found outside the plasma membrane of most bacteria and archaea and many eukaryotes.

cellular respiration A common metabolic pathway for production of ATP, involving transfer of electrons from compounds with high potential energy through an electron transport chain and ultimately to a final electron acceptor (often oxygen).

cellulase An enzyme that digests cellulose by catalyzing hydrolysis of the glycosidic linkages between glucose residues.

cellulose A structural polysaccharide composed of glucose monomers joined by β-1,4-glycosidic linkages. Found in the cell walls of algae, plants, and some bacteria and fungi.

Cenozoic era The most recent era of geologic time, beginning 66 million years ago, during which mammals and birds became the dominant vertebrates on land and angiosperms became the dominant plants on land.

central dogma The scheme for information flow in the cell: DNA → RNA → protein.

central nervous system (CNS) An aggregation of large numbers of neurons into clusters called ganglia in bilaterian animals. In vertebrates, the CNS consists of the brain and spinal cord. Compare with **nerve net** and **peripheral nervous system (PNS)**.

centriole One of two small cylindrical structures contained within the centrosome near the nucleus of a eukaryotic cell (found in animals but not in most plants). Consists of microtubule triplets and is structurally identical to a basal body.

centromere The region of a replicated chromosome where the two sister chromatids are joined most tightly and the kinetochore is formed during M phase.

centrosome A structure that serves as a microtubule-organizing center for the cell's cytoskeleton and for the spindle apparatus during cell division. Includes centrioles in animal cells and certain plants and fungi.

cephalization The formation in animals of a distinct anterior region (the head) where sense organs and a mouth are clustered.

cephalochordates One of the three major chordate lineages (Cephalochordata), comprising small, mobile organisms that live in marine sands and suspension feed; also called *lancelets* or *amphioxus*. Compare with **urochordates** and **vertebrates**.

cephalopods A lineage of mollusks including squid, octopuses, and cuttlefish. Distinguished by large brains, excellent vision, arms and tentacles, and (in living species) a reduced or absent shell.

cephalothorax In some arthropods, the tagma (body region) that is formed by the fusion of the head and thorax; anterior to the abdomen.

cerebellum A subdivision of the vertebrate hindbrain; involved in coordination of complex muscle movements, such as those required for locomotion and maintaining balance.

cerebrum A subdivision of the vertebrate forebrain responsible for integration of sensory information and coordination of motor responses; in mammals, the cerebrum is divided into left and right hemispheres and four lobes. See **frontal lobe, occipital lobe, parietal lobe**, and **temporal lobe**.

cervix The bottom portion of the uterus, containing an opening that leads to the vagina.

chaetae (singular: *chaeta*) Bristle-like extensions found in some annelids.

channel protein A transmembrane protein that forms a pore in a cell membrane, which may open or close in response to a signal. The structure of most channels allows them to admit just one or a few types of ions or molecules. Compare with **carrier protein**.

character Any heritable genetic, morphological, physiological, developmental, or behavioral characteristic of an organism to be studied. Also called a trait.

character displacement The evolutionary tendency for the traits of similar species that occupy overlapping ranges to change in a way that reduces interspecific competition.

chelicerae A pair of clawlike appendages found near the mouth of certain arthropods called chelicerates (spiders, mites, and relatives).

chelicerate A lineage of mostly terrestrial arthropods that include spiders, scorpions, ticks, mites, daddy longlegs, and horseshoe crabs; named for their clawlike appendages called chelicerae.

chemical bond An attractive force binding two atoms together. Covalent bonds, ionic bonds, and hydrogen bonds are types of chemical bonds.

chemical energy The potential energy stored in chemical bonds between atoms.

chemical equilibrium A dynamic but stable state of a reversible chemical reaction in which the forward reaction and reverse reactions proceed at the same rate, so that the concentrations of reactants and products remain constant.

chemical evolution The theory that simple chemical compounds in the early atmosphere and ocean combined via chemical reactions to form larger, more complex substances, eventually leading to the origin of life and the start of biological evolution.

chemical reaction Any process in which substances combine or are broken down into other substances; involves the making and/or breaking of chemical bonds.

chemiosmosis An energetic coupling mechanism whereby energy stored in an electrochemical proton gradient is used to drive an energy-requiring process such as production of ATP.

chemokine Any of a subset of cytokines that acts as a chemical signal attracting white blood cells to a site of tissue injury or infection.

chemolithotroph An organism (bacteria or archaea) that produces ATP by oxidizing inorganic molecules with high potential energy, such as ammonia (NH_3) or methane (CH_4). Also called *lithotroph*. Compare with **chemoorganotroph**.

chemoorganotroph An organism that produces ATP by oxidizing organic molecules with high potential energy such as sugars. Also called *organotroph*. Compare with **chemolithotroph**.

chemoreception The detection of specific chemicals in an animal's internal or external environment.

chemoreceptor A sensory cell specialized for detecting specific chemicals or classes of chemicals.

chiasma (plural: *chiasmata*) The X-shaped structure formed during meiosis by crossing over between non-sister chromatids in a pair of homologous chromosomes.

chitin A structural polysaccharide composed of *N*-acetyl-glucosamine (NAG) monomers joined end to end by β-1,4-glycosidic linkages. Found in the cell walls of fungi and many algae, and in external skeletons of insects and crustaceans.

chitons A lineage of marine mollusks that have a protective shell formed of eight calcium carbonate plates.

chlorophyll Any of several closely related green pigments, found in chloroplasts, that absorb light during photosynthesis.

chloroplast A chlorophyll-containing organelle, bounded by a double membrane, in which photosynthesis occurs; found in plants and algae. Also the location of starch, amino acid, fatty acid, purine, and pyrimidine synthesis.

choanocyte A specialized, flagellated feeding cell found in sponges (phylum Porifera).

cholecystokinin A peptide hormone secreted by cells in the lining of the small intestine. Stimulates the secretion of digestive enzymes from the pancreas and the release of bile by the gallbladder.

chordate Any member of the phylum Chordata. Chordates are deuterostomes distinguished by four morphological features present at some stage in their life cycles: a dorsal hollow nerve cord, pharyngeal slits or pouches, a notochord, and a post-anal tail. Includes vertebrates, cephalochordata, and urochordata.

chromatid One of the two identical double-stranded DNAs that are connected at the centromere and compose a replicated chromosome.

chromatin The complex of DNA and proteins, mainly histones, that compose eukaryotic chromosomes. Can be highly compact (heterochromatin) or loosely coiled (euchromatin).

chromatin remodeling The process by which the structure of chromatin is changed to allow or inhibit transcription. May involve chemical modification of histone proteins or reshaping of the chromatin by large multiprotein complexes in an ATP-requiring process.

chromatin-remodeling complex A group of proteins that use energy from ATP hydrolysis to shift nucleosomes on DNA so as to expose or hide DNA sequences from transcription factors.

chromosome Gene-carrying structure consisting of a single long molecule of double-stranded DNA and associated proteins (e.g., histones). Most prokaryotic cells contain a single circular chromosome; eukaryotic cells contain multiple noncircular (linear) chromosomes located in the nucleus.

chromosome-level mutation Any change in chromosome number, such as the loss of a chromosome (aneuploidy) or the gain of a set of chromosomes (polyploidy), or the change in the composition of individual chromosomes as a result of inversions, translocations, deletions, or duplications during cell division. Compare with **point mutation** and **lateral gene transfer**.

chromosome theory of inheritance The principle that genes are located on chromosomes and that patterns of inheritance are determined by the behavior of chromosomes during meiosis.

cilium (plural: *cilia*) One of many short, filamentous projections of some eukaryotic cells, containing a core of microtubules. Used to move the cell as well as to circulate fluid or particles around the surface of a stationary cell. See **axoneme**.

circadian clock An internal mechanism found in most organisms that regulates many body processes (sleep—wake cycles, hormonal patterns, etc.) in a roughly 24-hour cycle.

circulatory system The system responsible for moving oxygen, carbon dioxide, hormones, nutrients, wastes, and other materials within the animal body.

cisternae (singular: *cisterna*) Flattened, membrane-bound compartments that make up the Golgi apparatus.

citric acid cycle A series of eight chemical reactions that start with citrate (deprotonated citric acid) and ends with oxaloacetate. The cycle is formed when oxaloacetate reacts with acetyl CoA to form citrate as part of the pathway that oxidizes glucose to CO_2. Also known as the *Krebs cycle* or *tricarboxylic acid (TCA) cycle*.

clade See **monophyletic group**.

class I MHC protein A major histocompatibility protein that is present on the plasma membrane of virtually all nucleated cells of vertebrates. Functions in presenting epitopes of antigens to $CD8^+$ T cells.

class II MHC protein A major histocompatibility protein that is present on the plasma membrane of only certain cells involved in the immune response of vertebrates, including dendritic cells, macrophages, and B cells. Functions in presenting epitopes of antigens to $CD4^+$ T cells.

cleavage In animals, the series of rapid mitotic divisions, with little cell growth, that produces successively smaller cells (blastomeres) and transforms a zygote into a multicellular blastula.

cleavage furrow An indentation in the cell surface that occurs as the plasma membrane is pulled inward during cytokinesis in animal cells. The furrow deepens until the membrane fuses, dividing the cytoplasm into two daughter cells.

climate The prevailing, long-term weather conditions in a particular region.

climax community The stable, final community that develops from ecological succession.

clitoris A rod of erectile tissue in the external genitalia of female mammals. Is formed from the same embryonic tissue as the male penis and has a similar function in sexual arousal.

cloaca In a few mammals and many nonmammalian vertebrates, a body cavity into which the urinary, gastrointestinal, and reproductive tracts all empty.

clonal expansion The production of daughter lymphocytes that all arise originally from a single cell and that share the same antigen specificity.

clonal selection theory The dominant explanation for the generation of an adaptive immune response in vertebrates. According to the theory, the immune system retains a vast pool of inactive lymphocytes, each with a unique receptor for a unique epitope. Lymphocytes that encounter their complementary epitopes are stimulated to divide (selected and cloned), producing daughter cells that combat infection and confer immunity.

clone (1) An individual that is genetically identical to another individual. (2) A lineage of genetically identical individuals or cells. (3) As a verb, to make one or more genetic replicas of a cell or individual.

cloning vector A plasmid or other agent such as a virus used to transfer recombinant genes into cells. Also called a *vector*.

closed circulatory system A circulatory system in which the circulating fluid (blood) is confined to blood vessels and flows in a continuous circuit; found in vertebrates. Compare with **open circulatory system**.

cnidocyte A specialized stinging cell found in cnidarians (e.g., jellyfish, corals, and anemones) and used in capturing prey.

cochlea The organ of hearing in the inner ear of mammals, birds, and crocodilians. A coiled, fluid-filled tube containing specialized pressure-sensing cells (hair cells) that detect sounds of different pitches.

coding strand See **non-template strand**.

codominance An inheritance pattern in which heterozygotes exhibit both of the traits seen in each type of homozygous individual.

codon A sequence of three nucleotides in DNA or RNA that codes for an amino acid or a stop signal for protein synthesis.

coefficient of relatedness (*r*) A measure of how closely two individuals are related. Calculated as the probability that an allele in two individuals is inherited from the same ancestor.

coelom An internal, usually fluid-filled body cavity that is completely or partially lined with mesoderm.

coelomate An animal that has a true coelom, completely lined with mesoderm. Compare with **acoelomate** and **pseudocoelomate**.

coenocytic Containing many nuclei and a continuous cytoplasm through a filamentous body, without the body being divided into distinct cells. Some fungi are coenocytic.

coenzyme A small organic molecule that is required for an enzyme-catalyzed reaction. Often donates or receives electrons or functional groups during the reaction.

coenzyme A (CoA) A molecule that is required for many cellular reactions and that is often transiently linked to acetyl groups.

coenzyme Q A nonprotein molecule that shuttles electrons between membrane-bound complexes in the mitochondrial electron transport chain. Also called **ubiquinone** or *Q*.

coevolution A pattern of evolution in which two interacting species reciprocally influence each other's adaptations over time.

coevolutionary arms race A series of adaptations and counter-adaptations observed in species that interact closely over time and affect each other's fitness.

cofactor An inorganic ion, such as a metal ion, that is required for an enzyme to function normally. May be bound tightly to an enzyme or associate with it transiently during catalysis.

cohesion The tendency of certain like molecules (e.g., water molecules) to cling together due to attractive forces. Compare with **adhesion**.

cohesion-tension theory The theory that water movement upward through plant vascular tissues is due to loss of water from leaves (transpiration), which pulls a cohesive column of water upward.

cohort A group of individuals that are the same age and can be followed through time.

coleoptile A modified leaf that covers and protects the stems and leaves of grass seedlings.

collagen A fibrous, pliable glycoprotein that is a major component of the extracellular matrix of animal cells. Various subtypes differ in their tissue distribution, some of which are assembled into large fibrils and even larger cable-like fibers in the extracellular space.

collecting duct In the vertebrate kidney, a large, straight tube that receives filtrate from the distal tubules of several nephrons. Involved in the regulated reabsorption of water.

collenchyma Can refer to tissue or cell type. In plants, an elongated cell with cell walls thickened at the corners that provides support to growing plant parts; usually found as a simple tissue in strands along leaf veins and stalks. Compare with **parenchyma** and **sclerenchyma**.

colon The portion of the large intestine where feces are formed by compaction of wastes and reabsorption of water.

colony An assemblage of individuals. May refer to an assemblage of semi-independent cells or to a breeding population of multicellular organisms.

commensalism (adjective: *commensal*) A species relationship in which one organism (the commensal) benefits and the other (the host) is unaffected. Compare with **mutualism, parasitism, consumption**, and **competition**.

commitment The gradual process by which the developmental fate of an embryonic cell is specified, directing it to become a particular type of specialized cell. It is reversible in early stages.

communication In ecology, any process in which a signal from one individual modifies the behavior of another individual.

community All the populations of different species that interact with each other in a certain area.

companion cell In plants, a cell in the phloem that is connected via many plasmodesmata to adjacent sieve-tube elements. Companion cells provide materials to maintain sieve-tube elements and function in the loading and unloading of sugars into sieve-tube elements.

compass orientation A type of navigation in which movement occurs in a specific direction.

competition In ecology, the interaction of two species or two individuals trying to use the same limited resource (e.g., water, food, living space). May occur between individuals of the same species (intraspecific competition) or different species (interspecific competition). Compare with **mutualism, consumption**, and **commensalism**.

competitive exclusion principle The principle that two species cannot coexist in the same ecological niche in the same area because one species will outcompete the other.

competitive inhibition Inhibition of an enzyme's ability to catalyze a chemical reaction via a nonreactant molecule that competes with the substrate(s) for access to the active site.

complement system A set of proteins that circulate in the bloodstream and can destroy foreign cells by forming holes in their plasma membrane.

complementary base pairing The association between specific nitrogenous bases of nucleic acids stabilized by hydrogen bonding. Adenine pairs with thymine (in DNA) or uracil (in RNA), and guanine pairs with cytosine.

complementary DNA (cDNA) DNA produced either in the lab or by retroviruses, using an RNA transcript as a template and reverse transcriptase; corresponds to the transcribed region of a gene but lacks introns.

complementary strand A strand of RNA or DNA with a base sequence that forms via complementary base-pairing with the template strand.

complete digestive tract A digestive tract with two openings: a mouth and an anus.

complete dominance The form of dominance in which the phenotype of a heterozygote is indistinguishable from the phenotype of one of the homozygotes.

complete metamorphosis A type of metamorphosis in which the animal completely changes its form; includes a distinct larval stage.

complex tissue A tissue consisting of two or more cell types.

compound Any substance that consists of more than one different type of element chemically bonded together.

compound eye An eye formed of many independent lenses, each associated with a light-sensing columnar structure (ommatidium); occurs in arthropods. Compare with **simple eye**.

compound leaf A leaf consisting of two or more blades but defined by the presence of a single axillary bud where the petiole joins the branch or stem.

concentration gradient Difference across space (e.g., across a membrane) in the concentration of a dissolved substance.

condensation reaction A chemical reaction in which two molecules are joined covalently with the removal of an —OH from one and an —H from another to form water. In biology, most condensation reactions involve the joining of monomers into polymers. Also called a *dehydration reaction*. Compare with **hydrolysis**.

conduction (1) Direct transfer of heat between two objects that are in physical contact. Compare with **convection**. (2) Transmission of an electrical impulse along the axon of a nerve cell.

cone A vertebrate photoreceptor with a cone-shaped outer portion that is particularly sensitive to bright light of a certain color. Compare with **rod**.

conform To allow internal physiological conditions to vary with environmental conditions. Compare with **regulate**.

conidia Asexual spores produced by many fungi. Formed as a result of mitotic divisions; therefore, genetically identical to the parent fungus.

conjugation The transfer of genetic material between bacterial cells by direct cell-to-cell contact via formation of a bridge-like connection structure.

connective tissue An animal tissue consisting of loosely arranged cells in a liquid, jellylike, or solid extracellular matrix. Includes bone, cartilage, tendons, ligaments, and blood.

conservation biology The effort to study, preserve, and restore genetic diversity, species diversity, and diversity of ecosystem function in populations, communities, and ecosystems.

constant (*C*) region The invariant amino acid sequence in the polypeptides that comprise antibodies, B-cell receptors, and T-cell receptors. Compare with **variable (*V*) region**.

constitutive Always occurring or always present. Commonly used to describe proteins that are synthesized continuously or mutants in which one or more genetic loci are constantly expressed due to defects in gene control.

constitutive defense A defensive trait that is manifested even in the absence of a consumer (predator, herbivore, or parasite). Also called *standing defense*. Compare with **inducible defense**.

constitutive mutant An abnormal (mutated) strain that produces a product at all times, instead of under certain conditions.

consumer See **heterotroph**.

consumption In ecology, the interaction between species in which one organism eats or absorbs nutrients from another. Includes predation, herbivory, and parasitism. Compare with **mutualism, commensalism**, and **competition**.

continental shelf The portion of a geologic plate that extends under seawater from a continent.

continuous strand See **leading strand**.

contraception Any of several methods used to prevent pregnancy.

control In a scientific experiment, a group of organisms or samples that do not receive the experimental treatment but are otherwise identical to the group that does.

convection Transfer of heat by movement of large volumes of a gas or liquid. Compare with **conduction**.

convergent evolution The independent evolution of similar traits in different species due to adaptation to similar environmental conditions and a similar way of life.

cooperative binding The tendency of the protein subunits of hemoglobin to affect each other's oxygen binding such that each bound oxygen molecule increases the likelihood of further oxygen binding.

coprophagy The eating of feces.

copulation The act of sexual intercourse.

coral reef A large assemblage of colonial marine corals (Phylum Cnidaria) that usually serves as shallow-water, sunlit habitat for many other species as well.

core enzyme A general term for the enzyme within a multipart holoenzyme that is responsible for catalysis.

core promoter A DNA sequence in eukaryotes that binds general transcription factors to enable RNA polymerase to begin transcription. This sequence is sometimes simply called a promoter, although "core promoter" distinguishes it from other DNA sequences important for regulating transcription initiation.

co-receptor Any membrane protein that acts with some other membrane protein in a cell interaction or cell response.

co-repressor A molecule that binds to a bacterial repressor protein to induce an allosteric (shape) change in the repressor, thus allowing the repressor to bind to its operator and exert negative control of transcription.

Coriolis effect The tendency for moving air or water to be deflected from a straight path, swerving in a clockwise pattern in the Northern Hemisphere and in a counterclockwise direction in the Southern Hemisphere. Caused by the spinning of Earth on its axis.

cork The main tissue associated with tree bark; produced by, and exterior to, the cork cambium.

cork cambium One of two types of cylindrical meristems, consisting of a ring of undifferentiated plant cells in the stem and root of woody plants; produces **cork**. Compare with **vascular cambium**.

cork cell A cell in the protective outermost layer of a woody stem and root that produces and accumulates waxes that make the cell less permeable to water and gases.

corm A rounded, thick underground stem that can produce new plants via asexual reproduction.

cornea The transparent sheet of connective tissue at the very front of the eye in vertebrates and some other animals. Protects the eye and helps focus light.

corolla All of the petals of a flower.

corpus callosum A thick band of axons that connects the two hemispheres of the cerebrum in the mammalian brain.

corpus luteum A yellowish structure that secretes progesterone in an ovary. Formed from a follicle that has recently ovulated.

cortex (1) In animals, the outermost region of an organ, such as the kidney or adrenal gland. Compare with **medulla**. (2) In plants, a layer of ground tissue found outside the vascular bundles of roots and outside the pith of a stem.

corticotropin-releasing hormone (CRH) A peptide hormone, produced and secreted by the hypothalamus, that stimulates the anterior pituitary to release ACTH.

cortisol A steroid hormone, produced and secreted by the adrenal cortex, that increases blood glucose concentration and mediates the body's long-term stress response; the major glucocorticoid hormone in some mammals. Also called *hydrocortisone*.

cost–benefit analysis Decisions or analyses that weigh the fitness costs and benefits of a particular action.

cotransporter A transmembrane protein that facilitates diffusion of an ion down its previously established electrochemical gradient and uses the energy of that process to transport some other substance, in the same or opposite direction, *against* its concentration gradient. Also called *secondary active transporter*. See **antiporter** and **symporter**.

cotyledon The first leaf, or seed leaf, of a plant embryo. Used for storing and digesting nutrients and/or for early photosynthesis.

countercurrent exchanger In animals, any anatomical arrangement that allows the maximum transfer of heat or a soluble substance from one fluid to another. The two fluids must be flowing in opposite directions and have a heat or concentration gradient between them.

covalent bond A type of chemical bond in which two atoms share one or more pairs of electrons. Compare with **hydrogen bond** and **ionic bond**.

cranium A bony, cartilaginous, or fibrous case that encloses and protects the brain of vertebrates. Forms part of the skull. Also called *braincase*.

crassulacean acid metabolism (CAM) A variant type of photosynthesis that is related to the C_4 pathway in which CO_2 is fixed and stored in four-carbon organic acids. To reduce water and CO_2 loss by photorespiration, CO_2 is fixed at night when stomata are open and then released to feed the Calvin cycle during the day when stomata are closed.

CRISPR Acronym for a locus in bacteria or archaea—*C*lustered *R*egularly *I*nterspaced *S*hort *P*alindromic *R*epeats—that is comprised of repeat sequences separated by spacer sequences derived from viruses that have infected the cell or its ancestors. The CRISPR locus produces the crRNAs that, with a Cas protein, are the basis for current genome editing methods.

CRISPR-Cas system The current way to edit genomes based on an RNA (sgRNA) associated with a Cas protein that cuts DNA at sites complementary to the sgRNA.

cristae (singular: *crista*) Sac-like invaginations of the inner membrane of a mitochondrion. Location of the electron transport chain and ATP synthase.

critical night length The length of darkness that must be exceeded to initiate flowering in short-day plants or not exceeded to initiate flowering in long-day plants.

Cro-Magnons A prehistoric European population of modern humans (*Homo sapiens*) known from fossils, paintings, sculptures, and other artifacts.

crop In animal anatomy, a storage organ in the digestive tract of certain vertebrates.

cross A mating between two individuals that is used for genetic analysis.

cross-pollination Pollination of a flower by pollen from another individual, rather than by self-fertilization. Also called *crossing*.

crossing over The exchange of corresponding segments of non-sister chromatids between a pair of homologous chromosomes during meiosis I.

crosstalk Interactions among signaling pathways, triggered by different signals, that modify a cellular response.

crustaceans A lineage of arthropods that includes shrimp, lobsters, and crabs. Many have a carapace (a platelike portion of the exoskeleton covering the cephalothorax) and mandibles for biting or chewing.

cryptic species A species that cannot be distinguished from similar species by easily identifiable morphological traits.

culture In cell biology, a collection of cells or a tissue growing under controlled conditions, usually in suspension or on the surface of a dish of solid growth medium.

Cushing's disease A human endocrine disorder caused by loss of feedback inhibition by cortisol on ACTH secretion. Characterized by high ACTH and cortisol levels and wasting of body protein reserves.

cuticle A protective coating secreted by the outermost layer of cells of an animal or a plant; often functions to reduce evaporative water loss.

cyanobacteria A lineage of photosynthetic bacteria formerly known as blue-green algae. Likely the first life-forms to carry out oxygenic photosynthesis

cyclic AMP (cAMP) Cyclic adenosine monophosphate; a small molecule, derived from ATP, that is widely used by cells in signal transduction and transcriptional control.

cyclic electron flow Path of electrons in which excited electrons of photosystem I are transferred back to plastoquinone (PQ), the start of the electron transport chain normally associated with photosystem II. Instead of reducing $NADP^+$ to make NADPH, the electron energy is used to make ATP via photophosphorylation. Compare with **noncyclic electron flow**.

cyclin One of several regulatory proteins whose concentrations fluctuate cyclically throughout the cell cycle. Involved in control of the cell cycle via cyclin-dependent kinases.

cyclin-dependent kinase (Cdk) Any of several related protein kinases that are functional only when bound to a cyclin and are activated by other modifications. Involved in control of the cell cycle.

cytochrome *c* (cyt *c*) A soluble protein in the intermembrane space of mitochondria that shuttles electrons between membrane-bound complexes in the electron transport chain.

cytokine Any of a diverse group of signaling proteins, secreted largely by cells of the immune system, whose effects include increased white blood cell production, recruitment of cells to the site of an infection, tissue repair, and fever. Generally function to regulate the type, intensity, and duration of an immune response.

cytokinesis Division of the cytoplasm to form two daughter cells. In eukaryotic cells, typically occurs immediately after division of the nucleus by mitosis or meiosis.

cytokinins A class of plant hormones that stimulate cell division and retard aging.

cytoplasm All the contents of a cell bounded by the plasma membrane, excluding the nucleus (if present).

cytoplasmic determinant A regulatory molecule affecting development; it is distributed unevenly in a cell (typically an egg); therefore, if the plane of cleavage is controlled, the cytoplasmic determinant is transmitted to one daughter cell but not another, specifying the fate of the cell that obtains the determinant.

cytoplasmic streaming The directed flow of cytosol and organelles that facilitates distribution of materials within some large plant and fungal cells. Occurs along actin filaments and is powered by myosin.

cytoskeleton In eukaryotic cells, a network of protein fibers in the cytoplasm that are involved in cell shape, support, locomotion, and transport of materials within the cell. Prokaryotic cells have a similar but much less extensive network of fibers.

cytosol The fluid portion of the cytoplasm, excluding the contents of membrane-enclosed organelles.

cytotoxic T cell A type of $CD8^+$ effector T cell that induces apoptosis in infected and cancerous cells. Recognizes target cells via interactions with complementary class I MHC–peptide complexes. Also called *cytotoxic T lymphocyte (CTL)* and *killer T cell*. Compare with **helper T cell**.

dalton (Da) A unit of mass approximately equal to the mass of one proton or one neutron.

daughter strand The strand of DNA that is newly replicated from a template, or parental, strand of DNA.

day-neutral plant A plant whose flowering time is not affected by the relative length of day and night (the photoperiod). Compare with **long-day plant** and **short-day plant**.

dead space Air passages that are not involved in gas exchange with the blood; examples are the trachea and bronchi.

decomposer See **detritivore**.

decomposer food chain An ecological network of detritus, decomposers that eat detritus, and predators and parasites of the decomposers.

de-differentiation The process of reversing the normal course of cell specialization (differentiation). Can be induced to create induced pluripotent stem cells (*iPS cells*).

deep sequencing A method to learn the types of mRNAs or DNA sequences present in cells and their relative amounts. Involves preparing and extensively sequencing cDNAs prepared from the cell or tissue under investigation. Also known as *RNA-seq*.

dehydration reaction See **condensation reaction**.

deleterious In genetics, referring to any mutation, allele, or trait that reduces an individual's fitness. Compare with **beneficial** and **neutral**.

deletion In genetics, refers to a loss in a DNA sequence or chromosome.

demography The study of factors that determine the size and structure of populations through time.

denaturation (verb: *denature*) For a macromolecule, loss of its three-dimensional structure due to breakage of covalent and/or noncovalent chemical bonds, usually caused by exposure to heat, certain chemicals, or extreme pH conditions.

dendrite A short extension from a neuron's cell body that receives signals from other neurons.

dendritic cell A type of white blood cell that ingests and digests foreign antigens, moves to a lymph node, and presents the antigens' epitopes to $CD4^+$ and $CD8^+$ T cells in the form of MHC–peptide complexes on its plasma membrane.

dendrochronology The dating and study of annual growth rings in trees.

dense connective tissue A type of connective tissue distinguished by having an extracellular matrix dominated by collagen fibers. Found in tendons and ligaments.

density dependent In population ecology, referring to any characteristic that varies depending on population density.

density independent In population ecology, referring to any characteristic that does not vary with population density.

deoxyribonucleic acid (DNA) A nucleic acid composed of deoxyribonucleotides that carries the genetic information. Generally occurs as a double helix with two intertwined strands held together by noncovalent bonds. See also **double helix**.

deoxyribonucleoside triphosphate (dNTP) The monomer used by DNA polymerase to polymerize DNA. Consists of the sugar deoxyribose, a base (A, T, G, or C), and three phosphate groups.

deoxyribonucleotide A nucleotide consisting of a deoxyribose sugar, one or more phosphates, and one of four nitrogen-containing bases: adenine, guanine, cytosine, or thymine.

depolarization A change in membrane potential from its resting negative state to a less negative or a positive state; a normal phase in an action potential. Compare with **repolarization** and **hyperpolarization**.

deposit feeder An animal that eats organic matter in sediments.

derived trait A trait that is a modified form of an ancestral trait, found in a descendant.

dermal tissue system The tissue forming the outer layer of a plant; also called **epidermis**.

descent with modification The phrase used by Darwin to describe how species that lived in the past are the ancestors of species existing today, and that species change through time. See **evolution**.

desmosome A type of cell–cell attachment structure in animals, consisting of cadherins and other proteins anchored to intermediate filaments. Serves to link the cytoskeletons of adjacent cells and form strong intercellular attachments throughout a tissue. Compare with **gap junction** and **tight junction**.

determined The irrevocable (under normal conditions) commitment of an embryonic cell to a particular pathway of differentiation.

detritivore An organism whose diet consists mainly of dead organic matter (detritus). Various bacteria, fungi, protists, and animals are detritivores. Also called *decomposer*.

detritus A layer of dead organic matter that accumulates at ground level or on seafloors and lake bottoms.

deuterostomes A major lineage of bilaterian animals that includes echinoderms and chordates; named for the embryonic development of the anus before the mouth (literally, "second mouth"). Sister group to **protostomes**.

development The processes of forming a multicellular organism with specialized cells, tissues, and organs.

developmental biology The area of biology that considers the mechanisms and patterns of how organisms develop from a single cell into a multicellular adult.

developmental homology A similarity in embryonic form or developmental processes that is due to inheritance from a common ancestor.

diabetes mellitus A disease caused by a defect in insulin production (type 1) or in the response of cells to insulin (type 2). Characterized by abnormally high blood glucose levels and large volumes of glucose-containing urine.

diaphragm An elastic, sheetlike structure. In mammals, the muscular sheet that separates the chest and abdominal cavities. It contracts and moves downward during inhalation, expanding the chest cavity.

diastole The portion of the cardiac cycle during which the atria or ventricles of the heart are relaxed. Compare with **systole**.

diastolic blood pressure The force exerted by blood against artery walls during relaxation of the heart's left ventricle. Compare with **systolic blood pressure**.

dicot Any flowering plant (angiosperm) that has two cotyledons (embryonic leaves) upon germination. The dicots do not form a monophyletic group. Also called *dicotyledonous plant*. Compare with **eudicot** and **monocot**.

dideoxy sequencing A technique for determining the nucleotide sequence of DNA. Relies on the use of dideoxynucleoside triphosphates (ddNTPs), which terminate DNA replication.

diencephalon The part of the mammalian brain that relays sensory information to the cerebrum and functions in maintaining homeostasis.

differential centrifugation Procedure for separating cellular components according to their size and density by spinning a cell homogenate in a series of centrifuge runs. After each run, the supernatant is removed from the deposited material (pellet) and spun again at progressively higher speeds.

differential gene expression Expression of different sets of genes in cells with the same genome. Responsible for creating different cell types.

differentiation The process by which an unspecialized cell becomes a distinct specialized cell type (e.g., liver cell, brain cell), usually by changes in gene expression. Also called *cell differentiation*.

diffusion Spontaneous movement of a substance from one region to another, often with a net movement from a region of high concentration to one of low concentration (i.e., down a concentration gradient).

digestion The mechanical and chemical breakdown of food into molecules that can be absorbed into the body of an animal.

digestive tract In animals, a chamber or tube where digestion takes place. Also called *alimentary canal* or *gastrointestinal (GI) tract*.

dihybrid cross A mating between two homozygous parents that differ in alleles of two different genes. Produces dihybrid offspring. A dihybrid cross is usually continued through the F_2 generation.

dikaryotic Describing a fungal mycelium or hypha made up of cells containing two genetically distinct haploid nuclei.

dimer An association of two molecules that may be identical (homodimer) or different (heterodimer).

dioecious Describing an angiosperm species that has male and female reproductive structures on separate plants. Compare with **monoecious**.

diploblast (adjective: *diploblastic*) An animal whose body develops from two basic embryonic cell layers or tissues—ectoderm and endoderm. Compare with **triploblast**.

diploid (1) Having two sets of chromosomes ($2n$). (2) A cell or an organism with two sets of chromosomes, one set inherited from the mother and one set from the father. Compare with **haploid**.

direct sequencing A technique for identifying and studying microorganisms that cannot be grown in culture. Involves detecting and amplifying copies of specific genes in the microorganisms' DNA, sequencing these genes, and then comparing the sequences with the known sequences from other organisms.

directed-pollination hypothesis The hypothesis that flowers are adaptations that attract specific pollinators, increasing the likelihood that pollination will occur.

directional selection A mode of natural selection that favors one extreme phenotype with the result that the average phenotype of a population changes in one direction. Generally reduces overall genetic variation in a population. Compare with **disruptive selection** and **stabilizing selection**.

disaccharide A carbohydrate consisting of two monosaccharide sugar residues linked together.

discontinuous strand See **lagging strand**.

discrete trait An inherited trait that exhibits distinct phenotypes rather than the continuous variation of a quantitative trait such as body height.

dispersal The movement of individuals from their place of origin to the location where they live and breed as adults.

disruptive selection A mode of natural selection that favors extreme phenotypes at both ends of the range of phenotypic variation. Increases overall genetic variation in a population. Compare with **stabilizing selection** and **directional selection**.

distal tubule In the vertebrate kidney, the convoluted portion of a nephron into which filtrate moves from the loop of Henle. Involved in the regulated reabsorption of sodium ions and water. Compare with **proximal tubule**.

disturbance In ecology, any strong, short-lived disruption to a community that changes the distribution of living and/or nonliving resources.

disturbance regime The characteristic disturbances that affect a given ecological community.

disulfide bond A covalent bond between two sulfur atoms, typically in the side chains of certain amino acids (e.g., cysteine). Often contributes to tertiary and quaternary levels of protein structure.

DNA barcoding The use of well-characterized gene sequences to identify species.

DNA cloning Any of several techniques for producing many identical copies of a particular gene or other DNA sequence.

DNA fingerprinting Any of several methods for identifying individuals by unique features of their genomes. Commonly involves using PCR to produce many copies of short tandem repeats (microsatellites) and then analyzing their lengths. Also called *DNA profiling, DNA typing*.

DNA helicase An enzyme that breaks hydrogen bonds between nucleotides of DNA, "unzipping" a double-stranded DNA molecule.

DNA ligase An enzyme that joins pieces of DNA by catalyzing the formation of a phosphodiester linkage between the pieces.

DNA methylation The addition of a methyl group ($-CH_3$) to a DNA molecule.

DNA methyltransferase A class of enzymes that add a methyl group to bases in DNA. In many eukaryotes, methylation of cytosine is especially important. Methylation of DNA leads to chromatin condensation and is an important means of regulating gene expression in eukaryotes.

DNA microarray A set of single-stranded DNA fragments, representing thousands of different genes that are attached to a surface such as a glass slide. Can be used to determine which genes are expressed in different cell types, under different conditions, or at different developmental stages.

DNA polymerase Any enzyme that catalyzes synthesis of DNA.

DNA profiling See **DNA fingerprinting**.

DNA typing See **DNA fingerprinting**.

domain (1) A taxonomic category, based on similarities in basic cellular biochemistry, above the kingdom level. The three recognized domains are Bacteria, Archaea, and Eukarya. (2) A section of a protein that has a distinctive tertiary structure and function.

dominant Referring to a trait or allele whose phenotypic effect is observed when it is present in homozygous or heterozygous form. Dominant typically is used to mean **complete dominance**. Compare with **recessive**.

dormancy A temporary state of greatly reduced metabolic activity and growth in plants or plant parts (e.g., seeds, spores, bulbs, and buds).

dorsal Toward an animal's back and away from its belly. The opposite of **ventral**.

dorsal hollow nerve cord A bundle of nerves running the length of the body. A characteristic feature of chordates.

double fertilization A form of reproduction seen in flowering plants, in which one sperm cell fuses with an egg to form a zygote and the other sperm cell fuses with two polar nuclei to form the triploid endosperm.

double helix The secondary structure of DNA, consisting of two antiparallel DNA strands wound around each other. Some RNAs may also form a double helix in stem-and-loop secondary structures.

Down syndrome A human developmental disorder caused by trisomy (3 copies) of chromosome 21.

downstream In genetics, the direction in which RNA polymerase moves along a DNA strand. Compare with **upstream**.

duplication In genetics, refers to an additional copy of a DNA sequence or part of a chromosome.

duplication and divergence An evolutionary process in which new genes are created by duplication of an original gene followed by divergence of the copied gene sequences to allow one or both copies to acquire a new function.

dynein A class of motor proteins that uses the chemical energy of ATP to "walk" toward the minus end of a microtubule. Dyneins are responsible for bending of cilia and flagella, play a role in chromosome movement during mitosis, and can transport vesicles and organelles.

early endosome A small transient organelle that is formed by the accumulation of vesicles from receptor-mediated endocytosis and is an early stage in the formation of a lysosome.

ecdysone An insect hormone that triggers either molting (to a larger larval form) or metamorphosis (to the adult form), depending on the level of juvenile hormone.

ecdysozoans A major lineage of protostomes (Ecdysozoa) that grow by shedding their external skeletons (molting) and expanding their bodies. Includes arthropods, nematodes, and other groups. Compare with **lophotrochozoans**.

echinoderms A major lineage of deuterostomes (Echinodermata) distinguished by adult bodies with five-sided radial symmetry, a water vascular system, and tube feet. Includes sea urchins, sand dollars, and sea stars.

echolocation The use of echoes from vocalizations to obtain information about the locations of objects in the environment.

ecological selection Also known as *environmental selection*. A type of natural selection that favors individuals with heritable traits that enhance their ability to survive and reproduce in a certain physical and/or biological environment, excluding their ability to obtain a mate. Compare with **sexual selection**.

ecology The study of how organisms interact with each other and with their environment.

ecosystem All the organisms that live in a geographic area, together with the nonliving (abiotic) components that affect or exchange materials with the organisms; a biological community and its physical environment.

ecosystem diversity The structural and functional variety of biotic components in a region along with abiotic components, such as soil, water, and nutrients.

ecosystem function The sum of biological and chemical processes that are characteristic of a given ecosystem—such as primary production, nitrogen cycling, and carbon storage.

ecosystem services All of the benefits that humans derive, directly or indirectly, from ecosystem functions.

ecotourism Tourism that is based on observing wildlife or experiencing other aspects of natural areas.

ectoderm The outermost of the three basic cell layers (germ layers) in most animal embryos; gives rise to the outer covering and nervous system. Compare with **endoderm** and **mesoderm**.

ectomycorrhizal fungi (EMF) Fungi whose hyphae form a dense network that covers their host plant's roots but do not enter the root cells.

ectoparasite A parasite that lives on the outer surface of the host's body.

ectotherm An animal that gains most of its body heat from external sources as opposed to metabolic processes. Compare with **endotherm**.

effector Any cell, organ, or structure with which an animal can respond to external or internal stimuli. Usually functions, along with a sensor and an integrator, as part of a homeostatic system.

effector cells The activated lymphocytes that defend the body in an immune response. *Effector B cells* are called plasma cells and secrete antibodies. *Effector T cells* include cytotoxic T cells and helper T cells, which carry out cell-mediated responses.

efferent division The part of the nervous system that carries commands from the central nervous system to the rest of the body. Consists primarily of motor neurons. Compare with **afferent division**.

egg A mature female gamete and any associated external layers (such as a shell). In animals, also called *ovum*.

ejaculation The release of semen from the copulatory organ of a male animal.

ejaculatory duct A short duct through which sperm move during ejaculation; connects the vas deferens to the urethra.

electric current A flow of electric charge. Also called simply *current*.

electrical potential Potential energy created by a separation of electrical charges between two points. Also called **voltage**.

electrocardiogram (EKG) A recording of the electrical activity of the heart, obtained through electrodes on the skin.

electrochemical gradient The combined effect of an ion's concentration gradient and electrical (charge) gradient across a membrane that affects the diffusion of ions.

electrogenic fishes Any of various kinds of fishes that have specialized organs that generate electric fields in the water; these fields are used to detect objects.

electrolyte Any compound that dissociates into ions when dissolved in water. In nutrition, any of the major ions necessary for normal cell function.

electromagnetic spectrum The entire range of wavelengths of radiation extending from short wavelengths (high energy) to long wavelengths (low energy). Includes gamma rays, X-rays, ultraviolet, visible light, infrared, microwaves, and radio waves (from short to long wavelengths).

electron acceptor A reactant that gains an electron and is reduced in a reduction–oxidation reaction.

electron carrier Any molecule that readily accepts electrons from and donates electrons to other molecules. Protons may be transferred with the electrons in the form of hydrogen atoms.

electron donor A reactant that loses an electron and is oxidized in a reduction–oxidation reaction.

electron shell A group of electron orbitals with similar energies. Electron shells are arranged in roughly concentric layers around the nucleus of an atom, and electrons in outer shells have more energy than those in inner shells. Electrons in the outermost shell, the valence shell, often are involved in chemical bonding.

electron transport chain (ETC) Any set of membrane-bound protein complexes and mobile electron carriers involved in a coordinated series of redox reactions. The energy from the redox reactions is used to actively transport protons from one side of a membrane to the other.

electronegativity A measure of how strongly an atom pulls shared electrons toward itself in a bond.

electroreception The detection of electric fields.

electroreceptor A sensory cell specialized for detecting electric fields.

element A substance consisting of one type of atom with a specific number of protons. Elements preserve their identity in chemical reactions.

elevation Height of land above sea level.

elimination The removal of wastes from an animal's digestive tract.

elongation (1) The process by which RNA lengthens during transcription. (2) The process by which a polypeptide chain lengthens during translation.

elongation factor Any of a class of proteins involved in the elongation phase of translation that assist ribosomes in the synthesis of the growing peptide chain.

embryo An organism at an early stage of development; the stage after fertilization and zygote formation.

embryo sac The female gametophyte in flowering plants.

embryogenesis The production of an embryo from a zygote. Embryogenesis is an early event in development of animals and plants.

embryonic stem cell A cell from an early animal embryo that has the potential to differentiate into any cell type found in the adult.

Embryophyta An increasingly popular name for the lineage called land plants; reflects their retention of a fertilized egg.

embryophyte A plant that nourishes its embryos inside its own body. All land plants are embryophytes.

emergent property A property that stems from the interaction of simpler elements, is not exhibited by the simple elements, and is difficult or impossible to predict from the study of individual elements. For example, homeostasis is an emergent property of the collection of interacting molecules present in each cell.

emerging disease Any infectious disease, often a viral disease, that suddenly afflicts significant numbers of individuals for the first time; often due to changes in the host specificity of a pathogen or to radical changes in the genetic material of the pathogen.

emigration The migration of individuals away from one population to other populations. Compare with **immigration**.

emulsification (verb: *to emulsify*) The dispersion of fat into an aqueous solution. Usually requires the aid of an amphipathic substance such as a detergent or bile salts, which can break large fat globules into microscopic fat droplets.

endangered species A species whose numbers have decreased so much that it is in danger of extinction throughout all or part of its range.

endemic species A species that lives in one geographic area and nowhere else.

endergonic Referring to a chemical reaction that has a change in Gibbs free energy greater than zero ($\Delta G > 0$) and is nonspontaneous (i.e., requires an input of energy to occur). Compare with **exergonic**.

endocrine Relating to a chemical signal (hormone) that is released into the bloodstream by a cell and acts on a distant target cell.

endocrine disruptor An exogenous chemical that interferes with normal hormonal signaling.

endocrine gland A gland that secretes hormones directly into the bloodstream or interstitial fluid. Compare with **exocrine gland**.

endocrine system All of the glands, tissues, and cells that produce and secrete hormones into the bloodstream.

endocytosis General term for any pinching off of the plasma membrane that results in the uptake of material from outside the cell. Includes phagocytosis, pinocytosis, and receptor-mediated endocytosis. Compare with **exocytosis**.

endoderm The innermost of the three basic cell layers (germ layers) in most animal embryos; gives rise to the digestive tract and organs that connect to it (liver, lungs, etc.). Compare with **ectoderm** and **mesoderm**.

endodermis In plant roots, a cylindrical layer of cells that separates the cortex from the vascular tissue; location of the Casparian strip.

endomembrane system A system of organelles in eukaryotic cells that synthesizes, processes, transports, and recycles proteins and lipids. Includes the endoplasmic reticulum (ER), Golgi apparatus, and lysosomes.

endomycorrhizal fungi See **arbuscular mycorrhizal fungi (AMF)**.

endoparasite A parasite that lives inside the host's body.

endophyte (adjective: *endophytic*) A fungus that lives inside the tissues of a plant in a symbiotic relationship. Compare with **epiphyte**.

endoplasmic reticulum (ER) A network of interconnected membranous sacs and tubules found inside eukaryotic cells that functions in synthesizing lipids and proteins that reside in the endomembrane system and the plasma membrane or are secreted from the cell. See **rough** and **smooth endoplasmic reticulum**.

endoskeleton Bony and/or cartilaginous structures within the body that provide support. Examples are the spicules of sponges, the plates in echinoderms, and the bony skeleton of vertebrates. Compare with **exoskeleton** and **hydrostatic skeleton**.

endosperm A triploid ($3n$) tissue in the seed of a flowering plant (angiosperm) that serves as food for the plant embryo. Functionally analogous to the yolk in some animal eggs.

endospore A tough, resistant reproductive structure formed in certain bacteria in response to poor environmental conditions.

endosymbiont An organism that lives in a symbiotic relationship inside the body of its host.

endosymbiosis An association between organisms of two different species in which one lives inside the cell or cells of the other.

endosymbiosis theory The theory that mitochondria and chloroplasts evolved from prokaryotes that were engulfed by host cells and took up a symbiotic existence in those cells, a process termed primary endosymbiosis. In some eukaryotes, chloroplasts may have originated by secondary endosymbiosis; that is, when a cell engulfed a chloroplast-containing protist and retained its chloroplasts.

endotherm An animal that gains most of its body heat from internal metabolic processes. Compare with **ectotherm**.

endothermic Referring to a chemical reaction that absorbs heat. Compare with **exothermic**.

energetic coupling In cellular metabolism, the mechanism by which energy released from an exergonic reaction (commonly, hydrolysis of ATP) is used to make an endergonic reaction spontaneous.

energy The capacity to do work or to supply heat. May be stored (potential energy) or available in the form of motion (kinetic energy).

enhancer A set of regulatory sequences in eukaryotic DNA that may be located upstream or downstream of the start site of transcription, in nontranscribed regions, or in introns. Enhancers are often far from the gene they control. Binding of specific regulatory transcription factor proteins to an enhancer promotes the transcription of genes.

enrichment culture A method of detecting and obtaining cells with specific characteristics by placing a sample, containing many types of cells, under a specific set of conditions (e.g., temperature, salt concentration, available nutrients) and isolating those cells that grow rapidly in response.

enteric nervous system A network of neurons embedded in the wall of the gastrointestinal tract of vertebrates. Important in the regulation of digestion and possibly many other functions.

enteropeptidase An intestinal enzyme that converts pancreatic trypsinogen into its active form, trypsin, thereby triggering activation of a cascade of enzymes important in digesting nutrients in the small intestine.

enthalpy (H) A quantitative measure of the amount of potential energy, or heat content, of a system plus the pressure and volume it exerts on its surroundings.

entropy (S) A quantitative measure of the amount of disorder of any system, such as a group of molecules.

envelope (viral) A membrane that encloses the capsid of some viruses. Often includes specialized proteins that attach to host-cell surfaces.

environmental effect In genetics, any influence on phenotype that is not due to particular genotypes. Includes any aspect of the physical (e.g., temperature or humidity) or biological (e.g., population density, competition, or levels of sex hormones experienced during development) environment.

environmental selection Also known as ecological selection. A type of natural selection that favors individuals with heritable traits that enhance their ability to survive and reproduce in a certain physical and/or biological environment, excluding their ability to obtain a mate. Compare with **sexual selection**.

environmental sequencing See **metagenomics**.

enzyme A protein catalyst used by living organisms to increase the rate of biological reactions.

epicotyl In some embryonic plants, a portion of the embryonic stem that extends above the cotyledons.

epidemic The spread of an infectious disease throughout a population in a short time. Compare with **pandemic**.

epidermis The outermost layer of cells of any multicellular organism. See also **dermal tissue system** in plants.

epididymis A coiled tube on the outside of each testis in reptiles, birds, and mammals. The site of the final stages of sperm maturation.

epigenetic inheritance Any pattern of inheritance involving differences in phenotype that are not due to differences in the nucleotide sequence of genes.

epinephrine A catecholamine hormone, produced and secreted by the adrenal medulla, that triggers rapid responses related to the fight-or-flight response. Also called *adrenaline*.

epiphyte (adjective: *epiphytic*; noun: *epiphytism*) A nonparasitic plant that grows on the trunks or branches of other plants and is not rooted in soil.

epistasis A form of gene interaction in which the phenotype associated with a particular genotype of one gene can be masked, or hidden, by a specific genotype of another gene.

epithelial tissue See **epithelium**.

epithelium (plural: *epithelia*) An animal tissue consisting of sheetlike layers of tightly packed cells that line an organ, a gland, a duct, or a body surface. Also called *epithelial tissue*.

epitope A small region of a particular antigen that can bind to an antibody, B-cell receptor, or T-cell receptor.

equilibrium The sensory system located in the inner ear that detects orientation, motion, and acceleration.

equilibrium potential The membrane potential at which there is no net movement of a particular ion into or out of a cell.

ER signal sequence A short amino acid sequence that marks a polypeptide for transport to the endoplasmic reticulum, where synthesis of the polypeptide chain is completed and the signal sequence removed. See **signal recognition particle**.

erythropoietin (EPO) A peptide hormone, released by the kidney and other tissues in response to a low blood-oxygen level, that stimulates the bone marrow to produce more red blood cells.

esophagus The muscular tube that connects the mouth to the stomach.

essential amino acids Any amino acid that an animal cannot synthesize and must obtain from the diet. There are eight essential amino acids in adult humans: isoleucine, leucine, lysine, methionine, phenylalanine, threonine, tryptophan, and valine.

essential fatty acids Any fatty acid that an animal cannot synthesize and must obtain from the diet. There are two essential fatty acids in humans: linolenic acid and linoleic acid.

essential nutrient Any chemical element, ion, or compound that is required for normal growth, reproduction, and maintenance of a living organism and that cannot be synthesized by the organism.

ester linkage The covalent bond formed by a condensation reaction between a carboxyl group and a hydroxyl group. Ester linkages join fatty acids to glycerol to form a fat or phospholipid.

estivation Prolonged reduction in metabolic rate and activity to conserve energy and water during dry or hot weather. Compare with **hibernation** and **torpor**.

estradiol The major estrogen produced by the ovaries of female mammals and many other vertebrates. Stimulates development of the female reproductive tract, growth of ovarian follicles, and growth of breast tissue in mammals.

estrogen Any of a class of steroid hormones, including estradiol, estrone, and estriol, that generally promote female-like traits. Secreted by the gonads, fat tissue, and some other organs.

estrous cycle A female reproductive cycle in which the uterine lining is reabsorbed rather than shed in the absence of pregnancy and in which the female is sexually receptive only briefly during mid-cycle (estrus). Seen in all mammals except Old World monkeys and apes (including humans). Compare with **menstrual cycle**.

estuary An aquatic biome formed where streams and ocean meet, such that fresh water and salt water mix; highly productive because of abundant light and nutrients.

ethylene A gaseous plant hormone associated with senescence that induces fruits to ripen and flowers to fade.

etiolation The phenomenon in which plants grow long, thin, and spindly and tend to be pale yellow as a result of growing in shaded or dark conditions.

eudicot A member of a monophyletic group (lineage) of angiosperms that includes complex flowering plants and trees (e.g., roses, daisies, maples). All eudicots have two cotyledons, but not all dicots are members of this lineage. Compare with **dicot** and **monocot**.

Eukarya One of the three taxonomic domains of life, consisting of unicellular organisms (most protists, yeasts) and multicellular organisms (fungi, plants, animals) distinguished by a membrane-bound cell nucleus, numerous organelles, and an extensive cytoskeleton. Compare with **Archaea** and **Bacteria**.

eukaryote (adjective: *eukaryotic*) A member of the domain Eukarya; an organism whose cells contain a nucleus, numerous membrane-bound organelles, and an extensive cytoskeleton. May be unicellular or multicellular. Compare with **prokaryote**.

eusociality A complex social structure in which workers sacrifice most or all of their direct reproduction to help rear the queen's offspring. Common in insects such as ants, bees, wasps, and termites.

eutherians A lineage of mammals (Eutheria) whose young develop in the uterus and are not housed in an abdominal pouch. Also called *placental mammals*.

eutrophication The enrichment of water by nutrients such as nitrogen and/or phosphorous, leading to rapid growth of algae (algal bloom) that depletes the water of oxygen when the algae die and are eaten by bacteria; the resulting oxygen depletion can lead to the death of animals, causing a "dead zone."

evaporation The energy-absorbing phase change from a liquid state to a gaseous state. Many organisms evaporate water as a means of heat loss.

evo-devo Popular term for evolutionary developmental biology, a research field focused on how changes in developmentally important genes have led to the evolution of new phenotypes, especially body forms.

evolution (1) The theory that all organisms on Earth are related by common ancestry and that they have changed over time, and continue to change, via natural selection, genetic drift, gene flow, and mutation. (2) Any change in the genetic characteristics of a population over time, especially a change in allele frequencies.

evolutionary game theory A mathematical modeling approach to predicting the outcome of natural selection on behaviors when multiple "players" are interacting.

ex situ conservation Preserving species outside of natural areas (e.g., in zoos, aquaria, or botanical gardens).

excitable membrane A plasma membrane that is capable of generating an action potential. Neurons, muscle cells, and some other cells have excitable membranes.

excitatory postsynaptic potential (EPSP) A depolarization of a cell that makes the cell more likely to generate an action potential. Compare with **inhibitory postsynaptic potential (IPSP)**.

exergonic Referring to a chemical reaction that has a change in Gibbs free energy less than zero (ΔG)

exocrine gland A gland that secretes some substance through a duct into a space other than the circulatory system, such as the lumen of the digestive tract or the exterior of the body. Compare with **endocrine gland**.

exocytosis Secretion of intracellular molecules (e.g., hormones, collagen), contained within membrane-bound vesicles, to the outside of the cell by fusion of vesicles to the plasma membrane. Compare with **endocytosis**.

exon A transcribed region of a gene or region of a primary transcript that is retained in the mature RNA. Except for 5′ and 3′ UTRs, mRNA exons code for amino acids. Compare with **intron**.

exoskeleton A hard covering secreted on the outside of the body, used for body support, protection, and muscle attachment. Examples are the shell of mollusks and the outer covering (cuticle) of arthropods. Compare with **endoskeleton** and **hydrostatic skeleton**.

exothermic Referring to a chemical reaction that releases heat. Compare with **endothermic**.

exotic species A nonnative species that is introduced into a new area. Exotic species often are competitors, pathogens, or predators of native species.

expansin Any of a class of proteins that break hydrogen bonds between components in the primary cell wall of plants, allowing it to expand for cell growth.

experiment A powerful scientific tool in which researchers test the effect of a single, well-defined factor on a particular phenomenon.

exponential population growth The accelerating increase in the size of a population that occurs when the per capita growth rate is constant and density independent. Compare with **logistic population growth**.

expressed sequence tag (EST) A cDNA from a portion of a transcribed gene that is used to find the gene's physical location in the genome and to confirm that a region of DNA is transcribed.

extant species A species that is living today.

extensor A muscle that pulls two bones farther apart, increasing the angle of the joint between them, as in the extension of a limb or the spine. Compare with **flexor**.

extinct species A species that no longer exists.

extracellular digestion Digestion that takes place outside of an organism, as occurs in many fungi that make and secrete digestive enzymes.

extracellular matrix (ECM) A complex fiber composite in which animal cells are embedded, consisting of proteins (e.g., collagen, proteoglycan, and laminin) and polysaccharides produced by the cells.

extremophile An organism that thrives in an "extreme" environment (e.g., high-salt, high-temperature, low-temperature, or high-pressure).

F_1 generation First filial generation. The first generation of offspring produced from a mating (i.e., the offspring of the parental generation).

facilitated diffusion Passive movement (diffusion) of a substance across a membrane with the assistance of transmembrane carrier proteins or channel proteins.

facilitation In ecological succession, the phenomenon in which early-arriving species make conditions more favorable for later-arriving species. Compare with **inhibition** and **tolerance**.

facultative anaerobe Any organism that can survive and reproduce by performing aerobic respiration when oxygen is available or fermentation when it is not.

$FAD/FADH_2$ Oxidized and reduced forms, respectively, of flavin adenine dinucleotide. A nonprotein electron carrier that functions in the citric acid cycle and electron transport chain.

fallopian tube See **oviduct**.

fast muscle fiber A type of skeletal muscle fiber that is white, generates ATP by glycolysis, and contracts rapidly but fatigues easily. Also called *fast glycolytic*, or *Type IIb, fiber*.

fat A class of lipid consisting of three fatty acid molecules joined by ester linkages to a glycerol molecule. Also called *triacylglycerol* or *triglyceride*.

fate In the context of embryonic development, the type of adult cell or structure that normally forms from an embryonic cell or structure during the course of development.

fatty acid A lipid consisting of a hydrocarbon chain bonded at one end to a carboxyl group. Used by many organisms to store chemical energy; a major component of animal and plant fats and phospholipids.

fauna All the animal species characteristic of a particular region, period, or environment.

feather A specialized skin outgrowth, composed of keratin, present in all birds as well as in some nonavian dinosaurs. Used for flight, insulation, display, and other purposes.

feces The waste products of digestion.

fecundity The average number of female offspring produced by a single female in the course of her lifetime.

feedback inhibition A type of control in which high concentrations of the final product of a metabolic pathway inhibit one of the enzymes early in the pathway. A form of negative feedback.

fermentation Any of several metabolic pathways in the cytosol that regenerate oxidizing agents, such as NAD^+, by transferring electrons to an electron acceptor in the absence of an electron transport chain. Allows cells to continue making ATP in the absence of an electron transport chain.

ferredoxin In photosynthetic organisms, an iron- and sulfur-containing protein that is reduced by electrons from photosystem I. Can transfer electrons to the enzyme $NADP^+$ reductase, which catalyzes formation of NADPH.

fertility The average number of surviving children that each woman has during her lifetime.

fertilization Fusion of the nuclei of two gametes (often haploid) to form a zygote with a nucleus (often diploid). Exceptions to the haploid and diploid rule are seen in polyploid species.

fertilization envelope A physical barrier that forms around a fertilized egg in many animals. Prevents fertilization by more than one sperm (polyspermy).

fetal alcohol syndrome (FAS) A condition marked by hyperactivity, severe learning disabilities, and depression caused by exposure of an individual to high blood alcohol concentrations during embryonic development.

fetus In live-bearing (viviparous) animals, the unborn offspring after the embryonic stage. A fetus is usually developed sufficiently to be recognizable as belonging to a certain species.

fiber In plants, a type of elongated sclerenchyma cell that provides support to vascular tissue. Compare with **sclereid**.

fibroblast A type of cell that makes the fibers and extracellular matrix in loose connective tissue.

Fick's law of diffusion A mathematical relationship that describes the rates of diffusion of gases.

fight-or-flight response Rapid physiological changes that prepare the body for emergencies. Includes increased heart rate, increased blood pressure, and decreased digestion.

filament Any thin, threadlike structure, particularly (1) the threadlike extensions of a fish's gills or (2) part of a stamen; the slender stalk that bears the anthers in a flower.

filter feeder See **suspension feeder**.

filtrate Any fluid produced by filtration, in particular the fluid ("pre-urine") in the Malpighian tubules of insects and the nephrons of vertebrate kidneys.

filtration A process of removing large components from a fluid by forcing it through a filter. Occurs in renal corpuscles of the vertebrate kidney, allowing water and small solutes to pass from the blood into the nephron.

fimbria (plural: *fimbriae*) A long, needlelike projection from the cell membrane of bacteria that is involved in attachment to nonliving surfaces or other cells.

finite rate of increase (λ) The rate of increase of a population over a given period of time. Calculated as the ending population size divided by the starting population size. Compare with **intrinsic rate of increase** and **per capita rate of increase**.

first law of thermodynamics The principle of physics that energy is conserved in any process. Energy can be transferred and converted into different forms, but it cannot be created or destroyed.

fission (1) A form of asexual reproduction in which a prokaryotic cell divides to produce two genetically similar daughter cells by a process similar to mitosis of eukaryotic cells. Also called *binary fission*. (2) A form of asexual reproduction in which an animal splits into two or more individuals of approximately equal size.

fitness The ability of an individual to produce viable offspring relative to others of the same species.

fitness trade-off See **trade-off**.

flaccid Limp as a result of low internal (turgor) pressure (e.g., a wilted plant leaf). Compare with **turgid**.

flagellum (plural: *flagella*) A long, cellular projection that undulates (in eukaryotes) or rotates (in prokaryotes) to move the cell through an aqueous environment. See **axoneme**.

flatworms Members of the phylum Platyhelminthes. Distinguished by a broad, flat, unsegmented body that lacks a coelom. Flatworms belong to the lophotrochozoan branch of the protostome animals.

flavin adenine dinucleotide See **$FAD/FADH_2$**.

flexor A muscle that pulls two bones closer together, decreasing the angle of the joint between them, as in the flexion of a limb or the spine. Compare with **extensor**.

floral meristem A group of undifferentiated plant stem cells that can give rise to the four organs making up a flower.

florigen In plants, a protein hormone that is synthesized in leaves and transported to the shoot apical meristem, where it stimulates flowering.

flower In angiosperms, the part of a plant that contains reproductive structures. Typically includes a calyx, a corolla, and one or more stamens and/or carpels. See **perfect** and **imperfect flower**.

fluid connective tissue A type of connective tissue distinguished by having a liquid extracellular matrix; includes blood.

fluid feeder An animal that feeds by sucking or mopping up liquids such as nectar, plant sap, or blood.

fluid-mosaic model The widely accepted hypothesis that cellular membranes consist of proteins embedded in a fluid phospholipid bilayer.

fluorescence The spontaneous emission of light from an excited electron in a pigment falling back to its normal (ground) state.

follicle In a mammalian ovary, a sac of supportive cells containing an egg.

follicle-stimulating hormone (FSH) A peptide hormone produced and secreted by the anterior pituitary; it stimulates (in females) growth of eggs and follicles in the ovaries or (in males) sperm production in the testes.

follicular phase In a menstrual cycle, the phase during which follicles grow and the estradiol level increases; ends with ovulation.

food Any nutrient-containing material that can be consumed and digested by animals.

food chain A relatively simple pathway of energy and nutrient flow through a few species, each at a different trophic level, in an ecosystem. Might include, for example, a primary producer, a primary consumer, a secondary consumer, and a decomposer. A subset of a **food web**.

food web The complex network of consumption interactions among species in an ecosystem formed by the transfer of energy and nutrients among trophic levels. Consists of many food chains.

foot One of the three main parts of the mollusk body; a muscular appendage, used for movements such as crawling and/or burrowing into sediment.

foraging Searching for food.

forebrain One of the three main regions of the vertebrate brain; includes the olfactory bulb, cerebrum, thalamus, and hypothalamus. Compare with **hindbrain** and **midbrain**.

fossil Any physical trace of an organism that existed in the past. Includes tracks, burrows, fossilized bones, casts, and so on.

fossil record All of the fossils that have been found anywhere on Earth and that have been formally described in the scientific literature.

founder effect A change in allele frequencies that often occurs when a new population is established from a small group of individuals (founder event) due to sampling error (that is, the small group is not a representative sample of the source population).

frameshift mutation The addition or deletion of one or a few base pairs in a coding sequence that shifts the reading frame of the mRNA.

frequency The rate at which an event occurs per unit time. As applied to waves, the number of wave crests per second traveling past a stationary point. Determines the pitch of sound and the color of light.

frequency-dependent selection A pattern of selection in which certain alleles are favored only when they are rare; a form of balancing selection.

frontal lobe In the vertebrate brain, one of the four lobes of the cerebrum. Involved in complex decision making in humans.

fruit In flowering plants (angiosperms), a mature, ripened plant ovary (or group of ovaries), along with the seeds it contains and any adjacent fused parts; often functions in seed dispersal. See **aggregate fruit, multiple fruit**, and **simple fruit**.

fruiting body A structure formed in some prokaryotes, fungi, and protists for spore dispersal; usually consists of a base, a stalk, and a mass of spores at the top.

functional diversity In ecology, a measure of diversity in ecosystems based on the variety of functional traits present, such as primary productivity or nitrogen fixation.

functional genomics The study of how, when, and where specific genes are expressed within the genome and how their products interact to produce a functional organism.

functional group A small group of atoms bonded together in a precise configuration and exhibiting particular chemical properties that it imparts to any organic molecule in which it occurs.

fundamental niche The total theoretical range of environmental conditions that a species can tolerate. Compare with **realized niche**.

fungi A lineage of eukaryotes that typically have a filamentous body (mycelium) and obtain nutrients by absorption.

fungicide Any substance that can kill fungi or slow their growth.

G protein Any of various proteins that are activated by binding to guanosine triphosphate (GTP) and inactivated when GTP is hydrolyzed to GDP. In G-protein-coupled receptors, signal binding by a receptor directly triggers the activation of a G protein, leading to production of a second messenger or initiation of a phosphorylation cascade.

G_1 phase The phase of the cell cycle that constitutes the first part of interphase, before DNA synthesis (S phase).

G_2 phase The phase of the cell cycle between synthesis of DNA (S phase) and mitosis (M phase); the last part of interphase.

gallbladder A small pouch that stores bile from the liver and releases it as needed into the small intestine during digestion of fats.

gametangium (plural: *gametangia*) (1) The gamete-forming structure found in all land plants except angiosperms. Contains a sperm-producing antheridium and an egg-producing archegonium. (2) The gamete-forming structure of some chytrid fungi.

gamete A haploid reproductive cell that can fuse with another haploid reproductive cell of the opposite sex to form a diploid zygote. Exceptions to the haploid and diploid rule are seen in polyploid species. Most multicellular eukaryotes have two distinct forms of gametes: egg cells (ova) and sperm cells.

gametogenesis The production of gametes (eggs or sperm).

gametophyte In organisms undergoing alternation of generations, the multicellular haploid form that arises from a single haploid spore and produces gametes by mitosis and cell division. Compare with **sporophyte**.

ganglion (plural: *ganglia*) A mass of neurons in a central nervous system.

ganglion cell In the vertebrate retina, a neuron whose axon projects via the optic nerve to the brain.

gap junction A type of cell–cell attachment structure that directly connects the cytosol of adjacent animal cells, allowing passage of water, ions, and small molecules between the cells. Compare with **desmosome** and **tight junction**.

gastric lipase An enzyme produced by glands in the stomach that breaks down triglycerides into fatty acids and diglycerides.

gastrin A hormone produced by cells in the stomach lining in response to the arrival of food or to a neural signal from the brain. Stimulates other stomach cells to release hydrochloric acid.

gastropods A lineage of mollusks distinguished by a large, muscular foot and a unique feeding structure, the radula. Includes slugs and snails.

gastrula Stage of embryonic development that follows the blastula during which, in most vertebrate animals, the three distinct germ layers form and the anterior–posterior and dorsal-ventral axes are established. See **endoderm, mesoderm,** and **ectoderm**.

gastrulation The process of coordinated cell-shape changes and movements, including the movement of some cells from the outer surface of the embryo to the interior that results in the formation of the three germ layers (endoderm, mesoderm, and ectoderm) and establishes the axes of the embryo.

gated channel A channel protein that opens and closes in response to a specific stimulus, such as the binding of a particular substance or a change in voltage across the membrane.

gene (1) In a molecular context, a section of DNA (or RNA, for some viruses) that contains the regulatory sequences and coding information for the transcription of one or more related functional RNA molecules, some of which encode polypeptides. (2) In Mendelian genetics, the hereditary determinant of a trait, such as flower color or seed shape in pea plants.

gene duplication The formation of an additional copy of a gene, often by misalignment of chromosomes during crossing over. An important evolutionary process in creating new genes.

gene expression The entire set of processes, including transcription and translation, that convert information in DNA into a product of a gene, most commonly a protein, that contributes to the phenotype of a cell or organism.

gene family A set of genes whose DNA sequences are similar because they arose by duplication of an ancestral gene followed by divergence of the sequences due to accumulation of random mutations.

gene flow The movement of alleles between populations; occurs when individuals leave one population, join another, and breed.

gene interaction Two or more genes working together to control one trait.

gene pool All the alleles of all the genes in a certain population.

gene therapy The treatment of an inherited disease by introducing a normal form of the gene.

general transcription factor Any of a class of proteins, present in all eukaryotic cells, that assemble on promoters and help RNA polymerase initiate transcription. Sometimes called basal transcription factor. Compare with **regulatory transcription factor**.

generation time The average time between a mother's first offspring and her daughter's first offspring.

genetic bottleneck A reduction in the diversity of alleles in a population resulting from a sudden decrease in the size of that population (population bottleneck) due to a random event.

genetic code The set of all codons and their meanings.

genetic correlation A type of evolutionary constraint in which selection on one trait causes a change in another trait as well; may occur when the same gene(s) affect both traits.

genetic diversity The diversity of alleles or genes in a population, species, or group of species.

genetic drift Any change in allele frequencies due to chance. Causes allele frequencies to drift up and down randomly over time, and eventually can lead to the fixation or loss of alleles.

genetic equivalence When all different cell types of a multicellular individual possess the same genome.

genetic homology Similarity in DNA nucleotide sequences, RNA nucleotide sequences, or amino acid sequences due to inheritance from a common ancestor.

genetic map A diagram showing the relative positions and distances between genes along a chromosome. Also called a *linkage map*. Compare with **physical map**.

genetic marker A genetic locus that can be identified and traced in populations by molecular techniques or by a distinctive visible phenotype.

genetic recombination A change in the combination of alleles on a given chromosome or in a given individual. Also called *recombination*.

genetic regulatory cascade A set of regulatory genes that are linked in such a way that one initially activated gene triggers the expression of other regulatory genes, which in turn trigger the expression of yet more regulatory genes. Often used to control development.

genetic screen Any technique that identifies individuals with a particular type of mutation.

genetic variation (1) The number and relative frequency of alleles present in a particular population. (2) The proportion of phenotypic variation in a trait that is due to genetic rather than environmental influences in a certain population in a certain environment.

genetically modified organism (GMO) Plant or animal that has specific changes introduced into their DNA using genetic engineering methods. Most often refers to domestic species.

genetics The study of the inheritance of traits, the nature of genes, and how genes work.

genital (plural: *genitalia*) Any external reproductive structure of an animal.

genome All the hereditary material (DNA in cells, DNA or RNA in viruses) in a virus, cell, or organism, including but not confined to genes.

genome annotation The process of analyzing a genome sequence to identify key features such as genes, regulatory sequences, and splice sites.

genome-wide association study (GWAS) A method of locating genes by finding associations between a particular phenotype and a particular polymorphic genetic marker (almost always a SNP) in populations.

genomic DNA In the context of recombinant DNA studies or applications, DNA that is derived directly from cells, in contrast to cDNA (complementary DNA).

genomics The field of study concerned with sequencing, interpreting, and comparing whole genomes from different organisms.

genotype The alleles of a gene or genes present in a given individual. Compare with **phenotype**.

genus (plural: *genera*) In Linnaeus' system, a taxonomic category of closely related species. Always italicized and capitalized to indicate that it is a recognized scientific genus.

geologic time scale The sequence of eons, eras, and periods used to describe the geologic history of Earth.

germ layer In animals, one of the three embryonic tissue layers formed during gastrulation; gives rise to all other tissues and organs. See **endoderm, mesoderm**, and **ectoderm**.

germ theory of disease The theory that infectious diseases are caused by bacteria, viruses, and other microbes.

germination The process by which a seed becomes a young plant.

gestation In animal species with live birth (vivipary), the period of development inside the mother, from implantation of the embryo to birth.

gibberellins A class of hormones, found in plants and fungi, that stimulate growth. Gibberellic acid (GA) is one of the major gibberellins.

Gibbs free energy The energy of a system that can be converted into work. The amount of this energy that is available can be measured only by how it changes in a reaction (ΔG). See **standard free-energy change**.

gill Any organ in aquatic animals that exchanges gases and other dissolved substances between the blood or hemolymph and the surrounding water. Typically, a filamentous outgrowth of a body surface.

gill arch In aquatic vertebrates, a curved region of tissue between the gills. Gills are suspended from the gill arches.

gill filament In fishes, any of the many long, thin structures that extend from gill arches into the water and that function in gas exchange.

gill lamella (plural: *gill lamellae*) Any of the hundreds to thousands of sheetlike structures, each containing a capillary bed, that make up a gill filament.

gland An organ whose primary function is to secrete some substance, either into the blood (endocrine gland) or into some other space such as the interior of the gut or the outer surface of the body (exocrine gland).

glia Collective term for several types of cells in nervous tissue that are not neurons and do not conduct electrical signals but perform other functions, such as providing support, nourishment, or electrical insulation. Also called *glial cells*.

global carbon cycle See **carbon cycle, global**.

global climate change The global sum of all the local changes in temperature and precipitation patterns that accompany **global warming** (or in some past events, global cooling).

global gene regulation The regulation of multiple bacterial genes that are not part of one operon.

global nitrogen cycle See **nitrogen cycle, global**.

global phosphorus cycle See **phosphorus cycle, global**.

global warming A sustained increase in Earth's average surface temperature.

global water cycle See **water cycle, global**.

glomerulus (plural: *glomeruli*) (1) In the vertebrate kidney, a ball-like cluster of capillaries, surrounded by Bowman's capsule, at the beginning of a nephron. (2) In the vertebrate brain, a ball-shaped cluster of neurons in the olfactory bulb.

glucagon A peptide hormone produced by the pancreas in response to low blood glucose levels. Raises the blood glucose level by triggering breakdown of glycogen and stimulating gluconeogenesis. Compare with **insulin**.

glucocorticoid Any of a class of steroid hormones, produced and secreted by the adrenal cortex, that increase blood glucose concentration and mediate the body's long-term stress response. Glucocorticoids include cortisol and corticosterone. Compare with **mineralocorticoid**.

gluconeogenesis Synthesis of glucose, often from non-carbohydrate sources (e.g., proteins and fatty acids). In plants, used to produce glucose from products of the Calvin cycle. In animals, occurs in the liver in response to low insulin levels and high glucagon levels.

glucose A six-carbon monosaccharide that can be oxidized via cellular respiration or fermentation to produce ATP or serve as a source of carbon to synthesize other molecules.

glyceraldehyde-3-phosphate (G3P) The phosphorylated three-carbon carbohydrate formed from the fixation and reduction of CO_2 in the Calvin cycle.

glycerol A three-carbon molecule that forms the "backbone" of phospholipids and most fats.

glycogen A highly branched storage polysaccharide composed of glucose monomers joined by α-1,4- and α-1,6-glycosidic linkages. The major form of stored carbohydrate in animals.

glycolipid Any lipid molecule that is covalently bonded to one or more carbohydrates.

glycolysis A series of 10 chemical reactions in the cytosol that oxidize glucose to produce pyruvate, NADH, and ATP. Used by organisms as part of fermentation and/or cellular respiration.

glycoprotein Any protein with one or more covalently bonded carbohydrates, typically oligosaccharides.

glycosidic linkage The covalent bond formed by a condensation reaction between two sugar monomers; joins the residues of a polysaccharide. Also known as *glycosidic bond*.

glycosylation Addition of a carbohydrate group to a molecule.

glyoxysome Specialized type of peroxisome found in plant cells and packed with enzymes for processing the products of photosynthesis.

gnathostomes Vertebrate animals with jaws. Most vertebrates are gnathostomes.

Golgi apparatus A eukaryotic organelle, consisting of stacks of flattened membranous sacs (cisternae), that functions in processing and sorting proteins and lipids destined to be secreted or directed to other organelles. Also called *Golgi complex*.

gonad An animal organ (testis or ovary) that produces reproductive cells (gametes).

gonadotropin-releasing hormone (GnRH) A peptide hormone, produced and secreted by the hypothalamus, that stimulates release of follicle-stimulating hormone (FSH) and luteinizing hormone (LH) from the anterior pituitary.

Gram-negative Describing bacteria that look pink when treated with a Gram stain. These bacteria have a cell wall composed of a thin layer of peptidoglycan and an outer phospholipid layer. Compare with **Gram-positive**.

Gram-positive Describing bacteria that look purple when treated with a Gram stain. These bacteria have cell walls composed of a thick layer of peptidoglycan and no outer phospholipid later. Compare with **Gram-negative**.

Gram stain A staining technique used for the preliminary identification of prokaryotes based on cell wall composition. See **Gram-negative** and **Gram-positive**.

grana (singular: *granum*) Stacks of flattened thylakoid discs in chloroplasts where the light reactions of photosynthesis occur.

gravitropism The growth or movement of a plant in a particular direction in response to gravity.

grazing food chain The ecological network of primary producers, herbivores, and the predators and parasites that consume them.

great apes See **hominids**.

green algae A paraphyletic group of photosynthetic organisms that contain chloroplasts similar to those in land plants. Often classified as protists, green algae are the closest living relatives of land plants and form a monophyletic group with them.

greenhouse effect Selective energy absorption by greenhouse gases such as carbon dioxide (CO_2) and methane

(CH_4) in the atmosphere, which allows short-wavelength light energy to pass through but absorbs longer-wavelength infrared energy and reflects heat back to Earth. Causes a warming effect similar to that in a greenhouse.

greenhouse gas An atmospheric gas that absorbs and reflects infrared radiation, so that heat radiated from Earth is retained in the atmosphere instead of being lost to space. Includes carbon dioxide (CO_2), methane (CH_4), water vapor (H_2O), and nitrous oxides (N_2O).

gross primary productivity (GPP) In an ecosystem, the total amount of carbon fixed by photosynthesis (or more rarely, chemosynthesis), including that used for cellular respiration, over a given time period. Compare with **net primary productivity**.

ground meristem A primary meristem tissue that gives rise to the ground tissue system.

ground tissue An embryonic tissue layer that gives rise to parenchyma, collenchyma, and sclerenchyma—tissues other than the epidermis and vascular tissue. Also called *ground tissue system*.

ground tissue system Most common of the tissue systems; includes the following tissues/cell typesparenchyma, collenchyma, and sclerenchyma—tissues other than the epidermis and vascular tissue.

groundwater Any water below the land surface.

growth factor Any of a large number of signaling molecules that are secreted by certain cells and that stimulate other cells to grow, divide, or differentiate.

growth hormone (GH) A peptide hormone, produced and secreted by the mammalian anterior pituitary, that promotes lengthening of the long bones in children and muscle growth, tissue repair, and lactation in adults. Also called *somatotropin*.

guanosine triphosphate (GTP) A nucleotide consisting of guanine, a ribose sugar, and three phosphate groups. Can be hydrolyzed to release free energy. Commonly used in RNA synthesis and also functions in signal transduction in association with G proteins.

guard cell One of two specialized, crescent-shaped cells forming the border of a plant stoma. Guard cells can change shape to open or close the stoma. See also **stoma**.

gustation The sense of taste.

guttation Excretion of water droplets from plant leaves; visible in the early morning. Caused by root pressure.

gymnosperm A vascular plant that makes seeds but does not produce flowers. The gymnosperms include five lineages of green plants (cycads, ginkgoes, redwoods, pines, and gnetophytes). Compare with **angiosperm**.

gyre A large-scale ocean circulation or cycle that flows in the clockwise direction in the Northern Hemisphere and in the counterclockwise direction in the Southern Hemisphere.

H^+-ATPase See **proton pump**.

habitat degradation The human-caused reduction in the quality of a habitat.

habitat destruction Human-caused destruction of a natural habitat, often replaced by an urban, suburban, or agricultural landscape.

habitat fragmentation The breakup of a large region of a habitat into many smaller regions, often separated from each other by inhospitable areas of human development.

Hadley cell An atmospheric cycle of large-scale air movement in which warm equatorial air rises, moves north or south, and then descends at approximately 30°N or 30°S latitude.

hair cell A pressure-detecting mechanoreceptor cell in vertebrates that has tiny hairlike structures (stereocilia) jutting from its surface. Found in the inner ear, lateral line system, and ampullae of Lorenzini.

Hamilton's rule The proposition that an allele for altruistic behavior will be favored by natural selection only if $Br > C$, where B = the fitness benefit to the recipient, C = the fitness cost to the actor, and r = the coefficient of relatedness between recipient and actor.

haploid (1) Having one set of chromosomes ($1n$, or more commonly, n for short). (2) A cell or an individual organism with one set of chromosomes. Compare with **diploid**.

haploid number The number of distinct types of chromosomes in a cell. Symbolized as n.

hardwood Wood, typically from angiosperms, characterized by the presence of vessel elements (but may also contain tracheids).

Hardy–Weinberg equilibrium A state of agreement between observed allele frequencies in a population and allele frequencies predicted by the Hardy–Weinberg principle. See **Hardy–Weinberg principle**.

Hardy–Weinberg principle A principle of population genetics stating that genotype frequencies in a large population do not change from generation to generation in the absence of evolutionary processes (e.g., mutation, gene flow, genetic drift, and selection), and nonrandom mating.

haustorium (plural: *haustoria*) Highly modified stem or root of a parasitic plant. The haustorium penetrates the tissues of a host and absorbs nutrients and water.

head The anteriormost region of many bilaterian animals, usually containing specialized sensory structures and the brain.

hearing The sensation of the wavelike changes in air pressure called sound.

heart A muscular pump that circulates blood or hemolymph throughout an animal.

heart murmur A distinctive sound caused by backflow of blood through a defective heart valve.

heartwood The older xylem in the center of an older stem or root, containing protective compounds and no longer functioning in water transport.

heat Thermal energy that is transferred from an object at higher temperature to one at lower temperature.

heat of vaporization The energy required to change 1 gram of a liquid into a gas.

heavy chain The larger of the two types of polypeptide chains in an antibody or B-cell receptor; composed of a variable (V) region, which contributes to the antigen-binding site, and a constant (C) region. Differences in heavy-chain constant regions determine the different classes of immunoglobulins (IgA, IgE, etc.). Compare with **light chain**.

helper T cell A $CD4^+$ effector T cell that secretes cytokines and in other ways promotes the activation of other white blood cells. Activated by interacting with complementary class II MHC–peptide complexes on the surface of antigen-presenting cells such as dendritic cells.

heme A small molecule that binds to each of the four polypeptides in hemoglobin; contains an iron ion that can bind oxygen.

hemocoel A body cavity, present in arthropods and some mollusks, containing a pool of circulatory fluid (hemolymph) bathing the internal organs. Unlike a coelom, a hemocoel is not lined with mesodermally derived tissue.

hemoglobin An oxygen-binding protein consisting of four polypeptide subunits, each containing an oxygen-binding heme group. The major oxygen carrier in mammalian blood.

hemolymph The circulatory fluid of animals with open circulatory systems (e.g., insects), in which the fluid is not confined to blood vessels.

herbaceous Referring to a plant that is not woody.

herbivore (adjective: *herbivorous*) An animal that eats primarily plants or algae and rarely or never eats other animals. Compare with **carnivore** and **omnivore**.

herbivory The practice of eating plant or algal tissues. Compare with **predation** and **parasitism**.

heredity The transmission of traits from parents to offspring via genetic information.

heritable Referring to traits that can be transmitted from one generation to the next.

heterokaryotic Describing a fungal mycelium containing haploid nuclei that are genetically distinct.

heterospory (adjective: *heterosporous*) In seed plants, the production of two distinct types of spores: microspores, which become the male gametophyte, and megaspores, which become the female gametophyte. Compare with **homospory**.

heterotroph Any organism that cannot synthesize reduced organic compounds from inorganic sources and that must obtain them from other organisms. Some bacteria, some archaea, and virtually all fungi and animals are heterotrophs. Also called *consumer*. Compare with **autotroph**.

heterozygote advantage A pattern of natural selection that favors heterozygous individuals compared with homozygotes. Tends to maintain genetic variation in a population and thus is a form of balancing selection.

heterozygous Having two different alleles of a gene.

hexose A monosaccharide (simple sugar) containing six carbon atoms.

hibernation An energy-conserving physiological state, marked by a decrease in metabolic rate, body temperature, and activity, that lasts for a prolonged period (weeks to months), typically in response to winter cold and scarcity of food. Compare with **estivation** and **torpor**.

hindbrain One of the three main regions of the vertebrate brain, responsible for balance and sometimes hearing; includes the cerebellum and medulla oblongata. Compare with **forebrain** and **midbrain**.

hippocampus In the vertebrate brain, a paired structure at the inner edge of the temporal lobes of the cerebrum; involved in formation of new memories and in spatial navigation.

histamine A molecule released from mast cells during an inflammatory response that causes blood vessels to dilate and become more permeable, allowing circulating white blood cells to enter the area.

histone A member of a class of positively charged (basic) proteins associated with DNA in the chromatin of eukaryotic cells.

histone acetyltransferase (HAT) Any of a class of eukaryotic enzymes that loosen chromatin structure by adding acetyl groups to histone proteins.

histone code The hypothesis that specific combinations of chemical modifications of histone proteins contain information that influences chromatin condensation and gene expression.

histone deacetylase (HDAC) Any of a class of eukaryotic enzymes that condense chromatin by removing acetyl groups from histone proteins.

holoenzyme A general term for any multipart enzyme consisting of a core enzyme (which contains the active site for catalysis) along with other proteins required for full function.

homeobox A DNA sequence of about 180 base pairs that codes for a DNA-binding region in the resulting protein. Genes containing a homeobox usually play a role in controlling development of organisms.

homeostasis (adjective: *homeostatic*) The array of relatively stable chemical and physical conditions in an organism's cells, tissues, and organs. May be achieved by passively matching the conditions of a stable external environment (conformational homeostasis) or by active physiological processes (regulatory homeostasis) triggered by variations in the external or internal environment.

homeotherm An animal that has a constant or relatively constant body temperature. Compare with **poikilotherm**.

homeotic genes See ***Hox* genes**.

homeotic mutation A mutation that causes one body part to be substituted for another.

hominids Members of the family Hominidae; today's representatives are humans, chimpanzees, bonobos, gorillas, and orangutans. Distinguished by large body size, no tail, and an exceptionally large brain. Also called *great apes*.

hominin Any extinct or living species of bipedal ape, such as *Australopithecus africanus, Homo erectus*, and *Homo sapiens*.

homologous See **homology**.

homologous chromosome In diploid organisms, a member of a pair of chromosomes that are similar in size, shape, and gene content. Also called *homologous pair*.

homologous pair In diploid organisms, a pair of chromosomes that are the same size and shape and contain the same genes in the same positions along the chromosomes.

homologs The same type of chromosomes (carrying genes for the same traits) in a diploid or polyploid organism.

homology (adjective: *homologous*) Similarity among organisms of different species due to shared ancestry. Features that exhibit such similarity (e.g., DNA sequences, proteins, body parts) are said to be homologous. Compare with **homoplasy**.

homoplasy Similarity among organisms of different species due to reasons other than common ancestry, such as convergent evolution. Compare with **homology**.

homospory (adjective: *homosporous*) In seedless vascular plants, the production of just one type of spore. Compare with **heterospory**.

homozygous Having two identical alleles of a gene.

hormone Any of many different signaling molecules that circulate throughout the plant or animal body and can trigger characteristic responses in distant target cells at very low concentrations.

hormone receptor A protein to which a hormone binds in or on a target cell. Located intracellularly for lipid-soluble hormones such as steroids and in the plasma membrane for water-soluble hormones such as polypeptides.

hormone-response element A specific sequence in DNA to which a hormone–receptor complex can bind and affect gene transcription.

host An individual that has been invaded by an organism such as a parasite or virus, or that provides habitat or resources to a commensal organism.

host cell A cell that can be invaded by a parasitic organism or a virus and provides an environment that is conducive to the growth and reproduction of the organism or the replication of the virus.

***Hox* genes** A class of genes found in most animal phyla, including vertebrates, that are expressed in a distinctive pattern along body axes in development and control formation of specific structures. *Hox* genes code for transcription factors with a DNA-binding sequence called a homeobox.

human Any member of the genus *Homo*, which includes modern humans (*Homo sapiens*) and several extinct species.

human chorionic gonadotropin (hCG) A glycoprotein hormone produced by a human embryo and placenta from about week 3 to week 14 of pregnancy. Maintains the corpus luteum, which secretes hormones that preserve the uterine lining.

Human Genome Project The multinational research project that sequenced the human genome.

human immunodeficiency virus (HIV) A retrovirus that causes acquired immune deficiency syndrome (AIDS) in humans.

humoral (immune) response The type of immune response that is mediated through the production and secretion of antibodies, complement proteins, and other soluble factors that eliminate extracellular pathogens. Compare with **cell-mediated (immune) response**.

humus The decayed organic matter in soils.

hybrid The offspring of parents from two different strains, populations, or species.

hybrid zone A geographic area where interbreeding occurs between two species, sometimes producing fertile hybrid offspring.

hydrocarbon An organic molecule that contains only hydrogen and carbon atoms.

hydrogen bond A weak interaction between two molecules or different parts of the same molecule resulting from the attraction between a hydrogen atom with a partial positive charge and another atom (usually O or N) with a partial negative charge. Compare with **covalent bond** and **ionic bond**.

hydrogen ion (H^+) A single proton with a positive electric charge of +1; typically, one that is dissolved in solution or that is being transferred from one atom to another in a chemical reaction.

hydrolysis A chemical reaction in which a molecule is split into smaller molecules by reacting with water. In biology, most hydrolysis reactions involve the splitting of polymers into monomers. Compare with **condensation reaction**.

hydrophilic Interacting readily with water. Hydrophilic compounds are typically polar compounds containing partially or fully charged atoms. Compare with **hydrophobic**.

hydrophobic Not interacting readily with water. Hydrophobic compounds are typically nonpolar molecules. Compare with **hydrophilic**.

hydrophobic interactions Very weak interactions between nonpolar molecules, or nonpolar regions of the same molecule, when exposed to an aqueous solvent. The surrounding water molecules support these interactions by interacting with one another and encapsulating the nonpolar molecules.

hydroponic growth Growth of plants in liquid cultures instead of soil.

hydrostatic skeleton A system of body support involving a body wall in tension surrounding a fluid or soft tissue under compression. Also called *hydrostat*. Compare with **endoskeleton** and **exoskeleton**.

hydroxide ion (OH^-) An oxygen atom and a hydrogen atom joined by a single covalent bond and carrying a negative electric charge of −1.

hygiene hypothesis The hypothesis that allergies and autoimmune disorders arise in individuals who have had less exposure to pathogens and parasites, especially in early childhood. May explain the increased incidence of allergies and autoimmune disease in countries with high levels of sanitation.

hyperosmotic Referring to a solution that has a higher solute concentration, and therefore a lower water concentration, than another solution. Compare with **hyposmotic** and **isosmotic**.

hyperpolarization A change in membrane potential from its resting negative state to an even more negative state; a normal phase in an action potential. Compare with **depolarization** and **repolarization**.

hypersensitive reaction An intense allergic response by cells that have been sensitized by previous exposure to an allergen.

hypersensitive response (HR) In plants, the rapid death of a cell that has been infected by a pathogen, thereby reducing the potential for infection to spread throughout a plant. Compare with **systemic acquired resistance (SAR)**.

hypertension Abnormally high blood pressure.

hypertonic Comparative term designating a solution that, if outside a cell or vesicle, results in the loss of water and shrinkage of the membrane-bound structure. This solution has a greater solute concentration than the solution on the other side of the membrane. Used when the solute is unable to pass through the membrane. Compare with **hypotonic** and **isotonic**.

hypha (plural: *hyphae*) One of the long, branching strands of a fungal mycelium (the mesh-like body of a fungus). Also found in some protists.

hypocotyl The stem of a very young plant; the region between the cotyledon (embryonic leaf) and the radicle (embryonic root).

hyposmotic Referring to a solution that has a lower solute concentration, and therefore a higher water concentration, than another solution. Compare with **hyperosmotic** and **isosmotic**.

hypothalamic–pituitary axis The functional interaction of the hypothalamus and anterior pituitary gland, which are anatomically distinct but work together to regulate most of the other endocrine glands in the body.

hypothalamus A part of the brain that functions in maintaining the body's internal physiological state by regulating the autonomic nervous system, endocrine system, body temperature, water balance, and appetite.

hypothesis A testable statement that explains a phenomenon or a set of observations.

hypotonic Comparative term designating a solution that, if outside a cell or vesicle, results in the uptake of water and swelling or even bursting of the membrane-bound structure. This solution has a lower solute concentration than the solution on the other side of the membrane. Used when the solute is unable to pass through the membrane. Compare with **hypertonic** and **isotonic**.

immigration The migration of individuals into a particular population from other populations. Compare with **emigration**.

immobile nutrients nutrients that remain in position where they are first deposited, typically tied up in older leaves. Compare with **mobile nutrients**.

immune system The system whose primary function is to defend an organism against pathogens. Includes several types of cells (e.g., leukocytes). In vertebrates, also includes several organs where specialized cells mature or reside (e.g., thymus and lymph nodes).

immunity (adjective: *immune*) A state of being protected against infection by a pathogen.

immunization The conferring of immunity to a particular disease by artificial means.

immunoglobulin (Ig) An antibody or B-cell receptor.

immunological memory The ability of the immune system to mount a rapid, effective response to an antigen encountered years or decades earlier. In adaptive immunity, it is based on the formation of memory lymphocytes.

impact hypothesis The hypothesis that a collision between Earth and an asteroid caused the mass extinction that ended the Cretaceous, 66 million years ago.

imperfect flower A flower that contains male parts (stamens) *or* female parts (carpels) but not both. Compare with **perfect flower**.

implantation The process by which an embryo becomes embedded in the wall of a uterus (or oviduct), leading to the formation of a placenta. Occurs in mammals and some other viviparous vertebrates.

in situ hybridization A technique for detecting specific DNAs and mRNAs in cells and tissues by use of labeled complementary probes. Can be used to determine where and when particular genes are expressed in embryos.

inbreeding Mating between closely related individuals. Increases homozygosity of a population and often leads to a decline in the average fitness via selection (inbreeding depression).

inbreeding depression The decline in average fitness that takes place when homozygosity increases and heterozygosity decreases in a population due to inbreeding; results from the exposure of deleterious recessive alleles to selection.

inclusive fitness The combination of (1) direct production of offspring (direct fitness) and (2) extra production of offspring by relatives in response to help provided by the individual in question (indirect fitness).

incomplete digestive tract A digestive tract that has just one opening.

incomplete dominance An inheritance pattern in which the phenotype of a heterozygote is between the phenotypes of the homozygotes.

incomplete metamorphosis A type of metamorphosis in which the animal increases in size from one stage to the next, but typically does not dramatically change its body form.

independent assortment, principle of The concept that each pair of hereditary elements (alleles of the same gene) segregates (separates) independently of alleles of other genes during meiosis. One of Mendel's two principles of genetics. Compare with **principle of segregation**.

indeterminate growth A pattern of growth in which an individual continues to increase its overall body size throughout its life.

induced fit Change in the shape of the active site of an enzyme, as the result of initial weak binding of a substrate, so that it binds substrate more tightly.

induced pluripotent stem cell (iPS cell) An undifferentiated cell made in experimental settings by the de-differentiation of a specialized adult cell. Induced pluripotent cells have the potential to form any cell of the body if re-differentiated under appropriate conditions.

inducer A molecule that triggers transcription of a specific gene, often by binding to and inactivating a repressor protein.

inducible defense A physical, chemical, or behavioral defensive trait that is manifested only in response to the presence of a consumer (predator, or herbivore, or parasite) or pathogen. Compare with **constitutive defense**.

induction A mechanism of determining the fate of a cell during development that involves a signal from one cell determining what another cell will become during development.

infection thread An invagination of the plasma membrane of a root hair through which beneficial nitrogen-fixing bacteria enter the roots of their host plants (legumes).

infectious disease Disease caused by viruses, bacteria, fungi, or parasites that can be transmitted from one organism to another or acquired from the environment.

inflammatory response An aspect of the innate immune response, seen in most cases of infection or tissue injury, in which the affected tissue becomes swollen, red, warm, and painful.

ingestion The act of bringing food into the digestive tract.

inhibition In ecological succession, the phenomenon in which early-arriving species make conditions less favorable for the establishment of certain later-arriving species. Compare with **facilitation** and **tolerance**.

inhibitory postsynaptic potential (IPSP) A change in membrane potential, usually hyperpolarization, of a cell that makes the cell less likely to generate an action potential. Compare with **excitatory postsynaptic potential (EPSP)**.

initiation (1) In an enzyme-catalyzed reaction, the stage during which enzymes orient reactants precisely as they bind at specific locations within the enzyme's active site. (2) In DNA transcription, the stage during which RNA polymerase and other proteins assemble at the promoter sequence and open the strands of DNA to start transcription. (3) In translation, the stage during which a complex consisting of initiation factor proteins, a ribosome, an mRNA, and an aminoacyl tRNA corresponding to the start codon is formed to begin translation.

initiation factor Any of a class of proteins that assist ribosomes in binding to a messenger RNA molecule to begin translation.

innate behavior Behavior that is inherited genetically, does not have to be learned, and is typical of a species.

innate immunity A set of barriers to infection and generic defenses against broad types of pathogens. Produces an immediate response that involves many different leukocytes, which often activate an inflammatory response. Also called *innate immune response*. Compare with **adaptive immunity**.

inner cell mass (ICM) A cluster of cells, in the interior of a mammalian blastocyst, that develops into the embryo. Contrast with **trophoblast**.

inner ear The innermost portion of the mammalian ear, consisting of a fluid-filled system of tubes that includes the cochlea (which receives sound vibrations from the middle ear) and the semicircular canals (which function in balance).

insects A mostly terrestrial lineage of arthropods distinguished by three tagmata (head, thorax, abdomen), a single pair of antennae, and unbranched appendages.

insulin A peptide hormone produced by the pancreas in response to high levels of glucose (or amino acids) in the blood. Enables cells to absorb glucose and coordinates synthesis of fats, proteins, and glycogen. Compare with **glucagon**.

integral membrane protein Any membrane protein that spans the entire lipid bilayer. Also called *transmembrane protein*. Compare with **peripheral membrane protein**.

integrated pest management In agriculture or forestry, systems for managing insects or other pests that include carefully controlled applications of toxins, introduction of species that prey on pests, planting schemes that reduce the chance of a severe pest outbreak, and other techniques.

integrator A component of an animal's nervous system that functions as part of a homeostatic system by evaluating sensory information and triggering appropriate responses. See **effector** and **sensor**.

integrin Any of a class of cell-surface proteins that bind to proteins in the extracellular matrix, thus holding cells in place. Some integrins also function in cell–cell adhesions.

intercalated disc A type of specialized connection between adjacent heart muscle cells that contains gap junctions, allowing electrical signals to pass between the cells.

intermediate filament A long fiber, about 10 nm in diameter, composed of one of various proteins (e.g., keratins, lamins); one of the three types of cytoskeletal fibers. Used to form networks that help maintain cell shape and hold the nucleus in place. Compare with **actin filament** and **microtubule**.

intermediate muscle fiber A type of skeletal muscle fiber that is pink, generates ATP by both glycolysis and aerobic respiration, and has contractile properties that are intermediate between those of slow fibers and fast fibers. Also called fast oxidative/glycolytic fiber.

interneuron A neuron that passes signals from one neuron to another. Compare with **motor neuron** and **sensory neuron**.

internode The section of a plant stem between two nodes (sites where leaves attach).

interphase The portion of the cell cycle between one M phase and the next. Includes G_1 phase, S phase, and G_2 phase.

intersexual selection A type of sexual selection in which an individual of one sex choses a particular individual of the other sex for mating (often occurs through female choice).

interspecific competition Competition between members of different species for the same limited resource. Compare with **intraspecific competition**.

interstitial fluid In animals with a closed circulatory system, extracellular fluid that is not enclosed in blood vessels or lymphatic vessels.

intertidal zone The region between the low-tide and high-tide marks on a seashore.

intrasexual selection A type of sexual selection driven by competition among members of one sex (usually male–male) for an opportunity to mate.

intraspecific competition Competition between members of the same species for the same limited resource. Compare with **interspecific competition**.

intrinsic rate of increase (r_{max}) The rate of increase of a population when conditions are ideal (birth rates per individual are as high as possible and death rates per individual are as low as possible). Compare with **finite rate of increase** and **per capita rate of increase**.

intron A region of a gene that is transcribed into RNA but is later removed. Compare with **exon**.

invasive species An exotic (nonnative) species that, upon introduction to a new area, spreads rapidly and competes successfully with native species.

inversion A mutation in which a DNA sequence or a segment of a chromosome is flipped into the reversed orientation.

invertebrates A non-monophyletic group composed of animals without a backbone; includes about 95 percent of all animal species. Compare with **vertebrates**.

involuntary muscle A muscle that contracts without input from the nervous system or in response to stimulation by involuntary (parasympathetic or sympathetic), but not voluntary (somatic), neural stimulation. Compare with **voluntary muscle**.

ion An atom or a molecule that has lost or gained electrons and thus carries a full electric charge, either positive (cation) or negative (anion), respectively.

ion channel A type of channel protein that allows certain ions to diffuse across a plasma membrane down an electrochemical gradient.

ionic bond A chemical bond that is formed when an electron is completely transferred from one atom to another. Resulting ions remain associated due to their opposite electric charges. Compare with covalent bond and hydrogen bond.

iPS cell See **induced pluripotent stem cell**.

iris A ring of pigmented muscle just behind the cornea in the vertebrate eye that contracts or expands to control the amount of light entering the eye through the pupil.

isosmotic Referring to a solution that has the same solute concentration and water concentration as another solution. Compare with **hyperosmotic** and **hyposmotic**.

isotonic Comparative term designating a solution that, if inside a cell or vesicle, results in no net uptake or loss of water and thus no effect on the volume of the membrane-bound structure. This solution has the same solute concentration as the solution on the other side of the membrane. Compare with **hypertonic** and **hypotonic**.

isotope Any of several forms of an element that differ in the number of neutrons contained in their nuclei.

joint A place where two components (bones, cartilages, etc.) of a skeleton meet. May be movable (called an articulation) or immovable (e.g., skull sutures). Also called **articulation**.

juvenile An individual that has adult-like morphology but is not sexually mature.

juvenile hormone (JH) An insect hormone that prevents larvae from metamorphosing into adults.

karyogamy Fusion of two haploid nuclei to form a diploid nucleus. Occurs in many fungi, and in animals and plants during fertilization of gametes.

karyotype The distinctive appearance of all the metaphase or prometaphase chromosomes in an individual, including the number of chromosomes, their length, and their banding patterns.

keystone species A species that has an exceptionally large impact on the other species in its ecosystem relative to its abundance.

kidney In terrestrial vertebrates, one of a pair of organs situated at the back, behind the abdominal cavity, that filters the blood, produces urine, and secretes several hormones.

kilocalorie (kcal) A unit of energy often used to measure the energy content of food. A kcal of energy raises 1 kg of water 1°C.

kin selection A form of natural selection that favors traits that increase survival or reproduction of an individual's kin at the expense of the individual.

kinesin A class of motor proteins that uses the chemical energy of ATP to "walk" toward the plus end of a microtubule. Used to transport vesicles, particles, organelles, and chromosomes.

kinetic energy The energy of motion. Compare with **potential energy**.

kinetochore A protein complex that forms on a chromosome during M phase. Forms at the centromere and serves as a site for microtubule attachment. Contains motor proteins and microtubule-binding proteins that are involved in chromosome segregation during M phase.

kinetochore microtubule A microtubule in the spindle apparatus formed during mitosis or meiosis that is attached to the kinetochore on a chromosome.

kinocilium (plural: *kinocilia*) A single cilium that juts from the surface of many hair cells and functions in detecting pressure waves.

Koch's postulates Four criteria used to determine whether a suspected infectious agent causes a particular disease.

labium majus (plural: *labia majora*) One of two outer folds of skin that surround the labia minora, clitoris, and vaginal opening of female mammals.

labium minus (plural: *labia minora*) One of two folds of skin inside the labia majora and surrounding the opening of the urethra and vagina.

labor The strong muscular contractions of the uterus that expel the fetus from the uterus during birth.

***lac* operon** A set of three genes in *E. coli* that are transcribed into a single mRNA and required for metabolism of the sugar lactose. Studies of the *lac* operon revealed many insights about gene regulation.

lactation (verb: *lactate*) Production of milk by the mammary glands of mammals, to feed offspring.

lacteal A small lymphatic vessel extending into the center of a villus in the small intestine. Receives chylomicrons containing fat absorbed from food.

lactic acid fermentation Catabolic pathway in which pyruvate produced by glycolysis is converted to lactic acid in order to oxidize NADH to NAD^+.

lagging strand In DNA replication, the new strand of DNA that is synthesized discontinuously (as a series of short pieces that are later joined) in a direction moving away from the replication fork. Also called *discontinuous strand*. Compare with **leading strand**.

laminin An abundant protein in the extracellular matrix (ECM) of animals that binds to other ECM components and to integrins, helping to anchor cells. Predominantly found in the basal lamina. Many tissue-specific subtypes exist.

land plants A monophyletic group of organisms that contain chloroplasts similar to those in green algae. The vast majority of land plants are photosynthetic and live in terrestrial habitats. Includes mosses, hornworts, liverworts, and all vascular plants.

landscape ecology A field of ecological study that measures the structure and ecological function of landscapes that are a patchwork of fragments of natural ecosystems mixed with human-impacted areas such as roads, farms, and towns.

large intestine The posterior portion of the vertebrate digestive tract, consisting of the cecum, colon, and rectum. Its primary function is to form feces by absorbing water from and compacting the wastes delivered from the small intestine.

larva (plural: *larvae*). An immature form of an animal species in which the immature and adult stages have different body forms.

late endosome A membrane-bound vesicle that arises from an early endosome, accepts lysosomal enzymes from the Golgi, and matures into a lysosome.

latency In certain viruses that infect animals, a dormant state of coexistence with the host cell during which no new virions are produced. The viral genetic material is replicated as the host cell replicates. Genetic material may or may not be integrated in the host genome, depending on the virus. Compare with **lytic cycle** and **lysogeny**.

lateral bud A bud that forms at a node and may develop into a lateral (side) branch. Also called *axillary bud*.

lateral gene transfer Transfer of DNA between two different species. Also known as *horizontal gene transfer*. Compare with **chromosome-level mutation** and **point mutation**.

lateral line system A pressure-sensitive sensory organ found in many aquatic vertebrates.

lateral root A plant root that extends horizontally from another root.

leaching Loss of nutrients from soil via percolating water.

leading strand In DNA replication, the new strand of DNA that is synthesized in one continuous piece in a direction that follows the replication fork. Also called *continuous strand*. Compare with **lagging strand**.

leaf The main photosynthetic organ of vascular plants.

leak channel A type of ion channel that allows ions to leak across the membrane of a neuron in its resting state.

learning An enduring change in an individual's behavior that results from a specific experience.

leghemoglobin An iron-containing protein similar to hemoglobin. Found in infected cells of legume root nodules where it binds oxygen, preventing it from poisoning a bacterial enzyme needed for nitrogen fixation.

legumes Members of the pea plant family. Many form symbiotic associations with nitrogen-fixing bacteria in their roots.

lens A transparent structure that focuses incoming light onto a retina or other light-sensing apparatus of an eye.

lenticel Spongy segment in bark that allows gas exchange between cells in a woody stem and the atmosphere.

leukocyte Any of several types of blood cells, including neutrophils, macrophages, and lymphocytes, that reside in tissues and circulate in blood and lymph. Functions in tissue repair and defense against pathogens. Also called *white blood cell*.

lichen A mutualistic association of a fungus, often in the Ascomycota lineage, and a photosynthetic alga or cyanobacterium.

life cycle The sequence of developmental events and phases over the life span of an organism, from fertilization to offspring production.

life history The sequence of events in an individual's life from birth to reproduction to death, including how an individual allocates resources to growth, reproduction, and activities or structures that are related to survival.

life table A data set that summarizes the probability that an individual in a certain population will survive and reproduce in any given year over the course of its lifetime.

ligament A connective tissue that joins bones of an endoskeleton.

ligand Any molecule that binds to a specific site on a receptor molecule.

ligand-gated channel A type of ion channel that opens in response to binding by a certain molecule. Compare with **voltage-gated channel**.

light chain The smaller of the two types of polypeptide chains in an antibody or B-cell receptor; composed of a variable (*V*) region, which contributes to the antigen-binding site, and a constant (*C*) region. Compare with **heavy chain**.

light-harvesting complex A mobile accessory structure consisting of pigments that absorb light and transmit resonance energy to antenna pigments in photosystems.

lignin A complex polymer built from six-carbon rings and found in the secondary cell walls of some plants; it is exceptionally stiff and strong. Most abundant in woody plant parts.

limiting nutrient Any essential nutrient whose scarcity in the environment significantly reduces growth and reproduction of organisms.

limnetic zone Open water (not near shore) that receives enough sunlight to support photosynthesis but is too deep for plants to take root.

LINE See **long interspersed nuclear element**.

lineage See **monophyletic group**.

lingual lipase An enzyme produced by glands in the tongue that breaks down triglycerides into fatty acids and diglycerides.

linkage In genetics, a physical association between two genes because they are on the same chromosome; also the inheritance patterns showing dependent assortment of alleles that result from this association.

lipid Any organic substance that does not dissolve in water, but dissolves well in nonpolar organic solvents. Lipids include fatty acids, fats, oils, waxes, steroids, and phospholipids.

lipid bilayer The basic structural element of all cellular membranes; consists of a two-layer sheet of phospholipid molecules with their hydrophobic tails oriented toward the inside and their hydrophilic heads toward the outside. Also called *phospholipid bilayer*.

liposome An artificial vesicle formed by mixing amphipathic lipids, such as phospholipids, together in an aqueous solution.

littoral zone Shallow water near shore that receives enough sunlight to support photosynthesis and is shallow enough for plants to take root; often flowering plants are present.

liver A large, complex organ of vertebrates that performs many functions, including storage of glycogen, processing and conversion of food and wastes, and production of bile.

lobe-finned fishes A non-monophyletic group of bony fishes (including the living coelacanths and lungfish) that are more closely related to tetrapods than to ray-finned fishes; they have muscular fins with a single bone at the base.

locomotion Movement of an organism or cell from place to place under its own power.

locus (plural: *loci*) A gene's physical location on a chromosome.

logistic population growth The density-dependent decrease in growth rate as population size approaches the carrying capacity. Compare with **exponential population growth**.

long-day plant A plant that blooms in response to short nights (usually in late spring or early summer in the Northern Hemisphere). Compare with **day-neutral plant** and **short-day plant**.

long interspersed nuclear element (LINE) Any of the most abundant class of transposable elements in human genomes; can create copies of itself and insert them elsewhere in the genome.

long noncoding RNA (lncRNA) An abundant class of transcribed RNA molecules that are greater than 200 nucleotides in length and do not code for a protein.

loop of Henle In the kidneys of mammals and some birds, a long, U-shaped loop in a nephron that extends into the medulla. Functions as a countercurrent exchanger and multiplier in establishing and maintaining a medullary osmotic gradient that allows reabsorption of water from the collecting duct. Also called *nephron loop*.

loose connective tissue A type of connective tissue consisting of fibrous proteins in a soft matrix. Often functions as padding for organs.

lophophore A specialized feeding structure found in some lophotrochozoans and used in suspension (filter) feeding.

lophotrochozoans A major lineage of protostomes (Lophotrochozoa) that grow by extending their skeletons rather than by molting. Many phyla have a specialized feeding structure (lophophore) and/or ciliated larvae (trochophore). Includes rotifers, flatworms, segmented worms, and mollusks. Compare with **ecdysozoans**.

loss-of-function allele A mutant allele that does not produce a functional gene product. Also called **null allele**.

LUCA The *l*ast *u*niversal *c*ommon *a*ncestor of cells. This theoretical entity is proposed to be the product of chemical evolution and provided characteristics of life that are shared by all living organisms on Earth today.

lumen The interior space of any hollow structure (e.g., the rough ER) or organ (e.g., the stomach).

lung Any respiratory organ used for gas exchange between blood or hemolymph and air.

luteal phase In a menstrual cycle, the phase after ovulation, when the progesterone level is high and the body is preparing for a possible pregnancy.

luteinizing hormone (LH) A peptide hormone, produced and secreted by the anterior pituitary, that stimulates estrogen production, ovulation, and formation of the corpus luteum in females and testosterone production in males.

lymph The mixture of fluid and white blood cells that circulates through the ducts and lymph nodes of the lymphatic system in vertebrates.

lymph node Any of many small, oval structures that lymph moves through in the lymphatic system. Filters the lymph and screens it for pathogens and antigens. A major site of lymphocyte activation.

lymphatic system In vertebrates, a body-wide network of thin-walled, branching tubules and lymph nodes that is separate from the circulatory system. Collects excess fluid from body tissues and returns it to the blood; also functions as part of the immune system.

lymphocyte A cell that circulates through the bloodstream and lymphatic system and is responsible for the development of adaptive immunity. Key lymphocytes in the adaptive immune system are B cells and T cells.

lysogeny In certain viruses that infect bacteria (bacteriophages), a dormant state of coexistence with the host cell during which no new virions are produced. The viral genetic material is integrated in the host chromosome and replicated as the host cell replicates. Compare with **lytic cycle** and **lysogeny**.

lysosome A small, acidified organelle in animal cells and some plant cells (modified vacuoles) containing enzymes that catalyze hydrolysis reactions and can digest large molecules.

lysozyme An enzyme that functions in innate immunity by digesting bacterial cell walls. Occurs in lysosomes of phagocytes and is secreted in saliva, tears, and mucus.

lytic cycle A type of viral replicative growth in which the production and release of virions kills the host cell. Compare with **lysogeny** and **latency**.

M phase The phase of the cell cycle during which cell division occurs. Includes mitosis or meiosis and often cytokinesis.

M-phase-promoting factor (MPF) A complex of a cyclin and cyclin-dependent kinase that, when activated, phosphorylates a number of specific proteins that initiate mitosis or meiosis in eukaryotic cells.

macromolecular machine A group of proteins, and possibly other nonprotein macromolecules, that assemble to carry out a particular function.

macromolecule Generally, any large organic molecules made up of smaller molecules (monomers) joined together into a polymer. The main biological macromolecules are proteins, nucleic acids, and polysaccharides.

macronutrient Any element (e.g., nitrogen) that is required in large quantities for normal growth, reproduction, and maintenance of a living organism. Compare with **micronutrient**.

macrophage A type of white blood cell in the innate immune system that participates in the inflammatory response by secreting cytokines and phagocytizing invading pathogens and apoptotic cells. Also serves as an antigen-presenting cell in lymphocyte activation.

macropinocytosis A form of endocytosis in which a large vesicle containing fluid and small particles is pinched off from the cell membrane and brought into the interior of the cell.

macula (plural: *maculae*) A structure in the vestibule of the inner ear that detects orientation and linear acceleration.

magnetoreception The detection of magnetic fields.

magnetoreceptor A sensory cell specialized for detecting magnetic fields.

major histocompatibility protein See **class I and class II MHC proteins**.

malaria A human disease caused by five species of the protist *Plasmodium* and passed to humans by mosquitoes.

malignant tumor A tumor that is actively growing and disrupting local tissues or is spreading to other organs. Cancer consists of one or more malignant tumors. Compare with **benign tumor**.

Malpighian tubules A major excretory organ of insects, consisting of blind-ended tubes that extend from the gut into the insect's body cavity. They filter hemolymph and send the filtrate to the hindgut for further processing.

mammals One of the two lineages of amniotes (vertebrates that produce amniotic eggs) distinguished by hair (or fur) and mammary glands. Includes the monotremes (platypus and echidnas), marsupials, and eutherians (placental mammals).

mammary glands Specialized exocrine glands that produce and secrete milk for nursing offspring. A diagnostic feature of mammals.

mandibles Any mouthpart used in chewing. In vertebrates, the lower jaw. In insects, crustaceans, and myriapods, the first pair of mouthparts.

mantle One of the three main parts of the mollusk body; the thick outer tissue that protects the visceral mass and may secrete a calcium carbonate shell.

marsupials A lineage of mammals (Marsupiala) that nourish their young in an abdominal pouch after a very short period of development in the uterus.

mass extinction The extinction of a large number of diverse species around the world during a relatively short period of geologic time (about 1 million years). May occur due to sudden and extraordinary environmental changes. Compare with **background extinction**.

mass feeder An animal that ingests chunks of food.

mass number The total number of protons and neutrons in the nucleus of an atom.

mast cell A type of white blood cell that is stationary (embedded in tissue) and secretes factors, including histamine, that help trigger the inflammatory response. Particularly important in allergic responses and defense against parasites.

master regulator A gene product that can unleash a series of events that produce a specialized cell type, tissue, or body structure.

maternal chromosome A chromosome inherited from the mother.

maternal effect mutation A mutation in a mother that alters the phenotype of her offspring without causing an obvious mutant phenotype in the mother. Is explained by altered cytoplasmic determinants that control embryonic development that are placed in the egg by the mother.

mechanoreception The detection of changes in pressure.

mechanoreceptor A sensory cell specialized for detecting distortions caused by touch or pressure. Examples include hair cells in the cochlea and the lateral line system of vertebrates.

Mediator A large complex of proteins in eukaryotes that creates a physical link between regulatory transcription factors that are bound to DNA, the general transcription factors, and RNA polymerase.

medulla The innermost part of an organ (e.g., kidney or adrenal gland). Compare with **cortex**.

medulla oblongata In vertebrates, a region of the brain stem that, along with the cerebellum, forms the hindbrain.

medusa (plural: *medusae*) The free-floating stage in the life cycle of some cnidarians (e.g., jellyfish). Compare with **polyp**.

megapascal (MPa) A unit of pressure (force per unit area) equivalent to 1 million pascals (Pa).

megasporangium (plural: *megasporangia*) In heterosporous species of plants, a spore-producing structure that produces megaspores; these develop into female gametophytes.

megaspore In seed plants, a haploid (n) spore that is produced in a megasporangium by meiosis of a diploid ($2n$) megasporocyte; develops into a female gametophyte. Compare with **microspore**.

meiosis In sexually reproducing organisms, a special two-stage type of cell division in which one diploid ($2n$) parent cell produces haploid (n) cells (gametes); results in halving of the chromosome number. Also called *reduction division*.

meiosis I The first cell division of meiosis, in which synapsis, crossing over, and independent assortment occur. It separates homologous chromosomes from each other, producing daughter cells with half as many chromosomes as the parent cell.

meiosis II The second cell division of meiosis, in which sister chromatids are separated from each other. Meiosis II does not reduce chromosome number and is similar to **mitosis**.

melatonin A hormone, produced by the pineal gland, that regulates sleep–wake cycles and seasonal reproduction in vertebrates.

membrane potential A difference in electric charge across a cell membrane; a form of potential energy. Also called *membrane voltage*.

memory Retention of learned information.

memory cells A type of lymphocyte responsible for maintaining immunity for years or decades after an infection. Descended from activated B cells or T cells that responded to an infection or vaccination.

meniscus (plural: *menisci*) The concave boundary layer formed at most air–water interfaces due to adhesion and surface tension.

menstrual cycle A female reproductive cycle seen in Old World monkeys and apes (including humans) in which the uterine lining is shed (menstruation) if no pregnancy occurs. Compare with **estrous cycle**.

menstruation The periodic shedding of the uterine lining through the vagina that occurs in females of Old World monkeys and apes, including humans.

meristem (adjective: *meristematic*) In plants, a group of undifferentiated cells, including stem cells, that can divide and develop into various adult tissues throughout the life of a plant. See also **apical meristem** and **ground meristem**.

mesoderm The middle of the three basic cell layers (germ layers) in most animal embryos; gives rise to muscles, bones, blood, and some internal organs (kidney, spleen, etc.). Compare with **ectoderm** and **endoderm**.

mesoglea A gelatinous material, containing scattered ectodermal cells, that is located between the ectoderm and endoderm of cnidarians (e.g., jellyfish, corals, and anemones).

mesophyll cell A type of cell, found near the surfaces of plant leaves, that is specialized for the lightcapturing reactions of photosynthesis.

Mesozoic era The interval of geologic time, from 252 million to 66 million years ago, during which gymnosperms were the dominant plants on land and dinosaurs were the dominant vertebrates on land. Ended with extinction of the dinosaurs (except birds).

messenger RNA (mRNA) An RNA molecule transcribed from DNA that carries information (in codons) that specifies the amino acid sequence of a polypeptide.

metabolic pathway A linked series of biochemical reactions that sequentially changes an initial substrate to form a final product; the product of one reaction is the substrate of the next reaction.

metabolic rate The rate of energy use by all the cells of an individual. For aerobic organisms, often measured as the amount of oxygen consumed per hour.

metabolism The sum of all chemical reactions that occur within an organism.

metagenomics The inventory of all the genes in a community or ecosystem created by sequencing, analyzing, and comparing the genomes of the component organisms. Often refers to the study of microbial communities. Also called *environmental sequencing*.

metallothioneins Small plant proteins that bind to and prevent excess metal ions from acting as toxins.

metamorphosis A drastic transition from one developmental stage to another, such as from the larval to the adult form of an animal.

metaphase A stage in mitosis or meiosis during which chromosomes line up across the middle of the spindle.

metaphase plate An imaginary plane (between the spindle poles) along which chromosomes line up during metaphase of mitosis or meiosis.

metapopulation A population made up of many small, physically isolated populations connected by dispersal.

metastasis The spread of cancerous cells from their site of origin to distant sites in the body, where they may establish additional tumors.

methanogen A prokaryote that produces methane (CH_4) as a by-product of cellular respiration.

methanotroph A prokaryote that uses methane (CH_4) as its primary electron donor and source of carbon.

methyl salicylate (MeSA) A molecule that is hypothesized to function as a signal, transported among tissues, that triggers systematic acquired resistance in plants—a response to pathogen attack.

microbe Any microscopic organism, including bacteria, archaea, and various tiny eukaryotes.

microbiology The field of study concerned with microscopic organisms.

microbiome The ecological community of microbes, and their genomes, that share a particular space (e.g., the human gut).

microfibril A bundle of cellulose strands that serves as the fibrous component in plant cell walls.

microfilament See **actin filament**.

micronutrient Any element (e.g., iron, molybdenum, magnesium) that is required in very small quantities for normal growth, reproduction, and maintenance of a living organism. Compare with **macronutrient**.

micropyle The tiny pore in a plant ovule through which the pollen tube reaches the embryo sac.

microRNA (miRNA) A small, single-stranded RNA associated with proteins in an RNA-induced silencing complex (RISC). Processed from a longer miRNA gene transcript. Can bind to complementary sequences in mRNA molecules, allowing the associated proteins of RISC to degrade the bound mRNA or inhibit its translation. See **RNA interference**.

microsatellite See **short tandem repeat (STR)**.

microsporangium (plural: *microsporangia*) In heterosporous species of plants, a spore-producing structure that produces microspores; these develop into male gametophytes.

microspore In seed plants, a haploid (*n*) spore that is produced in a microsporangium by meiosis of a diploid (2*n*) microsporocyte; develops into a male gametophyte. Compare with **megaspore**.

microtubule A long tubular fiber, about 25 nm in diameter, formed by polymerization of tubulin protein dimers; one of the three types of cytoskeletal fibers. Involved in cell structure, movement, and transport of materials within the cell. Compare with **actin filament** and **intermediate filament**.

microtubule-organizing center (MTOC) General term for any structure (e.g., centrosome and basal body) where new microtubules originate in cells.

microvilli (singular: *microvillus*) Tiny protrusions from the surface of an epithelial cell that increase the surface area for absorption of substances.

midbrain One of the three main regions of the vertebrate brain; includes sensory integrating and relay centers. Compare with **forebrain** and **hindbrain**.

middle ear The air-filled middle portion of the mammalian ear, which contains three small bones that amplify sound and transmit it from the tympanic membrane to the inner ear.

migration (1) In ecology, a seasonal long-distance movement of large numbers of organisms from one geographic location or habitat to another. (2) In population genetics, movement of individuals from one population to another.

millivolt (mV) A unit of voltage equal to 1/1000 of a volt.

mimicry A phenomenon in which one species has evolved (or learns) to look or sound like another species. See **Batesian mimicry** and **Müllerian mimicry**.

mineralocorticoid Any of a class of steroid hormones, produced and secreted by the adrenal cortex, that regulate electrolyte levels and the overall volume of body fluids. Aldosterone is the principal mineralocorticoid in humans. Compare with **glucocorticoid**.

minerals Various inorganic substances that are important components of enzyme cofactors or structural materials in an organism.

mismatch repair A type of DNA repair used to correct mismatched base pairs in DNA that result from mistakes in DNA synthesis.

missense mutation A point mutation (change in a single base pair) that changes one amino acid for another within a protein.

mitochondrial DNA (mtDNA) A small circular (or linear in some species) chromosome that contains some of the genes that encode proteins and functional RNA molecules within mitochondria. There are normally many copies of this DNA in each mitochondrion.

mitochondrial matrix Central compartment of a mitochondrion, which is lined by the inner membrane; contains mitochondrial DNA, ribosomes, and the enzymes for pyruvate processing and the citric acid cycle.

mitochondrion (plural: *mitochondria*) A eukaryotic organelle that is surrounded by two membranes and is the site of aerobic respiration and ATP synthesis.

mitogen-activated protein kinase (MAPK) Any of a class of enzymes involved in signal transduction pathways that often activate cell division. In a cell, different types of MAPKs are organized in a series, where one activates another via phosphorylation. See also **phosphorylation cascade**.

mitosis In eukaryotic cells, the process of nuclear division that results in two daughter nuclei that have chromosomes and genes identical to the parent nucleus. Subsequent cytokinesis (division of the cytoplasm) yields two daughter cells.

mobile nutrients Nutrients that can be transported within plants, typically to regions where they are needed most (e.g., from older leaves to younger leaves). Compare with **immobile nutrients**.

mode of transmission The type of inheritance observed as a trait is passed from parent to offspring. Some common types are autosomal recessive, autosomal dominant, and X-linked recessive.

model organism An organism selected for scientific study based on features that make it easy to work with (e.g., body size, life span), in the hope that findings will apply to other species.

molarity The number of moles of a solute present in 1 liter of solution.

mole The amount of a substance that contains 6.022 $\times 10^{23}$ of its elemental entities (e.g., atoms, ions, or molecules). This number of molecules will have a mass equal to its molecular weight expressed in grams.

molecular biology Branch of biology that seeks to understand life through an underderstanding of the molecules that make up cells and the interactions between these molecules.

molecular chaperone A protein that facilitates the folding or refolding of a protein into its correct three-dimensional shape.

molecular formula A notation that indicates only the numbers and types of atoms in a molecule, such as H_2O for the water molecule. Compare with **structural formula**.

molecular weight The sum of the atomic weights of all the atoms in a molecule; roughly, the total number of protons and neutrons in the molecule.

molecule A substance made up of two or more atoms held together by covalent bonds.

mollusks (also spelled "molluscs") A lineage of protostomes with a body plan based on a muscular foot, visceral mass, and mantle. Includes chitons, clams, oysters, snails, slugs, squid, octopuses, and others.

molting A method of body growth, characteristic of ecdysozoans, that involves the shedding of an external protective cuticle (exoskeleton in arthropods), expansion of the soft body, and growth of a new external cuticle.

monocot Any flowering plant (angiosperm) that has a single cotyledon (embryonic leaf) upon germination. Monocots form a monophyletic group. Also called a monocotyledonous plant. Compare with **dicot**.

monoecious Describing an angiosperm species that has both male and female reproductive structures on each plant. Compare with **dioecious**.

monogamy A type of mating system where one male mates with one female per breeding season, forming a pair bond. Compare with **polygamy** and **promiscuity**.

monohybrid cross A mating between two homozygous parents that differ in alleles of one gene. Produces monohybrid offspring. A monohybrid cross is usually continued through the F_2 generation.

monomer A small molecule that can covalently bind to other similar molecules to form a larger macromolecule. Compare with **polymer**.

monophyletic group An evolutionary unit that includes an ancestral population and all of its descendants but no others. Also called a *clade* or *lineage*. Compare with **paraphyletic group** and **polyphyletic group**.

monosaccharide A molecule that has the molecular formula $(CH_2O)_n$ and cannot be hydrolyzed to form any smaller carbohydrates. Also called *simple sugar*. Compare with **disaccharide, oligosaccharide**, and **polysaccharide**.

monosomy The state of having only one copy of a particular type of chromosome in an otherwise diploid cell.

monotremes A lineage of mammals (Monotremata) that lay eggs and then nourish the young with milk. Includes just five living species: the platypus and four species of echidna, all with leathery beaks or bills.

morphogen A molecule that exists in a concentration gradient and provides spatial information to embryonic cells.

morphology The overall shape and appearance of an organism and its component parts.

morphospecies concept The definition of a species as a population or group of populations that have measurably different anatomical features from other groups. Also called *morphological species concept*. Compare with **biological species concept** and **phylogenetic species concept**.

motor neuron A nerve cell that carries signals from the central nervous system to effector cells in a muscle or gland. Compare with **interneuron** and **sensory neuron**.

motor protein A class of proteins whose major function is to convert the chemical energy of ATP into motion. Includes dynein, kinesin, and myosin.

mucosa-associated lymphoid tissue (MALT) Collective term for lymphocytes and other white blood cells associated with skin cells and mucus-secreting epithelial tissues in the digestive and respiratory tracts. Plays an important role in preventing entry of pathogens into the body.

mucous cell A type of cell, found in the epithelial layer of the stomach, that secretes mucus into the stomach.

mucus (adjective: *mucous*) A slimy mixture of glycoproteins (called mucins) and water that is secreted in many animal organs. Functions include providing lubrication and serving as a barrier to protect surfaces from infection.

Müllerian inhibitory substance A peptide hormone, secreted by the embryonic testis, that causes regression (withering away) of the female reproductive ducts.

Müllerian mimicry A type of mimicry in which two (or more) harmful species resemble each other. Compare with **Batesian mimicry**.

multicellularity The state of being composed of many cells that adhere to each other and do not all express the same genes, resulting in some cells having specialized functions.

multiple allelism The existence of more than two common alleles of the same gene.

multiple fruit A fruit (e.g., pineapple) that develops from many separate flowers and thus many carpels. Compare with **aggregate fruit** and **simple fruit**.

multiple sclerosis (MS) A human autoimmune disease in which the immune system attacks the myelin sheaths that insulate axons of neurons.

muscle fiber A muscle cell.

muscle tissue An animal tissue consisting of bundles of long, thin, contractile cells (muscle fibers). Functions primarily in movement.

mutant An individual that carries a mutation, particularly a new or rare mutation.

mutation Any permanent change in the hereditary material of an organism (DNA in most organisms, RNA in some viruses). The only source of new alleles in populations.

mutualism (adjective: *mutualistic*) A species relationship between two organisms (mutualists) that benefits both. Compare with **commensalism, consumption**, and **competition**.

mutualist An organism that is a participant and partner in a mutualistic relationship. See **mutualism**.

mycelium (plural: *mycelia*) A mass of underground filaments (hyphae) that form the body of a fungus. Also found in some protists and bacteria.

mycorrhiza (plural: *mycorrhizae*) A mutualistic association between certain fungi and the roots of most vascular plants, sometimes visible as nodules or nets in or around plant roots.

mycorrhizal Describes a fungus that lives symbiotically with the roots of vascular plants.

myelin sheath Multiple layers of myelin, derived from the cell membranes of certain glial cells, wrapped around the axon of a neuron and providing electrical insulation.

myocardial infarction Death of cardiac muscle cells when deprived of oxygen. Also called *heart attack*.

myofibril A long, slender structure composed of contractile proteins organized into repeating units (sarcomeres) in vertebrate cardiac muscle and skeletal muscle.

myosin Any one of a class of motor proteins that use the chemical energy of ATP to move along actin filaments in muscle contraction, cytokinesis, and vesicle transport.

myriapods A lineage of arthropods with long segmented trunks, each segment bearing one or two pairs of legs. Includes millipedes and centipedes.

NAD^+/NADH Oxidized and reduced forms, respectively, of nicotinamide adenine dinucleotide. A nonprotein electron carrier that functions in many of the redox reactions of metabolism.

$NADP^+$/NADPH Oxidized and reduced forms, respectively, of nicotinamide adenine dinucleotide phosphate. A nonprotein electron carrier that is reduced during the light-dependent reactions in photosynthesis and extensively used in biosynthetic reactions.

nastic movement A plant movement in which the direction of the response is not dependent on the direction of the stimulus.

natural experiment A situation in which a natural change in conditions enables comparisons of groups, rather than a manipulation of conditions by researchers.

natural selection The process by which individuals with certain heritable traits tend to produce more surviving offspring than do individuals without those traits, often leading to a change in the genetic makeup of the population. A major mechanism of evolution. The only evolutionary process that produces adaptation.

Neanderthal A recently extinct European species of hominin, *Homo neanderthalensis*, closely related to but distinct from modern humans.

nectar The sugary fluid produced by flowers that attracts and rewards pollinating animals.

nectary A nectar-producing structure in a flower.

negative control A type of regulation that works by slowing an ongoing process. In the context of gene transcription, the action of a regulatory protein to shut down transcription by binding to DNA in or near the gene.

negative feedback A self-limiting, corrective response in which a deviation in some variable (e.g., concentration of some compound) triggers responses aimed at returning the variable to a target value. Represents a means of maintaining homeostasis. Compare with **positive feedback**.

negative pressure ventilation Ventilation of the lungs by expanding the rib cage so as to "pull" air into the lungs. Compare with **positive pressure ventilation**.

negative-sense Describing a single strand of RNA that contains sequences complementary to those in the mRNA required to produce viral proteins. Compare with **ambisense** and **positive-sense**.

nematodes See **roundworms**.

nephron One of many tiny tubules inside the kidney that function in the formation of urine.

neritic zone Shallow marine waters beyond the intertidal zone, extending down to about 200 meters, where the continental shelf ends.

nerve A long strand of nervous tissue containing axons of neurons wrapped in connective tissue; carries signals between the central nervous system and some other part of the body.

nerve net A nervous system in which neurons are diffuse instead of being clustered into large ganglia or tracts; found in cnidarians and ctenophores.

nervous tissue An animal tissue consisting of nerve cells (neurons) and various supporting cells.

net primary productivity (NPP) In an ecosystem, the total amount of biomass generated by the fixation of carbon through photosynthesis per year minus the amount oxidized during cellular respiration. Compare with **gross primary productivity**.

net reproductive rate (R_0) The growth rate of a population per generation; equivalent to the average number of female offspring that each female produces over her lifetime.

neural tube A folded tube of ectoderm that forms along the dorsal side of an early vertebrate embryo; gives rise to the brain and the spinal cord.

neuroendocrine Relating to a chemical signal (hormone) that is released by a neuron and acts on a distant target cell.

neurogenesis The formation of new neurons from central nervous system stem cells.

neurohormone A hormone produced by a neuron.

neuron An animal cell that is specialized for the transmission of nerve impulses. Typically has dendrites, a cell body, and a long axon that forms synapses with other neurons. Also called *nerve cell.*

neurosecretory cell A nerve cell (neuron) that produces and secretes neurohormones into the bloodstream. In vertebrates, principally found in the hypothalamus. Also called *neuroendocrine cell.*

neurotoxin Any substance that specifically destroys or blocks the normal functioning of neurons.

neurotransmitter A molecule that transmits signals from one neuron to another neuron or to a muscle cell or gland cell. Examples are acetylcholine, dopamine, serotonin, and norepinephrine.

neutral In genetics, referring to any mutation, allele, or trait that has no effect on an individual's fitness. Compare with **deleterious** and **beneficial**.

neutrophil A type of white blood cell that is capable of moving through body tissues to engulf and digest pathogens and other foreign particles; also secretes various compounds that attack bacteria and fungi.

next-generation sequencing General term used to describe DNA sequencing technologies that are faster and cheaper than dideoxy sequencing. Millions of different DNAs can be amplified and sequenced in a single run.

niche The range of resources that a species can use and the range of conditions that it can tolerate. More broadly, the role that a species plays in its ecosystem.

niche differentiation The evolutionary change in resource use by competing species that occurs as the result of character displacement.

nicotinamide adenine dinucleotide See **NAD^+/NADH**.

nicotinamide adenine dinucleotide phosphate See **$NADP^+$/NADPH**.

nitrogen cycle, global The movement of nitrogen among abiotic and biotic reservoirs in terrestrial and aquatic ecosystems.

nitrogen fixation The incorporation of atmospheric nitrogen (N_2) into ammonia (NH_3), which can be used to make many organic compounds. Occurs in only a few lineages of prokaryotes.

nitrogenous wastes Compounds excreted by animals to rid their bodies of excess nitrogen. Main types include ammonia, urea, and uric acid.

nociceptor A sensory cell specialized for detecting tissue damage, usually resulting in the sensation of pain.

Nod factors Molecules produced by nitrogen-fixing bacteria that help them recognize and bind to roots of legumes.

node (1) In animals, any small thickening (e.g., a lymph node). (2) In plants, the part of a stem where leaves or leaf buds are attached. (3) In a phylogenetic tree, the point where two branches diverge, representing the point in time when an ancestral group split into two or more descendant groups; also called *fork.*

node of Ranvier One of the periodic gaps in the myelin sheath of a neuron's axon; serves as a site where an action potential can be regenerated.

nodule Globular structure on roots of legume plants that contain symbiotic nitrogen-fixing bacteria.

noncyclic electron flow Path of electron flow in which electrons originate from oxidizing water in photosystem II and then pass through an electron transport chain to photosystem I, where they ultimately reduce $NADP^+$ during the light-dependent reactions of photosynthesis. See also **Z scheme**.

nondisjunction An error that can occur during meiosis or mitosis; it results in one daughter cell receiving two copies of a particular chromosome while the other daughter cell receives none.

nonpolar covalent bond A covalent bond in which electrons are equally shared between two atoms of the same or similar electronegativity. Compare with **polar covalent bond**.

nonsense mutation A point mutation (change in a single base pair) that converts an amino-acid-specifying codon into a stop codon.

non-sister chromatids Any two chromatids that each come from a different member of a homologous chromosome pair. Crossing over occurs between non-sister chromatids. Compare with **sister chromatids**.

non-template strand The strand of DNA that is not transcribed during synthesis of RNA. With the exception of T's instead of U's, its base sequence corresponds to that of the RNA produced from the other strand. Also, in the case of protein-coding genes, called *coding strand.*

nonvascular plants A paraphyletic group of land plants that lack vascular tissue and reproduce using spores. The nonvascular plants include three lineages of green plants (liverworts, mosses, and hornworts).

norepinephrine A catecholamine used as a neurotransmitter in the sympathetic nervous system. Also produced by the adrenal medulla and functions as a hormone that triggers rapid physiological changes relating to the fight-or-flight response.

notochord A supportive, flexible rod that occurs in the back of a chordate embryo, ventral to the developing spinal cord. Replaced by vertebrae in most adult vertebrates. A defining feature of chordates.

nuclear envelope The double-layered membrane enclosing the nucleus of a eukaryotic cell.

nuclear lamina A lattice-like sheet of fibrous nuclear lamins, which are one type of intermediate filament. Lines the inner membrane of the nuclear envelope, stiffening the envelope and helping to organize the chromosomes.

nuclear lamins Intermediate filaments that make up the nuclear lamina layer—a lattice-like layer inside the nuclear envelope that stiffens the structure.

nuclear localization signal (NLS) A short amino acid sequence that marks a protein for delivery to the nucleus.

nuclear pore complex A large complex of about 30 proteins that form an opening in the nuclear envelope connecting the inside of the nucleus with the cytosol and allowing the free diffusion of small molecules and ions; also regulates transport of RNA and proteins.

nucleases Enzymes that digest RNA or DNA.

nucleic acid A macromolecule composed of nucleotide monomers. Generally used by cells to store or transmit hereditary information. Includes rib04onucleic acid and deoxyribonucleic acid.

nucleoid In prokaryotic cells, a dense, centrally located region that contains DNA but is not surrounded by a membrane.

nucleolus In eukaryotic cells, a specialized structure in the nucleus where ribosomal RNA processing occurs and ribosomal subunits are assembled.

nucleosome A group of eight histone proteins with about 200 nucleotides of DNA wrapped twice around it; the fundamental unit of chromatin.

nucleotide A molecule consisting of a five-carbon sugar (ribose or deoxyribose), one or more phosphate groups, and one of several nitrogen-containing bases. Equivalent to a nucleoside plus one or more phosphate groups.

nucleotide excision repair A type of DNA repair that removes a damaged region in one strand of DNA and replaces it with the correct sequence using the undamaged strand as a template.

nucleus (1) The center of an atom, containing protons and neutrons. (2) In eukaryotic cells, the large organelle containing the chromosomes and surrounded by a double membrane. (3) A discrete clump of neuron cell bodies in the brain, usually sharing a distinct function.

null allele See **loss-of-function allele**.

null hypothesis A hypothesis that specifies what the results of an experiment will be if the main hypothesis being tested is wrong. Often states that there will be no difference between experimental groups.

nutrient Any substance that an organism requires for normal growth, maintenance, or reproduction.

occipital lobe In the vertebrate brain, one of the four lobes of the cerebrum. Receives and interprets visual information.

oceanic zone The deeper waters of the open ocean beyond the continental shelf.

odorant Any airborne molecule that conveys information about food or the environment.

oil A polyunsaturated fat that is liquid at room temperature.

Okazaki fragment Short segment of DNA produced during replication of the lagging-strand template. The Okazaki fragments are eventually linked together to produce the lagging strand in newly synthesized DNA.

olfaction The sense of smell.

olfactory bulb A bulb-shaped projection of the vertebrate brain just above the nose. Receives and interprets odor information from the nose.

oligodendrocyte A type of glial cell that wraps around axons of some neurons in the vertebrate central nervous system, forming a myelin sheath that provides electrical insulation. Compare with **Schwann cell**.

oligopeptide A chain composed of fewer than 50 amino acid residues linked together by peptide bonds. Often referred to simply as *peptide.*

oligosaccharide A linear or branched carbohydrate chain generally consisting of fewer than 50 monosaccharides joined by glycosidic linkages. Compare with **monosaccharide** and **polysaccharide**.

ommatidium (plural: *ommatidia*) A light-sensing column in an arthropod's compound eye.

omnivore (adjective: *omnivorous*) An animal whose diet regularly includes a variety of organisms, including plants, animals, fungi, protists, archaea, and/or bacteria. Compare with **carnivore** and **herbivore**.

oncogene Any gene whose protein product stimulates cell division at all times and thus promotes cancer development. Often is a mutated form of a gene involved in regulating the cell cycle. See **proto-oncogene**.

one-gene, one-enzyme hypothesis The hypothesis that each gene is responsible for making one enzyme. This hypothesis has expanded to include genes that produce proteins other than enzymes, RNAs as final products, or two or more related products.

oogenesis The production of egg cells (ova).

oogonium (plural: *oogonia*) In an ovary, any of the diploid cells that can divide by mitosis to produce primary oocytes (which can undergo meiosis) and more oogonia.

open circulatory system A circulatory system in which the circulating fluid (hemolymph) is not confined to blood vessels; occurs in most invertebrates. Compare with **closed circulatory system**.

open reading frame (ORF) A sequence of DNA that, if converted into an RNA sequence and translated as a set of non-overlapping codons, would produce a polypeptide of substantial size. Discovery of an ORF identifies a protein-coding gene.

operator In bacterial DNA, a binding site for a repressor protein; located near the start of an operon.

operculum The stiff flap of tissue that covers the gills of teleost fishes.

operon A region of bacterial DNA that codes for a series of functionally related genes and is transcribed from a single promoter into one mRNA.

opsin A transmembrane protein that is covalently linked to retinal, the light-detecting pigment in the rods and cones of vertebrate retinas.

optic nerve A bundle of axons that runs from the eye to the brain in vertebrates.

optimal foraging The concept that animals forage in a way that maximizes the amount of usable energy they take in, given the costs of finding and ingesting their food and the risk of being eaten while they're at it.

optogenetics A technique in which specific types of cells are genetically engineered to express light-sensitive proteins. By stimulating or inhibiting the cells with light, researchers can determine the physiological roles of the cells.

orbital The region of space around an atomic nucleus in which an electron is present most of the time. Orbitals are grouped into electron shells.

organ A group of tissues organized into a functional and structural unit.

organ system A group of tissues and organs that work together to perform a function.

organelle Any discrete, often membrane-bound compartment within a cell (e.g., mitochondrion) that has a characteristic structure and function.

organic compounds A group of molecules that include at least one carbon atom; most have carbon–hydrogen bonds and carbon–carbon bonds. Organic compounds are widely used by living organisms.

organism Any living entity that contains one or more cells.

organogenesis A stage of embryonic development in which organs are formed. In animals, it follows gastrulation.

origin of replication The site on a chromosome where DNA replication begins.

osmoconformer An animal whose tissue osmolarity is isosmotic to its environment. Compare with **osmoregulator**.

osmolarity The concentration of dissolved solutes in a solution, measured in osmoles per liter.

osmoregulation The process by which an organism controls the concentration of water and solutes in its body.

osmoregulator An animal that actively regulates the osmolarity of its tissues. Compare with **osmoconformer**.

osmosis Diffusion of water across a selectively permeable membrane from a region of lower solute concentration (higher water concentration) to a region of higher solute concentration (lower water concentration). For osmosis to occur, the solute would not be able to pass through the membrane.

osteoblast A type of cell in bone that secretes calcium, proteins, and other substances that harden to form new bone tissue.

osteoclast A type of cell in bone that stimulates bone resorption by secreting acid, resulting in the release of calcium ions and other minerals into the blood.

osteoporosis A disease characterized by reduction in bone mass, leading to fragile bones susceptible to fracture.

ouabain A plant toxin that poisons Na^+/K^+-ATPases of animals.

out-of-Africa hypothesis The hypothesis that modern humans (*Homo sapiens*) evolved in Africa and spread to other continents, replacing other *Homo* species.

outcrossing Reproduction by fusion of gametes produced by different individuals. Compare with **self-fertilization**.

outer ear The outermost portion of the mammalian ear, consisting of the pinna (ear flap) and the ear canal. Funnels sound to the tympanic membrane.

outgroup Any taxon that is not part of the taxonomic group being studied; researchers often select one or more outgroups that are closely related to the ingroup to help root phylogenetic trees and to determine the direction of character changes.

oval window A membrane separating the air-filled middle ear from the fluid-filled cochlea in mammals; sound vibrations pass through it from the middle ear to the inner ear.

ovary The egg-producing organ of a female animal, or the fruit- and seed-producing structure in the female part of a flower.

overexploitation Unsustainable removal of wildlife from the natural environment for use by humans.

oviduct A narrow tube connecting the uterus to the ovary, through which an egg travels after ovulation. Site of fertilization and cleavage. Also called **fallopian tube**.

oviparous In animals, depositing fertilized eggs outside the body, where they develop and hatch. Compare with **ovoviviparous** and **viviparous**.

ovoviviparous In animals, producing eggs that are retained inside the body (nourished by yolk) until they are ready to hatch and released via live birth. Compare with **oviparous** and **viviparous**.

ovulation The release of an egg from an ovary of a female vertebrate. In humans, an ovarian follicle releases an egg at the end of the follicular phase of the menstrual cycle.

ovule In flowering plants, the structure inside an ovary that contains the female gametophyte and eventually (if fertilized) becomes a seed.

ovum (plural: *ova*) See **egg**.

oxidation The loss of electrons from an atom or molecule during a redox reaction, either by donation of an electron to another atom or molecule or by the shared electrons in covalent bonds moving farther from the atomic nucleus.

oxidative phosphorylation Production of ATP molecules by ATP synthase using the proton gradient established via redox reactions of an electron transport chain.

oxygen–hemoglobin equilibrium curve The graphed depiction of the percentage of hemoglobin in the blood that is saturated with oxygen at various partial pressures of oxygen.

oxygenic Referring to any process or reaction that produces oxygen. Photosynthesis in plants, algae, and cyanobacteria, which involves photosystem II, is oxygenic because it uses water as an electron source and produces O_2 as a by-product. Compare with **anoxygenic**.

oxytocin A peptide hormone, secreted by the posterior pituitary, that triggers labor and milk production in females and that stimulates pair bonding, parental care, and affiliative behavior in both sexes.

p53 protein A tumor suppressor protein (molecular weight of 53 kilodaltons) that responds to DNA damage by stopping the cell cycle, turning on DNA repair machinery, and, if necessary, triggering apoptosis. Encoded by the *p53* gene.

pacemaker cell Any of a group of specialized cardiac muscle cells in the sinoatrial (SA) node of the vertebrate heart that have an inherent rhythm and can generate an electrical impulse that spreads to other heart cells.

paleontologists Scientists who study the fossil record and the history of life.

Paleozoic era The interval of geologic time, from 541 million to 252 million years ago, during which fungi, land plants, and most animal lineages first appeared and diversified. Began with the Cambrian explosion and ended with the extinction of almost all multicellular life-forms at the end of the Permian period.

pancreas A large gland in vertebrates that has both exocrine and endocrine functions. Secretes digestive enzymes into a duct connected to the intestine and secretes several hormones (notably insulin and glucagon) into the bloodstream.

pancreatic amylase An enzyme produced by the pancreas that digests carbohydrates by catalyzing hydrolysis of the glycosidic linkages between glucose residues.

pancreatic lipase An enzyme produced by the pancreas that digests fats, releasing monoglycerides and fatty acids.

pandemic The spread of an infectious disease in a short time over a wide geographic area and often affecting a high proportion of a population. Compare with **epidemic**.

paracrine Relating to a chemical signal that is released by a cell and affects neighboring cells.

paraphyletic group A group that includes an ancestral population and *some* but not all of its descendants. Compare with **monophyletic group**.

parapodia (singular: *parapodium*) Appendages found in some annelids from which bristle-like structures (chaetae) extend.

parasite An organism that lives on a host species (ectoparasite) or in a host species (endoparasite) and that damages its host or takes resources from it.

parasitism (adjective: *parasitic*) A relationship between two organisms that is beneficial to one organism (the parasite) and detrimental, but usually not fatal, to the other (the host). Parasites are usually small relative to the host and consume relatively small amounts of tissue or nutrients while living inside the host (endoparasite) or on the surface of the host (ectoparasite). Compare with **herbivory** and **predation, commensalism** and **mutualism**.

parasitoid An organism that is free-living as an adult, but parasitic as a larva, and often kills the host. Most parasitoids are insects that lay eggs in the bodies of other insects.

parasympathetic nervous system The part of the autonomic nervous system that triggers responses for conserving or restoring energy, such as reduced heart rate and stimulated digestion. Compare with **sympathetic nervous system**.

parenchyma Can refer to tissue or cell type. In plants, a general type of cell with a relatively thin primary cell wall. These cells, which form a simple tissue found in leaves, the centers of stems and roots, and fruits, are involved in photosynthesis, storage, and transport. Compare with **collenchyma** and **sclerenchyma**.

parental care Any action by which an animal expends energy or assumes risks to benefit its offspring (e.g., building a nest, feeding and defending young).

parental generation The adults used in the first experimental cross of a breeding experiment.

parental strand A strand of preexisting DNA. The parental strand is used as a template during DNA synthesis and is also known as the *template strand*.

parietal cells A cell in the stomach lining that secretes hydrochloric acid.

parietal lobe In the vertebrate brain, one of the four lobes of the cerebrum. Involved in integrating sensory and motor functions.

parsimony The principle that the most likely explanation of a phenomenon is the most economical or simplest. When applied to comparison of alternative phylogenetic trees, it suggests that the one requiring the fewest character changes is most likely.

parthenogenesis A form of asexual reproduction in which offspring develop from unfertilized eggs.

partial pressure The pressure of one particular gas in a mixture of gases; the contribution of that gas to the overall pressure.

particulate inheritance The concept that inheritance is based on genes that do not blend together in offspring but instead remain separate, or particle-like.

pascal (Pa) A unit of pressure (force per unit area).

passive transport Diffusion of a substance across a membrane. When this event occurs with the assistance of membrane proteins, it is called *facilitated diffusion*.

patch clamping A technique for studying the electrical currents that flow through individual ion channels by applying the open tip of a microelectrode to a tiny patch of membrane.

paternal chromosome A chromosome inherited from the father.

pathogen (adjective: *pathogenic*) Any entity capable of causing disease, such as a microbe, virus, or prion.

pattern formation The series of events that determines the spatial organization of an entire embryo or parts of an embryo by setting the major body axes, for example, or setting up patterns in smaller groups of cells early in development.

pattern-recognition receptor On white blood cells, one of a class of membrane proteins, including Toll-like receptors, that bind to molecules commonly associated with foreign cells and viruses and signal innate immune responses against broad types of pathogens.

pectin A gelatinous polysaccharide found in the primary cell walls of plant cells. Attracts and holds water, forming a gel that resists compression forces and helps keep the cell wall moist.

pedigree A family tree of parents and offspring, showing inheritance of particular traits of interest.

pedipalps A pair of appendages found next to the claw-like chelicerae of certain arthropods called chelicerates (spiders, mites, and relatives); among other functions, used to manipulate food and transfer sperm.

penis The copulatory organ of male mammals, used to insert sperm into a female.

pentose A monosaccharide (simple sugar) containing five carbon atoms.

PEP carboxylase An enzyme that catalyzes addition of CO_2 to phosphoenolpyruvate, a three-carbon compound, forming a four-carbon organic acid. See also **C_4 pathway** and **crassulacean acid metabolism (CAM)**.

pepsin A protein-digesting enzyme secreted in inactive form (as pepsinogen) by chief cells in the stomach lining.

peptide See **oligopeptide**.

peptide bond The covalent bond formed by a condensation reaction between two amino acids.

peptidoglycan A complex structural polysaccharide found in bacterial cell walls.

per capita rate of increase (*r*) Also called *instantaneous rate of increase*. The difference between the birth rate per individual and the death rate per individual. Compare with **intrinsic rate of increase** and **finite rate of increase**.

perennial Describing a plant whose life cycle normally lasts for more than one year. Compare with **annual**.

perfect flower A flower that contains both male parts (stamens) and female parts (carpels). Compare with **imperfect flower**.

perforation In plants, a small hole in the primary and secondary cell walls of vessel elements that allows passage of water.

pericarp The part of a fruit, formed from the ovary wall, that surrounds the seeds and protects them. Corresponds to the flesh of most edible fruits and the hard shells of most nuts.

pericycle In plant roots, a layer of cells just inside the endodermis that give rise to lateral roots.

peripheral membrane protein Any membrane protein that does not span the entire lipid bilayer but instead binds to only one side of the bilayer. Compare with **integral membrane protein**.

peripheral nervous system (PNS) All the components of a nervous system that are outside the central nervous system. In vertebrates, includes the **somatic nervous system** and **autonomic nervous system**.

peristalsis Rhythmic waves of muscular contraction. In the digestive tract, pushes food along. In animals with hydrostatic skeletons, enables crawling.

permeability The tendency of a structure, such as a membrane, to allow a given substance to diffuse across it.

peroxisome An organelle found in most eukaryotic cells that contains enzymes for oxidizing fatty acids and other compounds, including many toxins, rendering them harmless. See **glyoxysome**.

petal Any of the leaflike organs arranged around the reproductive organs of a flower. Often the color and scent of petals attract pollinators.

petiole The stalk of a leaf.

pH A measure of the concentration of protons in a solution and thus of how acidic or basic the solution is. Defined as the negative of the base 10 logarithm of the proton concentration: $pH = -\log[H^+]$.

phagocytosis Uptake by a cell of small particles or cells by invagination and pinching off of the plasma membrane to form small, membrane-bound vesicles; one type of **endocytosis**.

pharyngeal jaw A secondary jaw in the back of the throat; found in some fishes, it aids in food processing. Derived from modified gill arches.

pharyngeal slits or pouches A set of parallel openings from the throat to the outside that function in feeding and/or gas exchange, or in vertebrate embryos, a set of throat pouches that are homologous to slits in other chordates. A characteristic trait of chordates.

phenology The yearly timing of events such as flowering and migration, in environments where seasonal changes occur.

phenotype The detectable traits of an individual. Compare with **genotype**.

phenotypic plasticity Variation in phenotype that is due to differences in environmental conditions. Occurs more commonly in plants than animals.

pheophytin The molecule in photosystem II that accepts excited electrons from the reaction center chlorophyll and passes them to an electron transport chain.

pheromone A chemical signal, released by an individual into the external environment, that can trigger changes in the behavior or physiology or both of another member of the same species.

phloem A plant vascular tissue that conducts sugars between roots and shoots; contain sieve-tube elements and companion cells. Primary phloem develops from the procambium of apical meristems; secondary phloem develops from the vascular cambium. Compare with **xylem**.

phosphatase An enzyme that removes phosphate groups from proteins or other molecules. Phosphatases often function in the inactivation of signaling pathways that involve the phosphorylation and activation of proteins.

phosphodiester linkage Chemical linkage between adjacent nucleotide residues in DNA and RNA. Forms when the phosphate group of one nucleotide condenses with the hydroxyl group on the sugar of another nucleotide. Also known as *phosphodiester bond*.

phosphofructokinase The enzyme that catalyzes synthesis of fructose-1,6-bisphosphate from fructose-6-phosphate, a key reaction in glycolysis (reaction 3). Also called *6-phosphofructokinase*.

phospholipid A class of lipid having a hydrophilic head (including a phosphate group) and a hydrophobic tail (consisting of two hydrocarbon chains). Major components of the plasma membrane and organelle membranes.

phosphorus cycle, global The movement of phosphorus among abiotic and biotic reservoirs in terrestrial and aquatic ecosystems.

phosphorylase An enzyme that breaks down glycogen by catalyzing hydrolysis of the α-glycosidic linkages between the glucose residues.

phosphorylation (verb*phosphorylate*) The addition of a phosphate group to a molecule.

phosphorylation cascade A series of enzyme-catalyzed phosphorylation reactions commonly used in signal transduction pathways to amplify and convey a signal inward from the plasma membrane.

photic zone In an aquatic habitat, water that is shallow enough to receive some sunlight (whether or not it is enough to support photosynthesis). Compare with **aphotic zone**.

photon A discrete packet of light energy; a particle of light.

photoperiod The amount of time per day (usually in hours) that an organism is exposed to light.

photoperiodism Any response by an organism to the relative lengths of day and night (i.e., photoperiod).

photophosphorylation Production of ATP molecules by ATP synthase using the proton-motive force generated either (1) during photosynthesis, as light-excited electrons flow through an electron transport chain, or (2) in some bacteria or archaea, as rhodopsin-like molecules use absorbed light energy to pump protons across the plasma membranes to create a chemiosmotic gradient.

photoreception The detection of light.

photoreceptor A molecule, a cell, or an organ that is specialized to detect light.

photorespiration A series of light-driven chemical reactions that "undoes" photosynthesis by consuming O_2 and releasing CO_2. Usually occurs when there are high O_2 and low CO_2 concentrations inside plant cells, which occurs in land plants when stomata must be kept closed to prevent dehydration.

photoreversibility A change in conformation that occurs in certain plant pigments when they are exposed to the particular wavelengths of light that they absorb; triggers responses by the plant.

photosynthesis The complex biological process that converts light energy to chemical energy stored in the carbohydrate G3P. Occurs in most plants, algae, and some bacteria.

photosystem One of two types of units, consisting of a central reaction center surrounded by antenna pigments, that is responsible for the light-capturing reactions of photosynthesis.

photosystem I Photosystem that contains antenna pigments and a reaction center with a pair of P700 chlorophyll molecules. Absorbed light energy is used to reduce $NADP^+$ to NADPH in noncyclic electron flow or ATP in cyclic electron flow.

photosystem II Photosystem that contains antenna pigments and a reaction center with a pair of P680 chlorophyll molecules. Absorbed light energy is used to reduce electron carriers in an electron transport chain that produces a proton-motive force for the synthesis of ATP. Oxygen is a by-product when water is split to obtain electrons.

phototroph An organism (most plants, algae, and some bacteria and archaea) that produces ATP through photophosphorylation.

phototropins A class of plant photoreceptors that detect blue light and initiate various responses.

phototropism Growth or movement of an organism in a particular direction in response to light.

phylogenetic diversity In ecology, a measure of evolutionary distinctiveness among species in a community, measured using branch lengths among species in a phylogeny.

phylogenetic species concept The definition of a species as the smallest monophyletic group in a phylogenetic tree. Compare with **biological species concept** and **morphospecies concept**.

phylogenetic tree A branching diagram that depicts the evolutionary relationships among species or other taxa.

phylogeny The evolutionary history of a group of organisms.

phylum (plural: *phyla*) In Linnaeus' system, a taxonomic category above the class level and below the kingdom level. In plants, sometimes called a *division*.

physical map A map of a chromosome that shows the number of base pairs between various genetic markers and genes. Compare with **genetic map**.

physiology How an organism's body functions.

phytochelatins Enzymatically synthesized peptides that bind to and prevent excess metal ions from acting as toxins.

phytochrome A specialized plant photoreceptor that exists in two shapes depending on the ratio of red to far-red light. This photoreceptor is involved in the timing of certain physiological processes, such as flowering, stem elongation, and germination.

picoplankton Plankton cells that are between 0.2 and 2.0 microns in diameter.

pigment Any molecule that absorbs certain wavelengths of visible light and reflects or transmits other wavelengths.

piloting A type of navigation in which animals use familiar landmarks to find their way.

pineal gland An endocrine gland, located in the brain, that secretes the hormone melatonin.

pioneering species Those species that appear first in recently disturbed areas.

pit In plants, a small hole in the secondary cell walls of tracheids and vessel elements that allows passage of water.

pitch The sensation produced by a particular frequency of sound. Low frequencies are perceived as low pitches; high frequencies, as high pitches.

pith In the shoot systems of plants, ground tissue located to the inside of the vascular bundles. Roots of some plants also have a pith.

pituitary gland A small gland located directly under the brain and physically and functionally connected to the hypothalamus. Produces and secretes an array of hormones that affect many other glands and organs.

placenta A structure that forms in the pregnant uterus from maternal and embryonic tissues. Delivers oxygen to the embryo/fetus, exchanges nutrients and wastes between the mother and the embryo/fetus, anchors the embryo/fetus to the uterine wall, and produces some hormones. Occurs in most mammals and in a few other vertebrates.

placental mammals See **eutherians**.

plankton Drifting small or microscopic organisms that serve as a food source in aquatic environments (includes animals, plants, protists, archaea, and bacteria).

plant body An entire plant, generally consisting of the root and shoot systems.

Plantae The monophyletic group that includes red, green, and glaucophyte algae as well as land plants.

plasma The liquid, noncellular portion of blood.

plasma cell An activated B cell that produces large numbers of antibodies against a specific antigen. Also called *effector B cell*.

plasma membrane A membrane that surrounds a cell, separating it from the external environment and selectively regulating passage of molecules and ions into and out of the cell. Also called *cell membrane*.

plasmid A small, usually circular, supercoiled DNA molecule that exists independently of the cell's main chromosome(s) in prokaryotes and some eukaryotes.

plasmodesmata (singular: *plasmodesma*) Physical connections between two plant cells, consisting of membrane-lined gaps in the cell walls through which the cells' plasma membranes, cytoplasm, and smooth ER can connect directly. Functionally similar to gap junctions in animal cells.

plasmogamy Fusion of the cytoplasm of two individuals. Occurs in many fungi.

plastocyanin (PC) A small protein that shuttles electrons originating from photosystem II to the reaction center of photosystem I during photosynthesis.

plastoquinone (PQ) A nonprotein electron carrier in the chloroplast electron transport chain. Receives excited electrons from photosystem II (noncyclic) or photosystem I (cyclic) and passes them through an electron transport chain. Also transports protons from the stroma to the thylakoid lumen, generating a proton-motive force.

platelet A small, membrane-bound cell fragment in vertebrate blood that functions in blood clotting. Derived from large cells in the bone marrow.

platyhelminthes See **flatworms**.

pleiotropy (adjective: *pleiotropic*) The ability of a single gene to affect more than one trait.

ploidy The number of complete chromosome sets present in a cell. *Haploid* refers to a ploidy of 1; *diploid*, a ploidy of 2; *triploid*, a ploidy of 3; and *tetraploid*, a ploidy of 4.

pluripotency (pluripotent) The abilty of a cell (called a pluriopent cell) to form any cell of the body.

pneumatophore An aerial root specialized for gas exchange.

poikilotherm An animal whose body temperature varies with the environmental temperature. Compare with **homeotherm**.

point mutation A mutation that results in a change in or an insertion or deletion of a single base pair in DNA. Compare with **chromosome-level mutation** and **lateral gene transfer**.

polar (1) Asymmetrical or unidirectional. (2) Carrying a partial positive charge on one side of a molecule and a partial negative charge on the other. Polar molecules are generally hydrophilic.

polar body Any of the tiny, nonfunctional cells that result from the unequal distribution of cytoplasm during meiosis of a primary oocyte.

polar covalent bond A covalent bond in which electrons are shared unequally between atoms differing in electronegativity, resulting in the more electronegative atom having a partial negative charge and the other atom having a partial positive charge. Compare with **nonpolar covalent bond**.

polar microtubule A microtubule that arises from one of the spindle poles in mitosis or meiosis and overlaps in the middle of the spindle apparatus with a similar microtubule from the opposite spindle pole.

polar nuclei In flowering plants, the nuclei in the female gametophyte that fuse with one sperm nucleus to produce the endosperm. Most species have two.

pollen grain In seed plants, a male gametophyte enclosed within a protective coat of sporopollenin.

pollen tube In flowering plants, a structure that grows out of a pollen grain after it reaches the stigma. The tube extends down the style, and two sperm cells travel through it to the ovule.

pollination The process by which pollen reaches the carpel of a flower: (1) in flowering plants, it is transferred from anther to stigma; (2) in gymnosperms, it reaches the ovule directly.

pollination syndrome Suites of flower characters that are associated with certain types of pollinators and that have evolved through natural selection imposed by the interaction between flowers and pollinators.

poly(A) signal In eukaryotes, a short sequence of nucleotides near the 3′ end of the future mRNA that signals cleavage of the pre-RNA and addition of the poly(A) tail.

poly(A) tail In eukaryotes, a sequence of about 100–250 adenine nucleotides added to the 3′ end of newly transcribed messenger RNA molecules.

polygamy A type of mating system where one male mates with two or more females (polygyny) or one female mates with more than one male (polyandry) during a breeding season. Compare with **monogamy** and **promiscuity**.

polygenic trait Any trait that is influenced by more than one gene. Polygenic traits are usually **quantitative traits**.

polymer Any large molecule composed of small repeating units (monomers) bonded together. The main biological polymers are proteins, nucleic acids, and polysaccharides.

polymerase chain reaction (PCR) A technique for rapidly generating millions of identical copies of a specific stretch of DNA. Works by multiple rounds of DNA replication, starting from primers designed by an investigator to amplify particular DNA sequences.

polymerization (verb: polymerize) The process by which many identical or similar small molecules (monomers) are covalently bonded to form a large molecule (polymer).

polymorphic species A species that has two or more distinct phenotypes in the same interbreeding population at the same time.

polymorphic trait A trait for which more than one form of an allele commonly appears in a population.

polyp The immotile (sessile) stage in the life cycle of some cnidarians (e.g., sea anemones). Compare with **medusa**.

polypeptide A chain typically consisting of 50 or more amino acids linked together by peptide bonds. Compare with **oligopeptide** and **protein**.

polyphyletic group A group based on convergent (homoplastic) characteristics that are not present in a common ancestor. Compare with monophyletic group.

polyploidy (adjective: *polyploid*) The state of having more than two full sets of chromosomes, either from the same species (autopolyploidy) or from different species (allopolyploidy).

polyribosome A messenger RNA molecule that is being translated by two or more ribosomes.

polysaccharide A linear or branched carbohydrate chain consisting of many monosaccharides joined by glycosidic linkages. Compare with **monosaccharide** and **oligosaccharide**.

polytomy A node in a phylogenetic tree that depicts an ancestral branch dividing into three or more descendant branches; usually indicates that insufficient data were available to resolve which taxa are more closely related.

population A group of individuals of the same species living in the same geographic area at the same time.

population density The number of individuals of a population per unit area.

population dynamics Changes in the size and other characteristics of populations through time and space.

population ecology The study of how and why the number of individuals in a population changes over time and space.

population genetics The branch of evolutionary biology responsible for investigating processes that cause changes in allele and genotype frequencies in populations.

population thinking A way of thinking that emphasizes the importance of variation among individuals in a population; the opposite of typological thinking, which ignores variation or considers it unimportant.

positive control A type of regulation that works by speeding up a process. In the context of gene transcription, the action of a regulatory protein to trigger expression by binding to DNA in or near the gene.

positive feedback A physiological mechanism in which a change in some variable stimulates a response that increases the change. Relatively rare in organisms but important in a few processes, such as generation of an action potential. Compare with **negative feedback**.

positive pressure ventilation Ventilation of the lungs by using increased pressure in the mouth to "push" air into the lungs. Compare with **negative pressure ventilation**.

positive-sense Describing a single strand of RNA that contains the same sequences as the mRNA required to produce viral proteins. Compare with **ambisense** and **negative-sense**.

posterior Toward an animal's tail and away from its head. The opposite of **anterior**.

posterior pituitary The part of the pituitary gland that contains the ends of hypothalamic neurosecretory cells and from which oxytocin and antidiuretic hormone are secreted. Compare with **anterior pituitary**.

postsynaptic cell A cell that receives signals, usually via neurotransmitters, from a neuron at a synapse. Compare with **presynaptic neuron**.

post-translational control Regulation of gene expression by modification of proteins (e.g., addition of a phosphate group or sugar residues) after translation.

postzygotic isolation Reproductive isolation resulting from mechanisms that operate after mating of individuals of two different species occurs. The most common mechanisms are the death of hybrid embryos or reduced fitness of hybrids.

potential energy Energy stored in matter as a result of its position or the position of electrons that form chemical bonds between atoms. Compare with **kinetic energy**.

Precambrian The interval between the formation of the Earth, about 4.6 billion years ago, and the appearance of most animal groups about 541 million years ago. Unicellular organisms were dominant for most of this era, and oxygen was virtually absent for the first 2 billion years.

predation The killing and eating of one organism (the prey) by another (the predator). Compare with **herbivory** and **parasitism**.

predator Any organism that kills another organism for food.

prediction A measurable or observable result of an experiment based on a particular hypothesis. A correct prediction provides support for the hypothesis being tested.

pre-mRNA In eukaryotes, the primary transcript of protein-coding genes. Pre-mRNA is processed to form mRNA.

pressure-flow hypothesis The hypothesis that sugar movement through phloem tissue is due to differences in the turgor pressure of phloem sap.

pressure potential (Ψ_P) A component of the potential energy of water caused by physical pressures on a solution. It can be positive or negative. Compare with **solute potential (Ψ_S)**.

presynaptic neuron A neuron that transmits signals, usually by releasing neurotransmitters, to another neuron or to a muscle cell or gland cell at a synapse. Compare with **postsynaptic cell**.

prey An organism that is consumed by another organism.

prezygotic isolation Reproductive isolation resulting from any combination of several mechanisms that prevent individuals of two different species from mating.

primary cell wall The outermost layer of a plant cell wall, made of cellulose fibers and gelatinous polysaccharides, that defines the shape of the cell and withstands the turgor pressure of the plasma membrane.

primary consumer An herbivore; an organism that eats plants, algae, or other primary producers. Compare with **secondary consumer**.

primary decomposer A decomposer (detritivore) that consumes detritus from plants.

primary growth In plants, an increase in the length of stems and roots due to the activity of apical meristems. Compare with **secondary growth**.

primary immune response An adaptive immune response to a pathogen that the immune system has not encountered before. Compare with **secondary immune response**.

primary meristem In plants, three types of partially differentiated cells that are produced by apical meristems: protoderm, ground meristem, and procambium. Compare with **apical meristem** and **cambium**.

primary oocyte Any of the large diploid cells in an ovarian follicle that can initiate meiosis to produce a haploid secondary oocyte and a polar body.

primary plant body All of the cells and tissues derived from apical meristems and primary meristems.

primary producer Any organism that creates its own food by photosynthesis (or more rarely, chemosynthesis) from reduced inorganic compounds and that is a food source for other species in its ecosystem. Also called *autotroph*.

primary spermatocyte Any of the diploid cells in a testis that can initiate meiosis I to produce two secondary spermatocytes.

primary structure The sequence of amino acid residues in a protein polymer; also the sequence of nucleotides in a single nucleic acid strand. Compare with **secondary, tertiary**, and **quaternary structure**.

primary succession The gradual colonization of a habitat of bare rock or gravel, usually after an environmental disturbance that removes all soil and previous organisms. Compare with **secondary succession**.

primary transcript In eukaryotes, a newly transcribed RNA molecule that has not yet been processed to a mature RNA. Called *pre-mRNA* when the processed RNA is an mRNA.

primase An enzyme that synthesizes a short stretch of RNA to use as a primer during DNA replication.

primates The lineage of mammals that includes prosimians (lemurs, lorises, etc.), monkeys, and great apes (including humans).

primer A short, single-stranded RNA molecule that base-pairs with a DNA template strand and is used as a starting point for DNA synthesis by DNA polymerase.

prion An infectious particle that consists entirely of protein. Prion proteins adopt two differently folded shapes: a normally folded shape and an infectious, often disease-causing shape. The infectious version can bind normally folded prion proteins and cause them to adopt the infectious shape. Also called *proteinaceous infectious particles.*

procambium A primary meristem tissue that gives rise to the vascular tissue.

product Any of the final materials formed in a chemical reaction.

progesterone A steroid hormone produced and secreted by the corpus luteum in the ovaries after ovulation and by the placenta during gestation; protects the uterine lining.

programmed cell death Regulated cell death that is used in development, tissue maintenance, and destruction of infected cells. Can occur in different ways; apoptosis is the best known.

prokaryote (adjective: *prokaryotic*) A member of the domain Bacteria or Archaea; a unicellular organism lacking a nucleus and containing relatively few organelles or cytoskeletal components. Compare with **eukaryote**.

prolactin A peptide hormone, produced and secreted by the anterior pituitary, that promotes milk production in female mammals and has a variety of effects on parental behavior and seasonal reproduction in other vertebrates.

prometaphase A stage in mitosis or meiosis during which the nuclear envelope breaks down and microtubules attach to kinetochores.

promiscuity A type of mating system where both males and females mate with multiple partners during a breeding season. Compare with **polygamy** and **monogamy**.

promoter A short nucleotide sequence in DNA that binds a sigma factor (in bacteria) or general transcription factors (in eukaryotes) to enable RNA polymerase to begin transcription. In bacteria, several contiguous genes are often transcribed from a single promoter. In eukaryotes, each gene generally has its own promoter.

promoter-proximal element In eukaryotic DNA, a regulatory sequence that is close to a promoter and can bind regulatory transcription factors.

proofreading A mechanism for error correction during DNA synthesis in which a DNA polymerase recognizes and removes a wrong deoxyribonucleotide added during DNA replication and then continues synthesis.

prop root A root that tends to grow from an aboveground portion of the stem, but its tip is embedded in soil and helps give extra support to the shoot system.

prophase The first stage in mitosis or meiosis, during which chromosomes condense and the spindle apparatus forms. Synapsis and crossing over occur during prophase of meiosis I.

prosimians A non-monophyletic group including all primates that are not arthropoids, including lemurs, lorises, pottos, and tarsiers. Compare with **anthropoids**.

prostate gland A gland in male mammals that surrounds the base of the urethra and secretes a fluid that is a component of semen.

prosthetic group A non-amino acid atom or molecule that is permanently attached to an enzyme or other protein and is required for its function.

protease An enzyme that can break up proteins by cleaving the peptide bonds between amino acid residues.

proteasome A macromolecular machine that destroys proteins that have been marked by the addition of ubiquitin.

protein A macromolecule consisting of one or more polypeptide chains. Each polypeptide has a unique sequence of amino acids and each protein generally possesses a characteristic three-dimensional shape.

protein kinase An enzyme that catalyzes the addition of a phosphate group to another protein, typically activating or inactivating the substrate protein.

proteinase inhibitors Defense compounds, produced by plants, that induce illness in herbivores by inhibiting digestive enzymes.

proteoglycan A type of highly glycosylated protein found in the extracellular matrix of animal cells; it attracts and holds water, forming a gel that resists compression forces.

proteome The complete set of proteins produced by a particular cell type.

proteomics The systematic study of the full set of proteins (the proteome) in a cell or tissue, including their interactions, localization, functions, regulation, and other features.

protist Any eukaryote that is not a land plant, animal, or fungus. Protists are a diverse paraphyletic group. Most are unicellular, but some are multicellular or form aggregations called colonies.

protocell A hypothetical pre-cell structure consisting of a membrane compartment that encloses replicating macromolecules, such as ribozymes.

protoderm The primary meristem tissue on the exterior layer of a young plant embryo that gives rise to the dermal tissue, or epidermis.

proton-motive force The combined effect of a proton gradient and an electric potential gradient across a membrane, which can drive protons across the membrane. Used by prokaryotes, mitochondria, and chloroplasts to power ATP synthesis or other energy-demanding processes via chemiosmosis.

proton pump A membrane protein that can hydrolyze ATP to power active transport of protons (H^+) across a membrane against an electrochemical gradient. Also called *H^+-ATPase*.

proto-oncogene Any gene that encourages cell division in a regulated manner, typically by triggering specific phases in the cell cycle. Mutation may convert it into an oncogene. See **oncogene**.

protostomes A major lineage of bilaterian animals including arthropods, mollusks, and annelids. Sister group to **deuterostomes**.

proximal tubule In the vertebrate kidney, the convoluted portion of a nephron into which filtrate moves from Bowman's capsule. Involved in the largely unregulated reabsorption of electrolytes, nutrients, and water. Compare with **distal tubule**.

proximate causation In biology, the immediate, mechanistic cause of a phenomenon (how it happens), as opposed to why it evolved. Also called *proximate explanation*. Compare with **ultimate causation**.

pseudocoelomate An animal having a coelom that is only partially lined with mesoderm. Compare with **acoelomate** and **coelomate**.

pseudogene A DNA sequence that closely resembles a functional gene but is not transcribed. Thought to have arisen by duplication of the functional gene followed by inactivation due to a mutation.

pseudopodium (plural: *pseudopodia*) A temporary bulge-like extension of certain protist cells; used in cell crawling and ingestion of food.

puberty The various physical and emotional changes that an immature human undergoes in reaching reproductive maturity. Also the period when such changes occur.

pulmonary artery A short, thick-walled artery that carries oxygen-poor blood from the heart to the lungs.

pulmonary circuit The part of the circulatory system that carries oxygen-poor blood from the heart to the lungs and returns oxygen-rich blood to the heart. It is separate from the rest of the circulatory system (the **systemic circuit**) in mammals and birds.

pulmonary vein A short, thin-walled vein that carries oxygen-rich blood from the lungs to the heart. Humans have four such veins.

pulse–chase experiment A type of experiment that involves using a labeled molecule (pulse) to mark a population of cells or molecules at a particular moment and then following their fate over time (chase).

pump Any membrane protein that uses energy (e.g., ATP) to change shape and power the active transport of a specific ion or molecule across a membrane in a single direction, often against its gradient. See **proton pump**.

pupa (plural: *pupae*) In insects undergoing complete metamorphosis, the life stage during which the juvenile form metamorphoses into the adult form.

pupil The hole in the center of the iris through which light enters a vertebrate or cephalopod eye.

pure line In genetics, a strain that produces offspring identical to the parents when self-fertilized or mated within the strain. Pure lines are homozygous for the gene under study.

purifying selection Selection that lowers the frequency of, or even eliminates, deleterious alleles.

purine A class of double-ringed nitrogenous bases [adenine (A), guanine (G)] found in nucleotides. Compare with **pyrimidine**.

pyrimidine A class of single-ringed nitrogenous bases [cytosine (C), uracil (U), thymine (T)] found in nucleotides. Compare with **purine**.

pyruvate dehydrogenase A large enzyme complex, located in the mitochondrial matrix of eukaryotes and the cytosol of prokaryotes, that is responsible for converting pyruvate to acetyl CoA during cellular respiration.

quantitative trait A trait that exhibits continuous phenotypic variation (as in human height) rather than the clearly separated phenotypes of a discrete trait (such as yellow versus green pea seeds).

quaternary structure In proteins, the overall three-dimensional shape formed from the combination of two or more polypeptide chains (subunits); determined by the number, relative positions, and interactions of the subunits. In nucleic acids, two or more distinct single strands will form this level of structure through hydrogen bonding between complementary bases and hydrophobic interactions with the aqueous environment. Compare with **primary, secondary**, and **tertiary structures**.

quorum sensing A type of cell signaling in unicellular organisms in which cells sense population density by detecting signaling molecules secreted by other cells. Cell activity often changes dramatically when the population reaches a threshold size, or quorum.

R-group Part of an amino acid's core structure that varies from a single hydrogen atom to large structures containing carbon rings. R-group variability is responsible for the variability in amino acid structure and function. Also called *side chains*.

radial axis An axis extending from the interior of a plant organ to the exterior.

radial symmetry An animal body pattern that has at least two planes of symmetry. Typically, the body is in the form of a cylinder or disk, and the body parts radiate from a central hub. Compare with **biradial symmetry** and **bilateral symmetry**.

radiation Transfer of heat between two bodies that are not in direct physical contact. More generally, the emission of electromagnetic energy of any wavelength.

radicle The root of a plant embryo.

radioactive isotope A version of an element that has an unstable nucleus, which will release radiation energy as it decays to a more stable form. Decay often results in the radioisotope losing protons and becoming a different element.

radioimmunoassay A competitive binding assay in which the quantity of hormone in a sample can be estimated. Uses radioactively labeled hormones that compete with the hormone in the sample to bind with an antibody.

radula A rasping feeding appendage in mollusks such as gastropods (snails, slugs).

rain shadow The dry region on the side of a mountain range away from the prevailing wind.

range The geographic distribution of a species.

Ras A small G protein that is often activated by enzyme-linked cell-surface receptors, including receptor tyrosine kinases, or G protein-coupled receptors. Activated Ras then initiates a phosphorylation cascade, culminating in a cell response.

ray In plant shoot systems with secondary growth, a lateral row of parenchyma cells produced by vascular cambium. Transports water and nutrients laterally across the stem.

ray-finned fishes Members of the Actinopterygii, a diverse group of fishes with fins supported by bony rods arranged in a ray pattern.

Rb protein A tumor suppressor protein that helps regulate progression of a cell from G_1 phase to S phase of the cell cycle. Defects in Rb protein are found in many types of cancer.

reactant Any of the starting materials in a chemical reaction.

reaction center Centrally located component of a photosystem containing proteins and a pair of specialized chlorophyll molecules. It is surrounded by antenna pigments that transmit resonance energy to excite the reaction center pigments.

reading frame A series of adjacent non-overlapping, three-base-long sequences (codons) in DNA or RNA that specify a sequence of amino acids in a polypeptide. The reading frame is set by the start codon.

realized niche The portion of the fundamental niche that a species actually occupies given limiting factors such as competition with other species. Compare with **fundamental niche**.

receptor-mediated endocytosis Uptake by a cell of certain extracellular macromolecules, bound to specific receptors in the plasma membrane, by pinching off the membrane to form small membrane-bound vesicles.

receptor tyrosine kinase (RTK) Any of a class of enzyme-linked cell-surface receptors that phosphorylate themselves after binding a signaling molecule. The activated, phosphorylated receptor then triggers a signal transduction pathway inside the cell.

recessive Referring to a trait or allele whose phenotypic effect is observed only in homozygous individuals. Compare with **dominant**.

reciprocal altruism Altruistic behavior that is exchanged between a pair of individuals at different times (i.e., sometimes individual A helps individual B, and sometimes B helps A).

reciprocal cross A cross in which the mother's and father's phenotypes are the reverse of those examined in a previous cross.

recombinant Possessing a new combination of alleles. May refer to a single DNA molecule, a chromosome, or an entire organism.

recombinant DNA technology A variety of techniques for isolating specific DNA fragments and linking them to different regions of DNA or adding them to a different host organism.

rectal gland A salt-excreting gland in the digestive system of sharks, skates, and rays.

rectum The most posterior portion of the digestive tract, where feces are held until they are expelled.

red blood cell A hemoglobin-containing cell that circulates in the blood and delivers oxygen from the lungs to the tissues.

redox potential A measure of the ability of a molecule to accept electrons, or become reduced, in a reduction-oxidation reaction. It is measured in volts, and larger values represent greater affinity for electrons. Also called *reduction potential*.

redox reaction Any chemical reaction that involves either the complete transfer of one or more electrons from one reactant to another, or a reciprocal shift in the position of shared electrons within one or more of the covalent bonds of two reactants. Also called *reduction–oxidation reaction*.

reduction The gain of electrons by an atom or molecule during a redox reaction, either by acceptance of an electron from another atom or molecule, or by the shared electrons in covalent bonds moving closer to the atomic nucleus.

reduction–oxidation reaction See **redox reaction**.

reference genome A genome sequence established for a species that serves as a point of comparison for genome sequences from other individuals of the same species.

restriction endonuclease Any of a class of bacterial

enzymes that cut DNA at a specific base-pair sequence (recognition site). Also called a *restriction enzyme*.

reflex An involuntary response to an environmental stimulus.

refractory No longer responding to stimuli that previously elicited a response. An example is the tendency of voltage-gated sodium channels to be less likely to reopen immediately after they have closed.

regulate To actively maintain homeostasis of internal physiological conditions. Compare with **conform**.

regulatory sequence Any segment of DNA or RNA that is involved in controlling the expression of a specific gene or process such as RNA splicing by binding a regulatory transcription factor protein or other protein.

regulatory transcription factor General term for a protein that binds to DNA regulatory sequences (eukaryotic enhancers, silencers, and promoterproximal elements), but not to the core promoter, leading to an increase or decrease in transcription of specific genes. Sometimes simply called a *transcription factor*. Compare with **general transcription factor**.

regulon A large set of genes distributed in bacterial DNA that are controlled by a single type of regulatory molecule. Regulon genes are transcribed in response to environmental cues and allow cells to respond to changing environments.

reinforcement In evolutionary biology, the natural selection for traits that prevent interbreeding between recently diverged species.

release factor Any of a class of proteins that trigger termination of translation when a ribosome reaches a stop codon.

renal corpuscle In the vertebrate kidney, the ball-like structure at the beginning of a nephron, consisting of a glomerulus and the surrounding Bowman's capsule. Acts as a filtration device.

replacement rate The number of offspring each female must produce over her entire life to "replace" herself and her mate, resulting in zero population growth. The actual number is slightly more than two because some offspring die before reproducing.

replication fork The Y-shaped site where a double-stranded molecule of DNA is separated into two single strands for replication and on which DNA synthesis occurs.

replicative growth The process by which cells reproduce or viruses produce new virions.

replisome The macromolecular machine that copies DNA; includes DNA polymerase, helicase, primase, and other enzymes.

repolarization Return to a resting potential after a membrane potential has changed; a normal phase in an action potential. Compare with **depolarization** and **hyperpolarization**.

repressor (1) In bacteria, a protein that binds to an operator sequence in DNA to prevent transcription when an inducer is not present and that comes off DNA to allow transcription when an inducer binds to the repressor protein. (2) In eukaryotes, a protein that binds to a silencer sequence in DNA to prevent or reduce gene transcription.

reproductive development The phase of plant development that involves development of the flower and reproductive cells. Follows vegetative development and occurs when a shoot apical meristem (SAM) transitions to a flower-producing meristem.

reptiles One of the two lineages of amniotes (vertebrates that produce amniotic eggs) distinguished by adaptations for life and reproduction on land. Living reptiles include turtles, snakes and lizards, crocodiles and alligators, and birds. Except for birds, all are typically ectotherms.

resilience In community ecology, a measure of how quickly a community recovers following a disturbance.

resistance In community ecology, a measure of the extent to which a community remains unchanged in the face of a disturbance.

resonance energy transfer Process of transferring energy from an excited donor pigment to an adjacent acceptor pigment, where another electron is excited in response.

respiratory system The collection of cells, tissues, and organs responsible for gas exchange between an animal and its environment.

resting potential The membrane potential of a cell in its resting, or unstimulated, state.

retina A thin layer of light-sensitive cells and neurons at the back of a simple eye, such as that of cephalopods and vertebrates.

retinal A light-absorbing pigment that is linked to an opsin protein in rods and cones of the vertebrate eye.

retrovirus A virus with an RNA genome that reverse-transcribes its RNA into double-stranded DNA, which is then inserted into the host's genome as part of the virus's replicative cycle.

reverse transcriptase An enzyme that can synthesize DNA from an RNA template.

rhizobia (singular: *rhizobium*) Members of the bacterial genus *Rhizobium;* nitrogen-fixing bacteria that live in root nodules of members of the pea family (legumes).

rhizoid The hairlike structure that anchors a nonvascular plant to the substrate.

rhizome A modified stem that runs horizontally underground and produces new plants at the nodes (a form of asexual reproduction). Compare with **stolon**.

rhodopsin A transmembrane complex that is instrumental in detection of light by rods of the vertebrate eye. It is composed of the transmembrane protein opsin and covalently linked to retinal, a light-absorbing pigment.

ribonucleic acid (RNA) A usually single-stranded nucleic acid composed of ribonucleotides. Functions include catalytic components of ribosomes (rRNA), transporters of amino acids (tRNA), and messages of the DNA code required for protein synthesis (mRNA), among others.

ribonucleotide A nucleotide consisting of a ribose sugar, one or more phosphates, and one of four nitrogen-containing bases: adenine, guanine, cytosine, or uracil.

ribosomal RNA (rRNA) An RNA molecule that forms part of the ribosome.

ribosome A large macromolecular machine that synthesizes proteins by using the genetic information encoded in messenger RNA. Consists of two subunits, each composed of ribosomal RNA and proteins.

ribosome binding site In a bacterial mRNA molecule, the sequence just upstream of the start codon to which a ribosome binds to initiate translation. Also called the *Shine–Dalgarno sequence*.

ribozyme Any RNA molecule that can act as a catalyst to increase the rate of a chemical reaction.

ribulose bisphosphate (RuBP) A five-carbon compound that combines with CO_2 in the first step of the Calvin cycle during photosynthesis or with O_2 in the first step of photorespiration.

RNA interference (RNAi) Degradation of an mRNA molecule or inhibition of its translation following its binding by a short RNA whose sequence is complementary to a portion of the mRNA.

RNA polymerase An enzyme that catalyzes the synthesis of RNA from ribonucleotides using a template usually consisting of DNA.

RNA processing In eukaryotes, the changes that a primary RNA transcript undergoes to become a mature RNA molecule. For pre-mRNA it includes the addition of a 5′ cap, a 3′ poly(A) tail, and splicing to remove introns.

RNA replicase A viral enzyme that can synthesize RNA from an RNA template. Also called an *RNA-dependent RNA polymerase*.

RNA-seq See **deep sequencing**.

RNA world hypothesis Proposal that at a particular stage in the evolution of life, RNA both stored genetic information and catalyzed its own replication and that RNA emerged before DNA and proteins during chemical evolution.

rod A vertebrate photoreceptor with a rod-shaped outer portion that is particularly sensitive to dim light but does not distinguish colors. Compare with **cone**.

root apical meristem (RAM) A group of undifferentiated cells at the tip of a plant root; these cells divide to produce new cells that differentiate into mature root tissue.

root cap A small group of cells that covers and protects the root apical meristem. Senses gravity and determines the direction of root growth.

root hair A long, thin outgrowth of the epidermal cells of plant roots, providing increased surface area for absorption of water and nutrients.

root pressure Positive pressure of xylem sap in the vascular tissue of roots. Generated during the night as a result of the accumulation of ions from the soil and subsequent osmotic movement of water into the xylem.

root system The belowground part of a plant.

rough endoplasmic reticulum (rough ER) The portion of the endoplasmic reticulum that is dotted with ribosomes. Involved in synthesizing plasma membrane proteins, secreted proteins, and proteins localized to the ER, Golgi apparatus, and lysosomes. Compare with **smooth endoplasmic reticulum**.

roundworms Members of the phylum Nematoda. Distinguished by an unsegmented body with a pseudocoelom and no appendages. Roundworms belong to the ecdysozoan branch of the protostome animals. Also called *nematodes*.

rubisco The enzyme that catalyzes the addition of a molecule of CO_2 (Calvin cycle) or O_2 (photorespiration) to ribulose bisphosphate. See also **carbon fixation**.

ruminants Member of a group of mammals (cattle, sheep, goats, deer, antelope, giraffe, and pronghorn) that have a four-chambered stomach specialized for digestion of plant cellulose. Ruminants regurgitate cud, a mixture of partially digested food and cellulose-digesting bacteria and protists, from the largest chamber (the rumen) for further chewing.

S phase The phase of the cell cycle during which DNA is synthesized for the replication of chromosomes.

salinity The proportion of solutes dissolved in water in natural environments, often designated in grams of solute per kilogram of water (cited as parts per thousand).

salivary amylase An enzyme produced by the salivary glands that digests starch and glycogen by catalyzing hydrolysis of the glycosidic linkages between glucose residues.

salivary gland In vertebrates, one of several glands that secrete saliva (a mixture of water, mucus-forming glycoproteins, and the enzyme salivary amylase) into the mouth.

sampling error The selection of a nonrepresentative sample from some larger population, due to chance; *not* a type of mistake in measurement.

saprophyte An organism that feeds primarily on dead plant material.

sapwood The younger xylem in the outer layer of wood of a stem or root, functioning primarily in water transport.

sarcomere The repeating contractile unit of a skeletal or cardiac muscle cell; the portion of a myofibril located between adjacent Z disks.

sarcoplasmic reticulum Sheets of smooth endoplasmic reticulum in a muscle cell. Stores calcium ions, which can be released into the cytoplasm to trigger contraction.

saturated Referring to lipids in which all the carbon-carbon bonds are single bonds. Such compounds have relatively high melting points. Compare with **unsaturated**.

scanning electron microscope (SEM) A microscope that produces surface images by reflecting electrons off a specimen coated with a layer of metal atoms. Compare with **transmission electron microscope (TEM)**.

scarify To scrape, rasp, cut, or otherwise damage the coat of a seed. Necessary in some species to trigger germination.

Schwann cells A type of glial cell that wraps around axons of some neurons in the vertebrate peripheral nervous system, forming a myelin sheath that provides electrical insulation. Compare with **oligodendrocyte**.

scientific name The unique, two-part name given to each species, with a genus name followed by a species name—as in *Homo sapiens*. Scientific names are always italicized. Also known as *Latin names*.

sclereid In plants, a relatively short type of sclerenchyma cell that usually functions in protection, such as in seed coats and nutshells. Compare with **fiber**.

sclerenchyma Can refer to tissue or cell type. In plants, a cell that has a thick secondary cell wall and provides support; forms a simple tissue that typically contains the tough structural polymer lignin and usually is dead at maturity. Includes fibers and sclereids. Compare with **collenchyma** and **parenchyma**.

scrotum A sac of skin that contains the testes and is suspended just outside the abdominal cavity of many male mammals.

second law of thermodynamics The principle of physics that the entropy of the universe or any closed system always increases.

second messenger A nonprotein signaling molecule (e.g., calcium) produced or activated inside a cell in response to stimulation by a chemical messenger (the "first" messenger) at the cell surface. Commonly used to relay the message of a hormone or other extracellular signaling molecule.

secondary active transport Transport of an ion or molecule in a defined direction (often against its gradient), made possible by the transport of another ion or molecule being moved along its gradient. Also called *cotransport*.

secondary cell wall The thickened inner layer of a cell wall formed by certain plant cells as they mature and after they have stopped growing; contains lignin in water-conducting cells. Provides support or protection.

secondary consumer A carnivore; an organism that eats herbivores. Compare with **primary consumer**.

secondary growth In plants, an increase in the width of stems and roots due to the activity of a cambium. Compare with **primary growth**.

secondary immune response An adaptive immune response to a pathogen that the immune system has encountered before. Normally much faster and more efficient than the primary response, due to immunological memory. Compare with **primary immune response**.

secondary metabolites Molecules that are closely related to compounds in key synthetic pathways and that often function in defense.

secondary oocyte A cell produced by meiosis I of a primary oocyte in an ovary. If fertilized, will complete meiosis II to produce an ootid (which develops into an ovum) and a polar body.

secondary phloem Phloem cells that are formed specifically from the vascular cambium and are therefore considered secondary growth.

secondary spermatocyte A cell produced by meiosis I of a primary spermatocyte in a testis. Can undergo meiosis II to produce two spermatids.

secondary structure In proteins, localized folding of a polypeptide chain into regular structures (i.e., alpha-helices and beta-pleated sheets) stabilized by hydrogen bonding between atoms of the peptide backbone. In nucleic acids, elements of structure (e.g., helices and hairpins) stabilized by hydrogen bonding between complementary bases and hydrophobic interactions with the aqueous environment. Compare with **primary, tertiary**, and **quaternary structures**.

secondary succession Gradual colonization of a habitat after an environmental disturbance (e.g., fire, windstorm, logging) that removes some or all previous organisms but leaves the soil intact. Compare with **primary succession**.

secondary xylem Xylem cells that are formed specifically from the vascular cambium and are therefore considered secondary growth.

secretin A peptide hormone secreted by cells in the small intestine in response to the arrival of food from the stomach. Stimulates secretion of bicarbonate ions (HCO_3^-) by the pancreas.

secretion Selective transport of solutes from the blood to the filtrate in the kidney tubules of vertebrates.

sedimentary rock A type of rock formed by gradual accumulation of sediment, particularly sand and mud, as in riverbeds and on the ocean floor. Most fossils are found in sedimentary rocks.

seed A plant reproductive structure consisting of an embryo, associated nutritive tissue (endosperm), and an outer protective layer (seed coat). In angiosperms, develops from the fertilized ovule of a flower.

seed banks A repository where seeds, representing many different varieties of domestic crops or wild species, are preserved.

seed coat A protective layer around a seed that encases both the embryo and the endosperm.

segment A well-defined body region that is repeated along the anterior–posterior body axis of an animal.

segmentation Division of the body or a part of it into a series of similar structures; exemplified by the body segments of insects, annelid worms, and vertebrates.

segregation, principle of The concept that each pair of hereditary elements (consisting of alleles of the same gene) separate from each other during meiosis. One of Mendel's two principles of genetics. Compare with **principle of independent assortment**.

selective adhesion The tendency of cells of one tissue type to specifically adhere to other cells of the same type.

selective permeability The property of a membrane that allows some substances to diffuse across it much more readily than other substances.

self-fertilization The fusion of two gametes produced by the same individual to form offspring. Also called *selfing*.

self-incompatible Plants that are unable to form viable seed when carpels are pollinated with pollen from the same plant.

self molecule A molecule that is synthesized by an organism and is a normal part of its cells and/or body, in contrast to nonself, or foreign, molecules.

self-renewal A process of cell division in which a cell divides to produce one copy of itself and another cell that specializes into a different cell type; usually used in the context of stem cell self-renewal.

semen The combination of sperm and accessory fluids that is released by male mammals and reptiles during ejaculation.

semicircular canals Three structures in the inner ear that sense rotation of the head.

semiconservative replication The way DNA replicates, with each strand of an existing DNA molecule serving as a template to create a new complementary DNA strand.

seminal vesicle In male mammals, one of a pair of reproductive glands that secrete a sugar-containing fluid into semen, which provides energy for sperm movement.

senescence The genetically programmed, active process of aging and eventual death.

sensor Any cell, organ, or structure with which an organism can sense some aspect of the external or internal environment. Usually functions, along with an integrator and an effector, as part of a homeostatic system.

sensory neuron A nerve cell that carries sensory signals to the central nervous system. Compare with **interneuron** and **motor neuron**.

sepal One of the protective leaflike organs enclosing a flower bud and, after blooming, the outermost portion of the flower.

septum (plural: *septa*) Any wall-like structure. In fungi, septa divide the filaments (hyphae) of mycelia into cell-like compartments.

serotonin A monoamine neurotransmitter; in vertebrates, involved in many brain functions, including sleep, pleasure, and mood.

serum The liquid that remains when cells and clot material are removed from clotted blood. Contains water, gases, growth factors, nutrients, and other dissolved substances. Compare with **plasma**.

sessile Permanently attached to a substrate; not capable of moving to another location.

set point A normal or target range of values for a regulated internal variable, such as body temperature or blood pH.

sex chromosome Chromosomes that differ in shape or in number in males and females. For example, the X and Y chromosomes of many animals. Compare with **autosome**.

sex-linked gene A gene located on either sex chromosome.

sex-linked inheritance Inheritance patterns observed in genes carried on sex chromosomes. In this case, females and males have different numbers of alleles of a gene. Often creates situations in which a trait appears more often in one sex. Also called *sex-linkage*.

sexual dimorphism Any trait that differs between males and females.

sexual reproduction Any form of reproduction in which genes from two parents are combined via fusion of gametes, producing offspring that are genetically distinct from both parents. Compare with **asexual reproduction**.

sexual selection A type of natural selection that favors individuals with traits that increase their ability to obtain mates or choose good mates. Compare with **ecological selection**.

shell A hard, protective outer structure.

Shine–Dalgarno sequence See **ribosome binding site**.

shoot apical meristem (SAM) A group of undifferentiated cells at the tip of a plant stem; these cells divide to produce new cells that can differentiate into mature shoot tissues.

shoot system The aboveground part of a plant that is made up of stems, leaves, and flowers (in angiosperms).

short-day plant A plant that blooms in response to long nights (usually in late summer or fall in the Northern Hemisphere). Compare with **day-neutral plant** and **long-day plant**.

short tandem repeat (STR) Short (2–6 base pairs) DNA sequence that is repeated sequentially along a chromosome. Also called *microsatellite* or *simple sequence repeat (SSR)*.

shotgun sequencing A method of sequencing genomes that is based on breaking the genome into small pieces, sequencing each piece separately, and then figuring out how the pieces are connected.

side chain See **R-group.**

sieve plate In plants, a pore-containing structure at each end of a sieve-tube element in phloem.

sieve-tube element In plants, an elongated sugar-conducting cell in phloem that lacks nuclei and has sieve plates at both ends, allowing sap to flow to adjacent cells.

sigma A bacterial protein that associates with the core RNA polymerase to allow recognition of promoters.

signal In behavioral ecology, any information-containing behavior or characteristic that has been shaped by natural selection.

signal receptor Any cellular protein that binds to a particular signaling molecule (e.g., a hormone or neurotransmitter) and triggers a response by the cell. Receptors for lipid-insoluble signals are transmembrane proteins in the plasma membrane; those for lipid-soluble signals (e.g., steroid hormones) are often located in the cytosol.

signal recognition particle (SRP) An RNA–protein complex that binds to the ER signal sequence in a polypeptide as it emerges from a ribosome and participates in transport of the ribosome–polypeptide complex to the ER membrane, where synthesis of the polypeptide is completed.

signal transduction The process by which a stimulus (e.g., a hormone, a neurotransmitter, or sensory information) outside a cell is converted to an intracellular signal required for a cellular response. Usually involves a specific sequence of molecular events called a signal transduction pathway, which may amplify the signal.

signal transduction cascade A series of enzyme-catalyzed phosphorylation reactions commonly used in signal transduction pathways to amplify and convey a signal inward from the plasma membrane.

silencer A regulatory sequence in eukaryotic DNA to which repressor proteins can bind, inhibiting gene transcription.

silent mutation A point mutation that changes the sequence of a codon without changing the amino acid that is specified.

simple eye An eye with only one light-collecting apparatus (e.g., one lens), as in vertebrates and cephalopods. Compare with **compound eye**.

simple fruit A fruit (e.g., apricot) that develops from a single flower that has a single carpel or several fused carpels. Compare with **aggregate fruit** and **multiple fruit**.

simple leaf A leaf consisting of a single blade defined by the presence of a single axillary bud where the petiole joins the branch or stem.

simple sequence repeat (SSR) See **short tandem repeat (STR)**.

simple tissue A tissue consisting of a single cell type.

single nucleotide polymorphism (SNP) A site on a chromosome where different versions of the chromosome have different nucleotides. Can be used as a genetic marker to help track the inheritance of nearby genes.

single-strand DNA–binding protein (SSBP) A protein that attaches to separated strands of DNA during replication, preventing them from re-forming a double helix.

sink Any tissue, site, or location where an element or a molecule is consumed or taken out of circulation (e.g., in plants, a tissue where sugar exits the phloem). Compare with **source**.

sinoatrial (SA) node In the right atrium of the vertebrate heart, a cluster of cardiac muscle cells that initiates the heartbeat and determines the heart rate. Compare with **atrioventricular (AV) node**.

siphon A tubelike appendage of many mollusks, often used for feeding or propulsion.

sister chromatids The paired, double-stranded DNA copies of a recently replicated chromosome. They are connected most tightly at the centromere and eventually separate during anaphase of mitosis or meiosis II. Compare with **non-sister chromatids**.

sister groups Two lineages that are each other's closest relatives, represented by two branches emerging from a node in a phylogenetic tree.

skeletal muscle The muscle tissue attached to the bones of the vertebrate skeleton. Consists of long, unbranched muscle fibers with a characteristic striped (striated) appearance; controlled voluntarily. Compare with **cardiac muscle** and **smooth muscle**.

sliding-filament model A model of muscle contraction in which thin (actin) filaments and thick (myosin) filaments slide past each other, thereby shortening the sarcomere. Shortening of all the sarcomeres in a myofibril results in contraction of the entire myofibril.

slow muscle fiber A type of skeletal muscle fiber that is red due to the abundance of myoglobin, generates ATP by oxidative phosphorylation, and contracts slowly but does not fatigue easily. Also called *slow oxidative*, or *Type I, fiber*.

small intestine The portion of the digestive tract between the stomach and the large intestine. The site of the final stages of digestion and of most nutrient absorption.

small nuclear ribonucleoprotein See **snRNP**.

smooth endoplasmic reticulum (smooth ER) The portion of the endoplasmic reticulum that lacks attached ribosomes. Involved in synthesizing lipids and removing toxic molecules. Compare with **rough endoplasmic reticulum**.

smooth muscle The unstriated muscle tissue that lines the intestine, blood vessels, and some other organs. Consists of tapered, unbranched cells that can sustain long contractions. Not voluntarily controlled. Compare with **cardiac muscle** and **skeletal muscle**.

snRNP (small nuclear riboprotein) Any of a class of small RNAs associated with proteins and found in the nucleus of eukaryotic cells, where they form the spliceosome and catalyze splicing. Pronounced *"snurp."*

sodium–potassium pump A transmembrane protein that uses the energy of ATP to move sodium ions out of the cell and potassium ions into the cell, normally against their electrochemical gradients. Also called Na^+/K^+*-ATPase*.

softwood Wood, typically from conifers, characterized by the presence of tracheids (but lacking vessel elements).

soil organic matter Organic (carbon-containing) compounds found in soil.

solute Any substance that is dissolved in a liquid.

solute potential (Ψ_S) A component of the potential energy of water caused by a difference in solute concentrations at two locations. Can be zero (pure water) or negative. Compare with **pressure potential (Ψ_P)**.

solution A liquid containing one or more dissolved solids or gases in a homogeneous mixture.

solvent Any liquid in which one or more solids or gases can dissolve.

soma See **cell body**.

somatic cell Any type of cell in a multicellular organism except eggs, sperm, and their precursor cells. Also called *body cell*.

somatic hypermutation Mutation that occurs in the variable regions of immunoglobulin genes when B cells are first activated and in memory cells, resulting in novel variation in the receptors that bind to antigens.

somatic nervous system The part of the vertebrate peripheral nervous system that controls skeletal muscles and is under voluntary control. Compare with **autonomic nervous system**.

somite One of the blocks of mesoderm that occur in pairs along both sides of the developing neural tube in a vertebrate embryo. Gives rise to muscle, vertebrae, ribs, and the dermis of the skin.

source Any tissue, site, or location where a substance is produced or enters circulation (e.g., in plants, the tissue where sugar enters the phloem). Compare with **sink**.

space-filling model A representation of a molecule where atoms are shown as balls that are color-coded and scaled to indicate the atom's identity and volume. Depicts spatial relationships between atoms more accurately than a ball-and-stick model.

speciation The evolution of two or more distinct species from a single ancestral species.

species An evolutionarily independent population or group of populations. Generally distinct from other species in appearance, behavior, habitat, ecology, genetic characteristics, and so on.

species–area relationship The mathematical relationship between the area of a certain habitat and the number of species that it can support.

species diversity The variety and relative abundance of the species present in a given ecological community.

species richness The number of species present in a given ecological community.

specific heat The amount of energy required to raise the temperature of 1 gram of a substance by 1°C; a measure of the capacity of a substance to absorb energy.

sperm A mature male gamete.

sperm competition Competition between the sperm of different males to fertilize eggs inside the same female.

spermatogenesis The production of sperm. Occurs continuously in a testis.

spermatogonium (plural: *spermatogonia*) In a testis, any of the diploid cells that can divide by mitosis to produce primary spermatocytes (which can undergo meiosis) and more spermatogonia.

spermatophore A gelatinous package containing sperm that is produced by males of species that have internal fertilization without copulation.

spermatozoan (plural: *spermatozoa*) The final product of spermatogenesis; a fully formed sperm cell.

sphincters A muscular valve that can close off a tube, as in a blood vessel or a part of the digestive tract.

spicule Stiff spike of silica or calcium carbonate that provides structural support in the body of many sponges.

spindle apparatus The array of microtubules responsible for moving chromosomes during mitosis and meiosis; includes kinetochore microtubules, polar microtubules, and astral microtubules.

spiracle In insects, a small opening in the exoskeleton that connects air-filled tracheae to the external environment, allowing for gas exchange.

spleen A dark red organ, found near the stomach of most vertebrates, that filters blood, stores extra red blood cells in case of emergency, and plays a role in immunity.

spliceosome In eukaryotes, a large, complex assembly of snRNPs (small nuclear ribonucleoproteins) and many proteins that catalyzes removal of introns from primary RNA transcripts.

splicing The process by which introns are removed from primary RNA transcripts and the remaining exons are connected together.

sporangium (plural: *sporangia*) A spore-producing structure found in seed plants, some protists, and some fungi (e.g., chytrids).

spore (1) In bacteria, a dormant form that generally is resistant to extreme conditions. (2) In eukaryotes, a single haploid cell produced by meiosis; it is distinct from a gamete, however, in being able to grow into a multicellular, haploid organism through mitotic divisions directly (no fertilization required).

sporophyte In organisms undergoing alternation of generations, the multicellular diploid form that develops by mitotic divisions after fertilization produces a zygote. Compare with **gametophyte**.

sporopollenin A watertight material that encases spores and pollen of modern land plants.

stabilizing selection A mode of natural selection that favors phenotypes near the middle of the range of phenotypic variation. Reduces overall genetic variation in a population. Compare with **disruptive selection** and **directional selection**.

stamen The male reproductive structure of a flower. Consists of an anther, in which pollen grains are produced, and a filament, which supports the anther. Compare with **carpel**.

standard free-energy change The equation used to measure the change in Gibbs free energy (ΔG) by calculating the changes in enthalpy (ΔH) and entropy (ΔS) that occur in a given chemical reaction. ΔG is less than 0 for spontaneous reactions and greater than 0 for nonspontaneous reactions.

stapes One of three small bones in the middle ear of vertebrates that receive vibrations from the tympanic membrane. The stapes contacts the oval window and passes the vibrations to the cochlea.

starch A mixture of two storage polysaccharides, amylose and amylopectin, both formed from α-glucose monomers. Amylopectin is branched, and amylose is unbranched. The major form of stored carbohydrate in plants.

start codon The AUG triplet in mRNA where protein synthesis begins; codes for the amino acid methionine.

statocyst A sensory organ of many arthropods that detects the animal's orientation in space (e.g., whether the animal is upside down).

statolith A tiny stone or dense particle found in specialized gravity-sensing organs in some animals such as lobsters, and in gravity-sensing tissues of plants.

statolith hypothesis The hypothesis that amyloplasts (dense, starch-storing plant organelles) serve as statoliths in gravity detection by plants.

stem cell Any relatively undifferentiated cell that can divide to produce a daughter cell that remains a stem cell and a daughter cell that can differentiate into specific cell types.

stems Vertical, aboveground structures that make up the shoot system of plants.

stereocilium (plural: *stereocilia*) One of many stiff projections from the surface of a vertebrate hair cell that are involved in detection of sound or of waterborne vibrations.

steroid A class of lipid with a characteristic four-ring hydrocarbon structure.

sticky ends The short, single-stranded ends of a DNA molecule cut by a restriction endonuclease; can form hydrogen bonds with complementary sticky ends.

stigma The sticky tip at the end of a flower carpel; pollen grains adhere to it.

stolon A modified stem that runs horizontally over the soil surface and produces new plants at the nodes (a form of asexual reproduction). Compare with **rhizome**.

stoma (plural: *stomata*) Generally, a pore or opening. In plants, a microscopic pore on the surface of a leaf or stem through which gas exchange occurs; typically surrounded by specialized cells that open the pore. See also **guard cell**.

stomach A tough, muscular pouch in the vertebrate digestive tract between the esophagus and small intestine. Physically breaks up food and begins digestion of proteins.

stomatal crypt Chamber located below the surface of leaves and containing numerous stomata and trichomes.

stop codon Any of three mRNA triplets (UAG, UGA, or UAA) that cause termination of protein synthesis. Also called a *termination codon*.

storage root A root specialized for storing carbohydrates or other nutrients that can be used by the shoot system when needed.

strain The lowest, most specific level of taxonomy; refers to a population of individuals that are genetically very similar or identical.

striated muscle Muscle tissue containing protein filaments organized into repeating structures that give the cells and tissues a banded appearance when viewed through a microscope.

stroma The fluid matrix of a chloroplast, enclosed inside a double-membrane envelope in which the thylakoids are embedded. Site of the Calvin cycle reactions.

structural formula A two-dimensional notation in which the chemical symbols for the constituent atoms are joined by straight lines representing single (—), double (=), or triple (≡) covalent bonds. Compare with **molecular formula**.

structural homology Similarities in adult organismal structures (e.g., limbs, flowers) that are due to inheritance from a common ancestor.

style The slender stalk of a flower carpel connecting the stigma and the ovary.

suberin Waxy substance found in the cell walls of cork tissue and in the Casparian strip of endodermal cells.

substrate (1) A reactant that interacts with a catalyst, such as an enzyme or ribozyme, in a chemical reaction. (2) A surface on which a cell or organism sits.

substrate-level phosphorylation Production of ATP or GTP by the transfer of a phosphate group from an intermediate substrate directly to ADP or GDP. Occurs in glycolysis and in the citric acid cycle.

succession In ecology, the gradual colonization of a habitat after an environmental disturbance (e.g., fire, flood), usually by a series of species. See **primary** and **secondary succession**.

sucrose A disaccharide formed from glucose and fructose. One of the two main products of photosynthesis.

sugar Synonymous with *carbohydrate*, though normally used in an informal sense to refer to small carbohydrates (monosaccharides and disaccharides).

summation The additive effect of multiple inhibitory and excitatory postsynaptic potentials on a neuron or muscle cell.

supporting connective tissue A type of connective tissue distinguished by having a firm extracellular matrix. Includes bone and cartilage.

surface tension The cohesive force that causes molecules at the surface of a liquid to stick together, thereby resisting deformation of the liquid's surface and minimizing its surface area.

survivorship On average, the proportion of offspring that survive to a particular age.

survivorship curve A graph depicting the percentage of a population that survives to different ages.

suspension feeder An animal that obtains food by filtering small particles or small organisms out of water. Also called *filter feeder*.

suspensor A multicellular structure that extends from the base of an embryo and transfers nutrients from endosperm to the developing embryo.

sustainability The planned use of environmental resources at a rate no faster than the rate at which they are naturally replaced.

sustainable agriculture Agricultural techniques that are designed to maintain long-term soil quality and productivity.

swim bladder A gas-filled organ of many ray-finned fishes; regulates buoyancy.

symbiosis (adjective: *symbiotic*) Any close and prolonged physical relationship between individuals of two different species. The association can increase, decrease, or have no effect on the fitness of the species involved. See **commensalism, mutualism**, and **parasitism**.

sympathetic nervous system The part of the autonomic nervous system that stimulates fight-or-flight responses, such as increased heart rate, increased blood pressure, and slowed digestion. Compare with **parasympathetic nervous system**.

sympatric speciation The divergence of populations living within the same geographic area into different species as the result of their genetic (not physical) isolation. Compare with **allopatric speciation**.

sympatry Condition in which two or more populations live in the same geographic area, or close enough to permit interbreeding. Compare with **allopatry**.

symplast In plants, the space inside the plasma membrane. The symplasts of adjacent cells are often connected through plasmodesmata. Compare with **apoplast**.

symporter A cotransport protein that allows an ion to diffuse down an electrochemical gradient, using the energy of that process to transport a different substance in the same direction *against* its concentration gradient. Compare with **antiporter**.

synapomorphy A shared, derived trait found in two or more taxa that is present in their most recent common ancestor but is missing in more distant ancestors. Useful for inferring evolutionary relationships.

synapse The interface between two neurons or between a neuron and an effector cell.

synapsis The physical pairing of corresponding regions of two homologous chromosomes during prophase I of meiosis. Crossing over occurs during synapsis.

synaptic cleft The space between two communicating nerve cells (or between a neuron and effector cell) at a synapse, across which neurotransmitters diffuse.

synaptic plasticity A change in the responsiveness or structure of a synapse that can occur after particular stimulation patterns. Thought to be the basis of learning and memory.

synaptic vesicle A small neurotransmitter-containing vesicle inside the end of an axon that releases neurotransmitters into the synaptic cleft by exocytosis.

synaptonemal complex A network of proteins that holds non-sister chromatids together during synapsis in meiosis I.

synergid One of two cells flanking the egg in the female gametophyte; releases chemical attractants that direct pollen-tube growth.

synthesis phase See S phase.

synthetic biology The branch of biology that uses biotechnology and engineering principles to design or redesign biological systems, often for useful purposes.

system A defined set of interacting chemical components under observation.

systematics The subdiscipline of biology that characterizes and classifies the relationships among all organisms on Earth.

systemic acquired resistance (SAR) A slow, widespread response of plants to a localized infection that protects healthy tissue from invasion by pathogens. Compare with **hypersensitive response (HR)**.

systemic circuit The part of the circulatory system that carries oxygen-rich blood from the heart to the rest of the body and returns oxygen-poor blood to the heart. It is separate from the **pulmonary circuit** in mammals and birds.

systemin A peptide hormone, produced by plant cells damaged by herbivores, that initiates a protective response in undamaged cells.

systems biology The study of the structure of networks of genes or proteins and how interactions between individual network components can lead to emergent biological properties.

systole The portion of the cardiac cycle during which the atria or ventricles of the heart are contracting. Compare with **diastole**.

systolic blood pressure The force exerted by blood against artery walls during contraction of the heart's left ventricle. Compare with **diastolic blood pressure**.

T cell A type of lymphocyte that, with B cells, is responsible for adaptive immunity in vertebrates. Originates in the bone marrow and matures in the thymus. Involved in activation of B cells ($CD4^+$ helper T cells) and destruction of infected cells ($CD8^+$ cytotoxic T cells). Also called *T lymphocyte*.

T-cell receptor (TCR) A transmembrane protein in the plasma membrane of a mature T cell that can bind to a specific antigen displayed on the surfaces of other cells. See **antigen presentation**.

T tubule Any of the membranous tubes that extend into the interior of a skeletal muscle cell, propagating action potentials throughout the cell and triggering the release of calcium ions from the sarcoplasmic reticulum.

tagmata (singular: *tagma*) Prominent body regions in arthropods, such as the head, thorax, and abdomen in insects.

taproot A large, vertical main root of a plant's root system.

taste bud A sensory structure, found mainly on the mammalian tongue, containing spindle-shaped cells that respond to chemical stimuli.

TATA-binding protein (TBP) A protein that binds to the TATA box in eukaryotic promoters and is a component of the transcription initiation complex.

TATA box A short DNA sequence in many eukaryotic promoters that is important for assembling general transcription factors and RNA polymerase at the core promoter; located about 30 base pairs upstream from the transcription start site.

taxon (plural: *taxa*) Any named group of organisms at any level of a classification system.

taxonomy The branch of biology concerned with describing, naming, and classifying goups of organisms.

tectorial membrane A membrane located in the vertebrate cochlea; takes part in the transduction of sound by bending the stereocilia of hair cells in response to sonic vibrations.

telomerase An enzyme that adds DNA to the ends of chromosomes (telomeres) to prevent their shortening by standard DNA synthesis; catalyzes DNA synthesis guided by an RNA template that is part of the enzyme.

telomere The end of a linear chromosome that contains a short repeated sequence of DNA.

telophase The final stage in mitosis or meiosis, during which daughter chromosomes (or homologous chromosomes in meiosis I) have moved to opposite poles and new nuclear envelopes begin to form around each set of chromosomes.

temperature A measurement of thermal energy present in an object or substance, reflecting how much the constituent matter is moving.

template strand A strand of RNA or DNA used to make a new, complementary strand via complementary base-pairing.

temporal lobe In the vertebrate brain, one of the four lobes of the cerebrum. Functions in memory, speech (in humans), and interpreting auditory information.

tendon A band of tough, fibrous connective tissue that connects a muscle to a bone.

tentacle A long, thin, muscular appendage typically used for sensing and feeding. Occurs in different forms in diverse animals, such as cephalopod mollusks and sea anemones.

termination (1) In enzyme-catalyzed reactions, the final stage in which the enzyme returns to its original conformation and products are released. (2) In transcription, the dissociation of the RNA and RNA polymerase from DNA. (3) In translation, the release of the polypeptide and dissociation of a ribosome from mRNA when the ribosome reaches a stop codon.

territory An area that is actively defended by an animal and that provides exclusive or semi-exclusive use of its resources by the owner.

tertiary consumer A carnivore; in a food chain or food web, an organism that feeds on secondary consumers. Compare with **primary consumer** and **secondary consumer**.

tertiary structure The overall three-dimensional shape of a single polypeptide chain, resulting from multiple interactions among the amino acid side chains and the peptide backbone. In nucleic acids, three-dimensional shape is formed by hydrogen bonding between complementary bases and other interactions. Compare with **primary, secondary**, and **quaternary structure**.

testcross The breeding of an individual that expresses a dominant phenotype but has an unknown genotype with an individual having only recessive alleles for the genes of interest. Used to infer the unknown genotype by observing the phenotypes seen in offspring.

testis (plural: *testes*) The sperm-producing organ of a male animal.

testosterone A steroid hormone, produced and secreted by the testes in male mammals and many other vertebrates, that stimulates sperm production and various male traits and reproductive behaviors.

tetrapod Any member of the vertebrate lineage with two pairs of limbs or that descended from a vertebrate with two pairs of limbs; includes amphibians, mammals, and reptiles (including birds).

texture A quality of soil, resulting from the relative abundance of different-sized particles making up the soil.

theory An explanation for a broad class of phenomena that is supported by a wide body of evidence. A theory serves as a framework for the development of new hypotheses.

thermal energy The total kinetic energy of a system that includes the motion of matter and is measured as temperature.

thermocline A steep gradient (cline) in environmental temperature, such as occurs in a thermally stratified lake or ocean.

thermophile A bacterium or archaean that thrives in very hot environments.

thermoreception The detection of heat energy.

thermoreceptor A sensory cell specialized for detecting changes in temperature.

thermoregulation Regulation of body temperature.

thick filament A filament composed of bundles of the motor protein myosin; anchored to the center of the sarcomere. Compare with **thin filament**.

thigmomorphogenesis The response by plants to mechanical stimuli such as wind or touch, in which plants alter their growth patterns.

thigmonastic movements Rapid plant movements that occur in response to touch or vibration and that are independent of the direction of the stimulus.

thigmotropism Growth or movement of an organism in response to contact with a solid object.

thin filament A filament composed of two coiled chains of actin and the regulatory proteins tropomyosin and troponin; anchored to the Z disk of the sarcomere. Compare with **thick filament**.

thorax A region of the body; in arthropods, one of the three prominent body regions (tagmata), located between the head and abdomen; in vertebrates, including humans, the chest.

thorn A modified plant stem shaped as a sharp, protective structure. Helps protect a plant against feeding by herbivores.

threshold potential The membrane potential that will trigger an action potential in a neuron or other excitable cell. Also called simply *threshold*.

thylakoid A membrane-bound network of flattened sac-like structures inside a plant chloroplast that functions in converting light energy to chemical energy. Stacks of thylakoid discs make up grana.

thymus An organ, located in the anterior chest or neck of vertebrates, in which immature T cells that originated in the bone marrow undergo maturation.

thyroid gland A gland in the neck that releases thyroid hormone (which increases metabolic rate) and calcitonin (which lowers blood calcium ion concentration).

thyroid hormone Either of two hormones, triiodothyronine (T_3) or thyroxine (T_4), produced by the thyroid gland. See **triiodothyronine** and **thyroxine**.

thyroid-stimulating hormone (TSH) A peptide hormone, produced and secreted by the anterior pituitary, that stimulates the release of thyroid hormones from the thyroid gland.

thyroxine (T_4) A lipid-soluble hormone, derived from the amino acid tyrosine, containing four iodine atoms and produced and secreted by the thyroid gland. Acts primarily to stimulate cellular metabolism. In mammals, T_4 is converted to the more active hormone triiodothyronine (T_3) in the liver.

tight junction A type of cell–cell attachment structure that links the plasma membranes of adjacent animal cells, forming a barrier that restricts movement of substances in the space between the cells. Most abundant in epithelia (e.g., the intestinal lining). Compare with **desmosome** and **gap junction**.

tissue A group of cells that function as a unit, such as muscle tissue in an animal or xylem tissue in a plant.

tissue system A broad category of plant tissues that includes tissues and cell types formed by the primary meristems (protoderm, ground meristem, procambium). Protoderm gives rise to the dermal tissue system, ground meristem to the ground tissue system, and procambium to the vascular tissue system.

tolerance In ecological succession, the phenomenon in which early-arriving species do not affect the probability that subsequent species will become established. Compare with **facilitation** and **inhibition**.

tonoplast The membrane surrounding a plant vacuole.

tool-kit genes A set of key developmental genes that establishes the body plan of animals and plants; present at the origin of the multicellular lineages and elaborated upon over evolutionary time by a process of duplication and divergence. Includes *Hox* genes.

topoisomerase An enzyme that prevents the twisting of DNA ahead of the advancing replication fork by cutting the DNA, allowing it to unwind, and rejoining it.

torpor An energy-conserving physiological state, marked by a decrease in metabolic rate, body temperature, and activity, that lasts for a short period (overnight to a few days or weeks). Occurs in some small mammals when the ambient temperature drops significantly. Compare with **hibernation**.

totipotent Capable of dividing and developing to form a complete, mature organism.

toxin A poison produced by a living organism, such as a plant, animal, or microorganism.

trachea (plural: *tracheae*) (1) In insects, any of the small, air-filled tubes that extend throughout the body and function in gas exchange. (2) In terrestrial vertebrates, the airway connecting the larynx to the bronchi. Also called *windpipe*.

tracheid In vascular plants, a long, thin, water-conducting cell that has pits where its lignin-containing secondary cell wall is absent, allowing water movement between adjacent cells. Compare with **vessel element**.

trade-off In evolutionary biology, an inescapable compromise between two traits that cannot be optimized simultaneously. Also called *fitness trade-off*.

trait Any observable characteristic at any level of observation of an individual.

transcription The process that uses a DNA template to produce a complementary RNA.

transcription factor General term for a protein that binds to a DNA regulatory sequence to influence transcription. It includes both **regulatory transcription factors** and **general transcription factors**.

transcriptional activator A eukaryotic regulatory transcription factor that binds to regulatory DNA sequences in enhancers or promoter-proximal elements to promote the initiation of transcription. Also called *activator*.

transcriptional control Regulation of gene expression by various mechanisms that change the rate at which genes are transcribed to produce messenger RNA. In negative transcriptional control, binding of a regulatory protein to DNA represses transcription; in positive transcriptional control, binding of a regulatory protein to DNA promotes transcription.

transcriptome The complete set of gene transcripts in a particular cell.

transduction (1) The conversion of information from one mode to another. For example, the process by which a stimulus outside a cell is converted into a response by the cell. (2) The transfer of DNA from one bacterial cell to another by a virus.

transfer RNA (tRNA) An L-shaped RNA molecule that has an anticodon at one end and an amino acid attachment site at the other. Each tRNA carries a specific amino acid and binds to the corresponding codon in messenger RNA during translation.

transformation (1) Incorporation of external DNA into a cell. Occurs naturally in some bacteria; can be induced in the laboratory. (2) Conversion of a normal mammalian cell to one that divides uncontrollably.

transgenic A plant or animal whose genome contains DNA introduced from another individual, often from a different species.

transition state A high-energy intermediate state of the reactants during a chemical reaction that must be achieved for the reaction to proceed. Compare with **activation energy**.

transitional feature A trait that is intermediate between a trait observed in ancestral (older) species and the homologous trait observed in derived (younger) species.

translation The process by which a polypeptide (a string of amino acids joined by peptide bonds) is synthesized from information in codons of messenger RNA.

translational control Regulation of gene expression by various mechanisms that alter the life span of messenger RNA or the efficiency of translation.

translocation (1) In plants, the movement of sugars and other organic nutrients through the phloem by bulk flow. (2) A type of mutation in which a piece of a chromosome moves to a nonhomologous chromosome. (3) During translation, the movement of a ribosome down a messenger RNA.

transmembrane protein See **integral membrane protein**.

transmission electron microscope (TEM) A microscope that produces images by passing a focused beam of electrons through extremely thin sections of a specimen stained with heavy metals. Compare with **scanning electron microscope**.

transpiration Loss of water vapor from aboveground plant parts. Occurs primarily through stomata.

transposable element Any of several kinds of DNA sequences that are capable of moving themselves, or copies of themselves, to other locations in the genome. Transposable elements include LINEs.

tree of life The phylogenetic tree that includes all organisms.

trichome A hairlike appendage that grows from epidermal cells in the shoot system of some plants. Trichomes exhibit a variety of shapes, sizes, and functions depending on species.

triiodothyronine (T_3) A lipid-soluble hormone, derived from the amino acid tyrosine, containing three iodine atoms and produced and secreted by the thyroid gland. Acts primarily to stimulate cellular metabolism. In mammals, T_3 has a stronger effect than does the related hormone thyroxine (T_4).

triose A monosaccharide (simple sugar) containing three carbon atoms.

triplet code A code in which a "word" of three letters encodes one piece of information. The genetic code is a triplet code because a codon is three nucleotides long and encodes one amino acid.

triploblast (adjective: *triploblastic*) An animal whose body develops from three basic embryonic cell layers or tissues: ectoderm, mesoderm, and endoderm. Compare with **diploblast**.

trisomy The state of having three copies of one type of chromosome in an otherwise diploid cell.

trochophore A larva with a ring of cilia around its middle that is found in some lophotrochozoans.

trophic cascade A series of changes in the abundance of species in a food web, usually caused by the addition or removal of a key predator.

trophic level A feeding level in an ecosystem, such as primary producer, primary consumer, and decomposer.

trophoblast The sheet of cells on the exterior of a blastocyst (the structure that results from cleavage during embryonic development in mammals).

tropism Directional movement in response to a directional stimulus.

tropomyosin A regulatory protein present in thin (actin) filaments that blocks the myosin-binding sites on these filaments in resting muscles, thereby preventing muscle contraction.

troponin A regulatory protein present in thin (actin) filaments that can move tropomyosin off the myosin-binding sites on these filaments, thereby allowing muscle contraction. Activated by a high intracellular calcium ion concentration.

true navigation The type of navigation by which an animal can reach a specific point on Earth's surface. Also called *map orientation*.

trypsin A protein-digesting enzyme that is secreted in inactive form (as trypsinogen) by the pancreas and activates several other protein-digesting enzymes.

tube feet One of the many small, mobile, fluid-filled extensions of the water vascular system of echinoderms; the part extending outside the body is called a podium, while the bulb within the body is the ampulla. Used in locomotion, feeding, and respiration.

tuber A modified plant rhizome that functions in the storage of carbohydrates.

tuberculosis (TB) A disease of the lungs caused by infection with the bacterium *Mycobacterium tuberculosis*.

tumor A mass of cells formed by uncontrolled cell division. Can be benign or malignant (cancerous).

tumor suppressor A protein (or gene) that prevents cell division when conditions are unfavorable, such as when the cell has DNA damage. Defects in the function of tumor suppressors, such as p53 or Rb, are associated with the uncontrolled cell replication in cancer.

turbidity Cloudiness of water caused by sediments and/or microscopic organisms.

turgid Swollen and firm as a result of high internal pressure (e.g., a plant cell containing enough water for the cytoplasm to press against the cell wall). Compare with **flaccid**.

turgor pressure The outward pressure exerted by the fluid contents of a living plant cell against its cell wall.

turnover In lake ecology, the complete mixing of upper and lower layers of water of different temperatures; occurs each spring and fall in temperate-zone lakes.

tympanic membrane (1) The membrane separating the middle ear from the outer ear in terrestrial vertebrates. Also called the *eardrum*. (2) A structure that functions in hearing in insects.

ubiquinone See **coenzyme Q**.

ultimate causation In biology, the reason that a trait or phenomenon is thought to have evolved; the adaptive advantage of that trait. Also called *ultimate explanation*. Compare with **proximate causation**.

umami The sensation triggered by glutamate, responsible for the taste of meat and of monosodium glutamate.

umbilical cord The cord that connects a developing mammalian embryo or fetus to the placenta and through which the embryo or fetus receives oxygen and nutrients from the mother.

uncoating A process in which the viral capsid is removed, leading to the release of the viral genome.

unequal crossing over An error in crossing over during meiosis I in which the two non-sister chromatids match up at different sites. Results in gene duplication in one chromatid and gene loss in the other.

unsaturated Referring to lipids in which at least one carbon-carbon bond is a double bond. Double bonds produce links in hydrocarbon chains and decrease the compound's melting point. Compare with **saturated.**

upstream In genetics, opposite to the direction in which RNA polymerase moves along a DNA strand. Compare with **downstream**.

urea The major nitrogenous waste of mammals, adult amphibians, and cartilaginous fishes. Compare with **ammonia** and **uric acid**.

ureter In vertebrates, a tube that transports urine from one kidney to the bladder.

urethra The tube that drains urine from the bladder to the outside environment. In male vertebrates, also used for passage of semen during ejaculation.

uric acid The major nitrogenous waste of birds, other reptiles, and most terrestrial arthropods. Compare with **ammonia** and **urea**.

urochordates One of the three major chordate lineages (Urochordata), comprising sessile or floating, filter-feeding animals that have a polysaccharide covering (tunic) and two siphons through which water enters and exits; include the ascidians, thalaceans, and larvaceans. Compare with **cephalochordates** and **vertebrates**.

uterus The organ in which embryos develop in mammals and some other viviparous vertebrates.

vaccination Artificial production of immunological memory against a pathogen. Uses isolated antigens or altered versions of the pathogen to stimulate an adaptive immune response in the absence of disease.

vaccine A preparation designed to stimulate an immune response against a particular pathogen without causing illness. May consist of inactivated pathogens, attenuated (weakened) pathogens, or parts of a pathogen (subunit vaccine).

vacuole A large organelle in plant and fungal cells that ordinarily is used for bulk storage of water, pigments, oils, or other substances. Some vacuoles contain enzymes and have a digestive function similar to lysosomes in animal cells.

vagina The birth canal of female mammals; a muscular tube that extends from the uterus through the pelvis to the exterior.

valence The number of unpaired electrons in the outermost electron shell of an atom; when an atom is involved in covalent bonding, valence often determines how many covalent bonds the atom can form.

valence electron An electron in the outermost electron shell, the valence shell, of an atom. Valence electrons tend to be involved in chemical bonding.

valence shell The outermost electron shell of an atom.

valve In circulatory systems, any of the flaps of tissue that prevent backward flow of blood, particularly in veins and in the heart.

van der Waals interaction A weak electrical attraction between two nonpolar molecules or parts of a molecule that have been brought together, often through hydrophobic interactions. In proteins, it contributes to tertiary and quaternary structures.

variable (*V*) region The amino acid sequence that varies in the polypeptides that comprise antibodies, B-cell receptors, and T-cell receptors. Forms the antigen-binding site. Compare with **constant (C) region**.

vas deferens (plural: *vasa deferentia*) A muscular tube that stores and transports semen from the epididymis to the ejaculatory duct. Also called the *ductus deferens*.

vasa recta In the vertebrate kidney, a network of blood vessels that runs alongside the loop of Henle of a nephron. Functions in reabsorption of water and solutes from the filtrate.

vascular bundle In a plant stem, a cluster of xylem and phloem strands that run the length of the stem.

vascular cambium One of two types of cylindrical meristems, consisting of a ring of undifferentiated plant cells in the stem and root of woody plants; produces secondary xylem (wood) and secondary phloem. Compare with **cork cambium**.

vascular tissue In plants, tissue that transports water, nutrients, and sugars. Made up of the complex tissues xylem and phloem, each of which contains several cell types. Also called *vascular tissue system*.

vascular tissue system In plants, tissues that transport water, nutrients, and sugars. Made up of the complex tissues xylem and phloem, each of which contains several cell types.

vector (1) A biting insect or other organism that transfers pathogens from one species to another. (2) A plasmid or other agent such as a virus used to transfer recombinant genes into cells. See **cloning vector**.

vegetative development The phase of plant development that involves growth and the production of all plant structures except the flower.

vein Any blood vessel that carries blood (oxygenated or not) under relatively low pressure from the rest of the body toward the heart. Compare with **artery**.

vena cava (plural: *venae cavae*) Either of two large veins that return oxygen-poor blood to the heart.

ventral Toward an animal's belly and away from its back. The opposite of **dorsal**.

ventricle (1) A thick-walled chamber of the heart that receives blood from an atrium and pumps it to the body or to the lungs or gills. (2) Any of several small, fluid-filled chambers in the vertebrate brain.

venule A small vessel that collects blood from a capillary bed.

vertebrae (singular: *vertebra*) The cartilaginous or bony elements that form the backbones of vertebrate animals.

vertebrates One of the three major chordate lineages (Vertebrata), comprising animals with a dorsal column of cartilaginous or bony structures (vertebrae) and a skull enclosing the brain. Includes fishes, amphibians, mammals, and reptiles (including birds). Compare with **cephalochordates** and **urochordates**.

vertigo A medical condition in which the sense of equilibrium is disrupted, usually due to dysfunction in the inner ear.

vesicle A membrane-enclosed compartment with an aqueous interior that is often used in cells to transport cargo between organelles or to the plasma membrane for secretion.

vessel element In vascular plants, a short, wide, water-conducting cell that has gaps through both the primary and secondary cell walls, allowing unimpeded passage of water between adjacent cells. Compare with **tracheid**.

vestibule The part of the inner ear housing the maculae, sensory structures responsible for the sense of equilibrium.

vestigial trait A reduced or incompletely developed structure that has no function, or reduced function, but is clearly similar to functioning organs or structures in ancestral or closely related species.

vicariance The physical splitting of a population into smaller, isolated populations by a geographic barrier.

villi (singular: *villus*) Small, fingerlike projections (1) of the lining of the small intestine or (2) of the fetal portion of the placenta adjacent to maternal arteries. Function to increase the surface area available for absorption of nutrients and gas exchange.

virion The infectious extracellular particle that is produced from a viral infection; used for transmitting the virus between hosts. It consists of a DNA or RNA genome that is often enclosed within a protein shell (capsid) that may be further enveloped in a phospholipid bilayer. Compare with **virus**.

virulence (adjective: *virulent*) The ability of a pathogen to cause severe disease in a susceptible host.

virus An obligate, intracellular parasite that is acellular but uses host-cell biosynthetic machinery to replicate. Compare with **virion**.

visceral mass One of the three main parts of the mollusk body; contains most of the internal organs and external gill.

visible light The range of wavelengths of electromagnetic radiation that humans can see, from about 400 to 710 nanometers.

vitamins Any of various organic micronutrients that usually function as coenzymes.

vitelline envelope A fibrous sheet of glycoproteins that surrounds mature eggs in many vertebrates. Surrounded by a thick, gelatinous matrix (the jelly layer) in some aquatic species. In mammals, called the *zona pellucida*.

viviparity A condition in which embryos and seeds begin to grow and germinate before they are released from the parent plant.

viviparous In animals, producing live young (instead of eggs) that develop within and are nourished by the body of the mother, typically via a placenta, before birth. Compare with **oviparous** and **ovoviviparous**.

volt (V) A unit of electrical potential (voltage).

voltage Potential energy created by a separation of electric charges between two points. Also called *electrical potential*.

voltage clamping A technique for imposing a constant membrane potential on a cell. Widely used to investigate ion channels.

voltage-gated channel A type of ion channel that opens or closes in response to a change in membrane voltage. Compare with **ligand-gated channel**.

voluntary muscle A muscle that contracts in response to stimulation by voluntary (somatic) but not involuntary (parasympathetic or sympathetic) neural stimulation. Compare with **involuntary muscle**.

vomeronasal organ (VMO) A sensory organ, located in the nasal cavity of tetrapod vertebrates, containing chemoreceptors that bind odorants and pheromones.

wall pressure The inward pressure exerted by a cell wall against the fluid contents of a living plant cell.

water cycle, global **The movement of water** among abiotic and biotic reservoirs in terrestrial and aquatic ecosystems.

water potential (Ψ) The potential energy of water in a certain environment compared with the potential energy of pure water at room temperature and atmospheric pressure. In living organisms, ψ equals the solute potential (ψ_S) plus the pressure potential (ψ_P).

water-potential gradient A difference in water potential in one region compared with that in another region. Determines the direction that water moves, always from regions of higher water potential to regions of lower water potential.

water table The upper limit of the underground layer of soil that is saturated with water.

water vascular system In echinoderms, a system of fluid-filled tubes and chambers that functions as a hydrostatic skeleton.

watershed The area drained by a single creek or river.

Watson–Crick pairing See **complementary base pairing**.

wavelength The distance between two successive crests or troughs in any regular wave, such as a light wave, sound wave, or wave in water.

wax A class of lipid with extremely long, saturated hydrocarbon tails. Harder and less greasy than fats

weather The short-term atmospheric conditions of temperature, moisture, sunlight, and wind at a specific place and time.

weed Any plant that is adapted for growth in disturbed soils.

white blood cell Any of several types of blood cells, including neutrophils, macrophages, and lymphocytes, that reside in tissues and circulate in blood and lymph. Functions in tissue repair and defense against pathogens. Also called *leukocyte*.

wild type The most common phenotype(s) seen in a wild population.

wildlife corridors Strips of wildlife habitat connecting populations that otherwise would be isolated by human-made development.

wilt To lose turgor pressure in a plant tissue.

wobble pairing Nonstandard base-pairing between the nucleotide in the third position of a codon and the corresponding nucleotide in the anticodon of a tRNA. Wobble pairing allows one anticodon to read more than one codon, which in turn allows roughly 40 different tRNAs to read all 61 amino-acid-specifying codons.

wood Xylem resulting from secondary growth; forms strong supporting material. Also called *secondary xylem*.

worm An animal with a long, soft body lacking limbs.

X-linkage See **X-linked inheritance**.

X-linked gene A gene located on the X chromosome.

X-linked inheritance Inheritance patterns for genes located on the X chromosome. Also called *X-linkage*.

X-ray crystallography A technique for determining the three-dimensional structure of large molecules, including proteins and nucleic acids, by analyzing the diffraction patterns produced by X-rays beamed at crystals of the molecule.

xenoestrogen Any foreign chemical that binds to estrogen receptors or otherwise induces estrogen-like effects.

xeroderma pigmentosum (XP) A human disease characterized by extreme sensitivity to ultraviolet light. Caused by autosomal recessive alleles that inactivate different genes involved in nucleotide excision DNA repair.

xerophyte A plant species with adaptations for survival in environments with little available water.

xylem A plant vascular tissue that conducts water and ions from roots to shoots; contains tracheids and/or vessel elements. Primary xylem develops from the procambium of apical meristems; secondary xylem, or wood, from the vascular cambium. Compare with **phloem**.

Y-linkage See **Y-linked inheritance**.

Y-linked gene A gene located on the Y chromosome.

Y-linked inheritance Inheritance patterns for genes located on the Y chromosome. Also called *Y-linkage*.

yeast Any fungus growing as a single-celled form. Also, a specific lineage of Ascomycota.

yolk The nutrient-rich cytoplasm inside an egg; used as food for the growing embryo.

Z disc The structure that forms each end of a sarcomere. Contains a protein that binds tightly to actin, thereby anchoring thin filaments.

Z-scheme Model for changes in the potential energy of electrons as they pass from photosystem II to photosystem I and ultimately to $NADP^+$ during the light--capturing reactions of photosynthesis. See also **noncyclic electron flow**.

zero population growth A state of stable population size due to fertility staying at the replacement rate for at least one generation.

zona pellucida The gelatinous layer around a mammalian egg. In other vertebrates, called the *vitelline envelope*.

zone of cellular division In plant roots, a group of apical meristematic cells just behind the root cap where cells are actively dividing.

zone of cellular elongation In plant roots, a group of young cells, derived from primary meristem tissues and located behind the apical meristem, that are increasing in length.

zone of (cellular) maturation In plant roots, a group of plant cells, located several millimeters behind the root cap, that are differentiating into mature tissues. Site of most nutrient uptake.

zygosporangium (plural: *zygosporangia*) The distinctive spore-producing structure in fungi that are members of the Zygomycota.

zygote The cell formed by the union of two gametes; a fertilized egg.

Credits

Photo Credits

Frontmatter FM00 George Ostertag/age fotostock/Alamy Stock Photo **FMUnit1** YOSHIKAZU TSUNO/Staff/Getty Images **FMUnit2** Robert Markus/Science Source **FMUnit3** Odophile.com **FMUnit5** Marc Volk/fStop/AGEfotostcok **FMUnit5** BIOPHOTO ASSOCIATES/Science Source/Getty images **FMUnit5** Wolstenholme Images/Alamy Stock Photo **FMUnit6** Goran Mihajlovski/Alamy Stock Photo **FMUnit8** ZUMA Press, Inc./Alamy Stock Photo

Chapter 1 Opener Dan Suzio **1.1a** Science Source/Science Source **1.1b** The Print Collector/Alamy Stock Photo **1.4** All Canada Photos/Alamy Stock Photo **1.5** Arvind Balaraman/Shutterstock **1.6** Rebecca Letz/Sipa Press/Newscom **1.8a** Steve Gschmeissner/Science Source **1.8b** Kwangshin Kim/Science Source **1.9** Corbis Premium RF/Alamy Stock Photo **1.12** The Company of Biologists, Ltd. **Case Study L** Ernesto Gianoli **Case Study R** Ernesto Gianoli **EOU1** Edmund Brodie, Ph.D **EOU1B** ZUMA Press, Inc./Alamy Stock Photo **EOU1T** Nordreisender/imageBROKER/Alamy Stock Photo

Bioskills Opener Ben Sadd/Frank Lane Picture Agency **B6.3** Lizabeth Allison **B6.5b** Rosalind Franklin/Science Source **B9.1a** Biology Pics/Science Source **B9.1b** Janice Carr/Center for Disease Control and Prevention (CDC) **B9.2a** Michael W. Davidson/ Florida State University Research Foundation **B9.2b** Michael W. Davidson/Florida State University Research Foundation **B9.2c** Peter Searson **B11.1aL** National Cancer Institute **B11.1aR** E.S. Anderson/Science Source **B11.1b** Sinclair Stammers/Science Source **B11.2a** Kwangshin Kim/Science Source **B11.2b** Richard L. (Larry) Blanton, Ph.D. **B11.2c** Nigel Cattlin/Science Source **B11.2d** SCIMAT/Science Source **B11.2e** Graphic Science/Alamy Stock Photo **B11.2f** Sinclair Stammers/Science Source **B11.2g** dra_schwartz/Getty Images

Chapter 2 Opener University of Delaware **2.1b** Ezra Shaw/Getty Images Sports/Getty Images **2.6c** Zedcor Wholly OwnedPhoto/Objects.net/Getty Images Plus/Getty Images **2.14b** Dietmar Nill/Picture Press/Getty Images **2.15c** John Sylvester/First Light/AGE Fotostock **2.19b** picture-alliance/Judaica-Samml/Newscom **2.20** SPL/Science Source **Case Study** Jerry Prezioso/NEFSC/NOAA

Chapter 3 3.7a Janice Carr/Center for Disease Control and Prevention (CDC) **3.7b** Janice Carr/Center for Disease Control and Prevention (CDC) **Case Study** ChameleonsEye/Shutterstock

Chapter 4 Opener SSPL/The Image Works **Case Study** SSPL/Science Museum/The Image Works

Chapter 5 Opener Dr. Keith Wheeler/Science Source **5.5a** Colloidial/Shutterstock **5.5b** Jim Occi/Fundamental Photographs, NYC **Case Study** Andrew Burton/Thomson Reuters (Markets) LLC

Chapter 6 6.2a Multiart/iStock/Getty Images Plus/Getty Images **6.2b** lepas2004/ iStock/Getty Images Plus/Getty Images **6.2c** nexus 7/Shutterstock **6.2d** Richard Megna/Fundamental Photographs **6.7L** Dr. rer. nat. Markus Drechsler **6.16** Janet Iwasa/Jack W. Szostak **6.19** Don W. Fawcett/Science Source **Case Study** Jan Otto/E+/ Getty Images

Chapter 7 Opener Dr. Torsten Witmann/Photo Researchers, Inc./Science Source **7.1** SPL/Science Source **7.2a** Dr. Gopal Murti/Science Source **7.2b** Biology Pics/Science Source **7.3** Eye of Science/Science Source **7.4** Eye of Science/Science Source **7.6** Don W. Fawcett/Science Source **7.7** Omikron Omikron/Science Source/Getty Images **7.8L** Don W. Fawcett/Science Source **7.8R** Don W. Fawcett/Science Source **7.9** Biophoto Associates/Science Source **7.10** Don W. Fawcett/Science Source **7.11** Biophoto Associates/Science Source **7.12** Don W. Fawcett/Science Source **7.13** Don W. Fawcett/ Science Source **7.14** Biophoto Associates/Science Source **7.15a** Don W. Fawcett/Science Source **7.15b** Don W. Fawcett/Science Source **7.15c** Omikron/Science Source **7.15d** Biophoto Associates/Science Source **7.16L** Don W. Fawcett/Science Source **7.17** Natalie Gassman/Karin Scarpinato Lab **7.24** Photograph by Dr. Conly L. Rieder, Wadsworth Center, Albany, New York 12201-0509 **7.24ab** Bruce J. Schnapp, Oregon Health Sciences University. **7.26a** SPL/Scienca Source **7.27a** SPL/Scienca Source **7.28** Don W. Fawcett/Science Source **Case Study** James Bannister/Alamy Stock Photo

Chapter 8 Opener Nature Production/Nature Picture Library **8.3a** Richard Megna/ Fundamental Photographs **8.3b** Richard Megna/Fundamental Photographs **8.4a** Richard Megna/Fundamental Photographs **8.4b** Richard Megna/Fundamental Photographs **8.4c** Richard Megna/Fundamental Photographs **8.6** SPL/Science Source **Case Study** Editorial Image, LLC/Alamy Stock Photo

Chapter 9 Opener Raiden32 **9.1a** Oliver Hoffmann/Shutterstock **9.1b** Gergo Orban/ Shutterstock **9.8** Frey, Terry **9.16** Delannoy, Michael **Case Study** Dimitri Otis/Stone/ Getty Images

Chapter 10 Opener R.B. Taylor/Science Source **10.3T** John Durham/Science Source **10.3M** Biophoto Associates/Science Source **10.3B** Biophoto Associates/Science Source **10-18** pernsanitfoto/Shutterstock **10-19a** Dr. Jeremy Burgess/Science Source **10-19b** Dr. Jeremy Burgess/Science Source **10.20** Bassham, James A. **Case Study** Santiago Urquijo/Moment/Getty Images

Chapter 11 Opener Alvin Telser/Science Source **11.1** Jochen Tack/imageBROKER/ Alamy Stock Photo **11.2b** Biophoto Associates/Science Source **11.3** Biophoto Associates/Science Source **11.4b** Joseph A. Buckwalter **11.5** David Goodsell **11.8L** Don W. Fawcett/Science Source **11.8R** Don W. Fawcett/Science Source **11.9a** Don W. Fawcett/Science Source **11.11a** Don W. Fawcett/Science Source **11.11b** Biophoto Associates/Science Source **11.18a** Bruno Bianco **11.18b** Quillin, Kim **Case Study** Nern, A. and Arkowitz, R.A. 2000. G proteins mediate changes in cell shape by stabilizing the axis of polarity. Molecular Cell 5(5): 853–864. Figure 2C

Chapter 12 Opener Pr. G. Giménez-Martín/Science Source **12.1a** Dr. Gopal Murti/ Science Source **12.1b** Biophoto Associates/Science Source **12.5-1** Photograph by Dr. Conly L. Rieder, Wadsworth Center, Albany, New York 12201-0509 **12.5-2** Photograph by Dr. Conly L. Rieder, Wadsworth Center, Albany, New York 12201–0509 **12.5-3** Photograph by Dr. Conly L. Rieder, Wadsworth Center, Albany, New York 12201–0509 **12.5-4** Photograph by Dr. Conly L. Rieder, Wadsworth Center, Albany, New York 12201-0509 **12.5-5** Photograph by Dr. Conly L. Rieder, Wadsworth Center, Albany, New York 12201-0509 **12.5-6** Photograph by Dr. Conly L. Rieder, Wadsworth Center, Albany, New York 12201-0509 **12.5-7** Photograph by Dr. Conly L. Rieder, Wadsworth Center, Albany, New York 12201-0509 **12.8a** Ed Reschke/Photolibrary/ Getty Images **12.8b** Michael Danilchik **Case Study** inga spence/Alamy Stock Photo **EOU2** Kankitti Chupayoong/Shutterstock **EOU3B** Jukka Palm/Shutterstock **EOU3T** Science Stock Photography/Science Source

Chapter 13 Opener David M. Phillips/Science Source **13.5** Jupiterimages/PHOTOS. com/Getty Images Plus/Getty Images **13.6-1** Warren Rosenberg/Fundamental Photographs **13.6-2-10** Warren Rosenberg/Fundamental Photographs **13.9** Look at Sciences/Science Source **13.13** Dr. Rebecca Shulte **Case Study** Karel Noppé/Alamy Stock Photo

Chapter 14 Opener Brian Johnston **14.17a** Carla Fernanda Reis **Case Study** Richard Tsong-Taatarii/Minneapolis Star Tribune/MCT

Chapter 15 Opener Dr. Gopal Murti/Science Source **15.1** Eye of Science/Science Source **15.5** Meselson, M., and F. W. Stahl. 1958. The replication of DNA in Escherichia coli. Proceedings of the National Academy of Sciences USA 44: 671–682 **15.7c** Dr Gopal Murti/Science Source **Case Study** Stephen Jaffe/AFP/Getty Images

Chapter 16 Opener Craig J. Venter Institute **16.4a** Rod Williams/Nature Picture Library **16.4b** Peromyscus Genetic Stock Center **16.9** Joanne Davidson, Mira Grigorova and Paul Edwards/University of Cambridge **Case Study** kazakovmaksim/istock/Getty Images

Chapter 17 Opener Professor Oscar L. Miller/Science Source **17.5a** Bert W. O'Malley, M.D. **17.8a** American Association for the Advancement of Science. Miller, O. L. et al. Visualization of bacterial genes in action. Science 169, 392–395 (1970) **Case Study** MAP/Jean-Yves Grospas/AGE Fotostock

Chapter 18 Opener Stephanie Schuller/Science Source **TYPS** Courtesy of EDVOTEK® (www.edvotek.com) **Case Study** Margaret Mcfall-Ngai

Chapter 19 19.2a Don Fawcett/Science Source **19.3** Bolzer A, Kreth G, Solovei I, Koehler D, Saracoglu K, Fauth C, et al. (2005) Three-Dimensional Maps of All Chromosomes in Human Male Fibroblast Nuclei and Prometaphase Rosettes. PLoS Biol 3(5): e157. https://doi.org/10.1371/journal.pbio.0030157 **19.5a** Barbara Hamkalo **19.5b** Dr. Victoria E. Foe **Case Study** Mageon/Shutterstock

Chapter 20 Opener Alfred Pasieka/Science Source **20.12** De Agostini Editorial/DEA PICTURE LIBRARY/Getty Images **20.16** ZUMA Press/Alamy Stock Photo **Case Study L** John Scrivener/E+/Getty Images **Case Study R** K-R. Jo et al. 2014. BMC Biotechnology 14: 50.

Chapter 21 Opener Francois Gohier/ardea.com/AGE Fotostock **21.1** Lee W. Wilcox **21.2a** John Chadwick/Newscom **21.6ab** Juan M. Hurle, MD PhD **21.6cd** Pierre Vanderhaeghen, Institute of Interdisciplinary Research (IRIBHM) **21.7a** Richard

Hutchings/Science Source **21.7b** Pierre Vanderhaeghen, Institute for Latoratory Animal Research ILAR/The National Academies **21.8** Steve Downer/ardea.com/Mary Evans Picture Library Ltd/AGE Fotostock **21.10a** Thomas Gregor **21.12a** Thomas Deerinck, NCMIR/Science Source **21.12b** Thomas Deerinck, NCMIR/Science Source **21.12c** Eye of Science/Science Source **Case Study L** Sherry Yates Young/Shutterstock **Case Study R** Devaux, Patricia **EOU4** blickwinkel/Trapp/Alamy Stock Photo

Chapter 22 **Opener** Michael Patrick O'Neill/Alamy Stock Photo **22.2a** Gary Crabbe/AGE Fotostock **22.2b** Michael Quinn/Grand Canyon NPS **22.2c** National Park Service U.S. Department of the Interior **22.2d** Michael Quinn/Grand Canyon NPS **22.3** Henry Neville Hutchinson/Biodiversity Heritage Library **22.5a** Bruna Dolores **22.5b** BSIP/Science Source **22.6b** Republished with permission of [Petren, K.,et al.], from [A phylogeny of Darwin's finches based on microsatellite DNA length variation. Proceedings of the Royal Society B: Biological Sciences, 266(1417), 321. (1999)] permission conveyed through Copyright Clearance Center, Inc. **22.8a** The Art Of Animals.co.uk/PetStockBoys/Alamy Stock Photo **22.8b** Reinhard, H/Arco Images/Alamy Stock Photo **22.8c** Usher, D./Arco Images GmbH/Alamy Stock Photo **22.8d** Juniors Bildarchiv/F360/Alamy Stock Photo **22.9** David Evans/National Geographic Creative/Alamy Stock Photo **22.11** Alison Wright/National Geographic Creative/Alamy Stock Photo **22.11 inset** D. & S. Tollerton/AGE Fotostock **22.12** Kim Quillin **22.14** ASSOCIATED PRESS/AP Images **22.T1-1** Michael K. Richardson **22.T1-2** Michael K. Richardson **22.T1-3** Michael K. Richardson **22.T3-1** dibrova/Shutterstock **22.T3-2** Jivko Kazakov/iStock/Getty Images **22.T3-3** Tony Garcia/Image Source Release info/Getty Images **Case Study** Karen Walzer/Moment/Getty Images

Chapter 23 **Opener** Chaiwat Subprasom/THAILAND ANIMALS SOCIETY/Reuters **23.9a** Eye of Science/Science Source **23.9b** Gary Meszaros/Science Source **23.10a** Sandra Matic/Alamy Stock Photo **23.11** Shaun Cunningham/Alamy Stock Photo **23.12a** Phil Savoie/Nature Picture Library **23.12b** Steve Mackay Photography/Getty Images **23.14** Westend61/Getty Images **23.15b** John Amato **23.17a** Mark Conlin/Visual&Written SL/Alamy Stock Photo **23.19a** Alasdair James/Getty Imgaes **Case Study** Westend61/Getty Images

Chapter 24 **Opener** Ron Sanford/Corbis Documentary/Getty Images **24.9-1** Michael Lustbader/Science Source **24.9-2** Cubo Images/Superstock **24.9-3** (c) 2008 Stephen Lynn Clancy **24.11-1** Tim Zurowski/All Canada Photos/Alamy Stock Photo **24.11-2** Jim Burns **24.11-3** Brian E. Small/Nature Photographers Ltd/Alamy Stock Photo **Case Study** ryflip/Fotolia

Chapter 25 **Opener** Indiana9 fossils and L & M Collectibles **25.7-1** Dr. Gerald D. Carr, PhD **25.7-2** Forest & Kim Starr **25.7-3** Forest & Kim Starr **25.8-1** Photo by O. Louis Mazzatenta/National Geographic/Getty Images **25.8-2** Sinclair Stammers/Science Source **25.8-3** Photo by O. Louis Mazzatenta/National Geographic/Getty Images **25.8-4** Verisimilus/Wikimedia **25.9** John Sibbick/Science Source **25.T4-1** Stefan Piasecki/GEUS **25.T4-2** Ismael Montero Verdu/iStock/Getty Images **25.T4-3** Richard Becker/Frank Lane Picture Agency **25.T4-4** Millard H. Sharp/Science Source **25.T4-5** Juan Carlos Cantero/Pixtal/AGE Fotostock **Case Study** Panther Media GmbH/Alamy Stock Photo **EOU5 AGE Fotostock America Inc.**

Chapter 26 **Opener** Ignacio Palacios/Lonely Planet Images/Getty Images **26.1** Courtesy of Doc Searls **26.1 inset** Priya DasSarma, University of Maryland Biotechnology Institute **26.2** Dr. Kari Lounatmaa/Science Source **26.3** S. V. Liu, J.Zhou, C. Zhang, et al. 1997. Thermophilic Fe(III)-reducing bacteria from the deep subsurface: The evolutionary implications. Science 277: 1106–1109. **26.8a B** Heide Schulz-Vogt/Max Planck Institute for Developmental Biology **26.8a T** Image Courtesy Thomas Deerinck and Mark Ellisman, NCMIR and J. Craig Venter, JCVI **26.8b B** Scimat/Science Source **26.8b T** Scimat/Science Source **26.8c T** USDA/Science Source **26.9** ASM/Science Source **26-12** Bruce J. Russell/BioMedia Associates **26.T4-1** The Society for Actinomycetes Japan **26.T4-2** David M. Phillips/Science Source **26.T4-3** David Dalton **26.T4-4** Andrew Syred/Science Source **26.T4-5** George L. Barron **26.T4-6** James Cavallini/Science Source **26.T5-1** Dr. Kenneth M. Stedman **26.T5-1 inset** Eye of Science/Science Source **26.T5-2** MARUM – Center for Marine Environmental Sciences, University of Bremen **26.T5-2 inset** Karl Stetter **26.T5-3** University of Washington

Chapter 27 **Opener** Martin Kreutz/Panther Media/AGE Fotostock **27.2a** Scenics & Science/Alamy Stock Photo **27.2b** Mark Conlin/Alamy Stock Photo **27.2c** Bob Gibbons/Frank Lane Picture Agency Limited **27.4a** JTB Photo/UIG Universal Images Group/Newscom **27.4b** Donald Anderson/Woods Hole Oceanographic Institution **27.6** Biophoto Associates/Science Source **27.12a** Steve Gschmeissner/Science Source **27.12b** Biophoto Associates/Science Source **27.12c** Eye of Science/Science Source **27.12d** Steve Gschmeissner/Science Source **27.13a** Biopphoto associates/Researchers, Inc./Science Source **27.13b** Gerd Guenther/Science Source **27.14** Eric V. Grave/Science Source **27.18a** Doug Perrine/Nature Picture Library/Alamy Stock Photo **27.T3-1** Alex Wild/alexanderwild.com **27.T3-3** Gerd Guenther/Science Source **27.T3-5** Oxford Scientific/Getty Images **27.T3-6** Wim Van Egmond/Science Source **27.T3-7** Andrew Syred/Science Source **Case Study L** Ray Simons/Science Source **Case Study R** National Academy of Science USA

Chapter 28 **Opener** RF Company/F285Alamy Stock Photo **28.1** Lynn Betts/USDA/Natural Resources Conservation Service **28.2b** Hugh Iltis/Doebley, John **28.3a** Dan McCoy/Rainbow/Alamy Stock Photo **28.3b** Lee W. Wilcox, Ph.D. **28.3c** Bob Gibbons/Alamy Stock Photo **28.4a** Lee W. Wilcox, Ph.D. **28.4b** Michael Sacco/E+/Getty Images **28.4c** Juniors Bildarchiv/AGEfotostock **28.4d** Radim Beznoska/Alamy Stock Photo **28.7** Lee W. Wilcox, Ph.D. **28.10b** ANP Photo/AGEfotostock **28.15a** Lee W. Wilcox, Ph.D. **28.15bL** Lee W. Wilcox, Ph.D. **28.15bR** Lee W. Wilcox, Ph.D. **28.17** M. I. Walker/Science Source **28.20a** Larry Deack **28.20b** Quest786/Shutterstock **28.20c** John Severns Severnjc **28.22a** Maxim Khytra/Shutterstock **28.22b** Andreas Nilsson/Shutterstock **28.25a** Dr Keith Wheeler/Science Source **28.25b** Dr Keith Wheeler/Science Source **28.25c** jonphoto/iStock/Getty Images **28.25d** Harrison, J. J. **28.25e** Garry DeLong/Science Source **28.25f** Dr. Keith Wheeler/Science Source **28.25g** Imo/Getty Images **28.25h** Shutterstock **28.26** Peter Bassett/Nature Picture Library **28.T2-1** Nick Upton/Nature Picture Library **28.T2-2** Willem Kolvoort/Nature Picture Library **28.T2-3** Graham, Linda E. **28.T2-4** M. I. Walker/Science Source **28.T3-1** Adrian Davies/Nature Picture Library **28.T3-2** Joan Edwards **28.T3-3** Daniel Vega/AGE Fotostock **28.T4-1** Andrea and Antonella Ferrari/NHPA/AGE Fotostock **28.T4-2** Science Stock Photography/Science Source **28.T4-3** Matt Meadows/Stockbyte/Getty Images **28.T4-4** F Hecker/Blickwinkel/AGE Fotostock **28.T5-1** Brigg/iStockphoto/Getty Images **28.T5-2** Peter Hallet/Science Source **28.T5-3** Peter Chadwick/Dorling Kindersley **28.T5-4** Lee W. Wilcox, Ph.D. **28.T5-5** Geoff Renner/Robert Harding World Imagery **28.T6-1** Scott Zona **28.T6-2** Elburg Botanic Media/Visions Pictures & Photography/AGE Fotostock **28.T6-3** John Beedle/Corbis Documentary/Getty Images **28.T6-4** GWI/Botanic Images In/AGE Fotostock **Case Study L** Premaphotos/Alamy Stock Photo **Case Study R** Nigel Cattlin/Alamy Stock Photo

Chapter 29 **Opener** Chaiwatphotos/Shutterstock **29.1a** Inga Spence/Science Source **29.1b** EGON/Alamy Stock Photo **29.2a** David Read/University of Sheffield **29.2b** Michael Amaranthus/Mycorrhizal Applications, Inc **29.3** B. G. Thomson/Science Source **29.4a** Eye of Science/Science Source **29.4b** Hecker/Sauer/Blickwinkel/Science Source **29.5aR** Susumu Nishinaga/Science Source **29.5bL** Barron, George L. **29.5bR** Biophoto Associates/Science Source **29.5c** Patrick Hickey/NIPHT LIMITED **29.6** N. Allen and G. Barron **29.7a** Melvin Fuller **29.7b** Ed Reschke/Photolibrary/Getty Images **29.7c** Biophoto Associates/Science Source **29.7d** Santamaria, Nino **29.8L** WebCat/Shutterstock **29.8R** E. Gueho/Science Source **29.12a** Mark Brundrett **29.12b** Mark Brundrett **29.13a** Marko König/imagebrok/imageBROKER/AGE Fotostock **29.13b** Wagner, Ralf **29.14** Thailand Wildlife/Alamy Stock Photo **29.T1-1** Dr. J. I. Ronny Larsson, Professor of Zoology **29.T1-2** Photo Researchers/Science History Images/Alamy Stock Photo **29.T1-3** Gregory G. Dimijian/Science Source **29.T1-4** Kristine Nichols, USDA Agricultural Research Service **29.T1-5** Zoonar/Jürgen Vogt/Zoonar GmbH RF/AGE Fotostock **29.T1-6** adrian hepworth/Alamy Stock Photo **Case Study** Darlyne A. Murawski/National Geographic/Getty Images

Chapter 30 **Opener** BILL CURTSINGER/National Geographic Creative **30.4** M. I. Walker/Photo Researchers, Inc./Science Source **30.5a** Steve Gschmeissner/Science Source **30.5b** Susumu Nishinaga/Science Source **30.6** Eric Röttinger **30.7** Source: Finnerty, J. R. (2004). Origins of Bilateral Symmetry: Hox and Dpp Expression in a Sea Anemone. Science 28 May 2004: Vol. 304, Issue 5675, pp. 1335–1337 DOI: 10.1126/science.1091946 **30-11L** Panganiban, G. et al. (1997). The origin and evolution of animal?appendages. Proceedings of the National Academy of Sciences, (May 13, 1997), 94 (10) 5162–5166. Copyright (1997) National Academy of Sciences, U.S.A. **30-11M** Panganiban, G. et al. (1997). The origin and evolution of animal?appendages. Proceedings of the National Academy of Sciences, (May 13, 1997), 94 (10) 5162–5166. Copyright (1997) National Academy of Sciences, U.S.A. **30-11R** Panganiban, G. et al. (1997). The origin and evolution of animal?appendages. Proceedings of the National Academy of Sciences, (May 13, 1997), 94 (10) 5162–5166. Copyright (1997) National Academy of Sciences, U.S.A. **30.T2-1** Perennou Nuridsany/Science Source **30.T2-2** Barry Mansell/Nature Picture Library **30.T2-3** Pascal Goetgheluck/Science Source **30.T2-4** Joze Maucec/Shutterstock **30.T3-1** Juniors Bildarchiv GmbH/Alamy Stock Photo **30.T3-2** Olga Khoroshunova/Fotolia **30.T3-3** Len Rue Jr./Science Source **30.T4-1** Joe Belanger/Alamy Stock Photo **30.T4-2** Todd Husman **30.T4-3** Heinz-jurgen Landshoe/AGE Fotostock America Inc. **30.T4-4** Four Oaks/Shutterstock **30.T5-1** Francesco Tomasinelli//Science Source **30.T5-2** Michael Zysman/Shutterstock **30.T5-3** PAUL R. STERRY/Nature Photographers Ltd/Alamy Stock Photo **30.T5-4** WaterFrame/Alamy Stock Photo **30.T5-5** Jeff Rotman/Science Source **30.T6-1** William J. Harrigan/Stephen Frink Collection/Alamy Stock Photo **30.T6-2** Brook Peterson/Stocktrek Images/Science Source **30.T6-3** Horia Bogdan/Shutterstock **30.T7-1** Andrew J. Martinez/Science Source **30.T7-2** Gregory G Dimijian/Science Source **30.T7-3** Alexander Semenov/Science Source **Case Study** Howard Chew/Alamy Stock Photo

Chapter 31 **Opener** Jochen Tack/Alamy Stock Photo **31.5a** Peter Parks/Image Quest Marine **31.5b** Carmel McDougall **31.7a** Michael Abbey/Science Source **31.7b** Sinclair Stammers/Science Source **31.7c** GAUDENCIO GARCIA/VWPICS/Visual&Written SL/Alamy Stock Photo **31.7d** blickwinkel/Fieber/Alamy Stock Photo **31.9** Jane Burton/Nature Picture Library **31.10** Adrian Costea/Fotolia **31.12a** Dave King/Dorling Kindersley **31.12b** Michael Mules/Alamy Stock Photo **31.13a** Eric Isselée/Shutterstock **31.13b** Holger Mette/Fotolia **31.16a** Nigel Cattlin/Alamy Stock Photo **31.16b** Dylan Beckshölt/Alamy Stock Photo **31.16b-inset** Mr.Smith Chetanachan/123RF **31.T1-1** blickwinkel/Fox/Alamy Stock Photo **31.T1-2** Franco Banfi/NHPA/AGE Fotostock **31.T1-3** Alexander Semenov/Science Source **31.T1-4** Sirajuddin Daud/EyeEm/Getty Images **31.T2-1** Martin Almqvist/Alamy Stock Photo **31.T2-2** Doug Wilson/Alamy Stock Photo

31.T2-3 WaterFrame_fba/WaterFrame/Alamy Stock Photo **31.T2-4** H. Schmidbauer/Blickwinkel/AGE Fotostock **31.T3-1** Sinclair Stammers/Science Source **31.T3-2** Steve Gschmeissner/Science Source **31.T3-3** George Grall/National Geographic Image Collection/Alamy Stock Photo **31.T3-4** Arts Illustrated Studios/Shutterstock **31.T4-1** Mark Smith/Science Source **31.T4-2** Juniors Bildarchiv GmbH/F300/Alamy Stock Photo **31.T4-3** Bennyartist/Shutterstock **31.T4-4** AlinaMD/Shutterstock **31.T5-1** Meul/ARCO/Nature Picture Library **31.T5-2** OGphoto/iStock/Getty Images Plus/Getty Images **31.T5-3** arlindo71/E+/Getty Images **31.T5-4** Trounce **31.T5-5** Yxowert/Shutterstock **31.T5-6** Marc Dubus/Fotolia **31.T5-7** Sergey Toronto/Fotolia **Case Study** Dr. Jürgen Schmidl

Chapter 32 Opener Fritz Wilhelm **32.2a** D.P. Wilson/FLPA/Science Source **32.2b** Juri Vinokurov/Fotolia **32.3b** Jeff Rotman/Nature Picture Library **32.7** Dr. Brian Choo Source: Zhu, M. et al. (2013). A Silurian placoderm with osteichthyan-like marginal jaw bones. Nature 502, 188–193 (10 October 2013) doi:10.1038/nature12617, **32.13** Radius Images/Alamy Stock Photo **32.14a** Holger Ehlers/Alamy Stock Photo **32.14b** J & C Sohns/Tier und Naturfotografie/AGE fotostock **32.18** Mark Thiessen/National Geographic Image Collection/Alamy Stock Photo **32.T1-1** Hal Beral/Corbis/Getty Images **32.T1-3** Jeff Rotman/Alamy Stock Photo **32.T1-4** Jeff Rotman/Science Source **32.T1-5** WaterFrame_fur/WaterFrame/Alamy Stock Photo **32.T2-1** Heather Angel/Natural Visions/Alamy Stock Photo **32.T2-2** Borut Furla/WaterF/AGE Fotostock **32.T3-1** Mark Conlin/Alamy Stock Photo **32.T3-2** Heather Angel/Natural Visions/Alamy Stock Photo **32.T4-1** Norbert Probst/imageBROKER/Alamy Stock Photo **32.T4-2** Charles Hood/Oceans Image/Photoshot **32.T4-3** Peter Scoones/Science Source **32.T4-4** Rudie Kuiter/Oceanwide Images **32.T5-1** Ondrej Prosicky/Shutterstock **32.T5-2** Papa Bravo/Shutterstock **32.T5-3** Photo Researchers, Inc./Science Source **32.T6-1** Nicole Duplaix/Getty images **32.T6-2** RF Pictures/Corbis **32.T6-3** Dennis Jacobsen/Alamy Stock Photo **32.T7-1** Francesco Puntiroli/Alamy Stock Photo **32.T7-2** Paul A. Souders/Corbis Documentary/Getty Images **32.T7-3** Aleksandar Mijatovic/Shutterstock **32.T7-4** W. Perry Conway/Corbis/Getty Images **32.T1-2** Alan James/Nature Picture Library **Case Study** Nacho Doce/Reuters

Chapter 33 Opener Jed Fuhrman **33.5a** Omikron/Science Source **33.5b** Biophoto Associates/Science Source **33.5c** Oliver Meckes/Science Source **33.5d** Eye of Science/Science Source **33.13a** Eye of Science/Science Source **33.13b** Biozentrum, University of Basel/Science Source **33.T2-1** SPL/Science Source **33.T2-2** Elizabeth Cerny-Chipman **33.T3-1** J A W Coetzer, by courtesy of the Department of Veterinary Tropical Diseases, Faculty of Veterinary Science, University of Pretoria, South Africa. **33.T3-2** Nigel Cattlin/Alamy Stock Photo **33.T3-3** mediacolor's/Alamy Stock Photo **33.T4-1** David Gray/Reuters **33.T4-2** Oktay Ortakcioglu/E+/Getty Images **Case Study** Xinhua/Alamy Stock Photo **EOU6-1** AGE Fotostock America Inc. **EOU6-2** Biophoto Associates/Science Source **EOU6-3** Wolstenholme Images/Alamy Stock Photo **EOU6-4** Eye of Science/Science Source

Chapter 34 Opener narvikk/E+/Getty Images **34.5a** Lee W. Wilcox **34.5b** Lee W. Wilcox **34.5c** David T. Webb **34.5d** David T. Webb **34.6a** Lee W. Wilcox **34.6b** Lee W. Wilcox **34.6c** Lee W. Wilcox **34.6d** Nigel Cattlin/Alamy Stock Photo **34.7** Eric Schrader/Pearson **34.9a** Ed Reschke/Photolibrary/Getty Images **34.9b** Ed Reschke/Photolibrary/Getty Images **34.10** Andrew Syred/Science Source **34.11a** John Durham/Science Source **34.11b** Lee W. Wilcox **34.12a** Lee W. Wilcox **34.12b** Lee W. Wilcox **34.12c** Lee W. Wilcox **34.13a** Paul Schulte **34.13b** Garry DeLong/Science Source **34.14a** Power and Syred/Science Source **34.14b** Mike Clayton, University of Wisconsin. **34.14c** Andrew Syred/Science Source **34.15a** Jim Bidlack **34.15b** Lee W. Wilcox **34.16a** Ed Reschke/Photolibrary/Getty Images **34.16b** Patrick J. Lynch/Science Source **34.18b** Dr. Jeremy Burgess/Science Source **34.19a** Dr. Keith Wheeler/Science Source **34.19b** Gary DeLong/Science Source **34.20** Garry DeLong/Science Source **34.22** Pascal Goetgheluck/Science Source **34.23a** Lee W. Wilcox **34.23b** John Durham/Science Source **34.23c** Adam Hart Davis/Science Source **34.T1-1** Photo Fun/Shutterstock **34.T1-2** PBouman/Shutterstock **34.T1-3** rukhsanaphotography/Fotolia **34.T1-4** Nigel Cattlin/Alamy Stock Photo **34.T2-1** Geoff Dann/Dorling Kindersley Limited **34.T2-2** Peter Anderson/DK Images **34.T2-3** Lee W. Wilcox **34.T2-4** Nigel Cattlin/Science Source **34.T2-5** PSW Gallery/Shutterstock **34.T3-1** eAlisa/Shutterstock **34.T3-2** Smit/Shutterstock **34.T3-3** Claude Nuridsany Marie Perennou/Science Source **34.T3-4** Filip Fuxa/Shutterstock **34.T3-5** Herman Brehm/Nature Picture Library **Case Study** R Kaufung/Blickwinkel/AGE Fotostock America Inc.

Chapter 35 Opener Marcus Siebert/imageBROKER/Alamy Stock Photo **35.3** CSP_dabjola/AGE Fotostock America Inc. **35.4** David Cook/blueshiftstudios/Alamy Stock Photo **35.6L** Lee W. Wilcox **35.6R** Lee W. Wilcox **35.8** Bruce Peter **35.10a** Steve Gschmeissner/Science Source **35.10b** Steve Gschmeissner/Science Source **35.18** Martin H. Zimmermann/Harvard Forest Archive/Harvard University **35.21ab** Source: Srivastava, A. C., Ganesan, S., Ismail, I. O., & Ayre, B. G. (2008). Functional Characterization of the Arabidopsis AtSUC2 Sucrose/H+ Symporter by Tissue-Specific Complementation Reveals an Essential Role in Phloem Loading But Not in Long-Distance Transport. Plant Physiology, 148(1), 200–211. http://doi.org/10.1104/pp.108.124776 **35.21b** Gottwald, J. et.al. (2000). Genetic evidence for the in planta role of phloem-specific plasma membrane sucrose transporters Proceedings of the National Academy of Sciences, (December 5, 2000) 97 (25) 13979–13984. Copyright (2000) National Academy of Sciences, U.S.A. **Case Study** Dr. Keith Wheeler/Science Source

Chapter 36 Opener Buena Vista Images/Getty Images **36.2a** Nigel Cattlin/Science Source **36.2b** Nigel Cattlin/Science Source **36.2c** Nigel Cattlin/Science Source **36.3** Emanuel Epstein **36.6a** Institute of Texan Cultures **36.6b** Tim McCabe/USDA Natural Resources Conservation Service **36.9a** Ed Reschke/Photolibrary/Getty Images **36.13b** Blumwald, Eduardo **36.14T** Hugh Spencer/Science Source **36.14BL** DR KEITH WHEELER/Science Source **36.14BR** Andrew Syred/Science Source **36.15** blickwinkel /Alamy Stock Photo **36.16a** Robert and Jean Pollock/Science Source **36.16b** Spike Walker/Wellcome Images - used under a Creative Commons License, https://creativecommons.org/licenses/by/4.0/36.16c Ross Frid/Alamy Stock Photo **36.16d** Frank Greenaway/Dorling Kindersley/Getty Images **36.17a** Ed Reschke/Stockbyte/Getty Images **36.17b** Florapix/Alamy Stock Photo **36.18** Maurice Ford/Getty Images **Case Study L** Scenics & Science/Alamy Stock Photo **Case Study R** 2012 Bazile et al./PLOS

Chapter 37 Opener Martin Shields/Alamy Stock Photo **37.4a** Prof. John Christie **37.4b** Prof. John Christie **37.8a** Nigel Cattlin/Alamy Stock Photo **37.8b** Nigel Cattlin/Alamy Stock Photo **37.9** Nigel Cattlin/Frank Lane Picture Agency Limited **37.12** Scheres Ben/The Company of Biologists, Ltd. **37.15** Wassim Chehab/Rice University **37.16** Rafael Campillo/AGE Fotostock **37.17a** I Weber/Blickwinkel/AGE Fotostock **37.17b** Blickwinkel/Weber/Alamy Stock Photo **37.18a** Lee W. Wilcox **37.18b** Lee W. Wilcox **37.19** Eiichi Tanimoto **37.23** B. Anthony Stewart/National Geographic Image Collection/Alamy Stock Photo **37.25** Peter Balint-Kurti/USDA Agricultural Research Service **37.28** Nigel Cattlin/Science Source **37.T2-1** Cathlyn Melloan/Stone/Getty Images **37.T2-2** Phil McLean/FLPA **37.T2-3** Michael P. Gadomski/Science Source **37.T2-4a** Ed Reschke/Photolibrary/Getty Images **37.T2-4b** Ed Reschke/Photolibrary/Getty Images **37.T2-5** Vnlit/Shutterstock **37.T2-6** Republished with permission of [American Society of Plant Biologists], from [Steven D. Clouse (2011). Brassinosteroid Signal Transduction: From Receptor Kinase Activation to Transcriptional Networks Regulating Plant Development, The Plant Cell Apr 2011, 23 (4) 1219–1230.] permission conveyed through Copyright Clearance Center, Inc. **Case Study** Kim Quillin

Chapter 38 Opener Walter Siegmund **38.1a** Dabjola/Shutterstock **38.1b** Jerome Wexler/Science Source **38.1c** Jerome Wexler/Science Source **38.1d** Carolina K. Smith MD/Shutterstock **38.3bT** Bernie Kasper **38.3bM** Rod Planck/Science Source **38.3bB** Rafael Ramirez Lee/Shutterstock **38.4a** Leonard Lessin/Science Source **38.4b** Leonard Lessin/Science Source **38.5a** Justin Voight/Getty Images **38.5b** blickwinkel/Jagel/Alamy Stock Photo **38.5c** Vince Bevan/Alamy Stock Photo **38.9L** GWI/Trevor Sims/AGEfotostock **38.9R** Nick Garbutt/Nature Picture Library **38.12** Lee W. Wilcox **38.13aL** Brian Johnston **38.13aR** Pablo Galán Cela/AGE Fotostock **38.13bL** Rafael Campillo/AGE Fotostock **38.13bR** Laurie Campbell/Photoshot License Limited **38.13cL** Nicholas and Sherry Lu Aldridge/FLPA **38.13cR** James Hardy/PhotoAlto Agency RF Collections/Getty Images **38.17R** Ed Reschke/Photolibrary/Getty Images **38.20abc** Source: Bar, M., Ori, Naomi (2014). Leaf development and morphogenesis. Development, 2014 141: 4219-4230 doi: 10.1242/dev.106195, Figure4. **38.22a** Photos by John L. Bowman **38.22b** Photos by John L. Bowman **38.22c** Photos by John L. Bowman **38.22d** Photos by John L. Bowman **Case Study L** blickwinkel/Jagel/Alamy Stock Photo **Case Study R** WILDLIFE GmbH/Alamy Stock Photo **EOU7B** Gilles MARTIN/Gamma-Rapho/Getty Images **EOU7T** Nigel Cattlin/Alamy Stock Photo

Chapter 39 Opener Gerard Lacz/AGE Fotostock **39.1a** Dave H. Funk **39.1b** David H. Funk **39.3** Republished with permission of [Petren, K.,et al.], from [A phylogeny of Darwin's finches based on microsatellite DNA length variation. Proceedings of the Royal Society B: Biological Sciences, 266(1417), 321. (1999)] permission conveyed through Copyright Clearance Center, Inc. **39.4** Deco Images II/Alamy Stock Photo **39.5a** Nina Zanetti/Pearson Education **39.5b** Ed Reschke/Photolibrary/getty Images **39.5c** Biophoto Associates/Science Source **39.6a** Ed Reschke/Photo Library/Getty Images **39.6b** Ed Reschke/Photo Library/Getty Images **39.8** Diego Cervo/Fotolia **39.10** Eric Isselee/Shutterstock **39.11a** Contributors are ES Pierson, G Kruitwagen and HPM Geurts **39.11b** Innerspace Imaging/SPL/Science Source **39.11c** G Macchiarelli A Caggiati P M Motta/Science Source **39.13** Derrick Hamrick/AGE Fotostock **39.14a** Scott Camazine/Science Source **39.14b** Masato Ono, Tamagawa University, Tokyo, Japan **39.14c** Masato Ono, Tamagawa University, Tokyo, Japan **39.15a** Corinne Heyning **39.T-1a** Alvin Telser/Science Source **39.T-1b** Nina Zanetti/Pearson Education **39.T-1c** Nina Zanetti/Pearson Education **39.T-1d** Nina Zanetti/Pearson Education **39.T-1e** Ed Reschke/Photo Library/Getty Images **Case Study** Emily Taylor/Pearson Education

Chapter 40 Opener Steve Winter/Getty Images **40.1** WaterFrame/Alamy Stock Photo **40.10a** Privatdozent Dr. med. habil. Holger Jast/Holger, Jastrow **Case Study** Paul Glendell/Alamy Stock Photo

Chapter 41 Opener Hari Venkatesh K R/Alamy Stock Photo **41.2a** wwing/E+Getty Images **41.2b** Danita Delimont/Gallo Images/Getty Images **41.3** Michael K. Oliver **41.4LMR** Photographs by Karel F. Liem, from Evolutionary strategies and morphological innovations: cichlid pharyngeal jaws. Systematic Zoology 22:425–441 (1974), p. 436, fig. 15B. Copyright 1974 by the Society of Systematic Biologists. **41-12** Biophoto Associates/Science Source **Case Study** Reinhard Dirscherl/Alamy Stock Photo

Chapter 42 Opener Natalie Behring/Bloomberg/Getty Images **42.4a** imageBROKER/SuperStock **42.4b** David Quillin **42.7** Chekaramit/Shutterstock **42.8a** Mark Westneat **42.8b** Mark Westneat **42.8c** Mark Westneat **42.18a** Science Source/Science Source

42.18b Ed Reschke/Photolibrary/Getty Images **42.21** Colleen G Farmer **Case Study** Marc Atkins/Mark Leech/Getty Images

Chapter 43 **Opener** Tom Barrick, Chris Clark, SGHMS/Science Source **43.8** Don W. Fawcett/Science Source **43.10** Thomas Deerinck, NCMIR/Science Source **43.19a** Allstar Picture Library/Alamy Stock Photo **43.19b** Kathy Hutchins/Hutchins/Newscom **43.19c** Peter Gudella/Shutterstock **43.19d** MARKA/Alamy Stock Photo **43.19e** WDG Photo/Shutterstock **43.20a** John B. Carnett/Popular Science/Getty Images **43.21a** Kevin Lee **43.23a** Nigel J. Dennis/Science Source **43.23b** Strand, Christine Rae **Case Study** Mary Evans Picture Library Ltd/Thomas Marent/AGEfotostock

Chapter 44 **Opener** Malcolm Schuyl/FLPA **44.1M** Dmitry Smirnov **44.3a** Dr. David Furness, Keele University/Science Source **44.7** Daniel Tuitert/ANP Photo/ AGE Fotostock **44.8** Amy Sanderson/ZUMA Wire/Alamy Stock Photo **44.9a** Piet Spaans Viridiflavus **44.10** Cuvillier Verlag **44.11a** Andrew Syred/Science Source **44.14a** David Quillin/Kim Quillin **44.14b** David Quillin/Kim Quillin **44.18a** Chase Swift/Corbis/Getty Images **44.18b** Ted Kinsman/Science Source **44.19** Jeff Rotman/Photolibrary/Getty Images **44.2** Christopher Zarnick **Case Study** Margaret Edge

Chapter 45 **Opener** Bence Mate/Nature Picture Library **45.1** Don W. Fawcett/Science Source **45.2** Don W. Fawcett/Science Source **45.9a** Life on white/Alamy Stock Photo **45.9b** Ted Kinsman/Science Source **45.9c** Alexandra Lande/Shuttertstock **45.12R** Glyn Kirk/ Stringer/AFP/Getty Images **45.15B** Eadweard Muybridge/Science Source **45.15T** LOC/ Science Source **45.16** Anders Hedenström **45.T-1a** Biophoto Associates/Science Source **45.T-1b** Ed Reschke/Photolibrary/Getty Images **45.T-1c** Nina Zanetti/Pearson Education **45.T-2a** Biophoto Associates/Science Source **Case Study** Lucas Jackson/Reuters

Chapter 46 **Opener** Anthony Mercieca/Science Source **46.8a** Harry Rogers/Science Source **46.8b** John Mitchell/Science Source **46.8c** John Cancalosi/Alamy Stock Photo **46.9a** Photo Researchers, Inc./Science Source **46.9b** Andrew Darrington/Alamy Stock Photo **46.9c** Perennou Nuridsany/Science Source **46.11** Kenneth M. Highfill/Science Source **46.13** Andrew Maguire/Aurora Open/Alamy Stock Photo **46.14** Living Art Enterprises/Science Source **Case Study** Martin Bernitti/Stringer/AFP/Getty Images

Chapter 47 **Opener** Jurgen & Christine Sohns/Frank Lane Picture Agency Limited **47.1a** Lutra/Photoshot **47.1b** Allison Gong **47.1c** Charles J. Cole **47.2** Bruce J. Russell/ BioMedia Associates **47.5a** Privatdozent Dr. med. habil. Holger Jast/Holger, Jastrow **47.10** Jurgen Freund/Nature Picture Library **47.11a** Jerry A. Coyne and Catherine Fernandez/Nature Magazine **47.11b** Jerry A. Coyne and Catherine Fernandez/National Institute of General Medical Sciences **47.12 inset** Michal Polak **47.12 inset** Johanna Rönn **47.14a-1** Michael Whitaker/Science Source **47.14a-2** Michael Whitaker/Science Source **47.14a-3** Michael Whitaker/Science Source **47.14b-1** Victor D. Vacquier **47.14b-2** Victor D. Vacquier **47.14b-3** Victor D. Vacquier **47.18a** David Shook **47.18b** David Shook **47.18c** David Shook **47.18d** David Shook **47.21a** Courtesy, Kathryn Tosney **47.21b** Courtesy, Kathryn Tosney **47.23** Biophoto Associates/Science Source **47.27a** John Cancalosi/Nature Picture Library **47.27b** John Cancalosi/Nature Picture Library **47.27c** John Cancalosi/Alamy Stock Photo **47.28a** Claude Edelmann/Science Source **47.28b** Claude Edelmann/Science Source **47.28c** Petit Format/Science Source **Case Study** naihei/Shutterstock

Chapter 48 **Opener** Eye of Science/Science Source **48.1** Susumo Nishinaga/Science Source **48.2** in Bruno Lemaitre, et.al. (1996). The Dorsoventral Regulatory Gene Cassette spätzle/Toll/cactus Controls the Potent Antifungal Response in Drosophila Adults, CELL Volume 86, Issue 6, 20 September 1996, Pages 973-983. **48.13a** Steve Gschmeissner/Science Source **48.13b** Steve Gschmeissner/Science Source **48.18** Pictorial Press Ltd/Alamy Stock Photo **Case Study L** Sciepro/Science Photo Library/ Getty Images **Case Study R** Cath Wadforth/Science Source **EOU8L** alslutsky/ Shutterstock **EOU8R** Don W. Fawcett/Science Source

Chapter 49 **Opener** Pulsar Imagens/Alamy Stock Photo **49.T1-1** Paul Nicklen/ National Geographic/Getty Images **49.T1-2** Sergey Gorshkov/Nature Picture Library **49.T1-3** Juniors Bildarchiv GmbH/Alamy Stock Photo **49.T1-4** Lynn M Stone/Nature Picture Library **49.T1-5** Russ Heinl/All Canada Photos/SuperStock **49.T2-1** Alida Latham/Danita Delimont/Alamy Stock Photo **49.T2-3** Pulsar Imagens/Alamy Stock Photo **49.11** Dr. Henrik Wahren, Research Centre for Applied Alpine Ecology, La Trobe University **49.12b** Stas Moroz/Shutterstock **49.13L** Reinhard Dirscherl/Alamy Stock Photo **49.13R** Doug Perrine/Alamy Stock Photo **49.14** Wild Wonders of Europe/ Maitland/Nature Picture Library/Alamy Stock Photo **49.17** NASA **49.T2-2** Science Source/Science Source **49.T3-1** John E Marriott/All Canada Photos/Glow Images **49. T3-2** Bill Brooks/Alamy Stock Photo **49.T3-3** Inga Nielsen/Shutterstock **49.T3-4** Anton Foltin/Shutterstock **49.T3-5** Stillfx/AGE Fotostock **49.T4-1** Pierre Leclerc/Shutterstock **49.T4-2** imageBROKER/SuperStock **49.T4-3** Yuriy Kulik/Shutterstock **49.T4-4** John Warburton Lee/SuperStock **49.T4-5** Alekss/Fotolia **49-UN-12** Clint Farlinger/Alamy Stock Photo **Case Study** Michael Patrick O'Neill/Alamy Stock Photo

Chapter 50 **Opener** Jane Gould/Alamy Stock Photo **50.1** Anup Shah/Nature Picture Library/Alamy Stock Photo **50.2L** The Africa Image Library/Alamy Stock Photo **50.2R** FLPA/Alamy Stock Photo **50.6a** walkingmoon/Getty Images **50.12** Suzanne L Collins Joseph T. Collins/Science Source **50.15** Adam Seward/Alamy Stock Photo **50.T1-1** jo Crebbin/Shutterstock **50.T1-2** Tony Fielding/Alamy Stock Photo **50.T1-3** Scott Bauer/U.S. Department of Agriculture **50.T1-4** Jane Gould/Alamy Stock Photo **Case Study** David Fleetham/Alamy Stock Photo

Chapter 51 **Opener** Reinhard Dirscherl/WaterFrame/AGE Fotostock **51.1R** Hecker/ Sauer/Blickwinkel/AGE Fotostock America Inc. **51.2** blickwinkel/McPHOTO/NBT/ Alamy Stock Photo **51.4a** blickwinkel/Alamy Stock Photo **51.6** Jeff Rotman/Oxford Scientific/Getty Images **51.12a** JC Schou/Biopix **51.13** Johner Images/Getty Images **51.15** Janine Wiedel/Janine Wiedel Photolibrary/Alamy Stock Photo **51.T1-1** Tracy Kenneth/Shutterstock **51.T1-2** Norbert Probst/imageBROKER/Alamy Stock Photo **51. T1-3** Adrian Weston/Alamy Stock Photo **51.T3-1** HamsterMan/Shutterstock **51.T3-2** Sorge/Agencja Fotograficzna Caro/Alamy Stock Photo **51.T3-3** Jeffrey Lepore/Science Source **51.T3-4** Denis Crawford/Alamy Stock Photo **51.T3-5** Peter Gilbert/Alamy Stock Photo **Case Study** Dan Callister/Alamy Stock Photo

Chapter 52 **Opener** Scott Tuason/Image Quest Marine **52.1** Kenneth Fink/Getty Images **52.9a** Photo credit: Stephen P. Yanoviak **52.9b** Photo credit: Stephen P. Yanoviak **52.9c** Photo credit: Stephen P. Yanoviak **52.10a** Merlin D. Tuttle/Science Source **52.10b** Reinhard Dirscherl/Alamy Stock Photo **52.17a** Neale Clark/robertharding/Alamy Stock Photo **52.17b** Tony C. Caprio **52.18-1** Elenamiv/Shutterstock **52.18-2** Michael P. Gadomski/Science Source **52.18-3** Michael P. Gadomski/Science Source **52.18-4** Aleksander Bolbot/age fotostock/Alamy Stock Photo **52.19L** Glacier Bay National Park and Preserve **52.19M** Christopher L. Fastie **52.19R** Glacier Bay National Park and Preserve **52.T1-1** Jeffrey Lepore/Science Source **52.T1-2** Olga Khoroshunova/Fotolia **52. T1-3** Juan Gaertner/Science Photo Library/Alamy Stock Photo **52.T1-4** ArtBoyMB/E+/ Getty Images **52.T1-5** Nigel Cattlin/Science Source **52.T2-1** Francesco Tomasinelli/Science Source **52.T2-2** Brian Bevan/ardea/AGE Fotostock **52.T2-3** Dirk Ercken/Shutterstock **52.T2-4** D J Mattaar/Shutterstock **52.T2-5** Shari L. Morris/AGE Fotostock **Case Study L** Michael Abbey/Science Source **Case Study R** Scott Camazine/Alamy Stock Photo

Chapter 53 **Opener** DVIDS/NASA https://svs.gsfc.nasa.gov/cgi-bin/details. cgi?aid=11719 **53.10a** Richard Hartnu **53.11** geogphotos/Alamy Stock Photo **53.12** John Campbell/USDA Forest Service **53.16** ChinaTopix/AP Images **53.T1-1** Gunter Marx/ Alamy Stock Photo **53.T1-2** Thorsten Milse/Robert Harding World Imagery **53.T1-3** Robert Pickett/Papilio/Alamy Stock Photo **53.T1-4** John Sullivan/Alamy Stock Photo **53. T1-5** Steve Ringman/National Geographic **Case Study** Alvaro Leiva/AGE Fotostock

Chapter 54 **Opener** Arco/C. Braun/Arco Images/AGE Fotostock **54.3** Jenkins, C.N. et.al. (2013). Global patterns of terrestrial vertebrate diversity and conservation, Proc Natl Acad Sci U S A. 2013 Jul 9 110(28): E2602–E2610, Figure 4. And adapted by Mannion, P.D, et.al. (2014). "The latitudinal biodiversity gradient through deep time", Trends in Ecology & Evolution Volume 29, Issue 1, January 2014, Pages 42-50. **54.6** Mark Richardson/Alamy Stock Photo **54.7a** NASA **54.7b** NASA **54.13** TPG/AP Images **54.T2-1** David Woodfall/Science Source **54.T2-2** Jeff Rotman/Science Source **54.T2-3** Mark Conlin/Alamy Stock Photo **54.T2-4** Simon Booth/Science Source **54.T2-5** Peter Scoones/Science Source **54.T3-1** Kondor83/Shutterstock **54.T3-2** Brian Brake/Science Source **54.T3-3** Specta/Shutterstock **54.T3-4** Koh Sze Kiat/Shutterstock **54.T4-1** Jeff Rotman/Photolibrary/Getty Images **54.T4-2** Mark Conlin/Alamy Stock Photo **54.T4-3** Chico Sanchez/AGE Fotostock **54.T4-4** Photo by Ela2007/Moment/Getty Images **54.T4-5** Stan Tekiela Author/Naturalist/Wildlife Photographer/Moment/Getty Images **54.T4-6** Michel Gunther/Biosphoto/Superstock **54.T4-7** Hans Blossey/Alamy Stock Photo **54.T4-8** Pete Niesen/Shutterstock **54.T4-9** ton koene/Alamy Stock Photo **54.T4-10** Fuse/Corbis/Getty Images **Case Study** Atlantide Phototravel/Corbis Documentary/ Getty Images **EOU9L** Alamy Stock Photo **EOU9R** Dan Suzio/Science Source

Illustration and Text Credits

Chapter 1 **1.11** Matthias Wittlinger, Rüdiger Wehner, Harald Wolf. **1.6** Darwin, C. (1859), *On the Origin of Species by Means of Natural Selection, or the Preservation of Favoured Races in the Struggle for Life* (1st ed.), London: John Murray. **1.9a** T. P. Young and L. A. Isbell (1991) Sex Differences in Giraffe Feeding Ecology: Energetic and Social Constraints, *Ethology* 87: 79–89. **EOC Case Study** S. N. Vignieri, J. G. Larson, and H. E. Hoekstra. 2010. *Evolution* 64: 2153–2158. **p. 2** François Jacob, 1965.

BioSkills **BS18.1** A. Crowe, C, Dicks, and M. P. Wenderoth (2008) Biology in bloom: Implementing Bloom's Taxonomy to enhance student learning in Biology. *CBE—Life Sciences Education* 7: 368–381, Table 3. **BS2 CYU** S. L. Potter, et al. (2012) Phase 2 clinical trial of intrathecal topotecan in children with refractory leptomeningeal leukemia: A Children's Oncology Group trial (P9962), *Pediatric Blood Cancer* 58: 362–365. **BS3 CYU** M. W. Baldwin, et al. (2014) Evolution of sweet taste perception in hummingbirds by transformation of the ancestral umami receptor, *Science,* 345: 929–933. **p. 45** Jacques Monod, 1965.

Chapter 2 **2.2** S. L. Miller. 1953. A production of amino acids under possible primitive Earth conditions. *Science* 117: 528–529. **2.21** S. L. Miller. 1953. A production of amino acids under possible primitive Earth conditions. *Science* 117: 528–529. **2T.1** D. R. Lide (ed.). 2008. Standard thermodynamic properties of chemical substances. In *CRC Handbook of Chemistry and Physics.* 89th ed. Boca Raton, FL: CRC Press. **EOC Case Study** J. G. Wallis, H. Wang, D. J. Guerra. 1997. *Plant Molecular Biology* 35: 323–330.

Chapter 3 **3.10(b)** S. Sheriff, W. A. Hendrickson, and J. L. Smith. 1987. Structure of myohemerythrin in the azidomet state at 1.7/1.3 resolution. *Journal of Molecular*

Biology 197: 273–296. **3.10(b2)** N. H. Haunerland, B. L. Jacobson, G. Wesenberg, et al. 1994. Three-dimensional structure of the muscle fatty-acid-binding protein isolated from the desert locust *Schistocerca gregaria. Biochemistry* 33: 12378–12385. **3.10(b3)** M. Baron, D. G. Norman, T. S. Harvey, et al. 1992. The three-dimensional structure of the first EGF-like module of human factor IX: Comparison with EGF and TGF-alpha. *Protein Science* 1: 81–90. **3.11a** P. B. Rupert, A. K. Mollah, M. C. Mossing, et al. (2000). The structural basis for enhanced stability and reduced DNA binding seen in engineered second-generation Cro monomers and dimers. *Journal of Molecular Biology* 2000 Mar 3; 296(4): 1079–90. **3.11b** S. Y. Park, T. Yokoyama, N. Shibayama, et al. 2006. 1.25 Å resolution crystal structures of human hemoglobin in the oxy, deoxy and carbonmonoxy forms. *Journal of Molecular Biology*, 360: 690–701. **3.6a** J. M. Chen, C. D. E. King, S. H. Feairheller, et al. 1991. An energetic evaluation of a Smith collagen microfibril model. *Journal of Protein Chemistry* 10: 535–552. **3.6b** Z. S. Juo, T. K. Chiu, P. M. Leiberman, et al. 1996. How proteins recognize the TATA box. *Journal of Molecular Biology* 261: 239–254. **3.6c** F. Korkmaz-Ozkan, S. Koster, W. Kuhlbrandt, et al. 2010. Correlation between the OmpG secondary structure and its pH-dependent alterations monitored by FTIR. *Journal of Molecular Biology* 401: 56–67. **3.6d** M. Marquart, J. Walter, J. Deisenhofer, et al. 1983. The geometry of the reactive site and of the peptide groups in trypsin, trypsinogen and its complexes with inhibitors. *Acta Crystallographica* Section B39: 480–490. **EOC Case Study C.** Mitea, et al. 2008. Efficient degradation of gluten by a prolyl endoprotease in a gastrointestinal model: implications for coeliac disease., *Journal of Medical Genetics*., vol. 57- 1 (January 2008).

Chapter 4 **4.10** F. Guo, A. R. Gooding, and T. R. Cech. 2004. Structure of the Tetrahymena ribozyme: Base triple sandwich and metal ion at the active site. *Molecular Cell* 16: 351–362. **4T.1** H. Shi and P. B. Moore. 2000. The crystal structure of yeast phenylalanine tRNA at 1.93 Å resolution: A classic structure revisited. *RNA* 6: 1091–1105. **EOC Case Study** Chargaff, F., Structure and function of nucleic acids as cell constituents, *Federation Proceedings*, vol. 10(3), (September 1951), pp. 654–659. **p. 102** Watson, J. D. and Crick F.H.C., A structure for Deoxyribose Nucleic Acid, *Nature*, VOL 171, (2 April 1953) page737, 1953.

Chapter 5 **EOC Case Study** R. Agrawal and F. G. Pinilla. 2012. *Journal of Physiology* 590: 2485–2499.

Chapter 6 **Opener** CHARMM-GUI Archive—Library of Pure Lipid Bi-layer (www.charmm-gui.org/?doc=archive&lib=lipid_pure), POPE Bilayer Library (pope_n256.pdb). Reference: S. Jo, T. Kim, and W. Im. 2007. Automated builder and database of protein/membrane complexes for molecular dynamics simulations. *PLoS ONE* 2 (9): e880. **6.11** J. de Gier, J. G. Mandersloot, L. L. M. Van Deenen. 1968. Lipid composition and permeability of liposomes. *Biochimica et Biophysica Acta* 150: 666–675. **6.22** K. Tani, T. Mitsuma, Y. Hiroaki, et al. 2009. Mechanism of aquaporin-4's fast and highly selective water conduction and proton exclusion. *Journal of Molecular* Biology 389: 694–706. **6.24** C. A. Bear, C. Li, N. Kartner, et al. (1992). Purification and functional reconstitution of the cystic fibrosis transmembrane conductance regulator (CFTR). *Cell* 68: 809–818. **6.23a** Y. Zhou, J. H. Morais-Cabral, A. Kaufman, et al. 2001. Chemistry of ion coordination and hydration revealed by a K^+ channel–Fab complex at 2.0 Å resolution. *Nature* 414: 43–48. **6.23b** L. G. Cuello, V. Jogini, D. M. Cortes, et al. 3FB7 Open KcsA potassium channel in the presence of Rb^+ ion. DOI: 10.2210/pdb3FB7/pdb. **EOC Case Study** C. M. C. Bassett, R. S. McCullough, A. L. Edel, et al. 2009. *Metabolism: Clinical and Experimental* 58: 1802–1808. **EOU Case Study Fig 2** R. P. Hartshorne et al. 1985. *PNAS | Neurobiology* 82: 240–244, Figure 1C. **EOU Case Study Fig 4** R. P. Hartshorne et al. 1985. *PNAS | Neurobiology* 82: 240–244, Figure 2B.

Chapter 7 **7.19b** J. D. Jamieson and G. E. Palade. 1967. Intracellular transport of secretory proteins in the pancreatic exocrine cell. *Journal of Cell Biology* 34: 597–615. **EOC Case Study** D. Murase et al. 2013. *Journal of Investigative Dermatology* 133: 2416–2424.

Chapter 8 **8.16** N. Nawani, B. P. Kapadnis, A. D. Das, et al. (2002) *Journal of Applied Microbiology* 93: 865–975. Also N. Nawani and B. P. Kapadnis. 2001. *Journal of Applied Microbiology* 90: 803–808. **8.4** Data courtesy of the Undergraduate Biotechnology Laboratory, California Polytechnic State University, San Luis Obispo. **8.8a** V. V. Lunin, Y. Li, J. D. Schrag, et al. (2004) Crystal structures of Escherichia coli ATP-dependent glucokinase and its complex with glucose. *Journal of Bacteriology* 186: 6915–6927. **8.8b** A. T. Cordeiro, A. J. Caceres, D. Vertommen, et al. (2007) The crystal structure of Trypanosoma cruzi glucokinase reveals features determining oligomerization and anomer specificity of hexose-phosphorylating enzymes. *Journal of Molecular Biology* 372: 1215–1226. **EOC Case Study** E. K. Jaffe et al. 2013. *Archives of Biochemistry and Biophysics* 530: 73–82.

Chapter 9 **9.13** X. Li, R. K. Dash, R. K. Pradhan, et al. 2010. *Journal of Physical Chemistry* B. 114: 16068–16082. **9.14** D. F. Wilson, M. Erecinska, and P. L. Dutton. 1974. *Annual Review of Biophysics and Bioengineering* 3: 203–230. Also V. D. Sled, N. I. Rudnitzky, Y. Hatefi, et al. 1994. *Biochemistry* 33: 10069–10075. **9.17** E. Racker and W. Stoeckenius. 1974. Reconstitution of purple membrane vesicles catalyzing light-driven proton uptake and adenosine triphosphate formation. *Journal of Biological Chemistry* 249: 662–663. **9.7** P. R. Evans and P. J. Hudson (1981) Phosphofructokinase: Structure and control, Philosophical Transactions of Royal Society of London, Series B: *Biological Sciences* 293: 53–62. **EOC Case Study** C. C. Caldeira da Silva et al. 2008. *Aging Cell* 7: 552–560.

Chapter 10 **10.11** R. Govindjee, Govindjee, and G. Hoch. 1964. *Plant Physiology* 39: 10–14. **10.22** T. C. Taylor and I. Andersson. 1997. The structure of the complex between rubisco and its natural substrate ribulose 1,5-bisphosphate. *Journal of Molecular Biology* 265: 432–444. **10.6** T. W. Engelmann. 1882. Oxygen excretion from plant cells in a microspectrum. *Botanische Zeitung* 40: 419–426. **10.7** G. S. Singhal et al. 1999. *Concepts in Photobiology: Photosynthesis and Photomorphogenesis*. Dordrecht: Kluwer Academic; co-published with Narosa Publishing House (New Delhi), 11–51. **10.20** A. A. Benson, J. A. Bassham, M. Calvin, et al. 1950. The path of carbon in photosynthesis. V. Paper chromatography and radioautography of the products. *Journal of the American Chemistry Society* 72: 1710–1718. **EOC Case Study** J. H. Mussgnug et al. 2007. *Plant Biotechnology Journal* 5: 802–814.

Chapter 11 **11.8b** B. Alberts, et al. (2002) *Molecular Biology of the Cell*, 4th ed. Fig. 19.5. New York: Garland Science. **EOC Case Study** M. Malleshaiah et al. 2010. *Nature* 456: 101–105.

Chapter 12 **12.13** National Cancer Institute (http://www.cancer.gov/types/common-cancers), Common Cancer Statistics, February 2017. **12.6** G. J. Gorbsky, P. J. Sammak, and G. G. Borisy. 1987. Chromosomes move poleward in anaphase along stationary microtubules that coordinately disassemble from their kinetochore ends. *Journal of Cell Biology* 104: 9–18. **12.9** J. L. Ptacin, et al. (2010) A spindle-like apparatus guides bacterial chromosome segregation, *Nature Cell Biology*, 12: 791–798, Fig. 5. **12.10** A. W. Murray and M. W. Kirschner. 1989. Cyclin synthesis drives the early embryonic cell cycle. *Nature* 339: 275–280. **EOC Case Study** C. Miceli, et al. 2013. *Cancer Gene Therapy* 20: 298–307. **EOU Case Study Fig 1** T. Noguchi et al. 2006. *Comparative Biochemistry and Physiology Part D: Genomics and Proteomics* 1: 145–152, Figure 2.

Chapter 13 **13.1** Adapted from Calvin Bridges, 1915, 1916, in the public domain. **13.11** www.ndss.org, National Down Syndrome Society (2012). **13.13** L. T. Morran, et al., Running with the Red Queen: Host-Parasite Coevolution Selects for Biparental Sex, *Science*, Vol. 333, Issue 6039, (08 Jul 2011) pp. 216–218. **EOC Case Study** T. Chiang, F. E. Duncan, K. Schindler, et al. 2010. *Current Biology* 20: 1522–1528.

Chapter 14 **14.11** T. H. Morgan, 1911. An attempt to analyze the constitution of the chromosomes on the basis of sex-limited inheritance in *Drosophila. Journal of Experimental Zoology* 11: 365–413. **14.12** T. H. Morgan. 1911. An attempt to analyze the constitution of the chromosomes on the basis of sex-limited inheritance in *Drosophila. Journal of Experimental Zoology* 11: 365–413. **14.19** Pan American Health Organization/WHO. 2004. Epidemiological Bulletin 25: 9–13, Graph 1. **14.2** G. Mendel, 1866. Versuche über Pflanzen-hybriden. *Verhandlungen des naturforschenden Vereines in Brünn* 4: 3–47. English translation available from ESP: Electronic Scholarly Publishing (www.esp.org). **14.3** G. Mendel, 1866. Versuche über Pflanzen-hybriden. *Verhandlungen des naturforschenden Vereines in Brünn* 4: 3–47. English translation available from ESP: Electronic Scholarly Publishing (www.esp.org).

Chapter 15 **15.14** R. C. Allsopp, et al. Telomere length predicts replicative capacity of human fibroblasts, *Proceedings of the National Academy of Sciences* USA, vol. 89, no. 21, (November 1, 1992) 10114–10118. **15.15** J. E. Cleaver. 1972. *Journal of Investigative Dermatology. Journal of Investigative Dermatology*. Volume 58, (3, March 1972), pp. 124–128. **15.2** A. D. Hershey and M. Chase. 1952. Independent functions of viral protein and nucleic acid in growth of bacteriophage. *Journal of General Physiology* 36: 39–56. **15.5** M. Meselson and F. W. Stahl. 1958. The replication of DNA in Escherichia coli. *Proceedings of the National Academy of Sciences* USA 44: 671–682. **EOC Question 10** P. Howard-Flanders and R. P. Boyce. 1966. *Radiation Research Supplement* 6: 156–184, Fig. 8. **EOC Case Study** B. Fournier et al. 2000. *Antimicrobial Agents and Chemotherapy* 44: 2160–2165.

Chapter 16 **16.2** A. M. Srb and N. H. Horowitz. 1944. The ornithine cycle in *Neurospora* and its genetic control. *Journal of Biological Chemistry* 154: 129–139. **EOC Case Study** R. L. Lamason et al. 2005. *Science* 310: 1782–1786. **p. 342** G. W. Beadle, The Gene of men and Molds, *Scientific American*, Vol. 179, No. 3, (September 1948), pp. 30–39. **p. 342** A. Srb and N. H. Horowitz. The ornithine cycle in *Neurospora* and its genetic control. *Journal of Biological Chemistry*, 154: 129–139.

Chapter 17 **17.12** I. Gruic-Sovulj, N. Uter, T. Bullock, et al. 2005. tRNA-dependent aminoacyl-adenylate hydrolysis by a nonediting class I aminoacylt-tRNA synthetase. *Journal of Biological Chemistry* 280: 23978–23986. **17.13** E. Villa, J. Sengupta, L. G. Trabuco, et al. 2009. Ribosome-induced changes in elongation factor Tu conformation control GTP hydrolysis. *Proceedings of the National Academy of Sciences* USA 106: 1063–1068. **17.10** M. B. Hoagland, M. L. Stephenson, J. F. Scott, et al. 1958. A soluble ribonucleic acid intermediate in protein synthesis. *Journal of Biological Chemistry* 231: 241–257. **17.2a** B. P. Hudson, J. Quispe, S. Lara-Gonzalez, et al. 2009. Three-dimensional EM structure of an intact activator-dependent transcription initiation complex. *Proceedings of the National Academy of Sciences* USA 106: 19830–19835. **EOC Case Study** T. J. Lindell, et al., Specific Inhibition of Nuclear RNA Polymerase II by α-Amanitin, *Science*, Vol. 170, Issue 3956, (23 Oct 1970), pp. 447–449. **p. 358** Kornberg, A., Ten Commandments: Lessons from the Enzymology of DNA Replication, *Journal of Bacteriology*, vol. 182, no. 13, (July 2000) pp: 3613–3618.

Chapter 18 **18.3** A. B. Pardee, F. Jacob, and J. Monod, The genetic control and cytoplasmic expression of "Inducibility" in the synthesis of ß-galactosidase by *E. coli*, *Journal of Molecular Biology*, Volume 1, Issue 2, (June 1959), Pages 165–178.

Chapter 19 **Opener** Schalch, T. et al. (2005) X-ray structure of a tetranucleosome and its implications for the chromatin fibre, *Nature* 436: 138–141. **19.11** G. J. Hannon (2002) RNA interference, *Nature*, vol. 418, p. 244–251. **19.6** I. N. H. Sandovici, H. Smith, and M. D. Nitert. 2011. Maternal diet and aging alter the epigenetic control of a promoter–enhancer interaction at the *Hnf4a* gene in rat pancreatic islets. *Proceedings of the National Academy of Sciences* USA 108: 5449–5454. **19.10** D. F. Wieczorek,, C. W. Smith, and B. Nadal-Ginard (1988) The rat alpha-tropomyosin gene generates a minimum of six different mRNAs coding for striated, smooth, and nonmuscle isoforms by alternative splicing, *Molecular and Cellular Biology*, vol. 8, no. 2, p. 679–694. **19.8b** P. C. Ma, et al. (1994) Crystal structure of MyoD bHLH domain–DNA complex: Perspectives on DNA recognition and implications for transcriptional activation, *Cell*, 77: 451–459. **EOC Case Study** C. A. Guenther et al. 2014. *Nature Genetics* 46: 748–752.

Chapter 20 **20.9** T. R. Gregory. 2005. *Nature Reviews Genetics* 6: 699–708. **20.5a** Y. Hou and S. Lin. 2009. *PLoS ONE* 4(9): e6978, Supplemental Table S1. **20.5b** KEGG: Kyoto Encyclopedia of Genes and Genomes, KEGG Organisms: Complete. **p. 418** U.S. President Bill Clinton at the Human Genome Announcement at the White House (20 Jun 2000).

Chapter 21 **21.3** H. Weintraub, S. J. Tapscott, R. L. Davis, et al. 1989. Activation of muscle-specific genes in pigment, nerve, fat, liver, and fibroblast cell lines by forced expression of MyoD. *Proceedings of the National Academy of Sciences* USA 86: 5434–5438. **EOU Case Study Fig 2** C. R. Feldman et al. 2016. *Heredity* 116: 84–91, Figure 3.

Chapter 22 **22.12** P. T. Boag and P. R. Grant. 1981. Intense natural selection in a population of Darwin's finches (*Geospizinae*) in the Galápagos. *Science* 214: 82–85. **22.13** P. R. Grant and B. R. Grant. 2014. 40 Years of Evolution. *Princeton* UP. **22.10** B. Borel 2017. *Nature* 543: 302–304. **22.7a** Based on an illustration by J. G. M. "Hans" Thewissen. **22.7b** Republished with permission of AAAS after P. D. Gingerich, M. ul Haq, I. S. Zalmout, et al. 2001. Origin of whales from early Artiodactyls: Hands and feet of Eocene Protocetidae from Pakistan. *Science* 293: 2239–2242, (http://www.sciencemag.org/content/293/5538/2239); permission conveyed through Copyright Clearance Center, Inc. **22.7c** Republished with permission of AAAS after P. D. Gingerich, M. ul Haq, I. S. Zalmout, et al. 2001. Origin of whales from early Artiodactyls: Hands and feet of Eocene Protocetidae from Pakistan. *Science* 293: 2239–2242, (http://www.sciencemag.org/content/293/5538/2239t); permission conveyed through Copyright Clearance Center, Inc. **22.7d** After Gingerich, P. D. et al., (2009). New Protocetid Whale from the Middle Eocene of Pakistan: Birth on Land, Precocial Development, and Sexual Dimorphism, *PLoS ONE* 4(2): e4366 (4 February 2009). **22.7e** Delphinapterus reproduced by permission of Skulls Unlimited International, Inc. (www.SkullsUnlimited.com). **EOC Case Study** S. N. Vignieri, J. G. Larson, and H. E. Hoekstra. 2010. Evolution 64: 2153–2158. **pp. 450, 455, 458** Darwin, C., *On the Origin of Species by Means of Natural Selection, or the Preservation of Favoured Races in the Struggle for Life* (London, John Murray 1875 (first publ. 1868)..

Chapter 23 **23.13** Reproduced by permission of Pearson Education, Inc., from S. Freeman and J. Herron. 2004. *Evolutionary Analysis*. 3rd ed., Figs. 6.15a, 6.15c. ©2004. **23.17** H. Araki, B. Cooper, and M. S. Blouin, Carry-over effect of captive breeding reduces reproductive fitness of wild-born descendants in the wild *Biology Letters*, Volume 5, issue 5, (23 October 2009), pp: 632–635. **23.18** O. Tenaillon, et al. 2016. Tempo and mode of genome evolution in a 50,000-generation experiment. *Nature*, 536: (11 August 2016) 165–170. **23.5** W. E. Johnson, et al., Genetic Restoration of the Florida Panther, *Science*, Vol. 329, Issue 5999,(24 Sep 2010), pp. 1641–1645. **23.6** P. T. Boag, and P. R. Grant, Intense Natural Selection in a Population of Darwin's Finches (*Geospizinae*) in the Galápagos, *Science*, Vol. 214, No. 4516 (Oct. 2, 1981), pp. 82–85. **23.7** Karn, M. N., H. Lang-Brown, J. J. MacKenzie, et al., Birth Weight, Gestation Time and Survival in Sibs, *Annuals of Human Genetics*, Volume 15, Issue 1 (January 1949), pp. 306–322. **23.8** S. Bhat, et al., Speciation Reversal in European Whitefish (*Coregonus lavaretus* (L.)) Caused by Competitor Invasion, *PLOS ONE*, (March 13, 2014). **23.10** J. D. Blount, et al., Carotenoid Modulation of Immune Function and Sexual Attractiveness in Zebra Finches, *Science*, Vol. 300, Issue 5616, (04 Apr 2003) pp. 125–127. **23.11bc** B. J. Le Boeuf and R. S. Peterson, Social Status and Mating Activity in Elephant Seals, *Science*, Vol. 163, Issue 3862, (03 Jan 1969) pp. 91–93. **23.19b** N. Moran and T. Jarvik, Lateral Transfer of Genes from Fungi Underlies Carotenoid Production in Aphids *Science* (30 April 2010) Vol. 328, no. 5978, pp. 624–627. **23T.1** W. C. Boyd, *Genetics and the Races of Man: an Introduction to Modern Physical Anthropology*, Little, Brown and Company, 453 pages. **EOC Case Study** M. Galetti, et al., Functional Extinction of Birds Drives Rapid Evolutionary Changes in Seed Size, *Science*, Vol. 340, Issue 6136, (31 May 2013), pp. 1086–1090.

Chapter 24 **24.11** S. Rohwer, E. Bermingham, and C. Wood. 2001. Plumage and Mitochondrial DNA Haplotype Variation Across a Moving Hybrid Zone, *Evolution* 55: 405–422. **24.2** C. Ribas et al. 2012. *Proceedings of the Royal Society* B 279: 681–689. **24.4** IUCN Elephant Specialist Group, 2017; A. L. Roca et al. 2001. *Science* 293: 1473–1477. **24.6** C. Ribas et al. 2012. *Proceedings of the Royal Society* B 279: 681–689. **24.10** S. Bhat et al. 2014. *PLOS ONE* 9: e91208. **24T.3** Artwork by Uko Gorter in R. Riesch et al. 2011. *Biological Journal of the Linnean Society* 106: 1–17; call data from A. D. Foote and J. A. Nystuen. 2007. *Journal of the Acoustic Society of America* 123: 1747–1752. **EOC Case Study** K. Prüfer et al. 2014. *Nature* 505: 43–49.

Chapter 25 **25.11** M. J. Benton. 1995. *Science* 268: 52–58. **25.12** R. Blakey. Colorado Plateau Geosystems, Arizona, USA. **25.5** International Stratigraphic Chart, 2009. International Commission on Stratigraphy (www.stratigraphy.org/column.php?id=Chart/TimeScale). The data on this site are modified from F. M. Gradstein and J. C. Ogg (eds.). 2004. *A Geologic Time Scale* 2004. Cambridge, UK: Cambridge University Press; and J. G. Ogg, G. Ogg, and F. M. Gradstein. 2008. *The Concise Geologic Time Scale*. Cambridge, UK: Cambridge University Press. **25.6** International Commission on Stratigraphy, 2017. **25.7** B. G. Baldwin, D. W. Kyhos, and J. Dvorak (1990) Chloroplast DNA evolution and adaptive radiation in the Hawaiian Silversword Alliance (Asteraceae-Madiinae), *Annals of the Missouri Botanical Garden*, 77 (1): 96–109, Fig. 2. **25.13a** G. Ceballos et al. 2015. *Science Advances* 1(5): e1400253; G. Ceballos et al. 2017. *Proceedings of the National Academy of Sciences* USA 114(30): e6089–e6096. **25.13b** G. Ceballos et al. 2015. *Science Advances* 1(5): e1400253; G. Ceballos et al. 2017. *Proceedings of the National Academy of Sciences* USA 114(30): e6089–e6096. **25.4c** DATA for (c): M. Nikaido, A. P. Rooney, and N. Okada. 1999. *Proceedings of the National Academy of Sciences* USA 96: 10261–10266. **25T.1** R. Agrawal and F. G. Pinilla. 2012. *Journal of Physiology* 590: 2485–2499. **EOC Case Study** W. Liu et al. 2013. *Nature* 467: 420–425. **EOU Case Study Fig 2** C. T. Hani n, E. D. Brodie Jr., E. D. Brodie III. 2008. *PLoS Biology* 6 (3): e60. **EOU Case Study Fig 3** C. R. Feldman et al. 2009. *PNAS* 106: 13415–3420.

Chapter 26 **26.13** D. F. Wilson, M. Erecinska, and P. L. Dutton. 1974. *Annual Review of Biophysics and Bioengineering* 3: 203–230; Tables 1 and 3. **26.3** S. V. Liu, J. Zhou, C. Zhang, et al. 1997. Thermophilic Fe(III)-reducing bacteria from the deep subsurface: The evolutionary implications. *Science* 277: 1106–1109. **p. 535** N. Pace, A Molecular View of Microbial Diversity and the Biosphere, *Science*, Vol. 276, Issue 5313, (02 May 1997), pp. 734–740.

Chapter 27 **27.18b** W. E. Esaias and H. C. Curl Jr. 1972. *Limnology and Oceanography* 17: 901–906. **EOC Case Study (Left)** Slime Mold Solves Maze in One Pass, Assisted by Gradient of Chemo-Attractants Andrew Adamatzky *IEEE Transactions On NanoBioscience*, Vol. 11, No. 2, June 2012 131 Figure 1d. **EOC Case Study (Right)** C. R. Reid, T. Latty, A. Dussutour, et al. 2012. *Proceedings of the National Academy of Sciences* USA 109: 17490–17494.

Chapter 28 **28.21** M. E. Hoballah, T. Gübitz, J. Stuurman, et al. 2007. Single gene-mediated shift in pollinator attraction in Petunia. *Plant Cell* 19: 779–790. **28.23** J.-F. Pombert, et al. (2005) The chloroplast genome sequence of the green alga *Pseudendoclonium akinetum* (Ulvophyceae) reveals unusual structural features and new insights into the branching order of chlorophyte lineages, *Molecular Biology and Evolution*, 22: 1903–1918. **28.24** P. S. Soltis, and D. E. Soltis (2004) The origin and diversification of angiosperms, *American Journal of Botany*, vol. 91, no. 10, p. 1614–1626, Figs. 1, 2, 3. **28.6** Y.-L. Qiu, et al. (2006) The deepest divergences in land plants inferred from phylogenomic evidence, *Proceedings of the National Academy of Sciences* USA, 103 (42): 15511–15516, Fig. 1. **28.9** Renzaglia, K. S., et al. (2007) Bryophyte phylogeny: Advancing the molecular and morphological frontiers, American Bryological and Lichenological Society, *The Bryologist*, vol, 110, no. 2 179–213. **EOC Case Study** N. Cronberg, R. Natcheva, and K. Hedlund. 2006. *Science* 313: 1255. **p. 578** Karl Niklas, Cornell University.

Chapter 29 **29.9** S. M. Adl, et al. (2005) The new higher level classification of eukaryotes with emphasis on the taxonomy of protists, *Journal of Eukaryotic Microbiology*, 52: 399–451. **29.10** T. Y. James, et al. (2006) Reconstructing the early evolution of fungi using a six-gene phylogeny, *Nature*, 443: 818–822, Fig. 1. **EOC Case Study** L. Yafetto, et al. (2008) The Fastest Flights in Nature: High-Speed Spore Discharge Mechanisms among Fungi, *PLoS ONE*, 3(9): e3237.

Chapter 30 **30.2** A. M.A. Aguinaldo, et al. Evidence for a clade of nematodes, arthropods, and other moulting animals. *Nature* (29 May 1997) Vol. 387, pp. 489–493. **30.7** J. R. Finnerty, K. Pang, P. Burton, et al. 2004. Origins of bilateral symmetry: *Hox* and *dpp* expression in a sea anemone. *Science* 304: 1335–1337. **EOC Case Study** T. Jezkova and J. J. Wiens. 2017. *American Naturalist* 189: 201–212.

Chapter 31 **31.1** IUCN 2017. The IUCN Red List of Threatened Species™. Version 2017-3. (December 5, 2017). **31.15** J. C. Regier, et al., Arthropod relationships revealed by phylogenomic analysis of nuclear protein-coding sequences. *Nature* (25 February 2010) Vol. 463, pp. 1079–1083. **EOC Case Study** Y. Basset, et. al., Arthropod Diversity in a Tropical Forest *Science*, Vol. 338, no. 6113 (December 14, 2012) pp. 1481–1484.

Chapter 32 **32.1** C. W. Dunn, A. Hejnol. D. Q. Matus, et al. 2008. Broad phylo-genomic sampling improves resolution of the animal tree of life. *Nature* 452: 745–750, Figs. 1, 2. **32.11** Color plate by Michael DiGiorgio. From Q. Li, K.-Q. Gao, J. Vinther, et al. 2010. Plumage color patterns of an extinct dinosaur. *Science* 327: 1371, Fig. 4. Reprinted by permission of the artist, Michael DiGiorgio. **32.15** A. B. Prasad, M. W. Allard, E. D. Green, et al. 2008. Confirming the phylogeny of mammals by use of large comparative sequence data sets. *Molecular Biology and Evolution* 25 (9): 1795–1808, Figs. 1, 2, 3. Oxford University Press. **32.16** C. Stringer, 2014. *Nature* 514: 427–429. **32.5** American Museum of Natural History; birdlife.org; fishbase.org; IUCN Red List 2010; Reptile-database.org. **32.6** J. E. Blair and S. B. Hedges. 2005. Molecular phylogeny and divergence times of deuterostome animals. *Molecular Biology and Evolution* 22: 2275–2284, Figs. 1, 3, 4. **32.8** E. B. Daeschler, N. H. Shubin, F. A. Jenkins, et al. 2006. A Devonian tetrapod-like fish and the evolution of the tetrapod body plan. *Nature* 440: 757–763, Fig. 6. **EOC Case Study** X-X. Luo. 2011. *Annual Review of Evolution, Ecology, and Systematics* 42: 355–350.

Chapter 33 **33.16** F. Gao, E. Bailes, D. L. Robertson, et al. 1999. Origin of HIV-1 in the chimpanzee *Pan troglodytes troglodytes*. *Nature* 397: 436–441, Fig. 2. **33.3** G. Pantaleo and A. S. Fauci. 1996. *Annual Review of Microbiology* 50: 825–854. **33.7** Data courtesy of the Undergraduate Biotechnology Laboratory, California Polytechnic State University, San Luis Obispo. **33.2a** R. Acuna-Soto, D. W. Stahle, M. K. Cleaveland, et al. 2002. Emerging Infectious Diseases 8: 360–362. **33.2b** E. Arias. 2010. United States Life Tables, 2006. *National Vital Statistics Reports* 58 (21): 1–40, Table 10. Hyattsville, MD: National Center for Health Statistics. **EOC Case Study** Q. Shao et al. 2016. *Development* 143: 4127–4136.

Chapter 34 **34.4** J. Clausen, D. D. Keck, and W. M. Hiesey. 1940. *Experimental studies on the nature of species. I. Effect of varied environments on western North American plants.* Publication 520. Washington, DC: Carnegie Institution. **EOC Case Study** H. B. Fan and Y. H. Wang. 2000. Effects of simulated acid rain on germination, foliar damage, chlorophyll contents and seedling growth of five hardwood species growing in China. *Forest Ecology and Management*, Vol 126, No. 3, (February 25, 2000) pp. 321–329.

Chapter 35 **35.12** C. Wei, Tyree, M.T. and Steudle, E., Direct Measurement of Xylem Pressure in Leaves of Intact Maize Plants. A Test of the Cohesion-Tension Theory Taking Hydraulic Architecture into Consideration, *Plant Physiology*, Vol. 121, no. 4 (Dec, 1999) pp. 1191–1205. **35.3** Cline, R. G. and Campbell, G. S., Seasonal and Diurnal Water Relations of Selected Forest Species, *Ecology*, Vol. 57, No. 2 (March 1, 1976) pp. 367–373. **35.21c** National Academy of Sciences. **EOC Case Study** E. I. Lammertsma, et al., Global CO_2 rise leads to reduced maximum stomatal conductance in Florida vegetation, *Proceedings of the National Academy of Sciences* USA, Vol. 108, no. 10 (October 19, 2010) pp. 4035–4040.

Chapter 36 **36.15** H. Fujikake et al. 2003. Quick and reversible inhibition of soybean root nodule growth by nitrate involves a decrease in sucrose supply to nodules. *Journal of Experimental Botany* 54 (386): 1379–1388. **36 36.3** D. I. Arnon and P. R. Stout. 1939. The essentiality of certain elements in minute quantity for plants with special reference to copper. *Plant Physiology* 14: 371–375. **EOC Case Study** V. Bazile et al. 2012. *PLoS ONE* 7: e36179.

Chapter 37 **37.11** A. Lang, M. K. Chailakhyan, and I. A. Frolova. 1977. Promotion and inhibition of flower formation in a day-neutral plant in grafts with a short-day plant and a long-day plant. *Proceedings of the National Academy of Sciences* USA 74: 2412–2416. **37.21** P. G. Blackman and W. J. Davies. 1985. Root to shoot communication in maize plants of the effects of soil drying. *Journal of Experimental Botany* 36: 39–48. **37.5** C. Darwin and F. Darwin. 1897. *The Power of Movement in Plants.* New York: D. Appleton & Co. **37T.1** H. A. Borthwick, S. B. Hendricks, M. W. Parker, et al. 1952. A reversible photoreaction controlling seed germination. *Proceedings of the National Academy of Sciences* USA 38: 662–666, Table 1. **EOC Case Study** A. G. Volkov et al. 2010. *Plant, Cell & Environment* 33: 163–173. **p. 790** Charles Darwin, *The Power of Movement in Plants* (1880).

Chapter 38 **38.14** J. J. Tewksbury and G. P. Nabhan. 2001. Directed deterrence by capsaicin in chilies. *Nature* 382: 403–404. **38.8** M. E. Hoballah T. Gübitz, J. Stuurman, et al. 2007. Single gene-mediated shift in pollinator attraction in Petunia. *Plant Cell* 19: 779–790. **EOC Case Study** A. Chautá-Mellizo, S. Campbell, M. Argenis Bonilla, et al. 2012. *Basic and Applied Ecology* 13: 524–532. **EOU Case Study Fig 1** A. Steppuhn et al. 2004. *PLoS Biology* 2(8): e217, Figure 4.

Chapter 39 **39.2** A. M. Kerr, S. N. Gershman, and S. K. Sakaluk. 2010. Experimentally induced spermatophore production and immune responses reveal a trade-off in crickets. *Behavioral Ecology* 21: 647–654. **39.10** K. Schmidt-Nielsen. 1984. *Scaling: Why Is Animal Size So Important?* Cambridge, UK: Cambridge University Press. **EOC Case Study** A. L. Jaffe, G. J. Slater, and M. E. Alfaro. 2011. *Biology Letters* 7: 558–561.

Chapter 40 **40.11** K. J. Ullrich, K. Kramer, et al. Present knowledge of the counter-current system in the mammalian kidney, *Progress in Cardiovascular Diseases*, Volume 3, Issue 5, March 1961, pp. 395–431. **40.16** J. R. Davis and D. F. DeNardo. The urinary bladder as a physiological reservoir that moderates dehydration in a large desert lizard, the Gila monster *Heloderma suspectum*, *Journal of Experimental Biology*, Volume 210, 2007, pp. 1472–1480. **40.2a** A. G. Becker et al. 2011. *Neotropical Ichthyology* 9: 895–900. **40.2b** W. W. Dowd et al. 2010. *Journal of Experimental Biology* 213: 210–224. **EOC Case Study** M. M. P. Camargo, M. N. Fernandes, and C. B. R. Martinez. 2009. *Aquatic Toxicology* 94: 40–46.

Chapter 41 **41.14** E.M. Wright, The intestinal Na^+/glucose cotransporter , *Annual Review of Physiology*, Volume 55, March 1993, pp. 575–589. **41.17** N. T. Nguyen, Xuan-Mai T. Nguyen, et al. Relationship Between Obesity and Diabetes in a US Adult Population: Findings from the National Health and Nutrition Examination Survey, 1999–2006, *Obesity Surgery*, Volume 21, Issue 3, March 2011, pp. 351–355. **EOC Case Study** D. P. German et al. Evolution of Herbivory in a Carnivorous Clade of Minnows (Teleostei: Cyprinidae): Effects on Gut Size and Digestive Physiology, *Physiological and Biochemical Zoology*, Volume 83, Number 1, January/February 2010, pp. 1–18.

Chapter 42 **42.19** E. H. Starling. 1896. On the Absorption of Fluids from the Connective Tissue Spaces. *The Journal of Physiology*. 19(4):312–326. **42.21** C. G. Farmer, T. J. Uriona, D. B. Olsen, M. Steenblik, and K. Sanders. 2008. The Right-to-Left Shunt of Crocodilians Serves Digestion. *Physiological and Biochemical Zoology* 81: 125–137. **42.22** J. E. Blair and S. B. Hedges. 2005. Molecular phylogeny and divergence times of deuterostome animals. *Molecular Biology and Evolution* 22: 2275–2284, Figs. 1, 3, 4. **42.9** R. A. Berner. 1999. Atmospheric oxygen over Phanerozoic time, *Proceedings of the National Academy of Sciences* USA 96: 10955–10957. **EOC Case Study** Source: L. M. Shapiro. 1984. *British Heart Journal* 52: 130–135.

Chapter 43 **43.19** R. Q. Quiroga, L. Reddy, G. Kreiman, et al. 2005. *Nature* 435: 1102–1107. **43.9** O. Loewi. 1921. Über humorale Übertragbarkeit der Herznervenwirkung. *Pflügers Archiv: European Journal of Physiology* 189: 239–242. **43.22b-c** K. C. Martin, A. Casadio, H. Zhu, et al. 1997. *Cell* 91: 927–938. **EOC Case Study** T. Narahashi, E. X. Albuquerque, and D. Deguchi. 1971. *Journal of General Physiology* 58: 54–70. **p. 937** W. Penfield, The Role of the Temporal Cortex in Recall of Past Experience and Interpretation of the Present, in *CIBA Foundation Symposium on the Neurological Basis of Behaviour*, G. E. W. Wolstenholme and C. M. O'Connor (Eds). Little, Brown & Company, 1958.

Chapter 44 **44.15** D. M. Hunt, et al. 1995. The chemistry of John Dalton's color blindness. *Science* 267: 984–988, Fig. 3. **44.2** J. E. Rose, J. E. Hind, D. J. Anderson, et al. 1971. *Journal of Neurophysiology* 34: 685–699. **44.10** K. Pohlmann, J. Atema, and T. Breithaupt. 2004. The importance of the lateral line in nocturnal predation of piscivorous catfish. *Journal of Experimental Biology* 207: 2971–2978. **EOC Case Study (Left)** M. K. McClintock. 1971. *Nature* 229: 244–245, Figure 1. **EOC Case Study (Right)** S. E. R. Hoover, C. I. Keeling, M. L. Winston, et al. 2003. *Naturwissenschaften* 90: 477–480. **p. 950** D. Martin, Robert Galambos, Neuroscientist Who Showed How Bats Navigate, Dies at 96, *The New York Times* (July 15, 2010).

Chapter 45 **45.3** PDB ID: 1KWO. D. M. Himmel, S. Gourinath, L. Reshetnikova, et al. 2002. Crystallographic findings on the internally uncoupled and near-rigor states of myosin: Further insights into the mechanics of the motor. *Proceedings of the National Academy of Sciences* USA 99: 12645–12650. **45-17** D. F. Hoyt and C. R. Taylor. 1981. Gait and the energetics of locomotion in horses. *Nature* 292: 239–240 **45-19** Schmidt-Nielsen, K. 1972. Locomotion: Energy cost of swimming, flying, and running. *Science* 177: 222–228. **EOC Case Study** W. J. Fink, D. L. Costill, and M. L. Pollock. 1977. *Annals of the New York Academy of Sciences* 301: 323–327.

Chapter 46 **46.3** A. A. Berthold. 1849. Transplantation der Hoden. *Arch. Anat. Physiol. Wissenschaft Med.* 16: 42–46. **46.6** T. W. Rall, E. W. Sutherland, and J. Berthet. 1957. *Journal of Biological Chemistry* 224: 463–475. **46.10** H. Illnerova, K. Hoffmann, and J. Vanecek. 1984. *Neuroendocrinology* 38: 226–231. **EOC Case Study (Left)** T. B. Hayes, A. Collins, M. Lee, et al. 2002. *Proceedings of the National Academy of Sciences* USA 99: 5476–5480. **EOC Case Study (Right)** B. G. Walker, P. D. Boersma, and J. C. Wingfield. 2006. *Conservation Biology* 20: 146–154.

Chapter 47 **47.11** C. S. C. Price, K. A. Dyer and J. A. Coyne. 1999. Sperm competition between *Drosophila* males involves both displacement and incapacitation. *Nature* 400: 449–452. **47.15** R. Shine and M. S. Y. Lee. 1999. A reanalysis of the evolution of viviparity and egg-guarding in squamate reptiles. The Herpetologists' League, Inc., *Herpetologica* 55 (4): 538–549, Figs. 1, 2, 3. **47.12b** C. Hotzy and G. Arnqvist. 2009. *Current Biology* 19: 404–407. **47.3a** R. G. Stross and J. C. Hill. 1965. *Science* 150: 1462–1464. **47.3b** O. T. Kleiven, P. Larsson, and A. Hoboek. 1992. *Oikos* 65: 197–206. **EOC Question 10** S. E. Anderson, G. E. Dallal, and A. Must. 2003. *Pediatrics* 111: 844–850. **EOC Case Study** U. J. Gaspard, M. A. Romus, D. Gillain, et al. 1983. *Contraception* 27: 577–590.

Chapter 48 **48.17** B. Alberts, et al. 2002. *Molecular Biology of the Cell*. 4th ed. New York: Garland Science, Taylor & Francis Group. Modified from Fig . 24-10. **48.7** M. Knossow, et al. 1984. Three-dimensional structure of an antigenic mutant of the influenza virus haemagglutinin. *Nature* 311 (5987): 678–680. **48.10** L. J. Stern, et al. 1994. Crystal structure of the human class II MHC protein HLA–DR1 complexed with an influenza virus peptide. *Nature* 368: 215–221. **48.2b** B. Lemaitre et al. 1996. *Cell* 86: 973–983. **48.5a** L. J. Harris, et al. 1997. Refined structure of an intact IgG2a monoclonal antibody. *Biochemistry* 36: 1581–1597. **48.5b** K. C. Garcia, et al. 1996. An $\alpha\beta$ T cell receptor structure at 2.5 Å and its orientation in the TCR-MHC complex. *Science* 274: 209–219. **EOC Case Study** P. K. Mishra et al. 2013. *Mucosal Immunology* 6: 297–308. **EOU Case Study Fig 1 and Fig 2** C. Feldman, N. Leblanc, R. del Carlo, and J. Reimche. 2018. Personal communication.

Chapter 49 **49.2** © 2016 Colorado Plateau Geosystems, Inc. **49.3** Ú. L. Vaz and J. C. Nabout. 2016. Using ecological niche models to predict the impact of global climate change on the geographical distribution and productivity of *Euterpe oleracea Mart.* (Arecaceae) in the Amazon. *Acta Botanica Brasilica* 30(2): 290–295. **49.9** D. M. Olson, E. Dinerstein, E. D. Wikramanayake, et al. 2001. *BioScience* 51: 933–938. **49.10** E. C. Ellis, K. Klein Goldewijk, S. Siebert, et al. 2010. *Global Ecology and Biogeography* 19: 589–606. **49.12a** M. B. Saffo. 1987. *BioScience* 37: 654–664. **49T.2** © 2016 Colorado Plateau Geosystems, Inc. **EOC Case Study** A. Q. Fogg, N. J. Brown-Peterson, and M. S. Peterson. 2017. *Bulletin of Marine Science* 93: 1–23. **p. 1055** Sean Carroll (2017) *The Serengeti Rules: The Quest to Discover How Life Works and Why It Matters*, Princeton University Press.

Chapter 50 **50.14** J. L. Hoogland. 1983. Nepotism and alarm calling in the black-tailed prairie dog (*Cynomys ludovicianus*). *Animal Behaviour* 31: 472–479. **50.3** S. Alem, C. J. Perry, et al. 2016. *PLOS Bioloy* 14: e1002564. **50.4** M. Sokolowski. 2001. *Drosophila*: Genetics meets behavior. *Nature Reviews Genetics* 2: 879–890, Fig. 2. **50.5** T-I. Yang and C-C. Chiao. 2016. Number sense and state-dependent valuation in cuttlefish. *Proceedings of the Royal Society* B 283: 2016.1379. **50.6** D. Crews. 1975. *Science* 189: 1059–1065. **50.7** T. Driessens, B. Vanhooydonck, and R. Van Damme. 2014. *Behavioral Ecology and Sociobiology* 68: 173–184. **50.8** K. J. Lohmann, C. M. F. Lohmann, L. M. Ehrhart, et al. 2004. Geomagnetic map used in sea-turtle navigation. *Nature* 428: 909–910. **50.9** C. Egevang, et al. 2010. Tracking of Arctic terns *Sterna paradisaea* reveals longest animal migration. *Proceedings of the National Academy of Sciences* USA 107(5): 2078–2081, Fig. 1B. **EOC Case Study** J. A. Goldbogen, B. L. Southall, S. L. DeRuiter, et al. 2013. *Proceedings of the Royal Society* B 280: 2013.0657.

Chapter 51 **51.1** T. Horvathova, et al. 2013. Length of activity season drives geographic variation in body size of a widely distributed lizard. *Ecology and Evolution* 3(8): 2424–2442, Figure 2. **51.11** C. J. Krebs, S. Boutin, R. Boonstra, et al. 1995. Impact of food and predation on the snowshoe hare cycle. *Science* 269: 1112–1115. **51.14** United Nations, *World Population Prospects: The 2017 Revision.* **51.16** United Nations, *World Population Prospects: The 2017 Revision.* **51.7** T. Fenchel. 1974. *Oecologia* 14: 317–326. **51.8** G. F. Gause. 1934. *The Struggle for Existence.* New York: Hafner Press. **51.9** V. B. Scheffer. 1951. *Scientific Monthly* 73: 356–362. **51.10** D. A. MacLulich. 1937. Fluctuations in the numbers of varying hare (*Lepus americanus*). *University of Toronto Studies: Biological Series* 43. **51.T2** H. Strijbosch and R. C. M. Creemers. 1988. Comparative demography of sympatric populations of *Lacerta vivipara* and *Lacerta agilis. Oecologia* 76: 20–26. **EOC Case Study** M. E. Dorcas, J. D. Willson, R. N. Reed, et al. 2012. *Proceedings of the National Academy of Sciences* USA 109: 2418–2422.

Chapter 52 **52.13** H. Toju et al. 2014. *Nature Communications* 5: 5273. **52.14** R. T. Paine. 1966. *American Naturalist* 100: 65–75. **52.16** D. G. Jenkins and A. L. Buikema. 1998. Do similar communities develop in similar sites? A test with zooplankton structure and function. *Ecological Monographs* 68: 421–443. **52.17** T. W. Swetnam. 1993. *Science* 262: 885–889. **52.2** G. F. Gause. 1934. *The Struggle for Existence.* New York: Hafner Press. **52.21** J. Rolland et al. 2014. *PLOS Biology* 12: e1001775; D. A. Duchêne and M. Cardillo. 2015. *Global Ecology and Biogeography* 24: 1203–1362. **52.4** J. H. Connell. 1961. The influence of interspecific competition and other factors on the distribution of the barnacle *Chthamalus stellatus. Ecology* 42: 710–723. **52.6** P. Grant. 1999. *Ecology and Evolution of Darwin's Finches*. Princeton, NJ: Princeton University Press. **52.8** G. H. Leonard, M. D. Bertness, and P. O. Yund. 1999. Crab predation, waterborne clues, and inducible defenses in the blue mussel, *Mytilus edulis. Ecology* 80: 1–14. **52.20** R. H. MacArthur and E. O. Wilson. 1963. An equilibrium theory of insular zoogeography. *Evolution* 17 (4): 373–387, Fig. 4. **EOC Case Study** F. Keesing, J. Brunner, S. Duerr, et al. 2009. *Proceedings of the Royal Society* B 276: 3911–3919.

Chapter 53 **53.1** J. R. Gosz et al. 2078. *Scientific American* 238: 93–102. **53.12** G. E. Likens, F. H. Bormann, N. M. Johnson, et al. 1970. Effects of forest cutting and herbicide treatment on nutrient. *Ecological Monographs* 40: 23–47. **53.13** U. Schneider et al. 2016. *Proceedings of the International Association of Hydrological Sciences* 374: 29–34. **53.14** V. L. McGuire, 2017, Water-level and recoverable water in storage changes, High Plains aquifer, predevelopment to 2015 and 2013–15, U.S. Geological Survey, *Scientific Investigations Report* 2017-5040. **53.15** D. Fowler et al. 2013. *Philosophical Transactions of the Royal Society* B 368(1621): 2013.0165. **53.17** B. Hönisch, A. Ridgwell, D. N. Schmidt, et al. 2012. *Science* 335: 1058–1063. **53.18** Carbon Dioxide Information Analysis Center. **53.21** *Emission Database for Global Atmospheric Research*, Version 4.3, 2016. **53.22** C. Moritz and R. Agudo. 2013. *Science* 341: 504–508; Intergovernmental Panel on Climate Change, 2014. *Climate Change 2014: Synthesis Report*. Core writing team, R. K. Pachauri and L. A. Meyer (eds.).IPCC, Geneva, Switzerland. **53.23** A. Dai. 2013. *Nature Climate Change* 3: 52–58. **53.8** http://earthobservatory.nasa.gov. **53.9** H. Leith and R. H. Whittaker (eds.). 1975. *Primary Productivity in the Biosphere*. New York: Springer-Verlag. **53.20** NOAA/ESRL; *Global Climate Change Impacts in the United States, 2009 Report*. T. R. Karl et al. (eds.). Cambridge UP; GISS/NASA. **EOC Case Study** P. M. Brando et al. 2014. *Proceedings of the National Academy of Sciences* USA 111: 6347–6352.

Chapter 54 **54.1** P. D. Mannion et al. 2014. The latitudinal biodiversity gradient through deep time. *Trends in Ecology and Evolution* 29: 42–50. **54.11** D Tilman, J. Knops, D. Wedin, et al. 1997. The influence of functional diversity and composition on ecosystem processes. *Science* 277: 1300–1302. **54.14** E. Dinerstein et al. 2017. *BioScience* 67: 534–545. **54.15** G. Eshel, A. Shepon, T. Makov et al. 2014. *Proceedings of the National Academy of Sciences* USA 111: 11996–12001. **54.4** S. Sloan et al. 2014. *Biological Conservation* 177: 12–24. **54.5** The International Union for Conservation of Nature (IUCN), 2017. *The IUCN Red List of Threatened Species.* Version 2017.3. **54.6** World Wildlife Fund (WWF), 2016. *Living Planet Report 2016: Risk and Resilience in a New Era*. **54.8** W. F. Laurance, S. G. Laurance, L. V. Ferreira, et al. 1997. Biomass collapse in Amazonian forest fragments. *Science* 278: 1117–1118. **54.9** J. M. Diamond. 1975. *Biological Conservation* 7: 129–146. **54.10** B. F. Sinervo et al. 2010. *Science* 328: 894–899. **EOC Case Study (Left)** D. S. Karp et al. 2013. *Ecology Letters* 16: 1339–1347. **EOC Case Study (Right)** D. S. Karp et al. 2013. *Ecology Letters* 16: 1339–1347. **EOU Case Study Fig 1** IUCN, Conservation International, and NatureServe. 2004. *Global Amphibian Assessment*.

Index

Boldface page numbers indicate a glossary entry; page numbers followed by an *f* indicate a figure; those followed by *t* indicate a table.

C

D

E

G

H

J

K

L

M

N

O

P

Q

R

S

U

V

X

Y

Z

✔ PUT IT ALL TOGETHER: Case Studies